HOUGHTON MIFFLIN

# COLLEGE DICTIONARY

## HOUGHTON MIFFLIN

# COLLEGE DICTIONARY

Houghton Mifflin Company Boston

All correspondence and inquiries should be directed to
Reference Division, Houghton Mifflin Company
One Beacon Street, Boston, MA 02108

**Library of Congress Cataloging in Publication Data**
Main entry under title:

Houghton Mifflin college dictionary.

1. English language—Dictionaries. I. Houghton Mifflin Company.
PE1628.H59 1986 423 85-19703
ISBN: 0-395-41031-2

Manufactured in the United States of America

# Table of Contents

# Dictionary Management and Staff

Anne H. Soukhanov • *Senior Editor*
Kaethe Ellis • *Coordinating Editor*

Howard Webber • *Publisher*
Margery S. Berube • *Director of Editorial Operations*

Dolores R. Harris • *Senior Editor, Usage*    Pamela B. DeVinne • *Editor, Pronunciation*
Marion Severynse • *Editor, Word Histories & Etymologies*

**Contributing Editors**

Kerry W. Metz • *Synonymies*
David R. Pritchard • *Biographical Names & Foreign Terms*

Caroline L. Becker
Elaine H. Brix
Ernest S. Hildebrand, Jr.
Rhonda L. Holmes
Rachel Lucas
Lawrence O. Masland
Ramona Michaelis
Trudy Nelson

Gretchen B. Armacost • *Special Consultant, Foreign Terms*

Christopher Leonesio • *Production Manager*

**Database Keyboarding Staff**

Brenda J. Bregoli • Celester Jackson • Ronald Perkins • Caren Raimondi

Sarah A. O'Reilly • *Art Research*
Laurel L. Cook • *Illustrations*
Donna L. Muise • *Production Assistant*
Geoffrey Hodgkinson • *Design*

**Special Assistants**

Francine D. Figelman
Annie B. Kelly
James Marciano
Patricia McTiernan
Ann L. Satchwill

# Preface to the Student

There are probably very few people who would claim the dictionary as their favorite book. All the words are there; it is the order that seems unexciting.

But to those of us at Houghton Mifflin Company whose work it is to prepare the materials that advance your learning and show you how to think for yourselves, the dictionary is the foundation of this entire effort.

The reason is that words stand for things and feelings, ideas and occurrences, memories and expectations, needs and possibilities, fears, desires, delights—the smallest and most ordinary issues of our lives as well as the largest and most powerful ones. Words are in some way the stuff of thoughts, and they are the only tools we have that permit us to share with others what is really going on inside our minds and hearts. They are the only instruments that reveal whether we are laughing from pleasure or embarrassment, whether our tears are signs of joy or sadness.

Not all of the words in the world are in any dictionary, even the biggest one. When you think of the special vocabulary of the scientist, which may be added to from month to month as what is discovered must be named and defined to fix it in human history, you will easily see that no printed dictionary can contain it all. Nor can such a dictionary contain all the words that speakers of English use, in every part of the world, as they talk about the experiences unique to them, as they joke and argue, and as they invent new words and meanings (for one of the pleasures of knowing language well is invention).

But in the *Houghton Mifflin College Dictionary* you will find all the words you need to become the kind of person who speaks and writes (and lives) not on a thin, barren, changeless level but with all the breadth and height and depth and movement you have a right to as a human being. As you discover those words, the dictionary can be the most important book you ever open.

We have designed the dictionary with you in mind. Probably its most common function is to help you get your spelling right, and of course your meaning. It shows you how to do term papers—from outline and notes to the final form of the paper. It tells you how to write letters that will be read seriously by those who receive them, including the President of the United States and the Secretary General of the United Nations. It shows you how to prepare a résumé when you're applying for a job. It gives you the advice you need to choose the correct uses of language and to avoid expressions that are trite or illogical. It helps you find a better word than the one you thought of. It helps you understand new words by understanding their parts, and the pronunciation code lets you say new words like a veteran, even though you may never have seen them before. It gives you basic information about important people and important places. It gives you a sense of how our words came to be in the first place.

As a matter of fact, you might easily become one of those people who *reads* dictionaries—including the pages that come before and after the alphabetical listings. Someone may have chided Ralph Waldo Emerson about reading a dictionary, because he wrote: "Neither is a dictionary a bad book to read. . . . It is full of suggestion,—the raw material of possible poems and histories. Nothing is wanting but a little shuffling."

Howard Webber, *Publisher*

# Explanatory Diagram: Major Elements of This Dictionary

MAIN-ENTRY WORD
SENSE NUMBER
SUBSENSE NUMBER
TEMPORAL LABEL
USAGE NOTE
PLURAL LABEL
INFLECTED FORMS, VERB
PRONUNCIATION
STYLISTIC LABEL
PRIMARY STRESS
SECONDARY STRESS
CROSS-REFERENCE
WORD HISTORY PARAGRAPH
SUBJECT LABEL
LANGUAGE LABEL
INFLECTED FORM, NOUN
INFLECTED FORMS, ADJECTIVE
VERBAL ILLUSTRATION

**quick** (kwĭk) *adj.* **-er, -est.** [ME *quicke*, swift, alive < OE *cwicu*, alive.] **1.** Moving or performing with speed and agility : FAST. **2.** Thinking or understanding rapidly and easily : BRIGHT <a *quick* learner> **3. a.** Perceiving or responding with speed and sensitivity : KEEN. **b.** Reacting immediately and sharply <a *quick* temper> **4. a.** Occurring or achieved in a relatively short time <a *quick* trip> **b.** Done or occurring immediately : PROMPT <*quick* service> **5.** Tending to react hastily or impulsively <*quick* to forgive> **6.** *Archaic.* Alive. **7.** *Archaic.* Pregnant. *—n.* **1.** Raw or sensitive exposed flesh, as under the fingernails. **2.** The most intimate and sensitive aspect of the emotions <pierced me to the *quick*> **3.** The living <the *quick* and the dead> **4.** The vital core of a thing : ESSENCE <the *quick* of the matter> *—adv.* Fast : promptly. ***usage:*** Both *quick* and *quickly* can be used as adverbs. In speech, *quick* is more frequent, as in *Come quick!* In writing, *quickly* is preferred, as in *They returned quickly when they heard the news.* **—quick′ly** *adv.* **—quick′ness** *n.*

**quick-and-dirt·y** (kwĭk′ən-dûr′tē) *adj.* Shoddily made or done : cheap <a *quick-and-dirty* construction project>

**quick assets** *pl.n.* Liquid assets, including cash on hand and assets readily convertible to cash.

**quick bread** *n.* A bread made with a leavening agent, as baking powder, that does not require a leavening period before baking.

**quick·en** (kwĭk′ən) *v.* **-ened, -en·ing, -ens.** *—vt.* **1.** To make more rapid : ACCELERATE. **2.** To make or bring alive : VITALIZE. **3.** To stimulate : stir <The good news *quickened* our interest.> **4.** To make steeper. *—vi.* **1.** To become more rapid. **2.** To come or return to life. **3.** To reach the stage of pregnancy when the fetus can be felt to move. **—quick′en·er** *n.*

**quick-freeze** (kwĭk′frēz′) *vt.* **-froze** (-frōz′), **-froz·en** (-frō′zən), **-freez·ing, -freez·es.** To freeze (food) by a process sufficiently rapid to retain desirable properties, as flavor and nutritional value.

**quick grass** *n.* [Var. of QUITCHGRASS.] Couch grass.

**quick·ie** (kwĭk′ē) *n. Informal.* Something done or made hastily.

**quick·lime** (kwĭk′līm′) *n.* [ME *quykke lyme*, transl. of Lat. *calx viva.*] Calcium oxide.

**quick·sand** (kwĭk′sănd′) *n.* A bed of loose sand mixed with water forming a soft, shifting mass that does not support heavy objects.

**quick·set** (kwĭk′sĕt′) *n. Chiefly Brit.* **1.** Cuttings or slips of a plant, as hawthorn, capable of rooting when set in the ground. **2.** A hedge grown from quickset.

**quick·sil·ver** (kwĭk′sĭl′vər) *n.* [ME < OE *cwicseolfor*, transl. of Lat. *argentum vivum.*] MERCURY 2. *—adj.* Unpredictable : mercurial.

▲ word history: The name *quicksilver* for the element mercury is a translation of Latin *argentum vivum*, literally "living silver." Mercury was so called because it is a silvery colored metal that is liquid at ordinary temperatures. In *quicksilver* the word *quick* preserves its original but now archaic sense "living, alive."

**quick·step** (kwĭk′stĕp′) *n. Mus.* A march for accompanying military quick time.

**quick-tem·pered** (kwĭk′tĕm′pərd) *adj.* Easily aroused to anger.

**quick time** *n.* A military marching pace of 120 steps per minute.

**quick-wit·ted** (kwĭk′wĭt′ĭd) *adj.* Mentally alert and sharp : CLEVER. **—quick′-wit′ted·ly** *adv.* **—quick′-wit′ted·ness** *n.*

**quid**[1] (kwĭd) *n.* [ME *quide*, cud < OE *cwidu.*] A cut of something to be chewed, esp. a plug of tobacco.

**quid**[2] (kwĭd) *n., pl.* **quid** or **quids.** [Orig. unknown.] *Chiefly Brit.* A pound sterling.

**quid·di·ty** (kwĭd′ĭ-tē) *n., pl.* **-ties.** [Med. Lat. *quidditas* < Lat. *quid*, what.] **1.** The inherent nature of a thing : ESSENCE. **2.** An unreasonably fine distinction : QUIBBLE.

**quid·nunc** (kwĭd′nŭngk′) *n.* [Lat. *quid nunc?* what now?] A prying or meddlesome person : BUSYBODY.

**quid pro quo** (kwĭd′ prō kwō′) *n.* [Lat., something for something.] An equal exchange or substitution.

**qui·es·cent** (kwī-ĕs′ənt, kwē-) *adj.* [Lat. *quiescens, quiescent-*, pr. part. of *quiescere*, to rest < *quies*, rest.] Inactive or still : DORMANT. **—qui·es′cence** *n.* **—qui·es′cent·ly** *adv.*

**qui·et** (kwī′ĭt) *adj.* **-et·er, -et·est.** [ME < OFr. *quiete* < Lat. *quietus*, p.part. of *quiescere*, to rest < *quies*, rest.] **1.** Making little or no sound : SILENT. **2.** Free of noise : HUSHED. **3.** Calm and unmoving : STILL <a *quiet* woodland pool> **4.** Free of agitation and turmoil : UNTROUBLED <a *quiet* life> **5.** Restful : soothing. **6.** Marked by tranquillity : PEACEFUL. **7.** Not showy or obtrusive : RESTRAINED. —*n.* The condition or quality of being quiet : TRANQUILLITY. —*v.* **-et·ed, -et·ing, -ets.** —*vt.* **1.** To cause to become quiet. **2.** *Law.* To make (a title) secure by freeing from all questions or claims. —*vi.* To become quiet <The heckler finally *quieted* down.> **—qui'et·ly** *adv.* **—qui'et·ness** *n.*

☆ ***syns:*** QUIET, INOBTRUSIVE, RESTRAINED, SUBDUED, UNOBTRUSIVE *adj. core meaning* : not showy or obtrusive <decor that is subtle and *quiet*> ***ant:*** gaudy, loud

**qui·et·ism** (kwī'ĭ-tĭz'əm) *n.* **1.** Christian mysticism calling for passive contemplation and the beatific annihilation of individual will. **2.** Quietness and passivity. **—qui'et·ist** *n.* **—qui'et·is'tic** *adj.*

**qui·e·tude** (kwī'ĭ-to͞od', -tyo͞od') *n.* [LLat. *quietudo* < Lat. *quietus,* p.part. of *quiescere,* to rest < *quies,* rest.] Quiet tranquillity.

**qui·e·tus** (kwī-ē'təs) *n.* [Short for Med. Lat. *quietus est,* he is discharged (of an obligation).] **1.** Something that suppresses, terminates, or allays. **2.** Release from life **:** DEATH. **3.** A final discharge, as of a debt or obligation.

**quiff** (kwĭf) *n.* [Orig. unknown.] *Chiefly Brit.* A tuft of hair, esp. a forelock.

**quill** (kwĭl) *n.* [ME *quil,* of Germanic orig.] **1.** The hollow stemlike main shaft of a feather. **2.** A large wing or tail feather. **3.** A writing pen made from a quill. **4.** A plectrum for a stringed musical instrument of the clavichord type. **5.** A toothpick made from the stem of a feather. **6.** One of the sharp hollow spines of a hedgehog or porcupine. **7.** A musical pipe having a hollow stem. **8.** A spool or bobbin for holding or winding yarn. **9.** A mechanical device consisting of a hollow cylinder that rotates on a solid shaft. —*vt.* **quilled, quill·ing, quills. 1.** To wind (yarn or thread) onto a quill. **2.** To imprint (fabric) with textured ridges.

**quill·back** (kwĭl'băk') *n., pl.* **-backs** *or* **quillback.** A North American freshwater fish, *Carpiodes cyprinus,* having a dorsal fin with one ray conspicuously extended.

**quill·wort** (kwĭl'wûrt', -wôrt') *n.* An aquatic plant of the genus *Isoetes,* having short, fleshy stems and grasslike leaves.

**quilt** (kwĭlt) *n.* [ME *quilte* < OFr. *cuilte* < Lat. *culcita,* mattress.] **1.** A bed covering consisting of two layers of fabric with a layer of batting or feathers between and stitched firmly together, usu. in a decorative pattern. **2.** A padded cover similar to a quilt. —*v.* **quilt·ed, quilt·ing, quilts.** —*vt.* **1.** To make into a quilt by stitching together (layers of fabric). **2.** To construct like a quilt <*quilt* a vest> **3.** To pad and stitch in decorative designs. —*vi.* **1.** To make a quilt. **2.** To do quilted work.

**quilt·ing** (kwĭl'tĭng) *n.* **1.** The act or process of doing quilted work. **2. a.** Material used to make quilts. **b.** Quilted material.

**quin-** *pref. var. of* QUINO-.

**quin·a·crine hydrochloride** (kwĭn'ə-krēn') *n.* [QUIN- + AC-R(ID)INE.] A bright yellow, bitter, crystalline compound, used chiefly to treat malaria.

**quin·a·liz·a·rin** (kwĭn'ə-lĭz'ə-rĭn) *n.* A reddish crystalline compound, $C_{14}H_8O_6$.

**qui·nate** (kwī'nāt') *adj.* [< Lat. *quini,* five each.] Arranged in groups of five <*quinate* leaflets>

**quince** (kwĭns) *n.* [ME *quynce,* pl. of *quyn,* quince < OFr. *coin* < Lat. *cotoneum,* var. of *cydoneum,* after *Cydonia,* an ancient town in Crete.] **1.** A tree, *Cydonia oblonga* orig. of western Asia, bearing white flowers and edible fruit. **2.** The yellow applelike fruit of the quince.

**quin·cun·cial** *also* **quin·cunx·ial** (kwĭn-kŭn'shəl) *adj.* Of, relating to, or forming a quincunx. **—quin·cun'cial·ly** *adv.*

**quin·cunx** (kwĭn'kŭngks') *n.* [Lat., five twelfths : *quinque,* five + *uncia,* twelfth < *unus,* unit.] An arrangement of five things with one at each corner of a square or rectangle and one at the center.

**quin·cunx·ial** (kwĭn-kŭn'shəl) *adj. var. of* QUINCUNCIAL.

**quin·de·cen·ni·al** (kwĭn'dĭ-sĕn'ē-əl) *adj.* [Lat. *quindecim,* fifteen (*quinque,* five + *decem,* ten) + *annus,* year.] **1.** Occurring once every 15 years. **2.** Lasting 15 years. —*n.* A 15th anniversary.

**qui·nel·la** (kwĭ-nĕl'ə, kē-) *or* **qui·nie·la** (kēn-yĕl'ə) *n.* [Am. Sp., a lottery-like game.] A system of betting in which the bettor wins by correctly picking the first two finishers of a race regardless of their order.

**quin·i·dine** (kwĭn'ĭ-dēn') *n.* A colorless crystalline alkaloid, $C_{20}H_{24}N_2O_2$, resembling quinine and used in treating certain heart disorders and malaria.

SUBENTRIES
SYNONYM PARAGRAPH
ANTONYMS
UNDEFINED RUN-ONS
BOLDFACE COLON
PART-OF-SPEECH LABEL FOR MAIN ENTRY
DEFINITION
PART-OF-SPEECH LABEL FOR SUBENTRY
VARIANT NOUN PLURALS
ETYMOLOGY
ANGLE BRACKETS
VARIANT CROSS-REFERENCE
SYLLABICATION DOTS
SCIENTIFIC BINOMIAL
SECONDARY VARIANT
EQUAL VARIANT
CHEMICAL FORMULA

# Explanatory Notes

Every user of this book should read these Explanatory Notes carefully, because a thorough understanding of the content and scope of the *Houghton Mifflin College Dictionary* is essential to its effective use.

## Guide Words

A pair of guide words is printed at the top of every numbered page as an aid to locating an entry on that page:

**Adlerian • adopt**

The word to the left of the bullet is the first entry on that page of the Dictionary. The word to the right of the bullet is the last entry on the page.

## Entry Order

Entry words are printed in boldface type and are positioned slightly to the left of the text columns. All entries—including compounds of two or more words—have been listed in strict alphabetical sequence, which accounts for each letter in turn through the full entry even if that entry consists of several words: **u·ra·ni·um, uranium 235, uranium 238, uranium dioxide, uranium enrichment, uranium trioxide, urano–.** Inclusion of numbers with a term (such as the *235* in *uranium 235*) has no effect on alphabetical order.

## Boldface Daggers: Americanisms

A boldface dagger appearing at the left of an entry word indicates that the entry or at least one of its senses is limited in use to a region or regions of the United States. The label *Regional* or a more narrowly specified geographic label also appears at such entries:

**†brew·is** (brōō′ĭs, brōōz) *n.* [ME < OFr. *brouetz*, dim. of *brouet*, broth, of Germanic orig.] *Regional.* Broth.

In the following example, the dagger signals an Americanism, and the *Regional* label at sense **2.** indicates that only sense **2.** is an Americanism:

**†an·y·more** (ĕn′ē-môr′, -mōr′) *adv.* **1.** At the present: from now on <can't go there *anymore*> **2.** *Regional.* Nowadays.

When a regional inflected form of an entry is shown at its own alphabetical place in the vocabulary with a cross-reference to the appropriate main entry, the cross-reference entry is also marked with a dagger and bears a label:

**†bought·en** (bôt′n) *v. Regional. var. p.p. of* BUY.

The above cross-reference indicates that the entry **boughten** is a regional variant of the past participle of the verb *buy.*

## Superscript Numerals

Two or more entry words that are identical in spelling but that have different etymologies are entered separately with superscript numerals immediately following them:

**bon·go**[1] (bŏng′gō, bông′-) *n., pl.* **-gos.** [Of Bantu orig.] An antelope, *Boocercus eurycerus* of central Africa, having a reddish-brown coat with white stripes and spirally twisted horns.
**bon·go**[2] (bŏng′gō, bông′-) *n., pl.* **-gos** *or* **-goes.** [Am. Sp. (West Indies) *bongó.*] One of a pair of connected tuned drums played with the hands.

## Syllabication

An entry and its inflected and derived forms (if there are any such forms) are syllabicated with boldface centered dots:

**u·ra·ni·nite** (yōō-rā′nə-nīt′) *n.*
**u·ra·ni·um** (yōō-rā′nē-əm) *n.*

In an entry for an open compound (i.e., a term composed of two or more words with a space between them), a word that appears elsewhere in the Dictionary as a separate entry is not syllabicated:

**uranium enrichment** *n.*

In the above example, *uranium* and *enrichment* are terms that are entered and syllabicated elsewhere in the Dictionary.

Pronunciations are also syllabicated but the syllabication of the phonetic form does not necessarily match the syllabication of the graphic form of the entry word. The former follows phonological rules, while the latter represents the traditional practice of typesetters and editors in breaking words at the ends of lines.

## Pronunciation

Pronunciations are given immediately after the boldface entry word, enclosed by parentheses. Variant pronunciations are given wherever necessary and follow the form to which they apply. A short key to the pronunciation symbols appears across the bottom of the inside column of each pair of facing pages. A full pronunciation key is given on page 64.

**Pronunciation symbols.** The pronunciation symbols have been selected so that the Dictionary user will be able to pronounce a word with only a quick reference to the key. These symbols have been so designed that they represent an acceptable pronunciation for each of the regional varieties of American English, since speakers of any one of these dialects will automatically apply the regional features of their speech in reading the key. All pronunciations are those of cultivated speech; when more than one pronunciation is given, all are acceptable in spite of the fact that there may be differences between them in frequency of occurrence.

**ə** This special symbol is called a *schwa.* It is used to represent a vowel that receives the weakest level of stress (reduced vowel) within a word and that almost always has a different quality than it would if stressed, as in the first syllables of **telemeter** (tĕl′ə-mē′tər) and **telemetry** (tə-lĕm′ĭ-trē).

**Stress.** There are three relative degrees of stress or loudness with which the syllables of a word are spoken. A syllable with primary, or strongest stress is signaled by a boldface mark (**′**) after it. A syllable with secondary, or weaker, stress has a lighter mark (′) after it. A syllable with no mark after it has the weakest stress in a multisyllabic word. Monosyllabic words have no stress mark because there is no other level of stress within the word with which the syllable can be compared.

**Spelling-sound correspondence table.** The English language employs over forty distinct speech sounds and uses an alphabet of only twenty-six characters to represent these sounds. This table, which is designed to help the user find a word known only through its pronunciation, lists the speech sounds of English and matches them with the possible spellings for each sound

| Pronunciation Symbol | Spelling | Example |
|---|---|---|
| (ă) | a | **pat** |
| | ai | **plaid** |
| | al | **half, halve** |
| | au | **laugh** |
| (ā) | a | **mane** |
| | ae | **maelstrom** |
| | ai | **aiguille, aim, maize, train** |
| | aig | **arraign** |
| | aigh | **straight** |
| | ao | **gaol** |
| | au | **gauge** |
| | ay | **pray, say** |
| | e | **bouquet, consomme, forte, sachet, suede** |
| | ea | **great, steak** |
| | ee | **matinee melee,** |
| | ei | **beige, rein, skein, veil** |
| | eig | **reign** |
| | eigh | **eight, neighbor, sleigh, weigh** |
| | ey | **dey, prey, they** |
| (âr) | a | **care** |
| | ae | **aerial, aerosol** |
| | ai | **air, bairn, pair** |
| | ay | **prayer** |
| | e | **berry, ere, there** |
| | ea | **pear, tear, wear** |
| | ei | **Eire, their** |
| | hei | **heiress** |
| (ä) | a | **father** |
| | ah | **ah, Mahdi, shah** |
| | al | **palm** |
| | e | **sergeant** |
| | ea | **heark, hearth** |
| | i | **lingerie** |
| (b) | b | **bib** |
| | bb | **cabbage, ebb, rubber** |
| | bh | **bhang** |
| | pb | **cupboard, raspberry** |

| Pronunciation Symbol | Spelling | Example |
|---|---|---|
| (ch) | c | **cello** |
| | ch | **church** |
| | cz | **Czech** |
| | tch | **catch, hitch, pitcher** |
| | te | **righteous** |
| | ti | **attention, bestial** |
| | tu | **denture, nature, pasture** |
| (d) | d | **deed** |
| | ed | **mailed, ringed, winged** |
| | dd | **gladden, ladder, saddle** |
| | dh | **dhow** |
| | ld | **would** |
| (ĕ) | a | **any, many** |
| | ae | **aesthete** |
| | ai | **again, said** |
| | ay | **says** |
| | e | **pet** |
| | ea | **bread, cleanse, steady, treasure** |
| | ei | **heifer, Seine** |
| | eo | **leopard** |
| | ie | **friend** |
| | oe | **oedipal** |
| | u | **bury** |
| (ē) | ae | **archaeology, paean** |
| | ay | **quay** |
| | e | **be** |
| | ea | **beach, eagle, leap, sea** |
| | ee | **beet, creep, peevish** |
| | ei | **deceit, receipt, receive, seize** |
| | eo | **people** |
| | ey | **covey, key** |
| | i | **bialy, miniskirt, piano, solarium, symposium** |
| | ie | **believe, grief, siege** |
| | oe | **amoeba, oenology, phoenix** |
| | y | **appendectomy, city, comedy** |
| (f) | f | **fame** |
| | fe | **rife** |
| | ff | **sniffle, tiff** |
| | gh | **enough, rough** |
| | lf | **half** |
| | ph | **alphabet, nymph, photo, sphinx** |
| (g) | g | **go** |
| | gg | **baggage, drugged, sluggish** |
| | gh | **ghastly, gherkin, ghetto, ghoul** |
| | gu | **guard, guest** |
| | gue | **catalogue, vague** |
| (h) | h | **house** |
| | wh | **who, whose** |
| | g | **Gila monster** |
| | j | **Jerez** |
| (hw) | wh | **wheat** |

| Pronunciation Symbol | Spelling | Example |
|---|---|---|
| (ĭ) | a | certificate, climate, village |
| | e | English, enough, rebuff, recite |
| | ee | been |
| | i | pin |
| | ia | carriage |
| | ie | sieve |
| | o | women |
| | u | busy |
| | ui | guilty |
| | y | cyst, hymn, symbol |
| (ī) | ai | aisle |
| | ay | aye, bayou |
| | ei | eider, height, stein |
| | ey | eying |
| | eye | eyeball |
| | ie | lie |
| | ig | align |
| | igh | high, right |
| | is | island, lisle |
| | uy | buy |
| | y | lyre, my, sky |
| | ye | rye |
| (îr) | e | cereal, here, series |
| | ea | beard, smear, tear |
| | ee | beer, steer |
| | ei | weird |
| | ie | pier |
| (j) | d | graduation, individual |
| | dg | judgment, trudging |
| | dge | bridge, lodge |
| | di | soldier |
| | dj | adjutant |
| | g | agile, angina, gem |
| | ge | diverge, stage |
| | gg | exaggerate |
| | j | jam |
| (k) | c | call, ecstasy, eczema |
| | cc | account |
| | cch | saccharin |
| | ch | alchemy, chorus, school |
| | ck | acknowledge, kick, pack |
| | cqu | lacquer |
| | cu | biscuit, circuit |
| | k | kind |
| | lk | walk |
| | qu | liquor, quay |
| | que | saque, torque |
| (kw) | ch | choir |
| | qu | quick |
| | cqu | acquit |
| (l) | l | lid |
| | le | lisle |
| | ll | llama, tall |
| | lle | faille |
| | lh | Lhasa |

| Pronunciation Symbol | Spelling | Example |
|---|---|---|
| (m) | chm | drachm |
| | gm | paradigm, phlegm |
| | lm | calm |
| | m | mum |
| | mb | comb, plumb |
| | mm | grammar, mummy |
| | mn | autumn, condemn, hymn |
| (n) | gn | align, gnat, gnome |
| | kn | knee, knife, know |
| | mn | mnemonic |
| | n | no |
| | nn | banner, bonny, inn |
| | pn | pneumatic |
| (ng) | n | anchor, congruent, ink, uncle |
| | ng | sing |
| | ngue | tongue |
| (ŏ) | a | water, what |
| | ho | honest |
| | o | pot |
| | ou | trough |
| (ō) | au | hautboy, mauve |
| | eau | beau, bureau, trousseau |
| | eo | yeoman |
| | ew | sew |
| | o | no |
| | oa | broach, croak, loan |
| | oe | foe |
| | oh | oh, ohm |
| | oo | brooch |
| | ou | boulder, shoulder |
| | ough | borough, though |
| | ow | low |
| | owe | owe, Marlowe |
| (ô) | a | all, water |
| | al | talk |
| | ah | Utah |
| | ar | warm |
| | as | Arkansas |
| | au | daughter, haunt |
| | aw | awe, awning, brawl, paw |
| | o | for, order |
| | oa | broad |
| | ough | thought, wrought |
| (oi) | oi | noise, voice |
| | oy | boy, royal |
| (ou) | au | sauerkraut |
| | aue | sauerbraten |
| | hou | hour |
| | ou | out |
| | ough | bough |
| | ow | now, scowl |
| (o͝o) | o | wolf, woman |
| | oo | took |
| | ou | should |
| | u | bush, full |

| Pronunciation Symbol | Spelling | Example |
|---|---|---|
| (o͞o) | eu | maneuver, rheumatic |
| | ew | drew |
| | ieu | lieutenant |
| | o | ado, move |
| | oe | canoe |
| | oo | boot |
| | ou | croup, troupe |
| | ough | through |
| | u | prudent, rule |
| | ue | ague, blue, sue |
| | ui | fruit, juice |
| (p) | p | pop |
| | pp | happen, snapper |
| (r) | r | barter, run |
| | rh | rhapsody, rheumatism, rhythm |
| | rr | cherry, marriage, porridge |
| | rrh | cirrhosis |
| | wr | wrap, wrench, write, wrong |
| (s) | c | cent, cite, cyst |
| | ce | mace, practice, sauce |
| | ps | psalm, pseudonym, psychology |
| | s | say |
| | sc | abscess, fascinate, scent |
| | sch | schism |
| | ss | mass, sassafras |
| | sth | isthmus |
| (sh) | ce | ocean |
| | ch | chandelier, marchioness, sachet |
| | ci | deficient, musician, special |
| | psh | pshaw |
| | s | sure |
| | sc | prescient |
| | sch | schist |
| | se | nauseous |
| | sh | ship |
| | si | mansion |
| | ss | fissure, mission |
| | ti | nation, partial |
| | xi | anxious |
| (t) | bt | debt |
| | ct | ctenophore |
| | ed | crashed, walked, stopped |
| | ght | bought, caught, wrought |
| | pt | ptarmigan, receipt |
| | t | take |
| | th | Thomas, thyme |
| | tt | better, mutton, sitter |
| | tw | two |

| Pronunciation Symbol | Spelling | Example |
|---|---|---|
| (th) | phth | phthisis |
| | th | thin |
| (*th*) | th | feather, this |
| (ŭ) | o | income, some, son |
| | oe | does |
| | oo | blood, flood |
| | ou | couple, doublet, trouble |
| | u | but |
| (yo͞o) | eau | beauty |
| | eu | eulogy, feud |
| | eue | queue |
| | ew | few, pewter |
| | ieu | adieu |
| | iew | view |
| | u | puce, use |
| | ue | cue, puerile |
| | you | you |
| | yu | yule |
| (ûr) | ear | pearl, yearn |
| | er | certain, hermit, stern |
| | eur | restaurateur |
| | ir | bird, first |
| | or | work |
| | our | journey, scourge |
| | yr | myrtle |
| | yrrh | myrrh |
| (v) | f | of |
| | ph | Stephen |
| | v | vain |
| | ve | save |
| (w) | o | one |
| | ou | Ouagadougou |
| | u | guanine, guano, guava |
| | w | with |
| (y) | i | minion, onion, opinion |
| | j | hallelujah |
| | y | yes |
| (z) | cz | czar |
| | s | hers, rise, yours |
| | se | cruise |
| | ss | dessert, hussar |
| | x | anxiety, xerography, xylophone |
| | z | zebra |
| | zz | buzz, fuzz |
| (zh) | ge | barrage |
| | s | measure, division |

*Note:* The letter *x* spells six sounds in English: ks, as in box, exit; gz, as in exact, exist; sh, as in anxious; gzh, as in luxurious, luxury; ksh (a variant of gzh), also as in luxurious, luxury; and z, as in anxiety, xerography.

## Part-of-speech Labels

The following italic part-of-speech labels appear after the pronunciation of an entry:

*adj.* adjective
*adv.* adverb
*aux. v.* auxiliary verb
*conj.* conjunction
*def. art.* definite article
*indef. art.* indefinite article
*interj.* interjection
*n.* noun
*prep.* preposition
*pron.* pronoun
*v.* verb

And these labels indicate inflected forms:

*compar.* comparative
*superl.* superlative

*p.t.* past tense
*p.p.* past participle

*pl.* plural
*sing.* singular

Words that occur only in the plural are labeled *pl.n.* Examples of these are *clothes* and *burnsides.*

The following italic labels classify verbs:

*vt.* transitive
*vi.* intransitive

The labels for word elements are:

*pref.* prefix
*suff.* suffix

Some entries are unlabeled; these include acronyms, contractions, symbols, trademarks, and certain word formatives that occur only in medial position:

**aren't** (ärnt, är'ənt). Are not.

**Day-Glo** (dā'glō'). A trademark for fluorescent materials.

**-'ll.** Shall : will <We*'ll* arrive later.>

## Inflected Forms

Inflected forms differ from main-entry forms by the addition of suffixes or by changes in the base form. The following inflected forms are given in the main entries in this Dictionary: (1) principal parts of all irregular and regular verbs, (2) all comparative and superlative degrees of adjectives and adverbs formed by inflection, and (3) irregular plurals of all nouns.

Inflected forms, syllabicated and set in boldface type, are given a pronunciation when necessary. When more than one inflected form is given, the forms are separated by commas. Inflected forms are usually cut back to the last syllable of the original entry word plus the inflected ending. Irregular inflections are written out to the extent that clarity requires. When inflected forms are cut back, the shortened forms are preceded by boldface hyphens:

**de·brief** (dē-brēf') *vt.* **-briefed, -brief·ing, -briefs.**

A single letter at the beginning or end of an entry word never stands alone, nor is it ever dropped in inflections:

**a·base** (ə-bās') *vt.* **a·based, a·bas·ing, a·bas·es.**

**Inflected forms: verbs.** The principal parts of verbs are entered in this order: past tense, past participle, present participle, and third person singular, present tense. When the past tense and past participle are identical, one form represents both:

**di·vulge** (dĭ-vŭlj') *vt.* **-vulged, -vulg·ing, -vulg·es.**

**do**[1] (do͞o) *v.* **did** (dĭd), **done** (dŭn), **do·ing, does** (dŭz)

**Inflected forms: adjectives and adverbs.** Adjectives and adverbs forming the comparative and superlative degrees by addition of *-er* and *-est* to unchanged entry words show these suffixes after their respective part-of-speech labels:

**blind** (blīnd) *adj.* **-er, -est.**

Irregular adjectives and adverbs are those whose forms change upon addition of *-er* and *-est.* These forms follow the general rules of style and presentation for all inflections:

**hot** (hŏt) *adj.* **hot·ter, hot·test.**

**wack·y** (wăk'ē) *also* **whack·y** (hwăk'ē, wăk'ē) *adj.* **-i·er, -i·est.**

**bad**[1] (băd) *adj.* **worse** (wûrs), **worst** (wûrst)

Remember that the existence of *-er* and *-est* forms does not preclude use of *more* and *most* with a simple adjective or adverb to express the comparative and superlative degrees.

**Inflected forms: nouns.** Plurals of nouns other than those formed by the simple addition of *-s* or *-es* are shown in this Dictionary and are labeled *pl.* A regular plural is also shown for a noun having both a regular and an irregular plural form as variants; both forms appear in such a case, and the convention for indicating primary and secondary variants applies to plurals as well:

**am·pul·la** (ăm-po͝ol'ə, -pŭl'ə) *n., pl.* **-pul·lae** (-po͝ol'ē, -pŭl'ē)

**am·pho·ra** (ăm'fər-ə) *n., pl.* **-pho·rae** (-fə-rē') *or* **-pho·ras.**

Regular plurals of nouns are also shown when their spelling might pose a problem for the user. Such nouns include those ending in *-o* and *-ey:*

**to·ma·to** (tə-mā'tō, -mä'-) *n., pl.* **-toes.**

**mon·ey** (mŭn'ē) *n., pl.* **mon·eys** *or* **mon·ies.**

A noun that is chiefly or exclusively plural both in form and in meaning is labeled *pl.n.*:

**clothes** (klōz, klōthz) *pl.n.* [ME < OE *clāðas* < *clāð,* cloth.] **1.** Wearing apparel : GARMENTS. **2.** Bedclothes.

A noun that is always plural in form but that does not always take a plural verb is entered as follows:

**pol·i·tics** (pŏl'ĭ-tĭks) *n.* **1.** *(sing. in number).* The art or science of government : POLITICAL SCIENCE. **2.** *(sing. in number).* The activities or affairs of a government, politician, or political party. **3.** *(sing. in number).* **a.** Conduct of or participation in political affairs, often professionally. **b.** The business, activities, or profession of one so involved. **4.** *(sing. in number).* The methods or tactics involved in managing a government or state. **5.** *(pl. in number).* Intrigue or maneuvering within a group <company *politics*>

**aer·o·bics** (â-rō'bĭks) *n.* [< AEROBIC.] *(sing. or pl. in number).* **1.** Conditioning of the cardiopulmonary system by means of vigorous exercise that seeks to increase efficiency of oxygen intake. **2.** Aerobic exercises, as running or calisthenics.

**cal·is·then·ics** (kăl′ĭs-thĕn′ĭks) *n.* [Gk. *kalli-* beautiful (< *kallos,* beauty) + *sthenos,* strength.] **1.** *(sing. or pl. in number).* Gymnastic exercises designed to develop muscular tone and promote physical well-being. **2.** *(sing. in number).* The practice of calisthenics.

**Inflected forms: separate entries.** Certain English verbs such as *be, do,* and *have,* have archaic inflected forms such as *art, dost,* and *hadst,* whose frequency of occurrence justifies their inclusion in this Dictionary. Such forms are entered separately at their own alphabetical places in the vocabulary:

**dost** (dŭst) *v. Archaic. 2nd person sing. present tense of* DO[1].

Also entered separately are the inflected forms of irregular verbs in which a vowel change indicates the change of tense. Such entries carry part-of-speech labels, and are syllabicated and pronounced:

**be·gan** (bĭ-găn′) *v. p.t. of* BEGIN.

Irregular plurals are entered separately in the Dictionary when such forms occur more than one entry away from the main entry in alphabetical order:

**a·phi·des** (ā′fĭ-dēz′, ăf′ĭ-) *n. pl. of* APHIS.

## Etymologies and Word History Paragraphs

**Etymologies.** The vocabulary of English is drawn from many different languages, and many words have had a long and eventful history. An *etymology* traces the history of a word from its most recent form to its earliest known source. In this Dictionary the etymology is enclosed in square brackets [ ] and is placed immediately before the definitions:

**bail·iff** (bā′lĭf) *n.* [ME *baillif* < OFr. < Med. Lat. *bajulivus* < Lat. *bajulus,* carrier.] **1.** A minor court officer whose duties include maintaining order in a courtroom during a trial. **2.** *Chiefly Brit.* An official who assists a sheriff and is empowered to execute writs, processes, and arrests. **3.** *Chiefly Brit.* An estate overseer : STEWARD.

The most recent stage before Modern English is given first, followed by the next earlier stage, until the earliest known stage is reached. The stages of the derivation are separated by the symbol < which means "borrowed from" or "derived from." For each stage a language name, linguistic form (appearing in *italic* type), and brief definition are given. In order to avoid repeating information, the language, form, or definition is not given for a particular stage if it is the same as the corresponding language, form, or definition of the immediately preceding stage. When an item is missing in the first stage of the etymology, it is presumed to be the same as that of the main entry word as it is defined in the first definition. If these principles are applied to the example above, the etymology shows that the word **bailiff** is first found in Middle English (ME) with the spelling *baillif* and the meaning "a minor court officer. . . ." This form was borrowed from Old French (OFr.), in which it had the same spelling and meaning. The immediate source of the Old French word was the Medieval Latin (Med. Lat.) *bajulivus,* still meaning "a minor court officer." This in turn was a development of the Latin (Lat.) noun *bajulus,* which meant simply "carrier." By these conventions a great deal of information can be fitted into a small space.

Sometimes a stage in the history of a word is inferred to have existed, but there is no written evidence of it. In such a case the assumed, but unrecorded, form is preceded by an asterisk (*):

**bal·ance** (băl′əns) *n.* [ME *balaunce* < OFr. < VLat. **bilancia* < Lat. *bilanx* : *bi-,* two + *lanx,* scale.] **1.** A weighing device composed of a rigid beam horizontally suspended by a low-friction support at its center, with identical weighing pans hung at either end, one of which holds an unknown weight while the effective weight in the other is increased by known amounts until the beam is level and motionless. **2.** Equilibrium or parity marked by cancellation of all forces by equal opposing forces. **3.** The means or power to decide. **4.** A state of bodily equilibrium. **5.** Emotional stability. **6.** A harmonious or satisfying arrangement of parts or elements.

Definitions are not given for such unrecorded forms.

The etymology of **balance** also shows the treatment of a compound word. The compound form (Lat. *bilanx*) is followed by a colon (:) that introduces the components making up the compound word. These components (*bi-, lanx*) are joined by a plus sign (+), and they belong to the same language as the compound word unless it is explicitly stated otherwise. Each component is given a definition and can be given a further etymology of its own:

**em·brace** (ĕm-brās′) *v.* **-braced, -brac·ing, -brac·es.** [ME *embracen* < OFr. *embracer* : *en-,* in (< Lat. *in-*) + *brace,* arms < Lat. *bracchium,* arm < Gk. *brakhiōn.*] *—vt.* **1.** To hold to one with the arms, usu. as a show of affection : HUG. **2. a.** To encircle. **b.** To twine around. **3.** To include, comprise, or contain : ENCOMPASS. **4.** To take up willingly or eagerly <*embrace* the fight for justice> **5.** To avail oneself of <*embrace* the chance> *—vi.* To join in an embrace. *—n.* **1.** An act of embracing : HUG. **2.** An enclosure. **3.** Eager acceptance.

**rec·om·mend** (rĕk′ə-mĕnd′) *vt.* **-mend·ed, -mend·ing, -mends.** [ME *recommenden* < Med. Lat. *recommendare* : Lat. *re-* (intensive) + Lat. *commendare,* to entrust (*com-,* together + *mandare,* to order).] **1.** To praise or commend to another as being desirable or worthy : ENDORSE <*recommended* you for the position> **2.** To make attractive or acceptable <Diligence *recommends* any person.> **3.** To commit to the charge of another : ENTRUST. **4.** To counsel or advise (that something be done). **—rec′om·mend′a·ble** *adj.* **—rec′om·mend′er** *n.*

The same conventions of style apply to these more complex derivations as to the simpler ones. A language name is not repeated if a form belongs to the same language as the form preceding. No language is specified for the components of a compound word if they are from the same language as the compound itself. In the etymology of **embrace,** *en-* and *brace* are Old French like *embracer,* but in the etymology of **recommend,** *re-* and *commendare* are Latin forms used to make the Medieval Latin compound *recommendare.* The further history of Old French *en-* and *brace* and Latin *re-* and *commendare* is also given. Latin *commendare* is itself a compound word. Its components are not introduced by a colon, but are enclosed within parentheses and joined by a plus sign.

In order to avoid repeating lengthy or complex derivations at one etymology it is often necessary to refer to another etymology for fuller information:

**heft** (hĕft) *n.* [Alteration of HEAVE.] Heaviness : weight. *—v.* **heft·ed, heft·ing, hefts.** *—vt.* **1.** To lift in order to determine the weight of. **2.** To hoist up : HEAVE. *—vi.* To weigh.

**hack·but** (hăk′bŭt′) *also* **hag·but** (hăg′-) *n.* [OFr. *haquebute* < MLG *hakebusse.* —see HARQUEBUS.] A harquebus. **—hack′but·eer′** (-bə-tîr′), **hack′but′ter** *n.*

**cor·ti·cos·ter·one** (kôr′tĭ-kŏs′tə-rōn′) *n.* [CORTICO- + STER(OL) + -ONE.] A corticoid, $C_{21}H_{30}O_4$, that induces hyperglycemia and deposition of glycogen in the liver.

Small capital letters are used to indicate the entries where a continuation of the etymologies of the words or, as in the case of **corticosterone,** their components can be found. Portions of words not used in a derivation, such as the last two letters of STEROL, are enclosed within parentheses.

Some words are not given etymologies. These include trademarks, interjections, words derived from geographic entries or from names of persons mentioned in the definition, and ethnic names that are the English equivalents of a group's own name for itself.

Words that are formed solely from other words, prefixes, and suffixes that are main entries in the Dictionary are also not given etymologies. Such words are considered to be combinations of easily recognizable components, and no explicit statement of their origins needs to be made. Examples are **nerve cell** (NERVE + CELL) and **inconsolable** (IN-[1] + CONSOLE[1] + -ABLE).

All the symbols and abbreviations used in the etymologies are listed on a separate page in the front of this Dictionary, and all language names are defined. Mandarin Chinese words are given in the Pinyin system of transliteration. Transliterations of Greek, Hebrew, Arabic, and Russian letters are given in the Alphabet Table.

**Word histories.** Word history paragraphs have been provided in addition to the etymologies for certain words whose derivations are especially interesting. Many of these words have been chosen because their development illustrates common and important linguistic processes and, it is hoped, an understanding of their development will provide a better appreciation of language as a living and ever-changing human phenomenon. An example of a word history paragraph is seen at the entry **bus:**

**bus** (bŭs) *n., pl.* **bus·es** *or* **bus·ses.** [Short for OMNIBUS.] **1.** A long motor vehicle for carrying passengers. **2.** *Informal.* An automobile. **3.** *Elect.* A bus bar. *—v.* **bused, bus·ing, bus·es** *or* **bussed, bus·sing, bus·ses.** *—vt.* **1.** To transport in a bus. **2.** To transport (schoolchildren) to achieve racial integration. *—vi.* **1.** To travel in a bus. **2.** To work as a bus boy.

▲ word history: *Bus,* as the etymology above shows, is short for *omnibus,* which was adopted into English from the French phrase *voiture omnibus,* a macaronic expression, part French, part Latin, meaning "vehicle for all." It appeared in French around 1830 and was both borrowed and shortened by the English before the end of the decade. *Omnibus* was also used to designate a waiter's assistant at a restaurant, though whether this usage derives from the use of *omnibus* to mean "vehicle" or if it was an independent use of the word is not clear. *Bus boy* is derived from the shortening of *omnibus* in this second sense. *Bus* meaning "to work as a bus boy" comes from bus boy and not directly from *omnibus.*

## Labels

The Dictionary has numerous subject and stylistic labels that provide the reader with specific topic or field orientation as well as guidance regarding various levels of English usage. The following paragraphs list and discuss these labels.

**Subject labels.** A subject label identifies the specific area to which a given definition applies. But such a label does not restrict any definition to a particular subject or field; it merely indicates that a definition has its primary application in the domain specified.

**disaster dump** *n. Computer Sci.* A printout that occurs as a result of a nonrecoverable program error.

**ax·il·lar·y** (ăk′sə-lĕr′ē) *adj.* **1.** *Anat.* Of, relating to, or near the axilla. **2.** *Bot.* Of, relating to, or situated in an axil.

**Stylistic and geographic labels.** Stylistic and geographic labels are restrictive: they indicate that an entry or a definition is limited to a particular level or style of usage or to a specific region. All senses of a term that are not restricted by such a label are appropriate for use in all contexts.

***Nonstandard.*** *Nonstandard,* the most restrictive label in this Dictionary, indicates that the word or sense to which it is applied is unacceptable to a broad group of educated native English speakers and writers:

**ir·re·gard·less** (ĭr′ĭ-gärd′lĭs) *adv. Nonstandard.* Regardless.

***Informal.*** Among educated native speakers and writers there are at least two levels of language—the language of formal discourse and that of conversation. While most words are equally appropriate in both contexts, many of them are more likely to occur in conversation than in formal prose. An example of such a term is **jim-dandy,** labeled *Informal:*

**jim-dan·dy** (jĭm′dăn′dē) [*Jim,* nickname for *James* + DANDY.] *Informal. n.* One that is very pleasing or excellent of its kind. *—adj.* Admirable : excellent.

Informal terms may, of course, appear in otherwise formal writing as a way of creating a particular effect.

***Slang.*** This label indicates a style of language rather than a level of formality or cultivation. The characteristic feature of slang is the user's attempt to achieve rhetorical effect by means of extravagant and often facetious figures of speech. Some forms of slang occur in the most cultivated speech but not in discourse intended to be formal. An example of a slang term is the verb *deep-six:*

**deep-six** (dēp′sĭks′) *vt.* **-sixed, -six·ing, -six·es.** *Slang.* **1.** To toss overboard. **2.** To toss out : get rid of <*deep-sixed* the leftovers>

***Obsolete.*** The label *Obs.* signals that a term or a sense is no longer in active use. That the object or situation to which a word refers may no longer exist or be in use does not make the word itself obsolete:

**brave** (brāv) . . . . . . . . . . . . . . . . . . *—n.* **1.** A North American Indian warrior. **2.** A courageous person. **3.** *Obs.* A bully. **4.** *Obs.* A boast or challenge.

Senses **3.** and **4.** of the noun **brave** have disappeared from current English use; hence they are labeled *Obs.* But although a caravel is no longer used on the high seas, the term itself is still used, especially in historical contexts; hence, it is not labeled:

**car·a·vel** *also* **car·a·velle** (kăr′ə-vĕl′) *or* **car·vel** (kär′vəl, -vĕl′) *n.* [OFr. *caravelle* < OPort. *caravela.*] A small light sailing ship used by the Spanish and Portuguese in the 15th and 16th cent.

***Archaic.*** This label is appended to words and senses that were once common but are now rare, although they occur in the literature of an earlier period:

**brede** (brēd) *n.* [Alteration of BRAID.] *Archaic.* An ornamental embroidered edging.

***Regional.*** When a term or expression is usually used in one area and little used in others, it carries an area label such as *Southern U.S.* or *New England.* When a word is common to several areas and yet is not in widespread use, it is labeled *Regional.*

**†corn pone** (pōn) *n. Southern U.S.* Corn bread made without milk or eggs.

**†ar·gu·fy** (är'gyə-fī') *v.* **-fied, -fy·ing, -fies.** *Regional.* —*vt.* To argue over. —*vi.* To argue stubbornly : WRANGLE. **—ar'gu·fi'er** *n.*

***Chiefly Brit.*** The label *Chiefly Brit.* reflects the fact that the distinction between British and American English is seldom exclusive. Furthermore, British terms are often used elsewhere in the world, as in Australia and New Zealand. An example of a term labeled *Chiefly Brit.* is:

**bubble and squeak** *n.* [Imit. of the sounds made as it cooks.] *Chiefly Brit.* Cabbage and potatoes fried together.

Other labels indicating that the occurrence of a particular term is confined to specific areas of the English-speaking community are:

| | |
|---|---|
| *Austral.* | Australian |
| *Ir.* | Irish |
| *Scot.* | Scots |
| *So. Afr.* | South African |

## Cross-references

Cross-references expand the information given at an entry and at the same time avoid unnecessary duplication.

**Synonymous cross-references.** When two terms are synonymous, a full definition appears at the primary, or most frequently occurring term; the secondary term is entered at its own place in the alphabet and is defined by the primary term, set in small capital letters and followed by the appropriate sense number indicating the sense of the main entry at which a full definition can be found. For instance, **afterword** is a cross-reference entry referring the reader to sense **2.** of **epilogue:**

**af·ter·word** (ăf'tər-wûrd') *n.* EPILOGUE 2.

And at sense **2.** of **epilogue,** the reader finds the full definition:

**ep·i·logue** *also* **ep·i·log** (ĕp'ə-lôg', -lŏg') *n.* [ME *epiloge* < OFr. *epilogue* < Lat. *epilogus* < Gk. *epilogos* < *epilegein,* to say more : *epi-,* in addition to + *legein,* to say.] **1. a.** A short speech spoken directly to the audience after the end of a play. **b.** The performer speaking such an epilogue. **2.** A short addition or concluding section at the end of a literary work, often discussing the future of its characters.

The entry **black diamond** contains two such cross-references, one referring the user to **carbonado**[2], and the other—a plural sense—referring the user to sense **1.a.** of **coal** for more complete information:

**black diamond** *n.* **1.** CARBONADO[2]. **2. black diamonds.** COAL 1a.

When the user turns to **carbonado**[2] and **coal 1.a.** the following information is available:

**car·bo·na·do**[2] (kär'bə-nā'dō, -nä'-) *n., pl.* **-dos.** [Port. < *carbone,* carbon < Fr. —see CARBON.] A form of chiefly Brazilian opaque or dark-colored diamond, used for drills.

**coal** (kōl) *n.* [ME *col* < OE.] **1. a.** A natural dark-brown to black solid used as fuel, formed from fossilized plants and consisting of amorphous carbon with various organic and some inorganic compounds.

**Variant cross-references.** The second kind of cross-reference is used to enter variant forms that appear more than one entry away from their primary forms:

**boulle** (bōōl) *n. var. of* BUHL.

A variant that applies to a specific sense of another entry is also cross-referenced to it:

**cas·si·no** (kə-sē'nō) *n. var. of* CASINO 3.

**ca·si·no** (kə-sē'nō) *n., pl.* **-nos.** [Ital. < *casa,* house < Lat., hut.] **1.** An Italian summer or country house. **2.** A public room or house esp. for gambling. **3.** *also* **cas·si·no.** A card game for two to four players in which cards on the table are matched by cards in the hand.

**Directional cross-references to tabular data.** The third kind of cross-reference directs the user from entry words to tabular data located elsewhere in the Dictionary. Such cross-references are used at entries for books of the Bible, letters of various alphabets, monetary units, and the months of three principal calendars:

**chron·i·cle** (krŏn'ĭ-kəl) *n.* . . . . . . . . **2. Chronicles** (*sing. in number*). —See table at BIBLE.

**be·ta** (bā'tə, bē'-) *n.* [Gk. *bēta,* of Phoenician orig.; akin to Heb. *bēth.*] **1.** The second letter of the Greek alphabet. —See table at ALPHABET.

**af·ghan·i** (ăf-găn'ē, -gä'nē) *n.* [Pashto.] —See table at CURRENCY.

**Av** (ôv, äb) *also* **Ab** (äb, äv, ôv) *n.* [Heb. *ābh* < Akkadian *abu.*] The 11th month of the Hebrew year. —See table at CALENDAR.

## Sense Order

When an entry word has more than one sense, the senses are arranged so that the word and its meanings can, to some extent, be perceived as a structured unit. Senses are arranged neither historically nor by frequency of occurrence. Instead, they are ordered according to central meaning clusters from which related subsenses and additional senses may evolve.

## Sense Division

Multisense entries have their meanings numbered sequentially in boldface:

**blear** (blîr) *vt.* **bleared, blear·ing, blears.** [ME *bleren.*] **1.** To blur (the eyes) with or as if with tears. **2.** To blur : dim. **—blear** *adj.*

When a numbered sense is composed of two or more closely related subsenses, the subsenses are introduced by the boldface letters **a., b., c.,** and so on:

**off·shore** (ôf'shôr', -shōr', ŏf'-) *adj.* **1.** Moving or directed away from the shore <an *offshore* current> **2. a.** Located or occurring at a distance from the shore <an *offshore* oil rig> **b.** Located or based outside the United States and not subject to U.S. tax laws <*offshore* corporations>

In a combined entry (i.e., an entry comprising more than one part of speech), the senses are numbered in separate sequences that begin with **1.** after each part-of-speech label:

**jam**[1] (jăm) *v.* **jammed, jam·ming, jams.** [Orig. unknown.] *—vt.* **1.** To drive or wedge forcibly into a tight position <*jammed* the lid on the box> **2.** To activate or apply suddenly, as automotive brakes. **3.** To cause to lock in an inoperable position <*jam* the typewriter keys> **4.** To fill to excess : pack tight <The fans *jammed* the arena.> **5.** To block or clog <The drain was *jammed* by debris.> **6.** To crush or bruise <*jam* a finger in the door> **7.** *Electron.* To interfere with or prevent the clear reception of (signals) by electronic means. *—vi.* **1.** To become wedged : STICK. **2.** To become inoperable because of jammed parts. **3.** To force into or through a tight space. **4.** *Mus.* To play jazz improvisations. *—n.* **1.** The act of jamming or state of being jammed. **2.** A crush or congestion in a limited space <a traffic *jam*> **3.** *Informal.* A difficult situation : PREDICAMENT.

**Boldface colon.** A boldface colon is used in this Dictionary to separate two or more definitional elements within a single sense or subsense:

**else** (ĕls) *adj.* [ME *elles* < OE.] **1.** Different **:** other <anybody *else*> **2.** More **:** additional <Do they require anything *else?*> *—adv.* **1.** In a different time,,place, or manner **:** DIFFERENTLY <How *else* could it be cooked?> **2.** If not **:** OTHERWISE <Watch out, or *else* you will make a big mistake.>

## Explanatory Phrases

Explanatory phrases introduced by dashes are sometimes used in certain entries such as those for function words, interjections, or intensives:

**on** (ŏn, ôn) *prep.* [ME < OE.] **1.** —Used to indicate: **a.** Position above and in contact with <The vase is *on* the bureau.> **b.** Contact with a surface, regardless of position <a painting *on* the wall> **c.** Location at or along <a cottage *on* the river> **d.** Proximity <a city *on* the Polish border> **e.** Attachment to or suspension from <pearls *on* a string> **2.** —Used to indicate: **a.** Movement or direction toward a position <threw the books *on* the ground> **b.** Movement toward, against, or onto <jump *on* the platform>

**gosh** (gŏsh) *interj.* [Alteration of GOD.] —Used to express mild surprise or delight.

**damned** (dămd) *adj.* **-er, -est. 1.** Condemned, esp. to eternal punishment. **2.** *Informal.* Deserving condemnation : DETESTABLE <this *damned* mud> **3.** —Used as an intensive <a *damned* idiot> *—adv. Informal.* Very <a *damned* poor manager>

## Verbal Illustrations

Thousands of verbal illustrations—many of them quotations—follow definitions. These illustrations, enclosed by angle brackets, show how a word is used in actual contexts, and are particularly helpful in illustrating figurative senses, transitive and intransitive verbs, and precise shades of meaning. Examples:

**bottom line** *n.* **1.** The lowest line in a financial statement, showing net income or loss. **2.** The final result or statement : UPSHOT <"the *bottom line*—for now—is that the city is heading toward default" *—New York*> **3.** The main or essential point.

**fug** (fŭg) *n.* [Orig. unknown.] A malodorous emanation <"In spite of the open windows the stench had become a reeking *fug*" —Colleen McCullough>

**fine-tune** (fīn'tyo͞on', -to͞on') *vt.* **-tuned, -tun·ing, -tunes.** To make precise, minute adjustments in <"advertising agencies kept *fine-tuning* the coolly calculated machinery of merchandising and hype" *—New Yorker*> **—fine'-tun'er** *n.*

## Variants

All variants in this Dictionary are acceptable in any context unless they are marked with a restrictive label such as *Nonstandard* or *Regional.* Variants are set in boldface type and are of two kinds—equal and secondary.

**Equal variants.** The italic word *or* between a main-entry term and a variant form indicates that the two forms are in almost equally frequent use:

**pole·ax** *or* **pole·axe** (pōl'ăks') [ME *pollax* : *poll,* head (< MLG *polle*) + *ax,* ax < OE *æxa.*] *—n.* **1.** A medieval battle-ax having an ax, or an ax, hammer, and pick combination, with a long shaft. **2.** An ax having a hammer face opposite the blade, used to slaughter cattle.

**Secondary variants.** When there is a distinct preference for one variant, the less preferred variant is introduced by the italic word *also*:

**am·bro·sial** (ăm-brō'zhəl, -zhē-əl) *also* **am·bro·sian** (-zhən, -zhē-ən) *adj.* **1.** Fragrant or delicious. **2.** Of or worthy of the gods.

**British variants.** A large group of terms have variants consisting of spellings preferred in British English and sometimes used in American English. Variants such as *litre* and *flavour* are labeled *Chiefly Brit.* Compounds or derivational forms of such terms do not repeat the variant there. For instance, the *Chiefly Brit.* **flavour** is entered as a variant for **flavor,** but **flavourful** and **flavouring** are not entered as variants of **flavorful** and **flavoring.**

## Phrasal Verbs and Idiomatic Expressions

**Phrasal verbs.** A phrasal verb is an expression consisting of a verb and an adverb or preposition with a unitary meaning equal to more than the sum of the separate meanings of its individual elements. Phrasal verbs, entered in boldface and introduced by dashes, are defined as subentries following the last verb definition of a main entry as shown in this example:

**lay**[1] (lā) *v.* **laid** (lād), **lay·ing, lays.** [ME *laien* < OE *lecgan.*] *—vt.* **1.** To cause to lie down <*lay* the baby in its bassinet> **2. a.** To bring to or place in a particular state or position. **b.** To bury. **3.** To put or set down : DEPOSIT. **4.** To produce and deposit <*lay* eggs> **5.** To cause to subside. **6.** To put up to or against <*lay* a finger to the nose> **7.** To put forward as a reproach or accusation <*laid* the blame on me> **8.** To put in order or readiness for use <*lay* places for six at dinner> **9.** To devise : make <*lay* plans> **10.** To spread over a surface <*lay* rail ties on a cinder bed> **11.** To place or give (importance) <*lay* emphasis on deportment> **12.** To impose as a burden or punishment <*lay* new taxes on the people> **13.** To present for examination <*lay* a case before a court> **14.** To put forward as a demand or assertion <*laid* claim to the treasure> **15.** To place (a bet) **:** WAGER. **16.** To aim (a gun or cannon). **17. a.** To place together (strands) to be twisted into rope. **b.** To make in this manner <*lay* up hawsers> *—vi.* **1.** To produce and deposit eggs. **2.** To bet **:** wager. **3.** *Nonstandard.* LIE[1] 1. **4.** To engage energetically in an action. **—lay aside. 1.** To give up **:** ABANDON <*lay aside* hope of success> **2.** To put aside for the future **:** SAVE. **—lay away. 1.** To reserve for the future **:** SAVE. **2.** To put aside and hold for future delivery. **—lay by.** To save. **—lay down. 1.** To store for the future. **2.** To specify as a guide or rule. **—lay in.** To store for future use. **—lay into.** To scold harshly. **—lay off. 1.** To terminate the employment of (a worker), esp. temporarily. **2.** To mark off. **3.** To give up **:** QUIT. **4.** To refrain from criticizing or annoying. **5.** To stop : cease. **—lay out. 1.** To make a detailed plan for. **2.** *Informal.* To speak. **3.** To prepare and clothe (a corpse) for burial. **4.** *Informal.* To knock to the ground or render unconscious by a blow. **—lay over.** To make a stopover in the course of a trip. **—lay to.** *Naut.* **1.** To bring (a ship) to a stop in open water. **2.** To remain stationary and face into the wind. **—lay up. 1.** To stock for future use. **2.** *Informal.* To confine with an injury or illness. **3.** To put (a ship) in dock, as for repairs. *—n.* **1. a.** The direction the strands of a rope or cable are twisted in <a left *lay*> **b.** The amount of such twist. **2.** The state of one that lays eggs <a hen coming into *lay*> **3. a.** *Chiefly Brit.* A line of activity. **b.** An occupation. **—lay down the law.** To assert positively and often arrogantly or harshly. **—lay it on thick. 1.** To exaggerate **:** overstate. **2.** To flatter effusively. **—lay of the land.** Nature, arrangement, or disposition.

The terms **lay aside, lay away, lay by, lay down, lay in, lay into, lay off, lay out, lay over, lay to,** and **lay up** are phrasal verbs.

**Idiomatic expressions.** Many entry words occur in phrases, the meanings of which cannot be derived from the meanings of the individual words making up the phrases. Such phrases, or idioms, introduced by boldface dashes, are entered and defined in alphabetical order within the entry for the most significant word element in the idiom, as at the end of the entry **lay**[1], where **lay down the law, lay it on thick,** and **lay of the land** are idioms.

The indefinite pronouns *one* and *someone,* when used parenthetically in idioms, indicate that the object or the possessive pronoun used with a particular idiom may vary according to context. The choice of *one* for the entry form means that the person referred to in the idiom can be identical to its subject; the pronoun *someone* means that the person referred to in the idiom cannot be identical to the subject. Thus, at the entry **breath,** the idiom **under (one's) breath** appears and its form of entry indicates that the sentence *He complained under his breath* is grammatical. At **pull,** however, the idiom **pull (someone's) leg** (i.e., to tease someone) indicates that the sentence *I pulled my leg* would be incorrect.

## Undefined Forms

**Undefined run-on entries.** A word formed from the main-entry word by the addition of a suffix is often found at the end of an entry. These run-on entries are related to the main-entry words at which they appear, but they have different grammatical functions as indicated by their italic part-of-speech labels. Run-on entries, set in boldface type and introduced by dashes, are syllabicated. Stress is indicated for all such undefined forms of more than one syllable, and pronunciation is indicated as required. For example, at the main entry **orchestra,** the undefined run-ons are:

**—or·ches'tral** (ôr-kĕs'trəl) *adj.* **—or·ches'tral·ly** *adv.*

When different run-on forms have the same grammatical function and the same meaning, they are separated by a comma and share a single part-of-speech label. For example, at **catarrh,** these adjectives have the same function and meaning and are run on together:

**—ca·tarrh'al, ca·tarrh'ous** *adj.*

In other instances, undefined run-ons have the same grammatical function but different meanings. These terms are run on separately as shown at the entry **catechize**:

**cat·e·chize** (kăt'ə-kīz') *vt.* **-chized, -chiz·ing, -chiz·es.** [ME *catecizen* < Med. Lat. *catechizare* < LGk. *katēkhizein* < Gk. *katēkhein,* to teach by word of mouth : *kata-,* according to + *ekhein,* to sound < *ekhē,* sound.] **1.** To teach esp. the principles of a religious creed by questions and answers. **2.** To question persistently or searchingly. **—cat'e·chi·za'tion** *n.* **—cat'e·chiz'er** *n.*

An entry word may appear unchanged in form at the end of the entry, but with a different part-of-speech label. This indicates that the word is used in a different grammatical function:

**An·glo-Nor·man** (ăng'glō-nôr'mən) *n.* **1.** One of the Normans who lived in England after the Norman conquest of England in 1066 or a descendant of these settlers. **2. a.** The dialect of Old French derived mainly from Norman French that was used by the Anglo-Normans. **b.** The form of this dialect used in English law until the 17th cent. **—An'glo-Nor'man** *adj.*

**Undefined self-explanatory lists.** Lists of undefined terms appear after entries for the prefixes **de–, dis–, non–, over–, re–,** and **un–.** These words—set in small boldface type—are syllabicated and the appropriate stress is indicated. Such words are considered "self-explanatory"; that is, the meaning equals the sum of the meaning of the prefix plus the meaning of the base form.

## Synonym Paragraphs

Synonymous terms judged to be of special interest to dictionary users are listed in clusters following the entry for the central term in the cluster. They are introduced by the italic boldface subheading ***syns.*** The synonyms are set in small capital letters followed by the appropriate part-of-speech label and the italic heading *core meaning* introducing a concise denotation of the sense shared by all of the synonyms in the cluster. At least one verbal illustration showing a typical context for the entry word and its synonyms appears in angle brackets after the core meaning. Many synonym paragraphs end with an antonym or antonyms, introduced by the italic boldface subheading ***ant.*** Synonym paragraphs may be short, relatively long, or discriminative:

**kin** (kĭn) *n.* [ME < OE *cyn.*] One's relatives **:** KINDRED. **—next of kin.** One's closet blood relatives.
☆ ***syns:*** KIN, KINDRED, KINSFOLK *n. core meaning* **:** one's relatives collectively <were finally united with their *kin*>

**fa·çade** *also* **fa·cade** (fə-säd') *n.* [Fr. < Ital. *facciata* < *faccia,* face < Lat. *facies.*] **1.** The face of a building. **2.** An artificial or deceptive outward appearance.
☆ ***syns:*** **1.** FAÇADE, FACE, FRONT, FRONTAL, FRONTISPIECE *n. core meaning* **:** the forward outer surface of a building <the famous *façade* of the Supreme Court building> **2.** FAÇADE, CLOAK, FACE, FRONT, GUISE, MASK, PRETENSE, PUT-ON, SHOW, VENEER, WINDOW-DRESSING *n. core meaning* **:** a deceptive outward appearance <a *façade* of respectability>

**ad·mon·ish** (ăd-mŏn'ĭsh) *vt.* **-ished, -ish·ing, -ish·es.** [ME *admonishen,* alteration of *amonesten* < OFr. *amonester* < VLat. **admonestare,* var. of Lat. *admonēre* : *ad,* to + *monēre,* to warn.] **1.** To reprove mildly or kindly but seriously. **2.** To warn against something **:** CAUTION. **3.** To point out obligations to, by means of a warning, reproof, or exhortation. **—ad·mon'ish·er** *n.* **—ad·mon'ish·ing·ly** *adv.* **—ad·mon'ish·ment** *n.*
☆ ***syns:*** ADMONISH, REBUKE, REPRIMAND, REPROACH, REPROVE *v. core meaning* **:** to address someone disapprovingly because of a fault or misdeed. While ADMONISH refers to mild, warning criticism <*admonished* me to slow down>, REPROVE and REPROACH imply somewhat stronger disapproval <*reproved* the mischievous child><*reproached* me for being late> REBUKE refers to sharp criticism <*rebuked* the insolent student> and REPRIMAND indicates very severe, often formal criticism of another <a general officer *reprimanded* by the President> ***ant:*** commend

## Usage Notes

Many readers look to the dictionary for guidance in matters of usage. Therefore, the editors of this Dictionary have made a selection from those terms that have traditionally presented problems to speakers and writers, and have included usage information in the form of a brief note at the appropriate entry. Each note is introduced by the heading ***usage*** at the definition or specific sense of the term to which it applies:

**da·ta** (dā'tə, dăt'ə, dä'tə) *pl.n.* [Lat., pl. of *datum.* —see DATUM.] *(sing. or pl. in number).* **1.** Information, esp. information organized for analysis or used as the basis for decision-making. **2.** Numerical information suitable for computer processing. **3.** *pl. of* DATUM 1. ***usage:*** *Data,* as the Latin plural of *datum,* in traditional use requires a plural verb, as in *These data are inconclusive.* However, the widespread occurrence of such sentences as *This data is inconclusive* indicates that *data* can now function as a singular form in English.

For the more complex problems not amenable to treatment in this form, see the front matter section "A Concise Guide to Style, Usage, and Diction," where a more extensive essay on problems of English usage is to be found.

# A Concise Guide to Style, Usage, and Diction

**This section of the Dictionary discusses and illustrates the basic conventions of American capitalization, punctuation, and italicization followed by a concise guide to American usage. The section concludes with a guide to better diction in which clichés and redundant expressions are listed and discussed as an aid to those who wish to achieve clear and effective prose style.**

## I. Style Guide

### Capitalization

Capitalize the following:

**1.** the first word of a sentence:

Some spiders are poisonous; others are not.

Are you my new neighbor?

**2.** the first word of a direct quotation, except when the quotation is split:

Joyce asked, "Do you think that the lecture was interesting?"

"No," I responded, "it was very boring."

Tom Paine said, "The sublime and the ridiculous are often so nearly related, that it is difficult to class them separately."

**3.** the first word of each line in a poem in traditional verse:

Half a league, half a league,
Half a league onward,
All in the valley of Death
Rode the six hundred.

—Alfred, Lord Tennyson

**4.** the names of people, of organizations and their members, of councils and congresses, and of historical periods and events:

Marie Curie
Benevolent and Protective Order of Elks
an Elk
Protestant Episcopal Church
an Episcopalian
the Democratic Party
a Democrat
the Nuclear Regulatory Commission
the U.S. Senate
the Middle Ages
World War I
the Battle of Britain

**5.** the names of places and geographic divisions, districts, regions, and locales:

| | |
|---|---|
| Richmond | Continental Divide |
| Vermont | Middle East |
| Argentina | Far North |
| Seventh Avenue | Gulf States |
| London Bridge | East Coast |
| Arctic Circle | the North |
| Eastern Hemisphere | the South Shore |

Do not capitalize words indicating compass points unless a specific region is referred to: Turn north onto Interstate 91.

**6.** the names of rivers, lakes, mountains, and oceans:

| | |
|---|---|
| Ohio River | Rocky Mountains |
| Lake Como | Atlantic Ocean |

**7.** the names of ships, aircraft, satellites, and space vehicles:

U.S.S. *Arizona*
*Spirit of St. Louis*
the spy satellite Ferret-D
Voyager II
the space shuttle Challenger

**8.** the names of nationalities, races, tribes, and languages:

| | |
|---|---|
| Spanish | Bantu |
| Maori | Russian |

**9.** words derived from proper names, except in their extended senses:

the Byzantine Empire

*but*

byzantine office politics

**10.** words indicating family relationships when used with a person's name as a title:

Aunt Toni and Uncle Jack

*but*

my aunt and uncle, Toni and Jack Walker

**11.** a title (i.e., civil, judicial, military, royal and noble, religious, and honorary) when preceding a name:

| | |
|---|---|
| Justice Marshall | Lord Mountbatten |
| General Jackson | Pope John Paul II |
| Mayor Daley | Professor Jacobson |
| Queen Victoria | Senator Byrd |

**12.** all references to the President and Vice President of the United States:

The President has entered the hall.

The Vice President presides over the Senate.

**13.** all key words in titles of literary, dramatic, artistic, and musical works:

the novel *The Old Man and the Sea*
the short story "Notes from Underground"
an article entitled "On Passive Verbs"
James Dickey's poem "In the Tree House at Night"
the play *Cat on a Hot Tin Roof*
Van Gogh's *Wheat Field and Cypress Trees*
Beethoven's *Emperor Concerto*

**14.** *the* in the title of a newspaper if it is a part of the title:

*The Wall Street Journal*
*but*
the New York *Daily News*

**15.** the first word in the salutation and in the complimentary close of a letter:

| | |
|---|---|
| My dear Carol, | Yours sincerely, |

**16.** epithets and substitutes for the names of people and places:

| | |
|---|---|
| Old Hickory | The Oval Office |
| Old Blood and Guts | the Windy City |

**17.** words used in personifications:

When is not Death at watch
Within those secret waters?
What wants he but to catch
Earth's heedless sons and daughters?
—Edmund Blunden

**18.** the pronoun *I*:

I told them that I had heard the news.

**19.** names for the Deity and sacred works:

| | |
|---|---|
| God | the Supreme Being |
| the Almighty | the Bible |
| Jesus | the Koran |
| Allah | the Talmud |

**20.** days of the week, months of the year, holidays, and holy days:

| | |
|---|---|
| Tuesday | Passover |
| May | Ramadan |
| Independence Day | Christmas |

**21.** the names of specific courts:

The Supreme Court of the United States
the Massachusetts Appeals Court
the United States Court of Appeals for the First Circuit

**22.** the names of treaties, accords, pacts, laws, and specific amendments:

Panama Canal Treaty
Treaty of Paris
Geneva Accords
Warsaw Pact countries
Sherman Antitrust Law
Labor Management Relations Act
took the Fifth Amendment

**23.** registered trademarks and service marks:

| | |
|---|---|
| Day-Glo | Comsat |

**24.** the names of geologic eras, periods, epochs, and strata and the names of prehistoric divisions:

| | |
|---|---|
| Paleozoic Era | Age of Reptiles |
| Precambrian | Bronze Age |
| Pleistocene | Stone Age |

**25.** the names of constellations, planets, and stars:

| | |
|---|---|
| Milky Way | Jupiter |
| Southern Crown | Uranus |
| Saturn | Polaris |

**26.** genus but not species names in binomial nomenclature:

*Rana pipiens*

**27.** New Latin names of classes, families, and all groups higher than genera in botanical and zoological nomenclature:

Nematoda

But do not capitalize derivatives from such names: nematodes.

**28.** many abbreviations and acronyms:

| | |
|---|---|
| Dec. | M.F.A. |
| Tues. | UNESCO |
| Lt. Gen. | MIRV |

## Italicization

Use italics to:

**1.** indicate titles of books, plays, and epic poems:

*War and Peace*
*The Importance of Being Earnest*
*Paradise Lost*

**2.** indicate titles of magazines and newspapers:

*New York* magazine
*The Wall Street Journal*
the New York *Daily News*

**3.** set off the titles of motion pictures and radio and television programs:

*Star Wars*
*All Things Considered*
*Masterpiece Theater*

4. indicate titles of major musical compositions:

Handel's *Messiah*
Adam's *Giselle*

5. set off the names of paintings and sculpture:

*Mona Lisa* *Pietà*

6. indicate words, letters, or numbers that are referred to:

The word *hiss* is onomatopoeic.
*Can't* means *won't* in your lexicon.
You form your *n*'s like *u*'s.
A 6 looks like an inverted *9*.

7. indicate foreign words and phrases not yet assimilated into English:

*C'est la vie* was the response to my complaint.

8. indicate the names of plaintiff and defendant in legal citations:

*Roe* v. *Doe*

9. emphasize a word or phrase:

When you appear on the national news, you are *somebody*.

Use this device sparingly.

10. distinguish New Latin names of genera, species, subspecies, and varieties in botanical and zoological nomenclature:

*Homo sapiens*

11. set off the names of ships and aircraft but not space vehicles:

U.S.S. *Arizona*
*Spirit of St. Louis*
Voyager II
the space shuttle Challenger
the spy satellite Ferret-D

# Punctuation

## Apostrophe

1. indicates the possessive case of singular and plural nouns, indefinite pronouns, and surnames combined with designations such as *Jr.*, *Sr.*, and *II*:

my sister's husband
my three sisters' husbands
anyone's guess
They answer each other's phones.
John Smith, Jr.'s car

2. indicates joint possession when used with the last of two or more nouns in a series:

Doe and Roe's report

3. indicates individual possession or authorship when used with each of two or more nouns in a series:

Smith's, Roe's, and Doe's reports

4. indicates the plurals of words, letters, and figures used as such:

60's and 70's *x*'s, *y*'s, and *z*'s

5. indicates omission of letters in contractions:

aren't that's o'clock

6. indicates omission of figures in dates:

the class of '63

## Brackets

1. enclose words or passages in quoted matter to indicate insertion of material written by someone other than the author:

A tough but nervous, tenacious but restless race [the Yankees]; materially ambitious, yet prone to introspection. . . . —Samuel Eliot Morison

2. enclose material inserted within matter already in parentheses:

(Vancouver [B.C.] January 1, 19–)

## Colon

1. introduces words, phrases, or clauses that explain, amplify, or summarize what has gone before:

Suddenly I realized where we were: Rome.

There are two cardinal sins from which all the others spring: impatience and laziness.
—Franz Kafka

2. introduces a long quotation:

In his original draft of the *Declaration of Independence,* Jefferson wrote: "We hold these truths to be sacred and undeniable; that all men are created equal and independent, that from that equal creation they derive rights inherent and inalienable. . ."

3. introduces a list:

We need the following items: pens, paper, pencils, blotters, and erasers.

4. separates chapter and verse numbers in Biblical references:

James 1:4

5. separates city from publisher in footnotes and bibliographies:

Boston: Houghton Mifflin, 1985.

6. separates hour and minute(s) in time designations:

9:30 a.m. a 9:30 meeting

7. follows the salutation in a business letter:

Gentlemen:

## Comma

1. separates the clauses of a compound sentence connected by a coordinating conjunction:

A difference exists between the musical works of Handel and Haydn, and it is a difference worth noting.

The comma may be omitted in short compound sentences:

I heard what you said and I am furious.

I got out of the car and I walked and walked.

**2.** separates *and* or *or* from the final item in a series of three or more:

Red, yellow, and blue may be mixed to produce all colors.

**3.** separates two or more adjectives modifying the same noun if *and* could be used between them without altering the meaning:

a solid, heavy gait

*but*

a polished mahogany dresser

**4.** sets off nonrestrictive clauses or phrases (i.e., those that if eliminated would not affect the meaning of the sentences):

The burglar, who had entered through the patio, went straight to the silver chest.

The comma should not be used when a clause is restrictive (i.e., essential to the meaning of the sentence):

The burglar who had entered through the patio went straight to the silver chest; the other burglar searched for the wall safe.

**5.** sets off words or phrases in apposition to a noun or noun phrase:

Plato, the famous Greek philosopher, was a student of Socrates.

The comma should not be used if such words or phrases precede the noun:

The Greek philosopher Plato was a student of Socrates.

**6.** sets off transitional words and short expressions that require a pause in reading or speaking:

Unfortunately, my friend was not well traveled.

Did you, after all, find what you were looking for?

I live with my family, of course.

**7.** sets off words used to introduce a sentence:

No, I haven't been to Paris.

Well, what do you think we should do now?

**8.** sets off a subordinate clause or a long phrase that precedes a principal clause:

By the time we found the restaurant, we were starved.

Of all the illustrations in the book, the most striking are those of the tapestries.

**9.** sets off short quotations and sayings:

The candidate said, "Actions speak louder than words."

"Talking of axes," said the Duchess, "chop off her head!" —Lewis Carroll

**10.** indicates omission of a word or words:

To err is human; to forgive, divine.

**11.** sets off the year from the month in full dates:

Nicholas II of Russia was shot on July 16, 1918.

But note that when only the month and the year are used, no comma appears:

Nicholas II of Russia was shot in July 1918.

**12.** sets off city and state in geographic names:

Atlanta, Georgia, is the transportation center of the South.

34 Beach Drive
Bedford, VA 24523

**13.** separates series of four or more figures into thousands, millions, etc.:

67,000 200,000

**14.** sets off words used in direct address:

I tell you, folks, all politics is applesauce. —Will Rogers

Thank you for your expert assistance, Dolores.

**15.** Separates a tag question from the rest of a sentence:

You forgot your keys again, didn't you?

**16.** sets off sentence elements that could be misunderstood if the comma were not used:

Some time after, the actual date for the project was set.

**17.** follows the salutation in a personal letter and the complimentary close in a business or personal letter:

Dear Jessica, Sincerely yours,

**18.** sets off titles and degrees from surnames and from the rest of a sentence:

Walter T. Prescott, Jr.
Gregory A. Rossi, S.J.
Susan P. Green, M.D., presented the case.

## Dash

**1.** indicates a sudden break or abrupt change in continuity:

"If—if you'll just let me explain—" the student stammered.

And the problem—if there really is one—can then be solved.

**2.** sets apart an explanatory, a defining, or an emphatic phrase:

Foods rich in protein—meat, fish, and eggs—should be eaten on a daily basis.

More important than winning the election, is governing the nation. That is the test of a political party—the acid, final test.
—Adlai E. Stevenson

3. sets apart parenthetical matter:

Wolsey, for all his faults—and he had many—was a great statesman, a man of natural dignity with a generous temperament. . .
—Jasper Ridley

4. marks an unfinished sentence:

"But if my bus is late—" he began.

5. sets off a summarizing phrase or clause:

The vital measure of a newspaper is not its size but its spirit—that is its responsibility to report the news fully, accurately, and fairly.
—Arthur H. Sulzberger

6. sets off the name of an author or source, as at the end of a quotation:

A poet can survive everything but a misprint.
—Oscar Wilde

## Ellipses

1. indicate, by three spaced points, omission of words or sentences within quoted matter:

Equipped by education to rule in the nineteenth century, . . . he lived and reigned in Russia in the twentieth century. —Robert K. Massie

2. indicate, by four spaced points, omission of words at the end of a sentence:

The timidity of bureaucrats when it comes to dealing with . . . abuses is easy to explain. . . .
—*New York*

3. indicate, when extended the length of a line, omission of one or more lines of poetry:

Roll on, thou deep and dark blue
ocean—roll!
. . . . . . . . . . . . . . . . . . . .
Man marks the earth with ruin—his
control
Stops with the shore.
—Lord Byron

4. are sometimes used as a device, as for example, in advertising copy:

To help you Move and Grow
with the Rigors of
Business in the 1980's. . .
and Beyond.
—*Journal of Business Strategy*

## Exclamation Point

1. terminates an emphatic or exclamatory sentence:

Go home at once!
You've got to be kidding!

2. terminates an emphatic interjection:

Encore!

## Hyphen

1. indicates that part of a word of more than one syllable has been carried over from one line to the next:

During the revolution, the nation was
beset with problems—looting, fighting, and famine.

2. joins the elements of some compounds:

great-grandparent
attorney-at-law
ne'er-do-well

3. joins the elements of compound modifiers preceding nouns:

high-school students
a fire-and-brimstone lecture
a two-hour meeting

4. indicates that two or more compounds share a single base:

four- and six-volume sets
eight- and nine-year olds

5. separates the prefix and root in some combinations; check the Dictionary when in doubt about the spelling:

| | |
|---|---|
| anti-Nazi | re-form/reform |
| re-elect | re-cover/recover |
| co-author | re-creation/recreation |

6. substitutes for the word *to* between typewritten inclusive words or figures:

pp. 145-155
the Boston-New York air shuttle

7. punctuates written-out compound numbers from 21 through 99:

forty-six years of age
a person who is forty-six
two hundred fifty-nine dollars

## Parentheses

1. enclose material that is not essential to a sentence and that if not included would not alter its meaning:

After a few minutes (some say less) the blaze was extinguished.

2. often enclose letters or figures to indicate subdivisions of a series:

A movement in sonata form consists of the following elements: (1) the exposition, (2) the development, and (3) the recapitulation.

3. enclose figures following and confirming written-out numbers, especially in legal and business documents:

The fee for my services will be two thousand dollars ($2,000.00).

4. enclose an abbreviation for a term following the written-out term, when used for the first time in a text:

The patient is suffering from acquired immune deficiency syndrome (AIDS).

## Period

1. terminates a complete declarative or mild imperative sentence:

   There could be no turning back as war's dark shadow settled irrevocably across the continent of Europe. —W. Bruce Lincoln

   Return all the books when you can.

   Would you kindly affix your signature here.

2. terminates sentence fragments:

   Gray clouds—and what looks like a veil of rain falling behind the East German headland. A pair of ducks. A tired or dying swan, head buried in its back feathers, sits on the sand a few feet from the water's edge. —Anthony Bailey

3. follows some abbreviations:

   Dec. Rev. St. Blvd. pp. Co.

## Question Mark

1. punctuates a direct question:

   Have you seen the new play yet?

   Who goes there?

   *but*

   I wonder who said "Nothing is easy in war."

   I asked if they planned to leave.

2. indicates uncertainty:

   Ferdinand Magellan (1480?–1521)
   Plato (427?–347 B.C.)

## Quotation Marks

1. Double quotation marks enclose direct quotations:

   "What was Paris like in the Twenties?" our daughter asked.

   "Ladies and Gentlemen," the Chief Usher said, "the President of the United States."

   Robert Louis Stevenson said that "it is better to be a fool than to be dead."

   When advised not to become a lawyer because the profession was already overcrowded, Daniel Webster replied, "There is always room at the top."

2. Double quotation marks enclose words or phrases to clarify their meaning or use or to indicate that they are being used in a special way:

   This was the border of what we often call "the West" or "the Free World."

   "The Windy City" is a name for Chicago.

3. Double quotation marks set off the translation of a foreign word or phrase:

   *die Grenze,* "the border"

4. Double quotation marks set off the titles of series of books, of articles or chapters in publications, of essays, of short stories and poems, of individual radio and television programs, and of songs and short musical pieces:

   "The Horizon Concise History" series

   an article entitled "On Reflexive Verbs in English"

   Chapter Nine, "The Prince and the Peasant"

   Pushkin's "The Queen of Spades"

   Tennyson's "Ode on the Death of the Duke of Wellington"

   "The Bob Hope Special"

   Schubert's "Death and the Maiden"

5. Single quotation marks enclose quotations within quotations:

   The blurb for the piece proclaimed, "Two years ago at Geneva, South Vietnam was virtually sold down the river to the Communists. Today the spunky little . . . country is back on its own feet, thanks to 'a mandarin in a sharkskin suit who's upsetting the Red timetable.' " —Frances FitzGerald

   Put commas and periods inside quotation marks; put semicolons and colons outside. Other punctuation, such as exclamation points and question marks, should be put inside the closing quotation marks only if part of the matter quoted.

## Semicolon

1. separates the clauses of a compound sentence having no coordinating conjunction:

   Do not let us speak of darker days; let us rather speak of sterner days. —Winston Churchill

2. separates the clauses of a compound sentence in which the clauses contain internal punctuation, even when the clauses are joined by conjunctions:

   Skis in hand, we trudged to the lodge, stowed our lunches, and donned our boots; and the rest of our party waited for us at the lifts.

3. separates elements of a series in which items already contain commas:

   Among those at the diplomatic reception were the Secretary of State; the daughter of the Ambassador to the Court of St. James's, formerly of London; and two United Nations delegates.

4. separates clauses of a compound sentence joined by a conjunctive adverb, such as *however, nonetheless,* or *hence*:

   We insisted upon a hearing; however, the Grievance Committee refused.

5. may be used instead of a comma to signal longer pauses for dramatic effect:

But I want you to know that when I cross the river my last conscious thought will be of the Corps; and the Corps; and the Corps.
—General Douglas MacArthur

## Virgule

**1.** separates successive divisions in an extended date:

fiscal year 1983/84

**2.** represents *per*:

35 km/hr    1,800 ft/sec

**3.** means *or* between the words *and* and *or*:

Take water skis and/or fishing equipment when you visit the beach this summer.

**4.** separates two or more lines of poetry that are quoted and run in on successive lines of a text:

The student actress had a memory lapse when she came to the lines "Double, double, toil and trouble/Fire burn and cauldron bubble/Eye of newt and toe of frog/Wool of bat and tongue of dog" and had to leave the stage in embarrassment.

# II. Problems in English Usage

This Dictionary has adopted a unique approach to problems of usage. At the entries for certain terms, points of usage are discussed and guidance is offered for matters which lend themselves to treatment in brief. Topics whose scope is so broad as to require extended discussion are dealt with in this guide. No attempt has been made to present an exhaustive handbook of usage and grammar; instead the guide treats only those problems which have traditionally been considered especially vexing by users of English. Neither does the guide dictate the choice of a single, supposedly correct usage. Its purpose is rather to instruct and inform, to present the evidence evenhandedly, and so permit the user to make a choice appropriate to a specific context and purpose.

It cannot be emphasized too strongly that in most cases the context of the situation and the intent of the speaker or writer will play the principal role in making this choice. For example, the language of a scholarly text or of an address given on a ceremonial occasion is usually in a measured and formal style, differing markedly from the more relaxed and informal discourse characteristic of a conversation among associates or a letter written to an old friend. Thus, for the purposes of the discussion that follows, it will be helpful to distinguish four levels of usage: the two levels, formal and informal, each subdivided into the two categories of spoken or written language.

**as, than**

The choice between the subjective form (e.g., *we, she, who*) and the objective form (e.g., *us, her, whom*) of English pronouns presents a problem for many. The circumstances in which such problems can arise are discussed below. In comparisons with *as* and *than,* a following pronoun is in the subjective case when it is the subject of an understood verb. In the following examples, the understood verb is in parentheses.

No one is as tall as he (is).

She can swim faster than he (can).

The pronoun is in the objective case, however, when it is the object of an understood verb or preposition.

It distressed her as much as (it distressed) me.

He talked to them more than (he talked to) us.

Similarly, after *as well as,* the case of the pronoun is determined by the way in which the pronoun functions in relationship to understood elements.

No one can play that cello sonata as well as she (can).

It was a present for her as well as (for) him.

**be**

After the forms of the verb *to be* the use of the subjective pronoun is traditionally prescribed when the pronoun refers to the subject.

It is I.

That can only be they at the door.

This use of the subjective can sometimes sound unnaturally formal, however, and for this reason the traditional rule is often violated. As an extreme example, almost all speakers will prefer

That picture isn't really her!

to the traditionally correct

That picture isn't really she!

Similarly, after a contraction, especially in speech, the objective form of the pronoun often occurs instead of the subjective form.

It's me.

This usage is entirely acceptable to all but the most staunch traditionalists.

**each**

When *each* precedes the noun it modifies, a pronoun or pronominal adjective referring to *each* is properly in the singular. If that noun is the subject of the sentence, the verb is also in the singular.

Each debater was allowed five minutes for his rebuttal.

When *each* follows the noun, however, the plural is usually correct.

The debaters were each allowed five minutes for their rebuttals.

—See also EVERYONE.

**everyone**

There is a great deal of uncertainty among users of English as to whether a singular or plural form is appropriate when referring to certain words and expressions. These terms may be divided into three categories. In the first are those words formed with *-body* or *-one,* such as *anybody, everyone, nobody,* and the term *no one,* which is written as two words. The second category includes such words as *each, either,* and *neither.* In the last category are the indefinite forms *whatever, whichever,* and *whoever.*

These terms are often considered to be plural in meaning or in reference although they are traditionally classified as singular. Therefore, to accord with the traditional view, any pronouns or pronominal adjectives making reference to these terms should also be in the singular.

*Everyone* is responsible for *his* own actions.
*Neither* of the two was able to sign *his* name.

The preceding examples are characteristic of the most formal level of expression.

Nevertheless, a plural form occurs very often in reference to these terms, especially in speech, and the problem, which has beset English speakers and even the best writers for centuries, persists to the present day.

Experience is the name that *everyone* gives to *their* mistakes.
—Oscar Wilde

This use of the plural is permissible in informal contexts. It should also be noted that when a pronoun or pronominal adjective referring to one of these constructions occurs outside the bounds of the same sentence, that pronoun cannot properly be in the singular.

*Everyone* rushed to the window. *They* were (not *He* was) amazed to see the size of the crowd that had gathered in the square.

This restriction, however, does not apply to constructions with *whoever.*

*Whoever* is responsible must be found. You must see to it that *his* punishment fits the crime.

An additional complication arises from the fact that the use of the masculine to refer to these terms is a practice that for one reason or another is objectionable to many. The choices facing the user, then, are on the one hand to follow the dictates of tradition by using the masculine forms or to attempt to satisfy the requirements of grammatical tradition, accuracy, and ideology by using both the masculine and feminine forms.

Each student will be expected to make his own travel arrangements.
Each student will be expected to make his or her own travel arrangements.

On the other hand, to avoid the seeming bias of the former sentence and the awkwardness of the latter by using the plural forms is a choice that may invite reproach. These considerations, however, lie outside the domain of grammar. Consequently, users must make this decision for themselves.

—See also EACH.

**if**

Depending upon the intended meaning, a clause that is introduced by *if* can contain either a subjunctive or an indicative verb. The subjunctive is used when the situation is known to be contrary to fact; the main clause of such a sentence will usually employ the modal verb *would* or *should.*

If he were elected, he would introduce radical programs to cure social ills.

The use of *would have been elected* instead of *were elected* in a sentence such as the above occurs with increasing frequency, but this usage is still considered to be incorrect.

An indicative verb occurs in a clause introduced by *if* when the situation is not known to be false or contrary to fact; the verb employed in the main clause will depend upon the intended meaning.

If my calculation is correct, then we have only two days left to comply with the law.

**that, which**

According to the traditional rule, *that* is used as the relative pronoun only in a restrictive clause, which is a clause that identifies or defines a particular individual or object. Such a clause is never separated from the rest of the sentence by a comma.

We need a saw that can cut through metal.

A nonrestrictive clause, by contrast, does not identify or define an entity, but merely provides additional information about an individual or object that is specified by other means. The pronouns *who* (or the inflected forms *whose* and *whom*) and *which,* as appropriate, occur in nonrestrictive clauses.

His paintings, which occupy a separate room in the National Gallery, are among the highest achievements of Western art.

My father, who was a superb cook, never followed a printed recipe in his life.

Some grammarians have argued that *which* should properly be employed only in nonrestrictive clauses and *that* should be confined only to restrictive clauses; however, such a neat division, while pleasing to logicians, is not supported by the facts of language. The two versions of the following sentence are equally correct.

A messenger brought the books that you asked for.

A messenger brought the books which you asked for.

The use of *which* is preferable on grounds of style when two or more clauses are linked by a conjunction.

It is a disease for which there is no cure and which is invariably fatal.

And, of course, *which* is obligatory whenever the antecedent is *that.*

He said only that which we all know to be the truth.

**there**

At the beginning of a sentence or clause, *there* before a linking verb such as *seem* or *be* serves to introduce the true subject, which comes after the verb in such constructions.

There has been a certain confusion about the proper role of government settling this dispute.

Whether the verb is singular or plural is usually determined by the true subject.

There is only one candidate that a rational person could support.

There are still many problems to be solved before the plan is put into effect.

It is sometimes the case, however, that a singular verb may occur before a compound subject, each of whose elements is singular.

There was (*or* were) weeping, wailing, and gnashing of teeth when we lost the championship.

And when such a compound subject contains both singular and plural elements, the verb can agree in number with the element that stands first, in the position closest to the verb.

There was a mother cat and her seven kittens in the basket.

There were seven kittens and a mother cat in the basket.

**whether, if**

After a verb such as *ask, doubt, know, learn* or *see,* either *if* or *whether* can introduce a clause indicating uncertainty.

Please see whether (*or* if) the papers are on my desk.

Sometimes such a sentence can be ambiguous when *if* introduces the clause.

Call and let me know if you plan to attend.

The first interpretation possible is that this sentence is a request to call, no matter what the decision is. In the second interpretation, the sentence is a request to call only in the event that the decision to attend has been made. In such cases rephrasing the sentence will eliminate the unclarity.

**who, whom**

According to traditional grammar, *who* is the appropriate form to use in contexts where other subjective pronouns, such as *we* or *he* would also occur.

Who is in charge of this department?
He who hesitates is lost.

As the subject of *is* and *hesitates* respectively, in the two sentences above *who* is quite properly in the subjective case.

In contexts requiring an objective form, such as *us* or *him,* then *whom* is appropriate to a formal style. And at all levels, *whom* is necessary when it is immediately preceded by a preposition.

To whom did you speak?
All the people whom we invited are planning to attend.

Although the rules are straightforward enough, problems can arise when the pronoun is at a distance from the elements of a phrase or sentence that determine the choice of case.

She interviewed the artist whom the committee had insisted that the mayor hire.

In constructing such a sentence, it is necessary to keep in mind the fact that *whom* will be the direct object of *hire.* Writing, which tends to be a more formal medium than speech, has the virtue of allowing a review of what has been produced in order to eliminate errors and inconsistencies. In speech or in the representation of speech, however, the distinction between *who* and *whom* is often not preserved.

Who are you speaking of?

—Thomas Hardy,
*Far from the Madding Crowd*

When a preposition and its pronoun object are separated from each other, the latter is often in the technically incorrect subjective case. This usage is a common occurrence in the works of the best writers and has been defended by many grammarians and students of language, including Noah Webster, on the grounds that rigid adherence to the rules, especially in informal contexts, yields sentences which sound stilted and pedantic. The same considerations as the foregoing apply to the choice between *whoever* and *whomever.*

—See also *that, who, which* at THAT.

# III. Clichés

This Dictionary defines *cliché* as "A trite or overused expression or idea." The English language abounds in clichés, many of which originated as metaphors, proverbs, or brief quotations. But historical changes in the language through the years have rendered many of these expressions meaningless. For instance, what does *fell* in *one fell swoop* mean? Others, such as *do one's thing* and *keep a low profile,* illustrate that such expressions age very fast through relentless use, and become stale. Since most clichés express rather clear meanings, the writer will have to determine whether it is a shade of meaning that is hard to convey by fresher wording. If the process of substitution is too difficult, use of some of the phrases that follow may be advisable; writing around the formulaic expression may produce something worse than hackneyed language, such as strained, wordy, or ambiguous discourse. But few on the following list are truly indispensable, and writers of fresh, original prose will avoid most of them.

**absence makes the heart grow fonder**
**add insult to injury**
**age before beauty**
**agonizing reappraisal**
**agree to disagree**
**albatross around one's neck**
**all in a day's work**
**all in the same boat**
**all over but the shouting**
**all things being equal**
**all things to all men**
**all work and no play**
**apple of one's eye**
**apple-pie order**
**armed to the teeth**
**arms of Morpheus**
**as luck would have it**
**at a loss for words** (*or* **never at a loss . . .**)
**at first blush**
**at sixes and sevens**
(an) **axe to grind**

**bag and baggage**
**bark up the wrong tree**
**bated breath**
**bathed in tears**
**beard the lion in his den**
**beat a dead horse**
**beat a hasty retreat**
**beat around the bush**
**beg to disagree**
**beggar description**
**bend** (*or* **lean**) **over backward**
**best foot forward**
**best-laid plans**
**best of all possible worlds**
**better late than never**
**between the devil and the deep blue sea**
**beyond the call of duty**
**beyond the pale**
**bigger than all outdoors**
**bigger** (*or* **larger**) **than life**
**bite off more than one can chew**
**bite the bullet**
**blushing bride**
**blush of shame**
**boggle the mind**
**bolt from the blue**
**bone of contention**
**born with a silver spoon**
**bosom of the family**
**brave the elements**
**breathe a sigh of relief**
**bright and early**
**bright as a button**
**bright-eyed and bushy-tailed**
**bright future**
**bring home the bacon**
**brown as a berry**
**budding genius**
**bull in a china shop**
**burn the midnight oil**
**busy as a bee**
**butter wouldn't melt in one's mouth**
**by leaps and bounds**
**by the same token**

**calm before the storm**
**can't see the forest for the trees**
**carry** (*or* **have**) **a chip on one's shoulder**
**carry coals to Newcastle**
(a) **case in point**
**caught on the horns of a dilemma**
**caught red-handed**
**chip off the old block**
**clear as mud**
(to) **coin a phrase**
**cold as ice**
**conspicuous by one's absence**
**cool as a cucumber**
**cross the Rubicon**
(a) **crying need**
**cut a long story short**
**cut off one's nose to spite one's face**
**cynosure of all eyes**

**daily repast**
**dead as a doornail**

defend to the death one's right to . . .
depths of despair
diamond in the rough
die in harness
die is cast
distaff side
do it up brown
do one's thing
dog in the manger
doom is sealed
doomed to disappointment
down in the dumps
down in the mouth
down one's alley
draw the line
drown one's sorrows
drunk as a lord (*or* skunk)
dull thud

early bird gets the worm
early to bed . . . to rise
ear to the ground
easier said than done
eat one's hat (*or* words)
epoch-making
eternal reward
eyes of the world

face the music
(the) fair sex
fall on deaf ears
far be it from me
(a) far cry
fast and loose
fate worse than death
fat's in the fire
feather in one's cap
feel one's oats
festive board
few and far between
few well-chosen words
fiddle while Rome burns
fight like a tiger
fill the bill
filthy lucre
fine and dandy
first and foremost
fit as a fiddle
flash in a the pan
flat as a flounder (*or* pancake)
flesh and blood
fly off the handle
fond farewell
food for thought
fools rush in
foot in one's mouth
foot the bill
foregone conclusion
forewarned is forearmed
free as a bird (*or* the air)
fresh as a daisy

generous to a fault
gentle as a lamb
get down to brass tacks
get one's back (*or* dander) up
(a) good time was had by all
goose that laid the golden egg
grain of salt
grand and glorious
graphic account
green-eyed monster
grin like a Cheshire cat
grind to a halt

hail fellow well met
hale and hearty
hand that rocks the cradle
handsome is as handsome does
handwriting on the wall
hapless victim
happy as a lark
happy pair
hard row to hoe
haughty stare
haul (*or* rake) over the coals
have a foot in the door
have a leg up
head over heels
heart of gold
heave a sigh of relief
hew to the line
high and dry
high as a kite
high on the hog
hit the nail on the head
hit the spot
hook, line, and sinker
hook or crook
hot as a firecracker (*or* pistol)
hue and cry
hungry as a bear (*or* lion)

if (the) truth be told
in full swing
in no uncertain terms
in on the ground floor
in seventh heaven
inspiring sight
in the final (*or* last) analysis
in the limelight
in the long run
in the nick of time
in this day and age
iron out a difficulty
irons in the fire
irony of fate
irreparable damage (*or* loss)
it goes without saying
it is interesting to note
it never rains but it pours
it's an ill wind
it's six of one and a half a dozen of the other
it stands to reason
it takes all kinds to make a world
it takes two to tango

jig is up
just deserts

keep a low profile
keep a stiff upper lip

**knock into a cocked hat**
**knock on wood**

**labor of love**
**land of milk and honey**
**land-office business**
**last but not least**
**last straw**
**law unto one's self**
**lead to the altar**
**lean and hungry look**
**lean over backward**
**leave in the lurch**
**leave no stone unturned**
**left-handed compliment**
**lend a helping hand**
**let one's hair down**
**let the cat out of the bag**
**let well enough alone**
**lick into shape**
**lid of secrecy**
**like a house afire (*or* on fire)**
**like a newborn babe**
**limp as a dish rag**
**lock, stock, and barrel**
**long arm of the law**
**look a gift horse in the mouth**
(as) **luck would have it**

**mad as a hatter**
**mad as a hornet (*or* wet hen)**
**mad as a March hare**
**mad dash**
**make a clean breast of**
**make a long story short**
**make a virtue of necessity**
**make bricks without straw**
**make ends meet**
**make hay while the sun shines**
**make no bones about**
**mantle of snow**
**matter or life and death**
**meek as Moses**
**meet one's Waterloo**
**meets the eye**
**method in one's madness**
**milk of human kindness**
**mince words**
**mind one's *p*'s and *q*'s**
**miss the boat**
**moment of truth**
**monarch of all one surveys**
**month of Sundays**
**moot question (*or* point)**
**more easily said than done**
**more sinned against than sinning**
**more than meets the eye**
(the) **more the merrier**
**motley crew**

**naked truth**
**name is legion**
**necessary evil**
**needle in a haysack**
**needs no introduction**
**neither fish nor fowl**
**never say die**
**nip in the bud**
**none the worse for wear**
**no sooner said than done**
**not to be sneezed (*or* sniffed) at**
**not wisely but too well**
**nothing new under the sun**

**of a high order**
**on cloud nine**
**on one's uppers**
**on the ball (*or* stick)**
**on the best (*or* unimpeachable) authority**
**on the bum (*or* the fritz)**
**on the lam**
**on the other hand**
**on the q.t.**
**on the wagon**
**once in a blue moon**
**one fell swoop**
**one's own worst enemy**
**open secret**
**opportunity knocks**
**other side of the coin**
**other things being equal**
**out of the frying pan into the fire**
**over a barrel**
**overcome with emotion**

**paint the town red**
**pandemonium reigned**
**part and parcel**
**pay the piper**
**penny for one's thoughts**
**penny wise, pound foolish**
**perfect gentleman**
**pet peeve**
**pillar of society**
**pillar to post**
**pinch pennies**
**play fast and loose**
**play it by ear**
**play second fiddle**
**play the Devil's advocate**
**plumb the depths**
(at this) **point in time**
**point with pride**
**poor but honest**
(the) **powers that be**
**pretty as a picture**
**pretty kettle of fish**
**pretty penny**
**psychological moment**
**pull the wool over one's eyes**
**pure as the driven snow**
**put on the dog**

**quick as lightning (*or* a flash)**
**quiet as a mouse**

**rack one's brains**
**rain cats and dogs**
**raise Cain**
**raise the roof**

red-letter day
reign supreme
render a decision
ring true
ripe old age
rub one the wrong way

sadder but wiser
sad to relate
save for a rainy day
seal one's fate (*or* doom)
second to none
seething mass
sell like hot cakes
separate the men from the boys
separate the sheep from the goats
shoot from the hip
(a) shot in the arm
shout from the rooftops
show one's hand
show the white feather
sick and tired
sight to behold
sing like a bird
skeleton in one's closet
small world
smell a rat
sour grapes
sow one's wild oats
stagger the imagination
start (*or* get) the ball rolling
steal one's thunder
stem to stern
stick in one's craw
stick out like a sore thumb
stick to one's guns
stick to one's knitting
stir up a hornet's nest
straight from the shoulder
straight and narrow
straw in the wind
straw that broke the camel's back
strong as an ox
stubborn as a mule
sweat of one's brow
sweet sixteen
sweet smell of success

take a dim view of
take a rain check
take it easy
take the bull by the horns
take up the cudgels
talk through one's hat
tell someone who cares
that is to say
that's for sure
throw caution to the winds
throw in the towel (*or* sponge)
throw one's hat in the ring
throw the book at
time hangs heavy
time immemorial
time of one's life
tip the scales
tired as a dog
tit for tat
to tell the truth
to the manner born
too funny for words
too little, too late
trip the light fantastic
true blue
turn over a new leaf

uncharted seas
up the creek without a paddle
usually reliable sources

vale of tears
view with alarm

wash one's hands of
wax poetic
wear two hats
wee (*or* small) hours
wet to the skin
what makes the world go round
when all is said and done
when you come (right) down to it
while ignorance is bliss
wide-open spaces
wise as an owl
without further ado
wolf in sheep's clothing
work one's fingers to the bone

# IV. Redundant Expressions

Redundancy—the needless repetition of ideas—is one of the principal obstacles to writing clear, precise prose. The list below gives some common redundant expressions. The elements repeated in the phrases and in the brief definitions are italicized. To eliminate redundancy, delete the italic elements in the phrases.

**anthracite** *coal*
(= a hard *coal* having a high carbon content)

*old* **antique**
(= an object having special value because of its *age*, esp. a work of art or handicraft more than 100 years *old*)

**ascend** *upward*
(= to go or move *upward*)

**assemble** *together*
(= to bring or gather *together*)

*pointed* **barb**
(= a sharp *point* projecting in reverse direction to the main point of a weapon or tool)

*first* **beginning**
(= the *first* part)

**big** *in size*
(= of considerable *size*)

**bisect** *in two*
(= to cut *into two* equal parts)

**blend** *together*
(= to combine, mix, or go well *together*)

**capitol** *building*
(= a *building* in which a legislative body meets)

**coalesce** *together*
(= to grow or come *together* so as to form a whole)

**collaborate** *together* or *jointly*
(= to work *together*, esp. in a joint effort)

*fellow* **colleague**
(=a *fellow* member of a profession, staff, or academic faculty)

**congregate** *together*
(= to bring or come *together* in a crowd)

**connect** *together*
(= to join or fasten *together*)

**consensus** *of opinion*
(= collective *opinion*)

**courthouse** *building*
(= a *building* in which judicial courts or county government offices are housed)

*habitual* **custom**
(= a *habitual* practice)

**descend** *downward*
(= to move, slope, extend, or incline *downward*)

**doctorate** *degree*
(= the *degree* or status of a doctor)

**endorse** (a check) *on the back*
(= to write one's signature *on the back of*, e.g., a check)

**erupt** *violently*
(= to emerge *violently* or to become *violently* active)

**explode** *violently*
(= to burst *violently* from internal pressure)

*real* **fact**
(= something with *real*, demonstrable existence)

*passing* **fad**
(= a *passing* fashion)

**few** *in number*
(= amounting to or made up of a *small number*)

**founder** *and sink*
(= to *sink* beneath the water)

*basic* **fundamental**
(= a *basic* or essential part)

**fuse** *together*
(= to mix *together* by or as if by melting)

*opening* **gambit**
(= a remark intended to *open* a conversation)

**gather** *together*
(= to come *together* or cause to come *together*)

*free* **gift**
(= something bestowed voluntarily and *without compensation*)

*past* **history**
(= a narrative of *past* events; something that took place *in the past*)

**hoist** *up*
(= to raise or to haul *up* with or as if with a mechanical device)

*current* or *present* **incumbent**
(= one *currently* holding an office)

*new* **innovation**
(= something *new* or unusual)

**join** *together*
(= to bring or put *together* so as to make continuous or form a unit)

**knots** *per hour*
(= a unit of speed, one nautical mile *per hour*, approx. 1.15 statute miles *per hour*)

**large** *in size*
(= greater than average *in size*)

**merge** *together*
(= to blend or cause to blend *together* gradually)

*necessary* **need**
(= something *necessary* or wanted)

*universal* **panacea**
(= a remedy for *all* diseases, evils, or difficulties)

*continue to* **persist**
(= to *continue* in existence)

*individual* **person**
(= an *individual* human being)

*advance* **planning**
(= detailed methodology, programs, or schemes worked out *beforehand* for the accomplishment of an objective)

*chief* or *leading* or *main* **protagonist**
(= the *leading* character in a Greek drama or other literary form; a *leading* or *principal* figure)

*original* **prototype**
(= an *original* type, form, or instance that is a model on which later stages are based or judged)

**protrude** *out*
(= to push or thrust *outward*)

**recall** *back*
(= to summon *back* to awareness; to bring *back*)

**recoil** *back*
(= to kick or spring *back*; to shrink *back* in fear or loathing; to fall *back*)

*new* **recruit**
(= a *new* member of a body or organization, esp. of a military force)

**recur** *again* or *repeatedly*
(= to occur *again* or *repeatedly*)

*temporary* **reprieve**
(= *temporary* relief, as from danger or pain)

**short** *in length* or *height*
(= having very little *length* or *height*)

**shuttle** *back and forth*
(= to move, go, or travel *back and forth*)

**skirt** *around*
(= to move or pass *around* rather than across or through)

**small** *in size*
(= characterized by relatively little *size* or slight *dimensions*)

**tall** *in height*
(having greater than average *height*)

*two* **twins**
(= one of *two* offspring born at the same birth; one of *two* identical or similar persons, animals, or things)

*completely* **unanimous**
(= being in *complete* harmony, accord, or agreement)

**visible** *to the eye*
(= perceptible *to the eye*)

*from* **whence**
(= *from* where; *from* what place; *from* what origin or source)

# Students' Guide to Typewriting Research Papers

This section of the Dictionary illustrates the proper way to prepare bibliography and note cards prior to writing a research paper. Facsimile illustrations exemplify the proper typewritten format of the paper itself. The major elements of a typical paper—the title page, the outline, the first text page together with two other text pages, the footnotes, and the bibliography—are accompanied by brief guidelines. The material is excerpted, adapted, and reprinted by permission from the *Practical English Handbook*, 6th ed., by Floyd C. Watkins and William B. Dillingham (Boston: Houghton Mifflin, 1982), pp. 271–284; pp. 294–325.

## Reference Works and Note Cards

Useful periodical indexes to be consulted when preparing to write a research paper include the following:

*Applied Science and Technology Index*, 1958—.
*Art Index*, 1929—.
*Biography Index*, 1946—.
*Biological and Agricultural Index*, 1964—.
*Book Review Digest*, 1905—.
*Book Review Index*, 1965—.
*British Humanities Index*, 1962—.
*Business Periodicals Index*, 1958—.
*Current Index to Journals in Education*, 1969—.
*Education Index*, 1929—.
*Essay and General Literature Index*, 1900–79.
*General Science Index*, 1978—.
*Humanities Index*, 1974—.
*Industrial Arts Index*, 1913–57.
*MLA International Bibliography of Books and Articles on the Modern Languages and Literature*, 1919—.
*Music Index*, 1949—.
*New York Times Index*, 1851—.
*New York Times Obituary Index*, 1858—.
*Nineteenth Century Readers' Guide to Periodical Literature*, 1890–99.
*Poole's Index to Periodical Literature*, 1802–1906.
*Public Affairs Information Service Bulletin*, 1915—.
*Readers' Guide to Periodical Literature*, 1900—.
*Social Sciences and Humanities Index*, 1965–74.
*Social Sciences Index*, 1974—.

Other useful reference books include:

*Articles on American Literature*, 1900–50, 1950–67, 1968–75.
*Cambridge Bibliography of English Literature*, 1941–57.
*Contemporary Authors*, 1962—.
*Current Biography*, 1940—.
*Dictionary of American Biography*, 1928–37. Supplements, 1944–80.
*Dictionary of American History*, 1976–81.
*Dictionary of National Biography*, 1885–1901. Supplements, 1912–71.
*Encyclopaedia Britannica*, 1984.
*Encyclopaedia Judaica*, 1972.
*Encyclopedia Americana*, 1983.
*Encyclopedia of Philosophy*, 1973.
*Encyclopedia of World Art*, 1959–68. Supplement, 1983.
*Encyclopedia of World History*, 1972.
*Information Please Almanac*, 1947—.
*International Encyclopedia of Social Sciences*, 1977. Biographical Supplement, 1979.
*McGraw-Hill Encyclopedia of Science and Technology*, 1982. *Yearbook of Science and Technology* Supplements.
*McGraw-Hill Encyclopedia of World Drama*, 1983.
*New Catholic Encyclopedia*, 1981.
*New Cambridge Bibliography of English Literature*, 1969—.
*New Cambridge Modern History*, 1957–79.
*New Grove Dictionary of Music and Musicians*, 1980.
*Oxford Classical Dictionary*, 1970.
*Oxford History of English Literature*, 1945—.
*Oxford Companion* Series (various subjects and dates).
*Princeton Encyclopedia of Poetry and Poetics*, 1974.
*Stateman's Yearbook: Statistical and Historical Annual of the States of the World*, 1864—.

**Bibliography Cards**

Keep a full record of all the primary and secondary sources you use in doing the research for your paper. Use one 3 × 5 inch card for each source. Keep the 3 × 5 cards in alphabetical order according to the authors' surnames.

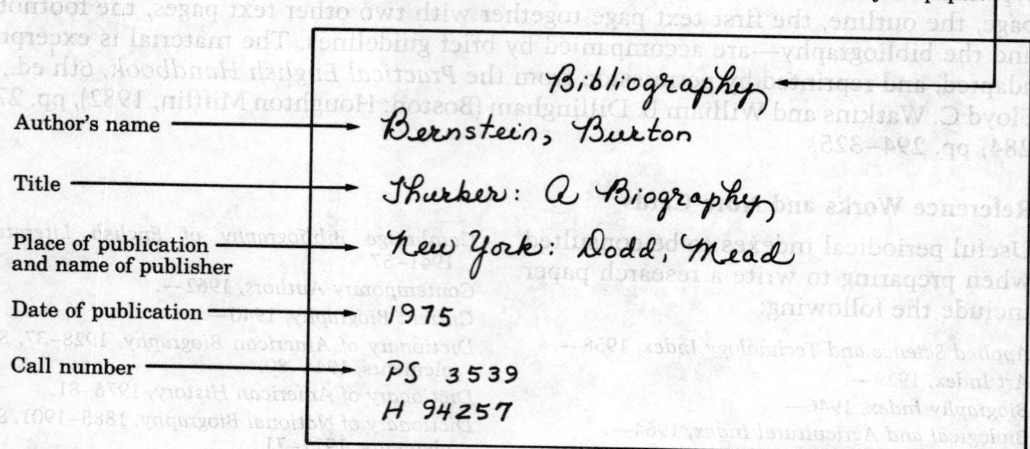


A label which indicates that this is a bibliography card. Keep these cards separate from those on which you take notes. They will eventually be used in making up the bibliography for your paper.

Place of publication and name of publisher

Date of publication

**Note Cards**

Keep note cards for the subjects you treat in your paper. Use one note card for each subject. Keep note cards for your paraphrased notes, all quotations, paraphrases, and quotations combined with paraphrases.

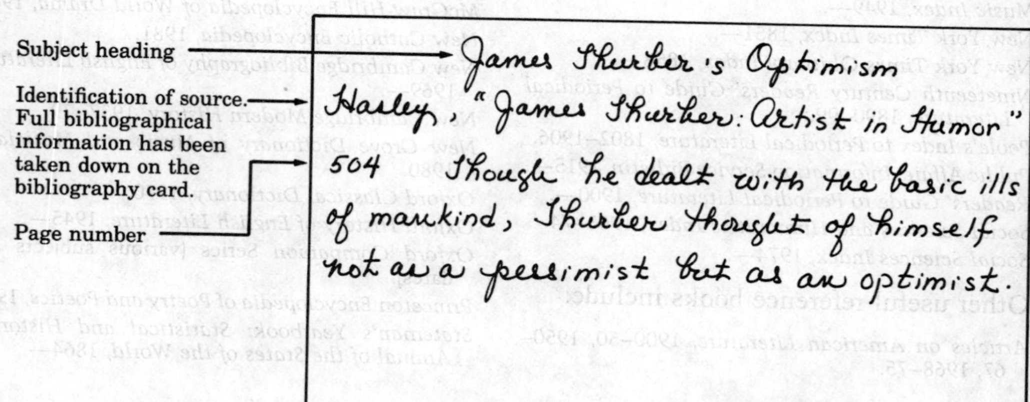

Paraphrased Notes
Subject heading
James Thurber's Optimism
Identification of source. Full bibliographical information has been taken down on the bibliography card.
Hasley, "James Thurber: Artist in Humor"
Page number
504 Though he dealt with the basic ills of mankind, Thurber thought of himself not as a pessimist but as an optimist.

**Quotation**

Subject heading → James Thurber's subjects

Abbreviated identification of source → Hasley

Page number → 564

Quotation → "With minor exceptions he did not explore the century's large social and political problems. War, religion, crime, poverty, civil rights—these were not his subjects. Instead he struck at the immemorial stupidities, cruelties, and perversities of men that lie at the root of our ills."

**Paraphrase**

Subject heading → Thurber's Cartoon of the Seal

Identification of source → Holmes, <u>Clocks</u>

Page number → 135

Paraphrased matter → Thurber's cartoon of the seal is difficult to analyze because it blends the real world with a world of fantasy and is therefore much like a dream.

**Quotation and Paraphrase**

Subject heading → Thurber's Cartoon of the Seal

Identification of source → Holmes, <u>Clocks</u>

Page number → 135

Quoted and paraphrased matter → "His most famous cartoon," of the seal, "has that mixture of the familiar and the strange" that makes it too much like a dream to be easily analyzed.

## The Title Page

Hope Through Fantasy in James Thurber's Drawings

By

Debra Warwick

English 101

Section 1

The material on the title page should be spaced so that it appears balanced on the page. Center the title and place it about one third from the top of the page. Include your name and the name and section number of the course as indicated.

## The Outline

Hope Through Fantasy in James Thurber's Drawings

THESIS STATEMENT: Although the comic drawings of James Thurber seem simple, they actually embody a complex world view with strong suggestions of pessimism but with an even stronger sense of hope.

I. Introduction--Thurber's career as a cartoonist
   A. His one-man exhibition at the Valentine Gallery
   B. Before the Valentine exhibit--his beginnings
      1. Collaboration with E. B. White
      2. Praise by British artist Paul Nash
      3. First collection of drawings: <u>The Seal</u>
   C. After the Valentine exhibit--his later years
II. The nature of Thurber's drawings
   A. Subject matter
   B. Technique
   C. Captions
III. The effects of Thurber's drawings
   A. Before careful analysis
   B. After careful analysis
IV. An examination of "The Seal"
   A. The woman
   B. The man
   C. The seal
V. Thurber's world view
   A. Reputation for bitterness
   B. His complexity and affirmation

If your instructor requests that you submit an outline with your paper, it should occupy a separate, unnumbered page following the title page and should follow the form for the outline illustrated above.

## The First Text Page

Hope Through Fantasy in James Thurber's Drawings

In December of 1934 the highly respected Valentine Gallery in Manhattan held one of the most unusual exhibitions in all of its history. Less traditional than many of the art galleries in New York City, the Valentine "devoted itself to the more advanced of the socially acceptable left-wing artists."[1] Nevertheless, this was a particularly odd one-man show: as viewers shuffled from one frame to another, they witnessed childlike reproductions of gentle dogs, aggressive women, and emasculated men all casually drawn as doodles on various sorts and sizes of sheets, even scratch paper.[2] Much puzzlement and disbelief ran through these lovers of art; but the exhibition was widely reviewed, and the artist, James Thurber, was in the end compared with such giants as Picasso and Matisse.[3] Since then, Thurber's position as a serious artist as well as a serious writer has been solidly established, although his cartoons, known best from the pages of The New Yorker, are deceptively simple. His fame has been accompanied by a reputation for bitterness, which is in some measure justified; but if pessimism is reflected in Thurber's drawings, a sense of hope is even stronger.

Thurber never took his role as an artist as seriously as other people did. He was completely self-taught. In fact, his brother William was supposedly the Thurber with talent for drawing. James' early doodles were not drastically different from the ones that brought him acclaim, but his family dismissed them, and he was advised not to waste his time. An urge so deep that he did not himself understand it kept him at these spare, undetailed, outline figures. He produced them so quickly and effortlessly that his respect for them was slight compared to his feelings for his writing,

Center the title on the page. Triple-space between the title and the first line of the text. The page number for the first page may be omitted or centered at the bottom.

## General Guidelines for All Text Pages

Allow ample and even margins.
Double-space the text.
Indent five spaces for paragraphs.
Leave two spaces after periods and other terminal punctuation.
Leave one space after other marks of punctuation.

## The Second Text Page: Placement of Footnote Numerals

2

which came slowly and painfully. Yet others saw depth in the drawings that seemed spontaneously dredged up from the unconscious mind of a genius. "Blending the worlds of reality and dream," as one critic has put it, the drawings, not the writings, "are the purest expression of his imagination."[4]

The person most responsible for getting Thurber's drawings into magazines and books was E.B. White. While sharing an office with him at The New Yorker, White noticed that Thurber produced dozens of drawings, which were promptly crumpled up and tossed into the trash. One of these turned out to be a seal resting on a large rock and looking dreamily off into the distant sea where there were two approaching specks. The caption read, "Hm, explorers." White was so amused and so convinced of a profound comic talent in Thurber that he tried to get the editor of The New Yorker to publish the cartoon.[5] This failed, but when they collaborated in 1929 on a satirical book, White insisted to the publishers that Thurber do the illustrations. Both for the writing and the drawings, the satirical book was a hit, and Thurber was on his way. Seeing how well the book was selling, the editor of The New Yorker then asked Thurber if he could publish the seal drawing. Thurber had thrown it away, however, and when he tried to reproduce it, it came out differently, and was destined to be the most famous of all his cartoons.[6]

The illustrations for the satirical book and the cartoons that steadily appeared in The New Yorker magazine thereafter brought Thurber much attention as a funny and peculiar scrawler, but he was surprised to realize in 1931 that he was being taken seriously by artists and critics of high reputation. When the respected British artist Paul Nash came to America, he sought out Thurber. "In the paintings of his American contemporaries he [Nash] found

Place the page number in the upper right-hand corner, two lines above the first line of text. Use Arabic numerals; do not put a period after the number. Place footnote numbers slightly above the line of type and after marks of punctuation. Do not leave a space before the number; do not place a period after the number. Number footnotes consecutively throughout the paper. Never repeat a number in the text even if the references are exactly the same.

## The Fourth Text Page: Placement of Quoted Matter

4

What the writer of one of Thurber's obituaries said about his dogs can also be said about most of the other animals: "His drawings of dogs, which he produced with abundance on the backs of envelopes, in telephone books and on tablecloths, had a quality that seemed to link them with no other beast on earth."[11]

If the general subject of Thurber's drawings is the war between men and women, with an animal of some sort frequently the innocent bystander, his technique is having no technique--that is, none that sophisticated training teaches. In answer to Alistair Cooke's question as to whether his cartoons had encouraged parents to submit their children's drawings, Thurber answered:

> It actually did--not only parents but very strange people. Some people thought my drawings were done under water; others that they were done by moonlight. But mothers thought that I was a little child or that my drawings were done by my grand-daughter. So they sent in their own children's drawings to the New Yorker, and I was told to write these ladies, and I would write them all the same letter: "Your son can certainly draw as well as I can. The only trouble is that he hasn't been through as much."[12]

With his usual tone of humility where his drawings were concerned, he was indulging in satire of others and himself, but he revealed nevertheless in his comment about how much he had "been through" an understanding of the relationship between his deepest inner self and his drawings. "No one understands," commented Dorothy Parker, "how he makes his boneless, loppy beings, with their shy kinship to the men and women of Picasso's later drawings," and likewise, no one knows "from what dark breeding-ground come the artist's ideas."[13] Like so many others, Dorothy Parker was intrigued, if not spellbound, by the drawing of the seal:

Prose quotations of about five lines or more generally should be set off as indicated here. Single-space within the quotation. Paragraphing within the passage should follow the source. Indent for paragraphs when you quote more than one. *Do not enclose blocked quotations in quotation marks.* When there is a quotation within a quotation, as here, use double quotation marks to indicate it.

## Footnotes: Listed Separately

10

Footnotes

1 "Morose Scrawler," Time, 31 Dec. 1934, p. 38.

2 "Morose Scrawler," p. 38.

3 In the previous year Smith College exhibited Thurber's drawings and those of George Grosz. Then after the one-man show at the Valentine Gallery, "it was clear that he had arrived as a comic artist," Charles S. Holmes, The Clocks of Columbus: The Literary Career of James Thurber (New York: Atheneum, 1972), p. 163.

4 Charles S. Holmes, Introd., Thurber: A Collection of Critical Essays, ed. Charles S. Holmes. Twentieth Century Views (Englewood Cliffs, N. J.: Prentice-Hall, 1974), p. 7.

5 Thurber recounted this incident in an interview with Alistair Cooke. "James Thurber in Conversation with Alistair Cooke," The Atlantic, Aug. 1956, p. 38.

6 Burton Bernstein calls it "one of the most celebrated and often-reprinted cartoons of the twentieth century." Thurber: A Biography (New York: Dodd, Mead, 1975), p. 190. The eminent humorist Robert Benchley was so taken with it that he sent Thurber a telegram that read as follows: "Thank you for the funniest drawing caption ever to appear in any magazine." See Holmes, Clocks, p. 136.

7 Margot Eates, Paul Nash: The Master of the Image, 1889-1946 (London: John Murray, 1973), p. 48.

8 His older brother accidently shot him in the eye with an arrow when they were children, causing the immediate loss of that eye. Thurber gradually lost the sight of the other eye because of complications from the accident and a cataract.

9 "That Thurber Woman," Newsweek, 22 Nov. 1943, p. 84.

10 Norris W. Yates, The American Humorist: Conscience of the Twentieth Century (Ames: Iowa Univ. Press, 1964), p. 278.

11 "James Thurber Is Dead at 66; Writer Was Also Comic Artist," New York Times, 3 Nov. 1961, p. 1, col. 3. An art critic remarked that Thurber's dog figure is "a fantasy dog, which arouses sympathy and laughter in many people and faintly depresses others." Alan Priest, "Mr. Thurber's Chinese Dog," The Metropolitan Museum of Art Bulletin, 4 (1946), 261.

Footnotes may be grouped together at the end of the paper as shown here, or they may be placed at the bottom of the pages on which they occur as illustrated in the forthcoming alternate style. The word *Footnotes* (or *Notes*) is centered on the page. Triple-space between the title and the first footnote; double-space between footnotes. Indent the first line of every footnote five spaces; do not indent succeeding lines. In footnotes (but not in the bibliography) the author's name is written in normal order, first name first.

## Footnotes: Alternate Styling at Bottom of Page

Hope Through Fantasy in James Thurber's Drawings

In December of 1934 the highly respected Valentine Gallery in Manhattan held one of the most unusual exhibitions in all of its history. Less traditional than many of the art galleries in New York City, the Valentine "devoted itself to the more advanced of the socially acceptable left-wing artists."[1] Nevertheless, this was a particularly odd one-man show: as viewers shuffled from one frame to another, they witnessed child-like reproductions of gentle dogs, aggressive women, and emasculated men all casually drawn as doodles on various sorts and sizes of sheets, even scratch paper.[2] Much puzzlement and disbelief ran through these lovers of art; but the exhibition was widely reviewed, and the artist, James Thurber, was in the end compared with such giants as Picasso and Matisse.[3] Since then, Thurber's position as a serious artist as well as a serious writer has been solidly established, although his cartoons, known best from the pages of The New Yorker, are deceptively simple. His fame has been accompanied by a reputation for bitterness, which is in some measure justified; but if pessimism is reflected in Thurber's drawings, a sense of hope is even stronger.

---

1 "Morose Scrawler," Time, 31 Dec. 1934, p. 38.

2 "Morose Scrawler," p. 38.

3 In the previous year Smith College exhibited Thurber's drawings and those of George Grosz. Then after the one-man show at the Valentine Gallery, "it was clear that he had arrived as a comic artist." Charles S. Holmes, The Clocks of Columbus: The Literary Career of James Thurber (New York: Atheneum, 1972), p. 163.

This sample shows how footnotes can be placed at the bottom of pages on which they occur. Follow your instructor's preference as to where you place the footnotes. Separate the footnotes from the text by a short ruled or typed line starting at the left-hand margin and placed far enough below the last line of the text so that it will not be mistaken for underlining to indicate italics.

## Bibliography

12

Works Cited

Auden, W. H. "The Icon and the Portrait." The Nation, 13 Jan. 1940, p. 48; rpt. in Thurber: A Collection of Critical Essays. Ed. Charles S. Holmes. Twentieth Century Views. Englewood Cliffs, N. J.: Prentice-Hall, 1974, pp. 59-61.

Becker, Stephen. Comic Art in America. Introd. Rube Goldberg. New York: Simon and Schuster, 1959.

Benson, E. M. "Phases of Fantasy." The American Magazine of Art, 28 (1935), 290-299.

Bernstein, Burton. Thurber: A Biography. New York: Dodd, Mead, 1975.

Black, Stephen A. James Thurber: His Masquerades. Studies in American Literature, 23. The Hague: Mouton, 1970.

Eates, Margot. Paul Nash: The Master of the Image, 1889-1946. London: John Murray, 1973.

Elias, Robert H. "James Thurber: The Primitive, the Innocent, and the Individual." The American Scholar, 27 (1958), 355-63.

Hasley, Louis. "James Thurber: Artist in Humor." The South Atlantic Quarterly, 73 (1975), 504-15.

Holmes, Charles S. The Clocks of Columbus: The Literary Career of James Thurber. New York: Atheneum, 1972.

________, introd. Thurber: A Collection of Critical Essays. Ed. Charles S. Holmes. Twentieth Century Views. Englewood Cliffs, N. J.: Prentice-Hall, 1974.

"James Thurber in Conversation with Alistair Cooke." The Atlantic, Aug. 1956, pp. 36-40.

"James Thurber Is Dead at 66; Writer Was Also Comic Artist." New York Times, 3 Nov. 1961, p. 1, cols. 3-4, p. 35, cols. 2-4.

"Morose Scrawler." Time, 31 Dec. 1934, p. 38.

Morsberger, Robert E. James Thurber. Twayne's United States Authors Series, 62. New York: Twayne, 1964.

Nugent, Elliott. "Notes on James Thurber the Man, or Men." New York Times, 22 Feb. 1940, Sec. 10, p. 3, cols. 1-4.

Parker, Dorothy, introd. James Thurber, The Seal in the Bedroom and Other Predicaments. New York: Harper & Row, 1932.

Priest, Alan. "Mr. Thurber's Chinese Dog." The Metropolitan Museum of Art Bulletin, 4 (1946), 260-61.

"That Thurber Woman." Newsweek, 22 Nov. 1943, pp. 84-86.

Thurber, James. "The Lady on the Bookcase." In The Beast in Me and Other Animals. New York: Harcourt, Brace, 1948, pp. 66-75.

Tobias, Richard C. The Art of James Thurber. Athens: Ohio Univ. Press, 1969.

Yates, Norris W. The American Humorist: Conscience of the Twentieth Century. Ames: Iowa Univ. Press, 1964.

Start the bibliography on a new page as the last section of your paper. Head the page Works Cited, centered. Triple-space below the heading. Do not indent the first line of an entry; indent succeeding lines five spaces. Double-space between entries, single-space within an entry. Authors are listed with surnames first. If a book has more than one author, however, the names of authors after the first one are put in normal order. List entries alphabetically. When more than one book by the same author is listed, use a line (about one inch) in place of the author's name in entries after the first. An entry without an author is listed alphabetically by the first word. List the inclusive pages of articles.

# Business Letter Styles

## Simplified Letter

**XYZ Company**

200 Enterprise Street, Chicago, IL 60631

**3 lines**

October 14, 19--

**3 lines**

Ms. Gail Hemingway
Vice President
KKKT Television
45 Brook Mountain Road
City, US 98765

**3 lines**

THE SIMPLIFIED LETTER

**3 lines**

Ms. Hemingway, this is an example of the Simplified Letter recommended by the Administrative Management Society. Its lean format lends itself readily to high-volume mailings. All of its internal elements are set tight with the left margin.

**2 lines**

The date is positioned three vertical line spaces below the last line of the printed letterhead. The inside address begins three lines below the date. There is no salutation. A capitalized subject line summarizes the thrust of the letter. The subject line is positioned three lines below the last line of the inside address.

The message, introduced by use of the recipient's name as shown in the first paragraph here, begins three lines below the subject line. All paragraphs, single-spaced internally, are set flush left. Double-spacing separates the paragraphs.

If a continuation sheet is used, at least three message lines must be carried over to it: at no time should the signature block stand alone there. The heading is typed six lines from the top of the page and is aligned with the left margin:

Ms. Hemingway
Page 2
October 14, 19--

Continue the message at least four lines below the last line of the heading. There is no complimentary close. The writer's name and title appear in capital letters at least five vertical line spaces below the last line of the message. A spaced hyphen separates the name from the title, all on the same line. Other notations may be typed two lines below the signature block.

David R. Ramsey

**5 lines**

DAVID R. RAMSEY - DIRECTOR/MEDIA RELATIONS

**2 lines**

ahs

## Block Letter

**XYZ Company**

200 Enterprise Street, Chicago, IL 60631

**3 lines**

October 14, 19--

**4 lines/variable**

Ms. Jennifer Stone
Vice President, Corporate Plans
CBA Corporation
43 Hunting Towers  Suite 100
City, US 98765

**2 lines**

Dear Ms. Stone:

**2 lines**

This is the Block Letter, all elements of which are aligned tight with the left margin. Spacing between letter parts is indicated by the key lines. Individual paragraphs in the message are single-spaced internally. Double-spacing separates each of the paragraphs one from another.

**2 lines**

If the letter exceeds one page, a continuation sheet is used. The heading begins six lines from the top edge of the page. The heading includes the name of the recipient, the page number, and the full date, blocked and aligned with the left margin as shown in the example here:

Ms. Jennifer Stone
Page 2
October 14, 19--

The message continues four lines below the heading. At least three message lines must be carried over to the continuation sheet: at no time should the complimentary close and the signature block stand alone there.

The writer's corporate title is shown under the typewritten signature line. Typist's initials, enclosure notations, and copy recipient listings are typewritten tight with the left margin and are spaced vertically as shown by the key lines.

**2 lines**

Sincerely yours,

**4 lines**

John B. Brown
Director of Marketing

**2 lines**

JBB:ahs
enclosures (2)

**2 lines**

cc: S.A. Longley

## Modified Block Letter

**XYZ Company**
200 Enterprise Street, Chicago, IL 60631

**4 lines/variable**

October 14, 19--

**3 lines/variable**

Mr. Irwin A. Levine
Treasurer
CCC Corporation
123 Westwood Blvd.
City, US 98765

**2 lines**

Dear Mr. Levine:

**2 lines**

Subject: Modified Block Letter

**2 lines**

This is an example of the Modified Block Letter, the distinctive features of which include the date line, the complimentary close, and the typewritten signature block aligned slightly to the right of center, as shown here. One also may align them tight with the right margin. A subject line (if included) is centered on the page two lines below the salutation and two lines above the message.

**2 lines**

The inside address, salutation, paragraphs of the message, typist's initials, enclosure notation, and list of copy recipients are all aligned flush with the left margin. If the message exceeds one page, a continuation sheet is used. The continuation sheet heading is typed six lines from the top edge of the page. It may be blocked and aligned tight with the left margin as shown within the Block Letter facsimile or it may be spaced across the page on a single line, as

Mr. Levine - 2 - October 14, 19--

At least three lines of the message must be carried over to the continuation sheet; at no time should the signature block and the complimentary close stand alone there.

Spacing between all elements of the Modified Block Letter is indicated by the key lines in this illustration.

**2 lines**

Very truly yours,

**4 lines** Nancy A. Martin

Nancy A. Martin
Comptroller

**2 lines**

NAM:ahs

encs: 5

**2 lines**

CC: L.A. Long
J.A. Matthews

## Modified Semi-block Letter

**XYZ Company**

200 Enterprise Street, Chicago, IL 60631

**4 lines/variable**

October 14, 19--

**4 lines/variable**

Ms. Cynthia Stanley
Personnel Director
Lexigrafix
123 Willow Run Parkway
City, US 98765

**2 lines**

Dear Ms. Stanley:

**2 lines**

This is the Modified Semi-block Letter. Its paragraphs are indented, and its date line, signature block, and complimentary close are vertically aligned slightly to the right of center, in the center of the page, or flush with the right margin. Placement of these elements depends on the writer's wishes.

**2 lines**

If a subject line is included, it is situated in the center of the page, two vertical line spaces below the salutation and two vertical line spaces above the first line of the message.

The paragraphs of the message are single-spaced internally but double-spacing is used between them to set them off one from another. If the message exceeds one page in length, a continuation sheet must be used. At least three message lines must be carried over onto the continuation sheet: at no time should the complimentary close and the signature block stand alone there.

Begin the heading of the continuation sheet six lines from the top edge of the page. Space the heading out on a single line as illustrated in the Modified Block facsimile herein.

Skip two lines from the last message line to the beginning of the complimentary close. Allow at least four lines for the written signature. Single-space the writer's name and title on two lines after the written signature space.

Typist's initials, an enclosure notation, or an indicator of copy recipients appear two spaces below the last line of the signature block, tight with the left margin.

If a postscript is included, it must be indented by the same number of spaces that the paragraphs of the message have been indented. The writer should initial any appended matter.

**2 lines**

Sincerely yours,

**4 lines** *Maria G. Godek*

Maria G. Godek
Senior Editor

MGG:ahs

## Hanging-indented Letter

**XYZ Company**
200 Enterprise Street, Chicago, IL 60631

**3 lines**

October 14, 19--

**2–3 lines/variable**

Mr. Robert V. Parker
Engineering Manager
VCC Corporation
P.O. Box 1234
City, US 98765

**2 lines**

Dear Mr. Parker:

**2 lines**

This is an example of the Hanging-indented Letter--a style usually reserved for direct mail advertising of new products and services. Its elegant, distinctive format lends itself well to this sort of business correspondence. Note that its paragraph alignment features a flush left first line with runovers indented and blocked by five character spaces. Paragraphs are single-spaced internally. Double-spacing separates the paragraphs.

**2 lines**

The date line, typed three lines below the letterhead, is set tight with the right margin. The salutation and inside address are set flush left. A subject line, if included, would also be aligned flush with the left margin. Spacing is indicated by the key lines.

A continuation sheet is used if the message exceeds one page. The continuation sheet heading is typewritten six lines below the top edge of the paper. It may be blocked tight with the left margin as shown in the Block Letter or it may be spaced across the top of the page on a single line as shown in the Modified Block Letter herein. Maintain margins and paragraph alignment matching those on the letterhead page.

Align the complimentary close and the typewritten signature block with the date line--i.e., tight with the right margin. The typist's initials, an enclosure notation, or a copy recipient notation may be typed two lines below the last line of the signature block, flush with the left margin. A postscript, if included, is also hanging-indented.

**2 lines**

Sincerely yours,

**4 lines** *Mary*

Mary A. Smythe
Public Relations

**2 lines**

MAS:ahs

**2 lines**

Encs: 6

**2 lines**

CC: Charles D. Rand
Judith T. Queensbury

## Job Application Letter

321 State Street Apt. 39
City, US 98765
December 2, 19--

Mr. Alexander R. Burke
Editor-in-Chief, College Texts
ABC Publishing Company, Inc.
89 Bacon Avenue
City, US 98765

Dear Mr. Burke:

Let me introduce myself in response to your advertisement regarding a Senior Editorship/Science Acquisitions, that appeared on page 98 of the November 23, 19-- issue of Publishing Weekly. As you can see from the enclosed Profile, I have sixteen years of uninterrupted college science textbook editing and publishing experience. This experience was supplemented recently by a series of courses on technical writing/editing for computerized publishing applications. My transcripts from the courses are available, should you wish to look them over.

I have a number of valuable potential authors at my disposal for the writing and editing of a college science series appropriate to the Eighties. I'm particularly interested in producing a series of biology textbooks with add-on diskettes--a project that might fit into your publishing plans.

Last week while attending the national meeting of the American Association for the Advancement of Science I ran into Joseph Hill, your Managing Editor. We had a nice visit. Joe, who has known me for over ten years, gave me a general idea of the directions in which ABC is moving in response to electronics and I was indeed impressed.

Although it would be a great wrench to leave Allied--a house in which I've enjoyed many challenges and successes--I nevertheless feel very much drawn to ABC because of its forward-looking management and editorial posture. Your recent success with the PhysSci Software/Textware series is a case in point. Any science editor in tune with the technology of the times would feel the same way, I'm sure.

Perhaps you might like to talk with me after you have read through my Profile. I can be reached at the number on the Profile after 5:00 p.m., weekdays. My office number is 678-2345.

I do hope that my background and experience are what you're looking for, and that my ideas parallel some of yours. I look forward to hearing from you at your convenience.

Sincerely yours,

Margaret E. Longford

Margaret E. Longford

Enclosure

## Résumé

PROFILE

Margaret E. Longford
321 State Street Apartment 39
City, US 98765
(800) 555-1212

Experience

| | |
|---|---|
| November 1975 - December 1983 | Allied College Publications, Inc.<br>College Textbook Division/Science<br>Associate Editor, Acquisitions |
| July 1969 - November 1975 | Howe & Row Publishers, Inc.<br>College Textbook Department<br>Assistant Editor, Life Sciences |
| March 1967 - July 1969 | LangData Incorporated<br>College Division<br>Editorial Assistant |

Experience Summary

Sixteen years' experience in line editing of college science textbooks . Author acquisitions . Production & Scheduling . Cost estimation . Budgets . Contracts . Revisions . Manuscript Review . Liaison with Typesetters, Artists, Designers, and Printers . Public Relations & Marketing Support . Liaison with Sales Personnel

Education

| | | |
|---|---|---|
| 1964 - 1967 | MBA | Taft Graduate School of Business Management |
| 1960 - 1964 | BA (cum laude) | Hartfield College<br>Major: Biology |
| 1956 - 1960 | diploma | Stonleigh School for Girls |

Languages

French (fluent)
German (fluent)
Russian (scientific only)

Publications List

Available upon request

References

Available upon request

# Forms of Address

Forms of address do not always follow set guidelines; the type of salutation is often determined by the relationship between correspondents or by the purpose and content of the letter. However, a general style applies to most occasions. In highly formal salutations, when the addressee is a woman, "Madam" should be substituted for "Sir." (Use "Madame" when the addressee is a foreign dignitary or diplomat.) When the salutation is informal, "Ms.," "Miss," or "Mrs." should be substituted for "Mr." If a woman addressee has previously indicated a preference for a particular form of address, that form should be used. In the following table, the slash (/) indicates alternative address stylings for women and men. ("Miss" or "Mrs." may be substituted for "Ms.")

| **Academics** | **Form of Address** | **Salutation** |
|---|---|---|
| assistant professor | Professor Joseph/Jane Stone<br>Mr./Ms./Dr. Joseph/Jane Stone | Dear Professor Stone:<br>Dear Mr./Ms. Stone:<br>Dear Dr. Stone: |
| associate professor | Professor Joseph/Jane Stone<br>Mr./Ms./Dr. Joseph/Jane Stone | Dear Professor Stone:<br>Dear Mr./Ms. Stone:<br>Dear Dr. Stone: |
| chancellor, university | Dr./Mr./Ms. Joseph/Jane Stone | Dear Chancellor Stone: |
| chaplain | The Reverend Joseph/Jane Stone | Dear Chaplain Stone:<br>Dear Mr./Ms. Stone:<br>Dear Father Stone: |
| dean, college or university | Dean Joseph/Jane Stone<br>*or*<br>Dr./Mr./Ms. Joseph/Jane Stone<br>Dean, School of ______ | Dear Dean Stone:<br><br>Dear Dr./Mr./Ms. Stone: |
| instructor | Mr./Ms./Dr. Joseph/Jane Stone | Dear Mr./Ms./Dr. Stone: |
| president | President Joseph/Jane Stone<br>*or*<br>Dr./Mr./Ms. Joseph/Jane Stone | Dear President Stone:<br><br>Dear Dr./Mr./Ms. Stone: |
| president/priest | The Reverend Joseph Stone, S.J.<br>President of ______ | Sir:<br>Dear Father Stone: |
| professor, college or university | Professor Joseph/Jane Stone<br>*or*<br>Dr./Mr./Ms. Joseph/Jane Stone | Dear Professor Stone:<br><br>Dear Dr./Mr./Ms. Stone: |

| **Clerical and Religious Orders** | **Form of Address** | **Salutation** |
|---|---|---|
| abbot, Roman Catholic | The Right Reverend Joseph Stone, O.S.B.<br>Abbot of ______ | Right Reverend Abbot:<br>Dear Father Abbot: |
| apostolic delegate | His Excellency<br>The Most Reverend Joseph Stone<br>Archbishop of ______<br>The Apostolic Delegate | Your Excellency:<br>My dear Archbishop: |
| archbishop, Greek Orthodox | The Most Reverend Joseph<br>Archbishop of ______ | Your Eminence: |
| archbishop, Roman Catholic | The Most Reverend Joseph Stone<br>Archbishop of ______ | Your Excellency: |
| archbishop, Russian Orthodox | The Most Reverend Joseph<br>Archbishop of ______ | Your Eminence: |
| archdeacon, Episcopal | The Venerable Joseph Stone<br>Archdeacon of ______ | Venerable Sir:<br>Dear Archdeacon Stone: |

| Clerical and Religious Orders | Form of Address | Salutation |
|---|---|---|
| archimandrite, Greek Orthodox | The Very Reverend Joseph Stone | Reverend Sir:<br>Dear Father Joseph: |
| archimandrite, Russian Orthodox | The Right Reverend Joseph Stone | Reverend Sir:<br>Dear Father Joseph: |
| archpriest, Greek Orthodox | The Reverend Joseph Stone | Dear Father Joseph: |
| archpriest, Russian Orthodox | The Very Reverend Joseph Stone | Dear Father Joseph: |
| bishop, Episcopal | The Right Reverend Joseph Stone<br>Bishop of ______ | Right Reverend Sir:<br>Dear Bishop Stone: |
| bishop, Greek Orthodox | The Right Reverend Joseph<br>Bishop of ______ | Your Grace: |
| bishop, Methodist | The Reverend Joseph Stone<br>Methodist Bishop | Dear Bishop Stone: |
| bishop, Roman Catholic | The Most Reverend Joseph Stone<br>Bishop of ______ | Your Excellency:<br>Dear Bishop Stone: |
| bishop, Russian Orthodox | The Most Reverend Joseph<br>Bishop of ______ | Your Grace: |
| brotherhood, Roman Catholic, member of | Brother Joseph Stone, C.F.C. | Dear Brother:<br>Dear Brother Joseph: |
| brotherhood, Roman Catholic, superior of | Brother Joseph, C.F.C.<br>Superior | Dear Brother Joseph: |
| canon, Episcopal | The Reverend Canon Joseph Stone | Dear Canon Stone: |
| cantor (man) | Cantor Joseph Stone | Dear Cantor Stone: |
| cantor (woman) | Cantor Jane Stone | Dear Cantor Stone: |
| cardinal | His Eminence<br>Joseph Cardinal Stone | Your Eminence: |
| clergyman, Protestant | The Reverend Joseph Stone<br>*or*<br>The Reverend Joseph Stone, D.D. | Dear Mr. Stone:<br>*or*<br>Dear Dr. Stone: |
| clergywoman, Protestant | The Reverend Jane Stone<br>*or*<br>The Reverend Jane Stone, D.D. | Dear Ms. Stone:<br>*or*<br>Dear Dr. Stone: |
| elder, Presbyterian | Elder Joseph/Jane Stone | Dear Elder Stone: |
| dean of a cathedral, Episcopal | The Very Reverend Joseph Stone<br>Dean of ______ | Dear Dean Stone: |
| metropolitan, Russian Orthodox | His Beatitude Joseph<br>Metropolitan of ______ | Your Beatitude: |
| moderator, Presbyterian | The Moderator of ______<br>*or*<br>The Reverend Joseph Stone<br>*or*<br>Dr. Joseph Stone | Reverend Sir:<br>My dear Sir:<br>Dear Mr. Moderator:<br><br>My dear Dr. Stone: |
| monsignor, Roman Catholic (domestic prelate) | The Right Reverend<br>Monsignor Joseph Stone | Right Reverend Monsignor:<br>Dear Monsignor:<br>Dear Monsignor Stone: |
| papal chamberlain | The Very Reverend<br>Monsignor Joseph Stone | Very Reverend and<br>Dear Monsignor Stone:<br>Dear Monsignor Stone: |
| patriarch, Greek Orthodox | His All Holiness<br>Patriarch Joseph | Your All Holiness: |
| patriarch, Russian Orthodox | His Holiness<br>the Patriarch of ______ | Your Holiness: |
| pope | His Holiness<br>The Pope | Your Holiness:<br>Most Holy Father: |
| president, Mormon Church | President Joseph Stone<br>Church of Jesus Christ of Latter-Day Saints | Dear President Stone: |

| Clerical and Religious Orders | Form of Address | Salutation |
|---|---|---|
| priest, Episcopal | The Reverend Joseph/Jane Stone<br>The Rev. Dr. Joseph/Jane Stone | Dear Mr./Ms. Stone:<br>Dear Dr. Stone: |
| priest, Greek Orthodox | The Reverend Joseph Stone | Dear Father Joseph: |
| priest, Roman Catholic | The Reverend Joseph Stone, S.J. | Dear Reverend Father:<br>Dear Father:<br>Dear Father Stone: |
| priest, Russian Orthodox | The Reverend Joseph Stone | Dear Father Joseph: |
| rabbi (man) | Rabbi Joseph Stone<br>*or*<br>Joseph Stone, D.D. | Dear Rabbi Stone:<br><br>Dear Dr. Stone: |
| rabbi (woman) | Rabbi Jane Stone<br>*or*<br>Jane Stone, D.D. | Dear Rabbi Stone:<br><br>Dear Dr. Stone: |
| sisterhood, Roman Catholic, member of | Sister Mary Viventia, C.S.J. | Dear Sister:<br>Dear Sister Viventia:<br>Dear Sister Mary: |
| sisterhood, Roman Catholic, superior of | The Reverend<br>Mother Superior, S.C. | Reverend Mother:<br>Dear Reverend Mother: |

| Diplomats | Form of Address | Salutation |
|---|---|---|
| ambassador, U.S. | The Honorable Joseph/Jane Stone<br>The Ambassador of the United States | Sir/Madam:<br>Dear Mr./Madam Ambassador: |
| ambassador to the U.S. | His/Her Excellency<br>Joseph/Jane Stone<br>The Ambassador of ______ | Excellency:<br>Dear Mr./Madame Ambassador: |
| chargé d'affaires, U.S. | The Honorable Joseph/Jane Stone<br>United States Chargé d'Affaires | Dear Mr./Ms. Stone: |
| chargé d'affaires to the U.S. | Joseph/Jane Stone, Esq.<br>Chargé d'Affaires of ______ | Dear Sir/Madame: |
| consul, U.S. | Joseph/Jane Stone, Esq.<br>United States Consul | Dear Mr./Ms. Stone: |
| consul to the United States | The Honorable Joseph/Jane Stone<br>Consul of ______ | Dear Mr./Ms. Stone: |
| minister, U.S. | The Honorable Joseph/Jane Stone<br>The Minister of the United States | Sir/Madam:<br>Dear Mr./Madam Minister: |
| minister to the U.S. | The Honorable Joseph/Jane Stone<br>The Minister of ______ | Sir/Madame:<br>Dear Mr./Madame Minister: |
| representative (foreign) to the United Nations (with rank of ambassador) | His/Her Excellency<br>Joseph/Jane Stone<br>Representative of ______ to the United Nations | Excellency:<br>My dear Mr./Madame Stone:<br>Dear Mr./Madame Ambassador: |
| secretary general, United Nations | His/Her Excellency<br>Joseph/Jane Stone<br>Secretary General of the United Nations | Dear Mr./Madam/Madame Secretary General: |
| undersecretary to the United Nations | The Honorable Joseph/Jane Stone<br>Undersecretary of the United Nations | Sir:/Madam: (if American)<br>Sir:/Madame: (if foreign)<br>My dear Mr./Ms. Stone:<br>Dear Mr./Ms. Stone: |
| U.S. representative to the United Nations | The Honorable Joseph/Jane Stone<br>United States Representative to the United Nations | Sir/Madam:<br>Dear Mr./Ms. Stone: |

| Federal, State, and Local Government Officials | Form of Address | Salutation |
|---|---|---|
| alderman | The Honorable Joseph/Jane Stone | Dear Mr./Ms. Stone: |
| assistant to the President | Mr./Ms. Joseph/Jane Stone | Dear Mr./Ms. Stone: |
| attorney general, U.S. | The Honorable Joseph/Jane Stone<br>Attorney General of the United States | Dear Mr./Madam Attorney General: |
| attorney general, state | The Honorable Joseph/Jane Stone<br>Attorney General<br>State of ________ | Dear Mr./Madam Attorney General: |
| assemblyman, state | The Honorable Joseph/Jane Stone | Dear Mr./Ms. Stone: |
| cabinet member | The Honorable Joseph/Jane Stone<br>Secretary of ________ | Sir/Madam:<br>Dear Mr./Madam Secretary: |
| cabinet member, former | The Honorable Joseph/Jane Stone | Dear Mr./Ms. Stone: |
| chief justice, U.S. Supreme Court | The Chief Justice of the United States | Dear Mr. Chief Justice:<br>Sir: |
| associate justice, U.S. Supreme Court | Mr./Madam Justice Stone | Dear Mr./Madam Justice:<br>Sir/Madam: |
| associate/chief justice, Supreme Court, former | The Honorable Joseph/Jane Stone | Dear Mr./Ms. Stone:<br>Dear Mr./Madam Justice Stone: |
| clerk, of a court | Joseph/Jane Stone, Esq.<br>Clerk of the Court of ________ | Dear Mr./Ms. Stone: |
| commissioner (federal, state, local) | The Honorable Joseph/Jane Stone | Dear Mr./Ms. Stone: |
| director, federal agency | The Honorable Joseph/Jane Stone<br>Director<br>________ Agency | Dear Mr./Ms. Stone: |
| district attorney | The Honorable Joseph/Jane Stone<br>District Attorney | Dear Mr./Ms. Stone: |
| governor | The Honorable Joseph/Jane Stone<br>Governor of ________ | Dear Governor Stone: |
| governor-elect | The Honorable Joseph/Jane Stone<br>Governor-elect of ________ | Dear Mr./Ms. Stone: |
| governor, former | The Honorable Joseph/Jane Stone | Dear Governor Stone:<br>Dear Mr./Ms. Stone: |
| judge, federal | The Honorable Joseph/Jane Stone<br>Judge of the United States District Court for the ________ District of ________ | Sir/Madam:<br>Dear Judge Stone: |
| judge, state or local | The Honorable Joseph/Jane Stone<br>Judge of the Court of ________ | Dear Judge Stone: |
| justice, Supreme Court, associate, chief, and former | **—See chief justice, supreme court** | |
| librarian of congress | The Honorable Joseph/Jane Stone<br>The Librarian of Congress | Sir:/Madam:<br>Dear Mr./Ms./Dr. Stone: |
| lieutenant governor | The Honorable Joseph/Jane Stone<br>Lieutenant Governor of ________ | Dear Mr./Ms. Stone: |
| mayor | The Honorable Joseph/Jane Stone<br>Mayor of ________ | Dear Mayor Stone: |
| postmaster general | The Honorable Joseph/Jane Stone<br>Postmaster General<br>United States Postal Service | Dear Mr./Madam Postmaster General: |
| president, U.S. | The President<br>The White House | Dear Mr. President: |
| president-elect, U.S. | The Honorable Joseph Stone<br>The President-elect of the United States | Dear Sir:<br>Dear Mr. Stone: |

| Federal, State, and Local Government Officials | Form of Address | Salutation |
|---|---|---|
| president, U.S., former | The Honorable Joseph Stone | Dear Mr. Stone:<br>Dear Mr. President:<br>Dear President Stone: |
| press secretary, to the President | Mr./Ms. Joseph/Jane Stone<br>Press Secretary to the President | Dear Mr./Ms. Stone: |
| representative, state | The Honorable Joseph/Jane Stone | Dear Mr./Ms. Stone: |
| representative, U.S. | The Honorable Joseph/Jane Stone<br>United States House of Represen-<br>tatives | Dear Mr./Ms. Stone: |
| secretary of state, for a state | The Honorable Joseph/Jane Stone<br>Secretary of State<br>State Capitol | Dear Mr./Madam Secretary: |
| senator, former (state or U.S.) | The Honorable Joseph/Jane Stone | Dear Senator Stone:<br>Dear Mr./Ms. Stone: |
| senator, state | The Honorable Joseph/Jane Stone<br>The State Senate<br>State Capitol | Dear Senator Stone: |
| senator, U.S. | The Honorable Joseph/Jane Stone<br>United States Senate | Dear Senator Stone: |
| speaker, U.S. House of Representatives | The Honorable Joseph/Jane Stone<br>Speaker of the House of Represen-<br>tatives | Dear Mr./Madam Speaker: |
| undersecretary, of cabinet department (applies to deputy and assistant secretaries also) | The Honorable Joseph/Jane Stone<br>Undersecretary of the Department<br>of ____________ | Dear Mr./Ms. Stone: |
| vice president, U.S. | The Vice President of the United<br>States<br>*or*<br>The Honorable Joseph Stone<br>Vice President of the United States | Sir:<br>My dear Mr. Vice President:<br>Dear Mr. Vice President:<br>Dear Mr. Vice President: |

| Military Rank—Officers* | Branch of Service | Form of Address | Salutation |
|---|---|---|---|
| admiral | USCG/USN | ADM Lee Stone, USCG/USN | Dear Admiral Stone: |
| brigadier general | USAF | Brig Gen Lee Stone, USAF | Dear General Stone: |
| | USA | BG Lee Stone, USA | Dear General Stone: |
| | USMC | BGen Lee Stone, USMC | Dear General Stone: |
| captain | USAF/USMC | Capt Lee Stone, USAF/USMC | Dear Captain Stone: |
| | USA | CPT Lee Stone, USA | Dear Captain Stone: |
| | USCG/USN | CAPT Lee Stone, USCG/USN | Dear Captain Stone: |
| chief warrant officer | USAF/USA | CWO Lee Stone, USAF/USA | Dear Mr./Ms. Stone: |
| colonel | USAF/USMC | Col Lee Stone, USAF/USMC | Dear Colonel Stone: |
| | USA | COL Lee Stone, USA | Dear Colonel Stone: |
| commander | USCG/USN | CDR Lee Stone, USCG/USN | Dear Commander Stone: |
| ensign | USCG/USN | ENS Lee Stone, USCG/USN | Dear Ensign Stone:<br>Dear Mr./Ms. Stone: |
| first lieutenant | USAF | 1st Lt Lee Stone, USAF | Dear Lt. Stone: |
| | USA | 1LT Lee Stone, USA | Dear Lt. Stone: |
| | USMC | 1stLt Lee Stone, USMC | Dear Lt. Stone: |

* The ranks and their abbreviations above are used with the names of military officers. The abbreviated rank is followed by the full name, a comma, and the appropriate abbreviation of the person's branch of service (USAF for U.S. Air Force, USA for U.S. Army, USCG for U.S. Coast Guard, USMC for U.S. Marine Corps, or USN for U.S. Navy). Example: ADM Lee Stone, USN. These forms of address apply to men and women, and the first name "Lee" is meant to cover both sexes. Subsequent tables treat cadet/midshipman and enlisted ranks.

| Military Rank—Officers | Branch of Service | Form of Address | Salutation |
|---|---|---|---|
| general | USAF/USMC | Gen Lee Stone, USAF/USMC | Dear General Stone: |
| | USA | GEN Lee Stone, USA | Dear General Stone: |
| lieutenant | USCG/USN | LT Lee Stone, USCG/USN | Dear Lt. Stone:<br>Dear Mr./Ms. Stone: |
| lieutenant colonel | USAF | Lt Col Lee Stone, USAF | Dear Colonel Stone: |
| | USA | LTC Lee Stone, USA | Dear Colonel Stone: |
| | USMC | LtCol Lee Stone, USMC | Dear Colonel Stone: |
| lieutenant commander | USCG/USN | LCDR Lee Stone, USCG/USN | Dear Commander Stone: |
| lieutenant general | USAF | Lt Gen Lee Stone, USAF | Dear General Stone: |
| | USA | LTG Lee Stone, USA | Dear General Stone: |
| | USMC | LtGen Lee Stone, USMC | Dear General Stone: |
| lieutenant (junior grade) | USCG/USN | LTJG Lee Stone, USCG/USN | Dear Lt. Stone:<br>Dear Mr./Ms. Stone: |
| major | USAF/USMC | Maj Lee Stone, USAF/USMC | Dear Major Stone: |
| | USA | MAJ Lee Stone, USA | Dear Major Stone: |
| major general | USAF | Maj Gen Lee Stone, USAF | Dear General Stone: |
| | USA | MG Lee Stone, USA | Dear General Stone: |
| | USMC | MajGen Lee Stone, USMC | Dear General Stone: |
| rear admiral | USCG/USN | RADM Lee Stone, USCG/USN | Dear Admiral Stone: |
| second lieutenant | USAF | 2d Lt Lee Stone, USAF | Dear Lt. Stone: |
| | USA | 2LT Lee Stone, USA | Dear Lt. Stone: |
| | USMC | 2dLt Lee Stone, USMC | Dear Lt. Stone: |
| vice admiral | USCG/USN | VADM Lee Stone, USCG/USN | Dear Admiral Stone: |
| warrant officer | USAF/USA | WO Lee Stone, USAF/USA | Dear Mr./Ms. Stone: |

**Cadets and Midshipmen**

| Rank | Form of Address | Salutation |
|---|---|---|
| cadet | Cadet Lee Stone | Dear Cadet Stone:<br>Dear Mr./Ms. Stone: |
| midshipman | Midshipman Lee Stone | Dear Midshipman Stone:<br>Dear Mr./Ms. Stone: |

**Enlisted Personnel: A Representative Listing**

| Rank | Branch of Service | Form of Address | Salutation |
|---|---|---|---|
| airman | USAF | AMN Lee Stone, USAF | Dear Airman Stone: |
| airman basic | USAF | AB Lee Stone, USAF | Dear Airman Stone: |
| airman first class | USAF | A1C Lee Stone, USAF | Dear Airman Stone: |
| chief petty officer | USCG/USN | CPO Lee Stone, USCG/USN | Dear Mr./Ms. Stone: |
| corporal | USA | CPL Lee Stone, USA | Dear Corporal Stone: |
| gunnery sergeant | USMC | GySgt Lee Stone, USMC | Dear Sergeant Stone: |
| lance corporal | USMC | L/Cpl Lee Stone, USMC | Dear Corporal Stone: |
| master sergeant | USAF | MSGT Lee Stone, USAF | Dear Sergeant Stone: |
| | USA | MSG Lee Stone, USA | Dear Sergeant Stone: |
| petty officer | USCG/USN | PO Lee Stone, USCG/USN | Dear Mr./Ms. Stone: |
| private | USA | PVT Lee Stone, USA | Dear Private Stone: |
| | USMC | Pvt Lee Stone, USMC | Dear Private Stone: |
| private first class | USA | PFC Lee Stone, USA | Dear Private Stone: |
| seaman | USCG/USN | SMN Lee Stone, USCG/USN | Dear Seaman Stone: |
| seaman first class | USCG/USN | S1C Lee Stone, USCG/USN | Dear Seaman Stone: |

**Enlisted Personnel: A Representative Listing**

| **Rank** | **Branch of Service** | **Form of Address** | **Salutation** |
|---|---|---|---|
| senior master sergeant | USAF | SMSGT Lee Stone, USAF | Dear Sergeant Stone: |
| sergeant | USAF | SGT Lee Stone, USAF | Dear Sergeant Stone: |
| | USA | SG Lee Stone, USA | Dear Sergeant Stone: |
| sergeant major (a title not a rank) | USA/USMC | SGM/Sgt. Maj. Lee Stone, USA/USMC | Dear Sergeant Major Stone: |
| specialist (as specialist 4th class) | USA | S4 Lee Stone, USA | Dear Specialist Stone: |
| staff sergeant | USAF | SSGT Lee Stone, USAF | Dear Sergeant Stone: |
| | USA | SSG Lee Stone, USA | Dear Sergeant Stone: |
| technical sergeant | USAF | TSGT Lee Stone, USAF | Dear Sergeant Stone: |

| **Professions** | **Form of Address** | **Salutation** |
|---|---|---|
| attorney | Mr./Ms. Joseph/Jane Stone<br>Attorney at Law<br>*or*<br>Joseph/Jane Stone, Esq. | Dear Mr./Ms. Stone: |
| dentist | Joseph/Jane Stone, D.D.S. | Dear Dr. Stone: |
| physician | Joseph/Jane Stone, M.D. | Dear Dr. Stone: |
| veterinarian | Joseph/Jane Stone, D.V.M. | Dear Dr. Stone: |

# Abbreviations and Labels in this Dictionary

**Subject Labels**

| Label | Full Form |
|---|---|
| *Aerospace.* | Aerospace |
| *Anat.* | Anatomy |
| *Astron.* | Astronomy |
| *Astrophysics.* | Astrophysics |
| *Baseball.* | Baseball |
| *Basketball.* | Basketball |
| *Biochem.* | Biochemistry |
| *Biol.* | Biology |
| *Bot.* | Botany |
| *Chem.* | Chemistry |
| *Christian Science.* | Christian Science |
| *Computer Sci.* | Computer Science |
| *Ecol.* | Ecology |
| *Elect.* | Electricity |
| *Electron.* | Electronics |
| *Engineer.* | Engineering |
| *Football.* | Football |
| *Genetics.* | Genetics |
| *Geol.* | Geology |
| *Gk. Myth.* | Greek Mythology |
| *Gk. & Rom. Myth.* | Greek & Roman Mythology |
| *Heraldry.* | Heraldry |
| *Law.* | Law |
| *Logic.* | Logic |
| *Math.* | Mathematics |
| *Med.* | Medicine |
| *Metallurgy.* | Metallurgy |
| *Meteorol.* | Meteorology |
| *Microbiol.* | Microbiology |
| *Mineral.* | Mineralogy |
| *Mormon Ch.* | Mormon Church |
| *Mus.* | Music |
| *Myth.* | Mythology |
| *Naut.* | Nautical |
| *Norse Myth.* | Norse Mythology |
| *Pathol.* | Pathology |
| *Philos.* | Philosophy |
| *Physics.* | Physics |
| *Physiol.* | Physiology |
| *Psychiat.* | Psychiatry |
| *Psychoanal.* | Psychoanalysis |
| *Psychol.* | Psychology |
| *Rom. Cath. Ch.* | Roman Catholic Church |
| *Rom. Myth.* | Roman Mythology |
| *Statistics.* | Statistics |
| *Zool.* | Zoology |

**Stylistic and Geographic Labels**

| | |
|---|---|
| *Archaic.* | Archaic |
| *Austral.* | Australian |
| *Chiefly Brit.* | Chiefly British |
| *Informal.* | Informal |

**Stylistic and Geographic Labels**

| | |
|---|---|
| *Ir.* | Irish |
| *Nonstandard.* | Nonstandard |
| *Obs.* | Obsolete |
| *Regional.* | Regional |
| *Scot.* | Scots |
| *Slang.* | Slang |
| *So. Afr.* | South African |

**Abbreviations & Symbols in Etymologies**

| | | | |
|---|---|---|---|
| abbr. | abbreviation | p. | past |
| adj. | adjective | part. | participle, participial |
| adv. | adverb | p.part. | past participle |
| aug. | augmentative | p.t. | past tense |
| cent. | century | perh. | perhaps |
| comp. | comparative | pl. | plural |
| conj. | conjunction | poss. | possibly |
| dial. | dialectal | pr. | present |
| dim. | diminutive | pr.part. | present participle |
| ety. | etymology | prob. | probably |
| fem. | feminine | redup. | reduplication |
| freq. | frequentative | sing. | singular |
| fut. | future | St. | Saint |
| gerund. | gerundive | superl. | superlative |
| imit. | imitative | tr. | transitive |
| imper. | imperative | transl. | translation |
| intr. | intransitive | ult. | ultimately |
| M | Middle | usu. | usually |
| masc. | masculine | v. | verb |
| Med. | Medieval | var. | variant |
| Mod. | Modern | | |
| n. | noun | * | unattested |
| naut. | nautical | < | derived from |
| O | Old | + | combined with |
| obs. | obsolete | ð | th in Old English and Old Norse |
| orig. | origin,originally | | |

**Language Abbreviations in Etymologies**

| Abbreviation | Language |
|---|---|
| Afr. | Afrikaans |
| AFr. | Anglo-French |
| Am. E. | American English |
| Am. Sp. | American Spanish |
| AN | Anglo-Norman |
| Ar. | Arabic |
| Aram. | Aramaic |
| Balt. | Baltic |
| Brit. | British |
| Brit. E. | British English |
| Canadian Fr. | Canadian French |
| Celt. | Celtic |
| Chin. | Chinese |

**Language Abbreviations in Etymologies**

| Abbreviation | Language |
|---|---|
| Dan. | Danish |
| Du. | Dutch |
| Egypt. | Egyptian |
| E. | English |
| Finn. | Finnish |
| Flem. | Flemish |
| Fr. | French |
| G. | German |
| Gael. | Gaelic |
| Gk. | Greek |
| Goth. | Gothic |
| Heb. | Hebrew |
| HG | High German |
| Hung. | Hungarian |
| Icel. | Icelandic |
| IE | Indo-European |
| Ir. | Irish |
| Iran. | Iranian |
| Ir. Gael. | Irish Gaelic |
| Ital. | Italian |
| J. | Japanese |
| Lat. | Latin |
| LGk. | Late Greek |
| Lith. | Lithuanian |
| Louisiana Fr. | Louisiana French |
| LG | Low German |
| MDu. | Middle Dutch |
| ME | Middle English |
| Med. Lat. | Medieval Latin |
| Mex. Sp. | Mexican Spanish |
| MHG | Middle High German |

**Language Abbreviations in Etymologies**

| Abbreviation | Language |
|---|---|
| MLG | Middle Low German |
| Mod. E. | Modern English |
| NLat. | New Latin |
| Norman Fr. | Norman French |
| Norw. | Norwegian |
| ODan. | Old Danish |
| OE | Old English |
| OFr. | Old French |
| OHG | Old High German |
| OIr. | Old Irish |
| OItal. | Old Italian |
| ON | Old Norse |
| ONFr. | Old North French |
| OProv. | Old Provençal |
| OSp. | Old Spanish |
| Pers. | Persian |
| Pidgin E. | Pidgin English |
| Pol. | Polish |
| Port | Portuguese |
| Prov. | Provençal |
| R. | Russian |
| Rum. | Rumanian |
| Sc. | Scottish |
| Scand. | Scandinavian |
| Sc. Gael. | Scottish Gaelic |
| Skt. | Sanskrit |
| Slav. | Slavic |
| Sp. | Spanish |
| Swed. | Swedish |
| Turk. | Turkish |
| VLat. | Vulgar Latin |

# Pronunciation Symbols

The pronunciation system in this Dictionary is explained in the section "Pronunciation" in the Explanatory Notes. The column below headed WEBSTER'S II is the pronunciation key used in this Dictionary. The right-hand column, labeled IPA, contains symbols from the International Phonetic Alphabet, a system widely used by scholars. The two systems do not precisely correspond because they were conceived for somewhat different purposes.

| Spellings | WEBSTER'S II | IPA |
|---|---|---|
| pat | ă | æ |
| pay | ā | e |
| care | âr | ɛr, er |
| father | ä | a:, a |
| bib | b | b |
| church | ch | tʃ |
| deed, milled | d | d |
| pet | ĕ | ɛ |
| bee | ē | i |
| fife, phase, rough | f | f |
| gag | g | g |
| hat | h | h |
| which | hw | hw (also ʍ) |
| pit | ĭ | ɪ |
| pie, by | ī | aɪ |
| pier | îr | ɪr, ir |
| judge | j | dʒ |
| kick, cat, pique | k | k |
| lid, needle | l (nēd′l) | l, l̩ [ˈnidl̩] |
| mum | m | m |
| no, sudden | n (sŭd′n) | n, n̩ [ˈsʌdn̩] |
| thing | ng | ŋ |
| pot, horrid | ŏ | ɑ |
| toe, hoarse | ō | o |
| caught, paw, for | ô | ɔ |
| noise | oi | ɔɪ |
| took | ŏŏ | ʊ |
| boot | o͞o | u |
| out | ou | aʊ |
| pop | p | p |
| roar | r | r |

| | WEBSTER'S II | IPA |
|---|---|---|
| sauce | s | s |
| ship, dish | sh | ʃ |
| tight, stopped | t | t |
| thin | th | θ |
| this | *th* | ð |
| cut | ŭ | ʌ |
| urge, term, firm, word, heard | ûr | ɝ, ɜr |
| valve | v | v |
| with | w | w |
| yes | y | j |
| zebra, xylem | z | z |
| vision, pleasure, garage | zh | ʒ |
| about, item, edible, gallop, circus | ə | ə |
| butter | ər | ɚ |

### Foreign

| | | |
|---|---|---|
| *French* feu, *German* schön | œ | œ |
| *French* tu, *German* über | ü | y |
| *German* ich, *Scottish* loch | KH | x |
| *French* bon | N | õ, æ̃, ã, œ̃ |

### Stress

| | | |
|---|---|---|
| Primary stress | ′ | **bi·ol′o·gy** (bī-ŏl′ə-jē) |
| Secondary stress | ′ | **bi′o·log′i·cal** (bī′ə-lŏj′ĭ-kəl) |

# A New Dictionary of American English

# Aa

**a** or **A** (ā) *n., pl.* **a's** or **A's. 1.** The first letter of the English alphabet. **2.** A speech sound represented by the letter *a.* **3.** The first in a series. **4.** *Mus.* **a.** The sixth tone in the scale of C major or the first note in the related minor scale. **b.** The key or scale in which A is the tonic. **c.** A written or printed note representing this tone. **d.** A string, key, or pipe tuned to the pitch of this tone. **5.** The highest grade in quality or rank. **6.** Something shaped like the letter A.

**a¹** (ə; ā *when stressed) indef. art.* [ME < OE *ān,* one.] **1.** —Used before nouns and noun phrases that denote a single, but unspecified, person or thing <*a* mountain> <*a* woman> **2.** —Used before terms, as *few* or *many,* denoting number < *a* hundred animals> <just *a* few of the players> **3.** The same <people of *a* kind> **4.** Any <not *a* bite to eat> **5. a.** —Used before a proper name to denote a type or a member of a class <the strength of *a* Hercules> **b.** —Used before a mass noun to indicate a single type or example <*a* dark beer>

▲ word history: The Old English adjective *ān,* "one," is the ancestor of both the indefinite article *a* or *an* and the number *one* in Modern English. These words illustrate the effect of stress on word development. In very early Middle English times *ān* began to be used as an indefinite article; as such it always preceded a noun and in that position lost its stress. The vowel consequently became short and the final *n* disappeared before consonants. As a number *ān* had a more independent existence and retained its stress, the quantity of its vowel, and its final consonant. By the middle of the 13th century, long *a* had become rounded to long o. The numeral *ān* became *one,* pronounced like *own.* The article, whose vowel was short, remained *a* or *an,* which it still is today.

**a²** (ə) *prep.* [ME *a* < OE *an,* in.] In or for each or every : PER <twice *a* day> <two dollars *a* yard>

**†a³** (ə; ā *when stressed) v.* [ME < *haven,* to have.] *Regional.* Have <We'd *a* come if it hadn't rained.>

**a-¹** or **an-** *pref.* [Gk.] Without : not <*a*moral><*a*typical>

**a-²** *pref.* [ME < OE < *an,* on.] **1.** On : in <*a*bed> **2.** In the act of <*a*borning> **3.** In the direction of <*a*stern> **4.** In a specified state or condition <*a*bloom><*a*buzz>

**aard·vark** (ärd′värk′) *n.* [Obs. Afr. : *aarde,* earth (< Du. < M.Du. *aerde*) + *vark,* pig (< M. Du. *varken*).] A burrowing mammal, *Orycteropus afer* of southern Africa, with a hairy, stocky body, large ears, a long, tubular snout, and powerful digging claws.

**aard·wolf** (ärd′wo͝olf′) *n.* [Afr. : *aarde,* earth + *wolf,* wolf (< M. Du.).] A hyenalike mammal, *Proteles cristatus* of southern and eastern Africa, with gray fur with black stripes, and feeding chiefly on termites and insect larvae.

**aardwolf**
*2½–4 feet long*

**Aa·ron·ic** (ā-rŏn′ĭk, ă-rŏn′-) *also* **Aa·ron·i·cal** (-ĭ-kəl) *adj.* **1.** Of, relating to, or characteristic of Aaron. **2.** Of or relating to the lower order of priests in the Mormon Church.

**Ab** (ăb, äv, ôv) *n. var. of* Av.

**ab-¹** *pref.* [Lat. < *ab.*] Away from <*ab*oral>

**ab-²** *pref.* [< ABSOLUTE.] —Used to indicate a centimeter-gram-second system electromagnetic unit <*ab*coulomb>

**AB** (ā′bē′) *n.* A human blood type of the ABO group.

**a·ba** (ə-bä′, ä-bä′) *n.* [Ar. *'abā'.*] **1.** A light fabric woven of camel or goat hair. **2.** A loose-fitting, often sleeveless garment made of aba, worn by Arabs.

**aba**
*Two types of abas worn by* (left) *a Palestinian fellah and* (right) *a Moslem from Yemen*

**ab·a·ca** (ăb′ə-kä′) *n.* [Sp. *abacá* < Tagolog *abaká.*] A Philippine plant, *Musa textilis,* related to the banana and whose leafstalks are the source of Manila hemp.

**a·back** (ə-băk′) *adv.* [ME *abak* < OE *on bæc* : *on,* to + *bæc,* back.] *Archaic.* Back : backward. **—take aback.** To startle or dumbfound.

**a·bac·te·ri·al** (ā′băk-tîr′ē-əl) *adj.* Not caused by or identified with bacteria.

**ab·a·cus** (ăb′ə-kəs, ə-băk′əs) *n., pl.* **ab·a·cus·es** or **ab·a·ci** (ăb′ə-sī′, ə-băk′ī′) [Lat. < Gk. *abax,* counting board.] **1.** A manual computing device consisting of a frame holding parallel rods strung with movable counters. **2.** A slab on the top of the capital of a column.

**a·baft** (ə-băft′) *adv.* [ME *on baft* : *on,* at + *baft,* rear < OE *beæftan,* behind (*be,* at + *æftan,* behind).] *Naut.* Toward the stern. **—a·baft′** *prep.*

**ab·a·lo·ne** (ăb′ə-lō′nē) *n.* [Am. Sp. *abulón.*] A large, edible marine gastropod of the genus *Haliotis,* with an ear-shaped shell with a row of holes and a pearly, colorful interior, often used for ornaments.

**ab·am·pere** (ăb-ăm′pîr′) *n.* A centimeter-gram-second electromagnetic unit of current, equal to the current that produces a force of two dynes per centimeter of length on each of two infinitely long straight parallel wires one centimeter apart, and equal to 10 amperes.

**a·ban·don** (ə-băn′dən) *vt.* **-doned, -don·ing, -dons.** [ME *abandounen* < OFr. *abandoner* < *a bandon* : *a,* at (< Lat. *ad*) + *bandon,* control.] **1.** To withdraw one's support or help from, esp. despite a duty, allegiance, or responsibility. **2.** To give up by leaving or ceasing to operate or inhabit, esp. as a result of impending threat <*abandon* an unsafe tenement> **3.** To surrender one's claim or right to : GIVE UP. **4.** To cease trying to continue : desist from <*abandon* a job-hunting effort> **5.** To yield (oneself) completely, as to emotion. —*n.* **1.** A complete surrender of inhibitions. **2.** Unlimited enthusiasm. **—a·ban′don·ment** *n.*

☆ ***syns:*** **1.** ABANDON, DESERT, FORSAKE, LEAVE, QUIT, THROW OVER *v. core meaning* : to give up without intending to return or claim again <*abandoned* their families> **2.** ABANDON, DESIST, DISCONTINUE, FORSWEAR, GIVE UP, LAY OFF, QUIT, RENOUNCE, STOP, SWEAR OFF *v. core meaning* : to cease trying to accomplish or continue <*abandoned* my studies>

**a·ban·doned** (ə-băn′dənd) *adj.* **1.** Having been given up and deserted : FORSAKEN. **2.** Recklessly unrestrained : SHAMELESS.

**ab·ap·i·cal** (ăb-ăp′ĭ-kəl, -ā′pĭ-) *adj.* Opposite to or directed away from the apex.

**a·base** (ə-bās′) *vt.* **a·based, a·bas·ing, a·bas·es.** [ME *abassen* <

OFr. *abaissir* : Lat. *ad-*, to + Med. Lat. *bassus*, low.] To lower in rank, prestige, or esteem : HUMILIATE. **—a·base′ment** *n.*

**a·bash** (ə-băsh′) *vt.* **a·bashed, a·bash·ing, a·bash·es.** [ME *abaishen*, to lose one's composure < OFr. *esbahier* : *es-* (intensive) + *baer*, to gape.] To embarrass or disconcert. **—a·bash′ment** *n.*

**a·ba·sia** (ə-bā′zhə) *n.* [NLat. : A-[1] + Gk. *basis*, step < *bainein*, to go.] Dysfunction of muscular coordination in walking.

**a·bate** (ə-bāt′) *v.* **a·bat·ed, a·bat·ing, a·bates.** [ME *abaten* < OFr. *abattre*, to beat : *a-*, to (< Lat. *ad-*) + *batre*, to beat < Lat. *battuere*.] —*vt.* **1.** To reduce in amount, degree, or intensity. **2.** To deduct from an amount : SUBTRACT. **3.** *Law.* **a.** To put an end to. **b.** To make void. —*vi.* **1.** To subside <The fever *abated.*> **2.** *Law.* To become void.

**a·bate·ment** (ə-bāt′mənt) *n.* **1. a.** Reduction in degree or intensity. **b.** The amount abated. **2.** *Law.* The act of abating.

**ab·at·toir** (ăb′ə-twär′) *n.* [Fr. < *abattre*, to strike down < OFr. —see ABATE.] A slaughterhouse.

**ab·ax·i·al** (ă-băk′sē-əl) *adj.* Situated away from the axis.

**ab·ba·cy** (ăb′ə-sē) *n.*, *pl.* **-cies.** [ME *abbatie* < LLat. *abbatia* < *abba*, father. —see ABBA.] The office, term, or jurisdiction of an abbot.

**ab·ba·tial** (ə-bā′shəl) *adj.* [ME *abbacyal* < LLat. *abbatialis* < *abbas*, abbot.] Of or relating to an abbey, abbot, or abbess.

**ab·bé** (ăb′ā′, ă-bā′) *n.* [Fr. < OFr. < LLat. *abbas*, abbot.] A French title orig. of the superior of an abbey, now given to any ecclesiastical figure.

**ab·bess** (ăb′ĭs) *n.* [ME *abesse* < OFr. < LLat. *abbatissa* < *abbas*, abbot.] The superior of a convent.

**Ab·be·vil·li·an** (ăb′ə-vĭl′ē-ən) *adj.* [After *Abbeville*, France.] Designating the earliest Paleolithic archaeological sites in Europe, marked by bifacial stone hand axes.

**ab·bey** (ăb′ē) *n.*, *pl.* **-beys.** [ME < OFr. *abaie* < LLat. *abbatia* < *abbas*, abbot.] **1.** A convent or monastery. **2.** An abbey church.

**ab·bot** (ăb′ət) *n.* [ME *abbod* < OE < LLat. *abbas* < LGk. *abbas* < Aram. *abbā*, father.] The superior of a monastery.

**ab·bre·vi·ate** (ə-brē′vē-āt′) *vt.* **-at·ed, -at·ing, -ates.** [ME *abbreviaten* < LLat. *abbreviare* : *ab-*, off + *breviare*, to shorten < *brevis*, short.] **1.** To shorten. **2.** To shorten (a word or phrase) to a form meant to represent the full form. **—ab·bre′vi·a′tor** *n.*

**ab·bre·vi·a·tion** (ə-brē′vē-ā′shən) *n.* **1.** The act or product of abbreviating. **2.** A shortened form of a word or phrase used chiefly in writing to represent the full form.

**ABC** (ā′bē′sē′) *n.*, *pl.* **ABC's. 1.** *often* **ABC's.** The alphabet. **2. ABC's.** The fundamentals of reading and writing.

**ABC art** *n.* Minimal art.

**ab·cou·lomb** (ăb-ko͞o′lŏm′, -lōm′) *n.* A centimeter-gram-second electromagnetic unit of charge, equal to ten coulombs or the charge passing in one second through any cross section of a conductor carrying a steady current of one abampere.

**ab·di·cate** (ăb′dĭ-kāt′) *v.* **-cat·ed, -cat·ing, -cates.** [Lat. *abdicare*, to disclaim : *ab-*, away + *dicare*, to proclaim.] —*vt.* To relinquish (power or responsibility) formally. —*vi.* To relinquish formally high office or responsibility. **—ab′di·ca·ble** (-kə-bəl) *adj.* **—ab′di·ca′tion** *n.* **—ab′di·ca′tor** *n.*

**ab·do·men** (ăb′də-mən, ăb-dō′mən) *n.* [Lat., belly.] **1.** The part of a mammal between the thorax and the pelvis that encloses the viscera. **2.** The major posterior part of the body of an arthropod. **—ab·dom′i·nal** (ăb-dŏm′ə-nəl) *adj.* **—ab·dom′i·nal·ly** *adv.*

**ab·du·cens** (ăb′do͞o′sənz, -dyo͞o′-) *n.*, *pl.* **ab·du·cen·tes** (ăb′do͞o-sĕn′tēz′, -dyo͞o-) [Lat. *abducens*, pr.part. of *abducere*, to take away. —see ABDUCT.] An abducens nerve.

**abducens nerve** *n.* Either of the sixth pair of cranial nerves that convey motor impulses to the eye muscles.

**ab·duct** (ăb-dŭkt′) *vt.* **-duct·ed, -duct·ing, -ducts.** [Lat. *abducere, abduct-* : *ab-*, away + *ducere*, to lead.] **1.** To carry off by force. **2.** *Physiol.* To draw away from the median line of a bone or muscle or from an adjacent part or limb. **—ab·duc′tion** *n.* **—ab·duc′tor** *n.*

**a·beam** (ə-bēm′) *adv.* At right angles to a ship's keel.

**a·be·ce·dar·i·an** (ā′bē-sē-dâr′ē-ən) *n.* [ME < Med. Lat. *abecedarium*, alphabet < LLat. *abecedarius*, alphabetical < the names of the letters A B C D.] **1.** One who studies or teaches the alphabet. **2.** A beginner. —*adj.* **1.** Relating to the alphabet. **2.** Arranged alphabetically. **3.** Elementary.

**a·bed** (ə-bĕd′) *adv.* In bed.

**a·bele** (ə-bēl′) *n.* [Du. *abeel* < OFr. *abel* < Med. Lat. *albellus*, dim. of Lat. *albus*, white.] The white poplar.

**A·be·lian group** (ə-bēl′yən, ə-bē′lē-ən) *n.* [After Niels Henrik *Abel* (1802–1829).] A commutative group.

**a·bel·mosk** (ā′bəl-mŏsk′) *n.* [NLat. *abelmoschus* < Ar. *ḥabbalmusk*, grain of musk : *ḥabb*, grain + *mosk*, musk < Pers. *mushk*.] A hairy plant, *Hibiscus abelmoschus* of tropical Asia, with large yellow flowers and musk-scented seeds used in perfumery.

**Ab·er·deen An·gus** (ăb′ər-dēn ăng′gəs) *n.* [After *Aberdeen* and *Angus*, former counties of Scotland.] Any of a breed of black, hornless beef cattle that originated in Scotland.

**ab·er·rant** (ă-bĕr′ənt) *adj.* **1.** Departing from the expected course. **2.** Deviating from the normal. **—ab·er′rance, ab·er′ran·cy** *n.*

**ab·er·ra·tion** (ăb′ə-rā′shən) *n.* [Lat. *aberratio*, diversion < *aberrare*, to go astray : *ab-*, from + *errare*, to stray.] **1.** A departure from the expected or proper course. **2.** A deviation from the normal or typical. **3.** Mental disorder or unsoundness. **4. a.** Blurring or distortion of an image. **b.** A physical defect in an optical element, as in a lens, causing such distortion. **5.** *Astron.* Apparent displacement of the position of a celestial body in the direction of motion of an observer on earth, caused by the motion of the earth and the finiteness of the velocity of light.

**a·bet** (ə-bĕt′) *vt.* **a·bet·ted, a·bet·ting, a·bets.** [ME *abetten* < OFr. *abeter*, to entice : *a*, to (< Lat. *ad-*) + *beter*, to bait, of Germanic orig.] To incite, encourage, or assist, esp. in wrongdoing. **—a·bet′ment** *n.* **—a·bet′tor, a·bet′ter** *n.*

**a·bey·ance** (ə-bā′əns) *n.* [OFr. *abeance*, desire < *abaer*, to gape at : *a-*, at (< Lat. *ad-*) + *baer*, to gape.] **1.** The condition of being temporarily set aside. **2.** *Law.* A condition of undetermined ownership, as of an estate that has not yet been assigned. **—a·bey′ant** *adj.*

☆ ***syns:*** ABEYANCE, DORMANCY, INTERMISSION, LATENCY, QUIESCENCE, REMISSION, SUSPENSION *n. core meaning* : temporary inactivity <hold a decision in *abeyance*> <cancer kept in *abeyance* by chemotherapy>

**ab·far·ad** (ăb-făr′ăd′, -əd) *n.* A centimeter-gram-second electromagnetic unit of capacitance, equal to one billion ($10^9$) farads or the capacitance of a capacitor having a charge of one abcoulomb and a potential difference of one abvolt.

**ab·hen·ry** (ăb-hĕn′rē) *n.*, *pl.* **-ries.** A centimeter-gram-second electromagnetic unit of inductance, equal to one-billionth ($10^{-9}$) henry or the inductance from a current variation of one abampere per second that produces an induced electromotive force of one abvolt.

**ab·hor** (ăb-hôr′) *vt.* **-horred, -hor·ring, -hors.** [ME *abhorren* < Lat. *abhorrēre*, to shrink from : *ab-*, from + *horrēre*, to shudder.] **1.** To consider with horror or disgust. **2.** To reject vehemently : SHUN. **—ab·hor′rence** (-hôr′əns, -hŏr′-) *n.* **—ab·hor′rer** *n.*

**ab·hor·rent** (ăb-hôr′ənt, -hŏr′-) *adj.* **1.** Disgusting. **2.** Arousing extreme repugnance. **3.** *Archaic.* Very strongly opposed. **—ab·hor′rent·ly** *adv.*

**A·bib** (ä-vēv′) *n.* [Heb. *'ābhībh*, (month of) fresh barley.] An earlier name for the month of Nisan in the ancient Hebrew calendar.

**a·bid·ance** (ə-bīd′ns) *n.* **1.** Continuance. **2.** Compliance.

**a·bide** (ə-bīd′) *v.* **a·bode** (ə-bōd′) *or* **a·bid·ed, a·bid·ing, a·bides.** [ME *abiden* < OE *ābīdan* : *a-* (intensive) + *bīdan*, to remain.] —*vt.* **1.** To wait patiently for : AWAIT. **2.** To persevere under : WITHSTAND. **3.** To accept or submit to. **4.** To put up with : TOLERATE <can't *abide* such negativity> —*vi.* **1. a.** To remain in one place or state. **b.** To dwell or sojourn. **2.** To continue or endure : LAST. **—abide by.** To comply with <*abide by* the rules> **—a·bid′er** *n.*

**a·bid·ing** (ə-bī′dĭng) *adj.* Long-lasting : ENDURING. **—a·bid′ing·ly** *adv.*

**A·bies** (ā′bēz, ä-bī′əs) *n.* [NLat. < Lat., silver fir.] A genus of firs in the pine family, with flattened, needlelike leaves and erect cones.

**ab·i·et·ic acid** (ăb′ē-ĕt′ĭk) *n.* [< Lat. *abies, abiet-*, silver fir.] A yellowish resinous powder, $C_{19}H_{29}COOH$, isolated from rosin and used in lacquers, varnishes, and soaps.

**a·bil·i·ty** (ə-bĭl′ĭ-tē) *n.*, *pl.* **-ties.** [ME *abilite* < OFr. *habilite* < Lat. *habilitas* < *habilis*, able.] **1.** Physical, mental, financial, or legal power to perform. **2.** A natural or acquired skill or talent.

☆ ***syns:*** **1.** ABILITY, ABLENESS, CAPABILITY, CAPACITY, COMPETENCE, FACULTY, MIGHT *n. core meaning* : physical or mental power to perform <the *ability* to learn math> **2.** ABILITY, ADEPTNESS, COMMAND, CRAFT, EXPERTISE, EXPERTNESS, KNACK, KNOW-HOW, MASTERY, PROFICIENCY, SKILL *n. core meaning* : natural or acquired facility in a specific activity <fine technical *ability*>

**-ability** or **-ibility** *suff.* [ME *-abilitie* < OFr. *-abilite* < Lat. *-abilitas*.] Ability, inclination, or suitability for a specified action or condition <teach*ability*><wear*ability*>

**ab in·i·ti·o** (ăb′ ĭ-nĭsh′ē-ō′) *adv.* [Lat.] From the beginning.

**a·bi·o·gen·e·sis** (ā′bī-ō-jĕn′ĭ-sĭs) *n.* Hypothetical development of living organisms from nonliving matter. **—a′bi·o·ge·net′ic** (-jə-nĕt′ĭk), **a′bi·o·ge·net′i·cal** *adj.* **—a′bi·og′e·nist** (-ŏj′ə-nĭst) *n.*

**a·bi·ot·ic** (ā′bī-ŏt′ĭk) *adj.* Not biotic. **—a′bi·o′sis** (-ō′sĭs) *n.* **—a′bi·ot′i·cal·ly** *adv.*

**ab·ject** (ăb′jĕkt′, ăb-jĕkt′) *adj.* [ME, outcast < Lat. *abjectus*, p.part. of *abicere*, to cast away : *ab-*, from + *jacere*, to throw.] **1.** Of the most contemptible kind : DESPICABLE <an *abject* coward> **2.** Of the most miserable kind : WRETCHED <lived in *abject* poverty> **—ab′ject′ly** *adv.* **—ab·ject′ness, ab·jec′tion** *n.*

**ab·jure** (ăb-jo͝or′) *vt.* **-jured, -jur·ing, -jures.** [ME *abjuren* < OFr. *abjurer* < Lat. *abjurare* : *ab-*, away + *jurare*, to swear.] **1.** To recant solemnly : REPUDIATE. **2.** To renounce under oath : FORSWEAR. **—ab′ju·ra′tion** (ăb′jo͞o-rā′shən) *n.* **—ab·jur′er** *n.*

**ab·la·tion** (ă-blā′shən) *n.* [LLat. *ablatio* < *ablatus*, p.part. of *auferre*, to carry away : *ab-*, away + *ferre*, to carry.] **1.** Surgical removal of a body part. **2.** The totality of erosive processes by which a glacier is reduced. **3.** *Aerospace.* Dissipation of heat generated by atmospheric friction, esp. in the atmospheric re-entry of a spacecraft or missile, by means of a melting heat shield.

ă **pat** ā **pay** âr **care** ä **father** ĕ **pet** ē **be** hw **which** ĭ **pit** ī **tie** îr **pier** ŏ **pot** ō **toe** ô **paw, for** oi **noise** o͝o **took**

**ab·la·tive** (ăb′lə-tĭv) *adj.* [ME < OFr. *ablatif* < Lat. *ablativus* < *ablatus,* p.part. of *auferre,* to carry away. —see ABLATION.] Designating a grammatical case indicating separation, direction away from, and sometimes manner or agency, found in some Indo-European languages. —*n.* **1.** The ablative case. **2.** A word in the ablative case.

**ablative absolute** *n.* An adverbial phrase in Latin syntactically independent from the rest of the sentence and containing two main elements both in the ablative case.

**ab·laut** (äb′lout′, äp′-) *n.* [G. *Ablaut* : *ab,* off (< OHG *aba*) + *Laut,* sound < MHG *lūt* < OHG *hlūt.*] A patterned change in root vowels of verb forms, characteristic of Indo-European languages, indicating alteration of tense, aspect, or function; e.g., *sing, sang, sung.*

**a·blaze** (ə-blāz′) *adj.* **1.** Being on fire. **2.** Radiantly bright : AGLOW.

**a·ble** (ā′bəl) *adj.* **a·bler, a·blest.** [ME < OFr. < Lat. *habilis* < *habēre,* to hold.] **1.** Having sufficient ability or resources. **2.** Capable or talented. **—a′bly** (ā′blē) *adv.*

**-able** *or* **-ible** *suff.* [ME < OFr. < Lat. *-abilis.*] **1.** Susceptible, capable, or worthy of a specified action <debat*able*> **2.** Inclined or given to a specified state or action <blam*able*>

**a·ble-bod·ied** (ā′bəl-bŏd′ēd) *adj.* Physically healthy and strong : SOUND.

**able-bodied seaman** *n.* An experienced seaman certified for all seaman's duties.

**able seaman** *n.* An able-bodied seaman.

**a·bloom** (ə-blo͞om′) *adj.* Being in bloom : FLOWERING.

**ab·lu·tion** (ə-blo͞o′shən, ă-blo͞o′-) *n.* [ME *ablucioun* < Lat. *ablutio* < *abluere,* to wash away : *ab,* away + *luere,* to wash.] **1.** A cleansing of the body, esp. in a religious ceremony. **2.** The liquid used in an ablution. **—ab·lu′tion·ar′y** (-shə-nĕr′ē) *adj.*

**ABM** (ā′bē-ĕm′) *n.* Antiballistic missile.

**Ab·na·ki** (ăb-nä′kē) *n., pl.* **Abnaki** *or* **-kis. 1. a.** A tribe of Indians of Maine, New Brunswick, and southern Quebec. **b.** A member of this tribe. **2.** The Algonquian language of the Abnaki.

**ab·ne·gate** (ăb′nĭ-gāt′) *vt.* **-gat·ed, -gat·ing, -gates.** [Lat. *abnegare, abnegat-,* to refuse : *ab-,* away + *negare,* to deny.] To deny to oneself : RENOUNCE.

**ab·ne·ga·tion** (ăb′nĭ-gā′shən) *n.* Self-denial : renunciation.

**ab·nor·mal** (ăb-nôr′məl) *adj.* [Lat. *abnormis* : *ab-,* away from + *norma,* rule.] Not normal : DEVIANT. **—ab·nor′mal·ly** *adv.*

☆ ***syns:*** ABNORMAL, ABERRANT, ANOMALOUS, ATYPICAL, DEVIANT, PRETERNATURAL, UNNATURAL *adj. core meaning* : departing from the normal <an *abnormal* interest in death>

**ab·nor·mal·i·ty** (ăb′nôr-măl′ĭ-tē) *n., pl.* **-ties. 1.** The condition of being abnormal. **2.** An abnormal phenomenon.

**abnormal psychology** *n.* Psychopathology.

**a·board** (ə-bôrd′, ə-bōrd′) *adv.* [ME *abōrd* : *a-,* on + *bōrd,* ship.] On board a passenger vehicle, as a ship, train, or aircraft. **—a·board′** *prep.*

**a·bode** (ə-bōd′) *n.* [ME *abod* < *abiden,* to wait. —see ABIDE.] **1.** A habitation. **2.** A sojourn.

**ab·ohm** (ă-bōm′) *n.* A centimeter-gram-second electromagnetic unit of resistance equal to one billionth ($10^{-9}$) of an ohm.

**a·bol·ish** (ə-bŏl′ĭsh) *vt.* **-ished, -ish·ing, -ish·es.** [ME *abolisshen* < OFr. *abolir, aboliss-* < Lat. *abolēre.*] **1.** To put an end to : ANNUL. **2.** To destroy completely : ANNIHILATE. **—a·bol′ish·a·ble** *adj.* **—a·bol′ish·er** *n.* **—a·bol′ish·ment** *n.*

**ab·o·li·tion** (ăb′ə-lĭsh′ən) *n.* [Fr. < Lat. *abolitio* < *abolitus,* p.part. of *abolēre,* to abolish.] **1.** An act of abolishing or the state of being abolished : ANNULMENT. **2.** The abolishment of slavery in the United States. **—ab′o·li′tion·ar′y** (-ə-nĕr′ē) *adj.*

**ab·o·li·tion·ism** (ăb′ə-lĭsh′ə-nĭz′əm) *n.* Advocacy of the abolition of slavery in the United States. **—ab′o·li′tion·ist** *n.*

**ab·o·ma·sum** (ăb′ō-mā′səm) *n., pl.* **-sa** (-sə). The fourth division of the stomach in ruminant animals, in which true digestion takes place. **—ab′o·ma′sal** (-səl) *adj.*

**A-bomb** (ā′bŏm′) *n.* An atomic bomb.

**a·bom·i·na·ble** (ə-bŏm′ə-nə-bəl) *adj.* [ME *abhominable* < OFr. < Lat. *abominabilis* < *abominari,* to abhor. —see ABOMINATE.] **1.** Utterly loathsome or detestable <*abominable* child abuse> **2.** Thoroughly unpleasant or disagreeable <*abominable* manners> **—a·bom′i·na·bly** *adv.*

**abominable snowman** *n.* A hirsute humanlike animal allegedly inhabiting the high Himalayas.

**a·bom·i·nate** (ə-bŏm′ə-nāt′) *vt.* **-nat·ed, -nat·ing, -nates.** [Lat. *abominari, abominat-,* to deprecate as a bad omen : *ab-,* away + *omen,* omen.] To regard as abominable. **—a·bom′i·na′tor** *n.*

**a·bom·i·na·tion** (ə-bŏm′ə-nā′shən) *n.* **1.** Extreme dislike or abhorrence. **2.** Something that elicits extreme dislike.

**ab·o·ral** (ă-bôr′əl, -bōr′-) *adj.* Opposite to or away from the mouth.

**ab·o·rig·i·nal** (ăb′ə-rĭj′ə-nəl) *adj.* **1.** Existing from the beginning or being the earliest of its type in an area : INDIGENOUS. **2.** Of or relating to aborigines. **—ab′o·rig′i·nal** *n.* **—ab′o·rig′i·nal·ly** *adv.*

**ab·o·rig·i·ne** (ăb′ə-rĭj′ə-nē) *n.* [< Lat. *aborigines,* original inhabitants : *ab-,* from + *origine,* beginning.] **1.** An indigenous inhabitant of a region. **2. aborigines.** The flora and fauna native to a geographic area.

**a·born·ing** (ə-bôr′nĭng) *adv.* While coming into being or getting started <research that nearly died *aborning*>

**a·bort** (ə-bôrt′) *v.* **a·bort·ed, a·bort·ing, a·borts.** [Lat. *abortare,* freq. of *aboriri,* to disappear, miscarry : *ab-,* away + *oriri,* to appear.] —*vi.* **1.** To terminate pregnancy prematurely. **2.** To fail to reach full development or maturation, as a bodily organ. **3.** To terminate an operation or procedure with a missile, a space vehicle, or an aircraft before completion, esp. because of equipment failure. —*vt.* **1.** To cause to terminate pregnancy prematurely. **2.** To interfere with the development of. **3.** To terminate before completion <*abort* a space launch><*abort* a takeoff> —*n.* **1.** The act of terminating an operation or procedure with a missile, space vehicle, or aircraft before completion, esp. because of equipment failure. **2.** *Computer Sci.* A procedure to terminate execution of a program when an unrecoverable error or malfunction occurs.

**a·bor·ti·fa·cient** (ə-bôr′tə-fā′shənt) *adj.* Inducing abortion. **—a·bor′ti·fa′cient** *n.*

**a·bor·tion** (ə-bôr′shən) *n.* **1.** Induced termination of pregnancy before the fetus is capable of independent survival. **2.** A fatally premature expulsion of an embryo or fetus from the womb. **3.** Cessation of normal growth, esp. of an organ, prior to full development or maturation. **4.** An aborted organism. **5.** One that is malformed or incompletely developed.

**a·bor·tion·ist** (ə-bôr′shə-nĭst) *n.* One who performs abortions.

**a·bor·tive** (ə-bôr′tĭv) *adj.* **1.** Failing to accomplish an objective : FUTILE. **2.** Imperfectly developed. **—a·bor′tive·ly** *adv.* **—a·bor′tive·ness** *n.*

**ABO system** (ā′bē′ō′) *n.* The antigenic system of human blood, which, by functioning genetically as an allelic unit, produces the four human blood groups or types, A, B, AB, and O.

**a·bou·li·a** (ə-bo͞o′lē-ə, ə-byo͞o′-) *n. var. of* ABULIA.

**a·bound** (ə-bound′) *vi.* **a·bound·ed, a·bound·ing, a·bounds.** [ME *abounden* < OFr. *abonder* < Lat. *abundare,* to overflow : *ab-,* from + *undare,* to flow < *unda,* wave.] **1.** To be plentiful in number or amount. **2.** To be fully supplied or filled : TEEM <lakes that *abound* with fish>

**a·bout** (ə-bout′) *adv.* [ME < OE *onbūtan* : *on,* in + *būtan,* outside.] **1.** Approximately <traveled *about* a year> **2.** Almost <just *about* gone> **3.** To a reversed position or direction. **4.** In no particular direction : here and there. **5.** All around : on every side <looked all *about*> **6.** In the area : NEAR <Flu is *about.*> **7.** In succession : one after another <Turn *about* is fair play.> —*prep.* **1.** On all sides of : SURROUNDING. **2.** In the vicinity of : AROUND. **3.** Almost the same as : close to <a child *about* your age> **4.** In reference to. **5.** In the possession of <kept my wits *about* me> **6.** Ready <*about* to begin> **7.** Engaged in <go *about* one's business> —*adj.* Astir <be up and *about*>

**a·bout-face** (ə-bout′fās′) *n.* **1.** The act of pivoting to face in the opposite direction from the original. **2.** A reversal of attitude or standpoint. —*vi.* **-faced, -fac·ing, -fac·es.** To reverse direction.

**a·bove** (ə-bŭv′) *adv.* [ME *aboven* < OE *abufan* : *a-,* on + *be,* by + *ufan,* up.] **1.** Overhead <the skies *above*> **2.** In or toward heaven. **3.** Upstairs <a noise from the apartment *above*> **4.** In or to a higher place. **5.** In an earlier part of a given text. **6.** In or to a higher rank or position. —*prep.* **1.** Over or higher than <hills rising *above* the plain> **2.** Superior to in rank, quality, or degree <honesty *above* profit> **3.** Beyond the level or reach of <a siren heard *above* the traffic> **4.** In preference to <takes water *above* wine> **5.** Too proud or honorable to stoop to <*above* cheating> **6.** More than <*above* average> —*n.* Something that is above. —*adj.* Appearing earlier in the same text. ***usage:*** The use of *above* in referring to earlier material in a text (*the above figures; take note of the above*) is more appropriate to business and legal contexts than to general writing. **—above all.** Of first importance.

**a·bove·board** (ə-bŭv′bôrd′, -bōrd′) *adv. & adj.* Without trickery or deceit.

▲ word history: *Aboveboard* is a simple compound of *above* and *board.* It is recorded as early as the 17th century as a gambler's term. If the gambler's hands were above the board or gaming table then presumably he could not surreptitiously change the cards or indulge in other forms of cheating.

**a·bove·ground** (ə-bŭv′ground′) *adj.* **1.** Situated on or above the surface of the ground. **2.** Existing or operating within the establishment or according to convention.

**ab o·vo** (ăb ō′vō) *adv.* [Lat., from the egg.] From the beginning.

**ab·ra·ca·dab·ra** (ăb′rə-kə-dăb′rə) *n.* [LLat.] **1.** A word once held to possess magical powers to ward off disease or disaster. **2.** Nonsensical talk : GIBBERISH.

**a·bra·chi·a** (ə-brā′kē-ə) *n.* [A-[1] + Gk. *brakhiōn,* arm.] Congenital absence of arms.

**a·brad·ant** (ə-brăd′nt) *n.* An abrasive. **—a·brad′ant** *adj.*

**a·brade** (ə-brād′) *vt.* **a·brad·ed, a·brad·ing, a·brades.** [Lat. *abradere,* to scrape off : *ab-,* off + *radere,* to scrape.] To wear away by friction : ERODE.

**a·bran·chi·ate** (ā-brăng′kē-ĭt, -āt′) *also* **a·bran·chi·al** (-kē-əl) *or*

o͞o **boot** ou **out** th **thin** *th* **this** ŭ **cut** û **urge** y **young**
yo͞o **abuse** zh **vision** ə **about,** item, edible, gallop, circus

**a·bran·chi·ous** (-kē-əs) *adj.* [Gk. *a-*, not + *brankhia*, gills.] Having no gills.

**a·bra·sion** (ə-brā′zhən) *n.* [Med. Lat. *abrasio* < Lat. *abradere*, to scrape off. —see ABRADE.] **1.** A wearing away by friction. **2.** A scraped or worn area.

**a·bra·sive** (ə-brā′sĭv, -zĭv) *adj.* Causing abrasion : ROUGH. **—a·bra′sive** *n.*

**ab·re·act** (ăb′rē-ăkt′) *vt.* **-act·ed, -act·ing, -acts.** [Transl. of G. *abreagieren* : *ab-*, away + *reagieren*, to react.] *Psychoanal.* To release (repressed emotions) by acting out, as in words, action, or the imagination, the situation causing the conflict. **—ab′re·ac′tion** *n.*

**a·breast** (ə-brĕst′) *adv.* [ME *abrest* < *on brest* : *on*, by + *brest*, breast.] **1.** Side by side. **2.** Up to date with <*abreast* of the news>

**a·bridge** (ə-brĭj′) *vt.* **a·bridged, a·bridg·ing, a·bridg·es.** [ME *abregen* < OFr. *abregier* < LLat. *abbreviare*, to shorten. —see ABBREVIATE.] **1.** To reduce the length of (a written text) : CONDENSE. **2.** To reduce or diminish <*abridge* civil rights> **3.** To cut short : CURTAIL. **—a·bridg′er** *n.*

**a·bridg·ment** *also* **a·bridge·ment** (ə-brĭj′mənt) *n.* **1.** The act of abridging or the state of being abridged. **2.** Condensation of a book.

**a·broach** (ə-brōch′) *adj.* [ME *abroche* < *a*, in + *broche*, a pointed object.] **1.** Opened or positioned so that a liquid, as wine, can be let out. **2.** Moving about : ASTIR.

**a·broad** (ə-brôd′) *adv.* [ME *abrod*, wide < *on brod* : *on*, at + *brod*, broad.] **1.** Out of one's own country. **2.** In a foreign country or countries. **3.** Away from one's home. **4.** In circulation : at large <Revolution is *abroad.*> **5.** Covering a large area : WIDELY. **6.** In error : ASTRAY.

**ab·ro·gate** (ăb′rə-gāt′) *vt.* **-gat·ed, -gat·ing, -gates.** [Lat. *abrogare*, *abrogat-* : *ab-*, away + *rogare*, to propose.] To abolish or annul by authority : NULLIFY. **—ab′ro·ga′tion** *n.*

**a·bro·sia** (ə-brō′zhə) *n.* [Gk. *abrōsia*, fasting : *a-*, not + *brōsis*, eating.] **1.** Abstinence from food : FASTING. **2.** A wasting away.

**a·brupt** (ə-brŭpt′) *adj.* [Lat. *abruptus*, p.part. of *abrumpere*, to break off : *ab-*, off + *rumpere*, to break.] **1.** Unexpectedly sudden. **2.** Rudely curt or brusque. **3.** Touching on one subject after another with sudden transitions <*abrupt*, nervous prose> **4.** Steeply inclined. **5.** *Biol.* Appearing to terminate abruptly : TRUNCATE. **—a·brupt′ly** *adv.* **—a·brupt′ness** *n.*

**a·brup·tion** (ə-brŭp′shən) *n.* A sudden breaking away or off.

**ab·scess** (ăb′sĕs′) *n.* [Lat. *abscessus*, absence < *abscedere*, to go away : *ab-*, away + *cedere*, to go.] A localized collection of pus, formed by tissue disintegration and surrounded by inflammation. —*vi.* **-scessed, -scess·ing, -scess·es.** To form an abscess.

**ab·scise** (ăb-sīz′) *v.* **-scised, -scis·ing, -scis·es.** [Lat. *abscindere*, *absciss-* : *ab-*, away + *caedere*, to cut.] —*vt.* To remove by cutting off. —*vi.* To shed by abscission.

**ab·scis·ic acid** *also* **ab·scis·sic acid** (ăb-sĭz′ĭk) *n.* A common abscisin that inhibits plant growth.

**ab·scis·in** *or* **ab·scis·sin** (ăb-sĭs′ĭn) *n.* [ABSCIS(SION) + -IN.] A group of plant hormones that may promote leaf abscission while inhibiting certain other growth mechanisms.

**ab·scis·sa** (ăb-sĭs′ə) *n., pl.* **-scis·sas** *or* **-scis·sae** (-sĭs′ē) [NLat. *(linea) abscissa*, (line) cut off < Lat. *abscissus*, p.part. of *abscindere*, to abcise.] The coordinate representing the distance of a point from the y-axis in a plane Cartesian coordinate system, measured along a line parallel to the x-axis.

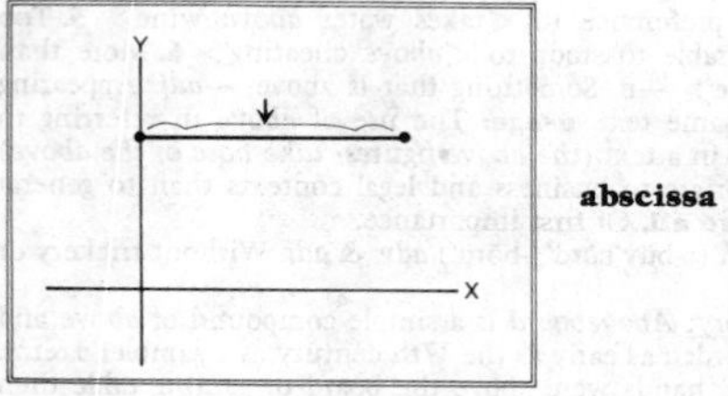


**abscissa**

**ab·scis·sion** (ăb-sĭzh′ən) *n.* **1.** An act of cutting off. **2.** The process by which plant parts, as leaves, are shed.

**ab·scond** (ăb-skŏnd′) *vi.* **-scond·ed, -scond·ing, -sconds.** [Lat. *abscondere*, to hide : *ab-*, away + *condere*, to put.] To leave quickly and secretly and hide oneself, esp. from the law. **—ab·scond′er** *n.*

**ab·sence** (ăb′səns) *n.* **1.** The state of being away. **2.** The time during which one is away. **3.** Lack <*absence* of curiosity>

**ab·sent** (ăb′sənt) *adj.* [ME < OFr. < Lat. *absens*, p.part. of *abesse*, to be away : *ab-*, away + *esse*, to be.] **1.** Not present : MISSING. **2.** Not existent : LACKING. **3.** Inattentive. —*vt.* (ăbsĕnt′) **-sent·ed, -sent·ing, -sents.** To keep (oneself) away. **—ab′sent·ly** *adv.*

**ab·sen·tee** (ăb′sən-tē′) *n.* One that is absent. —*adj.* **1.** Of or relating to one that is absent. **2.** Not in residence.

**absentee ballot** *n.* A ballot marked and mailed in advance by a voter away from the place where he or she is registered.

**ab·sen·tee·ism** (ăb′sən-tē′ĭz′əm) *n.* Habitual failure to appear, esp. for work or school.

**ab·sent-mind·ed** (ăb′sənt-mīn′dĭd) *adj.* **1.** Heedless of one's surroundings : PREOCCUPIED. **2.** Chronically forgetful. **—ab′sent-mind′ed·ly** *adv.* **—ab′sent-mind′ed·ness** *n.*

**absent without leave** *adj.* Absent without official permission from one's assigned military post or duties but without the intention to desert.

**ab·sinthe** *also* **ab·sinth** (ăb′sĭnth′) *n.* [Fr. < Lat. *absinthium*, wormwood.] **1.** A green liqueur having a bitter licorice flavor and made from wormwood. **2.** The wormwood.

**ab·so·lute** (ăb′sə-lo͞ot′, ăb′sə-lo͞ot′) *adj.* [ME *absolut* < Lat. *absolutus*, ended < *absoluere*, to finish : *ab-* from + *solvere*, to loose.] **1.** Perfect in nature or quality : COMPLETE. **2.** Not mixed : PURE <*absolute* alcohol> **3. a.** Not limited by restrictions or exceptions : UNCONDITIONAL <*absolute* freedom> **b.** Unqualified in extent or degree : TOTAL <*absolute* darkness> **4.** Not limited by constitutional provisions or other restraints <an *absolute* ruler> **5.** Unrelated to and independent of anything else <*absolute* music> **6.** Not to be doubted : POSITIVE <*absolute* truth> **7. a.** Denoting a construction in a sentence that is syntactically independent of the main clause; e.g., in *Their ship having sailed, we went home, Their ship having sailed* is an absolute phrase. **b.** Pertaining to a transitive verb when its object is implied but not stated; e.g., *inspires* in *We have a teacher who inspires.* **c.** Pertaining to an adjective or pronoun that stands alone, the noun it modifies being implied but not stated; e.g., *Theirs* and *best* in *Theirs were the best.* **8.** *Physics.* **a.** Pertaining to measurements or units of measurement derived from fundamental relationships of space, mass, and time. **b.** Pertaining to absolute temperature. **9.** *Law.* Complete and unconditional : FINAL. —*n.* **1.** Something absolute. **2. the Absolute.** *Philos.* **a.** Something considered to be the ultimate basis of all thought and being. **b.** Something considered to be independent of and unrelated to anything else. **—ab′so·lute′ness** *n.*

☆ ***syns:*** ABSOLUTE, ABSOLUTIST, ARBITRARY, AUTARCHIC, AUTOCRATIC, DESPOTIC, DICTATORIAL, MONOCRATIC, TOTALITARIAN, TYRANNICAL *adj. core meaning* : having and exercising supreme, unlimited political authority <an *absolute* ruler>

**absolute alcohol** *n.* Ethyl alcohol with no more than 1% water.

**absolute ceiling** *n.* The maximum altitude above sea level at which an aircraft or missile can maintain horizontal flight under standard atmospheric conditions.

**ab·so·lute·ly** (ăb′sə-lo͞ot′lē, ăb′sə-lo͞ot′lē) *adv.* **1.** Completely and definitely : without a doubt. **2.** In a manner that does not take a grammatical object.

**absolute magnitude** *n.* The intrinsic magnitude of a star computed as if viewed from a distance of 10 parsecs or 32.6 light-years.

**absolute music** *n.* Instrumental music that depends solely on its rhythmic, melodic, and contrapuntal structures.

**absolute pitch** *n.* **1.** The precise pitch of an isolated tone, as established by its rate of vibration measured on a standard scale. **2.** The ability to identify or sing any tone heard.

**absolute scale** *n.* A scale of temperature with absolute zero as the minimum and scale units equal in magnitude to centigrade degrees.

**absolute temperature** *n.* Temperature calculated or measured on the absolute scale.

**absolute value** *n.* **1.** The numerical value or magnitude of a quantity, as of a vector or of a negative integer, without regard to its sign. **2.** The modulus of a complex number, equal to the square root of the sum of the squares of the real and imaginary parts of the number.

**absolute zero** *n.* *Physics.* The temperature at which substances possess no thermal energy, equal to –273.15°C or –459.67°F.

**ab·so·lu·tion** (ăb′sə-lo͞o′shən) *n.* [ME < OFr. < Lat. *absolutio*, acquittal < *absolvere*, to absolve.] *Rom. Cath. Ch.* Formal remission of sin imparted by a priest as part of the sacrament of penance.

**ab·so·lut·ism** (ăb′sə-lo͞o′tĭz′əm) *n.* **1.** A form of government in which all power is vested in a single ruler : DESPOTISM. **2.** The political theory of absolutism. **3.** An absolute doctrine, principle, or opinion. **—ab′so·lut′ist** *n.* **—ab′so·lu·tis′tic** (-lo͞o-tĭs′tĭk) *adj.*

**ab·solve** (əb-zŏlv′, -sŏlv′) *vt.* **-solved, -solv·ing, -solves.** [ME *absolven* < Lat. *absolvere.* —see ABSOLUTE.] **1.** To clear of blame or guilt. **2.** To relieve of a requirement or obligation. **3. a.** To grant a remission of sin to. **b.** To pardon or remit (a sin). **—ab·solv′a·ble** *adj.* **—ab·solv′er** *n.*

**ab·sorb** (əb-sôrb′, -zôrb′) *vt.* **-sorbed, -sorb·ing, -sorbs.** [OFr. *absorber* < Lat. *absorbēre* : *ab-*, away + *sorbēre*, to suck.] **1.** To take in through or as if through pores or interstices : soak in or up <*absorbed* odd notions in childhood> **2.** To occupy the full attention of : ENGROSS. **3.** *Chem. & Physics.* To retain wholly, without reflection or transmission, that which is taken in. **4.** To assimilate <emigrants *absorbed* into the dominant culture> **5.** To receive the impact of without recoil or echo. **6.** To defray (costs). **7.** To accommodate

<tried to *absorb* the extra work> **—ab·sorb'a·bil'i·ty** (əb-sôr'bə-bĭl'ĭ-tē, -zôr'-) *n.* **—ab·sorb'a·ble** *adj.* **—ab·sorb'er** *n.* **—ab·sorb'ing·ly** *adv.*

**ab·sorbed** (əb-sôrbd', -zôrbd') *adj.* **1.** Engrossed. **2.** Sucked up or in. **3.** Assimilated. **—ab·sorb'ed·ly** (əb-sôr'bĭd-lē, -zôr'-) *adv.* **—ab·sorb'ed·ness** *n.*

**ab·sor·be·fa·cient** (əb-sôr'bə-fā'shənt, -zôr'-) *adj.* [ABSORB(E)(NT) + -FACIENT.] Inducing or causing absorption. **—ab·sor'be·fa'·cient** *n.*

**ab·sorb·ent** (əb-sôr'bənt, -zôr'-) *adj.* Capable of absorbing, as cotton. **—ab·sorb'ent** *n.* **—ab·sorb'en·cy** *n.*

**ab·sorp·tance** (əb-sôrp'təns, -zôrp'-) *n.* [ABSORPT(ION) + -ANCE.] The ratio of absorbed to incident radiation.

**ab·sorp·tion** (əb-sôrp'shən, -zôrp'-) *n.* [Lat. *absorptio* < *absorbēre,* to absorb. —see ABSORB.] **1.** The act or process of absorbing or the condition of being absorbed. **2.** Mental concentration. **—ab·sorp'·tive** (-tĭv) *adj.*

**absorption nebula** *n.* A nebula that absorbs all incident radiation without re-emission.

**absorption spectrum** *n. Physics.* The spectrum of dark lines and bands observed when radiation traverses an absorbing medium.

**ab·stain** (ăb-stān', əb-) *vi.* **-stained, -stain·ing, -stains.** [ME *absteinen,* to avoid < OFr. *abstenir* < Lat. *abstinēre,* to hold back : *ab-,* away + *tenēre,* to hold.] To refrain from something voluntarily. **—ab·stain'er** *n.*

**ab·ste·mi·ous** (ăb-stē'mē-əs, əb-) *adj.* [Lat. *abstemius* : *ab-,* away + *temetum,* liquor.] **1.** Consuming food and drink in moderation. **2.** Restricted to bare necessities. **—ab·ste'mi·ous·ly** *adv.* **—ab·ste'·mi·ous·ness** *n.*

**ab·sten·tion** (ăb-stĕn'shən, əb-) *n.* [LLat. *abstentio* < Lat. *abstinēre,* to hold back. —see ABSTAIN.] The act or habit of abstaining.

**ab·sti·nence** (ăb'stə-nəns) *n.* [ME < OFr. *abstenance* < Lat. *abstinentia* < *abstinēre,* to hold back. —see ABSTAIN.] **1.** Denial of the appetites. **2.** Habitual abstention from alcoholic beverages. **—ab'sti·nent** *adj.* **—ab'sti·nent·ly** *adv.*

**ab·stract** (ăb-străkt', ăb'străkt') *adj.* [ME < Lat. *abstractus,* p.part. of *abstrahere,* to draw away : *ab-,* away + *trahere,* to draw.] **1.** Considered apart from concrete existence <an *abstract* idea> **2.** Not applied or practical : THEORETICAL. **3.** Hard to understand : ABSTRUSE. **4.** Thought of or stated without reference to a specific instance <*abstract* words like "honesty" and "beauty"> **5.** Designating a genre of painting or sculpture whose intellectual and affective content depends solely on intrinsic form : NONOBJECTIVE. —*n.* (ăb'străkt'). **1.** A statement summarizing the important points of a given text. **2.** The concentrated essence of a larger whole. **3.** Something abstract, as a term. —*vt.* (ăb-străkt') **-stract·ed, -stract·ing, -stracts. 1.** To remove. **2.** To remove without permission : STEAL. **3.** To consider (e.g., a quality) without reference to a particular object or example. **4.** (ăb'străkt'). To summarize. **—ab·stract'er** *n.* **—ab·stract'ly** *adv.* **—ab·stract'ness** *n.*

**ab·stract·ed** (ăb-străk'tĭd, ăb'străk'-) *adj.* **1.** Removed or separated <metal *abstracted* from ore> **2.** Absent-minded : preoccupied <the *abstracted* look of a daydreamer> **—ab·stract'ed·ly** *adv.* **—ab·stract'ed·ness** *n.*

**abstract expressionism** *n.* A school of painting that flourished after World War II until the early 1960's, marked by the exclusion of representational content.

**ab·strac·tion** (ăb-străk'shən, əb-) *n.* **1.** The act or process of removing or separating. **2. a.** The act or process of separating the inherent qualities or properties of something from the actual physical object or concept to which they belong. **b.** A product of this process, as a general idea or word representing a physical concept. **3.** Preoccupation. **4.** An abstract work of art.

**ab·strac·tion·ism** (ăb-străk'shə-nĭz'əm) *n.* The theory and practice of abstract art. **—ab·strac'tion·ist** *n.*

**ab·strac·tive** (ăb-străk'tĭv, əb-) *adj.* Of or derived by abstraction.

**abstract of title** *n.* A brief history of the transfers of a piece of land, including all claims that could be made against it.

**ab·struse** (ăb-stro͞os', əb-) *adj.* [Lat. *abstrusus,* hidden, p.part. of *abstrudere,* to hide : *ab-,* away + *trudere,* to push.] Not easily understood : RECONDITE. **—ab·struse'ly** *adv.* **—ab·struse'ness** *n.*

**ab·surd** (əb-sûrd', -zûrd') *adj.* [Fr. *absurde* < Lat. *absurdus.*] **1.** Ridiculously incongruous or unreasonable. **2.** Of, relating to, or manifesting the view that there is no order or value in human life or in the universe. —*n.* **1.** The quality or condition of existing in a meaningless and irrational universe in which an individual's life has no meaning or purpose. **2.** The literary genre dealing with the theme of an absurd universe. **—ab·surd'i·ty** (-sûr'dĭ-tē, -zûr'-), **ab·surd'ness** *n.* **—ab·surd'ly** *adv.*

▲ word history: *Absurdus* in Latin meant "silly" and "irrational" just as *absurd* does in English, but its literal sense was "out of tune." It was used figuratively outside the realm of music and acoustics to mean "out of harmony with reason."

**absurd theater** *n.* Theater of the absurd.

**a·bu·li·a** *also* **a·bou·li·a** (ə-bo͞o'lē-ə, ə-byo͞o'-) *n.* [NLat. < Gk. *aboulia,* indecision : *a-,* without + *boulē,* will.] Impairment of the ability to act or decide independently. **—a·bu'lic** (-lĭk) *adj.*

**a·bun·dance** (ə-bŭn'dəns) *n.* **1.** A plentiful amount. **2.** Fullness to overflowing. **3.** Affluence.

**a·bun·dant** (ə-bŭn'dənt) *adj.* [ME *aboundant* < OFr. *abondant* < Lat. *abundans,* p.part. of *abundare,* to overflow. —see ABOUND.] **1.** Being in plentiful supply : AMPLE. **2.** Abounding : rich <an *abundant* harvest> <streams *abundant* in trout> **—a·bun'dant·ly** *adv.*

**a·buse** (ə-byo͞oz') *vt.* **a·bused, a·bus·ing, a·bus·es.** [ME *abusen* < OFr. *abuser* < *abus,* improper use < Lat. *abusus,* a using up, p.part. of *abuti,* to use up : *ab-,* away + *uti,* to use.] **1. a.** To use wrongly or improperly : MISUSE. **b.** *Obs.* To trick or deceive. **2.** To injure by maltreatment. **3.** To assail with abusive words : REVILE. —*n.* (ə-byo͞os'). **1.** Improper use or handling : MISUSE. **2.** A corrupt practice or custom. **3.** Physical maltreatment. **4.** Coarse or insulting language. **—a·bus'er** *n.*

☆ ***syns:*** **1.** ABUSE, ILL-USE, MALTREAT, MISHANDLE, MISTREAT, MISUSE *v. core meaning* : to hurt or injure by maltreatment <*abused* the child> **2.** ABUSE, MISAPPLY, MISAPPROPRIATE, MISUSE, PERVERT *v. core meaning* : to use wrongly and improperly <*abuse* a privilege>

**a·bu·sive** (ə-byo͞o'sĭv, -zĭv) *adj.* **1.** Of, relating to, or marked by abuse. **2.** Wrongly or incorrectly used or treated. **3.** Serving to abuse : INSULTING <*abusive* language> **—a·bu'sive·ly** *adv.*

**a·but** (ə-bŭt') *v.* **a·but·ted, a·but·ting, a·buts.** [ME *abutten* < OFr. *abouter,* to border on : *a-,* to + *bout,* end.] —*vi.* To touch at one end or side of something. —*vt.* To border on : ADJOIN. **—a·but'·ter** *n.*

**a·bu·ti·lon** (ə-byo͞ot'l-ŏn') *n.* [NLat. *Abutilon,* genus name < Ar. *aubūṭīlūn.*] A shrub or plant of the genus *Abutilon,* esp. the flowering maple.

**a·but·ment** (ə-bŭt'mənt) *n.* **1.** The act or process of abutting. **2. a.** Something that abuts. **b.** The point of contact of two abutting objects or parts. **3. a.** A structural part that bears the weight or pressure of an arch. **b.** A structure that supports the end of a bridge. **c.** A structure that anchors the cables of a suspension bridge.

**a·but·tals** (ə-bŭt'lz) *pl.n.* The parts of a piece of land that abut against other property.

**a·buzz** (ə-bŭz') *adj.* **1.** Filled with a buzzing sound. **2.** Filled or occupied with activity or talk <The conference room was *abuzz.*>

**ab·volt** (ăb'vōlt') *n.* A centimeter-gram-second electromagnetic unit of potential difference, equal to one hundred-millionth ($10^{-8}$) volt or the potential difference between two points such that one erg of work must be performed to move a one-abcoulomb charge from one of the points to the other.

**a·bysm** (ə-bĭz'əm) *n.* [ME *abīme* < OFr. < LLat. *abyssus.* —see ABYSS.] An abyss.

**a·bys·mal** (ə-bĭz'məl) *adj.* **1.** Immeasurably deep or extreme : FATHOMLESS. **2.** Of or like an abyss. **—a·bys'mal·ly** *adv.*

**a·byss** (ə-bĭs') *n.* [LLat. *abyssus* < Gk. *abussos,* bottomless : *a-,* without + *bussos,* bottom.] **1. a.** The primeval chaos. **b.** The bottomless pit : HELL. **2.** A yawning gulf. **3.** An immeasurably profound depth or void <the vast *abysses* of the galaxies>

**a·bys·sal** (ə-bĭs'əl) *adj.* **1.** Abysmal. **2.** Of or relating to the great depths of the oceans.

**Ab·ys·sin·i·an cat** (ăb'ĭ-sĭn'ē-ən) *n.* [After *Abyssinia,* former name for Ethiopia.] A short-haired cat orig. bred from Near Eastern stocks, with a reddish-brown coat tipped with small black markings.

**Ac** *symbol for* ACTINIUM.

**ac-** *pref. var. of* AD- 1. —Used before *c, k,* and *q.*

**-ac** *suff.* [NLat. *-acus,* adj. suffix < Gk. *-akos.*] —Used to form adjectives from nouns <ammoni*ac*>

**a·ca·cia** (ə-kā'shə) *n.* [Lat. < Gk. *akakia.*] **1.** A chiefly tropical tree of the genus *Acacia,* with compound leaves and tight small yellow or white flower clusters. **2.** Gum arabic.

**ac·a·deme** (ăk'ə-dēm') *n.* [< Lat. *Academia,* the Academy.] **1.** The academic environment. **2.** A scholar, teacher, or pedant. **3.** A college or university. **4.** Academic life.

**ac·a·de·mi·a** (ăk'ə-dē'mē-ə) *n.* [NLat. < Lat. *Academia,* the Academy.] The academic world : ACADEME.

**ac·a·dem·ic** (ăk'ə-dĕm'ĭk) *adj.* **1.** Of, relating to, or characteristic of a school. **2.** Pertaining to liberal or classical rather than technical or vocational education. **3.** Relating or belonging to a scholarly organization. **4.** Scholarly to the point of being oblivious to the outside world. **5.** Based on formal studies esp. at a college or university. **6.** Formalistic or conventional, as a style of painting or literature. **7. a.** Theoretical or speculative without a practical purpose or intention. **b.** Having no practical meaning or usefulness. —*n.* A student or teacher. **—ac'a·dem'i·cal·ly** *adv.*

**academic freedom** *n.* Freedom to pursue and teach relevant knowledge and to discuss it freely without interference, as from school or public officials.

**ac·a·de·mi·cian** (ăk'ə-də-mĭsh'ən, ə-kăd'ə-) *n.* A member of an art, literary, or scientific academy or society.

**ac·a·dem·i·cism** (ăk'ə-dĕm'ĭ-sĭz'əm) *also* **a·cad·e·mism** (ə-kăd'ə-mĭz'əm) *n.* Traditional formalism, esp. in art.

**a·cad·e·my** (ə-kăd'ə-mē) *n., pl.* **-mies.** [Lat. *Academia* < Gk. *Akadēmia,* the school where Plato taught.] **1.** A school for special instruction. **2.** A usu. private secondary or college-preparatory school. **3.** An association of scholars. **4. Academy.** A specified society of scholars or artists. **5. Academy. a.** Platonism. **b.** The disciples of Plato.

**A·ca·di·an** (ə-kā'dē-ən) *n.* **1.** One of the early French settlers of Acadia or their descendants. **2.** A dialect of French spoken by the Acadians. **—A·ca'di·an** *adj.*

**a·ca·jou** (ă-kə-zho͞o') *n.* [Fr., cashew < Port. *caju.*] Mahogany, esp. when used for making furniture.

**acantho-** *or* **acanth-** *pref.* [< Gk. *akanthos,* thorn plant < *akantha,* thorn.] Thorn <*acanth*oid>

**a·can·tho·ceph·a·lan** (ə-kăn'thə-sĕf'ə-lən) *also* **a·can·tho·ceph·a·lid** (-lĭd) *n.* [< NLat. *Acanthocephala,* phylum name : ACANTHO- + Gk. *kephalē,* head.] A parasitic worm of the phylum Acanthocephala, with a proboscis armed with hooked spines.

**a·can·thoid** (ə-kăn'thoid') *adj.* Resembling a thorn or spine.

**ac·an·thop·ter·yg·i·an** (ăk'ən-thŏp'tə-rĭj'ē-ən) *n.* [< NLat. *Acanthopterygii,* superorder name : ACANTHO- + Gk. *pterugion,* dim. of *pterus,* wing < *pteron,* feather.] A fish of the superorder Acanthopterygii, including fishes with spiny fins, as bass, perch, and mackerel. **—ac'an·thop'ter·yg'i·an** *adj.*

**a·can·thus** (ə-kăn'thəs) *n., pl.* **-thus·es** *or* **-thi** (-thī') [NLat. *Acanthus,* genus name < Gk. *akanthos,* thorn plant < *akantha,* thorn.] **1.** A plant of the genus *Acanthus* native to the Mediterranean region, with large, segmented, thistlelike leaves. **2.** An ornament patterned after acanthus leaves, used esp. on capitals of Corinthian columns.

**a·cap·ni·a** (ā-kăp'nē-ə) *n.* [NLat. < Lat. *acapnos,* without smoke (which contains carbon dioxide) < Gk. *akapnos : a-,* not + *kapnos,* smoke.] Absence of carbon dioxide in the blood and tissues.

**a cap·pel·la** (ä' kə-pĕl'ə) *adv.* [Ital. : *a,* as + *capella,* chapel.] *Mus.* Without instrumental accompaniment <sing *a cappella*>

**ac·a·ri** (ăk'ə-rī') *n. pl. of* ACARUS.

**ac·a·ri·a·sis** (ăk'ə-rī'ə-sĭs) *n.* [ACAR(ID) + -IASIS.] Infestation with mites.

**ac·a·rid** (ăk'ə-rĭd) *n.* [NLat. *Acaridae,* family name < *Acarus,* genus name < Gk. *akari,* a kind of mite.] An arachnid of the order Acarina, which includes the mites and ticks. **—ac'a·rid** *adj.*

**ac·a·roid resin** (ăk'ə-roid') *n.* [NLat. *acaroides.*] A yellow or reddish gum obtained from various Australian grass trees, used in varnishes, lacquers, and paper manufacturing.

**ac·a·rol·o·gy** (ăk'ə-rŏl'ə-jē) *n.* [ACAR(ID) + -LOGY.] The study of mites and ticks.

**ac·a·roph·i·ly** (ăk'ə-rŏf'ə-lē) *n.* [ACAR(ID) + -PHIL(O)- + -Y[2].] A symbiotic relationship to mites or ticks, usu. in plants.

**ac·a·ro·pho·bi·a** (ăk'ə-rə-fō'bē-ə) *n.* [ACAR(ID) + -PHOBIA.] Abnormal fear of mites or ticks.

**a·car·pous** (ā-kär'pəs) *adj. Bot.* Producing no fruit : STERILE.

**ac·a·rus** (ăk'ə-rəs) *n., pl.* **-ri** (-rī') [NLat. *Acarus,* genus name. —see ACARID.] A mite, esp. one of the genus *Acarus.*

**a·cat·a·lec·tic** (ā-kăt'l-ĕk'tĭk) *adj.* [LLat. *acatalecticus* < Gk. *akatalēktikos : a-,* not + *katalēktikos,* incomplete. —see CATALECTIC.] Designating a line of verse having the required number of syllables in the last foot. **—a·cat'a·lec'tic** *n.*

**a·cau·date** (ā-kô'dāt') *also* **a·cau·dal** (ā-kôd'l) *adj.* Having no tail.

**a·cau·les·cent** (ā'kô-lĕs'ənt) *adj.* Stemless or nearly so.

**ac·cede** (ăk-sēd') *vi.* **-ced·ed, -ced·ing, -cedes.** [ME *acceden,* to come near < Lat. *accedere,* to go near : *ad-,* to + *cedere,* to go.] **1.** To give consent. **2.** To come into an office or dignity. **3.** To become a party to an agreement or treaty. **—ac·ced'ence** (-sēd'ns) *n.* **—ac·ced'er** *n.*

**ac·cel·er·an·do** (ä-chĕl'ə-rän'dō) *adj.* [Ital. < Lat. *accelerandum* < *accelerare,* to hasten. —see ACCELERATE.] *Mus.* Gradually quickening in tempo. —Used as a direction. **—ac·cel'er·an'do** *adv.*

**ac·cel·er·ate** (ăk-sĕl'ə-rāt') *v.* **-at·ed, -at·ing, -ates.** [Lat. *accelerare : ad-* (intensive) + *celerare,* to quicken < *celer,* swift.] *—vt.* **1.** To increase the speed of. **2.** To bring about sooner than expected. **3.** To hasten the growth or progress of. **4.** *Physics.* To cause a change of velocity. *—vi.* To move or act faster. **—ac·cel'er·a'tive** *adj.*

**ac·cel·er·a·tion** (ăk-sĕl'ə-rā'shən) *n.* **1. a.** An act of accelerating. **b.** The process of being accelerated. **2.** *Physics.* The rate of change of velocity with respect to time.

**acceleration of gravity** *n.* Acceleration of freely falling bodies under the influence of terrestrial gravity, equal to 980.665 cm/sec$^2$ or approx. 32 ft/sec$^2$ at sea level.

**ac·cel·er·a·tor** (ăk-sĕl'ə-rā'tər) *n.* **1.** One that accelerates. **2.** A mechanical device, esp. the gas pedal of a motor vehicle, for increasing speed. **3.** *Chem.* A substance that increases the speed of a chemical reaction. **4.** *Physics.* A device, as an electrostatic generator, cyclotron, or linear accelerator, that accelerates charged subatomic particles or nuclei to energies useful for research.

**ac·cel·er·om·e·ter** (ăk-sĕl'ə-rŏm'ĭ-tər) *n.* [ACCELER(ATION) + -METER.] A device for measuring acceleration.

**ac·cent** (ăk'sĕnt') *n.* [ME < OFr. < Lat. *accentus,* accentuation : *ad-,* to + *cantus,* song < *canere,* to sing (transl. of Gk. *prosōidia,* voice modulations).] **1.** Relative prominence of a particular syllable of a word by greater intensity or by variation or modulation of pitch or tone. **2.** Vocal emphasis accorded a particular syllable, word, or phrase. **3.** A characteristic pronunciation, esp.: **a.** One determined by the regional or social background of the speaker. **b.** One determined by the phonetic habits of the speaker's native language transferred to his or her use of another language. **4.** A mark or symbol used to indicate the vocal quality to be given to a particular letter as in French. **5.** A mark or symbol used to indicate the stressed syllables of a spoken word <primary *accent*> **6.** Rhythmically significant stress in a line of verse. **7.** *Mus.* **a.** Special stress given to a note within a phrase. **b.** A mark representing this stress. **8.** *Math.* **a.** A mark or one of several marks used as a superscript to distinguish among variables represented by the same symbol. **b.** A mark used as a superscript to indicate the first derivative of a variable. **9.** A mark or one of several marks used as a superscript to indicate a unit, such as feet (') and inches (") in linear measurement. **10. a.** A distinctive quality or feature. **b.** Emphasis <an *accent* on fitness> *—vt.* (ăk'sĕnt', ăk-sĕnt') **-cent·ed, -cent·ing, -cents. 1.** To stress the pronunciation of. **2.** To mark with a printed accent. **3.** To call attention to : EMPHASIZE.

**ac·cen·tu·al** (ăk-sĕn'cho͞o-əl) *adj.* **1.** Of or relating to accent. **2.** Designating verse rhythm based on stress accents. **—ac·cen'tu·al·ly** *adv.*

**ac·cen·tu·ate** (ăk-sĕn'cho͞o-āt') *vt.* **-at·ed, -at·ing, -ates.** [Med. Lat. *accentuare, accentuat-* < Lat. *accentus,* accent.] **1.** To pronounce with a stress or accent. **2.** To mark with an accent. **3.** To emphasize. **—ac·cen'tu·a'tion** *n.*

**ac·cept** (ăk-sĕpt') *v.* **-cept·ed, -cept·ing, -cepts.** [ME *accepten* < Lat. *acceptare,* freq. of *accipere,* to receive : *ad-,* to + *capere,* to take.] *—vt.* **1.** To receive (something offered), esp. willingly. **2.** To admit to a group or place. **3. a.** To consider as usual, proper, or right. **b.** To consider as true. **c.** To understand as having a specific meaning, as a word or phrase. **4.** To endure <*accept* hardship> **5. a.** To answer affirmatively. **b.** To assume the duties or responsibilities of. **6.** To be able to take in or retain <a surface that won't *accept* oil paints> **7.** To receive officially, as a report. **8.** To consent to pay, as by a signed agreement. *—vi.* To receive something, esp. willingly.

**ac·cept·a·ble** (ăk-sĕp'tə-bəl) *adj.* **1.** Worthy of acceptance. **2.** Adequate : satisfactory <an *acceptable* report card> **3.** Designating an amount or level that can be allowed or endured. **—ac·cept'a·bil'i·ty, ac·cept'a·ble·ness** *n.* **—ac·cept'a·bly** *adv.*

**ac·cep·tance** (ăk-sĕp'təns) *n.* **1. a.** The act or process of accepting. **b.** The state of being accepted or acceptable. **2.** Approval. **3.** Belief in something : AGREEMENT. **4. a.** A formal indication by a debtor of willingness to pay a time draft or bill of exchange. **b.** A written instrument so accepted. **5.** *Law.* Compliance by one party with the terms and conditions of another's offer so that a contract becomes legally binding between them.

**ac·cep·tant** (ăk-sĕp'tənt) *adj.* Willingly accepting.

**ac·cep·ta·tion** (ăk'sĕp-tā'shən) *n.* **1.** The usual or accepted meaning, as of a word. **2.** Favorable reception. **3.** Ready belief.

**ac·cept·ed** (ăk-sĕp'tĭd) *adj.* That is gen. approved, believed, or recognized : RECEIVED.

☆ ***syns:*** ACCEPTED, CONVENTIONAL, ORTHODOX, RECEIVED, SANCTIONED *adj. core meaning* : generally approved or agreed on <*accepted* theories of evolution>

**ac·cep·tor** *also* **ac·cept·er** (ăk-sĕp'tər) *n.* **1.** One who signs a time draft or bill of exchange. **2.** *Chem.* **a.** The reactant in an induced reaction that has an increased rate of reaction in the presence of the inductor. **b.** The atom that contributes no atoms to a covalent bond.

**ac·cess** (ăk'sĕs') *n.* [ME *acces,* a coming to < OFr. < Lat. *accessus* < *accedere,* to arrive : *ad-,* to + *cedere,* to come.] **1.** A means of approaching : PASSAGE. **2.** The act of approaching. **3.** The right to enter or use <has *access* to official documents> **4.** The quality or state of being easy to approach or enter. **5.** An increase of growth. **6.** A sudden outburst <an *access* of weeping> *—vt.* **-cessed, -cess·ing, -cess·es.** To gain access to (e.g., computer information).

**ac·ces·sa·ry** (ăk-sĕs'ə-rē) *n. var. of* ACCESSORY.

**ac·ces·si·ble** (ăk-sĕs'ə-bəl) *adj.* **1.** Easily approached or entered. **2.** Easily obtained. **3.** Easy to get along with. **4.** Easily influenced : SUSCEPTIBLE <*accessible* to flattery> **—ac·ces'si·bil'i·ty, ac·ces'si·ble·ness** *n.* **—ac·ces'si·bly** *adv.*

**ac·ces·sion** (ăk-sĕsh'ən) *n.* **1.** The attainment of rank or dignity. **2. a.** An increase by means of something added <an *accession* of real estate> **b.** An addition. **3.** *Law.* **a.** The addition to or increase in value of property by means of improvements or natural growth. **b.** The right of a proprietor to ownership of such addition or increase. **4.** Agreement. **5.** Admittance. **6.** A sudden outburst <an *accession* of rage> *—vt.* **-sioned, -sion·ing, -sions.** To record as acquired. **—ac·ces'sion·al** *adj.*

**ac·ces·so·ry** *also* **ac·ces·sa·ry** (ăk-sĕs'ə-rē) *n., pl.* **-ries.** [ME *accessorie* < Med. Lat. *accessorius* < *accessor,* helper < Lat. *accessus,* approach. —see ACCESS.] **1. a.** A subordinate or supplementary part : ADJUNCT. **b.** Something nonessential but useful or decorative.

**2. a.** One who incites, aids, or abets a lawbreaker in the commission of a crime but is not present at the time of the crime <*accessory* before the fact> **b.** One who aids a criminal after the commission of a crime but was not present at the time of the crime <*accessory* after the fact> —*adj.* **1.** Having a supplementary function. **2.** Serving as an accessory to a lawbreaker. **—ac'ces·so'ri·al** (-sə-sôr'ē-əl, -sōr'-) *adj.* **—ac·ces'so·ri·ly** *adv.* **—ac·ces'so·ri·ness** *n.*

**accessory fruit** *n.* A fruit, as the pear, that contains fleshy tissue developed from floral parts as well as the ovary.

**accessory nerve** *n.* Either of the 11th pair of cranial nerves of higher vertebrate species that supply motor impulses to the upper thorax and pharynx.

**access road** *n.* A road giving access to a certain area.

**access time** *n. Computer Sci.* The time lag between a request for information stored in a computer and its delivery.

**ac·ciac·ca·tu·ra** (ä-chä'kə-to͝or'ə) *n.* [Ital. < *acciaccare*, to crush.] *Mus.* A short grace note one half step below a principal note, sounded immediately before or at the same time as the principal note to add sustained dissonance.

**ac·ci·dence** (ăk'sĭ-dəns, -dĕns') *n.* [LLat. *accidentia* < Lat. *accidens*, accident.] The section of morphology dealing with word inflections.

**ac·ci·dent** (ăk'sĭ-dənt, -dĕnt') *n.* [ME < OFr. < Lat. *accidens*, pr.part. of *accidere*, to happen : *ad-*, to + *cadere*, to fall.] **1.** An unexpected and undesirable event <a car *accident*><a skiing *accident*> **2.** Something that occurs unexpectedly or unintentionally. **3.** A nonessential circumstance or attribute. **4.** Fortune or chance <discovered by *accident*> **5.** *Geol.* An irregular or unusual natural formation or occurrence.

**ac·ci·den·tal** (ăk'sĭ-dĕn'tl) *adj.* **1.** Occurring unexpectedly and unintentionally. **2.** Not intrinsic : NONESSENTIAL. **3.** *Mus.* Of or denoting a sharp, flat, or natural not indicated in the key signature. —*n.* **1.** A nonessential factor or attribute. **2.** *Mus.* A chromatically altered note not belonging to the key signature. **—ac'ci·den'tal·ly** *adv.*

☆ ***syns:*** ACCIDENTAL, CASUAL, CHANCE, CONTINGENT, FLUKY, FORTUITOUS, INADVERTENT, ODD *adj. core meaning* : occurring unexpectedly <an *accidental* meeting> ***ant:*** planned

**accident insurance** *n.* Insurance against injury or death due to accident.

**ac·ci·dent-prone** (ăk'sĭ-dənt-prōn') *adj.* Susceptible to having a greater than average number of accidents.

**ac·cip·i·ter** (ăk-sĭp'ĭ-tər) *n.* [NLat. *Accipiter*, genus name < Lat., hawk.] A hawk of the genus *Accipiter*, characterized by a long tail and short wings. **—ac·cip'i·trine'** (-trīn', -trĭn) *adj.*

**accipiter**
*The sharp-shinned hawk, wingspan approximately 2 feet*

**ac·claim** (ə-klām') *v.* **-claimed, -claim·ing, -claims.** [Lat. *acclamare*, to shout at : *ad-*, at + *clamore*, to shout.] —*vt.* To salute or hail : APPLAUD. —*vi.* To shout approval. —*n.* Enthusiastic approval : ACCLAMATION. **—ac·claim'er** *n.*

**ac·cla·ma·tion** (ăk'lə-mā'shən) *n.* [Lat. *acclamatio* < *acclamare*, to shout at. —see ACCLAIM.] **1.** An expression of enthusiastic approval. **2.** An oral vote, esp. an enthusiastic vote of approval taken without formal ballot. **—ac·clam'a·to'ry** (ə-klăm'ə-tôr'ē, -tōr'ē) *adj.*

**ac·cli·mate** (ə-klī'mĭt, ăk'lə-māt') *vt. & vi.* **-mat·ed, -mat·ing, -mates.** [Fr. *acclimater* : *a-*, to (< Lat. *ad*) + *climat*, climate < OFr. —see CLIMATE.] To accustom or become accustomed to new surroundings or circumstances : ADAPT.

**ac·cli·ma·tion** (ăk'lə-mā'shən) *n.* **1.** The process of acclimating or of becoming acclimated. **2.** Adaptation of an organism to its natural climatic environment.

**ac·cli·ma·ti·za·tion** (ə-klī'mə-tĭ-zā'shən) *n.* **1.** The process of acclimatizing. **2.** Climatic adaptation of an organism, esp. a plant, that has been moved to a new environment.

**ac·cli·ma·tize** (ə-klī'mə-tīz') *vt. & vi.* **-tized, -tiz·ing, -tiz·es.** To acclimate. **—ac·cli'ma·tiz'er** *n.*

**ac·cliv·i·ty** (ə-klĭv'ĭ-tē) *n., pl.* **-ties.** [Lat. *acclivitas* < *acclivis*, uphill : *ad-*, to + *clivus*, slope.] An upward slope.

**ac·co·lade** (ăk'ə-lād', -läd') *n.* [Fr. < Prov. *acolada* < *acolar*, to embrace : *a-*, to (< Lat. *ad*) + *col*, neck < Lat. *collum*, neck.] **1.** An embrace of greeting or salutation. **2.** Approval : praise. **3.** The ceremonial bestowal of knighthood.

**ac·com·mo·date** (ə-kŏm'ə-dāt') *v.* **-dat·ed, -dat·ing, -dates.** [Lat. *accomodare, accomodat-*, to fit : *ad-*, to + *commodus*, suitable. —see COMMODIOUS.] —*vt.* **1.** To do a favor or service for : OBLIGE. **2.** To supply with (e.g., lodging). **3.** To have enough space for. **4.** To acclimate or adjust <*accommodated* myself to the new neighborhood> **5.** To reconcile, as differences. **6.** To allow for. —*vi.* To become adjusted, as the eye to focusing on distant objects. **—ac·com'mo·da'tive** *adj.* **—ac·com'mo·da'tive·ness** *n.* **—ac·com'mo·da'tor** *n.*

**ac·com·mo·dat·ing** (ə-kŏm'ə-dā'tĭng) *adj.* Willing to help. **—ac·com'mo·dat'ing·ly** *adv.*

**ac·com·mo·da·tion** (ə-kŏm'ə-dā'shən) *n.* **1.** The act of accommodating or state of being accommodated. **2.** Something that meets a need : CONVENIENCE. **3. accommodations. a.** Lodgings. **b.** A seat, compartment, or room on a public vehicle. **4.** Reconciliation of opposing views : COMPROMISE. **5.** *Physiol.* Adaptation or adjustment in an organism, organ, or part, as in the lens of the eye to permit retinal focus of images of objects at different distances. **6.** A financial favor, as a loan.

**accommodation ladder** *n.* A portable ladder or stairway hung from the side of a ship.

**accommodation paper** *n.* A note or bill drawn, accepted, or endorsed by one or more parties to enable another party to obtain credit or raise money without consideration or collateral.

**ac·com·pa·ni·ment** (ə-kŭm'pə-nē-mənt, ə-kŭmp'nē-) *n.* **1.** Something that accompanies : CONCOMITANT. **2.** An addition for embellishment, completeness, or symmetry : COMPLEMENT. **3.** A vocal or instrumental part that supports a solo part.

**ac·com·pa·nist** (ə-kŭm'pə-nĭst, ə-kŭmp'nĭst) *n.* A performer, as a pianist, who plays an accompaniment.

**ac·com·pa·ny** (ə-kŭm'pə-nē, ə-kŭmp'nē) *v.* **-nied, -ny·ing, -nies.** [ME *accompanien* < OFr. *acompagnier* : *a*, to (< Lat. *ad*) + *compaignon*, companion. —see COMPANION.] —*vt.* **1.** To go with as a companion. **2.** To supplement. **3.** To coexist or occur with. **4.** To perform an accompaniment to or for. —*vi.* To play a musical accompaniment.

**ac·com·plice** (ə-kŏm'plĭs) *n.* [< ME *a*, a + *complice*, companion. —see COMPLICE.] One who aids or abets a lawbreaker in a criminal act, either as a principal or an accessory.

**ac·com·plish** (ə-kŏm'plĭsh) *vt.* **-plished, -plish·ing, -plish·es.** [ME *accomplisshen* < OFr. *acomplir, accompliss-*, to complete : *a-*, to (< Lat. *ad*) + *complir*, to complete < Lat. *complere*, to fill out. —see COMPLETE.] **1.** To succeed in doing. **2.** To reach the end of : COMPLETE <*accomplish* a trip> **—ac·com'plish·a·ble** *adj.* **—ac·com'plish·er** *n.*

☆ ***syns:*** ACCOMPLISH, ACHIEVE, ATTAIN, GAIN, REACH, REALIZE *v. core meaning* : to succeed in doing <*accomplish* an objective>

**ac·com·plished** (ə-kŏm'plĭsht) *adj.* **1.** Completed : finished. **2.** Skilled : expert. **3.** Socially adept : POISED.

**ac·com·plish·ment** (ə-kŏm'plĭsh-mənt) *n.* **1.** The act of accomplishing or state of being accomplished. **2.** Something successfully completed. **3.** Social poise.

**ac·cord** (ə-kôrd') *vt.* **-cord·ed, -cord·ing, -cords.** [ME *accorden* < OFr. *acorder* < Med. Lat. *accordare* : Lat. *ad-*, to + Lat. *cor*, heart.] **1.** To cause to conform, agree, or harmonize. **2.** To grant <*accord* a favor>—*vi.* To be in agreement, unity, or harmony. —*n.* **1.** Agreement : harmony. **2. a.** A settlement or compromise of conflicting opinions. **b.** A settlement of points at issue between nations. **—of (one's) own accord.** Voluntarily.

**ac·cor·dance** (ə-kôr'dns) *n.* **1.** Agreement : conformity <in *accordance* with your wishes> **2.** The act of granting <*accordance* of civil rights>

**ac·cor·dant** (ə-kôr'dnt) *adj.* In agreement or harmony : CONSONANT. **—ac·cor'dant·ly** *adv.*

**according as** *conj.* **1.** Corresponding to the way in which : exactly as <gave me the new car *according as* I was promised> **2.** Depending on whether : IF <*according as* we approve your proposal>

**ac·cord·ing·ly** (ə-kôr'dĭng-lē) *adv.* **1.** Correspondingly. **2.** Consequently : therefore.

**according to** *prep.* **1.** On the authority of <*according to* informed sources> **2.** In keeping with <*according to* plan> **3.** As determined by <lined up *according to* height>

**ac·cor·di·on** (ə-kôr'dē-ən) *n.* [G. *Akkordion* < *Akkord*, chord < Fr. *accord*, harmony < OFr. *acorder*, to accord.] A portable musical instrument with a small keyboard and free metal reeds that sound when air is forced past them by pleated bellows operated by the player. —*adj.* Having folds or bends like the bellows of an accordion <a skirt with *accordion* pleats> **—ac·cor'di·on·ist** *n.*

**ac·cost** (ə-kôst', ə-kŏst') *vt.* **-cost·ed, -cost·ing, -costs.** [OFr. *acoster* < Med. Lat. *acostare*, to adjoin : Lat. *ad-*, to + Lat. *costa*, side.] **1.** To approach and address first. **2.** To solicit sexually.

**ac·couche·ment** (ä-ko͞osh-mäN') *n.* [Fr. < *accoucher*, to assist in childbirth : *a*, to (< Lat. *ad*) + *coucher*, to put to bed < OFr., to lay down. —see COUCH.] Parturition.

**†ac·count** (ə-kount') *n.* [ME < OFr. *acont* < *aconter*, to reckon : *a-*, to (< Lat. *ad*) + *cunter*, to count, ult. < Lat. *computare*, to sum up.

—see COMPUTE.] **1. a.** A narrative or record of events. **b.** A written or oral explanation. **2. a.** An exact list of monetary transactions. **b.** A detailed list. **3.** A business relationship involving the exchange of money or credit. **4.** Worth or importance <a citizen of some *account*> **5.** Profit or advantage <turned our talent to good *account*> —*vt.* **-count·ed, -count·ing, -counts.** To consider or esteem. **—account for. 1.** To make or render a reckoning <The missing funds haven't been *accounted for.*> **2.** To be the explanation or cause of. **3.** To be answerable for <*account for* one's misdeeds> **—call to account. 1.** To challenge or contest. **2.** To hold answerable for. **—give a good account of (oneself).** To act in a creditable way. **—on account. 1.** On credit. **2.** In partial payment of. **—on account of. 1.** Because of. **2.** *Regional.* Because. **—on no account.** Under no circumstances. **—on (one's) own account. 1.** On one's own behalf. **2.** On one's own : by oneself <decided to try it *on my own account*> **—take into account.** To take into consideration.

**ac·count·a·ble** (ə-koun'tə-bəl) *adj.* **1.** Required to render account : ANSWERABLE. **2.** Capable of being explained. **—ac·count'a·bil'i·ty, ac·count'a·ble·ness** *n.* **—ac·count'a·bly** *adv.*

**ac·count·ant** (ə-koun'tənt) *n.* One who keeps, audits, and inspects financial records and prepares financial and tax reports. **—ac·count'an·cy** (-tən-sē) *n.*

**account executive** *n.* An advertising executive who manages the account of one or more clients.

**ac·count·ing** (ə-koun'tĭng) *n.* The bookkeeping methods involved in making a financial record of business transactions and in the preparation of statements concerning the assets, liabilities, and operating results of a business.

**ac·cou·ter** (ə-kōō'tər) *vt.* **-tered, -ter·ing, -ters.** [Fr. *accoutrer* < OFr. *acoustrer.*] To outfit and equip, as for military duty.

**ac·cou·ter·ment** (ə-kōō'tər-mənt) *n.* **1.** The act of accoutering. **2.** *often* **accouterments.** The equipment other than weapons and clothing issued to a soldier. **3. accouterments.** The outward characteristics that serve to identify : TRAPPINGS.

**ac·cou·tre** (ə-kōō'tər) *v. Chiefly Brit. var. of* ACCOUTER.

**ac·cou·tre·ment** (ə-kōō'tər-mənt, -trə-mənt) *n. Chiefly Brit. var. of* ACCOUTERMENT.

**ac·cred·it** (ə-krĕd'ĭt) *vt.* **-it·ed, -it·ing, -its.** [Fr. *accréditer* : *a,* to (< Lat. *ad*) + *credit,* credit < OFr. —see CREDIT.] **1.** To attribute to. **2. a.** To authorize. **b.** To appoint as an ambassador to a foreign government. **3.** To certify as meeting a prescribed standard. **4.** To believe. **5.** To enter on the credit side of an account book.

**ac·cred·i·ta·tion** (ə-krĕd'ĭ-tā'shən) *n.* The act of accrediting or the state of being accredited, esp. the granting of approval to an institution of learning by an official review board after the school has met specific requirements.

**ac·crete** (ə-krēt') *v.* **-cret·ed, -cret·ing, -cretes.** [Back-formation < ACCRETION.] —*vt.* To make larger or greater, as by increased growth. —*vi.* **1.** To grow together : FUSE. **2.** To grow or increase gradually, as by addition.

**ac·cre·tion** (ə-krē'shən) *n.* [Lat. *accretio* < *accrescere,* to grow. —see ACCRUE.] **1. a.** Growth or increase in size by gradual external addition or accumulation. **b.** Something added externally to promote such growth or increase. **2.** *Biol.* The growing together of plant or animal tissues that are normally separate. **3.** *Geol.* Slow addition to land by deposition of water-borne sediment. **4.** An increase of land along the shores of a body of water, as by alluvial deposit. **5.** *Astron.* An increase in the mass of a celestial object by the collection of surrounding interstellar gases and objects by gravity. **—ac·cre'tion·ar'y** (-shə-nĕr'ē), **ac·cre'tive** *adj.*

**accretion disk** *n.* A ring of interstellar material surrounding a celestial object with an intense gravitational field, as a black hole.

**ac·cru·al** (ə-krōō'əl) *n.* **1.** The act or process of accruing : INCREASE. **2.** Something that increases or accrues.

**ac·crue** (ə-krōō') *vi.* **-crued, -cru·ing, -crues.** [ME *acreuen,* ult. < Lat. *accrescere,* to grow : *ad-,* to + *crescere,* to arise.] **1.** To come to someone or something as a gain or increment. **2.** To increase or accumulate by regular growth, as interest on capital. **3.** *Law.* To become an enforceable or permanent right. **—ac·crue'ment** *n.*

**ac·cul·tur·ate** (ə-kŭl'chə-rāt') *v.* **-at·ed, -at·ing, -ates.** —*vt.* To cause (e.g., a society) to change by the process of acculturation. —*vi.* To change by acculturation.

**ac·cul·tur·a·tion** (ə-kŭl'chə-rā'shən) *n.* **1.** Modification of one culture as a result of contact with a different, esp. a more advanced, culture. **2.** The process by which the culture of a particular society is instilled in a human being from infancy onward.

**ac·cum·bent** (ə-kŭm'bənt) *adj.* [Lat. *accumbens, accumbent-,* pr.part. of *accumbere,* to recline at table : *ad-,* near to + *cumbere,* to recline.] **1.** Reclining : leaning. **2.** *Bot.* Resting against another part, as a cotyledon.

**ac·cu·mu·late** (ə-kyōōm'yə-lāt') *v.* **-lat·ed, -lat·ing, -lates.** [Lat. *accumulare, accumulat-* : *ad-,* to + *cumulare,* to pile up < *cumulus,* heap.] —*vt.* To amass or gather. —*vi.* To grow or increase <bills *accumulating*> **—ac·cu'mu·la·ble** (-lə-bəl) *adj.*

**ac·cu·mu·la·tion** (ə-kyōōm'yə-lā'shən) *n.* **1.** The act of amassing or gathering. **2.** The process of growing into a large amount. **3.** A collected mass or heap. **4. a.** The growth of a principal sum by retention of interest or profit. **b.** The gradual purchase of securities in a depressed market in anticipation of rising prices. **c.** The difference between a bond's face value and its cost if purchased at a discount.

**ac·cu·mu·la·tive** (ə-kyōōm'yə-lā'tĭv, -lə-tĭv) *adj.* **1.** Marked by or showing the effects of accumulation : CUMULATIVE. **2.** Inclined to amass : ACQUISITIVE. **—ac·cu'mu·la·tive·ly** *adv.* **—ac·cu'mu·la·tive·ness** *n.*

**ac·cu·mu·la·tor** (ə-kyōōm'yə-lā'tər) *n.* **1.** One that accumulates. **2.** A register or electric circuit in a calculator or computer that stores figures. **3.** *Chiefly Brit.* An automobile storage battery.

**ac·cu·ra·cy** (ăk'yər-ə-sē) *n.* The quality or state of being accurate.

**ac·cu·rate** (ăk'yər-ĭt) *adj.* [Lat. *accuratus,* done with care, p.part. of *accurare,* to do with care : *ad-,* to + *curare,* to care for < *cura,* care.] **1.** In exact conformity to fact : ERRORLESS. **2.** Conforming closely to a standard <an *accurate* timepiece> **—ac'cu·rate·ly** *adv.* **—ac'cu·rate·ness** *n.*

☆ ***syns:*** ACCURATE, CORRECT, EXACT, FAITHFUL, PRECISE, PROPER, RIGHT, TRUE, VERACIOUS, VERICIDAL *adj. core meaning* : conforming to fact <an *accurate* description of the new house> ***ant:*** inaccurate

**ac·curs·ed** (ə-kûr'sĭd, ə-kûrst') *also* **ac·curst** (ə-kûrst') *adj.* [ME *acursed* < *acursen,* to put a curse on : *a-* (intensive) + OE *cursian,* to curse < *curs,* curse.] **1.** Under a curse : DOOMED. **2.** Abominable : damnable. **—ac·curs'ed·ly** *adv.* **—ac·curs'ed·ness** *n.*

**ac·cu·sa·tion** (ăk'yōō-zā'shən) *n.* **1.** An act of accusing. **2.** An allegation. **3.** *Law.* A formal charge brought before a court against a person, stating that he or she is guilty of a punishable offense.

☆ ***syns:*** ACCUSATION, CHARGE, DENOUNCEMENT, DENUNCIATION, IMPUTATION, INCRIMINATION, INDICTMENT *n. core meaning* : a charging of another with a misdeed <an *accusation* of murder>

**ac·cu·sa·tive** (ə-kyōō'zə-tĭv) *adj.* [ME *acusatif* < OFr. < Lat. *(casus) accusativus,* (case) of accusation < *accusare,* to accuse.] Of or relating to the case of a noun, pronoun, adjective, or participle that is the direct object of a verb or the object of certain prepositions. —*n.* The accusative case. **—ac·cu'sa·tive·ly** *adv.*

▲ word history: The connection between accusation and grammatical case is not immediately evident. The Latin verb *accusare,* the source of English *accuse,* means "to call to account," especially in the sense of answering a legal charge. *Accusativus* is an adjective derived from *accusare,* and it is the source of English *accusative. Accusativus* at bottom means "pertaining to an accusation," and it seems an odd choice for the name of a grammatical category. But *accusativus,* which has only the grammatical, not the legal, meaning in Latin, is a translation of the Greek word *aitiatikē,* which means both "causal" and "accusative," being itself derived from *aitia,* "accusation," "guilt," "cause." Thus the accusative case is the case of something caused or brought about by the verb.

**ac·cu·sa·to·ri·al** (ə-kyōō'zə-tôr'ē-əl, -tōr'-) *also* **ac·cu·sa·to·ry** (-tôr'ē, -tōr'ē) *adj.* Implying or containing accusation : ACCUSING.

**ac·cuse** (ə-kyōōz') *v.* **-cused, -cus·ing, -cus·es.** [ME *acusen* < Lat. *accusare* : *ad-,* to + *causa,* lawsuit.] —*vt.* **1.** To charge with a shortcoming or error. **2.** To bring charges against (someone) for a misdeed. —*vi.* To make an accusation against another. **—ac·cus'er** *n.* **—ac·cus'ing·ly** *adv.*

**ac·cused** (ə-kyōōzd') *n. Law.* The defendant or defendants in a criminal case.

**ac·cus·tom** (ə-kŭs'təm) *vt.* **-tomed, -tom·ing, -toms.** [ME *accustomen* < OFr. *acostumer* : *a,* to + *costume,* custom. —see CUSTOM.] To familiarize, as by constant practice, use, or habit.

**ac·cus·tomed** (ə-kŭs'təmd) *adj.* **1.** Usual, characteristic, or normal <began the job with my *accustomed* energy> **2.** In the habit of <*accustomed* to walking after dinner>

**AC/DC** (ā'sē-dē'sē) *adj.* [From the likening of a bisexual person to an appliance that works on either alternating or direct current.] *Slang.* BISEXUAL 3.

**ace** (ās) *n.* [ME *as* < OFr. < Lat., unit.] **1. a.** A single spot on a playing card, die, or domino. **b.** A playing card, die, or domino having one spot. **2.** In racket games: **a.** A serve that one's opponent fails to return. **b.** A point scored by serving an ace. **3.** The act of hitting a golf ball in the hole with one's first shot. **4.** *Informal.* A narrow margin. **5.** A military pilot who has shot down five or more enemy aircraft. **6.** An expert in a given field. —*adj. Informal.* First-rate <an *ace* swimmer> —*vt.* **aced, ac·ing, ac·es. 1.** To serve an ace against. **2.** To hit an ace in golf. **3.** To get the better of : SURPASS. **4.** To receive a grade of A on. **—ace in the hole. 1.** A hidden advantage. **2.** A hole in one in golf. **—within an ace of.** On the verge of.

**-acean** *suff.* [< NLat *-acea,* neuter pl. of *-aceus,* -aceous.] **1.** -ACEOUS. **2.** An organism belonging to a taxonomic group <cetacean>

**a·ce·di·a** (ə-sē'dē-ə) *n.* [LLat. < Gk. *akēdeia,* indifference : *a-,* without + *kēdos,* care.] Spiritual indifference : APATHY.

**A·cel·da·ma** (ə-sĕl'də-mə) *n.* [After *Aceldama,* a field bought with the money Judas received for betraying Jesus < Gk. *Akeldama* < Aram. *ḥăqēl dĕmā,* field of blood.] **1.** A place with dreadful associations. **2.** A place of bloodshed.

**a·cel·lu·lar** (ā-sĕl′yə-lər) *adj.* Containing no cells.

**-aceous** *suff.* [NLat. *-aceus* < Lat.] **1. a.** Of or relating to <amyl*aceous*> **b.** Resembling or having the nature of <ament*aceous*> **2.** Belonging to a specified taxonomic category <orchid*aceous*>

**a·ceph·a·lous** (ā-sĕf′ə-ləs) *adj.* [Med. Lat. *acephalus* < Gk. *akephalos* : *a-*, without + *kephalē*, head.] **1.** Headless or lacking a clearly defined head. **2.** Having no leader.

**†a·ce·qui·a** (ə-sā′kē-ə, ä-sä′-) *n.* [Sp. < Ar. *assaqīyāh*, irrigation ditch.] *Southwestern U.S.* An irrigation canal.

**ac·er·ate** (ăs′ə-rāt′) *also* **ac·er·at·ed** (-rā′tĭd) *adj.* [< Lat. *acer*, sharp.] *Biol.* Pointed at one end like a needle.

**a·cerb** (ə-sûrb′) *adj.* [Lat. *acerbus.*] **1.** Having a bitter or sour taste. **2.** Sharp or bitter in temper, mood, or expression. **—a·cer′bi·ty** (ə-sûr′bĭ-tē) *n.*

**ac·er·bate** (ăs′ər-bāt′) *vt.* **-bat·ed, -bat·ing, -bates.** [Lat. *acerbare, acerbat-*, to make harsh < *acerbus*, harsh.] To irritate or annoy.

**ac·er·ose** (ăs′ə-rōs′) *adj.* [NLat. *acerosus* < Lat. *acer*, sharp.] Sharp-pointed and slender, as a pine needle.

**a·cer·vate** (ə-sûr′vĭt, ăs′ər-vāt′) *adj.* [< Lat. *acervare, acervat-*, to heap up < *acervus*, heap.] *Bot.* Growing in small heaps or tight clusters. **—a·cer′vate·ly** *adv.*

**acet-** *pref. var. of* ACETO-.

**ac·e·tab·u·lum** (ăs′ĭ-tăb′yə-ləm) *n., pl.* **-la** (-lə) [Lat., vinegar cup < *acetum*, vinegar.] **1.** *Anat.* The cup-shaped cavity in the hipbone into which the head of the thighbone fits. **2.** *Zool.* A sucker, as that of an octopus or cuttlefish. **—ac′e·tab′u·lar** (-lər) *adj.*

**ac·e·tal** (ăs′ĭ-tăl′) *n.* [G. *Azetal* : *acet-*, aceto- + *Alkohol*, alcohol.] **1.** A colorless, flammable, volatile liquid, $CH_3CH(OC_2H_5)_2$, used as solvent and in cosmetics. **2.** Any of the class of compounds formed from aldehydes combined with alcohol.

**ac·et·al·de·hyde** (ăs′ĭ-tăl′də-hīd′) *n.* A colorless, flammable liquid, $C_2H_4O$, used to make acetic acid, perfumes, and drugs.

**a·cet·a·mide** (ə-sĕt′ə-mīd′, ăs′ĭt-ăm′ĭd′) *also* **a·cet·a·mid** (ə-sĕt′ə-mĭd, ăs′ĭt-ăm′ĭd) *n.* The crystalline amide of acetic acid, $CH_3CONH_2$, used in lacquers and explosives and as a solvent and wetting agent.

**a·cet·a·min·o·phen** (ə-sē′tə-mĭn′ə-fən, ăs′ə-) *n.* [ACET(O)- + AMIN(O)- + PHEN(OL).] A hydroxy derivative of acetanilide, $C_8H_9NO_2$, used in chemical synthesis and medicinally to reduce pain and fever.

**ac·et·an·i·lide** (ăs′ĭt-ăn′l-īd′) *also* **ac·et·an·i·lid** (-ĭd) *n.* [ACET(O)- + ANIL(INE) + -IDE.] A white crystalline compound, $C_6H_5NH(COCH_3)$, used medicinally to reduce pain and fever.

**ac·e·tate** (ăs′ĭ-tāt′) *n.* **1.** A salt or ester of acetic acid. **2.** Cellulose acetate or derivative products.

**a·ce·tic** (ə-sē′tĭk) *adj.* [< Lat. *acetum*, vinegar. —see ACETUM.] Of, relating to, or containing acetic acid or vinegar.

**acetic acid** *n.* A clear, colorless organic acid, $CH_3COOH$, with a distinctive pungent odor, used as a solvent and in making rubber, plastics, acetate fibers, pharmaceuticals, and photographic chemicals.

**acetic anhydride** *n.* An organic liquid, $(CH_3CO)_2O$, with a pungent odor, combining with water to produce acetic acid and used to make organic acetate derivatives.

**a·ce·ti·fy** (ə-sē′tə-fī′, ə-sĕt′ə-) *vt. & vi.* **-fied, -fy·ing, -fies.** To convert or become converted to acetic acid or vinegar. **—a·ce′ti·fi·ca′tion** *n.* **—a·ce′ti·fi·er** *n.*

**aceto-** *or* **acet-** *pref.* [< Lat. *acetum*, vinegar. —see ACETUM.] **1.** Acetic acid <*aceti*fy> **2.** Acetyl <*acet*anilide>

**ac·e·to·a·ce·tic acid** (ăs′ĭ-tō-ə-sē′tĭk, ə-sē′tō-) *n.* A syrupy, colorless acid, $CH_3COCH_2COOH$, excreted in the urine and found in abnormal quantities in the urine of diabetics.

**ac·e·tone** (ăs′ĭ-tōn′) *n.* A colorless, volatile, highly flammable liquid, $CH_3COCH_3$, widely used as an organic solvent. **—ac′e·ton′ic** (-tŏn′ĭk) *adj.*

**acetone body** *n.* A ketone body.

**ac·e·to·phe·net·i·din** (ăs′ĭ-tō-fə-nĕt′ĭ-dĭn, ə-sē′tō-) *n.* [ACETO- + PHEN(O) + ET(HYL) + -ID(E) + -IN.] A white powder or crystalline solid, $CH_3CONHC_6H_4OC_2H_5$, used medicinally to reduce pain and fever.

**a·ce·tous** (ə-sē′təs, ăs′ĭ-təs) *adj.* [ME, sour < Med. Lat. *acetus* < Lat. *acetum*, vinegar. —see ACETUM.] **1.** Of, relating to, or producing acetic acid or vinegar. **2.** Having an acetic taste.

**a·ce·tum** (ə-sē′təm) *n.* [Lat. *acetum* < *acēre*, to be sour < *acer*, sharp.] **1.** Vinegar. **2.** An acetic acid solution of a drug.

**a·ce·tyl** (ə-sēt′l, ăs′ĭ-tl) *n.* The acetic acid radical $CH_3CO$. **—ac′e·tyl′ic** (ăs′ĭ-tĭl′ĭk) *adj.*

**a·cet·y·late** (ə-sĕt′l-āt′) *vt.* **-lat·ed, -lat·ing, -ates.** To bring an acetyl group into (an organic molecule), using a reagent such as acetic anhydride. **—a·cet′y·la′tion** *n.*

**a·ce·tyl·cho·line** (ə-sēt′l-kō′lēn′) *n.* A white crystalline compound, $C_7H_{17}NO_3$, that transmits nerve impulses across intercellular gaps and forms salts used to lower blood pressure and increase peristalsis.

**a·ce·tyl·cho·lin·es·ter·ase** (ə-sēt′l-kō′lə-nĕs′tə-rās′, . -rāz′) *n.* Cholinesterase.

**a·cet·y·lene** (ə-sĕt′l-ēn′, -ən) *n.* A colorless, extremely flammable or explosive gas, $C_2H_2$, used as an illuminant and for metal welding and cutting. **—a·cet′y·len′ic** *adj.*

**acetylene series** *n.* A series of unsaturated aliphatic hydrocarbons, each containing at least one triple carbon bond, having chemical properties resembling acetylene and having the general formula $C_nH_{2n-2}$ with acetylene being the simplest member.

**a·ce·tyl·sal·i·cyl·ic acid** (ə-sēt′l-săl′ĭ-sĭl′ĭk) *n.* Aspirin.

**ace·y-deuc·y** (ā′sē-dōō′sē, -dyōō′-) *n.* [ACE + DEUCE.] A variation of backgammon.

**ach·a·la·sia** (ăk′ə-lā′zhə) *n.* [NLat. : A-[1] + Gk. *khalasis*, relaxation < *khalan*, to loosen.] Inability of a ring muscle or sphincter to relax.

**A·cha·tes** (ə-kā′tēz) *n.* [After *Achates*, the faithful companion of Aeneas, in the *Aeneid*, an epic poem by Virgil (70–19 B.C.).] A devoted friend.

**ache** (āk) *vi.* **ached, ach·ing, aches.** [ME *aken* < OE *acan.*] **1.** To suffer a dull, steady pain. **2.** To experience compassion or sympathy. **3.** *Informal.* To long for something **:** YEARN. *—n.* **1.** A dull, steady pain. **2.** *Informal.* A longing or desire.

**a·chene** (ə-kēn′) *n.* [NLat. *achenium* : Gk. *a-*, not + Gk. *khainein*, to yawn.] A small, dry, thin-walled fruit, as of the buttercup and dandelion, that does not split open when ripe. **—a·che′ni·al** (ə-kē′nē-əl) *adj.*

**A·cher·nar** (ā′kər-när′) *n.* [< Ar. *ākhir alnahr*, the end of the river.] A star in the constellation Eridanus that is one of the brightest stars in the sky and is 114 light-years from Earth.

**Ach·er·on** (ăk′ə-rŏn′, -rən) *n.* [Gk. *Akherōn.*] *Gk. Myth.* **1.** The river of woe over which Charon ferried the souls of the dead to Hades. **2.** Hades.

**A·cheu·li·an** *also* **A·cheu·le·an** (ə-shōō′lē-ən) *adj.* [Fr. *acheuléen*, after St. *Acheul*, France.] Designating a stage of culture of the European Lower Paleolithic Age between the second and third interglacial periods, characterized by symmetrical stone hand axes.

**a·chieve** (ə-chēv′) *v.* **a·chieved, a·chiev·ing, a·chieves.** [ME *acheven* < OFr. *achever* < *à chief (venir)*, (to come) to a head.] *—vt.* **1.** To accomplish successfully. **2.** To attain with effort <at last *achieved* the mountaintop> *—vi.* To accomplish something successfully. **—a·chiev′a·ble** *adj.* **—a·chiev′er** *n.*

**a·chieve·ment** (ə-chēv′mənt) *n.* **1.** The act of finishing or accomplishing. **2.** Something that has been accomplished successfully, esp. by means of persistent endeavor.

**A·chil·les** (ə-kĭl′ēz) *n.* [Lat. < Gk. *Akhilleus.*] *Gk. Myth.* The son of Peleus and Thetis and hero of Homer's *Iliad.*

**Achilles' heel** *n.* [From Achilles' being vulnerable only in the heel.] A vulnerable area **:** WEAKNESS.

**Achilles jerk** *n.* Reflex plantar flexion in response to a blow to the Achilles tendon.

**Achilles tendon** *n.* The large tendon extending from the heel bone to the calf muscle of the leg.

**ach·la·myd·e·ous** (ăk′lə-mĭd′ē-əs, ā′klə-) *adj. Bot.* Having no calyx or corolla.

**a·cho·li·a** (ā-kō′lē-ə) *n.* [NLat. : A-[1] + Gk. *kholē*, bile.] A decrease or absence of bile secretion in the small intestine.

**a·chon·drite** (ā-kŏn′drīt′) *n.* A stony meteorite that contains no chondrules. **—a′chon·drit′ic** (-drĭt′ĭk) *adj.*

**a·chon·dro·pla·sia** (ā-kŏn′drō-plā′zhə, -zhē-ə) *n.* Impaired development of cartilage at the ends of the long bones, resulting in congenital dwarfism. **—a·chon′dro·plas′tic** (-plăs′tĭk) *adj.*

**ach·ro·mat·ic** (ăk′rə-măt′ĭk) *adj.* [Gk. *akhrōmatos* : *a-*, not + *khrōma*, color.] **1.** Designating color perceived to have zero saturation and therefore no hue, as neutral grays. **2.** Refracting light without spectral color separation. **3.** *Biol.* Staining poorly with standard dyes. **4.** *Mus.* Having only the diatonic tones of the scale. **—ach′ro·mat′i·cal·ly** *adv.* **—a·chro′ma·tism** (ā-krō′mə-tĭz′əm), **ach′ro·ma·tic′i·ty** (ăk′rō-mə-tĭs′ĭ-tē) *n.*

**achromatic lens** *n.* A combination of lenses to produce images free of chromatic aberrations.

**a·chro·ma·tin** (ā-krō′mə-tĭn) *n.* [ACHROMAT(IC) + -IN.] The section of a cell nucleus that is relatively uncolored by stains or dyes. **—a·chro′ma·tin′ic** *adj.*

**a·chro·ma·tize** (ā-krō′mə-tīz′) *vt.* **-tized, -tiz·ing, -tiz·es.** To rid of color **:** make achromatic.

**a·chro·mic** (ā-krō′mĭk) *adj.* [A-[1] + CHROMO- + -IC.] Colorless.

**ach·y** (ā′kē) *adj.* **-i·er, -i·est.** Feeling aches. **—ach′i·ness** *n.*

**a·cic·u·la** (ə-sĭk′yə-lə) *n., pl.* **-lae** (-lē′) [NLat. < Lat., hairpin, dim. of *acus*, needle.] A needlelike bristle, spine, or crystal. **—a·cic′u·lar, a·cic′u·late** (-lĭt, -lāt′), **a·cic′u·lat′ed** (-lā′tĭd) *adj.*

**ac·id** (ăs′ĭd) *n.* [Lat. *acidus*, sour < *acēre*, to be sour < *acer*, sharp.] **1.** *Chem.* **a.** Any of a large class of substances whose aqueous solutions are capable of turning litmus indicators red, of reacting with and dissolving certain metals to form salts, of reacting with bases or alkalis to form salts, or having a sour taste. **b.** A substance that ionizes in solution to give the positive ion of the solvent. **c.** A substance capable of giving up a proton. **d.** A molecule or ion that can combine with another by forming a covalent bond with two electrons of the other. **2.** A substance having a sour taste. **3.** The quality of being

ōō **boot** ou **out** th **thin** *th* **this** ŭ **cut** ûr **urge** y **young**
yōō **abuse** zh **vision** ə **about, item, edible, gallop, circus**

sarcastic, incisive, or scornful. **4.** *Slang.* LSD. —*adj.* **1.** *Chem.* **a.** Of or relating to an acid. **b.** Having a high concentration of acid. **2.** Having a sour taste. **3.** Biting, sarcastic, or scornful <*acid* prose> **—ac'id·ly** *adv.* **—ac'id·ness** *n.*

**acid cell** *n.* A parietal cell of the stomach.

**ac·i·de·mi·a** (ăs'ĭ-dē'mē-ə) *n.* A condition in which blood pH is below normal.

**ac·id-fast** (ăs'ĭd-făst') *adj.* Not readily decolorized by acid <*acid-fast* bacteria> **—ac'id-fast'ness** *n.*

**ac·id·head** (ăs'ĭd-hĕd') *n. Slang.* A user of LSD.

**a·cid·ic** (ə-sĭd'ĭk) *adj.* **1.** Acid. **2.** Tending to form an acid.

**a·cid·i·fy** (ə-sĭd'ə-fī') *vt.* & *vi.* **-fied, -fy·ing, -fies.** To make or become acid. **—a·cid'i·fi'a·ble** *adj.* **—a·cid'i·fi·ca'tion** *n.* **—a·cid'i·fi'er** *n.*

**ac·i·dim·e·ter** (ăs'ĭ-dĭm'ĭ-tər) *n.* A hydrometer for determining the specific gravity of acid solutions. **—a·cid'i·met'ric** (ə-sĭd'ĭ-mĕt'rĭk) *adj.* **—ac'i·dim'e·try** *n.*

**a·cid·i·ty** (ə-sĭd'ĭ-tē) *n.* **1.** The quality, state, or degree of being acid. **2.** Hyperacidity.

**ac·i·do·phil·ic** (ăs'ĭ-dō-fĭl'ĭk) *also* **ac·i·doph·i·lus** (-dŏf'ə-ləs) *adj. Microbiol.* **1.** Thriving in an acid medium. **2.** Readily stained with acid dyes. **—a·cid'o·phil'** (ə-sĭd'ə-fĭl'), **a·cid'o·phile'** (-fīl') *n.*

**ac·i·doph·i·lus milk** (ăs'ĭ-dŏf'ə-ləs) *n.* [NLat. *acidophilus,* specific epithet of several species of bacteria : ACID + *-philus,* -philous.] Milk containing bacterial cultures that thrive in dilute acid, often used in treating gastrointestinal disorders.

**ac·i·do·sis** (ăs'ĭ-dō'sĭs) *n.* An abnormal increase in the acidity of bodily fluids, due to either acid accumulation or bicarbonate depletion. **—ac'i·dot'ic** (-dŏt'ĭk) *adj.*

**acid precipitation** *n.* Precipitation abnormally high in sulfuric and nitric acid content that is caused by industrial pollution.

**acid rain** *n.* Acid precipitation falling as rain.

**acid rock** *n.* Rock music with lyrics suggesting psychedelic and drug-related experiences.

**acid test** *n.* [From the testing of gold in nitric acid.] A decisive test.

**a·cid·u·late** (ə-sĭj'ə-lāt') *vt.* & *vi.* **-lat·ed, -lat·ing, -lates.** [ACIDUL(OUS) + -ATE[1].] To make or become somewhat acid. **—a·cid'u·la'tion** *n.*

**a·cid·u·lous** (ə-sĭj'ə-ləs) *adj.* [Lat. *acidulus,* sourish, dim. of *acidus,* sour. —see ACID.] Acid in taste or manner : CAUSTIC.

**ac·i·dur·i·a** (ăs'ĭ-dŏŏr'ē-ə, -dyŏŏr'-) *n.* A condition marked by excessive amounts of acid in the urine.

**ac·i·nar** (ăs'ĭ-nər, -när') *adj.* Of or relating to an acinus.

**ac·i·nus** (ăs'ə-nəs) *n., pl.* **-ni** (-nī') [Lat., berry.] **1.** *Bot.* One of the drupelets of an aggregate fruit such as the raspberry. **2.** A grape or a bunch of grapes. **3.** *Anat.* One of the small saclike dilations of a compound gland. **—a·cin'ic** (ə-sĭn'ĭk), **ac'i·nous** *adj.*

**ack-ack** (ăk'ăk') *n.* [British telephone code for AA, abbreviation for ANTIAIRCRAFT.] *Slang.* **1.** An antiaircraft gun. **2.** Antiaircraft fire.

**ac·knowl·edge** (ăk-nŏl'ĭj) *vt.* **-edged, -edg·ing, -edg·es.** [Blend of ME *knowledgen,* to acknowledge (< *knowen,* to know), and ME *aknouen,* to recognize < OE *oncnāwan,* to know.] **1. a.** To confess, avow, or admit the existence, reality, or truth of. **b.** To recognize the validity, force, or power of. **2. a.** To express recognition of <*acknowledged* our presence> **b.** To express gratitude for. **3.** To report the receipt of. **4.** *Law.* To accept or certify as legally binding. **—ac·knowl'edge·a·ble** *adj.*

☆ **syns:** ACKNOWLEDGE, ADMIT, ALLOW, AVOW, CONCEDE, CONFESS, GRANT, OWN UP *v. core meaning* : to recognize, often reluctantly, the reality or truth of <finally *acknowledged* the error> *ant:* deny

**ac·knowl·edged** (ăk-nŏl'ĭjd) *adj.* Commonly recognized or accepted.

**ac·knowl·edg·ment** *also* **ac·knowl·edge·ment** (ăk-nŏl'ĭj-mənt) *n.* **1.** The act of admitting to something. **2.** Recognition of another's existence, validity, authority, or right. **3.** A response in return for something done. **4.** An expression or token of thanks. **5.** A formal declaration made to authoritative witnesses to ensure legal validity.

**a·clin·ic line** (ā-klĭn'ĭk) *n.* The magnetic equator.

**ac·me** (ăk'mē) *n.* [Gk. *akmē.*] The point of utmost attainment : PEAK.

**ac·ne** (ăk'nē) *n.* [Poss. < Gk. *akmē,* point.] An inflammatory disease of the oil glands, causing pimples esp. on the face. **—ac'ned** *adj.*

**a·cock** (ə-kŏk') *adj.* & *adv.* In a cocked position.

**ac·o·lyte** (ăk'ə-līt') *n.* [ME *acolit* < OFr. < Med. Lat. *acolytus* < Gk. *akolouthos,* attendant < *akolouthein,* to follow.] **1.** One who assists a priest at Mass. **2.** An attendant or follower.

**ac·o·nite** (ăk'ə-nīt') *n.* [Fr. *aconit* < Lat. *aconitum* < Gk. *akoniton.*] **1.** The monkshood. **2.** The dried, poisonous root of a species of monkshood, *A. napellus,* sometimes used medicinally to reduce pain or fever.

**a·corn** (ā'kôrn', ā'kərn) *n.* [ME *akorn* < OE *æcern.*] The thick-walled nut of the oak tree, usu. set in a woody, cuplike base.

**acorn squash** *n.* A squash shaped somewhat like an acorn and having a longitudinally ridged rind.

**acorn tube** *n.* A small, acorn-shaped vacuum tube used in very high frequency devices.

**a·cot·y·le·don** (ā-kŏt'l-ēd'n) *n.* A plant, as a moss or fern, that does not have seed leaves. **—a·cot'y·le'don·ous** (-ēd'n-əs) *adj.*

**a·cous·tic** (ə-kōō'stĭk) *also* **a·cous·ti·cal** (-stĭ-kəl) *adj.* [Gk. *akoustikos,* pertaining to hearing < *akouein,* to hear.] **1.** Of or relating to sound, the sense of hearing, or the science of sound. **2. a.** Designed to carry sound or to aid in hearing. **b.** Controlling or absorbing sound <*acoustical* tile> **3. a.** Of, relating to, or being a musical instrument whose sound is not electronically modified. **b.** Being a musical performance that features acoustic instruments. —*n.* An acoustic musical instrument. **—a·cous'ti·cal·ly** *adv.*

**ac·ous·ti·cian** (ăk'ōō-stĭsh'ən) *n.* A specialist in acoustics.

**acoustic nerve** *n.* The eighth cranial nerve, consisting of the cochlear nerve, which conducts acoustic stimuli to the brain, and the vestibular nerve, which conducts stimuli related to bodily equilibrium to the brain.

**a·cous·tics** (ə-kōō'stĭks) *n.* **1.** *(sing. in number).* The scientific study of sound, esp. of its generation, transmission, and reception. **2.** *(pl. in number).* The total effect of sound, esp. as produced in an enclosed space.

**a·cous·to·e·lec·tric** (ə-kōō'stō-ĭ-lĕk'trĭk) *adj.* [ACOUST(IC) + ELECTRIC.] Pertaining to electroacoustics. **—a·cous'to·e·lec'tric·al·ly** *adv.*

**a·cous·to·op·tics** (ə-kōō'stō-ŏp'tĭks) *n.* [ACOUST(IC) + OPTICS.] *(sing. in number).* The science of the interaction of acoustic and optical phenomena. **—a·cous'to·op'ti·cal** *adj.* **—a·cous'to·op'ti·cal·ly** *adv.*

**ac·quaint** (ə-kwānt') *vt.* **-quaint·ed, -quaint·ing, -quaints.** [ME *aqueinten* < OFr. *acointier* < Med. Lat. *adcognitare* < Lat. *accognoscere,* to know perfectly : *ad-* (intensive) + *cognoscere,* to know. —see COGNITION.] **1. a.** To familiarize <*acquainted* myself with the route> **b.** To make known personally <Let me *acquaint* you with my parents.> **2.** To inform <*Acquaint* us with the facts.>

**ac·quain·tance** (ə-kwān'təns) *n.* **1.** Knowledge or information. **2.** Knowledge of a person acquired by a relationship less intimate than friendship. **3.** A person whom one knows. **—ac·quain'tance·ship'** *n.*

**ac·quaint·ed** (ə-kwān'tĭd) *adj.* **1.** Known by or familiar with another. **2.** Aware : informed <*acquainted* with the plan>

**ac·qui·esce** (ăk'wē-ĕs') *vi.* **-esced, -esc·ing, -esc·es.** [Lat. *acquiescere* : *ad-,* to + *quiescere,* to rest < *quies,* rest.] To consent or comply without protest. ***usage:*** *Acquiesce* usu. takes *to* (*acquiesced to their wishes*) or *in* (*acquiesced in the ruling*).

**ac·qui·es·cence** (ăk'wē-ĕs'əns) *n.* **1.** Unprotesting assent. **2.** The state of being acquiescent. **—ac'qui·es'cent** *adj.* **—ac'qui·es'cent·ly** *adv.*

**ac·quire** (ə-kwīr') *vt.* **-quired, -quir·ing, -quires.** [ME *acquere* < OFr. *aquerre* < Lat. *acquirere,* to add to : *ad-,* to + *quaerere,* to get.] **1.** To gain possession or control of : GET <*acquire* an education> **2.** To come to have <*acquire* immunity> **—ac·quir'a·ble** *adj.*

**acquired antibody** *n.* An antibody produced by an immune response as compared with one occurring in the system naturally.

**acquired immune deficiency syndrome** *n.* AIDS.

**acquired immunity** *n.* Immunity developed in the physiology during a lifetime.

**ac·qui·si·tion** (ăk'wĭ-zĭsh'ən) *n.* [ME *adquisicioun,* attainment < Lat. *acquisitio* < *acquirere,* to acquire : *ad-,* to + *quaerere,* to seek.] **1.** The act of acquiring. **2.** Something acquired. **3.** *Aerospace.* The process of locating a satellite, guided missile, or moving target so that its track or orbit can be determined.

**ac·quis·i·tive** (ə-kwĭz'ĭ-tĭv) *adj.* **1.** Marked by a strong desire to gain and possess : GREEDY. **2.** Tending to acquire and retain information, as the mind. **—ac·quis'i·tive·ly** *adv.* **—ac·quis'i·tive·ness** *n.*

**ac·quit** (ə-kwĭt') *vt.* **-quit·ted, -quit·ting, -quits.** [ME *aquiten* < OFr. *aquiter.*] **1.** To clear of a charge. **2.** To release from a duty. **3.** To repay (an obligation). **4.** To conduct (oneself) <*acquit* oneself honorably> **—ac·quit'ter** *n.*

**ac·quit·tal** (ə-kwĭt'l) *n. Law.* Judgment of a jury or judge that a defendant is not guilty of a crime as charged.

**ac·quit·tance** (ə-kwĭt'ns) *n.* A written release from an obligation.

**acr-** *pref. var. of* ACRO-.

**a·cra·sia** (ə-krā'zhə) *n.* [Gk. *akrasia,* incontinence < *akratēs,* powerless : *a-,* not + *kratos,* strength.] Lack of self-control.

**a·cre** (ā'kər) *n.* [ME *aker* < OE *æcer.*] **1.** A unit of area in the U.S. Customary System, used in land measurement and equal to 160 square rods, 4,840 square yards, or 43,560 square feet. **2. acres.** Property in the form of land : ESTATE. **3.** *often* **acres.** A wide expanse of land. **4.** *Archaic.* A field or plot of land.

▲ word history: The Old English word *æcer,* of which *acre* is the direct descendant, meant simply "field," especially a field of cultivated land. The word acquired the meaning of "a unit of land measurement" because it was used specifically to mean a field as large as one man could plow in one day. The size of such a field was fixed during medieval times at 4,840 square yards, which is still the size of a modern acre. Old English *æcer* is descended from the same Indo-European form as Latin *ager* and Greek *agros,* which both mean "field." The Modern English spelling comes from Old French *acre,* which is an alteration of the Latin form.

**a·cre·age** (ā′kər-ĭj, ā′krĭj) *n.* Area of land in acres.

**a·cre-foot** (ā′kər-fo͝ot′) *n.* The volume of water, 43,560 cubic feet, that will cover an area of 1 acre to a depth of 1 foot.

**a·cre-inch** (ā′kər-ĭnch′) *n.* One twelfth of an acre-foot, equal to 3,630 cubic feet.

**ac·rid** (ăk′rĭd) *adj.* [< Lat. *acer,* sharp.] **1.** Having a harsh taste or smell. **2.** Caustic in language or tone : ACRIMONIOUS. **—a·crid′i·ty** (ə-krĭd′ĭ-tē), **ac′rid·ness** *n.* **—ac′rid·ly** *adv.*

**ac·ri·dine** (ăk′rĭ-dēn′) *n.* A coal tar derivative, $C_{13}H_9N$, that has a strongly irritating odor and is used to make dyes and synthetics.

**ac·ri·fla·vine** (ăk′rə-flā′vēn′, -vĭn) *n.* [ACRI(DINE) + FLAVIN.] A brown or orange powder, $C_{14}H_{14}N_3Cl$, derived from acridine and used as an antiseptic.

**ac·ri·mo·ni·ous** (ăk′rə-mō′nē-əs) *adj.* Bitingly hostile in language or tone. **—ac′ri·mo′ni·ous·ly** *adv.* **—ac′ri·mo′ni·ous·ness** *n.*

**ac·ri·mo·ny** (ăk′rə-mō′nē) *n.* [Lat. *acrimonia,* sharpness < *acer,* sharp.] Ill-natured animosity : harshness of manner or tone.

**acro-** *or* **acr-** *pref.* [< Gk. *akros,* extreme.] **1. a.** Top : summit <*acro*petal> **b.** Height <*acro*phobia> **2. a.** Tip : beginning <*acro*nym> **b.** Extremity of the body <*acro*megaly>

**ac·ro·bat** (ăk′rə-băt′) *n.* [Fr. *acrobate* < Gk. *akrobatēs* < *akrobatein,* to walk on tiptoe : *akros,* high + *batos,* walker < *bainein,* to walk.] One skilled in feats of agility and balance. **—ac′ro·bat′ic** *adj.* **—ac′ro·bat′i·cal·ly** *adv.*

**ac·ro·bat·ics** (ăk′rə-băt′ĭks) *n.* (*sing. or pl. in number*). **1.** The actions of an acrobat. **2.** The art of an acrobat. **3.** A display of extraordinary agility.

**ac·ro·car·pous** (ăk′rō-kär′pəs) *adj.* [NLat. *acrocarpus* < Gk. *akrokarpos,* bearing fruit at the top : *akron,* top + *karpos,* fruit.] Having the spore-bearing capsule at the end or top of a leafy stem or stalk, as in many mosses.

**ac·ro·ceph·a·ly** (ăk′rə-sĕf′ə-lē) *or* **ac·ro·ce·phal·i·a** (-sə-fā′lē-ə) *n.* Oxycephaly. **—ac′ro·ce·phal′ic** (-făl′ĭk) *adj.*

**ac·ro·drome** (ăk′rə-drōm′) *also* **a·crod·ro·mous** (ə-krŏd′rə-məs) *adj.* Coming to a point and having the veins terminate at the tip. —Used of leaves.

**ac·ro·gen** (ăk′rə-jən) *n.* A flowerless plant, as a fern or moss, having a stem from whose tip all growth proceeds. **—ac′ro·gen′ic** (-jĕn′ĭk), **a·crog′e·nous** (ə-krŏj′ə-nəs) *adj.* **—a·crog′e·nous·ly** *adv.*

**a·cro·le·in** (ə-krō′lē-ĭn) *n.* [ACR(ID) + OLEIN.] A colorless, flammable, poisonous liquid, $CH_2CHCHO$, having an acrid odor and vapors harmful to the eyes.

**ac·ro·meg·a·ly** (ăk′rō-mĕg′ə-lē) *n.* [Fr. *acromégalie* : Gk. *akron,* extremity + Gk. *megas,* big.] Pathological enlargement of the bones of the hands, feet, and face, resulting from chronic overactivity of the pituitary gland. **—ac′ro·me·gal′ic** (-mĭ-găl′ĭk) *adj. & n.*

**ac·ro·mel·ic** (ăk′rō-mĕl′ĭk) *adj.* [ACRO- + Gk. *melos,* limb.] Of, relating to, or pertinent to the end of the extremities.

**a·cro·mi·on** (ə-krō′mē-ən) *n.* [NLat. < Gk. *akrōmion* : *akros,* extreme + *ōmion,* dim. of *ōmos,* shoulder.] The outer extremity of the scapula or shoulder blade.

**ac·ro·nym** (ăk′rə-nĭm′) *n.* A word formed from the initial letters of a name, as *WAC* for *W*omen's *A*rmy *C*orps, or by combining initial letters or parts of a series of words, as *radar* for *ra*dio *d*etecting *a*nd *r*anging. **—ac′ro·nym′ic, a·cron′y·mous** (ə-krŏn′ə-məs) *adj.*

**a·crop·e·tal** (ə-krŏp′ĭ-tl) *adj. Bot.* Developing upward from the base toward the apex, as some forms of inflorescence. **—a·crop′e·tal·ly** *adv.*

**ac·ro·pho·bi·a** (ăk′rə-fō′bē-ə) *n.* Abnormal fear of heights.

**a·crop·o·lis** (ə-krŏp′ə-lĭs) *n.* [Gk. *akropolis* : *akron,* top + *polis,* city.] The fortified height or citadel of an ancient Greek city, esp. the Athenian citadel.

**ac·ro·spire** (ăk′rə-spīr′) *n.* [Var. of dial. *akerspire* < ME : *acher,* ear of grain (< OE *æhher*) + *spire,* shoot < OE *spīr.*] The initial sprout from a germinating grain seed.

**a·cross** (ə-krôs′, ə-krŏs′) *prep.* [ME *acrois* < AN *an croiz* : *an,* in + *croiz,* cross.] **1.** On, at, or from the other side of <*across* the road> **2.** So as to cross : OVER <draw lines *across* the paper> **3.** From one side of to the other <a bridge *across* a river> **4.** Into contact with <came *across* my old roommate> *—adv.* **1.** From one side to the other <The street was empty and I ran *across.*> **2.** On or to the opposite side <came *across* by boat> **3.** Crossed : crosswise. **4.** Over <got my point *across*>

**a·cross-the-board** (ə-krôs′*th*ə-bôrd′, -bōrd′, ə-krŏs′-) *adj.* **1.** Designating a racing wager whereby equal amounts are bet on the same contestant to win, place, or show. **2.** Including all categories or members <*across-the-board* raises for the hospital staff>

**a·cros·tic** (ə-krô′stĭk, ə-krŏs′tĭk) *n.* [Fr. *acrostiche* < OFr. < Gk. *akrostikhis* : *akron,* end + *stikhos,* line.] **1.** A poem or series of lines in which certain letters, usu. the first in each line, form a name, motto, or message when read in sequence. **2.** A word square. **—a·cros′tic** *adj.* **—a·cros′ti·cal·ly** *adv.*

GRAB
RARE
ARTS
BEST

**acrostic**
*A word square*

**ac·ry·late resin** *also* **ac·ry·late** (ăk′rə-lāt′) *n.* Any of a class of acrylic resins used in emulsion paints, adhesives, plastics, and textile and paper finishes.

**a·cryl·ic** (ə-krĭl′ĭk) *n.* [ACR(OLEIN) + -YL + -IC.] **1.** Acrylic resin. **2.** Acrylic fiber. **3.** A paint containing acrylic resin. **—a·cryl′ic** *adj.*

**acrylic acid** *n.* A readily polymerized, colorless, corrosive liquid, $H_2C$:CHCOOH, used as a monomer for acrylate resins.

**acrylic fiber** *n.* A synthetic fiber polymerized from acrylonitrile.

**acrylic resin** *n.* Any of numerous thermoplastic or thermosetting polymers or copolymers of acrylic acid, methacrylic acid, esters of these acids, or acrylonitrile, used to produce synthetic rubbers and lightweight plastics.

**ac·ry·lo·ni·trile** (ăk′rə-lō-nī′trəl, -trēl′, -trīl) *n.* [ACRYL(IC RESIN) + NITRILE.] A colorless, liquid organic compound, $H_2C$:CHCN, used to make acrylic rubber and fibers.

**act** (ăkt) *n.* [ME < OFr. *acte* < Lat. *actus,* a doing, and *actum,* a thing done, both < *agere,* to do.] **1.** The process of doing : ACTION. **2.** Something done : DEED. **3.** An enactment, as of a judicial or legislative body. **4.** A formal written record of proceedings or transactions. **5. a.** One of the major divisions of a play or opera. **b.** A theatrical performance that forms part of a longer presentation. **6.** A manifestation of insincerity : POSE. ***usage:*** *Act* and *action* are distinct in meaning. An *act* is the deed accomplished by means of an *action.* Thus, a baseball pitcher may throw with an unnatural *action* (ie., a physical movement difficult for the human arm); a parent who abandons a child performs an unnatural *act* (the deed, but not the series of activities with which it is performed, is contrary to human nature). *—v.* **act·ed, act·ing, acts.** *—vt.* **1.** To assume the dramatic role of <*acted* Hamlet> **2.** To perform on the stage <*act* a comedy> **3.** To behave like : IMPERSONATE <*act* the villain> **4.** To behave as suitable for <Why can't you *act* your age?> *—vi.* **1.** To behave or conduct oneself. **2. a.** To perform in a dramatic role. **b.** To be suitable for theatrical performance, as a given play or role. **3.** To behave affectedly : PRETEND. **4.** To appear to be <The cat *acted* aloof.> **5.** To take action <We *acted* quickly.> **6.** To function or serve <umbrellas *acting* as shade> **7.** To take effect <hoped the pill would *act* quickly> **—act on (or upon). 1.** To act according to <*acted on* my suggestion> **2.** To produce an effect on <Water *acts on* stone.> **—act out. 1. a.** To perform in or as if in a play : DRAMATIZE <*act out* a song> **b.** To realize in action <*act out* one's ideals> **2.** *Psychoanal.* To express (e.g., unconscious impulses) in an overt way without understanding or awareness. **—act up.** To misbehave or malfunction. **—ac′ta·bil′i·ty** *n.* **—act′a·ble** *adj.*

**Ac·tae·on** (ăk-tē′ən) *n.* [Lat. < Gk. *Aktaiōn.*] *Gk. Myth.* A young hunter who was turned into a stag by Artemis, whom he had seen bathing, and who was killed by his own dogs.

**ACTH** (ā′sē′tē ăch′) *n.* [A(DRENO)C(ORTICO)T(ROPIC) H(ORMONE).] A pituitary hormone synthesized or extracted from mammalian pituitaries for use in stimulating secretion of adrenal cortex hormones, as cortisone.

**ac·tin** (ăk′tĭn) *n.* [Lat. *actus,* motion (< *agere,* to impel) + -IN.] A muscle protein, active with myosin in muscular contraction.

**actin-** *pref. var. of* ACTINO-.

**ac·ti·nal** (ăk′tĭ-nəl, ăk-tī′-) *adj.* Of or designating the part of a sea anemone or similar animal from which the tentacles or rays radiate. **—ac′ti·nal·ly** *adv.*

**act·ing** (ăk′tĭng) *adj.* **1.** Temporarily assuming the authority, duties, or function of another. **2. a.** Suitable for performing <an *acting* play> **b.** Containing directions for use in a dramatic performance. *—n.* **1.** The occupation or performance of an actor. **2.** Simulated behavior : PRETENSE.

**ac·tin·i·a** (ăk-tĭn′ē-ə) *also* **ac·tin·i·an** (-ən) *n., pl.* **-i·ae** (-ē-ē′) *also* **-i·ans.** [NLat. *Actinia,* sea anemone genus < Gk. *aktis,* ray.] A sea anemone or a related animal.

**ac·tin·ic** (ăk-tĭn′ĭk) *adj.* Of or relating to actinism. **—ac·tin′i·cal·ly** *adv.*

**actinic ray** *n.* Photochemically active radiation.

**ac·ti·nide** (ăk′tə-nīd′) *n.* Any of a series of chemically similar, mostly synthetic, radioactive elements with atomic numbers ranging from 89 (actinium) through 103 (lawrencium).

**ac·ti·nism** (ăk′tə-nĭz′əm) *n.* The intrinsic property in radiation that produces photochemical activity.

**ac·tin·i·um** (ăk-tĭn′ē-əm) *n. Symbol* **Ac** A radioactive metallic element found in uranium ores and used as a source of alpha rays; atomic number 89; longest-lived isotope Ac 227.

**actino-** *or* **actin-** *pref.* [< Gk. *aktis, aktin-*, ray.] **1.** Radial in form <*actinoid*> **2.** Actinic radiation <*actinometer*>

**ac·ti·no·car·pous** (ăk′tĭ-nō-kär′pəs) *adj.* Having flowers or fruit radiating from one point.

**ac·tin·o·gen** (ăk-tĭn′ə-jən) *n.* A radioactive element.

**ac·ti·noid**[1] (ăk′tə-noid′) *adj.* Having a radial form, as a starfish.

**ac·ti·noid**[2] (ăk′tə-noid′) *n.* An actinide.

**ac·tin·o·lite** (ăk-tĭn′ə-līt′) *n.* A greenish amphibole.

**ac·ti·no·mere** (ăk-tĭn′ə-mîr′) *n.* A segment composing part of the body of a radially symmetrical animal.

**ac·ti·nom·e·ter** (ăk′tə-nŏm′ĭ-tər) *n.* A radiometric instrument, as a pyrheliometer, used chiefly for meteorological measurements of terrestrial and solar radiation. **—ac′ti·no·met′ric** (-nō-mĕt′rĭk), **ac′ti·no·met′ri·cal** *adj.* **—ac′ti·nom′e·try** *n.*

**ac·ti·no·mor·phic** (ăk′tə-nō-môr′fĭk) *also* **ac·ti·no·mor·phous** (-fəs) *adj.* Divisible vertically through two or more planes into similar halves : radially symmetric, as a starfish.

**ac·ti·no·my·cete** (ăk′tə-nō-mī′sēt′, -mī-sēt′) *n.* Any of numerous gen. filamentous and often pathogenic microorganisms of the family Actinomycetaceae, resembling both bacteria and fungi.

**ac·ti·no·my·cin** (ăk′tə-nō-mī′sĭn) *n.* [< NLat. *Antinomyces*, a genus of soil bacteria : ACTINO- + Gk. *mukēs*, fungus.] Any of various often toxic antibiotic substances found in soil bacteria.

**ac·ti·no·my·co·sis** (ăk′tə-nō-mī-kō′sĭs) *n.* An inflammatory infection of cattle, swine, and occas. humans, caused by microorganisms of the genus *Actinomyces* and marked by lumpy tumors of the neck, chest, and abdomen. **—ac′ti·no·my·cot′ic** (-kŏt′ĭk) *adj.*

**ac·ti·non** (ăk′tə-nŏn′) *n.* A radioactive inert gaseous isotope of radon, with a half-life of 3.92 seconds.

**ac·ti·no·u·ra·ni·um** (ăk′tə-nō-yo͝o-rā′nē-əm) *n.* The isotope of uranium with mass number 235, fissionable with slow neutrons.

**ac·ti·no·zo·an** (ăk′tə-nō-zō′ən) *n.* An anthozoan.

**ac·tion** (ăk′shən) *n.* **1.** The process of acting or doing. **2.** An act or deed. **3.** A movement or a sequence of movements. **4.** Manner of movement. **5.** Habitual or vigorous activity or energy : INITIATIVE <someone of *action*> **6.** *often* **actions.** Behavior. **7. a.** The operating parts of a mechanism. **b.** The way in which such parts operate. **8.** A change that occurs in the body or in a bodily organ as a result of its functioning. **9.** A physical change, as in position, mass, or energy, undergone by an object or system. **10.** The plot of a story or play. **11.** The appearance of animation of a figure in a work of art. **12.** *Law.* **a.** A lawsuit. **b.** The right of an individual to exercise his or her privilege to legal process. **13.** Combat. **14.** Signficant or exciting activity <go where the *action* is>

**ac·tion·a·ble** (ăk′shə-nə-bəl) *adj.* Giving cause for legal action. **—ac′tion·a·bly** *adv.*

**action painting** *n.* A style of abstract painting using techniques such as dribbling or splashing to achieve an effect of spontaneity.

**action potential** *n.* A recorded change in electrical potential of a cell or tissue when stimulated.

**ac·ti·vate** (ăk′tə-vāt′) *vt.* **-vat·ed, -vat·ing, -vates. 1.** To set in motion. **2.** To create or organize (e.g., a military unit). **3.** To purify (sewage) by aeration. **4.** *Chem.* To accelerate a reaction in, as by heat. **5.** *Physics.* To make (a substance) radioactive. **—ac′ti·va′tion** *n.* **—ac′ti·va′tor** *n.*

**activated carbon** *or* **activated charcoal** *n.* Highly absorbent carbon obtained by heating granulated charcoal to exhaust contained gases, used in gas absorption, solvent recovery, or deodorization and as an antidote to certain poisons.

**activation analysis** *n.* A method for analyzing a material for its component chemical elements by bombarding it with nuclear particles or gamma rays and identifying the resultant radiations.

**ac·tive** (ăk′tĭv) *adj.* [ME *actif* < OFr. < Lat. *activus* < *actus*, moving, p.part. of *agere*, to impel.] **1.** In action : MOVING. **2.** Capable of functioning. **3.** Causing action or change. **4.** Participating <take an *active* role> **5.** In a state of action <an *active* geyser> **6.** Marked by energetic activity : LIVELY. **7.** Currently in operation or effect <an *active* disease> **8. a.** Denoting a verb inflection or voice indicating that the subject of the sentence is performing or causing the action expressed by the verb. **b.** Expressing action rather than a state of being. —Used of verbs such as *run, speak,* or *move.* **9.** Producing profit, interest, or dividends <*active* funds> **10.** Being on full military duty and full pay. **11.** *Mus.* Suggesting that something follows. —*n.* **1. a.** The active voice of a verb. **b.** A grammatical construction or form in the active voice. **2.** A participating member of an organization. **—ac′tive·ly** *adv.* **—ac′tive·ness** *n.*

**active immunity** *n.* A long-lasting immunity to disease due to antibody production by an organism.

**active transport** *n.* The movement of a chemical substance through a gradient of concentration or electrical potential in the direction opposite to normal diffusion, requiring the expenditure of energy.

**ac·tiv·ism** (ăk′tə-vĭz′əm) *n.* A theory or practice based on militant action. **—ac′tiv·ist** *n.*

**ac·tiv·i·ty** (ăk-tĭv′ĭ-tē) *n., pl.* **-ties. 1.** The state of being active. **2.** Energetic action. **3.** A specified form of supervised action or field of action, as in education or recreation. **4.** The intensity of a radioactive source.

**act of God** *n. Law.* An unforeseeable or inevitable event, as an earthquake or a flood, caused by nature.

**ac·to·my·o·sin** (ăk′tə-mī′ə-sĭn) *n.* [ACT(IN) + MYOSIN.] A system of actin and myosin that with other substances constitutes muscle fiber and is responsible for muscular expansion and contraction.

**ac·tor** (ăk′tər) *n.* [ME *actour* < Lat. *actor*, doer < *agere*, to do.] **1.** A theatrical performer. **2.** A participant.

**ac·tress** (ăk′trĭs) *n.* A woman who is a theatrical performer.

**Acts of the Apostles** *pl.n. (sing. in number).*—See table at BIBLE.

**ac·tu·al** (ăk′cho͞o-əl) *adj.* [ME < OFr. *actuel* < Lat. *actualis* < *actus*, acting < *agere*, to do.] **1.** Existing in fact or reality <*actual*, not ideal, conditions> **2.** Existing or acting at the present moment : CURRENT. **3.** Corresponding to all human facts. **4.** That is based on fact. **—ac′tu·al·ly** *adv.*

**ac·tu·al·i·ty** (ăk′cho͞o-ăl′ĭ-tē) *n., pl.* **-ties. 1.** The state or fact of being actual : REALITY. **2. actualities.** Actual facts or conditions.

☆ ***syns:*** ACTUALITY, BEING, EXISTENCE *n. core meaning* : the state or fact of having reality <ideas true in possibility but not in *actuality*>

**ac·tu·al·ize** (ăk′cho͞o-ə-līz′) *vt.* **-ized, -iz·ing, -iz·es. 1.** To realize in action. **2.** To describe realistically. **—ac′tu·al·i·za′tion** *n.*

**ac·tu·ar·y** (ăk′cho͞o-ĕr′ē) *n., pl.* **-ies.** [Lat. *actuarius*, secretary of accounts < *acta*, records < *actus*, p.part. of *agere*, to do.] A statistician who computes insurance risks and premiums. **—ac′tu·ar′i·al** *adj.* **—ac′tu·ar′i·al·ly** *adv.*

**ac·tu·ate** (ăk′cho͞o-āt′) *vt.* **-at·ed, -at·ing, -ates.** [Med. Lat. *actuare, actuat-* < Lat. *actus*, act < *agere*, to do.] **1.** To put into action or motion. **2.** To stimulate or motivate. **—ac′tu·a′tion** *n.*

**ac·tu·a·tor** (ăk′cho͞o-ā′tər) *n.* One that activates, esp. a device responsible for actuating another device, as one connected to a computer by a sensor link.

**ac·u·ate** (ăk′yo͞o-ăt′) *adj.* [ME *acuat* < Med. Lat. *acuatus* < Lat. *acus*, needle.] Pointed at the tip.

**a·cu·i·ty** (ə-kyo͞o′ĭ-tē) *n.* [ME *acuite* < OFr. < Lat. *acutus*, sharp.—see ACUTE.] Perceptual keenness : ACUTENESS.

**a·cu·le·ate** (ə-kyo͞o′lē-ĭt, -āt′) *adj.* [Lat. *aculeatus* < *aculeus*, sting, dim. of *acus*, needle.] *Biol.* Having a sting or prickles.

**a·cu·men** (ə-kyo͞o′mən) *n.* [Lat. *acumen* < *acuere*, to sharpen < *acus*, needle.] Accuracy and quickness of judgment : keen insight.

**a·cu·mi·nate** (ə-kyo͞o′mə-nĭt, -nāt′) *adj.* [Lat. *acuminatus*, p.part. of *acuminare*, to sharpen < *acumen*, acuteness. —see ACUMEN.] Tapering to a sharp point <an *acuminate* leaf> —*vt.* (-nāt′) **-nat·ed, -nat·ing, -nates.** To taper or sharpen. **—a·cu′mi·na′tion** *n.*

**ac·u·punc·ture** (ăk′yo͞o-pŭngk′chər) *n.* [Lat. *acus*, needle + PUNCTURE.] A traditional Chinese therapeutic technique whereby the body is punctured with fine needles. **—ac′u·punc′tur·ist** *n.*

**a·cute** (ə-kyo͞ot′) *adj.* [Lat. *acutus*, p.part. of *acuere*, to sharpen < *acus*, needle.] **1.** Having a sharp point. **2.** Keenly perceptive or discerning. **3.** Reacting readily to impressions : SENSITIVE. **4.** Extremely serious or significant : CRUCIAL <an *acute* food shortage> **5.** Extremely severe or sharp, as pain. **6.** *Med.* Reaching a crisis rapidly.—Used of a disease. **7.** *Mus.* High-pitched : shrill. **8.** *Geom.* Designating angles less than 90°. **—a·cute′ly** *adv.* **—a·cute′ness** *n.*

**acute accent** *n.* A mark (′) indicating: **a.** A raised pitch in certain languages such as Chinese and Ancient Greek. **b.** Primary stress of a spoken sound or syllable. **c.** Metrical stress in poetry. **d.** Sound quality or vowel length.

**a·cy·clic** (ā-sī′klĭk, ā-sĭk′lĭk) *adj.* **1.** *Bot.* Not having or forming cycles or whorls. **2.** *Chem.* Having an open-chain molecular structure rather than a ring-shaped structure.

**ac·yl** (ăs′əl) *n.* [AC(ID) + -YL.] *Chem.* A radical having the general formula RCO-, derived from an organic acid.

**ad**[1] (ăd) *n. Informal.* An advertisement.

**ad**[2] (ăd) *n.* ADVANTAGE 4.

**ad-** *pref.* [Lat. < *ad*, to.] **1.** *or* **ac-** *or* **af-** *or* **ag-** *or* **al-** *or* **ap-** *or* **as-** *or* **at-.** Toward : to <*ad*noun> **2.** Near : at <*ad*renal> **3.** In <*ad*umbrel>

**-ad** *suff.* Toward : in the direction of <cephal*ad*>

**ad·age** (ăd′ĭj) *n.* [Fr. < OFr. < Lat. *adagium*, proverb.] A short maxim : PROVERB.

**a·da·gio** (ə-dä′jō, -jē-ō′, -zhō, -zhē-ō) *adv.* [Ital. : *ad-*, at (< Lat.) + *agio*, ease < OProv. *aize* < Lat. *adjacens*, convenient. —see ADJACENT.] *Mus.* Slowly. —Used as a direction. —*adj. Mus.* Slower than andante in tempo. —*n., pl.* **-agios. 1.** *Mus.* A composition or movement played in an adagio tempo. **2.** A section of a pas de deux in which the ballerina and her partner perform steps requiring lyricism and great skill in lifting, balancing, and turning.

**Ad·am¹** (ăd'əm) *n.* [LLat. < Heb. *'adhām,* man < *'adhāmāh,* earth.] **1.** The first man and progenitor of humankind as described in the Old Testament. **2.** The unregenerate aspect of human nature. **—A·dam'ic** (ə-dăm'ĭk) *adj.*

**Ad·am²** (ăd'əm) *adj.* Of or relating to the neoclassic style of furniture and architecture originated by Robert and James Adam.

**Ad·am-and-Eve** (ăd'əm-ənd-ēv') *n.* [From its human-shaped corms.] The puttyroot.

**ad·a·mant** (ăd'ə-mənt, -mănt') *n.* [ME < OFr. *adamaunt* < Lat. *adamas* < Gk.] **1.** A legendary stone thought to be impenetrable. **2.** An extremely hard substance. *—adj.* Resolute : unyielding.

**ad·a·man·tine** (ăd'ə-măn'tēn', -tīn', -tĭn') *adj.* **1.** Made of or resembling adamant. **2.** Having the hardness or luster of a diamond. **3.** Unyielding : inflexible <an *adamantine* will>

**Ad·am's apple** (ăd'əmz) *n.* [Trans. of Heb. *tappūah hādām.*] The projection of the largest laryngeal cartilage at the front of the throat, esp. in men.

**Adam's needle** *n.* [From its spiny leaves.] SPANISH BAYONET 2.

**a·dapt** (ə-dăpt') *v.* **a·dapt·ed, a·dapt·ing, a·dapts.** [Lat. *adaptare* : *ad-,* to + *aptare,* to fit < *aptus,* fitting.] *—vt.* To adjust to a specified use or situation. *—vi.* To become adapted.

**a·dapt·a·ble** (ə-dăp'tə-bəl) *adj.* Able to adapt or to be adapted. **—a·dapt'a·bil'i·ty, a·dapt'a·ble·ness** *n.*

**ad·ap·ta·tion** (ăd'ăp-tā'shən) *n.* **1. a.** The state of being adapted. **b.** The act or process of adapting. **2.** Something that undergoes change to fit a new or special use or situation. **3.** An often hereditary alteration or adjustment by which a species or individual improves its condition in relationship to its environment. **4.** The responsive alteration of a sense organ to repeated stimuli. **5.** Behavioral change of an individual or group in adjustment to new or modified cultural surroundings. **—ad'ap·ta'tion·al** *adj.* **—ad'ap·ta'tion·al·ly** *adv.*

**a·dapt·er** *also* **a·dap·tor** (ə-dăp'tər) *n.* **1.** One that adapts. **2.** A device used to effect operative compatibility between different parts of one or more pieces of apparatus.

**a·dap·tion** (ə-dăp'shən) *n.* Adaptation.

**a·dap·tive** (ə-dăp'tĭv) *adj.* Capable of, suitable for, or tending toward adaptation. **—a·dap'tive·ly** *adv.* **—a·dap'tive·ness** *n.*

**adaptive radiation** *n.* The evolution of a relatively unspecialized species into several related species characterized by different specializations that fit them for life in various environments.

**A·dar** (ä-där') *n.* [ME < Heb. *Adhār* < Akkadian *addaru* < *adāru,* to be dark.] The sixth month of the Hebrew year. —See table at CALENDAR.

**Adar She·ni** (shā-nē') *n.* [Heb. *Adhār shēnī,* second Adar.] Veadar.

**ad·ax·i·al** (ăd-ăk'sē-əl) *adj.* Of, relating to, or being on the side toward the axis or stem.

**add** (ăd) *v.* **add·ed, add·ing, adds.** [ME *adden* < Lat. *addere* : *ad-,* to + *dare,* to give.] *—vt.* **1.** To unite or join so as to increase in size, quantity, or scope : APPEND. **2.** To combine (e.g., a column of figures) to form a sum. **3.** To say or write further. *—vi.* **1.** To create or constitute an addition <a skill that could *add* to your income> **2.** To find a sum in arithmetic. **—add up.** *Informal.* To be reasonable : make sense <What you say *adds up.*> **—add up to.** *Informal.* To indicate <The evidence *adds up to* homicide.> **—add'a·ble, add'i·ble** *adj.*

**ad·dax** (ăd'ăks') *n.* [Lat., of African orig.] An antelope, *Addax nasomaculatus* of northern Africa, with long, spirally twisted horns.

**add·ed-val·ue tax** (ăd'ĭd-văl'yo͞o) *n.* Value-added tax.

**ad·dend** (ăd'ĕnd', ə-dĕnd') *n.* [Short for ADDENDUM.] Any of a set of numbers to be added.

**ad·den·dum** (ə-dĕn'dəm) *n., pl.* **-da** (-də) [Lat. < *addere,* to add.] Something added, esp. a supplement to a book.

**add·er¹** (ăd'ər) *n.* One that adds, esp. a computer device that performs addition.

**ad·der²** (ăd'ər) *n.* [ME < *an addre,* alteration of *a naddre* : *a,* a + *naddre,* snake < OE *nædre.*] **1.** A venomous Old World snake of the family Viperidae, esp. the common viper, *Vipera berus* of Eurasia. **2.** Any of several nonvenomous snakes held to be harmful.

**ad·der's-mouth** (ăd'ərz-mouth') *n.* [From the resemblance of its flowers to the open mouths of snakes.] An orchid of the genus *Malaxis,* with small, usu. greenish flower clusters.

**ad·der's-tongue** (ăd'ərz-tŭng') *n.* [From the resemblance of the spike at the base of the frond to a snake's tongue.] **1.** A fern of the genus *Ophioglossum,* with a single sterile, leaflike frond and a spore-bearing stalk. **2.** The dogtooth violet.

**ad·dict** (ə-dĭkt') *vt.* **-dict·ed, -dict·ing, -dicts.** [< Lat. *addictus,* bondsman, p.part of *addicere,* to sentence : *ad-,* to + *dicere,* to adjudge.] To devote or give (oneself) habitually or compulsively to something (e.g., caffeine or alcohol). *—n.* (ăd'ĭkt). **1.** One who is addicted, esp. to narcotics. **2.** A devoted fan. **—ad·dic'tion** *n.* **—ad·dic'tive** *adj.*

**Ad·di·son's disease** (ăd'ĭ-sənz) *n.* [After Thomas *Addison* (1793–1860), its discoverer.] A disease caused by functional failure of the adrenal cortex, marked by a bronzelike skin pigmentation, anemia, and prostration.

**ad·di·tion** (ə-dĭsh'ən) *n.* [ME < OFr. < Lat. *additio* < *additus,* p.part. of *addere,* to add.] **1.** The act or process of adding. **2.** Something added, esp. a room or annex added to a building. **3.** The process of computing with sets of numbers so as to find their sum. **—in addition.** As well as : ALSO. **—in addition to.** Over and above : BESIDES. **—ad·di'tion·al** *adj.* **—ad·di'tion·al·ly** *adv.*

**ad·di·tive** (ăd'ĭ-tĭv) *adj.* Involving addition. *—n.* A substance added in small amounts to something else to alter it.

**additive identity** *n.* An identity element that in a given mathematical system leaves unchanged any element to which it is added.

**additive identity element** *n.* Zero.

**additive inverse** *n.* INVERSE 2b.

**ad·dle** (ăd'l) *v.* **-dled, -dling, -dles.** [< ME *adel,* muddled < OE, filth.] *—vt.* To confuse or muddle <"My brain is a bit *addled* by whiskey" —O'Neill> *—vi.* **1.** To become rotten. **2.** To become confused. *—adj.* **1.** Confused. **2.** Rotten <an *addled* egg>

**ad·dress** (ə-drĕs') *vt.* **-dressed, -dress·ing, -dress·es.** [ME *adressen* < OFr. *adresser* : *a-,* to (< Lat. *ad-*) + *dresser,* to arrange.—see DRESS.] **1.** To speak to. **2.** To make a formal speech to. **3.** To direct (a spoken or written message) to the attention of. **4.** To mark with a destination <*address* an envelope> **5. a.** To direct (oneself) in speech to. **b.** To direct the efforts or attention of (oneself) <*address* oneself to a job> **6.** To dispatch or consign (e.g., a ship) to an agent or factor. **7.** To adjust and aim the club at (a golf ball) in preparing for a stroke. *—n.* (ə-drĕs'). **1.** A formal spoken or written communication. **2.** A formal speech. **3.** *(also* ăd'rĕs'). The indication of destination, as on mail or parcels. **4.** *(also* ăd'rĕs').The location at which a person or an organization may be reached. **5.** *often* **addresses.** Courteous attentions esp. in courtship. **6.** A person's manner or bearing, esp. in conversation. **7.** Adroit handling of a situation. **8.** The act of dispatching or consigning a ship. **9.** *Computer Sci.* A number used in information storage or retrieval that is assigned to a specific memory location.

**ad·dress·a·ble** (ə-drĕs'ə-bəl) *adj.* Accessible through an address, as in computer memory.

**ad·dress·ee** (ăd'rĕ-sē', ə-drĕs'ē') *n.* One to whom something is addressed.

**ad·dress·er** *also* **ad·dres·sor** (ə-drĕs'ər) *n.* One that addresses.

**ad·duce** (ə-do͞os', ə-dyo͞os') *vt.* **-duced, -duc·ing, -duc·es.** [Lat. *adducere,* to bring to : *ad-,* to + *ducere,* to lead.] To bring forward and quote for formal consideration. **—ad·duce'a·ble, ad·duc'i·ble** *adj.*

☆ ***syns:*** ADDUCE, ADVANCE, CITE, LAY, PRESENT *v. core meaning* : to bring forward and quote for formal consideration <*adduced* strong, factual evidence before the court>

**ad·duct** (ə-dŭkt', ă-) *vt.* **-duct·ed, -duct·ing, -ducts.** [Back-formation from ADDUCTOR.] *Physiol.* To draw or pull toward the main axis. —Used of muscles. **—ad·duc'tion** *n.* **—ad·duc'tive** *adj.*

**ad·duc·tor** (ə-dŭk'tər) *n.* [NLat. < Lat. *adducere,* to bring to. —see ADDUCE.] A muscle that adducts.

**-ade** *suff.* [Fr., ult. < Lat. *-ata,* fem. of *-atus,* -ate.] A sweetened beverage of <lime*ade*> <orange*ade*>

**A·dé·lie penguin** (ə-dā'lē) *n.* [After the *Adélie* Coast, Antarctica.] A common medium-sized Antarctic penguin, *Pygoscelis adeliae,* that has a black back and head and white underparts and that lives and breeds in large exposed rookeries.

**-adelphous** *suff.* [NLat. *-adelphus* < Gk. *adelphos,* brother.] Having one or more groups of stamens <di*adelphous*>

**a·demp·tion** (ə-dĕmp'shən) *n.* [Lat. *ademptio,* a taking away < *adimere,* to take away : *ad-,* to + *emere,* to take.] *Law.* The disposal by a testator of specific property bequeathed in his or her will so as to invalidate the bequest.

**aden-** *pref. var. of* ADENO-.

**ad·e·nec·to·my** (ăd'n-ĕk'tə-mē) *n.* Surgical excision of a gland.

**ad·e·nine** (ăd'n-ēn') *n.* A purine derivative, $C_5H_5N_5$, that is a constituent of nucleic acid in organs, as the pancreas and spleen.

**ad·e·ni·tis** (ăd'n-ī'tĭs) *n.* Inflammation of a lymph gland or node.

**adeno-** *or* **aden-** *pref.* [< Gk. *adēn.*] Gland <*aden*ectomy>

**ad·e·no·car·ci·no·ma** (ăd'n-ō-kär'sə-nō'mə) *n.* A malignant tumor arising from glandular tissue. **—ad'e·no·car'ci·nom'a·tous** (-nŏm'ə-təs, -nō'mə-təs) *adj.*

**ad·e·no·hy·poph·y·sis** (ăd'n-ō-hī-pŏf'ĭ-sĭs) *n.* The anterior and intermediate glandular lobes of the pituitary gland. **—ad'e·no·hy·poph'y·se·al, ad'e·no·hy·poph'y·si·al** *adj.*

**ad·e·noid** (ăd'n-oid') *n. Often* **adenoids.** Lymphoid tissue growths in the nose above the throat that when swollen may obstruct nasal breathing, induce postnasal discharge, and impede speech.

**ad·e·noi·dal** (ăd'n-oid'l) *adj.* **1.** Of or resembling a gland. **2.** Of or relating to the adenoids. **3 a.** Having a nasal tone <an *adenoidal* vocalist> **b.** Breathing through the mouth because of enlarged adenoids.

**ad·e·no·ma** (ăd'n-ō'mə) *n.* An epithelial tumor of glandular origin and structure that is usu. of low-grade malignancy or benign. **—ad'·e·nom'a·tous** (-ŏm'ə-təs) *adj.*

**a·den·o·sine** (ə-dĕn'ə-sēn') *n.* [Blend of ADENINE and RIBOSE.] An organic compound, $C_{10}H_{13}N_5O_4$, that is a structural component of nucleic acids.

**adenosine di·phos·phate** (dī-fŏs′fāt′) *n.* ADP.

**adenosine mon·o·phos·phate** (mŏn′ō-fŏs′fāt′) *n.* **1.** Cyclic AMP. **2.** AMP.

**adenosine triphosphate** *n.* ATP.

**ad·e·no·vi·rus** (ăd′n-ō-vī′rəs) *n.* Any of various animal viruses that cause respiratory diseases in humans. **—ad′e·no·vi′ral** *adj.*

**a·den·yl·ate cy·clase** (ə-dĕn′l-ĭt sī′klās, ăd′n-ĭl′ĭt) *or* **ad·e·nyl cyclase** (ăd′n-ĭl) *n.* [ADEN(INE) + -YL + -ATE¹ + CYCL(O)- + -ASE.] The enzyme that catalyzes formation of cyclic AMP from ATP.

**a·dept** (ə-dĕpt′) *adj.* [Lat. *adeptus,* p.part. of *adipisci,* to arrive at.] Highly skilled : EXPERT. —*n.* (ăd′ĕpt′). A highly skilled person. **—a·dept′ly** *adv.* **—a·dept′ness** *n.*

**ad·e·quate** (ăd′ĭ-kwĭt) *adj.* [Lat. *adaequatus,* p.part. of *adaequare,* to equalize : *ad-,* to + *aequare,* to make equal < *aequus,* equal.] **1.** Able to satisfy a requirement. **2.** Barely sufficient of satisfactory. **—ad′e·qua·cy** (-kwə-sē), **ad′e·quate·ness** *n.* **—ad′e·quate·ly** *adv.*

**à deux** (ä′ dœ′) *adj.* [Fr.] Of or involving two individuals, esp. in private. —*adv.* Privately with only two individuals involved <picnicking *à deux*>

**ad·here** (ăd-hîr′) *vi.* **-hered, -her·ing, -heres.** [Fr. *adhérer* < Lat. *adhaerere,* to stick to : *ad-,* to + *haerere,* to stick.] **1.** To stick fast or together by or as if by being glued. **2.** To be devoted as a supporter or follower. **3.** To follow without deviation.

**ad·her·ence** (ăd-hîr′əns) *n.* **1.** The process or state of adhering. **2.** Faithful attachment or support : DEVOTION.

**ad·her·ent** (ăd-hîr′ənt) *adj.* **1.** Sticking or holding fast. **2.** *Bot.* Growing or fused together : ADNATE. —*n.* A supporter, as of a cause or individual. **—ad·her′ent·ly** *adv.*

**ad·he·sion** (ăd-hē′zhən) *n.* [Fr. *adhésion* < Lat. *adhaesio* < *adhaerere,* to adhere.] **1.** The act or state of adhering. **2.** Attachment or devotion. **3.** Assent. **4.** An abnormal condition in which bodily tissues that are ordinarily separate become united by fibrous tissue. **5.** Physical attraction or joining of two substances, esp. the macroscopically observable attraction of dissimilar substances. **6.** A fibrous band holding together normally separate anatomical structures. **7.** Pathological aggregation of dissimilar body materials to a visceral surface due to inflammation or trauma.

**ad·he·si·ot·o·my** (ăd-hē′zē-ŏt′ə-mē) *n., pl.* **-mies.** Surgical division of adhesions.

**ad·he·sive** (ăd-hē′sĭv, -zĭv) *adj.* **1.** Tending to adhere : STICKY. **2.** Gummed so as to adhere. **—ad·he′sive** *n.* **—ad·he′sive·ly** *adv.* **—ad·he′sive·ness** *n.*

**adhesive tape** *n.* Tape lined on one side with an adhesive.

**ad hoc** (ăd hŏk′, hōk′) *adj. & adv.* [Lat., to this.] For a specific purpose, case, or situation <formed an *ad hoc* committee>

**ad hom·i·nem** (ăd hŏm′ə-nĕm′) *adj. & adv.* [Lat., to the man.] Appealing to personal prejudices or emotions rather than to reason <an *ad hominem* debate>

**ad·i·a·bat·ic** (ăd′ē-ə-băt′ĭk, ā′dī-ə-) *adj.* [Gk. *adiabatos,* impassable : *a-,* not + *diabatos,* passable (*dia,* through + *batos,* passable < *bainein,* to go).] Of, pertaining to, or designating a reversible thermodynamic process executed at constant entropy. **—ad′i·a·bat′i·cal·ly** *adv.*

**a·dieu** (ə-dyōō′, ə-dōō′) *interj.* [ME < OFr. *a dieu,* (I commend you) to God : *a,* to (< Lat. *ad*) + *Dieu,* God < Lat. *deus.*] Good-by. —*n., pl.* **a·dieus** *or* **a·dieux** (ə-dyōōz′, ə-dōōz′). A farewell.

**ad in·fi·ni·tum** (ăd ĭn′fə-nī′təm) *adj. & adv.* [Lat., to infinity.] Without limit or end : FOREVER.

**ad in·ter·im** (ăd ĭn′tər-əm) *adj. & adv.* [Lat.] In the meantime.

**ad·i·os** (ăd′ē-ōs′, ä′dē-) *interj.* [Sp. *adíos* : *a,* to (< Lat. *ad*) + *Dios,* God < Lat. *deus.*] Good-by.

**ad·i·po·cere** (ăd′ə-pō-sîr′) *n.* [ADIPO(SE) + Lat. *cera,* wax.] A brown, fatty, waxlike substance that forms on dead animal tissues in response to moisture.

**ad·i·pose** (ăd′ə-pōs′) *adj.* [NLat. *adiposus* < Lat. *adeps,* lard.] Of or relating to animal fat : FATTY. —*n.* The fat found in adipose tissue. **—ad′i·pose′ness, ad′i·pos′i·ty** (-pŏs′ĭ-tē) *n.*

**adipose tissue** *n.* Bodily connective tissue that contains stored cellular fat.

**ad·it** (ăd′ĭt) *n.* [Lat. *aditus,* access < *adire,* to approach : *ad-,* toward + *ire,* to go.] An almost horizontal entrance to a mine.

**ad·ja·cent** (ə-jā′sənt) *adj.* [ME < Lat. *adjacens,* pr.part. of *adjacēre,* to lie near : *ad-,* near to + *jacēre,* to lie.] **1.** Close to : NEARBY <the house and *adjacent* pond> **2.** Next to : ADJOINING. **—ad·ja′cen·cy** *n.* **—ad·ja′cent·ly** *adv.*

☆ ***syns:*** ADJACENT, ABUTTING, ADJOINING, BORDERING, CONTERMINOUS, CONTIGUOUS, JUXTAPOSED, MEETING, TOUCHING *adj. core meaning* : sharing a common boundary <*adjacent* lots> ***ant:*** nonadjacent

**adjacent angle** *n.* Either of two angles having a common side and a common vertex.

**ad·jec·ti·val** (ăj′ĭk-tī′vəl) *adj.* Of, relating to, or functioning as an adjective. **—ad′jec·ti′val·ly** *adv.*

**ad·jec·tive** (ăj′ĭk-tĭv) *n.* [ME < OFr. *adjectif* < Lat. *adjectivus* < *adjicere,* to add to : *ad-,* to + *jacere,* to throw.] **1.** Any of a class of words used to modify a noun or other substantive by limiting, qualifying, or specifying. **2.** Any of a form class distinguished in English morphologically by one of several suffixes, as *-able, -ous, -er,* and *-est,* or syntactically by position in a phrase or sentence, as *white* in *a white house.* **3.** A subordinate or dependent. **—ad′jec·tive·ly** *adv.*

**adjective pronoun** *n.* A pronoun acting as an adjective, as *which* in *Which cars?* or *yourself* in *You yourself said so.*

**ad·join** (ə-join′) *v.* **-joined, -join·ing, -joins.** [ME *ajoinen* < OFr. *ajoindre* < Lat. *adjungere,* to join to : *ad-,* to + *jungere,* to join.] —*vt.* **1.** To be next to. **2.** To attach by joining. —*vi.* To be in or nearly in contact.

**ad·join·ing** (ə-joi′nĭng) *adj.* Bordering : contiguous.

**ad·journ** (ə-jûrn′) *v.* **-journed, -journ·ing, -journs.** [ME *ajournen* < OFr. *ajourner* : *a,* to (< Lat. *ad*) + *jour,* day < Lat. *diurnum.*] —*vt.* To suspend until a later stated time. —*vi.* **1.** To suspend proceedings to another time or location. **2.** *Informal.* To move from one location to another <*adjourned* to the den to read> **—ad·journ′ment** *n.*

**ad·judge** (ə-jŭj′) *vt.* **-judged, -judg·ing, -judg·es.** [ME *ajugen* < OFr. *ajuger* < Lat. *adjudicare.* —see ADJUDICATE.] **1.** To determine by judicial procedure : ADJUDICATE. **2.** To rule judicially. **3.** To award (e.g., damages) by law. **4.** To regard or consider.

**ad·ju·di·cate** (ə-jōō′dĭ-kāt′) *vt.* **-cat·ed, -cat·ing, -cates.** [Lat. *adjudicare, adjudicat-,* to award to (judicially) : *ad-,* to + *judicare,* to judge < *judex,* judge.] To hear and settle (a case) by judicial procedure. **—ad·ju′di·ca′tion** *n.* **—ad·ju′di·ca′tive** *adj.* **—ad·ju′di·ca′tor** *n.*

**ad·junct** (ăj′ŭngkt′) *n.* [Lat. *adjunctum* < *adjunctus,* p.part. of *adjungere,* to join to. —see ADJOIN.] **1.** One attached to another in a subordinate or dependent position. **2.** One associated with another in a duty or service in a subordinate or auxiliary capacity. **3.** A word or words added in order to clarify, qualify, or modify other words. **4.** *Logic.* A nonessential attribute. —*adj.* **1.** Added or connected in a subordinate or auxiliary capacity <an *adjunct* clause> **2.** Attached to a faculty or staff in a temporary or auxiliary capacity. **—ad·junc′tion** (ə-jŭngk′shən) *n.* **—ad·junc′tive** *adj.*

**ad·ju·ra·tion** (ăj′ə-rā′shən) *n.* **1.** A solemn command. **2.** An earnest appeal : ENTREATY. **—ad·jur′a·to′ry** (ə-jŏŏr′ə-tôr′ē, -tōr′ē) *adj.*

**ad·jure** (ə-jŏŏr′) *vt.* **-jured, -jur·ing, -jures.** [ME *adjuren* < Lat. *adjurare,* to swear to : *ad-,* to + *jurare,* to swear.] **1.** To command or enjoin solemnly, as under oath. **2.** To appeal to earnestly : ENTREAT. **—ad·jur′er, ad·ju′ror** *n.*

**ad·just** (ə-jŭst′) *v.* **-just·ed, -just·ing, -justs.** [Obs. Fr. *adjuster* < OFr. *ajoster* : Lat. *ad,* to + Lat. *juxta,* near.] —*vt.* **1.** To change so as to match or fit. **2.** To bring into proper relationship. **3.** To conform or adapt, as to new conditions. **4.** To make accurate by regulation. **5.** To decide how much is to be paid on (an insurance claim). **6.** To correct (the range and direction of a gun) in firing. —*vi.* To adapt oneself : CONFORM. **—ad·just′a·ble** *adj.* **—ad·just′a·bly** *adv.* **—ad·just′er, ad·jus′tor** *n.*

☆ ***syns:*** ADJUST, ATTUNE, FIX, REGULATE, SET, TUNE UP *v. core meaning* : to alter (parts of a device) for proper functioning <*adjust* the valves>

**ad·just·ment** (ə-jŭst′mənt) *n.* **1. a.** The act of making fit or conformable. **b.** The condition of being adjusted. **2.** A means for adjusting. **3.** The settlement of a debt or claim. **4.** A correction or modification <made an *adjustment* on the phone bill>

**ad·ju·tant** (ăj′ə-tənt) *n.* [Lat. *adjutans, adjutant-,* pr.part. of *adjutare,* freq. of *adjuvare,* to help : *ad-,* to + *juvare,* to help.] **1.** An administrative staff officer who assists a commanding officer. **2.** An assistant. **3.** The marabou. **—ad′ju·tan·cy** (-tən-sē) *n.*

**adjutant general** *n., pl.* **adjutants general. 1.** An adjutant of a military unit having a general staff. **2.** An officer in charge of the National Guard of one of the states of the United States. **3. Adjutant General.** The chief administrative officer of the U.S. Army.

**adjutant stork** *n.* The marabou.

**adjutant stork**
*Approximately 5 feet high*

**ad·ju·vant** (ăj′ə-vənt) *n.* [Lat. *adjuvans, adjuvant-,* pr.part. of *adjuvare,* to help. —see AID.] **1.** A pharmacological agent added to a drug to enhance its effect. **2.** An immunological agent that increases the antigenic response.

**Ad·le·ri·an** (ăd-lîr'ē-ən) *adj.* [After Alfred *Adler* (1870–1937).] Of or pertaining to a psychological school holding that behavior arises in subconscious efforts to compensate for inferiority or deficiency and that neurosis results from overcompensation.

**ad lib** (ăd lĭb') *adv.* [Short for AD LIBITUM.] In an improvisatory manner : SPONTANEOUSLY.

**ad-lib** (ăd-lĭb') *Informal.* —*v.* **-libbed, -lib·bing, -libs.** —*vt.* To improvise and deliver extemporaneously. —*vi.* To improvise or extemporize. —*n.* Something ad-libbed. —*adj.* Spoken or performed spontaneously <*ad-lib* comedy> **—ad-lib'ber** *n.*

**ad lib·i·tum** (ăd lĭb'ĭ-təm) *adj.* [Lat. *ad*, to + *libitum*, pleasure.] *Mus.* That may be performed or omitted. —Used as a direction.

**ad·man** (ăd'măn') *n. Informal.* A person in the advertising business.

**ad·meas·ure** (ăd-mĕzh'ər) *vt.* **-ured, -ur·ing, -ures.** [ME *amesuren* < OFr. *amesurer* : *a*, to (< Lat. *ad*) + *mesurer*, to measure.] To divide and distribute proportionally : APPORTION. **—ad·meas'ure·ment** *n.* **—ad·meas'ur·er** *n.*

**Ad·me·tus** (ăd-mē'təs) *n.* [Lat. < Gk. *Admētos.*] *Gk. Myth.* A king of Thessaly and husband of Alcestis.

**ad·min·is·ter** (ăd-mĭn'ĭ-stər) *v.* **-tered, -ter·ing, -ters.** [ME *administren* < OFr. *administrer* < Lat. *administrare* : *ad*, to + *ministrare*, to manage.] —*vt.* **1.** To have charge of : MANAGE. **2. a.** To give or apply formally, as a sacrament. **b.** To give as a remedy <*administer* a tranquilizer> **3.** To mete out <*administer* punishment> **4.** To manage or dispose of (a trust or estate) under a will or an official appointment. **5.** To tender (e.g., an oath). —*vi.* **1.** To manage as an administrator. **2.** To tend or minister <*administering* to our desires> **—ad·min'is·tra·ble** *adj.* **—ad·min'is·trant** *adj.* & *n.*

☆ ***syns:*** ADMINISTER, ADMINISTRATE, DIRECT, GOVERN, HEAD, MANAGE, RUN, SUPERINTEND *v. core meaning* : to have charge of (the affairs of others) <*administer* a colony>

**ad·min·is·trate** (ăd-mĭn'ĭ-strāt') *vt.* **-trat·ed, -trat·ing, -trates.** To administer.

**ad·min·is·tra·tion** (ăd-mĭn'ĭ-strā'shən) *n.* **1.** The management of affairs. **2.** The activity of a sovereign state in the exercise of its powers or duties. **3.** *often* **Administration.** The persons as a group who constitute the executive branch of a government. **4.** The management of a public or private institution. **5.** The term of office of an executive officer or body. **6.** *Law.* The management and disposal of a trust or estate, as by an executor. **7.** The dispensing, applying, or tendering of something, as medicine, a sacrament, or an oath. **—ad·min'is·tra'tive** (-strā'tĭv, -strə-) *adj.* **—ad·min'is·tra'tive·ly** *adv.*

**ad·min·is·tra·tor** (ăd-mĭn'ĭ-strā'tər) *n.* **1.** One who administers : EXECUTIVE. **2.** One appointed to administer an estate : EXECUTOR.

**ad·mi·ra·ble** (ăd'mər-ə-bəl) *adj.* Worthy of admiration. **—ad'mi·ra·ble·ness** *n.* **—ad'mi·ra·bly** *adv.*

☆ ***syns:*** ADMIRABLE, COMMENDABLE, ESTIMABLE, LAUDABLE, MERITORIOUS, PRAISEWORTHY *adj. core meaning* : worthy of admiration <*admirable* qualities of leadership>

**ad·mi·ral** (ăd'mər-əl) *n.* [ME *amiral* < OFr. < Ar. *'amīr al-*, commander of.] **1.** The commander in chief of a navy or fleet. **2.** An Admiral of the Fleet. **3.** In the U.S. Navy and U.S. Coast Guard: **a.** An officer of the next-to-the-highest rank. **b.** A rear admiral. **c.** A vice admiral. **4.** The ship carrying an admiral : FLAGSHIP. **5.** *Chiefly Brit.* The head of a fishing fleet. **6.** Any of various brightly colored butterflies of the genera *Limenitis* and *Vanessa*.

▲ word history: *Admiral* and *emir* are descended from the same Arabic word, *amir*, but since *admiral* has had a longer history in English it has undergone more alterations in both form and meaning than *emir*. The Arabic word *amir* means simply "commander," but its occurrence in phrases like *amir-al-bahr*, "commander of the sea," led Europeans to interpret *al* as part of the word for "commander." The Old French form of the Arabic title, for example, was *amiral*, which was the form borrowed into English. A *d* was inserted into the English word by scholars who knew that *am-* in French was frequently a simplification of Latin *adm-*, and they connected *amiral* with Latin *admirari*, "to wonder at." This spurious etymology caused the spelling with *d* to be adopted as the literary form among the learned in the 16th century, and it has persisted. *Emir* was borrowed into English in the 17th century as an Arabic or Moslem title, and, aside from having a variant form *ameer*, it has undergone little change.

**Admiral of the Fleet** *n.* The highest rank in the U.S. Navy, equivalent to General of the Army or field marshal.

**ad·mi·ral·ty** (ăd'mər-əl-tē) *n., pl.* **-ties. 1. a.** A court exercising jurisdiction over all maritime cases. **b.** Maritime law. **2. Admiralty.** The department of the British government controlling naval affairs.

**ad·mi·ra·tion** (ăd'mə-rā'shən) *n.* **1.** A feeling of wonder or delighted approval. **2.** An object of admiring regard.

**†ad·mire** (ăd-mīr') *v.* **-mired, -mir·ing, -mires.** [Fr. *admirer* < OFr. *amirer* < Lat. *admirari*, to wonder at : *ad-*, at + *mirari*, to wonder < *mirus*, wonderful.] —*vt.* **1.** To regard with admiration. **2.** To have a high opinion of : ESTEEM. **3.** *Archaic.* To marvel or wonder at. —*vi.* **1.** To feel or express admiration. **2.** *Regional.* To be pleased. **—ad·mir'er** *n.* **—ad·mir'ing·ly** *adv.*

**ad·mis·si·ble** (ăd-mĭs'ə-bəl) *adj.* **1.** Capable of being accepted or allowed. **2.** Worthy of being admitted. **—ad·mis'si·bil'i·ty, ad·mis'si·ble·ness** *n.* **—ad·mis'si·bly** *adv.*

**ad·mis·sion** (ăd-mĭsh'ən) *n.* [ME < Lat. *admissio* < *admittere*, to admit.] **1. a.** The act of admitting or allowing to enter. **b.** The state of being allowed to enter. **2.** The right to enter : ACCESS. **3.** An entrance fee. **4.** Appointment to a position or situation. **5.** A confession of wrongdoing. **6. a.** A voluntary acknowledgment that something is true. **b.** A fact or statement admitted or conceded. ***usage:*** *Admission* has a more general meaning than *admittance*, which is used only to refer to gaining physical access to a place. An *admission* to a theater (i.e., the price of entering) grants one the right to *admittance* (physical entry to the theater itself). **—ad·mis'sive** (-mĭs'ĭv) *adj.*

**ad·mit** (ăd-mĭt') *v.* **-mit·ted, -mit·ting, -mits.** [ME *admitten* < Lat. *admittere* : *ad*, to + *mittere*, to send.] —*vt.* **1.** To permit to enter. **2.** To serve as a means of entrance. **3.** To permit to join or exercise certain rights, functions, or privileges. **4.** To accommodate <The room *admits* 45.> **5.** To afford possibility for <The problem *admits* more than one solution.> **6.** To acknowledge or confess <*admit* the truth> **7.** To concede as true or valid <had to *admit* we were wrong> —*vi.* **1.** To allow or permit <a topic that *admits* of discussion> **2.** To afford entrance. **3.** To make admission <*admits* to the crime>

**ad·mit·tance** (ăd-mĭt'ns) *n.* **1.** The act of admitting or entering. **2.** Permission to enter. **3.** *Elect.* The reciprocal of impedance.

**ad·mit·ted·ly** (ăd-mĭt'ĭd-lē) *adv.* By general admission.

**ad·mix** (ăd-mĭks') *vt.* & *vi.* **-mixed, -mix·ing, -mix·es.** [Back-formation < obs. *admixt*, mixed into < Lat. *admixtus*, p.part. of *admiscēre*, to mix into : *ad*, to + *miscēre*, to mix.] To mix or blend.

**ad·mix·ture** (ăd-mĭks'chər) *n.* **1. a.** The act of mixing. **b.** The state of being mixed. **2.** A combination, mixture, or blend. **3.** Something added in mixing.

**ad·mon·ish** (ăd-mŏn'ĭsh) *vt.* **-ished, -ish·ing, -ish·es.** [ME *admonishen*, alteration of *amonesten* < OFr. *amonester* < VLat. **admonestare*, var. of Lat. *admonēre* : *ad*, to + *monēre*, to warn.] **1.** To reprove mildly or kindly but seriously. **2.** To warn against something : CAUTION. **3.** To point out obligations to, by means of a warning, reproof, or exhortation. **—ad·mon'ish·er** *n.* **—ad·mon'ish·ing·ly** *adv.* **—ad·mon'ish·ment** *n.*

☆ ***syns:*** ADMONISH, REBUKE, REPRIMAND, REPROACH, REPROVE *v. core meaning* : to address someone disapprovingly because of a fault or misdeed. While ADMONISH refers to mild, warning criticism <*admonished* me to slow down>, REPROVE and REPROACH imply somewhat stronger disapproval <*reproved* the mischievous child><*reproached* me for being late> REBUKE refers to sharp criticism <*rebuked* the insolent student> and REPRIMAND indicates very severe, often formal criticism of another <a general officer *reprimanded* by the President> ***ant:*** commend

**ad·mo·ni·tion** (ăd'mə-nĭsh'ən) *n.* [ME *amonicioun* < OFr. *amonition* < Lat. *admonitio* < *admonēre*, to admonish.] **1.** Mild reproof. **2.** Cautionary advice.

**ad·mon·i·to·ry** (ăd-mŏn'ĭ-tôr'ē, -tōr'ē) *adj.* Expressing admonition.

**ad·nate** (ăd'nāt') *adj.* [Lat. *adnatus*, var. of *agnatus*, connected by birth < *agnasci*, to be born in addition to. —see AGNATE.] *Biol.* Joined to or fused with another part or organ, as parts not usu. connected. **—ad·na'tion** (ăd-nā'shən) *n.*

**ad nau·se·am** (ăd nô'zē-əm) *adv.* [Lat., to the point of nausea.] To a disgusting or ridiculous degree <rambled on *ad nauseam*>

**ad·nex·a** (ăd-nĕk'sə) *n.* [NLat. < Lat. *adnexus*, *annexus*, p.part. of *annectere*, to bind to. —see ANNEX.] Accessory or subordinate anatomical parts, as ovaries and oviducts. **—ad·nex'al** *adj.*

**ad·noun** (ăd'noun') *n.* An adjective, specif. when used as a noun, as in *the bold and the brave.* **—ad·nom'i·nal** (ăd-nŏm'ə-nəl) *adj.*

**a·do** (ə-do͞o') *n.* [ME < the phrase *at do* : *at*, to + *do*, to do.] Fuss.

**a·do·be** (ə-dō'bē) *n.* [Sp. < Ar. *aṭṭoba* < *al- ṭoba*, the brick : *al*, the + *ṭōba*, brick.] **1.** A sun-dried, unburned brick of clay and straw. **2.** Clay or soil from which adobe is made. **3.** A structure of adobe.

**ad·o·les·cence** (ăd'l-ĕs'əns) *n.* **1.** The period of physical and psychological development from the onset of puberty to maturity. **2.** A transitional period of development between youth and maturity <the *adolescence* of a country>

**ad·o·les·cent** (ăd'l-ĕs'ənt) *adj.* [ME < OFr. < Lat. *adolescens*, pr.part. of *adolescere*, to grow up : *ad*, toward + *alescere*, to grow, inchoative of *alere*, to nourish.] Of, relating to, or undergoing adolescence. —*n.* An adolescent boy or girl : TEEN-AGER.

**Ad·o·nai** (ăd'n-ī', ăd'n-oi') *n.* [Heb. *adōnāi*, my lord < Phoenician *adōn*, lord.] Lord. —Used in Judaism as a spoken substitute for the name of God.

**A·don·is** (ə-dŏn'ĭs, ə-dō'nĭs) *n.* [Gk. *Adōnis* < Phoenician *adōn*, lord.] **1.** *Gk. Myth.* A youth loved by Aphrodite for his exceptional beauty. **2.** *often* **adonis.** A very handsome young man.

**a·dopt** (ə-dŏpt') *vt.* **a·dopt·ed, a·dopt·ing, a·dopts.** [ME *adopten* < OFr. *adopter* < Lat. *adoptare* : *ad*, to + *optare*, to choose.] **1.** To take into one's family legally and raise as one's own child. ***usage:*** One refers to an *adopted* child but *adoptive* parents. **2.** To take and follow by choice or assent, as a course of action. **3.** To take up and make one's own, as an idea. **4.** To take on : ASSUME <*adopted* an air of secrecy> **5.** To vote to accept <*adopt* a proposal> **6.** To select as a standard textbook or reference book in a course. **—a·dopt'a·ble** *adj.* **—a·dopt'er** *n.* **—a·dop'tion** *n.*

**a·dop·tee** (ə-dŏp'tē) *n.* One that is adopted.

**a·dop·tive** (ə-dŏp'tĭv) *adj.* **1. a.** Adopting or tending to adopt. **b.** Characteristic of adoption. **2.** Related or acquired by adoption. **—a·dop'tive·ly** *adv.*

**a·dor·a·ble** (ə-dôr'ə-bəl, ə-dōr'-) *adj.* **1.** *Informal.* Charming : delightful. **2.** *Archaic.* Worthy of adoration. **—a·dor'a·bil'i·ty, a·dor'a·ble·ness** *n.* **—a·dor'a·bly** *adv.*

**ad·o·ra·tion** (ăd'ə-rā'shən) *n.* **1.** The act of worship. **2.** Profound love, reverence, or admiration.

**a·dore** (ə-dôr', ə-dōr') *v.* **a·dored, a·dor·ing, a·dores.** [ME *adouren* < OFr. *adourer* < Lat. *adorare,* to pray to : *ad,* to + *orare,* to pray.] —*vt.* **1.** To worship as divine. **2.** To love or revere deeply. **3.** *Informal.* To like very much <*adores* parties> —*vi.* To worship. **—a·dor'er** *n.* **—a·dor'ing·ly** *adv.*

**a·dorn** (ə-dôrn') *vt.* **a·dorned, a·dorn·ing, a·dorns.** [ME *adournen* < OFr. *adourner* < Lat. *adornare* : *ad,* to + *ornare,* to decorate.] **1.** To decorate with or as if with ornaments. **2.** To enhance the beauty or distinction of : GRACE. **—a·dorn'er** *n.*

**a·dorn·ment** (ə-dôrn'mənt) *n.* **1.** The act of adorning. **2.** Something that adorns.

**ADP** (ā'dē'pē') *n.* An ester, $C_{10}H_{15}N_5O_{10}P_2$, that is an adenosine derivative formed in cells and converted to ATP for the storage of energy.

**ad rem** (ăd rĕm') *adj.* [Lat.] *Law.* Relevant to the point at issue : PERTINENT. **—ad rem'** *adv.*

**ad·re·nal** (ə-drē'nəl) *adj.* **1.** At, near, or on the kidneys. **2.** Of or relating to the adrenal glands or their secretions. —*n.* An adrenal gland. **—ad·re'nal·ly** *adv.*

**adrenal cortex** *n.* The three-zoned center of the adrenal glands.

**adrenal gland** *n.* Either of two small endocrine glands, one located above each kidney, consisting of the cortex, which secretes hormones, and the medulla, which secretes epinephrine.

**A·dren·a·lin** (ə-drĕn'ə-lĭn). A trademark for a preparation of adrenaline.

**a·dren·a·line** (ə-drĕn'ə-lĭn) *n.* Epinephrine.

**ad·re·ner·gic** (ăd'rə-nûr'jĭk) *adj.* [ADREN(ALINE) + Gk. *ergon,* work.] Of, relating to, or having chemical activity like that of epinephrine, as certain nerve fibers.

**a·dre·no·chrome** (ə-drē'nō-krōm', -nə-) *n.* [ADREN(ALINE) + CHROME.] A naturally occurring chemical formed during the oxidation of adrenaline.

**ad·re·no·cor·ti·co·trop·ic** (ə-drē'nō-kôr'tĭ-kō-trŏp'ĭk, -trō'pĭk) *also* **ad·re·no·cor·ti·co·troph·ic** (-trŏf'ĭk, -trō'fĭk) *adj.* [ADRE-N(AL) + CORTICO- + -TROPIC.] Stimulating or otherwise acting on the cortex of the adrenal gland.

**adrenocorticotropic hormone** *or* **ad·re·no·cor·ti·co·troph** (ə-drē'nō-kôr'tĭ-kō-trŏf) *or* **ad·re·no·cor·ti·co·tro·phin** (-kôr'tĭ-kō-trō'fən) *n.* ACTH.

**a·drift** (ə-drĭft') *adv.* & *adj.* **1.** Without being anchored. **2.** Without direction or purpose.

**a·droit** (ə-droit') *adj.* [Fr. < *a droit* : *a,* to (< Lat. *ad*) + *droit,* right < Lat. *directus.* —see DIRECT.] **1.** Dexterous : deft. **2.** Skillful and adept esp. in dangerous or difficult circumstances. **—a·droit'ly** *adv.* **—a·droit'ness** *n.*

**ad·sci·ti·tious** (ăd'sĭ-tĭsh'əs) *adj.* [< Lat. *ascitus,* assumed, p.part. of *asciscere,* to assume : *ad,* to + *sciscere,* to accept, inchoative of *scire,* to know.] Derived from something external : SUPPLEMENTAL.

**ad·sorb** (ăd-sôrb', -zôrb') *vt.* **-sorbed, -sorb·ing, -sorbs.** [AD- + Lat. *sorbēre,* to suck.] To take up by adsorption.

**ad·sor·bate** (ăd-sôr'bĭt, -bāt', ăd-zôr'-) *n.* An adsorbed substance.

**ad·sor·bent** (ăd-sôr'bənt, -zôr'-) *adj.* Capable of absorbing. **—ad·sor'bent** *n.*

**ad·sorp·tion** (ăd-sôrp'shən, -zôrp'-) *n.* [ADSORB + -TION.] Assimilation of gas, vapor, or dissolved matter by the surface of a solid or liquid. **—ad·sorp'tive** (-tĭv) *adj.*

**ad·u·lar·i·a** (ăj'ə-lâr'ē-ə, -lăr'-) *n.* [Ital. < Fr. *adulaire,* after *Adula,* a group of Swiss mountains.] A variety of orthoclase.

**ad·u·late** (ăj'ə-lāt') *vt.* **-lat·ed, -lat·ing, -lates.** [Back-formation < ADULATION.] To praise excessively or servilely. **—ad'u·la'tor** *n.* **—ad'u·la·to'ry** (-lə-tôr'ē, -tōr'ē) *adj.*

**ad·u·la·tion** (ăj'ə-lā'shən) *n.* [ME *adulacioun* < OFr. < Lat. *adulatio* < *adulari,* to flatter.] Excessive flattery or praise.

**a·dult** (ə-dŭlt', ăd'ŭlt') *n.* [Lat. *adultus,* p.part. of *adolescere,* to grow up. —see ADOLESCENT.] **1.** One who has attained maturity or legal age. **2.** A fully grown, mature organism. —*adj.* **1.** Fully developed and mature. **2.** Relating to, befitting, or intended for mature persons. **3. a.** Restricted to adults <*adult* movies> **b.** Relating to or dealing with explicitly sexual or pornographic material <*adult* bookshops> **—a·dult'hood'** (-hŏŏd') *n.*

**a·dul·ter·ant** (ə-dŭl'tər-ənt) *n.* A substance that adulterates. —*adj.* That adulterates.

**a·dul·ter·ate** (ə-dŭl'tə-rāt') *vt.* **-at·ed, -at·ing, -ates.** [Lat. *adulterare, adulterat-,* to pollute.] To make impure, spurious, or inferior by adding extraneous or improper ingredients. —*adj.* (-tər-ĭt). **1.** Spurious. **2.** Adulterous. **—a·dul'ter·a'tion** *n.* **—a·dul'ter·a'tor** *n.*

☆ ***syns:*** ADULTERATE, DOCTOR, LOAD *v. core meaning* : to make impure or defective by fraudulent addition of foreign matter <gasoline *adulterated* with water> ***ant:*** refine

**a·dul·ter·er** (ə-dŭl'tər-ər) *n.* One who commits adultery.

**a·dul·ter·ess** (ə-dŭl'trĭs, -tər-ĭs) *n.* A woman who commits adultery.

**a·dul·ter·ine** (ə-dŭl'tə-rīn', -rēn') *adj.* [Lat. *adulterinus* < *adulter,* adulterer < *adulterare,* to adulterate.] **1.** Marked by adulteration : SPURIOUS. **2.** Unauthorized by law : ILLEGAL. **3.** Born of adultery.

**a·dul·ter·ous** (ə-dŭl'tər-əs, -trəs) *adj.* Relating to, given to, or marked by adultery. **—a·dul'ter·ous·ly** *adv.*

**a·dul·ter·y** (ə-dŭl'tə-rē, -trē) *n., pl.* **-ries.** [ME < OFr. *avouterie* < Lat. *adulterium* < *adulter,* adulterer < *adulterare,* to adulterate.] Voluntary sexual intercourse between a married person and a partner other than the lawful spouse.

**ad·um·brate** (ăd'əm-brāt', ə-dŭm'-) *vt.* **-brat·ed, -brat·ing, -brates.** [Lat. *adumbrare, adumbrat-,* to overshadow : *ad,* to + *umbra,* shadow.] **1.** To give a sketchy outline of. **2.** To prefigure indistinctly : FORESHADOW. **3.** To disclose guardedly or partially. **—ad'um·bra'tion** *n.* **—ad·um'bra·tive** (ə-dŭm'brə-tĭv) *adj.* **—ad·um'bra·tive·ly** *adv.*

☆ ***syns:*** ADUMBRATE, FORESHADOW, PREFIGURE, PRESAGE *v. core meaning* : to give signs of in advance <events that *adumbrated* the revolution in Iran>

**a·dust** (ə-dŭst') *adj.* [ME < Lat. *adustus,* p.part. of *adurere,* to set fire to : *ad,* to + *urere,* to burn.] **1.** Burned. **2.** Melancholy.

▲ word history: *Adust,* which has etymologically nothing to do with *dust,* acquired the meaning "burned" directly from its Latin source but came to mean "melancholy" through its use in early medical writings. Dryness, heat, and a burnt color of the body and its components, like the blood, certain organs, and the skin, were considered symptoms of a melancholy temperament.

**ad va·lo·rem** (ăd' və-lôr'əm, -lōr'-) *adj.* [Lat.] According to the value <*ad valorem* taxes on imported goods>

**ad·vance** (ăd-văns') *v.* **-vanced, -vanc·ing, -vanc·es.** [ME *avauncen* < OFr. *avauncer* < VLat. **abantiare* < Lat. *abante,* from before : *ab,* from + *ante,* before.] —*vt.* **1.** To move or cause to move forward. **2.** To propose <*advance* an idea> **3.** To aid the growth or progress of. **4.** To promote <*advanced* me to sergeant> **5.** To cause to occur sooner : HASTEN. **6.** To raise in rate or amount. **7.** To pay (money or interest) before legally due. **8.** To supply or lend, esp. on credit. —*vi.* **1.** To move forward or onward. **2.** To improve : progress. **3.** To rise in rank, position, or value. **—advance on (***or* **upon).** To move against, as when attacking. —*n.* **1.** The act or process of moving forward. **2.** Progress or improvement. **3.** An increase in price or value. **4. advances.** Personal approaches to secure acquaintance, favor, or an agreement. **5. a.** The supplying of funds or goods on credit. **b.** The funds or goods so supplied. **6.** Payment of money before legally due. —*adj.* **1.** Made or given ahead of time : PRIOR <gave *advance* notice> **2.** Going before <an *advance* troop of police> **—in advance. 1.** In front. **2.** Ahead of time. **—ad·vanc'er** *n.*

☆ ***syns:*** **1.** ADVANCE, PROCEED, PROGRESS *v. core meaning* : to move forward. ADVANCE is most often limited to concrete instances of forward motion <The armored division *advanced* at top speed.><We *advanced* a step or two.>, but it can also refer to nonphysical movement <The editors *advanced* the deadline by one month.> PROCEED stresses continuing motion <We *proceeded* to the city by freeway.> PROGRESS suggests steady improvement or development <The patient is *progressing* nicely.> **2.** ADVANCE, FURTHER, PROMOTE *v. core meaning* : to cause to move forward or upward, as toward a goal <Medical research *advances* our knowledge of life itself.> PROMOTE and FURTHER stress active support and encouragement <an advertising campaign to *promote* a new product><*furthering* one's career through graduate study> ***ant:*** retard

**ad·vanced** (ăd-vănst') *adj.* **1.** Highly developed or complex <*advanced* technology> **2.** At a higher level than others <*advanced* studies> **3.** Ahead of the times : PROGRESSIVE. **4.** Far along in course or age <an *advanced* illness><at an *advanced* age>

**advanced degree** *n.* A university degree that is higher than a bachelor's.

**advanced standing** *n.* The status of a college student granted credit, usu. after testing, for courses omitted or taken elsewhere.

**advance guard** *n.* A detachment of troops sent ahead of the main force to reconnoiter and provide protection.

**advance man** *n.* **1.** An agent, as for a performing troupe, who makes advance business and publicity arrangements. **2.** An assistant, as to a political candidate, who makes advance arrangements, as for a public appearance.

**ad·vance·ment** (ăd-văns'mənt) *n.* **1.** The act of advancing. **2.** An improvement. **3.** Development : progress <the *advancement* of cancer research> **4.** A promotion, as in rank.

ă **pat** ā **pay** âr **care** ä **father** ĕ **pet** ē **be** hw **which** ĭ **pit** ī **tie** îr **pier** ŏ **pot** ō **toe** ô **paw, for** oi **noise** ŏŏ **took**

**ad·van·tage** (ăd-văn′tĭj) *n.* [ME *avauntage* < OFr. < *avant,* before < Lat. *abante,* from before. —see ADVANCE.] **1.** A factor conducive to success. **2.** Profit or benefit : GAIN. **3.** A relatively favorable position. **4. a.** The first point scored in tennis after deuce. **b.** The resulting score. —*vt.* **-taged, -tag·ing, -tag·es.** To benefit. **—take advantage of. 1.** To avail oneself of : put to good use. **2.** To profit selfishly by : EXPLOIT. **—to advantage.** So as to bring about a good effect.

**ad·van·ta·geous** (ăd′văn-tā′jəs, -vən-) *adj.* Affording benefit or gain. **—ad′van·ta′geous·ly** *adv.* **—ad′van·ta′geous·ness** *n.*

**ad·vect** (ăd-vĕkt′) *vt.* **-vect·ed, -vect·ing, -vects.** [Back-formation < ADVECTION.] **1.** To convey horizontally by advection. **2.** To transport (a substance) by advection.

**ad·vec·tion** (ăd-vĕk′shən) *n.* [Lat. *advectio,* conveyance < *advehere,* to carry to : *ad,* to + *vehere,* to carry.] A local change in a property of a system, as of atmospheric temperature, caused by motion of the fluid in a gradient of the property.

**ad·vent** (ăd′vĕnt′) *n.* [ME, the Advent season < OFr. < Lat. *adventus,* arrival < *advenire,* to come to : *ad,* to + *venire,* to come.] **1.** Arrival, esp. of something momentous <the *advent* of the space age> **2. Advent. a.** The coming or birth of Christ. **b.** The period of four Sundays before Christmas.

**Ad·vent·ist** (ăd′vĕn′tĭst) *n.* A member of any of several Christian denominations that believe Christ's second coming and the end of the world are imminent. **—Ad′vent′ism** *n.*

**ad·ven·ti·tia** (ăd′vĕn-tĭsh′ə, -vən-) *n.* [NLat. < Lat. *adventicius,* foreign. —see ADVENTITIOUS.] The external covering of an organ.

**ad·ven·ti·tious** (ăd′vĕn-tĭsh′əs, -vən-) *adj.* [Lat. *adventicious,* foreign < *adventus,* arrival. —see ADVENT.] **1.** Acquired by accident : not inherent. **2.** *Biol.* Appearing in an unusual place or in an irregular manner <*adventitious* shoots> **—ad′ven·ti′tious·ly** *adv.* **—ad′ven·ti′tious·ness** *n.*

**ad·ven·tive** (ăd-vĕn′tĭv) [< Lat. *adventus,* arrival. —see ADVENT.] *Biol.* —*adj.* Not native to and not fully established in a new habitat or environment <an *adventive* plant> —*n.* An adventive organism. **—ad·ven′tive·ly** *adv.*

**Advent Sunday** *n.* The first Sunday of Advent.

**ad·ven·ture** (ăd-vĕn′chər) *n.* [ME *aventure* < OFr. < Lat. *adventurus,* fut. part. of *advenire,* to arrive. —see ADVENT.] **1.** A hazardous undertaking. **2.** An unusual or suspenseful experience. **3.** Participation in hazardous or exciting experiences. **4.** A financial speculation or business venture. —*v.* **-tured, -tur·ing, -tures.** —*vt.* To venture or dare : RISK. —*vi.* To take risks. **—ad·ven′ture·some** (-səm) *adj.*

☆ ***syns:*** ADVENTURE, EMPRISE, ENTERPRISE, VENTURE *n. core meaning* : an exciting, often risky undertaking <our mountain-climbing *adventure*>

**ad·ven·tur·er** (ăd-vĕn′chər-ər) *n.* **1.** One who adventures. **2.** A soldier of fortune. **3.** A financial speculator. **4.** One who unscrupulously seeks wealth and social position.

**ad·ven·tur·ess** (ăd-vĕn′chər-ĭs) *n.* A woman who unscrupulously seeks social and financial advancement.

**ad·ven·tur·ous** (ăd-vĕn′chər-əs) *adj.* **1.** Inclined to undertake new and daring enterprises. **2.** Full of hazard or risk. **—ad·ven′tur·ous·ly** *adv.* **—ad·ven′tur·ous·ness** *n.*

**ad·verb** (ăd′vûrb′) *n.* [ME *adverbe* < OFr. < Lat. *adverbium* : *ad,* to + *verbum,* word.] **1.** A part of speech comprising a class of words that modify a verb, adjective, or other adverb. **2.** A word belonging to this class, as *slowly* in *The elephant moves slowly.*

**ad·ver·bi·al** (ăd-vûr′bē-əl) *adj.* Of, relating to, or used as an adverb. **—ad·ver′bi·al·ly** *adv.*

**ad ver·bum** (ăd vûr′bəm) *adv.* [Lat.] Word for word.

**ad·ver·sar·y** (ăd′vər-sĕr′ē) *n., pl.* **-ies.** [ME *adversarie* < Lat. *adversarius,* enemy < *adversus,* against. —see ADVERSE.] One who opposes another, esp. with animosity.

**ad·ver·sa·tive** (ăd-vûr′sə-tĭv) *adj.* [Lat. *adversativus* < *adversari,* to oppose < *adversus,* against. —see ADVERSE.] Expressing antithesis or opposition <the *adversative* conjunction "but"> **—ad·ver′sa·tive** *n.* **—ad·ver′sa·tive·ly** *adv.*

**ad·verse** (ăd-vûrs′, ăd′vûrs′) *adj.* [ME < OFr. *advers* < Lat. *adversus,* p.part. of *advertere,* to turn toward : *ad-,* toward + *vertere,* to turn.] **1.** Actively opposed : ANTAGONISTIC <an *adverse* reaction> **2.** Failing to promote one's interests or welfare <*adverse* living conditions> **3.** In an opposite or opposing direction or position. **4.** *Bot.* Facing the main stem. **—ad·verse′ly** *adv.* **—ad·verse′ness** *n.*

**ad·ver·si·ty** (ăd-vûr′sĭ-tē) *n., pl.* **-ties. 1.** A state of affliction or hardship : MISFORTUNE. **2.** An instance of misfortune.

**ad·vert¹** (ăd-vûrt′) *vi.* **-vert·ed, -vert·ing, -verts.** [ME *adverten* < OFr. *avertir,* to notice < Lat. *advertere,* to turn toward. —see ADVERSE.] To call attention : allude or refer.

**ad·vert²** (ăd′vûrt′) *n. Chiefly Brit.* An advertisement.

**ad·ver·tise** (ăd′vər-tīz′) *v.* **-tised, -tis·ing, -tis·es.** [ME *advertisen,* to notify < OFr. *avertir, avertiss-,* to notice. —see ADVERT¹.] —*vt.* **1.** To make public announcement of, esp. to proclaim the qualities or advantages of so as to increase sales <*advertise* a new product> **2.** To make generally known <*advertise* one's desire to bargain> **3.** *Archaic.* To warn or notify. —*vi.* **1.** To call the attention of the public to a product or business. **2.** To inquire or seek in a public notice, as in a newspaper. **—ad′ver·tis′er** *n.*

**ad·ver·tise·ment** (ăd′vər-tīz′mənt, ăd-vûr′tĭs-, -tĭz-) *n.* A notice designed to attract public attention or patronage.

**ad·ver·tis·ing** (ăd′vər-tī′zĭng) *n.* **1.** The act of calling public attention to a product or business. **2.** The business of preparing and distributing advertisements. **3.** Advertisements as a whole.

**ad·vice** (ăd-vīs′) *n.* [ME *avis* < OFr., view < Med. Lat. *advisus* : Lat. *ad-,* to + *visum,* something seen < *vidēre,* to see.] **1.** Opinion about a course of action : COUNSEL. **2.** *often* **advices.** Information or report, esp. when communicated from a distance : INTELLIGENCE.

**ad·vis·a·ble** (ăd-vī′zə-bəl) *adj.* Prudent : expedient <not *advisable* to travel now> **—ad·vis′a·bil′i·ty, ad·vis′a·ble·ness** *n.* **—ad·vis′a·bly** *adv.*

**ad·vise** (ăd-vīz′) *v.* **-vised, -vis·ing, -vis·es.** [ME *avisen* < OFr. *aviser* < *avis,* view. —see ADVICE.] —*vt.* **1.** To offer advice to. **2.** To recommend or suggest. **3.** To inform or notify. ***usage:*** The use of *advise* meaning "to inform" or "to notify" is best restricted to business or legal writing. —*vi.* **1.** To take counsel : CONSULT. **2.** To offer advice.

**ad·vised** (ăd-vīzd′) *adj.* **1.** Considered <ill-*advised*> **2.** Informed.

**ad·vis·ed·ly** (ăd-vī′zĭd-lē) *adv.* With careful consideration.

**ad·vi·see** (ăd-vī′zē) *n.* One that is advised.

**ad·vise·ment** (ăd-vīz′mənt) *n.* Careful consideration.

**ad·vis·er** *also* **ad·vi·sor** (ăd-vī′zər) *n.* **1.** One who advises. **2.** A person who offers advice, esp. professionally or officially. **3.** A teacher who advises students in academic and personal matters.

**ad·vi·so·ry** (ăd-vī′zə-rē) *adj.* **1.** Empowered to advise <an *advisory* council> **2.** Of, relating to, or containing advice <an *advisory* memo> —*n., pl.* **-ries.** A report, as on the weather, giving information and esp. a warning.

**ad·vo·ca·cy** (ăd′və-kə-sē) *n.* Active support, as of a cause.

**ad·vo·cate** (ăd′və-kāt′) *vt.* **-cat·ed, -cat·ing, -cates.** [< ME *advocat,* lawyer < OFr. *avocat* < Lat. *advocatus,* p.part. of *advocare,* to summon for counsel : *ad-,* to + *vocare,* to call.] To speak in favor of : RECOMMEND. —*n.* (-kĭt, -kāt′). **1.** One who supports or defends a cause <an *advocate* of equal rights> **2.** One who pleads in another's behalf, esp. a lawyer. **—ad′vo·ca′tor** *n.*

**ad·vow·son** (ăd-vou′zən) *n.* [ME *avouson* < OFr. *avoeson* < Med. Lat. *advocatia* < Lat. *advocatio,* a summoning < *advocare,* to summon. —see ADVOCATE.] The right to present a vacant benefice in English ecclesiastical law.

**ad·y·tum** (ăd′ĭ-təm) *n., pl.* **-ta** (-tə) [Lat. < Gk. *aduton* < *adutos,* not to be entered : *a-,* not + *duein,* to enter.] The sanctum in an ancient temple.

**adz** or **adze** (ădz) *n.* [ME *adese* < OE *adesa.*] An axlike tool with a curved blade at right angles to the handle, used for shaping wood.

**ad·zu·ki bean** (ăd-zo͞o′kē) *n.* [J. *azuki,* red bean.] A plant, *Phaseolus angularis,* with yellow flowers and pods yielding edible seeds, widely cultivated as a food crop in the Orient.

**ae·ci·o·spore** (ē′sē-ə-spôr′, -spōr′, -shē-) *n.* [AECI(UM) + SPORE.] A rust spore formed in a chainlike series in an aecium.

**ae·ci·um** (ē′sē-əm, ē′shē-) *n., pl.* **-ci·a** (-sē-ə, -shē-ə) [NLat. < Gk. *aikia,* injury < *aeikēs,* injurious.] A cuplike structure in rust fungi containing chains of aeciospores. **—ae′ci·al** (ē′sē-əl, -shē-) *adj.*

**a·e·des** (ā-ē′dēz) *n., pl.* **aedes.** [NLat. *Aedes,* genus name < Gk. *aēdēs,* unpleasant : *a-,* not + *ēdos,* pleasant.] A mosquito of the genus *Aëdes,* as *A. aegypti,* that transmits yellow fever and dengue.

**ae·dile** (ē′dīl′) *n.* [Lat. *aedilis* < *aedes,* house.] A magistrate in ancient Rome in charge of public works and games, police, and the grain supply.

**Ae·ge·an** (ĭ-jē′ən) *adj.* Of, relating to, or designating the Bronze Age civilization that flourished in the Aegean area.

**Ae·geus** (ē′jo͞os, ē′jē-əs) *n.* [Lat. < Gk. *Aigeus.*] *Gk. Myth.* A king of Athens and the father of Theseus.

**Aeg·ir** (ăg′ər) *n.* [ON.] *Norse Myth.* The god of the sea.

**ae·gis** (ē′jĭs) *n.* [Lat. < Gk. *aigis.*] **1.** *Gk. Myth.* The shield of Zeus, lent by him to Athena. **2.** Protection. **3.** Sponsorship.

**Ae·gis·thus** (ĭ-jĭs′thəs) *n.* [Lat. < Gk. *Aigisthos.*] *Gk. Myth.* The son of Thyestes and lover of Clytemnestra.

**-aemia** *suff. var. of* -EMIA.

**Ae·ne·as** (ĭ-nē′əs) *n.* [Lat. < Gk. *Aineias.*] The Trojan hero of Virgil's *Aeneid,* son of Anchises and Aphrodite and reputed ancestor of the Romans, who escaped the pillage of Troy and wandered for seven years before settling in Italy.

**a·e·ne·ous** or **a·e·ne·us** (ā-ē′nē-əs) *adj.* [Lat. *aeneus,* of bronze < *aes,* bronze.] Having a brassy or golden-green color.

**Ae·o·li·an** (ē-ō′lē-ən, -ōl′yən) *adj.* **1.** Of or relating to Aeolis or its people. **2.** *Gk. Myth.* Of or relating to Aeolus. **3. aeolian.** *var. of* EOLIAN. —*n.* **1.** A member of one of the major Greek tribes that settled in central Greece, Lesbos, and Aeolis. **2.** Aeolic.

**Aeolian harp** *n.* A boxlike musical instrument having stretched strings that sound when wind passes over them.

**Ae·ol·ic** (ē-ŏl′ĭk) *n.* A group of dialects of ancient Greek spoken by the Aeolians.

**ae·ol·i·pile** (ē-ŏl′ə-pīl′) *n.* [< Lat. *Aeoli pylae,* gates of Aeolus : *Aeolus,* Aeolus + *pylae,* gates < Gk. *pulē,* gate.] An ancient prototypal steam engine consisting of a cylindrical or spherical vessel fitted with circumferential exhaust jets and mounted to permit free rotation about the steam inlet axis.

**Ae·o·lus** (ē'ə-ləs) *n.* [Lat. < Gk. *Aiolos* < *aiolos,* nimble.] **1.** *Gk. Myth.* The god of the winds. **2.** A king of Thessaly and ancestor of the Aeolians.

**ae·on** (ē'ŏn', ē'ən) *n. var. of* EON.

**ae·o·ni·an** (ē-ō'nē-ən) *adj. var. of* EONIAN.

**ae·py·or·nis** (ē'pē-ôr'nĭs) *n.* [NLat. *Aepyornis,* genus name : Gk. *aipys,* high + Gk. *ornis,* bird.] Any of an extinct genus of large flightless birds of Madagascar.

**ae·quor·in** (ē-kwôr'ĭn, ē-kwŏr'-) *n.* [NLat. *Aequorea,* jellyfish genus + -IN.] A protein secreted by jellyfish that interacts with seawater to produce bioluminescent light.

**aer–** *pref. var. of* AERO–.

**aer·ate** (âr'āt') *vt.* **-at·ed, -at·ing, -ates. 1.** To supply or charge (liquid) with a gas, esp. to charge with carbon dioxide. **2.** To expose to the circulation of air for purification. **3.** To supply (blood) with oxygen.

**aer·a·tor** (âr'ā'tər) *n.* One that aerates.

**aer·i·al** (âr'ē-əl, ā-ĭr'ē-əl) *adj.* [< Lat. *aerius* < Gk. *aerios* < *aēr,* air.] **1.** Of, in, or caused by the air. **2.** Inhabiting the air. **3.** Imposingly high : LOFTY. **4.** Suggestive of air. **5.** Imaginary : unsubstantial. **6.** Of, for, or by aircraft. **7.** *Bot.* Growing in the air rather than in soil or under water <*aerial* roots> —*n.* (âr'ē-əl). An antenna.

**aer·i·al·ist** (âr'ē-ə-lĭst) *n.* An acrobat who performs high above the ground, as on a tightrope or trapeze.

**aerial ladder** *n.* An extensible ladder, esp. one mounted on a fire engine.

**aer·ie** *also* **aer·y** (âr'ē, ăr'ē, ĭr'ē) *n., pl.* **-ies.** [Med. Lat. *aeria* < OFr. *aire.*] **1.** The nest of a predatory bird, as an eagle, built on a high place. **2.** A house or stronghold built on a height.

**aero–** *or* **aer–** *pref.* [ME < OFr. < Lat. < Gk. < *aēr,* air.] **1. a.** Air : atmosphere <*aero*ballistics> **b.** Gas <*aerosol*> **2.** Aviation <*aero*nautics>

**aer·o·bal·lis·tics** (âr'ō-bə-lĭs'tĭks) *n. (sing. in number).* Ballistics, esp. of missiles. **—aer'o·bal·lis'tic** *adj.*

**aer·o·bat·ics** (âr'ə-băt'ĭks) *n. (sing. or pl. in number).* [AERO- + (ACRO)BATICS.] Performance of stunts, as rolls and loops, with an aircraft.

**aer·obe** (âr'ōb') *n.* [Fr. *aérobie* : Gk. *aēr,* air + Gk. *bios,* life.] An organism, as a bacterium, requiring molecular oxygen or air to live.

**aer·o·bic** (â-rō'bĭk) *adj.* **1.** Living or occurring only in the presence of oxygen. **2.** Of or relating to aerobics. **3.** Of or relating to aerobes.

**aer·o·bics** (â-rō'bĭks) *n.* [< AEROBIC.] *(sing. or pl. in number).* **1.** Conditioning of the cardiopulmonary system by means of vigorous exercise that seeks to increase efficiency of oxygen intake. **2.** Aerobic exercises, as running or calisthenics.

**aer·o·bi·ol·o·gy** (âr'ō-bī-ŏl'ə-jē) *n.* The branch of biology dealing with the atmospheric dispersion of materials of biologic significance, as microorganisms or pollen. **—aer'o·bi·o·log'i·cal** (-ə-lŏj'ĭ-kəl) *adj.* **—aer'o·bi·o·log'i·cal·ly** *adv.*

**aer·o·bi·um** (â-rō'bē-əm) *n.* [NLat. < AEROBE.] An aerobe.

**aer·o·drome** (âr'ə-drōm') *n. Chiefly Brit. var. of* AIRDROME.

**aer·o·dy·nam·ics** (âr'ō-dī-năm'ĭks) *n. (sing. in number).* The dynamics of gases, esp. of atmospheric interactions with moving objects. **—aer'o·dy·nam'ic** *adj.*

**aer·o·dyne** (âr'ə-dīn') *n.* [AERO- + Gk. *dunamis,* power < *dunasthai,* to be able.] A heavier-than-air aircraft that derives its lift from motion.

**aer·o·em·bo·lism** (âr'ō-ĕm'bə-lĭz'əm) *n.* **1.** The presence of air bubbles in the heart or blood vessels, often resulting from a neck wound. **2.** Caisson disease.

**aer·o·foil** (âr'ə-foil) *n. Chiefly Brit. var. of* AIRFOIL.

**aer·o·gram** *also* **aer·o·gramme** (âr'ə-grăm') *n.* AIR LETTER 2.

**aer·o·lite** (âr'ə-līt') *also* **aer·o·lith** (-lĭth') *n.* A chiefly silicious meteorite. **—aer'o·lit'ic** (-lĭt'ĭk) *adj.*

**aer·ol·o·gy** (â-rŏl'ə-jē) *n.* Total atmospheric meteorology as opposed to surface-based study. **—aer'o·log'ic** (âr'ə-lŏj'ĭk), **aer'o·log'i·cal** *adj.* **—aer·ol'o·gist** *n.*

**aer·o·mag·net·ics** (âr'ō-măg-nĕt'ĭks) *n. (sing. in number).* The science of magnetic characteristics associated with atmospheric conditions. **—aer'o·mag·net'ic** *adj.* **—aer'o·mag·net'i·cal·ly** *adv.*

**aer·o·me·chan·ics** (âr'ō-mə-kăn'ĭks) *n. (sing. in number).* The science of the motion and equilibrium of air and other gases, comprising aerodynamics and aerostatics. **—aer'o·me·chan'i·cal** *adj.* **—aer'o·me·chan'i·cal·ly** *adv.*

**aer·o·med·i·cine** (âr'ō-mĕd'ĭ-sĭn) *n.* The medical study and treatment of disturbances, disorders, and diseases resulting from or associated with atmospheric flight. **—aer'o·med'i·cal** *adj.*

**aer·o·me·te·or·o·graph** (âr'ō-mē'tē-ôr'ə-grăf', -ŏr'-) *n.* An aircraft instrument for simultaneously recording temperature, atmospheric pressure, and humidity.

**aer·om·e·ter** (â-rŏm'ĭ-tər) *n.* A device for determining the weight and density of air or other gas.

**aer·o·naut** (âr'ə-nôt') *n.* [AERO- + Gk. *nautēs,* sailor.] A pilot or navigator of a balloon or lighter-than-air craft.

**aer·o·nau·tics** (âr'ə-nô'tĭks) *n. (sing. in number).* **1.** Design and construction of aircraft. **2.** Aircraft navigation. **—aer'o·nau'tic, aer'o·nau'ti·cal** *adj.* **—aer'o·nau'ti·cal·ly** *adv.*

**aer·o·neu·ro·sis** (âr'ō-nŏŏ-rō'sĭs, -nyŏŏ-) *n.* Nervous exhaustion from prolonged piloting of aircraft.

**aer·on·o·my** (â-rŏn'ə-mē) *n.* The study of the upper atmosphere, esp. of regions of ionized gas.

**aer·o·pause** (âr'ō-pôz') *n.* The region of the atmosphere above which aircraft cannot fly.

**aer·o·pha·gia** (âr'ə-fā'jə) *n.* Abnormal spasmodic swallowing of air, esp. as a symptom of hysteria.

**aer·o·pho·bi·a** (âr'ə-fō'bē-ə) *n.* Abnormal fear of air, esp. drafts.

**aer·o·phore** (âr'ə-fôr', -fōr') *n.* A device to supply air to a non-breathing infant or to a person in an anaerobic environment, as a closed mine or an underwater area.

**aer·o·phyte** (âr'ə-fīt') *n. Bot.* An epiphyte.

**aer·o·plane** (âr'ə-plān') *n. Chiefly Brit. var. of* AIRPLANE.

**aer·o·shell** (âr'ə-shĕl') *n.* A protective all-covering shell for a spacecraft re-entering the atmosphere from space at high speeds.

**aer·o·sol** (âr'ə-sôl', -sōl') *n.* [AERO- + SOL(UTION).] **1.** A gaseous suspension of fine particles. **2. a.** A substance, as a detergent, insecticide, or paint, packaged under pressure in a dispenser. **b.** An aerosol bomb.

**aerosol bomb** *n.* A usu. hand-held container or dispenser from which an aerosol is released.

**aer·o·space** (âr'ō-spās') *adj.* **1.** Of or designating the earth's atmosphere and the space beyond. **2.** Of or relating to the science or technology of flight. **—aer'o·space'** *n.*

**aer·o·sphere** (âr'ō-sfîr') *n.* The lower part of the atmosphere in which unmanned and manned flight is possible.

**aer·o·stat** (âr'ō-stăt') *n.* [Fr. *aérostat* : Gk. *aēr,* air + Gk. *statos,* standing.] An aircraft, esp. a balloon or dirigible, deriving its lift from the buoyancy of surrounding air rather than from aerodynamic motion. **—aer'o·stat'ic, aer'o·stat'i·cal** *adj.*

**aer·o·stat·ics** (âr'ō-stăt'ĭks) *n. (sing. in number).* The science of gases in equilibrium and of the equilibrium of balloons or aircraft under changing atmospheric flight conditions.

**aer·o·ther·mo·dy·nam·ics** (âr'ō-thûr'mō-dī-năm'ĭks) *n. (sing. in number).* The study of the thermodynamics of gases, esp. at high relative velocities.

**aer·y**[1] (âr'ē, ā'ə-rē) *adj.* **-i·er, -i·est.** Ethereal.

**aer·y**[2] (âr'ē, ăr'ē, ĭr'ē) *n. var. of* AERIE.

**Aes·cu·la·pi·an** (ĕs'kyə-lā'pē-ən) *adj.* [After AESCULAPIUS.] Of or relating to medicine or the art of healing.

**Aes·cu·la·pi·us** (ĕs'kyə-lā'pē-əs) *n.* [Lat. < Gk. *Asklēpios.*] *Rom. Myth.* The god of medicine and healing.

**Ae·sir** (â'sîr') *pl.n.* [ON, pl. of *āss,* god.] *Norse Myth.* The gods.

**Ae·so·pi·an** (ē-sō'pē-ən) *also* **Ae·sop·ic** (ē-sŏp'ĭk) *adj.* **1.** Characteristic of Aesop's animal fables. **2.** Veiled in allegorical suggestions, hints, and euphemisms so as to elude political censorship <"they could express their views only in a diluted form, resorting to *Aesopian* hints and allusions" —Isaac Deutscher>

**aes·the·sia** (ĕs-thē'zhə) *n.* [Back-formation < ANESTHESIA.] The ability to perceive or feel.

**aes·thete** (ĕs'thēt') *n.* [Back-formation < AESTHETIC.] **1.** One who cultivates a superior appreciation of the beautiful. **2.** One whose pursuit and admiration of beauty is considered affected or excessive.

**aes·thet·ic** (ĕs-thĕt'ĭk) *adj.* [G. *ästhetisch* < NLat. *aestheticus* < Gk. *aisthētikos,* of sense perception < *aisthēta,* perceptible things < *aisthenasthai,* to perceive.] **1.** Of or relating to aesthetics. **2. a.** Of or relating to the sense of the beautiful. **b.** Artistic <an *aesthetic* failure> **3. a.** Having a love of beauty. **b.** *Informal.* Being in accordance with accepted notions of good taste. **—aes·thet'i·cal·ly** *adv.*

**aes·the·ti·cian** (ĕs'thĭ-tĭsh'ən) *n.* A critic concerned with the theory of beauty and the fine arts.

**aes·thet·i·cism** (ĕs-thĕt'ĭ-sĭz'əm) *n.* **1.** The pursuit of the beautiful. **2. a.** The belief that beauty is the basic principle from which all other principles are derived. **b.** A doctrine whereby art and artists are thought to have no obligation or responsibility other than that of striving for beauty.

**aes·thet·ics** (ĕs-thĕt'ĭks) *n. (sing. in number).* **1.** The branch of philosophy that provides a theory of the beautiful and of the fine arts. **2.** The theories and descriptions of the psychological response to beauty and artistic experiences. **3.** The branch of metaphysics concerned with the laws of perception in the philosophy of Kant.

**aes·ti·val** (ĕs'tə-vəl) *adj.* [ME *estival* < OFr. < Lat. *aestivalis* < *aestivus* < *aestas,* summer.] Of, relating to, or appearing in summer.

**aes·ti·vate** (ĕs'tə-vāt') *vi.* **-vat·ed, -vat·ing, -vates.** [Lat. *aestivare* < *aestivus,* summery. —see AESTIVAL.] **1.** *Zool.* To pass the summer, esp. in a state of dormancy. **2.** To spend the summer.

**aes·ti·va·tion** (ĕs'tə-vā'shən) *n.* **1.** The act of spending the summer. **2.** *Zool.* A state of dormancy during the summer or periods of drought. **3.** *Bot.* The arrangement of floral organs, as petals and sepals, in the unopened bud.

**Ae·ther** (ē'thər) *n.* [Lat. < Gk. *aithēr,* upper air.] *Gk. Myth.* The poetic personification of the upper air breathed by the Olympians.

**ae·ti·ol·o·gy** (ē'tē-ŏl'ə-jē) *n. var. of* ETIOLOGY.

**af-** *pref. var. of* AD- 1. —Used before *f.*

**a·far** (ə-fär') *adv.* [ME *afer* < *on fer,* far, and *of fer,* from afar < OE *feor,* far.] **1.** From a distance <traveling from *afar*> **2.** At or to a distance : far away <heard the sirens *afar* off>

**†a·feard** *also* **a·feared** (ə-fîrd') *adj.* [ME *afered* < OE *āfǣred,* p.part. of *āfǣran,* to frighten : *ā-* (intensive) + *fǣran,* to frighten < *fǣr,* fear.] *Regional & Archaic.* Afraid.

**a·fe·brile** (ā-fē'brəl, ā-fĕb'rəl) *adj.* Having no fever.

**af·fa·ble** (ăf'ə-bəl) *adj.* [OFr. < Lat. *affabilis* < *affari,* to speak to : *ad-,* to + *fari,* to speak.] **1.** Easy to converse with : AMIABLE. **2.** Marked by gentleness or graciousness <an *affable* manner> **—af'fa·bil'i·ty** *n.* **—af'fa·bly** *adv.*

**af·fair** (ə-fâr') *n.* [ME *afere* < OFr. *afaire* < *à faire,* to do.] **1.** Something done or to be done. **2. affairs.** Business transactions. **3. a.** An occurrence, event, or matter. **b.** An object or device <The couch was a shabby *affair.*> **c. affairs.** Personal belongings. **4.** A private matter. **5.** A matter causing scandal and controversy. **6.** A love affair, esp. a brief one.

**af·fect¹** (ə-fĕkt') *vt.* **-fect·ed, -fect·ing, -fects.** [Lat. *afficere, affect-* : *ad-,* to + *facere,* to do.] **1.** To bring about a change in : INFLUENCE. ***usage:*** As a verb *affect* is most commonly used in the sense "to influence" (*how smoking affects the health*), whereas *effect* means "to bring about" (*layoffs designed to effect savings*). **2.** To move emotionally. **3.** To attack or infect, as a disease. *—n.* (ăf'ĕkt'). **1.** *Psychol.* **a.** A feeling or emotion as distinguished from cognition, thought, or action. **b.** A strong feeling having active consequences. **2.** *Obs.* A disposition, feeling, or tendency. **—af'fect'less** *adj.*

☆ ***syns:*** AFFECT, GET TO, IMPACT, IMPRESS, INFLUENCE, MOVE, STRIKE, SWAY, TOUCH *v. core meaning* : to evoke a usu. strong mental or emotional response from <a play that deeply *affected* the audience><a generation *affected* by war>

**af·fect²** (ə-fĕkt') *vt.* **-fect·ed, -fect·ing, -fects.** [ME *affecten* < Lat. *affectare,* to strive after, freq. of *afficere,* to affect, influence.] **1.** To simulate or imitate in order to make a desired impression : ASSUME. **2. a.** To display a preference for <*affects* flashy clothes> **b.** *Archaic.* To fancy : love. **c.** To tend to assume by nature <*affect* crystalline form> **—af·fect'er** *n.*

☆ ***syns:*** AFFECT, ASSUME, COUNTERFEIT, FAKE, FEIGN, PRETEND, PUT ON, SIMULATE *v. core meaning* : to take on a false appearance of <*affected* concern but really did not care>

**af·fec·ta·tion** (ăf'ĕk-tā'shən) *n.* [Lat. *affectatio* < *affectare,* to strive after. —see AFFECT².] **1.** A show, pretense, or display. **2.** Artificial behavior designed to impress others.

**af·fect·ed¹** (ə-fĕk'tĭd) *adj.* **1.** Acted upon or influenced. **2.** Moved emotionally. **3.** Attacked by disease : INFECTED.

**af·fect·ed²** (ə-fĕk'tĭd) *adj.* **1.** Assumed or simulated to impress others. **2.** Speaking or behaving unnaturally to make an impression. **3.** Inclined : disposed. **—af·fect'ed·ly** *adv.* **—af·fect'ed·ness** *n.*

**af·fect·ing** (ə-fĕk'tĭng) *adj.* Evoking a usu. strong emotional response <an *affecting* scene of farewell> **—af·fect'ing·ly** *adv.*

**af·fec·tion** (ə-fĕk'shən) *n.* [ME *affeccioun* < OFr. *affection* < Lat. *affectio* < *afficere,* to affect, influence.] **1.** A fond or tender feeling toward another. **2.** *often* **affections.** Emotion. **3.** A pathological mental or physical condition. **4.** The act of influencing or acting on. **5.** The state of being influenced or acted on. **6.** An attribute. **7.** Mental tendency or disposition. **—af·fec'tion·al** *adj.* **—af·fec'tion·al·ly** *adv.*

**af·fec·tion·ate** (ə-fĕk'shə-nĭt) *adj.* **1.** Having or displaying affection. **2.** *Obs.* Strongly or favorably disposed. **—af·fec'tion·ate·ly** *adv.* **—af·fec'tion·ate·ness** *n.*

**af·fec·tive** (ə-fĕk'tĭv) *adj.* **1.** *Psychol.* Relating to or resulting from emotions or feelings rather than from thought. **2.** Relating to or evoking affection or emotion.

**af·fen·pin·scher** (ăf'ən-pĭn'chər) *n.* [G. *Affe,* monkey + *Pinscher,* terrier.] Any of a breed of small dogs of European origin, having dark, wiry, shaggy hair and a tufted muzzle.

**affenpinscher**
*10 inches high at shoulder*

**af·fer·ent** (ăf'ər-ənt) *adj.* [Lat. *afferens, afferent-,* pr.part. of *afferre,* to bring toward : *ad-,* toward + *ferre,* to bring.] Directed toward a central section or organ, as nerves that conduct impulses from the periphery of the body inward to the spinal cord.

**af·fi·ance** (ə-fī'əns) *vt.* **-anced, -anc·ing, -anc·es.** [OFr. *afiancer* < *affier,* to trust to < Med. Lat. *affidare* : Lat. *ad-,* to + Lat. *fidus,* faithful.] To betroth.

**af·fi·ant** (ə-fī'ənt) *n.* [OFr. pr.part. of *affier,* to trust to. —see AFFIANCE.] *Law.* One who makes an affidavit.

**af·fi·da·vit** (ăf'ĭ-dā'vĭt) *n.* [Med. Lat. *affidavit,* he has pledged < *affidare,* to pledge. —see AFFIANCE.] *Law.* A written declaration made under oath before an official, as a notary public.

**af·fil·i·ate** (ə-fĭl'ē-āt') *v.* **-at·ed, -at·ing, -ates.** [Med. Lat. *affiliare, affiliat-,* to adopt : *ad-,* to + *filius,* son.] *—vt.* **1.** To accept as a subordinate associate. **2.** To associate (oneself) with. **3.** To adopt as one's own child. **4.** *Law.* **a.** To determine the paternity of (an illegitimate child). **b.** To refer an illegitimate child to (its father). *—vi.* To associate or connect oneself <doctors who decided to *affiliate*> *—n.* (-ĭt). An associate or subordinate. **—af·fil'i·a'tion** *n.*

**af·fine** (ə-fīn') *adj.* [OFr. *affin,* closely related. —see AFFINED.] **1.** Of or pertaining to a mathematical transformation of coordinates that is equivalent to a translation, contraction, or expansion with respect to a fixed origin and fixed coordinate system. **2.** Of or relating to the geometry of affine transformations.

**af·fined** (ə-fīnd') *adj.* [Fr. *affiné* < OFr. *affin,* closely related < Lat. *affinis,* related by marriage : *ad-,* to + *fīnis,* boundary.] *Archaic.* **1.** Joined by kinship or affinity. **2.** Beholden : bound.

**af·fin·i·ty** (ə-fĭn'ĭ-tē) *n., pl.* **-ties.** [ME *affinite* < OFr. *afinite* < Lat. *affinitas* < *affinis,* related by marriage. —see AFFINED.] **1.** A natural personal attraction. **2.** Relationship by marriage. **3.** An inherent similarity between things. **4.** A resemblance or relationship between biological species implying a common origin. **5.** A chemical attraction or force that causes the atoms of certain elements to combine with atoms of another element and remain in the combined state.

☆ ***syns:*** AFFINITY, ATTRACTION, SYMPATHY *n. core meaning* : the relationship between individuals or things that are naturally or involuntarily drawn together <had an *affinity* for politics><a natural *affinity* for little children>

**af·firm** (ə-fûrm') *v.* **-firmed, -firm·ing, -firms.** [ME *affermen* < OFr. *afermer* < Lat. *affirmare* : *ad-,* to + *firmare,* to strengthen < *firmus,* strong.] *—vt.* **1.** To declare or maintain to be true. **2.** To confirm or ratify. *—vi. Law.* To declare formally and solemnly but not under oath. **—af·firm'a·ble** *adj.* **—af·firm'a·bly** *adv.* **—af·fir'mant** *adj. & n.* **—af'fir·ma'tion** (ăf'ər-mā'shən) *n.* **—af·firm'er** *n.*

**af·fir·ma·tive** (ə-fûr'mə-tĭv) *adj.* **1.** Affirming that something is true. **2.** *Logic.* Denoting a proposition in which the predicate states something about the subject to be true, as *apples have seeds.* *—n.* **1.** An affirmative word or phrase. **2.** The side in a debate that upholds a proposition. **—af·fir'ma·tive·ly** *adv.*

**affirmative action** *n.* Action taken to provide equal opportunity, as in admissions or employment, for minority groups and women.

**af·fix** (ə-fĭks') *vt.* **-fixed, -fix·ing, -fix·es.** [Med. Lat. *affiare,* freq. of Lat. *affigere* : *ad-,* to + *figere,* to fasten.] **1.** To attach <*affix* a heel to a shoe> **2.** To attribute or impute, as blame. **3.** To put at the end : append, as a postscript. *—n.* (ăf'ĭks'). **1.** Something added or attached. **2.** A word element, as a prefix or suffix, that can only occur as an attachment to a base, stem, or root. **—af·fix'a·ble** *adj.* **—af·fix'er** *n.*

**af·fla·tus** (ə-flā'təs) *n.* [Lat., p.part. of *afflare,* to breathe on : *ad-,* toward + *flare,* to blow.] A creative impulse : INSPIRATION.

**af·flict** (ə-flĭkt') *vt.* **-flict·ed, -flict·ing, -flicts.** [ME *afflighten* < Lat. *affligere,* to cast down : *ad-,* to + *fligere,* to strike.] To inflict physical or mental suffering on. **—af·flict'er** *n.* **—af·flic'tive** *adj.* **—af·flic'tive·ly** *adv.*

☆ ***syns:*** AFFLICT, AGONIZE, CURSE, EXCRUCIATE, PLAGUE, RACK, SCOURGE, STRIKE, TORMENT, TORTURE *v. core meaning* : to bring great harm or suffering to <a population *afflicted* with typhus>

**af·flic·tion** (ə-flĭk'shən) *n.* **1.** A condition of pain, suffering, or distress. **2.** A cause of pain, suffering, or distress.

**af·flu·ence** (ăf'lo͞o-əns) *n.* **1.** A plentiful supply of material goods : WEALTH. **2.** A plentiful supply : ABUNDANCE. **3.** A flowing toward.

**af·flu·en·cy** (ăf'lo͞o'ən-sē, ə-flo͞o'-) *n.* Affluence.

**af·flu·ent** (ăf'lo͞o-ənt) *adj.* [ME, abundant < OFr. < Lat. *affluens,* pr.part. of *affluere,* to abound in : *ad-,* to + *fluere,* to flow.] **1.** Wealthy. **2.** Abundant. **3.** Flowing freely. *—n.* A river or stream that flows into other water : TRIBUTARY. **—af'flu·ent·ly** *adv.*

**af·flux** (ăf'lŭks') *n.* [Lat. *affluxus,* p.part. of *affluere,* to flow to.—see AFFLUENT.] A flowing to or toward a particular area.

**af·ford** (ə-fôrd', ə-fōrd') *vt.* **-ford·ed, -ford·ing, -fords.** [ME *aforthen* < OE *geforðian,* to carry out < *forðian,* to further < *forð,* forward.] **1.** To have the financial means for. **2.** To be able to spare or give up <could *afford* two days for traveling> **3.** To be able to do with benefit or without harm <can *afford* to be honest><can't *afford* to lose sleep> **4.** To provide <a hobby that *affords* pleasure> **—af·ford'a·ble** *adj.*

**af·for·est** (ə-fôr'ĭst, ə-fŏr'-) *vt.* **-est·ed, -est·ing, -ests.** [Med. Lat.

*afforestare* : *ad-*, to + *forestare* < *foresta*, forest.] To convert (open land) into forest. **—af·for'es·ta'tion** (ə-fôr'ĭ-stā'shən, ə-fŏr'-) *n.*

**af·fran·chise** (ə-frăn'chīz') *vt.* **-chised, -chis·ing, -chis·es.** [ME < OFr. *afranchir, afranchiss-* : *a-*, to (< Lat. *ad-*) + *franchir*, to free < *franc*, free. —see FRANK.] To free, as from servitude or liabilities.

**af·fray** (ə-frā') *n.* [ME < OFr. *esfrei* < *esfreer*, to disturb.] A noisy brawl or quarrel. *—vt.* **-frayed, -fray·ing, -frays.** *Archaic.* To frighten.

**af·fri·cate** (ăf'rĭ-kĭt) *n.* [Lat. *affricare, affricat-*, to rub against : *ad-*, to + *fricare*, to rub.] A speech sound produced when the breath stream is completely stopped and then released at articulation, as the *t* plus *sh* sound in *clutch* or the *j* sound in *judge*.

**af·fric·a·tive** (ə-frĭk'ə-tĭv) *adj.* Of, relating to, or forming an affricate. **—af·fric'a·tive** *n.*

**af·fright** (ə-frīt') [ME *afrighten* < OE *afyrhtan.*] *Archaic.* *—vt.* **-fright·ed, -fright·ing, -frights.** To arouse fear in : TERRIFY. *—n.* **1.** Terror. **2.** A cause of terror. **3.** The act of frightening. **—af·fright'ment** *n.*

**af·front** (ə-frŭnt') *vt.* **-front·ed, -front·ing, -fronts.** [ME *afrounten* < OFr. *afronter* : Lat. *ad-*, to + Lat. *frons*, face.] **1.** To insult deliberately, esp. to the face : OFFEND. **2.** To meet face to face defiantly : CONFRONT. *—n.* **1.** A deliberate insult or offense. **2.** *Obs.* An encounter or meeting.

**af·fu·sion** (ə-fyōō'zhən) *n.* [LLat. *adfusio*, < *affundere*, to pour on : *ad-*, to + *fundere*, to pour.] A pouring on of liquid, as in baptism.

**Af·ghan** (ăf'găn', -gən) *n.* [Pashto *afghānī.*] **1.** A native of Afghanistan. **2.** Pashto. **3. afghan.** A wool coverlet knitted or crocheted in colorful geometric designs. **4.** An Afghan hound. *—adj.* Of or relating to Afghanistan, its people, or their language.

**Afghan hound** *n.* A large, slender dog of an ancient breed, with drooping ears, a pointed muzzle, and long, thick hair.

**af·ghan·i** (ăf-găn'ē, -gä'nē) *n.* [Pashto.] —See table at CURRENCY.

**a·fi·cio·na·do** (ə-fĭsh'ē-ə-nä'dō, ə-fĭs'ē-, ə-fē'sē-) *n.*, *pl.* **-dos.** [Sp. < p.part. of *aficionar*, to induce a liking for < *aficion*, liking < Lat. *affectio*, affection.] An enthusiastic admirer or follower : FAN.

**a·field** (ə-fēld') *adv.* **1.** Off the usual or desired track : ASTRAY. **2.** Away from one's usual environment : ABROAD.

**a·fire** (ə-fīr') *adj.* & *adv.* **1.** On fire. **2.** Intensely interested.

**a·flame** (ə-flām') *adj.* & *adv.* Afire.

**a·float** (ə-flōt') *adj.* & *adv.* **1.** Floating. **2.** At sea. **3.** In circulation : PREVAILING. **4.** Flooded. **5.** Moving without guidance : ADRIFT. **6.** Free of financial difficulty <tried to keep the company *afloat*>

**a·flut·ter** (ə-flŭt'ər) *adj.* **1.** Being in a flutter **2.** Filled with nervous excitement.

**a·foot** (ə-fŏŏt') *adj.* & *adv.* **1.** On foot. **2.** In progress.

**a·fore** (ə-fôr', ə-fōr') *adv.* & *prep.* & *conj.* [ME < OE *onforan* : *on*, at + *for*, fore.] *Archaic.* Before.

**a·fore·men·tioned** (ə-fôr'mĕn'shənd, ə-fōr'-) *adj.* Mentioned earlier or above.

**a·fore·said** (ə-fôr'sĕd', ə-fōr'-) *adj.* Spoken of earlier.

**a·fore·thought** (ə-fôr'thôt', ə-fōr'-) *adj.* Premeditated <The crime was commited with malice *aforethought.*>

**a·fore·time** (ə-fôr'tīm', ə-fōr'-) *Archaic.* *—adv.* At a former time : PREVIOUSLY. *—adj.* Earlier : former.

**a for·ti·o·ri** (ä fôr'tē-ôr'ē, ā fōr'tē-ō'rī') *adv.* [Lat.] For a stronger reason. —Used of a conclusion arrived at with greater logical necessity than another.

**a·foul of** (ə-foul') *prep.* In or into a condition of entanglement, collision, or conflict. **—run** (*or* **fall**) **afoul of.** To become entangled or in conflict with <*ran afoul of* the law>

**a·fraid** (ə-frād') *adj.* [ME *affraied*, p.part. of *affraien*, to frighten < OFr. *esfreer*, to disturb.] **1.** Filled with fear. **2.** Filled with dislike <Are you *afraid* of hard work?> **3.** Filled with regret <I'm *afraid* I have bad news.>

☆ ***syns:*** AFRAID, APPREHENSIVE, FEARFUL, FRIGHTENED, PETRIFIED, SCARED *adj. core meaning* : filled with fear <*afraid* of snakes><saw the mugger and was *afraid*>

**A-frame** (ā'frām') *n.* [From its being shaped like a capital *A.*] A structure, as a house, with steeply angled sides and a roof that reaches to the ground.

**af·reet** *also* **af·rit** (ăf'rēt', ə-frēt') *n.* [Ar. *'ifrīt.*] A gigantic and monstrous demon or powerful evil jinni in Arabic mythology.

**a·fresh** (ə-frĕsh') *adv.* Once more : AGAIN <start *afresh*>

**Af·ri·can** (ăf'rĭ-kən) *adj.* Of or relating to Africa or any of its peoples and languages. *—n.* **1.** A person born or living in Africa. **2.** A member of one of the indigenous peoples of Africa.

**African lily** *n.* A plant indigenous to southern Africa, *Agapanthus africanus*, bearing rounded blue, violet, or white flower clusters.

**African mahogany** *n.* **1. a.** An African tree of the genus *Khaya*, esp. *K. ivorensis*, with wood similar to that of true mahogany. **b.** The wood of African mahogany, used for furniture, musical instruments, and boat interiors. **2.** An African tree with wood resembling true mahogany.

**African marigold** *n.* A widely cultivated plant, *Tagetes erecta*, native to Mexico, with finely divided foliage and rounded yellow or orange flowers.

**African sleeping sickness** *n.* Sleeping sickness.

**African violet** *n.* A plant of the genus *Saintpaulia*, native to tropical Africa and widely cultivated as a house plant, esp. *S. ionantha*, bearing violet, white, or pink flowers.

**Af·ri·kaans** (ăf'rĭ-käns', -känz') *n.* [Afr. < Du. *Afrikaansch*, African.] A language that developed from 17th-cent. Dutch and is an official language of the Republic of South Africa.

**Af·ri·ka·ner** (ăf'rĭ-kä'nər) *n.* [Afr., African < Lat. *Africanus.*] An Afrikaans-speaking descendant of the Dutch settlers of South Africa.

**af·rit** (ăf'rēt', ə-frēt') *n. var. of* AFREET.

**Af·ro** (ăf'rō) *n.*, *pl.* **-ros.** [Perh. short for AFRO-AMERICAN.] A rounded, bushy hair style. *—adj.* **1.** Of or for an Afro hair style. **2.** African in style.

**Afro-** *pref.* [< Lat. *Afer*, an African.] African <*Afro*-Asiatic>

**Af·ro-A·mer·i·can** (ăf'rō-ə-mĕr'ĭ-kən) *adj.* Of or relating to American blacks of African ancestry, their history, or their culture. *—n.* An American black of African ancestry.

**Af·ro-A·si·at·ic** (ăf'rō-ā'zhē-ăt'ĭk, -zē-) *n.* A family of languages of southwestern Asia and northern Africa. **—Af'ro-A'si·at'ic** *adj.*

**aft** (ăft) *adv.* & *adj.* [Prob. shortening of ABAFT.] At, in, toward, or close to the stern of a vessel or tail of an aircraft.

**af·ter** (ăf'tər) *prep.* [ME < OE *æfter.*] **1.** Behind in place or order. **2.** In quest or pursuit of <seek *after* fame> **3.** Concerning <asked *after* you> **4.** At a later time than <come *after* dinner> **5.** Subsequent to and because of or regardless of <friends *after* all their differences> **6.** Following continually <year *after* year> **7.** Next to or lower than in order or importance. **8.** In the style of <satires *after* Horace> **9.** With the same or close to the same name as : in honor or commemoration of <named *after* a relative> **10.** According to the nature or desires of <someone *after* my own heart> **11.** Past the hour of <five minutes *after* three> *—adv.* **1.** Behind. **2.** Afterward. *—adj.* **1.** Subsequent in time or place : LATER **2.** Nearer the stern of a vessel or tail of an aircraft. *—conj.* Following or subsequent to the time that <I saw them *after* I arrived.> **—after all. 1.** Everything being considered. **2.** Eventually.

**af·ter·birth** (ăf'tər-bûrth') *n.* The placenta and fetal membranes expelled from the uterus after childbirth.

**af·ter·burn·er** (ăf'tər-bûr'nər) *n.* **1.** A device for augmenting the thrust of a jet engine by burning additional fuel with the uncombined oxygen in the hot exhaust gases. **2.** A device for burning or chemically altering unburned or partially burned carbon compounds in exhaust gases.

**af·ter·care** (ăf'tər-kâr') *n.* Treatment or special care given to convalescent patients, esp. after surgery.

**af·ter·clap** (ăf'tər-klăp') *n.* An unexpected, often disagreeable sequel to a supposedly concluded matter.

**af·ter·damp** (ăf'tər-dămp') *n.* An asphyxiating mixture of gases, primarily nitrogen and carbon dioxide gases in a mine after a fire or explosion.

**af·ter·deck** (ăf'tər-dĕk') *n.* *Naut.* The part of a ship's deck past amidships toward the stern.

**af·ter·ef·fect** (ăf'tĭr-ĭ-fĕkt') *n.* An effect following its cause after some delay, esp. a delayed or prolonged physiological or psychological response to a stimulus.

**af·ter·glow** (ăf'tər-glō') *n.* **1.** The light emitted or remaining after removal of a source of illumination, esp.: **a.** The atmospheric glow after sunset. **b.** The glow of an incandescent metal as it cools. **c.** Emission from a phosphor after removal of excitation. **2.** An agreeable feeling following a pleasant experience. **3.** A lingering impression of past glory or success.

**af·ter-hours** (ăf'tər-ourz') *adj.* **1.** Occurring after closing time, as consumption of liquor. **2.** Open after a legal or established closing time <an *after-hours* club for members only>

**af·ter·im·age** (ăf'tər-ĭm'ĭj) *n.* A visual image persisting after a visual stimulus ceases.

**af·ter·life** (ăf'tər-līf') *n.* **1.** A life held to follow death. **2.** A subsequent part of one's life.

**af·ter-mar·ket** (ăf'tər-mär'kĭt) *n.* The demand for goods or services associated with the upkeep of a previous purchase.

**af·ter·math** (ăf'tər-măth') *n.* [AFTER + obs. *math*, mowing, < OE *mǣð.*] **1.** A consequence, esp. of a misfortune or disaster. **2.** A period of time following a disastrous event <in the *aftermath* of the flood> **3.** A second crop of grass in the same season.

**af·ter·most** (ăf'tər-mōst') *adj.* **1.** *Naut.* Nearest the stern farthest aft. **2.** Nearest the end or rear : HINDMOST.

**af·ter·noon** (ăf'tər-nōōn') *n.* **1.** The day from noon until sunset. **2.** A later or closing period or part.

**af·ter·pains** (ăf'tər-pānz') *pl.n.* Pains or cramps from contraction of the uterus following childbirth.

**af·ter·piece** (ăf'tər-pēs') *n.* A short comic piece acted after a play.

**af·ter·sen·sa·tion** (ăf'tər-sĕn-sā'shən) *n.* A sensory impression that persists or recurs after removal of a stimulus.

**af·ter-shave** (ăf'tər-shāv') *n.* A usu. fragrant lotion for the face after shaving.

**af·ter·taste** (ăf′tər-tāst′) *n.* A taste or feeling persisting after the original stimulus is no longer present.

**af·ter·thought** (ăf′tər-thôt′) *n.* An idea, response, or explanation that occurs to one after an event or decision.

**af·ter·time** (ăf′tər-tīm′) *n.* The future.

**af·ter·ward** (ăf′tər-wərd) *also* **af·ter·wards** (-wərdz) *adv.* In or at a later time : SUBSEQUENTLY.

**af·ter·word** (ăf′tər-wûrd′) *n.* EPILOGUE 2.

**af·ter·world** (ăf′tər-wûrld′) *n.* A world believed to exist after death.

**aft·most** (ăft′mōst′) *adj. Naut.* Aftermost.

**Ag** [Lat. *argentum.*] *symbol for* SILVER.

**ag-** *pref. var. of* AD- 1. —Used before g.

**a·ga** *also* **a·gha** (ä′gə, ăg′ə) *n.* [Turk. *ağa.*] An important official of the Ottoman Empire.

**a·gain** (ə-gĕn′) *adv.* [ME < OE *ongeagn,* against.] **1.** Once more : ANEW. **2.** To a previous place, position, or state. **3.** Furthermore. **4.** On the other hand <might win and *again* might not> **5.** In response : in return. **—again and again.** Repeatedly. **—as much again. 1.** The same amount. **2.** Twice as much.

**a·gainst** (ə-gĕnst′) *prep.* [ME, alteration of *againes* < OE *ongeagn.*] **1.** In a direction opposite to <swam *against* the current> **2.** So as to come into forcible contact with <waves dashing *against* the cliffs> **3.** In contact with so as to rest or press on <leaned *against* the pillar> **4.** In hostile opposition to <struggle *against* fate> **5.** Contrary to <*against* my better judgment> **6.** In contrast or comparison with the setting or background of <black velvet *against* fair skin> **7.** In preparation for or anticipation of <food stored *against* winter> **8.** As a defense or safeguard from <protection *against* the cold> **9.** To the account or debt of <drew a check *against* my bank balance> **10.** Directly opposite to : FACING.

**a·ga·lac·ti·a** (ā-gə-lăk′tē-ə, -shē-ə, ăg′ə-) *n.* [NLat. < Gk. *agalaktia,* lack of milk : *a-,* not + *gala,* milk.] Nonsecretion or dysfunctional secretion of milk following childbirth.

**a·ga·ma** (ə-gä′mə, ăg′ə-) *n.* [Carib.] A small, long-tailed, insect-eating Old World lizard of the family Agamidae.

**Ag·a·mem·non** (ăg′ə-mĕm′nŏn′, -nən) *n.* [Gk. *Agamemnōn.*] *Gk. Myth.* The king of Mycenae, leader of the Greeks against Troy, husband of Clytemnestra, and father of Orestes, Electra, and Iphigenia.

**a·ga·mete** (ā′gə-mēt′, ā-găm′ēt′) *n.* An asexual reproductive cell that develops into an adult individual.

**a·gam·ic** (ā-găm′ĭk) *adj.* [< LLat. *agamus,* unmarried < Gk. *agamos* : *a-,* not + *gamos,* marriage.] *Biol.* Occurring or reproducing without the union of male and female cells. **—a·gam′i·cal·ly** *adv.*

**a·gam·o·gen·e·sis** (ā-găm′ə-jĕn′ĭ-sĭs, ăg′ə-mō-) *n.* [AGAM(IC) + -GENESIS.] Asexual reproduction, as by budding, cell division, or parthenogenesis.

**ag·a·mous** (ăg′ə-məs) *adj.* Agamic.

**ag·a·pan·thus** (ăg′ə-păn′thəs) *n.* [NLat. *Agapanthus,* genus name : Gk. *agapē,* love + Gk. *anthos,* flower.] A plant of the genus *Agapanthus,* which includes the African lily.

**a·gape**[1] (ə-gāp′, ə-găp′) *adv. & adj.* **1.** In a state of wonder or amazement. **2.** Wide open <stood with mouth *agape*>

**a·ga·pe**[2] (ä-gä′pā, ä′gə-pā′) *n.* [Gk. *agapē,* love.] **1.** Christian love. **2.** Spiritual, as opposed to sexual, love. **3.** The love feast accompanied by Eucharistic celebration in the early Christian church.

**a·gar** (ā′gär′, ä′gär′) *also* **a·gar-a·gar** (ā′gär-ā′gär′, ä′gär-ä′-) *n.* [Short for Malay *agar-agar.*] A gelatinous material prepared from certain marine algae and used as a base for bacterial culture media, as a laxative, and for gelling foods.

**ag·a·ric** (ăg′ər-ĭk, ə-găr′ĭk) *n.* [Lat. *agaricum,* a kind of fungus < Gk. *agarikon.*] **1.** A fungus of the family Agaricacae, including the common cultivated mushroom, *Agaricus campestris.* **2.** The dried fruiting body of the fungus *Fomes laricis,* once used medicinally.

**ag·ate** (ăg′ĭt) *n.* [ME *achate* < OFr. *acate* < Lat. *achates* < Gk. *akhatēs.*] **1.** A fine-grained, fibrous variety of chalcedony with color banding or irregular clouding. **2.** A playing marble made of agate or glass. **3.** A tool with agate parts, as a burnisher tipped with agate. **4.** A printer's type size, approx. 5½ points.

**agate line** *n.* A measure of space used in estimating printed advertising, usu. 1 column wide and 1/14 of an inch deep.

**a·ga·ve** (ə-gä′vē, ə-gā′-) *n.* [NLat. *Agave,* genus name < Gk. *agauē,* fem. of *agauos,* noble.] Any of numerous fleshy-leaved tropical American plants of the genus *Agave,* including the century plant, some species of which yield valuable fibers.

**age** (āj) *n.* [ME < OFr. *aage* < Lat. *aetas,* age.] **1.** A period of existence. **2.** A lifetime. **3.** The time in life when a person officially assumes certain rights and responsibilities : LEGAL AGE. **4.** A distinctive period or stage of life <the *age* of childhood> **5.** The latter portion of life. **6.** A geologic or historical period designated by distinctive characteristics <the *age* of enlightenment> **7.** *Informal.* A long time <haven't seen you in *ages*> **8.** *Psychol.* Mental age. *—v.* **aged, ag·ing, ag·es.** *—vt.* To cause to grow older or more mature. *—vi.* **1.** To become old. **2.** To manifest traits associated with old age. **3.** To become mature or ripe <wine *aging* in casks> **—ag′er** *n.*

☆ ***syns:*** AGE, DAY, EPOCH, ERA, PERIOD, TIME *n. core meaning* : a particular time notable for its distinctive characteristics <the Edwardian *age*>

**-age** *suff.* [ME < OFr. < Lat. *-aticum,* n. and adj. suffix.] **1.** Collection : mass <sewer*age*> **2.** Relationship : connection <parent*age*> **3.** Condition : state <vagabond*age*> **4. a.** An action <block*age*> **b.** Result of an action <break*age*> **5.** Residence or place of <vicar*age*> **6.** Charge or fee <cart*age*>

**ag·ed** (ā′jĭd) *adj.* **1.** Old. **2.** Of, pertaining to, or characteristic of old age. **3.** (ājd). Of the age of <*aged* seven> **4.** *Geol.* Near the base level of erosion. **—ag′ed·ly** *adv.* **—ag′ed·ness** *n.*

**age·ing** (ā′jĭng) *n. Chiefly Brit. var. of* AGING.

**age·ism** (ā′jĭz′əm) *n.* Discrimination against people of a certain age and esp. against the elderly. **—age′ist** *n.*

**age·less** (āj′lĭs) *adj.* **1.** Seeming not to show the effects of age. **2.** Existing forever : TIMELESS. **—age′less·ly** *adv.* **—age′less·ness** *n.*

**a·gen·cy** (ā′jən-sē) *n., pl.* **-cies.** [Med. Lat. *agentia* < *agens,* effective. —see AGENT.] **1.** Action : power. **2.** A mode of action : MEANS. **3.** A business or service officially acting for others. **4.** A governmental department of administration or regulations.

**a·gen·da** (ə-jĕn′də) *n.* [Lat. pl. of *agendum,* agendum.] *(sing. in number).* **1.** A list of things to be done, esp. the program for a meeting. **2.** *var. pl. of* AGENDUM.

**a·gen·dum** (ə-jĕn′dəm) *n., pl.* **-da** (-də) or **-das.** [Lat., neuter gerund. of *agere,* to do.] Something to be done, esp. an item on an agenda.

**a·gen·e·sis** (ā-jĕn′ĭ-sĭs) *n.* Failure of an organism, organ, or part to develop.

**a·gent** (ā′jənt) *n.* [ME < Lat. *agens,* effective, pr.part. of *agere,* to do.] **1.** One that acts or has the authority to act. **2.** One that acts as the representative of another. **3.** A means of doing something : INSTRUMENT. **4.** A force or substance that causes change <a leavening *agent*> **5.** A representative of a government or administrative department of a government <an IRS *agent*> **6.** A spy. **—a·gen′tial** (ā-jĕn′shəl) *adj.*

**a·gent pro·vo·ca·teur** (ä-zhäɴ′ prô-vô′kä-tœr′) *n., pl.* **a·gents pro·vo·ca·teurs** (ä-zhäɴ′ prô-vô′kä-tœr′) [Fr.] A secret agent who infiltrates an organization in order to incite its members to commit punishable acts.

**age of consent** *n. Law.* The age at which a female may legally choose to have sexual intercourse.

**ag·er·a·tum** (ăj′ə-rā′təm) *n.* [NLat. *Ageratum,* genus name < Gk. *agēratos,* ageless : *a-,* not + *gēras,* old age.] **1.** Any of various plants of the genus *Ageratum,* esp. *A. houstonianum,* a species bearing usu. violet-blue flower clusters. **2.** Any of several other plants with flower clusters similar to the ageratum.

**ag·gie**[1] (ăg′ē) *n.* [AG(ATE) + -IE.] A playing marble.

**ag·gie**[2] (ăg′ē) *n.* [AG(RICULTURAL) + -IE.] **1.** An agricultural college. **2.** A student enrolled at an agricultural college or in an agricultural program within a university.

**ag·gior·na·men·to** (ə-jôr′nə-mĕn′tō) *n., pl.* **-tos.** [Ital. < *aggiornare,* to update : *a-,* to (< Lat. *ad.*) + *giorno,* day < Lat. *diurnus,* daily < *dies,* day.] The process of updating an institution or organization.

**ag·glom·er·ate** (ə-glŏm′ə-rāt′) *vt. & vi.* **-at·ed, -at·ing, -ates.** [Lat. *agglomerare, agglomerat-,* to annex : *ad-,* to + *glomerare,* to form into a ball < *glomus,* ball.] To form into a rounded mass. *—adj.* (-ər-ĭt). Gathered into a rounded mass. *—n.* (-ər-ĭt). **1.** A mass of things clustered together : HEAP. **2.** A volcanic rock composed of rounded and angular fragments fused together. **—ag·glom′er·a′tive** (-ə-rā′tĭv, -ər-ə-tĭv) *adj.* **—ag·glom′er·a′tor** *n.*

**ag·glom·er·a·tion** (ə-glŏm′ə-rā′shən) *n.* **1. a.** The act or process of agglomerating. **b.** The state of being agglomerated. **2.** A confused or jumbled mass : AGGLOMERATE.

**ag·glu·ti·nate** (ə-glo͞ot′n-āt′) *v.* **-nat·ed, -nat·ing, -nates.** [Lat. *agglutinare, agglutinat* : *ad-,* to + *glutinare,* to glue < *gluten,* glue.] *—vt.* **1.** To join by adhesion. **2.** To combine (word elements) into a compound. **3.** *Physiol.* To cause (red blood cells or microorganisms) to clump together. *—vi.* **1.** To join together into a group or mass. **2.** To form words by agglutination. **3.** To undergo agglutination. **—ag·glu′ti·nant** *adj. & n.*

**ag·glu·ti·na·tion** (ə-glo͞ot′n-ā′shən) *n.* **1.** The process of agglutinating. **2.** A mass formed by agglutinating. **3.** Formation of words from morphemes that retain their original forms and meanings with little change during the combination process.

**ag·glu·ti·na·tive** (ə-glo͞ot′n-ā′tĭv) *adj.* **1.** Characteristic of agglutination : ADHESIVE. **2.** Designating a language in which words are formed primarily by agglutination.

**ag·glu·ti·nin** (ə-glo͞ot′n-ĭn) *n.* [AGGLUTIN(ATION) + -IN.] A substance inducing agglutination.

**ag·glu·tin·o·gen** (ăg′lo͞o-tĭn′ə-jən, ə-glo͞o′tn-) *n.* [AGGLUTIN(IN) + -GEN.] An antigen that stimulates the production of an agglutinin. **—ag′glu·tin′o·gen′ic** (ăg′lo͞o-tĭn′ə-jĕn′ĭk, ə-glo͞o′tn-) *adj.*

**ag·grade** (ə-grād′) *vt.* **-grad·ed, -grad·ing, -grades.** To fill and raise the level of (a streambed) by deposition of sediment. **—ag′gra·da′tion** (ăg′rə-dā′shən) *n.* **—ag′gra·da′tion·al** *adj.*

**ag·gran·dize** (ə-grăn′dīz′, ăg′rən-) *vt.* **-dized, -diz·ing, -diz·es.** [Fr. *aggrandir, aggrandiss-* : *a-*, to (< Lat. *ad-*) + *grandir*, to grow larger < Lat. *grandire* < *grandis*, large.] **1.** To increase the scope of : EXTEND. **2.** To make greater in power, influence, or reputation. **3.** To exaggerate the qualities of : EXALT. **—ag·gran′dize·ment** (ə-grăn′dĭz-mənt, -dīz′-) *n.* **—ag·gran′diz′er** *n.*

**ag·gra·vate** (ăg′rə-vāt′) *vt.* **-vat·ed, -vat·ing, -vates.** [Lat. *aggravare, aggravat-* : *ad*, to + *gravare*, to burden < *gravis*, heavy.] **1.** To make worse <bronchitis *aggravated* by smoking> **2.** *Informal.* To annoy <a talkative student who *aggravated* the teacher> **—ag′gra·vat′ing·ly** *adv.* **—ag′gra·va′tive** *adj.* **—ag′gra·va′tor** *n.*

**aggravated assault** *n. Law.* Any of various assaults that are more serious than a common assault, esp. one performed with intent to commit a crime.

**ag·gra·va·tion** (ăg′rə-vā′shən) *n.* **1.** The act of aggravating or state of being aggravated. **2.** One that irritates or makes worse. **3.** *Informal.* Annoyance : vexation.

**ag·gre·gate** (ăg′rĭ-gĭt) *adj.* [ME *aggregat* < Lat. *aggregare*, to add to : *ad-*, to + *gregare*, to collect < *grex*, flock.] **1.** Gathered together into a mass constituting a whole. **2.** *Bot.* Crowded or massed into a dense cluster. **3.** Composed of a mixture of minerals separable by mechanical means. *—n.* (-gĭt). **1.** A total or whole considered with reference to its constituent parts <an empire that was the *aggregate* of many states> **2.** The mineral materials, as sand or stone, used in making concrete. *—vt.* (-gāt′) **-gat·ed, -gat·ing, -gates. 1.** To gather into a mass, sum, or whole. **2.** To amount to. **—ag′gre·gate·ly** *adv.* **—ag′gre·ga′tion** *n.* **—ag′gre·ga′tive** *adj.* **—ag′gre·ga′tor** *n.*

**aggregate fruit** *n.* A fruit, as the raspberry, developed from the pistils of a single flower and consisting of a mass of drupelets.

**aggregate fruit**
*Two types of aggregate fruit: (left) a raspberry and (right) a strawberry*

**ag·gress** (ə-grĕs′) *vi.* **-gressed, -gress·ing, -gress·es.** [Fr. *aggresser* < Lat. *aggredi* : *ad-*, toward + *gradi*, to go.] To commit aggression.

**ag·gres·sion** (ə-grĕsh′ən) *n.* **1.** Initiation of forceful, usu. hostile action against another : ATTACK. **2.** The practice of attacking or encroaching, esp. in violation of territorial rights : INVASION. **3.** *Psychoanal.* Hostile action or behavior.

**ag·gres·sive** (ə-grĕs′ĭv) *adj.* **1.** Hostile : combative. **2. a.** Energetic and enterprising. **b.** Boldly assertive. **—ag·gres′sive·ly** *adv.* **—ag·gres′sive·ness** *n.*

**ag·gres·sor** (ə-grĕs′ər) *n.* One that engages in aggression.

**ag·grieve** (ə-grēv′) *vt.* **-grieved, -griev·ing, -grieves.** [ME *agreven* < OFr. *agrever* < Lat. *aggravare*, to make worse. —see AGGRAVATE.] **1.** To distress or afflict. **2.** To injure unjustly.

**ag·grieved** (ə-grēvd′) *adj.* **1.** Feeling distress or affliction. **2.** Treated wrongly : OFFENDED. **3.** *Law.* Treated unjustly, as by a decision of a court. **—ag·griev′ed·ly** (ə-grē′vĭd-lē) *adv.* **—ag·griev′ed·ness** *n.*

**a·gha** (ä′gə, ăg′ə) *n. var. of* AGA.

**a·ghast** (ə-găst′) *adj.* [ME *agast*, p.part. of *agasten*, to frighten : *a-* (intensive) + *gasten*, to frighten < OE *gæstan* < *gāst*, ghost.] Stricken with horror : APPALLED.

**ag·ile** (ăj′əl, ăj′īl′) *adj.* [OFr. < Lat. *agilis* < *agere*, to impel.] **1.** Able to move quickly and easily : NIMBLE. **2.** Mentally alert. **—ag′ile·ly** *adv.* **—ag′ile·ness** *n.* **—a·gil′i·ty** (ə-jĭl′ĭ-tē) *n.*

**†a·gin** (ə-gĭn′) *prep. Regional.* Against.

**ag·ing** (ā′jĭng) *n.* **1.** The process of becoming old or mature. **2.** An artificial process for imparting the characteristics and properties of age, as to wood.

**ag·i·o** (ăj′ē-ō′) *n., pl.* **-os.** [Ital. < Med. Gk. *allagion*, exchange < *allagē*, change < *allos*, other.] **1.** A premium paid for exchanging one currency for another. **2.** An allowance or premium for the difference in value between two currencies being exchanged.

**ag·i·tate** (ăj′ĭ-tāt′) *v.* **-tat·ed, -tat·ing, -tates.** [Lat. *agitare, agitat-*, freq. of *agere*, to impel.] *—vt.* **1.** To move with sudden forcefulness or violence <a hurricane *agitating* the trees> **2.** To upset emotionally. **3.** To try to arouse public interest in (e.g., a cause). **4.** *Archaic.* To ponder over. *—vi.* To stir up public interest in a cause. **—ag′i·tat′ed·ly** (-tā′tĭd-lē) *adv.* **—ag′i·ta′tive** *adj.*

**ag·i·ta·tion** (ăj′ĭ-tā′shən) *n.* **1. a.** The act of agitating. **b.** The state of being agitated. **2.** Extreme emotional disturbance : PERTURBATION. **3.** Arousal of public interest in a cause or controversial matter. **—ag′i·ta′tion·al** *adj.*

**ag·i·ta·to** (ăj′ĭ-tä′tō) *adj.* [Ital. < Lat. *agitare*, to agitate.] *Mus.* Fast and restless : AGITATED. —Used as a direction. **—ag′i·ta′to** *adv.*

**ag·i·ta·tor** (ăj′ĭ-tā′tər) *n.* **1.** One who agitates, esp. one who engages in political agitation. **2.** A mechanism that shakes or stirs, as in a washing machine.

**ag·it·prop** (ăj′ĭt-prŏp′) *n.* [R., department of agitation and propaganda : *agitatsiya*, agitation + *propaganda*, propaganda.] Communist-oriented political propaganda disseminated esp. through literature, drama, art, or music.

**A·gla·ia** (ə-glā′ə, ə-glī′ə) *n.* [Gk. < *aglaia*, splendor < *aglaos*, bright.] *Gk. Myth.* One of the Three Graces.

**a·gleam** (ə-glēm′) *adj. & adv.* Shining brightly : GLEAMING.

**ag·let** (ăg′lĭt) *n.* [ME < OFr. *aguillette*, dim. of *aguille*, needle < LLat. *acicula*, dim. of Lat. *acus*, needle.] **1.** A tag or metal sheath on the end of a lace, cord, or ribbon to facilitate its passing through eyelet holes. **2.** An ornamental device similar to the aglet.

**a·gley** (ə-glī′, ə-glā′, ə-glē′) *adv.* [Scottish : *a-*, on + *gley*, to squint < ME *glien*.] *Scot.* Awry : amiss.

**a·glim·mer** (ə-glĭm′ər) *adj. & adv.* Glimmering faintly.

**a·glit·ter** (ə-glĭt′ər) *adj.* Glittering : sparkling. **—a·glit′ter** *adv.*

**a·glow** (ə-glō′) *adj. & adv.* Glowing.

**a·gly·con** (ə-glī′kŏn) or **a·gly·cone** (-kōn′) *n.* A nonsugar component of a glycoside that is resolvable through hydrolysis.

**ag·mi·nate** (ăg′mə-nĭt, -nāt′) *also* **ag·mi·nat·ed** (-nā′tĭd) *adj.* [< Lat. *agmen, agmin-*, multitude.] *Bot.* Gathered in clusters.

**ag·nail** (ăg′nāl′) *n.* [ME *angnail*, corn < OE *angnægel*, a sore under the nail : *ang-*, tight + *nægel*, nail.] **1.** A hangnail. **2.** A painful swelling or sore around a fingernail or toenail.

**ag·nate** (ăg′nāt′) *adj.* [Lat. *agnatus*, a relation on the father's side < p.part. of *agnasci*, to be born in addition to : *ad-*, to + *nasci*, to be born.] **1.** Related on or descended from the male or father's side. **2.** From a common source : AKIN. *—n.* A relative on the male or father's side only. **—ag·nat′ic** (ăg-năt′ĭk) *adj.* **—ag·nat′i·cal·ly** *adv.* **—ag·na′tion** *n.*

**Ag·ni** (ŭg′nē) *n.* [Skt. *agniḥ*, fire.] The Vedic god of fire and guardian of humans.

**ag·no·men** (ăg-nō′mən) *n., pl.* **-nom·i·na** (-nŏm′ə-nə) [Lat. : *ad-*, to + *nomen*, name.] **1.** An additional cognomen given to a Roman citizen, often in honor of military victories. **2.** A nickname.

**ag·no·sia** (ăg-nō′zhə) *n.* [NLat. < Gk. *agnōsia*, ignorance : *a-*, not + *gnōsis*, knowledge < *gignōskein*, to know.] Pathologic loss of auditory, sensory, or visual comprehension.

**ag·nos·tic** (ăg-nŏs′tĭk) *n.* [< Gk. *agnōstos*, unknown : *a-*, not + *gnōstos*, known < *gignōskein*, to know.] One who believes that there can be no proof of the existence of God but does not deny the possibility that God exists. **—ag·nos′tic** *adj.* **—ag·nos′ti·cal·ly** *adv.*

**ag·nos·ti·cism** (ăg-nŏs′tĭ-sĭz′əm) *n.* **1.** *Philos.* The doctrines of the agnostics, holding that certainty or first or absolute truths are unattainable and that only perceptual phenomena are objects of exact knowledge. **2.** A theological theory that does not deny God but denies the possibility of knowing Him.

**Ag·nus De·i** (ăg′nəs dē′ī′, än′yo͞os dā′ē, ăg′no͞os′) *n.* [Lat.] **1.** The Lamb of God, an emblem of Christ. **2.** An iconographic representation of the Agnus Dei. **3.** A liturgical prayer to Christ.

**a·go** (ə-gō′) *adj. & adv.* [ME, p.part. of *agon*, to go away < OE *āgān* : *ā-* (intensive) + *gān*, to go.] Earlier than the present time : PAST <three months *ago*><died long *ago*>

**a·gog** (ə-gŏg′) *adv. & adj.* [ME *agogge* < OFr. *en gogue*, in merriment.] In a state of excitement and keen anticipation.

**-agog** *suff. var. of* -AGOGUE.

**à go·go** *also* **à·go-go** (ə-gō-gō′) *adv.* [Fr., galore. ] In a fast and lively manner : ENERGETICALLY.

**-agogue** *or* **-agog** *suff.* [LLat. *-agogus* < Gk. *-agogos* < *agein*, to lead.] A substance that stimulates the flow of <hem*agogue*>

**a·gone** (ə-gôn′, ə-gŏn′) *adj. & adv.* [ME *agon*, p.part. of *agon*, to go away. —see AGO.] *Archaic.* Gone by : PAST.

**a·gon·ic** (ā-gŏn′ĭk, ə-gŏn′-) *adj.* [< Gk. *agōnos* : *a-*, not + *gōnia*, angle.] Having no angle.

**agonic line** *n.* An imaginary line on the earth's surface connecting points where the magnetic declination is zero.

**ag·o·nist** (ăg′ə-nĭst) *n.* [Back-formation < ANTAGONIST.] **1.** *Physiol.* A muscle that contracts and is opposed by contraction in another muscle, the antagonist. **2.** One involved in a struggle or competition.

**ag·o·nis·tic** (ăg′ə-nĭs′tĭk) *also* **ag·o·nis·ti·cal** (-tĭ-kəl) *adj.* [Gk. *agōnistikos* < *agōnistēs*, combatant < *agōn*, contest.] **1.** Argumentative : combative. **2.** Struggling to achieve effect. **3.** Of or relating to athletic competitions, orig. those of the ancient Greeks. **—ag′o·nis′ti·cal·ly** *adv.*

**ag·o·nize** (ăg′ə-nīz′) *v.* **-nized, -niz·ing, -niz·es.** [OFr. *agoniser* < Med. Lat. *agonizare* < Gk. *agōnizesthai*, to struggle < *agōn*, contest.] *—vi.* **1.** To be in extreme physical or emotional pain : suffer intensely. **2.** To make a great effort : STRUGGLE. *—vt.* To cause great pain or anguish to. **—ag′o·niz′ing·ly** *adv.*

**ag·o·ny** (ăg′ə-nē) *n., pl.* **-nies.** [ME *agonie* < OFr. < Med. Lat. *agonia* < Gk. *agōnia* < *agōn*, struggle.] **1.** The suffering of intense physical or emotional pain. **2.** The struggle that precedes death. **3.** A sudden or intense emotion. **4.** A violent or intense struggle or effort.

**agony column** *n.* A newspaper column containing advertisements chiefly about missing relatives or friends.

**ag·o·ra¹** (ăg'ə-rə) *n., pl.* **-rae** (-rē') *or* **-ras.** [Gk.] An ancient Greek marketplace used as a gathering place for the populace.

**a·go·ra²** (ä'gə-rä') *n., pl.* **-rot** (-rōt') *or* **-roth** (-rōt') [Heb. *'agōrāh* < *āgōr*, to collect.] —See table at CURRENCY.

**ag·o·ra·pho·bi·a** (ăg'ə-rə-fō'bē-ə) *n.* [Gk. *agora*, open space + -PHOBIA.] Abnormal fear of open, esp. public, spaces. **—ag'o·ra·pho'bi·ac** *n.* **—ag'o·ra·pho'bic** (-fō'bĭk, -fŏb'ĭk) *adj.*

**a·go·rot** (ä'gə-rōt') *or* **a·go·roth** (-rōt') *n. var. pls. of* AGORA².

**a·gou·ti** (ə-gōō'tē) *n., pl.* **-tis** *or* **-ties.** [Fr. < Sp. (South America) *agutí* < Guarani *acutí*.] A tropical American burrowing rodent of the genus *Dasyprocta*, with grizzled brownish or dark-gray fur.

**agr-** *pref. var. of* AGRO-.

**a·graffe** *also* **a·grafe** (ə-grăf') *n.* [Fr. *agrafe* < OFr. *agrafer*, to hook onto : *a-*, to (< Lat. *ad*) + *grafer*, to hook < *grafe*, hook, of Germanic orig.] **1.** A hook and eye for fastening armor and clothing. **2.** A cramp iron for holding stones together in building.

**a·gran·u·lo·cy·to·sis** (ā-grăn'yə-lō-sī-tō'sĭs) *n.* A drug-induced disease marked by high fever, lesions of the mucous membranes, and a decrease in granular white blood corpuscles.

**ag·ra·pha** *also* **Ag·ra·pha** (ăg'rə-fə) *pl.n.* [Gk. < *agraphos*, unwritten : *a-* not + *graphein*, to write.] The sayings of Jesus not found in the Gospels.

**a·graph·i·a** (ā-grăf'ē-ə) *n.* [A-¹ + Gk. *graphein*, to write.] A disorder marked by the inability to write. **—a·graph'ic** *adj.*

**a·grar·i·an** (ə-grâr'ē-ən) *adj.* [< Lat. *agrarius* < *ager*, land.] **1.** Pertaining to land and its ownership. **2.** Pertaining to agricultural or rural matters. —*n.* One who favors equitable land distribution. **—a·grar'i·an·ly** *adv.*

**a·grar·i·an·ism** (ə-grâr'ē-ə-nĭz'əm) *n.* A movement for equitable distribution of land and for agrarian reform.

**a·grav·ic** (ā-grăv'ĭk) *adj.* [A-¹ + GRAV(ITY) + -IC.] Of or relating to a condition of no gravitation.

**a·gree** (ə-grē') *v.* **a·greed, a·gree·ing, a·grees.** [ME *agreen* < OFr. *agreer* < VLat. **aggratare* : Lat. *ad-*, to + Lat. *gratus*, pleasing.] —*vi.* **1.** To grant consent : ASSENT. **2.** To be in accord : MATCH <Our findings *agree* with yours.> **3.** To be of one opinion. **4.** To come to an understanding or to terms. **5.** To be appropriate or beneficial <Hot weather doesn't *agree* with me.> **6.** To correspond in gender, number, case, or person. —*vt.* To concede or grant : ADMIT <*agreed* that we should split the bill>

**a·gree·a·ble** (ə-grē'ə-bəl) *adj.* **1.** To one's liking : PLEASING. **2.** Being in accordance or harmony : SUITABLE. **3.** Ready to consent or submit : WILLING <*agreeable* to the suggestion> **—a·gree'a·bil'i·ty, a·gree'a·ble·ness** *n.* **—a·gree'a·bly** *adv.*

**a·greed** (ə-grēd') *adj.* **1.** Decided by mutual consent <the *agreed* time to meet> **2.** Of one opinion : in accord <My parents were *agreed*.> **3.** Granted : concurred.

**a·gree·ment** (ə-grē'mənt) *n.* **1.** The act of agreeing. **2.** Harmony of opinion : ACCORD. **3.** An arrangement between parties regarding a method of action : COVENANT. **4.** *Law.* **a.** A properly executed and legally binding compact. **b.** The writing or document embodying this compact. **5.** Correspondence between words in gender, number, case, or person.

**a·gres·tal** (ə-grĕs'təl) *also* **a·gres·tial** (-chəl) *adj.* [Lat. *agrestis* < *ager*, land.] Growing wild, esp. in cultivated areas.

**a·gres·tic** (ə-grĕs'tĭk) *also* **a·gres·ti·cal** (-tĭ-kəl) *adj.* **1.** Rustic : rural. **2.** Crude : unpolished.

**ag·ri·a** (ăg'rē-ə) *n.* [< Gk. *agria*, wild < *agros*, field.] Extensive or intense pustular eruption.

**ag·ri·busi·ness** (ăg'rə-bĭz'nĭs) *n.* [AGRI(CULTURE) + BUSINESS.] Farming engaged in as big business, embracing the production, processing, and distribution of farm products and the manufacture of farm equipment.

**ag·ri·cul·ture** (ăg'rĭ-kŭl'chər) *n.* [Lat. *agricultura* : *ager*, land + *cultura*, cultivation.] The science, art, and business of cultivating the soil, producing crops, and raising livestock : FARMING. **—ag'ri·cul'tur·al** *adj.* **—ag'ri·cul'tur·al·ly** *adv.* **—ag'ri·cul'tur·ist, ag'ri·cul'tur·al·ist** *n.*

**ag·ri·mo·ny** (ăg'rə-mō'nē) *n., pl.* **-nies.** [ME *agrimonie* < OFr. *aigremoine* < Lat. *agrimonia*, alteration of *argemonia* < Gk. *argemōnē*, poppy.] **1.** A plant of the genus *Agrimonia*, bearing compound leaves, long clusters of small yellow flowers, and bristly fruits. **2.** A plant related or similar to the agrimony, as the hemp agrimony.

**ag·ri·o·e·col·o·gy** (ăg'rē-ō-ī-kŏl'ə-jē) *n.* [Gk. *agrios*, wild (< *agros*, field) + ECOLOGY.] The ecology of cultivated or domestic plants.

**ag·ri·ol·o·gy** (ăg'rē-ŏl'ə-jē) *n.* [Gk. *agrios*, wild (< *agros*, field) + -LOGY.] The study of peoples whose cultures are at an early developmental stage. **—ag'ri·o·log'i·cal** (-ə-lŏj'ĭ-kəl) *adj.*

**agro-** *or* **agr-** *pref.* [< Gk. *agros*, field.] Field : soil <agrology>

**ag·ro·bi·ol·o·gy** (ăg'rō-bī-ŏl'ə-jē) *n.* The science of plant and animal growth and nutrition as related to soil variation and crop yield. **—ag'ro·bi'o·log'ic** (-ə-lŏj'ĭk), **ag'ro·bi'o·log'i·cal** *adj.* **—ag'ro·bi'o·log'i·cal·ly** *adv.* **—ag'ro·bi·ol'o·gist** *n.*

**a·grol·o·gy** (ə-grŏl'ə-jē) *n.* The applied science of soils in relation to crops. **—ag'ro·log'ic** (ăg'rə-lŏj'ĭk), **ag'ro·log'i·cal** *adj.* **—ag'ro·log'i·cal·ly** *adv.* **—a·grol'o·gist** *n.*

**a·gron·o·my** (ə-grŏn'ə-mē) *also* **ag·ro·nom·ics** (ăg'rə-nŏm'ĭks) *n.* The application of soil and plant sciences to soil management and crop production : scientific agriculture. **—ag'ro·nom'ic** (ăg'rə-nŏm'ĭk), **ag'ro·nom'i·cal** *adj.* **—a·gron'o·mist** *n.*

**ag·ros·tol·o·gy** (ăg'rə-stŏl'ə-jē) *n.* [Gk. *agrōstis*, a kind of wild grass (< *agros*, field) + -LOGY.] The botanical study of grasses.

**a·ground** (ə-ground') *adv. & adj.* Stranded in shallow water or on a reef or shoal <The ship ran *aground* in the storm.>

**a·gryp·ni·a** (ə-grĕp'nē-ə) *n.* [Gk. < *agrupnos*, wakeful.] Insomnia.

**a·gue** (ā'gyōō) *n.* [ME < OFr., short for *fievre ague*, sharp fever < Med. Lat. *febris acuta* : *febris*, fever + *acutus*, sharp.] **1.** A fever with periods of chill and sweating. **2.** A fit of shivering : CHILL. **—a'gu·ish** (-ĭsh) *adj.* **—a'gu·ish·ly** *adv.* **—a'gu·ish·ness** *n.*

**a·gue·weed** (ā'gyōō-wēd') *n.* **1.** A plant, *Gentiana quinquefolia* of eastern North America, with pale blue-violet or white flower clusters. **2.** Boneset.

**ah** (ä) *interj.* —Used to express various emotions, as surprise, pain, or satisfaction.

**a·ha** (ä-hä') *interj.* —Used to express surprise, triumph, or pleasure.

**a·head** (ə-hĕd') *adv.* **1.** At or to the front. **2.** In advance : BEFORE. **3.** Onward <forge *ahead*> **—be ahead.** *Informal.* To be gaining or winning. **—get ahead.** To attain success.

**ahead of** *prep.* In front of.

**a·hem** (ə-hĕm') *interj.* —Used to attract attention, as a warning, or to express doubt.

**a·him·sa** (ə-hĭm'sä') *n.* [Skt. *ahiṃsā* : *a-*, not + *hiṃsā*, injury < *hiṃsati*, he injures.] A Buddhist and Hindu doctrine of nonviolence expressing belief in the sacredness of all living creatures.

**a·hoy** (ə-hoi') *interj.* *Naut.* —Used to hail a ship or person or to attract attention.

**Ah·ri·man** (ä'rĭ-mən) *n.* [Pers. *Ahrīman*, prob. < Avestan *aṅra mainyu* : *aṅra*, evil + *mainyu*, spirit.] The spirit of evil in Zoroastrianism and the reputed arch rival of Ormazd.

**A·hu·ra Maz·da** (ä-hōōr'ə măz'də) *n.* [Avestan, wise god.] Ormazd.

**ai** (ī) *n.* [Port. < Tupi.] A three-toed sloth of the genus *Bradypus*.

**aid** (ād) *v.* **aid·ed, aid·ing, aids.** [ME *aiden* < OFr. *aider* < Lat. *adjutare*, freq. of *adjuvare*, to help : *ad-* (intensive) + *juvare*, to help.] —*vi.* To help or assist. —*vt.* To give help or assistance to. —*n.* **1.** The act or result of helping : ASSISTANCE. **2.** An assistant. **3.** An aide-de-camp. **4.** A money payment by a vassal to a feudal lord in medieval England. **—aid'er** *n.*

**aide** (ād) *n.* [Fr. < *aider*, to aid.] **1.** An aide-de-camp. **2.** An assistant.

**aide-de-camp** (ād'dĭ-kămp') *n., pl.* **aides-de-camp.** [Fr.] A military officer acting as secretary and confidential assistant to a superior officer.

**aide-mé·moire** (ād'măm-wär') *n.* [Fr.] **1.** Something, as a mnemonic device, that aids the memory. **2.** A memorandum outlining the major points of a proposed agreement or discussion, used esp. in diplomatic communications.

**AIDS** (ādz) *n.* [Acronym for ACQUIRED IMMUNE DEFICIENCY SYNDROME.] A disease of no known etiology in which the body's immunological system is destroyed.

**ai·grette** *or* **ai·gret** (ā-grĕt', ā'grĕt') *n.* [Fr., egret < OFr. —see EGRET.] **1.** An ornamental tuft of plumes, esp. the tail feathers of an egret. **2.** An ornament, as a spray of gems, resembling an aigrette.

**ai·guille** (ā-gwēl') *n.* [Fr., needle. —see AGLET.] **1.** A sharp, pointed mountain peak. **2.** A needle-shaped drill for boring holes in rock or masonry.

**ai·guil·lette** (ā'gwə-lĕt') *n.* [Fr. —see AGLET.] An ornamental cord or braid worn on the shoulder of a military uniform.

**ai·ki·do** (ī'kē-dō') *n.* [J. *aikidō* : *ai*, mutual + *ki*, spirit + *dō*, art.] A Japanese method of self-defense in which one's opponent's strength and weight are used against him.

**ail** (āl) *v.* **ailed, ail·ing, ails.** [ME *eilen* < OE *eglian* < *egle*, troublesome.] —*vi.* To feel ill. —*vt.* To make ill or uneasy.

**ai·lan·thus** (ā-lăn'thəs) *n.* [NLat. *Ailanthus*, genus name < Amboinese *ai lanto*, tree of heaven.] A deciduous tree, *Ailanthus altissima*, native to China but naturalized esp. in North American urban areas, with compound leaves and malodorous greenish flower clusters.

**ai·le·ron** (ā'lə-rŏn') *n.* [Fr., dim. of *aile*, wing < OFr. < Lat. *ala*.] Either of two movable flaps on the wings of an aircraft that can be used to control rolling and banking movements.

**ail·ment** (āl'mənt) *n.* A usu. mild physical or mental disorder.

**ai·lu·ro·phile** (ī-lōōr'ə-fīl', ā-lōōr'-) *n.* [Gk. *ailouros*, cat + -PHILE.] One who loves cats.

**ai·lu·ro·phobe** (ī-lōōr'ə-fōb', ā-lōōr'-) *n.* [Gk. *ailouros*, cat + -PHOBE.] One who hates or fears cats.

**aim** (ām) *v.* **aimed, aim·ing, aims.** [ME *amen* < OFr. *aesmer*, to guess at : *a-*, at (< Lat. *ad*) + *esmer*, to guess < Lat. *aestimare*, to estimate.] —*vt.* To direct (e.g., a weapon, remark, or blow). —*vi.*

1. To direct a weapon. 2. To direct one's efforts or purpose <*aim* at eliminating racism><*aim* to find the answer> —*n.* 1. The act of aiming. 2. The sighting or line of fire of something aimed. 3. *Obs.* An object or point aimed at : TARGET. 4. A purpose or intention. 5. *Obs.* A conjecture.

**aim·less** (ām'lĭs) *adj.* Lacking direction or purpose. **—aim'less·ly** *adv.* **—aim'less·ness** *n.*

**ain** (ān) *adj. Scot.* Own.

**ain't** (ānt). *Nonstandard.* 1. Am not. 2. —Used also as a contraction for *are not, is not, has not,* and *have not.* ***usage:*** Even though it would be useful as a contraction for *am not* and as an alternative form for *isn't, aren't, hasn't,* and *haven't, ain't* is still unacceptable in standard usage.

**Ai·nu** (ī'no͞o) *n., pl.* **Ainu** or **-nus.** 1. A member of an aboriginal Caucasian people inhabiting the northernmost islands of Japan. 2. The language of the Ainu.

**ai·o·li** (ī-ō'lē) *n.* [Prov. : *ai,* garlic (< Lat. *allium*) + *oli,* oil (< Lat. *oleum*).] A rich garlic-flavored mayonnaise.

**air** (âr) *n.* [ME < OFr. < Lat. *aer* < Gk. *aēr.*] 1. **a.** A colorless, odorless, tasteless gaseous mixture, mainly nitrogen (approx. 78%) and oxygen (approx. 21%) with lesser amounts of other gases, as argon, carbon dioxide, neon, and helium. **b.** This mixture with varying amounts of moisture, low-altitude pollutants, and particulate matter, enveloping the earth : ATMOSPHERE. **c.** The air or atmosphere in an enclosure <stale *air*> 2. The sky. 3. A breeze. 4. *Archaic.* Breath. 5. Public utterance <give *air* to one's outrage> 6. A characteristic impression : AURA <a house with an *air* of neglect> 7. Personal bearing, appearance, or manner <has an *air* of superiority> 8. **airs.** Affected manners <putting on *airs*> 9. *Mus.* A melody or tune, esp.: **a.** The soprano or treble part in a harmonized composition. **b.** A solo with or without accompaniment. 10. Air conditioning. —*vt.* **aired, air·ing, airs.** 1. To expose to the air : VENTILATE. 2. To give public utterance to. **—in the air.** 1. Prevalent. 2. Being thought out : UNCERTAIN. **—on (or off) the air.** Being (or not being) broadcast on radio or television. **—take the air.** To go outdoors for fresh air. **—up in the air.** 1. Not decided : UNCERTAIN. 2. Excited : agitated. **—walk on air.** To feel elated.

☆ ***syns:*** AIR, AMBIANCE, ATMOSPHERE, AURA, FEEL, FEELING, MOOD, SMELL, TONE *n. core meaning* : an impression caused by a predominant quality <exuded an *air* of fear>

**air bag** *n.* An automotive safety device designed to inflate upon collision and prevent passengers from pitching forward.

**air base** *n.* A base for military aircraft.

**air battery** *n.* A rechargeable battery in which current is generated as a result of oxidation of metal.

**air bladder** *n. Biol.* 1. An air-filled structure near the spinal column in many fishes that functions to maintain buoyancy or in some species as an aid in breathing or hearing. 2. An air-filled saclike structure, as one of the dilated parts of the thallus in some seaweeds.

**air·boat** (âr'bōt') *n.* A swamp boat.

**air·borne** (âr'bôrn', -bōrn') *adj.* 1. Carried by or through the air <*airborne* viruses> 2. Transported in aircraft, as military troops. 3. In flight : FLYING.

**air brake** *n.* A brake operated by compressed air.

**air·brush** also **air brush** (âr'brŭsh') *n.* An atomizer using compressed air to spray liquids, as paint, on a surface. —*vt.* **-brushed, -brush·ing, -brush·es.** To spray with an airbrush.

**air·burst** (âr'bûrst') *n.* An explosion of a projectile in the atmosphere.

**Air·bus** (âr'bŭs'). A trademark for a subsonic jet passenger aircraft.

**air chamber** *n.* 1. An enclosure filled with air for a particular purpose. 2. An air chamber, esp. in a hydraulic system, in which air elastically compresses and expands to regulate the flow of a fluid.

**air command** *n.* A unit of the U.S. Air Force that is larger than an air force.

**air-con·di·tion** (âr'kən-dĭsh'ən) *vt.* **-tioned, -tion·ing, -tions.** To furnish with or ventilate by air conditioning.

**air conditioner** *n.* An apparatus for controlling, esp. lowering, the temperature and humidity of an enclosure.

**air conditioning** *n.* 1. The state or condition produced by an air conditioner. 2. A system of air conditioners.

**air-cool** (âr'ko͞ol') *vt.* **-cooled, -cool·ing, -cools.** 1. To cool (e.g., an engine) by a flow of air. 2. To air-condition.

**air corridor** *n.* An air route established by international agreement.

**air cover** *n.* 1. Protective use of military aircraft during ground operations. 2. The aircraft used for air cover.

**air·craft** (âr'krăft') *n., pl.* **aircraft.** A machine or device, capable of atmospheric flight, esp. an airplane.

**aircraft carrier** *n.* A large naval ship with storage and service facilities for aircraft and a long flat deck on which they can take off and land at sea.

**air·crafts·man** (âr'krăfts'mən) also **air·craft·man** (-krăft'mən) *n.* A noncommissioned member of the British Royal Air Force or the Royal Canadian Air Force.

**air cushion** *n.* 1. An inflatable cushion. 2. An air spring.

**air-cush·ion** (âr'ko͝osh'ən) also **air-cush·ioned** (-ənd) *adj.* Of or relating to a ground-effect machine.

**air division** *n.* A unit of the U.S. Air Force larger than a wing and smaller than an air force.

**air·drome** (âr'drōm') *n.* 1. An airport. 2. A landing field. 3. A hangar for aircraft.

**air·drop** (âr'drŏp') *n.* A delivery, as of troops or supplies, by parachute from airborne aircraft. —*vt.* & *vi.* **-dropped, -drop·ping, -drops.** To drop from an airborne aircraft.

**air-dry** (âr'drī') *vt.* **-dried, -dry·ing, -dries.** To dry by exposure to the air. —*adj.* Sufficiently dry so that no moisture will be evaporated by further exposure to air.

**Aire·dale** (âr'dāl') *n.* [After *Airedale,* a valley in Yorkshire, England.] A large terrier of a long-legged breed having a wiry tan coat marked with black.

**air embolism** *n.* AEROEMBOLISM 1.

**air express** *n.* A system of transporting packages by air.

**air·fare** (âr'fâr') *n.* The charge for travel by aircraft.

**air·field** (âr'fēld') *n.* 1. An airport having hard-surfaced runways. 2. A landing strip.

**air·flow** (âr'flō') *n.* 1. A flow of air. 2. The air currents caused by the motion of a vehicle, as an automobile or aircraft.

**air·foil** (âr'foil') *n.* An aircraft part or surface, as a wing, propeller blade, or rudder, whose shape and orientation control stability, direction, lift, thrust, or propulsion.

**air force** *n.* 1. The aviation branch of a country's armed forces. 2. A unit of the U.S. Air Force larger than an air division and smaller than an air command.

**air·frame** (âr'frām') *n.* An aircraft without its power plant.

**air freight** *n.* 1. A system of transporting freight by air. 2. The charge for air freight.

**air gas** *n.* Producer gas.

**air·glow** (âr'glō') *n.* A low- or middle-latitude, relatively steady, faint photochemical luminescence in the upper atmosphere.

**air gun** *n.* A gun discharged by compressed air.

**air·head** (âr'hĕd') *n.* An area of hostile or enemy-controlled territory secured by paratroops.

**air hole** *n.* 1. An opening through which gas or air may pass. 2. A hole in the ice, as of a lake. 3. An air pocket.

**air hunger** *n.* The gasping, deep respiration typical of coma and diabetic acidosis.

**air·ing** (âr'ĭng) *n.* 1. Exposure to air, as for drying or freshening. 2. Exposure to open air esp. for exercise. 3. Public notice or expression. 4. A radio or television broadcast.

**air lane** *n.* An established route of travel for aircraft.

**air layering** *n.* A method of plant propagation in which a twig or shoot attached to the parent plant is wrapped in moist sphagnum moss or polyethylene plastic so that it will form roots and can later be removed and replanted.

**air·less** (âr'lĭs) *adj.* 1. Lacking air. 2. Lacking fresh air : STUFFY. 3. Lacking a breeze : STILL. **—air'less·ness** *n.*

**air letter** *n.* 1. An airmail letter. 2. A sheet of airmail paper that can be folded as an envelope with a message inside : AEROGRAM.

**air·lift** (âr'lĭft') *n.* A system of transporting troops or supplies by air when surface routes are obstructed or inaccessible. —*v.* **-lift·ed, -lift·ing, -lifts.** —*vt.* To transport by air, as when surface routes are blocked. —*vi.* To transport troops or supplies by air.

**air·line** (âr'līn') *n.* 1. **a.** A system for scheduled transport of passengers and freight by air. **b.** A business organization providing such a system. 2. An air lane. 3. The shortest distance between two geographic points.

**air·lin·er** (âr'lī'nər) *n.* A passenger-carrying aircraft operated by an airline.

**air lock** *n.* 1. An airtight chamber, usu. located between two regions of unequal pressure, in which air pressure can be regulated. 2. A blockage of flow, as in a radiator pipe, caused by trapped air.

**air·mail** (âr'māl') *vt.* **-mailed, -mail·ing, -mails.** To send (e.g., a letter) by air mail. —*adj.* Of, relating to, or for use with air mail.

**air mail** also **air·mail** (âr'māl') *n.* 1. The system of conveying mail by aircraft. 2. Mail conveyed by aircraft.

**air·man** (âr'mən) *n.* 1. An enlisted person in the U.S. Air Force. 2. An enlisted person working with aircraft in the U.S. Navy. 3. An aviator.

**airman basic** *n.* An enlisted person of the lowest rank in the U.S. Air Force.

**airman first class** *n.* An enlisted person in the U.S. Air Force ranking above an airman and below a sergeant.

**air mass** *n.* A large body of air with only small horizontal variations of pressure, temperature, and moisture.

**air mattress** *n.* An inflatable airtight sack used as a mattress.

**Air Medal** *n.* A decoration awarded by the U.S. Army, Air Force, or Navy for meritorious airborne conduct.

**air mile** *n.* A unit of distance in air navigation.

**air piracy** *n.* The hijacking of an aircraft in flight : SKYJACKING. **—air pirate** *n.*

**air·plane** (âr′plān′) *n.* [Alteration of *aeroplane,* prob. < Fr. *aéroplane* : *aéro-,* aero- + *planer,* to glide < *plan,* level < Lat. *planus,* flat.] A winged vehicle capable of flight, gen. heavier than air and propelled by jet engines or propellers.

**air plant** *n.* An epiphyte.

**air·play** (âr′plā′) *n.* Radio broadcasting of a record.

**air pocket** *n.* A downward air current that causes an aircraft to lose altitude abruptly.

**air police** *n.* Military police of an air force.

**air·port** (âr′pôrt′, -pōrt′) *n.* **1.** A cleared and leveled area where aircraft can take off and land, usu. having hard-surfaced landing strips, a control tower, hangars, passenger terminals, and accommodations for cargo. **2.** An installation similar to an airport in which the landing area is on water.

**air·proof** (âr′prōōf′) *adj.* Impermeable by air. *—vt.* **-proofed, -proof·ing, -proofs.** To make impermeable by air.

**air pump** *n.* Equipment for removing, compressing, or forcing a flow of air.

**air raid** *n.* An assault by hostile military aircraft.

**air rifle** *n.* A low-powered rifle that uses manually compressed air to discharge small pellets.

**air sac** *n.* Any of the air-filled spaces in a bird's body that connect the lungs and the bone cavities.

**air·screw** (âr′skrōō′) *n. Chiefly Brit.* The propeller of an aircraft.

**air-shed** (âr′shĕd′) *n.* [AIR + (WATER)SHED.] **1.** The air supply of a given area. **2.** The geographic area covered by an air supply.

**air·ship** (âr′shĭp′) *n.* A self-propelled lighter-than-air craft with directional control surfaces : DIRIGIBLE.

**air·sick·ness** (âr′sĭk′nĭs) *n.* Nausea and discomfort caused by nervous tension or changes in pressure or motion in an aircraft. **—air′sick** *adj.*

**air sock** *n.* A windsock.

**air·space** (âr′spās′) *n.* **1.** The section of the atmosphere above a particular land area, esp. that of a nation. **2.** The space occupied by an aircraft formation or used in a particular maneuver.

**air speed** *n.* The speed of an aircraft relative to the air.

**air splint** *n.* An inflatable cylinder used to immobilize fractures or sprains of extremities.

**air spray** *n.* **1.** A device that uses compressed air for spraying liquids. **2.** The liquid sprayed by an air spray.

**air spring** *n.* An enclosed, resilient volume of air that acts as a spring or shock absorber.

**air·strip** (âr′strĭp′) *n.* A cleared area serving a landing strip.

**airt** (ârt) *n.* [ME *art* < Sc. Gael. *aird.*] *Scot.* One of the cardinal points on the compass.

**air taxi** *n.* A small aircraft that makes short local flights to areas not serviced by regular airlines.

**air·tight** (âr′tīt′) *adj.* **1.** Impermeable by air or gas. **2.** Invulnerable : unassailable <an *airtight* excuse>

**air·time** (âr′tīm′) *n.* **1.** The time that a radio or television station is on the air. **2.** The time available for a particular broadcast.

**air-to-air missile** (âr′tə-âr′) *n.* A guided missile designed to be fired from aircraft at aircraft.

**air-to-sur·face missile** (âr′tə-sûr′fĭs) *n.* A guided missile designed to be fired from aircraft at targets on the ground.

**air vesicle** *n.* **1.** A terminal air sac in the lung where gas exchange occurs during respiration. **2.** An air-filled space in many water plants that aids in flotation.

**air·waves** (âr′wāvz′) *pl.n.* The medium used to transmit radio and television signals.

**air·way** (âr′wā′) *n.* **1.** A shaft or passageway, as in a mine, in which air circulates. **2.** An air lane.

**air·wor·thy** (âr′wûr′thē) *adj.* **-thi·er, -thi·est.** Fit to fly <an *airworthy* jet> **—air′wor′thi·ness** *n.*

**air·y** (âr′ē) *adj.* **-i·er, -i·est. 1.** Of, relating to, or like air. **2.** High in the air : LOFTY. **3.** Exposed to or open to the air : BREEZY. **4.** Performed in the air, as high leaps. **5.** Immaterial <an *airy* specter> **6.** Insubstantial : imaginative <*airy* speculations> **7.** Light as air : DELICATE. **8.** Nonchalant <an *airy* disregard for rules> **9.** Lighthearted : gay. **—air′i·ly** *adv.* **—air′i·ness** *n.*

**aisle** (īl) *n.* [ME *ele* < OFr., wing of a building < Lat. *ala.*] **1.** A section of a church divided laterally from the nave by a row of columns or pillars. **2.** A passageway between rows of seats, as in an auditorium. **3.** A passageway, as between shelves or counters in a grocery store.

**aitch** (āch) *n.* [Fr. *hache.*] The letter *h.*

**aitch·bone** (āch′bōn′) *n.* [ME *hach-boon* < the phrase *an hach-boon,* an aitchbone, alteration of *a nachebon* : *nache,* buttock (< OFr. < Lat. *natis*) + *bon,* bone < OE *bān.*] **1.** The rump bone in cattle. **2.** The cut of beef containing the aitchbone.

▲ word history: *Aitchbone* is a good example of a folk etymology, which is the refashioning of a word so that it resembles a more familiar, but unrelated, word. As the etymology shows, the incorrect division between the article and the noun in *a nachebon* resulted in a new noun whose first syllable was unrecognizable. This element was variously interpreted as *each, ash, ice,* and *aitch,* all familiar English words. The bone itself has no physical resemblance to the letter H.

**a·jar¹** (ə-jär′) *adv. & adj.* [ME *on char* : *on,* in + *char,* turn < OE *cierr.*] Partially opened.

**a·jar²** (ə-jär′) *adv. & adj.* Not in harmony : JARRING.

**A·jax** (ā′jăks′) *n.* [Lat. < Gk. *Aias.*] *Gk. Myth.* **1.** A son of Telamon of Salamis and warrior of great stature and prowess who fought against Troy. **2.** A son of Ileus of Locris and warrior of small stature and arrogant character who fought against Troy.

**a·kar·y·o·cyte** (ā-kăr′ē-ō-sīt′) *n.* A cell without a nucleus.

**ak·ee** (ăk′ē, ə-kē′) *n.* [Native word in Liberia.] **1.** A tropical African tree, *Blighia sapida,* with fragrant flowers and capsules containing black seeds. **2.** The edible aril surrounding the seeds of the akee.

**a·kim·bo** (ə-kĭm′bō) *adj. & adv.* [ME *in kenebowe.*] With hands on hips and elbows bowed outward.

**a·kin** (ə-kĭn′) *adj.* **1.** Related by blood. **2.** Similar in character or quality : ANALOGOUS. **3.** COGNATE 2.

**Ak·ka·di·an** (ə-kā′dē-ən) *n.* **1.** A native or inhabitant of ancient Akkad. **2.** The Semitic language of the Akkadians. **—Ak·ka′di·an** *adj.*

**Al** *symbol for* ALUMINUM.

**al-** *pref. var. of* AD-1. —Used before *l.*

**-al¹** *suff.* [ME < OFr. < Lat. *-alis.*] Of, pertaining to, or marked by <parent*al*>

**-al²** *suff.* [ME *-aille* < OFr. < Lat. *-alia,* neuter pl. of *-alis,* adj. suffix.] Action : process <retriev*al*>

**-al³** *suff.* [< ALDEHYDE.] Aldehyde <citronell*al*>

**a·la** (ā′lə) *n., pl.* **a·lae** (ā′lē) [Lat., wing.] A winglike structure or part, as an ear lobe or the membranous border of some seeds.

**à la** *also* **a la** (ä′lä, ä′lə, ăl′ə) *prep.* [Fr., short for *à la mode de,* in the manner of.] In the manner or style of <a story *à la* Poe>

**al·a·bas·ter** (ăl′ə-băs′tər) *n.* [ME *alabastre* < OFr. < Lat. *alabaster* < Gk. *alabastros,* poss. of Egypt. orig.] **1.** A dense, translucent, white or tinted fine-grained gypsum. **2.** A hard, translucent, occas. banded calcite. **3.** A pale yellowish pink to yellowish gray.

**à la carte** (ä′lə kärt′, ăl′ə) *adv. & adj.* [Fr., by the menu.] With a separate price for each item on the menu.

**a·lack** (ə-lăk′) *also* **a·lack·a·day** (ə-lăk′ə-dā′) *interj. Archaic.* —Used to express sadness, regret, or alarm.

**a·lac·ri·ty** (ə-lăk′rĭ-tē) *n.* [Lat. *alacritas* < *alacer,* lively.] **1.** Cheerful willingness. **2.** Speed or quickness. **—a·lac′ri·tous** (-təs) *adj.*

**A·lad·din** (ə-lăd′n) *n.* A boy in the *Arabian Nights* who acquires a magic lamp and a magic ring with which he can summon two jinn to grant his wishes.

**a·lae** (ā′lē) *n. pl. of* ALA.

**à la king** (ä′lə kĭng′, ăl′ə) *adj.* Cooked in a cream sauce with green pepper or pimiento and mushrooms <turkey *à la king*>

**†al·a·me·da** (ăl′ə-mē′də, -mā′-) *n.* [Sp. < *álamo,* poplar, alamo.] *Southwestern U.S.* A promenade or walk, esp. one shaded by alamos.

**†al·a·mo** (ăl′ə-mō′) *n., pl.* **-mos.** [Sp. *álamo* < Lat. *alnus,* alder, and *ulmus,* elm.] *Southwestern U.S.* A poplar tree, esp. a cottonwood.

**a·la·mode** (ä′lə-mōd′, ăl′ə-) *n.* [< À LA MODE.] A lustrous silk fabric for head coverings and scarfs.

**à la mode** (ä′lə mōd′, ăl′ə) *adj.* [Fr., in the fashion.] **1.** Stylish : fashionable. **2. a.** Served with ice cream, as pie. **b.** Braised with vegetables and served in a rich, brown sauce, as meats.

**al·a·nine** (ăl′ə-nēn′) *n.* [G. *Alanin,* ult. < *Aldehyd,* aldehyde.] An amino acid, $C_3H_7NO_2$, that is a constituent of most proteins.

**a·lar** (ā′lər) *adj.* [Lat. *alaris* < *ala,* wing.] **1.** Of, pertaining to, or having wings or alae. **2.** Resembling a wing, esp. in shape. **3.** *Anat.* Pertaining to the armpit : AXILLARY.

**a·larm** (ə-lärm′) *n.* [ME < OFr. *alarme* < OItal. *allarme* < *all' arme,* to arms : *alla,* to (< Lat. *ad illam,* to that) + *arme,* arms < Lat. *arma.*] **1.** Agitation and anxiety caused by the apprehension or realization of danger : FRIGHT. **2.** A warning of imminent danger. **3.** A device that warns of danger by means of a sound or signal. **4.** The sounding mechanism of an alarm clock or watch. **5.** A call to arms. *—vt.* **a·larmed, a·larm·ing, a·larms. 1.** To frighten by a sudden revelation of danger. **2.** To warn of or indicate imminent danger. **—a·larm′a·ble** *adj.* **—a·larm′ing·ly** *adv.*

☆ ***syns:*** ALARM, ALERT, TOCSIN, WARNING *n. core meaning* : a signal that warns of imminent danger <fighter pilots responding to an *alarm*>

**alarm clock** *n.* A clock that can be set to sound a bell or buzzer at a desired hour.

**a·larm·ist** (ə-lär′mĭst) *n.* One who needlessly and habitually alarms others. **—a·larm′ism** *n.*

**alarm reaction** *n.* An innate mechanism in animals and humans that provides a response to stressful circumstances.

**a·la·rum** (ə-lăr′əm, ə-lär′-) *n.* [ME *alarom,* var. of *alarm,* alarm.] *Archaic.* An alarm, esp. a call to arms.

**a·la·ry** (ā′lə-rē) *adj.* [Lat. *alarius* < *ala,* wing.] **1.** Of or pertaining to wings. **2.** Like a wing.

**a·las** (ə-lăs′) *interj.* —Used to express sadness, misery over loss, compassion, or apprehension of danger or evil.

**a·las·ka** (ə-lăs′kə) *n.* [After *Alaska.*] **1.** A heavy-duty rubberized overshoe. **2. a.** A heavy cotton and wool fabric for dresses and coats. **b.** A yarn made of cotton and wool.

**Alaska cedar** *n.* The Nootka cypress.

**A·las·kan malamute** (ə-lăs′kən) *n.* The malamute.

**a·las·tor** *also* **A·las·tor** (ə-lăs′tər, -tôr′) *n.* [Gk. *alastōr* < *alastos,* unforgettable : *a-,* not + *lathein,* to forget.] An avenging deity or spirit frequently evoked in Greek tragedy, the masculine personification of Nemesis.

**a·late** (ā′lāt′) *also* **a·lat·ed** (ā′lā′tĭd) *adj.* [Lat. *alatus* < *ala,* wing.] *Biol.* Having thin winglike extensions or parts.

**alb** (ălb) *n.* [ME *albe* < OE < Med. Lat. *alba* < Lat. *albus,* white.] A liturgical vestment consisting of a long white linen robe with tapered sleeves.

**al·ba·core** (ăl′bə-kôr′, -kōr′) *n., pl.* **albacore** or **-cores.** [Port. *albacor* < Ar. *al-bakrah* : *al,* the + *bakr,* young camel.] A large marine fish, *Thunnus alalunga* of warm seas, whose edible flesh is a major source of canned tuna.

**Al·ba·ni·an** (ăl-bā′nē-ən, -bān′yən, ôl-) *adj.* Of or relating to the People's Republic of Albania, its inhabitants, or their language. —*n.* **1.** A native or inhabitant of Albania. **2.** The Indo-European language of the Albanians.

**al·ba·tross** (ăl′bə-trôs′, -trŏs′) *n., pl.* **albatross** or **-tross·es.** [Prob. alteration of *alcatras,* pelican < Port. or Sp. *alcatraz,* of Ar. orig.] A large web-footed bird of the family Diomedeidae, chiefly of the oceans of the Southern Hemisphere, with long narrow wings and a hooked beak.

**al·be·do** (ăl-bē′dō) *n., pl.* **-dos.** [LLat., whiteness < Lat. *albus,* white.] The fraction of light or other electromagnetic radiation reflected by a surface.

**al·be·it** (ôl-bē′ĭt, ăl-) *conj.* [ME, although it be.] Although.

**al·bes·cent** (ăl-bĕs′ənt) *adj.* [Lat. *albescens, albescent-,* pr.part. of *albescere,* to become white < *albus,* white.] Becoming white or somewhat white.

**Al·bi·gen·ses** (ăl′bə-jĕn′sēz′) *pl.n.* [Med. Lat., pl. of *Albigensis,* inhabitant of Albiga, a town in southern France where the sect was dominant.] The members of a religious sect of southern France in the 12th and 13th cent., annihilated for heresy by the Inquisition. **—Al′bi·gen′sian** (-shən, -sē-ən) *adj.* **—Al′bi·gen′sian·ism** *n.*

**al·bi·nism** (ăl′bə-nĭz′əm) *n.* [Fr. *albinisme* < G. *Albinismus* < *Albino,* albino.] **1.** The congenital lack of normal pigmentation in a person, animal, or plant. **2.** The condition of being an albino.

**al·bi·no** (ăl-bī′nō) *n., pl.* **-nos.** [Port. < *albo,* white < Lat. *albus.*] An organism lacking normal pigmentation, as a person having inordinately pale skin and very light hair and lacking normal eye coloring or an animal having white hair or fur and red eyes.

**Al·bi·on** (ăl′bē-ən) *n.* [Lat.] A literary name for Britain.

**al·bite** (ăl′bīt′) *n.* [Swed. *albit* < Lat. *albus,* white.] A widely distributed white feldspar, $NaAlSi_3O_8$, one of the common rock-forming plagioclase group. **—al·bit′ic** (-bĭt′ĭk), **al·bit′i·cal** (-ĭ-kəl) *adj.*

**al·bum** (ăl′bəm) *n.* [Lat., blank tablet < *albus,* white.] **1.** A book or binder with blank pages for mounting a collection, as of stamps or photographs. **2. a.** A set of phonograph records stored together in jackets under one binding. **b.** The holder for such records. **c.** One or more long-playing records in a slipcase. **d.** A phonograph record. **e.** A recording of several musical pieces. **3.** An anthology of musical compositions, pictures, or literary selections. **4.** An oversize, handsomely printed book, esp. popular in the 19th cent., often having an abundance of illustrations and short, sentimental texts.

**al·bu·men** (ăl-byōō′mən) *n.* [Lat. < *albus,* white.] **1.** A nutritive substance surrounding a growing embryo, as the white of an egg or the material stored in a plant seed. **2.** Albumin.

**al·bu·min** (ăl-byōō′mĭn) *n.* [ALBUM(EN) + -IN.] A simple, water-soluble protein that is coagulated by heat, found in egg white, blood serum, milk, animal tissues, and many plant juices and tissues.

**al·bu·mi·noid** (ăl-byōō′mə-noid′) *also* **al·bu·mi·noi·dal** (-byōō′mə-noid′l) *adj.* Like albumin. —*n. Biochem.* Protein.

**al·bu·mi·nous** (ăl-byōō′mə-nəs) *adj.* Of, resembling, or relating to albumin or albumen.

**al·bu·mi·nu·ri·a** (ăl-byōō′mə-nŏŏr′ē-ə, -nyŏŏr-) *n.* The presence of albumin in the urine, occas. indicating kidney disease. **—al·bu′mi·nu′ric** (-nŏŏr′ĭk, -nyŏŏr′-) *adj.*

**al·bu·mose** (ăl′byə-mōs′, -mōz′) *n.* [Fr. : *albumine,* albumin + *-ose,* -ose.] Any of a class of albuminous substances formed by enzymatic action on proteins during digestion.

**al·bur·num** (ăl-bûr′nəm) *n.* [Lat. < *albus,* white.] Sapwood.

**Al·ca·ic** (ăl-kā′ĭk) *adj.* [LLat. *Alcaicus,* of Alcaeus < Gk. *Alkaïkos* < *Alkaios,* a Greek lyric poet of the 7th cent. B.C.] Of or describing a verse form used in Greek and Latin poetry, made up of strophes having four tetrametric lines. —*n.* Verse in Alcaic strophes.

**al·cai·de** *also* **al·cay·de** (ăl-kī′dē) *n.* [Sp. < Ar. *al-qā'id,* the commander < *qād,* to command.] The commander of a fortress in Spain or Portugal.

**al·cal·de** (ăl-käl′dē) *n.* [Sp. < Ar. *al-qāḍī* : *al,* the + *qāḍa,* to judge.] The chief governing or judicial official of a Spanish or Spanish-American town.

**al·cay·de** (ăl-kī′dē) *n. var. of* ALCAIDE.

**al·caz·ar** (ăl-kăz′ər, -kä′zər, ăl′kə-zär′) *n.* [Sp. *alcázar* < Ar. *alqaṣr* : *al,* the + *qaṣ,* camp.] A Spanish palace or fortress.

**Al·ces·tis** (ăl-sĕs′tĭs) *n.* [Lat. < Gk. *Alkēstis.*] *Gk. Myth.* The wife of King Admetus of Thessaly, who gave her life to save her husband's and was afterward rescued from Hades by Hercules.

**al·che·mist** (ăl′kə-mĭst) *n.* One who practices alchemy. **—al′che·mis′tic, al′che·mis′ti·cal** *adj.*

**al·che·mize** (ăl′kə-mīz′) *vt.* **-mized, -miz·ing, -miz·es.** To transmute by or as if by alchemy.

**al·che·my** (ăl′kə-mē) *n.* [ME *alkamie* < OFr. *alquemie* < Med. Lat. *alchymia* < Ar. *al-kīmiyā* : *al,* the + *kīmiyā,* alchemy < LGk. *khēmeia,* perh. < Gk. *Khēmia,* Egypt, of Egypt. orig.] **1.** A medieval chemical philosophy that had as its asserted aims the transmutation of base metals into gold and the search for the preparation of the elixir of longevity. **2.** An apparently magical power or transmutational process. **—al·chem′i·cal** (ăl-kĕm′ĭ-kəl), **al·chem′ic** *adj.* **—al·chem′i·cal·ly** *adv.*

**Al·cin·o·us** (ăl-sĭn′ō-əs) *n.* [Lat. < Gk. *Alkinoos.*] *Gk. Myth.* A king of Phaeacia who was hospitable to Odysseus.

**Alc·me·ne** (ălk-mē′nē) *n.* [Lat. < Gk. *Alkmēnē.*] *Gk. Myth.* Amphitryon's wife and mother by Zeus of Hercules.

**al·co·hol** (ăl′kə-hôl′) *n.* [Med. Lat., antimony < Ar. *al-koḥl* : *al,* the + *koḥl,* antimony.] **1.** A colorless volatile flammable liquid, $C_2H_5OH$, synthesized or derived from fermentation of sugars and starches and used, either pure or denatured, as a solvent and in drugs, cleaning solutions, explosives, and intoxicating beverages. **2.** Intoxicating liquor containing alcohol. **3.** Any of a series of hydroxyl compounds, the simplest of which are derived from saturated hydrocarbons, have the general formula $C_nH_{2n+1}OH$, and include ethanol and methanol.

**al·co·hol·ic** (ăl′kə-hô′lĭk, -hŏl′ĭk) *adj.* **1.** Of, relating to, or resulting from alcohol. **2.** Containing or preserved in alcohol. **3.** Suffering from alcoholism. —*n.* One who drinks alcoholic beverages habitually and excessively or who suffers from alcoholism.

**al·co·hol·ic·i·ty** (ăl′kə-hô-lĭs′ĭ-tē) *n.* Alcoholic content.

**al·co·hol·ism** (ăl′kə-hô′lĭz′əm) *n.* **1.** Excessive consumption of and psychophysiological dependence on alcoholic beverages. **2.** A chronic pathological condition, chiefly of the nervous and gastroenteric systems, caused by habitual excessive alcoholic consumption. **3.** Temporary mental disturbance, muscular incoordination, and paresis caused by excessive alcoholic consumption.

**al·co·hol·ize** (ăl′kə-hô-līz′) *vt.* **-ized, -iz·ing, -iz·es.** To mix, saturate, or treat with alcohol. **—al′co·hol′i·za′tion** *n.*

**al·co·hol·om·e·ter** (ăl′kə-hô-lŏm′ĭ-tər) *n.* A hydrometer for determining the percentage of alcohol in liquids.

**Al·co·ran** (ăl′kə-răn′) *n.* The Koran.

**al·cove** (ăl′kōv′) *n.* [Fr. *alcôve* < Sp. *alcoba* < Ar. *al-qubbah* : *al,* the + *qubbah,* vault.] **1.** A recess or partly enclosed extension connected to or forming part of a room **:** NOOK. **2.** A secluded enclosed structure in a garden, as a bower or gazebo.

**Al·cy·o·ne** (ăl-sī′ə-nē) *n.* [Lat. < Gk. *Alkuonē.*] **1.** *Gk. Myth.* The daughter of Aeolus who, grief-stricken over her dead husband Ceyx, threw herself into the sea and was changed into a kingfisher. **2.** *Gk. Myth.* A nymph, one of the Pleiades. **3.** *Astron.* The brightest star in the Pleiades, in the constellation Taurus.

**Al·deb·a·ran** (ăl-dĕb′ər-ən) *n.* [Med. Lat. < Ar. *al-dabarān* : *al,* the + *dabarān,* following < *dabar,* to follow.] A double star in the constellation Taurus, one of the brightest stars in the sky.

**al·de·hyde** (ăl′də-hīd′) *n.* [G. *Aldehyd* < NLat., short for *alcohol dehydrogenatum,* dehydrogenized alcohol.] **1.** Any of a class of highly reactive organic chemical compounds obtained by oxidation of primary alcohols, characterized by the common group CHO, and used to make resins, dyes, and organic acids. **2.** Acetaldehyde.

**al den·te** (ăl dĕn′tē) *adj. & adv.* [Ital.] Cooked enough to be firm but not soft, as pasta or vegetables.

**al·der** (ôl′dər) *n.* [ME < OE *alor.*] Any of various deciduous shrub or tree of the genus *Alnus,* growing in cool, moist places and yielding reddish wood used in cabinetwork.

**al·der·man** (ôl′dər-mən) *n.* [ME, a person of high rank < OE *ealdorman* : *ealdor,* chief (< *eald,* old) + *man,* man.] **1.** A member of the legislative body in many town and city governments. **2.** A member of the higher branch of a municipal or borough council in England and Ireland. **3.** In Anglo-Saxon England: **a.** A lord or prince. **b.** The primary officer of a shire. **—al′der·man·cy** (-sē) *n.* **—al′der·man′ic** (-măn′ĭk) *adj.*

**Al·der·ney** (ôl′dər-nē) *n., pl.* **-neys.** [After *Alderney,* one of the Channel Islands.] One of a breed of small dairy cattle orig. raised in the Channel Islands.

**al·dol** (ăl′dôl′, -dōl′) *n.* [ALD(EHYDE) + -OL¹.] A thick colorless to pale-yellow liquid, $C_4H_8O_2$, derived from acetaldehyde and used in perfumery and ore flotation.

**al·dol·ase** (ăl′də-lās′) *n.* An enzyme that promotes the breakdown of a fructose ester into triose sugars.

**al·dose** (ăl′dōs′, -dōz′) *n.* [ALD(EHYDE) + -OSE¹.] *Chem.* Any of a class of monosaccharide sugars having an aldehyde group.

**al·dos·ter·one** (ăl-dŏs′tə-rōn′) *n.* [ALD(EHYDE) + STER(OL) + -ONE.] A steroid hormone secreted by the adrenal cortex that regulates the salt and water balance in the body.

**al·drin** (ôl′drĭn) *n.* [After Kurt *Alder* (1902–1958).] An insecticide containing a naphthalene-derived compound, $C_{12}H_8Cl_6$.

**ale** (āl) *n.* [ME < OE *ealu.*] A fermented alcoholic beverage made of malt and hops, similar to but heavier than beer.

**a·le·a·to·ry** (ā′lē-ə-tôr′ē, -tōr′ē) *adj.* [Lat. *aleatorius* < *aleator*, gambler < *alea*, dice.] **1.** Dependent on chance or luck. **2.** Of or relating to gambling. **3.** *also* **a·le·a·to·ric** (ā′lē-ə-tôr′ĭk, -tōr′-). *Mus.* Using or made up of sound sequences played at random or arrived at by chance.

**A·lec·to** (ə-lĕk′tō) *n.* [Lat. < Gk. *Alēktō.*] *Gk. Myth.* One of the three Furies.

**a·lee** (ə-lē′) *adv. Naut.* Away from the wind.

**a·lef** (ä′lĕf, ä′ləf) *n. var. of* ALEPH.

**al·e·gar** (ăl′ĭ-gər, ā′lĭ-) *n.* [ME : *ale*, ale + *egre*, sharp < OFr. < Lat. *acer.*] Vinegar produced by fermenting ale.

**ale·house** (āl′hous′) *n.* An establishment where ale is sold and served.

**Al·e·man·ni** (ăl′ə-măn′ī) *pl.n.* [Lat., of Germanic orig.] A group of Germanic tribes that settled in Alsace and nearby areas during the 4th cent. A.D. and were conquered by the Franks in 496.

**Al·e·man·nic** (ăl′ə-măn′ĭk) *n.* A group of High German dialects spoken in Alsace, Switzerland, and parts of southern Germany. —*adj.* Of or relating to the Alemanni or their language.

**a·lem·bic** (ə-lĕm′bĭk) *n.* [ME *alambic* < OFr. < Med. Lat. *alembicus* < Ar. *al-anbīq* : *al*, the + *anbīq*, still < Gk. *ambix*, cup.] **1.** An apparatus formerly used for distilling. **2.** Something that purifies or transforms as if by distillation.

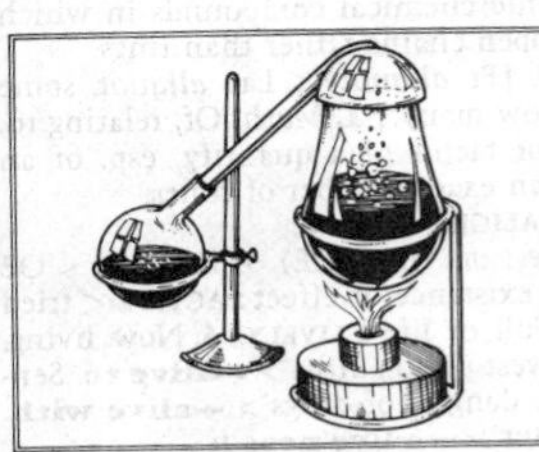

alembic

**a·leph** *also* **a·lef** (ä′lĕf, ä′ləf) *n.* [Heb. *ālеph* < *eleph*, ox.] The first letter of the Hebrew alphabet. —See table at ALPHABET.

**a·leph-null** (ä′lĕf-nŭl′, ä′ləf-) *n.* The first transfinite number.

**a·lert** (ə-lûrt′) *adj.* [Fr. *alerte* < Ital. *all' erta*, on the watch : *alla*, to (< Lat. *ad illam*, to that) + *erta*, watch, p.part. of *ergere*, to raise < Lat. *erigere.*] **1.** Vigilantly attentive : OBSERVANT. **2.** Mentally perceptive and responsive : QUICK. **3.** Lively or brisk. —*n.* **1.** A warning signal of attack or danger : ALARM. **2.** The time period during which an alert is in effect. —*vt.* **a·lert·ed, a·lert·ing, a·lerts.** To notify of imminent danger or risk : WARN. **—on the alert.** Watchful and prepared for danger or emergency.

☆ ***syns:*** ALERT, OBSERVANT, OPEN-EYED, VIGILANT, WAKEFUL, WARY, WATCHFUL, WIDE-AWAKE *adj. core meaning* : vigilantly attentive <*alert* to danger>

**al·eu·rone** (ăl′yə-rōn′) *also* **a·leu·ron** (-rŏn′) *n.* [G. *Aleuron* < Gk. *aleuron*, meal.] Protein consisting of minute granules forming the outermost layer of the endosperm in cereal grains. **—al′eu·ron′ic** (-rŏn′ĭk) *adj.*

**A·leut** (ə-lo͞ot′, ăl′ē-o͞ot′) *n., pl.* **Aleut** or **A·leuts.** [R.] **1.** An Eskimo native of the Aleutian Islands. **2.** The language of the Aleuts.

**A·leu·tian** (ə-lo͞o′shən) *adj.* Of or relating to the Aleuts or their language or culture.

**ale·wife¹** (āl′wīf′) *n.* [Perh. < ALEWIFE².] A fish, *Alosa pseudoharengus* of North American Atlantic waters and some inland lakes, closely related to the herrings.

**ale·wife²** (āl′wīf′) *n.* A woman who keeps an alehouse.

**al·ex·an·der** *also* **Al·ex·an·der** (ăl′ĭg-zăn′dər) *n.* [From the name *Alexander.*] A cocktail made with brandy or gin, crème de cacao, and sweet cream.

**Al·ex·an·dri·an** (ăl′ĭg-zăn′drē-ən) *adj.* **1.** Of or relating to Alexander the Great. **2.** Of or relating to Alexandria, Egypt. **3.** Of, typical of, or designating a school of Hellenistic literature, science, and philosophy located at Alexandria during the last three centuries B.C.

**al·ex·an·drine** *also* **Al·ex·an·drine** (ăl′ĭg-zăn′drĭn) *n.* [Fr. *alexandrin* < OFr. < *Alexandre*, title of a romance about Alexander the Great (356–323 B.C.) that was written in this meter.] A line of verse composed in iambic hexameter, usu. with a caesura after the third foot. —*adj.* Of, relating to, or written in alexandrines.

**al·ex·an·drite** (ăl′ĭg-zăn′drīt′) *n.* [G. *Alexandrit*, after Alexander I (1777–1825), czar of Russia.] A greenish variety of chrysoberyl that looks red in artificial light, used as a gemstone.

**a·lex·i·a** (ə-lĕk′sē-ə) *n.* [NLat. : A-¹ + Gk. *lexis*, speech < *legein*, to speak.] Loss of the ability to read as a result of cerebral lesions.

**a·lex·i·phar·mic** (ə-lĕk′sə-fär′mĭk) *adj.* [< Gk. *alexipharmakos* : *alexein*, to ward off + *pharmakon*, poison.] Preventing or counteracting effects of poison or infection. —*n.* An antidote.

**al·fal·fa** (ăl-făl′fə) *n.* [Sp. < Ar. *al-faṣfaṣah.*] A native Eurasian plant, *Medicago sativa*, bearing compound leaves with three leaflets and small purple flower clusters, widely grown for forage and as a commercial source of chlorophyll.

**al·fil·a·ri·a** or **al·fil·e·ri·a** (ăl-fĭl′ə-rē′ə) *n.* [Mex. Sp. *alfilerillo* < Sp., dim. of *alfiler*, pin < Ar. *al-khilāl*, the spine.] A native European plant, *Erodium cicutarium*, widely naturalized in North America, with finely divided leaves and small pink or purplish flowers.

**†al·for·ja** (ăl-fôr′wä) *n.* [Sp. < Ar. *al-khorj*, the supply.] *Western U.S.* A leather or canvas saddlebag.

**al·fres·co** (ăl-frĕs′kō) *adv.* [Ital. *al fresco*, in the fresh (air) : *a il*, in the + *fresco*, fresh.] Outdoors. —*adj.* Taking place outdoors.

**al·ga** (ăl′gə) *n., pl.* **-gae** (-jē) [Lat., seaweed.] Any of various chiefly aquatic, one-celled or multicellular plants without true stems, roots, and leaves but containing chlorophyll. **—al′gal** (ăl′gəl) *adj.*

**al·gar·ro·ba** or **al·ga·ro·ba** (ăl′gə-rō′bə) *n.* [Sp. < Ar. *al-kharrūbah.*] **1.** The mesquite. **2.** The carob. **3.** The edible pod of either the mesquite or carob tree.

**al·ge·bra** (ăl′jə-brə) *n.* [ME < Med. Lat. < Ar. *al-jebr*, the (science of) reuniting : *al*, the + *jabr*, reunification.] **1.** A generalization of arithmetic in which symbols, usu. letters of the alphabet, represent numbers or a specified set of numbers and are related by operations that hold for all numbers in the set. **2.** A set along with operations defined in the set that accord with specified laws. **—al′ge·bra′ist** (-brā′ĭst) *n.*

**al·ge·bra·ic** (ăl′jə-brā′ĭk) *adj.* **1.** Of, pertaining to, or designating algebra. **2.** Describing an expression, equation, or function in which only numbers, letters, and arithmetic operations are contained or used. **3.** Identifying or restricted to a finite number of algebraic operations. **—al′ge·bra′i·cal·ly** *adv.*

**algebraic language** *n.* A computer language whose statements resemble algebraic expressions.

**algebraic logic** *n.* The sequence of operations wherein a problem is entered into a calculator or computer in the order in which it would be written manually.

**algebraic number** *n.* A numerical root of a polynomial equation with rational coefficients.

**-algia** *suff.* [Gk. < *algos*, pain.] Pain <neur*algia*>

**al·gi·cide** (ăl′jə-sīd′) *n.* [ALG(A) + -CIDE.] A chemical agent that destroys algae in water.

**al·gid** (ăl′jĭd) *adj.* [Fr. *algide* < Lat. *algidus* < *algēre*, to be cold.] Somewhat cold : CHILLY. **—al·gid′i·ty** (-jĭd′ĭ-tē) *n.*

**al·gin** (ăl′jĭn) *n.* [ALG(A) + -IN.] A gelatinous substance derived from certain algae, esp. the giant kelp, and used as a thickener and emulsifier.

**algo-** *pref.* [Gk. < *algos*, pain.] Pain <*algo*meter>

**al·goid** (ăl′goid′) *adj.* Of or like algae.

**Al·gol** (ăl′gŏl′, -gôl′) *n.* [Ar. *al-ghūl*, the ghoul. —see GHOUL.] A double eclipsing variable star in the constellation Perseus.

**ALGOL** (ăl′gŏl′, -gôl′) *n.* [ALG(ORITHMIC) O(RIENTED) L(ANGUAGE).] A computer language by which numerical procedures may be precisely presented, used esp. for scientific problems.

**al·go·lag·ni·a** (ăl′gō-lăg′nē-ə) *n.* [NLat. : ALGO- + Gk. *lagneia*, lust.] Sexual pleasure derived from inflicting or experiencing pain. **—al′go·lag′nic** *adj.* **—al′go·lag′nist** *n.*

**al·gol·o·gy** (ăl-gŏl′ə-jē) *n.* [ALG(A) + -LOGY.] The study of algae. **—al′go·log′i·cal** (ăl′gə-lŏj′ĭ-kəl) *adj.* **—al′go·log′i·cal·ly** *adv.* **—al·gol′o·gist** *n.*

**al·gom·e·ter** (ăl-gŏm′ĭ-tər) *n.* An apparatus for determining sensitivity to pressure-induced pain. **—al′go·met′ric** (-gə-mĕt′rĭk), **al′go·met′ri·cal** *adj.* **—al·gom′e·try** *n.*

**Al·gon·ki·an** (ăl-gŏng′kē-ən) *n., pl.* **Algonkian** or **-ans.** [After the *Algonkin* (Algonquin) Indians.] **1.** *Geol.* Proterozoic. **2.** *var. of* ALGONQUIAN.

**Al·gon·kin** (ăl-gŏng′kĭn) *n. var. of* ALGONQUIN.

**Al·gon·qui·an** (ăl-gŏng′kwē-ən, -kē-ən) *also* **Al·gon·ki·an** (-kē-ən) *n., pl.* **Algonquian** or **-ans** *also* **Algonkian** or **-ans.** [< ALGONQUIN.] **1.** A family of Indian languages spoken from Labrador to the Carolinas between the Atlantic coast and the Rocky Mountains. **2.** A member of a tribe speaking an Algonquian language. **—Al·gon′qui·an** *adj.*

**Al·gon·quin** (ăl-gŏng′kwĭn, -kĭn) *also* **Al·gon·kin** (-kĭn) *n., pl.* **Algonquin** or **-quins** *also* **Algonkin** or **-kins.** [Canadian Fr.] **1. a.** Any of several Indian tribes formerly inhabiting the region along the Ottawa River and near the northern tributaries of the St. Lawrence River. **b.** A member of one of these tribes. **2.** The Algonquian language of the Algonquins.

**al·go·pho·bi·a** (ăl′gə-fō′bē-ə) *n.* Abnormal fear of pain.

**al·go·rism** (ăl′gə-rĭz′əm) *n.* [ME *algorisme* < OFr. < Med. Lat. *algorismus,* after Muhammad ibn-Musa *Al-Kharzimi* (780–850?).] The Arabic system of numeration : DECIMAL SYSTEM.

**al·go·ris·tic** (ăl′gə-rĭs′tĭk) *adj.* [< ALGORISM.] Yielding a precise answer, as a computational system guaranteeing accurate solution.

**al·go·rithm** (ăl′gə-rĭth′əm) *n.* [Var. of ALGORISM.] *Math.* A mathematical rule or procedure for solving a problem. **—al′go·rith′mic** (-rĭth′mĭk) *adj.*

▲ word history: *Algorithm* originated as a variant spelling of *algorism.* The spelling was probably influenced by the word *arithmetic* or its Greek source *arithm,* "number." With the development of sophisticated mechanical computing devices in the 20th century, however, *algorithm* was adopted as a convenient word for a recursive mathematical procedure, the computer's stock in trade. *Algorithm* has ceased to be used as a variant form of the older word.

**algorithmic language** *n.* A computer language presenting numerical procedures in standard form.

**a·li·as** (ā′lē-əs, āl′yəs) *n.* [Lat., otherwise < *alius,* other.] **1.** An assumed name. **2.** *Electron.* A false signal in telecommunication links from beats between signal frequency and sampling frequency. *—adv.* Otherwise named : also known as <Smith, *alias* Jones>

**A·li Ba·ba** (ä′lē bä′bə, ăl′ē) *n.* A poor woodcutter in the *Arabian Nights* who gains entrance to the treasure cave of the 40 thieves by saying the magic words "Open, Sesame!"

**al·i·bi** (ăl′ə-bī′) *n., pl.* **-bis.** [Lat., elsewhere : *alius,* other +*ubi,* where.] **1.** *Law.* A form of defense whereby a defendant tries to prove that he or she was elsewhere when the crime in question was committed. **2.** *Informal.* An excuse. *—vi.* **-bied, -bi·ing, -bis.** *Informal.* To offer an excuse for oneself.

**al·i·ble** (ăl′ə-bəl) *adj.* [Lat. *alibilis* < *alere,* to nourish.] Having nutritive value : NOURISHING.

**al·i·cy·clic** (ăl′ĭ-sī′klĭk, -sĭk′lĭk) *adj.* [ALI(PHATIC) + CYCLIC.] *Chem.* Of, relating to, or designating chemical compounds having both aliphatic and cyclic structures or characteristics.

**al·i·dade** (ăl′ĭ-dād′) *also* **al·i·dad** (-dăd′) *n.* [Fr. < Med. Lat. *allidada* < *al-'iḍāda,* revolving radius of a circle < *'aḍud,* upper arm.] **1.** An indicator or sighting apparatus on a plane table, used in angular measurement. **2.** A topographic surveying and mapping device with a telescope and graduated vertical circle.

**a·li·en** (ā′lē-ən, āl′yən) *adj.* [ME < OFr. < Lat. *alienus* < *alius,* other.] **1.** Owing political allegiance to another country or government : FOREIGN. **2.** Of, from, or typical of another person, place or thing : UNFAMILIAR. **3.** Inconsistent or opposed in nature or character <Stealing is *alien* to my nature.> *—n.* **1.** An unnaturalized foreign resident of a country. **2.** A member of another family, people, region, or country : FOREIGNER. **3.** An outsider. **4.** *Slang.* A creature from outer space. **5.** *Ecol.* A plant native to one region but naturalized in another. *—vt.* **-ened, -en·ing, -ens.** *Law.* To transfer (property) to a new owner.

**al·ien·a·ble** (āl′yə-nə-bəl, ā′lē-ə-) *adj. Law.* Capable of being transferred to a new owner. **—al′ien·a·bil′i·ty** *n.*

**al·ien·age** (āl′yə-nĭj, ā′lē-ə-) *n.* The legal status of being an alien.

**al·ien·ate** (āl′yə-nāt′, ā′lē-ə-) *vt.* **-at·ed, -at·ing, -ates.** [Lat. *alienare, alienat* < *alienus,* alien.] **1.** To cause to become indifferent or hostile : ESTRANGE <*alienate* one's family> **2.** To remove or dissociate (e.g., oneself). **3.** To cause to be withdrawn <alienate the affections of someone's spouse> **4.** *Law.* To transfer (property) to a new owner. **—al′ien·a′tor** *n.*

**al·ien·a·tion** (āl′yə-nā′shən, ā′lē-ə-) *n.* **1.** The condition of being alienated : ISOLATION. **2.** *Psychol.* A state of estrangement between the self and the objective world or between different aspects of the personality. **3.** *Law.* The act of transferring property or title to it to another.

**al·ien·ee** (āl′yə-nē′, ā′lē-ə-) *n. Law.* A person to whom ownership of property is transferred.

**al·ien·ism** (āl′yə-nĭz′əm, ā′lē-ə-) *n.* Alienage.

**al·ien·ist** (āl′yə-nĭst, ā′lē-ə-) *n.* [Fr. *aliéniste* < *aliéné,* insane < Lat. *alienatus,* p.part. of *alienare,* to deprive of reason. —see ALIENATE.] *Law.* A psychiatrist who has been accepted by a court as an expert on the mental competence of principals or witnesses.

**al·ien·or** (āl′yə-nôr′, ā′lē-ə-) *n. Law.* A person who transfers ownership of property to another.

**al·i·es·ter·ase** (ăl′ĭ-ĕs′tər-ās′, -āz′) *n.* [ALI(PHATIC) + ESTERASE.] An esterase contributing to ester-link hydrolysis, particularly in aliphatic esters.

**a·li·form** (ā′lə-fôrm′, ăl′ə-) *adj.* [Lat. *ala,* wing + -FORM.] Shaped like a wing : ALAR.

**a·light**[1] (ə-līt′) *vi.* **a·light·ed** *or* **a·lit** (ə-lĭt′), **a·light·ing, a·lights.** [ME *alighten* < OE *ālīhtan* : *ā-* (intensive) + *līhtan,* to relieve of a burden < *līht,* light.] **1.** To descend and settle, as after flight : LAND. **2.** To dismount. **3.** *Archaic.* To come upon by chance.

**a·light**[2] (ə-līt′) *adj.* [ME, p.part. of *alighten,* to set on fire < OE *alīhtan* : *a-* (intensive) + *līhtan,* to shine < *lēoht,* a light.] **1.** On fire : BURNING. **2.** Illuminated. **—a·light′** *adv.*

**a·lign** *also* **a·line** (ə-līn′) *v.* **a·ligned, a·lign·ing, a·ligns** *also* **a·lined, a·lin·ing, a·lines.** [Fr. *aligner* < OFr. : *a-,* to (< Lat. *ad-*) + *ligne,* line < Lat. *linea.*] *—vt.* **1.** To place in a line. **2.** To adjust (e.g., parts of a mechanism) to produce a proper condition or relationship. **3.** To ally (e.g., oneself) with one side of a dispute or cause. *—vi.* **1.** To fall into line. **2.** To be in correct adjustment. **—a·lign′er** *n.*

**a·lign·ment** *also* **a·line·ment** (ə-līn′mənt) *n.* **1.** Arrangement or position in a straight line. **2.** The process of adjusting a device or mechanism or the condition of a device or mechanism being adjusted. **3.** A ground plan. **4.** The act of aligning or the condition of being aligned.

**a·like** (ə-līk′) *adj.* [ME *ilike* < OE *gelīc.*] Having close resemblance : SIMILAR <"All good books are *alike*" —Hemingway> *—adv.* In the same manner or to the same degree <They all talk *alike.*> **—a·like′ness** *n.*

**al·i·ment** (ăl′ə-mənt) *n.* [ME < Lat. *alimentum* < *alere,* to nourish.] **1.** Something that nourishes : FOOD. **2.** Something that supports or sustains. *—vt.* (-mĕnt′) **-ment·ed, -ment·ing, -ments.** To supply with sustenance, as food. **—al′i·men′tal** (-mĕn′tl) *adj.* **—al′i·men′tal·ly** *adv.*

**al·i·men·ta·ry** (ăl′ə-mĕn′tə-rē, -trē) *adj.* **1.** Of or relating to food or nutrition. **2.** Giving sustenance or nourishment.

**alimentary canal** *n.* The mucous membrane-lined tube of the digestive system, extending from the mouth to the anus and including the pharynx, esophagus, stomach, and intestines.

**al·i·men·ta·tion** (ăl′ə-mĕn-tā′shən) *n.* **1.** The act or process of giving or receiving nourishment. **2.** Sustenance : support. **—al′i·men′ta·tive** (-tə-tĭv) *adj.*

**al·i·mo·ny** (ăl′ə-mō′nē) *n., pl.* **-nies.** [Lat. *alimonia,* sustenance < *alere,* to nourish.] **1.** *Law.* A court-ordered allowance for support, usu. given by a man to his former wife after a divorce or legal separation. **2.** A means of support : LIVELIHOOD.

**a·line** (ə-līn′) *v. var. of* ALIGN.

**A-line** (ā′līn′) *adj.* [From garments being shaped like a capital *A.*] Having a fitted top and a flared bottom <an *A-line* gown>

**a·line·ment** (ə-līn′mənt) *n. var. of* ALIGNMENT.

**al·i·phat·ic** (ăl′ə-făt′ĭk) *adj.* [< Gk. *aleiphar, aleiphat-,* oil.] Of, pertaining to, or indicating organic chemical compounds in which the carbon atoms are linked in open chains rather than rings.

**al·i·quot** (ăl′ĭ-kwŏt′, -kwət) *adj.* [Fr. *aliquote* < Lat. *aliquot,* some number : *alius,* some + *quot,* how many.] **1.** *Math.* Of, relating to, or indicating an exact divisor or factor of a quantity, esp. of an integer. **2.** Contained exactly or an exact number of times.

**a·lit** (ə-lĭt′) *v. var. p.t. & p.p. of* ALIGHT[1].

**a·live** (ə-līv′) *adj.* [ME < *on live* : *on,* in (< OE) + *live,* life < OE *līf.*] **1.** Having life : LIVING. **2.** In existence or effect : ACTIVE <tried to keep my dreams *alive*> **3.** Full of life : LIVELY. **4.** Now living. —Used as an intensive <the bravest person *alive*> **—alive to.** Sensitive to : aware of <*alive to* the dangers of drugs> **—alive with.** Full of <a room *alive with* chatter> **—a·live′ness** *n.*

☆ ***syns:*** ALIVE, ANIMATE, LIVE, LIVING, VITAL *adj. core meaning* : having life or existence <The patient is still *alive.*> ***ant:*** dead

**a·li·yah** (ä-lē′yä, ə-lē′yə) *n.* [Heb. *'alīyāh,* ascent.] Immigration of Jewish people into Israel.

**a·liz·a·rin** (ə-lĭz′ər-ĭn) *also* **a·liz·a·rine** (-ĭn, -ə-rēn′) *n.* [Fr. *alizarine* < *alizari,* madder root < Sp., prob. < Ar. *al-'aṣārah,* the juice pressed out.] An orange-red compound, $C_{14}H_8O_4$, used in dyes.

**al·ka·hest** (ăl′kə-hĕst′) *n.* [Med. Lat. *alchahest.*] The hypothetical universal solvent once sought by alchemists.

**al·ka·les·cent** (ăl′kə-lĕs′ənt) *adj.* [ALKAL(I) + -ESCENT.] Becoming alkaline or slightly alkaline. **—al′ka·les′cence, al′ka·les′cen·cy** *n.*

**al·ka·li** (ăl′kə-lī′) *n., pl.* **-lis** *or* **-lies.** [ME < Med. Lat. < Ar. *al-qualīy,* the ashes < *qalay,* to fry.] **1.** *Chem.* A carbonate or hydroxide of an alkali metal, whose aqueous solution is bitter, slippery, caustic, and typically basic in reactions. **2.** Any of various soluble mineral salts in natural water and arid soils. **3.** An alkali metal.

**al·ka·li·fy** (ăl-kăl′ə-fī′, ăl′kə-lə-fī′) *vt. & vi.* **-fied, -fy·ing, -fies.** To make or become alkaline.

**alkali metal** *n.* Any of a group of highly reactive metallic elements, including lithium, sodium, potassium, rubidium, cesium, and francium.

**al·ka·lim·e·ter** (ăl′kə-lĭm′ĭ-tər) *n.* A device for measuring alkalinity. **—al′ka·lim′e·try** *n.*

**al·ka·line** (ăl′kə-lĭn, -līn′) *adj.* **1.** Of, relating to, or containing an alkali. **2.** Having a pH greater than 7.

**alkaline earth** *n.* **1.** An oxide of an alkaline-earth metal. **2.** An alkaline-earth metal.

**alkaline-earth metal** *n.* Any of a group of metallic elements, esp. calcium, strontium, and barium, but gen. including beryllium, magnesium, and radium.

**al·ka·lin·i·ty** (ăl′kə-lĭn′ĭ-tē) *n.* The concentration of alkali or alkaline quality of a substance.

**al·ka·lize** (ăl′kə-līz′) *also* **al·ka·lin·ize** (-lə-nīz′) *v.* **-lized, -liz·ing, -liz·es** *also* **-ized, -iz·ing, -iz·es.** *—vt.* To make alkaline. *—vi.* To become an alkali. **—al′ka·li·za′tion** *n.*

**al·ka·loid** (ăl′kə-loid′) *n.* [ALKAL(I) + -OID.] Any of various physiologically active, nitrogen-containing organic bases obtained from plants, including nicotine, quinine, atropine, cocaine, and morphine. **—al'ka·loid'al** (-loid'l) *adj.*

**al·ka·lo·sis** (ăl'kə-lō'sĭs) *n.* [ALKAL(I) + -OSIS.] Abnormally high alkali content in the blood and tissues.

**al·kane series** (ăl'kān') *n.* [ALK(YL) + -ANE.] The paraffin series.

**al·ka·net** (ăl'kə-nĕt') *n.* [ME < Sp. *alcaneta*, dim. of *alcana*, henna < Med. Lat. *alchanna* < Ar. *al-hinnā'*, the henna.] **1. a.** A European plant, *Alkanna tinctoria*, whose roots yield a red dye. **b.** The root of the alkanet or a dye prepared from it. **2.** A hairy plant of the genus *Anchusa*, native to the Old World, with blue flower clusters. **3.** PUCCOON 1a.

**alkanet**
*Viper's bugloss*

**al·kene** (ăl'kēn') *n.* [ALK(YL) + -ENE.] An olefin.

**al·kine** (ăl'kīn') *n. var. of* ALKYNE.

**al·kyd** *or* **al·kyd resin** (ăl'kĭd) *n.* [ALKY(L) + (ACI)D.] A widely used durable synthetic resin derived from glycerol and phthalic anhydride and used in paints.

**al·kyl** (ăl'kəl) *n.* [G. *Alkohol*, alcohol + -YL.] *Chem.* A monovalent radical, as ethyl or propyl, with the general formula $C_nH_{2n+1}$.

**al·kyl·a·tion** (ăl'kə-lā'shən) *n. Chem.* A process in which an alkyl group is added to or substituted in a compound, as in the reaction of olefins with paraffin hydrocarbons to make high-octane fuels.

**al·kyne** *also* **al·kine** (ăl'kīn') *n.* [ALKY(L) + -(I)NE².] Any of a group of open-chain hydrocarbons with a triple bond and the general formula $C_nH_{2n-2}$.

**all** (ôl) *adj.* [ME *al* < OE *all*.] **1.** The total entity or extent of <*all* the West> **2.** The whole number, amount, or quantity of <*all* the guests> **3.** The utmost possible of <in *all* honesty> **4.** Every <*all* manner of trouble><*all* kinds of dogs> **5.** Any whatsoever <beyond *all* question> **6.** Nothing but : ONLY <*all* hair and teeth> *—pron.* **1.** Each and every one <*All* were lost.> **2.** Each and every thing <Ten cars raced and *all* crashed.> *—n.* **1.** Everything one has <gave our *all* to the cause> **2.** The whole number : TOTALITY <*all* of them> *—adv.* **1.** Wholly : entirely <*all* confused> **2.** Each : apiece <The score was seven *all*.> **3.** Exclusively <The mail is *all* for me.> ***usage:*** The phrase *all that* is often used in questions and negative sentences to mean "to the degree expected," as in *The news was not all that unexpected.* This usage is best limited to informal speech. **—all but.** Almost. **—all in.** *Informal.* Exhausted. **—all in all.** Everything being taken into account. **—and all.** And everything else, esp. of a specified kind <learning to drive *and all*> **—at all. 1.** In any and every way. **2.** To any extent. **—for all.** In spite of.

**all-** *pref. var. of* ALLO-.

**al·la breve** (ăl'ə brĕv', ä'lə brĕv'ā) *adv. & adj.* [Ital., according to the breve.] *Mus.* In duple or quadruple meter with the half note being the unit of time.

**Al·lah** (ăl'ə, ä'lə) *n.* [Ar. *Allāh* : *al*, the + *Ilāh*, god.] The Moslem supreme being.

**al·la·man·da** *also* **al·la·man·de** (ăl'ə-măn'də) *n.* [NLat. *Allamanda*, genus name, after Jean N. S. *Allamand* (1713–1787).] A tropical American woody vine of the genus *Allamanda*, with funnel-shaped yellow flowers.

**all-A·mer·i·can** (ôl'ə-mĕr'ĭ-kən) *adj.* **1. a.** Representative of the best or typical of its kind in the United States. **b.** Selected as the best amateur in the United States at a particular sports position or event. **2.** Composed of Americans or American materials exclusively. **3.** Being completely within the territorial limits of the United States. **4.** Of all the Americas. *—n. often* **All-American.** An all-American athlete.

**al·lan·toid** (ə-lăn'toid') *also* **al·lan·toid·al** (ăl'ən-toid'l) *adj.* **1.** Of or having an allantois. **2.** Sausage-shaped. *—n.* The allantois.

**al·lan·to·is** (ə-lăn'tō-ĭs) *n., pl.* **al·lan·to·i·des** (ăl'ən-tō'ĭ-dēz') [NLat. < Gk. *allantoeidēs*, sausage-shaped : *allas*, sausage + *eidos*, shape.] A membranous sac developing from the embryonic hindgut in mammals, birds, and reptiles. **—al'lan·to'ic** (ăl'ən-tō'ĭk) *adj.*

**all-a·round** (ôl'ə-round') *adj. var. of* ALL-ROUND.

**al·lay** (ə-lā') *vt.* **-layed, -lay·ing, -lays.** [ME *aleien* < OE *alecgan* : *a-* (intensive) + *lecgan*, to lay.] **1.** To relieve or lessen (e.g., grief or pain). **2.** To calm <*allay* one's fears> **—al·lay'er** *n.*



**all clear** *n.* **1.** A signal, usu. by siren, that an air raid is over. **2.** An expression signifying absence of immediate obstacles or imminent danger.

**al·le·ga·tion** (ăl'ĭ-gā'shən) *n.* [Fr. *allégation* < Lat. *allegatio* < *allegare*, to adduce : *ad-*, to + *legare*, to depute.] **1.** The act of alleging. **2.** Something alleged. **3.** A statement, as an excuse or plea, offered without proof. **4.** *Law.* An assertion that must be proved or supported with evidence.

**al·lege** (ə-lĕj') *vt.* **-leged, -leg·ing, -leg·es.** [ME *alleggen* < OFr. *alegier* < *esligier*, to disengage < LLat. **exlitigare* : Lat. *ex-*, out + Lat. *litigare*, to sue.] **1.** To state to be true : CLAIM. **2.** To assert without proof. **3.** To state (e.g., a plea or excuse) in support or denial of a claim or accusation. **4.** *Archaic.* To cite or quote, as in confirmation. **—al·lege'a·ble** *adj.* **—al·leg'er** *n.*

**al·leged** (ə-lĕjd', ə-lĕj'ĭd) *adj.* Represented as existing or as being as described but not so proved : SUPPOSED. **—al·leg'ed·ly** (ə-lĕj'ĭd-lē) *adv.*

**Al·le·ghe·ny spurge** (ăl'ĭ-gā'nē) *n.* [After the *Allegheny* Mountains.] A low-growing, shrubby plant, *Pachysandra procumbens* of the southeastern United States, bearing evergreen leaves and white or purplish flower spikes.

**Allegheny vine** *n.* The climbing fumitory.

**al·le·giance** (ə-lē'jəns) *n.* [ME *alligeaunce* < OFr. *ligeance* < *lige*, liege. —see LIEGE.] **1.** Loyalty or the obligation of loyalty, as to a person, nation, sovereign, or cause : FIDELITY. **2.** The obligations of a vassal to an overlord. **—al·le'giant** *adj.*

**al·le·gor·ic** (ăl'ĭ-gôr'ĭk, -gŏr'-) *also* **al·le·gor·i·cal** (-ĭ-kəl) *adj.* Of, pertaining to, or containing allegory. **—al'le·gor'i·cal·ly** *adv.*

**al·le·go·rize** (ăl'ĭ-gô-rīz', -gō-, -gə-) *v.* **-rized, -riz·ing, -riz·es.** *—vt.* **1.** To express as or in the form of an allegory. **2.** To interpret or treat as an allegory. *—vi.* To use or make allegory. **—al'le·go'ri·za'tion** *n.* **—al'le·go·riz'er** *n.*

**al·le·go·ry** (ăl'ĭ-gôr'ē, -gōr'ē) *n., pl.* **-ries.** [ME *allegorie* < Lat. *allegoria* < Gk. < *allēgorein*, to interpret allegorically : *allos*, other + *agoreuein*, to speak.] **1. a.** A literary, dramatic, or pictorial device in which each character, object, and event symbolically illustrates an idea or moral or religious principle. **b.** An instance of allegory. **2.** A symbolic representation. **—al'le·go'rist** *n.*

**al·le·gret·to** (ăl'ĭ-grĕt'ō, ä'lĭ-) [Ital., dim. of *allegro*, allegro.] *Mus.* *—adv.* Slower than allegro but faster than andante. —Used as a direction. *—n., pl.* **-tos.** An allegretto movement or passage. **—al'le·gret'to** *adj.*

**al·le·gro** (ə-lĕg'rō, ə-lā'-) [Ital., lively < Lat. *alacer*.] *Mus.* *—adv.* Faster than allegretto but slower than presto. —Used as a direction. *—n., pl.* **-gros.** An allegro movement or passage. **—al·le'gro** *adj.*

**al·lele** (ə-lēl') *n.* [G. *Allel*, short for *Allelomorph*, allelomorph.] Any of a group of possible mutational forms of a gene. **—al·le'lic** (ə-lē'lĭk, ə-lĕl'ĭk) *adj.* **—al·le'lism** *n.*

**al·le·lo·morph** (ə-lē'lə-môrf', ə-lĕl'ə-) *n.* [Gk. *allēlōn*, mutually (< *allos*, other) + -MORPH.] An allele. **—al·le'lo·mor'phic** (-môr'fĭk) *adj.* **—al·le'lo·mor'phism** (-môr'fĭz'əm) *n.*

**al·le·lu·ia** (ăl'ə-lōō'yə) *interj.* [ME < Med. Lat. *alleluja* < LGk. *allelouia* < Heb. *hallelūyāh*.] Hallelujah.

**al·le·mande** (ăl'ə-mănd', -mänd', ăl'ə-mănd', -mänd') *n.* [Fr., fem. of *allemand*, German < Lat. *Alemanni*, an ancient Germanic tribe.] **1. a.** A stately 16th-cent. dance in 2/2 time. **b.** A musical composition written for or as if for this dance, often used as the first movement of a suite. **2.** A lively 18th-cent. dance in 3/4 time.

**Al·len·ti·ac** (ə-lĕn'tē-ăk') *n., pl.* **Allentiac** *or* **-acs.** [Sp.] **1. a.** A tribe of Indians inhabiting west-central Argentina. **b.** A member of this tribe. **2.** The language of the Allentiac. **—Al·len'ti·ac'** *adj.*

**al·ler·gen** (ăl'ər-jən) *n.* [G. *Allergen* : *Allergie*, allergy + *-gen*, -gen.] Something that causes an allergy. **—al'ler·gen'ic** (-jĕn'ĭk) *adj.*

**al·ler·gic** (ə-lûr'jĭk) *adj.* **1.** Typical of or concerning allergy. **2.** Having an allergy. **3.** *Informal.* Strongly disinclined : AVERSE <*allergic* to housework>

**al·ler·gist** (ăl'ər-jĭst) *n.* A physician specializing in allergies.

**al·ler·gy** (ăl'ər-jē) *n., pl.* **-gies.** [G. *Allergie* : Gk. *allos*, other + Gk. *ergon*, effect.] **1.** Abnormal or pathological reaction to environmental substances, as pollens, foods, dust, or microorganisms, in amounts that do not affect most people. **2.** Anaphylaxis. **3.** *Informal.* An adverse sentiment : ANTIPATHY <an *allergy* to shopping>

**al·le·thrin** (ăl'ə-thrĭn') *n.* [ALL(YL) + (PYR)ETHRIN.] A synthetic insecticide, $C_{19}H_{26}O_3$, similar to pyrethrin.

**al·le·vi·ate** (ə-lē'vē-āt') *vt.* **-at·ed, -at·ing, -ates.** [LLat. *alleviare*, *alleviat-*, to lighten : Lat. *ad-*, to + *levis*, light.] To make less severe or more bearable : REDUCE <took an aspirin to *alleviate* my headache> **—al·le'vi·a'tion** *n.* **—al·le'vi·a'tor** *n.*

**al·le·vi·a·tive** (ə-lē'vē-ā'tĭv) *also* **al·le·vi·a·to·ry** (-ə-tôr'ē, -tōr'ē) *adj.* Helping to alleviate.

**al·ley¹** (ăl'ē) *n., pl.* **-leys.** [ME *alei* < OFr. *alée* < *aller*, to walk < Lat. *ambulare*.] **1.** A narrow street or passageway between or behind city buildings. **2.** A path between trees or flower beds in a park or garden. **3.** A bowling alley. **4.** The parallel lanes on either side of a tennis court used only in doubles matches. **—up (or down) one's alley.** *Slang.* Compatible with one's interests or qualifications.

**al·ley²** (ăl'ē) *n., pl.* **-leys.** [Short for ALABASTER.] A large playing marble, often used as the shooter.

**alley cat** *n.* **1.** A homeless cat : STRAY. **2.** A domestic cat with unknown ancestry.
**al·ley·way** (ăl′ē-wā′) *n.* A narrow passage between buildings.
**all-fired** (ôl′fīrd′) [Alteration of *hell-fired.*] *Slang.* —*adj.* Excessive : extreme. —*adv.* Excessively : extremely.
**All Fools' Day** *n.* April Fools' Day.
**all fours** *n.* **1.** All four limbs of an animal or person. **2.** Seven-up.
**All·hal·low·mas** (ôl′hăl′ō-məs) *also* **All·hal·lows** (ôl′hăl′ōz) *n.* All Saints' Day.
**all-heal** (ôl′hēl′) *n.* A plant, as the self-heal, thought to have healing powers.
**al·li·a·ceous** (ăl′ē-ā′shəs) *adj.* [Lat. *allium,* garlic + -ACEOUS.] Tasting or smelling of onions or garlic.
**al·li·ance** (ə-lī′əns) *n.* [ME < OFr. *aliance* < *alier,* to ally.] **1. a.** A formal pact of confederation between nations in a common cause. **b.** The nations so united. **2.** A union, relationship, or connection by kinship, marriage, or common interest. **3.** A union or conformity of quality or type : AFFINITY. **4.** The act of becoming allied or state of being allied.
☆ ***syns:*** ALLIANCE, ANSCHLUSS, COALITION, CONFEDERACY, CONFEDERATION, FEDERATION, LEAGUE, UNION *n. core meaning* : an association for a common cause <the Western economic *alliance*>
**al·lied** (ə-līd′, ăl′īd′) *adj.* **1.** Joined or united in a close relationship. **2.** Similar : related <music and *allied* interests>
**al·li·ga·tor** (ăl′ĭ-gā′tər) *n.* [< Sp. *el lagarto,* the lizard : *el,* the (< Lat. *ille,* that) + *lagarto,* lizard < Lat. *lacertus.*] **1.** Either of two large amphibious reptiles, *Alligator mississipiensis* of the southeastern United States or *A. sinensis* of China, with sharp teeth, powerful jaws, and a broad short snout. **2.** Leather made from the hide of an alligator. **3.** A tool or fastener having strong, often toothed, adjustable jaws.
**al·li·ga·tor·ing** (ăl′ĭ-gā′tər-ĭng) *n.* [From the resemblance of the cracks to the pattern of an alligator's scales.] The formation of cracks in paint.
**alligator pear** *n.* The avocado.
**alligator snapping turtle** *or* **alligator snapper** *n.* A large freshwater turtle, *Macroclemys temmincki* of the south-central United States, with a hooked beak and rough carapace.
**all-im·por·tant** (ôl′ĭm-pôr′tnt) *adj.* Very important.
**all-in·clu·sive** (ôl′ĭn-klo͞o′sĭv) *adj.* Including everything.
**al·lit·er·ate** (ə-lĭt′ə-rāt′) *v.* **-at·ed, -at·ing, -ates.** [Back-formation < ALLITERATION.] —*vi.* **1.** To use alliteration. **2.** To have or contain alliteration. —*vt.* To form or arrange with alliteration.
**al·lit·er·a·tion** (ə-lĭt′ə-rā′shən) *n.* [AD- + Lat. *littera,* letter.] The occurrence in a phrase or line of speech or writing of two or more words having the same initial sound; e.g., Whitman's line, "all summer in the sound of the sea."
**al·lit·er·a·tive** (ə-lĭt′ə-rā′tĭv, -ər-ə-) *adj.* Of, displaying, or characterized by alliteration. **—al·lit′er·a′tive·ly** *adv.* **—al·lit′er·a′tive·ness** *n.*
**al·li·um** (ăl′ē-əm) *n.* [NLat. *Allium,* genus name < Lat. *allium,* garlic.] A plant of the genus *Allium,* characterized by a pungent odor and including the onion, leek, chive, garlic, and shallot.
**all-night** (ôl′nīt′) *adj.* **1.** Continuing all night. **2.** Open all night.
**all·night·er** (ôl′nī′tər) *n. Slang.* The act or an instance of staying up all night, esp. to complete a project.
**allo-** *or* **all-** *pref.* [Gk. < *allos,* other.] **1.** Other : different <*allopatric*> **2.** Isomeric <*allocholesterol*>
**al·lo·cate** (ăl′ə-kāt′) *vt.* **-cat·ed, -cat·ing, -cates.** [Med. Lat. *allocare, allocat-* : Lat. *ad-,* to + Lat. *locare,* to place < *locus,* place.] **1.** To set aside for a special purpose. **2.** To distribute according to plan : ALLOT. **3.** To determine the location of. **—al′lo·ca·ble** (-kə-bəl) *adj.* **—al′lo·ca′tion** *n.*
**al·lo·cu·tion** (ăl′ə-kyo͞o′shən) *n.* [Lat. *allocutio* < *alloqui,* to speak to : *ad-,* to + *loqui,* to speak.] A formal, authoritative, esp. hortatory speech or address.
**al·log·a·my** (ə-lŏg′ə-mē) *n.* Cross-fertilization. **—al·log′a·mous** *adj.*
**al·lo·graft** (ăl′ə-grăft′) *n.* A homograft.
**al·lo·graph** (ăl′ə-grăf′) *n.* **1.** The configuration of a letter of an alphabet or of any unit in a system of writing. **2.** A letter or combination of letters that can represent one phoneme. **3.** Writing, esp. a signature, made by one person for another.
**al·lom·er·ism** (ə-lŏm′ə-rĭz′əm) *n.* Consistency in crystalline form with variation in chemical composition. **—al·lom′er·ous** *adj.*
**al·lom·e·try** (ə-lŏm′ĭ-trē) *n. Biol.* The study of the change in proportion of various parts of an organism as a consequence of development or growth. **—al′lo·met′ric** (ăl′ə-mĕt′rĭk) *adj.*
**al·lo·morph**[1] (ăl′ə-môrf′) *n.* A paramorph. **—al′lo·mor′phic** (-môr′fĭk) *adj.* **—al′lo·mor′phism** *n.*
**al·lo·morph**[2] (ăl′ə-môrf′) *n.* [ALLO- + MORPH(EME).] Any of the variant forms of a morpheme; e.g., the phonetic *s* of *cats, z* of *dogs,* and *iz* of *horses* are allomorphs of the English morpheme *s.* **—al′lo·mor′phic** (-môr′fĭk) *adj.* **—al′lo·mor′phism** *n.*
**al·lo·nym** (ăl′ə-nĭm′) *n.* [Fr. *allonyme* : Gk. *allos,* other + *onoma,* name.] The name of a usu. historical person assumed by a writer. **—al·lon′y·mous** (ə-lŏn′ə-məs) *adj.* **—al·lon′y·mous·ly** *adv.*
**al·lo·path** (ăl′ə-păth′) *also* **al·lop·a·thist** (ə-lŏp′ə-thĭst) *n.* One who advocates or practices allopathy.
**al·lop·a·thy** (ə-lŏp′ə-thē) *n.* [G. *Allopathie* : Gk. *allos,* other + *-pathie,* -pathy.] Therapy or treatment of disease with remedies that produce effects differing from those of the disease treated. **—al′lo·path′ic** (ăl′ə-păth′ĭk) *adj.* **—al′lo·path′i·cal·ly** *adv.*
**al·lo·pat·ric** (ăl′ə-păt′rĭk) *adj.* [ALLO- + Gk. *patra,* fatherland (< *patēr,* father) + -IC.] *Ecol.* Occurring in separate, widely differing geographic areas. **—al′lo·pat′ri·cal·ly** *adv.*
**al·lo·phane** (ăl′ə-fān′) *n.* [Gk. *allophanēs,* appearing otherwise : *allos,* other + *phainein,* to appear.] An amorphous, translucent, variously colored mineral, chiefly hydrous aluminum silicate.
**al·lo·phone** (ăl′ə-fōn′) *n.* Any of the variant forms of a phoneme; e.g., the aspirated *p* of *pit* and the unaspirated *p* of *spit* are allophones of the English phoneme *p.* **—al′lo·phon′ic** (-fŏn′ĭk) *adj.*
**al·lo·pu·ri·nol** (ăl′ō-pyo͝or′ə-nôl′) *n.* [ALLO- + PURIN(E) + -OL[2].] A drug, $C_5H_4N_4O_3$, used in treating gout.
**all-or-none** (ôl′ər-nŭn′) *adj.* Marked by either complete response or complete lack of response or effect, as in neurological action above a threshold.
**al·lo·steric** (ăl′ə-stĕr′ĭk) *adj.* Of or relating to molecular binding to an enzyme at a site other than the enzymatically active one.
**al·lot** (ə-lŏt′) *vt.* **-lot·ted, -lot·ting, -lots.** [ME *alotten* < OFr. *aloter* : *a-,* to (< Lat. *ad-*) + *lot,* portion, of Germanic orig.] **1.** To distribute by lot : APPORTION. **2.** To give or assign a portion for a particular purpose : ALLOCATE <*allot* five days for completing the report> **—al·lot′ter** *n.*
☆ ***syns:*** ALLOT, ADMEASURE, ALLOCATE, ALLOW, APPORTION, ASSIGN, GIVE, LOT, PORTION *v. core meaning* : to set aside or distribute as a share <*allotted* four ounces of grog to each sailor>
**al·lot·ment** (ə-lŏt′mənt) *n.* **1.** The act of allotting. **2.** Something allotted. **3.** A portion of a military person's pay set aside for his or her dependents or for insurance.
**al·lo·trope** (ăl′ə-trōp′) *n.* [Back-formation < ALLOTROPY.] A structurally different form of an allotropic element.
**al·lot·ro·py** (ə-lŏt′rə-pē) *n.* The existence, esp. in the solid state, of two or more crystalline or molecular structural forms of an element. **—al′lo·trop′ic** (ăl′ə-trŏp′ĭk), **al′lo·trop′i·cal** *adj.* **—al′lo·trop′i·cal·ly** *adv.*
**all' ot·ta·va** (ăl′ə-tä′və, äl′ō-) *adv.* [Ital., at the octave.] *Mus.* At an octave higher or lower than written. —Used as a direction.
**al·lot·tee** (ə-lŏt′ē′) *n.* One to whom something is allotted.
**all out** *adv.* With every possible effort or resource <went *all out*>
**all-out** (ôl′out′) *adj.* Using all one's resources <an *all-out* effort>
**all over** *adv.* **1.** Over the entire area or extent <covered *all over* with flour> **2.** Everywhere <looked *all over* for the missing book> **3.** In every respect <That's your mother *all over.*>
**all-o·ver** (ôl′ō′vər) *adj.* Covering an entire surface.
**al·low** (ə-lou′) *vt.* **-lowed, -low·ing, -lows.** [ME *allouen,* to approve, permit < OFr. *alouer* < both Lat. *allaudare,* to praise, and Med. Lat. *allocare,* to allocate.] **1.** To let do or happen : PERMIT. **2.** To acknowledge or admit : CONCEDE <*allow* the logic of an argument> **3.** To permit to have <*allow* oneself a drink before dinner> **4.** To make provision for <*allow* time for a shower> **5.** To sanction the presence of : let in. **6.** To provide (the needed amount). **7.** To have as one's opinion : BELIEVE. **8.** To grant as a discount or in exchange <*allowed* me $200 on my old car> **—allow for.** To consider <*allow for* their experience> **—allow of.** To admit of : be subject to <an accident that *allows of* several explanations> **—al·low′a·ble** *adj.* **—al·low′a·bly** *adv.*
**al·low·ance** (ə-lou′əns) *n.* **1.** The act or an instance of allowing. **2.** Something allowed. **3.** Something given, as money, at regular intervals or for a particular purpose. **4.** A reduction in price given in exchange for used merchandise : DISCOUNT. **5.** A consideration for possibilities or modifying circumstances <made *allowance* for mistakes> **6.** A permitted difference in dimension of closely mating machine parts. —*vt.* **-anced, -anc·ing, -anc·es. 1.** To restrict to a fixed allowance, as of food. **2.** To provide in a determined quantity.
**al·low·ed·ly** (ə-lou′ĭd-lē) *adv.* By general admission.
**al·loy** (ăl′oi′, ə-loi′) *n.* [OFr. *aloi* < *aloier,* to alloy < Lat. *alligare,* to bind to : *ad-,* to + *ligare,* to bind.] **1.** A homogeneous mixture or solid solution, usu. of two or more metals, the atoms of one taking the place of or occupying interstitial positions between the atoms of the other. **2.** The relative degree of mixture with a base metal : FINENESS. **3.** Something added that reduces purity or value. —*vt.* (ə-loi′, ăl′oi′) **-loyed, -loy·ing, -loys. 1.** To combine (metals) to form an alloy. **2.** To reduce purity or value of (a metal) by mixing with a cheaper metal. **3.** To debase by adding an inferior element.
**all-pur·pose** (ôl′pûr′pəs) *adj.* Useful in many ways.
**all right** *adv.* **1.** Satisfactory <For an old car, it's *all right.*> **2.** Correct <Your conclusions are *all right.*> **3.** Unhurt. **4.** Very well : YES. **5.** Without a doubt <It's a genuine antique, *all right.*> ***usage:*** It is not acceptable to write *all right* as the single form *alright.*

**all-right** (ôl'rīt') *adj. Slang.* **1.** Of sound character : DEPENDABLE. **2.** Good <an *all-right* concert>

**all-round** (ôl'round') *also* **all-a·round** (ôl'ə-round') *adj.* **1.** Including all aspects <*all-round* physical fitness> **2.** Capable of doing many things well : VERSATILE <an *all-round* musician>

**All Saints' Day** *n.* Nov. 1, a religious festival in honor of all saints.

**all·seed** (ôl'sēd') *n.* A plant, as knotgrass, that yields numerous seeds.

**All Souls' Day** *n.* Nov. 2, observed by the Roman Catholic Church as a day of prayer for souls in purgatory.

**all·spice** (ôl'spīs') *n.* **1.** A tropical American tree, *Pimenta officinalis,* bearing small white flowers and aromatic berries. **2.** The dried berries of the allspice, used as a seasoning.

**all-star** (ôl'stär') *adj.* Composed entirely of star performers <an *all-star* musical> —*n.* A player chosen for an all-star sports team.

**all-time** (ôl'tīm') *adj. Informal.* Of all time <an *all-time* favorite of mine>

**all told** *adv.* With everything considered : in all.

**al·lude** (ə-lo͞od') *vi.* **-lud·ed, -lud·ing, -ludes.** [Lat. *alludere,* to play with : *ad-,* to + *ludere,* to play < *ludus,* game.] To refer to something indirectly.

**al·lure** (ə-lo͝or') *v.* **-lured, -lur·ing, -lures.** [ME *aluren* < OFr. *alurer* : *a-,* to (< Lat. *ad-*) + *loirre,* bait, of Germanic orig.] —*vt.* To tempt with something desirable : ENTICE. —*vi.* To tempt or fascinate. —*n.* The power to tempt or entice : FASCINATION. **—al·lure'ment** *n.* **—al·lur'er** *n.* **—al·lur'ing·ly** *adv.*

**al·lu·sion** (ə-lo͞o'zhən) *n.* [LLat. *allusio,* a playing with < Lat. *alludere,* to play with. —see ALLUDE.] **1.** An indirect mention. **2.** An indirect but meaningful or pointed reference.

**al·lu·sive** (ə-lo͞o'sĭv) *adj.* Making or containing allusions : SUGGESTIVE. **—al·lu'sive·ly** *adv.* **—al·lu'sive·ness** *n.*

**al·lu·vi·a** (ə-lo͞o'vē-ə) *n. var. pl. of* ALLUVIUM.

**al·lu·vi·al** (ə-lo͞o'vē-əl) *adj.* Of, relating to, or made up of alluvium.

**alluvial fan** *n.* A fan-shaped deposit of alluvium at the mouth of a ravine.

**al·lu·vi·on** (ə-lo͞o'vē-ən) *n.* [Lat. *alluvio* < *alluere,* to wash against : *ad-,* to + *luere,* to wash.] **1.** Alluvium. **2.** The flow of water against a shore or bank. **3.** Inundation by water : FLOOD. **4.** *Law.* The increasing of land, esp. along a riverbed, by deposited alluvium.

**al·lu·vi·um** (ə-lo͞o'vē-əm) *n., pl.* **-vi·ums** *or* **-vi·a** (-vē-ə) [Lat. < *alluvius,* alluvial < *alluere,* to wash against. —see ALLUVION.] Sedimentary material deposited by flowing water, as in a riverbed or delta.

**al·ly** (ə-lī', ăl'ī') *v.* **-lied, -ly·ing, -lies.** [ME *allien* < OFr. *alier* < Lat. *alligare,* to bid to. —see ALLOY.] —*vt.* **1.** To unite in a formal relationship or bond, as by treaty. **2.** To unite in a personal alliance, as friendship or marriage. —*vi.* To enter into an alliance. —*n.* (ăl'ī', ə-lī'), *pl.* **-lies. 1.** One united with another in a formal or personal relationship. **2.** A friend or close associate.

**al·lyl** (ăl'əl) *n.* [Lat. *allium,* garlic + -YL (so called because it was first obtained from garlic).] The univalent organic radical $CH_2{:}CHCH_2$. **—al·lyl'ic** (ə-lĭl'ĭk) *adj.*

**Al·ma·gest** (ăl'mə-jĕst') *n.* [ME *almageste* < OFr. < Ar. *al-majisti* : *al,* the + Gk. *megistē (suntaxis),* greatest (composition), fem. of *megistos,* greatest, superl. of *megas,* great.] **1.** A detailed chronicle of astronomy and geography compiled by Ptolemy about A.D. 150. **2. almagest.** Any of several medieval treatises on astronomy or alchemy.

**al·ma ma·ter** or **Al·ma Ma·ter** (ăl'mə mä'tər, ăl'mə) *n.* [Lat., nourishing mother.] **1.** The school, college, or university one has attended. **2.** The anthem of an institution of higher learning.

**al·ma·nac** (ôl'mə-năk', ăl'-) *n.* [ME *almenak* < Med. Lat. *almanach,* perh. < Gk. *almenikhiaka,* calendars.] **1.** An annual publication including calendars with weather forecasts, astronomical information, tide tables, and other related tabular information. **2.** An annual publication of lists, charts, and tables of useful information.

**al·man·dine** (ăl'mən-dēn') *also* **al·man·dite** (-dīt') *n.* [ME *alabandine* < Lat. *alabandinus* < *Alabanda,* a town in ancient Asia Minor famous for its jewelry.] A deep violet-red garnet, chiefly $FeAl_2Si_3O_{12}$, found in metamorphic rocks and used as a gemstone.

**al·might·y** (ôl-mī'tē) *adj.* [ME *almighti* < OE *ealmihtig* : *eall,* all + *mihtig,* mighty < *miht,* might.] **1.** Having absolute power : OMNIPOTENT. **2.** *Informal.* Great or extreme <an *almighty* roar> —*adv. Slang.* Extremely <*almighty* thirsty> —*n.* **the Almighty.** GOD 1. **—al·might'i·ly** *adv.*

**al·mond** (ä'mənd, ăm'ənd) *n.* [ME *almande* < OFr. < LLat. *amandula,* alteration of Lat. *amygdala* < Gk. *amugdalē.*] **1.** A small tree native to the Mediterranean region, *Prunus amygdalus,* with pink flowers and fruit containing an edible nut. **2.** The ellipsoid-shaped nut of the almond tree, with a soft, yellowish-tan shell. **3.** Something shaped like an almond. **4.** A pale tan.

**al·mo·ner** (ăl'mə-nər, ä'mə-) *n.* [ME *aumener* < OFr. *aumonier,* ult. < Lat. *eleemosyna,* alms.] **1.** One who distributes alms. **2.** *Chiefly Brit.* A social worker in a hospital.

**al·most** (ôl'mōst', ôl-mōst') *adv.* [ME < OE *ealmæst* : *eall,* all + *mæst,* most.] Very nearly <*almost* finished>

**alms** (ämz) *pl.n.* [ME *almes* < OE *ælmesse* < LLat. *eleemosyna* < Gk. *eleēmosunē* < *eleēmōn,* pitiful < *eleos,* pity.] Money or goods given in charity.

**alms·house** (ämz'hous') *n.* A poorhouse.

**al·ni·co** (ăl'nĭ-kō') *n.* [AL(UMINUM) + NI(CKEL) + CO(BALT).] A strong, hard alloy of aluminum, cobalt, copper, iron, nickel, and occas. niobium or tantalum, used to make durable magnets.

**al·oe** (ăl'ō) *n.* [ME < OE *aluwe* < Lat. *aloe* < Gk. *aloē.*] **1.** Any of various plants of the genus *Aloe,* mostly native to southern Africa, bearing fleshy spiny-toothed leaves and red or yellow flowers. **2. aloes** *(sing. in number).* Bitter aloes. **3. aloes** *(sing. in number).* The fragrant wood of a tropical Asian tree, *Aquilaria agallocha.*

**a·loft** (ə-lôft', ə-lŏft') *adv.* [ME < ON *ā lopt* : *ā,* in + *lopt,* air.] **1.** In or into a high place. **2.** *Naut.* At or toward the upper rigging. —*prep.* On top of <aerialists performing *aloft* the high wire>

**a·lo·ha** (ə-lō'ə, -hə, ä-lō'ä', -hä') *interj.* [Hawaiian.] —Used as a greeting or farewell.

**al·o·in** (ăl'ō-ĭn) *n.* [ALO(E) + -IN.] A bitter crystalline compound derived from the aloe and used as a laxative.

**a·lone** (ə-lōn') *adj.* [ME < *al one,* all one.] **1.** Apart from anything or anyone else : SOLITARY. **2.** Excluding anyone else : ONLY <You *alone* are responsible.> **3.** With nothing further added <The painting *alone* costs $900.> **4.** Without equal : UNIQUE <You are *alone* in your ability to create dissension.> **—leave alone.** To refrain from interrupting or interfering with. **—let alone.** Not to speak of <I haven't one egg to spare, *let alone* a dozen.> **—let well enough alone.** To be satisfied with the status quo. **—stand alone.** To be without equal. **—a·lone'** *adv.* **—a·lone'ness** *n.*

☆ **syns:** ALONE, LONE, LONELY, LONESOME, SOLITARY *adj. core meaning* : apart from all others. ALONE, LONE, and SOLITARY all stress singleness <was *alone* on the beach><a *lone* pine on a hill><*solitary* confinement> LONELY and LONESOME convey a sense of isolation felt as a result of a lack of companionship <the *lonely* life of a widow><were *lonesome* for their old friends>

**a·long** (ə-lông', ə-lŏng') *adv.* [ME < OE *andlang* : *and,* against + *lang,* long.] **1.** In a line with : following the course of. **2.** With a forward motion <move *along*> **3.** In association : TOGETHER <disease *along* with poverty> **4.** As company or a companion <brought the dog *along*> **5.** *Informal.* Advanced to some degree <The morning was well *along.*> —*prep.* **1.** Over, through, or by the length of <roses growing *along* the fence> **2.** In accordance with <act *along* certain guidelines> **—all along.** From the very beginning <right *all along*> **—be along.** *Informal.* To arrive at a place <Will they *be along* soon?> **—get along. 1.** To go onward. **2.** To manage successfully : SURVIVE. **3.** To be compatible. **4.** *Slang.* To go away : get out.

**a·long·shore** (ə-lông'shôr', -shōr', ə-lŏng'-) *adv.* Along, near, or by the shore.

**a·long·side** (ə-lông'sīd', ə-lŏng'-) *adv.* Along, near, at, or to the side. —*prep.* Side by side with <marched *alongside* my friend>

**a·loof** (ə-lo͞of') *adj.* [Obs. *aloof,* toward the wind : A-[2] + obs. *loof,* luff.] Distant, esp. in one's social relations. —*adv.* At a distance, but within view. **—a·loof'ly** *adv.* **—a·loof'ness** *n.*

**al·o·pe·cia** (ăl'ə-pē'shə, -shē-ə) *n.* [Lat. *alopecia,* fox mange < Gk. *alōpekia* < *alōpēx,* fox.] Loss of hair. **—al'o·pe'cic** (-pē'sĭk) *adj.*

**a·loud** (ə-loud') *adv.* **1.** In a loud tone. **2.** With the voice : ORALLY.

**alp** (ălp) *n.* [Back-formation < the *Alps,* a group of mountains in Europe.] A high mountain.

**al·pac·a** (ăl-păk'ə) *n., plural* **alpaca** *or* **-as.** [Sp. < Aymara *allpaca.*] **1.** A domesticated South American mammal, *Lama pacos,* related to the llama, with long, fine, fleecy wool. **2. a.** The wool of the alpaca. **b.** Cloth made from alpaca. **3.** A glossy, usu. black, cotton or rayon and wool fabric.

**al·pen·glow** (ăl'pən-glō') *n.* [Partial transl. of G. *Alpenglühen* : *Alpen,* Alps + *glühen,* to glow.] A rosy glow suffusing snow-covered mountain peaks at sunrise or dusk on a clear day.

**al·pen·horn** (ăl'pən-hôrn') *n.* [G. *Alpenhorn* : *Alpen,* Alps + *Horn,* horn.] An extremely long curved wooden horn used by herdsmen in the Alps.

**alpenhorn**

**al·pen·stock** (ăl'pən-stŏk') *n.* [G. *Alpenstock* : *Alpen,* Alps + *Stock,* staff.] A long iron-tipped staff used by mountain climbers.

**al·pes·trine** (ăl-pĕs'trĭn) *adj.* [< Med. Lat. *alpestris* < *Alpes,* the Alps.] Growing at high altitudes.

**al·pha** (ăl′fə) *n.* [Gk., of Phoenician orig.; akin to Heb. *āleph*, aleph.] **1.** The first letter of the Greek alphabet. —See table at ALPHABET. **2.** The first part : BEGINNING. **3.** *Astron.* The brightest or largest star in a constellation. —*adj.* **1.** First in importance : PRIMARY. **2.** *Chem.* Closest to the functional group of atoms in a molecule. **3.** Alphabetical.

**alpha and omega** *n.* **1.** The first and the last. **2.** The most important element.

**al·pha·bet** (ăl′fə-bĕt′, -bĭt) *n.* [Lat. *alphabetum* < Gk. *alphabētos* : *alpha*, alpha + *beta*, beta, the first two letters of the Greek alphabet.] **1.** The letters of a given language, arranged in a traditional order. **2.** A system of characters or symbols representing sounds, objects, or concepts. **3.** The basic or elementary principles.

**al·pha·bet·i·cal** (ăl′fə-bĕt′ĭ-kəl) *also* **al·pha·bet·ic** (-bĕt′ĭk) *adj.* **1.** Arranged in the usual order of the letters of a language. **2.** Of, relating to, or expressed by an alphabet. **—al′pha·bet′i·cal·ly** *adv.*

**al·pha·bet·ize** (ăl′fə-bĭ-tīz′) *vt.* **-ized, -iz·ing, -iz·es. 1.** To put in alphabetical order. **2.** To express by or supply with an alphabet. **—al′pha·bet′i·za′tion** (ăl′fə-bĕt′ĭ-zā′shən) *n.* **—al′pha·bet·iz′er** *n.*

**alphabet soup** *n. Informal.* A confused mixture, esp. of governmental organizations identified by initials.

**Alpha Cen·tau·ri** (sĕn-tôr′ē) *n.* A double star in Centaurus, the brightest in the constellation.

**Alpha Cru·cis** (kro͞o′sĭs) *n.* A double star in the constellation Crux.

**alpha decay** *n.* Radioactive decay of an atomic nucleus by emission of an alpha particle.

**alpha helix** *n.* A common structure of proteins, characterized by a single chain of amino acids stabilized by hydrogen bonds. **—al′pha-hel′i·cal** (ăl′fə-hĕl′ĭ-kəl, -hē′lĭ-) *adj.*

**al·pha·nu·mer·ic** (ăl′fə-no͞o-mĕr′ĭk, -nyo͞o-) *also* **al·pha·mer·ic** (-fə-mĕr′ĭk) *adj.* [ALPHA(BETIC) + NUMERIC(AL).] **1.** Consisting of alphabetic and numerical symbols. **2.** *Computer Sci.* Consisting of alphabetic and numerical symbols and of punctuation marks, mathematical symbols, and other traditional symbols.

**alpha particle** *n.* A positively charged composite particle, indistinguishable from a helium atom nucleus and having two protons and two neutrons.

**alpha privative** *n.* The Greek negative prefix *a-* which occurs as *an-* before vowels.

**alpha ray** *n.* A stream of alpha particles.

**alpha rhythm** *also* **alpha wave** *n.* The most frequent electroencephalographic waveform found in recordings of the electrical activity of the adult cerebral cortex, usu. 8 to 12 smooth, regular oscillations per second in subjects at rest.

**al·pho·sis** (ăl-fō′sĭs) *n.* [Gk. *alphos*, leprosy + -OSIS.] Abnormal lack of skin pigmentation, as in albinism.

**al·pine** (ăl′pīn′) *adj.* [Lat. *Alpinus* < *Alpes*, the Alps.] **1. Alpine.** Of, relating to, or typical of the Alps or their inhabitants. **2.** Of or relating to high mountains. **3.** *Biol.* Living or growing on mountains above the timberline. **4.** Designed for or concerned with mountaineering. **5. Alpine.** Of or relating to competitive downhill racing and slalom skiing events. **6. Alpine.** Of or relating to a subdivision of the Caucasian race living in the Alps.

**al·pin·ist** *also* **Al·pin·ist** (ăl′pə-nĭst) *n.* A mountain climber : MOUNTAINEER. **—al′pin·ism** *n.*

**al·read·y** (ôl-rĕd′ē) *adv.* [ME *alredi* : *al*, all + *redi*, ready.] **1.** By this or a specified time : PREVIOUSLY. **2.** —Used as an intensive <Stop *already!*>

**al·right** (ôl-rīt′) *adv. Nonstandard.* All right.

**Al·sa·tian** (ăl-sā′shən) *adj.* Of or relating to Alsace, its inhabitants, or their culture. —*n.* **1.** A native or inhabitant of Alsace. **2.** *Chiefly Brit.* The German shepherd.

**al·sike clover** (ăl′sīk′) *n.* [After *Alsike*, Sweden.] A plant native to Eurasia, *Trifolium hybridum*, with compound leaves and pink or whitish flowers, widely grown for forage.

**al·so** (ôl′sō) *adv.* [ME < OE *ealswa* : *eall*, all + *swā*, so.] In addition : LIKEWISE. —*conj.* And in addition.

**al·so-ran** (ôl′sō-răn′) *n. Informal.* One that is defeated, as in a competition, election, or race.

**alt** (ălt) [Lat. *altus*, high.] *Mus.* —*adj.* Pitched in the first octave above the treble staff. —*n.* **1.** The first octave above the treble staff. **2.** A note or tone in the alt octave.

**Al·ta·ic** (ăl-tā′ĭk) *n.* [After the *Altai* Mountains.] A language family of Europe and Asia that includes the Turkic, Tungusic, and Mongolic subfamilies. —*adj.* **1.** Of or relating to the Altai Mountains. **2.** Of or relating to the Altaic language family.

**Al·tair** (ăl-tīr′, -târ′, ăl′tīr′, -târ′) *n.* [Ar. *al-ṭāir* < *al-nasr al-ṭāir*, the flying eagle.] A bright, double, variable star in the constellation Aquila.

**al·tar** (ôl′tər) *n.* [ME *auter* < OE *altar* < Lat. *altare.*] **1.** An elevated place or structure on which sacrifices may be offered or before which religious ceremonies may be enacted. **2.** A table before which the divine offices are recited and on which the Eucharist is celebrated in Christian churches.

**altar boy** *n.* An attendant to an officiating member of the clergy in the performance of a liturgical service : ACOLYTE.

**al·tar·piece** (ôl′tər-pēs′) *n.* Artwork, as a painting or carving, placed above and behind an altar.

**altar rail** *n.* A railing in front of the altar that separates the chancel from the rest of the church.

**altar stone** *n. Rom. Cath. Ch.* A stone slab containing relics that is incorporated into an altar.

**alt·az·i·muth** (ăl-tăz′ə-məth) *n.* [ALT(ITUDE) + AZIMUTH.] A mounting for astronomical telescopes that permits both horizontal and vertical rotation.

**al·ter** (ôl′tər) *v.* **-tered, -ter·ing, -ters.** [ME *alteren* < Med. Lat. *alterare* < Lat. *alter*, other.] —*vt.* **1.** To make different : MODIFY. **2.** To adjust (a garment) for a better fit. **3.** *Informal.* To spay or castrate. —*vi.* To become modified.

**al·ter·a·ble** (ôl′tər-ə-bəl) *adj.* Capable of being changed. **—al′ter·a·bil′i·ty, al′ter·a·ble·ness** *n.* **—al′ter·a·bly** *adv.*

**al·ter·a·tion** (ôl′tə-rā′shən) *n.* **1.** The act or process of altering. **2.** The condition of being altered : MODIFICATION.

**al·ter·a·tive** (ôl′tə-rā′tĭv, -tər-ə-tĭv) *adj.* **1.** Tending to bring about alteration. **2.** *Med.* Tending to restore normal health. —*n. Med.* An alterative treatment or medication.

**al·ter·cate** (ôl′tər-kāt′) *vi.* **-cat·ed, -cat·ing, -cates.** [Lat. *altercari, altercat-*, to quarrel < *alter*, other.] To argue vehemently.

**al·ter·ca·tion** (ôl′tər-kā′shən) *n.* A vehement quarrel.

**al·ter e·go** (ôl′tər ē′gō) *n.* [Lat., other I.] **1.** Another aspect of one's personality. **2.** An intimate friend or constant companion.

**al·ter·nate** (ôl′tər-nāt′, ăl′-) *v.* **-nat·ed, -nat·ing, -nates.** [Lat. *alternare, alternat-* < *alternus*, by turns < *alter*, other.] —*vi.* **1.** To occur in successive turns <Day *alternates* with night.> **2.** To change from one state, action, or place to another regularly <*alternates* between pitcher and catcher> —*vt.* **1.** To perform by turns. **2.** To cause to interchange regularly. —*adj.* (-nĭt). **1.** Happening or following successively. **2.** Designating or relating to every other one of a series <*alternate* rows> **3.** In place of another : SUBSTITUTE <an *alternate* method> **4.** *Bot.* **a.** Growing at alternating intervals on either side of a stem. **b.** Arranged alternately between other parts, as stamens between petals. —*n.* (-nĭt). One acting in the place of another : SUBSTITUTE. **—al′ter·nate·ly** *adv.* **—al′ter·nate·ness** *n.*

**alternate angle** *n.* An angle on one side of a transversal that cuts two lines, having one of the intersected lines as a side.

**alternating current** *n.* An electric current that reverses direction in a circuit at regular intervals.

**al·ter·na·tion** (ôl′tər-nā′shən, ăl′-) *n.* Successive change from one thing to another and back again.

**alternation of generations** *n.* Metagenesis.

**al·ter·na·tive** (ôl-tûr′nə-tĭv, ăl-) *n.* **1.** Choice between two mutually exclusive possibilities or either of these possibilities. **2.** One of a number of things from which one must be chosen. —*adj.* **1.** Necessitating or allowing a choice between two or more than two things. **2.** Existing outside traditional or conventional institutions or systems <an *alternative* church> **—al·ter′na·tive·ly** *adv.*

**alternative box** *n.* An element in a flow chart that signifies a decision to be made.

**alternative school** *n.* A school that is nontraditional, esp. in educational ideals or methods of teaching.

**al·ter·na·tor** (ôl′tər-nā′tər, ăl′-) *n.* An electric generator that produces alternating current.

**al·the·a** *also* **al·thae·a** (ăl-thē′ə) *n.* [Lat., mallows < Gk. *althaia* < *althein*, to heal.] **1.** The rose of Sharon. **2.** A plant of the genus *Althaea*, which includes the hollyhock.

**al·tho** (ôl-thō′) *conj. var. of* ALTHOUGH.

**alt·horn** (ălt′hôrn′) *n.* [G. : *alt*, alto + *Horn*, horn.] An alto saxhorn.

**al·though** *also* **al·tho** (ôl-thō′) *conj.* [ME : *al*, all + *though*, though.] Even though <*Although* I was ill, I went to work.>

**al·tim·e·ter** (ăl-tĭm′ĭ-tər) *n.* [Lat. *altus*, high + -METER.] An apparatus for determining elevation, esp. an aneroid barometer used in aircraft that senses pressure changes caused by changes in altitude. **—al·tim′e·try** *n.*

**al·ti·pla·no** (ăl′tĭ-plä′nō) *n.* [Am. Sp. : Lat. *altus*, high + Lat. *planum*, plain.] A high plateau.

**al·ti·tude** (ăl′tĭ-to͞od′, -tyo͞od′) *n.* [ME < Lat. *altitudo* < *altus*, high.] **1.** The elevation of an object above a reference level, esp. above sea level or above the earth's surface. **2.** *often* **altitudes.** A high area or location. **3.** *Astron.* The angular distance of a celestial object above the horizon. **4.** The perpendicular distance from the base of a geometric figure to the opposite vertex, parallel side, or parallel surface. **5.** A high rank or position. **—al′ti·tu′di·nal** (-to͞od′n-əl, -tyo͞od′-) *adj.*

**altitude sickness** *n.* Illness caused by an oxygen deficiency, as that encountered at high altitudes, and characterized by symptoms such as nausea, breathlessness, and nosebleed.

**al·to** (ăl′tō) *n., pl.* **-tos.** [Ital., high < Lat. *altus.*] *Mus.* **1.** A low female singing voice : CONTRALTO. **2.** A countertenor. **3.** The range between soprano and tenor. **4.** A singer whose voice is within the

# ALPHABET TABLE

## HEBREW

| Forms | Name | Sound |
|---|---|---|
| א | 'aleph<br>'alef | |
| ב | bēth | b (bh) |
| ג | gimel | g (gh) |
| ד | dāleth | d (dh) |
| ה | hē | h |
| ו | vav<br>waw | w |
| ז | zayin | z |
| ח | ḥeth | ḥ |
| ט | ṭeth | ṭ |
| י | yod<br>yodh | y |
| ך כ | kāph | k (kh) |
| ם מ | mēm | m |
| ל | lāmedh | l |
| ן נ | nūn | n |
| ס | samekh | s |
| ע | 'ayin | ' |
| ף פ | pē | p (ph) |
| ץ צ | sade<br>ṣadhe | ṣ |
| ק | qōph | q |
| ר | rēsh | r |
| שׂ | sin | s |
| שׁ | shin | sh |
| ת | tāv<br>tāw | t (th) |

Vowels are not represented in normal Hebrew writing, but for educational purposes they are indicated by a system of subscript and superscript dots. The transliterations shown in parentheses are used when the letter falls at the end of a word. The transliterations with subscript dots are pharyngeal consonants as in Arabic. The second forms shown are used when the letter falls at the end of a word.

## ARABIC

| Forms 1 | 2 | 3 | 4 | Name | Sound |
|---|---|---|---|---|---|
| ﺍ | ﺎ | | | 'alif | ' |
| ﺏ | ﺐ | ﺒ | ﺑ | bā | b |
| ﺕ | ﺖ | ﺘ | ﺗ | tā | t |
| ﺙ | ﺚ | ﺜ | ﺛ | thā | th |
| ﺝ | ﺞ | ﺠ | ﺟ | jīm | j |
| ﺡ | ﺢ | ﺤ | ﺣ | ḥā | ḥ |
| ﺥ | ﺦ | ﺨ | ﺧ | khā | kh |
| ﺩ | ﺪ | | | dāl | d |
| ﺫ | ﺬ | | | dhāl | dh |
| ﺭ | ﺮ | | | rā | r |
| ﺯ | ﺰ | | | zāy | z |
| ﺱ | ﺲ | ﺴ | ﺳ | sīn | s |
| ﺵ | ﺶ | ﺸ | ﺷ | shīn | sh |
| ﺹ | ﺺ | ﺼ | ﺻ | ṣād | ṣ |
| ﺽ | ﺾ | ﻀ | ﺿ | ḍād | ḍ |
| ﻁ | ﻂ | ﻄ | ﻃ | ṭā | ṭ |
| ﻅ | ﻆ | ﻈ | ﻇ | ẓā | ẓ |
| ﻉ | ﻊ | ﻌ | ﻋ | 'ayn | ' |
| ﻍ | ﻎ | ﻐ | ﻏ | ghayn | gh |
| ﻑ | ﻒ | ﻔ | ﻓ | fā | f |
| ﻕ | ﻖ | ﻘ | ﻗ | qāf | q |
| ﻙ | ﻚ | ﻜ | ﻛ | kāf | k |
| ﻝ | ﻞ | ﻠ | ﻟ | lām | l |
| ﻡ | ﻢ | ﻤ | ﻣ | mīm | m |
| ﻥ | ﻦ | ﻨ | ﻧ | nūn | n |
| ﻩ | ﻪ | ﻬ | ﻫ | hā | h |
| ﻭ | ﻮ | | | wāw | w |
| ﻯ | ﻰ | ﻴ | ﻳ | yā | y |

The different forms in the four numbered columns are used when the letters are in: (1) isolation; (2) juncture with a previous letter; (3) juncture with the letters on both sides; (4) juncture with a following letter.

Long vowels are represented by the consonants *'alif* (for ā), *wāw* (for ū), and *yā* (for ī). Short vowels are not usually written; they can, however, be indicated by the following signs: /*fatha* (for *a*), /*kesra* (for *i*), and ʼ*damma* (for *u*).

Transliterations with subscript dots represent "emphatic" or pharyngeal consonants, which are pronounced in the usual way except that the pharynx is tightly narrowed during articulation. When two dots are placed over the *hā*, the new letter thus formed is called *tā marbūta*, and is pronounced (t).

There are several other diacritical marks indicating such situations as the doubling of a consonant or the elision of a vowel.

## GREEK

| Forms | Name | Sound |
|---|---|---|
| Α α | alpha | a |
| Β β | beta | b |
| Γ γ | gamma | g (n) |
| Δ δ | delta | d |
| Ε ϵ | epsilon | e |
| Ζ ζ | zēta | z |
| Η η | ēta | ē |
| Θ θ | thēta | th |
| Ι ι | iota | i |
| Κ κ | kappa | k |
| Λ λ | lambda | l |
| Μ μ | mu | m |
| Ν ν | nu | n |
| Ξ ξ | xi | x |
| Ο ο | omicron | o |
| Π π | pi | p |
| Ρ ρ | rhō | r (rh) |
| Σ σ ς | sigma | s |
| Τ τ | tau | t |
| Υ υ | upsilon | u |
| Φ φ | phi | ph |
| Χ χ | chi<br>khi | kh |
| Ψ ψ | psi | ps |
| Ω ω | ōmega | ō |

The superscript ʽ on an initial vowel or *rhō*, called the rough breathing, represents an aspirate. Lack of aspiration on an initial vowel is indicated by the superscript ʼ, called the smooth breathing. When *gamma* precedes *kappa, xi, khi,* or another *gamma,* it has the value *n* and is so transliterated. The second lowercase form of *sigma* is used only in final position.

## RUSSIAN

| Forms | Sound |
|---|---|
| А а | a |
| Б б | b |
| В в | v |
| Г г | g |
| Д д | d |
| Е е | e |
| Ж ж | zh |
| З з | z |
| И и Й й | i, ĭ |
| К к | k |
| Л л | l |
| М м | m |
| Н н | n |
| О о | o |
| П п | p |
| Р р | r |
| С с | s |
| Т т | t |
| У у | u |
| Ф ф | f |
| Х х | kh |
| Ц ц | ts |
| Ч ч | ch |
| Ш ш | sh |
| Щ щ | shch |
| Ъ ъ | "[1] |
| Ы ы | y |
| Ь ь | '[2] |
| Э э | e |
| Ю ю | yu |
| Я я | ya |

[1]This letter, called the "hard sign," is very rare in modern Russian. It indicates that the previous consonant remains hard even when followed by a front vowel.

[2]This letter, called the "soft sign," indicates that the previous consonant is palatalized even when a front vowel does not follow.

alto range. **5.** An instrument that produces sound within the alto range. **6.** A part written for an alto voice or instrument.

▲ word history: *Alto* in Italian means "high." It is applied to the lowest female singing voice because the range of the female alto is the same as that of the highest male singing voice, which was originally called the alto.

**al·to·cu·mu·lus** (ăl'tō-kyo͞o'myə-ləs) *n.* [Lat. *altus*, high + CUMULUS.] A formation of roundish, fleecy, white or gray clouds.

**al·to·geth·er** (ôl'tə-gĕth'ər) *adv.* [ME *al togeder* : *al*, all + *togeder*, together.] **1.** Completely : entirely <started a new life *altogether*> **2.** With all included or counted <*Altogether* a dozen gifts arrived.> **3.** On the whole <*Altogether*, I'm sorry I went.> —*n.* A whole. **—in the altogether.** *Informal.* Nude.

**al·to-ri·lie·vo** *also* **al·to-re·lie·vo** (ăl'tō-rĭ-lē'vō, ăl'tō-rĕl-yā'vō) *n., pl.* **al·to-re·lie·vos** *also* **al·to-ri·lie·vi** (ăl'tō-rĕl-yā'vē) [Ital. *alto rilievo.*] High relief.

**al·to·stra·tus** (ăl'tō-strā'təs, -străt'əs) *n.* [Lat. *altus*, high + STRATUS.] An extended cloud formation of bluish or gray sheets or layers.

**al·tri·cial** (ăl-trĭsh'əl) *adj.* [< Lat. *altrix*, *altric-*, fem. of *altor*, nourisher < *alere*, to nourish.] Naked and helpless when hatched, as young pigeons.

**al·tru·ism** (ăl'tro͞o-ĭz'əm) *n.* [Fr. *altruisme*, prob. < Ital. *altrui*, someone else < Lat. *alter*, other.] Selfless regard or concern for the well-being of others. **—al'tru·ist** (-ĭst) *n.* **—al'tru·is'tic** *adj.* **—al'tru·is'ti·cal·ly** *adv.*

**al·u·la** (ăl'yə-lə) *n., pl.* **-lae** (-lē') [NLat., dim. of Lat. *ala*, wing.] The feathers attached to the part of a bird's wing corresponding to the thumb. **—al'u·lar** (-lər) *adj.*

**al·um** (ăl'əm) *n.* [ME < OFr. < Lat. *alumen.*] Any of various double sulfates of a trivalent metal such as aluminum, chromium, or iron and a univalent metal such as potassium or sodium, esp. aluminum potassium sulfate, $AlK(SO_4)_2 \cdot 12H_2O$, widely used industrially as clarifiers, hardeners, and purifiers and medicinally as topical astringents and styptics.

**a·lu·mi·na** (ə-lo͞o'mə-nə) *n.* [NLat. < Lat. *alumen*, alum.] Any of several forms of aluminum oxide, $Al_2O_3$, occurring naturally as corundum, in a hydrated form in bauxite, and with various impurities as ruby, sapphire, and emery, used in producing aluminum and in abrasives, refractories, ceramics, and electrical insulation.

**a·lu·mi·nate** (ə-lo͞o'mə-nāt', -nĭt) *n.* A chemical compound having aluminum as part of a negative ion.

**a·lu·mi·nif·er·ous** (ə-lo͞o'mə-nĭf'ər-əs) *adj.* [Lat. *alumen*, *alumin-* + -FEROUS.] Having or yielding aluminum, alumina, or alum.

**al·u·min·i·um** (ăl'yə-mĭn'ē-əm) *n.* [NLat. < *alumina*, alumina.] *Chiefly Brit. var. of* ALUMINUM.

**a·lu·mi·nize** (ə-lo͞o'mə-nīz') *vt.* **-nized, -niz·ing, -niz·es.** To cover or coat with aluminum or aluminum paint.

**a·lu·mi·nous** (ə-lo͞o'mə-nəs) *adj.* Of, relating to, or having aluminum or alum.

**a·lu·mi·num** (ə-lo͞o'mə-nəm) *n.* [ALUMIN(A) + -IUM.] *Symbol* **Al** A silvery-white, ductile metallic element used to form many hard, light, corrosion-resistant alloys; atomic number 13; atomic weight 26.98.

**aluminum oxide** *n.* Alumina.

**aluminum paste** *n.* Aluminum powder ground in oil, used in manufacturing aluminum paints.

**aluminum sulfate** *n.* A white crystalline compound, $Al_2(SO_4)_3$, used in papermaking, water purification, sanitation, and tanning.

**a·lum·na** (ə-lŭm'nə) *n., pl.* **-nae** (-nē') [Lat., fem. of *alumnus.*] A woman graduate or former student of a school, college, or university.

**a·lum·nus** (ə-lŭm'nəs) *n., pl.* **-ni** (-nī') [Lat., pupil < *alere*, to nourish.] A graduate or former student of a school, college, or university. ***usage:*** *Alumni* is gen. used to refer to both the alumni and alumnae of a coeducational institution.

**al·um·root** (ăl'əm-ro͞ot', -ro͝ot') *n.* **1.** A North American plant of the genus *Heuchera*, with small white, reddish, or green flower clusters and astringent roots. **2.** The wild geranium.

**A·lun·dum** (ə-lŭn'dəm). A trademark for a hard, artificial abrasive of fused alumina, used to make oilstones and grinding wheels.

**al·u·nite** (ăl'yə-nīt') *n.* [Fr. < *alun*, alum < Lat. *alumen.*] A gray mineral, essentially $K_2Al_3(OH)_6(SO_4)_3$, used in alum and fertilizer.

**al·ve·o·lar** (ăl-vē'ə-lər) *adj.* [Fr. *alvéolaire* < *alvéole*, alveolus < Lat. *alveolus.*] **1.** Of or relating to an alveolus. **2.** *Anat.* **a.** Relating to the portion of the jaw containing the tooth sockets. **b.** Relating to the alveoli of the lungs. **3.** Formed with the tip of the tongue touching or near the upper alveoli, as the English sounds for *t*, *d*, and *s*. —*n.* An alveolar sound. **—al·ve'o·lar·ly** *adv.*

**al·ve·o·late** (ăl-vē'ə-lĭt) *adj.* Honeycombed with alveoli. **—al·ve'o·la'tion** (-lā'shən) *n.*

**al·ve·o·lus** (ăl-vē'ə-ləs) *n., pl.* **-li** (-lī') [Lat., small hollow, dim. of *alveus*, a hollow < *alvus*, belly.] **1.** A small pit or cavity, as an individual cell of a honeycomb. **2.** A tooth socket in the jawbone. **3.** An air sac of the lungs at the termination of a bronchiole.

**al·ways** (ôl'wāz, -wĭz, -wēz) *adv.* [ME *alwei* < OE *ealne weg*, all the way : *ealne*, accusative of *eall*, all + *weg*, way.] **1.** At every instance : CONSISTENTLY <*always* late> **2.** For all time <We will love them *always.*> **3.** At any time : at will <You can *always* leave if you're dissatisfied.>

**a·lys·sum** (ə-lĭs'əm) *n.* [NLat. *Alyssum*, genus name < Gk. *alusson*, a plant believed to cure rabies : *a-*, not + *lussa*, rabies.] **1.** A plant of the genus *Alyssum*, with dense yellow or white flower clusters. **2.** Sweet alyssum.

**Alz·heim·er's disease** (älts'hī-mərz, ălts'-) *n.* [After Alois *Alzheimer* (1864–1915).] A severe neurological disorder marked by progressive dementia and cerebral cortical atrophy.

**am** (ăm; əm *when unstressed*) *v.* [ME < OE *eom.*] *1st person sing. present indicative of* BE.

**Am** *symbol for* AMERICIUM.

**a·mah** *also* **a·ma** (ä'mə, ä'mä) *n.* [Port. *ama*, nurse < Med. Lat. *amma*, mother.] An oriental maid, esp. a wet nurse.

**a·main** (ə-mān') *adv. Archaic.* **1.** Strongly and intensely. **2.** Speedily or hastily. **3.** Greatly : exceedingly.

**Am·a·lek·ite** (ăm'ə-lĕk'īt', ə-măl'ĭ-kīt') *n.* [Heb. *Amālēqī*, after *Amālēq*, Amalek.] A member of an ancient nomadic tribe thought to be descended from Esau's grandson Amalek.

**a·mal·gam** (ə-măl'gəm) *n.* [ME < OFr. *amalgame* < Med. Lat. *amalgama*, prob. ult. < Gk. *malagma*, soft mass.] **1.** Any of various alloys of mercury and other metals, as with tin or silver. **2.** A combination of diverse elements : MIXTURE.

**a·mal·ga·mate** (ə-măl'gə-māt') *v.* **-mat·ed, -mat·ing, -mates.** —*vt.* **1.** To mix so as to make a unified whole : BLEND. **2.** To mix or alloy (a metal) with mercury. —*vi.* **1.** To mix, unite, or consolidate. **2.** To unite or blend with another metal. **—a·mal'ga·ma'tive** *adj.* **—a·mal'ga·ma'tor** *n.*

**a·mal·ga·ma·tion** (ə-măl'gə-mā'shən) *n.* **1.** The act of amalgamating or condition of being amalgamated. **2.** A consolidation, as of several corporations. **3.** The dissolving of a metal in mercury to form an alloy.

**a·man·dine** (ä'mən-dēn', ăm'ən-) *adj.* [Fr. < *amande*, almond < OFr. *almande.* —see ALMOND.] Made or garnished with almonds.

**am·a·ni·ta** (ăm'ə-nī'tə, -nē'-) *n.* [NLat. *Amanita*, genus name < Gk. *amanitai*, a fungus.] Any of various usu. highly toxic mushrooms of the genus *Amanita*.

**a·man·ta·dine** (ə-măn'tə-dēn') *n.* [Alteration of E. *Adamantane*, a hydrocarbon + -INE[2].] An antiviral drug, $C_{10}H_{17}N{\cdot}HCl$, also used to treat Parkinson's disease.

**a·man·u·en·sis** (ə-măn'yo͞o-ĕn'sĭs) *n., pl.* **-ses** (-sēz') [Lat. *amanuensis* < the phrase *(servus) a manu*, (slave) at handwriting.] A secretary employed to take dictation or copy manuscript.

**am·a·ranth** (ăm'ə-rănth') *n.* [NLat. *Amaranthus*, genus name, alteration of Lat. *amarantus* < Gk. *amarantos*, unfading : *a-*, not + *marainein*, to wither.] **1.** A weedy plant of the genus *Amaranthus*, with small greenish or purplish flower clusters. **2.** An imaginary flower that never fades or dies. **3.** A deep reddish purple to dark or grayish purplish red. **4.** A dark red to purple azo dye.

**am·a·ran·thine** (ăm'ə-răn'thĭn, -thīn') *adj.* **1.** Of, relating to, or like the amaranth. **2.** Lastingly beautiful : UNFADING. **3.** Deep purple.

**am·a·relle** (ăm'ə-rĕl') *n.* [G. < Med. Lat. *amarellum* < Lat. *amarus*, bitter.] A variety of sour cherry with pale red fruit.

**am·a·ret·to** (ăm'ə-rĕt'ō) *n.* [Ital., dim. of *amaro*, bitter < Lat. *amarus.*] An almond-flavored liqueur.

**am·a·ryl·lis** (ăm'ə-rĭl'ĭs) *n.* [NLat. *Amaryllis*, genus name < Lat., name of a shepherdess < Gk. *Amarullis.*] **1.** A bulbous plant, *Amaryllis belladonna*, native to southern Africa, bearing large lilylike reddish or white flowers. **2.** Any of several similar or related plants. **3. Amaryllis.** A conventional name for a shepherdess in classical pastoral poetry.

**a·mass** (ə-măs') *vt.* **a·massed, a·mass·ing, a·mass·es.** [OFr. *amasser* : a-, to (< Lat. *ad-*) + *masse*, mass.] **1.** To gather or pile up : COLLECT. **2.** To accumulate, esp. for profit or pleasure. **—a·mass'a·ble** *adj.* **—a·mass'er** *n.* **—a·mass'ment** *n.*

**am·a·teur** (ăm'ə-tûr', -tər, -ə-cho͝or', -chər, -tyo͝or') *n.* [Fr. < Lat. *amator*, lover < *amare*, to love.] **1.** One who engages in an art, science, study, or athletic activity as a pastime rather than as a profession. **2.** An athlete who has never participated in competition for money or a livelihood. **3.** One lacking professional skill or ease in a given activity or area. —*adj.* **1.** Of, relating to, or performed by an amateur. **2.** Made up of amateurs. **3.** Unskillful or inexperienced. **—am'a·teur·ism** *n.*

☆ ***syns:*** AMATEUR, DILETTANTE, NONPROFESSIONAL *n. core meaning* : one lacking professional skill and ease in a given activity or area <foreign policy mismanaged by *amateurs*> AMATEUR and NONPROFESSIONAL additionally refer to one who engages in an activity for enjoyment rather than money <golfers who are *amateurs*><boxers who are *nonprofessionals*>, whereas DILETTANTE refers to one whose interest in an activity is merely superficial <a scholastic *dilettante*> ***ant:*** expert, professional

**am·a·teur·ish** (ăm'ə-tûr'ĭsh, -cho͝or'-, -tyo͝or'-) *adj.* Typical of an amateur. **—am'a·teur'ish·ly** *adv.* **—am'a·teur'ish·ness** *n.*

**A·ma·ti** (ä-mä'tē) *n.* A violin made by Nicolò Amati or the members of his family.

**am·a·tive** (ăm'ə-tĭv) *adj.* [Lat. *amare, amat-*, to love + -IVE.] Amatory. **—am'a·tive·ly** *adv.* **—am'a·tive·ness** *n.*

**am·a·tol** (ăm'ə-tôl', -tŏl') *n.* [< AM(MONIUM) + (TRINITRO)TOL(UENE).] A highly explosive mixture of ammonium nitrate and trinitrotoluene.

**am·a·to·ry** (ăm'ə-tôr'ē, -tōr'ē) *adj.* [Lat. *amatorius* < *amator*, lover < *amare*, to love.] Of, concerning, or promoting love, esp. sexual love.

**am·au·ro·sis** (ăm'ô-rō'sĭs) *n.* [Gk. *amaurōsis* < *amauros*, dark.] Total loss of vision : BLINDNESS. **—am'au·rot'ic** (-rŏt'ĭk) *adj.*

**a·maze** (ə-māz') *vt.* **a·mazed, a·maz·ing, a·maz·es.** [ME **amasen* < OE *āmasian*, to confound.] **1.** To affect with surprise or great wonder : ASTONISH. **2.** *Obs.* To bewilder. *—n. Archaic.* Wonder. **—a·maz'ed·ly** (ə-mā'zĭd-lē) *adv.* **—a·maz'ed·ness** (-nĭs) *n.*

**a·maze·ment** (ə-māz'mənt) *n.* **1.** A state of extreme surprise or wonder : ASTONISHMENT. **2.** *Obs.* Bewilderment : perplexity.

**Am·a·zon** (ăm'ə-zŏn', -zən) *n.* [ME < Lat. *Amazon* < Gk *Amazōn.*] **1.** *Gk. Myth.* A member of a nation of female warriors alleged to have lived in Scythia near the Black Sea. **2.** *often* **amazon.** A tall, vigorous, strong-willed woman.

▲ word history: The Greeks themselves devised what is probably a folk etymology for *Amazon.* According to Greek legends the Amazons cut off their right breasts to be able to use a bow more easily. The word thus appears to be formed in Greek of the prefix *a-*, "not," and *mazos*, "breast," in reference to the Amazons' mutilated condition. This explanation, however, is probably not the correct one. It is more likely that the word *Amazon* is a Greek spelling of a non-Greek tribal name.

**Am·a·zo·ni·an** (ăm'ə-zō'nē-ən) *adj.* **1.** Typical of or like an Amazon. **2.** *often* **amazonian.** Vigorous and strong-willed. —Used of women.

**am·a·zon·ite** (ăm'ə-zə-nīt') *n.* [After the *Amazon* River.] A green variety of microcline, often used as a semiprecious gemstone.

**amazon stone** *n.* Amazonite.

**am·bage** (ăm'bĭj) *n.* [Back-formation < ME *ambages*, equivocation < Lat. *ambages* : *ambi-*, around + *agere*, to drive.] *Archaic.* **1.** A circuitous pathway. **2.** *often* **ambages.** Roundabout ways. **—am·ba'gious** (ăm-bā'jəs) *adj.*

**am·bas·sa·dor** (ăm-băs'ə-dər, -dôr') *n.* [ME *ambassadour* < OFr. *ambassadeur* < Med. Lat. *ambactia*, mission, of Germanic orig.] **1.** A diplomatic official of the highest rank appointed and accredited as representative in residence by one government to another. **2.** Any of various diplomatic officials of the highest rank <an *ambassador* at large> **3.** A diplomatic official heading a country's permanent mission to an international organization, as the United Nations. **4. a.** An authorized representative or messenger. **b.** An unofficial representative <an *ambassador* of good will> **—am·bas'sa·do'ri·al** (-dôr'ē-əl, -dōr'-) *adj.* **—am·bas'sa·dor·ship'** *n.*

**am·ber** (ăm'bər) *n.* [ME *ambre* < OFr. < Med. Lat. *ambra* < Ar. *ambar*, ambergris.] **1.** A hard, translucent, yellow, orange, or brownish-yellow fossil resin, used for making ornamental objects, esp. jewelry. **2.** A brownish yellow. *—adj.* Brownish yellow.

**am·ber·gris** (ăm'bər-grĭs', -grēs') *n.* [ME < OFr. *ambre gris* : *ambre*, amber + *gris*, gray.] A waxy, grayish substance formed in the intestines of sperm whales and found floating at sea or washed ashore, used in perfumery as a fixative.

**am·ber·jack** (ăm'bər-jăk') *n., pl.* **amberjack** or **-jacks.** A food and game fish of the genus *Seriola*, of temperate and tropical marine waters.

**ambi-** *pref.* [Lat., around.] Both <*ambiversion*>

**am·bi·ance** *also* **am·bi·ence** (ăm'bē-əns, äɴ-byäɴs') *n.* [Fr. < *ambiant*, surrounding < Lat. *ambiens.* —see AMBIENT.] The distinctive atmosphere surrounding or suffusing a person, place, or thing.

**am·bi·dex·ter·i·ty** (ăm'bĭ-dĕk-stĕr'ĭ-tē) *n.* **1.** The quality or state of being ambidextrous. **2.** Hypocrisy or deceit.

**am·bi·dex·trous** (ăm'bĭ-dĕk'strəs) *adj.* [ME *ambidexter*, double dealing < Med. Lat. : Lat. *ambi-*, on both sides + Lat. *dexter*, right-handed.] **1.** Capable of using both hands with equal facility. **2.** Exceptionally dexterous. **3.** Hypocritical or misleading. **—am'bi·dex'trous·ly** *adv.*

**am·bi·ence** (ăm'bē-əns, äɴ-byäɴs') *n. var. of* AMBIANCE.

**am·bi·ent** (ăm'bē-ənt) *adj.* [Lat. *ambiens, ambient-*, pr.part. of *ambire*, to surround : *ambi-*, around + *ire*, to go.] Surrounding.

**am·bi·gu·i·ty** (ăm'bĭ-gyōō'ĭ-tē) *n., pl.* **-ties. 1.** The quality or state of being ambiguous. **2.** Something ambiguous.

**ambiguity error** *n. Computer Sci.* A gross, usu. transient error in the readout of an electronic device that is caused by imprecise synchronism, as in analog-to-digital conversion.

**am·big·u·ous** (ăm-bĭg'yōō-əs) *adj.* [Lat. *ambiguus*, uncertain < *ambigere*, to go about : *ambi-*, around + *agere*, to drive.] **1.** Liable to more than one interpretation. **2.** Uncertain or indefinite <paper of an *ambiguous* hue>

☆ ***syns:*** AMBIGUOUS, CLOUDY, EQUIVOCAL, NEBULOUS, OBSCURE, SIBYLLINE, UNCERTAIN, UNCLEAR, UNEXPLICIT *adj. core meaning* : liable to more than one interpretation <*ambiguous* wording> ***ant:*** clear, explicit

**am·bi·po·lar** (ăm'bĭ-pō'lər) *adj.* Applying equally to both positive ions and electrons in a plasma.

**am·bit** (ăm'bĭt) *n.* [Lat. *ambitus*, a going around < *ambire*, to go around. —see AMBIENT.] **1.** The external boundary : CIRCUIT. **2.** Scope or sphere : RANGE.

**am·bi·tion** (ăm-bĭsh'ən) *n.* [ME *ambicioun* < OFr. *ambition* < Lat. *ambitio* < *ambire*, to go around (for votes). —see AMBIENT.] **1. a.** An eager or strong desire to achieve something. **b.** The object or goal desired. **2.** A desire for exertion or activity : ENERGY <The heat killed my *ambition.*>

☆ ***syns:*** AMBITION, DRIVE, ENTERPRISE, INITIATIVE, PUSH *n. core meaning* : the wish, power, and ability to begin and follow through on a plan or task <young executives with *ambition*>

**am·bi·tious** (ăm-bĭsh'əs) *adj.* **1.** Full of, marked by, or prompted by ambition. **2.** Greatly desirous : EAGER. **3.** Requiring much effort : CHALLENGING. **—am·bi'tious·ly** *adv.* **—am·bi'tious·ness** *n.*

**am·biv·a·lence** (ăm-bĭv'ə-ləns) *n.* [G. *Ambivalenz* : Lat. *ambi-*, on both sides + Lat. *valens*, being strong < *valēre*, to be strong.] **1.** The existence of mutually conflicting emotions or thoughts about a person, object, or idea. **2.** Uncertainty as to what course to follow : INDECISION.

**am·biv·a·lent** (ăm-bĭv'ə-lənt) *adj.* Displaying ambivalence.

**am·bi·ver·sion** (ăm'bĭ-vûr'zhən, -shən) *n.* [AMBI- + (INTRO)VERSION or (EXTRO)VERSION.] A personality trait showing both introversion and extroversion. **—am'bi·vert'** (-vûrt') *n.*

**am·ble** (ăm'bəl) *vi.* **-bled, -bling, -bles.** [ME *amblen* < AN *aumbler* < Lat. *ambulare*, to walk.] **1.** To walk at a leisurely pace. **2.** To proceed smoothly by lifting first both legs on one side and then both on the other, as do horses. *—n.* **1.** An ambling gait, esp. that of a horse. **2.** A leisurely pace : SAUNTER. **—am'bler** *n.*

**am·blyg·o·nite** (ăm-blĭg'ə-nīt') *n.* [G. *Amblygonit* < Gk. *amblugōnios*, obtuse-angled : *amblus*, blunt + *gōnia*, angle.] A white or greenish mineral, $(Li,Na)Al(PO_4)(F,OH)$, a source of lithium.

**am·bly·o·pi·a** (ăm'blē-ō'pē-ə) *n.* [Gk. *ambluōpia* < *ambluōpos*, dim-sighted : *amblus*, dim. + *ōps*, *eye.*] Dimness of vision without apparent physical defect or disease of the eye : LAZY EYE. **—am'bly·o'pic** (-ō'pĭk, -ŏp'ĭk) *adj.*

**am·bo** (ăm'bō') *n., pl.* **am·bos** *or* **am·bo·nes** (ăm-bō'nēz) [Med. Lat. < Gk. *ambōn*, raised edge.] One of the two pulpits or raised stands in early Christian churches from which parts of the liturgy were chanted or read.

**am·boi·na** (ăm-boi'nə) *n. var. of* AMBOYNA.

**Am·boi·nese** (ăm'boi-nēz', -nēs') *n.* The language of Amboina.

**am·boy·na** *also* **am·boi·na** (ăm-boi'nə) *n.* [After *Amboina*, an island in the Moluccas, Indonesia.] The curly-grained, reddish-brown wood of a tree, *Pterocarpus indicus* of southeastern Asia, used for decorative cabinetwork.

**am·bro·sia** (ăm-brō'zhə, -zhē-ə) *n.* [Lat. < Gk. *ambrotos*, immortal, immortalizing : *a-*, not + *brotos*, mortal.] **1.** *Gk. & Rom. Myth.* The food of the gods, reputed to impart immortality. **2.** Something with an esp. delightful flavor or fragrance.

**am·bro·sial** (ăm-brō'zhəl, -zhē-əl) *also* **am·bro·sian** (-zhən, -zhē-ən) *adj.* **1.** Fragrant or delicious. **2.** Of or worthy of the gods. **—am·bro'sial·ly** *adv.*

**am·bro·type** (ăm'brō-tīp') *n.* [Gk. *ambrotos*, immortal + TYPE.] An early type of photograph made by imaging a negative on glass backed by a dark surface.

**am·bry** (ăm'brē) *n., pl.* **-bries.** [ME *aumeneri*, place where alms are distributed < OFr. *aumonerie* < Lat. *armarium*, closet < *arma*, tools.] **1.** A storeroom or cupboard : PANTRY. **2.** A niche near the altar in churches for keeping sacred vessels and vestments.

**ambs·ace** (ām'zās') *n.* [ME *ambes as* < OFr. < Lat. *ambas as*, : *ambo*, both + *as*, unit.] **1.** Double aces, the lowest throw at dice. **2.** Bad luck : MISFORTUNE. **3.** Something insignificant or worthless.

**am·bu·la·crum** (ăm'byə-lā'krəm) *n., pl.* **-cra** (-krə) [Lat., walk planted with trees < *ambulare*, to walk.] One of the five radial areas on the undersurface of an echinoderm, as a starfish, on which the tube feet are borne.

**am·bu·lance** (ăm'byə-ləns) *n.* [Fr. < *(hôpital) ambulant*, mobile (hospital) < Lat. *ambulans*, ambulant.] A vehicle equipped for transporting the sick or injured.

**ambulance chaser** *n. Slang.* **1.** A lawyer or a lawyer's agent who obtains clients by persuading accident victims to sue for damages. **2.** A lawyer avid for clients.

**am·bu·lant** (ăm'byə-lənt) *adj.* [Fr. < Lat. *ambulans*, pr.part. of *ambulare*, to walk.] Moving or walking about.

**am·bu·late** (ăm'byə-lāt') *vi.* **-lat·ed, -lat·ing, -lates.** [Lat. *ambulare, ambulat-*, to walk.] To walk or move about.

**am·bu·la·to·ry** (ăm'byə-lə-tôr'ē, -tōr'ē) *adj.* **1.** Of, relating to, or meant for walking. **2.** Capable of walking : not bedridden. **3.** Not stationary : moving about. **4.** *Law.* Capable of being changed or revoked, as a will during the lifetime of the testator. *—n., pl.* **-ries.** A sheltered place for walking, as a cloister.

**am·bus·cade** (ăm′bə-skād′, ăm′bə-skād′) *n.* [OFr. *embuscade* < Ital. *imboscata* < *imboscare*, to ambush < Med. Lat. **imbuscare* : Lat. *in-*, in + *buscus*, forest, of Germanic orig.] An ambush. —*vt.* **-cad·ed, -cad·ing, -cades.** To ambush. **—am′bus·cad′er** *n.*

**am·bush** (ăm′bo͝osh′) *n.* [ME *embushen*, to ambush < OFr. *embuschier* < Med. Lat. **imbuscare*. —see AMBUSCADE.] **1.** The act of lying in wait to attack by surprise. **2.** A surprise attack made from a hiding place. **3. a.** Those in hiding prepared to make an ambush. **b.** Their hiding place. **4.** A hidden danger or pitfall. —*vt.* **-bushed, -bush·ing, -bush·es. 1.** To hide and wait in ambush. **2.** To attack suddenly and without warning. **—am′bush′er** *n.*
☆ ***syns:*** AMBUSH, AMBUSCADE, BUSHWHACK, SURPRISE, WAYLAY *v. core meaning* : to attack suddenly and without warning <soldiers *ambushed* by partisans>

**a·me·ba** (ə-mē′bə) *n. var. of* AMOEBA.

**am·e·bi·a·sis** *also* **am·oe·bi·a·sis** (ăm′ə-bī′ə-sĭs) *n.* [AMEB(A) + -IASIS.] An infection caused by amoebas, esp. *Entamoeba histolytica.*

**a·me·bic dysentery** *also* **a·moe·bic dysentery** (ə-mē′bĭk) *n.* An acute inflammatory amebiasis of the colon, marked by severe pain and diarrhea and caused by the amoeba *Entamoeba histolytica.*

**a·me·bo·cyte** (ə-mē′bə-sīt′) *n. var. of* AMOEBOCYTE.

**a·meer** (ī-mîr′, ā-mîr′) *n. var. of* EMIR.

**a·me·lio·rate** (ə-mēl′yə-rāt′) *vt.* & *vi.* **-rat·ed, -rat·ing, -rates.** [Alteration of MELIORATE.] To make or become better : IMPROVE. **—a·me′lio·ra′tion** (-rā′shən) *n.*

**a·men** (ā-mĕn′, ä-mĕn′) *interj.* [ME < OE < LLat. *amen* < Gk. *amēn* < Heb. *āmēn*, certainly < *āman*, to strengthen.] —Used at the end of a prayer or a statement to express agreement or approval. —*adv.* Truly : certainly. —*n.* **1.** An utterance of "amen." **2.** An expression of conviction or assent.

**A·men** *also* **A·mon** (ä′mən) *n. Myth.* The Egyptian god of life and reproduction, represented as a man with a ram's head.

**a·me·na·ble** (ə-mē′nə-bəl, ə-mĕn′ə-) *adj.* [Alteration of ME *menable* < OFr. < *mener*, to lead < Lat. *minare*, to drive < *minari*, to threaten < *minae*, threats.] **1. a.** Willing to follow advice or suggestion. **b.** Obedient : tractable. **2.** Responsible to authority : ACCOUNTABLE. **3.** Open or liable to testing, criticism, or judgment <a theory *amenable* to proof>

**amen corner** *n.* A church seat reserved for persons leading responsive amens.

**a·mend** (ə-mĕnd′) *v.* **a·mend·ed, a·mend·ing, a·mends.** [ME *amenden* < OFr. < Lat. *emendare*, to correct : *ex-*, out of + *menda*, fault.] —*vt.* **1.** To improve. **2.** To remove the faults or errors of : CORRECT. **3.** To alter (e.g., a legislative measure) formally by adding, deleting, or rephrasing. —*vi.* To better one's conduct : REFORM.

**a·men·da·to·ry** (ə-mĕn′də-tôr′ē, -tōr′ē) *adj.* Serving or tending to amend : CORRECTIVE.

**a·mend·ment** (ə-mĕnd′mənt) *n.* **1.** An improvement. **2.** A correction. **3. a.** A revision. **b.** A formal statement of such a revision.

**a·mends** (ə-mĕndz′) *pl.n.* [ME *amendes* < OFr., pl. of *amende*, reparation < *amender*, to amend.] Reparation for insult, injury, or loss : RECOMPENSE <make *amends*>

**a·men·i·ty** (ə-mĕn′ĭ-tē, ə-mē′nĭ-) *n., pl.* **-ties.** [ME *amenite* < OFr. < Lat. *amoenitas* < *amoenus*, pleasant.] **1.** The quality of being pleasant or attractive. **2. a.** A feature that increases attractiveness or value, esp. of a piece of property. **b.** Something that increases physical or material comfort. **3. amenities.** Social courtesies : CIVILITIES.

**a·men·or·rhe·a** *or* **a·men·or·rhoe·a** (ā-mĕn′ə-rē′ə) *n.* [A-[1] + Gk. *mēn*, month + -RRHEA.] Abnormal suppression or absence of menstruation. **—a·men′or·rhe′ic** *adj.*

**am·ent**[1] (ăm′ənt, ā′mənt) *n.* [Lat. *amentum*, strap.] A catkin.

**a·ment**[2] (ā′mĕnt′, ā′mənt) *n.* [< Lat. *amens, ament-*, insane : *a-*, out of + *mens*, mind.] A mentally deficient person.

**am·en·ta·ceous** (ăm′ən-tā′shəs, ā′mən-) *adj.* **1.** Resembling or typical of an ament or catkin. **2.** Having aments or catkins.

**a·men·tia** (ā-mĕn′shə, -shē-ə) *n.* [Lat. *amentia* < *amens*, insane. —see AMENT[2].] Subnormal mental development.

**am·en·tif·er·ous** (ăm′ən-tĭf′ər-əs, ā′mən-) *adj. Bot.* Bearing aments or catkins.

**Am·er·a·sian** (ăm′ə-rā′zhən, -shən) *n.* [AMER(ICAN) + ASIAN.] A person of American and Asian descent. **—Am′er·a′sian** *adj.*

**a·merce** (ə-mûrs′) *vt.* **a·merced, a·merc·ing, a·merc·es.** [ME *amercen* < AN *amercier* < *à merci*, at the mercy of : *à*, to (< Lat. *ad*) + *merci*, mercy < Lat. *merces*, wages.] **1.** To punish by a fine imposed arbitrarily at the discretion of a court. **2.** To punish by imposing an arbitrary penalty.

**A·mer·i·can** (ə-mĕr′ĭ-kən) *adj.* **1.** Of, relating to, or typical of the United States of America, its people, culture, government, or history. **2.** Of, in, or relating to North or South America or the Western Hemisphere. **3.** Of or relating to the Indians inhabiting America. **4.** Indigenous to North or South America. —*n.* **1.** A native or inhabitant of America. **2.** A U.S. citizen. **—A·mer′i·can·ness** *n.*

**A·mer·i·ca·na** (ə-mĕr′ə-kä′nə, -kăn′ə, -kā′nə) *pl.n.* A collection of materials relating to American history, geography, or folklore.

**American Beauty** *n.* A long-stemmed rose bearing large purplish-red flowers.

**American cheese** *n.* A smooth, mild, white to yellowish cheddar cheese.

**American dream** *n.* An American ideal of social equality and esp. material success.

**American eagle** *n.* The bald eagle, esp. as it appears on the Great Seal of the United States.

**American elk** *n.* The wapiti.

**American English** *n.* The English language as used in the United States.

**American Indian** *n.* A member of any of the aboriginal peoples of North America (except the Eskimos), South America, and the West Indies, considered to belong to the Mongoloid ethnic division of the human species.

**A·mer·i·can·ism** (ə-mĕr′ĭ-kə-nĭz′əm) *n.* **1.** A custom, trait, or tradition peculiar to the United States. **2.** A usage of language typical of American English. **3.** Allegiance to the United States and its customs and institutions.

**A·mer·i·can·ist** (ə-mĕr′ĭ-kə-nĭst) *n.* **1.** One who studies a facet of America, as its history, geology, or culture. **2.** An anthropologist specializing in the study of American aboriginal culture. **3.** A person, other than a U.S. citizen, who is sympathetic to the United States and its policies.

**American ivy** *n.* Virginia creeper.

**A·mer·i·can·ize** (ə-mĕr′ĭ-kə-nīz′) *v.* **-ized, -iz·ing, -iz·es.** —*vt.* To assimilate into American culture. —*vi.* To become American in methods or spirit. **—A·mer′i·can·i·za′tion** *n.*

**American plan** *n.* A hotel plan whereby a guest is charged a fixed daily rate for room, meals, and service.

**American saddle horse** *n.* A three-gaited or five-gaited high-stepping saddle horse orig. bred in Kentucky.

**American sign language** *n.* An American system of communication for the deaf that uses manual signs.

**American Spanish** *n.* The Spanish language as used in the Western Hemisphere.

**American Standard Version** *n.* A revised version of the King James Bible published in the United States in 1901.

**am·er·i·ci·um** (ăm′ə-rĭsh′ē-əm) *n.* [After *America.*] A white metallic radioactive element used as a radiation source in research; atomic number 95; longest-lived isotope Am 243.

**Am·er·ind** (ăm′ə-rĭnd′) *also* **Am·er·in·di·an** (ăm′ə-rĭn′dē-ən) *n.* [AMER(ICAN) + IND(IAN).] An American Indian or an Eskimo. **—Am′er·in′di·an, Am′er·in′dic** *adj.*

**Am·es·lan** (ăm′ə-slăn′) *n.* American sign language.

**am·e·thop·ter·in** (ăm′ə-thŏp′tə-rən′) *n.* [A(MINO)- + METH- + *pter(oyl)*, a chemical radical + -IN.] Methotrexate.

**am·e·thyst** (ăm′ə-thĭst) *n.* [ME *amatiste* < OFr. < Lat. *amethystus* < Gk. *amethustos* : *a-*, not + *methuein*, to be intoxicated.] **1.** A purple or violet form of transparent quartz used as a gemstone. **2.** A purple variety of corundum used as a gemstone. **3.** A moderate purple to graying reddish purple. **—am′e·thys′tine** *adj.*

**am·e·tro·pi·a** (ăm′ĭ-trō′pē-ə) *n.* [Gk. *ametros*, without measure (*a-*, without + *metron*, measure) + -OPIA.] An eye abnormality, as nearsightedness, farsightedness, or astigmatism, caused by faulty refraction.

**Am·har·ic** (ăm-hăr′ĭk) *n.* [After *Amhara*, a former province of northern Ethiopia.] A Semitic language that is the official language in Ethiopia.

**a·mi·a·ble** (ā′mē-ə-bəl) *adj.* [ME < OFr. < LLat. *amicabilis*, amicable.] **1.** Pleasant in disposition : GOOD-NATURED. **2.** Cordial : congenial <an *amiable* group of people> **—a′mi·a·bil′i·ty, a′mi·a·ble·ness** *n.* **—a′mi·a·bly** *adv.*
☆ ***syns:*** AMIABLE, AFFABLE, AGREEABLE, COMPLAISANT, CORDIAL, EASY, EASYGOING, GENIAL, GOOD-NATURED *adj. core meaning* : pleasant and friendly <an *amiable* companion> ***ant:*** surly

**am·i·an·thus** (ăm′ē-ăn′thəs) *also* **am·i·an·tus** (-təs) *n.* [Lat. *amiantus* < Gk. *amiantos*, undefiled : *a-*, not + *mianein*, to defile.] An asbestos with fine, silky fibers.

**am·i·ca·ble** (ăm′ĭ-kə-bəl) *adj.* [ME < LLat. *amicabilis* < Lat *amicus*, friend.] **1.** Friendly. **2.** Harmonious. **—am′i·ca·bil′i·ty, am′i·ca·ble·ness** *n.* **—am′i·ca·bly** *adv.*

**am·ice** (ăm′ĭs) *n.* [ME, alteration of *amit* < OFr. < Lat. *amictus*, mantle < *amicio*, to wrap around : *ambi-*, around + *jacere*, to throw.] A liturgical vestment made of an oblong piece of white linen worn around the neck and shoulders and partly under the alb.

**a·mi·cus cu·ri·ae** (ə-mē′kəs kyo͝or′ē-ī′) *n., pl.* **a·mi·ci curiae** (ə-mē′kē) [Lat., friend of the court.] *Law.* A professional person or an organization that is not a party to a given litigation but is invited to advise a court on a matter of law in the case.

**a·mid** (ə-mĭd′) *also* **a·midst** (ə-mĭdst′) *prep.* [ME < OE *onmiddan* : *on*, in + *midde*, middle.] In the middle of : surrounded by.

**am·ide** (ăm′īd′, -ĭd) *n.* [AM(MONIA) + -IDE.] **1.** An organic compound, as acetamide, containing the $CONH_2$ group. **2.** A compound with a metal replacing hydrogen in ammonia, as sodium amide, $NaNH_2$. **—a·mid′ic** (ə-mĭd′ĭk, ă-mĭd′-) *adj.*

**am·i·dol** (ăm′ĭ-dôl′, -dōl′) *n.* [G. *Amidol*, a trademark.] A colorless crystalline compound $(NH_2)_2C_6H_3OH{\cdot}2HCl$, used in developing photographs.

**a·mid·ships** (ə-mĭd′shĭps′) *also* **a·mid·ship** (-shĭp′) *adv. Naut.* Halfway way between the bow and the stern.

**a·midst** (ə-mĭdst′) *prep. var. of* AMID.

**a·mi·go** (ə-mē′gō) *n., pl.* **-gos.** [Sp. < Lat. *amicus,* friend.] A friend.

**a·mine** (ə-mēn′, ăm′ĕn′) *n.* [AM(MONIUM) + -INE².] Any of a group of organic compounds of nitrogen, as ethylamine, $C_2H_5NH_2$, that may be regarded as ammonia derivatives in which one or more hydrogen atoms has been replaced by a hydrocarbon radical.

**-amine** *suff.* [< AMINE.] Amine <di*amine*>

**a·mi·no** (ə-mē′nō, ăm′ə-nō′) *adj.* [< AMINO-.] Relating to an amine or other chemical compound having $NH_2$ combined with a nonacid organic radical.

**amino-** *pref.* [< AMINE.] Having $NH_2$ combined with a nonacid organic radical <*amino*pyrine>

**amino acid** *n.* **1.** An organic compound having both an amino group $(NH_2)$ and a carboxylic acid group (COOH). **2.** A compound of the form $NH_2CHRCOOH$, found as essential components of the protein molecule.

**a·mi·no·ac·i·de·mi·a** (ə-mē′nō-ăs′ĭ-dē′mē-ə, ăm′ə-nō-) *n.* A condition marked by excess amino acids in the blood.

**a·mi·no·ac·i·du·ri·a** (ə-mē′nō-ăs′ĭ-dŏŏr′ē-ə, -dyŏŏr′-, ăm′ə-nō-) *n.* A condition marked by excess amino acids in the urine.

**a·mi·no·ben·zo·ic acid** (ə-mē′nō-bĕn-zō′ĭk, ăm′ə-nō-) *n.* Any of three benzoic acid derivatives, $C_7H_7NO_2$, esp. the yellowish para form, which is part of the vitamin B complex.

**a·mi·no·phe·nol** (ə-mē′nō-fē′nōl′, -nôl′, ăm′ə-nō-) *n.* One of three organic compounds with composition $C_6H_4NH_2OH$, used as photographic developers and dye intermediates.

**a·mi·no·py·rine** (ə-mē′nō-pī′rēn′, ăm′ə-nō-) *n.* [AMINO- + (ANTI)-PYRINE.] A colorless crystalline compound, $C_{13}H_{17}N_3O$, occas. used to reduce pain and fever.

**a·mir** (ĭ-mîr′, ā-mîr′) *n. var. of* EMIR.

**A·mish** (ä′mĭsh, ăm′ĭsh, ā′mĭsh) *pl.n.* [G. *amisch,* after Jacob *Amman,* 17th-cent. Swiss Mennonite bishop.] An orthodox Anabaptist sect that separated from the Mennonites in the late 17th cent. and exists today chiefly in southeastern Pennsylvania. *—adj.* Of or relating to the Amish.

**a·miss** (ə-mĭs′) *adj.* [ME *amis* < ON *a mis.*] Out of place or proper order <A patriotic song wouldn't be *amiss* here.> *—adv.* In a wrong or imperfect way <Plans had gone *amiss.*> **—take amiss.** To feel offended by **:** MISUNDERSTAND.

**a·mi·to·sis** (ā′mī-tō′sĭs) *n.* Cell division marked by simple nuclear cleavage without the formation of chromosomes. **—a′mi·tot′ic** (-tŏt′ĭk) *adj.* **—a′mi·tot′i·cal·ly** *adv.*

**am·i·trip·tyl·ine** (ăm′ə-trĭp′tə-lēn′) *n.* [AMI(NO)- + *tript-* (alteration and shortening of TRYPTOPHAN) + -YL + -INE².] An antidepressant drug, $C_{20}H_{23}N$.

**am·i·ty** (ăm′ĭ-tē) *n., pl.* **-ties.** [ME *amite* < OFr. < Med. Lat. **amicitas* < Lat. *amicus,* friend.] Peaceful relations **:** FRIENDSHIP.

**am·me·ter** (ăm′mē′tər) *n.* [AM(PERE) + -METER.] A device for measuring electric current.

**am·mine** (ăm′ēn′, ă-mēn′) *n.* [AMM(ONIA) + -INE².] Any of a class of chemical compounds, as aniline, obtained by replacing hydrogen atoms in ammonia with univalent hydrocarbon radicals. **—am′mi·no′** (ăm′ə-nō′, ə-mē′nō) *adj.*

**am·mo** (ăm′ō) *n.* Ammunition.

**am·mo·nia** (ə-mōn′yə) *n.* [NLat. < Lat. *(sal) ammoniacus,* (salt) of Amen < Gk. *Ammōniakos* < *Ammōn,* Amen (from its having been obtained from a region near the temple of Amen, in Libya).] **1.** A colorless pungent gas, $NH_3$, widely used to make fertilizers and nitrogen-containing organic and inorganic chemicals. **2.** Ammonium hydroxide.

**am·mo·ni·ac¹** (ə-mō′nē-ăk′) *also* **am·mo·ni·a·cal** (ăm′ə-nī′ə-kəl). *adj.* Of, containing, or like ammonia.

**am·mo·ni·ac²** (ə-mō′nē-ăk′) *n.* [ME *ammoniak* < *Ammōniacum* < Gk. *ammōniakon,* of Amen. —see AMMONIA.] A strong-smelling gum resin from the stems of a plant, *Dorema ammoniacum* of northern Asia, formerly used medicinally as an expectorant and stimulant.

**am·mo·ni·ate** (ə-mō′nē-āt′) *vt.* **-at·ed, -at·ing, -ates.** To treat or combine with ammonia. *—n.* A compound containing ammonia. **—am·mo′ni·a′tion** *n.*

**ammonia water** *n.* Ammonium hydroxide.

**am·mon·i·fi·ca·tion** (ə-mŏn′ə-fĭ-kā′shən, ə-mō′nə-) *n.* **1.** Treatment with ammonia or an ammonium compound. **2.** The generation of ammonia or ammonium compounds by the action of bacteria on nitrogenous organic matter in soil.

**am·mon·i·fy** (ə-mŏn′ə-fī′, ə-mō′nə-) *vt. & vi.* **-fied, -fy·ing, -fies.** To subject or be subjected to ammonification. **—am·mon′i·fi′er** *n.*

**am·mo·nite** (ăm′ə-nīt′) *also* **am·mo·noid** (-noid′) *n.* [NLat. *Ammonites* < Lat. *(cornu) Ammonis,* (horn) of Amen.] The coiled, flat, chambered shell of any of various extinct mollusks of the class *Cephalopoda,* found as fossils in Mesozoic formations.

**Am·mon·ite** (ăm′ə-nīt′) *n.* [LLat. *Ammonites,* the Ammonites < Heb. *'Ammōn,* city or people of Amman.] A member of a Semitic people living east of the Jordan River in Old Testament times.

**am·mo·ni·um** (ə-mō′nē-əm) *n.* [AMMON(IA) + -IUM.] The chemical ion $NH_4^+$.

**ammonium carbonate** *n.* A white powder, $(NH_4)HCO_3{\cdot}(NH_4)CO_2NH_2$, used in baking powders, smelling salts, and fire-extinguishing compounds.

**ammonium chloride** *n.* A somewhat hygroscopic white crystalline compound, $NH_4Cl$, used in dry cells, as a soldering flux, and as an expectorant.

**ammonium hydroxide** *n.* A colorless basic aqueous solution of ammonia, $NH_4OH$, used as a household cleanser and in making textiles, rayon, rubber, fertilizer, and plastics.

**ammonium nitrate** *n.* A colorless crystalline salt, $NH_4NO_3$, used in fertilizers, explosives, and solid rocket propellants.

**ammonium sulfate** *n.* A brownish-gray to white crystalline salt, $(NH_4)_2SO_4$, used in fertilizers and water purification.

**am·mo·noid** (ăm′ə-noid′) *n. var. of* AMMONITE.

**am·mu·ni·tion** (ăm′yə-nĭsh′ən) *n.* [OFr. *amunition,* from the phrase *la munition,* the provisioning < Lat. *munitio,* fortification. —see MUNITION.] **1. a.** Projectiles that can be propelled or discharged from guns. **b.** Nuclear, biological, chemical, or explosive material used as a weapon. **2.** A means of attack or defense.

**am·ne·sia** (ăm-nē′zhə) *n.* [NLat. < Gk. *amnēsia,* forgetfulness.] Partial or total loss of memory, esp. from shock, brain injury, illness, or psychological disturbance. **—am·ne′si·ac′** (-nē′zē-ăk′, -zhē-ăk′), **am·ne′sic** (-nē′zĭk, -sĭk) *n. & adj.* **—am·nes′tic** (-nĕs′tĭk) *adj.*

**am·nes·ty** (ăm′nĭ-stē) *n., pl.* **-ties.** [Lat. *amnestia* < Gk. *amnēstos,* not remembered.] A governmental pardon granted to a number of offenders, esp. for political offenses. *—vt.* **-tied, -ty·ing, -ties.** To grant amnesty to.

**am·ni·o·cen·te·sis** (ăm′nē-ō-sĕn-tē′sĭs) *n., pl.* **-ses** (-sēz′) [NLat. : AMNION + Gk. *kentēsis,* act of pricking < *kentein,* to prick.] Surgical removal of a sample of amniotic fluid from a pregnant woman, esp. for use in determining of sex or genetic disorder in the fetus.

**am·ni·og·ra·phy** (ăm′nē-ŏg′rə-fē) *n., pl.* **-phies.** [AMNIO(N) + -GRAPHY.] Radiography of the uterine cavity after injection of a radiopaque substance.

**am·ni·on** (ăm′nē-ən, -ŏn′) *n., pl.* **-ni·ons** *or* **-ni·a** (-nē-ə) [NLat. < Gk. *amniŏn,* plate to hold a sacrificial victim's blood.] A thin, tough, membranous sac of fluid containing the embryo of a mammal, bird, or reptile. **—am′ni·ot′ic** (-ŏt′ĭk), **am′ni·on′ic** (-ŏn′ĭk) *adj.*

**am·ni·os·co·py** (ăm′nē-ŏs′kə-pē) *n., pl.* **-pies.** [AMNIO(N) + -SCOPY.] Endoscopic examination of the amniotic cavity. **—am′ni·o·scope′** (-ə-skōp′) *n.*

**a·moe·ba** *also* **a·me·ba** (ə-mē′bə) *n., pl.* **-bas** *or* **-bae** (-bē) [NLat. < Gk. *amoibē,* change < *ameibein,* to change.] Any of various protozoans of the genus *Amoeba* and related genera, found in water and soil and as internal animal parasites, having a varying form, and moving by means of pseudopodia. **—a·moe′bic** (-bĭk) *adj.*

**am·oe·bi·a·sis** (ăm′ə-bī′ə-sĭs) *n. var. of* AMEBIASIS.

**amoebic dysentery** *n. var. of* AMEBIC DYSENTERY.

**a·moe·bo·cyte** *also* **a·me·bo·cyte** (ə-mē′bə-sīt′) *n.* [AMOEB(A) + -CYTE.] A cell, as a leucocyte, having amoebic form.

**a·moe·boid** (ə-mē′boid′) *adj.* Of or like an amoeba, esp. in changeability of form and means of locomotion.

**a·mok** (ə-mŭk′, ə-mŏk′) *adv. & adj. var. of* AMUCK.

**a·mo·le** (ə-mō′lē) *n.* [Mex. Sp. < Nahuatl *amolli.*] **1.** A chiefly southwestern North America plant with roots, bulbs, or other parts used as soap. **2.** The parts of the amole used as soap.

**A·mon** (ä′mən) *n. var. of* AMEN.

**a·mong** (ə-mŭng′) *also* **a·mongst** (ə-mŭngst′) *prep.* [ME < OE *on gemang* : *on,* in + *gemang,* throng.] **1.** Surrounded by **:** AMID <a house *among* the pines> **2.** In the group, number, or class of <We were *among* the fortunate.> **3.** In the company of <living *among* the poor> **4.** By or with many or most of <rebellion *among* the prisoners> **5.** By the joint action of <*Among* them, they finished the job.> **6.** With shares to each of <Divide this up *among* you.> **7.** With one another <arguing *among* themselves>

**a·mon·til·la·do** (ə-mŏn′tl-ä′dō) *n., pl.* **-dos.** [Sp. : *a-,* to (< Lat. *ad-*) + *Montilla,* a town in Spain.] A pale dry sherry.

**a·mor·al** (ā-môr′əl, ā-mŏr′-) *adj.* **1.** Outside or beyond the sphere of moral distinctions or judgments **:** neither moral nor immoral. **2.** Lacking moral judgment. **—a′mo·ral′i·ty** (ā′mō-răl′ĭ-tē, -mə-), **a·mor′al·ism** *n.* **—a·mor′al·ly** *adv.*

**am·o·ret·to** (ăm′ə-rĕt′ō, ä′mə-) *n., pl.* **-ti** (-tē) *or* **-tos.** [Ital., dim. of *Amore,* Cupid < Lat. *Amor* < *amor,* love < *amare,* to love.] A cupid.

**am·o·rist** (ăm′ə-rĭst) *n.* [Lat. *amor,* love + -IST.] One devoted to lovemaking or to writing about love.

**Am·o·rite** (ăm′ə-rīt′) *n.* [< Heb. *Emōrī.*] A member of a people inhabiting Canaan before the Israelites in Old Testament times.

**am·o·rous** (ăm′ər-əs) *adj.* [ME < OFr. *amoureus* < Med. Lat. *amorosus* < Lat. *amor,* love < *amare,* to love.] **1.** Strongly attracted to love, esp. sexual love. **2.** Indicative of love or sexual desire. **3.** Of or associated with love <*amorous* lyrics> **4.** In love. **—am′or·ous·ly** *adv.* **—am′or·ous·ness** *n.*

**a·mor·phous** (ə-môr′fəs) *adj.* [Gk. *amorphos* : *a-*, without + *morphē*, shape.] **1.** Lacking definite form : SHAPELESS. **2.** Of no particular sort : ANOMALOUS. **3.** Lacking organization or methodical arrangement. **4.** Lacking distinct crystalline structure. **—a·mor′phism** (-fiz′əm) *n.* **—a·mor′phous·ly** *adv.* **—a·mor′phous·ness** *n.*

**am·or·ti·za·tion** (ăm′ər-tĭ-zā′shən, ə-môr′tĭ-) *n.* **1.** The act or process of amortizing. **2.** Money set aside for the purpose of amortizing. **3.** In reckoning the yield of a bond bought at a premium, the periodic subtraction from its current yield of a proportionate share of the premium between the purchase date and the maturity date.

**am·or·tize** (ăm′ər-tīz′, ə-môr′tīz′) *vt.* **-tized, -tiz·ing, -tiz·es.** [ME *amortisen* < OFr. *amortir, amortiss-* < VLat. **admortire* : Lat. *ad-*, to + Lat. *mors*, death.] **1.** To liquidate (a debt) by installment payments or payment into a sinking fund. **2.** To write off (expenditures) by prorating over a certain period. **3.** *Law.* To sell or transfer (property) in mortmain. **—am′or·tiz′a·ble** *adj.*

**a·mor·tize·ment** (ə-môr′tĭz-mənt) *n.* Amortization.

**A·mos** (ā′məs) *n.* [Heb. *ʿAmōs.*] —See table at BIBLE.

**a·mount** (ə-mount′) *n.* [< ME *amounten*, to ascend < OFr. *amonter* < *amont*, upward < Lat. *ad montem*, to the hill < *mons*, hill.] **1.** The total of two or more quantities : AGGREGATE. **2.** A number : sum. **3.** A principal plus its interest, as in a loan. **4.** The total effect or meaning : IMPORT. **5.** Quantity. *—vi.* **a·mount·ed, a·mount·ing, a·mounts. 1.** To total in number or quantity. **2.** To be equivalent or tantamount <silence *amounting* to an admission of guilt>

**a·mour** (ə-mo͝or′) *n.* [ME < OFr. < Lat. *amor*, love < *amare*, to love.] A love affair, esp. an illicit one.

**a·mour-pro·pre** (ə-mo͝or′prôp′rə) *n.* [Fr.] Self-respect.

**A·moy** (ä-moi′, ə-moi′) *n.* [After *Amoy*, former name for Xiamen.] The dialect of Chinese spoken in and around the city of Xiamen and in southern Fujian, eastern Guandong, Hainan Island, and most of Taiwan.

**amp** (ămp) *n.* An ampere.

**AMP** (ā′ĕm-pē′) *n.* [A(DENOSINE) M(ONO)P(HOSPHATE).] A mononucleotide, $C_{10}H_{14}N_5O_7P$, found in animal cells, that is reversibly convertible to ADP and ATP.

**am·pe·lop·sis** (ăm′pə-lŏp′sĭs) *n.* [NLat. *Ampelopsis*, genus name : Gk. *ampelos*, grapevine + Gk. *opsis*, appearance.] A woody vine of the genus *Ampelopsis*, with small greenish or yellowish flowers.

**am·per·age** (ăm′pər-ĭj, ăm′pîr′ĭj) *n.* The strength of an electric current expressed in amperes.

**am·pere** (ăm′pîr′) *n.* [After André Marie *Ampère* (1775–1836).] **1.** A unit of electric current in the meter-kilogram-second system, the steady current that when flowing in straight parallel wires of infinite length, separated by a distance of one meter in free space, produces a force between the wires of $2 \times 10^{-7}$ newtons per meter of length. **2.** A unit in the International System specified as 1 International coulomb per second and equal to 0.999835 ampere.

**am·pere-hour** (ăm′pîr-our′) *n.* The electric charge transferred beyond a specified circuit point by a current of one ampere in one hour.

**am·pere-turn** (ăm′pîr-tûrn′) *n.* A unit of magnetomotive force in the meter-kilogram-second system equal to the magnetomotive force around a path linking one turn of a conducting loop carrying a current of one ampere.

**am·per·sand** (ăm′pər-sănd′) *n.* [Alteration of *and per se and*, "& (the sign) by itself (is the word) and."] The character or sign (&) representing *and.*

**am·phet·a·mine** (ăm-fĕt′ə-mēn′, -mĭn) *n.* [A(LPHA) + M(ETHYL) + PH(ENYL) + ET(HYL) + AMINE.] **1.** A colorless volatile liquid, $C_9H_{13}N$, used chiefly as a central nervous system stimulant. **2.** A phosphate or sulfate of amphetamine, used as a central nervous system stimulant.

**amphi-** *pref.* [Lat. < Gk. < *amphi*, on both sides.] **1. a.** Both <*amphi*biotic> **b.** On both sides <*amphi*stylar> **2.** Around <*amphi*thecium>

**am·phi·ar·thro·sis** (ăm′fē-är-thrō′sĭs) *n., pl.* **-ses** (-sēz′) [AMPHI- + ARTHROSIS.] A more or less immobile joint between bony surfaces connected by ligaments or elastic cartilage.

**am·phib·i·an** (ăm-fĭb′ē-ən) *n.* [< NLat. *Amphibia*, class name < Gk. *amphibion* < *amphibios*, amphibious.] **1.** Any of various cold-blooded, smooth-skinned vertebrate organisms of the class Amphibia, as a frog, that typically hatch as aquatic larvae that breathe by means of gills and metamorphose to an adult form with air-breathing lungs. **2.** An amphibious organism. **3.** An aircraft that can take off and land either on land or on water. **4.** A vehicle designed to travel over land and on water.

**am·phi·bi·ot·ic** (ăm′fə-bī-ŏt′ĭk) *adj.* Living in water during the early stage of development and on land during the adult stage.

**am·phib·i·ous** (ăm-fĭb′ē-əs) *adj.* [Gk. *amphibios* : *amphi-*, on both sides + *bios*, life.] **1.** Capable of living both on land and in water. **2.** Capable of operating on both land and water, as specially designed military vehicles. **3.** Of a mixed or twofold nature. **—am·phib′i·ous·ly** *adv.* **—am·phib′i·ous·ness** *n.*

**am·phi·bole** (ăm′fə-bōl′) *n.* [Fr. < LLat. *amphibolus*, ambiguous < Gk. *amphibolos*, doubtful < *amphiballein*, to throw on either side : *amphi-*, on both sides + *ballein*, to throw.] Any of a large group of structurally similar hydrated double silicate minerals, as hornblende, having mixtures of sodium, calcium, magnesium, iron, and aluminum. **—am′phi·bol′ic** (-bŏl′ĭk) *adj.*

**am·phib·o·lite** (ăm-fĭb′ə-līt′) *n.* A chiefly amphibole rock with minor plagioclase and a small amount of quartz.

**am·phib·o·lous** (ăm-fĭb′ə-ləs) *adj.* [Gk. *amphibolos.* —see AMPHIBOLE.] Having two meanings : AMBIGUOUS.

**am·phi·brach** (ăm′fə-brăk′) *n.* [Lat. *amphibrachys* < Gk. *amphibrakhus* : *amphi-*, on both sides + *brakhus*, short.] A trisyllabic metrical foot having one accented or long syllable between two unaccented or short syllables, as in the word *donation.*

**am·phic·ty·o·ny** (ăm-fĭk′tē-ə-nē) *n., pl.* **-nies.** [Gk. *Amphiktonia* < *amphiktuones*, neighbors : *amphi-*, on the periphery + *ktizein*, to settle.] A group of ancient Greek states sharing a common religious center or shrine. **—am·phic′ty·on′ic** (-ŏn′ĭk) *adj.*

**am·phim·a·cer** (ăm-fĭm′ə-sər) *n.* [Lat. *amphimacrus* < Gk. *amphimakros* : *amphi-*, on both sides + *makros*, long.] A trisyllabic metrical foot having an unaccented or short syllable between two accented or long syllables, as in the word *seventeen.*

**am·phi·mix·is** (ăm′fə-mĭk′sĭs) *n., pl.* **-mix·es** (-mĭk′sēz′) [NLat. : AMPHI- + Gk. *mixis*, a mingling < *mignunai*, to mingle.] True sexual reproduction, with fusion of sperm and egg nuclei. **—am′phi·mic′tic** (-mĭk′tĭk) *adj.*

**Am·phi·on** (ăm-fī′ən) *n.* [Gk. *Amphiōn.*] *Gk. Myth.* The son of Zeus and the twin brother of Zethus, with whom he conquered and fortified Thebes, building a wall around the city by charming the stones into place with the music of his magic lyre.

**am·phi·ox·us** (ăm′fē-ŏk′səs) *n.* [NLat. : AMPHI- + Gk. *oxus*, sharp.] The lancelet.

**am·phi·pod** (ăm′fə-pŏd′) *n.* [< NLat. *Amphipoda*, order name : AMPHI- + Gk. *pous*, foot.] Any of various small crustaceans of the order Amphipoda, which includes the beach fleas.

**am·phip·ro·style** (ăm-fĭp′rō-stīl′, ăm′fĭ-prō′stīl′) *adj.* [Lat. *amphiprostylos* < Gk. *amphiprostulos* : *amphi-*, on both sides + *prostulos*, with pillars in front. —see PROSTYLE.] Having a prostyle or set of columns at each end, but none on the sides. **—am·phip′ro·style′** *n.*

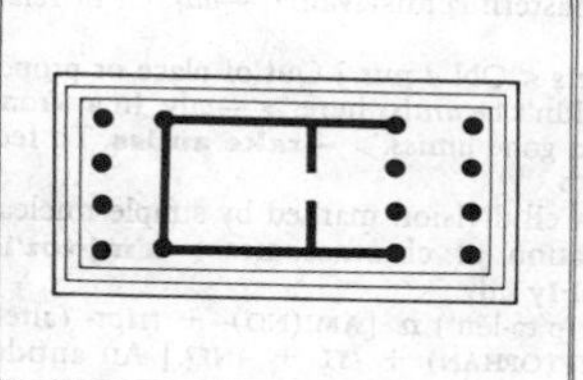

**amphiprostyle**
*Plan of the temple on the Ilissus, Athens*

**am·phis·bae·na** (ăm′fĭs-bē′nə) *n.* [Lat. < Gk. *amphisbaina* : *amphis*, both ways (< *amphi-*, on both sides) + *bainein*, to go.] A mythological serpent having a head at each end of its body and capable of moving in either direction. **—am′phis·bae′nic** *adj.*

**am·phi·sty·lar** (ăm′fĭ-stī′lər) *adj.* [AMPHI- + Gk. *stulos*, pillar.] Having architectural columns at both front and back or on each side.

**am·phi·the·a·ter** (ăm′fə-thē′ə-tər) *n.* [Lat. *amphitheatrum* < Gk. *amphitheatron* : *amphi-*, around + *theatron*, theater. —see THEATER.] **1.** An round or oval structure with tiers of seats rising gradually from a central space or arena. **2.** An arena where contests are held. **3.** A level area surrounded by upward sloping ground. **4.** An upper sloping gallery in a theater. **—am′phi·the·at′ric** (-ăt′rĭk), **am′phi·the·at′ri·cal** *adj.* **—am′phi·the·at′ri·cal·ly** *adv.*

**am·phi·the·a·tre** (ăm′fə-thē′ə-tər) *n. Chiefly Brit.* Amphitheater.

**am·phi·the·ci·um** (ăm′fĭ-thē′shē-əm, -sē-əm) *n., pl.* **-ci·a** (-shē-ə, -sē-ə) [NLat. : AMPHI- + Gk. *thēkion*, dim. of *thēkē*, case.] The external layer of cells of the spore-containing capsule of a moss.

**Am·phi·tri·te** (ăm′fĭ-trī′tē) *n.* [Gk. *Amphitritē.*] *Gk. Myth.* One of the Nereids and goddess of the sea.

**am·phit·ro·pous** (ăm-fĭt′rə-pəs) *adj. Bot.* Partially inverted so the point of attachment is near the middle. —Used of an ovule or seed.

**Am·phit·ry·on** (ăm-fĭt′rē-ən) *n.* [Gk. *Amphitruōn.*] *Gk. Myth.* A king of Thebes and the husband of Alcmene.

**am·pho·ra** (ăm′fər-ə) *n., pl.* **-pho·rae** (-fə-rē′) *or* **-pho·ras.** [Lat. < Gk. *amphoreus*, short for *amphiphoreus* : *amphi-*, on both sides + *phoreus*, bearer < *pherein*, to bear.] A two-handled jar with a narrow neck, used by the ancient Greeks and Romans for carrying oil or wine. **—am′pho·ral** (-fə-rəl) *adj.*

**am·pho·ter·ic** (ăm′fə-tĕr′ĭk) *adj.* [< Gk. *amphoterus*, either of two < *amphō*, both.] Capable of reacting either as an acid or a base.

**am·pi·cil·lin** (ăm′pə-sĭl′ən) *n.* [Blend of AMINO- and PENICILLIN.] An antibiotic related to penicillin that is effective against Gram-negative bacteria, used chiefly to treat urinary, intestinal, and respiratory tract infections.

**am·ple** (ăm′pəl) *adj.* **-pler, -plest.** [ME < OFr. < Lat. *amplus.*] **1. a.** Of large size, amount, extent, or capacity <an *ample* backyard> **b.** Large in degree or kind : generous or abundant. **2.** Sufficient for a particular need <*ample* supplies for the camping trip> **—am′ple·ness** *n.* **—am′ply** (-plē) *adv.*

**am·plex·i·caul** (ăm-plĕk′sĭ-kôl′) *adj.* [NLat. *amplexicaulis,* embracing stem : Lat. *amplexus,* an embracing < *amplector,* to embrace (*ambi-,* around + *plectere,* to twine) + Lat. *caulis,* stem.] Having a base that clasps or encircles the stem <*amplexicaul* leaves>

**am·pli·fi·ca·tion** (ăm′plə-fĭ-kā′shən) *n.* **1.** The act, an instance, or the result of amplifying. **2. a.** An addition to or expansion of a statement or idea. **b.** A statement with such an addition. **3.** *Physics.* **a.** The process of increasing the magnitude of a variable quantity, esp. the magnitude of a voltage or current, without changing any other quality. **b.** The result of such a process. **4.** *Electron.* GAIN[1] 4.

**am·pli·fi·er** (ăm′plə-fī′ər) *n.* **1.** One that amplifies, increases, or extends. **2.** *Electron.* A device to produce amplification, esp. one using transistors or electron tubes of an electrical signal.

**am·pli·fy** (ăm′plə-fī′) *v.* **-fied, -fy·ing, -fies.** [ME *amplifien* < OFr. *amplifier* < Lat. *amplificare* : *amplus,* large + *facere,* to make.] —*vt.* **1.** To make larger or more powerful : INCREASE. **2.** To expand, as by adding details <*amplify* a statement> **3.** To exaggerate. **4.** *Electron.* To produce amplification of. —*vi.* To express at length.

**am·pli·tude** (ăm′plĭ-to͞od′, -tyo͞od′) *n.* [Lat. *amplitudo* < *amplus,* large.] **1.** Greatness of size : MAGNITUDE. **2.** Fullness : plenitude. **3.** Breadth or range, as of intelligence. **4.** *Astron.* The angular distance along the horizon from true east or west to the intersection of the vertical circle of a celestial body with the horizon. **5.** *Physics.* The highest value of a periodically varying quantity. **6.** *Math.* **a.** The highest ordinate value of a periodic curve. **b.** The angle made with the positive horizontal axis by the vector representation of a complex number. **7.** *Electron.* The magnitude of a voltage or current waveform.

**amplitude modulation** *n.* *Electron.* **1.** The encoding of a carrier wave by alteration of its amplitude in accordance with an input signal. **2.** A broadcast system that uses amplitude modulation.

**am·poule** *also* **am·pule** *or* **am·pul** (ăm′po͞ol′, -pyo͞ol′) *n.* [Fr. < OFr. < Lat. *ampulla.*] A small, sealed glass vial used mainly as a container for a hypodermic injection solution.

**am·pul·la** (ăm-po͞ol′ə, -pŭl′ə) *n., pl.* **-pul·lae** (-po͞ol′ē, -pŭl′ē) [Lat.] **1.** A nearly round, two-handled bottle used by the ancient Romans for wine, oil, or perfume. **2.** A vessel for consecrated wine or holy oil. **3.** *Anat.* A small dilatation in a canal or duct, esp. one in the semicircular canal of the ear. **—am·pul′lar** (-ər) *adj.*

**am·pu·tate** (ăm′pyo͞o-tāt′) *vt.* **-tat·ed, -tat·ing, -tates.** [Lat. *amputare, amputat-,* to cut around : *ambi-,* around + *putare,* to cut.] To cut off (e.g., a limb), esp. by surgery. **—am′pu·ta′tion** *n.* **—am′pu·ta′tor** *n.*

**am·pu·tee** (ăm′pyo͞o-tē′) *n.* A person who has had one or more limbs removed by amputation.

**am·ri·ta** (ŭm-rē′tə) *n.* [Skt. *amrtam* : *a-,* without + *mrtam,* death.] **1.** The ambrosia prepared by the Hindu gods that confers immortality. **2.** The immortality achieved by drinking amrita.

**am·trac** *also* **am·track** (ăm′trăk′) *n.* [AM(PHIBIOUS) + TRAC(TOR).] A small, armed, amphibious vehicle first used in World War II to transport troops from ship to shore.

**a·muck** (ə-mŭk′) *also* **a·mok** (ə-mŭk′, ə-mŏk′) [Malay *amok.*] —*adv.* **1.** In a murderous frenzy <rioters running *amuck*> **2.** In a violent or uncontrolled manner <The disease had run *amuck.*> —*adj.* **1.** Crazed with murderous frenzy. **2.** Out of control.

**am·u·let** (ăm′yə-lĭt) *n.* [Lat. *amuletum.*] A charm worn esp. around the neck as a talisman against evil or injury.

**a·muse** (ə-myo͞oz′) *vt.* **a·mused, a·mus·ing, a·mus·es.** [OFr. *amuser,* to stupefy : *a,* to (< Lat. *ad* ) + *muser,* to stare stupidly.] **1.** To occupy in an agreeable or pleasing way : ENTERTAIN. **2.** To cause to laugh or smile by giving pleasure. **3.** *Archaic.* To divert the attentions of in order to deceive. **—a·mus′a·ble** *adj.* **—a·mus′er** *n.*

☆ ***syns:*** AMUSE, DISTRACT, DIVERT, ENTERTAIN *v. core meaning* : to occupy in an agreeable, pleasing way <*amused* myself with a good book>

**a·muse·ment** (ə-myo͞oz′mənt) *n.* **1.** The state of being amused. **2.** Something that amuses : DIVERSION

**amusement park** *n.* A commercial operation that offers entertainment, as rides and games.

**a·mus·ing** (ə-myo͞o′zĭng) *adj.* **1.** Pleasing or entertaining. **2.** Evoking laughter. **—a·mus′ing·ly** *adv.*

**a·mu·sive** (ə-myo͞o′zĭv, -sĭv) *adj.* Providing amusement.

**a·myg·da·la** (ə-mĭg′də-lə) *n., pl.* **-lae** (-lē) [NLat. < Lat., almond < Gk. *amugdalē.*] *Anat.* An almond-shaped mass of gray matter in the anterior portion of the temporal lobe.

**a·myg·dale** (ə-mĭg′dāl) *n.* [Gk. *amygdalē,* almond.] An amygdule.

**a·myg·da·lin** (ə-mĭg′də-lĭn) *n.* [Lat. *amygdala,* almond (< Gk. *amugdalē*) + -IN.] A glycoside obtained from the volatile essential oils of bitter almonds, peaches, and apricots, used in the production of Laetrile.

**a·myg·da·line** (ə-mĭg′də-lĭn, -līn′) *adj.* [Lat. *amygdalinus* < Gk. *amugdalinos* < *amugdalē,* almond.] Of, relating to, or similar to an almond.

**a·myg·da·loid** (ə-mĭg′də-loid′) *n.* [Lat. *amygdala,* almond (< Gk. *amugdalē*) + -OID.] A volcanic rock containing many amygdules. —*adj. also* **a·myg·da·loi·dal** (ə-mĭg′də-loi′dl). **1.** Almond-shaped. **2.** Of or relating to the amygdala. **3.** *Geol.* Like amygdaloid.

**a·myg·dule** (ə-mĭg′dyo͞ol) *n.* [Lat. *amygdala,* almond (from its shape) + (NOD)ULE.] A small gas bubble in igneous rock, as lava, filled with secondary minerals such as zeolite, calcite, or quartz.

**am·yl** (ăm′əl) *n.* [Lat. *amylum,* starch. —see AMYLUM.] **1.** The univalent organic radical $C_5H_{11}$, found in several isomeric forms in numerous organic compounds. **2.** *Slang.* Amyl nitrite.

**amyl-** *pref. var. of* AMYLO-.

**am·y·la·ceous** (ăm′ə-lā′shəs) *adj.* Of, relating to, or like starch : STARCHY.

**amyl acetate** *n.* An organic compound, $CH_3COOC_5H_{11}$, used commercially in isomeric mixtures as a flavoring agent, as a paint and lacquer solvent, and to make penicillin.

**amyl alcohol** *n.* Any of eight isomers of the composition $C_5H_{11}OH$, one of which is the principal constituent of fusel oil.

**am·y·lase** (ăm′ə-lās′, -lāz′) *n.* Any of various enzymes that convert starch to sugar.

**amyl nitrite** *n.* The nitrous acid ester of isoamyl alcohol, used medicinally as a vasodilator.

**amylo-** *or* **amyl-** *pref.* [< Lat. *amylum,* starch.] Starch <*amylose*>

**am·y·loid** (ăm′ə-loid′) *n.* **1.** A starchlike substance. **2.** *Pathol.* A hard protein deposit resulting from tissue degeneration. —*adj.* Starchlike.

**am·y·lol·y·sis** (ăm′ə-lŏl′ə-sĭs) *n.* Enzymatic conversion of starch to sugars. **—am′y·lo·lyt′ic** (-lō-lĭt′ĭk) *adj.*

**am·y·lop·sin** (ăm′ə-lŏp′sĭn) *n.* [AMYLO- + (TRY)PSIN.] The starch-digesting amylase produced by the pancreas.

**am·y·lose** (ăm′ə-lōs′, -lōz′) *n.* The relatively soluble portion of starch.

**am·y·lum** (ăm′ə-ləm) *n.* [Lat. < Gk. *amulon,* starch < *amulos,* not ground at a mill : *a-,* not + *mulē,* mill.] Starch.

**a·my·o·to·ni·a** (ā′mī-ə-tō′nē-ə) *n.* Lack of normal muscle tone.

**an[1]** (ən; ăn *when stressed*) *indef. art.* [ME < OE *ān,* one.] A[1]. —Used before words beginning with a vowel or with an unpronounced *h* <*an* emerald><*an* hour> ***usage:*** *An* should not be used before words such as *historical* and *hysterical* unless the *h* is not pronounced, a practice now uncommon in American speech.

**an[2]** *also* **an′** (ən; ăn *when stressed*) *conj.* [ME, short for *and,* and < OE.] *Archaic.* And if : IF <"*An* I may hide my face, let me play Thisby too" —Shakespeare>

**an-** *pref. var. of* A-[1]. —Used before vowels and frequently before *h.*

**-an[1]** *suff.* [ME < OFr. < Lat. *-anus,* adj. and n. suffix.] **1.** Of, relating to, or like <brachyur*an*> **2.** One relating to, belonging to, or like <librari*an*>

**-an[2]** *suff.* [Alteration of -ANE.] **1. a.** Unsaturated hydrocarbon <ureth*an*> **b.** Heterocyclic compound <fur*an*> **2.** Anhydride of a carbohydrate <dextr*an*>

**an·a[1]** (ăn′ə, ä′nə) *n., pl.* **ana** *or* **-as.** [< -ANA.] A collection of materials or an item in such a collection that reflects the character of a person or place.

**an·a[2]** (ăn′ə) *adv.* [ME < Med. Lat. < Gk., at the rate of.] Both in the same quantity. —Used to refer to prescription ingredients.

**ana-** *pref.* [Gk. < *ana,* up.] **1.** Upward : up <*ana*bolism> **2.** Backward : back <*ana*plasia> **3.** Again : anew <*ana*phylaxis>

**-ana** *or* **-iana** *suff.* [NLat. < Lat. *-ana,* neuter pl. of *-anus,* -an.] A collection of materials relating to a specified person or place <Americ*ana*><Shakespear*iana*>

**an·a·bae·na** (ăn′ə-bē′nə) *n.* [NLat. *Anabaena,* genus name < Gk. *anabainein,* to go up : *ana-,* up + *bainein,* to go.] Any of various freshwater algae of the genus *Anabaena,* sometimes causing a bad taste and odor in drinking water.

**An·a·bap·tist** (ăn′ə-băp′tĭst) *n.* [NLat. *anabaptista* : Gk. *ana-,* again + *baptizein,* to baptize < *baptein,* to dip.] A member of one of the radical movements of the Reformation of the 16th cent. that believed that only adult baptism was valid and that true Christians should not bear arms, use force, or hold government office. **—An′a·bap′tism** *n.* **—An′a·bap′tist** *adj.*

**an·a·bas** (ăn′ə-băs) *n.* [NLat. *Anabas,* genus name < Gk. *anabas,* climbing, pr.part. of *anabainein,* to go up. —see ANABATIC.] A member of the genus *Anabas,* which includes freshwater fishes of Africa and Asia.

**a·nab·a·sis** (ə-năb′ĭ-sĭs) *n., pl.* **-ses** (-sēz′) [Gk. < *anabainein,* to go up. —see ANABATIC.] A large-scale military advance.

**an·a·bat·ic** (ăn′ə-băt′ĭk) *adj.* [Gk. *anabatikos* < *anabainein,* to rise: *ana,* up + *bainein,* to go.] Of or relating to rising wind currents.

**an·a·bi·o·sis** (ăn′ə-bī-ō′sĭs) *n.* [NLat. < Gk. *anabiōsis* < *anabioun,* to return to life : *ana-,* back + *bioun,* to live < *bios,* life.] Restoration to life from a deathlike state. **—an′a·bi·ot′ic** (-ŏt′ĭk) *adj.*

**anabolic steroid** *n.* A synthetic hormone often used to increase muscle size and strength.

**a·nab·o·lism** (ə-năb′ə-lĭz′əm) *n.* [ANA- + (META)BOLISM.] The process by which simple substances are synthesized into the complex materials of living tissue. **—an′a·bol′ic** (ăn′ə-bŏl′ĭk) *adj.*

**a·nach·ro·nism** (ə-năk′rə-nĭz′əm) *n.* [Fr. *anachronisme* < Gk. *anakhronismos* < *anakronizein,* to be an anachronism : *ana-,* backward + *khronizein,* to belong in time < *khronos,* time.] **1.** The representation of something as existing or occurring at other than its proper or historical time. **2.** Something that is out of its proper time. **—a·nach′ro·nis′tic, a·nach′ro·nous** (-nəs) *adj.* **—a·nach′ro·nis′ti·cal·ly, a·nach′ro·nous·ly** *adv.*

**an·a·cli·sis** (ăn′ə-klī′sĭs, ə-năk′lĭ-sĭs) *n.* [Gk. *anaklisis,* a leaning back < *anaklinein,* to lean on : *ana-,* on + *klinein,* to lean.] Psychological dependence on others. **—an′a·clit′ic** (-klĭt′ĭk) *adj.*

**an·a·co·lu·thon** (ăn′ə-kə-lo͞o′thŏn′) *n., pl.* **-thons** *or* **-tha** (-thə) [LLat. < Gk. *anakolouthos,* inconsistent : *an-,* not + *akolouthos,* following.] An abrupt change within a sentence to a second grammatical construction inconsistent with the first, sometimes used for rhetorical effect; e.g., *I warned you that if you continue to smoke, what will become of you?* **—an′a·co·lu′thic** *adj.*

**an·a·con·da** (ăn′ə-kŏn′də) *n.* [Perh. alteration of Singhalese *henakandayā,* whip snake.] **1.** A large nonpoisonous, arboreal snake, *Eunectes murinus* of tropical South America, that constricts its prey in its coils. **2.** A snake similar or related to the anaconda.

**A·nac·re·on·tic** (ə-năk′rē-ŏn′tĭk) *adj.* Having a festive or romantic theme, as in the poetry of Anacreon. —*n.* An Anacreontic poem.

**an·a·cru·sis** (ăn′ə-kro͞o′sĭs) *n.* [NLat. < Gk. *anakrousis,* beginning of a tune < *anakrouein,* to begin : *ana-,* back + *krouein,* to push.] **1.** One or more unstressed syllables at the start of a line of verse, before the normal meter begins. **2.** *Mus.* An upbeat.

**an·a·dem** (ăn′ə-dĕm′) *n.* [Lat. *anadema* < Gk. *anadēma* < *anadein,* to bind up : *ana-,* up + *dein,* to bind.] *Archaic.* A garland or wreath for the head.

**an·a·di·plo·sis** (ăn′ə-də-plō′sĭs) *n.* [Lat. *anadiplosis* < Gk. *anadiplōsis* < *anadiploun,* to redouble : *ana-,* again + *diploun,* to double < *diplous,* double.] Rhetorical repetition of the word or phrase that ends one phrase at the start of the next phrase.

**a·nad·ro·mous** (ə-năd′rə-məs) *adj.* [Gk. *anadromos,* a running up : *ana-,* up + *dromos,* race.] Migrating up rivers from the sea to breed in fresh water, as salmon do.

**a·nae·mi·a** (ə-nē′mē-ə) *n. var. of* ANEMIA.

**a·nae·mic** (ə-nē′mĭk) *adj. var. of* ANEMIC.

**an·aer·obe** (ăn′ə-rōb′, ăn-âr′ōb′) *n.* A microorganism, as a bacterium, capable of living in the absence of free oxygen. **—an′aer·o′bic** (ăn′ə-rō′bĭk, -âr-ō′bĭk) *adj.* **—an′aer·o′bic·al·ly** *adv.*

**an·aes·the·sia** (ăn′ĭs-thē′zhə) *n. var. of* ANESTHESIA.

**an·aes·the·si·ol·o·gy** (ăn′ĭs-thē′zē-ŏl′ə-jē) *n. var. of* ANESTHESIOLOGY.

**an·aes·thet·ic** (ăn′ĭs-thĕt′ĭk) *adj. & n. var. of* ANESTHETIC.

**a·naes·the·tist** (ə-nĕs′thĭ-tĭst) *n. var. of* ANESTHETIST.

**a·naes·the·tize** (ə-nĕs′thĭ-tīz′) *v. var. of* ANESTHETIZE.

**an·a·glyph** (ăn′ə-glĭf′) *n.* [Gk. *anaglūphos,* carved in low relief : *ana-,* up + *gluphein,* to carve.] **1.** An ornament carved in low relief. **2.** A moving or still picture made up of two slightly different perspectives of the same subject in contrasting colors that are superimposed on each other, producing a three-dimensional effect when viewed through two correspondingly colored filters. **—an′a·glyph′ic, an′a·glyp′tic** (-glĭp′tĭk) *adj.*

**an·a·go·ge** *also* **an·a·go·gy** (ăn′ə-gō′jē) *n.* [LLat. < LGk. *anagōgē* : *ana,* up + *agein,* to lead.] A mystical interpretation of a word, passage, or text, esp. scriptural exegesis that discovers allusions to heaven or the afterlife. **—an′a·gog′ic** (-gŏj′ĭk), **an′a·gog′i·cal** *adj.* **—an′a·gog′i·cal·ly** *adv.*

**an·a·gram** (ăn′ə-grăm′) *n.* [Fr. *anagramme* : Gk. *ana-,* from bottom to top + Gk. *gramma,* letter < *graphein,* to write.] **1.** A word or phrase formed by rearranging the letters of another word or phrase. **2. anagrams** *(sing. in number).* A game whose object is to form words from a group of randomly picked letters. **—an′a·gram·mat′ic** (-grə-măt′ĭk) *adj.* **—an′a·gram·mat′i·cal·ly** *adv.*

**an·a·gram·ma·tize** (ăn′ə-grăm′ə-tīz′) *vt.* **-tized, -tiz·ing, -tizes.** To make an anagram of.

**a·nal** (ā′nəl) *adj.* [NLat. *analis* < Lat. *anus,* anus.] **1.** Of, relating to, or near the anus. **2.** *Psychoanal.* **a.** Of, relating to, or denoting the stage of infantile psychosexual development in which gratification is derived from sensations associated with the anus. **b.** Of, relating to, or denoting personality traits originating during toilet training and distinguished as anal-expulsive or anal-retentive.

**a·nal·cime** (ə-năl′sēm′) *also* **a·nal·cite** (-sīt′) *n.* [Fr. < Gk. *analkimos,* weak (from its weak electric power) : *an-,* not + *alkimos,* brave < *alkē,* strength.] A white or light-colored zeolite, occurring in diabase and certain basalts.

**an·a·lects** (ăn′ə-lĕkts′) *also* **an·a·lec·ta** (ăn′ə-lĕk′tə) *pl.n.* [Lat. *analecta* < Gk. *analekta* < *analegein,* to gather : *ana-,* up + *legein,* to gather.] Collected excerpts or selections from literary works. **—an′a·lec′tic** *adj.*

**an·a·lem·ma** (ăn′ə-lĕm′ə) *n.* [Lat., sundial < Gk. *analēmma,* support < *analambanein,* to take up. —see ANALEPTIC.] A graduated scale shaped like a figure eight indicating the sun's declination and the equation of time for every day of the year, usu. found on sundials and globes.

**an·a·lep·tic** (ăn′ə-lĕp′tĭk) *adj.* [Gk. *analēptikos* < *analambanein,* to take up : *ana-,* up + *lambanein,* to take.] Restorative or stimulating. —*n.* An analeptic medication.

**a·nal-ex·pul·sive** (ā′nəl-ĭk-spŭl′sĭv) *adj. Psychoanal.* Designating personality traits, as conceit, suspicion, ambition, and generosity, originating in habits, attitudes, or values associated with infantile pleasure in the expulsion of feces.

**an·al·ge·si·a** (ăn′əl-jē′zē-ə, -zhə) *n.* [NLat. < Gk. *analgēsia* : *an-,* without + *algos,* pain.] *Pathol.* Inability to feel pain although conscious. **—an′al·get′ic** *adj.*

**an·al·ge·sic** (ăn′əl-jē′zĭk, -sĭk) *n.* Something that relieves pain. —*adj.* Of or causing analgesia.

**an·a·log** (ăn′ə-lôg′, -lŏg′) *n. var. of* ANALOGUE.

**analog computer** *also* **analogue computer** *n.* A computer in which numerical data are represented by analogous physical magnitudes or electrical signals.

**analog data** *pl.n. (sing. or pl. in number).* Data collected or presented in continuous form, as voltage measurement or temperature variation.

**an·a·log·i·cal** (ăn′ə-lŏj′ĭ-kəl) *adj.* Of, relating to, composed of, or based on an analogy. **—an′a·log′i·cal·ly** *adv.*

**a·nal·o·gist** (ə-năl′ə-jĭst) *n.* One who looks for or reasons from analogies.

**a·nal·o·gize** (ə-năl′ə-jīz′) *v.* **-gized, -giz·ing, -giz·es.** —*vt.* To make an analogy to. —*vi.* To reason by analogy.

**a·nal·o·gous** (ə-năl′ə-gəs) *adj.* [Lat. *analogus* < Gk. *analogos,* proportionate : *ana-,* according to + *logos,* proportion < *legein,* to speak.] **1.** Corresponding in a way that allows the drawing of an analogy. **2.** *Biol.* Similar in function but not in evolutionary origin. **—a·nal′o·gous·ly** *adv.* **—a·nal′o·gous·ness** *n.*

**an·a·logue** *also* **an·a·log** (ăn′ə-lôg′, -lŏg′) *n.* [Fr. < Gk. *analogus,* proportionate —see ANALOGOUS.] **1.** One that bears an analogy to another. **2.** *Biol.* An organ or structure similar in function to one in another kind of organism but of dissimilar evolutionary origin. **3.** *Chem.* A structural derivative of a parent compound.

**a·nal·o·gy** (ə-năl′ə-jē) *n., pl.* **-gies.** [Lat. *analogia* < Gk. < *analogos,* proportionate. —see ANALOGOUS.] **1.** Correspondence in some respects between otherwise dissimilar things. **2.** *Biol.* Correspondence in function or position between organs of dissimilar evolutionary origin or structure. **3.** A form of logical inference, or an instance of it, based on the assumption that if two things are known to be alike in some respects, then they must be alike in other respects. **4.** The creation of forms on the basis of a proportion $a : b = c : x$.

**an·al·pha·bet·ic** (ăn-ăl′fə-bĕt′ĭk) *adj.* **1.** Not alphabetical. **2.** Unable to read : ILLITERATE. —*n.* An illiterate individual.

**a·nal-re·ten·tive** (ā′nəl-rĭ-tĕn′tĭv) *adj. Psychoanal.* Designating personality traits, as meticulousness, avarice, and obstinacy, originating in habits, attitudes, or values associated with infantile pleasure in retention of feces.

**a·nal·y·sand** (ə-năl′ĭ-sănd′) *n.* [< ANALYZE, by analogy with *multiplicand.*] One who is being psychoanalyzed.

**an·a·lyse** (ăn′ə-līz′) *v. Chiefly Brit. var. of* ANALYZE.

**a·nal·y·sis** (ə-năl′ĭ-sĭs) *n., pl.* **-ses** (-sēz′) [NLat. < Gk. *analusis,* a dissolving < *analuein,* to undo : *ana,* throughout + *luein,* to loosen.] **1.** Separation of an intellectual or substantial whole into its constituent parts for individual study. **2.** *Chem.* **a.** Separation of a substance into its constituent elements to determine either their nature (qualitative analysis) or their proportions (quantitative analysis). **b.** The stated findings of such separation or determination. **3.** *Math.* **a.** Methodology principally involving algebra and calculus as opposed to synthetic geometry, group theory, and number theory. **b.** The method of proof in which a known truth is sought as a consequence of reasoning from the thing to be proved. **4.** The use of two or more words instead of an inflected form to express a grammatical category. **5.** Psychoanalysis. **6.** Systems analysis.

**an·a·lyst** (ăn′ə-lĭst) *n.* **1.** One who analyzes. **2.** A licensed practitioner of psychoanalysis. **3.** A systems analyst.

**an·a·lyt·ic** (ăn′ə-lĭt′ĭk) *or* **an·a·lyt·i·cal** (-ĭ-kəl) *adj.* [LLat. *analyticus* < Gk. *analutikos* < *analuein,* to resolve. —see ANALYSIS.] **1.** Of or relating to analysis or analytics. **2.** Separating into elemental parts or basic principles. **3.** Reasoning from a perception of the parts and interrelations of a subject. **4.** Expert in or using analysis <an *analytic* intellect><an *analytic* method> **5.** *Logic.* Following necessarily. **6.** *Math.* **a.** Using, subjected to, or capable of being subjected to a methodology involving algebra and calculus. **b.** Proving a known truth by reasoning from the thing to be proved. **7.** Expressing a grammatical category by using two or more words instead of an inflected form. **8.** Psychoanalytic. **—an′a·lyt′i·cal·ly** *adv.*

**analytical balance** *n.* A balance for chemical analysis.

**analytic geometry** *n.* The analysis of geometric structures and properties mainly by algebraic operations on variables defined in terms of position coordinates.

**an·a·lyt·ics** (ăn'ə-lĭt'ĭks) *n. (sing. in number).* The branch of logic dealing with analysis.

**an·a·lyze** (ăn'ə-līz') *vt.* **-lyzed, -lyz·ing, -lyz·es.** [Prob. < Fr. *analyser* < *analyse,* analysis < Gk. *analusis.* —see ANALYSIS.] **1.** To separate into elemental parts or basic principles so as to determine the nature of the whole. **2.** To make a chemical analysis of. **3.** To make a mathematical analysis of. **4.** To psychoanalyze. **—an'a·lyz'a·ble** *adj.* **—an'a·ly·za'tion** *n.* **—an'a·lyz'er** *n.*

☆ ***syns:*** ANALYZE, ANATOMIZE, BREAK DOWN, DISSECT *v. core meaning* : to separate into parts for study <*analyzed* the ore and found iron>

**an·am·ne·sis** (ăn'ăm-nē'sĭs) *n., pl.* **-ses** (-sēz') [NLat. < Gk. *anamnēsis* < *anamimnēskein,* to remind : *ana-,* again, *mimnēskein,* to recall.] **1.** *Psychol.* Recollection : reminiscence. **2.** *Med.* A patient's complete case history. **—an'am·nes'tic** (-nĕs'tĭk) *adj.* **—an'am·nes'ti·cal·ly** *adv.*

**an·a·mor·phic** (ăn'ə-môr'fĭk) *adj.* Having or producing different optical magnification along mutually perpendicular radii.

**an·a·mor·pho·sis** (ăn'ə-môr'fə-sĭs) *n., pl.* **-ses** (-sēz') [Gk. *anamorphōsis,* re-formation : *ana-,* again + *morphē,* shape.] An image distorted so that it can be viewed correctly only from a special angle or with a special instrument.

**an·an·drous** (ăn-ăn'drəs) *adj. Bot.* Without stamens.

**An·a·ni·as** (ăn'ə-nī'əs) *n.* **1.** A liar in the New Testament who dropped dead when Peter rebuked him. **2.** A liar.

**an·an·thous** (ăn-ăn'thəs) *adj. Bot.* Having no flowers.

**an·a·pest** *also* **an·a·paest** (ăn'ə-pĕst') *n.* [Lat. *anapestus* < Gk. *anapaistos* : *ana-,* back + *paiein,* to strike (so called because an anapest is a reversed dactyl).] **1.** A metrical foot made up of two short syllables followed by one long one, as in the word *nonaligned.* **2.** A line of verse in anapest. **—an'a·pes'tic** *adj.*

**an·a·phase** (ăn'ə-fāz') *n.* The stage of mitosis in which the daughter chromosomes move toward the poles of the nuclear spindle.

**a·naph·o·ra** (ə-năf'ə-rə) *n.* [LLat. < Gk. < *anapherein,* to repeat : *ana-,* again + *pherein,* to carry.] The conscious repetition of a word or phrase at the beginning of several successive verses, clauses, or paragraphs.

**an·aph·ro·dis·i·a** (ăn-ăf'rə-dĭz'ē-ə, -dĭzh'ə) *n.* [Gk. : *an,* without + *aphrodisia,* sexual desire. —see APHRODISIAC.] Decrease or lack of sexual desire. **—an·aph'ro·dis'i·ac'** (ăn-ăf'rə-dĭz'ē-ăk') *adj.* & *n.*

**an·a·phy·lac·toid** (ăn'ə-fə-lăk'toid) *adj. Pathol.* **1.** Of or relating to an anaphylactic reaction that occurs without producing antibodies. **2.** Of or relating to a toxic reaction caused in an unsensitized person by an excessive dose of a substance that causes anaphylaxis in a sensitized person.

**an·a·phy·lax·is** (ăn'ə-fə-lăk'sĭs) *n.* [ANA- + (PRO)PHYLAXIS.] Hypersensitivity to a foreign substance, esp. in animals, induced by a small preliminary or sensitizing injection of the substance. **—an'a·phy·lac'tic** (-lăk'tĭk) *adj.* **—an'a·phy·lac'ti·cal·ly** *adv.*

**an·a·pla·sia** (ăn'ə-plā'zhə) *n.* Reversion of cells to a less differentiated or more primitive form.

**an·a·plas·tic** (ăn'ə-plăs'tĭk) *adj.* **1.** *Med.* Relating to the surgical restoration of a lost or absent part. **2.** Of or relating to anaplasia of cells.

**an·arch** (ăn'ärk') *n.* [Back-formation < ANARCHY.] A leader or advocate of anarchy.

**an·ar·chic** (ăn-är'kĭk) *or* **an·ar·chi·cal** (-kĭ-kəl) *adj.* **1.** Of, pertaining to, like, or promoting anarchy. **2.** Lacking order or control : LAWLESS. **—an·ar'chi·cal·ly** *adv.*

**an·ar·chism** (ăn'ər-kĭz'əm) *n.* **1.** The theory that all types of government are oppressive and undesirable and should be abolished. **2.** Active resistance and terrorism against the state. **3.** Rejection of all forms of coercive control and authority. **—an'ar·chis'tic** (ăn'ər-kĭs'tĭk) *adj.*

**an·ar·chist** (ăn'ər-kĭst) *n.* One who believes in, advocates, or engages in anarchism.

**an·ar·cho-syn·di·cal·ism** (ăn-är'kō-sĭn'dĭ-kəl-ĭz'əm) *n.* [ANARCH(Y) + SYNDICALISM.] Syndicalism.

**an·ar·chy** (ăn'ər-kē) *n., pl.* **-chies.** [Gk. *anarkhia* < *anarkhos,* without a ruler : *an-,* without + *arkhos,* ruler. —see -ARCH.] **1.** Absence of any form of political authority. **2.** Political disorder and confusion. **3.** Absence of any cohering principle, as a common purpose or standard.

**an·ar·thri·a** (ăn-är'thrē-ə) *n.* [NLat. < Gk. *anarthros,* not articulated. —see ANARTHROUS.] Loss of the ability to speak. **—an·ar'thric** (-thrĭk) *adj.*

**an·ar·throus** (ăn-är'thrəs) *adj.* [Gk. *anarthros,* not articulated : *an-,* without + *arthron,* joint.] *Zool.* Lacking joints.

**an·a·sar·ca** (ăn'ə-sär'kə) *n.* [NLat. : Gk. *ana,* throughout + *sarx,* flesh.] A general accumulation of serum in various tissues and bodily cavities. **—an'a·sar'cous** (-sär'kəs) *adj.*

**an·as·tig·mat** (ăn-ăs'tĭg-măt') *n.* An anastigmatic lens.

**an·as·tig·mat·ic** (ăn-ăs'tĭg-măt'ĭk) *adj.* Not astigmatic. —Used of a compound lens in which the separate elements compensate for the astigmatism of each.

**a·nas·to·mose** (ə-năs'tə-mōz', -mōs') *v.* **-mosed, -mos·ing, -mos·es.** [Back-formation < ANASTOMOSIS.] —*vt.* To unite or connect by anastomosis. —*vi.* To be connected by anastomosis, as blood vessels.

**a·nas·to·mo·sis** (ə-năs'tə-mō'sĭs) *n., pl.* **-ses** (-sēz') [NLat. < Gk. *anastomosis,* outlet < *anastomoun,* to furnish with a mouth : *ana-,* up + *stoma,* mouth.] **1.** The union or connection of branches, as of rivers, leaf veins, or blood vessels. **2.** A surgical connection of separate or severed hollow organs to form a continuous channel. **—a·nas'to·mot'ic** (-mŏt'ĭk) *adj.*

**a·nas·tro·phe** (ə-năs'trə-fē) *n.* [Gk. *anastrophē* < *anastrephein,* to turn upside-down : *ana-,* back + *strephein,* to turn.] Inversion of the normal syntactic order of words; e.g., *To church went we.*

**an·a·tase** (ăn'ə-tās', -tāz') *n.* [Fr. < Gk. *anatasis,* extension (from its long crystals) < *anateinein,* to extend : *ana-,* up + *teinein,* to stretch.] A rare blue or light-yellow to brown variety of titanium dioxide.

**a·nath·e·ma** (ə-năth'ə-mə) *n., pl.* **-mas.** [LLat., a person cursed < Gk. *anathēma,* an accursed thing < *anatithenai,* to dedicate : *ana-,* up + *tithenai,* to put.] **1.** A formal ecclesiastical ban, curse, or excommunication. **2.** A vehement denunciation : CURSE. **3.** One that is cursed or damned. **4.** One that is greatly detested.

**a·nath·e·ma·tize** (ə-năth'ə-mə-tīz') *vt.* **-tized, -tiz·ing, -tiz·es.** [Lat. *anathematizare* < Gk. *anathematizein* < *anathema,* anathema.] To proclaim an anathema on : CURSE.

**An·a·to·li·an** (ăn'ə-tō'lē-ən) *adj.* **1.** Of or relating to Anatolia or its inhabitants. **2.** Of or relating to a branch of the Indo-European language family that includes Hittite and other extinct languages of ancient Anatolia. —*n.* The Anatolian languages.

**an·a·tom·i·cal** (ăn'ə-tŏm'ĭ-kəl) *also* **an·a·tom·ic** (-tŏm'ĭk) *adj.* **1.** Of or relating to anatomy. **2.** Of or relating to dissection. **3.** Structural as opposed to functional. **—an'a·tom'i·cal·ly** *adv.*

**a·nat·o·mist** (ə-năt'ə-mĭst) *n.* An expert in or student of anatomy.

**a·nat·o·mize** (ə-năt'ə-mīz') *vt.* **-mized, -miz·ing, -miz·es. 1.** To dissect. **2.** To examine in minute detail. **—a·nat'o·mi·za'tion** *n.*

**a·nat·o·my** (ə-năt'ə-mē) *n., pl.* **-mies.** [ME *anatomie* < OFr. < LLat. *anatomia* < Gk. *anatomē,* dissection : *ana-,* up + *tomē,* a cutting < *temnein,* to cut.] **1. a.** The structure of a plant or animal or of any of its elements. **b.** The constituent structure of something. **2.** The science of the form and structure of organisms and their elements. **3.** A treatise on anatomical science. **4.** The dissection of a plant or animal to disclose the various elements and their positions, structure, and interrelation. **5.** A skeleton. **6.** A detailed analysis or examination. **7.** The human body.

☆ ***syns:*** ANATOMY, ARCHITECTURE, CONSTITUTION, FABRIC, MAKE-UP *n. core meaning* : constituent structure <studied the *anatomy* of the ancient civilization>

**a·nat·ro·pous** (ə-năt'rə-pəs) *adj.* Inverted so that the micropyle is next to the hilum and the embryonic root is at the other end. —Used of an ovule.

**-ance** *suff.* [ME < OFr. < Lat. *-antia,* n. suffix < *-āns,* -ant.] **1.** State or condition <absorpt*ance*> **2.** Action <continu*ance*>

**an·ces·tor** (ăn'sĕs'tər) *n.* [ME *auncestre* < OFr. < Lat. *antecessor* < *antecedere,* to precede : *ante-,* before + *cedere,* to go.] **1.** A person from whom one is descended, esp. if more remote than a grandparent : FOREFATHER. **2.** One that goes before : PREDECESSOR. **3.** *Law.* The person from whom an estate has been inherited. **4.** *Biol.* The actual or hypothetical organism or stock from which later kinds have evolved.

☆ ***syns:*** ANCESTOR, ANTECEDENT, ASCENDANT, FOREBEAR, FOREFATHER, PROGENITOR *n. core meaning* : a person from whom one is descended <My *ancestors* were farmers.> ***ant:*** descendant

**an·ces·tral** (ăn-sĕs'trəl) *adj.* Of, relating to, or evolved from an ancestor. **—an·ces'tral·ly** *adv.*

**an·ces·try** (ăn'sĕs'trē) *n., pl.* **-tries.** [ME *auncestrie* < OFr. *ancesserie* < *ancessour,* ancestor < Lat. *antecessor.* —see ANCESTOR.] **1.** Ancestral descent or lineage. **2.** Ancestors as a group.

**An·chi·ses** (ăn-kī'sēz') *n.* [Lat. < Gk. *Ankhisēs.*] *Gk.* & *Rom. Myth.* The father of Aeneas, rescued by his son from Troy.

**an·chor** (ăng'kər) *n.* [ME *anker* < OE *ancor* < Lat. *ancora* < Gk. *ankura.*] **1.** A heavy object attached to a vessel by a rope or cable and cast overboard to keep the vessel in place either by its weight or by flukes that grip the bottom. **2.** Something that provides a rigid point of support, as for securing a rope. **3.** Something that provides stability or security. **4.** An anchorman or anchorwoman. —*v.* **-chored, -chor·ing, -chors.** —*vt.* **1.** To secure firmly by or as if by an anchor. **2.** To narrate or coordinate (a newscast in which several correspondents give reports). —*vi.* To lie at anchor, as a ship. **—at anchor.** Anchored.

**an·chor·age** (ăng'kər-ĭj) *n.* **1.** A place for anchoring. **2.** A fee charged for the privilege of anchoring. **3.** A means of stabilizing or securing <My mother was the *anchorage* of our family.> **4.** The act of anchoring or condition of being at anchor.

**an·chor·ess** (ăng'kə-rĭs) *n.* A woman who lives as an anchorite.

**an·cho·rite** (ăng'kə-rīt') *also* **an·cho·ret** (-rĕt') *n.* [ME < Med. Lat. *anchorita* < LLat. *anachoreta* < Gk. *anakhorētēs* < *anakōrein,* to retire : *ana-*, back + *khōrein*, to withdraw.] One who has retired into seclusion for religious reasons. **—an'cho·rit'ic** (-rĭt'ĭk) *adj.*

**an·chor·man** (ăng'kər-măn') *n.* **1.** One heavily relied on. **2.** The usu. strongest member of a team who competes last in a relay race. **3.** The narrator or coordinator of a broadcast in which several correspondents give news reports.

**an·chor·wom·an** (ăng'kər-wo͝om'ən) *n.* A woman who narrates or coordinates a broadcast in which several correspondents give news reports.

**an·cho·vy** (ăn'chō'vē, ăn-chō'vē) *n., pl.* **anchovy** *or* **-vies.** [Sp. *anchova.*] Any of various small, herringlike marine fishes of the family Engraulidae, several species of which are widely used as food.

**an·chy·lose** (ăng'kə-lōs', -lōz') *v. var. of* ANKYLOSE.

**an·chy·lo·sis** (ăng'kə-lō'sĭs) *n. var. of* ANKYLOSIS.

**an·cien ré·gime** (äɴ-syăɴ' rā-zhēm') *n.* [Fr., old regime.] **1.** The sociopolitical system existing in France before the Revolution of 1789. **2.** A former sociopolitical or other system or mode.

**an·cient¹** (ān'shənt) *adj.* [ME *auncien* < OFr. < VLat. **anteanus* < Lat. *ante*, before.] **1.** Very old. **2.** Of, existing, or happening in remote times, esp. belonging to the historical period before the fall of the Western Roman Empire (A.D. 476). **3.** Antiquated : old-fashioned. **4.** Having qualities associated with age, wisdom, or long use : VENERABLE. —*n.* **1.** A very old person. **2.** A person who lived in ancient times. **3. ancients.** The peoples of the classical nations of antiquity. **4. ancients.** The ancient Greek and Roman authors. **—an'cient·ly** *adv.* **—an'cient·ness** *n.*

**an·cient²** (ān'shənt) *n.* [Var. of ENSIGN.] *Obs.* **1.** An ensign : FLAG. **2.** A flag-bearer or lieutenant.

**Ancient Chinese** *n.* The Chinese language as used until about the 13th cent. A.D.

**ancient history** *n.* **1.** The history of ancient times. **2.** *Informal.* Common knowledge, esp. of a recent event that has lost its original novelty or importance.

**an·cil·lar·y** (ăn'sə-lĕr'ē) *adj.* [Lat. *ancillaris*, pertaining to a maidservant < *ancilla*, maidservant, dim. of *ancula*, fem. of *anculus*, servant.] **1.** Subordinate <"For Degas, sculpture was never more than *ancillary* to his painting" —Herbert Read> **2.** Auxiliary <*ancillary* information> —*n., pl.* **-ies.** A servant.

**an·cip·i·tal** (ăn-sĭp'ĭ-təl) *adj.* [< Lat. *anceps, ancipit-*, two-headed : ambi-, two + *caput*, head.] Flattened and two-edged <*ancipital* plant stems>

**an·con** (ăng'kŏn') *n., pl.* **-an·con·es** (ăng-kō'nēz) [Lat. *ancon* < Gk. *ankōn*, elbow.] A projecting bracket used in classical architecture to support the upper elements of a cornice : CONSOLE.

**ancon**

**-ancy** *suff.* [Lat. *-antia.* —see -ANCE.] -ANCE.

**an·cy·lo·sto·mi·a·sis** (ăn'sə-lō-stō-mī'ə-sĭs, ăng'kə-lō-) *n.* [NLat. *Ancylostoma*, hookworm genus (Gk. *ankulos*, curved + Gk. *stoma*, mouth) + -IASIS.] A disease caused by hookworm infestation and characterized by progressive anemia.

**and** (ənd, ən; ănd *when stressed*) *conj.* [ME < OE.] **1.** Together or along with : as well as. —Used to connect words, phrases, or clauses that have the same grammatical function in a construction. **2.** Added to : PLUS <Two *and* two makes four.> **3.** As a result : in consequence <They invited me *and* here I am.> **4.** —Used between two verbs, esp. after such verbs as *go, come,* or *try* <try *and* find it><come *and* see> **5.** Then. —Used to begin a sentence. ***usage:*** The use of *and* to begin a sentence has a long and respectable history in English <"*And* it came to pass in those days" —Luke 2:1> **6.** *Archaic.* If.

**AND** (ănd) *n.* [< AND.] *Computer Sci.* A logic operator equivalent to the sentential connective "and."

**an·da·lu·site** (ăn'də-lo͞o'sīt') *n.* [Fr. *andalousite*, after *Andalusia*, where it was discovered.] A mineral aluminum silicate, $Al_2SiO_5$, usu. found in prisms of various colors.

**An·da·man·ese** (ăn'də-mə-nēz', -nēs') *n., pl.* **Andamanese. 1.** *also* **An·da·man** (ăn'də-mən). A member of a Negrito people inhabiting the Andaman Islands. **2.** The language of the Andamanese, of no known linguistic affiliation. **—An'da·man·ese'** *adj.*

**an·dan·te** (ăn-dăn'tā, ăn-dăn'tē) [Ital., pr.part. of *andare*, to walk, ult. < Lat. *ambulare*.] *Mus.* —*adv. & adj.* Moderate in tempo : faster than adagio, but slower than allegretto. —Used as a direction. —*n.* An andante movement or passage.

**an·dan·ti·no** (ăn'dăn-tē'nō, ăn'dăn-tē'nō) [Ital. dim. of *andante*, andante.] *Mus.* —*adv. & adj.* Slightly faster than andante in tempo. —Used as a direction. —*n., pl.* **-nos.** An andantino movement or passage.

**an·de·site** (ăn'dē-zīt') *n.* [G. *Andesit* : *Andes*, Andes (where it was found) + *-it*, -ite.] A fine-grained, gray volcanic rock, mainly plagioclase and feldspar.

**AND gate** *n. Computer Sci.* A signal circuit with two or more input wires that emits a signal only if all input wires receive coincident signals.

**and·i·ron** (ănd'ī'ərn) *n.* [ME *aundiren*, alteration of OFr. *andier.*] One of a pair of metal supports for fireplace logs.

**and/or** (ăn'dôr') *conj.* —Used to indicate that either *and* or *or* may be used to connect words, phrases, or clauses.

**andr-** *pref. var. of* ANDRO-.

**an·dra·dite** (ăn-drä'dīt') *n.* [After José B. de *Andrada* e Silva (d. 1838).] A green to brown or black calcium-iron garnet, $Ca_3Fe_2(SiO_4)_3$.

**andro-** *or* **andr-** *pref.* [Gk. < *anēr, andr-*, man.] **1.** Male : masculine <*androgen*> **2.** Stamen or anther <*androecium*>

**An·dro·cles** (ăn'drə-klēz') *n.* [Lat. < Gk. *Androklēs.*] A legendary Roman slave spared in the arena by a lion that identified him as the man who had once removed a thorn from its paw.

**an·droe·ci·um** (ăn-drē'shē-əm, -shəm) *n., pl.* **-ci·a** (-shē-ə, -shə) [NLat. : ANDR(O)- + Gk. *oikos*, house.] The stamens of a flower considered as a whole. **—an·droe'cial** (-shəl) *adj.*

**an·dro·gen** (ăn'drə-jən) *n.* A steroid hormone that develops and maintains masculine characteristics. **—an'dro·gen'ic** (-jĕn'ĭk) *adj.*

**an·drog·e·nize** (ăn-drŏj'ə-nīz') *vt.* **-nized, -niz·ing, -niz·es.** To treat with usu. massive doses of male hormones. **—an·drog'e·ni·za'tion** *n.*

**an·drog·e·nous** (ăn-drŏj'ə-nəs) *adj.* Of or relating to production of male offspring.

**an·drog·y·nous** (ăn-drŏj'ə-nəs) *adj.* [Lat. *androgynus*, hermaphrodite < Gk. *androgunos* : *anēr*, man + *gunē*, woman.] **1.** Having both female and male characteristics : HERMAPHRODITIC. **2.** *Bot.* Composed of staminate and pistillate flowers. —Used of the flower spikes of certain sedges. **3.** Being neither distinctly masculine nor feminine, as in dress, appearance, or behavior. **—an·drog'y·ny** (-ə-nē) *n.*

**an·droid** (ăn'droid') *adj.* Having human features. —*n.* A synthetic person created from biological materials.

**An·drom·a·che** (ăn-drŏm'ə-kē) *n.* [Lat. < Gk. *Andromakhē.*] *Gk. Myth.* The faithful wife of Hector, captured by the Greeks at the fall of Troy.

**An·drom·e·da** (ăn-drŏm'ə-də) *n.* [Lat. < Gk. *Andromedē.*] **1.** *Gk. Myth.* The daughter of Cepheus and Cassiopeia and wife of Perseus, who had rescued her from a sea monster. **2.** A constellation in the Northern Hemisphere. **3. andromeda.** Any of several hardy shrubs of the genus *Andromeda* or related genera.

**an·dros·ter·one** (ăn-drŏs'tə-rōn') *n.* [ANDRO- + STER(OL) + -ONE.] A male sex hormone, excreted in urine and produced synthetically from cholesterol.

**-androus** *suff.* [NLat. *-andrus* < Gk. *-andros*, having men < *anēr, andr-*, man.] Having a specified number or kind of stamens <*monandrous*>

**-andry** *suff.* [< Gk. *anēr, andr-*, man.] **1.** The condition of having a given kind or number of husbands <*monandry*> **2.** The condition of having a given kind or number of stamens <*polyandry*>

**-ane** *suff.* [Alteration of -ENE.] A saturated hydrocarbon <*hexane*>

**an·ec·dot·age** (ăn'ĭk-dō'tĭj) *n.* Anecdotes as a whole.

**an·ec·dot·al** (ăn'ĭk-dōt'l) *adj.* Relating to, marked by, or containing anecdotes.

**an·ec·dote** (ăn'ĭk-dōt') *n.* [Fr. < Gk. *anekdotos*, unpublished : *an-*, not + *ekdotos*, published < *ekdidonai*, to publish (*ek-*, out + *didonai*, to give).] **1.** A short, often oral, interesting or humorous account of a real or fictitious incident. **2.** *pl.* **-dotes** *or* **-do·ta** (-dō'tə). Secret or previously unknown details of history or biography.

☆ ***syns:*** ANECDOTE, FABLE, STORY, TALE, YARN *n. core meaning* : an entertaining, often oral account of a real or fictitious occurrence <told *anecdotes* about their travels>

**an·ec·dot·ic** (ăn'ĭk-dŏt'ĭk) *also* **an·ec·dot·i·cal** (-ĭ-kəl) *adj.* **1.** Anecdotal. **2.** Full of or inclined to telling anecdotes.

**an·ec·dot·ist** (ăn'ĭk-dō'tĭst) *n.* One who tells, collects, compiles, or publishes anecdotes.

**an·e·cho·ic** (ăn'ĕ-kō'ĭk) *adj.* Not having or producing echoes.

**a·nele** (ə-nēl') *vt.* **a·neled, a·nel·ing, a·neles.** [ME *anelen* : *an*, on (< OE) + *elen*, to anoint < *ele*, oil < OE < Lat. *oleum*. —see OIL.] *Archaic.* To anoint, esp. during extreme unction.

**a·ne·mi·a** *also* **a·nae·mi·a** (ə-nē'mē-ə) *n.* [NLat. < Gk. *anaimia* : *an-*, without + *haima*, blood.] A deficiency in the oxygen-carrying material of the blood, measured in unit volume concentrations of hemoglobin, red blood cell volume, and red blood cell number.

**a·ne·mic** *also* **a·nae·mic** (ə-nē'mĭk) *adj.* **1.** Of, relating to, or afflicted with anemia. **2. a.** Weak and listless. **b.** Pallid. **—a·ne'mi·cal·ly** *adv.*

**anemo-** *pref.* [< Gk. *anemos*, wind.] Wind <*anemometer*>

**a·nem·o·graph** (ə-nĕm′ə-grăf′) *n.* A recording anemometer.
**an·e·mog·ra·phy** (ăn′ə-mŏg′rə-fē) *n.* The science of making anemometric measurements.
**an·e·mom·e·ter** (ăn′ə-mŏm′ĭ-tər) *n.* An instrument for indicating and measuring wind force and velocity.
**an·e·mom·e·try** (ăn′ə-mŏm′ĭ-trē) *n.* The determination of wind force and velocity. **—an′e·mo·met′ric** (-mō-mĕt′rĭk), **an′e·mo·met′ri·cal** (-rĭ-kəl) *adj.*
**a·nem·o·ne** (ə-nĕm′ə-nē) *n.* [Lat. < Gk. *anemōnē.*] **1.** A plant of the genus *Anemone* of the North Temperate Zone, with white, purple, or red cup-shaped flowers. **2.** The sea anemone.
**anemone fish** *n.* A small, brightly colored marine fish of the genus *Amphiprion,* found near sea anemones.
**an·e·moph·i·lous** (ăn′ə-mŏf′ə-ləs) *adj.* Pollinated by wind-carried pollen.
**an·en·ceph·a·ly** (ăn′ən-sĕf′ə-lē) *n., pl.* **-lies.** Congenital absence of part or all of the brain. **—an′en·ce·phal′ic** (-sə-făl′ĭk) *adj.*
**†a·nent** (ə-nĕnt′) *prep.* [ME < OE *onefn,* near : *on,* on + *efn,* even.] **1.** Concerning : regarding **2.** *Regional.* Close to : OPPOSITE. **3.** *Obs.* & *Chiefly Brit. Regional.* On a level with : in a line with.
**an·er·oid** (ăn′ə-roid′) *adj.* [Fr. *anéroïde* : Gk. *a-,* not + Gk. *nēron,* water.] Not using fluid.
**aneroid barometer** *n.* A barometer in which variations of atmospheric pressure are indicated by the relative bulges of a thin elastic metal disk covering a chamber in which there is a partial vacuum.
**an·es·the·sia** *also* **an·aes·the·sia** (ăn′ĭs-thē′zhə) *n.* [NLat. < Gk. *anaisthēsia,* insensibility : *an,* without + *aisthēsis,* feeling < *aisthanesthai,* to feel.] **1.** Complete or partial loss of sensation, esp. tactile sensibility, caused by disease or an anesthetic. **2.** Artificially produced unconsciousness or local or general insensibility to pain.
**an·es·the·si·ol·o·gy** *also* **an·aes·the·si·ol·o·gy** (ăn′ĭs-thē′zē-ŏl′ə-jē) *n.* The branch of medicine concerned with the use of anesthetics. **—an′es·the′si·ol′o·gist** *n.*
**an·es·thet·ic** *also* **an·aes·thet·ic** (ăn′ĭs-thĕt′ĭk) [< Gk. *anaisthetos,* without feeling : *an-,* without + *aisthētos,* perceptible < *aisthanesthai,* to feel.] *—adj.* **1.** Relating to or like anesthesia. **2.** Inducing anesthesia. **3.** Lacking emotion : INSENSITIVE. *—n.* An agent that causes unconsciousness or insensitivity to pain.
**a·nes·the·tist** *also* **a·naes·the·tist** (ə-nĕs′thĭ-tĭst) *n.* One trained to administer anesthetics.
**a·nes·the·tize** *also* **a·naes·the·tize** (ə-nĕs′thĭ-tīz′) *vt.* **-tized, -tiz·ing, -tiz·es.** To bring about a condition of anesthesia in. **—a·nes′the·ti·za′tion** *n.*
**an·es·trus** (ăn-ĕs′trəs) *n.* An interval of sexual dormancy between two periods of estrus.
**an·eu·rysm** *also* **an·eu·rism** (ăn′yə-rĭz′əm) *n.* [Gk. *aneurusma* < *aneurein,* to dilate : *ana-,* throughout + *eurus,* wide.] **1.** A pathological blood-filled dilatation of a blood vessel. **2.** A similar dilatation of a cylindrical body.
**a·new** (ə-no͞o′, ə-nyo͞o′) *adv.* [ME *a new* : *a,* of + *new,* new.] **1.** Once more : AGAIN. **2.** In a new and different manner or form.
**an·frac·tu·os·i·ty** (ăn-frăk′cho͞o-ŏs′ĭ-tē) *n., pl.* **-ties. 1.** The condition or quality of being anfractuous. **2.** A winding channel, passage, or crevice. **3.** A complicated process, as of the intellect.
**an·frac·tu·ous** (ăn-frăk′cho͞o-əs) *adj.* [Fr. *anfractueux* < LLat. *anfractuosus,* < Lat. *anfractus,* winding < *ambi-,* around + *fractus,* broken, p.part, of *frangere,* to break.] Full of twists and turns : TORTUOUS.
**an·ga·ry** (ăng′gə-rē) *also* **an·gar·i·a** (ăng-gâr′ē-ə) *n.* [Fr. *angarie* < LLat. *angaria,* service to a lord < Gk. *angareia,* impressment for public service < *angaros,* courier.] The legal right of a belligerent state to seize, use, or destroy the property of a neutral state, provided that full compensation is made.
**an·gel** (ān′jəl) *n.* [ME < OE *engel* < LLat. *angelus* < LGk. *angelos,* < Gk. messenger.] **1. a.** An immortal spiritual being attendant upon God. **b.** The traditional representation of such a being as a human figure with a halo and wings. **2.** A guardian spirit or guiding influence. **3. a.** A kind and lovable person. **b.** One who manifests goodness, purity, and unselfishness. **4.** *Christian Science.* God's thoughts passing to man. **5.** *Informal.* A financial backer of an enterprise, esp. of a dramatic production. **6. a.** Enemy aircraft. **b.** A radar echo of unknown origin.
**angel cake** *n.* Angel food cake.
**angel dust** *n. Slang.* Phencyclidine.
**An·ge·le·no** (ăn′jə-lē′nō) *n., pl.* **-nos.** [Mex. Sp. *Angeleño,* after *Los Angeles,* California.] A native or inhabitant of Los Angeles.
**an·gel·fish** (ān′jəl-fĭsh′) *n., pl.* **angelfish** *or* **-fish·es. 1.** Any of several brightly colored fishes of the family Chaetodontidae of warm seas, with laterally compressed bodies. **2.** A freshwater fish indigenous to rivers of tropical South America, *Pterophyllum scalare,* with a flattened, usu. striped body, popular in aquariums.
**angel food cake** *n.* A white sponge cake made of egg whites, sugar, and flour.

**an·gel·ic** (ăn-jĕl′ĭk) *also* **an·gel·i·cal** (-ĭ-kəl) *adj.* **1.** Of, relating to, consisting of, or belonging to angels. **2. a.** Suggestive of an angel, as in goodness, purity, or beauty. **b.** *Informal.* Kind and lovable. **—an·gel′i·cal·ly** *adv.*
**an·gel·i·ca** (ăn-jĕl′ĭ-kə) *n.* [Med.Lat. *(herba) angelica,* angelic (herb) < LLat. < Gk. *angelikos* < *angelos,* messenger, angel.] **1.** A plant of the genus *Angelica,* bearing compound leaves and small white or greenish flower clusters, esp. *A. archangelica,* whose aromatic seeds, leaves, stems, and roots are used medicinally and as flavoring. **2.** The candied stem of the angelica. **3.** *often* **Angelica.** A sweet white wine or liqueur.
**angelica tree** *n.* Any of several spiny trees or shrubs, as the Hercules'-club.
**an·gel·ol·o·gy** (ān′jəl-ŏl′ə-jē) *n.* The branch of theology that deals with angels.
**angel shark** *n.* A raylike shark of the genus *Squatina,* with a broad, flat head and body.
**An·ge·lus** *also* **an·ge·lus** (ăn′jə-ləs) *n.* [Med. Lat., angel, first word of the prayer.] *Rom. Cath. Ch.* **1.** A devotional prayer recited in the morning, at noon, and at night to commemorate the Annunciation. **2.** A bell rung as a call to recite the Angelus.
**an·ger** (ăng′gər) *n.* [ME < ON *angr,* sorrow.] **1.** A feeling of great displeasure, hostility, indignation, or exasperation : WRATH. **2.** *Obs.* Trouble : affliction. **3.** *Chiefly Brit.* An inflammation or sore. *—v.* **-gered, -ger·ing, -gers.** *—vt.* **1.** To make angry. **2.** *Chiefly Brit. Regional.* To make sore or inflamed. *—vi.* To become angry.
**an·ger·ly** (ăng′gər-lē) *adv. Archaic.* In an angry manner : ANGRILY <"Again thou blushest *angerly*" —Tennyson>
**An·ge·vin** (ăn′jə-vĭn) *adj.* [Fr. < OFr. < Med. Lat. *Andegavinus* < *Andegavia,* Anjou.] **1.** Of or relating to the province of Anjou, France. **2.** Of or relating to the House of Anjou, esp. as represented by the Plantagenet kings of England.
**an·gi·na** (ăn-jī′nə, ăn′jə-) *n.* [Lat., quinsy < Gk. *ankonē,* a strangling.] **1.** A disease, as croup or diphtheria, marked by spasmodic and painful suffocation or spasms. **2.** Angina pectoris. **—an·gi′nal** *adj.* **—an′gi·nose′** (-jə-nōs′) *adj.*
**angina pec·to·ris** (pĕk′tə-rĭs) *n.* [NLat. : Lat. *angina,* quinsy + *pectus,* chest.] Severe paroxysmal chest pain associated with an insufficient supply of blood to the heart.
**an·gi·og·raph·y** (ăn′jē-ŏg′rə-fē) *n.* [Gk. *angeion,* vessel + -GRAPHY.] Roentgenography of the blood vessels after a radiopaque substance has been injected. **—an′gi·o·gram′** (-jē-ə-grăm′) *n.* **—an′gi·o·graph′ic** (ăn′jē-ə-grăf′ĭk) *adj.*
**an·gi·ol·o·gy** (ăn′jē-ŏl′ə-jē) *n.* [Gk. *angeion,* vessel + -LOGY.] The scientific study of blood and lymph vessels.
**an·gi·o·ma** (ăn′jē-ō′mə) *n., pl.* **-mas** *or* **-ma·ta** (-mə-tə) [Gk. *angeion,* vessel + -OMA.] A tumor made up of lymph and blood vessels.
**an·gi·op·a·thy** (ăn′jē-op′ə-thē) *n., pl.* **-thies.** [Gk. *angeion,* vessel + -PATHY.] Any of several diseases of the blood or lymph vessels.
**an·gi·o·sperm** (ăn′jē-ə-spûrm′) *n.* [Gk. *angeion,* vessel + SPERM[1].] A flowering plant of the class Angiospermae, having seeds enclosed in an ovary.
**an·gi·o·ten·sin** (ăn′jē-ō-tĕn′sĭn) *n.* [Gk. *angeion,* vessel + TENS(ION) + -IN.] Either of two polypeptide hormones that are powerful vasoconstrictors.
**an·gle[1]** (ăng′gəl) *vi.* **-gled, -gling, -gles.** [ME *anglen* < *angel,* fishhook < OE.] **1.** To fish with a hook and line. **2.** To attempt to get something, esp. by scheming or trickery. *—n. Obs.* A fishhook or fishing tackle.
**an·gle[2]** (ăng′gəl) *n.* [ME < OFr. < Lat. *angulus.*] **1.** *Math.* **a.** The figure formed by two lines diverging from a common point. **b.** The figure formed by two planes diverging from a common line. **c.** The rotation needed to superimpose either of two such lines or planes on the other. **d.** The space between such lines or surfaces. **e.** A solid angle. **2.** A sharp or projecting corner, as of a building. **3. a.** The place, position, or direction from which an object is presented to view <superb architecture from any *angle*> **b.** A particular part or phase, as of a problem : ASPECT. **4.** *Slang.* A devious method : SCHEME. *—v.* **-gled, -gling, -gles.** *—vt.* **1.** To move or turn at an angle. **2.** To hit (e.g., a ball or puck) at an angle. **3.** *Informal.* To impart a biased aspect or point of view to. *—vi.* To continue along or turn at or by angles <The road *angled* through the forest.>
**An·gle** (ăng′gəl) *n.* [< Lat. *Angli,* the Angles, of Germanic orig.] A member of a Germanic people that migrated to England from southern Denmark in the 5th cent. A.D., founded the kingdoms of Northumbria, East Anglia, and Mercia, and together with the Jutes and Saxons formed the Anglo-Saxon peoples.
**angle bracket** *n.* BRACKET 4b.
**an·gle·doz·er** (ăng′gəl-dō′zər) *n.* [Orig. a trademark.] A machine similar to a tractor, used to level or scrape ground and constructed so that the dirt is pushed off to one side.
**angle iron** *n.* A length of steel or iron bent at a right angle along its long dimension, used as a support or structural framework.
**angle of attack** *n.* The acute angle between the chord of an airfoil and a line representing the undisturbed relative airflow.
**angle of incidence** *n.* **1.** *Physics.* The angle formed by the path of a body or of radiation incident on a surface and a perpendicular to the surface at the point of impact. **2.** Angle of attack.

**angle of reflection** *n.* The acute angle formed by the path of a reflected body or reflected radiation with a perpendicular to the surface at the point of reflection.

**angle of refraction** *n.* The acute angle formed by the path of refracted radiation with a perpendicular to the refracting surface at the point of refraction.

**angle of yaw** *n.* The angle between an aircraft's longitudinal axis and its line of travel, as observed from above.

**angle plate** *n.* A right-angled metal bracket used on the faceplate of a lathe to hold pieces being worked.

**an·gle·pod** (ăng′gəl-pŏd′) *n.* A plant of the genus *Gonolobus* of the southern and central United States, bearing greenish or purple flowers and angular pods.

**an·gler** (ăng′glər) *n.* **1.** A person who fishes with a hook. **2.** A scheming person. **3.** An anglerfish.

**an·gler·fish** (ăng′glər-fĭsh′) *n., pl.* **anglerfish** *or* **-fish·es.** A marine fish of the order Lophiiformes or Pediculati, with a long dorsal fin ray that hangs above the mouth and serves as a lure for prey.

**an·gle·site** (ăng′glĭ-sīt′) *n.* [After *Anglesey,* Wales.] A lead sulfate mineral, found in colorless or tinted crystals.

**an·gle·worm** (ăng′gəl-wûrm′) *n.* A worm, as an earthworm, used as fishing bait.

**An·gli·an** (ăng′glē-ən) *adj.* Of or relating to the Angles. —*n.* **1.** An Angle. **2.** The Old English dialects of Mercian and Northumbrian.

**An·gli·can** (ăng′glĭ-kən) *adj.* [Med. Lat. *Anglicanus,* English < *Anglicus* < Lat. *Angli,* the Angles. —see ANGLE.] **1.** Of, relating to, or typical of the Church of England or any of the churches related to it in origin and communion, as the Protestant Episcopal Church. **2.** Of or relating to England or the English. —*n.* A member of the Church of England or of any of the churches related to it.

**Anglican Church** *n.* The Church of England and the churches in other nations that agree with it on doctrine and discipline and are in communion with the Archbishop of Canterbury.

**Anglican Communion** *n.* The Anglican Church.

**An·gli·can·ism** (ăng′glĭ-kən-ĭz′əm) *n.* The doctrine, system, and practice of the Anglican Church.

**An·gli·ce** (ăng′glĭ-sē′) *adv.* [Med. Lat. < *Anglicus,* English < Lat. *Angli,* the Angles. —see ANGLE.] In the English mode <Venezia, *Anglice* Venice>

**An·gli·cism** *also* **an·gli·cism** (ăng′glə-sĭz′əm) *n.* **1.** A word, phrase, or idiom peculiar to the English language, esp. as spoken in England : BRITICISM. **2.** A typically English quality.

**An·gli·cist** (ăng′glĭ-sĭst) *n.* A specialist in English linguistics.

**An·gli·cize** *also* **an·gli·cize** (ăng′glə-sīz′) *vt. & vi.* **-cized, -ciz·ing, -ciz·es.** To make or become English or similar to English in form, pronunciation, idiom, or character. **—An′gli·ci·za′tion** *n.*

**an·gling** (ăng′glĭng) *n.* The act, process, or art of fishing with a hook and line and usu. a rod.

**An·glo** (ăng′glō′) *n., pl.* **-glos.** [Short for ANGLO-AMERICAN.] *Informal.* An Anglo-American, esp. a white inhabitant of the United States who is not of Latin descent. **—An′glo′** *adj.*

**Anglo-** *pref.* [NLat. < Med. Lat. *Angli,* the English people < Lat., the Angles.] England : English <*Anglo*-Saxon>

**An·glo-A·mer·i·can** (ăng′glō-ə-mĕr′ĭ-kən) *adj.* **1.** Of, relating to, or between England and America, esp. the United States. **2.** Of or relating to Anglo-Americans. —*n.* An American, esp. a U.S. inhabitant whose language, ancestry, and culture are English.

**An·glo-Cath·o·lic** (ăng′glō-kăth′lĭk, -kăth′ə-lĭk) *n.* A member of the Anglican Communion whose religious convictions emphasize sacramental worship. **—An′glo-Cath′o·lic** *adj.*

**An·glo-French** (ăng′glō-frĕnch′) *adj.* Of, relating to, or between England and France or their peoples. —*n.* ANGLO-NORMAN 2.

**An·glo-In·di·an** (ăng′glō-ĭn′dē-ən) *n.* A person of English and Indian ancestry. **—An′glo-In′di·an** *adj.*

**An·glo-I·rish** (ăng′glō-ī′rĭsh) *n.* **1.** A native of England living in Ireland. **2.** A native of Ireland living in England. **3.** A person of Irish and English ancestry. **—An′glo-I′rish** *adj.*

**An·glo-Nor·man** (ăng′glō-nôr′mən) *n.* **1.** One of the Normans who lived in England after the Norman conquest of England in 1066 or a descendant of these settlers. **2. a.** The dialect of Old French derived mainly from Norman French that was used by the Anglo-Normans. **b.** The form of this dialect used in English law until the 17th cent. **—An′glo-Nor′man** *adj.*

**An·glo·phile** (ăng′glə-fīl′) *also* **An·glo·phil** (-fĭl) *n.* An admirer of England, the English, and English things. —*adj.* Of or like Anglophiles. **—An′glo·phil′i·a** (-fĭl′ē-ə) *n.*

**An·glo·phobe** (ăng′glə-fōb′) *n.* One who has an aversion to England, the English, or English things. —*adj.* Of or like Anglophobes. **—An′glo·pho′bi·a** *n.*

**An·glo·phone** (ăng′lə-fōn′) *n.* An English-speaking individual, esp. in a country where two or more languages are spoken. **—An′glo·phon′ic** (-fŏn′ĭk) *adj.*

**An·glo-Sax·on** (ăng′glō-săk′sən) *n.* **1.** A member of one of the Germanic peoples (Angles, Saxons, and Jutes) who settled in Britain in the 5th and 6th cent. **2.** Any of the descendants of the Anglo-Saxons who were dominant in England until the Norman Conquest of 1066. **3.** OLD ENGLISH 1. **4.** A person of English ancestry. —*adj.* Of, relating to, or typical of Anglo-Saxons or their descendants or their language or culture : ENGLISH.

**An·go·ra** (ăng-gôr′ə, -gōr′ə) *n.* [After *Angora* (Ankara), Turkey.] **1.** *often* **angora. a.** The long, silky hair of the Angora goat. **b.** The fine, light hair of the Angora rabbit, sometimes blended with wool in fabrics. **c.** A yarn or fabric made from either of these fibers. **2.** An Angora cat. **3.** An Angora goat. **4.** An Angora rabbit.

**Angora cat** *n.* A long-haired domestic cat.

**Angora goat** *n.* A domestic goat with long, silky hair.

**Angora goat**
*30 inches high at shoulder*

**Angora rabbit** *n.* A domestic rabbit with long, soft hair.

**an·gos·tu·ra bark** (ăng′gə-stŏŏr′ə, -tyŏŏr′ə) *n.* [After *Angostura,* former name of Ciudad Bolívar.] The bitter, aromatic bark of either of two Brazilian trees, *Galipea officinalis* or *Cusparia trifoliata,* used as a tonic and a flavoring agent.

**an·gry** (ăng′grē) *adj.* **-gri·er, -gri·est.** [ME *angri* < *anger,* anger.] **1.** Feeling or showing anger : IRATE. **2.** Indicative of or resulting from anger <an *angry* retort> **3.** Having a menacing or threatening aspect <an *angry* sky> **4.** Inflamed, as a sore. **—an′gri·ly** *adv.*

☆ ***syns:*** ANGRY, CHOLERIC, ENRAGED, FURIOUS, INDIGNANT, IRATE, MAD, SEETHING, SORE, WRATHFUL *adj. core meaning* : feeling or displaying anger <an *angry* customer><an *angry* look>

**angry young man** *also* **Angry Young Man** *n.* One of a group of English writers of the 1950's whose works reflect strong social protest.

**angst** (ängkst) *n.* [G.] A feeling of anxiety.

**ang·strom** *or* **ång·ström unit** (ăng′strəm) *n.* [After Anders Jonas *Ångström* (1814–1874).] A unit of length equal to one hundred millionth ($10^{-8}$) of a centimeter, used esp. to specify radiation wavelengths.

**an·guish** (ăng′gwĭsh) *n.* [ME *angwisshe,* < OFr. *anguisse* < Lat. *angustia,* narrowness < *angustus,* narrow.] Agonizing physical or mental pain : TORMENT. —*v.* **-guished, -guish·ing, -guish·es.** —*vt.* To cause to suffer anguish. —*vi.* To suffer anguish.

**an·gu·lar** (ăng′gyə-lər) *adj.* [Lat. *angularis* < *angulus,* angle.] **1.** Having, forming, or made up of an angle or angles. **2.** Measured by an angle or degrees of an arc. **3.** Bony and lean : GAUNT. **4.** Without grace or smoothness : AWKWARD. **5.** Unyielding in manner or disposition : STIFF. **—an′gu·lar·ly** *adv.* **—an′gu·lar·ness** *n.*

**angular acceleration** *n.* The rate of change of angular velocity with respect to time.

**an·gu·lar·i·ty** (ăng′gyə-lăr′ĭ-tē) *n., pl.* **-ties. 1.** The quality of being angular. **2. angularities.** Angular forms, outlines, or corners.

**angular momentum** *n.* **1.** The vector product of the position vector and linear velocity of a particle in motion relative to an axis. **2.** The sum of such products, one for each component particle of an extended body, expressed as the product of the angular velocity and the moment of inertia of the body.

**angular velocity** *n.* A vector quantity describing rotational motion, the magnitude of which is the time rate of change of angle and the direction of which is along the axis of rotation.

**an·gu·late** (ăng′gyə-lĭt, -lāt′) *adj.* Having angles or an angular shape. —*vt. & vi.* (-lāt′) **-lat·ed, -lat·ing, -lates.** To make or become angular. **—an′gu·late·ly** *adv.*

**an·gu·la·tion** (ăng′gyə-lā′shən) *n.* **1.** The formation of angles. **2.** An angular part, position, or formation.

**an·hin·ga** (ăn-hĭng′gə) *n.* [Port. < Tupi.] The water turkey.

**an·hy·dride** (ăn-hī′drīd′) *n.* [ANHYDR(OUS) + -IDE.] A chemical compound formed from another by the removal of water.

**an·hy·drite** (ăn-hī′drīt′) *n.* [ANHYDR(OUS) + -ITE[1].] A white to grayish or reddish mineral of anhydrous calcium sulfate, $CaSO_4$, occurring as layers in gypsum deposits.

**an·hy·drous** (ăn-hī′drəs) *adj.* [Gk. *anudros* : *an-,* without + *hudōr,* water.] Without water, esp. water of crystallization.

**a·ni** (ä-nē′) *n.* [Sp. (South America) *aní* < Tupi *ani.*] A chiefly tropical American bird of the genus *Crotophaga,* with black plumage and a long tail.

**an·il** (ăn′ĭl) *n.* [Fr. < Port. < Ar. *an-nīl,* the indigo < Pers. *nīl,* indigo.] The indigo plant or a blue dye derived from it.

ă **pat** ā **pay** âr **care** ä **father** ĕ **pet** ē **be** hw **which** ĭ **pit**
ī **tie** îr **pier** ŏ **pot** ō **toe** ô **paw, for** oi **noise** ŏŏ **took**

**an·ile** (ăn′īl′, ā′nīl′) *adj.* [Lat. *anilis* < *anus,* old woman.] Of, resembling, or similar to an old woman. **—a·nil′i·ty** (ə-nĭl′ĭ-tē) *n.*

**an·i·line** *also* **an·i·lin** (ăn′ə-lĭn) *n.* [G. *Anilin* : *Anil,* anil + *-in,* -ine.] A colorless, oily, poisonous benzene derivative, $C_6H_5NH_2$, used to make rubber, dyes, resins, pharmaceuticals, and varnishes. *—adj.* Derived from aniline.

**aniline black** *n.* A black dye that is produced on a fiber such as cotton by oxidation of aniline oil and is noted for its color intensity, fastness, and resistance to greening.

**aniline dye** *n.* Any of numerous synthetic dyes.

**an·i·ma** (ăn′ə-mə) *n.* [Lat.] The inner self : SOUL.

**an·i·mad·ver·sion** (ăn′ə-măd-vûr′zhən, -shən) *n.* [Lat. *animadversio* < *animadvertere,* to turn the mind toward —see ANIMADVERT.] **1.** Hostile criticism. **2.** A severely critical or censorious remark.

**an·i·mad·vert** (ăn′ə-măd-vûrt′) *vi.* **-vert·ed, -vert·ing, -verts.** [Lat. *animadvertere,* to turn the mind toward : *animus,* mind + *vertere,* to turn.] To remark or comment critically, usu. with strong disapproval.

**an·i·mal** (ăn′ə-məl) *n.* [Lat. < *animalis,* living < *anima,* soul.] **1.** An organism of the kingdom Animalia, distinguished from plants by certain characteristics, as the power of locomotion, fixed structure and limited growth, and nonphotosynthetic metabolism. **2.** An animal organism other than a human being, esp. a mammal. **3.** A person of bestial, brutish, or inhuman character. **4.** ANIMALITY 3. *—adj.* **1.** Of, relating to, or typical of animals. **2.** Relating to the physical or nonrational as distinct from the spiritual nature of humans.

**animal crackers** *pl.n.* Small cookies in various animal shapes.

**an·i·mal·cule** (ăn′ə-măl′kyōōl) *also* **an·i·mal·cu·lum** (-kyə-ləm) *n., pl.* **-cules** *also* **-cu·la** (-kyə-lə) [NLat. *animalculum,* dim. of Lat. *animal,* animal.] **1.** A microscopic or minute organism, as an amoeba or paramecium, usu. regarded as an animal. **2.** *Archaic.* A tiny animal, as a mosquito.

**animal heat** *n.* Heat generated in an animal's body.

**animal husbandry** *n.* The care and breeding of domestic animals, as cattle, hogs, sheep, and horses.

**an·i·mal·ism** (ăn′ə-mə-lĭz′əm) *n.* **1.** A state of enjoying good health and vigorous physical drives. **2.** A state of indifference to all but the physical appetites. **3.** The doctrine that the human being is purely animal with no spiritual nature. **—an′i·mal·ist** *n.* **—an′i·mal·is′tic** (-ĭs′tĭk) *adj.*

**an·i·mal·i·ty** (ăn′ə-măl′ĭ-tē) *n.* **1.** The nature or characteristics of an animal. **2.** The animal kingdom. **3.** The animal as distinct from the spiritual nature of humans.

**an·i·mal·ize** (ăn′ə-mə-līz′) *vt.* **-ized, -iz·ing, -iz·es.** **1.** To make coarse and brutal : DEHUMANIZE. **2.** To endow (a deity) with the attributes of an animal. **—an′i·mal·i·za′tion** *n.*

**animal kingdom** *n.* The category of living organisms that includes all animals.

**animal magnetism** *n.* **1.** Hypnotism or mesmerism. **2.** Magnetic personal presence. **3.** Sensualism. **4.** *Christian Science.* —Used to refer to <"The voluntary or involuntary action of error in all its forms" —Mary Baker Eddy>

**animal spirits** *pl.n.* Vigorous buoyancy of good health.

**animal starch** *n.* Glycogen.

**an·i·mate** (ăn′ə-māt′) *vt.* **-mat·ed, -mat·ing, -mates.** [Lat. *animare, animat-* < *anima,* soul.] **1.** To give life to. **2.** To impart interest or zest to : ENLIVEN <"The party was *animated* by all kinds of men and women" —René Dubos> **3.** To impart spirit, courage, or resolution to : ENCOURAGE. **4.** To inspire to action : PROMPT. **5.** To impart motion or activity to. **6.** To make, design, or produce (e.g., a cartoon) so as to create the illusion of motion. *—adj.* (ăn′ə-mĭt). **1.** Having life : ALIVE. **2.** Of or relating to animal life as distinct from plant life. **3.** Belonging to the class of nouns that stand for living things; e.g., the word "cat" is animate, the word "house" inanimate.

**an·i·mat·ed** (ăn′ə-mā′tĭd) *adj.* **1.** Filled with life, activity, vigor, or spirit. **2.** Made or designed so as to seem alive and moving. **3.** Containing or made of figures or objects that seem to move in a lifelike manner. **—an′i·mat′ed·ly** *adv.*

**animated cartoon** *n.* A usu. short motion picture having a photographed series of drawings.

**animated oat** *n.* A Mediterranean grass, *Avena sterilis,* bearing spikelets that move in response to changes in moisture.

**an·i·mat·er** (ăn′ə-mā′tər) *n. var. of* ANIMATOR.

**an·i·ma·tion** (ăn′ə-mā′shən) *n.* **1.** The act, process, or result of animating. **2.** The quality or condition of being animate. **3. a.** The art or process of designing, preparing, or producing animated cartoons or feature-length films. **b.** An animated cartoon.

**a·ni·ma·to** (ä′nē-mä′tō) *adv. & adj.* [Ital. < Lat. *animatus,* p.part. of *animare,* to animate.] *Mus.* In an animated or lively manner. —Used as a direction.

**an·i·ma·tor** *also* **an·i·mat·er** (ăn′ə-mā′tər) *n.* **1.** One that animates. **2.** One who designs, prepares, or produces an animated cartoon or feature-length film.

**an·i·mism** (ăn′ə-mĭz′əm) *n.* [G. *Animismus* : Lat. *anima,* soul + *-ismus,* -ism.] **1.** A primitive belief whereby natural phenomena and animate and inanimate things are held to possess an innate soul. **2.** A theory of psychic concepts or of spiritual beings. **3.** The hypothesis, first advanced by Pythagoras and Plato, of an immaterial force animating the universe. **4.** An 18th-cent. doctrine that viewed the soul as the vital principle and source of both the normal and the abnormal phenomena of life.

**an·i·mos·i·ty** (ăn′ə-mŏs′ĭ-tē) *n., pl.* **-ties.** [ME *animosite* < OFr. < LLat. *animositas,* courage < Lat. *animosus,* bold < *animus,* soul.] Long-standing or deep-seated hostility : ENMITY.

**an·i·mus** (ăn′ə-məs) *n.* [Lat., soul, mind.] **1.** An animating intention or purpose. **2.** A feeling of animosity : ILL WILL.

**an·i·on** (ăn′ī′ən) *n.* [Gk., (something) going up, pr.part. of *anienai,* to go up : *ana-,* up + *ienai,* to go.] A negatively charged ion that migrates to an anode, as in electrolysis.

**anis-** *pref. var. of* ANISO-.

**an·ise** (ăn′ĭs) *n.* [ME *anis* < OFr. < Lat. *anisum* < Gk. *anison.*] **1.** A plant native to the Mediterranean region, *Pimpinella anisum* with small yellowish-white flower clusters and licorice-flavored seeds. **2.** Aniseed.

**an·i·seed** (ăn′ĭ-sēd′) *n.* [ME *anis seed,* anise seed.] The licorice-flavored seed of the anise plant, used medicinally and as a flavoring.

**an·i·sei·ko·ni·a** (ăn-ī′sī-kō′nē-ə) *n.* [NLat. : ANIS(O)- + Gk. *eikōn,* image.] An ocular defect in which image, shape, and size differ in each eye.

**an·i·sette** (ăn′ə-sĕt′, -zĕt′) *n.* [Fr., dim. of *anis,* anise < OFr.] An anise-flavored liqueur.

**aniso-** *or* **anis-** *pref.* [NLat. < Gk. *anisos,* unequal: *an-,* not + *isos,* equal.] Unequal : dissimilar <*anisogamy*>

**an·i·sog·a·my** (ăn′ī-sŏg′ə-mē) *n.* A union between markedly different gametes. **—an′i·so·gam′ic** *adj.*

**an·i·som·er·ous** (ăn′ī-sŏm′ər-əs) *adj.* Having or designating floral whorls that have unequal numbers of parts.

**an·i·so·met·ric** (ăn-ī′sə-mĕt′rĭk) *adj.* Not isometric.

**an·i·so·me·tro·pi·a** (ăn-ī′sə-mĭ-trō′pē-ə) *n.* [ANISO- + Gk. *metros,* measure + -OPIA.] Difference in the refractive power of the eyes. **—an·i′so·me·trop′ic** *adj.*

**an·i·so·trop·ic** (ăn-ī′sə-trŏp′ĭk) *adj.* **1.** Not isotropic. **2.** *Physics.* Having properties that differ according to the direction of measurement. **—an·i′so·trop′i·cal·ly** *adv.* **—an′i·sot′ro·pism** (-sŏt′rə-pĭz′əm), **an′i·sot′ro·py** (-sŏt′rə-pē) *n.*

**an·ker·ite** (ăng′kə-rīt′) *n.* [G. *Ankerit,* after M.J. *Anker* (d. 1843).] A dolomitelike mineral in which iron partially replaces magnesium.

**ankh** (ăngk) *n.* [Of Egypt. orig.] An ansate cross.

**an·kle** (ăng′kəl) *n.* [ME *ancle,* prob. of ON orig.] **1.** The joint that connects the foot with the leg. **2.** The slender section of the leg just above the foot.

**an·kle·bone** (ăng′kəl-bōn′) *n.* The talus.

**an·klet** (ăng′klĭt) *n.* **1.** An ornament worn about the ankle. **2.** A sock reaching just above the ankle.

**an·ky·lose** *also* **an·chy·lose** (ăng′kə-lōs′, -lōz′) *v.* **-losed, -los·ing, -los·es.** [Back-formation < ANKYLOSIS.] *—vt.* To join or stiffen by ankylosis. *—vi.* To undergo ankylosis.

**an·ky·lo·sis** *also* **an·chy·lo·sis** (ăng′kə-lō′sĭs) *n.* [NLat. < Gk. *ankulōsis,* stiffening of the joints < *ankuloun,* bent.] **1.** *Anat.* The consolidation of bones or their parts forming a single unit. **2.** *Pathol.* The stiffening of a joint due to abnormal bone fusion.

**an·lace** (ăn′lĭs) *n.* [ME *anelas.*] A two-edged medieval dagger.

**an·la·ge** *also* **An·la·ge** (än′lä′gə) *n., pl.* **-ges** or **-gen** (-gən) [G., fundamental principle < MHG *anlāge,* request : *ane-,* on + *lāge,* act of laying.] **1.** The initial cell structure from which an embryonic part or organ develops : PRIMORDIUM. **2.** A fundamental principle.

**an·na** (ä′nə) *n.* [Hindi *ānā* < Skt. *aṇu-,* small.] A copper coin once used in India and Pakistan.

**an·nal·ist** (ăn′ə-lĭst) *n.* One who writes annals : CHRONICLER.

**an·nals** (ăn′əlz) *pl.n.* [Lat. (*libri*) *annales,* yearly (books) < *annalis,* yearly < *annus,* year.] **1.** A chronological record of the events of successive years. **2.** Chronicles. **3.** A periodical compiling the records and reports of an organization or a learned field.

**An·na·mese** (ăn′ə-mēz′, -mēs′) *also* **An·na·mite** (ăn′ə-mīt′). *—adj.* Of or relating to Annam, its inhabitants, their language, or their culture. *—n.* **1.** *pl.* **Annamese** *also* **-mites.** A native or inhabitant of Annam. **2.** Vietnamese.

**an·nat·to** (ə-nä′tō) *also* **ar·nat·to** (är-nä′tō) *n., pl.* **-tos.** [Of Cariban orig.] **1.** A small tropical American tree, *Bixa orellana,* with red or pinkish flowers and seeds used in cooking. **2.** A yellowish-red dyestuff obtained from the pulp of annatto seeds.

**an·neal** (ə-nēl′) *vt.* **-nealed, -neal·ing, -neals.** [ME *anelen* < OE *onælan,* to set fire to : *on,* on + *ælan,* to kindle < *āl,* fire.] **1.** To subject (glass or metal) to heat and slow cooling so as to toughen and reduce brittleness. **2.** To temper.

**an·ne·lid** (ăn′ə-lĭd) *also* **an·nel·i·dan** (ə-nĕl′ĭ-dən) [NLat. *Annelida,* phylum name < Fr. *annelé,* ringed < OFr. *annel,* ring < Lat. *anellus,* dim. of *anulus,* ring.] *—adj.* Of or belonging to the phylum Annelida, which includes worms having cylindrical segmented bodies, as the earthworms and leeches. *—n.* An annelid worm.

**an·nex** (ə-nĕks′, ăn′ĕks′) *vt.* **-nexed, -nex·ing, -nex·es.** [ME *an-*

*nexen* < OFr. *annexer* < Lat. *annexus*, p.part. of *annectere*, to connect : *ad-*, to + *nectere*, to bind.] **1.** To append or attach, esp. to a larger or more significant entity. **2.** To incorporate (territory) into an existing state or country. **3.** To add or attach, as an attribute, condition, or consequence. —*n.* (ăn′ĕks′, -ĭks) **1. a.** A structure added on to a larger one. **b.** An auxiliary structure situated near a main one. **2.** An addition to a record or document. **—an′nex·a′tion** (ăn′ĭk-sā′shən) *n.* **—an′nex·a′tion·al** *adj.* **—an′nex·a′tion·ism** *n.* **—an′nex·a′tion·ist** *n.*

**an·nexe** (ăn′ĭks′) *n. Chiefly Brit. var. of* ANNEX.

**An·nie Oak·ley** (ăn′ē ōk′lē) *n.* [After *Annie Oakley* (1860–1926).] *Slang.* A complimentary admission ticket.

▲ word history: Annie Oakley was a sharpshooter with Buffalo Bill's show, *The Wild West*, who sometimes used playing cards as targets. Free passes that were punched to prevent their being sold were thought to resemble Annie Oakley's handiwork.

**an·ni·hi·late** (ə-nī′ə-lāt′) *v.* **-lat·ed, -lat·ing, -lates.** [LLat. *annihilare, annihilat-* : Lat. *ad-*, to + Lat. *nihil*, nothing.] —*vt.* **1.** To destroy all traces of : OBLITERATE. **2.** To nullify or render void : ABOLISH. **3.** *Informal.* To overwhelm or vanquish completely. —*vi. Physics.* To participate in annihilation, as do an electron and a positron. **—an·ni′hi·la·bil′i·ty** (-lə-bĭl′ĭ-tē) *n.* **—an·ni′hi·la·ble** (-lə-bəl) *adj.* **—an·ni′hi·la′tor** *n.*

**an·ni·hi·la·tion** (ə-nī′ə-lā′shən) *n.* **1.** The act or process of annihilating. **2.** A condition of utter destruction. **3.** *Physics.* The phenomenon in which a particle and an antiparticle, as an electron and a positron, disappear with a resultant release of energy approx. equivalent to the sum of their masses.

**an·ni·ver·sa·ry** (ăn′ə-vûr′sə-rē) *n., pl.* **-ries.** [ME *anniversarie* < Med. Lat. *anniversarium* < Lat. *anniversarius*, returning yearly : *annus*, year + *versus*, p.part. of *vertere*, to turn.] **1.** The annual recurrence of the date of an earlier event. **2.** A commemorative celebration on the date of an anniversary.

**an·no Dom·i·ni** (ăn′ō dŏm′ə-nī′, dŏm′ə-nē) *adv.* [Lat., in the year of the Lord.] In a specified year of the Christian era. —Used chiefly in the abbreviated form <*A.D.* 495>

**an·no·tate** (ăn′ō-tāt′) *v.* **-tat·ed, -tat·ing, -tates.** [Lat. *annotare, annotat-*, to note down : *ad-*, to + *notāre*, to write < *nota*, note.] —*vt.* To furnish (a literary work) with critical commentary or explanatory notes : GLOSS. —*vi.* To gloss a text.

**an·no·ta·tion** (ăn′ō-tā′shən) *n.* **1.** The act or process of annotating. **2.** A critical or explanatory note : COMMENTARY.

**an·nounce** (ə-nouns′) *vt.* **-nounced, -nounc·ing, -nounc·es.** [ME *announcen* < OFr. *anoncier* < Lat. *annuntiare* : *ad-*, to + *nuntiare*, to report < *nuntius*, messenger.] **1.** To bring to public notice officially or formally <*announce* a marriage> **2.** To proclaim the presence or arrival of <*announce* a guest> **3.** To make aware or conscious of through the senses <Smells of cooking *announced* dinner.> **4.** To serve as an announcer.

**an·nounce·ment** (ə-nouns′mənt) *n.* **1.** The act of announcing. **2.** Something that is announced. **3.** A public statement or notice.

**an·nounc·er** (ə-noun′sər) *n.* **1.** One that announces. **2.** A radio or television performer who provides program continuity and delivers commercial announcements.

**an·noy** (ə-noi′) *v.* **-noyed, -noy·ing, -noys.** [ME *anoien* < OFr. *anoier* < LLat. *inodiare*, to make odious < Lat. *in odio*, odious : *in*, in + *odium*, hatred.] —*vt.* **1.** To bother or irritate. **2.** To disturb by repeated attacks : HARASS. —*vi.* To behave in an annoying way.

☆ **syns:** ANNOY, AGGRAVATE, BOTHER, BUG, CHAFE, DISTURB, EXASPERATE, FRET, GALL, GET, IRK, IRRITATE, NETTLE, PEEVE, PROVOKE, RUFFLE, VEX *v. core meaning* : to trouble (another) by repeated vexations <Their constant bickering *annoys* me.>

**an·noy·ance** (ə-noi′əns) *n.* **1.** Something that annoys : NUISANCE. **2.** The act of annoying. **3.** Irritation : vexation.

**an·noy·ing** (ə-noi′ĭng) *adj.* Causing annoyance : BOTHERSOME <an *annoying* tickle in my throat> **—an·noy′ing·ly** *adv.*

**an·nu·al** (ăn′yo͞o-əl) *adj.* [ME *annuel* < OFr. < LLat. *annualis* < Lat. *annus*, year.] **1.** Recurring, done, or performed every year : YEARLY <an *annual* checkup> **2.** Determined by a year's time <*annual* precipitation> **3.** *Bot.* Living and growing for only one year or season. —*n.* **1.** A periodical published yearly. **2.** A plant whose life cycle is completed in one year or season.

**annual ring** *n.* One of the concentric layers of wood, esp. in a tree trunk, indicating a year's growth in temperate climates and seasonal growth in regions of wet and dry seasons.

**an·nu·i·tant** (ə-no͞o′ĭ-tənt, ə-nyo͞o′-) *n.* One who is entitled to receive an annuity.

**an·nu·i·ty** (ə-no͞o′ĭ-tē, -nyo͞o′-) *n., pl.* **-ties.** [ME *annuite* < ANFr. < Med. Lat. *annuitas* < Lat. *annuus*, yearly < *annus*, year.] **1. a.** The annual payment of an allowance or income. **b.** The right to receive or the obligation to make this payment. **2.** An investment on which a person receives fixed payments for a lifetime or a specified number of years.

**an·nul** (ə-nŭl′) *vt.* **-nulled, -nul·ling, -nuls.** [ME *annullen* < OFr. *annuller* < LLat. *annullare* : Lat. *ad-*, to + Lat. *nullus*, none.] **1.** To make or declare void or invalid, as a marriage or a law : NULLIFY. **2.** To obliterate the existence or effect of.

**an·nu·lar** (ăn′yə-lər) *adj.* [OFr. *annulaire* < Lat. *anularis* < *anulus*, ring.] Shaped like or forming a ring.

**annular eclipse** *n.* A solar eclipse in which the moon covers all but a bright ring around the circumference of the sun.

**annular ligament** *n.* A ligament or fibrous band that rings the ankle or wrist joint.

**an·nu·late** (ăn′yə-lĭt, -lāt′) *also* **an·nu·lat·ed** (-lā′tĭd) *adj.* [Lat. *anulatus* < *anulus*, ring.] Having, made up of, or formed by rings or ringlike segments.

**an·nu·la·tion** (ăn′yə-lā′shən) *n.* **1.** The act or process of forming rings. **2.** A ringlike segment or structure.

**an·nu·let** (ăn′yə-lĭt) *n.* [Lat. *anulus*, ring + -ET.] A ringlike molding around the capital of a pillar.

**an·nu·li** (ăn′yə-lī′) *n. var. pl. of* ANNULUS.

**an·nul·ment** (ə-nŭl′mənt) *n.* **1.** The act of annulling. **2.** The retrospective and prospective invalidation of a marriage, as for nonconsummation, effected by means of a declaration stating that the marriage was never valid.

**an·nu·lus** (ăn′yə-ləs) *n., pl.* **-lus·es** *or* **-li** (-lī′) [Lat. *anulus*, ring.] **1.** A ringlike figure, part, structure, or marking. **2.** *Math.* The figure bounded by and containing the area between two concentric circles.

**an·nun·ci·ate** (ə-nŭn′sē-āt′) *vt.* **-at·ed, -at·ing, -ates.** [Lat. *annuntiare*. —see ANNOUNCE.] To announce or proclaim, esp. formally or officially.

**an·nun·ci·a·tion** (ə-nŭn′sē-ā′shən) *n.* **1.** The act of announcing. **2.** An announcement. **3. Annunciation. a.** The angel Gabriel's announcement of the Incarnation. **b.** The festival, on Mar. 25, celebrating of this event.

**Annunciation lily** *n.* The Madonna lily.

**an·nun·ci·a·tor** (ə-nŭn′sē-ā′tər) *n.* **1.** One that announces. **2.** An electrical signaling device used in hotels or offices to indicate the source of calls on a switchboard.

**an·nus mi·rab·i·lis** (ăn′əs mĭ-răb′ə-lĭs) *n.* [NLat., wondrous year.] An extraordinary year. <"Hungary's blood bath was the saddest event in that *annus mirabilis*" —C.L. Sulzberger>

**a·no·a** (ə-nō′ə) *n.* [Native word in Celebes.] A buffalo, *Anoa depressicornis* of Celebes and the Philippines, with short, pointed horns.

**anoa**
*6–7 feet long*

**an·ode** (ăn′ōd′) *n.* [Gk. *anodos*, a way up : *ana-*, up + *hodos*, way.] **1.** A positively charged electrode, as of an electrolytic cell, storage battery, or electron tube. **2.** The negatively charged terminal of a primary cell or of a storage battery that is supplying current.

**anode mud** *n.* The residue of electrolytic refining, esp. of copper, high in concentrations of inert metals such as platinum or gold.

**an·o·dize** (ăn′ə-dīz′) *vt.* **-dized, -diz·ing, -diz·es.** [ANOD(E) + -IZE.] To coat (a metallic surface) electrolytically with a protective oxide. **—an′o·di·za′tion** *n.*

**an·o·dyne** (ăn′ə-dīn′) *adj.* [Lat. *anodynus* < Gk. *anōdunos*, free from pain : *an-*, without + *odunē*, pain.] **1.** Capable of relieving pain. **2.** Serving to soothe or relax <*anodyne* hobbies> **3.** Watered-down : insipid. —*n.* **1.** A medicine that relieves pain. **2.** A soothing or comforting agent.

**a·noint** (ə-noint′) *vt.* **a·noint·ed, a·noint·ing, a·noints.** [ME *enointen* < OFr. *enoindre* < Lat. *inunguere* : *in-*, on + *ungere*, to smear.] **1.** To apply oil or ointment to. **2.** To place oil on as an indication of sanctification or consecration in a religious ceremony.

**anointing of the sick** *n. Rom. Cath. Ch.* The sacrament of anointing a critically ill person, praying for recovery, and asking for the absolution of sin.

**a·no·le** (ə-nō′lē) *n.* [Fr. *anolis* < Cariban.] A chiefly tropical New World lizard of the genus *Anolis*, having a distensible throat flap and the ability to change color.

**a·nom·a·lous** (ə-nŏm′ə-ləs) *adj.* [LLat. *anomalos* < Gk., uneven : *an-*, not + *homalos*, even < *homos*, same.] Deviating from the normal order, form, or rule.

**a·nom·a·ly** (ə-nŏm′ə-lē) *n., pl.* **-lies.** **1.** Deviation or departure from the normal order, form, or rule. **2.** Something irregular or abnormal. **3.** *Astron.* The angular deviation of a planet from its perihelion as observed from the sun. **—a·nom′a·lis′tic** (-lĭs′tĭk), **a·nom′a·lis′ti·cal** *adj.* **—a·nom′a·lis′ti·cal·ly** *adv.*

**an·o·mie** or **an·o·my** (ăn′ə-mē) *n.* [Gk. *anomia*, lawlessness < *anomos*, lawless : *a-*, without + *nomos*, law.] **1. a.** Collapse of social stability, as from erosion of standards. **b.** The state of alienation experienced by an individual or class in such a situation. **2.** Personal alienation resulting in unsocial behavior.

**a·non** (ə-nŏn′) *adv.* [ME, at once < OE *onān* : *on*, in + *ān*, one.] **1.** At another time : AGAIN. **2.** *Archaic.* In a short time : SOON <"Such good men as he which is *anon* to be interred" —Cotton Mather> **3.** *Archaic.* At once : IMMEDIATELY. **—ever (or now) and anon.** Time after time.

**an·o·nym** (ăn′ə-nĭm′) *n.* [Fr. *anonyme* < LLat. *anonymus*, anonymous.] **1.** An anonymous person. **2.** A pseudonym.

**an·o·nym·i·ty** (ăn′ə-nĭm′ĭ-tē) *n.* **1.** The quality or state of being unknown or obscure. **2.** One that is anonymous.

**a·non·y·mous** (ə-nŏn′ə-məs) *adj.* [LLat. *anonymus* < Gk. *anōnumos*, nameless : *an-*, without + *onoma*, name.] **1.** Having an unknown or unacknowledged name. **2.** Having an unknown or withheld authorship or agency <an *anonymous* donation>

**a·noph·e·les** (ə-nŏf′ə-lēz′) *n.* [NLat. *Anopheles*, genus name < Gk. *anōphelēs*, useless : *an-*, without + *ophelos*, advantage.] Any of various mosquitoes of the genus *Anopheles*, many of which carry the malaria parasite and transmit the disease to humans.

**an·o·rak** (ăn′ə-răk′) *n.* [Eskimo (Greenland) *ánorâq*.] **1.** A heavy, hooded jacket : PARKA. **2.** A light, waterproof, hooded jacket.

**an·o·rec·tic** (ăn′ə-rĕk′tĭk) or **an·o·ret·ic** (-rĕt′ĭk) *also* **an·o·rex·ic** (-rĕk′sĭk) [Gk. *anorektos* : *an-*, without + *oregein*, to reach out for.] *—adj.* **1.** Characterized by loss of appetite. **2.** Causing loss of or suppressing appetite. *—n.* **1.** A person who is anorectic. **2.** An anorectic drug.

**an·o·rex·i·a** (ăn′ə-rĕk′sē-ə) *n.* [Gk. : *an-*, without + *orexis*, appetite < *oregein*, to reach out for.] Loss of appetite, esp. as a result of disease or psychological disorder.

**anorexia nerv·o·sa** (nûr-vō′sə) *n.* [NLat., nervous anorexia.] A pathological condition occurring chiefly in young women that is marked by aversion to food and severe nutritional deficency and is thought to be psychological in origin.

**an·o·rex·ic** (ăn′ə-rĕk′sĭk) *adj.* & *n. var. of* ANORECTIC.

**an·or·thite** (ăn-ôr′thīt′) *n.* [Fr. : Gk. *an-*, not + Gk. *orthos*, straight (from its oblique crystals).] A rare plagioclase feldspar with high calcium oxide content, found in igneous rocks.

**an·or·tho·site** (ăn-ôr′thə-sīt′) *n.* [Fr. *anorthose*, a kind of feldspar (Gk. *an-*, not + Gk. *orthos*, straight) + -ITE.] A plutonic rock, mainly plagioclase.

**an·os·mi·a** (ăn-ŏz′mē-ə) *n.* [NLat. : Gk. *an-*, without + Gk. *osmē*, odor.] Loss of the sense of smell.

**an·oth·er** (ə-nŭth′ər) *adj.* **1.** One more : ADDITIONAL <*another* scoop of ice cream> **2.** Distinctly different from the first <tried *another* hair style> **3.** Some other <can do it *another* time> *—pron.* **1.** An additional one. **2.** A different one. **3.** One of an undetermined number or group <by one method or *another*>

**an·ov·u·lant** (ăn′ŏv′yə-lənt) *n.* [AN- + OVUL(ATION) + -ANT.] An anovulatory drug.

**an·o·vu·la·tion** (ăn-ō′vyə-lā′shən, -ŏv′yə-) *n.* Lack of ovulation.

**an·o·vu·la·to·ry** (ăn-ō′vyə-lə-tôr′ē, -tōr′ē, -ŏv′yə-) *adj.* Relating to the suppression of ovulation.

**an·ox·e·mi·a** (ăn′ŏk-sē′mē-ə) *n.* An abnormal decrease in the blood's oxygen content.

**an·ox·i·a** (ăn-ŏk′sē-ə) *n.* **1.** Absence of oxygen. **2.** A pathological deficiency of oxygen, esp. hypoxia. **—an·ox′ic** (-ŏk′sĭk) *adj.*

**an·sate** (ăn′sāt′) *also* **an·sat·ed** (-sā′tĭd) *adj.* [Lat. *ansatus* < *ansa*, handle.] Having a handle or a handlike part.

**ansate cross** *n.* A cross shaped like a T with a loop at the top.

**an·sat·ed** (ăn′sā′tĭd) *adj. var. of* ANSATE.

**An·schluss** (ăn′shlo͝os′) *n.* [G.] A union, esp. the political union of Nazi Germany and Austria in 1938.

**an·ser·ine** (ăn′sə-rīn′) *adj.* [NLat. *Anserinae*, subfamily name < Lat. *anserinus*, pertaining to geese < *anser*, goose.] **1.** Of or belonging to the subfamily Anserinae, which includes the geese. **2.** Gooselike. **3.** *also* **an·ser·ous** (-sər-əs). Stupid : silly.

**an·swer** (ăn′sər) *n.* [ME *answere* < OE *andswaru*.] **1.** A spoken or written reply, as to a question or demand. **2. a.** A solution or result, as to a problem. **b.** The correct response or solution. **3.** An act in response or retaliation. **4.** *Law.* The defense or reply to charges filed against a defendant. *—v.* **-swered, -swer·ing, -swers.** *—vi.* **1.** To respond in words or action. **2.** To be liable or accountable <had to *answer* for our mistakes> **3.** To serve the purpose : SUFFICE <"often I do use three words where one would *answer*" —Mark Twain> **4.** To match or correspond, as to a description. *—vt.* **1.** To reply to. **2.** To respond correctly to. **3.** To fulfill the demands of : SERVE. **4.** To conform or correspond to. **5.** To reply to an accusation <*answer* the charges>

☆ **syns:** ANSWER, REJOIN, REPLY, RESPOND, RETORT, RETURN *v. core meaning* : to speak or act in response to <*answer* a question> <*answer* a letter>

**an·swer·a·ble** (ăn′sər-ə-bəl) *adj.* **1.** Obligated to render account : ACCOUNTABLE <*answerable* for their hasty decisions> **2.** Capable of being answered. **3.** *Archaic.* Corresponding : suitable. **—an′swer·a·bil′i·ty, an′swer·a·ble·ness** *n.* **—an′swer·a·bly** *adv.*

**ant** (ănt) *n.* [ME *amte* < OE *ǣmete*.] Any of various social insects of the family Formicidae, typically having wings only in the males and fertile females and living in highly organized social colonies.

**ant-** *pref. var. of* ANTI-.

**-ant** *suff.* [ME < OFr. < Lat. *-ans*, *-ant-*, pr.part. suffix.] **1. a.** Performing, promoting, or causing a specified action <accept*ant*> **b.** Being in a specified state or condition <flipp*ant*> **2. a.** One that performs, promotes, or causes a specified action <deodor*ant*> **b.** One that undergoes a specified action <inhal*ant*>

**an·ta** (ăn′tə) *n., pl.* **-tae** (-tē) [Lat. *antae*, pilasters.] A thickening of the projecting end of the lateral wall of a Greek temple.

**ant·ac·id** (ănt-ăs′ĭd) *adj.* Neutralizing acids. *—n.* A substance that neutralizes acid.

**an·tae** (ăn′tē) *n. pl. of* ANTA.

**An·tae·us** (ăn-tē′əs) *n.* [Lat. < Gk. *Antaios*.] *Gk. Myth.* A giant, invincible while in contact with the ground, who was lifted into the air by Hercules and crushed to death.

**an·tag·o·nism** (ăn-tăg′ə-nĭz′əm) *n.* **1.** Mutual enmity : HOSTILITY. **2.** The condition of being an opposing or competing principle, force, or factor.

**an·tag·o·nist** (ăn-tăg′ə-nĭst) *n.* **1.** One who opposes and actively competes with another : ADVERSARY. **2.** *Anat.* A muscle that opposes another muscle. **3.** A drug that counteracts or neutralizes another drug. **—an·tag′o·nis′tic** *adj.* **—an·tag′o·nis′ti·cal·ly** *adv.*

**an·tag·o·nize** (ăn-tăg′ə-nīz′) *vt.* **-nized, -niz·ing, -niz·es.** [Gk. *antagōnizesthai*, to struggle against : *anti-*, against + *agōnizesthai*, to struggle < *agōn*, contest. —see AGONY.] **1.** To incur the dislike or antagonism of. **2.** To counteract.

**Ant·arc·tic** (ănt-ärk′tĭk, -är′tĭk) *adj.* [ME *Antartik* < OFr. *antartique* < Med. Lat. *antarcticus* < Lat., southern < Gk. *antarktikos* : *anti-*, opposite + *arktikos*, northern. —see ARCTIC.] Of or relating to the regions surrounding the South Pole.

**Antarctic Circle** *n.* A parallel of latitude, 66°33′ south, marking the limit of the South Frigid Zone.

**An·tar·es** (ăn-târ′ēz, -tăr′-) *n.* [Gk. *antarēs* : *anti-*, opposite + *Ares*, Mars.] A double and variable star in the constellation Scorpius, the brightest star in the southern sky.

▲ word history: The Greek name *Antares* can be interpreted in more than one way: "against Mars," "opposite Mars," or "instead of Mars." Since the planet Mars, which the Greeks called Ares, moves through the sky, it is not always opposite Antares in position. The name *Antares* more likely refers to the star's pronounced red color, which sometimes causes the star to be mistaken for the planet.

**ant cow** *n.* An aphid from which ants acquire honeydew.

**an·te** (ăn′tē) *n.* [< Lat., before.] **1.** The stake that each poker player must put into the pool before receiving new cards. **2.** *Slang.* An amount to be paid, esp. as one's share. *—vt.* **-ted** or **-teed, -te·ing, -tes. 1.** To put (one's stake) into the pool in poker. **2.** *Slang.* To pay <*ante* up>

**ante-** *pref.* [Lat. < *ante*, before.] **1.** Prior to : EARLIER <*ante*natal> **2.** In front of : BEFORE <*ante*room>

**ant·eat·er** (ănt′ē′tər) *n.* **1.** Any of several tropical American mammals of the family Myrmecophagidae that lack teeth and feed on ants and termites, esp. *Myrmecophaga tridactyla*, with a long, shaggy-haired tail, a long, narrow snout, and a long, sticky tongue. **2.** Any of several other animals, as the echidna, that feed on ants.

**an·te·bel·lum** (ăn′tē-bĕl′əm) *adj.* [Lat. *ante bellum*, before the war.] Belonging to the period prior to the American Civil War.

**an·te·cede** (ăn′tĭ-sēd′) *vt.* **-ced·ed, -ced·ing, -cedes.** [Lat. *antecedere* : *ante-*, before + *cedere*, to go.] To precede.

**an·te·ce·dence** (ăn′tĭ-sēd′ns) *n.* Precedence.

**an·te·ce·dent** (ăn′tĭ-sēd′nt) *adj.* Going before : PRECEDING. *—n.* **1.** One that precedes. **2.** An occurrence or event preceding another. **3. antecedents.** One's ancestors or ancestry. **4.** The word, phrase, or clause to which a relative pronoun refers. **5.** *Math.* The first term of a ratio. **6.** *Logic.* The conditional member of a hypothetical proposition. **—an′te·ce′dent·ly** *adv.*

**an·te·cham·ber** (ăn′tē-chām′bər) *n.* [Fr. *antichambre* : *anti-*, before (< Lat. *ante-*) + *chambre*, chamber.] A smaller room that is an entryway into a larger room.

**an·te·choir** (ăn′tĭ-kwīr′) *n.* A place in front of the choir reserved for the clergy and choir members.

**an·te·date** (ăn′tĭ-dāt′) *vt.* **-dat·ed, -dat·ing, -dates. 1.** To be of an earlier date than : precede in time. **2.** To give a date earlier than the actual date : BACKDATE. *—n.* A date given to an event or a document that is earlier than the actual date.

**an·te·di·lu·vi·an** (ăn′tĭ-də-lo͞o′vē-ən) *adj.* [ANTE- + Lat. *diluvium*, flood. —see DILUVIAL.] **1.** Taking place in or belonging to the era before the Biblical Flood. **2.** Very old : ANTIQUATED. **—an′te·di·lu′vi·an** *n.*

**an·te·fix** (ăn′tē-fĭks′) *n., pl.* **-fix·es** or **-fix·a** (-fĭk′sə) [< Lat. *antefixus*, fastened in front : *ante-*, before + *fixus*, fastened, p.part. of *fingere*, to fasten.] An upright ornament along the eaves of a tiled roof that conceals the joints between the rows of tiles. **—an'te·fix'·al** *adj.*

**an·te·lope** (ăn'tl-ōp') *n., pl.* **antelope** *or* **-lopes.** [ME < Med. Lat. *anthalopus* < LGk. *antholops.*] **1. a.** A slender, swift-running, long-horned ruminant of the family Bovidae of Africa and Asia. **b.** An animal, as the pronghorn, similar to a true antelope. **2.** Leather made from the hide of an antelope.

**an·te·me·rid·i·an** (ăn'tē-mə-rĭd'ē-ən) *adj.* [Lat. *antemeridianus* : *ante-*, before + *meridianus,* of noon. —see MERIDIAN.] Of, relating to, or occurring in the morning.

**an·te me·rid·i·em** (ăn'tē mə-rĭd'ē-əm) *adv. & adj.* [Lat. : *ante,* before + *meridies,* noon.] Before noon. —Used chiefly in the abbreviated form <10:30 *A.M.*><an *A.M.* conference> ***usage:*** Although *12 A.M.* denotes noon and *12 P.M.* denotes midnight, it is better to use *12 noon* and *12 midnight* to avoid confusion.

**an·te mor·tem** (ăn'tē môr'təm) *adj.* [Lat.] Before death.

**an·te·na·tal** (ăn'tē-nāt'l) *adj.* Before birth : PRENATAL.

**an·ten·na** (ăn-tĕn'ə) *n.* [Med. Lat. < Lat., sail yard.] **1.** *pl.* **-ten·nae** (-tĕn'ē). One of the paired, flexible, jointed sensory appendages on the head of an insect, myriapod, or crustacean. **2.** *pl.* **-nas.** A metallic device for transmitting and receiving electromagnetic waves : AERIAL. **—an·ten'nal** *adj.*

▲ word history: The classical Latin word *antenna* meant only "sail yard." The word was used in medieval writings for the sensory organs of insects because medieval naturalists thought that these organs resembled the parts of the yard that project beyond the sail.

**antenna loop** *n.* A flat coil or wire serving both as antenna and as part of the internal circuitry of a receiver.

**an·ten·nule** (ăn-tĕn'yo͞ol') *n.* [Fr., dim. of *antenne,* antenna < Lat. *antenna,* sail yard.] *Zool.* A small antenna, esp. one of the first pair in crustaceans.

**an·te·pen·di·um** (ăn'tē-pĕn'dē-əm) *n., pl.* **-di·a** (-dē-ə) [Med. Lat. : Lat. *ante-*, before + *pendere,* to hang.] A hanging placed in front of an altar or a pulpit.

**an·te·pe·nult** (ăn'tē-pē'nŭlt', -pĭ-nŭlt') *n.* [Short for LLat. *antepaenultima* < *antepaenultimus,* antepenultimate.] The third syllable from the end in a word, as *te* in *antepenult.*

**an·te·pe·nul·ti·mate** (ăn'tē-pĭ-nŭl'tə-mĭt) *adj.* [LLat. *antepaenultimus* : Lat. *ante-*, before + Lat. *paenultimus,* next to last. —see PENULT.] Third from the end in a series. —*n.* An antepenult.

**an·te·ri·or** (ăn-tîr'ē-ər) *adj.* [Lat., comp. of *ante,* before.] **1.** Located in front. **2.** Prior in time : EARLIER. **3.** *Zool.* **a.** Located near the head in lower animals. **b.** Located on or near the front of the body in higher animals. **4.** *Anat.* Located on or near the front of an organ or on the ventral surface of the body in humans. **5.** *Bot.* In front of and facing away from the axis or stem. **—an·te'ri·or·ly** *adv.*

**an·te·room** (ăn'tē-ro͞om', -ro͝om') *n.* **1.** An antechamber. **2.** A waiting room.

**ant·he·li·on** (ănt-hē'lē-ən, ăn-thē'-) *n., pl.* **-li·a** (-lē-ə) *or* **-ons.** [Gk. *anthēlion* < *anthēlios,* opposite the sun : *anti-*, opposite + *hēlios,* sun.] A luminous white, halolike area sometimes observed in the sky opposite the sun on the parhelic circle.

**ant·hel·min·tic** (ănt'hĕl-mĭn'tĭk, ăn'thĕl-) *also* **ant·hel·min·thic** (-thĭk) *adj.* [ANTI- + Gk. *helmins, helminth-*, worm.] Acting to expel or destroy intestinal worms. —*n.* A vermifuge.

**an·them** (ăn'thəm) *n.* [ME *anteme* < OE *antefn* < Med. Lat. *antiphona* < Gk. < *antiphōnos,* sounding in answer : *anti-*, in return + *phōnē,* voice.] **1.** A hymn of praise or allegiance. **2.** A devotional composition set to words from the Bible.

**an·the·mi·on** (ăn-thē'mē-ən) *n., pl.* **-mi·a** (-mē-ə) [Gk. < *anthos,* flower.] A motif used in Greek art having a pattern of honeysuckle or palm leaves in a radiating cluster.

**an·ther** (ăn'thər) *n.* [NLat. *anthera* < Med. Lat. *anthera,* pollen < Lat., a medicine extracted from flowers < Gk. *anthēros,* flowery < *anthos,* flower.] *Bot.* The organ at the upper end of a stamen that secretes and discharges pollen.

**an·ther·id·i·um** (ăn'thə-rĭd'ē-əm) *n., pl.* **-i·a** (-ē-ə) [NLat. < *anthera,* anther.] *Bot.* An organ producing male sex cells in algae, fungi, mosses, and ferns.

**an·ther·o·zo·id** (ăn'thər-ə-zō'ĭd) *n.* [ANTHER + ZO(O)ID.] *Bot.* A male sex cell produced by an antheridium.

**an·the·sis** (ăn-thē'sĭs) *n.* [Gk. *anthēsis,* flowering < *anthos,* flower.] *Bot.* The blooming period of a flower.

**ant·hill** (ănt'hĭl') *n.* A mound formed by ants or termites in digging or building a nest.

**antho-** *pref.* [< Gk. *anthos,* flower.] Flower <*antho*taxy>

**an·tho·cy·a·nin** (ăn'thō-sī'ə-nĭn) *also* **an·tho·cy·an** (-sī'ən, -ăn') *n.* [ANTHO- + CYANIN(E).] A water-soluble pigment that gives flowers and other plant parts colors ranging from blue to red.

**an·tho·di·um** (ăn-thō'dē-əm) *n., pl.* **-di·a** (-dē-ə) [NLat. < Gk. *anthōdēs,* flowerlike : *anthos,* flower + *-eidēs,* -oid.] The flower head of composite plants, as the aster or goldenrod.

**an·thol·o·gize** (ăn-thŏl'ə-jīz') *vt.* **-gized, -giz·ing, -giz·es.** To compile, include in, or publish an anthology. **—an·thol'o·gist** *n.*

**an·thol·o·gy** (ăn-thŏl'ə-jē) *n., pl.* **-gies.** [NLat. *anthologia* < Gk., flower gathering : *anthos,* flower + *logia,* collection < *legein,* to gather.] A collection of literary works, as poems, anecdotes, short stories, or plays. **—an'tho·log'i·cal** (ăn'thə-lŏj'ĭ-kəl) *adj.*

**an·tho·phore** (ăn'thə-fôr', -fōr') *n. Bot.* A stalklike part in certain flowers that supports the pistils and corolla.

**an·tho·tax·y** (ăn'thə-tăk'sē) *also* **an·tho·tax·is** (ăn'thə-tăk'sĭs) *n. Bot.* The arrangement of the parts of a flower.

**-anthous** *suff.* [NLat. *-anthus* < Gk. *anthos,* flower.] Having flowers <an*anthous*>

**an·tho·zo·an** (ăn'thə-zō'ən) *n.* Any of various marine organisms of the class Anthozoa, growing singly or in colonies and including the corals and sea anemones. **—an'tho·zo'an, an'tho·zo'ic** (-zō'ĭk) *adj.*

**an·thra·cene** (ăn'thrə-sēn') *n.* [Gk. *anthrax, anthrak-*, charcoal + *-ene.*] A crystalline hydrocarbon, $C_6H_4(CH_2)_2C_6H_4$, extracted from coal tar and used to make dyes and organic chemicals.

**an·thra·cite** (ăn'thrə-sīt') *n.* [Lat. *anthracites,* a kind of bloodstone < Gk. *anthrakitēs,* coallike < *anthrax,* charcoal.] A hard, clean-burning coal having a high carbon content and little volatile matter.

**an·thrac·nose** (ăn-thrăk'nōs') *n.* [Fr. : Gk. *anthrax,* carbuncle + Gk. *nosos,* disease.] A plant disease caused by fungi that results in dead spots on the leaves, twigs, or fruit.

**an·thrax** (ăn'thrăks') *n.* [ME *antrax,* malignant boil < Lat. *anthrax,* carbuncle < Gk. ] *Pathol.* **1.** An infectious, usu. fatal disease of cattle, sheep, and other warm-blooded animals caused by *Bacillus anthracis,* transmissible to humans, capable of affecting certain organs, and characterized esp. by malignant ulcers. **2.** *pl.* **-thra·ces** (-thrə-sēz'). A lesion caused by anthrax.

**an·throp·ic** (ăn-thrŏp'ĭk) *also* **an·throp·i·cal** (-ĭ-kəl) *adj.* [Gk. *anthrōpikos* < *anthropos,* humans.] Of or relating to humans or the era of human life.

**anthropo-** *pref.* [Gk. < *anthropos,* human being.] Human being <*anthropo*metry>

**an·thro·po·cen·tric** (ăn'thrə-pə-sĕn'trĭk) *adj.* **1.** Regarding the human being as the central fact or final aim of the universe. **2.** Interpreting reality solely in terms of human values and experience. **—an'thro·po·cen'tri·cal·ly** *adv.* **—an'thro·po·cen'trism** *n.*

**an·thro·po·gen·e·sis** (ăn'thrə-pə-jĕn'ĭ-sĭs) *n.* Scientific study of the origin of humans. **—an'thro·po·gen'ic** (-jĕn'ĭk) *adj.*

**an·thro·poid** (ăn'thrə-poid') *adj.* **1.** Resembling a human being, as the apes of the family Pongidae, which includes gorillas, chimpanzees, orangutans, and gibbons. **2.** Resembling or typical of an ape. —*n.* A member of the family Pongidae. **—an'thro·poid'al** *adj.*

**anthropoid ape** *n.* An anthropoid.

**an·thro·pol·o·gy** (ăn'thrə-pŏl'ə-jē) *n.* The scientific study of the origin and the physical, social, and cultural development and behavior of humans. **—an'thro·po·log'ic** (-pə-lŏj'ĭk), **an'thro·po·log'i·cal** *adj.* **—an'thro·po·log'i·cal·ly** *adv.* **—an'thro·pol'o·gist** *n.*

**an·thro·pom·e·try** (ăn'thrə-pŏm'ĭ-trē) *n.* The study and technique of human body measurement for use in anthropological classification and comparison. **—an'thro·po·met'ric** (-pə-mĕt'rĭk), **an'thro·po·met'ri·cal** (-rĭ-kəl) *adj.* **—an'thro·po·met'ri·cal·ly** *adv.* **—an'thro·pom'e·trist** *n.*

**an·thro·po·mor·phism** (ăn'thrə-pə-môr'fĭz'əm) *n.* The ascribing of human motivation, characteristics, or behavior to inanimate objects, animals, or natural phenomena. **—an'thro·po·mor'phic** *adj.* **—an'thro·po·mor'phi·cal·ly** *adv.*

**an·thro·po·mor·phize** (ăn'thrə-pə-môr'fīz') *vt.* **-phized, -phiz·ing, -phiz·es.** To ascribe human characteristics to.

**an·thro·po·mor·phous** (ăn'thrə-pə-môr'fəs) *adj.* [Gk. *anthropomorphos* : *anthropos,* human being + *morphē,* shape.] **1.** Having or suggesting human form and appearance. **2.** Anthropomorphic.

**an·thro·pop·a·thism** (ăn'thrə-pŏp'ə-thĭz'əm) *n.* [< Gk. *anthrōpopathēs,* with human feelings (*anthropos,* human being + *pathos,* feeling) + -ISM.] The ascribing of human sensibilities to nonhuman beings, objects, or phenomena. **—an'thro·po·path'ic** (-pə-păth'ĭk) *adj.*

**an·thro·poph·a·gus** (ăn'thrə-pŏf'ə-gəs') *n., pl.* **-gi** (-jī') [Lat. *anthropophagus* < Gk. *anthrōpophagos,* man-eating : *anthropos,* human being + *phagein,* to eat.] CANNIBAL 1. **—an'thro·po·phag'ic** (-pə-făj'ĭk), **an'thro·poph'a·gous** (-pŏf'ə-gəs) *adj.* **—an'thro·poph'a·gy** (-jē) *n.*

**an·thur·i·um** (ăn-tho͝or'ē-əm) *n.* [NLat. *Anthurium,* genus name : ANTH(O)- + Gk. *oura,* tail.] A tropical American plant of the genus *Anthurium.*

**an·ti** (ăn'tī', -tē) *n., pl.* **-tis.** [< ANTI-.] *Informal.* One who is opposed, as to a group, policy, proposal, or practice.

**anti-** *or* **ant-** *pref.* [Gk. < *anti,* opposite.] **1. a.** Opposite <*anti*mere> **b.** In opposition to : AGAINST <*anti*clerical> **c.** Counteracting : neutralizing <*anti*body> **2.** Reciprocal <*anti*logarithm>

**an·ti·a·bor·tion** (ăn'tē-ə-bôr'shən) *adj.* Opposed to abortion. **—an'ti·a·bor'tion·ist** *n.*

**an·ti·air·craft** (ăn'tē-âr'krăft') *adj.* Designed or used against aircraft or missile attack. —*n.* An antiaircraft weapon.

**an·ti·al·ler·gic** (ăn'tē-ə-lûr'jĭk) *also* **an·ti·al·ler·gen·ic** (ăn'tē-ăl'ər-jĕn'ĭk) *adj.* Alleviating or preventing allergies. **—an'ti·al'ler·gen'ic** *n.*

**an·ti·an·xi·e·ty** (ăn′tē-ăng-zī′ĭ-tē) *adj.* Alleviating or preventing anxiety <an *antianxiety* drug>

**an·ti·art** (ăn′tē-ärt′) *n.* Art that rejects traditional forms.

**an·ti·at·om** (ăn′tē-ăt′əm) *n.* An atom composed of antiparticles.

**an·ti·au·thor·i·tar·i·an** (ăn′tē-ə-thôr′ĭ-târ′ē-ən, -thŏr′-) *adj.* Rejecting authoritarians or authoritarianism. **—an′ti·au·thor·i·tar′i·an·ism** *n.*

**an·ti·bac·ter·i·al** (ăn′tē-băk-tîr′ē-əl) *adj.* Counteracting bacteria.

**an·ti·bal·lis·tic missile** (ăn′tē-bə-lĭs′tĭk) *n.* A defensive missile designed to intercept and destroy a ballistic missile in flight.

**an·ti·bar·y·on** (ăn′tē-băr′ē-ŏn′) *n.* The antiparticle of the baryon.

**an·ti·bi·o·sis** (ăn′tē-bī-ō′sĭs) *n.* An association between two or more organisms that is injurious to one of them.

**an·ti·bi·ot·ic** (ăn′tē-bī-ŏt′ĭk) *n.* A substance, as penicillin or streptomycin, that is produced by organisms such as fungi and bacteria, effective in the suppression or destruction of microorganisms, and widely used in the prevention and treatment of diseases. —*adj.* **1.** Of or relating to antibiotics. **2.** Of or relating to antibiosis. **—an′ti·bi·ot′i·cal·ly** *adv.*

**an·ti·bod·y** (ăn′tĭ-bŏd′ē) *n.* **1.** Any of various proteins in the blood that are generated in reaction to foreign proteins or polysaccharides, neutralize them, and so produce immunity against certain microorganisms or their toxins. **2.** An object made up of antimatter.

**an·ti·bus·ing** (ăn′tē-bŭs′ĭng) *adj.* Opposed to the busing of students as a means of achieving racial balance in the public schools.

**an·tic** (ăn′tĭk) *n.* [Ital. *antico,* ancient < Lat. *antiquus,* old.] **1.** A mischievous act or gesture : CAPER. **2.** *Archaic.* A clown. —*adj.* Ludicrous : far-fetched. **—an′ti·cal·ly** *adv.*

▲ word history: The source of *antic* is Italian *antico,* which means "ancient." The Italians used *antico* to refer to the grotesque designs found on some ancient Roman artifacts. *Antico* is derived from *antiquus,* the same Latin word that is the source of *antique,* which came into English from French. The two words are separate borrowings, and the English words *antic* and *antique* do not share any meanings.

**an·ti·can·cer** (ăn′tē-kăn′sər) *also* **an·ti·can·cer·ous** (-sər-əs) *adj.* Used or effective against cancer.

**an·ti·cat·a·lyst** (ăn′tē-kăt′l-ĭst) *n.* **1.** Something that retards or arrests a chemical reaction. **2.** Something that reduces or destroys the effectiveness of a catalyst.

**an·ti·cath·ode** (ăn′tē-kăth′ōd′) *n.* An electrode that is the target in a cathode-ray tube, esp. in an x-ray tube.

**an·ti·chlor** (ăn′tĭ-klôr′, -klōr′) *n.* [ANTI- + CHLOR(INE).] A substance, as sodium thiosulfate, that neutralizes the excess chlorine or hypochlorite left after bleaching textiles, fiber, or paper pulp. **—an′ti·chlo·ris′tic** (-klə-rĭs′tĭk) *adj.*

**an·ti·cho·lin·er·gic** (ăn′tē-kō′lə-nûr′jĭk) *adj.* Antagonistic to the physiological action of parasympathetic or other cholinergic nerve fibers. **—an′ti·cho′lin·er′gic** *n.*

**an·ti·cho·lin·es·ter·ase** (ăn′tē-kō′lə-nĕs′tə-rās′, -rāz′) *n.* A substance that inhibits a cholinesterase.

**an·ti·christ** (ăn′tĭ-krīst′) *n.* [ME *Antecrist* < OFr. < Med. Lat. *Antichristus* < LLat. < Gk. *Antikristos* : *anti-,* opposed to + *khristos,* anointed. —see CHRIST.] **1.** An enemy of Christ. **2. Antichrist.** The antagonist who was expected by the early Church to oppose Christ in the last days before the Second Coming. **3.** A false Christ.

**an·tic·i·pant** (ăn-tĭs′ə-pənt) *adj.* **1.** Acting or coming in advance. **2.** Anticipating : expectant. —*n.* One who anticipates.

**an·tic·i·pate** (ăn-tĭs′ə-pāt′) *vt.* **-pat·ed, -pat·ing, -pates.** [Lat. *anticipare, anticipat-,* to take before : *ante-,* before + *capere,* to take.] **1.** To feel or know beforehand : FORESEE. **2.** To look forward to, esp. eagerly : EXPECT. **3.** To act in advance so as to prevent : FORESTALL. **4.** To foresee and fulfill in advance. **5.** To cause to happen in advance : ACCELERATE. **6.** To use in advance, as income not yet available. **7.** To pay (a debt) before due. **—an·tic′i·pat′a·ble** *adj.* **—an·tic′i·pa′tor** *n.* **—an·tic′i·pa·to′ry** (-pə-tôr′ē, -tōr′ē) *adj.*

**an·tic·i·pa·tion** (ăn-tĭs′ə-pā′shən) *n.* **1.** The act of anticipating. **2.** Something anticipated. **3.** Foreknowledge : presentiment. **4.** The use or assignment of funds, esp. from a trust fund, before legitimately available. **5.** *Mus.* The introduction of one note of a new chord before the previous chord is resolved.

**an·tic·i·pa·tive** (ăn-tĭs′ə-pā′tĭv, -pə-tĭv) *adj.* Having or marked by anticipation : EXPECTANT. **—an·tic′i·pa′tive·ly** *adv.*

**an·ti·cler·i·cal** (ăn′tē-klĕr′ĭ-kəl) *adj.* Opposed to the clergy or the church's influence in political affairs. **—an′ti·cler′i·cal·ism** *n.*

**an·ti·cli·max** (ăn′tē-klī′măks′) *n.* **1.** A decline, as of a career, viewed in disappointing contrast with a previous rise. **2.** A commonplace or trivial event that comes at the end of a series of significant events. **3.** A sudden descent in speaking or writing from the impressive or significant to the unimpressive or inconsequential. **—an′ti·cli·mac′tic** (-klī-măk′tĭk) *adj.* **—an′ti·cli·mac′ti·cal·ly** *adv.*

**an·ti·cli·nal** (ăn′tē-klī′nəl) *adj.* Sloping downward in opposite directions, as an anticline.

**an·ti·cline** (ăn′tĭ-klīn′) *n. Geol.* A fold with layers sloping downward on both sides from a common crest.

**an·ti·clock·wise** (ăn′tē-klŏk′wīz) *adv.* & *adj.* Counterclockwise.

**an·ti·co·ag·u·lant** (ăn′tē-kō-ăg′yə-lənt) *n.* Something that counteracts coagulation, esp. of the blood. **—an′ti·co·ag′u·lant** *adj.*

**an·ti·co·in·ci·dence circuit** (ăn′tē-kō-ĭn′sĭ-dəns) *n.* A binary logic element designed to provide input signals to a device according to fixed rules.

**an·ti·con·vul·sant** (ăn′tē-kən-vŭl′sənt) *n.* A medicinal agent used to prevent convulsions. **—an′ti·con·vul′sive** *adj.*

**an·ti·cy·clone** (ăn′tē-sī′klōn′) *n.* An extensive system of winds spiraling outward from a high-pressure center, circling clockwise in the Northern Hemisphere and counterclockwise in the Southern Hemisphere. **—an′ti·cy·clon′ic** (-klŏn′ĭk) *adj.*

**an·ti·de·pres·sant** (ăn′tē-dĭ-prĕs′ənt) *n.* A drug that alleviates or prevents depression. **—an′ti·de·pres′sive** *adj.*

**an·ti·deu·ter·on** (ăn′tē-dōō′tə-rŏn′, -dyōō′-) *n.* The antimatter equivalent of deuteron.

**an·ti·dote** (ăn′tĭ-dōt′) *n.* [Lat. *antidotum* < Gk. *antidoton* < *antididonai,* to give as a remedy against : *anti-,* against + *didonai,* to give.] **1.** A remedy that counteracts the effects of a poison. **2.** An agent that counteracts an injurious effect. **usage:** *Antidote* may be followed by *to, for,* or *against* (*an antidote to snakebite; an antidote for poison; an antidote against rampant inflation*). **—an′ti·dot′al** (ăn′tĭ-dōt′l) *adj.* **—an′ti·dot′al·ly** *adv.*

**an·ti·e·lec·tron** (ăn′tē-ĭ-lĕk′trŏn′) *n.* A positron.

**an·ti·en·zyme** (ăn′tē-ĕn′zīm′) *n.* A substance that counteracts or neutralizes an enzyme. **—an′ti·en·zy·mat′ic** (-zī-măt′ĭk, -zĭ-), **an′ti·en·zy′mic** (-zī′mĭk) *adj.*

**an·ti·es·tab·lish·ment** (ăn′tē-ĭ-stăb′lĭsh-mənt) *adj.* Characterized by opposition to traditional social, political, or economic values or principles.

**an·ti·feb·rile** (ăn′tē-fĕb′rəl, -fē′brəl, -brĭl′) *adj.* Capable of reducing fever : ANTIPYRETIC. —*n.* An antifebrile drug or agent.

**an·ti·fed·er·al·ist** *also* **An·ti·fed·er·al·ist** (ăn′tē-fĕd′ər-ə-lĭst, -fĕd′rə-lĭst) *n.* One who was opposed to the ratification of the U.S. Constitution. **—an′ti·fed′er·al·ism** *n.*

**an·ti·fem·i·nist** (ăn′tē-fĕm′ə-nĭst) *adj.* Opposed to feminism. **—an′ti·fem′i·nism** *n.* **—an′ti·fem′i·nist** *n.*

**an·ti·fer·til·i·ty** (ăn′tĭ-fər-tĭl′ĭ-tē, ăn′tī-) *adj.* Capable of reducing or destroying fertility : CONTRACEPTIVE.

**an·ti·freeze** (ăn′tĭ-frēz′) *n.* A substance, as ethylene glycol or alcohol, mixed with a liquid, esp. water, to lower its freezing point.

**an·ti·gal·ax·y** (ăn′tē-găl′ək-sē) *n.* A galaxy made up of antimatter.

**an·ti·gen** (ăn′tĭ-jən) *n.* A substance that when introduced into the body stimulates the production of an antibody. **—an′ti·gen′ic** (-jĕn′ĭk) *adj.* **—an′ti·gen′i·cal·ly** *adv.* **—an′ti·ge·nic′i·ty** (-jə-nĭs′ĭ-tē) *n.*

**An·tig·o·ne** (ăn-tĭg′ə-nē) *n.* [Gk. *Antigonē.*] *Gk. Myth.* The daughter of Oedipus and Jocasta, who performed funeral rites over her brother's body in defiance of her uncle Creon.

**an·ti·grav·i·ty** (ăn′tē-grăv′ĭ-tē) *n.* The result of reducing or canceling a gravitational field. —*adj.* Canceling or reducing gravity or protecting against its effect.

**an·ti·he·li·um** (ăn′tē-hē′lē-əm) *n.* The antimatter equivalent of helium.

**an·ti·he·ro** (ăn′tĭ-hîr′ō) *n.* A protagonist, as in a literary work, marked by a lack of traditional heroic virtues and qualities.

**an·ti·his·ta·mine** (ăn′tē-hĭs′tə-mēn′, -mĭn) *n.* Any of various drugs used to reduce the physiological effects of histamine production in allergies and colds. **—an′ti·his′ta·min′ic** (-mĭn′ĭk) *adj.*

**an·ti·hy·dro·gen** (ăn′tē-hī′drə-jən) *n.* The antimatter equivalent of hydrogen.

**an·ti·in·flam·ma·to·ry** (ăn′tē-ĭn-flăm′ə-tôr′ē, -tōr′ē) *adj.* Counteracting inflammation.

**an·ti·knock** (ăn′tĭ-nŏk′) *n.* A substance, as tetraethyl lead, added to gasoline to reduce engine knock.

**an·ti·lep·ton** (ăn′tē-lĕp′tŏn′) *n.* The antiparticle of a lepton.

**an·ti·log** (ăn′tē-lôg′, -lŏg′) *n.* An antilogarithm.

**an·ti·log·a·rithm** (ăn′tē-lô′gə-rĭth′əm, -lŏg′ə-) *n.* The number for which a given logarithm stands; e.g., if log x equals *y,* the *x* is the antilogarithm of *y.* **—an′ti·log′a·rith′mic** *adj.*

**an·ti·ma·cas·sar** (ăn′tē-mə-kăs′ər) *n.* [ANTI- + *Macassar,* a brand of hair oil.] A protective covering for the backs of chairs and sofas.

**an·ti·mag·net·ic** (ăn′tē-măg-nĕt′ĭk) *adj.* Resistant or impervious to magnetization.

**an·ti·ma·lar·i·al** (ăn′tē-mə-lâr′ē-əl) *adj.* Effective against malaria. —*n.* An antimalarial drug.

**an·ti·mat·ter** (ăn′tĭ-măt′ər) *n.* Antiparticle matter having positron-surrounded nuclei that are composed of antiprotons and antineutrons.

**an·ti·mere** (ăn′tĭ-mîr′) *n. Biol.* A part or division corresponding to an opposite or similar part in an organism marked by bilateral or radial symmetry. **—an′ti·mer′ic** (-mĕr′ĭk) *adj.*

**an·ti·mi·cro·bi·al** (ăn′tē-mī-krō′bē-əl) *also* **an·ti·mi·cro·bic** (-bĭk) *adj.* Destroying or suppressing the growth of microorganisms. **—an′ti·mi·cro′bial** *n.*

**an·ti·mis·sile missile** (ăn′tē-mĭs′əl) *n.* A missile that can intercept and destroy another missile in flight.

**an·ti·mo·ni·al** (ăn′tə-mō′nē-əl) *adj.* Of or containing antimony. —*n.* A medicine containing antimony.
**an·ti·mo·ny** (ăn′tə-mō′nē) *n.* [ME *antimonie* < Med. Lat. *antimonium.*] *Symbol* **Sb** A metallic element used in a wide variety of alloys, esp. with lead in battery plates, and in paints, semiconductors, and ceramic products; atomic number 51; atomic weight 121.75.
**antimony glance** *n.* Stibnite.
**an·ti·ne·o·plas·tic** (ăn′tē-nē′ə-plăs′tĭk) *adj.* Suppressing the growth or expansion of neoplasms.
**an·ti·neu·tri·no** (ăn′tē-nōō-trē′nō, -nyōō-) *n., pl.* **-nos.** The antiparticle of the neutrino.
**an·ti·neu·tron** (ăn′tē-nōō′trŏn′, -nyōō′-) *n.* The antiparticle of the neutron.
**an·ti·node** (ăn′tĭ-nōd′) *n.* The point or region of greatest amplitude between adjacent nodes.
**an·ti·no·mi·an** (ăn′tĭ-nō′mē-ən) *n.* [Med. Lat. *antinomus* : Gk. *anti-*, against + *nomos*, law.] A member of a Christian sect believing that faith alone is necessary to salvation. **—an′ti·no′mi·an** *adj.* **—an′ti·no′mi·an·ism** *n.*
**an·tin·o·my** (ăn-tĭn′ə-mē) *n., pl.* **-mies.** [Lat. *antinomia* < Gk. *anti-*, against + *nomos*, law.] **1.** An apparent contradiction between valid principles or conclusions that seem equally necessary and reasonable. **2.** A contradiction, opposition, or conflict.
**an·ti·nov·el** (ăn′tē-nŏv′əl) *n.* A novel lacking traditional features of a work of fiction, as coherent structure and character development. **—an′ti·nov′el·ist** *n.*
**an·ti·nu·cle·on** (ăn′tē-nōō′klē-ŏn′, -nyōō′-) *n.* The antiparticle of a nucleon.
**an·ti·ox·i·dant** (ăn′tē-ŏk′sĭ-dənt) *n.* A chemical substance that inhibits oxidation.
**an·ti·par·ti·cle** (ăn′tē-pär′tĭ-kəl) *n.* A subatomic particle, as a positron, antiproton, or antineutron, having the same mass, average lifetime, spin, magnitude of magnetic moment, and magnitude of electric charge as the particle to which it corresponds but having the opposite sign of electric charge, opposite intrinsic parity, and opposite direction of magnetic moment.
**an·ti·pas·to** (ăn′tē-päs′tō) *n., pl.* **-tos** *or* **-ti** (-tē) [Ital. : *anti-*, before (< Lat. *ante*) + *pasto*, food < Lat. *pastus* < *pascere*, to feed.] An appetizer usu. including cheese, smoked meats, fish, and vegetables, served with oil and vinegar.
**an·tip·a·thet·ic** (ăn-tĭp′ə-thĕt′ĭk) *also* **an·tip·a·thet·i·cal** (-ĭ-kəl) *adj.* **1.** Having an inherent feeling of aversion, repugnance, or opposition. **2.** Causing antipathy. **—an·tip′a·thet′i·cal·ly** *adv.*
**an·tip·a·thy** (ăn-tĭp′ə-thē) *n., pl.* **-thies.** [Lat. *antipathia* < Gk. *antipatheia* : *anti-*, against + *pathos*, feeling.] **1.** A strong feeling of aversion, repugnance, or opposition. **2.** An object of aversion.
**an·ti·pe·ri·od·ic** (ăn′tē-pîr′ē-ŏd′ĭk) *adj.* Preventing regular recurrence of fever or disease. —*n.* An antiperiodic drug.
**an·ti·per·son·nel** (ăn′tē-pûr′sə-nĕl′) *adj.* Designed to injure or kill the military personnel or civilian population of an enemy country.
**an·ti·per·spi·rant** (ăn′tē-pûr′spər-ənt) *n.* A preparation applied esp. to the underarms to decrease or prevent excessive perspiration.
**an·ti·phlo·gis·tic** (ăn′tē-flə-jĭs′tĭk) *adj.* Reducing inflammation or fever. **—an′ti·phlo·gis′tic** *n.*
**an·ti·phon** (ăn′tə-fŏn′) *n.* [Fr. *antiphone* < Med. Lat. *antiphona*, sung responses. —see ANTHEM.] **1.** A devotional composition sung responsively as part of a liturgy. **2.** A short liturgical text chanted responsively before a psalm or canticle. **3.** A response or answer. **—an·tiph′o·nal** (-tĭf′ə-nəl) *adj.* **—an·tiph′o·nal·ly** *adv.*
**an·tiph·o·nar·y** (ăn-tĭf′ə-nĕr′ē) *n., pl.* **-ies.** A bound collection of antiphons.
**an·tiph·o·ny** (ăn-tĭf′ə-nē) *n., pl.* **-nies. 1.** Responsive singing or chanting. **2.** ANTIPHON 1. **3.** One that answers or echoes another.
**an·tip·o·dal** (ăn-tĭp′ə-dəl) *adj.* **1.** Of, relating to, or located on the opposite side or sides of the earth. **2.** Diametrically opposed.
**an·ti·pode** (ăn′tĭ-pōd′) *n.* [Back-formation from ANTIPODES.] A direct opposite.
**an·tip·o·des** (ăn-tĭp′ə-dēz′) *pl.n.* [ME < Lat. < Gk. < *antipous*, with the feet opposite : *anti-*, opposite + *pous*, foot.] **1.** Two places or regions on opposite sides of the earth. **2.** *(sing. or pl. in number).* One that is the exact opposite of another.
**an·ti·pol·lu·tion** (ăn′tē-pə-lōō′shən) *adj.* Intended to counteract or eliminate environmental pollution. **—an′ti·pol·lu′tion·ist** *n.*
**an·ti·pope** (ăn′tĭ-pōp′) *n.* [ME < Med. Lat. *antipapa* : Lat. *anti-*, opposed to + *papa*, pope.] One claiming to be pope in opposition to the one chosen by church law.
**an·ti·pov·er·ty** (ăn′tē-pŏv′ər-tē) *adj.* Intended to alleviate poverty.
**an·ti·pro·ton** (ăn′tē-prō′tŏn′) *n.* The antiparticle of the proton.
**an·ti·py·ret·ic** (ăn′tē-pī-rĕt′ĭk) *adj.* Reducing fever. —*n.* An antipyretic drug or agent. **—an′ti·py·re′sis** (-rē′sĭs) *n.*
**an·ti·py·rine** (ăn′tē-pī′rēn′) *n.* [Orig. a trademark.] A white powder, $C_{11}H_{12}N_2O$, used to reduce pain and fever.
**an·ti·quar·i·an** (ăn′tĭ-kwâr′ē-ən) *adj.* **1.** Of or relating to antiquaries or the study of antiquities. **2.** Dealing in or concerning rare old books. —*n.* An antiquary. **—an′ti·quar′i·an·ism** *n.*
**an·ti·quark** (ăn′tē-kwôrk′) *n.* The antiparticle of a quark.
**an·ti·quar·y** (ăn′tĭ-kwĕr′ē) *n., pl.* **-ies.** [Lat. *antiquarius* < *antiquus*, old.] A student of or dealer in antiquities.
**an·ti·quate** (ăn′tĭ-kwāt′) *vt.* **-quat·ed, -quat·ing, -quates.** [Lat. *antiquare, antiquat-*, to leave in an old state < *antiquus*, old.] To make old-fashioned or obsolete. **—an′ti·qua′tion** *n.*
**an·ti·quat·ed** (ăn′tĭ-kwā′tĭd) *adj.* **1.** So old as to be useless or unsuitable : OBSOLETE. **2.** Very old : AGED. **—an′ti·quat′ed·ness** *n.*
**an·tique** (ăn-tēk′) *adj.* [Fr. < Lat. *antiquus*, old.] **1.** Of, pertaining to, or belonging to ancient times. **2.** Belonging to, made in, or typical of an earlier period. **3.** Old-fashioned. —*n.* An object having special value because of its age, esp. a work of art or handicraft more than 100 years old. —*vt.* **-tiqued, -tiqu·ing, -tiques.** To give the appearance of an antique to. **—an·tique′ly** *adv.* **—an·tique′ness** *n.*
**an·tiqu·er** (ăn-tē′kər) *n.* One who treats or finishes new furniture so as to make it appear antique.
**an·tiq·ui·ty** (ăn-tĭk′wĭ-tē) *n., pl.* **-ties. 1.** Ancient times, esp. the times before the Middle Ages. **2.** The people, esp. the writers, of ancient times. **3.** The quality or state of being old or ancient. **4.** *often* **antiquities.** Something, as a relic, belonging to or dating from ancient times.
**an·ti·ra·chit·ic** (ăn′tē-rə-kĭt′ĭk) *adj.* Preventing or treating rickets. —*n.* An antirachitic drug or food.
**an·tir·rhi·num** (ăn′tə-rī′nəm) *n.* [NLat., genus name < Gk. *antirrhinon* : *anti-*, counterfeiting + *rhis*, nose.] A plant of the genus *Antirrhinum*, as a snapdragon.
**an·ti·scor·bu·tic** (ăn′tē-skôr-byōō′tĭk) *adj.* Preventing or treating scurvy. —*n.* An antiscorbutic food or drug.
**an·ti-Sem·ite** (ăn′tē-sĕm′īt′) *n.* One who discriminates against or is hostile to or prejudiced against Jews. **—an′ti-Se·mit′ic** (-sə-mĭt′ĭk) *adj.* **—an′ti-Sem′i·tism** *n.*
**an·ti·sep·sis** (ăn′tĭ-sĕp′sĭs) *n.* The destruction of microorganisms that cause disease, fermentation, or putrefaction.
**an·ti·sep·tic** (ăn′tĭ-sĕp′tĭk) *adj.* **1.** Of, relating to, or designating antisepsis. **2.** Capable of producing antisepsis. **3.** Entirely clean. **4.** Devoid of enlivening or enriching qualities. **5.** Impersonal. —*n.* An antiseptic drug or agent. **—an′ti·sep′ti·cal·ly** *adv.*
**an·ti·se·rum** (ăn′tĭ-sîr′əm) *n., pl.* **-rums** *or* **-ra** (-rə). Human or animal serum having antibodies for at least one antigen.
**an·ti·slav·er·y** (ăn′tē-slā′və-rē, -slāv′rē) *adj.* Opposed to slavery. **—an′ti·slav′er·y** *n.*
**an·ti·smog** (ăn′tē-smŏg′, -smôg′) *adj.* Counteracting or eliminating smog.
**an·ti·so·cial** (ăn′tē-sō′shəl) *adj.* **1.** Avoiding the society of others : UNSOCIABLE. **2. a.** Opposed to the established social order. **b.** Characterized by or engaging in behavior that violates conventional mores. **—an′ti·so′cial·ly** *adv.*
**an·ti·spas·mod·ic** (ăn′tē-spăz-mŏd′ĭk) *adj.* Alleviating or preventing spasms. —*n.* An antispasmodic drug.
**an·ti·stat·ic** (ăn′tē-stăt′ĭk) *also* **an·ti·stat** (-tē-stăt′) *adj.* Preventing or inhibiting the build-up of static electricity.
**an·tis·tro·phe** (ăn-tĭs′trə-fē) *n.* [LLat. < Gk. *antistrophē*, a turning back < *antistrephein*, to turn back : *anti-*, back + *strephein*, to turn.] **1.** The movement following and in the same meter as the strophe in ancient Greek choral poetry or drama, sung while the chorus moves in the opposite direction from that of the strophe. **2.** The second stanza in a poem having alternating stanzas in contrasting metric form. **—an′ti·stroph′ic** (ăn′tĭ-strŏf′ĭk) *adj.* **—an′ti·stroph′i·cal·ly** *adv.*
**an·ti·sub·ma·rine** (ăn′tē-sŭb′mə-rēn′, -sŭb′mə-rēn′) *adj.* Used against enemy submarines.
**an·ti·tank** (ăn′tē-tăngk′) *adj.* Designed for combat against armored vehicles, esp. tanks.
**an·tith·e·sis** (ăn-tĭth′ĭ-sĭs) *n., pl.* **-ses** (-sēz′) [LLat. < Gk., opposition < *antitithenai*, to oppose : *anti-*, against + *tithenai*, to set.] **1.** Direct contrast : OPPOSITION. **2.** The direct opposite. **3. a.** The juxtaposition of sharply contrasting ideas in balanced or parallel words, phrases, or grammatical structures; e.g., Milton's "He for God only, she for God in him." **b.** The second and contrasting part of such a juxtaposition. **4.** The second stage of the dialectic process.
**an·ti·thet·i·cal** (ăn′tĭ-thĕt′ĭ-kəl) *also* **an·ti·thet·ic** (-ĭk) *adj.* [LLat. *antitheticus*, Gk. *antithetikos* < *antitethenai*, to oppose. —see ANTITHESIS.] **1.** Relating to, like, or characterized by antithesis. **2.** Directly opposed. **—an′ti·thet′i·cal·ly** *adv.*
**an·ti·tox·ic** (ăn′tē-tŏk′sĭk) *adj.* **1.** Counteracting a toxin or poison. **2.** Of or relating to an antitoxin.
**an·ti·tox·in** (ăn′tē-tŏk′sĭn) *n.* **1.** An antibody formed in response to and capable of neutralizing a biological poison. **2.** An animal serum containing antitoxins.
**an·ti·trades** (ăn′tĭ-trādz′) *pl.n.* The westerly winds above the trade winds of the tropics, which become the westerly winds of the middle latitudes.
**an·ti·trust** (ăn′tē-trŭst′) *adj.* Opposing or regulating trusts, cartels, or similar business monopolies.
**an·ti·tu·mor** (ăn′tĭ-tōō′mər, -tyōō′-) *also* **an·ti·tu·mor·al** (-mər-əl) *adj.* Anticancer.

**an·ti·tus·sive** (ăn'tē-tŭs'ĭv) *adj.* Capable of relieving coughing. —*n.* An antitussive drug.

**an·ti·ven·in** (ăn'tē-vĕn'ĭn) *n.* [ANTI- + VEN(OM) + -IN.] **1.** An antitoxin for venom. **2.** An antiserum containing an antivenin.

**ant·ler** (ănt'lər) *n.* [ME *aunteler* < OFr. *antoillier.*] One of a pair of hard, bony, deciduous, usu. elongated and branched growths that grow on the heads of male deer and related animals. **—ant'lered** (ănt'lərd) *adj.*

**Ant·li·a** (ănt'lē-ə) *n.* [Lat. *antlia,* pump < Gk. *antlos,* bucket.] A constellation in the Southern Hemisphere.

**ant lion** *n.* **1.** Any insect of the family Myrmeleontidae, of which the adults resemble dragonflies. **2.** The larva of the ant lion, which digs holes to trap insects, as ants, for food.

**an·to·no·ma·sia** (ăn'tə-nə-mā'zhə) *n.* [Lat. < Gk. *antonomazein,* to name instead : *anti-,* instead of + *onomazein,* to name < *onoma,* name.] **1.** Substitution of a title or epithet for a proper name, as in calling an ambassador "Your Excellency." **2.** Substitution of a personal name for a common noun to designate a member of a group or class, as in calling a traitor a "Benedict Arnold."

**an·to·nym** (ăn'tə-nĭm') *n.* [ANT(I)- + -ONYM.] A word or expression having a meaning opposite to a meaning of another word or expression. **—an'to·nym'ic** (-nĭm'ĭk) *adj.* **—an·ton'y·mous** (ăn-tŏn'ə-məs) *adj.* **—an·ton'y·my** *n.*

**an·tre** (ăn'tər) *n.* [Fr. < Lat. *antrum,* cave.] A cave or cavern.

**an·trorse** (ăn'trôrs') *adj.* [NLat. *antrorsus,* perh. < Lat. *anterior,* before. —see ANTERIOR.] *Biol.* Directed forward and upward. **—an·trorse'ly** *adv.*

**an·trum** (ăn'trəm) *n., pl.* **-tra** (-trə) [LLat., cavity < Lat., cave < Gk. *antron.*] A cavity, usu. in bone, esp. either of the sinuses in the upper jaw opening into the nose. **—an'tral** *adj.*

**ant·sy** (ănt'sē) *adj. Slang.* Restless or fidgety.

**A·nu·bis** (ə-nōō'bĭs) *n.* [Lat. < Gk. *Anoubis,* of Egypt. orig.] *Myth.* A jackal-headed Egyptian god who conducted the dead to judgment.

Anubis

**a·nu·ran** (ə-nŏŏr'ən, ə-nyŏŏr'-) *adj.* [NLat. *Anura,* order of frogs and toads : AN- + Gk. *oura,* tail.] Of or relating to frogs and toads. **—a·nu'ran** *n.*

**an·u·re·sis** (ăn'yə-rē'sĭs) *n.* [AN- + Gk. *ourēsis,* urination < *ourein,* to urinate < *ouron,* urine.] Failure of the kidneys to secrete urine or failure of any urine that is secreted to reach the bladder. **—an'u·ret'ic** (-rĕt'ĭk) *adj.*

**a·nu·ri·a** (ə-nŏŏr'ē-ə, ə-nyŏŏr'-) *n.* Anuresis. **—a·nu'ric** (-ĭk) *adj.*

**a·nu·rous** (ə-nŏŏr'əs, ə-nyŏŏr'-) *adj.* Having no tail.

**a·nus** (ā'nəs) *n., pl.* **a·nus·es.** [Lat.] The excretory opening of the alimentary canal.

**an·vil** (ăn'vĭl) *n.* [ME *anvelt* < OE *anfilt.*] **1.** A heavy iron or steel block with a flat, smooth top on which metals are shaped by hammering. **2.** The fixed jaw in a set of calipers against which the object to be measured is placed. **3.** *Anat.* The incus.

**anx·i·e·ty** (ăng-zī'ĭ-tē) *n., pl.* **-ties.** [Lat. *anxietas* < *anxius,* anxious.] **1 a.** Uneasiness and distress about future uncertainties. **b.** A cause of uneasiness. **2.** *Psychiat.* Intense fear or dread lacking a clearly defined cause or a specific threat. **3.** Eagerness or earnestness, often accompanied by uneasiness.

☆ ***syns:*** ANXIETY, ANGST, CARE, CONCERN, DISQUIET, DISQUIETUDE, UNEASE, UNEASINESS, WORRY *n. core meaning* **:** a troubled state of mind <parental *anxiety*> ***ant:*** security

**anx·ious** (ăngk'shəs, ăng'shəs) *adj.* [Lat. *anxius* < *angere,* to torment.] **1. a.** Worried and distressed about an uncertainty. **b.** Attended with, exhibiting, or producing worry. **2.** Eagerly or earnestly desirous <*anxious* to get home> **—anx'ious·ly** *adv.* **—anx'ious·ness** *n.*

**an·y** (ĕn'ē) *adj.* [ME *ani* < OE *ænig.*] **1.** One or some, regardless of sort, quantity, or number <Wear *any* shirt you want.><Were there *any* phone calls for us?> **2. a.** One or another selected at random <*Any* parent would feel the same.> **b.** One or another without restriction or exception <will consider *any* solution> **3.** The whole amount of **:** ALL <should bank *any* extra money> **4.** An indeterminate number or amount <There isn't *any* ice.> —*pron.* **1.** Any one or ones among three or more. **2.** Any quantity or part. —*adv.* To any degree or extent **:** AT ALL. ***usage:*** The use of *any* without a following comparative adjective to mean "at all," as in *It didn't hurt any,* is considered inappropriate to formal style.

**an·y·bod·y** (ĕn'ē-bŏd'ē, -bŭd'ē) *pron.* Anyone. —*n.* A person of consequence.

**an·y·how** (ĕn'ē-hou') *adv.* **1.** In any way or by any means whatever **:** AT ALL. **2.** In any case. **3.** In a careless or haphazard way.

**†an·y·more** (ĕn'ē-môr', -mōr') *adv.* **1.** At the present **:** from now on <can't go there *anymore*> **2.** *Regional.* Nowadays.

**an·y·one** (ĕn'ē-wŭn', -wən) *pron.* Any person **:** ANYBODY.

**an·y·place** (ĕn'ē-plās') *adv.* To, in, or at any place **:** ANYWHERE.

**an·y·thing** (ĕn'ē-thĭng') *pron.* Any object, occurrence, or matter whatever. —*adv.* To any degree or extent **:** AT ALL. **—anything but.** By no means **:** not at all <is *anything but* lazy>

**an·y·time** (ĕn'ē-tīm') *adv.* At any time.

**an·y·way** (ĕn'ē-wā') *adv.* At any rate **:** NEVERTHELESS.

**an·y·ways** (ĕn'ē-wāz') *adv. Nonstandard.* Anyway.

**an·y·where** (ĕn'ē-hwâr', -wâr') *adv.* **1.** To, in, or at any place. **2.** To any extent or degree **:** AT ALL.

**an·y·wise** (ĕn'ē-wīz') *adv.* In any way or manner.

**An·zac** (ăn'zăk') *n.* [A(USTRALIAN AND) N(EW) Z(EALAND) A(RMY) C(ORPS).] A soldier from Australia or New Zealand.

**A-O·K** *also* **A-O·kay** (ā'ō-kā') *adj. & adv. Informal.* Perfectly OK.

**A-one** *also* **A-1** (ā'wŭn') *adj.* **1.** *Informal.* Excellent **:** first-rate. **2.** Having a hull and equipment in top condition, as a ship.

**a·o·rist** (ā'ər-ĭst) *n.* [< Gk. *aoristos,* indefinite : *a-,* not + *horistos,* definable < *horizein,* to define < *horos,* boundary.] A verb tense orig. used in classical Greek that usu. denotes past action without indicating completion, continuation, or repetition of the action. **—a'o·rist, a'o·ris'tic** *adj.* **—a'o·ris'ti·cal·ly** *adv.*

**a·or·ta** (ā-ôr'tə) *n., pl.* **-tas** *or* **-tae** (-tē) [NLat. < Gk. *aortē* < *aeirein,* to lift.] *Anat.* The main trunk of the systemic arteries, carrying blood away from the heart. **—a·or'tal, a·or'tic** *adj.*

**a·ou·dad** (ä'ōō-dăd', ou'dăd') *n.* [Fr. < Berber *audad.*] A wild sheep, *Ammotragus lervia* of northern Africa, with a beardlike growth of hair on the neck and chest and long, curved horns.

**ap-**[1] *pref. var. of* AD- 1. —Used before *p.*

**ap-**[2] *pref. var. of* APO-.

**a·pace** (ə-pās') *adv.* [ME *apas* < OFr. *à pas* : *à,* to (< Lat. *ad*) + *pas,* step. —see PACE.] At a swift pace **:** RAPIDLY.

**a·pache** (ə-păsh') *n., pl.* **a·paches** (ə-păsh') [Fr. < *Apache,* Apache Indian.] A member of the Parisian underworld.

**A·pach·e** (ə-păch'ē) *n., pl.* **Apache** *or* **-es.** [Mex. Sp., prob. < Zuñi *Apachu,* enemy.] **1. a.** A formerly nomadic tribe of Indians inhabiting the southwestern United States and northern Mexico. **b.** A member of this tribe. **2.** Any of the Athapascan languages of the Apache.

**ap·a·nage** (ăp'ə-nĭj) *n. var. of* APPANAGE.

**†ap·a·re·jo** (ăp'ə-rā'hō, -rā'ō) *n., pl.* **-jos.** [Mex. Sp. < Sp., equipment < *aparejar,* to prepare.] *Southwestern U.S.* A packsaddle made of a stuffed leather pad.

**a·part** (ə-pärt') *adv.* [ME < OFr. *à part,* to the side : *à,* to (< Lat. *ad*) + *part,* side < Lat. *pars.*] **1. a.** In pieces. **b.** To pieces. **2. a.** Separately or at a distance in time, place, or position. **b.** To one side **:** ASIDE. **3.** One from another. **4.** Separately or aside for a particular function or purpose. **5.** Considered or viewed separately. **6.** Excepted or excluded from consideration <Minor flaws *apart,* the play is superb.> —*adj.* Set apart **:** ISOLATED. —Used after a noun or in the predicate <a country *apart*>

**apart from** *prep.* With the exception of **:** BESIDES.

**a·part·heid** (ə-pärt'hīt', -hāt') *n.* [Afr. : Du. *apart,* separate < Fr. *à part,* apart + *-heid,* -hood.] An official policy of racial segregation in the Republic of South Africa.

**a·part·ment** (ə-pärt'mənt) *n.* [Fr. *appartement* < Ital. *appartemento* < *appartare,* to separate < *a parte,* apart : *a,* to (< Lat. *ad*) + *parte,* side < Lat. *pars.*] **1.** A room or suite of rooms designed to live in and gen. located in a building occupied by more than one household. **2.** A room.

**apartment house** *n.* A building divided into apartments.

**ap·a·tet·ic** (ăp'ə-tĕt'ĭk) *adj.* [Gk. *apatētikos,* deceptive < *apateuein,* to cheat < *apatē,* deceit.] Relating to or typical of coloration functioning as natural camouflage.

**ap·a·thet·ic** (ăp'ə-thĕt'ĭk) *also* **ap·a·thet·i·cal** (-ĭ-kəl) *adj.* [< APATHY.] **1.** Feeling or displaying little or no emotion. **2.** Feeling or displaying little or no interest. **—ap'a·thet'i·cal·ly** *adv.*

**ap·a·thy** (ăp'ə-thē) *n.* [Gk. *apatheia* < *apathēs,* without feeling : *a-,* without + *pathos,* feeling.] **1.** Lack of emotion or feeling **:** IMPASSIVITY. **2.** Lack of interest or regard **:** INDIFFERENCE.

**ap·a·tite** (ăp'ə-tīt') *n.* [G. *Apatit* < Gk. *apatē,* deceit (from its often being mistaken for other minerals).] A natural, variously colored calcium fluoride phosphate, $Ca_5F(PO_4)_3$, with chlorine, hydroxyl, or carbonate occas. replacing the fluoride, used as a source of phosphorus compounds.

**ape** (āp) *n.* [ME < OE *apa.*] **1. a.** Any of various large, tailless Old World primates of the family Pongidae, including the chimpanzee, gorilla, gibbon, and orangutan. **b.** A monkey. **2.** A mimic. **3.** *Informal.* A clumsy, ill-bred, coarse person. —*vt.* **aped, ap·ing, apes.** To imitate the actions of **:** MIMIC. **—ap'er** *n.*

**a·peak** (ə-pēk′) *adv.* & *adj.* [Fr. *à pic* : *à*, to (< Lat. *ad*) + *pic*, peak.] *Naut.* In a vertical or nearly vertical position or direction.

**ape·man** (āp′măn′) *n.* Any of several extinct primates held to be intermediate between apes and modern human beings.

**a·per·çu** (ä′pěr-sü′) *n., pl.* **-çus** (-sü′) [Fr.] A synopsis.

**a·pe·ri·ent** (ə-pîr′ē-ənt) *adj.* [Lat. *aperiens*, pr.part. of *aperire*, to open.] Gently laxative. **—a·pe′ri·ent** *n.*

**a·pe·ri·od·ic** (ā′pîr-ē-ŏd′ĭk) *adj.* Occurring without regularity. **—a′pe·ri·od′i·cal·ly** *adv.* **—a·pe′ri·o·dic′i·ty** (-ə-dĭs′ĭ-tē) *n.*

**a·pé·ri·tif** (ä-pĕr′ĭ-tēf′) *n.* [Fr. < OFr. *aperitif*, purgative < Med. Lat. *aperitivus* < Lat. *aperire*, to open.] A drink of liquor or wine taken before a meal to stimulate the appetite.

**ap·er·ture** (ăp′ər-chər) *n.* [Lat. *apertura* < *apertus*, open, p.part. of *aperire*, to open.] **1.** An opening, as a hole, gap, or slit : ORIFICE. **2.** A usu. adjustable opening in an optical instrument that limits the amount of light passing through a lens. **—ap′er·tur′al** *adj.*

**aperture card** *n.* A punched card on which a portion of a microfilmed document is mounted.

**a·pet·al·ous** (ā-pĕt′l-əs) *adj. Bot.* Lacking petals. **—a·pet′al·y** *n.*

**a·pex** (ā′pĕks′) *n., pl.* **a·pex·es** *or* **a·pi·ces** (ā′pĭ-sēz′, ăp′ĭ-) [Lat.] **1.** The highest point : VERTEX <the *apex* of the pyramid> **2.** The culmination, as of an activity or effort : CLIMAX <the *apex* of my professional life> **3.** A pointed end : TIP.

**a·phaer·e·sis** *or* **a·pher·e·sis** (ə-fĕr′ĭ-sĭs) *n., pl.* **-ses** (-sēz′) [LLat. < Gk. *aphairesis*, removal < *aphairein*, to take away from : *apo-*, away from + *hairein*, to take.] Loss of one or more letters or sounds from the beginning of a word, as in *'sides* for *besides*. **—aph′ae·ret′ic** (ăf′ə-rĕt′ĭk) *adj.*

**a·pha·gi·a** (ə-fā′jē-ə, -jə) *n.* Inability to swallow.

**aph·a·nite** (ăf′ə-nīt′) *n.* [Fr. < Gk. *aphanēs*, unseen : *a-*, not + *phainesthai*, to appear < *phainein*, to show.] Dense, homogeneous rock with constituents so fine that they are invisible to the naked eye. **—aph′a·nit′ic** (-nĭt′ĭk) *adj.*

**a·pha·sia** (ə-fā′zhə) *n.* [Gk. < *aphatos*, speechless : *a-*, not + *phanai*, to speak.] Partial or total loss of the ability to speak or comprehend speech, resulting from brain damage. **—a·pha′si·ac′** (-zē-ăk′) *n.* **—a·pha′sic** (-zĭk, -sĭk) *adj.* & *n.*

**a·phe·li·on** (ə-fē′lē-ən, ə-fēl′yən) *n., pl.* **-li·a** (-lē-ə) [NLat. : Gk. *apo-*, away from + Gk. *hēlios*, sun.] The orbital point on a planetary orbit farthest from the sun.

**a·phe·li·o·trop·ic** (ə-fē′lē-ə-trŏp′ĭk) *adj.* Turning away from the sun or another light source, as roots do. **—a·phe′li·o·trop′i·cal·ly** *adv.* **—a·phe′li·ot′ro·pism** (-ŏt′rə-pĭz′əm) *n.*

**a·pher·e·sis** (ə-fĕr′ĭ-sĭs) *n. var. of* APHAERESIS.

**aph·e·sis** (ăf′ĭ-sĭs) *n., pl.* **-ses** (-sēz′) [Gk., a release < *aphienai*, to let go : *apo-*, away + *hienai*, to send.] Loss of a short unstressed vowel from the beginning of a word, as in *possum* for *opossum*. **—a·phet′ic** (ə-fĕt′ĭk) *adj.* **—a·phet′i·cal·ly** *adv.*

**a·phid** (ā′fĭd, ăf′ĭd) *n.* [< NLat. *Aphis*, *Aphid-*, type genus.] Any of various small soft-bodied insects of the family Aphididae, which feed by sucking sap from plants. **—a·phid′i·an** (ə-fĭd′ē-ən) *adj.* & *n.*

**a·phi·des** (ā′fĭ-dēz′, ăf′ĭ-) *n. pl. of* APHIS.

**aphid lion** *n.* The larva of any of the insects of the family Chrysopidae, as the lacewing, that feed on aphids.

**a·phis** (ā′fĭs, ăf′ĭs) *n., pl.* **a·phi·des** (ā′fĭ-dēz′, ăf′ĭ-) [NLat. *Aphis*, genus name.] An aphid, esp. one of the genus *Aphis*.

**a·pho·ni·a** (ā-fō′nē-ə) *n.* [Gk. < *aphonos*, voiceless : *a-*, without + *phonē*, voice.] Loss of speech resulting from disease or injury to the speech organs. **—a·phon′ic** (ā-fŏn′ĭk, ā-fō′nĭk) *adj.*

**aph·o·rism** (ăf′ə-rĭz′əm) *n.* [Fr. *aphorisme* < OFr. < Med. Lat. *aphorismus* < Gk. *aphorismos* < *aphorizein*, to distinguish : *apo-*, off + *horos*, boundary.] **1.** A brief statement of a principle. **2.** A tersely phrased statement of an observation or truth : ADAGE. **—aph′o·ris′tic** (-rĭs′tĭk) *adj.* **—aph′o·ris′ti·cal·ly** *adv.*

**aph·o·rize** (ăf′ə-rīz′) *vi.* **-rized, -riz·ing, -riz·es.** To express in or as if in aphorisms.

**a·pho·tic** (ā-fō′tĭk) *adj.* Lacking light, as ocean depths.

**aph·ro·dis·i·ac** (ăf′rə-dĭz′ē-ăk′) *adj.* [Gk. *aphrodisiakos* < *aphrodisia*, sexual pleasures < *Aphroditē*, Aphrodite.] Stimulating or intensifying sexual desire. **—aph′ro·dis′i·ac** *n.* **—aph′ro·di·si′a·cal** (ăf′rə-dĭ-zī′ə-kəl) *adj.*

**Aph·ro·di·te** (ăf′rə-dī′tē) *n.* [Gk. *Aphroditē*.] **1.** *Gk. Myth.* The goddess of love and beauty. **2. aphrodite.** A brightly colored North American butterfly, *Argynnis aphrodite*.

**a·phyl·lous** (ā-fĭl′əs) *adj.* [Gk. *aphullos* : *a-*, without + *phullon*, leaf.] *Bot.* Bearing no leaves. **—a′phyl′ly** (ā′fĭl′ē) *n.*

**a·pi·an** (ā′pē-ən) *adj.* [Lat. *apianus* < *apis*, bee.] Of or relating to bees.

**a·pi·ar·i·an** (ā′pē-âr′ē-ən) *adj.* Relating to bees or to the keeping and care of bees. —*n.* A beekeeper.

**a·pi·a·rist** (ā′pē-ə-rĭst, ā′pē-ĕr′ĭst) *n.* A beekeeper.

**a·pi·ar·y** (ā′pē-ĕr′ē) *n., pl.* **-ies.** [Lat. *apiarium*, beehive < *apis*, bee.] A place where bees and beehives are kept.

**ap·i·cal** (ăp′ĭ-kəl, ā′pĭ-) *adj.* [NLat. *apicalis* < Lat. *apex*, top.] **1.** Of, relating to, or located at an apex. **2.** Relating to consonants articulated with the tip of the tongue, as *t*, *d*, and *s*. **—ap′i·cal·ly** *adv.*

**a·pi·ces** (ā′pĭ-sēz′, ăp′ĭ-) *n. var. pl. of* APEX.

**a·pic·u·late** (ə-pĭk′yə-lĭt) *adj.* [< NLat. *apiculus*, sharp point, dim. of Lat. *apex*, point.] Ending with a sharp tip, as certain leaves do.

**a·pi·cul·ture** (ā′pĭ-kŭl′chər) *n.* [Lat. *apis*, bee + CULTURE.] The raising and care of bees. **—a′pi·cul′tur·al** *adj.* **—a′pi·cul′tur·ist** *n.*

**a·piece** (ə-pēs′) *adv.* [ME *a pece* : *a*, by + *pece*, piece.] To or for each one <The twins got two new outfits *apiece*.>

**A·pis** (ā′pĭs) *n.* [Lat. < Gk., of Egypt. orig.] A sacred bull venerated by the ancient Egyptians.

**ap·ish** (ā′pĭsh) *adj.* **1.** Resembling an ape. **2.** Foolishly or slavishly imitative. **—ap′ish·ly** *adv.* **—ap′ish·ness** *n.*

**a·piv·o·rous** (ā-pĭv′ər-əs) *adj.* [Lat. *apis*, bee + -VOROUS.] Feeding on bees <*apivorous* birds>

**APL** (ā′pē-ĕl′) *n.* [A + P(ROGRAMMING) + L(ANGUAGE).] A computer-programming language designed for use at remote terminals.

**a·pla·cen·tal** (ā′plə-sĕn′tl) *adj.* Having no placenta, as marsupials.

**ap·la·nat·ic** (ăp′lə-năt′ĭk) *adj.* [< Gk. *aplanētos*, unable to go astray : *a-*, not + *planētos*, wandering < *planasthai*, to wander.] Of or relating to optical systems that correct for spherical aberration.

**a·pla·sia** (ə-plā′zhə, -zhē-ə) *n.* Defective development or congenital absence of tissue of an organ or an organic part.

**a·plas·tic** (ā-plăs′tĭk) *adj.* **1.** Lacking form. **2.** *Pathol.* Unable to form or regenerate tissue. **3.** *Pathol.* Of or pertaining to aplasia.

**a·plen·ty** (ə-plĕn′tē) *adj.* Being in abundance. —*adv.* **1.** In abundance. **2.** To an extreme degree.

**ap·lite** (ăp′līt′) *n.* [G. *Aplit* : Gk. *haplous*, single + *-it*, -ite.] A fine-grained, light-colored granitic rock consisting mainly of orthoclase and quartz. **—ap·lit′ic** (ă-plĭt′ĭk) *adj.*

**a·plomb** (ə-plŏm′, ə-plŭm′) *n.* [Fr., balance < OFr. *à aplomb*, perpendicularly : *à*, according to (< Lat. *ad*) + *plomb*, lead weight < Lat. *plumbum*, lead.] Confidence in oneself : POISE.

**ap·ne·a** *also* **ap·noe·a** (ăp-nē′ə, ăp′nē-ə) *n.* [NLat. < Gk. *apnoia* : *a-*, without + *pnoē*, breathing < *pnein*, to breathe.] Temporary suspension of respiration. **—ap·ne′ic** *adj.*

**apo-** *or* **ap-** *pref.* [Gk. < *apo*, away from.] **1. a.** Away from : OFF <*apheliotropic*> **b.** Separate <*apocarpous*> **2.** Without : not <*apogamy*> **3.** Related to : derived from <*apomorphine*> **4.** Metasomatic <*apophyllite*>

**A·poc·a·lypse** (ə-pŏk′ə-lĭps′) *n.* [ME *Apocalipse* < LLat. *Apocalipsis* < Gk. *apokalupsis*, revelation < *apokaluptein*, to uncover : *apo-*, away + *kaluptein*, to cover.] **1.** —See table at BIBLE. **2. apocalypse.** A prophetic revelation.

**a·poc·a·lyp·tic** (ə-pŏk′ə-lĭp′tĭk) *also* **a·poc·a·lyp·ti·cal** (-tĭ-kəl) *adj.* **1.** Of or relating to a prophetic revelation. **2.** Portending future disaster. **3.** Of or relating to a final decision or climax. **—a·poc′a·lyp′ti·cal·ly** *adv.*

**ap·o·carp** (ăp′ə-kärp′) *n.* [Back-formation < APOCARPOUS.] An apocarpous fruit.

**ap·o·car·pous** (ăp′ə-kär′pəs) *adj. Bot.* Having distinctly separated carpels. **—ap′o·car′py** (ăp′ə-kär′pē) *n.*

**ap·o·chro·mat·ic** (ăp′ə-krō-măt′ĭk) *adj.* Corrected for both chromatic and spherical aberration, as a lens.

**a·poc·o·pe** (ə-pŏk′ə-pē) *n.* [Lat. < Gk. *apokopē* < *apokoptein*, to cut off : *apo-*, off + *koptein*, to cut.] Loss of one or more letters or sounds from the end of a word, as in *doin'* for *doing*.

**ap·o·crine** (ăp′ə-krĭn, -krīn′, -krēn′) *adj.* [< Gk. *apokrinein*, to set apart : *apo-*, away + *krinein*, to separate.] Of or relating to a gland that loses part of its cytoplasm in secretion.

**A·poc·ry·pha** (ə-pŏk′rə-fə) *n.* [ME *Apocripha* < LLat. *apocryphus*, spurious < Gk. *apokruphos*, hidden : *apo-*, away + *kruptein*, to hide.] *(sing. or pl. in number).* **1.** The 14 Biblical books included in the Vulgate but considered uncanonical by Protestants because they are not part of the Hebrew Scriptures; 11 are accepted in the Roman Catholic canon. —See table at BIBLE. **2.** Early Christian writings proposed as additions to the New Testament but rejected by the major canons. **3. apocrypha.** Writings of questionable authorship or authenticity.

**a·poc·ry·phal** (ə-pŏk′rə-fəl) *adj.* **1.** Of questionable authorship or authenticity. **2.** Spurious : false. **3. Apocryphal.** Of or relating to the Apocrypha. **—a·poc′ry·phal·ly** *adv.*

**ap·o·dal** (ăp′ə-dəl) *also* **ap·o·dus** (-dəs) *adj.* [< Gk. *apous*, *apod-* : *a-*, without + *pous*, foot.] Having no feet or footlike appendages.

**ap·o·dic·tic** (ăp′ə-dĭk′tĭk) *adj.* [Lat. *apodicticus* < Gk. *apodeiktikos* < *apodeiknunai*, to demonstrate : *apo*, away from + *deiknunai*, to show.] Clearly demonstrated or proven : INCONTESTABLE. **—ap′o·dic′ti·cal·ly** *adv.*

**a·pod·o·sis** (ə-pŏd′ə-sĭs) *n., pl.* **-ses** (-sēz′) [Gk. < *apodidonai*, to give back : *apo-*, away + *didonai*, to give.] The clause stating the conclusion or consequence of a conditional sentence.

**ap·o·dous** (ăp′ə-dəs) *adj.* Apodal.

**ap·o·en·zyme** (ăp′ō-ĕn′zīm′) *n.* A protein requiring a coenzyme to function as an enzyme.

**a·pog·a·my** (ə-pŏg′ə-mē) *n.* Production of a new plant from a prothallus by budding without sexual reproduction, as in ferns. **—ap′o·gam′ic** (ăp′ə-găm′ĭk), **a·pog′a·mous** (ə-pŏg′ə-məs) *adj.*

**ap·o·gee** (ăp′ə-jē) *n.* [Fr. *apogée* < NLat. *apogaeum* < Gk. *apogaion* < *apogaios*, far from the earth : *apo-*, away from + *gaia*, earth.] **1.** The point in the orbit of the moon or of an artificial satellite most distant from the earth. **2.** The farthest or highest point : APEX. **—ap′o·ge′an** (-jē′ən) *adj.*

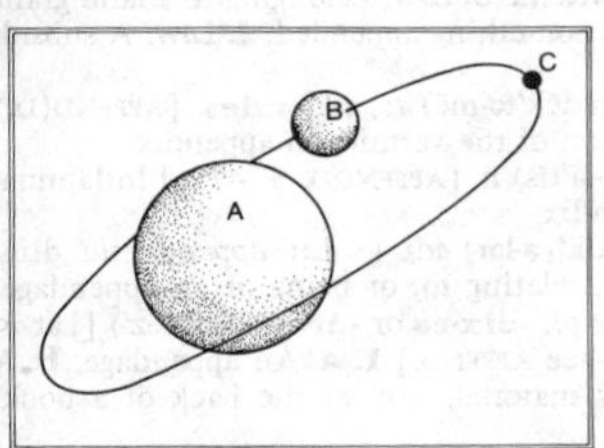


**apogee**
*A. earth, B. satellite, C. point of apogee*

**a·po·lit·i·cal** (ā′pə-lĭt′ĭ-kəl) *adj.* **1.** Having no involvement with or interest in politics. **2.** Having no political significance. **—a′po·lit′i·cal·ly** *adv.*

**A·pol·lo** (ə-pŏl′ō) *n.* [Lat. < Gk. *Apollōn.*] **1.** *Gk. Myth.* The god of the sun, prophecy, music, medicine, and poetry. **2. apollo.** A very handsome young man.

**Ap·ol·lo·ni·an** (ăp′ə-lō′nē-ən) *adj.* **1.** Of or relating to Apollo. **2. apollonian.** Clearly defined and well-ordered : HARMONIOUS. **3. apollonian.** Noble : serene.

**a·pol·o·get·ic** (ə-pŏl′ə-jĕt′ĭk) *adj.* **1.** Making or expressing an apology. **2.** Explaining or defending in speech or writing. —*n.* A formal defense. **—a·pol′o·get′i·cal·ly** *adv.*

**a·pol·o·get·ics** (ə-pŏl′ə-jĕt′ĭks) *n. (sing. in number).* The branch of theology dealing with the defense and proof of Christianity.

**ap·o·lo·gi·a** (ăp′ə-lō′jē-ə, -jə) *n.* [LLat., apology.] A formal defense or justification.

**a·pol·o·gist** (ə-pŏl′ə-jĭst) *n.* One who argues in defense or justification of another person or cause.

**a·pol·o·gize** (ə-pŏl′ə-jīz′) *vi.* **-gized, -giz·ing, -giz·es. 1.** To make regretful acknowledgment of or excuse for a fault or offense. **2.** To make a formal defense or justification. **—a·pol′o·giz′er** *n.*

**ap·o·logue** (ăp′ə-lôg′, -lŏg′) *n.* [Fr. < Lat. *apologus* < Gk. *apologos* : *apo-*, away + *logos*, speech.] A moral fable.

**a·pol·o·gy** (ə-pŏl′ə-jē) *n., pl.* **-gies.** [OFr. *apologie* < LLat. *apologia* < Gk. : *apo-*, away + *logos*, speech.] **1.** A statement of acknowledgment asking pardon or expressing regret for a fault or offense. **2.** A formal justification or defense. **3.** An inferior substitute.

**ap·o·lune** (ăp′ə-lo͞on′) *n.* [APO- + Lat. *luna*, moon.] The point of an orbit around the moon farthest from the moon's center.

**ap·o·mict** (ăp′ə-mĭkt′) *n.* [Back-formation < E. *apomictic*, produced by apomixis.] An organism produced by apomixis. **—ap′o·mic′tic** *adj.*

**ap·o·mix·is** (ăp′ə-mĭk′sĭs) *n.* [NLat. : APO- + Gk. *mixis*, sexual intercourse < *meignunai*, to have sexual intercourse.] A rare reproductive process in which a new individual is produced from a female cell or cells other than the egg cell, often in a way that imitates sexual reproduction.

**ap·o·mor·phine** (ăp′ə-môr′fēn′) *n.* A poisonous white crystalline alkaloid, $C_{17}H_{17}NO_2$, derived from morphine and used as an expectorant, emetic, and hypnotic.

**ap·o·neu·ro·sis** (ăp′ə-no͝o-rō′sĭs, -nyo͝o-) *n., pl.* **-ses** (-sēz′) [Gk. *aponeurōsis* < *aponeurousthai*, to become tendinous : *apo-*, away + *neuron*, sinew.] A sheetlike membrane resembling a flattened tendon that envelops a muscle or connects it to its insertion. **—ap′o·neu·rot′ic** (-rŏt′ĭk) *adj.*

**ap·o·phthegm** (ăp′ə-thĕm′) *n. var. of* APOTHEGM.

**a·poph·y·ge** (ə-pŏf′ə-jē) *n.* [Gk. *apophugē* : *apo-*, away + *phugē*, flight.] The curvature at the bottom or top of the shaft of an architectural column.

**a·poph·yl·lite** (ə-pŏf′ə-līt′, ăp′ə-fĭl′īt′) *n.* A white, pale-pink, or pale-green crystalline mineral, essentially $KCa_4FSi_4O_{10}{\cdot}8H_2O$.

**a·poph·y·sis** (ə-pŏf′ĭ-sĭs) *n., pl.* **-ses** (-sēz′) [NLat. < Gk. *apophusis* : *apo-*, off from + *phusis*, growth < *phuein*, to grow.] **1.** *Biol.* A swelling, projection, or outgrowth of an organ or part. **2.** *Geol.* A branch from a dike or vein. **—a·poph′y·sate′** (-sāt′), **a·poph′y·se′al** (ə-pŏf′ə-sē′əl) *adj.*

**ap·o·plec·tic** (ăp′ə-plĕk′tĭk) *adj.* **1.** Of, relating to, or causing apoplexy. **2.** Having or displaying symptoms of apoplexy. **3. a.** Apt to cause apoplexy. **b.** Verging on apoplexy. **—ap′o·plec′ti·cal·ly** *adv.*

**ap·o·plex·y** (ăp′ə-plĕk′sē) *n.* [ME *apoplexie* < OFr. < LLat. *apoplexia* < Gk. < *apoplēssein*, to cripple by a stroke : *apo-* (intensive) + *plēssein*, to strike.] Sudden loss of muscular control with diminution or loss of sensation and consciousness, caused by rupture or blocking of a cerebral blood vessel.

**a·port** (ə-pôrt′, ə-pōrt′) *adv. Naut.* On or toward the port side.

**ap·o·se·le·ne** (ăp′ō-sə-lē′nē) *n.* [APO- + Gk. *selene*, moon.] Apolune.

**ap·o·si·o·pe·sis** (ăp′ə-sī′ə-pē′sĭs) *n., pl.* **-ses** (-sēz′) [LLat. < Gk. *aposiōpēsis* < *aposiōpan*, to become silent : *apo-* (intensive) + *siōpan*, to be silent < *siōpē*, silence.] A sudden breaking off of a thought in midsentence. **—ap′o·si′o·pet′ic** pĕt′ĭk *adj.*

**a·pos·ta·sy** (ə-pŏs′tə-sē) *n., pl.* **-sies.** [ME *apostasie* < LLat. *apostasia*, defection < Gk. *aphistanai*, to revolt : *apo-*, away from + *histanai*, to stand.] Abandonment of one's religious faith, political party, cause, or principles : DEFECTION.

**a·pos·tate** (ə-pŏs′tāt′, -tĭt) *n.* [ME *apostata* < LLat. < Gk. *apostatēs* < *aphistanai*, to revolt. —see APOSTASY.] One guilty of apostasy. **—a·pos′tate′** *adj.*

**a·pos·ta·tize** (ə-pŏs′tə-tīz′) *vi.* **-tized, -tiz·ing, -tiz·es.** To be guilty of apostasy.

**a pos·te·ri·o·ri** (ä′ pŏ-stîr′ē-ôr′ē, -ôr′ī′, -ōr′ī′, ā′) *adj.* [Lat., from the subsequent.] *Logic.* Denoting reasoning from facts or particulars to general principles or from effects to causes : INDUCTIVE.

**a·pos·tle** (ə-pŏs′əl) *n.* [ME < OE *apostol* < LLat. *apostolus* < Gk. *apostulos*, messenger < *apostellein*, to send off : *apo-*, away from + *stellein*, to send.] **1. Apostle.** One of a group composed esp. of the 12 disciples chosen by Christ to preach His gospel. **2.** An early Christian missionary. **3.** A leader of the first Christian mission to a country or region. **4.** One of the 12 members of the Mormon administrative council. **5.** A leader or advocate of a cause or movement <an *apostle* of civil rights>

**Apostles' Creed** *n.* A Christian creed traditionally ascribed to the 12 Apostles.

**a·pos·to·late** (ə-pŏs′tə-lāt′, -lĭt) *n.* [LLat. *apostolatus* < *apostolus*, apostle.] **1.** The office, duties, or mission of an apostle. **2.** An association of individuals for the dissemination of a religion or a doctrine.

**ap·os·tol·ic** (ăp′ə-stŏl′ĭk) *adj.* **1.** Of, relating to, or contemporary with the 12 Apostles. **2.** Of or relating to the faith, teaching, or practice of the 12 Apostles. **3.** Of or relating to the pope as successor of Saint Peter.

**Apostolic Father** *n.* A church father who received personal instruction from the 12 Apostles or from their disciples.

**a·pos·tro·phe**[1] (ə-pŏs′trə-fē) *n.* [Fr. < LLat. *apostrophos* < Gk. < *apostrophein*, to turn away : *apo-*, away + *strephein*, to turn.] The superscript sign (′) indicating the omission of one or more letters from a word, the possessive case, and certain plurals, esp. those of numbers and letters. **—ap′os·troph′ic** (ăp′ə-strŏf′ĭk) *adj.*

**a·pos·tro·phe**[2] (ə-pŏs′trə-fē) *n.* [Lat. < Gk. < *apostrephein*, to turn away. —see APOSTROPHE[1].] A digression in discourse, esp. a turning away from an audience to address an absent or imaginary person. **—ap′os·troph′ic** (ăp′ə-strŏf′ĭk) *adj.*

**a·pos·tro·phize** (ə-pŏs′trə-fīz′) *vt. & vi.* **-phized, -phiz·ing, -phiz·es.** To address by or express in apostrophe.

**apothecaries' measure** *n.* A system of liquid volume measure used in pharmacy.

**apothecaries' weight** *n.* A system of weights used in pharmacy and based on an ounce equal to 480 grains and a pound equal to 12 ounces.

**a·poth·e·car·y** (ə-pŏth′ĭ-kĕr′ē) *n., pl.* **-ries.** [ME *apotecarie* < Med. Lat. *apothecarius* < LLat., clerk < Lat. *apotheca*, storehouse < Gk. *apothēke* < *apotithenai*, to store : *apo-*, away + *tithenai*, to put.] One who prepares and sells drugs and medicines : PHARMACIST.

**ap·o·the·ci·um** (ăp′ə-thē′sē-əm, -shē-) *n., pl.* **-ci·a** (-sē-ə, -shē-ə) [NLat. < Lat. *apotheca*, storehouse. —see APOTHECARY.] An open disk-shaped or cup-shaped fruiting body in certain fungi, lined with a spore-bearing layer. **—ap′o·the′cial** (-shəl) *adj.*

**ap·o·thegm** *also* **ap·o·phthegm** (ăp′ə-thĕm′) *n.* [Gk. *apophegma* < *apophthengesthai*, to speak plainly : *apo-*, away + *phthengesthai*, to speak.] A concise and witty instructive saying : MAXIM. **—ap′o·theg·mat′ic** (-thĕg-măt′ĭk), **ap′o·theg·mat′i·cal** (-ĭ-kəl) *adj.* **—ap′o·theg·mat′i·cal·ly** *adv.*

**ap·o·them** (ăp′ə-thĕm′) *n.* [APO- + Gk. *thema*, something laid down < *tithenai*, to put.] The perpendicular distance from the center of a regular polygon to any of the sides.

**a·poth·e·o·sis** (ə-pŏth′ē-ō′sĭs, ăp′ə-thē′ə-sĭs) *n., pl.* **-ses** (-sēz′) [LLat. < Gk. *apotheōsis* < *apotheoun*, to deify : *apo-*, from + *theos*, god.] **1.** Exaltation to divine rank or stature : DEIFICATION. **2.** An exalted or glorified example <the *apotheosis* of honesty>

**ap·o·the·o·size** (ăp′ə-thē′ə-sīz′, ə-pŏth′ē-ə-sīz′) *vt.* **-sized, -siz·ing, -siz·es.** To make the object of apotheosis.

**ap·o·tro·pa·ic** (ăp′ə-trō-pā′ĭk) *adj.* [Gk. *apotropaios* < *apotrepein*, to ward off : *apo-*, from + *trepein*, to turn.] Intended to ward off evil <an *apotropaic* amulet> **—ap′o·tro·pa′i·cal·ly** *adv.*

**Ap·pa·la·chian tea** (ăp′ə-lā′chən, -lā′chē-ən, -lăch′ən) *n.* [After the *Appalachian* Mountains.] The withe rod.

**ap·pall** (ə-pôl′) *vt.* **-palled, -pall·ing, -palls.** [ME *apallen*, to grow faint < OFr. *apalir* : *a-*, to (< Lat. *ad*) + *palir*, to grow pale < Lat. *pallēre.*] To fill with consternation or dismay.

**ap·pall·ing** (ə-pô′lĭng) *adj.* Causing consternation or dismay : FRIGHTFUL <an *appalling* car accident> **—ap·pall′ing·ly** *adv.*

**ap·pa·loo·sa** (ăp′ə-lo͞o′sə) *n.* [Prob. after the *Paloose* Indians, who bred the horse.] A horse orig. bred in northwestern North America, having a spotted rump.

**ap·pa·nage** *also* **ap·a·nage** (ăp'ə-nĭj) *n.* [Fr. *apanage* < OFr. < *apaner,* to make provisions for < Med. Lat. *appanare* : Lat. *ad-,* to + Lat. *panis,* bread.] **1.** A source of revenue, as land, given by a monarch for the maintenance of a member of the ruling family. **2.** Something extra offered or claimed as due : PERQUISITE. **3.** A natural accompaniment or adjunct.

**ap·pa·rat** (ăp'ə-răt', ä'pə-rät') *n.* [R.] **1.** APPARATUS 4. **2.** An underground political movement.

**ap·pa·ra·tchik** (ä'pə-rä'chĭk) *n.* [R. < *apparat,* apparat.] A member of a Communist apparat.

**ap·pa·ra·tus** (ăp'ə-răt'əs, -rā'təs) *n., pl.* **apparatus** *or* **-tus·es.** [Lat., preparation < *apparare,* to prepare : *ad-,* to + *parare,* to prepare.] **1.** The totality of means by which a designated function is performed or a specific task executed. **2. a.** A machine. **b.** A group of machines used together or in succession to accomplish a task. **3.** *Physiol.* A group of organs having a collective function <the digestive *apparatus*> **4.** A political organization.

**ap·par·el** (ə-păr'əl) *n.* [ME *appareil* < OFr. *apareil,* preparation < *apareillier,* to prepare < Lat. *apparare.* —see APPARATUS.] **1.** Clothing. **2.** Something that covers or adorns. —*vt.* **-eled, -el·ing, -els** *also* **-elled, -el·ling, -els. 1.** To clothe. **2.** To adorn.

**ap·par·ent** (ə-păr'ənt, ə-pâr'-) *adj.* [ME *apparaunt* < OFr. *aparant,* pr.part. of *aparoir,* to appear.] **1.** Easily seen : VISIBLE. **2.** Easily understood : OBVIOUS. **3.** Appearing as such but not necessarily so <an *apparent* thief> **—ap·par'ent·ly** *adv.* **—ap·par'ent·ness** *n.*

**apparent magnitude** *n.* MAGNITUDE 2.

**ap·pa·ri·tion** (ăp'ə-rĭsh'ən) *n.* [ME *apparicioun* < OFr. *apparition* < LLat. *apparitio,* an appearance < Lat. *apparēre,* to appear.] **1.** A ghostly figure : SPECTER. **2.** An unexpected or unusual sight. **3.** The act of appearing. **—ap'pa·ri'tion·al** *adj.*

**ap·par·i·tor** (ə-păr'ĭ-tər) *n.* [Lat. < *apparēre,* to appear.] An official once sent to carry out orders of a civil or ecclesiastical court.

**ap·peal** (ə-pēl') *n.* [ME *apel* < OFr. < *apeler,* to appeal < Lat. *appellare,* to entreat.] **1.** An earnest or urgent request, entreaty, or supplication. **2.** A resort or application, as to an acknowledged authority, for sanction, corroboration, or a decision. **3.** The power of attracting or of arousing interest <a new comedy with *appeal*> **4.** *Law.* **a.** The transfer of a case from a lower to a higher court for a new hearing. **b.** A request for a new hearing. **c.** A case so transferred. —*v.* **-pealed, -peal·ing, -peals.** —*vi.* **1.** To make an earnest or urgent request. **2.** To have recourse, as for corroboration : RESORT <*appeal* to reason> **3.** To be attractive or interesting. **4.** *Law.* To make or apply for an appeal. —*vt. Law.* To transfer or apply to transfer (a case) to a higher court for rehearing. **—ap·peal'a·ble** *adj.* **—ap·peal'er** *n.* **—ap·peal'ing·ly** *adv.*

**ap·pear** (ə-pîr') *vi.* **-peared, -pear·ing, -pears.** [ME *aperen* < OFr. *aparoir* < Lat. *apparēre* : *ad-,* to + *parēre,* to show.] **1.** To become visible. **2.** To come into existence <weeds *appearing* on the lawn> **3.** To look or seem to be. **4.** To seem likely <It *appears* we'll have to stay.> **5.** To be presented or published. **6.** *Law.* To present oneself formally before a court as defendant, plaintiff, or counsel.

☆ **syns:** APPEAR, LOOK, SEEM *v. core meaning* : to have the appearance of <*appeared* happy but really wasn't>

**ap·pear·ance** (ə-pîr'əns) *n.* **1.** The act or an instance of coming into sight. **2.** The act or an instance of coming into public view <put in a personal *appearance* at the benefit> **3.** Outward aspect <a kindly *appearance*> **4.** Something that appears : PHENOMENON. **5.** A superficial aspect : SEMBLANCE <keeping up an *appearance* of friendliness> **6. appearances.** Outward indications <to all *appearances,* a successful business>

**ap·pease** (ə-pēz') *vt.* **-peased, -peas·ing, -peas·es.** [ME *appesen* < OFr. *apaisier* : *a,* to (< Lat. *ad*) + *pais,* peace < Lat. *pax.*] **1.** To calm or pacify, esp. by giving what is demanded : PLACATE. **2.** To satisfy or relieve <*appease* hunger> **—ap·peas'a·ble** *adj.* **—ap·peas'a·bly** *adv.* **—ap·peas'er** *n.*

**ap·pease·ment** (ə-pēz'mənt) *n.* **1.** The act of appeasing or the state of being appeased. **2.** The policy of granting concessions to potential enemies to maintain peace.

**ap·pel** (ə-pĕl') *n.* [Fr., call < *appeler,* to call < OFr. *apeler,* to appeal.] A quick stamp of the foot used in fencing as a feint to produce an opening.

**ap·pel·lant** (ə-pĕl'ənt) *adj.* Of or relating to an appeal : APPELLATE. —*n.* One that appeals a court decision.

**ap·pel·late** (ə-pĕl'ĭt) *adj.* [Lat. *appellatus,* p.part. of *appellare,* to entreat.] Empowered to hear appeals and review lower court decisions <an *appellate* court>

**ap·pel·la·tion** (ăp'ə-lā'shən) *n.* [ME *appelacion* < Lat. *appellatio* < *appellare,* to entreat.] **1.** A name or title. **2.** The act of naming.

**ap·pel·la·tive** (ə-pĕl'ə-tĭv) *adj.* [ME < LLat. *appelativus* < *appellare,* to call upon, entreat.] **1.** Of or pertaining to the assignment of names. **2.** Designating a common noun. —*n.* A name or descriptive epithet. **—ap·pel'la·tive·ly** *adv.*

**ap·pel·lee** (ăp'ə-lē') *n.* [OFr. *apele* < *apeler,* to appeal.] One against whom an appeal is made.

**ap·pend** (ə-pĕnd') *vt.* **-pend·ed, -pend·ing, -pends.** [Lat. *appendere,* to hang upon : *ad-,* to + *pendere,* to hang.] **1.** To add as a supplement, as at the end of a book. **2.** To fix to : ATTACH.

**ap·pend·age** (ə-pĕn'dĭj) *n.* **1.** Something appended. **2.** *Biol.* A part or organ joined to an axis or trunk.

**ap·pen·dant** (ə-pĕn'dənt) *adj.* **1.** Attached as an appendage. **2.** Serving to accompany or attend. **3.** *Law.* Belonging to a land grant as a subsidiary right. —*n.* **1.** Something appended. **2.** *Law.* A subsidiary right.

**ap·pen·dec·to·my** (ăp'ən-dĕk'tə-mē) *n., pl.* **-mies.** [APPEND(IX) + -ECTOMY.] Surgical excision of the vermiform appendix.

**ap·pen·di·ci·tis** (ə-pĕn'dĭ-sī'tĭs) *n.* [APPENDIX + -ITIS.] Inflammation of the vermiform appendix.

**ap·pen·dic·u·lar** (ăp'ən-dĭk'yə-lər) *adj.* [< Lat. *appendicula,* dim. of *appendix,* appendix.] Of, relating to, or being of an appendage.

**ap·pen·dix** (ə-pĕn'dĭks) *n., pl.* **-dix·es** *or* **-di·ces** (-dĭ-sēz') [Lat. < *appendere,* to hang upon. —see APPEND.] **1. a.** An appendage. **b.** A collection of supplementary material, usu. at the back of a book. **2.** The vermiform appendix.

**ap·per·ceive** (ăp'ər-sēv') *vt.* **-ceived, -ceiv·ing, -ceives.** [ME *apperceiven,* to notice < OFr. *apercevoir* : *a,* toward (< Lat. *ad*) + *perceivre,* to perceive. —see PERCEIVE.] *Psychol.* To understand in terms of past perceptions.

**ap·per·cep·tion** (ăp'ər-sĕp'shən) *n.* [Fr. *aperception* < *apercevoir,* to notice < OFr. —see APPERCEIVE.] *Psychol.* **1.** Conscious perception with full awareness. **2.** The process of understanding by which newly observed qualities of an object are related to past experience. **—ap'per·cep'tive** (-sĕp'tĭv) *adj.*

**ap·per·tain** (ăp'ər-tān') *vi.* **-tained, -tain·ing, -tains.** [ME *appertenen* < OFr. *apartenir* < LLat. *appertinēre* : *ad-,* to + *pertinēre,* to belong. —see PERTAIN.] To belong as a proper function or part : PERTAIN.

**ap·pe·stat** (ăp'ĭ-stăt') *n.* [APPE(TITE) + -STAT.] The mechanism in the central nervous system that controls food intake.

**ap·pe·tence** (ăp'ĭ-təns) *n.* [Lat. *appetentia* < *appetere,* to strive after. —see APPETITE.] **1.** A strong desire or craving. **2.** A tendency or propensity.

**ap·pe·ten·cy** (ăp'ĭ-tən-sē) *n., pl.* **-cies.** Appetence.

**ap·pe·tite** (ăp'ĭ-tīt') *n.* [ME *appetit* < OFr. < Lat. *appetitus,* strong desire < *appetere,* to strive after : *ad-,* toward + *petere,* to seek.] **1.** A desire for food or drink. **2.** A physical desire. **3.** A strong desire or urge <an *appetite* for advancement> **—ap'pe·ti'tive** (ăp'ĭ-tī'tĭv, ə-pĕt'ĭ-tĭv) *adj.*

**ap·pe·tiz·er** (ăp'ĭ-tī'zər) *n.* A food or drink served usu. before a meal to whet the appetite.

**ap·pe·tiz·ing** (ăp'ĭ-tī'zĭng) *adj.* Appealing to or stimulating the appetite. **—ap'pe·tiz'ing·ly** *adv.*

**ap·plaud** (ə-plôd') *v.* **-plaud·ed, -plaud·ing, -plauds.** [Lat. *applaudere* : *ad-,* to + *plaudere,* to clap.] —*vi.* To express approval, esp. by clapping the hands. —*vt.* **1.** To express approval of, esp. by clapping the hands. **2.** To praise <*applauded* our successful accomplishment of the project> **—ap·plaud'er** *n.*

**ap·plause** (ə-plôz') *n.* [Med. Lat. *applausus* < Lat., p.part. of *applaudere,* to applaud.] **1.** Approval expressed esp. by the clapping of hands. **2.** Commendation : praise.

**ap·ple** (ăp'əl) *n.* [ME *appel* < OE *æppel.*] **1. a.** A tree, *Pyrus malus* of temperate regions, bearing fragrant pink or white flowers and edible fruit. **b.** The firm, rounded fruit of the apple tree or any of its varieties, with red, yellow, or green skin. **2. a.** A tree or plant having fruit resembling the apple, as the custard apple. **b.** The fruit of such a tree or plant. **3.** The hard wood of an apple tree.

**apple green** *n.* A moderate or vivid yellow green to light or strong yellowish green. **—ap'ple-green'** *adj.*

**ap·ple·jack** (ăp'əl-jăk') *n.* Brandy distilled from hard cider.

**apple of Peru** *n.* A plant native to tropical America, *Nicandra physalodes* bearing tubular blue or whitish flowers.

**ap·ple-pol·ish** (ăp'əl-pŏl'ĭsh) *vi.* **-ished, -ish·ing, -ish·es.** *Informal.* To curry favor by toadying. **—apple polisher** *n.*

**ap·ple·sauce** (ăp'əl-sôs') *n.* **1.** Apples stewed and sweetened and occas. spiced. **2.** *Slang.* Nonsense.

**Ap·ple·ton layer** (ăp'əl-tən) *n.* [After Sir Edward *Appleton* (1892–1965).] The F layer of the upper atmosphere.

**ap·pli·ance** (ə-plī'əns) *n.* [< APPLY.] A device, esp. one operated by electricity and designed for household use.

**ap·pli·ca·ble** (ăp'lĭ-kə-bəl, ə-plĭk'ə-) *adj.* Able to be applied : APPROPRIATE. **—ap'pli·ca·bil'i·ty** *n.* **—ap'pli·ca·bly** *adv.*

**ap·pli·cant** (ăp'lĭ-kənt) *n.* [Lat. *applicans,* pr.part. of *applicare,* to affix. —see APPLY.] One who applies, as for a job or admission.

**ap·pli·ca·tion** (ăp'lĭ-kā'shən) *n.* [ME *applicacioun* < Med. Lat. *applicatio* < Lat. *applicare,* to affix. —see APPLY.] **1.** The act of applying. **2.** Something applied, as a healing agent or cosmetic. **3. a.** The act of putting to a special use or purpose. **b.** A specific use to which something is put <the *application* of technology to education> **4.** Usability : relevance <found no *application* for the invention>

**5.** Close attention : DILIGENCE. **6. a.** A request, as for aid, employment, or admission. **b.** The form or document for such a request.

**ap·pli·ca·tive** (ăp′lĭ-kā′tĭv, ə-plĭk′ə-) *adj.* **1.** Marked by actual application : APPLIED. **2.** Practical. **—ap′pli·ca′tive·ly** *adv.*

**ap·pli·ca·tor** (ăp′lĭ-kā′tər) *n.* An instrument for applying something, as a healing agent or cosmetic.

**ap·pli·ca·to·ry** (ăp′lĭ-kə-tôr′ē, -tōr′ē, ə-plĭk′ə-) *adj.* Practical.

**ap·plied** (ə-plīd′) *adj.* Put into practice or to a particular use <*applied* engineering>

**ap·pli·qué** (ăp′lĭ-kā′) *n.* [Fr., p.part. of *appliquer,* to apply < Lat. *applicare,* to affix. —see APPLY.] A decoration or ornament, as in needlework, made by cutting pieces of one material and attaching them to the surface of another. —*vt.* **-quéd, -qué·ing, -qués.** To decorate with appliqué. **—ap′pli·qué′** *adj.*

**ap·ply** (ə-plī′) *v.* **-plied, -ply·ing, -plies.** [ME *applien* < OFr. *aplier* < Lat. *applicare,* to affix : *ad-,* to + *plicare,* to fold together.] —*vt.* **1.** To put on <*apply* paint to a wall> **2.** To put to or adapt for a special use <*applied* my earnings to tuition> **3.** To put (e.g., automotive brakes) into action. **4.** To devote (oneself or one's efforts) to something. —*vi.* **1.** To be pertinent or relevant <The law *applies to* everyone.> **2.** To ask or seek aid, employment, or admission.

**ap·pog·gia·tu·ra** (ə-pŏj′ə-to͝or′ə) *n.* [Ital. < *appoggiare,* to lean on < VLat. **appodiare* : Lat. *ad-,* to + *podium,* support < Gk. *podion,* base < *pous,* foot.] *Mus.* An embellishing note, usu. one step above or below the note it precedes and indicated by a small note or sign.

**ap·point** (ə-point′) *vt.* **-point·ed, -point·ing, -points.** [ME *appointen,* to decide < OFr. *apointier,* to arrange < *à point,* to the point : *à,* to (< Lat. *ad*) + *point,* point < Lat. *punctum* < *pungere,* to prick.] **1.** To name to fill an office or position. **2.** To fix or set, as a date or time, by authority or mutual agreement. **3.** To furnish or equip (e.g., an apartment), esp. appropriately. **4.** *Law.* To direct the disposition of (property) to a beneficiary in exercise of a power granted for this purpose by a preceding deed.

**ap·point·ee** (ə-poin′tē′, ăp′oin-) *n.* **1.** One named to an office or position. **2.** *Law.* One to whom a power of appointment of property is granted.

**ap·point·ive** (ə-poin′tĭv) *adj.* Relating to or filled by appointment.

**ap·point·ment** (ə-point′mənt) *n.* **1.** The act of appointing for an office or position. **2.** The office or position to which one has been appointed. **3.** An engagement <made an *appointment* with the dentist> **4. appointments.** Furnishings, fittings, or equipment. **5.** *Law.* The act of directing the disposition of property by virtue of a power granted for this purpose.

**ap·poin·tor** (ə-poin′tər, ə-poin′tôr′) *n. Law.* One that executes a power of appointment of property.

**ap·por·tion** (ə-pôr′shən, ə-pōr′-) *vt.* **-tioned, -tion·ing, -tions.** [OFr. *apportioner* : *à-* (< Lat. *ad*) + *portionner,* to divide into portions < *portion,* portion. —see PORTION.] To divide and distribute according to a plan or proportion : ALLOT.

**ap·por·tion·ment** (ə-pôr′shən-mənt, ə-pōr′-) *n.* **1.** The act of apportioning or state of being apportioned. **2. a.** Proportional distribution of the number of members of the U.S. House of Representatives on the basis of the population of each state. **b.** The allotment of direct taxes on the basis of state population.

**ap·pose** (ă-pōz′) *vt.* **-posed, -pos·ing, -pos·es.** [Back-formation < APPOSITION.] To place close to each other or side by side.

**ap·po·site** (ăp′ə-zĭt) *adj.* [Lat. *appositus,* p.part. of *apponere,* to put near : *ad-,* to + *ponere,* to put.] Suitable : appropriate. **—ap′po·site·ly** *adv.* **—ap′po·site·ness** *n.*

**ap·po·si·tion** (ăp′ə-zĭsh′ən) *n.* [ME *apposicioun* < Med. Lat. *appositio* < Lat. *apponere,* to put near. —see APPOSITE.] **1. a.** A construction in which a noun or noun phrase is placed with another as an explanatory equivalent, both having the same syntactic relation to the other elements in the sentence, as *Byron* and *the poet* in *The poet Byron died in Greece.* **b.** The relationship between such nouns or noun phrases. **2.** Placement side by side or close to each other. **3.** *Biol.* Growth of successive layers of a cell wall. **—ap′po·si′tion·al** *adj.* **—ap′po·si′tion·al·ly** *adv.*

**ap·pos·i·tive** (ə-pŏz′ĭ-tĭv) *adj.* Of or being in apposition. —*n.* A word or phrase in apposition. **—ap·pos′i·tive·ly** *adv.*

**ap·prais·al** (ə-prā′zəl) *n.* **1.** An act of appraising. **2.** An expert or official valuation as for taxation.

**ap·praise** (ə-prāz′) *vt.* **-praised, -prais·ing, -prais·es.** [ME *appreisen,* alteration of *apprisen* < OFr. *aprisier* < LLat. *appretiare* : Lat. *ad-,* to + Lat. *pretium,* price.] **1.** To determine the value of, esp. officially. **2.** To estimate the worth or features of : JUDGE <*appraise* a new product> **—ap·prais′a·ble** *adj.* **—ap·praise′ment** *n.* **—ap·prais′er** *n.*

**ap·pre·cia·ble** (ə-prē′shə-bəl) *adj.* Capable of being noticed, estimated, or measured : NOTICEABLE. **—ap·pre′cia·bly** *adv.*

**ap·pre·ci·ate** (ə-prē′shē-āt′) *v.* **-at·ed, -at·ing, -ates.** [LLat. *appretiare, appretiat-,* to appraise.] —*vt.* **1.** To recognize the quality, significance, or magnitude of. **2.** To be fully conscious of or sensitive to : REALIZE <I *appreciate* your difficulties.> **3.** To be thankful or show gratitude for. **4.** To perceive with aesthetic enjoyment. **5.** To raise in value or price. —*vi.* To increase in value or price. **—ap·pre′ci·a·tor** *n.* **—ap·pre′cia·to·ry** (-shə-tôr′ē, -tōr′ē) *adj.*

☆ ***syns:*** APPRECIATE, CHERISH, ESTEEM, PRIZE, RESPECT, SAVOR, TREASURE, VALUE *v. core meaning* : to recognize the worth, importance, or value of <a person who *appreciates* classical music>

**ap·pre·ci·a·tion** (ə-prē′shē-ā′shən) *n.* **1.** Recognition of the quality, value, significance, or magnitude of. **2.** A judgment or opinion, esp. a favorable one. **3.** An expression of gratitude. **4.** Awareness or perception, esp. of aesthetic qualities or values. **5.** An increase in value or price.

**ap·pre·cia·tive** (ə-prē′shə-tĭv, -shē-ā′tĭv) *adj.* Capable of or showing appreciation. **—ap·pre′cia·tive·ly** *adv.*

**ap·pre·hend** (ăp′rĭ-hĕnd′) *v.* **-hend·ed, -hend·ing, -hends.** [ME *apprehenden* < Lat. *apprehendere,* to seize : *ad-,* to + *prehendere,* to grasp.] —*vt.* **1.** To take into custody : ARREST. **2.** To grasp mentally : UNDERSTAND. **3.** To anticipate fearfully. —*vi.* To understand.

**ap·pre·hen·si·ble** (ăp′rĭ-hĕn′sə-bəl) *adj.* Capable of being apprehended or understood. **—ap′pre·hen′si·bly** *adv.*

**ap·pre·hen·sion** (ăp′rĭ-hĕn′shən) *n.* [LLat. *apprehensio,* understanding < *apprehendere,* to seize. —see APPREHEND.] **1.** A fearful or uneasy anticipation of the future : DREAD. **2.** The act of seizing or capturing : ARREST. **3.** Ability to understand.

**ap·pre·hen·sive** (ăp′rĭ-hĕn′sĭv) *adj.* **1.** Fearful or uneasy about the future : ANXIOUS. **2.** Capable of understanding : DISCERNING. **—ap′pre·hen′sive·ly** *adv.* **—ap′pre·hen′sive·ness** *n.*

**ap·pren·tice** (ə-prĕn′tĭs) *n.* [ME *apprentis* < OFr. < *aprendre,* to learn < Lat. *apprehendere,* to seize. —see APPREHEND.] **1.** One legally bound to work for another for a designated amount of time in return for instruction in a trade, art, or business. **2.** One who is learning a trade or occupation, esp. as a member of a labor union. **3.** One who lacks experience : BEGINNER. —*vt.* **-ticed, -tic·ing, -tic·es.** To place or take on as an apprentice. **—ap·pren′tice·ship′** *n.*

**ap·pressed** (ə-prĕst′) *adj.* [< Lat. *appressus,* p.part. of *apprimere,* to press down : *ad-,* to + *premere,* to press.] Lying flat or pressed closely against something, as leaves on a stem.

**ap·prise** (ə-prīz′) *vt.* **-prised, -pris·ing, -pris·es.** [Fr. *apprendre, appris-* < OFr. *aprendre,* to learn. —see APPRENTICE.] To inform.

**ap·prize** (ə-prīz′) *v. Chiefly Brit. var. of* APPRISE.

**ap·proach** (ə-prōch′) *v.* **-proached, -proach·ing, -proach·es.** [ME *approchen* < OFr. *aprochier* < LLat. *appropiare* : Lat. *ad-,* to + Lat. *propius,* nearer, comp. of *prope,* near.] —*vi.* To come close or closer, as in space or time. —*vt.* **1.** To move close or closer to. **2.** To approximate <enthusiam *approaching* fanaticism> **3.** To make overtures to <*approached* the star with a TV offer> **4.** To begin to deal with or work on. —*n.* **1.** The act of approaching. **2.** An approximation. **3.** A way or means of reaching something : ACCESS <an *approach* to the tower> **4.** The method used in dealing with or effecting something <an organized *approach*> **5.** An overture made by one person to another. **6.** The golf stroke following the drive from the tee with which a player tries to get the ball onto the putting green. **7.** Works, as trenches or bulwarks, for protecting troops who are besieging a fortified position.

**ap·proach·a·ble** (ə-prō′chə-bəl) *adj.* **1.** Capable of being approached. **2.** Not aloof : FRIENDLY. **—ap·proach′a·bil′i·ty** *n.*

**ap·pro·bate** (ăp′rə-bāt′) *vt.* **-bat·ed, -bat·ing, -bates.** [ME *approbaten* < Lat. *approbare,* to approve.] To approve, esp. officially. **—ap′pro·ba′tive, ap·pro′ba·to·ry** (ə-prō′bə-tôr′ē, -tōr′ē) *adj.*

**ap·pro·ba·tion** (ăp′rə-bā′shən) *n.* **1.** Commendation : praise. **2.** Official approval.

**ap·pro·pri·a·ble** (ə-prō′prē-ə-bəl) *adj.* Capable of being appropriated, as property.

**ap·pro·pri·ate** (ə-prō′prē-ĭt) *adj.* [ME *appropriat* < Med. Lat. *appropriatus,* p.part. of *appropriare,* to make one's own : Lat. *ad-,* to + Lat. *propius,* own.] Suitable : fitting. —*vt.* (-āt′) **-at·ed, -at·ing, -ates. 1.** To set apart for a specific use. **2.** To take possession of or make use of for oneself, often without permission or legal right. **—ap·pro′pri·ate·ly** *adv.* **—ap·pro′pri·ate·ness** *n.* **—ap·pro′pri·a′tive** (-ā′tĭv) *adj.* **—ap·pro′pri·a′tor** *n.*

**ap·pro·pri·a·tion** (ə-prō′prē-ā′shən) *n.* **1.** An act of appropriating. **2.** Something appropriated, esp. public funds set aside for a specific purpose. **3.** A legislative act authorizing the expenditure of a designated amount of public funds for a given purpose.

**ap·prov·al** (ə-pro͞o′vəl) *n.* **1.** An act of approving. **2.** Official approbation : SANCTION. **3.** Favorable regard : COMMENDATION. **—on approval.** For examination or trial by a potential customer without the obligation to buy.

**ap·prove** (ə-pro͞ov′) *v.* **-proved, -prov·ing, -proves.** [ME *approven* < OFr. *aprover* < Lat. *approbare* : *ad-,* to + *probare,* to test < *probus,* good.] —*vt.* **1.** To regard favorably <*approved* the new location> **2.** To confirm or agree to officially : SANCTION <The House *approved* the bill.> **3.** *Obs.* To prove or demonstrate. —*vi.* To feel, voice, or show approval <*approves* of fluoridation> **—ap·prov′a·ble** *adj.* **—ap·prov′ing·ly** *adv.*

**approved school** *n. Chiefly Brit.* A reform school.

**ap·prox·i·mate** (ə-prŏk′sə-mĭt) *adj.* [LLat. *approximatus,* p.part. of *approximare,* to approach : Lat. *ad-,* to + *proximare,* to come near < *proximus,* nearest.] **1.** Nearly exact, accurate, complete, or perfect.

2. Closely resembling : similar. **3.** Close together : NEAR. —*v.* (-māt′) **-mat·ed, -mat·ing, -mates.** —*vt.* **1.** To be nearly the same as. **2.** To bring near. —*vi.* To come near or close, as in nature, degree, or quality. **—ap·prox′i·mate·ly** *adv.*

**ap·prox·i·ma·tion** (ə-prŏk′sə-mā′shən) *n.* **1.** The act or result of approximating. **2.** *Math.* An inexact result adequate for a given purpose. **—ap·prox′i·ma′tive** *adj.* **—ap·prox′i·ma′tive·ly** *adv.*

**ap·pur·te·nance** (ə-pûr′tn-əns) *n.* [ME *appurtenaunce* < AN *apurtenance* < VLat. **appertinentia* < LLat. *appertinēre,* to appertain. ] **1.** Something added to a more important thing : ACCESSORY. **2. appurtenances.** Equipment, as clothing, tools, or instruments, used for a particular purpose or task : GEAR. **3.** *Law.* A right, privilege, or property considered incident to the principal property for purposes such as passage of title, conveyance, or inheritance. **—ap·pur′te·nant** *adj.*

**a·prax·i·a** (ā-prăk′sē-ə) *n.* [Gk., inaction : *a-,* without + *praxis,* action < *prassein,* to do.] Inability to perform coordinated movements due to lesions in the cerebral cortex. **—a·prac′tic** (ā-prăk′tĭk), **a·prax′ic** (ā-prăk′sĭk) *adj.*

**a·près-ski** (ä′prā-skē′, ăp′rā-) *n.* [Fr. : *après,* after + *ski,* skiing.] Social activities or events after skiing. **—a′près-ski′** *adj.*

**a·pri·cot** (ăp′rĭ-kŏt′, ā′prĭ-) *n.* [Alteration of earlier *abrecock* < Port. *albricoque* < Ar. *al-birqūq,* the apricot : *al,* the + LGk. *praikokion* < Lat. *praecoqquus,* ripe early (*prae-,* before + *coquere,* to ripen).] **1. a.** A tree native to western Asia and Africa, *Prunus armeniaca* widely cultivated for its edible fruit. **b.** The yellow-orange peachlike fruit of the apricot. **2.** A moderate, light, or strong orange to strong orange yellow.

**A·pril** (ā′prəl) *n.* [ME < Lat. *aprilis.*] The fourth month of the year according to the Gregorian calendar. —See table at CALENDAR.

**April fool** *n.* The victim of a trick played on April Fools' Day.

**April Fools' Day** *n.* Apr. 1, traditionally a day for playing practical jokes.

**a pri·o·ri** (ä′ prē-ôr′ē, -ōr′ē, ā′ prī-ôr′ī, -ōr′ī) *adj.* [Lat., from the former.] **1.** Proceeding from a known or assumed cause to a necessarily related effect : DEDUCTIVE. **2.** Based on a hypothesis or theory rather than on experiment or experience. **3.** Made before or without examination and not supported by factual study. **—a priori** *adv.* **—a′pri·or′i·ty** (-ôr′ĭ-tē, -ŏr′-) *n.*

**a·pron** (ā′prən) *n.* [ME < the phrase *a napron,* an apron < OFr. *naperon,* dim. of *nape,* tablecloth < Lat. *mappa,* napkin.] **1.** A garment, usu. tied or snapped in back, worn over the front of the body to protect clothing. **2.** Something resembling an apron, as a protective shield for a machine. **3.** The paved strip in front of and around airport hangars and terminal buildings. **4.** The part of a theatrical stage in front of the curtain. **5. a.** A platform at the entrance to a dock. **b.** A covering or structure along a shoreline for protection against erosion. **c.** A platform below a dam or in a sluiceway. **6.** A continuous conveyor belt. **7.** An area covered by sand and gravel deposited at the front of a glacial moraine. —*vt.* **a·proned, a·pron·ing, a·prons.** To cover, protect, or provide with an apron.

▲ word history: *An apron* was originally *a napron. Apron* is one of the numerous English words that have either lost or gained an initial *n* because of an incorrect division between the indefinite article and the noun. *Apron* is thus closely related to *napkin* and *napery.*

**apron string** *n.* **1.** The string of an apron. **2.** Complete control. —Used esp. in the phrase *tied to (someone's) apron strings.*

**ap·ro·pos** (ăp′rə-pō′) *adj.* [Fr. *à propos,* to the purpose.] Pertinent : appropriate. —*adv.* **1.** Pertinently : appropriately. **2.** By the way : INCIDENTALLY. —*prep.* Concerning : regarding <*apropos* your job application>

**apropos of** *prep.* [< APROPOS (adj.).] With reference to <a good story *apropos of* fishing>

**apse** (ăps) *n.* [LLat. *apsis.* —see APSIS.] **1.** A semicircular or polygonal, usu. domed projection of a building, esp. the altar or east end of a church. **2.** APSIS 2. **—ap′si·dal** (ăp′sĭ-dəl) *adj.*

**ap·sis** (ăp′sĭs) *n., pl.* **-si·des** (-sĭ-dēz′) [LLat. *apsis* < Lat., arch, vault < Gk. *hapsis* < *haptein,* to fasten.] **1.** APSE 1. **2.** *Astron.* The point in orbit of greatest or least distance of a celestial body from a center of attraction.

**apt** (ăpt) *adj.* [ME < Lat. *aptus,* p.part. of *apere,* to fasten.] **1.** Perfectly suitable : APPROPRIATE <an *apt* turn of phrase> **2.** Having probability : LIKELY <*apt* to be chilly tonight> **3.** Having a tendency : INCLINED <*apt* to talk too much> **4.** Quick to learn or understand.

*usage:* *Apt* is used instead of *likely* to indicate a natural tendency to error or undesirable behavior: *I am apt to lose my temper under severe stress* but *I am likely* (not *apt*) *to remain calm under severe stress.* **—apt′ly** *adv.* **—apt′ness** *n.*

**APT** (ā′pē-tē) *n.* [A(UTOMATICALLY) + P(ROGRAMMED) + T(OOL).] A computer language designed for programming numerically controlled machine tools.

**ap·ter·al** (ăp′tər-əl) *adj.* [< Gk. *apteros,* wingless : *a-,* without + *pteron,* wing.] Having no architectural columns along the sides.

**ap·ter·ous** (ăp′tər-əs) *adj.* **1.** *Zool.* Having no wings. **2.** *Bot.* Having no winglike parts or extensions.

**ap·ter·yx** (ăp′tə-rĭks′) *n.* [NLat. : A-[1] + Gk. *pterux,* wing.] The kiwi.

**ap·ti·tude** (ăp′tĭ-to͞od′, -tyo͞od′) *n.* [ME, tendency < Med. Lat. *aptitudo,* aptitude < *aptus,* apt.] **1.** A natural or acquired talent or ability : INCLINATION. **2.** Quickness in learning and understanding : INTELLIGENCE. **3.** The quality or condition of being appropriate.

**aptitude test** *n.* A standardized test for measuring the ability of an individual to acquire knowledge or develop skills.

**A·pus** (ā′pəs) *n.* [Lat. *apus,* a kind of swallow < Gk. *apous,* sand martin < *apous,* without feet. —see APODAL.] A constellation in the Southern Hemisphere.

**aq·ua** (ăk′wə, ä′kwə) *n., pl.* **aq·uae** (ăk′wē, ä′kwī′) or **aq·uas.** [Lat.] **1.** Water. **2.** An aqueous solution. **3.** A light bluish green to light greenish blue. **—aq′ua** *adj.*

**aq·ua·cade** (ăk′wə-kād′, ä′kwə-) *n.* [AQUA + (CAVAL)CADE.] An entertainment spectacle of swimmers and divers, often performing in unison to music.

**aq·ua for·tis** *also* **aq·ua·for·tis** (ăk′wə-fôr′tĭs, ä′kwə-) *n.* [NLat., strong water.] Nitric acid.

**Aqua Lung.** A trademark for an underwater breathing apparatus.

**aq·ua·ma·rine** (ăk′wə-mə-rēn′, ä′kwə-) *n.* [Lat. *aqua marina,* sea water.] **1.** A transparent blue-green variety of beryl, used as a gemstone. **2.** A pale blue to light greenish blue.

**aq·ua·naut** (ăk′wə-nôt′, ä′kwə-) *n.* [AQUA + Gk. *nautēs* sailor < *naus,* ship.] One trained and equipped to live in underwater installations and conduct, aid in, or be a subject of scientific research.

**aq·ua·plane** (ăk′wə-plān′, ä′kwə-) *n.* [AQUA + PLANE[1].] A board pulled by a motorboat and ridden by a person standing up on it. —*vi.* **-planed, -plan·ing, -planes.** To ride on an aquaplane.

**aqua re·gi·a** (rē′jē-ə, rē′jə) *n.* [Lat., royal water.] A corrosive, fuming, volatile mixture of hydrochloric and nitric acids, used for testing metals and dissolving platinum and gold.

**aq·ua·relle** (ăk′wə-rĕl′, ä′kwə-) *n.* [Fr. < obs. Ital. *acquarella,* water color, dim. of *acqua,* water < Lat. *aqua.*] A drawing in transparent water colors. **—aq′ua·rel′list** *n.*

**a·quar·i·a** (ə-kwâr′ē-ə) *n. var. pl. of* AQUARIUM.

**a·quar·ist** (ə-kwâr′ĭst) *n.* One who maintains an aquarium.

**a·quar·i·um** (ə-kwâr′ē-əm) *n., pl.* **-i·ums** or **-i·a** (-ē-ə) [< Lat. *aquarius,* of water < *aqua,* water.] **1.** A water-filled container in which fish or other aquatic animals and often plants are kept. **2.** A place for the public display of aquatic animals and plants.

**A·quar·i·us** (ə-kwâr′ē-əs) *n.* [ME < Lat., water carrier < *aqua,* water.] **1.** A constellation in the Southern Hemisphere. **2. a.** The 11th sign of the zodiac. **b.** One born under this sign.

**a·quat·ic** (ə-kwŏt′ĭk, ə-kwăt′-) *adj.* [OFr. *aquatique* < Lat. *aquaticus* < *aqua,* water.] **1.** Of or in water. **2.** Living or growing in or on the water. **3.** Taking place in or on the water, as sports. —*n.* **1.** An aquatic organism. **2. aquatics.** Sports performed in or on the water.

**aq·ua·tint** (ăk′wə-tĭnt′, ä′kwə-) *n.* [Fr. *aquatinte* < Ital. *acqua tinta,* dyed water : *acqua,* water (< Lat. *aqua*) + *tinta,* dyed < Lat. *tincta,* p.part. of *tingere,* to dye.] **1.** A process of etching capable of producing several tones by varying the etching time of different areas of a copper plate so that the resulting print resembles the flat tints of an ink or wash drawing. **2.** An etching made by aquatint. —*vt.* **-tint·ed, -tint·ing, -tints.** To etch in aquatint.

**a·qua·vit** (ä′kwə-vēt′) *n.* [Swed., Dan., and Norw. *akvavit* < Med. Lat. *aqua vitae,* water of life.] A clear, strong Scandinavian liquor distilled from potato or grain mash and flavored with caraway seed.

**aqua vi·tae** (vī′tē) *n.* [ME *aquavite* < Med. Lat. *aqua vitae,* water of life.] **1.** Alcohol. **2.** A strong liquor, as brandy.

**aq·ue·duct** (ăk′wĭ-dŭkt′) *n.* [Lat. *aquaeductus* : *aqua,* water + *ductus,* p.part. of *ducere,* to lead.] **1.** A pipe or channel for transporting water from a remote source, usu. by gravity. **2.** A bridgelike structure supporting a conduit or canal passing over a river or low ground. **3.** *Anat.* A fluid passage or channel.

**a·que·ous** (ā′kwē-əs, ăk′wē-) *adj.* [Med. Lat. *aqueus* < Lat. *aqua,* water.] **1.** Relating to, like, containing, or dissolved in water : WATERY. **2.** *Geol.* Formed from matter deposited by water, as certain sedimentary rocks.

**aqueous humor** *n.* A clear, lymphlike fluid in the chamber of the eye between the cornea and the lens.

**aqui-** *pref.* [Lat. < *aqua,* water.] Water <*aqui*culture>

**aq·ui·cul·ture** (ăk′wĭ-kŭl′chər, ä′kwĭ-) *n.* Hydroponics. **—aq′ui·cul′tur·al** *adj.*

**aq·ui·fer** (ăk′wə-fər, ä′kwə-) *n.* A water-bearing rock, rock formation, or group of rock formations. **—a·quif′er·ous** (ə-kwĭf′ər-əs) *adj.*

**Aq·ui·la** (ăk′wə-lə) *n.* [Lat. *aquila,* eagle.] A constellation in the Northern Hemisphere.

**aq·ui·le·gi·a** (ăk′wə-lē′jē-ə, -lē′jə) *n.* [NLat. *Aquilegia,* genus name < Med. Lat. *aquilegia,* columbine.] The columbine.

**aq·ui·line** (ăk′wə-līn′, -lĭn) *adj.* [Lat. *aquilinus* < *aquila,* eagle.] **1.** Of or like an eagle. **2.** Curved or hooked like an eagle's beak <had an elegant, *aquiline* nose>

**a·quiv·er** (ə-kwĭv′ər) *adj.* Marked by quivering.

**ar** (är) *n. var. of* ARE[2].

**Ar** *symbol for* ARGON.

**-ar** *suff.* [ME < OFr. *-er* < Lat. *-aris,* alteration of *-alis,* -al.] Of, pertaining to, or resembling <polar>

**A·ra** (âr′ə) *n.* [Lat. *ara,* altar.] A constellation in the Southern Hemisphere.

**Ar·ab** (ăr′əb) *n.* [Fr. *Arabe* < Lat. *Arabs* < Gk. *Araps* < Ar. *'Arab.*] **1.** A native or resident of Arabia. **2.** Any of a Semitic people orig. from Arabia but later widely scattered throughout the Near East, North Africa, and the Arabian Peninsula. **3.** Any of a nomadic people living in North African and Near Eastern desert regions. **4.** Any of a breed of swift, intelligent, graceful horses native to Arabia. **5.** A street Arab : WAIF. *—adj.* Arabian.

**ar·a·besque** (ăr′ə-bĕsk′) *n.* [Fr. < Ital. *arabesco,* in Arabian fashion < *Arab,* an Arab < Lat. *Arabs.*] **1.** An intricate and ornate design of intertwined floral, foliate, and geometric figures. **2.** A position in which a ballet dancer stands on one leg, the other leg extended backward with straight knee, and with the arms disposed in a conventional position. *—adj.* Formed as or in the style of an arabesque.

**A·ra·bi·an** (ə-rā′bē-ən) *adj.* Of or relating to Arabia or the Arabs. *—n.* **1.** A native or resident of Arabia. **2.** ARAB 4.

**Arabian camel** *n.* The dromedary.

**Ar·a·bic** (ăr′ə-bĭk) *adj.* Of or relating to Arabia, the Arabs, their language, or their culture. *—n.* A Semitic language consisting of numerous dialects that is the principal language of Arabia, Jordan, Syria, Iraq, Lebanon, Egypt, and parts of northern Africa.

**Arabic numeral** *n.* One of the numerical symbols 1, 2, 3, 4, 5, 6, 7, 8, 9, and 0.

**Ar·ab·ist** (ăr′ə-bĭst) *n.* One who specializes in the Arabic language or culture.

**ar·a·ble** (ăr′ə-bəl) *adj.* [ME < OFr. < Lat. *arabilis* < *arare,* to plow.] Suitable for cultivation <*arable* soil> *—n.* Arable land.

**A·rach·ne** (ə-răk′nē) *n.* [Lat. < Gk. *Arakhnē* < *arakhnē,* spider.] *Gk. Myth.* A maiden who was transformed into a spider by Athena for challenging her to a weaving contest.

**a·rach·nid** (ə-răk′nĭd) *n.* [NLat. *Arachnida,* class name < Gk. *arakhnē,* spider.] Any of various arthropods of the class Arachnida, as a spider, scorpion, mite, or tick, with four pairs of legs. **—a·rach′ni·dan** (-nĭ-dən) *adj. & n.*

**a·rach·noid** (ə-răk′noid′) *adj.* [NLat. *arachnoides* < Gk. *arakhnoeidēs,* cobweblike : *arakhnē,* spider + *-eidēs,* -oid.] **1.** Resembling a spider's web. **2.** Of or pertaining to the arachnids. **3.** Covered with or made up of thin, soft, entangled hairs. *—n.* **1.** An arachnid. **2.** A delicate membrane of the spinal cord and brain, lying between the pia mater and dura mater.

**Ar·a·go·nese** (ăr′ə-gə-nēz′, -nēs′) *adj.* Of or relating to Aragon, its inhabitants, their language, or their culture. *—n., pl.* **Aragonese.** A native or inhabitant of Aragon.

**a·rag·o·nite** (ə-răg′ə-nīt′, ăr′ə-gə-) *n.* [After *Aragon,* a region of Spain.] An orthorhombic mineral form of crystalline calcium carbonate, dimorphous with calcite.

**Ar·a·mae·an** (ăr′ə-mē′ən) *n. var. of* ARAMEAN.

**Ar·a·ma·ic** (ăr′ə-mā′ĭk) *n.* The Semitic language orig. of the ancient Arameans but widely used by non-Aramean peoples throughout southwest Asia from the 7th cent. B.C. to the 7th cent. A.D. **—Ar′a·ma′ic** *adj.*

**Ar·a·me·an** *also* **Ar·a·mae·an** (ăr′ə-mē′ən) *adj.* Of or relating to Aram, its inhabitants, their language, or their culture. *—n.* **1.** A native or resident of Aram. **2.** Aramaic.

**A·rap·a·ho** *also* **A·rap·a·hoe** (ə-răp′ə-hō′) *n., pl.* **Arapaho** *or* **-hos** *also* **Arapahoe** *or* **-hoes.** [Crow *aa-raxpé-ahu,* tattoo.] **1. a.** A tribe of Indians once living in the area of the Platte and Arkansas rivers, now settled in Oklahoma and Wyoming. **b.** A member of this tribe. **2.** The Algonquian language of the Arapaho.

**ar·a·pai·ma** (ăr′ə-pī′mə) *n.* [Sp. (South America), prob. < Tupi.] A very large South American freshwater food fish, *Arapaima gigas,* occas. reaching a length of 15 feet.

**ar·a·ro·ba** (ăr′ə-rō′bə) *n.* [Port., prob. of Tupian orig.] **1.** A Brazilian tree, *Andira araroba,* with yellowish wood from which a medicinal powder is obtained. **2.** The powder of the araroba.

**Ar·au·ca·ni·an** (ăr′ô-kā′nē-ən) *also* **A·rau·can** (ə-rô′kən) *n.* [Sp. *araucano,* after *Arauco,* a province in Chile.] **1.** A member of a group of Indian peoples of south-central Chile and the western pampas of Argentina. **2.** The language of the Araucanians, constituting an independent language family. **—Ar′au·ca′ni·an** *adj.*

**ar·au·car·i·a** (ăr′ô-kăr′ē-ə) *n.* [< Sp. *Araucaria,* (tree) of Arauco, a province in Chile.] Any of several evergreen trees of the genus *Araucaria,* as the monkey puzzle and Norfolk Island pine.

**Ar·a·wak** (ăr′ə-wăk′) *n., pl.* **Arawak** *or* **-waks. 1. a.** An Indian people now living chiefly in certain regions of the Guianas. **b.** A member of this people. **2.** The Arawakan language of the Arawak.

**Ar·a·wa·kan** (ăr′ə-wä′kən) *n., pl.* **Arawakan** *or* **-kans. 1.** A member of a group of Indian peoples living in a wide area of South America. **2.** A language family consisting of the languages spoken by the Arawakan peoples. **—Ar′a·wa′kan** *adj.*

**ar·ba·lest** *also* **ar·be·lest** (är′bə-lĭst) *n.* [ME *arblast* < OFr. *arbaleste* < LLat. *arcuballista* : Lat. *arcus,* bow + Lat. *ballista,* ballista. —see BALLISTA.] A medieval missile launcher designed on the crossbow principle. **—ar′ba·lest′er** (-lĕs′tər) *n.*

**arbalest**

**ar·bi·ter** (är′bĭ-tər) *n.* [ME *arbitour* < Lat. *arbiter.*] **1.** ARBITRATOR 1. **2.** One empowered to judge or ordain at will <an *arbiter* of good taste>

**ar·bi·tra·ble** (är′bĭ-trə-bəl) *adj.* **1.** Subject to judgment or decision by arbitration. **2.** Referrable to an arbitrator.

**ar·bi·trage** (är′bĭ-träzh′) *n.* [OFr., arbitration < *arbitrer,* to arbitrate < Lat. *arbitrari,* to give judgment. —see ARBITRATE.] Purchase of securities on one market for immediate resale on another so as to profit from a price discrepancy.

**ar·bit·ra·ment** (är-bĭt′rə-mənt) *n.* [ME *arbitrement* < OFr. < *arbitrer,* to judge < Lat. *arbitrari,* to give judgment. —see ARBITRATE.] **1.** An act of arbitrating. **2.** The judgment or award determined by an arbitrator or arbiter.

**ar·bi·trar·y** (är′bĭ-trĕr′ē) *adj.* [ME *arbitrarie* < Lat. *arbitrarius* < *arbiter,* arbiter.] **1.** Determined by impulse or whim. **2.** Based on or subject to individual judgment or discretion. **3.** Established by a court or judge rather than by a specific law or statute. **4.** Not limited by law : DESPOTIC. **—ar′bi·trar′i·ly** (-trâr′ə-lē) *adv.* **—ar′bi·trar′i·ness** *n.*

**ar·bi·trate** (är′bĭ-trāt′) *v.* **-trat·ed, -trat·ing, -trates.** [Lat. *arbitrari,* to give judgment < *arbiter,* arbiter.] *—vt.* **1.** To judge or decide as or in the manner of an arbitrator. **2.** To submit to settlement or judgment by arbitration. *—vi.* **1.** To serve as an arbitrator or arbiter. **2.** To refer or submit a dispute to arbitration.

**ar·bi·tra·tion** (är′bĭ-trā′shən) *n.* The process by which the parties to a dispute submit their differences to the judgment of an impartial person or group selected by mutual consent or statutory provision.

**ar·bi·tra·tor** (är′bĭ-trā′tər) *n.* **1.** One chosen to settle the issue between parties engaged in a dispute or controversy. **2.** ARBITER 2.

**ar·bor**[1] (är′bər) *n.* [ME *herber* < OFr. *herbier, erbier,* garden < *erbe,* herb. —see HERB.] **1.** A shady garden shelter or bower, often made of rustic work or latticework on which climbing plants, as vines and roses, are grown. **2.** *Obs.* An orchard or garden.

**ar·bor**[2] (är′bər) *n.* [Lat., tree.] **1.** An axis or shaft supporting a rotating part on a lathe. **2.** A bar for supporting cutting tools. **3.** A spindle of a wheel, as in watches and clocks. **4.** *Archaic.* A tree.

**Arbor Day** *n.* A day, usu. in the spring, observed by many U.S. states for the community planting of trees.

**ar·bo·re·al** (är-bôr′ē-əl, -bōr′-) *adj.* **1.** Relating to or like a tree. **2.** Inhabiting trees <*arboreal* primates> **—ar·bo′re·al·ly** *adv.*

**ar·bo·re·ous** (är-bôr′ē-əs, -bōr′-) *adj.* **1.** Having many trees. **2.** Resembling or typical of a tree : TREELIKE.

**ar·bo·res·cent** (är′bə-rĕs′ənt) *adj.* [Lat. *arborescens, arborescent-,* pr.part. of *arborescere,* to grow to be a tree < *arbor,* tree.] : ARBOREOUS 2. **—ar′bo·res′cence** *n.*

**ar·bo·re·tum** (är′bə-rē′təm) *n., pl.* **-tums** *or* **-ta** (-tə) [Lat., a place grown with trees < *arbor,* tree.] A place for the scientific study and public display of various species of trees and shrubs.

**ar·bo·ri·cul·ture** (är′bə-rĭ-kŭl′chər, är-bôr′ĭ-, -bōr′-) *n.* Cultivation of trees for study or for production of timber.

**ar·bo·ri·za·tion** (är′bər-ĭ-zā′shən) *n.* **1.** A treelike shape or arrangement, as that of certain minerals or fossils. **2.** Formation of an arborization.

**ar·bo·rize** (är′bə-rīz′) *vi.* **-rized, -riz·ing, -riz·es.** To have or form many branches.

**ar·bor·vi·tae** *also* **ar·bor vi·tae** (är′bər-vī′tē) *n.* [NLat. *arbor vitae,* tree of life.] **1. a.** An evergreen shrub or tree of the genus *Thuja,* bearing tiny scalelike leaves and egg-shaped cones. **b.** A similar tree of the genus *Thujopsis.* **2.** *Anat.* The white matter of the cerebellum, in cross section having the appearance of a tree.

**ar·bour** (är′bər) *n. Chiefly Brit. var. of* ARBOR[1].

**ar·bo·vi·rus** (är′bə-vī′rəs) *n.* [AR(THROPOD) + BO(RNE) + VIRUS.] Any of various arthropod-borne viruses that include the causative agent of encephalitis. **—ar′bo·vi′ral** *adj.* **—ar′bo·vi·rol′o·gy** (är′bō-vī-rŏl′ə-jē) *n.*

**ar·bu·tus** (är-byōō′təs) *n.* [NLat. *Arbutus,* genus name < Lat. *arbutus,* arbutus.] **1.** A broad-leaved evergreen tree of the genus *Arbutus,* with white or pinkish flower clusters. **2.** The trailing arbutus.

**arc** (ärk) *n.* [ME *ark* < OFr. *arc* < Lat. *arcus.*] **1.** Something shaped like a bow, curve, or arch. **2.** *Math.* A segment of a curve. **3.** *Elect.* A luminous discharge of electric current crossing a gap between two electrodes. —*adj. Math.* Denoting an inverse trigonometric function <the *arc* sine of a quantity> —*vi.* **arced** (ärkt) *or* **arcked, arc·ing** (är'kĭng) *or* **arck·ing, arcs.** To form an arc.

**ar·cade** (är-kād') *n.* [Fr., archway < Ital. *arcata* < *arco,* arch < Lat. *arcus.*] **1. a.** A series of arches supported by columns, piers, or pillars. **b.** An arched, roofed building or part of a building. **2.** A roofed passageway or lane, esp. one with shops on either side. —*vt.* **-cad·ed, -cad·ing, -cades.** To provide with or make into an arcade.

**Ar·ca·di·a** (är-kā'dē-ə) *n.* **1.** A region in ancient Greece held to be an ideal of rural simplicity and peacefulness. **2.** *often* **arcadia.** A region offering rural simplicity and tranquillity.

**Ar·ca·di·an** (är-kā'dē-ən) *adj.* **1.** Of, relating to, or typical of Arcadia. **2.** *often* **arcadian.** Rustic, tranquil, and simple : PASTORAL. —*n.* **1.** A native of Arcadia. **2.** *often* **arcadian.** One who leads, advocates, or prefers a simple, rural life. **3.** The dialect of ancient Greek used in Arcadia.

**Ar·ca·dy** (är'kə-dē) *n.* Arcadia.

**ar·ca·na** (är-kā'nə) *n. var. pl. of* ARCANUM.

**ar·cane** (är-kān') *adj.* [Lat. *arcanus,* secret < *arcere,* to shut up < *arca,* chest.] Known or understood only by a few : esoteric or secret.

**ar·ca·num** (är-kā'nəm) *n., pl.* **-na** (-nə) *or* **-nums.** [Lat. < *arcanus,* secret. —see ARCANE.] **1.** A profound secret : MYSTERY. **2.** The reputed secret of nature sought by alchemists. **3.** An elixir.

**arc-bou·tant** (är'bo͞o-täN') *n., pl.* **arcs-bou·tants** (är'bo͞o-täN') [Fr.] A flying buttress.

**arch¹** (ärch) *n.* [ME < OFr. *arche* < Lat. *arcus.*] **1.** A structural device, esp. of masonry, forming the curved, pointed, or flat upper edge of an opening or a support, as in a bridge or doorway. **2.** A structure, as a monument, that is similar to an arch. **3.** Something curved like an arch. **4.** *Anat.* An arch-shaped structure, esp. either of two such bony structures of the foot. —*v.* **arched, arch·ing, arch·es.** —*vt.* **1.** To supply with an arch. **2.** To cause to form an arch or similar curve. **3.** To span. —*vi.* To form an arch or archlike curve.

**arch²** (ärch) *adj.* [< ARCH-.] **1.** Principal : chief <an *arch*-criminal> **2.** Mischievous <an *arch* remark> **—arch'ly** *adv.* **—arch'ness** *n.*

**arch-** *pref.* [ME. —see ARCHI-.] **1.** *var. of* ARCHI-. **2.** The extreme or most typical example of its kind <*arch*conservative>

**-arch** *suff.* [ME *-arche* < OFr. < LLat. *-archa* < Lat. *-arches* < Gk. *-arkhēs* < *arkhos,* ruler < *arkhein,* to rule.] Leader <matri*arch*>

**Ar·chae·an** (är-kē'ən) *adj. var. of* ARCHEAN.

**archaeo-** *or* **archeo-** *pref.* [NLat. < Gk. *arkhaio-* < *arkhaios,* ancient < *arkhē,* beginning.] Ancient : earlier <*archaeo*pteryx>

**ar·chae·ol·o·gy** *or* **ar·che·ol·o·gy** (är'kē-ŏl'ə-jē) *n.* [Fr. *archéologie* < Gk. *arkhaiologia,* antiquarian lore : *arkhaios,* old + *-logia,* -logy.] Systematic recovery and examination of material evidence, as graves, buildings, tools, and pottery, remaining from past human life and culture. **—ar'chae·o·log'i·cal** (-ə-lŏj'ĭ-kəl), **ar'chae·o·log'ic** *adj.* **—ar'chae·ol'o·gist** *n.*

**ar·chae·op·ter·yx** (är'kē-ŏp'tər-ĭks) *n.* [NLat. : ARCHAEO- + Gk. *pterux,* bird < *pteron,* wing.] An extinct primitive bird of the genus *Archaeopteryx* of the Jurassic period, with lizardlike characteristics and representing a transitional form between reptiles and birds.

**Ar·chae·o·zo·ic** (är'kē-ə-zō'ĭk) *adj. & n. var. of* ARCHEOZOIC.

**ar·cha·ic** (är-kā'ĭk) *also* **ar·cha·i·cal** (-ĭ-kəl) *adj.* [Fr. *archaïque* < Gk. *arkhaikos,* old-fashioned < *arkhaios,* old < *arkhē,* beginning < *arkhein,* to begin.] **1.** Belonging to a much earlier time : ANCIENT. **2.** Not current or applicable : ANTIQUATED <*archaic* ideas about child-raising> **3.** Designating or typical of words and language that were once common but are now used chiefly to suggest an earlier style or period. **—ar·cha'i·cal·ly** *adv.*

**archaic smile** *n.* A representation of the human mouth with slightly upturned corners, featured in early Greek sculpture produced before the 5th cent. B.C.

**ar·cha·ism** (är'kē-ĭz'əm, -kā-) *n.* [NLat. *archaeismus* < Gk. *arkhaismos* < *arkhaios,* ancient. —see ARCHAIC.] **1.** An archaic word, phrase, expression or idiom. **2.** An archaic quality, style, or usage. **—ar'cha·ist** *n.* **—ar'cha·is'tic** (-ĭs'tĭk) *adj.*

**ar·cha·ize** (är'kē-īz', -kā-) *v.* **-ized, -iz·ing, -iz·es.** —*vt.* To make archaic. —*vi.* To express by archaisms. **—ar'cha·iz'er** *n.*

**arch·an·gel** (ärk'ān'jəl) *n.* [ME < OFr. < LLat. *archangelus* < LGk. *arkhangelos* : Gk. *arkh-,* archi- + *angelos,* angel.] **1.** A celestial being next in rank above an angel. **2. archangels.** The eighth of the nine orders of angels. **—arch'an·gel'ic** (-ăn-jĕl'ĭk) *adj.*

**arch·bish·op** (ärch-bĭsh'əp) *n.* [ME *arche-bishop* < OE *arce-bisceop* < LLat. *archiepiscopus* < LGk. *arkhiepiskopos* : Gk. *arkhi-,* archi- + *episkopos,* bishop. —see BISHOP.] A bishop of the highest rank, having jurisdiction over an archdiocese or province. **—arch·bish'op·ric** *n.*

**arch·dea·con** (ärch-dē'kən) *n.* [ME *arche-deken* < OE *arcediakon* < LLat. *archidiăconus* < LGk. *arkhidiakonos* : Gk. *arkhi,* archi- + *diakonos,* deacon. —see DEACON.] A church official in charge of temporal and ceremonial affairs in a diocese, with powers delegated from a bishop. **—arch·dea'con·ate** (-kə-nĭt) *n.* **—arch·dea'con·ship'** *n.*

**arch·dea·con·ry** (ärch-dē'kən-rē) *n., pl.* **-ries. 1.** The rank, office, or jurisdiction of an archdeacon. **2.** The territory or residence of an archdeacon.

**arch·di·o·cese** (ärch-dī'ə-sĭs, -sēs', -sēz') *n.* A diocese under an archbishop's jurisdiction. **—arch'di·oc'e·san** (-ŏs'ĭ-sən) *adj.*

**arch·du·cal** (ärch-do͞o'kəl, -dyo͞o'-) *adj.* Of or relating to an archduke or an archduchy.

**arch·duch·ess** (ärch-dŭch'ĭs) *n.* **1.** The wife or widow of an archduke. **2.** A woman having a rank equivalent to that of an archduke.

**arch·duch·y** (ärch-dŭch'ē) *n., pl.* **-ies.** The territory over which an archduke or archduchess has authority.

**arch·duke** (ärch-do͞ok', -dyo͞ok') *n.* A nobleman having a rank equivalent to that of a sovereign prince, esp. in imperial Austria.

**Ar·che·an** *also* **Ar·chae·an** (är-kē'ən) *adj.* [< Gk. *arkhaios,* ancient. —see ARCHAIC.] Of or relating to the oldest, predominantly igneous, rocks of the Precambrian era.

**ar·che·go·ni·um** (är'kĭ-gō'nē-əm) *n., pl.* **-ni·a** (-nē-ə) [NLat. < Gk. *arkhegonos,* original : *arkhe-,* archi- + *gonos,* race.] The multicellular female sex organ of mosses and related plants, producing a single egg. **—ar'che·go'ni·al** *adj.* **—ar'che·go'ni·ate** (-ĭt) *adj.*

**arch·en·e·my** (ärch-ĕn'ə-mē) *n.* **1.** A chief or principal enemy. **2.** *often* **Archenemy.** DEVIL 1.

**ar·chen·ter·on** (är-kĕn'tə-rŏn', -rən) *n.* The digestive cavity of the gastrula of an embryo. **—ar'chen·ter'ic** (är'kĕn-tĕr'ĭk) *adj.*

**archeo-** *pref. var. of* ARCHAEO-.

**ar·che·ol·o·gy** (är'kē-ŏl'ə-jē) *n. var. of* ARCHAEOLOGY.

**Ar·che·o·zo·ic** *also* **Ar·chae·o·zo·ic** (är'kē-ə-zō'ĭk) —*adj.* Of, belonging to, or designating the earlier of two gen. arbitrary divisions of the Precambrian era. —*n.* The Archeozoic era.

**arch·er** (är'chər) *n.* [ME < OFr. *archier* < Med. Lat. *arcarius* < Lat. *arcus,* bow.] **1.** One who shoots with a bow and arrow. **2. Archer.** Sagittarius.

**arch·er·fish** (är'chər-fĭsh') *n., pl.* **archerfish** *or* **-fish·es.** Any of several small Indo-Australian fishes of the family Toxotidae, having the ability to capture insects by squirting water at them.

**arch·er·y** (är'chə-rē) *n.* **1.** The art, skill, or sport of shooting with a bow and arrow. **2.** An archer's equipment. **3.** A group of archers.

**ar·che·spore** (är'kĭ-spôr', -spōr') *also* **ar·che·spo·ri·um** (-spôr'ē-əm, -spōr'-) *n., pl.* **-spores** *also* **-spo·ri·a** (-spôr'ē-ə, -spōr'-) [NLat. *archesporium* : *arche(gonium),* archegonium + *spora,* spore.] *Bot.* A spore-bearing cell or cell mass. **—ar'che·spo'ri·al** *adj.*

**ar·che·type** (är'kĭ-tīp') *n.* [Lat. *archetypum* < Gk. *arkhetupon* : *arkhe-,* archi- + *tupos,* model.] PROTOTYPE 1. **—ar'che·typ'al** (-tī'pəl), **ar'che·typ'ic** (-tĭp'ĭk), **ar'che·typ'i·cal** *adj.* **—ar'che·typ'i·cal·ly** *adv.*

**arch·fiend** (ärch-fēnd') *n.* **1.** A chief or principal fiend. **2. the Archfiend.** DEVIL 1.

**archi-** *or* **arch-** *pref.* [ME *arche-,* arch- < OE *ærce-* and OFr. *arche-,* both < Lat. *archi-* < Gk. *arkhi-* < *arkhein,* to begin, rule.] **1.** Most important : HIGHEST <*arch*duke> **2.** Earlier : primitive <*arch*enteron>

**ar·chi·di·ac·o·nal** (är'kĭ-dī-ăk'ə-nəl) *adj.* [< LLat. *archidiaconus,* archdeacon.] Of or relating to an archdeacon or archdiaconate.

**ar·chi·di·ac·o·nate** (är'kĭ-dī-ăk'ə-nĭt) *n.* [Med. Lat. *archidiaconatus* < LLat. *archidiăconus,* archdeacon.] The office or rank of an archdeacon.

**ar·chi·e·pis·co·pal** (är'kē-ĭ-pĭs'kə-pəl) *adj.* [Med. Lat. *archiepiscopālis* < LLat. *archiepiscopus,* archbishop.] Of or relating to an archbishop or an archbishopric. **—ar'chi·e·pis'co·pal'i·ty** (-păl'ĭ-tē) *n.* **—ar'chi·e·pis'co·pal·ly** *adv.* **—ar'chi·e·pis'co·pate** *n.*

**archiepiscopal cross** *n.* A processional crucifix mounted on a tall shaft and borne before an archbishop.

**ar·chil** (är'kĭl, -chĭl) *n. var. of* ORCHIL.

**ar·chi·mage** (är'kə-māj') *n.* [LGk. *arkhimagos* : Gk. *arkhe-,* archi- + Gk. *Magos,* wizard < OPers. *Maguus,* one of a tribe of Medes.] A chief wizard or great magician.

**ar·chi·man·drite** (är'kə-măn'drīt') *n.* [LLat. *archimandrita* < LGk. *arkhimandritēs* : Gk. *arkhi-,* archi- + *mandra,* monastery < Gk., cattle pen.] **1.** A cleric ranking below a bishop. **2.** The head of a Greek Orthodox monastery or group of monasteries.

**Ar·chi·me·de·an** (är'kə-mē'dē-ən, -mĭ-dē'-) *adj.* Of or relating to the Greek mathematician Archimedes or his inventions.

**Archimedean screw** *n.* An ancient device for raising water, having either a spiral tube around an inclined axis or an inclined tube containing a tight-fitting, broad-threaded screw.

**Ar·chi·me·des' screw** (är'kə-mē'dēz') *n.* An Archimedean screw.

**ar·chine** *also* **ar·shin** (är-shēn') *n.* [R. *arshin,* of Turkic orig.] A Russian unit of linear measure equivalent to 28 inches or approx. 71 centimeters.

**ar·chi·pel·a·go** (är'kə-pĕl'ə-gō') *n., pl.* **-goes** *or* **-gos.** [Ital. *Arcipelago,* the Aegean Sea : < Gk. *arkhi-,* archi- + Gk. *pelagos,* sea.] **1.** A large group of islands. **2.** A sea containing a large group of islands. **—ar'chi·pe·lag'ic** (-pə-lăj'ĭk) *adj.*

**ar·chi·tect** (är'kĭ-tĕkt') *n.* [OFr. *architecte* < Lat. *architectus* < Gk.

*arkhitektōn,* master builder : *arkhi-,* archi- + *tektōn,* builder.] **1.** One who designs and oversees the construction of large structures, as buildings, bridges, or ships. **2.** A planner or deviser <the chief *architect* of the peace treaty>

**ar·chi·tec·ton·ic** (är′kĭ-tĕk-tŏn′ĭk) *also* **ar·chi·tec·ton·i·cal** (-ĭ-kəl) *adj.* [Lat. *architectonicus,* architectural < Gk. *arkhitektonikos* < *arkhitektōn,* architect.] **1.** Of or relating to architecture or design. **2.** Having qualities typical of architectural structure and design. **3.** *Philos.* Relating to the scientific systematization of knowledge. **—ar′chi·tec·ton′i·cal·ly** *adv.*

**ar·chi·tec·ton·ics** (är′kĭ-tĕk-tŏn′ĭks) *n. (sing. in number).* **1.** The science of architecture. **2.** Structural design, as in a musical work. **3.** *Philos.* The scientific systematization of knowledge.

**ar·chi·tec·ture** (är′kĭ-tĕk′chər) *n.* [OFr. < Lat. *architectura* < *architectus,* architect.] **1.** The art and science of designing and erecting buildings. **2.** Architectural structures as a whole. **3.** A style and method of design and construction <Roman *architecture*> **4.** Design or system perceived by humans <the *architecture* of the solar system> **—ar′chi·tec′tur·al** *adj.* **—ar′chi·tec′tur·al·ly** *adv.*

**ar·chi·trave** (är′kĭ-trāv′) *n.* [OFr. < OItal. : *archi,* archi- (< Lat.) + *trave,* beam < Lat. *trabs.*] **1.** The lowermost part of an entablature, resting on top of a column in classical architecture. **2.** The molding around a door or window.

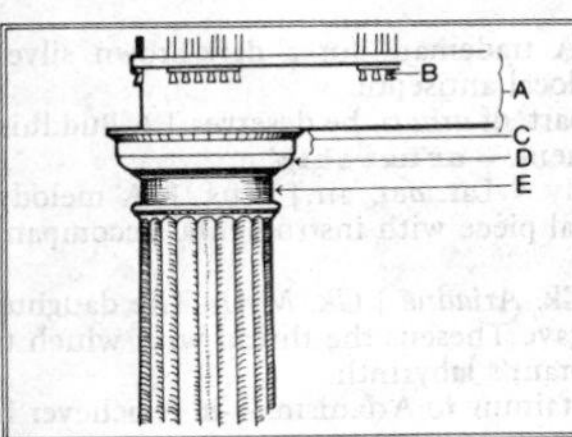


**architrave**
*On a Doric column:*
*A. architrave, B. guttae, C. abacus, D. echinus, E. necking*

**ar·chi·val** (är-kī′vəl) *adj.* Of, relating to, or kept in archives.

**ar·chi·vault** (är′kə-vôlt′) *n. var. of* ARCHIVOLT.

**ar·chives** (är′kīvz′) *pl.n.* [Fr. < LLat. *archium* < Gk. *arkheia* < *arkhē,* beginning < *arkhein,* to begin.] **1. a.** An organized body of records relating to an organization or institution. **b.** A place where such records are kept. **2.** A repository of data.

**ar·chi·vist** (är′kə-vĭst, -kī′-) *n.* One who is in charge of archives.

**ar·chi·volt** (är′kə-vōlt′) *also* **ar·chi·vault** (-vôlt′) *n.* [Ital. *archivolto* : *arco,* arch (< Lat. *arcus*) + *volta,* vault < Lat. *voluta,* fem. p.part. of *volvere,* to roll.] Decorative molding around an arched wall opening.

**ar·chon** (är′kŏn′, -kən) *n.* [Gk. *arkhōn,* ruler, pr.part. of *arkhein,* to rule.] **1.** One of the nine principal magistrates of ancient Athens. **2.** An official of the Byzantine Empire. **3.** One of several powers held to be superior to the angels in certain Gnostic systems. **—ar′chon·ship′** *n.*

**arch·priest** (ärch-prēst′) *n.* [ME *arche-prest* < OFr. *archeprestre* < LLat. *archipresbyter* < LGk. *arkhipresbuteros* : Gk. *arkhi-,* archi- + *presbuteros,* priest.] **1.** A priest formerly holding first rank among the members of a cathedral chapter, acting as chief assistant to a bishop. **2.** A rural dean. —Used as a title of honor. **3.** A high-ranking Russian Orthodox priest.

**arch·way** (ärch′wā′) *n.* **1.** A passageway under an arch. **2.** An arch covering or enclosing an entrance or passageway.

**-archy** *suff.* [ME *-archie* < OFr. < Lat. *-archia* < Gk. *-arkhia* < *-arkhēs,* ruler. —see -ARCH.] Rule : government <olig*archy*>

**ar·ci·form** (är′sə-fôrm′) *adj.* Formed like an arc.

**arc jet** *n.* An arc-jet engine.

**arc-jet engine** (ärk′jĕt′) *n.* A rocket engine that operates by heating the propellant gas with an electric arc.

**arcked** (ärkt) *v. var. p.t. & p.p. of* ARC.

**arck·ing** (är′kĭng) *v. var. present participle of* ARC.

**arc lamp** *also* **arc light** *n.* An electric lamp in which a current traverses a gas between two incandescent electrodes.

**arc·tic** (ärk′tĭk, är′tĭk) *adj.* [ME *artik* < Lat. *arcticus* < Gk. *arktikos* < *arktos,* bear, the northern constellation Ursa Major.] **1.** Extremely cold : FRIGID. **2. Arctic.** Of or pertaining to a geographic area extending from the North Pole to the northern timberline. —*n.* A warm, waterproof overshoe.

**Arctic Circle** *n.* The parallel of latitude 66°33′ north; the boundary between the North Temperate and North Frigid zones.

**arctic fox** *n.* A fox, *Alopex lagopus* of arctic regions, with fur that is white or light-gray in winter and brown or blue-gray in summer.

**Arc·tu·rus** (ärk-tŏŏr′əs, -tyŏŏr′-) *n.* [ME *Arctour* < Lat. *Arcturus* < Gk. *Arktouros* : *arktos,* bear + *ouros,* guard.] *Astron.* The brightest star in the constellation Boötes.

**ar·cu·ate** (är′kyōō-ĭt, -āt′) *also* **ar·cu·at·ed** (-ā′tĭd) *adj.* [Lat. *arcuatus,* p.part. of *arcuare,* to bend like a bow < *arcus,* bow.] Curved like a bow.

**ar·cu·a·tion** (är′kyōō-ā′shən) *n.* **1.** The process of curving or state of being curved. **2.** The use of arches or vaults in building.

**-ard** *or* **-art** *suff.* [ME < OFr., of Germanic orig.] One that habitually or excessively is in a given condition or performs a given action <drunk*ard*><bragg*art*>

**ar·deb** (är′dĕb′) *n.* [Ar. *ardabb* < Gk. *artabē.*] A unit of dry measure of several countries of the Near East, usu. equal to 5.6 U.S. bushels but with variations in different localities.

**ar·dent** (är′dnt) *adj.* [ME *ardaunt* < OFr. < Lat. *ardens,* pr.part. of *ardēre,* to burn.] **1. a.** Expressing or marked by emotional warmth, passion, or desire. **b.** Displaying or marked by strong enthusiasm or devotion : FERVENT <"an impassioned age, so *ardent* and serious in its pursuit of art" —Walter Pater> **2.** Fiery hot : BURNING. **—ar′den·cy** (-dn-sē) *n.* **—ar′dent·ly** *adv.* **—ar′dent·ness** *n.*

**ardent spirits** *pl.n.* Strong alcoholic liquors.

**ar·dor** (är′dər) *n.* [ME *ardour* < OFr. < Lat. *ardor* < *ardēre,* to burn.] **1. a.** Great warmth or intensity, as of emotion, passion, or desire. **b.** Strong enthusiasm or devotion : ZEAL <"the dazzling conquest of Mexico gave a new impulse to the *ardor* of discovery" —William H. Prescott> **2.** Intense heat.

**ar·dour** (är′dər) *n. Chiefly Brit. var. of* ARDOR.

**ar·du·ous** (är′jōō-əs) *adj.* [Lat. *arduus.*] **1.** Demanding great physical or mental care, effort, labor, or endurance <a ballet dancer's *arduous* training><a long, *arduous* journey> **2.** Hard to climb <an *arduous* mountain trail> **—ar′du·ous·ly** *adv.* **—ar′du·ous·ness** *n.*

**are¹** (är) *v.* [ME *aren* < OE *aron.*] *2nd person sing. & present tense indicative pl. of* BE.

**are²** (âr, är) *also* **ar** (är) *n.* [Fr. < Lat. *area,* open space.] A metric unit of area equal to 100 square meters.

**ar·e·a** (âr′ē-ə) *n.* [Lat. *area,* open space.] **1.** A flat piece of ground or open space. **2.** A part of the earth's surface : REGION. **3.** A distinct space or surface or one having a special function. **4.** Range or scope <the *area* of adult education> **5.** An areaway. **6.** The measure of a planar region or of the surface of a solid. **7.** *Computer Sci.* A section of computer storage set aside for a particular purpose.

**Area Code** *also* **area code** *n.* A number, usu. with three digits, assigned to a telephone area, as in the United States and Canada, and used when placing a call to another area.

**area rug** *n.* A rug covering limited floor space.

**area search** *n.* The systematic check of a large group of documents to select those belonging to a particular category.

**ar·e·a·way** (âr′ē-ə-wā′) *n.* **1.** A small sunken area allowing access or light and air to basement doors or windows. **2.** A passageway, often in close quarters between buildings.

**a·re·ca** (ə-rē′kə, ăr′ĭ-kə) *n.* [Port. < Malayalam *aṭekka.*] A tall palm of the genus *Areca* of Southeast Asia, with white flowers and red or orange egg-shaped nuts.

**a·re·na** (ə-rē′nə) *n.* [Lat. *harena.*] **1.** The area in the center of an ancient Roman amphitheater where contests and spectacles were held. **2.** A modern auditorium for sports events. **3.** A sphere of interest, activity, or conflict <the financial *arena*>

**ar·e·na·ceous** (ăr′ə-nā′shəs) *adj.* [Lat. *harenaceus* : *harena,* sand + *-aceus,* aceous.] **1.** Having a sandlike quality or appearance. **2.** Growing in sandy areas.

**arena stage** *n.* The stage of an arena theater.

**arena theater** *n.* A theater without a proscenium in which the stage is at the center of the auditorium, surrounded by seats.

**ar·e·nic·o·lous** (ăr′ə-nĭk′ə-ləs) *adj.* [Lat. *harena,* sand + -COLOUS.] Living or growing in sand.

**aren't** (ärnt, är′ənt). Are not.

**a·re·o·la** (ə-rē′ə-lə) *also* **ar·e·ole** (âr′ē-ōl′) *n., pl.* **-lae** (-lē′) *or* **-las** *also* **-oles** (-ōlz′) [Lat., small open space, dim. of *area,* open place.] **1.** *Biol.* A small space or interstice, as an area bounded by small veins in a leaf or an insect's wing. **2.** *Anat.* A small, dark-colored area around a center portion, as about a nipple or part of the iris of the eye. **—a·re′o·lar, a·re′o·late** (-lĭt) *adj.* **—a·re′o·la′tion** *n.*

**Ar·e·op·a·gite** (ăr′ē-ŏp′ə-jīt′, -gīt′) *n.* A member of the Areopagus. **—Ar′e·op′a·git′ic** *adj.*

**Ar·e·op·a·gus** (ăr′ē-ŏp′ə-gəs) *n.* The highest judicial council of ancient Athens.

**Ar·es** (âr′ēz) *n.* [Gk. *Arēs.*] *Gk. Myth.* The god of war.

**a·rête** (ə-rāt′) *n.* [Fr., fishbone < OFr. *areste* < Lat. *arista.*] A sharp, narrow mountain ridge or spur.

**Ar·e·thu·sa** (ăr′ə-thōō′zə, -sə) *n.* [Lat. < Gk. *Arethousa.*] **1.** *Gk. Myth.* A wood nymph who was changed into a fountain by Artemis. **2.** Any of several orchids of the genus *Arethusa,* esp. *A. bulbosa* of eastern North America, bearing a solitary rose-purple flower fringed with yellow.

**ar·gal** (är′gəl) *n. var. of* ARGOL.

**ar·ga·li** (är′gə-lē) *n., pl.* **argali** *or* **-lis.** [Mongolian.] A wild sheep, *Ovis ammon* of the mountains of central and northern Asia, with large, spirally curved horns.

**ar·gent** (är′jənt) *n.* [ME < OFr. < Lat. *argentum,* silver.] **1.** *Archaic.*

Silver or something like silver. **2.** *Heraldry.* The metal silver, represented by the color white.

**ar·gen·tic** (är-jĕn′tĭk) *adj.* Of or having silver.

**ar·gen·tif·er·ous** (är′jən-tĭf′ər-əs) *adj.* Bearing or producing silver.

**ar·gen·tine** (är′jən-tīn′, -tēn′) *adj.* [Fr. *argentin* < Lat. *argentinus* < *argentum,* silver.] Silvery. —*n.* **1.** Silver. **2.** A silvery metal.

**ar·gen·tite** (är′jən-tīt′) *n.* A valuable silver ore, $Ag_2S$, with a lustrous, lead-gray color.

**ar·gil** (är′jĭl) *n.* [ME *argilla* < Lat. < Gk. *argillos.*] Clay, esp. a white clay used by potters.

**ar·gil·la·ceous** (är′jə-lā′shəs) *adj.* Of, having, made of, or resembling clay.

**ar·gil·lite** (är′jə-līt′) *n.* A metamorphic rock, intermediate between shale and slate, that does not have true slaty cleavage.

**ar·gi·nine** (är′jə-nēn′) *n.* [G. *Arginin,* poss. < Gk. *arginoeis,* bright < *argos,* shining.] An amino acid, $C_6H_{14}N_4O_2$, necessary for nutrition and obtained from plant and animal protein or the digestive action of bacteria.

**Ar·give** (är′jīv′, -gīv′) *adj.* **1.** Of or designating Argos or Argolis. **2.** Greek. —*n.* A Greek, esp. an inhabitant of Argos or Argolis.

**Ar·go** (är′gō′) *n.* **1.** *Gk. Myth.* The ship in which Jason sailed in search of the Golden Fleece. **2.** A constellation in the Southern Hemisphere.

**ar·gol** (är′gôl′) *also* **ar·gal** (-gəl) *n.* [ME *argoil* < AN.] Crude potassium bitartrate, a by-product of wine-making.

**ar·gon** (är′gŏn′) *n.* [< Gk. *argos,* inert : *a-,* no + *ergon,* work.] *Symbol* **Ar** A colorless, odorless, inert gaseous element constituting approx. 1% of the earth's atmosphere, used in electric lamps, fluorescent tubes, and radio vacuum tubes; atomic number 18; atomic weight 39.94.

**Ar·go·naut** (är′gə-nôt′) *n.* [Lat. *Argonauta* < Gk. *Argonautēs* : *Argō,* the ship Argo + *nautēs,* sailor < *naus,* ship.] **1.** *Gk. Myth.* One who sailed with Jason on the *Argo* in search of the Golden Fleece. **2.** One who went to California in 1849 in search of gold. **3. argonaut.** The paper nautilus.

**ar·go·sy** (är′gə-sē) *n., pl.* **-sies.** [Alteration of obs. *ragusye* < Ital. *ragusea,* vessel of *Ragusa,* former name of the port of Dubrovnik, Yugoslavia.] **1.** A large merchant ship. **2.** A fleet of ships. **3.** A very rich supply <an *argosy* of adventure stories>

▲ word history: The derivation of *argosy* from *Ragusa* is not a case of simply switching around *r* and *a,* a process called metathesis. Early English forms of the city's name were *Aragouse* and *Arragouese,* and it is from these forms that *argosy* comes. There is no hint of a connection between *argosy* and the ship *Argo.*

**ar·got** (är′gō, -gət) *n.* [Fr.] The specialized vocabulary of a particular class or group.

**ar·gu·a·ble** (är′gyo͞o-ə-bəl) *adj.* Open to argument : DEBATABLE. **—ar′gu·a·bly** *adv.*

**ar·gue** (är′gyo͞o) *v.* **-gued, -gu·ing, -gues.** [ME *arguen* < OFr. *arguer* < Lat. *arguere,* to make clear.] —*vt.* **1.** To put forth reasons for or against : DEBATE. **2.** To attempt to prove by reasoning : MAINTAIN <*argued* that the Earth was flat> **3.** To give evidence of : INDICATE <"similarities cannot always be used to *argue* descent" —Isaac Asimov> **4.** To influence or persuade <*argued* us into staying> —*vi.* **1.** To present reasons for or against. **2.** To engage in a quarrel : DISPUTE. **—ar′gu·er** *n.*

☆ **syns:** ARGUE, BICKER, CONTEND, DISPUTE, FIGHT, HASSLE, QUARREL, QUIBBLE, SQUABBLE, TIFF, WRANGLE *v. core meaning* : to engage in a verbal exchange expressing conflict of opinions <The committee *argued* over expenditures.>

**†ar·gu·fy** (är′gyə-fī′) *v.* **-fied, -fy·ing, -fies.** *Regional.* —*vt.* To argue over. —*vi.* To argue stubbornly : WRANGLE. **—ar′gu·fi′er** *n.*

**ar·gu·ment** (är′gyə-mənt) *n.* [ME < OFr. < Lat. *argumentum* < *arguere,* to make clear.] **1. a.** A discussion in which disagreement is expressed : DEBATE. **b.** A quarrel. **c.** *Archaic.* A reason or matter for dispute or contention. **2. a.** A course of reasoning aimed at demonstrating the truth or falsehood of something. **b.** A fact or statement offered as proof or evidence. **3.** A summary of the plot or subject of a literary work. **4.** *Logic.* The minor premise in a syllogism. **5.** *Math.* **a.** The independent variable of a function. **b.** The amplitude of a complex number.

**ar·gu·men·ta** (är′gyə-mĕn′tə) *n. pl. of* ARGUMENTUM.

**ar·gu·men·ta·tion** (är′gyə-mĕn-tā′shən) *n.* **1.** Presentation and elaboration of an argument. **2.** Deductive reasoning in debate. **3.** A debate.

**ar·gu·men·ta·tive** (är′gyə-mĕn′tə-tĭv) *adj.* **1.** Given to arguing. **2.** Of or marked by argument. **—ar′gu·men′ta·tive·ly** *adv.* **—ar′gu·men′ta·tive·ness** *n.*

**ar·gu·men·tum** (är′gyə-mĕn′təm) *n., pl.* **-ta** (-tə) [Lat. —see ARGUMENT.] *Logic.* An argument, demonstration, or appeal to reason.

**argumentum ad hom·i·nem** (ăd hŏm′ə-nĕm′) *n.* [Lat., argument to the man.] An argument appealing to personal biases and emotions rather than to logic or reason.

**Ar·gus** (är′gəs) *n.* [Lat. < Gk. *Argos.*] **1.** *Gk. Myth.* A hundred-eyed giant who was made guardian of Io and was later slain by Hermes. **2.** A watchful person : GUARDIAN.

**Ar·gus-eyed** (är′gəs-īd′) *adj.* Highly observant and alert.

**argus pheasant** *n.* [After *Argus,* whose hundred eyes were given to a peacock's tail.] A large bird, *Argusianus argus* of southern Asia and the East Indies, bearing long tail feathers marked with brilliantly colored eyelike spots.

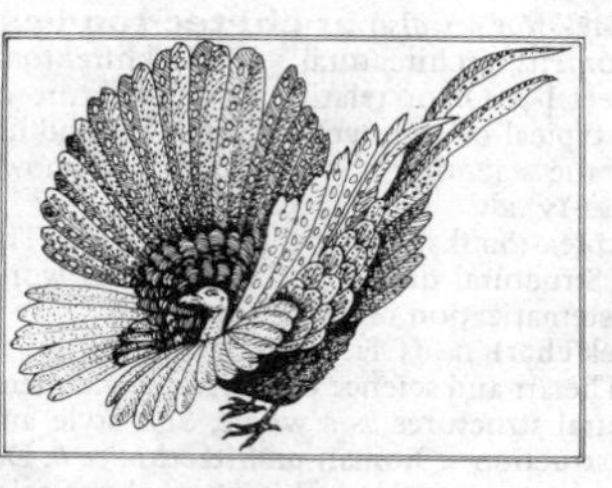

**argus pheasant**
*Size varies greatly, approximately 25 inches long*

**ar·gy-bar·gy** (är′gē-bär′gē) *n.* [Sc., prob. var. of E. *argle-bargle,* redup. of *argle,* to argue obstinately, prob. < ARGUE. —see ARGUE.] *Chiefly Brit.* A lively or disputatious discussion : WRANGLE.

**ar·gyle** *also* **ar·gyll** (är′gīl′) *n.* [After Campbell of *Argyle,* orig. from the pattern of their tartan.] **1.** A knitting pattern of varicolored, diamond-shaped areas on a solid color background. **2.** A stocking knit in an argyle pattern.

**Ar·gy·rol** (är′jə-rōl′, -rôl′). A trademark for a dark-brown silver-protein compound used as a local antiseptic.

**ar·hat** (är′hət) *n.* [Skt. < pr.part. of *arhati,* he deserves.] A Buddhist who has achieved enlightenment. **—ar′hat·ship′** *n.*

**a·ri·a** (är′ē-ə) *n.* [Ital., melody < Lat. *aer,* air.] *Mus.* **1.** A melody. **2.** A solo, often elaborate vocal piece with instrumental accompaniment, as in an opera.

**Ar·i·ad·ne** (ăr′ē-ăd′nē) *n.* [Gk. *Ariadnē.*] *Gk. Myth.* The daughter of Minos and Pasiphae who gave Theseus the thread with which to find his way out of the Minotaur's labyrinth.

**Ar·i·an** (âr′ē-ən, âr′-) *adj.* Pertaining to Arianism. —*n.* A believer in Arianism.

**-arian** *suff.* [Lat. *-arius,* -ary + -AN¹.] Advocate or promoter of <utilit*arian*>

**Ar·i·an·ism** (âr′ē-ə-nĭz′əm, âr′-) *n.* The doctrines of Arius, holding that Jesus was not of the same substance as God but was only the highest of created beings.

**ar·id** (ăr′ĭd) *adj.* [Lat. *aridus* < *arere,* to be dry.] **1.** Lacking moisture, esp. because of insufficient rainfall : DRY. **2.** Lacking interest or spirit : DULL. **—a·rid′i·ty** (ə-rĭd′ĭ-tē), **ar′id·ness** *n.*

**Ar·ies** (âr′ēz′, âr′ē-ēz′) *n.* [Lat. *aries,* ram.] **1.** A constellation in the Northern Hemisphere. **2. a.** The first sign of the zodiac. **b.** One born under this sign.

**a·ri·et·ta** (ä′rē-ĕt′ə) *also* **a·ri·ette** (-ĕt′) *n.* [Ital., dim. of *aria,* aria.] A short aria.

**a·right** (ə-rīt′) *adv.* [ME < OE *ariht* : *a-,* on + *riht,* right.] Rightly.

**ar·il** (ăr′əl) *n.* [Med. Lat. *arillus,* grape stone.] An outer covering or appendage of some seeds, arising at or near the hilum. **—ar′iled** *adj.*

**a·ri·o·so** (ä′rē-ō′sō, -zō) [Ital. < *aria,* aria.] *Mus.* —*adv.* In the style of an aria. —Used as a direction. —*adj.* Resembling an aria. —*n., pl.* **-sos.** An arioso passage or composition.

**a·rise** (ə-rīz′) *vi.* **a·rose** (ə-rōz′), **a·ris·en** (ə-rĭz′ən), **a·ris·ing, a·ris·es.** [ME *arisen* < OE *arīsan.*] **1.** To get up, as from a sitting or prone position. **2.** To move upward : ASCEND <smoke *arising* from chimneys> **3.** To come into being : ORIGINATE <customs that *arise* in antiquity> **4.** To result, issue, or proceed <discontent *arising* from poverty>

**a·ris·ta** (ə-rĭs′tə) *n., pl.* **-tae** (-tē) *or* **-tas.** [Lat., beard of grain.] A bristlelike part or process. **—a·ris′tate′** (-tāt′) *adj.*

**ar·is·toc·ra·cy** (ăr′ĭ-stŏk′rə-sē) *n., pl.* **-cies.** [OFr. *aristocratie,* government by the best < LLat. *aristocratia* < Gk. *aristokratia* : *aristos,* best + *kratos,* power.] **1.** A hereditary privileged ruling class or nobility. **2. a.** Government by the nobility or by a privileged upper class. **b.** A state or country having this form of government. **3. a.** Government by the best citizens. **b.** A state having this form of government. **4.** A group or class viewed as superior.

**a·ris·to·crat** (ə-rĭs′tə-krăt′, ăr′ĭs-) *n.* [Fr. *aristocrate* < *aristocratie,* aristocracy < OFr.] **1.** A member of the aristocracy. **2.** One having characteristics, as tastes, opinions, and manners, of the aristocracy. **3.** One advocating government by the aristocracy. **—a·ris′to·crat′ic, a·ris′to·crat′i·cal** *adj.* **—a·ris′to·crat′i·cal·ly** *adv.*

**Ar·is·to·te·li·an** *also* **Ar·is·to·te·le·an** (ăr′ĭ-stə-tē′lē-ən, -tēl′yən) *adj.* Of or relating to Aristotle or his philosophy. —*n.* **1.** A follower of Aristotle or his teachings. **2.** One who tends to be empirical or scientific in thought or methods. **—Ar′is·to·te′li·an·ism** *n.*

**Aristotelian logic** *n.* **1.** Aristotle's deductive method of logic, esp. the theory of the syllogism. **2.** The formal logic based on Aristotle's and dealing with the relations between propositions in terms of their form instead of their content.

**a·rith·me·tic** (ə-rĭth′mĭ-tĭk) *n.* [ME *arsmetrik,* alteration of OFr. *arismetique* < LLat. *arismetica,* alteration of Lat. *arithmetica* < Gk. *arithmētikē* (*tekhnē*), (the art) of counting < *arīthmein,* to count < *arithmos,* number.] **1.** The mathematics of integers under addition, subtraction, multiplication, division, involution, and evolution. **2.** Computation or problem solving involving real numbers and the arithmetic operations. **3.** A book on arithmetic. **—ar·ith·met·ic** (ăr′ĭth-mĕt′ĭk), **ar·ith·met·i·cal** (-ĭ-kəl) *adj.* **—ar′ith·met′i·cal·ly** *adv.* **—a·rith′me·ti′cian** (-tĭsh′ən) *n.*

**ar·ith·met·ic mean** (ăr′ĭth-mĕt′ĭk) *n.* The number obtained by dividing the sum of a given set of quantities by the number of quantities in the set.

**arithmetic progression** *n.* A sequence, as the odd integers 1, 3, 5, 7, . . . , in which each term after the first is formed by adding a constant to each preceding term.

**-arium** *suff.* [Lat., neuter of *-arius,* -ary.] A place or device containing or associated with <planet*arium*>

**ark** (ärk) *n.* [ME < OE *arc* < Lat. *arca,* chest.] **1.** The chest containing the Ten Commandments written on stone tablets, carried by the Hebrews during their desert wanderings. **2.** The Holy Ark. **3.** The boat God commanded Noah to build for shelter during the Flood. **4.** A large, commodious boat. **5.** A shelter or refuge.

**arm**[1] (ärm) *n.* [ME < OE *earm.*] **1.** An upper limb of the human body connecting the hand and wrist to the shoulder. **2.** An armlike part, as the forelimb of an animal or a long part projecting from a central support in a machine. **3.** Something meant to cover or support the human arm, as a sleeve on clothing or a projection on a piece of furniture. **4.** Something branching out from a large mass <an *arm* of the ocean> **5.** An administrative or functional branch, as of an organization. **6.** Power or authority. —Used esp. in the phrase *the long arm of the law.* **—an arm and a leg.** An extravagant amount <a vacation that cost *an arm and a leg*> **—with open arms.** Cordially. **—arm′like** *adj.*

**arm**[2] (ärm) *n.* [ME *armes* < OFr. < Lat. *arma.*] **1.** A weapon, esp. a firearm. **2.** A branch of a military force, as the infantry. **3. arms.** Warfare. **4. arms. a.** Heraldic bearings. **b.** Insignia, as of a state, official, family, or organization. *—v.* **armed, arm·ing, arms.** *—vi.* **1.** To supply or equip oneself with arms. **2.** To prepare oneself for or as if for warfare or conflict. *—vt.* **1.** To equip with weapons. **2.** To prepare for war. **3.** To provide with something that strengthens or promotes efficiency : FORTIFY <*armed* with righteous indignation> **4.** To prepare (e.g., a bomb) for detonation. **—(up) in arms.** Provoked and ready to fight : INDIGNANT <*up in arms* over higher taxes> **—arm′er** *n.*

**ar·ma·da** (är-mä′də, -mā′-) *n.* [Sp. < Med. Lat. *armata* < Lat. *armatus,* armed, p.part. of *armare,* to arm < *arma,* arms.] **1.** A fleet of warships. **2.** A large group of moving things.

**ar·ma·dil·lo** (är′mə-dĭl′ō) *n., pl.* **-los.** [Sp., dim. of *armado,* armored, p.part. of *armar,* to arm < Lat. *armare* < *arma,* arms.] Any of several omnivorous, burrowing mammals of the family Dasypodidae of southern North America and South America, with an armorlike covering of jointed, bony plates.

**Ar·ma·ged·don** (är′mə-gĕd′n) *n.* [LLat. *Armagedon* < Gk. < Heb. *har megiddōn,* mountain district of Megiddo, the site of several great battles mentioned in the Bible.] **1.** The scene of a final battle between the forces of good and evil, prophesied in the Bible to take place at the end of the world. **2.** A decisive conflict.

**Ar·ma·gnac** (är′mən-yăk′) *n.* [Fr., after *Armagnac,* a region in southwest France.] A dry brandy.

**ar·ma·ment** (är′mə-mənt) *n.* [Lat. *armamenta,* tools < *arma,* arms.] **1.** Weapons and equipment of war. **2.** *often* **armaments.** All the military forces and war materiel of a country. **3.** A military force equipped for war. **4.** The process of arming for war.

**ar·ma·men·tar·i·um** (är′mə-mĕn-târ′ē-əm) *n., pl.* **-i·ums** *or* **-i·a** (-ē-ə) [Lat., arsenal < *armamenta,* tools, armaments.] The equipment of a physician or medical institution, as books, supplies, and instruments.

**ar·ma·ture** (är′mə-cho͝or′, -chər) *n.* [Lat. *armatura,* equipment < *armare,* to arm < *arma,* arms.] **1.** *Elect.* **a.** The rotating part of a dynamo consisting of copper wire wound around an iron core. **b.** The moving part of an electromagnetic device such as a relay, buzzer, or loudspeaker. **c.** A piece of soft iron connecting the poles of a magnet. **2.** *Biol.* The protective structure or covering of a plant or animal. **3.** A framework used as a supporting core for clay sculpture. **4.** Armor.

**arm·chair** (ärm′chăr′) *n.* A chair with side structures to support the arms or elbows. *—adj.* Not actively involved <an *armchair* quarterback><an *armchair* voyager>

**armed** (ärmd) *adj.* **1.** Equipped with weapons. **2.** Having or marked by an arm or arms of a specified kind or number <thin-*armed*>

**armed forces** *pl.n.* A nation's military forces.

**Ar·me·ni·an** (är-mē′nē-ən, -mēn′yən) *n.* **1.** A native or resident of Armenia. **2.** The Indo-European language of the Armenians. **—Ar·me′ni·an** *adj.*

**ar·met** (är′mĕt′) *n.* [OFr.] A medieval light helmet with a neck guard and movable visor.

**arm·ful** (ärm′fo͝ol′) *n.* The amount an arm can hold.

**arm·hole** (ärm′hōl′) *n.* An opening for the arm in a garment.

**ar·mi·ger** (är′mə-jər) *n.* [Lat. : *arma,* arms + *gerere,* to carry.] **1.** An armorbearer for a knight : SQUIRE. **2.** One who is entitled to heraldic arms.

**ar·mil·lar·y sphere** (är′mə-lĕr′ē, är-mĭl′ə-rē) *n.* [Fr. *sphère armillaire* < Lat. *armilla,* bracelet < *armus,* shoulder.] An astronomical model with solid rings, all circles of a single sphere, used to show relationships among the principal celestial circles.

**Ar·min·i·an·ism** (är-mĭn′ē-ə-nĭz′əm) *n.* The doctrines of Jacobus Arminius and his followers, opposing the Calvinist doctrine of absolute predestination. **—Ar·min′i·an** *adj. & n.*

**ar·mi·stice** (är′mĭ-stĭs) *n.* [NLat. *armistitium* : Lat. *arma,* arms + Lat. *-stitium,* a stopping < *sistere,* to stop.] A temporary stoppage or suspension of hostilities by mutual consent : TRUCE.

**Armistice Day** *n.* Nov. 11, celebrated as the anniversary of the armistice of World War I in 1918 and called Veterans' Day since 1954.

**arm·let** (ärm′lĭt) *n.* **1.** An ornamental or identifying band worn on the arm. **2.** A small arm, as of the ocean.

**ar·moire** (ärm-wär′, är′mər) *n.* [OFr., var. of *armaire* < Lat. *armarium,* closet < *arma,* tools.] A large ornate wardrobe or cabinet.

**ar·mor** (är′mər) *n.* [ME *armure* < OFr. *armëure* < Lat. *armaturea* < *armare,* to arm < *arma,* arms.] **1.** A defensive covering, as chain mail, worn to protect the body against weapons. **2.** A tough protective covering, as the bony scales covering certain animals or metallic plates on tanks or warships. **3.** A safeguard or protection <an *armor* of indifference> **4. a.** The combat arms branch of an army specializing in the use of armored vehicles. **b.** The armored vehicles of an army. *—vt.* **-mored, -mor·ing, -mors.** To cover with armor.

**ar·mor·bear·er** (är′mər-bâr′ər) *n.* One who carries the armor of a warrior or knight.

**ar·mor-clad** (är′mər-klăd′) *adj.* ARMORED 1.

**ar·mored** (är′mərd) *adj.* **1.** Wearing or covered with armor <*armored* knights><an *armored* car> **2.** Equipped with armored vehicles, as a military unit.

**ar·mor·er** (är′mər-ər) *n.* **1.** One who makes or repairs armor. **2.** A weapons manufacturer. **3.** An enlisted person in charge of maintaining and reparing the small arms of a military unit.

**ar·mo·ri·al** (är-môr′ē-əl, -mōr′-) *adj.* Of or relating to heraldry or heraldic arms. *—n.* A treatise or book on heraldry.

**Ar·mor·ic** (är-môr′ĭk, -mŏr′-) *also* **Ar·mor·i·can** (-ĭ-kən) *adj.* Of or relating to Armorica or the people or language of Armorica. *—n.* **1.** A native or resident of Armorica. **2.** BRETON 2.

**armor plate** *n.* Hard steel plate used to cover warships, vehicles, and fortifications. **—ar′mor-plat′ed** *adj.*

**ar·mor·y** (är′mə-rē) *n., pl.* **-ies. 1.** An arsenal. **2.** A building for storing arms and military equipment, esp. one serving as headquarters for military reserve personnel. **3.** A munitions factory.

**ar·mour** (är′mər) *n. Chiefly Brit. var. of* ARMOR.

**arm·pit** (ärm′pĭt′) *n.* The hollow under the arm at the shoulder.

**arm·rest** (ärm′rĕst′) *n.* A support, as on a chair, for the arm.

**arm-twist·ing** (ärm′twĭs′tĭng) *n.* Use of esp. strong pressure to attain a goal <"subjecting vulnerable congressmen to some political *arm-twisting*" —James R. Gaines> **—arm′-twist′er** *n.*

**arm wrestling** *n.* A form of wrestling in which two opponents sit facing each other with usu. right hands interlocked and elbows firmly placed, as on a table surface, and attempt to force each other's arm down. **—arm′-wres′tle** *v.* **(-tled, -tling, -tles).**

**ar·my** (är′mē) *n., pl.* **-mies.** [ME *armee* < OFr. < Med. Lat. *armata.* —see ARMADA.] **1.** A large body of troops organized and trained for warfare on land. **2.** The military land forces of a country as a whole. **3.** A tactical and administrative military unit made up of a headquarters, two or more army corps, and auxiliary forces. **4.** A large group of people organized for a specific cause. **5.** A large multitude.

**Army Air Forces** *pl.n.* The aviation branch of the U.S. Army before the U.S. Air Force was established.

**army ant** *n.* Any of various chiefly tropical New World ants of the subfamily Dorylinae, forming large nomadic colonies.

**Army of the United States** *n.* A temporary organization of all military forces during a war, including the regular U.S. Army, the National Guard reserves, and Selective Service personnel.

**ar·my·worm** (är′mē-wûrm′) *n.* Any of various insect larvae that travel in large groups destroying vegetation, esp. the caterpillar of a New World moth, *Leucania,* or *Pseudaletia, unipuncta.*

**ar·nat·to** (är-nä′tō) *n. var. of* ANNATTO.

**ar·ni·ca** (är′nĭ-kə) *n.* [NLat. *Arnica,* genus name.] **1.** A plant of the genus *Arnica,* with bright-yellow, rayed flowers. **2.** A tincture of the dried flower heads of *A. montana,* used for sprains and bruises.

**ar·oid** (ăr′oid′, âr′-) *n.* [AR(UM) + -OID.] Any of various plants of the family Aracea, which includes the arums. **—ar′oid′** *adj.*

**a·roint** (ə-roint′) *vt.* **a·roint·ed, a·roint·ing, a·roints.** [Orig. unknown.] *Archaic.* Begone <"*Aroint* thee, witch!" —Shakespeare>

**a·ro·ma** (ə-rō′mə) *n.* [Lat. < Gk., *arōma,* aromatic herb.] **1.** A distinctive, usu. pleasant odor. **2.** A distinctive, intangible quality : AURA.

**ar·o·mat·ic** (ăr′ə-măt′ĭk) *adj.* **1.** Having a fragrant, sweet-smelling, or spicy aroma <*aromatic* incense> **2.** *Chem.* Of, relating to, or containing the six-carbon ring typical of the benzene series and related organic groups. —*n.* An aromatic plant or substance. **—ar′o·mat′i·cal·ly** *adv.* **—ar′o·mat′ic·ness** *n.*

**ar·o·ma·tic·i·ty** (ăr′ə-mə-tĭs′ĭ-tē, ə-rō′mə-) *n.* Aromatic quality or character, esp. the distinctive structure or properties of the aromatic chemical compounds.

**a·ro·ma·tize** (ə-rō′mə-tīz′) *vt.* **-tized, -tiz·ing, -tiz·es. 1.** To make aromatic. **2.** *Chem.* To subject to a reaction that produces an aromatic compound. **—a·ro′ma·ti·za′tion** *n.*

**a·rose** (ə-rōz′) *v. p.t. of* ARISE.

**a·round** (ə-round′) *adv.* [ME, in the round, in circumference.] **1.** On or to all sides or in all directions. **2.** In a circle or circular motion. **3.** To each member of a group <Will the cake go *around?*> **4.** In or toward the opposite direction, position, or attitude. **5.** From one place to another <travel *around*> **6.** *Informal.* In or close to one's present position <stood *around* for hours> **7.** *Informal.* To a specific place or area <Come *around* next week.> **8.** *Informal.* To a normal or desired state. **9.** *Informal.* Approximately <*around* two o'clock> —*prep.* **1.** On all sides of. **2.** So as to enclose, surround, or envelop. **3.** About the circumference or periphery of : ENCIRCLING. **4.** About the central point of <the moon's motion *around* the earth> **5.** In or to various places within or near <looking *around* the store> **6.** On or to the farther side of <the farmhouse *around* the bend> **7.** *Informal.* Approximately at : NEAR <vacations *around* the Catskills> **—get around.** *Informal.* **1.** To deal or cope with successfully. **2.** To succeed in evading or circumventing. **—get around to.** *Informal.* To find time to give one's attention to.

**a·round-the-clock** (ə-round′thə-klŏk′) *adj. var. of* ROUND-THE-CLOCK.

**a·rouse** (ə-rouz′) *v.* **a·roused, a·rous·ing, a·rous·es.** [< ROUSE, on the model of such pairs as *rise, arise.*] —*vt.* **1.** To awaken from or as if from sleep. **2.** To stir up : excite or provoke. —*vi.* To be or become aroused. **—a·rous′al** (ə-rou′zəl) *n.* **—a·rous′er** *n.*

**ar·peg·gi·o** (är-pĕj′ē-ō′, -pĕj′ō) *n., pl.* **-os.** [Ital. < *arpeggiare,* to play the harp < *arpa,* harp, of Germanic orig.] *Mus.* **1.** Production of the tones of a chord in rapid succession rather than simultaneously. **2.** A chord played or sung in arpeggio. **—ar·peg′gi·oed′** *adj.*

**ar·pent** (är-päɴ′) *n.* [Fr. < OFr. < Lat. *arepennis,* half acre, of Gaulish orig.] An old French unit of land measurement equivalent to approx. an acre.

**ar·que·bus** (är′kə-bəs, -kwə-) *n. var. of* HARQUEBUS.

**ar·rack** (ăr′ək, ə-răk′) *n.* [Ar. *'araq,* fruit juice.] A strong alcoholic drink of the Middle East and nearby regions of the Orient, usu. distilled from rice or molasses.

**ar·raign** (ə-rān′) *vt.* **-raigned, -raign·ing, -raigns.** [ME *arreinen* < OFr. *araisnier* < VLat. **adrationare* : Lat. *ad-,* to + Lat. *ratio,* account. —see REASON.] **1.** *Law.* To call before a court to answer to an indictment. **2.** Accuse : charge. **—ar·raign′er** *n.* **—ar·raign′ment** *n.*

**ar·range** (ə-rānj′) *v.* **-ranged, -rang·ing, -rang·es.** [ME *arengen* < OFr. *arengier* : *a-,* to (< Lat. *ad*) + *rengier,* to put in a line < *reng,* line, of Germanic orig.] —*vt.* **1.** To put into a specific order or relation : DISPOSE. **2.** To plan or prepare for. **3.** To agree about : SETTLE. **4.** *Mus.* To reset (music) for other instruments or voices or for another style of performance. —*vi.* **1.** To come to an agreement. **2.** To make preparations : PLAN. **—ar·rang′er** *n.*

**ar·range·ment** (ə-rānj′mənt) *n.* **1.** The act or process of arranging. **2.** The condition, manner, or result of being arranged : DISPOSAL. **3.** A collection of things that have been arranged. **4.** *often* **arrangements.** A provision or plan made in preparation for an undertaking. **5.** An agreement or settlement : DISPOSITION. **6.** *Mus.* **a.** An adaptation of a composition for other instruments or voices or to another style or level of difficulty. **b.** A composition so adapted.

**ar·rant** (ăr′ənt) *adj.* [Var. of ERRANT.] Completely such : THOROUGHGOING <an *arrant* liar> **—ar′rant·ly** *adv.*

**ar·ras** (ăr′əs) *n.* [ME, after *Arras,* France.] **1.** A tapestry. **2.** A wall hanging.

**ar·ray** (ə-rā′) *vt.* **-rayed, -ray·ing, -rays.** [ME *arraien* < OFr. *areer* < VLat. **arredare,* of Germanic orig.] **1.** To arrange or draw up, as troops in battle order. **2.** To clothe in finery : ADORN. —*n.* **1.** An orderly arrangement, esp. of troops. **2.** An impressive display of numerous persons or objects <"a heathenish *array* of monstrous clubs and spears" —Melville> **3.** Splendid attire : FINERY. **4.** *Math.* **a.** A rectangular arrangement of quantities in rows and columns, as in a matrix. **b.** Numerical data linearly ordered by magnitude. **5.** An arrangement of computer memory elements in one or several planes.

**ar·ray·al** (ə-rā′əl) *n.* **1.** The act or process of arraying. **2.** Something arrayed.

**ar·rear·age** (ə-rîr′ĭj) *n.* **1.** The state of being in arrears. **2.** An amount owed in payment.

**ar·rears** (ə-rîrz′) *pl.n.* [< ME *arrere,* behind < OFr. *arere* < LLat. *ad retro,* backward : Lat. *ad,* to + Lat. *retro,* behind.] **1.** An unpaid and overdue debt or unfulfilled obligation. **2.** The state of being behind in fulfilling contracted obligations or payments.

**ar·rest** (ə-rĕst′) *vt.* **-rest·ed, -rest·ing, -rests.** [ME *aresten* < OFr. *arester* < VLat. **arrestare* : Lat. *ad-,* to + Lat. *restare,* to stand still (*re-,* back + *stare,* to stand).] **1.** To stop or check the motion, progress, growth, or spread of <*arrest* a cold> **2.** To seize and hold under authority of the law. **3.** To capture and hold briefly (e.g., the attention) : ENGAGE. —*n.* **1.** The act of arresting or the state of being arrested. **2.** A device for arresting motion, esp. of a moving part. **—under arrest.** Detained in legal custody. **—ar·rest′er** *n.* **—ar·rest′ment** *n.*

☆ ***syns:*** ARREST, APPREHEND, BAG, BUST, COLLAR, DETAIN, NAB, PICK UP, PINCH, RUN IN, SEIZE *v. core meaning* : to take into legal custody <was *arrested* for car theft>

**ar·rest·ing** (ə-rĕs′tĭng) *adj.* Attracting and holding the attention : ENGAGING. **—ar·rest′ing·ly** *adv.*

**ar·rhyth·mi·a** (ə-rĭth′mē-ə) *n.* [Gk. *arruthmia,* lack of rhythm < *arruthmos,* unrhythmical : *a-,* without + *rhuthmos,* rhythm.] Irregularity in the force or rhythm of the heartbeat.

**ar·rhyth·mic** (ə-rĭth′mĭk) *also* **ar·rhyth·mi·cal** (-mĭ-kəl) *adj.* Lacking rhythm or rhythmic regularity. **—ar·rhyth′mi·cal·ly** *adv.*

**ar·ri·ère-ban** (ăr′ē-âr-bän′, -băn′) *n.* [Fr. < OFr. *ariere-ban,* alteration of *herban,* of Germanic orig.] **1.** A royal proclamation by which medieval French vassals were summoned to military service. **2.** The vassals summoned by an arrière-ban.

**ar·ri·ère-pen·sée** (ăr′ē-âr′päɴ-sā′) *n.* [Fr. : *arrière,* in back + *pensée,* thought.] An ulterior motive.

**ar·ris** (ăr′ĭs) *n., pl.* **arris** *or* **-ris·es.** [Alteration of OFr. *areste,* ridge. —see ARÊTE.] The sharp edge or ridge formed by two surfaces meeting at an angle, as in an architectural molding.

**ar·ri·val** (ə-rī′vəl) *n.* **1.** The act or process of arriving. **2.** One that arrives or has arrived. **3.** Attainment of a goal as a result of a process or effort.

**ar·rive** (ə-rīv′) *vi.* **-rived, -riv·ing, -rives.** [ME *ariven* < OFr. *ariver* < VLat. **arripare* : Lat. *ad-,* to + Lat. *ripa,* shore.] **1.** To reach a destination. **2.** To take place <The big day finally *arrived.*> **3.** To achieve success or recognition. **—arrive at.** To attain through a process or effort. **—ar·riv′er** *n.*

**ar·ri·viste** (ä-rē-vēst′) *n., pl.* **-vistes** (-vēst′) [Fr. < *arriver,* to arrive < OFr. *ariver.*] A social climber : UPSTART.

**ar·ro·ba** (ə-rō′bə) *n.* [Sp. and Port. < Ar. *ar-rub',* the quarter (of a quintal).] **1.** A former unit of weight in Spanish-speaking countries, equal to approx. 25 pounds. **2.** A former unit of weight in Portuguese-speaking countries, equal to approx. 32 pounds. **3.** A former liquid measure in Spanish-speaking countries, having varying value but approx. equal to 17 quarts when used to measure wine.

**ar·ro·gant** (ăr′ə-gənt) *adj.* [ME *arrogaunt* < OFr. < Lat. *arrogans,* pr.part. of *arrogare,* to arrogate.] **1.** Over convinced of one's own importance : HAUGHTY. **2.** Marked by or arising from haughty self-importance. **—ar′ro·gance** (-gəns) *n.* **—ar′ro·gant·ly** *adv.*

☆ ***syns:*** ARROGANT, CAVALIER, DISDAINFUL, HAUGHTY, HIGH-AND-MIGHTY, HOITY-TOITY, INSOLENT, LOFTY, LORDLY, OVERBEARING, OVERWEENING, PRESUMPTUOUS, PROUD, SUPERCILIOUS, SUPERIOR *adj. core meaning* : over convinced of one's own superiority and importance <an *arrogant,* selfish person>

**ar·ro·gate** (ăr′ə-gāt′) *vt.* **-gat·ed, -gat·ing, -gates.** [Lat. *arrogare, arrogat-* : *ad-,* to + *rogare,* to ask.] **1.** To claim, take, or assume for oneself without right. **2.** To attribute to another unjustifiably. **—ar′ro·ga′tion** *n.* **—ar′ro·ga′tive** *adj.* **—ar′ro·ga′tor** *n.*

**ar·ron·disse·ment** (ä-rôɴ′dēs-mäɴ′) *n.* [Fr. < *arrondir,* to round out : *a,* to (< Lat. *ad*) + *rondir,* to make round.] **1.** The chief administrative subdivision of a department in France. **2.** A municipal subdivision of some large French cities.

**ar·row** (ăr′ō) *n.* [ME *arwe* < OE *arewe.*] **1.** A thin, straight shaft for shooting from a bow, usu. made of light wood with a pointed head at one end and flight-stabilizing feathers at the other. **2.** Something similar to an arrow in form, function, or speed. **3.** A sign or symbol shaped like an arrow and used to indicate direction.

**ar·row·head** (ăr′ō-hĕd′) *n.* **1.** The pointed, removable striking tip of an arrow. **2.** Something shaped like an arrowhead, as a mark indicating a limit on a drawing. **3.** An aquatic or marsh plant of the genus *Sagittaria,* bearing arrowhead-shaped leaves and white flowers.

**ar·row·root** (ăr′ō-ro͞ot′, -ro͝ot′) *n.* [So called because it was used to draw poison from arrow wounds.] **1.** A tropical American plant, *Maranta arundinacea,* with roots that yield an edible starch. **2.** The starch from the arrowroot and from certain plants of the genera *Manihot, Curcuma,* and *Tacca.*

**ar·row·wood** (ăr′ō-wo͝od′) *n.* A small shrub of the genus *Viburnum,* having straight tough stems once used to make arrows.

**arrow worm** *n.* A small, slender marine worm of the phylum Chaetognatha, with prehensile bristles on each side of the mouth.

**ar·roy·o** (ə-roi′ō) *n., pl.* **-os.** [Sp., ult. < Lat. *arrugia,* mineshaft.] **1.** A deep gully cut by an intermittent stream. **2.** A brook or creek.

**ar·se·nal** (är′sə-nəl) *n.* [Ital. *arsenale* < Ar. *dār-aṣ-ṣināh* : *dār,* house + *aṣ-,* the + *ṣināh,* manufacture < *ṣanā,* he made.] **1.** A governmental establishment for the storing, manufacturing, or repairing of war materiel, as arms and ammunition. **2.** A stock of weapons. **3.** A store or supply <had an *arsenal* of circumstantial evidence>

**ar·se·nate** (är′sə-nĭt, -nāt′) *n.* A salt or ester of arsenic acid.

**ar·se·nic** (är′sə-nĭk) *n.* [ME *arsenik* < OFr. < Lat. *arrenicum* < Gk. *arsenikon,* yellow orpiment < Pers. *zarnīk* < *zar,* gold.] **1.** *Symbol* **As** A highly poisonous metallic element used in insecticides, weed killers, solid-state devices, and various alloys; atomic number 33; atomic weight 74.922. **2.** Arsenic trioxide. —*adj.* **ar·sen·ic** (är-sĕn′ĭk). Of or containing arsenic, esp. with valence 5.

**ar·sen·ic acid** (är-sĕn′ĭk) *n.* A poisonous white translucent crystalline compound, $H_3AsO_4$, used to make arsenates.

**ar·sen·i·cal** (är-sĕn′ĭ-kəl) *adj.* Of or containing arsenic. —*n.* A preparation or drug containing arsenic.

**ar·se·nic trioxide** (är′sə-nĭk) *n.* A poisonous white amorphous powder, $As_2O_3$, used in insecticides, rat poison, and weed killers.

**ar·se·nide** (är′sə-nīd′) *n.* A compound of arsenic with a more electropositive element.

**ar·se·ni·ous** (är-sē′nē-əs) *adj.* Of or containing arsenic, esp. with valence 3.

**ar·se·no·py·rite** (är′sə-nō-pī′rīt′) *n.* A silver-white to gray arsenic ore, essentially $FeS_2 \cdot FeAs_2$.

**ar·shin** (är-shēn′) *n. var. of* ARCHINE.

**ar·sine** (är-sēn′, är′sēn′) *n.* [ARS(ENIC) + -INE[2].] A colorless, flammable, very poisonous gas, $AsH_3$, used as a poison gas in the military, as a solid-state doping agent, and in organic synthesis.

**ar·sis** (är′sĭs) *n., pl.* **-ses** (-sēz′) [LLat., raising of the voice < Gk., upward beat < *aeirein,* to lift.] **1.** The unaccented or shorter part of a foot of quantitative verse. **2.** The accented or longer part of a foot of accentual verse. **3.** *Mus.* The unaccented part of a measure.

**ar·son** (är′sən) *n.* [AN < OFr. *arçun* < Med. Lat. *arsio* < Lat. *ardēre,* to burn.] The crime of maliciously setting fire to the property of another or of burning one's own property for an improper purpose, as to collect insurance. **—ar′son·ist** *n.*

**ars·phen·a·mine** (ärs-fĕn′ə-mēn′) *n.* [ARS(ENIC) + PHEN(YL) + AMINE.] A yellow hygroscopic powder, $C_{12}H_{12}N_2O_2As \cdot 2HCl \cdot 2H_2O$, once used to treat syphilis.

**art[1]** (ärt) *n.* [ME < OFr. < Lat. *ars.*] **1.** Human effort to imitate, supplement, alter, or counteract the work of nature. **2. a.** Conscious arrangement or production of sounds, colors, forms, movements, or other elements in a way that affects the aesthetic sense : production of the beautiful in a graphic or plastic medium. **b.** The study of these activities. **c.** The product of these activities. **3.** High quality of conception or execution, as found in works of beauty : aesthetic value. **4.** A field or category of art, as literature, music, or ballet. **5.** A nonscientific branch of learning, as one of the liberal arts. **6. a.** A system of principles and methods used in the performance of a set of activities <the *art* of baking> **b.** A trade or craft that applies such a system of principles and methods <pursuing the weaver's *art*> **7.** A specific skill in adept performance, held to require the exercise of intuitive faculties that cannot be learned solely by study <the *art* of storytelling> **8. a. arts.** Artful stratagems. **b.** Artfulness : cunning. **9.** Illustrative material in a printed work.

**art[2]** (ərt; ärt *when stressed*) *v.* [ME < OE *eart.*] *Archaic. 2nd person sing. present indicative of* BE.

**-art** *suff. var. of* -ARD.

**art dec·o** (dĕk′ō) *n.* [Fr. *Art Déco,* from *Exposition Internationale des Arts Décoratifs et Industriels Moderns,* a 1925 exposition in Paris, France.] An early 20th-cent. style of decorative art featuring geometric designs and bold colors.

**ar·te·fact** (är′tə-făkt′) *n. var. of* ARTIFACT.

**ar·tel** (är-tĕl′) *n.* [R. *artel′* < Ital. *artieri,* artisans < *arte,* work < Lat. *ars.*] A cooperative enterprise of agricultural or industrial workers in the U.S.S.R.

**Ar·te·mis** (är′tə-mĭs) *n.* [Gk.] *Gk. Myth.* The virgin goddess of the hunt and the moon and twin sister of Apollo.

**ar·te·mis·i·a** (är′tə-mĭzh′ē-ə, -mĭzh′ə, -mĭz′ē-ə) *n.* [ME *artemesie,* mugwort < OFr. < Lat. *artemisia* < Gk., wormwood, after *Artemis,* to whom it was sacred.] A plant of the genus *Artemisia,* as the sagebrush or wormwood.

**ar·te·ri·al** (är-tîr′ē-əl) *adj.* **1.** Of, like, or in an artery. **2.** Of or designating the bright-red blood in the arteries that has absorbed oxygen in the lungs. **3.** Of or designating a transportation route carrying a main flow with many branches. —*n.* A through road or street. **—ar·te′ri·al·ly** *adv.*

**ar·te·ri·al·ize** (är-tîr′ē-əl-īz′) *vt.* **-ized, -iz·ing, -iz·es.** To convert (venous blood) into arterial blood by absorption of oxygen in the lungs. **—ar·te′ri·al·i·za′tion** *n.*

**arterio-** *pref.* [Gk. *artērio-* < *artēria,* artery.] Artery <*arteriovenous*>

**ar·te·ri·og·ra·phy** (är-tîr′ē-ŏg′rə-fē) *n.* Roentgenography of the arteries following injection of a radiopaque dye. **—ar·te′ri·o·gram′** (är-tîr′ē-ə-grăm′) *n.* **—ar·te′ri·o·graph′ic** *adj.*

**ar·te·ri·ole** (är-tîr′ē-ōl′) *n.* [NLat. *arteriola,* dim. of Lat. *arteria,* windpipe < Gk.] A small terminal branch of an artery, esp. one that connects with a capillary. **—ar·te′ri·o′lar** (-ō′lər, -ə-lər) *adj.*

**ar·te·ri·o·scle·ro·sis** (är-tîr′ē-ō-sklə-rō′sĭs) *n.* A chronic disease in which thickening and hardening of arterial walls interferes with blood circulation. **—ar·te′ri·o·scle·rot′ic** (-rŏt′ĭk) *adj.*

**ar·te·ri·o·ve·nous** (är-tîr′ē-ō-vē′nəs) *adj.* Of, relating to, or connecting both arteries and veins.

**ar·te·ri·tis** (är′tə-rī′tĭs) *n.* Inflammation of an artery.

**ar·ter·y** (är′tə-rē) *n., pl.* **-ies.** [ME *arterie* < Lat. *arteria,* windpipe < Gk.] **1.** *Anat.* Any of a branching system of muscular tubes that carry blood away from the heart. **2.** A major transportation route into which local routes flow.

**ar·te·sian well** (är-tē′zhən) *n.* [Fr. *artésien,* of Artois, where such wells were first drilled.] A well drilled through impermeable strata to reach water capable of rising to the surface by internal hydrostatic pressure.

**art film** *n.* An often experimental motion picture meant to be a serious work of art and not intended for mass appeal.

**art·ful** (ärt′fəl) *adj.* **1.** Displaying art or skill. **2.** Clever and skillful. **3.** Deceitful or crafty. **4.** Not genuine : ARTIFICIAL. **—art′ful·ly** *adv.* **—art′ful·ness** *n.*

**arthr-** *pref. var. of* ARTHRO-.

**ar·thral·gia** (är-thrăl′jə, -jē-ə) *n.* Neuralgic pain in a joint. **—ar·thral′gic** (-jĭk) *adj.*

**ar·thri·tis** (är-thrī′tĭs) *n.* Inflammation of one or more joints. **—ar·thrit′ic** (-thrĭt′ĭk) *adj.* & *n.* **—ar·thrit′i·cal·ly** *adv.*

**arthro-** or **arthr-** *pref.* [< Gk. *arthron,* joint.] Joint <*arthropathy*><*arthroscopy*>

**ar·thro·mere** (är′thrə-mîr′) *n.* One of the typical body segments of an arthropod. **—ar′thro·mer′ic** (är′thrə-mĕr′ĭk, -mîr′ĭk) *adj.*

**ar·throp·a·thy** (är-thrŏp′ə-thē) *n.* A disease of a joint.

**ar·thro·pod** (är′thrə-pŏd′) *n.* [NLat. *Arthropoda,* phylum name : ARTHRO- + Gk. *pous,* foot.] Any of numerous invertebrate organisms of the phylum Arthropoda, including the insects, crustaceans, arachnids, and myriapods. **—ar·throp′o·dous** (är-thrŏp′ə-dəs), **ar·throp′o·dal** (-dəl) *adj.*

**ar·thros·co·py** (är-thrŏs′kə-pē) *n., pl.* **-pies.** Endoscopic examination of a joint, as the knee. **—ar′thro·scope′** (är′thrə-skōp′) *n.* **—ar′thro·scop′ic** (är′thrə-skŏp′ĭk) *adj.*

**ar·thro·sis** (är-thrō′sĭs) *n., pl.* **-ses** (-sēz′) [Gk. *arthrōsis* < *arthron,* joint.] **1.** A joint or connection between bones. **2.** A degenerative process in a joint.

**ar·thro·spore** (är′thrə-spôr′, -spōr′) *n.* A sporelike cell typical of segmented filamentous fungi or certain algae. **—ar′thro·spor′ic, ar′thro·spor′ous** *adj.*

**Ar·thur** (är′thər) *n.* [ME *Artur* < Med. Lat. *Artorius,* prob. of Celtic orig.] A legendary British hero, reputedly the king of the Britons in the 6th cent. A.D.

**Ar·thu·ri·an** (är-thŏŏr′ē-ən) *adj.* Of or relating to King Arthur and his Knights of the Round Table.

**ar·ti·choke** (är′tĭ-chōk′) *n.* [Dial. Ital. *articiocco,* alteration of *arcicioffo* < OSp. *alcarchofa* < Ar. *al-kharshūf.*] **1. a.** A thistlelike plant, *Cynara scolymus,* bearing a large flower head with numerous fleshy, scalelike bracts. **b.** The edible unopened flower head of an artichoke. **2.** The Jerusalem artichoke.

**ar·ti·cle** (är′tĭ-kəl) *n.* [ME < OFr. < Lat. *articulus,* part, dim. of *artus,* joint.] **1.** An individual thing in a class : ITEM <an *article* of apparel> **2.** A small thing. **3.** A particular section or item of a series in a written document, as a treaty or contract. **4.** A nonfictional literary composition, as a report or essay, that forms an independent part of a publication <a newspaper *article*> **5.** Any of a class of words, as the English terms *a, an,* and *the,* used to indicate nouns and specify their application. **6.** A specific point or matter. —*vt.* **-cled, -cling, -cles. 1.** To set forth or state in articles. **2.** To make specific or formal charges against : ACCUSE. **3.** To bind by the articles of a contract.

**ar·tic·u·lar** (är-tĭk′yə-lər) *adj.* [ME *articuler* < Lat. *articularis* < *articulus,* small joint. —see ARTICLE.] Of or relating to an anatomical joint. **—ar·tic′u·lar·ly** *adv.*

**ar·tic·u·late** (är-tĭk′yə-lĭt) *adj.* [Lat. *articulatus,* jointed, p.part. of *articulare,* to divide into joints < *articulus,* small joint. —see ARTICLE.] **1.** Capable of speaking. **2.** Spoken in or separated into clear and distinct syllables or words. **3.** Capable of, speaking in, or marked by clear, expressive language. **4.** *Biol.* Having joints or segments. —*v.* (är-tĭk′yə-lāt′) **-lat·ed, -lat·ing, -lates.** —*vt.* **1.** To utter (speech sounds) by moving the necessary organs of speech. **2.** To pronounce distinctly and carefully : ENUNCIATE. **3.** To express in coherent verbal form <couldn't *articulate* our problem> **4.** To unite by forming a joint. —*vi.* **1.** To utter speech sounds. **2.** To speak clearly and distinctly. **3.** To form a joint. **—ar·tic′u·la·ble** (-lə-bəl) *adj.* **—ar·tic′u·late·ly** *adv.* **—ar·tic′u·late·ness** *n.*

**ar·tic·u·la·tion** (är-tĭk′yə-lā′shən) *n.* **1.** The act or process of articulating : EXPRESSION <an *articulation* of my worst fears> **2. a.** The movements of speech organs used in making a particular speech sound. **b.** A speech sound, esp. a consonant. **3. a.** A jointing together or being jointed together. **b.** Method or manner of jointing. **4.** *Zool.* A joint between bones or between movable parts of an outside shell. **5.** *Bot.* **a.** A joint between two separable parts, as a leaf and a stem. **b.** A node or a space on a stem between two nodes. **—ar·tic'u·la·tive** (-lə-tĭv, -lā'tĭv), **ar·tic'u·la·to'ry** (-lə-tôr'ē, -tōr'ē) *adj.* **—ar·tic'u·la'tor** *n.*

**ar·ti·fact** *also* **ar·te·fact** (är'tə-făkt') *n.* [Lat. *ars, art-* + *factum*, something made, p.part. of *facere*, to make.] **1.** An object produced by human workmanship, esp. a tool, weapon, or ornament of archaeological or historical interest. **2.** *Biol.* A structure or substance not normally present but produced by an external agency or action. **—ar'ti·fac'tu·al** (-făk'cho͞o-əl) *adj.*

**ar·ti·fice** (är'tə-fĭs) *n.* [Fr. < OFr., craftsmanship < Lat. *artificium* < *artifex*, craftsman : *ars*, art + *-fex*, maker < *facere*, to make.] **1.** An artful device or stratagem. **2.** Subtle but base deception : TRICKERY. **3.** Cleverness : ingenuity.

**ar·tif·i·cer** (är-tĭf'ĭ-sər) *n.* **1.** A skilled worker at a craft. **2.** One adept at designing and constructing : INVENTOR.

**ar·ti·fi·cial** (är'tə-fĭsh'əl) *adj.* [ME < Lat. *artificialis* < *artificium*, craftsmanship. —see ARTIFICE.] **1.** Made by human beings rather than occurring in nature. **2.** Made in imitation of something natural. **3.** Not genuine or natural : FEIGNED <an *artificial* welcome> **—ar'ti·fi'ci·al'i·ty** (-fĭsh'ē-ăl'ĭ-tē) *n.* **—ar'ti·fi'cial·ly** *adv.*

☆ ***syns:*** ARTIFICIAL, MANMADE, SYNTHETIC *adj. core meaning* : made by human beings, not by nature <*artificial* snow><*artificial* furs> ***ant*** : natural

**artificial horizon** *n.* An instrument displaying a line on a flight indicator that lies within the horizontal plane and about which the pitching and banking movements of aircraft are shown.

**artificial insemination** *n.* The introduction of semen into the female reproductive organs without sexual contact.

**artificial intelligence** *n.* **1. a.** The characteristics of a machine programmed to imitate human intelligence functions. **b.** Research in methods of such programming. **2.** Research in methods of programming designed to supplement human intellectual abilities.

**artificial language** *n.* A language based on a set of prescribed rules established before the language is used.

**artificial respiration** *n.* A method used to restore normal breathing in an asphyxiated but living person, usu. by rhythmic forcing of air into and out of the lungs.

**ar·til·ler·y** (är-tĭl'ə-rē) *n.* [ME *artelrie* < OFr. *artillerie* < *artillier*, to fortify.] **1.** Large-caliber firing weapons, as howitzers and cannon, that are mounted and manned by crews. **2.** Troops armed with artillery. **3.** The combat arms branch of an army specializing in the use of artillery. **4.** The science of the use of guns : GUNNERY. **5.** A device for discharging missiles, as a catapult or crossbow.

**ar·til·ler·y·man** (är-tĭl'ə-rē-mən) *n.* A soldier in the artillery.

**artillery plant** *n.* A tropical American plant, *Pilea muscosa*, that releases its pollen with an explosive discharge.

**ar·ti·o·dac·tyl** (är'tē-ō-dăk'təl) *n.* [< NLat. *Artiodactyla*, order name : Gk. *artios*, even + *daktulos*, toe.] Any of various hoofed mammals of the order Artiodactyla, which includes cattle, deer, camels, and hippopotamuses, with an even number of toes on each foot. **—ar'ti·o·dac'tyl, ar'ti·o·dac'ty·lous** (-tə-ləs) *adj.*

**ar·ti·san** (är'tĭ-zən, -sən) *n.* [OFr. < OItal. *artigiano* < VLat. **artitianus* < Lat. *artitus*, skilled in the arts, p.part. of *artire*, to instruct in the arts < *ars*, art.] A person skilled in a particular craft.

**art·ist** (är'tĭst) *n.* [OFr. *artiste* < Ital. *artista* < *arte*, art < Lat. *ars*.] **1.** A creater of artistic works, esp. a painter, sculptor, or musician. **2.** One who performs one's work as if it were an art. **3.** An artiste.

**ar·tiste** (är-tēst') *n.* [Fr. < OFr., artist.] A public performer or entertainer, esp. a singer or dancer.

**ar·tis·tic** (är-tĭs'tĭk) *adj.* **1.** Of or pertaining to art or artists. **2.** Appreciative or highly aware of art or beauty. **3.** Showing skill and artistry <an *artistic* floral arrangement> **—ar·tis'ti·cal·ly** *adv.*

**art·ist·ry** (är'tĭ-strē) *n.* **1.** Artistic ability, quality, or workmanship. **2.** The practice or occupation of an artist.

**art·less** (ärt'lĭs) *adj.* **1.** Without guile, cunning, or deceit : INGENUOUS. **2.** Free of artificiality : NATURAL. **3.** Lacking art or skill : CRUDE. **4.** Lacking culture : IGNORANT. **—art'less·ly** *adv.* **—art'less·ness** *n.*

**art·mo·bile** (ärt'mə-bēl') *n.* A trailer for a traveling art exhibit.

**art nou·veau** (är' no͞o-vō', ärt') *n.* [Fr., new art.] A style of decoration and architecture of the late 19th and early 20th cent., marked by the use of flowing, sinuous lines.

**art song** *n.* A lyric song meant to be performed in recital and usu. accompanied by a piano.

**art·sy-craft·sy** (ärt'sē-krăft'sē) *adj. Informal.* Arty.

**art·work** (ärt'wûrk') *n.* **1.** Work in the graphic or plastic arts, esp. small, handmade decorative or artistic objects. **2.** The illustrative and decorative elements of printed materials.

**art·y** (är'tē) *adj.* **-i·er, -i·est.** *Informal.* Pretentiously or affectedly artistic. **—art'i·ly** *adv.* **—art'i·ness** *n.*

☆ ***syns:*** ARTY, ARTSY-CRAFTSY, CONTRIVED, PRECIOUS *adj. core meaning* : pretentiously artistic <*arty* movies>

**ar·um** (âr'əm, âr'-) *n.* [NLat. *Arum*, genus name < Lat., wake-robin < Gk. *aron*.] **1.** A plant of the genus *Arum*, bearing arrow-shaped leaves and small flowers on a spadix surrounded by or enclosed within a spathe. **2.** *also* **arum lily.** Any of several plants related to the arum, as the calla.

**a·rus·pex** (ə-rŭs'pĕks') *n. var. of* HARUSPEX.

**-ary** *suff.* [ME *-arie* < OFr. < Lat. *-arius*, adj. and n. suffix.] **1.** Of or pertaining to <bacill*ary*> **2.** One that pertains to or is connected with <bound*ary*>

**Ar·y·an** (âr'ē-ən, ăr'-) *n.* [Skt. *ārya-*, noble.] **1.** A member of an Indo-European-speaking people that invaded southwestern Asia and northwestern India in the second millennium B.C. **2.** A member of the people who spoke the parent language of the Indo-European languages. **3.** A member of any people speaking an Indo-European language. **4.** Indo-Iranian. *—adj.* **1.** Of or relating to the Indo-European languages or the hypothetical language from which they are derived. **2.** Of or relating to a speaker of an Indo-European language. **3.** Of or relating to Indo-Iranian. **4.** Of or relating to a presumed ethnic type exemplified by or descended from early speakers of Indo-European languages.

**ar·y·te·noid** (ăr'ĭ-tē'noid', ə-rĭt'n-oid') *adj.* [NLat. *arytaenoides* < Gk. *arutainoeidēs*, shaped like a ladle : *arutaina*, ladle (< *aruein*, to draw water) + *-eidēs*, shaped.] *Anat.* **1.** Of or relating to either of two small cartilages attached to the back of the larynx and to the vocal cords. **2.** Of or relating to any of three small muscles of the larynx. *—n.* An arytenoid cartilage or muscle. **—ar'y·te·noid'al** *adj.*

**†as¹** (ăz; əz *when unstressed*) *adv.* [ME < OE *ealswā*.] **1.** To the same extent or degree : EQUALLY. **2.** For instance <cooking spices, *as* cinnamon and cloves> *—conj.* **1.** To the same degree or quantity that. —Often used as a correlative after *so* or *as* <as good *as* gold><not so smart *as* you think> **2.** In the same manner or way that <Do *as* we suggest.> **3.** At the same time that : WHILE. **4.** For the reason that : BECAUSE. **5.** That the outcome is <was so gullible *as* to believe it> **6.** Though <Exciting *as* the party was, I'm glad it's over.> **7.** *Informal.* That <I don't know *as* I will.> *—pron.* **1.** That : which : who. —Used after *same* or *such* <found the same answer *as* you> **2.** A fact that <The class was canceled, *as* you know.> **3.** *Regional.* Who or which <Those *as* are willing can help clean up.> *—prep.* **1.** In the role, capacity, or function of <serving *as* hostess> **2.** In a manner similar to : LIKE <My hands were *as* ice.> **—as is.** *Informal.* Just the way it is : without making changes.

**as²** (ăs) *n., pl.* **as·ses** (ăs'ēz', ăs'ĭz) [Lat.] **1.** An ancient Roman coin of copper or copper alloy. **2.** A unit of weight in ancient Rome equal to about one troy pound.

**As** *symbol for* ARSENIC.

**as-** *pref. var. of* AD- 1. —Used before *s*.

**as·a·fet·i·da** *also* **as·a·foet·i·da** (ăs'ə-fĕt'ĭ-də) *n.* [ME < Med. Lat. : *asa*, gum (< Pers. *azā*) + Lat. *fetida*, stinking. —see FETID.] A yellow-brown, bitter, malodorous resinous material derived from the roots of several plants of the genus *Ferula* and formerly used medicinally.

**as·a·rum** (ăs'ə-rəm) *n.* [NLat. *Asarum*, genus name < Lat., wild spikenard < Gk. *asaron*.] The dried, strong-scented roots of the wild ginger, *Asarum canadense*, once used in medicine and as a flavoring.

**as·bes·tos** *also* **as·bes·tus** (ăs-bĕs'təs, ăz-) *n.* [Lat. < Gk., unquenchable : *a-*, not + *sbennunai*, to quench.] Either of two incombustible, chemical-resistant, fibrous mineral forms of impure magnesium silicate, used for fireproofing, electrical insulation, building materials, brake linings, and chemical filters. **—as·bes'tine** (-tĭn), **as·bes'tic** (-tĭk) *adj.* **—as·bes'tos, as·bes'tus** *adj.*

**as·bes·to·sis** (ăs'bĕs-tō'sĭs, ăz'-) *n.* [ASBEST(OS) + -OSIS.] Chronic lung inflammation resulting from prolonged inhalation of asbestos particles. **—as'bes·tot'ic** (-tŏt'ĭk) *adj.*

**as·bes·tus** (ăs-bĕs'təs, ăz-) *n. var. of* ASBESTOS.

**as·ca·ri·a·sis** (ăs'kə-rī'ə-sĭs) *n.* [ASCAR(ID) + -IASIS.] Infestation with nematode worms of the species *Ascaris lumbricoides*.

**as·ca·rid** (ăs'kə-rĭd) *n.* [NLat. *Ascaridae*, family name < Gk. *askaris*, intestinal worm.] A nematode worm of the family Ascaridae, as the common intestinal parasite *Ascaris lumbricoides*.

**as·cend** (ə-sĕnd') *v.* **-cend·ed, -cend·ing, -cends.** [ME *ascenden* < Lat. *ascendere* : *ad-*, toward + *scandere*, to climb.] *—vi.* **1.** To go or move upward : RISE. **2.** To slope upward. *—vt.* **1.** To move upward on or along : CLIMB. **2.** To come to occupy (a throne). **—as·cend'a·ble, as·cend'i·ble** *adj.*

**as·cen·dance** *also* **as·cen·dence** (ə-sĕn'dəns) *n.* Ascendancy.

**as·cen·dan·cy** *also* **as·cen·den·cy** (ə-sĕn'dən-sē) *n.* The state of being in the ascendant : DOMINATION <"Germany only awaits trade revival to gain an immense mercantile *ascendancy*" —Winston Churchill>

**as·cen·dant** *also* **as·cen·dent** (ə-sĕn'dənt) *—adj.* **1.** Moving or inclining upward : ASCENDING. **2.** Highest in position or influence : SUPERIOR. *—n.* **1.** The position or state of being dominant or in power. **2.** The section of the zodiac that rises in the east at the time of a particular event, as a person's birth. **3.** An ancestor.

**as·cen·dence** (ə-sĕn'dəns) *n. var. of* ASCENDANCE.

**as·cen·den·cy** (ə-sĕn'dən-sē) *n. var. of* ASCENDANCY.

**as·cen·dent** (ə-sĕn'dənt) *adj. & n. var. of* ASCENDANT.

**as·cend·er** (ə-sĕn'dər) *n.* **1.** One that ascends. **2. a.** The part of certain lower-case letters, as *b*, that extends above most other lower-case letters. **b.** A letter with an ascender.

**as·cend·ing** (ə-sĕn′dĭng) *adj.* Going, moving, or growing upward. **—as·cend′ing·ly** *adv.*

**as·cen·sion** (ə-sĕn′shən) *n.* [ME *ascensioun* < Lat. *ascensio* < *ascendere,* to ascend.] **1.** The act or process of ascending : ASCENT. **2.** The rising of a star above the horizon. **—as·cen′sion·al** *adj.*

**Ascension Day** *n.* The 40th day after Easter, observed in commemoration of Christ's ascension into heaven.

**as·cent** (ə-sĕnt′) *n.* [< ASCEND.] **1.** The act or process of ascending. **2.** An advancement, esp. in social status. **3.** An upward slope or incline. **4.** A going back in time or genealogical succession.

**as·cer·tain** (ăs′ər-tān′) *vt.* **-tained, -tain·ing, -tains.** [ME *acertainen,* to inform < OFr. *ascertener* : *a-,* to (< Lat. *ad-*) + *certain,* certain. —see CERTAIN.] **1.** To discover through experimentation or examination. **2.** *Archaic.* To make certain. **—as′cer·tain′a·ble** *adj.* **—as′cer·tain′a·ble·ness** *n.* **—as′cer·tain′a·bly** *adv.* **—as′cer·tain′ment** *n.*

**as·cet·ic** (ə-sĕt′ĭk) *n.* [Gk. *askētēs,* hermit < *askein,* to work.] One who leads a life of austere self-discipline, esp. as an act of religious devotion or penance. **—as·cet′ic** *adj.*

**as·ci** (ăs′ī′, -kī′) *n. pl. of* ASCUS.

**as·cid·i·an** (ə-sĭd′ē-ən) *n.* [< NLat. *Ascidia,* genus name < Gk. *askidion,* dim. of *askos,* wineskin.] Any of various saclike marine animals of the class Ascidiacea, including the sea squirts.

**as·cid·i·um** (ə-sĭd′ē-əm) *n., pl.* **-i·a** (-ē-ə) [NLat. < Gk. *askidion,* dim. of *askos,* wineskin.] *Bot.* A sac-shaped or bottle-shaped part or organ, as a leaf of a pitcher plant.

**ASCII** (ăs′kē) *n.* [Acronym for *American Standard Code Information Interchange.*] *Computer Sci.* **1.** A proposed standard for defining codes for information exchange between equipment produced by different manufacturers. **2.** A code following this proposed standard.

**as·ci·tes** (ə-sī′tēz) *n., pl.* **ascites.** [ME *aschites* < Med. Lat. *ascites* < Gk. *askitēs* < *askos,* belly.] Abnormal accumulation of serous fluid in the abdominal cavity. **—as·cit′ic** (-sĭt′ĭk) *adj.*

**As·cle·pi·us** (ə-sklē′pē-əs) *n.* [Gk. *Asklēpios.*] *Gk. Myth.* The god of medicine.

**asco-** *pref.* [NLat. < Gk. *askos,* bag.] Ascus <*ascospore*>

**as·co·carp** (ăs′kə-kärp′) *n. Bot.* A globular structure containing the spore sacs of ascomycetous fungi.

**as·co·go·ni·um** (ăs′kə-gō′nē-əm) *n., pl.* **-ni·a** (-nē-ə). *Bot.* A female reproductive structure of certain fungi.

**as·co·my·cete** (ăs′kō-mī′sēt′, -mī-sēt′) *n. Bot.* Any of numerous fungi that produce spores in an ascus.

**a·scor·bic acid** (ə-skôr′bĭk) *n.* [A-[1] + SCORB(UT)IC.] *Chem.* A white crystalline vitamin, $C_6H_8O_6$, occurring in citrus fruits, tomatoes, potatoes, and leafy green vegetables and used to prevent scurvy.

**as·co·spore** (ăs′kə-spôr′, -spōr′) *n.* A sexual spore formed in an ascus. **—as′co·spo′rous** (-spôr′əs, -spōr′-, ăs-kŏs′pər-əs), **as′co·spor′ic** (-spôr′ĭk, -spŏr′-) *adj.*

**as·cot** (ăs′kət, -kŏt′) *n.* [After *Ascot,* England.] A scarf knotted so its broad ends are laid flat, one over the other.

**as·cribe** (ə-skrīb′) *vt.* **-cribed, -crib·ing, -cribes.** [ME *ascriben* < Lat. *ascribere* : *ad-,* to + *scribere,* to write.] **1.** To attribute to a given cause, source, or origin <*ascribed* my bad mood to fatigue> **2.** To assign as an attribute <*ascribe* glory to God>

**as·crip·tion** (ə-skrĭp′shən) *n.* [Lat. *ascriptio* < *ascribere,* to ascribe.] **1.** The act of ascribing. **2.** A statement that ascribes.

**as·cus** (ăs′kəs) *n., pl.* **as·ci** (ăs′ī′, -kī′) [NLat. < Gk. *askos,* bag.] *Bot.* A membranous sac in certain fungi, containing ascospores.

**as·dic** (ăz′dĭk) *n.* [A(NTI)S(UBMARINE) + D(ETECTION) + I(NVESTIGATION) + C(OMMITTEE).] An antisubmarine sonar device.

**-ase** *suff.* [< DIASTASE.] Enzyme <amyl*ase*>

**a·sea** (ə-sē′) *adv.* Toward or on the sea or at sea.

**a·sep·sis** (ə-sĕp′sĭs, ā-) *n.* The state of being free from pathogenic organisms.

**a·sep·tic** (ə-sĕp′tĭk, ā-) *adj.* **1.** Of or relating to asepsis. **2. a.** Without animation or emotion. **b.** Impersonal. **—a·sep′ti·cal·ly** *adv.*

**a·sex·u·al** (ā-sĕk′sho͞o-əl) *adj.* **1.** Having no evident sex or sex organs. **2.** Relating to or characterizing reproduction involving a single individual and without male or female gametes, as in binary fission or budding. **—a·sex′u·al′i·ty** (-ăl′ĭ-tē) *n.* **—a·sex′u·al·ly** *adv.*

**as for** *prep.* With regard to <*As for* the punishment, it will fit the crime.>

**As·gard** (ăs′gärd′, ăz′-) *n.* [ON : *āss,* god + *garðr,* court.] *Norse Myth.* The heavenly residence of the gods and slain war heroes.

**ash[1]** (ăsh) *n.* [ME *asshe* < OE *asce.*] **1.** Grayish-white to black soft solid residue of combustion. **2.** *Geol.* Pulverized particulate matter ejected by volcanic eruption. **3. ashes.** Ruins. **4. ashes.** Human remains, esp. after cremation.

**ash[2]** (ăsh) *n.* [ME *asshe* < OE *æsc.*] **1.** A tree of the genus *Fraxinus,* with compound leaves, small greenish flower clusters, and winged seeds. **2.** The durable, close-grained, elastic wood of an ash.

**a·shamed** (ə-shāmd′) *adj.* [ME < OE *āsceamod,* p.part. of *āsceamian,* to feel shame.] **1.** Feeling shame or guilt. **2.** Reluctant through fear of shame <*ashamed* to say I was broke>

**A·shan·ti** (ə-shăn′tē, ə-shän′-) *n., pl.* **Ashanti** or **-tis. 1.** An inhabitant of Ashanti. **2.** Twi.

**ash can** *n.* **1.** A large, usu. metal receptacle for ashes or trash. **2.** *Slang.* A depth charge.

**ash·en[1]** (ăsh′ən) *adj.* **1.** Consisting of ashes. **2.** Like ashes, as in color <*ashen* skin>

**ash·en[2]** (ăsh′ən) *adj.* Of, relating to, or made from the wood of the ash tree.

**Ash·er** (ăsh′ər) *n.* [Heb. *Ashēr.*] The tribe of Israel descended from Asher, a son of Jacob.

**Ash·ke·naz·i** (ăsh′kə-năz′ē) *n., pl.* **-naz·im** (-năz′ĭm) [Heb. *Ashkĕnāzi.*] A gen. Yiddish-speaking, central or eastern European Jew. **—Ash′ke·naz′ic** (-năz′ĭk) *adj.*

**ash·lar** (ăsh′lər) *n.* [ME *assheler* < OFr. *aisselier,* beam < Lat. *axilla,* dim. of *axis,* plank.] **1. a.** A squared block of building stone. **b.** Masonry of such stones. **2.** A thin, dressed rectangle of stone for facing walls.

**a·shore** (ə-shôr′, ə-shōr′) *adv.* **1.** To or on the shore. **2.** Aground.

**ash·plant** (ăsh′plănt′) *n.* A walking stick or staff made from an ash sapling.

**ash·ram** (ăsh′rəm) *n.* [Skt. *aśrama.*] **1.** A Hindu philosopher or religious teacher's usu. secluded residence. **2.** A religious retreat or commune.

**Ash·to·reth** (ăsh′tə-rĕth′) *n.* [Heb. *'Ashtōreth.*] *Myth.* The ancient Syrian and Phoenician goddess of sexual love and fertility.

**ash·tray** (ăsh′trā′) *n.* A receptacle for tobacco ashes.

**A·shur** (ä′sho͝or′) *n.* [Assyrian *Ashūr.*] *Myth.* The principal Assyrian deity and god of war and empire.

**Ash Wednesday** *n.* The seventh Wednesday before Easter and the first day of Lent.

**ash·y** (ăsh′ē) *adj.* **-i·er, -i·est. 1.** Relating to, like, or covered with ashes. **2.** Having the color of ashes : PALLID. **—ash′i·ness** *n.*

**A·sian** (ā′zhən, ā′shən) *adj.* Of or relating to Asia or its people. *—n.* A native or resident of Asia.

**A·si·at·ic** (ā′zhē-ăt′ĭk, -shē-, -zē-) *adj.* Asian. *—n.* An Asian.

**Asiatic cholera** *n.* A form of cholera.

**a·side** (ə-sīd′) *adv.* **1.** On or to one side <move *aside*> **2.** Out of one's thoughts or mind <put our fears *aside*> **3.** Apart <an hour set *aside* for meditation> **4.** In reserve <funds put *aside* for school> **5.** Dispensed with <all kidding *aside*> *—n.* **1.** A line spoken by a character in a play that other actors on stage are supposed by dramatic convention not to hear. **2.** A parenthetical digression.

**aside from** *prep.* Except for : EXCLUDING.

**as if** *conj.* **1.** In the same way that it would be if <looked *as if* it would rain> **2.** That <seemed *as if* they'd never leave>

**as·i·nine** (ăs′ə-nīn′) *adj.* [Lat. *asininus,* of an ass < *asinus,* ass.] **1.** Marked by failure to exercise prudent judgment or common sense : SILLY <*asinine* behavior> **2.** Of, relating to, or like an ass. **—as′i·nine·ly** *adv.* **—as′i·nin′i·ty** (-nĭn′ə-tē) *n.*

**ask** (ăsk, äsk) *v.* **asked, ask·ing, asks.** [ME *asken* < OE *āscian.*] *—vt.* **1.** To put a question to. **2.** To seek information about : inquire about. **3.** To request of or for : SOLICIT <*asking* your help> **4. a.** To require or call for. **b.** To expect or demand <*ask* a lot of a friend> **5.** To invite. **6.** *Archaic.* To publish, as marriage banns. *—vi.* **1.** To inquire <*asked* about my job> **2.** To make a request <*asked* for more money>

**a·skance** (ə-skăns′) *adv.* [Orig. unknown.] **1.** With a side or oblique glance : SIDEWISE. **2.** With disapproval, skepticism, or distrust.

**a·skew** (ə-skyo͞o′) *adj. & adv.* [Prob. A-[2] + SKEW.] To one side : AWRY.

**a·slant** (ə-slănt′) *adj. & adv.* At a slant : OBLIQUELY. *—prep.* Obliquely over or across.

**a·sleep** (ə-slēp′) *adj.* **1.** In a condition of sleep. **2.** Inactive : dormant. **3.** Numb <My hand is *asleep.*> **4.** Dead. *—adv.* Into a state of sleep.

**as long as** *conj.* **1.** Since <*As long as* we're here, let's visit.> **2.** On the condition that <I'll go *as long as* I can drive.>

**a·slope** (ə-slōp′) *adv. & adj.* At a slope or slant.

**a·so·cial** (ā-sō′shəl) *adj.* **1.** Avoiding the company of others. **2.** Unable to interact adequately with others.

**as of** *prep.* On : at <in residence *as of* Sept. 15>

**asp** (ăsp) *n.* [ME *aspide* < Lat. *aspis* < Gk.] Any of several poisonous Old World snakes, as the small cobra, *Naja haje,* or the horned viper, *Cerastes cornutus,* both of Africa and Asia Minor.

**as·par·a·gus** (ə-spăr′ə-gəs) *n.* [Lat. < Gk. *asparagos.*] **1.** Any of several plants of the genus *Asparagus,* indigenous to Eurasia, with small scales or needlelike branchlets rather than true leaves, esp. the widely cultivated species *A. officinalis.* **2.** The edible, succulent young shoots of the asparagus.

**asparagus beetle** *n.* A small, spotted beetle, *Crioceris asparagi,* that infests and damages asparagus plants.

**asparagus fern** *n.* A vine indigenous to southern Africa *Asparagus plumosus,* with fernlike stems.

**as·par·tame** (ăs′pər-tām′, ə-spär′-) *n.* [ASPART(IC ACID) + (PHENYL)A(LANINE) + M(ETHYL) + E(STER).] An artificial sweetener, $C_{14}H_{18}N_2O_5$, derived from aspartic acid.

**as·par·tic acid** (ə-spär′tĭk) *n.* [< ASPARAGUS (from its being obtained by hydrolysis from an amino acid found in asparagus).] A nonessential amino acid, $C_4H_7NO_4$, occurring esp. in young sugar cane and sugar-beet molasses.

**as·par·to·ki·nase** (ə-spär′tō-kī′nās′) *n.* [ASPART(IC ACID) + KINASE.] An enzyme that catalyzes aspartic acid phosphorylation by ATP.

**as·pect** (ăs′pĕkt′) *n.* [ME < Lat. *aspectus,* a view, p.part. of *aspicere,* to look at : *ad-,* toward + *specere,* to look.] **1.** A particular facial expression : MIEN <a judge of stern *aspect*> **2.** Appearance to the eye, esp. when seen from a specific view. **3.** The way in which an idea, problem, or situation is considered mentally. **4.** A position facing or commanding a given direction : EXPOSURE. **5.** A side or surface facing in a particular direction <the dorsal *aspect* of the body> **6.** The configuration of the stars or planets in relation to one another or to an observer, held by astrologers to influence human affairs. **7.** A category of the verb denoting primarily the relation of the action to the passage of time, esp. in reference to completion, duration, or repetition. **8.** *Archaic.* A gaze : look.

**aspect ratio** *n.* The width-to-height ratio of a television image.

**as·pen** (ăs′pən) *n.* [ME *aspe* < OE *æsp.*] A tree of the genus *Populus,* bearing leaves attached by flattened leafstalks so that they flutter readily in the wind. *—adj.* **1.** Of or relating to an aspen. **2.** Trembling like the leaves of an aspen.

**as·per·ate** (ăs′pə-rāt′) *vt.* **-at·ed, -at·ing, -ates.** [Lat. *asperare, asperat-* < *asper,* rough.] To make uneven : ROUGHEN.

**as·per·ges** (ə-spûr′jēz) *n.* [Lat., from the phrase *asperges me,* you will sprinkle me, the first words of the rite.] *Rom. Cath. Ch.* A short rite preceding the High Mass on Sundays that consists of sprinkling the altar, clergy, and congregation with holy water.

**as·per·gill** (ăs′pər-jĭl) *also* **as·per·gil·lum** (ăs′pər-jĭl′əm) *n., pl.* **-gills** *also* **-gil·la** (-jĭl′ə) [NLat. *aspergillum* < Lat. *aspergere,* to sprinkle. —see ASPERSE.] *Rom. Cath. Ch.* An instrument, as a brush or a perforated container, used for sprinkling holy water.

**as·per·gil·li** (ăs′pər-jĭl′ī′) *n. pl. of* ASPERGILLUS.

**as·per·gil·lo·sis** (ăs′pər-jə-lō′sĭs) *n.* [ASPERGILL(US) + -OSIS.] An infectious disease esp. of the skin or lungs, caused by certain fungi of the genus *Aspergillus.*

**as·per·gil·lum** (ăs′pər-jĭl′əm) *n. var. of* ASPERGILL.

**as·per·gil·lus** (ăs′pər-jĭl′əs) *n., pl.* **-gil·li** (-jĭl′ī′) [NLat. < *aspergillum,* aspergill, from its resemblance to an aspergill brush.] Any of various fungi of the genus *Aspergillus,* which includes many common molds.

**as·per·i·ty** (ă-spĕr′ĭ-tē) *n.* [Lat. *asperitas* < *asper,* rough.] **1.** Roughness or harshness, as of surface, weather, or sound. **2.** Ill temper : IRRITABILITY.

**as·perse** (ə-spûrs′) *vt.* **-persed, -pers·ing, -pers·es.** [Lat. *aspergere, aspers-,* to sprinkle : *ad-,* toward + *spergere,* to strew.] **1.** To spread untrue charges or damaging insinuations against : DEFAME. **2.** To sprinkle, as with holy water.

**as·per·sion** (ə-spûr′zhən, -shən) *n.* **1.** A defamatory report or remark : CALUMNY. **2.** The act of defaming. **3.** A sprinkling, esp. with holy water.

**as·phalt** (ăs′fôlt′) *also* **as·phal·tum** (ăs-fôl′təm) *or* **as·phal·tus** (-təs) *n.* [ME *aspalt* < Med. Lat. *asphaltus* < Gk. *asphaltos.*] **1.** A brownish-black solid or semisolid mixture of bitumens obtained from native deposits or as a petroleum by-product and used in roofing, paving, and waterproofing. **2.** Mixed asphalt and crushed gravel or sand, used for roofing or paving. *—vt.* **-phalt·ed, -phalt·ing, -phalts.** To pave or cover with asphalt.

**as·phal·tite** (ăs′fôl-tīt′) *n.* A solid, dark-colored complex of hydrocarbons, found in natural veins and deposits.

**as·phal·tum** (ăs-fôl′təm) *or* **as·phal·tus** (-təs) *n. vars. of* ASPHALT.

**a·spher·ic** (ā-sfîr′ĭk, ā-sfĕr′-) *also* **a·spher·i·cal** (-ĭ-kəl) *adj.* Varying only slightly from sphericity and having only slight aberration, as a lens.

**as·pho·del** (ăs′fə-dĕl′) *n.* [Lat. *asphodelus* < Gk. *asphodelos.*] Any of several plants of the genera *Asphodeline* and *Asphodelus* of the Mediterranean region, with white or yellow flower clusters.

**as·phyx·i·a** (ăs-fĭk′sē-ə) *n.* [Gk. *asphuxia,* stopping of the pulse : *a-,* without + *sphuxis,* heartbeat < *sphuzein,* to throb.] Unconsciousness or death resulting from lack of oxygen.

**as·phyx·i·ant** (ăs-fĭk′sē-ənt) *adj.* Causing or tending to cause asphyxia. *—n.* An asphyxiant substance or condition.

**as·phyx·i·ate** (ăs-fĭk′sē-āt′) *v.* **-at·ed, -at·ing, -ates.** *—vt.* To cause asphyxia in : SMOTHER. *—vi.* To undergo asphyxia : SUFFOCATE. **—as·phyx′i·a′tion** *n.* **—as·phyx′i·a′tor** *n.*

**as·pic**[1] (ăs′pĭk) *n.* [Fr., asp (from the resemblance of the jelly's coloration to an asp's). —see ASPIC[2].] **1.** A molded dish of jellied meat, fish, vegetables, or fruit. **2.** A jellied garnish of meat or fish stock.

**as·pic**[2] (ăs′pĭk) *n.* [Fr. < OFr., alteration of *aspe* < Lat. *aspis* < Gk.] *Archaic.* An asp.

**as·pi·dis·tra** (ăs′pĭ-dĭs′trə) *n.* [NLat. *Aspidistra,* genus name < Gk. *aspis,* shield.] An Asian plant of the genus *Aspidistra,* esp. *A. lurida,* with long, tough, evergreen leaves and small brownish flowers, widely grown as a house plant.

**as·pi·rant** (ăs′pər-ənt, ə-spīr′-) *n.* One who aspires, esp. for advancement, honors, or a high position. *—adj.* Aspiring for position, recognition, or distinction.

**as·pi·rate** (ăs′pə-rāt′) *vt.* **-rat·ed, -rat·ing, -rates.** [Lat. *aspirare, aspirat,* to breathe on : *ad-* to + *spirare,* to breathe.] **1. a.** To pronounce (a vowel or word) with the initial release of breath associated with English *h,* as in *Halloween.* **b.** To follow (a consonant, esp. a stop consonant) with a puff of breath that is clearly audible before the next sound begins, as in English *p, t,* and *k* before vowels. **2.** *Med.* **a.** To remove (liquids or gases) with an aspirator. **b.** To draw (foreign matter, esp. food particles) into the lungs with the breath. *—n.* (-pər-ĭt). **1.** The speech sound represented by English *h.* **2.** The puff of air accompanying the release of a stop consonant. **3.** A speech sound followed by a puff of breath.

**as·pi·ra·tion** (ăs′pə-rā′shən) *n.* **1.** Expulsion of breath in speech. **2. a.** The pronunciation of a consonant with an aspirate. **b.** An aspirate. **3.** *Med.* **a.** Removal of liquids or gases with an aspirator. **b.** A drawing of foreign matter in the upper respiratory tract into the lungs with the breath. **4. a.** A strong desire for high achievement : AMBITION. **b.** An object of such desire : GOAL.

**as·pi·ra·tor** (ăs′pə-rā′tər) *n.* **1.** A device that removes liquids or gases from a space by suction, esp. one used medically to evacuate a bodily cavity. **2.** A suction pump used to create a partial vacuum.

**as·pi·ra·to·ry** (ə-spīr′ə-tôr′ē, -tōr′ē) *adj.* Of, relating to, or suited for breathing or suction.

**as·pire** (ə-spīr′) *vi.* **-pired, -pir·ing, -pires.** [ME *aspiren* < Lat. *aspirare,* to desire. —see ASPIRATE.] **1.** To have a fervent hope or ambition <*aspired* to be a prima ballerina> **2.** To strive toward an end <*aspiring* to great wealth> **3.** *Archaic.* To rise upward : SOAR. **—as·pir′er** *n.* **—as·pir′ing·ly** *adv.*

**as·pi·rin** (ăs′pə-rĭn, -prĭn) *n.* [Orig. a trademark.] **1.** A white crystalline compound of acetylsalicylic acid, $CH_3COOC_6H_4COOH$, used as an antipyretic and analgesic. **2.** Asprin in tablet or liquid form.

**a·squint** (ə-skwĭnt′) *adv. & adj.* [ME.] With a sidelong glance.

**ass** (ăs) *n., pl.* **ass·es** (ăs′ĭz) [ME *asse* < OE *assa.*] **1.** Any of several hoofed mammals of the genus *Equus,* closely related to the horses and zebras and including the domesticated donkey. **2.** A foolish or stupid person : DOLT.

**as·sa·gai** *or* **as·se·gai** (ăs′ə-gī′) *n.* [OFr. *azagaie,* prob. < OSp. *azagaya* < Ar. *az-zaghāyah* : *al,* the + Berber *zaghāyah,* spear.] **1.** A light spear or javelin, often with an iron tip, used by southern African tribesmen. **2.** A tree, *Curtisia faginea* of southern Africa, yielding wood used for making assagais.

**as·sai**[1] (ä-sī′) *n.* [Port. (Brazil) *assaí* < Tupi *assahi.*] **1.** A tropical South American palm tree of the genus *Euterpe,* with edible, fleshy purple fruit. **2.** A beverage made from the fruit of the assai.

**as·sai**[2] (ä-sī′) *adv.* [Ital. < VLat. **ad satis,* to sufficiency. —see ASSET.] *Mus.* Very. —Used in tempo directions.

**as·sail** (ə-sāl′) *vt.* **-sailed, -sail·ing, -sails.** [ME *assailen* < OFr. *assaillier* < VLat. **assalire,* var. of Lat. *assilire,* to jump on : *ad-,* onto + *salire,* to jump.] **1.** To attack with or as if with violent blows : ASSAULT. **2.** To attack verbally, as with ridicule. **—as·sail′a·ble** *adj.* **—as·sail′a·ble·ness** *n.* **—as·sail′er** *n.* **—as·sail′ment** *n.*

**as·sail·ant** (ə-sā′lənt) *n.* One who assails another.

**As·sam·ese** (ăs′ə-mēz′, -mēs′) *adj.* Of or relating to Assam, its people, or their language. *—n., pl.* **Assamese. 1.** A native or resident of Assam. **2.** The Indic language of the Assamese.

**as·sas·sin** (ə-săs′ĭn) *n.* [Fr. < Med. Lat. *assassinus* < Ar. *ḥashshāshīn,* user of hashish < *ḥashīsh,* hashish.] **1.** A murderer, esp. one who carries out a plot to kill a prominent person. **2. Assassin.** A member of a secret order of Moslem fanatics who terrorized and killed Christian Crusaders.

**as·sas·si·nate** (ə-săs′ə-nāt′) *vt.* **-nat·ed, -nat·ing, -nates. 1.** To murder (a prominent person). **2.** To destroy or injure (e.g., an opponent's character) treacherously. **—as·sas′si·na′tion** *n.* **—as·sas′si·na′tive** *adj.* **—as·sas′si·na′tor** *n.*

**assassin bug** *n.* Any of various predatory insects of the large family Reduviidae, with short, curved, powerful beaks adapted for sucking blood and capable of inflicting a painful bite.

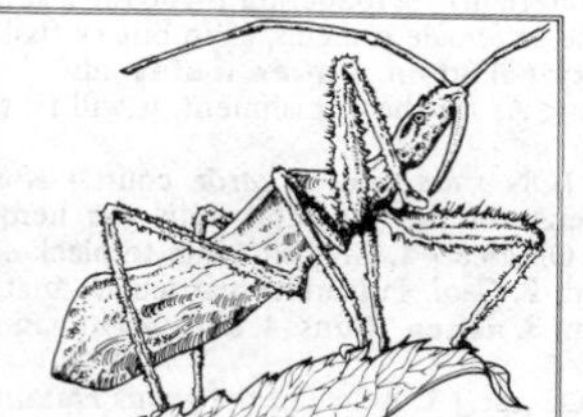

**assassin bug**
*Over 4,000 species, some up to one inch in length*

**as·sault** (ə-sôlt′) *n.* [ME *assaut* < OFr. < VLat. **assaltus,* var. of Lat. *assultus,* p.part. of *assilire,* to jump on. —see ASSAIL.] **1.** A violent physical or verbal attack. **2. a.** A military attack on a fortified place.

**b.** The final stage of an attack that includes close combat with the enemy. **3.** *Law.* An unlawful threat or attempt to injure another physically. **4.** RAPE[1] 1, 2. —*v.* **-sault·ed, -sault·ing, -saults.** —*vt.* To attack violently. —*vi.* To make an assault. **—as·sault'er** *n.*

**assault and battery** *n. Law.* The threat to use force upon another and the carrying out of the threat.

**as·say** (ăs'ā', ă-sā') *n.* [ME *assai* < OFr. —see ESSAY.] **1. a.** The qualitative or quantitative analysis of a substance, esp. of an ore or drug. **b.** A substance to be so analyzed. **c.** The result of such an analysis. **2.** An analysis or examination. **3.** *Obs.* An attempt : essay. —*v.* (ă-sā', ăs'ā') **-sayed, -say·ing, -says.** —*vt.* **1.** To subject to chemical analysis. **2.** To examine by trial or experiment : TEST <*assay* one's skill> **3.** To assess or evaluate. **4.** To attempt <*assay* skiing> —*vi.* To be shown by analysis to have a certain proportion, usu. of a precious metal. **—as·say'a·ble** *adj.* **—as·say'er** *n.*

**as·se·gai** (ăs'ə-gī') *n. var. of* ASSAGAI.

**as·sem·blage** (ə-sĕm'blĭj) *n.* **1.** ASSEMBLY 1. **2.** A group of people or things. **3.** A fitting together of manufactured parts, as of a machine. **4.** A sculpture consisting of an arrangement of miscellaneous objects, as scraps of metal, cloth, string, etc.

**as·sem·ble** (ə-sĕm'bəl) *v.* **-bled, -bling, -bles.** [ME *assemblen* < OFr. *assembler* < VLat. **assimulare* : Lat. *ad-*, to + Lat. *simul*, together.] —*vt.* **1.** To bring or gather together into a group or whole. **2.** To fit or join together the parts of. —*vi.* To come together : CONGREGATE.

**as·sem·bler** (ə-sĕm'blər) *n.* **1.** One that assembles. **2.** *Computer Sci.* A program operating on symbolic input data to produce the equivalent machine code.

**as·sem·bly** (ə-sĕm'blē) *n., pl.* **-blies. 1.** The act of assembling or state of being assembled. **2.** A group of persons gathered together for a common purpose : MEETING. **3. Assembly.** The lower house of the legislature in certain U.S. states. **4. a.** The combining of manufactured parts to make a completed product, esp. a machine. **b.** A set of parts so combined. **5.** The signal calling troops to form ranks.

**assembly language** *n. Computer Sci.* A programming language that is a close approximation of machine code.

**assembly line** *n.* A line of factory workers and equipment on which the product being assembled passes consecutively from operation to operation until completed.

**as·sem·bly·man** (ə-sĕm'blē-mən) *n.* A man who is a member of a legislative assembly.

**Assembly of God** *n.* A Pentecostal congregation founded in the United States in 1914.

**assembly time** *n. Computer Sci.* The time required for an assembler to translate symbolic language into machine instructions.

**as·sem·bly·wom·an** (ə-sĕm'blē-wo͝om'ən) *n.* A woman who is a member of a legislative assembly.

**as·sent** (ə-sĕnt') *vi.* **-sent·ed, -sent·ing, -sents.** [ME *assenten* < OFr. *assentir* < Lat. *assentari* : *ad-*, toward + *sentire*, to feel.] To express agreement : CONCUR. —*n.* **1.** Agreement, as to a plan or proposal. **2.** Acquiescence : consent. **—as·sent'er, as·sen'tor** *n.* **—as·sent'ing·ly** *adv.* **—as·sen'tive** *adj.* **—as·sen'tive·ness** *n.*

**as·sen·ta·tion** (ăs'ĕn-tā'shən) *n.* Servile or ill-considered agreement with another's opinions.

**as·sert** (ə-sûrt') *vt.* **-sert·ed, -sert·ing, -serts.** [Lat. *asserere, assert-*, : *ad-*, to + *serere*, to join.] **1.** To state or express positively : AFFIRM. **2.** To defend or maintain (e.g., one's rights). **—assert oneself.** To express oneself boldly or forcefully. **—as·sert'a·ble, as·sert'i·ble** *adj.* **—as·sert'er, as·ser'tor** *n.*

☆ ***syns:*** ASSERT, AFFIRM, AVER, AVOUCH, AVOW, DECLARE, HOLD, MAINTAIN, STATE *v. core meaning* : to put into words, positively and with conviction <*asserted* their innocence> ***ant:*** controvert, deny

**as·ser·tion** (ə-sûr'shən) *n.* **1.** The act of asserting. **2.** Something asserted. **—as·ser'tion·al** *adj.*

**as·ser·tive** (ə-sûr'tĭv) *adj.* Inclined to or displaying bold assertion : SELF-CONFIDENT. **—as·ser'tive·ly** *adv.* **—as·ser'tive·ness** *n.*

**assertiveness training** *n.* A method of training individuals to behave in a boldly self-confident manner.

**as·sess** (ə-sĕs') *vt.* **-sessed, -sess·ing, -sess·es.** [ME *assessen* < OFr. *assesser* < Lat. *assidēre*, to sit by (as an assistant judge) : *ad-*, near to + *sedēre*, to sit.] **1.** To estimate the value of (property) for taxation. **2.** To set or determine the amount of (e.g., a tax or fine). **3.** To charge (a person or property) with a special payment, as a tax or fine. **4.** To appraise or evaluate. **—as·sess'a·ble** *adj.*

**as·sess·ment** (ə-sĕs'mənt) *n.* **1.** The act, process, or an instance of assessing. **2.** An amount assessed.

**as·ses·sor** (ə-sĕs'ər) *n.* **1.** An official who makes assessments, as for taxation. **2.** An assistant to a judge, selected for his or her specialized knowledge. **—as'ses·so'ri·al** (ăs'ə-sôr'ē-əl, -sōr'-) *adj.*

**as·set** (ăs'ĕt') *n.* [Back-formation < E. *assets* < AN *asetz*, sufficient goods to settle a testator's debts and legacies < OFr. *asez*, enough < VLat. **ad satis* : Lat. *ad-*, to + *satis*, enough.] **1.** A useful or valuable quality or thing <Beauty can be a great *asset.*> **2.** A valuable material possession. **3. assets.** The entries on a balance sheet showing all properties and claims against others that may be directly or indirectly applied to cover liabilities.

▲ word history: *Asset* is an example of the process of back-formation. By this process a word is mistakenly analyzed as a base word augmented by an affix. *Asset* is a back-formation from the old legal term *assets*, which was not a plural noun (*asset* + *-s*); in fact, it was not a noun at all but an adjective. *Assets* was originally *asetz* or *asez*, an Old French word meaning simply "enough," as does *assez*, the modern French form. *Assets* was used as legal shorthand for "enough wealth to settle the claims made against a deceased person's estate." Because *assets* looked like a plural form and had a collective meaning, the word came to be treated grammatically as a plural. A singular form *asset* appeared in the 19th century to denote a single item in the "assets" column of a balance sheet, and from that usage the figurative meanings developed.

**as·sev·er·ate** (ə-sĕv'ə-rāt') *vt.* **-at·ed, -at·ing, -ates.** [Lat. *asseverare, asseverat-* : *ad-*, to + *severus*, serious.] To declare positively or seriously : AFFIRM. **—as·sev'er·a'tion** *n.*

**as·sib·i·late** (ə-sĭb'ə-lāt') *vt.* **-lat·ed, -lat·ing, -lates.** [AD- + SIBILATE.] To pronounce with a hissing sound. **—as·sib'i·la'tion** *n.*

**as·si·du·i·ty** (ăs'ĭ-do͞o'ĭ-tē, -dyo͞o'-) *n., pl.* **-ties. 1.** The quality or condition of being assiduous : DILIGENCE. **2.** *often* **assiduities.** Continuous personal attention : SOLICITUDE.

**as·sid·u·ous** (ə-sĭj'o͞o-əs) *adj.* [Lat. *assiduus* < *assidēre*, to attend to : *ad-*, near to + *sedēre*, to sit.] **1.** Constant in application or attention : DILIGENT <an *assiduous* employee> **2.** Persistent : unceasing <*assiduous* efforts> **—as·sid'u·ous·ly** *adv.* **—as·sid'u·ous·ness** *n.*

**as·sign** (ə-sīn') *vt.* **-signed, -sign·ing, -signs.** [ME *assignen* < OFr. *assignir* < Lat. *assignare* : *ad-*, to + *signare*, to mark < *signum*, sign.] **1.** To set aside for a particular purpose : DESIGNATE. **2.** To select for a duty or office : APPOINT. **3.** To give out as a task : ALLOT. **4.** To ascribe <*assigned* our failure to lack of planning> **5.** *Law.* To transfer (property, rights, or interests). **6.** To place (a unit or personnel) integrally into a military organization. —*n. Law.* An assignee. **—as·sign'a·bil'i·ty** *n.* **—as·sign'a·ble** *adj.* **—as·sign'a·bly** *adv.* **—as·sign'er** *n.*

**as·sig·nat** (ăs'ĭg-năt', ăs'ēn-yä') *n.* [Fr. < Lat. *assignatum*, to assign.] One of the notes of the paper currency issued in France (1789–96) by the revolutionary government and backed by the security of confiscated lands.

**as·sig·na·tion** (ăs'ĭg-nā'shən) *n.* **1.** The act of assigning. **2.** An assignment. **3.** An arrangement for a meeting between lovers : TRYST. **—as'sig·na'tion·al** *adj.*

**as·sign·ee** (ə-sī'nē', ăs'ĭ-nē') *n. Law.* **1.** One to whom a transfer of property, rights, or interest is made. **2.** One appointed to act for another : AGENT.

**as·sign·ment** (ə-sīn'mənt) *n.* **1.** The act of assigning. **2.** Something assigned. **3.** A position or post to which one is assigned. **4.** *Law.* **a.** The transfer of a claim, right, interest, or property. **b.** The document or deed by which such transfer is made.

**as·sign·or** (ə-sī'nôr', ə-sī'nər, ăs'ə-nôr') *n. Law.* One who makes an assignment.

**as·sim·i·late** (ə-sĭm'ə-lāt') *v.* **-lat·ed, -lat·ing, -lates.** [ME *assimilaten* < Lat. *assimilare*, to make similar to : *ad-*, to + *similis*, like.] —*vt.* **1.** *Physiol.* **a.** To consume and incorporate into the body : DIGEST. **b.** To transform (food) into living tissue. **2.** To absorb and incorporate (e.g., knowledge) mentally. **3.** To make or represent as similar. **4.** To alter (a speech sound) by assimilation. **5.** To absorb (an immigrant or culturally distinct group) into the prevailing culture. —*vi.* To become assimilated. **—as·sim'i·la·bil'i·ty** *n.* **—as·sim'i·la·ble** (-lə-bəl) *adj.* **—as·sim'i·la'tor** *n.*

**as·sim·i·la·tion** (ə-sĭm'ə-lā'shən) *n.* **1.** The act or process of assimilating. **2.** The condition or process of being assimilated. **3.** *Biol.* The process by which nourishment is changed into living tissue. **4.** The process by which a speech sound is modified to make it resemble an adjacent sound; e.g., the prefix *in-* in *intolerable* becomes *im-* in *impossible* by assimilation. **5.** The process whereby a group, as of minority or immigrant peoples, gradually adopts the characteristics of another culture.

**as·sim·i·la·tive** (ə-sĭm'ə-lā'tĭv) *also* **as·sim·i·la·to·ry** (-lə-tôr'ē, -tōr'ē) *adj.* Causing or characterized by assimilation.

**As·sin·i·boin** (ə-sĭn'ə-boin') *n., pl.* **Assiniboin** *or* **-boins.** [Fr. *Assiniboine*, of Ojibwa orig.] **1. a.** A tribe of Indians of northeastern Montana and adjacent regions of Canada. **b.** A member of this tribe. **2.** The Siouan language of the Assiniboin. **—As·sin'i·boin'** *adj.*

**as·sist** (ə-sĭst') *v.* **-sist·ed, -sist·ing, -sists.** [ME *assisten* < OFr. *assister* < Lat. *assistere* : *ad-*, near to + *sistere*, to stand.] —*vt.* **1.** To aid. **2.** To work with as an assistant. —*vi.* **1.** To give aid or support. **2.** To be present : ATTEND. —*n.* **1.** An act of giving aid : HELP. **2. a.** *Baseball.* A fielding and throwing of the ball that enables a teammate to put out a runner. **b.** A pass of the ball or puck to the teammate scoring a goal, as in basketball or ice hockey. **3.** A machine or mechanical device providing aid. **—as·sist'er** *n.*

**as·sis·tance** (ə-sĭs'təns) *n.* **1.** The act of assisting. **2.** Help : aid.

**as·sis·tant** (ə-sĭs'tənt) *n.* One that assists : AIDE. —*adj.* **1.** Having an auxiliary position : SUBORDINATE. **2.** Giving aid : AUXILIARY.

**assistant professor** *n.* A college teacher who ranks above an instructor and below an associate professor.

**as·sis·tant·ship** (ə-sĭs′tənt-shĭp′) *n.* An academic position that carries a stipend and usu. involves part-time teaching or research, granted to a qualified graduate student.

**as·size** (ə-sīz′) *n.* [ME *assise* < OFr., act of sitting < p.part. of *asseior*, to seat < Lat. *assidēre*, to sit beside : *ad-*, at + *sedēre*, to sit.] **1. a.** A session of a legislative body or court. **b.** An edict or decree made at such a session. **2. a.** An ordinance regulating weights and measures and the weights and prices of articles of consumption. **b.** The standards so set up. **3.** A judicial inquest, the writ by which it is instituted, or the verdict of the jurors. **4. assizes. a.** One of the periodic court sessions held in each of the counties of England and Wales for the trial of civil or criminal cases. **b.** The time or location of such sessions.

**as·so·ci·a·ble** (ə-sō′shē-ə-bəl, -shə-bəl) *adj.* Capable of being associated. **—as·so′ci·a·bil′i·ty, as·so′ci·a·ble·ness** *n.*

**as·so·ci·ate** (ə-sō′shē-āt′, -sē-) *v.* **-at·ed, -at·ing, -ates.** [ME *associaten* < Lat. *associare*, to join to : *ad-*, to + *socius*, companion.] —*vt.* **1.** To unite in a relationship. **2.** To connect or join together : LINK. **3.** To connect in the mind or imagination <*associates* tulips with springtime> —*vi.* **1.** To join in or form a league, union, or association. **2.** To keep company. —*n.* (ə-sō′shē-ĭt, -āt′, -sē-). **1.** One joined with another or others in an action, enterprise, or business : COLLEAGUE. **2.** A companion. **3.** Something that habitually accompanies or is associated with another. **4.** A member of an institution or society who is granted only partial status or privileges. —*adj.* (ə-sō′shē-ĭt, -āt′, -sē-). **1.** United with another or others and having equal or nearly equal status. **2.** Having partial status or privileges. **3.** Coming as a result of or accompanying : CONCOMITANT <unemployment and *associate* ills>

**associate professor** *n.* A college or university teacher who ranks above an assistant professor and below a full professor.

**as·so·ci·a·tion** (ə-sō′sē-ā′shən, -shē-) *n.* **1.** The act of associating or state of being associated. **2.** An organized group of people who share a common interest, activity, or purpose. **3.** A mental connection between thoughts, feelings, ideas, or sensations and someone or something specific. **4.** *Chem.* Any of various processes of chemical combination, as hydration, solvation, or complex-ion formation, depending on relatively weak chemical bonding. **5.** *Ecol.* A large number of organisms in a specific area with one or two dominant species. **—as·so′ci·a′tion·al** *adj.*

**association football** *n. Chiefly Brit.* Soccer.

**as·so·ci·a·tive** (ə-sō′shē-ā′tĭv, -sē-, -shə-tĭv) *adj.* **1.** Of, marked by, resulting from, or causing association. **2.** *Math.* Independent of the grouping of elements. —Used of mathematical operations; e.g., if $a + (b + c) = (a + b) + c$, the operation indicated by + is *associative*. **—as·so′ci·a′tive·ly** *adv.*

**as·soil** (ə-soil′) *vt.* **-soiled, -soil·ing, -soils.** [ME *assoilen* < OFr. *assoldre, assoil-* < Lat. *absolvere*, to set free : *ab-*, away + *solvere*, to loosen.] *Archaic.* **1.** To absolve or pardon. **2.** To atone for.

**as·so·nance** (ăs′ə-nəns) *n.* [Fr., < Lat. *assonans*, pr.part. of *assonare*, to respond to : *ad-*, to + *sonare*, to sound.] **1.** Resemblance in sound, esp. in the vowel sounds of words. **2.** A partial rhyme in which the accented vowel sounds correspond but the consonants differ, as in *save* and *main*. **—as′so·nant** *adj.* & *n.*

**as·sort** (ə-sôrt′) *v.* **-sort·ed, -sort·ing, -sorts.** [OFr. *assorter* : *a-*, to (< Lat. *ad-*) + *sorte*, kind < Lat. *sors*, chance, lot.] —*vt.* **1.** To distribute into groups according to kinds : CLASSIFY. **2.** To supply with a variety of goods. —*vi.* **1.** To be similar in kind : MATCH. **2.** To be in association. **—as·sort′a·tive** (ə-sôr′tə-tĭv) *adj.* **—as·sort′er** *n.*

**as·sort·ed** (ə-sôr′tĭd) *adj.* **1.** Consisting of a number of different kinds : VARIOUS. **2.** Organized in classes : CLASSIFIED. **3.** Matched or suited.

**as·sort·ment** (ə-sôrt′mənt) *n.* **1.** Separation into classes. **2.** A collection of various things : VARIETY.

**as·suage** (ə-swāj′) *vt.* **-suaged, -suag·ing, -suag·es.** [ME *asswagen* < OFr. *assuagier* < VLat. *assuaviare : Lat. *ad-*, to + *suavis*, sweet.] **1.** To make less burdensome or painful : EASE. **2.** To appease or satisfy, as thirst. **3.** To calm or pacify. **—as·suage′ment** *n.*

**as·sua·sive** (ə-swā′sĭv, -zĭv) *adj.* Soothing.

**as·sume** (ə-so͞om′) *vt.* **-sumed, -sum·ing, -sumes.** [ME *assumen* < Lat. *assumere*, to adopt : *ad-*, to + *sumere*, to take.] **1.** To put on (e.g., a garment). **2.** To take upon oneself : UNDERTAKE <*assume* the payments> **3.** To invest oneself formally with <*assume* the governorship> **4.** To take on : ADOPT <"The god *assumes* a human form" —Ruskin> **5.** To feign : affect <*assumed* an air of self-assurance> **6.** To take for granted : SUPPOSE. **7.** To take up or receive, as into heaven. **—as·sum′a·ble** *adj.* **—as·sum′a·bly** *adv.* **—as·sum′er** *n.*

**as·sumed** (ə-so͞omd′) *adj.* Fictitiously adopted, as a pseudonym. **—as·sum′ed·ly** (ə-so͞o′mĭd-lē) *adv.*

**as·sum·ing** (ə-so͞o′mĭng) *adj.* Presumptuous. **—as·sum′ing·ly** *adv.*

**as·sump·sit** (ə-sŭmp′sĭt) *n.* [Lat., he undertook.] *Law.* **1.** An agreement or promise not under seal : CONTRACT. **2.** A legal action to enforce or recover damages for a breach of assumpsit.

**as·sump·tion** (ə-sŭmp′shən) *n.* [ME < Lat. *assumptio*, adoption < *assumere*, to adopt. —see ASSUME.] **1.** The act of assuming. **2.** Something taken to be true without proof or demonstration. **3.** Presumption or arrogance. **4.** *Logic.* A minor premise. **5. Assumption. a.** The bodily taking up of the Virgin Mary into heaven after her death. **b.** A church feast on Aug. 15 that celebrates this event.

**as·sump·tive** (ə-sŭmp′tĭv) *adj.* **1.** Of or marked by assumption. **2.** Taken for granted. **3.** Presumptuous. **—as·sump′tive·ly** *adv.*

**as·sur·ance** (ə-sho͝or′əns) *n.* **1.** The act of assuring or state of being assured. **2.** A statement or indication that inspires confidence : GUARANTEE. **3. a.** Freedom from doubt : CERTAINTY. **b.** Self-confidence. **4.** Audacity. **5.** *Chiefly Brit.* Insurance.

**as·sure** (ə-sho͝or′) *vt.* **-sured, -sur·ing, -sures.** [ME *assuren* < OFr. *assurer* < Med. Lat. *assecurare*, to make sure : Lat. *ad-*, to + *securus*, secure.] **1.** To inform confidently, with a view to removing doubt. **2.** To cause to feel sure : CONVINCE. **3.** To give confidence to : REASSURE. **4.** To make certain : ENSURE <"Nothing in history *assures* the success of our civilization —Herbert J. Muller> **5.** To make safe or secure. **6.** *Chiefly Brit.* To insure, as against loss. ***usage:*** *Assure, ensure,* and *insure* all mean "to make secure or certain." But only *assure* is used to refer to a person in the sense of "to set the mind at rest." **—as·sur′a·ble** *adj.* **—as·sur′er** *n.*

**as·sured** (ə-sho͝ord′) *adj.* **1.** Guaranteed. **2.** Confident. **3.** Insured. **—as·sur′ed·ly** (ə-sho͝or′ĭd-lē) *adv.* **—as·sur′ed·ness** *n.*

**as·sur·gent** (ə-sûr′jənt) *adj.* [Lat. *assurgens, assurgent-*, pr.part. of *assurgere*, to rise up to : *ad-*, to + *surgere*, to rise. —see SURGE.] **1.** Rising or tending to rise. **2.** *Bot.* Slanting or curving upward : ASCENDING. **—as·sur′gen·cy** *n.*

**As·syr·i·an** (ə-sîr′ē-ən) *adj.* Of or relating to Assyria or its people, language, or culture. —*n.* **1.** A native or resident of Assyria. **2.** The Semitic language of the Assyrians.

**As·syr·i·ol·o·gy** (ə-sîr′ē-ŏl′ə-jē) *n.* The study of the ancient civilization of Assyria. **—As·syr′i·ol′o·gist** *n.*

**-ast** *suff.* [ME < Lat. *astes* < Gk. *-astēs*, n.suffix.] One associated with <ecdysi*ast*>

**a·sta·ble circuit** (ā′stā′bəl) *n. Computer Sci.* A circuit that alternates continuously between two unstable states.

**As·tar·te** (ə-stär′tē) *n.* [Lat. *Astarte* < Gk., of Semitic orig.] *Myth.* The Phoenician goddess of love and fertility.

**a·sta·sia** (ə-stā′zhə) *n.* [Gk., unsteadiness < *astatos*, unsteady : *a-*, not + *statos*, standing.] Inability to stand because of muscular incoordination.

**a·stat·ic** (ā-stăt′ĭk) *adj.* **1.** Unstable : unsteady. **2.** *Computer Sci.* Having no particular directional characteristics. **—a·stat′i·cal·ly** *adv.* **—a·stat′i·cism** *n.*

**as·ta·tine** (ăs′tə-tēn′, -tĭn) *n.* [Gk. *astatos*, unstable + -INE[2].] *Symbol* **At** A highly unstable radioactive element used in medicine as a radioactive tracer; atomic number 85; longest-lived isotope At 210.

**as·ter** (ăs′tər) *n.* [Lat., star < Gk. *astēr*.] **1.** A plant of the genus *Aster*, bearing rayed, daisylike flowers ranging in color from white to bluish, purple, or pink. **2.** The China aster. **3.** *Biol.* A star-shaped structure appearing in the cytoplasm of the cell and associated with the centrosome during mitosis.

**as·te·ri·at·ed** (ă-stîr′ē-ā′tĭd) *adj.* [< Gk. *asterios*, starry < *astēr*, star.] *Mineral.* Exhibiting asterism.

**as·ter·isk** (ăs′tə-rĭsk′) *n.* [LLat. *asteriscus* < Gk. *asteriskos*, dim. of *astēr*, star.] **1.** A star-shaped symbol (*) used in printing to indicate an omission or a reference to a footnote. **2.** An asterisk used to indicate an unattested linguistic form or entity. —*vt.* **-isked, -isk·ing, -isks.** To mark with an asterisk : STAR.

**as·ter·ism** (ăs′tə-rĭz′əm) *n.* [Gk. *asterismos*, constellation < *astēr*, star.] **1.** Three asterisks in triangular form used in a text to call attention to a following passage. **2.** *Astron.* **a.** A cluster of stars. **b.** A constellation. **3.** *Mineral.* A six-rayed starlike figure optically produced in some crystal structures by reflected or transmitted light. **—as′ter·is′mal** *adj.*

**a·stern** (ə-stûrn′) *adv. Naut.* **1.** Behind a vessel. **2.** Toward the rear of a vessel. **3.** To the rear : BACKWARD. **—a·stern′** *adj.*

**a·ster·nal** (ā-stûr′nəl) *adj. Anat.* **1.** Not connected to the sternum. **2.** Lacking a sternum.

**as·ter·oid** (ăs′tə-roid′) *n.* [Gk. *asteroeidēs*, starlike : *astēr*, star + *-eidēs*, like.] **1.** *Astron.* Any of numerous celestial bodies with typical diameters between one and several hundred miles and orbits lying chiefly between Mars and Jupiter. **2.** *Zool.* A starfish. —*adj. also* **as·ter·oid·al** (ăs′tə-roid′l). Star-shaped.

**As·ter·o·pe** (ă-stĕr′ə-pē) *n. var. of* STEROPE.

**as·the·ni·a** (ăs-thē′nē-ə) *n.* [NLat. < Gk. *astheneia* < *asthenēs*, weak : *a-*, without + *sthenos*, strength.] Loss or lack of bodily strength : WEAKNESS.

**as·then·ic** (ăs-thĕn′ĭk) *n.* A slender, lightly muscled human physique. —*adj. also* **as·then·i·cal** (-ĭ-kəl). Lacking bodily strength : WEAK.

**as·the·no·pi·a** (ăs′thə-nō′pē-ə) *n.* [ASTHEN(IA) + -OPIA.] Eyestrain, esp. accompanied by dimming of vision and headache. **—as′the·nop′ic** (-nŏp′ĭk) *adj.*

**as·then·o·sphere** (ăs-thĕn′ə-sfîr′) *n.* [Gk. *asthenēs*, weak (*a-*, not + *sthenos*, strength) + SPHERE.] A zone of the earth's mantle that lies below the lithosphere and consists of several hundred kilometers of deformable rock.

**asth·ma** (ăz′mə, ăs′-) *n.* [ME *asma* < Med. Lat. < Gk. *asthma.*] A chronic respiratory disease, often arising from allergies and accompanied by labored breathing, chest constriction, and coughing. **—asth·mat′ic** (-măt′ĭk) *adj.* & *n.* **—asth·mat′i·cal·ly** *adv.*

**as though** *conj.* As if <looked *as though* they had been crying>

**a·stig·ma·tism** (ə-stĭg′mə-tĭz′əm) *n.* [A-¹ + Gk. *stigma*, point < *stizein*, to tattoo.] **1.** A refractive defect of a lens that prevents focusing of sharply defined images. **2.** Faulty vision caused by defects in the lens of the eye. **—as′tig·mat′ic** (ăs′tĭg-măt′ĭk) *adj.* **—as′tig·mat′i·cal·ly** *adv.*

**a·stir** (ə-stûr′) *adj.* [Sc. *asteer* : *a*, on + *steer*, stir.] **1.** Moving about. **2.** Out of bed : AWAKE.

**as to** *prep.* **1.** With regard to <confused *as to* the actual events> **2.** According to <selected *as to* seniority>

**a·stom·a·tous** (ā-stŏm′ə-təs, ā-stō′mə-) *also* **as·tom·ous** (ăs′tə-məs) *or* **a·stom·a·tal** (ā-stŏm′ə-təl, ā-stō′mə-təl) *adj.* Having no mouth or stomata.

**a·ston·ied** (ə-stŏn′ēd) *adj.* [ME *astonied*, p.part. of *astonen*, to amaze.—see ASTONISH.] *Archaic.* Bewildered : dazed.

**a·ston·ish** (ə-stŏn′ĭsh) *vt.* **-ished, -ish·ing, -ish·es.** [Prob. alteration of obs. *astony*, to amaze < ME *astonen* < AN < VLat. **extonare* : Lat. *ex-*, out of + Lat. *tonare*, to thunder.] To fill with sudden wonder or amazement : SURPRISE. **—a·ston′ish·er** *n.* **—a·ston′ish·ing·ly** *adv.*

**a·ston·ish·ment** (ə-stŏn′ĭsh-mənt) *n.* **1.** Great surprise or amazement. **2.** A cause of amazement : MARVEL.

**a·stound** (ə-stound′) *vt.* **a·stound·ed, a·stound·ing, a·stounds.** [< Obs. *astoned*, p.part. of obs. *astony*, to amaze. —see ASTONISH.] To strike with sudden wonder. **—a·stound′ing·ly** *adv.*

**astr-** *pref. var. of* ASTRO-.

**as·tra·chan** (ăs′trə-kăn′, -kən) *n. var. of* ASTRAKHAN.

**As·tra·chan** (ăs′trə-kăn′, -kən) *n.* [After *Astrakhan*, a city in the U.S.S.R.] A tart red or yellow apple of Russian origin.

**a·strad·dle** (ə-străd′l) *adv.* In a straddling position : ASTRIDE. *—prep.* Upon or over and with a leg on each side of.

**As·trae·a** (ă-strē′ə) *n.* [NLat. < Gk. *astraios*, starry < *astron*, star.] *Gk. Myth.* The goddess of justice.

**as·tra·gal** (ăs′trə-gəl) *n.* [Lat. *astragalus* < Gk. *astragalos.*] A narrow, convex architectural molding, often having the form of beading.

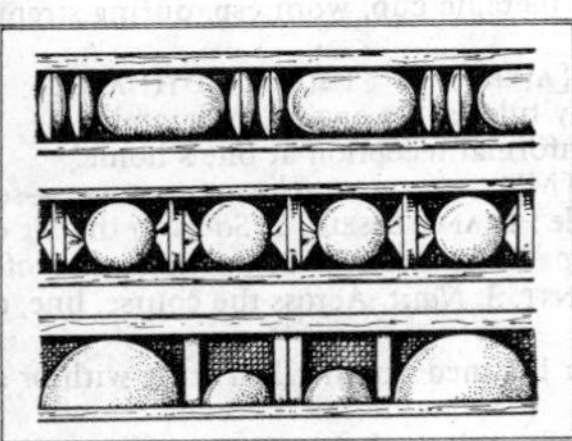

**astragal**
*Bead and reel pattern*

**as·trag·a·lus** (ə-străg′ə-ləs) *n., pl.* **-li** (-lī′) [NLat. < Gk. *astragalos*, vertebra.] The talus. **—as·trag′a·lar** *adj.*

**as·tra·khan** *also* **as·tra·chan** (ăs′trə-kăn′, -kən) *n.* **1.** The curly fur made from the skins of young lambs from the region of Astrakhan. **2.** A fabric with a looped, curly pile.

**as·tral** (ăs′trəl) *adj.* [LLat. *astralis* < Lat. *astrum*, star < Gk. *astron.*] **1.** Of, relating to, emanating from, or like the stars. **2.** *Biol.* Relating to or shaped like a cell aster : STAR-SHAPED. **—as′tral·ly** *adv.*

**as·tra·pho·bi·a** (ăs′trə-fō′bē-ə) *n.* [Gk. *astrapē*, lightning + -PHOBIA.] Abnormal fear of lightning and thunder.

**a·stray** (ə-strā′) *adv.* [ME < OFr. *estraie*, p.part. of *estraier*, to stray. —see STRAY.] **1.** Away from the correct path or direction. **2.** Away from the morally right or good. **—a·stray′** *adj.*

**as·trict** (ə-strĭkt′) *vt.* **-trict·ed, -trict·ing, -tricts.** [Lat. *astrictus*, p.part. of *astringere*, to bind fast. —see ASTRINGE.] To bind, esp. by moral or legal obligations. **—as·tric′tion** *n.*

**as·tric·tive** (ə-strĭk′tĭv) *adj.* Astringent. *—n.* An astringent. **—as·tric′tive·ly** *adv.* **—as·tric′tive·ness** *n.*

**a·stride** (ə-strīd′) *adv.* **1.** With a leg on each side of <riding *astride* the pony> **2.** With the legs wide apart. *—prep.* **1.** Upon or over and with a leg on each side of. **2.** On both sides of : SPANNING.

**astride of** *prep.* Astride.

**as·tringe** (ə-strĭnj′) *vt.* **-tringed, -tring·ing, -tring·es.** [Lat. *astringere*, to bind fast : *ad-*, to + *stringere*, to bind.] To draw together : CONSTRICT.

**as·trin·gent** (ə-strĭn′jənt) *adj.* **1.** *Med.* Tending to draw together or constrict tissue : STYPTIC. **2.** Harsh : severe. *—n.* An astringent substance or drug, as alum. **—as·trin′gen·cy** *n.* **—as·trin′gent·ly** *adv.*

**as·tri·on·ics** (ăs′trē-ŏn′ĭks) *n.* [ASTR(ONAUTIC) + (AV)IONICS.] *(sing. in number).* Electronics used in astronautics.

**astro-** *or* **astr-** *pref.* [ME < OFr. < Lat. < Gk. < *astron*, star.] **1 a.** Star <*astro*physics> **b.** Celestial body <*astro*metry> **c.** Outer space <*astro*naut> **2.** The aster of a cell <*astro*sphere>

**as·tro·bi·ol·o·gy** (ăs′trō-bī-ŏl′ə-jē) *n.* Exobiology.

**as·tro·bleme** (ăs′trə-blēm′) *n.* [ASTRO- + Gk. *blēma*, missile, wound < *ballein*, to throw.] A scar on the earth's surface left by meteorite impact.

**as·tro·chem·is·try** (ăs′trō-kĕm′ĭ-strē) *n.* The chemistry of stars and interstellar space.

**as·tro·cyte** (ăs′trə-sīt′) *n.* A star-shaped cell, esp. a neuroglial cell.

**as·tro·cy·to·ma** (ăs′trō-sī-tō′mə) *n., pl.* **-mas** *or* **-ma·ta** (-mə-tə). A malignant tumor of astrocytes.

**as·tro·dome** (ăs′trə-dōm′) *n.* **1.** A transparent dome on the top of an aircraft, through which celestial observations are made for navigation. **2.** A stadium covered by a dome.

**as·tro·dy·nam·ics** (ăs′trō-dī-năm′ĭks) *n.* *(sing. in number).* The dynamics of celestial bodies.

**as·tro·gate** (ăs′trə-gāt′) *vi.* **-gat·ed, -gat·ing, -gates.** [ASTRO- + (NAVI)GATE.] To navigate a spacecraft. **—as′tro·ga′tion** *n.* **—as′tro·ga′tor** *n.*

**as·tro·ge·ol·o·gy** (ăs′trō-jē-ŏl′ə-jē) *n.* The geologic study of celestial bodies.

**as·tro·labe** (ăs′trə-lāb′) *n.* [ME *astrelabie* < OFr. *astrelabe* < Med. Lat. *astrolabium* < Gk. *astrolabon*, planisphere : *astron*, star + *lambanein*, to take.] A medieval instrument used to determine the altitude of celestial bodies, as the sun.

**as·trol·o·gy** (ə-strŏl′ə-jē) *n.* [ME *astrologie* < OFr. < Lat. *astrologia*, astronomy < Gk. < *astrologos*, astronomer : *astron*, star + *logos*, speech.] The study of the positions and aspects of heavenly bodies in the belief that they have an influence on the course of human affairs. **—as·trol′o·ger** *n.* **—as′tro·log′ic** (ăs′trə-lŏj′ĭk), **as′tro·log′i·cal** *adj.* **—as′tro·log′i·cal·ly** *adv.*

**as·trom·e·try** (ə-strŏm′ĭ-trē) *n.* The scientific measurement of the positions and motions of celestial bodies. **—as′tro·met′ric** (ăs′trō-mĕt′rĭk), **as′tro·met′ri·cal** *adj.*

**as·tro·naut** (ăs′trə-nôt′) *n.* [ASTRO- + Gk. *nautēs*, sailor < *naus*, ship.] One trained to pilot, navigate, or otherwise take part in the flight of a spacecraft.

**as·tro·nau·tics** (ăs′trə-nô′tĭks) *n.* [ASTRO- + Lat. *nautica*, neuter pl. of *nauticus*, nautical. —see NAUTICAL.] *(sing. in number).* The science and technology of space flight. **—as′tro·nau′tic, as′tro·nau′ti·cal** *adj.* **—as′tro·nau′ti·cal·ly** *adv.*

**as·tro·nav·i·ga·tion** (ăs′trō-năv′ĭ-gā′shən) *n.* **1.** Navigation of spacecraft. **2.** Celestial navigation. **—as′tro·nav′i·ga′tor** *n.*

**as·tron·o·mer** (ə-strŏn′ə-mər) *n.* A scientist who specializes in astronomy.

**as·tro·nom·i·cal** (ăs′trə-nŏm′ĭ-kəl) *also* **as·tro·nom·ic** (-nŏm′ĭk) *adj.* **1.** Of or relating to astronomy. **2.** Inconceivably large : IMMENSE. **—as′tro·nom′i·cal·ly** *adv.*

**astronomical unit** *n.* A unit of length used in measuring astronomical distances, equal to the mean distance of the earth from the sun, approx. 93 million miles or 150 million kilometers.

**as·tron·o·my** (ə-strŏn′ə-mē) *n.* [ME *astronomie* < OFr. < Lat. *astronomia* < Gk. : *astron*, star + *-nomia*, -nomy.] The scientific study of the universe beyond the earth, esp. the observation, calculation, and interpretation of the positions, dimensions, distribution, composition, and evolution of celestial bodies and phenomena.

**as·tro·pho·tog·ra·phy** (ăs′trō-fə-tŏg′rə-fē) *n.* Astronomical photography. **—as′tro·pho′to·graph′ic** (-fō′tə-grăf′ĭk) *adj.*

**as·tro·phys·ics** (ăs′trō-fĭz′ĭks) *n.* *(sing. in number).* The physics of stellar phenomena. **—as′tro·phys′i·cal** *adj.* **—as′tro·phys′i·cist** (-fĭz′ĭ-sĭst) *n.*

**as·tro·sphere** (ăs′trō-sfîr′) *n.* **1.** The central portion of a cell aster : CENTROSPHERE. **2.** The entire cell aster with the exception of the centrosome.

**As·tro·turf** (ăs′trō-tûrf′). A trademark for an artificial grasslike ground covering.

**as·tute** (ə-sto͞ot′, ə-styo͞ot′) *adj.* [Lat. *astutus* < *astus*, craft.] Keen in judgment : SHREWD. **—as·tute′ly** *adv.* **—as·tute′ness** *n.*

**As·ty·a·nax** (ə-stī′ə-năks′) *n.* [Gk. *Astuanax.*] *Gk. Myth.* The young son of Hector and Andromache, killed by the conquering Greeks.

**a·sty·lar** (ā-stī′lər) *adj.* [A-¹ + Gk. *stulos*, pillar + -AR.] Having no columns or pilasters.

**a·sun·der** (ə-sŭn′dər) *adv.* [ME *asonder* < OE *onsundran* : *on*, on + *sundran*, separately < *sunder*, apart.] **1.** Into separate parts or groups. **2.** Apart from each other in direction or position. **—a·sun′der** *adj.*

**a·swarm** (ə-swôrm′) *adj.* Filled to overflowing : TEEMING.

**a·swirl** (ə-swûrl′) *adj.* Moving with a swirling motion.

**a·syl·lab·ic** (ā′sĭ-lăb′ĭk) *adj.* Not syllabic.

**a·sy·lum** (ə-sī′ləm) *n.* [ME *asilum* < Lat. *asylum* < Gk. *asulon*, sanctuary < *asulos*, inviolable : *a-*, without + *sulē*, right of seizure.] **1.** An institution for the care of the mentally ill or the aged. **2.** A place that provides protection or safety : REFUGE. **3.** A temple or church affording sanctuary for criminals or debtors. **4.** Protection and immunity from extradition granted by a government to a foreign political refugee.

**a·sym·met·ric** (ā'sĭ-mĕt'rĭk) *also* **a·sym·met·ri·cal** (-rĭ-kəl) *adj.* Not symmetrical. **—a'sym·met'ri·cal·ly** *adv.*

**a·sym·me·try** (ā-sĭm'ĭ-trē) *n.* Lack of symmetry.

**a·symp·to·mat·ic** (ā'sĭmp-tə-măt'ĭk) *adj.* Neither causing nor displaying symptoms. **—a'symp·to·mat'i·cal·ly** *adv.*

**as·ymp·tote** (ăs'ĭm-tōt', -ĭmp-) *n.* [NLat. *asymptota* < Gk. *asumptōtos*, not touching : *a-*, not + *sun-*, together + *ptōtos*, likely to fall.] *Math.* A line considered a limit to a curve in the sense that the perpendicular distance from a moving point on the curve to the line approaches zero as the point moves an infinite distance from the origin. **—as'ymp·tot'ic** (-tŏt'ĭk), **as'ymp·tot'i·cal** *adj.* **—as'ymp·tot'i·cal·ly** *adv.*

**a·syn·chro·nism** (ā-sĭng'krə-nĭz'əm) *n.* Lack of synchronism. **—a·syn'chro·nous** (-nəs) *adj.* **—a·syn'chro·nous·ly** *adv.*

**a·syn·de·ton** (ə-sĭn'dĭ-tŏn') *n.* [LLat. < Gk. *asundeton* < *asundetos*, without conjunctions : *a-*, not + *sundetos*, bound together < *sundein*, to bind together (*sun-*, together + *dein*, to bind).] Omission of conjunctions from constructions in which they would normally be used. **—as'yn·det'ic** (ăs'ĭn-dĕt'ĭk) *adj.* **—as'yn·det'i·cal·ly** *adv.*

**a·syn·tac·tic** (ā'sĭn-tăk'tĭk) *adj.* Not syntactic.

**at[1]** (ăt; ət *when unstressed*) *prep.* [ME < OE *æt.*] **1. a.** In the location of <*at* the lake> **b.** In the position of <*at* the end of the row> **2.** To or toward the direction of <threw it *at* me> **3.** Present : attending <*at* the picnic> **4.** In the state or condition of <*at* peace> **5.** In the manner of <*at* a gallop> **6.** To the extent or amount of <*at* 50¢ a quart> **7.** On or near the time or age of <*at* 10:30 A.M.> **8.** Because of <laugh *at* a joke> **9.** By way of : THROUGH. **10.** According to <*at* one's leisure> **11.** Dependent on <was *at* our mercy> **12.** Occupied with <*at* play>

**at[2]** (ät) *n., pl.* **at.** [Thai.] —See table at CURRENCY.

**At** *symbol for* ASTATINE.

**at-** *pref. var. of* AD- 1. —Used before *t.*

**At·a·brine** (ăt'ə-brĭn, -brēn'). A trademark for an antimalarial preparation, quinacrine hydrochloride.

**at·a·ghan** (ăt'ə-găn', -gən) *n. var. of* YATAGHAN.

**At·a·lan·ta** (ăt'ə-lăn'tə) *n.* [Lat. < Gk. *Atalantē.*] *Gk. Myth.* A maiden who agreed to marry any man who could outrun her and was defeated by Hippomenes when he dropped three golden apples that she paused to pick up.

**at all** *adv.* In any way whatsoever <can't hear it *at all*>

**at·a·man** (ăt'ə-măn') *n., pl.* **-mans.** [R., prob. < Pol. *hetman*, captain.] A Cossack chieftain.

**at·a·mas·co lily** (ăt'ə-măs'kō) *n.* [Algonquian (Virginia) *Attamusco.*] A plant, *Zephyranthes atamasco* of eastern North America, with funnel-shaped white or pinkish flowers.

**at·a·rac·tic** (ăt'ə-răk'tĭk) *also* **at·a·rax·ic** (-răk'sĭk) [Gk. *ataraktos*, undisturbed : *a-*, not + *taraktos*, disturbed < *tarassein*, to disturb.] *—adj.* Relating to or producing calmness and peace of mind. *—n.* A drug that reduces nervous tension : TRANQUILIZER.

**at·a·rax·i·a** (ăt'ə-răk'sē-ə) *n.* [Gk. < *ataraktos*, undisturbed. —see ATARACTIC.] Peace of mind : TRANQUILLITY.

**at·a·vism** (ăt'ə-vĭz'əm) *n.* [Fr. *atavisme* < Lat. *atavus*, ancestor : *atta*, father + *avus*, grandfather.] **1.** The reappearance of a characteristic in an organism after an absence of several generations, caused by a recessive gene or complementary genes. **2.** An individual or part displaying atavism. **—at'a·vist** *n.* **—at'a·vis'tic** *adj.* **—at'a·vis'ti·cal·ly** *adv.*

**a·tax·i·a** (ə-tăk'sē-ə) *also* **a·tax·y** (ə-tăk'sē) *n.* [Gk., disorder < *ataktos*, disorderly : *a-*, not + *taktos*, ordered < *tassein*, to arrange.] Inability to coordinate muscular movements.

**a·tax·ic** (ə-tăk'sĭk) *adj.* Of or pertaining to ataxia. *—n.* An individual displaying symptoms of ataxia.

**ate** (āt) *v. p.t of* EAT.

**-ate[1]** *suff.* [ME *-at* < OFr. < Lat. *-atus*, p.part. suffix of verbs in *-are.*] **1. a.** Having <nerv*ate*> **b.** Marked by <Latin*ate*> **c.** Resembling <lyr*ate*> **2. a.** One that is marked by <lamin*ate*> **b.** Rank : office <rabbin*ate*> **3.** To act upon in a specified manner <acidul*ate*>

**-ate[2]** *suff.* [NLat. *-atum* < Lat., neuter of *-atus*, p.part. suffix.] **1.** A derivative of a specified chemical compound or element <alumin*ate*> **2.** A salt or ester of a specified acid <acet*ate*>

**at·el·ier** (ăt'l-yā') *n.* [Fr. < OFr. *astelier*, carpenter's shop < *astele*, splinter < LLat. *astella*, alteration of Lat. *astula*, dim. of *assis*, board.] A workshop or studio, esp. an artist's studio.

**a tem·po** (ä tĕm'pō) *adv. & adj.* [Ital., in time.] *Mus.* Resuming the original tempo. —Used as a direction.

**a·tem·po·ral** (ā-tĕm'pər-əl) *adj.* Independent of time : TIMELESS.

**Ath·a·na·sian** (ăth'ə-nā'zhən) *adj.* Of or relating to Athanasius. *—n.* A follower of Athanasius and his doctrine of opposition to Arianism.

**Ath·a·pas·can** (ăth'ə-păs'kən) *also* **Ath·a·bas·can** (-băs'kən) *n.* [After Lake *Athabaska* in Canada < Cree *athapaskaaw*, there is scattered grass.] **1.** A group of related North American Indian languages including Navaho and Apache and languages of Alaska, northwestern Canada, and coastal Oregon and California. **2.** A member of an Athapascan-speaking tribe. **—Ath'a·pas'can** *adj.*

**a·the·ism** (ā'thē-ĭz'əm) *n.* [OFr. *atheisme* < *athee*, atheist < Gk. *atheos*, godless : *a-*, without + *theos*, god.] **1.** Disbelief in or denial of the existence of God. **2.** The quality or state of being godless.

**a·the·ist** (ā'thē-ĭst) *n.* One who denies the existence of God.

**a·the·is·tic** (ā'thē-ĭs'tĭk) *or* **a·the·is·ti·cal** (-tĭ-kəl) *adj.* **1.** Relating to or typical of atheism or of atheists. **2.** Inclined to atheism. **—a'the·is'ti·cal·ly** *adv.* **—a'the·is'tic·ness** *n.*

**ath·e·ling** (ăth'ə-lĭng, ăth'-) *n.* [ME < OE *ætheling*, prince.] An Anglo-Saxon nobleman or prince.

**A·the·na** (ə-thē'nə) *also* **A·the·ne** (-nē) *n.* [Lat. < Gk. *Athenē.*] *Gk. Myth.* The goddess of wisdom and the arts.

**ath·e·nae·um** *also* **ath·e·ne·um** (ăth'ə-nē'əm) *n.* [LLat. *Athenaeum*, a Roman school, after Gk. *Athenaion*, the temple of Athena.] **1.** An institution, as a literary club or scientific academy, for the promotion of learning. **2.** A library or reading room.

**ath·er·o·gen·e·sis** (ăth'ər-ō-jĕn'ĭ-sĭs) *n.* [ATHERO(MA) + -GENESIS.] The production of atheroma. **—ath'er·o·gen'ic** *adj.*

**ath·er·o·ma** (ăth'ə-rō'mə) *n., pl.* **-mas** *or* **-ma·ta** (-mə-tə) [Lat. < Gk. *atherōma*, tumor full of gruellike pus < *athēra*, gruel.] **1.** A deposit or degenerative accumulation of pulpy, acellular, lipid-containing materials, esp. in arterial walls. **2.** A form of arteriosclerosis induced and characterized by atheromas. **—ath'er·o·ma·to'sis** (-tō'sĭs) *n.* **—ath'er·om'a·tous** (-rŏm'ə-təs, -rō'mə-) *adj.*

**ath·er·o·scle·ro·sis** (ăth'ə-rō-sklə-rō'sĭs) *n.* [ATHERO(MA) + SCLEROSIS.] Atheromatous arteriosclerosis : ATHEROMA. **—ath'er·o·scle·rot'ic** (-rŏt'ĭk) *adj.*

**a·thirst** (ə-thûrst') *adj.* **1.** Strongly desirous : EAGER <*athirst* for adventure> **2.** *Archaic.* Thirsty.

**ath·lete** (ăth'lēt') *n.* [ME < Lat. *athleta* < Gk. *athlētēs*, contestant < *athlein*, to contend < *athlon*, prize.] **1.** One who participates in competitive sports. **2.** One having the natural aptitudes for physical exercises and sports, as strength, agility, and endurance.

**athlete's foot** *n.* A contagious skin infection caused by parasitic fungi usu. affecting the feet and sometimes the hands and resulting in itching, blisters, cracking, and scaling.

**ath·let·ic** (ăth-lĕt'ĭk) *adj.* **1.** Of, relating to, or appropriate to athletics or athletes. **2.** Physically strong. **—ath·let'i·cal·ly** *adv.* **—ath·let'i·cism** (-lĕt'ĭ-sĭz'əm) *n.*

**ath·let·ics** (ăth-lĕt'ĭks) *n.* **1.** *(pl. in number).* Athletic activities, as competitive sports. **2.** *(sing. in number).* The principles or system of athletic exercises and training.

**athletic supporter** *n.* An elastic support for the male genitals, sometimes employing a rigid metallic cup, worn esp. during strenuous athletic activity.

**ath·o·dyd** (ăth'ə-dĭd') *n.* [A(ERO)- + TH(ERM)ODY(NAMIC) + D(UCT).] A simple, essentially tubular jet engine, as a ramjet.

**at-home** (ət-hōm') *n.* An informal reception at one's home.

**a·thwart** (ə-thwôrt') *adv.* [ME : *a-*, on + *thwart*, across. —see THWART.] **1.** From side to side : TRANSVERSELY. **2.** So as to thwart or obstruct : PERVERSELY. *—prep.* **1.** From one side to the other of : ACROSS. **2.** Contrary to : AGAINST. **3.** *Naut.* Across the course, line, or length of.

**a·tilt** (ə-tĭlt') *adj.* **1.** Tilted or inclined upward. **2.** Tilting with or as if with a lance. **—a·tilt'** *adv.*

**-ation** *suff.* [ME *-acioun* < OFr. *-ation* < Lat. *-atio*, n. suffix < *-atus*, -ate.] **1. a.** Action or process <strangul*ation*> **b.** The result of an action or process <accultur*ation*> **2.** State, condition, or quality of <eburn*ation*>

**-ative** *suff.* [ME < OFr. *-atif* < Lat. *-ativus* < *-atus*, -ate.] Of, relating to, or associated with <talk*ative*>

**At·ka mackerel** (ăt'kə, ät'-) *n.* [After Atka Island, Alaska.] A food fish, *Pleurogrammus monopterygius* of northern Pacific waters.

**At·lan·te·an** (ăt'lăn-tē'ən, ăt-lăn'tē-) *adj.* [< Gk. *Atlas, Atlant-*, Atlas.] **1.** Of, relating to, or like Atlas. **2.** Of or relating to Atlantis.

**at·lan·tes** (ăt-lăn'tēz') *n. pl. of* ATLAS 6.

**At·lan·tic** (ăt-lăn'tĭk) *adj.* [Lat. *(mare) Atlanticum*, Atlantic (sea) < Gk. *(pelagos) Atlantikos* < *Atlas*, Atlas.] **1.** Of, in, near, on, or relating to the Atlantic Ocean. **2.** Of or relating to Atlas or the Atlas Mountains.

**Atlantic salmon** *n.* A food fish, *Salmo salar* of northern Atlantic waters.

**Atlantic Standard Time** *n.* Standard time as used in the region between the meridians at 52.5° and 67.5° west of Greenwich, England.

**At·lan·tis** (ăt-lăn'tĭs) *n.* [Gk. < *Atlas*, Atlas.] A legendary island in the Atlantic west of Gibraltar, said by Plato to have disappeared beneath the sea during an earthquake.

**At·las** (ăt'ləs) *n.* [Gk. *Atlas.*] **1.** *Gk. Myth.* A Titan condemned to support the heavens on his shoulders. **2.** A person bearing a great burden. **3. atlas.** A book or bound collection of maps. **4. atlas.** A volume of tables, charts, or plates that systematically illustrates a particular subject <an astronomical *atlas*> **5. atlas.** A large size of drawing paper, measuring 26 by 33 or 34 inches. **6. atlas,** *pl.* **at·lan·tes** (ăt-lăn'tēz'). A male figure used as a masonry column on a building. **7. atlas.** *Anat.* The first vertebra of the neck. **8.** An intercontinental ballistic missile developed by the U.S. Air Force.

▲ word history: The word *atlas* used to refer to a collection of maps is derived only indirectly from the name of the mythological figure. *Atlas* was the title of a work by the great 16th-century cartographer Mercator that consisted of two parts, a treatise on cosmology and a collection of maps. He named the entire work after Atlas, a legendary king of northern Africa who was sometimes either identified with, or considered descended from, the Titan Atlas from mythology. King Atlas was renowned for his learning, especially in astronomy. He was depicted on the title page of Mercator's work holding and measuring a globe; a globe already mapped lies at his feet. Mercator's successors retained the title *Atlas* for their collections of maps, and the word became generic for any work of this kind.

**at·man** (ät′mən) *n.* [Skt. *ātman,* breath, spirit.] **1.** The individual soul and principle of life in Hinduism. **2. Atman.** The supreme and universal soul, from which all individual souls arise in Hinduism.

**at·mom·e·ter** (ăt-mŏm′ĭ-tər) *n.* An instrument for measuring the rate of water evaporation. **—at′mo·met′ric** (-mō-mĕt′rĭk) *adj.* **—at·mom′e·try** *n.*

**at·mos·phere** (ăt′mə-sfîr′) *n.* [NLat. *atmosphaera* : Gk. *atmos,* vapor + Lat. *sphaera,* sphere.] **1.** The gaseous mass or envelope surrounding a celestial body, esp. that encompassing the earth, retained by the body's gravitational field. **2.** The atmosphere or climate in a given place. **3.** *Physics.* A unit of pressure equal to $1.01325 \times 10^5$ newtons per square meter. **4.** Environment regarded as having a physical or psychological influence <a cheerful *atmosphere*> **5.** The predominant tone or mood of a work of art. **6.** *Informal.* A quality or effect thought to be exotic or romantic.

**at·mos·pher·ic** (ăt′mə-sfĕr′ĭk, -sfîr′ĭk) *also* **at·mos·pher·i·cal** (-ĭ-kəl) *adj.* **1.** Of, relating to, or existing in the atmosphere. **2.** Produced by, relying on, or coming from the atmosphere. **—at′mos·pher′i·cal·ly** *adv.*

**atmospheric pressure** *n.* An exerted pressure of 1 atmosphere.

**at·mos·pher·ics** (ăt′mə-sfĕr′ĭks, -sfîr′-) *n. (sing. in number).* **1.** Electromagnetic radiation caused by natural phenomena, as lightning. **2.** Radio interference caused by electromagnetic radiation.

**at·mo·spher·i·um** (ăt′mə-sfîr′ē-əm) *n.* [ATMOSPHER(E) + (PLANETAR)IUM.] **1.** An optical device designed to project atmospheric phenomena, as clouds, on the inside of a dome. **2.** A room containing an atmospherium.

**a·toll** (ăt′ôl′, -ŏl′, ā′tôl′, ā′tŏl′) *n.* [Malayalam *atoḷu,* reef.] A ringlike coral island and reef that partially or totally encloses a lagoon.

**at·om** (ăt′əm) *n.* [ME *attome* < Lat. *atomus* < Gk. *atomos,* indivisible : *a-,* not + *temnein,* to cut.] **1.** *Physics & Chem.* **a.** A unit of matter, the smallest unit of an element, made up of a dense, central, positively charged nucleus surrounded by a system of electrons, equal in number to the number of nuclear protons, the entire structure having an approximate diameter of $10^{-8}$ centimeter and typically remaining undivided in chemical reactions except for limited removal, transfer, or exchange of certain electrons. **b.** This unit regarded as a source of nuclear energy. **2.** Something considered an irreducible constituent of a specified system. **3.** The irreducible, indestructive material unit of ancient atomism.

**atom bomb** *n.* An atomic bomb.

**a·tom·ic** (ə-tŏm′ĭk) *adj.* **1.** Of or relating to an atom or atoms. **2.** Of or employing atomic energy. **3.** Very small : INFINITESIMAL. **—a·tom′i·cal·ly** *adv.*

**atomic age** *also* **Atomic Age** *n.* The current era as marked by the discovery, technological applications, and sociopolitical consequences of atomic energy.

**atomic bomb** *n.* **1.** An explosive weapon of great destructive power derived from the rapid release of energy in the fission of heavy atomic nuclei, as of uranium 235. **2.** A bomb deriving its destructive power from the release of nuclear energy.

**atomic clock** *n.* An extremely accurate timekeeping device regulated in correspondence with a characteristic invariant frequency of an atomic or molecular system.

**atomic energy** *n.* **1.** The energy released from an atomic nucleus in fission or fusion. **2.** Atomic energy regarded as a source of practical power.

**at·o·mic·i·ty** (ăt′ə-mĭs′ĭ-tē) *n.* **1.** The state of being composed of atoms. **2.** *Chem.* **a.** The number of atoms in a molecule. **b.** Valence.

**atomic mass** *n.* The mass of an atomic system or constituent, usu. expressed in atomic mass units.

**atomic mass unit** *n.* A unit of mass equal to $1/_{12}$ the mass of the carbon isotope with mass number 12, approx. $1.6604 \times 10^{-24}$ gram.

**atomic number** *n.* The number of protons in an atomic nucleus.

**atomic pile** *n.* A nuclear reactor.

**atomic theory** *n.* **1.** The physical theory of the structure, properties, and behavior of the atom. **2.** Atomism.

**atomic weight** *n.* The average weight of an atom of an element, usu. expressed relative to one atom of the carbon isotope taken to have a standard weight of 12.

**at·om·ism** (ăt′ə-mĭz′əm) *n.* **1.** The ancient theory of Democritus, Epicurus, and Lucretius, according to which simple, indivisible, and indestructible atoms are the basic components of the entire universe. **2.** A theory according to which social institutions and processes arise solely from the acts of individual people. **3. a.** The division or tendency to divide into subclasses, groups, or units of a given society. **b.** Such a tendency accompanied by or arising from a strong subjective individualism. **—at′om·ist** *n.* **—at′om·is′tic** (-ĭs′tĭk), **at′om·is′ti·cal** *adj.* **—at′om·is′ti·cal·ly** *adv.*

**at·om·ize** (ăt′ə-mīz′) *vt.* **-ized, -iz·ing, -iz·es. 1.** To reduce or separate into atoms. **2. a.** To reduce (a liquid) to a spray. **b.** To spray (a liquid) in this form. **3.** To subject to bombardment with atomic weapons. **—at′om·i·za′tion** *n.*

**at·om·iz·er** (ăt′ə-mī′zər) *n.* A device for producing a fine spray, esp. one of medicine or perfume.

**atom smasher** *n.* An atomic particle accelerator.

**at·o·my¹** (ăt′ə-mē) *n., pl.* **-mies.** [< Lat. *atomi,* pl. of *atomus,* atom.] *Archaic.* **1.** A tiny particle. **2.** A tiny being.

**at·o·my²** (ăt′ə-mē) *n., pl.* **-mies.** [< *an atomy,* respelling of ANATOMY.] *Archaic.* A gaunt person or a skeleton.

**a·to·nal** (ā-tō′nəl) *adj. Mus.* Having apparently no key or tonality. **—a·to′nal·ly** *adv.*

**a·to·nal·ism** (ā-tō′nə-lĭz′əm) *n. Mus.* **1.** The lack of a tonal center or key as a principle of musical composition. **2.** The theory of atonal composition.

**a·to·nal·i·ty** (ā′tō-năl′ĭ-tē) *n.* A style of musical composition in which tonal center or key is disregarded.

**at once** *adv.* **1.** At one time : SIMULTANEOUSLY. **2.** Immediately <left *at once* when we heard the news>

**a·tone** (ə-tōn′) *v.* **a·toned, a·ton·ing, a·tones.** [ME *atonen,* to be reconciled < *at one,* in agreement : *at,* at + *one,* one.] *—vi.* **1.** To make amends, as for a fault or sin. **2.** *Archaic.* To agree. *—vt. Archaic.* **1.** To expiate. **2.** To reconcile or harmonize. **3.** To conciliate : APPEASE. **—a·ton′a·ble, a·tone′a·ble** *adj.* **—a·ton′er** *n.*

▲ word history: The derivation of *atone,* from *at* and *one,* has been obscured somewhat by the fairly recent change in the pronunciation of *one. One* used to be pronounced like *own,* but since the 17th century it has been pronounced like *won,* the past tense of *win.* The older pronunciation survives in *alone, lone, lonely,* and *only* in addition to *atone* and its derivatives; the new pronunciation occurs in *once.*

**a·tone·ment** (ə-tōn′mənt) *n.* **1.** Amends made for an injury or wrong : EXPIATION. **2.** In the Hebrew Scriptures man's reconciliation with God after having transgressed the covenant. **3. Atonement. a.** The redemptive life and death of Christ. **b.** The reconciliation of God and man thus brought about by Christ. **4.** *Christian Science.* The radical obedience and purification, exemplified in the life of Jesus, by which humanity finds oneness with God. **5.** *Archaic.* Reconciliation : concord.

**a·ton·ic** (ā-tŏn′ĭk) *adj.* [Fr. *atonique* < Gk. *atonos.* —see ATONY.] **1.** Not accented, as words and syllables. **2.** *Pathol.* Relating to, caused by, or marked by atony. *—n.* An unaccented word, syllable, or sound. **—at′o·nic′i·ty** (ăt′ə-nĭs′ĭ-tē) *n.*

**at·o·ny** (ăt′ə-nē) *n.* [LLat. *atonia* < Gk. < *atonos,* slack : *a-,* without + *tonos,* stretching, tone.] **1.** Insufficient muscular tone. **2.** Lack of accent or stress in phonetics.

**a·top** (ə-tŏp′) *adv.* On or at the top. *—prep.* On top of. **—a·top′** *adj.*

**-ator** *suff.* [ME *-atour* < OFr. < Lat. *-ator* : *-atus,* -ate + *-or,* -or.] One that acts in a given manner <radi*ator*>

**-atory** *suff.* [ME < Lat. *-atorius* : *-atus,* -ate + *-orius,* -ory.] **1. a.** Of or relating to <perspir*atory*> **b.** Tending to <amend*atory*> **2.** One that is connected with <reform*atory*>

**ATP** (ā′tē′pē′) *n.* [A(DENOSINE) T(RI)PHOSPHATE).] An adenosine-derived nucleotide, $C_{10}H_{16}N_5O_{13}P_3$, that supplies energy to cells through its conversion to ADP.

**ATP·ase** (ā′tē-pē′ās) *n.* An enzyme that hydrolyzes ATP.

**at·ra·bil·ious** (ăt′rə-bĭl′yəs) *also* **at·ra·bil·i·ar** (-bĭl′ē-ər) *adj.* [< Lat. *atra bilis,* black bile, transl. of Gk. *melankholia.* —see MELANCHOLY.] **1.** Inclined to melancholy. **2.** Ill-tempered : surly. **—at′ra·bil′ious·ness** *n.*

**A·treus** (ā′trōōs′, ā′trē-əs) *n.* [Gk.] *Gk. Myth.* A king of Mycenae, father of Agamemnon and Menelaus.

**a·tri·a** (ā′trē-ə) *n. var. pl. of* ATRIUM.

**a·tri·o·ven·tric·u·lar** (ā′trē-ō-vĕn-trĭk′yə-lər) *adj.* Of or relating to the atria and the ventricles of the heart.

**a·trip** (ə-trĭp′) *adj. & adv.* Just clear of the bottom, as an anchor.

**a·tri·um** (ā′trē-əm) *n., pl.* **a·tri·a** (ā′trē-ə) *or* **-ums.** [Lat. *atrium.*] **1. a.** A central courtyard, as in ancient Roman houses. **b.** A multistoried central court, as in a hotel, often having a skylight. **2.** A bodily cavity or chamber, as in the heart. **—a′tri·al** *adj.*

**a·tro·cious** (ə-trō′shəs) *adj.* [< Lat. *atrox, atroc-,* cruel.] **1.** Extremely evil or cruel : MONSTROUS <an *atrocious* felony> **2.** Exceptionally bad : ABOMINABLE <*atrocious* cooking><*atrocious* rudeness> **—a·tro′cious·ly** *adv.* **—a·tro′cious·ness** *n.*

**a·troc·i·ty** (ə-trŏs′ĭ-tē) *n., pl.* **-ties. 1.** Atrocious condition, quality, or behavior. **2.** An atrocious action, situation, or object : OUTRAGE.

**at·ro·phy** (ăt′rə-fē) *n., pl.* **-phies.** [LLat. *atrophia* < Gk. < *atrophos,* ill-nourished : *a-,* without + *trophē,* food.] **1.** *Pathol.* The emaciation or wasting away of bodily tissues or organs. **2.** A diminution or degeneration <moral *atrophy*> —*v.* **-phied, -phy·ing, -phies.** —*vt.* To affect with atrophy. —*vi.* To waste away : WITHER. **—a·troph′ic** (ā-trŏf′ĭk), **at′ro·phous** *adj.*

**at·ro·pine** (ăt′rə-pēn′, -pĭn) *also* **at·ro·pin** (-pĭn) *n.* [G. *Atropin* < NLat. *Atropa,* genus name of belladonna < Gk. *atropos,* unchangeable.] An extremely poisonous, bitter, crystalline alkaloid, $C_{17}H_{23}NO_3$, derived from belladonna and related plants and used to dilate the pupil of the eye and as an anesthetic and antispasmodic.

**At·ro·pos** (ăt′rə-pŏs′, -pəs) *n.* [Gk. < *atropos,* inexorable.] *Gk. Myth.* One of the three Fates.

**at·tach** (ə-tăch′) *v.* **-tached, -tach·ing, -tach·es.** [ME *attachen* < OFr. *attachier,* of Germanic orig.] —*vt.* **1.** To fasten on or affix to : connect or join. **2.** To connect as an adjunct or associated part. **3.** To add, as a signature. **4.** To ascribe or assign <*attached* no importance to the incident> **5.** To bind by personal ties, as of affection or loyalty <very *attached* to their pets> **6.** To appoint officially. **7.** To assign (personnel) to a military unit on a temporary basis. **8.** *Law.* To seize (persons or property) by legal writ. —*vi.* To adhere. **—at·tach′a·ble** *adj.* **—at·tach′er** *n.*

☆ ***syns:*** ATTACH, AFFIX, CLIP, CONNECT, COUPLE, FASTEN, FIX, MOOR, SECURE *v. core meaning* : to join one thing to another <the hinges to which the door is *attached*> ***ant*** : detach

**at·ta·ché** (ăt′ə-shā′, ă-tă′shā′) *n.* [Fr. < p.part. of *attacher,* to attach.] One officially assigned to the staff of a diplomatic mission to serve in a given capacity <a commercial *attaché*>

**attaché case** *n.* A briefcase resembling a small suitcase, with hinges and flat sides.

**at·tach·ment** (ə-tăch′mənt) *n.* **1.** The act of attaching or condition of being attached. **2.** Something, as a tie, band, or fastening, that joins one thing to another. **3.** A bond of affection or loyalty. **4.** A supplementary part : ACCESSORY <a vacuum cleaner with *attachments*> **5.** *Law.* **a.** The legal seizure of a person or property. **b.** The writ ordering an attachment.

**at·tack** (ə-tăk′) *v.* **-tacked, -tack·ing, -tacks.** [Fr. *attaquer* < OFr. < OItal. *attaccare,* of Germanic orig.] —*vt.* **1.** To set upon with violent force. **2.** To criticize strongly or in a hostile manner. **3.** To start work on with purpose and vigor <*attack* a backlog of orders> **4.** To begin to affect harmfully. —*vi.* To launch an attack. —*n.* **1.** The act of attacking : ASSAULT. **2.** The occurrence or onset of a disease. **3.** The initial movement in a task or undertaking. **4.** *Mus.* The way in which a passage or phrase is begun. **—at·tack′er** *n.*

☆ ***syns:*** ATTACK, ASSAIL, ASSAULT, BESET, HIT, STRIKE *v. core meaning* : to set upon with violent force <enemy troops *attacking* our positions>

**at·tain** (ə-tān′) *v.* **-tained, -tain·ing, -tains.** [ME *atteignen* < OFr. *ataindre,* to reach to < Lat. *attingere* : *ad-,* to + *tangere,* to touch.] —*vt.* **1.** To gain or accomplish by mental or physical effort <*attain* an objective> **2.** To arrive at <*attained* the mountaintop> —*vi.* To succeed in gaining or accomplishing <*attained* to the presidency> **—at·tain′a·bil′i·ty, at·tain′a·ble·ness** *n.* **—at·tain′a·ble** *adj.*

**at·tain·der** (ə-tān′dər) *n.* [ME *attendre,* conviction < OFr. *ataindre,* to convict, affect. —see ATTAIN.] *Law.* **1.** The loss of all civil rights legally consequent to a death sentence or to outlawry, esp. for treason. **2.** *Archaic.* Dishonor.

**at·tain·ment** (ə-tān′mənt) *n.* **1.** The act of attaining or condition of being attained. **2.** Something attained.

**at·taint** (ə-tānt′) *vt.* **-taint·ed, -taint·ing, -taints.** [ME *attaynten* < OFr. *ataint,* p.part. of *ataindre,* to affect. —see ATTAIN.] **1.** *Law.* To condemn by a sentence of attainder. **2.** *Archaic.* To disgrace. **3.** *Obs.* To accuse. —*n.* **1.** Attainder. **2.** *Archaic.* A disgrace : stigma.

**at·tar** (ăt′ər) *n.* [Pers. *′aṭir,* perfumed < Ar. *′uṭūr,* pl. of *′iṭr,* perfume.] A fragrant essential oil or perfume obtained from the petals of flowers, as roses.

**at·tempt** (ə-tĕmpt′) *vt.* **-tempt·ed, -tempt·ing, -tempts.** [ME *attempten* < OFr. *attempter* < Lat. *attemptare* : *ad-,* to + *temptare,* to test.] **1.** To try to do, make, or achieve. **2.** *Archaic.* To tempt. **3.** *Archaic.* To attack in order to subdue. —*n.* **1.** An effort or try. **2.** An attack or assault, as on one's life. **—at·tempt′a·ble** *adj.* **—at·tempt′er** *n.*

**at·tend** (ə-tĕnd′) *v.* **-tend·ed, -tend·ing, -tends.** [ME *attenden* < OFr. *atendre* < Lat. *attendere,* to heed : *ad-,* to + *tendere,* to stretch.] —*vt.* **1.** To be present at. **2.** To accompany as a circumstance or follow as a result <The announcement was *attended* by cheers.> **3. a.** To accompany or wait on as an attendant or servant. **b.** To take care of (e.g., a patient). **4.** To take charge of. **5.** To listen to : HEED. **6.** *Archaic.* To wait for : EXPECT. —*vi.* **1.** To be present. **2.** To apply or direct oneself <*attended* to the difficulty> **3.** To pay attention : heed. **4.** To remain ready to serve : WAIT <*attend* upon the queen> **5.** *Obs.* To delay or wait. **—at·tend′er** *n.*

**at·ten·dance** (ə-tĕn′dəns) *n.* **1.** The act of attending. **2.** Those that attend a function.

**at·ten·dant** (ə-tĕn′dənt) *n.* **1.** One who attends or serves another. **2.** One who is present. **3.** One that accompanies : CONCOMITANT. —*adj.* Accompanying or following as a result <the flu and *attendant* miseries> **—at·tend′ant·ly** *adv.*

**at·ten·tion** (ə-tĕn′shən) *n.* [ME *attencioun* < Lat. *attentio* < *attendere,* to heed. —see ATTEND.] **1.** Close or careful observation or heed : mental concentration. **2.** The ability or power to concentrate mentally. **3.** Observant consideration : NOTICE <Your complaint has come to my *attention.*> **4.** Courtesy or considerate regard, as for others' feelings. **5. attentions.** Acts of courtesy, consideration, or gallantry, esp. by a suitor. **6.** A military posture, with the body erect, eyes to the front, arms at the sides, and heels together. —Used as a command. **—at·ten′tion·al** *adj.*

**attention key** *n. Computer Sci.* A function key on terminals that interrupts program execution by the central processing unit.

**at·ten·tive** (ə-tĕn′tĭv) *adj.* **1.** Paying heed : OBSERVANT. **2.** Mindful of the well-being of others : CONSIDERATE. **—at·ten′tive·ly** *adv.* **—at·ten′tive·ness** *n.*

**at·ten·u·ate** (ə-tĕn′yo͞o-āt′) *v.* **-at·ed, -at·ing, -ates.** [Lat. *attenuare,* to make thin : *ad-,* to + *tenuis,* thin.] —*vt.* **1.** To make slender, fine, or small. **2.** To reduce in force, value, or amount : WEAKEN. **3.** To lessen the density of : RAREFY. **4.** *Biol.* To make less virulent. —*vi.* To become thin, weak, or fine. —*adj.* (ə-tĕn′yo͞o-ĭt). **1.** Weakened or reduced, as in strength or value. **2.** *Bot.* Gradually tapering to a slender point. **—at·ten′u·a·ble** (-ə-bəl) *adj.* **—at·ten′u·a′tion** *n.*

**at·ten·u·a·tor** (ə-tĕn′yo͞o-ā′tər) *n.* A device that reduces the amplitude of an electrical signal with minimal distortion.

**at·test** (ə-tĕst′) *v.* **-test·ed, -test·ing, -tests.** [Fr. *attester* < OFr. < Lat. *attestari* : *ad-,* to + *testis,* witness.] —*vt.* **1.** To assure the certainty or validity of : AFFIRM. **2. a.** To certify by signature or oath. **b.** To certify in an official capacity. **3.** To supply or be evidence of <a splendid mansion that *attested* their wealth> **4.** To constitute documentary proof of. **5.** To put under oath. —*vi.* To bear witness : give testimony <*attested* to my client's probity> —*n. Archaic.* An act or instance of attesting. **—at·test′ant** *n.* **—at′tes·ta′tion** (ăt′ĕs-tā′shən, ăt′ə-stā′-) *n.* **—at·test′er, at·tes′tor** *n.*

**at·tic** (ăt′ĭk) *n.* [< *Attic story,* a decorative structure over an order in a classical style.] **1.** A story or room directly beneath the roof of a house. **2.** A low wall or story above the cornice of a classical façade.

**At·tic** (ăt′ĭk) *adj.* **1.** Of, relating to, or typical of ancient Attica, Athens, or the Athenians. **2.** Marked by purity and simplicity, as a literary style. —*n.* The ancient Greek dialect used in Attica that became the literary language of ancient Greece.

**At·ti·cism** (ăt′ĭ-sĭz′əm) *n.* **1.** A typical feature of the Attic Greek language. **2.** An expression marked by conciseness and elegance.

**at·tire** (ə-tīr′) *vt.* **-tired, -tir·ing, -tires.** [ME *attiren* < OFr. *atirier,* to put in order : *a-,* to (< Lat. *ad-*) + *tire,* rank.] To dress or clothe, esp. in elaborate or splendid garments. —*n.* **1.** Clothes : array. **2.** *Heraldry.* The antlers of a deer.

**at·ti·tude** (ăt′ĭ-to͞od′, -tyo͞od′) *n.* [Fr. < Ital. *attitudine,* disposition < LLat. *aptitudo,* faculty < Lat. *aptus,* fit < *apere,* to fasten.] **1.** A position of the body or manner of carrying oneself <sat in an awkward *attitude*> **2.** A state of mind or feeling : DISPOSITION <a belligerent *attitude*> **3.** The orientation of an aircraft's axes relative to a reference line or plane, such as the horizon. **4.** *Aerospace.* The orientation of a spacecraft relative to its direction of motion. **5.** A position in which a ballet dancer stands on one leg with the other bent backward at the knee. **—at′ti·tu′di·nal** (ăt′ə-to͞od′n-əl, -tyo͞od′n-əl) *adj.*

**at·ti·tu·di·nize** (ăt′ĭ-to͞od′n-īz′, -tyo͞od′n-īz′) *vi.* **-nized, -niz·ing, -niz·es.** [Ital. *attitudine,* attitude, posture + -IZE.] To assume an affected attitude : POSTURE.

**atto-** *pref.* [< Dan. or Norw. *atten,* eighteen < ON *attjan.*] One quintillionth ($10^{-18}$) <*attotesla*>

**at·torn** (ə-tûrn′) *vi.* **-torned, -torn·ing, -torns.** [ME *attournen* < OFr. *atorner,* to assign to : *a-,* to (< Lat. *ad-*) + *torner,* to turn. —see TURN.] *Law.* To acknowledge a new owner as one's landlord. **—at·torn′ment** *n.*

**at·tor·ney** (ə-tûr′nē) *n., pl.* **-neys.** [ME *attourney* < OFr. *atorne,* p.part. of *atorner,* to attorn.] One legally appointed or retained to act for another, esp. an attorney at law. **—by attorney.** By proxy. **—at·tor′ney·ship′** *n.*

**attorney at law** *n.* One who is qualified to represent clients in a court of law and to advise them on legal matters : LAWYER.

**attorney general** *n., pl.* **attorneys general.** The chief law officer and legal counsel of a state or national government.

**at·to·tes·la** (ăt′ō-tĕs′lə) *n.* One quintillionth ($10^{-18}$) of a tesla.

**at·tract** (ə-trăkt′) *v.* **-tract·ed, -tract·ing, -tracts.** [ME *attracten* < Lat. *attrahere* : *ad-,* to + *trahere,* to draw.] —*vt.* **1.** To cause to draw near or adhere. **2.** To draw or direct to oneself by a quality or action. **3.** To evoke by arousing interest or admiration : ALLURE. —*vi.*

To have or use the power of attraction. **—at·tract'a·ble** *adj.* **—at·tract'er, at·trac'tor** *n.*

☆ ***syns:*** ATTRACT, ALLURE, APPEAL, DRAW, LURE, MAGNETIZE, PULL, TAKE *v. core meaning:* to direct or impel to oneself by a quality or action <a new celebrity who *attracted* the crowd's attention><a handsome face that *attracted* all eyes> ***ant:*** repel

**at·trac·tion** (ə-trăk'shən) *n.* **1.** The act or capability of attracting. **2.** The power or quality of attracting : CHARM. **3.** One that attracts. **4.** A public entertainment or spectacle.

**at·trac·tive** (ə-trăk'tĭv) *adj.* **1.** Having the power to attract. **2.** Pleasing to the eye or mind. **3.** Pleasingly suited to the wearer. **—at·trac'tive·ly** *adv.* **—at·trac'tive·ness** *n.*

**at·trib·ute** (ə-trĭb'yōōt) *vt.* **-ut·ed, -ut·ing, -utes.** [Lat. *attribuere, attribut- : ad-*, to + *tribuere*, to allot < *tribus*, tribe.] **1.** To assign to a cause or source : ASCRIBE. **2.** To consider or indicate as a creator or possessor <*attributed* the sculpture to Dégas> *—n.* (ăt'rə-byōōt'). **1.** A quality or characteristic : distinctive feature. **2.** An object associated with and serving to identify a character, personage, or office <A trident was the *attribute* of Neptune.> **3.** A word, esp. an adjective, used to ascribe a quality. **—at·trib'ut·a·ble** *adj.* **—at·trib'ut·er, at·trib'u·tor** *n.*

**at·tri·bu·tion** (ăt'rə-byōō'shən) *n.* **1.** The act of attributing. **2.** Something attributed.

**at·trib·u·tive** (ə-trĭb'yə-tĭv) *n.* A word or word group, as an adjective, placed adjacent to the noun it modifies without a linking verb, as *sick* in *the sick child.* *—adj.* **1.** Of or acting as an attributive <an *attributive* adjective> **2.** Of or having the nature of an attribution or attribute. **—at·trib'u·tive·ly** *adv.* **—at·trib'u·tive·ness** *n.*

**at·trit·ed** (ə-trī'tĭd) *adj.* Worn down by attrition.

**at·tri·tion** (ə-trĭsh'ən) *n.* [ME *atricioun* < Lat. *attritio*, act of rubbing against < *atterere*, to rub against : *ad-*, to + *terere*, to rub.] **1.** A rubbing away or wearing down by friction. **2.** A gradual lessening in number or strength due to constant stress. **3.** A gradual, natural reduction in membership or personnel, as through retirement. **4.** Repentance for sin motivated by fear of punishment rather than by love of God.

**at·tune** (ə-tōōn', ə-tyōōn') *vt.* **-tuned, -tun·ing, -tunes.** **1.** To tune. **2.** To bring into accord : HARMONIZE.

**a·twit·ter** (ə-twĭt'ər) *adj.* Nervously excited.

**a·typ·i·cal** (ā-tĭp'ĭ-kəl) *also* **a·typ·ic** (-ĭk) *adj.* Not conforming to type : UNUSUAL. **—a·typ'i·cal·ly** *adv.*

**Au** [Lat. *aurum*, gold.] *symbol for* GOLD.

**au·bade** (ō-bäd') *n.* [Fr. < OFr. < OSp. *albada* < *alba*, dawn.] A musical composition, as a love song, performed at dawn or in the early hours of the morning.

**au·ber·gine** (ō'bĕr-zhēn', ō'bər-jīn) *n.* [Fr. < Catalan *alberginia* < Ar. *albādinjān* < Pers. *bādin-gān.*] EGGPLANT 1b.

**au·burn** (ô'bərn) *n.* [ME < OFr. *auborne* < Lat. *alburnus*, whitish < *albus*, white.] A moderate reddish brown to brown.

**au cou·rant** (ō' kōō-räN') *adj.* [Fr.] Up-to-date : modern.

**auc·tion** (ôk'shən) *n.* [Lat. *auctio* < *augēre*, to increase.] **1.** A public sale in which property or items of merchandise are sold to the highest bidder. **2.** The bidding in the game of bridge. **3.** Auction bridge. *—vt.* **-tioned, -tion·ing, -tions.** To sell at or by an auction.

**auction bridge** *n.* A variety of the game of bridge in which tricks made in excess of the contract are scored toward game.

**auc·tion·eer** (ôk'shə-nîr') *n.* One who conducts an auction. *—vt.* **-eered, -eer·ing, -eers.** To auction.

**auctioneering device** *n. Computer Sci.* An instrument designed to automatically select either the highest or the lowest input signal from among two or more signals.

**auc·to·ri·al** (ôk-tôr'ē-əl, -tōr'-) *adj.* [< Lat. *auctor*, author.] Of or relating to an author.

**au·da·cious** (ô-dā'shəs) *adj.* [Lat. *audax, audaci-* < *audēre*, to dare.] **1.** Taking or willing to take risks : BOLD. **2.** Unrestrained by convention or propriety : INSOLENT. **—au·da'cious·ly** *adv.* **—au·da'cious·ness** *n.*

**au·dac·i·ty** (ô-dăs'ĭ-tē) *n., pl.* **-ties.** **1.** Willingness to take risks : BOLDNESS. **2.** Unrestrained impudence : BRASHNESS. **3.** An act or instance of audacity.

**au·di·ble** (ô'də-bəl) *adj.* [Med. Lat. *audibilis* < Lat. *audire*, to hear.] Capable of being heard. *—n. Football.* A new or substitute offensive play called by the quarterback or defensive formation called by a linebacker at the line of scrimmage as an adjustment to the opposing team's formation. **—au'di·bil'i·ty** *n.* **—au'di·ble·ness** *n.* **—au'di·bly** *adv.*

**au·di·ence** (ô'dē-əns) *n.* [ME < OFr. < Lat. *audientia* < *audiens*, pr.part. of *audire*, to hear.] **1.** A body of spectators, listeners, or readers of a work or performance. **2.** A formal hearing, as with a monarch or pope. **3.** An opportunity to be heard or to express one's views. **4.** The act of hearing or attending.

**au·di·ent** (ô'dē-ənt) *adj.* [Lat. *audiens, audient-*, pr.part. of *audire*, to hear.] Hearing : listening. **—au'di·ent** *n.*

**au·dile** (ô'dīl') *adj.* [< Lat. *audire*, to hear.] Capable of learning chiefly from auditory rather than visual or tactile stimuli. *—n.* An audile person.

**aud·ing** (ô'dĭng) *n.* [Lat. *audire*, to hear + -ING[2].] The process of hearing, acknowledging, and comprehending a spoken language.

**au·di·o** (ô'dē-ō') *adj.* [< AUDIO-.] **1.** Of or relating to audible sound. **2. a.** Of or relating to the broadcasting of sound. **b.** Of or relating to the high-fidelity reproduction of sound. *—n.* **1.** The part of television or motion-picture equipment that deals with sound. **2.** Audio broadcasting or reception. **3.** Audible sound.

**audio-** *pref.* [< Lat. *audire*, to hear.] **1.** Hearing <*audio*-lingual> **2.** Sound <*audio*phile>

**audio frequency** *n.* A set of frequencies, usu. ranging from 15 hertz to 20,000 hertz, typical of signals that are audible to the normal human ear.

**au·di·o-lin·gual** (ô'dē-ō-lĭng'gwəl) *adj.* Relating to or involving listening and speaking in learning a language.

**au·di·ol·o·gy** (ô'dē-ŏl'ə-jē) *n., pl.* **-gies.** The science of hearing defects and their treatment. **—au'di·o·log'i·cal** (-ə-lŏj'ĭ-kəl) *adj.* **—au'di·ol'o·gist** *n.*

**au·di·om·e·ter** (ô'dē-ŏm'ĭ-tər) *n.* An instrument for measuring hearing thresholds for pure tones of normally audible frequencies. **—au'di·o·met'ric** (-ō-mĕt'rĭk) *adj.* **—au'di·om'e·try** *n.*

**au·di·o·phile** (ô'dē-ə-fīl') *n.* One ardently interested in high-fidelity sound reproduction.

**au·di·o·typ·ing** (ô'dē-ō-tī'pĭng) *n.* Typing done directly from a tape recording. **—au'di·o·typ'ist** *n.*

**au·di·o-vis·u·al** (ô'dē-ō-vĭzh'ōō-əl) *adj.* **1.** Being both audible and visible. **2.** Of or relating to educational materials, as sound filmstrips, that present information in audible and visible form.

**au·di·o·vis·u·als** (ô'dē-ō-vĭzh'ōō-əlz) *pl.n.* Educational materials that use both sight and sound to present information.

**au·dit** (ô'dĭt) *n.* [ME < Lat. *auditus*, a hearing < p.part. of *audire*, to hear.] **1.** An examination of records or accounts to check their accuracy. **2.** An adjustment or correction of accounts. **3.** A verified account. *—v.* **-dit·ed, -dit·ing, -dits.** *—vt.* **1.** To examine, verify, or correct (e.g., accounts). **2.** To attend (a course) without receiving academic credit. *—vi.* To examine accounts.

**au·di·tion** (ô-dĭsh'ən) *n.* [Lat. *auditio* < *audire*, to hear.] **1.** The power or sense of hearing. **2.** An act of hearing. **3.** A usu. short performance, as by an actor or musician, to demonstrate talent or ability. *—vt. & vi.* **-tioned, -tion·ing, -tions.** To evaluate or present an audition.

**au·di·tive** (ô'dĭ-tĭv) *adj.* Auditory.

**au·di·tor** (ô'dĭ-tər) *n.* [ME < AN *auditour* < Lat. *auditor* < *audire*, to hear.] **1.** One who hears : LISTENER. **2.** One who audits accounts. **3.** One who audits an academic course.

**au·di·to·ri·um** (ô'də-tôr'ē-əm, -tōr'-) *n., pl.* **-ri·ums** *or* **-ri·a** (-tôr'ē-ə, -tōr'-) [Lat. < *audire*, to hear.] **1.** A room in a building such as a school or theater to accommodate a large audience. **2.** A building for public meetings, sports competitions, or artistic performances.

**au·di·to·ry** (ô'dĭ-tôr'ē, -tōr'ē) *adj.* [Lat. *auditorius* < *audire*, to hear.] Of or relating to the sense, organs, or experience of hearing.

**auditory nerve** *n. Anat.* The acoustic nerve.

**au fait** (ō fĕ') *adj.* [Fr., to the point.] Knowledgeable or skilled.

**Auf·klä·rung** (ouf'klā'rōōng) *n.* [G.] The Enlightenment.

**auf Wie·der·seh·en** (ouf vē'dər-zā'ən) *interj.* [G.] —Used to express a farewell.

**Au·ge·an** (ô-jē'ən) *adj.* [After *Augeas*, legendary Greek king who did not clean his stable for thirty years.] **1.** Exceedingly filthy from long neglect. **2.** Corrupt. **3.** Difficult : burdensome.

**au·gend** (ô'jĕnd') *n.* [LLat. *augendum* < *augendus*, gerund. of *augēre*, to increase.] A quantity to which the addend is added.

**au·ger** (ô'gər) *n.* [ME < *an auger*, alteration of *a nauger* < OE *nafogār*, auger.] **1.** A tool used to bore holes, as in ice or wood. **2.** A large tool for boring into the earth.

▲ word history: *An auger* was originally *a nauger.* It is one of the numerous English words that have lost an initial *n* as a result of the incorrect division between the indefinite article and the noun. The Old English word from which *auger* is derived, *nafogār*, is actually a compound word made up of *nafu*, "wheel hub" (related to *navel*), and *gār*, "spear." An *auger* is thus a tool for piercing and boring holes, like the hole in a wheel through which the axle passes.

**aught**[1] (ôt) [ME < OE *āuht.*] *Archaic. —pron.* Anything whatever : any least part. *—adv.* In any respect : at all.

**aught**[2] (ôt) *n.* [Alteration of A NAUGHT.] **1.** A cipher : zero. **2.** *Archaic.* Nothing.

**au·gite** (ô'jīt') *n.* [Lat. *augites*, a precious stone < Gk. *augitēs* < *augē*, brightness.] A dark-green to black pyroxene mineral containing large amounts of aluminum, iron, and magnesium.

**aug·ment** (ôg-mĕnt') *v.* **-ment·ed, -ment·ing, -ments.** [ME *augmentem* < OFr. *augmenter* < LLat. *augmentare*, to increase < *augmentum*, an increase < *augēre*, to increase.] *—vt.* **1.** To make greater, as in size, extent, or quantity : INCREASE. **2.** To add an augment to. *—vi.* To become greater. *—n.* (ôg'mĕnt'). **1.** An enlargement. **2.** The prefixation of a vowel or the lengthening of an initial vowel to indicate the past tense, esp. of Greek and Sanskrit verbs. **—aug·ment'a·ble** *adj.* **—aug·ment'er** *n.*

**aug·men·ta·tion** (ôg'mĕn-tā'shən) *n.* **1. a.** The act or process of augmenting. **b.** The condition of being augmented. **2.** Something that augments. **3.** *Mus.* The repetition of a theme in notes of usu. double time value.

**aug·men·ta·tive** (ôg-mĕn'tə-tĭv) *adj.* **1.** Capable of or tending to augment. **2.** Of, relating to, or designating esp. a word element that indicates an increase, as in size, force, or intensity, in the meaning of the original word. **—aug·men'ta·tive** *n.*

**aug·men·ted** (ôg-mĕn'tĭd) *adj. Mus.* Larger by a semitone than the corresponding major or perfect interval.

**au gra·tin** (ō grät'n, grăt'n) *adj.* [Fr.] Covered with bread crumbs or grated cheese and oven-browned <potatoes *au gratin*>

**au·gur** (ô'gər) *n.* [Lat.] **1.** One of a group of religious officials of ancient Rome who foretold events by observing and interpreting signs and omens. **2.** A prophet or seer : SOOTHSAYER. *—v.* **-gured, -gur·ing, -gurs.** *—vt.* **1.** To predict, as from signs or omens. **2.** To serve as an omen of : BETOKEN. *—vi.* **1.** To conjecture or foretell from signs or omens. **2.** To be a sign or omen <"traits that *augured* well for success" —Philip Horton> **—au'gu·ral** (ô'gyə-rəl) *adj.*

**au·gu·ry** (ô'gyə-rē) *n., pl.* **-ries.** [ME *augurie* < OFr. < Lat. *augurium* < *augur*, augur.] **1.** The art, ability, or practice of auguring. **2.** A sign or omen : PORTENT.

**au·gust** (ô-gŭst') *adj.* [Lat. *augustus*, venerable.] **1.** Inspiring admiration or awe. **2.** Venerable for reasons of age or high rank. **—au·gust'ly** *adv.* **—au·gust'ness** *n.*

**Au·gust** (ô'gəst) *n.* [ME < OE < Lat. *(mensis) Augustus*, (month) of Augustus, after *Augustus* Caesar.] The eighth month of the year according to the Gregorian calendar. —See table at CALENDAR.

**Au·gus·tan** (ô-gŭs'tən) *adj.* **1.** Relating to or typical of Augustus Caesar or his reign or times. **2.** Relating to or typical of English literature during the reign of Queen Anne. **—Au·gus'tan** *n.*

**Au·gus·tin·i·an** (ô'gə-stĭn'ē-ən) *adj.* **1.** Relating to Saint Augustine or his doctrines. **2.** Designating or belonging to any of several orders following or influenced by the rule of Saint Augustine. *—n.* **1.** A follower of the principles and doctrines of Saint Augustine. **2.** A monk belonging to an Augustinian order. **—Au'gus·tin'i·an·ism, Au·gus'tin·ism** *n.*

**au jus** (ō zhü') *adj.* [Fr.] Served with the natural juices or gravy <roast beef *au jus*>

**auk** (ôk) *n.* [Norw. *alk* < ON *alka.*] Any of several sea birds of the family Alcidae, as the razor-billed auk of northern regions, with a stocky body and short wings.

**auk·let** (ôk'lĭt) *n.* Any of various small auks of the genus *Aethia* and related genera, of northern Pacific coasts and waters.

**au lait** (ō lĕ') *adj.* [Fr.] With milk.

**auld** (ôld) *adj. Scot.* Old.

**auld lang syne** (ôld' lăng zīn', sīn') *n.* [Sc., old long since.] The good old days long past.

**au·lic** (ô'lĭk) *adj.* [Fr. *aulique*, Lat. *aulicus* < Gk. *aulikos* < *aulē*, court.] Relating to a royal court : COURTLY.

**au na·tu·rel** (ō' nä-tü-rĕl') *adj.* [Fr.] **1. a.** In a natural state. **b.** Nude. **2.** Cooked or served simply.

**aunt** (ănt, änt) *n.* [ME *aunte* < AN < OFr. *ante* < Lat. *amita*, paternal aunt.] **1.** The sister of one's father or mother. **2.** The wife of one's uncle.

**aunt·ie** *also* **aunt·y** (ăn'tē, än'-) *n. Informal.* Aunt.

**au pair** *also* **au pair girl** (ō' pâr') *n.* [Fr.] A foreign girl or woman who works for a family in exchange for room and board and an opportunity to learn the family's language.

**au·ra** (ôr'ə) *n., pl.* **au·ras** *or* **au·rae** (ôr'ē) [ME, gentle breeze < Lat. < Gk.] **1.** An invisible breath or emanation. **2.** A distinctive air or quality <an *aura* of suspense> **3.** *Pathol.* A sensation preceding the onset of certain nervous disorders.

**au·ral**[1] (ôr'əl) *adj.* [< Lat. *auris*, ear.] Of, relating to, or perceived by the ear.

**au·ral**[2] (ôr'əl) *adj.* Relating to or marked by an aura.

**au·rar** (ou'rär', œ'rär') *n. pl. of* EYRIR.

**au·re·ate** (ôr'ē-ĭt) *adj.* [ME *aureat* < Lat. *aureatus* < *aureus*, golden < *aurum*, gold.] **1.** Gold in color. **2.** Marked by florid and pompous literary style. **—au're·ate·ly** *adv.* **—au're·ate·ness** *n.*

**au·re·ole** (ôr'ē-ōl') *also* **au·re·o·la** (ô-rē'ə-lə) *n.* [ME < Med. Lat. *aureola (corona)*, golden (crown) < Lat. *aureolus*, golden < *aurum*, gold.] **1.** A circle of light or radiance surrounding the head or body of a representation of a deity or holy person : HALO. **2.** A bright band or region around a celestial body, as the sun or moon, esp. when observed through a haze or fog.

**Au·re·o·my·cin** (ôr'ē-ō-mī'sĭn). A trademark for chlortetracycline.

**au re·voir** (ō' rə-vwär') *interj.* [Fr.] —Used to express a farewell.

**au·ric** (ôr'ĭk) *adj.* [< Lat. *aurum*, gold.] Of, relating to, derived from, or having gold, esp. with valence 3.

**au·ri·cle** (ôr'ĭ-kəl) *also* **au·ric·u·la** (ô-rĭk'yə-lə) *n., pl.* **-cles** *also* **-las** *or* **-lae** (-lē') [Lat. *auricula*, dim. of *auris*, ear.] **1. a.** *Anat.* The outer part of the ear : PINNA. **b.** An atrium of the heart. **2.** *Biol.* An earlike part, process, or appendage, esp. at the base of an organ. **—au'ri·cled** (-kəld) *adj.*

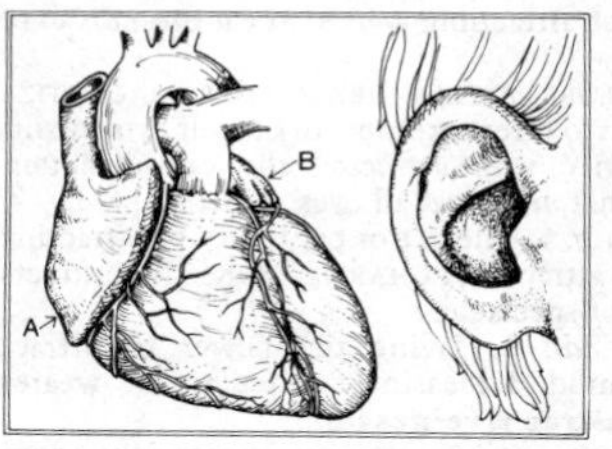

**auricle**
(Left) *of the heart showing A. right auricle, B. left auricle, and* (right) *of the ear*

**au·ric·u·la** (ô-rĭk'yə-lə) *n., pl.* **-las** *or* **-lae** (-lē') [NLat. < Lat., auricle.] **1.** A species of primrose indigenous to the Alps, *Primula auricula*, bearing variously colored flower clusters. **2.** *var. of* AURICLE.

**au·ric·u·lar** (ô-rĭk'yə-lər) *adj.* [Med. Lat. *auricularis* < *auricula*, auricle.] **1.** Of or relating to the sense or organs of hearing. **2.** Perceived by or spoken into the ear <an *auricular* report> **3.** Shaped like an ear. **4.** Of or relating to an auricle of the heart. *—n.* **au·riculars.** The feathers covering the opening of a bird's ear. **—au·ric'u·lar·ly** *adv.*

**au·ric·u·late** (ô-rĭk'yə-lĭt, -lāt') *also* **au·ric·u·lat·ed** (-lāt'ĭd) *adj.* [< Lat. *auricula*, auricle.] Having ears or earlike parts or extensions. **—au·ric'u·late·ly** *adv.*

**au·rif·er·ous** (ô-rĭf'ər-əs) *adj.* [Lat. *aurifer* : *aurum*, gold + *ferre*, to carry.] Containing or bearing gold.

**au·ri·form** (ôr'ə-fôrm') *adj.* [Lat. *auris*, ear + -FORM.] Ear-shaped.

**Au·ri·ga** (ô-rī'gə) *n.* [Lat. *auriga*, charioteer.] A constellation in the Northern Hemisphere.

**Au·rig·na·cian** (ôr'ĭg-nā'shən, ôr'ēn-yā'-) *adj.* [After *Aurignac*, a commune in France.] Of or relating to the Old World Upper Paleolithic culture between Mousterian and Solutrean, associated with Cro-Magnon man and marked by such artifacts as figures of stone and bone, graphic artwork, and the use of dress and adornment.

**au·rochs** (ou'rŏks', ôr'ŏks') *n.* [G.] **1.** The urus. **2.** The wisent.

**au·ro·ra** (ô-rôr'ə, ô-rōr'ə, ə-) *n.* [Lat.] **1. Aurora.** *Rom. Myth.* The goddess of the dawn. **2.** The dawn. **3. a.** Aurora borealis. **b.** Aurora australis. **—au·ro'ral, au·ro're·an** (-ē-ən) *adj.* **—au·ro'ral·ly** *adv.*

**aurora aus·tra·lis** (ô-strā'lĭs) *n.* [NLat., southern dawn.] A luminous phenomenon of the southern regions that corresponds to the aurora borealis of the northern regions.

**aurora bo·re·al·is** (bôr'ē-ăl'ĭs, bōr'-) *n.* [NLat., northern dawn.] Luminous bands or streamers of light that are sometimes visible in the night skies of the northern regions and are held to be caused by the ejection of charged particles into the magnetic field of the earth.

**au·rous** (ôr'əs) *adj.* [Lat. *aurum*, gold + -OUS.] Of, relating to, derived from, or having gold, esp. with valence 1.

**aus·cul·tate** (ô'skəl-tāt') *vt.* **-tat·ed, -tat·ing, -tates.** [Back-formation < AUSCULTATION.] *Med.* To examine (a person) by auscultation. **—aus'cul·ta'tive** *adj.* **—aus·cul'ta·to'ry** (ô-skŭl'tə-tôr'ē, -tōr'ē) *adj.*

**aus·cul·ta·tion** (ô'skəl-tā'shən) *n.* [Lat. *auscultatio* < *auscultare*, to listen to.] **1.** The act of listening. **2.** *Med.* Diagnostic monitoring of the sounds made by internal organs or an internal bodily part.

**aus·form** (ôs'fôrm') *vt.* **-formed, -form·ing, -forms.** [AUS(TENITIC) + (DE)FORM.] To subject a metal, esp. steel, to deformation, quenching, and tempering to improve its durability.

**aus·land·er** (ou'slĕn'dər, -slăn'-) *n.* [G. *Ausländer* : *aus*, out + *Land*, land.] A foreigner.

**aus·pex** (ô'spĕks') *n., pl.* **aus·pi·ces** (ô'spĭ-sēz') [Lat. —see AUSPICE.] An augur of ancient Rome, esp. one who interpreted omens observed in the actions of birds.

**aus·pi·cate** (ô'spĭ-kāt') *vt.* **-cat·ed, -cat·ing, -cates.** [Lat. *auspicari, aspicat-* < *auspex*, bird augur. —see AUSPICE.] To begin or inaugurate with a ceremony designed to bring good luck.

**aus·pice** (ô'spĭs) *n., pl.* **aus·pi·ces** (ô'spĭ-sēz', -sĭz) [Lat. *auspicium*, bird divination < *auspex*, bird augur : *avis*, bird + *-spex*, watcher < *specere*, to look.] **1. auspices.** Protection or support : PATRONAGE. **2.** A portent, omen, or augury, esp. when observed in the actions of birds. **3.** Observation of and divination from the actions of birds.

**aus·pi·cious** (ô-spĭsh'əs) *adj.* **1.** Accompanied by favorable circumstances : PROPITIOUS. **2.** Successful : prosperous. **—aus·pi'cious·ly** *adv.* **—aus·pi'cious·ness** *n.*

**Aus·sie** (ô'sē) *n.* [AUS(TRALIAN) + -IE.] *Informal.* A native or resident of Australia. **—Aus'sie** *adj.*

**aus·ten·ite** (ôs'tən-īt') *n.* [Fr., after Sir William Roberts-*Austen* (1843–1902).] A nonmagnetic solid solution of ferric carbide or carbon in iron, used in making corrosive-resistant steel. **—aus'ten·it'ic** (-ĭt'ĭk) *adj.*

**Aus·ter** (ôs'tər) *n.* [Lat.] The personification of the south wind.

**aus·tere** (ô-stîr') *adj.* [ME < OFr. < Lat. *austerus* < Gk. *austeros*, harsh.] **1. a.** Stern or severe. **b.** Somber. **2.** Ascetic. **3.** Lacking adornment or ornamentation : simple or plain. **—aus·tere'ly** *adv.* **—aus·tere'ness** *n.*

**aus·ter·i·ty** (ô-stĕr'ə-tē) *n., pl.* **-ties. 1.** The quality of being austere. **2.** Severe and rigid economy. **3.** An austere habit or practice.

**aus·tral** (ôs'trəl) *adj.* [< ME *auster,* south wind < Lat.] Of, relating to, or coming from the south.

**Aus·tra·lian** (ô-strāl'yən) *n.* **1.** A native or resident of the Commonwealth of Australia. **2.** An aborigine of Australia. **3.** Any of the languages of the Australian aborigines. *—adj.* **1.** Of or relating to Australia, its inhabitants, or their languages or cultures. **2.** *Ecol.* Of or designating the zoogeographic region that includes Australia and the islands adjacent to it, including New Guinea.

**Australian ballot** *n.* A printed ballot bearing the names of all candidates and the texts of propositions or referendums that is distributed to the voter at the polls and marked in secret.

**Australian crawl** *n.* A variation of the crawl that is executed with a flutter kick to each stroke.

**Australian terrier** *n.* A small dog orig. bred in Australia, with a coarse blackish coat marked with tan.

**Australian terrier**
*10 inches high at shoulder*

**Aus·tra·loid** (ôs'trə-loid') *adj.* [AUSTRAL(IAN) + -OID.] Of or relating to an ethnic group that includes the Australian aborigines. **—Aus'tra·loid'** *n.*

**aus·tra·lo·pith·e·cine** (ô-strā'lō-pĭth'ĭ-sīn') *n.* [< NLat. *Australopithecus* : Lat. *australis,* southern + *pithecus,* ape < Gk. *pithēkos.*] Any of several extinct humanlike primates of the genera *Australopithecus, Paranthropus,* or *Zinjanthropus,* known chiefly from Pleistocene fossil remains found in southern and eastern Africa. *—adj.* Of, relating to, or typical of the australopithecines.

**Austro-[1]** *pref.* [< Lat. *auster,* south.] Southern <*Austro*-Asiatic>

**Austro-[2]** *pref.* Austria : Austrian <*Austro*-Hungarian>

**Aus·tro-A·si·at·ic** (ôs'trō-ā'zhē-ăt'ĭk) *n.* A family of languages of southeastern Asia once dominant in northeastern India and Indochina. **—Aus'tro-A'si·at'ic** *adj.*

**Aus·tro·ne·sian** (ôs'trō-nē'zhən, -shən) *adj.* Of or relating to Austronesia, its peoples, or their languages. *—n.* A family of languages spoken in Austronesia that includes the Indonesian, Melanesian, Micronesian, and Polynesian subfamilies.

**au·ta·coid** *also* **au·to·coid** (ô'tə-koid') *n.* [AUT(O)- + Gk. *akos,* cure + -OID.] An organic substance, as a hormone, formed in an organ and secreted into the blood, lymph, or sap, from which it acts on other parts of the organism. **—au'to·coid'al** (-koid'l) *adj.*

**au·tar·chy** (ô'tär'kē) *n., pl.* **-chies.** [Gk. *autarkhia* < *autarkhos,* self-governing : *autos,* self + *arkhos,* ruler.] **1. a.** Complete control or power : AUTOCRACY. **b.** A country under autarchic rule. **2.** Autarky. **—au·tar'chic, au·tar'chi·cal** *adj.*

**au·tar·ky** (ô'tär'kē) *n., pl.* **-kies.** [Gk. *autarkeia,* self-sufficiency < *autarkēs,* self-sufficient : *autos,* self + *arkein,* to suffice.] **1.** A national policy of self-sufficiency and nonreliance on imports or economic aid. **2.** A self-sufficient country or region. **—au·tar'kic, au·tar'ki·cal** *adj.*

**au·te·col·o·gy** (ô'tĭ-kŏl'ə-jē) *n.* The ecology of a species or of individual organisms in relation to the environment. **—au'te·co·log'i·cal** *adj.*

**au·then·tic** (ô-thĕn'tĭk) *adj.* [ME *autentik* < OFr. *autentique* < LLat. *authenticus* < Gk. *authentikos* < *authentēs,* author.] **1. a.** Conforming to fact and therefore worthy of trust, reliance, or belief. **b.** Having an undisputed origin : GENUINE. **2.** *Law.* Executed with due process of law. **3.** *Mus.* **a.** Designating a medieval mode having a range from its final tone to the octave above it. **b.** Designating a cadence with the dominant chord immediately preceding the tonic chord. **4.** *Obs.* Authoritative. **—au·then'ti·cal·ly** *adv.* **—au'then·tic'i·ty** (ô'thĕn-tĭs'ə-tē) *n.*

**au·then·ti·cate** (ô-thĕn'tĭ-kāt') *vt.* **-cat·ed, -cat·ing, -cates.** To establish or prove as authentic. **—au·then'ti·ca'tion** *n.* **—au·then'ti·ca'tor** *n.*

**au·thor** (ô'thər) *n.* [ME *authour* < OFr. *autor* < Lat. *auctor,* creator < *augēre,* to create.] **1. a.** The original writer of a literary work. **b.** One who writes as a profession. **2. a.** An originator or creator. **b. Author.** GOD 1a. *—vt.* **-thored, -thor·ing, -thors. 1.** To be the author of : WRITE. **2.** To originate : create.

**au·thor·i·tar·i·an** (ə-thôr'ĭ-târ'ē-ən, ə-thŏr'-, ô-) *adj.* Marked by or advocating absolute obedience to authority. **—au·thor'i·tar'i·an** *n.* **—au·thor'i·tar'i·an·ism** *n.*

**au·thor·i·ta·tive** (ə-thôr'ĭ-tā'tĭv, ə-thŏr'-, ô-) *adj.* **1.** Having or arising from proper authority : OFFICIAL. **2.** Exercising authority : COMMANDING. **—au·thor'i·ta'tive·ly** *adv.* **—au·thor'i·ta'tive·ness** *n.*

**au·thor·i·ty** (ə-thôr'ĭ-tē, ə-thŏr'-, ô-) *n., pl.* **-ties.** [ME *autority* < OFr. *autorite* < Lat. *auctoritas* < *auctor,* creator.] **1. a.** The right and power to command, enforce laws, exact obedience, determine, or judge. **b.** A person or group invested with this right and power. **c.** The officials of a political unit having such right and power : GOVERNMENT. **2.** Freedom or right granted to another : AUTHORIZATION. **3.** A public agency or corporation with administrative powers in a specified field <a maritime *authority*> **4. a.** An accepted source of expert information or advice. **b.** A quotation or citation from such a source. **5.** An expert in a given field. **6.** Power to influence or persuade resulting from knowledge or experience <could speak with *authority*> **7.** Grounds for a course of action. **8.** An authoritative statement or decision that may be taken as a precedent.

**au·thor·i·za·tion** (ô'thər-ĭ-zā'shən) *n.* **1.** The act of authorizing. **2.** Something that authorizes : SANCTION.

**au·thor·ize** (ô'thə-rīz') *vt.* **-ized, -iz·ing, -iz·es.** [ME *autorisen* < OFr. *autoriser* < Med. Lat. *auctorizare* < Lat. *auctor,* author.] **1.** To give authority or power to. **2.** To approve or permit : SANCTION. **3.** To be sufficient grounds for : JUSTIFY. **—au'thor·iz'er** *n.*

**Authorized Version** *n.* The King James Bible.

**au·thor·ship** (ô'thər-shĭp') *n.* **1.** The profession or occupation of writing. **2.** An origin or source.

**au·tism** (ô'tĭz'əm) *n.* **1.** Acceptance of fantasy rather than reality : abnormal subjectivity. **2.** A form of childhood schizophrenia marked by acting out and withdrawal. **—au·tis'tic** (-tĭk) *adj.* & *n.*

**au·to** (ô'tō) [Short for AUTOMOBILE.] *Informal. —n., pl.* **-tos.** An automobile. *—vi.* **-toed, -to·ing, -tos.** To go by automobile.

**auto-** *pref.* [Gk. < *autos,* self.] **1.** Self : same <*auto*gamy> **2.** Automatic <*auto*pilot>

**au·to·an·ti·bod·y** (ô'tō-ăn'tĭ-bŏd'ē) *n., pl.* **-ies.** An antibody believed to act against cells of the organism in which it is formed.

**au·to·bahn** (ou'tō-bän') *n.* [G. : *auto,* automobile + *Bahn,* road.] A superhighway in Germany.

**au·to·bi·og·ra·phy** (ô'tō-bī-ŏg'rə-fē, -bē-ŏg'rə-fē) *n., pl.* **-phies.** The biography of a person written by himself or herself. **—au'to·bi·og'ra·pher** *n.* **—au'to·bi'o·graph'ic** (-bī'ə-grăf'ĭk), **au'to·bi'o·graph'i·cal** *adj.* **—au'to·bi'o·graph'i·cal·ly** *adv.*

**au·to·bus** (ô'tō-bŭs') *n., pl.* **-bus·es** *or* **-bus·ses.** BUS 1.

**au·to·ca·tal·y·sis** (ô'tō-kə-tăl'ĭ-sĭs) *n., pl.* **-ses** (-sēz'). Catalysis of a chemical reaction by one of the products of the reaction. **—au'to·cat'a·lyt'ic** *adj.*

**au·toch·thon** (ô-tŏk'thən) *n., pl.* **-thons** *or* **-tho·nes** (-thə-nēz') [Gk. *autokhthōn* : *autos,* self + *khthōn,* earth.] **1.** The first known or aboriginal inhabitant of a particular place. **2.** *Ecol.* An indigenous animal or plant.

**au·toch·tho·nous** (ô-tŏk'thə-nəs) *also* **au·toch·tho·nal** (-nəl) *or* **au·toch·thon·ic** (-thŏn'ĭk) *adj.* Native to a certain place : INDIGENOUS. **—au·toch'thon·ism, au·toch'tho·ny** *n.* **—au·toch'tho·nous·ly** *adv.*

**au·to·clave** (ô'tō-klāv') *n.* [Fr. : auto-, auto- + Lat. *clavis,* key.] A strong, pressurized, steam-heated vessel, as for sterilization.

**au·to·coid** (ô'tə-koid') *n. var. of* AUTACOID.

**au·toc·ra·cy** (ô-tŏk'rə-sē) *n., pl.* **-cies. 1.** Government by an individual having unlimited power : DESPOTISM. **2.** A country or state governed by autocracy.

**au·to·crat** (ô'tə-krăt') *n.* [Fr. *autocrate* < Gk. *autokratēs,* ruling by oneself : *auto-,* self + *kratos,* authority.] **1.** A ruler having absolute or unlimited power : DESPOT. **2.** A person with unrestricted power or authority. **—au'to·crat'ic, au'to·crat'i·cal** *adj.* **—au'to·crat'i·cal·ly** *adv.*

**au·to·cross** (ô'tō-krôs', -krŏs') *n.* A competition for automobiles that tests driving skill and speed.

**au·to-da-fé** (ou'tō-də-fā', ô'tō-) *n., pl.* **au·tos-da-fé** (ou'tōz-, ô'tōz-) [Port. *auto da fé,* act of the faith.] **1.** The public announcement of the sentences imposed on persons tried by the Inquisition and the public execution of these sentences by the secular authorities. **2.** The burning of heretics at the stake.

**au·to·di·dact** (ô'tō-dī'dăkt') *n.* [< Gk. *autodidaktos,* self-taught : *autos,* self + *didaktos,* taught. —see DIDACTIC.] A self-taught person. **—au'to·di·dac'tic** *adj.*

**au·to·dyne** (ô'tə-dīn') *n.* A heterodyne in which one tube serves simultaneously as oscillator and detector. **—au'to·dyne'** *adj.*

**au·toe·cious** (ô-tē'shəs) *adj.* [AUTO- + Gk. *oikos,* house.] *Biol.* Having all stages of a life cycle occur on the same host. **—au·toe'cism** *n.*

**au·to·er·o·tism** (ô'tō-ĕr'ə-tĭz'əm) *also* **au·to·e·rot·i·cism** (-ĭ-rŏt'ĭ-sĭz'əm) *n.* **1.** Self-gratification of sexual desire, as by masturbation. **2.** The arousal of sexual feeling without external stimulation. **—au'to·e·rot'ic** (-ĭ-rŏt'ĭk) *adj.*

**au·tog·a·my** (ô-tŏg′ə-mē) *n.* **1.** *Bot.* Fertilization of a flower by its own pollen : SELF-FERTILIZATION. **2.** *Biol.* The union of nuclei within and arising from a single cell, as in certain protozoans.
**au·to·gen·e·sis** (ô′tō-jĕn′ĭ-sĭs) *n.* Abiogenesis. **—au′to·ge·ne′tic** (-jə-nĕt′ĭk) *adj.* **—au′to·ge·net′i·cal·ly** *adv.*
**au·tog·e·nous** (ô-tŏj′ə-nəs) *also* **au·to·gen·ic** (ô′tə-jĕn′ĭk) *adj.* Self-produced : self-generated. **—au·tog′e·nous·ly** *adv.*
**au·to·gi·ro** *also* **au·to·gy·ro** (ô′tō-jī′rō) *n., pl.* **-ros.** [AUTO- + Gk. *guros,* circle.] An aircraft powered by a conventional propeller and provided with lift in flight by a freewheeling, horizontal rotor.
**au·to·graph** (ô′tə-grăf′) *n.* [Lat. *autographus,* written with one's own hand < Gk. *autographos* : *autos,* self + *graphein,* to write.] **1.** A person's own signature or handwriting. **2.** A manuscript in the author's own handwriting. —*vt.* **-graphed, -graph·ing, -graphs. 1.** To write one's name or signature on or in : SIGN. **2.** To write in one's own handwriting. **—au′to·graph′ic, au′to·graph′i·cal** *adj.* **—au′to·graph′i·cal·ly** *adv.*
**au·tog·ra·phy** (ô-tŏg′rə-fē) *n.* **1.** The writing of something in one's own handwriting. **2.** Autographs as a whole.
**au·to·gy·ro** (ô′tō-jī′rō) *n. var. of* AUTOGIRO.
**Au·to·harp** (ô′tō-härp′). A trademark for a musical instrument similar to a zither, on which a desired chord can be selected by depressing a particular damper.
**au·to·hyp·no·sis** (ô′tō-hĭp-nō′sĭs) *n.* **1.** The act or process of hypnotizing oneself. **2.** A self-induced hypnotic state. **—au′to·hyp·not′ic** (-nŏt′ĭk) *adj.*
**au·to·im·mune** (ô′tō-ĭ-myōōn′) *adj.* Related to or produced by autoantibodies. **—au′to·im·mun′i·ty** *n.* **—au′to·im·mun′i·za′·tion** (-ĭm′yə-nə-zā′shən) *n.* **—au′to·im′mun·ize′** *v.* **(-ized, -iz·ing, -i·zes).**
**au·to·in·dex** (ô′tō-ĭn′dĕks′) *n.* **1.** The procedure for preparation of an index to a body of material by means of a computer program. **2.** An index prepared by autoindex.
**au·to·in·fec·tion** (ô′tō-ĭn-fĕk′shən) *n.* Infection caused by bacteria, viruses, or parasites persisting on or in the body.
**au·to·in·oc·u·la·tion** (ô′tō-ĭn-ŏk′yə-lā′shən) *n.* **1.** Inoculation with a vaccine made from substances taken from the recipient's own body. **2.** A secondary infection caused by a disease already present in the body.
**au·to·in·tox·i·ca·tion** (ô′tō-ĭn-tŏk′sĭ-kā′shən) *n.* Self-poisoning resulting from endogenous microorganisms, metabolic wastes, or other toxins in the body.
**au·to·load·ing** (ô′tō-lō′dĭng) *adj.* Semiautomatic.
**au·tol·y·sate** (ô-tŏl′ĭ-sāt′, -zāt′) *n. Biochem.* An end product of autolysis.
**au·tol·y·sin** (ô-tŏl′ĭ-sĭn, ô′tə-lī′sĭn) *n.* A substance that results in autolysis.
**au·tol·y·sis** (ô-tŏl′ĭ-sĭs) *n.* Destruction of tissues or cells of an organism by autogenous substances, as enzymes. **—au′to·lyt′ic** (ô′tə-lĭt′ĭk) *adj.*
**au·to·mak·er** (ô′tō-mā′kər) *n.* An automobile manufacturer.
**Au·to·mat** (ô′tə-măt′). A trademark for a restaurant in which the customers obtain food esp. from coin-operated compartments.
**au·tom·a·ta** (ô-tŏm′ə-tə) *n. var. pl. of* AUTOMATON.
**au·to·mate** (ô′tə-māt′) *v.* **-mat·ed, -mat·ing, -mates.** [Back-formation from AUTOMATIC.] —*vt.* **1.** To convert to automation. **2.** To control or operate by automation. —*vi.* To convert to or make use of automation.
**automated teller** *n.* An unattended data system and related equipment activated by a bank customer to obtain banking services.
**au·to·mat·ic** (ô′tə-măt′ĭk) *adj.* [Gk. *automatos,* self-acting.] **1. a.** Acting or operating in a manner essentially independent of external influence or control. **b.** Self-regulating. **2. a.** Without volition or conscious control. **b.** Acting or performing in a mechanical or impersonal fashion. **3.** Capable of firing continuously until ammunition is exhausted. —*n.* **1.** An automatic firearm, esp. an automatic pistol. **2.** An automatic machine or device. **3.** *Football.* An audible. **—au′to·mat′i·cal·ly** *adv.* **—au′to·ma·tic′i·ty** (-mə-tĭs′ĭ-tē) *n.*
**automatic pilot** *n.* A device, as on an aircraft, that automatically maintains a preset course.
**au·to·ma·tion** (ô′tə-mā′shən) *n.* **1.** Automatic operation or control of a process or system or of equipment. **2.** The techniques and equipment used to bring about automatic operation or control. **3.** The condition of being automatically controlled or operated. **—au′to·ma′tive** *adj.*
**au·tom·a·tism** (ô-tŏm′ə-tĭz′əm) *n.* **1. a.** The quality or state of being automatic. **b.** Automatic mechanical action. **2.** *Philos.* The theory that the body is a machine whose functions are accompanied but not controlled by consciousness. **3.** *Physiol.* **a.** The automatic operation of cells and organs, as the beating of the heart. **b.** Performance of an act without conscious control, as in the operation of the reflexes. **4.** The suspension of consciousness in order to express subconscious ideas and feelings. **—au·tom′a·tist** *n.*
**au·tom·a·ti·za·tion** (ô-tŏm′ə-tĭ-zā′shən) *n.* Automation.
**au·tom·a·tize** (ô-tŏm′ə-tīz′) *vt.* **-tized, -tiz·ing, -tiz·es. 1.** To make automatic. **2.** To automate.
**au·tom·a·ton** (ô-tŏm′ə-tən, -tŏn′) *n., pl.* **-tons** or **-ta** (-tə) [Lat., self-operating machine < Gk. *automatos,* self-acting.] **1.** A robot. **2.** One that acts in a mechanical or impersonal way. **—au·tom′a·tous** *adj.*
**au·to·mo·bile** (ô′tə-mō-bēl′, -mō′bēl′) *n.* [Fr. : Gk. *autos,* self + *mobile,* mobile < OFr. —see MOBILE.] A self-propelled passenger vehicle used for land transport, usu. with four wheels and an internal-combustion engine. —*adj.* Automotive. **—au′to·mo·bil′ist** *n.*
**au·to·mo·tive** (ô′tə-mō′tĭv) *adj.* **1.** Containing its own means of propulsion. **2.** Of or relating to self-propelled vehicles.
**au·to·net·ics** (ô′tō-nĕt′ĭks) *n.* [AUTO- + (CYBER)NETICS.] *(sing. in number).* The study of automatic guidance and control systems.
**au·to·nom·ic** (ô′tə-nŏm′ĭk) *adj.* **1.** Autonomous : independent. **2.** *Physiol.* Of or relating to the autonomic nervous system. **3.** Resulting from internal causes : SPONTANEOUS. **—au′to·nom′i·cal·ly** *adv.*
**autonomic nervous system** *n.* The division of the vertebrate nervous system that regulates involuntary action, as of the intestines, heart, and glands, and comprises the sympathetic nervous system and the parasympathetic nervous system.
**au·ton·o·mous** (ô-tŏn′ə-məs) *adj.* [Gk. *autonomos,* self-ruling : *autos,* self + *nomos,* law.] **1. a.** Independent. **b.** Self-contained. **2. a.** Independent of the laws of another state or government : SELF-GOVERNING. **b.** Of or relating to an autonomy. **3.** Autonomic. **—au·ton′o·mous·ly** *adv.*
**au·ton·o·my** (ô-tŏn′ə-mē) *n., pl.* **-mies.** [Gk. *autonomia* < *autonomos,* autonomous.] **1.** The quality or condition of being self-governing. **2.** The right of self-government : INDEPENDENCE. **3.** A self-governing state, community, or group. **—au·ton′o·mist** *n.*
**au·to·phyte** (ô′tə-fīt′) *n.* An autotrophic plant. **—au′to·phyt′ic** (-fĭt′ĭk) *adj.*
**au·to·pil·er** (ô′tō-pī′lər) *n.* [AUTO- + (COM)PILER.] *Computer Sci.* A specific automatic compiler.
**au·to·pi·lot** (ô′tō-pī′lət) *n.* An automatic pilot.
**au·to·plas·ty** (ô′tō-plăs′tē) *n.* Surgical replacement or repair with tissue taken from the patient's own body. **—au′to·plas′tic** *adj.* **—au′to·plas′ti·cal·ly** *adv.*
**au·top·sy** (ô′tŏp′sē, ô′təp-) *n., pl.* **-sies.** [Gk. *autopsia,* a seeing for oneself : *auto-,* self + *opsis,* sight.] **1.** The examination of a corpse to determine the cause of death. **2.** A critical examination or assessment. **—au·top′sic, au·top′si·cal** *adj.* **—au·top′sist** *n.*
**au·to·route** (ô′tō-rōōt′) *n.* [Fr. : *auto,* automobile + *route,* road.] An expressway in France and French-speaking countries.
**au·to·some** (ô′tə-sōm′) *n.* A chromosome that is not a sex chromosome. **—au′to·so′mal** (-sō′məl) *adj.*
**au·to·stra·da** (ou′tō-strä′də, ô′tō-) *n.* [Ital. : *auto,* automobile + *strada,* street < LLat. *strata,* paved road.] An expressway in Italy.
**au·to·sug·ges·tion** (ô′tō-səg-jĕs′chən) *n. Psychol.* The process by which a person promotes self-acceptance of an opinion, belief, or plan of action : SELF-HYPNOSIS. **—au′to·sug·gest′i·bil′i·ty** *n.* **—au′to·sug·gest′i·ble** *adj.* **—au′to·sug·ges′tive** *adj.*
**au·tot·o·mize** (ô-tŏt′ə-mīz′) *vt. & vi.* **-mized, -miz·ing, -miz·es.** To cause the autotomy of or to undergo autotomy.
**au·tot·o·my** (ô-tŏt′ə-mē) *n. Zool.* The spontaneous separation of a body part, as the tail of certain lizards, for self-protection. **—au′to·tom′ic** (ô′tə-tŏm′ĭk) *adj.*
**au·to·tox·e·mi·a** *also* **au·to·tox·ae·mi·a** (ô′tō-tŏk-sē′mē-ə) *n. Pathol.* Autointoxication.
**au·to·tox·in** (ô′tō-tŏk′sĭn) *n.* A poison that acts on the organism in which it is generated. **—au′to·tox′ic** *adj.*
**au·to·trans·form·er** (ô′tō-trăns-fôr′mər) *n.* An electrical transformer in which the primary and secondary coils have some or all windings in common.
**au·to·troph** (ô′tə-trŏf′) *n.* [Back-formation < AUTOTROPHIC.] An autotrophic organism, as a green plant.
**au·to·troph·ic** (ô′tə-trŏf′ĭk, -trō′fĭk) *adj.* Designating or typical of plants or plantlike organisms capable of manufacturing their own food by synthesis of inorganic materials, as in photosynthesis. **—au′·to·troph′i·cal·ly** *adv.* **—au·tot′ro·phy** (ô-tŏt′rə-fē) *n.*
**au·to·work·er** (ô′tō-wûr′kər) *n.* One who works in the automobile industry.
**au·tumn** (ô′təm) *n.* [ME *autumpne* < OFr. < Lat. *autumnus.*] **1.** The season of the year between summer and winter. **2.** A period of maturity verging on decline. **—au·tum′nal** (-tŭm′nəl) *adj.* **—au·tum′nal·ly** *adv.*
**autumn crocus** *n.* A plant, indigenous to Europe and northern Africa, *Colchicum autumnale* with pink or purplish flowers that bloom in the autumn.
**au·tun·ite** (ō-tŭn′īt′, ô′tə-nīt′) *n.* [After *Autun,* France.] A yellowish fluorescent minor ore of uranium, $Ca(UO_2)_2(PO_4)_2 \cdot 10H_2O$.
**aux·e·sis** (ôg-zē′sĭs, ôk-sē′-) *n.* [Gk. *auxēsis,* growth < *auxanein,* to grow.] An increase in the size of a cell without cell division. **—aux·et′ic** (-zĕt′ĭk) *adj.* **—aux·et′i·cal·ly** *adv.*
**aux·il·ia·ry** (ôg-zĭl′yə-rē, -zĭl′ə-rē) *adj.* [Lat. *auxiliarius* < *auxilium,* help.] **1.** Giving or capable of giving assistance or support. **2.** Acting as a subsidiary : SUPPLEMENTARY. **3.** Held in or used as a reserve <*auxiliary* militia> **4.** *Naut.* Equipped with a motor as well as sails.

—*n., pl.* **-ries. 1.** An individual or group that assists or functions in an auxiliary capacity. **2.** A member of a foreign body of troops serving a country in war. **3.** An auxiliary verb. **4.** *Naut.* A sailing vessel equipped with a motor. **5.** A vessel, as a tug or supply ship, designed for and used in other than combat services.

**auxiliary verb** *n.* A verb, as *have, can,* or *will,* that accompanies particular forms of another verb of a clause to form a phrasal unit expressing person, number, tense, mood, voice, or aspect.

**aux·in** (ôk'sĭn) *n.* [< Gk. *auxein,* to grow.] Any of several plant hormones or similar substances produced synthetically that affect growth by causing the development of larger, elongated cells. **—aux·in'ic** *adj.* **—aux·in'i·cal·ly** *adv.*

**Av** (ôv, äb) *also* **Ab** (äb, äv, ôv) *n.* [Heb. *ābh* < Akkadian *abu.*] The 11th month of the Hebrew year. —See table at CALENDAR.

**a·vail** (ə-vāl') *v.* **a·vailed, a·vail·ing, a·vails.** [ME *availen* : *a-* (intensive) + OFr. *valoir, vail-,* to be worth < Lat. *valēre.*] —*vt.* To be of use or advantage to : HELP <Can anything *avail* us now?> —*vi.* To be of use, value, or advantage : SERVE. —*n.* Use, benefit, or advantage <struggled to no *avail*> **—avail (oneself) of.** To make use of <Please *avail* yourself *of* our services.> **—a·vail'ing·ly** *adv.*

**a·vail·a·ble** (ə-vā'lə-bəl) *adj.* **1.** Accessible for use : at hand. **2.** Having the qualities and the willingness to take on a responsibility <a list of *available* baby-sitters> **3.** *Archaic.* **a.** Capable of bringing about a desired end. **b.** Beneficial. **—a·vail'a·bil'i·ty, a·vail'a·ble·ness** *n.* **—a·vail'a·bly** *adv.*

**av·a·lanche** (ăv'ə-lănch') *n.* [Fr. < dial. Fr. *avalantse.*] **1.** A fall or slide of a large mass of material, as snow, rock, or earth, down a mountainside. **2.** Something resembling an avalanche <an *avalanche* of protests> —*v.* **-lanched, -lanch·ing, -lanch·es.** —*vi.* To fall, as an avalanche. —*vt.* To overwhelm.

**avalanche lily** *n.* [So called because it grows near the snow line and blooms when the snow begins to melt.] A plant, *Erythronium montanum* of western North America, with white flowers.

**Av·a·lon** (ăv'ə-lŏn') *n.* A legendary island paradise to which King Arthur went at his death.

**a·vant-garde** (ä'vänt-gärd') *n.* [Fr., vanguard.] A group active in the invention and application of new techniques in a given field, esp. in the arts. **—a'vant-garde'** *adj.*

**av·a·rice** (ăv'ə-rĭs) *n.* [ME < OFr. < Lat. *avaritia* < *avarus,* greedy < *avēre,* to desire.] Excessive desire for wealth : CUPIDITY.

**av·a·ri·cious** (ăv'ə-rĭsh'əs) *adj.* Excessively greedy, esp. for wealth. **—av'a·ri'cious·ly** *adv.* **—av'a·ri'cious·ness** *n.*

**a·vast** (ə-văst') *interj. Naut.* —Used as a command to stop or desist.

**av·a·tar** (ăv'ə-tär') *n.* [Skt. *avatāraḥ* : *ava,* down + *tarati,* he crosses.] **1. a.** An incarnation or embodiment, as of a quality or concept. **b.** A varying manifestation or aspect of a particular entity. **2.** The incarnation of esp. a Hindu deity in human form.

**a·vaunt** (ə-vônt', ə-vänt') *interj.* [ME < OFr. *avant.* —see VANGUARD.] *Archaic.* —Used as a command to be gone.

**a·ve** (ä'vā) *n.* [Lat., hail!] **1.** An expression of greeting or farewell. **2. Ave.** The Ave Maria.

**A·ve Ma·ri·a** (ä'vā mə-rē'ə) *n.* [ME < Med. Lat., hail, Mary.] The Hail Mary.

**a·venge** (ə-vĕnj') *vt.* **a·venged, a·veng·ing, a·veng·es.** [ME *avengen* < OFr. *avengier* : *a-,* to (< Lat. *ad-*) + *vengier,* to vindicate < Lat. *vindicare.*] **1.** To exact revenge or satisfaction for. **2.** To take vengeance on behalf of. **—a·veng'er** *n.* **—a·veng'ing·ly** *adv.*

☆ **syns:** AVENGE, REDRESS, REPAY, REQUITE, VINDICATE *v. core meaning* : to exact revenge for <*avenged* their child's murder>

**av·ens** (ăv'ənz) *n., pl.* **avens.** [ME *avence* < OFr.] **1.** A plant of the genus *Geum,* with irregularly shaped leaves, white, yellow, or reddish flowers, and plumed seed clusters. **2.** A plant of the genus *Dryas* of mountainous and arctic regions, related to the avens.

**a·ven·tu·rine** (ə-vĕn'chə-rēn', -rĭn) *also* **a·ven·tu·rin** (-rĭn) *n.* [Fr. < *aventure,* accident (so called because of its accidental discovery). —see ADVENTURE.] **1.** An opaque or semitranslucent brown glass flecked with small metallic particles, often of copper or chromic oxide. **2.** A variety of quartz or feldspar flecked with particles of mica, hematite, or other materials. **—a·ven'tu·rine'** *adj.*

**av·e·nue** (ăv'ə-nōō', -nyōō') *n.* [Fr. < OFr., p.part. of *avenir,* to approach < Lat. *advenire,* to come to. —see ADVENT.] **1.** A wide street or thoroughfare. **2. a.** A broad roadway lined with trees. **b.** *Chiefly Brit.* The drive leading from the main road up to a country house. **3.** A means of approach or access <new *avenues* of negotiation>

**a·ver** (ə-vûr') *vt.* **a·verred, a·ver·ring, a·vers.** [ME *averren* < OFr. *averer* < VLat. **adverare* : Lat. *ad-,* to + Lat. *verus,* true.] **1.** To declare positively : AFFIRM. **2.** *Law.* **a.** To assert formally as a fact. **b.** To prove or justify. **—a·ver'ment** *n.* **—a·ver'ra·ble** *adj.*

**av·er·age** (ăv'ər-ĭj, ăv'rĭj) *n.* [Obs. *averie,* shipping charges < OFr. *avarie,* damage to shipping < OItal. *avaria* < Ar. *'awārīyah,* damaged goods < *'awar,* blemish.] **1.** *Math.* **a.** A number that typifies a set of numbers of which it is a function. **b.** The arithmetic mean. **2.** A typical or usual level, degree, or kind. **3.** *Law.* **a.** The incurrence of and loss due to damage at sea to a ship or cargo. **b.** The equitable distribution of such a loss among concerned parties. **c.** Charges incurred through such a loss. **4.** Small expenses or charges that are usu. paid by the master of a ship. —*adj.* **1.** Of, relating to, or being a mathematical average. **2.** Typical : usual <of *average* height> **3.** Assessed in compliance with the laws of average. —*v.* **-aged, -ag·ing, -ag·es.** —*vt.* **1.** To calculate the average of. **2.** To accomplish or obtain an average of <*average* $20 a night in tips> **3.** To distribute proportionately. —*vi.* **1.** To be or amount to an average. **2.** To buy or sell more goods or shares to obtain more than an average price.

▲ word history: *Average* appears in English around 1500 as a maritime term referring in general to any expense, as a tax or loss from damage, over and above the cost of shipping freight. Such expenses were usually distributed proportionally among the interested parties in the venture. It is from the notion of the distribution of a sum to a number of persons that the idea of a mathematical average—the arithmetic mean—developed, and from this sense of a "mean" or "medium" figure the meanings "typical" and "usual" are derived.

**a·verse** (ə-vûrs') *adj.* [Lat. *aversus,* backward, p.part. of *avertere,* to avert.] **1.** Having a feeling of great distaste or aversion <was *averse* to being in crowds> **2.** *Bot.* Turned away from the central stem or axis. **—a·verse'ly** *adv.* **—a·verse'ness** *n.*

**aversion therapy** *n.* A therapy designed to modify antisocial habits or harmful addictions by creating a strong association with a disagreeable stimulus.

**a·ver·sive** (ə-vûr'sĭv, -zĭv) *adj.* Causing avoidance of an unpleasant or punishing stimulus, as in techniques of behavior modification. **—a·ver'sive·ly** *adv.*

**a·vert** (ə-vûrt') *vt.* **a·vert·ed, a·vert·ing, a·verts.** [ME *averten* < OFr. *avertir* < Lat. *avertere* : *ab-,* away from + *vertere,* to turn.] **1.** To turn away <*avert* one's face> **2.** To ward off or prevent <*avert* catastrophe> **—a·vert'i·ble, a·vert'a·ble** *adj.*

**A·ves·ta** (ə-vĕs'tə) *n.* [Pers. *apastāk,* text.] The sacred writings of the ancient Persians.

**A·ves·tan** (ə-vĕs'tən) *n.* The eastern dialect of Old Iranian and the language of the Avesta. —*adj.* Of or relating to the Avesta or Avestan.

**a·vi·an** (ā'vē-ən) *adj.* [< Lat. *avis,* bird.] Of, relating to, or typical of birds.

**a·vi·ar·y** (ā'vē-ĕr'ē) *n., pl.* **-ies.** [Lat. *aviarium* < *avis,* bird.] A large enclosure or cage for confining birds. **—a'vi·a·rist** (-ə-rĭst, -ĕr'ĭst) *n.*

**a·vi·a·tion** (ā'vē-ā'shən, ăv'ē-) *n.* [Fr. < Lat. *avis,* bird.] **1.** The operation of aircraft. **2.** The production of aircraft. **3.** Military aircraft.

**aviation medicine** *n.* The branch of medicine including aeromedicine and space medicine.

**a·vi·a·tor** (ā'vē-ā'tər, ăv'ē-) *n.* [Fr. *aviateur* < *aviation,* aviation.] One who operates an aircraft : PILOT.

**aviator glasses** *pl.n.* Tinted eyeglasses with a lightweight metal frame.

**a·vi·a·trix** (ā'vē-ā'trĭks, ăv'ē-) *n.* A woman who operates an aircraft.

**a·vi·cul·ture** (ā'vĭ-kŭl'chər, ăv'ĭ-) *n.* [Lat. *avis,* bird + -CULTURE.] The raising or keeping of birds. **—a'vi·cul'tur·ist** *n.*

**av·id** (ăv'ĭd) *adj.* [Fr. *avide* < Lat. *avidus* < *avēre,* to desire.] **1.** Ardently eager or greedy <*avid* for excitement> **2.** Marked by great enthusiasm <an *avid* bicyclist> **—av'id·ly** *adv.*

**av·i·din** (ăv'ĭ-dĭn) *n.* [AVID + -IN (from its affinity for biotin).] A protein in egg albumin capable of inactivating biotin and consequently inhibiting the growth of certain bacteria.

**a·vid·i·ty** (ə-vĭd'ĭ-tē) *n.* **1. a.** Eagerness. **b.** Excessive desire : GREED. **2.** *Chem.* **a.** The dissociation-dependent strength of an acid or base. **b.** Degree of affinity.

**a·vi·fau·na** (ā'və-fô'nə, ăv'ə-) *n.* [Lat. *avis,* bird + FAUNA.] All the birds of a specific region or time division. **—a'vi·fau'nal** *adj.*

**av·i·ga·tion** (ăv'ĭ-gā'shən) *n.* [AVI(ATION) + (NAVI)GATION.] Navigation of aircraft. **—av'i·ga'tor** *n.*

**a·vi·on·ics** (ā'vē-ŏn'ĭks, ăv'ē-) *n.* [AVI(ATION) + (ELECTR)ONICS.] *(sing. in number).* The science and technology of electronics applied to aeronautics and astronautics. **—a'vi·on'ic** *adj.*

**a·vir·u·lent** (ā-vîr'yə-lənt, ā-vîr'ə-) *adj.* Not virulent.

**a·vi·ta·min·o·sis** (ā-vī'tə-mĭ-nō'sĭs) *n.* A disease, as scurvy, caused by a deficiency of vitamins. **—a·vi'ta·min·ot'ic** (-nŏt'ĭk) *adj.*

**a·vo** (ä'vōō) *n., pl.* **a·vos.** [Port.] —See table at CURRENCY.

**av·o·ca·do** (ăv'ə-kä'dō, ä'və-) *n., pl.* **-dos.** [Mex. Sp. *aguacate* < Nahuatl *ahuactl.*] **1.** A tropical American tree, *Persea americana,* grown for its edible fruit. **2.** The oval or pear-shaped fruit of the avocado, with leathery green or blackish skin, a large seed, and greenish-yellow pulp.

**av·o·ca·tion** (ăv'ō-kā'shən) *n.* [Lat. *avocatio,* diversion < *avocare,* to call away : *ab-,* away + *vocare,* to call.] **1.** An activity pursued in addition to one's regular work or profession, usu. for enjoyment : HOBBY. **2.** *Archaic.* One's regular work or profession.

**av·o·cet** (ăv'ə-sĕt') *n.* [Fr. *avocette* < Ital. *avocetta.*] A long-legged shore bird of the genus *Recurvirostra,* with a long, slender, upturned beak.

**A·vo·ga·dro number** (ä'və-gä'drō, ăv'ə-) *n.* [After Amadeo Avogadro (1776–1856).] The number of molecules in a mole of a substance, approx. $6.0225 \times 10^{23}$.

**A·vo·ga·dro's law** (ä'və-gä'drōz, ăv'ə-) *n.* The principle that equal volumes of different gases under identical conditions of pressure and temperature have the same number of molecules.

**a·void** (ə-void′) *vt.* **a·void·ed, a·void·ing, a·voids.** [ME *avoiden* < AN *avoider,* to empty out < OFr. *esvuidier* : *es-,* out (< Lat. *ex-*) + *vuidier,* to empty < *voide,* empty. —see VOID.] **1.** To keep away from : SHUN. **2.** To keep from happening. **3.** *Law.* To annul or make void : INVALIDATE. **4.** *Obs.* To void. **—a·void′a·ble** *adj.* **—a·void′a·bly** *adv.* **—a·void′er** *n.*

☆ ***syns:*** AVOID, BY-PASS, DODGE, DUCK, ELUDE, ESCAPE, ESCHEW, EVADE, GET AROUND, SHUN *v. core meaning* : to keep away from <*avoided* their responsibilities><tried to *avoid* a confrontation> *ant:* face

**a·void·ance** (ə-void′ns) *n.* **1.** The act or an instance of avoiding or shunning. **2.** *Law.* An annulment.

**av·oir·du·pois** (ăv′ər-də-poiz′) *n.* [ME *avoir de pois,* commodities sold by weight, alteration of OFr. *aveir de peis,* goods of weight.] **1.** Avoirdupois weight. **2.** *Informal.* Weight or heaviness, esp. personal weight.

**avoirdupois weight** *n.* A system of weights and measures based on a pound containing 16 ounces, 7,000 grains or 453.59 grams.

**a·vouch** (ə-vouch′) *vt.* **a·vouched, a·vouch·ing, a·vouch·es.** [ME *avouchen,* to cite as a warrant < OFr. *avochier* < Lat. *advocare,* to summon : *ad-,* to + *vocare,* to call.] **1.** To take responsibility for : GUARANTEE. **2. a.** To declare positively : AFFIRM. **b.** To prove : establish. **3. a.** To acknowledge one's responsibility for. **b.** To avow.

**a·vow** (ə-vou′) *vt.* **a·vowed, a·vow·ing, a·vows.** [ME *avowen* < OFr. *avouer* < Lat. *advocare,* to call upon. —see AVOUCH.] **1.** To acknowledge openly (e.g., guilt). **2.** To declare oneself to be. **—a·vow′a·ble** *adj.* **—a·vow′a·bly** *adv.* **—a·vow′ed·ly** (-ĭd-lē) *adv.* **—a·vow′er** *n.*

**a·vow·al** (ə-vou′əl) *n.* An acknowledgment or admission.

**a·vulse** (ə-vŭls′) *vt.* **a·vulsed, a·vuls·ing, a·vuls·es.** [Lat. *avellere, avuls-,* to tear off : *ab-,* away + *vellere,* to pull.] To tear off forcibly.

**a·vul·sion** (ə-vŭl′shən) *n.* **1. a.** A forcible separation. **b.** A part removed by avulsion. **2.** *Law.* The removal of a piece of land from one property onto another as a result of a shift in the course of a boundary stream.

**a·vun·cu·lar** (ə-vŭng′kyə-lər) *adj.* [< Lat. *avunculus,* maternal uncle.] **1.** Of or relating to an uncle. **2.** Resembling an uncle, esp. in kindness or benevolence.

**aw** (ô) *interj.* —Used to express sympathy, disgust, or disbelief.

**a·wait** (ə-wāt′) *v.* **a·wait·ed, a·wait·ing, a·waits.** [ME *awaiten* < ONFr. *awaitier* : *a-,* to (< Lat. *ad-*) + *waitier,* to watch, of Germanic orig.] —*vt.* **1.** To wait for. **2.** To be in store for <don't know what *awaits* us> **3.** *Obs.* To lie in ambush for. —*vi.* To wait.

**a·wake** (ə-wāk′) *v.* **a·woke** (ə-wōk′), **a·waked, a·wak·ing, a·wakes.** [ME *awaken* < OE *āwacian.*] —*vt.* **1.** To rouse from sleep : WAKEN. **2.** To stir the interest of : EXCITE. **3.** To arouse (e.g., fears or memories). —*vi.* **1.** To wake up. **2.** To become alert. **3.** To become aware or cognizant <*awoke* to the truth> —*adj.* **1.** Not asleep. **2.** Alert : vigilant.

**a·wak·en** (ə-wā′kən) *vi.* & *vt.* **-ened, -en·ing, -ens.** [ME *awakenen* < OE *āwæcnian* : *ā-,* on + *wæcnian,* to waken.] To wake up or cause to wake up. **—a·wak′en·er** *n.*

**a·ward** (ə-wôrd′) *vt.* **a·ward·ed, a·ward·ing, a·wards.** [ME *awarden* < AN *awarder,* to decide (a legal question), of Germanic orig.] **1.** To grant as due or merited. **2.** To declare as legally due <*awarded* damages to the defendant> —*n.* **1.** A decision, as one made by a judge or arbitrator. **2.** Something awarded or granted, as for merit. **—a·ward′a·ble** *adj.* **—a·ward′er** *n.*

**a·ward·ee** (ə-wôr-dē′) *n.* One that receives an award.

**a·ware** (ə-wâr′) *adj.* [ME < OE *gewær.*] **1.** Having knowledge or cognizance. **2.** *Obs.* Vigilant : watchful. **—a·ware′ness** *n.*

☆ ***syns:*** AWARE, COGNIZANT, CONVERSANT, HIP, KNOWING, MINDFUL, SENSIBLE, SENTIENT, WISE (to) *adj. core meaning* : being conscious and understanding of <*aware* of our own shortcomings><*aware* of the trickery>

**a·wash** (ə-wŏsh′, ə-wôsh′) *adj.* **1.** Washed by waves. **2.** Flooded. **3.** Floating on waves. **—a·wash′** *adv.*

**a·way** (ə-wā′) *adv.* [ME < OE *aweg.*] **1.** From a particular place <walked *away* from me> **2.** At a distance. **3. a.** In a different direction : ASIDE <looked *away*> **b.** In a secure or proper place <put the toys *away*> **4.** Out of existence <The snow melted *away.*> **5.** From one's possession <gave the secret *away*> **6.** Continuously <labored *away* at the task> **7.** Immediately <Fire *away!*> —*adj.* **1.** Absent <*away* for ten days> **2.** At a distance <a month *away*> **3.** Played on an opponent's field or grounds. **4.** *Baseball.* Out <bases loaded, two *away*> **—away with. 1.** Take away. **2.** Go away. —Often used imperatively.

**awe** (ô) *n.* [ME < ON *agi.*] **1. a.** An emotion of mixed reverence, dread, and wonder. **b.** Fearful veneration or respect. **2.** *Archaic.* The power to inspire reverence or fear. **3.** *Obs.* Dread. —*vt.* **awed, aw·ing, awes.** To inspire with awe.

**a·wea·ry** (ə-wîr′ē) *adj.* Weary : tired.

**a·weath·er** (ə-wĕth′ər) *adv. Naut.* To windward.

**a·weigh** (ə-wā′) *adj. Naut.* Hanging just clear of the bottom. —Used of an anchor.

**awe·some** (ô′səm) *adj.* **1.** Inspiring awe. **2.** Displaying or marked by awe. **—awe′some·ly** *adv.* **—awe′some·ness** *n.*

**awe·struck** (ô′strŭk′) *also* **awe·strick·en** (-strĭk′ən) *adj.* Full of awe.

**aw·ful** (ô′fəl) *adj.* [ME *aweful,* awe-inspiring < OE *egefull.*] **1.** Extremely bad or unpleasant : TERRIBLE. **2.** Commanding awe. **3.** Filled with awe. **4.** Immense <an *awful* responsibility> —*adv. Informal.* Very <was *awful* tired> **—aw′ful·ly** *adv.* **—aw′ful·ness** *n.*

**a·while** (ə-hwīl′) *adv.* For a short time.

**awk·ward** (ôk′wərd) *adj.* [ME *awkeward,* in the wrong way < *awke,* wrong < ON *ǒfugr,* backward.] **1.** Marked by a lack of dexterity and grace, esp. in physical movement. **2. a.** Clumsily lacking in the ability to do or perform : UNSKILLFUL. **b.** Clumsily or unskillfully performed. **3.** Difficult to handle or manage <an *awkward* carton to move> **4. a.** Causing embarrassment and distress <an *awkward* comment> **b.** Marked by embarrassment or unease <an *awkward* pause> **5.** Requiring tact and discretion <an *awkward* circumstance> **—awk′ward·ly** *adv.* **—awk′ward·ness** *n.*

☆ ***syns:*** AWKWARD, GAWKY, GRACELESS, INEPT, KLUTZY, LUMBERING, LUMPISH, UNGAINLY, UNGRACEFUL *adj. core meaning* : lacking physical dexterity and grace <an *awkward* dancer>

**awl** (ôl) *n.* [ME *aul* < OE *æl.*] A pointed tool for boring holes, as in leather or wood.

**awl·wort** (ôl′wûrt′, -wôrt′) *n.* [From its awl-shaped leaves.] A small aquatic plant, *Subularia aquatica* of the Northern Hemisphere, bearing a tuft of narrow, pointed leaves and minute white flowers.

**awn** (ôn) *n.* [ME *awne* < ON *ǒgn.*] A slender, bristlelike terminal process, as those found at the tips of the spikelets in many grasses. **—awned** *adj.* **—awn′less** *adj.*

**awn·ing** (ô′nĭng) *n.* [Orig. unknown.] A protective, rooflike covering, as over a window or door.

**a·woke** (ə-wōk′) *v. p.t. of* AWAKE.

**a·wok·en** (ə-wō′kən) *v. Chiefly Brit. p.p. of* AWAKE.

**AWOL** *or* **awol** (ā′wôl′) *adj.* Absent without leave, esp. from military service. —*n.* One that is AWOL.

**a·wry** (ə-rī′) *adv.* **1.** Turned or twisted toward one side : ASKEW. **2.** Away from the correct course : AMISS <plans went *awry*> **—a·wry′** *adj.*

**ax** *or* **axe** (ăks) *n., pl.* **ax·es** (ăk′sĭz) [ME < OE *æxa.*] **1.** A tool with a bladed head mounted on a handle, used to fell or split lumber. **2.** A similar tool or weapon, as a battle-ax. **3.** *Informal.* **a.** A sudden termination of employment <gave me the *ax*> **b.** A sudden or ruthless removal. —*vt.* **axed, ax·ing, ax·es. 1.** To use an ax. **2.** To remove ruthlessly or abruptly. **—have an ax to grind.** To pursue a subjective or selfish aim.

**ax·el** (ăk′səl) *n.* [After *Axel* Paulsen, 19th-cent. Norwegian figure skater.] A jump in figure skating with 1½ turns in the air.

**a·xen·ic** (ā-zĕn′ĭk, ā-zē′nĭk) *adj.* Free from symbionts or parasites : UNCONTAMINATED. **—a·xen′i·cal·ly** *adv.*

**ax·es** *n.* **1.** (ăk′sēz′). *pl. of* AXIS. **2.** (ăk′sĭz). *pl. of* AX.

**ax·i·al** (ăk′sē-əl) *adj.* **1.** Relating to or forming an axis. **2.** Located on, around, or in the direction of an axis. **—ax′i·al′i·ty** (-ăl′ĭ-tē) *n.* **—ax′i·al·ly** *adv.*

**ax·il** (ăk′sĭl) *n.* [Lat. *axilla,* armpit.] The angle between the upper surface of a leafstalk, flower stalk, branch, or similar part, and the stem or axis from which it arises.

**ax·il·la** (ăk-sĭl′ə) *n., pl.* **-il·lae** (-sĭl′ē) [Lat.] The armpit or an analogous part.

**ax·il·lar** (ăk-sĭl′ər, ăk′sə-lər) *adj.* Axillary. —*n.* One of the feathers in the axilla of a bird's wing.

**ax·il·lar·y** (ăk′sə-lĕr′ē) *adj.* **1.** *Anat.* Of, relating to, or near the axilla. **2.** *Bot.* Of, relating to, or situated in an axil. —*n., pl.* **-ies.** An axillar.

**ax·i·ol·o·gy** (ăk′sē-ŏl′ə-jē) *n.* [Gk. *axios,* worth + -LOGY.] *Philos.* The study of the nature of values and value judgments. **—ax′i·o·log′i·cal** (-ə-lŏj′ĭ-kəl) *adj.* **—ax′i·o·log′i·cal·ly** *adv.* **—ax′i·ol′o·gist** (-ŏl′ə-jĭst) *n.*

**ax·i·om** (ăk′sē-əm) *n.* [Lat. *axioma* < Gk. *axiōma* < *axios,* worthy.] **1.** A self-evident or universally recognized truth : MAXIM. **2.** An established rule, principle, or law. **3.** *Math.* & *Logic.* **a.** An undemonstrated proposition concerning an undefined set of elements, properties, functions, and relationships : POSTULATE. **b.** A self-evident or accepted principle.

**ax·i·o·mat·ic** (ăk′sē-ə-măt′ĭk) *also* **ax·i·o·mat·i·cal** (-ĭ-kəl) *adj.* Of, relating to, or like an axiom : SELF-EVIDENT. **—ax′i·o·mat′i·cal·ly** *adv.*

**ax·is** (ăk′sĭs) *n., pl.* **ax·es** (ăk′sēz′) [ME < Lat.] **1.** A straight line about which a body or geometric object rotates or may be thought to rotate. **2.** *Math.* **a.** An unlimited line, half-line, or line segment serving to orient a space or a geometric object, esp. a line about which the object is symmetric. **b.** A reference line from which distances or angles are measured in a coordinate system. **3.** A center line to which parts of a structure or body may be referred. **4.** An imaginary line to which elements of a work of art are referred for measurement or symmetry. **5.** *Anat.* **a.** The second cervical vertebra on which the head pivots. **b.** Any of various central structures or standard abstract

lines used as a positional referent. **6.** *Bot.* The main stem or central part about which organs or plant parts are arranged.
**axis deer** *n.* [Lat. *axis,* a kind of animal.] A deer, *Axis axis* of central Asia, with a white-spotted brown coat.

**axis deer**
*4–7 feet long*

**ax·ite** (ăk'sīt') *n.* [AX(ON) + -ITE.] *Anat.* One of the terminal fibers of an axon.
**ax·le** (ăk'səl) *n.* [ME *axel* < OE *eaxl.*] **1.** A supporting shaft or member on which a wheel or pair of wheels revolves. **2. a.** The spindle of an axletree. **b.** Either end of an axletree.
**ax·le·tree** (ăk'səl-trē') *n.* A crossbar or rod supporting a vehicle, as a cart, and having terminal spindles on which the wheels revolve.
**ax·man** (ăks'mən) *n.* One who wields an ax, esp. a worker who fells trees or chops logs.
**Ax·min·ster** (ăks'mĭn'stər) *n.* [After *Axminster,* England.] A carpet with stiff jute backing and long, soft cut-wool pile.
**ax·o·lotl** (ăk'sə-lŏt'l) *n.* [Nahuatl : *atl,* water + *xolotl,* servant.] Any of several western North American and Mexican salamanders of the genus *Ambystoma,* which, unlike most amphibians, often retain their external gills and become sexually mature without undergoing metamorphosis.
**ax·on** (ăk'sŏn') *also* **ax·one** (-sōn') *n.* [Gk. *axōn,* axis.] The core of a nerve fiber that usu. conducts impulses away from the body of a nerve cell.
**ax·seed** (ăks'sēd') *n.* [From its ax-shaped pods.] The crown vetch.
**ay¹** (ī) *interj. Archaic.* —Used to express surprise or distress.
**ay²** (ī) *n. & adv. var. of* AYE¹.
**ay³** (ā) *adv. var. of* AYE².
**a·yah** (ä'yə, ä'ə, ī'ə) *n.* [Hindi *āyā* < Port. *aia,* nursemaid < Lat. *avia,* grandmother.] A native maid or nursemaid in India.
**a·ya·tol·lah** (ī'ə-tō'lə, -tŏl'ə) *n.* [Pers. : Ar. *ayat,* sign + *allāh,* God.] An Islamic religious leader of the Shiite sect.
**aye¹** *also* **ay** (ī) [Prob. alteration of I.] *—n.* An affirmative vote or voter. *—adv.* Yes.
**aye²** *also* **ay** (ā) *adv.* [ME *ai* < ON *ei.*] Always : ever.
**aye-aye** (ī'ī') *n.* [Fr. < Malagasay *aiay,* prob. imit. of its cry.] A lemur, *Daubentonia madagascariensis* of Madagascar, with large ears, a long, bushy tail, and rodentlike teeth.
**a·yin** (ī'ĭn) *n.* [Heb. *'ayin.*] The 16th letter of the Hebrew alphabet. —See table at ALPHABET.
**Ay·ma·ra** (ī'mä-rä') *n., pl.* **Aymara** *or* **-ras. 1. a.** An Indian people inhabiting Bolivia and Peru. **b.** A member of this people. **2. a.** The language of the Aymara. **b.** A language family consisting of Aymara. **—Ay'ma·ran'** *adj. & n.*
**Ayr·shire** (âr'shîr, -shər) *n.* One of a breed of brown and white dairy cattle orig. bred in Ayr, Scotland.
**az-** *pref. var. of* AZO-.
**a·zal·ea** (ə-zāl'yə) *n.* [NLat. < Gk. *azaleos,* dry (so called because it grows in dry soil).] Any of a group of deciduous or evergreen shrubs of the genus *Rhododendron* of the North Temperate Zone, cultivated for their variously colored flowers.
**a·zan** (ä-zän') *n.* [Ar. *adhān* < *adhina,* to proclaim.] The Moslem summons to prayer, called by the muezzin from a minaret of a mosque five times a day.
**A·za·zel** (ə-zā'zəl, ăz'ə-zĕl') *n.* [Heb. *'azāzēl.*] In ancient Hebrew tradition the rebel leader of the angels who seduced mankind.
**A·zer·bai·ja·ni** (ä'zər-bī-jä'nē, ăz'ər-) *n., pl.* **Azerbaijani** *or* **-nis. 1.** A native or resident of Azerbaijan. **2.** The Turkic language of Azerbaijan.
**A·zil·ian** (ə-zĭl'yən) *adj.* [After le Mas d'Azil, a village in France.] Of or describing a western European culture following the Magdalenian era and preceding the Neolithic.
**az·i·muth** (ăz'ə-məth) *n.* [ME *azimut* < OFr. < Ar. *as-sumūt* : *as,* the + *sumūt,* pl. of *samt,* compass bearing.] **1.** The horizontal angular distance from a fixed reference direction to a position, object, or object referent, as to a great circle intersecting a celestial body, usu. measured clockwise in degrees along the horizon from a point due south. **2.** The lateral deviation of a projectile or bomb. **—az'i·muth'·al** (-mŭth'əl) *adj.* **—az'i·muth'al·ly** *adv.*
**azimuthal equidistant projection** *n.* A map projection of the earth designed so that a straight line from a given point on the map to any other point gives the shortest distance between the two points.
**az·ine** (ăz'ēn', ā'zēn') *n.* A six-membered heterocyclic compound, as pyridine, containing one or more atoms of nitrogen.
**azine dye** *n.* Any of various dyes obtained from phenazine.
**az·o** (ăz'ō, ā'zō) *adj.* [< AZO-.] Having a nitrogen group.
**azo-** *or* **az-** *pref.* [< Fr. *azote,* nitrogen : Gk. *a-,* not + Gk. *zoē,* life.] Having a nitrogen group, esp. one attached at both ends in a covalent bond to other groups <azole>
**azo dye** *n.* Any of various red, brown, or yellow acidic or basic dyes derived from nitrobenzene in an alkaline solution.
**a·zo·ic** (ā-zō'ĭk) *adj.* Of or relating to geologic periods that precede the appearance of life.
**az·ole** (ăz'ōl', ā'zōl') *n.* An organic compound having a five-membered heterocyclic ring with two double bonds.
**a·zon·ic** (ā-zŏn'ĭk, ā-zō'nĭk) *adj.* Not confined to a particular zone or region : not local.
**az·o·te·mi·a** (ăz'ə-tē'mē-ə, ā'zə-) *n.* [Fr. *azote,* nitrogen + -EMIA.] Uremia. **—az'o·te'mic** (-mĭk) *adj.*
**az·oth** (ăz'ŏth', -ôth') *n.* [Ar. *az-zā'ūq.*] Mercury regarded in alchemy as the primary source of all metals.
**a·zo·to·bac·ter** (ā-zō'tə-băk'tər) *n.* [Fr. *azote,* nitrogen + BACTER(IA).] Any of various nitrogen-fixing bacteria of the family Azotobacteraceae.
**az·o·tu·ri·a** (ăz'ə-tŏŏr'ē-ə, -tyŏŏr'-) *n.* [Fr. *azote,* nitrogen + -URIA.] Increase of nitrogenous substances in the urine.
**Az·ra·el** (ăz'rā-ĕl') *n.* [Ar. *Azrā'īl* < Heb. *'Ăzar'ēl,* God has helped.] The angel who separates the soul from the body at death in Moslem and Jewish legend.
**Az·tec** (ăz'tĕk') *n.* [Sp. *Azteca* < Nahuatl *Aztecatl.*] **1.** A member of an Indian people of Central Mexico noted for their advanced civilization before Cortés invaded Mexico in 1519. **2.** Nahuatl. *—adj. also* **Az·tec·an** (-tĕk'ən). Of the Aztecs or their language, culture, or empire.
**az·ure** (ăzh'ər) *n.* [ME < OFr. *azur* < Med. Lat. *azura* < Ar. *al-lāzaward* < Pers. *lājwārd,* lapis lazuli.] **1.** A light purplish blue. **2.** An azure pigment. **3.** The blue sky.
**az·ur·ite** (ăzh'ə-rīt') *n.* An azure-blue vitreous mineral of basic copper carbonate, $2CuCO_3 \cdot Cu(OH)_2$, used as a copper ore and as a gemstone.
**a·zy·gous** (ā-zī'gəs) *adj.* Occurring singly : not paired.

# Bb

**b** *or* **B** (bē) *n., pl.* **b's** *or* **B's. 1.** The second letter of the English alphabet. **2.** A speech sound represented by the letter *b.* **3.** The second in a series. **4.** *Mus.* **a.** The seventh tone in the scale of C major or the second tone in the relative minor scale. **b.** The key or a scale in which B is the tonic. **c.** A written or printed note representing this tone. **d.** A string, key, or pipe tuned to the pitch of this tone. **5.** The second highest grade in quality or rank. **6.** A human blood type of the ABO group.
**B** *symbol for* BORON.
**Ba** *symbol for* BARIUM.
**baa** (bă, bä) *vi.* **baaed, baa·ing, baas.** [Imit.] To make a bleating sound, as that of a sheep. **—baa** *n.*
**Ba·al** (bā'əl) *n., pl.* **-als** *or* **-al·im** (-ə-lĭm) [Heb. *bá'al,* lord.] **1.** Any of various local fertility and nature gods of the ancient Semitic peoples regarded by the Hebrews as false gods. **2.** *often* **baal.** A false god.
**ba·ba** (bä'bə) *n.* [Fr. < Pol.] A leavened rum cake.

**ba·bas·su** (bä'bə-sōō') *n.* [Port. *babaçú.*] A Brazilian palm tree, *Orbignya martiana* or *O. speciosa,* having hard nuts from which an oil similar to coconut oil is produced.

**Bab·bitt¹** (băb'ĭt) *n.* [After George F. *Babbitt,* the main character in the novel *Babbitt* by Sinclair Lewis.] A member of the American middle class whose unthinking attachment to its business and social ideals is such as to make him a model of narrow-mindedness and self-satisfaction. **—Bab'bitt·ry** *n.*

**Bab·bitt²** (băb'ĭt). A trademark for a soft, silvery antifriction alloy composed of tin with small amounts of copper and antimony.

**bab·ble** (băb'əl) *v.* **-bled, -bling, -bles.** [ME *babelen.*] *—vi.* **1.** To utter meaningless, confused words or sounds. **2.** To talk foolishly : CHATTER. **3.** To make a continuous low, murmuring sound <a *babbling* brook> *—vt.* **1.** To utter rapidly and incoherently. **2.** To blurt out impulsively <*babbled* their excuses> *—n.* **1.** Inarticulate or meaningless talk or sounds. **2.** Foolish talk. **3.** A continuous low, murmuring sound. **—bab'bler** *n.*

☆ ***syns:*** BABBLE, BLATHER, BLATHERSKITE, DOUBLE TALK, GABBLE, GIBBERISH, JABBER, JABBERWOCKY, NONSENSE, PRATE, PRATTLE, TWADDLE *n. core meaning* : unintelligible or foolish talk <a speech that was merely *babble*>

**babe** (bāb) *n.* [ME.] **1.** An infant. **2.** An innocent or naive person. **3.** *Slang.* **a.** A young woman. **b.** A person.

**Ba·bel** (bā'bəl, băb'əl) *n.* [Heb. *Bābhēl* < Akkadian *Bāb-ilu,* gate of God.] **1.** The site of a tower reaching to heaven whose construction was interrupted by the confusion of tongues, according to the Old Testament. **2. babel. a.** A confusion of voices or sounds. **b.** Noise and confusion.

**ba·be·sia** (bə-bē'zhə) *n.* [NLat. *Babesia,* genus name, after Victor *Babes* (1854–1926).] Any of the family Babesiidae of parasitic sporozoans that affect the blood of vertebrates, as dogs and sheep.

**bab·e·si·a·sis** (băb'ĭ-zī'ə-sĭs) *also* **ba·be·si·o·sis** (bə-bē'ē-ō'sĭs) *n.* A disease or infection caused by babesia.

**ba·bies'-breath** (bā'bēz-brĕth') *n. var. of* BABY'S-BREATH

**bab·i·ru·sa** *also* **bab·i·rus·sa** *or* **bab·i·rous·sa** (băb'ə-rōō'sə, bä'bə-) *n.* [Malay *bābīrūsa* : *bābī,* hog + *rūsa,* deer.] A wild pig, *Babyrousa babyrussa* of the East Indies, the male of which has long, upward-curving tusks.

**babirusa**
*4–5 feet long*

**bab·ka** (bäb'kə) *n.* [Pol.] A coffee cake flavored with orange rind, rum, almonds, and raisins.

**ba·boon** (bă-bōōn') *n.* [ME *babewyne* < OFr. *babuin.*] **1.** A chiefly African monkey of the genus *Papio* or *Chaeropithecus* or related genera, with an elongated, doglike muzzle. **2.** *Slang.* A boor. **—ba·boon'ish** *adj.*

**ba·bu** (bä'bōō) *n.* [Hindi *bābū,* father.] **1.** A Hindu form of address equivalent to *Mr.* **2.** A Hindu clerk literate in English.

**ba·bul** (bə-bōōl') *n.* [Pers. *babūl.*] A tree, *Acacia arabica* of northern Africa and India, that is a source of gum arabic, of a hardwood used in carving, and of tannin.

**ba·bush·ka** (bə-bōōsh'kə) *n.* [R., grandmother, dim. of *baba,* old woman.] A woman's head scarf, folded triangularly and worn tied under the chin.

**ba·by** (bā'bē) *n., pl.* **-bies.** [ME *babi.*] **1. a.** A very young child : INFANT. **b.** The youngest member of a family or group. **c.** A very young animal. **2.** A person who behaves in an infantile way. **3.** *Slang.* A young woman or girl. **4.** *Slang.* An object of personal concern or interest. *—vt.* **-bied, -by·ing, -bies.** To treat oversolicitously : CODDLE. **—ba'by·hood'** *n.* **—ba'by·ish** *adj.*

☆ ***syns:*** BABY, CATER (to), CODDLE, COSSET, INDULGE, MOLLYCODDLE, OVERINDULGE, PAMPER, SPOIL *v. core meaning* : to treat with excessive indulgence <*babied* their child>

**baby blue** *n.* A very light to very pale greenish or purplish blue.

**ba·by-blue-eyes** (bā'bē-blōō'īz') *n. (sing. or pl. in number).* A low-growing plant, *Nemophila menziesii* of California, with bell-shaped blue flowers.

**baby bond** *n.* A bond issued in an amount less than $1,000.

**baby carriage** *n.* A small four-wheeled carriage for an infant.

**baby grand** *n.* A small grand piano approx. five feet long.

**Bab·y·lon** (băb'ə-lən, -lŏn') *n.* [After *Babylon,* an ancient city of Babylonia, noted for its luxury.] A place characterized by great luxury and often corruption.

**Bab·y·lo·ni·an** (băb'ə-lō'nē-ən) *adj.* **1.** Of or relating to ancient Babylonia or Babylon or their people, culture, or language. **2.** Characterized by a luxurious, pleasure-seeking, and often immoral way of life. *—n.* **1.** A native or inhabitant of ancient Babylon or Babylonia. **2.** The form of Akkadian used in ancient Babylonia.

**ba·by's-breath** *or* **ba·bies'-breath** (bā'bēz-brĕth') *n.* **1.** A plant of the genus *Gypsophila,* esp. *G. paniculatum,* with many small white flowers in branching clusters. **2.** Any of several plants similar to the baby's breath, having small, sweet-smelling flowers.

**ba·by-sit** (bā'bē-sĭt') *v.* **-sat** (-săt'), **-sit·ting, -sits.** *—vi.* To act as a baby sitter. *—vt.* To take care of.

**ba·by-sit·ter** (bā'bē-sĭt'ər) *n.* One hired to care for one or more children when the parents are away.

**ba·by-tears** (bā'bē-tîrz') *also* **ba·by's-tears** (-bēz-) *n. (sing. or pl. in number).* A creeping plant native to Corsica, *Helxine soleirolii,* bearing many leaves and tiny green flowers.

**bac·ca·lau·re·ate** (băk'ə-lôr'ē-ĭt) *n.* [Med. Lat. *baccalaureatus* < *baccalarius,* bachelor.] **1.** The degree of bachelor, conferred on graduates of most U.S. colleges and universities. **2.** A farewell address or sermon delivered to a graduating class.

**bac·ca·rat** (bä'kə-rä', băk'ə-) *n.* [Fr. *baccara.*] A card game in which the winner holds two or three cards totaling closest to nine.

**bac·cate** (băk'āt') *adj.* [< Lat. *bacca,* berry.] **1.** Bearing berries. **2.** Resembling a berry in form or texture.

**Bac·chae** (băk'ē) *pl.n.* [Lat. < Gk. *bakkhai.*] The priestesses and women followers of Bacchus.

**bac·cha·nal** (băk'ə-năl', -näl', băk'ə-nəl) *n.* [Lat. *bacchanalis,* of Bacchus < *Bacchus,* Bacchus < Gk. *Bakkhos.*] **1.** A participant in the Bacchanalia. **2.** *often* **bacchanals.** BACCHANALIA 1. **3.** A drunken or riotous celebration. **4.** A reveler. **—bac'cha·nal'** *adj.*

**bac·cha·na·lia** (băk'ə-nāl'yə, -nā'lē-ə) *n., pl.* **bacchanalia.** [Lat. < *bacchanalis,* of Bacchus < *Bacchus,* Bacchus < Gk. *Bakkhos.*] **1. Bacchanalia.** The ancient Roman festival held in honor of Bacchus. **2.** A drunken or riotous festivity : ORGY. **—bac'cha·na'lian** (-nāl'yən, -nā'lē-ən) *adj.* & *n.*

**bac·chant** (bə-kănt', -känt', băk'ənt) *n., pl.* **bac·chants** *or* **bac·chan·tes** (bə-kăn'tēz, -kän'-, -kănts', -känts') [Lat. *bacchans, bacchant-,* pr.part. of *bacchari,* to celebrate the festival of Bacchus < *Bacchus,* Bacchus < Gk. *Bakkhos.*] **1.** A priest or votary of Bacchus. **2.** A boisterous, drunken carouser. **—bac·chan·tic** (-kăn'tĭk) *adj.*

**bac·chan·te** (bə-kăn'tē, -kăn'-, -kănt', -känt') *n.* [Fr. < Lat. *bacchans,* bacchant.] A priestess or woman votary of Bacchus.

**bac·chan·tes** (bə-kăn'tēz, -kän'-, kănts', -känts') *n. var. pl. of* BACCHANT.

**Bac·chic** (băk'ĭk) *adj.* **1.** Of or relating to Bacchus. **2. bacchic.** Drunken and carousing.

**Bac·chus** (băk'əs) *n.* [Lat. < Gk. *Bakkhos.*] *Gk. Myth.* The god of grape-growing, wine, and revelry.

**bac·cif·er·ous** (băk-sĭf'ər-əs) *adj.* [Lat. *baccifer* : *bacca,* berry + *ferre,* to carry.] Bearing berries.

**bac·ci·form** (băk'sə-fôrm') *adj.* [Lat. *bacca,* berry + -FORM.] Shaped like a berry.

**bach** *also* **batch** (băch) *vi.* **bached, bach·ing, bach·es** *also* **batched, batch·ing, batch·es.** [Short for BACHELOR.] *Slang.* To live alone as a bachelor. **—bach** *n.*

**bach·e·lor** (băch'ə-lər, băch'lər) *n.* [ME *bacheler* < OFr., squire < Med. Lat. *baccalarius.*] **1.** An unmarried man. **2.** A young feudal knight in the service of another knight. **3. a.** A college or university degree signifying completion of the undergraduate curriculum and graduation. **b.** One holding such a degree. **4.** A young male fur seal that is kept from the breeding territory by older males. **—bach'e·lor·dom, bach'e·lor·hood', bach'e·lor·ship'** *n.*

**bach·e·lor's-but·ton** (băch'ə-lərz-bŭt'n, băch'lərz-) *n.* **1.** The cornflower. **2.** DAISY 2. **3.** A plant similar to the bachelor's-button, having buttonlike flowers or flower heads.

**bac·il·lar·y** (băs'ə-lĕr'ē, bə-sĭl'ə-rē) *also* **ba·cil·lar** (bə-sĭl'ər, băs'ə-lər) *adj.* [< BACILLUS.] **1.** Rod-shaped. **2.** Of, relating to, or caused by bacilli.

**ba·cil·li** (bə-sĭl'ī') *n. pl. of* BACILLUS.

**ba·cil·lo·pho·bi·a** (bə-sĭl'ə-fō'bē-ə) *n.* [BACILL(US) + -PHOBIA.] A psychologically aberrant fear of bacilli.

**bac·il·lu·ri·a** (băs'ə-lōōr'ē-ə) *n.* [BACILL(US) + -URIA.] Presence of bacilli in the urine.

**ba·cil·lus** (bə-sĭl'əs) *n., pl.* **-cil·li** (-sĭl'ī') [NLat. *Bacillus,* genus name < LLat., little rod, dim. of *baculum,* rod.] **1.** Any of various rod-shaped, aerobic bacteria of the genus *Bacillus,* often occurring in chainlike formations. **2.** Any bacterium, esp. a rod-shaped one.

**bac·i·tra·cin** (băs'ĭ-trā'sĭn) *n.* [BACI(LLUS) + Margaret *Trac(y),* an American child in whose blood it was first isolated + -IN.] An antibiotic obtained from the bacterium *Bacillus subtilis* and used externally as a salve.

**back¹** (băk) *n.* [ME *bak* < OE *bæc.*] **1. a.** The region of the vertebrate body located nearest the spine, in humans composed of the rear area from the neck to the pelvis. **b.** The analogous dorsal region in other animals. **2.** The backbone or spine. **3.** The area or part farthest from the front. **4.** The part opposite the front. ***usage:*** The expression *back of* is an informal variant of *in back of* and is best avoided in formal contexts. **5.** The reverse side, as of a coin. **6.** A part that supports or strengthens from the rear. **7. a.** The part of a book where the pages are stitched or glued together into the binding. **b.** A book binding. **8. a.** A player, as in football, who takes a position behind the front line. **b.** The position of back, as in football. *—v.* **backed, back·ing, backs.** *—vt.* **1.** To cause to move backward or in a reverse direction. **2.** To provide or strengthen with a back or backing. **3.** To furnish support, help, or encouragement for <contributors who *backed* the community theater> **4.** To adduce evidence in support of : SUBSTANTIATE. **5.** To bet on. **6.** To form the back or background of. **7.** To endorse by signing on the back of. *—vi.* **1.** To move backward. **2.** To shift to a counterclockwise direction. — Used of the wind. **—back down.** To withdraw from a position, opinion, or commitment. **—back off. 1.** To retreat. **2.** To draw away. **—back out. 1.** To withdraw from (a project or plan) before its completion. **2.** To fail to keep a commitment or promise. **—back up.** To accumulate in a clogged state <Rush-hour traffic was *backed up* for miles.> *—adj.* **1.** Situated in the rear. **2.** Distant from a center of activity : REMOTE. **3.** Of a past date : not current. **4.** Owing or due from an earlier time <*back* payments> **5.** Being in a backward direction. **6.** Articulated with the tongue positioned toward the rear of the mouth. *—adv.* **1.** BACKWARD 1. **2.** In, to, or toward a former location. **3.** In, to, or toward a former state. **4.** In, to, or toward a past time. **5.** In reserve or concealment. **6.** Under restraint or in check. **7.** In reply or return. **—go back on. 1.** To fail to keep (a promise or commitment). **2.** To betray or desert (a person). **—turn (one's) back on. 1.** To reject, as in anger. **2.** To forsake. **—backed** *adj.*

**back$^{2}$** (băk) *n.* [Du. *bak* < Fr. *bac,* ferry boat < OFr.] A shallow vat or tub used chiefly by brewers.

**back·ache** (băk'āk') *n.* Discomfort localized in the region of the spine or back.

**back·bench·er** (băk'bĕn'chər) *n.* [So-called because such members sit in the back.] *Chiefly Brit.* A junior member of the House of Commons.

**back·bite** (băk'bīt') *v.* **-bit** (-bĭt'), **-bit·ten** (-bĭt'n), **-bit·ing, -bites.** *—vt.* To use spiteful language about (a person). *—vi.* To speak spitefully about a person. **—back'bit'er** *n.*

**back·board** (băk'bôrd', -bōrd') *n.* **1.** A board placed under or behind an object to provide firmness or support. **2.** *Basketball.* The elevated vertical board from which the basket projects.

**back·bone** (băk'bōn') *n.* **1.** The vertebrate spine or spinal column. **2.** Something, as the keel of a ship, resembling a backbone. **3.** A main support or sustaining factor. **4.** Strength of character : DETERMINATION. **—back'boned'** *adj.*

**back·break·ing** (băk'brā'kĭng) *adj.* Demanding great physical exertion : ARDUOUS. **—back'break'er** *n.*

**back burner** *n. Informal.* A position of low priority <Let's put that project on the *back burner.*>

**back·comb** (băk'kōm') *vt.* **-combed, -comb·ing, -combs.** To tease (hair).

**back·coun·try** (băk'kŭn'trē) *n.* A sparsely inhabited rural region.

**back·court** (băk'kôrt', -kōrt') *n.* **1.** The part of a court between the service line and the base line in tennis and other net games. **2.** The part of the playing area farthest from the goal or target wall in some court games, as handball. **3.** *Basketball.* **a.** The half of the court that a team defends. **b.** The part of a team that comprises the two guard positions. **c.** The players in these positions.

**back·cross** (băk'krôs', -krŏs') *vt.* **-crossed, -cross·ing, -cross·es.** To mate (a first-generation hybrid) with a parent or member of the parental stock. **—back'cross'** *n.*

**back·date** (băk'dāt') *vt.* **-dated, -dat·ing, -dates.** To date prior to the true date : ANTEDATE.

**back·door** (băk'dôr', -dōr') *adj.* Surreptitious : clandestine.

**back·drop** (băk'drŏp') *n.* **1.** A painted curtain hung at the back of a stage set. **2.** The setting, as of a historical event.

**back·er** (băk'ər) *n.* A supporter or helper.

**back·field** (băk'fēld') *n. Football.* **1. a.** The players stationed behind the line of scrimmage. **b.** The positions filled by these players. **2.** The area in which the backfield lines up.

**back·fire** (băk'fīr') *n.* **1.** A fire started to extinguish an oncoming fire, as on a prairie, by clearing an area in its path. **2.** An explosion of prematurely ignited fuel or of unburned exhaust gases in an internal-combustion engine. *—vi.* **-fired, -fir·ing, -fires. 1.** To explode in or make the sound of a backfire. **2.** To produce an unexpected and undesired result <a plot that *backfired* on the perpetrators>

**back-for·ma·tion** (băk'fôr-mā'shən) *n.* **1.** A new word created by removing from an existing word what is mistakenly thought to be an affix, as *laze* from *lazy* or *edit* from *editor.* **2.** The process of forming words by back-formation.

**back·gam·mon** (băk'găm'ən) *n.* [BACK + GAMMON$^{1}$.] A board game for two persons, played with pieces whose moves are determined by throws of dice.

**back·ground** (băk'ground') *n.* **1.** The scenery or ground located behind something. **2.** The part of a painting or picture depicting what lies behind the objects in the foreground. **3.** An inconspicuous or unimportant position. **4.** Attendant circumstances and events. **5.** One's total experience, education, and knowledge <an extensive academic *background*> **6.** Subdued music played esp. as an accompaniment to dialogue in a dramatic performance. **7.** Sound or radiation present at a relatively constant low level at a specific location. **—back'ground'** *adj.*

**back·ground·er** (băk'groun'dər) *n. Slang.* An informal meeting at which an official provides background information, as to the news media, about a government issue.

**back·hand** (băk'hănd') *n.* **1.** A stroke or motion, as of a racket, made with the back of the hand facing outward and the arm moving forward. **2.** Handwriting with letters slanting to the left. *—vt.* **-hand·ed, -hand·ing, -hands.** To perform, catch, or hit backhand. **—back'hand** *adv.*

**back·hand·ed** (băk'hăn'dĭd) *adj.* **1.** Made with or using a backhand. **2.** Oblique : roundabout <*backhanded* compliments> **—back'hand'ed·ly** *adv.* **—back'hand'ed·ness** *n.*

**back·ing** (băk'ĭng) *n.* **1.** Something that forms a back. **2. a.** Support : aid. **b.** Approval or endorsement.

**back·lash** (băk'lăsh') *n.* **1.** A sudden or violent backward whipping motion. **2.** An antagonistic reaction to a prior action <conservative *backlash*> **3.** A snarl in the part of a fishing line wound around the reel. **4.** The play resulting from loose connections between mechanical elements, as gears.

**back·less** (băk'lĭs) *adj.* Lacking a back <a *backless* bench><a *backless* gown>

**back·log** (băk'lôg', -lŏg') *n.* **1.** A reserve supply. **2.** An accumulation, as of unfinished work or unfilled orders. **3.** A large log at the back of a fire in a fireplace. *—vt. & vi.* **-logged, -log·ging, -logs.** To acquire as or become a backlog.

**back matter** *n.* Material, as an appendix, that follows the main body of a book.

**back mutation** *n.* A reversal process whereby a gene that has undergone mutation returns to its previous state. **—back'-mu'tate'** (băk'myōō'tāt', -myōō-tāt') *v.* **(-tat·ed, -tat·ing, -tates).**

**back·pack** (băk'păk') *n.* **1.** A knapsack, often mounted on a lightweight frame, that is worn on the back to carry camping supplies. **2.** A piece of equipment made for use while being carried on the back. *—v.* **-packed, -pack·ing, -packs.** *—vi.* To hike while carrying a backpack. *—vt.* To carry in a backpack. **—back'pack'er** *n.* **—back'pack'ing** *n.*

**back·ped·al** (băk'pĕd'l) *vi.* **-aled, -al·ing, -als** *or* **-alled, -al·ling, -als.** To move backward or withdraw <The spokesperson later *backpedaled* on the issue.>

**back·rest** (băk'rĕst') *n.* A rest for the back.

**back·rush** (băk'rŭsh') *n.* The seaward return of water after the landward motion of a wave.

**back·saw** (băk'sô') *n.* A saw reinforced by a metal band along its back edge.

**back·scat·ter** (băk'skăt'ər) *n.* Deflection of waves or particles through angles greater than 90° by electromagnetic or nuclear forces. **—back'scat'ter** *v.* **(-tered, -ter·ing, -ters).**

**back seat** *n.* **1.** A seat in the back, esp. of a vehicle. **2.** *Informal.* A subordinate position.

**back-seat driver** (băk'sēt') *n. Informal.* **1.** A passenger in a motor vehicle who constantly advises, corrects, or nags the driver. **2.** One who persists in giving unsolicited advice.

**back·set** (băk'sĕt') *n.* SETBACK 1.

**back·side** (băk'sīd') *n. Informal.* The buttocks.

**back·slap** (băk'slăp') *v.* **-slapped, -slap·ping, -slaps.** *—vt.* To demonstrate jocular good will for. *—vi.* To demonstrate jocular good will. **—back'slap'per** *n.*

**back·slide** (băk'slīd') *vi.* **-slid** (-slĭd'), **-slid** *or* **-slid·den** (-slĭd'n), **-slid·ing, -slides.** To lapse into sin or wrongdoing, esp. in religious practice. **—back'slid'er** *n.*

**back·spin** (băk'spĭn') *n.* A spin that tends to retard, arrest, or reverse the linear motion of an object, esp. of a ball.

**back·stage** (băk'stāj') *adv.* **1.** In or toward the part of the stage behind the performing area in a theater. **2.** Secretly : privately. *—adj.* (băk'stāj'). **1.** Of, relating to, occurring in, or situated behind the performing area of a theater. **2.** Not known to the public : PRIVATE.

**back·stairs** (băk'stârz') *also* **back·stair** (-stâr') *adj.* Furtively carried on : CLANDESTINE <*backstairs* gossip>

**back·stay** (băk'stā') *n.* **1.** *Naut.* A rope or shroud extending from the top of a mast aft to the ship's side or stern to help support the mast. **2.** A supporting device at or for the back of an object.

**back·stitch** (băk'stĭch') *n.* A stitch made by inserting the needle at the midpoint of a preceding stitch so that the stitches overlap by half lengths. *—vt. & vi.* **-stitched, -stitch·ing, -stitch·es.** To sew with backstitches.

**back·stop** (băk'stŏp') *n.* **1.** A screen or fence that prevents a ball from being thrown or hit far out of a playing area. **2.** *Baseball.* A catcher. *—vt.* **-stopped, -stop·ping, -stops. 1.** To serve as a backstop for. **2. a.** To support. **b.** To substitute for (another) in an emergency.

**back·street** (băk′strēt′) *adj.* Backstairs.
**back·stretch** (băk′strĕch′) *n.* The part of an oval racecourse farthest from the spectators and opposite the homestretch.
**back·stroke** (băk′strōk′) *n.* **1.** A backhanded stroke. **2.** A swimming stroke executed with the swimmer on his or her back.
**back·swept** (băk′swĕpt′) *adj.* Swept or angled backward.
**back·swim·mer** (băk′swĭm′ər) *n.* Any of various insects of the family Notonectidae that swim or float on their backs.
**back·sword** (băk′sôrd′, -sōrd′) *n.* **1.** A sword with only one cutting edge. **2.** SINGLESTICK 1.
**back talk** *n.* An insolent retort.
**back·track** (băk′trăk′) *vi.* **-tracked, -track·ing, -tracks. 1.** To go back over the course by which one has come. **2.** To reverse one's policy or position.
**back·up** (băk′ŭp′) *n.* **1.** A reserve or substitute. **2. a.** Support or backing. **b.** A background accompaniment, as for a musical performer. *—adj.* Extra : standby <a *back-up* driver>
**back·ward** (băk′wərd) *adv.* **1.** At, to, or toward the back. **2.** With the back leading. **3.** In a reverse manner or order. **4.** To, toward, or into the past. **5.** Toward a worse condition. ***usage:*** The adverb forms *backward* and *backwards* are interchangeable: *leaned backward; moved the chair backwards.* Only *backward* is an adjective: *a backward view.* *—adj.* **1.** Directed or facing toward the back. **2.** Performed, carried out, or arranged in a reverse manner or order. **3.** Behind others in progress or development. **—back′ward·ly** *adv.* **—back′ward·ness** *n.* **—back′wards** *adv.*
**back·wash** (băk′wŏsh′, -wôsh′) *n.* **1.** A backward flow of water, as from the action of oars. **2.** A backward flow of air, as from an aircraft propeller. **3.** A result of an event : AFTERMATH.
**back·wa·ter** (băk′wô′tər, -wŏt′ər) *n.* **1. a.** Water held or pushed back by or as if by a dam or current. **b.** A body of water thus formed. **2.** A place or situation considered stagnant or backward <a cultural and economic *backwater*>
**back·woods** (băk′wŏŏdz′, -wŏŏdz′) *pl.n.* Heavily wooded, uncultivated, thinly settled areas. **—back′woods′man** *n.*
**back yard** *also* **back·yard** (băk′yärd′) *n.* A yard behind a house.
**ba·con** (bā′kən) *n.* [ME < OFr., of Germanic orig.] The salted and smoked meat from the back and sides of a pig.
**Ba·co·ni·an** (bā-kō′nē-ən) *adj.* **1.** Of, relating to, or characteristic of the works or thought of Francis Bacon. **2.** Of or designating the theory that Francis Bacon was the author of the plays attributed to Shakespeare.
**bacter-** *pref. var. of* BACTERIO-.
**bac·te·re·mi·a** (băk′tə-rē′mē-ə) *n.* Presence of viable bacteria in the blood. **—bac′te·re′mic** (-mĭk) *adj.* **—bac′te·re′mi·cal·ly** *adv.*
**bacteri-** *pref. var. of* BACTERIO-.
**bac·te·ri·a** (băk-tîr′ē-ə) *n. pl. of* BACTERIUM.
**bacterial capsule** *n.* A dense mucous protective layer that envelops some bacteria.
**bac·te·ri·cide** (băk-tîr′ĭ-sīd′) *n.* An agent that destroys bacteria. **—bac·te′ri·cid′al** (-sīd′l) *adj.*
**bac·ter·in** (băk′tə-rĭn) *n.* A vaccine prepared from dead bacteria.
**bacterio-** *or* **bacteri-** *or* **bacter-** *pref.* [NLat. *bacterium*, bacterium.] Bacteria : bacterial <*bacterioscopy*>
**bac·te·ri·o·cin** (băk-tîr′ē-ə-sĭn′) *n.* [BACTERIO- + (COLI)CIN.] A bacterially produced antibacterial agent.
**bac·te·ri·o·gen·ic** (băk-tîr′ē-ə-jĕn′ĭk) *adj.* Caused by bacteria.
**bac·te·ri·ol·o·gy** (băk-tîr′ē-ŏl′ə-jē) *n.* The study of bacteria. **—bac·te′ri·o·log′ic** (-ə-lŏj′ĭk), **bac·te′ri·o·log′i·cal** *adj.* **—bac·te′ri·o·log′i·cal·ly** *adv.* **—bac·te′ri·ol′o·gist** *n.*
**bac·te·ri·ol·y·sis** (băk-tîr′ē-ŏl′ĭ-sĭs) *n., pl.* **-ses** (-sēz′). Dissolution of bacteria. **—bac·te′ri·o·lyt′ic** (-ə-lĭt′ĭk) *adj.*
**bac·te·ri·o·phage** (băk-tîr′ē-ə-fāj′) *n.* A submicroscopic, usu. viral organism that destroys bacteria. **—bac·te′ri·o·phag′ic** (-făj′ĭk), **bac·te′ri·oph′a·gous** (-ŏf′ə-gəs) *adj.* **—bac·te′ri·o·phag′i·cal·ly** *adv.*
**bac·te·ri·os·co·py** (băk-tîr′ē-ŏs′kə-pē) *n.* Microscopic study of bacteria. **—bac·te′ri·o·scop′ic** (-ə-skŏp′ĭk), **bac·te′ri·o·scop′i·cal** *adj.* **—bac·te′ri·o·scop′i·cal·ly** *adv.* **—bac·te′ri·os′co·pist** *n.*
**bac·te·ri·o·sta·sis** (băk-tîr′ē-ō-stā′sĭs) *n., pl.* **-ses** (-sēz′). Arrestment or inhibition of bacterial growth without killing the bacteria. **—bac·te′ri·o·stat′ic** (-stăt′ĭk) *adj.* **—bac·te′ri·o·stat′i·cal·ly** *adv.*
**bac·te·ri·o·stat** (băk-tîr′ē-ə-stăt′) *n.* An agent producing bacteriostasis.
**bac·te·ri·um** (băk-tîr′ē-əm) *n., pl.* **-ri·a** (ē-ə) [NLat. < Gk. *baktērion*, little rod, dim. of *baktron*, rod.] Any of numerous unicellular microorganisms of the class Schizomycetes, occurring in many forms, existing either as free-living organisms or as parasites, and having a broad range of biochemical, often pathogenic properties. **—bac·te′ri·al** *adj.* **—bac·te′ri·al·ly** *adv.*
**bac·te·ri·u·ri·a** (băk-tîr′ē-yŏŏr′ē-ə) *n.* Presence of bacteria in urine.
**bac·te·rize** (băk′tə-rīz′) *vt.* **-rized, -riz·ing, -riz·es.** To change or cause a change in by means of bacteria. **—bac′te·ri·za′tion** *n.*
**bac·te·roid** (băk′tə-roid′) *adj.* Like bacteria. *—n.* Any of various structurally modified bacteria, as those occurring on the roots of leguminous plants.
**Bac·tri·an camel** (băk′trē-ən) *n.* A two-humped camel, *Camelus bactrianus*, native to central and southwestern Asia.

**Bactrian camel**
*4–10 feet long*

**bac·u·li·form** (băk′yə-lə-fôrm′, bə-kyōō′lə-) *adj.* [Lat. *baculum*, stick + -FORM.] Rod-shaped.
**bad**[1] (băd) *adj.* **worse** (wûrs), **worst** (wûrst) [ME *badde*, prob. < OE *bæddel*, an effeminate person.] **1.** Not reaching an adequate standard : POOR. **2.** Evil : wicked. **3.** Disobedient : naughty. **4.** Disagreeable, unpleasant, or disturbing. **5.** Unfavorable <a play that received *bad* reviews> **6.** Not fresh : SPOILED <a *bad* peach> **7.** Injurious <had *bad* habits> **8.** Not functioning properly : DEFECTIVE <a *bad* phone connection> **9.** Faulty : incorrect <*bad* grammar> **10.** Not valid or genuine <passed a *bad* check> **11.** Severe : intense <had a *bad* cold> **12. a.** Being in poor health or in pain. **b.** Being in poor condition : DISEASED <a child with *bad* teeth> **13.** Sorry : regretful. **14.** *compar.* **bad·der,** *superl.* **bad·dest.** *Slang.* Very good : GREAT. *—n.* **1.** Something bad <had to decide between the good and the *bad*> **2.** A wicked or unhappy state. *—adv. Informal.* Badly. ***usage:*** The use of *bad* as an adverb, while common in informal speech, should be avoided in writing. Formal usage requires: *My tooth hurts badly* (not *bad*). **—bad′ness** *n.*
☆ ***syns:*** BAD, BUM, DEFICIENT, UNSATISFACTORY *adj. core meaning* : being below par <a really *bad* writer>
**bad**[2] (băd) *v. Archaic. var. p.t. of* BID.
**bad blood** *n.* Bitterness between persons.
**bad·der·locks** (băd′ər-lŏks′) *n.* [Orig. unknown.] *(sing. or pl. in number).* An edible seaweed, *Alaria esculenta*, with long, yellowish-green fronds.
**bade** (băd, bād) *v. var. p.t. of* BID.
**badge** (băj) *n.* [ME *bagge.*] **1. a.** An emblem worn as an insignia of rank, office, or membership in a group. **b.** An emblem presented as an award or honor. **2.** A characteristic mark or sign. **—badge** *v.* **(badged, badg·ing, badg·es).**
**badg·er** (băj′ər) *n.* [Orig. unknown.] **1. a.** A carnivorous burrowing animal of the family Mustelidae, as *Meles meles* of Eurasia or *Taxidea taxus* of North America, with short legs, long claws on the front feet, and a heavy, grizzled coat. **b.** The fur or hair of a badger. **c.** A mammal, as the honey badger, related or similar to the badger. **2. Badger.** *Slang.* A native or inhabitant of Wisconsin. *—vt.* **-ered, -er·ing, -ers.** To pester.
**bad·i·nage** (băd′n-äzh′) *n.* [Fr. < *badin*, joker < Prov. *badar*, to gape < LLat. *badare.*] Light, playful banter.
**bad·lands** (băd′lăndz′) *pl.n.* An area of barren land having roughly eroded ridges, peaks, and mesas.
**bad·ly** (băd′lē) *adv.* **1.** In a bad manner. **2.** Very much.
**bad·min·ton** (băd′mĭn′tən) *n.* [After *Badminton*, the Duke of Beaufort's country seat in Gloucestershire, England.] A game played by volleying a shuttlecock back and forth over a high, narrow net by a light, long-handled racket.
**bad·mouth** *also* **bad-mouth** (băd′mouth′, -mouth′) *vt.* **-mouthed, -mouth·ing, -mouths.** *Slang.* To criticize or disparage, often spitefully.
**bad news** *n. Slang.* An unpleasant, unwelcome person or situation.
**Bae·de·ker** (bā′dĭ-kər) *n.* [After Karl *Baedeker* (1801–1859), publisher of guidebooks to Europe.] A guidebook.
**baf·fle** (băf′əl) *vt.* **-fled, -fling, -fles.** [Perh. < Fr. *bafouer*, to ridicule.] **1.** To frustrate (a person) as by confusing or perplexing : STYMIE. **2.** To impede the movement or force of. *—n.* **1.** A usu. static device that regulates light or the flow of a fluid. **2.** A partition preventing interference between sound waves in a loudspeaker. **—baf·fle·ment** *n.* **—baf′fler** *n.*
**bag** (băg) *n.* [ME *bagge* < ON *baggi.*] **1. a.** A usu. flexible container <a grocery *bag*> **b.** A handbag : purse. **c.** A piece of hand luggage. **d.** An organic sac or pouch, as a cow's udder. **2.** Something resembling a bag. **3.** *Naut.* The bulging part of a sail. **4.** A bagful. **5.** An amount of game taken or legally permitted to be taken. **6.** *Baseball.* A base. **7.** *Slang.* An area of interest or skill <Painting is not my *bag.*> *—v.* **bagged, bag·ging, bags.** *—vt.* **1.** To put into a bag. **2.** To cause to bulge like a bag. **3.** To capture or kill as game. **4.** *Informal.* To gain possession of : CAPTURE <*bagged* the criminal> *—vi.* **1.** To hang loosely. **2.** To swell out : BULGE. **—bag and baggage. 1.** With all one's belongings. **2.** Entirely : completely. **—in the bag.** *Slang.* **1.** Certain to be successful. **2.** Intoxicated.

**ba·gasse** (bə-găs′) *n.* [Fr. < Sp. *bagazo*, dregs < Lat. *baca*, berry.] The dry pulp remaining from sugar cane after extraction of the juice.
**bag·a·telle** (băg′ə-tĕl′) *n.* [Fr. < Ital., *bagatella*, poss. < Lat. *baca*, berry.] **1.** TRIFLE 1. **2.** A game played on an oblong table with a cue and balls.
**ba·gel** (bā′gəl) *n.* [Yiddish *beygel* < MHG *bouc*, ring < OHG *boug*.] A glazed ring-shaped roll having a chewy, tough texture.
**bag·ful** (băg′fo͝ol′) *n.*, *pl.* **-fuls.** The amount that a bag can hold.
**bag·gage** (băg′ĭj) *n.* [ME *bagage* < OFr. *bague*, bundle.] **1.** The containers, as trunks, bags, parcels, and suitcases, in which one carries one's belongings while traveling : LUGGAGE. **2.** The movable equipment and supplies of an army. **3.** Superfluous or burdensome practices, regulations, or ideas. **4. a.** A wanton or immoral woman. **b.** An impudent or saucy girl or woman.
**bag·ging** (băg′ĭng) *n.* Material used for making bags.
**bag·gy** (băg′ē) *adj.* **-gi·er, -gi·est.** Hanging loosely : BULGING. **—bag′gi·ly** *adv.* **—bag′gi·ness** *n.*
**bag lady** *n.* A homeless woman esp. one in a large city, who carries all her possessions in a shopping bag.
**bag·man** (băg′mən) *n.* **1.** *Slang.* A person who receives and collects illicit payments, as for racketeers. **2.** *Chiefly Brit.* A traveling sales representative.
**ba·gnio** (băn′yō) *n.*, *pl.* **-gnios.** [Ital. *bagno*, bath < Lat. *balneum* < Gk. *balaneion*.] **1.** A brothel. **2.** *Obs.* An Oriental slave prison.
▲ word history: Public bathhouses, at least in England, at one time had such an unsavory reputation as places for illicit sexual encounters that *bagnio*, which was borrowed from Italian with the meaning "bath," eventually lost that meaning and acquired the sense "brothel."
**bag·pipe** (băg′pīp′) *n.* *often* **bagpipes.** A musical instrument having a flexible bag inflated either by a tube with valves or by bellows, a double-reed melody pipe, and from one to four drone pipes. **—bag′pip′er** *n.*
**ba·guette** (bă-gĕt′) *n.* [Fr., rod < Ital. *bacchetta*, dim. of *bacchio*, rod < Lat. *baculum*, stick.] **1. a.** A gem cut into a narrow rectangle. **b.** The form of such a gem. **2.** A narrow, convex molding.
**bag·wig** (băg′wĭg′) *n.* A wig with the back hair encased in a small silk sack, worn in the 18th cent.
**bag·worm** (băg′wûrm′) *n.* The larva of any of several moths of the family Psychidae that encloses itself in a fibrous case and eats and destroys tree foliage.
**bah** (bä, bă) *interj.* —Used to express impatient rejection or contempt.
**Ba·ha·i** (bä-hä′ē, -hī′) *adj.* [Pers. *bahā′ī*, of glory < *bahā*, glory.] Of, relating to, or designating a religion founded in 1863 in Iran and emphasizing the spiritual unity of all humankind. **—Ba·ha′i** *n.* **—Ba·ha′ism** (bə-hä′ĭz′əm, -hī′-) *n.* **—Ba·ha′ist** *n.*
**Ba·ha·sa Indonesia** (bä-hä′sə) *n.* [Indonesian, Indonesian language.] A dialect of Malay that is the official language of the Republic of Indonesia.
**baht** (bät) *n.*, *pl.* **bahts** or **baht.** [Thai *bāt*.] —See table at CURRENCY.
**bail**[1] (bāl) *n.* [ME *baile*, custody < OFr. *bail* < *baillier*, to take charge of < Lat. *bajulare*, to carry a load < *bajulus*, carrier of a burden.] **1.** Security, usu. a sum of money, exchanged for the release of an arrested person as a guarantee of his or her appearance for trial. **2.** Release from imprisonment provided by the payment of bail. **3.** One providing bail. *—vt.* **bailed, bail·ing, bails. 1.** To secure the release of by posting bail. **2.** To release (a person) for whom bail has been paid. **3.** *Informal.* To extricate from a difficult situation <had to *bail* their teen-agers out of trouble> **4.** To transfer (property) to another for a special purpose but without permanent transference of ownership.
**bail**[2] (bāl) *v.* **bailed, bail·ing, bails.** [< ME *baille*, bucket < OFr., poss. < Lat. *bajulus*, porter.] *—vt.* **1.** To remove (water) from a boat by repeatedly filling a container and emptying it over the side. **2.** To empty (a boat) of water by bailing. *—vi.* To empty a boat of water by bailing. **—bail out. 1.** To parachute from an aircraft : EJECT. **2.** *Slang.* To abandon a project or enterprise. **—bail** *n.* **—bail′er** *n.*
**bail**[3] (bāl) *n.* [ME *baill* prob. < ON *beygla*.] **1.** The arched, hooplike handle of a container, as a pail. **2.** An arch or hoop, such as those used to support the top of a covered wagon. **3.** A hinged bar on a typewriter that holds the paper against the platen.
**bail**[4] (bāl) *n.* [ME, bailey.] **1.** *Chiefly Brit.* A pole or bar used to confine or separate animals. **2.** One of the two crossbars that form the top of a wicket in the game of cricket.
**bail·a·ble** (bā′lə-bəl) *adj.* **1.** Eligible for bail. **2.** Allowing or admitting of bail <*bailable* offenses>
**bail·ee** (bā-lē′) *n.* One to whom property is bailed.
**bail·er** (bā′lər) *n.* *var. of* BAILOR.
**bai·ley** (bā′lē) *n.*, *pl.* **-leys.** [ME *baille* < OFr.] **1.** The outer wall of a castle. **2.** The space enclosed by a bailey.
**Bai·ley bridge** (bā′lē) *n.* [After Sir Donald *Bailey* (b. 1901).] A steel bridge designed to be shipped in parts and assembled fast.
**bail·ie** (bā′lē) *n.* [ME *baillie* < OFr. *baillif*, bailiff.] **1.** A Scottish municipal officer corresponding to an English alderman. **2.** *Obs.* A bailiff.
**bail·iff** (bā′lĭf) *n.* [ME *baillif* < OFr. < Med. Lat. *bajulivus* < Lat. *bajulus*, carrier.] **1.** A minor court officer whose duties include maintaining order in a courtroom during a trial. **2.** *Chiefly Brit.* An official who assists a sheriff and is empowered to execute writs, processes, and arrests. **3.** *Chiefly Brit.* An estate overseer : STEWARD.
**bail·i·wick** (bā′lə-wĭk′) *n.* [ME *bailliwik* : *baillif*, bailiff + *wik*, town < OE *wīc* < Lat. *vicus*.] **1.** One's specific area of interest, skill, or authority. **2.** The office or district of a bailiff.
**bail·ment** (bāl′mənt) *n.* *Law.* **1.** The process of posting bail for an accused person. **2.** The act of delivering goods or personal property to another in trust.
**bail·or** (bā′lər, bā-lôr′) *also* **bail·er** (bā′lər) *n.* *Law.* One who bails property to another.
**bail·out** (bāl′out′) *n.* A rescue from financial difficulties.
**bails·man** (bālz′mən) *n.* *Law.* One who posts bail for another.
**bain-ma·rie** (băn′mə-rē′) *n.*, *pl.* **bains-ma·rie** (băn′mə-rē′) [Fr. < Med. Lat. *balneum Mariae*, bath of Mary.] A device consisting of a large pan containing hot water in which smaller pans may be set to cook slowly or keep warm.
**bairn** (bârn) *n.* [ME *barn* < OE *bearn*.] *Scot.* A child.
**bait**[1] (bāt) *n.* [ME, partly < ON *beita*, to hunt with dogs, and partly < ON *beita*, food.] **1.** Lure, as food, that is placed on a hook or in a trap and used in the taking of animals, as fish or birds. **2.** An enticement. **3.** *Archaic.* A stop for food or rest during a trip. *—v.* **bait·ed, bait·ing, baits.** *—vt.* **1.** To place bait in (a trap) or on (a fishing hook). **2.** To entice or lure, esp. by trickery. **3.** To set dogs on (e.g., a chained animal) for sport. **4.** To torment, esp. with persistent insults, criticisms, or ridicule. **5.** TEASE 3. **6.** To feed and water (an animal) esp. on a journey. *—vi.* *Archaic.* To stop for food or rest during a trip. ***usage:*** *Bait* is sometimes used improperly for *bate* in the phrase *bated breath.* **—bait′er** *n.*
☆ ***syns:*** BAIT, BADGER, HECKLE, HECTOR, HOUND, NEEDLE, RIDE, TAUNT *v. core meaning* : to torment persistently with insults or ridicule <a hostile reporter who continually *baited* the President>
▲ word history: It is probable that *bait*[1] is a blend of three Old Norse words: *beit*, "pasturage," *beita*, "food," and *beita*, a verb, "to graze, hunt, put bait on a hook." The verb *beita* is derived from *bīta*, "to bite," a cognate of English *bite*. *Beita* is a causative verb, that is, its basic meaning is "to cause to bite." Some of the senses of the English word *bait*[1] come directly from Old Norse, but others arose as the word developed in English.
**bait**[2] (bāt) *v.* *var. of* BATE[2].
**bait and switch** *n.* A sales tactic in which a bargain-priced item is used to attract customers who are then urged to buy a similar, but more expensive item.
**bai·za** (bī′zä) *n.* [Ar. < Hindi *paisā*.] —See table at CURRENCY.
**baize** (bāz) *n.* [Fr. *baies* < *bai*, bay-colored < Lat. *badius*.] Cotton or woolen material napped to imitate felt.
**bake** (bāk) *v.* **baked, bak·ing, bakes.** [ME *baken* < OE *bacan*.] *—vt.* **1.** To cook (food) with dry heat, esp. in an oven. **2.** To dry or harden by subjecting to heat in or as if in an oven. *—vi.* **1.** To cook food by baking. **2.** To become baked. *—n.* **1. a.** The act or process of baking. **b.** An amount that is baked. **2.** A social gathering at which food is baked and served.
**Ba·ke·lite** (bā′kə-līt′). A trademark for any of a group of thermosetting plastics having high chemical and electrical resistance and used in a variety of manufactured articles.
**bak·er** (bā′kər) *n.* **1.** One who bakes. **2.** A portable oven.
**baker's dozen** *n.* [From the former custom of bakers to add an extra roll as a safeguard against the possibility of 12 weighing light.] A group of 13.
**bak·er·y** (bā′kə-rē) *n.*, *pl.* **-ies.** An establishment where baked goods are prepared or sold.
**bake·shop** (bāk′shŏp′) *n.* A bakery.
**bak·ing** (bā′kĭng) *n.* **1.** BAKE 1a. **2.** BAKE 1b.
**baking powder** *n.* A mixture of baking soda, starch, and at least one slightly acidic compound such as cream of tartar, used as a leavening agent in baking.
**baking soda** *n.* Sodium bicarbonate.
**ba·kla·va** (bä′klə-vä′) *n.* [Turk.] A dessert made of paper-thin layers of pastry, chopped nuts, and honey.
**bak·sheesh** (băk′shēsh′, băk-shēsh′) *n.*, *pl.* **baksheesh.** [Pers. *bakhshīsh* < *bakhshīdan*, to give.] A gratuity : tip.
**BAL**[1] (băl) *n.* [B(RITISH) + A(NTI-) + L(EWISITE).] A colorless, oily, viscous liquid, $C_3H_5(SH)_2(OH)$, used as an antidote for poisoning caused by lewisite, organic arsenic compounds, or heavy metals.
**BAL**[2] (bē′ā-ĕl′) *n.* [B(ASIC) A(SSEMBLY) L(ANGUAGE).] *Computer Sci.* A low-level assembly language.
**Ba·laam** (bā′ləm) *n.* [Gk. < Heb. *Bil′ām*.] An Old Testament prophet who was commanded to curse the Israelites but blessed them instead after being rebuked by the ass he rode.

**bal·a·lai·ka** (băl'ə-lī'kə) *n.* [R.] A three-stringed Russian musical instrument with a triangular body.

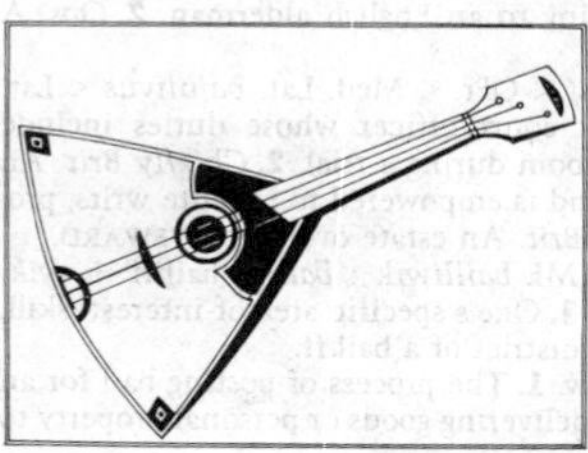
**balalaika**

**bal·ance** (băl'əns) *n.* [ME *balaunce* < OFr. < VLat. **bilancia* < Lat. *bilanx* : *bi-*, two + *lanx*, scale.] **1.** A weighing device composed of a rigid beam horizontally suspended by a low-friction support at its center, with identical weighing pans hung at either end, one of which holds an unknown weight while the effective weight in the other is increased by known amounts until the beam is level and motionless. **2.** Equilibrium or parity marked by cancellation of all forces by equal opposing forces. **3.** The means or power to decide. **4.** A state of bodily equilibrium. **5.** Emotional stability. **6.** A harmonious or satisfying arrangement of parts or elements. **7.** A force or influence tending to produce equilibrium : COUNTERPOISE. **8.** The difference in magnitude between opposing forces or influences. **9. a.** Equality of totals in the debit and credit sides of an account. **b.** The difference between such totals. **10.** A leftover : remainder. **11.** *Chem.* Equality of the number, kinds, and net electric charge of reacting species on each side of a chemical equation. **12.** *Math.* Equality as to the net number of reduced symbolic quantities on each side of an equation. **13.** A balance wheel. **14. Balance.** Libra. —*v.* **-anced, -anc·ing, -anc·es.** —*vt.* **1.** To weigh in or as if in a balance. **2.** To compare by or as if by turning over in the mind. **3.** To bring into or maintain in a state of equilibrium. **4.** To act as an offsetting weight or force to : COUNTERBALANCE. **5. a.** To compute the difference between the debits and credits of (an account). **b.** To reconcile or equalize the sums of the debits and credits of (an account). **c.** To settle by paying what is owed. **6.** To bring into or keep in equal or satisfying harmony or proportion. **7.** *Math.* To bring (an equation) into mathematical balance. **8.** *Chem.* To bring (a chemical equation) into chemical balance. **9.** To move toward and then away from (a dance partner). —*vi.* **1.** To be in or come into a state of equilibrium. **2.** To be equal or equivalent. **3.** To sway or waver as if losing or regaining equilibrium. **4.** To move toward and then away from a dance partner. **—in the balance.** In an undetermined position <the future of our nation hanging *in the balance*>

☆ ***syns:*** BALANCE, COUNTERPOISE, EQUILIBRIUM, EQUIPOISE, STASIS *n. core meaning* : a stable state marked by cancellation of all forces by equal opposing forces <the *balance* of geopolitical power>

**balance beam** *n.* **1.** A narrow horizontal wooden beam raised about four feet above the floor and used for balancing exercises in gymnastics. **2.** A competitive gymnastics event in which balancing feats are performed on the balance beam.

**balance of payments** *n.* A record of a nation's total payments to foreign countries, including the price of imports, the outflow of capital and gold, and the total receipts from abroad, including the price of exports and the inflow of capital and gold.

**balance of power** *n.* Distribution of power whereby no single nation is able to dominate or interfere with others.

**balance of trade** *n.* The difference in value between a nation's total exports and total imports.

**bal·anc·er** (băl'ən-sər) *n.* **1.** One that balances. **2.** HALTER[2].

**balance sheet** *n.* A statement of financial assets and liabilities at a given date.

**balance wheel** *n.* A wheel regulating the rate of movement in machine parts, as in a watch.

**bal·as** (băl'əs) *n.* [ME < OF *balais* < Med. Lat. *balascus* < Ar. *badakhsh* < Pers. *Badhakhshān*, a region in Afghanistan.] A rose-red to orange spinel.

**ba·la·ta** (bə-lä'tə) *n.* [Am. Sp. (West Indies), of Cariban orig.] **1.** A tropical American tree, *Manilkara bidentata* or *Mimusops balata*, that yields a latexlike sap. **2.** A tough, nonelastic, rubberlike gum obtained from the sap of the balata and used for golf-ball covers, industrial belting, and gaskets.

**bal·bo·a** (băl-bō'ə) *n.* [After Vasco Núñez de *Balboa* (1475–1519).] —See table at CURRENCY.

**bal·brig·gan** (băl-brĭg'ən) *n.* [After *Balbriggan*, Ireland.] A knitted unbleached-cotton underwear fabric.

**bal·co·ny** (băl'kə-nē) *n., pl.* **-nies.** [Ital. *balcone* < OItal., scaffold, of Germanic orig.] **1.** A platform projecting from the wall of a building and surrounded by a railing, balustrade, or parapet. **2.** A gallery projecting over the main floor in a theater or auditorium.

**bald** (bôld) *adj.* **-er, -est.** [ME *balled.*] **1.** Having no hair on the head. **2.** Lacking a natural or the usual covering <a *bald* spot in the yard> **3.** Having white feathers or markings on the head. **4.** Unadorned. **5.** Undisguised <*bald* lies> **—bald'ly** *adv.*

**bal·da·chin** (bôl'də-kĭn, băl'-) *also* **bal·da·chi·no** (băl'də-kē'nō) *n., pl.* **-chins** *also* **-chi·nos.** [Ital. *baldacchino* < OItal. < *Baldacco*, Baghdad.] **1.** A rich silk and gold brocade fabric. **2.** A canopy of fabric carried in church processions or placed over an altar, throne, or dais. **3.** A stone or marble structure in the form of a canopy, esp. over a church altar.

**bald cypress** *n.* A cone-bearing but deciduous tree, *Taxodium distichum* of the southeastern United States, growing in swamps and damp ground.

**bald eagle** *n.* A North American eagle, *Haliaeetus leucocephalus*, having a dark body and a white head and tail.

**bal·der·dash** (bôl'dər-dăsh') *n.* [Orig. unknown.] Nonsense.

**bald-faced** (bôld'fāst') *adj.* **1.** Brash. **2.** BALD 5.

**bald·head** (bôld'hĕd') *n.* **1.** One having a bald head. **2.** Any of several birds with white markings on the head.

**bald·ing** (bôl'dĭng) *adj. Informal.* Becoming bald.

**bald·pate** (bôld'pāt') *n.* **1.** BALDHEAD 1. **2.** The widgeon.

**bal·dric** (bôl'drĭk) *n.* [ME *baudrik*, prob. < OFr. *baudre.*] A usu. ornamented leather belt worn across the chest to support a sword or bugle.

**Bald·win** (bôl'dwĭn) *n.* [After Loammi *Baldwin* (1740–1807).] A red-skinned American variety of apple.

**bale[1]** (bāl) *n.* [ME, perh. < OFr., of Germanic orig.] A large bound, often wrapped package of raw or finished material. —*vt.* **baled, bal·ing, bales.** To wrap in bales. **—bal'er** *n.*

**bale[2]** (bāl) *n.* [ME < OE *balu.*] **1.** Evil. **2.** Mental anguish.

**ba·leen** (bə-lēn') *n.* [ME *balene* < OFr. *baleine* < Lat. *ba-laena*, whale < Gk. *phalaina.*] WHALEBONE 1.

**bale·ful** (bāl'fəl) *adj.* **1.** Harmful or malignant in intent or effect. **2.** Portending evil : OMINOUS. ***usage:*** Although *baleful* and *baneful* overlap in meaning, *baleful* usu. applies to something that threatens or foreshadows evil (*a baleful look*), while *baneful* most often applies to something that is actually harmful or destructive (*a baneful influence*). **—bale'ful·ly** *adv.* **—bale'ful·ness** *n.*

**Ba·li·nese** (bä'lə-nēz', -nēs') *n.* **1.** A native or inhabitant of Bali. **2.** The Indonesian language of Bali. **—Ba'li·nese'** *adj.*

**balk** *also* **baulk** (bôk) [ME *balken*, to plow up in ridges < *balk*, ridge < OE *balc.*] —*v.* **balked, balk·ing, balks** *also* **baulked, baulk·ing, baulks** —*vi.* **1.** To stop short and refuse to go on. **2.** To refuse obstinately or abruptly. **3. a.** *Baseball.* To make an illegal motion before pitching. **b.** To make an incomplete or misleading motion in a sport. —*vt.* **1.** To put obstacles in the way of. **2.** *Archaic.* To let go by : MISS. **3.** *Baseball.* To move one or more runners ahead one base by balking. —*n.* **1.** A hindrance or check. **2.** An incomplete or misleading motion, esp. an illegal move made by a baseball pitcher. **3. a.** An unplowed strip of land. **b.** A ridge between furrows. **4.** A wooden beam or rafter. **5.** One of the spaces between the cushion and the balk line on a billiard table. **—balk'er** *n.*

**Bal·kan** (bôl'kən) *adj.* **1.** Of or relating to the Balkan Peninsula or the Balkan Mountains. **2.** Of or relating to the Balkan States or their inhabitants.

**Bal·kan·ize** *also* **bal·kan·ize** (bôl'kə-nīz') *vt.* **-ized, -iz·ing, -iz·es.** [From the political division of the Balkans in the early 20th cent.] To divide (a region or territory) into small, often hostile units. **—Bal'kan·i·za'tion** *n.*

**balk line** *also* **balk·line** (bôk'līn') *n.* A line parallel to one end of a billiard table from behind which opening shots with the cue ball are made.

**balk·y** (bô'kē) *adj.* **-i·er, -i·est.** Tending to balk : OBSTINATE <a *balky* horse><a *balky* child> **—balk'i·ness** *n.*

**ball[1]** (bôl) *n.* [ME *bal* < ON *böllr.*] **1. a.** A spherical or almost spherical body. **b.** A spherical entity <a *ball* of flames> **2. a.** A rounded movable object used in various sports and games. **b.** A game, esp. baseball, played with such an object. **c.** A ball moving, thrown, hit, or kicked in a certain way. **d.** *Baseball.* A pitched ball not swung at by the batter that does not pass through the strike zone. **3. a.** A solid spherical or pointed projectile, as that shot from a cannon. **b.** Such projectiles as a whole. **4.** A rounded part or protuberance, esp. of the body. —*v.* **balled, ball·ing, balls.** To form or become formed into a ball. **—ball up.** *Slang.* To confuse or bungle : mess up. **—on the ball.** *Slang.* **1.** Alert, competent, or efficient. **2.** Having qualities assuring success.

**ball[2]** (bôl) *n.* [Fr. *bal* < OFr. < *baller*, to dance < LLat. *ballare* < Gk. *ballizein.*] **1.** A formal gathering for social dancing. **2.** *Slang.* An extremely enjoyable time.

**bal·lad** (băl'əd) *n.* [ME < OFr. *ballade* < OProv. *balada*, song sung while dancing < *balar*, to dance < LLat. *ballare*, to dance. —see BALL[2].] **1. a.** A narrative poem, often of folk origin and intended to be sung, consisting of simple stanzas and usu. having a recurrent refrain. **b.** The music for a ballad. **2.** A popular, esp. romantic or sentimental song. **—bal·lad'ic** (bə-lăd'ĭk, bă-) *adj.*

**bal·lade** (bə-läd', bă-) *n.* [ME < OFr., ballad.] **1.** A verse form usu. consisting of three stanzas of eight or ten lines each, with the same concluding line in each stanza and a brief final stanza, ending with the same last line as that of the preceding stanzas. **2.** A musical composition, usu. for the piano, having the romantic or dramatic quality of a ballad.

**bal·lad·eer** (băl'ə-dîr') *n.* A ballad singer.

**bal·lad·ist** (băl'ə-dĭst') *n.* A singer or writer of ballads.

**bal·lad·ry** (băl'ə-drē) *n.* Ballads as a whole.

**ballad stanza** *n.* A four-line stanza often used in ballads, rhyming in the second and fourth lines and having four metrical feet in the first and third lines and three in the second and fourth.

**ball-and-sock·et joint** (bôl'ən-sŏk'ĭt) *n.* **1.** A joint consisting of a spherical knob or knoblike part fitted into a socket so that some degree of motion is possible in nearly any direction. **2.** *Physiol.* An articulation (e.g., the hip joint) in which the rounded head of one bone fits into the cap-shaped cavity of the other and allows movement in any direction.

**bal·last** (băl'əst) *n.* [Perh. < OSwed. or ODan. *barlast* : *bar*, bare + *last*, load.] **1.** Heavy material put into the hold of a ship or the gondola of a balloon to enhance stability. **2.** Coarse gravel or crushed rock laid to form a bed for roads or railroads. **3.** A factor or element providing stability, esp. in character. *—vt.* **-last·ed, -last·ing, -lasts. 1.** To stabilize or provide with ballast. **2.** To fill (a railroad bed) or as if with ballast.

**ball bearing** *n.* **1.** A friction-reducing bearing consisting of a ring-shaped track containing freely revolving hard metal balls against which a rotating shaft or other part turns. **2.** A hard ball used in a ball bearing.

**ball carrier** *n. Football.* The player carrying the ball on an offensive play.

**ball cock** *n.* A self-regulating device controlling the supply of water in a tank, cistern, or toilet by a float connected to a valve that opens or closes with a change in water level.

**ball control** *n.* An offensive strategy, esp. in football and basketball, in which a team attempts to retain possession of the ball for long periods or deliberately slows the pace of the game.

**bal·le·ri·na** (băl'ə-rē'nə) *n.* [Ital. < *ballare*, to dance < LLat. < Gk. *ballizein*.] A woman who is a principal ballet dancer in a troupe.

**bal·let** (bă-lā', băl'ā') *n.* [Fr. < Ital. *balletto*, dim. of *ballo*, dance < *ballare*, to dance. —see BALLERINA.] **1.** An artistic dance form characterized by grace and precision of movement and an elaborate formal technique. **2.** A theatrical presentation of group or solo dancing to a musical accompaniment, usu. using costume and scenic effects and conveying a story, theme, or atmosphere. **3.** A musical composition written or used for ballet. **4.** A troupe that performs ballet. **—bal·let'ic** (bă-lĕt'ĭk) *adj.*

**bal·let·o·mane** (bă-lĕt'ə-măn') *n.* [Back-formation < BALLETOMANIA.] An admirer of ballet. **—bal·let'o·ma'ni·a** (-ə-mā'nē-ə, -măn'yə) *n.*

**ball·flow·er** (bôl'flou'ər) *n.* An ornament shaped like a ball cupped in the petals of a circular flower.

**ball game** *n.* **1.** A game, esp. baseball, that is played with a ball. **2.** *Informal.* **a.** A competition, as a political race. **b.** A particular condition or situation.

**bal·lis·ta** (bə-lĭs'tə) *n., pl.* **-tae** (-tē') [Lat. < Gk. *ballein*, to throw.] An ancient or medieval military engine used to hurl heavy projectiles during battles.

**bal·lis·tic** (bə-lĭs'tĭk) *adj.* [< BALLISTA.] **1.** Of or relating to ballistics. **2.** Of or relating to projectiles, their motion, or their effects. **—bal·lis'ti·cal·ly** *adv.*

**ballistic missile** *n.* A projectile that assumes a free-falling trajectory after an internally guided, self-powered ascent.

**bal·lis·tics** (bə-lĭs'tĭks) *n. (sing. in number).* **1. a.** The study of the dynamics of projectiles. **b.** The study of the flight characteristics of projectiles. **2. a.** The study of the functioning of firearms. **b.** The study of the firing, flight, and effect of ammunition. **—bal'lis·ti'cian** (băl'ĭ-stĭsh'ən) *n.*

**bal·lis·to·car·di·o·gram** (bə-lĭs'tō-kär'dē-ə-grăm') *n.* [BALLIST(IC) + CARDIOGRAM.] A recording made by a ballistocardiograph.

**bal·lis·to·car·di·o·graph** (bə-lĭs'tō-kär'dē-ə-grăf') *n.* [BALLIST(IC) + CARDIOGRAPH.] A device for measuring the volume of blood passing through the heart in a specific time period by measurement of the body's recoil against the ejection movements of the heart's ventricular muscles.

**bal·lis·to·pho·bi·a** (bə-lĭs'tə-fō'bē-ə) *n.* [BALLIST(IC) + -PHOBIA.] An abnormal fear of projectiles.

**ball lightning** *n.* A phenomenon associated with thunderstorms, usu. held to consist of a moving luminous sphere of ionized gas.

**ball of fire** *n.* A highly energetic or dynamic person.

**ball of wax** *n. Informal.* An unspecified set of items or circumstances <"wants international fame and fortune, the whole *ball of wax*" —*Newsweek*>

**bal·lo·net** (băl'ə-nā') *n.* [Fr. *ballonnet*, dim. *ballon*, balloon.] One of several small auxiliary gasbags placed inside a balloon or a nonrigid airship that can be inflated or deflated in flight to control and maintain shape and buoyancy.

**bal·loon** (bə-lo͞on') *n.* [Fr. *ballon* < Ital. *pallone*, aug. of *palla*, ball, of Germanic orig.] **1. a.** A spherical, flexible, nonporous bag inflated with a lighter-than-air gas, as helium, that causes it to rise and float in the atmosphere. **b.** Such a bag with sufficient capacity to lift a suspended gondola. **2.** Any of variously shaped, brightly colored inflatable rubber bags used as toys. **3.** A rounded or irregularly shaped outline containing the words that a cartoon character is represented as saying. *—v.* **-looned, -loon·ing, -loons.** *—vi.* **1.** To ascend or ride in a balloon. **2.** To expand or swell. **3.** To increase quickly. *—vt.* To cause to expand by or as if by inflating. *—adj.* **1.** Relating to or like of a balloon <a gown with *balloon* sleeves> **2.** Having periodic payments that are insufficient to pay back a note, thus requiring a large final payment <a *balloon* mortgage> **—bal·loon'ist** *n.*

**balloon sail** *n. Naut.* A rather large foresail used to supplement or replace a jib when going before the wind.

**balloon tire** *n.* A pneumatic tire with a wide tread, inflated to low pressure, and designed for maximum cushioning.

**bal·lot** (băl'ət) *n.* [Ital. *ballotta*, dim. of *balla*, ball, of Germanic orig.] **1.** A sheet of paper used to cast a vote, esp. a secret vote. **2.** An act, process, or method of voting, esp. by use of secret ballots. **3.** A list of candidates running for office : TICKET. **4.** The total votes cast in an election. **5.** Right to vote : FRANCHISE. **6.** A small ball used to register a vote. *—vi.* **-lot·ed, -lot·ing, -lots. 1.** To cast a ballot : VOTE. **2.** To draw lots. **—bal'lot·er** *n.*

▲ word history: The earliest meaning of *ballot* in English is "a small ball used to register a vote," which was also the meaning of *ballotta* in Italian. The ball was dropped into a box or other container, and in this manner it was possible to vote secretly. When yea and nay votes were recorded with different color balls, the negative was often expressed by a black ball. This practice gave rise to the verb *blackball*, meaning "to exclude someone from membership by a negative vote."

**bal·lotte·ment** (bə-lŏt'mənt) *n.* [Fr. < *ballotter*, to toss < *ballotte*, dim. of *balle*, ball, of Germanic orig.] A technique for detecting or examining a floating object in the body, as: **a.** The use of a finger to push sharply against the uterus and detect the presence or position of a fetus by its return impact. **b.** A test for a floating kidney in which the kidney is moved by alternating external digital pressures.

**ball·park** *also* **ball park** (bôl'pärk') *n.* **1.** A park or stadium in which ball games are played. **2.** *Informal.* One's own sphere of influence or territory <"the coalition was playing behind the scenes, in its own *ballpark*" —Hugh Gardner> **—in the ballpark.** *Informal.* Within the proper range <sales projections *in the ballpark*> **—ball'park'** *adj.*

**ball-point pen** (bôl'point') *n.* A pen whose writing point is a small ball bearing that transfers ink stored in a cartridge onto a writing surface.

**ball·room** (bôl'ro͞om', -ro͝om') *n.* A large room for dancing.

**ball valve** *n.* A valve regulated by the position of a free-floating ball that moves in response to fluid or mechanical pressure.

**bal·ly·hoo** (băl'ē-ho͞o') [Orig. unknown.] *Informal.* *—n., pl.* **-hoos. 1.** Sensational, clamorous advertising. **2.** A noisy uproar. *—vt.* **-hooed, -hoo·ing, -hoos.** To advertise by sensational methods.

**bal·ly·rag** (băl'ē-răg') *v. var. of* BULLYRAG.

**balm** (bäm) *n.* [ME *baume* < OFr. *basme* < Lat. *balsamum*, balsam.] **1.** An aromatic, oily resin exuded by various chiefly tropical trees and shrubs and used in medicine. **2.** An aromatic ointment, oil, or unguent. **3. a.** An aromatic herb native to Eurasia, *Melissa officinalis*, with small, fragrant white flower clusters. **b.** An aromatic plant similar to the balm. **4.** A pleasing fragrance. **5.** A soothing or healing agent : SALVE.

**bal·ma·caan** (băl'mə-kăn', -kän') *n.* [After *Balmacaan*, an estate near Inverness, Scotland.] A loose, full overcoat with raglan sleeves, orig. made of rough woolen cloth.

**balm of Gil·e·ad** (gĭl'ē-əd, -ăd') *n.* [After *Gilead*, region of ancient Palestine known for its balm.] **1.** An aromatic evergreen tree of the genus *Commiphora*, esp. *C. opobalsamum* of Africa and Asia Minor. **2.** A fragrant resin obtained from the balm of Gilead. **3.** A North American deciduous tree, *Populus candicans*, with broad, heart-shaped leaves. **4.** A fragrant resin obtained from the balsam fir.

**Bal·mor·al** (băl-môr'əl, -mŏr'-) *n.* [After *Balmoral* Castle, Scotland.] **1.** A brimless Scottish cap with a flat, round top. **2.** *often* **balmoral.** A heavy, laced walking shoe.

**balm·y** (bä'mē) *adj.* **-i·er, -i·est. 1.** Like balm in quality or fragrance. **2.** Mild <a *balmy* day> **3.** *Slang.* Eccentric in behavior. **—balm'i·ly** *adv.* **—balm'i·ness** *n.*

**bal·ne·ol·o·gy** (băl'nē-ŏl'ə-jē) *n.* [Lat. *balneum*, bath + -LOGY.] Medical therapy utilizing mineral baths.

**ba·lo·ney** (bə-lō'nē) *n., pl.* **-neys. 1.** *Informal. var. of* BOLOGNA. **2.** *Slang.* Nonsense.

o͞o **boot** ou **out** th **thin** *th* **this** ŭ **cut** ûr **urge** y **young**
yo͞o **abuse** zh **vision** ə **about**, **item**, **edible**, **gallop**, **circus**

**bal·sa** (bôl′sə) *n.* [Sp.] **1.** A tree, *Ochroma lagopus* of tropical America, whose wood is unusually lightweight. **2.** The wood of the balsa tree. **3.** A raft made of a frame fastened to buoyant cylinders of wood or metal.

**bal·sam** (bôl′səm) *n.* [Lat. *balsamum* < Gk. *balsamon.*] **1. a.** An oily or gummy oleoresin, usu. containing benzoic or cinnamic acids, obtained from the exudations of various trees and shrubs and used as a base for cough syrups and perfumes. **b.** A fragrant ointment used as medication. **2.** A tree yielding an aromatic, resinous substance, esp. the balsam fir. **3.** A plant of the genus *Impatiens,* esp. *I. balsamina,* cultivated for its double, variously colored flowers. **—bal·sam′ic** (-săm′ĭk) *adj.*

**balsam fir** *n.* An evergreen tree, *Abies balsamea* of northeastern North America, with small needles and cones approx. 2½ inches in length.

**balsam of Pe·ru** (pə-ro͞o′) *n.* The aromatic resin of a tropical American tree, *Myroxylon pereirae,* used in making perfumes and other products.

**balsam of To·lu** (tə-lo͞o′) *n.* The aromatic resin of a tropical American tree, *Myroxylon toluiferum,* used in cough remedies and in perfumery.

**balsam pear** *n.* An Old World tropical vine, *Momordica charantia,* bearing yellow-orange fruit.

**balsam poplar** *n.* A North American tree, *Populis balsamifera,* having large buds coated with a gummy, fragrant resin.

**Balt** (bôlt) *n.* A member of the Baltic-speaking people inhabiting the southeastern shores of the Baltic Sea and at one time occupying a wide area bounded by Danzig (Gdańsk), Riga, Moscow, and Kiev.

**Bal·tic** (bôl′tĭk) *adj.* **1.** Of or relating to the Baltic Sea or to the Baltic States and their inhabitants or cultures. **2.** Of or relating to the branch of the Indo-European language family that contains Latvian, Lithuanian, and Old Prussian. —*n.* The Baltic language branch.

**Bal·ti·more oriole** (bôl′tə-môr′, -mōr′) *n.* [After George Calvert (1580?–1632), 1st Lord *Baltimore.*] An American songbird, *Icterus galbula,* the male of which has brilliant orange, black, and white plumage.

**Bal·to-Sla·vic** (bôl′tō-slä′vĭk, -slăv′ĭk) *n.* A subfamily of the Indo-European language family that consists of the Baltic and Slavic branches. **—Bal′to-Sla′vic** *adj.*

**Ba·lu·chi** (bə-lo͞o′chē) *n., pl.* **Baluchi** or **-chis. 1.** A native or inhabitant of Baluchistan. **2.** The Iranian language of the Baluchi.

**bal·us·ter** (băl′ə-stər) *n.* [Fr. *balustre* < Ital. *balaustro* < *balaustra,* pomegranate flower (from a resemblance to the post) < Lat. *balaustium* < Gk. *balaustion.*] A post or support of a handrail.

**bal·us·trade** (băl′ə-strād′) *n.* [Fr. < Ital. *balustrade* < *balaustro,* baluster.] A rail and the row of posts that support it, as along the edge of a staircase.

**Bam·ba·ra** (băm-bä′rä) *n., pl.* **Bambara** or **-ras. 1 a.** A Negroid people of the upper Niger River valley. **b.** A member of this people. **2.** The Mande language of the Bambara.

**bam·bi·no** (băm-bē′nō, bäm-) *n., pl.* **-nos** or **-ni** (-nē) [Ital., dim. of *bambo,* child.] **1.** An infant : baby. **2.** A representation of the infant Jesus.

**bam·boo** (băm-bo͞o′) *n., pl.* **-boos.** [Orig. unknown.] **1. a.** Any of various mostly tropical grasses of the genus *Bambusa,* having hard-walled stems with ringed joints. **b.** The hollow woody stems of the bamboo. **2.** A tall, bamboolike grass of the genera *Arundinaria, Phyllostachys,* or *Dendrocalamus.*

**Bamboo Curtain** *n.* A political, esp. ideological barrier in Asia, esp. China.

**bam·boo·zle** (băm-bo͞o′zəl) *vt.* **-zled, -zling, -zles.** [Orig. unknown.] *Informal.* To deceive : hoodwink. **—bam·boo′zle·ment** *n.*

**ban¹** (băn) *vt.* **banned, ban·ning, bans.** [ME *bannen,* to summon, banish, curse, partly < OE *bannan,* to summon, and partly < ON *banna,* to prohibit, curse.] **1.** To prohibit, esp. by official decree. **2.** *Archaic.* To curse : execrate. —*n.* **1.** An excommunication or condemnation by church officials. **2.** A prohibition imposed by law or official decree. **3.** Censure, esp. through public opinion. **4.** A summons to arms in feudal times. **5.** A curse : imprecation.

**ban²** (bän) *n., pl.* **ba·ni** (bä′nē) [Rum. < Serbo-Croatian *bān,* warlord.] —See table at CURRENCY.

**ba·nal** (bə-näl′, -năl′, bā′nəl) *adj.* [Fr. < OFr., shared by tenants in a feudal jurisdiction < *ban,* summons to military service, of Germanic orig.] Lacking originality or freshness : HACKNEYED. **—ba·nal′i·ty** (bə-năl′ĭ-tē, bā-) *n.* **—ba·nal′ly** *adv.*

**ba·nan·a** (bə-năn′ə) *n.* [Port. and Sp., of African orig.] **1.** A treelike tropical or subtropical plant of the genus *Musa,* esp. *M. sapientum,* a widely cultivated species having long, broad leaves and hanging clusters of edible fruit. **2.** The crescent-shaped fruit of a banana plant, having pulpy, white flesh and thick, easily removed yellow or red-dish skin.

**banana oil** *n.* **1.** A liquid mixture of nitrocellulose and amyl acetate, or a similar solvent, having a bananalike odor. **2.** Amyl acetate.

**ba·nan·as** (bə-năn′əz) *adj. Slang.* Crazy.

**banana seat** *n.* [From its shape.] An elongated bicycle seat that usu. curves upward in the back.

**banana split** *n.* Several scoops of ice cream and usu. flavored syrups or sauces, nuts, fruit, and whipped cream served on a banana split lengthwise.

**band¹** (bănd) *n.* **1.** A thin strip of flexible material for encircling and binding one object or for holding several objects together. **2.** A strip contrasting with something else, as in color. **3. a. bands.** *Archaic.* A physical restraint : FETTER. **b.** A moral or legal restraint. **4.** A simple ungrooved ring, esp. a wedding ring. **5.** A narrow strip of fabric used to trim, finish, or reinforce garments. **6.** A collar or neckband. **7.** *Biol.* A chromatically or functionally differentiated strip or stripe in or on an organism. **8.** *Physics.* **a.** A range of a physical variable, as of radiation wavelength or frequency. **b.** A range of very closely spaced electron energy levels in solids, the distribution and nature of which determine the electrical properties of a material. **9.** Any of the distinct grooves on a long-playing phonograph record that contains an individual selection or separate section of a whole. **10.** The cords across the back of a book to which the quires or sheets are attached. —*vt.* **band·ed, band·ing, bands. 1.** To tie, bind, or encircle with or as if with a band. **2.** To mark or identify with or as if with a band.

**band²** (bănd) *n.* [OFr., prob. of Germanic orig.] **1.** A group of people, animals, or objects. **2.** A group of musicians playing as an ensemble. —*v.* **band·ed, band·ing, bands.** —*vt.* To assemble or unite in a group. —*vi.* To form a group : UNITE.

**band·age** (băn′dĭj) *n.* [Fr. < *bande,* band, strip.] A strip of material used to cover and protect an injury. —*vt.* **-aged, -ag·ing, -ag·es.** To apply a bandage to. **—band′ag·er** *n.*

**Band-Aid** (bănd′ād′). A trademark for an adhesive bandage with a gauze pad in the center, used for protecting minor wounds.

**ban·dan·na** or **ban·dan·a** (băn-dăn′ə) *n.* [Prob. < Port. < Hindi *bāndhnū,* a dyeing process in which cloth is knotted < *bāndhnā,* to tie < Skt. *badhnāti,* he ties.] A large handkerchief, usu. brightly colored and figured.

**band·box** (bănd′bŏks′) *n.* A lightweight cylindrical box for holding small garments and accessories.

**ban·deau** (băn-dō′) *n., pl.* **-deaux** (-dōz′) or **-deaus.** [Fr. < OFr. *bandel,* dim. of *bande,* band, strip.] **1.** A narrow band for the hair. **2.** A brassiere.

**ban·de·ri·lla** (băn′də-rē′ə, -rēl′yə) *n.* [Sp., dim. of *bandera,* banner.] A decorated barbed dart that a banderillero thrusts into the bull's neck or shoulder muscles during a bullfight.

**ban·de·ri·lle·ro** (băn′də-rē-âr′ō, -rēl-yâr′ō) *n., pl.* **-ros.** [Sp. < *banderilla,* banderilla.] One who implants the banderillas in a bullfight.

**ban·de·role** or **ban·de·rol** (băn′də-rōl′) *also* **ban·ne·rol** (băn′ə-rōl′) *n.* [Fr. < Ital. *banderoula,* dim. of *bandiera,* banner.] **1.** A narrow forked streamer or flag. **2.** A long inscribed ribbon or scroll.

**ban·di·coot** (băn′dĭ-ko͞ot′) *n.* [Telegu *pandi-kokku* : *pandi,* pig + *kokku,* rat.] **1.** Any of several ratlike marsupials of the family Peramelidae, of Australia and adjacent islands, with long tapering snouts and long hind legs. **2.** Any of several large rats of the genera *Bandicota* and *Nesokia* of southeastern Asia.

**ban·dit** (băn′dĭt) *n., pl.* **-dits** or **ban·dit·ti** (băn-dĭt′ē) [Ital. *bandito* < *bandire,* to band together, prob. of Germanic orig.] **1.** A robber. **2.** An outlaw. **3.** One who cheats or exploits others. **—ban′dit·ry** *n.*

**band·mas·ter** (bănd′măs′tər) *n.* The conductor of a musical band.

**ban·dog** (băn′dôg′, -dŏg′) *n.* [ME *band-dogge.*] A dog kept chained as a watchdog or because of its ferocity.

**ban·do·leer** or **ban·do·lier** (băn′də-lîr′) *n.* [Fr. *bandoulière* < Sp. *bandolera,* dim. of *banda,* band, prob. of Germanic orig.] A belt worn across a soldier's chest and fitted with small pockets or loops for carrying cartridges.

**ban·dore** (băn′dôr′, -dōr′) *also* **ban·do·ra** (băn-dôr′ə, -dōr′ə) *n.* [Port. *bandurra* < LLat. *pandura* < Gk. *pandoura.*] An ancient musical instrument similar to a guitar.

**band-pass filter** (bănd′păs′) *n.* An electric filter that blocks all signals but those within a selected frequency range.

**band saw** *n.* A power saw consisting of a toothed metal band coupled to and continuously driven around two wheels.

**band shell** *n.* A bandstand with a concave, almost hemispheric wall at the rear that serves as a sounding board.

**bands·man** (băndz′mən) *n.* A musician in a band.

**band·stand** (bănd′stănd′) *n.* An often roofed platform or stand for a band or orchestra.

**band·wag·on** (bănd′wăg′ən) *n.* **1.** An elaborately decorated wagon used to transport musicians in a parade. **2.** A cause or party that attracts increasing numbers of adherents.

**band·width** (bănd′wĭdth′, -wĭth′) *n.* The range of consecutive frequencies comprising a band.

**ban·dy** (băn′dē) *vt.* **-died, -dy·ing, -dies.** [Orig. unknown.] **1. a.** To toss back and forth. **b.** To hit (e.g., a ball) back and forth. **2. a.** To exchange (words). **b.** To discuss casually or frivolously. **3.** To use lightly or indiscriminately. —*adj.* Bowed or bent in an outward curve <*bandy* legs> —*n., pl.* **-dies. 1.** A game similar to modern field hockey. **2.** A stick, bent at one end, used in playing bandy.

**ban·dy-leg·ged** (băn′dē-lĕg′ĭd) *adj.* Bowlegged.
**bane** (bān) *n.* [ME, destroyer < OE *bana.*] **1.** *Archaic.* Fatal injury or ruin. **2.** A cause of death, destruction, or ruin. **3.** A deadly poison.
**bane·ber·ry** (bān′bĕr′ē) *n.* **1.** A plant of the genus *Actaea,* with white flower clusters and red or white poisonous berries. **2.** The berry of a baneberry plant.
**bane·ful** (bān′fəl) *adj.* Causing death, destruction, or ruin : HARMFUL. **—bane′ful·ly** *adv.*
**bang**[1] (băng) *n.* [Prob. of Scand. orig.] **1.** A sudden loud noise, as of an explosion. **2.** A sudden loud blow. **3.** *Informal.* A sudden burst of activity. **4.** *Slang.* A sense of excitement : THRILL <got a *bang* out of the party> *—v.* **banged, bang·ing, bangs.** *—vt.* **1.** To strike heavily, often repeatedly : BUMP. **2.** To close suddenly and loudly : SLAM. **3.** To handle noisily or violently. *—vi.* **1.** To make a sudden loud, explosive noise. **2.** To crash noisily against or into something. **—bang away. 1.** To attack persistently, esp. with questions. **2.** To work diligently. **—bang up.** To damage extensively. *—adv.* Exactly : precisely.
**bang**[2] (băng) *n.* [Perh. short for BANGTAIL.] *often* **bangs.** A fringe of hair cut short and straight across the forehead. *—vt.* **banged, bang·ing, bangs.** To cut in a bang.
**bang**[3] (băng) *n. var. of* BHANG.
**ban·ga·lore torpedo** (băng′gə-lôr′) *n.* [After *Bangalore,* India.] A piece of metal pipe filled with an explosive, chiefly used to clear a path through barbed wire or to detonate land mines.
**bang·er** (băng′ər) *n. Chiefly Brit.* A sausage.
**bang·kok** (băng′kŏk′, băng-kŏk′) *n.* [After *Bangkok,* Thailand.] A hat made of a fine straw.
**ban·gle** (băng′gəl) *n.* [Hindi *baṅgrī,* glass bracelet.] **1.** A rigid bracelet or anklet, esp. one with no clasp. **2.** An ornament that hangs from a bracelet or necklace.
**Bang's disease** (băngz) *n.* [After Bernhard L. F. *Bang* (1848–1932).] Brucellosis.
**bang·tail** (băng′tāl′) *n.* [BANG[1] + TAIL.] *Slang.* A racehorse.
**bang-up** (băng′ŭp′) *adj. Slang.* Very good : EXCELLENT.
**ba·ni** (bä′nē) *n. pl. of* BAN[2].
**ban·ian** (băn′yən) *n.* [Port. < Gujarati, *vāṇiyo* < Skt. *vāṇoija,* a merchant.] **1.** A Hindu merchant or trader belonging to a caste whose members eat no meat. **2.** *var. of* BANYAN.
**ban·ish** (băn′ĭsh) *vt.* **-ished, -ish·ing, -ish·es.** [ME *banishen* < OFr. *banir, baniss-,* of Germanic orig.] **1.** To force to leave a country or place by official decree : EXILE. **2.** To drive away : EXPEL <*banished* my doubts and fears> **—ban′ish·er** *n.* **—ban′ish·ment** *n.*
☆ ***syns:*** BANISH, DEPORT, EXILE, EXPATRIATE, EXPEL, OSTRACIZE *v. core meaning* : to force to leave a place by decree <was *banished* from the country for inciting a rebellion>
**ban·is·ter** *also* **ban·nis·ter** (băn′ĭ-stər) *n.* [Var. of BALUSTER.] **1.** A baluster. **2.** The handrail of a staircase.
**ban·jo** (băn′jō) *n., pl.* **-jos** or **-joes.** [Prob. of African orig.] A fretted stringed instrument having a narrow neck and a hollow circular body with a stretched diaphragm of vellum on which the bridge rests. **—ban′jo·ist** *n.*
**bank**[1] (băngk) *n.* [ME, of Scand. orig.] **1.** A piled-up mass. **2.** A steep natural incline. **3.** An artificial embankment. **4.** A slope of land that adjoins a body of water. **5.** A large elevated area of a sea floor. **6.** The cushion of a billiard or pool table. **7.** The lateral tilting of an aircraft executed in a turn. *—v.* **banked, bank·ing, banks.** *—vt.* **1.** To border or protect with a ridge or embankment. **2.** To pile up : AMASS. **3.** To cover (a fire), as with ashes or fresh fuel, to ensure continued low burning. **4.** To construct with a slope rising to the outside edge. **5.** To tilt (an aircraft) laterally in flight. **6.** To strike (a billiard ball) so that it rebounds from the table's cushion. *—vi.* **1.** To rise in or take the form of a bank. **2.** To tilt an aircraft laterally when turning.
**bank**[2] (băngk) *n.* [Fr. *banque* < Ital. *banca,* moneychanger's table, of Germanic orig.] **1. a.** An establishment where money is stored for saving or commercial purposes or is invested, supplied for loans, or exchanged. **b.** The offices or building in which such an establishment is located. **2. a.** The funds of a gambling casino. **b.** The funds held by a dealer or banker in a gambling game. **c.** The reserve pieces, cards, chips, or play money in a game, as poker, from which the players may draw. **3.** A supply for emergency use <an eye *bank*> **4.** A place of safekeeping or storage <a computer's memory *bank*> **5.** *Obs.* A moneychanger's table or place of business. *—v.* **banked, bank·ing, banks.** *—vt.* To deposit in a bank. *—vi.* **1.** To transact business with a bank or maintain a bank account. **2.** To operate a bank. **—bank on.** *Informal.* To rely on : DEPEND <*banked on* their acceptance>
**bank**[3] (băngk) *n.* [ME, bench < OFr. *banc,* of Germanic orig.] **1.** A set of similar or matched things arranged in a row, esp.: **a.** A set of elevators. **b.** A row of typewriter keys. **2.** A bench for rowers in a galley. **3.** The lines of printed type under a headline. *—vt.* **banked, bank·ing, banks.** To arrange or set up in a row.
▲ word history: The history of *bank* illustrates how the same word can acquire different meanings by being borrowed from different languages at different times. All three homographs of *bank* are derived from the same Germanic ancestor. *Bank*[1] was borrowed in the early Middle English period, probably from Old Norwegian, which used the word to refer to natural objects like a bank of clouds or the bank of a river. *Bank*[1] retains these senses and has extended them to objects of human manufacture. *Bank*[2] appeared in English in the late 15th century as a borrowing of Italian *banca,* "moneychanger's table," via French *banque.* Italian *banca* is derived from German *bank,* "bench," though in the process of borrowing the meaning had become more specialized. *Bank*[3] appears in Middle English with the meaning "bench" in the judicial sense; it was borrowed from Old French *banc,* a derivative of Late Latin *bancus,* which is ultimately derived from the same Germanic form that gave rise to the Old Norwegian ancestor of *bank*[1] and the German ancestor of *bank*[2].
**bank·a·ble** (băng′kə-bəl) *adj.* **1.** Acceptable to or at a bank. **2.** Guaranteed to bring profit <a *bankable* film star>
**bank acceptance** *n.* A draft or bill of exchange drawn on and accepted by a bank.
**bank account** *n.* Funds deposited in a bank that are credited to and subject to withdrawal by the depositor.
**bank annuities** *pl.n. Chiefly Brit.* Consols.
**bank·book** (băngk′bo͝ok′) *n.* A depositor's book in which the bank records his or her deposits and withdrawals.
**bank·card** (băngk′kärd′) *n.* A credit card issued by a bank.
**bank discount** *n.* The interest on a loan computed in advance and deducted at the time the loan is made.
**bank·er**[1] (băng′kər) *n.* **1.** An officer or owner of a bank. **2.** The player in charge of the bank in a gambling game.
**bank·er**[2] (băng′kər) *n.* One engaged in cod fishing on the Newfoundland banks.
**bank·er**[3] (băng′kər) *n.* [< BANK[3], bench (obs.).] A mason's or sculptor's workbench.
**banker's acceptance** *n.* A bank acceptance.
**bank holiday** *n.* **1.** A day on which banks are legally closed. **2.** *Chiefly Brit.* A legal holiday when banks are ordered to remain closed.
**bank·ing** (băng′kĭng) *n.* The business of a bank or the occupation of a banker.
**bank note** *n.* A note issued by a bank representing its promise to pay a specific sum to the bearer on demand and acceptable as money.
**bank paper** *n.* **1.** Bank notes. **2.** Commercial paper, as securities, drafts, and bills of exchange, that are acceptable by a bank.
**bank rate** *n.* The discount rate established by a country's central bank.
**bank·roll** (băngk′rōl′) *n.* **1.** A roll of paper money. **2.** *Informal.* Ready cash. *—vt.* **-rolled, -roll·ing, -rolls.** To underwrite the expense of (e.g., a business venture). **—bank′roll′er** *n.*
**bank·rupt** (băngk′rŭpt′, -rəpt) *n.* [Fr. *banqueroute* < Ital. *bancarotta* : *banca,* moneychanger's table + *rotta,* p.part. of *rompere,* to break < Lat. *rumpere.*] **1.** *Law.* A debtor who, upon voluntary petition or one invoked by his or her creditors, is judged legally insolvent and whose remaining property is administered for those creditors or distributed among them. **2.** One lacking a given resource or quality <an intellectual *bankrupt*> *—adj.* **1.** Legally declared a bankrupt : INSOLVENT. **2.** Financially ruined. **3.** Sterile : depleted <a *bankrupt* society> **4.** Destitute of a given quality <*bankrupt* of mercy> *—vt.* **-rupt·ed, -rupt·ing, -rupts.** To cause to become bankrupt. **—bank′rupt′cy** (-rŭpt′sē, -rəp-sē) *n.*
**ban·ner** (băn′ər) *n.* [ME *banere* < OFr. *baniere* < VLat. **bandaria* < LLat. *bandum,* of Germanic orig.] **1. a.** A piece of material attached to a staff and used as a standard by a sovereign, knight, or military commander. **b.** The flag of a nation, state, armed force, or sovereign. **2.** A piece of fabric bearing a legend or motto, as of a club. **3.** A headline spanning the width of a newspaper page. **4.** STANDARD 6a. *—adj.* Exceptionally good : OUTSTANDING <This was a *banner* year for car sales>
**ban·ner·et**[1] (băn′ər-ĭt, -ə-rĕt′) *also* **ban·ner·ette** (băn′ə-rĕt′) *n.* [ME *baneret* < OFr. *banerete,* dim. of *baniere,* banner.] A small banner <*bannerets* fluttering in a light breeze>
**ban·ner·et**[2] (băn′ər-ĭt, -ə-rĕt′) *n.* [ME *baneret* < OFr. < *baniere,* banner.] A feudal knight entitled to lead men into battle under his own standard and ranking between knight bachelor and baron.
**ban·ner·ette** (băn′ə-rĕt′) *n. var. of* BANNERET[1].
**ban·ner·ol** (băn′ə-rōl′) *n. var. of* BANDEROLE.
**ban·nis·ter** (băn′ĭ-stər) *n. var. of* BANISTER.
**ban·nock** (băn′ək) *n.* [ME *bannok* < OE *bannuc* < Gael. *bannach.*] *Chiefly Brit. Regional.* A usu. unleavened griddlecake made of oatmeal, barley, or wheat flour.
**banns** (bănz) *pl.n.* [ME *banes,* pl. of *ban,* proclamation, partly < OE *gebann,* and partly < OFr. *ban,* of Germanic orig.] An announcement, esp. in a church, of a forthcoming marriage.
**ban·quet** (băng′kwĭt) *n.* [OFr., dim. of *banc,* bench, of Germanic orig.] **1.** An elaborate, sumptuous repast. **2.** A ceremonial dinner honoring a guest or an occasion. *—vt. & vi.* **-quet·ed, -quet·ing, -quets.** To entertain at or partake of a banquet. **—ban′quet·er** *n.*
**banquet room** *n.* A large room suitable for banquets.

**†ban·quette** (băng-kĕt′) *n.* [Fr. < Prov. *banqueta*, dim. of *banca*, bench, of Germanic orig.] **1.** A platform lining a trench or parapet wall on which soldiers may stand when firing. **2.** *Southern U.S.* A sidewalk. **3.** A long upholstered bench placed against or built into a wall. **4.** A ledge or shelf, as on a buffet.

**ban·shee** (băn′shē) *n.* [Ir. Gael. *bean sídhe*, woman of the fairies.] A female spirit in Gaelic folklore held to presage a death in the family by wailing.

**ban·tam** (băn′təm) *n.* [After *Bantam*, Indonesia.] **1.** Any of various breeds of diminutive domestic fowl. **2.** A small, aggressive person. **—ban′tam** *adj.*

**ban·tam·weight** (băn′təm-wāt′) *n.* A boxer in the weight class of 112 to 118 pounds.

**ban·ter** (băn′tər) *n.* [Orig. unknown.] Good-humored, teasing conversation. *—v.* **-tered, -ter·ing, -ters.** *—vt.* To speak to in a playful or teasing way. *—vi.* To exchange mildly teasing remarks. **—ban′ter·er** *n.* **—ban′ter·ing·ly** *adv.*

**bant·ling** (bănt′lĭng) *n.* [Prob. alteration of G. *Bänkling*, bastard < *Bank*, bench < OHG.] A young child.

**Ban·tu** (băn′to͞o) *n., pl.* **Bantu** *or* **-tus. 1.** A member of a Negro tribe of central and southern Africa. **2.** A group of Niger-Congo languages spoken in central and southern Africa. **—Ban′tu** *adj.*

**ban·yan** *also* **ban·ian** (băn′yən) *n.* [Var. of BANIAN.] A tree, *Ficus benghalensis* of tropical India and the East Indies, with large oval leaves, reddish fruit, and many aerial roots that develop into additional trunks.

**ban·zai** (bän-zī′) *n.* [J., (may you live) ten thousand years < Chin. (Mandarin) *wan$^4$ sui$^4$* : *wan$^4$*, ten thousand + *sui$^4$*, year.] A Japanese battle cry or patriotic cheer.

**ba·o·bab** (bā′ə-băb′, bou′-) *n.* [Prob. a native word in central Africa.] A tree, *Adansonia digitata* of tropical Africa, with a trunk up to 30 feet in diameter, large pendulous white flowers, and hard-shelled fleshy fruit.

**baobab**
*Approximately 35 feet high, diameter of trunk to 30 feet*

**bap·tism** (băp′tĭz′əm) *n.* [ME *bapteme* < OFr. < LLat. *baptisma* < Gk. *baptismos* < *baptizein*, to baptize.] **1.** A Christian sacrament marked by symbolic use of water to rid the recipient of original sin and resulting in admission into Christianity. **2.** A ceremony, trial, or experience by which one is initiated, purified, or given a name. **3.** *Christian Science.* A submergence in Spirit or purification by Spirit. **—bap·tis′mal** (băp-tĭz′-məl) *adj.* **—bap·tis′mal·ly** *adv.*

**baptism of fire** *n.* **1.** A soldier's first combat experience. **2.** A severe ordeal experienced for the first time.

**Bap·tist** (băp′tĭst) *n.* **1.** A member of a Protestant denomination believing that the sacrament of baptism should be given only to adult members upon a profession of faith and usu. by immersion. **2. baptist.** One who baptizes. **—Bap′tist** *adj.*

**bap·tis·ter·y** *also* **bap·tis·try** (băp′tĭ-strē) *n., pl.* **-ies** *also* **-tries. 1.** A part of a church or a separate building used for baptizing. **2.** A font used for baptism.

**bap·tize** (băp-tīz′, băp′tīz′) *v.* **-tized, -tiz·ing, -tiz·es.** [ME *baptizen* < OFr. *baptiser* < LLat. *baptizare* < Gk. *baptizein* < *baptein*, to dip.] *—vt.* **1.** To admit into Christianity by baptism. **2. a.** To cleanse or purify. **b.** To initiate. **3.** To give a first name to. *—vi.* To administer baptism. **—bap·tiz′er** *n.*

**bar$^1$** (bär) *n.* [ME *barre* < OFr.] **1.** A rather long, straight, rigid piece of solid material used as a support, fastener, barrier, or structural or mechanical member. **2. a.** A solid oblong block of a substance <a *bar* of soap> **b.** A rectangular block of a precious metal. **3.** An impediment : obstacle. **4.** A ridge, as of sand or gravel, on a shore or streambed that is formed by tides or currents. **5.** A narrow marking, as a stripe or band. **6.** *Heraldry.* A pair of horizontal parallel lines across a shield. **7.** *Law.* **a.** Nullification, defeat, or prevention of a claim or action. **b.** The process by which this is accomplished. **8.** The railing in a courtroom enclosing the area where the judges and attorneys sit, witnesses are heard, and defendants are tried. **9.** A system of law courts. **10.** A place of judgment : TRIBUNAL. **11. a.** Lawyers as a group. **b.** The legal profession. **12.** *Mus.* **a.** A vertical line dividing a staff into equal measures. **b.** A measure. **13. a.** A counter at which drinks, esp. alcoholic drinks, and occas. food are served. **b.** An establishment or part of a building having such a counter. *—vt.* **barred, bar·ring, bars. 1.** To fasten securely with a bar. **2.** To shut in or out with or as if with bars. **3.** To obstruct or impede : BLOCK. **4.** To keep out : EXCLUDE. **5.** To mark with bars or stripes. **6.** *Law.* To stop (a claim or action) by legal objection. *—prep.* Except for : EXCLUDING <the best concert, *bar* none other>

**bar$^2$** (bär) *n.* [G. < Gk. *baros*, weight.] A unit of pressure equal to $10^5$ newtons per square meter or 0.98697 standard atmosphere.

**bar-** *pref. var. of* BARO-.

**bar·a·the·a** (băr′ə-thē′ə) *n.* [Orig. unknown.] A soft fabric of silk and cotton or silk and wool.

**barb$^1$** (bärb) *n.* [ME *barbe* < OFr., beard < Lat. *barba.*] **1.** A sharp point projecting in reverse direction to the main point of a weapon or tool, as on an arrow. **2.** A cutting remark. **3.** *Bot.* A hooked bristle or hairlike projection. **4.** One of the parallel filaments projecting from the main shaft of a feather. **5.** An Old World freshwater fish of the genus *Barbus* or *Puntius* and related genera. **6.** A linen covering for a woman's head, throat, and chin worn in medieval times. *—vt.* **barbed, barb·ing, barbs.** To furnish with a barb.

**barb$^2$** (bärb) *n.* [Fr. *barbe* < *Barbarie*, Barbary States.] **1.** A breed of horse introduced into Spain from northern Africa by the Moors. **2.** One of a breed of domestic pigeons with dark plumage.

**Bar·ba·dos cherry** (bär-bā′dōs, -dəs) *n.* A tropical and semitropical American shrub, *Malpighia glabra*, yielding edible, acid red fruit.

**bar·bar·i·an** (bär-bâr′ē-ən) *n.* [Fr. *barbarien* < Lat. *barbaria*, foreign country < *barbarus*, barbarous.] **1.** A member of a people regarded by those of another nation or group as having a primitive civilization. **2.** A fierce, brutal, or cruel person. **3.** An insensitive, uncultured person : BOOR. *—adj.* **1.** Of or relating to primitive people. **2.** Lacking refinement. **—bar·bar′i·an·ism** *n.* **—bar·bar′i·cal·ly** *adv.*

☆ **syns:** BARBARIAN, BARBARIC, BARBAROUS *adj.* All three terms can mean uncivilized and are interchangeable in <*barbarian* tribes> BARBARIAN, BARBARIC, and BARBAROUS also mean lacking refinement <*barbaric* (or *barbarian* or *barbarous*) eating habits> BARBARIC and BARBAROUS can additionally describe what is savagely violent or cruel <a *barbarous* (or *barbaric*) massacre>

**bar·bar·ic** (bär-băr′ĭk) *adj.* **1.** Of, relating to, or characteristic of barbarians. **2.** Crude or unrestrained in taste, style, or behavior.

**bar·ba·rism** (bär′bə-rĭz′əm) *n.* [OFr. *barbarisme* < Lat. *barbarismus* < Gk. *barbarismos*, foreign speech < *barbaros*, foreign.] **1.** An instance, act, trait, or custom marked by coarseness or brutality. **2. a.** The use of words or forms felt to be incorrect or nonstandard. **b.** A specific word or form so used.

**bar·bar·i·ty** (bär-băr′ĭ-tē) *n., pl.* **-ties. 1.** Harsh or cruel behavior. **2.** An inhuman act. **3.** Crudity and coarseness.

**bar·ba·rize** (bär′bə-rīz′) *vt.* & *vi.* **-rized, -riz·ing, -riz·es.** To make or become barbarous. **—bar′ba·ri·za′tion** *n.*

**bar·ba·rous** (bär′bər-əs) *adj.* [Lat. *barbarus* < Gk. *barbaros.*] **1.** Primitive in culture and customs : UNCIVILIZED. **2.** Savage : brutal. **3.** Lacking culture or refinement : COARSE. **4.** Marked by the use or occurrence of barbarisms in language. **—bar′ba·rous·ly** *adv.* **—bar′ba·rous·ness** *n.*

**Bar·ba·ry ape** (bär′bə-rē) *n.* A tailless monkey, *Macaca sylvana* of Gibraltar and northern Africa.

**bar·bas·co** (bär-băs′kō) *n., pl.* **-cos.** [Am. Sp.] Any of several tropical American trees of the genus *Lonchocarpus*, used locally as the source of a poison for killing fish.

**bar·bate** (bär′bāt′) *adj.* [Lat. *barbatus* < *barba*, beard.] Having a beard or tufted hairs similar to a beard.

**bar·be·cue** (bär′bĭ-kyo͞o′) *n.* [Am. Sp. *barbacoa* < Haitian, framework of sticks < Taino.] **1.** A grill, pit, or outdoor fireplace for roasting meat. **2. a.** A whole animal carcass or section thereof roasted or broiled over an open fire or on a spit. **b.** A usu. outdoor social gathering at which food is prepared in this way. *—vt.* **-cued, -cu·ing, -cues.** To roast, broil, or grill (meat) over live coals or an open fire.

**barbed** (bärbd) *adj.* **1.** Having barbs. **2.** Cutting : stinging <*barbed* remarks intended to hurt> **—barb′ed·ness** (bär′bĭd-nĭs) *n.*

**barbed wire** *n.* Twisted strands of fence wire with barbs at regular intervals.

**bar·bel** (bär′bəl) *n.* [Obs. Fr. < OFr., dim. of *barbe*, beard < Lat. *barba.*] **1.** One of the slender, whiskerlike sensory organs on the head of certain fishes, as catfish. **2.** Any of several Old World freshwater fish of the genus *Barbus.*

**bar·bell** (bär′bĕl′) *n.* A bar with adjustable weights at each end, lifted for sport or exercise.

**bar·bel·late** (bär′bə-lāt′, bär-bĕl′ĭt, -āt′) *adj.* [< NLat. *barbella*, dim. of Lat. *barbula*, little beard < *barba*, beard.] Having tiny, hooked bristles or hairs.

**bar·ber** (bär′bər) *n.* [ME *barbour* < OFr. < Med. Lat. *barbator* < *barba*, beard < Lat.] One who cuts hair and shaves or trims beards as an occupation. *—v.* **-bered, -ber·ing, -bers.** *—vt.* **1.** To cut the hair of. **2.** To shave or trim the beard of. *—vi.* To work as a barber.

**bar·ber·ry** (bär′bĕr′ē) *n.* [ME *berberie* < OFr. *berberis* < Ar. *barbārīs.*] Any of various shrubs of the genus *Berberis*, with small leaves, yellow flower clusters, and small orange or red berries.

**bar·ber·shop** (bär′bər-shŏp′) *n.* A barber's place of business. *—adj.*

Of, consisting of, or relating to the performance of sentimental songs in four-part harmony.

**barber's itch** *n.* A skin eruption on the neck, esp. ringworm.

**bar·bet** (bär'bĭt) *n.* [Fr. *barbu* < Lat. *barbatus*, barbate < *barba*, beard.] Any of various tropical birds of the family Capitonidae, having a broad bill bristled at the base and brightly colored plumage and related to the toucans.

**bar·bette** (bär-bĕt') *n.* [Fr., dim. of *barbe*, beard < Lat. *barba*.] **1.** A platform or mound within a fort from which guns are fired over the parapet. **2.** An armored protective cylinder around a revolving turret on a warship.

**bar·bi·can** (bär'bĭ-kən) *n.* [ME < OFr. *barbacane* < Med. Lat. *barbacana*, perh. of Ar. or Pers. orig.] A fortification, as a tower, on the approach to a castle or town, esp. one at a gate or drawbridge.

**bar·bi·cel** (bär'bĭ-sĕl') *n.* [NLat. *barbicella*, dim. of Lat. *barba*, beard.] One of many minute projections that fringe the edges of the barbules of feathers and interlock with those on adjacent barbules.

**bar·bi·tal** (bär'bĭ-tôl') *n.* [BARBIT(URIC ACID) + *-al* (as in *veronal*).] A white crystalline compound, $C_8H_{12}N_2O_3$, used as a sedative.

**bar·bi·tu·rate** (bär-bĭch'ər-ĭt, -ə-rāt', bär'bĭ-tŏŏr'ĭt, -āt', -tyŏŏr'-) *n.* [BARBITUR(IC ACID) + -ATE.] **1.** A salt or ester of barbituric acid. **2.** Any of a group of barbituric acid derivatives used as sedatives or hypnotics.

**bar·bi·tu·ric acid** (bär'bĭ-tŏŏr'ĭk, -tyŏŏr'-) *n.* [Partial transl. of G. *Barbitursäure*.] An organic acid, $C_4H_4N_2O_3$, used in making barbiturates and some plastics.

**bar·bule** (bär'byōōl) *n.* [Lat. *barbula*, dim. of *barba*, beard.] A small pointed projection, esp. one of the small projections fringing the edges of the barbs of feathers.

**barb·wire** (bärb'wīr') *n.* Barbed wire.

**bar·ca·role** *also* **bar·ca·rolle** (bär'kə-rōl') *n.* [Fr. < Ital. *barcaruola* < *barcaruolo*, gondolier < *barca*, boat < LLat.] **1.** A Venetian gondolier's song with a rhythm suggestive of rowing. **2.** A musical composition imitating a barcarole.

**bar chart** *n.* A bar graph.

**bard[1]** (bärd) *n.* [ME < Ir. Gael. *bàrd* and Welsh *bardd*.] **1.** One of an ancient Celtic order of singing poets who composed and recited verses on the legends and history of their tribes. **2.** An exalted national poet. **—bard'ic** (bär'dĭk) *adj.*

**bard[2]** *also* **barde** (bärd) *n.* [ME *barde* < OFr., prob. < OItal. *barda* < Ar. *barda'ah*, stuffed packsaddle.] Armor used to protect or decorate a horse. *—vt.* **bard·ed, bard·ing, bards.** To equip with bards.

**bare[1]** (bâr) *adj.* **bar·er, bar·est.** [ME *bar* < OE *bær*.] **1.** Lacking the appropriate or usual covering or clothing : NAKED. **2.** Exposed to view : UNDISGUISED. **3.** Lacking the usual equipment, furnishings, or ornamentation <walls *bare* of paintings> **4.** Lacking addition, adornment, or qualification : PLAIN <Give me the *bare* facts.> **5.** Just sufficient <the *bare* necessities> **6.** *Obs.* Bareheaded. *—vt.* **bared, bar·ing, bares.** To make bare. **—bare'ness** *n.*

☆ **syns:** BARE, BALD, NAKED, NUDE *adj. core meaning*: without the usual covering. BARE, BALD, NAKED, and NUDE can apply to persons or things <a *bare* arm><*bare* fields><a *bald* head><*bald* wintery hills><*naked* feet><*naked* tree branches><a *nude* cherub><a *nude* statue> BARE, BALD, and sometimes NAKED also describe what is blunt or without qualification <the *bare* facts><the *bald* truth><a *naked* lie>

**bare[2]** (bâr) *v. Archaic. var. p.t. of* BEAR[1].

**bare·back** (bâr'băk') *also* **bare·backed** (-băkt') *adj. & adv.* On an animal, as a horse or pony, without a saddle <*bareback* riding><riding *bareback*>

**bare bones** *pl.n.* The most basic elements <outlined the *bare bones* of the plan> **—bare'-bones'** (bâr'bōnz') *adj.*

**bare·faced** (bâr'fāst') *adj.* **1. a.** Having no covering over the face. **b.** Having no beard. **2.** Without disguise : UNCONCEALED. **3.** BALD 5. **—bare'fac'ed·ly** (-fā'sĭd-lē, -fāst'lē) *adv.* **—bare'fac'ed·ness** *n.*

**bare·foot** (bâr'fŏŏt') *also* **bare·foot·ed** (-fŏŏt'ĭd) *adj. & adv.* Wearing nothing on the feet.

**ba·rege** *also* **ba·rège** (bə-rĕzh') *n.* [Fr. *barège* < *Barèges*, a town in France.] A sheer fabric woven of silk or cotton and wool.

**bare·hand·ed** (bâr'hăn'dĭd) *adj. & adv.* **1.** Having no covering on the hands. **2.** Unaided by tools or weapons. **—bare'hand'ed·ness** *n.*

**bare·head·ed** (bâr'hĕd'ĭd) *adj. & adv.* Having no covering on the head. **—bare'head'ed·ness** *n.*

**bare·leg·ged** (bâr'lĕg'ĭd, -lĕgd') *adj. & adv.* Having no covering over the legs. **—bare'leg'ged·ness** *n.*

**bare·ly** (bâr'lē) *adv.* **1.** By a very little : HARDLY <could *barely* see the way> **2.** In a scanty way : SPARSELY <a *barely* furnished efficiency apartment>

**barf** (bärf) *vi.* **barfed, barf·ing, barfs.** [Orig. unknown.] *Slang.* To vomit.

**bar·fly** (bär'flī') *n. Slang.* A habitué of bars.

**bar·gain** (bär'gĭn) *n.* [ME *bargaine* < OFr., of Germanic orig.] **1. a.** An agreement or contract, esp. one involving purchase and sale of goods or services. **b.** The terms of such an agreement. **c.** The property acquired or services rendered as a result of such an agreement. **2.** Something offered or acquired at a price advantageous to the buyer. *—v.* **-gained, -gain·ing, -gains.** *—vi.* **1.** To negotiate the terms of an agreement. **2.** To arrive at an agreement. *—vt.* To exchange : trade. **—bargain for.** To count on : EXPECT. **—into** (*or* **in**) **the bargain.** More than what is expected. **—bar'gain·er** *n.*

**barge** (bärj) *n.* [ME, boat < OFr., poss. < Lat. *barca*.] **1.** A long, large, usu. flat-bottomed boat that is unpowered and towed by other craft, used for transporting freight. **2.** A large pleasure boat. **3.** A flag officer's powerboat. *—v.* **barged, barg·ing, barg·es.** *—vt.* To transport by barge. *—vi.* **1.** To move about clumsily. **2.** To enter rudely and abruptly : INTRUDE.

**barge·board** (bärj'bôrd', -bōrd') *n.* [Orig. unknown.] An often ornately carved board that is attached along the projecting edge of a gable roof.

**bar·gee** (bär-jē') *n. Chiefly Brit.* A bargeman.

**bar·gel·lo** (bär-zhĕl'ō) *n., pl.* **-los.** [After the *Bargello*, a museum in Florence, Italy, which contains chairs upholstered in fabric worked in this stitch.] A needlepoint stitch producing zigzag lines.

**barge·man** (bärj'mən) *n.* A crew member or the master of a barge.

**bar graph** *n.* A graph of parallel, usu. vertical, bars or rectangles with lengths proportional to specified quantities in a set of data.

**bar·hop** (bär'hŏp') *vi.* **-hopped, -hop·ping, -hops.** To go from one bar to another during an evening.

**bar·i·at·rics** (băr'ē-ăt'rĭks) *n.* (*sing. in number*). A branch of medicine dealing with the treatment of obesity. **—bar'i·at'ric** *adj.* **—bar'i·a·tri'cian** (-ə-trĭsh'ən) *n.*

**ba·ril·la** (bə-rēl'yə, -rē'yə) *n.* [Sp. *barrilla*.] **1.** An Old World plant, *Salsola kali* or *S. soda*, or a plant similar to it, *Halogeton soda*, burned to generate a form of sodium carbonate. **2.** The sodium carbonate obtained from a barilla.

**bar·ite** (bâr'īt', băr'-) *n.* [Gk. *barus*, heavy + -ITE.] A colorless crystalline mineral of barium sulfate that is the primary source of barium chemicals.

**bar·i·tone** *also* **bar·y·tone** (băr'ĭ-tōn') *n.* [Ital. *baritono* < Gk. *barutonos*, deep sounding : *barus*, heavy + *tonus*, tone.] **1. a.** A male singer or voice having a range higher than a bass and lower than a tenor. **b.** A part written for a voice with such a range. **2.** A brass wind instrument with a range similar to that of a baritone.

**bar·i·um** (bâr'ē-əm, băr'-) *n.* [BAR(YTA) + -IUM.] *Symbol* **Ba** A soft, silvery-white metal, used to deoxidize copper, in various alloys, and in rat poison; atomic number 56; atomic weight 137.34. **—bar'ic** (-ĭk) *adj.*

**barium sulfate** *n.* A fine white powder, $BaSO_4$, used as a pigment, as a filler for textiles, rubbers, and plastics, and as an indicator in x-ray photography of the digestive tract.

**barium yellow** *n.* A pigment made of barium chromate, $BaCrO_4$.

**bark[1]** (bärk) *n.* [< ME *berken*, to bark < OE *beorcan*.] **1.** The harsh, abrupt sound made by a dog. **2.** A sound, as a cough, that is similar to a bark. *—v.* **barked, bark·ing, barks.** *—vi.* **1.** To utter a bark. **2.** To make a sound similar to a bark. **3.** To speak sharply : SNAP. **4.** *Informal.* To work as a barker. *—vt.* To utter in a loud, harsh voice.

**bark[2]** (bärk) *n.* [ME < ON *bǫrkr*.] **1.** The outer covering of the woody stems, branches, roots, and main trunks of trees and other woody plants. **2.** A specific kind of bark used for a special purpose, as in tanning or medicine. *—vt.* **barked, bark·ing, barks. 1.** To remove bark from (a tree or log). **2.** To rub off the skin of : ABRADE. **3.** To treat medically, tan, or dye using bark.

**bark[3]** *also* **barque** (bärk) *n.* [ME *barke*, boat < OFr. *barque* < OItal. *barca* < LLat.] **1.** A sailing ship with from three to five masts, all of them square-rigged except the after mast which is fore-and-aft rigged. **2.** A small sailing vessel.

**bark beetle** *n.* Any of various small insects of the family Scolytidae that damage trees by boring along the surface of the wood.

**bar·keep·er** (bär'kē'pər) *also* **bar·keep** (-kēp') *n.* **1.** An owner or manager of a bar selling alcoholic beverages. **2.** A bartender.

**bar·ken·tine** *also* **bar·quen·tine** (bär'kən-tēn') *n.* [Prob. a blend of BARK[3] and BRIGANTINE.] A sailing ship with from three to five masts of which only the foremast is square-rigged, the others being fore-and-aft rigged.

**bark·er[1]** (bär'kər) *n.* **1.** One that barks. **2.** *Informal.* An employee who stands before the entrance to a show and solicits customers with a loud, colorful sales pitch.

**bark·er[2]** (bär'kər) *n.* One that removes bark from trees or logs or prepares it for tanning.

**bark·y** (bär'kē) *adj.* **-i·er, -i·est.** Covered with, containing, or resembling bark.

**bar-le-duc** *also* **Bar-le-Duc** (bär'lĭ-dŏŏk') *n.* [After *Bar-le-Duc*, France.] A savory preserve made of white currants or gooseberries.

**bar·ley** (bär'lē) *n.* [ME *barli* < OE *bærlic*.] **1.** A cereal grass, *Hordeum vulgare*, having bearded flower spikes with edible seeds. **2.** The grain of barley, used as food and in making beer, ale, and whiskey.

**bar·ley·corn** (bär'lē-kôrn') *n.* **1.** The grain of barley. **2.** A unit of measure equal to the width of a grain of barley, or approx. $^1/_3$ inch.

**barley sugar** *n.* A clear, hard candy made by boiling down sugar.

**barm** (bärm) *n.* [ME *berme* < OE *beorma*.] The yeasty foam that rises to the surface of fermenting malt liquors.

**bar·maid** (bär'mād') *n.* A woman who serves drinks in a bar.

**bar·man** (bär′mən) *n.* A bartender.

**Bar·me·cid·al** (bär′mĭ-sīd′l) *also* **Bar·me·cide** (bär′mĭ-sīd′) *adj.* [After *Barmecide,* a nobleman in *The Arabian Nights,* who served a beggar an imaginary feast.] Apparently plentiful or abundant <a *Barmecidal* feast>

**bar mitz·vah** (bär mĭts′və) *n.* [Heb. *bar mitzvāh* : *bar,* son + *mitzvah,* commandment.] **1.** A 13-year-old Jewish male, considered an adult and responsible for his moral and religious duties. **2.** The ceremony that initiates and recognizes a boy as a bar mitzvah. —*vt.* **-vahed, -vah·ing, -vahs.** To confirm by bar mitzvah.

**barm·y** (bär′mē) *adj.* **-i·er, -i·est. 1.** Full of barm : FOAMY. **2.** *Chiefly Brit.* Insane : crazy.

**barn** (bärn) *n.* [ME *bern* < OE *berern* : *bere,* barley + *ern,* house.] **1.** A farm building used for storing farm products and sheltering livestock. **2.** A shed for housing vehicles, as railroad cars. **3.** *Physics.* A unit of area equal to $10^{-24}$ square centimeter, used to express nuclear cross sections.

**bar·na·cle** (bär′nə-kəl) *n.* [ME *bernak,* a kind of goose (from the belief that the geese were produced from the shellfish) < Med. Lat. *bernaca.*] **1.** Any of various marine crustaceans of the order Cirripedia that in the adult stage form a hard shell and remain attached to a submerged surface. **2.** The barnacle goose. **—bar′na·cled** *adj.*

▲ word history: The history of *barnacle* is a tangled story. The word does not appear earlier than the 13th century and its ultimate etymology is unknown. The word was originally applied only to the bird now called the *barnacle goose;* its application to the crustacean is of considerably later date. Because the bird, which summers in the arctic, was never observed to breed, fantastic theories were elaborated to account for its genesis. One theory held that the bird grew from the little shells found on trees or driftwood by the seashore. These shells are what are now known as *barnacles.* The heat of the sun, which Shakespeare tells us bred crocodiles from the mud of the Nile, was also supposed to cause the small crustaceans to grow into the large birds.

**barnacle goose** *n.* A waterfowl, *Branta leucopsis* of northern Europe and Greenland, having black, white, and gray plumage.

**barn dance** *n.* A social gathering, often held in a barn, with music and square dancing.

**barn owl** *n.* A predatory nocturnal bird, *Tyto alba,* with light-brown and white plumage, that often frequents barns.

**barn·storm** (bärn′stôrm′) *vi.* **-stormed, -storm·ing, -storms. 1.** To travel around the countryside presenting plays, lecturing, or making political speeches. **2.** To appear at county fairs and carnivals in exhibitions of stunt flying and parachute jumping. **—barn′storm′er** *n.*

**barn swallow** *n.* A widely distributed bird, *Hirundo rustica,* with a deeply forked tail, a dark-blue back, and tan underparts.

**barn·yard** (bärn′yärd′) *n.* The often fenced-in area surrounding a barn. —*adj.* Smutty <*barnyard* humor>

**baro-** or **bar-** *pref.* [< Gk. *baros,* weight.] Weight : pressure <*barometer*>

**bar·o·gram** (băr′ə-grăm′) *n.* A graphic record produced by a barograph.

**bar·o·graph** (băr′ə-grăf′) *n.* A self-registering barometer. **—bar′o·graph′ic** *adj.*

**ba·rom·e·ter** (bə-rŏm′ĭ-tər) *n.* **1.** An instrument for measuring atmospheric pressure, used in weather forecasting and in determining elevation. **2.** An indicator of fluctuations <economic *barometers* such as interest rates> **—bar′o·met′ric** (băr′ə-mĕt′rĭk), **bar′o·met′ri·cal** *adj.* **—bar′o·met′ri·cal·ly** *adv.* **—ba·rom′e·try** *n.*

**bar·on** (băr′ən) *n.* [ME < OFr., prob. of Germanic orig.] **1. a.** A feudal tenant holding his rights and title directly from the monarch or another feudal superior. **b.** A lord or nobleman : PEER. **2. a.** A member of the lowest rank of nobility in Great Britain, certain European countries, and Japan. **b.** The rank or title of such a nobleman. **3.** One with great wealth, power, and influence in a specified sphere <oil *barons* and other captains of industry>

**bar·on·age** (băr′ə-nĭj) *n.* **1.** The rank, title, or dignity of a baron. **2.** The peers of a kingdom.

**bar·on·ess** (băr′ə-nĭs) *n.* **1.** The wife or widow of a baron. **2.** A woman holding a barony in her own right.

**bar·on·et** (băr′ə-nĭt, băr′ə-nĕt′) *n.* [ME, dim. of *baron,* baron.] **1.** A British hereditary title of honor, ranking next below a baron, held by a commoner. **2.** The bearer of a baronet.

**bar·on·et·age** (băr′ə-nĭ-tĭj, -nĕt′ĭj) *n.* **1.** The rank or dignity of a baronet : BARONETCY. **2.** Baronets as a group.

**bar·on·et·cy** (băr′ə-nĭt-sē, -nĕt′sē) *n.* BARONETAGE 1.

**ba·rong** (bə-rông′, -rŏng′) *n.* [Native word in the Philippines.] A large, broad-bladed knife used by the Moros of the Philippines.

**ba·ro·ni·al** (bə-rō′nē-əl) *adj.* **1.** Of or relating to a baron or barony. **2.** Befitting a baron : STATELY <the *baronial* mansions of Newport>

**bar·o·ny** (băr′ə-nē) *n., pl.* **-nies. 1.** The domain of a baron. **2.** BARONAGE 1.

**ba·roque** (bə-rōk′) *adj.* [Fr. < Ital. *barocco.*] **1.** Of, relating to, or characteristic of an artistic and architectural style developed in Europe from about 1550 to 1700 and typified by elaborate ornamentation, as scrolls and curves. **2.** Of, relating to, or characteristic of a style of musical composition that flourished in Europe from about 1600 to 1750, marked by chromaticism, strict forms, and elaborate ornamentation. **3.** Ornate or flamboyant in style <a *baroque* luxuriance of language> **4.** Irregular in shape <a strand of *baroque* pearls> **—ba·roque′** *n.* **—ba·roque′ly** *adv.*

**bar·o·re·cep·tor** (băr′ə-rĭ-sĕp′tər) *n.* A sensory nerve ending, as in the carotid sinus, that is sensitive to pressure change.

**ba·rouche** (bə-ro͞osh′) *n.* [G. *Barutsche* < Ital. *biroccio* < LLat. *birotus,* two-wheeled : *bi-,* two + *rota,* wheel.] A four-wheeled carriage with a collapsible top, two double seats inside opposite each other, and a box seat outside in front for the driver.

**barque** (bärk) *n. var. of* BARK[3].

**bar·quen·tine** (bär′kən-tēn′) *n. var. of* BARKENTINE.

**bar·rack[1]** (băr′ək) *vt.* **-racked, -rack·ing, -racks.** To house in barracks.

**bar·rack[2]** (băr′ək) *vt.* **-racked, -rack·ing, -racks.** [Orig. unknown.] *Chiefly Brit.* To shout or jeer at. **—bar′rack·er** *n.*

**bar·racks** (băr′əks) *n. (sing. or pl. in number).* [Fr. *baraques* < Sp. *barraca,* soldier's tent < Catalan.] **1.** A building or group of buildings used to house soldiers. **2.** A large unadorned building used for temporary occupancy.

**barracks bag** *n.* A soldier's cloth bag, usu. with a drawstring, for the storage of clothing or laundry.

**bar·ra·coon** (băr′ə-ko͞on′) *n.* [Sp. *barracón,* aug. of *barraca,* hut.] A barracks in which slaves or convicts were temporarily confined.

**bar·ra·cu·da** (băr′ə-ko͞o′də) *n., pl.* **barracuda** or **-das.** [Mex. Sp.] Any of various voracious, chiefly tropical marine fishes of the genus *Sphyraena,* having long, narrow bodies and projecting jaws with fanglike teeth.

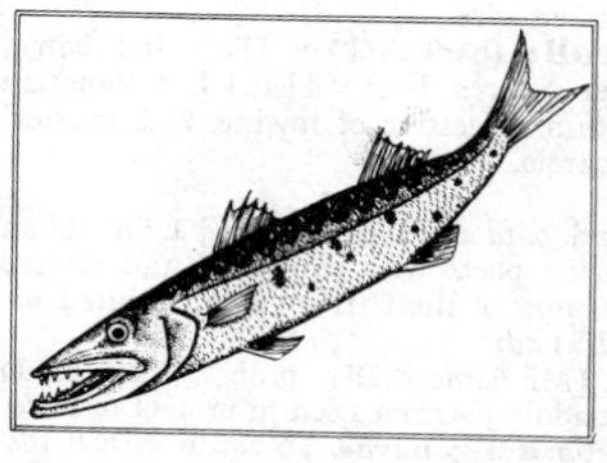

**barracuda**
*5–6 feet long*

**bar·rage[1]** (bär′ĭj) *n.* [Fr. < *barrer,* to bar < *barre,* bar < OFr.] An artificial obstruction in a watercourse, esp. one built to promote irrigation : DAM.

**bar·rage[2]** (bə-räzh′) *n.* [Fr. *(tir de) barrage,* barrier (fire).] **1.** A heavy curtain of artillery fire directed in front of friendly troops to screen and protect them. **2.** A rapid, concentrated discharge of projectiles, as from small arms. **3.** A concentrated outpouring. —*vt.* **-raged, -rag·ing, -rag·es.** To direct a barrage at.

☆ ***syns:*** BARRAGE, BOMBARDMENT, BURST, CANNONADE, FUSILLADE, HAIL, SALVO, SHOWER, STORM, VOLLEY *n. core meaning* : a concentrated outpouring <a *barrage* of bullets><a *barrage* of complaints>

**barrage balloon** *n.* A balloon anchored singly or in series over a military objective to block passage of enemy aircraft.

**bar·ra·mun·da** (băr′ə-mŭn′də) *also* **bar·ra·mun·di** (-dē) *n., pl.* **barramunda** or **-das** *also* **barramundi** or **-dis.** [Native word in Australia.] An Australian food fish, as the river fish *Scleropages leichhardtdii* or the lungfish *Neoceratodus forsteri.*

**†bar·ran·ca** (bə-răng′kə) *n.* [Sp., prob. < Iberian.] *Southwestern U.S.* A deep ravine : GORGE.

**bar·ra·tor** *also* **bar·ra·ter** (băr′ə-tər) *n.* [ME *baratour* < OFr. *barateor,* swindler < *barater,* to cheat.] *Law.* One that commits barratry.

**bar·ra·try** (băr′ə-trē) *n., pl.* **-tries.** [ME *barratrie,* the sale of church offices < OFr. *baraterie,* deception < *barater,* to cheat.] **1.** *Law.* Persistent incitement of lawsuits. **2.** An unlawful breach of duty on the part of a ship's master or crew resulting in injury to the ship's owner. **3.** Sale or purchase of positions in church or state. **—bar′ra·trous** (-trəs) *adj.* **—bar′ra·trous·ly** *adv.*

**barred owl** *n.* A North American owl, *Strix varia,* having barred, brownish plumage, a streaked belly, and a strident hoot.

**bar·rel** (băr′əl) *n.* [ME *barel* < OFr. *baril.*] **1.** A large cylindrical container, usu. made of wooden staves bound together with hoops, having a flat top and bottom of equal diameter. **2.** The quantity that a barrel will hold. **3.** A unit of capacity or volume. **4.** The cylindrical part or hollow shaft of a mechanism, as: **a.** The metal, cylindrical part of a firearm through which the bullet travels. **b.** A cylinder containing a movable piston. **c.** The drum of a capstan. **d.** The cylinder within the mechanism of a timepiece that contains the mainspring. **5.** *Informal.* A large quantity <had a *barrel* of fun> —*v.* **-reled, -rel·ing, -rels** or **-relled, -rel·ling, -rels.** —*vt.* To put or pack in a barrel. —*vi. Slang.* To move at a high speed <*barreling* down the road at 90 m.p.h.>

**barrel chair** *n.* A large, upholstered chair having a high, rounded back resembling a half barrel.

**bar·rel·house** (băr′əl-hous′) *n.* **1.** A disreputable drinking establishment. **2.** An early style of jazz characterized by free group improvisation and an accented two-beat rhythm.

**barrel organ** *n.* A portable musical instrument operated by the action of a revolving barrel with pegs or pins that open air valves leading from a bellows to a series of pipes.

**barrel roll** *n.* A flight maneuver in which an aircraft makes a complete rotation on its longitudinal axis while approx. maintaining its original direction.

**bar·ren** (băr′ən) *adj.* [ME *barreine* < OFr. *baraigne.*] **1. a.** Producing no offspring : CHILDLESS. **b.** Incapable of producing offspring : STERILE. **2.** Lacking vegetation <*barren* fields> **3.** Unproductive of gains or results : UNPROFITABLE <*barren* efforts> **4.** Lacking a specified quality <writing *barren* of original thoughts> **5.** Not lively or interesting : DULL. —*n. often* **barrens.** A tract of unproductive land, often with scrubby trees. **—bar′ren·ly** *adv.* **—bar′ren·ness** *n.*

**barren strawberry** *n.* A low-growing plant, *Waldsteinia fragarioides* of eastern North America, with yellow flowers and small, dry, inedible fruit.

**bar·rette** (bə-rĕt′, bä-) *n.* [Fr. < dim. of *barre,* bar < OFr.] A small clasp used by women and girls for holding the hair in place.

**bar·ri·cade** (băr′ĭ-kād′, băr′ĭ-kād′) *n.* [Fr. < OFr. *barrique,* barrel < Sp. *barrica.*] **1.** A structure set up across a route of access for defense or the obstruction of passage. **2.** An obstruction : barrier. —*vt.* **-cad·ed, -cad·ing, -cades. 1.** To block with a barricade. **2.** To keep in or out by a barricade. **—bar′ri·cad′er** *n.*

**bar·ri·er** (băr′ē-ər) *n.* [ME *barrer* < OFr. *barriere* < LLat. *barraria* < *barra,* bar.] **1.** A structure, as a fence or wall, built to bar passage. **2.** Something that hinders or restricts <spoke different languages, a real *barrier* to mutual understanding> **3.** A boundary : limit. **4.** Something immaterial that separates or holds apart. **5.** A movable gate that keeps racehorses in line before the start of a race. **6. barriers.** The palisades or fences enclosing the lists of a medieval tournament. **7.** *Geol.* A section of the Antarctic ice shelf that extends beyond the coastline, resting partly on the ocean floor.

**barrier reef** *n.* A long, narrow ridge of coral or rock parallel to and relatively close to a coastline, separated from it by a lagoon too deep for coral growth.

**bar·ring** (bär′ĭng) *prep.* Apart from the occurrence of : EXCEPTING.

**bar·ri·o** (bä′rē-ō′, băr′-) *n., pl.* **-os.** [Sp. < Ar. *barrī,* of an open area < *barr,* open area.] **1.** An enclave, ward, or district in a Spanish-speaking country. **2.** A chiefly Spanish-speaking community or neighborhood in a U.S. city.

**bar·ris·ter** (băr′ĭ-stər) *n.* [Prob. < BAR[1].] *Chiefly Brit.* An attorney admitted to plead at the bar in the superior courts.

**bar·room** (bär′ro͞om′, -ro͝om′) *n.* A room or building in which alcoholic beverages are sold at a counter or bar.

**bar·row[1]** (băr′ō) *n.* [ME *barowe* < OE *bearwe.*] **1.** A flat, rectangular tray or cart with handles at each end. **2.** A wheelbarrow.

**bar·row[2]** (băr′ō) *n.* [ME *borewe* < OE *beorg.*] A large mound of earth or stones placed over an ancient burial site.

**bar·row[3]** (băr′ō) *n.* [ME *barow* < OE *bearg.*] A pig castrated before reaching sexual maturity.

**bar sinister** *n.* **1.** *Heraldry.* A charge held to signify bastardy. **2.** A hint or proof of illegitimate birth.

**bar·tend·er** (bär′tĕn′dər) *n.* One who mixes and serves alcoholic drinks at a bar.

**bar·ter** (bär′tər) *v.* **-tered, -ter·ing, -ters.** [ME *barteren,* prob. < OFr. *berator.*] —*vi.* To trade goods or services without exchanging money. —*vt.* To trade (goods or services) without exchanging money. —*n.* **1.** An act or instance of bartering. **2.** Something bartered. **—bar′ter·er** *n.*

**Bar·tho·lin's gland** (bär′tl-ĭnz, -thə-lĭnz) *n.* [After Kaspar *Bartholin* (1655–1738).] Either of two small compound racemose glands located on either side of the lower vagina that secrete mucus during coitus.

**bar·ti·zan** *also* **bar·ti·san** (bär′tĭ-zən, bär′tĭ-zăn′) *n.* [Alteration of *bratticing,* timberwork < BRATTICE.] A small overhanging turret on a tower or wall. **—bar′ti·zaned** *adj.*

**Bart·lett** (bärt′lĭt) *n.* [After Enoch *Bartlett* (1779–1860).] A widely grown variety of pear with large, juicy yellow fruit.

**bar·y·cen·ter** (băr′ĭ-sĕn′tər) *n.* [Gk. *barus,* heavy + CENTER.] *Physics.* Center of mass.

**bar·y·on** (băr′ē-ŏn′) *n.* [Gk. *barus,* heavy + -ON.] Any of a family of subatomic particles, including the nucleon and hyperon multiplets, that participate in strong interactions, have half-integral spins, and are gen. more massive than mesons. **—bar′y·on′ic** *adj.*

**baryon number** *n.* A quantum number equal to the difference between the number of baryons and the number of antibaryons in a system of subatomic particles.

**bar·y·sphere** (băr′ə-sfîr′) *n.* CENTROSPHERE 2.

**ba·ry·ta** (bə-rī′tə) *n.* [NLat. < Gk. *barutēs,* weight < *barus,* heavy.] Any of several barium compounds, as barium sulfate.

**ba·ry·tes** (bə-rī′tēz′) *n.* [Gk. *barutēs,* weight < *barus,* heavy.] Barite.

**bar·y·tone** (băr′ĭ-tōn′) *n. var. of* BARITONE.

**bas·al** (bā′səl, -zəl) *adj.* **1.** Of, relating to, located at, or forming a base. **2.** Fundamental : basic. **—bas′al·ly** *adv.*

**basal metabolic rate** *n.* The rate at which energy is used by an organism at complete rest, measured in humans by the heat given off per unit time.

**basal metabolism** *n.* The least amount of energy required to maintain vital functions in an organism at complete rest.

**ba·salt** (bə-sôlt′, bā′sôlt′) *n.* [Lat. *basaltes,* alteration of *basanites (lapis),* touchstone < Gk. *basanitēs* < *basanos,* of Egypt. orig.] A hard, dense, dark, often glassy volcanic rock composed chiefly of plagioclase, augite, and magnetite. **—ba·sal′tic** (-sôl′tĭk) *adj.*

**bas·cule** (băs′kyo͞ol) *n.* [Fr., seesaw : *bas,* low + *cul,* bottom.] A device, as a drawbridge, counterbalanced so that when one end is lowered the other is raised.

**base[1]** (bās) *n.* [ME < OFr. < Lat. *basis* < Gk.] **1.** The lowest or bottom part. **2.** A supporting layer or part : FOUNDATION <a *base* of solid granite supporting the tower> **3.** A fundamental principle or underlying concept : BASIS. **4.** A chief constituent. **5.** The fact, observation, or premise from which a reasoning process is begun. **6.** *Baseball.* Any one of the four corners of an infield, marked by a bag or plate, that must be touched by a runner before he or she can score a run. **7.** A center of organization, supply, or activity : HEADQUARTERS. **8. a.** A fortified center of operations. **b.** A supply center for a large force. **9.** The lowest part of a structure, as a wall, regarded as a separate architectural unit. **10.** *Heraldry.* The lower part of a shield. **11.** A morpheme or morphemes considered as a form to which affixes or other bases may be added. **12.** *Math.* **a.** The side or face of a geometric figure to which an altitude is or is thought to be drawn. **b.** The number that is raised to various powers to generate the principal counting units of a number system. **c.** The number raised to the logarithm of a designated number in order to produce that designated number. **13.** A line used as a reference for computations or measurement. **14.** *Chem.* **a.** Any of a large class of compounds, including the hydroxides and oxides of metals, having a bitter taste, a slippery solution, the ability to turn litmus blue, and the ability to react with acids to form salts. **b.** A molecular or ionic substance capable of combining with a proton to form a new substance. **c.** A substance that provides a pair of electrons for a covalent bond with an acid. —*adj.* **1.** Forming or serving as a base. **2.** Located at or close to the base. —*vt.* **based, bas·ing, bas·es. 1.** To form or make a base for. **2.** To find a basis for : ESTABLISH. **—off base. 1.** Badly mistaken. **2.** Unprepared : unawares.

**base[2]** (bās) *adj.* **bas·er, bas·est.** [ME *bas* < OFr., low < Med. Lat. *bassus.*] **1.** Having or resulting from low moral standards <a *base* act><*base* instincts> **2.** Inferior in quality or value. **3.** Containing inferior substances <*base* metals> **4.** Valueless or greatly depreciated in value : DEBASED. **5.** *Archaic.* Of low birth or rank. **6.** *Obs.* Short in stature. —*n. Obs.* BASS[2]. **—base′ly** *adv.* **—base′ness** *n.*

**base·ball** (bās′bôl′) *n.* **1.** A game played with a bat and ball by two opposing teams of nine players, each team playing alternately in the field and at bat, and the players at bat having to run a course of four bases laid out in a diamond pattern in order to score. **2.** The ball used in baseball.

**base·board** (bās′bôrd′, -bōrd′) *n.* A molding that conceals the joint between an interior wall and a floor.

**base·born** (bās′bôrn′) *adj.* **1.** Being of humble birth. **2.** Born of unwed parents : ILLEGITIMATE. **3.** Contemptible : ignoble.

**base·burn·er** (bās′bûr′nər) *n.* A stove or furnace that automatically replenishes consumed coal or other fuel from above.

**Base Exchange.** A trademark for a Post Exchange on a naval or air force base.

**base hit** *n. Baseball.* A hit by which the batter reaches base safely without an error or force play being made.

**base·less** (bās′lĭs) *adj.* Devoid of basis or foundation in fact : UNFOUNDED.

**base level** *n.* The lowest level to which a land surface can be reduced by the action of running water.

**base line** *n.* **1.** A line serving as a base, as for measurement. **2.** *Baseball.* An area within which a base runner must stay when running between successive bases. **3.** A line bounding each back end of a tennis court.

**base·man** (bās′mən) *n. Baseball.* A player assigned to first, second, or third base.

**base·ment** (bās′mənt) *n.* **1.** The substructure or foundation of a building. **2.** The lowest habitable story of a building, usu. below ground level.

**basement membrane** *also* **basement lamina** *n.* A thin, primarily collagenous, delicate layer of connective tissue underlying the epithelium.

**ba·sen·ji** (bə-sĕn′jē) *n.* [Bantu.] A dog orig. bred in Africa that has a short smooth coat and does not bark.

**base pair** *n.* One of the pairs of compounds, as adenine and thymine, that along with a hydrogen bond form the connections between the complementary strands of DNA.

**base runner** *n. Baseball.* A member of the team at bat who has safely reached or is trying to reach a base.

**ba·ses** (bā'sēz') *n. pl. of* BASIS.

**bash** (băsh) *vt.* **bashed, bash·ing, bash·es.** [Orig. unknown.] *Informal.* To strike with a heavy blow. —*n.* **1.** *Informal.* A heavy blow. **2.** *Slang.* A large, often lavish party.

**ba·shaw** (bə-shô') *n. Obs.* A pasha.

**bash·ful** (băsh'fəl) *adj.* [(A)BASH + -FUL.] **1.** Timid : shy <a *bashful* child> **2.** Marked by, showing, or resulting from social shyness or self-consciousness <a *bashful* smile> **—bash'ful·ly** *adv.* **—bash'ful·ness** *n.*

**basi-** or **baso-** *pref.* [< Lat. *basis,* base < Gk.] **1.** Base : lower part <*basi*petal> **2.** Chemical base : chemically basic <*baso*phil>

**ba·sic** (bā'sĭk) *adj.* **1.** Of, relating to, or forming a base : FUNDAMENTAL <*basic* principles of grammar> **2.** Of, being, or functioning as a starting point. **3.** *Chem.* **a.** Producing, resulting from, or relating to a base. **b.** Containing a base, esp. in excess of acid. **4.** *Geol.* Containing little silica, as igneous rocks. **—ba'sic** *n.* **—ba'si·cal·ly** *adv.* **—ba·sic'i·ty** (-sĭs'ĭ-tē) *n.*

**BA·SIC** (bā'sĭk) *n.* [B(EGINNER'S) + A(LL-PURPOSE) + S(YMBOLIC) + I(NSTRUCTION) + C(ODE).] *Computer Sci.* A common programming language often used with remote or time-sharing centers.

**ba·si·chro·mat·ic** (bā'sĭ-krō-măt'ĭk) *adj.* That is easily stained with basic dye.

**basic oxide** *n.* A metallic oxide that is a base or that forms a hydroxide if combined with water.

**basic process** *n.* A method of manufacturing steel that uses a furnace lined with a basic refractory material.

**basic training** *n.* The initial training period of a recruit in the armed forces.

**ba·sid·i·a** (bə-sĭd'ē-ə) *n. pl. of* BASIDIUM.

**ba·sid·i·o·my·cete** (bə-sĭd'ē-ō-mī'sēt', -mī-sēt') *n.* [NLat. *Basidiomycetes,* class name : BASIDIUM + Gk. *mukēs,* fungus.] A fungus of the class Basidiomycetes, including the mushrooms, puffballs, and other fungi that bear spores on a basidium. **—ba·sid'i·o·my·ce'tous** (-mī-sē'təs) *adj.*

**ba·sid·i·o·spore** (bə-sĭd'ē-ə-spôr', -spōr') *n.* A spore formed on a basidium. **—ba·sid'i·o·spo'rous** *adj.*

**ba·sid·i·um** (bə-sĭd'ē-əm) *n., pl.* **-i·a** (-ē-ə) [NLat. < Lat. *basis,* base < Gk.] A club-shaped cell characteristic of basidiomycetous fungi on which usu. four sexual spores are borne at the tip. **—ba·sid'i·al** *adj.*

**ba·si·fy** (bā'sə-fī') *vt.* **-fied, -fy·ing, -fies.** *Chem.* To make basic. **—ba'si·fi·ca'tion** *n.* **—ba'si·fi'er** *n.*

**bas·il** (băz'əl, bā'zəl) *n.* [ME *basile* < OFr. < Med. Lat. *basilico* < Gk. *basilikon,* royal.] **1.** An Old World herb, *Ocimum basilicum,* having small white flower spikes and aromatic leaves used as seasoning. **2.** A European plant related to basil, *Satureja vulgaris,* widely naturalized in North America and having small, dense pink or purplish flower clusters.

**bas·i·lar** (băs'ə-lər) *also* **bas·i·lar·y** (-lĕr'ē) *adj.* [NLat. *basilaris* < Lat. *basis,* base < Gk.] Of, relating to, or located at or near the base, esp. the base of the skull.

**ba·sil·i·ca** (bə-sĭl'ĭ-kə) *n.* [Lat. < Gk. *basilikē (stoa),* royal (portico) < *basileus,* king.] **1. a.** An oblong ancient Roman building with a semicircular apse at one end, used as a court or place of assembly. **b.** A building of this design used as a Christian church. **2.** *Rom. Cath. Ch.* A church or cathedral accorded certain ceremonial rights by the pope. **—ba·sil'i·can** (-kən) *adj.*

**bas·i·lisk** (băs'ə-lĭsk', băz'-) *n.* [ME < Lat. *basiliscus* < Gk. *basiliskos,* dim. of *basileus,* king.] **1.** A legendary serpent or dragon with lethal breath and glance. **2.** Any of various tropical American lizards of the genus *Basiliscus,* with an erectile crest at the back of the head.

**basilisk**
*29–32 inches long*

**ba·sin** (bā'sĭn) *n.* [ME *bacin* < OFr.] **1. a.** A round, open container used esp. for holding liquids. **b.** The amount that a basin will hold. **2.** A washbowl : sink. **3. a.** An artificially enclosed section of a river or harbor designed so that the water level remains unaffected by tidal changes. **b.** A small enclosed or partly enclosed body of water. **4.** A region drained by one river system. **5.** *Geol.* **a.** Land in which the rock strata are tilted toward a common center. **b.** A bowl-shaped depression in the land or in an ocean floor. **—ba'sin·al** *adj.*

**bas·i·net** (băs'ə-nĕt', băs'ə-nĭt) *n.* [ME *bacinet* < OFr., dim. of *bacin,* basin.] A light, often visored medieval helmet.

**ba·sip·e·tal** (bā-sĭp'ĭ-tl, -zĭp'-) *adj. Bot.* Developing or growing from the top toward the base, as certain forms of inflorescence. **—ba·sip'e·tal·ly** *adv.*

**ba·sis** (bā'sĭs) *n., pl.* **-ses** (-sēz') [Lat. < Gk.] **1.** A supporting element : FOUNDATION. **2.** The chief component or fundamental ingredient. **3.** An essential principle.

**bask** (băsk) *vi.* **basked, bask·ing, basks.** [ME *basken.*] **1.** To expose oneself to pleasant warmth. **2.** To thrive or take pleasure.

**bas·ket** (băs'kĭt) *n.* [ME.] **1. a.** A container made from interwoven material, as rushes or twigs. **b.** The amount a basket will hold. **2.** Something like a basket. **3.** *Basketball.* **a.** Either of the two goals, each consisting of a metal hoop from which an open-bottomed circular net is suspended. **b.** The score, usu. worth two points, made by throwing the ball through the basket. **—bas'ket·ful'** (-fo͝ol') *n.*

**bas·ket·ball** (băs'kĭt-bôl') *n.* **1.** A game played between two teams of five players each, the object being to throw the ball through an elevated basket on the opponent's side of the rectangular court. **2.** The ball used in basketball.

**basket case** *n. Informal.* One that is in a completely hopeless or useless condition <"Psychological *basket cases* desperately in search of love and identity" —*Village Voice*>

**basket fern** *n.* **1.** A tropical American sword fern, *Nephrolepis pectinata.* **2.** A male fern.

**basket hilt** *n.* A sword hilt with a basket-shaped guard serving to cover and protect the hand.

**bas·ket·ry** (băs'kĭ-trē) *n.* **1.** The process or craft of making baskets. **2.** Baskets at a whole.

**basket star** *n.* Any of various marine organisms of the class Ophiuroidea, related to the starfishes and having slender, branched arms.

**basket weave** *n.* A textile weave of double threads interlaced to produce a checkered pattern similar to that of a woven basket.

**basking shark** *n.* A very large shark, *Cetorhinus maximus,* that feeds on plankton and often floats near the surface of water.

**bas mitz·vah** or **bas miz·vah** (bäs mĭts'və) *n. vars. of* BAT MITZVAH.

**baso-** *pref. var. of* BASI-.

**ba·so·phil** (bā'sə-fĭl, -zə-) *n.* A cell, esp. a white blood cell, having granules that exhibit an affinity for basic dyes. **—ba'so·phil'ic, ba·soph'i·lous** (bə-sŏf'ə-ləs) *adj.*

**ba·so·pho·bi·a** (bā'sə-fō'bē-ə, -zə-) *n.* An abnormal fear of standing erect or walking.

**basque** (băsk) *n.* [Fr. < Prov. *basta,* perh. of Germanic orig.] A close-fitting bodice.

**Basque** (băsk) *n.* [Fr. < Lat. *Vasco.*] **1.** One of a people of unknown origin inhabiting the western Pyrenees in France and Spain. **2.** The language of the Basques, of no known linguistic affiliation. **—Basque** *adj.*

**bas-re·lief** (bä'rĭ-lēf') *n.* [Fr. < Ital. *bassorilievo* : *basso,* low (< Med. Lat. *bassus*) + *rilievo,* relief < *rilievare,* to raise < Lat. *relevare.* —see RELIEF.] Low relief.

**bass**[1] (băs) *n., pl.* **bass** or **bass·es.** [ME *bace,* var. of dial. *barse* < OE *bærs.*] **1.** Any of several North American freshwater fishes of the family Centrarchidae, related to but larger than the sunfishes. **2.** A marine fish of the family Serranidae, as the sea bass.

**bass**[2] (bās) *n.* [ME *bas.*] **1.** A low-pitched tone or sound. **2.** The tones in the lowest register of a musical instrument. **3.** The lowest part in vocal or instrumental part music. **4. a.** A male singing voice of the lowest range. **b.** A man who has a singing voice in the lowest range. **5.** A musical instrument, esp. a double bass, that produces tones in a low register. —*adj.* **1.** Having a deep tone. **2.** Low in pitch.

**bass**[3] (băs) *n.* [Var. of BAST.] Bast.

**bass clef** (bās) *n. Mus.* A clef that designates F below middle C as being on the fourth line above the bottom of the staff.

**bass drum** (bās) *n.* A large drum with a cylindrical body and two heads that produces a low, resonant sound.

**bas·set** (băs'ĭt) *n.* [Fr. < OFr., dim of *basse,* fem. adj. of *bas,* low.] The basset hound.

**basset horn** *n.* A tenor clarinet in F, having a range of 3½ octaves pitched between the range of an alto clarinet and that of a bass clarinet.

**basset hound** *n.* A short-haired dog orig. bred in France and having a long body, short legs, and long, drooping ears.

**bass horn** (bās) *n.* A tuba.

**bas·si** (bä'sē) *n. var. pl. of* BASSO.

**bassi pro·fun·di** (prə-fo͝on'dē) *n. var. pl. of* BASSO PROFUNDO.

**bas·si·net** (băs'ə-nĕt', băs'ə-nĕt') *n.* [Fr., small basin.] An oblong basketlike bed for an infant.

**bass·ist** (bā'sĭst) *n.* A double bass player.

**bas·so** (băs'ō, bä'sō) *n., pl.* **bas·sos** or **bas·si** (bä'sē) [Ital. < Med. Lat. *bassus,* low.] An operatic bass.

**bas·soon** (bə-so͞on', bă-) *n.* [Fr. *basson* < Ital. *bassone,* augmentative of *basso,* basso.] A low-pitched woodwind instrument with a double reed and a long wooden body attached to a lateral tube leading to the mouthpiece. **—bas·soon'ist** *n.*

**bas·so pro·fun·do** (băs'ō prə-fŭn'dō, bä'sō prə-fo͞on'dō) *n., pl.* **basso pro·fun·dos** *or* **bas·si pro·fun·di** (bä'sē prə-fo͞on'dē) [Ital.] **1.** A bass voice of the lowest range. **2.** A singer having a basso profundo.

**bas·so-re·lie·vo** (băs'ō-rĭ-lē'vō) *n., pl.* **-vos.** [Ital. *bassorilievo.* — see BAS-RELIEF.] Low relief.

**bass viol** (bās) *n. Mus.* **1.** A double bass. **2.** A viola da gamba.

**bass·wood** (băs'wo͝od') *n.* **1.** A linden tree of eastern North America, esp. *Tilia americana,* with fragrant yellowish flower clusters. **2.** The soft, light-colored wood of a basswood.

**bast** (băst) *n.* [ME, inner bark of the linden tree < OE *bæst.*] **1.** The fibrous or somewhat woody outer layer of the stems of plants such as flax, hemp, and ramie. **2.** Fibrous material that is obtained chiefly from plants or from certain trees and used to make cordage and textiles.

**bas·tard** (băs'tərd) *n.* [ME < OFr., perh. < *(fils de) bast,* packsaddle (child).] **1.** An illegitimate child. **2.** Something of irregular, inferior, or dubious origin. *—adj.* **1.** Born of unwed parents : ILLEGITIMATE. **2.** Not genuine : SPURIOUS. **3.** Of inferior breed or kind. **4.** Resembling a known kind or species but not truly such. **—bas'tard·ly** *adj.*

▲ word history: In Old French *fils de bast* literally meant "child of a packsaddle"; the phrase refers to the unsanctified circumstances in which the child was conceived. Travelers used packsaddles as beds—often, no doubt, as impromptu marriage beds. The word *bastard* was formed in Old French from *bast,* "packsaddle," and the pejorative suffix *-ard.*

**bas·tard·ize** (băs'tər-dīz') *vt.* **-ized, -iz·ing, -iz·es.** To debase : corrupt. **—bas'tard·i·za'tion** *n.*

**bastard toadflax** *n.* A plant of the genus *Comandra,* esp. *C. umbellata* of eastern North America, with small, rounded greenish flower clusters.

**bastard wing** *n.* An alula.

**bas·tard·y** (băs'tər-dē) *n.* **1.** Illegitimate birth : ILLEGITIMACY. **2.** Procreation of a bastard.

**baste¹** (bāst) *vt.* **bast·ed, bast·ing, bastes.** [ME *basten* < OFr. *bastir.*] To sew loosely with large running stitches so as to hold together temporarily. **—bast'er** *n.*

**baste²** (bāst) *vt.* **bast·ed, bast·ing, bastes.** [Orig. unknown.] To moisten (e.g., meat) periodically with a liquid, as butter or sauce, esp. while cooking. **—bast'er** *n.*

**baste³** (bāst) *vt.* **bast·ed, bast·ing, bastes.** [Orig. unknown.] **1.** To beat vigorously : THRASH. **2.** To berate.

**bas·tille** *also* **bas·tile** (bă-stēl') *n.* [ME *bastel* < OFr. *bastille* < Med. Lat. *bastile* < *bastire,* to build.] A prison : jail.

**Bastille Day** *n.* Jul. 14 observed in France in commemoration of the destruction of the Bastille in 1789.

**bas·ti·na·do** (băs'tə-nā'dō, -nä'-) *also* **bas·ti·nade** (-nād', -näd') [Sp. *bastonada* < *baston,* stick < LLat. *bastum.*] *—n., pl.* **-does** *also* **-nades. 1.** A beating with a stick or cudgel, esp. on the soles of the feet. **2.** A stick or cudgel. *—vt.* **-doed, -do·ing, -does** *also* **-nad·ed, -nad·ing, -nades.** To subject to a beating : THRASH.

**bas·tion** (băs'chən, -tē-ən) *n.* [Fr. < Ital. *bastione* < *bastire,* to build.] **1.** A projecting part of a fortification. **2.** A well-fortified area or position. **3.** Something likened to a defensive stronghold <the *bastion* of freedom> **—bas'tioned** *adj.*

**bast·naes·ite** (băst'nĭ-sīt') *n.* [Swed. *bastnäsit,* after *Bastnäs,* a region of Sweden.] A yellowish to reddish-brown mineral fluorocarbonate, used as a rare-earth ore.

**bat¹** (băt) *n.* [ME < OE *batt.*] **1.** A stout wooden club or stick : CUDGEL. **2.** A blow, as with a stick. **3. a.** A rounded, usu. wooden club, wider and heavier at the hitting end and tapering at the handle, used to strike a baseball. **b.** A wooden club having a broad, flat-surfaced hitting end and a distinct, narrow handle, used in the game of cricket. **c.** The racket used in various other games, as squash. **4.** *Slang.* A binge : spree. *—v.* **bat·ted, bat·ting, bats.** *—vt.* **1.** To hit with or as if with a club or bat. **2.** *Baseball.* To have (a certain percentage) as a batting average. **3.** *Informal.* To produce hurriedly <*bat* out a speech in an hour> **4.** *Informal.* To discuss or consider at length <*bat* an idea around> *—vi.* **1.** *Baseball.* **a.** To use a bat. **b.** To have a turn at bat. **2.** *Slang.* To go from place to place aimlessly : WANDER. **—at bat.** Taking one's turn to bat, as in baseball. **—go to bat for.** *Informal.* To support or defend. **—right off the bat.** *Informal.* Immediately.

**bat²** (băt) *n.* [Alteration of ME *bakke,* of Scand. orig.] Any of various nocturnal flying mammals of the order Chiroptera, with membranous wings extending from the forelimbs to the hind limbs or tail.

**bat³** (băt) *vt.* **bat·ted, bat·ting, bats.** [Prob. a var. of BATE².] To wink : flutter <*bat* one's eyelashes>

**batch¹** (băch) *n.* [ME *bacche* < OE **bæcce* < *bacan,* to bake.] **1.** An amount of food produced at one baking. **2.** A quantity produced as the result of one operation <a *batch* of asphalt> **3.** The quantity of material needed for one operation. **4.** A group of individuals or objects. **5.** *Computer Sci.* A set of data or jobs to be processed in a single program run.

**batch²** (băch) *v. var. of* BACH.

**bate¹** (bāt) *vt.* **bat·ed, bat·ing, bates.** [ME *baten,* short for *abaten.* —see ABATE.] **1.** To lessen the force of : MODERATE. **2.** To take away : SUBTRACT.

**bate²** *also* **bait** (bāt) *vi.* **bat·ed, bat·ing, bates** *also* **bait·ed, bait·ing, baits.** [ME *baten* < OFr. *batre,* to beat.] To flap the wings wildly in impatience. —Used of a falcon.

**ba·teau** (bă-tō') *n., pl.* **-teaux** (-tōz') [Fr., boat < OFr. *batel,* prob. < OE *bāt.*] A light, flat-bottomed riverboat.

**bateau bridge** *n.* A pontoon bridge.

**Bates·i·an mimicry** (bāt'sē-ən) *n.* [After Henry W. *Bates* (1825–1892).] The resemblance of a harmless species to a species whose defense against predation is based on repellent traits.

**bat·fish** (băt'fĭsh') *n., pl.* **batfish** *or* **-fish·es.** Any of various marine anglerfishes of the family Ogcocephalidae, with a retractable appendage above the mouth.

**bat·fowl** (băt'foul') *vi.* **-fowled, -fowl·ing, -fowls.** To catch roosting birds at night by blinding them with a light.

**bath¹** (băth, bäth) *n., pl.* **baths** (băthz, bäthz, băths, bäths) [ME < OE *bæð.*] **1.** The act of washing or soaking the body. **2.** The water used for a bath. **3. a.** A bathtub. **b.** A bathroom. **4. a.** A building equipped for bathing. **b.** *often* **baths.** A spa. **5.** A liquid or a liquid and its container in which matter is dipped or soaked in order to process it <a *bath* of acid>

**bath²** (băth) *n.* [Heb.] An ancient Hebrew unit of liquid measure, equal to approx. ten U.S. gallons.

**Bath chair** (băth, bäth) *n.* [After *Bath,* England.] A hooded wheelchair used esp. for invalids, as at a spa.

**bathe** (bāth) *v.* **bathed, bath·ing, bathes.** [ME *bathen* < OE *baðian.*] *—vi.* **1.** To take a bath. **2.** To go into the water esp. for swimming. **3.** To become immersed in or as if in liquid. *—vt.* **1.** To immerse in liquid. **2.** To wash in a liquid. **3.** To apply a liquid to for soothing or therapeutic purposes. **4.** To seem to wash or pour over : SUFFUSE <a room *bathed* in sunlight> **—bath'er** *n.*

**ba·thet·ic** (bə-thĕt'ĭk) *adj.* [Prob. a blend of BATHOS and PATHETIC.] Marked by bathos.

**bath·house** (băth'hous', bäth'-) *n.* **1.** A building with bathing facilities. **2.** A building with dressing rooms for swimmers.

**bathing suit** *n.* A swimsuit.

**batho-** *pref. var. of* BATHY-.

**bath·o·lith** (băth'ə-lĭth') *n.* Igneous rock that has melted and intruded surrounding strata at great depths. **—bath'o·lith'ic** *adj.*

**ba·thom·e·ter** (bə-thŏm'ĭ-tər) *n.* An instrument for measuring the depth of water.

**bath·o·pho·bi·a** (băth'ə-fō'bē-ə) *n.* Abnormal fear of depths.

**ba·thos** (bā'thŏs') *n.* [Gk., depth.] **1. a.** A ludicrously abrupt transition from an elevated to a commonplace style. **b.** An anticlimax. **2. a.** Insincere or grossly sentimental pathos. **b.** Gross triteness.

**bath·robe** (băth'rōb', bäth'-) *n.* A loose-fitting robe.

**bath·room** (băth'ro͞om', -ro͝om', bäth'-) *n.* A room with facilities for bathing or showering and usu. containing a washbasin and toilet.

**bath salts** *pl.n.* A perfumed crystalline substance for softening bath water.

**bath·tub** (băth'tŭb', bäth'-) *n.* A tub for bathing.

**bathy-** *or* **batho-** *pref.* [< Gk. *bathus,* deep.] **1.** Deep : depth <*batho*lith> **2.** Deep-sea <*bathy*sphere>

**ba·thym·e·try** (bə-thĭm'ĭ-trē) *n.* Measurement of the depth of large bodies of water. **—bath'y·met'ric** (băth'ə-mĕt'rĭk), **bath'y·met'ri·cal** *adj.* **—bath'y·met'ri·cal·ly** *adv.*

**bath·y·pe·lag·ic** (băth'ə-pə-lăj'ĭk) *adj.* Of, relating to, or living in the depths of the ocean, esp. below 2,000 feet.

**bath·y·scaph** (băth'ĭ-skăf') *also* **bath·y·scaphe** (-skăf', -skāf') *n.* [Fr. *bathyscaphe* : Gk. *bathus,* deep + Gk. *scaphē,* boat.] A free-diving, self-contained deep-sea research vessel consisting of a large flotation hull with a manned observation capsule attached to its underside.

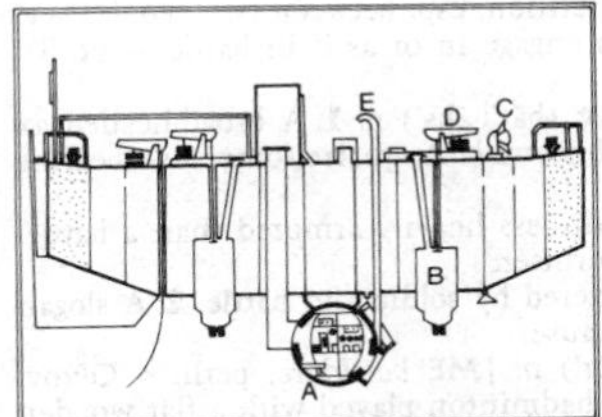


**bathyscaph**
*A. observation gondola, B. pellet ballast hopper, C. propeller, D. release magnet, E. snorkel*

**bath·y·sphere** (băth'ĭ-sfîr') *n.* A reinforced spherical deep-diving chamber.

**ba·tik** (bə-tēk', băt'ĭk) *n.* [Malay < Javanese, painted.] **1. a.** A method of dyeing print into a fabric in which parts of the cloth not

intended to be dyed are covered with removable wax. **b.** A design dyed into cloth by this method. **2.** A cloth dyed by batik.
**ba·tiste** (bə-tēst′, bă-) *n.* [Fr. < OFr.] A fine, plain-woven fabric made from various fibers and used esp. for clothing.
**bat·man** (băt′mən) *n.* [Obs. *bat,* packsaddle < ME *batt* < OFr. *bat.*] A British army orderly.
**bat mitz·vah** *or* **bat miz·vah** (bät mĭts′və) *also* **bas mitz·vah** *or* **bas miz·vah** (bäs) *n.* [Heb. *baṭ mitzvāh* : *baṭ,* daughter + *mitzvāh,* commandment.] **1.** A Jewish girl aged 12 to 14 years who assumes her Jewish duties and responsibilities. **2.** The ceremony that initiates and recognizes a girl as a bat mitzvah.
**ba·ton** (bə-tŏn′, bă-, băt′n) *n.* [Fr. *bâton* < OFr. *baston.*] **1.** A short staff carried by some public officials as a symbol of office. **2.** A slender wooden stick or rod used by a conductor to direct a band or an orchestra. **3.** A hollow metal rod with heavy rubber tips twirled by a drum major or majorette.
**ba·tra·chi·an** (bə-trā′kē-ən) *adj.* [< NLat. *Batrachia,* former order name < Gk. *batrakhos,* frog.] Of or relating to frogs and toads. —*n.* A frog or toad.
**bats** (băts) *adj. Slang.* Insane : crazy.
**bats·man** (băts′mən) *n.* BATTER².
**batt** (băt) *n.* A mass of cotton fibers.
**bat·tal·ion** (bə-tăl′yən) *n.* [OFr. *battaillon* < OItal. *battaglione* < *battaglia,* a body of troops < VLat. **battalia.* —see BATTLE.] **1.** A tactical military unit typically consisting of a headquarters company and four infantry companies or a headquarters battery and four artillery batteries. **2.** A large body of military troops. **3.** A great number.
**bat·ten¹** (băt′n) *vi.* **-tened, -ten·ing, -tens.** [Ult. < ON *batna,* to improve.] **1.** To become fat. **2.** To thrive and prosper, esp. at another's expense.
**bat·ten²** (băt′n) *n.* [Alteration of BATON.] **1.** A narrow strip of wood used esp. for flooring. **2.** One of several flexible strips of wood placed in pockets at the outer edge of a sail to keep it flat. —*vt.* **-tened, -ten·ing, -tens. 1.** To furnish with battens. **2.** To fasten or secure with battens.
**bat·ter¹** (băt′ər) *v.* **-tered, -ter·ing, -ters.** [ME *bateren* < OFr. *battre* < Lat. *battuere.*] —*vt.* **1.** To beat heavily and repeatedly so as to hurt, bruise, or destroy. **2.** To damage, as by heavy wear. —*vi.* To hit heavily and repeatedly. —*n.* A damaged area on the face of printing type or on a printing plate.
**bat·ter²** (băt′ər) *n.* The player at bat in baseball and cricket.
**bat·ter³** (băt′ər) *n.* [ME *bater,* prob. < *bateren,* to batter, beat.] A thick, beaten liquid mixture, as of flour, milk, and eggs, used in cooking.
**bat·ter⁴** (băt′ər) *n.* [Orig. unknown.] A slope, as of the outer side of a wall, that recedes from bottom to top. —*vt.* **-tered, -ter·ing, -ters.** To construct in a batter.
**battered child syndrome** *n.* A combination of serious physical injuries, as bruises, scratches, hematomas, burns, or malnutrition, inflicted on a child through gross abuse usu. by parents or guardians.
**bat·ter·ing-ram** *also* **battering ram** (băt′ər-ĭng-răm′) *n.* **1.** A heavy beam used in ancient warfare to batter down walls and gates. **2.** A heavy metal bar used by firefighters to break down doors and walls.
**bat·ter·y** (băt′ə-rē) *n., pl.* **-ies.** [Fr. *batterie* < *battre,* to batter < OFr.] **1. a.** An act of pounding or beating. **b.** *Law.* The unlawful beating or other use of force on another person. **2. a.** An emplacement for a piece of artillery. **b.** A set of heavy guns, as on a warship. **c.** The basic tactical artillery unit corresponding to the company in the infantry. **3.** An array or grouping of like things to be used together. **4.** *Baseball* The pitcher and catcher on a team. **5.** The percussion section of an orchestra. **6.** A device for generating an electric current by chemical reaction.
**bat·ting** (băt′ĭng) *n.* **1.** The act of one who bats. **2.** Cotton or wool fiber wadded into rolls or sheets and used for stuffing furniture and mattresses and lining quilts.
**bat·tle** (băt′l) *n.* [ME *bataille* < OFr. < VLat. **battalia* < LLat. *battualia,* fighting and fencing exercises < Lat. *battuere,* to batter.] **1. a.** Large-scale combat between two armed forces. **b.** Armed fighting : COMBAT. **2.** Intense competition, esp. between two people. —*v.* **-tled, -tling, -tles.** —*vi.* To engage in or as if in battle. —*vt.* To fight against. **—bat′tler** *n.*
**bat·tle-ax** *or* **bat·tle-axe** (băt′l-ăks′) *n.* **1.** A broad-headed ax once used as a weapon. **2.** *Informal.* A quarrelsome, overbearing woman.
**battle cruiser** *n.* A warship less heavily armored than a battleship and with the speed of a cruiser.
**battle cry** *n.* **1.** A shout uttered by soldiers in battle. **2.** A slogan used by the proponents of a cause.
**bat·tle·dore** (băt′l-dôr′, -dōr′) *n.* [ME *batildore,* perh. < OProv. *batedor.*] **1.** An early form of badminton played with a flat wooden paddle and a shuttlecock. **2.** The paddle used in battledore.
**battle fatigue** *n.* Combat fatigue.
**bat·tle·field** (băt′l-fēld′) *n.* **1.** A field where a battle is fought. **2.** A sphere of contention.
**bat·tle·front** (băt′l-frŭnt′) *n.* The area where opponents meet in battle.
**bat·tle·ground** (băt′l-ground′) *n.* A battlefield.
**bat·tle·ment** (băt′l-mənt) *n.* [ME *batelment* < OFr. *battillement.*] A parapet built on top of a wall, with defensive or decorative indentations. **—bat′tle·ment′ed** (-mĕn′tĭd) *adj.*
**battle royal** *n., pl.* **battles royal. 1.** A battle involving many combatants. **2.** A fight to the finish. **3.** An intense altercation.
**bat·tle·ship** (băt′l-shĭp′) *n.* A large modern warship carrying the greatest number of guns and clad with the heaviest armor.
**battleship gray** *n.* A medium gray. **—bat′tle·ship-gray′** *adj.*
**bat·tle·wag·on** (băt′l-wăg′ən) *n. Slang.* A battleship.
**bat·ty** (băt′ē) *adj.* **-ti·er, -ti·est.** *Slang.* Insane. **—bat′ti·ness** *n.*
**bau·bee** (bô-bē′, bô′bē) *n. var. of* BAWBEE.
**bau·ble** (bô′bəl) *n.* [ME *babel* < OFr., plaything.] **1.** A small, showy, cheap ornament. **2.** *Archaic.* A baton carried by a court jester as a mock scepter of his office.
**baud** (bôd) *n.* [After J.M.E. *Baud* (d. 1903).] *Computer Sci.* A unit of speed in data transmission, as one bit per second for binary signals.
**Bau·haus** (bou′hous′) *adj.* [G., an architecture school founded by Walter Gropius (1883–1969).] Of, relating to, or characteristic of a 20th-cent. school of design whose aesthetic was influenced by and derived from techniques and materials employed esp. in industrial fabrication and manufacture.
**baulk** (bôk) *v. & n. var. of* BALK.
**Bau·mé scale** (bō-mā′) *n.* [After Antoine *Baumé* (1728–1804).] A hydrometer scale that separately covers liquids with specific gravities greater and less than 1.
**baux·ite** (bôk′sīt′) *n.* [Fr., after Les *Baux,* France.] The principal ore of aluminum, 30–75% $Al_2O_3{\cdot}nH_2O$, with ferric oxide and silica as impurities.
**Ba·var·i·an** (bə-vâr′ē-ən) *n.* **1.** A native or inhabitant of Bavaria. **2.** The High German dialect of Bavaria and Austria. **—Ba·var′i·an** *adj.*
**baw·bee** *also* **bau·bee** (bô-bē′, bô′bē) *n.* [After Alexander Orrok, Laird of *Sillbawby,* 16th-cent. Scottish master of the mint.] *Scot.* A halfpenny.
**bawd** (bôd) *n.* [ME *bawde,* prob. < OFr. *baude,* bold < OHG *bald.*] **1.** A woman brothel keeper : MADAM. **2.** A prostitute.
**bawd·ry** (bô′drē) *n.* [ME *bawdery* < *bawde,* bawd.] Coarse, risqué, or obscene language.
**bawd·y** (bô′dē) *adj.* **-i·er, -i·est. 1.** Humorously coarse : RISQUÉ. **2.** Vulgar : lewd. **—bawd′i·ly** *adv.* **—bawd′i·ness** *n.*
**bawd·y·house** (bô′dē-hous′) *n.* A house of prostitution.
**bawl** (bôl) *v.* **bawled, bawl·ing, bawls.** [ME *baulen,* to bark, of Scand. orig.] —*vi.* **1.** To sob loudly : WAIL. **2.** To shout vehemently. —*vt.* To utter in a loud, vehement voice <a sergeant major *bawling* orders> **—bawl out.** *Informal.* To reprimand harshly or loudly. **—bawl** *n.* **—bawl′er** *n.*
**bay¹** (bā) *n.* [ME *baye* < OFr. *baie* < LLat. *baia,* perh. < Iberian.] **1.** A body of water partially enclosed by land but with a wide outlet to the sea. **2. a.** A broad stretch of low land between hills. **b.** An arm of prairie partially enclosed by woodland.
**bay²** (bā) *n.* [ME < OFr. *baee,* an opening < *baer,* to gape.] **1.** A part of a structure, as a building, that is marked off by vertical elements. **2. a.** A bay window. **b.** A recess or opening in a wall. **3.** An extension of a building : WING. **4.** A compartment in a barn for storing hay or grain. **5.** A ship's sickbay. **6.** An aircraft bomb bay.
**bay³** (bā) *adj.* [ME < OFr. *bai* < Lat. *badius.*] Reddish-brown. —*n.* **1.** A reddish brown. **2.** A reddish-brown animal, esp. a horse.
**bay⁴** (bā) *n.* [ME < *baien,* to bark < OFr. *baiier.*] **1.** A deep, prolonged bark <the *bay* of hounds> **2.** The position of one cornered by pursuers and forced to turn and fight at close quarters. **3.** The position of one checked or kept at a safe distance <a mob kept at *bay* by riot police> —*v.* **bayed, bay·ing, bays.** —*vi.* To utter a bay. —*vt.* **1.** To pursue with barking. **2.** To utter by or as if by barking. **3.** To bring to bay.
**bay⁵** (bā) *n.* [ME *bai,* laurel berry < OFr. *baie* < Lat. *baca,* berry.] **1.** LAUREL 1. **2.** A crown or wreath made esp. of the leaves and branches of the bay and awarded as a sign of honor or victory. **3.** *often* **bays.** Renown : honor.
**ba·ya·dere** (bī′ə-dîr′, -dâr′) *n.* [Fr. *bayadère* < Port. *bailadeira,* dancer < *bailar,* to dance < LLat. *ballare* < Gk. *ballizein.*] A fabric having contrasting horizontal stripes.
**bay·ber·ry** (bā′bĕr′ē) *n.* **1.** An aromatic shrub or small tree of the genus *Myrica,* esp. *M. pensylvanica* of eastern North America, having gray, waxy berries. **2.** A tropical American tree, *Pimenta acris,* yielding an oil used in making bay rum. **3.** The fruit of a bayberry.
**bay leaf** *n.* The dried aromatic leaf of the bay, *Laurus nobilis,* or of the bayberry, *Pimenta acris,* used as seasoning in cooking.
**bay lynx** *n.* The bobcat.
**bay·o·net** (bā′ə-nĭt, -nĕt′, bā′ə-nĕt′) *n.* [Fr. *baionnette* < *Bayonne,* a city in France.] A knife adapted to fit the muzzle end of a rifle and used in hand-to-hand combat. —*vt.* **-net·ed, -net·ing, -nets** *or* **-net·ted, -net·ting, -nets.** To prod or stab with a bayonet.

**bay·ou** (bī′o͞o, bī′ō) *n.* [Louisiana Fr. < Choctaw *bayuk.*] A marshy, sluggish body of water tributary to a lake or river.
**bay rum** *n.* An aromatic liquid obtained by distilling the leaves of the bayberry tree, *Pimenta acris,* with rum or synthesized from alcohol, water, and various oils.
**bay rum tree** *n.* The bayberry.
**bay window** *n.* **1.** A large window projecting from the outer wall of a building and forming a recess within. **2.** *Slang.* A protruding belly : PAUNCH.
**ba·zaar** *also* **ba·zar** (bə-zär′) *n.* [Prob. < Ital. *bazarro* < Pers. *bāzār.*] **1.** An Oriental market consisting of a street lined with shops and stalls. **2.** A shop or part of a store for the sale of miscellaneous merchandise. **3.** A fair or sale at which miscellaneous merchandise is sold, often for charity.
**ba·zoo·ka** (bə-zo͞o′kə) *n.* [After the *bazooka,* a crude wind instrument made of pipes, invented and named by Bob Burns (1896–1956).] A portable weapon consisting of a long metal smoothbore tube for firing small, armor-piercing rockets at short range.
**BB** (bē′bē) *n.* [Perh. from the letter *b.*] A standard size of lead shot measuring about .46 centimeter, or .18 inch in diameter.
**BB gun** *n.* A small air rifle for firing BB shot.
**B cell** *n.* [B(ONE-MARROW-DERIVED) + CELL.] A lymphocyte that arises from the bone marrow of humans, produces antibodies, and is vital in the body's defense against pyogenic bacteria.
**bdel·li·um** (dĕl′ē-əm) *n.* [Lat. < Gk. *bdellion,* of Semitic orig.] An aromatic gum resin similar to myrrh, produced by various trees of the genus *Commiphora* of western Asia and Africa.
**be** (bē) *vi.* **was** (wŏz, wŭz; wəz *when unstressed*) *or* **were** (wûr), **be·ing** (bē′ĭng), **been** (bĭn) [ME *been* < OE *bēon.*] **1.** To exist in actuality : have reality or life <I think, therefore I *am.*> **2.** To occupy a given position <The paper *is* on the desk.> **3.** To take place : OCCUR. **4.** To go. —Used chiefly in the past and perfect tenses <Have you ever *been* to France?> **5.** *Archaic.* To belong : befall. **6.** —Used as a copula linking a subject and a predicate nominative, adjective, or pronoun, in senses such as: **a.** To equal in meaning or identity <"To be a Christian *was* to be a Roman" —James Bryce> **b.** To symbolize : signify <A *is* excellent, B *is* good.> **c.** To belong to a given class or group <The human being *is* a primate.> **d.** To have or exhibit a given quality or characteristic <The coat *is* beautiful.> <All of us *are* human.> —*aux.* **1.** Used with the past participle of a transitive verb to form the passive voice <The election *is* held annually.> **2.** —Used with the present participle of a verb to express a continuing action <We *are* trying to meet our production quotas.> **3.** —Used with the infinitive of a verb to express intention, obligation, or future action <They *were* to call before they left.> <I *am* to take the responsibility.> **4.** *Archaic.* —Used with the past participle of certain intransitive verbs of motion to form the perfect tense <"Where *be* those roses gone which sweetened so our eyes?" —Philip Sidney>
**Be** *symbol for* BERYLLIUM.
**be-** *pref.* [ME < OE *bi-.*] **1.** Completely : thoroughly : excessively. —Used as an intensive <*be*moan> **2.** On : around : over <*be*smear> **3.** —Used to form transitive verbs from nouns, adjectives, and intransitive verbs, as: **a.** Make : cause to become <*be*sot> **b.** Affect or provide with <*be*spangle>
**beach** (bēch) *n.* [Orig. unknown.] **1.** The sandy, pebbly, or rocky shore of a body of water. **2.** The sand, pebbles, or rocks on a shore. —*vt.* **beached, beach·ing, beach·es.** To haul or drive ashore.
**beach buggy** *n.* A dune buggy.
**beach·comb·er** (bēch′kō′mər) *n.* **1.** One who lives on matter found on beaches or in wharf areas. **2.** A long wave rolling in toward a beach.
**beach flea** *n.* Any of various small, jumping crustaceans of the family Orchestiidae, living on sandy beaches at or near the tide line.
**beach grass** *n.* A grass of the genus *Ammophila,* growing on sandy shores and dunes and having spikelets in long, crowded clusters.
**beach·head** (bēch′hĕd′) *n.* **1.** A position on an enemy shoreline captured by troops in advance of an invading force. **2.** A position opening the way for further development : FOOTHOLD.
**beach pea** *n.* A North American plant, *Lathyrus maritimus* of the Atlantic coast or *L. littoralis* of the Pacific coast, with purplish flowers and sprawling stems.
**beach plum** *n.* A seacoast shrub, *Prunus maritima* of northeastern North America, with white flowers and edible, plumlike fruit.
**beach wormwood** *n.* A seacoast plant native to Asia, *Artemisia stelleriana,* covered with dense white down and having small yellow flowers.
**bea·con** (bē′kən) *n.* [ME *beken* < OE *bēacen.*] **1.** A signal fire, esp. one used to warn of an enemy's approach. **2.** A coastal signaling or guiding device. **3.** A radio transmitter that emits a signal as a warning or guide. **4.** Something that warns or guides. —*vt. & vi.* **-coned, -con·ing, -cons.** To furnish with or function as a beacon.
**bead** (bēd) *n.* [ME *bede,* rosary bead < OE *gebed,* prayer.] **1. a.** A small ball-shaped piece of material pierced for stringing or threading. **b. beads.** A necklace made of such pieces. **c. beads.** A rosary. **2.** A small round object, esp.: **a.** A small drop of moisture. **b.** A bubble of gas in a liquid. **c.** A small knob of metal on the muzzle of a firearm, used for sighting. **3.** A strip of material, usu. wood, with one molded edge placed flush against the inner part of a door or window frame. —*vt. & vi.* **bead·ed, bead·ing, beads.** To furnish with or collect into beads.

▲ word history: The connection between a small round object and a prayer lies in the medieval Christian practice of keeping count of prayers by means of beads threaded on a string. Telling one's beads—saying one's prayers—with the aid of a rosary was such a common way of praying that the word for "prayer" gradually became the word for the counter. By modern times *bead* no longer meant "prayer" at all, but had been extended to signify other small round objects, such as drops of water.

**bead·ing** (bē′dĭng) *n.* **1.** Beads or material used for beads. **2.** Ornamentation with beads. **3.** A narrow half-rounded molding. **4.** A narrow piece of openwork lace through which ribbon may be run.
**bea·dle** (bēd′l) *n.* [ME *bedele,* herald < OE *bydel.*] A minor parish official in an English church whose duties include ushering and keeping order during services.
**bead·work** (bēd′wûrk′) *n.* **1.** BEADING 2. **2.** BEADING 3.
**bead·y** (bē′dē) *adj.* **-i·er, -i·est. 1.** Small, round, and shiny <*beady* eyes> **2.** Decorated or covered with beads.
**bea·gle** (bē′gəl) *n.* [ME *begle.*] One of a breed of small hounds having short legs, drooping ears, and a smooth coat with white, black, and tan markings.
**beak** (bēk) *n.* [ME *bek* < OFr. *bec* < Lat. *beccus,* of Celt. orig.] **1. a.** The horny projecting structure forming the mandibles of a bird : BILL. **b.** A part or organ resembling a bird's bill, as in some turtles, insects, or fish. **2.** A hard, cone-shaped, or pointed part or structure. **3.** *Informal.* The human nose. **—beaked** (bēkt) *adj.*
**beak·er** (bē′kər) *n.* [ME *biker* < ON *bikarr,* prob. < Med. Lat. *bicarius* < Gk. *bikos,* jug.] **1.** A large wide-mouthed drinking cup. **2.** An open glass cylinder with a pouring lip, used as a laboratory container.
**beam** (bēm) *n.* [ME < OE *bēam.*] **1.** A squared-off log or large oblong piece of timber, metal, or stone used esp. in construction. **2.** The breadth of a ship at the widest point. **3.** *Informal.* The width across a person's hips. **4.** A steel tube or wooden roller on which the warp is wound in a loom. **5.** An oscillating lever connected to an engine piston rod and used to transmit power to the crankshaft. **6.** The bar of a balance from which weighing pans are suspended. **7.** One of the main stems of a deer's antlers. **8.** The main horizontal bar on a plow to which the share, colter, and handles are attached. **9. a.** A ray of light. **b.** A group of particles traveling together in close parallel trajectories. **10.** A radio beam. —*v.* **beamed, beam·ing, beams.** —*vi.* **1.** To emit light : SHINE. **2.** To smile broadly. —*vt.* To emit or transmit. **—on the beam. 1.** Following a radio beam, as an aircraft. **2.** *Informal.* On the right track.
**beam-ends** (bēm′ĕndz′) *pl.n.* The ends of a ship's beams.
**beam·y** (bē′mē) *adj.* **-i·er, -i·est. 1.** Broad in the beam. **2.** Emitting beams, as of light : RADIANT.
**bean** (bēn) *n.* [ME *bene* < OE *bēan.*] **1. a.** Any of several plants of the genus *Phaseolus,* having compound leaves, white or yellow flowers, and seed-bearing pods. **b.** The edible seed or pod of the bean. **c.** Any of several plants related to the bean and bearing similar pods and seeds. **d.** Any of various other seeds or pods resembling beans, as the coffee bean. **2.** *Slang.* The human head. **3. beans.** *Slang.* A small amount <doesn't know *beans* about dictionary editing> **4.** *Chiefly Brit.* A fellow : chap. —*vt.* **beaned, bean·ing, beans.** *Slang.* To hit on the head with a thrown object, esp. a pitched baseball. **—full of beans. 1.** Energetic. **2.** Full of nonsense. **—spill the beans.** To disclose a secret.
**bean·bag** (bēn′băg′) *n.* A small bag filled with dried beans and used for throwing in games.
**bean ball** *n.* *Baseball.* A pitch aimed at the batter's head.
**bean blight** *n.* A disease of the bean caused by the bacterium *Xanthomonas phaseoli,* which results in yellow-brown blotches on all parts of the plant.
**bean caper** *n.* A plant of the genus *Zygophyllum,* esp. *Z. fabago,* a Middle Eastern shrub bearing edible buds used as capers.
**bean curd** *n.* [Transl. of Chin. (Mandarin) *dou$^4$ fu$^3$* : *dou$^4$,* bean + *fu$^3$,* curdled.] A soft, cheeselike food made from puréed soy beans.
**bean·ie** (bē′nē) *n.* A small brimless cap.
**bean·o** (bē′nō) *n., pl.* **-os.** Bingo in which beans are used as markers.
**bean·pole** (bēn′pōl′) *n.* **1.** A thin pole for supporting bean vines. **2.** *Slang.* A very tall, thin person.
**bean sprout** *n.* A young, tender shoot of certain beans, as the soybean, used in cooking.
**bean·stalk** (bēn′stôk′) *n.* The stem of a bean plant.
**bean tree** *n.* A tree, as the catalpa, that bears beanlike fruit.
**bear[1]** (bâr) *v.* **bore** (bôr, bōr), **borne** *or* **born** (bôrn, bōrn), **bear·ing, bears.** [ME *beren* < OE *beran.*] —*vt.* **1.** To hold up : SUPPORT. **2.** To move while supporting : CARRY. **3.** To hold in the mind : HARBOR <*bearing* a grudge> **4.** To relate <*bearing* good news> **5.** To have as a visible characteristic <*bore* a scar on the forehead> **6.** To have as a quality : EXHIBIT. **7.** To carry (oneself) in a particular way : CONDUCT. **8.** To be accountable for : ASSUME <*bore* heavy responsi-

o͞o **boot** ou **out** th **thin** *th* **this** ŭ **cut** û r **urge** y **young**
yo͞o ab**use** zh vi**si**on ə **a**bout, it**e**m, ed**i**ble, gall**o**p, circ**u**s

bilities> **9.** To have a tolerance for: ENDURE. **10.** To admit of <This case *bears* further investigation.> **11.** To give birth to. **12.** To produce: YIELD <plants *bearing* flowers> **13.** To give as testimony <*bearing* witness> **14.** To move by steady pressure: PUSH <boats *borne* by the tides> —*vi.* **1.** To yield a product: PRODUCE. **2.** To be relevant: APPLY <how this theory *bears* on the course of our experimentation> **3.** To exert pressure. **4.** To exert oneself determinedly: FORGE. **5.** To proceed in a given direction <*Bear* left at the next corner.> ***usage:*** The past participle *born* occurs only in constructions with the verb *to be*, as in *That writer was born in 1757. Borne*, said of the act of birth, refers only to the mother's role, as in *She has borne three children.* In all other senses of *bear*, *borne* is the past participle. **—bear down. 1.** To overwhelm: vanquish. **2.** To apply maximum effort and concentration. **—bear down on.** To affect in a harmful way. **—bear out.** To prove right or justified: CONFIRM. **—bear up.** To withstand stress, difficulty, or attrition <*bore up* well during the crisis> **—bear with.** To be patient with.

**bear²** (bâr) *n.* [ME *bere* < OE *bera*.] **1. a.** Any of various usu. omnivorous mammals of the family Ursidae, having a shaggy coat and a short tail and walking with the entire lower surface of the foot touching the ground. **b.** An animal, as the koala, that resembles a bear. **2.** One who is awkward, clumsy, or ill-mannered. **3.** One who sells securities or commodities in the expectation that prices will fall.

**bear·a·ble** (bâr′ə-bəl) *adj.* Capable of being endured. **—bear′a·bil′i·ty** *n.* **—bear′a·bly** *adv.*

**bear·bait·ing** (bâr′bā′tĭng) *n.* The sport of setting dogs to attack or torment a chained bear.

**bear·ber·ry** (bâr′bĕr′ē) *n.* A trailing shrub, *Arctostaphylos uva-ursi* of northern regions, having small evergreen leaves, white or pink flowers, and red berries.

**beard** (bîrd) *n.* [ME *berd* < OE *beard*.] **1.** The hair on a man's chin, cheeks, and throat. **2.** A hairy or hairlike growth such as that on or near the face of certain mammals. **3.** A tuft or group of bristles on certain plants: AWN. **4.** The part of a piece of type between the face and the shoulder: NECK. —*vt.* **beard·ed, beard·ing, beards. 1.** To furnish with a beard. **2.** To grasp by the beard. **3.** To confront boldly. **—beard′ed** *adj.* **—beard′ed·ness** *n.* **—beard′less** *adj.* **—beard′less·ness** *n.*

**bearded iris** *n.* An iris having a beardlike growth at the base of the three lower, recurved petals.

**bearded vulture** *n.* The lammergeier.

**beard-tongue** (bîrd′tŭng′) *n.* Any of various plants of the chiefly North American genus *Penstemon*, having variously colored, tubular, two-lipped flowers.

**bear·er** (bâr′ər) *n.* One that bears, as: **a.** A porter. **b.** One who presents a check or other redeemable note for payment. **c.** A fruit-bearing plant. **d.** A pallbearer.

**bearer bond** *n.* A bond payable to the holder.

**bear grass** *n.* **1.** A tall plant, *Xerophyllum tenax* of northwestern North America, having narrow grasslike leaves and white flowers in a large terminal cluster. **2.** Any of several plants similar or related to the bear grass, esp. any of several species of yucca.

**bear hug** *n.* A rough hug.

**bear·ing** (bâr′ĭng) *n.* **1.** The manner in which one carries or conducts oneself <the poise and *bearing* of a winner> **2. a.** A part supporting another machine part or structure. **b.** A device that supports, guides, and reduces the friction of motion between fixed and moving machine parts. **3.** Something that bears weight or acts as a support. **4.** The part of an arch or beam that rests on a support. **5. a.** The act, power, or period of producing fruit or offspring. **b.** The quantity produced: YIELD. **6.** Direction, esp. angular direction measured from one position to another using geographic or celestial reference lines. **7.** *often* **bearings.** Awareness of one's position or situation relative to one's surroundings. **8.** Relevancy, relationship, or connection <testimony with no *bearing* on the case> **9.** *Heraldry.* A charge or device borne on a field.

**bearing rein** *n.* A rein for a horse: CHECKREIN.

**bear·ish** (bâr′ĭsh) *adj.* **1.** Like a bear. **2. a.** Causing, expecting, or characterized by falling stock-market prices. **b.** Pessimistic, negative, or skeptical. **—bear′ish·ly** *adv.* **—bear′ish·ness** *n.*

**bé·ar·naise sauce** (bā′är-nāz′, -ər-) *n.* [Fr. *béarnaise*, fem. of *béarnais*, of Béarn, region of southwestern France.] A sauce similar to hollandaise but flavored with tarragon, shallots, and chervil.

**bear·skin** (bâr′skĭn′) *n.* **1.** Something made from the skin of a bear. **2.** A tall military headdress made of black fur.

**beast** (bēst) *n.* [ME *beste* < OFr. < Lat. *bestia*.] **1. a.** An animal. **b.** A large four-footed animal. **2.** Animal nature. **3.** A brutal person.

**beast epic** *n.* A long verse narrative in which the characters are animals with human feelings and motives.

**beast·ings** (bē′stĭngz) *n.* *var. of* BEESTINGS.

**beast·ly** (bēst′lē) *adj.* **-li·er, -li·est. 1.** Of or like a beast. **2.** Disagreeable: nasty. —*adv. Chiefly Brit.*—Used as an intensive <a *beastly* hot summer> **—beast′li·ness** *n.*

**beast of burden** *n.* A pack animal, as a mule.

**beat** (bēt) *v.* **beat, beat·en** (bēt′n) *or* **beat, beat·ing, beats.** [ME *beten* < OE *bēaten*.] —*vt.* **1.** To strike repeatedly. **2.** To punish by hitting: FLOG. **3.** To pound or strike against repeatedly <waves *beating* the rocky coastline> **4.** To shape by repeated blows: FORGE. **5.** To make flat by pounding or trampling. **6.** To mix rapidly with an instrument. **7.** To flap <birds *beating* their wings> **8.** To strike so as to produce a signal or music. **9.** To mark or count (time or rhythm) with the hands or with a baton. **10.** To defeat, as in a contest. **11.** *Informal.* To be superior: SURPASS <Driving *beats* walking.> **12.** *Slang.* To perplex: baffle. **13.** *Informal.* To avoid or counter the effects of: CIRCUMVENT <*beat* the rush-hour traffic> —*vi.* **1.** To inflict repeated blows. **2.** To pulsate rhythmically. **3.** *Physics.* To cause beating by superposing waves of different frequencies. **4.** To emit sound when struck. **5.** To sound a signal <drums *beating*> **6.** To flap repeatedly. **7.** To hunt through woods or underbrush in search of game. **8.** *Naut.* To progress against the wind by tacking. **—beat back.** To force to retreat. **—beat down.** To force or persuade (a seller) to accept a lower price. **—beat off.** To drive away. **—beat out.** *Baseball.* To make a hit on a ground ball by fast running to first base. —*n.* **1.** A stroke or blow, esp. one producing a sound or acting as a signal. **2.** A periodic pulsation. **3.** *Physics.* An amplitude pulse produced by beating. **4.** *Mus.* **a.** A regular and rhythmical unit of time. **b.** The gesture used by a conductor to indicate this unit of time. **c.** The symbol representing this unit of time. **5.** The measured and rhythmical sound of verse: METER. **6.** The area regularly covered by a police officer, sentry, or newspaper reporter. **7.** *Slang.* Reportage of a news item obtained ahead of one's competitors. **8.** A beatnik —*adj.* **1.** *Informal.* Fatigued. **2.** Of or relating to beatniks. **—beat around (or about) the bush.** To approach a subject in a roundabout way. **—beat it.** *Slang.* To get going.

**beat·en** (bēt′n) *adj.* **1.** Made thin or formed by hammering <*beaten* gold> **2.** Worn by continuous use: familiar and much traveled. **3.** BEAT 1.

**beat·er** (bē′tər) *n.* **1.** One that beats, esp. an instrument for beating food. **2.** One who drives wild game from under cover for a hunter.

**be·a·tif·ic** (bē′ə-tĭf′ĭk) *adj.* [LLat. *beatificus*: Lat. *beatus*, p.part. of *beare*, to bless + *facere*, to make.] Exhibiting or producing exalted joy or blessedness. **—be′a·tif′i·cal·ly** *adv.*

**be·at·i·fy** (bē-ăt′ə-fī′) *vt.* **-fied, -fy·ing, -fies.** [Fr. *beatifier* < LLat. *beatificare*: Lat. *beatus*, p.part. of *beare*, to bless + Lat. *facere*, to make.] **1.** To make blessedly happy. **2.** *Rom. Cath. Ch.* To proclaim (a deceased person) to be one of the blessed and worthy of public religious honor. **3.** To exalt above all others. **—be·at′i·fi·ca′tion** *n.*

**beat·ing** (bē′tĭng) *n.* **1.** Punishment by whipping. **2.** A defeat. **3.** A pulsation, as of the heart. **4.** *Physics.* Periodic alternation of amplitude maxima and minima produced by interference between two waves of different frequency.

**be·at·i·tude** (bē-ăt′ĭ-to͞od′, -tyo͞od′) *n.* [Fr. *béatitude* < Lat. *beatitudo* < *beare*, to bless.] **1.** Supreme blessedness or happiness. **2. Beatitude.** One of nine declarations of blessedness made by Jesus in the Sermon on the Mount.

**beat·nik** (bēt′nĭk) *n.* [< *beat generation*, a group of unconventional young people of the 1950's + -NIK.] One acting and dressing with pointed, often exaggerated disregard for what is deemed proper, and given to radical and extravagant social criticism or self-expression.

**beau** (bō) *n.*, *pl.* **beaus** *or* **beaux** (bōz) [Fr., handsome < Lat. *bellus*.] **1.** The sweetheart of a girl or woman. **2.** A man who is excessively interested in fine clothes and social etiquette: DANDY.

**Beau Brum·mell** (bō brŭm′əl) *n.* [After George Bryan ("*Beau*") *Brummell* (1778–1840).] BEAU 2.

**Beau·fort scale** (bō′fərt) *n.* [After Sir Francis *Beaufort* (1774–1857).] A scale on which successive ranges of wind velocities are assigned code numbers from 0 to 12 or from 0 to 17.

**beau geste** (bō zhĕst′) *n.*, *pl.* **beaux gestes** (bō zhĕst′) *or* **beau gestes** (bō zhĕst′) [Fr.] **1.** A gracious gesture. **2.** A seemingly noble, yet meaningless gesture.

**beau i·de·al** (bō′ ī-dē′əl) *n.*, *pl.* **beau i·de·als.** [Fr. *beau idéal*, ideal beauty.] **1.** The concept of perfect beauty. **2.** An idealized model.

**Beau·jo·lais** (bō′zhō-lā′) *n.* [After *Beaujolais*, a region of central France.] A red table wine of French origin.

**beau monde** (bō mŏnd′, môɴd′) *n.*, *pl.* **beaux mondes** (bō môɴd′) *or* **beau mondes** (bō mŏndz′) [Fr., fine world.] Fashionable society.

**beaut** (byo͞ot) *n.* [Short for BEAUTY.] *Slang.* BEAUTY 4.

**beau·te·ous** (byo͞o′tē-əs) *adj.* Beautiful. **—beau′te·ous·ly** *adv.* **—beau′te·ous·ness** *n.*

**beau·ti·cian** (byo͞o-tĭsh′ən) *n.* One skilled in cosmetic treatments, as in a beauty parlor.

**beau·ti·ful** (byo͞o′tə-fəl) *adj.* Having qualities that delight the eye. **—beau′ti·ful·ly** *adv.* **—beau′ti·ful·ness** *n.*

☆ ***syns:*** BEAUTIFUL, BEAUTEOUS, BONNY, COMELY, FAIR, GOOD-LOOKING, GORGEOUS, LOVELY, PRETTY, PULCHRITUDINOUS, RAVISHING *adj. core meaning*: having qualities that delight the eye <*a beautiful* face> ***ant***: ugly

**beautiful people** *also* **Beautiful People** *pl.n.* Prominent people, esp. in international society.

**beau·ti·fy** (byōō′tə-fī′) *vt.* & *vi.* **-fied, -fy·ing, -fies.** To make or become beautiful. **—beau′ti·fi·ca′tion** *n.* **—beau′ti·fi′er** *n.*
**beau·ty** (byōō′tē) *n., pl.* **-ties.** [ME *beaute* < OFr. *bealte* < VLat. **bellitas* < Lat. *bellus,* beautiful.] **1.** A pleasing quality associated with harmony of form or color, excellence of craftsmanship, truthfulness, originality, or another often unspecifiable property. **2.** One that is beautiful, esp. a beautiful woman. **3.** A quality or feature that is most effective, gratifying, or telling <The *beauty* of this venture is that we risk nothing.> **4.** An outstanding or conspicuous example <a *beauty* of a blunder>
**beau·ty·bush** (byōō′tē-bŏŏsh′) *n.* A shrub native to China, *Kolkwitzia amabilis,* grown for its profusely blooming pink flowers.
**beauty parlor** *n.* An establishment providing women with services including hair treatment, manicures, and facials.
**beauty salon** *n.* A beauty parlor.
**beauty shop** *n.* A beauty parlor.
**beauty spot** *n.* **1.** A small black mark penciled or glued on a woman's face or shoulders to accentuate the fairness of her skin or conceal an imperfection. **2.** A mole or freckle.
**beaux** (bōz) *n. var. pl. of* BEAU.
**beaux-arts** (bō-zär′) *pl.n.* [Fr.] The fine arts.
**beaux es·prits** (bō′zĕ-sprē′) *n. pl. of* BEL ESPRIT.
**beaux gestes** (bō zhĕst′) *n. var. pl. of* BEAU GESTE.
**beaux mondes** (bō môNd′) *n. var. pl. of* BEAU MONDE.
**bea·ver**[1] (bē′vər) *n.* [ME *bever* < OE *beofor.*] **1. a.** A large aquatic rodent of the genus *Castor,* having thick brown fur, webbed hind feet, a paddlelike hairless tail, and chisellike front teeth adapted for gnawing bark and felling trees used to build dams. **b.** The fur of a beaver. **2.** A top hat orig. made of the beaver's underfur. **3.** A napped wool fabric, similar to felt, used for outer garments.
**bea·ver**[2] (bē′vər) *n.* [ME *baviere* < OFr., child's bib < *bave,* saliva.] **1.** Armor attached to a helmet or breastplate to protect the mouth and chin. **2.** The visor on a helmet.
**bea·ver·board** (bē′vər-bôrd′, -bōrd′) *n.* [Orig. a trademark.] A light, semirigid building material of compressed wood pulp, used for walls and partitions.
**be·bop** (bē′bŏp′) *n.* [Imit. of a two-beat phrase in this music.] BOP[2].
**be·calm** (bĭ-käm′) *vt.* **-calmed, -calm·ing, -calms. 1.** To render motionless for lack of wind. **2.** To make calm or still.
**be·came** (bĭ-kām′) *v. p.t. of* BECOME.
**be·cause** (bĭ-kôz′, -kŭz′) *conj.* [ME.] For the reason that : SINCE. ***usage:*** *Because* sometimes occurs in informal speech to mean "just because," as in *Because they work hard they think they should be promoted.* This use should be avoided in formal style.
**because of** *prep.* By reason of : on account of.
**bec·ca·fi·co** (bĕk′ə-fē′kō) *n., pl.* **-cos.** [Ital. : *beccare,* to peck (< *becco,* beak < Lat. *beccus*) + *fico,* fig (< Lat. *ficus*).] A small songbird eaten as a delicacy in Italy.
**bé·cha·mel sauce** (bā′shə-mĕl′) *n.* [Fr. *sauce béchamelle,* after Louis de *Béchamel* (d. 1703), its inventor.] A white sauce of butter, flour, and milk or cream.
**be·chance** (bĭ-chăns′) *vi.* & *vt.* **-chanced, -chanc·ing, -chanc·es.** *Archaic.* To happen or happen to.
**bêche-de-mer** (bĕsh′də-mâr′) *n., pl.* **bêches-de-mer** (bĕsh′də-mâr′) [Fr.] **1.** The trepang. **2.** A lingua franca that combines Malay and English, spoken in the southwest Pacific.
**Bech·u·a·na** (bĕch′ōō-ä′nə) *n., pl.* **Bechuana** *or* **-nas. 1.** A member of a Bantu people inhabiting Botswana in south-central Africa. **2.** The Bantu language of the Bechuana.
**beck**[1] (bĕk) *n.* [ME *bek,* an order < *bekken,* to beckon.] A gesture of beckoning or summons. **—at (one's) beck and call.** Ready to comply with any wish or command.
**beck**[2] (bĕk) *n.* [ME < ON *bekkr.*] *Chiefly Brit.* A small brook.
**beck·et** (bĕk′ĭt) *n.* [Orig. unknown.] *Naut.* A device, as a looped rope, hook and eye, strap, or grommet, for holding or fastening loose ropes, spars, or oars in position.
**beck·on** (bĕk′ən) *v.* **-oned, -on·ing, -ons.** [ME *beknen* < OE *bēcnan.*] *—vt.* **1.** To signal or summon, as by waving. **2.** To attract as if with gestures <"a lovely, sunny country that seemed to *beckon* them on to the Emerald City" —L. Frank Baum>. *—vi.* **1.** To make a signaling or summoning gesture. **2.** To have a strong attraction. **—beck′on** *n.* **—beck′on·er** *n.* **—beck′on·ing·ly** *adv.*
**be·cloud** (bĭ-kloud′) *vt.* **-cloud·ed, -cloud·ing, -clouds.** To darken with or as if with clouds : OBSCURE.
**be·come** (bĭ-kŭm′) *v.* **-came** (-kām′), **-come, -com·ing, -comes.** [ME *becomen* < OE *becuman.*] *—vi.* To grow or come to be. *—vt.* **1.** To be appropriate or suitable to. **2.** To look good with <a new suit that *becomes* you> **—become of.** To be the fate or subsequent condition of.
**be·com·ing** (bĭ-kŭm′ĭng) *adj.* **1.** Suitable : appropriate. **2.** Pleasing : attractive. **—be·com′ing·ly** *adv.* **—be·com′ing·ness** *n.*
**Bec·que·rel ray** (bĕ-krĕl′, bĕk′ə-rĕl′) *n.* [After Antoine Henri *Becquerel* (1852–1908).] *Obs.* Radiation associated with radioactivity.



**bed** (bĕd) *n.* [ME < OE.] **1. a.** A piece of furniture for reclining and sleeping, typically including a flat, rectangular frame and a mattress resting on springs. **b.** A bedstead. **c.** A mattress. **2.** A place where one may sleep : LODGING. **3.** A time at which one goes to sleep. **4.** A small plot of cultivated or planted land. **5.** The bottom of a body of water, as a watercourse. **6.** A supporting, underlying, or securing part, esp.: **a.** A layer of food surmounted by another kind of food. **b.** A foundation of crushed rock or a similar substance for a road or railroad : ROADBED. **c.** A layer of mortar upon which stones or bricks are laid. **7.** *Geol.* **a.** A rock mass of large horizontal extent bounded, esp. above, by physically different material. **b.** A deposit, as of ore, parallel to the local stratification. *—v.* **bed·ded, bed·ding, beds.** *—vt.* **1.** To furnish with a bed or sleeping quarters <*bedded* the guests down in the livingroom> **2.** To put to bed. **3.** To plant in a prepared bed of soil. **4.** To lay flat or arrange in layers. **5.** To embed. *—vi.* **1.** To go to bed. **2.** To form strata.
**be·daub** (bĭ-dôb′) *vt.* **-daubed, -daub·ing, -daubs. 1.** To smear : soil. **2.** To ornament in a vulgar and showy fashion <*bedaubed* the performers' costumes with gold sequins>
**be·daz·zle** (bĭ-dăz′əl) *vt.* **-zled, -zling, -zles.** To dazzle so much as to confuse or blind. **—be·daz′zle·ment** *n.*
**bed·bug** *also* **bed bug** (bĕd′bŭg′) *n.* A wingless, bloodsucking insect, *Cimex lectularius,* with a flat reddish body and a disagreeable odor, that often infests human dwellings.
**bed·cham·ber** (bĕd′chām′bər) *n.* A bedroom.
**bed·clothes** (bĕd′klōz′, -klōthz′) *pl.n.* Coverings, such as sheets and blankets, used on a bed.
**bed·ding** (bĕd′ĭng) *n.* **1.** Bedclothes. **2.** Straw or similar material for animals to sleep on. **3.** A foundation or bottom layer. **4.** *Geol.* Stratification of rocks into beds.
**be·deck** (bĭ-dĕk′) *vt.* **-decked, -deck·ing, -decks.** To adorn in a showy way.
**bedes·man** (bēdz′mən) *n.* [ME *bedeman* : *bede,* prayer (< OE *gebed*) + *man,* man (< OE mann).] An almsman.
**be·dev·il** (bĭ-dĕv′əl) *vt.* **-iled, -il·ing, -ils** *or* **-illed, -il·ling, -ils. 1.** To torment devilishly : PLAGUE. **2.** To worry, annoy, or frustrate. **3.** To possess with or as if with a devil : BEWITCH. **4.** To ruin : spoil. **—be·dev′il·ment** *n.*
**be·dew** (bĭ-dōō′, -dyōō′) *vt.* **-dewed, -dew·ing, -dews.** To wet with or as if with dew.
**bed·fel·low** (bĕd′fĕl′ō) *n.* **1.** A person with whom one shares a bed : BEDMATE. **2.** A collaborator, associate, or ally.
**Bed·ford cord** (bĕd′fərd) *n.* [After *Bedford,* England.] A heavy fabric in a ribbed weave with wide or narrow raised cords very similar to corduroy.
**be·dight** (bĭ-dīt′) *vt.* **-dight** *or* **-dight·ed, -dight·ing, -dights.** *Archaic.* [ME *bidighten.*] To dress or adorn.
**be·dim** (bĭ-dĭm′) *vt.* **-dimmed, -dim·ming, -dims.** To make dim.
**be·di·zen** (bĭ-dī′zən, -dĭz′ən) *vt.* **-zened, -zen·ing, -zens.** To bedeck. **—be·di′zen·ment** *n.*
**bed·lam** (bĕd′ləm) *n.* [ME *Bedlem,* Hospital of St. Mary of *Bethlehem,* London, an insane asylum.] **1.** A place or situation of noisy uproar and confusion. **2.** *Archaic.* A lunatic asylum.
**bed·lam·ite** (bĕd′lə-mīt′) *n.* A madman : lunatic.
**Bed·ling·ton terrier** (bĕd′lĭng-tən) *n.* [After *Bedlington,* England.] A dog orig. bred in England, with a woolly grayish or brownish coat.

**Bedlington terrier**
*16½ inches high at shoulder*

**bed·mate** (bĕd′māt′) *n.* BEDFELLOW 1.
**bed molding** *n.* **1.** The molding between the corona and frieze of an entablature. **2.** A molding below a projecting part.
**bed of roses** *n.* A state of idyllic comfort or luxury.
**Bed·ou·in** *also* **Bed·u·in** (bĕd′ōō-ĭn, bĕd′wĭn) *n.* [ME *Bedoin* < OFr. *beduin* < Ar. *badāwīn,* pl. of *badāwī* < *badw,* desert.] An Arab of one of the nomadic North African, Arabian, and Syrian tribes.
**bed·pan** (bĕd′păn′) *n.* **1.** A metal, glass, or plastic receptacle for the excreta of bedridden persons. **2.** A warming pan.
**bed·plate** (bĕd′plāt′) *n.* A plate, frame, or platform functioning as a base or support for a machine.
**bed·post** (bĕd′pōst′) *n.* A vertical post at the corner of a bed.
**be·drag·gle** (bĭ-drăg′əl) *vt.* **-gled, -gling, -gles.** To cause to become wet and limp.
**be·drag·gled** (bĭ-drăg′əld) *adj.* **1. a.** Wet and limp. **b.** Soiled by or as if by dragging in the mud. **2.** Deteriorated : dilapidated.

**bed·rid·den** (bĕd′rĭd′n) *also* **bed·rid** (-rĭd′) *adj.* [ME *bedreden* < OE *bedrida,* bedridden person : *bed,* bed + *rīda,* rider (< *rīdan,* to ride).] Confined to one's bed due to illness or infirmity.

**bed·rock** (bĕd′rŏk′) *n.* **1.** The solid rock underlying all soil, sand, clay, gravel, and loose material on the earth's surface. **2.** The lowest or bottom level. **3.** Fundamental principles.

**bed·roll** (bĕd′rōl′) *n.* A portable roll of bedding used esp. by campers and others who sleep outdoors.

**bed·room** (bĕd′ro͞om′, -ro͝om′) *n.* A room for sleeping. —*adj.* **1.** Dealing with or suggestive of sexual relations <*bedroom* eyes> **2.** Relating to or inhabited by commuters <*bedroom* communities>

**bed·side** (bĕd′sīd′) *n.* The space alongside a bed. **—bed′side′** *adj.*

**bedside manner** *n.* The attitude and conduct of a physician in the presence of a patient.

**bed·sit·ter** (bĕd′sĭt′ər) *or* **bed-sit** (-sĭt′) *n. Chiefly Brit.* A bed-sitting-room.

**bed-sit·ting-room** (bĕd-sĭt′ĭng-ro͞om′, -ro͝om′) *n. Chiefly Brit.* A one-room apartment serving as a bedroom and a sitting room.

**bed·so·ni·a** (bĕd-sō′nē-ə) *n.* [NLat., after Sir Samuel P. *Bedson* (d. 1969).] Any of a group of intracellular parasites, as of the genus *Chlamydia,* including the causative agents of diseases such as trachoma and psittacosis.

**bed·sore** (bĕd′sôr′, -sōr′) *n.* A pressure-induced skin ulceration occurring during long confinement to bed.

**bed·spread** (bĕd′sprĕd′) *n.* A usu. decorative bed cover.

**bed·spring** (bĕd′sprĭng′) *n.* The springs that support the mattress of a bed.

**bed·stead** (bĕd′stĕd′) *n.* The frame that supports a bed.

**bed·straw** (bĕd′strô′) *n.* Any of various plants of the genus *Galium,* with whorled leaves, small white or yellow flowers, and prickly burrs.

**bed·time** (bĕd′tīm′) *n.* The time when one goes to bed.

**Bed·u·in** (bĕd′o͞o-ĭn, bĕd′wĭn) *n. var. of* BEDOUIN.

**bed-wet·ting** (bĕd′wĕt′ĭng) *n.* Enuresis, esp. when occurring in bed at night. **—bed wetter** *n.*

**bee**[1] (bē) *n.* [ME < OE *bēo.*] **1.** Any of various winged, hairy-bodied, usu. stinging insects of the order Hymenoptera, including many solitary species as well as the social members of the family Apidae, characterized by specialized structures for sucking nectar and gathering pollen from flowers. **2.** A social gathering where people combine work, competition, and amusement <a quilting *bee*>

**bee**[2] (bē) *n.* [ME *bei,* a metal ring < OE *bēag.*] A bee block.

**bee**[3] (bē) *n.* The letter *b.*

**bee balm** *n.* Oswego tea.

**bee block** *n. Naut.* A piece of hardwood on either side of a bowsprit through which forestays are reeved.

**bee·bread** (bē′brĕd′) *n.* A brownish substance consisting of a mixture of pollen and nectar, fed by bees to their larvae.

**beech** (bēch) *n.* [ME *beche* < OE *bēce.*] **1.** A tree of the genus *Fagus,* characterized by smooth, light-colored bark and edible nuts partly enclosed in a prickly husk, esp. *F. grandifolia* of eastern North America and *F. sylvatica* of Europe. **2.** The wood of a beech.

**beech·drops** (bēch′drŏps′) *n., pl.* **beechdrops.** A leafless plant, *Epifagus virginiana* of eastern North America, that has brownish or purplish flowers and is parasitic on the roots of the beech tree.

**beech·nut** (bēch′nŭt′) *n.* The small edible nut of the beech tree.

**bee-eat·er** (bē′ē′tər) *n.* Any of various chiefly tropical Old World birds of the family Meropidae, with brightly colored plumage, that feed mostly on bees.

**beef** (bēf) *n., pl.* **beeves** (bēvz) [ME < OFr. *boef* < Lat. *bos.*] **1. a.** A full-grown steer, bull, ox, or cow, esp. one intended for use as meat. **b.** The meat of a slaughtered full-grown steer, bull, ox, or cow. **2.** *Informal.* Human muscle : BRAWN. **3.** *pl.* **beefs.** *Slang.* A complaint. —*vi.* **beefed, beef·ing, beefs.** *Slang.* To complain. **—beef up.** *Slang.* To make stronger, as by addition or reinforcement.

**beef·a·lo** (bē′fə-lō′) *n., pl.* **-loes** *or* **-los** *or* **-lo.** [BEEF + (BUFF)ALO.] A hybrid resulting from a cross between the American buffalo and domestic cattle.

**beef bour·gui·gnon** (bo͝or′gēn-yôn′) *n.* [Fr. *boeuf bourguignon* < *Bourgogne,* Burgundy, a region of France.] Braised beef cubes simmered in a red-wine sauce with mushrooms, carrots, and onions.

**beef·burg·er** (bēf′bûr′gər) *n.* A hamburger.

**beef·cake** (bēf′kāk′) *n.* [BEEF + (CHEESE)CAKE.] *Informal.* Photographs of minimally clad, muscular men.

**beef·eat·er** (bēf′ē′tər) *n.* **1.** A yeoman of the royal guard in England. **2.** A warder of the Tower of London. **3.** *Slang.* An Englishman.

**bee fly** *n.* Any of various flies of the family Bombyliidae, resembling bees and having larvae that are parasitic on the young of bees, wasps, and other insects.

**beef·steak** (bēf′stāk′) *n.* A slice of beef suitable for broiling or frying.

**beefsteak fungus** *n.* An edible fungus, *Fistularia hepatica,* growing on decaying wood and having a large, irregularly shaped reddish cap.

**beef stro·ga·noff** (strō′gə-nôf′, -nŏf′) *n.* [After Count Paul *Stroganoff* (1744?–1817).] Thinly sliced sautéed beef served with mushrooms and sour cream, often with flat noodles.

**beef·wood** (bēf′wo͝od′) *n.* [Perh. from its reddish color.] Any of various trees of the genus *Casuarina,* mostly native to Australia, with small scalelike leaves and flowers.

**beef·y** (bē′fē) *adj.* **-i·er, -i·est. 1.** Like beef. **2.** Muscular in physique : BRAWNY. **—beef′i·ness** *n.*

**bee gum** *n.* **1.** A hollow gum tree in which bees hive. **2.** A beehive, esp. one in a hollow gum tree.

**bee·hive** (bē′hīv′) *n.* **1.** A hive for bees. **2.** A place teeming with activity.

**bee·keep·er** (bē′kē′pər) *n.* One who keeps bees : APIARIST.

**bee·line** (bē′līn′) *n.* [From the belief that a pollen-laden bee flies straight to its hive.] A straight, direct course.

**Be·el·ze·bub** (bē-ĕl′zə-bŭb′) *n.* [LLat. *Beëlzebub* < Gk. *Beelzeboub* < Heb. *bá'al zbūb,* lord of the flies.] DEVIL 1.

**bee moth** *n.* A moth, *Galleria mellonella* or *Achroia grisella,* that lays its eggs in beehives where the larvae feed on the honeycombs and the young bees.

**been** (bĭn) *v. p.p. of* BE.

**beep** (bēp) *n.* [Imit.] A signaling or warning sound, as from a horn or an electronic device. —*vi.* & *vt.* **beeped, beep·ing, beeps.** To make or cause to make a beep.

**beep·er** (bē′pər) *n.* One that beeps, esp. a small portable electronic device that emits a beeping signal when the person carrying it is paged.

**bee plant** *n.* Any of various fragrant, nectar-bearing plants that attract bees.

**beer** (bîr) *n.* [ME *bere* < OE *bēor.*] **1.** A fermented alcoholic beverage brewed from malt and flavored with hops. **2.** A drink made from extracts of various roots and plants.

**beer·y** (bîr′ē) *adj.* **-i·er, -i·est. 1.** Tasting or smelling of beer. **2.** Affected or produced by beer <a *beery* grin>

**beest·ings** *also* **beast·ings** (bē′stĭngz) *n.* [ME *besting* < OE *bēost,* beestings.] *(sing. or pl. in number).* The first milk given by a mammal, as a cow, after parturition : COLOSTRUM.

**bees·wax** (bēz′wăks′) *n.* **1.** The yellowish to dark-brown wax secreted by the honeybee for making honeycombs. **2.** Commercial wax obtained by processing and purifying the crude wax of the honeybee and used in making crayons, candles, and polishes.

**beet** (bēt) *n.* [ME *bete* < OE *bēte* < Lat. *beta.*] **1.** A widely cultivated plant of the genus *Beta,* esp. *B. vulgaris,* having leaves occas. eaten as greens and a thickened, fleshy root. **2.** The dark red bulbous root of the beet, eaten as a vegetable.

**beet armyworm** *n.* An armyworm, *Spodoptera exigua,* that feeds mostly on alfalfa, beets, and other vegetables.

**bee·tle**[1] (bēt′l) *n.* [ME *bityl* < OE *bitela* < *bītan,* to bite.] **1.** Any of many insects of the order Coleoptera, having biting mouth parts and front wings modified to form horny wing covers that overlie the membranous rear wings when at rest. **2.** An insect like a beetle.

**bee·tle**[2] (bēt′l) *adj.* [ME *bitel-(brouwed),* having shaggy or protruding (eyebrows).] Overhanging : jutting <*beetle* brows> —*vi.* **-tled, -tling, -tles.** To overhang : jut.

**bee·tle**[3] (bēt′l) *n.* [ME *betel* < OE *bīetel.*] **1.** A heavy mallet having a large wooden head. **2.** A small wooden household mallet. **3.** A heavy wooden club used in stamping and finishing handmade linen. **4.** A cloth-finishing machine that stamps cloth with revolving wooden hammers.

**bee·tle·bung** (bēt′l-bŭng′) *n.* The sour gum.

**bee·tle·weed** (bēt′l-wēd′) *n.* The galax.

**bee tree** *n.* **1.** A hollow tree in which bees live. **2.** A tree, as the basswood, having flowers rich in nectar.

**beet·root** (bēt′ro͞ot′, -ro͝ot′) *n. Chiefly Brit.* The root of the beet.

**beetroot purple** *n.* A deep to very deep purplish red.

**beeves** (bēvz) *n. pl. of* BEEF.

**be·fall** (bĭ-fôl′) *v.* **-fell** (-fĕl′), **-fall·en** (-fô′lən), **-fall·ing, -falls.** [ME *befallen* < OE *befeallen,* to fall.] —*vi.* To happen, esp. by chance. —*vt.* To happen to.

**be·fit** (bĭ-fĭt′) *vt.* **-fit·ted, -fit·ting, -fits.** To be suitable to or appropriate for.

**be·fit·ting** (bĭ-fĭt′ĭng) *adj.* **1.** Suitable : appropriate. **2.** Proper. **—be·fit′ting·ly** *adv.*

**be·fog** (bĭ-fôg′, -fŏg′) *vt.* **-fogged, -fog·ging, -fogs. 1.** To cover or obscure with or as if with fog. **2.** To cause confusion in : MUDDLE.

**be·fool** (bĭ-fo͞ol′) *vt.* **-fooled, -fool·ing, -fools. 1.** To make a fool of. **2.** To play a trick on : HOODWINK.

**be·fore** (bĭ-fôr′, -fōr′) *adv.* [ME < OE *beforan.*] **1.** In the past : EARLIER. **2.** In front : AHEAD. —*prep.* **1.** Previous to in time. **2.** In front of. **3.** In prospect for : AWAITING <Your entire future lies *before* you.> **4.** In the presence of. **5.** Under the consideration or jurisdiction of <the case *before* the judge> **6.** Superior to or in advance of, as in rank, state, or development. —*conj.* **1.** In advance of the time when <*before* we met> **2.** Rather than <I would die *before* I would desert my family.>

**be·fore·hand** (bĭ-fôr′hănd′, -fōr′-) *adv.* & *adj.* **1.** In anticipation. **2.** In advance : EARLY.

**be·fore·time** (bĭ-fôr′tīm′, -fōr′-) *adv. Archaic.* Formerly.

**be·foul** (bĭ-foul′) *vt.* **-fouled, -foul·ing, -fouls. 1.** To make dirty: SOIL. **2.** To cast aspersions on: SLANDER.

**be·friend** (bĭ-frĕnd′) *vt.* **-friend·ed, -friend·ing, -friends.** To behave as a friend to.

**be·fud·dle** (bĭ-fŭd′l) *vt.* **-dled, -dling, -dles. 1.** To perplex: confuse. **2.** To stupefy with or as if with alcoholic liquor.

**beg** (bĕg) *v.* **begged, beg·ging, begs.** [ME *beggen*, prob. < OFr. *begart*, beggar, ult. < MDu. *beggaert*.] —*vt.* **1.** To ask for as charity <*begged* a quarter> **2.** To ask earnestly for or of: ENTREAT <*begged* me to stay> —*vi.* **1.** To solicit alms. **2.** To make a humble or urgent plea. **—beg off.** To ask to be released from (e.g., an obligation). **—beg the question. 1.** To employ an argument that assumes as valid the very same argument that one is trying to prove. **2.** To dodge an issue.

**be·gan** (bĭ-găn′) *v. p.t. of* BEGIN.

**be·gat** (bĭ-găt′) *v. Archaic. var. p.t. of* BEGET.

**be·get** (bĭ-gĕt′) *vt.* **-got** (-gŏt′), **-got·ten** (-gŏt′n) *or* **-got, -get·ting, -gets.** [ME *begete* < OE *begietan*, to obtain.] **1.** To father <*beget* children> **2.** To cause to exist: PRODUCE. **—be·get′ter** *n.*

**beg·gar** (bĕg′ər) *n.* [ME < OFr. *begart*, ult. < MDu. *beggaert*.] **1.** One who begs, esp. for livelihood. **2.** An impoverished person. **3.** A rascal. —*vt.* **-gared, -gar·ing, -gars. 1.** To impoverish. **2.** To exhaust the resources of <enchantment that *beggars* all description>

**beg·gar·ly** (bĕg′ər-lē) *adj.* Of or relating to a beggar: POOR. **—beg′gar·li·ness** *n.*

**beg·gar's-lice** (bĕg′ərz-līs′) *n.* **1.** *(sing. or pl. in number).* A plant, as the stickseed, bearing small, prickly fruit that clings to clothing or animal fur. **2.** A seed of a beggar's-lice.

**beg·gar-ticks** (bĕg′ər-tĭks′) *n.* **1.** *(sing. or pl. in number).* A plant, esp. the bur marigold and the tick trefoil, having seeds that cling to clothing, often by barbed bristles. **2.** The seeds of a beggar-ticks.

**beg·gar·weed** (bĕg′ər-wēd′) *n.* A West Indian plant, *Desmodium purpureum*, grown in the southern United States as forage.

**beg·gar·y** (bĕg′ə-rē) *n.* **1.** Extreme poverty. **2.** Beggars as a group.

**be·gin** (bĭ-gĭn′) *v.* **-gan** (-găn′), **-gun** (-gŭn′), **-gin·ning, -gins.** [ME *beginnen* < OE *biginnan*.] —*vi.* **1.** To start to do something: COMMENCE. **2.** To come into being <when life on earth *began*> **3.** To do or be in the least degree <Your remark doesn't even *begin* to answer my question.> —*vt.* **1.** To start to do: COMMENCE. **2.** To cause to come into being: ORIGINATE.

☆ ***syns:*** BEGIN, ARISE, COMMENCE, ORIGINATE, START *v. core meaning:* to come into existence <an uprising that *began* with labor unrest>

**be·gin·ner** (bĭ-gĭn′ər) *n.* **1.** A person who begins something. **2.** A person who is just starting to learn or do something: NOVICE.

**be·gin·ning** (bĭ-gĭn′ĭng) *n.* **1.** The act or process of bringing or being brought into being: START. **2.** The time at which something begins or is begun. **3.** The location where something begins or is begun. **4.** A source: origin. **5.** The first part. **6.** *often* **beginnings.** An early, rudimentary phase.

**beginning rhyme** *n.* **1.** Rhyme at the beginning of consecutive lines of verse. **2.** Alliteration.

**be·gird** (bĭ-gûrd′) *vt.* **-girt** (-gûrt′) *or* **-gird·ed, -girt, -gird·ing, -girds.** To gird or encircle: SURROUND.

**be·gone** (bĭ-gôn′, -gŏn′) *interj.* [ME *begone* : *be*, imper. of *been*, to be + *gone*, gone.] —Used as an order of dismissal.

**be·go·nia** (bĭ-gōn′yə) *n.* [NLat. *Begonia*, genus name, after Michel *Bégon* (1638–1710).] Any of various chiefly tropical plants of the genus *Begonia*, having often brightly colored or veined and irregular leaves and variously colored waxy flowers.

**be·gor·ra** (bĭ-gôr′ə, -gŏr′ə) *interj.* [Alteration of *by God*.] *Ir.* —Used as a mild swearword.

**be·got** (bĭ-gŏt′) *v. p.t. & var. p.p. of* BEGET.

**be·got·ten** (bĭ-gŏt′n) *v. var. p.p. of* BEGET.

**be·grime** (bĭ-grīm′) *vt.* **-grimed, -grim·ing, -grimes.** To smear or soil with or as if with grime.

**be·grudge** (bĭ-grŭj′) *vt.* **-grudged, -grudg·ing, -grudg·es. 1. a.** To envy the possession or enjoyment of. **b.** To envy for a possession. **2.** To give with reluctance. **—be·grudg′er** *n.* **—be·grudg′ing·ly** *adv.*

**be·guile** (bĭ-gīl′) *vt.* **-guiled, -guil·ing, -guiles. 1.** To mislead by guile: DECEIVE. **2.** To deprive of by guile: CHEAT. **3.** To distract the attention of: DIVERT. **4.** To pass (time) pleasantly. **5.** To charm: delight. **—be·guile′ment** *n.* **—be·guil′er** *n.*

**be·guine** (bĭ-gēn′) *n.* [Fr. (West Indies) *béguine* < Fr. *béguin*, flirtation.] **1.** A ballroom dance based on a native dance of Martinique and St. Lucia. **2.** The music for the beguine.

**Beg·uine** (bā′gēn′, bā-gēn′) *n.* [OFr.] A member of any of several Roman Catholic lay sisterhoods existing in the Netherlands since the 12th cent.

**be·gum** (bā′gəm, bē′-) *n.* [Urdu < Turk. *begim*, possessive of *beg*, bey.] A high-ranking Moslem lady.

**be·gun** (bĭ-gŭn′) *v. p.p. of* BEGIN.

**be·half** (bĭ-hăf′, -häf′) *n.* [ME *bihalve* : *bi*, by + *half*, side. —see HALF.] Interest, support, or benefit. **—in behalf of.** In the interest of. **—on behalf of.** On the part of: speaking for.

**be·have** (bĭ-hāv′) *v.* **-haved, -hav·ing, -haves.** [ME *behaven*.] —*vi.* **1.** To act, react, function, or perform in a specified way. **2. a.** To conduct oneself in a specified way. **b.** To conduct oneself in a proper way. —*vt.* **1.** To conduct (oneself) properly. **2.** To conduct (oneself) in a specified way.

**be·hav·ior** (bĭ-hāv′yər) *n.* **1.** The manner in which one behaves. **2.** One's actions or reactions under specified circumstances. **3.** The manner in which a machine operates. **—be·hav′ior·al** *adj.* **—be·hav′ior·al·ly** *adv.*

**behavioral psychophysics** *n.* A branch of psychology dealing primarily with the measurement of sensory capacities in nonaberrant animal specimens.

**behavioral science** *n.* A science, as sociology or anthropology, that attempts to discover general truths about human social behavior. **—behavioral scientist** *n.*

**be·hav·ior·ism** (bĭ-hāv′yə-rĭz′əm) *n.* The psychological school holding that objectively observable organismic behavior constitutes the essential or exclusive scientific basis of psychological data and investigation and stressing the role of environment as a determinant of behavior. **—be·hav′ior·ist** *n.* **—be·hav′ior·is′tic** *adj.*

**behavior modification** *n.* Modification of behavioral traits through psychological means, as reinforcement and aversion therapy. **—behavior modifier** *n.*

**be·hav·iour** (bĭ-hāv′yər) *n. Chiefly Brit. var. of* BEHAVIOR.

**be·head** (bĭ-hĕd′) *vt.* **-head·ed, -head·ing, -heads.** [ME *biheveden* < OE *behēafdian* : *be-*, away from + *hēafod*, head.] To decapitate <*beheaded* many aristocrats during the revolution>

**be·held** (bĭ-hĕld′) *v. p.t & p.p.* of BEHOLD.

**be·he·moth** (bĭ-hē′məth, bē′ə-məth) *n.* [Heb. *bəhēmōth*, pl. of *bəhēmāh*, beast.] **1.** A huge animal, possibly the hippopotamus, described in the Old Testament. **2.** Something enormous.

**be·hest** (bĭ-hĕst′) *n.* [ME *bihest*, promise < OE *behæs*.] **1.** An authoritative command: ORDER. **2.** An urgent request.

**be·hind** (bĭ-hīnd′) *adv.* [ME *bihinde* < OE *behindan*.] **1.** In, to, or toward the rear <ran *behind*> **2.** In a place or condition that has been left or passed <I left my coat *behind*.> **3.** In arrears: LATE <*behind* in your mortgage payments> **4.** Below standard: in or into an inferior position <fell far *behind* in class> **5.** Slow <My watch is running *behind*.> **6.** *Archaic.* In reserve: yet to come. —*prep.* **1.** At the back of or in the rear of <I sat *behind* them.> **2.** On the farther side or other side of: BEYOND <*behind* the wall> **3.** In a place or time that has been passed or left by <Our troubles are *behind* us.> **4.** Later than <The train is *behind* schedule.> **5.** Less advanced than <*behind* us in high technology> **6. a.** Hidden or concealed by <*behind* the hand> **b.** In the background of: UNDERLYING <*Behind* their every action was selfishness.> **7.** Serving to support <had the entire army *behind* them> **8.** In pursuit of <the posse hard *behind* them> —*n. Informal.* The buttocks.

**be·hind·hand** (bĭ-hīnd′hănd′) *adv. & adj.* **1.** In arrears. **2.** Behind time: SLOW. **3.** In a backward state.

**be·hind-the-scenes** (bĭ-hīnd′thə-sēnz′) *adj.* Carried out in secret <*behind-the-scenes* negotiations>

**be·hold** (bĭ-hōld′) *vt.* **-held** (-hĕld′), **-hold·ing, -holds.** [ME *biholden* < OE *behealdan*.] To look upon. —Often used in the imperative. **—be·hold′er** *n.*

**be·hold·en** (bĭ-hōl′dən) *adj.* [ME *biholden*.] Obliged, as for a favor.

**be·hoof** (bĭ-ho͞of′) *n.* [ME *bihove* < OE *behōfe*.] Benefit: advantage.

**be·hoove** (bĭ-ho͞ov′) *v.* **-hooved, -hoov·ing, -hooves.** [ME *behǫven* < OE *behōfian*.] —*vt.* To be necessary or proper for <It *behooves* you to try to be polite.> —*vi.* To be necessary or proper.

**beige** (bāzh) *n.* [Fr.] **1.** A soft fabric of undyed and unbleached wool. **2.** A light grayish brown or yellowish brown to grayish yellow. —*adj.* Light grayish brown or yellowish brown to grayish yellow.

**be·ing** (bē′ĭng) *n.* **1.** The quality or state of having existence. **2. a.** An object, idea, or symbol that exists, is held to exist, or is depicted as existing. **b.** A person <"The artist after all is a solitary *being*" —Virginia Woolf> **3.** Basic or essential nature. **4.** *Philos.* **a.** That which can be conceived as existing. **b.** Absolute existence in its perfect and unqualified state: the essence of existence.

**Be·ja** (bā′jə) *n., pl.* **Beja. 1. a.** A pastoral nomadic tribe living between the Nile River and the Red Sea. **b.** A member of this tribe. **2.** The Cushitic language of the Beja.

**bel** (bĕl) *n.* [After Alexander Graham *Bell* (1847–1922).] The logarithm to the base 10 of the ratio of two levels of power, used to measure voltage or sound intensity and equal to 10 decibels.

**be·la·bor** (bĭ-lā′bər) *vt.* **-bored, -bor·ing, -bors. 1.** To attack with blows. **2.** To assail verbally. **3.** To harp on <*belabored* the point>

**be·la·bour** (bĭ-lā′bər) *v. Chiefly Brit. var. of* BELABOR.

**be·lat·ed** (bĭ-lā′tĭd) *adj.* Being too late: TARDY <a *belated* sympathy card> **—be·lat′ed·ly** *adv.* **—be·lat′ed·ness** *n.*

**be·lay** (bĭ-lā′) *v.* **-layed, -lay·ing, -lays.** [ME *beleggen*, to surround < OE *belecgan*.] —*vt.* **1.** *Naut.* To secure or make fast (e.g., a rope) by winding on a cleat or pin. **2.** To secure (a mountain climber) at the end of a length of rope. **3.** To cause to stop. —*vi.* **1.** To be made secure. **2.** To stop <*Belay* there!> —*n.* The securing of a rope on a rock or other projection in mountain climbing.

**belaying pin** *n.* A short, removable wooden or metal pin fitted in a hole in the rail of a boat and used for securing running gear.

**bel can·to** (bĕl kän′tō) *n.* [Ital., beautiful singing.] A style of operatic singing characterized by rich tonal lyricism and brilliant display of vocal technique.

**belch** (bĕlch) *v.* **belched, belch·ing, belch·es.** [ME *belchen* < OE *bealcan.*] —*vi.* **1.** To expel gas noisily from the stomach through the mouth. **2.** To erupt violently. **3.** To gush forth. —*vt.* **1.** To expel (gas) noisily from the stomach through the mouth : ERUCT. **2.** To eject violently <a volcano *belching* fire and ash> **—belch** *n.*

**bel·dam** *also* **bel·dame** (bĕl′dəm) *n.* [ME, grandmother : OFr. *bel,* beautiful (< Lat. *bellus*) + OFr. *dame,* woman. —see DAME.] An old, esp. ugly or loathsome woman.

**be·lea·guer** (bĭ-lē′gər) *vt.* **-guered, -guer·ing, -guers.** [Du. *belegeren: be-,* around + *leger,* camp.] **1.** To besiege by surrounding with troops. **2.** To harass : beset <*beleaguered* by complex problems and issues>

**bel·em·nite** (bĕl′əm-nīt′) *n.* [NLat. *belemnites* < Gk. *belemnon,* dart.] A cigar-shaped fossilized internal shell of an extinct cephalopod related to the cuttlefish.

**bel es·prit** (bĕl′ĕ-sprē′) *n., pl.* **beaux es·prits** (bō′zĕ-sprē′) [Fr., fine mind.] An intelligent, cultivated person.

**bel·fry** (bĕl′frē) *n., pl.* **-fries.** [ME *berfrei* < OFr., portable siege tower, of Germanic orig.] **1.** A bell tower, esp. one attached to a building. **2.** The part of a tower in which bells are suspended. **—bel′fried** *adj.*

**Bel·gae** (bĕl′gī′, -jē′) *pl.n.* [Lat.] An ancient Gallic people once inhabiting what is now Belgium and northern France.

**Bel·gian** (bĕl′jən) *n.* A native or inhabitant of Belgium. **—Bel′gian** *adj.*

**Belgian hare** *n.* A large, reddish-brown domestic rabbit orig. bred from Belgian stock in England.

**Bel·gic** (bĕl′jĭk) *adj.* **1.** Of or relating to Belgium or the Belgians. **2.** Of or relating to the Netherlands. **3.** Of or relating to the Belgae.

**Be·li·al** (bē′lē-əl, bēl′yəl) *n.* [Heb. *blīyya′al,* uselessness: *blīy,* without + *ya′al,* use.] **1.** A satanic personification of wickedness and ungodliness alluded to in the New Testament. **2.** One of the fallen angels in Milton's *Paradise Lost* who rebelled against God.

**be·lie** (bĭ-lī′) *vt.* **-lied, -ly·ing, -lies.** [ME *belīen* < OE *belēogan.*] **1.** To picture falsely : MISREPRESENT. **2.** To show to be false. **3.** To disappoint or leave unfulfilled. **4.** *Archaic.* To tell lies about.

**be·lief** (bĭ-lēf′) *n.* [ME *bileve,* alteration of OE *gelēafa.*] **1.** The mental act, condition, or habit of placing trust or confidence in a person or thing. **2.** Mental acceptance of or conviction in the truth or actuality of something. **3.** Something believed or accepted as true, esp. a particular tenet or a body of tenets accepted by a group of persons.
☆ ***syns:*** BELIEF, CONVICTION, FEELING, IDEA, NOTION, OPINION, PERSUASION, POSITION, SENTIMENT, VIEW *n. core meaning* : something believed or accepted as true <a *belief* in life after death><liberal political *beliefs*>

**be·lieve** (bĭ-lēv′) *v.* **-lieved, -liev·ing, -lieves.** [ME *bileven* < OE *belēfan.*] —*vt.* **1.** To accept as true or real. **2.** To credit with veracity. **3.** To expect or suppose : THINK. —*vi.* **1.** To have faith, esp. religious faith. **2.** To have faith or confidence : TRUST. **3.** To have confidence in the truth, value, or existence of something <*believed* in free will> **4.** To think : judge. **—be·liev′a·ble** *adj.* **—be·liev′er** *n.*
☆ ***syns:*** **1.** BELIEVE, ACCEPT, BUY, SWALLOW *v. core meaning* : to regard as true or real <*believed* the story> ***ant:*** disbelieve **2.** BELIEVE, CONSIDER, DEEM, FIGURE, HOLD, THINK *v. core meaning* : to have an opinion <*believe* that smoking is bad for one's health>

**be·like** (bĭ-līk′) *adv. Archaic.* Perhaps : probably.

**be·lit·tle** (bĭ-lĭt′l) *vt.* **-tled, -tling, -tles. 1.** To represent or speak of as unimportant : DISPARAGE. **2.** To cause to seem less or little. **—be·lit′tle·ment** *n.* **—be·lit′tler** *n.*

**bell**[1] (bĕl) *n.* [ME *belle* < OE.] **1.** A hollow metal instrument, usu. cup-shaped with a flared opening, that emits a metallic tone when struck. **2.** Something shaped like a bell, as: **a.** The round, flared mouth of certain musical wind instruments. **b.** The corolla of a flower. **c.** A hollow, usu. inverted vessel, as a diving bell. **3.** *Naut.* **a.** A stroke on a bell to mark the hour. **b.** The time indicated by the striking of a bell, divided into half hours. **4. bells.** Bell-bottoms. —*v.* **belled, bell·ing, bells.** —*vt.* **1.** To put a bell on. **2.** To shape or cause to flare like a bell. —*vi.* To take on the form of a bell. **—bell the cat.** To perform a daring act.

**bell**[2] (bĕl) *n.* [< ME *bellen,* to bellow < OE *bellan.*] The bellowing or baying cry of some animals, as a deer in rut or a beagle on the hunt. —*vi.* **belled, bell·ing, bells.** To bellow : bay.

**bel·la·don·na** (bĕl′ə-dŏn′ə) *n.* [Ital. : *bella,* beautiful + *donna,* lady.] **1.** A poisonous Eurasian plant, *Atropa belladonna,* with purplish-red, bell-shaped flowers and small black poisonous berries. **2.** An atropine powder or tincture derived from the leaves and roots of the belladonna and used to treat colic, asthma, and hyperacidity.

**belladonna lily** *n.* AMARYLLIS 1.

**bell·bird** (bĕl′bûrd′) *n.* Any of various tropical American birds of the family Cotingidae, with a bell-like call.

**bell-bot·tom** (bĕl′bŏt′əm) *adj.* Having legs flaring out at the bottom <*bell-bottom* jeans>

**bell-bot·toms** (bĕl′bŏt′əmz) *pl.n.* Trousers with flared legs.

**bell·boy** (bĕl′boi′) *n.* A boy or man employed by a hotel to assist guests, as by carrying luggage and running errands.

**bell buoy** *n.* A buoy fitted with a warning bell that is activated by the movement of the waves.

**belle** (bĕl) *n.* [Fr., beautiful < OFr. *bel* < Lat. *bella,* fem. of *bellus.*] An attractive, much-admired girl or woman.

**belle é·poque** (ā-pŭk′) *n.* [Fr., beautiful age.] An era of artistic and cultural refinement in a society, esp. in France at the turn of the century.

**Bel·ler·o·phon** *n.* (bə-lĕr′ə-fən, -fŏn′) *n.* [Lat. < Gk. *Bellerophōn.*] *Gk. Myth.* The Corinthian hero who, with the aid of the winged horse Pegasus, killed the Chimera.

**belles-let·tres** (bĕl-lĕt′rə) *n.* [Fr. : *belles,* fine + *lettres,* letters.] *(sing. in number).* Literature regarded for its aesthetic value rather than for its didactic or informative content.

**bel·let·rist** (bĕl-lĕt′rĭst) *n.* A writer of belles-lettres. **—bel·let′rism** *n.* **—bel′le·tris′tic** (bĕl′ĭ-trĭs′tĭk) *adj.*

**bell·flow·er** (bĕl′flou′ər) *n.* Any of various plants of the genus *Campanula,* having blue bell-shaped flowers.

**bell·hop** (bel′hŏp′) *n.* A bellboy.

**bel·li·cose** (bĕl′ĭ-kōs′) *adj.* [ME < Lat. *bellicosus* < *bellicus,* of war < *bellum,* war.] Pugnacious : warlike. **—bel′li·cose′ly** *adv.* **—bel′li·cos′i·ty** (-kŏs′ĭ-tē), **bel′li·cose′ness** *n.*

**bel·lig·er·ent** (bə-lĭj′ər-ənt) *adj.* [Lat. *belligerans, belligerant-,* pr.part. of *belligerare,* to wage war : *bellum,* war + *gerere,* to make.] **1.** Inclined or eager to fight. **2.** Of, relating to, or carrying on warfare. —*n.* **1.** One that is belligerent. **2.** One that is engaged in war. **—bel·lig′er·ence, be·lig′er·en·cy** *n.* **—bel·lig′er·ent·ly** *adv.*
☆ ***syns:*** BELLIGERENT, BELLICOSE, COMBATIVE, CONTENTIOUS, HOSTILE, PUGNACIOUS, TRUCULENT *adj. core meaning* : having a very aggressive attitude, ready to do battle <a *belligerent* drunk> ***ant:*** friendly

**bell jar** *n.* A cylindrical glass vessel with a rounded top and an open base used to protect and display fragile objects or to establish a controlled atmosphere or environment in scientific experiments.

**bell·man** (bĕl′mən) *n.* A town crier.

**bell metal** *n.* A tin and copper alloy used to make bells.

**Bel·lo·na** (bə-lō′nə) *n.* [Lat. < *bellum,* war.] *Rom. Myth.* The goddess of war.

**bel·low** (bĕl′ō) *v.* **-lowed, -low·ing, -lows.** [ME *belwen* < OE *belgan,* to be enraged.] —*vi.* **1.** To roar, as a large animal. **2.** To shout in a deep voice. —*vt.* To utter in a loud, powerful voice. —*n.* **1.** The roar of a large animal. **2.** A very loud utterance <*a bellow* of pain and rage> **—bel′low·er** *n.*

**bel·lows** (bĕl′ōz, -əz) *n.* [ME *belows,* pl. of *below,* a bellows < OE *belg,* bag, belly.] *(sing. or pl. in number).* **1.** A device for producing a strong current of air, as for sounding a pipe organ or increasing the draft to a fire, composed of a flexible, valved air chamber that is contracted and expanded by pumping to force the air through a nozzle. **2.** Something resembling a bellows, as the pleated windbag of an accordion. **3.** The lungs.

**bell pepper** *n.* **1.** A pepper plant, *Capsicum frutescens grossum,* cultivated for its edible fruit. **2.** The mild-flavored, bell-shaped fruit of the bell pepper, usu. red when ripe but often eaten when green.

**Bell's Law** *n.* [After Sir Charles *Bell* (1774–1842).] *Anat.* **1.** The law that in the spinal cord the dorsal roots are of sensory function and the ventral roots are of major function. **2.** The neurological law that in a reflex arc nerve impulses are conducted in only one direction.

**Bell's palsy** *n.* [After Sir Charles *Bell* (1774–1842).] A suddenly occurring unilateral facial paralysis of unknown etiology, presumed to be caused by virally induced swelling of the seventh nerve.

**bell·weth·er** (bĕl′wĕth′ər) *n.* **1.** A usu. castrated male sheep that wears a bell hung from its neck and is followed by a flock of sheep. **2.** A leader, esp. of a passive group of followers.

**bell·wort** (bĕl′wûrt′, -wôrt′) *n.* A plant of the genus *Uvularia* of eastern North America, with yellow bell-shaped flowers.

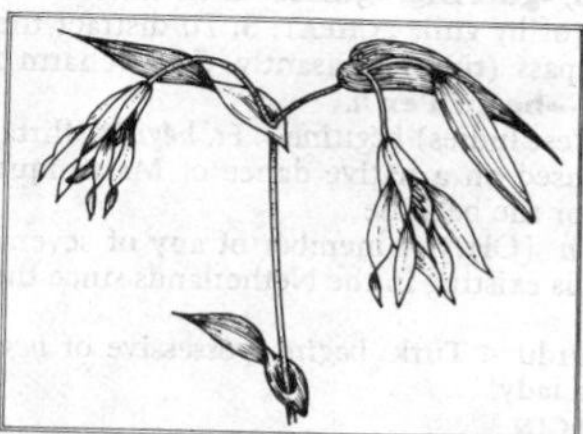

**bellwort**

**bel·ly** (bĕl′ē) *n., pl.* **-lies.** [ME *beli* < OE *belg.*] **1.** The part of the body of mammals between the rib cage and the pelvis that contains the intestines : ABDOMEN. **2.** The underside of certain vertebrate bodies, as those of snakes and fish. **3. a.** The stomach. **b.** Appetite for food. **4.** A protruding part <the *belly* of a sail> **5.** The womb : uterus. **6.** A deep hollow interior : BOWEL <a ship's *belly*> **7.** The bulging part of a muscle. **8.** The front part of the body of a stringed musical instrument : TABLE. —*vi. & vt.* **-lied, -ly·ing, -lies.** To bulge or cause to bulge.

**bel·ly·ache** (bĕl′ē-āk′) *n.* **1.** An ache in the stomach or abdomen. **2.** *Slang.* A complaint. —*vi.* **-ached, -ach·ing, -aches.** *Slang.* To grumble or complain, esp. in a whining way. **—bel′ly·ach′er** *n.*

**bel·ly·band** (bĕl′ē-bănd′) *n.* **1.** A band passed around the belly of an animal to secure something, as a saddle. **2.** An encircling cloth band for holding in a baby's protruding navel.

**bel·ly·but·ton** (bĕl′ē-bŭt′n) *n. Informal.* The navel.

**belly dance** *n.* A dance in which the performer makes sinuous movements of the belly. **—bel′ly-dance′** *v.* **(-danced, -danc·ing, -danc·es). —belly dancer** *n.*

**belly flop** *n.* A dive in which the front of the body hits flat against the surface of the water. **—bel′ly-flop′** *v.* **(-flopped, -flop·ping, -flops).**

**bel·ly·ful** (bĕl′ē-fo͝ol′) *n. Informal.* An amount exceeding what one desires or can endure.

**bel·ly-land** (bĕl′ē-lănd′) *vi.* **-land·ed, -land·ing, -lands.** To land an aircraft on its underside without deployment of landing gear. **—belly landing** *n.*

**belly laugh** *n.* A deep laugh.

**bel·o·ne·pho·bi·a** (bĕl′ə-nə-fō′bē-ə) *n.* [Gk. *belonē*, needle + -PHOBIA.] Abnormal fear of sharply pointed objects.

**be·long** (bĭ-lông′, -lŏng′) *vi.* **-longed, -long·ing, -longs.** [ME *belongen.*] **1.** To have a proper, appropriate, or suitable place. **2. a.** To be naturally associated with something. **b.** To fit into a group naturally <a kid who just didn't *belong*> **—belong to. 1.** To be the property or concern of. **2.** To be a member of (a group).

**be·long·ing** (bĭ-lông′ĭng, -lŏng′-) *n.* **1. belongings.** Personal possessions. **2.** Close and secure relationship <a sense of *belonging*>

**be·lov·ed** (bĭ-lŭv′ĭd, -lŭvd′) *adj.* [ME, p.part. of *beloven*, to love.] Dearly loved. **—be·lov′ed** *n.*

**be·low** (bĭ-lō′) *adv.* [ME *biloghe* : *bi*, by + *loogh*, low.] **1.** In or to a lower place : BENEATH. **2. a.** On or to a lower floor : DOWNSTAIRS. **b.** *Naut.* On or to a lower deck. **3.** Following or lower down on a page. **4.** Farther down, as along a slope or valley. **5.** In or to hell or Hades. **6.** On earth. **7.** In a lower rank or class. —*prep.* **1.** Underneath : beneath. **2.** Lower than, as on a graduated scale. **3.** Unworthy of or unsuitable to the rank or dignity of.

**Bel·shaz·zar** (bĕl-shăz′ər) *n.* [Heb. *Bēlshassar.*] The son of Nebuchadnezzar II and the last king of Babylon, who was warned of his downfall and death by the handwriting on the wall.

**belt** (bĕlt) *n.* [ME < OE < Lat. *balteus.*] **1.** A band, as of leather or cloth, worn around the waist to support garments, secure tools or weapons, or serve as decoration. **2.** Something resembling a belt <a *belt* of outbuildings> **3. a.** A beltway. **b.** A belt line. **4.** A strap that holds a person securely in a seat : SEAT BELT. **5.** A continuous band or chain for transferring motion or power or conveying materials from one wheel or shaft to another. **6.** A band of tough reinforcing material beneath the tread of a tire. **7.** A distinctive geographic region. **8.** *Slang.* A powerful blow : PUNCH. **9.** *Slang.* A strong emotional reaction. **10.** *Slang.* A drink of liquor <a *belt* of scotch> —*vt.* **belt·ed, belt·ing, belts. 1.** To encircle : gird. **2.** To attach with or as if with a belt. **3.** To mark with or as if with a belt. **4.** To strike with a belt. **5.** *Slang.* To strike forcefully : PUNCH. **6.** *Slang.* To sing in a loud and forceful way <*belt* out a love song> **—below the belt.** Not according to the rules : UNFAIRLY. **—tighten (one's) belt.** To become frugal. **—under (one's) belt.** Having been added to one's knowledge and experience.

**Bel·tane** (bĕl′tən) *n.* [ME < Sc. Gael. *bealltainn.*] **1.** May Day in the old Scottish calendar. **2.** The ancient Celtic May Day celebration.

**belt highway** *n.* A beltway.

**belt line** *n.* A transportation line, as of trains, that makes a complete circuit of an urban area.

**belt tightening** *n.* A decrease in spending.

**belt·way** (bĕlt′wā′) *n.* A highway skirting an urban area.

**be·lu·ga** (bə-lo͞o′gə) *n.* [R. *byelukha* < *byelii*, white.] **1.** The white whale. **2.** A sturgeon, *Huso huso* of the Black and Caspian seas, whose roe is used for caviar.

**bel·ve·dere** (bĕl′vĭ-dîr′) *n.* [Ital. : *bel*, beautiful (< Lat. *bellus*) + *vedere*, view (< Lat. *vidēre*, to see).] A structure, as an open roofed gallery, located so as to command a view.

**be·ma** (bē′mə) *n., pl.* **-ma·ta** (-mə-tə) [LLat. < Gk. *bēma*, platform.] **1.** The platform from which services are conducted in a synagogue. **2.** The enclosed area about the altar of an Eastern Orthodox church.

**be·mire** (bĭ-mīr′) *vt.* **-mired, -mir·ing, -mires. 1.** To soil with mud. **2.** To cause to bog down in mud.

**be·moan** (bĭ-mōn′) *v.* **-moaned, -moan·ing, -moans.** —*vt.* **1.** To mourn over : LAMENT. **2.** To express pity or grief for. —*vi.* To mourn.

**be·muse** (bĭ-myo͞oz′) *vt.* **-mused, -mus·ing, -mus·es. 1.** To cause to be bewildered : CONFUSE. **2.** To cause to be lost in thought. **—be·mus′ed·ly** (-myo͞o′zĭd-lē) *adv.*

**ben**[1] (bĕn) [ME *binne*, within < OE *binnan.*] *Scot.* —*n.* The inner room or parlor of a house. —*adv.* Inside : within. —*prep.* Within.

**ben**[2] (bĕn) *n.* [Dial. Ar. *bēn* < Ar. *bān.*] Any of several Asiatic trees of the genus *Moringa*, bearing winged seeds that yield an oil used in perfumes and cosmetics.

**Bence-Jones protein** (bĕns′jōnz′) *n.* [After Henry *Bence-Jones* (1814–1873).] An abnormal globulin group that can appear in serum and in urine in association with multiple myeloma.

**bench** (bĕnch) *n.* [ME < OE *benc.*] **1.** A long, often backless seat for two or more people. **2.** A thwart. **3. a.** The seat for judges in a courtroom. **b.** The office or position of a judge. **c.** The judge or judges constituting a court. **4. a.** A seat occupied by an official. **b.** The office of an official. **5.** A strong worktable <a carpenter's *bench*> **6.** A platform on which animals, esp. dogs, are exhibited. **7. a.** The place where the players on a team sit while not participating in a game. **b.** A team's reserve players. **8. a.** A narrow, level stretch of land interrupting a declivity. **b.** A level elevation of land along a shore, esp. one marking a former shoreline. —*vt.* **benched, bench·ing, bench·es. 1.** To provide with benches. **2.** To seat on a bench. **3.** To show (dogs) in a bench show. **4.** To keep out of or remove from a game <*benched* the quarterback>

**bench·er** (bĕn′chər) *n.* **1.** One who sits on a bench. **2.** *Chiefly Brit.* A member of the inner or higher bar who acts as a governor of one of the Inns of Court. **3.** One, as a magistrate, who occupies an official bench.

**bench mark** *also* **bench·mark** (bĕnch′märk′) *n.* **1.** A surveyor's mark made on a stationary object of previously determined position and elevation and used as a reference point in tidal observations and surveys. **2. benchmark.** A standard of measurement or evaluation.

**bench show** *n.* An indoor exhibition of small animals, esp. a competitive dog show.

**bench warrant** *n.* A warrant issued by a judge or court of law ordering the arrest of an offender.

**bend**[1] (bĕnd) *v.* **bent** (bĕnt), **bend·ing, bends.** [ME *benden* < OE *bendan.*] —*vt.* **1.** To bring into tension, as by drawing in with a string <*bend* a bow> **2.** To force to assume a curved or angular shape. **3.** To cause to swerve from a straight line : DEFLECT. **4.** To turn (e.g., one's attention) in a given direction. **5.** To influence coercively : SUBDUE. **6.** To apply (the mind) closely : CONCENTRATE. **7.** *Naut.* To fasten <*bend* a mainsail onto the boom> —*vi.* **1. a.** To turn or be altered from straightness or from an initial shape or position <This wire *bends* easily.> **b.** To assume a curved, crooked, or angular form or direction <The small trees *bent* in the wind.> **2.** To take a new direction : SWERVE. **3.** To incline the body : STOOP. **4.** To bow in submission : YIELD. **5.** To apply oneself closely : CONCENTRATE <*bend* to the task> —*n.* **1.** The act or fact of bending. **2.** The state of being bent. **3.** Something bent. **4.** *Naut.* **a. bends.** The thick planks in a ship's side : WALES. **b.** A knot that joins a rope to a rope or to another object. **5. bends** *(sing. or pl. in number).* Caisson disease. **—bend over backward.** To make more than the required effort.

**bend**[2] (bĕnd) *n.* [ME, prob. < OFr. *bende*, band.] *Heraldry.* A band running from the upper dexter corner of an escutcheon to the lower sinister corner.

**Ben Da·vis** (bĕn dā′vĭs) *n.* A variety of large red winter cooking apple grown in western North America.

**Ben Day** *also* **ben·day** *or* **Ben·day** (bĕn-dā′) *n.* [After *Benjamin Day* (1838–1916), its inventor.] A method of adding a tone to a printed image by imposing a transparent sheet of patterns, as dots, on the image at a given stage of a photographic reproduction process.

**bend·ed** (bĕn′dĭd) *v. Archaic. var. p.t. & p.p.* of BEND[1].

**bend·er** (bĕn′dər) *n.* **1.** One that bends. **2.** *Slang.* A drinking spree.

**bend sinister** *n. Heraldry.* A band running from the upper sinister corner of an escutcheon to the lower dexter corner.

**be·neath** (bĭ-nēth′) *adv.* [ME *binethe* < OE *bineoðan.*] **1.** In a lower place : BELOW. **2.** Underneath. —*prep.* **1.** Lower than : BELOW. **2.** Covered or concealed by <The ground lay *beneath* the snow.> **3.** Under the control, force, or influence of. **4. a.** Lower than, as in rank or station. **b.** Unworthy of <was *beneath* me to beg>

**ben·e·dict** (bĕn′ĭ-dĭkt′) *n.* [After *Benedick*, a character in *Much Ado About Nothing* by Shakespeare.] A newly married man who has previously been considered a confirmed bachelor.

**Ben·e·dic·tine** (bĕn′ĭ-dĭk′tĭn, -tēn′) *adj.* Of or pertaining to Saint Benedict of Nursia or the order he founded. **—Ben′e·dic′tine** *n.*

**ben·e·dic·tion** (bĕn′ĭ-dĭk′shən) *n.* [ME *benediccioun* < OFr. *benediction* < Lat. *benedictio* < *benedicere*, to bless : *bene*, well + *dicere*, to say.] **1.** A blessing. **2.** An invocation of divine blessing, usu. at the end of a church service. **3.** *often* **Benediction.** *Rom. Cath. Ch.* A short service consisting of prayers, the singing of a Eucharistic hymn, and the blessing of the congregation with the host. **4.** A blessed state. **—ben′e·dic′tive, ben′e·dic′to·ry** (-dĭk′tə-rē) *adj.*

**Ben·e·dic·tus** (bĕn′ĭ-dĭk′təs) *n.* [Lat., blessed.] **1.** A canticle that begins *Benedictus qui venit in nomine Domini*, "Blessed is he that cometh in the name of the Lord." **2.** A canticle that begins *Benedictus Dominus Deus Israel,* "Blessed be the Lord God of Israel."

**ben·e·fac·tion** (bĕn′ə-făk′shən, bĕn′ə-făk′-) *n.* [LLat. *benefactio* < Lat. *bene facere,* to do well.] **1.** Conferment of a benefit. **2.** A charitable act or gift.

**ben·e·fac·tor** (bĕn′ə-făk′tər) *n.* One who gives financial or other aid. **—ben′e·fac′tress** *n.*

**be·nef·ic** (bə-nĕf′ĭk) *adj.* [Lat. *beneficus* < *bene facere,* to do well.] Beneficent.

**ben·e·fice** (bĕn′ə-fĭs) *n.* [ME < OFr. < Med. Lat. *beneficium* < Lat., benefit < *beneficus,* benefic.] **1. a.** A church office endowed with fixed capital assets that provide a living. **b.** The revenue from such assets. **2.** A landed estate granted in feudal tenure. —*vt.* **-ficed, -fic·ing, -fic·es.** To endow or provide with a benefice.

**be·nef·i·cence** (bə-nĕf′ĭ-səns) *n.* [OFr. < Lat. *beneficentia* < *beneficus,* benefic.] **1.** The quality or state of being beneficent. **2.** BENEFACTION 2.

**be·nef·i·cent** (bə-nĕf′ĭ-sənt) *adj.* **1.** Marked by or performing kind or charitable acts. **2.** BENEFICIAL 1. **—be·nef′i·cent·ly** *adv.*

**ben·e·fi·cial** (bĕn′ə-fĭsh′əl) *adj.* [Fr. *bénéficial* < LLat. *beneficialis* < Lat. *beneficium,* benefit < *beneficus,* benefic.] **1.** Promoting a favorable result : ADVANTAGEOUS. **2.** *Law.* Receiving or having the right to receive proceeds or other advantages. **—ben′e·fi′cial·ly** *adv.* **—ben′e·fi′cial·ness** *n.*

☆ ***syns:*** BENEFICIAL, ADVANTAGEOUS, BENEFIC, BENEFICIENT, BENIGNANT, FAVORABLE, GOOD, PROPITIOUS, SALUTARY *adj. core meaning* : affording benefit <a *beneficial* climate> ***ant:*** detrimental

**ben·e·fi·ci·ar·y** (bĕn′ə-fĭsh′ē-ĕr′ē, -fĭsh′ə-rē) *n., pl.* **-ies. 1.** One that receives a benefit. **2.** *Law.* The recipient of benefits, as funds or property, as from an insurance policy or will. **3.** The holder of a church benefice. —*adj.* Relating to or holding a feudal benefice.

**ben·e·fit** (bĕn′ə-fĭt) *n.* [ME < OFr. *bienfait* < Lat. *benefactum,* good deed < *bene facere,* to do well.] **1.** Something promoting or enhancing well-being : ADVANTAGE. **2.** *Archaic.* A kindly act. **3.** Payments made or entitlements available in accord with a wage agreement, insurance contract, or public assistance program. **4.** A public entertainment, performance, or social event held to raise funds for a person or a worthy cause. —*v.* **-fit·ed, -fit·ing, -fits** *also* **-fit·ted, -fit·ting, -fits.** —*vt.* To be helpful or useful to. —*vi.* To gain advantage : PROFIT <*benefit* from past experience>

**benefit of clergy** *n.* **1.** Exemption from trial or punishment in a civil court given to the clergy in the Middle Ages. **2.** Authorized sanction of a religious rite.

**benefit of the doubt** *n.* A favorable judgment granted in the absence of complete evidence.

**be·nev·o·lence** (bə-nĕv′ə-ləns) *n.* **1.** An inclination or tendency to do kind or charitable acts. **2.** A kindly act. **3.** A compulsory tax or payment exacted by some English monarchs without parliamentary consent.

**be·nev·o·lent** (bə-nĕv′ə-lənt) *adj.* [ME < Lat. *benevolens,* well-wishing : *bene,* well + *volens,* pr.part. of *velle,* to wish.] **1.** Marked by or suggesting benevolence : KINDLY. **2.** Of, concerned with, or organized for the benefit of charity <a *benevolent* fund> **—be·nev′o·lent·ly** *adv.*

**Ben·ga·li** (bĕn-gô′lē, bĕng-) *n.* **1.** An inhabitant of Bengal. **2.** An inhabitant of Bangladesh. **3.** The modern Indic language of West Bengal and Bangladesh. —*adj.* Of or characteristic of Bengal, its inhabitants, or its language.

**ben·ga·line** (bĕng′gə-lēn′) *n.* [Fr. < *Bengal,* Bengal.] A fabric having a crosswise ribbed effect made of silk, wool, or synthetic fibers.

**Ben·gal light** (bĕn-gôl′, bĕng-) *n.* A firework that burns with a brilliant, sustained blue light, once used for signaling.

**be·night·ed** (bĭ-nī′tĭd) *adj.* **1.** Overtaken by darkness or night. **2.** Morally or intellectually unenlightened. **—be·night′ed·ly** *adv.* **—be·night′ed·ness** *n.*

**be·nign** (bĭ-nīn′) *adj.* [ME *benigne* < OFr. < Lat. *benignus* : *bene,* well + *genus,* born.] **1.** Having a kind disposition. **2.** Exhibiting gentleness and mildness. **3.** Tending to promote well-being : BENEFICIAL. **4.** *Pathol.* Not malignant <a *benign* tumor> **—be·nign′ly** *adv.*

**be·nig·nan·cy** (bĭ-nĭg′nən-sē) *n., pl.* **-cies.** Benignity.

**be·nig·nant** (bĭ-nĭg′nənt) *adj.* **1.** Favorable : beneficial. **2.** Kind and gracious, occas. in a patronizing way. **—be·nig′nant·ly** *adv.*

**be·nig·ni·ty** (bĭ-nĭg′nĭ-tē) *n., pl.* **-ties. 1.** The quality or state of being benign. **2.** A kindly or gracious act.

**ben·i·son** (bĕn′ĭ-zən, -sən) *n.* [ME *benisoun* < OFr. *beneisson* < Lat. *benedictio.* —see BENEDICTION.] A benediction : blessing.

**ben·ja·min** (bĕn′jə-mĭn) *n.* [Var. of *benjoin,* benzoin.] Benzoin.

**Ben·ja·min** (bĕn′jə-mĭn) *n.* [Heb. : *ben,* son + *yāmīn,* right hand.] The tribe of Israel descended from Benjamin, the youngest son of Jacob and Rachel. **—Ben′ja·mite′** (-mīt′) *n.*

**benjamin bush** *n.* SPICEBUSH 1.

**ben·ne** *or* **ben·ni** (bĕn′ē) *n.* [Of African orig.] **1.** The sesame plant. **2.** The seeds or oil of the sesame plant.

**ben·net** (bĕn′ĭt) *n.* Herb bennet.

**ben·ni** (bĕn′ē) *n. var. of* BENNE.

**Ben·ning·ton** *or* **Ben·ning·ton ware** (bĕn′ĭng-tən) *n.* Ceramic ware, as earthenware with a brown mottled glaze, made in Bennington, Vermont.

**ben·ny** (bĕn′ē) *n., pl.* **-nies.** [< BENZEDRINE.] *Slang.* An amphetamine tablet.

**bent[1]** (bĕnt) *adj.* [< p.p. of BEND.] **1.** Not straight : CROOKED. **2.** On a fixed course of action : DETERMINED <was *bent* on going to the concert> —*n.* **1.** A tendency, disposition, or inclination. **2.** Limit of endurance. **3.** A structural member used for strengthening a bridge or trestle transversely.

**bent[2]** (bĕnt) *n.* [ME, grassy plain < OE *beonet.*] **1.** Any of several grasses of the genus *Agrostis,* some species of which are used in lawn mixtures and for hay. **2.** The stiff stalk of various grasses. **3.** A moor.

**Ben·tham·ism** (bĕn′thə-mĭz′əm) *n.* The utilitarian philosophy of Jeremy Bentham, holding that pleasure is the chief end of life and that the greatest happiness for the greatest number should be the ultimate goal of people. **—Ben′tham·ite′** (-mīt′) *n.*

**ben·thos** (bĕn′thŏs′) *n.* [Gk.] **1.** The bottom of a sea or a lake. **2.** The organisms living on sea or lake bottoms. **—ben′thic** (bĕn′thĭk), **ben·thon′ic** (bĕn-thŏn′ĭk) *adj.*

**ben·ton·ite** (bĕn′tə-nīt′) *n.* [After Fort *Benton,* Montana.] Either of two principally aluminum silicate clays, containing some magnesium and iron, distinguished by sodium or calcium content with corresponding high or low swelling capacity, and used in various adhesives, cements, and ceramic fillers. **—ben′ton·it′ic** (-nĭt′ĭk) *adj.*

**bent·wood** (bĕnt′wŏŏd′) *n.* Wood that has been steamed until pliable and then bent and shaped. **—bent′wood′** *adj.*

**be·numb** (bĭ-nŭm′) *vt.* **-numbed, -numb·ing, -numbs.** [< ME *binomen,* p.part. of *benimen,* to take away < OE *beniman.*] **1.** To make numb, esp. by cold. **2.** To make inactive : DULL <"the anesthetic afternoon *benumbs,* sickens our senses" —Karl Shapiro> **—be·numb′ment** *n.*

**benz-** *pref. var. of* BENZO-.

**benz·al·de·hyde** (bĕn-zăl′də-hīd′) *n.* A colorless or yellowish, strongly reactive volatile oil, $C_6H_5CHO$, used as a solvent and flavoring and in perfumery.

**Ben·ze·drine** (bĕn′zĭ-drēn′). A trademark for a brand of amphetamine.

**ben·zene** (bĕn′zēn′, bĕn-zēn′) *n.* A clear, colorless, highly refractive flammable liquid, $C_6H_6$, derived from petroleum and used in the manufacture of chemical products such as DDT, detergents, insecticides, and motor fuels.

**benzene ring** *n.* The hexagonal ring structure in the benzene molecule and its substitutional derivatives, each vertex of which is occupied and distinguished by a carbon atom.

**benzene series** *n.* A series of chemically related aromatic hydrocarbons containing the benzene ring, the simplest member of which is benzene.

**ben·zi·dine** (bĕn′zĭ-dēn′) *n.* [Prob. BENZ(ENE) + -ID(E) + -INE.] A yellowish, white, or reddish-gray crystalline powder, $C_{12}H_{12}N_2$, used in making dyes and to detect blood stains.

**ben·zine** (bĕn′zēn′, bĕn-zēn′) *also* **ben·zin** (bĕn′zĭn) *n.* **1.** Ligroin. **2.** Benzene.

**benzo-** *or* **benz-** *pref.* [< BENZOIN.] Benzene : benzoic acid <*benzophenone*>

**ben·zo·ate** (bĕn′zō-āt′) *n.* A salt or ester of benzoic acid.

**benzoate of soda** *n.* Sodium benzoate.

**ben·zo·caine** (bĕn′zə-kān′) *n.* A white, odorless, tasteless crystalline ester, $C_9H_{11}NO_2$, used as a local anesthetic.

**ben·zo·di·az·e·pine** (bĕn′zō-dī-ăz′ə-pēn′, -pĭn) *n.* [BENZO- + DIAZEP(AM) + -INE.] Any of several chemical compounds used as sedatives and muscle relaxants.

**ben·zo·ic acid** (bĕn-zō′ĭk) *n.* A white crystalline acid, $C_6H_5COOH$, used to season tobacco and in perfumes, dentifrices, and germicides.

**ben·zo·in** (bĕn′zō-ĭn, -zoin′) *n.* [Fr. *benjoin* < Ital. *benzoino* < Ar. *lubān jāwī,* frankincense of Java.] **1.** Any of several resins containing benzoic acid, obtained as a gum from various trees of the genus *Styrax* and used in ointments, perfumes, and medicine. **2.** Any of various aromatic shrubs and trees of the genus *Lindera,* including the spicebush. **3.** A white or yellowish crystalline compound, $C_{14}H_{12}O_2$, derived from benzaldehyde and used as an antiseptic.

**ben·zol** (bĕn′zôl′, -zōl′) *n.* Benzene.

**ben·zo·phe·none** (bĕn′zō-fĭ-nōn′, -fē′nōn′) *n.* A white crystalline compound, $(C_6H_5)_2CO$, used in perfumery and in medicine.

**ben·zo·py·rine** (bĕn′zō-pī′rēn′, -pī-rēn′) *n.* A yellow, crystalline, aromatic hydrocarbon, $C_{20}H_{12}$, that is a carcinogen found in coal tar and cigarette smoke.

**ben·zo·yl** (bĕn′zō-ĭl′) *n.* The univalent radical $C_6H_5CO$, derived from benzoic acid.

**benzoyl peroxide** *n.* A flammable white granular solid, $(C_6H_5CO)_2O_2$, used as a bleaching agent for flour, fats, waxes, and oils, as a polymerization catalyst, and in pharmaceuticals.

**ben·zyl** (bĕn′zĭl, -zēl′) *n.* The univalent radical $C_6H_5CH_2$, derived from toluene.

**Be·o·wulf** (bā′ə-wŏŏlf′) *n.* [OE *Bēowulf.*] The legendary hero of an anonymous Old English epic poem.

**be·queath** (bĭ-kwēth′, -kwēth′) *vt.* **-queathed, -queath·ing, -queaths.** [ME *bequethen* < OE *becweðan.*] **1.** *Law.* To leave or give by will. **2.** To pass on : hand down. **—be·queath′al** (-kwē′thəl, -thəl), **be·queath′ment** *n.* **—be·queath′er** *n.*

**be·quest** (bĭ-kwĕst′) *n.* [ME *biqueste* < *bequethen,* to bequeath.] **1.** The act of bequeathing. **2.** Something bequeathed : LEGACY.

**be·rate** (bĭ-rāt′) *vt.* **-rat·ed, -rat·ing, -rates.** To scold harshly.

**Ber·ber** (bûr′bər) *n.* [Ar. *Barbar.*] **1.** A member of one of several Moslem tribes of North Africa. **2.** Any of the Afro-Asiatic languages of the Berbers. **—Ber′ber** *adj.*

**ber·ber·ine** (bûr′bə-rēn′) *n.* [NLat. *Berberis,* barberry genus < OFr. *berberis,* barberry + -INE.] A bitter-tasting yellow alkaloid, $C_{20}H_{19}NO_5$, obtained from plants such as the barberry and from the root of a North American plant, *Hydrastis canadensis,* and used in medicine.

**ber·ceuse** (bĕr-sœz′) *n., pl.* **-ceuses** (-sœz′) [Fr. < *bercer,* to rock.] **1.** A lullaby. **2.** A soothing musical composition.

**be·reave** (bĭ-rēv′) *vt.* **-reaved** *or* **-reft** (-rĕft′), **-reav·ing, -reaves.** [ME *bireven* < OE *berēafian.*] *Archaic.* **1.** To deprive of (something valued). **2.** To deprive of (a loved one) by death. **—be·reave′ment** *n.* **—be·reav′er** *n.*

**be·reft** (bĭ-rĕft′) *adj.* [< p.part. of BEREAVE.] **1. a.** Deprived of <*bereft* of their self-worth> **b.** Lacking a needed or expected element. **2.** Suffering the death of a loved one : BEREAVED.

**Ber·e·ni·ce's Hair** (bĕr′ə-nī′sēz) *n.* Coma Berenices.

**be·ret** (bə-rā′, bĕr′ā′) *n.* [Fr. *béret* < OProv. *birret.* —see BIRETTA.] A round visorless cloth cap.

**berg** (bûrg) *n.* An iceberg.

**ber·ga·mot** (bûr′gə-mŏt′) *n.* [Fr. *bergamote* < Ital. *bergamotta,* prob. < Turk. *beg-armûdĭ,* prince's pear.] **1. a.** A small spiny tree, *Citrus aurantium bergamia,* bearing pear-shaped fruit whose rind yields an aromatic oil. **b.** The oil of the bergamot, used in perfumery. **2.** A plant of the genus *Monarda,* esp. the wild bergamot.

**Berg·mann's rule** (bûrg′mənz) *n.* [After Karl Georg L.C. *Bergmann* (1814–1865).] An axiom stating that in any warm-blooded animal species that is polytypic and wide-ranging the body size of each geographic group varies with the average environmental temperature.

**Berg·son·ism** (bĕrg′sə-nĭz′əm) *n.* The philosophy of Henri Bergson, which contends that all living forms arise from a persisting natural force, the élan vital. **—Berg·so′ni·an** (-sō′nē-ən) *n.*

**ber·i·ber·i** (bĕr′ē-bĕr′ē) *n.* [Singhalese, redup. of *beri,* weakness.] A thiamine deficiency disease of the peripheral nervous system, endemic in eastern and southern Asia and characterized by partial paralysis of the extremities, emaciation, and anemia.

**Be·ring time** (bîr′ĭng, bâr′-) *n.* [After the *Bering* Sea.] The time in western Alaska and the Aleutian Islands, which lie in the 11th time zone west of Greenwich, England.

**Berke·le·ian·ism** (bär′klē-ə-nĭz′əm, bûr′-) *n.* The philosophy of George Berkeley, which holds that material objects have no independent being but exist only as concepts of a human or divine mind. **—Berke′le·ian** *n.*

**ber·ke·li·um** (bər-kē′lē-əm, bûrk′lē-əm) *n.* [After *Berkeley,* California.] *Symbol* **Bk** A synthetic radioactive element; atomic number 97; longest-lived isotope Bk 247.

**Berk·shire** (bûrk′shîr′, -shər) *n.* [After *Berkshire,* a county in England.] One of a domestic breed of black hogs with white feet and faces.

**ber·lin** (bər-lĭn′) *n.* [After *Berlin,* Germany.] **1.** A light wool used in making clothing, esp. gloves. **2.** A four-wheeled covered carriage.

berlin

**ber·line** *also* **ber·lin** (bər-lĭn′) *n.* [Fr. < *Berlin,* Berlin, Germany.] A limousine with a glass window between the front and rear seats.

**berm** *also* **berme** (bûrm) *n.* [Fr. *berme* < MDu.] **1. a.** A narrow ledge or shelf, as along a slope. **b.** A shoulder of a road. **2.** A ledge between a parapet and a moat in a fortification.

**Ber·mu·da grass** (bər-myōō′də) *n.* A grass, *Cynodon dactylon,* that has wiry creeping rootstocks and is used for lawns and pasturage in warm regions.

**Bermuda lily** *n.* A lily, *Lilium longiflorum,* widely cultivated in Bermuda, with large white trumpet-shaped flowers.

**Bermuda onion** *n.* A large mild-flavored yellow-skinned onion.

**Bermuda rig** *n. Naut.* A fore-and-aft rig distinguished by a tall triangular mainsail and widely used on cruising and racing vessels.

**Ber·mu·das** (bər-myōō′dəz) *pl.n.* Bermuda shorts.

**Bermuda shorts** *pl.n.* Shorts ending just above the knees.

**Ber·noul·li distribution** (bər-nōō′lē) *n.* [After Jakob *Bernoulli* (1654–1705).] *Statistics.* Binomial distribution.

**Bernoulli effect** *n.* [After Daniel *Bernoulli* (1700–1782).] The phenomenon of internal pressure reduction with increased stream velocity in a fluid.

**Bernoulli's law** *n.* [Statistics law after Jakob *Bernoulli;* physics law after Daniel *Bernoulli.*] **1.** *Statistics.* The probability theorem stating that for a very large number of independent repeated Bernoulli trials the observed relative frequency of successes will approximate the probability of success on each trial. **2.** *Physics.* The relationship between internal fluid pressure and fluid velocity, essentially a statement of the conservation of energy, that has as a consequence the Bernoulli effect.

**Bernoulli trial** *n.* [After Jakob *Bernoulli* (1654–1705).] *Statistics.* An experiment having just two possible results, usu. denoted *success* and *failure,* with the property that the occurrence of one excludes the occurrence of the other in any given trial.

**ber·ry** (bĕr′ē) *n., pl.* **-ries.** [ME *berye* < OE *berie.*] **1.** A usu. fleshy edible fruit, as the strawberry, blackberry, or raspberry. **2.** *Bot.* A fleshy fruit, as the grape, blueberry, or tomato, developed from a single ovary and having few or many seeds but not a single stone. **3.** A seed or dried kernel, as that of the coffee plant. **4.** The small dark egg of certain crustaceans or fishes. *—vi.* **-ried, -ry·ing, -ries.** **1.** To hunt for or gather berries. **2.** To produce or bear berries.

**ber·seem** (bər-sēm′) *n.* [Ar. *birsīm* < Coptic *bersīm.*] A clover, *Trifolium alexandrinum,* native to northern Africa and southwestern Asia, grown for soil improvement in dry regions of southwestern North America.

**ber·serk** (bər-sûrk′, -zûrk′) *adj.* **1.** Destructively or frenziedly violent. **2.** Deranged. *—n.* A berserker. **—ber·serk′** *adv.*

**ber·serk·er** (bər-sûr′kər, -zûr′-) *n.* [ON *berserkr* : *björn,* bear + *serkr,* shirt.] An ancient Scandinavian warrior who fought in battle with frenzied violence and fury.

**berth** (bûrth) *n.* [Prob. < BEAR[1].] **1.** A built-in bed or bunk on a ship or train. **2.** A space at a wharf for a ship to dock or anchor. **3.** Sufficient space for a ship to maneuver. **4.** A position of employment, esp. on a ship. **5.** A place to sleep. **6.** A space where a vehicle can be parked. *—v.* **berthed, berth·ing, berths.** *—vt.* **1.** To bring (a ship) to a berth. **2.** To provide with a berth. *—vi.* To come to a berth : DOCK. **—give a wide berth to.** To keep well clear of : AVOID.

**ber·tha** (bûr′thə) *n.* [Fr. *berthe* < *Berthe,* the name Bertha.] A wide, deep, often lace collar covering the shoulders.

**Ber·til·lon system** (bûr′tl-ŏn′, bĕr′tē-yôN′) *n.* [After Alphonse *Bertillon* (1853–1914).] A system once used for identifying people by means of a record of various body measurements, coloring, and markings.

**ber·yl** (bĕr′əl) *n.* [ME < OFr. < Lat. *beryllus* < Gk. *bērullos.*] A mineral, essentially aluminum beryllium silicate, $Be_3Al_2Si_6O_{18}$, occurring in hexagonal prisms and being the main source of beryllium. **—ber′yl·line** (-ə-lĭn, -līn′) *adj.*

**be·ryl·li·um** (bə-rĭl′ē-əm) *n.* [< BERYL.] *Symbol* **Be** A high-melting, lightweight, corrosion-resistant, rigid, steel-gray metallic element used as an aerospace structural material, as a moderator and reflector in nuclear reactors, and in a copper alloy used for springs, electrical contacts, and nonsparking tools; atomic number 4; atomic weight 9.0122.

**be·seech** (bĭ-sēch′) *vt.* **-sought** (-sôt′) *or* **-seeched, -seech·ing, -seech·es.** [ME *besechen* < OE *besēcan.*] **1.** To address an earnest or urgent request to : IMPLORE. **2.** To request earnestly <*besought* help in vain> **—be·seech′er** *n.*

**be·seem** (bĭ-sēm′) *vt.* **-seemed, -seem·ing, -seems.** [ME *besemen.*] *Archaic.* To be appropriate for : BEFIT.

**be·set** (bĭ-sĕt′) *vt.* **-set, -set·ting, -sets.** [ME *besetten* < OE *besettan.*] **1.** To attack from all sides : ASSAIL. **2.** To trouble persistently : HARASS <"*beset* by a ghostly band of doubts" —Sherwood Anderson> **3.** To hem in : SURROUND. **4.** To stud, as with jewels. **—be·set′ment** *n.*

**be·shrew** (bĭ-shrōō′) *vt.* **-shrewed, -shrew·ing, -shrews.** [ME *beshrewen.*] *Archaic.* To invoke evil on : CURSE.

**be·side** (bĭ-sīd′) *prep.* [ME *biside* < OE *be sīdan* : *be,* by + *sīde,* side.] **1.** Next to : at the side of. **2.** In comparison with. **3.** Except for. **4.** Wide of : apart from <a comment that was *beside* the point> *—adv. Archaic.* **1.** In addition. **2.** Nearby. **—beside (oneself).** In a state of great agitation.

**be·sides** (bĭ-sīdz′) *adv.* [ME *bisides,* genitive of *biside,* beside.] **1.** In addition : ALSO. **2.** Furthermore : moreover. **3.** Else : otherwise. *—prep.* **1.** In addition to. **2.** Other than : EXCEPT. ***usage:*** In modern usage the senses "in addition to" and "except for" are more often expressed by *besides* than *beside,* as in *They had few friends besides us.*

**be·siege** (bĭ-sēj′) *vt.* **-sieged, -sieg·ing, -sieg·es.** [ME *besegen.*] **1.** To encircle with troops : lay siege to. **2.** To crowd around : hem in. **3.** To importune or harass, as with requests. **4.** To cause to feel distressed. **—be·sieg′er** *n.*

**be·smear** (bĭ-smîr′) *vt.* **-smeared, -smear·ing, -smears.** To smear <a lab coat *besmeared* with dark stains>
**be·smirch** (bĭ-smûrch′) *vt.* **-smirched, -smirch·ing, -smirch·es.** To soil : sully. **—be·smirch′er** *n.* **—be·smirch′ment** *n.*
**be·som** (bē′zəm) *n.* [ME *besum* < OE *besema.*] **1.** A bundle of twigs attached to a handle and used as a broom. **2.** The broom plant.
**be·sot** (bĭ-sŏt′) *vt.* **-sot·ted, -sot·ting, -sots.** To stupefy or muddle with or as if with liquor.
**be·sought** (bĭ-sôt′) *v. var. p.t. & p.p. of* BESEECH.
**be·spake** (bĭ-spāk′) *v. Archaic. var. p.t. of* BESPEAK.
**be·spat·ter** (bĭ-spăt′ər) *vt.* **-tered, -ter·ing, -ters.** To spatter.
**be·speak** (bĭ-spēk′) *vt.* **-spoke** (-spōk′), **-spo·ken** (-spō′kən) *or* **-spoke, -speak·ing.** [ME *bespeken,* to speak out < OE *besprecan.*] **1.** To be or give a sign of : SIGNIFY. **2.** *Archaic.* To speak to : ADDRESS. **3.** To claim or engage in advance : RESERVE. **4.** To foretell : portend.
**be·spec·ta·cled** (bĭ-spĕk′tə-kəld) *adj.* Wearing eyeglasses.
**be·spoke** (bĭ-spōk′) *or* **be·spo·ken** (-spō′kən) *adj. Chiefly Brit.* **1.** Made-to-order. **2.** Dealing in custom-made articles.
**be·sprent** (bĭ-sprĕnt′) *adj.* [ME *bespreynt,* p.part. of *besprengen,* to besprinkle < OE *besprengan.*] *Archaic.* Sprinkled over.
**be·sprin·kle** (bĭ-sprĭng′kəl) *vt.* **-kled, -kling, -kles.** To sprinkle.
**Bes·se·mer converter** (bĕs′ə-mər) *n.* A large pear-shaped container used in the Bessemer process.
**Bessemer process** *n.* [After Henry *Bessemer* (1813–1898).] A method for making steel by blasting compressed air through molten iron, burning out excess carbon and other impurities.
**best** (bĕst) *adj.* [ME < OE *betst.*] **1.** Exceeding all others in excellence, achievement, or quality : most excellent <the *best* politician> **2.** Most satisfactory, suitable, or useful : most desirable <the *best* procedure> **3.** Greatest : largest <the *best* part of a day> *—adv.* **1.** In the best way. **2.** To the greatest extent or degree : MOST. *—n.* **1.** Something that is best. **2.** The best condition or quality <act your *best*> **3.** One's best clothing. **4.** The best effort one can make. **5.** One's warmest wishes. *—vt.* **best·ed, best·ing, bests.** To get the better of : SURPASS. **—at best.** Under the most favorable circumstances <an unfortunate remark *at best*>
**be·stead** (bĭ-stĕd′) *Archaic.* *—vt.* **-stead·ed** *or* **-stead, -stead·ing, -steads.** To be of use to : AVAIL.
**bes·tial** (bĕs′chəl, bĕst′yəl) *adj.* [ME < OFr. < LLat. *bestialis* < Lat. *bestia,* beast.] **1.** Of, relating to, or like an animal. **2.** Having the manners or qualities of a brute : SAVAGE. **3.** Lacking in intelligence. **—bes′ti·al′i·ty** *n.* **—bes′tial·ly** *adv.*
**bes·tial·ize** (bĕs′chə-līz′, bĕs′-) *vt.* **-ized, -iz·ing, -iz·es.** To make bestial : BRUTALIZE.
**bes·ti·ar·y** (bĕs′chē-ĕr′ē, bēs′-) *n., pl.* **-ies.** [Med. Lat. *bestiarium* < Lat. *bestia,* beast.] A medieval collection of allegoric fables about real and imaginary animals, each fable having a moral.
**be·stir** (bĭ-stûr′) *vt.* **-stirred, -stir·ring, -stirs.** [ME *bestiren* < OE *bestyrian,* to pile up.] To cause to become active : ROUSE.
**best man** *n.* A bridegroom's chief attendant.
**be·stow** (bĭ-stō′) *vt.* **-stowed, -stow·ing, -stows.** [ME *bestowen.*] **1.** To present as a gift or honor : CONFER. **2.** To apply : USE. **3.** *Archaic.* To store or house. **—be·stow′a·ble** *adj.* **—be·stow′al** *n.*
**be·strew** (bĭ-strōō′) *vt.* **-strewed, -strewed** *or* **-strewn** (-strōōn′), **-strew·ing, -strews. 1.** To strew (a surface) with things so as to cover it. **2.** To lie scattered over or about.
**be·stride** (bĭ-strīd′) *vt.* **-strode** (strōd′), **-strid·den** (-strĭd′n), **-strid·ing, -strides.** [ME *bestriden* < OE *bestrīdan.*] **1.** To stand or sit on with the legs astride. **2.** To step over. **3.** To tower over.
**best seller** *n.* A product, as a book, that is among those sold in the greatest numbers. **—best′-sell′ing** *adj.*
**bet** (bĕt) *n.* [Orig. unknown.] **1.** An agreement usu. between two parties that the one who has made a wrong prediction about an uncertain outcome will forfeit something to the other : WAGER. **2.** A fact, event, or outcome on which a wager is made. **3.** An object or amount risked in a wager : STAKE. **4.** One on which a stake is placed. *—v.* **bet** *or* **bet·ted, bet·ting, bets.** *—vt.* **1.** To stake (e.g., an amount) in a bet. **2.** To make a bet with. **3.** To make a bet on (an outcome or a contestant). **4.** To maintain confidently, as if making a bet. *—vi.* To make or place a bet. **—you bet.** *Informal.* Of course : SURELY.
**be·ta** (bā′tə, bē′-) *n.* [Gk. *bēta,* of Phoenician orig.; akin to Heb. *bēth.*] **1.** The second letter of the Greek alphabet. —See table at ALPHABET. **2.** The second item in a series or system of classification. **3.** *Physics.* **a.** A beta particle. **b.** A beta ray.
**be·ta-ad·re·ner·gic** (bā′tə-ăd′rə-nûr′jĭk, bē′-) *adj.* Of, relating to, or being a beta-receptor.
**be·ta-block·er** (bā′tə-blŏk′ər, bē′-) *n.* A drug inhibiting absorption of adrenalin by interference with beta-receptor action.
**beta cell** *n.* **1.** Any of the cells in the islands of Langerhans that secrete insulin. **2.** Any of the basophilic chemophiles in the anterior lobe of the adenohypophysis.
**be·ta-en·dor·phin** (bā′tə-ĕn-dôr′fĭn, bē′-) *n.* An endorphin that is a powerful pain suppressant produced by the pituitary gland.
**beta globulin** *n.* Any of several globulins intermediate in their particulate motility response to electrophoresis as compared with alpha and gamma globulins.
**be·ta·ine** (bē′tə-ēn′) *n.* [Lat. *beta,* beet + -INE.] A sweet crystalline alkaloid, $C_5H_{11}NO_2$, found in sugar beets and other plants and used to treat muscular degeneration.
**be·take** (bĭ-tāk′) *vt.* **-took** (-tŏŏk′), **-tak·en, -tak·ing, -takes. 1.** To cause (oneself) to go or move. **2.** *Archaic.* To commit : APPLY.
**be·ta-ox·i·da·tion** (bā′tə-ŏk′sĭ-dā′shən, bē′-) *n.* Fatty-acid catabolism in which two-carbon fragments are removed successively from the carboxyl end of the chain.
**beta particle** *n.* A high-speed electron or positron, esp. one given off in radioactive decay.
**beta ray** *n.* A stream of beta particles, esp. of electrons.
**be·ta-re·cep·tor** (bā′tə-rĭ-sĕp′tər, bē′-) *n.* A site in the autonomic nervous system that is activated by or strongly reacts to adrenergic agents, as epinephrine, by generating inhibitory action.
**beta rhythm** *also* **beta wave** *n.* The second most common waveform observed in electroencephalograms of the adult brain, having a frequency from 18 to 30 cycles per second and associated with an alert waking state.
**be·ta·tron** (bā′tə-trŏn′, bē′-) *n.* A fixed-radius magnetic induction electron accelerator capable of accelerating electrons to energies ranging from a few million to a few hundred million electron volts.
**be·tel** (bēt′l) *n.* [Port. < Malayalam *veṭṭila.*] A climbing Asiatic plant, *Piper betle,* whose leaves are chewed with the betel nut and lime by many people of southeastern Asia.
**Be·tel·geuse** (bēt′l-jōōz′, bĕt′l-jœz′) *n.* [Fr. *Bételgeuse,* prob. < Ar. *bīt aljauzā.*] A bright-red intrinsic variable star, 527 light-years from Earth, in the constellation Orion.
**betel nut** *also* **be·tel·nut** (bēt′l-nŭt′) *n.* The seed of the fruit of the betel palm.
**betel palm** *n.* A palm tree, *Areca catechu* of tropical Asia, with featherlike leaves and orange or scarlet fruit.
**bête noire** (bĕt nwär′) *n.* [Fr. : *bête,* beast + *noire,* black.] One particularly disliked or to be avoided.
**beth** (bĕt) *n.* [Heb. *bēth.*] The second letter of the Hebrew alphabet. —See table at ALPHABET.
**beth·el** (bĕth′əl) *n.* [Heb. *bēth ′Ēl,* house of God.] **1.** A holy or hallowed place. **2.** A chapel for sailors.
**be·think** (bĭ-thĭngk′) *v.* **-thought** (-thôt′), **-think·ing, -thinks.** [ME *bethinken* < OE *beðencan.*] *—vt.* **1.** *Archaic.* To reflect on : think about. **2.** To remind (oneself) : REMEMBER. *—vi. Archaic.* To meditate : PONDER.
**be·tide** (bĭ-tīd′) *vt. & vi.* **-tid·ed, -tid·ing, -tides.** [ME *betiden* : *be-,* thoroughly + *tiden,* to happen < OE *tīdan.*] To happen to or to take place.
**be·times** (bĭ-tīmz′) *adv.* [ME.] **1.** In good time : EARLY. **2.** *Archaic.* Soon : quickly.
**bê·tise** (bā-tēz′) *n., pl.* **-tises** (-tēz′) [Fr. < *bête,* foolish < *bête,* beast < OFr. *beste* < Lat. *bestia.*] **1.** Foolishness : stupidity. **2.** A foolish or stupid act.
**be·to·ken** (bĭ-tō′kən) *vt.* **-kened, -ken·ing, -kens.** [ME *betokenen* : *be-,* thoroughly + *toknen,* to signify < OE *tacnian.*] To be or give a portent or sign of.
**bet·o·ny** (bĕt′n-ē) *n., pl.* **-nies.** [ME *betone* < OFr. *betoine* < Lat. *betonica,* prob. < *Vettones,* an ancient Iberian tribe.] **1.** A plant of the genus *Stachys,* esp. *S. officinalis,* having a reddish-purple flower spike. **2.** The lousewort.
**be·took** (bĭ-tŏŏk′) *v. p.t. of* BETAKE.
**be·tray** (bĭ-trā′) *vt.* **-trayed, -tray·ing, -trays.** [ME *betrayen* : *be-,* thoroughly + *trayen,* to betray < OFr. *trair* < Lat. *tradee.* —see TRADITION.] **1.** To be a traitor to or commit treason against. **2.** To reveal in a breach of confidence. **3.** To make known accidentally <shifty eyes *betraying* guilt> **4.** To reveal : indicate. **5.** To lead astray : DECEIVE. **—be·tray′al** *n.* **—be·tray′er** *n.*
**be·troth** (bĭ-trōth′, -trôth′) *vt.* **-trothed, -troth·ing, -troths.** [ME *betrouthen* : *be-,* in relation to + *trouthe,* troth.] To promise to give in marriage or marry.
**be·troth·al** (bĭ-trō′thəl, -trô′thəl) *n.* **1.** An act of betrothing or the state of being betrothed. **2.** A mutual promise to marry.
**be·trothed** (bĭ-trōthd′, -trôtht′) *n.* A person to whom one is engaged to be married.
**bet·ta** (bĕt′ə) *n.* [NLat. *Betta,* genus name.] Any of a genus, *Betta,* of small, long-finned freshwater fishes with striking coloration, found in southeastern Asia.
**bet·ter[1]** (bĕt′ər) *adj.* [ME < OE *betera.*] **1.** Greater in excellence or higher in quality. **2.** More useful, desirable or suitable. **3.** Larger : greater <the *better* part of the artist's work> **4.** Healthier than before. *—adv.* **1.** In a more excellent way. **2. a.** To a greater degree or extent. **b.** To greater advantage or use. **3.** More <*better* than a mile> ***usage:*** Using *better* to mean "more," as in *They live better than 11 miles from town,* is considered by many to be unacceptable in formal style. *—n.* **1.** Something better. **2.** A superior, as in position or intelligence. *—v.* **-tered, -ter·ing, -ters.** *—vt.* **1.** To improve <*bettered* my position in life> **2.** To surpass : exceed. *—vi.* To become better.

ă **pat** ā **pay** âr **care** ä **father** ĕ **pet** ē **be** hw **which** ĭ **pit**
ī **tie** îr **pier** ŏ **pot** ō **toe** ô **paw, for** oi **noise** ŏŏ **took**

**bet·ter[2]** (bĕt′ər) *n. var. of* BETTOR.

**bet·ter·ment** (bĕt′ər-mənt) *n.* **1.** An improvement. **2.** An improvement that repairs real property and adds to its value.

**bet·ter-off** (bĕt′ər-ôf′, -ŏf′) *adj.* Being in a better or more prosperous condition.

**bet·tor** *also* **bet·ter** (bĕt′ər) *n.* One who bets.

**be·tween** (bĭ-twēn′) *prep.* [ME *betwene* < OE *betwēonum.*] **1. a.** In the interval or position separating <*between* the buildings><*between* Thanksgiving and Christmas> **b.** Intermediate to, as in quantity, amount, or degree <measures *between* four and five feet> **2.** Connecting spatially <a route *between* Dallas and Fort Worth> **3. a.** By the combined effect or effort of <*Between* the sun and wind, the wash dried.> **b.** In the combined ownership of <They had a controlling interest *between* them.> **4.** As measured against. —Used often to express a reciprocal relationship <choose *between* swimming and tennis> **—between you and me.** In strictest confidence. **—in between.** In an intermediate situation. **—be·tween′** *adv.*

**be·tween·times** (bĭ-twēn′tīmz′) *adv.* At or during pauses.

**be·twixt** (bĭ-twĭkst′) *adv.* & *prep.* [ME < OE *betwyx.*] *Archaic.* Between. **—betwixt and between.** In an intermediate position.

**Beu·lah** (byo͞o′lə) *n.* **1.** The land of Israel in the Old Testament. **2.** The land of peace in Bunyan's *Pilgrim's Progress.*

**bev·a·tron** (bĕv′ə-trŏn′) *n.* [B(ILLION) + E(LECTRON) + V(OLTS) + -TRON.] *Physics.* A proton synchrotron.

**bev·el** (bĕv′əl) *n.* [OFr. **bevel* < *baif,* open-mouthed < *bayer,* to gape.] **1.** The angle or inclination of a surface or line that meets another at any angle but 90°. **2.** A rule having an adjustable arm used to draw or measure angles or to fix a surface at an angle. *—v.* **-eled, -el·ing, -els** *or* **-elled, -el·ling, -els.** *—vt.* To cut at an inclination that forms an angle other than a right angle. *—vi.* To be inclined : SLOPE.

**bevel gear** *n.* Either of a pair of gears having teeth surfaces cut so that the gear shafts are not parallel.

**bev·er·age** (bĕv′ər-ĭj, bĕv′rĭj) *n.* [ME *beverege* < OFr. *bevrage* < *beivre,* to drink < Lat. *biber.*] A liquid for drinking, usu. excluding water.

**bev·y** (bĕv′ē) *n., pl.* **-ies.** [ME.] **1.** A group of birds or animals, esp. larks or quail : FLOCK. **2.** A group : assemblage.

**be·wail** (bĭ-wāl′) *vt.* **-wailed, -wail·ing, -wails. 1.** To express sorrow or regret over. **2.** To cry about. **—be·wail′er** *n.*

**be·ware** (bĭ-wâr′) *v.* **-wared, -war·ing, -wares.** [ME *be ware* : *be,* imper. of *been,* to be + *ware,* on one's guard (< OE *wær*).] *—vt.* To be cautious of. *—vi.* To be cautious.

**be·whis·kered** (bĭ-hwĭs′kərd, -wĭs′-) *adj.* Having whiskers.

**be·wil·der** (bĭ-wĭl′dər) *vt.* **-dered, -der·ing, -ders. 1.** To befuddle or confuse, esp. with a variety of conflicting situations, objects, or statements. **2.** To cause to lose one's bearings. **—be·wil′dered·ly** *adv.* **—be·wil′dered·ness** *n.* **—be·wil′der·ing·ly** *adv.* **—be·wil′der·men** *n.*

**be·witch** (bĭ-wĭch′) *vt.* **-witched, -witch·ing, -witch·es.** [ME *bewicchen* : *be-,* thoroughly + *wicchen,* to enchant < OE *wiccian* < *wicca,* witch and *wicce,* wizard.] To place under one's power by or as if by magic : captivate totally. **—be·witch′er·y** *n.* **—be·witch′ing** *adj.* **—be·witch′ing·ly** *adv.*

**be·witch·ment** (bĭ-wĭch′mənt) *n.* **1.** The act of bewitching or the state of being bewitched. **2.** A spell that bewitches.

**be·wray** (bĭ-rā′) *vt.* **-wrayed, -wray·ing, -wrays.** [ME *bewreien* : *be-,* thoroughly + *wreien,* to accuse < OE *wregan.*] *Archaic.* To disclose, esp. unintentionally : BETRAY.

**bey** (bā) *n.* [Turk.] **1.** A provincial governor in the Ottoman Empire. **2.** A ruler of the former kingdom of Tunis. **3.** A Turkish title of honor and respect.

**be·yond** (bē-ŏnd′, bĭ-yŏnd′) *prep.* [ME < OE *begeondan.*] **1.** On the far side of : PAST. **2.** Later than. **3.** Past the understanding, reach, or scope of <cruelty *beyond* words> **4.** To a degree or amount greater than <rich *beyond* my wildest hopes> **5.** In addition to <sought nothing *beyond* bread and shelter> **—be·yond′** *adv.*

**bez·ant** (bĕz′ənt, bə-zănt′) [ME *besant* < OFr. < Lat. *Byzantius,* of Byzantium.] **1.** A gold coin issued in Byzantium : SOLIDUS. **2.** A flat disk used as an architectural decoration.

**bez·el** (bĕz′əl) *n.* [Orig. unknown.] **1.** A slanting surface or bevel on the edge of a cutting tool. **2.** The upper, faceted portion of a cut gem above the girdle. **3.** A groove or flange that holds a beveled edge, as of a watch crystal or a gem.

**be·zique** (bə-zēk′) *n.* [Fr. *bésique.*] A card game similar to pinochle that is played with a deck of 64 cards.

**be·zoar** (bē′zôr′, -zōr′) *n.* [ME *bezear* < OFr. *bezar* < Ar. *bāzahr* < Pers. *pād-zahr* : *pād,* protecting against + *zahr,* poison.] A hard gastric or intestinal mass found chiefly in ruminants and once regarded as a magical antidote to poison.

**B-girl** (bē′gûrl′) *n.* [B(AR) + GIRL.] A woman who works in a bar and encourages customers to spend money freely.

**Bha·ga·vad-Gi·ta** (bä′gə-väd-gē′tə) *n.* [Skt. *bhagavad-gītā,* song of the blessed one (Krishna).] A sacred Hindu text that is part of the *Mahabharata,* an ancient Sanskrit epic.

**bhang** *also* **bang** (băng) *n.* [Hindi *bhāṅg* < Skt. *bhaṅgā.*] **1.** The hemp plant. **2.** Any of several narcotics made from the dried flowers and leaves of hemp.

**Bhu·tan·ese** (bo͞o′tə-nēz′, -nēs′) *n., pl.* **Bhutanese. 1.** A native or inhabitant of Bhutan. **2.** The Sino-Tibetan language of Bhutan. *—adj.* Of or characteristic of Bhutan, its people, or their culture and language.

**bi-[1]** *or* **bin-** *pref.* [Lat. < *bis,* twice.] **1. a.** Two <*bi*form> **b.** Both <*bin*aural> **c.** Both sides, parts, or directions <*bi*concave> **2. a.** Occurring at intervals of two <*bi*centennial> **b.** Occurring twice during <*bi*weekly> **3. a.** Containing twice the proportion of a specified chemical element or group necessary for stability <*bi*carbonate> **b.** Containing two chemical atoms, radicals, or groups <*bi*phenyl>

**bi-[2]** *pref. var. of* BIO-.

**Bi** *symbol for* BISMUTH.

**bi·a·ly** (bē-ä′lē) *n., pl.* **-lys.** [After *Bialystok,* Poland.] A round, flat baked roll with onion flakes on top.

**bi·an·nu·al** (bī-ăn′yo͞o-əl) *adj.* Happening twice each year : SEMIANNUAL. **—bi·an′nu·al·ly** *adv.*

**bi·as** (bī′əs) *n.* [OFr. *biais,* oblique.] **1.** A line cutting diagonally across the grain of fabric. **2. a.** An inclination or preference, esp. one that interferes with impartial judgment : PREJUDICE. **b.** A specified instance of this. **3. a.** An irregularity or weight in a ball that causes it to swerve, as in lawn bowling. **b.** The tendency of such a ball to swerve. **4.** The fixed voltage applied to an electrode. *—vt.* **-ased, -as·ing, -as·es** *or* **-assed, -as·sing, -as·ses. 1.** To cause to have a prejudiced view. **2.** To apply a small voltage to (a grid).

☆ ***syns:*** BIAS, PARTIALITY, PREJUDICE, PREPOSSESSION *n. core meaning* : an inclination for or against that inhibits impartial judgment <a decision influenced by personal *bias*>

**bi·ath·lon** (bī-ăth′lən, -lŏn′) *n.* [BI- + Gk. *athlon,* contest.] An athletic competition that combines cross-country skiing and rifle shooting.

**bi·ax·i·al** (bī-ăk′sē-əl) *adj.* Having two axes. **—bi·ax′i·al·ly** *adv.*

**bib** (bĭb) *n.* [Prob. < ME *bibben,* to drink, perh. < Lat. *bibere.*] **1.** A napkin tied under the chin and worn, esp. by young children, to protect the clothing while eating. **2.** The part of an apron or overalls covering the chest. *—v.* **bibbed, bib·bing, bibs.** *—vt.* To drink : imbibe. *—vi.* To indulge in drinking : TIPPLE.

**bib and tucker** *n. Informal.* Clothing.

**bibb** (bĭb) *n.* [Alteration of BIB.] **1.** A bracket supporting the trestletrees on a ship's mast. **2.** A bibcock.

**bib·ber** (bĭb′ər) *n.* [< BIB.] A tippler.

**Bibb lettuce** (bĭb) *n.* [After Jack *Bibb,* 19th-cent. American vegetable grower.] A lettuce forming a small, loose head and having tender, dark-green leaves.

**bib·cock** (bĭb′kŏk′) *n.* A faucet with a nozzle that bends downward.

**bi·be·lot** (bē′bə-lō′, bē-blō′) *n.* [Fr. < OFr. *beubelet,* from a redup. of *bel,* beautiful < Lat. *bellus,* handsome.] A small ornamental object : TRINKET.

**Bi·ble** (bī′bəl) *n.* [ME < OFr. < Med. Lat. *biblia* < Gk., pl. of *biblion,* book < *biblos,* papyrus < *Bublos,* a Phoenician port.] **1. a.** The sacred book of Christianity, including both the Old Testament and the New Testament. **b.** The Old Testament, the sacred book of Judaism. **c.** A specific copy of a Bible <their own *Bible*> **d.** A book or collection of writings comprising the sacred text of a religion. **2. bible.** A book held to be authoritative in its field <the *bible* of Japanese cooking>

**Bible Belt** *n.* Sections of the United States, esp. in the South and Middle West, where Protestant fundamentalism prevails.

**bib·li·cal** *also* **Bib·li·cal** (bĭb′lĭ-kəl) *adj.* [Med. Lat. *biblicus* < *biblia,* Bible.] **1.** Of, relating to, or contained in the Bible. **2.** Being in keeping with the nature of the Bible, esp.: **a.** Suggestive of the people or times depicted in the Bible. **b.** Suggestive of the prose or narrative style of the King James Bible. **—Bib′li·cal·ly** *adv.*

**Bib·li·cist** (bĭb′lĭ-sĭst) *n.* **1.** An expert on the Bible. **2.** A literal interpreter of the Bible. **—Bib′li·cism** *n.*

**biblio-** *pref.* [< Gk. *biblion,* book. —see BIBLE.] Book <*biblio*phile>

**bib·li·o·film** (bĭb′lē-ō-fĭlm′) *n.* Microfilm used esp. to photograph book pages.

**bib·li·og·ra·pher** (bĭb′lē-ŏg′rə-fər) *n.* **1.** An expert in the description and cataloguing of printed matter. **2.** A compiler of a bibliography or bibliographies.

**bib·li·og·ra·phy** (bĭb′lē-ŏg′rə-fē) *n., pl.* **-phies. 1. a.** A list of the works of a specific author or publisher. **b.** A list of writings on a single subject. **2.** Description and identification of the editions, dates of issue, authorship, and typography of written material, as books. **—bib′li·o·graph′i·cal** (-ə-grăf′ĭ-kəl), **bib′li·o·graph′ic** (-ĭk) *adj.* **—bib′li·o·graph′i·cal·ly** *adv.*

**bib·li·ol·a·try** (bĭb′lē-ŏl′ə-trē) *n.* **1.** Excessive reverence for a literal interpretation of the Bible. **2.** Extreme devotion to books. **—bib′li·ol′a·ter** *n.* **—bib′li·ol′a·trous** *adj.*

# BOOKS OF THE BIBLE

## HEBREW SCRIPTURES

| | | | | |
|---|---|---|---|---|
| Genesis | II Samuel | Joel | Haggai | Lamentations |
| Exodus | I Kings | Amos | Zechariah | Ecclesiastes |
| Leviticus | II Kings | Obadiah | Malachi | Esther |
| Numbers | Isaiah | Jonah | Psalms | Daniel |
| Deuteronomy | Jeremiah | Micah | Proverbs | Ezra |
| Joshua | Ezekiel | Nahum | Job | Nehemiah |
| Judges | THE TWELVE | Habakkuk | Song of Songs | I Chronicles |
| I Samuel | Hosea | Zephaniah | Ruth | II Chronicles |

## OLD TESTAMENT

| Jerusalem Version | King James Version | Jerusalem Version | King James Version |
|---|---|---|---|
| Genesis | Genesis | Song of Solomon | Song of Solomon |
| Exodus | Exodus | Wisdom | |
| Leviticus | Leviticus | Ecclesiasticus | |
| Numbers | Numbers | Isaiah | Isaiah |
| Deuteronomy | Deuteronomy | Jeremiah | Jeremiah |
| Joshua | Joshua | Lamentations | Lamentations |
| Judges | Judges | Baruch | |
| Ruth | Ruth | Ezekiel | Ezekiel |
| I Samuel | I Samuel | Daniel | Daniel |
| II Samuel | II Samuel | Hosea | Hosea |
| I Kings | I Kings | Joel | Joel |
| II Kings | II Kings | Amos | Amos |
| I Chronicles | I Chronicles | Obadiah | Obadiah |
| II Chronicles | II Chronicles | Jonah | Jonah |
| Ezra | Ezra | Micah | Micah |
| Nehemiah | Nehemiah | Nahum | Nahum |
| Tobit | | Habakkuk | Habakkuk |
| Judith | | Zephaniah | Zephaniah |
| Esther | Esther | Haggai | Haggai |
| Job | Job | Zechariah | Zechariah |
| Psalms | Psalms | Malachi | Malachi |
| Proverbs | Proverbs | I Maccabees | |
| Ecclesiastes | Ecclesiastes | II Maccabees | |

## NEW TESTAMENT

| | | | |
|---|---|---|---|
| Matthew | I Corinthians | II Thessalonians | I Peter |
| Mark | II Corinthians | I Timothy | II Peter |
| Luke | Galatians | II Timothy | I John |
| John | Ephesians | Titus | II John |
| Acts | Philippians | Philemon | III John |
| | Colossians | Hebrews | Jude |
| Romans | I Thessalonians | James | Revelation |

**bib·li·o·man·cy** (bĭb′lē-ə-măn′sē) *n., pl.* **-cies.** Divination by interpretation of a random selection from a book, esp. the Bible.

**bib·li·o·ma·ni·a** (bĭb′lē-ə-mā′nē-ə, -mān′yə) *n.* An exaggerated liking for acquiring books. **—bib′li·o·ma′ni·ac′** (-ăk′) *n.* **—bib′li·o·ma·ni′a·cal** (-mə-nī′ə-kəl) *adj.*

**bib·li·o·phile** (bĭb′lē-ə-fīl′) *also* **bib·li·o·phil** (-fĭl′) *or* **bib·li·oph·i·list** (bĭb′lē-ŏf′ə-lĭst) *n.* A book lover or collector. **—bib′li·oph′i·lism** *n.* **—bib′li·oph′i·lis′tic** *adj.*

**bib·li·o·pho·bi·a** (bĭb′lē-ə-fō′bē-ə) *n.* Abnormal fear of books. **—bib′li·o·phobe′** *n.* **—bib′li·o·pho′bic** *adj.*

**bib·li·o·pole** (bĭb′lē-ə-pōl′) *also* **bib·li·op·o·list** (bĭb′lē-ŏp′ə-lĭst) *n.* [Lat. *bibliopola* < Gk. *bibliopōlēs* : *biblion*, book + *pōlein*, to sell.] A dealer in rare books. **—bib′li·o·pol′ic** (-pŏl′ĭk), **bib′li·o·pol′i·cal** *adj.*

**bib·li·o·the·ca** (bĭb′lē-ə-thē′kə) *n.* [Lat. < Gk. *bibliothēkē* : *biblion*, book + *thēkē*, case.] **1.** A book collection **:** LIBRARY. **2.** A catalogue of books. **—bib′li·o·the′cal** *adj.*

**bib·li·ot·ics** (bĭb′lē-ŏt′ĭks) *n.* *(sing. in number).* Examination of written documents to ascertain authorship or authenticity.

**bib·u·lous** (bĭb′yə-ləs) *adj.* [Lat. *bibulus* < *bibere*, to drink.] **1.** Given to or marked by drinking alcoholic beverages. **2.** Very absorbent. **—bib′u·lous·ly** *adv.* **—bib′u·lous·ness** *n.*

**bi·cam·er·al** (bī-kăm′ər-əl) *adj.* Composed of two legislative chambers or branches.

**bi·cap·su·lar** (bī-kăp′sə-lər) *adj. Bot.* Having two capsules or one capsule with two cells.

**bi·car·bon·ate** (bī-kär′bə-nāt′, -nĭt) *n.* The radical group $HCO_3$ or a compound, as sodium bicarbonate, containing it.

**bicarbonate of soda** *n.* Sodium bicarbonate.

**bi·cau·dal** (bī-kôd′l) *adj.* Having two tails.

**bi·cel·lu·lar** (bī-sĕl′yə-lər) *adj.* Having two cells.

**bi·cen·ten·a·ry** (bī′sĕn-tĕn′ə-rē, bī-sĕn′tə-nĕr′ē) *n., pl.* **-ries.** A bicentennial. **—bi′cen·ten′a·ry** *adj.*

**bi·cen·ten·ni·al** (bī′sĕn-tĕn′ē-əl) *adj.* **1.** Happening once every 200 years. **2.** Lasting for 200 years. **3.** Relating to a 200th anniversary. **—bi′cen·ten′ni·al** *n.*

**bi·cen·tric** (bī-sĕn′trĭk) *adj.* Having two centers. **—bi′cen·tric′i·ty** (-trĭs′ĭ-tē) *n.*

**bi·ceph·a·lous** (bī-sĕf′ə-ləs) *adj.* Having two heads.

**bi·ceps** (bī′sĕps′) *n., pl.* **biceps** *or* **-ceps·es** (-sĕp′sĭz) [NLat. < Lat., two-headed : *bi-*, two + *caput*, head.] A muscle with two points of origin, esp.: **a.** The large muscle at the front of the upper arm that flexes the elbow joint. **b.** The large muscle at the back of the thigh that flexes the knee joint.

**bi·chlo·ride** (bī-klôr′īd′, -klōr′-) *n.* Dichloride.

**bi·chro·mate** (bī-krō′māt′, -mĭt) *n.* Dichromate.

**bi·cil·i·ate** (bī-sĭl′ē-ĭt, -āt′) *adj.* Having two cilia.

**bi·cip·i·tal** (bī-sĭp′ĭ-tl) *adj.* [< NLat. *biceps, bicipit-*, biceps.] Of or relating to the biceps.
**bick·er** (bĭk′ər) *vi.* **-ered, -er·ing, -ers.** [ME *bikeren*, to attack.] **1.** To engage in a petty quarrel : SQUABBLE. **2.** To flicker : quiver. —*n.* A petty quarrel : SQUABBLE. **—bick′er·er** *n.*
**bi·col·or** (bī′kŭl′ər) *also* **bi·col·ored** (-ərd) *adj.* Having two colors <a *bicolor* illustration>
**bi·con·cave** (bī′kŏn-kāv′, bī-kŏn′kāv′) *adj.* Concave on both sides or surfaces. **—bi′con·cav′i·ty** (-kăv′ĭ-tē) *n.*
**bi·con·vex** (bī′kŏn-vĕks′, bī-kŏn′vĕks′) *adj.* Convex on both sides or surfaces. **—bi′con·vex′i·ty** (-vĕk′sĭ-tē) *n.*
**bi·corn** (bī′kôrn′) *also* **bi·cor·nu·ate** (bī-kôr′nyōō-ĭt, -āt′) *adj.* [Lat. *bicornis* : *bi-*, two + *cornu*, horn.] **1.** Having two horns or horn-shaped parts. **2.** Shaped like a crescent.
**bi·cor·po·ral** (bī-kôr′pər-əl) *also* **bi·cor·po·re·al** (bī′kôr-pôr′ē-əl, -pōr′-) *adj.* Having two distinct bodies or main parts.
**bi·cos·tate** (bī-kŏs′tāt′) *adj.* [BI- + COST(A) + -ATE.] *Bot.* Having two main longitudinal ribs <a *bicostate* leaf>
**bi·cul·tur·al** (bī-kŭl′chər-əl) *adj.* Of or pertaining to two distinct cultures in one nation or geographic region. **—bi·cul′tur·al·ism** *n.*
**bi·cus·pid** (bī-kŭs′pĭd) *also* **bi·cus·pi·date** (-pĭ-dāt′) *adj.* [NLat. *bicuspis, bicuspid-* : *bi-* two + *cuspis*, cusp.] Having two points or cusps, as the crescent moon. —*n.* A bicuspid tooth, esp. a premolar.

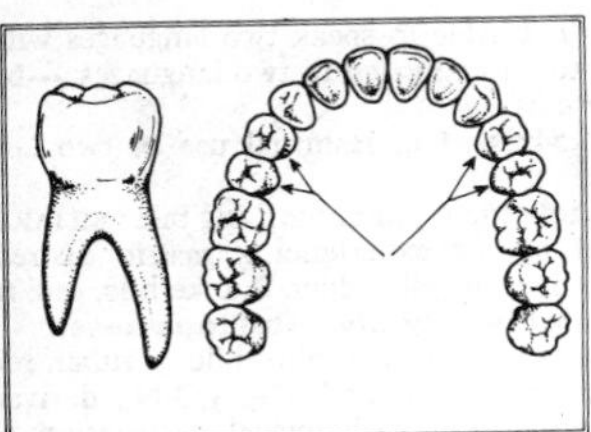

**bicuspid**

**bicuspid valve** *n.* The cardiac valve, made up of two triangular flaps, situated in the orifice linking the left auricle and ventricle.
**bi·cy·cle** (bī′sĭk′əl, -sĭ′kəl) *n.* [Fr. : *bi-*, two + Gk. *kuklos*, wheel.] A vehicle consisting of a metal frame on two spoked wheels one behind the other and having a seat, handlebars for steering, and two pedals or a small motor by which it is driven. —*vi.* **-cled, -cling, -cles.** To ride on a bicycle. **—bi′cy·clist** *n.*
**bi·cy·clic** (bī-sī′klĭk, -sĭk′lĭk) *also* **bi·cy·cli·cal** (-sī′klĭ-kəl, -sĭk′lĭ-) *adj.* **1.** Consisting of or having two cycles. **2.** *Bot.* Made up of or arranged in two distinct whorls, as the petals of a flower. **3.** *Chem.* Containing molecules with two fused rings.
**bid** (bĭd) *v.* **bade** (băd, bād) *or* **bid, bid·den** (bĭd′n) *or* **bid, bid·ding, bids.** [Partly < ME *bidden*, to ask, command (< OE *biddan*), and partly < ME *beden*, to offer (< OE *bēodan*).] —*vt.* **1.** To order : command. **2.** To utter (a greeting or salutation). **3.** To invite to attend : SUMMON. **4.** *p.t.* & *p.p.* **bid.** To state one's intention to take (tricks of a certain number or suit in cards) <*bid* four spades> **5.** *p.t.* & *p.p.* **bid.** To offer (an amount) as a price. —*vi.* **1.** *p.t.* & *p.p.* **bid.** To make an offer to accept or pay a specified price. **2.** *p.t.* & *p.p.* **bid.** To try to attain something : STRIVE. —*n.* **1. a.** An offer of a price. **b.** The amount offered. **2.** An invitation, esp. one offering membership in a group or club. **3. a.** The act of bidding in card games. **b.** The number of tricks or points declared. **c.** The trump or no-trump declared. **d.** A player's turn to bid. **4.** A sincere effort to attain something. **—bid fair.** To appear likely. **—bid′der** *n.*
**bi·dar·ka** (bī-där′kə) *n.* [R. *baidarka.*] A hide-covered canoe used by Eskimos of Alaska.
**bid·da·ble** (bĭd′ə-bəl) *adj.* **1.** Capable of being bid. **2.** Easily controlled : TRACTABLE.
**bid·den** (bĭd′n) *v. var. p.p. of* BID.
**bid·ding** (bĭd′ĭng) *n.* **1.** A demand : command. **2.** A request to appear : SUMMONS. **3.** Bids as a whole, as at an auction.
**bid·dy**[1] (bĭd′ē) *n., pl.* **-dies.** [Orig. unknown.] A hen or young chicken.
**bid·dy**[2] (bĭd′ē) *n., pl.* **-dies.** [Nickname for *Bridget.*] *Slang.* A woman, esp. a talkative old one.
**Biddy Basketball** *n.* [Alteration of dial. *bitty*, small (BIT + -Y) + BASKETBALL.] Basketball for youngsters with baskets at a height of 8½ feet.
**bide** (bīd) *v.* **bid·ed** *or* **bode** (bōd), **bid·ed, bid·ing, bides.** [ME *biden* < OE *bīdan.*] —*vi.* **1.** To remain in a particular state or condition. **2. a.** To wait : tarry. **b.** To be left : REMAIN. —*vt.* To await <*bide* one's time>
**bi·den·tate** (bī-dĕn′tāt′) *adj.* Having two teeth or toothlike processes.

ōō **boot** ou **out** th **thin** *th* **this** ŭ **cut** ûr **urge** y **young**
yōō **abuse** zh **vision** ə **about**, **item**, **edible**, **gallop**, **circus**

**bi·det** (bē-dā′) *n.* [Fr.] A basinlike bathroom fixture designed to be straddled for bathing the genitals and the posterior parts.
**bi·di·a·lec·tal** (bī′dī-ə-lĕk′təl) *adj.* Using two dialects of the same language. **—bi′di·a·lec′tal·ism** *n.*
**bi·don·ville** (bē′dôN-vēl′) *n.* [Fr. : *bidon*, tin can (< OFr., canteen, prob. of Scand. orig.) + *ville*, town. —see VILLAGE.] A shantytown on the outskirts of a city, as in France.
**Bie·der·mei·er** (bē′dər-mī′ər) *adj.* [After Gottlieb *Biedermeier*, the imaginary author of poems written by Ludwig Eichroth (1827–1892).] Of or relating to a heavy style of furniture developed in Germany during the first half of the 19th cent. and modeled after French Empire styles.
**bi·en·ni·a** (bī-ĕn′ē-ə) *n. var. pl. of* BIENNIUM.
**bi·en·ni·al** (bī-ĕn′ē-əl) *adj.* **1.** Living or lasting for two years. **2.** Happening every second year. **3.** Having a normal life cycle of two years. —*n.* **1.** An event that occurs once every two years. **2.** A plant that normally needs two years to reach maturity, producing leaves in the first year, blooming and yielding fruit in the second year, and then dying. **—bi·en′ni·al·ly** *adv.*
**bi·en·ni·um** (bī-ĕn′ē-əm) *n., pl.* **-ni·ums** *or* **-ni·a** (-ē-ə) [Lat : *bi-*, two + *annus*, year.] A two-year period.
**bier** (bîr) *n.* [ME *bere* < OE *bēr.*] A stand on which a corpse or a coffin containing a corpse is placed before burial.
**bi·fa·cial** (bī-fā′shəl) *adj.* **1.** Having two faces or fronts. **2.** *Bot.* Having upper and lower surfaces dissimilar and distinct. **3.** Having two opposing surfaces that are alike.
**biff** (bĭf) *vt.* **biffed, biff·ing, biffs.** [Imit.] *Slang.* To punch : sock. **—biff** *n.*
**bi·fid** (bī′fĭd) *adj.* Divided into two equal parts or lobes by a median cleft. **—bi·fid′i·ty** (-fĭd′ĭ-tē) *n.* **—bi′fid·ly** *adv.*
**bi·fi·lar** (bī-fī′lər) *adj.* Fitted with or involving the use of two wires or threads. **—bi·fi′lar·ly** *adv.*
**bi·flag·el·late** (bī-flăj′ə-lĭt, -lāt′) *adj. Zool.* Having two flagella.
**bi·fo·cal** (bī-fō′kəl) *adj.* **1.** Having two different focal lengths. **2.** Having one section that corrects for distant vision and another that corrects for close vision.
**bi·fo·cals** (bī-fō′kəlz) *pl.n.* Eyeglasses with bifocal lenses.
**bi·fo·li·ate** (bī-fō′lē-ĭt, -āt′) *adj. Bot.* Having two leaves.
**bi·fo·li·o·late** (bī-fō′lē-ə-lāt′, -lĭt) *adj. Bot.* Having two leaflets.
**bi·fo·rate** (bī-fôr′āt′, -fōr′-, bī′fə-rāt′) *adj.* [BI- + Lat. *forare, forat-*, to pierce.] *Biol.* Having two openings.
**bi·form** (bī′fôrm′) *adj.* Having a combination of the features or qualities of two distinct individuals.
**bi·fur·cate** (bī′fər-kāt′, bī-fûr′-) *v.* **-cat·ed, -cat·ing, -cates.** [Med. Lat. *bifurcare, bifurcat-*, to divide < Lat. *bifurcus*, two-pronged : *bi-*, two + *furcus*, fork.] —*vt.* To divide into two branches. —*vi.* To separate into two branches : FORK. **—bi′fur·cate′** *adj.* **—bi′fur·cate′ly** *adv.* **—bi′fur·ca′tion** *n.*
**big** (bĭg) *adj.* **big·ger, big·gest.** [ME, prob. of Scand. orig.] **1.** Of considerable size, number, quantity, magnitude, or extent : LARGE. **2. a.** *Obs.* Of great force or violence. **b.** Of great intensity : STRONG. **3.** Adult : grown-up. **4.** Pregnant <*big* with child> **5.** Filled up : brimming over. **6.** Having or exercising considerable authority or influence. **7.** Conspicuous in position, importance or wealth : PROMINENT. **8.** Of great significance : MOMENTOUS. **9.** Loud and firm : RESOUNDING. **10.** Generous : bountiful. **11.** *Informal.* Self-important : boastful. —*adv.* **1.** In a boastful or pretentious way <talked *big* about their plans> **2.** With much success. **—big on.** Enthusiastic about <*big on* science fiction> **—make it big.** To be very successful. **—big′gish** *adj.* **—big′ly** *adv.* **—big′ness** *n.*
☆ ***syns:*** BIG, CONSIDERABLE, EXTENSIVE, GOOD, GREAT, HEALTHY, LARGE, SIZABLE, TIDY *adj. core meaning* : notably above average in amount, size, or extent <a *big* inheritance><a *big* stadium> ***ant:*** little, small
**big·a·mous** (bĭg′ə-məs) *adj.* **1.** Involving bigamy. **2.** Guilty of bigamy. **—big′a·mous·ly** *adv.*
**big·a·my** (bĭg′ə-mē) *n., pl.* **-mies.** [ME *bigamie* < OFr. < *bigame*, bigamous < LLat. *bigamus* : Lat. *bi-*, two + Gk. *gamos*, marriage.] *Law.* Entry into marriage with one person while still legally married to another. **—big′a·mist** *n.*
**big·ar·reau** (bĭg′ə-rō′) *n.* [Fr. < *bigarrer*, to variegate.] Any of several varieties of sweet cherry with firm, often light-colored flesh.
**big bang** *n.* The cosmic explosion that marked the origin of the universe according to the big bang theory.
**big bang theory** *n.* A cosmological theory that the universe originated billions of years ago from the violent eruption of a point source.
**Big Bertha** *n.* [After *Bertha* Krupp von Bohlen und Halbach (1886–1957).] A huge cannon used by Germany in World War I.
**big brother** *n.* [After *Big Brother*, a character in the novel *Nineteen Eighty-Four* by George Orwell (1903–1950).] **1.** An older brother. **2.** A man who befriends a disadvantaged boy. **3. Big Brother.** A vague, threatening figure representing the all-seeing, ruthless power of a totalitarian government.
**Big Broth·er·ism** (brŭth′ə-rĭz′əm) *n.* Authoritarian efforts at complete control, as of an individual or a country.
**Big Dipper** *n.* A cluster of seven stars in the constellation Ursa Major, forming the shape of a dipper.

**bi·gem·i·nal** (bī-jĕm′ə-nəl) *adj.* [LLat. *bigeminus,* doubled : Lat. *bi-,* two + Lat. *geminus,* paired.] Occurring in pairs : TWINNED.

**bi·gem·i·ny** (bī-jĕm′ə-nē) *n.* [LLat. *bigeminus,* doubled + -Y.] A cardiovascular condition wherein the pulse occurs in groups of two rapid beats with a pause after each pair of beats.

**big·eye** (bĭg′ī′) *n.* Any of several tropical marine fishes of the family Priacanthidae, with large eyes and rough reddish scales.

**Big·foot** (bĭg′fo͝ot′) *n.* [From the size of the footprints believed to belong to it.] The sasquatch.

**big game** *n.* **1.** Large animals hunted or caught for sport. **2.** *Slang.* An important objective, esp. when dangerous. **—big′-game′** *adj.*

**big·ge·ty** (bĭg′ĭ-tē) *adj. var. of* BIGGITY.

**big·gie** (bĭg′ē) *n. Informal.* **1.** A big wheel. **2.** Something, as a corporation, that is considered big or important.

**big·gi·ty** *also* **big·ge·ty** (bĭg′ĭ-tē) *adj.* [< BIG.] *Informal.* Conceited : self-important.

**big·head** (bĭg′hĕd′) *n.* **1.** *Informal.* Conceit : egotism. **2.** Any of various diseases of animals typified by swelling of the head. **—big′-head′ed** *adj.* **—big′head′ed·ness** *n.*

**big-heart·ed** (bĭg′här′tĭd) *adj.* Kind : generous. **—big′-heart′ed·ly** *adv.* **—big′-heart′ed·ness** *n.*

**big·horn** (bĭg′hôrn′) *n.* A wild sheep, *Ovis canadensis* of western North American mountains, the male of which has massive curved horns.

**big house** *n. Slang.* A penitentiary.

**bight** (bīt) *n.* [ME, bend < OE *byht.*] **1.** A loop or slack part in a rope. **2. a.** A curve or bend, esp. in a shoreline. **b.** A wide bay formed by such a curve or bend. —*vt.* **bight·ed, bight·ing, bights.** To secure or tie in with a bight.

**big league** *n.* **1.** A major league. **2.** Big time. **—big leaguer** *n.*

**big-league** (bĭg′lēg′) *adj.* **1.** Major-league. **2.** Outstanding in one's field <a *big-league* publisher>

**big·mouth** (bĭg′mouth′) *n.* **1.** Any of various fishes with unusually large mouths. **2.** *Slang.* A gossipy or loud-mouthed person.

**big·mouthed** (bĭg′mouthd′, -moutht′) *adj.* **1.** Having a large mouth. **2.** Speaking indiscreetly or loudly : LOUD-MOUTHED.

**big-name** (bĭg′nām′) *adj.* **1.** Of superior rank in popular acknowledgment. **2.** Of or involving one that is big-name.

**big·no·ni·a** (bĭg-nō′nē-ə) *n.* [NLat. *Bignonia,* genus name, after Jean-Paul Bignon (1662–1743).] A plant of the genus *Bignonia,* esp. *B. capreolata,* a woody vine.

**big·ot** (bĭg′ət) *n.* [Fr. < OFr.] One fanatically devoted to one's own group, religion, race, or politics and intolerant of those who differ. **—big′ot·ed** *adj.* **—big′ot·ed·ly** *adv.* **—big′ot·ed·ness** *n.*

**big·ot·ry** (bĭg′ə-trē) *n.* The attitude, state of mind, or behavior characteristic of a bigot : INTOLERANCE.

**big shot** *n. Slang.* An influential or important person.

**big-tick·et** (bĭg′tĭk′ĭt) *adj. Informal.* Having a high price : EXPENSIVE <*big-ticket* items such as cars and private planes>

**big time** *n. Slang.* The most prestigious level of achievement in a competitive field. **—big′-time′** *adj.* **—big′-tim′er** *n.*

**big toe** *n.* The largest toe of the human foot.

**big top** *n. Informal.* **1.** CIRCUS 1c. **2.** CIRCUS 1a, b.

**big tree** *n.* The giant sequoia.

**big wheel** *n. Slang.* A person of importance or authority.

**big·wig** (bĭg′wĭg′) *n. Slang.* A big wheel.

**Bi·ha·ri** (bĭ-hä′rē) *n., pl.* **-ris. 1.** A native or inhabitant of Bihar. **2.** The Indic language of the Biharis.

**bi·jou** (bē′zho͞o′) *n., pl.* **-joux** (-zho͞o′, -zho͞oz′) [Fr. < Breton *bizou,* ring with a stone < *biz,* finger.] A small, exquisitely crafted trinket.

**bi·jou·te·rie** (bē-zho͞o′tə-rē) *n.* [Fr.] A collection of trinkets.

**bike** (bīk) *n.* [Short for BICYCLE.] **1.** A bicycle. **2.** A motorcycle. **3.** A motorbike. —*vi.* **biked, bik·ing, bikes.** To ride a bike.

**bik·er** (bī′kər) *n.* A motorcyclist, esp. a member of a motorcycle gang.

**bike·way** (bīk′wā′) *n.* A roadway for bicycles.

**bi·ki·ni** (bĭ-kē′nē) *n.* [Fr. < *Bikini,* an atoll in the Marshall Islands.] **1.** A very brief two-piece bathing suit for women. **2.** Brief underpants encircling the hips rather than the waist.

**bi·la·bi·al** (bī-lā′bē-əl) *adj.* **1.** Pronounced or articulated with both lips, as the consonants *b, p, m,* and *w.* **2.** Relating to both lips. —*n.* A bilabial sound or consonant. **—bi·la′bi·al·ly** *adv.*

**bi·la·bi·ate** (bī-lā′bē-ĭt, -āt′) *adj. Bot.* Having two lips, as a flower or corolla.

**bil·an·der** (bĭl′ən-dər, bī′lən-) *n.* [Du. *bijlander*: *bij,* by (< MDu. *bie*) + *land,* land (< MDu.).] A small two-masted sailing vessel, used esp. along the coasts of, and on canals in the Low Countries.

**bi·lat·er·al** (bī-lăt′ər-əl) *adj.* **1.** Of, relating to, or having two sides : TWO-SIDED. **2.** Having two symmetric sides. **3.** Affecting or undertaken by two sides equally : binding on both parties <a *bilateral* trade agreement> **—bi·lat′er·al·ism** *n.* **—bi·lat′er·al·ly** *adv.* **—bi·lat′er·al·ness** *n.*

**bil·ber·ry** (bĭl′bĕr′ē) *n.* [Prob. of Scand. orig.] **1.** Any of several shrubby plants of the genus *Vaccinium,* having edible blue or blackish berries. **2.** The fruit of a bilberry plant.

**bil·bo** (bĭl′bō) *n., pl.* **-boes.** [Poss. after *Bilbao,* Spain.] An iron bar with sliding fetters once used to shackle the feet of prisoners.

**bile** (bīl) *n.* [Fr. < Lat. *bilis.*] **1.** A bitter, alkaline, brownish-yellow or greenish-yellow liquid secreted by the liver, stored in the gallbladder, discharged into the duodenum, that helps digestion chiefly by saponifying fats. **2.** Ill temper : IRASCIBILITY.

**bile acid** *n.* Any of the liver-generated steroid acids that appear in the bile as sodium salts.

**bile duct** *n.* Any of the passages in the liver that convey bile from the liver to the hepatic duct, uniting with the cystic duct to form the common bile duct.

**bile salt** *n.* **1.** Any of the sodium salts bile. **2.** A mixture of ox-gall salts used medicinally as a laxative or hepatic stimulant.

**bilge** (bĭlj) *n.* [Prob. alteration of BULGE.] **1.** The lowest inner part of a ship's hull. **2.** Bilge water. **3.** The bulging part of a cask or barrel. **4.** *Slang.* NONSENSE. —*v.* **bilged, bilg·ing, bilg·es.** —*vi.* **1.** To spring a leak in the bilge. **2.** To swell or bulge. —*vt.* To break open the bilge of. **—bilg′y** *adj.*

**bilge keel** *n.* Either of two beams or fins fastened lengthwise along the outside of a ship's bilge to prevent heavy rolling.

**bilge water** *n.* Water that seeps into a ship's bilge.

**bil·har·zi·a·sis** (bĭl′här-zī′ə-sĭs) *n.* [< NLat. *Bilharzia,* genus name, after Theodor *Bilharz* (1825–1862).] Schistosomiasis.

**bil·i·ar·y** (bĭl′ē-ĕr′ē) *adj.* Of or relating to bile.

**biliary cirrhosis** *n.* Progressive inflammatory disease of the liver caused by obstruction of the bile duct.

**bi·lin·e·ar** (bī-lĭn′ē-ər) *adj. Math.* Linear with respect to each of two variables or positions.

**bi·lin·gual** (bī-lĭng′gwəl) *adj.* **1.** Able to speak two languages with equal facility. **2.** Of, relating to, or expressed in two languages. **—bi·lin′gual** *n.* **—bi·lin′gual·ly** *adv.*

**bi·lin·gual·ism** (bī-lĭng′gwə-lĭz′əm) *n.* Habitual use of two languages, esp. in speaking.

**bil·ious** (bĭl′yəs) *adj.* **1.** Of, relating to, or containing bile : BILIARY. **2.** Relating to, characterized by, or experiencing gastric distress caused by sluggishness of the liver or gallbladder. **3.** Like bile, esp. in color. **4.** Irascible : peevish. **—bil′ious·ly** *adv.* **—bil′ious·ness** *n.*

**bil·i·ru·bin** (bĭl′ĭ-ro͞o′bĭn, bĭl′ĭ-ro͞o′-) *n.* [Lat. *bilis,* bile + *ruber,* red + -IN.] A reddish-yellow organic compound, $C_{33}H_{36}O_6N_4$, derived from hemoglobin during normal and pathological destruction of erythrocytes.

**bil·i·ver·din** (bĭl′ĭ-vûr′dĭn, bĭl′ĭ-vûr′-) *n.* [Swed.: Lat. *bilis* + OFr. *verd,* green. —see VERDANT.] A green compound, $C_{33}H_{34}O_6N_4$, occurring in bile, occas. formed by oxidation of bilirubin.

**bilk** (bĭlk) *vt.* **bilked, bilk·ing, bilks.** [Perh. an alteration of BALK.] **1.** To defraud, swindle, or cheat. **2.** To evade payment of. **3.** To thwart or frustrate. **4.** To elude. —*n.* **1.** A cheat. **2.** A swindle or hoax. **—bilk′er** *n.*

**bill**[1] (bĭl) *n.* [ME *bille* < Norman Fr. < Med. Lat. *billa,* alteration of *bulla,* seal on a document < Lat., bubble.] **1.** An itemized list of fees or charges. **2.** A list of items, as a menu. **3.** Theatrical entertainment. **4.** An advertising poster or public notice. **5.** A piece of legal paper money. **6.** A commercial note, as a bill of exchange. **7.** A draft of a proposed law presented to a legislative body for approval. **8.** *Law.* A document presented to a court and containing a formal statement of a case, petition, or complaint. —*vt.* **billed, bill·ing, bills. 1.** To present a statement of charges or costs to. **2.** To enter on a statement of costs or on an itemized list. **3.** To announce, advertise, or schedule by public notice or as part of a program. **—fill the bill.** *Informal.* To meet all requirements. **—foot the bill.** *Informal.* To pay the cost of in full.

**bill**[2] (bĭl) *n.* [ME < OE *bile.*] **1.** The horny beak of a bird. **2.** A beaklike mouth part, as of a turtle. **3.** The visor of a cap. **4.** The tip of an anchor fluke. —*vi.* **billed, bill·ing, bills.** To touch beaks together.

**bill**[3] (bĭl) *n.* [ME *bil* < OE.] **1.** A billhook. **2.** A halberd or similar weapon with a long handle and hooked blade.

**bil·la·bong** (bĭl′ə-bông′, -bŏng′) *n.* [Native word in Australia.] *Austral.* **1.** A dead-end channel extending from the main stream of a river. **2.** A streambed filled with water only in the rainy season. **3.** A backwater or stagnant pool.

**bill·board** (bĭl′bôrd′, -bōrd′) *n.* A structure that displays advertisements in public places or alongside highways.

**bill·er** (bĭl′ər) *n.* One, as a clerk or a machine, that makes out bills.

**bil·let**[1] (bĭl′ĭt) *n.* [ME, official register < OFr. *billette,* dim. of *bulle,* document < Med. Lat. *bulla,* document, seal < Lat., bubble.] **1. a.** Board and lodging for troops, esp. in a civilian building. **b.** A written order to provide a billet. **2.** *Informal.* A position of employment : JOB. **3.** *Archaic.* A short letter : NOTE. —*v.* **-let·ed, -let·ing, -lets.** —*vt.* **1. a.** To quarter (soldiers), esp. in civilian buildings. **b.** To serve (a person) with a written order to provide a billet. **2.** To assign lodging to. —*vi.* To be quartered : LODGE.

**bil·let**[2] (bĭl′ĭt) *n.* [ME < OFr. *billette,* dim. of *bille,* log < Med. Lat. *billus,* poss. of Celt. orig.] **1.** A short thick piece of wood, as firewood. **2.** One of a series of log-shaped or square decorations on a molding. **3. a.** A bar of steel or iron in an intermediate manufacturing stage. **b.** A small ingot of nonferrous metal. **4. a.** The section of a harness strap that passes through a buckle. **b.** A loop or pocket for holding the tongue of a harness strap.

**bil·let-doux** (bĭl′ā-dōō′) *n., pl.* **bil·lets-doux** (-dōōz′) [Fr.: *billet,* short note + *doux,* sweet.] A love letter.

**bill·fish** (bĭl′fĭsh′) *n., pl.* **billfish** *or* **-fish·es. 1.** A fish of the family Istiophoridae, as a marlin or sailfish, with an elongated spearlike or swordlike snout and upper jaw. **2.** Any of various other fishes with long pointed jaws.

**bill·fold** (bĭl′fōld′) *n.* A folding pocket-sized case for money and personal papers.

**bill·head** (bĭl′hĕd′) *n.* A sheet of paper with a business name and address printed at the top, used for invoicing.

**bill·hook** (bĭl′hŏŏk′) *n.* A tool with a curved blade attached to a handle, used esp. for clearing brush and for rough pruning.

**bil·liard** (bĭl′yərd) *n.* A shot in billiards : CAROM. **—bil′liard** *adj.*

**bil·liards** (bĭl′yərdz) *pl.n.* [Fr. *billard.*] *(sing. in number).* **1.** A game played on an oblong cloth-covered table with raised cushioned edges, in which a long tapered cue is used to hit three small balls. **2.** A game similar to billiards, as one that is played on a table with pockets.

**bill·ing** (bĭl′ĭng) *n.* **1.** Relative importance of theatrical performers as indicated by the position and type size in which their names appear on marquees, programs or advertisements. **2. a.** Advertising. **b.** *often* **billings.** The total amount of business done in a particular period, as by a company.

**bil·lings·gate** (bĭl′ĭngz-gāt′, -gĭt) *n.* [After *Billingsgate,* a fish market in London, England.] Foul, abusive language.

**bil·lion** (bĭl′yən) *n.* [Fr., a million million : *bi-,* second power + *million,* million.] **1.** The cardinal number equal to $10^9$. **2.** *Chiefly Brit.* The cardinal number equal to $10^{12}$. **3.** An indefinitely large number. **—bil′lion** *adj. & pron.*

**bil·lion·aire** (bĭl′yə-nâr′) *n.* [BILLION + (MILLION)AIRE.] A person whose wealth amounts to at least a billion, as of dollars or pounds.

**bil·lionth** (bĭl′yənth) *n.* **1.** The ordinal number matching the number one billion in a series. **2.** One of a billion equal parts. **—bil′lionth** *adj. & adv.*

**bill of attainder** *n.* A former legislative act declaring a person guilty of a crime, usu. treason, without trial and subjecting him or her to capital punishment and attainder.

**bill of exchange** *n.* A written order directing that a specified sum of money be paid to a specified person.

**bill of fare** *n.* A menu.

**bill of goods** *n.* A consignment of goods.

**bill of health** *n.* A certificate stating whether there is infectious disease in a port of departure or aboard a ship, given to the ship's master.

**bill of lading** *n.* A document listing and acknowledging receipt of goods for shipment.

**bill of rights** *n.* **1.** A formal summary of the rights and liberties guaranteed to a people <a consumer *bill of rights*> **2. Bill of Rights.** The first ten amendments to the U.S. Constitution. **3. Bill of Rights.** A declaration of rights limiting the power of the crown, enacted by the English Parliament in 1689.

**bill of sale** *n.* A document that attests a transfer by sale of the ownership of personal property.

**bil·lon** (bĭl′ən) *n.* [Fr. < OFr., ingot, log. —see BILLET.] **1.** An alloy of silver or gold with a greater proportion of another metal, as copper, used in making coins. **2.** An alloy of silver with a high percentage of copper, used in making medals and tokens.

**bil·low** (bĭl′ō) *n.* [Prob. < ON *bylgja.*] **1.** A large ocean wave. **2.** A great surge, as of smoke. *—v.* **-lowed, -low·ing, -lows.** *—vi.* **1.** To surge in billows. **2.** To swell out. *—vt.* To cause to billow. **—bil′low·i·ness** *n.* **—bil′low·y** *adj.*

**bill·post·er** (bĭl′pō′stər) *n.* One who posts notices, posters, or advertisements. **—bill′post′ing** *n.*

**bil·ly[1]** (bĭl′ē) *n., pl.* **-lies.** [Prob. < *Billy,* nickname for William.] A billy club.

**bil·ly[2]** (bĭl′ē) *n., pl.* **-lies.** [Prob. short for *billycan* : *billa,* water (native word in Australia) + CAN[2].] *Austral.* A metal kettle or pot used in camp cooking.

**billy club** *n.* A short wooden club, used esp. by police officers.

**bil·ly·cock** (bĭl′ē-kŏk′) *n.* [Orig. unknown.] *Chiefly Brit.* A man's low-crowned felt hat similar to a derby.

**billy goat** *n. Informal.* A male goat.

**bi·lo·bate** (bī-lō′bāt′) *also* **bi·lo·bat·ed** (-bā′tĭd) *or* **bi·lobed** (bī′lōbd′) *adj.* Divided into or having two lobes.

**bi·lob·u·lar** (bī-lŏb′yə-lər, -lō′byə-) *adj.* Bilobate.

**bi·loc·u·lar** (bī-lŏk′yə-lər) *also* **bi·loc·u·late** (-lĭt, -lāt′) *adj.* Divided into or containing two cavities, cells, or chambers.

**Bi·lox·i** (bə-lŭk′sē, -lŏk′-) *n., pl.* **Biloxi** *or* **-is.** One of a tribe of Indians orig. living in the lower Mississippi River valley.

**bil·sted** (bĭl′stĕd′) *n.* [Orig. unknown.] SWEET GUM 1.

**bil·tong** (bĭl′tŏng′, -tông′) *n.* [Afr. : *bil,* buttock (< MDu. *bille*) + *tong,* tongue (< MDu. *tonghe*).] *So. Afr.* Narrow strips of meat dried in the sun.

**bi·man·u·al** (bī-măn′yōō-əl) *adj.* Using or requiring the use of two hands. **—bi·man′u·al·ly** *adv.*

**bi·max·il·lar·y** (bī-măk′sə-lĕr′ē) *adj.* Relating to both halves of the maxilla.

**bi·mes·tri·al** (bī-mĕs′trē-əl) *adj.* [Lat. *bimestris* : *bi-,* two + *mensis,* month.] Bimonthly.

**bi·me·tal·lic** (bī′mə-tăl′ĭk) *adj.* **1.** Composed of two metals. **2.** Of, based on, or using the principles of bimetallism.

**bi·met·al·lism** (bī-mĕt′l-ĭz′əm) *n.* The use of two metals, esp. gold and silver, as a monetary standard of currency. **—bi·met′al·list** *n.* **—bi·met′al·lis′tic** *adj.*

**bi·mod·al** (bī-mōd′l) *adj.* Having two distinct statistical modes. **—bi′mo·dal′i·ty** (bī′mō-dăl′ĭ-tē) *n.*

**bi·mo·lec·u·lar** (bī′mə-lĕk′yə-lər) *adj.* Relating to, consisting of, or affecting two molecules. **—bi′mo·lec′u·lar·ly** *adv.*

**bi·month·ly** (bī-mŭnth′lē) *adj.* **1.** Occurring every two months. **2.** Occurring twice a month : SEMIMONTHLY. *—adv.* **1.** Once every two months. **2.** Twice a month : SEMIMONTHLY. *—n., pl.* **-lies.** A bimonthly publication.

**bi·mor·phe·mic** (bī′môr-fē′mĭk) *adj.* That is composed of two morphemes.

**bin** (bĭn) *n.* [ME *binne* < OE.] A container or enclosed storage space. *—vt.* **binned, bin·ning, bins.** To store in a bin.

**bin-** *pref. var. of* BI-[1].

**bi·nal** (bī′nəl) *adj.* [NLat. *binalis,* twin < Lat. *bini,* two by two.] Double : twofold.

**bi·na·ry** (bī′nə-rē) *adj.* [LLat. *binarius* < Lat. *bini,* two by two.] **1.** Characterized by or composed of two different parts or components : TWOFOLD. **2.** Of or based on the number 2 or the binary numeration system <a *binary* digit> **3.** *Chem.* Comprised of or containing only molecules consisting of two kinds of atoms. **4.** *Mus.* Having two sections or subjects. *—n., pl.* **-ries.** Something binary, esp. a binary star.

**binary digit** *n.* Either of the digits 0 or 1, used to represent numbers in the binary numeration system.

**binary fission** *n.* Fission, esp. of a cell or of an atomic nucleus, resulting in two approx. equal products.

**binary numeration system** *n.* A system of numeration, based on 2, in which numerals are represented as sums of powers of 2 and in which all numerals can be written with the symbols 0 and 1.

**binary operation** *n.* An operation, as addition, that is applied to two elements of a set to produce a single element of the set.

**binary star** *n.* A stellar system composed of two stars orbiting about a common center of mass and often appearing as a single visual or telescopic object.

**bi·nate** (bī′nāt′) *adj.* Consisting of two parts or divisions : growing in pairs <a *binate* leaf> **—bi′nate′ly** *adv.*

**bi·na·tion·al** (bī-năsh′ə-nəl, -năsh′nəl) *adj.* Of, pertaining to, or involving two nations.

**bin·au·ral** (bī-nôr′əl, bĭn-ôr′-) *adj.* **1. a.** Having or related to two ears. **b.** Hearing with both ears. **2.** Of or relating to sound transmission from two sources with varying acoustics, as in tone or pitch, relative to a listener. **—bin·au′ral·ly** *adv.*

**bind** (bīnd) *v.* **bound** (bound), **bind·ing, binds.** [ME *binden* < OE *bindan.*] *—vt.* **1.** To secure, as with a rope or cord. **2.** To fasten by encircling, as with a belt. **3.** To bandage. **4.** To restrain with or as if with bonds. **5.** To obligate, as with a sense of moral duty. **6.** *Law.* To place under legal obligation by contract or oath. **7.** To make certain or irrevocable <*bind* a bargain> **8.** To employ as an apprentice : INDENTURE. **9.** To cause to stick together in a mass. **10.** To enclose and fasten (pages of a book) between protective covers. **11.** To furnish with a border or edge for reinforcement or ornamentation. **12.** To constipate. *—vi.* **1.** To tie up something. **2.** To be tight and uncomfortable. **3.** To become solid or compact : COHERE. **4.** To be compulsory. **—bind off.** To cast off in knitting. **—bind over.** *Law.* To hold on bail or place under bond. *—n.* **1. a.** Something that binds. **b.** An act of binding or the state of being bound. **2.** *Informal.* A difficult situation <caught in a *bind*>

**bind·er** (bīn′dər) *n.* **1.** One who binds books : BOOKBINDER. **2.** Something, as a cord, used to bind. **3.** A notebook cover with rings or clamps for holding paper. **4.** Something, as eggs in batter, that causes uniform consistency, solidification, or cohesion. **5. a.** An attachment on a reaping machine that ties grain in bundles. **b.** A machine that reaps and ties grain. **6.** *Law.* A payment or written statement making an agreement legally binding until the completion of a formal contract.

**bind·er·y** (bīn′də-rē) *n., pl.* **-ies.** A place where books are bound.

**bind·ing** (bīn′dĭng) *n.* **1.** The act of one that binds. **2.** Something that binds or is used as a binder. **3.** The cover that holds together the pages of a book. **4.** A strip attached over or along the edge of something for reinforcement or ornamentation. *—adj.* **1.** Serving to bind. **2.** Uncomfortably tight. **3.** Obligatory. **—bind′ing·ly** *adv.* **—bind′ing·ness** *n.*

**binding energy** *n.* **1.** The energy released in binding a group of particles into a single system, esp. a group of nucleons into an atomic

nucleus. **2.** The work required to remove an atomic electron to an infinitely remote position from its orbit.
**binding force** *n.* A strong interaction.
**bin·dle·stiff** (bĭn'dl-stĭf') *n.* [E. *bindle,* alteration of BUNDLE + STIFF.] *Slang.* A hobo or migrant worker.
**bind·weed** (bīnd'wēd') *n.* **1.** Any of several trailing or twining plants of the genus *Convolvulus,* with pink or white trumpet-shaped flowers. **2.** Any of various plants similar to a bindweed.
**bine** (bīn) *n.* [Dial. of *bind.*] **1.** The flexible stem of a climbing or twining plant, as the hop, woodbine, or bindweed. **2.** A plant whose stem is a bine.
**Bi·net age** (bĭ-nā') *n.* A person's mental age as determined by the Binet-Simon scale.
**Bi·net-Si·mon scale** (bĭ-nā'sē-mōN', -sī'mən) *n.* [After Alfred *Binet* (1857–1911) and Théodore *Simon* (1873–1961).] Any of a series of early psychological tests of childhood intelligence.
**binge** (bĭnj) [Dial. *binge,* to soak.] *Slang.* —*n.* **1.** A drunken spree. **2.** A period of uncontrolled self-indulgence <an eating *binge*> —*vi.* **binged, bing·ing, bing·es.** To be uncontrolled and self-indulgent.
**bin·go** (bĭng'gō) *n., pl.* **-gos.** [Orig. unknown.] A game of chance in which players put markers on a pattern of numbered squares according to numbers drawn and announced by a caller.
**bin·na·cle** (bĭn'ə-kəl) *n.* [Alteration of ME *bitakil* < OSp. *bitácula* or OPort. *bitácola* < Lat. *habitaculum,* habitation < *habitare,* to inhabit.] The nonmagnetic stand which supports a ship's compass case.
**bin·oc·u·lar** (bə-nŏk'yə-lər, bī-) *adj.* **1.** Relating to, used by, or involving both eyes simultaneously. **2.** Having two eyes arranged to produce stereoscopic vision. —*n. often* **binoculars.** An optical device, esp. a pair of field glasses, designed for use by both eyes simultaneously. **—bin·oc'u·lar'i·ty** (-lăr'ĭ-tē) *n.* **—bin·oc'u·lar·ly** *adv.*
**bi·no·mi·al** (bī-nō'mē-əl) *adj.* [< NLat. *binomius:* BI- + Gk. *nomos,* part.] Composed of or relating to two names or terms. —*n.* **1.** *Math.* An expression composed of two terms joined by a plus or minus sign. **2.** *Biol.* A taxonomic name. **—bi·no'mi·al·ly** *adv.*
**binomial distribution** *n.* The frequency distribution of the probability of a specified number of successes in an arbitrary number of repeated independent Bernoulli trials.
**binomial nomenclature** *n.* A system of classifying plants and animals by a double name, the first being the name of the genus and the second that of the species.
**binomial theorem** *n.* A mathematical theorem that specifies the expansion of a binomial to any power without requiring the explicit multiplication of the binomial terms.
**bi·nu·cle·ate** (bī-nōō'klē-ĭt, -āt', -nyōō'-) *also* **bi·nu·cle·at·ed** (-ā'tĭd) *adj.* Having two nuclei.
**bio-** *or* **bi-** *pref.* [Gk. < *bios,* life.] **1. a.** Life <*bio*lysis> **b.** Living organism <*bio*me> **2.** Biology : biological <*bio*physics>
**bi·o·ac·tiv·i·ty** (bī'ō-ăk-tĭv'ĭ-tē) *n.* The effect of a given agent, as a vaccine, on a living organism.
**bi·o·as·say** (bī'ō-ăs'ā', -ă-sā') *n.* Evaluation of a drug by comparison of its effect with that of a standard on a test organism.
**bi·o·as·tro·nau·tics** (bī'ō-ăs'trə-nô'tĭks) *n.* *(sing. in number).* Study of the medical and biological effects of space flight.
**bi·o·a·vail·a·bil·i·ty** (bī'ō-ə-vā'lə-bĭl'ĭ-tē) *n.* The degree to which an agent, as a nutrient or drug, becomes available at the physiological site of activity.
**bi·o·cat·a·lyst** (bī'ō-kăt'l-ĭst) *n.* A substance that modifies and initiates the rate of a biological process. **—bi'o·cat'a·lyt'ic** (-ĭt'ĭk) *adj.*
**bi·o·ce·nol·o·gy** (bī'ō-sə-nŏl'ə-jē) *n.* *Ecol.* Study of communities and member interactions in nature.
**biochemical oxygen demand** *n.* The amount of dissolved oxygen needed to meet the metabolic requirements of microorganisms in a water environment rich in organic matter, as sewage.
**bi·o·chem·is·try** (bī'ō-kĕm'ĭ-strē) *n.* The chemistry of biological processes and substances. **—bi'o·chem'i·cal** (-ĭ-kəl) *adj.* **—bi'o·chem'i·cal·ly** *adv.* **—bi'o·chem'ist** *n.*
**bi·o·cide** (bī'ə-sīd') *n.* A substance, as an antibiotic or pesticide, capable of destroying living organisms. **—bi'o·cid'al** (-sīd'l) *adj.*
**bi·o·cli·ma·tol·o·gy** (bī'ō-klī'mə-tŏl'ə-jē) *n.* Study of the effects of climate on organic life. **—bi'o·cli·mat'ic** (-klī-măt'ĭk) *adj.*
**bi·o·de·grad·a·ble** (bī'ō-dĭ-grā'də-bəl) *adj.* Capable of being decomposed by natural biological processes.
**bi·o·eth·ics** (bī'ō-ĕth'ĭks) *n.* *(sing. or pl. in number).* Study of the moral and ethical questions involved in applying new biological and medical findings, as in genetic engineering, neurobiology, and drug research. **—bi'o·eth'ic** *adj.* **—bi'o·eth'i·cist** (-ĭ-sĭst) *n.*
**bi·o·feed·back** (bī'ō-fēd'băk') *n.* A technique in which an attempt is made to consciously control a bodily function believed to be involuntary, as heartbeat or blood pressure, by using an instrument to monitor the function and to signal changes in it.
**bi·o·fla·vo·noid** (bī'ō-flā'və-noid') *n.* Any of a group of biologically active substances widely found in plants and functioning in the maintenance of the walls of small blood vessels.
**bi·o·gas** (bī'ō-găs') *n.* A mixture of carbon dioxide and methane produced through bacterial action.
**bi·o·gen·e·sis** (bī'ō-jĕn'ĭ-sĭs) *also* **bi·og·e·ny** (bī-ŏj'ə-nē) *n.* **1.** The doctrine that living organisms develop only from other living organisms and not from nonliving matter. **2.** Generation of living organisms from other living organisms. **—bi'o·ge·net'ic** (-jə-nĕt'ĭk), **bi'o·ge·net'i·cal** *adj.* **—bi'o·ge·net'i·cal·ly** *adv.*
**bi·o·ge·og·ra·phy** (bī'ō-jē-ŏg'rə-fē) *n.* Biological study of the geographic distribution of animals and plants. **—bi'o·ge'o·graph'ic** (-jē'ə-grăf'ĭk), **bi'o·ge'o·graph'i·cal** *adj.*
**bi·og·ra·pher** (bī-ŏg'rə-fər, bē-) *n.* A writer of a biography.
**bi·o·graph·i·cal** (bī'ə-grăf'ĭ-kəl) *also* **bi·o·graph·ic** (-grăf'ĭk) *adj.* **1.** Containing, composed of, or relating to the facts or events in a person's life. **2.** Of or relating to biography as a literary form. **—bi'o·graph'i·cal·ly** *adv.*
**bi·og·ra·phy** (bī-ŏg'rə-fē, bē-) *n., pl.* **-phies.** [Med. Gk. *biographia* : *bios,* life + *graphia,* -graphy.] **1.** A written account of a person's life. **2.** Biographies as a whole.
**bi·o·haz·ard** (bī'ō-hăz'ərd) *n.* A biological material, esp. if infective, that poses a threat to humans or their environment.
**bi·o·in·stru·men·ta·tion** (bī'ō-ĭn'strə-mĕn-tā'shən) *n.* **1.** Use of instruments for recording or transmitting physiological data. **2.** Instruments used in bioinstrumentation.
**bi·o·log·i·cal** (bī'ə-lŏj'ĭ-kəl) *also* **bi·o·log·ic** (-lŏj'ĭk) *adj.* **1.** Of or relating to biology. **2.** Of, relating to, caused by, or affecting life or living organisms. —*n.* **biologic.** A drug derived from a biological source. **—bi'o·log'i·cal·ly** *adv.*
**biological clock** *n.* An intrinsic biological mechanism responsible for the periodicity or other time-dependent aspects of certain classes of behavior in living organisms.
**biological half-life** *n.* HALF-LIFE 2a.
**biological warfare** *n.* Warfare in which disease-producing microorganisms or organic biocides are used to destroy human life, livestock, or crops.
**bi·ol·o·gy** (bī-ŏl'ə-jē) *n.* [G. *Biologie* : Gk. *bios,* life + Gk. *logos,* reckoning.] **1.** The science of living organisms and life processes, including the study of structure, functioning, growth, origin, evolution, and distribution of living organisms. **2.** The life processes or characteristic phenomena of a group or category of living organisms. **3.** The animal and plant life of a region or place. **—bi·ol'o·gist** *n.*
**bi·o·lu·mi·nes·cence** (bī'ō-lōō'mə-nĕs'əns) *n.* Emission of visible light by living organisms such as the firefly and various fish, fungi, and bacteria. **—bi'o·lu'mi·nes'cent** *adj.*
**bi·ol·y·sis** (bī-ŏl'ĭ-sĭs) *n.* Death caused or accompanied by lysis. **—bi'o·lyt'ic** (bī'ə-lĭt'ĭk) *adj.*
**bi·o·mass** (bī'ō-măs') *n.* The total mass of living matter within a given volume of environment.
**bi·ome** (bī'ōm') *n.* An entire community of living organisms in a single major ecological region.
**bi·o·med·i·cine** (bī'ō-mĕd'ĭ-sĭn) *n.* **1.** The branch of medicine concerned with human survival and functioning in abnormally stressful environments and with the medical aspects of protective modification of those environments. **2.** The study of medicine as it relates to all biological systems. **—bi'o·med'i·cal** *adj.*
**bi·o·met·rics** (bī'ō-mĕt'rĭks) *n.* *(sing. in number).* The statistical study of biological data. **—bi'o·met'ric, bi'o·met'ri·cal** *adj.* **—bi'o·met'ri·cal·ly** *adv.*
**bi·om·e·try** (bī-ŏm'ĭ-trē) *n.* Biometrics.
**bi·on·ic** (bī-ŏn'ĭk) *adj.* [BIO- + (ELECTR)ONIC.] **1.** Consisting of or enhanced by or as if by electronic or mechanical devices or components. **2.** Of or relating to bionics.
**bi·on·ics** (bī-ŏn'ĭks) *n.* [BI(O)- + (ELECTR)ONICS.] *(sing. in number).* Application of biological principles to the design and study of engineering systems, esp. electronic systems.
**bi·o·nom·ics** (bī'ə-nŏm'ĭks) *n.* [< Fr. *bionomique,* pertaining to ecology < *bionomie,* ecology : Gk. *bios,* life + Gk. *nomos,* law.] *(sing. in number).* Ecology. **—bi'o·nom'ic, bi'o·nom'i·cal** *adj.* **—bi'o·nom'i·cal·ly** *adv.*
**bi·ont** (bī'ŏnt') *n.* A living organism. **—bi·on'tic** (bī-ŏn'tĭk) *adj.*
**bi·o·phys·ics** (bī'ō-fĭz'ĭks) *n.* *(sing. in number).* The physics of biological processes. **—bi'o·phys'i·cal** *adj.* **—bi'o·phys'i·cal·ly** *adv.* **—bi'o·phys'i·cist** *n.*
**bi·o·plasm** (bī'ō-plăz'əm) *n.* Living protoplasm.
**bi·op·sy** (bī'ŏp'sē) *n., pl.* **-sies.** The study of tissue taken from a living organism, esp. in examination for the presence of disease. **—bi·op'sic** (bī-ŏp'sĭk) *adj.*
**bi·o·rhythm** (bī'ō-rĭth'əm) *n.* **1.** An intrinsically patterned cyclical biological function or process. **2.** The determining factor in a biorhythm. **—bi'o·rhyth'mic, bi'o·rhyth'mi·cal** *adj.*
**bi·o·sci·ence** (bī'ō-sī'əns) *n.* Life science.
**bi·os·co·py** (bī-ŏs'kə-pē) *n., pl.* **-pies.** Medical examination of a body to determine whether it is dead.
**-biosis** *suff.* [NLat. < Gk. *biōsis,* way of life < *bioun,* to live < *bios,* life.] A way of living <para*biosis*>
**bi·o·sphere** (bī'ə-sfîr') *n.* The part of the earth and its atmosphere in which living things exist.
**bi·o·syn·the·sis** (bī'ō-sĭn'thĭ-sĭs) *n.* Production of complex sub-

stances from simple ones by or with living organisms. **—bi'o·syn·thet'ic** (-thĕt'ĭk) *adj.* **—bi'o·syn·thet'i·cal·ly** *adv.*

**bi·o·ta** (bī-ō'tə) *n.* [NLat. < Gk. *biotē*, way of life.] The animal and plant life of a particular region.

**bi·o·tech·nol·o·gy** (bī'ō-tĕk-nŏl'ə-jē) *n.* The engineering and biological study of relationships between human beings and machines. **—bi'o·tech'no·log'i·cal** (-nə-lŏj'ĭ-kəl) *adj.*

**bi·ot·ic** (bī-ŏt'ĭk) *adj.* [Gk. *biōtikos* < *bios*, life.] Relating to life or specific life conditions.

**biotic potential** *n.* **1.** The chance of survival of a specific organism in a specific, esp. an unfavorable, environment. **2.** The growth rate of a population that maintains a stable age distribution.

**bi·o·tin** (bī'ə-tĭn) *n.* [Gk. *biotos*, life + -IN.] A colorless crystalline vitamin, $C_{10}H_{16}N_2O_3S$, often considered in the vitamin B complex and found in liver, egg yolk, milk, and yeast.

**bi·o·tite** (bī'ə-tīt') *n.* [G. *Biotit*, after Jean Baptiste *Biot* (1774–1862).] A dark-brown to black mica, $K(Mg, Fe)_3AlSi_3O_{10}(OH)_2$, found in metamorphic and igneous rocks. **—bi'o·tit'ic** (-tĭt'ĭk) *adj.*

**bi·o·tope** (bī'ə-tōp') *n.* [BIO- + Gk. *topos*, place.] A limited ecological region in which the environment suits certain forms of life.

**bi·o·trans·for·ma·tion** (bī'ō-trăns'fər-mā'shən) *n.* Chemical transformation within a living system.

**bi·o·tron** (bī'ə-trŏn') *n.* A climate-control chamber for studying a living organism's responses to specific environmental conditions.

**bi·o·type** (bī'ə-tīp') *n.* A group of organisms with identical genetic but varying physical characteristics. **—bi'o·typ'ic** (-tĭp'ĭk) *adj.*

**bi·o·vu·lar** (bī-ō'vyə-lər, -ŏv'yə-) *adj.* Derived from two ova, as fraternal twins.

**bip·a·rous** (bĭp'ər-əs) *adj.* **1.** *Biol.* Producing two offspring in a single birth. **2.** *Bot.* Having two axes or branches.

**bi·par·ti·san** (bī-pär'tĭ-zən, -sən) *adj.* Consisting of or supported by members of two political parties. **—bi·par'ti·san·ism** *n.* **—bi·par'ti·san·ship'** *n.*

**bi·par·tite** (bī-pär'tīt') *adj.* [< Lat. *bipartire, bipartit-*, to divide into two parts : *bi-*, two + *partire*, to part < *pars*, share.] **1.** Having two parts. **2.** Having two corresponding parts, one for each party <a *bipartite* pact> **3.** *Bot.* Divided into two, almost to the base <*bipartite* leaves> **—bi·par'tite'ly** *adv.* **—bi'par·ti'tion** (-tĭsh'ən) *n.*

**bi·ped** (bī'pĕd') *n.* [Lat. *bipes, biped-*, two-footed : *bi-*, two + *pes*, foot.] A two-footed animal. **—bi'ped'** *adj.* **—bi·ped'al** (bī-pĕd'l) *adj.*

**bi·pet·al·ous** (bī-pĕt'l-əs) *adj.* Having two petals : DIPETALOUS.

**bi·phen·yl** (bī-fĕn'əl, -fē'nəl) *n.* A colorless crystalline compound, $C_{12}H_{10}$, used as a heat-transfer agent, in fungicides, and in organic synthesis.

**bi·pin·nate** (bī-pĭn'āt') *adj.* Having opposite leaflets subdivided into opposite leaflets, as compound leaves. **—bi·pin'nate'ly** *adv.*

**bi·plane** (bī'plān') *n.* An aircraft with single or paired wings fixed at two different levels, esp. one above and one below the fuselage.

**bi·pod** (bī'pŏd') *n.* A supporting stand with two legs.

**bi·po·lar** (bī-pō'lər) *adj.* **1.** Relating to or having two poles. **2.** Relating to or involving both of the earth's poles. **3.** Having or expressing two contradictory ideas or qualities. **—bi'po·lar'i·ty** (-lăr'ĭ-tē) *n.*

**bi·po·ten·ti·al·i·ty** (bī'pə-tĕn'shē-ăl'ĭ-tē) *n.* Capacity to perform both the male and the female sexual functions.

**bi·pro·pel·lant** (bī'prə-pĕl'ənt) *n.* A two-component rocket propellant, as liquid oxygen and liquid hydrogen, combined as fuel and oxidizer.

**bi·quad·rat·ic** (bī'kwŏ-drăt'ĭk) *Math.* —*adj.* Of or relating to the fourth degree. —*n.* An algebraic equation of the fourth degree.

**bi·quar·ter·ly** (bī-kwôr'tər-lē) *adj.* Occurring or appearing two times during each three-month period of the year <*biquarterly* meetings> **—bi·quar'ter·ly** *adv.*

**bi·ra·cial** (bī-rā'shəl) *adj.* Of, for, or composed of members of two races. **—bi·ra'cial·ism** *n.*

**bi·ra·di·al** (bī-rā'dē-əl) *adj.* Bilaterally and radially symmetrical.

**bi·ra·mous** (bī-rā'məs) *adj.* Bearing two branches, as in an arthropod appendage.

**birch** (bûrch) *n.* [ME < OE *birc.*] **1.** Any of several deciduous trees of the genus *Betula*, common in the Northern Hemisphere and having white, yellowish, or gray bark separable from the wood in sheets. **2.** The close-grained, hard wood of a birch tree. **3.** A rod from a birch tree, used as a whip. —*vt.* **birched, birch·ing, birch·es.** To whip with or as if with a birch rod.

**Birch·er** (bûr'chər) *also* **Birch·ite** (-chīt') *or* **Birch·ist** (-chĭst) *n.* A member or supporter of the John Birch Society. **—Birch'ism** *n.*

**bird** (bûrd) *n.* [ME < OE, young bird.] **1.** A member of the class Aves, including warm-blooded, egg-laying, feathered vertebrates with forelimbs modified to form wings. **2.** A bird hunted as game. **3.** *Slang.* **a.** An aircraft. **b.** A guided missile or rocket. **4.** A clay pigeon. **5.** A shuttlecock. **6.** *Slang.* One who is odd or remarkable <a strange *bird*>. **7.** *Chiefly Brit.* A young woman. **8.** *Slang.* RASPBERRY 4. —*vi.* **bird·ed, bird·ing, birds.** **1.** To observe and identify birds in their natural surroundings. **2.** To trap, shoot, or catch birds. **—for the birds.** Worthless or objectionable. **—bird'er** *n.*

**bird·bath** (bûrd'băth', -bäth') *n.* A water-filled basin in which birds may bathe.

**bird·brain** (bûrd'brān') *n.* *Slang.* A silly person.

**bird·cage** (bûrd'kāj') *n.* A cage for birds.

**bird·call** (bûrd'kôl') *n.* **1.** The song of a bird. **2. a.** An imitation of the song of a bird. **b.** A small device for producing this sound.

**bird cherry** *n.* A cherry tree native to Eurasia, *Prunus padus*, with white flower clusters and small black fruit.

**bird colonel** *n.* [From the eagle of the insignia.] *Slang.* A full colonel in the U.S. Army, Air Force, or Marine Corps.

**bird dog** *n.* **1.** A dog used to hunt game birds : GUN DOG. **2.** *Slang.* One who seeks out something for another.

**bird-dog** (bûrd'dôg', -dŏg') *v.* **-dogged, -dog·ging, -dogs.** —*vi.* To watch closely. —*vt.* To seek out : FOLLOW.

**bird·house** (bûrd'hous') *n.* **1.** An aviary. **2.** A box made as a nesting place for birds.

**bird·ie** (bûr'dē) *n.* **1.** *Informal.* A small bird. **2.** One stroke under par for a hole in golf. **3.** A shuttlecock. —*vt.* **-ied, -ie·ing, -ies.** To shoot (a hole in golf) in one stroke under par.

**bird·lime** (bûrd'līm') *n.* **1.** A sticky substance smeared on twigs to capture small birds. **2.** Something that captures and ensnares. —*vt.* **-limed, -lim·ing, -limes.** To smear or catch with birdlime.

**bird·man** (bûrd'mən) *n.* **1.** (*also* -măn'). One who works with birds, as an ornithologist. **2.** *Slang.* An aviator.

**bird of paradise** *n.* Any of various birds of the family Paradisaeidae native to New Guinea and adjacent areas, the male of which usu. has long tail feathers and brilliant plumage.

**bird-of-par·a·dise flower** (bûrd'əv-păr'ə-dīs', -dīz') *n.* A perennial plant, *Strelitzia reginae*, having purple bracts and large orange or yellow flowers with blue tongues.

**bird of passage** *n.* **1.** A migratory bird. **2.** One who moves from place to place frequently.

**bird of prey** *n.* A predatory carnivorous bird, as the eagle or hawk.

**bird pepper** *n.* **1.** A plant, *Capsicum frutescens*, the probable ancestor of the mild peppers and many of the pungent peppers. **2.** The narrow, very pungent fruit of the bird pepper.

**bird·seed** (bûrd'sēd') *n.* A mixture of seeds used for feeding birds, esp. caged birds.

**bird's-eye** (bûrdz'ī') *n.* **1.** Any of various plants bearing small, brightly colored flowers. **2. a.** A fabric woven with a pattern of small diamonds, each with a dot in the center. **b.** The pattern of such a fabric. —*adj.* **1.** Dappled or patterned with spots held to resemble birds' eyes. **2.** Seen from high above or from a distance <a *bird's-eye* view of the great city>

**bird's-foot** (bûrdz'fo͝ot') *n.*, *pl.* **bird's-foots.** A plant, as the bird's-foot trefoil, that bears flowers, leaves, or pods resembling a bird's foot or claw.

**bird's-foot fern** *n.* A fern native to California, *Pellaea mucronata*, bearing fronds with wiry leaves grouped to resemble a bird's foot.

**bird's-foot trefoil** *n.* A sprawling plant, *Lotus corniculatus*, bearing yellow flowers and seed pod clusters resembling the claws of a bird.

**bird's-nest fungus** (bûrdz'nĕst') *n.* A fungus of the family Nidulariaceae, whose cuplike fruiting body contains several round egglike structures that enclose the spores.

**bird watcher** *n.* One who observes and identifies birds in their natural habitats. **—bird watching** *n.*

**bird·y·back** (bûr'dē-băk') *n.* [BIRD + -Y + (PIGGY)BACK.] Transport of loaded truck trailers by aircraft.

**bi·re·frin·gence** (bī'rĭ-frĭn'jəns) *n.* The splitting or resolution of a light wave into two waves with mutually perpendicular vibration directions by an optically anisotropic medium such as calcite or quartz. **—bi're·frin'gent** *adj.*

**bi·reme** (bī'rēm') *n.* [Lat. *biremis* : *bi-*, two + *remus*, oar.] An ancient galley with two tiers of oars on each side.

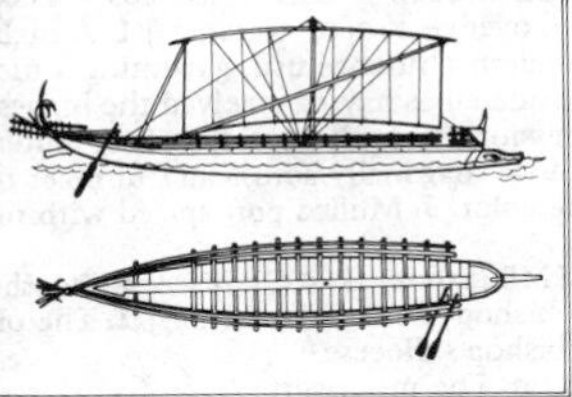

**bireme**
*An ancient Greek bireme*

**bi·ret·ta** (bə-rĕt'ə) *n.* [Ital. *berretta* < OProv. *birret*, cap < LLat. *birrus*, hooded cloak.] A stiff square cap worn esp. by Roman Catholic clergy.

**birl** (bûrl) *v.* **birled, birl·ing, birls.** [Blend of BIRR and WHIRL.] —*vt.* To cause (a floating log) to spin rapidly by rotating with the feet. —*vi.* **1.** To participate in birling. **2.** To whirr or hum. —*n.* A whirring noise : HUM. **—birl'er** *n.*

**birl·ing** (bûr′lĭng) *n.* A game of skill, esp. among lumberjacks, in which two competitors try to balance on a floating log while spinning it with their feet.

**birr¹** (bûr) *n.* [ME *bir,* strong wind < ON *byrr,* favorable wind.] A whirring sound.

**birr²** (bîr) *n., pl.* **birr** *or* **birrs.** [Native word in Ethiopia.] —See table at CURRENCY.

†**birth** (bûrth) *n.* [ME, of ON orig.] **1.** The beginning of existence : the fact of being born. **2.** A beginning : origin. **3. a.** The act of bearing young : PARTURITION. **b.** Passage of a child from the uterus. **4.** Ancestry : parentage. **5.** Origin : lineage. —*vt.* **birthed, birth·ing, births.** *Chiefly Regional.* **1.** To deliver (a baby). **2.** To bear (a child). **—give birth to.** To bring into being.

**birth canal** *n.* The cavity of the vagina and the uterus traversed by the fetus in parturition.

**birth certificate** *n.* An official record of a person's parentage and the date, place, and time of birth.

**birth control** *n.* Voluntary limitation or control of the number of children conceived, esp. by planned use of contraceptive techniques.

**birth·day** (bûrth′dā′) *n.* **1.** The day of one's birth. **2.** The anniversary of one's birth.

**birthday suit** *n.* The state of being naked.

**birthing room** *n.* A hospital delivery room furnished like a private bedroom and equipped for natural childbirth.

**birth·mark** (bûrth′märk′) *n.* A blemish or mole present on the body from birth : NEVUS.

**birth pang** *n.* **1.** *often* **birth pangs.** One of the repetitive pains occurring in childbirth. **2. birth pangs.** Turmoil or tumult associated esp. with a major social change.

**birth·place** (bûrth′plās′) *n.* A place of birth or origin.

**birth·rate** (bûrth′rāt′) *n.* The number of births in a specified population per unit time, esp. per year.

**birth·right** (bûrth′rīt′) *n.* **1.** A privilege granted a person by virtue of his or her birth. **2.** A special privilege accorded a first-born.

**birth·root** (bûrth′ro͞ot′, -ro͝ot′) *n.* A North American plant of the genus *Trillium,* esp. *T. erectum,* bearing purplish flowers with an unpleasant odor and tuberlike roots, used at one time as an aid in childbirth.

**birth·stone** (bûrth′stōn′) *n.* A jewel associated with the month of one's birth.

**birth trauma** *n.* An injury or emotional shock sustained by an infant during birth.

**birth·wort** (bûrth′wûrt′, -wôrt′) *n.* Any of several tropical woody vines of the genus *Aristolochia,* bearing reddish or brownish, usu. unpleasantly scented flowers.

**bis** (bĭs) *adv.* [Fr. < Lat., twice.] Encore : again. —Used esp. as a direction in music.

**bis·cuit** (bĭs′kĭt) *n., pl.* **-cuits** *or* **biscuit.** [ME *biscute* < OFr. *biscuit* < Med. Lat. *biscoctus* : Lat. *bis,* twice + Lat. *coquere,* to cook.] **1.** A small cake of shortened bread leavened with baking powder or soda. **2.** *Chiefly Brit.* **a.** A cookie. **b.** A thin, crisp cracker. **3.** A pale brown : BEIGE. **4.** Clay that has been fired once but not glazed.

**bise** (bēz) *n.* [ME < OFr., of Germanic orig.] A cold north wind of the Swiss Alps, France, and Italy.

**bi·sect** (bī′sĕkt′, bī-sĕkt′) *v.* **-sect·ed, -sect·ing, -sects.** —*vt.* To cut into two equal parts. —*vi.* To split : fork. **—bi·sec′tion** *n.* **—bi·sec′tion·al** *adj.* **—bi·sec′tion·al·ly** *adv.*

**bi·sec·tor** (bī′sĕk′tər, bī-sĕk′-) *n.* Something that bisects, esp. a straight line bisecting an angle.

**bi·ser·rate** (bī-sĕr′āt′) *adj. Biol.* **1.** Having serrations that are themselves serrated : doubly serrate. **2.** Serrated on both sides <*biserrate* antennae>

**bi·sex·u·al** (bī-sĕk′sho͞o-əl) *adj.* **1.** Of or relating to both sexes. **2.** Having both male and female organs : HERMAPHRODITIC. **3.** Sexually attracted to members of both sexes. —*n.* **1.** A bisexual organism : HERMAPHRODITE. **2.** A person sexually attracted to members of both sexes. **—bi′sex·u·al′i·ty** (-ăl′ĭ-tē) *n.* **—bi·sex′u·al·ly** *adv.*

**bish·op** (bĭsh′əp) *n.* [ME < OE *bisceop* < LLat. *episcopus* < LGk. *episkopos* < Gk., overseer : *epi,* over + *skopos,* watcher.] **1.** A high-ranking Christian cleric, in modern churches usu. governing a diocese and in some churches considered as having received the highest ordination in unbroken succession from the apostles. **2.** A miter-shaped chess piece that can move diagonally across any number of unoccupied spaces of the same color. **3.** Mulled port spiced with oranges, sugar, and cloves.

**bish·op·ric** (bĭsh′ə-prĭk) *n.* [ME *bishop-rik* < OE *bisceop-rīce,* the diocese of a bishop : *bisceop,* bishop + *rīce,* dominion.] **1.** The office or rank of a bishop. **2.** A bishop's diocese.

**bish·op's-cap** (bĭsh′əps-kăp′) *n.* The miterwort.

**bis·muth** (bĭz′məth) *n.* [Obs. G. *Bismut,* alteration of *Wismut* : *Wise,* meadow + *Mut,* claim to a mine.] *Symbol* **Bi** A white, crystalline, brittle metallic element used in alloys to form sharp castings for objects sensitive to high temperatures and in various low-melting alloys for fire-safety devices; atomic number 83; atomic weight 208.980. **—bis′muth·al** *adj.*

**bis·na·ga** (bĭs-nä′gə) *n.* [Sp. *biznaga,* alteration of *vitznauac* (< Nahuatl *huitznahuac* : *huitztli,* spine + *nahuac,* around.] Any of several spiny, globe-shaped or barrel-shaped cacti of the southwestern United States and Mexico.

**bi·son** (bī′sən, -zən) *n.* [Fr. < Lat., of Germanic orig.] **1.** A hoofed mammal, *Bison bison* of western North America, with a dark-brown coat, a shaggy mane, and short curved horns : BUFFALO. **2.** An animal, *B. bonasus* of Europe, similar to but somewhat smaller than the bison : WISENT.

**bisque¹** (bĭsk) *n.* [Fr.] **1. a.** A thick, rich soup made from meat, fish, or shellfish. **b.** A thick cream soup made of puréed vegetables. **2.** Ice cream mixed with crushed macaroons or nuts.

**bisque²** (bĭsk) *n.* [< BISCUIT.] **1.** BISCUIT 4. **2. a.** A pale orange yellow to yellowish gray. **b.** A color ranging from moderate yellowish pink to grayish yellow.

**bisque³** (bĭsk) *n.* [Fr.] An advantage given an inferior player in certain games, esp. a free point taken when desired in a tennis set.

**bis·sex·tile** (bī-sĕk′stĭl, -stīl′, bĭ-) *adj.* [LLat. *bissextilis* < Lat. *bissextus,* an intercalary day : *bi-,* twice + *sextus,* sixth.] **1.** Of or relating to a leap year. **2.** Of or relating to the extra day falling in a leap year. —*n.* A leap year.

▲ word history: A leap year is called a *bissextile* year because it originally contained a month with two "sixths." The extra sixth day was the intercalary day inserted in the Julian calendar between the sixth and fifth days before the calends of March. There were thus two days that could be considered the sixth day—the first being *sextus,* the second *bissextus.*

**bi·sta·ble** (bī-stā′bəl) *n. Computer Sci.* Of or pertaining to hardware having the ability to assume only two stable states, such as *on* or *off,* or *1* or *0.*

**bi·state** (bī′stāt′) *adj.* Of, pertaining to, or involving two states.

**bis·ter** *also* **bis·tre** (bĭs′tər) *n.* [Fr. *bistre.*] **1.** A water-soluble, yellowish-brown pigment made from soot obtained from beech or other wood. **2.** A grayish to yellowish brown. **—bis′tered** *adj.*

**bis·tort** (bĭs′tôrt′) *n.* [OFr. *bistorte* < Med. Lat. *bistorta* : Lat. *bis,* twice + Lat. *torquēre,* to twist.] Any of several plants of the genus *Polygonum,* esp.: **a.** A Eurasian plant, *P. bistorta,* bearing small pinkish flowers in pointed clusters. **b.** A similar plant, *P. bistortoides* of the mountains of western North America, bearing pink or white oval flower clusters.

**bis·tou·ry** (bĭs′tə-rē) *n., pl.* **-ries.** [Fr. *bistouri* < OFr. *bistoric,* dagger.] A surgical knife for minor incisions.

**bis·tre** (bĭs′tər) *n. var. of* BISTER.

**bis·tro** (bē′strō, bĭs′trō) *n., pl.* **-tros.** [Fr.] A small bar or nightclub.

**bi·sul·cate** (bī-sŭl′kāt′) *adj.* Cleft or cloven <a *bisulcate* hoof>

**bi·sul·fate** (bī-sŭl′fāt′) *n.* The inorganic acid group $HSO_4$ or a compound containing it.

**bi·sul·fide** (bī-sŭl′fīd′) *n.* A disulfide.

**bi·sul·fite** (bī-sŭl′fīt′) *n.* The inorganic acid group $HSO_3$ or a compound containing it.

**bit¹** (bĭt) *n.* [ME *bite,* morsel < OE *bita.*] **1. a.** A small piece, portion, or amount. **b.** The tiniest amount. **2.** A brief amount of time : MOMENT. **3. a.** An entertainment routine given regularly by a performer : ACT <a comedy *bit*> **b.** A short scene or episode in a theatrical performance. **4.** A small and insignificant role, as in a play or movie. **5.** *Informal.* **a.** A particular kind of action, situation, or behavior <We did the whole tourist *bit.*> **b.** A matter under discussion or consideration <What's this *bit* about condiminiums?> **6.** *Informal.* An amount equal to ⅛ of a dollar <two *bits*> **7.** *Chiefly Brit.* A small coin <a threepenny *bit*> **—bit by bit.** Little by little.

☆ ***syns:*** BIT, GRAIN, HOOT, IOTA, JOT, MINIM, MODICUM, MOLECULE, ORT, OUNCE, PARTICLE, SCRAP, SHRED, SMIDGEN, SMITCH, SPECK, TITTLE, WHIT *n. core meaning* : the tiniest amount <not a *bit* of truth to that story>

**bit²** (bĭt) *n.* [ME *bite* < OE, cut.] **1.** The sharp part of a tool. **2.** A pointed and threaded tool for boring and drilling that is secured in a brace, bitstock, or drill press. **3.** The part of a key that enters the lock and engages the bolt or tumblers. **4.** The metal mouthpiece of a bridle that controls and curbs an animal. **5.** Something that controls or curbs. —*vt.* **bit·ted, bit·ting, bits. 1.** To place a bit in the mouth of (a horse). **2.** To control or check as if with a bit. **3.** To make or grind a bit on (a key).

**bit³** (bĭt) *n.* [B(INARY) (DIG)IT.] *Computer Sci.* **1.** A single character of a language having just two characters, as either of the binary digits 0 or 1. **2.** A unit of information equivalent to the choice of either of two equally likely alternatives. **3.** A unit of information storage capacity, as of a computer memory.

**bit⁴** (bĭt) *v. p.t. & var. p.p. of* BITE.

**bi·tar·trate** (bī-tär′trāt′) *n.* The tartrate of an acid.

**bitch** (bĭch) *n.* [ME *bicche* < OE *bicce.*] **1.** A female canine. **2.** *Slang.* A complaint. —*v.* **bitched, bitch·ing, bitch·es.** —*vi. Slang.* To complain : GRUMBLE.

**bit decay** *n. Slang.* A mythical disease of computer software in which programs are believed to stop working properly or completely if they are not used for a long period of time.

**bite** (bīt) *v.* **bit** (bĭt), **bit·ten** (bĭt′n) *or* **bit, bit·ing, bites.** [ME *biten* < OE *bītan.*] —*vt.* **1.** To cut, tear, or grip with or as if with the teeth. **2.** To pierce the skin of, esp. with the teeth or fangs. **3.** To cut into with a sharp instrument. **4.** To grip, grab, or seize. **5.** To eat into : CORRODE. **6.** To cause to sting or smart. —*vi.* **1.** To cut, tear, or grip something with or as if with the teeth. **2.** To have a stinging effect. **3.** To have a sharp taste. **4.** To swallow or take bait <The fish are *biting* today.> **5.** To be taken in by a ploy or deception. —*n.* **1.** An act of biting. **2.** A wound resulting from biting. **3. a.** A stinging or smarting sensation. **b.** Incisive, penetrating quality. **4.** An amount of food taken into the mouth at one time. **5.** *Informal.* A light meal : SNACK. **6.** A secure grip or hold applied by a tool or machine on a working surface. **7.** The angle at which the upper and lower teeth meet. **8.** The corrosive action of acid on an etcher's metal plate. —**bite the bullet.** To face an unpleasant situation bravely and stoically. —**bite the dust. 1.** To fall dead. **2.** To be defeated. **3.** To become useless. —**put the bite on.** *Slang.* To borrow money from. —**bit′a·ble, bite′a·ble** *adj.* —**bit′er** *n.*

**bite·wing** (bīt′wĭng′) *n.* A dental x-ray plate with a central flap for the patient to bite, holding the plate in position.

**bit·ing** (bī′tĭng) *adj.* **1.** Causing a stinging sensation. **2.** Penetrating : incisive. —**bit′ing·ly** *adv.*

**bit·stock** (bĭt′stŏk′) *n.* A handle or brace in which a boring or drilling bit is secured.

**bitt** (bĭt) *n.* [Orig. unknown.] A vertical post set on the deck of a ship for securing cables. —*vt.* **bitt·ed, bitt·ing, bitts.** To wind (a cable) around a bitt.

**bit·ten** (bĭt′n) *v. var. p.p. of* BITE.

**bit·ter** (bĭt′ər) *adj.* **-er, -est.** [ME < OE.] **1.** Having or being a taste that is sharp, acrid, and unpleasant. **2.** Causing sharp physical or mental pain or discomfort : HARSH <a *bitter* cold spell><*bitter* regrets> **3.** Difficult or distasteful to accept, admit, or bear <the *bitter* reality> **4.** Caused by or exhibiting strong animosity <*bitter* enemies> **5.** Resulting from or expressive of severe grief, anguish, or disappointment <cried *bitter* tears> **6.** Marked by anguished resentfulness or rancor <"He was already a *bitter* elderly man with a gray face" —John Dos Passos> —*vt.* & *vi.* **-tered, -ter·ing, -ters.** To make or become bitter. —*n.* **bitters.** A bitter, usu. alcoholic liquid made with roots or herbs and used in cocktails or as a tonic. —**bit′ter·ly** *adv.* —**bit′ter·ness** *n.*

**bitter almond** *n.* A variety of the common almond, *Prunus amygdalus amara*, bearing bitter kernels that yield a highly poisonous oil.

**bitter aloes** *n.* (*sing in number*). A cathartic drug derived from the juice of the fleshy leaves of a tropical plant, *Aloe barbadensis.*

**bitter apple** *n.* **1.** The colocynth. **2.** The fruit of the colocynth.

**bitter end** *n.* **1.** *Naut.* The end of a rope or cable that is wound around a bitt. **2.** A final, painful, or disastrous extremity.

**bit·tern**[1] (bĭt′ərn) *n.* [ME *biture* < OFr. *butor.*] A wading bird of the genera *Botaurus* or *Ixobrychus*, with mottled, brownish plumage and a deep, resonant cry.

**bit·tern**[2] (bĭt′ərn) *n.* [< BITTER.] The solution of bromides, magnesium, and calcium salts left after sodium chloride is crystallized out of sea water.

**bit·ter·nut** (bĭt′ər-nŭt′) *n.* A hickory tree, *Carya cordiformis* of eastern North America, bearing nuts with bitter kernels.

**bitter principle** *n.* Any of a large number of bitter substances, frequently of vegetable origin.

**bit·ter·root** (bĭt′ər-rōōt′, -rŏŏt′) *n.* A plant, *Lewisia rediviva* of western North America, with pink or white flowers and a starchy, edible root.

**bitter rot** *n.* A destructive fungal disease of fruit caused by *Glomerella cingulata.*

**bit·ter·sweet** (bĭt′ər-swēt′) *n.* **1.** A North American woody vine, *Celastrus scandens*, bearing orange or yellowish fruits that split open to expose seeds enclosed in fleshy scarlet arils. **2.** A sprawling vine, *Solanum dulcamara* of Eurasia, bearing purple flowers and poisonous scarlet berries. **3.** A dark to deep reddish orange. —*adj.* **1.** Both bitter and sweet. **2.** Engendering both pain and pleasure. **3.** Dark to deep reddish orange.

**bit·ter·weed** (bĭt′ər-wēd′) *n.* A plant that contains a bitter principle, as the ragweed or a plant of the genus *Picris.*

**bi·tu·men** (bĭ-tōō′mən, -tyōō′-, bī-) *n.* [ME *bithumen*, a mineral pitch from the Near East < Lat. *bitumen.*] Any of various mixtures of hydrocarbons and other substances, occurring naturally or distilled from coal or petroleum, found in asphalt and tar, and used for surfacing roads and waterproofing. —**bi·tu′mi·noid′** (-mə-noid′) *adj.*

**bi·tu·mi·nize** (bĭ-tōō′mə-nīz′, -tyōō′-, bī-) *vt.* **-nized, -niz·ing, -niz·es.** To treat with bitumen. —**bi·tu′mi·ni·za′tion** *n.*

**bi·tu·mi·nous** (bĭ-tōō′mə-nəs, -tyōō′-, bī-) *adj.* **1.** Like or containing bitumen. **2.** Of or relating to bituminous coal.

**bituminous coal** *n.* A mineral coal that burns with a smoky, yellow flame, yielding volatile bituminous constituents.

**bi·va·lent** (bī-vā′lənt) *adj.* **1.** *Chem.* Having valence 2. **2.** Composed of two homologous chromosomes or two sets of such chromosomes. —**bi·va′lence, bi·va′len·cy** *n.*

**bi·valve** (bī′vălv′) *n.* A mollusk, as an oyster or clam, with a two-hinged shell. —*adj.* **1.** Having a shell composed of two hinged parts. **2.** Composed of two similar separable parts. —**bi′valved′** *adj.*

**bi·var·i·ate** (bī-vâr′ē-ĭt, -āt′) *adj.* Having two variables.

**biv·ou·ac** (bĭv′ōō-ăk′, bĭv′wăk′) *n.* [Fr., prob. < dial. G. *beiwacht*, supplementary night watch.] A temporary, often open-air encampment. —*vi.* **-acked, -ack·ing, -acks** or **-acs.** To camp in a bivouac.

**bi·week·ly** (bī-wēk′lē) *adj.* **1.** Happening every two weeks. **2.** Happening twice a week : SEMIWEEKLY. —*n., pl.* **-lies.** A publication issued every two weeks. —*adv.* **1.** Every two weeks. **2.** Twice a week : SEMIWEEKLY.

**bi·year·ly** (bī-yîr′lē) *adj.* **1.** Happening every two years. **2.** Happening twice a year : SEMIYEARLY. —*adv.* **1.** Every two years. **2.** Twice a year : SEMIYEARLY.

**bi·zarre** (bĭ-zär′) *adj.* [Fr. < Sp. *bizarro*, brave, prob. < Basque *bizar*, beard.] Strikingly unconventional or far-fetched <*bizarre* behavior and attire> —**bi·zarre′ly** *adv.* —**bi·zarre′ness** *n.*

**bi·zon·al** (bī-zō′nəl) *adj.* Of or relating to the affairs of a zone under the joint administration or government of two powers.

**Bk** *symbol for* BERKELIUM.

**blab** (blăb) *v.* **blabbed, blab·bing, blabs.** [ME *blabben*, to talk foolishly.] —*vt.* To reveal (secret matters) esp. through indiscreet or outspoken talk. —*vi.* **1.** To reveal secret matters. **2.** To chatter indiscreetly. —*n.* **1.** One who blabs. **2.** Lengthy chatter. —**blab′by** *adj.*

**blab·ber** (blăb′ər) *vi.* **-bered, -ber·ing, -bers.** [ME *blaberen.*] To chatter. —*n.* **1.** Idle chatter. **2.** BLAB 1.

**blab·ber·mouth** (blăb′ər-mouth′) *n. Slang.* A person who chatters indiscreetly and at length.

**black** (blăk) *adj.* **-er, -est.** [ME *blak* < OE *blæc.*] **1.** Being of the darkest achromatic visual value : producing or reflecting comparatively little light and having no predominant hue. **2.** Having little or no light <a *black* and stormy night> **3.** *often* **Black.** Belonging to an ethnic group having dark skin, esp. Negroid. **4.** Dark in color. **5.** Soiled, as from soot : DIRTY. **6.** Wicked : evil <*black* deeds> **7.** Cheerless and depressing <a *black* mood> **8.** Marked by anger or sullenness <gave me a *black* look> **9.** *often* **Black.** Attended with disaster : CALAMITOUS <a *black* day in American history> **10.** Deserving of, indicating, or incurring censure or dishonor <"man . . . has written one of his *blackest* records as a destroyer on the oceanic islands" —Rachel Carson> **11.** Wearing black clothing. **12.** Served without milk or cream <*black* coffee> —*n.* **1.** An achromatic color value of minimum lightness or maximum darkness : one extreme of the neutral gray series, the opposite being white. **2.** Complete or almost complete absence of light : DARKNESS. **3.** Black clothing, esp. such clothing worn for mourning. **4.** *often* **Black.** A member of a Negroid people : NEGRO. ***usage:*** The term *black* is preferred instead of *Negro*, and it is often but not always uncapitalized <"Together, *blacks* and whites can move our country beyond racism" —Whitney Young, Jr.> **5.** A black-colored chess or checker piece. —*vt.* & *vi.* **blacked, black·ing, blacks.** To make or become black. —**black out. 1.** To extinguish or conceal during an air raid all lights that might help enemy aircraft find a target. **2.** To lose memory or consciousness temporarily. **3.** To prohibit the dissemination of, esp. by censorship <*blacked out* all information about the air strikes> —**in the black.** On the credit side of a ledger. —**black′ish** *adj.* —**black′ly** *adv.* —**black′ness** *n.*

☆ ***syns:*** BLACK, EBON, EBONY, INKY, JET, JETTY, ONYX, PITCH-BLACK, PITCHY, SABLE, SOOTY *adj. core meaning* : of the darkest achromatic visual hue <*black* eyes><a *black* hat> ***ant:*** white

**black alder** *n.* **1.** A deciduous holly, *Ilex verticillata* of eastern North America, bearing bright-scarlet berries. **2.** A tree, *Alnus glutinosa* of Eurasia, with dark bark.

**black·a·moor** (blăk′ə-mŏŏr′) *n.* [BLACK + MOOR.] A dark-skinned person, esp. an African Negro.

**black-and-blue** (blăk′ən-blōō′) *adj.* Discolored from coagulation of blood below the surface of the skin.

**Black and Tan** *n.* [From the color of their uniforms.] A member of the Royal Irish Constabulary, a force of British soldiers sent to Ireland to suppress the Sinn Fein disturbances of 1919–21.

**black-and-tan terrier** (blăk′ən-tăn′) *n.* A Manchester terrier.

**black and white** *n.* **1.** Writing or print. **2.** A picture or photograph in tones of black and white.

**black-and-white** (blăk′ənd-hwīt′, -wīt′) *adj.* **1.** Being in writing or print. **2.** Being partially black and partially white. **3.** Drawn or painted in dark pigment on a light background or vice versa. **4.** Reproducing visual images in tones of gray <*black-and-white* television> **5.** Evaluating things as either entirely good or entirely bad.

**black art** *n.* Black magic.

**black bag job** *n. Slang.* Illegal entry and search carried out by a law enforcement agency.

**black·ball** (blăk′bôl′) *n.* **1.** A negative vote that prevents admission of an applicant to an organization. **2.** A small black ball used as a negative ballot. —*vt.* **-balled, -ball·ing, -balls. 1.** To vote against, esp. to block the admission of. **2.** To exclude from a social group. —**black′ball′er** *n.*

**black bass** *n.* A North American freshwater game fish of the genus *Micropterus.*
**black bear** *n.* Either of two black or dark-brown bears, *Euarctos* or *Ursus americanus* of North America or *Selenarctos thibetanus* of Asia.
**black belt** *n.* **1. a.** The rank of expert in a system of self-defense such as karate or judo. **b.** The black-colored belt symbolizing the rank of expert. **c.** A person holding a black belt. **2.** A region of rich, black soil. **3.** A part of a region, state, or city having a chiefly black population.
**black·ber·ry** (blăk′bĕr′ē) *n.* **1.** A woody plant of the genus *Rubus,* with canelike, usu. thorny stems and glossy black, edible berries. **2.** The fruit of a blackberry.
**black bile** *n.* One of the four humors of medieval physiology, supposedly causing melancholia.
**black·bird** (blăk′bûrd′) *n.* **1.** Any of various New World birds of the family Icteridae, as the redwing or grackle, of which the male has black or mostly black plumage. **2.** An Old World songbird, *Turdus merula,* of which the male is black with a yellow bill.
**black·board** (blăk′bôrd′, -bōrd′) *n.* A smooth, hard, often dark-colored panel for writing on with chalk.
**black·bod·y** (blăk′bŏd′ē) *n.* A theoretically perfect absorber of all incident radiation.
**blackbody radiation** *n.* Thermal radiation emitted by a blackbody at a specific temperature.
**black book** *n.* A record of people to punish or blacklist.
**black box** *n.* **1.** A device or theoretical construct, esp. an electric circuit, with known or specified performance characteristics but unknown or unspecified constituents and means of operation. **2.** A removable unit of electronic equipment on an aircraft, as an in-flight voice recorder.
**black bryony** *n.* A climbing European plant, *Tamus communis,* bearing small greenish flowers and poisonous red berries.
**black·buck** (blăk′bŭk′) *n.* An antelope, *Antilope cervicapra* of India, of which the male has spiral horns and a dark back.
**black·cap** (blăk′kăp′) *n.* **1.** The black raspberry. **2.** A small European bird, *Sylvia atricapilla,* of which the male is gray with a black crown.
**black·cock** (blăk′kŏk′) *n.* A male black grouse.
**black cohosh** *n.* A tall plant, *Cimicifuga racemosa* of eastern North America, bearing small whitish flowers in long clusters.
**black comedy** *n.* Comedy that uses black humor.
**black crappie** *n.* An edible North American fish, *Pomoxis nigromaculatus,* with dark, mottled coloring.
**black·damp** (blăk′dămp′) *n.* A gas composed of a mixture of carbon dioxide and nitrogen, encountered in mines after fires and explosions.
**Black Death** *n.* [From the dark splotches it causes on its victims.] A form of plague caused by the bacillus *Yersina pestis,* pandemic throughout Europe and most of Asia for several years after 1353.
**black diamond** *n.* **1.** CARBONADO[2]. **2. black diamonds.** COAL 1a.
**black·en** (blăk′ən) *v.* **-ened, -en·ing, -ens.** —*vt.* **1.** To make dark or black. **2.** To defame or sully. —*vi.* To become black or dark. **—black′en·er** *n.*
**Black English** *n.* A distinctive nonstandard dialect of English spoken by many American blacks.
**black eye** *n.* **1.** A bruised discoloration of the flesh around the eye. **2.** A bad name.
**black-eyed pea** (blăk′īd′) *n.* COWPEA 2.
**black-eyed Susan** *n.* **1.** Any of several North American plants of the genus *Rudbeckia,* esp. *R. hirta,* with hairy stems and leaves and flowers with orange-yellow rays and dark-brown centers. **2.** A vine native to tropical Africa, *Thunbergia alata,* bearing purple-throated white or orange-yellow flowers.
**black·face** (blăk′fās′) *n.* **1.** Make-up for a conventionalized comic travesty of blacks, as in a minstrel show. **2.** An actor in a minstrel show. **3.** Boldface printing type.
**black·fish** (blăk′fĭsh′) *n., pl.* **blackfish** or **-fish·es. 1.** Any of various dark-colored fishes, as: **a.** A freshwater fish, *Dallia pectoralis* of far northern regions. **b.** The tautog. **2.** The pilot whale.
**black flag** *n.* A Jolly Roger.
**black fly** *n.* Any of various small, dark-colored biting flies of the family Simuliidae.
**Black·foot** (blăk′fo͝ot′) *n., pl.* **Blackfoot** or **-feet** (-fēt′). **1. a.** Any of three Indian tribes once inhabiting the regions of Montana, Alberta, and Saskatchewan. **b.** A member of any of these tribes. **2.** The Algonquian language of the Blackfoot. **—Black′foot′** *adj.*
**black gold** *n.* Petroleum.
**black grouse** *n.* A Eurasian game bird, *Lyrurus tetrix,* of which the male is black with white markings.
**black·guard** (blăg′ərd, -ärd′) *n.* **1.** An unprincipled person : SCOUNDREL. **2.** A foul-mouthed person. —*vt.* **-guard·ed, -guard·ing, -guards.** To abuse verbally. **—black′guard·ism** *n.* **—black′guard·ly** *adj.* & *adv.*
**black gum** *n.* The sour gum.
**Black Hand** *n.* A secret society that engaged in acts of terrorism and blackmail in the United States in the early 20th cent.
**black haw** *n.* A shrub, *Viburnum prunifolium,* with white flower clusters and purple-black fruit.
**black·head** (blăk′hĕd′) *n.* **1.** A plug of dried fatty matter capped with blackened dust and epithelial debris that clogs a pore of the skin. **2.** An infectious, often fatal liver and intestinal disease of turkeys and some wildfowl. **3.** A bird with dark head markings.
**black·heart** (blăk′härt′) *n.* A plant disease in which the inner tissues darken.
**black-heart·ed** (blăk′här′tĭd) *adj.* Wicked : evil.
**black hole** *n.* A small celestial body with an intense gravitational field believed to be a collapsed star.
**black horehound** *n.* A strong-smelling plant, *Ballota nigra* of Europe, bearing purple flower clusters.
**black humor** *n.* Morbid and absurd humor, esp. in its development as a literary genre. **—black humorist** *n.*
**black·ing** (blăk′ĭng) *n.* **1.** Lampblack. **2.** A black paste or liquid for polishing shoes.
**black·jack**[1] (blăk′jăk′) *n.* **1.** A leather-covered bludgeon with a short flexible strap or shaft, used as a weapon. **2.** An oak tree, *Quercus marilandica* of the southeastern United States, with blackish bark. **3.** A card game in which the object is to accumulate cards with a total count nearer to 21 than that of the dealer. **4.** Sphalerite. —*vt.* **-jacked, -jack·ing, -jacks. 1.** To hit with a blackjack. **2.** To coerce by threats.
**black·jack**[2] (blăk′jăk′) *n.* [BLACK + ME *jakke,* leather container < OFr. *jacque.*] A tankard made of tarred leather.
**black knot** *n.* A disease of plum and cherry trees caused by the fungus *Dibotryon morbosa* and resulting in black, knotlike swellings on the branches.
**black lead** (lĕd) *n.* Graphite.
**black·leg** (blăk′lĕg′) *n.* **1.** An infectious, usu. fatal gas gangrene affecting the heavily muscled upper parts of the legs of cattle and sheep. **2.** A bacterial or fungous plant disease that causes the stems of plants to turn black. **3.** One who cheats at cards, esp. a professional gambler : CARDSHARP. **4.** *Chiefly Brit.* A strikebreaker : scab.
**black letter** *n.* A heavy typeface with broad counters and thick, ornamental serifs.
**black light** *n.* Invisible infrared or ultraviolet radiation.
**black·list** (blăk′lĭst′) *n.* A list of organizations or persons that are disapproved, boycotted, or suspected of disloyalty. —*vt.* **-list·ed, -list·ing, -lists.** To place (a name) on a blacklist. **—black′list′er** *n.*
**black lung** *n.* Pneumoconiosis caused by long-term inhalation of coal dust.
**black magic** *n.* Magic in league with the Devil : WITCHCRAFT.
**black·mail** (blăk′māl′) *n.* [BLACK + MAIL[3].] **1. a.** Extortion of money or something valuable from a person by threatening to expose a past criminal act or discreditable information. **b.** Something valuable extorted in this way. **2.** Tribute once paid to freebooters along the Scottish border for protection from plunder. **—black′mail′** *v.* **(-mailed, -mail·ing, -mails). —black′mail′er** *n.*
▲ word history: *Blackmail* has nothing to do with the post office. *Black* is used in the figurative sense of "evil" or "wicked." *Mail* is a Scottish word meaning "rent" or "tribute." The term *blackmail* originated in Scotland, where Highland chiefs at one time extorted tribute from Lowlanders and Englishmen on the Scottish border in return for protection from being plundered.
**Black Ma·ri·a** (mə-rī′ə) *n.* [BLACK + the name *Maria.*] A patrol wagon.
**black market** *n.* **1.** The illegal business of selling or buying goods in violation of such restrictions as rationing or price controls. **2.** A place where black market operations are carried on.
**black-mar·ket** (blăk′mär′kĭt) *v.* **-ket·ed, -ket·ing, -kets.** —*vt.* To sell in the black market. —*vi.* To buy or sell in the black market. **—black marketer, black marketeer** *n.*
**black mass** *n.* A travesty of the Roman Catholic Mass as part of the alleged observances of Satanism.
**black measles** *n.* A severe form of measles, typified by a dark rash.
**black medic** or **black medick** *n.* A cloverlike plant native to Europe, *Medicago lupulina,* with compound leaves, small yellow flower heads, and black pods.
**black money** *n.* Income, as from illegal activities, not reported to the government for tax purposes.
**Black Muslim** *n.* A member of a black religious group, the Nation of Islam, that advocates segregation of blacks and whites and the establishment of a black nation.
**black mustard** *n.* A plant native to Eurasia, *Brassica nigra,* with yellow flower clusters and pungent seeds that are a source of the condiment mustard.
**Black Nationalist** *n.* A member of a group of militant blacks who urge separatism from whites and the establishment of self-governing black communities. **—Black Nationalism** *n.*
**black oak** *n.* A deciduous tree, *Quercus velutina* of eastern North America, with durable, hard wood.

**black·out** (blăk′out′) *n.* **1.** The concealing or extinguishing of lights that might be seen by enemy aircraft crews during an air raid. **2.** Lack of illumination due to an electrical power failure. **3. a.** The abrupt extinguishing of all stage lights in a theater to mark the passage of time or the end of an act or a scene. **b.** A short comic vaudeville skit ending with lights off. **4.** Temporary loss of consciousness or memory. **5.** A suppression or stoppage <a wartime news *blackout*>

**Black Panther** *n.* A member of an organization of militant black Americans.

**black pepper** *n.* PEPPER 1b.

**Black Plague** *n.* Black Death.

**black·poll** (blăk′pōl′) *n.* A North American warbler, *Dendroica striata,* of which the male has a black cap.

**black poplar** *n.* A shade tree native to Eurasia *Populus nigra,* with spreading branches and pointed triangular leaves.

**Black Power** *n.* A movement among black Americans emphasizing social equality and racial pride through the creation of black political and cultural institutions.

**black raspberry** *n.* **1.** A prickly shrub, *Rubus occidentalis* of eastern North America, bearing black fruit. **2.** The fruit of the black raspberry.

**Black Rod** *n.* The chief usher of the House of Lords and other British institutions.

**black rot** *n.* A plant disease caused by bacteria or fungi that results in darkening of the leaves and decay.

**black sheep** *n.* One considered disgraceful or undesirable by a respectable group or family.

**Black Shirt** *n.* A member of a fascist party organization having a black shirt as part of its uniform.

**black·smith** (blăk′smĭth′) *n.* [From the color of iron.] **1.** One who shapes and forges iron with a hammer and anvil. **2.** One who makes, repairs, and fits horseshoes. **—black′smith′ing** *n.*

**black·snake** (blăk′snāk′) *n.* **1.** A dark-colored, chiefly nonvenomous snake, as the black racer, *Coluber constrictor,* or the black rat snake, *Elaphe obsoleta,* both of North America. **2.** A long, tapering braided leather or rawhide whip with a snapper on the end.

**black spot** *n.* A plant disease caused by bacteria or fungi that results in small black spots on the leaves.

**black spruce** *n.* An evergreen tree, *Picea mariana* of northern North America, growing mostly in bogs.

**black·strap** (blăk′străp′) *n.* A dark, very thick molasses used in manufacturing industrial alcohol and as an ingredient in cattle feed.

**black studies** *pl.n.* Studies dealing with Afro-American culture.

**black-tailed deer** (blăk′tāld′) *also* **black·tail deer** (-tāl′) *n.* The mule deer.

**black tea** *n.* A dark tea, the leaf of which is fully fermented before drying.

**black·thorn** (blăk′thôrn′) *n.* A thorny Eurasian shrub, *Prunus spinosa,* with white flower clusters and bluish-black, plumlike fruit.

**black thorn** *n.* The pear haw.

**black tie** *n.* **1.** A black bow tie worn with a dinner jacket. **2.** Semiformal evening wear for men, typically requiring a dinner jacket. **—black′-tie′** *adj.*

**black tip** *n.* Any of several plant diseases typified by dark necrotic areas at the tip of the fruit or seed.

**black·top** (blăk′tŏp′) *n.* A bituminous material, as asphalt, used to pave roads. *—vt.* **-topped, -top·ping, -tops.** To pave with blacktop.

**black vomit** *n.* **1.** Vomit composed of bloody matter. **2.** Severe yellow fever with symptomatic regurgitation of black vomit.

**black vulture** *n.* A carrion-eating bird, *Coragyps atratus* of central North America and South America, with black plumage and a bald, black head.

**black walnut** *n.* **1.** A deciduous tree, *Juglans nigra* of eastern North America, with dark, hard wood and edible nuts. **2.** The wood of the black walnut. **3.** The nut of the black walnut.

**black·wash** (blăk′wŏsh′, -wôsh′) *vt.* **-washed, -wash·ing, -wash·es.** [BLACK + (WHITE)WASH.] To bring from concealment : DISCLOSE.

**black·wa·ter fever** (blăk′wô′tər, -wŏt′ər) *n.* A severe, often fatal malaria with symptomatic excretion of blood in the urine.

**black widow** *n.* [From the fact that the female eats its mate.] A New World spider, *Latrodectus mactans,* of which the extremely venomous female is black with red markings on the underside.

**blad·der** (blăd′ər) *n.* [ME *bladdre* < OE *blǣdre.*] **1.** *Anat.* Any of various distensible membranous sacs found in most animals, esp. the urinary bladder. **2.** Something resembling a bladder. **3.** *Bot.* An inflated, hollow structure, as the air sac in certain seaweeds. **4.** *Pathol.* A blister, cyst, or postule filled with air or fluid.

**blad·der·nose** (blăd′ər-nōz′) *n.* The hooded seal.

**blad·der·nut** (blăd′ər-nŭt′) *n.* Any of several shrubs or small trees of the genus *Staphylea,* of the North Temperate Zone, bearing small whitish flowers and inflated seed pods.

**bladder worm** *n.* The bladderlike, encysted larva of the tapeworm.

**blad·der·wort** (blăd′ər-wûrt′, -wôrt′) *n.* Any of various aquatic plants of the genus *Utricularia,* with yellow or violet flowers and in most species small bladders that trap minute aquatic animals.

**bladder wrack** *n.* A rockweed, *Fucus vesiculosus,* with forked brownish-green fronds having air-filled bladders.

**blade** (blād) *n.* [ME < OE *blæd.*] **1.** The flat-edged cutting part of a sharpened tool or weapon. **2. a.** A sword. **b.** A swordsman. **3.** A dashing young man. **4.** A flat, thin part or section <the *blade* of a propeller> **5.** The metal part of an ice skate. **6.** *Anat.* The scapula. **7.** *Bot.* **a.** The leaf of a grass or similar plant. **b.** The expanded, usu. green part of a leaf. **8.** The upper surface of the tongue just behind the tip. **—blad′ed** *adj.*

**blade-ap·ple** (blād′ăp′əl) *n.* A spiny, vinelike tropical American cactus, *Pereskia aculeata,* bearing true leaves, white flowers, and pulpy yellow fruit.

**blah** (blä) [Imit. of meaningless talk.] *Slang.* *—n.* **1.** Nonsense. **2. blahs.** A general feeling of psychological or physical dissatisfaction or discomfort. *—adj.* Dull and uninteresting.

**blain** (blān) *n.* [ME < OE *blegen.*] A skin swelling or sore.

**blam·a·ble** *also* **blame·a·ble** (blā′mə-bəl) *adj.* Deserving blame : CULPABLE. **—blam′a·ble·ness** *n.* **—blam′a·bly** *adv.*

**blame** (blām) *vt.* **blamed, blam·ing, blames.** [ME *blamen* < OFr. *blasmer* < alteration of LLat. *blasphemare,* to reproach. —see BLASPHEME.] **1.** To hold (someone or something) at fault. **2.** To find fault with. **3.** To place responsibility for (something) on a person. *—n.* **1.** Responsibility for a fault or error. **2.** Condemnation : censure. **—blam′er** *n.*

☆ ***syns:*** BLAME, CENSURE, CONDEMN, CRITICIZE, DENOUNCE, DENUNCIATE, FAULT, PAN, RAP *v. core meaning* : to find fault with <*blamed* the Administration for the recession>

**blame·a·ble** (blā′mə-bəl) *adj. var. of* BLAMABLE.

**blamed** (blāmd) *adj & adv.* Damned.

**blame·ful** (blām′fəl) *adj.* Blameworthy. **—blame′ful·ly** *adv.* **—blame′ful·ness** *n.*

**blame·less** (blām′lĭs) *adj.* Free from blame : INNOCENT. **—blame′less·ly** *adv.* **—blame′less·ness** *n.*

**blame·wor·thy** (blām′wûr′thē) *adj.* **-thi·er, -thi·est.** Deserving blame : REPREHENSIBLE. **—blame′wor′thi·ness** *n.*

**blanc fixe** (blängk′ fēks′) *n.* [Fr. : *blanc,* white + *fixe,* fixed.] Powdered barium sulfate used as a base for water-color pigments.

**blanch** (blănch) *also* **blench** (blĕnch) *v.* **blanched, blanch·ing, blanch·es** *also* **blenched, blench·ing, blench·es.** [ME *blaunchen,* to make white < OFr. *blanchir* < *blanche,* fem. of *blanc,* white.] *—vt.* **1.** To take the color from : BLEACH. **2.** To whiten (a growing food plant, as celery) by covering to cut off direct light. **3.** To whiten (a metal) by soaking in acid or by coating with tin. **4. a.** To scald so as to loosen the skin of (e.g., almonds). **b.** To scald (food) briefly, as before freezing. **5.** To cause to turn white or become pale. *—vi.* To turn white or pale. **—blanch′er** *n.*

**blanc·mange** (blə-mänj′, -mänzh′) *n.* [ME *blankmanger,* white meat with rice < OFr. *blanc manger,* white food.] A flavored, sweetened milk pudding thickened with cornstarch.

**bland** (blănd) *adj.* **-er, -est.** [Lat. *blandus,* caressing.] **1.** Marked by a moderate, undisturbing, or tranquil quality, esp.: **a.** Pleasant in manner. **b.** Not irritating or stimulating <a *bland* concoction> **c.** Balmy : mild. **2.** Lacking distinctive character : INSIPID. **—bland′ly** *adv.* **—bland′ness** *n.*

**blan·dish** (blăn′dĭsh) *vt.* **-dished, -dish·ing, -dish·es.** [ME *blandishen* < OFr. *blandir* < Lat. *blandiri* < *blandus,* flattering.] To coax by wheedling or flattery : CAJOLE. **—blan′dish·er** *n.* **—blan′dish·ment** *n.*

**blank** (blăngk) *adj.* **-er, -est.** [ME < OFr. *blanc,* white.] **1.** Bearing no writing, print, or marking. **2.** Not completed or filled in <a *blank* form of application> **3.** Lacking certain features <a *blank* wall> **4.** Expressing nothing : VACANT <"Although his gestures were elaborate, his face was *blank*" —Nathanael West> **5.** Appearing dazed or confused : BEWILDERED. **6.** Devoid of character or activity : EMPTY. **7.** Barren : fruitless <*blank* efforts> **8.** Absolute : utter <a *blank* denial> *—n.* **1.** An empty place : VOID. **2. a.** An empty space, as one to be filled in on a document. **b.** A document with one or more such spaces. **3.** An unfinished material, part, or article, as a key form, that is prepared and stored for eventual finishing. **4.** A gun cartridge with a charge of powder but no bullet. **5.** A losing lottery ticket. **6.** A mark, usu. a dash (—), indicating the omission of a word or letter. **7.** The center white circle of a target : BULL'S-EYE. *—vt.* **blanked, blank·ing, blanks.** **1.** To remove : obliterate <a view *blanked* by the strong glare of the sun> **2.** To delete : invalidate. **3.** To prevent (an opponent) from scoring in a sports contest. **4.** To stamp or punch from flat stock, esp. with a die. **—draw a blank.** To fail to come up with an idea, answer, or solution. **—blank′ly** *adv.* **—blank′ness** *n.*

**blank cartridge** *n.* BLANK 4.

**blank check** *n.* **1.** A signed check with no amount filled in. **2.** Utter freedom of action : CARTE BLANCHE.

**blank endorsement** *n.* An endorsement on a check or negotiable note that names no payee, thereby making it payable to the bearer.

**blan·ket** (blăng′kĭt) *n.* [ME < OFr. < *blanc,* white.] **1.** A piece of woven material used as a covering for warmth, esp. on a bed. **2.** A thick covering or enclosing layer <a *blanket* of wet leaves> —*adj.* Applying to or covering all conditions or requirements <a *blanket* tax bill> —*vt.* **-ket·ed, -ket·ing, -kets. 1.** To cover with or as if with a blanket. **2.** To cover so as to suppress or inhibit. **3.** To apply to uniformly and with no exceptions.

**blan·ket·flow·er** (blăng′kĭt-flou′ər) *n.* The gaillardia.

**blanket stitch** *n.* The buttonhole stitch used for edging around a blanket.

**blan·ket-stitch** (blăng′kĭt-stĭch′) *vt.* **-stitched, -stitch·ing, -stitch·es.** To sew with a blanket stitch.

**blank verse** *n.* Verse consisting of unrhymed lines, usu. of iambic pentameter.

**blare** (blâr) *v.* **blared, blar·ing, blares.** [ME *bleren.*] —*vi.* To sound loudly and stridently. —*vt.* **1.** To cause to sound loudly and stridently. **2.** To utter loudly. —*n.* **1.** A loud, strident noise. **2.** Flamboyance.

**blar·ney** (blär′nē) *n.* [After the *Blarney* Stone, in Blarney Castle, Blarney, Ireland, said to give skill in flattery to those who kiss it.] **1.** Smooth, flattering talk. **2.** Nonsensical or deceptive talk. **—blar′ney** *v.* **(-neyed, -ney·ing, -neys).**

**bla·sé** (blä-zā′) *adj.* [Fr., p.part. of *blaser,* to cloy.] **1.** Uninterested because of frequent exposure or indulgence. **2.** Extremely sophisticated and worldly.

**blas·pheme** (blăs-fēm′) *v.* **-phemed, -phem·ing, -phemes.** [ME *blasfemen* < OFr. *blasfemer* < LLat. *blasphemare* < Gk. *blasphēmien* < *blasphēmos,* blasphemous.] —*vt.* **1.** To speak of (God or something sacred) irreverently or impiously. **2.** To revile : execrate. —*vi.* To speak blasphemy. **—blas·phem′er** *n.*

**blas·phe·mous** (blăs′fə-məs) *adj.* [LLat. *blasphemus* < Gk. *blasphēmos.*] Profane or irreverent. **—blas′phe·mous·ly** *adv.* **—blas′phe·mous·ness** *n.*

**blas·phe·my** (blăs′fə-mē) *n., pl.* **-mies. 1. a.** A contemptuous or profane act, utterance, or writing concerning God. **b.** Claiming for oneself the attributes and rights of God. **2.** An irreverent act, attitude, or utterance regarding something inviolable or sacrosanct.

**blast** (blăst) *n.* [ME < OE *blǣst.*] **1. a.** A strong gust of wind. **b.** The effect of such a gust. **2.** A forcible stream of air, gas, or steam from an opening, esp. one in a blast furnace to aid combustion. **3. a.** The act of blowing a wind instrument or whistle. **b.** The sound or noise produced by this. **4. a.** An explosion <a *blast* of dynamite> **b.** The effect of an explosion. **c.** An explosive charge. **5.** A plant disease that results in failure of flowers to open or of fruit or seeds to mature. **6.** A damaging or destructive influence. **7.** A violent verbal assault or outburst. **8.** *Slang.* A satisfyingly exciting event or experience <had a *blast* at the party> —*v.* **blast·ed, blast·ing, blasts.** —*vt.* **1.** To fragment by or as if by explosion : SMASH. **2.** To ruin : frustrate <"Our hopes were all *blasted* at one blow" —Frederick Douglass> **3.** To cause to shrivel, wither, or mature imperfectly by or as if by blast or blight. **4.** To make or open by or as if by explosion <*blast* a tunnel through the mountain> **5.** *Informal.* To attack or criticize vigorously. —*vi.* **1.** To detonate explosives. **2.** To emit a loud, strident noise. **3.** To wither, shrivel, or mature imperfectly. **4.** *Informal.* To criticize or attack vigorously. **5.** *Slang.* To shoot. **6.** *Electron.* To distort sound recording or transmission by overloading a microphone or loudspeaker. **—blast off.** To take off, as a rocket or space vehicle. **—full blast.** Full speed, volume, or capacity <a machine running *full blast*> **—blast′er** *n.*

**blast-** *pref. var. of* BLASTO-.

**-blast** *suff.* [< Gk. *blastos,* bud.] Bud : germ : cell : cell layer <endo*blast*>

**blast·ed** (blăs′tĭd) *adj.* **1.** Blighted : withered. **2.** *Informal.* Extremely annoying : OBNOXIOUS.

**blas·te·ma** (blă-stē′mə) *n., pl.* **-mas** *or* **-ma·ta** (-mə-tə) [Gk. *blastēma,* offspring < *blastos,* bud.] A segregated region of embryonic cells from which a specific organ develops. **—blas·te′mal** (blă-stē′məl), **blas′te·mat′ic** (blăs′tə-măt′ĭk), **blas·te′mic** (blă-stē′mĭk) *adj.*

**blast furnace** *n.* A furnace in which a blast of air intensifies combustion.

**-blastic** *suff.* [< -BLAST.] Having a specified number or kind of buds, germs, cells, or cell layers <mero*blastic*>

**blasting gelatin** *n.* A dynamite containing nitrocellulose as well as nitroglycerin.

**blasto-** *or* **blast-** *pref.* [< Gk. *blastos,* bud.] Bud : germ <*blasto*cyst><*blasto*coel>

**blas·to·coel** *or* **blas·to·coele** (blăs′tə-sēl′) *n.* [BLASTO- + -COEL.] The cavity of a blastula. **—blas′to·coe′lic** *adj.*

**blas·to·cyst** (blăs′tə-sĭst′) *n.* The germinal vesicle. **—blas′to·cys′tic** (-sĭs′tĭk) *adj.*

**blas·to·derm** (blăs′tə-dûrm′) *n.* The layer of cells surrounding the blastocoel and generating the germinal disc from which the embryo develops in most placental vertebrates. **—blas′to·der·mat′ic** (-dər-măt′ĭk), **blas′to·der′mic** (-dûr′mĭk) *adj.*

**blas·to·disc** (blăs′tə-dĭsk′) *n.* The germinal disc.

**blast·off** *also* **blast-off** (blăst′ôf′) *n.* A rocket or space vehicle launch.

**blas·to·gen·e·sis** (blăs′tə-jĕn′ĭ-sĭs) *n.* **1.** The theory that inherited characteristics are transmitted from parent to offspring by germ plasm. **2.** Reproduction by budding. **—blas′to·ge·net′ic** (-jə-nĕt′ĭk), **blas′to·gen′ic** (-jĕn′ĭk) *adj.*

**blas·to·ma** (blă-stō′mə) *n., pl.* **-mas** *or* **-ma·ta** (-mə-tə). A neoplasm comprised of immature and undifferentiated cells.

**blas·to·mere** (blăs′tə-mîr′) *n.* A cell formed during cleavage of a fertilized ovum. **—blas′to·mer′ic** (-mîr′ĭk, -mĕr′-) *adj.*

**blas·to·pore** (blăs′tə-pôr′, -pōr′) *n.* [BLASTO- + PORE².] The mouthlike opening into the primitive intestinal cavity of the gastrula. **—blas′to·por′al** (-pôr′əl, -pōr′-) *adj.*

**blas·tu·la** (blăs′chə-lə) *n., pl.* **-las** *or* **-lae** (-lē′) [NLat. < Gk. *blastos,* bud.] An early embryonic form consisting of a hollow cellular sphere. **—blas′tu·lar** (-lər) *adj.* **—blas′tu·la′tion** (-lā′shən) *n.*

**blat** (blăt) *v.* **blat·ted, blat·ting, blats.** [Imit.] —*vt.* To utter unthinkingly : BLURT. —*vi.* **1.** To make the characteristic cry of a sheep : BLEAT. **2.** To make a harsh, raucous noise. **—blat** *n.*

**bla·tant** (blāt′nt) *adj.* [Prob. < Lat. *blatire,* to blab.] **1.** Unpleasantly and often vulgarly loud and noisy. **2.** Conspicuous, often to an offensive degree <a *blatant* lie> **—bla′tan·cy** *n.* **—bla′tant·ly** *adv.*

**blath·er** (blă*th*′ər) *also* **bleth·er** (blĕ*th*′-) [ME < ON *blaðra* < *blaðr,* nonsense.] —*vi.* **-ered, -er·ing, -ers.** To talk nonsensically. —*n.* Nonsensical talk. **—blath′er·er** *n.*

**blath·er·skite** (blă*th*′ər-skīt′) *n.* [BLATHER + dial. *skite,* contemptible person.] **1.** A foolish, babbling person. **2.** Blather.

**blaze¹** (blāz) *n.* [ME *blase* < OE *blæse.*] **1.** A brilliant burst of fire : FLAME. **2.** A bright, steady light or glare <the *blaze* of the summer sun> **3.** A destructive fire. **4.** A brilliant, striking display <a *blaze* of fall colors> **5.** A sudden outburst <a *blaze* of fury> **6. blazes.** *Slang.* Hell. —*v.* **blazed, blaz·ing, blaz·es.** —*vi.* **1.** To burn with a bright flame. **2.** To shine brightly. **3.** To be resplendent <a garden *blazing* with red roses> **4.** To flare up suddenly <hot tempers *blazing*> **5.** To shoot rapidly and continuously <guns *blazing*> —*vt.* **1.** To cause to blaze : BURN. **2.** To shine or be resplendent with <windows *blazing* with lights> **—blaz′ing·ly** *adv.*

**blaze²** (blāz) *n.* [Of Germanic orig.] **1.** A white or light-colored spot on the face of an animal. **2.** A mark cut or painted on a tree to indicate a trail. —*vt.* **blazed, blaz·ing, blaz·es. 1.** To mark (a tree) with or as if with blazes. **2.** To indicate (a trail) by marking trees with blazes.

**blaze³** (blāz) *vt.* **blazed, blaz·ing, blaz·es.** [ME *blasen* < MDu. *blāsen,* to blow.] To make known publicly : PROCLAIM.

**blaz·er** (blā′zər) *n.* **1.** One that blazes. **2.** A lightweight, informal sports jacket.

**blazing star** *n.* **1.** A North American plant, *Chamaelirium luteum,* bearing small white flowers in long clusters. **2.** A North American plant of the genus *Liatris,* with tuftlike purple or pinkish flower clusters. **3.** A plant, *Mentzelia laevicaulis* of western North America, with large pale yellow flowers.

**bla·zon** (blā′zən) *vt.* **-zoned, -zon·ing, -zons.** [ME *blasoun,* shield < OFr. *blason.*] **1.** *Heraldry.* **a.** To describe (a coat of arms) in proper terms. **b.** To paint or depict (a coat of arms) with accurate detail. **2.** To adorn or embellish with or as if with blazons. **3.** To announce : proclaim. —*n.* **1.** *Heraldry.* **a.** A coat of arms. **b.** The description or representation of a coat of arms. **2.** An ostentatious display. **—bla′zon·er** *n.* **—bla′zon·ment** *n.*

**bla·zon·ry** (blā′zən-rē) *n., pl.* **-ries. 1.** *Heraldry.* **a.** The art of accurately describing or representing armorial bearings. **b.** A coat of arms. **2.** A brilliant display.

**bleach** (blēch) *v.* **bleached, bleach·ing, bleach·es.** [ME *blechen* < OE *blǣcan.*] —*vt.* **1.** To remove the color from, as by chemical agents. **2.** To make white or colorless. —*vi.* To become white or colorless. —*n.* **1.** A chemical bleaching agent. **2. a.** The act of bleaching. **b.** A degree of bleaching obtained.

**bleach·er** (blē′chər) *n.* **1.** One that bleaches. **2.** *often* **bleachers.** An often unroofed outdoor grandstand.

**bleaching powder** *n.* A powder, as chlorinated lime or calcium hypochlorite, used as a bleach in solution.

**bleak¹** (blēk) *adj.* **-er, -est.** [ME *bleik* < ON *bleikr.*] **1.** Exposed to the elements : BARREN. **2.** Cold and cutting : RAW. **3. a.** Gloomy and somber : DREARY <"Life in the Aran Islands has always been *bleak* and difficult" —John M. Synge> **b.** Discouraging : depressing <a *bleak* outlook for the future> **—bleak′ly** *adv.* **—bleak′ness** *n.*

**bleak²** (blēk) *n.* [ME *bleke* < OE *blǣc,* bright.] A European freshwater fish of the genus *Alburnus,* whose silvery scales are used to make artificial pearls.

**blear** (blîr) *vt.* **bleared, blear·ing, blears.** [ME *bleren.*] **1.** To blur (the eyes) with or as if with tears. **2.** To blur : dim. **—blear** *adj.*

**blear-eyed** (blîr′īd′) *adj. var. of* BLEARY-EYED.

**blear·y** (blîr′ē) *adj.* **-i·er, -i·est. 1.** Blurred or dimmed by or as if by tears. **2.** Vaguely outlined : INDISTINCT. **3.** Worn-out : exhausted. **—blear′i·ly** *adv.* **—blear′i·ness** *n.*

**blear·y-eyed** (blîr′ē-īd′) *also* **blear-eyed** (blîr′īd′) *adj.* **1.** Blurred by or as if by tears. **2.** Mentally dull : UNPERCEPTIVE.

**bleat** (blēt) *n.* [ME *blet* < *bleten,* to bleat < OE *blǣtan.*] **1.** The cry of a goat or sheep. **2.** A sound similar to a bleat. —*v.* **bleat·ed,**

**bleat·ing, bleats.** —*vi.* **1.** To utter a bleat. **2.** To utter a sound like a bleat : WHINE. —*vt.* To utter in a whining voice. —**bleat'er** *n.*

**bleb** (blĕb) *n.* [Prob. var. of BLOB.] **1.** A small blister or pustule. **2.** An air bubble. —**bleb'by** *adj.*

**bleed** (blēd) *v.* **bled** (blĕd), **bleed·ing, bleeds.** [ME *bleden* < OE *blēdan.*] —*vi.* **1.** To emit or lose blood. **2.** To be wounded, esp. in battle. **3.** To feel sympathetic grief or anguish <My heart *bleeds* over their loss.> **4.** To exude sap or a similar fluid, as a bruised plant does. **5.** *Slang.* To pay out money, esp. an exorbitant amount. **6.** To become mixed or run <Madras *bleeds* when washed.> **7.** To show through a layer of paint, as resin or a stain in wood. **8.** To be printed so as to go off the edge or edges of a page after trimming. —*vt.* **1. a.** To take blood from <*bleed* a patient> **b.** To extract sap or juice from. **2. a.** To draw liquid or gaseous contents from : DRAIN. **b.** To draw off (liquid or gaseous matter) from a container. **3.** *Slang.* To obtain money from, esp. by improper means. **4. a.** To cause (e.g., an illustration) to bleed. **b.** To trim (e.g., a page) so closely as to mutilate the printed or illustrative matter. —*n.* **1.** Illustrative matter that bleeds. **2. a.** A page trimmed so as to bleed. **b.** The part of a page that is trimmed off.

**bleed·er** (blē'dər) *n.* **1.** A hemophiliac. **2.** *Slang.* A blood vessel severed by trauma or surgery that requires attention for the arrest of blood flow.

**bleed·ing-heart** (blē'dĭng-härt') *n.* **1.** A plant of the genus *Dicentra,* bearing nodding pink flowers, esp. the widely cultivated species *D. spectabilis.* **2.** A person regarded as excessively sympathetic toward the alleged underprivileged. —**bleed'ing-heart'** *adj.*

**bleep** (blēp) *n.* [Imit.] A brief high-pitched sound, as from an electronic device. —*vt.* **bleeped, bleep·ing, bleeps.** To blip.

**blem·ish** (blĕm'ĭsh) *vt.* **-ished, -ish·ing, -ish·es.** [ME *blemisshen* < OFr. *blemir,* to make pale, of Germanic orig.] To impair or spoil by a flaw : MAR. —*n.* A flaw or defect, esp. one adversely affecting appearance. —**blem'ish·er** *n.*

**blench**[1] (blĕnch) *vi.* **blenched, blench·ing, blench·es.** [ME *blenchen* < OE *blencan,* to deceive.] To recoil out of fear : QUAIL. —**blench'er** *n.*

**blench**[2] (blĕnch) *v. var. of* BLANCH.

**blend** (blĕnd) *v.* **blend·ed** *or* **blent** (blĕnt), **blend·ing, blends.** [ME *blenden* < OE *blandan.*] —*vt.* **1.** To combine so as to render the constituent parts indistinguishable. **2.** To mix (different varieties or grades) thoroughly so as to obtain a new mixture of a particular quality or consistency <*blend* tobaccos> —*vi.* **1.** To form a uniform mixture : INTERMINGLE <"The smoke *blended* easily into the odor of the other fumes" —Norman Mailer> **2.** To become merged into one : UNITE. **3.** To go well together : HARMONIZE. —*n.* **1.** Something blended : MIXTURE. **2.** The act of blending. **3.** A word produced by combining parts of other words, as *smog* from *smoke* and *fog.*

**blende** (blĕnd) *n.* [G. < *blenden,* to deceive < OHG *blenten,* to blind.] **1.** Any of various shiny minerals composed chiefly of metallic sulfides. **2.** Sphalerite.

**blended whiskey** *n.* Whiskey that is either a blend of two or more straight whiskeys or a blend of whiskey and neutral spirits.

**blend·er** (blĕn'dər) *n.* One that blends, esp. an electrical appliance with whirling blades that chops, mixes, or liquefies foods.

**blending inheritance** *n.* Inheritance of characters intermediate between those of parents widely divergent in those characters.

**blen·ny** (blĕn'ē) *n., pl.* **-nies.** [Lat. *blennius,* a kind of sea fish < Gk. *blennos.*] A small, elongated marine fish, chiefly of the families Blenniidae or Clinidae.

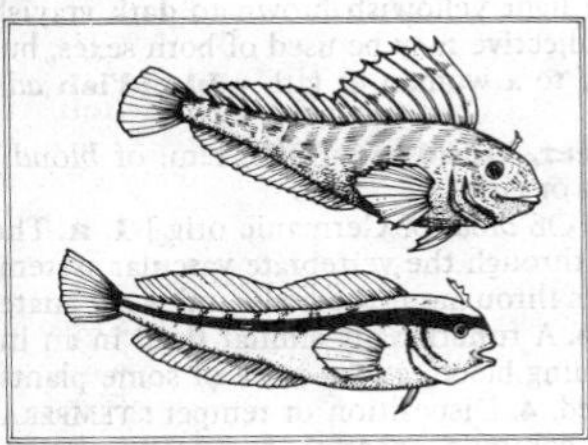

**blenny**
(Top) *butterfly blenny,* (bottom) *long-striped blenny. Both types up to 10 inches long*

**blent** (blĕnt) *v. var. p.t. & p.p. of* BLEND.

**bleph·a·ri·tis** (blĕf'ə-rī'tĭs) *n.* [Gk. *blepharon,* eyelid + -ITIS.] Inflammation of the eyelid.

**bleph·a·ro·spasm** (blĕf'ə-rō-spăz'əm) *n.* [NLat. *blepharospasmus* : Gk. *blepharon,* eyelid + Gk. *spasmos,* cramp.] Uncontrollable winking caused by involuntary contraction of an eyelid muscle.

**bles·bok** (blĕs'bŏk') *n., pl.* **blesbok** *or* **-boks.** [Afr. : *bles,* white mark on an animal's face (< MDu.) + *bok,* buck (< MDu. *boc*).] An African antelope, *Damaliscus albifrons,* with curved horns and a face marked with white.

**bless** (blĕs) *vt.* **blessed** *or* **blest** (blĕst), **bless·ing, bless·es.** [ME *blessen* < OE *blētsian.*] **1.** To make holy by religious rite : SANCTIFY. **2.** To make the sign of the cross over so as to sanctify. **3.** To invoke divine favor on. **4.** To honor as holy : GLORIFY. **5.** To confer well-being or prosperity on. **6.** To endow, as with talent. —**bless'er** *n.*

**bless·ed** (blĕs'ĭd) *also* **blest** (blĕst) *adj.* **1. a.** Worthy of being worshiped : HOLY. **b.** Held in veneration : REVERED. **2.** *Rom. Cath. Ch.* Enjoying the eternal happiness of heaven. —Used as a title for those who have been beatified. **3.** Enjoying happiness : FORTUNATE. **4.** Bringing happiness : PLEASURABLE. **5.** *Slang.* —Used as an intensive <without a *blessed* crumb of bread> —**bless'ed·ly** *adv.* —**bless'ed·ness** *n.*

**Blessed Sacrament** *n. Rom. Cath. Ch.* The consecrated Host.

**Blessed Virgin** *n.* The Virgin Mary.

**bless·ing** (blĕs'ĭng) *n.* **1.** The act of one who blesses. **2.** Something promoting happiness, well-being, or prosperity : BOON. **3.** Approval. **4.** A short prayer before or after a meal.

**blest** (blĕst) *v. var. p.t. & p.p. of* BLESS. —*adj. var. of* BLESSED.

**bleth·er** (blĕth'ər) *v. & n. var. of* BLATHER.

**bleu cheese** *n.* Blue cheese.

**blew**[1] (blo͞o) *v. p.t. of* BLOW[1].

**blew**[2] (blo͞o) *v. p.t. of* BLOW[3].

**blight** (blīt) *n.* [Orig. unknown.] **1.** Any of several plant diseases that destroy leaves, growing tips, or an entire plant. **2.** An adverse environmental condition, as water or air pollution. **3.** One that withers hopes or ambitions, impairs growth, or halts prosperity. **4.** The state of being affected by or as if by blight. —*v.* **blight·ed, blight·ing, blights.** —*vt.* **1.** To cause (e.g., a plant) to be affected with blight. **2.** To cause to decline or decay. **3.** To ruin : destroy. —*vi.* To suffer blight.

**blight·er** (blī'tər) *n.* **1.** One that blights. **2.** *Chiefly Brit.* A worthless fellow.

**blimp**[1] (blĭmp) *n.* [Orig. unknown.] A nonrigid, buoyant aircraft.

**blimp**[2] (blĭmp) *n.* [After Colonel *Blimp,* a cartoon character invented by David Low (1891–1963).] *Chiefly Brit.* One whose views blend ultraconservative jingoism and misinformation.

**blind** (blīnd) *adj.* **-er, -est.** [ME < OE.] **1.** Being without sight. **2.** Of, relating to, or intended for sightless persons. **3.** Performed by instruments without the use of sight <a *blind* approach to the airport> **4.** Performed without preparation, forethought, or knowledge <a *blind* effort> **5.** Unable or unwilling to understand <*blind* to their defects> **6.** Not based on reason or evidence <*blind* belief> **7.** *Slang.* Drunk. **8.** Independent of human control <*blind* luck> **9. a.** Difficult to comprehend or see : ILLEGIBLE <*blind* inscriptions> **b.** Illegibly or incompletely addressed <*blind* mail> **10. a.** Hidden from sight <a *blind* seam> **b.** Screened from the view of an oncoming driver <a *blind* curve> **11.** Closed at one end <a *blind* passageway> **12.** Having no opening <a *blind* wall> **13.** *Bot.* Failing to flower. —*n.* **1.** Something that hinders vision or shuts out light. **2.** A shelter for concealing esp. duck hunters. **3.** Something intended to conceal the true nature, esp. of an activity : SUBTERFUGE. —*adv.* **1.** Without seeing : BLINDLY <fly the plane *blind*> **2.** *Informal.* Into a stupor <quickly drank themselves *blind*> —*vt.* **blind·ed, blind·ing, blinds. 1.** To deprive of sight. **2.** To dazzle. **3.** To deprive of judgment or perception. **4.** To deprive of light. —**blind'ing·ly** *adv.* —**blind'ly** *adv.* —**blind'ness** *n.*

**blind alley** *n. Informal.* An erroneous or futile undertaking.

**blind date** *n. Informal.* **1.** A social engagement between a woman and a man who have not previously met. **2.** Either of the persons on a blind date.

**blind·er** (blīn'dər) *n.* **1.** One that blinds. **2. blinders.** A pair of flaps attached to a horse's bridle to curtail side vision. **3. blinders.** An obstacle to the view or to perception.

**blind·fish** (blīnd'fĭsh') *n., pl.* **blindfish** *or* **-fish·es.** A fish having rudimentary, nonfunctioning eyes, esp. the cavefish.

**blind·fold** (blīnd'fōld') *vt.* **-fold·ed, -fold·ing, -folds.** [ME *blindfellen* < OE *geblindfellian,* to strike blind.] **1.** To cover the eyes of with or as if with a bandage. **2.** To prevent from seeing and esp. from comprehending. —*n.* A bandage to cover the eyes. —*adj.* **1.** With eyes covered. **2.** Reckless.

**blind gut** *n.* **1.** A digestive cavity with an opening only at one end. **2.** CECUM 2.

**blind hinge** *n.* A hinge constructed to allow the hinged piece to swing shut by its own weight unless held open.

**blind·man's buff** (blīnd'mănz') *n.* [*Buff,* short for *buffet,* a blow.] A game in which a blindfolded player tries to catch and identify another of the players.

**blind pig** *n. Slang.* A speakeasy.

**blind side** *n.* The side away from where one is looking.

**blind spot** *n.* **1.** *Anat.* The small, optically insensitive region where the optic nerve enters the retina. **2.** A part of an area not directly observable. **3.** An area of weak radio reception. **4.** A subject about which one is very ignorant or prejudiced.

**blind staggers** *n. (sing. in number).* STAGGER 3.

**blind tiger** *n. Slang.* A speakeasy.

**blind·worm** (blīnd'wûrm') *n.* [From its small eyes.] The slowworm.

**bli·ni** (blē'nē) *pl.n.* [R., pl. of *blin,* pancake.] Small pancakes served with caviar, sour cream, melted butter, and herring.

**blink** (blĭngk) *v.* **blinked, blink·ing, blinks.** [ME *blinken.*] —*vi.* **1.** To close and open the eyes rapidly. **2.** To look through half-closed eyes, as in a bright glare : SQUINT. **3.** To shine with intermittent gleams. **4.** To be startled. —*vt.* **1.** To close and open (the eyes) rapidly. **2.** To refuse to recognize or face : IGNORE <*blink* unpleasant facts> **3.** To signal (a message) with a flashing light. **—blink at.** To pretend not to see <*blink at* drug smuggling> —*n.* **1.** The act or an instance of closing and opening the eyes rapidly. **2.** A quick look or glimpse. **3.** The time required to blink. **4.** A flash of light : TWINKLE. **5.** ICEBLINK 1. **—on the blink.** *Slang.* Out of order.

**blink·er** (blĭng'kər) *n.* **1.** A light that conveys a message or warning by blinking. **2.** *Slang.* EYE 2a, b. **3. blinkers.** Goggles. **4. blinkers.** BLINDERS 2.

**blintz** (blĭnts) *also* **blin·tze** (blĭn'tsə) *n.* [Yiddish *blintse* < R. *blinyets,* dim. of *blin,* pancake.] A thin, rolled pancake usu. stuffed with cottage cheese and often served with sour cream.

**blip** (blĭp) *vt.* **blipped, blip·ping, blips.** [Alteration of BLEEP.] To interrupt recorded sounds, as on a videotape <*blipped* the curse from the TV script> —*n.* **1.** A spot of light on a radar screen. **2.** A brief interruption of the sound received in a television program as a result of blipping.

**bliss** (blĭs) *n.* [ME *blisse* < OE *bliss.*] **1.** Extreme happiness : JOY. **2.** Ecstasy of salvation : spiritual joy. **—bliss'ful** *adj.* **—bliss'ful·ly** *adv.* **—bliss'ful·ness** *n.*

**blis·ter** (blĭs'tər) *n.* [ME < OFr. *blestre,* boil, of Germanic orig.] **1.** A thin, rounded swelling of the skin, containing watery matter, caused by burning or irritation. **2.** A swelling on a plant similar to a blister. **3.** An air bubble resembling a blister, as on a painted surface. **4.** A rounded, often transparent protuberance, as one on an aircraft <an observation *blister*> —*v.* **-tered, -ter·ing, -ters.** —*vt.* **1.** To cause a blister to form on. **2.** To reprove harshly. —*vi.* To break out in blisters. **—blis'ter·y** *adj.*

**blister beetle** *n.* Any of various beetles of the family Meloidae that secrete a substance capable of blistering the skin.

**blister copper** *n.* [From its blistered appearance.] An almost pure copper produced in an intermediate stage of copper refining.

**blis·ter·ing** (blĭs'tər-ĭng) *adj.* **1.** Intensely hot. **2.** Severe : harsh. **3.** Very rapid <a *blistering* pace>

**blister rust** *n.* Any of several diseases of pine trees, caused by various fungi of the genus *Cronartium* and producing cankers and blisters on the bark.

**blite** (blīt) *n.* [ME, an herb < Lat. *blitum,* spinach < Gk. *bliton.*] Strawberry blite.

**blithe** (blīth, blīth) *adj.* **blith·er, blith·est.** [ME < OE *blīðe*] **1.** Filled with gaiety : CHEERFUL. **2.** Carefree : casual. **—blithe'ly** *adv.* **—blithe'ness** *n.*

**blith·er** (blĭth'ər) *vi.* **-ered, -er·ing, -ers.** [Alteration of BLATHER.] To blather.

**blithe·some** (blīth'səm, blīth'-) *adj.* Merry : cheerful. **—blithe'some·ly** *adv.* **—blith'some·ness** *n.*

**blitz** (blĭts) *n.* [Short for BLITZKRIEG.] **1.** A blitzkrieg. **2.** An intensive air raid or series of air raids. **3.** An intense campaign <an advertising *blitz*> **—blitz** *v.* **(blitzed, blitz·ing, blitz·es).**

**blitz·krieg** (blĭts'krēg') *n.* [G. : *Blitz,* lightning + *Krieg,* war.] **1.** A sudden, swift military offensive. **2.** A swift, concerted effort.

**bliz·zard** (blĭz'ərd) *n.* [Orig. unknown.] **1.** A violent windstorm accompanied by intense cold and driving snow. **2.** An extremely heavy snowstorm with high winds. **3.** A torrent <a *blizzard* of congratulations>

**bloat** (blōt) *v.* **bloat·ed, bloat·ing, bloats.** [< ME *blout,* soft, puffed < ON *blauvt.*] —*vt.* **1.** To cause to swell up, as with gas or liquid. **2.** To puff up, as with vanity. **3.** To cure (fish) by soaking in brine and half-drying in smoke. —*vi.* To become swollen or inflated. —*n.* A swelling of the rumen or intestinal tract of a domestic animal, caused by the gases of fermentation of green forage.

**bloat·er** (blō'tər) *n.* A lightly salted and briefly smoked herring or mackerel.

**blob** (blŏb) *n.* [ME *blober,* bubble.] **1.** A soft, amorphous mass. **2.** A shapeless splotch or daub, esp. of color. —*vt.* **blobbed, blob·bing, blobs.** To splash or mark with blobs : SPLOTCH.

**bloc** (blŏk) *n.* [Fr. < OFr., block.] **1.** A group united for common action. **2.** An often bipartisan coalition of U.S. legislators acting together for a common purpose <the oil *bloc*>

**block** (blŏk) *n.* [ME *blok* < OFr. < MDu., of Germanic orig.] **1. a.** A solid piece, as of wood, with one or more flat sides. **b.** A solid piece used as a structural member or as a support. **c.** A solid piece on which chopping or cutting is done. **d.** A solid piece on which persons are beheaded. **2.** A stand from which auctioned articles are sold. **3.** A mold or form on which something is shaped or displayed. **4.** A piece of a substance, as of wood or stone, that is prepared for engraving. **5. a.** A pulley or a system of pulleys set in a casing. **b.** An engine block. **6.** A bloc. **7.** A set of like items, as shares of stock, sold or handled as a unit. **8.** A group of four or more unseparated postage stamps. **9.** A group of Canadian townships in an unsurveyed area. **10. a.** A section of a city or town bounded on each side by consecutive streets. **b.** A segment of a street bounded by successive cross streets and including its buildings and inhabitants. **11.** A large building divided into separate units, as apartments. **12.** A length of railroad track controlled by signals. **13.** An act of obstructing or hindering. **14.** An obstacle : hindrance. **15.** An act of bodily obstruction, esp. legal interference with an opposing football player to clear the path of the ball carrier. **16.** *Med.* Interruption, esp. obstruction of a neural or digestive process. **17.** *Psychol.* Sudden cessation of a thought process without an immediate observable cause, occas. regarded as a consequence of repression. **18.** *Slang.* A person's head <I'll knock your *block* off.> **19.** A blockhead. —*v.* **blocked, block·ing, blocks.** —*vt.* **1.** To shape into a block. **2.** To support, strengthen, or retain in place by a block. **3.** To shape with or on a block. **4.** To stop or impede the passage of or movement through : OBSTRUCT. **5.** To indicate broadly without great detail : SKETCH. **6.** To impede the movement of (an opposing team member or a ball) by physical interference. **7.** *Med.* To interrupt the proper functioning of (a physiological process). **8.** *Psychol.* To fail to remember. **9.** To run (trains) on the block system. —*vi.* To obstruct the movement of an opposing team member in a sport. **—on the block.** Up for sale, esp. at an auction. **—block'er** *n.*

**block·ade** (blŏ-kād') *n.* **1.** Closure of an area, as a city or harbor, by hostile forces so as to prevent entrance and exit of traffic and communication. **2.** The forces used in a blockade. **—block·ade'** *v.* **(-ad·ed, -ad·ing, -ades). —block·ad'er** *n.*

**block·ade-run·ner** (blŏ-kād'rŭn'ər) *n.* One, as a ship, that runs a blockade. **—block·ade'-run'ning** *n.*

**block·age** (blŏk'ĭj) *n.* **1.** The act of obstructing. **2.** An obstruction.

**block and tackle** *n.* Pulley blocks and ropes or cables for hauling and hoisting heavy objects.

**block·bust·er** (blŏk'bŭs'tər) *n.* *Informal.* **1.** A bomb capable of destroying a city block. **2.** Something notably effective or successful. **3.** One that engages in blockbusting.

**block·bust·ing** (blŏk'bŭs'tĭng) *n.* *Informal.* The practice of persuading whites to sell their homes quickly and usu. at a loss by appealing to the fear that minority groups and esp. blacks will move into the neighborhood, with a resulting decline in property values.

**block·head** (blŏk'hĕd') *n.* A stupid person : DOLT.

**block·house** (blŏk'hous') *n.* **1.** A military fortification with loopholes for defensive firing or observation. **2.** A heavily reinforced building used for missile and space launch operations and for observation of weapons testing. **3.** A fort made of squared timbers with a projecting upper story.

**block·ish** (blŏk'ĭsh) *adj.* **1.** Resembling a block. **2.** Stupid : dull. **—block'ish·ly** *adv.* **—block'ish·ness** *n.*

**block letter** *n.* **1.** A letter printed or written sans serif. **2.** A sans-serif type style.

**block plane** *n.* A small plane for cutting across the grain of wood.

**block printing** *n.* Printing from engraved or carved wooden or linoleum blocks.

**block signal** *n.* A fixed signal at the entrance to a railroad block, indicating whether trains may enter.

**block system** *n.* A system for controlling and safeguarding the flow of railway trains in which track is divided into blocks, each controlled by automatic signals.

**block·y** (blŏk'ē) *adj.* **-i·er, -i·est.** Resembling a block : STOCKY.

**bloke** (blōk) *n.* [Orig. unknown.] *Chiefly Brit.* FELLOW 1a.

**blond** (blŏnd) *adj.* **-er, -est.** [OFr.] **1.** Having fair hair and skin and usu. light eyes. **2.** Of a flaxen or golden color or of any light shade of auburn or pale yellowish brown. **3.** Light-colored <*blond* furniture> —*n.* **1.** A blond person. **2.** A light yellowish brown to dark grayish yellow. ***usage:*** *Blond* as an adjective may be used of both sexes, but *blonde* as a noun refers only to a woman or girl. **—blond'ish** *adj.* **—blond'ness** *n.*

**blonde** (blŏnd) *adj.* **blond·er, blond·est.** [OFr., fem. of *blond.*] Blond. —*n.* A blonde woman or girl.

**blood** (blŭd) *n.* [ME *blod* < OE *blōd,* of Germanic orig.] **1. a.** The fluid circulated by the heart through the vertebrate vascular system, carrying oxygen and nutrients throughout the body and waste materials to excretory channels. **b.** A functionally similar fluid in an invertebrate. **c.** A fluid resembling blood, as the juice of some plants. **2.** Life : lifeblood. **3.** Bloodshed. **4.** Disposition or temper : TEMPERAMENT. **5. a.** Descent from a common ancestor. **b.** Family relationship. **c.** Descent from noble or royal lineage <a princess of the *blood*> **d.** Recorded descent from purebred stock. **e.** Racial or national ancestry. **6.** Personnel <new *blood* in the department> **7.** A dashing young man. —*vt.* **blood·ed, blood·ing, bloods. 1. a.** To give (a hunting dog) its first taste of blood. **b.** To initiate (a novice hunter) by marking his face with the blood of the prey. **2.** To subject (recruits) to baptism of fire. —*adj.* Purebred. **—in cold blood.** Dispassionately and deliberately.

**blood agar** *n.* A culture medium containing whole blood for cultivation of bacteria.

**blood bank** *n.* **1.** A place where whole blood or plasma is identi-

ă pat ā pay âr care ä father ĕ pet ē be hw which ĭ pit ī tie îr pier ŏ pot ō toe ô paw, for oi noise ŏŏ took

fied by type, processed, and stored for future transfusion. **2.** Blood or plasma in a blood bank.

**blood bath** *n.* A massacre.

**blood-brain barrier** (blŭd′brān′) *n.* **1.** A physiologic differentiation phenomenon that deaccelerates the penetration of certain substances into the brain while allowing more rapid perfusion into other tissues. **2.** A membrane that separates brain tissues from circulating blood.

**blood brother** *n.* **1.** A brother by birth. **2.** One of two individuals pledged to mutual fidelity and trust by a ceremony involving the mingling of each other's blood. **—blood brotherhood** *n.*

**blood clot** *n.* A solidified mass of blood caused by polymerization of fibrin molecules and trapped cells.

**blood count** *n.* **1.** The number of red and white corpuscles in a specific volume of blood. **2.** Determination of a blood count.

**blood·cur·dling** (blŭd′kûrd′lĭng) *adj.* Causing great horror. **—blood′cur′dling·ly** *adv.*

**blood·ed** (blŭd′ĭd) *adj.* **1.** Having blood or a temperament of a specified kind <a cold-*blooded* monster><hot-*blooded* young street toughs> **2.** Thoroughbred.

**blood fluke** *n.* A schistosome.

**blood group** *n.* Any of several immunologically distinct, genetically determined classes of human blood, clinically identified by agglutination reactions.

**blood·guilt** (blŭd′gĭlt′) *n.* Guilt caused by murder or bloodshed.

**blood heat** *n.* The usual temperature (98.6°F) of human blood.

**blood·hound** (blŭd′hound′) *n.* **1.** A hound with a smooth coat, drooping ears, sagging jowls, and a keen sense of smell. **2.** *Informal.* A relentless pursuer.

**blood·less** (blŭd′lĭs) *adj.* **1.** Deficient in or lacking blood. **2.** Pale and anemic. **3.** Achieved without bloodshed <a *bloodless* victory> **4.** Lacking vivacity. **—blood′less·ly** *adv.* **—blood′less·ness** *n.*

**blood·let·ting** (blŭd′lĕt′ĭng) *n.* **1.** Bleeding of a vein as therapy. **2.** A draining away, as of lifeblood. **3.** Bloodshed. **—blood′let′ter** *n.*

**blood·line** (blŭd′līn′) *n.* **1.** Direct line of descent. **2.** Pedigree.

**blood·mo·bile** (blŭd′mə-bēl′) *n.* [BLOOD + (AUTO)MOBILE.] A motor vehicle equipped for collecting blood from donors.

**blood money** *n.* **1.** Compensation paid to the next of kin of a murder victim. **2.** Money gained at the cost of another's life or livelihood. **3.** Money paid to a hired killer.

**blood plasma** *n.* The pale-yellow or gray-yellow, protein-containing fluid portion of the blood in which the corpuscles are suspended.

**blood platelet** *n.* A platelet.

**blood poisoning** *n.* **1.** Toxemia. **2.** Septicemia.

**blood pressure** *n.* Pressure of the blood within the arteries, primarily maintained by contraction of the left ventricle.

**blood pudding** *n.* A sausage of cooked swine's blood and suet.

**blood red** *n.* A moderate to vivid red. **—blood′-red′** (blŭd′rĕd′) *adj.*

**blood relation** *n.* A person related by birth rather than by marriage. **—blood relationship** *n.*

**blood·root** (blŭd′ro͝ot′, -ro͞ot′) *n.* A woodland plant, *Sanguinaria canadensis* of eastern North America, with a fleshy rootstock, red juice, and a single pale flower.

**blood sausage** *n.* Blood pudding.

**blood serum** *n.* Blood plasma with the fibrin removed.

**blood·shed** (blŭd′shĕd′) *n.* Shedding of blood, esp. the injury or killing of human beings.

**blood·shot** (blŭd′shŏt′) *adj.* Red and irritated.

**blood·stain** (blŭd′stān′) *n.* A stain caused by blood. **—blood′-stain′** *v.* **(-stained, -stain·ing, -stains).**

**blood·stone** (blŭd′stōn′) *n.* A deep-green chalcedony flecked with red jasper.

**blood·stream** *also* **blood stream** (blŭd′strēm′) *n.* The stream of blood in the circulatory system of a living body.

**blood·suck·er** (blŭd′sŭk′ər) *n.* **1.** An animal, as a leech, that sucks blood. **2.** *Informal.* One who clings to another : PARASITE. **—blood′-suck′ing** *adj.*

**blood test** *n.* A usu. diagnostic examination of a blood sample, esp. to detect syphilis.

**blood·thirst·y** (blŭd′thûr′stē) *adj.* Eager for or marked by bloodshed. **—blood′thirst′i·ly** *adv.* **—blood′thirst′i·ness** *n.*

**blood type** *n.* Blood group.

**blood vessel** *n.* An elastic, tubular canal, as an artery, vein, or capillary, through which blood circulates.

**blood·worm** (blŭd′wûrm′) *n.* A segmented worm of the genera *Polycirrus* or *Enoplobranchus,* with a bright-red body and often used for bait.

**blood·wort** (blŭd′wûrt′, -wôrt′) *n.* Any of various chiefly South American plants of the family Haemodoraceae, with roots that contain a red juice.

**blood·y** (blŭd′ē) *adj.* **-i·er, -i·est. 1.** Stained with blood. **2.** Of, typical of, or containing blood. **3.** Accompanied by or giving rise to bloodshed. **4.** Cruel : bloodthirsty. **5.** Suggesting the color of blood : BLOOD-RED. **6.** —Used as an intensive <a *bloody* fool> —*adv.* —Used as an intensive <*bloody* well right> —*vt.* **-ied, -y·ing, -ies.** To stain or color with or as if with blood. **—blood′i·ly** *adv.* **—blood′i·ness** *n.*

**bloody mary** *also* **Bloody Mary** *n.* A cocktail usu. made of vodka, tomato juice, and seasonings.

**bloom[1]** (blo͞om) *n.* [ME *blom* < ON *blōm.*] **1.** The flower or blossoms of a plant. **2. a.** The state or a time of being in flower. **b.** A state or time of vigor, freshness, and beauty : PRIME <"the radiant *bloom* of Greek genius" —Edith Hamilton> **3.** A fresh, rosy complexion. **4. a.** *Bot.* A delicate, powdery coating, as that on some fruits or on some leaves and stems. **b.** A similar coating, as on newly minted coins. **5.** An excessive planktonic growth in a body of water. —*v.* **bloomed, bloom·ing, blooms.** —*vi.* **1.** To bear flowers. **2.** To shine with health and vigor : GLOW. **3.** To grow or flourish. —*vt.* **1.** *Obs.* To cause to flower. **2.** To cause to flourish. **—bloom′y** *adj.*

☆ ***syns:*** BLOOM, BLOSSOM, BURGEON, EFFLORESCE, FLOWER *v. core meaning* : to bear flowers <roses *blooming* on the fence>

**bloom[2]** (blo͞om) *n.* [ME *blome,* lump of metal < OE *blōma.*] **1.** A steel bar, usu. more than 36 square inches or 232 square centimeters in cross section, prepared for rolling. **2.** A mass of wrought iron ready for further working.

**bloom·er[1]** (blo͞o′mər) *n.* **1. a.** A plant that blooms. **b.** One who attains full development of his or her abilities or talents <an early *bloomer*> **2.** *Slang.* A blunder.

**bloom·er[2]** (blo͞o′mər) *n.* [After Amelia *Bloomer* (1818–1894).] **1.** A costume, once worn by women and girls, of loose trousers gathered about the ankles and worn under a short skirt. **2. bloomers. a.** Wide, loose trousers gathered at the knee, once worn by women and girls for athletics. **b.** Women's underpants of similar design.

**bloom·ing** (blo͞o′mĭng) *adj.* [Prob. a euphemism for BLOODY.] *Chiefly Brit.* —Used as an intensive <a *blooming* fool>

**bloop·er** (blo͞o′pər) *n.* [< *bloop,* imit. of the sound of such a hit.] **1.** *Baseball.* A short, weakly hit fly ball that carries just beyond the infield. **2.** *Informal.* An embarrassing error.

**blos·som** (blŏs′əm) *n.* [ME < OE *blōstm.*] **1.** A flower or mass of flowers, esp. of a plant that yields edible fruit. **2.** The state or time of flowering. —*vi.* **-somed, -som·ing, -soms. 1.** To come into flower : BLOOM. **2.** To develop : flourish. **—blos′som·y** *adj.*

**blot[1]** (blŏt) *n.* [ME.] **1.** A spot : stain. **2.** A moral stigma : DISGRACE. —*v.* **blot·ted, blot·ting, blots.** —*vt.* **1.** To stain or spot. **2.** *Obs.* To bring moral disgrace to. **3.** To obliterate : cancel. **4.** To make obscure : DARKEN. **5.** To soak up or dry with absorbent material. —*vi.* **1.** To spill or spread in a blot. **2.** To become blotted. **—blot out.** To destroy utterly : ANNIHILATE.

**blot[2]** (blŏt) *n.* [Orig. unknown.] **1.** An exposed backgammon piece. **2.** *Archaic.* A weak point.

**blotch** (blŏch) *n.* [Prob. a blend of BLOT and BOTCH.] **1.** A spot or blot. **2.** A discoloration on the skin : BLEMISH. **3.** A plant disease caused by fungi that results in black or brown dead areas on fruit or leaves. —*vt.* & *vi.* **blotched, blotch·ing, blotch·es.** To mark or become marked with blotches. **—blotch′i·ness** *n.* **—blotch′y** *adj.*

**blot·ter** (blŏt′ər) *n.* **1.** A piece or pad of blotting paper. **2.** A book containing daily records.

**blotting paper** *n.* Absorbent paper used to soak up excess ink.

**blouse** (blous, blouz) *n.* [Fr.] **1.** A loosely fitting shirt extending to the waist or slightly below. **2.** The service coat or tunic worn by members of the U.S. Army. —*vi.* & *vt.* **bloused, blous·ing, blous·es.** To hang or cause to hang loose and full.

**blou·son** (blou′sŏn′, blo͞o′zŏn′) *n.* [Fr., dim. of *blouse,* blouse.] A woman's garment with a fitted waistband over which material blouses.

**blow[1]** (blō) *v.* **blew** (blo͞o), **blown** (blōn), **blow·ing, blows.** [ME *blowen* < OE *blāwan.*] —*vi.* **1.** To be in motion, as air. **2.** To be carried by or as if by the wind. **3.** To expel a current of air, as from the mouth. **4.** To produce a sound by expelling a current of air, as in playing a musical wind instrument. **5.** To breathe hard : PANT. **6.** To burn out or melt, as a fuse. **7.** To burst suddenly <The tire *blew.*> **8.** To spout air and water, as a whale. **9.** *Slang.* To boast. **10.** *Slang.* To go away : DEPART. —*vt.* **1.** To cause to move by means of a current of air. **2.** To expel (air) from the mouth. **3.** To cause air to be expelled from. **4.** To drive a current of air on, in, or through. **5.** To clear out or free of obstruction by forcing air through. **6.** To shape or form (e.g., glass) by forcing air or gas through at the end of a pipe. **7. a.** To cause (a wind instrument) to sound. **b.** To sound <a bugle *blowing* retreat> **8.** To cause (a horse) to be out of breath. **9.** To cause to disintegrate due to an explosion <Dynamite *blew* the mountain to pieces.> **10.** To lay or deposit eggs in. —Used of a fly. **11.** To melt or disable (a fuse). **12.** *Slang.* To spend (money) freely. **13.** *Slang.* To handle ineptly. **14.** *Slang.* To depart in a great hurry <*blew* town> **—blow away.** *Slang.* **1.** To kill by shooting. **2.** To affect intensely : OVERWHELM. **—blow in.** *Slang.* To arrive. **—blow out. 1.** To extinguish or be extinguished by blowing <*blew* out all the candles> **2.** To fail, as an electrical apparatus. **—blow over.** To pass away without effect <a storm that soon *blew over*> **—blow up. 1.** To come into being <A squall *blew up.*> **2.** To fill with air : INFLATE. **3.** To enlarge (a photographic image or print). **4.** *Slang.* To lose one's temper. —*n.* **1. a.** A blast of air or wind. **b.** A storm. **2.** An act or instance of blowing. **3.** *Slang.* An act of bragging. **—blow hot and cold.** To change one's mind often : VACILLATE. **—blow off steam.** To give vent to pent-up emotion. **—blow (one's) mind.** *Slang.* To affect with intense emotion.

**blow²** (blō) *n.* [ME *blaw.*] **1.** A sudden hard hit, as with the fist or an instrument. **2.** A sudden unexpected shock. **3.** A sudden unexpected attack. **—come to blows.** To begin to fight.

**blow³** (blō) *n.* [ME *blowen,* to bloom < OE *blōwan.*] A mass of blossoms. —*vi. & vt.* **blew** (blo͞o), **blown** (blōn), **blow·ing, blows.** To bloom or cause to bloom.

**blow·ball** (blō'bôl') *n.* A fluffy seed ball, as that of a dandelion.

**blow-by-blow** (blō'-bī-blō') *adj.* Displaying careful attention to details <a *blow-by-blow* account>

**blow-dry** (blō'drī') *vt.* **-dried, -dry·ing, -dries.** To dry and often style with a hand-held hair dryer. **—blow'-dry'er** *n.*

**blow·er** (blō'ər) *n.* **1.** One that blows, esp. a fan. **2.** *Slang.* A braggart. **3.** *Chiefly Brit.* A telephone.

**blow·fish** (blō'fĭsh') *n., pl.* **blowfish** or **-fish·es.** PUFFER 2.

**blow·fly** (blō'flī') *n.* Any of several flies of the family Calliphoridae that deposit their eggs in carcasses or carrion or in open sores and wounds.

**blow·gun** (blō'gŭn') *n.* A long narrow pipe through which darts or pellets may be blown.

**blow·hard** (blō'härd') *n. Slang.* A braggart : boaster.

**blow·hole** (blō'hōl') *n.* **1.** A nostril at the highest point on the head of whales and other cetaceans. **2.** A hole in ice to which aquatic mammals, as dolphins, come to breathe. **3.** A vent permitting the escape of air or other gas.

**blown¹** (blōn) *adj.* [< p.part. of BLOW¹.] **1.** Swollen : distended. **2.** Out of breath. **3.** Flyblown. **4.** Formed by blowing.

**blown²** (blōn) *adj.* [< p.part. of BLOW³.] Completely opened <a full-*blown* rose>

**blow·out** (blō'out') *n.* **1. a.** A sudden bursting, as of a tire. **b.** The hole made by such a bursting. **2.** A sudden escape of a confined gas. **3.** *Slang.* A large party.

**blow·pipe** (blō'pīp') *n.* **1.** A metal tube in which a flow of gas and a controlled flow of air are mixed to concentrate the heat of a flame. **2.** A blowgun. **3.** A long narrow iron pipe used to gather, work, and blow molten glass.

**blow·sy** (blou'zē) *adj. var. of* BLOWZY.

**blow·torch** (blō'tôrch') *n.* A usu. portable gas burner that generates a flame hot enough to melt soft metals.

**blow·up** (blō'ŭp') *n.* **1.** An explosion. **2.** A violent outburst of temper. **3.** A photographic enlargement.

**blow·y** (blō'ē) *adj.* **-i·er, -i·est.** Breezy : windy.

**blow·zy** *also* **blow·sy** (blou'zē) *adj.* **-zi·er, -zi·est** *also* **-si·er, -si·est.** [< Obs. *blowze,* beggar wench.] **1.** Having a coarsely ruddy, bloated appearance. **2.** Disheveled and frowzy : UNKEMPT.

**BLT** (bē'ĕl-tē') *n.* A bacon, lettuce, and tomato sandwich.

**blub·ber¹** (blŭb'ər) *v.* **-bered, -ber·ing, -bers.** [ME *bluberen,* to bubble < *bluber,* foam.] —*vi.* To sob noisily. —*vt.* **1.** To utter while sobbing. **2.** To make wet and swollen by sobbing. —*n.* Loud sobbing. **—blub'ber·er** *n.* **—blub'ber·ing·ly** *adv.*

**blub·ber²** (blŭb'ər) *n.* [ME *bluber,* foam.] **1.** The thick layer of fat between the skin and the muscle layers of whales and other marine mammals. **2.** Excessive body fat. **3.** A large invertebrate sea nettle or medusa. —*adj.* Swollen and protruding. **—blub'ber·y** *adj.*

**blu·cher** (blo͞o'chər, -kər) *n.* [After Gebhard L. von *Blücher* (1742–1819).] **1.** A high shoe or half boot. **2.** A shoe with a one-piece vamp and tongue and a top that laps over the vamp.

**bludg·eon** (blŭj'ən) *n.* [Orig. unknown.] A short, heavy, usu. wooden club with one end loaded or thicker than the other. —*vt.* **-eoned, -eon·ing, -eons. 1.** To hit with or as if with a club. **2.** To bully or threaten. **—bludg'eon·er, bludg'eon·eer'** (-ə-nîr') *n.*

**blue** (blo͞o) *n.* [ME < OFr. *bleu,* of Germanic orig.] **1.** Any of a group of colors that may vary in lightness and saturation, whose hue is that of a clear sky; the hue of that portion of the spectrum lying between green and violet; one of the additive or light primaries; one of the psychological primary hues, evoked in the normal observer by radiant energy of wavelength approx. 475 nanometers. **2. a.** A pigment or dye imparting the color blue. **b.** Bluing. **3. a.** An object of the color blue. **b.** Blue clothing. **4.** One who wears a blue uniform. **5.** *often* **Blue. a.** A member of the Union Army in the Civil War. **b.** The Union Army. **6. blues. a.** The blue uniform of the U.S. Navy. **b.** The dress uniform of the U.S. Army. **7.** A small blue butterfly of the family Lycaenidae. **8. a.** The sky. **b.** The sea. —*adj.* **blu·er, blu·est. 1.** Of the color blue. **2.** Bluish or having blue or bluish parts. **3.** Having a gray or purplish color, as from cold or contusion. **4.** Wearing blue. **5. a.** Depressed : gloomy. **b.** Dreary : dismal. **6.** Puritanical : severe. **7.** Aristocratic. **8.** Risqué : indecent <a *blue* joke> —*vt. & vi.* **blued, blu·ing, blues.** To make or become blue. **—once in a blue moon.** Very seldom : RARELY. **—out of the blue. 1.** From an unexpected source. **2.** At an unexpected time. **—blue'ly** *adv.* **—blue'ness** *n.*

**blue baby** *n.* An infant born with bluish skin caused by inadequate oxygenation of the blood, a symptom of a congenital cardiac or pulmonary defect.

**blue·beard** (blo͞o'bîrd') *n.* [< *Bluebeard,* a character in a folk tale.] A man who first marries and then murders one wife after another.

**blue·bell** (blo͞o'bĕl') *n.* A plant bearing bell-shaped, blue flowers, esp.: **a.** A European plant, *Scilla nonscripta,* with grasslike leaves and fragrant, blue-violet flower clusters. **b.** The harebell. **c.** A plant of the genus *Mertensia.*

**blue·ber·ry** (blo͞o'bĕr'ē) *n.* **1.** Any of several North American shrubs of the genus *Vaccinium,* bearing small urn-shaped flowers and edible berries. **2.** The juicy blue, purplish, or blackish berry of a blueberry shrub.

**blue·bill** (blo͞o'bĭl') *n.* The scaup.

**blue·bird** (blo͞o'bûrd') *n.* Any of several North American birds of the genus *Sialia,* with blue plumage and in the male of most species a rust-colored breast.

**blue-black** (blo͞o'blăk') *adj.* Very dark blue in color.

**blue blood** *n.* [Transl. of Sp. *sangre azul;* prob. from the visible veins of fair-complexioned aristocrats.] **1.** Aristocratic lineage. **2.** A member of the aristocracy. **—blue'-blood'ed** *adj.*

**blue·blos·som** (blo͞o'blŏs'əm) *n.* A shrub, *Ceanothus thyrsiflorus* of the U.S. west coast, with profuse clusters of small blue flowers.

**blue·bon·net** (blo͞o'bŏn'ĭt) *n.* **1. a.** A plant, *Lupinus subcarnosus* of Texas and adjacent regions, with compound leaves and blue flower clusters. **b.** Any of several other plants similar to the bluebonnet. **2. a.** A broad, blue woolen cap worn in Scotland. **b.** A person wearing a bluebonnet.

**blue book** *also* **blue·book** (blo͞o'bo͝ok') *n.* **1.** An official list of persons employed by the U.S. government. **2.** *Informal.* A book listing the names of socially prominent people. **3.** A blank notebook with blue covers in which to write college tests and examinations.

**blue·bot·tle** (blo͞o'bŏt'l) *n.* **1.** Any of several flies of the genus *Calliphora,* with a bright metallic-blue body and breeding in decaying organic matter. **2.** The cornflower.

**blue cheese** *n.* A semisoft cheese made of cow's milk, with a flavor similar to Roquefort cheese and greenish-blue mold.

**blue chip** *n.* **1.** A stock selling at a high price due to public confidence in its consistently steady earnings over the long term. **2.** A highly valuable asset or property.

▲ word history: Blue chips in poker are usually the chips with the highest value. It is from this usage that the term *blue chip* is applied to highly regarded stocks.

**blue·coat** (blo͞o'kōt') *n.* One, esp. a police officr, who wears a blue uniform.

**blue cohosh** *n.* A plant, *Caulophyllum thalictroides* of eastern North America, bearing compound leaves and greenish or purplish flower clusters.

**blue-col·lar** (blo͞o'kŏl'ər) *adj.* Of or relating to wage earners, esp. as a class, whose tasks are carried out in work clothes and usu. involve manual labor.

**blue·curls** *also* **blue curls** (blo͞o'kûrlz') *n.* (*sing. or pl. in number*). Any of several North American plants of the genus *Trichostema,* having two-lipped blue flowers with long, curved stamens.

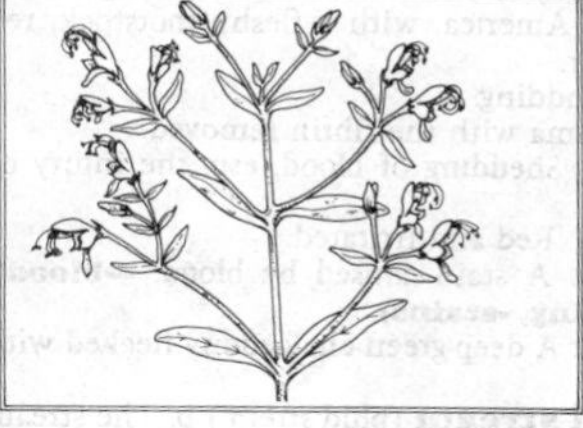

**bluecurls**

**blue devils** *pl.n. Informal.* A feeling of depression.

**blue-eyed grass** (blo͞o'īd') *n.* Any of various plants of the genus *Sisyrinchium,* chiefly of North America, with grasslike leaves and small, starlike blue flowers.

**blue-eyed Mary** *n.* A plant, *Collinsia verna* of eastern North America, with two-lipped white and blue flowers.

**blue·fish** (blo͞o'fĭsh') *n., pl.* **bluefish** or **-fish·es. 1.** A voracious food and game fish, *Pomatomus saltatrix* of temperate and tropical waters of the Atlantic and Indian oceans. **2.** Any of various fishes that are mainly blue.

**blue flag** *n.* Any of several wild irises bearing blue flowers, esp. *Iris versicolor* of eastern North America.

**blue flu** *n.* [So called because police officers' uniforms are usually blue.] A sickout, esp. by uniformed police officers.

**blue fox** *n.* **1.** The arctic fox during its summer color phase, when its pelt is bluish gray. **2.** The fur of a blue fox.

**blue·gill** (blo͞o'gĭl') *n.* A common, edible sunfish, *Lepomis macrochirus* of North American lakes and streams.

**blue·grass** (blo͞oʹgrăsʹ) *n.* **1.** A grass of the genus *Poa*, esp. *P. pratensis*, native to Eurasia but naturalized throughout North America. **2.** Folk music that originated in the southern United States, usu. played on banjos and guitars and with rapid tempos and jazzlike improvisation.

**blue-green alga** (blo͞oʹgrēnʹ) *n.* An alga of the division Cyanophyta or Myxophyceae, thought to be among the simplest plant forms.

**blue grouse** *n.* A wildfowl, *Dendragapus obscurus* of western North America, with mostly gray plumage.

**blue gum** *n.* A tall timber tree native to Australia, *Eucalyptus globulus*, bearing aromatic leaves and outer bark that peels off in shreds.

**blue·head** (blo͞oʹhĕdʹ) *n.* A marine fish, *Thalassoma bifasciatum* of tropical Atlantic waters, of which the male has a green body and a blue head.

**blue·hearts** (blo͞oʹhärtsʹ) *n.* *(sing. or pl. in number).* A hairy plant, *Buchnera americana* of central North America, bearing deep-purple flower spikes.

**blue heron** *n.* A variety of heron with blue or blue-gray plumage.

**blue·ing** (blo͞oʹĭng) *n. var. of* BLUING.

**blue·ish** (blo͞oʹĭsh) *adj. var. of* BLUISH.

**blue·jack** (blo͞oʹjăkʹ) *n.* [BLUE + (BLACK)JACK[1]] An oak tree, *Quercus cinerea* of the southern United States, having narrow unlobed leaves.

**blue·jack·et** (blo͞oʹjăkʹĭt) *n.* An enlisted person in the U.S. or British Navy.

**blue jay** *n.* A North American bird, *Cyanocitta cristata*, having predominantly blue plumage and a crested head.

**blue jeans** *pl.n.* JEANS 2.

**blue law** *n.* **1.** One of a body of laws in colonial New England designed to enforce strict moral standards. **2.** A law aimed at regulating commercial and other activities on Sunday.

**blue mold** *n.* Any of several fungi of the genus *Penicillium*, forming a bluish growth on food and other surfaces.

**blue·nose** (blo͞oʹnōzʹ) *n.* A puritanical person.

**blue note** *n.* [From its use in blues music.] *Mus.* A flatted note, esp. the third or seventh note of a chord, in place of an expected major interval.

**blue-pen·cil** (blo͞oʹpĕnʹsəl) *vt.* **-ciled, -cil·ing, -cils** *also* **-cilled, -cil·ling, -cils.** To edit, correct, or revise with or as if with a blue pencil.

**blue pe·ter** (pēʹtər) *n.* A blue flag with a white square in the center, flown to signal that a ship is prepared to sail.

**blue-plate** (blo͞oʹplātʹ) *adj.* Being a main course of a restaurant meal usu. offered at a special price.

**blue point** *n.* An edible oyster found mainly off Blue Point, Great South Bay, Long Island, New York.

**blue·print** (blo͞oʹprĭntʹ) *n.* **1.** A photographic reproduction, as of architectural plans or technical drawings, rendered as white lines on a blue background. **2.** A carefully designed plan. —*vt.* **-print·ed, -print·ing, -prints. 1.** To make a blueprint of. **2.** To lay a plan for.

**blue ribbon** *n.* **1.** A blue ribbon awarded as the first prize in a competition. **2.** An award for excellence. **—blueʹ-ribʹbon** *adj.*

**blue-rib·bon jury** (blo͞oʹrĭbʹən) *n.* A jury whose members have been selected for their special qualifications, as a high degree of intelligence, enabling them to deal with complex legal issues.

**blues** (blo͞oz) *pl.n.* [Short for BLUE DEVILS.] *(sing. or pl. in number).* **1.** A state of depression. **2.** A style of jazz evolved from southern American Negro secular songs and usu. distinguished by flatted thirds and sevenths and a slow tempo.

**blue shift** *n.* An apparent decrease in the wavelength of radiation emitted by an approaching celestial body as a consequence of the Doppler effect.

**blue-sky law** (blo͞oʹskīʹ) *n.* A law designed to protect the public from buying fraudulent securities.

**blue spruce** *n.* An evergreen tree, *Picea pungens* of the Rocky Mountain region, bearing bluish-green needles.

**blues-rock** (blo͞ozʹrŏkʹ) *n.* A style of music combining blues and rock 'n' roll.

**blue·stock·ing** (blo͞oʹstŏkʹĭng) *n.* [After the *Blue Stocking* Society, a nickname for a chiefly female literary club of 18th-cent. London.] A scholarly or pedantic woman. **—blueʹstockʹing** *adj.*

▲ word history: The term *bluestocking* seems always to have been one of contempt and derision, for it originally signified one who was informally and unfashionably dressed in blue worsted rather than black silk stockings. Such informal wear was common at literary and intellectual gatherings in 18th-century London, which were scornfully dubbed "bluestocking" societies by those who preferred parties where they could play cards and indulge in other idle amusements in their best and most fashionable clothes. Since the literary gatherings were organized and attended primarily by women, the term *bluestocking* was transferred, sneer and all, to any woman with pretensions or aspirations to literature and learning.

**blue·stone** (blo͞oʹstōnʹ) *n.* **1.** A bluish-gray sandstone used for building and paving. **2.** A stone similar to the bluestone.

**blue streak** *n. Informal.* **1.** Something moving very fast. **2.** A rapid and seemingly interminable stream of words.

**blu·ets** (blo͞oʹĭts) *n.* [< Fr. *bluet*, cornflower.] *(sing. or pl. in number).* A slender, low-growing plant, *Houstonia caerulea* of eastern North America, with small light-blue flowers having yellow centers.

**blue vitriol** *n.* Copper sulfate.

**blue·weed** (blo͞oʹwēdʹ) *n.* Viper's bugloss.

**blue whale** *n.* A large whale, *Sibbaldus musculus*, with a bluish-gray back and longitudinal grooves along the belly and throat.

**bluff**[1] (blŭf) *v.* **bluffed, bluff·ing, bluffs.** [Perh. < Du. *bluffen*, to boast.] —*vt.* **1.** To mislead or deceive. **2.** To impress, deter, or intimidate by showing more confidence than the facts support. **3.** To try to mislead (opponents) in poker by heavy betting on a poor hand or by little or no betting on a good one. —*vi.* To engage in a false display of confidence or strength. —*n.* **1.** An act or the practice of bluffing. **2.** One who bluffs. **—call (someone's) bluff.** To challenge or expose a false display of confidence or strength. **—bluffʹa·ble** *adj.* **—bluffʹer** *n.*

**bluff**[2] (blŭf) *n.* [Orig. unknown.] A steep headland, promontory, river bank, or cliff. —*adj.* **-er, -est. 1.** Having a steep, broad front. **2.** Having a blunt and rough but not unkind manner. **—bluffʹly** *adv.* **—bluffʹness** *n.*

**blu·ing** *also* **blue·ing** (blo͞oʹĭng) *n.* **1.** A coloring agent for counteracting the yellowing of laundered fabrics. **2.** A rinsing agent for giving a silver tint to graying hair.

**blu·ish** *also* **blue·ish** (blo͞oʹĭsh) *adj.* Somewhat blue. **—bluʹish·ness** *n.*

**blun·der** (blŭnʹdər) *n.* [ME *blunderen*, to go blindly, prob. < ON *blunda*, to doze.] A serious mistake usu. caused by ignorance, stupidity, or confusion. —*v.* **-dered, -der·ing, -ders.** —*vi.* **1.** To move unsteadily or clumsily. **2.** To make a blunder. —*vt.* **1.** To bungle or botch. **2.** To say thoughtlessly or stupidly. **—blunʹder·er** *n.* **—blunʹder·ing·ly** *adv.*

**blun·der·buss** (blŭnʹdər-bŭsʹ) *n.* [Alteration of Du. *donderbus* : *donder*, thunder (< MDu. *doner*) + *bus*, gun (< MDu. *busse*, tube < Lat. *buxis*, box).] **1.** A short musket of flaring muzzle and wide bore, once used to scatter shot at close range. **2.** A clumsy or stupid person : BLUNDERER.

**blunt** (blŭnt) *adj.* **-er, -est.** [ME.] **1.** Having a dull edge or end. **2.** Abrupt and frank : BRUSQUE. **3.** Slow to understand : DULL. **4.** Lacking in feeling : INSENSITIVE. —*v.* **blunt·ed, blunt·ing, blunts.** —*vt.* **1.** To dull the edge of. **2.** To make less effective : WEAKEN <"Division of purpose *blunted* our offensive spirit"—T.E. Lawrence> —*vi.* To become blunt. **—bluntʹly** *adv.* **—bluntʹness** *n.*

**blur** (blûr) *v.* **blurred, blur·ring, blurs.** [Orig. unknown.] —*vt.* **1.** To make hazy and indistinct : OBSCURE. **2.** To smear or stain : SMUDGE. **3.** To lessen the perception of : DIM. —*vi.* **1.** To become indistinct. **2.** To make blurs. —*n.* **1.** A smear or blot : SMUDGE. **2.** Something hazy and indistinct. **—blurʹri·ness** *n.* **—blurʹry** *adj.*

**blurb** (blûrb) *n.* [Coined by Gelett Burgess (1866–1951).] A brief publicity notice, as on a book jacket.

**blurt** (blûrt) *vt.* **blurt·ed, blurt·ing, blurts.** [Prob. imit.] To utter suddenly and impulsively <*blurted* out the truth>

**blush** (blŭsh) *vi.* **blushed, blush·ing, blush·es.** [ME *blushen* < OE *blyscan*.] **1.** To become suddenly red in the face, as from modesty, embarrassment, or shame : FLUSH. **2.** To become red or rosy. **3.** To feel ashamed or embarrassed. —*n.* **1.** A sudden reddening of the face, as from modesty, embarrassment, or shame. **2.** A red or rosy color. **—at (or on) first blush.** At first glance. **—blushʹful** *adj.* **—blushʹing·ly** *adv.*

**blush·er** (blŭshʹər) *n.* **1.** One that blushes. **2.** Make-up for giving a rosy tint to the face and esp. the cheekbones.

**blus·ter** (blŭsʹtər) *v.* **-tered, -ter·ing, -ters.** [ME *blusteren* < MLG *blüsteren*.] —*vi.* **1.** To blow in loud, violent gusts, as wind in a storm. **2.** To act or speak with noisy threats or boasts. —*vt.* To bully with swaggering threats. —*n.* **1.** A violent gusty wind. **2.** Noisy confusion : TURBULENCE. **3.** Noisily boastful or threatening talk. **—blusʹter·er** *n.* **—blusʹter·y, blusʹter·ous** *adj.*

**bo** (bō) *n., pl.* **bos.** [Prob. short for BOY.] *Slang.* A fellow : pal.

**bo·a** (bōʹə) *n.* [NLat. *Boa*, genus name < Lat. *boa*, a large water snake.] **1.** Any of various large, nonvenomous, chiefly tropical snakes of the family Boidae, including the anaconda, python, boa constrictor, and other snakes that coil around and suffocate their prey. **2.** A long fluffy scarf of soft material such as feathers.

**boa constrictor** *n.* A large nonvenomous snake, *Constrictor constrictor* of tropical America, with brown markings.

**boar** (bôr, bōr) *n.* [ME *bor* < OE *bār*.] **1.** An uncastrated male pig. **2.** A wild pig, *Sus scrofa* of Eurasia and northern Africa, that has dark dense bristles and is the ancestor of the domestic hog. **—boarʹish** *adj.* **—boarʹish·ness** *n.*

**board** (bôrd, bōrd) *n.* [ME *bord* < OE.] **1.** A long flat slab of sawed lumber : PLANK. **2.** A flat piece of rigid material, as wood adapted for a special use. **3.** A flat surface on which a game is played. **4.** The hard cover of a book. **5. boards.** A theater stage. **6. a.** A table, esp. one set for serving food. **b.** Food or meals as a whole <room and *board*> **7.** A table at which official meetings are held. **8.** An organized body of administrators or investigators <a *board* of directors><a *board* of inquiry> **9.** An electrical equipment panel <a circuit *board*> **10.** *Obs.* A border : edge. **11. boards.** The wooden structure surrounding an ice-hockey rink. **12.** *Naut.* **a.** The side of a ship. **b.** A leeboard. **c.** A centerboard. —*v.* **board·ed, board·ing, boards.** —*vt.* **1.** To close or cover with boards. **2. a.** To furnish with meals in return for pay. **b.** To house <*boarded* students at a hostel> **3.** To enter or go aboard (a vehicle or ship). **4.** To come alongside (a ship). **5.** *Obs.* To approach. —*vi.* To receive meals in return for pay. **—across the board. 1.** Designating a bet that a dog or horse will win, place, or show. **2.** Affecting all members, divisions, or categories equally <raises *across the board*> **—go by the board. 1.** To be ruined, unnoticed, or ignored. **2.** To be swept overboard. **—on board.** Aboard.

**board·er** (bôr′dər, bōr′-) *n.* One who pays a stipulated amount for regular meals or for meals and lodging.

**board foot** *n., pl.* **board feet.** A unit of lumber measurement equal to one foot square by one inch thick.

**board game** *n.* A game of strategy played by moving pieces on a board.

**boarding house** *also* **board·ing·house** (bôr′dĭng-hous′, bōr′-) *n.* A house providing lodging and meals.

**boarding school** *n.* A school where pupils are supplied with lodging and meals.

**board measure** *n.* Measurement in board feet.

**board of education** *n.* A school board.

**board of trade** *n.* An association of bankers and business people to promote common commercial interests.

**board rule** *n.* A measuring stick for ascertaining board feet.

**board·walk** (bôrd′wôk′, bōrd′-) *n.* **1.** A walk of wooden planks. **2.** A wooden promenade along a beach or waterfront.

**boast**[1] (bōst) *v.* **boast·ed, boast·ing, boasts.** [ME *bosten* < *bost,* brag.] —*vi.* To speak with excessive pride about one's own accomplishments, abilities, or possessions : BRAG. —*vt.* **1.** To speak about with excessive pride. **2.** To take pride in or be enhanced by the possession of <The city *boasts* a fine new art gallery.> —*n.* **1.** The act or an instance of bragging. **2.** A source of pride. **—boast′er** *n.* **—boast′ful** *adj.* **—boast′ful·ly** *adv.* **—boast′ful·ness** *n.*

☆ ***syns:*** BOAST, BRAG, CROW, GASCONADE, RODOMONTADE, VAUNT *v. core meaning:* to talk with excessive pride <*boasted* about their accomplishments>

**boast**[2] (bōst) *vt.* **boast·ed, boast·ing, boasts.** [Orig. unknown.] To form or shape (stone) roughly with a broad chisel.

**boat** (bōt) *n.* [ME *boot* < OE *bāt.*] **1.** A rather small, usu. open craft. **2.** A ship. **3.** A dish shaped like a boat <a sauce *boat*> —*v.* **boat·ed, boat·ing, boats.** —*vi.* To travel by boat, as for pleasure. —*vt.* **1.** To transport by boat. **2.** To put into a boat. **—in the same boat.** In the same situation or condition.

**boat·bill** (bōt′bĭl′) *n.* A tropical American wading bird, *Cochlearius cochlearius,* with a large boat-shaped bill.

**boat·er** (bō′tər) *n.* **1.** One who boats. **2.** A stiff straw hat with a flat crown.

**boat hook** *n.* A pole with a metal point and hook at one end used esp. for maneuvering logs, rafts, and boats.

**boat·house** (bōt′hous′) *n.* A house for storing boats.

**boat·load** (bōt′lōd′) *n.* The load that a boat can carry.

**boat·man** (bōt′mən) *n.* One who works on, deals with, or operates boats. **—boat′man·ship′** *n.*

**boat·swain** *also* **bo's'n** *or* **bos'n** *or* **bo·sun** (bō′sən) *n.* [ME *botswein* < OE *bātswān.*] A warrant officer or petty officer in charge of a ship's deck crew, rigging, cables, and anchors.

**boat-tailed grackle** (bōt′tāld′) *n.* A bird, *Cassidix mexicanus* of the southern United States and Mexico, with a long tail and glossy black plumage on the male.

**boat train** *n.* A train that regularly takes passengers between a city and a port.

**bob**[1] (bŏb) *v.* **bobbed, bob·bing, bobs.** [ME *bobben.*] —*vt.* **1.** To hit lightly and quickly : TAP. **2.** To cause to move up and down. —*vi.* **1.** To move up and down <heads *bobbing* in agreement> **2.** To grab at hanging or floating objects with the teeth. **3.** To bow or curtsy. **—bob up.** To appear or arise suddenly. —*n.* **1.** A light blow or tap. **2.** A quick, jerky movement of the head or body.

**bob**[2] (bŏb) *n.* [ME *bobbe.*] **1.** A small knoblike pendent object, as a plumb bob. **2.** A fishing float or cork. **3.** A small curl or lock of hair. **4.** A short haircut. **5.** The docked tail of a horse. **6. a.** A bobsled. **b.** A bob skate. —*v.* **bobbed, bob·bing, bobs.** —*vi.* To fish with a bob. —*vt.* To cut short, as hair. **—bob′ber** *n.*

**bob**[3] (bŏb) *n., pl.* **bob.** [Orig. unknown.] *Chiefly Brit.* A shilling.

**bob·bin** (bŏb′ĭn) *n.* [Fr. *bobine.*] **1.** A spool or reel that holds thread or yarn for sewing, spinning, weaving, knitting, or lace making. **2.** Narrow braid used as trimming.

**bob·bi·net** (bŏb′ə-nĕt′) *n.* [BOBBI(N) + NET.] A machine-woven net fabric with hexagonal meshes.

**bobbin lace** *n.* An elaborate handmade lace made by interlacing thread around small notched pins or bobbins stuck into a pillow.

**bob·ble** (bŏb′əl) *v.* **-bled, -bling, -bles.** [Freq. of BOB[1].] —*vi.* To bob up and down. —*vt.* To fumble (e.g., a ball). —*n.* A mistake : blunder.

**bob·by** (bŏb′ē) *n., pl.* **-bies.** [After Sir *Robert* Peel (1788–1850).] *Chiefly Brit.* A police officer.

**bobby pin** *n.* [< BOB[2].] A small usu. metal hair clip with the ends pressed tightly together.

**bobby socks** *also* **bobby sox** *pl.n. Informal.* Ankle socks worn by girls or women.

**bob·by·sox·er** *also* **bobby sox·er** (bŏb′ē-sŏk′sər) *n. Informal.* A teen-age girl of the 1940's who followed current fads.

**bob·cat** (bŏb′kăt′) *n.* [From its short tail.] A wild cat, *Lynx rufus* of North America, with spotted reddish-brown fur, tufted ears, and a short tail.

**bob·o·link** (bŏb′ə-lĭngk′) *n.* [Imit. of its cry.] An American migratory songbird, *Dolichonyx oryzivorus,* of which the male has black, white, and yellowish plumage.

**bob skate** *n.* A skate with two parallel bearing edges.

**bob·sled** (bŏb′slĕd′) *n.* **1.** A long racing sled with a steering mechanism to control the front runners. **2. a.** A long sled made of two shorter sleds joined in tandem. **b.** Either of these two smaller sleds. **—bob′sled′** *v.* **(-sled·ded, -sled·ding, -sleds).**

**bob·stay** (bŏb′stā′) *n.* [Prob. < BOB[1].] A rope or chain for steadying a ship's bowsprit.

**bob·tail** (bŏb′tāl′) *n.* **1.** A short or shortened tail. **2.** An animal, as a horse, having a bobtail. **3.** Something that has been cut short or abbreviated. **—bob′tailed′** *adj.*

**bob·white** (bŏb-hwīt′, -wīt′) *n.* [Imit. of its cry.] A small North American quail, *Colinus virginianus,* having brown plumage with white markings.

**bo·cac·cio** (bə-kä′chō, -chē-ō′) *n., pl.* **-cios.** [Mex. Sp., prob. < Sp. *bocachón,* big mouth, aug. of *boca,* mouth.] A rockfish, *Sebastodes paucispinis* of American Pacific waters.

**boc·cie** *or* **boc·ci** *or* **boc·ce** (bŏch′ē) *n.* [Ital. *bocce,* pl. of *boccia,* ball.] A game originating in Italy that is played with wooden balls on a long narrow clay or dirt court and that resembles bowling.

**bock beer** *n.* [G. *Bockbier,* alteration of *Einbecker Bier,* after *Einbeck,* West Germany.] A dark strong beer, the first that is drawn from the vats in springtime.

**bod** (bŏd) *n. Slang.* BODY 1a.

**bode**[1] (bōd) *vt.* **bod·ed, bod·ing, bodes.** [ME *boden* < OE *bodian,* to announce < *boda,* messenger.] **1.** To be an omen of. **2.** *Archaic.* To predict.

**bode**[2] (bōd) *v. var. p.t. of* BIDE.

**bo·de·ga** (bō-dā′gə) *n.* [Sp. < Lat. *apotheca,* storehouse. —see APOTHECARY.] **1.** A small grocery store, occas. combined with a wineshop. **2.** A warehouse for storing wine.

**bo·dhi·satt·va** (bō′dĭ-sŭt′və) *n.* [Skt. : *bodhiḥ,* perfect knowledge + *sattvam,* reality.] One who, out of compassion, forgoes nirvana for the sake of saving others.

**bod·ice** (bŏd′ĭs) *n.* [Var. of *bodies,* pl. of BODY.] **1.** The fitted part of a dress from the waist to the shoulder. **2.** A woman's laced outer garment, worn like a vest over a blouse. **3.** *Obs.* A corset.

▲ word history: *Bodice* is a specialized use of *bodies,* the plural of *body.* The spelling represents the old pronunciation of the plural with voiceless *s.* A *body* was the part of a woman's dress that covered the torso. A "pair of bodies" was an inner corsetlike garment made of two "bodies" quilted and stiffened to provide support. *Stays,* likewise a plural, was also used for a similar garment. Since a *body* and a *bodice* both covered the torso, the words were used synonymously for a time until *body* died out and *bodice* took its place.

**bod·ied** (bŏd′ēd) *adj.* Having a body of a specified kind <strong-*bodied*><weak-*bodied*>

**bod·i·less** (bŏd′ē-lĭs) *adj.* Having no body, form, or substance.

**bod·i·ly** (bŏd′l-ē) *adj.* **1.** Of, relating to, situated within, or displayed by the body <*bodily* organs> **2.** Physical rather than mental or spiritual. —*adv.* **1.** In the flesh : in person. **2.** As a complete physical entity <carried the child *bodily* from the room>

**bod·ing** (bō′dĭng) *n.* A foreboding, esp. of evil.

**bod·kin** (bŏd′kĭn) *n.* [ME *boidekyn.*] **1.** A small, sharply pointed instrument for making holes in leather or fabric. **2.** A blunt needle for pulling ribbon or tape through a series of loops or a hem. **3.** A long hairpin, usu. with an ornamental head. **4.** An awl or pick for extracting letters from set printing type. **5.** *Archaic.* A dagger or stiletto.

**bod·y** (bŏd′ē) *n., pl.* **-ies.** [ME < OE *bodig.*] **1. a.** The entire material structure and substance of an organism, esp. of a human being or an animal. **b.** The physical part of a person as opposed to the mind or spirit. **c.** A corpse or carcass. **2. a.** The torso of a human being or animal. **b.** The part of a garment covering the torso. **3.** *Law.* **a.** A person. **b.** A group of individuals considered to be an entity : CORPORATION. **4.** A number of persons, concepts, or things regarded as a whole : GROUP <a legislative *body*> **5.** The main or central part, as: **a.** A church nave. **b.** The content of a book or document exclusive of prefatory matter, codicils, indexes, or appendices. **c.** The passenger- and cargo-carrying section of an aircraft, ship, or vehicle. **d.** The sound box of a musical instrument. **6.** A bounded aggregate of matter

ă **pat** ā **pay** âr **care** ä **father** ĕ **pet** ē **be** hw **which** ĭ **pit**
ī **tie** îr **pier** ŏ **pot** ō **toe** ô **paw, for** oi **noise** ŏŏ **took**

<a *body* of water> **7.** Consistency of substance <a Chablis with *body*><hair with *body*> **8.** The part of a block of printing type underlying the impression surface. —*vt.* **-ied, -y·ing, -ies. 1.** To furnish with a body. **2.** To give shape to.

☆ ***syns:*** BODY, BULK, CORPUS, SUBSTANCE *n. core meaning* : the main part <the *body* of the thesis>

**body bag** *n.* A zippered usu. rubber bag for transporting a human corpse.

**body building** *n.* The art or practice of developing the body through exercise and diet, esp. for competitive exhibition. **—body builder** *n.*

**body cavity** *n.* The coelom.

**body corporate** *n. Law.* CORPORATION 1a.

**body count** *n.* A count of individual bodies, as those killed in military operations.

**body English** *n.* **1.** The natural or instinctive tendency of a person to try to influence the movement of a propelled object, as a ball, by twisting his or her body toward the desired goal. **2.** The usu. irregular movement or spin of a propelled object as if it were influenced by body English.

**bod·y·guard** (bŏd'ē-gärd') *n.* A usu. armed person or group of persons responsible for the protection of one or more other persons.

**body language** *n.* The postures, gestures, and facial expressions by which one communicates nonverbally with others.

**body louse** *n.* A parasitic louse, *Pediculus humanus*, infesting human beings and their clothing.

**body politic** *n.* The people as a group of a politically organized unit, as a nation or state.

**body shirt** *n.* **1.** A woman's top made with a sewn-in or snapped crotch. **2.** A tight-fitting blouse or shirt.

**body shop** *n.* A shop for the repair of automotive bodies.

**body snatcher** *n.* One who steals corpses from graves.

**body stocking** *n.* A tight-fitting usu. one-piece garment covering the torso and occas. having sleeves and legs.

**body suit** *n.* A tight-fitting one-piece garment for the torso.

**bod·y·surf** (bŏd'ē-sûrf') *vi.* **-surfed, -surf·ing, -surfs.** To ride flat on a wave without a surfboard. **—bod'y·surf'er** *n.*

**body wall** *n.* The external animal body surface composed of ectoderm and mesoderm, which encloses the body cavity.

**body work** *n.* The process or act of repairing automotive bodies.

**boehm·ite** (bā'mīt', bō'-) *n.* [G. *Böhmit*, after J. *Böhm*, 20th-cent. German scientist.] A white to dark reddish brown orthorhombic mineral, AlO(OH), found in bauxite.

**Boer** (bōr, bôr, bo͝or) *n.* [Du., farmer < MDu. *gheboer.*] A Dutch colonist or descendant of a Dutch colonist in South Africa.

**bof·fo** (bŏf'ō) *adj.* [Short for slang *boffola*, hit, success.] *Slang.* Very successful : EXCELLENT.

**Bo·fors gun** (bō'fôrz', bo͞o'-) *n.* [After *Bofors*, Sweden.] An automatic, double-barreled antiaircraft gun.

**bog** (bôg, bŏg) *n.* [Ir. Gael. *bogach* < *bog*, soft.] Soft, waterlogged ground : MARSH. —*v.* **bogged, bog·ging, bogs.** —*vt.* To cause to sink in or as if in a bog <*bogged* down in a mass of regulations> —*vi.* To be hindered and slowed. **—bog'gy** *adj.*

**bog asphodel** *n.* Either of two related bog plants, *Narthecium americanum* of the southeastern United States or *N. ossifragum* of Europe, with a yellow flower cluster.

**bo·gey** (bō'gē) *n., pl.* **-geys. 1.** (*also* bo͝og'ē, bo͞o'gē). *var. of* BOGY[1]. **2. a.** An estimated standard golf score. **b.** One golf stroke over par on a hole. **3.** *Slang.* An unidentified flying aircraft.

**bog·ey·man** (bo͝og'ē-măn', bō'gē-, bo͞o'gē-) *n. var. of* BOOGIEMAN.

**bog·gle** (bŏg'əl) *v.* **-gled, -gling, -gles.** [Prob. < *boggle*, dial. var. of BOGLE.] —*vi.* **1.** To hesitate or evade as if in fear or doubt. **2.** To shy away with fright or astonishment. **3.** To botch or bungle. —*vt.* To cause to be overcome, as with astonishment or fright. **—bog'gler** *n.*

**bo·gie[1]** *also* **bo·gy** (bō'gē) *n., pl.* **-gies.** [Orig. unknown.] **1.** A railroad car or locomotive undercarriage with two, four, or six wheels that swivels so curves can be negotiated. **2.** Any of several wheels or supporting and aligning rollers inside the tread of a tractor or tank.

**bo·gie[2]** (bō'gē, bo͝og'ē, bo͞o'gē) *n. var. of* BOGY[1].

**bo·gle** (bō'gəl) *n.* [Sc. *bogill.*] A hobgoblin : bogy.

**bog rosemary** *n.* A low-growing evergreen shrub, *Andromeda glaucophylla* of northern regions, growing in wet ground and bearing small pink flowers.

**bog·trot·ter** (bôg'trŏt'ər, bŏg'-) *n.* One who lives in or frequents bogs.

**bo·gus** (bō'gəs) *adj.* [< E. *bogus*, a device for making counterfeit money.] Counterfeit : fake <*bogus* $20 bills>

**bog·wood** (bôg'wo͝od', bŏg'-) *n.* Wood preserved in a peat bog.

**bo·gy[1]** *also* **bo·gey** *or* **bo·gie** (bō'gē, bo͝og'ē, bo͞o'gē) *n., pl.* **-gies** *also* **-geys** *or* **-gies.** [Orig. unknown.] **1.** An evil or mischievous spirit : HOBGOBLIN. **2.** Something causing annoyance or harassment.

**bo·gy[2]** (bō'gē) *n. var. of* BOGIE[1].

**bo·hea** (bō-hē') *n.* [Chin. (Fujian) *bu-i*, after *Wu-i Shan*, a range of hills on the border of Jiangxi and Fujan Provinces.] A black tea produced in China.

**bo·he·mi·a** *also* **Bo·he·mi·a** (bō-hē'mē-ə) *n.* [Back-formation < BOHEMIAN.] **1.** A community of persons with artistic or literary tastes who adopt manners and mores markedly different from those expected or approved of by the majority of society. **2.** The district in which bohemians live.

**Bo·he·mi·an** (bō-hē'mē-ən) *n.* **1.** A native or inhabitant of Bohemia. **2.** A Gypsy. **3.** The Czech dialects of Bohemia. **4. bohemian.** One with artistic or literary interests who disregards conventional standards of behavior. **—Bo·he'mi·an** *adj.*

▲ word history: The nomadic people called *Bohemians* who roamed Europe from late medieval times had in fact come from the borderlands of India and Iran, not from Bohemia. The Europeans gave them such names as *Bohemian* and *Gypsy* (from *Egyptian*) because they did not know where their original homelands had been. The French word *bohémien* was also used of "social gypsies"—artists and writers who led unconventional and irregular lives, abandoning their own class of society for a much lower one. Thackeray introduced this sense of the word into English in *Vanity Fair.* The noun *bohemia*, meaning the artistic community or its dwelling place, is derived from *bohemian* and is not directly from the name of the district in Czechoslovakia.

**Bohr theory** (bôr, bōr) *n.* [After Niels *Bohr* (1885–1962).] A model of atomic structure in which electrons travel around the nucleus in orbits determined by quantum conditions.

**boil[1]** (boil) *v.* **boiled, boil·ing, boils.** [ME *boillen* < OFr. *boilir* < Lat. *bullire.*] —*vi.* **1. a.** To vaporize a liquid by application of heat. **b.** To reach the boiling point. **c.** To undergo boiling. **2.** To be agitated like boiling water : CHURN <water *boiling* over the rocks> **3.** To be stirred up or greatly excited <Their impertinence made me *boil.*> —*vt.* **1.** To heat to the boiling point. **2.** To cook or clean by boiling. **3.** To form or separate, as sugar, by the process of boiling. **—boil away.** To evaporate by boiling. **—boil down. 1.** To reduce in bulk or size by boiling. **2.** To condense or be condensed : REDUCE. **—boil over. 1.** To overflow while boiling. **2.** To lose one's temper or explode in passsion. **—boil** *n.*

☆ ***syns:*** **1.** BOIL, PARBOIL, SIMMER, STEW *v. core meaning* : to cook (food) in liquid heated to the point of steaming <*boil* potatoes> **2.** BOIL, BUBBLE, BURN, CHURN, FERMENT, MOIL, SEETHE, SIMMER, SMOLDER *v. core meaning* : to be in a state of emotional turmoil <was silently *boiling* over the delay>

**boil[2]** (boil) *n.* [ME *bile* < OE *bȳl.*] A painful, localized pus-filled swelling of the skin and subcutaneous tissue resulting from bacterial infection.

**boil·er** (boi'lər) *n.* **1.** An enclosed vessel in which water is heated and circulated, either as hot water or steam, for heating or power. **2.** A container, as a kettle, for boiling liquids. **3.** A storage tank for hot water.

**boil·er·mak·er** (boi'lər-mā'kər) *n.* **1.** A person who makes or repairs boilers. **2.** *Slang.* A drink of whiskey with a beer chaser.

**boil·er·plate** (boi'lər-plāt') *n.* **1.** A steel plate used to make the shells of steam boilers. **2.** Journalistic material, as syndicated features, available in plate or mat form.

**boil·er-room** (boi'lər-ro͞om', -ro͝om') *adj. Informal.* Of, relating to, or involving usu. illegal, high-pressure telephone sales tactics.

**boiling point** *n.* **1.** The temperature at which a liquid boils, esp. under standard atmospheric conditions. **2.** *Informal.* **a.** The point at which one loses one's temper. **b.** A point of crisis : TURNING POINT.

**boil·off** (boil'ôf', -ŏf') *n.* Vaporization of liquid, as a rocket fuel.

**bois·ter·ous** (boi'stər-əs, -strəs) *adj.* [ME *boistres*, var. of *boistous*, rude.] **1.** Rough and stormy : VIOLENT. **2.** Loud, noisy, unrestrained, and undisciplined. **—bois'ter·ous·ly** *adv.* **—bois'ter·ous·ness** *n.*

**bok choy** (bŏk choi') *n. var. of* PAK CHOI.

**Bok·mål** (bo͝ok'môl', bōk'-) *n.* [Norw. : *bok*, book + *mål*, language.] Riksmål.

**bo·la** (bō'lə) *also* **bo·las** (-ləs) *n.* [Sp. (South America) *bolas*, pl. of Sp. *bola*, ball.] A rope with weights attached used esp. in South America to catch cattle or game by entangling the legs.

**bola tie** *n. var. of* BOLO TIE.

**bold** (bōld) *adj.* **-er, -est.** [ME < OE *bald.*] **1.** Fearless and daring : COURAGEOUS. **2.** Requiring or showing courage and bravery. **3.** Unduly forward : BRAZEN. **4.** Clear and distinct to the eye : CONSPICUOUS. **5.** Steep, as a cliff : ABRUPT. **6.** Designating boldface type. **—bold'ly** *adv.* **—bold'ness** *n.*

**bold·face** (bōld'fās') *n.* Type with thick, heavy lines. —*vt.* **-faced, -fac·ing, -fac·es. 1.** To mark (copy) for printing in boldface. **2.** To print or set in boldface. **—bold'faced'** *adj.*

**bold-faced** (bōld'fāst') *adj.* Impudent : brazen.

**bole** (bōl) *n.* [ME < ON *bolr.*] A tree trunk.

**bo·le·ro** (bō-lâr'ō, bə-) *n., pl.* **-ros.** [Sp.] **1.** A short jacket worn open in the front. **2. a.** A Spanish dance in triple meter. **b.** The music for a bolero.

**bo·le·tus** (bō-lē'təs) *n., pl.* **-tus·es** *or* **-ti** (-tī') [NLat. *Boletus*, genus name < Lat. *boletus*, mushroom.] A fungus of the genus *Boletus*, with an umbrella-shaped cap and spore-bearing tubules on the underside, of which some species are poisonous and others edible.

**bo·li·var** (bō-lē′vär′, bŏl′ə-vər) *n., pl.* **bo·li·vars** *or* **bo·li·var·es** (bō′lē-vä′räs′) [After Simón *Bolívar* (1783–1830).] —See table at CURRENCY.
**boll** (bōl) *n.* [ME.] The rounded seed pod or capsule of a plant, as flax or cotton.
**bol·lard** (bŏl′ərd) *n.* [ME, prob. < *bole,* tree trunk < ON *bolr.*] A thick post on a ship or wharf for securing ropes and hawsers.
**bol·lix** (bŏl′ĭks) *vt.* **-lixed, -lix·ing, -lix·es.** [Alteration of *ballocks,* testicles < ME *balloks* < OE *beallucas.*] *Slang.* To throw into confusion : BOTCH <managed to *bollix* the whole project>
**boll weevil** *n.* **1.** A small, grayish, long-snouted beetle, *Anthonomus grandis* of Mexico and the southern United States, with destructive larvae that damage cotton bolls. **2.** *Informal.* A conservative Southern Democrat in the U.S. House of Representatives.
**boll·worm** (bōl′wûrm′) *n.* **1.** The larva of a moth, *Pectinophora gossypiella,* that is damaging to cotton. **2.** The corn earworm.
**bo·lo** (bō′lō) *n., pl.* **-los.** [Sp.] A long, heavy, single-edged machete orig. used in the Philippines.
**bo·lo·gna** (bə-lō′nē, -nə, -nyə) *also* **ba·lo·ney** *or* **bo·lo·ney** (-nē) *n.* [After *Bologna,* Italy.] A seasoned smoked sausage made of mixed meats.
**bo·lom·e·ter** (bō-lŏm′ĭ-tər) *n.* [Gk. *bolē,* ray + -METER.] An instrument for measuring radiant heat by correlating the radiation-induced change in electrical resistance of a blackened metal foil with the amount of radiation that is absorbed. **—bo′lo·met′ric** (bō′lə-mĕt′rĭk) *adj.*
**bo·lo·ney** (bə-lō′nē) *n. var. of* BOLOGNA.
**bolo tie** *also* **bola tie** *n.* [Alteration of BOLA + TIE.] A necktie made a piece of cord fastened with an ornamental clasp or bar.
**Bol·she·vik** (bōl′shə-vĭk′, bŏl′-) *n., pl.* **-viks** *or* **-vi·ki** (-vē′kē) [R. *Bol′shevik* < *bol′shii,* comp. of *bol′shoi,* large.] **1. a.** A participant in the Russian Revolution who belonged to the Communist Party of the Soviet Union. **b.** A member of the left-wing majority group of the Russian Social Democratic Party who adopted Lenin's theses on party organization (1903). **2.** *often* **bolshevik.** An extreme radical.
**Bol·she·vism** *also* **bol·she·vism** (bōl′shə-vĭz′əm, bŏl′-) *n.* **1.** The strategy developed by the Bolsheviks between 1903 and 1917 with a view to seizing state power and establishing the dictatorship of the proletariat. **2.** Soviet Communism.
**Bol·she·vist** *also* **bol·she·vist** (bōl′shə-vĭst, bŏl′-) *n.* A Bolshevik.
**bol·ster** (bōl′stər) *n.* [ME < OE.] A long, narrow cushion or pillow. —*vt.* **-stered, -ster·ing, -sters. 1.** To support or prop up with or as if with a bolster. **2.** To buoy up <the compliments that *bolstered* our morale> **—bol′ster·er** *n.*
**bolt¹** (bōlt) *n.* [ME < OE, heavy arrow.] **1.** A wooden or metal bar that slides into a socket and is used to fasten doors and gates. **2.** A metal bar or rod in a lock mechanism that is thrown or withdrawn by turning the key. **3.** A fastener having a threaded pin or rod with a head at one end, designed to be inserted through holes in assembled parts and secured by a mated nut that is tightened by application of torque. **4. a.** A sliding metal bar that positions the cartridge in breechloading rifles, closes the breech, and ejects the spent cartridge. **b.** A similar device in any other breech mechanism. **5.** A short, heavy arrow with a thick head, used esp. with a crossbow. **6.** A flash of lightning : THUNDERBOLT. **7.** A sudden or unexpected event. **8.** A sudden movement to or away. **9.** A large roll of cloth <bought a *bolt* of calico> —*v.* **bolt·ed, bolt·ing, bolts.** —*vt.* **1.** To secure or lock with or as if with a bolt. **2.** *Archaic.* To shoot or discharge (a projectile). **3.** To arrange or roll (e.g., lengths of cloth) on a bolt. **4.** To eat hurriedly and with little chewing : GULP. **5.** To desert or withdraw support from (a political party). **6.** To utter impulsively : blurt out. —*vi.* **1.** To move or spring suddenly. **2.** To break from a rider's control and run away, as a horse. **3.** To run off suddenly. **4.** To break away from a political party or its policies. **5.** To flower or produce seeds prematurely. **—bolt from the blue.** A sudden, often shocking surprise.
**bolt²** (bōlt) *vt.* **bolt·ed, bolt·ing, bolts.** [ME *bulten* < OFr. *buleter,* of Germanic orig.] To pass through a sieve : SIFT.
**bolt·er¹** (bōl′tər) *n.* **1.** A horse given to bolting. **2.** A person who gives up membership in or withdraws support from a political party.
**bolt·er²** (bōl′tər) *n.* **1.** A machine used for sifting, esp. for sifting flour. **2.** A person who operates a bolter.
**bol·to·ni·a** (bōl-tō′nē-ə) *n.* [NLat., genus name after James Bolton (d. 1799).] One of several North American plants of the genus *Boltonia,* bearing daisylike flowers with white, violet, or pinkish rays.
**bolt·rope** (bōlt′rōp′) *n.* A rope sewn into the outer edge of a sail to prevent the sail from tearing.
**bo·lus** (bō′ləs) *n., pl.* **-lus·es.** [Med. Lat. *bolus* < Gk. *bōlos,* lump of earth.] **1.** A small round mass. **2.** A large pill or tablet. **3.** A soft mass of chewed food.
**bomb** (bŏm) *n.* [Fr. *bombe* < Ital. *bomba,* prob. < Lat. *bombus,* a booming sound < Gk. *bombos.*] **1. a.** An explosive weapon detonated by a predetermined means, as impact, proximity to an object, or a timing mechanism. **b.** An atomic bomb. **c. the bomb.** Nuclear weapons in general. **2.** A weapon detonated to release destructive material, as smoke or gas. **3.** *Football.* A long forward pass designed to achieve much yardage in a single play. **4. a.** A container capable of withstanding high internal pressure. **b.** A vessel for storing compressed gas. **c.** A portable, manually operated container that ejects a spray, foam, or gas under pressure. **5.** *Slang.* A dismal failure : FIASCO. **6.** A mass of lava exploded from a volcano. **7.** *Chiefly Brit.* A large amount of money. **8.** *Chiefly Brit.* An old car. —*v.* **bombed, bomb·ing, bombs.** —*vt.* To attack, damage, or destroy with or as if with bombs. —*vi.* **1.** To drop bombs. **2.** *Slang.* To fail miserably <The new play *bombed.*>
**bom·bard** (bŏm′bärd′) *n.* [ME < OFr. *bombarde* < Med. Lat. *bombarda* prob. < Lat. *bombus,* a booming sound < Gk. *bombos.*] An early cannon that fired stone balls. —*vt.* (bŏm-bärd′) **-bard·ed, -bard·ing, -bards. 1.** To attack with bombs, shells, or missiles. **2.** To attack persistently, as with questions or insults. **3.** To irradiate (an atom). **4.** To attack with a bombard. **—bom·bard′er** *n.* **—bom·bard′ment** *n.*
**bom·bar·dier** (bŏm′bər-dîr′) *n.* [Fr. < OFr. *bombarde,* bombard.] **1.** An aircraft crew member who operates the bombing equipment. **2.** A noncommissioned British artillery officer. **3.** A soldier who operated a bombard.
**bombardier beetle** *n.* Any of various beetles of the genus *Brachinus* and related genera that expel an acrid secretion from the posterior end of the abdomen.
**bom·bar·don** (bŏm′bər-dōn′, bŏm-bär′dn) *n.* [Fr. < Ital. *bombardone,* aug. of *bombardo,* alteration of *bombarda,* bombard < Med. Lat.] **1.** A brass musical instrument resembling a tuba but with a lower pitch : bass or contrabass tuba. **2.** A 16-foot reed stop on an organ.
**bom·bast** (bŏm′băst′) *n.* [OFr. *bombace,* cotton padding < LLat. *bombax,* cotton < Lat. *bombyx,* silk < Gk. *bombux,* silkworm.] Flamboyant, pompous writing or speech. **—bom′bast′er** *n.* **—bom·bas′tic** (-băs′tĭk) *adj.* **—bom·bas′ti·cal·ly** *adv.*
**Bom·bay duck** (bŏm-bā′) *n.* **1.** A food fish, *Harpodon nehereus* of India. **2.** The dried flesh of the Bombay duck or a dish or relish prepared from it.
**bom·ba·zine** (bŏm′bə-zēn′) *n.* [Fr. *bombasin* < LLat. *bombacinum,* silken texture < Lat. *bombycinum* < *bombyx,* silk < Gk. *bombux,* silkworm.] A fine twilled fabric of silk and worsted or cotton, often dyed black.
**bomb bay** *n.* The compartment in the fuselage of a military aircraft from which bombs are dropped.
**bombe** (bŏm, bôNb) *n.* [Fr.] A dessert consisting of two or more layers of ice cream of different flavors frozen in a round or melon-shaped mold.
**bombed** (bŏmd) *adj. Slang.* Drunk.
**bomb·er** (bŏm′ər) *n.* One that bombs, esp. a military aircraft designed to carry and drop bombs.
**bomb·proof** (bŏm′pro͞of′) *adj.* Designed and constructed to resist destruction by bombs.
**bomb·shell** (bŏm′shĕl′) *n.* **1.** BOMB 1a, 2. **2.** A sudden, usu. shocking surprise.
**bomb·sight** (bŏm′sīt′) *n.* A device in an aircraft for aiming bombs.
**bom·by·cid** (bŏm′bĭ-sĭd) *n.* [< NLat. *Bombycidae,* family name < Lat. *bombyx,* silkworm < Gk. *bombux.*] A moth of the family Bombycidate, which includes the silkworms.
**bo·na fide** (bō′nə fīd′, fī′dē, bŏn′ə) *adj.* [Lat., in good faith.] **1.** Performed or made in good faith : SINCERE <a *bona fide* verbal agreement> **2.** Authentic : genuine <a *bona fide* signature>
**bo·na fides** (bō′nə fīdz′, fī′dēs, bŏn′ə) *pl.n.* Authentic credentials <a defector whose *bona fides* could not be checked>
**bo·nan·za** (bə-năn′zə) *n.* [Sp., exaggerated good, aug. of *bueno,* good < Lat. *bonus.*] **1.** A rich mine, vein, or pocket of ore. **2.** A source of great wealth or prosperity.
**Bo·na·part·ist** (bō′nə-pär′tĭst) *n.* A follower or supporter of the French emperors Napoleon I, Napoleon III, or their dynasty. **—Bo′na·part′ism** *n.*
**bon·bon** (bŏn′bŏn′) *n.* [Fr., redup. of *bon,* good < Lat. *bonus.*] A candy coated with fondant or chocolate and often with a fruit, nut, or fondant center.
**bon·bon·nière** (bŏn′bŏn-yâr′) *n.* [Fr. < *bonbon,* bonbon.] **1.** A small ornate candy dish or box. **2.** A confectioner's store.
**bond** (bŏnd) *n.* [ME < ON *band.*] **1.** Something, as a fetter, cord, or band, that binds, ties, or fastens together. **2.** *often* **bonds.** *Archaic.* Captivity : confinement. **3.** A uniting tie or force : LINK <*bonds* of friendship> **4.** A binding agreement : COVENANT. **5.** A duty, promise, or obligation by which one is bound. **6. a.** An agent that causes two or more objects or parts to cohere <a thermoplastic *bond*> **b.** Cohesion or union brought about by such an agent. **7.** A chemical bond. **8.** *Law.* **a.** A written and sealed obligation, esp. one requiring payment of a stipulated amount of money on or before a given day. **b.** A sum of money paid as bail or surety. **c.** One who acts as bail : BONDSMAN. **9.** A certificate of debt issued by a government or corporation guaranteeing payment of the original investment plus interest by a specified future date. **10.** The condition of warehousing taxable goods until the duties or taxes due on them are paid. **11.** An insur-

ance contract in which an agency guarantees payment to an employer in the event of unforeseen financial loss through the actions of an employee. **12.** An overlapping arrangement of masonry components in a wall or on a flat surface, as a walkway. **13.** Bond paper. —*v.* **bond·ed, bond·ing, bonds.** —*vt.* **1.** To mortgage or place a guaranteed bond on. **2.** To furnish bond or surety for. **3.** To place (e.g., an employee) under bond or guarantee. **4.** To join securely, as with cement or glue. **5.** To lay (e.g., bricks) in an overlapping pattern. —*vi.* To hold together or secure with or as if with a bond. **—bond'a·ble** *adj.* **—bond'er** *n.*

**bond·age** (bŏn'dĭj) *n.* [ME < AN < ME *bonde*, serf < OE *bōnda*, husbandman < ON *bondi*.] **1.** The condition of a slave or serf : SERVITUDE. **2.** Subjection to a power, force, or influence. **3.** Villeinage.

☆ ***syns:*** BONDAGE, SERVITUDE, SLAVERY *n. core meaning* : the condition of being involuntarily under the power of another <the Israelites who toiled in Egyptian *bondage*> BONDAGE can also be used figuratively <addicts held in *bondage* to heroin> SERVITUDE stresses subjection or submission to a master <involuntary *servitude* in Siberia> SLAVERY implies being owned as a possession or being treated as property <the abolition of *slavery* after the Civil War> ***ant:*** freedom, liberty

**bond·hold·er** (bŏnd'hōl'dər) *n.* The owner of a bond or bonds.

**bond·ing** (bŏn'dĭng) *n.* Formation of close, specialized human relationships, as those that link parent and child.

**bond·maid** (bŏnd'mād') *n.* A woman who is a bondservant.

**bond·man** (bŏnd'mən) *n.* A man who is a bondservant.

**bond paper** *n.* A superior grade of strong paper used esp. in business correspondence.

**bond·ser·vant** (bŏnd'sûr'vənt) *n.* One obligated to work without wages : SLAVE.

**bonds·man** (bŏndz'mən) *n.* **1.** A bondman. **2.** One who provides bond or surety for another.

**bond·wom·an** (bŏnd'wo͝om'ən) *n.* A bondmaid.

**bone** (bōn) *n.* [ME *bon* < OE *bān*.] **1. a.** The dense, semirigid, porous, calcified connective tissue of the skeleton of most vertebrates. **b.** One of numerous anatomically distinct skeletal structures made of bone. **c.** A piece of bone. **2. bones. a.** The skeleton. **b.** The body. **c.** Mortal remains. **3.** Material, as ivory, resembling bone. **4.** Something made of bone or similar material, esp.: **a.** A piece of material, as whalebone, used as a corset stay. **b. bones.** *Informal.* Dice. **5. bones.** The basic design or plan <the bare *bones* of the plot> **6. a. bones.** Flat clappers made of bone or wood used by the end person in a minstrel show. **b. Bones** *(sing. in number).* The end person in a minstrel show. **7.** Something given to soothe or placate <a Thanksgiving turkey as a *bone* to unhappy workers> —*v.* **boned, bon·ing, bones.** —*vt.* **1.** To remove the bones from. **2.** To stiffen (clothing) with stays, as of whalebone. —*vi. Slang.* To study intensely, usu. at the last minute <*boned* up on physics for the test> **—bone of contention.** The subject of an argument. **—feel in (one's) bones.** To have an intuition of. **—have a bone to pick.** To have grounds for a complaint. **—make no bones about. 1.** To be straightforward about. **2.** To offer no argument or objection.

**bone ash** *n.* The white, powdery calcium phosphate ash of burned bones, used as a fertilizer, in making ceramics, and in cleaning and polishing compounds.

**bone·black** *also* **bone black** (bōn'blăk') *n.* A black pigment containing approx. 10% charcoal, made by roasting bones in an airtight container and used in polishes, as a filtering medium, and in decolorizing sugar.

**bone charcoal** *n.* Boneblack.

**bone china** *n.* White, translucent china made from clay mixed with bone ash.

**bone conduction** *n.* Transmission of sound by bone, esp. to the inner ear by the bones of the skull.

**bone-dry** (bōn'drī') *adj.* **1.** Without a trace of moisture. **2.** Opposed to or prohibiting the use or sale of alcoholic beverages.

**bone·fish** (bōn'fĭsh') *n., pl.* **bonefish** *or* **-fish·es.** A warm water, marine game fish, *Albula vulpes*, with silvery scales.

**bone·head** (bōn'hĕd') *n. Slang.* A stupid person : DUNCE. **—bone'head'ed** *adj.* **—bone'head'ed·ness** *n.*

**bone meal** *n.* Crushed or ground bones used as plant fertilizer and animal feed.

**bon·er** (bō'nər) *n. Slang.* A blunder.

**bone·set** (bōn'sĕt') *n.* [From its use as a folk medicine.] Any of various plants of the genus *Eupatorium*, esp. *E. perfoliatum* of eastern North America, with broad white flower clusters.

**bon·fire** (bŏn'fīr') *n.* [ME *bonfir* : *bon*, bone + *fir*, fire.] A large outdoor fire.

**bong** (bŏng, bông) *n.* [Imit.] A deep ringing sound, as of a bell. —*vt.* & *vi.* **bonged, bong·ing, bongs.** To sound with a bong : RING.

**bon·go**[1] (bŏng'gō, bông'-) *n., pl.* **-gos.** [Of Bantu orig.] An antelope, *Boocercus eurycerus* of central Africa, having a reddish-brown coat with white stripes and spirally twisted horns.

**bongo**[1]
*4½ feet high at shoulder*

**bon·go**[2] (bŏng'gō, bông'-) *n., pl.* **-gos** *or* **-goes.** [Am. Sp. (West Indies) *bongó*.] One of a pair of connected tuned drums played with the hands.

**bon·ho·mie** (bŏn'ə-mē') *n.* [Fr. < *bonhomme*, good-natured man : *bon*, good (< Lat. *bonus*) + *homme*, man < Lat. *homo*.] A pleasant, affable disposition : GENIALITY.

**bon·i·face** (bŏn'ə-fəs, -fās') *n.* [After *Boniface*, an innkeeper in *The Beaux' Strategem* by George Farquhar (1678–1707).] An innkeeper.

**boning knife** *n.* A narrow-bladed, sharp-pointed knife for boning poultry, meat, and fish.

**bo·ni·to** (bə-nē'tō) *n., pl.* **bonito** *or* **-tos.** [Sp. < *bonito*, pretty < Lat. *bonus*, good.] **1.** One of several marine food and game fishes of the genus *Sarda*, related to the tuna. **2.** One of several fishes similar to the bonito.

**bon·kers** (bŏng'kərz) *adj.* [Orig. unknown.] *Slang.* Crazy <That loud music is driving me *bonkers*.>

**bon mot** (bôN mō') *n., pl.* **bons mots** (bôN mō', mōz') [Fr. : *bon*, good + *mot*, word.] A clever saying : WITTICISM.

**bon·net** (bŏn'ĭt) *n.* [ME *bonet*, cap < OFr.] **1. a.** A hat secured on the head by ribbons tied under the chin, worn by women and girls. **b.** *Scot.* A brimless cap worn by men. **c.** A feather headdress worn by some American Indians. **2.** A removable metal plate over a machine part, as a valve. **3.** *Chiefly Brit.* HOOD[1] 4c. **4.** A wind screen for a chimney. **5.** A strip of canvas laced to a fore-and-aft sail to increase sail area. —*vt.* **-net·ed, -net·ing, -nets.** To put a bonnet on.

**bon·ny** *also* **bon·nie** (bŏn'ē) *adj.* **-ni·er, -ni·est.** [Orig. unknown.] *Chiefly Scot.* Pleasing or attractive : PRETTY. **—bon'ni·ly** *adv.* **—bon'ni·ness** *n.*

**bon·ny·clab·ber** (bŏn'ē-klăb'ər) *n.* [Ir. Gael. *bainne*, milk + *claba*, thick.] *Chiefly Brit.* Sour clotted milk.

**bon·sai** (bŏn-sī') *n., pl.* **bonsai.** [J., potted plant : *bon*, basin (< Chin. *pen*[2]) + *sai*, to plant (< Chin. *zai*[1]).] **1.** The craft of growing dwarfed, ornamentally shaped shrubs or trees in small, shallow pots. **2.** A tree or shrub grown by bonsai.

**bon·spiel** (bŏn'spēl') *n.* [Prob. of Du. orig.] *Scot.* A tournament or match in curling.

**bon·te·bok** (bŏn'tə-bŏk') *n.* [Afr. : *bont*, spotted (< MDu.) + *bok*, buck < MDu. *boc*).] A nearly extinct South African antelope, *Damaliscus pygargus*, with a reddish coat, a white rump, and a white mark on the face.

**bon ton** (bŏn tŏn') *n.* [Fr. : *bon*, good + *ton*, tone.] **1.** Sophisticated style or manner. **2.** Fashionable society.

**bo·nus** (bō'nəs) *n., pl.* **-nus·es.** [< Lat., good.] **1.** Something given or paid in addition to the usual or expected. **2.** Money paid to a state by a company in return for a corporate charter. **3.** A grant from a government to military veterans. **4.** An extra dividend paid to stockholders of a company from profits. **5.** A premium paid for a loan.

**bon vi·vant** (bôN' vē-väN') *n., pl.* **bons vi·vants** (bôN' vē-väN') [Fr. : *bon*, good + *vivant*, pr.part. of *vivre*, to live.] One who has refined tastes, esp. one who enjoys good food and drink.

**bon voy·age** (bôN' vwä-yäzh') *interj.* [Fr. : *bon*, good + *voyage*, journey.] —Used to wish a departing traveler a pleasant journey.

**bon·y** (bō'nē) *adj.* **-i·er, -i·est. 1.** Of, relating to, like, or composed of bone. **2.** Having an internal skeleton of bones. **3.** Full of bones. **4.** Having prominent bones : GAUNT. **—bon'i·ness** *n.*

**bonze** (bŏnz) *n.* [Fr. < Port. *bonzo* < J. *bonsō*, of Chin. orig.] A Buddhist monk.

**bon·zer** (bŏn'zər) *adj.* [Poss. alteration of BONANZA.] *Austral.* Very good : EXCELLENT.

**boo**[1] (bo͞o) *n., pl.* **boos.** [Imit.] A vocal sound uttered to show contempt, scorn, or disapproval. —*interj.* —Used to frighten or surprise or to express contempt, scorn, or disapproval. —*v.* **booed, boo·ing, boos.** —*vi.* To utter a boo. —*vt.* To express contempt, scorn, or disapproval of by booing.

**boo**[2] (bo͞o) *n.* [Orig. unknown.] *Slang.* Marijuana.

**boob** (bo͞ob) *n.* [Short for BOOBY.] *Slang.* A foolish, stupid person : DOLT.

**boob·oi·sie** (bo͞ob'wä-zē') *n.* [BOOB[1] + (BOURGE)OISIE.] A class of people made up of the gullible and stupid.

**boo-boo** (bo͞o'bo͞o) *n., pl.* **-boos.** [Perh. alteration of *boohoo*, to weep noisily.] *Slang.* **1.** A blunder. **2.** A slight physical injury.

**boob tube** *n. Slang.* Television.

**boo·by** (bo͞o'bē) *n., pl.* **-bies.** [Prob. Sp. *bobo* < Lat. *balbus*, stammering.] *Slang.* **1.** A stupid person : DUNCE. **2.** One of several tropical sea birds of the genus *Sula*, related to the gannets.

**booby hatch** *n.* **1.** *Naut.* A raised covering over a small hatchway. **2.** *Slang.* A hospital or asylum for the insane.

**booby prize** *n.* An insignificant or comical award given to the one with the worst score in a game or contest.

**booby trap** *n.* **1.** A usu. concealed, often explosive device triggered by the movement of an apparently harmless object. **2.** A device or situation that catches a person off guard : TRAP.

**boo·dle** (bo͞od′l) [Du. *boedel*, estate < MDu. *bōdel*.] *Slang.* —*n.* **1. a.** Money, esp. counterfeit money. **b.** Money accepted as a bribe. **2.** Stolen goods : SWAG. **3.** A crowd of people : CABOODLE. —*vt.* & *vi.* **-dled, -dling, -dles.** To bribe or accept a bribe. **—boo′dler** *n.*

**boog·ie** (bo͝og′ē) *vi.* **-ied, -ie·ing, -ies.** [Short for BOOGIE-WOOGIE.] *Slang.* **1.** To dance to rock 'n' roll music. **2.** To be able to accept or go along with someone or something.

**boog·ie·man** *also* **boog·y·man** *or* **bog·ey·man** (bo͝og′ē-măn′, bō′gē-, bo͞o′gē-) *n.* [Var. of BOGY + MAN.] A terrifying person or thing : HOBGOBLIN.

**boog·ie-woog·ie** (bo͝og′ē-wo͝og′ē, bo͞o′gē-wo͞o′gē) *n.* [Orig. unknown.] A style of jazz piano marked by a repeated rhythmic and melodic pattern in the bass.

**boog·y·man** (bo͝og′ē-măn′, bō′gē-, bo͞o′gē-). *n. var. of* BOOGIEMAN.

**book** (bo͝ok) *n.* [ME *bok* < OE *bōc*.] **1.** A set of written or printed pages fastened on one side and enclosed between protective covers. **2.** A printed or written literary work. **3.** A bound volume of blank or ruled pages. **4. a.** A volume for recording financial transactions. **b. books.** Such records as a whole. **5.** A main division of a larger written or printed work <a *book* of the New Testament> **6.** A libretto. **7.** The script of a play. **8. Book.** The Bible. **9.** A set of rigid, prescribed rules <always did things by the *book*> **10.** Something held to be a source of knowledge. **11.** A small packet of similar items bound together <a *book* of stamps> **12.** A record of bets placed on a race. **13.** The number of tricks needed in a card game before any tricks can have scoring value, as the first six tricks taken by the declaring side in bridge. **14.** A bundle of tobacco leaves sliced lengthwise. —*vt.* **booked, book·ing, books. 1.** To register or list in or as if in a book. **2.** To record charges against (a person) on a police blotter. **3.** To arrange for (e.g., tickets) in advance : RESERVE. **4.** To hire (e.g., entertainers). **—in (one's) book.** In one's opinion. **—like a book.** Completely : thoroughly <knows me *like a book*> **—one for the books.** *Informal.* Something noteworthy. **—on the books.** Registered or recorded. **—throw the book at.** *Slang.* **1.** To make all possible charges against (e.g., a lawbreaker). **2.** To punish or reprimand severely. **—book′er** *n.*

**book·bind·ing** (bo͝ok′bīn′dĭng) *n.* The art, trade, or profession of binding books. **—book′bind′er** *n.* **—book′bind′er·y** *n.*

**book·case** (bo͝ok′kās′) *n.* A piece of furniture with shelves for holding books.

**book·end** *also* **book end** (bo͝ok′ĕnd′) *n.* A support at the end of a row of books to keep them upright.

**book·ie** (bo͝ok′ē) *n. Slang.* BOOKMAKER 2.

**book·ing** (bo͝ok′ĭng) *n.* **1.** An engagement, as for a performance by an entertainer. **2.** A reservation, as for lodging at an inn.

**book·ish** (bo͝ok′ĭsh) *adj.* **1.** Of, relating to, or like a book. **2.** Enjoying books : STUDIOUS. **3.** Relying on book learning. **4.** Dull : pedantic. **—book′ish·ly** *adv.* **—book′ish·ness** *n.*

**book jacket** *n.* DUST JACKET 1.

**book·keep·ing** (bo͝ok′kē′pĭng) *n.* The art or practice of recording business accounts and transactions. **—book′keep′er** *n.*

**book learning** *n.* Knowledge gained from books rather than from practical experience.

**book·let** (bo͝ok′lĭt) *n.* A small, usu. paperbound book or pamphlet.

**book·louse** (bo͝ok′lous′) *n.* One of various small, often wingless insects of the order Psocoptera or Corrodentia, some species of which damage books.

**book lung** *n.* A sacculate respiratory organ in some arachnids, consisting of a group of membranous folds resembling book pages.

**book·mak·er** (bo͝ok′mā′kər) *n.* **1.** A person who edits, prints, publishes, or binds books. **2.** A person who takes and pays off bets, as on a sporting event. **—book′mak′ing** *n.*

**book·man** (bo͝ok′mən) *n.* **1.** One fond of books and reading. **2.** A bookseller.

**book·mark** (bo͝ok′märk′) *n.* A marker, as a ribbon, placed between the pages of a book to mark the reader's place.

**book·mo·bile** (bo͝ok′mō-bēl′) *n.* [BOOK + (AUTO)MOBILE.] A truck or van equipped to serve as a mobile lending library.

**Book of Common Prayer** *n.* The book of services and prayers used in the Anglican Church.

**Book of Mormon** *n.* The sacred text of the Mormon Church.

**book·plate** (bo͝ok′plāt′) *n.* A label bearing the owner's name that is usu. pasted on the inside cover of a book.

**book·rack** (bo͝ok′răk′) *n.* **1.** A shelf or rack for books. **2.** A frame or rack for supporting an open book.

**book review** *n.* A critical appraisal of a book.

**book·sell·er** (bo͝ok′sĕl′ər) *n.* One who sells books.

**book·shelf** (bo͝ok′shĕlf′) *n.* A shelf or set of shelves for holding books.

**book·shop** (bo͝ok′shŏp′) *n.* A bookstore.

**book·stall** (bo͝ok′stôl′) *n.* A stall where books are sold.

**book·stand** (bo͝ok′stănd′) *n.* **1.** A small counter where books are sold. **2.** A bookrack.

**book·store** (bo͝ok′stôr′, -stōr′) *n.* A store where books are sold.

**book·tell·er** (bo͝ok′tĕl′ər) *n.* A person who reads books aloud for reproduction on phonograph records or tape recordings.

**book·worm** (bo͝ok′wûrm′) *n.* **1.** The larva of any of various insects that infest books and feed on the paste in the bindings. **2.** A person who spends much time reading or studying.

**Bool·e·an** (bo͞o′lē-ən) *adj.* [After George *Boole* (1815–1864).] Of or relating to an algebraic combinatorial system treating variables, as propositions and computer logic elements, through the operators AND, OR, NOT, IF, THEN, and EXCEPT.

**boom**[1] (bo͞om) *v.* **boomed, boom·ing, booms.** [ME *bummen*.] —*vi.* **1.** To make a deep, resonant sound. **2.** To grow or develop rapidly : FLOURISH <The oil business is *booming*.> —*vt.* **1.** To give forth with a deep, resonant sound. **2.** To cause to grow or flourish : BOOST. —*n.* **1.** A booming sound, as of an explosion. **2.** A period of economic prosperity. **3.** A sudden increase, as in wealth.

**boom**[2] (bo͞om) *n.* [Du., pole < MDu.] **1.** *Naut.* A long spar extending from a mast to hold or extend the foot of a sail. **2.** A long pole extending upward at an angle from the mast of a derrick to support or guide objects lifted or suspended. **3. a.** A barrier made of a chain of floating logs enclosing other free-floating logs. **b.** The area enclosed by such a barrier. **4.** A floating barrier extending across a river, lake, or harbor to obstruct navigation. **5.** A long, movable arm for maneuvering a microphone.

**boo·mer·ang** (bo͞o′mə-răng′) *n.* [Native word in Australia.] **1.** A flat, curved, usu. wooden missile that returns to the thrower when it is hurled. **2.** A statement or course of action that backfires against its originator. —*vi.* **-anged, -ang·ing, -angs.** To have the opposite effect from the one intended : BACKFIRE.

**boom·let** (bo͞om′lĭt) *n.* A small boom, as in business.

**boon**[1] (bo͞on) *n.* [ME *bone* < ON *bōn*, prayer.] **1.** Something beneficial that is bestowed : BLESSING. **2.** *Archaic.* A request or favor.

**boon**[2] (bo͞on) *adj.* [ME *bon*, good < OFr. < Lat. *bonus*.] **1.** Convivial : jolly <a *boon* companion> **2.** *Archaic.* Generous : kind.

**boon·docks** (bo͞on′dŏks′) *pl.n.* [Tagalog *bundok*, mountain.] *Slang.* **1.** Wild and dense brush : JUNGLE. **2.** Rural country.

**boon·dog·gle** (bo͞on′dô′gəl, -dŏg′əl) [< *boondoggle*, a plaited leather cord worn by Boy Scouts (coined by R.H. Link, 20th-cent. American scoutmaster).] *Informal.* —*n.* Pointless, unnecessary work. —*vi.* **-gled, -gling, -gles.** To waste time or money on boondoggle. **—boon′dog′gler** *n.*

**boon·ies** (bo͞o′nēz) *pl.n.* [Shortening and alteration of BOONDOCKS.] *Slang.* BOONDOCKS 2.

**boor** (bo͝or) *n.* [Du. *boer* < MDu. *gheboer*.] **1.** A peasant. **2.** A coarse, rude person.

☆ ***syns:*** BOOR, BARBARIAN, BOUNDER, CHUFF, CHURL, PHILISTINE, YAHOO *n. core meaning* : an unrefined, rude person <a *boor* who insulted me at lunch>

**boor·ish** (bo͝or′ĭsh) *adj.* Like a boor. **—boor′ish·ly** *adv.* **—boor′ish·ness** *n.*

**boost** (bo͞ost) *vt.* **boost·ed, boost·ing, boosts.** [Orig. unknown.] **1.** To lift by or as if by pushing up from below. **2.** To increase : RAISE. **3.** To stir up enthusiasm for. **4.** *Slang.* To shoplift. —*n.* **1.** A push upward or ahead. **2.** An increase <a *boost* in pay>

**boost·er** (bo͞o′stər) *n.* **1.** A device for increasing power or effectiveness. **2.** An enthusiastic promoter. **3.** *Electron.* A radio-frequency amplifier. **4. a.** A rocket that assists the main propulsive system of an aircraft or spacecraft. **b.** A rocket used to launch a missile or space vehicle. **5.** A supplementary dose of a vaccine injected to maintain immunity. **6.** *Slang.* A shoplifter.

**booster cable** *n.* An electric cable used to connect a discharged automotive battery to a power source for charging.

**boot**[1] (bo͞ot) *n.* [ME *bote* < OFr.] **1.** A protective piece of usu. leather footgear covering the foot and part or all of the leg. **2.** A protective sheath for a horse's leg. **3.** A torture instrument used to crush the foot and leg. **4.** A protective covering or sheath, esp. a patch for the inner casing of an automotive tire to protect a weak spot or break. **5.** *Chiefly Brit.* TRUNK 6. **6.** A scabbard on a saddle or vehicle to hold a gun. **7. a.** A kick. **b.** *Slang.* A quick, pleasurable feeling : THRILL. **8.** A marine or navy recruit in basic training. **9.** *Slang.* A summary, usu. discourteous dismissal, as from work. —*vt.* **boot·ed, boot·ing, boots. 1.** To put boots on. **2.** To kick. **3.** *Slang.* To discharge : DISMISS.

**†boot**[2] (bo͞ot) *vi.* **boot·ed, boot·ing, boots.** [ME *boten* < OE *bōt*, help.] *Archaic.* To be of help or advantage : AVAIL. —*n.* **1.** *Regional.* Something given in addition. **2.** *Archaic.* Avail : advantage. **—to boot.** In addition : BESIDES.

**boot·black** (bo͞ot′blăk′) *n.* One who cleans and polishes shoes for a living.

**boot camp** *n.* A training camp for marine or navy recruits.

**boo·tee** *also* **boo·tie** (bo͞o′tē) *n.* A soft, usu. knitted baby shoe.

**Bo·ö·tes** (bō-ō′tēz) *n.* [Lat. *Bootes* < Gk. *Boōtēs* < *boōtēs,* plowman < *boōtein,* to plow < *bous,* ox.] A constellation in the Northern Hemisphere.

**booth** (bo͞oth) *n., pl.* **booths** (bo͞othz, bo͞oths) [ME *bothe,* prob. of Scand. orig.] **1.** A small enclosed compartment, usu. accommodating only one person and providing privacy <a phone *booth*> **2.** A seating area in a restaurant having a table and seats whose backs serve as partitions. **3.** A small stall or stand for the display and sale of goods.

**boo·tie** (bo͞o′tē) *n. var. of* BOOTEE.

**boot·jack** (bo͞ot′jăk′) *n.* A forked device for holding a boot secure while the foot is being withdrawn.

**boot·leg** (bo͞ot′lĕg′) *v.* **-legged, -leg·ging, -legs.** [From a smuggler's practice of carrying liquor in the legs of boots.] *—vt.* To produce, sell, or transport (e.g., alcoholic liquor) for sale illegally. *—vi.* **1.** To engage in bootlegging. **2.** *Football.* To fake a handoff, conceal the ball on the hip, and roll out in order to pass or esp. to rush around the end. —Used of a quarterback. *—n.* **1.** Goods smuggled or illicitly produced or sold. **2.** The part of a boot above the instep. **3.** *Football.* A bootleg play. *—adj.* Produced, sold, or transported illegally. **—boot′leg′ger** *n.*

**boot·less** (bo͞ot′lĭs) *adj.* Without benefit or advantage : USELESS. **—boot′less·ly** *adv.* **—boot′less·ness** *n.*

**boot·lick** (bo͞ot′lĭk′) *v.* **-licked, -lick·ing, -licks.** *—vt.* To treat servilely or obsequiously. *—vi.* To behave servilely or obsequiously : FAWN. **—boot′lick′er** *n.*

**boot·strap** (bo͞ot′străp′) *n.* **1.** A leather or cloth loop sewn at the side or the top rear of a boot to help in pulling the boot on. **2.** *Computer Sci.* A subroutine used to establish the full routine or another routine. *—vt.* **-strapped, -strap·ping, -straps.** To establish (a computer program) with a bootstrap. *—adj.* **1.** Undertaken or accomplished with minimal resources or assistance. **2.** Being or relating to a self-initiating or self-sustaining process, as in a computer. **—by one's (own) bootstraps.** By one's own efforts.

**bootstrap loader** *n. Computer Sci.* A hardware device that stores instructions for loading and transferring control to a bootstrap.

**boot tree** *n.* A shoetree.

**boo·ty** (bo͞o′tē) *n., pl.* **-ties.** [ME *bottyne* < OFr. *butin* < MLG *būte,* exchange.] **1.** Plunder taken from an enemy in wartime. **2.** Stolen or seized goods. **3.** A valuable prize, award, or gain.

**booze** (bo͞oz) [ME *bous* < MDu. *būse,* drinking vessel.] *Slang. —n.* **1.** Alcoholic beverages. **2.** A drinking spree. *—vi.* **boozed, booz·ing, booz·es.** To drink alcoholic beverages chronically or excessively. **—booz′er** *n.* **—booz′y** *adj.*

**bop¹** (bŏp) *vt.* [Imit.] *Informal.* **bopped, bop·ping, bops.** To hit or strike. **—bop** *n.*

**bop²** (bŏp) *n.* [Short for BEBOP.] A style of jazz marked by rhythmic, harmonic complexity and innovations.

**bo·ra** (bôr′ə, bōr′ə) *n.* [Dial. Ital. < Lat. *Boreas,* Boreas.] A strong, cold northerly wind of the Adriatic Sea.

**bo·rac·ic** (bə-răs′ĭk) *adj.* [< Med. Lat. *borax, borac-,* borax.] *var. of* BORIC.

**bor·age** (bôr′ĭj, bŏr′-) *n.* [ME < OFr. *bourage* < Med. Lat. *borago,* perh. < Ar. *abū 'āraq,* father of sweat (from its use as a sudorific).] A plant indigenous to southern Europe and northern Africa, *Borago officinalis,* with blue star-shaped flowers and leaves sometimes used for salads.

**bo·rane** (bôr′ān′, bōr′-) *n.* [BOR(ON) + -ANE.] One of a series of boron-hydrogen compounds.

**bo·rate** (bôr′āt′, bōr′-) *n.* A salt of boric acid.

**bo·rax** (bôr′ăks′, -əks, bōr′-) *n.* [ME *boras* < OFr. *boreis* < Med. Lat. *borax* < Ar. *būraq* < Pers. *būrah.*] **1.** A hydrated sodium borate. **2.** An anhydrous sodium borate used in making glass and ceramics.

**Bo·ra·zon** (bôr′ə-zŏn′, bōr′-). A trademark for an extremely hard boron nitride formed at very high pressures and temperatures.

**Bor·deaux** (bôr-dō′) *n., pl.* **Bordeaux** (bôr-dōz′). A red or white wine produced in the region surrounding Bordeaux, France.

**Bordeaux mixture** *n.* [Transl. of Fr. *bouillie bordelaise.*] A mixture of copper sulfate, lime, and water used as a fungicide.

**bor·der** (bôr′dər) *n.* [ME *bordure* < OFr. *bordëure* < *border,* to border < *bort,* border, of Germanic orig.] **1.** A margin, rim, or edge. **2.** A decorative strip around the edge or rim of something, as a fabric. **3.** A strip of ground containing ornamental plants or shrubbery. **4.** The line or frontier area separating political or geographic boundaries. *—vt.* **-dered, -der·ing, -ders. 1.** To put a border on. **2.** To lie along or adjacent to the border of. **—border on (***or* **upon). 1.** To adjoin. **2.** To approach or come near to in character <Their behavior *bordered* on insolence.> **—bor′der·er** *n.*

**bor·der·land** (bôr′dər-lănd′) *n.* **1. a.** Land located on or near a border or frontier. **b.** Outskirts. **2.** An indeterminate or uncertain area or situation.

**bor·der·line** *also* **border line** (bôr′dər-līn′) *n.* **1.** A line that establishes or marks a border. **2.** An intermediate area between two qualities or conditions. *—adj.* **1.** Located near or at a border. **2. a.** Being between two conditions or points : INTERMEDIATE. **b.** Being just below the average, standard, or normal <a child of *borderline* intelligence> **c.** Verging on the obscene <told a *borderline* joke> **d.** Being of marginal validity <a *borderline* theory at best>

**bor·de·tel·la per·tus·sis** (bôr′də-tĕl′ə pər-tŭs′ĭs) *n.* [NLat. : *Bordetella,* genus name (after Jules *Bordet,* 1870–1961) + *pertussis,* whooping cough.] The coccobacillus causing whooping cough.

**bor·dure** (bôr′jər) *n.* [ME. —see BORDER.] *Heraldry.* A border around a shield.

**bore¹** (bôr, bōr) *v.* **bored, bor·ing, bores.** [ME *boren* < OE *borian.*] *—vt.* **1.** To make a hole in or through, as with a drill. **2.** To form (e.g., a tunnel) by drilling, digging, or burrowing. *—vi.* **1.** To make a hole in or through something by or as if by boring. **2.** To advance steadily or laboriously. *—n.* **1.** A hole or passage made by or as if by boring. **2.** The interior diameter of a hole, tube, or cylinder. **3.** The caliber of a firearm. **4.** A drilling tool.

**bore²** (bôr, bōr) *vt.* **bored, bor·ing, bores.** [Orig. unknown.] To make weary with repetition, tedium, or dullness. *—n.* One that arouses boredom. **—bore′dom** (bôr′dəm, bōr′-) *n.*

☆ ***syns:*** BORE, PALL, TIRE, WEARY *v. core meaning* : to fatigue with dullness or tedium <was *bored* by the play> ***ant:*** interest

**bore³** (bôr, bōr) *n.* [ME *bare,* wave < ON *bāra.*] A high and often dangerous tidal wave in a narrow estuary.

**bore⁴** (bôr, bōr) *v. p.t. of* BEAR¹.

**bo·re·al** (bôr′ē-əl, bōr′-) *adj.* [ME < LLat. *Boreālis* < Lat. *Boreas,* Boreas.] **1.** Of or relating to the north : NORTHERN. **2.** Of or concerning the north wind. **3. Boreal.** Of or pertaining to the forest areas and tundras of the North Temperate Zone and Arctic region.

**Bo·re·as** (bôr′ē-əs, bōr′-) *n.* [ME < Lat. *Boreas* < Gk.] **1.** The north wind personified. **2.** *Gk. Myth.* The god of the north wind.

**bore·cole** (bôr′kōl′, bōr′-) *n.* [Du. *boerenkool* : *boer,* peasant + *kool,* cabbage.] Kale.

**bor·er** (bôr′ər, bōr′-) *n.* **1. a.** A tool for boring or drilling. **b.** A person who works with such a tool. **2.** An insect or insect larva, as the corn borer, that bores into plants. **3.** One of various mollusks that bore into soft rock or wood.

**bo·ric** (bôr′ĭk, bōr′-) *also* **bo·rac·ic** (bə-răs′ĭk) *adj.* Of, relating to, derived from, or containing boron.

**boric acid** *n.* A white or colorless crystalline compound, $H_3BO_3$, used as an antiseptic and preservative and in fireproofing compounds, cosmetics, cements, and enamels.

**bo·ride** (bôr′īd′, bōr′-) *n.* [BOR(ON) + -IDE.] A binary compound of boron with a more electropositive element or radical.

**bor·ing** (bôr′ĭng, bōr′-) *adj.* Dull : uninteresting. **—bor′ing·ly** *adv.*

**born** (bôrn) *adj.* [< p.part. of BEAR¹.] **1.** Brought into life or being. **2.** Having from birth a particular quality or talent <a *born* writer> **3.** Derived, resulting, or arising from <inventions *born* of necessity> **4.** Being a native of a given country <Russian-*born*> **5.** Destined from birth <*born* to lead others>

**born-a·gain** (bôrn′ə-gĕn′) *adj.* **1. a.** Of, relating to, or being an individual who has made a conversion or renewed a commitment to Jesus Christ as personal savior <a *born-again* Christian> **b.** Of or relating to evangelical Christianity <a *born-again* religious experience> **2.** Marked by renewal, resurgence, or return <*born-again* enthusiasm><a *born-again* conservative>

**borne** (bôrn) *v. var. p.p. of* BEAR¹.

**born·ite** (bôr′nīt′) *n.* [After Ignaz von *Born* (1742–1791).] A brownish-bronze copper ore with composition $Cu_5FeS_4$.

**bo·ron** (bôr′ŏn′, bōr′-) *n.* [BOR(AX) + (CARB)ON.] *Symbol* **B** A soft, brown, amorphous or crystalline nonmetallic element used in flares, nuclear reactor control elements, abrasives, and hard metallic alloys; atomic number 5; atomic weight 10.811.

**boron carbide** *n.* A very hard, black crystalline compound or solid solution, $B_4C$, used as an abrasive, in control rods for nuclear reactors, and as a reinforcing filament in structural materials.

**bo·ro·sil·i·cate glass** (bôr′ō-sĭl′ĭ-kĭt, bōr′-, -kāt′) *n.* A strong heat-resistant glass containing a minimum of 5% boric oxide.

**bor·ough** (bûr′ō, bŭr′-ō) *n.* [ME *burgh,* city < OE *burg.*] **1.** A self-governing incorporated town in some U.S. states. **2.** One of the five administrative units of New York City. **3.** *Chiefly Brit.* **a.** A town with a municipal corporation and certain rights, as self-government. **b.** A town that sends one or more representatives to Parliament.

**bor·ough-Eng·lish** (bûr′ō-ĭng′glĭsh, bŭr′-) *n.* A former custom in certain parts of England whereby the right to inherit an estate went to the youngest son.

**Bor·rel·i·a** (bə-rĕl′ē-ə, -rē′lē-ə) *n.* [NLat., after Amédée *Borrel* (1867–1936).] A genus of locomotive helical bacteria of the family Spirochaetaceae, some of which cause relapsing fever in humans.

**bor·row** (bŏr′ō, bôr′ō) *v.* **-rowed, -row·ing, -rows.** [ME *borwen* < OE *borgian.*] *—vt.* **1.** To receive (something) on loan with the understanding of returning it or its equivalent. **2.** To adopt or use as one's own <*borrowed* another author's ideas> **3.** To take (a unit of ten) from the next larger denomination in the minuend so as to make a number larger than the number to be subtracted. **4.** To take from one language and assimilate into another. *—vi.* To obtain or receive something. **—bor′row·er** *n.*

**bor·row·ing** (bŏr′ō-ĭng, bôr′-) *n.* Something borrowed, esp. a word borrowed from one language into another.

**borscht** *also* **borsht** (bôrsht) *or* **borsch** (bôrsh) *n.* [R. *borshch.*] A beet soup served hot or cold, usu. with sour cream.

**borscht circuit** *n.* [From the popularity of BORSCHT in their cuisine.] *Slang.* The predominantly Jewish resort hotels of the Catskill Mountains that employ entertainers.

**borsht** (bôrsht) *n. var. of* BORSCHT.

**bort** (bôrt) *n.* [Prob. < Du. *boort.*] **1.** Poorly crystallized diamonds used for industrial cutting and abrasion. **2.** An imperfect diamond : CARBONADO. **—bort′y** *adj.*

**bor·zoi** (bôr′zoi′) *n.* [R. *borzaya.*] A large, slenderly built dog orig. bred in Russia, with a narrow, pointed head and silky coat.

**bos·cage** *also* **bos·kage** (bŏs′kĭj) *n.* [ME *boskage* < OFr. *boscage* < *bosc,* forest, of Germanic orig.] A mass of shrubs or trees : THICKET.

**bosh** (bŏsh) *n.* [Turk. *boş,* useless.] *Informal.* Nonsense.

**bosk** (bŏsk) *n.* [Back-formation < *bosky.*] A small wooded area.

**bos·kage** (bŏs′kĭj) *n. var. of* BOSCAGE.

**Bos·kop man** (bŏs′kŏp′) *n.* [After *Boskop,* a region in the Transvaal where the remains were first found.] A Stone Age man of southern Africa believed to be an ancestor of the Bushmen and Hottentots.

**bosk·y** (bŏs′kē) *adj.* **-i·er, -i·est.** [< ME *bosk,* bush < Med. Lat. *bosca,* of Germanic orig.] **1.** Abounding in trees and shrubs : WOODED. **2.** Shaded by trees or bushes. **—bosk′i·ness** *n.*

**bo′s′n** *or* **bos′n** (bō′sən) *n. vars. of* BOATSWAIN.

**Bos·ni·an** (bŏz′nē-ən) *also* **Bos·ni·ac** (-nē-ăk′) *adj.* Of or relating to Bosnia. —*n.* **1.** A native of Bosnia. **2.** The Serbo-Croatian language of the Bosnians.

**bos·om** (bo͝oz′əm, bo͞o′zəm) *n.* [ME < OE *bōsm.*] **1.** The chest of a human being, esp. the female breasts. **2.** The part of a garment covering the chest. **3.** The center : heart <in the *bosom* of our family> **4.** The chest considered as the source of emotion.

**bo·son** (bō′sŏn) *n.* [After Jagadis Chandra *Bose* (1858–1937).] A particle, as a photon, pion, or alpha particle, having zero or integral spin and obeying statistical rules that permit any number of identical particles to occupy the same quantum state.

**boss**[1] (bôs, bŏs) *n.* [Du. *baas,* master.] **1. a.** An employer or supervisor of workers. **b.** One who makes decisions or exercises authority. **2.** A politician who controls a party or political machine. —*v.* **bossed, boss·ing, boss·es.** —*vt.* **1.** To supervise or control. **2.** To give orders to, esp. in an arrogant or domineering way <*bossed* the children around> —*vi.* To be or act as a boss. —*adj.* **1.** *Slang.* Topnotch : first-rate.

**boss**[2] (bôs, bŏs) *n.* [ME *boce* < OFr.] **1.** A circular protuberance. **2.** A raised area used as ornamentation. **3.** A raised ornament, as at the intersection of the ribs in vaulted roofs. **4. a.** An enlarged part of a shaft to which another shaft is coupled or to which a wheel or gear is keyed. **b.** A hub, esp. of a propeller. **5.** A metal ornament for protecting the corners or centers of books. —*vt.* **bossed, boss·ing, boss·es. 1.** To decorate with bosses. **2.** To emboss.

**boss**[3] (bôs, bŏs) *n.* [Orig. unknown.] A calf or cow.

**bos·sa no·va** (bŏs′ə nō′və) *n.* [Port. : *bossa,* trend + *nova,* new.] **1.** A lively Brazilian dance similar to the samba. **2.** Music that is a blend of jazz and samba.

**boss·ism** (bô′sĭz′əm, bŏs′ĭz′-) *n.* Domination of a political organization by a political boss.

**boss·y**[1] (bô′sē, bŏs′ē) *adj.* **-i·er, -i·est.** Given to ordering others around : DOMINEERING. **—boss′i·ly** *adv.* **—boss′i·ness** *n.*

**boss·y**[2] (bô′sē, bŏs′ē) *adj.* Decorated with studs or raised ornaments.

**boss·y**[3] (bô′sē, bŏs′ē) *n., pl.* **-ies.** *Informal.* A calf or cow.

**Bos·ton bag** (bô′stən) *n.* A handbag or satchel with handles on both sides of the top opening.

**Boston bull** *n.* A Boston terrier.

**Boston cream pie** *n.* A cake with custard filling.

**Boston fern** *n.* A fern, *Nephrolepis exaltata bostoniensis,* with arching or drooping fronds having opposite leaflets.

**Boston ivy** *n.* A widely cultivated climbing woody vine, *Parthenocissus tricuspidata,* native to Asia, that has three-lobed leaves.

**Boston lettuce** *n.* A cultivated lettuce forming a rounded head with soft-textured, yellow-green leaves.

**Boston terrier** *n.* A small dog orig. bred in New England as a cross between a bulldog and a bull terrier.

**bo·sun** (bō′sən) *n. var. of* BOATSWAIN.

**Bos·well** (bŏz′wĕl′, -wəl) *n.* [After James *Boswell* (1740–1795).] A person who meticulously records the words and deeds of a contemporary.

**bot** *also* **bott** (bŏt) *n.* [ME.] The parasitic larva of a botfly.

**bo·tan·i·cal** (bə-tăn′ĭ-kəl) *also* **bo·tan·ic** (-tăn′ĭk) *adj.* [Fr. *botanique* < LLat. *botanicus* < Gk. *botanikos* < *botanē,* plant.] **1.** Of or relating to plants or plant life. **2.** Of or relating to botany. —*n.* A medicinal preparation obtained from a plant or plants. **—bo·tan′i·cal·ly** *adv.*

**bot·a·nist** (bŏt′n-ĭst) *n.* A specialist in the study of plants.

**bot·a·nize** (bŏt′n-īz′) *v.* **-nized, -niz·ing, -niz·es.** —*vi.* **1.** To secure plants for botanical study. **2.** To examine plants scientifically. —*vt.* To investigate (an area) for botanical study. **—bot′a·niz′er** *n.*

**bot·a·ny** (bŏt′n-ē) *n., pl.* **-nies. 1.** The science of plants. **2.** The flora of a particular region or district. **3.** The characteristics and phenomena of a plant group or category. **4. a.** A scholarly work on botany. **b.** A specific system of botany.

**botch** (bŏch) *vt.* **botched, botch·ing, botch·es.** [ME *bocchen,* to mend.] **1.** To ruin by carelessness or clumsiness. **2.** To make or perform clumsily : BUNGLE. **3.** To repair or mend haphazardly. —*n.* **1.** A defective or ruined piece of work. **2.** A badly repaired part or flaw. **—botch′er** *n.* **—botch′y** *adj.*

☆ ***syns:*** BOTCH, BOLLIX, BUNGLE, FOUL UP, FUMBLE, GOOF (up), LOUSE (up), MESS (up), MISHANDLE, MISMANAGE, MUDDLE, MUFF, SCREW UP, SNAFU, SPOIL *v. core meaning* : to harm severely through inept handling <a project *botched* by incompetent management>

**bot·fly** *also* **bot fly** (bŏt′flī′) *n.* One of various winged insects, chiefly of the genera *Gasterophilus* and *Oestrus,* with larvae that are parasitic on humans and other animals.

**both** (bōth) *adj.* [ME *bothe* < ON *bāðir.*] Relating to or being two in conjunction : being one and the other <*Both* students arrived.> —*pron.* The one and the other <*Both* of those soldiers are patriots.> —*conj.* —Used with *and* to indicate that each of two things in a coordinated phrase or clause is included <*both* men and women>

***usage:*** When *both* is used with *and* to link elements in a sentence, it is stylistically desirable that the linked elements be grammatically parallel. Thus, phrases such as *in both India and China* or *both in India and in China* are preferable to *both in India and China.*

**both·er** (bŏth′ər) *v.* **-ered, -er·ing, -ers.** [Orig. unknown.] —*vt.* **1.** To irritate, esp. by petty annoyances. **2. a.** To make agitated or nervous : FLUSTER. **b.** To make confused or perplexed : PUZZLE. **3.** To intrude without invitation or warrant : DISTURB. **4.** To give trouble to <a knee that *bothers* me constantly> —*vi.* **1.** To take the trouble : concern oneself. **2.** To cause trouble. —*n.* A cause or state of disturbance. —*interj. Chiefly Brit.* —Used to express annoyance or mild irritation.

**both·er·a·tion** (bŏth′ə-rā′shən) *n.* A vexation.

**both·er·some** (bŏth′ər-səm) *adj.* Causing bother.

**both·ri·um** (bŏth′rē-əm) *n.* [NLat. < Gk. *bothrion,* dim. of *bothros,* pit.] A suction groove on the scolex of the pseudophyllidean tapeworm.

**bo tree** (bō) *n.* [Singhalese *bo* < Pali *bodi(taru),* (tree of) wisdom < Skt. *bodhi,* enlightenment < *bodhati,* he awakes.] The peepul.

**bot·ry·oid·al** (bŏt′rē-oid′l) *also* **bot·ry·oid** (bŏt′rē-oid′) *adj.* [Gk. *botruoeidēs* < *botrus,* bunch of grapes.] Shaped like a bunch of grapes. **—bot′ry·oid′al·ly** *adv.*

**bots** (bŏts) *n.* [Perh. < Sc. Gael. *boiteag,* maggot.] (*sing. in number*). An ailment in mammals, as cattle and horses, caused by alimentary infestations of parasitic botfly larvae.

**bott** (bŏt) *n. var. of* BOT.

**bot·tle** (bŏt′l) *n.* [ME *botel* < OFr. *botele* < Med. Lat. *buticula,* dim. of LLat. *butis,* cask.] **1.** A usu. glass receptacle having a comparatively narrow neck and a mouth that can be capped or corked. **2.** The quantity a bottle contains. **3. a.** Formula or bottled milk fed in place of mother's milk. **b.** Alcoholic drink <hit the *bottle*> —*vt.* **-tled, -tling, -tles.** To place in a bottle. **—bottle up. 1.** To hold in : RESTRAIN <*bottled up* my anger> **2.** To seal up : BLOCK. **—bot′tler** *n.*

**bot·tle·brush** (bŏt′l-brŭsh′) *n.* A shrub or tree of the genera *Callistemon* or *Melaleuca,* native to Australia, having dense flower spikes with protruding stamens suggestive of a brush for cleaning bottles.

bottlebrush

**bottle club** *n.* A private establishment where patrons may purchase bottles of liquor for consumption after legal closing hours.

**bottled gas** *n.* Gas, as butane or propane, stored in pressurized portable tanks.

**bot·tle-feed** (bŏt′l-fēd′) *vt.* **-fed** (-fĕd′), **-feed·ing, -feeds.** To feed, as a baby, with a bottle.

**bottle gentian** *n.* A plant, *Gentiana andrewsii* of eastern and central North America, with deep-blue flowers that remain closed.

**bottle gourd** *n.* **1.** The calabash vine. **2.** The fruit of the calabash vine.

**bottle green** *n.* A dark to grayish green. **—bot′tle-green′** (bŏt′l-grēn) *adj.*

ă **pat** ā **pay** âr **care** ä **father** ĕ **pet** ē **be** hw **wh**ich ĭ **pit** ī **tie** îr **pier** ŏ **pot** ō **toe** ô **paw, for** oi **noise** o͝o **took**

**bot·tle·neck** (bŏt′l-nĕk′) *n.* **1.** The narrow part of a bottle near the top. **2.** A narrow or obstructed section of a highway or pipeline. **3.** A hindrance to production or progress. **4.** *Mus.* A style of guitar playing in which an object, as a piece of glass or metal, is pressed against the strings to achieve a gliding effect. —*vt.* **-necked, -neck·ing, -necks.** To impede or slow down by creating a bottleneck.

**bot·tle-nosed dolphin** (bŏt′l-nōzd′) *n.* A marine mammal of the genus *Tursiops*, with a short, protruding beak.

**bottle tree** *n.* Any of several trees of the genus *Stercula* or *Brachychiton*, native to Australia, having a bottlelike swelling of the trunk.

**bot·tom** (bŏt′əm) *n.* [ME *botme* < OE *botm.*] **1. a.** The lowest or deepest part. **b.** The last place or position <at the *bottom* of the class> **c.** *Baseball.* The second half of an inning. **2.** The underside. **3.** The supporting part : FOUNDATION. **4.** The basic underlying quality : ESSENCE. **5.** The solid surface under a body of water. **6.** *often* **bottoms.** Low-lying alluvial land adjacent to a river : BOTTOMLAND. **7.** The part of a ship's hull below the water line. **8.** A ship. **9.** *often* **bottoms.** Pajama trousers. **10.** *Informal.* The buttocks. **11.** The seat of a chair. **12.** Staying power, as of a horse : STAMINA. —*v.* **-tomed, -tom·ing, -toms.** —*vt.* **1.** To provide with an underside or foundation. **2.** To establish : found. **3.** To grasp the meaning of : FATHOM. —*vi.* **1.** To be or become based or grounded. **2.** To rest on or touch the bottom. **—bottom out.** To descend, as securities, to the lowest point possible, after which only a rise may occur. **—bot′tom·er** *n.*

**bottom break** *n.* A branch arising from the stem base of a plant.

**bottom fauna** *n.* Marine vegetation growing in the benthic region of the ocean depths.

**bot·tom·land** (bŏt′əm-lănd′) *n.* Low-lying land along a river.

**bot·tom·less** (bŏt′əm-lĭs) *adj.* **1.** Having no bottom. **2.** Too deep to be measured. **3.** Difficult or impossible to understand : UNFATHOMABLE. **4.** Limitless. **—bot′tom·less·ly** *adv.*

**bottom line** *n.* **1.** The lowest line in a financial statement, showing net income or loss. **2.** The final result or statement : UPSHOT <"the *bottom line*—for now—is that the city is heading toward default" —*New York*> **3.** The main or essential point.

**bot·tom-line** (bŏt′əm-līn′) *adj.* Relating to or concerned chiefly with costs and profits. **—bot′tom-lin′er** *n.*

**bottom round** *n.* A cut of meat, as a steak or roast, taken from the outer section of a round of beef.

**bot·tom·ry** (bŏt′əm-rē) *n.* [Alteration of Du. *bodemerij* < *bodem*, ship.] A contract by which a shipowner borrows money to finance a voyage, pledging the ship as security.

**bot·u·lin** (bŏch′ə-lĭn) *n.* [Lat. *botulus*, sausage + -IN.] A nerve toxin produced by botulinum and found in improperly canned or improperly smoked foods.

**bot·u·li·num** (bŏch′ə-lī′nəm) *n.* [NLat. < Lat. *botulus*, sausage.] A bacterium, *Clostridium botulinum*, that secretes botulin.

**bot·u·lism** (bŏch′ə-lĭz′əm) *n.* [G. *Botulismus* < Lat. *botulus*, sausage.] A virulent, often fatal food poisoning caused by botulin and marked by vomiting, abdominal pain, muscular weakness, and visual disturbance.

**bou·clé** or **bou·cle** (bo͞o-klā′) *n.* [Fr., p.part. of *boucler*, to curl < OFr. *boucle*, curl of hair.] **1.** A yarn, usu. three-ply and with one thread looser than the others, that produces a rough-textured cloth. **2.** Fabric woven or knitted from bouclé.

**bou·doir** (bo͞o′dwär′, -dwôr′) *n.* [Fr. < OFr. *bouder*, to sulk.] A woman's private sitting room, dressing room, or bedroom.

**bouf·fant** (bo͞o-fänt′) *adj.* [Fr., pr.part. of *bouffer*, to puff up < OFr.] Puffed out : FULL <a *bouffant* bridal veil>

**bouffe** (bo͞of) *n.* Comic opera.

**bou·gain·vil·le·a** *also* **bou·gain·vil·lae·a** (bo͞o′gən-vĭl′ē-ə, -vĭl′yə) *n.* [NLat. *Bougainvillea*, genus name, after Louis Antoine de *Bougainville* (1729–1811).] A woody tropical American vine of the genus *Bougainvillea*, with inconspicuous flowers surrounded by red, purple, or orange bracts.

**bough** (bou) *n.* [ME < OE *bōh.*] A large tree branch.

**bought** (bôt) *v. p.t. & p.p. of* BUY.

**†bought·en** (bôt′n) *v. Regional. var. p.p. of* BUY.

**bou·gie** (bo͞o′zhē, -jē) *n.* [Fr. < OFr., a fine wax < *Bougie*, Bejaïa, Algeria.] **1.** A wax candle. **2.** *Med.* **a.** A slender, cylindrical, pliable implement inserted into a bodily canal, as the urethra or rectum. **b.** A suppository.

**bouil·la·baisse** (bo͞o′yə-bās′) *n.* [Fr. < Prov. *bouiabaisso* : *boui*, imper. of *bouie*, to boil (< Lat. *bullire*) + *abaisso*, imper. of *abeissa*, to lower.] A highly seasoned fish stew made of several kinds of fish and shellfish.

**bouil·lon** (bo͞o′yŏn′, bo͝ol′-, -yən) *n.* [Fr. < OFr. < *boulir*, to boil < Lat. *bullire.*] A clear, thin broth made usu. from beef or chicken.

**bouillon cube** *n.* A small cube of evaporated seasoned meat, poultry, or vegetable stock.

**boul·der** (bōl′dər) *n.* [ME *bulder*, prob. of Scand. orig.] A large rounded rock mass that is on the surface of the ground or is imbedded in the soil.

**bou·le**[1] (bo͞o′lē, bo͞o-lā′) *n.* [Gk. *boulē.*] **1. a.** The senate of 400 founded in ancient Athens by Solon. **b.** A legislative assembly in one of the ancient Greek states. **2.** The lower house of the modern Greek legislature.

**boule**[2] (bo͞ol) *n.* [Fr., ball < Lat. *bulla.*] A pear-shaped synthetic sapphire, ruby, or other alumina-based gem, produced by fusing and tinting alumina.

**boule**[3] (bo͞ol) *n. var. of* BUHL.

**boul·e·vard** (bo͝ol′ə-värd′, bo͞o′lə-) *n.* [Fr. < OFr. *boloart*, rampart converted to a promenade < MDu. *bolwerc*, bulwark < MHG.] A wide city street.

**bou·le·vard·ier** (bo͞o′lə-vär-dyā′, bo͝ol′ə-, -dîr′) *n.* [Obs. Fr. < *boulevard*, boulevard.] A man about town.

**bou·le·ver·se·ment** (bo͞o′lə-vĕr′sə-mäN′) *n.* [Fr. < OFr. *bouleverser*, to overturn : *boule*, ball + *verser*, to overturn < Lat. *versare*, to turn.] **1.** A reversal. **2.** An uproar : tumult.

**boulle** (bo͞ol) *n. var. of* BUHL.

**bounce** (bouns) *v.* **bounced, bounc·ing, bounc·es.** [ME *bounsen*, to beat.] —*vi.* **1. a.** To rebound elastically from a collision. **b.** To collide and rebound elastically several times in succession. **2.** To bound thumpingly <*bounced* into the room> **3.** *Informal.* To be sent back by a bank as worthless <a check that *bounced*> —*vt.* **1.** To cause to collide and rebound. **2.** *Slang.* **a.** To expel forcibly. **b.** To dismiss from employment. **3.** To write on an overdrawn bank account. —*n.* **1.** A bound or rebound. **2.** A sudden spring or leap. **3.** Capacity to bounce : SPRING <a tennis ball with lots of *bounce*> **4.** Liveliness : spirit. **5.** *Slang.* Expulsion. **6.** *Chiefly Brit.* Impudent bluster.

**bounc·er** (boun′sər) *n.* **1.** One that bounces. **2.** *Slang.* One employed to expel disorderly persons from a public place.

**bounc·ing** (boun′sĭng) *adj.* **1.** Healthy : vigorous. **2.** Lively : spirited. **—bounc′ing·ly** *adv.*

**bouncing Bet** (bĕt) *n.* [< *Bet*, nickname for *Elizabeth.*] An Old World plant, *Saponaria officinalis*, with fragrant pink or white flower clusters.

**bounc·y** (boun′sē) *adj.* **-i·er, -i·est. 1.** Tending to bounce. **2.** Elastic : springy. **3.** Energetic : lively. **—bounc′i·ly** *adv.*

**bound**[1] (bound) *vi.* **bound·ed, bound·ing, bounds.** [Fr. *bondir*, to bounce < OFr., to resound, perh. < Lat. *bombitare*, to hum < *bombus*, a humming sound < Gk. *bombos.*] **1.** To leap forward or upward : SPRING. **2.** To advance by bounds. —*n.* **1.** A leap : jump. **2.** A bounce.

**bound**[2] (bound) *n.* [ME < OFr. *bunde* < Med. Lat. *bodina*, of Celtic orig.] **1.** *often* **bounds.** The limit of something <went beyond the *bounds* of credibility><joy that knew no *bounds*> **2. bounds.** The territory on, within, or near limiting lines. —*v.* **bound·ed, bound·ing, bounds.** —*vt.* **1.** To provide a limit to. **2.** To form the boundary or limit of. **3.** To identify and set the boundaries of : DEMARCATE. —*vi.* To border on another country, state, or place : ADJOIN. **—out of bounds.** Beyond boundaries or limits.

**bound**[3] (bound) *adj.* [< p.part. of BIND.] **1.** Confined by bonds : TIED <*bound* and gagged> **2.** Under legal or moral obligation or contract <*bound* by a promise> **3.** Indentured <a *bound* apprentice> **4.** Equipped with a cover or binding <*bound* books> **5.** Predetermined : certain <We are *bound* to be late.> **6.** Determined : resolved <They are *bound* to lose.> **7.** Constipated.

**bound**[4] (bound) *adj.* [ME *boun*, ready < ON *būinn*, p.part. of *būa*, to get ready.] On the way : HEADED <*bound* for the office>

**bound·a·ry** (boun′də-rē, -drē) *n., pl.* **-ries. 1.** Something indicating a border or limit. **2.** The border or limit indicated.

**boundary layer** *n.* The almost motionless fluid layer located immediately adjacent to a boundary, as the surface of a solid, past which the fluid flows.

**bound·en** (boun′dən) *adj.* [ME, p.part. of *binden*, to bind < OE *bindan.*] **1.** *Archaic.* Under obligation : OBLIGED. **2.** Obligatory <our *bounden* duty>

**bound·er** (boun′dər) *n.* **1.** One that bounds. **2.** *Chiefly Brit.* A vulgar, cocksure person.

**bound form** *n.* A linguistic element that always occurs as part of another word, as *-ly* in *lovely.*

**bound·less** (bound′lĭs) *adj.* Being without limits : INFINITE. **—bound′less·ly** *adv.* **—bound′less·ness** *n.*

**boun·te·ous** (boun′tē-əs) *adj.* [ME *bountevous* < OFr. *bontive*, benevolent < *bonte*, bounty.] **1.** Giving freely. **2.** Plentiful : copious. **—boun′te·ous·ly** *adv.* **—boun′te·ous·ness** *n.*

**boun·ti·ful** (boun′tə-fəl) *adj.* Bounteous. **—boun′ti·ful·ly** *adv.* **—boun′ti·ful·ness** *n.*

**boun·ty** (boun′tē) *n., pl.* **-ties.** [ME *bounte* < OFr. *bonte* < Lat. *bonitas*, goodness < *bonus*, good.] **1. a.** Liberality in giving. **b.** Something liberally given. **2.** A reward, inducement, or payment, esp. one given by a government for acts beneficial to the state, as the killing of predatory animals.

**bounty hunter** *n.* A person who hunts predatory animals or outlaws for a bounty.

**bou·quet** (bō-kā′, bo͞o-) *n.* [Fr. < OFr. *bosquet*, thicket, dim. of *bosc*, forest, of Germanic orig.] **1.** A cluster of flowers : NOSEGAY. **2.** The fragrance characteristic of a wine or a liqueur.

**bou·quet gar·ni** (bō-kā′ gär-nē′, bo͞o-) *n., pl.* **bou·quets gar·nis** (bō-kāz′ gär-nē′, bo͞o-) [Fr. : *bouquet*, bunch + *garni*, p.part. of *gar-*

*nir,* to garnish.] A bunch of herbs wrapped in cheesecloth or tied together and immersed in a soup or stew as seasoning.

**bour·bon** (bûr′bən) *n.* [After *Bourbon* County, Kentucky.] A whiskey distilled from a fermented mash having not less than 51% corn.

**bour·don** (bŏŏr′dn) *n.* [ME *burdoun* < OFr. *bourdon.*] **1.** The drone bass of a bagpipe. **2.** An organ stop, usu. of the 16-foot pipes.

**bourg** (bŏŏrg) *n.* [ME < OFr. < Lat. *burgus,* fortress, of Germanic orig.] **1.** A medieval village situated near a castle. **2.** A French market town.

**bour·geois** (bŏŏr-zhwä′, bŏŏr′zhwä′) *n., pl.* **bourgeois.** [Fr. < OFr. *burgeis* < *bourg,* bourg.] **1.** A member of the bourgeoisie. **2.** One whose attitudes and behavior conform to middle-class conventions and standards. **3.** A member of the property-owning class according to Marxist theory : CAPITALIST. —*adj.* **1.** Characteristic of the middle class. **2.** Preoccupied with respectability and material values.

**bour·geoise** (bŏŏr-zhwäz′, bŏŏr′zhwäz′) *n., pl.* **bour·geois·es** (bŏŏr-zhwä′zĭz, bŏŏr′zhwä′-) [Fr., fem. of *bourgeois,* bourgeois.] A woman member of the bourgeoisie. **—bour·geoise′** *adj.*

**bour·geoi·sie** (bŏŏr′zhwä-zē′) *n.* [Fr. < *bourgeois,* bourgeois.] **1.** The middle class. **2.** The social group in Marxist theory opposed to the proletariat in the class struggle : capitalist class.

**bour·geon** (bûr′jən) *v. var. of* BURGEON.

**bourn[1]** *also* **bourne** (bôrn, bōrn, bŏŏrn) *n.* [ME < OE *burna.*] A small brook.

**bourn[2]** *also* **bourne** (bôrn, bōrn, bŏŏrn) *n.* [Fr. *bourne* < OFr. *bodne,* limit < Med. Lat. *bodina,* of Celt. orig.] *Archaic.* **1.** The terminal point of a trip or action : GOAL. **2.** A boundary.

**bour·rée** (bŏŏ-rā′, bŏŏ-) *n.* [Fr.] **1.** A 17th-cent. French dance resembling the gavotte, usu. in quick duple time beginning with an upbeat. **2.** The music for a bourrée.

**bourse** (bŏŏrs) *n.* [Fr. < LLat. *bursa,* bag. —see PURSE.] The stock exchange of a continental European city, esp. Paris.

**bouse** *also* **bowse** (bouz) *v.* **boused, bous·ing, bous·es** *also* **bowsed, bows·ing, bows·es.** [Orig. unknown.] *Naut.* —*vt.* To hoist with a tackle. —*vi.* To hoist.

**bou·stro·phe·don** (bŏŏ′strə-fēd′n, -fē′dŏn′) *n.* [< Gk. *boustrophēdon,* turning like an ox while plowing: *bous,* ox + *strephein,* to turn.] An ancient writing method in which the lines were inscribed alternately from right to left and from left to right. **—bou·stroph′e·don′ic** (-strŏf′ĭ-dŏn′ĭk) *adj.*

**bout** (bout) *n.* [ME, bend.] **1.** A contest between antagonists : MATCH. **2.** A period spent in a particular way : SPELL <"His tremendous *bouts* of drinking had wrecked his health" —Thomas Wolfe>

**bou·tique** (bŏŏ-tēk′) *n.* [Fr. < OProv. *botica* < Lat. *apotheca,* storehouse. —see APOTHECARY.] A small retail shop specializing in gifts, fashionable clothes, and accessories.

▲ word history: *Boutique* is one of three English words that are derived from the Greek word *apothēkē,* "storehouse." The others are *apothecary* and *bodega.* All three came into English by way of Latin, which borrowed the Greek word as *apotheca.* The Latin suffix *-arius* added to this noun formed a new word, *apothecarius,* meaning "storekeeper"; the neuter plural, *apothecaria,* meant "the things pertaining to a storekeeper." In postclassical times the Latin words came to refer primarily to the storing and selling of medicines and drugs, and with such meanings the word *apothecary* was borrowed into English. *Boutique* and *bodega* are both derived from *apotheca* and come to English from French and Spanish, respectively. *Boutique* originally meant a small shop of any kind, but is now restricted to a small, fashionable retail store. *Bodega* in Spanish means "wine cellar" or "tavern"; only in American Spanish has it been extended to mean "grocery store" as well.

**bou·ton** (bŏŏ-tôn′) *n.* [Fr., button.] A club-shaped enlargement at the terminus of a nerve fiber.

**bou·ton·niere** *also* **bou·ton·nière** (bŏŏ′tə-nîr′, -tən-yâr′) *n.* [Fr. *boutonnière,* buttonhole < OFr. < *bouton,* button.] A flower or small bunch of flowers worn in a buttonhole.

**bou·var·di·a** (bŏŏ-vär′dē-ə) *n.* [NLat. *Bouvardia,* genus name, after Charles *Bouvard* (1572–1658).] A tropical American shrub of the genus *Bouvardia,* with red or white, often fragrant flower clusters.

**Bou·vier des Flan·dres** (bŏŏ-vyā′ də flăn′dərz, fläN′drə) *n.* [Fr. : *bouvier,* cowherd + *des,* of + *Flandres,* Flanders.] A large rough-coated dog orig. used in Belgium for herding and guarding cattle.

**bou·zou·ki** (bŏŏ-zŏŏ′kē, bə-) *n.* [Mod. Gk. *mpouzouki.*] A Greek stringed instrument similar to a mandolin.

**bo·vid** (bō′vĭd) *adj.* [< NLat. *Bovidae,* family name < Lat. *bos,* cow.] Of or belonging to the family Bovidae, which includes hoofed, hollow-horned ruminants such as cattle, sheep, goats, and buffaloes. **—bo′vid** *n.*

**bo·vine** (bō′vīn′, -vēn′) *adj.* [LLat. *bovinus* < Lat. *bos,* cow.] **1.** Of, relating to, or like an animal of the genus *Bos,* as an ox or cow. **2.** Dull : sluggish. —*n.* A bovine animal.

**bow[1]** (bou) *n.* [Poss. of LG orig.] **1.** The front section of a ship or boat. **2.** The oar or oarsman nearest to the bow of a boat.

**bow[2]** (bou) *v.* **bowed, bow·ing, bows.** [ME *bowen* < OE *būgan.*] —*vi.* **1.** To bend or curve downward : STOOP. **2.** To bend the body, head, or knee, as in greeting, courtesy, or acknowledgment. **3.** To comply or yield : SUBMIT. —*vt.* **1.** To bend (the head, knee, or body) so as to express greeting, consent, courtesy, submission, or veneration. **2.** To convey (e.g., a greeting) by bowing. **3.** To escort in or out with bows <*bowed* us into the theater> **4.** To cause to acquiesce : SUBMIT. **5.** To cause to bend downward : OVERBURDEN <Sorrow *bowed* them down.> **—bow out.** To remove oneself : RESIGN. —*n.* An inclination of the head or body, as in greeting, acknowledgment, or submission.

**bow[3]** (bō) *n.* [ME *bowe* < OE *boga.*] **1.** Something bent, curved, or arched. **2.** A weapon composed of a curved, flexible strip of material, as wood, strung taut from end to end and used to launch arrows. **3. a.** An archer. **b.** Archers as a group. **4.** A rod with usu. horsehair drawn tightly between its two raised ends, used to play stringed instruments of the violin and viol families. **5.** A knot usu. having two loops and two ends : BOWKNOT. **6. a.** A frame for the lenses of eyeglasses. **b.** The part of an eyeglass frame passing over the ear. **7.** A rainbow. **8.** An oxbow. —*v.* **bowed, bow·ing, bows.** —*vt.* **1.** To bend (an object) into the shape of a bow. **2.** To play (a stringed instrument) with a bow. —*vi.* **1.** To bend into a bow or curve. **2.** To play a stringed instrument with a bow.

**bowd·ler·ize** (bōd′lə-rīz′, boud′-) *vt.* **-ized, -iz·ing, -iz·es.** [After Thomas Bowdler (1754–1825).] To expurgate (e.g., a publication) prudishly. **—bowd′ler·ism** *n.* **—bowd′ler·i·za′tion** *n.* **—bowd′ler·iz′er** *n.*

**bow·el** (bou′əl, boul) *n.* [ME < OFr. *bouele* < LLat. *botellus,* small intestine, dim. of *botulus,* sausage.] **1. a.** An intestine, esp. of a human being. **b.** *often* **bowels.** The digestive tract below the stomach. **2. bowels.** The interior <in the *bowels* of the city> **3. bowels.** *Archaic.* The seat of pity or the gentler emotions.

**bow·er[1]** (bou′ər) *n.* [ME *bour,* a dwelling < OE *būr.*] **1.** A shaded, leafy recess : ARBOR. **2.** A woman's private chamber in a medieval castle. **3.** A rustic cottage or country retreat. —*vt.* **-ered, -er·ing, -ers.** To enclose in or as if in a bower. **—bow′er·y** (-ə-rē) *adj.*

**bow·er[2]** (bou′ər) *n.* A ship's heaviest anchor, carried at the bow.

**bow·er·bird** (bou′ər-bûrd′) *n.* Any of various birds of the family Ptilonorhynchidae of Australia and New Guinea, of which the males of many species build bowers of grasses, twigs, and colored materials to attract females.

**bow·er·y** (bou′ə-rē, bou′rē) *n., pl.* **-ies.** [Du. *bouwerij* < *bouwen,* to cultivate < MDu.] **1.** A farm or plantation owned by one of the early Dutch settlers of New York. **2.** A section of a city marked by flophouses, cheap bars, and vagrants.

**bow·fin** (bō′fĭn′) *n.* A freshwater fish, *Amia calva* of central and eastern North America, the only known extant species of the family Amidae.

**bow·front** (bō′frŭnt′) *adj.* Having an outward-curving front <a *bowfront* bureau>

**bow·head** (bō′hĕd′) *n.* A large-headed whale, *Balaena mysticetus* of Arctic seas.

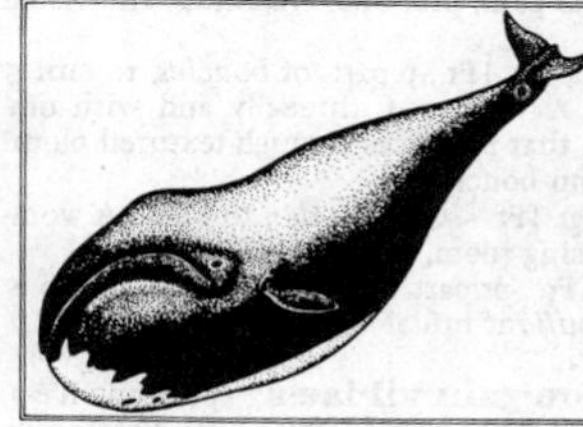

**bowhead**
*50–60 feet long*

**bow·ie knife** (bō′ē, bŏŏ′ē) *n.* [After James *Bowie* (1790?–1836).] A large thick-bladed single-edged hunting knife.

**bow·knot** (bō′nŏt′) *n.* A knot with large decorative loops.

**bowl[1]** (bōl) *n.* [ME *bowle* < OE *bolla.*] **1. a.** A hemispherical vessel, wider than it is deep, for food or fluids. **b.** The contents of a bowl. **2.** A drinking goblet. **3. a.** A bowl-shaped part, as of a spoon. **b.** The receptacle in a toilet. **4. a.** A bowl-shaped stadium or outdoor theater. **b.** A football game played between selected teams after the regular season. **5.** A bowl-shaped topographic depression.

**bowl[2]** (bōl) *n.* [ME *boule* < OFr. < Lat. *bulla,* bubble.] **1.** A large, wooden ball weighted or slightly flattened so as to roll with a bias. **2.** A roll or throw of the ball, as in bowling. **3.** A revolving cylinder or drum in a machine. **4. bowls** *(sing. in number).* Lawn bowling. —*v.* **bowled, bowl·ing, bowls.** —*vi.* **1.** To take part in a game of bowling. **2.** To throw or roll a ball in bowling. **3.** To move smoothly and rapidly <The train *bowled* along.> **4.** To hurl a cricket ball from one end of the pitch toward the batsman at the other in a way distinguished from throwing. —*vt.* **1.** To throw or roll (a ball) in bowling. **2.** To make or achieve by bowling. **—bowl out.** To retire (a batsman in cricket) with a bowled ball that knocks the bails off the wicket. **—bowl over.** To take by surprise : ASTOUND.

**bow·leg** (bō'lĕg') *n.* A leg with an outward curvature in the region of the knee. **—bow'leg'ged** (bō'lĕg'ĭd, -lĕgd') *adj.*

**bowl·er¹** (bō'lər) *n.* One that bowls.

**bowl·er²** (bō'lər) *n. Chiefly Brit.* [After John Bowler, a 19th-cent. London hatmaker.] DERBY 3.

**bow·line** (bō'lĭn, -līn') *n.* [ME *bouline.*] **1.** *Naut.* A rope leading forward from the leech of a square sail to hold the leech forward when sailing close-hauled. **2.** A knot forming a loop that does not slip.

**bowl·ing** (bō'lĭng) *n.* **1.** A game played by rolling a ball down a wooden alley in order to knock down a triangular group of ten pins. **2.** A game similar to bowling, as skittles or ninepins.

**bowling alley** *n.* **1.** A smooth, level wooden alley used in bowling. **2.** A place containing bowling alleys.

**bow·man¹** (bō'mən) *n.* An archer.

**bow·man²** (bou'mən) *n.* An oarsman who is stationed at the bow of a boat.

**Bow·man's capsule** (bo'mənz) *n.* [After Sir William *Bowman* (1816–1892).] The renal or malpighian corpuscle, which acts as a filter in urine formation in the kidney.

**Bowman's glands** *pl.n.* [After Sir William *Bowman* (1816–1892).] The olfactory glands, which keep the olfactory surface moist.

**bow·man's root** (bō'mənz) *n.* A plant, *Gillenia trifoliata* of eastern North America, with compound leaves and small white or pinkish flowers.

**bowse** (bouz) *v. var. of* BOUSE.

**bow·sprit** (bou'sprĭt', bō'-) *n.* [ME *bousprět.*] A spar extending forward from the stem of a ship.

**bow·string** (bō'strĭng') *n.* The string of a bow.

**bowstring hemp** *n.* **1.** A plant of the genus *Sansevieria,* with thick, erect leaves. **2.** The fiber from the leaves of bowstring hemp, used esp. for cordage.

**bow tie** (bō) *n.* A small necktie tied in the shape of a bow.

**bow window** (bō) *n.* A curved bay window.

**bow-wow** (bou'wou') *n.* [Imit.] **1.** The bark of a dog. **2.** A dog.

**bow·yer** (bō'yər) *n.* **1.** An archer. **2.** A person who makes bows.

**box¹** (bŏks) *n.* [ME < OE < LLat. *buxis* < Gk. *puxis* < *puxos,* box tree.] **1. a.** A rectangular container usu. having a lid or cover. **b.** The quantity a box will hold. **2.** A separated compartment in a public place, as a theater, for the accommodation of a small group. **3.** A small structure serving as a shelter. **4.** A box stall. **5.** The raised driver's seat of a coach or carriage. **6.** *Baseball.* One of various designated areas on the diamond where the pitcher, catcher, batter, or coaches stand. **7.** Featured printed matter enclosed by hairlines, a border, or white space and placed within or between text columns. **8.** A cut in the side of a tree through which sap is collected. **9.** An insulating, enclosing, or protective casing or part in a machine. **10.** An awkward or perplexing situation : PREDICAMENT. **11.** A penalty box. *—vt.* **boxed, box·ing, box·es. 1.** To pack in a box. **2.** To confine in or as if in a box. **3.** *Naut.* To boxhaul.

**box²** (bŏks) *n.* [ME.] A blow or slap with the hand. *—v.* **boxed, box·ing, box·es.** *—vt.* **1.** To hit with the fist or hand. **2.** To take part in a boxing match with. *—vi.* To fight with the fists : SPAR.

**box³** (bŏks) *n., pl.* **box** or **box·es.** [ME < OE < Lat. *buxus* < Gk. *puxos.*] **1. a.** An evergreen tree or shrub of the genus *Buxus,* esp. *B. sempervirens,* used for hedges, borders, and garden mazes. **b.** The yellow wood of the box, used to make musical instruments, rulers, inlays, and engraving blocks. **2.** A tree whose timber or foliage is similar to that of box.

**box calf** *n.* [After Joseph *Box,* a 19th-cent. London bootmaker.] Calfskin treated with chromium salts and with square markings on the grain.

**box·car** (bŏks'kär') *n.* An enclosed railway car usu. with movable side doors for transporting freight.

**box coat** *n.* [BOX¹ + COAT.] **1.** A heavy overcoat once worn by coachmen. **2.** A coat designed to hang loosely from the shoulders.

**box-el·der** (bŏks'ĕl'dər) *n.* A maple tree, *Acer negundo* of North America, with compound leaves and lobed leaflets.

**box·er¹** (bŏk'sər) *n.* A person who boxes, esp. a prizefighter.

**box·er²** (bŏk'sər) *n.* [G. < BOXER¹.] A short-haired dog orig. bred in Germany, with a brownish coat and a short square-jawed muzzle.

**Box·er** (bŏk'sər) *n.* [Approximate transl. of Chin. (Mandarin) *yi⁴ he² quan²,* righteous harmonious fists, alteration of *yi⁴ he² tuan²,* righteous harmonious society.] A member of a secret society in China that attempted to drive foreigners from the country in 1900 by violence and to force Chinese Christians to give up their religion.

**boxer shorts** *pl.n.* Full-cut undershorts.

**box·fish** (bŏks'fĭsh') *n., pl.* **boxfish** or **-fish·es.** The trunkfish.

**box·haul** (bŏks'hôl') *vt.* **-hauled, -haul·ing, -hauls.** *Naut.* To turn (a square-rigged ship) about on the heel by bracing the sails aback.

**box·ing¹** (bŏk'sĭng) *n.* Material for making boxes.

**box·ing²** (bŏk'sĭng) *n.* The sport of fighting with the fists.

**Boxing Day** *n.* The first weekday after Christmas, observed in parts of the British Commonwealth as a holiday, when Christmas gifts or boxes are traditionally given to service workers.

**boxing glove** *n.* A heavily padded leather glove worn in boxing.

**box kite** *n.* A tailless, rectangular, box-shaped kite.

**box lunch** *n.* A lunch packed in a container esp. for traveling.

**box office** *n.* **1.** A ticket office, as of a theater or stadium. **2.** The drawing power of a theatrical entertainment or performer.

**box pleat** *n.* A double pleat formed by two facing folds.

**box score** *n.* A printed summary, as of a baseball game, in the form of a table listing each player and his performance statistics.

**box seat** *n.* **1.** A seat in a box at a theater, concert hall, or stadium. **2.** An area or location favorable for observing something.

**box spring** *n. often* **box springs.** A bedspring consisting of a frame enclosed with cloth and containing rows of coil springs.

**box stall** *n.* An enclosed stall for a single animal.

**box·thorn** (bŏks'thôrn') *n.* The matrimony vine.

**box turtle** *n.* A North American turtle of the genus *Terrapene,* with a high-domed shell.

**box·wood** (bŏks'wŏŏd') *n.* **1.** BOX³.

**box·y** (bŏk'sē) *adj.* **-i·er, -i·est.** Like a box. **—box'i·ness** *n.*

**boy** (boi) *n.* [ME *boi,* poss. < OFr. *embuié,* p.part. of *embuier,* to fetter.] **1.** A male child or youth. **2.** *Informal.* A grown man : FELLOW. **3.** A manservant. *—interj.* —Used to express mild astonishment, elation, or disgust. **—boy'hood'** *n.*

**bo·yar** *also* **bo·yard** (bō-yär') *n.* [R. *boyarin* < Old R., of Turkish orig.] A member of a Russian aristocratic class abolished by Peter I.

**boy·cott** (boi'kŏt') *vt.* **-cott·ed, -cott·ing, -cotts.** [After Charles C. *Boycott* (1832–1897).] To abstain from using, buying, or dealing with to express protest or to coerce. **—boy'cott'** *n.* **—boy'cott'er** *n.*

**boy·friend** *also* **boy friend** (boi'frĕnd') *n.* **1.** A male friend. **2.** *Informal.* A favored male companion or sweetheart.

**boy·ish** (boi'ĭsh) *adj.* Typical of or suitable for a boy. **—boy'ish·ly** *adv.* **—boy'ish·ness** *n.*

**Boyle's law** (boilz) *n.* [After Robert *Boyle* (1627–1691), its formulator.] The principle that at a fixed temperature the pressure of a confined ideal gas varies inversely with its volume.

**Boy Scout** *n.* **1.** A member of a worldwide organization of young men and boys, founded in England in 1908, for character development and citizenship training. **2.** One who helps others.

**boy·sen·ber·ry** (boi'zən-bĕr'ē) *n.* [After Rudolph *Boysen* (d. 1950).] **1.** A prickly bramble hybridized from the loganberry and various blackberries and raspberries. **2.** The large, wine-red, edible berry of the boysenberry.

**bo·zo** (bō'zō) *n., pl.* **-zos.** [Orig. unknown.] *Slang.* **1.** A fellow : guy. **2.** A dunce : fool.

**Br** *symbol for* BROMINE.

**bra** (brä) *n.* A brassiere.

**brab·ble** (brăb'əl) *vi.* **-bled, -bling, -bles.** [Poss. < M. Du. *brabbelen,* to jabber.] To quarrel noisily : WRANGLE. **—brab'ble** *n.* **—brab'bler** *n.*

**brace** (brās) *n.* [ME < OFr., two arms < Lat. *bracchia,* pl. of *bracchium,* arm < Gk. *brakhiōn.*] **1.** A device that holds or fastens two or more parts together or in place : CLAMP. **2.** A device, as a supporting beam in a building, that steadies or holds something erect. **3. braces.** *Chiefly Brit.* A pair of suspenders. **4.** An appliance used to support a bodily part <a knee *brace*> **5.** *often* **braces.** An arrangement of bands and wires fixed to the teeth to correct irregular alignment. **6.** *Naut.* A rope by which a yard is swung and secured on a square-rigged ship. **7.** A protective pad strapped to an archer's bow arm. **8.** A leather loop that slides to change the tension on the cords of a drum. **9.** A set of connected musical staves. **10.** A cranklike handle with an adjustable aperture at one end for securing and turning a bit. **11.** One of two symbols, { }, used to connect written or printed lines. **12.** *Math.* Either of a pair of symbols, { }, used to indicate aggregation or clarify the grouping of quantities when parentheses and square brackets have already been used. **13.** *pl.* **brace.** A pair of like things <a *brace* of partridges><a *brace* of dueling pistols> **14.** A very stiff, erect posture. *—v.* **braced, brac·ing, brac·es.** *—vt.* **1.** To furnish with a brace. **2.** To support or hold steady with or as if with a brace. **3.** To prepare or position so as to be ready for impact or danger. **4.** To invigorate : stimulate <mountain air *bracing* the hikers> **5.** To turn (the yards of a ship) by the braces. *—vi.* **1.** To get ready. **2.** To assume a stiff, erect posture.

**brace·let** (brās'lĭt) *n.* [ME < OFr., dim. of *bracel,* armlet < Lat. *bracchiale* < *bracchium,* arm < Gk. *brakhiōn.*] **1.** An ornamental band encircling the wrist. **2.** Something, as handcuffs, resembling a bracelet.

**brac·er¹** (brā'sər) *n.* **1.** One that braces. **2.** *Informal.* A stimulating drink : TONIC.

**bra·cer²** (brā'sər) *n.* [ME < OFr. *braceüre* < *bras,* arm < Lat. *bracchium* < Gk. *brakhiōn.*] An arm or wrist guard worn by archers and fencers.

**bra·ce·ro** (brə-sâr'ō) *n., pl.* **-ros.** [Sp., laborer < *brazo,* arm < Lat. *bracchium* < Gk. *brakhiōn.*] A Mexican laborer allowed to enter the United States and work for a limited time.

**brach** (brăch) *n.* [ME *brache,* back-formation < OFr. *brachez,* pl. of *brachet,* hunting dog < OHG *bracco.*] *Obs.* A bitch hound.

**bra·chi·a** (brā′kē-ə, brăk′ē-ə) *n. pl. of* BRACHIUM.
**bra·chi·al** (brā′kē-əl, brăk′ē-) *adj.* [Lat. *bracchialis* < *bracchium,* arm < Gk. *brakhiōn.*] Of, relating to, or like the arm or a similar or homologous part.
**bra·chi·ate** (brā′kē-ĭt, -āt′, brăk′ē-) *adj.* [Lat. *bracchiatus* < *bracchium,* arm < Gk. *brakhiōn.*] Having widely spreading branches arranged in pairs. —*vi.* (-āt′) **-at·ed, -at·ing, -ates.** To swing by the arms from branch to branch, as certain apes do. **—bra′chi·a′tion** *n.*
**brach·i·o·pod** (brăk′ē-ə-pŏd′, brā′kē-) *n.* [BRACHI(UM) + -POD.] A marine invertebrate of the phylum Brachiopoda, with bivalve dorsal and ventral shells and a pair of tentacled, armlike structures on either side of the mouth. **—brach′i·o·pod′** *adj.*
**bra·chi·um** (brā′kē-əm, brăk′ē-) *n., pl.* **bra·chi·a** (brā′kē-ə, brăk′ē-ə) [Lat. *bracchium,* arm < Gk. *brakhiōn.*] An arm or homologous anatomical structure, as a flipper or wing.
**brachy-** *pref.* [< Gk. *brakhus,* short.] Short <*brachy*dactylic>
**brach·y·ce·phal·ic** (brăk′ĭ-sə-făl′ĭk) *also* **brach·y·ceph·a·lous** (-sĕf′ə-ləs) *adj.* Having a short, almost round head, the width of which is at least 80% as great as the length. **—brach′y·ceph′a·ly** (-sĕf′ə-lē), **brach′y·ceph′a·lism** (-sĕf′ə-lĭz′əm) *n.*
**brach·y·dac·tyl·ic** (brăk′ĭ-dăk-tĭl′ĭk) *also* **brach·y·dac·ty·lous** (-dăk′tə-ləs) *adj.* Having abnormally short fingers or toes. **—brach′y·dac′ty·ly** (-dăk′tə-lē) *n.*
**bra·chyl·o·gy** (bră-kĭl′ə-jē) *n., pl.* **-gies.** [Med. Lat. *brachylogia* < Gk. *brakhulogia* : *brakhus,* short + *logos,* speech.] **1.** Brief, concise speech. **2.** A condensed phrase or expression.
**bra·chyp·ter·ous** (bră-kĭp′tər-əs) *adj.* [Gk. *brakhupteros* : *brakhus,* short + *pteron,* wing.] Having short wings <*brachypterous insects*> **—bra·chyp′ter·ism** (-tə-rĭz′əm) *n.*
**brach·y·u·ran** (brăk′ē-yŏŏr′ən) *also* **brach·y·u·ral** (-əl) *or* **brach·y·u·rous** (-əs) *adj.* [< NLat. *Brachyura,* name suborder : Gk. *brakhus,* short + Gk. *oura,* tail.] Of or belonging to the Brachyura, a group of crustaceans characterized by a short abdomen concealed under the cephalothorax and including the true crabs. **—brach′y·u′ran** *n.*
**brac·ing** (brā′sĭng) *adj.* Invigorating. **—brac′ing·ly** *adv.*
**brack·en** (brăk′ən) *n.* [ME *braken,* prob. of Scand. orig.] **1.** A fern, *Pteridium aquilinum,* with tough stems and finely divided fronds. **2.** An area overgrown with bracken. **3.** A large, coarse fern.
**brack·et** (brăk′ĭt) *n.* [OFr. *braguette,* codpiece, dim. of *brague,* breeches < OProv. *braga* < Lat. *bracae.*] **1.** A simple rigid L-shaped structure, one arm of which is fixed to a vertical surface, the other projecting horizontally to support a weight, as a shelf. **2.** A wall-anchored fixture adapted to support a load. **3.** A small shelf or shelves supported by brackets. **4. a.** Either of a pair of symbols, [ ], used to enclose written or printed material. **b.** Either of a pair of symbols, < >, similarly used and in mathematics used esp. together to indicate the average of a contained quantity. **c.** BRACE 12. **5.** A classification or grouping, esp. of taxpayers according to income. **6.** The space between two rounds of artillery, the first aimed beyond a target and the second aimed short of it, used to determine range. —*vt.* **-et·ed, -et·ing, -ets. 1.** To support with a bracket or brackets. **2.** To place within or as if within brackets. **3.** To classify or group together. **4.** To fire beyond and short of (a target) in order to determine range.
**bracket fungus** *n.* Any of various fungi that form shelflike growths on tree trunks and wood structures.
**brack·ish** (brăk′ĭsh) *adj.* [Du. *brak.*] **1.** Containing some salt : BRINY. **2.** Unpalatable : distasteful. **—brack′ish·ness** *n.*
**bract** (brăkt) *n.* [NLat. *bractea* < Lat. gold leaf.] A leaflike plant part, usu. small but occas. showy and sometimes brightly colored, located either below a flower or on the stalk of a flower cluster. **—brac′te·al** (brăk′tē-əl) *adj.*
**brac·te·ate** (brăk′tē-ĭt, -āt′) *adj.* [NLat. *bracteatus* < *bractea,* gold leaf.] Having bracts.
**brac·te·o·late** (brăk′tē-ə-lĭt, -lāt′) *adj.* Having bracteoles.
**brac·te·ole** (brăk′tē-ōl′) *n.* [NLat. *bracteola* < Lat., dim. of *bractea,* gold leaf.] A small or secondary bract.
**brad** (brăd) *n.* [ME < ON *broddr,* spike.] A tapered nail with a small head or a slight side projection instead of a head.
**brad·awl** (brăd′ôl′) *n.* A small awl with a chisel edge, used in making holes in wood for brads or screws.
**brady-** *pref.* [NLat. < Gk. *bradus,* slow.] Slow <*brady*cardia>
**brad·y·car·di·a** (brăd′ĭ-kär′dē-ə) *n.* [BRADY- + Gk. *kardia,* heart.] Abnormally slow heartbeat. **—brad′y·car′dic** (-dĭk) *adj.*
**brad·y·lex·i·a** (brăd′ĭ-lĕk′sē-ə) *n.* [BRADY- + Gk. *lexis,* speech < *legein,* to speak.] A slowness of reading not attributable to lack of intelligence.
**brad·y·lo·gia** (brăd′ə-lō′jə, -jē-ə) *n.* [NLat. : BRADY- + Gk. *-logia,* -logy.] Abnormally slow speech.
**brae** (brā) *n.* [ME *bra* < ON *brā,* eyelid.] *Scot.* A hillside : slope.
**brag** (brăg) *v.* **bragged, brag·ging, brags.** [ME *braggen* < *brag,* ostentatious.] —*vi.* To talk boastfully. —*vt.* To assert boastfully. —*n.* **1.** Boastful or arrogant speech or manner. **2.** Something boasted of. **3.** A braggart. **4.** A card game similar to poker. —*adj.* **brag·ger, brag·gest.** Exceptionally fine. **—brag′ger** *n.*
**brag·ga·do·ci·o** (brăg′ə-dō′sē-ō′, -shē-ō′, -shō) *n., pl.* **-os.** [Alteration of *Braggadocchio,* the personification of vainglory in *The Fairie Queene* by Sir Edmund Spenser (1552–1599).] **1.** A braggart. **2. a.** Empty, pretentious bragging. **b.** Swaggering manner.
**Bragg angle** (brăg) *n.* [After William Henry *Bragg* (1862–1942) and William Lawrence *Bragg* (1890–1971).] The angle between an incident x-ray beam and a set of crystal planes for which the secondary radiation displays maximum intensity as a result of constructive interference.
**brag·gart** (brăg′ərt) *n.* [Fr. *bragard* < *braguer,* to brag, perh. < ME *braggen.*] A person given to empty, pretentious boasting : BRAGGER. —*adj.* Boastful.
**Bragg's law** (brăgz) *n.* [After William Henry *Bragg* (1862–1942) and William Lawrence *Bragg* (1890–1971).] The fundamental law of x-ray crystallography, $n\lambda = 2d\sin\Theta$, where $n$ is an integer, $\lambda$ is the wavelength of a beam of x-rays incident on a crystal with lattice planes separated by distance $d$, and $\Theta$ is the Bragg angle.
**Brah·ma**[1] (brä′mə) *n.* [Skt. *brahman.*] **1.** The personification of divine reality in its creative aspect as a member of the Hindu triad. **2.** *var. of* BRAHMAN 1, 3.
**Brah·ma**[2] *also* **brah·ma** (brä′mə, brā′-) *n.* [After the *Brahmaputra* River in southern Asia.] A large domestic fowl orig. bred in Asia that has feathered legs.
**Brah·man** (brä′mən) *n.* [Skt.] **1.** *also* **Brah·ma** (-mə). The essential divine reality of the universe in Hinduism, the eternal spirit from which all being originates and to which all returns. **2.** *also* **Brah·min** (-mĭn). A member of the highest Hindu caste, orig. composed only of priests. **3.** *also* **Brah·ma** (-mə) *or* **Brah·min** (-mĭn). One of a breed of domestic cattle developed in the southern United States from stock orig. bred in India and having a hump between the shoulders and a pendulous dewlap. **—Brah·man′ic** (-măn′ĭk), **Brah·man′i·cal** *adj.*
**Brah·man·ism** (brä′mə-nĭz′əm) *also* **Brah·min·ism** (brä′mĭ-) *n.* **1.** The religious practices and beliefs of ancient India as reflected in the Vedas. **2.** The social caste system of the Brahmans of India. **—Brah′man·ist** *n.*
**Brah·min** (brä′mĭn) *n.* **1.** A highly cultured and socially exclusive person, esp. a member of one of the old New England families. **2.** *var. of* BRAHMAN 2, 3. **—Brah·min′ic** (-mĭn′ĭk), **Brah·min′i·cal** *adj.*
**Brah·min·ism** (brä′mĭ-nĭz′əm) *n.* **1.** The behavior or attitudes characteristic of a cultural or social elite. **2.** *var. of* BRAHMANISM.
**braid** (brād) *vt.* **braid·ed, braid·ing, braids.** [ME *braiden* < OE *bregdan,* to weave.] **1.** To interweave three or more strands of : PLAIT. **2.** To decorate or edge with an ornamental trim. **3.** To produce by interweaving. **4.** To fasten or decorate (hair) with a band or ribbon. —*n.* **1.** A narrow length of plaited material, as hair or fabric. **2.** A thin, flat woven strip of cloth used for binding or decorating fabrics. **3.** A ribbon or band used to fasten the hair. **—braid′er** *n.*
**braid·ed** (brā′dĭd) *adj.* Flowing in an interconnected network of channels <a *braided* stream>
**braid·ing** (brā′dĭng) *n.* Braided embroidery.
**brail** (brāl) *n.* [ME *brayle* < OFr. *brail,* belt < Med. Lat. *bracale* < Lat. *bracae,* breeches.] A line used to furl loose-footed sails. —*vt.* **brailed, brail·ing, brails.** To gather in (a sail) with brails.
**Braille** *also* **braille** (brāl) *n.* [After Louis *Braille* (1809–1852).] A system of writing and printing for the blind, in which varied arrangements of raised dots representing letters and numerals can be identified by touch.

1 a, 2 b, 3 c, 4 d, 5 e, 6 f, 7 g, 8 h, 9 i, 0 j, k, l, m, n, o, p, q, r, s, t, u, v, w, x, y, z

Braille

**brain** (brān) *n.* [ME < OE *brægen.*] **1. a.** The part of the central nervous system in the vertebrate cranium that is responsible for interpretation of sensory impulses, coordination and control of bodily activities, and exercise of emotion and thought. **b.** A functionally similar portion of the invertebrate nervous system. **2. brains.** Intellectual capacity. **3.** *Informal.* A highly intelligent person. **4.** *often* **brains.** The supreme planner, as of a movement. **5.** An automatic device, as a computer, central to a computation or control process. —*vt.* **brained, brain·ing, brains. 1.** To smash in the skull of. **2.** *Slang.* To hit on the head. **—on the brain.** Obsessively in mind. **—rack (or beat) (one's) brains.** To think as hard as one can.
**brain case** *n.* The brainpan.

**brain child** *n. Informal.* An original plan or idea attributed to a specific individual or group.
**brain coral** *n.* A coral of the genus *Meandrina,* forming rounded colonies that resemble the human brain.
**brain death** *n.* Death as shown by absence of central-nervous-system activity. **—brain'-dead'** (brān'dĕd') *adj.*
**brain drain** *n.* Defection of professionals, as scientists or scholars, to countries offering higher salaries and better living conditions.
**brain fever** *n.* Encephalitis.
**brain·less** (brān'lĭs) *adj.* Lacking intelligence : STUPID. **—brain'less·ly** *adv.* **—brain'less·ness** *n.*
**brain·pan** (brān'păn') *n.* The part of the skull containing the brain : CRANIUM.
**brain-pick·ing** (brān'pĭk'ing) *n.* The act of probing another's mind for information. **—brain'-pick'er** *n.*
**brain·pow·er** (brān'pou'ər) *n.* **1.** Intellectual ability or power. **2.** People with accellerated mental ability.
**brain scanner** *n.* A CAT scanner used to x-ray the brain. **—brain scan** *n.*
**brain·sick** (brān'sĭk') *adj.* Of, relating to, or caused by insanity : MAD. **—brain'sick'ly** *adv.* **—brain'sick'ness** *n.*
**brain·stem** (brān'stĕm') *n.* The part of the brain consisting of the medulla oblongata, pons, and mesencephalon and connecting the spinal cord to the forebrain and cerebrum.
**brain·storm** (brān'stôrm') *n.* **1.** A sudden and violent disturbance in the brain. **2. a.** A spontaneous clever idea. **b.** A foolish idea.
**brain·storm·ing** (brān'stôr'mĭng) *n.* A method of problem solving in which all members of a group spontaneously contribute ideas. **—brain'storm'** *v.* **(-stormed, -storm·ing, -storms). —brain'storm'er** *n.*
**brain trust** *n.* A group of experts who serve as unofficial advisers and policy planners, esp. in a government. **—brain truster** *n.*
**brain·wash** (brān'wŏsh', -wôsh') *vt.* **-washed, -wash·ing, -wash·es.** [Back-formation < BRAINWASHING.] To subject to brainwashing.
**brain·wash·ing** (brān'wŏsh'ĭng, -wô'shĭng) *n.* [Transl. of Chin. (Mandarin) *xi*³ *nao*³ : *xi*³, to wash + *nao*³, brain.] **1.** Intensive, usu. political indoctrination aimed at changing a person's basic convictions and attitudes and replacing them with a fixed and unquestioned set of beliefs. **2.** Indoctrination or persuasion by advertising or salesmanship.
**brain wave** *n.* **1.** A rhythmic fluctuation of electric potential between parts of the brain. **2.** BRAINSTORM 2a.
**brain·y** (brā'nē) *adj.* **-i·er, -i·est.** *Informal.* Intellectually advanced : SMART. **—brain'i·ly** *adv.* **—brain'i·ness** *n.*
**braise** (brāz) *vt.* **braised, brais·ing, brais·es.** [Fr. *braiser* < *braise,* hot charcoal < OFr. *brese,* of Germanic orig.] To cook (meat, poultry, or vegetables) by browning in fat and then simmering in a small amount of liquid in a covered pan.
**brake¹** (brāk) *n.* [ME.] **1.** A device for separating the fibers of flax or hemp by crushing or beating. **2.** A heavy harrow for breaking clods of earth. **3.** A handle on a machine, as a pump. **4.** A machine for bending and folding sheet metal. *—vt.* **braked, brak·ing, brakes. 1.** To crush (flax or hemp) in a brake. **2.** To break up (clods of earth) with a harrow.
**brake²** (brāk) *n.* [Perh. ME *breake,* bridle, curb.] **1.** A device for reducing or stopping motion, as of a vehicle, esp. by contact friction. **2.** Something that reduces or stops action. *—v.* **braked, brak·ing, brakes.** *—vt.* To reduce the speed of with or as if with a brake. *—vi.* To operate or apply a brake.
**brake³** (brāk) *n.* [ME.] A fern, esp. bracken.
**brake⁴** (brāk) *n.* [ME.] An area overgrown with dense brushwood, briers, and undergrowth : THICKET.
**brake⁵** (brāk) *n. var. of* BREAK 19.
**brake⁶** (brāk) *v. Archaic. var. p.t. of* BREAK.
**brake·age** (brā'kĭj) *n.* Braking capacity or action.
**brake band** *n.* A flexible belt tightened around a brake drum to reduce or stop the motion of a shaft or wheel.
**brake drum** *n.* A metal cylinder to which pressure is applied by a braking mechanism so as to arrest rotation of the wheel or shaft to which the cylinder is attached.
**brake fluid** *n.* Liquid used in a hydraulic brake cylinder.
**brake horsepower** *n.* The actual or useful horsepower of an engine, usu. determined from the force exerted on a dynamometer connected to the drive shaft.
**brake·man** (brāk'mən) *n.* A railroad employee who assists the conductor and inspects the train.
**brake pad** *n.* A flat block that presses against the disk of a disc brake.
**brake shoe** *n.* A curved block that presses against and reduces or stops the rotation of a wheel or shaft.
**braking rocket** *n.* A retrorocket.
**bra·less** (brä'lĭs) *adj. & adv.* Wearing no brassiere.
**bram·ble** (brăm'bəl) *n.* [ME *brembel* < OE *brǣmbel.*] **1.** A prickly plant or shrub of the genus *Rubus,* esp. the blackberry or the raspberry. **2.** A prickly bush or shrub. **—bram'bly** *adj.*
**bram·bling** (brăm'blĭng) *n.* [BRAMB(LE) + -LING¹.] A finch, *Fringilla montifringilla* of northern Eurasia, with black, white, and rust-brown plumage.
**bran** (brăn) *n.* [ME < OFr., of Celt. orig.] **1.** The seed husk of cereals, as wheat, rye, and oats, separated from the flour by sifting or bolting. **2.** Cereal by-products used as a food.
**branch** (brănch) *n.* [ME < OFr. *branche* < LLat. *branca,* paw.] **1.** A secondary woody stem or limb growing from the trunk or main stem of a tree, bush, or shrub or from another secondary limb. **2.** A part like or structurally analogous to a branch. **3.** A limited part of a larger or more complex body, esp.: **a.** An academic or vocational field of specialization. **b.** A local unit of a business. **c.** A division of a group, as a tribe or family, held to stem from a common ancestor. **4.** A subdivision of a language family. **5. a.** A tributary of a river. **b.** A small creek, stream, or brook. **6.** *Math.* A part of a curve that is separated, as by discontinuities or extreme points. **7.** *Computer Sci.* **a.** A program instruction causing a departure from the normal instructional sequence. **b.** The instructions executed as the result of such a departure. *—v.* **branched, branch·ing, branch·es.** *—vi.* **1.** To spread out in branches. **2.** To separate into subdivisions : DIVERGE. **3.** *Computer Sci.* To depart from an instructional sequence as a result of a branch. *—vt.* **1.** To separate into or as if into branches. **2.** To embroider with a design of foliage or flowers. **—branch off.** To diverge from the main part or course. **—branch out.** To expand one's interests or activities. **—branched** *adj.* **—branch'less** *adj.* **—branch'y** *adj.*
**bran·chi·a** (brăng'kē-ə) *n., pl.* **-chi·ae** (-kē-ē) [Lat. < Gk. *brankhia,* gills.] A breathing organ, as a gill. **—bran'chi·al** *adj.*
**bran·chi·ate** (brăng'kē-ĭt, -āt') *adj.* Having branchiae or gills.
**bran·chi·o·pod** (brăng'kē-ə-pŏd') *n.* [NLat. *Branchiopoda,* subclass name : Lat. *branchia,* gill + Gk. *pous,* foot.] Any of various crustaceans of the subclass Branchiopoda, having a segmented body and flattened, limblike appendages. **—bran'chi·o·pod', bran'chi·op'o·dan** (-ŏp'ə-dən), **bran'chi·op'a·dous** (-ŏp'ə-dəs) *adj.*
**branch water** *n.* [< *branch water,* water from a stream.] Plain water, esp. when mixed with liquor.
**brand** (brănd) *n.* [ME, torch < OE.] **1. a.** A trademark or distinctive name of a product or manufacturer. **b.** The make of a product thus marked. **2.** A mark indicating identity or ownership burned on the hide of an animal with a hot iron. **3.** A mark once burned into the skin of criminals. **4.** A mark of notoriety or shame : STIGMA. **5.** An iron heated and used for branding. **6.** Burning or charred wood. **7.** *Archaic.* A sword. *—vt.* **brand·ed, brand·ing, brands.** To mark with or as if with a brand <*branded* the cattle><was *branded* as a liar> **—brand'er** *n.*
**branding iron** *n.* A heated metal rod used for branding.
**bran·dish** (brăn'dĭsh) *vt.* **-dished, -dish·ing, -dish·es.** [ME *brandissen* < OFr. *brandir, bandiss-* < *brand,* sword, of Germanic orig.] **1.** To wave threateningly, as a weapon. **2.** To display ostentatiously. *—n.* A threatening or defiant gesture. **—bran'dish·er** *n.*
**brand·ling** (brănd'lĭng) *n.* [BRAND (from its markings) + -LING.] A reddish-brown earthworm, *Eisenia foetida,* often used as bait.
**brand name** *n.* TRADE NAME 1.
**brand-new** (brănd'no͞o', -nyo͞o') *adj.* Totally new.
**bran·dy** (brăn'dē) *n., pl.* **-dies.** [Short for *brandy-wine* < Du. *brandewijn* : *branden,* to distill + *wijn,* wine.] An alcoholic liquor distilled from fermented fruit juice or wine. *—vt.* **-died, -dy·ing, -dies.** To blend, flavor, or preserve with brandy.
**branks** (brăngks) *n.* [Orig. unknown.] *(sing. or pl. in number).* A metal bridle with a bit to restrain the tongue, once used to punish scolds.
**bran·ni·gan** (brăn'ĭ-gən) *n.* [Prob. from the name *Brannigan.*] *Slang.* **1.** A raucous argument : BRAWL. **2.** A drinking spree : BINGE.
**brant** (brănt) *also* **brent** (brĕnt) *n., pl.* **brant** or **brants** *also* **brent** or **brents.** [Orig. unknown.] A wild goose of the genus *Branta* that breeds in arctic regions, esp. *B. bernicla,* with a black head and neck.
**brash¹** (brăsh) *adj.* **-er, -est.** [Orig. unknown.] **1.** Rash and unthinking : HASTY. **2.** Saucy : impudent. **3.** Brittle <*brash* timbers> **—brash'ly** *adv.* **—brash'ness** *n.*
**brash²** (brăsh) *n.* [Perh. an alteration of Fr. *brèche,* breach.] An accumulation of rubble or fragments, as of floating ice.
**brass** (brăs) *n.* [ME *bras* < OE *bræs.*] **1.** An alloy of copper and zinc. **2.** Ornaments, objects, or utensils made of brass. **3.** *often* **brasses.** Wind instruments, as the French horn and trombone, made of brass. **4.** A brass memorial plaque or tablet. **5.** A bushing sleeve or similar lining for a bearing, made from a copper alloy. **6.** *Informal.* Blatant self-assurance : EFFRONTERY. **7.** *Slang.* Brass hats. **8.** *Chiefly Brit. Slang.* Money. **—brass** *adj.*
**bras·sage** (brăs'ĭj) *n.* [Fr., act of stirring, coining money, brassage < *brasser,* to brew, mix < OFr. *bracier* < Lat. *braces,* a kind of grain, of Celtic orig.] A charge issued by a government to cover the cost of converting bullion to coins.
**bras·sard** (brə-särd', brăs'ärd') *n.* [Fr. < Prov. *brassal* < *bras,* arm < Lat. *brācchium* < Gk. *brakhion.*] **1.** A cloth badge worn around the upper arm. **2.** Armor for the arm.

**brass·bound** (brăs′bound′) *adj.* Firmly established : RIGID.
**brass-col·lar** (brăs′kŏl′ər) *adj.* Voting the straight party ticket.
**bras·se·rie** (brăs′ə-rē′, brăs-rē′) *n.* [Fr. < *brasser,* to brew. —see BRASSAGE.] A restaurant offering alcoholic beverages, esp. wine or beer.
**brass hat** *n.* [From the gold braid on his hat.] *Slang.* A high-ranking military officer or civilian official.
**brass·ie** *also* **brass·y** (brăs′ē) *n., pl.* **-ies.** A wooden golf club with a brass-plated sole, used for long low shots.
**bras·siere** *or* **bras·sière** (brə-zîr′) *n.* [Fr. < OFr. *braciere,* armguard < *bras,* arm < Lat. *brācchium* < Gk. *brakhion.*] A woman's undergarment that supports and contours the breasts.
**brass knuckles** *pl.n.* A weapon with a metal strip or chain having holes or links into which the fingers fit.
**brass tacks** *pl.n. Informal.* Essential facts : BASICS.
**brass·y**[1] (brăs′ē) *adj.* **-i·er, -i·est. 1.** Made of or adorned with brass. **2.** Resembling brass in color. **3.** Like or marked by the sound of brass instruments. **4.** Showy and cheap : FLASHY. **5.** *Informal.* Brazenly insolent. **—brass′i·ly** *adv.* **—brass′i·ness** *n.*
**brass·y**[2] (brăs′ē) *n. var. of* BRASSIE.
**brat** (brăt) *n.* [Poss. < *brat,* coarse garment < ME < OE *bratt,* of Celt. orig.] A spoiled, bad-mannered child. **—brat′ty** *adj.*
**brat·tice** (brăt′ĭs) *n.* [ME *bretice,* defensive structure < OFr. *breteche* < Med. Lat. *brittisca.*] A partition, esp. one that is erected to ventilate a mine. *—vt.* **-ticed, -tic·ing, -tic·es.** To equip with a brattice.
**brat·tle** (brăt′l) *n.* [Imit.] *Scot.* A clattering or rattling sound. **—brat′tle** *v.* **(-tled, -tling, -tles).**
**brat·wurst** (brăt′wûrst′, -vŏŏrst′) *n.* [G. < OHG *brātwurst* : *brāto,* meat + *wurst,* sausage.] A sausage of chopped, seasoned fresh pork.
**Braun·schwei·ger** (broun′shwī′gər) *n.* [G. < *Braunschweig,* Brunswick, West Germany.] A smoked liver sausage.
**bra·va** (brä′vä, brä-vä′) *interj.* [Ital., fem. of *bravo,* bravo.] —Used to express approval in applauding a woman. **—bra′va** *n.*
**bra·va·do** (brə-vä′dō) *n., pl.* **-does** *or* **-dos.** [Sp. *bravada* < *bravo,* brave.] **1.** Arrogant behavior. **2.** A pretense of courage.
**brave** (brāv) *adj.* **brav·er, brav·est.** [OFr. < OItal. *bravo.*] **1.** Possessing or displaying courage : VALIANT. **2.** Making a fine display : SPLENDID. **3.** *Archaic.* Excellent. *—n.* **1.** A North American Indian warrior. **2.** A courageous person. **3.** *Obs.* A bully. **4.** *Obs.* A boast or challenge. *—v.* **braved, brav·ing, braves.** *—vt.* **1.** To undergo or face courageously. **2.** To defy : challenge. **3.** *Obs.* To make splendid. *—vi. Obs.* To boast. **—brave′ly** *adv.* **—brave′ness** *n.*
☆ ***syns:*** BRAVE, AUDACIOUS, BOLD, COURAGEOUS, DAUNTLESS, FEARLESS, GALLANT, GUTSY, HEROIC, INTREPID, METTLESOME, PLUCKY, STOUTHEARTED, UNAFRAID, UNDAUNTED, VALIANT, VALOROUS *adj. core meaning* : having or displaying courage <a *brave* soldier><a *brave* rescue attempt> ***ant:*** cowardly
**brav·er·y** (brā′və-rē, brāv′rē) *n., pl.* **-ies. 1.** The quality or state of being brave : COURAGE. **2.** Splendor, as of attire : SHOW.
**bra·vis·si·mo** (brä-vĭs′ə-mō′) *interj.* [Ital., superl. of *bravo,* fine.] —Used to express great approval.
**bra·vo**[1] (brä′vō, brä-vō′) *interj.* [Ital., fine.] —Used to express approval. **—bra′vo** *n.*
**bra·vo**[2] (brä′vō) *n., pl.* **-voes** *or* **-vos.** [Ital., brigand.] A hired killer.
**bra·vu·ra** (brə-vŏŏr′ə, -vyŏŏr′ə) *n.* [Ital. < *bravo,* fine.] **1.** *Mus.* Brilliant technique or style in performance. **2.** Showy display or behavior. **—bra·vu′ra** *adj.*
**braw** (brô) *adj.* **-er, -est.** [Sc., alteration of BRAVE.] *Scot.* Fine or splendid : EXCELLENT.
**brawl** (brôl) *n.* [ME *brall* < *brallen,* to quarrel.] **1.** A noisy argument or fight. **2.** *Slang.* A loud party. *—vi.* **brawled, brawl·ing, brawls. 1.** To argue noisily. **2.** To flow noisily, as water. **—brawl′er** *n.* **—brawl′ing·ly** *adv.*
**brawn** (brôn) *n.* [ME, muscle < OFr. *braon,* meat, of Germanic orig.] **1.** Stout, well-developed muscles. **2.** Muscular strength and power. **3.** *Chiefly Brit.* **a.** A pig. **b.** A pickled or preserved preparation made from meat of the head or feet of a pig.
**brawn·y** (brô′nē) *adj.* **-i·er, -i·est.** Strong and muscular. **—brawn′i·ly** *adv.* **—brawn′i·ness** *n.*
**bray**[1] (brā) *v.* **brayed, bray·ing, brays.** [ME *brayen* < OFr. *braire,* prob. of Celt. orig.] *—vi.* **1.** To utter the loud, harsh cry of a donkey. **2.** To sound loudly and harshly. *—vt.* To utter loudly and harshly. **—bray** *n.*
**bray**[2] (brā) *vt.* **brayed, bray·ing, brays.** [ME *brayen* < OFr. *breier,* of Germanic orig.] **1.** To crush and pound in or as if in a mortar. **2.** To spread (ink) thinly over a surface.
**bray·er**[1] (brā′ər) *n.* One that brays.
**bray·er**[2] (brā′ər) *n.* A small hand roller for spreading ink thinly and evenly over printing type.
**braze**[1] (brāz) *vt.* **brazed, braz·ing, braz·es.** [ME *brasen* < OE *brasian* < *bræs,* brass.] **1.** To make of or decorate with brass. **2.** To make hard like brass.
**braze**[2] (brāz) *vt.* **brazed, braz·ing, braz·es.** [Prob. < Fr. *braser* < OFr. < *brese,* hot coals.] To solder (two pieces of metal) together using a hard solder with a high melting point. **—braz′er** *n.*
**bra·zen** (brā′zən) *adj.* [ME *brasen* < OE *bræsen* < *bræs,* brass.] **1.** Made of brass. **2.** Resembling brass in color, quality, or hardness. **3.** Having a loud, resonant sound like that of a brass trumpet. **4.** Bold : impudent. *—vt.* **-zened, -zen·ing, -zens.** To face or undergo with bold self-assurance <*brazened* out the crisis> **—bra′zen·ly** *adv.* **—bra′zen·ness** *n.*
**bra·zen·faced** (brā′zən-fāst′) *adj.* BRAZEN 4.
**bra·zier**[1] (brā′zhər) *n.* [ME *brasier* < *bras,* brass.] A person who works with brass.
**bra·zier**[2] (brā′zhər) *n.* [Fr. *brasier* < *braise,* hot coals.] A metal pan for holding burning coals.
**braz·i·lin** (brăz′ə-lĭn, brə-zĭl′ən) *n.* [Fr. *brésiline* < *brésil,* brazilwood < OFr. *bresil.*] A crystalline compound, $C_{16}H_{14}O_5$, obtained from brazilwood and used as a dye.
**Bra·zil nut** (brə-zĭl′) *n.* [After *Brazil.*] **1.** A tree, *Bertholletia excelsa* of tropical South America, having hard, round, woody pods that contain approx. 20 to 30 nuts. **2.** The edible nut of the Brazil nut.
**bra·zil·wood** (brə-zĭl′wŏŏd′) *n.* [Obs. *brazil,* brazilwood (< ME *brasil* < OFr. *bresil*) + WOOD.] The red wood of any of several tropical trees of the genus *Caesalpinia,* used for cabinetwork and as the source of a red or purple dye.
**breach** (brēch) *n.* [ME *breche* < OE *brēc.*] **1.** A violation or infraction, as of a law, obligation, or promise. **2.** A gap or rift in a solid structure, as a dike or fortification. **3.** A disruption of friendly relations : ESTRANGEMENT. **4.** A leap of a whale from the water. **5.** The breaking of waves. **6.** *Obs.* A wound : injury. *—v.* **breached, breach·ing, breach·es.** *—vt.* To make a hole or gap in : break through. *—vi.* To leap from the water <watched the whale *breach*>
☆ ***syns:*** BREACH, BREAK, CLEFT, GAP, HOLE, PERFORATION, RENT, RUPTURE *n. core meaning* : an opening in a solid structure <a *breach* in the dike>
**breach of promise** *n.* Nonfulfillment of a promise, esp. a promise to marry.
**bread** (brĕd) *n.* [ME < OE *brēad.*] **1.** A usu. leavened staple food made from a flour or meal mixture that is shaped into loaves and baked. **2.** Food in general. **3. a.** The necessities of life : LIVELIHOOD. **b.** Something that nourishes : SUSTENANCE. **c.** *Slang.* Money. *—vt.* **bread·ed, bread·ing, breads.** To coat with bread crumbs, as before cooking.
▲ word history: Old English *brēad* does occur (in the northern dialects) with the modern meaning, but the more common and probably older sense of the Old English word is "crumb," "morsel," or "fragment." The usual word for "bread" in Old English was *hlāf,* which has come down to Modern English as *loaf*[1].
**bread and butter** *n. Informal.* Livelihood.
**bread-and-butter** (brĕd′n-bŭt′ər) *adj.* **1.** Influenced by or undertaken out of necessity <a *bread-and-butter* job> **2.** Showing gratitude for hospitality <wrote a *bread-and-butter* note>
**bread·bas·ket** (brĕd′băs′kĭt) *n.* **1.** A basket for serving bread. **2.** A geographic region that is a principal source of grain. **3.** *Slang.* The stomach.
**bread·board** (brĕd′bôrd′, -bōrd′) *n.* **1.** A board on which bread is sliced. **2.** An experimental model, esp. of an electric circuit : PROTOTYPE. *—vt.* **-board·ed, -board·ing, -boards.** To construct an experimental model of (e.g., an electric circuit). **—bread′board′ing** *n.*
**bread·fruit** (brĕd′frōōt′) *n.* **1.** A tree, *Artocarpus communis* or *A. incisa* of Polynesia, with deeply lobed leaves and round, usu. seedless fruit. **2.** The edible fruit of the breadfruit.
**bread mold** *n.* A fungus, *Rhizopus nigricans,* that forms a dense, cottony growth on foods, as bread.
**bread·nut** (brĕd′nŭt′) *n.* **1.** A tree, *Brosimum alicastrum* of Central America and the West Indies, bearing round, nutlike fruit. **2.** The fruit of the breadnut, ground to produce a substitute for wheat flour.
**bread·root** (brĕd′rōōt′, -rŏŏt′) *n.* A plant, *Psoralea esculenta* of the central North American plains, with an edible, starchy root.
**bread·stuff** (brĕd′stŭf′) *n.* **1.** Bread. **2.** Flour, meal, or grain used to make bread.
**breadth** (brĕdth) *n.* [ME *brede* < OE *brǣd.*] **1.** The measure or dimension from side to side. **2.** A piece usu. produced in a standard width <a *breadth* of canvas> **3.** Wide scope or extent. **4.** Freedom from narrowness, as of views or interests.
**breadth·wise** (brĕdth′wīz′) *also* **breadth·ways** (-wāz′) *adv. & adj.* In the direction of the breadth.
**bread·win·ner** (brĕd′wĭn′ər) *n.* One whose earnings support a family or household.
**break** (brāk) *v.* **broke** (brōk), **bro·ken** (brō′kən), **break·ing, breaks.** [ME *breken* < OE *brecan.*] *—vt.* **1.** To cause to come apart into pieces suddenly or violently : SMASH. **2.** To crack without separating into pieces. **3.** To render unusable or inoperative. **4.** To part or pierce the surface of <*break* ground> **5.** To cause to burst. **6.** To fracture a bone of. **7.** To force or make a way through : PENETRATE. **8.** To force one's way out of : escape from <*break* jail> **9.** To put an end to by force or strong opposition <*broke* the union> **10.** To fail to conform to : VIOLATE. **11.** To disrupt abruptly : INTERRUPT <con-

centration *broken* by a noise> **12.** To cause to give up a habit. **13.** To train to obey : TAME. **14.** To disrupt the order or regularity of <*break* ranks> **15.** To destroy the completeness of <*break* a set of books> **16.** To lessen in force or effect <*break* a fall> **17.** To weaken or destroy, as in spirit or health. **18.** To overwhelm with grief. **19.** To reduce to bankruptcy. **20.** To reduce in rank : DEMOTE. **21.** To reduce to or exchange for smaller monetary units <*break* a ten-dollar bill> **22.** To exceed or outdo <*break* a speed record> **23.** To make known <*break* a news story> **24.** To find the key or solution to. **25.** *Law.* To invalidate (a will) by judicial action. **26.** *Elect.* To open <*break* a circuit> —*vi.* **1.** To become separated into pieces or fragments. **2.** To become inoperative. **3.** To give way : COLLAPSE. **4.** To diminish or discontinue abruptly <The drought finally *broke.*> **5.** To disperse or scatter. **6.** To move away or escape suddenly. **7.** To change direction suddenly. **8.** To come into being or notice, esp. suddenly. **9.** To emerge above the surface of water. **10.** To be overwhelmed with grief. **11.** To begin abruptly to produce or utter something <*broke* into laughter> **12.** To plummet <Stock prices *broke* at the news.> **13.** To collapse or crash into surf or spray, as waves. **14.** To change from one tonal quality or musical register to another. **15.** *Baseball.* To curve near or over a plate <The pitch *broke* sharply.> **16.** *Informal.* To occur in a specified way <Things *broke* well for the new company.> **—break down. 1.** To fail to function. **2.** To have a mental or physical collapse. **3.** To become or cause to become distressed or upset. **4.** To consider in parts : ANALYZE. **5.** To undergo chemical decompositon. **6.** To undergo electrical breakdown. **—break in. 1.** To train or adapt for a purpose. **2.** To reduce the stiffness of (a new product) <*break in* new shoes><*break in* a new car> **3.** To enter forcibly or suddenly. **—break in on** (or **upon).** To interrupt. **—break into. 1.** To enter forcibly, suddenly, or illegally. **2.** To interrupt. **3.** To begin suddenly. **—break off. 1.** To stop suddenly, as in speaking. **2.** To discontinue (a relationship). **—break out. 1.** To become affected with a skin eruption. **2.** To develop suddenly and forcefully <Fighting *broke out* along the DMZ.> **3. a.** To ready for action or use <*break out* the ammunition> **b.** To produce for consumption <*break out* the beer> **—break up. 1.** To bring or come to an end. **2.** *Informal.* To burst or cause to burst into laughter. —*n.* **1.** The act of breaking. **2.** The result of breaking. **3.** A beginning <the *break* of day> **4.** A dash, esp. to escape. **5.** A disruption in continuity or regularity. **6.** A pause, as from work. **7.** A sudden or marked change. **8.** *Informal.* A chance occurrence, esp. an unexpected opportunity. **9.** Severance of ties. **10.** A sudden decline in prices. **11.** A caesura. **12. a.** The space between two paragraphs. **b.** A series of three dots ( . . . ) used to indicate an omission in a text. **13.** *Elect.* Interruption of a flow of current. **14.** *Mus.* **a.** The point at which a register or a tonal quality changes to another register or tonal quality. **b.** The change itself. **c.** A solo jazz cadenza played during the pause between the regular phrases or choruses of a melody. **15.** The swerve of a ball from a straight path of flight when thrown, as in baseball. **16.** The opening shot in billiards. **17.** A run or unbroken series of successful shots, as in billiards or croquet. **18.** Failure to score a strike or a spare in a given bowling frame. **19.** *also* **brake.** A high, open horse-drawn carriage with four wheels. **—break camp.** To pack up equipment and leave a campsite.

**break·a·ble** (brā′kə-bəl) *adj.* Capable of being broken. —*n. often* **breakables.** Easily broken articles. **—break′a·ble·ness** *n.*

**break·age** (brā′kĭj) *n.* **1.** The act or process of breaking. **2.** An amount broken. **3. a.** Loss or damage due to breaking. **b.** An allowance for such loss or damage.

**break·a·way** (brāk′ə-wā′) *adj.* **1.** Designating a theatrical prop designed to fall apart easily. **2.** Capable of or favoring independent action <a *breakaway* consumer group> —*n.* **1. a.** One that breaks away. **b.** The act or process of breaking away. **2.** A breakaway object.

**break·bone fever** (brāk′bōn′) *n.* Dengue.

**break·down** (brāk′doun′) *n.* **1. a.** The act or process of breaking down. **b.** The state of being broken down. **2.** *Elect.* Failure of an insulator or insulating medium to prevent discharge or current flow. **3.** A collapse in mental or physical health. **4.** An analysis, outline, or summary composed of itemized data or essentials. **5.** Disintegration or decomposition into parts or elements.

**break·er**[1] (brā′kər) *n.* **1.** One that breaks. **2.** A machine or plant for breaking up a substance, as rock. **3.** *Elect.* A circuit breaker. **4.** A wave that breaks into foam, esp. against a shoreline.

**break·er**[2] (brā′kər) *n.* [Alteration of Sp. *barrica.*] A small water cask used on a lifeboat.

**break·fast** (brĕk′fəst) *n.* [ME *brekfast* : *breken,* to break + *faste,* a fast < ON *fasta.*] The first meal of the day. **—break′fast** *v.* **(-fast·ed, -fast·ing, -fasts). —break′fast·er** *n.*

**break·front** (brāk′frŭnt′) *n.* A high, wide cabinet or bookcase with a central section projecting beyond the end sections.

**break-in** (brāk′ĭn′) *n.* **1.** Forcible entry, as into a dwelling, for an illegal purpose, as theft. **2.** A testing or training period intended to improve performance.

**break·ing** (brā′kĭng) *n.* [Transl. of G. *Brechung.*] Change of a simple vowel to a diphthong, often caused by the influence of neighboring consonants.

**breaking and entering** *n. Law.* The unauthorized forcible entry into another's premises for the purpose of committing a crime.

**breaking point** *n.* **1.** The point at which one breaks down under stress. **2.** A critical juncture.

**break·neck** (brāk′nĕk′) *adj.* Extremely fast and dangerous.

**break·out** (brāk′out′) *n.* **1.** A forceful break from a restrictive situation. **2.** A military offensive undertaken to escape encirclement.

**break·through** (brāk′thrōō′) *n.* **1.** The act of breaking through an obstacle or restriction. **2.** A military offensive penetrating an enemy defensive line. **3.** A major accomplishment or success that permits further progress, as in technology.

**break·up** (brāk′ŭp′) *n.* **1.** The act of breaking up or the state of having been broken up. **2.** *Informal.* A loss of composure.

**break·wa·ter** (brāk′wô′tər, -wŏt′ər) *n.* A barrier protecting a harbor or shore from the impact of waves.

**bream**[1] (brēm) *n., pl.* **bream** or **breams.** [ME *breme* < OFr., of Germanic orig.] **1.** A European freshwater fish of the genus *Abramis,* with a flattened body and silvery scales. **2.** A fish similar or related to the bream.

**bream**[2] (brēm) *vt.* **breamed, bream·ing, breams.** [Orig. unknown.] To clean (a wooden ship's hull) by heating and scraping.

**breast** (brĕst) *n.* [ME *brest* < OE *brēost.*] **1.** The mammary gland, esp. the human mammary gland. **2.** The surface of the body extending from the neck to the abdomen. **3.** The seat of affection and emotion. **4.** Something likened to the breast. —*vt.* **breast·ed, breast·ing, breasts.** To advance against resolutely : CONFRONT.

**breast-beat·ing** (brĕst′bē′tĭng) *n.* A noisy demonstration, esp. of self-accusation.

**breast·bone** (brĕst′bōn′) *n.* The sternum.

**breast-feed** (brĕst′fēd′) *vt.* **-fed** (-fĕd′), **-feed·ing, -feeds.** To feed (a baby) mother's milk from the breast : NURSE.

**breast·plate** (brĕst′plāt′) *n.* **1.** A piece of armor plate covering the breast. **2.** A square cloth set with 12 precious stones representing the 12 tribes of Israel, formerly worn by a Jewish high priest. **3.** The plastron of a turtle's or tortoise's shell.

**breast stroke** *n.* A swimming stroke performed by lying face down in the water, extending the arms in front of the head, then sweeping them both back laterally under the surface of the water while doing a frog kick. **—breast′strok′er** *n.*

**breast·work** (brĕst′wûrk′) *n.* A hastily constructed, temporary fortification that is usu. breast high.

**breath** (brĕth) *n.* [ME *breth* < OE *brǣð.*] **1.** The air inhaled and exhaled in respiration. **2.** The act or process of breathing : RESPIRATION. **3.** Capacity to breathe. **4.** A single respiration. **5.** Exhaled air, as evidenced by vapor, odor, or heat. **6.** A momentary rest or pause. **7. a.** A momentary stirring of air. **b.** A slight gust of fragrant air. **8.** A suggestion or trace. **9.** A soft-spoken sound : WHISPER. **10.** Exhalation of air without vibration of the vocal cords, as in the articulation of *p* and *s.* **—hold (one's) breath.** To await anxiously. **—in the same breath.** At or almost at the same time. **—out of breath.** Breathless, as from exertion. **—take (one's) breath away.** To leave one in awe or surprise. **—under (one's) breath.** In a whisper or muted voice.

**breathe** (brēth) *v.* **breathed, breath·ing, breathes.** [ME *brethen* < *breth,* breath.] —*vi.* **1.** To inhale and exhale air. **2.** To be alive : LIVE. **3.** To move or blow gently, as air. **4.** To be exhaled or emanated, as a fragrance. **5.** To pause to rest or regain breath. —*vt.* **1.** To inhale and exhale during respiration. **2.** To impart as if by breathing : INSTILL <*breathed* life into the old building> **3.** To exhale : emit. **4.** To utter, esp. quietly : WHISPER <never *breathed* a word of scandal> **5.** To make apparent : MANIFEST. **6.** To allow (a person or animal) to rest or regain breath. **—breathe (one's) last.** To die. **—breath′a·ble** *adj.*

**breath·er** (brē′thər) *n.* **1.** One that breathes, esp. in a given way. **2.** *Informal.* A short rest period <take a *breather*>

**breath·ing** (brē′thĭng) *n.* **1.** The act or process of respiration. **2.** Either of two marks used in Greek to indicate aspiration of an initial sound or the absence of such aspiration.

**breathing space** *n.* **1.** Sufficient space to permit ease of breathing or movement. **2.** An opportunity to rest or think.

**breathing spell** *n.* BREATHING SPACE 2.

**breath·less** (brĕth′lĭs) *adj.* **1. a.** Without breath : not breathing. **b.** Dead. **2.** Having no air : STILL. **3.** Out of breath. **4.** Holding the breath from excitement or suspense. **5.** Marked by sudden excitement : INSPIRING **—breath′less·ly** *adv.* **—breath′less·ness** *n.*

**breath·tak·ing** (brĕth′tā′kĭng) *adj.* Very exciting : AWESOME. **—breath′tak′ing·ly** *adv.*

**breath·y** (brĕth′ē) *adj.* **-i·er, -i·est.** Marked by audible or noisy breathing. **—breath′i·ly** *adv.* **—breath′i·ness** *n.*

**brec·ci·a** (brĕch′ē-ə, brĕch′ə) *n.* [Ital., of Germanic orig.] Rock composed of sharp-angled fragments cemented in a fine matrix.

**brec·ci·ate** (brĕch′ē-āt′) *vt.* **-ci·at·ed, -ci·at·ing, -ci·ates.** To form (rock) into breccia. **—brec′ci·a′tion** *n.*

**bred** (brĕd) *v. p.t. & p.p. of* BREED.

**brede** (brēd) *n.* [Alteration of BRAID.] *Archaic.* An ornamental embroidered edging.

**bred-in-the-bone** (brĕd′ĭn-*th*ə-bōn′) *adj.* **1.** Deeply instilled <*bred-in-the-bone* allegiance> **2.** DYED-IN-THE-WOOL 2.

**†breech** (brēch) *n.* [ME *brech* < OE *brēc*, pl. of *brōc*, leg covering.] **1.** The lower rear portion of the human trunk : BUTTOCKS. **2.** The lower part of a pulley. **3.** The part of a firearm to the rear of the bore. **3. breeches. a.** Trousers extending to or just below the knee. **b.** *Informal.* Trousers. —*vt.* **breeched, breech·ing, breech·es.** *Archaic & Regional.* To clothe with breeches.

**breech·block** (brēch′blŏk′) *n.* The metal block that closes and secures the breech end of a breechloading gun.

**breech·cloth** (brēch′klôth′, -klŏth′) *also* **breech·clout** (-klout′) *n.* A loincloth.

**breech delivery** *n.* Delivery of a fetus with the buttocks or feet appearing first.

**breeches buoy** *n.* An apparatus used for rescues and transfers at sea, consisting of sturdy canvas breeches attached at the waist to a ring buoy that is suspended from a pulley running along a rope from ship to shore or from ship to ship.

**breech·ing** (brē′chĭng, brĭch′ĭng) *n.* The strap of a harness that passes behind a draft animal's haunches.

**breech·load·er** (brēch′lō′dər) *n.* A gun loaded at the breech. **—breech′load′ing** *adj.*

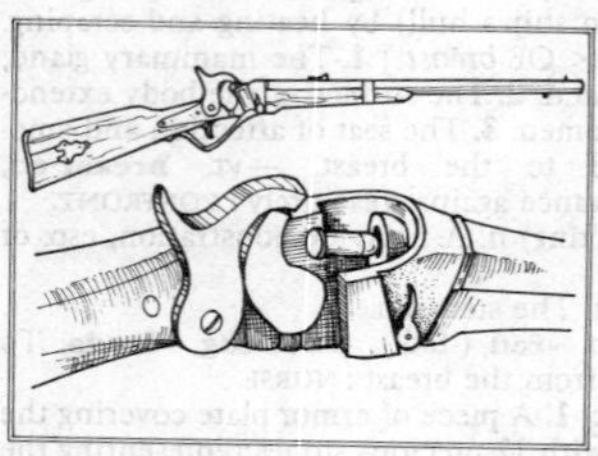

**breechloader**
*A breechloading rifle*

**breech presentation** *n.* The position of a fetus during labor in which the buttocks or feet appear first.

**breed** (brēd) *v.* **bred** (brĕd), **breed·ing, breeds.** [ME *breden* < OE *brēdan.*] —*vt.* **1.** To give birth to (offspring). **2.** To bring about : ENGENDER. **3. a.** To cause to reproduce : RAISE. **b.** To develop new or improved strains in (animals or plants). **4.** To bring up : REAR. —*vi.* **1.** To produce offspring. **2.** To originate and thrive <Revolution *breeds* in hunger.> —*n.* **1.** A genetic strain or type of organism, usu. a domestic animal, having consistent and recognizable inherited characteristics, esp. such a strain developed and maintained by humans. **2.** A type : kind <a new *breed* of criminal>

**breed·er** (brē′dər) *n.* **1.** One who breeds animals or plants. **2.** An animal kept to produce offspring. **3.** A cause or source. **4.** *Physics.* A breeder reactor.

**breeder reactor** *n.* A nuclear reactor that generates as well as consumes fissionable material, esp. one that generates more fissionable material than it consumes.

**breed·ing** (brē′dĭng) *n.* **1.** Line of descent : ANCESTRY. **2.** Training in the proper forms of social and personal conduct. **3.** Production of offspring or young. **4.** Propagation of plants or animals.

**breeding ground** *n.* **1.** A place to which animals go to breed. **2.** A place or set of circumstances that encourages the development of ideas or conditions <a *breeding ground* for revolution>

**breeks** (brēks) *pl.n.* [ME, var. of *breeches.*] *Scot.* Breeches.

**breeze¹** (brēz) *n.* [Perh. < OSp. *briza*, northeast wind.] **1.** A light air current : gentle wind. **2.** *Meteorol.* A wind of from 4 to 31 miles or 6.4 to 49.6 kilometers per hour. **3.** *Chiefly Brit.* A commotion or disturbance : ARGUMENT. **4.** *Informal.* An easily accomplished task. —*vi.* **breezed, breez·ing, breez·es. 1.** To blow lightly. **2.** *Informal.* To progress swiftly and effortlessly <*breezed* through the exam>

**breeze²** (brēz) *n.* [Prob. < Fr. *braise*, hot coals < OFr. *brese.*] The residue left when coke or charcoal is made.

**breeze·way** (brēz′wā′) *n.* A roofed, open-sided or screened passageway connecting two structures, as a house and a garage.

**breez·y** (brē′zē) *adj.* **-i·er, -i·est. 1.** Exposed to breezes : WINDY. **2.** Fresh : lively. **—breez′i·ly** *adv.* **—breez′i·ness** *n.*

**breg·ma** (brĕg′mə) *n., pl.* **-ma·ta** (-mə-tə) [NLat. < LLat., top of head < Gk.] The junction of the sagittal and coronal sutures at the top of the skull. **—breg·mat′ic** (-măt′ĭk) *adj.*

**brems·strah·lung** (brĕms′shträ′ləng) *n.* [G. : *Bremse*, brake + *Strahlung*, radiation < *Strahl*, ray.] Electromagnetic radiation produced by an electrically charged subatomic particle subjected to sudden deceleration in the electric field of an atomic nucleus.

**Bren gun** (brĕn) *n.* [Blend of *Brno*, Czechoslovakia, and *Enfield*, England.] A light, air-cooled submachine gun used by the British Army in World War II.

**brent** (brĕnt) *n. var. of* BRANT.

**†br'er** (brûr, brĕr) *n. Southern U.S.* BROTHER 1, 2.

**breth·ren** (brĕth′rən) *n. Archaic. pl. of* BROTHER.

**Bret·on** (brĕt′n) *n.* [ME < OFr. < Lat. *Briton.*] **1.** A native or resident of Brittany. **2.** The Celtic language of Brittany. **—Bret′on** *adj.*

**breve** (brĕv, brēv) *n.* [ME, short, var. of *bref* < OFr. < Lat. *brevis.*] **1.** A symbol placed over a vowel to show that it has a short sound, as the ă in *pat.* **2.** A symbol similar to a breve used to indicate that a syllable of verse is short or unstressed. **3.** *Mus.* A single note equivalent to two whole notes.

**bre·vet** (brə-vĕt′, brĕv′ĭt) *n.* [ME, official letter < AN, dim. of *bref*, letter < OFr. *brief* < Lat. *breve*, dispatch < *brevis*, short.] An often honorary commission promoting a military officer without increasing pay or authority. —*vt.* **-vet·ted, -vet·ting, -vets** *or* **-vet·ed, -vet·ing, -vets.** To promote by brevet. **—bre·vet′cy** (brə-vĕt′sē) *n.*

**bre·vi·ar·y** (brē′vē-ĕr′ē, brĕv′ē-) *n., pl.* **-ies.** [ME *breviarie* < Med. Lat. *breviarium*, abridgment, ult. < Lat. *brevis*, short.] A book containing the hymns, offices, and prayers for the canonical hours.

**brev·i·ty** (brĕv′ĭ-tē) *n.* [Lat. *brevitas* < *brevis*, short.] **1.** Briefness of duration. **2.** Concise expression : TERSENESS.

**brew** (broo) *v.* **brewed, brew·ing, brews.** [ME *brewen* < OE *brēowan.*] —*vt.* **1.** To make (ale or beer) from malt and hops by infusion, boiling, and fermentation. **2.** To make (a beverage) by boiling, steeping, or mixing various ingredients. **3. a.** To instigate : incite <*brew* discontent> **b.** To concoct : devise. —*vi.* **1.** To make ale or beer as an occupation. **2.** To be imminent : IMPEND <"in spite of storms *brewing* on every frontier" —John Dos Passos> —*n.* **1. a.** A beverage made by brewing. **b.** A cup of coffee or tea. **c.** A glass of beer. **2.** The amount of beverage brewed at one time. **—brew′age** *n.* **—brew′er** *n.*

**brew·er's yeast** (broo′ərz) *n.* A yeast, *Saccharomyces cerevisiae*, used in brewing and as a source of B-complex vitamins.

**brew·er·y** (broo′ə-rē, broor′ē) *n., pl.* **-ies.** An establishment for making malt liquors.

**brew·house** (broo′hous′) *n.* A brewery.

**†brew·is** (broo′ĭs, brooz) *n.* [ME < OFr. *brouetz*, dim. of *brouet*, broth, of Germanic orig.] *Regional.* Broth.

**bri·ar¹** *also* **bri·er** (brī′ər) *n.* [Fr. *bruyère*, heath.] **1.** A shrub or small tree, *Erica arborea* of southern Europe, with a hard woody root used to make tobacco pipes. **2.** A pipe made from briarroot or from a similar wood.

**bri·ar²** (brī′ər) *n. var. of* BRIER¹.

**bri·ard** (brē-är′, -ärd′) *n.* [Fr. < *Brie*, a region of France.] A sturdily built, rough-coated dog of an ancient French breed.

**Bri·ar·e·us** (brī-âr′ē-əs) *n.* [Lat. < Gk. *Briareōs* < *briaros*, strong.] *Gk. Myth.* A giant who assisted Zeus and the Olympians against the Titans.

**bri·ar·root** (brī′ər-root′, -root′) *n.* The hard woody root of the briar.

**bri·ar·wood** (brī′ər-wood′) *n.* Wood from the root of the briar.

**bribe** (brīb) *n.* [ME < OFr., alms.] **1.** Something, as money or a favor, offered or given to someone in a position of trust to induce him or her to act dishonestly. **2.** Something that serves to influence or persuade. —*v.* **bribed, brib·ing, bribes.** —*vt.* **1.** To give, offer, or promise a bribe to. **2.** To gain influence over or corrupt by bribery. —*vi.* To give, offer, or promise bribes. **—brib′a·ble** *adj.* **—brib′er** *n.*

☆ ***syns:*** BRIBE, BOODLE, PAYOFF, PAYOLA *n. core meaning:* something (as money, property, or a favor) given, offered, or promised as inducement to dishonest behavior on the part of the intended recipient. BRIBE and PAYOFF are interchangeable in this sense <business executives who used *bribes* to gain foreign contracts><a Vice President who accepted *payoffs* for political favors> but BOODLE refers strictly to money used to obtain an end <The lobbyist passed the *boodle* on to the senator.> PAYOLA, on the other hand, refers primarily to inducing disc jockeys to promote records through illicit payments or expensive gifts <radio stations charged with *payola*>

**brib·er·y** (brī′bə-rē) *n., pl.* **-ies.** The act or practice of offering, giving, or taking a bribe.

**bric-a-brac** (brĭk′ə-brăk′) *n.* [Fr. *bric-à-brac.*] Small usu. ornamental objects valued for their rarity, antiquity, or curiosity.

**brick** (brĭk) *n.* [ME *brike* < MDu. *bricke.*] **1.** A molded, rectangular block of baked clay used as a building and paving material. **2.** An object shaped like a brick <a *brick* of gold> **3.** *Informal.* A splendid person. —*vt.* **bricked, brick·ing, bricks. 1.** To construct, line, or pave with brick. **2.** To close or wall with brick. **—brick′y** *adj.*

**brick·bat** (brĭk′băt′) *n.* **1.** A piece of brick, esp. one used as a weapon or missile. **2.** A critical remark.

**brick·kiln** (brĭk′kĭln′, -kĭl′) *n.* A kiln in which bricks are baked.

**brick·lay·er** (brĭk′lā′ər) *n.* One skilled in building with bricks. **—brick′lay′ing** *n.*

**brick red** *n.* **1.** A moderate reddish brown. **2.** A moderate to strong brown. **—brick′-red′** (brĭk′rĕd′) *adj.*

**brick·work** (brĭk′wûrk′) *n.* **1.** A structure of bricks. **2.** Construction with bricks.

**brick·yard** (brĭk′yärd′) *n.* A place where bricks are made.

**bri·dal** (brīd′l) *n.* [ME *bridale*, wedding feast < OE *brȳdeala* : *brȳd*, bride + *ealu*, ale.] A marriage ceremony : WEDDING. —*adj.* Of or relating to a bride or a marriage ceremony : NUPTIAL.

▲ word history: As the etymology above shows, *bridal* is a compound of two Old English words that mean "bride" and "ale." The compound originally meant "wedding feast," but by late medieval times was used to refer to the ceremony itself. *Bridal* as an adjective probably arose from the interpretation of *-al* as the common adjectival suffix.

**bridal wreath** *n.* Either of two related shrubs, *Spiraea prunifolia* or *S. vanhouttei,* cultivated for their profuse white flowers.

**bride** (brīd) *n.* [ME < OE *brȳd.*] A woman recently married or about to be married.

**bride·groom** (brīd'gro͞om', -gro͝om') *n.* [Alteration of ME *bridegome* < OE *brȳdguma* : *brȳd,* bride + *guma,* man.] A man recently married or about to be married.

**brides·maid** (brīdz'mād') *n.* A woman attendant of a bride at a marriage ceremony.

**bridge¹** (brĭj) *n.* [ME *brigge* < OE *brycge.*] **1.** A structure spanning and providing passage over an obstacle, as a waterway. **2.** Something resembling or analogous to a bridge. **3. a.** The upper bony ridge of the human nose. **b.** The part of a pair of eyeglasses that rests against this ridge. **4.** *Mus.* **a.** A thin upright piece of wood in some stringed instruments that supports the strings above the sounding board. **b.** A transitional passage between two subjects or movements. **5.** A fixed or removable replacement for one or several but not all of the natural teeth, usu. anchored at each end to a natural tooth. **6.** A crosswise platform above the main deck of a ship from which the ship is controlled. **7. a.** A device used to steady the cue in billiards. **b.** The hand used in place of this device. **8.** *Elect.* Any of various circuits containing a branch that connects two points of equal potential and consequently carries no current when the circuit is suitably adjusted. —*vt.* **bridged, bridg·ing, bridg·es. 1.** To build a bridge over. **2.** To cross by or as if by a bridge. **—bridge'a·ble** *adj.*

**bridge²** (brĭj) *n.* [Orig. unknown.] One of several card games derived from whist and played with one deck of cards divided equally among four people.

**bridge·board** (brĭj'bôrd', -bōrd') *n.* A notched board at either side of a staircase that supports the treads and risers.

**bridge·head** (brĭj'hĕd') *n.* [Transl. of Fr. *tête de pont.*] A military position established by advance troops on the enemy's side of a river or pass to afford protection for the main attacking force.

**bridge·work** (brĭj'wûrk') *n.* **1.** A dental bridge. **2.** Prosthetics.

**bridg·ing** (brĭj'ĭng) *n.* Wooden braces between beams, as of a roof, that provide reinforcement and distribution of stress.

**bri·dle** (brīd'l) *n.* [ME *bridel* < OE *brīdel.*] **1.** A harness consisting of a headstall, bit, and reins, which fits a horse's head and is used as a restraint or guide. **2.** A curb : check. **3.** *Naut.* A span of chain, wire, or rope securable at both ends to an object and slung from its center point. —*v.* **-dled, -dling, -dles.** —*vt.* **1.** To put a bridle on. **2.** To control or restrain with or as if with a bridle. —*vi.* **1.** To lift the head and draw in the chin as an expression of scorn or resentment. **2.** To show resentment or anger : take offense <*bridled* at the criticism> **—bri'dler** *n.*

**bridle path** *n.* A trail for riding horses.

**Brie** (brē) *n.* [After *Brie,* a region of France.] A soft, mold-ripened, whole-milk cheese.

**brief** (brēf) *adj.* **-er, -est.** [ME *bref* < OFr. *bref* < Lat. *brevis,* short.] **1.** Short in time, length, or extent. **2.** Condensed in expression : SUCCINCT. **3.** Abrupt : curt. —*n.* **1.** A short or condensed statement. **2.** A condensation of a lengthy document or series of documents. **3.** *Law.* **a.** An abstract of all of the documents affecting the title of real property. **b.** A document containing all facts and points of law pertinent to a specific case, filed by an attorney before arguing the case in court. **4.** *Rom. Cath. Ch.* A papal letter relating to disciplinary matters. **5.** A briefing. **6. briefs.** Short tight-fitting underpants. —*vt.* **briefed, brief·ing, briefs. 1.** To summarize. **2.** To give concise preparatory instructions or advice to. **3.** *Chiefly Brit.* To send a legal brief to. **4.** *Chiefly Brit.* To hire (an attorney) as counsel. **—in brief.** In a few words. **—brief'ly** *adv.* **—brief'ness** *n.*

☆ ***syns:*** BRIEF, SHORT *adj. core meaning* : not long in time or duration <a *brief* interval> ***ant*** : lengthy, long

**brief·case** (brēf'kās') *n.* [< BRIEF (document).] A portable rectangular case for carrying papers or books.

**brief·ing** (brē'fĭng) *n.* **1.** The act or process of giving or receiving concise preparatory instructions, information, or advice. **2.** The information conveyed during a briefing.

**brief·less** (brēf'lĭs) *adj.* Having no brief and thus no clients <a *briefless* lawyer>

**bri·er¹** *also* **bri·ar** (brī'ər) *n.* [ME *breir* < OE *brǣr.*] A thorny plant or bush, esp. a prickly-stemmed rosebush. **—bri'er·y** *adj.*

**bri·er²** (brī'ər) *n. var. of* BRIAR¹.

**brig** (brĭg) *n.* [Short for BRIGANTINE.] **1.** A two-masted, square-rigged sailing ship. **2.** A ship's prison. **3.** A guardhouse.

**bri·gade** (brĭ-gād') *n.* [Fr. < OFr., company < OItal. *brigata* < *brigare,* to fight < LLat. *briga,* strife.] **1. a.** A military unit consisting of a large number of troops. **b.** A unit of the U.S. Army composed of two or more regiments, once commanded by a brigadier general, but now by a colonel <an airborne *brigade*> **2.** A group of persons organized for a specific purpose. —*vt.* **-gad·ed, -gad·ing, -gades.** To form into a brigade.

**brig·a·dier** (brĭg'ə-dîr') *n.* [Fr. < *brigade,* brigade.] A brigadier general.

**brigadier general** *n., pl.* **brigadier generals.** An officer ranking above a colonel and below a major general in the U.S. Army, Air Force, and Marine Corps.

**brig·and** (brĭg'ənd) *n.* [ME *brigaunt* < OFr., prob. < OItal. *brigante,* skirmisher < *brigare,* to fight. —see BRIGADE.] One who plunders and robs : BANDIT. **—brig'and·age** (-ən-dĭj), **brig'and·ism** *n.*

**brig·an·tine** (brĭg'ən-tēn') *n.* [Fr. *brigantin* < OFr. *brigandin* < OItal. *brigantino,* skirmishing ship < *brigante,* skirmisher. —see BRIGAND.] A two-masted sailing ship, square-rigged on the foremast and having a fore-and-aft mainsail with square maintopsails.

**bright** (brīt) *adj.* **-er, -est.** [ME < OE *beorht.*] **1. a.** Reflecting or emitting light readily : SHINING. **b.** Comparatively high on the scale of brightness. **2.** Brilliant in color : VIVID. **3.** Resplendent : splendid. **4.** Full of hope : PROMISING. **5.** Joyful : happy. **6.** Mentally agile : SMART. —*n.* **1.** A thin, flat paintbrush used to highlight. **2. brights.** High-beam headlights. —*adv.* In a bright way. **—bright'ly** *adv.*

☆ ***syns:*** BRIGHT, BRILLIANT, EFFULGENT, INCANDESCENT, LUMINOUS, LUSTROUS, RADIANT, REFULGENT *adj. core meaning* : giving off or reflecting light readily <*bright* street lamps><*bright* sparkling diamonds>

**bright·en** (brīt'n) *vt.* & *vi.* **-ened, -en·ing, -ens. 1.** To make or become bright or brighter. **2.** To make or become more cheerful. **—bright'en·er** *n.*

**bright·ness** (brīt'nĭs) *n.* **1.** The quality or state of being bright. **2.** The effect or sensation by means of which an observer is able to ascertain differences in luminance. **3.** The dimension of a color that represents its similarity to one of a series of achromatic colors ranging from very dim to very bright.

**Bright's disease** (brīts) *n.* [After Richard *Bright* (1789–1858).] Chronic nephritis.

**bright·work** (brīt'wûrk') *n.* Highly polished metal fixtures.

**brill** (brĭl) *n., pl.* **brill** or **brills.** [Orig. unknown.] An edible flatfish, *Scophthalmus rhombus* of European waters.

**bril·liant** (brĭl'yənt) *adj.* [Fr. *brillant,* pr.part. of *briller,* to shine < Ital. *brillare,* perh. < *brillo,* beryl < Lat. *beryllus.*] **1.** Full of light : SHINING. **2. a.** Brightly vivid in color. **b.** Designating a color that has a combination of strong saturation and high lightness. **3.** *Mus.* Sharp and clear in tone. **4.** Magnificent : glorious. **5.** Superb : extraordinary <a *brilliant* recital> **6.** Exceptionally intelligent or inventive. —*n.* A precious gem finely cut with numerous facets. **—bril'liance, bril'lian·cy** *n.* **—bril'liant·ly** *adv.* **—bril'liant·ness** *n.*

**bril·lian·tine** (brĭl'yən-tēn') *n.* [Fr. *brillantine* < *brillant,* brilliant.] **1.** An oily, perfumed hairdressing. **2.** A glossy fabric made from cotton and worsted or cotton and mohair.

**Brill's disease** (brĭlz) *n.* [After Nathan E. *Brill* (1860–1925).] A form of epidemic typhus thought to be a recurrence of an earlier infection.

**brim** (brĭm) *n.* [ME *brimme,* of Germanic orig.] **1.** The rim or uppermost edge of a vessel, as a cup. **2.** A projecting rim or edge <the *brim* of a hat> **3.** A border or edge, esp. one surrounding a body of water. —*vt.* & *vi.* **brimmed, brim·ming, brims.** To fill or be full to the brim. **—brim over.** To overflow.

**brim·ful** (brĭm'fo͝ol') *adj.* Full to the brim.

**brim·stone** (brĭm'stōn') *n.* [ME *brimston* < OE *brynstān.*] Sulfur.

**brin** (brĭn) *n.* [Fr.] One of the ribs of a fan.

**brin·dle** (brĭn'dl) *n.* [Back-formation < BRINDLED.] **1.** A brindled color. **2.** A brindled animal.

**brin·dled** (brĭn'dld) *adj.* [Alteration of ME *brended,* perh. < *brende,* p.part. of *brennen,* to burn < ON *brenna.*] Tawny or grayish with streaks or spots of a darker color.

**brine** (brīn) *n.* [ME < OE *brīne.*] **1.** Water saturated with or containing large amounts of a salt, esp. of sodium chloride. **2. a.** The water of a sea or an ocean. **b.** A large body of salt water. **3.** Salt water for preserving and pickling foods. —*vt.* **brined, brin·ing, brines.** To immerse, preserve, or pickle in brine. **—brin'er** *n.*

**Bri·nell hardness** (brĭ-nĕl') *n.* [After Johann A. *Brinell* (1849–1925).] The relative hardness of metals and alloys, determined by forcing a steel ball into a test piece under standard conditions and measuring the surface area of the resulting indentation.

**Brinell number** *n.* The numerical value assigned to the Brinell hardness of metals and alloys.

**brine shrimp** *n.* A small crustacean of the genus *Artemia.*

**bring** (brĭng) *vt.* **brought** (brôt), **bring·ing, brings.** [ME *bringen* < OE *bringan.*] **1.** To take with oneself to a place : convey or carry along <*brought* their cameras> **2.** To carry as an attribute or contribution <*brought* specialized expertise to the job> **3.** To lead or force into a specified state, situation, or location <Gambling *brought* them to ruin.> **4.** To succeed in persuading : INDUCE. **5.** To cause to occur as a consequence or concomitant <The rains *brought* mud slides to the canyon.> **6.** To cause to become apparent to the mind : RECALL <*bring* back memories of old times> **7.** *Law.* To advance or set forth (e.g., charges) in a court. **8.** To sell for : FETCH. **—bring about.** To cause to happen. **—bring around** (*or* **round**). **1.** To cause to adopt an opinion or course of action. **2.** To cause to

o͞o **boot** ou **out** th **thin** *th* **this** ŭ **cut** ûr **urge** y **young**
yo͞o **abuse** zh **vision** ə **about,** it**e**m, ed**i**ble, gall**o**p, circ**u**s

recover consciousness. **—bring down. 1.** To cause to collapse or fail. **2.** To kill. **—bring forth. 1.** To give rise to : PRODUCE. **2.** To bear (fruit or young). **—bring forward. 1.** To present : PRODUCE. **2.** To carry (a sum) from one page or column to another in bookkeeping. **—bring in. 1.** To give or submit (a verdict). **2.** To produce or yield (profits or income). **—bring off.** To carry out successfully. **—bring on. 1.** To result in : CAUSE. **2.** To cause to appear. **—bring out. 1.** To reveal or expose. **2.** To publish or produce. **—bring over.** To win over. **—bring to. 1.** To cause to recover consciousness. **2.** To cause (a ship) to turn into the wind and lose way. **—bring to light.** To uncover : reveal. **—bring up. 1.** To take care of and educate (a child) : REAR. **2.** To introduce into discussion : MENTION. **3.** To vomit. **—bring'er** *n.*

**bring·down** (brĭng'doun') *n.* Something that disappoints, dismays, or disturbs.

**brink** (brĭngk) *n.* [ME, prob. of Scand. orig.] **1. a.** The upper edge, as of a steep slope. **b.** The margin of land bordering a body of water. **2.** VERGE[1] 3.

**brink·man·ship** (brĭngk'mən-shĭp') *n.* The practice of following a perilous course of action stopping just short of the point of no return.

**brin·y** (brī'nē) *adj.* **-i·er, -i·est.** Of, relating to, or resembling brine : SALTY. *—n. Slang.* The sea. **—brin'i·ness** *n.*

**bri·o** (brē'ō) *n.* [Ital., of Celt. orig.] Liveliness : vigor.

**bri·oche** (brē-ôsh', -ŏsh') *n.* [Fr. < OFr. < dial. *brier,* var. of *broyer,* to knead, of Germanic orig.] A soft, light-textured roll made from eggs, butter, flour, and yeast.

**bri·o·lette** (brē'ə-lĕt') *n.* [Fr.] A pear-shaped gem, esp. a diamond, cut with long triangular facets.

**bri·quette** *also* **bri·quet** (brĭ-kĕt') *n.* [Fr. *briquette,* dim. of *brique,* brick < MDu. *bricke.*] A block of compressed coal dust or charcoal, used for kindling and fuel.

**bri·sance** (brĭ-zäns', -zäNs') *n.* [Fr. < *brisant,* pr.part. of *briser,* to break, of Celt. orig.] The shattering effect of a sudden release of energy, as in an explosion. **—bri·sant'** (-zänt', -zäNt') *adj.*

**brisk** (brĭsk) *adj.* **-er, -est.** [Orig. unknown.] **1.** Lively : vigorous <a *brisk* jog> **2.** Keen or sharp in speech or manner. **3.** Stimulating and invigorating <*brisk* autumn air> **4.** Pleasantly zestful <a *brisk* tea> **—brisk'ly** *adv.* **—brisk'ness** *n.*

**bris·ket** (brĭs'kĭt) *n.* [ME *brusket,* poss. of Scand. orig.] **1.** The chest of an animal. **2.** The ribs and meat from the brisket.

**bris·ling** (brĭz'lĭng, brĭs'-) *n.* [Norw. < LG *bretling* < *bret,* broad.] SPRAT 1.

**bris·tle** (brĭs'əl) *n.* [ME *bristel.*] A short, coarse, stiff hair or hairlike part. *—v.* **-tled, -tling, -tles.** *—vi.* **1.** To erect the bristles, as an angry, excited, or frightened animal. **2.** To react with agitation to anger, excitement, or fear. **3.** To stand erectly on end like bristles. **4.** To be covered or thick with or as if with bristles <a thicket *bristling* with briers> *—vt.* **1.** To cause to stand erect like bristles. **2.** To furnish with bristles. **3.** To disturb : ruffle. **—bris'tly** *adj.*

**bris·tle·cone pine** (brĭs'əl-kōn') *n.* A small pine, *Pinus aristata* of the western United States, that is the longest-living conifer known.

**bris·tle·tail** (brĭs'əl-tāl') *n.* One of various wingless insects of the order Thysanura, with bristlelike posterior appendages.

**Bris·tol board** (brĭs'təl) *n.* [After *Bristol,* England.] A smooth, heavy pasteboard of fine quality.

**brit** *also* **britt** (brĭt) *n.* [Perh. < Cornish *brythel,* mackerel.] **1.** The young of herring and similar fish. **2.** Minute marine organisms, as crustaceans of the genus *Calanus,* that are a major source of food for many fish and whales.

**bri·tan·nia metal** *also* **Bri·tan·nia metal** (brĭ-tăn'yə, -tăn'ē-ə) *n.* [< Lat. *Brittania,* Britain.] A white alloy of tin with copper, antimony, and occas. bismuth and zinc.

**Bri·tan·nic** (brĭ-tăn'ĭk) *adj.* British.

**britch·es** (brĭch'ĭz) *pl.n.* [Alteration of BREECHES.] *Informal.* Breeches. **—too big for (one's) britches.** *Informal.* Cocky : overconfident.

**Brit·i·cism** (brĭt'ĭ-sĭz'əm) *also* **Brit·ish·ism** (-shĭz'əm) *n.* A word, phrase, or idiom typical of or peculiar to English as it is spoken in Great Britain.

**Brit·ish** (brĭt'ĭsh) *adj.* [ME *Brittish* < OE *Bryttisc,* pertaining to ancient Britons.] **1.** Of, relating to, or typical of Great Britain, the United Kingdom, or the British Empire. **2.** Of, relating to, or typical of the ancient Britons. *—n.* **1.** *(pl. in number).* The people of Great Britain. **2.** British English. **3.** The Celtic language of the ancient Britons.

**British English** *n.* The English language as used in England as distinguished from that used elsewhere.

**Brit·ish·er** (brĭt'ĭ-shər) *n. Informal.* BRITON 1.

**Brit·ish·ism** (brĭt'ĭ-shĭz'əm) *n. var. of* BRITICISM.

**British thermal unit** *n.* The quantity of heat required to raise the temperature of one pound of water by one degree Fahrenheit.

**Brit·on** (brĭt'n) *n.* [ME *Breton* < OFr. < Lat. *Britto,* Celt, of Celtic orig.] **1.** A native or inhabitant of Britain. **2.** One of a Celtic people who inhabited ancient Britain before the Roman invasion.

**britt** (brĭt) *n. var. of* BRIT.

**Brit·ta·ny spaniel** (brĭt'n-ē) *n.* [After *Brittany,* a region of northwestern France.] A large spaniel orig. bred in France.

**brit·tle** (brĭt'l) *adj.* **-tler, -tlest.** [ME *britel.*] **1.** Likely to break : FRAGILE <*brittle* figurines> **2.** Difficult to deal with : SNAPPISH <a *brittle* temper> *—n.* A caramelized sugar candy with nuts in it. **—brit'tle·ly** (brĭt'l-ē) *adv.* **—brit'tle·ness** *n.*

**brittle star** *n.* One of various marine organisms of the class Ophiuroidea, related to the starfish but with long, slender, whiplike arms.

**Brix scale** (brĭks) *n.* [After Adolf F. *Brix* (d. 1870), its inventor.] A hydrometer scale for measuring the sugar content of a solution at a given temperature.

**broach[1]** (brōch) *n.* [ME *broche* < OFr., spit.] **1. a.** A tapered tool for shaping or enlarging a hole. **b.** The hole made by a broach. **2.** A spit for roasting meat. **3.** A mason's narrow chisel. **4.** A tool for broaching casks. **5.** *var. of* BROOCH. *—vt.* **broached, broach·ing, broach·es. 1. a.** To bring up (a subject) for discussion : talk about. **b.** To announce <*broached* the vacation schedule> **2.** To pierce in order to draw off liquid <*broached* a cask of wine> **3.** To shape or enlarge (a hole) with a broach. **—broach'er** *n.*

**broach[2]** (brōch) *vi.* & *vt.* **broached, broach·ing, broach·es.** [Poss. < BROACH[1].] *Naut.* To veer or cause to veer broadside to the wind and waves <kept the yacht from *broaching* to>

**broad** (brôd) *adj.* **-er, -est.** [ME *brod* < OE *brād.*] **1.** Wide from side to side. **2.** Large in expanse : SPACIOUS. **3.** Clear and open <*broad* daylight> **4.** General in scope <a *broad* restriction> **5.** Tolerant or liberal. **6.** Main : essential. **7.** Apparent : obvious <a *broad* clue> **8.** Bordering on indelicacy : RIBALD. **9.** Heavily regional <a *broad* drawl> **10.** Indicating a vowel pronounced with the tongue placed low and flat and the oral cavity wide open, as the *a* in *bath* when pronounced like the *a* in *bard.* *—n.* **1.** A broad part. **2.** *Slang.* A woman or girl. **—broad'ly** *adv.* **—broad'ness** *n.*

☆ ***syns:*** **1.** BROAD, WIDE *adj. core meaning* : extending from side to side over a large area <the *broad* shoulders of a fullback> ***ant:*** narrow **2.** BROAD, AMPLE, EXPANSIVE, EXTENSIVE, SPACIOUS *adj. core meaning* : large in expanse <a *broad,* velvety lawn>

**broad arrow** *n.* **1.** An arrow with a wide, barbed head. **2.** A wide arrowhead mark identifying British government property.

**broad·ax** *also* **broad·axe** (brôd'ăks') *n.* A short-handled ax with a wide, flat head : BATTLE-AX.

**broad·band** (brôd'bănd') *adj.* Of, having, or relating to a wide band of electromagnetic frequencies. **—broad'band'** *n.*

**broad bean** *n.* **1.** An Old World plant, *Vicia faba,* cultivated for its edible pods and seeds. **2.** The flattened seed of the broad bean.

**broad·bill** (brôd'bĭl') *n.* **1. a.** One of various birds of the family Eurylaimidae of Africa and tropical Asia, with brightly colored plumage and a short, wide bill. **b.** A broad-billed bird, as the shoveler. **2.** The swordfish.

**broad·cast** (brôd'kăst') *v.* **-cast** *or* **-cast·ed, -cast·ing, -casts.** *—vt.* **1.** To transmit (a program) by radio or television. **2.** To make known over a wide area <*broadcast* gossip> **3.** To throw (seed) about : SCATTER. *—vi.* **1.** To transmit a radio or television program. **2.** To participate in a radio or television program. *—n.* **1.** Transmission of a radio or television signal. **2.** A radio or television program. *—adj.* **1.** Scattered over a wide area. **2.** Of or relating to radio or television broadcasting. *—adv.* Over a wide area. **—broad'cast'er** *n.*

**Broad-Church** (brôd'chûrch') *adj.* Of or relating to members of the Anglican Communion advocating liberalism of ritual and doctrine. **—Broad'-Church'man** *n.*

**broad·cloth** (brôd'klôth', -klŏth') *n.* **1.** A densely textured woolen cloth with a lustrous finish and a plain or twill weave. **2.** A closely woven cotton, silk, or synthetic fabric with a narrow crosswise rib.

**broad·en** (brôd'n) *vt.* & *vi.* **-ened, -en·ing, -ens.** To make or become broad or broader. **—broad'en·er** *n.*

**broad gauge** *n.* A railroad track wider than standard gauge.

**broad-gauge** (brôd'gāj') *adj.* **1.** Having a broad gauge. **2.** *Informal.* Having a wide scope : LIBERAL.

**broad·head** (brôd'hĕd') *n.* **1.** A steel arrowhead with usu. two sharp edges. **2.** An arrow with a broadhead.

**broad jump** *n.* A long jump.

**broad·leaf** (brôd'lēf') *n.* One of various tobacco plants with broad leaves. *—adj.* Broad-leaved.

**broad-leaved** (brôd'lēvd') *also* **broad-leafed** (-lēft') *adj.* Having broad leaves, as the rhododendron, rather than needles.

**broad·loom** (brôd'lo͞om') *adj.* Woven on a wide loom. *—n.* A broadloom carpet.

**broad-mind·ed** (brôd'mīn'dĭd) *adj.* Liberal and tolerant. **—broad'mind'ed·ly** *adv.* **—broad'-mind'ed·ness** *n.*

**broad·sheet** (brôd'shēt') *n.* BROADSIDE 4.

**broad·side** (brôd'sīd') *n.* **1.** The side of a ship above the water line. **2. a.** All the guns on one side of a warship. **b.** The simultaneous discharge of all the guns on one side of a warship. **3.** An explosive verbal denunciation or attack. **4.** A large sheet of paper printed on one side. **5.** A broad unbroken surface. *—adv.* **1.** With the side turned to a given object. **2.** In a random manner.

**broad-spec·trum** (brôd'spĕk'trəm) *adj.* Effective against many microorganisms or insects <*broad-spectrum* insecticides>

**broad·sword** (brôd′sôrd′, -sōrd′) *n.* A cutting sword with a wide blade.

**broad·tail** (brôd′tāl′) *n.* **1.** The karakul. **2.** The black pelt of a prematurely born karakul sheep, with wavy markings and a flat surface.

**Broad·way** (brôd′wā′) *n.* **1.** The principal theater district of New York City, located on or near Broadway, a Manhattan thoroughfare. **2.** The American legitimate stage <went from *Broadway* to motion pictures> —*adj.* Of, relating to, or produced on Broadway < a *Broadway* musical>

**bro·cade** (brō-kād′) *n.* [Sp. or Port. *brocado* < Ital. *brocato* < *brocco*, twisted thread < VLat. ***brocca*, spike < Lat. *brocchus*, of Celt. orig.] A heavy fabric interwoven with a rich, raised design. —*vt.* **-cad·ed, -cad·ing, -cades.** To weave brocade.

**broc·a·tel** *also* **broc·a·telle** (brŏk′ə-tĕl′) *n.* [Fr. *brocatelle* < Ital. *broccatello*, dim. of *broccato*, brocade.] A fabric resembling brocade but with a more highly raised design.

**broc·co·li** *also* **broc·o·li** (brŏk′ə-lē) *n.* [Ital., pl. of *broccolo*, cabbage top, dim. of *brocco*, shoot. —see BROCADE.] **1.** A plant, *Brassica oleracea italica*, closely related to the cauliflower, with a branched greenish flower head. **2.** The flower head of the broccoli, eaten as a vegetable before the tightly clustered green buds have opened.

**bro·chette** (brō-shĕt′) *n.* [Fr. < OFr., dim. of *broche*, spit.] A small skewer.

**bro·chure** (brō-shŏŏr′) *n.* [Fr. < *brocher*, to stitch < *broche*, knitting needle < OFr., spit.] A small booklet or pamphlet.

**brock** (brŏk) *n.* [ME < OE *broc*, of Celt. orig.] *Chiefly Brit. & Scot.* A badger.

**brock·et** (brŏk′ĭt) *n.* [ME *broket* < OFr. *brocard* < *broque*, animal's horn, var. of *broche*, spit.] **1.** A two-year-old male red deer. **2.** One of several small deer of the genus *Mazama* of South America, with short, unbranched horns.

**broc·o·li** (brŏk′ə-lē) *n. var. of* BROCCOLI.

**bro·gan** (brō′gən) *n.* [Ir. Gael. *brōgan*, dim. of *brōg*, brogue.] A heavy ankle-high work shoe.

**brogue** (brōg) *n.* [Ir. and Sc. Gael. *brōg*, OIr. *brōc*, shoe.] **1. a.** A rough, heavy shoe, once worn in Scotland and Ireland. **b.** A strong oxford shoe, usu. with ornamental perforations. **2.** A strong dialectal accent, esp. a strong Irish accent.

**broi·der** (broi′dər) *vt.* **-dered, -der·ing, -ders.** *Obs.* To embroider. **—broi′der·y** (-də-rē) *n.*

**broil**[1] (broil) *v.* **broiled, broil·ing, broils.** [ME *broilen* < OFr. *bruler*.] —*vt.* To cook by direct radiant heat, as over hot coals. —*vi.* To become boiled. —*n.* **1.** The act or state of broiling. **2.** Something that is broiled.

**broil**[2] (broil) *n.* [< obs. *broil*, to brawl < ME *broilen* < OFr. *brouiller*.] A raucous quarrel : BRAWL. —*vi.* **broiled, broil·ing, broils.** To engage in a broil.

**broil·er** (broi′lər) *n.* **1.** One that broils. **2. a.** A small electric oven for broiling. **b.** The part of a stove used for broiling. **3.** A young chicken suitable for broiling.

**broke** (brōk) *adj.* [Alteration of BROKEN.] *Informal.* Completely without money : POOR.

**bro·ken** (brō′kən) *adj.* [< p.part. of BREAK.] **1.** Forcibly fractured into pieces : SHATTERED. **2.** Not upheld : BREACHED <*broken* promises> **3.** Incomplete : fragmentary. **4.** Disorganized : routed <*broken* troops> **5.** Intermittently stopping and starting : DISCONTINUOUS. **6.** Varying abruptly, as in pitch <*broken* cries> **7.** Spoken or written imperfectly <*broken* Italian> **8.** Irregular and uneven <the *broken* lunar surface> **9.** Humbled : subdued. **10.** Tamed and trained. **11.** Exhausted : weakened. **12.** Crushed by grief <a *broken* will to live> **13.** Bankrupt. **14.** Not functioning : INOPERATIVE. **—bro′ken·ly** *adv.* **—bro′ken·ness** *n.*

**bro·ken-down** (brō′kən-doun′) *adj.* **1.** Not in working order. **2.** In poor condition, as from old age : INFIRM.

**bro·ken·heart·ed** (brō′kən-här′tĭd) *adj.* Inconsolably sad.

**broken home** *n.* A family in which the parents are divorced or separated.

**broken wind** *n.* HEAVES 5.

**bro·ker** (brō′kər) *n.* [ME < AN *brocour*.] **1.** One that acts as an agent for others in negotiating contracts, purchases, or sales. **2.** A stockbroker.

**bro·ker·age** (brō′kər-ĭj) *n.* **1.** The business of a broker. **2.** A fee or commission paid to a broker.

**brom-** *pref. var. of* BROMO-.

**bro·mate** (brō′māt′) *n.* A salt of bromic acid. —*vt.* **-mat·ed, -mat·ing, -mates. 1.** To treat (a substance) chemically with a bromate. **2.** To combine (a substance) chemically with bromine.

**brome** (brōm) *n.* [NLat. *Bromus*, genus name < Lat. *bromos*, oats < Gk.] A grass of the genus *Bromus*, with spikelets in loose, often drooping clusters.

**brome grass** *n.* Brome.

**bro·me·li·ad** (brō-mē′lē-ăd′) *n.* [< NLat. *Bromelia*, type genus, after Olaf *Bromelius* (1639–1705).] One of various mostly epiphytic plants of the family Bromeliaceae, including the pineapple, Spanish moss, and many species cultivated as house plants.

**bro·mic acid** (brō′mĭk) *n.* A corrosive, colorless liquid, $HBrO_3$, used to make pharmaceuticals and dyes.

**bro·mide** (brō′mīd′) *n.* **1.** A binary compound of bromine. **2.** Potassium bromide. **3. a.** A worn-out notion or remark : PLATITUDE. **b.** A wearisome person : BORE. **—bro·mid′ic** (-mĭd′ĭk) *adj.*

▲ word history: Several bromine compounds, especially potassium bromide, have been used medicinally as sedatives. In 1906 Gelett Burgess wrote a book entitled *Are You a Bromide?* in which he used *bromide* to mean a tiresome person of unoriginal thought and trite conversation. *Bromide* was soon after extended to denote the kind of commonplace remarks made by such persons.

**bro·mi·nate** (brō′mə-nāt′) *vt.* **-nat·ed, -nat·ing, -nates.** To combine (a substance) with bromine or a bromine compound.

**bro·mine** (brō′mēn′) *n.* [Fr. *brome* < Gk. *brōmos*, stench + -INE.] *Symbol* **Br** A heavy, corrosive, reddish-brown, nonmetallic liquid element used in gasoline antiknock mixtures, fumigants, and photographic chemicals; atomic weight 79.909; atomic number 35.

**bro·mism** (brō′mĭz′əm) *also* **bro·min·ism** (brō′mə-nĭz′əm) *n.* Poisoning due to overuse of bromides.

**bromo-** *or* **brom-** *pref.* [Prob. < Fr. *brome* < Gk. *brōmos*, stench.] Bromine <*bromide*>

**bronch-** *pref. var. of* BRONCHO-.

**bron·chi** (brŏng′kī) *n. pl. of* BRONCHUS.

**bron·chi·a** (brŏng′kē-ə) *n. pl. of* BRONCHIUM.

**bron·chi·al** (brŏng′kē-əl) *adj.* Of or relating to the bronchi, the bronchia, or the bronchioles. **—bron′chi·al·ly** *adv.*

**bronchial asthma** *n.* A usu. allergic asthma of the bronchi.

**bronchial pneumonia** *n.* Broncho pneumonia.

**bronchial tube** *n.* A bronchus or any of its branches.

**bron·chi·ec·ta·sis** (brŏng′kē-ĕk′tə-sĭs) *n.* [Gk. *bronkhia*, bronchial tubes < *bronkhos*, windpipe + Gk. *ektasis*, extension (*ek-*, out + *teinein*, to stretch).] Chronic dilatation of the bronchial tubes, with cough and formation of mucopurulent matter.

**bron·chi·ole** (brŏng′kē-ōl′) *n.* Any of the fine, thin-walled, tubular extensions of a bronchus. **—bron′chi·o′lar** (-ō′lər) *adj.*

**bron·chi·tis** (brŏn-kī′tĭs, brŏng-) *n.* Acute or chronic inflammation of the mucous membrane of the bronchial tubes. **—bron·chit′ic** (-kĭt′ĭk) *adj.*

**bron·chi·um** (brŏng′kē-əm) *n., pl.* **-chi·a** (-kē-ə) [NLat., sing. of LLat. *bronchia*, bronchial tubes < Gk. *bronkhia* < *bronkhos*, windpipe.] A bronchial tube smaller than a bronchus and larger than a bronchiole.

**broncho-** *or* **bronch-** *pref.* [LLat. < Gk. *bronkho-* < *bronkhos*, windpipe.] Bronchus : bronchial <*broncho*scope>

**bron·cho·con·stric·tion** (brŏng′kō-kən-strĭk′shən) *n.* Constriction of the bronchial tubes. **—bron′cho·con·stric′tor** *adj.*

**bron·cho·pneu·mon·ia** (brŏng′kō-nŏŏ-mōn′yə, -nyŏŏ-) *n.* Inflammation of the lungs following infection of the bronchi.

**bron·cho·scope** (brŏng′kə-skōp′) *n.* A slender tubular instrument with a small light on the end for inspecting the interior of the bronchi. **—bron′cho·scop′ic** (-skŏp′ĭk) *adj.* **—bron′cho·scop′i·cal·ly** *adv.* **—bron·chos′co·pist** (brŏn-kŏs′kə-pĭst, brŏng-) *n.* **—bron·chos′co·py** (-kə-pē) *n.*

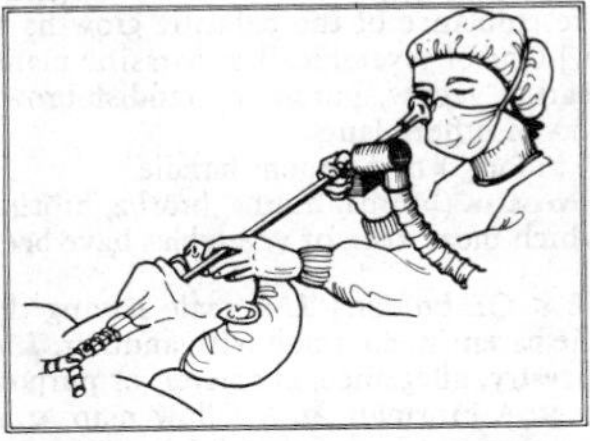

**bronchoscope**

**bron·chus** (brŏng′kəs) *n., pl.* **-chi** (-kī′) [NLat. < Gk. *bronkhos*, windpipe.] Either of two main branches of the trachea, leading directly to the lungs.

**bron·co** (brŏng′kō) *n., pl.* **-cos.** [Mex. Sp. < Sp., wild.] A wild or semiwild horse of western North America.

**bron·co·bust·er** (brŏng′kō-bŭs′tər) *n.* A cowboy who breaks wild horses.

**bron·to·saur** (brŏn′tə-sôr′) *also* **bron·to·sau·rus** (brŏn′tə-sôr′əs) *n.* [NLat. *Brontosaurus*, genus name : Gk. *brontē*, thunder + Gk. *sauras*, lizard.] A large herbivorous dinosaur of the genus *Apatosaurus* or *Brontosaurus*, of the Jurassic period.

**Bronx cheer** (brŏngks) *n.* [After the *Bronx*, a borough of New York City.] *Slang.* RASPBERRY 4.

**bronze** (brŏnz) *n.* [Fr. < Ital. *bronzo*.] **1. a.** An alloy of copper and tin, occas. with traces of other metals. **b.** An alloy of copper, with or without tin, and antimony, phosphorus, or other components. **2.** A work of art made of bronze. **3. a.** A moderate yellowish to olive brown. **b.** A pigment of this color. —*adj.* Of the color bronze. —*vt.*

**bronzed, bronz·ing, bronz·es.** To give a bronze appearance to. **—bronz'y** *adj.*
**Bronze Age** *n.* A period of human culture between the Stone Age and the Iron Age, marked by the use of bronze implements and weapons.
**bronz·er** (brŏn'zər) *n.* A cosmetic imparting a tanned appearance to the skin.
**Bronze Star** *n.* A U.S. Army decoration awarded for heroism or meritorious achievement in ground combat.
**brooch** (brōch, bro͞och) *also* **broach** (brōch) *n.* [ME *broche,* pointed tool. —see BROACH[1].] A decorative clasp or pin.
**brood** (bro͞od) *n.* [ME < OE *brōd.*] **1.** The young of certain animals, esp. a group of young birds or fowl hatched at one time and cared for together. **2.** The children in one family. *—v.* **brood·ed, brood·ing, broods.** —*vt.* **1.** To sit on or hatch (eggs). **2.** To protect (young) by or as if by covering with the wings. —*vi.* **1.** To sit on or hatch eggs. **2.** To hover envelopingly. **3.** To ponder at length and unhappily : WONDER. —*adj.* Kept for breeding. **—brood'ing·ly** *adv.*
☆ ***syns:*** BROOD, MOPE, STEW, WORRY *v. core meaning* : to dwell on moodily and at length <*brooded* about their poor health>
**brood·er** (bro͞o'dər) *n.* **1.** One that broods. **2.** A heated enclosure used for raising fowl.
**brood·mare** (bro͞od'mâr') *n.* A mare kept for breeding.
**brood·y** (bro͞o'dē) *adj.* **-i·er, -i·est. 1.** Meditative : moody. **2.** Inclined to sit on eggs to hatch them. **—brood'i·ness** *n.*
**brook[1]** (bro͝ok) *n.* [ME < OE *brōc.*] A small natural freshwater stream.
**brook[2]** (bro͝ok) *vt.* **brooked, brook·ing, brooks.** [ME *brouken* < OE *brūcan,* to use, enjoy.] To put up with : TOLERATE.
**brook·ite** (bro͝ok'īt') *n.* [After Henry J. *Brooke* (1771–1857).] A red-brown to black titanium dioxide mineral with orthorhombic crystals.
**brook lamprey** *n.* One of several usu. small lampreys that live mostly in brooks.
**brook·let** (bro͝ok'lĭt) *n.* A small brook.
**brook·lime** (bro͝ok'līm') *n.* [ME *broke-lemok* : *broke,* brook + *lemok,* a kind of brooklime < OE *hleomoce.*] Either of two closely related trailing plants, *Veronica americana* of North America and *V. beccabunga,* native to Eurasia, growing in moist places and bearing small blue flowers.
**brook trout** *n.* A freshwater game fish, *Salvelinus fontinalis* of eastern North America.
**brook·weed** (bro͝ok'wēd') *n.* Either of two related plants, *Samolus valerandi* of Europe and *S. floribundus* of North America, bearing small white flowers and growing in moist areas.
**broom** (bro͞om, bro͝om) *n.* [ME < OE *brōm.*] **1.** A brush of stiff fibers or bristles bound together on a long stick and used for sweeping. **2. a.** A shrub of the genus *Cytisus,* native to Eurasia, with compound leaves and yellow or white flowers. **b.** One of several similar or related shrubs, esp. of the genus *Genista.* —*vt.* **broomed, broom·ing, brooms.** To sweep with or as if with a broom. **—broom'y** *adj.*
**broom·corn** (bro͞om'kôrn', bro͝om'-) *n.* A grass, *Sorghum vulgare technicum,* whose stiff branching stalks are used to make brooms and brushes.
**broom moss** *n.* A moss of the genus *Dicranum,* esp. *D. scoparium,* with leaves turned to one side along the stem.
**broom·rape** (bro͞om'rāp', bro͝om'-) *n.* [Transl. of Lat. *rapum genistae,* broom tuber, from the resemblance of the parasitic growths to tubers on the roots of broom.] One of several leafless parasitic plants of the genus *Orobanche,* bearing yellow, purple, or reddish-brown flowers and living on the roots of other plants.
**broom·stick** (bro͞om'stĭk', bro͝om'-) *n.* A broom handle.
**broth** (brôth, brŏth) *n., pl.* **broths** (brôths, brŏths, brôthz, brŏthz) [ME < OE *broð.*] Water in which meat, fish, or vegetables have been boiled : STOCK.
**broth·er** (brŭ*th*'ər) *n.* [ME < OE *brōðor.*] **1.** A male having the same parents as another or one parent in common with another. **2.** A person sharing a common ancestry, allegiance, character, or purpose with another or others, esp.: **a.** A kinsman. **b.** A fellow man. **c.** A fellow member, as of a fraternity. **d.** A close male friend : COMRADE. **3. a.** A member of a men's religious order who is not in holy orders but engages in the work of the order. **b.** A lay member of a men's religious order.
**broth·er·hood** (brŭ*th*'ər-ho͝od') *n.* **1.** The quality or state of being brothers. **2.** An association, as a fraternity, united for common purposes. **3.** All the members of a specific profession or trade.
**broth·er-in-law** (brŭ*th*'ər-ĭn-lô') *n., pl.* **broth·ers-in-law. 1.** The brother of one's spouse. **2.** The husband of one's sister. **3.** The husband of the sister of one's spouse.
**broth·er·ly** (brŭ*th*'ər-lē) *adj.* Characteristic of brothers : FRATERNAL. **—broth'er·li·ness** *n.* **—broth'er·ly** *adv.*
**brougham** (bro͞om, bro͞o'əm, brōm, brō'əm) *n.* [After Henry P. *Brougham* (1778–1868).] **1.** A closed four-wheeled carriage with an open driver's seat in front. **2.** An automobile with an open driver's seat. **3.** An electrically powered automobile resembling a coupé.
**brought** (brôt) *v. p.t.* & *p.p. of* BRING.
**brou·ha·ha** (bro͞o'hä-hä') *n.* [Fr.] A noisy wrangle : UPROAR.
**brow** (brou) *n.* [ME < OE *brū.*] **1. a.** The superciliary ridge over the eyes. **b.** The eyebrow. **c.** The forehead. **2.** Countenance : mien. **3.** The edge of a steep place.
**brow·beat** (brou'bēt') *vt.* **-beat, -beat·en** (-bēt'n), **-beat·ing, -beats.** To intimidate : bully. **—brow'beat'er** *n.*
**brown** (broun) *n.* [ME < OE *brūn.*] One of a group of colors between red and yellow in hue that are medium to low in lightness and low to moderate in saturation. —*adj.* **-er, -est. 1.** Of the color brown. **2.** Deeply suntanned. —*v.* **browned, brown·ing, browns.** —*vt.* To make brown, esp. to cook until brown. —*vi.* To become brown. **—brown'ish** *adj.* **—brown'ness** *n.*
**brown alga** *n.* A brownish, chiefly marine alga of the division Phaeophyta, including the rockweeds and the kelps.
**brown bagging** *n.* **1.** The practice of taking one's own liquor into a public establishment, as a restaurant, where setups are available. **2.** The practice of taking one's lunch to work, usu. in a brown paper bag. **—brown bagger** *n.*
**brown bear** *n.* A large bear, *Ursus arctos* of Alaska and northern Eurasia, with brown to yellowish fur.
**brown belt** *n.* **1.** A rank of proficiency in karate or judo ranking next below black belt. **2.** The brown-colored belt symbolizing this rank. **3.** A person holding a brown belt.
**brown Bet·ty** (bĕt'ē) *n.* [*Betty,* nickname for *Elizabeth.*] Baked pudding made with apples, bread crumbs, raisins, sugar, and spices.
**brown bread** *n.* **1.** A bread made of a dark flour. **2.** A steamed bread usu. made of cornmeal, flour, and molasses.
**brown coal** *n.* Lignite.
**brown fat** *n.* Adipose tissue whose oxidation is a major source of heat in mammals.
**Brown·i·an motion** (brou'nē-ən) *n.* [After Robert *Brown* (1773–1858).] Random motion of microscopic particles suspended in a liquid or gas, caused by collision with molecules of the surrounding medium.
**Brownian movement** *n.* Brownian motion.
**brown·ie** (brou'nē) *n.* **1.** A kindly sprite believed to do helpful work at night. **2. Brownie.** A member of the Girl Scouts from seven to nine years old. **3.** A square or bar of moist, rich, often nut-filled chocolate cake.
**Brownie point** *n.* [From the practice of awarding points for achievement by Brownies in the Girl Scouts.] Credit considered as earned, esp. by impressing a superior favorably.
**Brown·ing automatic rifle** (brou'nĭng) *n.* [After John M. *Browning* (1855–1926).] A .30 caliber air-cooled, automatic or semi-automatic, gas-operated, magazine-fed rifle.
**Browning machine gun** *n.* [After John M. *Browning* (1855-1926).] A .30 or .50 caliber belt-fed, water-cooled machine gun.
**brown lung disease** *n.* Byssinosis.
**brown·out** (broun'out') *n.* [BROWN + (BLACK)OUT.] A reduction or cutback in electric power, esp. as a result of a shortage.
**brown patch** *n.* A disease of grasses caused by a fungus, *Pellicularia filamentosa,* that results in circular dying areas.
**brown rat** *n.* The Norway rat.
**brown rice** *n.* Unpolished rice, retaining the germ and the yellowish outer layer containing the bran.
**brown rot** *n.* **1.** A disease of fruit, as peaches, caused by fungi of the genus *Monolinia.* **2.** A disease of citrus trees, caused by fungi of the genus *Phytophthora.*
**Brown Shirt** *n.* [Transl. of G. *Braunhemd.*] A storm trooper.
**brown·stone** (broun'stōn') *n.* **1.** A brownish-red sandstone used as a building material. **2.** A house built or faced with brownstone.
**brown study** *n.* A state of reverie or deep thought.
**brown sugar** *n.* Sugar whose crystals retain a thin coating of dark syrup.
**Brown Swiss** *n.* One of a hardy breed of dairy cattle orig. bred in Switzerland.
**brown-tail moth** (broun'tāl') *n.* A small white and brown moth, *Euproctis phaeorrhoea,* whose larvae damage shade-tree foliage and cause an irritating skin rash.
**brown thrasher** *n.* A North American bird, *Toxostoma rufum,* with a dark-streaked breast and reddish-brown back.
**brown trout** *n.* A widely naturalized European freshwater game fish, *Salmo trutta,* with speckled sides.
**browse** (brouz) *v.* **browsed, brows·ing, brows·es.** [Perh. < Fr. *broust,* young shoot < OFr. *brost,* of Germanic orig.] —*vi.* **1. a.** To inspect leisurely and casually. **b.** To read superficially : SKIM. **2.** To feed on vegetation, as leaves or young shoots. —*vt.* **1.** To look over casually. **2. a.** To nibble : crop. **b.** To graze on. —*n.* **1.** Young twigs, leaves, and tender shoots of plants or shrubs that are fit for animals to eat. **2.** An act of browsing. **—brows'er** *n.*
**bru·cel·lo·sis** (bro͞o'sə-lō'sĭs) *n.* [NLat. *Brucella,* genus name, after Sir David *Bruce,* (1855–1931) + -OSIS.] **1.** Undulant fever. **2.** A disease of cattle caused by the bacillus *Brucella abortus* and resulting in abortions in newly infected animals.
**bru·cine** (bro͞o'sēn', -sĭn) *n.* [After James *Bruce* (1730–1794).] A

poisonous white crystalline alkaloid, $C_{23}H_{26}O_4N_2 \cdot 2H_2O$, derived from nux vomica seeds.

**bru·in** (broo'ĭn) *n.* [Du., brown.] A bear.

**bruise** (brooz) *v.* **bruised, bruis·ing, bruis·es.** [ME *bruisen* < OE *brȳsan,* to crush, and OFr. *bruisier,* to crush.] *—vt.* **1.** To injure without breaking or rupturing. **2.** To mar or dent. **3.** To pound into fragments : CRUSH. **4.** To hurt, esp. psychologically. *—vi.* To experience or undergo bruising. *—n.* **1.** An injury in which the skin is not broken : CONTUSION. **2.** An injury, esp. to one's feelings.

**bruis·er** (broo'zər) *n. Slang.* A large, powerfully built man.

**bruit** (broot) *vt.* **bruit·ed, bruit·ing, bruits.** [< ME, noise < OFr., p.part. of *bruire,* to roar.] To spread news of : REPEAT. *—n.* **1.** *Archaic.* **a.** A rumor. **b.** A clamor. **2.** *Med.* An abnormal sound heard during auscultation.

**bru·mal** (broo'məl) *adj.* [Lat. *brumalis* < *bruma,* winter.] Of, relating to, or occurring in winter.

**brume** (broom) *n.* [Fr., ult. < Lat. *bruma,* winter.] Heavy mist or fog. **—bru'mous** (broo'məs) *adj.*

**brum·ma·gem** (brŭm'ə-jəm) *adj.* [Alteration of *Birmingham,* England (from the counterfeit coins made there in the 17th cent.).] Gaudy : tawdry. *—n.* Something cheap and showy.

**brunch** (brŭnch) *n.* [BR(EAKFAST) + (L)UNCH.] A meal eaten as a combination of breakfast and lunch.

**bru·net** (broo-nĕt') *adj.* [Fr. < OFr. < *brun,* brown, of Germanic orig.] **1.** Of a dark coloring or complexion. **2.** Having dark brown or black hair or eyes. *—n.* A person with brown hair.

**bru·nette** (broo-nĕt') *adj.* Having dark or brown hair. *—n.* A girl or woman with dark or brown hair.

**Brun·hild** (broon'hilt') *n.* [G.] A queen of Iceland in Germanic legend who is won as a bride by Gunther.

**Bruns·wick stew** (brŭnz'wĭk) *n.* [After *Brunswick* County, Virginia.] A vegetable stew that usu. contains chicken and rabbit or squirrel meat.

**brunt** (brŭnt) *n.* [ME.] **1.** The main impact, force, or burden, as of a blow. **2.** *Obs.* A violent attack.

**brush[1]** (brŭsh) *n.* [ME *brusshe* < OFr. *brosse,* perh. < *brosse,* brushwood.] **1.** A device consisting of bristles fastened into a handle, for scrubbing, polishing, or painting. **2.** The act of using a brush. **3.** A light touch in passing : GRAZE. **4.** A brief encounter <a *brush* with the law> **5.** A bushy tail. **6.** A sliding connection completing a circuit between a fixed and a moving conductor. *—v.* **brushed, brush·ing, brush·es.** *—vt.* **1.** To use a brush on so as to clean, polish, or groom. **2.** To apply with or as if with motions of a brush. **3.** To remove with or as if with motions of a brush. **4.** To dismiss abruptly. **5.** To touch lightly in passing. *—vi.* **1.** To use or apply a brush. **2.** To move past something so as to touch it lightly. **—brush up.** To refresh one's memory. **—brush'er** *n.* **—brush'y** *adj.*

☆ ***syns:*** BRUSH, FLICK, GLANCE, GRAZE, SKIM *n. core meaning* : light and momentary contact with another < the *brush* of a bird's wing against a leaf>

**brush[2]** (brŭsh) *n.* [ME *brusshe* < OFr. *brosse,* brushwood.] **1. a.** A thicket of shrubs or bushes. **b.** Land covered by brush. **2.** Cut or broken branches. **—brush'y** *adj.*

**brush discharge** *n.* A faintly visible, relatively slow crackling discharge of electricity without sparking.

**brushed** (brŭsht) *adj.* Of or designating fabrics that have a nap produced by brushing.

**brush·fire** (brŭsh'fīr') *n.* A fire in low-growing scrubby trees and brush.

**brushfire war** *n.* A relatively insignificant and limited military action, usu. for harrassment.

**brush-off** (brŭsh'ôf', -ŏf') *n. Slang.* An abrupt dismissal.

**brush·wood** (brŭsh'wood') *n.* BRUSH[2].

**brush·work** (brŭsh'wûrk') *n.* **1.** Work done with a brush. **2.** The way in which an artist applies paint.

**brusque** *also* **brusk** (brŭsk) *adj.* [Fr. < Ital. *brusco.*] Rudely abrupt : BLUNT. **—brusque'ly** *adv.* **—brusque'ness** *n.*

**brus·que·rie** (brŭs'kə-rē') *n.* [Fr. < *brusque,* brusque.] Curtness : brusqueness.

**Brus·sels carpet** (brŭs'əlz) *n.* [After *Brussels,* Belgium.] A machine-made carpet consisting of small colored woolen loops that form a heavy patterned pile.

**Brussels lace** *n.* Net lace with an appliqué design.

**Brussels sprout** *n.* **1.** A variety of cabbage, *Brassica oleracea gemmifera,* having a stout stem studded with budlike heads. **2. Brussels sprouts.** The small edible heads of the Brussels sprout.

**brut** (broot) *adj.* [Fr. < OFr., rough < Lat. *brutus,* heavy.] Very dry, as champagne.

**bru·tal** (broot'l) *adj.* **1.** Characteristic of a brute : MEAN <a *brutal* attack> **2.** Disconcertingly accurate <the *brutal* statistics> **3.** Harsh <a *brutal* wind> **4.** Cruelly insensitive <a *brutal* lie> **—bru·tal'i·ty** *n.* **—bru'tal·ly** *adv.*

**bru·tal·ize** (broot'l-īz') *vt.* **-ized, -iz·ing, -iz·es. 1.** To make brutal. **2.** To treat brutally. **—bru'tal·i·za'tion** *n.*

**brute** (broot) *n.* [< ME, nonhuman < OFr. *brut* < Lat. *brutus,* stupid.] **1.** An animal other than a human being : BEAST. **2.** A brutal person. *—adj.* **1.** Of or relating to beasts : ANIMAL. **2.** Typical of a brute, esp.: **a.** Completely instinctive or physical <*brute* strength> **b.** Displaying a lack of reason or intelligence <an impulsive, *brute* reaction> **c.** Savage : cruel. **—brut'ism** *n.*

**bru·ti·fy** (broo'tə-fī') *vt.* & *vi.* **-fied, -fy·ing, -fies.** To brutalize or become brutalized.

**brut·ish** (broo'tĭsh) *adj.* **1.** Of or befitting a brute. **2.** Showing a lack of reason or intelligence. **3.** Coarse : crude. **4.** Carnal : sensual. **—brut'ish·ly** *adv.* **—brut'ish·ness** *n.*

**Bryn·hild** (brĭn'hĭld') *n.* [ON *Brynhildr.*] A Valkyrie who is revived from an enchanted sleep by Sigurd.

**bryo-** *pref.* [NLat. < Gk. *bruon,* moss.] Moss <*bryology*>

**bry·ol·o·gy** (brī-ŏl'ə-jē) *n.* The botany of bryophytes. **—bry'o·log'i·cal** (-ə-lŏj'ĭ-kəl) *adj.*

**bry·o·ny** (brī'ə-nē) *n., pl.* **-nies.** [Lat. *bryonia* < Gk. *bruōnia.*] **1.** The black bryony. **2.** The white bryony.

**bry·o·phyte** (brī'ə-fīt') *n.* A plant of the major botanical division Bryophyta, including the true mosses, peat mosses, and liverworts. **—bry'o·phyt'ic** (-fĭt'ĭk) *adj.*

**bry·o·zo·an** (brī'ə-zō'ən) *n.* [< NLat. *Bryozoa,* phylum name : Gk. *bruon,* moss + Gk. *zōa,* pl. of *zōon,* animal.] Any of various small aquatic animals of the phylum Bryozoa that reproduce by budding and form mosslike or branching colonies. *—adj.* Of or belonging to the Bryozoa.

**Bryth·on** (brĭth'ən, -ŏn') *n.* [Welsh.] **1.** An ancient Celtic Briton of Cornwall, Wales, or Cumbria. **2.** A person who speaks a Brythonic language.

**Bry·thon·ic** (brĭ-thŏn'ĭk) *adj.* Of, relating to, or typical of the Brythons or their language. *—n.* The branch of the Celtic languages that includes Welsh, Breton, and Cornish.

**B-school** (bē'skool') *n.* A school of business administration.

**bub·ble** (bŭb'əl) *n.* [< ME *bubelen,* to bubble.] **1.** A light, thin-walled sphere of liquid containing air or gas. **2.** A small globule of gas trapped in a liquid or solid, as in a carbonated beverage. **3.** A sound like something bubbling. **4.** Something insubstantial, groundless, or ephemeral, as an abortive scheme. **5.** A usu. transparent glass or plastic dome. *—v.* **-bled, -bling, -bles.** *—vi.* **1.** To form or give off bubbles. **2.** To move or flow with a gurgling sound. **3.** To express lively animation or excitement <*bubbling* with joy> *—vt.* To cause to form bubbles. **—bub'bly** *adj.*

**bubble and squeak** *n.* [Imit. of the sounds made as it cooks.] *Chiefly Brit.* Cabbage and potatoes fried together.

**bubble chamber** *n.* An apparatus for detecting the paths of charged particles or inferring the paths of electrically neutral particles by examination of trails of bubbles that form on ions produced in a superheated liquid.

**bubble gum** *n.* Chewing gum that can be blown into bubbles.

**bubble memory** *n.* A computer memory in which binary digits are represented by the presence or absence of magnetic bubbles.

**bub·bler** (bŭb'lər) *n.* A drinking fountain in which the water flows upward through a small nozzle.

**bub·bly** (bŭb'lē) *n., pl.* **-blies.** *Informal.* Champagne.

**bub·by** (boob'ē, boob'ē) *n., pl.* **-bies.** [Orig. unknown.] *Slang.* A woman's breast.

**bu·bo** (boo'bō, byoo'-) *n., pl.* **-boes.** [ME < Med. Lat. < Gk. *boubōn.*] An inflamed swelling of a lymphatic gland, esp. in the armpit or groin. **—bu·bon'ic** (-bŏn'ĭk) *adj.*

**bubonic plague** *n.* A contagious, usu. fatal epidemic disease caused by bacteria of the genus *Pasteurella,* transmitted by fleas from infected rats and marked by chills, fever, vomiting, diarrhea, and buboes.

**buc·cal** (bŭk'əl) *adj.* [< Lat. *bucca,* cheek.] Of or relating to the cheeks or mouth cavity.

**buc·ca·neer** (bŭk'ə-nîr') *n.* [Fr. *boucanier* < *boucaner,* to cure meat < *boucan,* barbecue frame < Tupi *mocaen.*] PIRATE 1. **—buc'ca·neer'** *v.* **(-neered, -neer·ing, -neers).**

**buck[1]** (bŭk) *n.* [ME *bukke* < OE *buc.*] **1.** The adult male of some animals, as the deer or rabbit. **2.** *Informal.* **a.** A robust or high-spirited young man. **b.** A fop. *—v.* **bucked, buck·ing, bucks.** *—vi.* **1.** To rear up or leap forward suddenly, as a horse. **2.** To move rapidly forward with the head lowered : BUTT. **3.** To move forward with sudden jerks : JOLT. **4.** To resist stubbornly and obstinately : BALK. **5.** *Informal.* To strive doggedly <*bucking* for a raise> *—vt.* **1.** To throw (a rider or burden) by bucking. **2.** To butt against with the head. **3.** *Football.* To charge into (an opponent's line) carrying the ball. **4.** To oppose stubbornly and directly : struggle against. **—buck up.** *Informal.* To summon one's courage. *—adj.* Of the lowest rank in a specified military category <a *buck* private><a *buck* sergeant> **—buck'er** *n.*

**buck[2]** (bŭk) *n.* [Short for SAWBUCK.] **1.** A sawhorse. **2.** A leather-covered frame for vaulting in gymnastics.

**buck[3]** (bŭk) *n.* [Short for BUCKSKIN.] *Slang.* A dollar.

**buck[4]** (bŭk) *n.* [Short for *buckhorn knife.*] A marker once placed before a poker player next in line to deal. **—pass the buck.** To shift accountability or blame to another.

oo b**oo**t ou **ou**t th **th**in *th* **th**is ŭ c**u**t ûr **ur**ge y **y**oung
yoo ab**u**se zh vi**s**ion ə **a**bout, it**e**m, edib**i**le, gall**o**p, circ**u**s

**buck and wing** *n.* A fast solo tap dance with much springing of the legs and heel clicking.

**buck·a·roo** *also* **buck·er·oo** (bŭk'ə-ro͞o') *n., pl.* **-roos** *also* **-oos.** [Sp. *vaquero* < *vaca,* cow < Lat. *vacca.*] A cowboy.

**buck·bean** (bŭk'bēn') *n.* [Transl. of Flem. *bocks boonen.*] A marsh plant, *Menyanthes trifoliata,* with a creeping rootstock and white or reddish flowers.

**buck·board** (bŭk'bôrd', -bōrd') *n.* [Obs. *buck,* body of a wagon (< ME *buke,* belly < OE *būc*) + BOARD.] An open, four-wheeled carriage with the seat attached to a flexible board extending from the front to the rear axle.

buckboard

**buck·er·oo** (bŭk'ə-ro͞o') *n. var. of* BUCKAROO.

**buck·et** (bŭk'ĭt) *n.* [ME < OFr. *buket.*] **1. a.** A cylindrical vessel for holding or carrying liquids or solids : PAIL. **b.** The amount a bucket holds. **2.** A machine compartment that receives and conveys material, as the scoop of a power shovel. *—v.* **-et·ed, -et·ing, -ets.** *—vt.* **1.** To hold, carry, or put in a bucket. **2.** To ride (a horse) hard. *—vi.* **1.** To proceed jerkily and rapidly <*bucketing* over a dirt road> **2.** To make haste.

**bucket brigade** *n.* A line of people passing buckets of water hand to hand to put out a fire.

**bucket seat** *n.* A seat with a rounded or molded back for one person, as in a sports car.

**bucket shop** *n.* [< *bucket shop,* a saloon selling small amounts of liquor in buckets, from its resemblance to the forerunner of such brokerage operations, which dealt in small units of stocks and commodities.] A dishonest brokerage operation that accepts orders to buy or sell securities or commodities but delays executing the orders on the gamble that prices will change adversely to the interests of the customer so that the broker can then pocket what the customer thinks has been lost.

**buck·eye** (bŭk'ī') *n.* [From the seed's appearance.] **1.** One of several North American trees of the genus *Aesculus,* with compound leaves and erect white or reddish flower clusters. **2.** The glossy brown nut of a buckeye.

**buck fever** *n. Informal.* The nervous excitement of a novice hunter at the first sight of game.

**buck·hound** (bŭk'hound') *n.* A hound for hunting deer.

**buck·le**[1] (bŭk'əl) *n.* [ME *bocle* < OFr. *boucle* < Lat. *buccula,* cheek strap of a helmet, dim. of *bucca,* cheek.] **1.** A metal frame with one or more movable tongues used esp. for fastening two strap or belt ends. **2.** An ornament resembling a buckle. *—vt. & vi.* **-led, -ling, -les.** To fasten or be fastened with a buckle. **—buckle down.** To apply oneself with determination.

**buck·le**[2] (bŭk'əl) *v.* **-led, -ling, -les.** [ME *boclen* < OFr. *boucler,* to fasten with a buckle < *boucle,* buckle.] *—vi.* **1.** To warp, bend, or crumple, as under heat or pressure. **2.** To give way : COLLAPSE <supports *buckling* from the load> **3.** To surrender : yield <*buckled* from the constant harangue> *—vt.* To cause to warp, bend, or crumple. *—n.* A distortion, as a bend or bulge.

**buck·ler** (bŭk'lər) *n.* [ME *bokler* < OFr. *bocler* < *boucle,* boss on a shield < Lat. *buccula.*] **1.** A small round shield either carried or worn on the arm. **2.** A means of protection : DEFENSE. *—vt.* **-lered, -ler·ing, -lers.** To shield with or as if with a buckler : PROTECT.

**buck·o** (bŭk'ō) *n., pl.* **-oes. 1.** A bully. **2.** *Chiefly Ir.* A lad.

**buck passer** *n.* A person who tries to shift accountability to another.

**buck·ram** (bŭk'rəm) *n.* [ME *bokeram,* fine linen < OFr. *boquerant,* poss. after *Bukhara,* U.S.S.R.] **1.** A coarse, heavily sized cotton fabric used in bookbinding and for stiffening garments. **2.** *Obs.* Formality : stiffness. *—adj.* Resembling buckram in stiffness. *—vt.* **-ramed, -ram·ing, -rams.** To stiffen with or as if with buckram.

**buck·saw** (bŭk'sô') *n.* [< BUCK[2].] A wood-cutting saw, usu. set in an H-shaped frame.

**buck·shee** (bŭk'shē) *n.* [Var. of BAKSHEESH.] *Chiefly Brit.* **1.** A gratuity or windfall. **2.** An extra ration.

**buck·shot** (bŭk'shŏt') *n.* A large lead shot for shotgun shells.

**buck·skin** (bŭk'skĭn') *n.* **1. a.** The skin of a male deer. **b.** A strong, soft, grayish-yellow leather. **2. buckskins.** Breeches or shoes made from buckskin. **3.** A person wearing clothes of buckskin, esp. a backwoodsman of early America. **4.** A buckskin-colored horse.

**buck·thorn** (bŭk'thôrn') *n.* [Transl. of NLat. *cervi spina.*] A shrub or tree of the genus *Rhamnus,* esp. *R. cathartica* of Eurasia, with spine-tipped branches and small greenish flowers.

**buck·tooth** (bŭk'to͞oth') *n.* A prominent, projecting front tooth. **—buck'toothed'** (-to͞otht') *adj.*

**buck·wheat** (bŭk'hwēt', -wēt') *n.* [Partial transl. of MDu. *boecweite* : *boek,* beech + *weite,* wheat.] **1.** A plant of the genus *Fagopyrum,* esp. *F. esculentum,* native to Asia, having fragrant white flowers and small triangular seeds. **2.** The edible seeds of buckwheat, often ground into flour.

**bu·col·ic** (byo͞o-kŏl'ĭk) *adj.* [Lat. *bucolicus* < Gk. *boukolikos* < *boukolos,* cowherd < *bous,* cow.] **1.** Of or relating to shepherds or flocks : PASTORAL. **2.** Of or relating to the countryside or its people : RUSTIC. *—n.* A pastoral poem. **—bu·col'i·cal·ly** *adv.*

**bud** (bŭd) *n.* [ME *budde.*] **1.** *Bot.* **a.** A small protuberance on a stem or branch, often enclosed in protective scales, containing an undeveloped shoot, leaves, or flowers. **b.** The state or stage of having buds. **2.** *Biol.* **a.** An asexually produced protuberance, as on a polyp, that develops into a mature, complete organism. **b.** A small, rounded organic part resembling a plant bud. **3.** One that is not yet fully developed. *—v.* **bud·ded, bud·ding, buds.** *—vi.* **1.** To put forth or produce buds. **2.** To begin to develop or grow from or as if from a bud. **3.** To be in an undeveloped stage or condition. *—vt.* **1.** To cause to put forth buds. **2.** To graft a bud onto (a plant). **—bud'der** *n.*

**Bud·dha** (bo͞o'də, bŏŏd'ə) *n.* [Skt., enlightened < *bodhati,* he awakes.] **1.** A Buddhist sage who has achieved a state of perfect illumination in accordance with the teachings of Gautama Buddha. **2.** A representation of Gautama Buddha.

**Bud·dhism** (bo͞o'dĭz'əm, bŏŏd'ĭz'-) *n.* **1.** The doctrine, attributed to Gautama Buddha, that suffering is inseparable from existence but that inward extinction of the self and of the senses culminates in a state of illumination beyond both suffering and existence. **2.** The religion of eastern and central Asia represented by the many differing sects that profess Buddhism and venerate Gautama Buddha. **—Bud'dhist** *n.* **—Bud·dhis'tic** *adj.*

**bud·dle** (bŭd'l) *n.* [Orig. unknown.] An inclined trough for separating ore from waste by washing with running water.

**bud·dle·ia** (bŭd'lē-ə, bŭd-lē'ə) *n.* [After Adam *Buddle* (d. 1715).] The butterfly bush.

**bud·dy** (bŭd'ē) *n., pl.* **-dies.** *Informal.* [Prob. alteration of BROTHER.] **1.** A good friend : COMRADE. **2.** Fellow : friend.

**buddy system** *n.* An informal arrangement in which persons are paired, as for mutual assistance or safety.

**budge**[1] (bŭj) *v.* **budged, budg·ing, budg·es.** [OFr. *bouger* < VLat. **bullicare,* to bubble < Lat. *bullire,* to boil.] *—vi.* **1.** To move or stir slightly <never *budged* at all> **2.** To change a position or attitude : YIELD <refused to *budge* on the decision> *—vt.* **1.** To cause to move slightly. **2.** To cause to change a position or attitude.

**budge**[2] (bŭj) *n.* [ME *buge.*] Fur, usu. lambskin, once used to trim academic robes. *—adj. Archaic.* Extremely formal : SOLEMN.

**budg·er·i·gar** (bŭj'ə-rē-gär', bŭj'ə-rē'-) *n.* [Native word in Australia.] A parakeet native to Australia, *Melopsittacus undulatus,* with green, yellow, or blue plumage.

**budg·et** (bŭj'ĭt) *n.* [ME *bouget,* wallet < OFr. *bougette,* dim. of *bouge,* leather bag < Lat. *bulga,* of Celt. orig.] **1. a.** An itemized summary of probable expenditures and income for a given period. **b.** A systematic plan for meeting expenses in a given period. **c.** The total sum of money allocated for a particular purpose or time period. **2.** A stock or collection with definite limits. *—vt.* **-et·ed, -et·ing, -ets. 1.** To plan in advance the expenditure of (e.g., money). **2.** To enter or plan for in a budget. **—budg'et·ar'y** (bŭj'ĭ-tĕr'ē) *adj.* **—budg'et·er** *n.*

**budg·ie** (bŭj'ē) *n. Informal.* A budgerigar.

**buff**[1] (bŭf) *n.* [OFr. *buffle,* buffalo.] **1.** A soft, thick, undyed leather made chiefly from the skins of buffalo, elk, or oxen. **2. a.** A moderate or yellow light. **b.** A moderate orange yellow. **c.** A military coat made of buff. **3.** *Informal.* The bare skin <swam in the *buff*> **4.** A polishing implement covered with a soft material, as velvet. *—adj.* Of the color of buff. *—vt.* **buffed, buff·ing, buffs. 1.** To polish or shine with a buff. **2.** To make the color of buff.

**buff**[2] (bŭf) *vt.* **buffed, buff·ing, buffs.** [Prob. imit.] To deaden the shock of.

**buff**[3] (bŭf) *n.* [From the buff-colored uniform worn by New York volunteer firemen at one time.] *Informal.* A fan : enthusiast <model railroad *buffs*>

**buf·fa·lo** (bŭf'ə-lō') *n., pl.* **-loes** *or* **-los** *or* **buffalo.** [Ital. or Port. *bufalo* < VLat. **bufalus* < Lat. *bubalus* < Gk. *boubalos.*] **1. a.** One of several oxlike Old World mammals of the family Bovidae, as *Syncerus caffer* of Africa, with massive, downward-curving horns. **b.** The bison. **2.** The buffalo fish. *—vt.* **-loed, -lo·ing, -loes.** *Slang.* **1.** To intimidate. **2.** To bewilder : confuse.

**buffalo berry** *n.* **1.** A North American shrub, *Shepherdia argentea* or *S. canadensis,* with small yellowish flowers and red or yellowish berries. **2.** The berry of a buffalo berry.

**buffalo bug** *n.* The carpet beetle.

**buffalo fish** *n.* One of several North American humped-back freshwater fishes of the genus *Ictiobus.*

**buffalo grass** *n.* A short grass, *Buchloe dactyloides* of the plains east of the Rocky Mountains.

**buffalo robe** *n.* The dressed skin of the North American bison, used as a lap robe, cape, or blanket.

**buff·er**[1] (bŭf'ər) *n.* A device used for shining or polishing.

**buff·er**[2] (bŭf'ər) *n.* [Prob. < BUFF[2].] **1.** Something that lessens or absorbs the shock of an impact. **2.** One that protects by intercepting or moderating adverse pressures or influences. **3.** Something that separates the entities, as a neutral area between two conflicting powers. **4. a.** *Chem.* A substance capable of maintaining the relative concentrations of hydrogen and hydroxyl ions in a solution by neutralizing, within limits, added acids or bases. **b.** *Computer Sci.* A device or area used to store data temporarily and deliver at a rate different from that at which it was received. —*vt.* **-ered, -er·ing, -ers.** *Chem.* To treat (a solution) with a buffer.

**buf·fet**[1] (bə-fā', bo͞o-) *n.* [Fr.] **1.** A large sideboard with drawers and cupboards. **2. a.** A counter from which meals or refreshments are served. **b.** A restaurant having such a counter. **3.** A meal at which guests serve themselves from various dishes displayed on a table or sideboard.

**buf·fet**[2] (bŭf'ĭt) *n.* [ME < OFr. *bufet,* dim. of *buffe,* blow.] A blow or cuff with or as if with the hand. —*v.* **-fet·ed, -fet·ing, -fets.** —*vt.* **1.** To club or hit, esp. with the hand. **2.** To strike against forcefully : BATTER <waves *buffeting* the pier> **3.** To force (one's way) with or as if with crude blows. —*vi.* To force one's way, esp. with difficulty. **—buf'fet·er** *n.*

**buf·fi** (bo͞o'fē) *n. var. pl. of* BUFFO.

**buffing wheel** *n.* A wheel covered with a soft material, as velvet or leather, for shining and polishing metal.

**buf·fle·head** (bŭf'əl-hĕd') *n.* [Obs. *buffle,* buffalo (< OFr. < VLat. **bufalus*) + HEAD.] A small North American duck, *Bucephala albeola,* with black and white plumage and a densely feathered head.

**buf·fo** (bo͞o'fō) *n., pl.* **-fi** (-fē) or **-fos.** [Ital. < *buffare,* to puff.] A male comic opera singer.

**buf·foon** (bə-fo͞on') *n.* [Fr. *bouffon* < OItal. *buffone* < *buffa,* jest < *buffare,* to puff.] **1.** A jester : clown. **2.** A person who makes coarse jokes. **—buf·foon'er·y** (bə-fo͞o'nə-rē) *n.*

**bug** (bŭg) *n.* [Orig. unknown.] **1.** Any of various wingless or four-winged insects of the order Hemiptera, and esp. the suborder Heteroptera, with mouth parts adapted for piercing and sucking. **2.** An insect. **3.** *Informal.* A disease-producing microorganism. **4.** A mechanical, electrical, or other systemic defect or difficulty. **5.** *Slang.* BUFF[3]. **6.** A small hidden device, as a microphone used for eavesdropping. —*vt.* **bugged, bug·ging, bugs. 1.** *Slang.* To annoy : PESTER. **2.** To equip (e.g., a telephone circuit) with a concealed electronic listening device. **—bug'ger** *n.*

**bug·a·boo** (bŭg'ə-bo͞o') *n., pl.* **-boos.** [Perh. of Celt. orig.] **1.** A bugbear. **2.** A steady source of concern.

**bug·bane** (bŭg'bān') *n.* Any of several plants of the genus *Cimicifuga,* esp. *C. americana* of eastern North America, with small white flower clusters believed to repel insects.

**bug·bear** (bŭg'bâr') *n.* [Obs. *bug,* hobgoblin (< ME *bugge,* poss. < MWelsh *bwga,* ghost) + BEAR.] **1.** An object of obsessive dread. **2.** *Archaic.* A hobgoblin.

**bug-eyed** (bŭg'īd') *adj. Slang.* Agog, as with amazement.

**bug·gy**[1] (bŭg'ē) *n., pl.* **-gies.** [Orig. unknown.] **1.** A small, light, four-wheeled horse-drawn carriage. **2.** A baby carriage.

**bug·gy**[2] (bŭg'ē) *adj.* **-gi·er, -gi·est. 1.** Infested with bugs. **2.** *Slang.* Crazy. **—bug'gi·ness** *n.*

**bug·house** (bŭg'hous') [< BUGGY[2].] *Slang.* —*n.* An insane asylum. —*adj.* Mentally unsound : INSANE.

**bu·gle**[1] (byo͞o'gəl) *n.* [ME < OFr. < Lat. *buculus,* steer, dim. of *bos,* ox.] A brass wind instrument shorter than a trumpet and lacking keys or valves. —*vi.* **-gled, -gling, -gles. 1.** To play a bugle. **2.** To give forth a deep, prolonged sound similar to the bay of a hound. **—bu'gler** *n.*

**bu·gle**[2] (byo͞o'gəl) *n.* [Orig. unknown.] A tubular bead used to trim clothing.

**bu·gle**[3] (byo͞o'gəl) *n.* [ME < OFr. < LLat. *bugula* < Lat. *bugillo.*] A plant of the genus *Ajuga,* native to Eurasia, with spikes or small, dense, blue or white flower clusters.

**bu·gle·weed** (byo͞o'gəl-wēd') *n.* [Perh. from its tubular flowers.] **1.** A plant of the genus *Lycopus,* esp. *L. virginicus,* with small, whitish flowers and an aromatic odor. **2.** BUGLE[3].

**bu·gloss** (byo͞o'glŏs', -glôs') *n.* [ME *buglosse* < OFr. < Lat. *buglossa* < Gk. *bouglōssos* : *bous,* ox + *glōssa,* tongue.] Any of several plants of the genera *Lycopsis, Echium,* or *Anchusa,* with hairy stems and leaves and blue flower clusters.

**bug moss** *n.* A moss of the genus *Buxbaumia,* having a flattened, asymmetric capsule borne on a rough stalk.

**bug·seed** (bŭg'sēd') *n.* A low-growing plant of the genus *Corispermum,* with narrow leaves and flat seeds.

**bug·sha** (bo͝og'shä') *n.* [Ar.] —See table at CURRENCY.

**buhl** *also* **boule** *or* **boulle** (bo͞ol) *n.* [After André C. *Boule* (1642–1732).] Inlaid furniture decoration of elaborate designs in tortoiseshell, ivory, and metals of various colors.

**buhr·stone** *also* **burr·stone** (bûr'stōn') *n.* [Var. of BURR[1] + STONE.] **1.** A tough limestone impregnated with silica, used for millstones. **2.** A millstone made from buhrstone.

**build** (bĭld) *v.* **built** (bĭlt), **build·ing, builds.** [ME *bilden* < OE **byldan* < *bold,* a dwelling.] —*vt.* **1.** To form by combining materials or parts : CONSTRUCT. **2.** To give form to according to a definite plan or process : CREATE. **3.** To establish, create, or strengthen <*build* a bank account> **4.** To establish a basis for : FOUND <*built* a case from evidence> —*vi.* **1.** To engage in construction. **2.** To be a builder. **3.** To develop in extent or magnitude <Excitement began to *build.*> **—build in.** To construct as an integral or permanent part of. **—build up. 1.** To develop by degrees or in stages <*built up* their strength> <*building up* a gun collection> **2.** To enhance the value or reputation of <trying to *build up* a worthless product> —*n.* Physical make-up, as of a person.

**build·er** (bĭl'dər) *n.* **1.** One that builds, esp. a person who contracts for and supervises the construction of a building. **2.** An abrasive or filler used in detergent or soap.

**build·ing** (bĭl'dĭng) *n.* **1.** A structure that is built. **2.** The act, process, art, or occupation of constructing.

**build-up** *also* **build·up** (bĭld'ŭp') *n.* **1.** The act or process of amassing or increasing. **2.** *Informal.* Widely favorable publicity, esp. by a systematic campaign.

**built** (bĭlt) *v. p.t. & p.p. of* BUILD.

**built-in** (bĭlt'ĭn') *adj.* **1.** Constructed as part of a larger unit : not detachable. **2.** Forming a permanent or essential quality or element <a *built-in* penalty clause> **—built'-in'** *n.*

**built-up** (bĭlt'ŭp') *adj.* **1.** Made by fastening several layers or sections together. **2.** Filled with buildings <a *built-up* section of town>

**bulb** (bŭlb) *n.* [Lat. *bulbus* < Gk. *bolbos,* bulbous plant.] **1. a.** *Bot.* A modified underground stem, as that of the onion or tulip, usu. surrounded by scalelike modified leaves and containing stored food for the undeveloped shoots of the new plant enclosed within it. **b.** An underground stem or root resembling this, as a corm, rhizome, or tuber. **c.** A plant that grows from a bulb. **2.** A rounded projection or part. **3.** An incandescent lamp or its glass housing. **4.** *Anat.* Any of various rounded, enlarged, or bulb-shaped structures, esp. the medulla oblongata.

**bul·bar** (bŭl'bər, -bär') *adj.* Of, relating to, or typical of a bulb, esp. of the medulla oblongata.

**bul·bil** (bŭl'bəl, -bĭl') *n.* [Fr. *bulbille,* dim. of *bulbe,* bulb < Lat. *bulbus.*] A small bulblike part growing above ground on a flower stalk or in a leaf axil.

**bul·bous** (bŭl'bəs) *adj.* **1.** Resembling a bulb : ROUNDED. **2.** *Bot.* Bearing bulbs or growing from a bulb. **—bul'bous·ly** *adv.*

**bul·bul** (bo͝ol'bo͝ol') *n.* [Pers. < Ar.] **1.** Any of various chiefly tropical Old World songbirds of the family Pycnonotidae, with grayish or brownish plumage. **2.** A songbird believed to be a nightingale, often mentioned in Persian poetry.

**Bul·gar** (bŭl'gär', bo͝ol'-) *n.* BULGARIAN 1.

**Bul·gar·i·an** (bŭl-gâr'ē-ən, bo͝ol-) *adj.* Of, relating to, or typical of Bulgaria, its inhabitants, or their language. —*n.* **1.** A native or resident of Bulgaria. **2.** The Slavic language of the Bulgarians.

**bulge** (bŭlj) *n.* [ME, pouch < OFr. *bouge* < Lat. *bulga,* bag, of Celt. orig.] **1.** A protruding part, as an outward curve or swelling. **2.** The rounded lower section of a ship's hull. **3.** *Slang.* An advantage. —*v.* **bulged, bulg·ing, bulg·es.** —*vt.* To cause to curve outward. —*vi.* To grow larger : SWELL. **—bulg'i·ness** *n.* **—bulg'y** *adj.*

**bul·gur** (bo͞ol-go͞or', bŭl'gər) *n.* [Turk.] Dried cracked wheat prepared for food.

**bu·lim·i·a** (byo͞o-lĭm'ē-ə) *n.* [NLat. < Gk. *boulima* : *bous,* ox + *limos,* hunger.] Insatiable appetite.

**bulk**[1] (bŭlk) *n.* [ME < ON *bulki,* cargo.] **1.** Great size, mass, or volume. **2. a.** A distinct portion of matter, esp. when large. **b.** The body of a human being, esp. when obese. **3.** The greater part <"the great *bulk* of necessary work can never be anything but painful" —Bertrand Russell> **4.** Thickness of paper or cardboard in relation to weight. **5.** A ship's hold or the cargo stowed therein. —*v.* **bulked, bulk·ing, bulks.** —*vi.* **1.** To be or appear to be massive in size, volume, importance, or consequence : LOOM. **2.** To increase in importance or size. —*vt.* **1.** To cause to swell or expand. **2.** To form into a mass : COHERE. **—in bulk. 1.** Loose : unpackaged. **2.** In large numbers, qualities.

☆ ***syns:*** BULK, AMPLITUDE, MAGNITUDE, MASS, SIZE, VOLUME *n. core meaning* : a great amount, extent, or dimension <the monstrous *bulk* of a supertanker>

**bulk**[2] (bŭlk) *n.* [Orig. unknown.] A frame structure, as a stall or booth, jutting from the front of a building.

**bulk·age** (bŭl'kĭj) *n.* A substance that stimulates peristalsis by increasing the bulk of material in the intestine.

**bulk·head** (bŭlk'hĕd') *n.* [BULK[2] + HEAD.] **1.** One of the upright partitions dividing a ship into compartments and preventing the spread of leakage or fire. **2.** A wall or embankment constructed in a mine or tunnel to protect against earth slides, fire, water, or gas. **3.** A horizontal or sloping structure providing access to a cellar stairway or an elevator shaft.

**bulk·y** (bŭl'kē) *adj.* **-i·er, -i·est. 1.** Extremely large : MASSIVE. **2.** Unwieldy : clumsy. **—bulk'i·ly** *adv.* **—bulk'i·ness** *n.*

**bull[1]** (bŏŏl) *n.* [ME *bule* < OE **bulla* < ON *boli.*] **1. a.** An adult male bovine mammal. **b.** The uncastrated adult male of domestic cattle. **c.** The male of certain other mammals, as the elephant and moose. **2.** An exceptionally powerful and aggressive person. **3.** A person who buys commodities or securities in anticipation of a rise in prices or who tries by speculative purchases to effect such a rise. **4. Bull.** Taurus. **5.** *Slang.* A police officer. **6.** *Slang.* Nonsense. *—v.* **bulled, bull·ing, bulls.** *—vt.* **1.** To engage in speculative buying so as to raise the price of (stocks) or prices in (a market). **2.** To push : force <*bulled* their way through the crowd> *—vi.* **1.** To rise in price. **2.** To push through or ahead forcefully. *—adj.* **1.** Male <a *bull* elephant> **2.** Resembling a bull : POWERFUL. **3.** Marked by rising prices <a *bull* market>

**bull[2]** (bŏŏl) *n.* [ME *bulle* < OFr. < Med. Lat. *bulla.*] **1.** An official document issued by the pope. **2.** The bulla with which a bull is sealed.

**bull[3]** (bŏŏl) *n.* [Orig. unknown.] *Informal.* A blunder.

**bul·la** (bŏŏl'ə) *n., pl.* **bul·lae** (bŏŏl'ē) [Med. Lat. < Lat., bubble, seal.] **1.** A round seal affixed to a papal bull. **2.** *Pathol.* A large blister or vesicle.

**bul·lace** (bŏŏl'ĭs) *n.* [ME *bolas* < OFr. *buloce.*] DAMSON.

**bul·lae** (bŏŏl'ē) *n. pl. of* BULLA.

**bul·late** (bŏŏl'āt', bŭl'-) *adj.* [< Lat. *bulla,* bubble.] Puckered or blistered in appearance <*bullate* leaves>

**bull·bait·ing** (bŏŏl'bā'tĭng) *n.* The sport of baiting bulls with dogs.

**bull·bat** (bŏŏl'băt') *n.* [From its roaring sound while in flight.] NIGHTHAWK 1a.

**†bull·dog** (bŏŏl'dôg', -dŏg') *n.* **1.** A short-haired dog of a breed having a large head, strong, square jaws with dewlaps, and a stocky body. **2.** A short-barreled, large-caliber revolver. **3.** A heat-resistant material used to line puddling furnaces. **4.** *Chiefly Brit.* A proctor's assistant at Oxford or Cambridge. *—adj.* Stubborn. *—vt.* **-dogged, -dog·ging, -dogs.** *Western U.S.* To throw (a steer) by seizing its horns and twisting its neck until the animal falls. **—bull'dog'ger** *n.*

**bulldog edition** *n.* The early morning edition of a newspaper.

**bull·doze** (bŏŏl'dōz') *vt.* **-dozed, -doz·ing, -doz·es.** [Poss. < obs. *bulldose,* severe beating : BULL[1] + DOSE.] **1.** *Slang.* To bully or coerce : INTIMIDATE. **2.** To clear, dig up, or move with a bulldozer.

**bull·doz·er** (bŏŏl'dō'zər) *n.* **1.** A tractor with a horizontal blade in front used esp. for clearing or grading land. **2.** *Slang.* BULLY[1] 1.

**bul·let** (bŏŏl'ĭt) *n.* [Fr. *boulette,* dim. of *boule,* ball < OFr. < Lat. *bulla.*] **1. a.** A spherical or pointed cylindrical metallic projectile fired from a pistol or rifle. **b.** Such a projectile in a metal casing : CARTRIDGE. **2.** An object resembling a bullet in shape, action, or effect. **3.** A heavy dot (•) used to call attention to a printed passage.

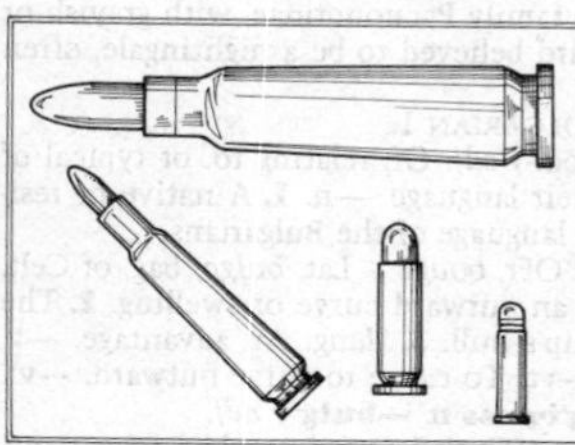

**bullet**
*Various calibers of bullets: (top) 7.62 mm. M60 machine gun, (bottom) .17 rifle, .45 pistol, and .22 pistol*

**bul·le·tin** (bŏŏl'ĭ-tn, -tĭn) *n.* [Fr., prob. < OFr. *bullette* < *bulle,* bull. —see BULL[2].] **1.** A printed or broadcast statement on a matter of public interest. **2.** A periodical, esp. one published by an organization. *—vt.* **-tined, -tin·ing, -tins.** To inform by bulletin.

**bulletin board** *n.* A board on which notices are posted.

**bul·let·proof** (bŏŏl'ĭt-prŏŏf') *adj.* Impenetrable by bullets <a *bulletproof* car> *—vt.* **-proofed, -proof·ing, -proofs.** To make bulletproof.

**bull fiddle** *n.* A double bass.

**bull·fight** (bŏŏl'fīt') *n.* A public spectacle, esp. in Spain and Mexico, in which a fighting bull is engaged in a series of traditional maneuvers culminating usu. with the matador's ceremonial execution of the bull by sword. **—bull'fight'er** *n.* **—bull'fight'ing** *n.*

**bull·finch** (bŏŏl'fĭnch') *n.* **1.** A European bird, *Pyrrhula pyrrhula,* with a short, thick bill and, in the male, a red breast. **2.** One of several similar finches.

**bull·frog** (bŏŏl'frôg', -frŏg') *n.* A large frog, chiefly of the genus *Rana,* esp. *R. catesbeiana* of North America, with a deep, resonant croak.

**bull·head** (bŏŏl'hĕd') *n.* **1.** A North American freshwater catfish of the genus *Ictalurus.* **2.** A fish of the family Cottidae, as the sculpin and the miller's thumb.

**bull·head·ed** (bŏŏl'hĕd'ĭd) *adj.* Extremely stubborn : OBSTINATE. **—bull'head'ed·ly** *adv.* **—bull'head'ed·ness** *n.*

**bull·horn** (bŏŏl'hôrn') *n.* An electric megaphone that amplifies the volume, esp. of a voice.

**bul·lion** (bŏŏl'yən) *n.* [ME, an ingot of precious metal, partly < OFr. *billon* < *bille,* stick, bubble < *boilir,* to boil, and partly < OFr. *boillon,* molten metal.] **1. a.** Silver or gold regarded with respect to quantity rather than value. **b.** Gold or silver in bars, ingots, or plates. **2.** A heavy lace trimming of twisted gold or silver threads.

**bull·ish** (bŏŏl'ĭsh) *adj.* **1.** Like a bull. **2. a.** Causing, expecting, or marked by rising stock-market prices. **b.** Confident. **—bull'ish·ly** *adv.* **—bull'ish·ness** *n.*

**bull·mas·tiff** (bŏŏl'măs'tĭf) *n.* A heavy-set dog of a breed orig. developed from the bulldog and the mastiff.

**Bull Moose** *n.* [From the party's emblem.] A member or supporter of the Bull Moose Party.

**Bull Moose Party** *n.* PROGRESSIVE PARTY 1.

**bull·necked** (bŏŏl'nĕkt') *adj.* Having a short thick neck.

**bull·nose** (bŏŏl'nōz') *n.* A contagious disease of swine caused by the bacillus *Actinomyces necrophorus* and characterized by an infection and swelling of the snout.

**bul·lock** (bŏŏl'ək) *n.* [ME *bullok* < OE *bulluc.*] **1.** A castrated bull : STEER. **2.** A young bull.

**bull·pen** (bŏŏl'pĕn') *n.* **1.** An enclosure for confining bulls. **2.** *Informal.* A place for the temporary detention of prisoners. **3.** *Baseball.* **a.** An area where relief pitchers warm up during a game. **b.** The relief pitchers of a team considered as a group.

**bull·ring** (bŏŏl'rĭng') *n.* A circular arena for bullfights.

**bull session** *n.* *Informal.* A rambling informal group discussion.

**bull's-eye** *also* **bull's eye** (bŏŏlz'ī') *n.* **1. a.** The small central circle on a target. **b.** A shot that hits this circle. **2.** Something that precisely achieves a desired goal. **3.** A thick circular piece of glass set, as in a roof, to admit light. **4.** A circular opening or window. **5. a.** A planoconvex lens used to concentrate light. **b.** A lantern or lamp having such a lens. **6.** A piece of hard, round candy.

**bull snake** *n.* A nonvenomous North American snake of the genus *Pituophis,* with yellow and brown or black markings.

**bull terrier** *n.* A short-haired dog of a breed orig. developed by crossing a bulldog and a terrier.

**bull thistle** *n.* [From its large head.] A coarse Eurasian weed, *Cirsium vulgare,* with spiny stems and leaves and purple flowers.

**bull tongue** *n.* A heavy plow with a single shovel, used primarily in cotton fields.

**bull·whip** (bŏŏl'hwĭp', -wĭp') *n.* A long, plaited rawhide whip with a knotted end. *—vt.* **-whipped, -whip·ping, -whips.** To whip with a bullwhip.

**bul·ly[1]** (bŏŏl'ē) *n., pl.* **-lies.** [Poss. < MDu. *boele,* sweetheart.] **1.** One who is habitually cruel, esp. to smaller or weaker people. **2.** *Archaic.* A hired ruffian. **3.** *Archaic.* A pimp. **4.** *Obs.* A fine fellow. **5.** *Obs.* A sweetheart. *—v.* **-lied, -ly·ing, -lies.** *—vt.* To intimidate with superior size or strength. *—vi.* To behave like a bully. *—adj. Informal.* Splendid. *—interj.* —Used to express approval.

**bul·ly[2]** (bŏŏl'ē) *n.* [Perh. Fr. *bouilli,* boiled meat < *bouiller,* to boil.] Canned or pickled beef.

**bul·ly·rag** (bŏŏl'ē-răg') *also* **bal·ly·rag** (băl'ē-) *vt.* **-ragged, -rag·ging, -rags.** To intimidate by bullying.

**bul·rush** (bŏŏl'rŭsh') *n.* [ME *bulrish.*] **1.** A grasslike sedge of the genus *Scirpus,* growing in wet places. **2.** A marsh plant, as the cattail.

**bul·wark** (bŏŏl'wərk, -wôrk', bŭl'-) *n.* [ME *bulwerk* < MDu. *bolwerk* < MHG *bolwerc* : *bole,* plank + *werc,* work < OHG.] **1.** A structure, as a wall, raised as a defensive fortification : RAMPART. **2.** Something serving as a principal defense. **3.** A breakwater. **4.** *often* **bulwarks.** The part of a ship's side above the upper deck. *—vt.* **-warked, -wark·ing, -warks. 1.** To fortify with a bulwark. **2.** To provide defense or protection for.

**bum[1]** (bŭm) *n.* [Short for *bummer,* lazy person, prob. < G. *Bummler* < *bummeln,* to loaf.] **1.** A vagrant : hobo. **2.** One who avoids work and tries to live off others. **3.** One who performs poorly : INCOMPETENT. *—v.* **bummed, bum·ming, bums.** *Informal. —vi.* **1.** To live by begging or sponging, often while moving from place to place. **2.** To loaf. *—vt.* To acquire by begging or sponging. *—adj. Slang.* **1.** Worthless : invalid. **2.** Not functioning properly : CRIPPLED. **—on the bum.** *Slang.* **1.** Living as a tramp or hobo. **2.** Out of order : BROKEN.

**bum[2]** (bŭm) *vi.* **bummed, bum·ming, bums.** [ME *bummen.*] *Chiefly Brit.* To make a humming sound : DRONE.

**bum[3]** (bŭm) *n.* [ME *bom.*] *Chiefly Brit.* The buttocks.

**bum·ble[1]** (bŭm'bəl) *v.* **-bled, -bling, -bles.** [Prob. alteration of BUNGLE.] *—vi.* To speak or behave clumsily or falteringly. *—vt.* To botch up : BUNGLE.

**bum·ble[2]** (bŭm'bəl) *vi.* **-bled, -bling, -bles.** [ME *bomblen.*] To make a humming or droning sound : BUZZ. **—bum'ble** *n.*

**bum·ble·bee** (bŭm'bəl-bē') *n.* [BUMBLE[2] + BEE.] A large hairy bee of the genus *Bombus.*

**bum·boat** (bŭm′bōt′) *n.* [Perh. BUM[3] + BOAT.] A small boat used for peddling provisions and small wares to ships anchored offshore.

**bum·mer** (bŭm′ər) *n. Slang.* **1. a.** A bad reaction to a hallucinogenic drug. **b.** A disagreeable person, event, or situation. **2.** A failure.

**bump** (bŭmp) *v.* **bumped, bump·ing, bumps.** [Imit.] —*vt.* **1.** To collide with forcefully. **2.** To cause to knock against an obstacle. **3.** To knock to a new position : DISPLACE. **4.** *Informal.* To displace by right of seniority or authority <was *bumped* from the day shift> —*vi.* **1.** To knock against something forcefully. **2.** To proceed with jolts and jerks. **—bump into.** To meet by chance. **—bump off.** *Slang.* To murder. —*n.* **1.** A forceful blow, collision, or jolt. **2.** A slight swelling or lump. **3.** One of the natural protuberances on the human skull. **4.** A suggestive forward thrust of the pelvis, as in a striptease act.

**bump·er**[1] (bŭm′pər) *n.* **1.** One that bumps. **2. a.** Either of two metal structures, typically horizontal bars, attached to the front and rear of a car to absorb the impact of a collision. **b.** A protective device used to absorb shocks.

**bump·er**[2] (bŭm′pər) *n.* [Perh. < BUMP.] **1.** A drinking vessel filled to the brim. **2.** Something extraordinarily large. —*adj.* Unusually full or abundant <a *bumper* crop>

**bumper sticker** *n.* A sticker bearing a printed message for display on a vehicle's bumper.

**bump·kin** (bŭmp′kĭn, bŭm′-) *n.* [Orig. unknown.] **1.** An unsophisticated, simple country person : YOKEL. **2.** A short spar projecting from the deck of a ship, used for extending a sail or securing a block or stay.

**bump·tious** (bŭmp′shəs) *adj.* [Perh. a blend of BUMP and FRACTIOUS.] Obtrusively forward and crudely assertive : PUSHY. **—bump′tious·ly** *adv.* **—bump′tious·ness** *n.*

**bump·y** (bŭm′pē) *adj.* **-i·er, -i·est.** Covered with or full of bumps. **—bump′i·ly** *adv.* **—bump′i·ness** *n.*

**bun**[1] (bŭn) *n.* [ME *bunne*, prob. < OFr. *bugne*, boil, of Celt. orig.] **1.** A small bread roll, often sweetened or spiced. **2.** A tight roll of hair worn at the back of the head.

**bun**[2] (bŭn) *n.* [Orig. unknown.] A drunken spree.

**Bu·na** (bo͞o′nə, byo͞o′-). A trademark for synthetic rubber made by polymerization of butadiene and sodium.

**bunch** (bŭnch) *n.* [ME *bunche*, prob. < OFr. *bonge*, bundle.] **1.** A group of like items growing, fastened, or placed together : CLUSTER. **2.** *Informal.* A group of people. **3.** A swelling or lump. —*v.* **bunched, bunch·ing, bunch·es.** —*vt.* To gather or form into a bunch. —*vi.* **1.** To form a cluster or group. **2.** To protrude : swell. **—bunch′y** *adj.*

**bunch·ber·ry** (bŭnch′bĕr′ē) *n.* The dwarf cornel.

**bunch·flow·er** (bŭnch′flou′ər) *n.* A bog plant, *Melanthium virginicum* of the eastern United States, with narrow leaves and a branching greenish flower cluster.

**bun·co** *also* **bun·ko** (bŭng′kō) [Sp. *banca*, card game < Ital. *banca*, bank.] *Informal.* —*n. pl.* **-cos** *also* **-kos.** A swindle or confidence game. —*vt.* **-coed, -co·ing, -cos** *also* **-koed, -ko·ing, -kos.** To swindle.

**bun·combe** (bŭng′kəm) *n. var. of* BUNKUM.

**bund**[1] (bŭnd) *n.* [Hindi *band* < Pers.] **1.** An embankment or dike, esp. in India. **2.** A street running along a harbor or waterway, esp. in the Orient.

**bund**[2] (bo͝ond, bŭnd) *n.* [G. < MHG *bunt.*] **1.** An association, esp. a political association. **2.** *often* **Bund.** A pro-Nazi German-American organization of the 1930's. **—bund′ist** *n.*

**Bun·des·rat** *also* **Bun·des·rath** (bo͝on′dəs-rät′) *n.* [G. : *Bundes*, genitive of *Bund*, confederation + *Rat*, council.] **1.** The upper house of the federal legislature of West Germany. **2.** The federal council of certain countries, as of Switzerland and Austria.

**Bun·des·tag** (bo͝on′dəs-täg′) *n.* [G. : *Bund*, confederation + *-tag*, meeting < MHG *tagen*, to meet < *tag*, day < OHG *tac.*] The lower house of the federal legislature of West Germany.

**bun·dle** (bŭn′dl) *n.* [ME *bundel*, prob. < MDu. *bondel.*] **1.** A group of objects fastened together. **2.** Something tied up or wrapped for carrying : PACKAGE. **3.** *Biol.* A cluster or strand of specialized cells. **4.** *Bot.* A vascular bundle. **5.** *Slang.* A large amount of money. —*v.* **-dled, -dling, -dles.** —*vt.* **1.** To tie, wrap, or fasten together. **2.** To hustle away and with little fuss. **3.** To dress warmly <*bundled* them up in winter clothes> —*vi.* **1.** To leave abruptly and unceremoniously. **2.** To sleep in the same bed while fully clothed, a custom practiced by engaged couples in early New England. **—bun′dler** *n.*

**bung** (bŭng) *n.* [ME *bunge* < MDu. *bonge*, perh. < LLat. *puncta*, hole < Lat. *pungere*, to prick.] **1.** A stopper for the bunghole of a keg or barrel. **2.** A bunghole. —*vt.* **bunged, bung·ing, bungs.** To close with or as if with a cork or stopper.

**bun·ga·low** (bŭng′gə-lō′) *n.* [Hindi *baṅglā* < *Bengal*, a region in eastern India and Bangladesh.] A small, usu. one-story cottage.

**bung·hole** (bŭng′hōl′) *n.* The hole for filling or draining a keg or barrel.

**bun·gle** (bŭng′gəl) *v.* **-gled, -gling, -gles.** [Perh. of Scand. orig.] —*vi.* To work or act ineptly. —*vt.* To manage (a task) badly : MISHANDLE. **—bun′gler** *n.* **—bun′gling·ly** *adv.*

**bun·ion** (bŭn′yən) *n.* [Poss. alteration of obs. *bunny*, swelling.] A painful, inflamed swelling at the bursa of the big toe.

**bunk**[1] (bŭngk) *n.* [Orig. unknown.] **1.** A narrow bed, as on a ship, built like a shelf against a wall. **2.** A double-decker bed. **3.** *Informal.* A sleeping place. —*v.* **bunked, bunk·ing, bunks.** —*vi.* **1.** To sleep in a bunk. **2.** To sleep, esp. in makeshift quarters. —*vt.* To provide with sleeping quarters.

**bunk**[2] (bŭngk) *n.* [Short for BUNKUM.] *Slang.* Nonsense.

**bunk bed** *n.* BUNK[1] 2.

**bun·ker** (bŭng′kər) *n.* [Sc. *bonker*, chest.] **1.** A bin or tank for fuel storage, as on a ship. **2.** A sand trap on a golf course. **3.** A fortified earthwork, as for the protection of a gun emplacement. —*vt.* **-kered, -ker·ing, -kers.** To store or place in a bunker.

**bunk·house** (bŭngk′hous′) *n.* Sleeping quarters, as on a ranch.

**bunk·mate** (bŭngk′māt′) *n.* One with whom one shares sleeping quarters.

**bun·ko** (bŭng′kō) *n. var. of* BUNCO.

**bun·kum** *also* **bun·combe** (bŭng′kəm) *n.* [After *Buncombe* County, North Carolina.] Meaningless or empty talk : CLAPTRAP.

**bun·ny** (bŭn′ē) *n., pl.* **-nies.** [Dial. *bun*, rabbit.] *Informal.* A rabbit.

**Bun·ra·ku** (bo͞on-rä′ko͞o, bo͞on′rä′-) *n.* [J. : *bun*, literary composition + *raku*, easy.] A traditional Japanese puppet theater featuring large wooden puppets.

**buns** (bŭnz) *pl.n.* [< dial. *bun*, hind part of a rabbit or squirrel < Sc. Gael., stump, bottom.] *Slang.* The buttocks.

**Bun·sen burner** (bŭn′sən) *n.* [After Robert W. *Bunsen* (1811–1899).] A small laboratory burner having a vertical metal tube connected to a gas source and producing a hot flame from a mixture of gas and air let in through adjustable holes at the base.

**bunt**[1] (bŭnt) *v.* **bunt·ed, bunt·ing, bunts.** [Orig. unknown.] —*vt.* **1.** To push or strike with or as if with the horns or head : BUTT. **2.** *Baseball.* To bat (a pitched ball) with a half swing so that the ball rolls slowly in front of the infielders. —*vi. Baseball.* To bunt a pitch. —*n.* **1.** A butt with or as if with the horns or head. **2.** *Baseball.* **a.** The act of bunting. **b.** A bunted ball.

**bunt**[2] (bŭnt) *n.* [Orig. unknown.] **1.** *Naut.* The midsection of a square sail. **2.** The sagging middle part of a fishnet.

**bunt**[3] (bŭnt) *n.*, [Orig. unknown.] A disease of cereal grasses, as rye or wheat, caused by fungi of the genus *Tilletia* and resulting in sooty black spores in place of normal seeds.

**bunt·ing**[1] (bŭn′tĭng) *n.* [Orig. unknown.] **1.** A light cotton or woolen cloth for making flags. **2.** Flags as a group. **3.** Long colored strips of cloth used for festive decoration.

**bunt·ing**[2] (bŭn′tĭng) *n.* [ME.] Any of various birds of the family Fringillidae, with short, cone-shaped bills.

**bunt·ing**[3] (bŭn′tĭng) *n.* [Orig. unknown.] A hooded, snug-fitting sleeping bag for infants.

**bunt·line** (bŭnt′lĭn, -līn′) *n. Naut.* A rope that keeps a square sail from bellying when it is being hauled up for furling.

**bun·ya** (bŭn′yə) *also* **bun·ya·bun·ya** (bŭn′yə-bŭn′yə) *n.* [Native word in Australia.] An evergreen tree native to Australia, *Araucaria bidwilli* with sharp-pointed, close-set leaves and large cones.

**Bun·yan·esque** (bŭn′yə-nĕsk′) *adj.* **1.** Of, relating to, or suggestive of the allegorical writings of John Bunyan. **2. a.** Of, relating to, or suggestive of the stories about Paul Bunyan. **b.** Of very large size.

**buoy** (bo͞o′ē, boi) *n.* [ME *boie* < OFr.] **1.** A float, often having a bell or light, moored in water as a warning of danger or as a marker for a channel. **2.** A ring-shaped life preserver. —*vt.* **buoyed, buoy·ing, buoys. 1.** To mark with or as if with a buoy. **2.** To keep afloat. **3.** To elevate the spirits of : HEARTEN.

**buoy·ance** (boi′əns, bo͞o′yəns) *n.* Buoyancy.

**buoy·an·cy** (boi′ən-sē, bo͞o′yən-) *n.* **1.** The tendency or capacity to remain afloat in a liquid or to rise in air or gas. **2.** The upward force that a fluid exerts on an object less dense than itself. **3.** Ability to recover rapidly from setbacks. **4.** Lightness of spirit : CHEERFULNESS.

**buoy·ant** (boi′ənt, bo͞o′yənt) *adj.* [Sp. *boyante*, pr.part. of *boyar*, to refloat a boat < *boya*, buoy < OFr. *boie.*] **1.** Capable of floating or keeping things afloat. **2.** Sprightly : cheerful. **—buoy′ant·ly** *adv.*

**bu·pres·tid** (byo͞o-prĕs′tĭd) *n.* [NLat. *Buprestidae*, family name < *Buprestis*, type genus < Lat. *buprestis*, beetle harmful to cattle < Gk. *bouprestis* : *bous*, ox + *prēthein*, to swell up.] Any of various often brightly colored beetles of the family Buprestidae, many of which are destructive wood borers as larvae.

**bur**[1] (bûr) *also* **burr** *n.* [ME *burre*, of Scand. orig.] **1. a.** The rough, prickly, or spiny fruit husk, seed pod, or flower of various plants, as the chestnut or the burdock. **b.** A plant producing burs. **2.** One that is persistently clinging or nettlesome. **3.** A rough protuberance, esp. a burl on a tree. **4.** *var. of* BURR[1]. —*vt. var. of* BURR[1].

**bur**[2] (bûr) *n. & v. var. of* BURR[2].

**bur**[3] (bûr) *n. var. of* BURR[3].

**bur·ble** (bûr′bəl) *n.* [ME *burblen*, to bubble.] **1.** A bubbling or rushing sound. **2.** Meaningless sounds : BABBLE. **3.** A separation in the boundary layer of air about a moving streamlined body, as the wing of an aircraft, causing a breakdown in smooth airflow and resulting in turbulence. —*vi.* **-bled, -bling, -bles. 1.** To bubble : gurgle. **2.** To babble : prattle.

**bur·bot** (bûr′bət) *n., pl.* **burbot** *or* **-bots.** [ME < OFr. *borbote.*] A freshwater fish, *Lota lota* of the Northern Hemisphere, related to the cod.

**bur cucumber** *n.* **1.** A climbing vine, *Sicyos angulatus* of eastern North America, with lobed leaves, small greenish flowers, and bristly, egg-shaped fruit. **2.** The fruit of the bur cucumber.

**bur·den**[1] (bûr′dn) *n.* [ME < OE *byrðen.*] **1. a.** Something carried. **b.** Something difficult to bear emotionally or physically. **2.** A responsibility : duty. **3. a.** The amount of cargo a vessel can carry. **b.** The weight of the cargo carried by a vessel at one time. **4.** Transport of heavy loads. —*vt.* **-dened, -den·ing, -dens. 1.** To load or overload. **2.** To weigh down : OPPRESS.

**bur·den**[2] (bûr′dn) *n.* [Var. of BOURDON.] **1.** The bass accompaniment to a song. **2.** The refrain or chorus of a musical composition. **3.** The drone of a bagpipe. **4.** A theme or recurring idea.

**burden of proof** *n.* Responsibility of proving a disputed charge or allegation.

**bur·den·some** (bûr′dn-səm) *adj.* Being or imposing a burden. **—bur′den·some·ly** *adv.* **—bur′den·some·ness** *n.*

☆ ***syns:*** BURDENSOME, ARDUOUS, BACKBREAKING, DEMANDING, DIFFICULT, EXIGENT, FORMIDABLE, HARD, HEAVY, LABORIOUS, ONEROUS, OPPRESSIVE, RIGOROUS, SEVERE, TAXING, TOUGH, TRYING, WEIGHTY *adj. core meaning* : imposing a severe test of physical or spiritual strength <*burdensome* farm chores><*burdensome* responsibilities>

**bur·dock** (bûr′dŏk′) *n.* [BUR[1] + DOCK[4].] A coarse, weedy plant of the genus *Arctium,* native to Eurasia, with large heart-shaped leaves and purplish flowers surrounded by hooked bristles.

**bu·reau** (byŏŏr′ō) *n., pl.* **-reaus** *or* **-reaux** (-ōz) [Fr., desk, cloth cover for desks < OFr. *burel,* woolen cloth, ult. < LLat. *burra,* shaggy garment.] **1.** A chest of drawers, esp. one with a mirror. **2.** *Chiefly Brit.* A writing desk with drawers. **3. a.** A government department or subdivision within a department. **b.** An office, usu. of a large organization, that performs a specific duty. **c.** A business that offers a specified kind of information <a travel *bureau*>

**bu·reauc·ra·cy** (byŏŏ-rŏk′rə-sē) *n., pl.* **-cies.** [Fr. *bureaucratie* : *bureau,* office (see BUREAU) + Gk. *-kratia,* rule < *kratos,* strength.] **1. a.** Administration of a government chiefly through bureaus staffed with nonelective officials. **b.** The departments and their officials as a whole. **2.** Government marked by diffusion of authority among numerous offices and adherence to inflexible rules of operation. **3.** An administrative system in which the need to follow complex procedures impedes effective action.

**bu·reau·crat** (byŏŏr′ə-krăt′) *n.* **1.** An official who adheres rigidly to regulations, forms, and procedures. **2.** An official of a bureaucracy. **—bu′reau·crat′ic** *adj.* **—bu′reau·crat′i·cal·ly** *adv.*

**bu·reau·crat·ese** (byŏŏr′ə-kră-tēz′, -tēs′) *n.* A style of language used esp. by bureaucrats, marked by euphemism and jargon.

**bu·reau·cra·tize** (byŏŏ-rŏk′rə-tīz′) *vt.* **-tized, -tiz·ing, -tiz·es.** To bring under bureaucratic control or influence.

**bu·reaux** (byŏŏr′ōz) *n. var. pl. of* BUREAU.

**bu·rette** *also* **bu·ret** (byŏŏ-rĕt′) *n.* [Fr., dim. of *buire,* vase for liquors.] A uniform-bore glass tube with fine gradations and a stopcock at the bottom, used esp. in laboratory procedures for accurate fluid dispensing and measurement.

**burg** (bûrg) *n.* [ME *burgh* < OE *burg.*] **1.** A fortified or walled town. **2.** *Informal.* A city or town.

**bur·gage** (bûr′gĭj) *n.* [ME < Med. Lat. *burgagium* < *burgus,* fortified town, of Germanic orig.] A tenure in England and Scotland under which property of the king or a lord in a town was held for an annual rent.

**bur·gee** (bər-jē′, bûr′jē) *n.* [Perh. < dial. Fr. *bourgeais,* shipowner < OFr. *burgeis,* citizen < *bourg,* bourg.] A small flag displayed by ships for identification or signals.

**bur·geon** *also* **bour·geon** (bûr′jən) *vi.* **-geoned, -geon·ing, -geons.** [ME *burgeonen* < OFr. *borjoner.*] **1. a.** To put forth new buds, leaves, or greenery : SPROUT. **b.** To begin to blossom or grow. **2.** To develop rapidly : FLOURISH.

**burg·er** (bûr′gər) *n. Informal.* **1.** A hamburger. **2.** A sandwich similar to a hamburger but with a nonbeef filling.

**bur·gess** (bûr′jĭs) *n.* [ME *burgeis* < OFr. < LLat. *burgensis.*] **1.** A freeman or citizen of an English borough. **2.** A member of the British Parliament formerly representing a town, borough, or university. **3.** A member of the lower house of the legislature of colonial Virginia or Maryland.

**burgh** (bûrg) *n.* [Sc., var. of BOROUGH.] A chartered town or borough in Scotland.

**burgh·er** (bûr′gər) *n.* [Ult. < MHG *burgaere* < OHG *burgārī* < *burg,* city.] **1.** A member of the mercantile class of a medieval city. **2.** A citizen of a town or borough.

**bur·glar** (bûr′glər) *n.* [AN *burgler* < Med. Lat. *burgulator,* var. of *burgator* < LLat. *burgus,* fortified town, of Germanic orig.] A person who commits burglary. **—bur·glar′i·ous** (bər-glâr′ē-əs) *adj.*

**bur·glar·ize** (bûr′glə-rīz′) *vt.* **-ized, -iz·ing, -iz·es. 1.** To break into, enter, and steal from. **2.** To commit burglary against.

**bur·glar·proof** (bûr′glər-prōōf′) *adj.* Secure against burglary.

**bur·gla·ry** (bûr′glə-rē) *n., pl.* **-ries.** The crime of breaking into and entering a building with intention to steal.

**bur·gle** (bûr′gəl) *vt.* **-gled, -gling, -gles.** [Back-formation < BURGLAR.] *Informal.* To burglarize.

**bur·go·mas·ter** (bûr′gə-măs′tər) *n.* [Partial transl. of Du. *burgemeester* : *burg,* town (< MDu. *burch*) + *meester,* master.] The principal magistrate of a city or town in the Netherlands, Flanders, Austria, and Germany : MAYOR.

**bur·go·net** (bûr′gə-nĭt, bûr′gə-nĕt′) *n.* [OFr. *bourguignotte,* prob. < *Bourgogne,* Burgundy, a region of southeastern France.] A 16th-cent. helmet.

**bur·goo** (bûr′gōō′, bər-gōō′) *n., pl.* **-goos.** [Orig. unknown.] **1.** Thick oatmeal gruel. **2. a.** A thick, spicy soup or stew of meat and vegetables. **b.** A picnic or gathering where burgoo is served.

**Bur·gun·dy** (bûr′gən-dē) *n., pl.* **-dies. 1. a.** Any of various red or white wines produced in Burgundy, France. **b.** Any of various similar wines produced elsewhere. **2. burgundy.** A dark grayish or blackish purple to dark purplish red or reddish brown.

**bur·i·al** (bĕr′ē-əl) *n.* [ME *buriel* < OE *byrgels.*] The act or process of burying a corpse.

**bu·rin** (byŏŏr′ĭn, bûr′-) *n.* [Fr.] A pointed steel cutting tool used in engraving or in carving stone.

**burke** (bûrk) *vt.* **burked, burk·ing, burkes.** [After William *Burke* (1792–1829).] **1.** To kill by suffocation or strangulation so as to leave the body intact and suitable for dissection. **2.** To suppress or get rid of quietly.

**burl** (bûrl) *n.* [ME *burle* < OFr. *bourle,* tuft of wool, dim. of *bourre,* coarse wool < LLat. *burra,* shaggy garment.] **1.** A knot, lump, or slub in yarn or cloth. **2. a.** A large rounded outgrowth on a tree trunk or branch. **b.** The strongly marked wood from such an outgrowth used as veneer. —*vt.* **burled, burl·ing, burls.** To finish (cloth) by removing loose threads or burls. **—burl′er** *n.*

**bur·lap** (bûr′lăp′) *n.* [Orig. unknown.] **1.** A coarsely woven cloth made of jute, flax, or hemp fibers and used esp. in the manufacture of bagging and wrapping. **2.** A lightweight fabric similar to burlap, used for clothing.

**bur·lesque** (bər-lĕsk′) *n.* [< Fr., comical < Ital. *burlesco* < *burla,* joke < VLat. **burrula,* dim. of LLat. *burrae,* nonsense.] **1.** A literary or dramatic work that ridicules a subject by absurd exaggeration or ludicrous imitation. **2.** An absurd or mocking imitation : TRAVESTY. **3.** Vaudeville entertainment characterized by ribald comedy, dancing, and nudity. —*v.* **-lesqued, -lesqu·ing, -lesques.** —*vt.* To imitate mockingly or humorously <"always bringing junk . . . home, as if he were *burlesquing* his role as provider" —John Updike> —*vi.* To use the techniques or methods of burlesque. **—bur·lesque′ly** *adv.* **—bur·lesqu′er** *n.*

**bur·ley** (bûr′lē) *n., pl.* **-leys.** [Prob. < the name *Burley.*] A light-colored tobacco grown primarily in Kentucky.

**bur·ly** (bûr′lē) *adj.* **-li·er, -li·est.** [ME *burlich.*] Heavy, strong, and muscular : HUSKY. **—bur′li·ly** *adv.* **—bur′li·ness** *n.*

**bur marigold** *n.* A plant of the genus *Bidens,* with yellow flowers and pointed seeds that cling to fur and clothing.

**Bur·mese** (bər-mēz′, -mēs′) *adj.* Of, relating to, or typical of Burma, its natives and inhabitants, their language, or their culture. —*n., pl.* **Burmese. 1.** A native or inhabitant of Burma. **2.** The Sino-Tibetan language of Burma.

**burn**[1] (bûrn) *v.* **burned** *or* **burnt** (bûrnt), **burn·ing, burns.** [ME *burnen* < OE *beornan* and *bærnan.*] —*vt.* **1. a.** To cause to undergo combustion. **b.** To destroy with fire. **2.** *Physics.* To cause to undergo nuclear fission or fusion. **3.** To damage or injure by fire, heat, or a heat-producing agent. **4. a.** To kill with fire. **b.** *Slang.* To execute, esp. to electrocute. **5.** To produce by fire or heat <*burned* a hole in the couch> **6.** To use as a fuel. **7.** To impart a sensation of intense heat to. **8.** To brand (an animal). **9.** To harden or impart a finish to by subjecting to intense heat : FIRE <*burn* coal into coke> **10.** To make angry <a comment that *burned* me up> **11.** *Slang.* **a.** To defeat in a contest, esp. by a narrow margin. **b.** To deceive or swindle : CHEAT. —*vi.* **1.** To undergo combustion. **2.** To be on fire : FLAME. **3.** To emit heat or light by or as if by fire. **4.** To be destroyed, injured, damaged, or changed by or as if by fire. **5.** To feel or look hot. **6.** To be consumed with strong emotion, esp.: **a.** To be or become angry. **b.** To be very eager <was *burning* to hear the news> **7.** To be imprinted by or as if by burning. **8.** *Slang.* To be electrocuted. **9.** To impart a sensation of heat. **10.** To become sunburned. **—burn out. 1.** To stop burning from lack of fuel. **2.** To wear out or become inoperative as a result of heat or friction. **3.** To become exhausted, esp. as a result of long-term stress. —*n.* **1.** An injury produced by fire, heat, or a heat-producing agent. **2.** A burned place or area. **3.** The process or result of burning, as in manufacturing bricks. **4.** A stinging sensation. **5.** A sunburn. **6.** *Aerospace.* A single firing of a rocket. **7.** *Slang.* A swindle. **—burn (one's) bridges.** To eliminate the possibility of retreat.

☆ ***syns:*** BURN, BLAZE, FLAME, FLARE *v. core meaning* : to undergo combustion <Wood shavings *burn* easily.>

**burn[2]** (bûrn) *n.* [ME < OE.] *Chiefly Scot.* A small stream : BROOK.

**burn bag** *n.* A container for holding classified documents to be burned.

**burn·er** (bûr′nər) *n.* **1.** One that burns. **2.** The part of a stove, furnace, or lamp that is lighted to produce a flame. **3.** A device, as a furnace, in which matter is burned.

**bur·net** (bər-nĕt′, bûr′nĭt) *n.* [ME < OFr. *burnette,* dark brown, dim. of *brun,* brown, of Germanic orig.] A plant of the genus *Sanguisorba,* with cucumber-flavored leaves and small white, brownish-red, or dark brown flower clusters.

**burn·ing** (bûr′nĭng) *adj.* **1.** Marked by intense heat. **2.** Passionate : intense <a *burning* desire for revenge> **3.** Of immediate import : URGENT <*burning* issues> **—burn′ing·ly** *adv.*

**burning bush** *n.* **1.** A plant or shrub having foliage that turns bright red, as the wahoo and the summer cypress. **2.** The gas plant.

**burning glass** *n.* A convex lens used to focus the sun's rays and produce heat, esp. for ignition.

**bur·nish** (bûr′nĭsh) *vt.* **-nished, -nish·ing, -nish·es.** [ME *burnishen* < OFr. *burnir, burniss-,* var. of *brunir* < *brun,* shining, of Germanic orig.] **1.** To make glossy by or as if by rubbing : POLISH. **2.** To rub with a tool that serves esp. to polish. *—n.* A smooth, glossy finish or appearance : LUSTER. **—bur′nish·er** *n.*

**bur·noose** *also* **bur·nous** (bər-no͞os′) *n.* [Fr. *burnous* < Ar. *bournous* < Gk. *birros,* cloak < LLat. *birrus.*] A hooded cloak worn esp. by Arabs.

**burn·out** (bûrn′out′) *n.* **1.** A failure in a device attributable to burning, excessive heat, or friction. **2.** *Aerospace.* **a.** Termination of rocket or jet-engine operation because of fuel exhaustion or shutoff. **b.** The point at which this termination occurs. **3. a.** Physical or emotional exhaustion, esp. as a result of long-term stress. **b.** A person who is burned out.

**burn·sides** (bûrn′sīdz′) *pl.n.* [After Ambrose E. *Burnside* (1824–1881).] Heavy side whiskers and a moustache, worn with the chin clean-shaven.

**burnt** (bûrnt) *v. var. p.t. & p.p. of* BURN[1].

**bur oak** *n.* A timber tree, *Quercus macrocarpa* of eastern North America, with acorns enclosed within a deep, fringed cup.

**burp** (bûrp) *n.* [Imit.] A belch. *—v.* **burped, burp·ing, burps.** *—vi.* To belch. *—vt.* To cause (a baby) to belch, esp. after feeding.

**burp gun** *n.* A portable lightweight submachine gun.

**burr[1]** *also* **bur** (bûr) [ME *burre,* of Scand. orig.] *—n.* **1.** A rough edge or area remaining from the cutting or shaping of material. **2.** A rotary cutting tool designed to be attached to a drill. **3.** *var. of* BUR[1]. *—vt.* **burred, burr·ing, burrs** *also* **burs. 1.** To form a burr on. **2.** To remove burrs from.

**burr[2]** *also* **bur** (bûr) [Imit.] *—n.* **1.** A rough trilling of the letter *r,* as in Scottish pronunciation. **2.** A buzzing or whirring sound. *—v.* **burred, burr·ing, burrs** *also* **burs.** *—vt.* To pronounce with a burr. *—vi.* **1.** To speak with a burr. **2.** To make a whirring or buzzing sound.

**burr[3]** *also* **bur** (bûr) *n.* [Var. of obs. *burrow* < ME *burwhe.*] **1.** A washer that fits around the smaller end of a rivet. **2.** A blank punched from a sheet of metal.

**bur reed** *n.* A marsh plant of the genus *Sparganium,* with narrow leaves and round, prickly fruit.

**bur·ri·to** (bo͞o-rē′tō, bə-) *n., pl.* **-tos.** [Am. Sp. < Sp., little donkey, dim. of *burro,* burro.] A flour tortilla wrapped around a filling, as beef, beans, or cheese.

**bur·ro** (bûr′ō, bo͝or′ō, bŭr′ō) *n., pl.* **-ros.** [Sp. < *borrico,* donkey < LLat. *burricus,* small horse.] A small donkey.

**bur·row** (bûr′ō, bŭr′ō) *n.* [ME *borow.*] A hole or tunnel dug in the ground by a small animal, as a rabbit or a mole, for habitation or refuge. *—v.* **-rowed, -row·ing, -rows.** *—vi.* **1.** To dig a burrow. **2.** To live or hide in a burrow. **3.** To move or progress by or as if by digging or tunneling <"Suddenly the train is *burrowing* through the pinewoods" —William Styron> *—vt.* **1.** To make by or as if by tunneling. **2.** To dig a burrow in or through. **3.** *Archaic.* To hide in a burrow. **—bur′row·er** *n.*

**burr·stone** (bûr′stōn′) *n. var. of* BUHRSTONE.

**bur·ry** (bûr′ē) *adj.* **-ri·er, -ri·est. 1.** Like a bur : PRICKLY. **2.** Full of or covered with burs.

**bur·sa** (bûr′sə) *n., pl.* **-sae** (-sē) *or* **-sas.** [NLat. < Med. Lat., purse < Gk.] A saclike bodily cavity, esp. one located between joints or at points of friction between moving structures. **—bur′sal** (-səl) *adj.*

**bur·sar** (bûr′sər, -sär′) *n.* [Med. Lat. *bursarius* < *bursa,* purse.] An official in charge of funds, as at a college or university : TREASURER.

**bur·sa·ry** (bûr′sə-rē) *n., pl.* **-ries.** [Med. Lat. *bursaria* < *bursa,* purse.] **1.** A treasury, esp. of a public institution or religious order. **2.** A scholarship granted to a student, esp. at a Scottish university. **—bur·sar′i·al** (bər-sâr′ē-əl) *adj.*

**burse** (bûrs) *n.* [Med. Lat. *bursa* < Gk., purse.] **1.** A purse. **2.** A flat cloth case for carrying the corporal used in celebrating the Eucharist.

**bur·seed** (bûr′sēd′) *n.* The stickseed.

**bur·si·form** (bûr′sə-fôrm′) *adj. Anat.* Shaped like a pouch.

**bur·si·tis** (bər-sī′tĭs) *n.* Inflammation of a bursa, esp. in the shoulder, elbow, or knee joint.

**burst** (bûrst) *v.* **burst, burst·ing, bursts.** [ME *bursten* < OE *berstan.*] *—vi.* **1. a.** To come open or fly apart suddenly or violently, esp. from internal pressure. **b.** To explode. **2.** To be or seem to be full to the point of breaking open. **3.** To emerge or come forth suddenly <*burst* through the doorway> **4.** To give way from overwhelming emotion <thought my heart would *burst* from grief> **5.** To give sudden utterance or expression <*burst* into laughter><*burst* into tears> *—vt.* **1.** To cause to burst. **2.** *Computer Sci.* To separate (a continuous roll of printout) into individual sheets. *—n.* **1.** A sudden outbreak or outburst : EXPLOSION. **2.** The result of bursting. **3.** A sudden, intense expression of emotion <a *burst* of anger> **4.** An abrupt intense increase : RUSH <a *burst* of power> **5.** The explosion of a projectile or bomb on impact or in the air. **6.** A volley of shots.

**burst·er** (bûr′stər) *n. Computer Sci.* An offline device used to burst computer printout.

**bur·then** (bûr′thən) *n. Archaic. var. of* BURDEN[1].

**bur·ton** (bûr′tn) *n.* [Orig. unknown.] *Naut.* A light tackle having double or single blocks, used to hoist or tighten rigging.

**bur·weed** (bûr′wēd′) *n.* A plant, as the burdock, that bears burs.

**bur·y** (bĕr′ē) *vt.* **-ied, -y·ing, -ies.** [ME *burien* < OE *byrgan,* to inter.] **1.** To conceal by or as if by covering with earth. **2.** To place (a corpse) in a grave, a tomb, or the sea : INTER. **3.** To cover from view : HIDE <*buried* their faces in their hands> **4.** To occupy (oneself) with deep concentration : ABSORB. **5.** To put an end to : ABANDON <Let's *bury* our quarrel and be friends.> **—bur′i·er** *n.*

**burying beetle** *n.* Any of various black or black and orange beetles of the genus *Necrophorus* that bury small dead animals, as mice, on which they feed and lay their eggs.

**bus** (bŭs) *n., pl.* **bus·es** *or* **bus·ses.** [Short for OMNIBUS.] **1.** A long motor vehicle for carrying passengers. **2.** *Informal.* An automobile. **3.** *Elect.* A bus bar. *—v.* **bused, bus·ing, bus·es** *or* **bussed, bus·sing, bus·ses.** *—vt.* **1.** To transport in a bus. **2.** To transport (schoolchildren) to achieve racial integration. *—vi.* **1.** To travel in a bus. **2.** To work as a bus boy.

▲ word history: *Bus,* as the etymology above shows, is short for *omnibus,* which was adopted into English from the French phrase *voiture omnibus,* a macaronic expression, part French, part Latin, meaning "vehicle for all." It appeared in French around 1830 and was both borrowed and shortened by the English before the end of the decade. *Omnibus* was also used to designate a waiter's assistant at a restaurant, though whether this usage derives from the use of *omnibus* to mean "vehicle" or if it was an independent use of the word is not clear. *Bus boy* is derived from the shortening of *omnibus* in this second sense. *Bus* meaning "to work as a bus boy" comes from bus boy and not directly from *omnibus.*

**bus bar** *n. Elect.* A conducting bar that carries heavy currents to supply several electric circuits.

**bus boy** *n.* A restaurant employee who clears away dirty dishes and serves as a waiter's assistant.

**bus·by** (bŭz′bē) *n., pl.* **-bies.** [Poss. < the name *Busby.*] A tall full-dress fur hat worn in certain British army regiments.

**bush[1]** (bo͝osh) *n.* [ME.] **1.** A low branching woody plant, usu. smaller than a tree : SHRUB. **2.** A thick growth of shrubs. **3. a.** Land covered with a dense growth of shrubs. **b.** Land remote from settlement : BACKLAND. **4.** A shaggy mass, as of hair. **5.** A fox's tail. **6. a.** *Archaic.* A clump of ivy used as the sign of a tavern. **b.** *Obs.* A tavern. *—v.* **bushed, bush·ing, bushes.** *—vi.* **1.** To grow or branch out like a bush. **2.** To extend in a bushy growth. *—vt.* To decorate, protect, or support with bushes.

**bush[2]** (bo͝osh) *vt.* **bushed, bush·ing, bush·es.** [< E. *bush,* bushing.] To furnish or line with a bushing.

**bush baby** *n.* The galago.

**bush bean** *n.* A shrubby plant, *Phaseolus vulgaris humilis,* a variety of the string bean.

**bush·buck** (bo͝osh′bŭk′) *n.* [Transl. of Afr. *bosbok.*] A small African antelope, *Tragelaphus scriptus,* with white markings and twisted horns.

**bush clover** *n.* A plant or shrub of the genus *Lespedeza,* with compound leaves and purple or yellowish flower clusters.

**bushed** (bo͝osht) *adj.* [Orig. unknown.] Very tired.

**bush·el[1]** (bo͝osh′əl) *n.* [ME < OFr. *boissel* < *boisse,* one sixth of a bushel, of Celt. orig.] **1. a.** A unit of volume or capacity in the U.S. Customary System, used in dry measure and equal to 4 pecks, 35.24 liters, or 2,150.42 cubic inches. **b.** A unit of volume or capacity in the British Imperial System, used in dry and liquid measure and equal to 2,219.36 cubic inches. **2.** A container with the capacity of a bushel. **3.** *Informal.* A large amount.

**bush·el[2]** (bo͝osh′əl) *vt.* **-eled, -el·ing, -els** *or* **-elled, -el·ling, -els.** [Prob. < G. *bosseln,* to do odd jobs.] To alter (clothing). **—bush′el·man** *n.*

**bush honeysuckle** *n.* A North American shrub of the genus *Diervilla,* with yellow flowers that turn reddish.

**Bu·shi·do** *also* **bu·shi·do** (bo͝osh′ĭ-dō′, bo͞o′shĭ-) *n.* [J. *bushidō* : *bushi,* warrior (< Chin. *wu[3]shi[4]*) + *dō,* way (< Chin. *dao[4]*).] The traditional Japanese samurai code prizing honor above life.

o͞o **boot** ou **out** th **thin** *th* **this** ŭ **cut** ûr **urge** y **young**
yo͞o **abuse** zh **vision** ə **about,** it**e**m, ed**i**ble, gall**o**p, circ**u**s

**bush·ing** (bo͝osh'ĭng) *n.* [MDu. *busse.*] **1.** A fixed or removable lining used to constrain, guide, or reduce friction. **2.** An insulating lining for an aperture through which a conductor passes. **3.** An adapter threaded to permit joining of pipes with different diameters.
**bush jacket** *n.* A long cotton shirtlike jacket usu. with four flat pockets and a belt.
**bush league** *n. Slang.* A minor league, esp. in baseball. **—bush leaguer** *n.*
**bush-league** (bo͝osh'lēg') *adj. Slang.* **1.** Of or belonging to a bush league. **2.** Inferior : second-rate.
**Bush·man** (bo͝osh'mən) *n.* [Transl. of Afr. *boschjeman.*] **1.** A member of a race of nomadic people of southwestern Africa. **2.** Any of the Khoisan languages spoken by the Bushmen. **3. bushman.** *Austral.* A backwoodsman.
**bush·mas·ter** (bo͝osh'măs'tər) *n.* A large, poisonous snake, *Lachesis muta* of tropical America, with brown and grayish markings.
**bush pig** *n.* [Transl. of Afr. *bosvark.*] A wild hog, *Potamochoerus porcus* of southern Africa, with long tufts of hair on the face and ears.
**bush pilot** *n.* A pilot who flies a small airplane to and from areas inaccessible to larger aircraft or other means of transportation.
**bush·rang·er** (bo͝osh'rān'jər) *n.* **1.** A backwoodsman. **2.** *Austral.* An outlaw living in the bush.
**bush·tit** (bo͝osh'tĭt') *n.* Either of two small, long-tailed birds, *Psaltriparus minimus* or *P. melanotis* of western North America, with predominantly gray plumage.
**bush·whack** (bo͝osh'hwăk', -wăk') *v.* **-whacked, -whack·ing, -whacks.** —*vi.* **1.** To make one's way through thick woods by cutting away bushes and branches. **2.** To travel through or live in the woods. **3.** To fight as a guerrilla in the bush. —*vt.* To ambush. **—bush'whack'er** *n.*
**bush·y** (bo͝osh'ē) *adj.* **-i·er, -i·est. 1.** Overgrown with bushes. **2.** Thick and shaggy. **—bush'i·ly** *adv.* **—bush'i·ness** *n.*
**busi·ness** (bĭz'nĭs) *n.* [ME *businesse* < *bisi,* busy.] **1. a.** The occupation, work, or trade in which one is engaged. **b.** A specific pursuit or occupation. **2.** Commercial, industrial, or professional dealings. **3.** A commercial enterprise or establishment. **4.** Volume or amount of commercial trade <*Business* was increasing.> **5.** Commercial dealings : PATRONAGE. **6.** One's rightful or proper interest or concern. **7.** Serious work or endeavor <promised to get down to *business*> **8.** An affair or matter <a distressing *business*> **9.** An incidental action performed by an actor on the stage to fill a pause between lines or to provide interesting detail. **10.** *Informal.* A verbal reprimand : SCOLDING. **11.** *Obs.* The condition of being busy.
☆ ***syns:*** BUSINESS, COMMERCE, INDUSTRY, TRADE, TRAFFIC *n. core meaning* : commercial, industrial, or professional activity <when *business* is slow> BUSINESS applies broadly to all gainful activity. INDUSTRY is the production and manufacture of goods <the plastics *industry*>, while COMMERCE and TRADE are the exchange and distribution of commodities <interstate *commerce*> <the publishing *trade*> TRAFFIC may suggest illegal trade <*traffic* in narcotics>
**business administration** *n.* A college or university course offering instruction in general business principles and practices.
**business card** *n.* A small card conveying information about a business or a business representative.
**busi·ness·like** (bĭz'nĭs-līk') *adj.* **1.** Displaying characteristics thought to be advantageous in business. **2.** Purposeful : earnest.
**busi·ness·man** (bĭz'nĭs-măn') *n.* A man engaged in business.
**business person** *n.* A person engaged in business.
**busi·ness·wom·an** (bĭz'nĭs-wo͝om'ən) *n.* A woman engaged in business.
**bus·ing** *also* **bus·sing** (bŭs'ĭng) *n.* Transportation of children by bus to schools outside their neighborhoods, esp. as a way of achieving racial integration.
**busk** (bŭsk) *vt.* **busked, busk·ing, busks.** [ME *busken* < ON *būask,* reflexive of *būa,* to prepare.] *Chiefly Scot.* To make ready : PREPARE.
**bus·kin** (bŭs'kĭn) *n.* [OFr. *bouzequin.*] **1.** A laced boot reaching halfway to the knee. **2. a.** A thick-soled laced half boot worn by actors of Greek and Roman tragedies. **b.** Tragedy.
**bus·man** (bŭs'mən) *n.* A person who runs a bus.
**bus·man's holiday** (bŭs'mənz) *n. Informal.* A vacation on which a person engages in activity similar to his or her usual work.
**buss** (bŭs) *vt.* & *vi.* **bussed, buss·ing, buss·es.** [Prob. imit.] To kiss. **—buss** *n.*
**bust¹** (bŭst) *n.* [Fr. *buste* < Ital. *busto,* poss. < Lat. *bustum,* sepulchral monument.] **1. a.** A woman's bosom. **b.** *Archaic.* The human chest. **2.** A piece of sculpture representing a person's head, shoulders, and upper chest.
**bust²** (bŭst) *v.* **bust·ed, bust·ing, busts.** [Var. of BURST.] —*vt.* **1.** To smash or break, esp. forcefully. **2.** To cause to come to an end : BREAK UP <*busted* the union><*busted* up their marriage> **3.** To break or tame (a horse). **4.** To cause to become short of money or bankrupt. **5.** To reduce the rank of : DEMOTE. **6.** To hit or punch. **7. a.** To place under arrest. **b.** To make a raid on. —*vi.* **1.** To break or burst. **2.** To become short of money or bankrupt. —*n.* **1.** A flop : failure. **2.** Bankruptcy. **3.** A period of widespread financial depression. **4.** A punch : blow. **5.** A spree. **6. a.** An arrest <a narcotics *bust*> **b.** A raid <a gambling *bust*>
**bus·tard** (bŭs'tərd) *n.* [ME < blend of OFr. *bistarde* and *oustarde,* both < Lat. *avis tarda.*] Any of various large Old World birds of the family Otididae, frequenting open, grassy regions.

**bustard**
*40 inches long*

**bust·er** (bŭs'tər) *n.* **1.** One who bursts or breaks up <a gang *buster*> **2.** One who breaks horses : BRONCOBUSTER. **3. a.** Something remarkable or very large. **b.** A particularly husky child. **4.** A spree. **5.** Fellow. —Used as a term of familiar address <Got the time, *buster*?>
**bus·tle¹** (bŭs'əl) *vi.* & *vt.* **-tled, -tling, -tles.** [Prob. var. of obs. *buskle,* freq. of BUSK³.] To hurry or cause to hurry energetically and busily. **—bus'tle** *n.*
**bus·tle²** (bŭs'əl) *n.* [Orig. unknown.] A frame or pad to support and expand the fullness of the back of a woman's skirt.
**bust·y** (bŭs'tē) *adj.* **-i·er, -i·est.** *Informal.* Full-bosomed. **—bust'i·ness** *n.*
**bu·sul·fan** (byo͞o-sŭl'fən) *n.* [Blend of BUTANE and SULFONYL.] An antineoplastic drug, $C_6H_{14}O_6S_2$.
**bus·way** (bŭs'wā') *n.* A road or lane of a road such as an expressway set aside for buses, usually during rush hours.
**bus·y** (bĭz'ē) *adj.* **-i·er, -i·est.** [ME < OE *bisig.*] **1.** Actively engaged in work : OCCUPIED. **2.** Crowded with activity <a *busy* day> **3.** Nosy : meddlesome. **4.** Being in use, as a telephone line. **5.** Cluttered with detail to the point of being distracting. —*vt.* **-ied, -y·ing, -ies.** To make busy. **—bus'i·ly** *adv.* **—bus'y·ness** *n.*
☆ ***syns:*** BUSY, ENGAGE, ENGROSS, OCCUPY *v. core meaning* : to make or become involved in an activity <*busied* themselves with legal matters>
**bus·y·bod·y** (bĭz'ē-bŏd'ē) *n.* A nosy, meddlesome person.
**bus·y·work** (bĭz'ē-wûrk') *n.* Usu. unproductive activity that takes up time.
**but** (bŭt; bət *when unstressed*) *conj.* [ME < OE *būtan.*] **1.** On the contrary <caused not success *but* failure> **2.** Contrary to expectation : YET <worked hard *but* accomplished nothing> **3.** Save : except <Everyone *but* I knew the secret.> **4.** With the exception that : except that. —Used to introduce a dependent clause <would have fought back *but* that they lacked weapons> **5.** Without the result that <It never rains *but* it pours.> **6.** Other than <I have no desire *but* to end poverty.> **7.** That. —Often used after a negative <No doubt *but* justice will prevail.> **8.** That . . . not. —Used after a negative or question <There never is a program presented *but* someone will dislike it.> **9.** *Archaic.* If not : UNLESS <"Beshrew me *but* I love her heartily" —Shakespeare> **10. a.** Before : when. **b.** *Archaic* & *Nonstandard.* Than. —*prep.* **1.** With the exception of : BARRING. **2.** Other than <The accusations are nothing *but* lies.> ***usage:*** In informal speech *but* is often used with a negative in sentences like *It won't take but an hour.* This construction should be avoided in formal style; instead, use *It will only take an hour.* —*adj.* Merely : just <joy that lasted *but* a moment> —*n.* An objection, restriction, or exception <no *buts* about it> **—but what.** *Informal.* That not <don't know *but what* I'll be there>
**but-** *pref.* [< BUTYRIC.] Containing four carbon atoms <*butyl*>
**bu·ta·di·ene** (byo͞o'tə-dī'ēn', -dī-ēn') *n.* [BUTA(NE) + DI- + -ENE.] A colorless, highly flammable hydrocarbon, $C_4H_6$, obtained from petroleum and used to make synthetic rubber.
**bu·tane** (byo͞o'tān') *n.* Either of two isomers of a gaseous hydrocarbon, $C_4H_{10}$, produced synthetically from petroleum and used as a household fuel, refrigerant, and aerosol propellant and in manufacturing synthetic rubber.
**bu·ta·no·ic acid** (byo͞o'tə-nō'ĭk) *n.* Butyric acid.
**bu·ta·nol** (byo͞o'tə-nôl', -nōl') *n.* Either of two butyl alcohols derived from butane and used as solvents and in organic synthesis.
**bu·ta·none** (byo͞o'tə-nōn') *n.* A colorless flammable ketone, $C_4H_8O$, used in lacquers, paint removers, cements and adhesives, celluloid, and cleaning fluids.
**butch·er** (bo͝och'ər) *n.* [ME *bucher* < OFr. *bouchier* < *boc,* he-goat.] **1. a.** A person who slaughters and dresses animals commercially. **b.** A meat vendor. **2.** One who kills brutally or sadistically.

3. A vendor, as of candy and magazines, on a train. 4. A bungler : botcher. —*vt.* **-ered, -er·ing, -ers. 1.** To slaughter or prepare (animals) for market. **2.** To kill brutally or pointlessly. **3.** To spoil by botching : BUNGLE. **—butch'er·er** *n.*

**butch·er·bird** (bŏŏch'ər-bûrd') *n.* A bird, esp. the shrike, that impales its prey on thorns.

**butcher knife** *n.* A heavy-duty knife, approx. 8 inches or 20.3 centimeters long, with a broad blade.

**butcher's broom** *n.* A shrub native to Europe, *Ruscus aculeatus*, with stiff, prickle-tipped, flattened stems resembling true leaves.

**butch·er·y** (bŏŏch'ə-rē) *n.*, *pl.* **-ies. 1.** The trade of a butcher. **2.** *Chiefly Brit.* A slaughterhouse. **3.** Wanton or cruel killing : CARNAGE. **4.** Something botched.

**bu·te·o** (byōō'tē-ō') *n.*, *pl.* **-os.** [NLat., genus name < Lat. *buteo*, a kind of hawk or falcon.] A hawk of the genus *Buteo*, having broad wings and broad, rounded tails.

**but·ler** (bŭt'lər) *n.* [ME < OFr. *bouteillier*, bottle-bearer < *bouteille, botele*, bottle.] The chief male servant in a household.

**butler's pantry** *n.* A service and storage room between a kitchen and dining room.

**butt**[1] (bŭt) *v.* **butt·ed, butt·ing, butts.** [ME *butten* < OFr. *bouter*, to strike, of Germanic orig.] —*vt.* To hit or push against with the head or horns : RAM. —*vi.* **1.** To hit or push something with the head or horns. **2.** To project out or forward. **—butt in.** *Informal.* To meddle in other people's affairs. —*n.* A push or blow with the head or horns. **—butt'er** *n.*

**butt**[2] (bŭt) *vt.* & *vi.* **butt·ed, butt·ing, butts.** [ME *butten* < OFr. *bouter*, to adjoin.] To join or be joined end to end : ABUT. —*n.* **1.** A butt joint. **2.** A butt hinge.

**butt**[3] (bŭt) *n.* [ME *butte* < OFr.] **1.** One that serves as an object of scorn or contempt <was the *butt* of their insults> **2. a.** A target. **b. butts.** A target range. **c.** An obstacle behind a target for stopping the shot. **3.** *Obs.* A limit : goal.

**butt**[4] (bŭt) *n.* [ME *butte.*] **1.** The larger or thicker end. **2. a.** An unburned end, as of a cigarette. **b.** *Informal.* A cigarette. **3.** A short or broken remnant : STUB. **4.** *Informal.* The buttocks.

**butt**[5] (bŭt) *n.* [ME < OFr. *boute* < LLat. *buttis.*] **1.** A large cask. **2.** A unit of volume equal to 126 U.S. gallons or approx. 477 liters.

**butte** (byōōt) *n.* [Fr. < OFr. *butt*, mound behind targets.] A hill that rises abruptly from the surrounding area and has sloping sides and a flat top.

**but·ter** (bŭt'ər) *n.* [ME *butere* < OE < Lat. *butyrum* < Gk. *bouturon*, cow cheese : *bous*, cow + *turos*, cheese.] **1.** A soft yellowish or whitish emulsion of butterfat, water, air, and often salt, churned from milk or cream and processed for food. **2.** A substance similar to butter, esp.: **a.** A food spread made from fruit or nuts. **b.** A vegetable fat with a nearly solid consistency at ordinary temperatures. **3.** *Informal.* Flattery. —*vt.* **-tered, -ter·ing, -ters.** To put butter on or in. **—butter up.** To praise or flatter excessively.

**but·ter-and-eggs** (bŭt'ər-ən-ĕgz') *n.* (*sing. or pl. in number*). A North American plant, *Linaria vulgaris*, with numerous narrow leaves and a spike of spurred pale-yellow and orange flowers.

**but·ter·ball** (bŭt'ər-bôl') *n.* **1.** A ball of butter. **2.** *Informal.* A fat or chubby person. **3.** The bufflehead.

**†butter bean** *n.* **1.** The wax bean. **2.** *Regional.* The lima bean.

**but·ter·bur** (bŭt'ər-bûr') *n.* A plant of the genus *Petasites*, with woolly leaves and stems and fragrant whitish or purple flowers.

**but·ter·cup** (bŭt'ər-kŭp') *n.* A plant of the genus *Ranunculus*, with glossy yellow flowers.

**but·ter·fat** (bŭt'ər-făt') *n.* The oily content of milk from which butter is made, consisting chiefly of the glycerides of oleic, stearic, and palmitic acids.

**but·ter·fin·gers** (bŭt'ər-fĭng'gərz) *n.* (*sing. in number*). An awkward or clumsy person apt to drop things. **—but'ter·fin'gered** *adj.*

**but·ter·fish** (bŭt'ər-fĭsh') *n.*, *pl.* **butterfish** *or* **-fish·es.** [From its slippery mucous coating.] **1.** A marine food fish, *Poronotus triacanthus* of the North American Atlantic coast, with a flattened body. **2.** One of various fishes similar or related to the butterfish.

**but·ter·fly** (bŭt'ər-flī') *n.* [ME *butterflye* < OE *butorflēoge.*] **1.** Any of various insects of the order Lepidoptera, with slender bodies, knobbed antennae, and four broad, usu. colorful wings. **2.** One principally interested in whimsical pleasure. **3.** The butterfly stroke. **4. butterflies.** An uneasy feeling or mild nausea caused esp. by nervous anticipation. —*vt.* **-flied, -fly·ing, -flies.** To cut and spread open and flat, as shrimp.

**butterfly bush** *n.* A shrub of the genus *Buddleia*, cultivated for their purplish or white clusters.

**butterfly fish** *n.* Any of various tropical marine fishes of the family Chaetodontidae, most of which are brightly colored.

**butterfly pea** *n.* A twining vine, *Clitoria mariana* of the eastern United States, with compound leaves and pale-blue flowers.

**butterfly stroke** *n.* A swimming stroke in which both arms are drawn upward out of the water and forward with a simultaneous up-and-down kick of the feet.

**butterfly valve** *n.* **1.** A disk turning on a diametrical axis inside a pipe, used as a throttle valve or damper. **2.** A valve composed of two semicircular plates hinged on a common spindle, used to permit flow in one direction only.

**butterfly weed** *n.* A North American plant, *Asclepias tuberosa*, with flat-topped, bright-orange flower clusters.

**but·ter·milk** (bŭt'ər-mĭlk') *n.* **1.** The sour liquid that remains after the butterfat has been removed from whole milk or cream by churning. **2.** A cultured sour milk made by adding certain microorganisms to sweet milk.

**but·ter·nut** (bŭt'ər-nŭt') *n.* [From the nut's oiliness.] **1. a.** A tree, *Juglans cinerea* of eastern North America, with compound leaves and egg-shaped nuts. **b.** The edible, oily nut of the butternut. **c.** The bark of the butternut. **d.** A brownish color or dye obtained from butternut bark. **e. butternuts.** Clothing dyed with butternut extract. **2.** *Informal.* A Confederate soldier or partisan in the Civil War. **3.** The souari nut.

**but·ter·scotch** (bŭt'ər-skŏch') *n.* A syrup, sauce, candy, or flavoring made of butter, brown sugar, and occas. artificial flavorings.

**but·ter·weed** (bŭt'ər-wēd') *n.* **1.** A plant, *Senecio glabellus* of the southern and central United States, with yellow flowers. **2.** The horseweed.

**but·ter·wort** (bŭt'ər-wûrt', -wôrt') *n.* [From the leaves' oiliness.] A plant of the genus *Pinguicula*, esp. *P. vulgaris* of wet places, with fleshy, greasy leaves and spurred violet-blue flowers.

**but·ter·y**[1] (bŭt'ə-rē) *adj.* **1.** Resembling, containing, or spread with butter. **2.** *Informal.* Marked by flattery. **—but'ter·i·ness** *n.*

**but·ter·y**[2] (bŭt'ə-rē, bŭt'rē) *n.*, *pl.* **-ies.** [ME *buttrie* < OFr. *boterie* < LLat. *botaria* < *bota*, var. of *butta*, cask.] *Chiefly Brit.* **1.** A pantry or wine cellar. **2.** A place in colleges and universities where students may buy provisions.

**butt hinge** *n.* [< BUTT[2].] A hinge composed of two plates attached to abutting surfaces of a door and door jamb and joined by a pin.

**butt·in·sky** (bŭ-tĭn'skē) *n.*, *pl.* **-skies.** [BUTT + IN + *-sky*, last syllable in many Slavic surnames.] *Slang.* A meddler.

**butt joint** *n.* [< BUTT[2].] A joint formed by two abutting surfaces placed squarely together.

**but·tock** (bŭt'ək) *n.* [ME < OE *buttuc*, end.] **1.** Either of the two rounded, fleshy parts of the rump. **2. buttocks.** The rump.

**but·ton** (bŭt'n) *n.* [ME < OFr. *bouton* < *bouter*, to thrust, of Germanic orig.] **1. a.** A gen. disk-shaped fastener used to join two parts of a garment by fitting through a loop or buttonhole. **b.** Such an object used for decoration. **2.** An object resembling a button, esp.: **a.** A push-button switch. **b.** The tip of a fencing foil. **c.** A fused metal or glass globule. **3.** A knoblike organic structure, esp.: **a.** The head of a small mushroom. **b.** The tip of a rattlesnake's tail. **4.** A round flat badge bearing a design or printed information. **5.** *Slang.* The end of the chin. —*v.* **-toned, -ton·ing, -tons.** —*vt.* **1.** To fasten with buttons. **2.** To furnish or decorate with buttons. —*vi.* To have buttons for closing or fastening <a blouse that *buttons* in the back> **—on the button.** *Informal.* Exactly : precisely. **—but'ton·er** *n.* **—but'ton·y** *adj.*

**but·ton·ball** (bŭt'n-bôl') *n.* [So called because of its button-shaped fruit.] SYCAMORE 1.

**but·ton·bush** (bŭt'n-bŏŏsh') *n.* A North American shrub, *Cephalanthus occidentalis*, with small white, spherical flower clusters.

**but·ton-down** (bŭt'n-doun') *adj.* **1.** Having the ends of the collar fastened down by buttons <a *button-down* dress shirt> **2.** *also* **but·toned-down.** Conventional, conservative, or unimaginative <*button-down* politicians>

**but·ton·hole** (bŭt'n-hōl') *n.* **1.** A slit in a garment or piece of fabric for fastening a button. **2.** *Chiefly Brit.* A boutonniere. —*vt.* **-holed, -hol·ing, -holes. 1.** To make a buttonhole in. **2.** To sew with a buttonhole stitch. **3.** To accost and detain (a person) in conversation. **—but'ton·hol'er** *n.*

**buttonhole stitch** *n.* A loop stitch that forms a reinforced edge, as around a buttonhole.

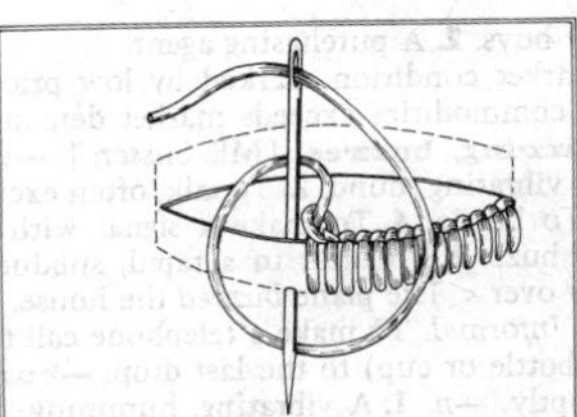

**buttonhole stitch**

**but·ton·hook** (bŭt'n-hŏŏk') *n.* A small hook for pulling buttons through buttonholes.

**but·ton·quail** (bŭt'n-kwāl') *n.* Any of various small, quaillike Old World birds of the family Turnicidae.

**button snakeroot** *n.* **1.** BLAZING STAR 2. **2.** Rattlesnake master.

**but·ton·wood** (bŭt'n-wo͝od') *n.* [From its button-shaped fruit.] SYCAMORE 1.

**but·tress** (bŭt'rĭs) *n.* [ME *buteras* < OFr. *bouteret* < *bouter,* to strike against, of Germanic orig.] **1.** A usu. brick or stone reinforcing structure built against a wall or building. **2.** Something resembling a buttress. **3.** A horny growth on the heel of a horse's hoof. **4.** Something that supports or reinforces. —*vt.* **-tressed, -tress·ing, -tress·es. 1.** To support or reinforce with a buttress. **2.** To bolster or strengthen <a theory *buttressed* by statistics>

**butt shaft** *n.* A blunt, unbarbed arrow.

**butt weld** *n.* A welded butt joint.

**butt-weld** (bŭt'wĕld') *vt.* **-weld·ed, -weld·ing, -welds.** To join by a butt weld.

**bu·tut** (bo͞o'to͞ot') *n., pl.* **butut** *or* **-tuts.** [Native word in Gambia.] —See table at CURRENCY.

**bu·tyl** (byo͞ot'l) *n.* A hydrocarbon radical, $C_4H_9$, with the structure of butane and valence 1.

**butyl alcohol** *n.* One of four isomeric alcohols, $C_4H_9OH$, used as solvents and in organic synthesis.

**bu·tyl·ate** (byo͞ot'l-āt') *vt.* **-at·ed, -at·ing, -ates.** To bring a butyl group into (a compound). **—bu'tyl·a'tion** *n.*

**bu·tyl·ene** (byo͞ot'l-ēn') *n.* One of three gaseous isomeric ethylene hydrocarbons, $C_4H_8$, used chiefly to make synthetic rubbers.

**butyl rubber** *n.* A synthetic rubber produced by copolymerization of a butylene with isoprene, used in tires, inner tubes, and insulation.

**bu·ty·ra·ceous** (byo͞o'tə-rā'shəs) *adj.* [Lat. *butyrum,* butter + -ACEOUS.] Like butter.

**bu·tyr·al·de·hyde** (byo͞o'tə-răl'də-hīd') *n.* [BUTYR(IC) + ALDEHYDE.] A transparent, highly flammable liquid, $C_4H_8O$, used in synthesizing resins.

**bu·ty·rate** (byo͞o'tə-rāt') *n.* [BUTYR(IC) + -ATE.] A salt or ester of butyric acid.

**bu·tyr·ic** (byo͞o-tĭr'ĭk) *adj.* [< Lat. *butyrum,* butter.] **1.** Of, relating to, containing, or derived from butter. **2.** Of, relating to, or derived from butyric acid.

**butyric acid** *n.* Either of two colorless isomeric acids, $C_3H_7COOH$, occurring in animal milk fats and used in disinfectants, emulsifying agents, and pharmaceuticals.

**bu·ty·rin** (byo͞o'tər-ĭn) *n.* [Fr. *butyrine* < Lat. *butyrum,* butter.] One of three isomeric glyceryl esters of butyric acid, naturally present in butter.

**bux·om** (bŭk'səm) *adj.* [ME, obedient < OE **būhsum* < *būgan,* to bow, submit.] **1.** Healthily plump and ample of figure, as a large-bosomed woman. **2.** *Archaic.* Vivacious : lively. **3.** *Obs.* Obedient : yielding. **—bux'om·ly** *adv.* **—bux'om·ness** *n.*

▲ word history: A *buxom* woman in modern parlance is "healthily plump" and most likely large-bosomed as well. It is hard to believe that such a meaning grew out of the original sense "obedient," but an intermediate stage is probably to be found in another obsolete sense, "gracious, courteous, obliging." This usage developed into the sense "lively, jolly," and from there it was a short step to using *buxom* to indicate someone who looked lively and jolly, as ample-figured persons are reputed to be.

**buy** (bī) *v.* **bought** (bôt), **buy·ing, buys.** [ME < OE *bycgan.*] —*vt.* **1.** To acquire in exchange for money or its equivalent : PURCHASE. **2.** To be capable of purchasing. **3.** To acquire by sacrifice, exchange, or trade. **4.** To bribe <*buy* a judge> **5.** *Slang.* To believe in <wouldn't *buy* their excuses> —*vi.* To purchase goods. **—buy in. 1.** To purchase stock or interest, as in a company. **2.** To purchase back for the original owner, as at an auction when the bidding is low. **—buy off.** To bribe in order to proceed without interference or be freed from an obligation or prosecution. **—buy out.** To purchase the controlling stock, business rights, or interests of. **—buy up.** To purchase all that is available of. —*n.* **1.** The act of buying : PURCHASE. **2.** *Informal.* A bargain. **—buy'a·ble** *adj.*

☆ ***syns:*** BUY, PURCHASE *v. core meaning* : to acquire in exchange for money or something of equal value <*buy* a used car> ***ant:*** sell

**buy·er** (bī'ər) *n.* **1.** One that buys. **2.** A purchasing agent.

**buyers' market** *n.* A market condition marked by low prices occurring when a supply of commodities exceeds market demand.

**buzz** (bŭz) *v.* **buzzed, buzz·ing, buzz·es.** [ME *bussen.*] —*vi.* **1.** To make a low, droning or vibrating sound. **2.** To talk, often excitedly, in subdued tones. **3.** To bustle. **4.** To make a signal with a buzzer. —*vt.* **1.** To cause to buzz. **2.** To utter in a rapid, subdued voice. **3.** *Informal.* To fly low over <The plane *buzzed* the house.> **4.** To signal with a buzzer. **5.** *Informal.* To make a telephone call to. **6.** *Chiefly Brit.* To drink (a bottle or cup) to the last drop. **—buzz off.** *Informal.* To leave abruptly. —*n.* **1.** A vibrating, humming, or droning sound. **2.** A low murmur. **3.** *Informal.* A telephone call <Give me a *buzz* at noon.> **4.** *Slang.* A pleasant intoxication, as from alcohol.

**buz·zard** (bŭz'ərd) *n.* [ME *busard* < OFr. < Lat. *buteo.*] **1.** A North American vulture, as the turkey buzzard. **2.** *Chiefly Brit.* A hawk of the genus *Buteo.* **3.** An unpleasant, often greedy person.

**buzz bomb** *n.* ROBOT BOMB.

**buzz·er** (bŭz'ər) *n.* An electric signaling device, as a doorbell, that makes a buzzing sound.

**buzz saw** *n.* A circular saw.

**buzz word** *n.* An important-sounding technical word or phrase used primarily to impress lay people.

**bwa·na** (bwä'nə) *n.* [Swahili < Ar. *abūna,* our father.] Master : sir. —Used chiefly in eastern Africa as a form of address.

**by**[1] (bī) *prep.* [ME < OE *bi.*] **1.** Next to : close to <the light *by* the window> **2.** With the help or use of : THROUGH <We came *by* the main road.> **3.** Up to and beyond : PAST <drove *by* the entrance> **4.** In the period of : DURING <working *by* night> **5.** Not later than <*by* noon> **6. a.** In the amount of <requests *by* the thousands> **b.** To the extent of <closer *by* a foot> **7. a.** According to. **b.** With respect to. **8.** In the name of <swore *by* my honor> **9.** Through the agency or action of <peace *by* negotiation> **10.** —Used to indicate a succession of specified units of measure <filed out one *by* one><Little *by* little they made headway.> **11. a.** —Used in multiplication and division <3 *by* 5> **b.** —Used with measurements <a section 4 *by* 8 feet> —*adv.* **1.** On hand : NEARBY <stay *by*> **2.** Aside : away <set it *by* for now> **3.** Up to, alongside, and past <The runner raced *by.*> **4.** Into the past <as time goes *by*> **—by and by.** A little later. **—by and large.** For the most part.

**by**[2] (bī) *n. var. of* BYE.

**by-** *or* **bye-** *pref.* **1.** By <*by*gone> **2.** Secondary : incidental <*by*way>

**by-and-by** (bī'ən-bī') *n.* **1.** A time or occasion in the future. **2.** The hereafter.

**by-blow** (bī'blō') *n.* An indirect blow.

**bye** *also* **by** (bī) *n.* [< BY.] **1.** A secondary matter : SIDE ISSUE. **2.** The position of one who draws no opponent for a round in a tournament, thus advancing to the next round. **—by the bye.** By the way : INCIDENTALLY.

**bye-** *pref. var. of* BY-.

**bye-bye** (bī'bī', bī-bī') *interj.* [Redup. of *(good-)bye,* var. of GOOD-BY.] *Informal.* —Used to express farewell.

**by-e·lec·tion** *also* **bye-e·lec·tion** (bī'ĭ-lĕk'shən) *n.* A special election held between general elections to fill a vacancy in a legislature or parliament.

**by·gone** (bī'gôn', -gŏn') *adj.* Gone by : PAST <*bygone* times> —*n.* A past occurrence. **—let bygones be bygones.** To let past differences be forgotten.

**by·law** (bī'lô') *n.* [ME *bilawe,* local regulations, poss. of Scand. orig.] **1.** A secondary law. **2.** A rule or law governing the internal affairs of an organization.

**by-line** *also* **by·line** (bī'līn') —*n.* A line at the head of a newspaper or magazine article with the author's name. —*vt.* **-lined, -lin·ing, -lines.** To write (an article) under a by-line. **—by'-lin'er** *n.*

**by-name** (bī'nām') *n.* **1.** A surname. **2.** A nickname.

**by-pass** *also* **by·pass** (bī'păs') —*n.* **1.** A road or highway that passes around or to one side of a congested area or obstruction. **2.** A conduit for channeling a liquid or gas around another pipe or a fixture. **3.** A means of circumvention. **4.** *Elect.* SHUNT 3. **5.** *Med.* **a.** An alternative passage created surgically between two blood vessels, esp. to avoid an obstruction. **b.** An operation to create a by-pass. —*vt.* **-passed, -pass·ing, -pass·es. 1.** To avoid by using a by-pass. **2.** To ignore or be heedless of. **3.** To cause (e.g., piped liquid) to follow a by-pass.

**by·past** (bī'păst') *adj.* Past : bygone.

**by·path** (bī'păth', -päth') *n.* An indirect or little-used path.

**by-play** (bī'plā') *n.* Secondary action or speech taking place while the main action proceeds, esp. on a theater stage.

**by-prod·uct** (bī'prŏd'əkt) *n.* **1.** Something produced in the making of something else. **2.** A secondary result : SIDE EFFECT.

**byre** (bīr) *n.* [ME < OE *bȳre.*] *Chiefly Brit.* A cowshed or barn.

**by·road** (bī'rōd') *n.* A side road.

**By·ron·ic** (bī-rŏn'ĭk) *adj.* Of or typical of the poet Byron or his works. **—By·ron'i·cal·ly** *adv.*

**bys·si·no·sis** (bĭs'ĭ-nō'sĭs) *n.* [LLat. *byssinum,* linen garment (< Lat. *byssus,* a kind of cloth < Gk. *bussos,* flax, of Semitic orig.) + -OSIS.] Pneumoconiosis caused by the long-term inhalation of cotton dust and marked by chronic bronchitis.

**bys·sus** (bĭs'əs) *n., pl.* **bys·sus·es** *or* **bys·si** (bĭs'ī') [Lat., linen cloth < Gk. *bussos,* linen, of Semitic orig.] **1.** *Zool.* A mass of filaments by means of which certain bivalve mollusks, as mussels, attach themselves to fixed surfaces. **2.** A fine-textured linen used by the ancient Egyptians as wrapping for mummies.

**by·stand·er** (bī'stăn'dər) *n.* A witness to an event.

**by·street** (bī'strēt') *n.* A side street.

**byte** (bīt) *n.* [Alteration and blend of BIT[1] and BITE.] *Computer Sci.* A sequence of adjacent binary digits operated on as a unit.

**by·way** (bī'wā') *n.* **1.** A byroad. **2.** A secondary or overlooked field of study.

**by·word** (bī'wûrd') *n.* [ME *byworde* < OE *bīwORD* < transl. of Lat. *proverbium.*] **1.** A well-known saying : PROVERB. **2.** One that prover-

ă **pat** ā **pay** âr **care** ä **father** ĕ **pet** ē **be** hw **which** ĭ **pit** ī **tie** îr **pier** ŏ **pot** ō **toe** ô **paw, for** oi **noise** o͝o **took**

bially represents a type, class, or quality. **3.** Something noteworthy or notorious. **4.** An epithet or nickname.

**Byz·an·tine** (bĭz'ən-tēn', -tīn', bĭ-zăn'tĭn) *adj.* **1.** Of, relating to, or typical of the ancient city of Byzantium, its inhabitants, or their culture. **2.** Of or designating the architectural style developed from the 5th cent. A.D. in Byzantium, marked by round arches, massive domes, intricate minarets and spires, and extensive use of mosaic. **3.** Of or referring to the style of painting and design developed in Byzantium, marked by formal design, frontal, stylized presentation of figures, rich use of color, esp. gold, and gen. religious subject matter. **4.** Of or relating to the Eastern Orthodox Church or its rites. **5. a.** Of, relating to, or marked by intrigue: DEVIOUS <"A fine hand for *Byzantine* deals and cozy arrangements" —*New York*> **b.** Highly complicated: INTRICATE <"A financial empire of *Byzantine* complexity" —*Newsweek*> —*n.* A native or inhabitant of Byzantium.

# Cc

**c** or **C** (sē) *n., pl.* **c's** or **C's. 1.** The third letter of the English alphabet. **2.** A speech sound represented by the letter *c.* **3.** The third in a series. **4. C** *Mus.* **a.** The first tone in the scale of C major. **b.** The third tone in the relative minor scale. **c.** The key or a scale in which C is the tonic. **d.** A written or printed note representing this tone. **5.** The third highest grade in quality or rank. **6. C** The Roman numeral for 100.

**C** *symbol for* CARBON.

**Ca** *symbol for* CALCIUM.

**cab**[1] (kăb) *n.* [Short for TAXICAB.] **1.** A taxicab. **2.** A one-horse vehicle for public hire. **3.** The covered compartment of a heavy vehicle, as a truck or locomotive, in which the driver or operator sits.

**cab**[2] (kăb) *n.* [Heb. *qabh.*] A Hebrew measure equal to approx. two quarts.

**ca·bal** (kə-băl') *n.* [Fr. *cabale* < Med. Lat. *cabala.* —see CABALA.] **1.** A conspiratorial group of plotters. **2.** A secret plot or scheme. —*vi.* **-balled, -bal·ling, -bals.** To form a cabal.

**cab·a·la** or **cab·ba·la** (kăb'ə-lə, kə-bä'-) *n.* [Med. Lat. < Heb. *qabbālāh,* received doctrine < *qābal,* he received.] **1.** *often* **Cabala.** An occult theosophy of rabbinical origin, widely disseminated in medieval Europe, based on an esoteric interpretation of the Hebrew Scriptures. **2.** A secret, esoteric, or occult doctrine. **—cab'a·lism** *n.* **—cab'a·list** *n.* **—cab'a·lis'tic** *adj.* **—cab'a·lis'ti·cal·ly** *adv.*

**†cab·al·le·ro** (kăb'ə-lâr'ō, -əl-yâr'ō) *n., pl.* **-ros.** [Sp. < LLat. *caballarius,* horse groom < Lat. *caballus,* horse.] **1.** A Spanish gentleman: CAVALIER. **2.** *Southwestern U.S.* A skilled equestrian.

**ca·ban·a** *also* **ca·ba·ña** (kə-băn'ə, -băn'yə) *n.* [Sp. *cabaña* < LLat. *capanna,* hut.] A shelter on a beach or at a swimming pool that is used as a bathhouse.

**cab·a·ret** (kăb'ə-rā') *n.* [Fr. < ONFr., liquor store, perh. < LLat. *camera,* room. —see CHAMBER.] **1.** A restaurant or nightclub offering live entertainment. **2.** A floor show presented by a cabaret.

**cab·bage** (kăb'ĭj) *n.* [ME *caboche* < OFr. *caboce,* head.] **1.** An edible plant, *Brassica oleracea capitata,* grown in temperate climates worldwide and having a short, thick stalk and a large head formed by tightly overlapping green or reddish leaves. **2.** An edible leaf bud of the cabbage palm. **3.** *Slang.* Money. —*vi.* **-baged, -bag·ing, -bag·es.** To form or grow in a head, as cabbage does.

**cabbage butterfly** *n.* A white butterfly of the genus *Pieris,* with larvae that feed on cabbage.

**cabbage palm** *n.* A tropical American palm tree, *Roystonea oleracea,* with leaf buds that are edible when young.

**cabbage palmetto** *n.* A palmetto.

**cabbage rose** *n.* A prickly shrub native to the Caucasus, *Rosa centifolia,* with large, fragrant, many-petaled pink flowers.

**cab·bage·worm** (kăb'ĭj-wûrm') *n.* Any of several caterpillars that feed on and destroy cabbage, esp. the bright-green larva of the cabbage butterfly.

**cabbage yellow** *n.* A disease of cabbage marked by the yellowing of leaves and caused by the fungus *Fusarium conglutinans.*

**cab·ba·la** (kăb'ə-lə, kə-bä'-) *n. var. of* CABALA.

**cab·by** (kăb'ē) *n., pl.* **-bies.** *Informal.* A driver of a taxicab.

**ca·ber** (kā'bər, kä'-) *n.* [Sc. Gael. *cabar.*] *Scot.* A heavy wooden pole thrown as a demonstration of strength in a sporting contest.

**cab·er·net** (kăb'ər-nā') *n.* [Fr.] A dry red wine made from the grape variety *Cabernet sauvignon.*

**cab·e·zon** (kăb'ĭ-zŏn') *n.* [Sp. *cabezón,* aug. of *cabeza,* head < VLat. **capitia* < Lat. *caput,* head.] A large edible fish, *Scorpaenichthys marmoratus* of North American Pacific coastal waters.

**cab·in** (kăb'ĭn) *n.* [ME *caban* < OFr. *cabane* < LLat. *capanna.*] **1.** A small, roughly constructed house: COTTAGE. **2. a.** A compartment in a ship used as living quarters by an officer or a passenger. **b.** An enclosed compartment in a boat used as a shelter or living quarters. **c.** The enclosed space in an aircraft for the crew, passengers, or cargo. —*vt.* & *vi.* **-ined, -in·ing, -ins.** To confine or live in or as if in a cabin.

**cabin boy** *n.* A boy employed as a servant aboard a ship.

**cabin class** *n.* A class of accommodations on some passenger ships, lower than first class and higher than tourist class.

**cabin cruiser** *n.* A powerboat with a cabin.

**cab·i·net** (kăb'ə-nĭt) *n.* [OFr., small room, dim. of ONFr. *cabine,* gambling house.] **1.** An upright cupboardlike repository with shelves, drawers, or compartments, as for the safekeeping or display of a collection of objects. **2.** *often* **Cabinet.** A body of people appointed by a chief of state or a prime minister to head the executive departments of the government and act as official advisers. **3.** *Archaic.* A private room set aside for a specific activity.

**cab·i·net·mak·er** (kăb'ə-nĭt-mā'kər) *n.* A craftsman specializing in making fine wooden furniture. **—cab'i·net·mak'ing** *n.*

**cab·i·net·ry** (kăb'ĭ-nĭ-trē) *n.* Cabinetwork.

**cab·i·net·work** (kăb'ə-nĭt-wûrk') *n.* Finished woodwork made by a cabinetmaker.

**cabin fever** *n.* Anxiety or distress caused by a lack of environmental stimulation, as when living in a remote, sparsely populated region or a small enclosed space.

**ca·ble** (kā'bəl) *n.* [ME < Norman Fr. < LLat. *capulum,* lasso < Lat. *capere,* to seize.] **1. a.** A strong, large-diameter heavy steel or fiber rope. **b.** Something resembling a cable. **2.** *Elect.* A bound or sheathed group of mutually insulated conductors. **3. a.** *Naut.* A heavy rope or chain for mooring or anchoring a ship. **b.** A unit of nautical length equal to 720 feet or approx. 220 meters in the United States and 608 feet or approx. 185 meters in Britain. **4.** A cablegram. **5.** Cable television. —*v.* **-bled, -bling, -bles.** —*vt.* **1. a.** To send a cablegram to. **b.** To transmit (a message) by telegraph. **2.** To furnish or fasten with a cable or cables. —*vi.* To send a cablegram.

**cable car** *n.* A car designed to operate on a cableway or cable railway.

**ca·ble·cast** (kā'bəl-kăst') *n.* A telecast transmitted via cable television. **—ca'ble·cast'** *v.* **(-cast** or **-cast·ed, -cast·ing, -casts). —ca'ble·cast'er** *n.*

**ca·ble·gram** (kā'bəl-grăm') *n.* A telegram sent by submarine cable.

**ca·ble-laid** (kā'bəl-lād') *adj.* Constructed of three ropes of three strands each, twisted together counterclockwise.

**cable railway** *n.* A railroad on which the cars are moved by an endless cable driven by a stationary engine.

**cable stitch** *n.* A knitting stitch producing a twisted rope design.

**ca·blet** (kā'blĭt) *n.* A cable-laid rope with a circumference of less than 10 inches or 25.4 centimeters.

**cable television** *also* **cable TV** *n.* A television distribution system in which station signals, picked up by elevated antennas, are delivered via cable to subscribers' receivers.

**ca·ble·vi·sion** (kā'bəl-vĭzh'ən) *n.* Cable television.

**ca·ble·way** (kā'bəl-wā') *n.* A suspended cable used as a track for a cable car.

**cab·man** (kăb'mən) *n.* The driver of a cab.

**cab·o·chon** (kăb'ə-shŏn') *n.* [OFr., aug. of *caboche,* head.] **1.** A highly polished, convex-cut, unfaceted gem. **2.** The cabochon style of cutting. **—cab'o·chon'** *adv.*

**ca·boo·dle** (kə-bōōd'l) *n.* [Perh. alteration of *kit and boodle.*] *Informal.* The entire lot.

**ca·boose** (kə-bōōs') *n.* [Perh. < Du. *kabuis.*] **1.** The last car on a freight train, having kitchen and sleeping facilities for the train crew. **2.** *Obs.* **a.** A ship's galley. **b.** A cast-iron range used in a ship's galley in the early 19th cent. **c.** An outdoor oven or fireplace.

**cab·o·tage** (kăb′ə-täzh′) *n.* [Fr. < *caboter*, to coast.] Coastal trade or navigation.

**Cab·ot's ring** (kăb′əts) *n.* [After Richard C. *Cabot* (1868–1939).] Ringlike inclusions in the red blood cells occas. found in cases of severe anemia.

**ca·bret·ta** (kə-brĕt′ə) *n.* [Sp. and Port. *cabra*, goat (< Lat. *capra*, she-goat) + Sp. *-etta*, -ette.] A soft kidlike leather made from sheepskin having coarse, hairlike wool.

**ca·bril·la** (kə-brē′yə, -brĭl′ə) *n.* [Sp., dim. of *cabra*, goat < Lat. *capra*, she-goat.] A sea bass, esp. *Epinephelus guttatus* of tropical waters.

**cab·ri·ole** (kăb′rē-ōl′) *n.* [Fr., caper (from its resemblance to the foreleg of a capering animal). —see CABRIOLET.] A style of furniture leg, typical of Queen Anne and Chippendale furniture, that curves outward and narrows downward into an ornamental foot.

**cab·ri·o·let** (kăb′rē-ə-lā′) *n.* [Fr., dim. of *cabriole*, caper < OFr. < OItal. *capriola* < *capriolo*, roebuck < Lat. *capreolus*, wild goat < *caper*, he-goat.] **1.** A two-wheeled, one-horse carriage with two seats and a folding top. **2.** An automobile with a folding top.

**cab·stand** (kăb′stănd′) *n.* A place for taxicabs waiting for hire.

**ca·ca·o** (kə-kä′ō, -kā′ō) *n., pl.* **-os.** [Sp. < Nahuatl *cacahuatl*, cacao beans.] **1.** An evergreen tropical American tree, *Theobroma cacao*, with yellowish flowers and reddish-brown seedpods. **2.** The seed of the cacao tree, used in making chocolate, cocoa, and cocoa butter.

**cacao bean** *n.* The seed of the cacao tree.

**cacao butter** *n.* Cocoa butter.

**cach·a·lot** (kăsh′ə-lŏt′, -ə-lō′) *n.* [Fr.] The sperm whale.

**cache** (kăsh) *n.* [Fr. < *cacher*, to hide < OFr. < VLat. **coactiare* < Lat. *coactare*, to constrain, freq. of *cogere*, to drive together. —see COGENT.] **1.** A hiding place for storing provisions and other necessities. **2.** A safe place for concealment, as of valuables. **3.** A store of goods hidden in a cache. **4.** *Computer Sci.* A fast storage buffer in the central processing unit of a computer. —*vt.* **cached, cach·ing, cach·es.** To hide or store in a cache.

**ca·chec·tic** (kə-kĕk′tĭk) *adj.* [Fr. *cachectique* < Lat. *cachecticus* < Gk. *kakhektikos* < *kakhexia*, bad condition of the body. —see CACHEXIA.] Relating to or marked by cachexia.

**cache·pot** (kăsh′pŏt′, -pō′) *n.* [Fr. : *cacher*, to hide + *pot*, pot.] An ornamental container for a flowerpot.

**ca·chet** (kă-shā′) *n.* [OFr. < *cacher*, to hide. —see CACHE.] **1.** A seal on a letter or document. **2.** A quality or feature that confers distinction or prestige <a profession with a *cachet* of glamour> **3. a.** A commemorative design stamped on an envelope to mark a postal or philatelic event. **b.** A motto forming part of a postal cancellation. **4.** A wafer capsule once used by pharmacists for presenting an unpleasant-tasting drug.

**ca·chex·i·a** (kə-kĕk′sē-ə) *n.* [LLat. < Gk. *kakhexia* : *kakos*, bad + *hexis*, condition < *ekhein*, to have.] A general wasting of the body during a chronic disease.

**cach·in·nate** (kăk′ə-nāt′) *vi.* **-nat·ed, -nat·ing, -nates.** [Lat. *cachinnare, cachinnat-*.] To laugh loudly or convulsively : GUFFAW. **—cach′in·na′tion** *n.*

**ca·chou** (kă-sho͞o′, kăsh′o͞o) *n.* [Fr. < Port. *cachu* < Malayalam *kāccu*.] **1.** Catechu. **2.** A pastille used to sweeten the breath.

**ca·chu·cha** (kə-cho͞o′chə) *n.* [Sp.] An Andalusian solo dance in 3/4 time.

**ca·cique** (kə-sēk′) *n.* [Sp., of Arawakan orig.] **1.** An Indian chief, esp. in the Spanish West Indies and other parts of Latin America during colonial and postcolonial times. **2.** A local Spanish or Latin American political boss. **3.** A tropical oriole.

**cack·le** (kăk′əl) *v.* **-led, -ling, -les.** [ME *cakelen*.] —*vi.* **1.** To make the shrill cry typical of a hen after laying an egg. **2.** To laugh or speak in a way similar to a hen's cackle. —*vt.* To utter in cackles. —*n.* **1.** The act or sound of cackling. **2.** Shrill, brittle laughter. **3.** Foolish chatter. **—cack′ler** *n.*

**caco-** *pref.* [Gk. *kako-* < *kakos*, bad.] Bad <cacography>

**cac·o·dyl** (kăk′ə-dĭl′) *n.* [Gk. *kakōdēs*, bad-smelling (*kakos*, bad + *ozein*, to smell) + -YL.] **1.** The arsenic group $As(CH_3)_2$. **2.** A poisonous oil, $As_2(CH_3)_4$, with a garlicky smell. **—cac′o·dyl′ic** *adj.*

**cac·o·ë·thes** (kăk′ō-ē′thēz) *n.* [Lat. *cacoethus* < Gk. *kakoēthes* : *kakos*, bad + *ēthos*, disposition.] An irresistible compulsion.

**cac·o·gen·ics** (kək′ə-jĕn′ĭks) *n.* (*sing. in number*). Dysgenics. **—cac′o·gen′ic** *adj.*

**ca·cog·ra·phy** (kə-kŏg′rə-fē) *n.* **1.** Bad handwriting. **2.** Bad spelling.

**cac·o·mis·tle** (kăk′ə-mĭs′əl) *n.* [Mex. Sp. < Nahuatl *tlacomiztli* : *tlaco*, half + *miztli*, mountain lion.] Either of two small, carnivorous mammals, *Bassariscus astutus* of the southwestern United States or *Jentinkia sumichrasti* of Central America, with grayish or brownish fur and a black-banded tail.

**ca·coph·o·nous** (kə-kŏf′ə-nəs) *adj.* [Gk. *kakophōnos* : *kakos*, bad + *phōnē*, sound.] Having a harsh, unpleasant sound : DISCORDANT <*cacophonous* laughter> **—ca·coph′o·nous·ly** *adv.*

**ca·coph·o·ny** (kə-kŏf′ə-nē) *n., pl.* **-nies.** [Fr. *cacophonie* < Gk. *kakophonia* < *kakophōnos*, cacophonous.] **1.** Harsh, discordant sound : DISSONANCE. **2.** Harsh or unharmonious use of language.

**cac·tus** (kăk′təs) *n., pl.* **-ti** (-tī′) *or* **-tus·es.** [NLat. *Cactus*, type genus < Lat., cardoon < Gk. *kaktos*.] Any of a large group of plants of the family Cactaceae, chiefly native to arid regions of the New World and having thick, fleshy, often prickly stems that function as leaves and in some species showy flowers and edible fruit.

**ca·cu·mi·nal** (kə-kyo͞o′mə-nəl) *adj.* [< Lat. *cacumen*, summit.] Articulated with the tip of the tongue turned back and up toward the roof of the mouth : RETROFLEX.

**cad** (kăd) *n.* [Short for CADDIE.] An ungentlemanly man. **—cad′dish** *adj.* **—cad′dish·ly** *adv.* **—cad′dish·ness** *n.*

**ca·das·ter** *also* **ca·das·tre** (kə-dăs′tər) *n.* [Fr. < Ital. *cadastro*, var. of OItal. *catastico* < LGk. *katastikhon*, register : *kata-*, by + *stikhos*, line.] A public record, survey, or map of the value, extent, and ownership of land as a basis of taxation. **—ca·das′tral** *adj.*

**ca·dav·er** (kə-dăv′ər) *n.* [Lat. < *cadere*, to die.] A corpse, esp. one intended for dissection. **—ca·dav′er·ic** (-ə-rĭk) *adj.*

**ca·dav·er·ine** (kə-dăv′ə-rēn′) *n.* A syrupy, colorless fuming ptomaine, $NH_2(CH_2)_5NH_2$, formed by carboxylation of lysine in decaying animal flesh.

**ca·dav·er·ous** (kə-dăv′ər-əs) *adj.* **1.** Suggestive of death. **2. a.** Of deathly pallor <a *cadaverous* face> **b.** Emaciated : gaunt <a *cadaverous* stray dog> **—ca·dav′er·ous·ly** *adv.* **—ca·dav′er·ous·ness** *n.*

**cad·dice** (kăd′ĭs) *n. var. of* CADDIS.

**caddice fly** *n. var. of* CADDIS FLY.

**caddice worm** *n. var. of* CADDIS WORM.

**cad·die** *also* **cad·dy** (kăd′ē) [French *cadet*, caddie, cadet.] —*n., pl.* **-dies. 1.** One hired to serve as an attendant to a golfer, esp. by carrying the clubs. **2.** *Scot.* A boy who does odd jobs. —*vi.* **-died, -dy·ing, -dies.** To serve as a caddie.

**cad·dis** *also* **cad·dice** (kăd′ĭs) *n.* [ME *cadace* < AN *cadaz* < OProv. *cadarz*.] A coarse woolen fabric, yarn, or ribbon binding.

**caddis fly** *also* **caddice fly** *n.* [< obs. *cad*, var. of COD[2] (from the tube in which the larva lives).] Any of various four-winged insects of the order Trichoptera, found near lakes and streams.

**caddis worm** *also* **caddice worm** *n.* The aquatic, wormlike larva of the caddis fly, enclosed in a cylindrical case covered with grains of sand, fragments of shell, etc.

**Cad·do** (kăd′ō) *n., pl.* **Caddo** *or* **-dos.** [Prob. < Caddoan *Kädohädächo*.] A member of an Indian confederacy of Caddoan linguistic stock, once inhabiting Arkansas, Louisiana, and eastern Texas.

**Cad·do·an** (kăd′ō-ən) *n.* A family of Indian languages once spoken in the Dakotas, Nebraska, Kansas, Oklahoma, Arkansas, Texas, and Louisiana.

**cad·dy**[1] (kăd′ē) *n., pl.* **-dies.** [Malay *kati*, catty.] A small container, esp. for holding tea.

**cad·dy**[2] (căd′ē) *n.* & *v. var. of* CADDIE.

**cade**[1] (kād) *adj.* [ME.] Having been left by its mother and then reared by hand <a *cade* calf>

**cade**[2] (kād) *n.* [Fr. < OProv. < Med. Lat. *catanus*.] A shrub, *Juniperus oxycedrus* of the Mediterranean region, whose wood yields an oily brown liquid used in treating skin ailments.

**-cade** *suff.* [< CAVALCADE.] Procession <motor*cade*>

**ca·delle** (kə-dĕl′) *n.* [Fr. < Prov. *cadello* < Lat. *catella*, fem. of *catellus*, puppy < *catulus*, the young of animals.] A small blackish beetle, *Tenebroides mauritanicus*, both the larval and adult forms of which damage stored grain and packaged foods.

**ca·dence** (kād′ns) *also* **ca·den·cy** (kād′n-sē) *n., pl.* **-denc·es** *also* **-den·cies.** [ME < OFr. < OItal. *cadenza* < *cadere*, to fall < Lat.] **1.** Balanced, rhythmic flow, as of verse or oratory. **2.** The measure or beat of movement, as in marching or dancing. **3. a.** A falling inflection of the voice, as at the end of a sentence. **b.** General inflection or modulation of the voice. **4.** *Mus.* A progression of chords moving to a harmonic close or point of rest. **—ca′denced** *adj.*

**ca·dent** (kād′nt) *adj.* [Lat. *cadens, cadent-*, pr.part. of *cadere*, to fall.] **1.** Having cadence. **2.** *Archaic.* Falling <*cadent* tears>

**ca·den·za** (kə-dĕn′zə) *n.* [Ital. < OItal., cadence.] **1.** An elaborate ornamental melodic flourish interpolated into a vocal piece, as an aria. **2.** An extended virtuosic section for a soloist near the end of a movement of a concerto.

**ca·det** (kə-dĕt′) *n.* [Fr. < dial. *capdet*, captain < LLat. *capitellum*, dim. of Lat. *caput*, head.] **1.** A student at a military school. **2.** A younger son or brother. **3.** *Slang.* A pimp. **—ca·det′ship′** *n.*

**cadge** (kăj) *vi.* & *vt.* **cadged, cadg·ing, cadg·es.** [Perh. back-formation < obs. *cadger*, peddler < ME *cadgear*.] *Informal.* To beg or obtain by begging. **—cadg′er** *n.*

**cad·mi·um** (kăd′mē-əm) *n.* [NLat. < Lat. *cadmia*, calamine (from its being found with calamine in zinc ore).] *Symbol* **Cd** A soft, bluish-white metallic element used in low-friction alloys, solders, dental amalgams, and nickel-cadmium storage batteries; atomic number 48; atomic weight 112.40. **—cad′mic** (-mĭk) *adj.*

**cadmium sulfate** *n.* A compound, $CdSO_4$, that forms colorless crystals and is used as an antiseptic.

**Cad·mus** (kăd′məs) *n.* [Lat. < Gk *Kadmos*.] *Gk. Myth.* A Phoenician prince who killed a dragon and sowed its teeth, from which sprang up an army of men who fought one another until only five survived and with these five Cadmus founded the city of Thebes.

**cad·re** (kăd′rē) *n.* [Fr. < Ital. *quadra* < Lat. *quadrum*, a square.] **1.** A framework. **2.** A small group of trained personnel around which a larger organization can be built and trained.

**ca·du·ce·us** (kə-do͞o′sē-əs, -shəs, -dyo͞o′-) *n., pl.* **-ce·i** (-sē-ī′) [Lat. *caduceus,* alteration of Gk. *karukeion* < *karux,* herald.] **1.** A herald's wand or staff, esp. in ancient times. **2. a.** *Myth.* A winged staff with two serpents twined around it, carried by Hermes. **b.** An insignia bearing a caduceus and used as the symbol of the medical profession. **—ca·du′ce·an** (-sē-ən, -shən) *adj.*

**ca·du·ci·ty** (kə-do͞o′sĭ-tē, -dyo͞o′-) *n.* [Fr. *caducité* < *caduc,* frail < Lat. *caducus.* —see CADUCOUS.] **1.** Frailty of old age : SENILITY. **2.** The quality or state of being perishable : IMPERMANENCE.

**ca·du·cous** (kə-do͞o′kəs, -dyo͞o′-) *adj.* [Lat. *caducus,* falling < *cadere,* to fall.] *Biol.* Falling off or shedding at an early stage of development, as the gills of amphibians or the leaves of some plants.

**cae·cil·ian** (sə-sĭl′yən, -sĭl′ē-ən, -sēl′-) *n.* [< Lat. *caecilia,* a kind of lizard < *caecus,* blind (from its small eyes).] A legless, burrowing, wormlike, tropical amphibian of the order Gymnophiona.

**cae·cum** (sē′kəm) *n. var. of* CECUM.

**Cae·lum** (sē′ləm) *n.* [Lat. *caelum,* sculptor's chisel < *caedere,* to cut.] A constellation in the Southern Hemisphere.

**caer·phil·ly** (kär-fĭl′ē) *n.* [After *Caerphilly,* a district in Wales.] A mild white cheese orig. made in Wales.

**Cae·sar** (sē′zər) *n.* [Lat., after Gaius Julius *Caesar* (100–44 B.C.).] **1.** A surname of the early Roman emperors that after Hadrian became the title of the junior imperial colleague of the Augustus. **2.** *often* **caesar.** A dictator or autocrat.

**Cae·sar·e·an** *also* **Cae·sar·i·an** *or* **Ce·sar·e·an** *or* **Ce·sar·i·an** (sĭ-zâr′ē-ən) *n.* A Caesarean section.

**Caesarean section** *also* **caesarean section** *n.* [From the belief that Julius *Caesar* (100–44 B.C.) was delivered by this operation.] Surgical incision through the abdominal wall and uterus, performed to extract a fetus.

**Cae·sar·ism** (sē′zə-rĭz′əm) *n.* Military or imperial dictatorship. **—Cae′sar·ist** *n.* **—Cae′sar·is′tic** *adj.*

**caesar salad** *n.* [After *Caesar's,* a restaurant in Tijuana, Mexico.] A tossed green salad with anchovies, croutons, and grated cheese and a dressing of olive oil, lemon juice, and a raw or coddled egg.

**cae·si·um** (sē′zē-əm) *n. var. of* CESIUM.

**caes·pi·tose** (sĕs′pĭ-tōs′) *adj. var. of* CESPITOSE.

**caes·tus** (sĕs′təs) *n. var. of* CESTUS².

**cae·su·ra** *also* **ce·su·ra** (sĭ-zho͝or′ə, -zo͝or′ə) *n., pl.* **-su·ras** *or* **-su·rae** (-zho͝or′ē, -zo͝or′ē) [Lat. < *caedere,* to cut off.] **1.** A pause in a line of verse dictated by sense or natural speech rhythm rather than by metrics. **2.** A break in a line of Latin and Greek prosody caused by the ending of a word within a foot, esp. when it coincides with a sense division. **3.** *Mus.* A pause or breathing at a point of rhythmic division in a melody. **—cae·su′ral, cae·su′ric** *adj.*

**ca·fé** *also* **ca·fe** (kă-fā′, kə-) *n.* [Fr., café, coffee < Turk. *kahve.* —see COFFEE.] A coffeehouse, restaurant, or bar.

**ca·fé au lait** (kă-fā′ ō lā′) *n.* [Fr., coffee with milk.] **1.** Coffee with hot milk. **2.** A light coffee color.

**café fil·tre** (fĭl′trə) *n.* [Fr., filter coffee.] Coffee made by passing hot water through ground coffee and a filtering device underneath.

**café noir** (nwär′) *n.* [Fr., black coffee.] Coffee without cream or milk.

**caf·e·te·ri·a** (kăf′ĭ-tîr′ē-ə) *n.* [Sp. *cafetería,* coffee shop < *café,* coffee < Turk. *kahve.* —see COFFEE.] A restaurant in which customers are served at a counter and carry their meals on trays to tables.

**caf·feine** *also* **caf·fein** (kă-fēn′, kăf′ēn′, kăf′ē-ĭn) *n.* [G. *Kaffein* < *Kaffee,* coffee < Fr. *café.* —see CAFÉ.] A bitter white alkaloid, $C_8H_{10}N_4O_2{\cdot}H_2O$, found in coffee, tea, and kola nuts and used as a stimulant and diuretic.

**caf·tan** (kăf′tăn′, kăf-tăn′) *n.* [R. *kaftan* < Turk.] A full-length, longsleeved, often loose-fitting tunic worn chiefly in the Near East.

**cage** (kāj) *n.* [ME < OFr. < Lat. *cavea* < *cavus,* hollow.] **1.** A structure for confining birds or animals, enclosed on at least one side by a wire or barred grating in order to let in air and light. **2.** An enclosure for confining prisoners. **3.** Something resembling a cage, as a framework having a cagelike appearance or construction. **4.** An elevator car. **5.** *Baseball.* **a.** A backstop used for batting practices. **b.** A catcher's mask. **6. a.** *Basketball.* BASKET 3a. **b.** A hockey goal, made of a network frame. —*vt.* **caged, cag·ing, cag·es.** To lock up or confine in or as if in a cage.

**cage·ling** (kāj′lĭng) *n.* A pet bird kept in a cage.

**cag·ey** *also* **ca·gy** (kā′jē) *adj.* **-i·er, -i·est.** [Orig. unknown.] **1.** Careful : wary. **2.** Shrewd : crafty. **—cag′i·ly** *adv.* **—cag′i·ness** *n.*

**ca·hier** (kä-yā′) *n.* [Fr. < OFr. *caier* < Lat. *quaterni,* group of four < *quattuor,* four.] **1.** A number of pages gathered together, as in a loose-leaf binder : NOTEBOOK. **2.** A report, as of the proceedings of a meeting or convention.

**ca·hoots** (kə-ho͞ots′) *pl.n.* [Perh. < Fr. *cahute,* cabin < OFr.] *Informal.* Questionable or illegal collaboration <in cahoots with a dishonest attorney>

**Ca·hui·lla** (kə-wē′ə) *n., pl.* **Cahuilla** *or* **-llas.** [Sp., of Am. Indian orig.] A Shoshonean language of southeastern California.

**cai·man** *also* **cay·man** (kā′mən, kā-măn′, kī-măn′) *n., pl.* **-mans.** [Sp. *caimán* < Carib *acayuman.*] A tropical American crocodilian of the genus *Caiman* and related genera, resembling the alligator.

**Cain** (kān) *n.* [After *Cain,* the son of Adam and Eve, who murdered his brother Abel.] A murderer.

**-caine** *suff.* [< COCAINE.] A synthetic alkaloid anesthetic <*eucaine*>

**cai·no·to·pho·bi·a** (kā-nō′tə-fō′bē-ə) *n.* [Gk. *kainotēs,* newness + PHOBIA.] Abnormal fear of newness.

**ca·ïque** (kä-ēk′) *n.* [Fr. < Ital. *caicco* < Turk. *kayiuk.*] **1.** A long narrow rowboat used in the Middle East. **2.** A small sailing vessel used in the eastern Mediterranean.

caïque

**caird** (kârd) *n.* [Sc. Gael. *ceard,* craftsman < OIr. *cerd,* artist.] *Scot.* An itinerant tinker or handyman.

**cairn** (kârn) *n.* [ME *carne* < Sc. Gael. *carn.*] A mound of stones erected as a memorial or landmark. **—cairned** (kârnd) *adj.*

**cairn·gorm** (kârn′gôrm′) *n.* [After the *Cairngorm* Mountains, Scotland.] A smoky-brown or yellow variety of quartz, used as a semiprecious gemstone.

**Cairn terrier** *n.* [So called because it hunts among cairns.] A small dog orig. bred in Scotland, with a broad head and a rough, shaggy coat.

**cais·son** (kā′sŏn′, -sən) *n.* [Fr., aug. of *caisse,* box < OProv. *caisa* < Lat. *capsa* < *capere,* to hold.] **1.** A watertight structure in which underwater construction is effected. **2.** CAMEL 2. **3.** A floating structure used for closing off the entrance to a dock or canal lock. **4.** A large box open at the top and one side, designed to fit against the side of a ship and used to repair damaged hulls under water. **5. a.** A large box used to hold ammunition. **b.** A large, usu. two-wheeled horse-drawn vehicle used at one time to transport ammunition.

**caisson disease** *n.* A disorder, esp. in divers and underwater workers, caused by release of nitrogen bubbles in the tissues and blood upon too rapid a return from high pressure to atmospheric pressure, marked by pains in the joints, cramps, paralysis, and eventual death unless treated by gradual decompression.

**cai·tiff** (kā′tĭf) *n.* [ME *caitif* < Norman Fr. < Lat. *captivus,* prisoner. —see CAPTIVE.] A base, contemptible coward. **—cai′tiff** *adj.*

**Ca·jan** (kā′jən) *n. var. of* CAJUN.

**ca·jole** (kə-jōl′) *vt.* **-joled, -jol·ing, -joles.** [Fr. *cajoler.*] To coax gently and persistently : WHEEDLE <*cajoled* me into attending the party> **—ca·jol′er** *n.* **—ca·jol′er·y** (-jō′lə-rē) *n.* **—ca·jol′ing·ly** *adv.*

**Ca·jun** *also* **Ca·jan** (kā′jən) *n.* [Alteration of ACADIAN.] A native of Louisiana held to be a descendant of French exiles from Acadia.

**cake** (kāk) *n.* [ME < ON *kaka.*] **1.** A sweet baked mixture of flour, liquid, eggs, and other ingredients shaped into a loaf or rounded layers. **2.** A flat, thin mass of dough or batter that is baked or fried, as a pancake. **3.** A patty of fried food <a fish *cake*> **4.** A shaped or molded piece, as of soap. —*vt.* & *vi.* **caked, cak·ing, cakes.** To form into a cake.

**cake·walk** (kāk′wôk′) *n.* **1.** A promenade, once performed by American blacks as an entertainment in which those executing the most complex steps won cakes as prizes. **2. a.** A strutting dance based on the cakewalk. **b.** The music for this dance. **—cake′walk′** *v.* **(-walked, -walk·ing, -walks). —cake′walk′er** *n.*

**Cal·a·bar bean** (kăl′ə-bär) *n.* [After *Calabar,* Nigeria.] The dark-brown, poisonous seed of a woody vine, *Physostigma venenosum* of tropical Africa, that is the source of the drug physostigmine

**cal·a·bash** (kăl′ə-băsh′) *n.* [Fr. *calabasse,* gourd < Sp. *calabaza.*] **1.** An Old World vine, *Lagenaria siceraria* yielding large, hard-shelled gourds. **2.** A tropical American tree, *Crescentia cujete,* bearing large, rounded fruit. **3.** The fruit of a calabash. **4.** A utensil, as a dish or ladle, made from the fruit of a calabash.

**†cal·a·boose** (kăl′ə-bo͞os′) *n.* [Louisiana Fr. *calabouse* < Sp. *calabozo,* dungeon.] *Regional.* A jail.

**ca·la·di·um** (kə-lā′dē-əm) *n.* [NLat. *Caladium,* genus name < Malay *kĕladi,* an aroid.] A tropical plant of the genus *Caladium,* widely cultivated as house plants for their colorful foliage.

**cal·a·man·co** (kăl′ə-măng′kō) *n., pl.* **-coes.** [Sp. *calamaco,* perh. < Lat. *calamancus,* felt cap.] A glossy woolen fabric with a checked pattern on one side.

**cal·a·man·der** (kăl′ə-măn′dər) *n.* [Prob. < Du. *kalamander.*] The hard black-and-brown-striped wood of certain tropical Asiatic trees of the genus *Diospyros,* used in furniture-making.

**cal·a·mi** (kăl′ə-mī′) *n. pl. of* CALAMUS.

**cal·a·mine** (kăl′ə-mīn′, -mĭn) *n.* [Fr. < Med. Lat. *calamina,* alteration of Lat. *cadmia* < Gk. *kadmeia* < *kadmeios,* Theban < *Kadmos,* Cadmus.] **1.** A white or occas. iron- or copper-stained mineral, essentially $Zn_4Si_2O_7(OH)_2 \cdot H_2O$. **2.** A pink, odorless, tasteless powder of zinc oxide with a small amount of ferric oxide, dissolved in mineral oils and used in skin lotions.

**cal·a·mint** (kăl′ə-mĭnt′) *n.* [ME *calaminte* < OFr. *calamente* < Med. Lat. *calamentum* < Lat. *calaminthe* < Gk. *kalaminthē.*] Any of several aromatic plants of the genus *Satureja,* esp. the Eurasian variety *S. calamintha,* with purplish or pink flower clusters.

**cal·a·mite** (kăl′ə-mīt′) *n.* [NLat. *Calamites,* genus name < Lat. *calamus,* reed < Gk. *kalamos.*] Any of various extinct treelike plants of the genus *Calamites,* similar to but much larger than the horsetails and known only as fossils.

**ca·lam·i·tous** (kə-lăm′ĭ-təs) *adj.* Causing or involving calamity. **—ca·lam′i·tous·ly** *adv.* **—ca·lam′i·tous·ness** *n.*

**ca·lam·i·ty** (kə-lăm′ĭ-tē) *n., pl.* **-ties.** [ME *calamite* < OFr. < Lat. *calamitas.*] **1.** An extremely serious event fraught with terrible loss and affliction. **2.** A state of dire distress or misfortune.

**cal·a·mon·din** (kăl′ə-mŏn′dĭn) *n.* [Tagalog *kalamunding.*] **1.** A citrus tree, *Citrus mitis* of the Philippine Islands. **2.** The acid, globular fruit of the calamondin, resembling a small orange.

**cal·a·mus** (kăl′ə-məs) *n., pl.* **-mi** (-mī′) [Lat., reed < Gk. *kalamos.*] **1. a.** The sweet flag. **b.** The aromatic root of the sweet flag. **2.** Any of various tropical Asiatic palms of the genus *Calamus,* from some of which rattan is obtained. **3.** QUILL 1.

**ca·lan·do** (kə-län′dō) *adj.* [Ital. < *calare,* to let down < Lat. *chalare* < Gk. *khalan.*] *Mus.* Gradually decreasing in tempo and volume. **—ca·lan′do** *adv.*

**ca·lash** (kə-lăsh′) *also* **ca·lèche** (-lĕsh′) *n.* [Fr. *calèche* < G. *Kalesche,* of Slavic orig.] **1. a.** A carriage with low wheels and a collapsible top. **b.** The top of such a carriage. **2.** A folding bonnet worn in the late 18th cent. by women.

**cal·a·thus** (kăl′ə-thəs) *n., pl.* **-thi** (-thī′) [Lat. < Gk. *kalathos.*] A vase-shaped basket represented in Greek painting and sculpture.

**calc-** *pref. var. of* CALCI-.

**cal·ca·ne·o·cu·boid ligament** (kăl-kā′nē-ō-kyo͞o′boid′) *n.* The ligament connecting the calcaneus and the cuboid bones.

**cal·ca·ne·us** (kăl-kā′nē-əs) *also* **cal·ca·ne·um** (-nē-əm) *n., pl.* **-ne·i** (-nē-ī′) *also* **-ne·a** (-nē-ə) [LLat., heel < Lat. *calcaneum* < *calx.*] The quadrangular bone at the back of the tarsus. **—cal·ca′ne·al** *adj.*

**cal·car**[1] (kăl′kär′) *n., pl.* **cal·car·i·a** (kăl-kâr′ē-ə) [Lat., spur < *calx,* heel.] An anatomical spur or spurlike projection.

**cal·car**[2] (kăl′kär′) *n.* [Ital. *calcara* < Lat. *calcaria* < *calx,* lime. —see CALX.] A furnace once used in glassmaking for preparing frit.

**cal·car·e·ous** (kăl-kâr′ē-əs) *adj.* [Lat. *calcarius* < *calx,* lime. —see CALX.] Made up of, having, or typical of calcium carbonate, calcium, or limestone : CHALKY.

**cal·ca·rine fissure** (kăl′kə-rīn′) *n.* A calcarine sulcus.

**calcarine sulcus** *n.* [CALCAR[1] + -INE.] A sulcus on the occipital lobe of the brain.

**cal·ce·i·form** (kăl′sē-ə-fôrm′) *adj.* [Lat. *calceus,* shoe + -FORM.] *Bot.* Slipper-shaped : calceolate.

**cal·ce·o·lar·i·a** (kăl′sē-ə-lâr′ē-ə) *n.* [NLat. *Calceolaria,* genus name < Lat. *caleolus,* small shoe. —see CALCEOLATE.] A plant of the genus *Calceolaria,* native to tropical America and having yellow, speckled, slipper-shaped flowers.

**cal·ce·o·late** (kăl′sē-ə-lāt′) *adj.* [< Lat. *calceolus,* dim. of *calceus,* shoe.] Shaped like a slipper, as the blossoms of some orchids.

**cal·ces** (kăl′sēz′) *n. var. pl. of* CALX.

**calci-** *or* **calc-** *pref.* [< Lat. *calx, calc-,* lime < Gk. *khalix,* pebble.] Calcium : calcium salt <*calciferous*>

**cal·cic** (kăl′sĭk) *adj.* Made up of, having, derived from, or relating to calcium or lime.

**cal·ci·cole** (kăl′sĭ-kōl′) *n.* [CALCI- + Lat. *colere,* to inhabit.] A plant thriving in lime-rich soil. **—cal·cic′o·lous** (-sĭk′ə-ləs) *adj.*

**cal·ci·co·sis** (kăl′sĭ-kō′sĭs) *n.* [CALCI- + *-cosis* (as in *silicosis*).] A pneumoconiosis caused by inhalation of calcium carbonate dust.

**cal·cif·er·ol** (kăl-sĭf′ə-rôl′, -rōl′) *n.* [CALCIF(EROUS) + (ERGOST)EROL.] Vitamin $D_2$

**cal·cif·er·ous** (kăl-sĭf′ər-əs) *adj.* Of, forming, or having calcium or calcium carbonate.

**cal·cif·ic** (kăl-sĭf′ĭk) *adj.* Producing salts of lime, as in the formation of eggshells.

**cal·ci·fi·ca·tion** (kăl′sə-fĭ-kā′shən) *n.* **1. a.** Impregnation with calcium or calcium salts, as with calcium carbonate. **b.** Hardening, as of tissue, by such impregnation. **2.** A calcified substance or part.

**cal·ci·fuge** (kăl′sə-fyo͞oj′) *n.* A plant that does not thrive in lime-rich soil. **—cal·cif′u·gal** (-sĭf′yə-gəl), **cal·cif′u·gous** (-yə-gəs) *adj.*

**cal·ci·fy** (kăl′sə-fī′) *vt. & vi.* **-fied, -fy·ing, -fies.** To make or become stony or chalky by deposition of calcium salts.

**cal·ci·mine** (kăl′sə-mīn′) *n.* [Orig. unknown.] A white or tinted liquid containing zinc oxide, water, glue, and coloring matter, used as a coating wash for walls and ceilings. *—vt.* **-mined, -min·ing, -mines.** To cover or wash with calcimine.

**cal·cine** (kăl-sīn′, kăl′sīn′) *v.* **-cined, -cin·ing, -cines.** [ME *calcinen* < OFr. *calciner* < Med. Lat. *calcinare* < Lat. *calx,* lime. —see CALX.] *—vt.* To heat (a substance) to a high temperature but below the melting or fusing point, causing loss of moisture, reduction, or oxidation. *—vi.* To undergo calcination. **—cal′ci·na′tion** (-sə-nā′shən) *n.*

**cal·ci·no·sis** (kăl′sə-nō′sĭs) *n.* [CALCI- + -OSIS.] An abnormal condition in which calcium salts are deposited in a tissue of the body, as the skin.

**cal·cite** (kăl′sīt′) *n.* A common crystalline form of natural calcium carbonate, the basic constituent of marble, limestone, and chalk. **—cal·cit′ic** (-sĭt′ĭk) *adj.*

**cal·ci·to·nin** (kăl′sĭ-tō′nĭn) *n.* [CALCI- + TON(IC) + -IN.] A hormone that functions in calcium metabolism, produced by the thyroid gland.

**cal·ci·um** (kăl′sē-əm) *n.* [NLat. < Lat. *calx,* lime. —see CALX.] *Symbol* **Ca** A silvery metallic element that occurs in bone, shells, limestone, and gypsum and forms compounds used to make plaster, quicklime, cement, and metallurgic and electronic materials; atomic number 20; atomic weight 40.08.

**calcium carbide** *n.* A grayish-black crystalline compound, $CaC_2$, obtained by heating pulverized limestone or quicklime with carbon and used to generate acetylene gas, as a dehydrating agent, and in making graphite and hydrogen.

**calcium carbonate** *n.* A colorless or white crystalline compound, $CaCO_3$, occurring naturally as chalk, limestone, marble, and other forms and used in manufactured products including commercial chalk, medicines, and dentifrices.

**calcium chloride** *n.* A white deliquescent compound, $CaCl_2$, used primarily as a drying agent, refrigerant, and preservative.

**calcium cyanamide** *n.* A gray-black compound, $Ca(CN)_2$, used as a fertilizer and weed killer.

**calcium fluoride** *n.* A white powder, $CaF_2$, used in emery wheels, carbon electrodes, and cements.

**calcium hydroxide** *n.* A soft white powder, $Ca(OH)_2$, used in manufacturing mortar, cements, calcium salts, paints, hard rubber products, and petrochemicals.

**calcium hypochlorite** *n.* A white crystalline solid, $Ca(OCl)_2$, used as a bactericide, fungicide, and bleaching agent.

**calcium light** *n.* LIMELIGHT 1.

**calcium oxide** *n.* A white caustic lumpy powder, CaO, used as a refractory, a flux, and an industrial alkali and in the manufacture of steel and glass, in waste treatment, and in insecticides.

**calcium phosphate** *n.* Any of several phosphate compounds, esp.: **a.** A white crystalline powder, $CaHPO_4$ or $CaHPO_4 \cdot 2H_2O$, used as a food, as a plastic stabilizer, and in glass. **b.** A colorless deliquescent powder, $CaH_4(PO_4)_2 \cdot H_2O$, used in baking powders, as a plant food and plastic stabilizer, and in glass. **c.** A white amorphous powder, $Ca_3(PO_4)_2$, used in ceramics, rubber, fertilizers, plastic stabilizers, and as a food supplement.

**calc·spar** *or* **calc-spar** (kălk′spär′) *n.* [Partial transl. of Swed. *kalkspat* : *kalk,* lime < OSwed. < MLG < Lat. *calx,* lime. + *spat,* spar (mineral).] Calcite.

**calc-tu·fa** (kălk′to͞o′fə, -tyo͞o′fə) *also* **calc-tuff** (-tŭf′) *n.* [CALC(AREOUS) + TUFA.] A porous or spongy deposit of calcium carbonate found in calcareous mineral springs.

**cal·cu·la·ble** (kăl′kyə-lə-bəl) *adj.* **1.** Capable of being calculated or estimated. **2.** Dependable. **—cal′cu·la·bil′i·ty** *n.*

**†cal·cu·late** (kăl′kyə-lāt′) *v.* **-lat·ed, -lat·ing, -lates.** [Lat. *calculare, calculat-* < *calculus,* small stone used in reckoning, dim. of *calx,* small stone for gaming. —see CALX.] *—vt.* **1.** To ascertain by computation : RECKON. **2.** To estimate : evaluate. **3.** To fit or make suitable for a purpose <a motorcycle that is *calculated* to go 90 miles an hour> **4.** *Regional.* **a.** To intend. **b.** To think : suppose. *—vi.* **1.** To execute a mathematical process. **2.** *Regional.* **a.** To suppose. **b.** To count on <We're *calculating* on your help.>

☆ ***syns:*** CALCULATE, CAST, COMPUTE, FIGURE (out), RECKON *v. core meaning* : to ascertain by mathematics <astronomers *calculating* the positions of the planets>

**cal·cu·lat·ed** (kăl′kyə-lā′tĭd) *adj.* **1.** Undertaken after careful estimation of the likely outcome <took *calculated* business risks> **2.** Likely <a strategy *calculated* to succeed> **3.** Determined by mathematical calculation. **—cal′cu·lat′ed·ly** *adv.*

**cal·cu·lat·ing** (kăl′kyə-lā′tĭng) *adj.* **1.** Performing calculations. **2. a.** Crafty : shrewd. **b.** Coldly scheming.

**cal·cu·la·tion** (kăl′kyə-lā′shən) *n.* **1.** The act, process, or result of calculating. **2.** An estimate based on probabilities. **3.** Deliberation : foresight. **—cal′cu·la′tive** *adj.*

**cal·cu·la·tor** (kăl′kyə-lā′tər) *n.* **1.** One who performs calculations. **2.** A keyboard machine that automatically performs mathematical operations. **3.** A set of mathematical tables used in calculating.

**cal·cu·li** (kăl′kyə-lī′) *n. var. pl. of* CALCULUS.

**cal·cu·lous** (kăl′kyə-ləs) *adj.* Relating to, caused by, or having a calculus or calculi.

**cal·cu·lus** (kăl′kyə-ləs) *n., pl.* **-li** (-lī′) *or* **-lus·es.** [Lat., small stone used in reckoning. —see CALCULATE.] **1.** *Pathol.* An abnormal concretion, as a kidney stone, usu. formed of mineral salts. **2.** *Math.* **a.** Analysis or calculation using a special symbolic notation. **b.** The combined mathematics of differential and integral calculus.

**calculus of variations** *n.* The mathematical analysis of the maxima and minima of definite integrals, the integrands of which are functions of independent variables, dependent variables, and the derivatives of one or more dependent variables.

**cal·de·ra** (kăl-dâr′ə, -dîr′ə, kôl-) *n.* [Sp., caldron < LLat. *caldaria.*] A large crater formed by volcanic explosion or the collapse of a volcanic cone.

**cal·dron** *also* **caul·dron** (kôl′drən) *n.* [ME *caudron* < Norman Fr. < LLat. *caldaria* < Lat. *caldarius,* suitable for warming < *calidus,* warm.] A large vessel, as a kettle or vat, used for boiling.

▲ word history: The Latin word *caldaria,* "kettle," has three descendants in Modern English that arrived by three different routes. *Caldron,* spelled *caudron,* is the earliest form, appearing in Middle English in the 14th century. It was borrowed from Norman French, which preserved the original *k* sound of *c* that occurred in the Latin form. The other forms of *caldaria* came from the central French variant of *caudron* and reflect the change of Latin *c* to *ch* characteristic of this dialect. *Chaldron,* from Old French *chauderon,* was at first used in English as a synonym of *caldron* meaning "kettle," but by the 19th century it meant a unit of dry measure used only for coal. The third form is *chowder,* from the modern French word *chaudière,* "pot." It was borrowed in the 18th century with the meaning "fish stew," a meaning it probably developed in the French spoken in Brittany and the Maritime Provinces of Canada. *Chowder* is not the only word for a cooking vessel that has come to denote the contents instead: *lasagna* and *casserole* are two others.

**ca·lèche** (kə-lĕsh′) *n. var. of* CALASH.

**cal·en·dar** (kăl′ən-dər) *n.* [ME *calender* < OFr. *calendier* < Med. Lat. *calendra* < Lat., account book < *kalendae,* calends (from the fact that monthly interest was due on the calends).] **1.** A system of reckoning time in which the beginning, length, and divisions of a year are arbitrarily defined or otherwise established. **2.** A table showing the months, weeks, and days in at least one specific year. **3.** A list or schedule, esp. one arranged chronologically, as of cases on a court docket. **4.** *Obs.* A guide: example. —*vt.* **-dared, -dar·ing, -dars.** To enter on a calendar.

## THREE PRINCIPAL CALENDARS

| GREGORIAN | | HEBREW | | MOSLEM | |
|---|---|---|---|---|---|
| | | Months correspond approximately to those in parentheses. | | Beginning of year retrogresses through the solar year of the Gregorian calendar. | |
| **Name** | **Days** | **Name** | **Days** | **Name** | **Days** |
| January | 31 | Tishri<br>(September–October) | 30 | Muharram | 30 |
| February<br>in leap year | 28<br>29 | Heshvan<br>in some years<br>(October–November) | 29<br>30 | Safar | 29 |
| March | 31 | Kislev<br>in some years<br>(November–December) | 29<br>30 | Rabi I | 30 |
| April | 30 | Tevet<br>(December–January) | 29 | Rabi II | 29 |
| May | 31 | Shevat<br>(January–February) | 30 | Jumada I | 30 |
| June | 30 | Adar*<br>in leap year<br>(February–March) | 29<br>30 | Jumada II | 29 |
| July | 31 | Nisan<br>(March–April) | 30 | Rajab | 30 |
| August | 31 | Iyar<br>(April–May) | 29 | Sha'ban | 29 |
| September | 30 | Sivan<br>(May–June) | 30 | Ramadan | 30 |
| October | 31 | Tammuz<br>(June–July) | 29 | Shawwal | 29 |
| November | 30 | Av<br>(July–August) | 30 | Dhu'l-Qa'dah | 30 |
| December | 31 | Elul<br>(August–September) | 29 | Dhu'l-Hijja<br>in leap year | 29<br>30 |

*Adar is followed in leap year by the intercalary month Veadar or Adar Sheni, having 29 days.

**calendar month** *n.* MONTH 1.

**calendar year** *n.* YEAR 1.

**cal·en·der** (kăl′ən-dər) *n.* [Fr. *calendre* < Med. Lat. *calendra* < Lat. *cylindrus,* roller < Gk. *kulindros* < *kulindein,* to roll.] A machine in which paper or cloth is made smooth and glossy by being pressed through rollers. —*vt.* **-dered, -der·ing, -ders.** To press in a calender. **—cal′en·der·er** *n.*

**ca·len·dri·cal** (kə-lĕn′drĭ-kəl) *also* **ca·len·dric** (-drĭk) *adj.* Of, relating to, or used in a calendar.

**cal·ends** (kăl′əndz, kā′ləndz) *n., pl.* **calends.** [ME *kalendes* < Lat. *kalendae.*] The day of the new moon and first day of the month in the ancient Roman calendar. **—ca·len′dal** (kə-lĕn′dəl) *adj.*

**ca·len·du·la** (kə-lĕn′jə-lə) *n.* [NLat. *Calendule,* genus name < Med. Lat., marigold < Lat. *kalendae,* calends.] A plant of the genus *Calendula,* with orange-yellow flowers.

**cal·en·ture** (kăl′ən-chŏŏr′) *n.* [Sp. *calentura* < *calentar,* to heat < Lat. *calens,* pr.part. of *calēre,* to be warm.] A tropical fever once thought to be caused by the heat.

**calf[1]** (kăf, käf) *n., pl.* **calves** (kăvz, kävz) [ME < OE *cealf.*] **1. a.** A young cow or bull. **b.** The young of certain other mammals, as the elephant or whale. **2.** Calfskin leather. **3.** A large floating chunk of ice split from a glacier, iceberg, or floe. **4.** An awkward, callow youth.

**calf[2]** (kăf, käf) *n., pl.* **calves** (kăvz, kävz) [ME < ON *kalfi.*] The fleshy, muscular back part of the human leg between the knee and ankle.

**calf's-foot jelly** *also* **calves'-foot jelly** (kăvz′fŏŏt′, kăfs′-, kävz′-, käfs′-) *n.* A gelatinous food made by boiling calves' feet.

**calf·skin** (kăf′skĭn′, käf′-) *n.* **1.** The hide of a calf. **2.** Fine leather made from the hide of a calf.

**Cal·i·ban** (kăl′ə-băn′) *n.* The grotesque, brutish slave in Shakespeare's play *The Tempest.*

**cal·i·ber** (kăl′ə-bər) *n.* [OFr. *calibre* < OItal. *calibro* < Ar. *qālib,* mold.] **1. a.** The diameter of the inside of a tube. **b.** The diameter of the bore of a firearm. **c.** The diameter of a bullet or shell. **2.** Degree of worth or distinction <an executive of a high *caliber*>

**cal·i·brate** (kăl′ə-brāt′) *vt.* **-brat·ed, -brat·ing, -brates. 1.** To check, adjust, or standardize systematically the graduations of a quantitative measuring instrument. **2.** To determine the caliber of (a tube). **—cal′i·bra′tion** *n.* **—cal′i·bra′tor** *n.*

**cal·i·bre** (kăl′ə-bər) *n. Chiefly Brit. var. of* CALIBER.

**ca·li·ces** (kă′lĭ-sēz′, kăl′ĭ-) *n. pl. of* CALIX.

**ca·li·che** (kə-lē′chē) *n.* [Sp., pebble in a brick < Lat. *calx,* pebble. —see CALX.] **1. a.** A crude sodium nitrate occurring naturally in Chile, Peru, and the southwestern United States, used as fertilizer. **b.** Sodium nitrate. **2.** A hard soil layer cemented by calcium carbonate and found in deserts and other arid or semiarid regions.

**cal·i·co** (kăl′ĭ-kō′) *n., pl.* **-coes** or **-cos.** [After *Calicut* (Kozhikode), India.] **1.** A coarse, usu. brightly printed cloth. **2.** *Chiefly Brit.* White cotton cloth. **3.** An animal having a mottled coat.

**calico bass** *n.* [From the colored spots on its body.] The black crappie.

**calico bush** *n.* The mountain laurel.

**ca·lif** (kā′lĭf, kăl′ĭf) *n. var. of* CALIPH.

**Cal·i·for·nia laurel** (kăl′ə-fôr′nyə) *n.* An aromatic evergreen tree, *Umbellularia californica* of the North American Pacific Coast, with yellowish-green flower clusters and yellowish-green fleshy fruit.

**California nutmeg** *n.* An evergreen tree, *Torreya californica,* with spiny, pointed leaves and purple-streaked, greenish fruit.

**California poppy** *n.* A plant, *Eschscholtzia californica* of the North American Pacific Coast, with finely divided bluish-green leaves and orange-yellow flowers.

**California quail** *n.* A plump, chunky bird, *Lophortyx californicus* of western North America, with brown and gray plumage and a curving black plume on the crown of the head.

**cal·i·for·ni·um** (kăl′ə-fôr′nē-əm) *n.* [After *California.*] *Symbol* **Cf** A synthetic radioactive element produced in trace quantities by helium isotope bombardment of curium; atomic number 98; longest-lived isotope Cf 25.

**ca·lig·i·nous** (kə-lĭj′ə-nəs) *adj.* [< Lat. *caliginosus* < *caligo,* darkness.] *Archaic.* Dark and gloomy.

**cal·i·pash** (kăl′ə-păsh′, kăl′ə-păsh′) *n.* [Prob. alteration of Sp. *carapacho,* carapace.] An edible, gelatinous, greenish substance lying beneath a turtle's upper shell.

**cal·i·pee** (kăl′ə-pē′, kăl′ə-pē′) *n.* [Prob. alteration of CALIPASH.] An edible, gelatinous, yellowish substance lying above a turtle's lower shell.

**cal·i·per** *also* **cal·li·per** (kăl′ə-pər) [Alteration of CALIBER.] —*n.* **1.** *often* **calipers.** An instrument composed of two curved hinged legs, used for measuring internal and external dimensions. **2.** A large instrument having a fixed and a movable arm on a graduated stock, used for measuring the diameters of logs and similar objects. **3.** A vernier caliper. —*vt. & vi.* **-pered, -per·ing, -pers.** To measure or determine dimensions with calipers.

**ca·liph** *also* **ca·lif** (kā'lĭf, kăl'ĭf) *n.* [ME *calife* < OFr. < Ar. *khalīfa* < *khalafa,* he succeeded.] The secular and religious head of a Moslem state. **—ca'liph·ate'** (kā'lĭ-fāt', kăl'ĭ-, -fĭt) *n.*

**cal·i·sa·ya** (kăl'ĭ-sā'ə) *n.* [Sp., prob. < *Calisaya,* an area of Bolivia where a species of quinine bark was first discovered.] The bark of a tree of the genus *Cinchona,* esp. *C. calisaya,* from which quinine is obtained.

**cal·is·then·ics** (kăl'ĭs-thĕn'ĭks) *n.* [Gk. *kalli-* beautiful (< *kallos,* beauty) + *sthenos,* strength.] **1.** *(sing. or pl. in number).* Gymnastic exercises designed to develop muscular tone and promote physical well-being. **2.** *(sing. in number).* The practice of calisthenics. **—cal'is·then'ic** *adj.*

**ca·lix** (kā'lĭks, kăl'ĭks) *n., pl.* **ca·li·ces** (kā'lĭ-sēz', kăl'ĭ-) [Lat. *calix, calic-,* cup.] CHALICE 2.

**calk¹** (kôk) *n.* [Short for obs. *calkin* < ME *kakun* < MDu. *kalkoen,* hoof < OFr. *calcain,* heel < Lat. *calcaneum* < *calx.*] **1.** A pointed extension on the toe or heels of a horseshoe, designed to prevent slipping. **2.** A spiked plate attached to the bottom of a shoe to prevent slipping and preserve the sole. *—vt.* **calked, calk·ing, calks. 1.** To supply with or fasten on calks. **2.** To cut or injure with a calk.

**calk²** (kôk) *v. var. of* CAULK.

**call** (kôl) *v.* **called, call·ing, calls.** [ME *callen* < ON *kalla.*] *—vt.* **1.** To cry out loudly : PROCLAIM <*called* my name> **2.** To summon <*call* them to supper> **3.** To convoke or convene (a meeting). **4.** To summon to a given career or pursuit <was *called* to the priesthood> **5.** To awaken. **6.** To telephone (someone). **7. a.** To make a characteristic cry, as a bird. **b.** To lure by imitating such a cry. **8.** To name <What will you *call* the child?> **9.** To estimate as being : CONSIDER <I *call* that unfair.> **10.** To designate as such : LABEL <Nobody *calls* me a cheat.> **11.** To bring to action or under consideration <*call* a case to trial> **12.** To demand payment of (e.g., a loan). **13. a.** To stop or postpone (a game) because of inclement weather, darkness, or other adverse conditions. **b.** To indicate a decision in regard to (e.g., a foul or play). **c.** *Baseball.* To indicate a decision in regard to (a pitch). **d.** To choose or select (plays to be run). **14. a.** To predict (the outcome of a billiard shot) before playing. **b.** To ask (another billiard player) to do so. **15.** To predict accurately <You've *called* the game's outcome every time.> **16.** To demand to see the hand of (a poker opponent) by equaling his or her bet. **17.** To demand that a person make good a boast or support a statement with facts <*call* one's bluff> **18.** To shout (directions) in rhythm for square dances. *—vi.* **1.** To telephone. **2.** To pay a short visit. **3.** To attract attention by shouting. **—call back.** To telephone again. **—call down. 1.** To invoke, as from heaven. **2.** To find fault with : BERATE. **—call for. 1.** To go and get or stop for. **2.** To be appropriate for : WARRANT <This *calls for* a toast.> **3.** To require <work that *calls for* accuracy> **—call forth.** To evoke. **—call in. 1.** To take out of circulation <*calling in* silver dollars> **2.** To summon for help or advice. **—call off. 1.** To cancel or postpone. **2.** To restrain or recall. **—call on** (or **upon).** To request or order (someone) to do something. **—call out.** To cause to assemble : SUMMON <*call out* the troops> **—call up. 1.** To summon into military service <was *called up* for active duty> **2.** To remember or cause to remember. **3.** To bring forth for action or discussion. *—n.* **1.** A shout. **2. a.** The characteristic cry of an animal, esp. a bird. **b.** An instrument or sound made to imitate such a cry, used as a lure <a duck *call*> **3.** Need or occasion <no *call* for that remark> **4.** Demand <not much *call* for manual typewriters today> **5.** A claim on a one's time or life <the *call* of duty> **6.** A short visit, esp. one made as a formality or for business purposes. **7.** A summons or invitation. **8. a.** A signal, as made by a horn or bell. **b.** The sounding of a horn to encourage hounds during a hunt. **9.** A vocation, as to the ministry. **10.** A roll call. **11.** A notice of rehearsal times posted in a theater. **12.** The decision of a sports umpire or referee. **13.** An instruction in square dancing to begin a different step or set. **14.** A demand or request for payment of a debt, as by redeeming bonds. **15. a.** An agreement in which a trader may, for a fee, buy a certain quantity of a stock or commodity for a specified price within a limited period of time. **b.** A demand for payment due on stock bought on margin when the value has shrunk. **—call into question.** To raise doubts about. **—call to mind.** To remind of. **—close call.** A narrow escape. **—on call. 1.** Payable on demand. **2.** Available whenever summoned <doctors *on call*> **—within call.** Close enough to come if summoned.

☆ ***syns:*** CALL, CONVENE, CONVOKE, MUSTER, SUMMON *v. core meaning* : to demand to appear, come, or assemble <*call* a meeting of the legislature>

**cal·la** (kăl'ə) *n.* [NLat. *Calla,* genus name < Gk. *kallaia,* wattle of a cock.] **1.** A tropical or semitropical plant of the genus *Zantedeschia,* esp. *Z. aethiopica,* widely cultivated for its large white spathe that encloses a yellow spadix. **2.** A marsh plant, *Calla palustris* of the North Temperate Zone, with small, densely clustered greenish flowers partially enclosed in a spreading white spathe.

**call·back** (kôl'băk') *n.* RECALL 6.

**call·board** (kôl'bôrd', -bōrd') *n.* A bulletin board backstage in a theater for posting instructions and notices.

**call·boy** (kôl'boi') *n.* **1.** One who tells theatrical performers when it is time for them to go on stage. **2.** A bellboy.

**call·er¹** (kô'lər) *n.* One that calls.

**cal·ler²** (kăl'ər) *adj.* [ME *calour.*] *Scot.* **1.** Fresh. —Used of food, esp. fish. **2.** Cool and refreshing, as a breeze.

**call girl** *n.* A prostitute hired by telephone.

**call house** *n.* A house of prostitution.

**cal·lig·ra·phy** (kə-lĭg'rə-fē) *n.* [Fr. *calligraphie* < Gk. *kalligraphia,* beautiful writing : *kalli-,* beautiful (< *kallos,* beauty) + *graphein,* to write.] **1.** The art of fine handwriting. **2.** Handwriting. **—cal·lig'ra·pher, cal·lig'ra·phist** *n.* **—cal'li·graph'ic** (kăl'ĭ-grăf'ĭk) *adj.*

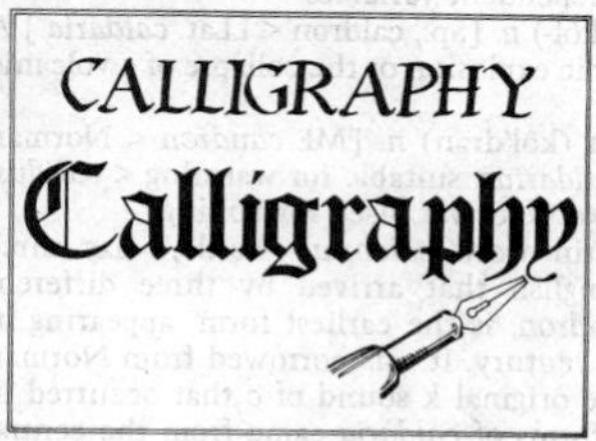

calligraphy

**call-in** (kôl'ĭn') *adj.* Inviting listeners to make broadcast telephone calls <a *call-in* talk show>

**call·ing** (kô'lĭng) *n.* **1.** A strong inner urge or impulse. **2.** An occupation, profession, or career <the *calling* of the priesthood>

**calling card** *n.* A card bearing one's name and often one's address and telephone number, used for social or business purposes.

**Cal·li·o·pe** (kə-lī'ə-pē') *n.* [Lat. < Gk. *Kalliopē* : *kalli-,* beautiful (< *kallos,* beauty) + *ops,* voice.] **1.** *Gk. Myth.* The Muse of epic poetry. **2. calliope** (*also* kăl'ē-ōp'). A musical instrument equipped with steam whistles, played from a keyboard.

**cal·li·op·sis** (kăl'ē-ŏp'sĭs) *n.* The coreopsis.

**cal·li·per** (kăl'ə-pər) *n.* & *v. var. of* CALIPER.

**cal·li·pyg·i·an** (kăl'ə-pĭj'ē-ən) *also* **cal·li·py·gous** (-pī'gəs) *adj.* [Gk. *kallipugos* : *kalli-,* beautiful (< *kallos,* beauty) + *pugē,* buttocks.] Having beautifully proportioned buttocks.

**Cal·lis·to** (kə-lĭs'tō) *n.* [Lat. < Gk. *Kallistō* < *kallistos,* superl. of *kalos,* beautiful.] **1.** *Gk. Myth.* A nymph loved by Zeus and hated by Hera, who changed her into a bear, after which Zeus placed her in the sky as the constellation Ursa Major. **2.** One of the moons of Jupiter, the largest known moon of any planet.

**call letters** *pl.n.* The code letters or numbers used to identify a radio or television transmitting station.

**call loan** *n.* A loan repayable on demand at any time.

**call market** *n.* The market for call money.

**call money** *n.* Money lent by banks, usu. to stockbrokers, subject to repayment on demand at any time.

**call number** *n.* A library number used to classify a book and indicate its placement on the shelves.

**cal·lose** (kăl'ōs') *n.* [< Lat. *callosus,* callous.] A complex branched carbohydrate component of plant cell walls.

**cal·los·i·ty** (kə-lŏs'ĭ-tē-) *n., pl.* **-ties.** [ME *callosite* < OFr. *callosité* < Lat. *callositas* < *callosus,* callous.] **1.** The state of being calloused. **2.** Hardheartedness. **3.** A callus.

**cal·lous** (kăl'əs) *adj.* [ME < OFr. *cailleuse* < Lat. *callosus* < *callum,* hard skin.] **1.** Having calluses. **2.** Emotionally hardened. *—vt.* & *vi.* **-loused, -lous·ing, -lous·es.** To make or become callous. ***usage:*** Although the verb and adjective forms are spelled *callous,* the noun form is spelled *callus.* **—cal'lous·ly** *adv.* **—cal'lous·ness** *n.*

**cal·low** (kăl'ō) *adj.* [ME *calwe,* bald < OE *calu.*] **1.** Immature and inexperienced <a *callow* youth> **2.** Not yet having feathers, as a bird : UNFLEDGED. **—cal'low·ness** *n.*

**call rate** *n.* The rate of interest charged on call loans.

**cal·lus** (kăl'əs) *n., pl.* **-lus·es.** [Lat.] **1. a.** A localized thickening and enlargement of the horny layer of the skin. **b.** The hard bony tissue surrounding the ends of a fractured bone. **2.** *Bot.* Hardened tissue that develops over a wound or the cut end of a woody stem. *—vi.* **-lused, -lus·ing, -lus·es.** To form or develop a callus.

**calm** (käm) *adj.* **-er, -est.** [ME *calme* < OFr. < OItal. *calma* < LLat. *cauma,* heat of the day < Gk. *kauma* < *kaiein,* to burn.] **1.** Almost or utterly motionless : UNDISTURBED. **2.** Not excited or agitated : COMPOSED. *—n.* **1.** Absence or cessation of motion. **2.** Serenity : peace. **3.** Little or no wind. *—vt.* & *vi.* **calmed, calm·ing, calms.** To make or become calm. **—calm'ly** *adv.* **—calm'ness** *n.*

☆ ***syns:*** CALM, PEACEFUL, PLACID, SERENE, TRANQUIL *adj. core meaning* : free of movement, noise, or disturbing emotion <spent a *calm* night asleep> While CALM is the most general term implying freedom from agitation, PEACEFUL and TRANQUIL stress undisturbed quietude <*peaceful* times><a *tranquil* lifestyle> PLACID suggests not being easily shaken physically or emotionally <a *placid*

lake><a *placid* disposition> SERENE suggests a lofty, almost spiritual calm <a *serene* smile>

**cal·ma·tive** (kă'mə-tĭv, käl'mə-) *adj.* Having relaxing or pacifying properties : SEDATIVE. **—calm'a·tive** *n.*

**cal·o·mel** (kăl'ə-mĕl', -məl) *n.* [Fr. : Gk. *kalos,* beautiful + *melas,* black.] A white, tasteless compound, $Hg_2Cl_2$, used as a purgative.

**cal·o·re·cep·tor** (kăl'ə-rĭ-sĕp'tər) *n.* [Lat. *calor,* heat + RECEPTOR.] A sensory receptor that detects warmth.

**ca·lor·ic** (kə-lôr'ĭk, -lŏr'-) *adj.* [Fr. *calorique* < Lat. *calor,* heat.] Of or relating to heat or calories. —*n.* A hypothetically indestructible, highly elastic, self-repellent, all-pervading fluid once believed to be responsible for production, possession, and transfer of heat. **—ca·lor'i·cal·ly** *adv.*

**cal·o·rie** (kăl'ə-rē) *n.* [Fr. < Lat. *calor,* heat.] **1.** Any of several approx. equal units of heat, each measured as the quantity of heat required to raise the temperature of 1 gram of water by 1°C from a standard initial temperature, esp. from 3.98°C, 14.5°C, or 19.5°C, at 1 atmosphere pressure. **2.** The unit of heat equal to 1/100 the quantity of heat required to raise the temperature of 1 gram of water from 0°C to 100°C at 1 atmosphere pressure. **3.** The unit of heat equal to the amount of heat required to raise the temperature of 1 kilogram of water by 1°C at 1 atmosphere pressure. **4.** The unit of heat equal to 4.184 joules.

**cal·o·rif·ic** (kăl'ə-rĭf'ĭk) *adj.* [Fr. *calorifique* < Lat. *calorificus* : *calor,* heat + *facere,* to make.] Of or producing heat or calories.

**cal·o·rim·e·ter** (kăl'ə-rĭm'ĭ-tər) *n.* [Lat. *calor,* heat + -METER.] **1.** A device for measuring heat. **2.** The part of a calorimeter, usu. a container for holding a sample, in which the heat measured causes a change of state. **—ca·lor'i·met'ric** (kə-lôr'ə-mĕt'rĭk, -lŏr'-) *adj.* **—ca·lor'i·met'ri·cal·ly** *adv.*

**cal·o·rim·e·try** (kăl'ə-rĭm'ĭ-trē) *n.* [Lat. *calor,* heat + -METRY.] Measurement of the amount of heat evolved or absorbed in a chemical reaction, change of state, or formation of a solution.

**ca·lotte** (kə-lŏt') *n.* [Fr. dim. of OFr. *cale,* cap, of Germanic orig.] **1.** A skullcap, esp. one worn by members of the Roman Catholic clergy. **2.** *Biol.* A caplike structure.

**ca·loy·er** (kə-loi'ər, kăl'ə-yər) *n.* [Fr. < obs. Ital. *caloiero* < LGk. *kalogēros,* venerable : Gk. *kalos,* beautiful + *gēros,* old age.] An Eastern Orthodox monk.

**cal·pac** *also* **cal·pack** (kăl'păk', kăl-păk') *n.* [Turk. *kalpāk.*] A large black, usu. felt or sheepskin cap worn in Turkey, Armenia, and other Near Eastern regions.

**calque** (kălk) *n.* [Fr. < *calquer,* to copy < Ital. *calcare,* to press < Lat., to tread on < *calx,* heel.] **1.** Semantic borrowing in which a word is given an extended meaning by analogy with that of a word having the same basic meaning in another language. **2.** A loan translation. —*vt.* **calqued, calqu·ing, calques.** To model (a word's meaning) on that of an analogous word in another language.

**cal·trop** *also* **cal·trap** (kăl'trəp, kôl'-) *n.* [ME *calketrappe,* partly < Norman Fr., and partly OE < *calcatrippe,* thistle, both < Med. Lat. *calcatrippa,* thistle.] **1.** An iron ball with four projecting spikes so arranged that when three of the spikes are on the ground, the fourth points upward, once used to delay the advance of cavalry and infantry. **2.** Any of several plants with spiny burs or bracts, as members of the genera *Tribulus* and *Kallstroemia.*

**cal·u·met** (kăl'yə-mĕt', -mĭt, kăl'yə-mĕt') *n.* [Canadian Fr. < dial. Fr., straw < LLat. *calamellus,* dim. of Lat. *calamus,* reed < Gk. *kalamos* straw.] A long-stemmed ornamented pipe used in ceremonies by North American Indians.

**ca·lum·ni·ate** (kə-lŭm'nē-āt') *vt.* **-at·ed, -at·ing, -ates.** [Lat. *calumniari, calumniat-* < *calumnia,* calumny.] To make malicious, false statements about : SLANDER. **—ca·lum'ni·a'tion** *n.* **—ca·lum'ni·a'tor** *n.*

**ca·lum·ni·ous** (kə-lŭm'nē-əs) *adj.* Containing or implying calumny : SLANDEROUS. **—ca·lum'ni·ous·ly** *adv.*

**cal·um·ny** (kăl'əm-nē) *n., pl.* **-nies.** [ME *calumnie* < OFr. *calomnie* < Lat. *calumnia.* < *calvi,* to deceive.] **1.** A false statement maliciously made to injure someone. **2.** Utterance of calumny : SLANDER.

**cal·va·dos** (kăl'və-dōs') *n.* [Fr., after *Calvados,* a department of France.] A French brandy made from apples.

**cal·var·i·um** (kăl-vâr'ē-əm) *n., pl.* **-i·ums** *or* **-i·a** (-ē-ə) [NLat. < Lat. *calvaria,* skull < *calvus,* bald.] The superior part of the skull that lacks the lower jaw and facial parts.

**cal·va·ry** (kăl'və-rē) *n., pl.* **-ries.** [Fr. *calvaire* < *Calvaire,* Calvary, the hill outside Jerusalem where Jesus Christ was crucified.] **1.** A sculptured depiction of the Crucifixion. **2.** ORDEAL 1.

**Calvary cross** *n. Heraldry.* A Latin cross set on three steps.

**calve** (kăv, käv) *v.* **calved, calv·ing, calves.** [ME *calven* < OE *calfian.*] —*vi.* **1.** To give birth to a calf. **2.** To break up and lose a portion of itself. —Used of a glacier or an iceberg. —*vt.* **1.** To give birth to (a calf). **2.** To set loose (a mass of ice).

**calves[1]** (kăvz, kävz) *n. pl. of* CALF[1].

**calves[2]** (kăvz, kävz) *n. pl. of* CALF[2].

**calves'-foot jelly** (kăvz'fo͝ot', kăfs'-, kävz'-, käfs'-) *n. var. of* CALF'S-FOOT JELLY.

**Cal·vin·ism** (kăl'vĭ-nĭz'əm) *n.* The religious doctrines of John Calvin, stressing the omnipotence of God and the salvation of the elect by God's grace alone. **—Cal'vin·ist** *n.* **—Cal'vin·is'tic** *adj.* **—Cal'vin·is'ti·cal·ly** *adv.*

**calx** (kălks) *n., pl.* **calx·es** *or* **cal·ces** (kăl'sēz') [Lat., lime < Gk. *khalix,* pebble.] **1.** The crumbly residue left after a mineral or metal has been calcined or roasted. **2.** Lime : chalk. **3.** Calcium oxide.

**ca·ly·ces** (kā'lĭ-sēz', kăl'ĭ-) *n. var. pl. of* CALYX.

**ca·ly·cine** (kā'lĭ-sĭn', kăl'ĭ-) *adj.* Of, relating to, or like a calyx.

**ca·ly·cle** (kā'lĭ-kəl, kăl'ĭ-) *n.* [Fr. *calicule* < Lat. *calyculus,* dim. of *calyx,* calyx.] *Bot.* Epicalyx. **—ca·lyc'u·late'** (kə-lĭk'yə-lāt', -lĭt) *adj.*

**ca·lyc·u·lus** (kə-lĭk'yə-ləs) *n., pl.* **-li** (-lī') [Lat. —see CALYCLE.] *Biol.* A small cup-shaped structure. **—ca·lyc'u·lar** *adj.*

**Ca·lyp·so** (kə-lĭp'sō) *n.* [Lat. < Gk. *Kalupso* < *kaluptein,* to conceal.] **1.** *Gk. Myth.* A sea nymph who delayed Odysseus on her island, Ogygia, for seven years. **2. calypso** *pl.* **-sos.** An orchid, *Calypso bulbosa* of the North Temperate Zone, having a pinkish flower with a slipper-shaped lip. **3. calypso** *pl.* **-sos** *or* **-soes.** Music that originated in the West Indies, particularly in Trinidad, and is marked by improvised lyrics on topical or humorous subjects.

**ca·lyp·tra** (kə-lĭp'trə) *n.* [Gk. *kaluptra,* vein < *kaluptein,* to cover.] **1.** The protective cap covering the spore case of a moss or related plant. **2.** A hoodlike or caplike structure. **—ca·lyp'trate'** (-trāt') *adj.*

**ca·lyx** (kā'lĭks, kăl'ĭks) *n., pl.* **ca·lyx·es** *or* **ca·ly·ces** (kā'lĭ-sēz', kăl'ĭ-) [Lat. *calyx, calic-* < Gk. *kalux.*] **1.** The outer protective covering of a flower, composed of a series of leaflike, usu. green sepals. **2.** A cuplike or funnel-shaped animal structure. **3.** A collecting structure in the kidney.

**cam** (kăm) *n.* [Du. *cam,* comb, of Germanic orig.] An eccentric wheel mounted on a rotating shaft and used to produce variable or reciprocating motion in another engaged or contacted part.

**ca·ma·ra·der·ie** (kä'mə-rä'də-rē, kăm'ə-răd'ə-) *n.* [Fr. < *camarade,* comrade < OFr., roommate. —see COMRADE.] Good will and lighthearted rapport between or among friends.

**cam·a·ril·la** (kăm'ə-rĭl'ə, -rē'yə) *n.* [Sp. < *cámara,* room < LLat. *camera* < Lat., arched roof < Gk. *kamara,* vault.] **1.** A secret, unofficial adviser to an early Spanish king. **2. a.** A group of confidential advisers. **b.** CABAL 1.

**cam·as** *also* **cam·ass** (kăm'əs) *n.* [Chinook Jargon *kamass.*] **1.** A North American plant of the genus *Camassia,* esp. *C. quamash* of western North America, with a blue or white flower cluster and an edible bulb. **2.** The death camas.

**cam·ber** (kăm'bər) *n.* [ME *caumber,* curved < OFr. *cambre* < *cambrer,* to arch < *caerare,* to vault < *camera,* vault < Lat. —see CHAMBER.] **1. a.** A slightly arched surface, as of a road, a ship's deck, or an airfoil. **b.** The state of having an arched surface. **2.** A setting of automotive wheels closer together at the bottom than at the top. —*vi.* & *vt.* **-bered, -ber·ing, -bers.** To arch or cause to arch slightly.

**cam·bist** (kăm'bĭst) *n.* [Fr. *cambiste,* exchange broker < Ital. *cambista* < *cambio,* exchange < LLat. *cambire,* to exchange.] **1.** A manual providing exchange rates of different currencies and equivalents of different weights and measures. **2.** A dealer in or expert on international exchange. **—cam'bism, cam'bis·try** *n.*

**cam·bi·um** (kăm'bē-əm) *n.* [Med. Lat., change < LLat. *cambire,* to exchange.] A layer of cells in the stems and roots of vascular plants that generates phloem and xylem. **—cam'bi·al** *adj.*

**Cam·bri·an** (kăm'brē-ən) *adj.* [< Med. Lat. *Cambria,* Wales < Welsh *Cymry.*] **1.** Of or relating to Wales : WELSH. **2.** Of, belonging to, or designating the geologic time, system of rocks, and sedimentary deposits of the first period of the Paleozoic era, characterized by warm seas and desert land areas. —*n.* **1.** A Welshman. **2.** The Cambrian period.

**cam·bric** (kām'brĭk) *n.* [Obs. *cameryk* < obs. Flem. *kameryk* < *Kameryk,* Cambrai, a city in France.] A finely woven white linen or cotton fabric.

**cambric tea** *n.* [So called because it is thin and white like cambric.] A drink for children, made of hot water, milk, sugar, and usu. a small amount of tea.

**came[1]** (kām) *n.* [Orig. unknown.] A slender, grooved lead bar that holds together the panes in stained glass or latticework windows.

**came[2]** (kām) *v. p.t. of* COME.

**cam·el** (kăm'əl) *n.* [ME < Lat. *camelus* < Gk. *kamēlos,* of Semitic orig.] **1.** A humped, long-necked ruminant mammal of the genus *Camelus,* domesticated in Old World desert regions as a beast of burden and a source of wool, milk, and meat. **2.** A device used to raise a sunken vessel.

**cam·el·back** (kăm'əl-băk') *adj.* Humped or arching in a curve.

**cam·el·eer** (kăm'ə-lîr') *n.* One who drives or rides a camel.

**ca·mel·lia** (kə-mēl'yə) *n.* [NLat. *Camellia,* genus name, after Georg Josef Kamel (1661–1706).] **1.** A native Asian shrub or tree of the genus *Camellia,* esp. *C. japonica,* with shiny evergreen leaves and variously colored flowers. **2.** The flower of a camellia.

**ca·mel·o·pard** (kə-mĕl'ə-pärd') *n.* [Med. Lat. *camelopardus* < Lat. *camelopardalis* < Gk. *kamēlopardalis* : *kamēlos,* camel + *pardalis,* var. of *pardos,* pard (so called because the giraffe has a head like a camel's and the spots of a leopard).] **1.** *Archaic.* A giraffe. **2.** *Her-*

*aldry.* A bearing resembling a giraffe but represented with long curved horns.

**Ca·mel·o·par·da·lis** (kə-mĕl'ō-pär'dl-ĭs) *n.* [Lat. *camelopardalis,* camelopard.] A constellation in the Northern Hemisphere.

**Cam·e·lot** (kăm'ə-lŏt') *n.* **1.** The legendary town where King Arthur had his court. **2.** A place, time, or circumstance characterized by ideal beauty, peacefulness, and enlightenment.

**camel's hair** *n.* **1.** The soft, fine hair of a camel or a substitute for it. **2.** A soft, heavy cloth, usu. light tan, made chiefly of camel's hair.

**Cam·em·bert** (kăm'əm-bâr') *n.* [Fr., after *Camembert,* France.] A creamy, mold-ripened cheese that softens on the inside as it ages.

**cam·e·o** (kăm'ē-ō') *n., pl.* **-os.** [ME *cameu* < OFr. *camaieu* < OSp. *camafeo.*] **1. a.** A technique of engraving in relief on a gem or other stone, esp. with layers of different hues, cut so the raised design is of one color and the background of another. **b.** A gem so cut. **c.** A medallion with a profile cut in raised relief. **2.** A brief but dramatic appearance of a prominent performer in a single scene on a television show or in a film. —*vt.* **-oed, -o·ing, -os. 1.** To make into or as if into cameo. **2.** To depict in sharp, delicate relief, as in a literary composition.

**cam·er·a** (kăm'ər-ə, kăm'rə) *n.* [LLat., room. —see CHAMBER.] **1.** A device for taking photographs, gen. composed of a lightproof enclosure having an aperture with a shuttered lens through which the image of an object is focused and recorded on a photosensitive film or plate. **2.** The part of a television transmitting apparatus that receives the primary image on a light-sensitive cathode tube and transforms it into electrical impulses. **3.** A camera obscura. **4.** *pl.* **-er·ae** (-ə-rē). A judge's private chamber.

**cam·er·al** (kăm'ər-əl) *adj.* [Med. Lat. *cameralis* < *camera,* office < LLat., room. —see CHAMBER.] **1.** Relating to a judge's chamber and to the judicial business transacted therein. **2.** Relating to public finance and state business or to a council that manages such matters.

**camera lu·ci·da** (lo͞o'sĭ-də) *n.* [NLat., light chamber.] An optical device that projects a virtual image of an object onto a plane surface, esp. for tracing.

**cam·er·a·man** (kăm'ər-ə-măn', kăm'rə-) *n.* One who operates a motion-picture or television camera.

**camera ob·scu·ra** (əb-skyo͝or'ə) *n.* [NLat., dark chamber.] A darkened chamber in which the real image of an object is received through a small opening or lens and focused in natural color onto another surface.

**cam·er·lin·go** (kăm'ər-lĭng'gō) *also* **cam·er·len·go** (-lĕng'gō) *n., pl.* **-gos.** [Ital. *camarlingo.*] *Rom. Cath. Ch.* The cardinal who manages the pope's secular activities.

**cam·i·on** (kăm'ē-ən, kăm-yôN') *n.* [Fr. < OFr. *chamion.*] **1.** A low, sturdy wagon. **2. a.** TRUCK[1] 1. **b.** BUS 1.

**†ca·mi·sa** (kə-mē'sə) *n.* [Sp. < LLat. *camisia,* shirt.] *Southwestern U.S.* A shirt or chemise.

**cam·i·sa·do** (kăm'ĭ-sā'dō, -sä'-) *n., pl.* **-dos.** [Prob. < Sp. *encamisado,* shirted (so called because the attackers wore white shirts over their armor for identification) < *camisa,* shirt. —see CAMISA.] *Archaic.* A sudden, surprise attack by night.

**ca·mise** (kə-mēz', -mēs') *n.* [Ar. *qamīṣ* < LLat. *camisia,* shirt.] A loose shirt, shift, or tunic.

**cam·i·sole** (kăm'ĭ-sōl') *n.* [Fr. < OProv. *camisolla,* dim. of *camisa,* shirt < LLat. *camisia.*] **1.** A woman's sleeveless undergarment. **2.** A short negligee.

**Cam·lan** (kăm'lən) *n.* The legendary battlefield where King Arthur was mortally wounded.

**cam·let** (kăm'lĭt) *n.* [ME *chamelet* < OFr. *chamelot* < Ar. *ḥamlat.*] **1.** A rich cloth of Oriental origin, held to have been made orig. of camel's hair and silk and later of goat's hair and silk or other combinations. **2.** A garment made from camlet.

**cam·o·mile** (kăm'ə-mīl') *n. var. of* CHAMOMILE.

**Ca·mor·ra** (kə-môr'ə, -mŏr'ə) *n.* [Ital., perh. < *camorra,* a kind of smock, said to have been worn by members of the society.] **1.** A Neapolitan secret society organized about 1820, notorious for its violence and blackmail. **2.** An unscrupulous, clandestine group.

**cam·ou·flage** (kăm'ə-fläzh', -fläj') *n.* [Fr. < *camoufler,* to disguise < Ital. *camuffare.*] **1.** Concealment of personnel or materiel from an enemy by making them appear to be part of the natural surroundings. **2.** A means of concealment. —*vt. & vi.* **-flaged, -flag·ing, -flag·es.** To conceal by or utilize camouflage. **—cam'ou·flag'er** *n.*

**camp[1]** (kămp) *n.* [OFr. < Lat. *campus,* field.] **1. a.** A place where a group of people, as soldiers, are temporarily billeted in makeshift shelters, as tents or huts. **b.** The shelters in a campsite. **c.** The persons using shelters in a campsite. **2.** A place composed of more or less permanent shelters, used for recreation. **3.** Military service. **4.** A group of persons, parties, or states favorable to a common cause, doctrine, or political system <the liberal *camp*> —*v.* **camped, camp·ing, camps.** —*vt.* To shelter or lodge in a camp : ENCAMP. —*vi.* **1.** To make or set up a camp. **2.** To live in or as if in a camp : SETTLE <*camped* in the new house until the furniture arrived>

**camp[2]** (kămp) [Orig. unknown.] *Slang.* —*n.* **1. a.** An affectation or appreciation of manners and tastes usu. considered outlandish, vulgar, or banal. **b.** Behavior displaying such affectation or appreciation. **2.** Banality, vulgarity, or artificiality when appreciated for its humor. —*vi.* **camped, camp·ing, camps.** To act in an outlandish, vulgar, or banal manner. **—camp, camp'y** *adj.*

**cam·paign** (kăm-pān') *n.* [Fr. *campagne* < OFr., battlefield < OItal. *campagna* < LLat. *campania,* open country < *campus,* field.] **1.** A series of military operations launched to achieve a specific objective within a given area. **2.** An operation undertaken, as by propaganda, to reach a political, social, or commercial goal. **—cam·paign'** *v.* **(-paigned, -paign·ing, -paigns). —cam·paign'er** *n.*

**cam·pa·ni·le** (kăm'pə-nē'lē) *n., pl.* **-les** (-lēz) or **-li** (-lē) [Ital. < *campana,* bell < LLat.] A bell tower, esp. one close to but unattached to a building.

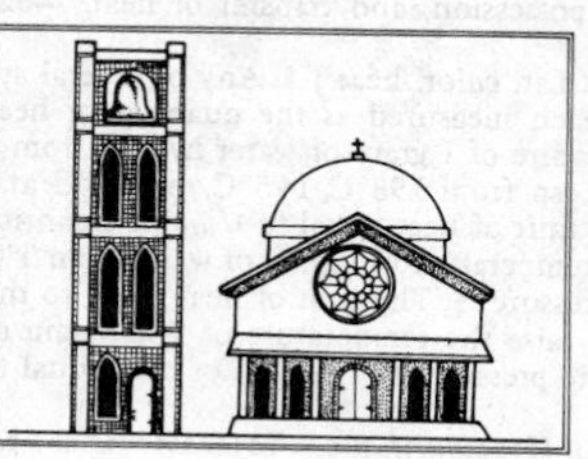

campanile

**cam·pa·nol·o·gy** (kăm'pə-nŏl'ə-jē) *n.* [LLat. *campana,* bell + -LOGY.] The art or study of bell casting and ringing. **—cam'pa·nol'o·gist** *n.*

**cam·pan·u·la** (kăm-păn'yə-lə) *n.* [NLat. *Campanula,* genus name, dim. of LLat. *campana,* bell.] Any of various plants of the genus *Campanula,* including the bellflowers.

**cam·pan·u·late** (kăm-păn'yə-lĭt, -lāt') *adj.* [NLat. *campanula,* dim. of LLat. *campana,* bell + -ATE[1]] *Bot.* Bell-shaped.

**camp·er** (kăm'pər) *n.* **1.** One that camps. **2. a.** A compact, vanlike vehicle similar to an automobile and trailer combination, intended to function as a dwelling and used for camping or on long motor trips. **b.** A portable shelter resembling the top part of a trailer, made to be mounted on a pickup truck to form a recreational vehicle.

**cam·pes·tral** (kăm-pĕs'trəl) *adj.* [Lat. *campester,* of a field < *campus,* field.] Of, relating to, or growing in uncultivated land or open fields.

**cam·pes·tri·an** (kăm-pĕs'trē-ən) *adj.* [< Lat. *campestria,* a plain < *campus,* field.] Relating to the northern Great Plains.

**camp·fire** (kămp'fīr') *n.* **1.** An outdoor fire in a camp, used for warmth or cooking. **2.** A meeting held around a campfire.

**Camp Fire Girl** *n.* A member of the Camp Fire Girls, an organization for girls aged 7 through 18 that strives to inculcate good values and character and develop practical skills.

**camp follower** *n.* **1.** A civilian follower of an army who sells goods or services to it. **2.** One who follows but does not belong to a main body or group.

**camp·ground** (kămp'ground') *n.* An area used for setting up a camp or holding a camp meeting.

**cam·phene** (kăm'fēn') *n.* [CAMPH(OR) + -ENE.] A colorless crystalline compound, $C_{10}H_{16}$, used in making synthetic camphor and insecticides.

**cam·phor** (kăm'fər) *n.* [ME *caumfre* < OFr. *camphre* < Med. Lat. *camphora* < Ar. *kāfūr.*] A volatile crystalline compound, $C_{10}H_{16}O$, obtained from the wood of the camphor tree or synthesized and used as an insect repellent, in manufacturing film, plastics, lacquers, and explosives, and medicinally as a stimulant, expectorant, and diaphoretic. **—cam·phor'ic** (-fôr'ĭk, -fŏr'-) *adj.*

**cam·phor·ate** (kăm'fə-rāt') *vt.* **-at·ed, -at·ing, -ates.** To treat or impregnate with camphor.

**camphor ice** *n.* A skin ointment composed of camphor, white wax, spermaceti, and castor oil.

**camphor oil** *n.* The oil obtained from the wood of the camphor tree.

**camphor tree** *n.* An evergreen tree indigenous to eastern Asia, *Cinnamomum camphora,* having aromatic wood that is a source of camphor.

**cam·pi·on** (kăm'pē-ən) *n.* [Orig. unknown.] Any of various plants of the genus *Lychnis* or related genera, with small red, pink, or white flowers.

**camp meeting** *n.* An evangelistic gathering held in a tent or outdoors.

**cam·po** (kăm'pō, käm'-) *n., pl.* **-pos.** [Am. Sp. < Sp., field < Lat. *campus.*] A large, grassy South American plain with occasional bushes and small trees.

**camp·o·ree** (kăm'pə-rē') *n.* [CAMP + (JAMB)OREE.] A gathering of Boy Scouts on a local or district level.

**camp robber** *n.* The Canada jay.

**camp·site** (kămp'sīt') *n.* An area suitable or used for camping.

**camp·stool** (kămp'sto͞ol') *n.* A light folding stool.

**cam·pus** (kăm′pəs) *n., pl.* **-pus·es.** [Lat., field.] **1.** The grounds of a school, college, or university. **2.** A field in ancient Rome used for various events, as military exercises.

**cam·py·lot·ro·pous** (kăm′pə-lŏt′rə-pəs) *adj.* [Gk. *kampulos*, curved + -TROPOUS.] *Bot.* With a partially inverted, curved ovule.

**cam·shaft** (kăm′shăft′) *n.* An engine shaft fitted with a cam or cams.

**can**[1] (kăn; kən *when unstressed*) *aux.v. p.t.* **could** (ko͝od) [ME < OE, first and third person pr. indicative of *cunnan*, to know how.] **1.** —Used to indicate: **a.** Mental or physical ability <I *can* see you today.> **b.** Possession of a given power, right, or privilege <The President *can* veto congressional bills.> **c.** Possession of a given skill or capacity <I *can* tune the piano as well as play it.> **2.** —Used to indicate possibility or probability <I wonder if they *can* still be alive.> **3.** —Used to request or grant permission <*Can* I be excused?><Yes, you *can.*> ***usage:*** In speech *can* is used more often than *may* to express permission, even though traditionalists insist that *can* should be used only to express the capacity to do something. Technically, correct usage therefore requires: *May I have the car tonight?* But because the contraction *mayn't* is felt to be stilted, the negative form is usu. expressed as *Can't I have the car tonight?*

**can**[2] (kăn) *n.* [ME *canne*, a water container < OE.] **1.** A usu. cylindrical metal container. **2. a.** An airtight container, usu. made of tin-coated iron, in which foods or beverages are preserved. **b.** The contents of such a container. **3.** *Slang.* A prison. **4.** *Slang.* **a.** A toilet. **b.** A rest room. **5.** *Slang.* The buttocks. —*vt.* **canned, can·ning, cans. 1.** To seal in a can or jar for future use **:** PRESERVE. **2.** *Slang.* To make a recording of <*can* the audience's laughter> **3.** *Slang.* **a.** To dismiss from school or employment. **b.** To quit or dispense with <Let's *can* the chatter.> **—can′ner** *n.*

**Can·a·da balsam** (kăn′ə-də) *n.* **1.** The balsam fir. **2.** A viscous, yellowish, transparent resin obtained from the balsam fir and used as a mounting cement for microscopic specimens.

**Canada goose** *n.* A common wild goose, *Branta canadensis* of North America, with grayish plumage, a black neck and head, and a white face patch.

**Canada jay** *n.* A bird, *Perisoreus canadensis* of North American conifer forests, with gray plumage and a black-capped head.

**Canada thistle** *n.* A weedy Eurasian plant, *Cirsium arvense*, with prickly leaves and purplish flower clusters.

**Ca·na·di·an bacon** (kə-nā′dē-ən) *n.* Cured rolled bacon from the loin of a hog.

**Canadian French** *n.* The language of the French-Canadians.

**ca·naille** (kə-nī′, -nāl′) *n.* [Fr. < Ital. *canaglia* < *cane*, dog < Lat. *canis.*] The masses of common people **:** RABBLE.

**ca·nal** (kə-năl′) *n.* [Partly < Fr., channel, and partly < ME, tube, both < Lat. *canalis*, tube, channel.] **1.** An artificial waterway or artificially improved river used for irrigation, shipping, or travel. **2.** *Anat.* A tube **:** duct. **3.** *Astron.* One of the faint, hazy markings resembling straight lines on the surface of Mars. —*vt.* **-nalled, -nal·ling, -nals** or **-naled, -nal·ing, -nals. 1.** To dig an artificial waterway through. **2.** To provide with a canal or canals.

**can·a·lic·u·late** (kăn′ə-lĭk′yə-lĭt, -lāt′) *adj.* [Lat. *canaliculatus* < *canaliculus*, dim. of *canalis*, channel.] Having grooves or channels.

**can·a·lic·u·lus** (kăn′ə-lĭk′yə-ləs) *n., pl.* **-li** (-lī′) [Lat., dim. of *canalis*, conduit.] A small bodily channel, as a tear duct. **—can′a·lic′u·lar** (-lər) *adj.*

**can·a·li·za·tion** (kăn′ə-lĭ-zā′shən) *n.* **1.** The act or an instance of canalizing. **2.** A canal system.

**can·a·lize** (kăn′ə-līz′) *vt.* **-lized, -liz·ing, -liz·es. 1.** To furnish with, build, or convert into a canal or canals. **2.** To channel into a specific direction.

**can·a·pé** (kăn′ə-pā′, -pē) *n.* [Fr. < *canapé*, couch < Med. Lat. *canapeum*, mosquito net. —see CANOPY.] A cracker or small thin piece of bread or toast spread with cheese, meat, or relish and served as an appetizer.

▲ word history: The Middle English spelling of *canopy* was *canope* or *canape*, which makes its relationship to *canapé* easier to see. The two words actually represent earlier and later borrowings of the same word from French. The Latin and Greek forms of the word signified a bed or couch hung about with mosquito netting. The English word *canopy* preserves the idea of a covering; the modern French word *canapé* preserves the idea of a seat and means "couch." French *canapé* also means "appetizer, canapé." The appetizer was apparently visualized by the French who invented it as a little seat for a savory tidbit.

**ca·nard** (kə-närd′) *n.* [Fr., prob. < the phrase *vendre un canard à moitié*, to half-sell a duck, to swindle.] An unfounded, false, and deliberately misleading story.

**ca·nar·y** (kə-nâr′ē) *n., pl.* **-ies.** [Fr. *canari* < OSp. *canario* < (*Islas*) *Canarias*, Canary (Islands) < LLat. *Canariae* (*Insulae*), (islands) of dogs < Lat. *canis*, dog.] **1.** A songbird native to the Canary Islands, *Serinus canaria*, greenish to yellow in color and long bred as a cage bird. **2.** *Slang.* A stool pigeon **:** informer. **3.** A sweet white wine from the Canary Islands. **4.** A lively 16th-cent. court dance. **5.** A light to moderate or vivid yellow.

**canary grass** *n.* A grass native to Europe, *Phalaris canariensis*, with straw-colored seeds used to feed birds.

**ca·nas·ta** (kə-năs′tə) *n.* [Sp. < *canasto*, basket < Lat. *canistrum*. —see CANISTER.] A card game for two to six players, played with two decks of cards and related to rummy.

**can·can** (kăn′kăn′) *n.* [Fr.] An exuberant dance performed by women and marked by high kicks.

**can·cel** (kăn′səl) *v.* **-celed, -cel·ing, -cels** *also* **-celled, -cel·ling, -cels.** [ME < Norman Fr. *canceler* < Lat. *cancellare*, to cross out < *cancelli*, lattice, dim. of *cancer*, lattice.] —*vt.* **1.** To cross out with markings such as lines. **2.** To annul or invalidate <*cancel* an invitation> **3.** To mark or perforate (e.g., a postage stamp) to indicate that it may not be used again. **4.** To equalize or make up for **:** OFFSET. **5.** *Math.* **a.** To remove a common factor from the numerator and denominator of a fractional expression. **b.** To remove a common factor or term from both members of an equation or inequality. **6.** To omit or delete (printed material). —*vi.* To balance or neutralize one another. —*n.* **1. a.** Omission or deletion of printed matter. **b.** The matter omitted or deleted or its replacement. **2.** A part of a book used as a substitute for an original part of the book. **—can′cel·a·ble** *adj.* **—can′cel·er** *n.* **—can′cel·la′tion, can′ce·la′tion** *n.*

☆ ***syns:*** CANCEL, COUNTERACT, NEUTRALIZE *v. core meaning* **:** to make ineffective by applying an opposing force or amount <Two opposing votes *cancel* each other.>

**can·cel·late** (kăn-sĕl′ĭt, kăn′sə-lāt′) *also* **can·cel·lat·ed** (-lā′tĭd) *adj.* [Lat. *cancellatus*, p.part. of *cancellare*, to make like a lattice. —see CANCEL.] *Anat.* Cancellous.

**can·cel·lous** (kăn-sĕl′əs, kăn′sə-ləs) *adj. Anat.* Having a coarse net-like or spongy structure. —Used of bone.

**can·cer** (kăn′sər) *n.* [Lat. *cancer, cancr-.*] **1. a.** Any of various malignant neoplasms that manifest invasiveness and a tendency to metastasize to new sites. **b.** The pathological condition characterized by such growths. **2.** A pernicious, spreading evil <A *cancer* of bigotry spread through the country.> **3. Cancer.** A constellation in the Northern Hemisphere. **4. Cancer. a.** The fourth sign of the zodiac. **b.** One born under this sign. **—can′cer·ous** (-sər-əs) *adj.*

**can·croid** (kăng′kroid′) *adj.* **1.** Like a cancer. **2.** Similar to a crab. —*n.* A skin cancer.

**can·del·a** (kăn-dĕl′ə) *n.* [Lat. *candela*, candle.] A unit of luminous intensity equal to $1/_{60}$ of the luminous intensity per square centimeter of a blackbody radiating at the temperature of solidification of platinum (2,046°K).

**can·de·la·bra** (kăn′dl-ä′brə, -ăb′rə, -ā′brə) *n.* A candelabrum.

**can·de·la·brum** (kăn′dl-ä′brəm, -ăb′rəm, -ā′brəm) *n., pl.* **-bra** (-brə) or **-brums.** [Lat. < *candela*, candle.] A large multibranched decorative candlestick.

**can·dent** (kăn′dənt) *adj.* [Lat. *candens, candent-*, pr.part. of *candēre*, to shine.] Having a white-hot glow **:** INCANDESCENT.

**can·des·cence** (kăn-dĕs′əns) *n.* [< Lat. *candescens*, pr.part. of *candescere*, inceptive of *candēre*, to shine.] The state of being white hot **:** INCANDESCENCE. **—can·des′cent** *adj.* **—can·des′cent·ly** *adv.*

**can·did** (kăn′dĭd) *adj.* [Fr. *candide* < Lat. *candidus* < *candēre*, to shine.] **1.** Devoid of prejudice **:** IMPARTIAL. **2.** Lacking pretense or reserve **:** STRAIGHTFORWARD. **3.** Not rehearsed or posed <a *candid* wedding picture> —*n.* An unposed informal photograph. **—can′did·ly** *adv.* **—can′did·ness** *n.*

**can·di·da** (kăn′dĭ-də) *n.* [NLat. *Candida*, genus name < Lat., fem. of *candidus*, white.] Any of the pathogenic yeastlike imperfect fungi of the genus *Candida*.

**can·di·date** (kăn′dĭ-dāt′, -dĭt) *n.* [Lat. *candidatus*, clothed in white (from the white togas worn by Romans seeking office) < *candidus*, white. —see CANDID.] **1.** One who seeks or is nominated for an office, prize, or honor. **2.** One apt to gain a certain position or come to a certain fate <a *candidate* for a heart attack> **—can′di·da·cy** (-də-sē), **can′di·da·ture′** (-də-cho͝or′, -chər) *n.*

**candid camera** *n.* A small, easily operated camera with a fast lens for taking unposed or informal photographs.

**can·di·di·a·sis** (kăn′dĭ-dī′ə-sĭs) *n.* A fungous infection caused by a member of the genus *Candida*.

**can·died** (kăn′dēd) *adj.* Permeated, covered, encrusted, or cooked with sugar <*candied* apples>

**can·dle** (kăn′dl) *n.* [ME *candel* < OE < Lat. *candela* < *candēre*, to shine.] **1.** A solid, usu. cylindrical mass of a fatty substance, as tallow or wax, with an axially embedded wick that is burned to provide light. **2.** Something like a candle. **3.** *Physics.* **a.** An obsolete unit of luminous intensity, orig. defined in terms of a wax candle with standard composition and equal to 1.02 candelas. **b.** A candela. —*vt.* **-dled, -dling, -dles.** To examine (an egg) for freshness in front of a light. **—burn (one's) candle at both ends.** To expend one's energy by doing too many things at once. **—not hold a candle to.** To be not nearly as good as (another). **—can′dler** *n.*

**can·dle·ber·ry** (kăn′dl-bĕr′ē) *n.* The wax myrtle or its fruit.

**can·dle·fish** (kăn′dl-fĭsh′) *n., pl.* **candlefish** or **-fish·es.** An oily, edible fish, *Thaleichthys pacificus* of northern Pacific waters, once dried by Indians and used as a torch.

**can·dle-foot** (kăn′dl-fo͝ot′) *n.* A foot-candle.

**can·dle·hold·er** (kăn′dl-hōl′dər) *n.* A candlestick.
**can·dle·light** (kăn′dl-līt′) *n.* **1.** Illumination from a candle or candles. **2.** Dusk : twilight.
**Can·dle·mas** (kăn′dl-məs) *n.* [ME *candelmasse* < OE *candelmæsse* : *candel,* candle + *mæsse,* mass (from the blessing of candles at the feast).] A church festival celebrated on Feb. 2 as the feast of the purification of the Virgin Mary and the presentation of the infant Christ in the temple.
**can·dle·nut** (kăn′dl-nŭt′) *n.* **1.** A tree, *Aleurites moluccana* of tropical Asia and Polynesia, bearing nuts that yield an oil used in paints and varnishes. **2.** The nut of the candlenut.
**can·dle·pin** (kăn′dl-pĭn′) *n.* **1.** A slender bowling pin used in a variation of the game of tenpins. **2. candlepins** *(sing. in number).* A bowling game using a ball smaller than that used in tenpins.
**can·dle·pow·er** (kăn′dl-pou′ər) *n.* Luminous intensity expressed in standard candles.
**can·dle·stick** (kăn′dl-stĭk′) *n.* [ME *candelstikke* < OE *candelsticca.*] An often ornamental holder for securing a candle or candles.
**can·dle·wick** (kăn′dl-wĭk′) *n.* The wick of a candle.
**can·dle·wick·ing** (kăn′dl-wĭk′ĭng) *n.* **1.** Soft, heavy cotton thread similar to that used to make wicks for candles. **2.** Embroidery made of tufts of candlewicking.
**can·dle·wood** (kăn′dl-wŏŏd′) *n.* **1.** The ocotillo. **2.** The resinous wood of the ocotillo or plants similar to it.
**can·dor** (kăn′dər) *n.* [Lat. < *candēre,* to shine.] **1.** Frankness of expression : SINCERITY. **2.** Freedom from prejudice : IMPARTIALITY.
**can·dour** (kăn′dər) *n. Chiefly Brit. var. of* CANDOR.
**can·dy** (kăn′dē) *n., pl.* **-dies.** [Short for *sugar candy* < ME *sugre candie* < OFr. *sucre candi* < Ital. *zucchero candi* < Ar. *sukkar qandī* : *sukkar,* sugar + *qandī,* candied < *qand,* cane sugar, of Dravidian orig.] **1.** A sweet, rich confection made with sugar or corn syrup, often combined with chocolate, dairy products, fruits, or nuts. **2.** A single piece of a rich, sweet confection. —*v.* **-died, -dy·ing, -dies.** —*vt.* **1.** To reduce to sugar crystals. **2.** To cook, preserve, or coat with sugar or syrup. **3.** To make pleasant : SWEETEN. —*vi.* **1.** To crystallize, as sugar. **2.** To become coated with sugar or syrup.
▲ word history: The simple noun *candy* does not occur in English until the 18th century; the word first appears in the 15th century in the compound *sugar candy. Sugar candy* signified a particular kind of sugar that resulted from boiling and crystallization. In this form sugar was often served as a confection, and in British English today *candy* is largely restricted to sweets of sugar candy. In the United States *candy* includes confections made of other ingredients besides sugar, such as chocolate, fruit, and nuts. The original sense is retained in the verb *candy,* which means "to coat with sugar" and not with "candy" in the wider sense.
**candy striper** *n.* [From the resemblance of the volunteer's red and white striped uniform to a candy cane.] A usu. teen-age volunteer nurse's aide.
**can·dy·tuft** (kăn′dē-tŭft′) *n.* [Obs. *Candy,* var. of *Candia,* Crete + TUFT.] Any of various plants of the genus *Iberis,* with white, red, or purplish flower clusters.
**cane** (kān) *n.* [ME < Norman Fr. < Med. Lat. *canna* < Gk. *kanna,* reed.] **1. a.** A slender, jointed, woody but usu. flexible stem, as of bamboo, rattan, or some palms. **b.** A plant having such a stem. **c.** Such stems or strips of such stems used for wickerwork. **2.** A grass, *Arundinaria gigantea* of the southeastern United States, with long stiff stems and often forming canebrakes. **3.** The long, woody stem of a raspberry, blackberry, certain roses, or similar plants. **4.** Sugar cane. **5.** A stick used as an aid in walking. **6.** A rod used for flogging. —*vt.* **caned, can·ing, canes. 1.** To make, supply, or repair with cane. **2.** To strike or beat with a cane. **—can′er** *n.*
**cane·brake** (kān′brāk′) *n.* A dense thicket of cane.
**ca·nes·cent** (kə-nĕs′ənt) *adj.* [Lat. *canescens, canescent-,* p.part. of *canescere,* inchoative of *canēre,* to be white < *canus,* white.] **1.** *Biol.* Covered with whitish or grayish down : HOARY. **2.** Turning grayish or white. **—ca·nes′cence** *n.*
**cane sugar** *n.* Sucrose.
**Ca·nes Ve·nat·i·ci** (kā′nēz vĭ-năt′ĭ-sī′) *n.* [Lat., hunting dogs.] A constellation in the Northern Hemisphere.
**Ca·nic·u·la** (kə-nĭk′yə-lə) *n.* [Lat. < dim. of *canis,* dog.] Sirius.
**ca·nic·u·lar** (kə-nĭk′yə-lər) *adj.* **1.** Of or relating to the Dog Star. **2.** Relating to the dog days of Jul. and Aug.
**ca·nine** (kā′nīn′) *adj.* [Lat. *caninus* < *canis,* dog.] **1.** Of, relating to, or typical of a member of the family Canidae, including dogs, wolves, and foxes. **2.** Of or designating one of the conical teeth located between the incisors and the first bicuspids. —*n.* **1.** A canine animal. **2.** A canine tooth.
**Ca·nis Ma·jor** (kā′nĭs mā′jər, kăn′ĭs) *n.* [Lat., the larger dog.] A constellation in the Southern Hemisphere.
**Canis Mi·nor** (mī′nər) *n.* [Lat., the smaller dog.] A constellation in the Southern Hemisphere.
**can·is·ter** (kăn′ĭ-stər) *n.* [Lat. *canistrum,* basket < Gk. *kanastron* < *kanna,* reed.] **1.** A usu. thin metal container for holding dry foods. **2.** A metallic cylinder that when fired from a gun bursts and scatters the shot packed inside it. **3.** The part of a gas mask having a filter for removing poison gas from the surrounding air.
**can·ker** (kăng′kər) *n.* [ME < OE *cancer,* cancer, and < OFr. *cancre,* cancer, both < Lat. *cancer.*] **1.** An ulcerous sore of the mouth and lips. **2.** A necrotic area in a plant surrounded by healthy wood or bark. **3.** Any of several animal diseases marked by chronic inflammatory processes. **4.** A source of spreading corruption or decay. —*v.* **-kered, -ker·ing, -kers.** —*vt.* **1.** To attack or infect with canker. **2.** To cause to decay or become corrupt. —*vi.* To become infected with or as if with canker.
**can·ker·ous** (kăng′kər-əs) *adj.* **1.** Of or infected with a canker : ULCEROUS. **2.** Causing canker.
**canker sore** *n.* A small, painful ulcer usu. of the mouth.
**can·ker·worm** (kăng′kər-wûrm′) *n.* The larva of either of two moths, *Paleacrita vernata* or *Alsophila pometaria,* that destroys fruit and shade trees.
**can·na** (kăn′ə) *n.* [NLat. *Canna,* genus name < Lat. *canna,* cane.] Any of various tropical plants of the genus *Canna,* with broad leaves and yellow or red flowers.
**can·na·bi·di·ol** (kăn′ə-bĭ-dī′ôl′, -ōl′) *n.* [CANNABI(S) + DI-[1] + -OL.] A chemical constituent of cannabis, $C_{21}H_{28}(OH)_2$.
**can·na·bin** (kăn′ə-bĭn) *n.* [CANNAB(IS) + -IN.] A resinous material extracted from cannabis.
**can·na·bis** (kăn′ə-bĭs) *n.* [Lat. < Gk. *kannabis.*] **1.** The hemp plant. **2.** The dried flower buds of the hemp plant. **—can′na·bic** *adj.*
**canned** (kănd) *adj.* **1.** Preserved and sealed in a can or jar. **2.** *Informal.* Recorded or taped <*canned* laughter>
**canned heat** *n.* Alcohol or paraffin fuel packed in small cans and used to heat food.
**can·nel** *or* **can·nel coal** (kăn′əl) *n.* [Perh. short for *cannel coal,* var. of *candle coal* (from its bright flame).] A bituminous coal that burns brightly with a great deal of smoke.
**can·nel·lo·ni** (kăn′ə-lō′nē) *n.* [Ital., pl. of *cannellone,* tubular soup noodle < *cannello,* small tube, dim. of *canna,* reed < Lat. < Gk. *kanna.*] An Italian pasta dish of large stuffed macaroni that is baked and served with a tomato or cream sauce.
**can·ner·y** (kăn′ə-rē) *n., pl.* **-ies.** An establishment where foods, as meat or vegetables, are canned.
**can·ni·bal** (kăn′ə-bəl) *n.* [< Sp. *Caníbalis,* the name recorded by Christopher Columbus of the man-eating Caribs of Cuba and Haiti, of Arawakan orig.] **1.** One who eats the flesh of other human beings. **2.** An animal that feeds on others of its own kind. **—can′ni·bal·ism** *n.* **—can′ni·bal·is′tic** *adj.*
**can·ni·bal·ize** (kăn′ə-bə-līz′) *vt.* **-ized, -iz·ing, -iz·es. 1.** To remove serviceable parts from (e.g., damaged airplanes or tanks) for use in the repair of other equipment. **2.** To deprive (an organization) of personnel or equipment for use in another organization. **—can′ni·bal·i·za′tion** *n.*
**can·ni·kin** (kăn′ĭ-kĭn) *n.* [Prob. < Du. *kanneken* < MDu. *canneken,* dim. of *canne,* can, of Germanic orig.] **1.** A little can or cup. **2.** A wooden bucket.
**can·no·li** (kə-nō′lē, kă-) *n.* [Ital., pl. of *cannolo,* tube, dim. of *canna* < Lat., reed. —see CANE.] A fried pastry roll with a creamy, usu. sweet filling.
**can·non** (kăn′ən) *n., pl.* **cannon** *or* **-nons.** [ME *canon* < OFr. < OItal. *cannone* < *canna,* tube < Lat., reed. —see CANE.] **1.** A weapon for firing projectiles, having a heavy metal tube mounted on a carriage. **2.** A heavy firearm larger than 0.60 caliber. **3.** The loop at the top of a bell by which the bell is suspended. **4.** A round bit for a horse. **5.** The section of leg containing the cannon bone. **6.** *Chiefly Brit.* A carom made in billiards. —*vt.* & *vi.* **-noned, -non·ing, -nons. 1.** To bombard with or fire cannon. **2.** *Chiefly Brit.* To carom or cause to carom, as in billiards.
**can·non·ade** (kăn′ə-nād′) *vt.* & *vi.* **-ad·ed, -ad·ing, -ades.** [Prob. < Fr. *canonade,* discharge of artillery < Ital. *cannonata* < *cannone,* cannon. < OItal.] To assault with or deliver heavy artillery fire. **—can′non·ade′** *n.*
**can·non·ball** *also* **cannon ball** (kăn′ən-bôl′) *n.* **1.** A round projectile fired from a cannon. **2.** A jump into water made with the arms grasping the upraised knees. **3.** Something that is fast-moving, as a train. —*vi.* **-balled, -ball·ing, -balls. 1.** To travel rapidly. **2.** To make a cannonball jump into water.
**cannon bone** *n.* A supporting leg bone in some hoofed mammals.
**can·non·eer** (kăn′ə-nîr′) *n.* [OFr. *canonier* < *canon,* cannon.] A gunner : artilleryman.
**cannon fodder** *n.* [Transl. of G. *Kanonenfutter.*] Soldiers considered as expendable resources in warfare.
**can·non·ry** (kăn′ən-rē) *n., pl.* **-ries. 1.** Artillery. **2.** Artillery fire.
**cannon shot** *n.* **1.** Ammunition for a cannon. **2.** A shot or shots fired by cannon. **3.** A cannon's firing distance.
**can·not** (kăn′ŏt, kə-nŏt′, kă-) *aux.v.* Can not. ***usage:*** In the phrase *cannot but,* which is sometimes criticized as a double negative, *but* is used in the sense of "except." *One cannot but admire your courage* is therefore acceptable.
**can·nu·la** *also* **can·u·la** (kăn′yə-lə) *n., pl.* **-las** *or* **-lae** (-lē′) [Lat., dim. of *canna,* reed. —see CANE.] A tube inserted into a bodily cavity to drain fluid or insert medication.

**can·nu·lar** (kăn′yə-lər) *adj.* Cannulate.
**can·nu·late** *also* **can·u·late** (kăn′yə-lāt′). —*vt.* **-lat·ed, -lat·ing, -lates.** To insert a cannula in. —*adj.* Tubular : hollow. **—can′nu·la′tion** *n.*
**can·ny** (kăn′ē) *adj.* **-ni·er, -ni·est.** [< CAN[1].] **1.** Careful and shrewd. **2.** Thrifty : frugal. **3.** *Chiefly Scot.* **a.** Pleasant : attractive. **b.** Gentle : mild. **—can′ni·ly** *adv.* **—can′ni·ness** *n.*
**ca·noe** (kə-no͞o′) *n.* [Obs. *canoa* < Sp., of Cariban orig.] A light, slender boat with pointed ends, propelled by paddles. —*v.* **-noed, -noe·ing, -noes.** —*vt.* To carry or send by canoe. —*vi.* **1.** To travel in a canoe. **2.** To propel a canoe. **—ca·noe′ist** *n.*
**can of worms** *n. Informal.* A complicated, troublesome situation.
**can·on**[1] (kăn′ən) *n.* [ME *canoun,* partly < OE, and partly < OFr. *canon,* both < Lat. *canon,* rule < Gk. *kanōn.*] **1.** A law or code of laws established by a church council. **2.** A secular law, rule, or code of law. **3.** A basis for judgment : CRITERION. **4.** The books of the Bible officially recognized as the Holy Scripture. **5.** *Rom. Cath. Ch.* The part of the Mass beginning after the Sanctus and ending just before the Lord's Prayer. **6.** *Rom. Cath. Ch.* The calendar of accepted saints. **7.** An authoritative list, as of the works of an author. **8.** *Mus.* A composition or passage in which the same melody is repeated by one or more voices, overlapping in time in the same or a related key.
**can·on**[2] (kăn′ən) *n.* [ME *canoun* < Norman Fr. *canun* < LLat. *canonicus* < *canon,* rule.] **1.** A priest serving in a cathedral or collegiate church. **2.** A member of a religious community living under common rules and bound by vows.
**ca·ñon** (kăn′yən) *n. var. of* CANYON.
**can·on·ess** (kăn′ə-nĭs) *n.* A member of a religious community of women living under a common rule but not bound by vows.
**ca·non·i·cal** (kə-nŏn′ĭ-kəl) *also* **ca·non·ic** (-ĭk) *adj.* **1.** Relating to, required by, or abiding by canon law. **2.** Of or appearing in the Biblical canon. **3.** Officially approved : ORTHODOX. **—ca·non′i·cal·ly** *adv.* **—can′on·ic′i·ty** (kăn′ə-nĭs′ĭ-tē) *n.*
**canonical hours** *pl.n.* **1. a.** A special form of prayer, prescribed by canon law, usu. to be recited at specified times of the day, either in common or individually. **b.** The times of day set aside for these prayers. **2.** *Chiefly Brit.* The hours between 8:00 A.M. and 3:00 P.M., during which marriages may legally take place in parish churches.
**ca·non·i·cals** (kə-nŏn′ĭ-kəlz) *pl.n.* The dress prescribed by canon for officiating clergy.
**ca·non·i·cate** (kə-nŏn′ĭ-kāt′, -kĭt) *n.* [Med. Lat. *canonicatus* < LLat. *canonicus,* canon.] CANONRY 1.
**can·on·ist** (kăn′ə-nĭst) *n.* An expert in canon law. **—can′on·is′tic, can′on·is′ti·cal** *adj.*
**can·on·ize** (kăn′ə-nīz′) *vt.* **-ized, -iz·ing, -iz·es. 1.** *Rom. Cath. Ch.* To declare (a deceased person) to be a saint and entitled to be fully honored. **2.** To include in the Biblical canon. **3.** To approve as being within canon law. **—can′on·i·za′tion** *n.* **—can′on·iz′er** *n.*
**canon law** *n.* The body of officially established rules governing the faith and practice of the members of a Christian church.
**can·on·ry** (kăn′ən-rē) *n., pl.* **-ries. 1.** The position or benefice of a canon. **2.** Canons as a group.
**Ca·no·pic** (kə-nō′pĭk, -nŏp′ĭk) *adj.* [After *Canopus,* a city of ancient Egypt.] Designating an ancient Egyptian vessel, as a vase, urn, or jar, used to hold the remains of the dead.
**Ca·no·pus** (kə-nō′pəs) *n.* [Lat. < Gk. *kanōpos.*] A star in the constellation Carina, the second-brightest star in the sky.
**can·o·py** (kăn′ə-pē) *n., pl.* **-pies.** [ME *canape* < Med. Lat. *canapeum,* mosquito net < Lat. *conopeum* < Gk. *kōnōpion* < *Kanōpos,* Canopus, a city of ancient Egypt.] **1.** A cloth fastened or held horizontally above a person or an object for covering, protection, or ornamentation. **2.** A rooflike ornamental architectural structure. **3.** A high, overarching covering <a vast *canopy* of green leaves> **4. a.** The transparent, movable enclosure over an aircraft's cockpit. **b.** The hemispherical surface of a parachute. —*vt.* **-pied, -py·ing, -pies.** To cover with or as if with a canopy.
**ca·no·rous** (kə-nôr′əs, -nōr′-, kăn′ər-əs) *adj.* [Lat. *canorus* < *canor,* tune < *canere,* to sing.] Pleasing to the ears : MELODIOUS. **—ca·no′rous·ly** *adv.* **—ca·no′rous·ness** *n.*
**canst** (kănst) *v. Archaic. 2nd person sing. present tense of* CAN[1].
**cant**[1] (kănt) *n.* [ME, side < Norman Fr.] **1.** Angular deviation from a vertical or horizontal surface or plane : INCLINATION. **2. a.** A tilting thrust or motion. **b.** The tilt caused by such a thrust or motion. **3.** An outer corner, as of a building. **4.** A slanted edge or surface. —*v.* **cant·ed, cant·ing, cants.** —*vt.* **1.** To set at an oblique angle. **2.** To give a slanting edge to : BEVEL. **3.** To change the direction of suddenly. —*vi.* **1.** To tilt to one side : SLANT. **2.** To take an oblique direction or course : swing around, as a ship.
**cant**[2] (kănt) *n.* [Prob. < Norman Fr., singing < *canter,* to sing < Lat. *cantare,* freq. of *canere,* to sing.] **1.** Whining, affected speech. **2.** Monotonous or mechanical discourse. **3.** Hypocritically pious language. **4.** The special vocabulary peculiar to the members of an underworld group : ARGOT. **5.** The special terminology understood by the members of a profession, discipline, or class but obscure to the general population : JARGON. —*vi.* **cant·ed, cant·ing, cants. 1.** To speak in a whining, pleading tone. **2.** To speak tediously or sententiously : MORALIZE. **3.** To use specialized jargon or argot. **—cant′ing·ly** *adv.* **—cant′ing·ness** *n.*
**can't** (kănt, känt). Cannot.
**can·ta·bi·le** (kän-tä′bə-lā′) *adj.* & *adv.* [Ital. < LLat. *cantabilis,* worthy of being sung < Lat. *cantare,* freq. of *canere,* to sing.] *Mus.* Smooth, lyrical, and flowing in style. —Used as a direction. **—can·ta′bi·le′** *n.*
**Can·ta·brig·i·an** (kăn′tə-brĭj′ē-ən) *adj.* [< Med. Lat. *Cantabrigia,* Cambridge.] Of or relating to Cambridge University. —*n.* A student or graduate of Cambridge University.
**can·ta·la** (kăn-tä′lə) *n.* [Orig. unknown.] **1.** A tropical American century plant, *Agave cantula* cultivated for its coarse, tough fiber. **2.** The fiber of the cantala.
**can·ta·loupe** *also* **can·ta·loup** (kăn′tl-ōp′) *n.* [Fr. *cantaloup* < Ital. *cantalupo* < *Cantalupo,* a former papal villa near Rome.] **1.** A variety of melon, *Cucumis melo cantalupensis,* having fruit with a ribbed, rough rind and aromatic orange flesh. **2.** Any of several melons similar to the cantaloupe. **3.** The fruit of a cantaloupe.
**can·tan·ker·ous** (kăn-tăng′kər-əs) *adj.* [Perh. ult. < ME *contek,* contentious.] Ill-tempered : disagreeable. **—can·tan′ker·ous·ly** *adv.* **—can·tan′ker·ous·ness** *n.*
**can·ta·ta** (kən-tä′tə) *n.* [Ital. *(aria) cantata,* sung (aria) < *cantare,* to sing < Lat., freq. of *canere.*] A vocal and instrumental composition made up of choruses, solos, and recitatives.
**can·teen** (kăn-tēn′) *n.* [Fr. *cantine* < Ital. *cantina,* wine cellar < *canto,* corner.] **1. a.** A store for on-base military personnel. **b.** *Chiefly Brit.* A club for troops. **2.** An institutional recreation hall or cafeteria. **3.** A temporary or mobile eating place, esp. one set up in an emergency <a Red Cross *canteen*> **4. a.** A mess kit. **b.** A box divided into compartments containing a set of cooking gear. **5.** A flask for drinking water of the kind carried by soldiers and hikers.
**can·ter** (kăn′tər) *n.* [Short for *Canterbury gallop,* after *Canterbury,* England, toward which pilgrims rode at an easy pace.] A gait slower than a gallop but faster than a trot. —*v.* **-tered, -ter·ing, -ters.** —*vi.* To move or ride at a canter. —*vt.* To cause (a horse) to go at a canter.
**Can·ter·bur·y bells** (kăn′tər-bĕr′ē) *n.* [From the association of the flowers with the bells on the horses of Canterbury pilgrims.] *(sing. or pl. in number).* A plant native to Europe, *Campanula medium,* widely cultivated for its violet-blue bell-shaped flowers.
**can·tha·ris** (kăn′thər-ĭs) *n., pl.* **can·thar·i·des** (kăn-thăr′ĭ-dēz′) [Lat., a kind of beetle < Gk. *kantharis.*] A toxic preparation of the crushed, dried bodies of the beetle *Lytta vesicatoria* or *Cantharis vesicatoria,* once used as a counterirritant for skin blisters and as an aphrodisiac.
**can·thi** (kăn′thī′) *n. pl. of* CANTHUS.
**can·thi·tis** (kăn-thī′tĭs) *n.* Inflammation of the canthus.
**cant hook** *n.* [< CANT[1].] A peavey.
**can·thus** (kăn′thəs) *n., pl.* **-thi** (-thī′) [LLat. < Gk. *kanthos.*] The corner at either side of the eye, formed by the meeting of the upper and lower eyelids.
**can·ti·cle** (kăn′tĭ-kəl) *n.* [ME < Lat. *canticulum,* dim. of *cantus,* song < *canere,* to sing.] A song or chant, esp. a nonmetrical hymn with words from a Biblical text.
**can·ti·le·ver** (kăn′tl-ē′vər, -ĕv′ər) *n.* [Perh. CANT[1] + LEVER.] **1.** A projecting structure, as a beam, supported at one end. **2.** A structural member, as a beam, that projects beyond a fulcrum and is supported by a balancing member or a downward force behind the fulcrum. **3.** A bracket or block supporting a balcony or cornice. —*vt.* **-vered, -ver·ing, -vers.** To extend outward or build as a cantilever.
**cantilever bridge** *n.* A bridge formed by two projecting beams or trusses that are joined in the center by a connecting member and are supported on piers and anchored by counterbalancing members.
**can·til·late** (kăn′tl-āt′) *vt.* & *vi.* **-lat·ed, -lat·ing, -lates.** [Lat. *cantillare, cantillat-,* to hum < *cantare,* to sing, freq. of *canere.*] To chant or recite in a musical monotone. **—can′til·la′tion** *n.*
**†can·ti·na** (kăn-tē′nə) *n.* [Sp., canteen < Ital., wine cellar < *canto,* corner.] *Southwestern U.S.* A bar that serves liquor.
**can·tle** (kăn′tl) *n.* [ME *cantel,* corner < Norman Fr., dim. of *cant,* corner.] **1.** The rear part of a saddle. **2.** A corner or portion, esp. when cut off.
**can·to** (kăn′tō) *n., pl.* **-tos.** [Ital. < Lat. *cantus,* song. —see CANTICLE.] A principal division of a long poem.
**can·ton** (kăn′tən, -tŏn′) *n.* [Fr. < OFr. < Ital. *cantone,* aug. of *canto,* corner.] **1. a.** A small territorial division of a country, esp. one of the states of Switzerland. **b.** A subdivision of an arrondissement in France. **2.** *Heraldry.* A small, square division of a shield, usu. in the upper right corner. **3.** A usu. rectangular division of a flag in the upper corner next to the staff. —*vt.* **-toned, -ton·ing, -tons. 1.** To divide into cantons or territorial districts. **2.** (kăn-tŏn′, -tōn′). To assign quarters to (troops). **—can′ton·al** (kăn′tə-nəl, kăn-tŏn′əl) *adj.*
**Can·ton crepe** (kăn′tŏn′) *n.* [After *Canton* (Guangzhou), China.] A soft fabric with a finely crinkled texture, similar to crêpe de Chine but heavier.
**Can·ton·ese** (kăn′tə-nēz′, -nēs′) *n.* The dialect of Chinese spoken in and around Guangzhou (formerly Canton), China. **—Can′ton·ese′** *adj.*

**Canton flannel** (kăn′tŏn′) *n.* [After *Canton* (Guangzhou), China.] Flannelette.

**can·ton·ment** (kăn-tōn′mənt, -tŏn′-) *n.* **1.** A group of more or less temporary billets for troops. **2.** Assignment of troops to temporary quarters.

**Canton ware** *n.* [After *Canton* (Guangzhou), China.] Ceramic ware including blue-and-white enameled porcelain exported from China, esp. during the 18th and 19th cent.

**can·tor** (kăn′tər) *n.* [Lat., singer < *canere,* to sing.] **1.** The official soloist or chief singer of the liturgy in a synagogue. **2.** One who leads a church choir or congregation in singing.

**can·trip** (kăn′trəp *n.* [Orig. unknown.] *Scot.* **1.** A witch's trick. **2.** A mischievous trick : PRANK.

**can·tus fir·mus** (kăn′təs fîr′məs, fûr′-) *n.* [Med. Lat., fixed melody.] A plainsong serving as the basis of a polyphonic composition by the addition of contrapuntal voices, as in 15th-cent. polyphony.

**can·u·la** (kăn′yə-lə) *n. var. of* CANNULA.

**can·u·late** (kăn′yə-lāt′) *v. & adj. var. of* CANNULATE.

**can·vas** (kăn′vəs) *n.* [ME *canevas* < Norman Fr. *canevaz* < Lat. *cannabis,* hemp. —see CANNABIS.] **1.** A heavy, coarse, closely woven fabric of cotton, hemp, or flax, used for making tents and sails. **2. a.** A piece of canvas on which a painting is done, esp. an oil painting. **b.** A painting of this kind. **3.** Sailcloth. **4.** Sails as a whole. **5. a.** Tents as a whole. **b.** A circus tent. **6.** A fabric of coarse open weave, used as a foundation for needlework. **7.** The floor of a ring in which boxing or wrestling takes place.

**can·vas·back** (kăn′vəs-băk′) *n.* A North American duck, *Aythya valisneria,* with a reddish-brown head and neck and a whitish back.

**canvas duck** *n.* A fabric of lightweight cotton or linen.

**can·vass** (kăn′vəs) *v.* **-vassed, -vass·ing, -vass·es.** [< obs. *canvass,* to toss in a canvas sheet as punishment < CANVAS.] —*vt.* **1.** To examine carefully or discuss thoroughly. **2. a.** To go through (a region) or go to (persons) to solicit votes, orders, or subscriptions. **b.** To conduct a survey of (public opinion) on a given subject : POLL. —*vi.* **1.** To make a thorough examination or carry on a detailed discussion. **2.** To solicit political support, sales orders, or opinions. —*n.* **1.** A thorough examination or detailed discussion. **2. a.** A solicitation of votes, sales orders, or opinions. **b.** A public opinion survey. **—can′vass·er** *n.*

**can·yon** *also* **ca·ñon** (kăn′yən) *n.* [Sp. *cañon,* aug. of *caña,* tube < Lat. *canna,* reed < Gk. *kanna.*] A narrow chasm with steep walls, formed by running water : GORGE.

**can·zo·ne** (kăn-zō′nē, känt-sō′nä) *n., pl.* **-nes** (-nēz, -näz) *or* **-ni** (-nē) [Ital. < Lat. *cantio,* song < *canere,* to sing.] **1.** A lyric 13th-cent. Italian or Provençal poetic form. **2.** A polyphonic song form evolving from the canzone and resembling the madrigal.

**can·zo·net** (kăn′zə-nĕt′) *n.* [Ital. *canzonetta,* dim. of *canzone.* —see CANZONE.] A short, lighthearted air or song.

**caou·tchouc** (kou′cho͝ok′, -cho͝ok′) *n.* [Fr. < obs. Sp. *cauchuc* < Quechua.] RUBBER¹ 1.

**cap** (kăp) *n.* [ME *cappe* < OE *cæppe* < LLat. *cappa.*] **1.** A usu. soft, close-fitting, brimless or visored head covering. **2. a.** A special head covering worn to indicate occupation, rank, or membership in a specific group. **b.** MORTARBOARD 2. **3. a.** An object similar to a cap <a bottle *cap*> **b.** A limit or restraint <put a *cap* on government spending> **4.** The capital of an architectural column. **5.** The pileus of a fungus. **6. a.** A percussion cap. **b.** A small explosive charge enclosed in paper for use in a toy gun. **7.** Any of several sizes of writing paper, as foolscap. —*vt.* **capped, cap·ping, caps. 1.** To put a cap on. **2.** To lie over or on top of : COVER <Snow *capped* the Alps.> **3.** To apply the finishing touch to : COMPLETE <*cap* a meal with champagne> **4.** To surpass.

**ca·pa·bil·i·ty** (kā′pə-bĭl′ĭ-tē) *n., pl.* **-ties. 1.** The quality or state of being capable : ABILITY. **2.** Potential ability <has the *capability* to excel> **3.** The capacity to be used, treated, or developed for a particular purpose.

**ca·pa·ble** (kā′pə-bəl) *adj.* [Fr. < OFr. < LLat. *capabilis,* spacious < *capere,* to take.] **1.** Having ability or capacity : EFFICIENT. **2.** Having the requisite mental or physical capacity : QUALIFIED. **3.** Open or susceptible to <an error *capable* of correction> **—ca′pa·ble·ness** *n.* **—ca′pa·bly** *adv.*

**ca·pa·cious** (kə-pā′shəs) *adj.* [< Lat. *capax, capac-* < *capere,* to take.] Spacious. **—ca·pa′cious·ly** *adv.* **—ca·pa′cious·ness** *n.*

**ca·pac·i·tance** (kə-păs′ĭ-təns) *n.* [CAPACIT(Y) + -ANCE.] **1.** The ratio of charge to potential on an electrically charged, isolated conductor. **2.** The ratio of the electric charge transferred from one to the other of a pair of conductors to the resulting potential difference between them. **3. a.** The property of a circuit element that permits it to store charge. **b.** The part of the circuit exhibiting capacitance. **—ca·pac′i·tive** (-tĭv) *adj.* **—ca·pac′i·tive·ly** *adv.*

**ca·pac·i·tate** (kə-păs′ĭ-tāt′) *vt.* **-tat·ed, -tat·ing, -tates.** [CAPACIT(Y) + -ATE¹.] To render fit or qualified. **—ca·pac′i·ta′tion** *n.*

**ca·pac·i·tor** (kə-păs′ĭ-tər) *n.* An electric circuit element composed of two metallic plates separated by a dielectric, used to store a charge temporarily.

**ca·pac·i·ty** (kə-păs′ĭ-tē) *n., pl.* **-ties.** [ME *capacite* < OFr. < Lat. *capacitas* < *capax,* spacious. —see CAPACIOUS.] **1. a.** Ability to receive, hold, or absorb. **b.** A measure of this ability : VOLUME. **2.** The maximum amount that can be contained. **3.** The maximum or optimum amount of production. **4.** The ability to learn or retain information. **5.** The ability to do something : FACULTY. **6.** Suitability for or receptiveness to specified treatment <the *capacity* of plastic to be molded> **7.** The position in which one functions : ROLE. **8.** Legal qualification or authority. **9.** *Obs.* Capacitance.

**cap-a-pie** *or* **cap-à-pie** (kăp′ə-pē′) *adv.* [OFr. *(de) cap a pie,* (from) head to foot < OProv. *(de) cap a pe.*] From head to foot.

**ca·par·i·son** (kə-păr′ĭ-sən) *n.* [Obs. Fr. *caparasson* < Sp. *caparazón.*] **1.** A usu. ornamental covering for a horse's saddle or harness. **2.** Richly decorated clothing : FINERY. —*vt.* **-soned, -son·ing, -sons.** To outfit with a caparison.

**cape¹** (kāp) *n.* [Fr., partly < OProv. *cape,* and partly < Sp. *capa,* both < Med. Lat. *cappa,* cloak.] A sleeveless garment fastened at the throat and worn hanging over the shoulders.

**cape²** (kāp) *n.* [ME *cap* < OFr. < OProv. < Lat. *caput,* head.] A point of land projecting into a sea or other body of water : PROMONTORY.

**Cape Cod cottage** *n.* [After *Cape Cod,* Massachusetts.] A compact one- or one-and-a-half-story house with a central chimney and gabled roof.

**Cape cowslip** *n.* [After *Cape* of Good Hope, a province of the Republic of South Africa.] A bulbous South African plant of the genus *Lachenalia,* with drooping red or yellow flower clusters that are widely cultivated as potted plants.

**Cape gooseberry** *n.* A tropical American plant, *Physalis peruviana,* with edible yellow berries and yellow flowers.

**Cape jasmine** *n.* The gardenia.

**cap·e·lin** (kăp′ə-lĭn, kăp′lĭn) *also* **cap·lin** (kăp′lĭn) *n.* [Canadian Fr. *capelan* < Fr., codfish < OProv.] A small, edible marine fish, *Mallotus villosus* of northern Atlantic and Pacific waters, related to and resembling the smelts.

**Ca·pel·la** (kə-pĕl′ə) *n.* [Lat., dim. of *caper,* goat.] A double star in Auriga, the brightest star in the constellation.

**ca·per¹** (kā′pər) *n.* [Alteration of CAPRIOLE.] **1.** A playful leap or hop. **2.** A wild escapade. **3.** A criminal plot or undertaking. —*vi.* **-pered, -per·ing, -pers.** To leap or frisk about : FROLIC.

**ca·per²** (kā′pər) *n.* [ME *caperis* < Lat. *capparis* < Gk. *kapparis.*] **1.** A spiny, trailing shrub, *Capparis spinosa* of the Mediterranean region. **2. a.** A pickled flower bud of the caper, used as a condiment. **b.** A similar pickled bud or pod.

**cap·er·cail·lie** (kăp′ər-kāl′yē, -kā′lē) *also* **cap·er·cail·zie** (-kāl′zē) *n.* [Sc. Gael. *capull coille* : *capull,* horse (prob. < Lat. *caballus*) + *coille,* forest.] A large grouse, *Tetrao urogallus* of northern Europe, with dark plumage and a fanlike tail.

**cape·skin** (kāp′skĭn′) *n.* [After *Cape* of Good Hope, a province of the Republic of South Africa.] Soft sheepskin leather used esp. for gloves.

**Ca·pe·tian** (kə-pē′shən) *adj.* Relating or belonging to the French dynasty (987–1328) founded by Hugh Capet. —*n.* A member of the Capetian dynasty.

**cap·ful** (kăp′fo͝ol′) *n., pl.* **-fuls.** The amount a cap will hold.

**cap gun** *n.* A cap pistol.

**ca·pi·as** (kā′pē-əs) *n.* [ME < Med. Lat. < Lat., you may arrest (the first word of the writ) < *capere,* to seize.] *Law.* A writ authorizing an officer to arrest the person specified therein.

**cap·il·lar·i·ty** (kăp′ə-lăr′ĭ-tē) *n., pl.* **-ties.** The interaction between contacting surfaces of a liquid and a solid that distorts the liquid surface from a planar shape.

**cap·il·lar·o·scope** (kăp′ə-lăr′ə-skōp′) *n.* [CAPILLAR(Y) + -SCOPE.] A microscope used in capillaroscopy.

**cap·il·la·ros·co·py** (căp′ə-lə-rŏs′kə-pē) *n.* [CAPILLAR(Y) + -SCOPY.] Diagnostic examination of the capillaries.

**cap·il·lar·y** (kăp′ə-lĕr′ē) *adj.* [Lat. *capillaris* < *capillus,* hair.] **1.** Relating to or resembling a hair : fine and slender. **2.** Having a small internal diameter, as a tube. **3.** *Anat.* In, of, or relating to the capillaries. **4.** *Physics.* Of or relating to capillarity. —*n., pl.* **-ies. 1.** *Anat.* One of the minute blood vessels that connect the arteries and veins. **2.** A tube with a small internal diameter.

**capillary attraction** *n.* The force that results from greater adhesion of a liquid to a solid surface than internal cohesion of the liquid itself and causes the liquid to be raised against a vertical surface, as water is in a clean glass tube.

**capillary bed** *n.* The capillary network in a particular area or organ of the body.

**cap·i·tal¹** (kăp′ĭ-tl) *n.* [< ME, principal < OFr. < Lat. *capitalis* < *caput,* head.] **1. a.** A city or town that is the official seat of government in a state or nation. ***usage:*** A city or town serving as the seat of a government is a *capital;* the building in which a legislature meets is a *capitol.* **b.** A city or region that is the center of a particular activity. **2. a.** Wealth, as money or property, owned, used, or accumulated in business by an individual, partnership, or corporation. **b.** Material wealth used or available for use in the production of more wealth. **3. a.** The net worth of a business after all liabilities have been deducted. **b.** The funds contributed to a business by the

ă **pat** ā **pay** âr **care** ä **father** ĕ **pet** ē **be** hw **which** ĭ **pit** ī **tie** îr **pier** ŏ **pot** ō **toe** ô **paw, for** oi **noise** o͝o **took**

owners or stockholders. **4.** Capitalists as a group or class. **5.** An asset: advantage. **6.** A capital letter. —*adj.* **1.** First and foremost: PRINCIPAL <decisions of *capital* importance> **2.** Of or relating to a political capital. **3.** First-rate <a *capital* fellow> **4.** Very serious <a *capital* error> **5.** Involving death or calling for the death penalty <a *capital* crime such as premeditated murder> **6.** Of or relating to monetary capital. **7.** Designating an upper-case letter.

**cap·i·tal²** (kăp'ĭ-tl) *n.* [ME < Norman Fr. < LLat. *capitellum,* dim. of *caput,* head.] The upper part of a pillar or column.

**capital account** *n.* **1.** An account stating the amount of funds and assets invested in a business by the owners or stockholders, including retained earnings. **2.** A statement of the net worth of a business enterprise at a given time.

**capital assets** *pl.n.* Long-term assets, as land or buildings.

**capital expenditure** *n.* Funds expended for additions or improvements to plant or equipment.

**capital gain** *n.* Profit from the sale of capital assets.

**capital gains distribution** *n.* A payment to shareholders realized from the sale of capital assets, as securities.

**capital goods** *pl.n.* Goods used in producing commodities.

**cap·i·tal-in·ten·sive** (kăp'ĭ-tl-ĭn-tĕn'sĭv) *adj.* Requiring or having a large expenditure of capital in comparison to labor <a *capital-intensive* industry>

**cap·i·tal·ism** (kăp'ĭ-tl-ĭz'əm) *n.* **1.** An economic system, marked by open competition in a free market, in which the means of production and distribution are privately or corporately owned and development is proportionate to increasing accumulation and reinvestment of profits. **2.** A political or social system considered to be based on capitalism.

**cap·i·tal·ist** (kăp'ĭ-tl-ĭst) *n.* **1.** An investor of capital in business, esp. one having a major interest in an important enterprise. **2.** A very wealthy person. **3.** A supporter of capitalism.

**cap·i·tal·is·tic** (kăp'ĭ-tl-ĭs'tĭk) *adj.* Of or relating to capitalism or capitalists. **—cap'i·tal·is'ti·cal·ly** *adv.*

**cap·i·tal·i·za·tion** (kăp'ĭ-tl-ĭ-zā'shən) *n.* **1.** The act, practice, or result of capitalizing. **2. a.** The total value of owners' shares in a business firm. **b.** The authorized or outstanding stock or bonds in a corporation. **3.** Conversion of anticipated future income into present value. **4.** Use of upper-case letters in printing or writing.

**cap·i·tal·ize** (kăp'ĭ-tl-īz') *v.* **-ized, -iz·ing, -iz·es.** —*vt.* **1.** To use as or convert into capital. **2.** To supply with capital or investment funds. **3.** To authorize a certain amount of capital stock of (a business). **4.** To convert (debt) into capital stock or shares. **5.** To estimate the present value of (an asset). **6.** To include (expenditures) in business accounts as assets instead of expenses. **7. a.** To write or print in upper-case letters. **b.** To begin a word with an upper-case letter. —*vi.* To turn to advantage <*capitalize* on a competitor's error> **—cap'i·tal·iz'a·ble** *adj.*

**capital letter** *n.* A letter written or printed in a size larger than and usu. in a form differing from its equivalent lower-case letter.

**capital levy** *n.* A tax on capital assets or real property.

**cap·i·tal·ly** (kăp'ĭ-tl-ē) *adv.* In an excellent way.

**capital punishment** *n.* Infliction of the death penalty for the commission of certain extremely serious crimes.

**capital ship** *n.* A warship, as a battleship, of the largest class.

**capital stock** *n.* **1.** The total amount of stock authorized for issue by a corporation. **2.** The total stated or par value of the permanently invested capital of a corporation.

**cap·i·tate** (kăp'ĭ-tāt') *adj.* [Lat. *capitatus,* having a head < *caput,* head.] **1.** *Zool.* Enlarged or globular at an end, as some tentacles. **2.** *Bot.* Forming a headlike mass or dense cluster, as the inflorescence of certain flowers.

**cap·i·ta·tion** (kăp'ĭ-tā'shən) *n.* [Lat. *capitatio,* poll tax < *caput,* head.] A tax fixed at an equal sum per person, as a per capita or poll tax. **—cap'i·ta'tive** *adj.*

**cap·i·tol** (kăp'ĭ-tl) *n.* [ME *Capitol,* Jupiter's temple in Rome < Lat. *Capitolium.*] **1.** The building in which a state legislature meets. **2. Capitol.** The building in which the U.S. Congress meets.

**Capitol Hill** *n.* The U.S. Congress.

**ca·pit·u·lar** (kə-pĭch'ə-lər) *adj.* [Med. Lat. *capitularis* < *capitulum,* chapter. —see CHAPTER.] Of or relating to a chapter, esp. an ecclesiastical one. **—ca·pit'u·lar·ly** *adv.*

**ca·pit·u·lar·y** (kə-pĭch'ə-lĕr'ē) *n., pl.* **-ies. 1.** A member of an ecclesiastical chapter. **2. a.** An ecclesiastical or civil ordinance. **b.** A set of such ordinances, esp. those promulgated by Charlemagne and his successors.

**ca·pit·u·late** (kə-pĭch'ə-lāt') *vi.* **-lat·ed, -lat·ing, -lates.** [Med. Lat. *capitulare, capitulat-,* to draw up in chapters < *capitulum,* chapter. —see CHAPTER.] **1.** To surrender under conditional terms. **2.** To cease resisting: ACQUIESCE. **—ca·pit'u·lant** *n.* **—ca·pit'u·la'tor** *n.*

**ca·pit·u·la·tion** (kə-pĭch'ə-lā'shən) *n.* **1.** The act of capitulating. **2.** An instrument containing the terms of surrender. **3.** An enumeration of the central parts of a subject: SUMMARY. **—ca·pit'u·la·to'ry** (-lə-tôr'ē, -tōr'ē) *adj.*

**ca·pit·u·lum** (kə-pĭch'ə-ləm) *n., pl.* **-la** (-lə) [NLat. < Lat., dim. of *caput,* head.] **1.** *Bot.* A dense headlike cluster of stalkless flowers. **2.** A small knob, as the end of a bone or the knoblike tip of an insect's antenna.

**cap·lin** kăp'lĭn) *n. var. of* CAPELIN.

**ca·po¹** (kā'pō) *n., pl.* **-pos.** [Short for *capo tasto* < Ital. *capotasto* : *capo,* head + *tasto,* fret.] A small movable bar placed across the fingerboard of a stringed instrument, as a guitar, for changing simultaneously the pitch of all the strings.

**ca·po²** (kä'pō, kăp'ō) *n., pl.* **-pos.** [Ital., head < Lat. *caput.*] The head of an organized crime family or a branch of it.

**ca·pon** (kā'pŏn', -pən) *n.* [ME *capoun,* partly < OE *capun,* and partly < OFr. *chapon,* both < Lat. *capo.*] A castrated rooster.

**cap·o·ral** (kăp'ər-əl, kăp'ə-răl') *n.* [Fr., short for *tabac de caporal,* corporal's tobacco.] A strong, dark tobacco.

**ca·pote** (kə-pōt') *n.* [Fr., dim. of *cape,* cloak. —see CAPE¹.] A long, usu. hooded cloak or coat.

**cap·per** (kăp'ər) *n.* **1.** One that caps or makes caps. **2.** *Informal.* **a.** A climactic point or ending. **b.** A highlight. **3.** *Slang.* A decoy, as in a confidence game: SHILL.

**cap pistol** *n.* A toy pistol with a hammer action that detonates a mildly explosive cap.

**cap·puc·ci·no** (kăp'ə-chē'nō, kä'pə-) *n., pl.* **-nos.** [Ital., Capuchin (from the resemblance of its color to the color of the monk's habit).] Espresso coffee mixed or topped with steamed milk or cream.

**cap·re·o·late** (kăp'rē-ə-lāt', kə-prē'-) *adj.* [< Lat. *capreolus,* tendril, wild goat. —see CAPRIOLE.] *Biol.* Resembling or having tendrils.

**cap·ric acid** (kăp'rĭk) *n.* [< Lat. *caper, capr-,* goat (from the acid's nasty smell).] A white crystalline compound, $CH_3(CH_2)_8COOH$, obtained by distilling coconut oil and used in manufacturing perfumes and fruit flavors.

**ca·pric·cio** (kə-prē'chō, -chē-ō') *n., pl.* **-cios.** [Ital. —see CAPRICE.] **1.** *Mus.* An instrumental work with an improvised style and a free form. **2.** A prank: caper. **3.** A fanciful whim.

**ca·pric·cio·so** (kə-prē-chō'sō) *adj.* [Ital. < *capriccio.* —see CAPRICE.] *Mus.* Lively and free. —Used as a direction.

**ca·price** (kə-prēs') *n.* [Fr. < Ital. *capriccio* : *capo,* head (< Lat. *caput*) + *riccio,* curly (< Lat. *ericius,* hedgehog).] **1.** An impulsive change of mind: WHIM. **2.** An inclination to change one's mind impulsively. **3.** *Mus.* A capriccio.

**ca·pri·cious** (kə-prĭsh'əs, -prē'shəs) *adj.* Characterized by or subject to whim: UNPREDICTABLE. **—ca·pri'cious·ly** *adv.* **—ca·pri'cious·ness** *n.*

☆ ***syns:*** CAPRICIOUS, CHANGEABLE, ERRATIC, FICKLE, INCONSISTENT, INCONSTANT, MERCURIAL, TEMPERAMENTAL, UNPREDICTABLE, UNSTABLE, UNSTEADY, VARIABLE, VOLATILE, WHIMSICAL *adj. core meaning:* following no predictable pattern <*capricious* winds><a *capricious* flirt>

**Cap·ri·corn** (kăp'rĭ-kôrn') *n.* [ME *Capricorne* < Lat. *Capricornus* : *caper,* goat + *cornu,* horn.] **1.** A constellation in the Southern Hemisphere. **2. a.** The tenth sign of the zodiac. **b.** One born under this sign.

**cap·ri·fi·ca·tion** (kăp'rə-fĭ-kā'shən) *n.* [Lat. *caprificatio* < *caprificare,* to ripen figs by caprification < *caprificus,* caprifig.] A method of assuring pollination of the edible fig by having certain wasps carry pollen from the flowers of the caprifig to those of the edible variety.

**cap·ri·fig** (kăp'rə-fĭg') *n.* [ME < Lat. *caprificus* : *caper,* goat + *ficus,* fig.] A wild variety of fig, *Ficus carica sylvestris* of the eastern Mediterranean region, used in caprification of the edible fig.

**cap·ri·ole** (kăp'rē-ōl') *n.* [Fr. < Ital. *capriola,* somersault < *capriolo,* wild goat < Lat. *capreolus,* dim. of *caper,* goat.] **1.** An upward leap made by a trained horse without going forward and with all feet off the ground. **2.** A leap: jump. —*vi.* **-oled, -ol·ing, -oles.** To perform a capriole.

**ca·pri pants** (kä'prē, kə-prē') *also* **ca·pris** (kä'prēz, kə-prēz') *pl.n.* [After *Capri,* Italy.] Tight-fitting, nearly ankle-length pants with a slit on the outside of the leg bottoms.

**ca·pro·ic acid** (kə-prō'ĭk, kă-) *n.* [< Lat. *caper, capr-,* goat.] A liquid fatty acid, $C_6H_{12}O_2$, occurring in animal fats and oils and used in manufacturing pharmaceuticals and flavors.

**ca·pryl·ic acid** (kə-prĭl'ĭk, kă-) *n.* [< E. *capryl,* a radical found in caprylic acid.] A liquid fatty acid, $C_8H_{16}O_2$, with a rancid taste, used in manufacturing dyes and perfumes.

**cap·sa·i·cin** (kăp-sā'ĭ-sĭn) *n. var. of* CAPSICIN.

**cap screw** *n.* A long-threaded, usu. square-headed bolt used in fastening machine parts.

**Cap·si·an** (kăp'sē-ən) *adj.* [Fr. *capsien* < Lat. *Capsa,* Gafsa, Tunisia, near where remains of this culture were found.] Of or designating a Paleolithic culture of northern Africa and southern Europe.

**cap·si·cin** (kăp'sĭ-sĭn) *also* **cap·sa·i·cin** (kăp-sā'ĭ-sĭn) *n.* [CAPSIC(UM) + -IN.] A peppery, reddish-brown liquid, $C_{18}H_{27}O_3N$, obtained from plants of the genus *Capsicum* and used in flavoring vinegar and pickles and medicinally as an irritant.

**cap·si·cum** (kăp'sĭ-kəm) *n.* [NLat. *Capsicum,* genus name, perh. < Lat. *capsa,* box (from its podlike fruit).] **1.** A tropical plant of the genus *Capsicum.* **2.** The dried fruit of pungent varieties of *C. frutescens,* used as a gastric stimulant and counterirritant.

**cap·sid** (kăp′sĭd) *n.* [< Lat. *capsa,* box.] The proteinaceous covering of a virus particle.

**cap·size** (kăp′sīz′, kăp-sīz′) *vi.* & *vt.* **-sized, -siz·ing, -siz·es.** [Orig. unknown.] To overturn or cause to overturn.

**cap·so·mere** (kăp′sə-mîr′) *n.* [CAPS(ID) + -MERE.] An individual subunit comprising a capsid.

**cap·stan** (kăp′stən, -stăn′) *n.* [ME < Norman Fr. < OProv. *cabestan* < Lat. *capistrum,* halter < *capere,* to seize.] **1.** *Naut.* A device consisting of a vertical cylinder rotated manually or by motor, used for hoisting weights by winding in a cable. **2.** A small cylindrical pulley used for regulating the speed of magnetic tape in a tape recorder.

**cap·stone** (kăp′stōn′) *also* **cope·stone** (kōp′-) *n.* **1.** The top stone of a wall or other structure. **2.** The final stroke : CULMINATION <the *capstone* of a brilliant career>

**cap·su·late** (kăp′sə-lāt′, -lĭt) *also* **cap·su·lat·ed** (-lā′tĭd) *adj.* Being in or having been formed into a capsule. **—cap′su·la′tion** *n.*

**cap·sule** (kăp′səl, -so͞ol) *n.* [Fr. < Lat. *capsula,* dim. of *capsa,* box.] **1.** A soluble container, usu. of gelatin, enclosing an oral dose of medication. **2.** *Anat.* A fibrous, membranous, or fatty envelope, as the sac surrounding the kidney, that encloses an organ or part. **3.** *Microbiol.* A mucopolysaccharide layer enveloping certain bacteria. **4.** *Bot.* **a.** A fruit that contains two or more seeds and that dries and splits open. **b.** The spore case of a bryophyte, as a moss. **5.** A pressurized modular compartment of an aircraft or spacecraft, esp. one designed to accommodate a crew or to be ejected. **6.** A brief summary <a *capsule* of the evening news> *—vt.* **-suled, -sul·ing, -sules. 1.** To enclose in or furnish with a capsule. **2.** To condense or summarize <*capsuled* the news story> **—cap′su·lar** *adj.*

**cap·tain** (kăp′tən) *n.* [ME *capitain* < OFr. < LLat. *capitaneus* < Lat. *caput,* head.] **1.** A commander, leader, or guide, esp.: **a.** The officer in command of a ship. **b.** A precinct chief in a police or fire department. **c.** The designated leader of a sports team. **2. a.** A commissioned officer in the U.S. Army, Air Force, or Marine Corps ranking below a major and above a first lieutenant. **b.** A commissioned officer in the U.S. Navy ranking below a rear admiral and above a commander. **3.** A leading figure <a *captain* of the steel industry> *—vt.* **-tained, -tain·ing, -tains.** To command or direct. **—cap′tain·cy** *n.* **—cap′tain·ship′** *n.*

**cap·tan** (kăp′tən, -tăn′) *n.* [Short for MERCAPTAN.] An agricultural fungicide, $C_9H_8Cl_3NO_2S$.

**cap·tion** (kăp′shən) *n.* [< obs. *caption,* an arrest < Lat. *captio* < *capere,* to seize.] **1.** A title, brief explanation, or description accompanying an illustration or picture. **2.** A film subtitle. **3.** A title or heading, as of a document. **4.** *Law.* The part of a legal document that states the time, place, and authority of its execution. *—vt.* **-tioned, -tion·ing, -tions.** To furnish a caption for.

**cap·tious** (kăp′shəs) *adj.* [ME *capcious* < Lat. *captiosus* < *captio,* sophism < *capere,* to take.] **1.** Inclined to make petty criticisms : CARPING. **2.** Intended to entrap or confuse : DECEPTIVE <*captious* questions> **—cap′tious·ly** *adv.* **—cap′tious·ness** *n.*

**cap·ti·vate** (kăp′tĭ-vāt′) *vt.* **-vat·ed, -vat·ing, -vates.** [LLat. *captivare, captivat-,* to capture < Lat. *captivus,* prisoner < *capere,* to seize.] **1.** To fascinate : enrapture. **2.** *Archaic.* To capture. **—cap′ti·va′tion** *n.* **—cap′ti·va′tor** *n.*

**cap·tive** (kăp′tĭv) *n.* [ME *catif* < OFr. < Lat. *captivus* < *capere,* to seize.] **1.** One, as a prisoner, who is forcibly confined, restrained, or subjugated. **2.** One who is enslaved by a strong emotion. *—adj.* **1.** Held as prisoner <*captive* POW's> **2.** Under restraint or control <*captive* nations> **3.** Captivated : enraptured. **4.** Obliged to be present <a *captive* audience> **—cap·tiv′i·ty** *n.*

**cap·tor** (kăp′tər, -tôr′) *n.* [LLat., hunter < Lat. *capere,* to seize.] One who takes or keeps another as a captive.

**cap·ture** (kăp′chər) *vt.* **-tured, -tur·ing, -tures.** [Fr. < OFr. < Lat. *captura,* a catching of animals < *capere,* to seize.] **1.** To take captive. **2.** To win possession or control of, as in a contest <*captured* the pennant> **3.** To succeed in preserving in a permanent form <*captured* the model's likeness in the painting> *—n.* **1.** The act of capturing : SEIZURE. **2.** One that is seized, caught, or won. **3.** *Physics.* The phenomenon in which an atomic nucleus absorbs a subatomic particle, often with the subsequent emission of radiation.

**ca·puche** (kə-po͞och′, -po͞osh′) *n.* [Ital. *capuccio* < *cappa,* hood.] A hood on a cloak, esp. the long, pointed cowl worn by a Capuchin monk.

**cap·u·chin** (kăp′yə-chĭn, kə-pyo͞o′-, -shĭn) *n.* [Fr. < Ital. *cappuccino* < *cappuccio,* hood. —see CAPUCHE.] **1. Capuchin.** A monk belonging to the Order of Friars Minor Capuchins, a branch that broke away from the Franciscans in 1525. **2.** A woman's hooded cloak. **3.** A long-tailed monkey of the genus *Cebus* of Central and South America, many of which have hoodlike tufts of hair on the head.

**cap·y·ba·ra** (kăp′ə-bär′ə, -băr′ə) *n.* [Port. *capibara* < Tupi.] A large short-tailed semiaquatic rodent, *Hydrochoerus hydrochaeris* of tropical South America, that often attains a length of more than 4 feet or approx. 1.2 meters.

**car** (kär) *n.* [ME *carre,* cart < Norman Fr., ult. < Lat. *carrus,* cart.] **1.** An automobile. **2.** A conveyance, as a streetcar, with wheels that run along tracks. **3.** *Archaic.* A chariot. **4.** A boxlike enclosure for passengers or freight on a conveyance, as an elevator car.

▲ word history: *Car* was a moribund word in English until modern technology brought it back to life. By the middle of the 19th century *car* was only a poetic synonym for *chariot,* to which it is related. With the advent of the railroad, *car* received new life by being used in the United States for what the English call a "railway carriage." The word was also used for the part of an elevator that carries passengers or freight, but *car* sprang to life after the term *motorcar* was adopted late in the 19th century for the automobile. *Motorcar* was soon shortened to *car,* which lost all its archaic and poetic associations in the brave new era.

**car·a·bao** (kăr′ə-bou′, kä′rə-) *n., pl.* **-baos.** [Sp. < Visayan *karabáw.*] The water buffalo.

**car·a·bid** (kăr′ə-bĭd, kə-răb′ĭd) *n.* [NLat. *Carabidae,* family name < Gk. *karabos,* horned beetle.] Any of various black carnivorous beetles of the family Carabidae.

**car·a·bi·neer** *or* **car·a·bi·nier** (kăr′ə-bə-nîr′) *n. vars. of* CARBINEER.

**car·a·cal** (kăr′ə-kăl′) *n.* [Fr. < Turk. *kara kūlāk* : *kara,* black + *kulak,* ear.] A wild cat, *Lynx caracal* of Africa and southern Asia, with short fawn-colored fur and long tufted ears.

**car·a·car·a** (kăr′ə-kăr′ə, -kə-rä′) *n.* [Sp., and Port. *caracará,* both < Tupi *caracara.*] Any of several large, carrion-eating or predatory birds of the subfamily Caracarinae of South and Central America and the southern United States, related to the hawks and falcons.

**car·ack** (kăr′ək) *n. var. of* CARRACK.

**car·a·cole** (kăr′ə-kōl′) *also* **car·a·col** (-kŏl′) *n.* [Fr. < Sp. *caracol,* snail.] A half turn to either side performed by an equestrian. *—vi.* **-coled, -col·ing, -coles.** To perform a caracole.

**car·a·cul** (kăr′ə-kəl) *n.* **1.** The loosely curled fur of a karakul lamb. **2.** *var. of* KARAKUL.

**ca·rafe** (kə-răf′) *n.* [Fr. < Ital. *caraffa* < Sp. *garaffa* < Ar. *gharrāf* < *gharafa,* he dipped.] A glass bottle, often with a flared lip, used for serving water or wine.

**car·a·geen** (kăr′ə-gēn′) *n. var. of* CARRAGEEN.

**car·a·mel** (kăr′ə-məl, -mĕl′, kär′məl) *n.* [Fr. < Sp. *caramelo.*] **1.** A smooth, chewy candy made with sugar, butter, cream or milk, and flavoring. **2.** Burnt sugar used for coloring and sweetening foods.

**car·a·mel·ize** (kăr′ə-mə-līz′, kär′mə-līz′) *vt.* & *vi.* **-ized, -iz·ing, -iz·es.** To convert into or be converted into caramel. **—car′a·mel·i·za′tion** *n.*

**ca·ran·gid** (kə-răn′jĭd, -răng′gĭd) *n.* [NLat. *Carangidae,* family name < Fr. *carangue,* mackerel < Sp. *caranga.*] Any of various fishes of the family Carangidae, including the jacks and pompanos. **—ca·ran′gid** *adj.*

**car·a·pace** (kăr′ə-pās′) *n.* [Fr. < Sp. *carapacho.*] **1.** *Zool.* A hard bony outer covering, as the fused dorsal plates of a turtle or the part of the exoskeleton covering the head and thorax of a crustacean. **2.** A protective covering resembling a carapace.

**car·at** (kăr′ət) *n.* [Fr. < OFr. < Med. Lat. *carratus* < Ar. *qīrāṭ,* weight of four grains < Gk. *keration,* a weight, dim. of *keras,* horn.] **1.** A unit of weight for precious stones, equal to 200 milligrams. **2.** *var. of* KARAT.

**car·a·van** (kăr′ə-văn′) *n.* [Fr. *caravane* < Pers. *kārwān.*] **1.** A group of travelers journeying together, esp. across a desert. **2.** A single file of vehicles or pack animals. **3.** A large covered vehicle : VAN. **4.** *Chiefly Brit.* A house trailer.

**car·a·van·sa·ry** (kăr′ə-văn′sə-rē) *also* **car·a·van·se·rai** (-rī′) *n., pl.* **-ries** *also* **-rais.** [Pers. *kārwānsarāī* : *kārwān,* caravan + *sarāī,* palace.] **1.** An inn built around a large court for accommodating caravans in the Near or Far East. **2.** A large inn.

**car·a·vel** *also* **car·a·velle** (kăr′ə-vĕl′) *or* **car·vel** (kär′vəl, -vĕl′) *n.* [OFr. *caravelle* < OPort. *caravela.*] A small light sailing ship used by the Spanish and Portuguese in the 15th and 16th cent.

**car·a·way** (kăr′ə-wā′) *n.* [ME *carewei,* prob. < Med. Lat. *carvi* < Ar. *karawyā* < Gk. *karon.*] **1.** A Eurasian plant, *Carum carvi,* having finely divided leaves and small whitish flower clusters. **2.** The pungent, aromatic seeds of the caraway, used in cooking.

**carb-** *pref. var. of* CARBO-.

**car·ba·mate** (kär′bə-māt′, kär-băm′āt′) *n.* [CARBAM(IC ACID) + -ATE.] A salt or ester of carbamic acid, esp. one that is used as an insecticide.

**car·bam·ic acid** (kär-băm′ĭk) *n.* [CARB(O)- + AM(IDE) + -IC.] An acid, $CH_3NO_2$, used in the form of its esters and salts, as urea.

**car·ba·mide** (kär′bə-mīd′, kär-băm′ĭd) *n.* [CARB(O)- + AMIDE.] Urea.

**carb·an·i·on** (kär-băn′ī′ən, -ī′ŏn′) *n.* A negatively charged ion located at a carbon position.

**car·ba·ryl** (kär′bə-rĭl′) *n.* [CARB(AMATE) + AR(OMATIC) + -YL.] A carbamate, $C_{11}H_{11}NO_2$, used as a general purpose insecticide.

**car·ben·i·cil·lin** (kär-bĕn′ĭ-sĭl′ĭn) *n.* [CAR(BOXYL) + BEN(ZYL) + (PEN)ICILLIN.] A broad-spectrum antibiotic of the penicillin group.

**car·bide** (kär′bīd′) *n.* A binary carbon compound, esp. calcium carbide, composed of carbon and a more electropositive element.

**car·bine** (kär′bīn′, -bēn′) *n.* [Fr. *carabine* < OFr. *carabin,* soldier

armed with a musket.] A light shoulder rifle with a short barrel, orig. used by cavalry.
**car·bi·neer** (kär′bə-nîr′) *also* **car·a·bi·neer** *or* **car·a·bi·nier** (kăr′ə-) *n.* A soldier armed with a carbine.
**car·bi·nol** (kär′bə-nôl′, -nōl′) *n.* **1.** Methanol. **2.** An alcohol derived from methanol.
**carbo-** *or* **carb-** *pref.* [Fr. < *carbone*, carbon.] Carbon <*carbohydrate*>
**car·bo·cy·clic** (kär′bō-sī′klĭk, -sĭk′lĭk) *adj. Chem.* Having a ring made up of carbon atoms, as benzene.
**car·bo·hy·drase** (kär′bō-hī′drās′, -drāz′) *n.* Any of various enzymes that catalyze the hydrolysis of a carbohydrate.
**car·bo·hy·drate** (kär′bō-hī′drāt′) *n.* Any of a group of chemical compounds, including sugars, starches, and cellulose, containing carbon, hydrogen, and oxygen only, with the ratio of hydrogen to oxygen atoms usu. 2:1.
**car·bo·lat·ed** (kär′bə-lā′tĭd) *adj.* Containing or treated with carbolic acid.
**car·bol·ic acid** (kär-bŏl′ĭk) *n.* [CARB- + Lat. *oleum*, oil + -IC.] PHENOL 1.
**car·bon** (kär′bən) *n.* [Fr. *carbone* < Lat. *carbo*, charcoal.] **1.** *Symbol* **C** A naturally abundant nonmetallic element that occurs in many inorganic and in all organic compounds, exists in amorphous, graphitic, and diamond allotropes, and is capable of chemical self-bonding to form a large number of chemically, biologically, and commercially important long-chain molecules; atomic number 6; atomic weight 12.01115. **2. a.** A sheet of carbon paper. **b.** A copy made by using carbon paper. **3.** *Elect.* **a.** Either of two rods through which current flows to form an arc in lighting or in welding. **b.** A carbonaceous electrode in an electric cell. **—car′bon·ous** (-bə-nəs) *adj.*
**carbon 14** *n.* A naturally radioactive carbon isotope with atomic mass 14 and half-life 5,700 years, used in dating ancient carbon-containing objects.
**car·bon-14 dating** (kär′bən-fôr-tēn′, -fōr-) *n.* Carbon dating.
**car·bo·na·ceous** (kär′bə-nā′shəs) *adj.* Composed of, containing, relating to, or yielding carbon.
**car·bo·na·do**[1] (kär′bə-nā′dō, -nä′-) *n., pl.* **-dos** *or* **-does.** [Sp. *carbonada* < *carbón*, charcoal < Lat. *carbo.*] A piece of scored and broiled fish, fowl, or meat. —*vt.* **-doed, -do·ing, -dos. 1.** To score and broil (fish, fowl, or meat). **2.** *Archaic.* To slice or cut.
**car·bo·na·do**[2] (kär′bə-nā′dō, -nä′-) *n., pl.* **-dos.** [Port. < *carbone*, carbon < Fr. —see CARBON.] A form of chiefly Brazilian opaque or dark-colored diamond, used for drills.
**car·bon·ate** (kär′bə-nāt′) *vt.* **-at·ed, -at·ing, -ates. 1.** To charge with carbon dioxide gas, as a beverage. **2.** CARBONIZE 1. **3.** To change into a carbonate. —*n.* (-nāt′, -nĭt). A salt or ester of carbonic acid. **—car′bon·a′tion** *n.* **—car′bon·a′tor** *n.*
**carbonated water** *n.* SODA WATER 1.
**carbon bisulfide** *n.* Carbon disulfide.
**carbon black** *n.* Any of various finely divided forms of carbon derived from the incomplete combustion of natural gas or petroleum oil and used primarily in ink and rubber.
**carbon copy** *n.* **1.** A replica, as of a letter, made by using carbon paper. **2.** *Informal.* A copy of another : DUPLICATE <a *carbon copy* of one's parent>
**carbon cycle** *n.* **1.** The carbon-nitrogen cycle. **2.** *Biol.* The cycle of natural processes in which atmospheric carbon in the form of carbon dioxide is converted to carbohydrates by photosynthesis, metabolized by animals, and ultimately returned to the atmosphere as a carbon dioxide waste or decomposition product.
**carbon dating** *n.* Determination of the approximate age of carbon-containing objects by use of the radiation rate of carbon 14.
**carbon dioxide** *n.* A colorless, odorless, incombustible gas, $CO_2$, formed during respiration, combustion, and organic decomposition and used in food refrigeration, carbonated beverages, inert atmospheres, fire extinguishers, and aerosols.
**carbon disulfide** *n.* A clear, flammable liquid, $CS_2$, used to make viscose rayon and cellophane, as a solvent for fats, rubber, resins, waxes, and sulfur, and in matches, fumigants, and pesticides.
**carbon fiber** *n.* A very strong light fiber made by pyrolyzing synthetic fibers, as rayon, and used in high-strength composites.
**car·bon·ic acid** (kär-bŏn′ĭk) *n.* A weak, unstable acid, $H_2CO_3$, present in solutions of carbon dioxide in water.
**carbonic acid gas** *n.* Carbon dioxide.
**car·bon·if·er·ous** (kär′bə-nĭf′ər-əs) *adj.* **1.** Producing, containing, or relating to carbon or coal. **2. Carboniferous.** Of, belonging to, or designating a geologic division of the Paleozoic era following the Devonian and preceding the Permian, including the Mississippian and Pennsylvanian periods and marked esp. in the Pennsylvanian by swamp formation and deposition of plant remains later hardened into coal. —*n.* **Carboniferous.** The Carboniferous period.
**car·bo·ni·um** (kär-bō′nē-əm) *n.* A positively charged organic ion, as $H_3C$, having one less electron than a corresponding free radical and behaving chemically as if the positive charge were localized on the carbon atom.
**car·bon·i·za·tion** (kär′bə-nĭ-zā′shən) *n.* **1.** The process of carbonizing. **2.** Destructive distillation of bituminous coal to obtain coke and other fractions.
**car·bon·ize** (kär′bə-nīz′) *vt.* **-ized, -iz·ing, -iz·es. 1.** To reduce or convert to carbon, as by partial burning. **2.** To coat or combine with carbon. **—car′bon·iz′er** *n.*
**carbon monoxide** *n.* A colorless, odorless, extremely poisonous gas, CO, formed by incomplete combustion of carbon or a carbonaceous material, including gasoline.
**car·bon-ni·tro·gen cycle** (kär′bən-nī′trə-jən) *n.* A chain of thermonuclear reactions in which nitrogen isotopes are formed in intermediate stages and carbon acts as a catalyst to convert four protons into one helium nucleus, the entire sequence thought to generate significant amounts of energy in certain classes of stars.
**carbon paper** *n.* A lightweight paper coated on one side with a dark waxy pigment that is transferred by the impact of typewriter keys or by writing pressure onto paper.
**carbon process** *n.* A photographic printing process using permanent pigments, as carbon, in a sensitized tissue or film of gelatin.
**carbon star** *n.* Any of a class of carbon-rich stars with primarily low temperatures.
**carbon tetrachloride** *n.* A poisonous, nonflammable, colorless liquid, $CCl_4$, used in fire extinguishers and as a solvent.
**car·bon·yl** (kär′bə-nĭl′, -nēl′) *n.* **1.** The bivalent radical CO. **2.** A metal compound, as $Ni(CO)_4$, containing the CO group. **—car′bon·yl′ic** (-nĭl′ĭk) *adj.*
**carbonyl chloride** *n.* Phosgene.
**car·bo·rane** (kär′bə-rān′) *n.* [Blend of CARBON and BORANE.] Any of a class of stable compounds containing carbon, hydrogen, and boron.
**Car·bo·run·dum** (kär′bə-rŭn′dəm). A trademark for a silicon carbide abrasive.
**car·box·yl** (kär-bŏk′səl) *n.* [CARB(O)- + OX(Y)- + -YL.] A univalent radical, COOH, typical of all organic acids. **—car′box·yl′ic** (-sĭl′ĭk) *adj.*
**car·box·yl·ase** (kär-bŏk′sə-lās′, -lāz′) *n.* A plant enzyme producing acetaldehyde and carbon dioxide from pyruvic acid.
**car·box·yl·a·tion** (kär-bŏk′sə-lā′shən) *n.* Introduction of a carboxyl group into a compound or molecule.
**car·box·yl·ic acid** (kär′bŏk-sĭl′ĭk) *n.* An organic acid containing one or more carboxyl groups.
**car·boy** (kär′boi′) *n.* [Pers. *qarāba.*] A large glass or plastic bottle, usu. encased in a protective basket or crate and often used for holding corrosive liquids.
**car·bun·cle** (kär′bŭng′kəl) *n.* [ME < OFr. < Lat. *carbunculus*, dim. of *carbo*, coal.] **1.** A painful, localized pus-producing infection of the skin and subcutaneous tissue. **2.** *Obs.* A deep-red, unfaceted, convex garnet. **—car′bun′cled** *adj.* **—car·bun′cu·lar** (-kyə-lər) *adj.*
**car·bu·ret** (kär′bə-rāt′, -byə-, -rĕt′) *vt.* **-ret·ed, -ret·ing, -rets** *or* **-ret·ted, -ret·ting, -rets.** [< obs. *carburet*, carbide < Fr. *carbure* < Lat. *carbo*, carbon.] To mix with carbon or hydrocarbons so as to increase available fuel energy.
**car·bu·re·tor** (kär′bə-rā′tər, -byə-) *n.* [< CARBURET.] A device in gasoline engines that mixes fuel vapor and air before combustion.
**car·bu·ret·tor** (kär′bə-rĕt′ər, -byə-) *n. Chiefly Brit. var. of* CARBURETOR.
**car·bu·rize** (kär′bə-rīz′, -byə-) *vt.* **-rized, -riz·ing, -riz·es.** [CARBUR(ET) + -IZE.] **1.** To treat with carbon. **2.** To treat with hydrocarbons. **—car′bu·ri·za′tion** *n.*
**car·ca·jou** (kär′kə-jōō′, -zhōō′) *n.* [Canadian Fr. < Algonquian *karkajou.*] The wolverine.
**car·ca·net** (kär′kə-nĕt′, -nĭt) *n.* [OFr. *carcan*, collar.] A jeweled necklace, collar, or headband.
**car·case** (kär′kəs) *n. Archaic. var. of* CARCASS.
**car·cass** (kär′kəs) *n.* [Fr. *carcasse* < OFr. *carcois.*] **1.** The dead body of a slaughtered animal. **2.** The living body of a human being. **3.** Worthless remains <the *carcass* of a once powerful and glorious empire> **4.** A framework or basic structure, as of a ruined building.
**carcino-** *pref.* [Gk. *karkino-* < *karkinos*, crab, cancer.] Cancer : cancerous <*carcinogen*>
**car·cin·o·gen** (kär-sĭn′ə-jən, kär′sə-nə-jĕn′) *n.* A cancer-causing agent. **—car′ci·no·gen′e·sis** (kär′sə-nə-jĕn′ə-sĭs) *n.* **—car′cin·o·gen′ic** (kär′sə-nə-jĕn′ĭk) *adj.* **—car′ci·no·ge·nic′i·ty** (kär′sə-nə-jə-nĭs′ĭ-tē) *n.*
**car·ci·no·ma** (kär′sə-nō′mə) *n., pl.* **-mas** *or* **-ma·ta** (-mə-tə) [Lat., cancerous ulcer < Gk. *karkinōma* < *karkinos.*] A malignant tumor derived from epithelial tissue. **—car′ci·nom′a·tous** (-nŏm′ə-təs, -nō′mə-) *adj.* **—car′ci·no′ma·toid** (-nō′mə-toid′) *adj.*
**car·ci·no·ma·to·sis** (kär′sə-nō′mə-tō′sĭs) *n.* Existence of carcinomas at many anatomical sites.
**car coat** *n.* A three-quarter length coat.
**card**[1] (kärd) *n.* [ME *carde*, OFr. *carte* < Lat. *charta*, leaf of papyrus < Gk. *khartēs.*] **1.** A small, flat, usu. rectangular piece of stiff paper or thin pasteboard, esp.: **a.** A playing card. **b.** A greeting card. **c.** A postcard. **d.** One bearing a person's name and other information used for identification or classification. **2. cards** *(sing. or pl. in number).*

**a.** A game played with cards. **b.** The playing of a game with cards. **3.** A program, esp. for a sports event. **4.** A compass card. **5.** *Informal.* An amusing or eccentric person : CHARACTER. —*vt.* **card·ed, card·ing, cards. 1.** To furnish with or attach to a card. **2.** To list on a card : CATALOGUE. **3.** *Informal.* To check the identification of, esp. so as to verify legal age. **—have a card up (one's) sleeve.** To have a secret resource or plan in reserve. **—in the cards.** Apt or destined to happen. **—put (**or **lay) (one's) cards on the table.** To reveal one's intentions or motives frankly and clearly.

**card²** (kärd) *n.* [ME *carde* < Med. Lat. *cardus* < Lat. *cardus,* thistle.] **1.** A wire-toothed brush or a machine fitted with rows of wire teeth used to disentangle fibers, as of wool, before spinning. **2.** A device for raising the nap on a fabric. —*vt.* **card·ed, card·ing, cards.** To comb out or brush with a card. **—card'er** *n.*

**car·da·mom** *or* **car·da·mum** (kär'də-məm) *also* **car·da·mon** (-mən) *n.* [Lat. *cardamomum* < Gk. *kardamōmon* : *kardamon,* cress + *amōmon,* an Indian spice.] **1.** A tropical Asiatic perennial plant, *Elettaria cardamomum,* with large hairy leaves and capsular fruit whose seeds are used as a spice and in medicine. **2.** An East Indian plant, *Amomum cardamomum,* whose seeds are used as an inferior substitute for true cardamom seed.

**card·board** (kärd'bôrd', -bōrd') *n.* A thin stiff pasteboard.

**card-car·ry·ing** (kärd'kăr'ē-ĭng) *adj.* [From the assumption that such a person carries a membership card.] **1.** Being an enrolled member of an organization, esp. the Communist party. **2.** Being strongly identified with or devoted to a cause or ideal <a *card-carrying* conservative>

**card catalog** *n.* An alphabetical listing, esp. of books in a library, made with a separate card for each item.

**cardi-** *pref. var. of* CARDIO-.

**car·di·a** (kär'dē-ə) *n.* [Gk. *kardia,* heart, cardiac orifice of the stomach.] The opening of the esophagus into the stomach.

**car·di·ac** (kär'dē-ăk') *adj.* [Lat. *cardiacus* < Gk. *kardiakos* < *kardia,* heart.] **1.** Of, near, or relating to the heart. **2.** Of or relating to the cardia. —*n.* One having a heart disorder.

**cardiac massage** *n.* A resuscitative procedure characterized by the rhythmic compression of the chest in an effort to restore proper circulation and respiration.

**cardiac muscle** *n.* The striated muscle of the heart.

**car·di·al·gia** (kär'dē-ăl'jə, -jē-ə) *n.* [Gk. *kardialgia* : *kardia,* heart + *algos,* pain.] Heartburn.

**car·di·gan** (kär'dĭ-gən) *n.* [After the Seventh Earl of *Cardigan;* James Thomas Brudenell (1797–1868).] A sweater or knitted jacket opening down the front.

**car·di·nal** (kär'dn-əl, kärd'nəl) *adj.* [ME < Lat. *cardinalis,* principal, pertaining to a hinge < *cardo,* hinge.] **1.** Being of prime importance : PIVOTAL. **2.** Of a dark to deep or vivid red color. —*n.* **1.** *Rom. Cath. Ch.* A member of the College of Cardinals who is appointed by and ranks just below the pope. **2.** A dark to deep or vivid red. **3.** A North American bird, *Richmondena cardinalis,* with a crested head, a short thick bill, and bright-red plumage in the male. **4.** A short hooded cloak, orig. of scarlet cloth, worn by women in the 18th cent. **5.** A cardinal number. **—car'di·nal·ship'** *n.*

**car·di·nal·ate** (kär'dn-ə-lĭt, kärd'nə-, -lāt') *n. Rom. Cath. Ch.* **1.** The College of Cardinals. **2.** The position, rank, dignity, or term of a cardinal.

**cardinal flower** *n.* A plant, *Lobelia cardinalis* of eastern North America, with a terminal cluster of brilliant scarlet flowers.

**cardinal number** *n.* A number, as 3 or 11 or 412, indicating quantity but not order.

**cardinal point** *n.* One of the four main directions on a compass.

**cardinal virtues** *pl.n.* The four qualities of justice, prudence, fortitude, and temperance.

**cardio-** *or* **cardi-** *pref.* [Gk. *kardio-* < *kardia,* heart.] Heart <*cardio*vascular>

**car·di·o·ac·cel·er·a·tor** (kär'dē-ō'ăk-sĕl'ə-rā'tər) *n.* An agent that increases the heart rate. **—car'di·o'ac·cel'er·a'tion** *n.*

**car·di·o·gen·ic** (kär'dē-ō-jĕn'ĭk, -jē'nĭk) *adj.* Originating in a cardiac condition.

**car·di·o·gram** (kär'dē-ə-grăm') *n.* The curve traced by a cardiograph, used in diagnosis of heart defects.

**car·di·o·graph** (kär'dē-ə-grăf') *n.* A device for recording the mechanical movements of the heart. **—car'di·og'ra·phy** (-ŏg'rə-fē) *n.*

**car·di·oid** (kär'dē-oid') *n.* A heart-shaped plane curve, the locus of a fixed point on a circle that rolls on the circumference of another circle with the same radius.

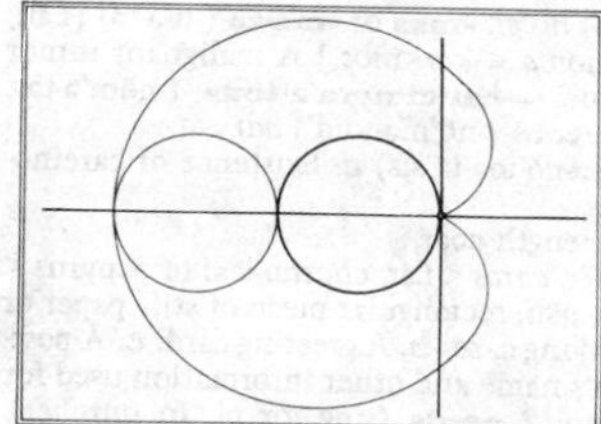

**cardioid**

**car·di·ol·o·gy** (kär'dē-ŏl'ə-jē) *n.* The medical study of the diseases and functioning of the heart. **—car'di·ol'o·gist** *n.*

**car·di·o·meg·a·ly** (kär'dē-ō-mĕg'ə-lē) *n.* Megalocardia.

**car·di·o·pul·mo·nar·y** (kär'dē-ō-pŏŏl'mə-nĕr'ē) *adj.* Of or relating to the heart and the lungs.

**cardiopulmonary resuscitation** *n.* A procedure employed after cardiac arrest in which cardiac massage, drugs, and mouth-to-mouth resuscitation are used to restore breathing.

**car·di·o·res·pi·ra·to·ry** (kär'dē-ō-rĕs'pər-ə-tôr'ē, -rĭ-spīr'ə-tôr'ē, -tōr'ē) *adj.* Of or relating to the heart and the respiratory system.

**car·di·o·vas·cu·lar** (kär'dē-ō-văs'kyə-lər) *adj.* Of, relating to, or involving the heart and the blood vessels.

**car·di·tis** (kär-dī'tĭs) *n.* Inflammation of the heart.

**car·doon** (kär-dōōn') *n.* [Fr. *cardon* < Prov. < Lat. *carduus,* wild thistle.] A plant, *Cynara cardunculus* of southern Europe, having spiny leaves, purple flowers, and an edible leafstalk.

**card·sharp** (kärd'shärp') *also* **card·sharp·er** (-shär'pər) *n.* An expert cheat at cards. **—card'sharp'ing** *n.*

**care** (kâr) *n.* [ME < OE *cearu.*] **1.** A troubled, distressed state of mind : WORRY. **2.** Mental suffering : GRIEF. **3.** A source of worry, attention, or solicitude. **4.** Caution in avoiding harm or danger. **5.** The function of watching, guarding, or overseeing : CHARGE. **6.** Attentiveness to detail : painstaking application <read the manuscript with *care*> —*v.* **cared, car·ing, cares.** —*vi.* **1.** To be concerned or interested. **2.** To object : mind <I won't *care* if you borrow my book.> **3.** To have a liking or attachment. —*vt.* **1.** To be inclined : WISH <We don't *care* to go.> **2.** To be concerned to the degree of <I don't *care* one bit what you think.>

☆ ***syns:*** CARE, CHARGE, CUSTODY, GUARDIANSHIP, SUPERVISION *n. core meaning* : the function of watching, guarding, or overseeing <left the children in the *care* of their grandparents>

**ca·reen** (kə-rēn') *v.* **-reened, -reen·ing, -reens.** [< Fr. *(en) carene,* (on) the keel < OFr. *carene* < OItal. *carena* < Lat. *carina.*] —*vi.* **1.** To swerve or lurch while in motion <a car *careening* around a corner> **2.** *Naut.* To lean to one side, as a ship sailing in the wind. **3.** *Naut.* To turn a ship on its side for cleaning, caulking, or repairing. —*vt. Naut.* **1.** To cause (a ship) to lean to one side : TILT. **2. a.** To lean (a ship) on one side for cleaning, caulking, or repairing. **b.** To clean, caulk, or repair (a ship in this position). —*n. Naut.* **1.** The act or process of careening a ship. **2.** The position of a careened ship. **—ca·reen'er** *n.*

**ca·reer** (kə-rîr') *n.* [Fr. *carrière* < OFr., racecourse < OProv. *carriera,* street < Med. Lat. *(via) carraria,* (road) for carts < LLat. *carra,* cart. —see CAR.] **1. a.** A chosen profession or occupation <a *career* in the foreign service> **b.** The general progression of one's life, esp. in one's profession <a professor with a distinguished *career*> **2.** *Archaic.* **a.** A path or course. **b.** A rapid course or swift progression, as of the sun through the heavens. **3. a.** Speed <"My hasting days fly on with full *career*" —Milton> **b.** The moment of highest pitch or peak activity <The empire was now in the full *career* of its triumphs.> —*vi.* **-reered, -reer·ing, -reers.** To rush headlong. —*adj.* Having undertaken a given occupation <*career* diplomats>

**ca·reer·ism** (kə-rîr'ĭz'əm) *n.* The practice of seeking one's professional advancement by all possible means. **—ca·reer'ist** *n.*

**care·free** (kâr'frē') *adj.* Free from all cares.

**care·ful** (kâr'fəl) *adj.* **1.** Cautious in thought, speech, or action : CIRCUMSPECT. **2.** Thorough and painstaking : CONSCIENTIOUS. **3.** Showing care : SOLICITOUS. **4.** *Archaic.* Full of cares or anxiety. **—care'ful·ly** *adv.* **—care'ful·ness** *n.*

☆ ***syns:*** **1.** CAREFUL, CAUTIOUS, CHARY, CIRCUMSPECT, PRUDENT *adj. core meaning* : trying attentively to avoid danger or risk <a *careful* driver> ***ant:*** reckless **2.** CAREFUL, METICULOUS, PAINSTAKING, SCRUPULOUS *adj. core meaning* : marked by attention to all aspects or details <*careful* writing> ***ant:*** careless

**care·less** (kâr'lĭs) *adj.* **1.** Not taking sufficient care : NEGLIGENT. **2.** Marked by or resulting from lack of forethought or thoroughness. **3.** Showing a lack of consideration. **4.** Indifferent : unconcerned <*careless* about your health> **5.** Effortless : unstudied <*careless* elegance> **6.** Carefree. **—care'less·ly** *adv.* **—care'less·ness** *n.*

☆ ***syns:*** **1.** CARELESS, FECKLESS, HEEDLESS, INATTENTIVE, THOUGHTLESS *adj. core meaning* : lacking or marked by a lack of care <a *careless* remark> ***ant:*** careful **2.** CARELESS, MESSY, SLAPDASH, SLIPSHOD, SLOPPY, SLOVENLY, UNTIDY *adj. core meaning* : indifferent to accuracy or neatness <a *careless* writer> ***ant:*** careful

**ca·ress** (kə-rĕs') *n.* [Fr. *caresse* < Ital. *carezza* < *caro,* dear < Lat. *carus.*] A gentle touch or gesture of fondness, tenderness, or love. —*vt.* **-ressed, -ress·ing, -ress·es. 1. a.** To touch or stroke affectionately or lovingly. **b.** To touch or move as if with a caress <a warm, soft breeze that *caressed* us> **2.** To treat fondly, kindly, or favorably. **—ca·ress'er** *n.* **—ca·ress'ing·ly** *adv.* **—ca·res'sive** *adj.*

☆ ***syns:*** CARESS, CUDDLE, FONDLE, PET *v. core meaning* : to touch or stroke affectionately <patted and *caressed* the puppy>

**car·et** (kăr′ĭt) *n.* [Lat., there is lacking < *carere,* to lack.] A proofreading symbol (‸) used to indicate where material is to be inserted in a written or printed line.
**care·tak·er** (kâr′tā′kər) *n.* One employed to look after goods, property, or another person : CUSTODIAN.
**care·worn** (kâr′wôrn′, -wōrn′) *adj.* Showing the effects of care <a tired, *careworn* face>
**car·fare** (kär′fâr′) *n.* Fare charged a passenger.
**car·go** (kär′gō) *n., pl.* **-goes** or **-gos.** [Sp. < *cargar,* to load < LLat. *carricare* < Lat. *carrus,* cart, of Celtic orig.] Freight carried by a ship, aircraft, or other transport vehicle.
**car·hop** (kär′hŏp′) *n.* One who waits on customers at a drive-in restaurant.
**Car·ib** (kăr′ĭb) *n., pl.* **Carib** or **-ibs.** [Sp. *Caribe,* of Cariban orig.] **1.** *also* **Car·i·ban** (kăr′ə-bən, kə-rē′bən). **a.** A group of American Indian peoples of northern South America and the Lesser Antilles. **b.** A member of one of these peoples. **2.** Any of the languages of the Carib. **—Car′ib** *adj.*
**Car·i·ban** (kăr′ə-bən, kə-rē′bən) *n., pl.* **Cariban** or **-bans. 1.** *var. of* CARIB 1. **2.** A language family comprising the Carib languages. **—Car′i·ban** *adj.*
**Car·ib·be·an** (kăr′ə-bē′ən, kə-rĭb′ē-ən) *n.* A Carib Indian. *—adj.* **1.** Of or relating to the Caribbean Sea and its islands. **2.** Of or relating to the Carib or their language.
**ca·ri·be** (kə-rē′bē) *n.* [Am. Sp. < Sp. *Caribe,* Carib.] The piranha.
**car·i·bou** (kăr′ə-bo͞o′) *n., pl.* **caribou** or **-bous.** [Canadian Fr., of Algonquian orig.] A deer, *Rangifer tarandus* of arctic regions of the New World, having antlers in both sexes.
**car·i·ca·ture** (kăr′ĭ-kə-cho͝or′) *n.* [Fr. < Ital. *caricatura* < *caricare,* to exaggerate, to load < Lat. *carrus,* cart, of Celtic orig.] **1.** An esp. pictorial representation in which the subject's distinctive features or peculiarities are intentionally distorted or exaggerated to generate a comic or grotesque effect. **2.** The process or art of creating caricatures. **3.** An imitation so inferior as to be absurd. *—vt.* **-tured, -turing, -tures.** To represent in or as if in a caricature : SATIRIZE. **—car′i·ca·tur′ist** *n.*
**car·ies** (kâr′ēz) *n., pl.* **caries.** [Lat.] Bone or tooth decay.
**car·il·lon** (kăr′ə-lŏn′, -lən) *n.* [Fr., alteration of OFr. *quarregnon* < LLat. *quaternio,* set of four. —see QUATERNION.] **1.** A stationary set of chromatically tuned bells in a tower, usu. played via a keyboard. **2.** A composition arranged or written for a carillon. *—vi.* **-lonned, -lon·ning, -lons.** To play a carillon.
**car·il·lon·neur** (kăr′ə-lə-nûr′) *n.* A player of a carillon.
**ca·ri·na** (kə-rī′nə, -rē′-) *n., pl.* **-nae** (-nē′) [NLat. < Lat., keel.] A keel-shaped ridge, as that on the breastbone of a bird or in the petals of certain flowers.
**Ca·ri·na** (kə-rī′nə) *n.* [Lat. < *carina,* keel.] A constellation in the Southern Hemisphere.
**car·i·nate** (kăr′ə-nāt′, -nĭt) *also* **car·i·nat·ed** (-nā′tĭd) *adj. Biol.* Having or shaped like a keel : RIDGED.
**car·i·ole** *also* **car·ri·ole** (kăr′ē-ōl′) *n.* [Fr. *carriole* < OProv. *carriola,* dim. of *carri,* chariot < Lat. *carrus,* cart, of Celtic orig.] **1.** A small, open one-horse carriage. **2.** A light covered cart.
**car·i·ous** (kâr′ē-əs) *adj.* Having caries : DECAYED. **—car′i·os′i·ty** (-ŏs′ĭ-tē), **car′i·ous·ness** *n.*
**†carl** or **carle** (kärl) *n.* [ME < ON *karl,* man.] **1.** *Archaic.* A peasant or farmer. **2.** *Obs.* A bondman : serf. **3.** *Regional.* A churl.
**car·line** or **car·lin** (kär′lĭn) *n.* [ME *kerling* < ON < *karl,* man.] *Scot.* A woman, esp. an old woman.
**car·ling** (kär′lĭng, -lĭn) *n.* [Fr. *carlingue* < OFr. *calingue,* prob. < ON *kerling,* old woman < *karl,* man.] *Naut.* One of the short fore and aft timbers that connect the transverse beams supporting a ship's deck.
**Carl·ist** (kär′lĭst) *n.* A supporter of Don Carlos, the pretender to the Spanish throne, or his heirs. **—Carl′ism** *n.*
**car·load** (kär′lōd′) *n.* **1.** The load that a car can carry. **2.** The official minimum weight necessary to ship freight at the carload rate.
**Car·lo·vin·gian** (kär′lə-vĭn′jən, -jē-ən) *adj.* & *n. var. of* CAROLINGIAN.
**car·man** (kär′mən) *n.* **1.** A conductor, as of a streetcar. **2.** A driver of a car or cart.
**Car·mel·ite** (kär′mə-līt′) *n.* **1.** A monk or mendicant friar belonging to the order of Our Lady of Mount Carmel, founded in 1155. **2.** A member of a community of nuns of the Carmelite order, founded in 1452. **—Car′mel·ite′** *adj.*
**car·min·a·tive** (kär-mĭn′ə-tĭv, kär′mə-nā′-) *adj.* [ME *carminatif* < Lat. *carminare,* to card wool < *carmen,* card for wool.] Inducing expulsion of gas from the stomach and intestines. *—n.* A carminative drug.
**car·mine** (kär′mĭn, -mīn′) *n.* [Fr. *carmin* < Med. Lat. *carminium,* prob. a blend of Ar. *qirmiz,* kermes, and Lat. *minium,* cinnabar.] **1.** A strong to vivid red color. **2.** A crimson pigment derived from cochineal. **—car′mine** *adj.*
**car·nage** (kär′nĭj) *n.* [OFr. < OItal. *carnaggio* < LLat. *carnactium,* meat < Lat. *caro.*] **1.** Massive slaughter : MASSACRE <battlefield *carnage*> **2.** *Obs.* Corpses, esp. of troops killed in battle.
**car·nal** (kär′nəl) *adj.* [ME < Lat. *carnalis* < Lat. *caro,* flesh.] **1.** Relating to sensual desires and appetites. **2.** Earthly : temporal. **—car·nal′i·ty** (kär-năl′ĭ-tē) *n.* **—car′nal·ly** *adv.*
**carnal knowledge** *n.* Sexual intercourse.
**car·nall·ite** (kär′nə-līt′) *n.* [G. *Carnallit,* after Rudolf von *Carnall* (1804–1874).] A white, brownish, or reddish mineral, $KCl{\cdot}MgCl_2{\cdot}6H_2O$, used to make potash salts.
**car·nas·si·al** (kär-năs′ē-əl) *adj.* [Fr. *carnassier,* carnivorous < Lat. *caro,* flesh.] Adapted for tearing apart flesh. —Used of teeth. *—n.* The last upper premolar and the first lower molar teeth in carnivorous mammals.
**car·na·tion** (kär-nā′shən) *n.* [Prob. < OFr., flesh-colored < OItal. *carnagione* < *carne,* flesh < Lat. *caro.*] **1. a.** A Eurasian plant, *Dianthus caryophyllus,* widely cultivated for its fragrant white, pink, or red flowers with fringed petals. **b.** The flower of the carnation. **2.** A flesh-colored tint formerly used in painting.
**car·nau·ba** (kär-nô′bə, -nou′-) *n.* [Port., prob. of Tupi origin.] **1.** A palm tree, *Copernica cerifera* of tropical South America. **2.** *also* **carnauba wax.** A hard wax obtained from the leaves of the carnauba, used as a polish and in candles.
**car·nel·ian** (kär-nēl′yən) *also* **cor·nel·ian** (kôr-) *n.* [ME *corneline* < OFr.] A pale to deep red or reddish-brown variety of clear chalcedony, used in jewelry.
**car·net** (kär-nā′) *n.* [Fr., notebook < OFr. *quernet* < Lat. *quaterni,* group of four < *quattour,* four.] **1.** An official, esp. international permit to cross borders. **2.** A book of bus or subway tickets. **3.** A book of postage stamps.
**car·ney** (kär′nē) *n. var. of* CARNY.
**car·ni·tine** (kär′nĭ-tēn′) *n.* [G. *Karnitin* < *Karnin,* a basic substance derived from meat.] A betaine commonly occurring in the liver and in skeletal muscle.
**car·ni·val** (kär′nə-vəl) *n.* [Ital. *carnevale* < OItal. *carnelevare,* Shrovetide : *carne,* meat (< Lat. *caro*) + *levare,* to remove < Lat.] **1.** The season just before Lent, marked by merrymaking and feasting. **2.** A time of revelry : FESTIVAL. **3.** A traveling amusement show featuring exhibits and rides.
**car·ni·vore** (kär′nə-vôr′, -vōr′) *n.* [Fr. < Lat. *carnivorus,* carnivorous.] **1.** An animal belonging to the order Carnivora, including predominantly flesh-eating mammals. **2.** A flesh-eating or predatory organism, as a bird of prey or an insectivorous plant.
**car·niv·o·rous** (kär-nĭv′ər-əs) *adj.* [Lat. *carnivorus* : *caro,* flesh + *vorare,* to swallow up.] **1.** Belonging or relating to the order Carnivora. **2.** Flesh-eating : predatory. **3.** *Bot.* Capable of trapping and absorbing insects or other small organisms : INSECTIVOROUS <*carnivorous* plants such as the Venus's-flytrap> **—car·niv′o·rous·ly** *adv.* **—car·niv′o·rous·ness** *n.*
**car·no·tite** (kär′nə-tīt′) *n.* [Fr., after M.A. *Carnot* (d. 1920), French inspector general of mines.] A yellow uranium ore with composition $K_2(UO_2)_2(VO_4)_2{\cdot}3H_2O$.
**car·ny** *also* **car·ney** (kär′nē) *n., pl.* **-nies** *also* **-neys.** *Slang.* **1.** CARNIVAL 3. **2.** A carnival worker.
**car·ob** (kăr′əb) *n.* [OFr. *carobe* < Med. Lat. *carrubium* < Ar. *kharrūbah.*] An evergreen tree, *Ceratonia siliqua* of the Mediterranean region, with compound leaves and edible pods.
**ca·roche** (kə-rōch′, -rōsh′) *n.* [OFr. *carroche* < OItal. *carroccio,* aug. of *carro,* cart < Lat. *carrus,* of Celtic orig.] A stately 16th- and 17th-cent carriage.
**car·ol** (kăr′əl) *v.* **-oled, -ol·ing, -ols** *also* **-olled, -ol·ling, -ols.** [ME *carolen* < OFr. *caroler* < *carole,* a carol < LLat. *choraula,* choral song < Lat. *choraules,* accompanist < Gk. *khoraulēs* : *khoros,* choral dance + *aulos,* reed instrument.] *—vt.* **1.** To celebrate in song. **2.** To sing joyously. *—vi.* **1.** To sing joyously. **2.** To go from house to house singing Christmas songs. *—n.* **1.** A song of praise or joy, esp. for Christmas. **2.** An old round dance often accompanied by singing. **—car′ol·er** *n.*
**Car·o·le·an** (kăr′ə-lē′ən) *adj.* [< Med. Lat. *Carolus,* Charles.] Of or relating to Charles I or Charles II of England.
**Car·o·line** (kăr′ə-līn′, -lĭn) *adj.* [NLat. *Carolinius* < Med. Lat. *Carolus,* Charles.] Relating to the life and times of Charles I or Charles II of England.
**Car·o·lin·gian** (kăr′ə-lĭn′jən, -jē-ən) *also* **Car·lo·vin·gian** (kär′lə-vĭn′jən, -jē-ən) *adj.* [Fr. *Carolingien,* alt. of *Carlovingian,* ult. < OHG *Karl,* Charles.] Related to or belonging to the Frankish dynasty that was founded by Pepin the Short in A.D. 751 and that lasted until A.D. 987 in France and A.D. 911 in Germany. **—Car′o·lin′gian, Car′lo·vin′gian** *n.*
**Car·o·lin·i·an** (kăr′ə-lĭn′ē-ən) *adj.* **1.** Caroline. **2.** Of or relating to Charlemagne and his times.
**car·om** (kăr′əm) *n.* [Obs. *carambole* < Sp. *carambola* < Fr. *carambole.*] **1. a.** A billiards shot in which the cue ball successively strikes two other balls : BILLIARD. **b.** A similar shot in related games, as pool. **2.** A collision followed by a rebound. *—v.* **-omed, -om·ing, -oms.** *—vi.* **1.** To collide with and rebound. **2.** To make a carom, as in billiards. *—vt.* To cause to carom.
**car·o·te·nase** (kăr′ə-tē-nās, -nāz) *n.* An enzyme that catalyzes the hydrolysis of a carotenoid.

**ca·ro·tene** (kăr′ə-tēn′) *also* **ca·ro·tin** (-tĭn) *n.* [G. *Karotin* < Lat. *carota*, carrot.] An orange-yellow to red hydrocarbon, $C_{40}H_{56}$, existing in three isomeric forms, occurring in many plants as a pigment, and converted to vitamin A in the animal liver.

**car·o·te·ne·mi·a** (kăr′ə-tə-nē′mē-ə) *n.* A condition in which there is carotene in the blood, sometimes marked by yellowing of the skin.

**ca·rot·e·noid** (kə-rŏt′n-oid′) *n.* Any of a class of yellow- to deep-red pigments, as the carotenes, occurring in many vegetable oils and some animal fats.

**ca·rot·id** (kə-rŏt′ĭd) *n.* [Fr. *carotide* < Gk. *karōtides* < *karoun*, to stupefy.] Either of the two major arteries in the neck that carry blood to the head. **—ca·rot′id** *adj.*

**carotid body** *n.* A chemoreceptor in the bifurcations of the carotid arteries that contains cells that are responsive to changes in oxygen in the blood and help control respiratory activity.

**carotid sinus** *n.* An arterial enlargement located at the bifurcations of the carotid arteries that contains numerous baroreceptors that function in arterial pressure control.

**car·o·tin** (kăr′ə-tĭn) *n. var. of* CAROTENE.

**ca·rous·al** (kə-rou′zəl) *n.* A loud, riotous drinking party.

**ca·rouse** (kə-rouz′) *n.* [G. *garaus (trinken)*, (to drink) all out.] Loud, drunken merrymaking. *—vi.* **-roused, -rous·ing, -rous·es. 1.** To drink excessively. **2.** To go on a drinking spree. **—ca·rous′er** *n.*

**car·ou·sel** *or* **car·rou·sel** (kăr′ə-sĕl′, -zĕl′) *n.* [Fr. *carrousel* < Ital. *carosello*, a kind of tournament.] **1.** MERRY-GO-ROUND 1, 2. **2.** A tournament in which knights or horsemen engaged in exercises and races.

**carp**[1] (kärp) *vi.* **carped, carp·ing, carps.** [ME *carpen* < ON *karpa*, to boast.] To complain constantly : NAG. **—carp′er** *n.*

**carp**[2] (kärp) *n., pl.* **carp** *or* **carps.** [ME *carpe*, partly < OFr. *carpe*, and partly < Med. Lat. *carpa*, both of Germanic orig.] **1.** An edible freshwater fish, *Cyprinus carpio*, frequently bred in ponds and lakes. **2.** Any of various fishes of the family Cyprinidae.

**-carp** *suff.* [NLat. *-carpium* < Gk. *-karpion* < *karpos*, fruit.] Fruit : fruitlike structure <cysto*carp*>

**car·pal** (kär′pəl) *adj.* [NLat. *carpalis* < Gk. *karpos*, wrist.] Of, relating to, or near the carpus. *—n.* A bone of the carpus.

**car·pe di·em** (kär′pĕ dē′ĕm′, -əm, dī′-) *n.* [Lat., seize the day.] The admonition to enjoy the pleasures of the moment without considering the future.

**car·pel** (kär′pəl) *n.* [NLat. *carpellum* < Gk. *karpos*, fruit.] The central ovule-bearing female organ of a flower, composed of a modified leaf forming one or more sections of the pistil. **—car′pel·lar′y** (-pə-lĕr′ē) *adj.*

**car·pel·late** (kär′pə-lāt′, -lĭt) *adj. Bot.* Having carpels.

**car·pen·ter** (kär′pən-tər) *n.* [ME < Norman Fr. *carpentier* < LLat. *carpentarius (artifex)*, (maker) of a carriage < *carpentum*, a two-wheeled carriage, of Celtic orig.] One whose occupation is constructing, finishing, and repairing wooden objects and structures. *—v.* **-tered, -ter·ing, -ters.** *—vt.* To build or repair (wooden structures). *—vi.* To work as a carpenter. **—car′pen·try** (-trē) *n.*

**carpenter ant** *n.* Any of various ants of the genus *Camponotus* that nest in and are destructive to wood.

**carpenter bee** *n.* Any of various bees of the families Xylocopidae and Ceratinidae that do not live in colonies and that bore tunnels into wood to lay their eggs.

**car·pet** (kär′pĭt) *n.* [ME < OFr. *carpite* < OItal. *carpita* < *carpire*, to pluck < Lat. *capere*.] **1. a.** A thick, heavy floor covering, usu. woven of wool or synthetic fibers : RUG. **b.** The fabric used for carpets. **2.** A surface similar to a carpet <a *carpet* of green moss> *—vt.* **-pet·ed, -pet·ing, -pets.** To cover with or as if with a carpet. **—on the carpet. 1.** In the position of being reprimanded by a superior. **2.** Under consideration or discussion.

**car·pet·bag** (kär′pĭt-băg′) *n.* An old-fashioned traveling bag made of carpet fabric.

**car·pet·bag·ger** (kär′pĭt-băg′ər) *n.* [So called because they carried their belongings in a carpetbag.] **1.** A Northerner who went to the South after the American Civil War for political or financial advantage. **2.** A nonresident politician who represents or seeks to represent a locality for political self-interest. **—car′pet·bag′ger·y** *n.*

**carpet beetle** *n.* A small beetle of the genera *Anthrenus* or *Attagenus*, the larva of which is injurious to fabrics and furs.

**car·pet·ing** (kär′pĭ-tĭng) *n.* **1.** Material used for carpets. **2. a.** A carpet. **b.** Carpets as a whole.

**carpet knight** *n.* A soldier, orig. a knight, who has spent his life away from battle.

**car·pet·weed** (kär′pĭt-wēd′) *n.* A low-growing weedy plant, *Mollugo verticillata*, forming dense mats and having whorled leaves and small greenish-white flowers.

**car·pi** (kär′pī′) *n. pl. of* CARPUS.

**-carpic** *suff.* [Prob. < NLat. *-carpicus* < Gk. *karpos*, fruit.] -CARPOUS.

**carp·ing** (kär′pĭng) *adj.* Naggingly critical. **—carp′ing·ly** *adv.*

**carpo-** *pref.* [< Gk. *karpos*, fruit.] Fruit <*carpo*phore>

**car·po·go·ni·um** (kär′pə-gō′nē-əm) *n., pl.* **-ni·a** (-nē-ə). *Bot.* The female structure producing sex cells in certain red algae. **—car′po·go′ni·al** (-əl) *adj.*

**car·pol·o·gy** (kär-pŏl′ə-jē) *n.* The branch of botany concerned with the study of fruit and seeds. **—car·pol′o·gist** *n.*

**car-pool** (kär′pōōl′) *n.* **1.** An arrangement whereby several commuters travel together in one car, sharing the costs and usu. taking turns providing the car. **2.** A group of commuters participating in a car-pool. *—vi.* **-pooled, -pool·ing, -pools.** To travel in a car-pool. *—vt.* To take turns driving (passengers) in a car-pool <*car-pooled* the children to their football games> **—car′-pool′er** *n.*

**car·poph·a·gous** (kär-pŏf′ə-gəs) *adj.* [Gk. *karpophagos* : *karpos*, fruit + *phagein*, to eat.] Feeding on fruit.

**car·po·phore** (kär′pə-fôr′, -fōr′) *n.* **1.** The elongated part of the axis of certain flowers to which the carpels are attached. **2.** A fruiting body or the stalk of a fruiting body in a fungus.

**car·port** (kär′pôrt′, -pōrt′) *n.* A roof projecting from the side of a building, used to shelter an automobile.

**car·po·spo·ran·gi·um** (kär′pə-spə-răn′jē-əm) *n., pl.* **-gi·a** (-jē-ə). A specialized sporangium in red algae in which carpospores are formed.

**car·po·spore** (kär′pə-spôr′, -spōr′) *n.* A nonmotile haploid or diploid spore formed within the carpogonium of red algae.

**-carpous** *suff.* [NLat. *-carpus* < Gk. *karpos*, fruit.] A given number or kind of fruit <poly*carpous*>

**car·pus** (kär′pəs) *n., pl.* **-pi** (-pī′) [NLat. < Gk. *karpos*, wrist.] **1. a.** The wrist. **b.** The bones in the wrist. **2.** A joint corresponding to the wrist in quadrupeds.

**car·rack** *also* **car·ack** (kăr′ək) *n.* [ME *carik* < OFr. *caraque* < OSp. *carraca* < Ar. *qarāqīr*, pl. of *qurqūr*, merchant ship.] A large galleon used in the 14th, 15th, and 16th cent.

**car·ra·geen** *also* **car·ra·gheen** *or* **car·a·geen** (kăr′ə-gēn′) *n.* [After *Carragheen*, near Waterford, Ireland.] Irish moss.

**car·ra·geen·in** *or* **car·ra·geen·an** *also* **car·ra·gheen·in** (kăr′ə-jē′nən) *n.* A colloid derived from carrageen, used as a clarifying and stabilizing agent in foods.

**car·re·four** (kăr′ə-fōōr′) *n.* [Fr. < OFr. *carrefor* < Lat. *quadrifurcus*, four forked : *quadri-*, four + *furca*, fork.] **1.** A crossroads. **2.** A public square : PLAZA.

**car·rel** *also* **car·rell** (kăr′əl) *n.* [OFr. *carole* and Med. Lat. *carola*.] A nook near or in the stacks in a library, used for private study.

**car·riage** (kăr′ĭj) *n.* [ME *cariage* < Norman Fr. < *carier*, to carry. —see CARRY.] **1.** A four-wheeled, usu. horse-drawn passenger vehicle. **2.** *Chiefly Brit.* A railroad car for passengers. **3.** A baby carriage. **4.** A wheeled support or frame for moving a heavy object, as a cannon. **5.** A machine part for holding or shifting another part. **6. a.** The act or process of transporting or carrying. **b.** (kăr′ē-ĭj). The cost of or charge for transporting. **7.** Manner of holding and moving one's head and body : POSTURE. **8.** *Archaic.* Management.

**carriage dog** *n.* A Dalmatian.

**carriage trade** *n.* Wealthy patrons, as of a business.

**car·rick bend** (kăr′ĭk) *n.* [< Obs. *carrick*, carrack < ME *carik*.] *Naut.* A knot used to fasten two cables or hawsers together.

**carrick bitt** *n.* [Prob. < obs. *carrick*, carrack. —see CARRICK BEND.] *Naut.* Either of the two posts supporting the windlass on a ship's deck.

**car·ri·er** (kăr′ē-ər) *n.* **1.** One that transports or conveys. **2.** One that deals in transporting passengers or goods. **3.** A mechanism by which something is conveyed or conducted. **4.** *Med.* One that is at least temporarily immune to a pathogen that it transmits directly or indirectly to others <a *carrier* of encephalitis> **5.** *Electron.* **a.** A carrier wave. **b.** A charge-carrying entity, esp. an electron or a hole in a semiconductor. **6.** An aircraft carrier.

**carrier pigeon** *n.* A homing pigeon trained to carry messages.

**carrier wave** *n.* An electromagnetic wave that can be modulated, as in frequency, amplitude, or phase to transmit speech, music, images, or signals.

**car·ri·ole** (kăr′ē-ōl′) *n. var. of* CARIOLE.

**car·ri·on** (kăr′ē-ən) *n.* [ME *careine* < AN < VLat. *_caronia_ < Lat. *caro*, flesh.] Dead, decaying flesh.

**carrion crow** *n.* A common European black crow, *Corvus corone*.

**carrion flower** *n.* **1.** A climbing vine, *Smilax herbacea* of eastern North America, having small, greenish flower clusters with an odor of decaying flesh. **2.** Any of several plants similar to the carrion flower whose flowers emit an unpleasant odor.

**car·rot** (kăr′ət) *n.* [OFr. *carotte* < Lat. *carota* < Gk. *karōton*.] **1.** A widely cultivated plant, *Daucus carota sativa*, having finely divided leaves, small flat white flower clusters, and an edible, yellow-orange root. **2.** The long tapering root of the carrot, eaten as a vegetable. **3.** Something offered as an attractive but often illusory inducement.

**car·rot-and-stick** (kăr′ət-ən-stĭk′) *adj.* Combining a promised reward with a threat of punishment <a *carrot-and-stick* approach to diplomacy>

**car·rot·y** (kăr′ə-tē) *adj.* **1.** Having the color of a carrot. **2.** Having carrot-colored hair.

**car·rou·sel** (kăr′ə-sĕl′, -zĕl′) *n. var. of* CAROUSEL.

**car·ry** (kăr′ē) *v.* **-ried, -ry·ing, -ries.** [ME *carien* < Norman Fr.

*carier* < *carre,* cart. —see CAR.] —*vt.* **1.** To bear from one place to another : TRANSPORT <*carry* books><*carry* passengers> **2.** To make known, take, bring, or communicate (e.g., a message) <*carried* the good news> **3.** To serve as a means for the conveyance or transmission of. **4.** To hold or bear while moving. **5.** To hold or be capable of holding <This ladder *carries* only one person.> **6.** To support or sustain the responsibility of. **7.** To have or keep on one's person <*carry* a gun> **8.** To be pregnant with <*carry* a child to term> **9. a.** To hold and move (the body or a part of it) in a given way. **b.** To behave or conduct (oneself) in a given way. **10.** To extend or continue in a certain direction or to a given point or degree. **11.** To cause to move : IMPEL. **12.** To take or seize, esp. by force : CAPTURE. **13.** To gain victory, support, or acceptance for, esp. to secure the adoption of. **14.** To win most of the votes of <Their candidate *carried* the South.> **15.** To be successful in : WIN. **16.** To win over the interest of <The comedian's humor *carried* the audience.> **17.** To support or corroborate (e.g., a claim). **18. a.** To have as a necessary or customary feature or adjunct <an appliance *carrying* a two-year guarantee> **b.** To involve necessarily as a condition, consequence, or effect <The crime *carried* a ten-year sentence.> **19.** To transfer from one place, as a column, page, or book, to another. **20.** To keep in stock : offer for sale <a store that *carries* hardware> **21.** To keep in one's accounts as a debtor <*carried* the customers during their unemployment> **22.** To place before the public, as through a mass medium <The concert was *carried* by the local station.> **23.** To produce as a crop. **24.** To support or sustain (livestock). **25.** To sing (e.g., a melody) on key <couldn't *carry* a tune> **26.** To cover (a distance) or advance beyond (a point or object) in one golf stroke. —*vi.* **1.** To act as a bearer <taught the puppy to fetch and *carry*> **2. a.** To be transmitted or conveyed <a voice that *carries* poorly> **b.** To have or exert propulsive force. **3.** To admit of being transported in a certain manner <perishables that do not *carry* well> **4.** To hold the neck and head in a certain way. —Used of a horse. **5.** To be accepted or approved <The motion *carries.*> **—carry away.** To move or excite greatly <*carried away* by enthusiasm> **—carry forward. 1.** To progress with <*carry forward* the project> **2.** To transfer (an entry) to the next column, page, or book or to another account. **—carry off. 1.** To cause the death of <was *carried off* by plague> **2.** To seize and detain (a person) unlawfully. **3.** To handle (e.g., a situation) successfully. **4.** To win, as a prize, award, or honor <*carry off* first place> **—carry on. 1.** To conduct : administer. **2.** To engage in <*carry on* a love affair> **3.** To continue without halting <*carry on* in the face of trouble> **4.** To behave in an excited, improper, or silly way. **—carry out. 1.** To put into practice or effect. **2.** To obey (orders). **3.** To bring to a conclusion : ACCOMPLISH. **—carry over. 1.** CARRY FORWARD 2. **2.** To continue at another time : PUT OFF. **—carry through. 1.** To accomplish : complete. **2.** To enable to endure : SUSTAIN. —*n., pl.* **-ries. 1.** The act or process of carrying. **2.** A portage, as between two navigable bodies of water. **3. a.** The range of a gun or projectile. **b.** The distance traveled by a golf ball. **4.** *Football.* An act or instance of rushing with the ball.

☆ **syns:** CARRY, BEAR, CONVEY, LUG *v. core meaning* : to support while moving <*carry* groceries home>

**car·ry·all** (kăr'ē-ôl') *n.* **1.** A large tote bag or pocketbook. **2. a.** A covered one-horse carriage with two seats. **b.** A closed car with two lengthwise seats facing each other.

**carrying capacity** *n.* The maximum number of inhabitants that an environment can support without detrimental effects.

**carrying charge** *n.* The interest charged on the balance owed when paying in installments.

**car·ry·on** (kăr'ē-ŏn') *n.* A luggage item compact enough to be carried aboard an aircraft by a passenger.

**car·ry·out** (kăr'ē-out') *adj.* Take-out <*carry-out* food> **—car'ry·out'** *n.*

**car·ry-o·ver** (kăr'ē-ō'vər) *n.* **1.** A quantity, as of goods or commodities, left over or held for future use. **2.** A sum transferred to a new column, page, book, or business account.

**car·sick** (kär'sĭk') *adj.* Nauseated from vehicular motion. **—car'sick'ness** *n.*

**cart** (kärt) *n.* [ME, prob. < ON *kartr.*] **1.** A two-wheeled vehicle drawn by an animal, as a horse, and used for carrying goods. **2.** An open two-wheeled carriage. **3.** A small, light vehicle moved by hand <a tea *cart*> —*vt.* **cart·ed, cart·ing, carts. 1. a.** To transport in a cart. **b.** To convey laboriously, as if in a cart : LUG. **2.** To remove or transport unceremoniously or forcibly <*carted* the suspect off to jail> **—cart'a·ble** *adj.* **—cart'er** *n.*

**cart·age** (kär'tĭj) *n.* **1.** The act or process of transporting by cart. **2.** The cost of transporting by cart.

**carte blanche** (kärt blänsh', blänch') *n., pl.* **cartes blanches** (blänsh', blän'shĭz, blänch', blän'chĭz) [Fr. : *carte,* document + *blanche,* blank.] Unrestricted power to act at one's own discretion.

**car·tel** (kär-tĕl') *n.* [G. *Kartell* < Fr. *cartel* < Ital. *cartello,* placard, dim. of *carta,* card < Lat. *charta,* leaf of papyrus. —see CARD[1].] **1.** A combination of independent businesses formed to regulate production, pricing, and marketing of goods by the members <an oil *cartel*> **2.** An official agreement between governments at war, esp. one concerning exchange of prisoners. **3.** A political group in some countries united in a common cause.

**Car·te·sian** (kär-tē'zhən) *adj.* [NLat. *Cartesianus* < *Cartesius,* René Descartes (1596–1650).] Of or relating to the philosophy or methods of Descartes. **—Car·te'sian·ism** *n.*

**Cartesian coordinate** *n.* A coordinate in a Cartesian coordinate system.

**Cartesian coordinate system** *n.* **1.** A two-dimensional coordinate system in which the coordinates of a point are its distances from two intersecting, often perpendicular straight lines, the distance from each being measured along a straight line parallel to the other. **2.** A three-dimensional coordinate system in which the coordinates of a point are its distances from each of three intersecting, often mutually perpendicular planes along lines parallel to the intersection of the other two.

**Cartesian plane** *n.* A plane having all points described by Cartesian coordinates.

**Cartesian product** *n.* A set of element pairs (x, y) constructed from given sets such that x belongs to X and y to Y.

**Car·thu·sian** (kär-thōō'zhən) *n.* [Med. Lat. *Cartusiensis* < *Chartreuse,* the place in France where the order's first monastery was built.] *Rom. Cath. Ch.* A member of a contemplative order founded during the 11th cent. by St. Bruno. **—Car·thu'sian** *adj.*

**car·ti·lage** (kär'tl-ĭj) *n.* [Lat. *cartilago.*] A tough white fibrous connective tissue attached to the articular surfaces of bones, a major constituent of the fetal and young vertebrate skeleton that is largely converted to bone with maturation.

**cartilage bone** *n.* A bone generated from cartilage.

**car·ti·lag·i·nous** (kär'tl-ăj'ə-nəs) *adj.* **1.** Of or relating to cartilage. **2.** Having a skeleton consisting mostly of cartilage.

**cartilaginous fish** *n.* A fish whose skeleton is composed chiefly of cartilage.

**cart·load** (kärt'lōd') *n.* The amount that a cart can carry.

**car·to·gram** (kär'tə-grăm') *n.* [Fr. *cartogramme* : *carte,* map (< OFr., card) + *-gramme,* -gram.] A presentation of statistical data in geographic distribution on a map.

**car·tog·ra·phy** (kär-tŏg'rə-fē) *n.* [Fr. *cartographie* : *carte,* map (< OFr., card) + *-graphie,* -graphy.] The art or technique of making maps. **—car·tog'ra·pher** *n.* **—car'to·graph'ic** (kär'tə-grăf'ĭk), **car'to·graph'i·cal** *adj.*

**car·ton** (kär'tn) *n.* [Fr., pasteboard < Ital. *cartone,* pasteboard, aug. of *carta,* card < Lat. *charta,* leaf of papyrus. —see CARD[1].] **1. a.** A cardboard box. **b.** A container made from paper products. **2.** The contents of a carton. —*vt.* **-toned, -ton·ing, -tons.** To place or pack in a carton.

▲ word history: *Carton* and *cartoon* were borrowed from the same French word but at different times. The French word *carton* originally meant "pasteboard, cardboard" and was extended to mean a drawing, especially an artist's sketch, made on such material. *Carton* in the sense "drawing" was borrowed into English in the 17th century, at a time when the French syllable *-on* was rendered as *-oon* in English. The English word therefore is *cartoon.* The English word *carton* represents a 19th-century borrowing of the French word. By this time the original sense of "pasteboard" had been extended in French to mean a container made from stiff paper material. Other pairs of English words related in the same way as *carton* and *cartoon* are *dragon/dragoon, marron/maroon*[2], and *salon/saloon.*

**car·toon** (kär-tōōn') *n.* [Fr. *carton* < Ital. *cartone,* pasteboard. —see CARTON.] **1.** A drawing depicting a humorous situation, often accompanied by a caption. **2.** A pictorial satire or comment on a current event, usu. accompanied by words : CARICATURE. **3.** A preliminary sketch similar in size to the work, as a fresco, that is to be copied from it. **4.** An animated cartoon. **5.** A comic strip. —*v.* **-tooned, -toon·ing, -toons.** —*vt.* To sketch a humorous or satirical representation of : CARICATURE. —*vi.* To draw satirical or humorous sketches. **—car·toon'ist** *n.*

**car·touche** or **car·touch** (kär-tōōsh') *n.* [Fr. < Ital. *cartoccio* < *carta,* card. —see CARTON.] **1.** A scroll-like tablet used either to provide space for an inscription or to serve as an ornament. **2.** An oval or oblong figure in ancient Egyptian hieroglyphics that encloses characters listing the names or epithets of royal or divine personages. **3.** *Obs.* A heavy paper cartridge case.

**car·tridge** (kär'trĭj) *n.* [Earlier *cartage,* var. of Fr. *cartouche.* —see CARTOUCHE.] **1. a.** A tubular metal or cardboard and metal case containing the powder and primer of small arms ammunition or shotgun shells. **b.** An ammunition case fitted with a projectile, as a bullet. **2.** A small modular unit of equipment designed to be inserted into a larger piece of equipment. **3.** A removable case containing the stylus and electric conversion circuitry in a phonograph pickup. **4.** A case containing reeled magnetic tape, a pickup reel, and guide and feed mechanisms, used instead of separate reels in some tape recorders and players. **5.** A case with photographic film that can be loaded directly into a camera.

**cartridge belt** *n.* A belt having pockets or loops for carrying ammunition.

**cartridge clip** *n.* A metal container or frame for holding cartridges to be loaded into an automatic firearm.

**car·tu·lar·y** *also* **char·tu·lar·y** (kär′chə-lĕr′ē) *n., pl.* **-ies.** [Med. Lat. *cartularium* < Lat. *cartula*, dim. of *carta*, leaf of papyrus. —see CARD[1].] A collection of deeds or charters.

**cart·wheel** (kärt′hwēl′, -wēl′) *n.* **1.** A tumbling maneuver in which the body turns over sideways with the arms and legs spread like the spokes of a wheel. **2.** *Slang.* A large coin, as a silver dollar.

**ca·run·cle** (kə-rŭng′kəl, kăr′ŭng′-) *n.* [Obs. Fr. *caruncule* < Lat. *caruncula*, dim. of *caro*, flesh.] **1.** A fleshy, naked outgrowth, as a fowl's wattles. **2.** *Bot.* An excrescence on a seed at or near the hilum. **—ca·run′cu·lar** (-lər) *adj.* **—ca·run′cu·late** (-lĭt, -lāt′), **ca·run′cu·lat′ed** (-lā′tĭd) *adj.*

**car·va·crol** (kär′və-krôl′, -krōl′) *n.* [NLat. *carvi*, specific epithet of *Carum carvi*, caraway + Lat. *acer, acr-*, sharp + -OL.] A liquid phenol, $C_{10}H_{14}O$, used in flavorings and fungicides.

**carve** (kärv) *v.* **carved, carv·ing, carves.** [ME *kerven* < OE *ceorfan.*] —*vt.* **1. a.** To slice into pieces. **b.** To divide by parceling out <*carve* up a large estate among heirs> **2.** To cut into a desired shape. **3.** To make or form by or as if by cutting <*carved* my initials in the bark><*carved* out an empire> **4.** To decorate by carving. —*vi.* **1.** To engrave or cut figures as a hobby or trade. **2.** To disjoint, slice, and serve meat or poultry. **—carv′er** *n.*

**car·vel** (kär′vəl, -vĕl′) *n. var. of* CARAVEL.

**car·vel-built** (kär′vəl-bĭlt′, -vĕl′-) *adj.* Built with the hull planks lying flush or edge to edge rather than overlapping, as a ship.

**carv·en** (kär′vən) *v. Archaic. var. p.t. & p.p. of* CARVE.

**carv·ing** (kär′vĭng) *n.* **1.** The cutting of wood, stone, etc., to form a figure or design. **2.** A figure or design formed by carving.

**car wash** *n.* A place equipped for washing cars.

**car·y·at·id** (kăr′ē-ăt′ĭd) *n., pl.* **-ids** *or* **-i·des** (-ĭ-dēz′) [< Lat. *Caryatides*, caryatids, maidens of Caryae < Gk. *Karuatides* < *Karuai*, Caryae, a village in Greece.] A supporting column sculptured into a woman's figure. **—car′y·at′i·dal** (-ĭ-dəl), **car′y·at′i·de′an** (-ĭ-dē′ən), **car′y·a·tid′ic** (-ə-tĭd′ĭk) *adj.*

**caryatid**

**caryo-** *pref. var. of* KARYO-.

**car·y·op·sis** (kăr′ē-ŏp′sĭs) *n., pl.* **-op·ses** (-ŏp′sēz′) *or* **-op·si·des** (-ŏp′sĭ-dēz′) [CARY(O)- + -OPSIS.] A one-celled, one-seeded dry fruit, as a grain of wheat, having its outer coat fused to its surface.

**ca·sa·ba** *also* **cas·sa·ba** (kə-sä′bə) *n.* [After *Kasaba*, former name for Turgutlu, Turkey.] A winter melon having a yellow rind and sweet, whitish flesh.

**Cas·a·no·va** (kăs′ə-nō′və, kăz′-) *n.* [After G.J. *Casanova* de Seingalt (1725–1798).] A promiscuous, unscrupulous man : LIBERTINE.

**ca·sa·va** (kə-sä′və) *n. var. of* CASSAVA.

**Cas·bah** (kăz′bä′, käz′-) *n.* [Fr. < dial. Ar. *qaṣbah* < Ar. *qasabah*, fortress.] **1.** The citadel and palace of a northern African sovereign. **2.** The native quarter in a northern African city.

**cas·cade** (kă-skād′) *n.* [Fr. < Ital. *cascata* < *cascare*, to fall < VLat. **casicare* < Lat. *cadere.*] **1.** A waterfall or a series of small waterfalls over steep rocks. **2.** Something as a fall of lace, resembling a cascade. **3.** A succession of processes, operations, or units. **4.** *Elect.* A series of components or networks the output of each of which serves as the input for the next. —*vi. & vt.* **-cad·ed, -cad·ing, -cades.** To fall or cause to fall from one level to another in a continuous series.

**cas·car·a** (kă-skăr′ə) *n.* [Sp. *cáscara*, bark < *cascar*, to break < VLat. **quassicare* < Lat. *quassare* < *quatere*, to shake.] **1.** The cascara buckthorn. **2.** Cascara sagrada.

**cascara buckthorn** *n.* A shrub or tree, *Rhamnus purshiana* of northwestern North America, the bark of which is the source of cascara sagrada.

**cascara sa·gra·da** (sə-grä′də) *n.* [Am. Sp. *cáscara sagrada* : Sp. *cáscara*, bark + Sp. *sagrada*, sacred.] The dried bark of the cascara buckthorn, used medicinally as a stimulant, cathartic, and laxative.

**cas·ca·ril·la** (kăs′kə-rĭl′ə) *n.* [Sp., dim. of *cáscara*, bark. —see CASCARA.] **1.** A shrub, *Croton eluteria* of the West Indies, with bitter, aromatic bark. **2.** *also* **cascarilla bark.** The bark of the cascarilla, used as a tonic.

**cascarilla oil** *n.* An oil obtained from cascarilla bark, used as a flavoring.

**case[1]** (kās) *n.* [ME *cas* < OFr. < Lat. *casus* < p.part. of *cadere*, to fall.] **1.** An instance of the existence or occurrence of something. **2. a.** An occurrence of disease. **b.** A client, as of a physician or attorney. **3.** A set of circumstances : SITUATION. **4.** A set of reasons, arguments, or supporting facts offered in justification of a statement, action, or situation. **5.** A matter <a *case* of honor> **6.** A situation requiring investigation, esp. by an official body. **7.** *Law.* **a.** An action or suit or just grounds for an action. **b.** The facts or evidence offered in support of a claim. **8.** *Informal.* An eccentric person. **9. a.** The syntactic relationship of a noun, pronoun, or adjective to the other words of a sentence, indicated in inflected languages by the assumption of declensional endings and in noninflected languages by the position of the words within the sentence. **b.** The form or position of a word that indicates a syntactic relationship. **c.** Such forms, positions, or relationships as a whole. **10. a.** A pattern of inflection of nouns, pronouns, and adjectives to express different syntactic functions in a sentence. **b.** The form of such an inflected word. —*vt.* **cased, cas·ing, cas·es.** *Slang.* To examine with care, as in planning a crime <*cased* the bank before robbing it> **—in any case.** Regardless of what occurred or will occur. **—in case.** If it happens that : IF.

▲ word history: The grammatical sense of *case* did not develop in Latin but was imported along with the concept from Greek. Latin *casus* basically meant "a falling," and it was used to translate Greek *ptōsis*, which meant both "a falling" and "grammatical case." To the Greek grammarians any difference from the nominative form of nouns and the present indicative form of verbs was a "falling" from the basic form. Cases thus originally included derived adverbs, the different genders of adjectives, and other moods and tenses of verbs, as well as the oblique cases of nouns.

**case[2]** (kās) *n.* [ME < Norman Fr. *casse* < Lat. *capsa.*] **1.** A container. **2.** A decorative or protective covering. **3.** A box with its contents. **4.** A set or pair, as of pistols : BRACE. **5.** The framework of a window, door, or stairway. **6.** A shallow, compartmented tray for storing printing type or type matrices. —*vt.* **cased, cas·ing, cas·es.** To put into, cover, or protect with a case.

**ca·se·ate** (kā′sē-āt′) *vi.* **-at·ed, -at·ing, -ates.** [< Lat. *caseus*, cheese.] To undergo caseation.

**ca·se·a·tion** (kā′sē-ā′shən) *n.* [< CASEATE.] Necrotic degeneration of bodily tissue into a cheeselike substance.

**case·book** (kās′bŏŏk′) *n.* A book containing source materials in a specific area, used in teaching and as a reference.

**case·hard·en** (kās′här′dn) *vt.* **-ened, -en·ing, -ens. 1.** To harden the surface of (iron or steel) by high-temperature shallow infusion of carbon followed by quenching. **2.** To make insensitive <a person *casehardened* by poverty>

**case history** *n.* An organized set of facts pertinent to the development of an individual or group condition under study or treatment, esp. in sociology, psychiatry, or medicine.

**ca·sein** (kā′sēn′, kā′sē-ĭn) *n.* [Ult. < Lat. *caseus*, cheese.] A white, tasteless, odorless milk and cheese protein, used to manufacture plastics, adhesives, and paints and in making foods.

**case knife** *n.* **1.** A knife held in a sheath or case. **2.** A table knife.

**case law** *n.* Law based on judicial decision and precedent rather than statute.

**case load** *n.* The number of cases handled in a given period, as by a clinic or social services agency.

**case·mate** (kās′māt′) *n.* [OFr. < OItal. *casamatta.*] **1.** A fortified enclosure for artillery on board a warship. **2.** An armored compartment for artillery on a rampart. **—case′mat′ed** *adj.*

**case·ment** (kās′mənt) *n.* [ME < Med. Lat. *casamentum* < *casa*, dwelling < Lat.] **1. a.** A window sash that opens outward by hinges. **b.** A window having casement sashes. **2.** A case or covering. **—case′ment·ed** *adj.*

**ca·se·ous** (kā′sē-əs) *adj.* [< Lat. *caseus*, cheese.] Like cheese.

**ca·sern** *also* **ca·serne** (kə-zûrn′) *n.* [Fr. *caserne.*] A barracks for soldiers.

**case shot** *n.* **1. a.** CANISTER 2. **b.** The shot in a canister. **2.** A shrapnel shell.

**case study** *n.* A detailed analysis of an individual or group, esp. as an exemplary model of medical, psychological, or social phenomena.

**case system** *n.* A method of teaching law that emphasizes the study of selected cases rather than legal textbooks.

**case·work** (kās′wûrk′) *n.* The part of a social worker's duties treating the problems of a specific case. **—case′work′er** *n.*

**cash[1]** (kăsh) *n.* [OFr. *casse*, money box < Lat. *capsa*, case.] **1.** Ready money : currency or coins. **2.** Payment for goods or services in money or by check. —*vt.* **cashed, cash·ing, cash·es.** To exchange for or convert into ready money. **—cash in. 1.** To withdraw from an enterprise by or as if by settling one's account. **2.** *Slang.* To die. **—cash in on.** To take advantage of <*cash in on* an opportunity to get rich>

**cash[2]** (kăsh) *n., pl.* **cash.** [Port. *caixa* < Tamil *kācu*, a small coin < Skt. *karṣa*, a weight.] An Oriental coin of small denomination, esp. a copper and lead coin with a square hole in its center.

**cash·book** (kăsh′bŏŏk′) *n.* A book wherein a record of cash receipts and expenditures is kept.

**cash crop** *n.* A crop grown esp. for sale and usu. being an important source of income.

ă **pat** ā **pay** âr **care** ä **father** ĕ **pet** ē **be** hw **which** ĭ **pit**
ī **tie** îr **pier** ŏ **pot** ō **toe** ô **paw, for** oi **noise** ŏŏ **took**

**cash discount** *n.* A reduction in the price of an item for sale allowed if payment is made within a stipulated period.
**cash·ew** (kăsh′o͞o, kə-sho͞o′) *n.* [Port. *acajú* < Tupi.] **1.** A tropical American evergreen tree, *Anacardium occidentale*, bearing kidney-shaped nuts that protrude from a fleshy receptacle. **2.** *also* **cashew nut.** The nut of the cashew.
**cash flow** *n.* The cash receipts or net income after taxes and other disbursements from one or more assets for a given period, often used as an indicator of corporate worth.
**cash·ier**[1] (kă-shîr′) *n.* [Du. *cassier* < Fr. *caissier* < *caisse*, money box < OFr. *casse.* —see CASH[1]] **1.** The officer of a bank or business in charge of disbursing and receiving money. **2.** One who handles cash transactions for a business operation such as a supermarket.
**ca·shier**[2] (kă-shîr′) *vt.* **-shiered, -shier·ing, -shiers.** [Du. *casseren* < OFr. *casser*, to dismiss. —see QUASH.] To dismiss from a position of command or responsibility <a colonel *cashiered* from the army>
**cashier's check** *n.* A check drawn by a bank on its own funds and signed by the bank's cashier.
**cash·mere** (kăzh′mîr′, kăsh′-) *n.* [After *Kashmir*, a region in India.] **1.** Fine, downy wool growing beneath the outer hair of the Cashmere goat. **2.** A soft fabric made of wool from the Cashmere goat or of similar fibers.
**Cashmere goat** *n.* A goat native to the Himalayan regions of India and Tibet, greatly valued for its wool.
**cash register** *n.* A machine that tabulates the amount of sales transactions, makes a permanent and cumulative record of them, and has a drawer in which cash may be kept.
**cas·i·mere** (kăz′ə-mîr′, kăs′-) *n. var. of* CASSIMERE.
**cas·ing** (kā′sĭng) *n.* **1.** The act of encasing. **2.** An outer cover <a shell *casing*> **3.** The cleaned intestines of cattle, sheep, or hogs, used to contain processed meat. **4.** The framework for a window or door. **5.** A metal pipe or tube used as a lining for water, oil, or gas wells.
**ca·si·no** (kə-sē′nō) *n., pl.* **-nos.** [Ital. < *casa*, house < Lat., hut.] **1.** An Italian summer or country house. **2.** A public room or house esp. for gambling. **3.** *also* **cas·si·no.** A card game for two to four players in which cards on the table are matched by cards in the hand.
**cask** (kăsk) *n.* [Sp. *casco*, helmet < *cascar*, to break. —see CASCARA.] **1.** A barrel. **2.** The quantity that a cask will hold.
**cas·ket** (kăs′kĭt) *n.* [ME, poss. alteration of OFr. *cassette.* —see CASSETTE.] **1.** A small case or chest, as for jewels. **2.** A coffin. —*vt.* **-ket·ed, -ket·ing, -kets.** To enclose in a casket.
**casque** (kăsk) *n.* [Fr. < Sp. *casco.* —see CASK.] **1.** Armor, as a helmet, worn on the head. **2.** *Zool.* A helmetlike structure or protuberance. **—casqued** (kăskt) *adj.*
**cas·sa·ba** (kə-sä′bə) *n. var. of* CASABA.
**Cas·san·dra** (kə-săn′drə) *n.* [Lat. < Gk. *Kassandra.*] **1.** *Gk. Myth.* A daughter of Priam, the king of Troy, endowed with the gift of prophecy but fated by Apollo never to be believed. **2.** One who makes unheeded prophecies.
**cas·sa·tion** (kă-sā′shən) *n.* [OFr. < *casser*, to annul. —see QUASH.] Abrogation or annulment by a higher authority.
**cas·sa·va** *also* **ca·sa·va** (kə-sä′və) *n.* [Sp. *cazabe*, cassava bread < Taino *caçábi.*] **1.** A tropical American plant of the genus *Manihot*, with a large, starchy root. **2.** A starch derived from the root of the cassava, used in tapioca and as a staple food in the tropics.
**cas·se·role** (kăs′ə-rōl′) *n.* [Fr., saucepan, dim. of OFr. *casse*, roasting pan < OHG *chezi.*] **1. a.** A usu. earthenware, glass, or cast-iron dish in which food is both baked and served. **b.** Food prepared and served in a casserole. **2.** *Chem.* A small-handled, deep porcelain crucible for heating and evaporating substances.
**cas·sette** (kə-sĕt′, kă-) *n.* [Fr., small box < OFr., dim. of *casse*, case < Lat. *capsa.*] **1.** A light-proof camera cartridge for daylight loading of photographic film. **2.** *Electron.* **a.** A small cartridge containing unlooped magnetic tape for use in certain recorders and players. **b.** A recorder or player designed to use cassettes.
**cas·sia** (kăsh′ə) *n.* [ME < Lat., a kind of plant < Gk. *kassia*, of Semitic orig.] **1.** Any of various chiefly tropical trees, shrubs, and plants of the genus *Cassia*, with compound leaves, usu. yellow flowers, and long pods. **2. a.** A tree, *Cinnamomum cassia* of tropical Asia, whose bark is similar to cinnamon but is of inferior quality. **b.** The bark of this tree, used as a spice.
**cassia oil** *n.* An oil derived from the bark of the tree *Cinnamomum cassia.*
**cas·si·mere** *also* **cas·i·mere** (kăz′ə-mîr′, kăs′-) *n.* [Obs. *Cassimere*, var. of *Kashmir*, a region in India.] A plain or twilled woolen cloth.
**cas·si·no** (kə-sē′nō) *n. var. of* CASINO 3.
**Cas·si·o·pe·ia** (kăs′ē-ə-pē′ə) *n.* [Lat. < Gk. *Kassiopeia.*] A constellation in the Northern Hemisphere.
**cas·sis** (kə-sēs′) *n.* [Fr. < Lat. *cassia*, a kind of plant. —see CASSIA.] **1.** A European bush, *Ribes nigrum*, that bears black currants. **2.** A cordial made from the berries of the cassis.

**cas·sit·er·ite** (kə-sĭt′ə-rīt′) *n.* [Fr. *casitérite* < Gk. *kasseritos*, tin.] A yellow, red-brown, or black mineral, $SnO_2$, an important tin ore.
**cas·sock** (kăs′ək) *n.* [OFr. *casaque*, long coat, perh. < Pers. *kazagand*, padded jacket.] A long garment reaching to the feet and worn by the clergy and some laypersons assisting in church services.
**cas·so·war·y** (kăs′ə-wĕr′ē) *n., pl.* **-ies.** [Malay *kĕsuari.*] A large, flightless bird of the genus *Casuarius* of New Guinea and adjacent areas, having a large, bony projection on the top of the head and brightly colored wattles.
**cast** (kăst) *v.* **cast, cast·ing, casts.** [ME *casten* < ON *kasta.*] —*vt.* **1.** To throw violently or forcefully : FLING. **2.** To shed <a snake *casting* its skin> **3.** To throw forth : DROP <*cast* anchor> **4.** To throw on the ground, as in wrestling. **5.** To deposit or give (a ballot). **6.** To turn or direct <*cast* our eyes on the orator> **7.** To put forth : DIRECT <torches *casting* light><*cast* aspersions on my character> **8.** *Archaic.* To bestow : confer <"The government I *cast* upon my brother" —Shakespeare> **9.** To throw (dice). **10.** To give birth to prematurely <The cow *cast* a calf.> **11.** To cause (hounds) to scatter and circle in search of a lost scent. **12. a.** To choose theatrical performers for (e.g., a play). **b.** To assign a certain role to (a theatrical performer). **c.** To assign a theatrical performer to (a part). **13.** To form (e.g., liquid metal) into a given shape by pouring into a mold. **14.** To arrange in a system : FORMULATE. **15.** To contrive : devise. **16.** To add up (a column of figures) : COMPUTE. **17.** To calculate astrologically : FORECAST. **18.** To warp : twist. **19.** *Obs.* To consider : ponder. **20.** To stereotype or electroplate. **21.** *Naut.* To turn (a ship) : change to the opposite tack. —*vi.* **1. a.** To throw out a lure or bait at the end of a fishing line. **b.** To throw. **2. a.** To add a column of figures. **b.** To conjecture or forecast. **3.** To receive form or shape in a mold. **4.** To search for a lost scent in hunting with hounds. **5.** *Naut.* **a.** To veer to leeward from a former course : FALL OFF. **b.** To put about : TACK. **6.** To choose the performers for a play, movie, or other theatrical presentation. **7.** *Obs.* To turn over something in the mind : PONDER. **—cast about. 1.** To look for <*cast about* for lost keys> **2.** To devise means : CONTRIVE. **—cast aside (**or **away). 1.** To discard as useless : REJECT. **2.** To squander or waste. **—cast off. 1.** CAST ASIDE 1. **2.** To let go : set loose. **3.** To make the last row of stitches in knitting. **4.** To estimate the space a manuscript will occupy when set into type. **—cast on.** To make the first row of stitches in knitting. **—cast out.** To drive out by force : EXPEL. **—cast up. 1.** To vomit. **2.** To add up. —*n.* **1. a.** The act of casting. **b.** The distance cast. **2. a.** The throwing of a fishing line or net into the water. **b.** The fishing line or net thrown. **c.** *Chiefly Brit.* The fishing leader with flies or baited hooks attached. **3. a.** A throw of dice. **b.** The number thrown. **c.** A stroke of fortune or fate : LOT. **4. a.** A turning of the eye in a certain direction. **b.** A slight squint. **5.** Something thrown off, out, or away, as the skin shed by an insect. **6. a.** Addition of a column of figures : CALCULATION. **b.** A conjecture or forecast. **7. a.** The act of founding. **b.** The amount of molten material poured into a mold at a single operation. **c.** Something formed by this means. **8.** An impression formed in a mold or matrix : MOLD. **9.** The configuration in which something is made or constructed : ARRANGEMENT. **10.** The performers in a theatrical presentation. **11.** A rigid dressing, usu. made of gauze and plaster of Paris, as for a broken bone. **12.** A slight trace of color. **13.** Outward form or look : APPEARANCE. **14.** Sort : type <consider oneself to be of a macho *cast*> **15.** An inclination : tendency. **16.** Distortion of shape. **17.** The circling of hounds to pick up a scent in hunting. **—cast lots.** To draw lots in order to determine something by chance. **—cast (one's) lot (in) with.** To join or side with for better or worse.
**cas·ta·nets** (kăs′tə-nĕts′) *pl.n.* [Sp. *castañeta* < *castaña*, chestnut < Lat. *castanea.* —see CHESTNUT.] A rhythm instrument composed of a pair of slightly concave shells of ivory or hardwood, held in the palm of the hand by a connecting cord over the thumb and clapped together with the fingers.
**cast·a·way** (kăst′ə-wā′) *adj.* **1.** Cast adrift or ashore : SHIPWRECKED. **2.** Discarded. —*n.* **1.** A shipwrecked person. **2.** One that has been discarded or rejected.
**caste** (kăst) *n.* [Port. *casta*, race < *casto*, pure < Lat. *castus.*] **1.** One of the four major hereditary classes of Hindu society, each caste distinctly separated from the others by restrictions placed on occupation and marriage. **2. a.** A social class separated from others by hereditary, professional, or financial distinctions. **b.** A social system based on caste. **c.** The social status conferred by a system based on caste. **3.** A specialized level in a colony of social insects, as ants, in which its members, as workers, carry out a specific function.
**cas·tel·lan** (kăs′tə-lən) *n.* [ME *castelain* < Norman Fr. < Lat. *castellanus*, of a castle < *castellum*, castle.] The governor of a castle.
**cas·tel·la·ny** (kăs′tə-lā′nē) *n., pl.* **-nies. 1.** The jurisdiction of a castellan. **2.** The lands appertaining to a castle.
**cas·tel·lat·ed** (kăs′tə-lā′tĭd) *adj.* [Med. Lat. *castellatus*, p.part of *castellare*, to fortify as a castle < Lat. *castellum*, castle.] **1.** Having turrets and battlements. **2.** Having a castle. **—cas′tel·la′tion** *n.*
**cast·er** (kăs′tər) *n.* **1.** One that casts. **2.** *also* **cas·tor.** A small wheel on a swivel, attached under a piece of furniture or other heavy object to facilitate moving it. **3.** *also* **cas·tor. a.** A small bottle or cruet for condiments. **b.** A stand for holding a set of these bottles.

**cas·ti·gate** (kăs′tĭ-gāt′) *vt.* **-gat·ed, -gat·ing, -gates.** [Lat. *castigare, castigat-* < *castus,* pure.] **1.** To chastise or punish. **2.** To criticize severely. **—cas′ti·ga′tion** *n.* **—cas′ti·ga′tor** *n.*

**Cas·tile soap** *also* **cas·tile soap** (kă-stēl′) *n.* [After *Castile,* a province of Spain, where the soap was first made.] A fine, hard, white, odorless soap made with olive oil and sodium hydroxide.

**Cas·til·ian** (kă-stĭl′yən) *n.* **1. a.** The Spanish dialect of Castile. **b.** The standard literary and official form of Spanish, based on this dialect. **2.** A native or inhabitant of Castile. *—adj.* Of or relating to Castile, its people, or their language and culture.

**cast·ing** (kăs′tĭng) *n.* **1. a.** The act or process of making casts or molds. **b.** Something cast in a mold. **2.** CAST 2a. **3.** Selection of theatrical performers. **4.** Something cast off or out.

**casting vote** *n.* The vote of a presiding officer in an assembly or council, cast to break a tie.

**cast iron** *n.* A hard, brittle, nonmalleable iron-carbon alloy containing 2.0–4.5% carbon, 0.5–3% silicon, and lesser amounts of sulfur, manganese, and phosphorus.

**cast-i·ron** (kăst′ī′ərn) *adj.* **1.** Made of cast iron. **2.** Rigid : inflexible <*cast-iron* regulations>

**cas·tle** (kăs′əl) *n.* [ME *castel,* partly < OE and partly < Norman Fr., both < Lat. *castellum,* dim. of *castrum,* fort.] **1. a.** A fort or fortified cluster of buildings usu. dominating the surrounding country and held by a vassal of a ruler in feudal societies. **b.** A fortified stronghold converted to residential use. **2.** A building resembling a castle. **3.** A place of privacy, security, or refuge. **4.** A rook in the game of chess. *—v.* **-tled, -tling, -tles.** *—vt.* **1.** To place in or as if in a castle. **2.** To move (the chess king) from his own square two squares to one side and then, in the same move, bring the rook from that side to the square immediately past the new position of the king. *—vi.* To move the chess king and rook by castling.

**cas·tled** (kăs′əld) *adj.* Castellated.

**cast·off** (kăst′ôf′, -ŏf′) *n.* **1.** One that has been discarded. **2.** Calculation of the amount of space a manuscript will occupy when typeset.

**cast-off** (kăst′ôf′, -ŏf′) *adj.* Discarded : rejected <*cast-off* clothes>

**cas·tor¹** (kăs′tər) *n.* [ME, beaver < Lat. < Gk. *kastōr* < *Kastōr,* Castor.] **1.** An oily, brown, odorous substance obtained from glands in the groin of the beaver and used as a perfume fixative. **2.** A beaver hat. **3.** A heavy wool fabric used esp. for overcoats.

**cas·tor²** (kăs′tər) *n. var. of* CASTER 2, 3.

**Cas·tor** (kăs′tər) *n.* [Lat. < Gk. *Kastōr.*] **1.** *Gk. Myth.* One of the Dioscuri. **2.** A double star in the constellation Gemini, the brightest star in the group.

**castor bean** *n.* [CASTOR (OIL) + BEAN.] **1.** The castor-oil plant. **2.** The poisonous seed of the castor bean.

**castor oil** *n.* [Poss. from a former use as a substitute for castor in medicine.] A colorless or yellowish oil extracted from castor-oil plant seeds and used as a cathartic and lubricant.

**cas·tor-oil plant** (kăs′tər-oil′) *n.* A large plant native to tropical Africa and Asia, *Ricinus communis,* grown for ornament and for the extraction of castor oil from its poisonous seeds.

**cas·trate** (kăs′trāt′) *vt.* **-trat·ed, -trat·ing, -trates.** [Lat. *castrare, castrat-.*] **1.** To remove the testicles of : GELD. **2.** To remove the ovaries of : SPAY. **3.** To remove the vitality or force of, esp. by expurgating. **—cas·tra′tion** *n.*

**Cas·tro·ism** (kăs′trō-ĭz′əm) *n.* The political, governmental, and socioeconomic policies and principles of the Cuban leader Fidel Castro. **—Cas′tro·ite′** (-īt′) *n.*

**ca·su·al** (kăzh′o͞o-əl) *adj.* [ME *casuel* < OFr. < LLat. *casualis* < Lat. *casus,* event. —see CASE¹.] **1.** Occurring by chance : ACCIDENTAL. **2.** Taking place at irregular intervals : OCCASIONAL. **3.** Showing little interest : NONCHALANT. **4. a.** Devoid of ceremony or formality. **b.** Suited for everyday wear or use : INFORMAL <*casual* clothes> **5.** Not serious or thorough : SUPERFICIAL <a *casual* inspection of the documents> **6.** Not close or intimate <*casual* friends> *—n.* **1.** *Chiefly Brit.* One who receives temporary welfare relief. **2.** One who works at irregular intervals. **3.** A soldier temporarily attached to a unit while awaiting permanent assignment. **—ca′su·al·ly** *adv.* **—ca′su·al·ness** *n.*

**ca·su·al·ty** (kăzh′o͞o-əl-tē) *n., pl.* **-ties.** [ME *casuelte* < OFr. < Lat. *casualis,* casual.] **1.** A serious accident, esp. one involving loss of life. **2.** One who is injured or killed in an accident. **3.** One injured, killed, captured, or missing in military action.

**ca·su·a·ri·na** (kăzh′o͞o-ə-rī′nə) *n.* [NLat. *Casuarina,* genus name < Malay *kĕsuari,* cassowary (from the resemblance of its twigs to the drooping feathers of the cassowary).] Any of various trees of the genus *Casuarina,* including the beefwoods.

**ca·su·ist** (kăzh′o͞o-ĭst) *n.* [Fr. *casuiste* < Sp. *casuista* < Lat. *casus,* case. —see CASE¹.] One who is highly skilled in or given to casuistry.

**ca·su·is·tic** (kăzh′o͞o-ĭs′tĭk) *also* **ca·su·is·ti·cal** (-tĭ-kəl) *adj.* Of or relating to casuists or casuistry. **—ca′su·is′ti·cal·ly** *adv.*

**ca·su·ist·ry** (kăzh′o͞o-ĭ-strē) *n.* [< CASUIST.] **1.** Determination of right and wrong in questions of conduct or conscience by the application of general principles of ethics. **2.** A subtle but misleading or false application of ethical principles.

**ca·sus bel·li** (kā′səs bĕl′ī′, kä′səs bĕl′ē′) *n.* [NLat., occasion of war.] An act or event provoking, justifying, or used as an excuse for a declaration of war.

**cat** (kăt) *n.* [ME < OE *catt* < Lat. *cattus.*] **1. a.** A carnivorous mammal, *Felis catus* or *F. domesticus,* domesticated since early times as a catcher of rats and mice and as a pet and existing in several breeds and varieties. **b.** Any of the other animals of the family Felidae, including the lion, tiger, leopard, and lynx. **c.** The fur of a domestic cat. **2.** *Slang.* A spiteful woman. **3.** A cat-o′-nine-tails. **4.** A catfish. **5.** *Naut.* **a.** A cathead. **b.** A device for raising an anchor to the cathead. **c.** A catboat. **6.** *Slang.* FELLOW 1a. *—vt.* **cat·ted, cat·ting, cats.** To hoist an anchor to the cathead. **—let the cat out of the bag.** To disclose a secret : spill the beans.

▲ word history: In ancient times the cat, now kept as a pet, was not a domestic animal in Europe. The Latin word *felis,* which has been adopted as the scientific name for the cat genus, denoted only a kind of wild cat. The Greek word *ailouros* did denote the domestic cat as well as other kinds of cats, but the only domestic cats mentioned by the early Greek writers lived in Egypt, not Greece. By the 4th century A.D., however, the domestic cat had been introduced into Europe, and a new word appeared to denote this creature: *cattus* in Latin and *katta* in Greek. The Latin form is the source of the word for "cat" in all the Romance languages. It was also borrowed into West and North Germanic at an early date and is the source of the Modern English word *cat.* Some form of the word, perhaps the ancestor of the Greek and Latin forms, was borrowed into the Celtic and Slavic languages by at least the very early medieval period. Although a common European word, the word for "cat" represented by Latin *cattus* is not of Indo-European origin; it was most likely borrowed from a language of northern Africa.

**CAT** (kăt) *n.* Computerized axial tomography.

**cata-** *pref.* [Gk. *kata-* < *kata,* down.] **1.** Down <*cata*dromous> **2.** Reverse : backwards : degenerative <*cata*plasia>

**ca·tab·o·lism** (kə-tăb′ə-lĭz′əm) *n.* [< Gk. *katabolē,* a throwing down < *kataballein,* to throw down : *kata-,* down + *ballein,* to throw.] Metabolic change of complex molecules into simple molecules. **—cat′a·bol′ic** (kăt′ə-bŏl′ĭk) *adj.* **—cat′a·bol′i·cal·ly** *adv.*

**ca·tab·o·lite** (kə-tăb′ə-līt′) *n.* [CATABOL(ISM) + -ITE.] A substance produced in catabolism.

**ca·tab·o·lize** (kə-tăb′ə-līz′) *vi.* & *vt.* **-lized, -liz·ing, -liz·es.** To undergo or cause to undergo catabolism.

**cat·a·chre·sis** (kăt′ə-krē′sĭs) *n., pl.* **-ses** (-sēz′) [Lat. *catachresis* < Gk. *katakhrēsis,* excessive use < *katakhrēsthai,* to use up : *kata-,* completely + *khrēsthai,* to use.] **1. a.** Strained use of a word or phrase, as for rhetorical effect. **b.** An intentionally paradoxical figure of speech. **2.** Use of a wrong word in a context. **—cat′a·chres′tic** (-krĕs′tĭk) *adj.*

**cat·a·clysm** (kăt′ə-klĭz′əm) *n.* [Fr. *cataclysme* < Lat. *cataclysmos,* deluge < Gk. *kataklusmos* < *katakluzein,* to inundate : *kata-,* down + *kluzein,* to wash.] **1.** A sudden violent change in the earth's crust. **2.** A violent upheaval or disaster. **3.** A devastating flood. **—cat′a·clys′mic** (-klĭz′mĭk), **cat′a·clys′mal** (-klĭz′məl) *adj.*

**cat·a·comb** (kăt′ə-kōm′) *n.* [OFr. *catacombe* < LLat. *catacumbae* (pl.).] *often* **catacombs.** An underground chamber with recesses for graves.

**ca·tad·ro·mous** (kə-tăd′rə-məs) *adj.* Migrating down river to breed in marine waters <*catadromous* fish>

**cat·a·falque** (kăt′ə-fălk′, -fôlk′) *n.* [Fr. < Ital. *catafalco.*] The raised structure on which a coffin rests during a state funeral.

**Cat·a·lan** (kăt′l-ăn′, -ən) *adj.* Of or relating to Catalonia or to its people, language, or culture. *—n.* **1.** A native or resident of Catalonia. **2.** The Romance language of Catalonia.

**cat·a·lase** (kăt′l-ās′, -āz′) *n.* [CATAL(YSIS) + -ASE.] An enzyme in the blood and tissues that catalyzes the decomposition of hydrogen peroxide into water and oxygen.

**cat·a·lec·tic** (kăt′l-ĕk′tĭk) *adj.* [LLat. *catalecticus* < Gk. *katalektikos,* incomplete < *katalēgein,* to leave off : *kata-,* off + *lēgein,* to stop.] Designating a verse lacking part of the last foot.

**cat·a·lep·sy** (kăt′l-ĕp′sē) *n., pl.* **-sies.** [ME *catalempsi* < Med. Lat. *catalempsia* < Gk. *katalēpsis* < *katalambanein,* to seize upon : *kata-* (intensive) + *lambanein,* to seize.] Muscular rigidity, lack of awareness of environment, and lack of response to external stimuli, often associated with epilepsy, schizophrenia, and hysteria. **—cat′a·lep′tic** (-ĕp′tĭk) *adj.*

**cat·a·logue** *also* **cat·a·log** (kăt′l-ôg′, -ŏg′) [ME *cathaloge* < OFr. *catalogue* < LLat. *catalogus,* enumeration < Gk. *catalogos* < *katalegein,* to list : *kata-* (intensive) + *legein,* to count.] *—n.* **1. a.** A systematized list, often featuring descriptions of the listed items. **b.** A publication containing such a list <a mail-order *catalogue*> **2.** A card catalog. *—vt.* & *vi.* **-logued, -logu·ing, -logues** *also* **-loged, -log·ing, -logs.** To list in or make a catalogue. **—cat′a·logu′er, cat′a·log′er** *n.*

**ca·tal·pa** (kə-tăl′pə, -tôl′-) *n.* [Creek *kutuhlpa.*] A chiefly North American tree of the genus *Catalpa,* having large leaves, whitish flower clusters, and long, slender pods.

**ca·tal·y·sis** (kə-tăl′ĭ-sĭs) *n., pl.* **-ses** (-sēz′) [Gk. *katalusis,* dissolution < *katalueín,* to dissolve : *kata-,* intensive + *luein,* to loosen.]

The action of a catalyst, esp. modification of the rate of a chemical reaction by a catalyst. **—cat'a·lyt'ic** (kăt'l-ĭt'ĭk) *adj.* **—cat'a·lyt'i·cal·ly** *adv.*

**cat·a·lyst** (kăt'l-ĭst) *n.* [< CATALYSIS.] **1.** *Chem.* A substance, usu. present in small amounts relative to the reactants, that modifies and esp. increases the rate of a chemical reaction without being consumed in the process. **2.** One that precipitates a process or event <The President's visit was the *catalyst* for riots.>

**catalytic converter** *n.* A reaction chamber typically containing a finely divided platinum-iridium catalyst into which exhaust gases from an automotive engine are passed together with excess air so that carbon monoxide and hydrocarbon pollutants are oxidized to carbon dioxide and water.

**catalytic cracker** *n.* An oil refinery unit in which the cracking of petroleum takes place in the presence of a catalyst.

**cat·a·lyze** (kăt'l-īz') *vt.* **-lyzed, -lyz·ing, -lyz·es.** [< CATALYSIS.] To modify the rate of (a chemical reaction) as a catalyst. **—cat'a·lyz'er** *n.*

**cat·a·ma·ran** (kăt'ə-mə-răn') *n.* [Tamil *kaṭṭumaram* : *kaṭṭu-*, to tie + *maram*, tree.] **1.** A sailboat with two parallel hulls. **2.** A raft of logs or floats lashed together.

**cat·a·me·ni·a** (kăt'ə-mē'nē-ə) *n.* [Gk. *katamēnia* : *kata-*, according to + *mēn*, month.] *Physiol.* Menses. **—cat'a·me'ni·al** *adj.*

**cat·a·mite** (kăt'ə-mīt') *n.* [Lat. *catamitus* < *Catamitus*, Ganymede < Etruscan *Catmite* < Gk. *Ganumēdēs*.] A boy kept by a pederast.

**cat·a·mount** (kăt'ə-mount') *also* **cat·a·moun·tain** (kăt'ə-moun'tən) *n.* [Short for *catamountain*, var. of *cat of the mountain*.] A wild cat, as the lynx or mountain lion.

**cat-and-mouse** (kăt'n-mous') *adj.* Of or employing continuous torment and teasing while probing the vulnerabilities of an opponent and awaiting an opportunity to attack.

**cat·a·pho·re·sis** (kăt'ə-fə-rē'sĭs) *n. Chem.* Electrophoresis. **—cat'a·pho·ret'ic** (-rĕt'ĭk) *adj.* **—cat'a·pho·ret'i·cal·ly** *adv.*

**cat·a·phyll** (kăt'ə-fĭl') *n. Bot.* A modified or rudimentary leaf.

**cat·a·pla·sia** (kăt'ə-plā'zhə, -zhē-ə) *n.* Degenerative reversion of cells or tissue to a less differentiated form. **—cat'a·plas'tic** (-plăs'tĭk) *adj.*

**cat·a·plasm** (kăt'ə-plăz'əm) *n.* [OFr. *cataplasme* < LLat. *cataplasma* < Gk. *kataplasma* < *kataplassein*, to plaster over : *kata-* (intensive) + *plassein*, to mold.] *Med.* A poultice.

**cat·a·plex·y** *n.* [Gk. *kataplēxis*, fixation (of the eyes in paralysis) < *kataplessein*, to amaze, terrify : *kata-*, down + *plessein*, to strike.] A sudden state of immobility with loss of muscle tone usu. caused by extreme emotional stress.

**cat·a·pult** (kăt'ə-pŭlt', po͝olt') *n.* [OFr. *catapulte* < Lat. *catapulta* < Gk. *katapaltēs* : *kata-*, down + *pallein*, to hurl.] **1.** An ancient military machine for hurling missiles, as large stones or arrows. **2.** A mechanism for launching aircraft without a runway or with a short runway, as from an aircraft carrier. **3.** A slingshot. —*v.* **-pult·ed, -pult·ing, -pults.** —*vt.* To hurl or launch from or as if from a catapult. —*vi.* To become catapulted <*catapulted* into stardom>

**cat·a·ract** (kăt'ə-răkt') *n.* [Lat. *cataracta* < Gk. *katauraktēs* < *katarassein*, to dash down : *kata-*, down + *rassein*, to strike.] **1.** A huge waterfall. **2.** An enormous downpour. **3.** *Pathol.* Opacity of the lens or capsule of the eye, causing partial or total blindness.

**ca·tarrh** (kə-tär') *n.* [OFr. *catarrhe* < LLat. *catarrhus* < Gk. *katarrhous* < *katarrein*, to flow down : *kata-*, down + *rhein*, to flow.] Inflammation of mucous membranes, esp. of the nose and throat. **—ca·tarrh'al, ca·tarrh'ous** *adj.*

**ca·tas·ta·sis** (kə-tăs'tə-sĭs) *n., pl.* **-ses** (-sēz') [Gk. *katastasis*, settlement < *kathistanai*, to settle : *kata-*, down + *histanai*, to set.] **1.** The intensified part of the action of a classical tragedy directly preceding the catastrophe. **2.** The climax of a play.

**ca·tas·tro·phe** (kə-tăs'trə-fē) *n.* [Gk. *katastrophē*, an overturning < *katastrephein*, to overturn : *kata-*, down + *strephein*, to turn.] **1.** A sudden, terrible calamity : DISASTER. **2.** A total failure : FIASCO. **3.** CATACLYSM 1. **4.** The climax of a play, esp. a classical tragedy. **—cat'a·stroph'ic** (kăt'ə-strŏf'ĭk) *adj.* **—cat'a·stroph'i·cal·ly** *adv.*

**cat·a·to·ni·a** (kăt'ə-tō'nē-ə) *n.* [NLat. < G. *Katatonie* < Gk. *katatonos*, stretching down : *kata-*, down + *teinein*, to stretch.] A schizophrenic disorder marked by plastic immobility of the limbs, stupor, negativism, and mutism. **—cat'a·ton'ic** (-tŏn'ĭk) *adj.* & *n.*

**Ca·taw·ba** (kə-tô'bə) *n., pl.* **Catawba** or **-bas. 1. a.** A tribe of Indians once living along the Catawba River in the Carolinas. **b.** A member of this tribe. **c.** The Siouan language of the Catawba. **2. a.** *pl.* **-bas.** A light-red North American grape developed from the fox grape, *Vitis labrusca*. **b.** Wine made from the Catawba grapes.

**cat·bird** (kăt'bûrd') *n.* [From the resemblance of one of its calls to the mewing of a cat.] A North American songbird, *Dumetella carolinensis*, with chiefly slate-gray plumage.

**catbird seat** *n.* A powerful position.

**cat·boat** (kăt'bōt') *n. Naut.* A broad-beamed sailboat carrying a single sail on a mast stepped well forward.

**cat·bri·er** (kăt'brī'ər) *n.* A thorny vine of the genus *Smilax*, esp. *S. rotundifolia*, with heart-shaped leaves, small green flowers, and blackish berries.

**cat·call** (kăt'kôl') *n.* A harsh or shrill call or whistle expressing derision or extreme disapproval. **—cat'call'** *v.* **(-called, -call·ing, -calls).**

**catch** (kăch, kĕch) *v.* **caught** (kôt), **catch·ing, catch·es.** [ME *cacchen* < AN *cachier*, to chase < Lat. *captare*, freq. of *capere*, to seize.] —*vt.* **1.** To seize or capture, esp. after a chase. **2.** To take by trapping or snaring <*catch* a mouse> **3.** To come upon suddenly, unexpectedly, or accidentally <*caught* me unaware> **4. a.** To lay hold of forcibly or suddenly : GRASP. **b.** To grab so as to stop the motion of <*catch* a baseball> **5. a.** To overtake <*caught* me on the steep hill> **b.** To reach in time to board <*caught* the train at five> **6. a.** To entangle : grip. **b.** To cause to become suddenly or accidentally hooked, entangled, or fastened. **7.** To strike : hit <a blow that *caught* me in the chest> **8.** To check (oneself) during an action. **9.** To become subject to or contract, as by exposure or contagion <*catch* the flu> **10.** To become affected by or imbued with <*caught* the happy mood of the party> **11.** To perceive suddenly, momentarily, or quickly <*caught* a glimpse of the President> **12.** To grasp mentally : APPREHEND <didn't *catch* the meaning of that sentence> **13.** To apprehend and reproduce accurately by or as if by artistic means. **14.** To attract and fix <tried to *catch* your attention> **15.** To charm : captivate. **16.** *Informal.* To see (e.g., a theatrical performance) <*caught* their act in Las Vegas> —*vi.* **1.** To become held, entangled, or fastened. **2.** To act or move so as to hold someone or something. **3.** To be communicable : SPREAD. **4.** To take fire : KINDLE. **5.** *Baseball.* To act as catcher. **—catch on.** *Informal.* **1.** To understand. **2.** To become popular <a melody that *caught on* quickly> **—catch up. 1.** To detect (someone) in a mistake. **2.** To come up from behind : OVERTAKE. **3.** To become involved with, often unwillingly <*caught up* in the plot> **4.** To become up to date <*catch up* on the evening news> **5.** To absorb completely : ENGROSS. —*n.* **1.** The act of catching. **2.** Something that catches, esp. a device for fastening or for checking motion. **3.** Something caught <a *catch* of catfish> **4.** An amount caught. **5.** A choking or stoppage of the breath or voice. **6.** A stop or break in a mechanism. **7.** *Informal.* One worth catching. **8.** *Informal.* A tricky or unsuspected condition or drawback. **9.** A snatch : fragment. **10.** *Mus.* A canonical, often rhythmically intricate composition for three or more voices, popular esp. in the 17th and 18th cent. **11. a.** The seizure and holding of a thrown, kicked, or batted ball before it hits the ground. **b.** A game of throwing and catching a ball. **—catch it.** *Informal.* To receive punishment or scolding. **—catch (one's) breath.** To take a brief rest.

☆ ***syns:*** CATCH, CLUTCH, GRAB, NAB, SEIZE, SNATCH *v. core meaning* : to lay hands on (something moving) <The outfielder *caught* the fly ball.>

▲ word history: Nowadays if a hunter "chases" an animal, he does not necessarily "catch" it, but it was not always so. *Catch* and *chase* are doublets, ultimately the same word but derived from two different dialects of Old French. *Catch* descends from the Norman form *cachier*, which meant only "to chase" and not "to catch." It is first recorded as an English word in the early 13th century. *Chase* descends from the central, or Parisian, form *chacier*, which, like *cachier*, meant only "to chase." This is still the meaning of modern French *chasser*. In English *catch* was first used to mean "capture," and after the adoption of *chacier* in the 14th century it came to be used exclusively in that sense.

**catch·all** (kăch'ôl', kĕch'-) *n.* A receptacle for odds and ends.

**catch-as-catch-can** (kăch'əz-kăch-kăn', kĕch'əz-kĕch-) *adj.* Seizing any opportunity or utilizing any available means to achieve an end.

**catch·er** (kăch'ər, kĕch'-) *n.* One that catches, esp. the baseball player whose position is behind home plate and who signals for and receives pitches.

**catch·fly** (kăch'flī', kĕch'-) *n.* A plant of the genus *Silene* and related genera, having white, pink, or red flowers with sticky stems and calyxes.

**catch·ing** (kăch'ĭng, kĕch'-) *adj.* **1.** Infectious. **2.** Attractive.

**catch·ment** (kăch'mənt, kĕch-) *n.* **1.** A catching or collecting of water. **2. a.** A structure, as a basin, for collecting or draining water. **b.** The amount of water so collected.

**catch·pen·ny** (kăch'pĕn'ē, kĕch-) *adj.* Designed to sell without concern for quality : CHEAP. —*n., pl.* **-nies.** A catchpenny item.

**catch phrase** *n.* An often repeated word or slogan.

**catch·pole** *also* **catch·poll** (kăch'pōl', kĕch'-) *n.* [ME *cacchepol* < Norman Fr. *cachepol*, prob. < OFr. *chacepol* : *chacier*, to hunt (< Lat. *captare*, freq. of *capere*, to seize.) + *poul*, rooster < Lat. *pullus*, chicken.] A sheriff's officer, esp. one who arrests debtors.

**Catch-22** (kăch'twĕn-tē-to͞o', kĕch'-) *n.* [After *Catch-22*, a novel by Joseph Heller (b. 1923).] A paradox in which seeming alternatives actually cancel each other out, leaving no means of escape from a dilemma.

**catch·up** (kăch'əp, kĕch'-) *n. var. of* KETCHUP.

**catch-up** (kăch'ŭp') *adj.* Designed or intended to catch up to a standard <"*catch-up* increases in prices to restore profit margins" —*Newsweek*>

**catch·word** (kăch′wûrd′, kĕch′-) *n.* **1.** A catch phrase. **2. a.** A guideword. **b.** The first word of a page printed at the bottom of the preceding page.

**catch·y** (kăch′ē, kĕch′ē) *adj.* **-i·er, -i·est. 1.** CATCHING 2. **2.** Easily remembered. **3.** Tricky : deceptive. **4.** Fitful or spasmodic.

**cate** (kāt) *n.* [Short for *acate*, purchased food < ME < Norman Fr. *acat* < *acater*, to buy. —see CATER.] *Archaic.* A choice or dainty food : DELICACY.

**cat·e·che·sis** (kăt′ĭ-kē′sĭs) *n., pl.* **-ses** (-sēz′) [Gk. *katakhēsis* < *katekhein*, to teach by word of mouth. —see CATECHIZE.] Instruction of catechumens. **—cat′e·chet′i·cal** (-kĕt′ĭ-kəl) *adj.*

**cat·e·chin** (kăt′ĭ-kĭn′) *n.* [CATECH(U) + -IN.] A substance, $C_{15}H_{14}O_6$, derived from catechu and used in tanning and dyeing.

**cat·e·chism** (kăt′ĭ-kĭz′əm) *n.* [LLat. *catechismus* < LGk. *katēkhismos* < *katēkhizein*, to teach by word of mouth. —see CATECHIZE.] **1.** A short book giving a brief summary of the basic principles of a religion in question-and-answer form. **2.** A book in question-and-answer form giving basic instruction in a subject. **3.** A question-and-answer examination, as of a political figure.

**cat·e·chist** (kăt′ĭ-kĭst) *n.* [LLat. *catechista* < LGk. *katēchistēs* < *katēkhizein*, to teach by word of mouth. —see CATECHIZE.] One who catechizes, esp. one who instructs catechumens for baptism. **—cat′e·chis′tic** (-kĭs′tĭk), **cat′e·chis′ti·cal** (-tĭ-kəl) *adj.*

**cat·e·chize** (kăt′ə-kīz′) *vt.* **-chized, -chiz·ing, -chiz·es.** [ME *catecizen* < Med. Lat. *catechizare* < LGk. *katēkhizein* < Gk. *katēkhein*, to teach by word of mouth : *kata-*, according to + *ekhein*, to sound < *ekhē*, sound.] **1.** To teach esp. the principles of a religious creed by questions and answers. **2.** To question persistently or searchingly. **—cat′e·chi·za′tion** *n.* **—cat′e·chiz′er** *n.*

**cat·e·chol** (kăt′ĭ-kôl′, -kōl′) *n.* [CATECH(U) + -OL.] **1.** Catechin. **2.** Pyrocatechol.

**cat·e·cho·la·mine** (kăt′ĭ-kō′lə-mēn′, -kô′-) *n.* Any of a group of amines, which include epinephrine, norepinephrine, and dopamine, that are derived from tyrosine and have a hormonal function.

**cat·e·chu** (kăt′ə-cho͞o′) *n.* [Prob. < Malay *kachu*, of Dravidian orig.] Any of several water-soluble, resinous, astringent substances used in tanning and dyeing, as that obtained from a tree, *Acacia catechu* of southern Asia, or from a woody vine, *Uncaria gambier* of Malaya.

**cat·e·chu·men** (kăt′ĭ-kyo͞o′mən) *n.* [ME *cathecumine* < LLat. *catechumenus* < Gk. *katēkhoumenos*, pr. passive part. of *katēkhein*, to instruct. —see CATECHIZE.] **1.** One who is being taught the principles of Christianity. **2.** One who is being instructed in a subject at an elementary level.

**cat·e·gor·i·cal** (kăt′ĭ-gôr′ĭ-kəl, -gŏr′-) *or* **cat·e·gor·ic** (-ĭk) *adj.* **1.** Utterly without exception or qualification : ABSOLUTE <a *categorical* refusal> **2.** Of, concerning, or included in a category. **—cat′e·gor′i·cal·ly** *adv.* **—cat′e·gor′i·cal·ness** *n.*

**categorical imperative** *n.* An absolute and universally binding moral law in Kant's ethical system.

**cat·e·go·rize** (kăt′ĭ-gə-rīz′) *vt.* **-rized, -riz·ing, -riz·es.** To place into a category. **—cat′e·go·ri·za′tion** *n.*

**cat·e·go·ry** (kăt′ĭ-gôr′ē, -gōr′ē) *n., pl.* **-ries.** [LLat. *categoria*, class of predicables < Gk. *katēgoria* < *katēgorein*, to predicate : *kata-*, against + *agora*, assembly.] **1.** A specifically defined division in a classification system : CLASS. **2.** *Logic.* Any of the basic classifications into which all knowledge can be placed.

**ca·te·na** (kə-tē′nə) *n., pl.* **-nae** (-nē′) *or* **-nas.** [Med. Lat., chain < Lat.] A closely linked series.

**cat·e·nar·y** (kăt′n-ĕr′ē, kə-tē′nə-rē) *n., pl.* **-ies.** [< Lat. *catenarius*, of a chain < *catena*, chain.] **1.** The curve theoretically formed by a perfectly flexible, uniformly dense and thick, inextensible cable suspended from two points. **2.** Something shaped like a catenary. **—cat′e·nar′y** *adj.*

**cat·e·nate** (kăt′n-āt′) *vt.* **-nat·ed, -nat·ing, -nates.** [Lat. *catenare*, *catenat-* < *catena*, chain.] To form into a series of chainlike links. **—cat′e·na′tion** *n.*

**ca·ten·u·late** (kə-tĕn′yə-lĭt) *adj.* [< Lat. *catenula*, dim. of *catena*, chain.] Made up of chainlike links.

**ca·ter** (kā′tər) *v.* **-tered, -ter·ing, -ters.** [< obs. *cater*, a buyer of provisions < ME *catour*, short for *acatour* < Norman Fr. < *acater*, to buy < VLat.* *accaptare* : Lat. *ad-*, to + Lat. *captare*, to catch < *capere*, to take.] *—vi.* **1.** To provide food, services, or entertainment. **2.** To provide something wished for or needed <*catered* to my every whim> *—vt.* To provide food service for <*catering* a banquet> **—ca′ter·er** *n.*

**cat·er·an** (kăt′ər-ən) *n.* [ME *ketharan*, prob. < Sc. Gael. *ceathairneach.*] A former robber of the Scottish Highlands.

**cat·er-cor·nered** (kăt′ər-kôr′nərd, kăt′ē-) *also* **cat·er-cor·ner** (-nər) *or* **cat·ty-cor·nered** (kăt′ē-kôr′nərd) *adj.* [< obs. *cater*, four at dice < ME < OFr. *catre*, four < Lat. *quattuor*.] Diagonal. **—cat′er-cor′nered, cat′er-cor′ner, cat′ty-cor′nered** *adv.*

**cat·er·cous·in** (kăt′ər-kŭz′ĭn) *n.* [Orig. unknown.] A very close friend.

**cat·er·pil·lar** (kăt′ər-pĭl′ər, kăt′ə-) *n.* [ME *catirpel*, prob. < OFr. *chatepelose* : *chate*, cat + *pelose*, hairy < Lat. *pilosus* < *pilus*, hair.] **1.** The wormlike, often brightly colored, hairy or spiny larva of a butterfly or moth. **2.** Any of various insect larvae similar to those of the caterpillar.

**cat·er·waul** (kăt′ər-wôl′) *vi.* **-wauled, -waul·ing, -wauls.** [ME *caterwawen.*] **1.** To make a discordant cry or screech like a cat in heat. **2.** To have a noisy argument. **—cat′er·waul′** *n.*

**cat·fish** (kăt′fĭsh′) *n., pl.* **catfish** *or* **-fish·es.** Any of numerous scaleless, chiefly freshwater fishes of the order Siluriformes, with whiskerlike barbels extending from the upper jaw.

**cat·gut** (kăt′gŭt′) *n.* A tough, thin cord or thread made from the dried intestines of certain animals, used for stringing musical instruments and tennis rackets and for surgical ligatures.

**ca·thar·sis** (kə-thär′sĭs) *n., pl.* **-ses** (-sēz′) [Gk. *katharsis* < *kathairein*, to purge < *katharos*, pure.] **1.** *Med.* Purgation, esp. for the digestive system. **2.** A purifying release of the emotions or of tension, esp. through art. **3.** *Psychoanal.* **a.** A technique used to relieve tension and anxiety by bringing repressed material to consciousness. **b.** Abreaction.

**ca·thar·tic** (kə-thär′tĭk) *adj.* [LLat. *catharticus* < Gk. *kathartikos* < *kathairein*, to purge. —see CATHARSIS.] Inducing catharsis : PURGATIVE. *—n.* A cathartic agent, esp. a laxative.

**cat·head** (kăt′hĕd′) *n. Naut.* A beam projecting outward from the bow and used as a support to raise the anchor.

**ca·the·dra** (kə-thē′drə) *n., pl.* **-drae** (-drē) [Lat., chair < Gk. *kathedra* : *kata-*, down + *hedra*, seat.] **1.** The official throne of a bishop. **2.** The office or see of a bishop. **3.** The official chair of an office or position, as of a professor.

**ca·the·dral** (kə-thē′drəl) *n.* [Obs. *cathedral church* < ME *cathedral*, of a diocese < OFr. < Med.Lat. *cathedralis* < Lat. *cathedra*, chair. —see CATHEDRA.] **1.** The principal church of a bishop's see and one that contains his official throne. **2.** A large or important church. *—adj.* **1.** Of, relating to, or containing a bishop's throne. **2.** Relating to or issuing from a chair of office or authority : AUTHORITATIVE. **3.** Of or relating to a cathedral.

**ca·thep·sin** (kə-thĕp′sĭn) *n.* [G. *Kathepsin* < Gk. *kathepsein*, to digest : *kata-*, down + *hepsein*, to boil.] Any of various proteolytic enzymes that catalyze the hydrolysis of proteins into polypeptides.

**cath·er·ine wheel** (kăth′ər-ĭn, kăth′rĭn) *n.* [After St. *Catherine* of Alexandria (d. 307 A.D.).] PINWHEEL 2.

**cath·e·ter** (kăth′ĭ-tər) *n.* [LLat. < Gk. *kathetēr*, surgical instrument for emptying the bladder < *kathienai*, to drop : *kata-*, down + *hienai*, to send.] A slender, flexible tube inserted into a bodily channel, as a vein, to distend or maintain an opening to an internal cavity.

**cath·e·ter·ize** (kăth′ĭ-tə-rīz′) *vt.* **-ized, -iz·ing, -iz·es.** To introduce a catheter into. **—cath′e·ter·i·za′tion** *n.*

**ca·thex·is** (kə-thĕk′sĭs) *n., pl.* **-thex·es** (-thĕk′sēz′) [Gk. *kathexis*, a holding < *katekhein*, to hold fast : *kata-*, down + *ekhein*, to hold.] Concentration of emotional energy on an object or idea. **—ca·thec′tic** (-tĭk) *adj.*

**cath·ode** (kăth′ōd′) *n.* [Gk. *kathodos*, descent : *kata-*, down + *hodos*, way.] **1.** A negatively charged electrode, as of an electrolytic cell, storage battery, or electron tube. **2.** The positively charged terminal of a primary cell or of a storage battery that is supplying current. **—ca·thod′ic** (kă-thŏd′ĭk) *adj.* **—ca·thod′i·cal·ly** *adv.*

**cathode ray** *n.* **1.** A stream of electrons emitted by the cathode in electrical discharge tubes. **2.** One of the electrons emitted in a stream from a cathode-ray tube.

**cath·ode-ray tube** (kăth′ōd-rā′) *n.* A vacuum tube in which a hot cathode emits electrons that are accelerated as a beam through a relatively high voltage anode, further focused or deflected electrostatically or electromagnetically, and allowed to fall on a fluorescent screen.

**cath·o·lic** (kăth′ə-lĭk, kăth′lĭk) *adj.* [OFr. *catholique* < LLat. *catholicus* < Gk. *katholikos* < *katholou*, in general : *kata-*, according to + *holos*, whole.] **1.** Broad or general in scope : UNIVERSAL. **2.** Broad and comprehensive in interests and sympathies : LIBERAL. **3. Catholic. a.** Of or relating to the universal Christian church. **b.** Of or relating to the ancient undivided Christian church. **c.** Of or designating those churches that have claimed to be representatives of the ancient undivided church. **d.** Of or concerning the Roman Catholic Church. *—n.* **Catholic.** A member of a Catholic church, esp. a Roman Catholic. **—ca·thol′i·cal·ly** (kə-thŏl′ĭ-kə-lē, -ĭk-lē) *adv.*

**Catholic Church** *n.* The Roman Catholic Church.

**Ca·thol·i·cism** (kə-thŏl′ĭ-sĭz′əm) *n.* The faith, doctrine, system, and practice of a Catholic church, esp. the Roman Catholic Church.

**cath·o·lic·i·ty** (kăth′ə-lĭs′ĭ-tē) *n.* **1.** The quality or state of being catholic : BROAD-MINDEDNESS. **2.** General acceptance : UNIVERSALITY. **3. Catholicity.** Roman Catholicism.

**ca·thol·i·cize** (kə-thŏl′ĭ-sīz′) *vt.* & *vi.* **-cized, -ciz·ing, -ciz·es.** **1.** To make or become catholic. **2.** To convert or be converted to Catholicism.

**ca·thol·i·con** (kə-thŏl′ĭ-kŏn′) *n.* [Fr. < Med. Lat. < Gk. *katholikon*, neuter of *katholikos*, universal. —see CATHOLIC.] A universal panacea.

**cat·house** (kăt′hous′) *n. Slang.* A house of prostitution.
**cat·i·on** (kăt′ī′ən) *n.* [Gk. *kation,* (a thing) going down : *kata-,* down + *ienai,* to go.] A positively charged ion that in an electrolyte moves toward a negative electrode. **—cat′i·on′ic** (kăt′ī-ŏn′ĭk) *adj.*
**cation exchange** *n.* A chemical process used in water softening in which cations of like charge are exchanged equally between a solid, as zeolite, and a solution, as water.
**cat·kin** (kăt′kĭn′) *n.* [< obs. Du. *katteken,* dim. of *katte,* cat (from its resemblance to a cat's tail).] *Bot.* A dense, often drooping flower cluster, as that of a birch, composed of small scalelike flowers.
**cat·like** (kăt′līk′) *adj.* Like a cat : STEALTHY <walked with *catlike* grace>
**cat·nap** (kăt′năp′) *n.* A short nap. **—cat′nap′** *v.* **(-napped, -nap·ping, -naps).**
**cat·nip** (kăt′nĭp′) *n.* A hairy aromatic plant native to Eurasia, *Nepeta cataria* to which cats are strongly attracted.
**cat-o′-nine-tails** (kăt′ə-nīn′tālz′) *n.* [So called because it leaves marks like the scratches of a cat.] A whip having nine knotted cords fastened to a handle.
**ca·top·tric** (kə-tŏp′trĭk) *also* **ca·top·tri·cal** (-trĭ-kəl) *adj.* [Gk. *katoptrikos* < *katoptron,* mirror.] Of or relating to mirrors and reflected images. **—ca·top′trics** *n.*
**cat rig** *n. Naut.* The rig of a catboat.
**CAT scan** *n.* A cross-sectional picture produced by a CAT scanner.
**CAT scanner** *n.* A device that produces cross-sectional x-rays of the body using computerized axial tomography.
**CAT scanning** *n.* The act or process of using a CAT scanner.
**cat's cradle** *n.* **1.** A child's game in which an intricately looped string is transferred from the hands of one player to the next, resulting in a succession of different loop patterns. **2.** Something likened to a cat's cradle <a *cat's cradle* of intrigues>
**cat scratch disease** *or* **cat scratch fever** *n.* A disease in humans marked by fever and lymphadenitis and believed to be transmitted by cats.
**cat's-eye** (kăts′ī′) *n.* **1.** Any of various semiprecious gems displaying a band of reflected light that shifts position as the gem is turned. **2.** A colored reflector attached to the back of a vehicle to indicate its presence on the road at night. **3.** A marble with eyelike circles.
**cat's-paw** *also* **cats·paw** (kăts′pô′) *n.* [From a fable about a monkey that used a cat's paw to pull chestnuts out of a fire.] **1.** One used by another as a dupe or tool. **2.** A light breeze that ruffles small areas of a water surface. **3.** *Naut.* A hitch in the bight of a rope, on which a tackle is hooked.
**cat·sup** (kăt′səp, kăch′əp, kĕch′-) *n. var. of* KETCHUP.
**cat·tail** (kăt′tāl′) *n.* A marsh plant of the genus *Typha,* esp. *T. latifolia,* having long, straplike leaves and a dense cylindrical head of minute brown flowers.
**cat·tie** (kăt′ē) *n. var. of* CATTY[1].
**cat·tle** (kăt′l) *pl.n.* [ME *catel,* livestock, property < Norman Fr. < Med. Lat. *capitale,* property < Lat. *capitalis,* principal. —see CAPITAL[1].] **1.** Animals of the genus *Bos,* esp. those of the domesticated species *B. taurus,* raised in many breeds for dairy products and meat. **2.** Human beings, esp. when regarded as an unthinking mob.

▲ word history: *Cattle* and *chattel* are doublets from medieval French, both derived from Latin *capitale,* which is also the ancestor of English *capital*[1]. All three words refer to wealth in one form or another. *Cattle* is derived from Norman French *catel* and in medieval times meant "movable property" in general. The word became restricted to that sense because livestock was such an important form of property. The further narrowing of *cattle* to refer only to bovine animals occurred in the 19th century. *Chattel* represents the central French form *chatel.* It was adopted in medieval England as a legal term for "movable property," supplanting *cattle. Capital*[1] was also borrowed from French in medieval times; it represents not the regular French development of Latin *capitale* but instead a French reborrowing of the Latin word. Its original use in English was as an adjective, first meaning "relating to the head" and later meaning "principal" or "chief." *Capital*[1] came to denote "wealth" by its use to mean someone's principal substance or property.

**cat·tle·man** (kăt′l-mən, -măn′) *n.* One who raises or tends cattle.
**cattle prod** *n.* An electrified prod for controlling cattle.
**cat·tle·ya** (kăt′lē-ə) *n.* [NLat. *Cattleya,* genus name, after William *Cattley* (d. 1832).] An orchid of the genus *Cattleya,* with large rose-purple or white flowers.
**cat·ty**[1] *also* **cat·tie** (kăt′ē) *n., pl.* **-ties.** [Malay *kati.*] An Asian unit of weight gen. equivalent to $1\frac{1}{3}$ pounds avoirdupois.
**cat·ty**[2] (kăt′ē) *adj.* **-ti·er, -ti·est. 1.** Catlike. **2.** Subtly cruel or malicious : SPITEFUL <*catty* remarks> **—cat′ti·ly** *adv.* **—cat′ti·ness** *n.*
**cat·ty-cor·nered** (kăt′ē-kôr′nərd) *adj.* & *adv. var. of* CATER-CORNERED.
**cat·walk** (kăt′wôk′) *n.* A narrow platform or pathway, as on the sides of a bridge.
**Cau·ca·sian** (kô-kā′zhən, -kăzh′ən) *n.* [After the *Caucasus,* a mountain range in the Soviet Union.] **1.** A native or resident of the Caucasus. **2.** A member of the Caucasoid ethnic division. *—adj.* **1.** Of or relating to the Caucasus region, its people, or their culture. **2.** Caucasoid.
**Cau·ca·soid** (kô′kə-soid′) *adj.* **1.** Of, relating to, or designating a major ethnic division of the human species having certain distinctive physical characteristics such as skin color varying from very light to brown and fine hair ranging from straight to wavy or curly and regarded as including groups of peoples indigenous to or inhabiting Europe, northern Africa, southwestern Asia, and the Indian subcontinent and persons of this ancestry in other parts of the world. **2.** Of, relating to, or typical of Caucasoids. *—n.* A member of the Caucasoid ethnic division.
**cau·cus** (kô′kəs) *n., pl.* **-cus·es** *or* **-cus·ses.** [Poss. of Algonquian orig.] **1.** A closed meeting of the members of a political party to make policy decisions and select candidates for office. **2.** *Chiefly Brit.* A committee within a political party charged with setting policy. *—vi.* **-cused, -cus·ing, -cus·es** *or* **-cussed, -cus·sing, -cus·ses.** To assemble in or hold a caucus.
**cau·dad** (kô′dăd′) *adv.* [Lat. *cauda,* tail + -AD.] *Anat.* Toward the tail or posterior part of the body.
**cau·dal** (kôd′l) *adj.* [NLat. *caudalis* < Lat. *cauda,* tail.] **1.** *Anat.* Of, at, or near the tail or hind parts : POSTERIOR. **2.** *Zool.* Taillike. **—cau′dal·ly** *adv.*
**caudal fin** *n.* The tail fin of a fish.
**cau·date** (kô′dāt′) *also* **cau·dat·ed** (-dā′tĭd) *adj.* [NLat. *caudatus* < Lat. *cauda,* tail.] Having a tail. **—cau·da′tion** *n.*
**caudate nucleus** *n.* A large ganglion in the lateral ventricle of the brain that functions in motor control.
**cau·dex** (kô′dĕks′) *n., pl.* **-di·ces** (-dĭ-sēz′) *or* **-dex·es.** [Lat. *caudex, caudic-,* tree trunk.] **1.** The thickened base of the stem of certain perennials. **2.** A woody trunklike stem, as that of a tree fern.
**cau·di·llo** (kou-*thē*l′yō, -*thē*′yō) *n., pl.* **-llos.** [Sp., leader < LLat. *capitellum,* small head, dim. of Lat. *caput,* head.] A military dictator in Spanish-speaking countries.
**cau·dle** (kôd′l) *n.* [ME *caudel* < Norman Fr. < *chaud,* warm < Lat. *calidus.*] A warm beverage of wine or ale mixed with sugar, eggs, bread, and spices, given to sick people.
**caught** (kôt) *v. p.t. & p.p. of* CATCH.
**caul** (kôl) *n.* [ME *calle* < OE *cawl,* basket.] **1.** A part of the membrane that surrounds a fetus and that occas. covers its head at birth. **2.** The large omentum.
**caul·dron** (kôl′drən) *n. var. of* CALDRON.
**cau·les·cent** (kô-lĕs′ənt) *adj.* [Lat. *caulis,* stem + -ESCENT.] *Bot.* Having a stem showing above the ground.
**cau·li·cle** (kô′lĭ-kəl) *n.* [Lat. *cauliculus,* dim. of *caulis,* stem.] *Bot.* A small stem.
**cau·li·flow·er** (kô′lĭ-flou′ər, kŏl′ĭ-) *n.* [Prob. < Ital. *cavolofiore* : *cavolo,* cabbage (< LLat. *caulus* < Lat. *caulis*) + *fiore,* flower < Lat. *flos.*] **1.** A plant, *Brassica oleracea botrytis,* related to the cabbage and broccoli and having an enlarged, edible flower head. **2.** The compact whitish flower head of the cauliflower.
**cauliflower ear** *n.* An ear deformed by repeated blows.
**cau·line** (kô′līn′) *adj.* [NLat. *caulinus* < Lat. *caulis,* stem.] *Bot.* Of, having, or growing on a stem.
**caulk** *also* **calk** (kôk) *vt.* **caulked, caulk·ing, caulks** *also* **calked, calk·ing, calks.** [OFr. *cauquer,* to press < Lat. *calcare,* to tread < *calx,* heel.] **1.** To make (a boat) watertight by packing seams with oakum or tar. **2.** To make (e.g., pipes) watertight or airtight by filling in cracks. **—caulk′er** *n.*
**caus·al** (kô′zəl) *adj.* **1.** Relating to, involving, or being a cause <the *causal* factors of the recession> **2.** Indicating or expressing a cause. **3.** Originating from a cause. *—n.* A word or grammatical element expressing a cause or reason. **—caus′al·ly** *adv.*
**cau·sal·i·ty** (kô-zăl′ĭ-tē) *n., pl.* **-ties. 1.** The relationship between cause and effect. **2.** A causal force, agency, or quality.
**cau·sa·tion** (kô-zā′shən) *n.* **1.** The act or process of causing. **2.** CAUSE 1a. **3.** Causality.
**caus·a·tive** (kô′zə-tĭv) *adj.* **1.** Functioning as an agent or cause. **2.** Designating a verb or verbal affix that expresses causation. **—caus′a·tive** *n.* **—caus′a·tive·ly** *adv.*
**cause** (kôz) *n.* [ME < OFr. < Lat. *causa,* reason.] **1. a.** Something that produces an effect, result, or consequence. **b.** The person, event, or state responsible for an action or result. **2.** A basis for an action or decision : REASON. **3.** A goal or principle served with dedication and zeal <"the *cause* of freedom versus tyranny" —Hannah Arendt> **4.** The interests of a person or group engaged in a struggle. **5.** *Law.* **a.** A ground for legal action. **b.** A lawsuit. **6.** A subject under discussion or debate. *—vt.* **caused, caus·ing, caus·es.** To be the cause of. **—caus′a·ble** *adj.* **—cause′less** *adj.* **—caus′er** *n.*

☆ ***syns:*** CAUSE, ANTECEDENT, REASON *n. core meaning* : that which produces a result or effect <scientists studying the *cause* of sunspots> ***ant:*** effect

**cause cé·lè·bre** (kōz′ sā-lĕb′rə) *n., pl.* **causes cé·lè·bres** (kōz′ sā-lĕb′rə) [Fr. : *cause,* case + *célèbre,* celebrated.] **1.** A notorious legal case. **2.** A highly controversial issue.
**cau·se·rie** (kōz-rē′) *n.* [Fr. < *causer,* to talk < Lat. *causari,* to discuss < *causa,* case.] **1.** A chat. **2.** A short, informal piece of writing.

**cause·way** (kôz′wā′) *n.* [ME *caucewei* < *cauce,* raised road < Norman Fr. *caucie* < VLat. **calciata* < Lat. *calx,* limestone. —see CALX.] **1.** A raised roadway, as across water. **2.** A paved road.

**caus·tic** (kô′stĭk) *adj.* [Lat. *causticus* < Gk. *kaustikos* < *kaiein,* to burn.] **1.** Capable of corroding, burning, dissolving, or otherwise eating away by chemical action. **2.** Sharp and bitter : SARCASTIC. **3.** Of or relating to light emitted from a point source and reflected or refracted from a curved surface. —*n.* A caustic substance or material. **—caus′ti·cal·ly** *adv.* **—caus·tic′i·ty** (kô-stĭs′ĭ-tē) *n.*

**caustic potash** *n.* Potassium hydroxide.

**caustic soda** *n.* Sodium hydroxide.

**cau·ter·ize** (kô′tə-rīz′) *vt.* **-ized, -iz·ing, -iz·es.** [OFr. *cauteriser* < LLat. *cauterizare,* to brand < Gk. *kautēriazein* < *kautērion,* branding iron. —see CAUTERY.] To burn or sear with a cautery. **—cau′ter·i·za′tion** *n.*

**cau·ter·y** (kô′tə-rē) *n., pl.* **-ies.** [Lat. *cauterium,* branding iron < Gk. *kautērion* < *kaiein,* to burn.] **1.** A caustic agent or a very hot or cold instrument used to destroy aberrant tissue. **2.** Cauterization.

**cau·tion** (kô′shən) *n.* [ME *caucioun* < OFr. *caution* < Lat. *cautio* < *cavēre,* to take care.] **1.** Careful forethought. **2.** A warning : admonishment. **3.** *Informal.* One that is striking or alarming. —*vt.* **-tioned, -tion·ing, -tions.** To warn against danger. **—cau′tion·ar′y** (-shə-nĕr′ē) *adj.*

☆ ***syns:*** CAUTION, CALCULATION, CAREFULNESS, CIRCUMSPECTION, PRECAUTION, WARINESS *n. core meaning* : careful forethought <climbed the icy steps with *caution*> <proceed with *caution*>

**cau·tious** (kô′shəs) *adj.* Exhibiting or practicing caution : CAREFUL. **—cau′tious·ly** *adv.* **—cau′tious·ness** *n.*

**cav·al·cade** (kăv′əl-kād′, kăv′əl-kād′) *n.* [Fr. < OFr. < OItal. *cavalcata* < *cavalcare,* to ride on horseback < VLat. **caballicare* < Lat. *caballus,* horse.] **1.** A ceremonial procession, esp. of riders or horse-drawn carriages. **2.** A colorful procession or display.

**cav·a·lier** (kăv′ə-lîr′) *n.* [OFr. < OItal. *cavaliere* < LLat. *caballarius,* horseman < Lat. *caballus,* horse.] **1.** An armed horseman : KNIGHT. **2.** A chivalrous gentleman. **3. Cavalier.** A supporter of Charles I of England in his struggles against Parliament : ROYALIST. —*adj.* **1.** Arrogant or disdainful : HAUGHTY. **2.** Carefree and gay : OFFHAND.

**ca·val·la** (kə-văl′ə) *n., pl.* **-las** or **cavalla.** [Sp. *caballa,* horse mackerel < LLat. *caballus,* horse.] **1.** Any of various tropical marine food fishes of the family Carangidae. **2.** The king mackerel.

**cav·al·ry** (kăv′əl-rē) *n., pl.* **-ries.** [OFr. *cavallerie* < OItal. *cavalleria* < *cavaliere,* cavalier.] Troops trained to fight in armored vehicles or on horseback. **—cav′al·ry·man** *n.*

**cave** (kāv) *n.* [ME < OFr. < Lat. *cava* < *cavus,* hollow.] An underground hollow often having an opening in the side of a hill or cliff. —*vt.* **caved, cav·ing, caves.** To hollow out. **—cave in.** *Informal.* **1.** To collapse, as from being undermined <a tunnel that *caved in*> **2.** To give up all opposition : YIELD <*caved in* when they heard our threats>

**ca·ve·at** (kā′vē-ăt′, kăv′ē-, kä′vē-ät′) *n.* [Lat., let him beware < *cavēre,* to beware.] **1.** *Law.* A formal notice filed by an interested party with a court or officer requesting the postponement of a proceeding until he or she is heard. **2.** A warning or caution. —*v.* **-at·ed, -at·ing, -ats.** —*vi. Law.* To enter a caveat. —*vt. Slang.* To do or say (something) with an accompanying warning.

**caveat emp·tor** (ĕmp′tôr′) *n.* [Lat., let the buyer beware.] The axiom that one who buys something does so at one's own peril.

**cave·fish** (kāv′fĭsh′) *n., pl.* **cavefish** or **-fish·es.** Any of various freshwater fishes of the family Amblyopsidae of subterranean waters, with rudimentary eyes.

**cave-in** (kāv′ĭn′) *n.* **1.** An act of caving in. **2.** A place where the ground has caved in.

**cave man** *n.* **1.** A prehistoric human being who lived in caves. **2.** *Informal.* One whose behavior is crude or brutal, esp. toward women. **—cave′man′** *adj.*

**cav·ern** (kăv′ərn) *n.* [ME *caverne* < OFr. < Lat. *caverna* < *cavus,* hollow.] A large cave. —*vt.* **-erned, -ern·ing, -erns. 1.** To enclose in or as if in a cavern. **2.** To hollow out.

**cav·er·nic·o·lous** (kăv′ər-nĭk′ə-ləs) *adj.* Living in caverns or caves.

**cav·ern·ous** (kăv′ər-nəs) *adj.* **1.** Filled with caverns. **2.** Like a cavern. **3.** Filled with cavities : POROUS. **—cav′ern·ous·ly** *adv.*

**ca·vet·to** (kə-vĕt′ō) *n., pl.* **-vet·ti** (-vĕt′ē) or **-vet·tos.** [Ital. < *cavo,* hollow < Lat. *cavus.*] A concave architectural molding shaped like a circular quadrant.

**cav·i·ar** *also* **cav·i·are** (kăv′ē-är′, kä′vē-) *n.* [Prob. < OItal. *caviaro* < Turk. *havyār.*] The roe esp. of a sturgeon, salted, seasoned, and eaten as an appetizer or garnish.

**cav·il** (kăv′əl) *v.* **-iled, -il·ing, -ils** *also* **-illed, -il·ling, -ils.** [OFr. *caviller* < Lat. *cavillari,* to criticize < *cavilla,* a jeering.] —*vi.* To find fault unnecessarily. —*vt.* To quibble about. —*n.* A captious or trivial objection. **—cav′il·er** *n.*

**cav·i·ta·tion** (kăv′ĭ-tā′shən) *n.* [< CAVITY.] **1.** Sudden formation and collapse of low-pressure bubbles in liquids by mechanical forces, as those resulting from rotation of a marine propeller. **2.** Formation of cavities in tissue or an organ, esp. as a result of disease.

**cav·i·ty** (kăv′ĭ-tē) *n., pl.* **-ties.** [Fr. *cavité* < OFr. *cavete* < LLat. *cavitas* < Lat. *cavus,* hollow.] **1.** A hollow or hole. **2.** A hollow area within the body. **3.** A pitted area in a tooth caused by caries.

**ca·vort** (kə-vôrt′) *vi.* **-vort·ed, -vort·ing, -vorts.** [Perh. var. of CURVET.] **1.** To bound or prance about in a sprightly way : CAPER. **2.** To frolic.

**ca·vy** (kā′vē) *n., pl.* **-vies.** [Prob. < Galibi *cabiai.*] Any of various short-tailed or apparently tailless South American rodents of the family Caviidae, including the guinea pig and the capybara.

**caw** (kô) *n.* [Imit.] The hoarse, raucous sound of a crow or similar bird. —*vi.* **cawed, caw·ing, caws.** To utter a caw.

**cay** (kē, kā) *n.* [Sp. *cayo,* prob. < OFr. *quai,* quay.] A small low islet formed chiefly of coral or sand.

**cay·enne pepper** (kī-ĕn′, kā-) *n.* [Obs. *kian* < Tupi *kyinha.*] A condiment made from the pungent fruit of a variety of the plant *Capsicum frutescens.*

**cay·man** (kā′mən, kā-măn′, kī-măn′) *n. var. of* CAIMAN.

**Ca·yu·ga** (kā-yōō′gə, kī-) *n., pl.* **Cayuga** or **-gas. 1. a.** A tribe of Indians once living around Cayuga and Seneca lakes in central New York. **b.** A member of this tribe. **2.** The Iroquoian language of the Cayuga. **—Ca·yu′ga** *adj.*

**†Cay·use** (kī-yōōs′, kī′yōōs′) *n., pl.* **Cayuse** or **-us·es. 1. a.** A tribe of Indians living in Oregon. **b.** A member of this tribe. **c.** The Sahaptin language of the Cayuse. **2.** *Western U.S.* A horse, esp. an Indian pony. **—Cay·use′** *adj.*

**Cb** *symbol for* COLUMBIUM.

**CB** (sē-bē′) *n., pl.* **CB's.** Citizens band.

**CB·er** (sē-bē′ər) *n.* One who uses a CB radio.

**C clef** *n. Mus.* A clef sign used to form any of three clefs, soprano, alto, or tenor, by locating the tone C (261.7 cycles per second) on, respectively, the lowest line of the staff, the middle line, or the fourth (next to the highest) line.

**Cd** *symbol for* CADMIUM.

**Ce** *symbol for* CERIUM.

**cease** (sēs) *v.* **ceased, ceas·ing, ceas·es.** [ME *cesen* < OFr. *cesser* < Lat. *cessare,* to stop, freq. of *cedere,* to yield.] —*vt.* To put an end to : DISCONTINUE <The assembly line *ceased* production.> —*vi.* To come to an end : STOP <noise that finally *ceased*> —*n.* Cessation <pressure without *cease*>

**cease-fire** (sēs′fīr′) *n.* **1.** An order to cease firing. **2.** Suspension of active hostilities : TRUCE.

**cease·less** (sēs′lĭs) *adj.* Never stopping : ENDLESS <*ceaseless* nagging> **—cease′less·ly** *adv.* **—cease′less·ness** *n.*

**ce·cec·to·my** (sē-sĕk′tə-mē) *n., pl.* **-mies.** [CEC(UM) + -ECTOMY.] Surgical excision of the cecum.

**ce·cro·pi·a moth** (sĭ-krō′pē-ə) *n.* [NLat. *cecropia,* specific name < Lat., fem. of *Cecropius,* Athenian < Gk. *Kekropios* < *Kekrops,* Cecrops, a legendary Athenian king.] A large North American moth, *Hyalophora cecropia,* with wide wings having red, white, and black markings.

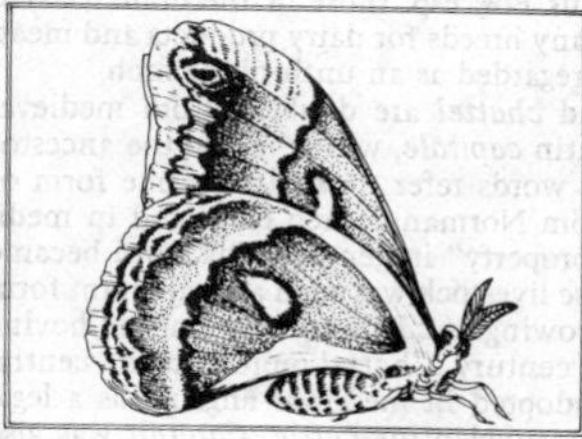

**cecropia moth**
*Wingspan up to 6 inches*

**ce·cum** *also* **cae·cum** (sē′kəm) *n., pl.* **-ca** (-kə) [NLat. < Lat. *(intestinum) caecum,* blind (intestine) < *caecus,* blind.] **1.** A cavity with only one opening. **2.** *Anat.* The large blind pouch forming the beginning of the large intestine. **—ce′cal** (sē′kəl) *adj.*

**ce·dar** (sē′dər) *n.* [ME *cedre* < OFr. < Lat. *cedrus* < Gk. *kedros.*] **1. a.** An Old World coniferous evergreen tree of the genus *Cedrus, as the cedar of Lebanon.* **b.** A similar evergreen tree, chiefly of the genera *Thuja, Chamaecyparis,* or *Juniperus.* **2.** The durable, aromatic, often reddish wood of a cedar.

**ce·dar·bird** (sē′dər-bûrd′) *n.* A cedar waxwing.

**cedar of Lebanon** *n.* A tall evergreen tree, *Cedrus libani* of Asia Minor, with short dark needles and fragrant hard wood.

**cedar waxwing** *n.* [Prob. so called because it eats the berries of the red cedar.] A North American bird, *Bombycilla cedrorum,* with a crested head and chiefly brown plumage.

**cede** (sēd) *vt.* **ced·ed, ced·ing, cedes.** [OFr. *ceder* < Lat. *cedere.*] **1.** To surrender possession of formally or officially. **2.** To yield or grant, as by a treaty.

**ce·di** (sā′dē) *n., pl.* **cedi** or **-dis.** [Native word in Ghana.] —See table at CURRENCY.

**ce·dil·la** (sĭ-dĭl′ə) *n.* [Obs. Sp., dim. of *ceda,* the letter *z* < LLat. *zeta,* zeta (so called because a small *z* was formerly used to make a hard *c* sibilant).] A mark ( ̧ ) placed beneath the letter *c,* as in the spelling of the French word *garçon,* to indicate that the letter is to be pronounced (s).

**cee** (sē) *n.* The letter *c.*

**cei·ba** (sā′bə) *n.* [Sp., prob. < Arawakan.] Any of various large tropical trees of the genus *Ceiba,* including the silk-cotton tree, which is the source of kapok.

**ceil** (sēl) *vt.* **ceiled, ceil·ing, ceils.** [ME *celen* < OFr. *celer* < Lat. *caelare,* to carve < *caelum,* chisel. —see CAELUM.] **1.** To make a ceiling for. **2.** To provide (a ship) with interior planking.

**ceil·ing** (sē′lĭng) *n.* [ME *celing* < *celen,* to ceil.] **1. a.** The interior upper surface of a room. **b.** Material used to cover this surface. **2.** Something like a ceiling <a *ceiling* of green leaves over the arbor> **3.** Planking applied to the interior framework of a ship. **4.** A maximum limit, esp. as prescribed by law <wage and price *ceilings*> **5.** Any of various vertical boundaries, esp. of atmospheric visibility, cloud cover altitude, or operable aircraft altitude. **—ceil′inged** *adj.*

**ceil·om·e·ter** (sē-lŏm′ĭ-tər) *n.* [CEIL(ING) + -METER.] A photoelectric instrument for determining cloud heights.

**cel·a·don** (sĕl′ə-dŏn′) *n.* [Fr., after *Céladon,* a character in *Astrée,* a romance by Honoré d'Urfé (1568–1625).] **1.** A pale to very pale green. **2.** A pale to very pale blue.

**cel·a·don·ite** (sĕl′ə-dn-īt′) *n.* A soft mica with a green hue and a high iron content.

**Ce·lae·no** (sĭ-lē′nō) *n.* [Lat. < Gk. *Kelainō.*] **1.** *Gk. Myth.* One of the Pleiades. **2.** One of the six stars in the Pleiades cluster visible to the naked eye.

**cel·an·dine** (sĕl′ən-dīn′, -dēn′) *n.* [ME *celidoine* < OFr. < Med. Lat. *celidonia* < Lat. *chelidonia* < Gk. *khelidonion* < *khelidōn,* swallow (from the association by ancient writers of the plant with the habits of the swallow).] **1.** A Eurasian plant, *Chelidonium majus,* having deeply divided leaves, yellow flowers, and yellow-orange juice. **2.** The lesser celandine.

**-cele**[1] *suff.* Tumor : hernia <cysto*cele*>

**-cele**[2] *suff. var. of* -COEL.

**cel·e·brant** (sĕl′ə-brənt) *n.* **1.** The priest officiating at the Eucharist. **2.** A participant in a celebration. ***usage:*** The use of *celebrant* to mean "a participant in a celebration," as in *a New Year's Eve celebrant,* has recently gained currency, but traditionalists who are reluctant to adopt this new usage may prefer *celebrator.*

**cel·e·brate** (sĕl′ə-brāt′) *v.* **-brat·ed, -brat·ing, -brates.** [Lat. *celebrare, celebrat-* < *celeber,* famous.] *—vt.* **1.** To observe (a day or event) with ceremonies of respect, festivity, or rejoicing. **2.** To perform (a religious ceremony) <*celebrate* Mass> **3.** To praise publicly : HONOR <a sonnet *celebrating* love> *—vi.* **1.** To observe an occasion with ceremony or festivity. **2.** To perform a religious ceremony. **3.** To take part in festivities <*celebrated* after graduation> **—cel′e·bra′tion** *n.* **—cel′e·bra′tor** *n.*

☆ ***syns:*** CELEBRATE, COMMEMORATE, KEEP, OBSERVE, SOLEMNIZE *v. core meaning* : to mark (a day or event) with ceremony <We *celebrate* Thanksgiving with a traditional turkey dinner.>

**cel·e·brat·ed** (sĕl′ə-brā′tĭd) *adj.* Well-known : famous.

**ce·leb·ri·ty** (sə-lĕb′rĭ-tē) *n., pl.* **-ties.** [Lat. *celebritas,* fame < *celeber,* famous.] **1.** A famous person. **2.** Renown : fame.

☆ ***syns:*** CELEBRITY, LUMINARY, NAME, NOTABLE, PERSONAGE *n. core meaning* : a famous person <*celebrities* of stage, screen, and television>

**ce·le·ri·ac** (sə-lĭr′ē-ăk′, -lĕr′-) *n.* [Alteration of CELERY.] A variety of celery, *Apium graveolens rapaceum,* cultivated for its edible, turniplike root.

**ce·ler·i·ty** (sə-lĕr′ĭ-tē) *n.* [ME *celerite* < OFr. < Lat. *celeritas* < *celer,* swift.] Speed : swiftness <moved with *celerity*>

**cel·er·y** (sĕl′ə-rē) *n., pl.* **-ies.** [Fr. *céleri* < dial. Ital. *seleri,* pl. of *selero,* alteration of Lat. *selinon,* parsley < Gk.] A Eurasian plant, *Apium graveolens dulce,* cultivated for its edible stalks.

**celery cabbage** *n.* Chinese cabbage.

**ce·les·ta** (sə-lĕs′tə) *also* **ce·leste** (sə-lĕst′) *n.* [Fr. *célesta* < *céleste,* celestial < Lat. *caelestis.*] A musical instrument with a keyboard and metal plates struck by hammers that produce bell-like tones.

**ce·les·tial** (sə-lĕs′chəl) *adj.* [ME < OFr. < Lat. *caelestis* < *caelum,* sky.] **1.** Of or relating to the sky or the heavens. **2.** Of, from, or resembling heaven : SPIRITUAL <*celestial* beings> **3.** Supreme in nature or kind <*celestial* bliss> **4. Celestial.** Of or relating to the Chinese people or to the former Chinese Empire. *—n.* A heavenly being. **—ce·les′tial·ly** *adv.*

**celestial equator** *n.* A great circle on the celestial sphere in the same plane as the earth's equator.

**celestial globe** *n.* A model of the celestial sphere showing the stars and other celestial bodies.

**celestial horizon** *n.* HORIZON 2c.

**celestial mechanics** *n. (sing. in number).* The science of the motion of celestial bodies under the influence of gravitational forces.

**celestial navigation** *n.* Ship or aircraft navigation based on the positions of celestial bodies.

**celestial pole** *n.* Either of two diametrically opposite points at which the extensions of the earth's axis intersect the celestial sphere.

**celestial sphere** *n.* An imaginary sphere of infinite extent with the earth at its center on which the stars, planets, and other heavenly bodies appear to be located.

**cel·es·tite** (sĕl′ĭ-stīt′, sə-lĕs′-) *n.* [G. *Zölestin* < Lat. *caelestis,* celestial.] An important white, red-brown, or light-blue strontium ore, primarily strontium sulfate, $SrSO_4$.

**ce·li·ac** *also* **coe·li·ac** (sē′lē-ăk′) *adj.* [Lat. *coeliacus* < Gk. *koiliakos* < *koilia,* abdomen < *koilos,* hollow.] Of or relating to the abdomen. *—n.* One afflicted with celiac disease.

**celiac disease** *n.* A chronic nutritional disturbance of infants and young children, caused by improper absorption of fats and resulting in malnutrition, distended abdomen, and diarrhea.

**cel·i·ba·cy** (sĕl′ə-bə-sē) *n.* [Lat. *caelibatus* < *caelebs,* unmarried.] **1.** The state of being unmarried, esp. by reason of religious vows. **2.** Abstinence from sexual intercourse.

**cel·i·bate** (sĕl′ə-bĭt) *n.* [< Lat. *caelebs, caelib-,* unmarried.] **1.** One who remains unmarried, esp. by reason of religious vows. **2.** One who abstains from sexual intercourse. **—cel′i·bate** *adj.*

**cell** (sĕl) *n.* [ME *celle,* partly < OE *cell,* and partly < OFr., both < Lat. *cella,* chamber.] **1.** A narrow, confining room, as in a prison or convent. **2.** A small one-room abode. **3.** A small religious house dependent on a larger one, as a priory within an abbey. **4.** The primary organizational unit of a movement <a Communist *cell*> **5.** *Biol.* The smallest structural unit of an organism that is capable of independent functioning, composed of one or more nuclei, cytoplasm, various organelles, and inanimate matter, all surrounded by a semipermeable plasma membrane. **6.** *Biol.* A small enclosed cavity or space, as a compartment in a honeycomb or within a plant ovary or an area bordered by veins in an insect's wing. **7.** *Elect.* **a.** A single unit for electrolysis or for conversion of chemical into electric energy, usu. composed of a container with electrodes and an electrolyte. **b.** A single unit that converts radiant energy into electric energy <fuel *cells*> **8.** *Computer Sci.* A basic unit of storage in a computer memory that can hold one unit of information, as a character or word. *—v.* **celled, cell·ing, cells.** *—vt.* To store in a honeycomb. *—vi.* To live in a cell.

**cel·la** (sĕl′ə) *n., pl.* **cel·lae** (sĕl′ē′) [Lat.] The inner room of an ancient Greek or Roman temple.

**cel·lar** (sĕl′ər) *n.* [ME *celer* < OFr. < LLat. *cellarium,* pantry < Lat. *cella,* storeroom.] **1.** A room for storage, usu. below ground or beneath a building. **2. a.** A cool, dark room for storing wines. **b.** A stock of wines. **3.** *Informal.* The lowest level, esp. in the relative standing of athletic teams. *—vt.* **-lared, -lar·ing, -lars.** To store in a cellar.

**cel·lar·age** (sĕl′ər-ĭj) *n.* **1.** A fee charged for cellar storage. **2.** Cellars as a whole.

**cel·lar·er** (sĕl′ər-ər) *n.* [ME *celerer* < OFr. < Lat. *cellarius,* steward < *cella,* storeroom.] One, as in a monastic community, who is responsible for maintaining adequate supplies of food and drink.

**cel·lar·ette** *also* **cel·lar·et** (sĕl′ə-rĕt′) *n.* A cabinet for storing wine bottles.

**cell membrane** *n.* Plasma membrane.

**cel·lo** (chĕl′ō) *n., pl.* **-los.** [Short for VIOLONCELLO.] A four-stringed instrument of the violin family, pitched lower than the viola but higher than the double bass. **—cel′list** (chĕl′ĭst) *n.*

**cel·lo·bi·ose** (sĕl′ə-bī′ōs′, -ōz′) *n.* [CELL(ULOSE) + BI- + -OSE.] A disaccharide obtained from the partial hydrolysis of cellulose.

**cel·loi·din** (sə-loid′n) *n.* [CELL(ULOSE) + -OID + -IN.] A pure pyroxylin in which specimen sections are embedded for microscopic examination.

**cel·lo·phane** (sĕl′ə-fān′) *n.* [Orig. a trademark.] A thin, flexible, transparent cellulose material made from wood pulp, used as a moistureproof wrapping.

**cel·lu·lar** (sĕl′yə-lər) *adj.* **1.** Relating to or resembling a cell. **2.** Made of or containing a cell or cells. **—cel′lu·lar′i·ty** (-lăr′ĭ-tē) *n.* **—cel′lu·lar·ly** *adv.*

**cel·lu·lase** (sĕl′yə-lās′, -lāz′) *n.* [CELLUL(OSE) + -ASE.] Any of several enzymes found in fungi, bacteria, and lower animals, that hydrolyze cellulose.

**cel·lule** (sĕl′yo͞ol) *n.* [Fr. < Lat. *cellula,* dim. of *cella,* chamber.] A small cell.

**cel·lu·lite** (sĕl′yə-līt′) *n.* [CELLUL(OSE) + -ITE.] A fatty deposit, as around the thighs and buttocks.

**cel·lu·li·tis** (sĕl′yə-lī′tĭs) *n.* [Lat. *cellula,* cellule + -ITIS.] Inflammation of subcutaneous tissue.

**cel·lu·loid** (sĕl′yə-loid′) *n.* [Orig. a trademark.] A colorless, flammable material made from nitrocellulose and camphor.

**cel·lu·lo·lyt·ic** (sĕl′yə-lō-lĭt′ĭk) *adj.* [CELLULO(SE) + -LYTIC.] Of, relating to, or bringing about hydrolysis of cellulose.

**cel·lu·lose** (sĕl′yə-lōs′, -lōz′) *n.* [Fr. < *cellule,* biological cell, cellule.] An amorphous carbohydrate polymer, $(C_6H_{10}O_5)_x$, the main constituent of all plant tissues and fibers, used in making paper, textiles, and explosives. **—cel'lu·lo'sic** (-lō'sĭk, -zĭk) *adj.*

**cellulose acetate** *n.* A cellulose resin used in lacquers, photographic film, transparent sheeting, and cigarette filters.

**cellulose nitrate** *n.* Nitrocellulose.

**cell wall** *n.* The permeable, rigid outermost layer of a plant cell composed chiefly of cellulose.

**ce·lom** (sē'ləm) *n. var. of* COELOM.

**Cel·o·tex** (sĕl'ə-tĕks'). A trademark for a building board made of compressed bagasse, used for insulation.

**Cel·si·us** (sĕl'sē-əs, -shəs) *adj.* [After Anders *Celsius* (1701–1744), its inventor.] Of or relating to a temperature scale that registers the freezing point of water as 0°C and the boiling point as 100°C under normal atmospheric pressure.

**celt** (sĕlt) *n.* [LLat. *celtis,* chisel.] A prehistoric axlike tool.

**Celt** (kĕlt, sĕlt) *n.* [Fr. *Celte* < Lat. *Celta,* sing. of *Celtae.*] **1.** One of an ancient people of western and central Europe, including the Britons and the Gauls. **2.** A speaker or a descendant of a speaker of a Celtic language.

**Celt·ic** (kĕl'tĭk, sĕl'-) *n.* A subfamily of the Indo-European language family comprising the Brythonic and the Goidelic branches. *—adj.* Of or relating to the Celtic people and languages.

**Celtic cross** *n.* An upright cross superimposed on a circle.

**Celt·i·cism** (kĕl'tĭ-sĭz'əm, sĕl'-) *n.* **1.** A Celtic custom. **2.** A fondness for Celtic customs. **3.** A Celtic idiom.

**Celt·i·cist** (kĕl'tĭ-sĭst, sĕl'-) *n.* A specialist in Celtic culture or languages.

**cem·ba·lo** (chĕm'bə-lō') *n., pl.* **-los.** [Ital., short for *clavicembalo* < Med. Lat. *clavicymbalum* : Lat. *clavis,* key + Lat. *cymbalum,* cymbal. —see CYMBAL.] A harpsichord. **—cem'ba·list** *n.*

**ce·ment** (sĭ-mĕnt') *n.* [ME *ciment* < OFr. < Lat. *caementum,* rough stone < *caedere,* to cut.] **1.** Any of various construction adhesives, consisting of powdered, calcined rock and clay materials, that form a paste with water and can be molded or poured to set as a solid mass. **2.** A substance that hardens to function as an adhesive **:** GLUE. **3.** Something, as a concern or feeling, that unites <Love was the *cement* that saved their marriage.> **4.** *Geol.* A chemically precipitated substance that binds particles of clastic rocks. **5.** Cementum. *—v.* **-ment·ed, -ment·ing, -ments.** *—vt.* **1.** To bind with or as if with cement. **2.** To cover or coat with cement. *—vi.* To become cemented. **—ce·ment'er** *n.*

**ce·men·ta·tion** (sē'mĕn-tā'shən) *n.* **1.** The act, process, or result of cementing. **2.** A metallurgical coating process in which iron or steel is immersed in a powder of another metal, as zinc, chromium, or aluminum, and heated to a temperature below the melting point of either.

**ce·ment·ite** (sĭ-mĕn'tīt') *n.* [< CEMENT.] A hard, brittle iron carbide, $Fe_3C$, found in steel with more than 0.85% carbon.

**ce·men·ti·tious** (sē'mĕn-tĭsh'əs) *adj.* Of, relating to, or having the characteristics of cement.

**cement mixer** *n.* A concrete mixer.

**ce·men·tum** (sĭ-mĕn'təm) *n.* [NLat. <Lat. *caementum,* rough stone < *caedere,* to cut.] A bony substance covering the root of a tooth.

**cem·e·ter·y** (sĕm'ĭ-tĕr'ē) *n., pl.* **-ies.** [ME *cimiterie* < Med. Lat. *cimiterium* < LLat. *coemeterium* < Gk. *koimētērion* < *koiman,* to put to sleep.] A place for burying the dead **:** GRAVEYARD.

**cen·a·cle** (sĕn'ə-kəl) *n.* [ME < OFr. < LLat. *cenaculum* < Lat. *cena,* dinner.] A small dining room, usu. on an upper floor.

**-cene** *suff.* [< Gk. *kainos,* new.] Recent. —Used in names of geologic periods <Oligo*cene*>

**ceno-** *pref. var. of* COENO-.

**cen·o·bite** *also* **coen·o·bite** (sĕn'ə-bīt', sē'nə-) *n.* [LLat. *coenobita* < *coenobium,* convent < Gk. *koinobion* < *koinobios,* living in community : *koinos,* common + *bios,* life.] A member of a religious community. **—cen'o·bit'ic** (-bĭt'ĭk), **cen'o·bit'i·cal** *adj.*

**ce·no·gen·e·sis** *also* **coe·no·gen·e·sis** (sē'nō-jĕn'ĭ-sĭs, sĕn'ō-) *n.* [Gk. *kainos,* new + GENESIS.] The environmentally determined development of characteristics or structures in an organism. **—ce'no·ge·net'ic** (-jə-nĕt'ĭk) *adj.* **—ce'no·ge·net'i·cal·ly** *adv.*

**ce·no·spe·cies** (sē'nə-spē'shēz) *n.* A group of species capable of interbreeding.

**cen·o·taph** (sĕn'ə-tăf') *n.* [OFr. *cenotaphe* < Lat. *cenotaphium* < Gk. *kenotaphion* : *kenos,* empty + *taphos,* tomb.] A monument erected in honor of a deceased person whose remains are interred elsewhere. **—cen'o·taph'ic** *adj.*

**Ce·no·zo·ic** (sē'nə-zō'ĭk, sĕn'ə-) *adj.* [Gk. *kainos,* new + -ZOIC.] Of, belonging to, or designating the latest era of geologic time, which includes the Tertiary and Quaternary periods and is marked by the evolution of mammals, birds, plants, modern continents, and glaciation. *—n.* The Cenozoic era.

**cense** (sĕns) *vt.* **censed, cens·ing, cens·es.** [ME *censen,* short for *encensen* < *encens,* incense. —see INCENSE[2].] **1.** To perfume with incense. **2.** To offer incense to.

**cen·ser** (sĕn'sər) *n.* [ME, short for *encenser* < Norman Fr. *encensier* < *encens,* incense < OFr. —see INCENSE[2].] An incense vessel.

**cen·sor** (sĕn'sər) *n.* [Lat., Roman censor < *censēre,* to assess.] **1.** One authorized to examine material, as literature or plays, and remove or suppress anything considered objectionable. **2.** An official, as in the military, who examines personal mail and official dispatches to expunge classified information. **3.** One who condemns or censures. **4.** One of two ancient Roman officials responsible for supervising the public census, public behavior, and morals. **5.** *Psychoanal.* The agent responsible for censorship. *—vt.* **-sored, -sor·ing, -sors.** To examine and expurgate <*censored* the dissidents' newsletter> **—cen·so'ri·al** (sĕn-sôr'ē-əl, -sōr'-) *adj.*

☆ ***syns:*** CENSOR, BAN, STIFLE, SUPPRESS *v. core meaning* **:** to keep from being published or otherwise disseminated <*censored* the news story>

**cen·so·ri·ous** (sĕn-sôr'ē-əs, -sōr'-) *adj.* [Lat. *censorius,* of a censor < *censor,* Roman censor.] **1.** Tending to censure. **2.** Expressing censure. **—cen·so'ri·ous·ly** *adv.* **—cen·so'ri·ous·ness** *n.*

**cen·sor·ship** (sĕn'sər-shĭp') *n.* **1.** The act, process, or policy of censoring. **2.** The office or authority of a Roman censor. **3.** *Psychoanal.* Inhibition by either ego or superego of conscious awareness of painful feelings or ideas.

**cen·sur·a·ble** (sĕn'shər-ə-bəl) *adj.* Deserving of or apt to censure. **—cen'sur·a·ble·ness, cen'sur·a·bil'i·ty** *n.* **—cen'sur·a·bly** *adv.*

**cen·sure** (sĕn'shər) *n.* [Lat. *censura,* severe judgment < *censor,* Roman censor.] **1.** An expression of blame or disapproval. **2.** An official rebuke. *—vt.* **-sured, -sur·ing, -sures.** To criticize severely **:** REBUKE. **—cen'sur·er** *n.*

**cen·sus** (sĕn'səs) *n.* [Lat., registration of citizens < *censēre,* to assess.] An official, usu. periodic enumeration of a population.

**cent** (sĕnt) *n.* [OFr., hundred < Lat. *centum.*] —See table at CURRENCY.

**cen·tal** (sĕn'təl) *n.* [< Lat. *centum,* hundred.] A hundredweight.

**cen·taur** (sĕn'tôr') *n.* [ME < Lat. *Centaurus* < Gk. *Kentauros.*] *Gk. Myth.* One of a race of monsters having the head, arms, and trunk of a man and the body and legs of a horse.

**Cen·tau·rus** (sĕn-tôr'əs) *n.* [Lat. *Centaurus,* centaur.] A constellation in the Southern Hemisphere.

**cen·tau·ry** (sĕn'tôr'ē, -tōr'ē) *n., pl.* **-ries.** [ME *centorie* < Lat. *centaureum* < Gk. *kentaureion* < *Kentauros,* centaur (from the legend that the plant's medicinal properties were discovered by a centaur).] **1.** A Eurasian, plant of the genus *Centaurium* esp. *C. umbellatum,* with rose-purple flower clusters. **2.** A plant of the genus *Centaurea,* as the cornflower.

**cen·ta·vo** (sĕn-tä'vō) *n., pl.* **-vos.** [Sp. < Lat. *centum,* hundred.] —See table at CURRENCY.

**cen·te·nar·i·an** (sĕn'tə-nâr'ē-ən) *n.* [< Lat. *centenarius,* of a hundred. —see CENTENARY.] A person 100 years old or older. **—cen'te·nar'i·an** *adj.*

**cen·ten·a·ry** (sĕn-tĕn'ə-rē, sĕn'tə-nĕr'ē) *adj.* [Lat. *centenarius,* of a hundred < *centum,* hundred.] **1.** Of or relating to a 100-year period. **2.** Happening once every 100 years. *—n., pl.* **-ries. 1.** A 100-year period. **2.** A centennial.

**cen·ten·ni·al** (sĕn-tĕn'ē-əl) *adj.* [Lat. *centum,* hundred + (BI)ENNIAL.] **1.** Of or relating to an age or period of 100 years. **2.** Happening once every 100 years. **3.** Of or relating to a 100th anniversary. *—n.* A 100th anniversary or a celebration of it. **—cen·ten'ni·al·ly** *adv.*

**cen·ter** (sĕn'tər) *n.* [ME *centre* < OFr. < Lat. *centrum* < Gk. *kentron.*] **1.** A point equidistant or at the average distance from all points on the sides or outer boundaries of something **:** MIDDLE. **2. a.** A point equidistant from the vertexes of a regular polygon. **b.** A point equidistant from all points on the circumference of a circle or on the surface of a sphere. **3.** A point around which something revolves **:** AXIS. **4.** A part of an object surrounded by the rest of it **:** CORE. **5. a.** A hub of activity or influence <a communications *center*> **b.** A place of dense population. **6.** One that is the object of attention, interest, activity, or emotion <I was the *center* of controversy.> **7.** One occupying a middle position. **8.** A political policy or group representing a compromise between the right and the left. **9.** A team player occupying a middle position on a field, court, or forward line. *—v.* **-tered, -ter·ing, -ters.** *—vt.* **1.** To place in or on a center. **2.** To concentrate at a center <*centered* my arguments on the key issues> **3.** *Football.* To pass (the ball) from the line to a back. *—vi.* **1.** To be concentrated **:** CLUSTER. **2.** To have a central theme or concern.

☆ ***syns:*** CENTER, MIDDLE, MIDPOINT, MIDST *n. core meaning* **:** a point more or less equidistant from all sides of something <flowers in the *center* of a round table>

**center bit** *n.* A bit with a sharp center point, used for boring holes.

**cen·ter·board** (sĕn'tər-bôrd', -bōrd') *n.* A flat board or metal plate that can be lowered through the bottom of a sailboat to prevent drifting and provide stability.

**center field** *n. Baseball.* **1.** The middle part of the outfield behind second base. **2.** The position of center field. **—center fielder** *n.*

**center of gravity** *n.* The point in or near a body at which the gravitational potential energy of the body is equal to that of a single particle of the same mass located at that point and through which the resultant of the gravitational forces on the component particles of the body acts.

**center of mass** *n.* The point about which the sum of all the linear moments of mass of the particles in a body is zero.
**cen·ter·piece** (sĕn′tər-pēs′) *n.* A decorative object or arrangement placed at the center of a table.
**center punch** *n.* A sharp-pointed tool used in metalworking to mark centers or center lines on pieces to be drilled.
**cen·tes·i·mal** (sĕn-tĕs′ə-məl) *adj.* [< Lat. *centesimus* < *centum,* hundred.] **1.** Hundredth. **2.** Relating to or divided into hundredths. **—cen·tes′i·mal·ly** *adv.*
**cen·tes·i·mo¹** (sĕn-tĕs′ə-mō′) *n., pl.* **-mos** or **-mi** (-mē) [Ital. < Lat. *centesimus,* centesimal.] —See table at CURRENCY.
**cen·tes·i·mo²** (sĕn-tĕs′ə-mō′) *n., pl.* **-mos.** [Sp. < Lat. *centesimus,* centesimal.] —See table at CURRENCY.
**cen·te·sis** (sĕn-tē′sĭs) *n., pl.* **-ses** (-sēz′) [Gk. *kentēsis,* act of pricking < *kentein,* to prick.] Surgical puncture of a membrane or bodily cavity usu. for diagnostic purposes.
**centi-** *pref.* [Fr. < Lat. *centum,* hundred.] One hundredth ($10^{-2}$) <*centi*liter>
**cen·ti·grade** (sĕn′tĭ-grăd′) *adj.* [Fr. : Lat. *centum-,* hundred + Lat. *gradus,* degree.] Celsius.
**cen·ti·gram** (sĕn′tĭ-grăm′) *n.* One hundredth ($10^{-2}$) of a gram.
**cen·ti·li·ter** (sĕn′tə-lē′tər) *n.* One hundredth ($10^{-2}$) of a liter.
**cen·time** (sän′tēm′, sĕn′-) *n.* [Fr. < *cent,* hundred < Lat. *centum.*] —See table at CURRENCY.
**cen·ti·me·ter** *also* **cen·ti·me·tre** (sĕn′tə-mē′tər, sän′-) *n.* A unit of length equal to 1/100 of a meter or 0.3937 inch.
**cen·ti·me·ter-gram-sec·ond system** (sĕn′tə-mē′tər-grăm′-sĕk′ənd) *n.* A coherent system of units for mechanics, electricity, and magnetism, in which the basic units of length, mass, and time are the centimeter, gram, and second.
**cen·ti·mo** (sĕn′tə-mō′) *n., pl.* **-mos.** [Sp. *céntimo* < Fr. *centime,* centime.] —See table at CURRENCY.
**cen·ti·pede** (sĕn′tə-pēd′) *n.* [Lat. *centipeda* : *centum,* hundred + *pes,* foot.] Any of various wormlike arthropods of the class Chilopoda, having numerous body segments, each with a pair of legs, the front pair modified into venomous biting organs.
**cen·ti·poise** (sĕn′tə-poiz′) *n.* One hundredth ($10^{-2}$) of a poise.
**cent·ner** (sĕnt′nər) *n.* [G. *Zentner* < Lat. *centenarius,* of a hundred. —see CENTENARY.] **1.** A unit of weight corresponding to the hundredweight, equal to 110.23 pounds, used in several European countries. **2.** An assaying unit equal to one dram.
**cen·to** (sĕn′tō) *n., pl.* **-tos.** [LLat. < Lat., patchwork.] A literary work pieced together from the works of several authors.
**centr-** *pref. var. of* CENTRO-.
**cen·tral** (sĕn′trəl) *adj.* [Lat. *centralis* < *centrum,* center.] **1.** At, in, near, or being the center. **2.** Having dominant power, influence, or control <the company's *central* office> **3.** Of great importance : ESSENTIAL <the *central* theme of the book> **4.** Easily reached from various points <a *central* artery through town> **5.** Of or constituting a single source controlling all components of a system <*central* heating> **6.** *Anat. & Physiol.* **a.** Denoting that part of the nervous system constituted by the brain and spinal cord. **b.** Relating to a centrum. **7.** Pronounced with the tongue in a neutral position, as *e* in *mister.* —*n.* **1.** A telephone exchange. **2.** A telephone exchange operator. **—cen′tral·ly** *adv.*
**central angle** *n.* An angle having radii as sides and the center of a circle as its vertex.
**cen·tral·ism** (sĕn′trə-lĭz′əm) *n.* Assignment of power and authority to a central leadership in an organization, as a political system. **—cen′tral·ist** *n.* **—cen′tral·is′tic** *adj.*
**cen·tral·i·ty** (sĕn-trăl′ĭ-tē) *n.* **1.** The quality or state of being central. **2.** The tendency to be or remain at the center.
**cen·tral·ize** (sĕn′trə-līz′) *v.* **-ized, -iz·ing, -iz·es.** —*vt.* **1.** To draw into or toward a center : CONSOLIDATE. **2.** To bring under a single, central authority. —*vi.* To come together at a center : CONCENTRATE. **—cen′tral·i·za′tion** *n.* **—cen′tral·iz′er** *n.*
**central nervous system** *n.* The portion of the vertebrate nervous system composed of the brain and spinal cord.
**central processing unit** *n. Computer Sci.* The part of a computer that interprets and executes instructions.
**Central Standard Time** *n.* The local civil time of the 90th meridian west of Greenwich, England, six hours earlier than Greenwich time, observed in the central United States.
**Central Time** *n.* Central Standard Time.
**cen·tre** (sĕn′tər) *n.* & *v. Chiefly Brit. var. of* CENTER.
**centri-** *pref. var. of* CENTRO-.
**cen·tric** (sĕn′trĭk) *also* **cen·tri·cal** (-trĭ-kəl) *adj.* [Gk. *kentrikos* < *kentron,* center.] **1.** Located at, relating to, or having a center. **2.** *Physiol.* Of or originating at a nerve center. **—cen′tri·cal·ly** *adv.* **—cen·tric′i·ty** (sĕn-trĭs′ĭ-tē) *n.*
**-centric** *suff.* [< Lat. *centrum,* center.] **1.** Having a given kind or number of centers <poly*centric*> **2.** Having a given object as the center <geo*centric*>
**cen·trif·u·gal** (sĕn-trĭf′yə-gəl, -trĭf′ə-) *adj.* [< NLat. *centrifugus* : Lat. *centrum,* center + *fugere,* to flee.] **1.** Moving or directed away from a center or axis. **2.** Operated by centrifugal force. **3.** *Physiol.* Relating to impulses transmitted away from a nerve center : EFFERENT. **4.** *Bot.* Developing outward from a center or axis. **5.** Tending away from centralization <*centrifugal* urban societal trends> **—cen·trif′u·gal·ly** *adv.*
**centrifugal force** *n.* The component of apparent force on a body in curvilinear motion, as observed from that body, that is directed away from the center of curvature or axis of rotation.
**cen·tri·fuge** (sĕn′trə-fyo͞oj′) *n.* [Fr. < NLat. *centrifugus,* centrifugal.] A device consisting of a compartment spun about a central axis to separate materials of different density or to simulate gravity with centrifugal force. —*vt.* **-fuged, -fug·ing, -fug·es.** To separate, dehydrate, or test by means of a centrifuge. **—cen·trif′u·ga′tion** (sĕn-trĭf′yə-gā′shən, -trĭf′ə-) *n.*
**cen·tri·ole** (sĕn′trē-ōl′) *n.* [G. *Zentriol* < NLat. *centriolum,* dim. of Lat. *centrum,* center.] A tiny cylindrical organelle, considered a pole of the mitotic figure and located at the center of a centrosome.
**cen·trip·e·tal** (sĕn-trĭp′ĭ-tl) *adj.* [Lat. *centrum,* center + -PETAL.] **1.** Directed or moving toward a center or axis. **2.** Operated by centripetal force. **3.** *Physiol.* Relating to nerve impulses transmitted toward the central nervous system : AFFERENT. **4.** *Bot.* Developing inward toward the center or axis, as some forms of inflorescence do. **5.** Tending to centralize. **—cen·trip′e·tal·ly** *adv.*
**centripetal force** *n.* The component of force acting on a body in curvilinear motion that is directed toward the center of curvature or axis of rotation.
**cen·trism** (sĕn′trĭz′əm) *n.* A political philosophy of avoiding extremes by taking a midway position. **—cen′trist** *n.*
**centro-** or **centr-** or **centri-** *pref.* [< Gk. *kentron,* center.] Center <*centro*id>
**cen·tro·bar·ic** (sĕn′trə-băr′ĭk) *adj.* [Gk. *kentrobarikē,* theory of the center of gravity : *kentron,* center + *baros,* weight.] Of or relating to the center of gravity.
**cen·troid** (sĕn′troid′) *n.* **1.** The center of mass of an object having constant density. **2.** The point in a system of masses each of whose coordinates is a weighted mean of coordinates of the same dimension of points within the system, the weights being determined by the density function of the system.
**cen·tro·lec·i·thal** (sĕn′trə-lĕs′ə-thəl) *adj.* [CENTRO- + LECITH(IN) + -AL.] Having the yolk concentrated in the center of the egg cell.
**cen·tro·mere** (sĕn′trə-mîr′) *n.* The region of a chromosome to which the spindle fiber is attached during mitosis.
**cen·tro·some** (sĕn′trə-sōm′) *n.* A small mass of differentiated cytoplasm containing the centriole. **—cen′tro·so′mic** (-sō′mĭk) *adj.*
**cen·tro·sphere** (sĕn′trə-sfîr′) *n.* **1.** The mass of cytoplasm surrounding the centriole in a centrosome. **2.** The earth's central core.
**cen·trum** (sĕn′trəm) *n., pl.* **-trums** or **-tra** (-trə) [Lat., center.] The major part of a vertebra, exclusive of the bases of the neural arch.
**cen·tum** (kĕn′təm) *adj.* [Lat., hundred (a word whose initial sound illustrates the preservation of Indo-European velar *k*).] Of, relating to, or comprising the group of Indo-European languages that retained the velar *k* and the labiovelar *kw* of primitive Indo-European.
**cen·tu·ri·on** (sĕn-to͝or′ē-ən, -tyo͝or′-) *n.* [ME *centurioun* < OFr. *centurion* < Lat. *centurio* < *centuria,* century.] An officer who commanded a century in the Roman army. **—cen·tu′ri·al** *adj.*
**cen·tu·ry** (sĕn′chə-rē) *n., pl.* **-ries.** [Lat. *centuria,* a group of a hundred < *centum,* hundred.] **1. a.** A period of 100 years. **b.** Each of the successive periods of 100 years before or since the advent of the Christian era. **2. a.** A unit of the Roman army orig. composed of 100 men. **b.** One of the 193 electoral divisions of the Roman people. **3.** A group of 100 things.
**century plant** *n.* A fleshy plant of the genus *Agave,* some species of which bloom only once in 10 to 20 years and then die, esp. *A. americana,* having grayish leaves and greenish flowers.
**ce·orl** (chā′ôrl) *n.* [OE.] An Anglo-Saxon freeman of the lowest class.
**cephal-** *pref. var. of* CEPHALO-.
**ceph·a·lad** (sĕf′ə-lăd′) *adv.* Toward the head or anterior section.
**ceph·al·al·gia** (sĕf′ə-lăl′jə, -jē-ə) *n.* Pain concentrated in the region of the head.
**ce·phal·ic** (sə-făl′ĭk) *adj.* [OFr. *cephalique* < Lat. *cephalicus* < Gk. *kephalikos* < *kephalē,* head.] **1.** Of or relating to the head or skull. **2.** Located on, in, or near the head. **—ce·phal′i·cal·ly** *adv.*
**-cephalic** *suff.* [-CEPHAL(OUS) + -IC.] -CEPHALOUS.
**cephalic index** *n.* The ratio of the maximum width of the head to its maximum length, multiplied by 100.
**ceph·a·lin** (sĕf′ə-lĭn) *n.* A phosphatide derived from the brain and spinal cord, usu. of cattle, and used as a hemostatic agent.
**ceph·a·li·za·tion** (sĕf′ə-lĭ-zā′shən) *n.* Gradually increasing concentration of the brain and sensory organs in the head during animal evolution.
**cephalo-** or **cephal-** *pref.* [Lat. < Gk. *kephalo-* < *kephalē,* head.] Head <*cephalo*thorax>
**ceph·a·lo·chor·date** (sĕf′ə-lə-kôr′dāt′) *adj.* [NLat. *Cephalochordata,* subphylum name : CEPHALO- + *Chordata,* chordate phylum name.] Of or belonging to the subphylum Cephalochordata, includ-

ing primitive forerunners of the vertebrates such as the lancelet. *—n.* A member of the Cephalochordata.

**ceph·a·lo·pod** (sĕf′ə-lə-pŏd′) *n.* [NLat. *Cephalapoda,* class name : CEPHALO- + Gk. *pous,* foot.] A mollusk of the class Cephalopoda, as an octopus, having a beaked head, an internal shell in some species, and prehensile tentacles. *—adj.* Of, relating to, or belonging to the Cephalopoda. **—ceph′a·lop′o·dan** (sĕf′ə-lŏp′ə-dən) *n.* & *adj.*

**ceph·a·lo·spo·rin** (sĕf-ə-lə-spôr′ĭn, -spōr′-) *n.* [< NLat. *Cephalosporium,* genus name : CEPHALO- + *spora,* spore.] Any of various antibiotics produced by an imperfect fungus of the genus *Cephalosporium.*

**ceph·a·lo·tho·rax** (sĕf′ə-lə-thôr′ăks′, -thōr′-) *n.* The anterior section of arachnids and many crustaceans, composed of the fused head and thorax.

**-cephalous** *suff.* [Gk. *-kephalos* < *kephalē,* head.] Having a specified kind of head or number of heads <di*cephalous*>

**-cephaly** *suff.* [< Gk. *kephalē,* head.] A specified condition of the head <micro*cephaly*>

**Ce·phe·id** (sē′fē-ĭd, sĕf′ē-) *n.* [< CEPHEUS.] Any of a class of intrinsically variable stars with highly regular periods of light pulsation.

**Ce·pheus** (sē′fyo͞os′, -fē-əs, sĕf′ē-) *n.* [Lat. *Cepheus* < Gk. *Kēpheus.*] A constellation in the Northern Hemisphere.

**ce·ra·ceous** (sə-rā′shəs) *adj.* [Lat. *cera,* wax + -ACEOUS.] Waxy.

**ce·ram·al** (sə-răm′əl) *n.* [CERAM(IC) + AL(LOY).] Cermet.

**ce·ram·ic** (sə-răm′ĭk) *n.* [< Gk. *keramikos,* of pottery < *keramos,* potter's clay.] **1.** Any of various hard, brittle, heat- and corrosion-resistant materials made by shaping and then firing a nonmetallic mineral, as clay, at a high temperature. **2. a.** An object made of ceramic. **b. ceramics** *(sing. in number).* The art or technique of making objects of ceramic, esp. from fired clay or porcelain. **—ce·ram′ic** *adj.* **—ce·ram′ist** *n.*

**ce·ras·tes** (sə-răs′tēz) *n., pl.* **cerastes.** [ME < Lat. < Gk. *kerastēs,* horned serpent < *keras,* horn.] A venomous snake of the genus *Cerastes,* as the horned viper.

**ce·rate** (sîr′āt′) *n.* [Lat. *ceratum* < *cera,* wax.] A hard, unctuous, fat- or wax-based solid, occas. medicated, at one time applied to the skin directly or on dressings.

**ce·rat·ed** (sîr′ā′tĭd) *adj.* [< Lat. *ceratus,* p.part. of *cerare,* to wax < *cera,* wax.] **1.** Coated with wax. **2.** Possessing a cere.

**cerato-** or **cerat-** *pref. vars. of* KERATO-.

**ce·rat·o·dus** (sə-răt′ə-dəs) *n., pl.* **-dus·es.** [NLat. *Ceratodus,* genus name : Gk. *keras,* horn + Gk. *odous,* tooth.] Any of various extinct lungfishes of the genus *Ceratodus,* of the Triassic and Cretaceous periods.

**cer·a·toid** (sĕr′ə-toid′) *adj.* Hornlike.

**Cer·ber·us** (sûr′bər-əs) *n.* [Lat. < Gk. *Kerberos.*] *Gk.* & *Rom. Myth.* A three-headed dog guarding the entrance of Hades. **—Cer′ber·e′an** (sûr′bə-rē′ən) *adj.*

**cer·car·i·a** (sər-kâr′ē-ə) *n., pl.* **-i·ae** (-ē-ē′) or **-i·as.** [NLat. < Gk. *kerkos,* tail.] The parasitic larva of a trematode worm, having a tail that disappears in the adult stage. **—cer·car′i·al** *adj.*

**cer·co·pi·the·coid** (sûr′kə-pĭ-thē′koid′, -pĭth′ĭ-koid′) *adj.* [< NLat. *Ceropithecidae,* family name < Lat. *cercopithecus,* long-tailed ape < Gk. *kerkopithēkos* : *kerkos,* tail + *pithēkos,* ape.] Of or belonging to the family Cercopithecidae, including monkeys such as the baboons, mandrills, macaques, and langurs. *—n.* A member of the Cercopithecidae.

**cere**[1] (sîr) *vt.* **cered, cer·ing, ceres.** [ME *seren* < OFr. *cirer,* to cover with wax < Lat. *cerare* < *cera,* wax.] To wrap in or as if in cerecloth.

**cere**[2] (sîr) *n.* [ME *sere* < OFr. *cire* < Med. Lat. *cera* < Lat., wax.] A fleshy or waxlike swelling at the base of the upper part of the beak in some birds, such as parrots. **—cered** *adj.*

**ce·re·al** (sîr′ē-əl) *n.* [Lat. *cerealis,* of grain < *Ceres,* Ceres.] **1.** An edible grain, as wheat, oats, or corn. **2.** A grain-producing grass. **3.** A food prepared from grain. *—adj.* Of or relating to cereals.

**cer·e·bel·lum** (sĕr′ə-bĕl′əm) *n., pl.* **-bel·lums** or **-bel·la** (-bĕl′ə) [Med. Lat. < Lat., dim. of *cerebrum,* brain.] The brain structure responsible for coordination and regulation of complex voluntary muscular movement, lying posterior to the pons and medulla oblongata and inferior to the occipital lobes of the cerebral hemispheres. **—cer′e·bel′lar** (-bĕl′ər) *adj.*

**cerebr-** *pref. var. of* CEREBRO-.

**cer·e·bral** (sə-rē′brəl, sĕr′ə-brəl) *adj.* **1.** Of or relating to the brain or cerebrum. **2.** Appealing to or marked by the workings of the intellect <a *cerebral* novel> **—cer′e′bral·ly** *adv.*

**cerebral cortex** *n.* The extensive outer layer of gray tissue of the cerebral hemispheres, responsible for higher nervous functions.

**cerebral hemisphere** *n.* Either hemisphere of the cerebrum of the brain, divided by the longitudinal cerebral fissure.

**cerebral palsy** *n.* Impaired muscular power and coordination due to brain damage usu. at or prior to birth. **—cer′e·bral-pal′sied** *adj.*

**cer·e·brate** (sĕr′ə-brāt′) *vi.* **-brat·ed, -brat·ing, -brates.** [< Lat. *cerebrum,* brain.] To use the power of reason : THINK. **—cer′e·bra′tion** *n.*

**cerebro-** or **cerebr-** *pref.* [< CEREBRUM.] Brain : cerebrum <*cerebro*side>

**cer·e·bro·side** (sĕr′ə-brə-sīd′, sə-rē′-) *n.* [CEREBR(O)- + -OS(E) + -IDE.] Any of various compounds found in the brain and other nerve tissue, yielding upon decomposition a fatty acid, an unsaturated amino-alcohol, and a sugar.

**cer·e·bro·spi·nal** (sĕr′ə-brō-spī′nəl, sə-rē′brō-) *adj.* Of or relating to the brain and spinal cord.

**cerebrospinal fluid** *n.* The serumlike fluid that bathes the lateral ventricles of the brain and the cavity of the spinal cord.

**cerebrospinal meningitis** *n.* An acute, often fatal, infectious epidemic meningitis.

**cer·e·bro·vas·cu·lar** (sĕr′ə-brō-văs′kyə-lər, sə-rē′brō-) *adj.* Of or relating to the blood vessels supplying the brain.

**cer·e·brum** (sĕr′ə-brəm, sə-rē′brəm) *n., pl.* **-brums** or **-bra** (-brə) [Lat., brain.] The large rounded structure of the brain occupying most of the cranial cavity, divided into two cerebral hemispheres by a deep median sagittal groove and joined at the bottom by the corpus callosum.

**cere·cloth** (sîr′klôth′, -klŏth′) *n.* Wax-coated cloth once used for wrapping corpses.

**cer·e·ment** (sĕr′ə-mənt, sîr′mənt) *n.* Cerecloth.

**cer·e·mo·ni·al** (sĕr′ə-mō′nē-əl) *adj.* Of, marked by, or involved in ceremony. ***usage:*** As an adjective *ceremonial* is applicable only to things and means "having to do with ceremony" (*ceremonial occasions*). By contrast, *ceremonious,* when used to characterize persons, means "devoted to forms and ritual" (*ceremonious courtiers*). *—n.* **1.** A set of ceremonies prescribed for an occasion : RITE. **2.** A ceremony. **—cer′e·mo′ni·al·ism** *n.* **—cer′e·mo′ni·al·ist** *n.* **—cer′e·mo′ni·al·ly** *adv.*

**cer·e·mo·ni·ous** (sĕr′ə-mō′nē-əs) *adj.* **1.** Of or relating to a ceremony. **2.** Fond of ceremony. **3.** Marked by ceremony. **—cer′e·mo′ni·ous·ly** *adv.* **—cer′e·mo′ni·ous·ness** *n.*

☆ ***syns:*** CEREMONIOUS, COURTLY, FORMAL, POLITE, PUNCTILIOUS *adj. core meaning* : fond of or given to ceremony <The Japanese are a *ceremonious* people.>

**cer·e·mo·ny** (sĕr′ə-mō′nē) *n., pl.* **-nies.** [ME *cerimonie* < Lat. *caerimonia,* religious rite.] **1.** A formal act or set of acts performed as prescribed by ritual, custom, or etiquette. **2.** A conventional social gesture or act with no intrinsic purpose. **3.** Strict observance of formalities or etiquette.

**Če·ren·kov effect** (chə-rĕng′kôf) *n.* [After P.A. *Čerenkov* (1904–1958), its discoverer.] Emission of light by a particle passing through a transparent nonconducting solid at a speed greater than the speed of light in that solid.

**Čerenkov radiation** *n.* Light emitted in the Čerenkov effect.

**Ce·res** (sîr′ēz) *n.* [Lat.] **1.** *Rom. Myth.* The goddess of agriculture. **2.** The first asteroid to be discovered, having an orbit between Mars and Saturn.

**ce·re·us** (sîr′ē-əs) *n.* [NLat. *Cereus,* genus name < Lat. *cereus,* taper (from the cacti's shape) < *cera,* wax.] A tall tropical American cactus, esp. of the genus *Cereus,* as the night-blooming cereus.

**ce·ric** (sîr′ĭk, sĕr′-) *adj.* [CER(IUM) + -IC.] Of, relating to, or containing cerium, esp. with valence 4.

**ce·ric oxide** (sîr′ĭk, sĕr′-) *n.* A pale yellow-white powder, $CeO_2$, used in ceramics, to polish glass, and to sensitize photosensitive glass.

**cer·iph** (sĕr′ĭf) *n. Chiefly Brit. var. of* SERIF.

**ce·rise** (sə-rēs′, -rēz′) *n.* [Fr. < OFr., cherry. —see CHERRY.] A deep to vivid purplish red.

**ce·ri·um** (sîr′ē-əm) *n.* [After CERES.] *Symbol* **Ce** A lustrous, iron-gray, malleable metallic element, used in various metallurgical and nuclear applications; atomic number 58; atomic weight 140.12.

**cer·met** (sûr′mĕt′) *n.* [CER(AMIC) + MET(AL).] A material composed of processed ceramic particles bonded with metal and used in high-strength and high-temperature applications.

**cer·nu·ous** (sûr′nyo͞o-əs) *adj.* [Lat. *cernuus,* bowing forward.] *Bot.* Hanging downward : DROOPING.

**ce·ro** (sîr′ō, sĕr′ō) *n., pl.* **-ros** or **cero.** [< Sp. *sierra,* sawfish. —see SIERRA.] **1.** An edible fish, *Scomberomorus regalis* of western Atlantic waters, with a dark-blue back. **2.** The king mackerel.

**ce·ro·tic acid** (sə-rō′tĭk, -rŏt′ĭk) *n.* [< Lat. *ceratum,* cerate.] An acid, $C_{25}H_{51}COOH$, found in waxes, as beeswax.

**ce·ro·type** (sîr′ə-tīp′, sĕr′ə-) *n.* [Gk. *kēros,* wax + -TYPE.] The process of preparing a printing surface for electrotyping by first engraving on a wax-coated metal plate.

**ce·rous** (sîr′əs) *adj.* [CER(IUM) + -OUS.] Of, relating to, or containing cerium, esp. with valence 3.

**cer·tain** (sûr′tn) *adj.* [ME < OFr. < VLat. *certanus < Lat. *certus,* p. part. of *cernere,* to determine.] **1.** Definite : fixed <a *certain* sum of money for food> **2.** Sure to happen : INEVITABLE <*certain* disaster> **3.** Established beyond question or doubt <What is *certain* is that every effect has a cause.> **4.** Unfailing : sure <a *certain* cure> **5.** Confident : assured. **6. a.** Not specified or named but assumed to be known <a *certain* individual> **b.** Named but not familiar or well-known <a *certain* Ms. Doe> **7.** Appreciable but unspecified <a *certain* degree><a *certain* charm> ***usage:*** Although some consider *certain* to be an absolute term, it may often be qualified by an adverb, as in *Nothing is more certain than death and taxes.* —*pron.* An indefinite but limited number : SOME. **—for certain.** Without doubt : SURELY. **—cer'tain·ly** *adv.*

**cer·tain·ty** (sûr'tn-tē) *n., pl.* **-ties. 1.** Freedom from doubt. **2.** A clearly established fact.

☆ ***syns:*** CERTAINTY, CERTITUDE, CONFIDENCE, CONVICTION, SURENESS, SURETY *n. core meaning* : freedom from doubt <denied the rumor with absolute *certainty*> ***ant:*** uncertainty

**cer·tes** (sûr'tēz, sûrts) *adv.* [ME < OFr. < *cert* < Lat. *certus,* certain.] *Archaic.* Certainly : truly.

**cer·tif·i·cate** (sər-tĭf'ĭ-kĭt) *n.* [ME *certificat* < Med. Lat. *certificatum* < LLat. *certificare,* to certify.] **1.** A document testifying to accuracy or truth. **2.** A document issued to a person completing a course of study not leading to a diploma. **3.** A document certifying that a person may officially practice a specific profession. **4.** A document certifying ownership. —*vt.* (-kāt') **-cat·ed, -cat·ing, -cates.** To furnish with, testify to, or authorize by a certificate. **—cer·tif'i·ca·to'·ry** (-kə-tôr'ē, -tōr'ē) *adj.*

**certificate of deposit** *n.* A bank certificate stating that the named person has a specified sum of money on deposit.

**cer·ti·fi·ca·tion** (sûr'tə-fĭ-kā'shən) *n.* **1.** The act of certifying or state of being certified. **2.** A certified statement.

**certified check** *n.* A check guaranteed by a bank to be covered by sufficient funds on deposit.

**certified mail** *n.* Uninsured first-class mail whose delivery is recorded by having the addressee sign for it.

**certified public accountant** *n.* A public accountant who has received a certificate stating that he or she has met a state's legal requirements to practice.

**cer·ti·fy** (sûr'tə-fī') *v.* **-fied, -fy·ing, -fies.** [ME *certifien* < OFr. *certifier* < LLat. *certificare* : Lat. *certus,* certain + Lat. *facere,* to make.] —*vt.* **1. a.** To confirm formally as true, accurate, or genuine, esp. in writing. **b.** To guarantee as meeting a standard. **2.** To acknowledge in writing on the face of (a check) that the signature of the maker is genuine and that the depositor has sufficient funds on deposit for its payment. **3.** To declare legally insane. **4.** To make certain : ASSURE. **5.** To issue a license or certificate to. —*vi.* To testify <*certify* to the truth of the matter> **—cer'ti·fi'a·ble** *adj.* **—cer'ti·fi'a·bly** *adv.* **—cer'ti·fi'er** *n.*

**cer·ti·o·rar·i** (sûr'shē-ə-râr'ē, -rä'rē) *n.* [ME < Lat., to be certified (from the word's occurrence in the writ) < *certiorare,* to certify < *certion,* comp. of *certus,* certain.] *Law.* A writ from a higher court to a lower one requesting a transcript of the proceedings of a case for review.

**cer·ti·tude** (sûr'tĭ-to͞od', -tyo͞od') *n.* [ME < LLat. *certitudo* < Lat. *certus,* certain.] **1.** Complete assurance. **2.** Inevitability.

**ce·ru·le·an** (sə-ro͞o'lē-ən) *adj.* [Lat. *caeruleus,* dark blue.] Sky-blue.

**ce·ru·lo·plas·min** (sə-ro͞o'lō-plăz'mĭn) *n.* [CERUL(EAN) + PLASM(A) + -IN.] A blood glycoprotein that functions in the storage of copper.

**ce·ru·men** (sə-ro͞o'mən) *n.* [NLat. < Lat. *cera,* wax.] A yellowish waxy secretion of the external ear : EARWAX.

**ce·ru·mi·nous gland** (sə-ro͞o'mə-nəs) *n.* [< NLat. *cerumen, cerumin-,* cerumen.] Any of the specialized sweat glands located in the external auditory canal that produce cerumen.

**ce·ruse** (sə-ro͞os', sîr'o͞os') *n.* [ME < OFr. < Lat. *cerussa.*] White lead.

**ce·rus·site** (sə-rŭs'īt') *n.* [G. *Zerussit* < Lat. *cerussa,* ceruse.] Natural lead carbonate, $PbCO_3$, a lead ore.

**cer·vi·cal** (sûr'vĭ-kəl) *adj.* [NLat. *cervicalis* < Lat. *cervix, cervic-,* neck.] Of or relating to the neck or cervix <a *cervical collar*>

**cer·vi·ci·tis** (sûr'vĭ-sī'tĭs) *n.* Inflammation of the cervix of the uterus.

**cer·vine** (sûr'vīn') *adj.* [Lat. *cervinus* < *cervus,* deer.] Relating to, resembling, or typical of a deer.

**cer·vix** (sûr'vĭks) *n., pl.* **cer·vix·es** *or* **cer·vi·ces** (sûr'vĭ-sēz', sər-vī'sēz) [Lat. *cervix, cervic-* neck.] **1.** The neck. **2.** A neck-shaped anatomical structure, as the narrow outer end of the uterus.

**Ce·sar·e·an** *or* **Ce·sar·i·an** (sĭ-zâr'ē-ən) *n. vars. of* CAESAREAN.

**ce·si·um** *also* **cae·si·um** (sē'zē-əm) *n.* [NLat. < Lat. *caesius,* bluish gray (from its blue spectral lines).] *Symbol* **Cs** A soft, silvery-white ductile metal, liquid at room temperature, the most electropositive and alkaline of the elements, used in photoelectric cells; atomic number 55; atomic weight 132.905.

**ces·pi·tose** *also* **caes·pi·tose** (sĕs'pĭ-tōs') *adj.* [NLat. *caespitosus* < Lat. *caespes,* turf.] Growing in dense tufts or turflike clumps. **—ces'pi·tose'ly** *adv.*

**cess¹** (sĕs) *n.* [Alteration of obs. *assess,* assessment < ASSESS.] A tax.

**cess²** (sĕs) *n.* [Poss. short for SUCCESS.] *Ir.* Luck.

**ces·sa·tion** (sĕ-sā'shən) *n.* [ME *cessacioun* < OFr. < Lat. *cessatio* < *cessare,* to stop. —see CEASE.] The act or fact of ceasing : HALT.

**ces·sion** (sĕsh'ən) *n.* [ME < OFr. < Lat. *cessio* < *cedere,* to yield.] **1.** An act of ceding, as of territory to another country by treaty. **2.** A ceded territory.

**ces·sion·ar·y** (sĕsh'ə-nĕr'ē) *n., pl.* **-ies.** One to whom a cession is made : ASSIGNEE.

**cess·pit** (sĕs'pĭt') *n.* [CESS(POOL) + PIT.] A cesspool.

**cess·pool** (sĕs'po͞ol') *n.* [Perh. alteration of obs. *cesperalle,* drainpipe < ME *suspiral* < OFr. *souspiral,* breathing hole < *souspirer,* to breathe < Lat. *suspirare,* to sigh. —see SUSPIRE.] A covered hole or pit for waste or sewage.

**ces·tode** (sĕs'tōd') *n.* [NLat. *Cestoda,* class name < Lat. *cestus,* belt < Gk. *kestos.*] A flatworm of the class Cestoda, including the tapeworms.

**ces·tus¹** (sĕs'təs) *n., pl.* **-ti** (-tī') [Lat., belt < Gk. *kestos.*] A belt or girdle.

**ces·tus²** *also* **caes·tus** (sĕs'təs) *n., pl.* **-tus·es.** [Lat. *caestus* < *caedere,* to strike.] A covering for the hand, made of leather straps weighted with iron or lead, worn by ancient Roman boxers.

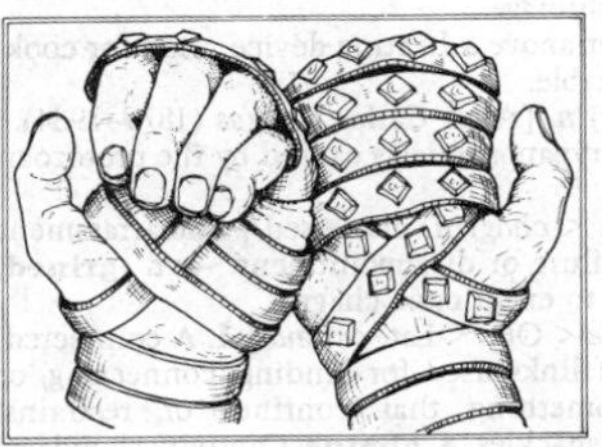
**cestus²**

**ce·su·ra** (sĭ-zho͝or'ə, -zo͝or'ə) *n. var. of* CAESURA.

**ce·ta·ce·an** (sĭ-tā'shən) *adj.* [< NLat. *Cetacea,* order name < Lat. *cetus,* whale < Gk. *kētos.*] Of or belonging to the order Cetacea, including fishlike aquatic mammals such as the whale and porpoise. —*n.* An aquatic mammal of the order Cetacea.

**ce·ta·ceous** (sĭ-tā'shəs) *adj.* [Lat. *cetus,* whale (< Gk. *kētos*) + -ACEOUS.] Cetacean.

**ce·tane** (sē'tān') *n.* [Lat. *cetus,* whale (so called because it is found in sperm whale oil) + -ANE.] A colorless liquid, $C_{16}H_{34}$, used as a solvent and in standardized hydrocarbons.

**cetane number** *also* **cetane rating** *n.* The performance rating of a diesel fuel, expressed as the percentage of cetane that must be mixed with liquid methylnaphthalene to produce the same ignition performance as the diesel fuel being rated.

**ce·ter·is pa·ri·bus** (kā'tər-ĭs păr'ə-bəs) *adv.* [NLat, with other things equal.] With all other factors or things being the same.

**ce·tol·o·gy** (sĭ-tŏl'ə-jē) *n.* [Lat. *cetus,* whale (< Gk. *kētos*) + -LOGY.] The zoology of whales and related aquatic mammals. **—ce'·to·log'i·cal** (sēt'l-ŏj'ĭ-kəl) *adj.* **—ce·tol'o·gist** *n.*

**Ce·tus** (sē'təs) *n.* [Lat. *cetus,* whale < Gk. *kētos.*] A constellation in the Southern Hemisphere.

**ce·tyl alcohol** (sēt'l) *n.* [Lat. *cetus,* whale + -YL.] A waxy alcohol, $C_{16}H_{34}O$, used in cosmetics and pharmaceuticals.

**Cey·lon moss** (sĭ-lŏn') *n.* [After *Ceylon,* former name of Sri Lanka.] A red seaweed, *Gracilaria lichenoides* of the East Indies, used for making agar.

**Cf** *symbol for* CALIFORNIUM.

**Cha·blis** (shă-blē', shä-, shăb'lē) *n.* [After *Chablis,* France.] A very dry white Burgundy wine.

**cha-cha** (chä'chä') *n.* [Sp. (Carribean) *chachachá.*] A rhythmic ballroom dance that originated in Latin America. **—cha'-cha'** *v.* **(-chaed, -cha·ing, -chas).**

**chac·ma** (chăk'mə) *n.* [Hottentot.] A grayish-black baboon, *Chaeropithecus ursinus* or *Papio ursinus* of southern and eastern Africa.

**cha·conne** (shă-kôn', -kŏn') *n.* [Fr. < *chacona,* a kind of dance.] **1. a.** A slow stately 18th-cent. dance. **b.** The music for the chaconne. **2.** A musical form with variations based on a reiterated harmonic pattern.

**chad** (chăd) *n.* [Orig. unknown.] The small pieces of paper resulting from the formation of holes in punched tape or data cards. **—chad'·less** *adj.*

**chae·ta** (kē'tə) *n., pl.* **-tae** (-tē') [NLat. < Gk. *khaitē,* long hair.] *Zool.* A bristle or seta of certain worms.

**chae·tog·nath** (kē'tŏg-năth') *n.* [NLat. *Chaetognatha,* phylum name : CHAETA + Gk. *gnathos,* jaw.] Any of various marine worms of the phylum Chaetognatha, including the arrow worms.

**chafe** (chāf) *v.* **chafed, chaf·ing, chafes.** [ME *chafen* < OFr. *chaufer,* to warm < VLat. **calefare,* alteration of Lat. *calefacere* : *calēre,* to be warm + *facere,* to make.] —*vt.* **1.** To irritate or wear away by friction. **2.** To annoy : vex. **3.** To heat or warm by rubbing. —*vi.* **1.** To cause friction : RUB. **2.** To become worn or sore from friction. **3.** To be or become annoyed <*chafed* at the delay> —*n.* **1.** Warmth, wear, or soreness produced by friction. **2.** Annoyance : irritation.

☆ ***syns:*** CHAFE, ABRADE, EXCORIATE, FRET, GALL *v. core meaning* : to make (the skin) raw by or as if by friction <*chafed* my knee against the concrete>

**cha·fer** (chā′fər) *n.* [ME, a kind of beetle < OE *ceafer.*] A beetle of the family Scarabaeidae, as the cockchafer.

**chaff**[1] (chăf) *n.* [ME *chaf* < OE *ceaf.*] **1.** Husks of grain remaining after separation from the seed. **2.** Finely cut straw or hay used as fodder. **3.** Trivial or worthless matter. **4.** Strips of metal foil released in the atmosphere to inhibit radar.

**chaff**[2] (chăf) *vt.* & *vi.* **chaffed, chaff·ing, chaffs.** [Perh. var. of CHAFE.] To subject to or engage in good-natured banter. —*n.* Good-natured banter. **—chaff′er** *n.*

**chaf·fer** (chăf′ər) *v.* **-fered, -fer·ing, -fers.** [ME *chaffaren* < *chaffare*, bargaining : *chep*, trade (< OE *cēap*) + *fare*, business < OE *faru.*] —*vi.* **1.** *Chiefly Brit.* To bandy words. **2.** To bargain or haggle. —*vt.* **1.** To bargain or haggle for. **2.** To barter. —*n. Archaic.* Bargaining : haggling. **—chaf′fer·er** *n.*

**chaf·finch** (chăf′ĭnch) *n.* [ME *chaffinche* : *chaf*, chaff, husk + *finch*, finch.] A small European songbird, *Fringilla coelebs*, with predominantly reddish-brown plumage.

**chafing dish** *n.* A dish set above a heating device, used for cooking or warming food at the table.

**Cha·gas disease** (shä′gəs) *n.* [After Carlos *Chagas* (1879–1934).] A South American form of trypanosomiasis caused by the protozoan *Trypanosoma cruzi.*

**cha·grin** (shə-grĭn′) *n.* [Fr. < *chagrin*, distressed.] Embarrassment or humiliation caused by failure or disappointment. —*vt.* **-grined, -grin·ing, -grins.** To cause to experience chagrin.

**chain** (chān) *n.* [ME *chaine* < OFr. < Lat. *catena.*] **1.** A connected, flexible, usu. metal series of links used for binding, connecting, or transmitting motion. **2.** Something that confines or restrains. **3. chains.** Bonds, fetters, or shackles. **4. chains.** Captivity : bondage. **5.** A series of connected or related things <a *chain* of uncanny coincidences> **6.** A number of establishments, as stores or theaters, commonly owned or managed. **7.** A mountain range. **8.** *Chem.* A group of atoms bonded in a chainlike spatial configuration. **9. a.** A measuring instrument for surveying, composed of 100 linked pieces of iron or steel. **b.** A unit of length, equal to 100 links, 66 feet, or approx. 20 meters. **10. a.** A measuring instrument used in engineering. **b.** A unit of length, equal to 100 feet or 30.5 meters. —*vt.* **chained, chain·ing, chains. 1.** To bind or make fast with a chain or chains. **2.** To bind or fetter : CONFINE.

**Chain-A-Matic** (chā′nə-măt′ĭk). A trademark for a balance with an adjustable calibrated chain suspended from the beam.

**chain gang** *n.* A group of convicts chained together for outdoor work.

**chain letter** *n.* A letter directing the recipient to send out multiple copies so that its circulation increases in a geometric progression as long as the instructions are carried out.

**chain·like** (chān′līk′) *adj.* Resembling a chain.

**chain mail** *n.* Flexible armor of joined metal links or scales.

**chain·man** (chān′mən) *n.* Either of the two persons who hold a measuring chain in surveying.

**chain pump** *n.* A pump that lifts water by means of containers that are attached to an endless chain and pass under water and up over a wheel.

**chain-re·act** (chān′rē-ăkt′) *vi.* **-act·ed, -act·ing, -acts.** To undergo a chain reaction.

**chain-reacting pile** *n.* A nuclear reactor.

**chain reaction** *n.* **1.** A series of events each of which induces or influences its successor. **2.** *Physics.* A multistage nuclear reaction, esp. a self-sustaining series of fissions in which the average number of neutrons produced per unit of time exceeds the number absorbed or lost. **3.** *Chem.* A series of reactions in which one product of a reacting set is a reactant in the following set.

**chain saw** *n.* A power saw with teeth linked in an endless chain.

**chain-smoke** (chān′smōk′) *v.* **-smoked, -smok·ing, -smokes.** —*vi.* To smoke esp. cigarettes in succession. —*vt.* To smoke (e.g., cigarettes) in continuing succession. **—chain smoker** *n.*

**chain stitch** *n.* A decorative stitch in which loops are connected in a chainlike fashion.

**chain store** *n.* One of several retail stores under the same ownership.

**chair** (châr) *n.* [ME *chaiere* < OFr. < Lat. *cathedra.* —see CATHEDRA.] **1.** A piece of furniture composed of a seat, legs, back, and often arms, intended to seat a single person. **2.** A seat of office, authority, or dignity, as that of a bishop. **3. a.** The position or office of a person in authority. **b.** One holding such an office or position, esp. one presiding over a meeting. **4.** *Slang.* The electric chair. **5.** A seat carried about on poles : SEDAN CHAIR. **6.** A metal block for supporting and holding railroad track in position. —*vt.* **chaired, chair·ing, chairs. 1.** To place or seat in a chair. **2.** To install in an authoritative position, esp. as a presiding officer. **3.** To preside over (a meeting). **4.** *Chiefly Brit.* To carry (a person) aloft in triumph, usu. in a chair.

**chair car** *n.* A parlor car.

**chair lift** *n.* A cable-suspended, power-driven chair assembly for transporting people, as skiers, up or down mountains.

**chair·man** (châr′mən) *n.* **1.** The presiding officer of an assembly, meeting, committee, academic department, or board. **2.** One who carries or wheels others in a chair. —*vt.* **-manned, -man·ning, -mans.** CHAIR 3.

**chair·man·ship** (châr′mən-shĭp′) *n.* The office or term of a chairman or chairwoman.

**chair·per·son** (châr′pûr′sən) *n.* A person who presides over an assembly, meeting, committee, academic department, or board.

**chair·wom·an** (châr′wo͝om′ən) *n.* A woman who presides over an assembly, meeting, committee, academic department, or board.

**chaise** (shāz) *n.* [Fr., chair, alteration of OFr. *chaiere.* —see CHAIR.] **1.** A light, open carriage, often with a collapsible hood, esp. a two-wheeled, horse-drawn carriage. **2.** A post chaise.

**chaise longue** (shāz lông′) *n.*, *pl.* **chaise longues** *or* **chaises longues** (shāz lông′) [Fr. : *chaise*, chair + *longue*, long.] A reclining chair with a seat that supports the sitter's outstretched legs.

**chak·ra** (chŭk′rə) *n.* [Skt. *cakram*, wheel.] One of the seven centers of spiritual energy in the human body according to yoga.

**cha·lah** (KHä′lə) *n. var. of* CHALLAH.

**cha·la·za** (kə-lā′zə, -lăz′ə) *n.*, *pl.* **-zae** (-zē′) *or* **-zas.** [Gk. *khalaza*, hailstone.] **1.** *Zool.* One of two spiral bands of tissue in an egg, connecting the yolk to the lining membrane. **2.** *Bot.* The part of an ovule opposite the micropyle that serves as a point of attachment for the integuments and the nucellus.

**cha·la·zi·on** (kə-lā′zē-ən, -ŏn′) *n.* [NLat. < Gk. *khalazion*, dim. of *khalaza*, hard lump.] A small hard tumor of the eyelid.

**chal·ced·o·ny** (kăl-sĕd′n-ē) *n.*, *pl.* **-nies.** [ME *calcedoine* < LLat. *chalcedonius* < Gk. *khalkēdōn*, a mystical stone, perh. < *Khalkēdōn*, an ancient town in Asia Minor.] A translucent to transparent milky or grayish quartz with microscopic crystals arranged in slender fibers in parallel bands. **—chal′ce·don′ic** (kăl′sĭ-dŏn′ĭk) *adj.*

**chal·cid** (kăl′sĭd) *n.* [< NLat. *Chalcis*, genus name < Gk. *khalkos*, copper (from the wasp's metallic color).] Any of various minute wasps of the superfamily Chalcidoidea, of which the larvae of many species are parasitic on the larval stages of other insects.

**chal·co·cite** (kăl′kə-sīt′) *n.* [Alteration of obs. *chalcosine* : Gk. *khalkos*, copper.] An important copper ore, essentially $CuS_2$.

**chal·co·py·rite** (kăl′kə-pī′rīt′) *n.* [Gk. *khalkos*, copper + PYRITES.] An important copper ore, essentially $CuFeS_2$.

**chal·co·sis** (kăl-kō′sĭs) *n.* [Gk. *khalkos*, copper + -OSIS.] Copper poisoning, occas. marked by copper deposits forming in the tissues.

**Chal·de·an** *also* **Chal·dae·an** (kăl-dē′ən) *or* **Chal·dee** (kăl′-dē′) *n.* [< Lat. *Chaldaeus* < Gk. *Khaldaios* < *Khaldaia*, Chaldea, a region of ancient Babylonia.] **1.** A member of an ancient Semitic people who ruled in Babylonia. **2.** The Semitic language of the Chaldeans. **3.** One versed in the occult. **—Chal·da′ic** (-dā′ĭk) *n.* & *adj.* **—Chal·de′an** *adj.*

**chal·dron** (chôl′drən) *n.* [OFr. *chauderon* < *chaudiere*, kettle < LLat. *caldaria.* —see CALDRON.] A unit of dry measure, as for coal, equal to 32 to 36 bushels, once used in England.

**cha·let** (shă-lā′, shăl′ā) *n.* [Fr.] **1.** A dwelling with a gently sloping overhanging roof, common to Alpine regions. **2.** The hut of a herdsman in the Alps.

**chal·ice** (chăl′ĭs) *n.* [ME < OFr. < Lat. *calix.*] **1.** A cup or goblet. **2.** A cup for the Eucharistic wine. **3.** A cup-shaped blossom.

**chal·i·co·sis** (kăl′ĭ-kō′sĭs) *n.* [Gk. *khalix*, *khalik-*, pebble + -OSIS.] Pneumoconiosis caused by inhalation of stone dust.

**chal·i·co·there** (kăl′ĭ-kə-thîr′) *n.* [NLat. *Chalicotherium*, genus name : Gk. *khalix*, stone + Gk. *thērion*, dim. of *thēr*, beast.] Any of various extinct ungulate mammals of the Eocene to Pleistocene epochs, with distinctive three-clawed, three-toed feet.

**chalk** (chôk) *n.* [ME < OE *cealk* < Lat. *calx*, lime. —see CALX.] **1.** A soft compact calcium carbonate, $CaCO_3$, with varying amounts of silica, quartz, feldspar, or other mineral impurities, gen. gray-white or yellow-white and derived chiefly from fossil seashells. **2.** A piece of often colored chalk or chalklike substance used for marking on a blackboard. **3.** A mark or picture made with chalk. **4.** A reckoning, as of credit given : TALLY. —*vt.* **chalked, chalk·ing chalks. 1.** To mark, draw, or write with chalk. **2.** To smear or cover with chalk. **3.** To make pale : WHITEN. **4.** To treat (e.g., soil) with chalk. **—chalk up. 1.** To earn : score <*chalk up* points> **2.** To credit <*Chalk* this *up* to experience.> **—chalk′i·ness** *n.* **—chalk′y** *adj.*

**chalk·board** (chôk′bôrd′, -bōrd′) *n.* A usu. green or black panel for writing on with chalk : BLACKBOARD.

**chalk·stone** (chôk′stōn′) *n. Pathol.* A tophus.

**chalk talk** *n.* An often informal lecture illustrated with diagrams chalked on a blackboard.

**chal·lah** *also* **cha·lah** (KHä′lə) *n.* [Heb. *ḥallāh.*] A loaf of yeast-leavened, usu. braided white egg bread traditionally eaten by Jews on the Sabbath, holidays, and other ceremonial occasions.

**chal·lenge** (chăl′ənj) *n.* [ME *chalenge* < OFr. < *chalanger*, to accuse < Lat. *calumniari*, to accuse falsely < *calumnia*, calumny < *calvi*, to deceive.] **1.** A call to engage in a contest or fight. **2.** A demand for an explanation. **3.** A sentry's demand for identification. **4.** Requirement for full use of one's abilities, or resources <a job that offers me a *challenge*> **5.** A claim that a vote is invalid or that a voter is unqualified. **6.** *Law.* A formal objection, esp. to the qualifications of a juror or jury. —*v.* **-lenged, -leng·ing, -leng·es.** —*vt.*

**1.** To call to engage in a contest or fight. **2.** To take exception to : DISPUTE. **3.** To order to halt and be identified. **4.** *Law.* To take formal objection to (e.g., a juror). **5.** To question the qualifications or validity of. **6.** To have due claim to. **7.** To summon to action, effort, or use <a problem that *challenges* the imagination> —*vi.* **1.** To make or utter a challenge. **2.** To begin barking upon picking up a scent, as hunting dogs do. **—chal'lenge·a·ble** *adj.* **—chal'leng·er** *n.*

**chal·lis** (shăl'ē) *n.* [Poss. < the surname *Challis.*] A light clothing fabric of wool, cotton, or rayon.

**cha·lone** (kā'lōn', kăl'ōn') *n.* [< Gk. *khalōn,* pr.part. of *khalan,* to slacken.] A hormone inhibiting a metabolic process.

**cha·lyb·e·ate** (kə-lĭb'ē-ĭt, -lē'bē-) *adj.* [NLat. *chalybeatus* < Lat. *chalybs,* steel < Gk. *khalups* < *Khalups,* Chalybes, a people of Asia Minor famous for their steel.] **1.** Impregnated with or containing salts of iron. **2.** Tasting like iron, as mineral-spring water. —*n.* Water or medicine containing iron in solution.

**cham** (kăm) *n.* [Fr. < Turk *khān.*] *Archaic.* KHAN[1] 2.

**Cha·mae·leon** *also* **Cha·me·leon** (kə-mēl'yən, -mē'lē-ən) *n.* [Lat. *chamaeleon,* chameleon.] A constellation in the Southern Hemisphere.

**cham·ae·phyte** (kăm'ə-fīt') *n.* [Gk. *khamai,* on the ground + -PHYTE.] A perennial plant bearing its winter buds very close to the soil surface.

**cham·ber** (chām'bər) *n.* [ME *chaumbre* < OFr. < LLat. *camera,* chamber < Lat., vault < Gk. *kamara.*] **1. a.** A room in a house, esp. a bedroom. **b. chambers.** *Chiefly Brit.* A suite of rooms : APARTMENT. **2.** *often* **chambers.** A judge's office. **3.** A formal reception room in a palace or official residence. **4.** A hall for the meeting of an assembly, esp. a legislative assembly. **5.** A legislative, judicial, or deliberative assembly. **6.** A board or council. **7.** A place where governmental funds are received and held : TREASURY. **8.** An enclosed space or compartment : CAVITY. **9. a.** An enclosed space at the bore of a gun that holds the charge. **b.** The part of a cylinder in a revolver that receives the cartridge. —*vt.* **-bered, -ber·ing, -bers. 1.** To put in or as if in a chamber : CONFINE. **2.** To provide with a chamber.

**chambered nautilus** *n.* A cephalopod mollusk, *Nautilus pompilius* of the Pacific and Indian oceans, having a coiled and partitioned shell lined with a pearly layer.

**cham·ber·lain** (chām'bər-lən) *n.* [ME *chaumberlein* < OFr. *chamberlene* < LLat. *camera,* chamber.] **1.** An official who manages the household of a sovereign or nobleman. **2.** A high-ranking officer in a royal court. **3.** An official who receives the rents and fees of a municipality : TREASURER. **4.** *Rom. Cath. Ch.* An often honorary papal attendant.

**cham·ber·maid** (chām'bər-mād') *n.* A maid who cleans and cares for bedrooms, as in hotels.

**chamber music** *n.* Music appropriate for performance in a private room or small concert hall and written for an ensemble such as a trio or quartet.

**chamber of commerce** *n.* An association of business people for the promotion of commercial interests in the community.

**chamber pot** *n.* A portable vessel used as a toilet.

**cham·bray** (shăm'brā') *n.* [After *Cambrai,* France.] A fine, lightweight gingham woven with white threads across a colored warp.

**cha·me·leon** (kə-mēl'yən, -mē'lē-ən) *n.* [ME *camelioun* < Lat. *chamaeleon* < Gk. *khamaileōn* : *khamai,* on the ground + *leōn,* lion.] **1.** Any of various tropical Old World lizards of the family Chamaeleonidae that are able to change color. **2.** The anole. **3.** A changeable, inconstant person. **4. Chameleon.** *var. of* CHAMAELEON. **—cha·me'le·on'ic** (-lē-ŏn'ĭk) *adj.*

**cham·fer** (chăm'fər) *vt.* **-fered, -fer·ing, -fers.** [Prob. ult. < OFr. *chanfreindre* : *chant,* edge + *fraindre,* to break < Lat. *frangere.*] **1.** To cut off the edge or corner of : BEVEL. **2.** To cut a groove in : FLUTE. —*n.* **1.** A flat surface made by cutting off an edge or corner. **2.** A furrow or groove, as in a column.

**cham·fron** (chăm'frən) *n.* [ME *shamfron* < OFr. *chanfrein.*] Medieval armor for the front of a horse's head.

**cha·mi·so** (chə-mē'sō) *n., pl.* **-sos.** [Sp. *chamiza,* wild brush < *chamizo,* dry brush gathered for firewood.] A shrub, *Adenostoma fasciculatum* of California, having small white flower clusters and forming dense thickets.

**cham·ois** (shăm'ē) *n., pl.* **cham·ois** (shăm'ēz, shăm-wä') [OFr.] **1.** A hoofed mammal, *Rupricapra rupricapra* of mountainous regions of Europe, having upright horns with backward-hooked tips. **2.** *also* **cham·my** *pl.* **-mies.** Soft leather made from the hide of the chamois or other animals, such as deer or sheep. **3.** A moderate to grayish yellow. —*vt.* **-oised, -ois·ing, -ois·es. 1.** To dress or prepare like chamois. **2.** To polish or dry with chamois leather.

**cham·o·mile** or **cam·o·mile** (kăm'ə-mīl') *n.* [ME *camomille* < LLat. *chamomilla* < Lat. *chamaemelon* < Gk. *khamaimēlon* : *khamai,* on the ground + *mēlon,* apple.] **1.** A plant of the genus *Anthemis,* esp. *A. nobilis,* an aromatic plant native to Eurasia, with finely dissected leaves and white flowers. **2.** A similar plant of the genus *Matricaria,* esp. *M. chamomilla.*

**champ[1]** (chămp) *also* **chomp** (chŏmp) [Perh. imit.] —*v.* **champed, champ·ing, champs** *also* **chomped, chomp·ing, chomps.** —*vt.* **1.** To bite on impatiently or restlessly. **2.** To chew on noisily. **3.** *Scot.* To crush or trample. —*vi.* To work the jaws vigorously. **—champ at the bit.** To be very impatient. **—champ** *n.*

**champ[2]** (chămp) *n. Informal.* A champion.

**cham·pagne** (shăm-pān') *n.* **1. a.** A sparkling white wine produced in Champagne, a region of France. **b.** A similar wine made elsewhere. **2.** A pale orange yellow to grayish yellow or yellowish gray.

**cham·paign** (shăm-pān') *n.* [ME *champain* < OFr. *champaigne* < LLat. *campania,* open country. —see CAMPAIGN.] Level and open country. **—cham·paign'** *adj.*

**cham·pak** *also* **cham·pac** (chăm'păk', chŭm'pŭk) *n.* [Hindi *campak* < Skt. *campakaḥ.*] A tree, *Michelia champaca* of India and the East Indies, bearing yellow flowers and yielding a camphorlike substance and an oil used in perfumery.

**cham·per·ty** (chăm'pər-tē) *n., pl.* **-ties.** [ME *champartie* < OFr. *champart,* the lord's share of the tenant's crop : *champ,* field (< Lat. *campus*) + *part,* share (< Lat. *pars*).] *Law.* An illegal sharing in the proceeds of a lawsuit by an outside party that has promoted the litigation. **—cham'per·tous** (-təs) *adj.*

**cham·pi·gnon** (shăm-pĭn'yən) *n.* [Fr. < OFr. *champigneul,* prob. ult. < Lat. *campus,* field.] An edible mushroom, esp. the common species *Agaricus campestris.*

**cham·pi·on** (chăm'pē-ən) *n.* [ME *champioun* < OFr. *champion* < Med. Lat. *campio.*] **1.** The holder of first place or the winner of first prize in a contest, esp. in sports. **2.** A defender, advocate, or supporter of a cause or another person. **3.** One who fights : WARRIOR. —*vt.* **-oned, -on·ing, -ons. 1.** To fight as champion of : DEFEND <"*championed* the government and defended the system of taxation" —Samuel Chew> **2.** *Obs.* To defy or challenge. —*adj.* **1.** Holding first place or prize. **2.** Exceeding all others : ARCH <"the *champion* playboy of the Western World" —John Millington Synge>

**cham·pi·on·ship** (chăm'pē-ən-shĭp') *n.* **1.** The position or title of a champion. **2.** Defense or support : ADVOCACY. **3.** A competition or series of competitions held to determine a winner.

**chance** (chăns) *n.* [ME, unexpected event < OFr. < VLat. **cadentia* < Lat. *cadere,* to happen.] **1. a.** The abstract nature or quality shared by unexpected, random, or unpredictable events : CONTINGENCY. **b.** This quality considered as causing or determining the outcome of such events : LUCK. **2.** The likelihood of occurrence of an event : PROBABILITY. **3. a.** An unexpected, random, or unpredicted event. **b.** A fortuitous event. **4. a.** An opportunity <a *chance* to better oneself> **b.** A risk or hazard : GAMBLE. **c.** A raffle or lottery ticket. **5.** *Baseball.* An opportunity to make a putout or an assist that counts as an error if unsuccessful. —*adj.* **1.** Taking place unexpectedly <a *chance* encounter> **2.** Determined or marked by whim or caprice : ARBITRARY. —*v.* **chanced, chanc·ing, chanc·es.** —*vi.* To happen by chance. —*vt.* To take the risk or hazard of. **—chance on** (*or* **upon).** To discover or come upon accidentally.

**chance·ful** (chăns'fəl) *adj.* **1.** Full of chance : EVENTFUL. **2.** *Archaic.* Dependent on chance. **3.** *Obs.* Risky.

**chan·cel** (chăn'səl) *n.* [ME *chauncel* < OFr. *chancel.*] The space around the altar of a church for the clergy and choir, often enclosed by a lattice.

**chan·cel·ler·y** (chăn'sə-lə-rē, -slə-rē, chän'-) *n., pl.* **-ies.** [ME *chancelrie* < OFr. *chancelerie* < *chancelier,* chancellor.] **1.** The position, rank, office, or department of a chancellor. **2.** The building in which the office of a chancellor is located. **3.** The official place of business of an embassy or consulate.

**chan·cel·lor** (chăn'sə-lər, -slər) *n.* [ME *chaunceler* < OFr. *chancelier* < LLat. *cancellarius,* doorkeeper < Lat. *cancelli,* lattice. —see CANCEL.] **1.** A high-ranking official, esp.: **a.** A king's or nobleman's secretary. **b.** *Chiefly Brit.* The chief secretary of an embassy. **c.** The chief minister of state in some European countries. **2. a.** *Chiefly Brit.* The honorary or titular head of a university. **b.** The president of some U.S. universities. **3.** The presiding judge of a court of chancery or equity in some U.S. states. **—chan'cel·lor·ship'** *n.*

**Chancellor of the Exchequer** *n.* A cabinet member and the highest finance minister in the British government.

**chance-med·ley** (chăns'mĕd'lē) *n.* [ME *chaunce medley* < Norman Fr. *chance medlee,* mixed chance.] **1.** *Law.* A sudden quarrel resulting in an unpremeditated homicide. **2.** A random action.

**chan·cer·y** (chăn'sə-rē) *n., pl.* **-ies.** [ME *chauncerie* < OFr. *chancelerie,* chancellery < *chancelier,* chancellery.] **1. a.** A court with jurisdiction in equity. **b.** The proceedings and practice of a court of chancery : EQUITY. **2.** An office of archives. **3.** One of the five divisions of the High Court of Justice of Great Britain, presided over by the Lord High Chancellor. **4.** CHANCELLERY 3. **—in chancery. 1.** *Law.* In litigation or pending in a court of chancery. **2.** *Informal.* In a hopeless predicament.

**chan·cre** (shăng'kər) *n.* [Fr. < Lat. *cancer,* ulcer.] **1.** A dull-red, hard, insensitive lesion that is the first manifestation of syphilis. **2.** An ulcer located at the initial point of entry of a pathogen. **—chan'crous** (-krəs) *adj.*

**chan·croid** (shăng'kroid') *n.* [Fr. *chancroïde: chancre,* chancre +

*-oide*, -oid.] A soft, nonsyphilitic, usu. venereal lesion of the genital region. **—chan'croid'al** (-kroid'l) *adj.*

**chanc·y** (chăn'sē, chän'-) *adj.* **-i·er, -i·est. 1.** Uncertain : risky <a *chancy* proposition> **2.** *Scot.* Lucky : propitious.

**chan·de·lier** (shăn'də-lîr') *n.* [Fr. < OFr., prob < VLat. *candela*rum* < Lat. *candelabrum*, candelabrum. —see CANDELABRUM.] A branched light fixture holding a number of bulbs or candles and usu. suspended from a ceiling.

**chan·delle** (shăn-dĕl') *n.* [Fr. < *chandelle*, candle < OFr. —see CHANDLER.] A sudden, steep climbing turn of an aircraft, executed to change flight direction and gain altitude simultaneously.

**chan·dler** (chănd'lər) *n.* [ME *chaundeler* < OFr. *chandelier* < *chandelle*, candle < Lat. *candela*. —see CANDLE.] **1.** A maker or seller of candles. **2.** A dealer in specified equipment or goods <a ship *chandler*>

**Chan·dler Wobble** (chănd'lər) *or* **Chan·dler's Wobble** (chănd'lərz) *n.* [After Seth C. *Chandler* (1846–1913).] An oscillation in the earth's rotational axis having a period of approx. 14 months.

**chan·dler·y** (chănd'lə-rē) *n.*, *pl.* **-ies. 1.** The stock or business of a chandler. **2.** A place for storing candles.

**change** (chānj) *v.* **changed, chang·ing, chang·es.** [ME *chaungen* < Norman Fr. *chaunger* < OFr. *changier* < LLat. *cambiare*, prob. of Celtic orig.] *—vt.* **1. a.** To make different : ALTER. **b.** To give a totally different form or appearance to : TRANSFORM <irrigation that *changed* the arid land to fertile soil> **2.** To give and receive reciprocally : INTERCHANGE <Let's *change* places.> **3.** To exchange for or replace by another, usu. of the same kind or class <*change* one's name> **4.** To lay aside, abandon, or leave for another : SWITCH <*change* planes><*change* procedures> **5.** To give or receive the equivalent of (money) in lower denominations or in foreign currency. **6.** To put fresh clothes or coverings on <*change* a baby's diapers><*changed* the bed> *—vi.* **1.** To become different <*changed* as they matured> **2.** To go from one phase to another, as the moon. **3.** To make an exchange. **4.** To transfer from one vehicle to another. **5.** To put on other clothing <*change* for dinner> **6.** To become deeper in tone. —Used of the voice. **—change off. 1.** To alternate with another person in performing a task. **2.** To alternate between two different tasks or between a task and a period of rest. *—n.* **1. a.** The act, process, or result of changing. **b.** Substitution of one thing for another <went to France for a *change* of scene> **2.** Transition from one state or phase to another <the *change* of seasons> **3.** Something different : VARIETY <came home early for a *change*> **4.** A different or fresh set of clothing. **5. a.** Money of smaller denomination given or received in exchange for money of higher denomination. **b.** The balance of money returned when an amount given is more than what is due. **c.** Coins. **6.** *Mus.* A pattern or order in which bells are rung. **7.** A market or exchange where business is transacted. **—change hands.** To pass from one owner to another. **—change (one's) mind.** To reverse an opinion or a decision. **—change'less** *adj.* **—chang'er** *n.*

☆ ***syns:*** CHANGE, ALTER, MODIFY, MUTATE, TURN, VARY *v. core meaning* : to make or become different <an event that *changed* the world><a face that had *changed* with age><*change* a liquid into a gaseous or solid form>

**change·a·ble** (chān'jə-bəl) *adj.* **1.** Apt to change : CAPRICIOUS <*changeable* moods> **2.** Able to undergo alteration <*changeable* habits> **3.** Changing color or appearance when seen from different angles <*changeable* taffeta> **—change'a·bil'i·ty, change'a·ble·ness** *n.* **—change'a·bly** *adv.*

**change·ful** (chānj'fəl) *adj.* Tending or able to change : VARIABLE. **—change'ful·ly** *adv.* **—change'ful·ness** *n.*

**change·ling** (chānj'lĭng) *n.* **1.** A child secretly exchanged for another. **2.** *Archaic.* A fickle person. **3.** *Archaic.* An idiot.

**change of life** *n.* Menopause.

**change·o·ver** (chānj'ō'vər) *n.* Conversion to a different purpose or from one system to another, esp. in production techniques or type of equipment.

**change ringing** *n.* The ringing of a set of bells with all possible unrepeated variations.

**chan·nel**[1] (chăn'əl) *n.* [ME *chanel* < OFr. < Lat. *canalis*.] **1.** A stream or riverbed. **2.** The deeper part of a river or harbor, esp. a deep navigable passage. **3.** A broad strait. **4.** A tubular passage for liquids. **5.** A course through which something may be directed or moved <a *channel* of thought> **6. channels.** Official communication routes. **7.** *Electron.* A specified frequency band for transmitting and receiving electromagnetic signals, as for television. **8.** A trench, furrow, or groove. **9.** A rolled metal bar with a bracket-shaped section. *—vt.* **-neled, -nel·ing, -nels** *also* **-nelled, -nel·ling, -nels. 1.** To make or cut channels in. **2.** To form a channel or flute in. **3.** To direct or guide along a desired course <*channel* one's efforts into worthwhile projects>

**chan·nel**[2] (chăn'əl) *n.* [Alteration of obs. *chainwale* : CHAIN + WALE.] *Naut.* A wood or steel ledge projecting from a sailing ship's sides to spread the shrouds and keep them clear of the gunwales.

**channel bass** *n.* The red drum.

**channel black** *n.* [< CHANNEL[1].] A finely divided carbon black, formed on iron plate by direct exposure to a natural gas flame and used in inks, paints, typewriter ribbons, crayons, and polishes.

**chan·nel-lag deposit** (chăn'əl-lăg') *n.* Residue deposited in a channel as a stream runs its natural course.

**chan·son** (shäN-sôN') *n.*, *pl.* **-sons** (-sôN', -sôNz') [Fr. < OFr. < Lat. *cantio* < *cantare*, to sing, freq. of *canere*.] A song, esp. a French cabaret song.

**chan·son de geste** (shäN-sôN' də zhĕst') *n.*, *pl.* **chan·sons de geste** (-sôN', -sôNz) [Fr.: *chanson*, song + *de*, of + *geste*, heroic exploit.] An Old French epic poem genre of the 11th to the 13th cent.

**chant** (chănt) *n.* [Prob. < Fr., song < OFr. < Lat. *cantus* < *canere*, to sing.] **1. a.** A short, simple melody in which a number of syllables or words are sung on the same note. **b.** A psalm or canticle sung in this manner. **2.** A song or melody. **3.** A monotonous rhythmic call or shout <the *chant* of demonstrators at a rally> *—v.* **chant·ed, chant·ing, chants.** *—vt.* **1.** To sing or intone to a chant. **2.** To celebrate in song. **3.** To say in the manner of a chant. *—vi.* **1.** To sing, esp. in the manner of a chant. **2.** To speak monotonously. **—chant'ing·ly** *adv.*

**chant·er** (chăn'tər) *n.* **1.** One who chants, as a chorister. **2.** A priest who sings in a chantry. **3.** The pipe of a bagpipe on which the melody is played.

**chan·te·relle** (shăn'tə-rĕl', shän'-) *n.* [Fr. < NLat. *cantharellus*, dim. of Lat. *cantharus*, cup (from the mushroom's shape) < Gk. *kantharos*.] An edible yellow mushroom, *Cantharellus cibarius*, with a pleasant fruity odor.

**chan·teuse** (shäN-tœz') *n.* [Fr., fem. of *chanteur*, singer < *chanter*, to sing < OFr. < Lat. *cantare*, freq. of *canere*.] A woman singer, esp. a nightclub singer.

**chan·tey** *also* **chan·ty** (shăn'tē, chăn'-) *n.*, *pl.* **-teys** *also* **-ties.** [Prob. < Fr. *chantez*, imper. of *chanter*, to sing < OFr. < Lat. *cantare*, freq. of *canere*.] A song sung by sailors to the rhythm of their movements while working.

**chan·ti·cleer** (chăn'tĭ-klîr', shăn'-) *n.* [ME *chauntecler* < OFr. *chantecler* : *chanter*, to sing + *cler*, clear < Lat. *clarus*.] ROOSTER 1a.

**chan·try** (chăn'trē) *n.*, *pl.* **-tries.** [ME *chaunterie* < OFr. < *chanter*, to sing. —see CHANTEY.] **1.** An endowment to cover expenses for the saying of masses and prayers, usu. for the soul of the founder of the endowment. **2.** An endowed altar or chapel.

**chan·ty** (shăn'tē, chăn'-) *n. var. of* CHANTEY.

**Cha·nu·kah** (KHä'nə-kə, hä'-) *n.* [Heb. *ḥanukkāh*, consecration.] An 8-day Jewish festival beginning on the 25th day of the month of Kislev and commemorating the victory of the Maccabees over the Syrians in 165 B.C. and the rededication of the Temple at Jerusalem.

**cha·os** (kā'ŏs') *n.* [Lat., formless matter < Gk. *khaos*.] **1.** A state or place of total confusion or disorder. **2.** *often* **Chaos.** The disordered state of unformed matter and infinite space believed, according to some religious cosmological views, to have existed prior to the ordered universe. **3.** *Obs.* A vast chasm or abyss. **—cha·ot'ic** (-ŏt'ĭk) *adj.* **—cha·ot'i·cal·ly** *adv.*

**chap**[1] (chăp) *v.* **chapped, chap·ping, chaps.** [ME *chappen*.] *—vt.* To cause (the skin) to split or roughen, esp. due to cold or exposure. *—vi.* To split or become rough and sore <skin that *chaps* easily> **—chap** *n.*

**chap**[2] (chăp) *n.* [Short for CHAPMAN.] FELLOW 1a.

†**cha·pa·re·jos** *also* **cha·pa·ra·jos** (shăp'ə-rā'ōs) *pl.n.* [Mex. Sp. *chaparreras*.] *Southwestern U.S.* Chaps.

**chap·ar·ral** (shăp'ə-răl') *n.* [Sp. < *chaparro*, evergreen oak < Basque *txapar*, dim. of *saphar*, thicket.] A dense thicket of shrubs and small trees, esp. in the southwestern United States and Mexico.

**chaparral cock** *n.* The roadrunner.

**chaparral pea** *n.* A thorny shrub, *Pickeringia montana* of California, having reddish-purple flowers and forming dense thickets.

**chap·book** (chăp'bo͝ok') *n.* [CHAP(MAN) + BOOK (so called because it was orig. sold by chapmen).] A small book of poems, ballads, stories, or religious tracts.

**chape** (chāp, chăp) *n.* [ME < OFr., covering < LLat. *cappa*.] A metal trimming or mounting on a scabbard or sheath.

**cha·peau** (shă-pō') *n.*, *pl.* **-peaux** (-pōz') *or* **-peaus** (-pōz') [Fr. < OFr. *chapel* < Med. Lat. *cappellus* < LLat. *cappa*.] HAT 1.

**chap·el** (chăp'əl) *n.* [ME *chapele* < OFr. < Med. Lat. *capella*, chapel, cape (from a shrine containing the cape of St. Martin of Tours), dim. of *cappa*, cloak.] **1.** A place of worship smaller than and subordinate to a church. **2.** A place of worship in an institution, as a university or hospital. **3.** The services held at a chapel. **4.** A recess or room in a church reserved for small or special services. **5.** A place of worship for those not affiliated with or not members of an established church. **6.** A choir or orchestra connected with a chapel or royal court. **7. a.** An association of workers in a print shop. **b.** *Obs.* A printing house or print shop.

**chap·er·on** *also* **chap·er·one** (shăp'ə-rōn') [Fr. < *chaperon*, hood < OFr. < *chape*, covering. —see CHAPE.] *—n.* **1.** A person, esp. an older or married woman, who accompanies a young unmarried woman in public. **2.** An older person who attends and supervises a social gathering for young people. *—vt.* **-oned, -on·ing, -ons** *also*

**-oned, -on·ing, -ones.** To act as chaperon to or for. **—chap′er·on′·age** (-rō′nĭj) *n.*

**chap·fall·en** (chăp′fô′lən, chŏp′-) *also* **chop·fall·en** (chŏp′-) *adj.* [< obs. *chaps,* alteration of CHOPS.] Dispirited.

**chap·i·ter** (chăp′ĭ-tər) *n.* [ME *chapitre,* chapter, chapiter < OFr. < *chapitle* < Lat. *capitulum,* dim. of *caput,* head.] The capital of an architectural column.

**chap·lain** (chăp′lĭn) *n.* [ME *chapelein* < OFr. *chapelain* < Med. Lat. *cappellanus* < *capella,* chapel. —see CHAPEL.] **1.** A member of the clergy attached to a chapel. **2.** A minister or layperson who conducts religious services, as for a legislative assembly. **3.** A member of the clergy attached to a military unit. **—chap′lain·cy, chap′lain·ship′** *n.*

**chap·let** (chăp′lĭt) *n.* [ME *chapelet* < OFr., dim. of *chapel,* hat. —see CHAPEAU.] **1.** A wreath or garland for the head. **2.** *Rom. Cath. Ch.* **a.** A string of prayer beads with one third the number of a rosary's beads. **b.** The prayers counted on such beads. **3.** A string of beads. **4.** A small molding carved so as to resemble a string of beads. **—chap′let·ed** *adj.*

**chap·man** (chăp′mən) *n.* [ME < OE *cēapman* : *cēap,* trade (ult. < Lat. *caupo,* tradesman) + *mann,* man.] **1.** *Chiefly Brit.* A peddler. **2.** *Archaic.* A dealer or merchant.

**chaps** (chăps, shăps) *pl.n.* [Short for Mex. Sp. *chaparreras.*] Heavy leather trousers without a seat, worn over ordinary trousers by cowboys to protect their legs.

**chap·ter** (chăp′tər) *n.* [ME *chapitre,* chapter, chapiter. —see CHAPITER.] **1.** A division of a book or other writing, usu. numbered or titled. **2.** A period or sequence of events marking a distinct change of pattern <a sad *chapter* in the history of our nation> **3.** A local branch of an organization, as a club or fraternity. **4. a.** An assembly of the canons of a church. **b.** The canons as a group. **5.** An assembly of the members or representatives of a religious house, community, or order. **6.** A meeting of a sorority or fraternity. **7.** A short Scriptural passage read after the psalms in some church services.

**chapter house** *n.* **1.** A building in which the chapter of a cathedral or monastery assembles. **2.** A house in which a chapter of a fraternity or sorority lives and meets.

**char**[1] (chär) *vt.* & *vi.* **charred, char·ring, chars.** [Back-formation < CHARCOAL.] **1.** To scorch or become scorched. **2.** To reduce or become reduced to charcoal by incomplete combustion. *—n.* A charred substance.

**char**[2] *also* **charr** (chär) *n., pl.* **char** *or* **chars** *also* **charr** *or* **charrs.** [Orig. unknown.] A fish of the genus *Salvelinus,* related to the trout, esp. the widely distributed species *S. alpinus.*

**char**[3] (chär) *n.* [ME, a piece of work < OE *cierr,* a turning.] **1.** A chore, esp. a household task. **2.** A charwoman. *—vi.* **charred, char·ring, chars. 1.** To do tasks or chores. **2.** To work as a charwoman.

**char·a·banc** (shăr′ə-băng′) *n.* [Fr. *char à bancs* : *char,* carriage + *à,* with + *bancs,* benches.] *Chiefly Brit.* A large bus, often used for sightseeing.

**char·a·cin** (kăr′ə-sĭn) *also* **char·a·cid** (-sĭd) *n.* [NLat. *Characinidae,* former family name < Gk. *kharax,* a kind of fish.] Any of numerous chiefly tropical freshwater fishes of the family Characidae, many of which are popular aquarium fish.

**char·ac·ter** (kăr′ək-tər) *n.* [ME *carecter* < Lat. *character* < Gk. *kharaktēr* < *kharassein,* to inscribe.] **1.** The combination of emotional, intellectual, and moral qualities distinguishing one person or group from another. **2.** A distinctive feature or attribute : CHARACTERISTIC. **3.** Moral or ethical strength : INTEGRITY <a person with *character*> **4.** Public estimation : REPUTATION. **5.** Status, capacity, or role <in my *character* as a single parent> **6.** *Informal.* An eccentric person. **7.** An important, influential person. **8.** A person portrayed in an artistic piece, as a drama or novel. **9.** A description of one's attributes, traits, or abilities. **10.** A formal written statement as to competency and dependability, given by an employer to a former employee : RECOMMENDATION. **11.** A symbol or mark used in a writing system. **12.** *Computer Sci.* **a.** One of a set of symbols, as letters or numbers, arranged to express information. **b.** The multibit code representing such a character. **13.** A style of printing or writing. **14.** A symbol used in secret writing. **15.** *Genetics.* A structure, function, or attribute determined by a gene or group of genes. *—adj.* **1.** Having the ability to act in roles emphasizing traits markedly different from those of the performer <a *character* actress> **2.** Of or requiring the abilities of a character actor or actress <a *character* part in the play> *—vt.* **-tered, -ter·ing, -ters. 1.** To write, print, engrave, or inscribe. **2.** *Archaic.* To portray, describe, or represent. **—in character.** Consistent with one's character or behavior. **—out of character.** Inconsistent with one's character or behavior.

☆ ***syns:*** CHARACTER, COMPLEXION, DISPOSITION, MAKE-UP, NATURE *n. core meaning* : the combination of emotional, intellectual, and moral qualities that distinguishes a person <Goodness is the foundation of a truly great person's *character.*>

**char·ac·ter·is·tic** (kăr′ək-tə-rĭs′tĭk) *adj.* Serving to identify or set apart a person or group : DISTINCTIVE. *—n.* **1.** A distinguishing attribute or element. **2.** *Math.* The integral part of a logarithm as distinguished from the mantissa <6 is the *characteristic* of the logarithm 6.3214.> **—char′ac·ter·is′ti·cal·ly** *adv.*

☆ ***syns:*** CHARACTERISTIC, DISTINCTIVE, INDIVIDUAL, PECULIAR, TYPICAL, VINTAGE *adj. core meaning* : serving to identify or set apart <the zebra's *characteristic* stripes><behavior *characteristic* of psychopaths><a *characteristic* remark>

**char·ac·ter·i·za·tion** (kăr′ək-tər-ĭ-zā′shən) *n.* **1.** The act of characterizing. **2.** A representation of a person's attributes or peculiarities. **3.** Creation or delineation of a character or characters on the stage or in writing, esp. by imitating or describing actions, gestures, or speeches.

**char·ac·ter·ize** (kăr′ək-tə-rīz′) *vt.* **-ized, -iz·ing, -iz·es. 1.** To describe the qualities or peculiarities of <*characterized* that person as weak> **2.** To be a distinguishing trait or mark of <a seaport *characterized* by its charm> **—char′ac·ter·iz′er** *n.*

**char·ac·ter·y** (kăr′ək-tə-rē, kə-răk′-) *n., pl.* **-ies.** A system of characters or symbols used to express or convey thought and meaning.

**cha·rade** (shə-rād′) *n.* [Fr.] **1. a. charades.** A game in which words or phrases are represented in pantomime, often syllable by syllable, until they are guessed by the other players. **b.** An episode in this game or the word so represented. **2.** An easily perceivable pretense : TRAVESTY.

**char·broil** (chär′broil′) *vt.* **-broiled, -broil·ing, -broils.** [CHAR[1] + BROIL.] To broil (meat) over charcoal.

**char·coal** (chär′kōl′) *n.* [ME *charcol.*] **1.** A black, porous carbonaceous material produced by the destructive distillation of wood and used as a fuel, filter, and absorbent. **2.** A drawing pencil or crayon made from charcoal. **3.** A drawing executed with a charcoal pencil or crayon. **4.** A dark grayish brown to black or dark purplish gray. *—vt.* **-coaled, -coal·ing, -coals.** To draw, write, or blacken with charcoal.

**charcoal rot** *n.* A plant disease caused by a fungus, *Macrophomina phaseoli,* that results in black, decayed tissue.

**Char·cot's disease** (shär-kōz′) *n.* Multiple sclerosis.

**chard** (chärd) *n.* [Fr. *carde* < OProv. *cardon,* cardoon. —see CARDOON.] A variety of beet, *Beta vulgaris cicla,* having large succulent leaves eaten as a vegetable.

**charge** (chärj) *v.* **charged, charg·ing, charg·es.** [ME < OFr. *chargier,* to load < LLat. *carricare* < Lat. *carrus,* cart, of Celtic orig.] *—vt.* **1.** To entrust with a responsibility, duty, or obligation <*charged* me with the task of teaching the beginners> **2.** To give orders to <*charged* them not to reveal classified data> **3.** To make an accusation against : ACCUSE <*charged* them with the crime> **4.** To set or ask (a given amount) as a price <*charged* $100 for the dress> **5.** To hold financially liable : demand payment from <*charged* me for the balance due> **6.** To postpone payment on (a service or purchase) by recording as a debt <*charging* a new sofa> **7.** To attack violently. **8. a.** To direct or put (a weapon) into position for use. **b.** To load (a firearm). **9.** To instruct (a jury) about legal points and weight of evidence. —Used of a judge. **10.** *Elect.* **a.** To cause formation of a net electric charge on or in (e.g., a conductor). **b.** To energize (a storage battery). **11. a.** To furnish or fill to capacity. **b.** To cause to be saturated : IMPREGNATE <air *charged* with perfume><an atmosphere *charged* with tension> **12.** To place a heraldic bearing on. *—vi.* **1.** To rush forward in or as if in a violent attack. **2.** To demand or ask payment <*charged* for the passports> **3.** To make an entry to one's debit. **—charge off.** To consider as a loss. *—n.* **1.** Custody or supervision. **2.** An obligation : responsibility. **3.** A person or thing entrusted to one's care or management. **4.** An order, command, or injunction. **5.** Instructions given by a judge to a jury about points of law and weight of evidence. **6.** An accusation : indictment <a *charge* of murder> **7. a.** Expense : cost. **b.** The price set or asked <no *charge* for the repairs> **8.** A financial burden, as a tax or lien. **9.** A debt or an entry in an account recording a debt. **10. a.** A rushing, forceful attack. **b.** A command to attack. **11.** *Informal.* A feeling of pleasurable excitement : THRILL <got a *charge* out of the new game> **12.** The maximum quantity that an apparatus or container can hold at one time. **13.** A quantity of explosive to be set off at one time. **14. a.** The intrinsic property of matter responsible for all electric phenomena, esp. for the force of the electromagnetic interaction, occurring in two forms arbitrarily designated negative and positive. **b.** A measure of this property. **c.** The net measure of this property possessed by a body or contained in a bounded region of space. **15.** A heraldic bearing or figure.

**charge·a·ble** (chär′jə-bəl) *adj.* **1.** Suitable to be charged, as to an account. **2.** Liable to be accused or indicted. **—charge′a·ble·ness** *n.*

**charge account** *n.* A credit arrangement in which the customer receives purchases or services prior to payment.

**charge conjugation** *n.* **1.** A mathematical operator that changes the sign of the charge and of the magnetic moment of every particle in the system to which it is applied. **2.** Theoretical conversion of matter to antimatter or of antimatter to matter.

**char·gé d'af·faires** (shär-zhā′ də-fâr′) *n., pl.* **char·gés d'affaires** (-zhā′, -zhāz′) [Fr. : *chargé,* charged + *d'affaires,* with affairs.] **1.** A governmental official temporarily put in charge of diplomatic affairs in the absence of the ambassador or minister. **2.** The lowest-ranking diplomatic representative accredited by a government to the minister of foreign affairs of another.

**charge density** *n.* The electric charge per unit area or per unit volume of a body or of a region of space.

**charge nurse** *n.* A nurse in charge of a hospital or nursing-home ward.

**charg·er¹** (chär′jər) *n.* **1.** One that charges. **2.** A horse trained for battle. **3.** An instrument that recharges storage batteries.

**char·ger²** (chär′jər) *n.* [ME *chargeour* < OFr. < *chargier,* to load. —see CHARGE.] *Archaic.* A large shallow dish : PLATTER.

**char·i·ness** (châr′ē-nĭs) *n.* **1.** The quality of being chary : FRUGALITY. **2.** *Obs.* Strict integrity.

**char·i·ot** (chăr′ē-ət) *n.* [ME < OFr. *charriote* < *char,* cart < Lat. *carras,* of Celtic orig.] **1.** An ancient horse-drawn two-wheeled vehicle used in war, races, and processions. **2.** A light four-wheeled carriage used for ceremonial occasions or for pleasure. —*vt.* & *vi.* **-ot·ed, -ot·ing, -ots.** To transport or ride in a chariot.

**char·i·o·teer** (chăr′ē-ə-tîr′) *n.* **1.** A driver of a chariot. **2. Charioteer.** Auriga.

**cha·ris·ma** (kə-rĭz′mə) *also* **char·ism** (kăr′ĭz′əm) *n., pl.* **-ma·ta** (-mə-tə) *also* **-isms.** [Gk. *kharisma,* divine gift < *kharizesthai,* to favor < *kharis,* favor.] **1. a.** The power or quality of winning the devotion of large numbers of people. **b.** Great personal magnetism : CHARM. **2.** A divinely inspired gift or power, as the ability to perform miracles.

**char·is·mat·ic** (kăr′ĭz-măt′ĭk) *adj.* **1.** Of, relating to, or marked by charisma. **2.** Of, relating to, or being a Christian religious movement emphasizing divinely inspired powers or gifts, as of healing or prophecy. —*n.* A member of a charismatic group or movement.

**char·i·ta·ble** (chăr′ĭ-tə-bəl) *adj.* **1.** Generous in giving financial or other aid to the needy. **2.** Mild or tolerant in judging others : LENIENT. **3.** Of, for, or concerned with charity. **—char′i·ta·ble·ness** *n.* **—char′i·ta·bly** *adv.*

**char·i·ty** (chăr′ĭ-tē) *n., pl.* **-ties.** [ME *charite* < OFr., Christian love < Lat. *caritas,* affection < *carus,* dear.] **1.** Provision of aid to the poor : ALMSGIVING. **2.** Something given to help the needy : ALMS. **3.** An institution, organization, or fund set up to help the needy. **4.** An act or feeling of benevolence, good will, or affection. **5.** Forbearance in judging others : LENIENCY. **6. a.** God's benevolence toward people. **b.** Brotherly love.

**cha·ri·va·ri** *also* **chiv·a·ree** (shĭv′ə-rē′, shĭv′ə-rē′) *n., pl.* **-ris** *also* **-rees.** [Fr. < LLat. *caribaria,* headache < Gk. *karēbaria* : *karē,* head + *barus,* heavy.] A noisy mock serenade to newlyweds.

**char·kha** *also* **char·ka** (chûr′kə, chär′-) *n.* [Hindi *carkha.*] A spinning wheel, esp. one used in India for cotton.

**char·la·tan** (shär′lə-tən) *n.* [Fr. < Ital. *ciarlatano,* perh. < *ciarlare,* to prattle.] One who is not what he or she claims to be : QUACK. **—char′la·tan′ic** (-tăn′ĭk), **char′la·tan′i·cal** *adj.* **—char′la·tan·ism, char′la·tan·ry** *n.*

**Charles's law** (chärl′zĭz) *n.* [After Jacques *Charles* (1746–1823), its formulator.] The physical law that the volume of a fixed mass of gas held at a constant pressure varies directly with the absolute temperature.

**Charles's Wain** (wān) *n.* [After *Charlemagne* (742–814).] The Big Dipper.

**Charles·ton** (chärl′stən) *n.* [After *Charleston,* South Carolina.] A fast dance in 4/4 time, popular during the 1920's.

**char·ley horse** (chär′lē) *n.* [Orig. unknown.] *Informal.* A muscle cramp esp. in the arm or leg, caused by injury or excessive exertion.

**char·lock** (chär′lək, -lŏk′) *n.* [ME *cherlok* < OE *cerlic.*] A weedy Eurasian plant, *Brassica kaber,* with hairy stems and yellow flowers.

**char·lotte** (shär′lət) *n.* [Fr.< *Charlotte,* Charlotte.] A mold of sponge cake or bread with a filling, as of fruits, whipped cream, or custard, and served as a dessert.

**charlotte russe** (ro͞os′) *n.* [Fr. : *charlotte,* charlotte + *russe,* Russian.] A cold dessert of Bavarian cream set in a mold lined with ladyfingers.

**charm** (chärm) *n.* [ME *charme,* magic spell < OFr. < Lat. *carmen,* incantation.] **1.** The power or quality of attracting or fascinating. **2.** A fascinating or attractive feature or quality. **3.** A small ornament worn on another piece of jewelry, as a bracelet. **4.** A small object worn for its supposed magical effect, as in warding off evil : AMULET. **5.** An act or formula held to have magical power. **6.** Incantation of a magic word or verse. **7.** *Physics.* A quantum property of one of the quarks whose conservation explains the absence of certain strange-particle decay modes and that accounts for the longevity of the J particle. —*v.* **charmed, charm·ing, charms.** —*vt.* **1.** To attract or delight greatly or irresistibly : FASCINATE. **2.** To act on with or as if with magic : BEWITCH. —*vi.* **1.** To be alluring. **2.** To act as an amulet or charm. **3.** To employ spells. **—charm′ing·ly** *adv.*

⭑ ***syns:*** CHARM, BEWITCH, ENCHANT, ENTHRALL, ENTRANCE, SPELLBIND, WITCH *v. core meaning* : to act on with or as if with magic <*charmed* the authorities into waiving the regulations>

**charm·er** (chär′mər) *n.* **1.** A person, esp. an attractive woman, who charms or has the power to charm. **2.** A sorcerer.

**char·mo·ni·um** (chär-mō′nē-əm) *n.* [< CHARM.] Any of various elementary particles having a charm quark and antiquark.

**char·nel** (chär′nəl) *n.* [ME < OFr. < LLat. *carnale* < Lat. *caro,* flesh.] A charnel house. —*adj.* Resembling, suggesting, or suitable for receiving the dead.

**charnel house** *n.* A building, room, or vault in which the bones or bodies of the dead are placed.

**Char·on** (kâr′ən) *n.* [Lat. < Gk. *Kharōn.*] *Gk. Myth.* The ferryman who conveyed the dead to Hades over the river Styx.

**char·qui** (chär′kē) *n.* [Sp. < Quechua *ch'arki.*] JERKY².

**charr** (chär) *n. var. of* CHAR².

**chart** (chärt) *n.* [OFr. *charte* < Lat. *charta,* leaf of papyrus. —see CARD¹.] **1.** A map showing coastlines, water depths, or other data useful to navigators. **2.** An outline map on which special data, as weather information, can be plotted. **3.** A sheet presenting data in graph or tabular form. **4.** GRAPH 2. —*vt.* **chart·ed, chart·ing, charts. 1.** To make a chart of. **2.** To plan in detail.

**char·ta·ceous** (kär-tā′shəs) *adj.* [LLat. *chartaceus* < Lat. *chata,* leaf of papyrus. —see CARD¹.] Like paper : PAPERY.

**char·ter** (chär′tər) *n.* [ME *chartre* < OFr. < Lat. *chartula,* dim. of *charta,* leaf of papyrus. —see CARD¹.] **1.** A document issued by a monarch, legislative body, or other authority, creating a public or private corporation, as a city, college, or bank, and delineating its privileges and purposes. **2.** A written grant from the sovereign power of a country conferring certain rights and privileges on a person, a corporation, or the populace. **3.** A document setting forth the principles, functions, and organization of a corporate body. **4.** An authorization from a central organization to establish a local branch or chapter. **5.** Special privilege or immunity. **6.** A contract for the commercial leasing of a vessel or space on a vessel. **7.** The hiring or leasing of an aircraft, vessel, or transport vehicle. **8.** A written instrument given as evidence of agreement, transfer, or contract : DEED. —*vt.* **-tered, -ter·ing, -ters. 1.** To grant a charter to : establish by charter. **2.** To engage the temporary use of, for a fee <*charter* a plane> **—char′ter·er** *n.*

**chartered accountant** *n. Chiefly Brit.* A member of one of the institutes of accountants granted a royal charter.

**char·ter·house** (chär′tər-hous′) *n.* [By folk ety. < OFr. *chartrouse.* —see CHARTREUSE.] A Carthusian monastery.

**charter member** *n.* An original member or founder of an organization or group.

**Chart·ism** (chär′tĭz′əm) *n.* The principles and practices of a 19th-cent. party of social and political reformers, chiefly workingmen, active in England.

**chart·ist** (chär′tĭst) *n.* **1.** A stock-market specialist who uses charts and graphic records to interpret market action, predict trends, or forecast price movements of individual stocks. **2. Chartist.** An advocate of Chartism.

**char·treuse** (shär-tro͞oz′, -tro͞os′, -trœz′) *n.* [< CHARTREUSE.] A strong to brilliant greenish yellow to moderate or strong yellow green.

**Chartreuse.** A trademark for a green or yellow liqueur.

**char·treu·sin** (shär-tro͞o′zĭn, -sĭn) *n.* [NLat. *chartreusis,* specific epithet of *Streptomyces chartreusis* + -IN.] An antibiotic, $C_{18}H_{18}O_{18}$, effective against some Gram-positive microorganisms and derived from a strain of *Streptomyces chartreusis.*

**char·tu·lar·y** (kär′chə-lĕr′ē) *n. var. of* CARTULARY.

**char·wom·an** (chär′wo͝om′ən) *n. Chiefly Brit.* A woman hired to do cleaning usu. in a large building.

**char·y** (châr′ē) *adj.* **-i·er, -i·est.** [ME *chari,* diligent < OE *cearig,* sorrowful.] **1.** Very cautious : WARY. **2.** Shy <*chary* of meeting new people> **3.** Careful in the use of financial or material resources. **—char′i·ly** *adv.*

**Cha·ryb·dis** (kə-rĭb′dĭs) *n.* [Lat. < Gk. *Kharubdis.*] *Gk. Myth.* A whirlpool off the Sicilian coast, opposite the cave of Scylla.

**chase¹** (chās) *v.* **chased, chas·ing, chas·es.** [ME *chacen,* to hunt < OFr. *chacier* < Lat. *captare.* —see CATCH.] —*vt.* **1.** To follow in order to catch or overtake : PURSUE. **2.** To follow (game) so as to capture or kill. **3.** To follow earnestly or regularly. **4.** To put to flight <*chased* the dog away> —*vi.* **1.** To go in pursuit. **2.** *Informal.* To go hurriedly <*chased* around looking for us> —*n.* **1.** The act of chasing : PURSUIT. **2. a. the chase.** The sport of hunting. **b.** QUARRY¹. **3. a.** An unenclosed game preserve. **b.** The right to hunt or keep game within certain boundaries of land.

**chase²** (chās) *n.* [Perh. < Fr. *chasse,* case < Lat. *capsa.*] A rectangular steel or iron frame into which pages or columns of type are locked for printing or plate-making.

**chase³** (chās) *n.* [Perh. < OFr. *chas,* enclosure < Lat. *capsa,* box.] **1. a.** A groove cut in an object : SLOT. **b.** A trench or channel for drainpipes or wiring. **c.** A longitudinal groove for a tenon or tongue. **2.** The part of a gun in front of the trunnions. —*vt.* **chased, chas·ing, chas·es. 1.** To decorate (metal) by engraving or embossing. **2. a.** To groove : indent. **b.** To cut (the thread of a screw).

**chas·er¹** (chā′sər) *n.* **1.** One that pursues or chases. **2.** *Informal.* A drink, as of beer, taken after a drink of hard liquor.

**chas·er²** (chā′sər) *n.* **1.** One who decorates metal by engraving or embossing. **2.** A steel tool for cutting or finishing screw threads.

**chasm** (kăz′əm) *n.* [Lat. *chasma* < Gk. *khasma.*] **1.** A deep crack in the earth's surface. **2.** A sudden interruption of continuity. **3.** A pronounced difference of opinion, interests, or loyalty. **—chas′mal** (kăz′məl) *adj.*
**chas·sé** (shă-sā′) *n.* [Fr. < *chasser,* to chase < OFr. *chacier.* —see CHASE[1].] A dance movement consisting of one or more quick, gliding steps with the same foot always leading. —*vi.* **-séd, -sé·ing, -sés.** To make or perform a chassé.
**chasse·pot** (shăs′pō′) *n.* [Fr., after Antoine *Chassepot* (1833–1905), its inventor.] A breech-loading rifle introduced into the French army in 1866 that fires a paper cartridge.
**chas·seur** (shă-sûr′) *n.* [Fr. < OFr. *chaceor* < *chacier,* to pursue. —see CHASE[1].] **1.** A soldier, esp. one of certain light cavalry or infantry troops trained for rapid maneuvers. **2.** A huntsman. **3.** A uniformed footman.
**Chas·sid** (KHä′sĭd) *n., pl.* **Chas·si·dim** (KHä-sē′dĭm) [Heb. *hasīdh,* pious.] A member of a mystical Jewish sect founded in Poland about 1750 in opposition to the formalistic Judaism of the period and to ritual laxity. **—Chas·si′dic** *adj.* **—Chas·si′dism** *n.*
**chas·sis** (shăs′ē, chăs′ē) *n., pl.* **chas·sis** (-ēz) [Fr. *châssis* < OFr. < VLat. **capsicium* < Lat. *capsa,* box.] **1.** The rectangular steel frame, supported on springs and attached to the axles, that holds the body and engine of an automotive vehicle. **2.** The landing gear of an aircraft, including the wheels, floats, and other structures that support the aircraft on land or water. **3.** The frame on which a casement gun carriage moves forward and backward. **4.** The framework to which the functioning parts of a radio, television, etc., are attached.
**chaste** (chāst) *adj.* **chast·er, chast·est.** [ME < OFr. < Lat. *castus.*] **1.** Morally pure. **2. a.** Not having experienced sexual intercourse : VIRGINAL. **b.** Abstaining from unlawful sexual intercourse. **c.** Abstaining from sexual intercourse : CELIBATE. **3.** Simple in design or style. **—chaste′ly** *adv.* **—chaste′ness** *n.*
☆ ***syns:*** CHASTE, DECENT, MODEST, PURE, VIRTUOUS *adj. core meaning* : morally beyond reproach <a *chaste* nun>
**chas·ten** (chā′sən) *vt.* **-tened, -ten·ing, -tens.** [Alteration of obs. *chaste* < ME *chasten* < OFr. *chastiier, chastiss-* < Lat. *castigare* < *castus,* pure.] **1.** To chastise : punish. **2.** To restrain : moderate. **3.** To refine : purify <*chasten* one's speaking style> **—chas′ten·er** *n.*
**chas·tise** (chăs-tīz′) *vt.* **-tised, -tis·ing, -tis·es.** [ME *chastisen* < OFr. *chastiier, chastiss-,* to chasten.] **1.** To punish, usu. by beating. **2.** To criticize. **3.** *Archaic.* To purify. **—chas·tis′a·ble** *adj.* **—chas·tise′ment** (chăs-tīz′mənt, chăs′tĭz-) *n.* **—chas·tis′er** *n.*
**chas·ti·ty** (chăs′tĭ-tē) *n.* [ME *chastite* < OFr. *chastete* < Lat. *castitas* < *castus,* pure.] **1.** The quality or state of being chaste. **2. a.** Virginity. **b.** Virtuousness. **c.** Celibacy.
**chastity belt** *n.* A device worn in medieval times by women to prevent sexual intercourse.
**chas·u·ble** (chăz′ə-bəl, chăzh′ə-, chăs′ə-) *n.* [Fr. < OFr. < LLat. *casubla,* hooded garment.] A long, sleeveless vestment worn over the alb by the celebrant at Mass.
**chat** (chăt) *vi.* **chat·ted, chat·ting, chats.** [ME *chatten,* to jabber, short for *chateren.*] To converse in an informal or familiar way. —*n.* **1.** An informal or familiar conversation. **2.** A bird known for it's chattering call, as of the genera *Saxicola* or *Icteria.*
**cha·teau** *also* **châ·teau** (shă-tō′) *n., pl.* **-teaux** (-tōz′) [Fr. *château* < OFr. *chastel* < Lat. *castellum,* castle. —see CASTLE.] **1.** A French castle or manor house. **2.** A large country house.
**Châ·teau·bri·and** *also* **châ·teau·bri·and** (shă-tō′-brē-äɴ′) *n.* [After Vicomte de *Châteaubriand,* François René (1768–1848).] A double-thick tender center cut of beef tenderloin that is usu. broiled or grilled and served with a sauce.
**chat·e·lain** (shăt′l-ān′) *n.* [ME *chatelein* < OFr. *chastelain* < Lat. *castellanus* < *castellum,* castle. —see CASTLE.] The keeper of a castle.
**chat·e·laine** (shăt′l-ān′) *n.* [Fr. *châtelaine,* fem. of *châtelain,* chatelain < OFr. *chastelain.*] **1. a.** The mistress of a castle or chateau. **b.** The mistress of a large, fashionable household. **2.** A clasp or chain worn at the waist for holding keys, a purse, or a watch.
**cha·toy·ant** (shə-toi′ənt) *adj.* [Fr., pr.part. of *chatoyer,* to shimmer < *chat,* cat.] Having a changeable luster. —*n.* A chatoyant stone or gemstone, as the cat's-eye. **—cha·toy′an·cy** (-ən-sē) *n.*
**chat·tel** (chăt′l) *n.* [ME *chatel,* property < OFr. < Med. Lat. *capitale.* —see CATTLE.] **1.** An article of movable, personal property. **2.** A slave.
**chattel mortgage** *n.* A mortgage on personal property as security for an obligation or debt.
**chat·ter** (chăt′ər) *v.* **-tered, -ter·ing, -ters.** [ME *chateren.*] —*vi.* **1.** To utter a rapid series of short, inarticulate sounds <birds *chattering*> **2.** To talk rapidly, incessantly, and about trivia : JABBER. **3.** To click quickly and repeatedly <teeth *chattering*><machine guns *chattering*> **4.** To vibrate or rattle while in operation, as a power tool. —*vt.* To utter rapidly and incessantly. —*n.* **1.** Idle, incessant talk. **2.** Sharp, rapid sounds made by a bird or animal. **3.** A series of quick rattling or clicking sounds. **—chat′ter·er** *n.*
**chat·ter·box** (chăt′ər-bŏks′) *n.* A very talkative person.
**chatter mark** *also* **chat·ter·mark** (chăt′ər-märk′) *n.* **1.** A riblike marking on wood or metal, caused by vibration of a cutting tool. **2.** *Geol.* One of a series of short scars on a glaciated rock surface.
**chat·ty** (chăt′ē) *adj.* **-ti·er, -ti·est. 1.** Given to informal conversation <a *chatty* neighbor> **2.** Informal : familiar <a *chatty* letter> **—chat′ti·ly** *adv.* **—chat′ti·ness** *n.*
**chaud·froid** (shō-frwä′) *n.* [Fr. : *chaud,* hot + *froid,* cold.] **1.** A jellied white or brown sauce used as an aspic for cold meats or fish. **2.** Molded cold meat or fish dishes garnished with chaudfroid.
**chauf·feur** (shō′fər, shō-fûr′) *n.* [Fr., stoker < *chauffer,* to heat < OFr. *chaufer.* —see CHAFE.] One employed to drive a private car. —*v.* **-feured, -feur·ing, -feurs.** —*vt.* **1.** To serve as a driver for (someone). **2.** To convey as a chauffeur does <*chauffeured* our house guests around the city> —*vi.* To serve as a chauffeur.
**chaul·moo·gra** (chôl-mōō′grə) *n.* [Bengali *cāulmugrā* : *cāul,* rice + *mugrā,* hemp.] A tree of tropical Asia, esp. *Taraktogenos kurzii* and any one of the genus *Hydnocarpus,* having seeds that yield an oil used in treating leprosy.
**chaunt** (chônt, chänt) *n.* & *v. Archaic. var. of* CHANT.
**chausses** (shōs) *pl.n.* [ME *chauces* < OFr. < Lat. *calceus,* shoe.] Medieval armor for the legs and feet, made of mail.

**chausses**

**chau·vin·ism** (shō′və-nĭz′əm) *n.* [Fr. *chauvinisme,* after Nicholas *Chauvin,* a legendary French soldier.] **1.** Militant devotion to and glorification of one's country. **2.** Prejudiced belief in the superiority of one's own group <male *chauvinism*> **—chau′vin·ist** *n.* **—chau′vin·is′tic** (-nĭs′tĭk) *adj.* **—chau′vin·is′ti·cal·ly** *adv.*
**†chaw** (chô) —*vi.* & *vt.* [Var. of CHEW.] *Regional.* **chawed, chaw·ing, chaws.** To chew. **—chaw** *n.*
**cha·yo·te** (chä-yō′tā) *n.* [Sp. < Nahuatl *chayotli.*] **1.** A tropical American vine, *Sechium edule,* bearing edible squashlike fruit. **2.** The fruit of the chayote.
**cha·zan** *or* **chaz·zen** (KHä′zən) *n.* [Heb. *ḥazzān.*] CANTOR 1.
**cheap** (chēp) *adj.* **-er, -est.** [< ME *chep,* price, purchase < OE *cēap,* trade < Lat. *caupo,* tradesman.] **1.** Relatively low in cost : INEXPENSIVE. **2.** Charging low prices. **3.** Requiring little effort <a *cheap* victory> **4.** Of or considered of small value <Life was very *cheap.*> **5.** Being of poor quality : INFERIOR <*cheap* rhinestone jewelry> **6.** Undeserving of respect : VULGAR <*cheap* jokes> **7.** Stingy : miserly <was too *cheap* to pay the employees well> **8. a.** Obtainable at a low rate of interest. **b.** Devalued, as in buying power <*cheap* dollars> —*adv.* Inexpensively. **—cheap′ly** *adv.* **—cheap′ness** *n.*
☆ ***syns:*** **1.** CHEAP, INEXPENSIVE, LOW *adj. core meaning* : not having a high price <*cheap* vegetables> ***ant:*** costly, expensive **2.** CHEAP, CHEESY, POOR, SCHLOCKY, SHODDY *adj. core meaning* : of decidedly inferior quality <*cheap* merchandise> ***ant:*** precious
**cheap·en** (chē′pən) *v.* **-ened, -en·ing, -ens.** —*vt.* **1.** To make cheap or cheaper. **2.** To deprecate or belittle. —*vi.* To become cheap or cheaper. **—cheap′en·er** *n.*
**cheap shot** *n.* An unjust action or statement directed esp. at a vulnerable target, as a public figure.
**cheap·skate** *also* **cheap skate** (chēp′skāt′) *n. Slang.* A miser.
**cheat** (chēt) *v.* **cheat·ed, cheat·ing, cheats.** [ME *cheten,* to confiscate, short for *acheten,* var. of *escheten* < *eschete,* escheat. —see ESCHEAT.] —*vt.* **1.** To deceive by trickery : SWINDLE. **2.** To mislead. **3.** To deprive by trickery : DEFRAUD <*cheated* the Indians of their land> **4.** To elude <*cheat* death> —*vi.* **1.** To practice fraud. **2.** *Informal.* To be sexually unfaithful. —*n.* **1.** A swindle or fraud. **2.** A swindler. **3.** *Law.* Fraudulent acquisition of another's property. **4.** A grass, *Bromus secalinus,* with rough blades and wheatlike ears. **—cheat′er** *n.* **—cheat′ing·ly** *adv.*
☆ ***syns:*** CHEAT, BILK, COZEN, DEFRAUD, FLIMFLAM, GULL, GYP, MULCT, ROOK, SWINDLE, TAKE, VICTIMIZE *v. core meaning* : to get money or something else from by deceitful trickery <*cheated* them out of their inheritance>
**check** (chĕk) *n.* [ME *chek,* check in chess < OFr. *eschec* < Ar. *shāh* < Pers., king, king in chess.] **1.** An abrupt halt. **2.** A restraint or control <keep a *check* on one's emotions> **3.** Supervised control, as of efficiency or accuracy. **4.** A standard of comparison to verify accuracy : TEST. **5.** A mark to show verification or approval. **6.** A ticket or identification slip <a baggage *check*> **7.** A bill at an eating establishment or bar. **8.** A chip or counter used in gambling games. **9.** A written order to a bank to pay the amount specified from funds on deposit : DRAFT. **10. a.** A pattern of small squares, as on a chessboard.

**b.** One of the squares of such a pattern. **c.** A fabric patterned with small squares. **11.** A small crack. **12. a.** A move in chess that directly attacks one's opponent's king but is not a checkmate. **b.** The position or tactical condition of a king so attacked. **13.** The act of impeding an opponent in control of the puck in ice hockey, either by blocking progress with the body or by jabbing at the puck with the stick. —*interj.* **1.** —Used to declare to a chess opponent that his or her king is in check. **2.** *Informal.* —Used to express agreement or understanding. —*v.* **checked, check·ing, checks.** —*vt.* **1.** To arrest the motion of abruptly : HALT. **2.** To restrain : curb. **3.** To slow the growth of : RETARD. **4.** To rebuke or rebuff. **5.** To test or examine, as for accuracy or efficiency : VERIFY. **6.** To put a check mark on or next to <*checked* off each item> **7.** To deposit for temporary safekeeping <*check* your coat> **8.** To make cracks or chinks in. **9.** To move in chess so as to put (an opponent's king) under direct attack. **10.** To impede an opponent in control of the puck in ice hockey, either by using the body to block the opponent or by jabbing at the puck with the hockey stick. —*vi.* **1.** To come to a sudden halt : STOP. **2.** To have item-for-item correspondence : AGREE <All the figures *check.*> **3.** To make an investigation to ascertain accuracy <*check* on an arrival time> **4.** To write a check on a bank account. **5.** To crack in a pattern of checks, as paint. **6.** To pause to relocate a scent. —Used of hunting dogs. **7.** To place a chess opponent's king in check. **8.** To abandon the proper game and follow baser prey in falconry. **—check in.** To register, as at a hotel. **—check out. 1.** To leave after going through a required procedure, as after paying a hotel bill. **2.** To take possession after a recording procedure <*check out* a book from the library> **—in check.** Under restraint <kept my emotions *in check*> **—check'a·ble** *adj.*

**check·book** (chĕk'bo͝ok') *n.* A book that contains blank checks issued by a bank.

**checked** (chĕkt) *adj.* **1.** Patterned with checks. **2.** Held in check : RESTRAINED. **3.** Included in a stopped or closed syllable <a *checked* vowel>

**check·er** (chĕk'ər) *n.* [ME *cheker,* chessboard < OFr. *eschequier* < *eschec,* check in chess. —see CHECK.] **1. a. checkers** *(sing. in number).* A game played on a checkerboard by 2 players, each using 12 pieces. **b.** One of the round flat pieces used in this game. **2. a.** A checked pattern. **b.** One of the checks in such a pattern. **3.** One who checks, examines, or supervises. **4.** One who receives items for temporary storage or safekeeping <a coat *checker*> **5.** A cashier. —*vt.* **-ered, -er·ing, -ers. 1.** To mark with a checked pattern. **2.** To diversify in color, shading, or character : VARIEGATE.

**check·er·ber·ry** (chĕk'ər-bĕr'ē) *n.* [CHECKER, the wild service tree (dial.) + BERRY.] The wintergreen or its red, edible, spicy berry.

**check·er·bloom** (chĕk'ər-blo͞om') *n.* [CHECKER, the wild service tree (dial.) + BLOOM.] A plant, *Sidalcea malvaeflora* of California, with long rose-pink flower clusters.

**check·er·board** (chĕk'ər-bôrd', -bōrd') *n.* A game board that is divided into 64 squares of 2 alternating colors on which chess and checkers are played.

**check·ered** (chĕk'ərd) *adj.* **1.** Divided into squares. **2.** Marked by light and dark patches. **3.** Marked by great changes or shifts in fortune, some of them negative <a *checkered* acting career>

**checking account** *n.* A bank account in which checks may be written against deposits.

**check list** *n.* A list in which items can be compared, scheduled, verified, or identified.

**check·mate** (chĕk'māt') *vt.* **-mat·ed, -mat·ing, -mates.** [ME *chekmat* < OFr. *eschec mat* < Ar. *shāh māt,* the king is dead < Pers.] **1.** To attack (a chess opponent's king) in such a way that no escape or defense is possible, thus ending the game. **2.** To defeat decisively. —*n.* **1. a.** A move that constitutes an inescapable and indefensible attack on a chess opponent's king. **b.** The position or strategic condition of a king so attacked. **2.** Decisive defeat. —*interj.* —Used to declare the checkmate of an opponent's king in chess.

**check·off** (chĕk'ôf', -ŏf') *n.* Collection of dues from union members by authorized deductions from their wages.

**check·out** (chĕk'out') *n.* **1.** The act, time, or place of checking out, as at a hotel, library, or supermarket. **2.** A test, as of a machine, for proper functioning. **3.** An investigation or inspection.

**check·point** (chĕk'point') *n.* A place, as between two international borders, where surface traffic is stopped for inspection.

**check·rein** (chĕk'rān') *n.* **1.** A short rein connected from a horse's bit to the saddle to keep the horse from lowering its head. **2.** A rein joining the bit of one of a span of horses to the driving rein of the other horse.

**check·room** (chĕk'ro͞om', -ro͝om') *n.* A room in which items, as coats, can be stored temporarily.

**check·row** (chĕk'rō') *n.* A row, as of corn, in which the distance between plants is the same as the distance between adjacent rows to allow cross-cultivation. —*vt.* **-rowed, -row·ing, -rows.** To plant in checkrows.

**checks and balances** *pl.n.* Maintenance of the balance of power between various branches of a government.

**check·up** (chĕk'ŭp') *n.* **1.** A thorough examination, as for verification. **2.** A physical examination.

**Ched·dar** *also* **ched·dar** (chĕd'ər) *n.* [After *Cheddar,* England.] A type of smooth, hard cheese varying from mild to extra sharp in flavor.

**cheek** (chēk) *n.* [ME *cheke* < OE *cēace.*] **1.** The fleshy part of either side of the face beneath the eye and between the nose and ear. **2.** Something like the cheek in position or shape. **3.** Either of the buttocks. **4.** Arrogant self-confidence. —*vt.* **cheeked, cheek·ing, cheeks.** *Informal.* To speak impudently to. **—cheek by jowl.** Side by side and in intimate contact.

**cheek·bone** (chēk'bōn') *n.* The zygomatic bone.

**cheek pouch** *n.* An enlargement in the cheeks of some rodents that functions as a means of carrying food.

**cheek·y** (chē'kē) *adj.* **-i·er, -i·est.** Saucy or impudent : BRAZEN. **—cheek'i·ly** *adv.* **—cheek'i·ness** *n.*

**cheep** (chēp) *n.* [Imit.] A faint, shrill sound like that of a young bird : CHIRP. —*vi.* **cheeped, cheep·ing, cheeps.** To utter a faint, shrill sound : CHIRP. **—cheep'er** *n.*

**cheer** (chîr) *n.* [ME *chere,* mood < OFr. *chiere,* face < LLat. *cara* < Gk. *kara,* head.] **1.** Gaiety or happiness : JOY. **2.** Something providing gaiety or happiness. **3.** A shout of approval, encouragement, or congratulation. **4.** Food or drink : REFRESHMENT <a cup of *cheer*> —*interj.* **cheers.** —Used as a toast. —*v.* **cheered, cheer·ing, cheers.** —*vt.* **1.** To make cheerful or more cheerful. **2.** To encourage with or as if with cheers. **3.** To salute or acclaim with cheers. —*vi.* **1.** To shout cheers. **2.** To become cheerful <had a good rest and soon *cheered* up> **—cheer'er** *n.* **—cheer'ing·ly** *adv.*

**cheer·ful** (chîr'fəl) *adj.* **1. a.** Being in good spirits : HAPPY. **b.** Exhibiting good spirits. **2.** Producing good spirits. **3.** Willing : good-humored <*cheerful* assistance> **—cheer'ful·ly** *adv.* **—cheer'ful·ness** *n.*

☆ ***syns:*** CHEERFUL, BRIGHT, CHEERY, CHIPPER, HAPPY, LIGHTHEARTED, SUNNY *adj. core meaning* : being in or displaying good spirits <a *cheerful* person><a *cheerful* smile> ***ant:*** cheerless

**cheer·i·o** (chîr'ē-ō') *interj.* [< CHEER.] *Chiefly Brit.* —Used in greeting or parting.

**cheer·lead·er** (chîr'lē'dər) *n.* One who leads the cheering of fans and spectators, as at a football game.

**cheer·less** (chîr'lĭs) *adj.* Devoid of cheer. **—cheer'less·ly** *adv.* **—cheer'less·ness** *n.*

**cheer·y** (chîr'ē) *adj.* **-i·er, -i·est.** Cheerful : bright. **—cheer'i·ly** *adv.* **—cheer'i·ness** *n.*

**cheese¹** (chēz) *n.* [ME *chese* < OE *cyse* < Lat. *caseus.*] **1. a.** A solid food made from the pressed curd of milk, often seasoned and aged. **b.** A molded mass of this substance. **2.** Something like cheese. **3. cheeses. a.** The common mallow, *Malva rotundifolia,* a creeping plant with pale-lavender or white flowers and flat, round, ridged fruits. **b.** The fruit of the cheese plant.

**cheese²** (chēz) *vt.* **cheesed, chees·ing, chees·es.** [Orig. unknown.] *Slang.* To stop. **—cheese it.** *Slang.* —Used imperatively as a warning to watch out or get away fast.

**cheese³** (chēz) *n.* [Perh. < Urdu *chīz,* thing < Pers.] *Slang.* An important person.

**cheese·burg·er** (chēz'bûr'gər) *n.* A hamburger topped with melted cheese.

**cheese·cake** (chēz'kāk') *n.* **1.** *also* **cheese cake.** A cake made of sweetened cottage cheese or cream cheese, eggs, milk, sugar, and flavorings. **2.** *Slang.* **a.** A photograph of a scantily clothed girl. **b.** Such photographs as a whole.

**cheese·cloth** (chēz'klôth', -klŏth') *n.* A coarse, loosely woven cotton gauze, orig. used for wrapping cheese.

**cheese-par·ing** (chēz'pâr'ĭng) *n.* **1.** Something valueless. **2.** Stinginess. —*adj.* Miserly.

**chees·y** (chē'zē) *adj.* **-i·er, -i·est. 1. a.** Like cheese. **b.** Containing cheese. **2.** *Slang.* Of inferior quality. **—chees'i·ness** *n.*

**chee·tah** *also* **che·tah** (chē'tə) *n.* [Hindi *cītā* < Skt. *citrakāya,* tiger : *citra,* variegated + *kāyaḥ,* body.] A long-legged, swift-running wild cat, *Acinonyx jubatus* of Africa and southwestern Asia, that has black-spotted, tawny fur and nonretractile claws and is occas. trained to pursue game.

**chef** (shĕf) *n.* [Fr., short for *chef de cuisine,* head of the kitchen.] A cook, esp. the chief cook of a large kitchen staff.

**chef-d'oeu·vre** (shā-dœ'vrə, -dûrv') *n., pl.* **chefs-d'oeuvre** (-dœ'vrə, -dûrv', -dûrvz') [Fr. : *chef,* head + *d'oeuvre,* of work.] A literary or artistic masterpiece.

**chef's salad** *n.* A tossed green salad usu. including raw vegetables, hard-boiled eggs, and julienne strips of cheese and meat.

**chei·lo·sis** (kī-lō'sĭs) *n.* [Gk. *kheilos,* lip + -OSIS.] Inflammation and fissuring of the lips with cracking at the angles of the mouth due to a riboflavin deficiency.

**cheiro-** *pref. var. of* CHIRO-.

**chei·ro·plas·ty** *also* **chi·ro·plas·ty** (kī'rō-plăs'tē) *n.* [CHIRO- + -PLASTY.] Plastic surgery of the hand.

**che·la** (kē'lə) *n., pl.* **-lae** (-lē) [NLat. < Gk. *khēlē,* claw.] A pincerlike claw of a crustacean, as of a crab or lobster.

**che·late** (kē'lāt') *adj.* **1.** *Zool.* Having or typical of a chela. **2.** *Chem.* Of or relating to a heterocyclic ring containing a metal ion attached by coordinate bonds to at least two nonmetal ions in the same molecule. —*vt.* **-lat·ed, -lat·ing, -lates.** To form a ring compound by joining a chelating agent to a metal ion. **—che'late'** *n.* **—che·la'tion** *n.* **—che'la'tor** *n.*

**che·lic·er·a** (kĭ-lĭs'ər-ə) *n., pl.* **-erae** (-ə-rē') [NLat. : CHELA + Gk. *keras,* horn.] Either of the first pair of appendages near the mouth of an arachnid, as a spider, often modified for grasping.

**che·li·form** (kē'lə-fôrm') *adj.* Shaped like a chela : PINCERLIKE.

**Chel·li·an** *or* **Chel·le·an** (shĕl'ē-ən) *adj.* [After *Chelles,* a commune in France.] Abbevillian.

**che·lo·ni·an** (kĭ-lō'nē-ən) *adj.* [< NLat. *Chelonia,* order name < Gk. *khelōnē,* tortoise.] *Zool.* Of or belonging to the order Chelonia, including the turtles and tortoises. —*n.* A member of the Chelonia.

**chem-** *or* **chemi-** *pref. vars. of* CHEMO-.

**chem·ic** (kĕm'ĭk) *adj. Archaic.* **1.** Chemical. **2.** Alchemic.

**chem·i·cal** (kĕm'ĭ-kəl) *adj.* [Obs. *chimical* < *chimic,* alchemist < NLat. *chimicus* < Med. Lat. *alchimicus* < *alchymia,* alchemy. —see ALCHEMY.] **1.** Of or relating to chemistry. **2.** Of or relating to the properties or actions of chemicals. —*n.* A substance produced by or used in a chemical process. **—chem'i·cal·ly** *adv.*

**chemical bond** *n.* Any of several forces or mechanisms, esp. the ionic bond, covalent bond, and metallic bond, by which atoms or ions are bound in a molecule or crystal.

**chemical engineering** *n.* **1.** The profession or science involving application of chemistry to industrial uses. **2.** The technology of large-scale chemical and chemical materials production. **—chemical engineer** *n.*

**Chemical Mace.** A trademark for a mixture of organic chemicals used in aerosol form as a weapon to disable with intense burning eye pain, blepharospasm, acute bronchitis, and respiratory irritation.

**chemical warfare** *n.* Warfare using chemicals, as direct weapons, esp. irritants, asphyxiants, contaminants, poisons, defoliants, and incendiaries.

**chem·i·lu·mi·nes·cence** (kĕm'ə-lo͞o'mə-nĕs'əns) *n.* Emission of light resulting from a chemical reaction at environmental temperatures. **—chem'i·lu'mi·nes'cent** *adj.*

**che·min de fer** (shə-măN' də fâr') *n.* [Fr. : *chemin,* road + *de,* of + *fer,* iron.] A gambling game similar to baccarat.

**che·mise** (shə-mēz') *n.* [ME < OFr., shirt < LLat. *camisia.*] **1.** A woman's loose, shirtlike undergarment. **2.** SHIFT 7a.

**chem·i·sette** (shĕm'ĭ-zĕt') *n.* [Fr., dim. of *chemise,* shirt < OFr. —see CHEMISE.] **1.** A short, sleeveless bodice, worn at one time by women. **2.** A blouse front worn at one time by women.

**chem·i·sorb** (kĕm'ĭ-sôrb') *also* **chem·o·sorb** (-ə-sôrb') *vt.* **-sorbed, -sorb·ing, -sorbs.** [CHEMI- + (AB)SORB.] To take up and chemically bind (a substance) on the surface of another substance. **—chem'i·sorp'tion** (-sôrp'shən) *n.*

**chem·ist** (kĕm'ĭst) *n.* [Obs. *chimist* < NLat. *chimista* < Med. Lat. *alchymista,* alchemist < *alchymia,* alchemy. —see ALCHEMY.] **1.** A scientist specializing in chemistry. **2.** *Chiefly Brit.* A pharmacist. **3.** *Obs.* An alchemist.

**chem·is·try** (kĕm'ĭ-strē) *n., pl.* **-tries. 1.** The science of the composition, structure, properties, and reactions of matter, esp. of atomic and molecular systems. **2.** The composition, structure, properties, and reactions of a substance.

**chemo-** *or* **chemi-** *or* **chem-** *pref.* [NLat. < LGk. *khēmeia,* alchemy. —see ALCHEMY.] Chemicals : chemical <*chemo*therapy>

**che·mo·au·to·troph** (kē'mō-ô'tə-trŏf', -trôf', kĕm'ō-) *n.* [CHEMO- + AUTO- + Gk. *trophos,* feeder < *trephein,* to cause to grow.] An organism that obtains its nutritive substances through inorganic chemical oxidation. **—che'mo·au'to·tro'phic** (-trō'fĭk) *adj.* **—che'mo·au'to·tro'phi·cal·ly** *adv.* **—che'mo·au·tot'ro·phy** (-ô-tŏt'rə-fē) *n.*

**che·mo·pro·phy·lax·is** (kē'mō-prō'fə-lăk'sĭs, kĕm'ō-) *n.* Use of chemicals in the prevention of infectious disease. **—che'mo·pro'phy·lac'tic** (-tĭk) *adj.*

**che·mo·re·cep·tion** (kē'mō-rĭ-sĕp'shən, kĕm'ō-) *n.* Reaction of a sense organ to a chemical stimulus. **—che'mo·re·cep'tive** *adj.* **—che'mo·re'cep·tiv'i·ty** (-rē'sĕp-tĭv'ĭ-tē) *n.*

**che·mo·re·cep·tor** (kē'mō-rĭ-sĕp'tər, kĕm'ō-) *n.* A nerve ending or sense organ, as of smell or taste, sensitive to chemical stimuli.

**che·mo·sen·so·ry** (kē'mō-sĕn'sə-rē, kĕm'ō-) *adj.* Of or relating to the sensory reception of a chemical stimulus.

**chem·os·mo·sis** (kĕm'ŏz-mō'sĭs, -ŏs-) *n.* Chemical action between substances separated by a semipermeable membrane. **—chem'os·mot'ic** (-mŏt'ĭk) *adj.*

**chem·o·sorb** (kĕm'ə-sôrb') *v. var. of* CHEMISORB.

**che·mo·sphere** (kē'mə-sfîr', kĕm'ə-) *n.* The region of the atmosphere between 20 and 120 miles, or approx. 32 and 193 kilometers, altitude in which photochemical reactions initiated by solar radiation occur.

**che·mo·sur·ger·y** (kē'mō-sûr'jə-rē, kĕm'ō-) *n.* Selective destruction of tissue by use of chemicals.

**che·mo·syn·the·sis** (kē'mō-sĭn'thĭ-sĭs, kĕm'ō-) *n.* Synthesis of organic substances, as food nutrients, using the energy of chemical reactions. **—che'mo·syn·thet'ic** (-sĭn-thĕt'ĭk) *adj.* **—che'mo·syn·thet'i·cal·ly** *adv.*

**che·mo·tax·is** (kē'mō-tăk'sĭs, kĕm'ō-) *n.* Orientation or motion of a freely moving living organism relative to a chemical substance. **—che'mo·tac'tic** (-tăk'tĭk) *adj.* **—che'mo·tac'ti·cal·ly** *adv.*

**che·mo·tax·on·o·my** (kē'mō-tăk-sŏn'ə-mē, kĕm'ō-) *n.* Classification of organisms according to biochemical criteria. **—che'mo·tax'o·nom'ic** (-tăk'sə-nŏm'ĭk) *adj.* **—chemo·tax'o·nom'i·cal·ly** *adv.* **—che'mo·tax·on'o·mist** *n.*

**che·mo·ther·a·py** (kē'mō-thĕr'ə-pē, kĕm'ō-) *n.* Prevention or treatment of disease with chemical agents. **—che'mo·ther'a·peu'tic** (-pyo͞o'tĭk) *adj.*

**che·mot·ro·pism** (kĭ-mŏt'rə-pĭz'əm) *n.* Movement or growth of an organism, esp. a plant, in response to chemical stimuli. **—che'mo·trop'ic** (kē'mō-trŏp'ĭk, kĕm'ō-) *adj.*

**chem·ur·gy** (kĕm'ər-jē, kĭ-mûr'-) *n.* Development of new industrial chemical products from organic raw materials, esp. materials of agricultural origin. **—che·mur'gic** (kĭ-mûr'jĭk), **che·mur'gi·cal** *adj.*

**che·nille** (shə-nēl') *n.* [Fr. *chenille,* caterpillar < Lat. *canicula,* dim. of *canis,* dog.] **1.** A soft tufted silk, cotton, or worsted cord used in embroidery or for fringing. **2.** Fabric made of chenille.

**che·no·pod** (kē'nə-pŏd', kĕn'ə-) *n.* [NLat. *Chenopodiaceae,* family name < *Chenopodium,* type genus : Gk. *khēn,* goose + Gk. *podion,* dim. of *pous,* foot.] A plant of the goosefoot family, Chenopodiaceae, including spinach and beets as well as many common weeds.

**cheque** (chĕk) *n. Chiefly Brit. var. of* CHECK 9.

**cheq·uer** (chĕk'ər) *n. Chiefly Brit. var. of* CHECKER.

**cher·i·moy·a** (chĕr'ə-moi'ə) *n.* [Am. Sp. *chirimoya* < Quechua *chirimuya.*] **1.** A tropical American tree, *Annona cherimola,* bearing yellow flowers and edible fruit with white, soft, aromatic pulp. **2.** The fruit of the cherimoya.

**cher·ish** (chĕr'ĭsh) *vt.* **-ished, -ish·ing, -ish·es.** [ME *cherishen* < OFr. *cherir, cheriss-* < *cher,* dear < Lat. *carus.*] **1.** To hold dear and treat affectionately. **2.** To keep fondly in mind <*cherish* a childhood memory> **—cher'ish·er** *n.* **—cher'ish·ing·ly** *adv.*

**cher·no·zem** (chĕr'nə-zĕm', chĕr'nə-zhôm') *n.* [R., short for *chĕrnaya zemlya,* black earth.] A black topsoil, rich in humus, typical of cool to temperate semiarid regions, as the grasslands of European Russia.

**Cher·o·kee** (chĕr'ə-kē', chĕr'ə-kē') *n., pl.* **Cherokee** *or* **-kees.** [< Cherokee *tsalaki.*] **1. a.** A tribe of Indians once living in North Carolina and northern Georgia and now settled in Oklahoma. **b.** A member of this tribe. **2.** The Iroquoian language of the Cherokee. **—Cher'o·kee'** *adj.*

**Cherokee rose** *n.* A climbing rose, *Rosa laevigata* of Chinese origin, with large, white, fragrant flowers.

**che·root** (shə-ro͞ot', chə-) *n.* [Tamil *śurruṭṭu* < *śuruḷ,* curl.] A cigar whose ends are square-cut.

**cher·ry** (chĕr'ē) *n., pl.* **-ries.** [ME *cheri* < Norman Fr. *cherise,* var. of OFr. *cerise* < Med. Lat. *ceresia* < Lat. *cerasus* < Gk. *kerasos.*] **1. a.** A tree of the genus *Prunus,* bearing small globe-shaped or heart-shaped fruit with a small hard stone, esp. *P. avium,* the common sweet cherry, and *P. cerasus,* the sour cherry. **b.** The fruit or wood of the cherry tree. **2.** A moderate or strong red to purplish red.

**cherry birch** *n.* Blackberry.

**cherry laurel** *n.* A European shrub, *Prunus laurocerasus,* with evergreen foliage and white flowers.

**cherry leaf spot** *n.* A disease of the cherry caused by the fungus *Coccomyces hiemalis* that hinders the growth of the tree and causes spotting of the leaves.

**cherry pepper** *n.* The pungent red, yellow, or purplish fruit of a tropical plant, *Capsicum frutescens cerasiforme.*

**cherry picker** *n.* Any of various large, usu. mobile cranes with a long maneuverable vertical boom often supporting a work platform.

**cherry plum** *n.* MYROBALAN 1.

**cher·ry·stone** (chĕr'ē-stōn') *n.* The quahog clam when half-grown and comparatively small.

**cherry tomato** *n.* A common tomato, *Lycospermum esculentum cerasiforme,* with small red or yellow fruit.

**cher·so·nese** (kûr'sə-nēz', -nēs') *n.* [Lat. *chersonesus* < Gk. *khersonēsos* : *khersos,* dry land + *nēsos,* island.] A peninsula.

**chert** (chûrt) *n.* [Orig. unknown.] **1.** Any of various microscopically crystalline mineral varieties of silica. **2.** A siliceous rock of chalcedonic or opaline silica occurring in limestone.

**cher·ub** (chĕr'əb) *n., pl.* **cher·u·bim** (chĕr'ə-bĭm', -yə-bĭm') [Heb. *kărūbh.*] **1. a.** A winged celestial being. **b.** One of the second order of angels. **2.** A representation of an angelic cherub, portrayed as a winged child with a chubby, rosy face. **3.** *pl.* **cherubs.** A person, esp. a child, with an innocent chubby face. **—che·ru'bic** (chə-ro͞o'bĭk) *adj.* **—che·ru'bi·cal·ly** *adv.*

**cher·vil** (chûr'vəl) *n.* [ME *chervel* < OE *cerfille* < Lat. *chaerephylla* < Gk. *khairephullon.*] **1.** An aromatic Eurasian plant, *Anthriscus* *cerefolium,* having leaves used in soups and salads. **2.** Any of several related plants, esp. *Chaerophyllum bulbosum,* with an edible root.

**Ches·a·peake Bay retriever** (chĕs′ə-pēk′) *n.* A hunting dog orig. bred in the United States, having a thick, short, brownish coat.

**chesh·ire cheese** (chĕsh′ər) *n.* [After *Cheshire,* England.] A hard, yellow English cheese made from cow's milk.

**chess**[1] (chĕs) *n.* [ME *ches,* short for OFr. *esches,* pl. of *eschec,* check in chess. —see CHECK.] A board game for two players, each having an initial force of a king, a queen, two bishops, two knights, two rooks, and eight pawns, all maneuvered according to individual movement rules with the objective being the checkmate of the opposing king.

**chess**[2] (chĕs) *n.* [Orig. unknown.] A weedy grass, esp. cheat.

**chess**[3] (chĕs) *n., pl.* **chess** or **chess·es.** [ME *ches,* tier < OFr. *chasse,* frame < Lat. *capsa,* box.] One of the floorboards of a pontoon bridge.

**chess·board** (chĕs′bôrd′, -bōrd′) *n.* A board marked with 64 squares, used in playing chess.

**chess·man** (chĕs′măn′, -mən) *n.* A piece used in the game of chess.

**chest** (chĕst) *n.* [ME < OE *cest.*] **1.** The part of the body between the neck and the abdomen, enclosed by the ribs and the breastbone. **2. a.** A sturdy box with a lid and often a lock, used esp. for storage. **b.** A small shelved closet or cabinet used for storing supplies. **3. a.** A public institution's treasury. **b.** The funds kept there. **4. a.** A box for shipping goods, as tea. **b.** The quantity packed in such a shipping box. **5.** A sealed receptacle for liquid, gas, or steam. **6.** A bureau : dresser. **—chest′ed** (chĕs′tĭd) *adj.*

**ches·ter·field** (chĕs′tər-fēld′) *n.* [After a 19th-cent. Earl of *Chesterfield.*] **1.** A single-breasted or double-breasted overcoat, usu. with concealed buttons and a velvet collar. **2.** A large overstuffed sofa with upright armrests.

**Ches·ter White** (chĕs′tər) *n.* A white hog orig. bred in Chester County, Pennsylvania.

**chest·nut** (chĕs′nŭt′, -nət) *n.* [ME *chesteine* < OFr. *chastaigne* < Lat. *castanea* < Gk. *kastenea.*] **1. a.** Any of several trees of the genus *Castanea* of the Northern Hemisphere, bearing nuts enclosed in a prickly bur. **b.** The nut of any of these trees, edible when cooked. **c.** The hard wood of these trees, used in furniture and as a building material. **2.** The horse chestnut. **3.** A grayish brown to moderate reddish brown. **4.** A reddish-brown horse. **5.** A small hard callus on the inner surface of a horse's foreleg. **6. a.** A stale joke. **b.** Something, as a story, lacking freshness or originality. *—adj.* Of a grayish brown to moderate reddish brown.

**chestnut blight** *n.* A disease of the native American chestnut tree caused by a fungus, *Endothia parasitica,* and resulting in cankers on the trunk and branches and eventual death.

**chestnut oak** *n.* A tree, *Quercus prinus* of eastern and central North America, having leaves like those of the chestnut.

**chest·y** (chĕs′tē) *adj.* **-i·er, -i·est.** *Informal.* **1.** Having a large or well-developed chest. **2.** Arrogant : conceited. **—chest′i·ness** *n.*

**che·tah** (chē′tə) *n. var. of* CHEETAH.

**chet·rum** (chē′trəm, chĕt′rəm) *n.* [Native word in Bhutan.] —See table at CURRENCY.

**che·val-de-frise** (shə-văl′də-frēz′) *n., pl.* **che·vaux-de-frise** (shə-vō′-) [Fr., Frisian horse, from its use by Frisians to compensate for a lack of cavalry.] **1.** Jagged glass or a spiked obstacle set in the masonry on the top of a wall to prevent trespassing or escape. **2.** A spiked obstacle attached to a wooden frame, once used to hinder the advance of enemy cavalry.

**che·val·et** (shə-văl′ā, shĕv′ə-lā′) *n.* [Fr. < dim. of *cheval,* horse < Lat. *caballus.*] The bridge of a stringed musical instrument.

**che·val glass** (shə-văl′) *n.* [< Fr. *cheval,* support, horse. —see CHEVALET.] A long mirror mounted on swivels in a frame.

**chev·a·lier** (shĕv′ə-lîr′) *n.* [ME *chevaler* < OFr. *chevalier* < LLat. *caballarius,* horseman < *caballus,* horse.] **1.** A member of an order of knighthood or merit, as the French Legion of Honor. **2.** A French nobleman of the lowest rank in former times. **3.** *Archaic.* A knight. **4.** A chivalrous man.

**cheve·lure** (shəv-lür′) *n.* [ME *cheveler* < OFr. *cheveleure* < Lat. *capillatura* < *capillus,* hair.] A head of hair, esp. a coiffure.

**Chev·i·ot** (shĕv′ē-ət, chĕv′-) *n.* [After the *Cheviot* Hills, a range of hills between England and Scotland.] **1.** One of a breed of sheep with short thick wool, orig. raised on the border of England and Scotland. **2. cheviot.** A woolen fabric with a coarse twill weave, used mainly for suits and overcoats and orig. made from the wool of the Cheviot sheep.

**chev·ron** (shĕv′rən) *n.* [ME *cheveroun* < OFr., rafter (from the meeting of rafters at an angle) < Lat. *capra,* goat.] **1.** An insignia of stripes meeting at an angle, worn on the sleeve of a military or police uniform to indicate rank, merit, or length of service. **2.** *Heraldry.* A device shaped like an inverted V. **3.** A V-shaped pattern, esp. an architectural fret.

**chev·ro·tain** (shĕv′rə-tān′) *n.* [Fr. *chevrotin* < OFr., dim. of *chevrot,* kid, dim. of *chevre,* goat < Lat. *capra,* she-goat.] A small hornless ruminant of the genera *Hyemoschus* or *Tragulus* of central Africa and southeast Asia.

**chew** (cho͞o) *v.* **chewed, chew·ing, chews.** [ME *cheuen* < OE *cēowan.*] *—vt.* **1.** To grind and bite with the teeth : MASTICATE. **2.** To ponder <*chew* our problems over> *—vi.* **1.** To make a crushing and grinding motion with the teeth. **2.** To ponder. **3.** *Informal.* To use chewing tobacco. **—chew out.** *Slang.* To scold. *—n.* **1.** An act of chewing. **2.** Something chewed <a *chew* of tobacco> **—chew the fat** (or **rag**). *Slang.* To talk idly or casually. **—chew′er** *n.*

**chewing gum** *n.* A sweetened, flavored preparation for chewing, usu. made of chicle.

**che·wink** (chĭ-wĭngk′) *n.* [Imit. of its song.] The towhee.

**Chey·enne** (shī-ăn′, -ĕn′) *n., pl.* **Cheyenne** or **-ennes.** [Canadian Fr. < Dakota *šahíyena.*] **1. a.** A tribe of Indians once living in central Minnesota and North and South Dakota and now settled in Montana and Oklahoma. **b.** A member of this tribe. **2.** The Algonquian language of the Cheyenne. **—Chey·enne′** *adj.*

**Cheyne-Stokes respiration** (chān′stōks′, chā′nē-stōks′) *n.* [After John *Cheyne* (1777–1836) and William *Stokes* (1804–1878).] Abnormal breathing marked by slow rhythmic waxing and waning of respiratory depth.

**chi** (kī) *n.* [Gk. *khi.*] The 22nd letter of the Greek alphabet. —See table at ALPHABET.

**Chi·an·ti** (kē-än′tē, -ăn′-) *n.* [After the *Chianti* Mountains, Italy.] A dry, usu. red table wine.

**chi·ao** (jē′ou′) *n., pl.* **chiao.** [Chin. (Mandarin) *jiao*[3].] The jiao.

**chi·a·ro·scu·ro** (kē-är′ə-sko͝or′ō, -skyo͝or′ō) *n., pl.* **-ros.** [Ital. : *chiaro,* light (< Lat. *clarus,* clear) + *oscuro,* dark < Lat. *obscurus.*] **1.** The technique of using light and shade in pictorial representation. **2.** Arrangement of light and dark elements in a pictorial work of art. **—chi·a′ro·scu′rist** *n.*

**chi·as·ma** (kī-ăz′mə) *also* **chi·asm** (kī′ăz′əm) *n., pl.* **-ma·ta** (-mə-tə) or **-mas** *also* **-asms.** [Gk. *khiasma, khiasmat-,* crosspiece < *khiazein,* to mark with an X < *khi,* chi, from the letter's shape.] **1.** *Anat.* A crossing or intersection of two tracts, as of nerves or ligaments. **2.** A point of contact between homologous chromosomes, regarded as the cytological manifestation of crossing over. **—chi·as′mal, chi·as′mic, chi′as·mat′ic** (-măt′ĭk) *adj.*

**chi·as·ma·ty·py** (kī-ăz′mə-tī′pē) *n.* [CHIASMA + -TYP(E) + -Y[2].] Meiotic twisting between pairs of homologous chromosomes that produces chiasmata.

**chi·as·mus** (kī-ăz′məs) *n., pl.* **-mi** (-mī′) [NLat. < Gk. *khiasmos* < *khiazein,* to mark with an X. —see CHIASMA.] A rhetorical inversion of the second of two parallel structures; e.g., *He went to the office, but home went she.*

**chi·as·to·lite** (kī-ăs′tə-līt′) *n.* [< Gk. *chiastos,* crossed, p.part. of *khiazein,* to mark with an X. —see CHIASMA.] A mineral variety of andalusite with carbonaceous impurities regularly arranged along the longer axis of the crystal.

**chiaus** (chous, choush) *n.* [Turk. *çavuş* < *çav,* news.] An official Turkish messenger, emissary, or sergeant.

**Chib·cha** (chĭb′chə) *n., pl.* **Chibcha** or **-chas. 1. a.** An extinct tribe of Indians once inhabiting Colombia. **b.** A member of this tribe. **2.** The extinct language of the Chibcha.

**Chib·chan** (chĭb′chən) *n.* **1.** A Central or South American Indian ethnic stock including the Chibcha. **2.** A language stock of Central and South America that includes Chibcha.

**chi·bouk** *also* **chi·bouque** (chĭ-bo͝ok′, shĭ-) *n.* [Fr. *chibouque* < Turk. *çibuk.*] A Turkish tobacco pipe with a long stem and a red clay bowl.

**chic** (shēk) *adj.* **-er, -est.** [Fr.] **1.** Sophisticated and stylish. **2.** Smartly attired. *—n.* **1.** Elegant sophistication in dress and manner. **2.** Stylishness. **—chic′ly** *adv.* **—chic′ness** *n.*

**chi·ca·lo·te** (chē′kə-lō′tē) *n.* [Sp. < Nahuatl *chicalotl.*] A prickly poppy, *Argemone platyceras* of the southwestern United States and tropical America, with grayish foliage and large white flowers.

**Chi·ca·na** (chĭ-kä′nə, shĭ-) *n.* [Alteration of Sp. *Mejicana,* fem. of *Mejicano,* Mexican.] A Mexican-American woman. **—Chi·ca′na** *adj.*

**chi·cane** (shĭ-kān′, chĭ-) *v.* **-caned, -can·ing, -canes.** [Fr. *chicaner* < OFr., to quibble.] *—vt.* & *vi.* To trick or to use tricks. *—n.* Chicanery. **—chi·can′er** *n.*

**chi·can·er·y** (shĭ-kā′nə-rē, chĭ-) *n., pl.* **-ies. 1.** Deception by trickery. **2.** A trick : subterfuge.

**Chi·ca·no** (chĭ-kä′nō, shĭ-) *n., pl.* **-nos.** [Mex. Sp., var. of Sp. *Mejicano,* a Mexican < *Méjico,* Mexico.] A Mexican-American. **—Chi·ca′no** *adj.*

**chi·chi** (shē′shē) *adj.* [Fr.] Ostentatiously stylish : SHOWY.

**chick** (chĭk) *n.* [ME *chike,* short for *chiken,* chicken < OE *cicen.*] **1. a.** A young chicken. **b.** The young of a bird. **2.** A child. **3.** *Slang.* A young woman or girl.

**chick·a·dee** (chĭk′ə-dē′) *n.* [Imit. of its song.] Any of several small, plump North American birds of the genus *Parus,* with predominantly gray plumage and a dark-crowned head.

**chick·a·ree** (chĭk′ə-rē′) *n.* [Imit. of its song.] A squirrel, *Tamiascurus douglasi* of northwestern North America, resembling the red squirrel.

**Chick·a·saw** (chĭk′ə-sô′) *n., pl.* **Chickasaw** or **-saws. 1. a.** A tribe of Indians orig. of Mississippi and later removed to Oklahoma.

**b.** A member of this tribe. **2.** The Muskhogean language of the Chickasaw. **—Chick'a·saw'** *adj.*

**chick·en** (chĭk'ən) *n.* [ME *chiken* < OE *cicen.*] **1. a.** The common domestic fowl or its young. **b.** A bird similar or related to the chicken. **c.** The flesh of the common domestic fowl. **2.** *Slang.* A young woman. *—adj. Slang.* **1.** Afraid. **2.** Cowardly. *—vi.* **-ened, -en·ing, -ens.** *Slang.* To act in a cowardly way <*chickened* out at the last minute>

**chicken breast** *n.* A chest deformity marked by a projecting sternum, caused by rickets. **—chick'en-breast'ed** *adj.*

**chicken feed** *n. Slang.* A small amount of money.

**chicken hawk** *n.* Any of various hawks that prey on or have the reputation of preying on chickens.

**chick·en-heart·ed** (chĭk'ən-här'tĭd) *adj.* CHICKEN 2. **—chick'en·heart'ed·ness** *n.*

**chick·en-liv·ered** (chĭk'ən-lĭv'ərd) *adj.* CHICKEN 2.

**chicken pox** *n.* An acute contagious viral disease, usu. of young children, characterized by skin eruption and a slight fever.

**chicken wire** *n.* A light-gauge galvanized wire fencing, usu. made with hexagonal mesh.

**chick·pea** (chĭk'pē') *n.* [Obs. *chichpease* : ME *chiche,* chickpea (< OFr. < Lat. *cicer*) + *pease,* pea.] **1.** A bushy plant, *Cicer arietenum,* grown in the Mediterranean region and central Asia and bearing edible seeds. **2.** A pealike seed of the chickpea, used as food.

**chick·weed** (chĭk'wēd') *n.* [So called because it is eaten by chickens.] A plant of the genera *Cerastium* or *Stellaria,* esp. *S. media,* a weedy plant with white flowers.

**chic·le** (chĭk'əl) *n.* [Sp. < Nahuatl *chictli.*] The coagulated milky juice of the sapodilla, used as the chief ingredient in chewing gum.

**chi·co** (chē'kō) *n., pl.* **-cos.** [Short for CHICALOTE.] GREASEWOOD 1.

**chic·o·ry** (chĭk'ə-rē) *n., pl.* **-ries.** [ME *cicoree* < OFr. < Lat. *cichorium* < Gk. *kikhora.*] **1.** A plant, *Cichorium intybus,* with usu. blue flowers and leaves used in salads. **2.** The root of the chicory, roasted and ground for mixing with coffee or as a coffee substitute.

**chide** (chīd) *v.* **chid·ed** *or* **chid** (chĭd), **chid·ed** *or* **chid** *or* **chid·den** (chĭd'n), **chid·ing, chides.** [ME *chiden* < OE *cīdan.*] *—vt.* To reprimand so as to improve <*chided* them for bad table manners> *—vi.* To express disapproval. **—chid'er** *n.* **—chid'ing·ly** *adv.*

**chief** (chēf) *n.* [ME *chef* < OFr. < Lat. *caput,* head.] **1.** One of the highest rank or authority : LEADER. **2.** *often* **Chief. a.** A ship's chief engineering officer. **b.** A chief petty officer. **3.** *Slang.* A boss. **4.** The upper section of a heraldic shield. **5.** The highest or most important part. *—adj.* **1.** Being the highest in rank, authority, or office. **2.** Most important : PRINCIPAL <my *chief* reason for objecting> *—adv. Archaic.* Chiefly.

**chief cell** *n.* Any of the secretory cells of the gastric glands that produce pepsin.

**chief justice** *also* **Chief Justice** *n.* The presiding judge of a high court having several judges, esp. of the U.S. Supreme Court.

**chief·ly** (chēf'lē) *adv.* **1.** Above all : ESPECIALLY. **2.** Mostly : mainly. *—adj.* Of or similar to a chief.

**chief of staff** *n.* **1.** *often* **Chief of Staff.** The ranking officer of the U.S. Army, Navy, or Air Force, responsible to the secretary of his or her branch and to the President. **2.** The senior military staff officer at the division level or higher.

**chief of state** *n.* The head of a nation.

**chief·tain** (chēf'tən) *n.* [ME *chevetain* < OFr. < LLat. *capitaneus* < Lat. *caput,* head.] The leader of a group, esp. a clan or tribe.

**chiff·chaff** (chĭf'chăf') *n.* [Imit. of its song.] A small European warbler, *Phylloscopus collybita,* having yellowish plumage.

**chif·fon** (shĭ-fŏn', shĭf'ŏn') *n.* [Fr., chiffon, rag < *chiffe,* old rag.] **1.** A sheer fabric. **2. chiffons.** Ribbons, laces, or other ornamental accessories for women's clothing. *—adj.* **1.** Of, relating to, or like chiffon. **2.** Having a light and fluffy consistency.

**chif·fo·nier** (shĭf'ə-nîr') *n.* [Fr. < *chiffon,* rag. —see CHIFFON.] A narrow, high chest of drawers, often with an attached mirror.

**chig·ger** (chĭg'ər) *n.* [Var. of CHIGOE.] **1.** Any of various small six-legged larvae of mites of the family Trombidiidae, causing intensely irritating itching when lodged on the skin. **2.** CHIGOE 1.

**chi·gnon** (shēn-yŏn', shēn'yŏn') *n.* [Fr. < OFr. *chaignon,* chain < Lat. *catena.*] A roll or knot of hair worn at the back of the head or nape of the neck.

**chig·oe** (chĭg'ō, chē'gō) *n.* [Of Cariban orig.] **1.** A small tropical flea, *Tunga penetrans,* of which the fertile female burrows under the skin, causing intense irritation and sores that may become severely infected. **2.** CHIGGER 1.

**Chi·hua·hua** (chĭ-wä'wä, -wə) *n.* [After *Chihuahua,* Mexico.] A tiny dog orig. bred in Mexico, with pointed ears and a smooth coat.

**chil·blain** (chĭl'blān') *n.* [CHIL(L) + BLAIN.] An inflammation followed by itchy irritation on the hands, feet, or ears, caused by exposure to moist cold. **—chil'blained'** *adj.*

**child** (chīld) *n., pl.* **chil·dren** (chĭl'drən) [ME < OE *cild.*] **1.** A person between birth and puberty. **2. a.** An unborn infant : FETUS. **b.** A baby : infant. **3.** A childish or immature person. **4.** A son or daughter : OFFSPRING. **5.** *often* **children.** Members of a tribe : DESCENDANTS <*children* of Abraham> **6.** Figurative offspring <a *child* of nature> **—with child.** Pregnant. **—child'less** *adj.* **—child'less·ness** *n.*

▲ word history: The plural of *child* is really a double plural, and neither plural suffix belongs to the original declension of the word. The earliest Old English form *cild* was a neuter noun that formed the plural by adding no suffix, like the Modern English words *sheep* and *deer.* Other neuter nouns, however, formed plurals by suffixing *-ru,* and in later Old English times a new plural, *cildru,* was used for *cild.* This form developed into *childer.* In Old English still another class of nouns formed the plural with the suffix *-an,* which survived in Middle English as *-en. Oxen* and *brethren* are modern plurals that show this suffix. In some dialects of Middle English *-en* was the usual plural suffix, and it was added to *childer* to make it conform to other nouns. *Childeren* was probably pronounced *children,* which became the modern spelling of the plural of *child.*

**child abuse** *n.* Maltreatment of a child by a parent, guardian, or other adult that includes intentional acts resulting in physical injury, toleration of and complicity in conditions injurious to the child's health, or sexual assault upon the child.

**child·bear·ing** (chīld'bâr'ĭng) *n.* Parturition.

**child·bed** (chīld'bĕd') *n.* The state of a woman in childbirth.

**childbed fever** *n.* Puerperal fever.

**child·birth** (chīld'bûrth') *n.* Parturition.

**childe** (chīld) *n.* [ME *child,* child.] *Archaic.* A child of noble birth.

**child·hood** (chīld'ho͝od') *n.* The time or state of being a child.

**child·ish** (chīl'dĭsh) *adj.* **1.** Of, similar to, or for a child. **2.** Lacking maturity : PUERILE. **—child'ish·ly** *adv.* **—child'ish·ness** *n.*

☆ ***syns:*** CHILDISH, BABYISH, IMMATURE, INFANTILE, JUVENILE, PUERILE *adj. core meaning* : of or characteristic of a child, esp. in immaturity <*childish* temper tantrums> ***ant:*** adult

**child labor** *n.* Full-time employment of children under a minimum legal age.

**child·like** (chīld'līk') *adj.* Like or befitting a child <a *childlike,* innocent smile> **—child'like'ness** *n.*

**child neglect** *n.* Failure on the part of a parent or parental substitute to supervise a child and provide requisite care and protection.

**child·proof** (chīld'pro͞of') *adj.* Designed to resist tampering by children <a *childproof* aspirin bottle>

**chil·dren** (chĭl'drən) *n. pl. of* CHILD.

**child's play** *n.* **1.** An easy task. **2.** A trivial matter.

**chil·e** (chĭl'ē) *n. var. of* CHILI.

**chil·e con car·ne** *also* **chil·i con car·ne** (chĭl'ē kŏn kär'nē) *n.* [Sp.: *chile,* chili + *con,* with + *carne,* meat.] A highly spiced dish of red peppers, meat, and often beans.

**Chile saltpeter** *n.* Sodium nitrate.

**chil·i** *also* **chil·e** *or* **chil·li** (chĭl'ē) *n., pl.* **-ies** *also* **-es** *or* **-lies.** [Sp. *chile* < Nahuatl *chilli.*] **1. a.** The very pungent fruit of several varieties of a woody plant, *Capsicum frutescens.* **b.** A condiment made from the dried fruits of the chili. **2.** Chile con carne.

**chil·i·ad** (kĭl'ē-ăd', -əd) *n.* [LLat. *chilias, chiliad-* < Gk. *khilias* < *khilioi,* thousand.] **1.** A group containing 1,000 elements. **2.** One thousand years.

**chil·i·asm** (kĭl'ē-ăz'əm) *n.* [NLat. *chiliasmus* < LLat. *chiliastes,* a chiliast < *chilias,* chiliad.] The belief that Christ will reign on earth for 1,000 years. **—chil'i·ast'** (-ăst', -əst) *n.* **—chil'i·as'tic** *adj.*

**chil·i·bur·ger** (chĭl'ē-bûr'gər) *n.* A hamburger covered with chili.

**chil·i con car·ne** (chĭl'ē kŏn kär'nē) *n. var. of* CHILE CON CARNE.

**chil·i·dog** (chĭl'ē-dôg', -dŏg') *n.* A hot dog covered with chili.

**chili sauce** *n.* A spiced sauce of chilies and tomatoes.

**chill** (chĭl) *n.* [ME *chele* < OE *cēle.*] **1.** A moderate but penetrating cold. **2.** A feeling of coldness, as with a fever. **3.** A dampening of enthusiasm, spirit, or joy <bad news that put a *chill* on the party> **4.** A sudden numbing fear or dread. *—adj.* **1.** Chilly. **2.** Dispiriting : discouraging. *—v.* **chilled, chill·ing, chills.** *—vt.* **1.** To affect with cold. **2.** To discourage. **3.** To lower in temperature. **4.** *Metallurgy.* To harden (a metallic surface) by rapid cooling. *—vi.* **1.** To be seized with cold. **2.** To become cold. **3.** *Metallurgy.* To become hard by rapid cooling. **—chill'ing·ly** *adv.* **—chill'ness** *n.*

**chill·er** (chĭl'ər) *n.* One that chills or frightens : THRILLER.

**chil·li** (chĭl'ē) *n. var. of* CHILI.

**chill·y** (chĭl'ē) *adj.* **-i·er, -i·est. 1.** Cold enough to cause shivering. **2.** Seized with cold : SHIVERING. **3.** Distant and cool : UNFRIENDLY <a *chilling* greeting> **—chill'i·ly** *adv.* **—chill'i·ness** *n.*

**chi·lo·pod** (kī'lə-pŏd') *n.* [NLat. *Chilopoda,* class name : Gk. *kheilos,* lip + Gk. *pous,* foot (so called because the foremost pair of legs are jawlike appendages).] Any of various arthropods of the class Chilopoda, including the centipedes.

**chi·mae·ra** (kī-mîr'ə, kĭ-) *n.* [NLat. *Chimaera,* type genus < Lat., *chimera.*] **1.** One of the noncommercial fish of the family Chimaeridae. **2.** *var. of* CHIMERA.

**chime[1]** (chīm) *n.* [ME *chimbe* < OFr., var. of *cimble,* cymbal < Lat. *cymbalum.* —see CYMBAL.] **1.** An apparatus for striking a bell or bells to make a musical sound. **2.** *often* **chimes.** A set of bells tuned to the musical scale and used as an orchestral instrument. **3.** A single bell. **4.** The musical sound made by a bell or bells. **5.** Agreement : accord <Their views are in *chime* with mine.> *—v.* **chimed,**

**chim·ing, chimes.** —*vi.* **1.** To sound with a harmonious ring when struck. **2.** To make a musical sound by striking a chime. **3.** To agree : harmonize <Their opinions *chimed* with ours.> —*vt.* **1.** To produce (music) by striking bells. **2.** To strike (a bell) to produce music. **3.** To make known (the hour) by ringing bells. **4.** To call, send, or welcome by ringing bells. **—chime in. 1.** To break into, as a conversation : INTERRUPT <*chimed in* with an extraneous comment> **2.** To join in harmoniously. **—chim'er** *n.*

**chime²** (chīm) *n.* [ME *chimb.*] The rim of a cask.

**chi·me·ra** *also* **chi·mae·ra** (kī-mîr'ə, kĭ-) *n.* [ME *chimere* < OFr. < Lat. *chimaera* < Gk. *khimaira.*] **1. Chimera.** *Gk. Myth.* A fire-breathing she-monster usu. depicted as a composite of a lion, a goat, and a serpent. **2.** A foolish fancy. **3.** *Biol.* An organism, esp. a plant, with tissues from at least two genetically distinct parents.

**Chimera**

**chi·mer·i·cal** (kī-mĕr'ĭ-kəl, -mîr'-, kĭ-) *also* **chi·mer·ic** (-mĕr'ĭk, -mîr'-) *adj.* **1.** Like a chimera : IMAGINARY. **2.** Given to unrealistic fantasies : FANCIFUL. **—chi·mer'i·cal·ly** *adv.*

**chim·ney** (chĭm'nē) *n., pl.* **-neys.** [ME *chimene* < OFr. *cheminee* < LLat. *caminata,* fireplace < Lat. *caminus,* furnace < Gk. *kaminos.*] **1. a.** A structural passage through which smoke and gases escape from a fire or furnace : FLUE. **b.** The usu. vertical structure containing a chimney. **c.** The part of such a structure rising above a roof. **2.** *Chiefly Brit.* A smokestack, as of a locomotive. **3.** A glass tube for enclosing a lamp's flame. **4.** Something, as a narrow cleft in a cliff, resembling a chimney.

**chim·ney·piece** (chĭm'nē-pēs') *n.* **1.** The mantel of a fireplace. **2.** A decoration over a fireplace.

**chimney pot** *n.* A pipe put on top of a chimney to improve draft.

**chimney sweep** *n.* One who cleans soot from chimneys.

**chimney swift** *n.* A small, dark, swallowlike New World bird, *Chaetura pelagica,* that frequently nests in chimneys.

**chimp** (chĭmp) *n. Informal.* A chimpanzee.

**chim·pan·zee** (chĭm'păn-zē', -pən-, chĭm-păn'zē) *n.* [Of Bantu orig.] An anthropoid ape, *Pan troglodytes* of tropical Africa, having gregarious habits and a high degree of intelligence.

**chin** (chĭn) *n.* [ME < OE *cin.*] The central forward portion of the lower jaw. —*v.* **chinned, chin·ning, chins.** —*vt.* **1.** To pull (oneself) up with the arms while grasping an overhead horizontal bar until one's chin is level with the bar. **2.** To place (a violin) under the chin. —*vi.* **1.** *Informal.* To chatter. **2.** To chin oneself.

**chi·na** (chī'nə) *n.* **1.** High-quality porcelain or ceramic ware, orig. made in China. **2.** Porcelain ware.

**China aster** *n.* A plant native to China, *Callistephus chinensis,* widely cultivated for its variously colored flowers.

**chi·na·ber·ry** (chī'nə-bĕr'ē) *n.* **1.** A spreading tree native to Asia, *Melia azedarach,* widely grown for its white or purple flower clusters. **2.** A soapberry tree, *Sapindus marginatus* or *S. saponaria* of the West Indies, Mexico, and the southwestern United States. **3.** The fruit of a chinaberry.

**China rose** *n.* A shrub, *Rosa chinensis,* that has fragrant red or pink flowers and is the ancestor of many cultivated hybrid roses.

**Chi·na·town** (chī'nə-toun') *n.* A neighborhood inhabited by Chinese people.

**China tree** *n.* CHINABERRY 1.

**chi·na·ware** (chī'nə-wâr') *n.* Tableware made of china.

**†chinch** (chĭnch) *n.* [Sp. *chinche* < Lat. *cimex,* bug.] *Regional.* A bedbug.

**chinch bug** *n.* A small black-and-white insect, *Blissus leucopterus,* very destructive to grains and grasses.

**chin·che·rin·chee** (chĭn'chə-rĭn-chē', chĭng'kə-) *n.* [Orig. unknown.] A bulbous plant, *Ornithogalum thyrsoides* of southern Africa, with long white or yellow flower clusters or spikes.

**chin·chil·la** (chĭn-chĭl'ə) *n.* [Sp., prob. < Aymara.] **1. a.** A squirrellike rodent, *Chinchilla laniger,* native to the mountains of South America and widely raised in captivity for its soft pale-gray fur. **b.** The fur of the chinchilla. **2.** A thick twilled cloth of wool and cotton, used for overcoats.

**Chin·co·teague pony** (shĭng'kə-tēg', chĭng'-) *n.* [After *Chincoteague* Island, Virginia.] A small inbred North American horse that runs wild on certain islands off the Virginia coast.

**chine** (chīn) *n.* [ME < OFr. *eschine,* of Germanic orig.] **1. a.** The backbone : spine. **b.** A cut of meat containing part of the backbone. **2.** A ridge or crest. **3.** The line of intersection between the side and bottom of a flat-bottomed or V-bottom boat.

**Chi·nese** (chī-nēz', -nēs') *adj.* Of or relating to China or its culture, people, or languages. —*n., pl.* **Chinese. 1. a.** A native or resident of China. **b.** A person of Chinese ancestry. **2. a.** A branch of the Sino-Tibetan language family consisting of the various dialects spoken by the Chinese people. **b.** Any of the dialects spoken by the Chinese people.

**Chinese anise** *n.* Star anise.

**Chinese cabbage** *n.* A Chinese plant, *Brassica pekinensis,* with a cylindrical head of crisp, edible leaves.

**Chinese calendar** *n.* The lunar calendar of the Chinese people.

**Chinese Chippendale** *n.* Chippendale furniture reflecting certain Oriental stylistic influences.

**Chinese date** *n.* **1.** The jujube. **2.** The fruit of the jujube.

**Chinese evergreen** *n.* A plant, *Aglaonema simplex* of tropical Asia, that has glossy pointed green leaves and is widely grown as a house plant.

**Chinese fire drill** *n. Slang.* Utter chaos.

**Chinese houses** *n. (sing. or pl. in number).* A plant, *Collinsia bicolor* of California, with white and rose-purple flowers.

**Chinese ink** *n.* India ink.

**Chinese lantern** *n.* **1.** A decorative collapsible lantern of thin brightly colored paper. **2.** One of the papery inflated seed cases of the winter cherry.

**Chinese lantern plant** *n.* The winter cherry.

**Chinese puzzle** *n.* **1.** An intricate puzzle. **2.** A difficult problem.

**Chinese red** *n.* Vermilion.

**Chinese restaurant syndrome** *n.* A complex of symptoms, including facial pressure, perspiration, dizziness, and headache, that can occur after eating food prepared with large amounts of monosodium glutamate.

**Chinese sacred lily** *n.* A variety of the polyanthus narcissus, *Narcissus tazetta orientalis,* having fragrant yellow and white flowers and frequently grown as a house plant.

**Chinese white** *n.* Zinc oxide.

**Chinese wood oil** *n.* Tung oil.

**chink¹** (chĭngk) *n.* [Perh. alteration of obs. *chine* < ME, crack < OE *cine.*] A narrow crack or fissure. —*vt.* **chinked, chink·ing, chinks. 1.** To make chinks in. **2.** To fill chinks in. **—chink'y** *adj.*

**chink²** (chĭngk) *n.* [Imit.] A short metallic sound. —*vi.* & *vt.* **chinked, chink·ing, chinks.** To make or cause to make a chink.

**chi·no** (chē'nō, shē'-) *n., pl.* **-nos.** [Am. Sp., toasted (from its original tan color).] **1.** A coarse twilled cotton fabric used for uniforms and sports clothes. **2. chinos.** Trousers made of chino.

**chi·noi·se·rie** (shēn'wäz-rē') *n.* [Fr. < *chinois,* Chinese < *Chine,* China.] **1.** A style in art reflecting Chinese influence through use of elaborate decoration and intricate patterns. **2.** An object reflecting Chinese artistic influence.

**Chi·nook** (shĭ-no͝ok', chĭ-) *n., pl.* **Chinook** or **-nooks.** [Salish *c'inuk.*] **1. a.** A tribe of Indians once inhabiting the Columbia River basin in Oregon. **b.** A member of this tribe. **c.** The Chinookan language of this tribe. **2. chinook.** A moist warm wind blowing from the sea on the Oregon and Washington coasts. **3. chinook.** A warm dry wind that descends from the eastern slopes of the Rocky Mountains, causing a rapid rise in temperature.

**Chi·nook·an** (shĭ-no͝ok'ən, chĭ-) *n.* A North American Indian language family of Washington and Oregon. —*adj.* Of or relating to the Chinook Indians, their language, or their culture.

**Chinook Jargon** *n.* A pidgin language combining English, French, Chinook, and other Indian dialects that was once used by Indians and fur traders in the Pacific Northwest.

**Chinook salmon** *n.* A salmon, *Oncorhynchus tshawytscha* of northern Pacific waters, valued as a food fish.

**chin·qua·pin** (chĭng'kə-pĭn') *n.* [Of Algonquian orig.] **1.** A small shrubby tree, *Castanea pumila* of the eastern United States. **2.** A large evergreen tree, *Castanopsis chrysophella* of the North American Pacific Coast. **3.** The nut of a chinquapin.

**chintz** (chĭnts) *n.* [Obs. *chints,* pl. of *chint,* calico cloth < Hindi *chīṇt* < Skt. *citra-,* variegated.] A printed, glazed, usu. brightly colored cotton fabric.

**chintz·y** (chĭnt'sē) *adj.* **-i·er, -i·est. 1.** Of, relating to, or decorated with chintz. **2.** Gaudy : cheap <*chintzy* furnishings>

**chip¹** (chĭp) *n.* [ME < OE *cyp,* beam.] **1.** A small piece broken or cut off. **2.** A mark, as a crack, caused by chipping. **3. a.** A small disk or counter used in poker and other games to represent money. **b. chips.** *Slang.* Money. **4.** *Electron.* A minute square of a thin semiconducting material, as silicon or germanium, doped and otherwise processed to have specified electrical characteristics, esp. such a square before attachment of electrical leads and packaging as an electronic component or integrated circuit. **5.** A thin, brittle slice of a food <a potato *chip*> **6. chips.** *Chiefly Brit.* French-fried potatoes. **7.** Material, as wood, palm leaves, or straw, cut and dried for weaving. **8.** A fragment of dried animal dung, used as fuel. **9.** A worthless item. **10.** A chip shot. —*v.* **chipped, chip·ping, chips.** —*vt.* **1.** To break a small piece from. **2.** To chop or cut with a sharp implement. **3.** To shape or carve by cutting or chopping. —*vi.* To become broken off. **—chip in.** *Informal.* **1.** To contribute money or labor. **2.** To chime in. **3.** To put up chips or money as one's bet in poker and other games. **—chip off the old block.** One who resembles either of one's parents. **—chip on (one's) shoulder.** Persistent bitterness or resentment.

**chip[2]** (chĭp) *vi.* **chipped, chip·ping, chips.** [Imit.] To cheep. **—chip** *n.*

**chip[3]** (chĭp) *n.* [Orig. unknown.] A trick method of throwing one's opponent in wrestling.

**Chip·e·wy·an** (chĭp'ə-wī'ən) *n., pl.* **Chipewyan** or **-ans.** [Cree *čīpwayān,* parka wearer : *cīpw-,* pointed + *-ayān,* skin.] **1. a.** A tribe of Indians living in Canada in the area between Great Slave Lake and Lake Athabasca on the west and Hudson Bay on the east. **b.** A member of this tribe. **2.** The Athapascan language of the Chipewyan.

**chip·munk** (chĭp'mŭngk') *n.* [Alteration of obs. *chitmunk,* of Algonquian orig.] A small rodent, *Tamias striatus* of eastern North America, or any of several similar rodents of the genus *Eutamias* of western North America and northern Asia, resembling a squirrel but smaller and having a striped back.

**chipped beef** *n.* Smoked and thinly sliced dried beef, usu. served with a cream sauce.

**Chip·pen·dale** (chĭp'ən-dāl') *adj.* [After Thomas *Chippendale* (1718–1779).] Of, relating to, or designating furniture styled with flowing lines and rococo ornamentation.

**chip·per[1]** (chĭp'ər) *n.* One that chips or cuts.

**chip·per[2]** (chĭp'ər) *vi.* **-pered, -per·ing, -pers.** [< CHIP[2].] **1.** To chirp or twitter, as a bird. **2.** To chatter nonsensically.

**chip·per[3]** (chĭp'ər) *adj.* [Poss. alteration of Brit. dial. *kipper,* lively.] *Informal.* Cheerful : pert <a *chipper* grin>

**Chip·pe·wa** (chĭp'ə-wô', -wä', -wā', -wə) *n., pl.* **Chippewa** or **-was.** Ojibwa.

**chipping sparrow** *n.* A small North American sparrow, *Spizella passerina,* with a reddish-brown crown.

**chip·py** (chĭp'ē) *n., pl.* **-pies.** [< CHIP[2].] **1.** The chipping sparrow. **2.** *Slang.* A prostitute.

**chip shot** *n.* A short, lofted golf stroke, used in approaching the green.

**chi·ral** (kī'rəl) *adj.* [CHIRO- + -AL.] Of or relating to the handedness of an asymmetric molecule. **—chi·ral'i·ty** (kī-răl'ĭ-tē) *n.*

**chi-rho** (kī'rō', kē'-) *n.* [CHI + RHO, first two letters of Greek *Khristos,* Christ.] A monogram and symbol for Christ, composed of the superimposed Greek letters chi (X) and rho (P), often embroidered on altar cloths and clerical vestments.

**chiro-** or **cheiro-** *pref.* [Lat. < Gk. *kheir,* hand.] Hand <*chiropractic*>

**chi·rog·ra·phy** (kī-rŏg'rə-fē) *n.* Penmanship. **—chi·rog'ra·pher** *n.* **—chi'ro·graph'ic** (kī'rə-grăf'ĭk), **chi'ro·graph'i·cal** *adj.*

**chi·ro·man·cy** (kī'rə-măn'sē) *n.* [ME *ciromancie* < Med. Lat. *chiromantia* < Gk. *kheiromantis,* palmist : *kheir,* hand + *mantis,* diviner.] Palmistry. **—chi'ro·man'cer** *n.*

**Chi·ron** (kī'rŏn') *n.* [Lat. < Gk. *Kheirōn.*] *Gk. Myth.* The wise centaur who tutored Achilles, Hercules, and Asclepius.

**chi·ro·plas·ty** (kī'ro-plăs'tē) *n. var. of* CHEIROPLASTY.

**chi·rop·o·dy** (kī-rŏp'ə-dē, shī-) *n.* [CHIR(O)- + -POD + -Y.] Podiatry. **—chi·rop'o·dist** *n.*

**chi·ro·prac·tic** (kī'rə-prăk'tĭk) *n.* [CHIRO- + Gk. *praktikos,* effective. —see PRACTICAL.] A therapeutic system in which disease is regarded as the result of neural malfunction and manipulation of the spinal column and other structures is the preferred treatment method. **—chi'ro·prac'tor** *n.*

**chi·rop·ter·an** (kī-rŏp'tər-ən) *n.* [NLat. *Chiroptera,* order name : CHIRO- + -PTER.] A flying mammal of the order Chiroptera, including the bats. **—chi·rop'ter·an** *adj.*

**chirp** (chûrp) *n.* [Imit.] A short, high-pitched sound, as that made by a bird. **—chirp** *v.* **(chirped, chirp·ing, chirps).**

**chirr** (chûr) *n.* [Imit.] A harsh, trilling sound, as that made by a cricket. **—chirr** *v.* **(chirred, chir·ring, chirs).**

**chir·rup** (chûr'əp, chĭr'-) *v.* **-ruped, -rup·ing, -rups.** [Var. of CHIRP.] —*vi.* **1.** To utter a series of chirps. **2.** To make clicking, clucking sounds with the lips, as in urging on a horse. —*vt.* **1.** To sound with chirps. **2.** To make clucking sounds to. —*n.* **1.** A series of chirps. **2.** A series of clucks or clicking sounds, as those made to urge on a horse.

**chi·rur·geon** (kī-rûr'jən) *n.* [ME *cirurgien* < OFr. *cirurgiien* < Lat. *chirurgia,* surgery. —see SURGERY.] *Archaic.* A surgeon.

**chis·el** (chĭz'əl) *n.* [ME < OFr., prob. ult. < Lat. *caedere,* to cut.] A metal tool with a sharp beveled edge, used to cut and shape stone, wood, or metal. —*v.* **-eled, -el·ing, -els** *also* **-elled, -el·ling, -els.** —*vt.* **1.** To shape or cut with a chisel. **2.** *Slang.* **a.** To cheat : swindle. **b.** To obtain by deception. —*vi.* **1.** To use a chisel. **2.** *Slang.* To use unethical methods to gain an end : CHEAT. **—chis'el·er** *n.*

**chi-square** (kī'skwâr') *n.* A statistic used in testing a hypothesis concerning the discrepancy between observed and expected results, that is calculated as the sum of the squares of observed values minus expected values divided by the expected values.

**chit[1]** (chĭt) *n.* [Obs. *chitty* < Hindi *ciṭṭhi,* note, letter < Skt. **ciṣṭa,* message.] **1.** CHECK 7. **2.** *Chiefly Brit.* A short letter : NOTE.

**chit[2]** (chĭt) *n.* [ME, young animal.] A child, esp. a pert girl.

**chit·chat** (chĭt'chăt') *n.* [Redup. of CHAT.] **1.** Small talk. **2.** Gossip. —*vi.* **-chat·ted, -chat·ting, -chats.** To indulge in chitchat.

**chi·tin** (kīt'n) *n.* [Fr. *chitine* < NLat. *chiton,* mollusk < Gk. *khitōn,* chiton.] A semitransparent horny substance, primarily a mucopolysaccharide, forming the chief component of crustacean shells, insect exoskeletons, and the cell walls of certain fungi. **—chi'tin·ous** *adj.*

**chit·lins** or **chit·lings** (chĭt'lĭnz) *pl.n. vars. of* CHITTERLINGS.

**chi·ton** (kīt'n, kī'tŏn') *n.* [Gk. *khitōn,* tunic, of Semitic orig.] **1.** A tunic worn by ancient Greek men and women. **2.** Any of various marine mollusks of the class Amphineyra, living on rocks and having shells made up of eight overlapping transverse plates.

**chit·tam·wood** (chĭt'əm-wŏŏd') *n.* [Prob. of Muskhogean orig.] A North American tree, esp. a smoke tree.

**chit·ter** (chĭt'ər) *vi.* **-tered, -ter·ing, -ters.** [ME *chiteren.*] To twitter, as birds do.

**chit·ter·lings** *also* **chit·lins** or **chit·lings** (chĭt'lĭnz) *pl.n.* [ME *chiterling.*] The small intestines of pigs, cooked and eaten as food.

**chi·val·ric** (shĭ-văl'rĭk, shĭv'əl-) *adj.* Chivalrous.

**chiv·al·rous** (shĭv'əl-rəs) *adj.* **1.** Having the qualities of gallantry and honor attributed to an ideal knight. **2.** Of or relating to chivalry. **3.** Courteous and considerate, esp. toward women. **—chiv'al·rous·ly** *adv.* **—chiv'al·rous·ness** *n.*

**chiv·al·ry** (shĭv'əl-rē) *n., pl.* **-ries.** [ME *chevalrie* < OFr. *chevalerie* < *chevalier,* knight. —see CHEVALIER.] **1. a.** Medieval knighthood. **b.** The principles and customs of medieval knighthood. **2. a.** The qualities, as bravery, courtesy, and honor, that were idealized by knighthood. **b.** The manifestation of any of these qualities. **3.** A group of knights or gallant gentlemen.

**chiv·a·ree** (shĭv'ə-rē', shĭv'ə-rē') *n. var. of* CHARIVARI.

**chive** (chīv) *n.* [ME *cive* < OFr. < Lat. *cepa,* onion.] **1.** A plant native to Eurasia, *Allium schoenoprasum,* with purplish flowers and hollow, grasslike leaves. **2.** *often* **chives.** The leaves of the chive, used as a seasoning.

**chiv·vy** or **chiv·y** (chĭv'ē) *vt.* **-vied, -vy·ing, -vies** *also* **-ied, -y·ing, -ies.** [< *chevy,* a hunting cry, prob. short for *chevy chase,* pursuit < *Chevy Chase,* title of a ballad about a border skirmish.] **1.** To harass or annoy. **2.** To get, direct, or manipulate by persistent, often petty maneuvering.

**chlam·y·date** (klăm'ĭ-dāt') *adj.* [Lat. *chlamydatus,* cloaked < *chlamys,* cloak < Gk. *khlamus.*] *Zool.* Having a mantle. —Used of mollusks.

**chla·myd·e·ous** (klə-mĭd'ē-əs) *adj.* [Lat. *chlamys, chlamyd-,* cloak + -EOUS.] *Bot.* Having a floral envelope.

**chla·myd·o·spore** (klə-mĭd'ə-spôr', -spōr') *n.* [Lat. *chlamys, chlamyd-,* cloak + SPORE.] A thick-walled fungus spore derived from a hyphal cell.

**chlam·ys** (klăm'ĭs, klā'mĭs) *n., pl.* **chlam·ys·es** or **chlam·y·des** (klăm'ĭ-dēz') [Lat. < Gk. *khlamus.*] A short mantle fastened at the shoulder, worn by ancient Greek men.

**chlo·as·ma** (klō-ăz'mə) *n., pl.* **-ma·ta** (-mə-tə) [NLat. < Gk. *khloasma,* greenness < *khloazein,* to be green < *kloos,* green color.] A patchy brown or black skin discoloration, usu. on the face.

**chlor-** *pref. var. of* CHLORO-.

**chlor·ac·ne** (klôr-ăk'nē, klōr-) *n.* A skin condition resembling acne caused by exposure to chlorinated hydrocarbons.

**chlo·ral** (klôr'əl, klōr'-) *n.* A colorless, mobile oily liquid, $CCl_3CHO$, a penetrating lung irritant, used to make DDT and chloral hydrate.

**chloral hydrate** *n.* A colorless crystalline compound, $CCl_3CH(OH)_2$, used medicinally as a sedative and hypnotic.

**chlo·ra·mine** (klôr'ə-mēn', klōr'-) *n.* A compound containing nitrogen and chlorine, esp. an unstable colorless liquid, $NH_2Cl$, used to manufacture hydrazine.

**chlo·ram·phen·i·col** (klôr'ăm-fĕn'ĭ-kôl', klōr'-, -kōl') *n.* [CHLOR(O)- + AM(IDE) + PHE(NO)- + NI(TRO)- + (GLY)COL.] An antibiotic, $C_{11}H_{12}Cl_2N_2O_5$, derived from the soil bacterium *Streptomyces venezuelae* or synthesized.

**chlo·rate** (klôr'āt', klōr'-) *n.* The inorganic group $ClO_3$ or a compound containing it.

**chlor·dane** (klôr'dān', klōr'-) *also* **chlor·dan** (-dăn') *n.* [CHLOR(O)- + (IN)D(ENE) + -ANE.] A colorless, odorless viscous liquid, $C_{10}H_6Cl_8$, used as an insecticide.

**chlor·di·az·e·pox·ide** (klôr'dī-ăz'ə-pŏk'sīd', klōr'-) *n.* [CALOR(O)- + DI- + AZ(O) + EP(I)- + OXIDE.] A compound, $C_{16}H_{14}ClN_3O$, whose hydrochloride is used as a tranquilizer.

**chlo·rel·la** (klə-rĕl'ə) *n.* [NLat. *Chlorella,* genus name < Gk. *khlōros,* green.] Any of various green algae of the genus *Chlorella,* widely used in studies of photosynthesis.

**chlo·ren·chy·ma** (klə-rĕng'kə-mə) *n.* [CHLOR(OPHYLL) + -ENCHYMA.] Plant tissue, esp. stem tissue, containing chlorophyll.

**chlo·ric** (klôr'ĭk, klōr'-) *adj.* Of, relating to, or containing chlorine.

**chloric acid** *n.* A strongly oxidizing, unstable acid, $HClO_3 \cdot 7H_2O$.

**chlo·ride** (klôr′īd′, klōr′-) *n.* A binary compound of chlorine. **—chlo·rid′ic** (klə-rĭd′ĭk) *adj.*

**chloride of lime** *n.* Chlorinated lime.

**chlo·ri·nate** (klôr′ə-nāt′, klōr′-) *vt.* **-nat·ed, -nat·ing, -nates.** To treat or combine with chlorine or a chlorine compound. **—chlo′ri·na′tion** *n.* **—chlo′ri·na′tor** *n.*

**chlorinated lime** *n.* A white powder of varying composition, as $CaCl(ClO) \cdot 4H_2O$, produced by chlorinating slaked lime and used as a bleach.

**chlo·rine** (klôr′ēn′, klōr′-, -ĭn) *n. Symbol* **Cl** A highly irritating, greenish-yellow gaseous element, used in water purification, as a disinfectant, a bleaching agent, and in making chloroform and carbon tetrachloride; atomic number 17; atomic weight 35.45.

**chlo·rite¹** (klôr′īt′, klōr′-) *n.* [Lat. *chloritis*, a green precious stone < Gk. *khlōritis* < *khlōros*, green.] A gen. green or black secondary mineral, $(Mg, Fe, Al)_6(Si, Al)_4O_{10}(OH)_8$, often formed by metamorphic alteration of primary dark rock minerals.

**chlo·rite²** (klôr′īt′, klōr′-) *n.* The inorganic group $ClO_2$ or a compound containing it.

**chloro-** *or* **chlor-** *pref.* [< Gk. *khlōros*, green.] **1.** Green <*chlorosis*> **2.** Chlorine <*chloroform*>

**chlo·ro·ben·zene** (klôr′ō-bĕn′zēn′, -bĕn-zēn′, klōr′-) *n.* A colorless, volatile flammable liquid, $C_6H_5Cl$, used to prepare phenol, DDT, and aniline and as a general solvent.

**chlo·ro·car·bon** (klôr′ō-kär′bən, klōr′-) *n.* A compound consisting of chlorine and carbon.

**chlo·ro·fluor·o·car·bon** (klôr′ō-floŏr′ō-kär′bən, -flôr′-, -flōr′-, -floō′ər-ō-, klōr′-) *n.* Any of various compounds consisting of carbon, hydrogen, chlorine, and fluorine, once used as aerosol propellants and refrigerants.

**chlo·ro·form** (klôr′ə-fôrm′, klōr′-) *n.* [CHLORO- + FORM(YL).] A clear, colorless, heavy liquid, $CHCl_3$, used in refrigerants, propellants, and resins and as an anesthetic. *—vt.* **-formed, -form·ing, -forms. 1.** To anesthetize or kill with chloroform. **2.** To apply chloroform to.

**chlo·ro·hy·drin** (klôr′ō-hī′drĭn, klōr′-) *n.* An aliphatic organic chemical compound that is both an alkyl chloride and an alcohol, often containing a single chlorine atom and a single hydroxyl group on adjacent carbon atoms.

**Chlo·ro·my·ce·tin** (klôr′ō-mī-sēt′n, klōr′-). A trademark for chloramphenicol.

**chlo·ro·phyll** *also* **chlo·ro·phyl** (klôr′ə-fĭl, klōr′-) *n.* Any of a group of related green pigments found in photosynthetic organisms, esp.: **a.** Chlorophyll a. **b.** Chlorophyll b.

**chlorophyll a.** A waxy blue-black microcrystalline green-plant pigment, $C_{55}H_{72}MgN_4O_5$, with a blue-green alcohol solution.

**chlorophyll b.** A green-plant pigment similar to chlorophyll a, $C_{55}H_{70}MgN_4O_6$, having a brilliant green alcohol solution.

**chlo·ro·pic·rin** (klôr′ə-pĭk′rĭn, klōr′-) *n.* [CHLORO- + PICR(O)- + -IN.] An oily colorless liquid, $CCl_3NO_2$, used to make dyestuffs, disinfectants, insecticides, fumigants, and poison gas.

**chlo·ro·plast** (klôr′ə-plăst′, klōr′-) *also* **chlo·ro·plas·tid** (klôr′ə-plăs′tĭd, klōr′-) *n. Bot.* A plastid in photosynthetic plants that contains chlorophyll.

**chlo·ro·prene** (klôr′ə-prēn′, klōr′-) *n.* [CHLORO- + (ISO)PRENE.] A colorless liquid, $C_4H_5Cl$, used as the monomer of neoprene rubber.

**chlo·ro·quine** (klôr′ə-kwīn′, -kwēn′, klōr′-) *n.* [Blend of CHLORO- and QUINOLINE.] A compound, $C_{18}H_{26}ClN_3$, used in treating malaria and occas. in treating lupus erythematosus.

**chlo·ro·sis** (klə-rō′sĭs) *n.* **1.** *Bot.* An abnormal condition of plants, marked by absence of or deficiency in green pigment and caused by lack of light, mineral deficiency, or genetic disorders. **2.** *Pathol.* An iron-deficiency anemia chiefly affecting girls at puberty and marked by greenish skin coloration. **—chlo·rot′ic** (-rŏt′ĭk) *adj.* **—chlo·rot′i·cal·ly** *adv.*

**chlor·prom·a·zine** (klôr-prŏm′ə-zēn′, -prō′mə-, klōr-) *n.* [CHLOR(O)- + PRO(PYL) + METH(YL) + AZINE.] An oily liquid, $C_{17}H_{19}ClN_2S$, derived from phenothiazine and used as a sedative, tranquilizer, and antiemetic.

**chlor·tet·ra·cy·cline** (klôr′tĕt-rə-sī′klēn′, klōr′-) *n.* An antibiotic, $C_{22}H_{23}ClN_2O_8$, obtained from the soil bacterium *Streptomyces aureofaciens.*

**cho·a·na** (kō′ə-nə) *n., pl.* **-nae** (-nē′) [Gk. *khoanē*, funnel < *khein*, to pour.] A funnel-shaped opening, esp. one of the internal nares.

**cho·an·o·cyte** (kō-ăn′ə-sīt′) *n.* [Gk. *khoanē*, funnel (< *khein*, to pour) + -CYTE.] *Biol.* One of the flagellated cells that line the body cavity of a sponge.

**chock** (chŏk) *n.* [Orig. unknown.] **1.** A block or wedge placed under something, as a wheel, to keep it from moving. **2.** *Naut.* A heavy metal or wooden fitting with two inward curving jaws through which a rope or cable can be run. *—vt.* **chocked, chock·ing, chocks. 1.** To fit with or secure by a chock. **2.** To place (a boat) on chocks. *—adv.* As close as possible <*chock* up against the rail>

**chock-a-block** (chŏk′ə-blŏk′) *adj.* **1.** Drawn so close as to have the blocks touching. —Used of a ship's hoisting tackle. **2.** Crowded <a house *chock-a-block* with guests> **—chock′-a-block′** *adv.*

**chock-full** (chŏk′foŏl′, chŭk′-) *adj.* [ME *chokkeful.*] Completely filled <*chock-full* of food>

**choc·o·late** (chô′kə-lĭt, chôk′lĭt, chŏk′-) *n.* [Sp. < Aztec *xocolatl* : *xococ*, bitter + *atl*, water.] **1.** Husked, roasted, and ground cacao seeds, often combined with a sweetener or flavoring agent. **2.** A beverage or candy made from chocolate. **3.** A grayish to deep reddish brown to deep grayish brown. *—adj.* **1.** Flavored with chocolate. **2.** Of a grayish to deep reddish brown to deep grayish brown.

**chocolate tree** *n.* CACAO 1.

**Choc·taw** (chŏk′tô) *n., pl.* **Choctaw** *or* **-taws.** [Choctaw *Chahta.*] **1. a.** A tribe of Indians once inhabiting southern Mississippi and Alabama and now settled in Oklahoma. **b.** A member of this tribe. **2.** The Muskhogean language of the Choctaw.

**choice** (chois) *n.* [ME *chois* < OFr. < *choisir*, to choose, of Gothic orig.] **1.** An act of choosing : SELECTION. **2.** Power, right, or liberty to choose : OPTION. **3.** One chosen. **4.** A number or variety from which to choose <a wide *choice* of shoes> **5.** Something best or preferable. **6.** An alternative. *—adj.* **choic·er, choic·est. 1. a.** Of fine quality : SELECT. **b.** Appealing to refined taste. **2.** Selected carefully. **3.** Of the U.S. Government grade of meat higher than good and lower than prime. **—choice′ly** *adv.* **—choice′ness** *n.*

☆ ***syns:*** CHOICE, ELECTION, OPTION, PREFERENCE, SELECTION *n. core meaning* : the act of choosing <a price that influenced my *choice* of rugs><had no *choice* in the matter>

**choir** (kwīr) *n.* [ME *quer* < OFr. *cuer* < Lat. *chorus.* —see CHORUS.] **1. a.** An organized group of singers, esp. one performing church music. **b.** The part of a church used by such singers. **2.** The part of a cruciform church between the nave and the main altar : CHANCEL. **3. a.** A musical group or band. **b.** A section of a musical group or band. *—vi.* **choired, choir·ing, choirs.** To sing in chorus.

**choir·boy** (kwīr′boi′) *n.* A boy who is a member of a choir.

**choir loft** *n.* A gallery for a church choir.

**choir·mas·ter** (kwīr′măs′tər) *n.* The director of a choir.

**choke** (chōk) *v.* **choked, chok·ing, chokes.** [ME *choken* < OE *ācēocian.*] *—vt.* **1.** To interfere with or terminate the normal respiration of, esp. by constricting or breaking the windpipe or by polluting the air. **2.** To stop by or as if by strangling : SUPPRESS <*choke* back tears><*choked* off all discussion> **3.** To reduce the air intake of (a carburetor), thus enriching the fuel mixture. **4.** To slow down the movement, growth, or action of. **5.** To obstruct by filling or crowding : CLOG <streets *choked* with traffic> **6.** To fill completely : JAM. **7.** To grip (a bat, racket, or club) at a point nearer the hitting surface. *—vi.* **1. a.** To become suffocated. **b.** To have difficulty in breathing, swallowing, or speaking. **2.** To be obstructed. **—choke up.** *Informal.* **1.** To be unable to speak, due to strong emotion. **2.** To fail to perform effectively, due to nervous tension. *—n.* **1.** The act or sound of choking. **2. a.** Something that chokes. **b.** A narrow part, as the chokebore of a gun. **3.** A device used in an internal-combustion engine to enrich the fuel mixture by reducing air flow to the carburetor.

**choke·ber·ry** (chōk′bĕr′ē) *n.* [From its bitter fruit.] **1.** Any of various North American shrubs of the genus *Aronia*, having bitter-tasting red, black, or purple fruit. **2.** The fruit of the chokeberry.

**choke·bore** (chōk′bôr′, -bōr′) *n.* **1.** A shotgun bore that narrows toward the muzzle to prevent wide scattering of the shot. **2.** A gun equipped with a chokebore.

**choke·cher·ry** (chōk′chĕr′ē) *n.* [From its bitter fruit.] **1.** A North American shrub or tree, *Prunus virginiana*, with long white flower clusters and extremely astringent dark-red or blackish fruit. **2.** The fruit of the chokecherry.

**choke·damp** (chōk′dămp′) *n.* [So called because it causes suffocation in mines.] Blackdamp.

**chok·er** (chō′kər) *n.* **1.** One that chokes. **2. a.** A necklace fitting closely around the throat. **b.** A tight, high collar. **c.** A narrow fur neckpiece.

**chol-** *pref. var. of* CHOLE-.

**cho·lan·gi·og·ra·phy** (kō-lăn′jē-ŏg′rə-fē) *n.* [CHOL(E)- + Gk. *angeion*, vessel + -GRAPHY.] Roentgenographic examination of the bile ducts. **—cho·lan′gi·o·graph′ic** (-ə-grăf′ĭk) *adj.*

**chole-** or **chol-** *pref.* [< Gk. *kholē*, bile.] Bile <*cholesterol*>

**cho·le·cyst** (kō′lĭ-sĭst′) *n.* The gallbladder.

**cho·le·cys·tec·to·my** (kō′lĭ-sĭ-stĕk′tə-me) *n., pl.* **-mies.** Surgical removal of the gallbladder.

**cho·le·li·thi·a·sis** (kō′lə-lĭ-thī′ə-sĭs) *n.* Presence of gallstones in the gallbladder.

**chol·er** (kŏl′ər, kō′lər) *n.* [ME *colre* < OFr. < Lat. *cholera*, jaundice < Gk. *kholera.*] **1.** Anger. **2.** *Archaic.* **a.** One of the four humors of the body thought in the Middle Ages to cause anger and bad temper when present in excess : BILE. **b.** Biliousness.

**chol·er·a** (kŏl′ər-ə) *n.* [Lat., jaundice. —see CHOLER.] An acute, often fatal, infectious epidemic disease caused by the microorganism *Vibrio comma* and characterized by watery diarrhea, vomiting, cramps, suppression of urine, and collapse. **—chol′e·ra′ic** (-ə-rā′ĭk) *adj.* **—chol′e·roid′** (-ə-roid′) *adj.*

**cholera mor·bus** (môr′bəs) *n.* [NLat. : Lat. *cholera*, jaundice + Lat. *morbus*, disease.] Acute gastroenteritis occurring in summer and autumn and marked by severe cramps, diarrhea, and vomiting.

**holera nos·tras** (nŏs′trəs) *n.* Cholera morbus.
**hol·er·ic** (kŏl′ə-rĭk, kə-lĕr′ĭk) *adj.* **1.** Easily angered. **2.** Exhibiting or expressing anger. **—chol′er·i·cal·ly, chol′er·ic·ly** *adv.*
**ho·le·sta·sis** (kō′lĭ-stā′sĭs) *n.* Suppression of biliary flow.
**ho·les·ter·in** (kə-lĕs′tər-ĭn) *n.* Cholesterol.
**ho·les·ter·ol** (kə-lĕs′tə-rôl′, -rōl′) *n.* [Gk. *kholē*, bile + Gk. *stereos*, solid + -OL (so called because it was first found in gallstones).] A glistening white soapy crystalline substance, $C_{27}H_{45}OH$, the most common animal sterol, a precursor of a form of vitamin D and a universal tissue constituent, occurring notably in bile, gallstones, the brain, blood cells, plasma, egg yolk, and seeds.
**ho·lic acid** (kō′lĭk) *n.* [Gk. *kholikos*, bilious < *kholē*, bile.] An abundant crystalline bile acid, $C_{24}H_{40}O_5$.
**ho·line** (kō′lēn′) *n.* A natural amine, $C_5H_{15}NO_2$, often classed in the vitamin B complex and a precursor of acetylcholine.
**ho·lin·er·gic** (kō′lə-nûr′jĭk) *adj.* [(ACETYL)CHOLIN(E) + Gk. *ergon*, work.] **1.** Activated by or capable of liberating the acetylcholine. **2.** Having physiological effects similar to acetylcholine.
**ho·lin·es·ter·ase** (kō′lə-nĕs′tə-rās′, -rāz′) *n.* [CHOLIN(E) + ESTERASE.] An enzyme that hydrolyzes acetylcholine to form acetic acid and choline.
**hol·la** (choi′ə, -yə) *n.* [Mex. Sp. < obs. Sp. *cholla*, upper part of the head, poss. < OFr. *cholle*, head, of Germanic orig.] A spiny cactus of the genus *Opuntia*, having cylindrical rather than flattened stem segments.
**homp** (chŏmp) *v.* & *n.* *var. of* CHAMP[1].
**hon** (chŏn) *n.*, *pl.* **chon.** [Korean.] —See table at CURRENCY.
**hondr-** *or* **chondri-** *pref. vars. of* CHONDRO-.
**hon·dri·fy** (kŏn′drə-fī′) *vt.* & *vi.* **-fied, -fy·ing, -fies.** To change into cartilage. **—chon′dri·fi·ca′tion** *n.*
**hon·dri·o·some** (kŏn′drē-ə-sōm′) *n.* A mitochondrion.
**hon·drite** (kŏn′drīt′) *n.* A stone of meteoric origin characterized by chondrules. **—chon·drit′ic** (-drĭt′ĭk) *adj.*
**hon·dri·tis** (kŏn-drī′tĭs) *n.* Inflammation of cartilage.
**hondro-** *or* **chondri-** *or* **chondr-** *pref.* [< Gk. *khondros*, granule, cartilage.] **1.** Cartilage <*chondro*cranium> **2.** Granule <*chondr*ite>
**hon·dro·cra·ni·um** (kŏn′drō-krā′nē-əm) *n.*, *pl.* **-ni·ums** *or* **-ni·a** (-nē-ə). The embryonic cranium.
**hon·dro·i·tin** (kŏn-drō′ĭ-tĭn) *n.* [< *chondroitic acid*, an acid occurring in cartilage.] A mucopolysaccharide found in cartilage in its sulfated form.
**hon·dro·ma** (kŏn-drō′mə) *n.*, *pl.* **-mas** *or* **-ma·ta** (-mə-tə). A cartilaginous growth.
**hon·dro·ma·la·cia** (kŏn′drō-mə-lā′shə) *n.* [CHONDRO- + Gk. *malakia*, softness < *malakos*, soft.] Abnormal softening of cartilage.
**hon·drule** (kŏn′drōōl′) *n.* *Geol.* A small round granule of extraterrestrial origin found embedded in some meteorites.
**hoose** (chōōz) *v.* **chose** (chōz), **chos·en** (chō′zən), **choos·ing, choos·es.** [ME *chesen* < OE *cēosan*.] *—vt.* **1.** To select from several possible alternatives. **2.** To prefer above others. **3.** To want : desire. *—vi.* To make a choice : SELECT. **—choos′er** *n.*
☆ **syns: 1.** CHOOSE, CULL, ELECT, OPT, PICK OUT, SELECT, SINGLE (out) *v. core meaning* : to make a choice from several alternatives <*chose* only the best> ***ant:*** reject **2.** CHOOSE, DESIRE, LIKE, PLEASE, WANT, WILL, WISH *v. core meaning* : to have the desire or inclination to <thought they could do as they *chose*>
**choos·y** *also* **choos·ey** (chōō′zē) *adj.* **-i·er, -i·est.** Hard to please : PERSNICKETY. **—choos′i·ness** *n.*
**chop[1]** (chŏp) *v.* **chopped, chop·ping, chops.** [ME *choppen*.] *—vt.* **1.** To cut by striking with a heavy, sharp tool, as an axe. **2.** To form or shape by chopping. **3.** To cut into small bits : MINCE. **4.** To cut short <*chop* off a sentence> **5.** To hit or hit at with a short, swift downward stroke in a sport. *—vi.* **1.** To make heavy, cutting strokes. **2.** To move roughly or suddenly. *—n.* **1.** An act of chopping. **2.** A swift, short, cutting blow or stroke. **3.** A chopped-off piece, esp. a cut of meat, usu. taken from the rib, shoulder, or loin and containing a bone. **4.** A short, irregular motion of waves.
**chop[2]** (chŏp) *vi.* **chopped, chop·ping, chops.** [*chop*, to exchange (obs.) < ME *choppen*.] To change direction suddenly, as a ship in the wind.
**chop[3]** (chŏp) *n.* [Hindi *chhāp*, seal.] **1.** An official stamp or permit in the Far East. **2.** Quality <first *chop*>
**chop·fall·en** (chŏp′fô′lən) *adj. var. of* CHAPFALLEN.
**chop·house** (chŏp′hous′) *n.* A restaurant specializing in steaks and chops.
**chop·per** (chŏp′ər) *n.* **1.** One that chops. **2.** A device that interrupts an electric current or beam of radiation. **3.** *Slang.* A helicopter. **4. choppers.** *Slang.* Teeth, esp. a set of false teeth. **5.** A motorcycle, esp. a customized one.
**chopping block** *n.* A wooden block on which food is prepared, as by chopping.
**chop·py[1]** (chŏp′ē) *adj.* **-pi·er, -pi·est. 1.** Rough with many small waves <*choppy* seas> **2.** Marked by abrupt transitions : JERKY <a *choppy* prose style> **—chop′pi·ness** *n.*
**chop·py[2]** (chŏp′ē) *adj.* **-pi·er, -pi·est.** Abruptly shifting : VARIABLE. —Used of the wind.
**chops** (chŏps) *pl.n.* [Orig. unknown.] The jaws, cheeks, or jowls of animals or human beings.
**chop·sticks** (chŏp′stĭks′) *pl.n.* [Pidgin E. *chop*, fast (prob. < Cantonese *kap*) + STICKS.] A pair of slender wooden, ivory, or plastic sticks used as eating utensils primarily in Oriental countries and restaurants.
**chop su·ey** (chŏp sōō′ē) *n.* [Chin. (Cantonese) *tsap[2] sui[4]*, mixed pieces : *tsap[2]*, mixed + *sui[4]*, to break up.] A Chinese-American dish made of small pieces of meat cooked with bean sprouts and other vegetables, served with rice.
**cho·ra·gus** (kə-rā′gəs) *n.*, *pl.* **-gi** (-jī′) [Lat. < Gk. *khoragos* : *khoros*, chorus + *agein*, to lead.] The leader of a chorus in Greek drama. **—cho·rag′ic** (-răj′ĭk) *adj.*
**cho·ral** (kôr′əl, kōr′-) *adj.* [Med. Lat. *choralis* < *chorus*, choral dance < Lat. —see CHORUS.] **1.** Of or relating to a chorus or choir. **2.** Written for performance by a chorus or choir. *—n. var. of* CHORALE. **—cho′ral·ly** *adv.*
**cho·rale** *also* **cho·ral** (kə-răl′, -räl′) *n.* [G. *Choral(gesang)*, choral (song) < Med. Lat. *choralis*, choral.] **1.** A Protestant hymn tune. **2.** A harmonized hymn, esp. one for the organ. **3.** A chorus or choir.
**chorale prelude** *n.* A chiefly baroque musical composition for the organ, with an elaborate contrapuntal structure based on the melody of a hymn or chorale.
**choral speaking** *n.* The recitation of poetry or prose by a chorus.
**chord[1]** (kôrd, kōrd) *n.* [ME *cord* < *accord*, agreement < OFr. *acorde* < *acorder*, to agree. —see ACCORD.] **1.** A combination of three or more usu. concordant tones sounded simultaneously. **2.** Harmony, as of color. **3.** An emotional feeling or response <a crisis that struck a sympathetic *chord*> *—vt.* & *vi.* **chord·ed, chord·ing, chords.** To furnish with or form a chord : HARMONIZE.
**chord[2]** (kôrd, kōrd) *n.* [Alteration of CORD.] **1.** A line segment joining two points on a curve. **2.** A straight line connecting the leading and trailing edges of an airfoil. **3.** *Archaic.* The string of a musical instrument. **4.** *var. of* CORD 5.
▲ word history: The spelling *chord* for *cord* was introduced in the 16th century as etymologically more correct by scholars who knew Latin and Greek. The Greek origin of both words is *khordē*, meaning "gut," "string," "musical note," and "sausage." Both instrument strings and sausage casings were made of animal guts. Latin borrowed the Greek word as *chorda* and used it to mean primarily "instrument string" and "cord, rope." English borrowed the word from French, which had preserved the Latin meanings. The native English cognate of Greek *khordē* is *yarn*.
**chord·al** (kôr′dl) *adj.* *Mus.* **1.** Relating to or consisting of a harmonic chord. **2.** Giving prominence to harmonic rather than contrapuntal structure.
**chor·date** (kôr′dāt′, -dĭt) *n.* [NLat. *Chordata*, phylum name < Lat. *chorda*, cord. —see CORD.] Any of numerous animals of the phylum Chordata, including all vertebrates and certain marine animals, as the lancelets, having a notochord.
**chord organ** *n.* An electronic or reed organ equipped with buttons for producing chords.
**chore** (chôr, chōr) *n.* [Var. of CHAR[3].] **1.** A routine or minor duty. **2. chores.** Daily or routine domestic tasks, esp. of a farmer. **3.** An unpleasant task. *—vi.* **chored, chor·ing, chores.** To work at chores.
**-chore** *suff.* [< Gk. *khōrein*, to move.] A plant distributed by a specified agency <*zoo*chore>
**cho·re·a** (kô-rē′ə, kō-) *n.* [Lat., dance < Gk. *khoreia*, choral dance < *khoros*.] A nervous disorder, esp. of children, characterized by irregular and uncontrollable movements of the muscles of the arms, legs, and face.
**cho·re·o·graph** (kôr′ē-ə-grăf′, kōr′-) *vt.* & *vi.* **-graphed, -graph·ing, -graphs.** To create the choreography of or specialize in choreography. **—cho′re·og′ra·pher** (kôr′ē-ŏg′rə-fər, kōr′-) *n.*
**cho·re·og·ra·phy** (kôr′ē-ŏg′rə-fē, kōr′-) *n.* [Fr. *chorégraphie* : Gk. *khoreia*, choral dance (< *khoros*) + *-graphie*, -graphy.] **1.** The art of creating and arranging dances or ballets. **2.** The art and technique of dance notation. **3.** The art of dancing. **—cho′re·o·graph′ic** (-ə-grăf′ĭk) *adj.* **—cho′re·o·graph′i·cal·ly** *adv.*
**cho·ri·amb** (kôr′ē-ămb′, -ăm′, kōr′-) *n.* [LLat. *choriambus* < Gk. *khoriambos* : *khoreios*, of a chorus (< *khoros*) + *iambos*, iamb.] **1.** A metrical foot consisting of a trochee followed by an iamb. **2.** A foot of verse used in lyric poetry having two unstressed syllables flanked by the two rhythmic stresses marking the first and last syllables of the foot. **—cho′ri·am′bic** (-ăm′bĭk) *adj.*
**cho·ric** (kôr′ĭk, kōr′-, kŏr′-) *adj.* [Gk. *khorikos* < *khoros*, choral dance.] Of or relating to a chorus, esp. a Greek chorus.
**cho·rine** (kôr′ēn′, kōr′-) *n.* [CHOR(US) + -INE[1].] A chorus girl.
**cho·ri·o·al·lan·to·is** (kôr′ē-ō′ə-lăn′tō-ĭs, kōr′-) *n.* [CHORIO(N) + ALLANTOIS.] The vascular fetal membrane consisting of the fused chorion and allantois. **—cho′ri·o·al′lan·to′ic** (-ō-ăl′ən-tō′ĭk) *adj.*
**cho·ri·oid** (kôr′ē-oid′, kōr′-) *adj.* [Gk. *khorioeidēs*, like an afterbirth : *khorion*, afterbirth + *-eidēs*, -oid.] Choroid.

**cho·ri·on** (kôr′ē-ŏn′, kōr′-) *n.* [Gk. *khorion.*] The outer membrane enclosing the embryo in reptiles, birds, and mammals. **—cho′ri·on′ic** (-ŏn′ĭk) *adj.*
**cho·ris·ter** (kôr′ĭ-stər, kŏr′-, kōr′-) *n.* [ME *queristre* < Norman Fr. **cueristre* < Med. Lat. *chorista* < *chorus,* chorus < Lat., choral dance. —see CHORUS.] **1.** A choir singer, esp. a choirboy. **2.** A choir leader.
**cho·ri·zo** (chə-rē′zō, -sō) *n.* [Sp.] A spicy pork sausage.
**cho·rog·ra·phy** (kə-rŏg′rə-fē) *n.* [Lat. *chorographia* < Gk. *khōrographia* : *khōros,* place + *-graphia,* writing < *graphein,* to write.] **1.** The technique of mapping a region. **2.** A map or description of a region. **—cho·rog′ra·pher** *n.* **—cho′ro·graph′ic** (kôr′ə-grăf′ĭk, kōr′-), **cho′ro·graph′i·cal** *adj.* **—cho′ro·graph′i·cal·ly** *adv.*
**cho·roid** (kôr′oid′, kōr′-) *also* **cho·roi·de·a** (kô-roi′dē-ə, kō-) *n.* [< Gk. *khoroeidēs,* like an afterbirth, alteration of *khorioeidēs.* —see CHORIOID.] *Anat.* The dark-brown vascular coat of the eye between the retina and the sclera. —*adj.* **1.** Resembling the chorion. **2.** Resembling the corium. **3.** Of or relating to the choroid.
**chor·tle** (chôr′tl) *n.* [Blend of CHUCKLE and SNORT.] A snorting, joyful chuckle. —*vi.* & *vt.* **-tled, -tling, -tles.** To express or utter with a chortle. **—chor′tler** *n.*
**cho·rus** (kôr′əs, kōr′-) *n., pl.* **-rus·es.** [Lat., choral dance < Gk. *khoros.*] **1.** *Mus.* **a.** A composition in four or more parts written for numerous singers. **b.** A song refrain in which the audience joins the soloist. **c.** A repeat of the opening statement of a popular song played by the whole group. **d.** A solo section based on the primary melody of a popular song and played by a member of the group. **e.** A group of singers who perform choral compositions. **f.** A group of vocalists and dancers who support the soloists and leading actors, as in operas and musical comedies. **2. a.** A group of persons who speak or sing a given dramatic or poetic part or composition in unison. **b.** An actor who recites the prologue and epilogue to an Elizabethan drama and sometimes comments on the action. **3.** In Greek poetry and drama: **a.** A ceremonial dance performed to the singing of odes. **b.** The part of a drama consisting of choric dance and ode. **c.** The body of actors whose choric performance comments on and accompanies the action of the play. **4. a.** A song, speech, or other utterance made in concert by many people. **b.** A simultaneous utterance by a number of people. —*vt.* & *vi.* **-rused, -rus·ing, -rus·es** or **-russed, -rus·sing, -rus·ses.** To sing or utter in chorus.
**chorus girl** *n.* A young woman who sings or dances in a theatrical chorus.
**chose**[1] (chōz) *v. p.t. of* CHOOSE.
**chose**[2] (shōz) *n.* [Fr. < Lat. *causa,* thing.] *Law.* An item of personal property : CHATTEL.
**cho·sen** (chō′zən) *adj.* [P.part. of CHOOSE.] **1.** Selected from or preferred above others. **2.** Elect. —*n., pl.* **chosen. 1.** One of the elect. **2.** The elect as a group.
**chott** (shŏt) *n.* [Fr. < Ar. *shaṭṭ.*] **1.** The depression surrounding a salt marsh, esp. in North Africa. **2.** The bed of a dried salt marsh.
**chough** (chŭf) *n.* [ME.] A crowlike Old World bird of the genus *Pyrrhocorax,* with red legs and black plumage.
**chow**[1] (chou) *also* **chow chow** *n.* [Pidgin E., prob. < Chin. (Mandarin) *kou*[3], dog.] A heavy-set dog orig. bred in China, with a long, dense, black or reddish-brown coat and a blackish tongue.
**chow**[2] (chou) [Pidgin E., prob < Chin. (Mandarin) *chao*[3], to stir-fry.] *Slang.* —*n.* Food. —*vi.* **chowed, chow·ing, chows.** To eat.
**chow-chow** (chou′chou′) *n.* [Pidgin E., prob. < Chin. (Mandarin) *chao*[3], to stir-fry.] A relish made of chopped vegetables pickled in mustard.
**chow chow** *n. var. of* CHOW[1].
**chow·der** (chou′dər) *n.* [Fr. *chaudière,* stew pot < OFr. < LLat. *caldaria.* —see CALDRON.] A thick soup or stew containing shellfish or fish, esp. clams, and vegetables, often in a milk base.
**chow mein** (chou′ mān′) *n.* [Chin. (Mandarin) *chao*[3] *mian*[4] : *chao*[3], to stir-fry + *mian*[4], noodles.] A Chinese-American dish combining meat and stewed vegetables, usu. served over fried noodles.
**chres·ard** (krĕs′ərd) *n.* [Gk. *khrēsis,* use (< *khrēsthai,* to use) + Gk. *ardein,* to water.] Water present in the soil and available for plant absorption.
**chres·tom·a·thy** (krĕ-stŏm′ə-thē) *n., pl.* **-thies.** [Gk. *khrēstomatheia* : *khrēstos,* useful (< *khrēsthai,* to use) + *-matheia,* learning < *manthanein,* to learn.] A selection of literary passages, used to study a language or literature. **—chres′to·math′ic** (krĕs′tə-măth′ĭk) *adj.*
**chrism** (krĭz′əm) *n.* [ME *crisme,* chrism, chrisom < OE *crisma* < LLat. *chrisma* < Gk. *khrisma,* ointment < *khriein,* to anoint.] **1.** A mixture of oil and balsam consecrated by a bishop and used for anointing in church sacraments, as baptism. **2.** A sacramental anointing, esp. upon confirmation into the Eastern Orthodox Church. **—chris′mal** (krĭz′məl) *adj.*
**chris·om** (krĭz′əm) *n.* [ME *crisom,* var. of *crisme,* chrism, chrisom, chrism.] **1.** A white robe or cloth worn by an infant at baptism. **2.** *Archaic.* An infant wearing a baptismal robe : BABY.
**Christ** (krīst) *n.* [ME *Crist* < OE *Crīst* < Lat. *Christus* < Gk. *Khristos* < *khristos,* anointed < *khriein,* to anoint.] **1.** The Messiah. **2.** Jesus. **3.** *Christian Science.* —Used to refer to <"The divine manifestation of God, which comes to the flesh to destroy incarnate error" —Mary Baker Eddy> **—Christ′li·ness** *n.* **—Christ′ly** *adj.*
**chris·ten** (krĭs′ən) *vt.* **-tened, -ten·ing, -tens.** [ME *cristnen* < OE *cristnian* < *Cristen,* Christian.] **1.** To baptize into a Christi[an] church. **2.** To name at baptism. **3.** To name and dedicate ceremo[ni]ally <*christen* a battleship> **4.** *Informal.* To use for the first tim[e].
**Chris·ten·dom** (krĭs′ən-dəm) *n.* [ME *Cristendom* < OE *Criste[n]dōm* : *Cristen,* Christian + *-dom,* -dom.] **1.** Christians as a who[le]. **2.** The Christian world. **3.** Christianity.
**chris·ten·ing** (krĭs′ə-nĭng) *n.* The Christian sacrament of baptis[m].
**Chris·tian** (krĭs′chən) *adj.* [ME *Cristen* < OE < Lat. *Christianus* < Gk. *Khristianos* < *Khristos,* Christ.] **1.** Declaring belief in Jesus [as] Christ or following the religion based on His teachings. **2.** Relating [to] or derived from Jesus or His teachings. **3.** Manifesting the qualities [or] spirit of Christ : CHRISTLIKE. **4.** Relating to or typical of Christiani[ty] or its adherents. **5.** *Informal.* Kind : decent. —*n.* **1.** One who profess[es] belief in Jesus as Christ or follows the religion based on His teac[h]ings. **2.** One who lives according to the teachings of Jesus. **—Chris[′]tian·ly** *adv.*
**Christian era** *n.* The period beginning with the birth of Jesu[s].
**chris·ti·an·i·a** (krĭs′tē-ăn′ē-ə, -ä′nē-ə, krĭs′chē-) *n.* [Norw. < *Chr[is]tiania,* the former name for Oslo, Norway.] A ski turn in which th[e] body is swung from a crouching position in order to make a stop [or] change direction.

**christiania**

**Chris·ti·an·i·ty** (krĭs′chē-ăn′ĭ-tē, krĭs′tē-) *n., pl.* **-ties. 1.** Th[e] Christian religion. **2.** Christians as a group : CHRISTENDOM. **3.** Th[e] state or fact of being a Christian.
**Chris·tian·ize** (krĭs′chə-nīz′) *vi.* & *vt.* **-ized, -iz·ing, -iz·es.** T[o] adopt Christianity or convert (another) to Christianity. **—Chris[′]tian·i·za′tion** *n.* **—Chris′tian·iz′er** *n.*
**Christian Science** *n.* The church and religious system founde[d] by Mary Baker Eddy, emphasizing healing through spiritual means a[s] an important element of Christianity and teaching pure divine goo[d]ness as underlying the scientific reality of existence. **—Christia[n] Scientist** *n.*
**Christ·like** (krīst′līk′) *adj.* Having the spiritual qualities or attr[i]butes of Christ. **—Christ′like′ness** *n.*
**Christ·mas** (krĭs′məs) *n.* [ME *Cristemas* < OE *Crīstes mæsse* : *Crīst,* Christ + *mæsse,* mass. —see MASS.] **1.** Dec. 25, celebrated b[y] Christians as the anniversary of the birth of Jesus. **2.** Christmastid[e].
**Christmas berry** *n.* Toyon.
**Christmas cactus** *n.* A spineless epiphytic cactus, *Zygocactu[s] truncatus* of South America, cultivated as a house plant for its re[d,] pink, or white flowers.
**Christmas disease** *n.* [After Stephen *Christmas,* the first p[a]tient in whom the disease was diagnosed and studied.] A hemophili[a] caused by a deficiency of the plasma thromboplastin componen[t].
**Christmas Eve** *n.* The night before Christmas.
**Christmas fern** *n.* Dagger fern.
**Christmas rose** *n.* An evergreen plant native to Europe, *Helle[borus] niger* with a poisonous root and white or pinkish-green flower[s] that bloom in late fall or winter.
**Christ·mas·tide** (krĭs′məs-tīd′) *n.* The Christian church festiva[l] extending from Dec. 24 through Jan. 6.
**Christmas tree** *n.* An evergreen or artificial tree decorated wit[h] ornaments and lights during the Christmas season.
**Chris·tol·o·gy** (krĭ-stŏl′ə-jē) *n., pl.* **-gies. 1.** The study of Christ['s] person and qualities. **2.** A doctrine or theory based on Christ or Hi[s] teachings. **—Chris′to·log′i·cal** (krĭs′tə-lŏj′ĭ-kəl) *adj.*
**Christ's-thorn** (krīsts′thôrn′) *n.* A plant, as the jujube, of th[e] Near East, bearing spiny thorns and thought to have been used fo[r] Christ's crown of thorns.
**chris·ty** (krĭs′tē) *n., pl.* **-ties.** A christiania.
**chrom-** *pref. var. of* CHROMO-.
**chro·ma** (krō′mə) *n.* [Gk. *khrōma,* color.] An aspect of color b[y] which a sample appears to differ from a gray of the same brightnes[s] or lightness and that corresponds to saturation of the perceived color.
**chro·maf·fin** (krō′mə-fĭn) *adj.* [CHROMO- + Lat. *affinis,* related.] Capable of being stained with chromium salts.
**chromat-** *pref. var. of* CHROMATO-.
**chro·mate** (krō′māt′) *n.* A salt or ester of chromic acid.

ă **pat** ā **pay** âr **care** ä **father** ĕ **pet** ē **be** hw **which** ĭ **pi[t]**
ī **tie** îr **pier** ŏ **pot** ō **toe** ô **paw, for** oi **noise** ŏŏ **too[k]**

**chro·mat·ic** (krō-măt′ĭk) *adj.* [Gk. *khrōmatikos* < *khrōma,* color.] **1. a.** Relating to color or colors. **b.** Relating to color perceived to have a saturation greater than zero. **2.** *Mus.* **a.** Of, relating to, or based on the chromatic scale. **b.** Relating to chords or harmonies based on nonharmonic tones. **—chro·mat′i·cal·ly** *adv.* **—chro·mat′i·cism** *n.*

**chromatic aberration** *n.* Color distortion in an image produced by a lens because of the dependence of lens refractivity on the wavelength of light and marked by a variation in the focusing of colors.

**chro·ma·tic·i·ty** (krō′mə-tĭs′ĭ-tē) *n.* The aspect of color including consideration of its dominant wavelength and purity.

**chro·mat·ics** (krō-măt′ĭks) *n. (sing. in number).* Scientific study of color. **—chro′ma·tist** (-mə-tĭst) *n.*

**chromatic scale** *n. Mus.* A scale of 12 semitones.

**chro·ma·tid** (krō′mə-tĭd) *n.* Either of two daughter strands of a duplicated chromosome while still joined by a single centromere.

**chro·ma·tin** (krō′mə-tĭn) *n.* A complex of proteins and nucleic acids present in chromosones and marked by intense staining with basic dyes. **—chro′ma·tin′ic** *adj.*

**chromato-** or **chromat-** *pref.* [< Gk. *khrōma, khrōmat-,* color.] **1.** Color <*chromatics*> **2.** Chromatin <*chromatolysis*>

**chro·mat·o·gram** (krō-măt′ə-grăm′) *n.* The absorbent column or strip of material containing the stratographically differentiated constituents separated from a solution or mixture by chromatography.

**chro·ma·tog·ra·phy** (krō′mə-tŏg′rə-fē) *n.* Separation of complex mixtures by percolation through a selectively adsorbing medium, as through a column of magnesia, gelatin, or starch, yielding stratified, sometimes chromatically distinct constituent layers. **—chro·mat′o·graph′** (krō-măt′ə-grăf′) *n.* **—chro′ma·tog′ra·pher** *n.* **—chro·mat′o·graph′ic** *adj.* **—chro·mat′o·graph′i·cal·ly** *adv.*

**chro·ma·tol·y·sis** (krō′mə-tŏl′ĭ-sĭs) *n.* Solution and disintegration of stainable material, as of chromatin, within a cell. **—chro·mat′o·lyt′ic** (-măt′l-ĭt′ĭk) *adj.*

**chro·mat·o·phore** (krō-măt′ə-fôr′, -fōr′) *n. Biol.* A pigment-containing or pigment-producing cell, esp. a pigment-containing animal cell, as in lizards, that by contraction or expansion can change the color of the skin. **—chro·mat′o·phor′ic** (-fôr′ĭk, -fōr′-) *adj.*

**chrome** (krōm) *n.* [Fr. < Gk. *khrōma,* color (from the brilliant colors of chromium compounds).] **1. a.** Chromium. **b.** An object plated with a chromium alloy. **2.** A pigment containing chromium. —*vt.* **chromed, chrom·ing, chromes. 1.** To plate with chromium. **2.** To tan or dye with a chromium compound.

**-chrome** *suff.* [< Gk. *khrōma,* color.] **1.** Colored <auto*chrome*> **2.** Color: pigment <uro*chrome*>

**chrome alum** *n.* A violet-red crystalline compound, $CrK(SO_4)_2 \cdot 12H_2O$, used in tanning and photography and as a mordant.

**chrome green** *n.* **1.** Any of a class of green pigments consisting of chrome yellow and iron blue in various proportions. **2.** A very dark yellowish green to moderate or strong green.

**chrome red** *n.* A light orange to red pigment consisting of basic lead chromate with varying proportions of $PbCrO_4$ and PbO.

**chrome yellow** *n.* Lead chromate, $PbCrO_4$, a yellow pigment often combined with lead sulfate, $PbSO_4$, for lighter hues.

**chro·mic** (krō′mĭk) *adj.* Of, relating to, or containing chromium, esp. with valence 3.

**chromic acid** *n.* **1.** A corrosive, oxidizing acid, $H_2CrO_4$, known only in solution. **2.** The anhydride of chromic acid, $CrO_3$, a purplish crystalline material that reacts explosively with reducing agents and is used in chromium plating, as an oxidizing agent, and to color glass and rubber.

**chromic oxide** *n.* A bright-green crystalline powder, $Cr_2O_3$, used in metallurgy and as a paint pigment.

**chro·mite** (krō′mīt′) *n.* A widely distributed black to brownish-black chromium ore, $FeCr_2O_4$.

**chro·mi·um** (krō′mē-əm) *n.* [NLat. < Fr. *chrome.* —see CHROME.] *Symbol* **Cr** A lustrous, hard, steel-gray metallic element for hardening steel alloys, for producing stainless steels, and for use in corrosion-resistant decorative platings; atomic number 24; atomic weight 51.996.

**chromo-** or **chrom-** *pref.* [< Gk. *khrōma,* color.] **1.** Color <*chromo*plast> **2.** Chromium <*chrom*ous>

**chro·mo·dy·nam·ics** (krō′mō-dī-năm′ĭks) *n. (sing. in number).* The physics of the relationship between quarks and esp. the nature of the strong interaction, color, and the exchange of gluons.

**chro·mo·gen** (krō′mə-jən) *n.* **1.** *Chem.* A substance capable of chemical conversion into a pigment or dye. **2.** *Biol.* A strongly pigmented or pigment-generating organ or organelle. **—chro′mo·gen′ic** (-jĕn′ĭk) *adj.*

**chro·mo·lith·o·graph** (krō′mə-lĭth′ə-grăf′) *n.* A colored print produced by chromolithography.

**chro·mo·li·thog·ra·phy** (krō′mə-lĭ-thŏg′rə-fē) *n.* The process or art of printing color pictures from a series of zinc or stone plates by lithography. **—chro′mo·li·thog′ra·pher** *n.* **—chro′mo·lith′o·graph′ic** (-lĭth′ə-grăf′ĭk) *adj.*

**chro·mo·mere** (krō′mə-mîr′) *n.* One of the serially aligned chromatin granules forming a chromosome.

**chro·mo·ne·ma** (krō′mə-nē′mə) *n., pl.* **-ma·ta** (-mə-tə) [CHROMO- + Gk. *nēma,* thread.] The coiled filamentous core of a chromosome. **—chro′mo·ne′mal** (-nē′məl), **chro·mo′ne·mat′al** (-nē-măt′l), **chro′mo·ne·mat′ic** (-nə-măt′ĭk), **chro′mo·ne′mic** (-nē′mĭk) *adj.*

**chro·mo·phil** (krō′mə-fĭl′) *adj.* Capable of being stained readily with dyes.

**chro·mo·phore** (krō′mə-fôr′, -fōr′) *n.* A molecular group capable of selective light absorption resulting in coloration of aromatic compounds. **—chro′mo·phor′ic** (-fôr′ĭk, -fōr′-) *adj.*

**chro·mo·plast** (krō′mə-plăst′) *n.* A colored plastid containing a pigment other than or in addition to chlorophyll.

**chro·mo·pro·tein** (krō′mə-prō′tēn′, -tē-ĭn) *n.* A substance made up of a protein and a chromophore or pigment.

**chro·mo·some** (krō′mə-sōm′) *n.* A DNA-containing linear body of the cell nuclei of plants and animals, responsible for determination and transmission of hereditary characteristics. **—chro′mo·so′mal, chro′mo·so′mic** *adj.* **—chro′mo·so′mal·ly** *adv.*

**chro·mo·sphere** (krō′mə-sfîr′) *n.* **1.** An incandescent, transparent layer of gas, chiefly hydrogen, several thousand miles in depth, that lies above and surrounds the photosphere of the sun but is distinctly separate from the corona. **2.** A gaseous layer similar to a chromosphere around a star. **—chro′mo·spher′ic** (-sfîr′ĭk, -sfĕr′-) *adj.*

**chro·mous** (krō′məs) *adj.* Of, relating to, or containing chromium, esp. with valence 2.

**chron-** *pref. var. of* CHRONO-.

**chro·nax·y** *also* **chro·nax·ie** (krō′năk′sē, krŏn′ăk′-) *n., pl.* **-ies.** [Fr. *chronaxie* : Gk. *khronos,* time + Gk. *axia,* value < *axios,* worthy.] The time interval needed to stimulate a nerve or muscle fiber electrically with twice the minimum current necessary to elicit a threshold response.

**chron·ic** (krŏn′ĭk) *adj.* [Fr. *chronique* < Lat. *chronicus* < Gk. < *khronos,* time.] **1.** Of long duration. **2.** Subject to a habit or disease for a lengthy period : INVETERATE. **—chron′i·cal·ly** *adv.* **—chro·nic′i·ty** (krŏ-nĭs′ĭ-tē) *n.*

☆ ***syns:* 1.** CHRONIC, CONTINUING, LINGERING, PERSISTENT, PROLONGED, PROTRACTED *adj. core meaning* : of long duration <*chronic* feelings of doubt> **2.** CHRONIC, HABITUAL, HABITUATED, INVETERATE *adj. core meaning* : subject to a disease or habit for a long time <a *chronic* invalid><a *chronic* drug user>

**chron·i·cle** (krŏn′ĭ-kəl) *n.* [ME *cronicle* < Norman Fr., var. of OFr. *cronique* < Lat. *chronica* < Gk. *khronika,* annals < *khronikos,* of time < *khronos,* time.] **1.** A chronological record of historical events. **2. Chronicles** (*sing. in number*). —See table at BIBLE. —*vt.* **-cled, -cling, -cles.** To record in or in the form of a chronicle. **—chron′i·cler** (-klər) *n.*

**chrono-** or **chron-** *pref.* [< Gk. *khronos,* time.] Time <*chrono*meter>

**chron·o·bi·ol·o·gy** (krŏn′ō-bī-ŏl′ə-jē) *n.* The study of biological rhythms.

**chron·o·gram** (krŏn′ə-grăm′, krō′nə-) *n.* **1.** The record produced by a chronograph. **2.** An inscribed phrase in which certain letters can be read as Roman numerals indicating a specific date. **—chron′o·gram·mat′ic** (-grə-măt′ĭk) *adj.* **—chron′o·gram·mat′i·cal·ly** *adv.*

**chron·o·graph** (krŏn′ə-grăf′, krō′nə-) *n.* An instrument that shows or graphically records time intervals, as the duration of an event. **—chron′o·graph′ic** *adj.* **—chron′o·graph′i·cal·ly** *adv.*

**chron·o·log·i·cal** (krŏn′ə-lŏj′ĭ-kəl, krō′nə-) *also* **chron·o·log·ic** (-lŏj′ĭk) *adj.* **1.** Arranged in order of time of occurrence. **2.** Being in accordance with or pertaining to chronology. **—chron′o·log′i·cal·ly** *adv.*

**chronological age** *n.* The number of years a person has lived, used in psychometrics as a comparison standard for various performance measures.

**chro·nol·o·gy** (krə-nŏl′ə-jē) *n., pl.* **-gies. 1.** Determination of dates and the sequence of events. **2.** Arrangement of events in time. **3.** A chronological table or list. **—chro·nol′o·gist** *n.*

**chro·nom·e·ter** (krə-nŏm′ĭ-tər) *n.* An extremely precise timepiece, as a clock or watch. **—chron′o·met′ric** (krŏn′ə-mĕt′rĭk, krō′nə-), **chron′o·met′ri·cal** *adj.* **—chron′o·met′ri·cal·ly** *adv.*

**chro·nom·e·try** (krə-nŏm′ĭ-trē) *n.* Scientific measurement of time.

**chron·o·scope** (krŏn′ə-skōp′, krō′nə-) *n.* An instrument for measuring minute time intervals. **—chron′o·scop′ic** (-skŏp′ĭk) *adj.*

**chrys-** *pref. var. of* CHRYSO-.

**chrys·a·lid** (krĭs′ə-lĭd) *n.* A chrysalis. **—chrys′a·lid** *adj.*

**chrys·a·lis** (krĭs′ə-lĭs) *n., pl.* **chrys·a·lis·es** *or* **chry·sal·i·des** (krĭ-săl′ĭ-dēz′) [Lat. *chrysallis* < Gk. *khrusallis,* gold-colored pupa of a butterfly < *khrusos,* gold.] The third stage in the development of an insect, esp. a moth or butterfly, enclosed in a case or cocoon.

**chry·san·the·mum** (krĭ-săn′thə-məm, -zăn′-) *n.* [Lat. *chrysanthemum* < Gk. *khrusanthemon,* a kind of flower : *khrusos,* gold + *anthemon,* flower < *anthos.*] **1.** Any of several plants of the genus

*Chrysanthemum,* the cultivated forms esp. bearing flowers of various colors and sizes. **2.** The flower of a chrysanthemum.

**chrys·a·ro·bin** (krĭs′ə-rō′bĭn) *n.* [CHRYS(O)- (from its golden color) + (AR)AROB(A) + -IN.] A medicine derived from a deposit found in the wood of the araroba tree and used for treating certain chronic skin conditions.

**chrys·el·e·phan·tine** (krĭ-sĕl′ə-făn′tēn′, -tīn′) *adj.* [Gk. *khruselephantinos* : *khrusos,* gold + *elephas,* ivory.] Made of gold and ivory.

**chryso-** *or* **chrys-** *pref.* [< Gk. *khrusos,* gold.] Gold : golden <*chryso*therapy>

**chrys·o·ber·yl** (krĭs′ə-bĕr′əl) *n.* [Lat. *chrysoberyllus* < Gk. *khrusobērullos* : *khrusos,* gold + *bērullos,* beryl.] A green to yellow vitreous mineral, $BeAl_2O_4$, used as a gemstone.

**chrys·o·lite** (krĭs′ə-līt′) *n.* [ME *crisolite* < OFr. < Lat. *chrysolithus* < Gk. *khrusolithos* : *khrusos,* gold + *lithos,* stone.] Olivine.

**chrys·o·prase** (krĭs′ə-prāz′) *n.* [ME *crisopase* < OFr. < Lat. *chrysoprasus* < Gk. *khrusoprasos* : *khrusos,* gold + *prason,* leek.] Apple-green chalcedony used as a gemstone.

**chrys·o·ther·a·py** (krĭs′ō-thĕr′ə-pē) *n.* Treatment of disease with gold compounds.

**chrys·o·tile** (krĭs′ə-tĭl′) *n.* [G. *Chrysotil* : Gk. *khrusos,* gold + Gk. *tilos,* something plucked < *tillein,* to pluck.] A fibrous mineral variety of serpentine forming part of commercial asbestos.

**chthon·ic** (thŏn′ĭk) *also* **chtho·ni·an** (thō′nē-ən) *adj.* [Gk. *khthonios,* under the earth < *khthōn,* earth.] Relating to the gods and spirits of the underworld.

**chub** (chŭb) *n., pl.* **chub** *or* **chubs.** [ME *chubbe.*] **1.** A freshwater fish of the family Cyprinidae, related to the carp and the minnow, esp. a Eurasian species, *Leuciscus cephalus.* **2.** A fish, as a whitefish of the genus *Coregonus* or a marine fish of the genus *Kyphosus,* not closely akin to the true chub.

**chub·by** (chŭb′ē) *adj.* **-bi·er, -bi·est.** [Prob. < CHUB (from the plumpness of the fish).] Plump. **—chub′bi·ness** *n.*

**chuck¹** (chŭk) *vt.* **chucked, chuck·ing, chucks.** [Orig. unknown.] **1.** To pat or squeeze fondly or playfully, esp. under the chin. **2.** To throw. **3.** *Informal.* To throw out : DISCARD. **4.** *Informal.* To force out : EJECT. —*n.* **1.** An affectionate pat or squeeze under the chin. **2.** A throw, toss, or pitch.

**†chuck²** (chŭk) *n.* [Dial. *chuck,* lump.] **1.** A cut of beef running from the neck to the ribs and including the shoulder blade. **2.** *Western U.S.* Food. **3.** A clamp that holds a tool or material being worked in a machine such as a lathe.

**chuck³** (chŭk) *vi.* **chucked, chuck·ing, chucks.** [Imit.] To make a clucking sound. **—chuck** *n.*

**chuck-a-luck** (chŭk′ə-lŭk′) *n.* [Prob. CHUCK¹ + LUCK.] A gambling game in which players bet on possible combinations of three thrown dice.

**†chuck·hole** (chŭk′hōl′) *n.* [Prob. < CHUCK¹.] *Regional.* A rut or pothole in a road.

**chuck·le** (chŭk′əl) *vi.* **-led, -ling, -les.** [Prob. freq. of CHUCK³.] **1.** To laugh quietly or to oneself. **2.** To cluck, as a hen. —*n.* A quiet laugh of mild amusement or satisfaction. **—chuck′ler** *n.*

**chuck·le·head** (chŭk′əl-hĕd′) *n. Informal.* A stupid and gauche person : BLOCKHEAD. **—chuck′le·head′ed** *adj.*

**chuck wagon** *n.* A wagon having food and cooking utensils.

**chuck·wal·la** (chŭk′wŏl′ə) *n.* [Mex. Sp. *chacahuala* < Cahuilla *tcáxxwal.*] A lizard, *Sauromalus obesus* of the southwestern United States and Mexico, related to the iguana.

**chuck-will's-wid·ow** (chŭk′wĭlz-wĭd′ō) *n.* [Imit. of its song.] A bird, *Caprimulgus carolinensis* of the southern and central United States, resembling the whippoorwill.

**chu·fa** (cho͞o′fə) *n.* [Sp. *chufar,* to make fun of < VLat. **sufilare,* var. of Lat. *sibilare,* to whistle at.] An Old World sedge, *Cyperus esculentus,* with edible, nutlike tubers.

**chuff** (chŭf) *n.* [ME *chuffe.*] A rude, insensitive person.

**chug** (chŭg) *n.* [Imit.] A usu. short, repetitive, dull, explosive sound made by or as if by a laboring engine. —*vi.* **chugged, chug·ging, chugs. 1.** To make chugs. **2.** To move while making chugs.

**chug·a·lug** (chŭg′ə-lŭg′) *v.* **-lugged, -lug·ging, -lugs.** [Imit.] *Slang* —*vt.* To swallow the contents of (a container, as of beer) without pausing. —*vi.* To swallow liquid, as beer, without pausing.

**chu·kar** (chə-kär′) *n.* [Hindi *cakor* < Skt. *cakoraḥ.*] An Old World partridge, *Alectoris graeca,* introduced into western North America.

**Chuk·chi** *also* **Chuk·chee** (cho͝ok′chē) *n., pl.* **Chukchi** *or* **-chis** *also* **Chukchee** *or* **-chees. 1. a.** A Mongoloid people of northeastern Siberia. **b.** A member of this people. **2.** The language of the Chukchi, noted for being pronounced differently by men and women.

**chuk·ka** (chŭk′ə) *n.* [Alteration of CHUKKER (so called because polo players wear a similar boot).] A short ankle-length usu. suede boot with two pairs of eyelets.

**chuk·ker** *also* **chuk·kar** (chŭk′ər) *n.* [Hindi *cakkar,* circle < Skt. *cakram.*] One of the periods of play, lasting 7½ minutes, in a polo match.

**chum¹** (chŭm) *n.* [Perh. short for *chamber fellow,* roommate.] An intimate friend. —*vi.* **chummed, chum·ming, chums. 1.** To be an intimate friend. **2.** To share the same room.

**chum²** (chŭm) *n.* [Orig. unknown.] Bait usu. consisting of oily fish ground up and scattered on the water. —*vi.* **chummed, chum·ming, chums.** To fish with chum.

**chum·my** (chŭm′ē) *adj.* **-mi·er, -mi·est.** *Informal.* Friendly : intimate. **—chum′mi·ly** *adv.* **—chum′mi·ness** *n.*

**chump¹** (chŭmp) *n.* [Perh. a blend of CHUNK and LUMP or STUMP.] A dolt : blockhead.

**chump²** (chŭmp) *vt.* & *vi.* **chumped, chump·ing, chumps.** [Var. of CHAMP¹.] To chew or make a chewing movement.

**chunk** (chŭngk) *n.* [Perh. var. of CHUCK².] **1.** A thick piece or mass. **2.** A substantial amount <a *chunk* of money>

**chunk·y** (chŭng′kē) *adj.* **-i·er, -i·est. 1.** Short and thick : STOCKY. **2.** In chunks <*chunky* stew> **—chunk′i·ness** *n.*

**church** (chûrch) *n.* [ME *chirche* < OE *cirice* < LGk. *kuriakon* < Gk. *kuriakos,* of the lord < *kurios,* lord.] **1.** *often* **Church.** The company of all Christians considered as a mystic spiritual body. **2.** A building for public, esp. Christian, worship. **3.** A congregation. **4.** Public divine worship in a church. **5.** *often* **Church.** A specified Christian denomination <the Methodist *Church*> **6.** Ecclesiastical power <no union of *church* and state> **7.** The clerical profession : CLERGY. **8.** *Christian Science.* —Used to refer to <"The structure of Truth and Love" —Mary Baker Eddy> —*vt.* **churched, church·ing, church·es.** To conduct a church service for, esp. to perform a religious service for (a woman after childbirth). —*adj.* Of or relating to the church : ECCLESIASTICAL <*church* doctrine and dogma>

**church·go·er** (chûrch′gō′ər) *n.* One who attends church. **—church′go′ing** *adj.* & *n.*

**church key** *n. Informal.* A bottle or can opener with a triangular pointed head.

**church·ly** (chûrch′lē) *adj.* Of, relating to, or befitting a church <*churchly* attire> **—church′li·ness** *n.*

**church·man** (chûrch′mən) *n.* **1.** A clergyman : priest. **2.** A member of a church. **—church′man·ly** *adj.* **—church′man·ship′** *n.*

**Church of Christ, Scientist** *n.* The official name of the Christian Science Church.

**Church of England** *n.* The episcopal and liturgical national church of England.

**Church of Jesus Christ of Latter-day Saints** *n.* The official name of the Mormon Church.

**Church of Rome** *n.* The Roman Catholic Church.

**Church Slavonic** *n.* The literary language of Slavic manuscripts written after the early 11th cent. and still used as a liturgical language by several churches of the Eastern Orthodoxy with Slavic congregations.

**church·war·den** (chûrch′wôr′dn) *n.* **1.** A lay officer in the Anglican Church chosen annually by the vicar or the congregation to handle the secular and legal affairs of the parish. **2.** One of two elected chief lay officers of the vestry in the Episcopal Church.

**church·wom·an** (chûrch′wo͝om′ən) *n.* A woman who is a member of a church.

**church·yard** (chûrch′yärd′) *n.* A yard adjoining to a church, often used for burial.

**churl** (chûrl) *n.* [ME < OE *ceorl,* peasant.] **1.** A rude boor. **2.** A miser : niggard. **3. a.** A ceorl. **b.** A medieval English peasant.

**churl·ish** (chûr′lĭsh) *adj.* **1. a.** Of or like a churl : BOORISH. **b.** Befitting a churl : VULGAR. **2.** Difficult to work with : INTRACTABLE. **—churl′ish·ly** *adv.* **—churl′ish·ness** *n.*

**churn** (chûrn) *n.* [ME *chirne* < OE *cyrn.*] A vessel or device that agitates milk or cream to separate the oily globules from the caseous and serous parts, used to make butter. —*v.* **churned, churn·ing, churns.** —*vt.* **1. a.** To stir or agitate (milk or cream) in a churn. **b.** To make by agitating milk or cream. **2.** To shake or agitate vigorously <Wind *churned* up the bay.> —*vi.* **1.** To make butter by operating a churn. **2.** To move with or produce great agitation <waves *churning* in a northeast gale> **—churn out.** To produce automatically and esp. abundantly. **—churn′er** *n.*

**churr** (chûr) *n.* [Imit.] The sharp, whirring or trilling sound made by some insects and birds. —*vi.* **churred, churr·ing, churrs.** To make a churring sound.

**Chur·ri·gue·resque** (cho͝or′ĭ-gə-rĕsk′) *adj.* [Sp. *churrigueresco,* after José *Churriguera* (1650–1723).] Of or pertaining to a style of baroque architecture of Spain and its Latin-American colonies, marked by extravagant and elaborate decoration.

**chute** (sho͞ot) *n.* [Fr., a fall < OFr. *cheoite,* p.part. of *cheoir,* to fall < Lat. *cadere.*] **1.** An inclined trough, passage, or channel through or down which things may pass. **2.** A waterfall or rapid. **3.** *Informal.* A parachute.

**chut·ney** (chŭt′nē) *n.* [Hindi *caṭnī.*] A pungent relish of fruits, spices, and herbs.

**chutz·pah** (KHo͝ot′spə) *n.* [Yiddish.] *Slang.* Gall : brazenness.

**Chu·vash** (cho͞o-väsh′) *n., pl.* **Chuvash** *or* **-vash·es.** [R. < Chuvash *čǎvaš.*] **1.** One of a Turkic-speaking Tartar people living chiefly in the Chuvash A.S.S.R. **2.** The Turkic language of the Chuvash.

**chyle** (kīl) *n.* [Lat. *chylus* < Gk. *khulos,* juice < *khein,* to pour.] Thick white or pale-yellow fluid, consisting of lymph and finely emulsified fat, taken up by the lacteals from the intestine in digestion. **—chy·la'ceous** (kī-lā'shəs), **chy'lous** (kī'ləs) *adj.*

**chy·lo·mi·cron** (kī'lō-mī'krŏn') *n.* [CHYL(E) + Gk. *mikron,* small thing < neuter of *mikros,* small.] One of the microscopic fat particles in the blood that are formed during digestion of fat.

**chyme** (kīm) *n.* [LLat. *chymus* < Gk. *khumos,* juice < *khein,* to pour.] The thick semifluid mass of partly digested food that is passed from the stomach to the duodenum. **—chy'mous** (kī'məs) *adj.*

**chy·mo·sin** (kī'mə-sĭn) *n.* [CHYM(E) + -OS(E) + -IN.] Rennin.

**chy·mo·tryp·sin** (kī'mə-trĭp'sĭn) *n.* [CHYM(E) + TRYPSIN.] A pancreatic digestive enzyme.

**ciao** (chou) *interj.* [Ital. < dial. *schiavo,* (I am your) slave.] —Used to express greeting or farewell.

**ci·bo·ri·um** (sĭ-bôr'ē-əm, -bōr'-) *n., pl.* **-bo·ri·a** (-bôr'ē-ə, -bōr'-) [Med. Lat. *ciborium* < Lat., a drinking cup < Gk. *kibōrion.*] **1.** A vaulted canopy over an altar. **2.** A covered receptacle for holding the consecrated wafers of the Eucharist.

**ci·ca·da** (sĭ-kā'də, -kä'-) *n., pl.* **-das** *or* **-dae** (-dē') [NLat. *Cicada,* type genus < Lat *cicada,* cicada.] Any of various insects of the family Cicadidae, with a broad head, membranous wings, and in the male a pair of resonating organs that produce a high-pitched, drone.

**cicada killer** *n.* A wasp, *Sphecius speciosus,* that preys on cicadas.

**cic·a·trix** (sĭk'ə-trĭks', sĭ-kā'trĭks) *n., pl.* **cic·a·tri·ces** (sĭk'ə-trī'sēz, sĭ-kā'trĭ-sēz') [ME *cicatrice* < Lat. *cicatrix.*] **1.** Recently formed connective tissue on a healing wound **:** SCAR TISSUE. **2.** *Bot.* A scar left where a leaf or a branch has been detached. **—cic'a·tri'cial** (sĭk'ə-trĭsh'əl), **ci·cat'ri·cose'** (sĭ-kăt'rĭ-kōs') *adj.*

**cic·e·ly** (sĭs'ə-lē) *n., pl.* **-lies.** [ME *seseli* < Lat. *seselis* < Gk.] Sweet cicely.

**cic·e·ro·ne** (sĭs'ə-rō'nē, chē'chə-) *n., pl.* **-nes** *or* **-ni** (-nē) [Ital., after *Cicerone,* Marcus Tullius Cicero (106–43 B.C.).] A guide for sightseers.

**cich·lid** (sĭk'lĭd) *n.* [NLat. *Cichlidae,* family name < Gk. *kikhlē,* a kind of fish.] Any of various tropical freshwater fishes of the family Cichlidae, with spiny fins, many of which are popular as aquarium fish. **—cich'lid** *adj.*

**-cide** *suff.* [Fr., partly < Lat. *-cida,* killer, and partly < Lat. *-cidium,* killing, both < *caedere,* to kill.] **1.** Killer <*bactericide*> **2.** Act of killing <*ecocide*>

**ci·der** (sī'dər) *n.* [ME *sidre* < OFr. < LLat. *sicera,* intoxicating drink < Gk. *sikera* < Heb. *shēkār* < *shākar,* he drank deeply.] Juice pressed from fruits, esp. apples, and used for making vinegar or as a fermented or unfermented beverage.

**ci·gar** (sĭ-gär') *n.* [Sp. *cigarro,* poss. < Yucatec *sik,* to shred.] A compact roll of tobacco leaves prepared for smoking.

**cig·a·rette** *also* **cig·a·ret** (sĭg'ə-rĕt', sĭg'ə-rĕt') *n.* [Fr., dim. of *cigare,* cigar < Sp. *cigarro.*] A small roll of finely cut tobacco for smoking, enclosed in a wrapper of thin paper.

**cig·a·ril·lo** (sĭg'ə-rĭl'ō) *n., pl.* **-los.** [Sp., dim. of *cigarro,* cigar.] A small, narrow cigar.

**ci·gar-store Indian** (sĭ-gär'stôr', -stōr') *n.* A wooden effigy of an American Indian brave holding a cluster of cigars and once used as the emblem of a tobacconist.

**ci·lan·tro** (sĭ-lăn'trō) *n.* [Sp., alteration of LLat. *coliandrum* < Lat. *coriandrum.* —see CORIANDER.] The parsleylike leaves of fresh coriander used in cooking.

**cil·i·a** (sĭl'ē-ə) *n. pl. of* CILIUM.

**cil·i·ar·y** (sĭl'ē-ĕr'ē) *adj.* **1.** Of, relating to, or like cilia. **2.** Of or relating to the ciliary body.

**ciliary body** *n.* The thickened part of the vascular tunic of the eye that connects the choroid with the iris.

**ciliary movement** *n.* Cellular motion marked by rhythmic beating of cilia along the cell surface.

**cil·i·ate** (sĭl'ē-ĭt, -āt') *adj.* Having cilia. *—n.* Any of various protozoans of the class Ciliata, having numerous cilia. **—cil'i·ate·ly** *adv.*

**cil·i·a·ted** (sĭl'ē-ā'tĭd) *adj.* Ciliate.

**cil·ice** (sĭl'ĭs) *n.* [Fr. < Lat. *cilicium,* a covering made of Cilician goat's hair < *Cilicia,* a province in Asia Minor.] A coarse cloth **:** HAIRCLOTH.

**cil·i·o·late** (sĭl'ē-ə-lāt') *adj.* [< NLat. *ciliolum,* dim. of *cilium,* cilium.] Having minute cilia.

**cil·i·um** (sĭl'ē-əm) *n., pl.* **-i·a** (-ē-ə) [NLat. < Lat., eyelid.] **1.** A microscopic hairlike process extending from a cell surface and often capable of rhythmical motion. **2.** An eyelash.

**ci·met·i·dine** (sĭ-mĕt'ĭ-dēn', -dīn') *n.* [E. *ci-* (alteration of CYANO-) + MET(HYL) + -IDINE.] A drug that acts as a histamine receptor antagonist and is used for treating gastrointestinal diseases.

**ci·mex** (sī'mĕks') *n., pl.* **cim·i·ces** (sĭm'ĭ-sēz') [NLat. *Cimex,* genus name < Lat. *cimex,* bug.] An insect of the genus *Cimex,* including the bedbug.

**Cim·me·ri·an** (sĭ-mĭr'ē-ən) *adj.* [< Lat. *Cimmerri,* the Cimmerians < Gk. *Kimmerioi.*] Very gloomy **:** DARK. *—n. Myth.* One of a people described by Homer as inhabiting a land of eternal darkness.

**cinch** (sĭnch) *n.* [Sp. *cincha* < Lat. *cingula* < *cingere,* to gird.] **1.** A girth for a saddle or pack. **2.** *Informal.* A firm grip. **3.** *Slang.* **a.** Something easy to do. **b.** A sure thing. *—v.* **cinched, cinch·ing, cinch·es.** *—vt.* **1.** To put a saddle girth on. **2.** *Informal.* To get a tight grip on. **3.** *Slang.* To make certain of <*cinch* the deal> *—vi.* To tighten a saddle girth.

**cin·cho·na** (sĭng-kō'nə, sĭn-chō'-) *n.* [NLat. *Cinchona,* genus name, after Francisca Henriquez de Ribera (1576–1639), Countess of *Chinchón.*] **1.** Any of various South American trees and shrubs of the genus *Cinchona,* whose bark yields quinine. **2.** The dried bark of a cinchona tree. **—cin·chon'ic** (sĭng-kŏn'ĭk, sĭn-) *adj.*

**cin·cho·nine** (sĭng'kə-nēn', sĭn'chə-) *n.* [CINCHON(A) + -INE.] An alkaloid, $C_{19}H_{22}N_2O$, derived from the bark of various cinchona trees and used as an antimalarial agent.

**cin·cho·nism** (sĭng'kə-nĭz'əm, sĭn'chə-) *n.* A pathological condition characterized by deafness, headache, giddiness, and dimming eyesight, that results from an overdose of cinchona.

**cinc·ture** (sĭngk'chər) *n.* [Lat. *cinctura* < *cingere,* to gird.] **1.** A belt **:** girdle. **2.** Something that surrounds or encompasses. *—vt.* **-tured, -tur·ing, -tures.** To encompass or gird.

**cin·der** (sĭn'dər) *n.* [ME *sinder* < OE, dross.] **1.** A burned or partially burned substance, as coal, that is not reduced to ashes but is incapable of further combustion. **2.** A partially charred substance that can burn further but without flame. **3. cinders.** Ashes. **4.** SCORIA 1. **5.** SLAG 1. *—vt.* **-dered, -der·ing, -ders.** To burn or reduce to cinders. **—cin'der·y** *adj.*

**Cin·der·el·la** (sĭn'də-rĕl'ə) *n.* [< CINDER.] **1.** The fairy-tale heroine who escapes from a life of drudgery through the intervention of a fairy godmother and marries a handsome prince. **2.** One who gains affluence or recognition after obscurity and neglect.

**cin·e·ast** (sĭn'ē-ăst) *also* **cin·é·aste** (sĭn'ā-ăst') *n.* [Fr. *cinéaste* < *ciné,* cinema.] A film enthusiast.

**cin·e·ma** (sĭn'ə-mə) *n.* [Short for CINEMATOGRAPH.] **1.** A motion picture. **2.** A motion-picture theater. **3. a.** Motion pictures as a whole. **b.** The film industry. **4.** The art of making films. **—cin'e·mat'ic** (sĭn'ə-măt'ĭk) *adj.* **—cin'e·mat'i·cal·ly** *adv.*

**cin·e·mat·o·graph** (sĭn'ə-măt'ə-grăf') *n.* [Fr. *cinématographe* **:** Gk. *kinēma,* motion (< *kinein,* to move) + *-graphe,* -graph.] *Chiefly Brit.* A movie camera or projector. **—cin'e·mat'o·graph'ic** *adj.* **—cin'e·mat'o·graph'i·cal·ly** *adv.*

**cin·e·ma·tog·ra·phy** (sĭn'ə-mə-tŏg'rə-fē) *n.* The technique of making movies. **—cin'e·ma·tog'ra·pher** *n.*

**cin·e·mat·o·ra·di·og·ra·phy** (sĭn'ə-măt'ə-rā'dē-ŏg'rə-fē) *n.* [Gk. *kinēma, kinēmat-,* movement (< *kinein,* to move) + RADIOGRAPHY.] *Med.* Radiography of a body part or organ in motion. **—cin'e·mat'o·ra'di·o·graph'ic** (-ə-grăf'ĭk) *adj.* **—cin'e·mat'o·ra'di·o·graph'i·cal·ly** *adv.*

**ci·né·ma vé·ri·té** (sē'nā-mä' vā'rē-tā') *n.* [Fr. *cinéma-vérité* **:** *cinéma,* cinema + *vérité,* truth.] Filmmaking emphasizing unbiased realism.

**cin·e·ol** *also* **cin·e·ole** (sĭn'ē-ōl') *n.* [NLat. *cina,* wormseed + Lat. *oleum,* oil.] Eucalyptol.

**Cin·er·am·a** (sĭn'ə-răm'ə, -rä'mə). A trademark for a motion-picture process designed to produce a realistic effect.

**cin·e·rar·i·a** (sĭn'ə-râr'ē-ə) *n.* [NLat. *Cineraria,* genus name < Lat. *cinerarius,* of ashes (from the ash-colored down on its leaves) < *cinis,* ashes.] A plant native to the Canary Islands, *Senecio cruentis,* bearing variously colored daisylike flowers in flat clusters and widely grown as a house plant.

**cin·e·rar·i·um** (sĭn'ə-râr'ē-əm) *n., pl.* **-i·a** (ē-ə) [Lat. < *cinerarius,* of ashes < *cinis,* ashes.] A place for keeping the ashes of a cremated corpse. **—cin'er·ar'y** (sĭn'ə-rĕr'ē) *adj.*

**ci·ne·re·ous** (sĭ-nĭr'ē-əs) *adj.* [Lat. *cinereus* < *cinis,* ashes.] **1.** Consisting of or like ashes. **2.** Of the color of ashes.

**cin·er·in** (sĭn'ər-ĭn) *n.* [Lat. *cinis, ciner-,* ashes + -IN.] A compound, $C_{20}H_{28}O_3$ or $C_{21}H_{28}O_5$, used in insecticides.

**cin·gu·lum** (sĭng'gyə-ləm) *n., pl.* **-la** (-lə) [NLat. < Lat., girdle < *cingere,* to gird.] *Biol.* A girdlelike structure, band, or marking. **—cin'gu·late** (-lĭt), **cin'gu·la'ted** (-lā'tĭd) *adj.*

**cin·na·bar** (sĭn'ə-bär') *n.* [ME *cinabare* < Lat. *cinnabaris* < Gk. *kinnabari.*] **1.** A heavy reddish mercuric sulfide, HgS, the principal source of mercury. **2.** Red mercuric sulfide used as a pigment. **3.** VERMILION 1.

**cin·na·mon** (sĭn'ə-mən) *n.* [ME *cinamome* < OFr. < Lat. *cinnamomum* < Gk. *kinnamōmon.*] **1.** Either of two trees, *Cinnamomum zeylanicum* or *C. loureirii* of tropical Asia, with very aromatic bark. **2.** The yellowish-brown bark of a cinnamon tree, dried and often ground, used as a spice. **3.** Any of several trees yielding a spice similar to cinnamon. **4.** A light yellowish brown. *—adj.* Of a light yellowish brown. **—cin·nam'ic** (sĭ-năm'ĭk) *adj.*

**cinnamon bear** *n.* The American black bear during the phase when its color is reddish-brown.

**cinnamon stone** *n.* Essonite.

**cin·quain** (sĭng'kān', săng'-) *n.* [Fr. **:** *cinq,* five (< Lat. *quinque*) + *quatrain,* quatrain.] A five-line stanza.

**cinque** (sĭngk, săngk) *n.* [ME *cink* < OFr. < Lat. *quinque,* five.] The number five in cards or dice.

**cin·que·cen·to** (chĭng'kwĭ-chĕn'tō) *n.* [Ital., short for *millecinquecento,* one thousand five hundred.] The 16th cent., esp. in Italian art and architecture.

**cinque·foil** (sĭngk'foil', săngk'-) *n.* [ME *cinkfoil* < OFr. < Lat. *quinquefolium* : *quinque,* five + *folium,* leaf.] **1.** Any of various plants of the genus *Potentilla,* bearing compound leaves, often having five lobes. **2.** A design with five sides composed of converging arcs, usu. used as a panel or a frame for glass.

**ci·on** (sī'ən) *n. var. of* SCION 2.

**ci·pher** *also* **cy·pher** (sī'fər) [ME *cifre* < OFr. < Med. Lat. *cifra* < Ar. *ṣifr.*] *—n.* **1.** The mathematical symbol (0) denoting absence of quantity : ZERO. **2.** An Arabic numeral or figure : NUMBER. **3.** The Arabic system of numerical notation. **4.** One without influence or value : NONENTITY. **5. a.** A cryptographic system in which units of plain text of regular length, usu. letters, are arbitrarily substituted or transposed according to a predetermined key. **b.** The key to such a system. **c.** A message in code. **6.** A design combining or interweaving letters or initials : MONOGRAM. *—v.* **-phered, -pher·ing, -phers.** *—vi.* To solve problems in arithmetic : CALCULATE. *—vt.* **1.** To put in secret writing : ENCIPHER. **2.** To solve by means of arithmetic.

**cir·ca** (sûr'kə) *prep.* [Lat. < *circum,* around < *circus,* circle.] In approximately <written *circa* 1600>

**cir·ca·di·an** (sər-kā'dē-ən, -kăd'ē-, sûr'kə-dī'ən, -dē'-) *adj.* [Lat. *circa,* about + Lat. *dies,* day.] *Biol.* Exhibiting approx. 24-hour periodicity.

**circadian rhythm** *n.* A daily rhythmic activity cycle, based on 24-hour intervals, that is exhibited by many organisms.

**Cir·cas·sian** (sər-kăsh'ən, -kăsh'ē-ən) *n.* **1.** An inhabitant of Circassia, esp. a member of a Caucasian people inhabiting Circassia. **2.** The non-Indo-European language of the Circassians. *—adj.* Of or relating to the people, language, or region of Circassia.

**Circassian walnut** *n.* The mottled or veined light-brown wood of the English walnut used esp. in ornamental cabinetwork.

**Cir·ce** (sûr'sē) *n.* [Lat. < Gk. *Kirkē.*] *Myth.* An enchantress described in the *Odyssey* who detained Odysseus for a year and turned his men into swine. **—Cir·ce·an** (sûr'sē-ən, sər-sē'ən) *adj.*

**cir·ci·nate** (sûr'sə-nāt') *adj.* [Lat. *circinatus,* p.part. of *circinare,* to make circular < *circinus,* pair of compasses < *circus,* circle.] **1.** Ring-shaped. **2.** Rolled up from the tip, as a young fern frond. **—cir'ci·nate'ly** *adv.*

**Cir·ci·nus** (sûr'sə-nəs) *n.* [Lat. *circinus,* pair of compasses < *circus,* circle.] A constellation in the Southern Hemisphere.

**cir·cle** (sûr'kəl) *n.* [ME *cercle* < OFr. < Lat. *circulus,* dim. of *circus,* circle.] **1.** A plane curve everywhere equidistant from a fixed center. **2.** A planar region bounded by a circle. **3.** Something, as a ring, shaped like a circle. **4.** A circular course, circuit, or orbit. **5.** A curved section or tier of seats in a theater. **6.** A series or process that finishes at its starting point or continuously repeats itself : CYCLE. **7.** A group of individuals sharing a common interest, activity, or achievement. **8.** A territorial or administrative division, esp. of a European province. **9.** A sphere of influence or interest : DOMAIN <well-known in tennis *circles*> **10.** *Logic.* A fallacy in reasoning in which the premise is used to prove the conclusion, and the conclusion used to prove the premise. *—v.* **-cled, -cling, -cles.** *—vt.* **1.** To make or form a circle around : ENCLOSE. **2.** To move in a circle around. *—vi.* To move in circles : REVOLVE <The eagles *circled* slowly overhead.> **—cir'cler** (-klər) *n.*

☆ **syns:** CIRCLE, GYRE, ORB, RING *n. core meaning* : a closed plane curve everywhere equidistant from a fixed point <a bracelet shaped like a *circle* of diamonds>

**cir·clet** (sûr'klĭt) *n.* [ME *cerclet* < OFr., dim. of *cercle,* circle.] A small circle <a *circlet* of gold>

**cir·cuit** (sûr'kĭt) *n.* [ME, circumference < OFr. < Lat. *circuitus,* a going around < *circumire,* to go around : *circum,* around + *ire,* to go.] **1. a.** A closed, usu. circular curve. **b.** The area enclosed by such a curve. **2. a.** A route or path the complete traversal of which without local change of direction requires returning to the starting point. **b.** The act of following such a route or path. **c.** A journey made on such a route or path. **3.** *Elect.* **a.** A closed path followed or capable of being followed by an electric current. **b.** A configuration of electrically or electromagnetically connected devices or components. **4. a.** A regular or accustomed course from place to place, as that of a salesperson : ROUND. **b.** The area or district thus covered, esp. a territory under the jurisdiction of a judge, in which he or she holds periodic court sessions. **5.** An association of theaters in which plays, acts, or films move from theater to theater for performance. **6.** An association of clubs, teams, or arenas of competition. *—vi. & vt.* **-cuit·ed, -cuit·ing, -cuits.** To make a circuit or circuit of.

**circuit breaker** *n.* An automatic switch that stops the flow of electric current in an overloaded or otherwise abnormally stressed electric circuit.

**circuit court** *n.* The lowest court of record in some U.S. states, occas. holding sessions in different places.

**cir·cu·i·tous** (sər-kyo͞o'ĭ-təs) *adj.* [Med. Lat. *circuitosus* < Lat. *circuitus,* a going round. —see CIRCUIT.] Being or taking a roundabout course. **—cir·cu'i·tous·ly** *adv.* **—cir·cu'i·ty, cir·cu'i·tous·ness** *n.*

**circuit rider** *n.* A member of the clergy who travels from church to church in a district, esp. a rural district.

**cir·cuit·ry** (sûr'kĭ-trē) *n.* **1.** The design of or detailed plan for an electric circuit. **2.** Electric circuits as a whole.

**cir·cu·lar** (sûr'kyə-lər) *adj.* **1.** Of or relating to a circle. **2.** Shaped like or almost like a circle : ROUND. **3.** Moving in or forming a circle. **4.** Indirect : circuitous. **5.** Marked by reasoning in a circle <a *circular* discussion> **6.** Addressed or distributed to a large number of persons. *—n.* A printed advertisement, directive, or notice for mass distribution. **—cir'cu·lar'i·ty** (-lăr'ĭ-tē) *n.* **—cir'cu·lar·ly** *adv.*

**circular file** *n. Slang.* A wastebasket.

**circular function** *n.* Trigonometric function.

**cir·cu·lar·ize** (sûr'kyə-lə-rīz') *vt.* **-ized, -iz·ing, -iz·es.** To publicize with circulars. **—cir'cu·lar·i·za'tion** *n.* **—cir'cu·lar·iz'er** *n.*

**circular measure** *n.* The measure of an angle in radians.

**circular saw** *n.* An electric saw composed of a toothed disk rotated at high speed.

**cir·cu·late** (sûr'kyə-lāt') *v.* **-lat·ed, -lat·ing, -lates.** [Lat. *circulare, circulat-,* to make round < *circulus,* dim. of *circus,* circle.] *—vi.* **1.** To move in or flow through a circle or circuit <electricity *circulating* through the building> **2.** To move around, as from person to person or place to place <a candidate *circulating* through the crowd> **3.** To move about or flow freely, as air. **4.** To spread widely among persons or places : DISSEMINATE <Bad news tends to *circulate* quickly.> *—vt.* To cause to move about or be distributed. **—cir'cu·la'tive** (-lā'tĭv) *adj.* **—cir'cu·la'tor** *n.* **—cir'cu·la·tor'y** (-lə-tôr'ē, -tōr'ē) *adj.*

**circulating decimal** *n.* A repeating decimal.

**circulating library** *n.* A lending library.

**circulating medium** *n.* Currency or coin exchangeable for goods without endorsement.

**cir·cu·la·tion** (sûr'kyə-lā'shən) *n.* **1.** Movement in a circle or circuit. **2.** Movement of blood through bodily vessels due to the heart's pumping action. **3.** Movement or passage through a system of vessels, as of water through pipes : FLOW. **4.** Free movement or passage. **5.** Passage of something, as money or news, from place to place or person to person. **6.** The condition of being passed about and widely known : DISTRIBUTION. **7. a.** Distribution of printed material, esp. copies of periodicals and newspapers, among readers. **b.** The number of copies of a publication sold or distributed.

**circulatory system** *n.* The system of structures by which blood and lymph are circulated throughout the body.

**circum-** *pref.* [Lat. < *circum,* around < *circus,* circle.] Around : about <*circum*lunar>

**cir·cum·am·bi·ent** (sûr'kəm-ăm'bē-ənt) *adj.* Enclosing. **—cir'cum·am'bi·ence, cir'cum·am'bi·en·cy** *n.* **—cir'cum·am'bi·ent'ly** *adv.*

**cir·cum·cise** (sûr'kəm-sīz') *vt.* **-cised, -cis·ing, -cis·es.** [ME *circumcisen* < Lat. *circumcidere,* to cut around : *circum-,* around (< *circus,* circle) + *caedere,* to cut.] To remove the prepuce of (a male) or clitoris of (a female). **—cir'cum·cis'er** *n.*

**cir·cum·ci·sion** (sûr'kəm-sĭzh'ən) *n.* **1.** The act of circumcising. **2.** A religious ceremony in which one is circumcised. **3. Circumcision.** A church festival celebrated on Jan. 1 commemorating the circumcision of Jesus.

**cir·cum·duc·tion** (sûr'kəm-dŭk'shən) *n.* [Lat. *circumductio,* act of leading around < *circumducere,* to lead around: *circum-,* around (< *circus,* circle) + *ducere,* to lead.] The movement of a limb such that the distal end of the limb delineates an arc.

**cir·cum·fer·ence** (sər-kŭm'fər-əns) *n.* [ME < OFr. < Lat. *circumferentia* < *circumferre,* to carry around : *circum-,* around (< *circus,* circle) + *ferre,* to carry.] **1.** The boundary line of a circle. **2. a.** The boundary line of a closed curvilinear figure : PERIMETER. **b.** The length of such a boundary. **—cir·cum'fer·en'tial** (-fə-rĕn'shəl) *adj.*

☆ **syns:** CIRCUMFERENCE, AMBIT, CIRCUIT, COMPASS, PERIMETER *n. core meaning* : a line around a closed figure or area <walked around the *circumference* of the lake>

**cir·cum·flex** (sûr'kəm-flĕks') *n.* [Lat. *circumflexus,* a bending around < p.part. of *circumflectere,* to bend around : *circum-,* around (< *circus,* circle) + *flectere,* to bend.] A mark (ˆ) used over a vowel in certain languages or in phonetic keys to indicate quality of pronunciation. *—adj.* **1.** Marked with a circumflex. **2.** Curving around <a *circumflex* blood vessel>

**cir·cum·fuse** (sûr'kəm-fyo͞oz') *vt.* **-fused, -fus·ing, -fus·es.** [Lat. *circumfundere, circumfus-* : *circum-,* around (< *circus,* circle) + *fundere,* to pour.] **1.** To pour or diffuse around : SPREAD. **2.** To surround, as with liquid : SUFFUSE. **—cir'cum·fu'sion** *n.*

**cir·cum·lo·cu·tion** (sûr'kəm-lō-kyo͞o'shən) *n.* [ME *circumlocucioun* < Lat. *circumlocutio* : *circum-,* around (< *circus,* circle) + *loqui,* to speak.] **1.** Use of indirect and wordy language. **2.** Written or spoken evasion. **3.** A roundabout expression. **—cir'cum·loc'u·to'ri·ly** (-lŏk'yə-tôr'ə-lē, -tōr'-) *adv.* **—cir'cum·loc'u·to'ry** (-lŏk'yə-tôr'ē, -tōr'ē) *adj.*

**cir·cum·lu·nar** (sûr′kəm-lo͞o′nər) *adj.* Surrounding or orbiting the moon.
**cir·cum·nav·i·gate** (sûr′kəm-năv′ĭ-gāt′) *vt.* **-gat·ed, -gat·ing, -gates.** To sail completely around <*circumnavigate* the globe> **—cir′cum·nav′i·ga′tion** *n.* **—cir′cum·nav′i·ga′tor** *n.*
**cir·cum·nu·tate** (sûr′kəm-no͞o′tāt′, -nyo͞o′-) *vi.* **-tat·ed, -tat·ing, -tates.** [CIRCUM- + Lat. *nutare, nutat-*, to sway.] *Bot.* To grow or move with an irregular elliptical or spiral motion. **—cir′cum·nu·ta′tion** *n.*
**cir·cum·po·lar** (sûr′kəm-pō′lər) *adj.* **1.** Located in one of the polar regions. **2.** *Astron.* Designating a star that from a given observer's latitude does not go below the horizon.
**cir·cum·ro·tate** (sûr′kəm-rō′tāt′) *vi.* **-tat·ed, -tat·ing, -tates.** To turn like a wheel : REVOLVE. **—cir′cum·ro·ta′tion** *n.* **—cir′cum·ro′ta·to′ry** (-tə-tôr′ē, -tōr′ē) *adj.*
**cir·cum·scis·sile** (sûr′kəm-sĭs′əl, -īl′) *adj.* [CIRCUM- + Lat. *scissilis*, able to be split < *scindere*, to split.] *Bot.* Splitting or opening along a transverse circular line <a *circumscissile* seed capsule>
**cir·cum·scribe** (sûr′kəm-skrīb′) *vt.* **-scribed, -scrib·ing, -scribes.** [ME *circumscriben* < Lat. *circumscribere* : *circum-*, around (< *circus*, circle) + *scribere*, to write.] **1.** To draw a line around : ENCIRCLE. **2.** To confine within bounds : RESTRICT. **3.** To determine the boundaries of : DEFINE. **4. a.** To enclose (a polygon or polyhedron) within a configuration of lines, curves, or surfaces so that every vertex of the enclosed object is incident on the enclosing configuration. **b.** To be erected as such an enclosing configuration. **—cir′cum·scrib′a·ble** *adj.* **—cir′cum·scrib′er** *n.*
**cir·cum·scrip·tion** (sûr′kəm-skrĭp′shən) *n.* **1.** The act of circumscribing or state of being circumscribed. **2.** Something that circumscribes. **3.** A circumscribed space. **4.** A circular inscription, as on a medal. **—cir′cum·scrip′tive** *adj.* **—cir′cum·scrip′tive·ly** *adv.*
**cir·cum·so·lar** (sûr′kəm-sō′lər) *adj.* Surrounding or orbiting the sun.
**cir·cum·spect** (sûr′kəm-spĕkt′) *adj.* [ME < Lat. *circumspectus*, p.part. of *circumspicere*, to take heed : *circum-*, around (< *circus*, circle) + *specere*, to look.] Heedful of circumstances or future consequences. **—cir′cum·spec′tion** *n.* **—cir′cum·spect′ly** *adv.*
**cir·cum·stance** (sûr′kəm-stăns′) *n.* [ME < OFr. < Lat. *circumstantia* < *circumstare*, to stand around : *circum-*, around (< Lat. *circus*, circle) + *stare*, to stand.] **1.** A fact or condition attending an event and having bearing on it. **2.** A fact or condition that determines or must be considered in the determining of a course of action. **3.** The sum of determining factors beyond willful control <a victim of *circumstance*> **4.** *often* **circumstances.** Financial status or means <in straitened *circumstances*> **5.** Additional or accessory information : DETAIL. **6.** Formal display : CEREMONY <marched in pomp and *circumstance* through the church> *—vt.* **-stanced, -stanc·ing, -stanc·es.** To place in particular circumstances or conditions : SITUATE. **—under no circumstances.** In no case : NEVER. **—under (or in) the circumstances.** Such being the case.
**cir·cum·stan·tial** (sûr′kəm-stăn′shəl) *adj.* **1.** Of, relating to, or dependent on circumstances. **2.** Of no primary significance : INCIDENTAL. **3.** Complete and full of detail <a *circumstantial* account of the campaign> **—cir′cum·stan′tial·ly** *adv.*
**circumstantial evidence** *n. Law.* Evidence not bearing directly on the fact in dispute but on various attendant circumstances from which the judge or jury might infer the occurrence of the fact in dispute.
**cir·cum·stan·ti·al·i·ty** (sûr′kəm-stăn′shē-ăl′ĭ-tē) *n., pl.* **-ties.** **1.** The quality of being fully detailed. **2.** A particular detail.
**cir·cum·stan·ti·ate** (sûr′kəm-stăn′shē-āt′) *vt.* **-at·ed, -at·ing, -ates.** [Lat. *circumstantia*, circumstance + -ATE[1].] To set forth or verify with circumstances : give detailed proof or description of. **—cir′cum·stan′ti·a′tion** *n.*
**cir·cum·ter·res·tri·al** (sûr′kəm-tə-rĕs′trē-əl) *adj.* Surrounding or orbiting the earth.
**cir·cum·val·late** (sûr′kəm-văl′āt′) *vt.* **-lat·ed, -lat·ing, -lates.** [Lat. *circumvallare, circumvallat-* : *circum-*, around (< *circus*, circle) + *vallum*, rampart with palisades.] To surround with or as if with a defensive barrier, as a rampart. *—adj.* (-āt′, -ĭt). Surrounded with or as if with a defensive barrier, as a rampart. **—cir′cum·val·la′tion** *n.*
**cir·cum·vent** (sûr′kəm-vĕnt′) *vt.* **-vent·ed, -vent·ing, -vents.** [Lat. *circumvenire, circumvent-* : *circum-*, around (< *circus*, circle) + *venire*, to come.] **1.** To surround and entrap by devious means. **2.** To overcome by clever maneuvering. **3.** To avoid by or as if by passing around. **—cir′cum·vent′er, cir′cum·ven′tor** *n.* **—cir′cum·ven′tion** *n.* **—cir′cum·ven′tive** *adj.*
**cir·cum·vo·lu·tion** (sər-kŭm′və-lo͞o′shən, sûr′kəm-vō-) *n.* [ME *circumvolucioun* < Med. Lat. *circumvolutio* < Lat. *circumvolvere*, to roll around. —see CIRCUMVOLVE.] **1.** An act of turning, coiling, or folding about a core, center, or axis. **2.** A single coil, turn, or fold : CONVOLUTION.
**cir·cum·volve** (sûr′kəm-vŏlv′) *vi.* & *vt.* **-volved, -volv·ing, -volves.** [Lat. *circumvolvere* : *circum-*, around (< *circus*, circle) + *volvere*, to roll.] To revolve or cause to revolve.
**cir·cus** (sûr′kəs) *n.* [Lat., circle.] **1. a.** Entertainment consisting usu. of varied performances by acrobats, clowns, and trained animals. **b.** A traveling company that presents such entertainments. **c.** A circular arena, surrounded by tiers of seats and sometimes covered by a tent, in which such shows are presented. **2.** A roofless, oval enclosure surrounded by tiers of seats and used in ancient times for public spectacles. **3.** *Chiefly Brit.* An open circular place where several streets intersect. **4.** *Informal.* A place or activity marked by confused or noisy disorder <The office is a *circus.*>
**cirque** (sûrk) *n.* [Fr. < Lat. *circus*, circle.] A steep hollow, often containing a small lake, at the upper end of a mountain valley.
**cir·rate** (sĭr′āt′) *adj.* [Lat. *cirratus*, curled < *cirrus*, curl.] *Biol.* Having or of the nature of a cirrus or cirri.
**cir·rho·sis** (sĭ-rō′sĭs) *n.* [Gk. *kirros*, tawny (from the color of the diseased liver) + -OSIS.] A chronic disease of the liver characterized by progressive destruction and regeneration of liver cells and increased connective tissue formation that ultimately results in blockage of portal circulation, portal hypertension, liver failure, and death. **—cir·rhot′ic** (-rŏt′ĭk) *adj.*
**cirri-** *pref. var. of* CIRRO-.
**cir·ri·ped** (sĭr′ə-pĕd′) *also* **cir·ri·pede** (-pēd′) *n.* [NLat. *Cirripedia*, order name : CIRRUS + Lat. *pes*, foot.] Any of various crustaceans of the order Cirripedia, including the barnacles and similar organisms that fasten themselves to objects or become parasitic in the adult stage. **—cir′ri·ped′** *adj.*

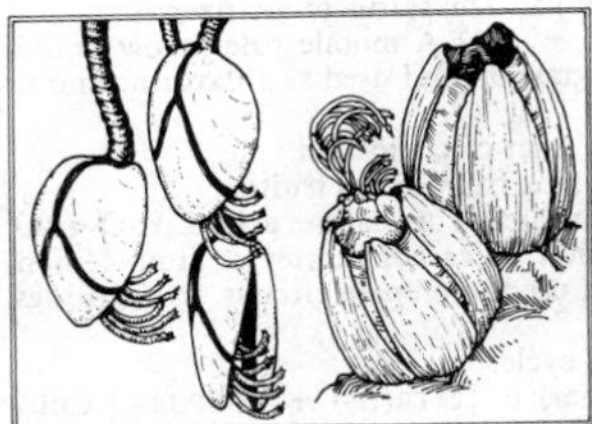

**cirriped**
*Two types of cirripeds:* (left) *goose barnacles and* (right) *acorn barnacles*

**cirro-** or **cirri-** *pref.* [< CIRRUS.] Cirrus cloud <*cirro*stratus>
**cir·ro·cu·mu·lus** (sĭr′ō-kyo͞om′yə-ləs) *n.* A high-altitude cloud made up of a series of small, regularly arranged cloudlets shaped like ripples or grains.
**cir·ro·stra·tus** (sĭr′ō-strā′təs, -străt′əs) *n.* A high-altitude, thin hazy cloud, usu. covering the sky and often generating a halo effect.
**cir·rus** (sĭr′əs) *n., pl.* **cir·ri** (sĭr′ī′) [Lat., curl of hair.] **1.** A high-altitude cloud made up of narrow bands or patches of thin, gen. white, fleecy parts. **2.** *Bot.* A tendril. **3.** *Zool.* A slender, flexible appendage, as a tentacle.
**cis-** *pref.* [Lat. < *cis*, on this side of.] On this side <*cis*atlantic>
**cis·at·lan·tic** (sĭs′ət-lăn′tĭk) *adj.* Being on this side of the Atlantic.
**cis·co** (sĭs′kō) *n., pl.* **-coes** or **-cos.** [Canadian Fr. *ciscoette* < Ojibwa *pemitewiskawet*, oily-skinned fish.] A North American freshwater fish of the genus *Coregonus* or *Leucichthys*, related to the whitefish.
**cis·lu·nar** (sĭs-lo͞o′nər) *adj.* Being between the earth and the moon.
**cis·mon·tane** (sĭs-mŏn′tān′) *adj.* [Fr. *cismontain* < Lat. *cismontanus* : *cis-*, cis- + *montanus*, of the mountains < *mons*, mountain.] Being on this side of the mountain.
**cist**[1] (sĭst) *n.* [Lat. *cista* < Gk. *kistē*.] An ancient Roman wicker receptacle used for carrying sacred utensils in processions.
**cist**[2] (sĭst, kĭst) *n.* [Welsh, chest < Lat. *cista* < Gk. *kistē*.] A Neolithic stone coffin.
**Cis·ter·cian** (sĭ-stûr′shən) *n.* [Fr. *Cistertien* < Med. Lat. *Cistercium*, Cîteaux, France, site of an abbey.] A member of a contemplative monastic order founded in France in 1098 by reformist Benedictines. **—Cis·ter′cian** *adj.*
**cis·tern** (sĭs′tərn) *n.* [ME *cisterne* < Lat. *cisterna* < *cista*, box < Gk. *kistē*.] **1.** A receptacle for holding liquid, esp. a tank for catching and storing rainwater. **2.** *Anat.* A cisterna. **—cis·tern′al** (sĭ-stûr′nəl) *adj.*
**cis·ter·na** (sĭ-stûr′nə) *n., pl.* **-nae** (-nē′) [NLat. < Lat., cistern.] **1.** A fluid-containing sac or space in the body of an organism. **2.** One of the saclike vesicles that make up the endoplasmic reticulum.
**cis·tron** (sĭs′trŏn′) *n.* [CIS- + TR(ANS)- + -ON[1].] A subunit of a gene that is complementary to another subunit on the same gene such that they form a single functional unit. **—cis·tron′ic** (sĭ-strŏn′ĭk) *adj.*
**cit·a·del** (sĭt′ə-dəl, -dĕl′) *n.* [OFr. *citadelle* < OItal. *citadella*, dim. of *cittade*, city < Lat. *civitas*.] **1.** A fortress in a commanding position in or near a city. **2.** A stronghold or fortified place : BULWARK.
**ci·ta·tion** (sī-tā′shən) *n.* **1.** The act of citing. **2. a.** A quoting of a source for substantiation. **b.** A source so cited : QUOTATION. **3.** *Law.* A reference to previous court decisions or authoritative writings. **4.** An official commendation for meritorious action, esp. in military

service. **5.** Enumeration or mention, as of facts. **6.** An official summons, esp. one calling for appearance in court <a traffic *citation*> **—ci·ta'tion·al** *adj.* **—ci'ta·to'ry** (sī'tə-tôr'ē, -tōr'ē) *adj.*

**cite** (sīt) *vt.* **cit·ed, cit·ing, cites.** [ME *citen,* to summon < OFr. *citer* < Lat. *citare,* freq. of *ciēre,* to call.] **1.** To quote as an authority or example. **2.** To bring forward or mention as support, illustration, or proof. **3.** To commend for meritorious action. **4.** To bring to another's attention : MENTION. **5.** To summon before a court of law. **—cit'a·ble** *adj.*

**cith·a·ra** (sĭth'ər-ə, kĭth'-) *n.* [Lat. < Gk. *kithara.*] An ancient musical instrument similar to the lyre.

**cith·er** (sĭth'ər, sĭth'-) *also* **cith·ern** (-ərn) *n.* [Fr. *cithare* < Lat. *cithara,* cithara < Gk. *kithara.*] A cittern.

**cit·ied** (sĭt'ēd) *adj.* Having a city or cities.

**cit·i·fied** (sĭt'ĭ-fīd') *adj.* Characterized by or having customs, manners, fashions, or other qualities attributed to city dwellers.

**cit·i·fy** (sĭt'ĭ-fī') *vt.* **-fied, -fy·ing, -fies. 1.** To cause to become urban. **2.** To give the styles and manners of a city to. **—cit'i·fi·ca'tion** *n.*

**cit·i·zen** (sĭt'ĭ-zən) *n.* [ME *citisein* < AN *citesein,* prob. alteration of OFr. *citeain* < *cite,* city.] **1.** A person owing loyalty to and entitled by birth or naturalization to the protection of a particular state. **2.** A resident of a city or town, esp. one permitted to vote and enjoy other privileges there. **3.** A civilian as distinguished from an employee of a state. **—cit'i·zen·ly** *adj.*

**cit·i·zen·ry** (sĭt'ĭ-zən-rē) *n., pl.* **-ries.** Citizens as a whole.

**citizen's arrest** *n.* Arrest by a citizen who is authorized to do so by his or her status as a citizen.

**citizens band** *n.* A radio-frequency band set aside for private use.

**cit·i·zen·ship** (sĭt'ĭ-zən-shĭp') *n.* The status of a citizen.

**cit·ral** (sĭt'răl') *n.* [CITR(US) + -AL.] A mobile pale-yellow liquid, $C_{10}H_{16}O$, derived from lemon-grass oil and used as a flavoring and in perfume.

**cit·rate** (sĭt'rāt') *n.* A salt or ester of citric acid.

**cit·ric** (sĭt'rĭk) *adj.* Of or derived from citrus fruits.

**citric acid** *n.* A colorless translucent crystalline acid, $C_6H_8O_7{\cdot}H_2O$, chiefly derived by fermentation of carbohydrates or from lemon, lime, and pineapple juices and used to prepare citrates, in flavorings, and in metal polishes.

**citric acid cycle** *n.* Krebs cycle.

**cit·ri·cul·ture** (sĭt'rĭ-kŭl'chər) *n.* [CITR(US) + CULTURE.] Cultivation of citrus fruits. **—cit'ri·cul'tur·ist** *n.*

**ci·trine** (sĭ-trēn', sĭt'rēn') *n.* [ME, yellow < OFr. *citrin* < Med. Lat. *citrinus* < Lat. *citrus,* citrus tree.] **1.** A pale-yellow quartz resembling topaz. **2.** A light to moderate olive. **—ci·trine'** *adj.*

**cit·ron** (sĭt'rən) *n.* [Fr. < OFr. < Lat. *citrus.*] **1. a.** A tree native to Asia, *Citrus medica,* bearing lemonlike fruit with a thick, aromatic rind. **b.** The fruit of the citron. **2.** A variety of watermelon, *Citrullus vulgaris citroides,* bearing fruit gen. considered inedible and with a hard rind used as flavoring. **3.** The preserved or candied rind of a citron, used esp. in baking. **4.** A grayish green yellow. **—cit'ron** *adj.*

**cit·ro·nel·la** (sĭt'rə-nĕl'ə) *n.* [NLat. < Fr. *citronnelle,* lemon oil, dim. of *citron,* citron.] **1.** A tropical Eurasian grass, *Cymbopogon nardus,* with bluish-green, lemon-scented leaves. **2.** A light-yellow, aromatic oil derived from citronella and used in insect repellents and perfumery.

**cit·ro·nel·lal** (sĭt'rə-nĕl'ăl') *n.* [CITRONELL(A) + -AL.] A colorless mixture of isomeric liquids, $C_9H_{17}CHO$, the principal constituent of citronella oil.

**cit·rul·line** (sĭt'rə-lēn') *n.* [NLat. *Citrullus,* plant genus + -INE.] An amino acid, $C_6H_{13}N_3O_3$, produced as an intermediate during urea formation in the liver.

**cit·rus** (sĭt'rəs) *adj.* [NLat., *Citrus,* genus name < Lat. *citrus,* citrus tree.] **1.** Of or relating to trees or shrubs of the genus *Citrus,* many of which bear edible fruit such as the orange, lemon, lime, and grapefruit. **2.** Of or typical of the fruits of citrus trees or shrubs. *—n., pl.* **-rus·es** *or* **citrus.** A citrus tree or shrub.

**cit·tern** (sĭt'ərn) *n.* [Ult. < Lat. *cithara,* cithara < Gk. *kithara.*] A 16th-cent. pear-shaped guitar.

**cit·y** (sĭt'ē) *n., pl.* **-ies.** [ME *cite* < OFr. < Lat. *civitas* < *civis,* citizen.] **1.** A center of population, commerce, and culture. **2. a.** An incorporated U.S. municipality with definite boundaries and legal powers set forth in a charter granted by the state. **b.** A high-ranking Canadian municipality, usu. determined by population but varying by province. **c.** A large incorporated town in Great Britain, usu. the seat of a bishop, with its title conferred by the Crown. **3.** Inhabitants of a city as a whole. **4.** An ancient Greek city-state.

**city council** *n.* The governing body of a city.

**city desk** *n.* A department in a newspaper handling local news.

**city editor** *n.* **1.** A newspaper editor responsible for local news and reporters' assignments. **2.** *Chiefly Brit.* The editor in charge of commercial and financial news.

**city father** *n.* A municipal official, as a council member.

**city hall** *n.* **1.** The building containing administrative offices of a municipal government. **2.** Municipal government, esp. its officials regarded as a group.

**city manager** *n.* An administrator appointed by a city council to manage city affairs.

**cit·y·scape** (sĭt'ē-skāp') *n.* **1.** An artistic representation, as a painting or photograph, of a city. **2.** A city regarded as a scene <"the vast *cityscape* of lower Manhattan" —*The New Yorker*>

**city slicker** *n. Informal.* A person displaying the sophisticated and smart style traditionally associated by rural people with the manners and customs of the city.

**cit·y-state** (sĭt'ē-stāt') *n.* A sovereign state composed of an independent city and its surrounding territory.

**civ·et** (sĭv'ĭt) *n.* [Fr. *civette* < OFr. < OItal. *zibetto* < Ar. *zabād.*] **1.** Any of various catlike mammals of the family Viverridae of Asia and Africa, with anal scent glands that secrete a fluid with a musky odor. **2.** Fluid secreted by a civet, used in perfumery. **3.** The fur of a civet.

**civ·ic** (sĭv'ĭk) *adj.* [Lat. *civicus* < *civis,* citizen.] Of, relating to, or belonging to a city, a citizen, or citizenship.

**civ·ics** (sĭv'ĭks) *n. (sing. in number).* The branch of political science concerned with civic affairs.

**civ·ies** (sĭv'ēz) *pl.n. var. of* CIVVIES.

**civ·il** (sĭv'əl) *adj.* [ME < Lat. *civilis* < *civis,* citizen.] **1.** Of, relating to, or befitting a citizen or citizens. **2.** Of or relating to citizens and their relations with the state. **3.** Having to do with ordinary citizens or ordinary community life as opposed to the military or the ecclesiastical. **4.** Of or in accordance with organized society : CIVILIZED. **5.** Observing or befitting accepted social usages : POLITE <ask a *civil* question> **6.** Designating or according to legally recognized divisions of time <a *civil* year> **7.** *Law.* Relating to the rights of private individuals and to legal proceedings involving these rights as distinguished from criminal, military, or international courts, proceedings, or rules. **—civ'il·ly** *adv.*

**civil death** *n. Law.* Total deprivation of civil rights resulting from conviction for treason or another serious offense.

**civil defense** *n.* The emergency measures to be taken by an organized body of civilian volunteers for the protection of life and property in case of a natural disaster, enemy attack, or invasion.

**civil disobedience** *n.* Refusal to obey civil laws regarded as unjust, usu. by employing methods of passive resistance.

**civil engineer** *n.* An engineer trained to design and construct public works. **—civil engineering** *n.*

**ci·vil·ian** (sĭ-vĭl'yən) *n.* **1.** A person following the pursuits of civil life. **2.** A student of or specialist in Roman or civil law. *—adj.* Of or relating to civilians or civil life : NONMILITARY.

**ci·vil·i·ty** (sĭ-vĭl'ĭ-tē) *n., pl.* **-ties. 1.** Courtesy. **2.** A courteous act or utterance.

**civ·i·li·za·tion** (sĭv'ə-lĭ-zā'shən) *n.* **1.** An advanced stage of development in the arts and sciences accompanied by corresponding political, social, and cultural complexity. **2.** The type of culture and society developed by a particular nation or region or in a particular epoch <the *civilization* of ancient Rome> **3.** The act or process of civilizing or of reaching a civilized state. **4.** *Informal.* Modern society with its conveniences <left *civilization* to live in the jungle>

**civ·i·lize** (sĭv'ə-līz') *vt.* **-lized, -liz·ing, -liz·es. 1.** To bring out of a savage or primitive state. **2.** To educate or enlighten : REFINE. **—civ'i·liz'a·ble** *adj.* **—civ'i·liz'er** *n.*

**civ·i·lized** (sĭv'ə-līzd') *adj.* **1. a.** Having a highly developed culture and society. **b.** Of, relating to, or typical of a people or nation so developed. **2.** Polite or cultured : REFINED <My colleagues are a *civilized* group of people.>

**civil law** *n.* **1.** The body of law concerned with the rights of private citizens in a specific state or nation as distinguished from criminal, military, or international law. **2.** Ancient Roman law, esp. that which applied to private citizens. **3.** A system of law originating in Roman law as distinguished from common or canon law.

**civil liberty** *n.* Liberty legally guaranteeing to the individual the rights of free speech, thought, and action, limited only insofar as their use does not affect the rights of others. **—civil libertarian** *n.*

**civil marriage** *n.* A marriage performed by a civil official.

**civil rights** *pl.n.* Rights belonging to a person by virtue of his or her status as a citizen or as a member of civil society. **—civil'-rights'** *adj.*

**civil servant** *n.* A person employed in a nation's civil service.

**civil service** *n.* **1.** All branches of public service that are not legislative, judicial, naval, or military. **2.** The employees of the civil branches of a government.

**civil war** *n.* War between factions or regions of a single nation.

**civ·vies** *also* **civ·ies** (sĭv'ēz) *pl.n.* [Shortening and alteration of CIVILIAN.] *Slang.* Civilian clothes.

**Cl** *symbol for* CHLORINE.

**clab·ber** (klăb'ər) *n.* [Short for obs. *bonnyclabber* < Ir. Gael. *bainne clabair* : *bainne,* milk + *clabar,* thick sour milk.] Sour, curdled milk. *—vt. & vi.* **-bered, -ber·ing, -bers.** To curdle.

**clack** (klăk) *v.* **clacked, clack·ing, clacks.** [ME *clakken* < ON *klaka.*] *—vi.* **1.** To make an abrupt, dry sound, as of the impact of two hard surfaces. **2.** To chatter aimlessly or at length. **3.** To cackle or cluck, as a hen. *—vt.* To cause to make an abrupt, dry sound. *—n.*

**1.** A clacking sound. **2.** An object that makes a clacking sound. **3.** Aimless, prolonged talk : CHATTER. **—clack'er** *n.*

**clack valve** *n.* A hinged valve that allows fluids to flow only in one direction and clacks when the valve closes.

**Clac·to·ni·an** (klăk-tō'nē-ən) *adj.* [After *Clacton*-on-Sea, England.] Of or relating to a lower Paleolithic culture of northwestern Europe.

**clad[1]** (klăd) *vt.* **clad, clad·ding, clads.** [< CLAD[2].] To sheathe (a metal) with another metal.

**clad[2]** (klăd) *v. var. p.t. & p.p. of* CLOTHE.

**clad·ding** (klăd'ĭng) *n.* Metal coating bonded onto another metal.

**cla·doc·er·an** (klə-dŏs'ər-ən) *n.* [NLat. *Cladocera,* order name : Gk. *klados,* branch + Gk. *keras,* horn.] Any of various small aquatic crustaceans of the order Cladocera, including the water fleas. *—adj.* Of or belonging to the Cladocera.

**clad·ode** (klăd'ōd') *n.* [NLat. *cladodium* < Gk. *klados,* branch.] A cladophyll. **—cla·do'di·al** (klə-dō'dē-əl) *adj.*

**clad·o·gen·e·sis** (klăd'ə-jĕn'ĭ-sĭs) *n.* [Gk. *klados,* branch + -GENESIS.] An evolutionary pattern in which the differentiation of two or more groups within a population generates additional taxa. **—clad'o·ge·net'ic** (-jə-nĕt'ĭk) *adj.* **—clad'o·ge·net'i·cal·ly** *adv.*

**clad·o·phyll** (klăd'ə-fĭl') *n.* [Gk. *klados,* twig + -PHYLL.] A branch or part of a stem that resembles and functions as a leaf.

**clag** (klăg) *vt. & vi.* **clagged, clag·ging, clags.** [Perh. < dial. *clag,* to cover with mud, of Scand. orig.] To clog. **—clag** *n.*

**claim** (klām) *vt.* **claimed, claim·ing, claims.** [ME *claimen* < OFr. *clamer* < Lat. *clamare,* to call.] **1.** To demand or ask for as one's own or one's due <*claim* a share of an estate> **2.** To state to be true : ASSERT. **3.** To deserve or call for : REQUIRE <a situation that *claims* the efforts of many> *—n.* **1.** A demand for something as one's rightful due. **2.** A basis for demanding : TITLE. **3.** Something claimed, esp.: **a.** A tract of land staked out by a homesteader or miner. **b.** Money demanded in accordance with an insurance policy or other formal arrangement. **4.** A statement of fact : assertion of truth <advertising that makes too many *claims*> **—claim'a·ble** *adj.* **—claim'ant, claim'er** *n.*

**claiming race** *n.* A horse race in which each entry may be purchased at a previously fixed price, the right to buy often being limited to those people entering horses in that race.

**clair de lune** (klâr' də lo͞on') *n.* [Fr. : *clair,* light + *de,* of + *lune,* moon.] **1.** Pale, grayish-blue glaze applied to various kinds of Chinese porcelain. **2.** The color of clair de lune.

**clair·voy·ance** (klâr-voi'əns) *n.* [Fr. : *clair,* clear (< Lat. *clarus*) + *voyant,* pr.part. of *voir,* to see < Lat. *vidēre.*] **1.** Power to perceive things that are naturally beyond the range of human senses. **2.** Acute intuitive insight. **—clair·voy'ant** *n.*

**clam[1]** (klăm) *n.* [Obs. *clam-shell,* clam < CLAM[2].] **1.** Any of various usu. burrowing marine and freshwater bivalve mollusks of the class Pelecypoda, including members of the genera *Venus* and *Mya,* many of which are edible. **2.** *Informal.* A close-mouthed person. *—vi.* **clammed, clam·ming, clams.** To hunt for clams. **—clam up.** To cease talking or remain silent.

**clam[2]** (klăm) *n.* [ME < OE, bond.] A clamp or vise.

**cla·mant** (klā'mənt, klăm'ənt) *adj.* [Lat. *clamans, clamant-,* pr.part. of *clamare,* to cry out.] **1.** Clamorous : loud. **2.** Urgent : compelling. **—cla'mant·ly** *adv.*

**clam·bake** (klăm'bāk') *n.* **1.** A seashore picnic where clams, fish, corn, and other foods are baked in layers on buried hot stones. **2.** *Informal.* A party or gathering, esp. a noisy and lively one.

**clam·ber** (klăm'bər, klăm'ər) *vt. & vi.* **-bered, -ber·ing, -bers.** [ME *clambren.*] To climb with difficulty, esp. on all fours. *—n.* **—clam'ber** *n.* **—clam'ber·er** *n.*

**clam chowder** *n.* Any of various soups made of shucked clams, salt pork, onions, and potatoes.

**clam·my** (klăm'ē) *adj.* **-mi·er, -mi·est.** [ME *clammi,* sticky < *clam,* of MLG orig.] Disagreeably moist, sticky, and usu. cold <a clammy atmosphere on a foggy day> **—clam'mi·ly** *adv.* **—clam'mi·ness** *n.*

**clam·or** (klăm'ər) *n.* [ME *clamour* < OFr. < Lat. *clamor,* shout < *clamare,* to cry out.] **1.** Loud outcry : HUBBUB. **2.** A vehement expression of discontent or protest : public outcry. **3.** Loud and sustained noise : DIN. *—v.* **-ored, -or·ing, -ors.** *—vi.* **1.** To make a clamor. **2.** To make insistent demands or complaints. *—vt.* **1.** To exclaim noisily and insistently. **2.** To drive or influence by clamor. **—clam'or·er** *n.*

**clam·or·ous** (klăm'ər-əs) *adj.* Making, full of, or marked by clamor. **—clam'or·ous·ly** *adv.* **—clam'or·ous·ness** *n.*

**clam·our** (klăm'ər) *n. & v. Chiefly Brit. var. of* CLAMOR.

**clam·our·ous** (klăm'ər-əs) *adj. Chiefly Brit. var. of* CLAMOROUS.

**clamp** (klămp) *n.* [ME < MDu. *klampe.*] A device for joining, gripping, supporting, or compressing structural or mechanical parts. *—vt.* **clamped, clamp·ing, clamps.** To fasten, grip, or support with or as if with a clamp. **—clamp down.** *Informal.* To become repressive.

**clamp·er** (klăm'pər) *n.* A spiked plate attached to the sole of a shoe to prevent slipping on ice.

**clam·shell** (klăm'shĕl') *n.* **1.** The shell of a clam. **2.** A dredging bucket made of two hinged jaws.

**clam·worm** (klăm'wûrm') *n.* Any of various segmented marine worms of the genus *Nereis.*

**clan** (klăn) *n.* [Sc. Gael. *clann,* family < OIr. *cland,* offspring < Lat. *planta,* sprout.] **1.** A traditional social unit in the Scottish Highlands, made up of a number of families claiming a common ancestor and following the same hereditary chieftain. **2.** A division of a tribe tracing descent from a common ancestor. **3.** A large group of relatives, friends, or associates.

**clan·des·tine** (klăn-dĕs'tĭn) *adj.* [Lat. *clandestinus* < *clam,* secretly.] Concealed or kept secret : SURREPTITIOUS. **—clan·des'tine·ly** *adv.* **—clan·des'tine·ness** *n.*

**clang** (klăng) *vi. & vt.* **clanged, clang·ing, clangs.** [Lat. *clangere.*] To make or cause to make a loud, metallic, resonant sound. *—n.* **1.** A clanging sound. **2.** The strident call of a bird such as a crane or goose.

**clang·er** (klăng'ər) *n. Chiefly Brit.* A faux pas.

**clan·gor** (klăng'ər, klăng'gər) *n.* [Lat. < *clangere,* to clang.] A clang or repeated clanging : DIN. *—vi.* **-gored, -gor·ing, -gors.** To make a clangor. **—clan'gor·ous** *adj.* **—clan'gor·ous·ly** *adv.*

**clan·gour** (klăng'ər, klăng'gər) *n. & v. Chiefly Brit. var. of* CLANGOR.

**clank** (klăngk) *n.* [Imit.] A sharp and hard metallic sound. *—vi.* **clanked, clank·ing, clanks.** To make a clanking sound.

**clan·nish** (klăn'ĭsh) *adj.* **1.** Of, relating to, or typical of a clan. **2.** Inclined to stick together and exclude outsiders <*clannish* mountain people> **—clan'nish·ly** *adv.* **—clan'nish·ness** *n.*

**clans·man** (klănz'mən) *n.* One who belongs to a clan.

**clans·wom·an** (klănz'wo͝om'ən) *n.* A woman who belongs to a clan.

**clap** (klăp) *v.* **clapped, clap·ping, claps.** [ME *clappen* < OE *clappan.*] *—vi.* **1.** To strike the palms of the hands together with a sudden explosive sound, as in applauding. **2.** To make a sudden sharp noise. *—vt.* **1. a.** To strike (the hands) together repeatedly with an abrupt, loud sound. **b.** To applaud in this manner. **2.** To tap with the open hand, as in greeting <*clapped* me on the back> **3.** To put, move, or send promptly or suddenly <*clapped* the lions into cages> **4.** *Informal.* To put together hastily <*clap* together a budget> *—n.* **1. a.** The act or sound of clapping the hands. **b.** A loud, sharp, or explosive noise. **2.** A sharp blow with the open hand : SLAP. **3.** *Obs.* A sudden stroke, esp. of bad luck.

**clap·board** (klăb'ərd, klăp'bôrd', -bōrd') *n.* [Partial transl. of MDu. *clapholt* : *clappen,* to split + *holt,* board.] A long, narrow board with one edge thicker than the other, overlapped to cover the outer walls of frame houses. *—vt.* **-board·ed, -board·ing, -boards.** To cover with clapboards.

**clap·per** (klăp'ər) *n.* **1.** One that claps, esp. the tongue of a bell. **2. clappers.** Two flat pieces of wood that are held between the fingers and struck together rhythmically. **3.** *Slang.* The tongue of a talkative person.

**clap·per·claw** (klăp'ər-klô') *vt.* **-clawed, -claw·ing, -claws.** *Archaic.* **1.** To scratch or claw. **2.** To revile or berate.

**clapper rail** *n.* A North American marsh bird, *Rallus longirostris,* with brownish plumage, a long bill, and a clattering cry.

**clap·trap** (klăp'trăp') *n.* [Obs. *claptrap,* a theatrical trick to win applause : CLAP + TRAP[1].] Pretentious or insincere language.

**claque** (klăk) *n.* [Fr. < *claquer,* to clap.] **1.** A group of persons hired to applaud at a performance. **2.** A group of fawning admirers.

**clar·ence** (klăr'əns) *n.* [After the Duke of *Clarence* (1765–1837), later William IV of England.] A four-wheeled closed carriage seating four passengers.

**clar·et** (klăr'ĭt) *n.* [Short for ME *claret win,* light-colored wine < OFr. *vin claret* : *vin,* wine (< Lat. *vinum*) + *claret,* light-colored < *clair,* clear < Lat. *clarus.*] **1. a.** The dry red wine of Bordeaux, France. **b.** A similar wine made elsewhere. **2.** A dark or grayish purplish red to dark purplish pink.

**claret cup** *n.* A chilled mixed drink of red wine variously combined with soda and fruit juices.

**Cla·re·tian** (klə-rē'shən, klă-) *n.* [After St. Anthony *Claret* (1807–1870).] A member of the Congregation of the Missionary Sons of the Immaculate Heart of Mary. **—Cla·re'tian** *adj.*

**clar·i·fy** (klăr'ə-fī') *v.* **-fied, -fy·ing, -fies.** [ME *clarifien* < OFr. *clarifier* < LLat. *clarificare* : Lat. *clarus,* clear + Lat. *facere,* to make.] *—vt.* **1.** To make clear or easier to understand : ELUCIDATE. **2.** To make clear by removing impurities, often by heating gently <*clarify* butter> *—vi.* To become clear. **—clar'i·fi·ca'tion** *n.* **—clar'i·fi'er** *n.*

☆ ***syns:*** CLARIFY, CLEAR (up), ELUCIDATE, ILLUMINATE *v. core meaning* : to make clear or clearer <*clarified* the law>

**clar·i·net** (klăr'ə-nĕt') *also* **clar·i·o·net** (klăr'ē-ə-nĕt') *n.* [Fr. *clarinette,* dim. of *clarine,* cattle bell < *clair,* clear < Lat. *clarus.*] A woodwind instrument having a straight, cylindrical tube with a flaring bell and a single-reed mouthpiece, played by means of finger holes and keys. **—clar'i·net'ist, clar'i·net'tist** *n.*

**clar·i·on** (klăr′ē-ən) *n.* [ME *clarioun* < OFr. *clarion* < Med. Lat. *clario* < Lat. *clarus,* clear.] **1.** A medieval trumpet with a clear, shrill tone. **2.** The sound made by a clarion or a sound resembling it. —*adj.* Shrill and clear <heard the *clarion* call to arms>

**clar·i·o·net** (klăr′ē-ə-nĕt′) *n. var. of* CLARINET.

**clar·i·ty** (klăr′ĭ-tē) *n.* [ME *clerte,* brightness < OFr. *clarte* < Lat. *claritas,* clearness < *clarus,* clear.] The quality or state of being clear : LUCIDITY.

**clark·i·a** (klär′kē-ə) *n.* [NLat. *Clarkia,* genus name, after William *Clark* (1770–1838).] A plant of the genus *Clarkia* of western North America, bearing red, purple, or pink flowers.

**clar·y** (klâr′ē) *n., pl.* **-ies.** [ME *clare,* partly < Med. Lat. *sclarea* and partly < OE *slaria.*] A European plant of the genus *Salvia,* esp. *S. sclarea,* an aromatic herb with bluish-white flowers.

**clash** (klăsh) *v.* **clashed, clash·ing, clash·es.** [Imit.] —*vi.* **1.** To collide with a loud, harsh noise. **2.** To conflict <members of Congress *clashing* over the budget> —*vt.* To strike together with a harsh, metallic noise. —*n.* **1.** A loud, resounding metallic noise, as that made by two objects colliding. **2.** A state of disharmony or disagreement. **3.** An often heated argument.

**clas·mat·o·cyte** (klăz-măt′ə-sīt′) *n.* [Gk. *klasma, klasmat-,* fragment (< *klan,* to break) + -CYTE.] Histiocyte. **—clas·mat′o·cyt′ic** (-sĭt′ĭk) *adj.*

**clasp** (klăsp) *n.* [ME *claspe,* prob. of OE orig.] **1.** A fastening, as a hook, used to hold two parts or objects together. **2. a.** An embrace : hug. **b.** A grip or grasp of the hand. **3.** A small metal bar attached to a military decoration indicating the action for which it was awarded. —*vt.* **clasped, clasp·ing, clasps. 1.** To fasten with or as if with a clasp. **2.** To hold in a tight grasp : EMBRACE. **3.** To grip firmly in or with the hand. **—clasp′er** *n.*

**clasp·er** (klăs′pər) *n.* A modified part of the pelvic fin of male elasmobranch fishes that aids sperm transmission during copulation.

**clasp knife** *n.* A pocketknife with a folding blade.

**class** (klăs) *n.* [Fr. *classe* < Lat. *classis,* class of citizens.] **1. a.** A set, group, collection, or configuration containing members having or believed to have at least one attribute in common : KIND. **b.** *Statistics.* An interval in a frequency distribution. **2.** A division by quality, grade, or rank. **3. a.** A social stratum whose members share similar political, economic, and cultural characteristics. **b.** Social rank, esp. high rank. **4. a.** A group of alumni or students graduated in the same year. **b.** A group of students studying the same subject. **c.** The period during which such a group meets. **5. a.** *Biol.* A taxonomic category ranking below a phylum and above an order. **b.** A subdivision of a larger group. **6. a.** A grade of mail. **b.** The quality of accommodation on a public vehicle <travel tourist *class*> **7.** *Slang.* Great style or quality <a violinist with *class*> —*vt.* **classed, class·ing, class·es.** To arrange, group, or rate according to qualities or characteristics : CLASSIFY.

☆ ***syns:*** CLASS, CATEGORY, CLASSIFICATION, ORDER, SET *n. core meaning* : a subdivision of a larger group <the *class* of concertgoers who enjoy chamber music>

▲ word history: *Class* is from Latin *classis,* "a division of the Roman people." *Classis* had an adjective, *classicus,* that originally meant "belonging to a *classis*" but later meant "belonging to the highest-ranking *classis,*" from which the sense "first-class, superior" developed. English *classic* and *classical* are derived from *classicus* in the more general sense. Because the ancient Greek and Roman authors were considered models of excellence and were universally studied in schools and universities throughout Europe until recent times, the terms *classic* and *classical* came to be associated first with ancient Greek and Roman literature of all degrees of excellence and then with anything pertaining to ancient Greece or Rome.

**class action** *n.* A lawsuit in which the plaintiff or plaintiffs bring suit both on their own behalf and on behalf of others who have the same claim against the defendant.

**class-con·scious** (klăs′kŏn′shəs) *adj.* Aware of belonging to a specific socioeconomic class. **—class′-con′scious·ness** *n.*

**clas·sic** (klăs′ĭk) *adj.* **1.** Being of the highest class or rank. **2. a.** Serving as an outstanding representative of a kind : MODEL. **b.** Well-known and typical <a *classic* Western with ambushes and shootouts> **3.** Having lasting significance or recognized worth. **4.** CLASSICAL 1a. **5.** Of or in accordance with established principles and methods in the arts and sciences. **6.** Of lasting literary or historical significance. —*n.* **1.** An artist, author, or work gen. held to be of the highest rank or excellence. **2.** A literary work of ancient Greece or Rome. **3.** Something typical or traditional.

**clas·si·cal** (klăs′ĭ-kəl) *adj.* **1. a.** Of or relating to the culture of ancient Greece and Rome, esp. the art, architecture, and literature. **b.** Relating to or versed in studies of antiquity. **2. a.** Relating to or designating European music of the latter half of the 18th cent. **b.** Designating music in the educated European tradition. **3.** Standard and authoritative <*classical* methods of building construction> **4.** Of or relating to nonrelativistic or nonquantum physics <*classical* mechanics> **—clas′si·cal′i·ty** (-kăl′ĭ-tē), **clas′si·cal·ness** *n.* **—clas′si·cal·ly** *adv.*

**clas·si·cism** (klăs′ĭ-sĭz′əm) *n.* **1.** Aesthetic attitudes and principles based on the culture, art, and literature of ancient Greece and Rome and characterized by emphasis on form, simplicity, proportion, and restrained emotion. **2.** Classical scholarship. **3.** A Greek or Latin e pression or idiom.

**clas·si·cist** (klăs′ĭ-sĭst) *n.* A specialist in or student of the classic

**clas·si·fi·ca·tion** (klăs′ə-fĭ-kā′shən) *n.* **1.** The act or result of cla sifying. **2.** *Biol.* The systematic grouping of organisms into categori based on shared characteristics or traits : TAXONOMY. **—clas′si·fi·c to′ry** (klăs′ə-fĭ-kə-tôr′ē, -tōr′ē, klə-sĭf′ĭ-, klăs′ə-fĭ-kā′tə-rē) *adj.*

**classified advertisement** *n.* A usu. brief advertisemen printed in a newspaper or magazine along with others of the sam category.

**clas·si·fy** (klăs′ə-fī′) *vt.* **-fied, -fy·ing, -fies. 1.** To organize or a range according to class or category. **2.** To designate (e.g., a doc ment) as secret and available only to authorized person **—clas′si·fi′a·ble** *adj.* **—clas′si·fi′er** *n.*

**clas·sis** (klăs′ĭs) *n., pl.* **-ses** (-ēz′) [Lat., class of citizens.] **1.** A go erning body of pastors and elders in certain Reformed churches ha ing jurisdiction over local churches. **2.** The district or church governed by a classis.

**class mark** *n. Statistics.* MARK[1] 22.

**class·mate** (klăs′māt′) *n.* A member of the same academic clas

**class·room** (klăs′ro͞om′, -ro͝om′) *n.* A room in which classes mee

**class·y** (klăs′ē) *adj.* **-i·er, -i·est.** *Slang.* Elegant : stylish <a *class* paint job> **—class′i·ness** *n.*

**clast** (klăst) *n.* [< Gk. *klastos,* broken < *klan,* to break.] A roc fragment.

**clas·tic** (klăs′tĭk) *adj.* [< Gk. *klastos,* broken < *klan,* to break **1.** Separable into parts or having removable sections <a *clastic* an tomical model> **2.** *Geol.* Made up of fragments : FRAGMENTA **—clas′tic** *n.*

**clath·rate** (klăth′rāt′) *adj.* [Lat. *clathratus,* p.part. of *clathrare,* furnish with a lattice < Gk. *klēithron,* door bar < *kleiein,* to close **1.** *Biol.* Latticelike in structure or appearance. **2.** *Chem.* Of or rela ing to inclusion complexes in which molecules of one substance a completely enclosed within the crystal structure of another. —*n.* clathrate compound.

**clat·ter** (klăt′ər) *v.* **-tered, -ter·ing, -ters.** [ME *clateren.*] —*v* **1.** To make a rattling sound. **2.** To move with a rattling sound. **3.** T talk noisily and rapidly : CHATTER. —*vt.* To cause to make a rattlin sound. —*n.* **1.** A rattling sound. **2.** A loud disturbance : COMMOTIO **3.** Noisy talk : CHATTER. **—clat′ter·er** *n.*

**clau·di·ca·tion** (klô′dĭ-kā′shən) *n.* [ME *claudicacioun* < La *claudicatio* < *claudicare,* to limp < *claudus,* lame.] A halt in a pe son's walk : LIMP.

**clause** (klôz) *n.* [ME < OFr. < Med. Lat. *clausa,* close of a rhetoric period < Lat. *claudere,* to close.] **1.** A group of words having a subje and predicate and forming part of a compound or complex sentenc **2.** A distinct article, stipulation, or provision in a documen **—claus′al** (klô′zəl) *adj.*

**claus·tral** (klô′strəl) *adj. var. of* CLOISTRAL.

**claus·tro·pho·bi·a** (klô′strə-fō′bē-ə) *n.* [Lat. *claustrum,* enclose place (< *claudere,* to close) + -PHOBIA.] Abnormal fear of confine spaces. **—claus′tro·pho′bic** (-fō′bĭk) *adj.*

**cla·vate** (klā′vāt′) *adj.* [NLat. *clavatus* < Lat. *clava,* club.] Havin one end thickened : CLUB-SHAPED. **—cla′vate′ly** *adv.*

**clave[1]** (klāv) *v. Archaic. var. p.t. of* CLEAVE[1].

**clave[2]** (klāv) *v. Archaic. var. p.t. of* CLEAVE[2].

**cla·ver** (klā′vər) **-vered, -ver·ing, -vers.** [Perh. of Celt. orig.] *Sco* —*vi.* To gossip or talk idly. —*n.* Idle talk : GOSSIP.

**clav·i·chord** (klăv′ĭ-kôrd′) *n.* [Med. Lat. *clavichordium* : Lat. *cla vis,* key + Lat. *chorda,* string < Gk. *khordē.*] An early musical ke board instrument with a soft sound produced by tangents strikin horizontal strings.

**clav·i·cle** (klăv′ĭ-kəl) *n.* [NLat. *clavicula* < Lat., dim. of *clavis,* ke (from its shape).] A bone linking the sternum and the scapul **—cla·vic′u·lar** (klə-vĭk′yə-lər) *adj.* **—cla·vic′u·late′** (-lāt′) *adj.*

**clav·i·corn** (klăv′ĭ-kôrn′) *adj.* [NLat. *Clavicornia,* family name Lat. *clava,* club + Lat. *cornu,* horn.] Belonging to or designating group of beetles of the suborder Polyphaga, with clavate antenna

**cla·vier** (klə-vîr′, klā′vē-ər, klăv′ē-) *n.* [Fr. < OFr., key-bearer < La *clavis,* key.] **1.** A keyboard. **2.** A stringed keyboard instrument, as harpsichord.

**clav·i·form** (klăv′ə-fôrm′) *adj.* [Lat. *clava,* club + -FORM.] Clavate

**cla·vus** (klā′vəs, klä′-) *n.* [Lat., nail.] CORN[2].

**claw** (klô) *n.* [ME *clave* < OE *clawu.*] **1.** A sharp, often curved na on the toe of a mammal, bird, or reptile. **2. a.** A pincerlike structure as a chela, on the limb of a crustacean or other arthropod. **b.** A lim terminating in such a structure. **3.** Something resembling a claw, a the cleft end of a hammerhead. **4.** *Bot.* The narrowed basal part c certain petals or sepals. —*vt. & vi.* **clawed, claw·ing, claws.** To di or scratch or make digging or scratching movements with or as i with claws.

**claw hammer** *n.* **1.** A hammer with one end of the head forke for removing nails. **2.** A swallow-tailed coat.

**claw hatchet** *n.* A hatchet with one end of the head forked

**clay** (klā) *n.* [ME *clei* < OE *clæg.*] **1.** A fine-grained, firm natural material, plastic when wet, that consists primarily of hydrated silicates of aluminum and is widely used to make bricks, tiles, and pottery. **2.** An earth that forms a paste with water and hardens when heated, esp. one with grains smaller than 0.002 millimeters in diameter. **3.** Moist earth : MUD. **4.** The human body as opposed to the spirit. **—clay′ey** (klā′ē), **clay′ish** (klā′ĭsh) *adj.*

**clay mineral** *n.* Any of various hydrous silicates that have a crystalline structure and are components of clay.

**clay·more** (klā′môr′, -mōr′) *n.* [Gael. *claidheamh mōr* : *claidheamh*, sword + *mōr*, great.] **1.** A large, double-edged broadsword once used by Scottish Highlanders. **2.** A claymore mine.

**claymore mine** *n.* A lens-shaped, ground-emplaced antipersonnel mine whose blast is focused only in the direction of the advancing enemy.

**clay pigeon** *n.* A clay disk thrown as a flying target for skeet and trapshooting.

**clay·to·ni·a** (klā-tō′nē-ə) *n.* [NLat. *Claytonia*, herb genus, after John *Clayton* (1693–1773).] The spring beauty.

**clean** (klēn) *adj.* **-er, -est.** [ME *clene* < OE *clǣne.*] **1.** Free from dirt, stain, or impurities : UNSOILED. **2.** Free from foreign matter : UNADULTERATED. **3.** Producing little radioactive fallout or contamination <a *clean* nuclear bomb> **4.** Without imperfections or blemishes. **5.** Free from clumsiness <a *clean* hit> **6.** Without restrictions or encumbrances. **7.** Complete : entire <a *clean* break> **8.** Having few alterations or corrections : LEGIBLE <*clean* typing> **9.** Blank <a *clean* page> **10.** Morally pure : SINLESS <led a *clean* life> **11.** Not obscene or ribald <a *clean* joke> **12.** Honest and fair, as in sports <a *clean* competitor> **13.** *Slang.* **a.** Not carrying concealed weapons. **b.** Free from narcotics addiction, use, or possession. **c.** Innocent of a suspected crime. *—adv.* **1. a.** So as to be clean <streets washed *clean* by the rain> **b.** In a clean way <played the entire match *clean*> **2.** *Informal.* Wholly : entirely <an arrow that went *clean* through the target> *—v.* **cleaned, clean·ing, cleans.** *—vt.* **1. a.** To rid of dirt. **b.** To free of impurities. **2.** To prepare (fowl or other food) for cooking. **3.** To remove the contents from : EMPTY <*cleaned* the platter> *—vi.* To undergo or perform the act of ridding of dirt and impurities. **—clean house.** To eliminate or discard what is undesirable <After the election the President decided to *clean house.*> **—clean out. 1.** To drive or force out <*cleaned out* all opponents of government policies> **2.** To deprive completely, as of money <The expensive vacation *cleaned* me *out.*> **—clean up. 1.** To dispose of : SETTLE <*cleaned up* old debts> **2.** *Informal.* To make a large profit, often in a short time <speculated on computer stocks and *cleaned up*> **—clean′a·ble** *adj.* **—clean′ness** *n.*

☆ ***syns:*** CLEAN, ANTISEPTIC, CLEANLY, IMMACULATE, SPOTLESS, STAINLESS *adj. core meaning* : free from dirt <*clean* clothes><*clean* instruments> ***ant*** : dirty

**clean-cut** (klēn′kŭt′) *adj.* **1.** Clearly and sharply outlined or defined. **2.** Trim and neat <*clean-cut* cadets>

**clean·er** (klē′nər) *n.* **1.** One whose work or business is cleaning. **2.** A machine or substance used in cleaning.

**clean-hand·ed** (klēn′hăn′dĭd) *adj.* Guiltless : innocent.

**clean-limbed** (klēn′lĭmd′) *adj.* With well-formed limbs.

**clean·ly** (klĕn′lē) *adj.* **-li·er, -li·est.** Habitually and carefully clean and neat. *—adv.* (klēn′lē). In a clean way. **—clean′li·ness** (klĕn′lē-nĭs) *n.*

**clean room** *n.* A room that is kept contaminant-free, esp. for the handling of precision parts.

**cleanse** (klĕnz) *vt.* **cleansed, cleans·ing, cleans·es.** [ME *clensen* < OE *clǣnsian.*] To free from dirt, defilement, or guilt : CLEAN.

**cleans·er** (klĕn′zər) *n.* One that cleans, as soap or detergent.

**clean-shav·en** (klēn′shā′vən) *adj.* **1.** With the beard or hair shaved off. **2.** Having recently shaved.

**clean·up** (klēn′ŭp′) *n.* **1.** A thorough cleaning. **2.** *Informal.* A very large profit.

**clear** (klîr) *adj.* **-er, -est.** [ME *cler* < OFr. < Lat. *clarus.*] **1.** Free from clouds, mist, or haze <a *clear* sky> **2.** Free from what dims, obscures, or darkens : TRANSPARENT <*clear* glass> **3.** Free from flaw, blemish, or impurity <a *clear* skin> **4.** Free from impediment, obstruction, or hindrance : OPEN <a *clear* passage> **5.** Plain or evident to the mind : UNMISTAKABLE <a *clear* example of fraud> **6.** Easily perceptible to the eye or ear : DISTINCT. **7.** Discerning or perceiving easily : KEEN <a *clear* thinker> **8.** Free from doubt or confusion : CERTAIN. **9.** Free from limitation or qualification : ABSOLUTE. **10.** Free from guilt <a *clear* conscience> **11.** Freed from contact or connection : DISENGAGED. **12.** Without charges or deductions : NET <won a *clear* $40,000> **13.** Containing nothing. *—adv.* **1.** Clearly : distinctly <spoke loud and *clear*> **2.** *Informal.* All the way : ENTIRELY <climbed *clear* to the top> *—v.* **cleared, clear·ing, clears.** *—vt.* **1.** To make clear, light, or bright. **2.** To rid of blemishes, impurities, muddiness, or foreign matter. **3.** To make plain or intelligible <*cleared* up the mystery of the child's disappearance> **4. a.** To rid of obstructions or entanglements <*clear* the path of booby traps> **b.** To remove or get rid of (obstructions or entanglements) <*clear* fallen rocks from the road> **5.** To free from a legal charge or imputation of guilt : ACQUIT <was *cleared* of blame in the accident> **6.** To pass by, under, or over without contact <The horses *cleared* all the jumps.> **7.** To settle (a debt) by paying it. **8.** To gain (a given amount) as net earnings or profit. **9.** To pass (a bill of exchange, as a check) through a clearing-house. **10.** To free (a ship or cargo) from legal detention at a harbor by fulfilling harbor and customs requirements. **11.** To give (an aircraft) clearance, as for takeoff. **12.** To free (the throat) of phlegm by coughing. *—vi.* **1.** To become clear. **2.** To pass through a clearing-house. **3.** To comply with harbor and customs requirements in discharging a cargo or in entering or leaving a port. **—clear out.** *Informal.* To leave a place, often in haste. *—n.* **1.** A clear or open space. **2.** Clearance. **—clear the air.** To dispel emotional tensions or differences. **—in the clear. 1.** Free from dangers or burdens. **2.** Not subject to suspicion or accusations of guilt <Testimony of witnesses put me *in the clear.*> **—clear′a·ble** *adj.* **—clear′er** *n.* **—clear′ly** *adv.* **—clear′ness** *n.*

**clear-air turbulence** (klîr′âr′) *n.* Severe atmospheric turbulence that occurs under otherwise tranquil conditions and subjects aircraft to strong updrafts and downdrafts.

**clear·ance** (klîr′əns) *n.* **1.** The act or process of clearing. **2.** A space cleared : CLEARING. **3.** The amount by which a moving object clears something. **4.** An intervening space or distance allowing free play, as between machine parts. **5.** Permission for a vehicle to proceed, as after inspection of cargo or equipment. **6.** Official certification of blamelessness, reliability, or suitability. **7.** A sale of old merchandise, gen. at reduced prices. **8.** The passage of bills of exchange, as checks, through a clearing-house.

**clear-cut** (klîr′kŭt′) *adj.* **1. a.** Distinctly and sharply outlined or defined. **b.** Marked by or having keen intellect. **2.** Evident : plain.

**clear-eyed** (klîr′īd′) *adj.* **1.** With sharp, bright eyes : KEEN-SIGHTED. **2.** Mentally acute or perceptive.

**clear-head·ed** (klîr′hĕd′ĭd) *adj.* Having a clear, orderly mind. **—clear′-head′ed·ly** *adv.* **—clear′-head′ed·ness** *n.*

**clear·ing** (klîr′ĭng) *n.* **1.** Land from which the trees and other obstructions have been removed. **2. a.** Exchange among banks of checks, drafts, and notes and the settlement of differences arising from it. **b. clearings.** The total of claims presented daily at a clearing-house.

**clear·ing-house** *also* **clear·ing·house** (klîr′ĭng-hous′) *n.* An office where banks exchange checks and drafts and settle accounts.

**clear-sight·ed** (klîr′sī′tĭd) *adj.* **1.** Having sharp, clear vision. **2.** Perceptive : discerning. **—clear′-sight′ed·ly** *adv.* **—clear′-sight′ed·ness** *n.*

**clear·sto·ry** (klîr′stôr′ē, -stōr′ē) *n. var. of* CLERESTORY.

**clear·weed** (klîr′wēd′) *n.* A plant, *Pilea pumila* of eastern North America, with small green flowers and translucent stems and leaves.

**clear·wing** (klîr′wĭng′) *n.* Any of various moths of the family Aegeriidae, with scaleless, transparent wings.

**cleat** (klēt) *n.* [ME *clete*, of OE orig.] **1.** A strip of wood or iron for strengthening or supporting the surface to which it is attached. **2.** A piece of iron, rubber, or leather attached to the underside of a shoe to preserve the sole or prevent slipping. **3.** A piece of metal or wood with projecting arms or ends on which a rope can be wound or secured. **4.** A wedge-shaped piece of material fastened onto something, as a spar, for support or to prevent slipping. **5.** A spurlike device used to grip a tree or pole in climbing. *—vt.* **cleat·ed, cleat·ing, cleats.** To supply, support, secure, or strengthen with a cleat.

**cleav·age** (klē′vĭj) *n.* **1.** The act of splitting or cleaving. **2.** The state of being split or cleft : FISSURE. **3.** *Mineral.* The splitting or tendency to split of a crystallized substance along definite crystalline planes, yielding smooth surfaces. **4.** *Biol.* The process of or any of various stages in cell division that produce a blastula from a fertilized ovum. **5.** *Informal.* The separation between a woman's breasts.

**cleave**[1] (klēv) *v.* **cleft** (klĕft) *or* **cleaved** *or* **clove** (klōv), **cleft** *or* **cleaved** *or* **clo·ven** (klō′vən), **cleav·ing, cleaves.** [ME *cleven* < OE *clēofan.*] *—vt.* **1.** To split or separate, as with an ax. **2.** To make or accomplish by or as if by cutting <*cleave* their way through the underbrush> **3.** To penetrate or pierce. *—vi.* **1.** To split or separate, esp. along a natural line of division. **2.** To make one's way : PASS. **—cleav′a·ble** *adj.*

**cleave**[2] (klēv) *vi.* **cleaved** *or* **clove** (klōv), **cleaved, cleav·ing, cleaves.** [ME *cleven* < OE *cleofian.*] **1.** To adhere or cling. **2.** To be faithful.

**cleav·er** (klē′vər) *n.* A heavy, axlike knife or hatchet used esp. by butchers.

**cleav·ers** (klē′vərz) *n.* [ME *cliver* < OE *clīfe.*] *(sing. or pl. in number).* A plant of the genus *Galium*, esp. *G. aparine*, with small white flowers and prickly stems and seeds.

**cleek** (klēk) *n.* [ME *cleike*, large hook.] **1.** A number-one golf iron, with very little loft to the club face. **2.** *Scot.* A large hook.

**clef** (klĕf) *n.* [Fr. < Lat. *clavis*, key.] *Mus.* A symbol on a staff showing the pitch of the notes.

**cleft** (klĕft) *adj.* [P.part of CLEAVE[1].] **1.** Split : divided. **2.** *Bot.* Having deeply divided lobes or divisions <a *cleft* leaf> *—n.* **1.** A crack or crevice. **2.** A split or indentation between two parts, as of the chin.

**cleft palate** *n.* A congenital fissure in the roof of the mouth.

**cleis·tog·a·mous** (klī-stŏg'ə-məs) *also* **cleis·to·gam·ic** (klī'stə-găm'ĭk) *adj.* [Gk. *kleistos,* closed (< *kleiein,* to close) + -GAMOUS.] *Bot.* Marked by self-fertilization in an unopened budlike state. **—cleis·tog'a·mous·ly** *adv.* **—cleis·tog'a·my** (-mē) *n.*

**cleis·to·the·ci·um** (klī'stə-thē'sē-əm) *n., pl.* **-ci·a** (-sē-ə) [NLat. : Gk. *kleistos,* closed + Gk. *thēkion,* small case, dim. of *thēkē,* chest.] A closed spherical ascocarp.

**clem·a·tis** (klĕm'ə-tĭs) *n.* [NLat. *Clematis,* genus name < Lat. *clematis,* a creeping plant < Gk. *klēmatis* < *klēma,* twig.] A plant or vine of the genus *Clematis* of eastern Asia and North America, bearing white or variously colored flowers and plumelike seeds.

**clem·en·cy** (klĕm'ən-sē) *n., pl.* **-cies. 1.** Mercy, esp. toward an offender or enemy : LENIENCY. **2.** A kind, merciful, or lenient act. **3.** Mildness, esp. of weather.

**clem·ent** (klĕm'ənt) *adj.* [ME < Lat. *clemens.*] **1.** Lenient or merciful. **2.** Mild <*clement* April days> **—clem'ent·ly** *adv.*

**clench** (klĕnch) *vt.* **clenched, clench·ing, clench·es.** [ME *clenchen* < OE *(be)clencan.*] **1.** To bring together (hands or teeth) tightly : CLOSE. **2.** To grasp or grip tightly. **3.** To clinch (e.g., a bolt). **4.** *Naut.* To fasten with a clinch. —*n.* **1.** A tight grip or grasp. **2.** A device that clenches or holds fast. **3.** *Naut.* CLINCH 4.

**cle·o·me** (klē-ō'mē) *n.* [NLat. *Cleome,* genus name.] Any of various mostly tropical plants of the genus *Cleome,* esp. *C. spinosa,* cultivated for its white or purplish flower clusters with long, conspicuous stamens.

**clepe** (klēp) *vt.* **cleped** (klēpt, klĕpt), **cleped** *or* **clept** *or* **y·cleped** (ĭ-klēpt', ĭ-klĕpt') *or* **y·clept, clep·ing, clepes.** [ME *clepen* < OE *cleopian,* to cry out.] *Archaic.* To call by the name of.

**clep·sy·dra** (klĕp'sĭ-drə) *n., pl.* **-dras** *or* **-drae** (-drē') [Lat. < Gk. *klepsudra* : *kleps-,* secretly (< *kleptein,* to steal) + *hudōr,* water.] An ancient device that measured time by marking the regulated flow of water through a small opening.

**clept** (klĕpt, klēpt) *v. var. p.p. of* CLEPE.

**clere·sto·ry** *also* **clear·sto·ry** (klîr'stôr'ē, -stōr'ē) *n., pl.* **-ries.** [ME *clerestorie,* perh. : *cler,* giving light, clear + *storie,* tier. —see STORY.] **1.** The upper part of the nave, transepts, and choir of a church, containing windows. **2.** A windowed wall or construction similar to a clerestory, used for light and ventilation.

**cler·gy** (klûr'jē) *n., pl.* **-gies.** [ME *clergie* < OFr. < *clerc,* cleric. —see CLERK.] The body of people ordained for religious service.

**cler·gy·man** (klûr'jē-mən) *n.* A man who is a member of the clergy.

**cler·gy·wo·man** (klûr'jē-wo͝om'ən) *n.* A woman who is a member of the clergy.

**cler·ic** (klĕr'ĭk) *n.* [LLat. *clericus.* —see CLERK.] A member of the clergy.

**cler·i·cal** (klĕr'ĭ-kəl) *adj.* **1.** Of or relating to clerks or office workers. **2.** Of, relating to, or typical of the clergy or a member of it. —*n.* **1.** A cleric. **2.** One advocating clericalism. **—cler'i·cal·ly** *adv.*

**cler·i·cal·ism** (klĕr'ĭ-kə-lĭz'əm) *n.* A policy of supporting the influence of the clergy in political or secular matters. **—cler'i·cal·ist** *n.*

**cler·i·hew** (klĕr'ə-hyo͞o') *n.* [After Edmund *Clerihew* Bentley (1875–1956), its inventor.] A humorous quatrain about a person who is gen. named in the first line.

**cler·i·sy** (klĕr'ĭ-sē) *n.* [G. *klerisei,* clergy < Med. Lat. *clericia* < LLat. *clericus,* priest. —see CLERK.] Educated people considered as a class : LITERATI.

**clerk** (klûrk; *Brit.* klärk) *n.* [ME, clergyman, secretary < OE and OFr. *clerc,* clergyman, both < LLat. *clericus* < Gk. *klērikos,* belonging to the clergy < *klēros,* inheritance.] **1.** A person who performs such office tasks as keeping records, handling correspondence, or filing. **2.** A person who keeps the records and performs the regular business of a legislative body or court. **3.** A salesperson in a store. **4.** *Archaic.* A clergyman. **5.** *Archaic.* **a.** A literate person. **b.** A scholar. —*vi.* **clerked, clerk·ing, clerks.** To work or serve as a clerk. **—clerk'dom** *n.* **—clerk'ship'** *n.*

**clerk·ly** (klûrk'lē) *adj.* **-li·er, -li·est. 1.** Of or relating to a clerk or clerks. **2.** *Archaic.* Scholarly. **—clerk'li·ness** *n.*

**†clev·er** (klĕv'ər) *adj.* **-er, -est.** [Perh. < ME *cliver,* expert in seizing, perh. of ON orig.] **1.** Mentally quick and original : BRIGHT. **2.** Nimble or skillful with the hands : DEXTEROUS. **3.** Showing quick-wittedness <a *clever* alibi> **4.** *Regional.* Suitable : handy. **—clev'er·ly** *adv.* **—clev'er·ness** *n.*

☆ ***syns:*** CLEVER, ALERT, BRIGHT, INTELLIGENT, KEEN, SHARP, SMART *adj. core meaning* : mentally quick and original <a *clever* college student>

**clev·is** (klĕv'ĭs) *n.* [Obs. *clevi,* prob. of Scand. orig.] A U-shaped metal piece with holes in each end through which a pin or bolt is run, used for attaching a drawbar to a plow.

**clew** (klo͞o) *n.* [ME *cleve* < OE *clewe.*] **1.** A ball of yarn or thread. **2. clews.** The cords by which a hammock is suspended. **3.** *Naut.* **a.** One of the two lower corners of a square sail. **b.** The lower aft corner of a fore-and-aft sail. **4.** *var. of* CLUE. —*vt.* **clewed, clew·ing, clews. 1.** To roll or coil into a ball. **2.** *Naut.* To raise the lower corners of (a square sail) by means of clew lines. **3.** *var. of* CLUE.

**clew line** *n. Naut.* A rope for raising the clew of a sail up to the yard or mast.

**cli·ché** (klē-shā') *n.* [Fr. < *clicher,* to stereotype.] **1.** A trite expression or idea. **2.** A stereotype or electrotype printing plate.

**click** (klĭk) *n.* [Imit.] **1.** A brief, sharp, nonresonant sound. **2.** A mechanical device, as a detent, that snaps into position. **3.** An implosive speech sound, common in some African languages, produced by drawing air into the mouth and clicking the tongue. —*v.* **clicked, click·ing, clicks.** —*vi.* **1.** To make a click or series of clicks. **2.** *Slang.* **a.** To become a success. **b.** To function well together. —*vt.* To cause to click. **—click'er** *n.*

**click beetle** *n.* A beetle of the family Elateridae, able to right itself from an overturned position by flipping into the air with a clicking sound.

**cli·ent** (klī'ənt) *n.* [ME < OFr. < Lat. *cliens.*] **1.** One for whom professional services are rendered, as by an attorney. **2.** A customer : patron. **3.** One dependent on the patronage or protection of another <Cuba—a *client* of the Soviet Union> **—cli'ent·age** (-ən-tĭj) *n.* **—cli·en'tal** (klī-ĕn'tl, klī'ən-) *adj.*

**cli·en·tele** (klī'ən-tĕl', klē'än-, klē-äɴ-) *n.* [Fr. *clientèle* < Lat. *clientela,* clientship < *cliens,* client.] **1.** The clients of a professional person as a group. **2.** Customers or patrons as a group.

**cliff** (klĭf) *n.* [ME *clif* < OE.] A high, steep, or overhanging rock face. **—cliff'y** *adj.*

**cliff dweller** *n.* **1.** A member of certain prehistoric Indian tribes of the southwestern United States who lived in caves in the sides of cliffs. **2.** *Informal.* One who lives in an apartment house, esp. in a city. **—cliff dwelling** *n.*

**cliff·hang·er** (klĭf'hăng'ər) *n.* **1.** A melodramatic serial in which each episode ends in suspense. **2.** A contest so closely matched that the outcome is doubtful until the end. **—cliff'hang'ing** *adj.*

**cliff swallow** *n.* A North American swallow, *Petrochelidon pyrrhonota,* that builds a bottle-shaped mud nest on the face of a cliff or bluff or under the eaves of a roof.

**cli·mac·ter·ic** (klī-măk'tər-ĭk, klī'măk-tĕr'ĭk) *n.* [Lat. *climactericus,* of a dangerous period in life < Gk. *klimatērikos* < *klimaktēr,* dangerous point, rung of a ladder < *klimax,* ladder.] **1.** A period of life when physiological changes take place in the body. **2.** The menopause. **3.** A crucial period. —*adj.* Relating to a crucial period. **—cli'mac·ter'i·cal** (klī'măk-tĕr'ĭ-kəl) *adj.*

**cli·mac·tic** (klī-măk'tĭk) *also* **cli·mac·ti·cal** (-tĭ-kəl) *adj.* Relating to or constituting a climax. **—cli·mac'ti·cal·ly** *adv.*

**cli·ma·graph** (klī'mə-grăf') *n. var. of* CLIMOGRAPH.

**cli·mate** (klī'mĭt) *n.* [ME *climat* < OFr. < LLat. *clima* < Gk. *klima,* region of the earth.] **1.** Meteorological conditions, including temperature, precipitation, and wind, that prevail in a region. **2.** A region manifesting particular meteorological conditions. **3.** A prevailing condition or atmosphere <a *climate* of suspicion> **—cli·mat'ic** (-măt'ĭk), **cli·mat'i·cal** *adj.* **—cli·mat'i·cal·ly** *adv.*

**cli·ma·tol·o·gy** (klī'mə-tŏl'ə-jē) *n.* Meteorological study of climate. **—cli'ma·to·log'ic** (-tə-lŏj'ĭk), **cli'ma·to·log'i·cal** *adj.* **—cli'ma·tol'o·gist** *n.*

**cli·max** (klī'măks') *n.* [Lat., rhetorical climax < Gk. *klimax,* ladder.] **1.** The point of greatest intensity in a series of events : CULMINATION. **2.** Orgasm. **3. a.** A series of ideas stated in an ascending order of rhetorical force or intensity. **b.** The final statement in such a series. **4.** The stage in ecological development or evolution in which a community of organisms becomes stable and starts to perpetuate itself. —*vi.* & *vt.* **-maxed, -max·ing, -max·es.** To reach or bring to a climax.

☆ ***syns:*** CLIMAX, ACME, APEX, APOGEE, CREST, CROWN, CULMINATION, HEIGHT, MERIDIAN, PEAK, PINNACLE, SUMMIT, ZENITH *n. core meaning* : the highest point or state < The *climax* of his career was election to the Presidency.>

**climb** (klīm) *v.* **climbed, climb·ing, climbs.** [ME *climben* < OE *climban.*] —*vt.* To move up or mount, esp. by using the hands and feet : ASCEND <*climbed* a hill> —*vi.* **1.** To rise or move up, esp. by using the hands and feet <*climbed* up the telephone pole> **2.** To reach a higher status, rank, or condition. **3.** To slant or slope upward <a trail *climbing* gradually to the peak> **4.** To grow in an upward direction, as some plants do, by twining around an object for support. **—climb down.** To descend, esp. by means of the hands and feet <*climbed down* the rope ladder> —*n.* **1.** The act of climbing : ASCENT. **2.** A place to be climbed. **—climb'a·ble** (klī'mə-bəl) *adj.*

**climb·er** (clī'mər) *n.* **1.** One that climbs. **2.** One who seeks to gain a higher professional or social position. **3.** A plant that grows upward by twining around an object. **4.** A climbing iron.

**climbing fumitory** *n.* A weak-stemmed climbing vine, *Adlumia fungosa* of eastern North America, bearing spurred pinkish or white flowers.

**climbing hempweed** *n.* A vine, *Mikania scandens* of eastern and central North America, bearing small white flower clusters.

**climbing iron** *n.* An iron bar with spurs or spikes attached, which can be strapped to a boot or shoe used in climbing.

**climbing perch** *n.* A freshwater fish, *Anabas testudineus* of tropical Asia, able to move along the ground with the aid of its gill covers and pectoral fins.

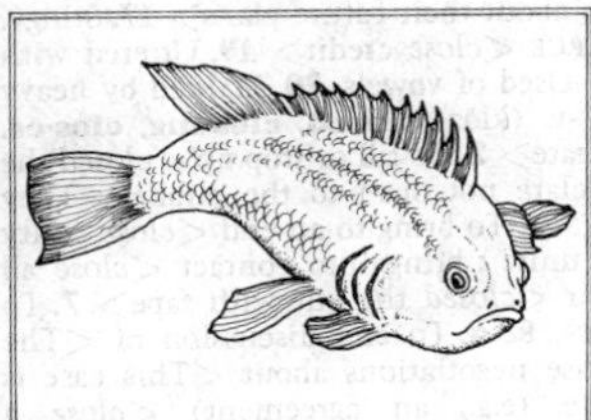
**climbing perch**
*Up to 12 inches long*

**clime** (klīm) *n.* [ME, region of the earth < LLat. *clima* < Gk. *klima.*] CLIMATE 1, 2.
**cli·mo·graph** *also* **cli·ma·graph** (klī′mə-grăf′) *n.* [CLIM(ATE) + GRAPH.] A representation of climatic data in which one climatic factor, as temperature, is plotted against another, as moisture.
**clin-** *pref. var. of* CLINO-.
**-clinal** *suff.* [< Gk. *klinein,* to lean.] Sloping <syn*clinal*>
**cli·nan·dri·um** (klĭ-năn′drē-əm) *n., pl.* **-dri·a** (-drē-ə) [NLat. : Gk. *klinē,* couch (< *klinein,* to recline) + NLat. *-andrium,* stamen < Gk. *anēr,* man.] *Bot.* A hollow containing the anther in the upper part of the column of an orchid.
**clinch** (klĭnch) *v.* **clinched, clinch·ing, clinch·es.** [Alteration of CLENCH.] —*vt.* **1. a.** To secure or fix (e.g., a nail) by bending down or flattening the protruding end. **b.** To fasten together in this way. **2.** To settle conclusively <*clinched* the agreement> **3.** *Naut.* To fasten with a clinch. —*vi.* **1.** To be held together securely. **2.** To hold a boxing opponent's body with one or both arms to prevent or hinder punches. **3.** *Slang.* To embrace. —*n.* **1.** Something, as a clamp, that clinches. **2.** The clinched part of a nail, bolt, or rivet. **3.** The act or an instance of clinching in boxing. **4.** *Naut.* A knot in a rope made by a half hitch with the end of the rope fastened back by seizing. **5.** *Slang.* An embrace.
**clinch·er** (klĭn′chər) *n.* **1.** One who clinches. **2.** A nail or bolt for clinching. **3.** A tool for clinching nails or bolts. **4.** *Informal.* A decisive point, fact, or remark.
**cline** (klīn) *n.* [< Gk. *klinein,* to lean.] *Ecol.* A series of differing characteristics within members of a species or population, caused by gradual environmental changes or transitions. **—clin′al** (klī′nəl) *adj.*
**-cline** *suff.* [< Gk. *klinein,* to lean.] Slope <anti*cline*>
**cling** (klĭng) *vi.* **clung** (klŭng), **cling·ing, clings.** [ME *clingen* < *clingan.*] **1.** To hold tight or adhere to something, as by grasping, sticking, or entwining. **2. a.** To remain close. **b.** To fit closely, as to the body <a wet suit that *clings*> **c.** To resist separation. **3.** To hold on, often doggedly <*cling* to outmoded customs> —*n.* A clingstone. **—cling′er** *n.*
**cling·fish** (klĭng′fĭsh′) *n., pl.* **clingfish** *or* **-fish·es.** One of various small marine fishes of the family Gobiesocidae, with an adhesive disk under the front part of the body by which it fastens itself to rocks and seaweed.
**cling·stone** (klĭng′stōn′) *n.* A fruit, esp. a peach, whose pulp adheres partially to the stone. **—cling′stone′** *adj.*
**clin·ic** (klĭn′ĭk) *n.* [Fr. *clinique* < Lat. *clinicus,* physician < Gk. *klinikos* < *klinē,* bed < *klinein,* to recline.] **1. a.** A training session for medical students in which they observe the examination and treatment of patients. **b.** A class receiving this instruction. **2.** An institution associated with a hospital or medical school that deals chiefly with outpatients. **3.** A medical establishment run by several specialists working together. **4.** A center that offers counsel or instruction <a divorce *clinic*><a remedial reading *clinic*>
**-clinic** *suff.* [< Gk. *klinein,* to lean.] **1.** Sloping <iso*clinic*> **2.** Having a specified number of oblique axial intersections <tri*clinic*>
**clin·i·cal** (klĭn′ĭ-kəl) *adj.* **1.** Of, relating to, or connected with a clinic. **2.** Of or relating to direct observation and treatment of patients. **3.** Highly objective and devoid of emotion : ANALYTICAL <a *clinical* account of their difficulties> **—clin′i·cal·ly** *adv.*
**clinical pathology** *n.* Scientific study of the diagnosis and treatment of disease through laboratory analysis of clinical specimens, as tissue.
**clinical thermometer** *n.* A small self-registering glass thermometer for measuring body temperature.
**cli·ni·cian** (klĭ-nĭsh′ən) *n.* [Fr. *clincien* < *clinique,* clinic. —see CLINIC.] A physician or psychologist specializing in clinical studies or practice.
**clink[1]** (klĭngk) *vi.* & *vt.* **clinked, clink·ing, clinks.** [ME *clinken* < MDu.] To make or cause to make a soft, sharp, ringing sound. —*n.* **1.** A clinking sound. **2.** *Chiefly Brit.* The shrill cry of a bird.
**clink[2]** (klĭngk) *n.* [After *Clink,* a prison in London, England.] *Slang.* A prison.
**clink·er** (klĭng′kər) *n.* [Obs. Du. *klinckaerd* < MDu. *klinken,* to clink.] **1.** Incombustible residue, fused into an irregular lump, remaining after combustion of coal. **2.** A partially vitrified brick or a mass of bricks fused together. **3.** A very hard burned brick. **4.** Vitrified matter expelled by a volcano. **5.** *Slang.* A mistake or fault, esp. in a musical performance. —*vi.* **-ered, -er·ing, -ers.** To create clinkers in burning, as coal does.
**clink·er-built** (klĭng′kər-bĭlt′) *adj.* [< obs. *clinker,* clinch nail < ME *clinken,* prob. var. of *clenchen,* to clench < OE *(be)clencan.*] Built with overlapping planks or boards, as a ship.
**clink·stone** (klĭngk′stōn′) *n.* Phonolite.
**clino-** *or* **clin-** *pref.* [NLat. < Gk. *klinein,* to slope.] Slope : slant <*clino*meter>
**cli·nom·e·ter** (klī-nŏm′ĭ-tər) *n.* An instrument for measuring the angle of an incline, as of an embankment. **—cli′no·met′ric** (-nə-mĕt′rĭk), **cli′no·met′ri·cal** *adj.* **—cli·nom′e·try** *n.*
**clin·quant** (klĭng′kənt, klăN-kăN′) *adj.* [Fr., glistening, tinkling < obs. *clinquer,* to clink, perh. < MDu. *klinken.*] Ornamented with gold or silver : TINSELED. —*n.* Imitation gold leaf : TINSEL.
**clin·to·ni·a** (klĭn-tō′nē-ə) *n.* [NLat. *Clintonia,* genus name, after DeWitt *Clinton* (1769–1828).] A plant of the genus *Clintonia,* with broad leaves, white, greenish-yellow, or purplish flowers, and usu. blue berries.
**Cli·o** (klī′ō) *n.* [Lat. < Gk. *Kleiō* < *kleiein,* to tell.] **1.** *Gk. Myth.* The Muse of history. **2.** A statuette awarded annually for outstanding achievement in radio and television advertising.
**cli·o·met·rics** (klī′ə-mĕt′rĭks) *n.* [CLIO + -METRICS.] *(sing. in number).* The study of history using advanced mathematical methods of data processing and analysis. **—cli′o·met′ric** *adj.* **—cli′o·me·tri′cian** (-mĭ-trĭsh′ən) *n.*
**clip[1]** (klĭp) *v.* **clipped, clip·ping, clips.** [ME *clippen* < ON *klippa.*] —*vt.* **1.** To cut, cut off, or cut out with scissors or shears. **2.** To make shorter by cutting : TRIM. **3.** To cut off the edge of. **4.** To cut short : CURTAIL. **5.** To fail to pronounce or write fully <*clipped* their words> **6.** *Informal.* To hit with a sharp blow. **7.** *Slang.* To cheat or overcharge. —*vi.* **1.** To cut something. **2.** *Informal.* To move rapidly. —*n.* **1.** The act of clipping. **2.** Something clipped off, as a sequence from a motion picture. **3. a.** The wool shorn at one shearing. **b.** A season's shearing. **4.** *Informal.* A quick, sharp blow. **5.** *Informal.* A brisk pace. **6. clips.** A pair of shears.
**clip[2]** (klĭp) *n.* [ME, hook < *clippen,* to embrace < OE *clyppan.*] **1.** A device for gripping : CLASP. **2.** A piece of jewelry attached by a clasp or clip. **3.** A flange on the top of a horseshoe. **4.** A cartridge clip. —*vt.* **clipped, clip·ping, clips. 1.** To grip securely : FASTEN. **2.** To join (one thing) to another. **3.** *Football.* To block (an opponent not carrying the ball) illegally from the rear.
**clip·board** (klĭp′bôrd′, -bōrd′) *n.* A small writing board with a spring clip for holding papers or a writing pad.
**clip joint** *n. Slang.* A restaurant or place of public entertainment where customers are overcharged.
**clip·per** (klĭp′ər) *n.* **1.** One who cuts, shears, or clips. **2.** *often* **clippers.** A tool for cutting, clipping, or shearing. **3.** A sharp-bowed mid-19th cent. sailing vessel, having tall masts and sharp lines and built for speed. **4.** A fast-moving vehicle.
**clip·ping** (klĭp′ĭng) *n.* Something, esp. an item from a newspaper, that is cut off or out.
**clip·sheet** (klĭp′shēt′) *n.* A sheet of paper containing news items and other material, usu. printed on only one side for convenience in clipping and reprinting.
**clique** (klēk, klĭk) *n.* [Fr.] An exclusive group of friends or associates. —*vi.* **cliqued, cliqu·ing, cliques.** *Informal.* To form, associate in, or act as a clique. **—cliqu′ey, cliqu′y, cliqu′ish** *adj.* **—cliqu′ish·ly** *adv.* **—cliqu′ish·ness** *n.*
**cli·tel·lum** (klī-tĕl′əm) *n., pl.* **-tel·la** (-tĕl′ə) [NLat., alteration of Lat. *clitellae,* packsaddle.] A swollen, glandular, saddlelike region in the epidermis of certain annelid worms, as the earthworm.
**clit·o·ris** (klĭt′ər-ĭs, klī′tər-) *n.* [Gk. *kleitoris,* poss. < *kleiein,* to shut.] A small, erectile organ at the upper end of the vulva, homologous with the penis. **—clit′o·ral** (-əl) *adj.*
**clo·a·ca** (klō-ā′kə) *n., pl.* **-cae** (-sē′) [Lat.] **1.** A sewer. **2.** A latrine. **3.** *Zool.* **a.** The cavity into which the intestinal, genital, and urinary tracts open in vertebrates such as fish, reptiles, birds, and some primitive mammals. **b.** The posterior part of the intestinal tract in various invertebrates. **—clo·a′cal** (-kəl) *adj.*
**cloak** (klōk) *n.* [ME *cloke* < OFr., var. of *cloche,* cloak, bell < Med. Lat. *clocca,* bell, from its shape.] **1.** A usu. sleeveless, loose outer garment. **2.** Something that covers or conceals <a *cloak* of secrecy> —*vt.* **cloaked, cloak·ing, cloaks. 1.** To cover with or as if with a cloak. **2.** To hide : conceal.
**cloak-and-dagger** (klōk′ən-dăg′ər) *adj.* Marked by melodramatic intrigue and often espionage.
**cloak·room** (klōk′ro͞om′, -ro͝om′) *n.* **1.** A room where coats and other items may be left temporarily, as in a theater. **2.** An anteroom of a legislative chamber where delegates often confer with colleagues.

**clob·ber** (klŏb′ər) *vt.* **-bered, -ber·ing, -bers.** [Orig. unknown.] *Slang.* **1.** To strike violently and repeatedly : BATTER. **2.** To defeat decisively.

**cloche** (klōsh) *n.* [Fr., bell < OFr. < Med. Lat. *clocca.*] **1.** A bell-shaped glass vessel used to cover plants or food. **2.** A woman's close-fitting bell-shaped hat.

**clock[1]** (klŏk) *n.* [ME *clokke* < OFr. *cloke,* var. of *cloche,* bell < Med. Lat. *clocca.*] **1.** An instrument other than a watch for measuring or indicating time, esp. a mechanical device with a numbered dial and moving hands or pointers or one with a digital display. **2.** A time clock. **3.** A metering device, as a speedometer. **4.** *Bot.* The downy flower head of a dandelion that has gone to seed. —*v.* **clocked, clock·ing, clocks.** —*vt.* To record the time or speed of, as with a stopwatch <*clocked* the racer at over 150 miles an hour> —*vi.* To record working hours with a time clock <I *clocked* in at 8:30 A.M.> **—clock′er** *n.*

**clock[2]** (klŏk) *n.* [Perh. < CLOCK[1], bell (obs.), from an original bell-shaped appearance.] An embroidered or woven decoration on the side of a sock or stocking.

**clock radio** *n.* A radio with a built-in clock that can be set to turn the radio on automatically.

**clock·wise** (klŏk′wīz′) *adv.* In the same direction as the rotating hands of a clock. **—clock′wise′** *adj.*

**clock·work** (klŏk′wûrk′) *n.* The mechanism of a clock or a similar mechanism. **—like clockwork.** With machinelike regularity and precision. **—clock′work′** *adj.*

**clod** (klŏd) *n.* [ME < OE.] **1.** A lump or chunk, esp. of earth or clay. **2.** Earth or soil. **3.** An ignorant or stupid person : DOLT. **—clod′dish** *adj.* **—clod′dish·ly** *adv.* **—clod′dish·ness** *n.*

**clod·hop·per** (klŏd′hŏp′ər) *n.* **1.** A clumsy, coarse person : BUMPKIN. **2.** A big, heavy shoe.

**clog** (klŏg) *n.* [ME, block attached to an animal's leg.] **1.** An obstacle or hindrance. **2.** A weight, as a block, attached to the leg of an animal to hinder movement. **3.** A heavy, usu. wooden-soled shoe. —*v.* **clogged, clog·ging, clogs.** —*vt.* **1.** To block up : OBSTRUCT <evening traffic *clogging* the bridges> **2. a.** To impede or encumber (an animal) with a cog. **b.** To impede or hamper. —*vi.* **1.** To become obstructed or choked up. **2.** To thicken or stick together so as to obstruct. **3.** To do a clog dance.

**clog dance** *n.* A dance performed while wearing clogs and characterized by heavy, stamping steps.

**cloi·son·né** (kloi′zə-nā′, klə-wä′zə-) *n.* [Fr., p.part. of *cloisonner,* to partition < *cloison,* partition < VLat. **clausio* < Lat. *claudere,* to close.] **1.** Enamelware in which the different colors of the surface decoration are separated by thin strips of metal set on edge. **2.** The process or method of producing cloisonné. **—cloi′son·né′** *adj.*

**clois·ter** (kloi′stər) *n.* [ME *cloistre* < OFr. < Lat. *claustrum,* enclosed place < *claudere,* to close.] **1.** A covered walk with an open colonnade on one side, running along the inside walls of buildings that face a quadrangle. **2. a.** A place, esp. a monastery or convent, devoted to religious seclusion. **b.** Monastery or convent life. —*vt.* **-tered, -ter·ing, -ters.** **1.** To confine in or as if in a cloister : SECLUDE. **2.** To furnish (a building) with a cloister.

**clois·tral** (kloi′strəl) *also* **claus·tral** (klô′-) *adj.* **1.** Of, resembling, or like a cloister : SECLUDED. **2.** Living in a cloister.

**clomb** (klōm) *v. Archaic. var. p.t. & p.p. of* CLIMB.

**clom·i·phene** (klōm′ə-fēn′, klō′mə-) *n.* [C(H)LO(RO)- + (A)MI(NE) + PHEN(YL).] A drug, $C_{26}H_{28}ClNO$, used in its citrate form to stimulate ovulation.

**clomp** (klŏmp) *vi.* **clomped, clomp·ing, clomps.** [Imit.] To walk heavily and noisily.

**clone** (klōn) *n.* [Gk. *klōn,* twig.] **1.** A group of genetically identical cells descended from a single common ancestor. **2.** One or more organisms descended asexually from a single ancestor. **3.** One that is an exact replica of another. —*v.* **cloned, clon·ing, clones.** —*vi.* To create a genetic duplicate of an individual organism through asexual reproduction, as by stimulating a single cell. —*vt.* **1.** To duplicate (an organism) asexually by cloning. **2.** To create (a new organism) asexually by cloning. **—clon′al** *adj.* **—clon′al·ly** *adv.* **—clon′er** *n.*

**clo·nor·chi·a·sis** (klō′nôr-kī′ə-sĭs) *n.* [NLat. *Clonorchis,* former genus name + -IASIS.] A parasitic infection of mammals usu. caused by ingestion of raw fish infected with the trematode *Opisthorchis sinensis.*

**clo·nus** (klō′nəs) *n., pl.* **-nus·es.** [< Gk. *klonos,* turmoil.] A convulsion marked by rapidly alternating muscular contraction and relaxation. **—clo′nic** (klō′nĭk, klŏn′ĭk) *adj.* **—clo·nic′i·ty** (klō-nĭs′ĭ-tē, klŏ-), **clo′nism** (klō′nĭz′əm, klŏn′ĭz′əm) *n.*

**clop** (klŏp) *n.* [Imit.] The drumming sound of a horse's hoof striking pavement. —*vi.* **clopped, clop·ping, clops.** To make a clop.

**close** (klōs) *adj.* **clos·er, clos·est.** [ME *clos,* closed < OFr. < Lat. *clausus,* p.part. of *claudere,* to close.] **1.** Near in space or time. **2. a.** Near in relationship <*close* relatives> **b.** Bound by mutual interests, loyalties, or affections : INTIMATE <*close* companions> **3.** Having little or no space between elements or parts : COMPACT <a *close* mesh> **4.** Being near to the surface <a *close* haircut> **5.** Being almost even, as in tally or score <a *close* contest> **6.** Very similar to an original <a *close* copy of the painting> **7.** Strict : rigorous <paid *close* attention> **8.** Shut or shut in : not open. **9.** Enclosed or almost enclosed. **10.** Confining or narrow : CROWDED. **11.** Tightly fitting <*close* sleeves> **12.** Lacking fresh air : STUFFY. **13.** Confined to specific persons or groups : RESTRICTED. **14.** Strictly confined or confining <*close* direction> **15.** Hidden from view : SECLUDED. **16.** Taciturn : reticent <*close* about their future plans> **17.** Stingy. **18.** Not easy to acquire : SCARCE <*close* credit> **19.** Uttered with the tongue near the palate. —Used of vowels. **20.** Marked by heavy punctuation, esp. commas. —*v.* (klōz) **closed, clos·ing, clos·es.** —*vt.* **1.** To shut <*closed* the gate> **2.** To fill or stop up <*closed* the gap with cement> **3.** To declare not open to the public <They *closed* the dump for cleanup.> **4.** To bring to an end <*close* treaty negotiations> **5.** To join or unite : bring into contact <*close* all circuits> **6.** To draw together <*closed* the cut with tape> **7.** To enclose on all sides : SHUT IN. **8. a.** To end discussion of <The matter is *closed.*> **b.** To cease negotiations about <This case is *closed.*> **c.** To consummate (e.g., an agreement) <*close* a deal><*close* a house sale> —*vi.* **1.** To become shut <The window *closed.*> **2.** To come to an end <a play that *closed* quickly> **3.** To discontinue operation <The business *closed* last year.> **4.** To engage in a one-on-one struggle : GRAPPLE <troops *closed* in combat> **5.** To reach an agreement. **6.** To come together. **—close in. 1.** To surround with an oppressive, isolating effect <grim reality *closing in*> **2.** To surround and advance upon so as to cut off all chance of escape <The tanks *closed in* on the village.> **3.** To enshroud to such a degree that entrance and exit are impossible <a mountain pass *closed in* by a blizzard> **—close out.** To dispose of merchandise, usu. at reduced prices. —*n.* (klōz). **1.** The act of closing. **2.** A conclusion : finish <steered the conference to a successful *close*> **3.** An enclosed place, esp. land surrounding or beside a cathedral. **4.** *Archaic.* A fight at close quarters. **5.** (klōs). *Scot. & Brit. Regional.* A narrow lane or alley. —*adv.* (klōs). In a close position : NEAR <hugged *close* to the wall> **—close′ly** *adv.* **—close′ness** *n.* **—clos′er** (klō′zər) *n.* **—clos′ing** (klō′zĭng) *n.*

**close call** (klōs) *n. Informal.* A narrow escape.

**closed** (klōzd) *adj.* **1.** Having boundaries : ENCLOSED. **2.** Blocked or barred to entry or passage. **3.** Explicitly limited : RESTRICTED <a *closed* fraternity> **4.** Self-contained. **5.** Carried on in secrecy <*closed* rites of imitation> **6.** *Math.* **a.** Of or relating to a curve, as a circle, having no end points. **b.** Of or relating to a surface having no boundary curves. **c.** Marked by or having the property by which an operation acting on an element in a set produces an element within the set.

**closed-cap·tioned** (klōzd′kăp′shənd) *adj.* Being a telecast with captions visible only on a specially equipped receiver <a *closed-captioned* television movie for the deaf>

**closed chain** *n. Chem.* RING[1] 14.

**closed circuit** *n.* **1.** A television transmission circuit with a limited number of reception stations and no broadcast facilities. **2.** An electric circuit providing an uninterrupted endless path for the flow of current.

**closed corporation** *n.* A corporation in which owned stock is owned by relatively few people and is rarely bought or sold on the open market.

**closed-end investment company** (klōzd′ĕnd′) *n.* A company with fixed capitalization whose shares are bought and sold by investors and whose capital is invested in other companies.

**closed gentian** *n.* The bottle gentian.

**closed interval** *n.* INTERVAL 3b.

**closed shop** *n.* A union shop.

**close-fist·ed** (klōs′fĭs′tĭd) *adj.* Stingy.

**close-grained** (klōs′grānd′) *adj.* Compact or dense in structure or texture <*close-grained* wood>

**close-hauled** (klōs′hôld′) *adv. Naut.* With sails trimmed flat for sailing as close to the wind as possible. **—close′-hauled′** *adj.*

**close-mind·ed** (klōs′mīn′dĭd) *adj.* **1.** Intolerant of the beliefs and opinions of others. **2.** Stubborn.

**close-mouthed** (klōs′mouthd′, -moutht′) *adj.* Not inclined to talk : RETICENT.

**close-or·der drill** (klōs′ôr′dər) *n.* Military drill in marching, maneuvering, and handling of arms in which the participants perform at close intervals.

**close-out** (klōz′out′) *n.* A sale in which all goods are disposed of, usu. at reduced prices.

**close shave** (klōs) *n. Informal.* A narrow escape.

**clos·et** (klŏz′ĭt, klô′zĭt) *n.* [ME, private room < OFr., dim. of *clos,* enclosure < Lat. *clausum* < *clausus,* enclosed. —see CLOSE.] **1.** A small room, cabinet, or recess for storage. **2.** A small private chamber, as for studying. **3.** A water closet : TOILET. —*vt.* **-et·ed, -et·ing, -ets.** To enclose or shut up in a private room, as for discussion <*closet* oneself with special consultants> —*adj.* **1. a.** Confidential : private. **b.** Hidden : secret <a *closet* drug addict> **2.** Based on theory and speculation rather than practice.

**closet drama** *n.* A play to be read rather than performed.

ă **pat** ā **pay** âr **care** ä **father** ĕ **pet** ē **be** hw **which** ĭ **pit**
ī **tie** îr **pier** ŏ **pot** ō **toe** ô **paw,** **for** oi **noise** ŏŏ **took**

**close-up** (klōs′ŭp′) *n.* **1.** A picture, as a motion-picture or television shot, taken at close range. **2.** A close or intimate look.

**clos·trid·i·um** (klŏ-strĭd′ē-əm) *n., pl.* **-i·a** (-ē-ə) [NLat. *Clostridium,* genus name < Gk. *klōstēr,* spindle < *klōthein,* to spin.] Any of various rod-shaped, spore-forming, chiefly anaerobic bacteria of the genus *Clostridium,* as the nitrogen-fixing bacteria found in soil and those causing botulism. **—clos·trid′i·al** (-əl) *adj.*

**clo·sure** (klō′zhər) *n.* [ME < OFr. < Lat. *clausura* < *clausus,* enclosed. —see CLOSE.] **1.** The act of closing or the condition of being closed. **2.** Something that closes. **3.** A finish : conclusion. **4.** *var. of* CLOTURE. **5.** The property of being mathematically closed. *—v.* **-sured, -sur·ing, -sures.** To end by cloture.

**clot** (klŏt) *n.* [ME < OE.] A thick, viscous, or coagulated lump. *—v.* **clot·ted, clot·ting, clots.** *—vi.* To form into clots. *—vt.* **1.** To cause to clot. **2.** To fill or cover with clots.

**cloth** (klôth, klŏth) *n., pl.* **cloths** (klôths, klôthz, klŏths, klŏthz) [ME < OE *clāð.*] **1.** Material formed by weaving, knitting, pressing, or felting of natural or synthetic fibers. **2.** A piece of material for a particular purpose, as a tablecloth. **3.** *Naut.* **a.** Canvas. **b.** A sail. **4.** Distinctive professional attire. **5. a.** The dress of the clergy. **b.** The clergy.

**cloth·bound** (klôth′bound′, klŏth′-) *adj.* Describing a book bound in thick paper boards covered with cloth.

**clothe** (klō*th*) *vt.* **clothed** or **clad** (klăd), **cloth·ing, clothes.** [ME *clothen* < OE *clāðian* < *clāð,* cloth.] **1.** To put clothes on. **2.** To provide clothes for. **3.** To cover as if with clothes : INVEST.

**clothes** (klōz, klō*th*z) *pl.n.* [ME < OE *clāðas* < *clāð,* cloth.] **1.** Wearing apparel : GARMENTS. **2.** Bedclothes.

**clothes·horse** (klōz′hôrs′, klō*th*z′-) *n.* **1.** A frame on which clothes are hung. **2.** One regarded as being overly concerned with dress and fashion.

**clothes·line** (klōz′līn′, klō*th*z′-) *n.* A cord, rope, or wire on which clothes are hung.

**clothes moth** *n.* Any of various moths of the family Tineidae, whose larvae feed on wool, hair, fur, and feathers.

**clothes·pin** (klōz′pĭn′, klō*th*z′-) *n.* A clip for holding clothes on a clothesline.

**clothes·press** *also* **clothes press** (klōz′prĕs′, klō*th*z′-) *n.* A chest, closet, or wardrobe for clothes.

**clothes tree** *n.* An upright pole or stand with pegs or hooks on which to hang clothes.

**cloth·ier** (klō*th*′yər, klō′*th*ē-ər) *n.* A maker or seller of clothing or cloth.

**cloth·ing** (klō′*th*ĭng) *n.* **1.** CLOTHES 1. **2.** A covering.

**Clo·tho** (klō′thō) *n.* [Lat. < Gk. *Klōthō* < *klōthein,* to spin.] *Gk. Myth.* One of the three Fates, spinner of the thread of destiny.

**cloth yard** *n.* The standard unit of cloth measurement, equal to 36 inches or 0.9144 meters.

**clo·ture** (klō′chər) *also* **clo·sure** (-zhər) *n.* [Fr. *clôture,* alteration of OFr. *closure.* —see CLOSURE.] A parliamentary procedure by which debate is ended and an immediate vote is taken on the matter being discussed. *—vt.* **-tured, -tur·ing, -tures.** To close (a parliamentary debate) by cloture.

**cloud** (kloud) *n.* [ME, hill, cloud < OE *clūd,* hill.] **1. a.** A visible body of fine droplets of water or particles of ice dispersed in the atmosphere above the earth's surface at various altitudes up to several miles. **b.** A visible airborne mass, as of steam, smoke, or dust. **2.** A large moving body of things on the ground or in the air : SWARM. **3.** Something that darkens or fills with gloom. **4.** A dark blemish or spot, as on a polished stone. **5.** Something that obscures. **6.** A suspicion or charge affecting a reputation. **7.** A collection of charged particles <an electron *cloud*> *—v.* **cloud·ed, cloud·ing, clouds.** *—vt.* **1.** To cover with or as if with clouds. **2.** To make gloomy or troubled. **3.** To cast aspersions on : SULLY. *—vi.* To become cloudy or overcast. **—cloud′less** *adj.*

▲ word history: The weather in medieval England must not have been very sunny, because two modern English words meaning "sky" meant "cloud" in earlier times. *Welkin* (from Old English *wolcen*) meant "cloud" until the 12th century, as the related German word *Wolke* still does. The word *sky* itself is an Old Norse word for "cloud." It was borrowed into English with that meaning in the 13th century and was not used to mean "sky" until about a century later. The word *cloud* (from Old English *clūd*) meant "hill" or "rock" until about the 14th century; it is related to *clod* and *clot.* The ordinary Old and Middle English word for "sky" was *heaven,* which retains that sense in Modern English only in the plural *heavens.* The use of *heaven* to denote the abode of God is recorded as early as the word itself.

**cloud·ber·ry** (kloud′bĕr′ē) *n.* A creeping plant, *Rubus chamaemorus* of northern regions, bearing white flowers and edible, reddish-orange fruit.

**cloud·burst** (kloud′bûrst′) *n.* A sudden rainstorm : DOWNPOUR.

**cloud chamber** *n.* A device in which the formation of chains of droplets on ions generated by the passage of charged subatomic particles through a supersaturated vapor is used to detect such particles, to infer the presence of neutral particles, and to study certain nuclear reactions.

**cloud nine** *n.* A state of elation.

**cloud seeding** *n.* A technique of stimulating rainfall, esp. by distributing quantities of dry ice crystals or silver iodide smoke through clouds : RAINMAKING.

**cloud·y** (klou′dē) *adj.* **-i·er, -i·est. 1.** Full of or covered with clouds : OVERCAST. **2.** Of or resembling a cloud or clouds. **3.** Marked with indistinct masses or streaks. **4.** Not transparent <*cloudy* water> **5. a.** Liable to more than one interpretation. **b.** Not clearly perceived or perceptible. **6.** Gloomy : troubled. **—cloud′i·ly** *adv.* **—cloud′i·ness** *n.*

**†clout** (klout) *n.* [ME, prob. < OE *clūt,* patch.] **1.** A blow, esp. with the fist. **2.** *Baseball.* A long, powerful hit. **3.** *Informal.* **a.** Influence : pull <economic *clout*> **b.** Power : muscle. **4.** An archery target. **5.** *Archaic & Regional.* A piece of cloth used for mending : PATCH. *—vt.* **clout·ed, clout·ing, clouts. 1.** To hit, esp. with the fist. **2.** *Archaic & Regional.* To bandage or patch.

**clove**[1] (klōv) *n.* [ME *clow,* clove spice < OFr. *clou (de girofle),* nail (of the clove tree) < Lat. *clavus,* nail.] **1.** An East Indian evergreen tree, *Eugenia aromatica,* whose aromatic unopened flower buds are used as a spice. **2.** *often* **cloves.** A spice consisting of dried clove flower buds.

▲ word history: It may seem odd that a segment of a garlic bulb and the spice called "cloves" in English should share the same name, but *clove*[1] and *clove*[2] are not at all related. *Clove*[2], meaning "bulb section," comes from Old English *clufu,* a noun related to *clēofan,* "to split," the ancestor of *cleave*[1]. *Clove*[1] as a name for the spice is really a misnomer. The full name of the spice in Old French was *clou de girofle,* literally "nail of the clove tree." *Clou* is from Latin *clavus,* "nail." The dried flower bud of the clove tree, which is the part used as a spice, somewhat resembles a small nail or tack. The English gradually shortened the full name of the spice to *clow,* whose modern form is *clove.*

**clove**[2] (klōv) *n.* [ME < OE *clufu.*] A small section of a separable bulb, as that of garlic.

**clove**[3] (klōv) *v. var. p.t. & archaic p.p. of* CLEAVE[1].

**clove**[4] (klōv) *v. var. p.t. & archaic var. p.p. of* CLEAVE[2].

**clove hitch** *n.* [ME *clove,* split, p.part. of *cleven,* to split < OE *clēofan.*] *Naut.* A knot for securing a line to a spar, post, or other object, consisting of two turns with the second held under the first.

**clo·ven** (klō′vən) *adj.* [< P.part. of CLEAVE[1].] Divided : split.

**cloven foot** *n.* A cloven hoof. **—clo′ven-foot′ed** *adj.*

**cloven hoof** *n.* **1.** A cleft or divided hoof, as in cattle or deer. **2.** Evil, based on the image of Satan as a figure with cloven hoofs.

**clo·ven-hoofed** (klō′vən-ho͞oft′, -ho͝oft′, -ho͞ovd′, -ho͝ovd′) *adj.* **1.** Having cloven hoofs, as cattle do. **2.** Devilish : satanic.

**clove oil** *n.* An aromatic oil distilled from the dried flower buds of the clove tree, used as an antiseptic.

**clove pink** *n.* A variety of the carnation, *Dianthus caryophyllus,* bearing flowers with a spicy fragrance.

**clo·ver** (klō′vər) *n.* [ME < OE *clæfre.*] **1.** A plant of the genus *Trifolium,* bearing compound leaves with three leaflets and tight heads of small flowers. **2.** A plant related to the clover, as the bush clover. **—in clover.** Living a carefree life of comfort or prosperity.

**clo·ver·leaf** (klō′vər-lēf′) *n.* A highway interchange at which two highways cross each other on different levels and have curving access and exit ramps enabling vehicles to go in any of four directions.

**clown** (kloun) *n.* [Perh. of LG orig.] **1.** A comic performer who entertains by jokes, antics, and tricks in a circus or play. **2.** A boor. **3.** A peasant. *—vi.* **clowned, clown·ing, clowns. 1.** To behave like a clown. **2.** To perform as a clown. **—clown′ish** *adj.* **—clown′ish·ly** *adv.* **—clown′ish·ness** *n.*

**clox·a·cil·lin** (klŏk′sə-sĭl′ĭn) *n.* [C(H)L(ORO)- + OX- + A(ZO) + (PENI)CILLIN.] A synthetic antibiotic of the penicillin group, effective against staphylococci.

**cloy** (kloi) *v.* **cloyed, cloy·ing, cloys.** [Obs. *accloy* < ME *acloien.*] *—vt.* To supply with too much of something, esp. with something too rich or sweet : SURFEIT. *—vi.* To cause to feel surfeited. **—cloy′ing·ly** *adv.* **—cloy′ing·ness** *n.*

**cloze** (klōz) *n.* [Alteration of CLOSURE.] A test of reading comprehension in which the test taker is asked to supply words that have been systematically removed from the text. **—cloze** *adj.*

**club** (klŭb) *n.* [ME < ON *klubba.*] **1.** A stout, heavy stick, usu. thicker at one end than at the other, suitable as a weapon : CUDGEL. **2.** A bat or stick used to drive a ball in certain games, esp. a stick with a curved head used in golf and hockey. **3. a.** A black figure on a playing card, shaped like a clover or trefoil leaf. **b.** A card marked with such figures. **c. clubs.** The suit so marked. **4.** A group of people organized for a common purpose, esp. one that holds regular meetings. **5.** The facilities used for the meetings of a club. *—v.* **clubbed, club·bing, clubs.** *—vt.* **1.** To beat or strike with or as if with a club. **2.** To use (a firearm) as a club by holding the barrel and hitting with the butt end. **3.** *Archaic.* To gather or combine (e.g., hair) into a clublike mass. **4.** To contribute for a joint or common purpose. *—vi.*

**1.** *Archaic.* To form or gather into a mass. **2.** To join or combine for a common purpose.

**club·ba·ble** *also* **club·a·ble** (klŭb′ə-bəl) *adj. Informal.* Suited to membership in a social club : SOCIABLE.

**club·by** (klŭb′ē) *adj.* **-bi·er, -bi·est. 1.** Typical of a club or club members. **2.** Sociable : friendly. **3.** Exclusive : clannish <the *clubby* atmosphere in our company> **—club′bi·ness** *n.*

**club car** *n.* A railroad passenger car typically having lounge chairs, tables, and a buffet or bar.

**club chair** *n.* An upholstered easy chair with arms and a low back.

**club·foot** (klŭb′fo͝ot′) *n.* **1.** Congenital deformity of the foot, characterized by a misshapen appearance often looking like a club. **2.** A foot so deformed. **—club′foot′ed** *adj.*

**club·house** (klŭb′hous′) *n.* **1.** A building occupied by a club. **2.** The locker room for an athletic team.

**club·man** (klŭb′mən, -măn′) *n.* A man who belongs to a club or clubs, esp. one who is active in club life.

**club moss** *n.* [From the club-shaped strobiles on some species of this plant.] An evergreen, erect or creeping, mosslike plant of the genus *Lycopodium*, with tiny, scalelike, overlapping leaves and reproducing by spores.

**club root** *n.* A disease of cabbage and related plants, caused by the fungus *Plasmodiophora brassicae* and resulting in large, distorted swellings on the roots.

**club sandwich** *n.* A sandwich, usu. three slices of bread, filled with various meats, tomato, lettuce, and dressing.

**club soda** *n.* An effervescent, unflavored water used in various alcoholic and nonalcoholic drinks.

**club steak** *n.* Delmonico steak.

**club·wom·an** (klŭb′wo͝om′ən) *n.* A woman who belongs to a club or clubs, esp. one who is active in club life.

**cluck** (klŭk) *n.* [Imit.] **1. a.** The sound made by a hen when brooding or calling her chicks. **b.** A sound like a cluck. **2.** *Informal.* A stupid person. *—v.* **clucked, cluck·ing, clucks.** *—vi.* **1.** To utter a cluck. **2.** To make a sound like a cluck, as in coaxing a horse. *—vt.* **1.** To call by making a cluck. **2.** To express by clucking <The teacher *clucked* a mild reproof.>

**clue** *also* **clew** (klo͞o) [Var. of CLEW (from Theseus' use of a thread as a guide through the Cretan labyrinth).] *—n.* Something that leads to the solution of a problem or mystery. *—vt.* **clued, clue·ing** *or* **clu·ing, clues** *also* **clewed, clew·ing, clews.** To give (someone) guiding information <*Clue* me in on what's really happening.>

▲ word history: *Clue* and *clew* were at one time simply spelling variants, both meaning "ball," especially a ball of yarn or thread. The meaning "guide to a solution" developed from the story of Theseus and the Minotaur. Theseus, the great Athenian hero, had the task of killing the Minotaur, a monster half man and half bull, who lived in the Labyrinth, King Minos' maze on Crete. Finding the Minotaur was no problem, but finding the way out of the Labyrinth again would have been impossible if Ariadne, Minos' daughter, had not provided Theseus with a clew—a ball—of string. Theseus unwound the ball as he entered and wound it up as he returned, thus following a sure path out of the maze. Allusions to this "clew of thread" or a "clew to a maze" were very common from Chaucer's day to modern times and appeared in contexts that referred to various kinds of difficulties. As a result, the figurative import of the word *clew* was gradually lost, and all associations with a ball of twine were broken. In very recent times, especially since the advent of detective fiction, the spelling *clue* has primarily signified the meaning "guide to a solution." The older spelling *clew* survives for the obsolescent sense "ball of thread" and the nautical senses derived from it.

**Clum·ber spaniel** *also* **clum·ber spaniel** (klŭm′bər) *n.* [After *Clumber*, an estate in Nottinghamshire, England.] A dog orig. bred in England, with short legs and a silky, chiefly white coat.

**clump** (klŭmp) *n.* [Prob. LG *klump* < MLG *klumpe.*] **1.** A clustered mass : LUMP. **2.** A thick grouping, as of bushes or trees. **3.** A dull, heavy sound : THUD. *—v.* **clumped, clump·ing, clumps.** *—vi.* **1.** To form clumps. **2.** To walk with a dull, heavy sound. *—vt.* To gather into or form clumps of. **—clump′y** *adj.*

**clum·sy** (klŭm′zē) *adj.* **-si·er, -si·est.** [< obs. *clumse*, to be numb with cold < ME *clomsen*, of ON orig.] **1.** Lacking physical coordination, skill, or grace : AWKWARD. **2.** Awkwardly made : UNWIELDY <*clumsy* boots> **3.** Inept : gauche <a *clumsy* explanation> **—clum′si·ly** *adv.* **—clum′si·ness** *n.*

**clung** (klŭng) *v. p.t. & p.p. of* CLING.

**clunk** (klŭngk) *n.* [Imit.] **1.** A dull sound : THUMP. **2.** A hefty blow. *—v.* **clunked, clunk·ing, clunks.** *—vi.* **1.** To make or move with a clunk. **2.** To strike something with a clunk. *—vt.* To strike with a clunk.

**clunk·er** (clŭng′kər) *n.* **1.** A broken-down car. **2.** A flop : failure.

**clu·pe·id** (klo͞o′pē-ĭd) *n.* [NLat. *Clupeidae*, family name < Lat. *clupea*, a kind of small fish.] Any of various oily, soft-finned fishes of the family Clupeidae, including the herrings and menhadens. *—adj.* Of or belonging to the Clupeidae.

**clus·ter** (klŭs′tər) *n.* [ME < OE *clyster.*] **1.** A group of the same or similar elements occurring closely together : BUNCH. **2.** Two or more successive consonants in a word, as *cl* and *st* in the word *cluster.* *—vi & vt.* **-tered, -ter·ing, -ters.** To gather or grow into clusters or to cause to grow or form into clusters.

**cluster headache** *n.* A severe headache like a migraine that can occur several times daily for a period of weeks.

**clutch**[1] (klŭch) *v.* **clutched, clutch·ing, clutch·es.** [ME *clucchen*, var. of *clicchen* < OE *clyccan.*] *—vt.* **1.** To grasp and hold tightly. **2.** To snatch or seize. *—vi.* To attempt to seize or grasp. *—n.* **1.** A hand, claw, talon, or paw in the act of grasping. **2.** A tight grasp. **3.** *often* **clutches.** Power or control <in the *clutches* of a blackmailer> **4.** A device for gripping and holding. **5. a.** A device for engaging and disengaging two working parts of a shaft or of a shaft and a driving mechanism. **b.** The lever, pedal, or other apparatus that activates such a device. **6.** A tense or critical situation <always played well in the *clutch*>

**clutch**[2] (klŭch) *n.* [Var. of dial. *cletch*, perh. < *cleck*, to hatch < ME *clekken* < ON *klekja.*] **1.** The number of eggs produced or incubated at one time. **2.** A brood of chickens. *—vt.* **clutched, clutch·ing, clutch·es.** To hatch (chicks).

**clut·ter** (klŭt′ər) *n.* [Prob. < ME *cloteren*, to clot.] **1.** A confused or disordered collection or state : JUMBLE. **2.** A confused noise : CLATTER. *—v.* **-tered, -ter·ing, -ters.** *—vt.* To litter or pile in a disordered state. *—vi.* **1.** To run or move with confusion and bustle. **2.** To make a clatter.

**Clydes·dale** (klīdz′dāl′) *n.* A large, powerful draft horse bred in the Clyde valley, Scotland.

**clyp·e·ate** (klĭp′ē-ĭt) *also* **clyp·e·at·ed** (-ā′tĭd) *adj.* **1.** Shaped like a round shield. **2.** Having a clypeus.

**clyp·e·us** (klĭp′ē-əs) *n., pl.* **-e·i** (-ē-ī′) [NLat. < Lat. *clipeus*, round shield.] *Biol.* A shieldlike structure, esp. a plate on the front of the head of an insect. **—clyp′e·al** *adj.*

**clys·ter** (klĭs′tər) *n.* [ME *clister* < Lat. *clyster* < Gk. *klustēr*, clyster pipe < *kluzein*, to wash out.] *Med.* An enema.

**Cly·tem·nes·tra** *also* **Cly·taem·nes·tra** (klī′təm-nĕs′trə) *n.* [Lat. < Gk. *Klutaimnēstra.*] *Gk. Myth.* The wife of Agamemnon.

**Cm** *symbol for* CURIUM.

**cni·do·blast** (nī′də-blăst′) *n.* [Gk. *knidē*, nettle + -BLAST.] A modified interstitial cell in coelenterates that produces a nematocyst.

**Co** *symbol for* COBALT.

**co-** *pref.* [ME < Lat. < *com-*, com-.] **1.** With or together : JOINT <*co*education> **2. a.** Partner or associate in an activity <*co*-author> **b.** Subordinate or assistant <*co*pilot> **3.** To the same extent or degree <*co*extend> **4.** Complement of an angle <*co*tangent>

**co·ac·er·vate** (kō-ăs′ər-vāt′) *n.* [< Lat. *coacervatus*, p.part. of *coacervare*, to heap together : *co(m)-*, together + *acervare*, to heap < *acervus*, a heap.] *Chem.* A cluster of droplets separated out of a lyophilic colloid. **—co·ac′er·vate** *adj.* **—co·ac′er·va′tion** *n.*

**coach** (kōch) *n.* [Fr. *coche*, ult. < Hung. *kocsi*, after *Kocs*, Hungary, where such carriages were first made.] **1.** A large closed carriage with four wheels. **2.** A closed automobile, usu. with two doors. **3.** A motorbus. **4.** A railroad passenger car. **5.** Low-priced passenger accommodations on a train or aircraft. **6.** One who trains athletes or athletic teams. **7.** One who gives private instruction, as in acting or singing. **8.** A private tutor who prepares a student for an examination. *—vt. & vi.* **coached, coach·ing, coach·es. 1.** To train or act as a coach. **2.** To transport by or ride in a coach. **—coach′er** *n.*

▲ word history: *Coach* was used to mean "a tutor" or "a trainer" in allusion to the speed of stagecoaches and railway coaches. In the days before automobiles and airplanes, the fastest method of travel was by coach—at first horsedrawn and later steam-powered. A *coach* in university parlance was an instructor who brought his students along at the fastest possible rate.

**coach dog** *n.* [So called because it was trained to run behind a coach.] A Dalmatian.

**coach·man** (kōch′mən) *n.* **1.** One who drives a coach. **2.** An artificial fishing fly with white wings, a multi-colored feathered body with a brown hackle, and a gold tag.

**co·ac·tion** (kō-ăk′shən) *n.* [ME *coaccioun* < Lat. *coactio*, a collecting < *cogere*, to collect : *co(m)-*, together + *agere*, to drive.] **1.** Impelling or restraining force : COMPULSION. **2.** Joint action. **—co·ac′tive** *adj.* **—co·ac′tive·ly** *adv.*

**co·ad·ju·tant** (kō-ăj′ə-tənt) *n.* An assistant : helper.

**co·ad·ju·tor** (kō′ə-jo͞o′tər, kō-ăj′ə-tər) *n.* [ME *coadjutour* < Lat. *coadjutor* : *co(m)-* (intensive) + *adjutor*, assistant < *adjutare*, to aid.] **1.** An assistant : coworker. **2.** A bishop's assistant.

**co·ad·u·nate** (kō-ăj′ə-nĭt, -nāt′) *adj.* [LLat. *coadunatus*, p.part. of *coadunare*, to combine : Lat. *co(m)-*, together + Lat. *adunare*, to unite (*ad-*, to + *unus*, one).] Grown together : closely joined. **—co·ad′u·na′tion** (-nā′shən) *n.* **—co·ad′u·na′tive** *adj.*

**co·ag·u·lant** (kō-ăg′yə-lənt) *n.* An agent causing coagulation. **—co·ag′u·lant** *adj.*

**co·ag·u·lase** (kō-ăg′yə-lās′, -lāz′) *n.* [COAGUL(ATE) + -ASE.] A blood-clotting enzyme, as thrombin or rennin.

**co·ag·u·late** (kō-ăg′yə-lāt′) *v.* **-lat·ed, -lat·ing, -lates.** [ME *coagulaten* < Lat. *coagulare* < *coagulum,* coagulator. —see COAGULUM.] —*vt.* To cause transformation of (a liquid or sol) into a soft, semi-solid, or solid mass. —*vi.* To become coagulated. **—co·ag′u·la·bil′i·ty** *n.* **—co·ag′u·la·ble** *adj.* **—co·ag′u·la′tion** *n.* **—co·ag′u·la′tor** *n.*

**co·ag·u·lum** (kō-ăg′yə-ləm) *n., pl.* **-la** (-lə) [Lat., coagulator < *cogere,* to condense : *co(m)-,* together + *agere,* to drive.] A coagulated mass.

**coal** (kōl) *n.* [ME *col* < OE.] **1. a.** A natural dark-brown to black solid used as fuel, formed from fossilized plants and consisting of amorphous carbon with various organic and some inorganic compounds. **b.** A piece of coal. **2.** A glowing or charred piece of solid fuel, as coal or wood. **3.** Charcoal. —*v.* **coaled, coal·ing, coals.** —*vt.* **1.** To burn (a combustible solid) to a charcoal residue. **2.** To provide with coal. —*vi.* To take on coal.

**coal·er** (kō′lər) *n.* Something, esp. a train or ship, for carrying coal.

**co·a·lesce** (kō′ə-lĕs′) *vi.* **-lesced, -lesc·ing, -lesc·es.** [Lat. *coalescere* : *co(m)-,* together + *alescere,* to grow, inceptive of *alere,* to nourish.] **1.** To grow together : FUSE. **2.** To come together so as to form one whole : UNITE <Neighborhood groups *coalesced* into one large organization.> **—co′a·les′cence** *n.* **—co′a·les′cent** *adj.*

**coal gas** *n.* **1.** A gaseous mixture produced by the destructive distillation of bituminous coal and used as commercial fuel. **2.** The gaseous mixture released by burning coal.

**coal·i·fi·ca·tion** (kō′lə-fĭ-kā′shən) *n.* The process by which coal is formed from plant materials. **—coal′i·fy** *v.* **(-fied, -fy·ing, -fies).**

**co·a·li·tion** (kō′ə-lĭsh′ən) *n.* [Med. Lat. *coalitio* < Lat. *coalescere,* to grow together. —see COALESCE.] **1.** An esp. temporary alliance of factions, parties, or nations. **2.** Combination into one body : UNION. **—co′a·li′tion·ist** *n.*

**coal measures** *pl.n. Geol.* **1. Coal Measures.** A stratigraphic unit equivalent to the Pennsylvanian or Upper Carboniferous periods. **2.** Strata of the Carboniferous period containing coal deposits.

**coal oil** *n.* Kerosene.

**Coal·sack** (kōl′săk′) *n.* **1.** A dark nebula near the Southern Cross. **2.** A dark region of the sky, the Northern Coalsack, near the Northern Cross.

**coal tar** *n.* A viscous black liquid derived from the destructive distillation of coal, used in many dyes, drugs, and organic chemicals and for waterproofing, paints, roofing, and insulation materials.

**coam·ing** (kō′mĭng) *n.* [Orig. unknown.] A raised curb or rim around an opening, as in a ship's deck, designed to keep out water.

**co-an·chor** (kō-ăng′kər) *n.* Either of two news commentators who are anchorpersons during a broadcast. **—co-an′chor** *v.* **(-chored, -chor·ing, -chors).**

**co·arc·tate** (kō-ärk′tāt′) *adj.* [Lat. *coarctatus,* p.part. of *coarctare,* to compress : *co(m)-,* together + *artare,* to compress < *artus,* confined.] **1.** Designating an insect pupa compressed in the larval shell. **2.** Having a constricted separation between the abdomen and thorax. **—co′arc·ta′tion** *n.*

**coarse** (kôrs, kōrs) *adj.* **coars·er, coars·est.** [ME *cors,* prob. < *course,* custom. —see COURSE.] **1.** Of low, common, or inferior quality. **2.** Lacking in refinement or delicacy. **3.** Consisting of large particles. **4.** Harsh : rough <a *coarse* tweed fabric> **—coarse′ly** *adv.* **—coarse′ness** *n.*

☆ ***syns:*** COARSE, BOORISH, CHURLISH, CRASS, CRUDE, GROSS, PHILISTINE, RAW, ROUGH, RUDE, TASTELESS, UNCOUTH, VULGAR *adj. core meaning* : lacking delicacy or refinement <*coarse* language and manners> ***ant:*** refined

**coarse-grained** (kôrs′grānd′, kōrs′-) *adj.* **1.** With a coarse texture. **2.** Not refined : INDELICATE.

**coars·en** (kôr′sən, kōr′-) *vt. & vi.* **-ened, -en·ing, -ens.** To make or become coarse.

**coast** (kōst) *n.* [ME *coste* < OFr. < Lat. *costa,* side.] **1.** The land next to the sea : SEASHORE. **2.** *Obs.* The border or frontier of a country. **3.** A slope down which one may coast, as on a sled. **4.** The act of sliding or coasting : SLIDE. **5. Coast.** The U.S. Pacific Coast. —*v.* **coast·ed, coast·ing, coasts.** —*vi.* **1. a.** To slide down an inclined slope, as on a sled. **b.** To move smoothly and effortlessly. **2.** To move without further acceleration. **3.** To sail along or near a coast. **4.** To move or act aimlessly or with little effort. —*vt.* To sail or move along the coast or border of. **—coast′al** *adj.*

**coast artillery** *n.* Artillery for protecting coastal areas.

**coast·er** (kō′stər) *n.* **1.** One that coasts. **2.** A vessel engaged in coastal trade. **3.** A coasting sled or toboggan. **4.** A disk placed under a bottle, pitcher, or drinking glass to protect the surface below. **5.** A small tray, often on wheels, for passing something, as a wine decanter, around a table.

**coaster brake** *n.* A brake and clutch operating on the rear wheel and drive mechanism of a bicycle when pedaling is reversed.

**coast guard** *also* **Coast Guard** *n.* **1.** The naval or military coastal patrol of a nation, responsible for the protection of life and property at sea, coastal defense, and enforcement of customs, immigration, and navigation laws. **2.** A coastguardsman.

**coast·guards·man** (kōst′gärdz′mən) *n.* A member of a coast guard.

**coast·line** (kōst′līn′) *n.* The shape or boundary of a coast.

**coast rhododendron** *n.* An evergreen shrub, *Rhododendron californicum* or *R. macrophyllum* of the Pacific coast of North America, bearing rose-purple flowers.

**coast·ward** (kōst′wərd) *adv. & adj.* Toward or directed toward the coast. **—coast′wards** (-wərdz) *adv.*

**coast·ways** (kōst′wāz′) *adv.* Coastwise.

**coast·wise** (kōst′wīz′) *adj. & adv.* Following, by way of, or along the coast.

**coat** (kōt) *n.* [ME *cote* < OFr.] **1. a.** An outer garment covering the body from the shoulders to the waist or below. **b.** A garment extending to just below the waist and usu. forming the top part of a suit. **2.** A natural integument or outer covering, as the fur of an animal. **3.** A layer of covering material : COATING <a thick *coat* of varnish> —*vt.* **coat·ed, coat·ing, coats. 1.** To provide or cover with a coat. **2.** To cover with a layer, as of paint. **—coat′ed** *adj.*

**co·a·ti** (kō-ä′tē) *n.* [Port. *coatí* < Tupi : *cua,* belt + *tim,* nose.] An omnivorous mammal of the genus *Nasua* of South and Central America and southwestern United States, resembling the raccoon but with a longer snout and tail.

**co·a·ti·mun·di** *also* **co·a·ti·mon·di** (kō-ä′tē-mŭn′dē) *n.* [Tupi.] The coati.

**coat·ing** (kō′tĭng) *n.* **1.** COAT 3. **2.** Cloth for making coats.

**coat of arms** *n.* **1.** A surcoat or tabard blazoned with heraldic bearings. **2.** A representation of a coat of arms.

**coat of mail** *n., pl.* **coats of mail.** An armored coat made of chain mail, interlinked rings, or overlapping metal plates : HAUBERK.

**coat·tail** (kōt′tāl′) *n.* **1.** The loose back part of a coat below the waist. **2. coattails.** The skirts of a formal or dress coat.

**co·au·thor** (kō-ô′thər) *n.* A joint author. —*vt.* **-thored, -thor·ing, -thors.** To be a co-author of.

**coax** (kōks) *v.* **coaxed, coax·ing, coax·es.** [Obs. *cokes,* to fool < *cokes,* fool.] —*vt.* **1.** To persuade or try to persuade by flattery or pleading : WHEEDLE. **2.** To obtain by persistent persuasion. **3.** *Obs.* To fondle or caress. —*vi.* To use persuasion or inducement. **—coax′er** *n.* **—coax′ing·ly** *adv.*

☆ ***syns:*** COAX, BLANDISH, CAJOLE, SOFT-SOAP, SWEET-TALK, WHEEDLE *v. core meaning* : to try to persuade by gentle, persistent urging or flattery <*coaxed* me into attending the cocktail party>

**co·ax·i·al** (kō-ăk′sē-əl) *adj.* Having or mounted on a common axis.

**coaxial cable** *n.* High-frequency telephone, telegraph, and television transmission cable consisting of a conducting outer metal tube enclosing and insulated from a central conducting core.

**cob** (kŏb) *n.* [Prob. < obs. *cob,* round object.] **1.** The central core of an ear of corn : CORNCOB. **2.** A male swan. **3.** A thickset short-legged horse. **4.** A small lump or mass, as of coal.

**co·bal·a·min** (kō-băl′ə-mĭn) *also* **co·bal·a·mine** (-mēn′) *n.* [COBAL(T) + (VIT)AMIN.] Vitamin $B_{12}$.

**co·balt** (kō′bôlt′) *n.* [G. *Kobalt* < MHG *Kobolt,* goblin (from the trouble it gave silver miners).] *Symbol* **Co** A hard, brittle metallic element, used for magnetic alloys, high-temperature alloys, and glass and ceramic pigments; atomic number 27; atomic weight 58.9332.

**cobalt 60** *n.* A radioactive isotope of cobalt with mass number 60 and exceptionally intense gamma-ray activity, used in radiotherapy, metallurgy, and materials testing.

**cobalt blue** *n.* **1.** A blue to green pigment composed of a variable mixture of cobalt and aluminum oxides. **2.** A deep to vivid blue or strong greenish blue.

**co·balt·ite** (kō′bôl-tīt′) *also* **co·balt·ine** (-tēn′) *n.* A silver-white to gray mineral, CoAsS, an important cobalt ore used in ceramics.

**cob·ber** (kŏb′ər) *n.* [Orig. unknown.] *Austral.* A comrade.

**cob·ble**[1] (kŏb′əl) *n.* [Back-formation < COBBLESTONE.] **1.** A cobblestone. **2.** COB COAL 2. —*vt.* **-bled, -bling, -bles.** To pave with cobblestones.

**cob·ble**[2] (kŏb′əl) *vt.* **-bled, -bling, -bles.** [Prob. back-formation < COBBLER.] **1.** To mend or make (boots or shoes). **2.** To put together clumsily : BUNGLE.

**cob·bler**[1] (kŏb′lər) *n.* [ME *cobeler.*] **1.** One who mends shoes and boots. **2.** *Archaic.* One who is clumsy at work : BUNGLER.

**cob·bler**[2] (kŏb′lər) *n.* [Orig. unknown.] **1.** A deep-dish fruit pie with a thick top crust. **2.** An iced drink of wine or liqueur, sugar, and citrus fruit.

**cob·ble·stone** (kŏb′əl-stōn′) *n.* [ME *cobelston.*] A naturally rounded stone once used for paving streets.

**cob coal** *n.* **1.** Rounded lumps of coal in various sizes. **2.** A lump of coal approx. the size of a cobblestone.

**co·bel·lig·er·ent** (kō′bə-lĭj′ər-ənt) *n.* A nation allied with another or others in waging war.

**co·bi·a** (kō′bē-ə) *n.* [Orig. unknown.] A large game fish, *Rachycentron canadum* of tropical and subtropical seas.

**co·ble** (kō′bəl) *n.* [ME *cobel,* ult. < Lat. *caupulus,* a kind of small ship.] **1.** *Chiefly Brit.* A small, flat-bottomed fishing boat with a lugsail on a raking mast. **2.** *Scot.* A flat-bottomed rowboat.

**cob·nut** (kŏb′nŭt′) *n.* **1.** A tree, *Corylus avellana grandis,* related to the hazel. **2.** The large, edible nut of the cobnut.

**CO·BOL** *or* **Co·bol** (kō′bôl′) *n.* [CO(MMON) B(USINESS) O(RIENTED) L(ANGUAGE).] A language based on English words and phrases, used to program computers for business applications.

**co·bra** (kō′brə) *n.* [Short for Port. *cobra (de capello)*, snake (with a hood) < Lat. *colubra*, snake.] **1.** Any of several venomous snakes of the genus *Naja* and related genera of Asia and Africa, able to expand the skin of the neck to form a flattened hood. **2.** Leather made from the skin of a cobra.

**cob·web** (kŏb′wĕb′) *n.* [ME *coppeweb* : *coppe*, spider (short for *attercoppe* < OE *ātorcoppe* : *ātor*, poison + *copp*, head) + *web*, web < OE.] **1. a.** The web spun by a spider to catch its prey. **b.** A single thread of a spider's web. **2.** Something resembling a cobweb in gauziness or flimsiness. **3.** An intricate plot : SNARE <a *cobweb* of plotting and scheming> **4. cobwebs.** Disorder : confusion. —*vt.* **-webbed, -web·bing, -webs.** To cover with or as if with cobwebs. **—cob′web′by** *adj.*

**co·ca** (kō′kə) *n.* [Sp. < Quechua *kúka.*] **1.** A South American tree, *Erythroxylon coca*, bearing leaves that contain cocaine and related alkaloids. **2.** The dried leaves of the coca or related plants, chewed as a stimulant by people of the Andes.

**co·caine** *also* **co·cain** (kō-kān′, kō′kān′) *n.* A colorless or white crystalline narcotic alkaloid, $C_{17}H_{21}NO_4$, extracted from coca leaves and used medically as a local anesthetic.

**co·cain·ism** (kō-kā′nĭz′əm) *n.* Habitual use of cocaine.

**co·cain·ize** (kō-kā′nīz′) *vt.* **-ized, -iz·ing, -iz·es.** To anesthetize (a body part) with cocaine. **—co·cain′i·za′tion** *n.*

**co·car·box·y·lase** (kō′kär-bŏk′sə-lās′, -lāz′) *n.* A coenzyme that functions in the catalysis of the decarboxylation of pyruvic acid in the Krebs cycle.

**coc·ci** (kŏk′sī, kŏk′ī′) *n. pl. of* COCCUS.

**coc·cid** (kŏk′sĭd) *n.* [NLat. *Coccidae*, family name < *Coccus*, genus name < Gk. *kokkos*, grain.] An insect of the family Coccidae, including the scale insects and mealybugs.

**coc·cid·i·oi·do·my·co·sis** (kŏk-sĭd′ē-oi′dō-mī-kō′sĭs) *n.* [NLat. *Coccidiodes*, genus name (< *Coccidia*, order name < COCCUS) + MYCOSIS.] A fungus disease that usu. affects the lungs of humans and other animals, caused by the fungus *Coccidioides immitis*.

**coc·cid·i·o·sis** (kŏk-sĭd′ē-ō′sĭs) *n.* [NLat. *Coccidia*, order name (< COCCUS) + -OSIS.] A disease mainly of animals and rarely of humans, caused by an infection of the digestive tract by parasitic protozoa of the order Coccidia.

**coc·co·ba·cil·lus** (kŏk′ō-bə-sĭl′əs) *n., pl.* **-cil·li** (-sĭl′ī′) [COCC(US) + -O- + BACILLUS.] A short, oval bacillus.

**coc·cus** (kŏk′əs) *n., pl.* **coc·ci** (kŏk′sī′, kŏk′ī′) [NLat. < Gk. *kokkos*, grain.] **1.** A spherical or spheroidal bacterium. **2.** *Bot.* A division that contains a single seed and splits apart from a many-lobed fruit. **—coc′coid′** (kŏk′oid′), **coc′cal** (kŏk′əl) *adj.*

**-coccus** *suff.* [NLat. < Gk. *kokkos*, berry.] A microorganism of spherical or spheroidal shape <strepto*coccus*>

**coc·cyg·e·al** (kŏk-sĭj′ē-əl) *adj.* [< Gk. *kokkyx*, *kokkyg-*, coccyx.] Of or relating to the coccyx.

**coc·cyx** (kŏk′sĭks) *n., pl.* **coc·cy·ges** (kŏk-sī′jēz, kŏk′sĭ-jēz′) [Gk. *kokkux*, cuckoo, coccyx (from the resemblance of the bone to a cuckoo's beak).] A small bone at the base of the spinal column, composed of several fused rudimentary vertebrae. **—coc·cyg′e·al** (kŏk-sĭj′ē-əl) *adj.*

**Co·chin China** (kō′chĭn, kŏch′ĭn) *n.* [After *Cochin China*, a former name for a region of Vietnam.] A large domestic fowl orig. bred in Asia, with thickly feathered legs.

**coch·i·neal** (kŏch′ə-nēl′, kŏch′ə-nēl′, kō′chə-, kō′chə-) *n.* [Fr. *cochenille* < Sp. *cochinilla*, prob. < Lat. *coccineus*, scarlet < Gk. < *kokkos*, kermes berry (from its use in making scarlet dye).] **1.** A brilliant red dye made by drying and pulverizing the bodies of the females of a tropical American scale insect, *Dactylopius coccus*, that feeds on certain species of cacti. **2.** A vivid red. **—coch′i·neal′** *adj.*

**cochineal insect** *n.* A tropical American insect, *Dactylopius coccus*, that feeds on certain species of cacti.

**coch·le·a** (kŏk′lē-ə, kō′klē-ə) *n., pl.* **-le·ae** (-lē-ē′) [NLat. < Lat., snail shell < Gk. *kokhlias*, snail < *kokhlos*, land snail.] A spiral tube of the inner ear resembling a snail shell and having nerve endings essential for hearing. **—coch′le·ar** *adj.*

**cochlear nerve** *n.* A division of the acoustic nerve.

**coch·le·ate** (kŏk′lē-ĭt, -āt′, kō′klē-) *also* **coch·le·at·ed** (-ā′tĭd) *adj.* [Lat. *cochleatus* < *cochlea*, snail shell. —see COCHLEA.] Being spirally twisted like a snail shell.

**cock¹** (kŏk) *n.* [ME *cok* < OE *coc.*] **1. a.** The adult male of the domestic fowl : ROOSTER. **b.** A male bird. **2.** A weather vane shaped like a rooster : WEATHERCOCK. **3.** A leader or chief. **4.** A faucet or valve for regulating the flow of a liquid or gas. **5. a.** The hammer in a firearm. **b.** The position of a hammer when ready for firing. **6.** A tilting or jaunty turn upward <the *cock* of your cap> —*v.* **cocked, cock·ing, cocks.** —*vt.* **1.** To set the hammer of (a firearm) in position for firing. **2.** To tilt or turn up or to one side, usu. in a jaunty or alert way. **3.** To raise in preparation to throw or hit. —*vi.* **1.** To cock the hammer of a firearm. **2.** To turn or stick up.

**cock²** (kŏk) *n.* [ME *cok.*] A cone-shaped pile of straw or hay. —*vt.* **cocked, cock·ing, cocks.** To arrange (straw or hay) in a cock.

**cock·ade** (kŏ-kād′) *n.* [Alteration of obs. *cockard* < Fr. *cocarde* < OFr. *coquarde*, vain < *coq*, cock.] A rosette worn esp. on a hat as a badge. **—cock·ad′ed** (kŏ-kā′dĭd) *adj.*

**cock-a-hoop** (kŏk′ə-ho͝op′, -ho͞op′) *adj.* [From the phrase *to set cock on hoop*, to drink festively.] **1.** Elated : jubilant. **2.** Boastful. **3.** Askew. **—cock′-a-hoop′** *adv.*

**Cock·aigne** (kŏ-kān′) *n.* [ME *cokaigne* < OFr. < (*pais de*) *cokaigne*, land of plenty, prob. < MLG *kōkenje*, small cake, dim. of *kōke*, cake.] An imaginary land of luxury and ease.

**cock-a-leek·ie** *also* **cock·a·leek·ie** (kŏk′ə-lē′kē) *n.* [Alteration of *cockie*, dim. of COCK¹ + *leekie*, dim. of LEEK.] A cream soup of leeks and chicken.

**cock·a·lo·rum** (kŏk′ə-lôr′əm, -lōr′-) *n.* [Perh. alteration of obs. Flem. *kockeloeren*, to crow.] **1.** A little man with an unduly good opinion of himself. **2.** Boastful talk.

**cock·a·ma·mie** *also* **cock·a·ma·my** (kŏk′ə-mā′mē) *adj.* [Prob. alteration of DECALCOMANIA.] *Slang.* **1.** Nearly valueless : TRIFLING. **2.** Nonsensical : ludicrous <a *cockamamie* excuse for a car>

**cock-and-bull story** (kŏk′ən-bo͝ol′) *n.* A highly improbable tale passed off as true.

**cock·a·tiel** *also* **cock·a·teel** (kŏk′ə-tēl′) *n.* [Du. *kaketielje*, prob. ult. < Malay *kakatua*, cockatoo.] A crested parrot, *Nymphicus hollandicus* of Australia, with gray and yellow plumage.

**cock·a·too** (kŏk′ə-to͞o′) *n., pl.* **-toos.** [Du. *kaketoe*, < Malay *kakatua.*] Any of various parrots of the genus *Kakatoe* and related genera of Australia and adjacent islands, having a long, erectile crest.

**cock·a·trice** (kŏk′ə-trĭs, -trīs′) *n.* [ME *cocatrice*, basilisk < OFr. *cocatris* < Med. Lat. *calcatrix* (< Lat. *calcare*, to track < *calx*, heel), transl. of Gk. *ikhneumōn*, tracker. —see ICHNEUMON.] A mythical serpent hatched from a cock's egg and having the power to kill by its glance.

**cock·boat** (kŏk′bōt′) *n.* [ME *cokbote* : *cok*, cockboat (< AN *coque*) + *bot*, boat < OE *bāt.*] A small rowboat, esp. one used as a tender and kept on a ship.

**cock·chaf·er** (kŏk′chā′fər) *n.* An Old World beetle of the family Scarabaeidae, esp. *Melolontha melolontha*, a species destructive to plants.

**Cock·croft-Wal·ton accelerator** (kŏk′krôft-wôl′tən) *n.* [After Sir John Douglas *Cockcroft* (1897–1967) and Ernest Thomas Sinton *Walton* (b. 1903), its inventors.] *Physics.* A positive-ion accelerator, composed of several stages of a voltage-doubling circuit together with an ion source and a discharge tube, used in the first purely artificial disintegration of an atomic nucleus.

**cock·crow** (kŏk′krō′) *n.* DAWN 1.

**cocked hat** *n.* A hat with the brim turned up in two or three places, esp. a three-cornered hat : TRICORN.

**cock·er¹** (kŏk′ər) *n.* **1.** A cocker spaniel. **2. a.** A keeper or trainer of gamecocks. **b.** One who promotes or attends cockfights.

**cock·er²** (kŏk′ər) *vt.* **-ered, -er·ing, -ers.** [ME *cokeren.*] To pamper or spoil.

**cock·er·el** (kŏk′ər-əl) *n.* [ME *cokerel*, dim. of *cok*, cock < OE *coc.*] A young rooster.

**cocker spaniel** *n.* [From its original use in hunting woodcocks.] A dog orig. bred in England, with long, drooping ears and a variously colored silky coat.

**cock·eye** (kŏk′ī′) *n.* A squinting eye.

**cock·eyed** (kŏk′īd′) *adj.* **1.** Cross-eyed. **2.** *Slang.* **a.** Being askew : CROOKED. **b.** Ridiculous : foolish. **c.** Drunk.

**cock·fight** (kŏk′fīt′) *n.* A fight between gamecocks often fitted with metal spurs. **—cock′fight′ing** *adj. & n.*

**cock·le¹** (kŏk′əl) *n.* [ME *cokel* < OFr. *coquille*, shell < VLat. **conchillia* < Lat. *conchyllium* < Gk. *konkhulion*, dim. of *konkhē*, mussel.] **1.** Any of various bivalve mollusks of the family Cardiidae, with rounded or heart-shaped shells with radiating ribs. **2.** COCKLESHELL 1a, 2. **3.** A wrinkle : pucker. —*vi. & vt.* **-led, -ling, -les.** To become or cause to become wrinkled or puckered. **—cockles of (one's) heart.** One's innermost feelings.

**cock·le²** (kŏk′əl) *n.* [ME *cokkel* < OE *coccel.*] Any of several plants often growing as weeds in grain fields.

**cock·le·bur** (kŏk′əl-bûr′) *n.* **1.** Any of several coarse weeds of the genus *Xanthium*, bearing prickly burs. **2.** The bur of a cocklebur.

**cock·le·shell** (kŏk′əl-shĕl′) *n.* **1. a.** The shell of a cockle. **b.** A shell similar to that of a cockle. **2.** A small, light boat.

**cock·loft** (kŏk′lôft′, -lŏft′) *n.* [Prob. from its use as a roosting place.] A small garret or loft.

**cock·ney** (kŏk′nē) *n., pl.* **-neys.** [ME *cokenei*, pampered child, prob. : *cok*, cock (< OE *coc*) + *ei*, egg < OE *ǣg.*] **1.** *often* **Cockney.** A native of London's East End. **2.** The dialect or accent of cockneys. **—cock′ney** *adj.*

**cock-of-the-rock** (kŏk′əv-thə-rŏk′) *n., pl.* **cocks-of-the-rock.** [From its habit of nesting on rocks.] A South American bird, *Rupicola rupicola* or *R. peruviana*, with a distinctive crest and bright-orange or reddish plumage in the male.

cock-of-the-rock
*12 inches long*

**cock·pit** (kŏk'pĭt') *n.* **1. a.** The space in the fuselage of a small aircraft with seats for the pilot, copilot, and sometimes passengers. **b.** The space for the pilot and crew in a large airliner. **2.** A pit or enclosed space for cockfights. **3.** A place where many battles have been fought. **4. a.** An apartment in an old warship below the water line, used as quarters for junior officers and as a station for the wounded during a battle. **b.** An area near the stern in small decked vessels.

**cock·roach** (kŏk'rōch') *n.* [By folk ety. < Sp. *cucaracha.*] Any of various oval, flat-bodied insects of the family Blattidae, several species of which are common household pests.

**cocks·comb** (kŏks'kōm') *n.* **1.** The comb of a rooster. **2.** The cap of a jester, decorated to resemble the comb of a rooster. **3.** Any of several plants of the genus *Celosia*, esp. *C. argentea cristata*, bearing a showy crested or rolled flower cluster. **4.** *also* **cox·comb.** A pretentious fop.

**cock·shy** (kŏk'shī') *n., pl.* **-shies.** [From an old game in which sticks were shied at a cock.] *Chiefly Brit.* **1.** A mark aimed at in throwing contests. **2.** The throw in a throwing contest.

**cock·spur thorn** (kŏk'spûr') *n.* [From the resemblance of its thorn to a cock's spur.] A small, thorny North American tree, *Crataegus crus-galli*, bearing white flowers and small red fruit.

**cock·sure** (kŏk'shŏŏr') *adj.* Completely, often arrogantly sure. **—cock'sure'ly** *adv.* **—cock'sure'ness** *n.*

**cock·tail** (kŏk'tāl') *n.* **1.** A mixed alcoholic drink consisting usu. of brandy, whiskey, vodka, or gin combined with fruit juices or other liquors and usu. served chilled. **2.** An appetizer, as a juice or seafood served with a sharp sauce <shrimp *cocktail*> *—adj.* **1.** Of or relating to cocktails. **2.** Suitable for semiformal wear.

**cock·y** (kŏk'ē) *adj.* **-i·er, -i·est.** *Informal.* Self-assertive or self-confident : CONCEITED. **—cock'i·ly** *adv.* **—cock'i·ness** *n.*

**co·co** (kō'kō) *n., pl.* **-cos.** [Sp. < Port., goblin, from the face suggested by holes on the inner coconut shell.] **1.** The coconut palm. **2.** The coconut. *—adj.* Made of fibers from the coconut shell.

**co·coa** (kō'kō) *n.* [Var. of CACAO.] **1. a.** Powder made from cacao seeds after they have been roasted, ground, and freed of most of their fatty oil. **b.** A hot drink made by combining cocoa powder with water or milk and sugar. **2.** A moderate brown to reddish brown. **—co'coa** *adj.*

▲ word history: The confusion of *cocoa* with *coco* can be traced to Samuel Johnson's great dictionary, published in 1755. Johnson himself maintained the distinction between the two words in his own writing, but by some editorial or printing error the definitions for *coco* and *cocoa* were printed together under the word *cocoa.* That was unfortunate, because *coco* and *cocoa* are two different words that refer to two different trees. The cacao tree of tropical America produces both cocoa and chocolate. The name *cacao* comes from Nahuatl (Aztec) *cachuatl*, and *chocolate* comes from *xocolatl*, an unrelated word in the same language. The word *cocoa* is simply a variant spelling of *cacao.* The coconut or coco palm originated in the East Indies. Its name is not a native name like *cacao* but comes from Portuguese *coco*, "goblin," referring to the facelike appearance of the three holes at the bottom of the fruit.

**cocoa butter** *n.* A yellowish-white, waxy solid obtained from cacao seeds and used in manufacturing pharmaceuticals, confections, and soap.

**co·coa·nut** (kō'kə-nŭt', -nət) *n. var. of* COCONUT.

**co·co·bo·lo** (kō'kə-bō'lō) *n., pl.* **-los.** [Sp. < Arawakan *kakabali.*] **1.** A tropical American tree, *Dahlbergia retusa*, with hard, dark wood banded with light streaks. **2.** The wood of the cocobolo, used in cabinetwork.

**co·con·scious** (kō'kŏn'shəs) *adj.* Aware or conscious of the same things. *—n. Psychiat.* Mental processes outside the realm of conscious awareness or activity, as with schizophrenics. **—co·con'scious·ness** *n.*

**co·co·nut** *also* **co·coa·nut** (kō'kə-nŭt', -nət) *n.* The fruit of the coconut palm, a large seed with a thick, hard shell that encloses edible white meat and has a milky fluid filling the hollow center.

**coconut oil** *n.* An oil extracted from coconuts used in foods and in manufacture of soaps.

**coconut palm** *n.* A tall palm tree native to the East Indies, *Cocos nucifera*, bearing coconuts as fruit.

**co·coon** (kə-kōōn') *n.* [Fr. *cocon* < Prov. *coucoun*, dim. of *coco*, shell.] **1. a.** A covering of silk or similar fibrous material spun by the larvae of moths and other insects to protect their pupal stage. **b.** A similar protective covering or structure, as that of a spider or earthworm. **2.** A protective plastic coating over mothballed military or naval equipment.

**co·cotte** (kô-kôt') *n.* [Fr.] A prostitute.

**Co·cy·tus** (kō-kī'təs) *n.* [Lat. < Gk. *kōkutos* < *kōkutos*, lamentation < *kōkuein*, to wail.] *Gk. Myth.* One of the six rivers of Hades.

**cod**[1] (kŏd) *n., pl.* **cod** *or* **cods.** [ME.] Any of various marine fishes of the family Gadidae, esp. *Gadus morhua* or *G. callarias*, an important food fish of Northern Atlantic waters.

**†cod**[2] (kŏd) *n.* [ME < OE *codd.*] **1.** *Regional.* A husk or pod. **2.** *Obs.* A bag. **3.** *Archaic.* The scrotum.

**co·da** (kō'də) *n.* [Ital. < Lat. *cauda*, tail.] *Mus.* A passage bringing a movement or composition to a formal close.

**cod·dle** (kŏd'l) *vt.* **-dled, -dling, -dles.** [Poss. < CAUDLE.] **1.** To cook in water just below the boiling point. **2.** To treat overindulgently. **—cod'dler** *n.*

**code** (kōd) *n.* [ME < OFr. < Lat. *codex.*] **1.** A comprehensive and systematically arranged collection of laws. **2.** A systematic collection of regulations and rules of conduct or procedure <a *code* of ethics> **3. a.** A system of signals used to represent letters or numbers in sending messages. **b.** A system of symbols, letters, or words given certain arbitrary meanings, used for sending messages requiring secrecy or brevity. *—vt.* **cod·ed, cod·ing, codes. 1.** To systematize and arrange (laws and regulations) into a code. **2.** To convert (e.g., a message) into code.

**co·deine** (kō'dēn', -dē-ĭn) *n.* [Fr. *codéine* : Gk. *kōdeia*, poppy head + *-ine*, -ine.] An alkaloid narcotic, $C_{18}H_{21}NO_3$, derived from opium or morphine, used for cough relief, as an analgesic, and as a hypnotic.

**Code Na·po·lé·on** (kōd' nä-pō-lā-ôN') *n.* [Fr.] The code of French civil law, prepared between 1804 and 1807 under the direction of Napoleon Bonaparte.

**co·dex** (kō'dĕks') *n., pl.* **co·di·ces** (kō'dĭ-sēz', kŏd'ĭ-) [Lat.] **1.** A manuscript volume, esp. of a classic work or of the Scriptures. **2.** *Obs.* A code of laws.

**cod·fish** (kŏd'fĭsh') *n., pl.* **codfish** *or* **-fish·es.** COD[1].

**codg·er** (kŏj'ər) *n.* [Perh. alteration of obs. *cadger*, peddler. —see CADGE.] *Informal.* An old man.

**co·di·ces** (kō'dĭ-sēz', kŏd'ĭ-) *n. pl. of* CODEX.

**cod·i·cil** (kŏd'ə-sĭl) *n.* [ME < Lat. *codicilis*, dim. of *codex*, codex.] **1.** *Law.* A supplement or appendix to a will. **2.** A supplement or appendix. **—cod'i·cil'la·ry** (kŏd'ə-sĭl'ə-rē) *adj.*

**cod·i·fy** (kŏd'ĭ-fī', kō'də-) *vt.* **-fied, -fy·ing, -fies. 1.** To reduce (e.g., laws) to a code. **2.** To arrange or systematize. **—cod'i·fi·ca'tion** *n.* **—cod'i·fi'er** *n.*

**cod·ling**[1] (kŏd'lĭng) *also* **cod·lin** (-lĭn) *n.* [ME *querdlyng.*] *Chiefly Brit.* **1.** A long tapering apple. **2.** An unripe apple.

**cod·ling**[2] (kŏd'lĭng) *n., pl.* **-lings** *or* **codling.** A young cod.

**codling moth** *also* **codlin moth** *n.* A small grayish moth, *Carpocapsa pomonella*, whose larvae are destructive to various fruits, esp. apples.

**cod-liv·er oil** (kŏd'lĭv'ər) *n.* Oil obtained from the liver of cod and rich in vitamins A and D.

**co·dom·i·nance** (kō-dŏm'ə-nəns) *n.* The condition in which two different alleles in a gene pair appear together in a heterozygote. **—co·dom'i·nant** *adj.*

**co·don** (kō'dŏn') *n.* [COD(E) + -ON[1].] A sequence of three adjacent nucleotides specifying the insertion of an amino acid in a specific structural position during protein synthesis.

**cod·piece** (kŏd'pēs') *n.* [ME *codpece* : *cod*, scrotum (< OE *codd*, bag) + *pece*, piece. —see PIECE.] A pouch at the crotch of the tight-fitting breeches worn by 15th- and 16th-cent. men.

**cods·wal·lop** (kŏdz'wŏl'əp) *n.* [Orig. unknown.] *Chiefly Brit.* Nonsense.

**co·ed** *or* **co-ed** (kō'ĕd') [Short for *coeducational student.*] *Informal. —n.* A woman student at a coeducational university or college. *—adj.* Coeducational.

**co·ed·u·ca·tion** *also* **co-ed·u·ca·tion** (kō-ĕj'ə-kā'shən) *n.* The system of education in which men and women attend the same institution or classes. **—co·ed'u·ca'tion·al** *adj.* **—co·ed'u·ca'tion·al·ly** *adv.*

**co·ef·fi·cient** (kō'ə-fĭsh'ənt) *n.* **1.** *Math.* **a.** A numerical factor of an elementary algebraic term, as 2 in the term 2x. **b.** The product of all but one of the factors of an expression, the product being considered a distinct entity with respect to the excluded factor and to a designated operation. **2.** A numerical measure of a physical or chemical property that is constant for a system under given conditions.

**coefficient of correlation** *n.* Correlation coefficient.

**-coel** *or* **-coele** *or* **-cele** *suff.* [NLat. *-coela* < Gk. *koilos*, hollow.] Chamber : cavity <blasto*coel*>

**coe·la·canth** (sē'lə-kănth') *n.* [NLat. *Coelecanthus*, former genus name : Gk. *koilos*, hollow + Gk. *akantha*, spine.] Any of various fishes of the order Coelacanthiformes, previously known only in fossil form until a living species, *Latimeria chalumnae* of African marine waters, was identified in 1938. **—coe'la·can'thine'** (-kăn'thīn', -thĭn) *adj.* **—coe'la·can'thous** (-thəs) *adj.*

**-coele** *suff. var. of* COEL.

**coe·len·te·ra** (sĭ-lĕn'tər-ə) *n. pl. of* COELENTERON.

**coe·len·ter·ate** (sĭ-lĕn'tə-rāt', -tər-ĭt) *n.* [NLat. *Coelenterata*, phylum name : Gk. *koilos*, hollow + Gk. *enteron*, intestines.] An invertebrate animal of the phylum Coelenterata, having a radially symmetric body with a saclike internal cavity and including the jellyfishes, hydras, sea anemones, and corals. *—adj.* Of or belonging to the Coelenterata. **—coe·len·ter'ic** (sĭ-lĕn-tĕr'ĭk) *adj.*

**coe·len·ter·on** (sĭ-lĕn'tə-rŏn', -tər-ən) *n., pl.* **-te·ra** (-tər-ə) [NLat. : Gk. *koilos*, hollow + Gk. *enteron*, intestine.] A coelenterate's saclike body cavity.

**coe·li·ac** (sē'lē-ăk') *adj. var. of* CELIAC.

**coe·lom** *also* **ce·lom** *or* **coe·lome** (sē'ləm) *n.* [Gk. *koilōma*, cavity < *koilos*, hollow.] The body cavity in all animals higher than the coelenterates and some primitive worms, formed by a splitting of the mesoderm into two layers.

**coeno-** *or* **ceno-** *pref.* [NLat. < Gk. *koino-* < *koinos*, common.] Common <*coenocyte*>

**coen·o·bite** (sĕn'ə-bīt', sē'nə-) *n. var. of* CENOBITE.

**coe·no·cyte** (sē'nə-sīt') *n.* An organism made up of a multinucleate protoplasmic mass resulting from nuclear division without the formation of a new cell wall or membrane, as in slime molds and certain algae and fungi. **—coe'no·cyt'ic** (-sĭt'ĭk) *adj.*

**coe·no·gen·e·sis** (sē'nō-jĕn'ĭ-sĭs, sĕn'ō-) *n. var. of* CENOGENESIS.

**coe·nu·rus** (sĭ-nŏŏr'əs, -nyŏŏr'-) *n., pl.* **-nu·ri** (-nŏŏr'ī', -nyŏŏr'ī') [COEN(O)- + -UR(O)US.] The larval stage of a tapeworm, *Multiceps multiceps* or *Taenia multiceps*, that attacks the central nervous system of ruminant animals.

**co·en·zyme** (kō-ĕn'zīm') *n.* A heat-stable organic molecule that must be associated with an enzyme for the enzyme to function. **—co·en·zy·mat·ic** (-zə-măt'ĭk) *adj.* **—co·en'zy·mat'i·cal·ly** *adv.*

**coenzyme A** *n.* A coenzyme in all living cells that functions as an acetylating agent and is also required in fatty acid metabolism.

**co·e·qual** (kō-ē'kwəl) *adj.* Equal with one another. **—co·e'qual** *n.* **—co'e·qual'i·ty** (-kwŏl'ĭ-tē) *n.* **—co·e'qual·ly** *adv.*

**co·erce** (kō-ûrs') *vt.* **-erced, -erc·ing, -erc·es.** [Lat. *coercēre*, to confine : *co(m)-*, together + *arcēre*, to restrain.] **1.** To force to act or think in a given way by pressure, threats, or intimidation : COMPEL. **2.** To dominate or restrain forcibly. **3.** To bring about by force <efforts to *coerce* agreement> **—co·erc'er** *n.* **—co·erc'i·ble** *adj.*

☆ ***syns:*** COERCE, DRAGOON, FORCE, STRONG-ARM *v. core meaning* : to compel by pressure or threats <*coerced* them into the car at gunpoint>

**co·er·cion** (kō-ûr'zhən, -shən) *n.* **1.** The act or practice of coercing. **2.** Coercive power. **—co·er'cion·ar'y** (-zhə-nĕr'ē, -shə-) *adj.*

**co·er·cive** (kō-ûr'sĭv) *adj.* Marked by or tending toward coercion. **—co·er'cive·ly** *adv.* **—co·er'cive·ness** *n.*

**co·es·sen·tial** (kō'ĭ-sĕn'shəl) *adj.* Having identical nature or essence. **—co'es·sen'ti·al'i·ty** (-shē-ăl'ĭ-tē), **co'es·sen'tial·ness** *n.* **—co'es·sen'tial·ly** *adv.*

**co·e·ta·ne·ous** (kō'ĭ-tā'nē-əs) *adj.* [< LLat. *coaetaneus*, a contemporary : Lat. *co(m)-*, same + Lat. *aetas*, age.] Coeval. **—co'e·ta'ne·ous·ly** *adv.* **—co'e·ta'ne·ous·ness** *n.*

**co·e·ter·nal** (kō'ĭ-tûr'nəl) *adj.* Equally eternal <believed that the three persons of the Trinity are *coeternal*> **—co'e·ter'nal·ly** *adv.*

**co·e·ter·ni·ty** (kō'ĭ-tûr'nĭ-tē) *n.* Existence for eternity with another or others.

**co·e·val** (kō-ē'vəl) *adj.* [Lat. *coaevus* : *co(m)-*, same + *aevum*, age.] Originating or existing during the same period of time : CONTEMPORARY. *—n.* One of the same era or period. **—co·e'val·ly** *adv.*

**co·ex·ist** (kō'ĭg-zĭst') *vi.* **-ist·ed, -ist·ing, -ists. 1.** To exist together in the same place or at the same time. **2.** To live in peace together despite differences, esp. as a matter of diplomatic policy. **—co'ex·is'tence** *n.* **—co'ex·is'tent** *adj.*

**co·ex·tend** (kō'ĭk-stĕnd') *v.* **-tend·ed, -tend·ing, -tends.** *—vt.* To cause to extend through the same space or time. *—vi.* To reach to or attain the same limit in space or time. **—co'ex·ten'sion** *n.*

**co·ex·ten·sive** (kō'ĭk-stĕn'sĭv) *adj.* Having the same limits, boundaries, or scope. **—co'ex·ten'sive·ly** *adv.*

**co·fac·tor** (kō'făk'tər) *n.* A substance, as an inorganic ion, coenzyme, or vitamin, that activates an enzyme.

**cof·fee** (kô'fē, kŏf'ē) *n.* [Turk. *kahve* < Ar. *qahwah*.] **1. a.** A tree of the genus *Coffea*, native to eastern Asia and Africa, yielding berries containing beans used to prepare a beverage, esp. *C. arabica*, the chief commercial source of the beans. **b.** Seeds or beans of the coffee tree. **c.** An aromatic, mildly stimulating beverage prepared from coffee beans. **2.** A moderate brown to dark or dark grayish yellowish brown.

**coffee break** *n.* A short work break during which coffee or other refreshments may be consumed.

**coffee cake** *n.* A cake made of sweetened yeast dough, often containing raisins or nuts and topped with icing or powdered sugar.

**cof·fee·house** *also* **coffee house** (kô'fē-hous', kŏf'ē-) *n.* A place serving coffee and other refreshments to customers.

**coffee klatch** *or* **coffee klatsch** (klăch, kläch) *n.* [Partial transl. of G. *Kaffeeklatsch* : *Kaffee*, coffee (< Ital. *caffè* < Turk. *kahve* < Ar. *qahwah*) + *Klatsch*, chat.] A gathering for coffee an[d] casual conversation.

**coffee mill** *n.* A device for grinding roasted coffee beans.

**cof·fee·pot** (kô'fē-pŏt', kŏf'ē-) *n.* A pot used for brewing or servin[g] coffee.

**coffee shop** *n.* A small restaurant serving light meals.

**coffee table** *n.* A long, low table, often placed before a sofa.

**cof·fee-ta·ble book** (kô'fē-tā'bəl, kŏf'ē-) *also* **cof·fee-ta·ble** (-tā'blər) *n.* An oversized, elaborately designed book often used f[or] display, as on a coffee table.

**coffee tree** *n.* **1.** A tree of the genus *Coffea*, producing coffe[e] beans. **2.** The Kentucky coffee tree.

**cof·fer** (kô'fər, kŏf'ər) *n.* [ME *cofre* < OFr. < Lat. *cophinus*, basket < Gk. *kophinos*.] **1.** A strongbox. **2.** *often* **coffers.** Financial resource[s] : FUNDS <union *coffers*> **3.** A decorative sunken panel in a soffi[t], ceiling, dome, or vault. **4.** A canal lock. **5.** A cofferdam. *—vt.* **-fered -fer·ing, -fers. 1.** To put in a coffer. **2.** To supply with decorativ[e] sunken panels.

**cof·fer·dam** (kô'fər-dăm', kŏf'ər-) *n.* **1.** A temporary watertight en[-]closure built in the water and pumped dry to expose the bottom s[o] that construction, as of piers, may be undertaken. **2.** A watertigh[t] chamber attached to a ship's side to facilitate underwater repairs.

**cof·fin** (kô'fĭn, kŏf'ĭn) *n.* [ME *cofin*, basket < OFr. < Lat. *cophinu[s]* < Gk. *kophinos*.] **1.** An oblong box in which a corpse is buried. **2.** A horse's hoof. *—vt.* **-fined, -fin·ing, -fins.** To place in or as if in a coffin.

**coffin bone** *n.* The bone inside a horse's hoof.

**coffin corner** *n.* *Football.* A corner within 10 yards of the defend[-]ing team's goal line, in which area the ball may be punted out of bounds, thus placing the receiving team very close to its goal line.

**coffin nail** *n.* [From the unhealthful effects of smoking ciga[-]rettes.] *Slang.* A cigarette.

**cof·fle** (kô'fəl, kŏf'əl) *n.* [Ar. *qāfilah*, caravan.] A file of animals prisoners, or slaves, chained together in transit. **—cof'fle** *v.* **(-fled -fling, -fles).**

**cog¹** (kŏg) *n.* [ME *cogge*.] **1.** One of a series of teeth on the rim of a wheel that by engagement transmit motive force to a corresponding wheel. **2.** A cogwheel. **3.** A minor functionary in an organization. **—cogged** *adj.*

**cog²** (kŏg) *v.* **cogged, cog·ging, cogs.** [Orig. unknown.] *—vt.* To load or manipulate (dice) fraudulently. *—vi.* To cheat, esp. at dice. *—n.* A swindle.

**cog³** (kŏg) *n.* [Orig. unknown.] A tenon projecting from a wooden beam and fitting into an opening in another beam to form a joint. **—cog** *v.* **(cogged, cog·ging, cogs).**

**co·gen·er·a·tion** (kō-jĕn'ə-rā'shən) *n.* Utilization of its own waste energy by an industrial facility to produce electricity.

**co·gent** (kō'jənt) *adj.* [Lat. *cogens*, *cogent-*, pr.part. of *cogere*, to force : *co(m)-*, together + *agere*, to drive.] Appealing strongly to the intellect or reasoning powers : CONVINCING <a *cogent* presentation of the facts> **—co'gen·cy** (-jən-sē) *n.* **—co'gent·ly** *adv.*

**cog·i·tate** (kŏj'ĭ-tāt') *vi. & vt.* **-tat·ed, -tat·ing, -tates.** [Lat. *cogitare*, *cogitat-* : *co(m)-* (intensive) + *agitare*, to consider.] To take careful thought or think carefully about. **—cog'i·ta·ble** (kŏj'ĭ-tə-bəl) *adj.* **—cog'i·ta'tor** *n.*

**cog·i·ta·tion** (kŏj'ĭ-tā'shən) *n.* **1.** Thoughtful consideration. **2.** A serious thought.

**cog·i·ta·tive** (kŏj'ə-tā'tĭv) *adj.* **1.** Of or relating to cogitation. **2.** Inclined to or capable of cogitation. **—cog'i·ta'tive·ly** *adv.* **—cog'i·ta'tive·ness** *n.*

**co·gnac** (kōn'yăk', kŏn'-, kôn'-) *n.* **1.** A brandy from the vicinity of Cognac in western France. **2.** A fine brandy.

**cog·nate** (kŏg'nāt') *adj.* [Lat. *cognatus* : *co(m)*, together + *gnatus*, born, var. of *natus*, p.part. of *nasci*, to be born.] **1.** Related by blood : having a common ancestor. **2.** Related in origin, as certain words in different languages derived from the same root. **3.** Related or analogous in character, nature, or function. *—n.* One cognate with another, esp.: **a.** A person sharing a common ancestor with another. **b.** A word related to one in another language. **—cog·na'tion** *n.*

**cog·ni·tion** (kŏg-nĭsh'ən) *n.* [ME *cognicioun* < Lat. *cognitio* < *cognoscere*, to learn : *co(m)-* (intensive) + *gnoscere*, to know.] **1.** The mental faculty or process by which knowledge is acquired. **2.** Knowledge gained, as through perception, reasoning, or intuition. **—cog·ni'tion·al, cog'ni·tive** (kŏg'nĭ-tĭv) *adj.*

**cognitive dissonance** *n.* *Psychol.* A condition of conflict resulting from inconsistency between one's beliefs and one's actions.

**cog·ni·za·ble** (kŏg'nĭ-zə-bəl, kŏg-nī'-) *adj.* **1.** Capable of being known or perceived. **2.** *Law.* Capable of being tried before a given court. **—cog'ni·za·bly** *adv.*

**cog·ni·zance** (kŏg'nĭ-zəns) *n.* [ME *conissaunce* < OFr. *conoissance* < *connoistre*, to know < Lat. *cognoscere*, to learn. —see COGNITION.] **1.** Conscious knowledge or recognition : AWARENESS. **2.** Range of knowledge or understanding. **3.** Notice : observance <took *cognizance* of their objections> **4.** *Law.* **a.** Examination of a case by a court. **b.** The right or power of a court's jurisdiction. **c.** Admission of an action or fact : CONFESSION. **5.** A heraldic crest or badge worn to distinguish the bearer.

**cog·ni·zant** (kŏg'nĭ-zənt) *adj.* [< COGNIZANCE.] Fully informed : AWARE <*cognizant* of all the difficulties>

**cog·no·men** (kŏg-nō'mən) *n., pl.* **-mens** *or* **-nom·i·na** (-nŏm'ə-nə) [Lat. : *co(m)-*, together + *gnomen, nomen,* name.] **1.** A family name : SURNAME. **2.** The third and usu. last name of a citizen of ancient Rome; e.g., *Caesar* in *Caius Julius Caesar.* **3.** A name, esp. a descriptive nickname. **—cog·nom'i·nal** (-nŏm'ə-nəl) *adj.*

**co·gno·scen·te** (kŏn'yə-shĕn'tē, kŏg'nə-) *n., pl.* **-ti** (-tē) [Obs. Ital. < Lat. *cognoscens,* pr.part. of *cognoscere,* to know. —see COGNITION.] A person of superior taste or knowledge.

**cog·no·vit** (kŏg-nō'vĭt) *n.* [Lat., he has acknowledged < *cognoscere,* to recognize. —see COGNITION.] *Law.* Written admission by a defendant of liability, made to avoid the expense of a trial.

**co·gon** (kō-gōn') *n.* [Sp. *cogón* < Tagalog *kugon.*] A tall tropical grass of the genus *Imperata,* esp. *I. cylindrica* or *I. exaltata* of the Philippines and adjacent islands, used for thatch.

**cog railway** *n.* A railway designed to climb steep slopes, having a locomotive with a center cogwheel that engages with a cogged center rail to provide traction.

**Cogs·well chair** (kŏgz'wĕl', -wəl) *n.* [Prob. < the name *Cogswell.*] An upholstered easy chair, open under the armrests, with a sloping back and cabriole front legs.

**cog·wheel** (kŏg'hwēl', -wēl') *n.* One of a set of cogged wheels within a mechanism.

**co·hab·it** (kō-hăb'ĭt) *vi.* **-it·ed, -it·ing, -its.** [LLat. *cohabitare* : Lat. *co(m)-*, together + Lat. *habitare,* to dwell.] **1.** To live together as spouses. **2.** To live together in a sexual relationship when not legally married. **—co·hab'i·tant** *n.* **—co·hab'i·ta'tion** *n.*

**co·heir** (kō-âr') *n.* A joint heir.

**co·heir·ess** (kō-âr'ĭs) *n.* A joint heiress.

**co·here** (kō-hîr') *v.* **-hered, -her·ing, -heres.** [Lat. *cohaerēre* : *co(m)-*, together + *haerēre,* to cling.] *—vi.* **1.** To stick together in a mass. **2.** To be logically connected. *—vt.* To cause to form a united or orderly whole.

**co·her·ent** (kō-hîr'ənt, -hĕr'-) *adj.* **1.** Sticking together : COHERING. **2.** Marked by an orderly or logical relation of parts that affords comprehension or recognition <*coherent* arguments> **3.** *Physics.* Of or relating to waves with a continuous relationship among phases. **4.** Of or relating to a system of units of measurement in which a small number of basic units are defined from which all others in the system are derived by multiplication or division only. **—co·her'ence, co·her'en·cy** *n.* **—co·her'ent·ly** *adv.*

**co·he·sion** (kō-hē'zhən) *n.* [Fr. *cohésion,* ult. < Lat. *cohaerēre,* to cling together. —see COHERE.] **1.** The act, process, or state of cohering. **2.** *Physics.* Mutual attraction by which the elements of a body are held together. **3.** *Bot.* Congenital joining of two parts. **—co·he'sive** (-sĭv, -zĭv) *adj.* **—co·he'sive·ly** *adv.* **—co·he'sive·ness** *n.*

**co·he·sion·less** (kō-hē'zhən-lĭs) *adj.* Of or relating to a soil whose constituent particles do not cohere.

**co·hort** (kō'hôrt') *n.* [ME < Lat. *cohors.*] **1.** One of the 10 divisions of a Roman legion, having 300–600 men. **2.** A group united in a struggle. **3.** *Informal.* A companion or associate. ***usage:*** Although *cohort* in the sense of "a companion or associate" occurs at all levels, many consider this usage unacceptable in formal contexts.

**co·ho salmon** (kō'hō) *n.* [Orig. unknown.] A food and game fish, *Oncorhyncus kisutch,* orig. of Pacific waters.

**co·hosh** (kō'hŏsh') *n.* [Prob. of Algonquian orig.] BANEBERRY 1.

**co·hune** (kō-hōōn') *n.* [Sp. (Central America) < Mosquito *ókhún.*] A tropical American palm tree, *Attalea cohune,* with long featherlike leaves and oily nuts.

**cohune palm** *n.* The cohune.

**coif** (koif) *n.* [ME < OFr. *coife* < LLat. *cofea,* helmet.] **1.** A tight-fitting cap worn under a veil, as by nuns. **2.** A white skullcap worn at one time by English lawyers. **3.** A heavy steel or leather skullcap worn at one time under a helmet or mail hood. **4.** (*also* kwäf). A coiffure. *—vt.* (koif) **coifed, coif·ing, coifs. 1.** To cover with or as if with a coif. **2.** (*also* kwäf). To arrange or dress (the hair).

**coif·feur** (kwä-fûr') *n.* [Fr. < *coiffer,* to coif < OFr. *coife,* coif. —see COIF.] A hairdresser.

**coif·feuse** (kwä-fûrz', -fyōōz') *n.* [Fr., fem. of *coiffeur,* coiffeur.] A woman hairdresser.

**coif·fure** (kwä-fyŏŏr') *n.* [Fr. < *coiffer,* to coif. —see COIFFEUR.] A hair style. *—vt.* **-fured, -fur·ing, -fures.** To arrange or dress (the hair).

**coil[1]** (koil) *n.* [Prob. < OFr. *coillir,* to gather < Lat. *colligere.* —see COLLECT.] **1. a.** A series of connected spirals or concentric rings formed by winding or gathering. **b.** A single ring or spiral within a series. **2.** A spiral pipe or series of spiral pipes, as in a radiator. **3.** *Elect.* **a.** A wound spiral of two or more turns of insulated wire, used to introduce inductance into a circuit. **b.** A device of which such a spiral is the major part. *—v.* **coiled, coil·ing, coils.** *—vt.* **1.** To wind in spirals or concentric rings. **2.** To wind into a shape resembling a coil. *—vi.* **1.** To form coils. **2.** To move in a spiral course. **—coil'er** *n.*

**coil[2]** (koil) *n.* [Orig. unknown.] A disturbance : fuss.

**coin** (koin) *n.* [ME < OFr., die for stamping coins, wedge < Lat. *cuneus,* wedge.] **1.** A small, usu. flat and circular piece of metal issued and authorized by a government for use as money. **2.** COINAGE 2a. **3.** A corner or cornerstone of a building. *—vt.* **coined, coin·ing, coins. 1.** To make (coins) from metal : MINT <*coin* half dollars> **2.** To make coins from (metal) <*coin* silver> **3.** To invent (a word or expression). *—adj.* Requiring one or more coins for operation <a *coin* vending machine> **—coin'a·ble** *adj.* **—coin'er** *n.*

**coin·age** (koi'nĭj) *n.* **1.** The right or process of making coins. **2. a.** Metal currency. **b.** A system of metal currency. **3. a.** A coined word or phrase. **b.** Invention of new words.

**co·in·cide** (kō'ĭn-sīd') *vi.* **-cid·ed, -cid·ing, -cides.** [Med. Lat. *coincidere* : Lat. *co(m)-*, together + Lat. *incidere,* to occur. —see INCIDENT.] **1. a.** To occupy the same position simultaneously. **b.** To have the same dimensions. **2.** To take place at the same time. **3.** To correspond exactly.

**co·in·ci·dence** (kō-ĭn'sĭ-dəns, -dĕns') *n.* **1.** The state or fact of coinciding. **2.** A seemingly planned sequence of accidentally occurring events.

**coincidence gate** *n. Electron.* GATE[1] 7.

**co·in·ci·dent** (kō-ĭn'sĭ-dənt) *adj.* **1.** Occupying the same position. **2.** Taking place at the same time. **3.** Matching point for point : COINCIDING <*coincident* circles>

**co·in·ci·den·tal** (kō-ĭn'sĭ-dĕn'təl) *adj.* Happening as or due to coincidence. **—co·in'ci·den'tal·ly** *adv.*

**co·in·sur·ance** (kō'ĭn-shŏŏr'əns) *n.* **1.** Insurance held jointly by two or more insurers. **2.** Insurance in which a party insures property for less than its assessed value and agrees to be responsible for the difference.

**co·in·sure** (kō'ĭn-shŏŏr') *vt.* **-sured, -sur·ing, -sures. 1.** To insure jointly. **2.** To insure with coinsurance.

**coir** (koir) *n.* [Malayam *kāyar,* cord < *kāyaru,* to be twisted.] Fiber obtained from a coconut husk, used to make rope and matting.

**co·i·tus** (kō'ĭ-təs, kō-ē'-) *n.* [Lat. < *coire,* to copulate : *co(m)-*, together + *ire,* to go.] Physical union of male and female sexual organs, leading to orgasm and ejaculation of semen. **—co'i·tal** *adj.*

**coitus in·ter·rup·tus** (ĭn'tə-rŭp'təs) *n.* [Lat. : *coitus,* copulation + *interruptus,* interrupted.] Sexual intercourse purposely interrupted by withdrawal of the male prior to ejaculation.

**coke[1]** (kōk) *n.* [Perh. < ME *colk,* core.] Solid carbonaceous residue obtained from bituminous coal after removal of volatile material by destructive distillation, used as fuel and in making steel. *—vt. & vi.* **coked, cok·ing, cokes.** To convert or change into coke.

**coke[2]** (kōk) *n. Slang.* Cocaine.

**Coke** (kōk). A trademark for a soft drink.

**col** (kŏl) *n.* [Fr. < OFr., neck < Lat. *collum.*] A pass between two mountain peaks or a gap in a ridge.

**col-[1]** *pref. var. of* COM-. —Used before *l.*

**col-[2]** *pref. var. of* COLO-.

**co·la[1]** (kō'lə) *n.* A carbonated soft drink containing an extract prepared from kola nuts.

**co·la[2]** (kō'lə) *n. pl. of* COLON[1] 2.

**co·la[3]** (kō'lə) *n. var. pl. of* COLON[2].

**co·la[4]** (kō'lə) *n. var. of* KOLA.

**col·an·der** (kŭl'ən-dər, kŏl'-) *also* **cul·len·der** (kŭl'-) *n.* [ME *colyndore,* prob. alteration of OProv. *colador* < VLat. **colator* < Lat. *colare,* to strain < *colum,* sieve.] A perforated bowl-shaped kitchen utensil for draining off liquids and rinsing food.

**cola nut** *n. var. of* KOLA NUT.

**col·can·non** (kŏl-kăn'ən) *n.* [Ir. Gael. *cal ceannan* : *cal,* cabbage (< OIr. < Lat. *caulis*) + *ceannan,* white-headed (*ceann,* head + *fionn,* white).] An Irish dish of mashed potatoes and cabbage.

**col·chi·cine** (kŏl'chĭ-sēn', kŏl'kĭ-) *n.* [COLCHIC(UM) + -INE.] A poisonous alkaloid, $C_{22}H_{25}NO_6$, used experimentally for inducing chromosome doubling and medicinally for treating gout.

**col·chi·cum** (kŏl'chĭ-kəm, kŏl'kĭ-) *n.* [Lat., a plant with a poisonous root < Gk. *Kolkhikon,* meadow saffron, after *Kolkhos,* Colchis, a region east of the Black Sea.] **1.** A bulbous plant of the genus *Colchicum,* as the autumn crocus. **2.** The dried seeds or corms of *C. autumnale,* a source of colchicine.

**col·co·thar** (kŏl'kə-thər, -thär') *n.* [Sp. *colcotar* < Ar. *qolqoṭār.*] A brownish-red iron oxide obtained as a residue after heating ferrous sulfate, used in glass polishing and as a pigment.

**cold** (kōld) *adj.* **-er, -est.** [ME < OE *ceald.*] **1. a.** Having a low temperature <a *cold* climate> **b.** Having a body temperature lower than normal <a *cold,* clammy forehead> **c.** Feeling no warmth : CHILLY <I'm *cold.*> **2. a.** Marked by a lack of heat <a *cold* basement> **b.** Being at less than optimum warmth <*cold* spaghetti> **c.** Chilled by refrigeration or ice <*cold* drinks> **3.** Not marked or affected by emotion : OBJECTIVE <*cold* analysis> **4.** Without appeal to the senses or feelings : DEPRESSING <a *cold* picture> **5. a.** Not affectionate or friendly : ALOOF. **b.** Unenthusiastic or uninterested <a *cold* welcome> **c.** Without sexual desire : FRIGID. **6.** Designating a color or tone, as pale gray, that suggests little warmth. **7. a.** *Infor-*

*mal.* Insensible : unconscious <knocked *cold* in the first round> **b.** Dead <*cold* in the tomb> **8.** *Slang.* Not stolen or suspected of use in illegal activities <a *cold* car> **9.** *Informal.* Marked by unqualified certainty <knew the lines *cold*> **10.** So intense as to be almost uncontrollable <*cold* rage> *—adv. Informal.* **1.** Totally : unqualifiedly <was turned down *cold*> **2.** Without advance preparation <took the test *cold* and failed> *—n.* **1. a.** Relative lack of warmth. **b.** The sensation resulting from lack of warmth : CHILL. **2.** A viral infection marked by inflammation of the mucous membranes of the respiratory passages and accompanying fever, chills, coughing, and sneezing. **3.** Chilly weather. **—in cold blood.** Without feeling, passion, or remorse. **—(out) in the cold.** Lacking benefits given to others : NEGLECTED. **—cold′ly** *adv.* **—cold′ness** *n.*

☆ ***syns:*** **1.** COLD, CHILL, CHILLY, COOL, NIPPY *adj. core meaning* : marked by a low temperature <a *cold* night> **2.** COLD, CHILL, FRIGID, GLACIAL, ICY *adj. core meaning* : lacking all friendliness and warmth <a *cold* stare>

**cold agglutinin** *n.* An agglutinin in human blood serum that induces agglutination only at low temperatures.

**cold-blood·ed** (kōld′blŭd′ĭd) *adj.* **1. a.** Lacking feeling or emotion <a *cold-blooded* criminal> **b.** Carried out without feeling or emotion <*cold-blooded* slaughter> **2.** *Zool.* Poikilothermous. **—cold′-blood′ed·ly** *adv.* **—cold′-blood′ed·ness** *n.*

**cold chisel** *n.* A chisel made of tempered, hardened steel and used for cutting cold metal.

**cold cream** *n.* An emulsion for softening and cleansing the skin.

**cold cuts** *pl.n.* Slices of assorted cold meats.

**cold duck** *n.* [Transl. of G. *Kalte Ente,* a drink made from a mixture of wines.] A beverage of champagne and sparkling burgundy.

**cold feet** *n. Slang.* Fearfulness that interferes with completing a course of action.

**cold frame** *n.* A structure having a wooden frame and a glass top for protecting young plants from the cold.

**cold front** *n.* The leading edge of a cold air mass moving against and eventually replacing a warm air mass.

**cold-heart·ed** (kōld′här′tĭd) *adj.* Without sympathy or feeling. **—cold′-heart′ed·ly** *adv.* **—cold′-heart′ed·ness** *n.*

**cold light** *n.* **1.** Light producing little or no heat. **2.** Light emitted by a process other than incandescence.

**cold pack** *n.* **1.** *Med.* A therapeutic pack consisting of a cold, damp sheet or a sealed plastic bag containing a cold-transmitting agent. **2.** A canning process in which uncooked food is packed in jars or cans, then sterilized by heat.

**cold rubber** *n.* A durable, strong synthetic rubber polymerized at low temperatures.

**cold-shoul·der** (kōld′shōl′dər) *vt.* **-dered, -der·ing, -ders.** *Informal.* To give (someone) the cold shoulder : SNUB.

**cold shoulder** *n. Informal.* Deliberate coldness or disregard.

**cold sore** *n.* A small sore on the lips, often accompanying a fever or cold : FEVER BLISTER.

**cold storage** *n.* Protective storage, as of furs or foods, in a refrigerated place.

**cold sweat** *n.* Simultaneous chill and perspiration, usu. induced by fear, pain, or shock.

**cold turkey** *n. Informal.* Immediate, total withdrawal from something on which one has become dependent, as an addictive drug. **—talk cold turkey.** To speak frankly and bluntly.

**cold type** *n.* Typesetting done without the casting of metal.

**cold war** *n.* A state of political tension and military rivalry between nations that stops short of actual full-scale war. **—cold warrior** *n.*

**cold-wa·ter** (kōld′wô′tər, -wŏt′ər) *adj.* Lacking modern plumbing or heating facilities <*cold-water* tenements>

**cold wave** *n.* **1.** An abrupt onset of cold weather. **2.** A permanent wave in which the hair is set by chemicals rather than by heat.

**cold welding** *n.* Welding of two materials under high pressure or vacuum without using heat. **—cold′-weld′** *v.* **(-weld·ed, -weld·ing, -welds).**

**cole** (kōl) *n.* [ME *col* < OE *cāl* < Lat. *caulis,* cabbage.] A plant of the genus *Brassica,* as the cabbage or rape.

**co·lec·to·my** (kə-lĕk′tə-mē) *n., pl.* **-mies.** Surgical removal of part or all of the colon.

**cole·man·ite** (kōl′mə-nīt′) *n.* [After William T. *Coleman* (1824–1893).] A natural white or colorless hydrated calcium borate, $Ca_2B_6O_{11}{\cdot}5H_2O$, a principal source of borax.

**co·le·op·ter·a** (kō′lē-ŏp′tər-ə) *n.* [NLat. *Coleoptera,* order name < Gk. *koleopteros,* sheath-winged : *koleon,* sheath + *pteron,* wing.] The beetles and weevils. **—co′le·op′ter·ist** *n.*

**co·le·op·ter·an** (kō′lē-ŏp′tər-ən, kŏl′ē-) *also* **co·le·op·ter·on** (-tə-rŏn′) *n.* [NLat. *Coleoptera,* order name < Gk. *koleopteros,* sheath-winged : *koleon,* sheath + *pteron,* wing.] An insect of the order Coleoptera, having forewings modified to form tough protective covers for the hind wings, and including the beetles and weevils. *—adj.* Of or belonging to the Coleoptera. **—co′le·op′ter·ous** *adj.*

**co·le·op·tile** (kō′lē-ŏp′tĭl, kŏl′ē-) *n.* [NLat. *coleoptilum* : Gk. *koleon,* sheath + Gk. *ptilon,* plume.] The first seedling leaf in grasses and similar monocotyledons, forming a protective sheath around the plumule.

**co·le·o·rhi·za** (kō′lē-ə-rī′zə, kŏl′ē-) *n., pl.* **-zae** (-zē) [Gk. *koleon,* sheath + Gk. *rhiza,* root.] A protective sheath surrounding the embryonic root of grasses and similar monocotyledons.

**cole·slaw** *also* **cole slaw** (kōl′slô′) *n.* [Du. *koolsla* : *kool,* cabbage (< MDu. *cōle* < Lat. *caulis*) + *sla,* short for *salade,* salad < Fr. < OFr. —see SALAD.] A salad of shredded raw cabbage with a dressing.

**co·le·us** (kō′lē-əs) *n.* [NLat. *Coleus,* genus name < Gk. *koleos,* sheath (from the way its filaments are joined.] Any of various plants of the genus *Coleus* of Eurasia and Africa, cultivated for their showy leaves, often marked with red, yellow, or white.

**cole·wort** (kōl′wûrt′, -wôrt′) *n.* Cole.

**coli-** *pref. var. of* COLO-.

**col·ic** (kŏl′ĭk) *n.* [ME *colik,* suffering with colic < OFr. *colique* < Lat. *colicus* < Gk. *kōlikos* < *kōlon,* colon.] **1.** Acute, paroxysmal pain in the abdomen, caused by obstruction, spasm, or distention of any of the hollow viscera. **2.** Severe abdominal pain in infants, usu. resulting from gas in the alimentary canal. **—col′ick·y** (kŏl′ĭ-kē) *adj.*

**col·i·cin** (kŏl′ĭ-sĭn, kō′lĭ-) *n.* [COL(ON) + -IC + -IN.] A protein produced by certain strains of the colon bacillus that is lethal to other strains of bacteria of the same species.

**col·i·ci·no·ge·nic·i·ty** (kŏl′ĭ-sə-nə-jə-nĭs′ĭ-tē) *n.* The ability to produce colicin. **—col′i·ci·no·gen′ic** (-jĕn′ĭk) *adj.*

**col·ic·root** (kŏl′ĭk-ro͞ot′, -ro͝ot′) *n.* **1.** A plant, *Aletris farinosa* of the eastern United States, having a cluster of tubular white flowers and a bitter root once used in medicine. **2.** A plant believed to relieve or cure colic.

**col·ic·weed** (kŏl′ĭk-wēd′) *n.* A plant of the genera *Dicentra* or *Corydalis,* as the squirrel corn.

**co·li·form** (kō′lə-fôrm′, kŏl′ə-) *adj.* Of, relating to, or like the colon bacillus. **—co′li·form′** *n.*

**co·lin·e·ar** (kō-lĭn′ē-ər) *adj. Genetics.* Containing elements corresponding to one another and arranged in the same linear sequence. **—co·lin′e·ar′i·ty** (-ăr′ĭ-tē) *n.*

**col·i·se·um** *also* **col·os·se·um** (kŏl′ĭ-sē′əm) *n.* [Med. Lat., an amphitheater in Rome, Italy < Lat. *colosseum,* neuter of *colosseus,* gigantic < *colossus,* huge statue < Gk. *kolossos.*] A large amphitheater or sports arena.

**co·lis·tin** (kə-lĭs′tĭn, kō-) *n.* [NLat. *Colistinus,* specific epithet of the bacterium that produces it.] An antibiotic produced by the bacterium *Bacillus colistinus,* effective against a wide range of Gram-positive microorganisms.

**co·li·tis** (kō-lī′tĭs) *n.* Inflammation of the mucous membrane of the colon.

**coll-** *pref. var. of* COLLO-.

**col·lab·o·rate** (kə-lăb′ə-rāt′) *vi.* **-rat·ed, -rat·ing, -rates.** [LLat. *collaborare, collaborat-* : Lat. *com-,* together + Lat. *laborare,* to work < *labor,* work.] **1.** To work together, esp. in a joint intellectual effort <*collaborated* on a biography> **2.** To cooperate treasonably, as with an enemy occupying one's country. **—col·lab′o·ra′tion** *n.* **—col·lab′o·ra′tive** *adj.* **—col·lab′o·ra′tor** *n.*

**col·lab·o·ra·tion·ist** (kə-lăb′ə-rā′shə-nĭst) *n.* One who cooperates treasonably with an enemy occupying one's country. **—col·lab′o·ra′tion·ism** *n.*

**col·lage** (kō-läzh′, kə-) *n.* [Fr. < *coller,* to glue < *colle,* glue, ult. < Gk. *kolla.*] An artistic composition of objects and materials pasted over a surface, often with unifying lines and color.

**col·la·gen** (kŏl′ə-jən) *n.* [Gk. *kolla,* glue + -GEN.] The fibrous albuminoid constituent of bone, cartilage, and connective tissue. **—col′la·gen′ic** (kŏl′ə-jĕn′ĭk), **col·lag′e·nous** (kə-lăj′ə-nəs) *adj.*

**col·lag·e·nase** (kə-lăj′ə-nās′, -nāz′, kŏl′ə-jə-) *n.* Any of various enzymes that catalyze the breakdown of collagen and gelatin.

**col·lap·sar** (kə-lăp′sär′) *n.* [COLLAPSE + *-ar,* as in *quasar.*] A black hole.

**col·lapse** (kə-lăps′) *v.* **-lapsed, -laps·ing, -laps·es.** [Lat. *collabi, collaps-,* to fall together : *com-,* together + *labi,* to fall.] *—vi.* **1.** To fall down or inward suddenly : cave in. **2.** To break down suddenly in strength or health and cease to function <a government that *collapsed*><a patient who *collapsed*> **3.** To fold compactly <temporary fencing that *collapses*> *—vt.* To cause to collapse. *—n.* **1.** The act of falling down or inward, as from loss of supports. **2.** An abrupt failure of function, strength, or health : BREAKDOWN. **—col·laps′i·bil′i·ty** *n.* **—col·laps′i·ble, col·laps′a·ble** *adj.*

**col·lar** (kŏl′ər) *n.* [ME *coler* < OFr. *colier* < Lat. *collare* < *collum,* neck.] **1.** The part of a garment around the neck. **2.** A necklace. **3.** An identifying or restraining band put around the neck of an animal. **4.** The cushioned part of a harness that presses against the shoulders of a draft animal. **5.** *Informal.* An arrest. **6.** *Biol.* An encircling structure or bandlike marking resembling a collar. **7.** A ringlike device used to limit, secure, or guide a machine part. *—vt.* **-lared, -lar·ing, -lars. 1.** To furnish with a collar. **2.** *Informal.* **a.** To seize : detain. **b.** To arrest.

**col·lar·bone** (kŏl′ər-bōn′) *n. Anat.* The clavicle.

**collar cell** *n. Biol.* A choanocyte.

**col·lard** (kŏl′ərd) *n.* [Var. of COLEWORT.] **1.** A variety of kale, *Bras-*

*sica oleracea acephala,* with a crown of edible leaves. **2. collards.** Leaves of the collard used as a vegetable.

**col·late** (kə-lāt′, kŏl′āt′, kō′lāt′) *vt.* **-lat·ed, -lat·ing, -lates.** [Lat. *conferre, collat-,* to collect. —see CONFER.] **1.** To examine and compare carefully so as to note points of disagreement. **2.** To assemble in proper logical or numerical sequence. **3.** To examine (gathered sheets) in order to arrange them in proper sequence before binding. **4.** To verify the order and completeness of (the pages of a volume). **5.** To admit (a cleric) to a benefice. **—col·la′tor** *n.*

**col·lat·er·al** (kə-lăt′ər-əl) *adj.* [ME < Med. Lat. *collateralis* : Lat. *co(m)-,* together + Lat. *latus,* side.] **1.** Located or running side by side : PARALLEL. **2.** Coinciding in tendency or effect : CONCOMITANT. **3.** Serving to corroborate or support <*collateral* facts> **4.** Secondary in nature : SUBORDINATE. **5.** Of, designating, or guaranteed by a security pledged against the performance of an obligation <a *collateral* loan> **6.** Having an ancestor in common but descended from a different line. —*n.* **1.** Property acceptable as security for a loan or other obligation. **2.** A collateral relative. **—col·lat′er·al·ly** *adv.*

**col·la·tion** (kə-lā′shən, kŏ-, kō-) *n.* **1.** The act or process of collating. **2. a.** A light meal permitted on fast days. **b.** A light meal.

**col·league** (kŏl′ēg′) *n.* [OFr. *collegue* < Lat. *collega* : *com-,* together + *legare,* to depute.] A fellow member of a profession, staff, or academic faculty. **—col′league·ship′** *n.*

**col·lect[1]** (kə-lĕkt′) *v.* **-lect·ed, -lect·ing, -lects.** [ME *collecten* < Lat. *colligere* : *com-,* together + *legere,* to gather.] —*vt.* **1.** To bring together in a group : ASSEMBLE. **2.** To accumulate as a hobby or for study <*collect* butterflies> **3.** To call for and obtain payment of <*collect* unpaid debts> **4.** To recover control of <*collect* one's thoughts> —*vi.* **1.** To gather together : CONGREGATE. **2.** To take in payments or donations <*collecting* for the library fund> —*adj.* With payment to be made by the recipient <a *collect* phone call> —*adv.* So that the recipient is charged <make a phone call *collect*> **—col·lect′i·ble, col·lect′a·ble** *adj.*

**col·lect[2]** (kŏl′ĭkt, -ĕkt′) *n.* [ME *collecte* < OFr. < Med. Lat. *collecta,* short for *oratio ad collectam,* prayer at the gathering.] A brief formal prayer used in various Western liturgies before the epistle at Mass and varying with the day.

**col·lec·ta·ne·a** (kŏl′ĕk-tā′nē-ə) *pl.n.* [Lat. < *collectaneus,* gathered < *collectus,* p.part. of *colligere,* to gather. —see COLLECT[1].] A selection of passages from one or more authors : ANTHOLOGY.

**col·lect·ed** (kə-lĕk′tĭd) *adj.* **1.** Self-possessed : composed. **2.** Brought or placed together from various sources <the *collected* short stories of O. Henry> **—col·lect′ed·ly** *adv.* **—col·lect′ed·ness** *n.*

**col·lec·tion** (kə-lĕk′shən) *n.* **1.** The act or process of collecting. **2.** A group of objects or works kept together, esp. to be viewed or studied. **3.** An accumulation : deposit. **4. a.** A collecting of money, as in church. **b.** The sum collected.

**col·lec·tive** (kə-lĕk′tĭv) *adj.* **1.** Assembled or accumulated into a whole. **2.** Of, relating to, typical of, or made by a number of individuals taken or acting as a group <a *collective* plan> —*n.* **1.** An undertaking or business set up on the principle of control and ownership of the means of production and distribution by the workers involved, usu. under government supervision. **2.** A collective noun. **—col·lec′tive·ly** *adv.* **—col·lec′tive·ness** *n.*

**collective bargaining** *n.* Negotiation between the representatives of organized workers and their employer to determine wages, working hours, rules, and working conditions.

**collective farm** *n.* A farm or a group of farms organized as a unit and managed and worked cooperatively by a group of laborers under government supervision.

**collective fruit** *n. Bot.* A multiple fruit.

**collective mark** *n.* A trademark or service mark used by members of a cooperative, association, or other collective group or organization, including marks used to indicate membership in a union, association, or other organization.

**collective noun** *n. Gram.* A noun that denotes a collection of persons or things regarded as a unit. ***usage:*** A collective noun takes a singular verb when the reference is to a group as a whole and a plural verb when the reference is to members of a group as single individuals: *The orchestra was playing. The orchestra have all gone home.*

**col·lec·tiv·ism** (kə-lĕk′tə-vĭz′əm) *n.* The system or principle of ownership and control of the means of production and distribution by the people collectively. **—col·lec′tiv·ist** *n.* **—col·lec′tiv·is′tic** *adj.* **—col·lec′tiv·is′ti·cal·ly** *adv.*

**col·lec·tiv·i·ty** (kŏl′ĕk-tĭv′ĭ-tē, kə-lĕk′-) *n.* **1.** The quality or state of being collective. **2.** The people as a whole.

**col·lec·tiv·ize** (kə-lĕk′tə-vīz′) *vt.* **-ized, -iz·ing, -iz·es.** To organize (an economy, industry, or enterprise) on the basis of collectivism. **—col·lec′tiv·i·za′tion** *n.*

**col·lec·tor** (kə-lĕk′tər) *n.* **1.** One that collects. **2.** One employed to collect taxes, duties, or other payments. **3.** One who makes a collection, as of paintings. **4. a.** *Elect.* A conducting contact between parts of an electric circuit in relative motion. **b.** *Electron.* The output terminal of a three-terminal semiconducting device, esp. of a transistor. **—col·lec′tor·ship′** *n.*

**col·leen** (kŏ-lēn′, kŏl′ēn′) *n.* [Ir. Gael. *cailīn,* dim. of *caile,* girl.] An Irish girl.

**col·lege** (kŏl′ĭj) *n.* [ME < OFr. < Lat. *collegium,* association < *collega,* colleague.] **1. a.** A school of higher learning that grants the bachelor's degree in liberal arts or science or both. **b.** Any of the undergraduate schools or divisions of a university offering courses and granting degrees in a specific field. **c.** A professional or technical school, often affiliated with a university, offering the bachelor's or master's degree. **d.** The building or buildings of such a school. **e.** *Chiefly Brit.* A self-governing society of scholars for instruction or study, incorporated within a university. **f.** An institution in France for secondary education not supported by the state. **2.** A company or assemblage, esp. a body of persons having a common purpose or common duties <a *college* of physicians> **3.** A body of clergy living together on an endowment.

**College of Cardinals** *n. Rom. Cath. Ch.* The body comprising all the cardinals that elects the pope, assists him in governing the church, and administers the Holy See when vacant.

**col·le·gi·a** (kə-lē′jē-ə, -lĕg′ē-ə) *n. var. pl. of* COLLEGIUM.

**col·le·gi·al** (kə-lē′jē-əl, -jəl) *adj. var. of* COLLEGIATE.

**col·le·gi·al·i·ty** (kə-lē′jē-ăl′ĭ-tē) *n.* [< Lat. *collegiate,* of colleagues < *collegium,* association < *collega,* colleague.] **1.** Shared authority among colleagues. **2.** *Rom. Cath. Ch.* The doctrine that bishops collectively share collegiate authority.

**col·le·gian** (kə-lē′jən, -jē-ən) *n.* A college student or recent college graduate.

**col·le·giate** (kə-lē′jĭt, -jē-ĭt) *also* **col·le·gi·al** (-jē-əl, -jəl) *adj.* [Med. Lat. *collegiatus* < Lat. *collegium,* association. —see COLLEGE.] **1.** Of, relating to, or like a college. **2.** Of, for, or typical of college students. **3.** Of or relating to a collegiate church.

**collegiate church** *n.* **1.** An Anglican or Roman Catholic church other than a cathedral, having a chapter of canons and presided over by a provost or dean. **2. a.** A U.S. church associated with others under a common body of ministers. **b.** An association of such churches. **3.** A church in Scotland served by two or more ministers at the same time.

**col·le·gi·um** (kə-lē′jē-əm, -lĕg′ē-) *n., pl.* **-le·gi·a** (-lē′jē-ə, -lĕg′ē-ə) *or* **-le·gi·ums.** [R. *kollegya* < Lat. *collegium,* association < *collega,* colleague.] An executive council or committee of equally empowered members, esp. one supervising a commissariat, industry, or other organization in the Soviet Union.

**col·lem·bo·lan** (kə-lĕm′bə-lən) *n.* [COLL(O)- + Gk. *embolos,* peg.] The springtail.

**col·len·chy·ma** (kə-lĕng′kə-mə) *n.* [COLL(O)- + -ENCHYMA.] Supportive tissue of plants, composed of elongated, approx. rectangular cells with cell walls thickened at the corners. **—col′len·chym′a·tous** (kŏl′ən-kĭm′ə-təs) *adj.*

**col·len·chyme** (kŏl′ən-kīm′) *n.* [< COLLENCHYMA.] A gelatinous mesenchyme that makes up a layer in the body wall of many coelenterates and ctenophores.

**col·let** (kŏl′ĭt) *n.* [Fr., dim. of *col,* collar < Lat. *collum,* neck.] **1.** A cone-shaped sleeve used in a lathe for holding circular or rodlike pieces. **2.** A metal collar used in watchmaking for joining one end of a balance spring to the balance staff. **3.** A circular flange or rim, as in a ring, into which a gem is set. —*vt.* **-let·ed, -let·ing, -lets.** To set in or supply with a collet.

**col·lide** (kə-līd′) *vi.* **-lid·ed, -lid·ing, -lides.** [Lat. *collidere* : *com-,* together + *laedere,* to strike.] **1.** To come together with direct, violent impact. **2.** To meet in opposition : CLASH <dissimilar cultures *colliding*>

**col·lie** (kŏl′ē) *n.* [Sc.] A large dog orig. bred in Scotland as a sheep dog, with long hair and a long, narrow muzzle.

**col·lier** (kŏl′yər) *n.* [ME *colier* < *col,* coal < OE.] *Chiefly Brit.* **1.** A coal miner. **2.** A coal ship.

**col·lier·y** (kŏl′yə-rē) *n., pl.* **-ies.** *Chiefly Brit.* A coal mine.

**col·li·gate** (kŏl′ĭ-gāt′) *vt.* **-gat·ed, -gat·ing, -gates.** [Lat. *colligare, colligat-* : *com-,* together + *ligare,* to tie.] **1.** To tie or group together. **2.** *Logic.* To bring (isolated observations) together by an explanation or hypothesis that applies to them all. **—col′li·ga′tion** *n.*

**col·li·ga·tive** (kŏl′ĭ-gā′tĭv) *adj.* [G. *kolligativ* < Lat. *colligatus,* p.part of *colligare,* to bind : *com-,* together + *ligare,* to tie.] Depending on the quantity of molecules but not on their chemical nature.

**col·li·mate** (kŏl′ə-māt′) *vt.* **-mat·ed, -mat·ing, -mates.** [NLat. *collimare, collimat-,* alteration of Lat. *collineare,* to aim : *com-* (intensive) + *lineare,* to make straight < *linea,* line.] **1.** To make parallel : LINE UP. **2.** To adjust the line of sight of (an optical device). **—col′li·ma′tion** *n.*

**col·li·ma·tor** (kŏl′ə-mā′tər) *n.* A device capable of collimating radiation, as a long narrow tube in which strongly absorbing or reflecting walls permit only radiation traveling parallel to the tube axis to traverse the entire length.

**col·lin·e·ar** (kə-lĭn′ē-ər, kŏ-) *adj.* **1.** Lying on the same line. **2.** Containing a common line : COAXIAL. **—col·lin′e·ar′i·ty** (-ăr′ĭ-tē) *n.*

**col·lins** (kŏl′ənz) *n.* [Prob. from the name *Collins.*] A tall iced drink made with liquor, as gin, and lemon or lime juice.

**col·lin·si·a** (kə-lĭn′zē-ə) *n.* [NLat. *Collinsia,* genus name, after Zaccheus *Collins* (1764–1831).] A North American plant of the genus *Collinsia,* bearing blue-and-white or purplish flowers.
**col·li·sion** (kə-lĭzh′ən) *n.* [ME < Lat. *collisio* < *collidere,* to strike. —see COLLIDE.] **1.** The act or process of colliding : CRASH. **2.** *Physics.* A dynamic event consisting of the interaction between two or more bodies, usu. of very brief duration, resulting in a change of momentum of at least one participating body. **—col·li′sion·al** *adj.*
**collision course** *n.* A course, as of individuals, opposing philosophies, or moving objects that will end in conflict or collision if continued unchanged <revolutionaries on a *collision course* with the government>
**collo-** or **coll-** *pref.* [NLat. < Gk. *kolla,* glue.] **1.** Glue <*collen*chyma> **2.** Colloid <*collo*type>
**col·lo·cate** (kŏl′ə-kāt′) *vt.* **-cat·ed, -cat·ing, -cates.** [Lat. *collocare, collocat-* : *com-,* together + *locare,* to place < *locus,* place.] To place together or in proper order : ARRANGE.
**col·lo·ca·tion** (kŏl′ō-kā′shən) *n.* **1.** The act of collocating or the state of being collocated. **2.** Arrangement or juxtaposition, esp. of words. **—col′lo·ca′tion·al** *adj.*
**col·lo·di·on** (kə-lō′dē-ən) *also* **col·lo·di·um** (-əm) *n.* [NLat. *collodium* < Gk. *kollōdēs,* glutinous < *kolla,* glue.] A highly flammable, colorless or yellowish syrupy solution of pyroxylin in alcohol and ether, used to hold surgical dressings, as a coating for some skin diseases, and for making photographic plates.
**col·logue** (kə-lōg′) *vi.* **-logued, -logu·ing, -logues.** [Orig. unknown.] *Chiefly Brit. Regional.* To confer secretly : CONSPIRE.
**col·loid** (kŏl′oid′) *n.* **1.** *Chem.* **a.** A suspension of finely divided particles in a continuous medium, esp. a gaseous, liquid, or solid substance, as an atmospheric fog, a paint, or foam rubber, containing suspended particles that are approx. 5 to 5,000 angstroms in size, do not settle out of the substance rapidly, and are not readily filtered. **b.** The particulate matter so suspended. **2.** *Physiol.* A clear gelatinous secretion of the thyroid gland. **3.** *Pathol.* Gelatinous material resulting from colloid degeneration or colloid carcinoma. **—col′loid′, col·loid′al** (kə-loid′l, kō-) *adj.* **—col·loid′al·ly** *adv.*
**col·lop** (kŏl′əp) *n.* [ME.] **1.** A small piece or slice, esp. of meat. **2.** A roll of bodily flesh.
**col·lo·qui·a** (kə-lō′kwē-ə) *n. var. pl. of* COLLOQUIUM.
**col·lo·qui·al** (kə-lō′kwē-əl) *adj.* [< COLLOQUY.] **1.** Typical of or appropriate to the spoken language or to writing that seeks the effect of speech : INFORMAL. **2.** Relating to conversation : CONVERSATIONAL. **—col·lo′qui·al·ly** *adv.* **—col·lo′qui·al·ness** *n.*
**col·lo·qui·al·ism** (kə-lō′kwē-ə-lĭz′əm) *n.* **1.** Colloquial quality or style. **2.** A colloquial expression.
**col·lo·qui·um** (kə-lō′kwē-əm) *n., pl.* **-qui·ums** *or* **-qui·a** (-kwē-ə) [Lat., conversation < *colloqui,* to talk together : *com-,* together + *loqui,* to speak.] **1.** An informal meeting to exchange views. **2.** An academic seminar on a broad field of study, each meeting of which is usu. led by a different lecturer.
**col·lo·quy** (kŏl′ə-kwē) *n., pl.* **-quies.** [Lat. *colloquium,* conversation. —see COLLOQUIUM.] **1.** A conversation, esp. a formal one. **2.** A written dialogue.
**col·lo·type** (kŏl′ə-tīp′) *n.* **1.** A printing process in which a glass plate with a gelatin surface carries the image to be reproduced. **2.** A print made by the collotype process.

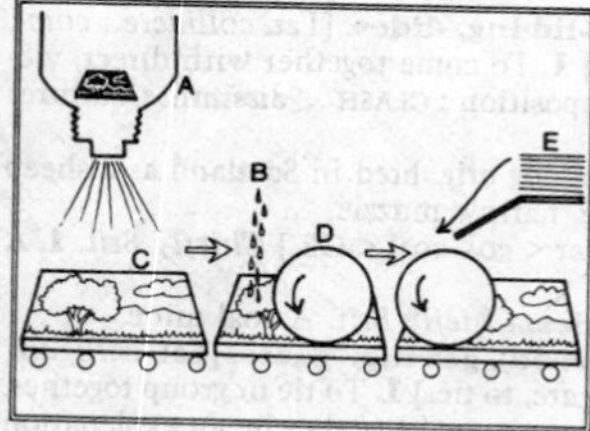


**collotype**
*The process of collotype printing showing:*
*A. negative, B. ink,*
*C. printing plates,*
*D. impression cylinder,*
*E. paper*

**col·lude** (kə-lo͞od′) *vi.* **-lud·ed, -lud·ing, -ludes.** [Lat. *colludere* : *com-,* together + *ludere,* to play < *ludus,* play.] To act together secretly to achieve a fraudulent or deceitful purpose : CONNIVE. **—col·lud′er** *n.*
**col·lu·sion** (kə-lo͞o′zhən) *n.* [ME < Lat. *collusio* < *colludere,* to collude.] A secret agreement between two or more persons for a fraudulent or deceitful purpose **—col·lu′sive** (-lo͞o′sĭv, -zĭv) *adj.* **—col·lu′sive·ly** *adv.* **—col·lu′sive·ness** *n.*
**col·lu·vi·um** (kə-lo͞o′vē-əm) *n., pl.* **-vi·a** (-vē-ə) *or* **-vi·ums.** [Lat., collection of washings < *colluere,* to wash thoroughly : *com-* (intensive) + *lavere,* to wash.] A loose deposit of rock debris accumulated at the base of a cliff or slope. **—col·lu′vi·al** *adj.*
**col·lyr·i·um** (kə-lîr′ē-əm) *n., pl.* **-i·ums** *or* **-i·a** (-ē-ə) [Lat. < Gk. *kollurion,* dim. of *kollura,* roll of bread.] A medicinal lotion applied to the eye : EYEWASH.
**col·ly·wob·bles** (kŏl′ē-wŏb′əlz) *pl.n.* [*Colly,* alteration of COLIC + WOBBLE.] (*sing. or pl. in number*). *Informal.* A pain in the stomach or bowels : BELLYACHE.
**colo-** or **coli-** or **col-** *pref.* [NLat. < Lat. *colon* < Gk. *kolon,* large intestine.] Colon <*colo*stomy>
**col·o·bo·ma** (kŏl′ə-bō′mə) *n., pl.* **-ma·ta** (-mə-tə) [NLat. *coloboma, colobomat-* < Gk. *kolobōma,* part removed in mutilation < *koloboun,* to mutilate < *kolobos,* maimed.] A lesion or fissure of the eye or eyelid.
**col·o·cynth** (kŏl′ə-sĭnth′) *n.* [Lat. *colocynthis* < Gk. *kolokunthis.*] **1.** A vine, *Citrullus colocynthis* of the Mediterranean region, bearing a small, bitter fruit. **2.** The fruit of the colocynth, used as a cathartic.
**co·logne** (kə-lōn′) *n.* [Short for Fr. *eau de Cologne,* water of Cologne, after *Cologne,* Germany.] A scented liquid made of alcohol and fragrant oils.
**co·lon**[1] (kō′lən) *n., pl.* **-lons.** [Lat., part of a verse < Gk. *kōlon,* metrical unit.] **1. a.** A punctuation mark (:) used after a word introducing a quotation, explanation, example, or series and after the salutation of a business letter. **b.** The sign (:) used between numbers or groups of numbers in expressions of time and ratios <2:30 A.M.><1:2> **2.** *pl.* **co·la** (-lə). A section of a rhythmical period in Greek and Latin verse, composed of two to six feet and having one principal accent.
**co·lon**[2] (kō′lən) *n., pl.* **-lons** *or* **-la** (-lə) [ME < Lat. < Gk. *kolon.*] The section of the large intestine extending from the cecum to the rectum. **—co·lon′ic** (kə-lŏn′ĭk) *adj.*
**co·lon**[3] (kō-lōn′) *n., pl.* **-lons** *or* **-lo·nes** (-lō′nās′) [Sp. *colón* < *Christóbal Colón,* Christopher Columbus (1451–1506).] —See table at CURRENCY.
**colon bacillus** *n.* A bacillus, *Escherichia coli,* normally occurring in all vertebrate intestinal tracts and occas. virulent, causing pyelitis or infantile diarrhea.
**colo·nel** (kûr′nəl) *n.* [Alteration of obs. *coronel* < Fr. < OItal. *colonello,* dim. of *colonna,* column of soldiers < Lat. *columna,* pillar.] **1. a.** An officer in the U.S. Army, Air Force, or Marine Corps ranking immediately above a lieutenant colonel and below a brigadier general. **b.** An officer of similar rank in other military or paramilitary organizations. **2.** An honorary title awarded by some U.S. states. **—colo′nel·cy, colo′nel·ship′** *n.*

▲ word history: The improbable pronunciation "kernel" for the word spelled *colonel* represents the triumph of popular speech over learned tinkering. The French form actually borrowed in the 16th century was *coronel,* from the Italian form *colonello.* The substitution of *r* for *l* in the French form is an example of dissimilation. By this process two similar or identical sounds, like the two *l* sounds in *colonello,* become less alike. In English usage *coronel* was respelled as *colonel,* with the original *l* restored, but the pronunciation based on this spelling did not win out over "kernel," which was the pronunciation of the older English form.

**Colonel Blimp** (blĭmp) *n.* [After *Colonel Blimp,* a cartoon character by David Low (1891–1963).] An elderly, pompous, short-sighted reactionary, esp. a government official or army officer.
**co·lo·nes** (kō-lō′nās′) *n. var. pl. of* COLON³.
**co·lo·ni·al** (kə-lō′nē-əl) *adj.* **1.** Of, relating to, possessing, or inhabiting a colony or colonies. **2.** *often* **Colonial. a.** Of or relating to the 13 British colonies that became the original United States of America. **b.** Of or relating to the U.S. colonial period. **3.** *often* **Colonial.** Designating an architectural style common in the American colonies just before and during the Revolution. **4.** Living in, forming, or consisting of a colony <*colonial* organisms> —*n.* An inhabitant of a colony. **—co·lo′ni·al·ly** *adv.*
**co·lo·ni·al·ism** (kə-lō′nē-ə-lĭz′əm) *n.* A policy by which a nation maintains or extends its control over foreign dependencies. **—co·lo′ni·al·ist** *n.*
**col·o·nist** (kŏl′ə-nĭst) *n.* **1.** An original founder or settler of a colony. **2.** An inhabitant of a colony.
**co·lon·i·tis** (kō′lə-nī′tĭs) *n.* Colitis.
**col·o·nize** (kŏl′ə-nīz′) *v.* **-nized, -niz·ing, -niz·es.** —*vt.* **1. a.** To establish a colony or colonies in. **b.** To migrate to, settle in, and occupy as a colony. **2.** To establish in a new settlement : form a colony of. —*vi.* **1.** To set up or form a colony. **2.** To settle in a colony or colonies. **—col′o·ni·za′tion** (-nĭ-zā′shən) *n.* **—col′o·niz′er** *n.*
**col·on·nade** (kŏl′ə-nād′) *n.* [Fr. < Ital. *colonnato* < *colonna,* column < Lat. *columna.*] A series of columns situated at regular intervals. **—col′on·nad′ed** *adj.*
**col·o·ny** (kŏl′ə-nē) *n., pl.* **-nies.** [ME *colonie* < Lat. *colonia* < *colonus,* settler < *colere,* to cultivate.] **1. a.** A group of emigrants or their descendants who settle in a distant land but remain subject to or closely connected with the parent country. **b.** A territory thus settled. **2.** A region politically controlled by a distant country. **3. a.** A group of people with the same ethnic origin or interests concentrated in a particular area <the Russian *colony* in San Francisco> **b.** The area or place occupied by such a group. **4.** A group of the same kind of animals, plants, or one-celled organisms living or growing together. **5.** A visible growth of microorganisms in a nutrient medium.

**col·o·phon** (kŏl'ə-fŏn', -fən) *n.* [LLat. < Gk. *kolophōn*, finishing touch.] **1.** An inscription placed usu. at the end of a book, giving facts relating to its publication. **2.** A publisher's emblem or trademark placed usu. on the title page of a book.

**col·or** (kŭl'ər) *n.* [ME *colour* < OFr. < Lat. *color.*] **1.** That aspect of things caused by differing qualities of the light reflected or emitted by them, which may be defined in terms of the observer or of the light: **a.** The appearance of objects or light sources described in terms of the individual's perception of them, involving hue, lightness, and saturation for objects and hue, brightness, and saturation for light sources. **b.** The characteristics of light by which the individual is made aware of objects or light sources through the ocular receptors, described in terms of dominant wavelength, luminance, and purity. **2.** A substance, as a dye, pigment, or paint, that imparts color. **3. a.** General appearance of the skin : COMPLEXION. **b.** A ruddy complexion. **c.** A reddening of the face : BLUSH. **4.** The skin pigmentation of a person not classed as a Caucasian, esp. that of a black. **5. colors.** A flag or banner, as of a country or military unit. **6. colors.** A distinguishing symbol, badge, ribbon, or mark <the *colors* of a college> **7. colors.** One's beliefs or position <Stick to your *colors.*> **8.** *often* **colors.** Character : nature <revealed your true *colors*> **9. a.** Outward, often deceptive appearance. **b.** Appearance of truth or authenticity : PLAUSIBILITY. **10.** Variety of expression or effect. **11.** Vivid and picturesque detail. **12.** Traits of behavior or personality that attract interest. **13.** Use or effect of color in painting. **14.** *Mus.* Tonal quality. **15.** *Law.* An apparent or prima-facie right, pretext, or ground. **16. colors.** The salute made during the ceremony of raising or lowering the flag. **17.** A particle or bit of gold found in auriferous gravel or sand. **18.** *Physics.* A quantum characteristic of quarks that determines their role in the strong interaction. —*v.* **-ored, -or·ing, -ors.** —*vt.* **1.** To impart color to or change the color of. **2.** To give a distinctive quality or character to. **3.** To misrepresent, esp. by exaggeration or distortion. —*vi.* **1.** To take on color or become colored. **2.** To change color. **3.** To become red in the face. **—col'or·er** *n.*

**col·or·a·ble** (kŭl'ər-ə-bəl) *adj.* **1.** Intended to deceive : SPECIOUS. **2.** Seemingly genuine or true : PLAUSIBLE. **—col'or·a·bil'i·ty, col'or·a·ble·ness** *n.* **—col'or·a·bly** *adv.*

**col·o·ra·do** (kŏl'ə-rä'dō) *adj.* [Sp. reddish < *colorar, colorear*, to color < Lat. *colorare* < *color*, color.] Of medium strength and color. —Used of cigars.

**Col·o·ra·do potato beetle** (kŏl'ə-rä'dō, -răd'ə) *n.* The potato beetle.

**col·or·ant** (kŭl'ər-ənt) *n.* Something, esp. a dye, pigment, paint, or ink, that imparts color or modifies color.

**col·or·a·tion** (kŭl'ə-rā'shən) *n.* **1.** Arrangement of colors. **2.** The sum of the principles or beliefs of a person, group, or institution.

**col·or·a·tu·ra** (kŭl'ər-ə-tŏŏr'ə, -tyŏŏr'ə) *n.* [Obs. Ital. < LLat., coloring < Lat. *colorare*, to color < *color*, color.] *Mus.* **1.** Florid, ornamental vocal trills and runs. **2.** Music characterized by coloratura. **3.** A soprano specializing in coloratura.

**color bar** *n.* Color line.

**col·or·blind** (kŭl'ər-blīnd') *adj.* **1.** Partially or totally unable to distinguish certain colors. **2. a.** Not subject to racial prejudices. **b.** Not recognizing racial distinctions <"Our Constitution is *colorblind*, and neither knows nor tolerates classes among citizens"—John Marshall Harlan> **—col'or·blind'ness** *n.*

**col·or·breed** (kŭl'ər-brēd') *vt.* **-bred** (-brĕd'), **-breed·ing, -breeds.** To breed (plants or animals) selectively so as to produce new or desired colors.

**col·or·cast** (kŭl'ər-kăst') *v.* **-cast** or **-cast·ed, -cast·ing, -casts.** [COLOR + (BROAD)CAST.] —*vt.* To broadcast (a television program) in color. —*vi.* To televise in color. **—col'or·cast'** *n.*

**col·or-code** (kŭl'ər-kōd') *vt.* **-cod·ed, -cod·ing, -codes.** To color, as papers or wires, according to a code for easy identification.

**col·ored** (kŭl'ərd) *adj.* **1.** Having color. **2. a.** Of an ethnic group not regarded as Caucasian, esp. Negro. **b.** Of mixed racial strains. **3.** Distorted or biased. **—col'ored** *n.*

**col·or·fast** (kŭl'ər-făst') *adj.* Having color that will not fade or run with wear or washing. **—col'or·fast'ness** *n.*

**color filter** *n.* A photographic filter for increasing contrast or taking photographs through haze.

**col·or·ful** (kŭl'ər-fəl) *adj.* **1.** Full of color or colors. **2.** Richly varied : VIVID. **—col'or·ful·ly** *adv.* **—col'or·ful·ness** *n.*

☆ ***syns:*** COLORFUL, BRIGHT, GAY, SHOWY, VIVID *adj. core meaning* : full of color <*colorful* roses> ***ant:*** colorless

**color guard** *n.* The ceremonial escort for a flag.

**col·or·if·ic** (kŭl'ə-rĭf'ĭk) *adj.* Producing or imparting color.

**col·or·im·e·ter** (kŭl'ə-rĭm'ĭ-tər) *n.* **1.** An instrument for determining or specifying colors, as by comparison with spectroscopic or visual standards. **2.** An instrument that measures the concentration of a known solution constituent by comparison with colors of standard solutions of that constituent. **—col'or·i·met'ric** (-ər-ə-mĕt'rĭk) *adj.* **—col'or·i·met'ri·cal·ly** *adv.* **—col'or·im'e·try** *n.*

**col·or·ing** (kŭl'ər-ĭng) *n.* **1.** The art, manner, or process of applying color. **2.** A substance for coloring something. **3.** Appearance with regard to color. **4.** Characteristic aspect, style, or tone. **5.** False or misleading appearance.

**col·or·ist** (kŭl'ər-ĭst) *n.* **1.** A painter skilled in achieving special effects with color. **2.** A hairdresser specializing in dyeing hair. **—col'or·is'tic** *adj.*

**col·or·less** (kŭl'ər-lĭs) *adj.* **1.** Devoid of color. **2.** Weak or dull in color : PALLID. **3.** Lacking animation, variety, or distinction : DULL <a *colorless* speaker> **—col'or·less·ly** *adv.* **—col'or·less·ness** *n.*

**color line** *n.* A barrier, created by custom, law, or economic differences, separating nonwhite persons from whites.

**co·los·sal** (kə-lŏs'əl) *adj.* [Fr. < Lat. *colossus*, colossus.] Enormous in size, extent, or degree : GIGANTIC. **—co·los'sal·ly** *adv.*

**col·os·se·um** (kŏl'ĭ-sē'əm) *n. var. of* COLISEUM.

**co·los·si** (kə-lŏs'ī') *n. var. pl. of* COLOSSUS.

**Co·los·sians** (kə-lŏsh'ənz, -lŏs'ē-ənz) *pl.n. (sing. in number).* —See table at BIBLE.

**co·los·sus** (kə-lŏs'əs) *n., pl.* **-los·si** (-lŏs'ī') *or* **-los·sus·es.** [Lat. < Gk. *kolossos.*] **1.** A huge statue. **2.** Something resembling a colossus, as in importance or size.

**co·los·to·my** (kə-lŏs'tə-mē) *n., pl.* **-mies.** Surgical construction of an artificial excretory opening from the colon.

**co·los·trum** (kə-lŏs'trəm) *n.* [Lat.] The first milk secreted by the mammary glands just after childbirth, lasting for a few days.

**col·our** (kŭl'ər) *n.* & *v. Chiefly Brit. var. of* COLOR.

**-colous** *suff.* [< Lat. *-cola*, inhabitant.] Having a given kind of habitat <arenicolous>

**col·pi·tis** (kŏl-pī'tĭs) *n.* [Gk. *kolpos*, vagina + -ITIS.] Inflammation of the vaginal mucous membrane.

**col·por·teur** (kŏl'pôr'tər, -pōr'-) *n.* [Fr., alteration of OFr. *comporteur* < *comporter*, to peddle, conduct. —see COMPORT.] A peddler of devotional literature. **—col'por'tage** (-pôr'tĭj, -pōr'-) *n.*

**col·po·scope** (kŏl'pə-skōp') *n.* [Gk. *kolpos*, vagina, womb + -SCOPE.] A speculum used in diagnostic examination of the cervical and vaginal tissues.

**col·pos·co·py** (kŏl-pŏs'kə-pē) *n., pl.* **-pies.** [Gk. *kolpos*, vagina, womb + -SCOPY.] Examination of the cervical and vaginal tissues by means of a colposcope.

**colt** (kōlt) *n.* [ME < OE.] **1.** A young male horse. **2.** A youthful, inexperienced person. **3.** A rope whip once used for disciplining sailors aboard ship.

**col·ter** *also* **coul·ter** (kōl'tər) *n.* [ME < OE *culter* ult. < Lat.] A blade or wheel on a plow for making vertical cuts in the sod.

**colt·ish** (kōl'tĭsh) *adj.* **1.** Of, relating to, or like a colt. **2.** Lively and playful : FRISKY. **—colt'ish·ly** *adv.* **—colt'ish·ness** *n.*

**colts·foot** (kōlts'fŏŏt') *n., pl.* **-foots.** [From the shape of its leaves.] An Old World plant, *Tussilago farfara*, bearing yellow flowers that appear before the heart-shaped leaves.

**col·u·brid** (kŏl'ə-brĭd, kŏl'yə-) *n.* [NLat. *Colubridae*, family name < Lat. *coluber*, snake.] Any of numerous chiefly nonvenomous snakes of the family Colubridae, including the king snakes and garter snakes. —*adj.* Of or belonging to the Colubridae.

**col·u·brine** (kŏl'ə-brīn', kŏl'yə-) *adj.* Colubrid.

**co·lu·go** (kə-lōō'gō) *n., pl.* **-gos.** [Malay.] The flying lemur.

**Co·lum·ba** (kə-lŭm'bə) *n.* [NLat. < Lat. *columba*, dove.] A constellation in the Southern Hemisphere.

**col·um·bar·i·um** (kŏl'əm-bâr'ē-əm) *also* **col·um·bar·y** (kŏl'əm-bĕr'ē) *n., pl.* **-i·a** (-ē-ə) *also* **-ies.** [Lat., sepulchre for urns, dovecote < *columba*, dove.] **1. a.** A vault with niches for urns containing ashes of the dead. **b.** A niche in such a vault. **2. a.** A dovecote. **b.** A pigeonhole in a dovecote.

**Co·lum·bi·a** (kə-lŭm'bē-ə) *n.* [After Christopher COLUMBUS (1451–1506).] Feminine personification of the United States.

**Co·lum·bi·an** (kə-lŭm'bē-ən) *adj.* **1.** Of or relating to the United States. **2.** Of or relating to Christopher Columbus.

**col·um·bine** (kŏl'əm-bīn') *n.* [ME < Med. Lat. *columbina* < Lat. *columbinus*, dovelike (from the resemblance of the inverted flower to a cluster of doves) < *columba*, dove.] A plant of the genus *Aquilegia*, having variously colored flowers with five conspicuously spurred petals. —*adj.* Dovelike.

**co·lum·bite** (kə-lŭm'bīt') *n.* [COLUMB(IUM) + -ITE.] A black mineral, essentially $(Fe, Mn)(Nb, Ta)_2O_6$, used as a source of niobium and tantalum.

**co·lum·bi·um** (kə-lŭm'bē-əm) *n.* [< NLat. *Columbia*, the United States.] *Symbol* **Cb** Niobium.

**Columbus Day** *n.* Oct. 12, a holiday officially celebrated in the United States on the second Monday in Oct. to honor Christopher Columbus.

**col·u·mel·la** (kŏl'yə-mĕl'ə, kŏl'ə-) *n., pl.* **-mel·lae** (-mĕl'ē) [Lat., dim. of *columna*, column.] One of several small columnlike structures in some animals and plants. **—col'u·mel'lar** (-mĕl'ər), **col'u·mel'late'** (-mĕl'āt') *adj.*

**col·umn** (kŏl'əm) *n.* [ME *columne* < Lat. *columna.*] **1.** A supporting pillar consisting of a base, a cylindrical shaft, and a capital. **2.** Something like a column in function or form. **3. a.** One of two or more vertical sections of typed or printed lines lying side by side on a page, separated by a blank space or rule. **b.** A feature article that appears regularly, as in a publication. **4.** A formation, as of troops or vehicles, in which all elements follow one behind the other. **5.** *Bot.* An organ formed by the fusion of stamens or of stamens and pistils, as in the orchid. **—col'umned** (kŏl'əmd) *adj.*

**co·lum·nar** (kə-lŭm'nər) *adj* **1.** Shaped like a column. **2.** Constructed with or having columns.

**co·lum·ni·a·tion** (kə-lŭm'nē-ā'shən) *n.* Arrangement or use of columns in a building.

**col·um·nist** (kŏl'əm-nĭst, -ə-mĭst) *n.* A writer of a newspaper or magazine column.

**col·za** (kŏl'zə, kōl'-) *n.* [Fr. < Du. *koolzaad* : *kool,* cabbage (< MDu. *côle* < Lat. *caulis*) + *zaad,* seed < MDu. *saet.*] RAPE[2].

**com-** *or* **col-** *or* **con-** *pref.* [ME < OFr. < Lat.] Together : with : joint : jointly <*commingle*>

**co·ma**[1] (kō'mə) *n., pl.* **-mas.** [Gk. *kōma, komat-,* deep sleep.] A deep, prolonged unconsciousness, usu. the result of disease, injury, or poison.

**co·ma**[2] (kō'mə) *n., pl.* **-mae** (-mē) [Lat., hair < Gk. *komē.*] **1.** *Astron.* The nebulous luminescent cloud containing the nucleus and constituting the major portion of the head of a comet. **2.** *Bot.* A tuft of hairs, as on some seeds. **3.** *Optics.* A diffuse pear-shaped image of a point source. **—co'mal** *adj.*

**Coma Ber·e·ni·ces** (bĕr'ə-nī'sēz') *n.* [Lat., hair of Berenice.] A constellation in the Northern Hemisphere.

**co·mae** (kō'mē) *n. pl. of* COMA[2].

**Co·man·che** (kə-măn'chē) *n., pl.* **Comanche** *or* **-ches.** [Sp. < Ute *kimanči.*] **1. a.** A tribe of Indians once ranging over the western plains from Wyoming to Texas and now living in Oklahoma. **b.** A member of this tribe. **2.** The Uto-Aztecan language of the Comanche. **—Co·man'che** *adj.*

**Co·man·che·an** (kə-măn'chē-ən) *adj.* [After *Comanche,* a county in Texas.] Of, pertaining to, or designating the geologic time, system of rocks, or sedimentary deposits of the Mesozoic era between the Jurassic and the Upper Cretaceous. —*n.* The Comanchean period.

**co·mate**[1] (kō'māt') *also* **co·mose** (-mōs') *adj.* [Lat. *comatus,* having long hair < *coma,* hair < Gk. *komē.*] Having or resembling a tuft of hair.

**co·mate**[2] (kō-māt', kō'māt') *n.* A mate : companion.

**co·ma·tose** (kō'mə-tōs', kŏm'ə-) *adj.* **1.** Of, relating to, or affected with coma : UNCONSCIOUS. **2.** Marked by lethargy : TORPID. **—co'ma·tose'ly** *adv.*

**co·mat·u·lid** (kə-măch'ə-lĭd) *also* **co·mat·u·la** (-lə) *n., pl.* **-lids** *also* **-lae** (-lē) [NLat. *Comatulidae,* former family name < LLat. *comatulus,* having neatly curled hair < Lat. *comatus,* having long hair.

—see COMATE.] Any of several marine invertebrates of the order Crinoidea that are attached to a surface by a stalk when young but are free-swimming as adults.

**comb** (kōm) *n.* [ME < OE.] **1. a.** A thin, toothed strip for smoothing, arranging, or fastening the hair. **b.** Something resembling a comb, as a card for dressing and cleansing fiber, as wool. **c.** A currycomb. **2. a.** The fleshy crest or ridge on the crown of the head of domestic fowl and other birds, most prominent in the male. **b.** Something resembling a fowl's comb. **3.** A honeycomb. —*v.* **combed, comb·ing, combs.** —*vt.* **1.** To dress or arrange with or as if with a comb. **2.** To card (e.g., wool). **3.** To search thoroughly. —*vi.* To roll and break <The surf *combed* with a great roar.>

**com·bat** (kəm-băt', kŏm'băt') *v.* **-bat·ed, -bat·ing, -bats** *also* **-bat·ted, -bat·ting, -bats.** [OFr. *combattre* < VLat. **combattere* : Lat. *com-,* with + Lat. *battuere,* to beat.] —*vt.* **1.** To fight against in battle. **2.** To oppose vigorously : RESIST <drugs that *combat* infection> —*vi.* To engage in fighting : STRUGGLE. —*n.* (kŏm'băt'). Fighting, esp. armed conflict : BATTLE.

**com·bat·ant** (kəm-băt'nt, kŏm'bə-tnt) *n.* One engaging in fighting or armed combat. —*adj.* Engaging in combat.

**combat fatigue** *n.* A nervous disorder, usu. temporary but sometimes leading to a permanent neurosis, caused by the stress of prolonged combat or similar situations and marked by severe anxiety, depression, irritability, and other related symptoms.

**com·bat·ive** (kəm-băt'ĭv) *adj.* Having or showing an eagerness to fight : BELLIGERENT. **—com·bat'ive·ly** *adv.* **—com·bat'ive·ness** *n.*

**comb·er** (kō'mər) *n.* **1.** One that combs. **2.** A long ocean wave that has reached its highest point or broken into foam : BREAKER.

**com·bi·na·tion** (kŏm'bə-nā'shən) *n.* **1.** The act or process of combining or state of being combined. **2.** Something that results from combining two or more things : COMPOUND. **3.** An alliance or association for a common purpose : COALITION. **4.** A sequence of numbers or letters needed to open a combination lock. **5.** A one-piece undergarment having underpants and an undershirt or chemise. **6.** *Math.* One or more elements chosen from a set without regard to order of selection. **—com'bi·na'tion·al** *adj.*

**combination lock** *n.* A lock that can be opened only by turning its dial through a predetermined sequence of positions identified on the dial face by numbers or letters.

**com·bi·na·tive** (kŏm'bə-nā'tĭv, kəm-bī'nə-tĭv) *adj.* **1.** Of, relating to, or stemming from combination. **2.** Tending, serving, or able to combine.

**com·bi·na·to·ri·al** (kŏm'bə-nə-tôr'ē-əl, -tōr'-, kəm-bī'nə-) *adj.* **1.** Of, relating to, or involving combinations. **2.** Relating to the arrangement and manipulation of mathematical elements in sets.

**com·bi·na·tor·ics** (kŏm'bə-nə-tôr'ĭks, -tōr'-, kəm-bī'nə-) *pl.n.* *(sing. in number).* Combinatorial mathematics.

**com·bine** (kəm-bīn') *v.* **-bined, -bin·ing, -bines.** [ME *combinen* < OFr. *combiner* < LLat. *combinare* : Lat. *com-,* together + *bine,* two at a time.] —*vt.* **1.** To bring into a state of unity : MERGE. **2.** To join (two or more substances) to make a single substance, as a chemical compound : MIX. **3.** To have or display in combination. —*vi.* **1.** To become united : COALESCE. **2.** To join forces for a common purpose. **3.** *Chem.* To form a chemical compound. —*n.* (kŏm'bīn'). **1.** A power-driven harvesting machine for cutting, threshing, and cleaning grain. **2.** An alliance of persons or groups joined together for the furtherance of commercial or political interests. **3.** A combination. **—com·bin'er** *n.*

**comb·ings** (kō'mĭngz) *pl.n.* Loose hairs, wool, or other material removed with a comb.

**combining form** *n.* A word element that joins with other word forms to create compounds; e.g., *-logy* in *gynecology, macro-* in *macrochemistry,* or *Sino-* in *Sino-Soviet.*

**combining weight** *n.* Equivalent weight.

**comb jelly** *n.* A ctenophore.

**com·bo** (kŏm'bō) *n., pl.* **-bos.** [Short for COMBINATION.] **1.** *Informal.* A small musical ensemble. **2.** *Slang.* The consequence or product of combining : COMBINATION.

**comb plate** *n.* A locomotive apparatus in ctenophores made up of a plate of fused cilia.

**com·bus·ti·ble** (kəm-bŭs'tə-bəl) *adj.* **1.** Capable of catching fire and burning. **2.** Easily aroused or stirred to action. —*n.* A combustible substance. **—com·bus'ti·bil'i·ty** *n.* **—com·bus'ti·bly** *adv.*

**com·bus·tion** (kəm-bŭs'chən) *n.* [ME *combustion* < LLat. *combustio* < Lat. *comburere,* to burn up : *com-* (intensive) + *urere,* to burn.] **1.** The process of burning. **2.** *Chem.* A rapid chemical change, esp. oxidation, that produces heat and light. **3.** Violent anger or agitation. **—com·bus'tive** (-tĭv) *adj.*

**combustion chamber** *n.* An enclosure in which combustion, esp. of a fuel or propellant, is started and controlled.

**come** (kŭm) *vi.* **came** (kām), **come, com·ing, comes.** [ME *comen* < OE *cuman.*] **1. a.** To proceed toward the speaker or toward a specified place : APPROACH <*Come* to me.> **b.** To proceed in a specified way <*came* willingly> **2.** To make progress : ADVANCE <has *come* an incredibly long way> **3.** To arrive at a specific point in a series or as a consequence of orderly progression <Daybreak is *coming.*> **4.** To move into view : APPEAR. **5.** To happen in time. **6. a.** To arrive at a specific result or end <*come* to an agreement> **b.** To arrive at or reach a specific state or condition <*came* to my wits' end> **c.** To move or be brought to a specific position <*came* to a sudden stop> **7.** To extend : reach <water that *came* only to their ankles> **8. a.** To exist at a specific point or place <The number 2 *comes* before 3.> **b.** To have priority or greater importance : RANK <Your health *comes* first.> **9. a.** To happen <How did we *come* to be invited?> **b.** To take place as a result <This *comes* of your meddling.> **10.** To fall to one <No good can *come* of this.> **11.** To take place in the mind <A new notion *came* to me.> **12. a.** To spring from : DESCEND <*comes* from a wealthy background> **b.** To be derived : ORIGINATE <Plants *come* from seeds.> **13.** To be a native or inhabitant of <*comes* from Dallas> **14.** To add up : TOTAL <Dinner *came* to more than $40.> **15.** To become <The latch *came* open.> **16.** To be available or obtainable <sweaters that *come* in all sizes and colors> **17.** To turn out to be : PROVE <a dream that can't *come* true> **—come about. 1.** To take place : OCCUR. **2.** To turn around. **3.** *Naut.* To change tack. **—come across. 1.** To find or meet accidentally. **2.** *Slang.* To do or give what is wanted. **3.** To give an impression <*came across* as an officious boor> **—come again.** *Slang.* To say something again : REPEAT. **—come along.** To achieve success : PROGRESS. **—come around** (or **round**). **1.** To recover : revive. **2.** To alter one's opinion or position <They *came around* after further discussion of the problem.> **—come at. 1.** To obtain : get. **2.** To rush at : ASSAIL. **—come back. 1.** To return to past success or former status after a period of misfortune. **2.** To retort : ANSWER. **3.** To recur to the memory. **—come between.** To cause hard feelings or separation. **—come by. 1.** To come into possession of : ACQUIRE. **2.** To pay a visit. **—come down. 1.** To lose wealth or status <has really *come down* in the world> **2.** To be descended or passed down <a tradition that *comes down* from ancient times> **—come down with.** To become ill <*came down with* the flu> **—come forward.** To volunteer one's services. **—come in. 1.** To arrive <New recruits will be *coming in* soon.> **2.** To arrive among those who complete a competition or race <*came in* second> **—come in for.** *Informal.* To receive or get <a performance that *came in for* savage criticism> **—come into.** To inherit <*came into* a great deal of money> **—come off. 1.** To take place : OCCUR. **2.** To acquit oneself <is sure to *come off* badly if put

ă **pat** ā **pay** âr **care** ä **father** ĕ **pet** ē **be** hw **which** ĭ **pit** ī **tie** îr **pier** ŏ **pot** ō **toe** ô **paw, for** oi **noise** ŏŏ **took**

to the test> **3.** To succeed <an experiment that *came off*> **—come on. 1.** To find or meet by accident. **2.** To impart a particular personal image <*comes on* as a well-meaning bungler> **—come out. 1.** To become known. **2.** To be issued or published <The new budget figures just *came out.*> **3.** To make a formal entry, as into society. **4.** To end up : RESULT. **—come out with. 1.** To disclose publicly <*came out with* a new party platform> **2.** To put into words : UTTER <always *comes out with* an amusing comment> **—come over. 1.** To happen to : POSSESS <An urge to cry *came over* me.> **2.** To switch sides, as in a controversy. **3.** *Informal.* To pay a visit. **—come through. 1.** To turn out well : SUCCEED. **2.** *Informal.* To do as expected <I asked for help, and you *came through.*> **3.** To exist in the fact of adversity : SURVIVE. **—come to. 1.** To recover consciousness. **2.** *Naut.* **a.** To bring the blow into the wind. **b.** To anchor. **3.** To be a matter of. **—come up.** To manifest itself : ARISE <The problem never *came up* before.> **—come upon.** To find or meet by accident. **—come up to.** To be the equal of <acting talent that doesn't *come up to* yours> **—come up with.** *Informal.* To propose, esp. as a solution to a problem : PRODUCE. *—interj.* —Used to express anger, impatience, or remonstrance <*Come,* that's quite enough!> **—come a cropper.** To fail utterly. **—come alive.** To become receptive and animated. **—come apart.** To disintegrate mentally or physically. **—come clean.** To confess. **—come off it.** *Slang.* To stop acting or speaking in a foolish or pompous way. **—come to grips with.** To face squarely <*came to grips with* the disaster>

**come·back** (kŭm′băk′) *n.* **1.** A return to former success or status. **2.** A return to popularity. **3.** *Informal.* A quick, witty reply : RETORT. **4.** The act of making up a deficit, as in a contest or game.

**co·me·di·an** (kə-mē′dē-ən) *n.* **1.** A professional entertainer who tells amusing stories or performs various other comic acts. **2.** An actor in comedy. **3.** A comedy writer. **4.** One who amuses or attempts to be amusing : CLOWN.

**co·me·dic** (kə-mē′dĭk) *adj.* Of or pertaining to comedy.

**co·me·di·enne** (kə-mē′dē-ĕn′) *n.* [Fr. *comédienne,* fem. of *comédien,* comedian < *comédie,* comedy < Lat. *comoedia.*] A woman who performs professionally as a comedian.

**com·e·do** (kŏm′ĭ-dō′) *n., pl.* **-dos** or **-do·nes** (-dō′nēz) [Lat., glutton < *comedere,* to eat up : *com-* (intensive) + *edere,* to eat.] A blackhead.

**come·down** (kŭm′doun′) *n.* **1.** A decline or sudden drop in position, status, or prestige. **2.** *Informal.* **a.** A feeling of disappointment or depression. **b.** Something causing disappointment or depression <The sad play was a real *comedown.*>

**com·e·dy** (kŏm′ĭ-dē) *n., pl.* **-dies.** [ME *comedie* < Lat. *comoedia* < Gk. *kōmōidia* < *kōmōidos,* comedian : *kōmos,* revel + *aoidos,* singer < *aeidein,* to sing.] **1. a.** A dramatic work that has a humorous theme and characters and a happy ending. **b.** The branch of drama comprising such works. **2.** A literary composition having a comic theme or using the methods of comedy. **3.** The branch of literature dealing with comedies. **4.** The art or technique of composing comedy. **5.** A comic element of literature or life. **6.** A comic event.

▲ word history: Comedy, like so many of the literary genres of Europe, originated in ancient Greece. Its exact beginnings are lost, but the word *comedy* itself provides a clue to the development of the form. Greek *kōmōidia* is derived from *kōmōidos,* a compound of *kōmos,* "revel, carousal," and *aoidos,* "singer." The earliest kind of comedy thus seems to have been a performance by a singer at the public revels. These celebrations were part of certain important religious festivals in Greek life and contained elements like mime, masquerades, and choruses that were later stylized and incorporated into a literary and dramatic form. Comedies have characteristically happy endings probably because the Greek religious revels ended in a feast.

**comedy of manners** *n.* A comedy satirizing the ways of fashionable society.

**come-hith·er** (kŭm-hĭth′ər) *adj.* Alluring : beguiling.

**come·ly** (kŭm′lē) *adj.* **-li·er, -li·est.** [ME *comli* < OE *cymlic,* lovely.] **1.** Having a pleasing appearance : ATTRACTIVE. **2.** Conforming to accepted standards : PROPER. **—come′li·ness** *n.*

**come-on** (kŭm′ŏn′, -ôn′) *n.* Something offered to allure or attract : ENTICEMENT.

**com·er** (kŭm′ər) *n.* **1.** One that arrives or comes. **2.** *Informal.* One that shows much promise of fame or success.

**co·mes·ti·ble** (kə-mĕs′tə-bəl) *adj.* [OFr. < LLat. *comestibilis* < Lat. *comedere,* to eat up. —see COMEDO.] Fit to be eaten : EDIBLE. *—n.* Something that can be eaten as food.

**com·et** (kŏm′ĭt) *n.* [ME *comete* < OE *comēta* < Lat. *cometa* < Gk. *komētēs* < *(astēr) komētēs,* long-haired (star) < *koman,* to grow hair long < *komē,* hair.] *Astron.* A celestial body, observable only in the part of its orbit that is relatively close to the sun, having a head made up of a solid nucleus surrounded by a nebulous coma up to 1.5 million miles or 2.4 million kilometers in diameter and an elongated curved vapor tail arising from the coma when sufficiently close to the sun, and thought to consist mainly of ammonia, methane, carbon dioxide, and water. **—com′et·ar′y** (-ĭ-tĕr′ē), **co·met′ic** (kə-mĕt′ĭk) *adj.*

**come·up·pance** *also* **come·up·ance** (kŭm′ŭp′əns) *n. Informal.* A justly deserved punishment or retribution.

**com·fit** (kŭm′fĭt, kŏm′-) *n.* [ME *confit* < OFr. < Lat. *conficere,* to prepare : *com-* (intensive) + *facere,* to make.] A candy : confection.

**com·fort** (kŭm′fərt) *vt.* **-fort·ed, -fort·ing, -forts.** [ME *comforten* < OFr. *conforter,* to strengthen < LLat. *confortare* : Lat. *com-* (intensive) + Lat. *fortis,* strong.] **1.** To soothe in time of grief or fear : CONSOLE. **2.** To make less severe or more bearable : RELIEVE. *—n.* **1.** A condition of ease or well-being. **2.** Consolation in time of grief or fear. **3.** Help : assistance <gave *comfort* to the enemy> **4.** One that brings comfort. **5.** Capacity to give physical ease and well-being. **—com′fort·ing·ly** *adv.*

▲ word history: *Comfort,* as noun and verb, is ultimately derived from Latin *com–,* an intensive prefix, and *fortis,* "strong." The Latin verb *confortare* and its descendents in the medieval Romance languages meant "to strengthen." The verb and the noun were borrowed into English from French in the 13th century. The senses "to strengthen" for the verb and "strengthening" and "solace" for the noun both occur at that early date. The meaning "ease" for the noun is not recorded until the early 19th century, although the meaning "at ease" for *comfortable* developed during the 18th century. These weakened senses are now the normal ones and the older meanings are obsolete.

**com·fort·a·ble** (kŭm′fər-tə-bəl, kŭmf′tə-bəl) *adj.* **1.** Affording pleasurable ease <a *comfortable* house> **2.** Being in a state of comfort. **3.** *Informal.* Sufficient : adequate <a *comfortable* salary> **—com′fort·a·ble·ness** *n.* **—com′fort·a·bly** *adv.*

**com·fort·er** (kŭm′fər-tər) *n.* **1.** One that comforts. **2. Comforter.** The Holy Spirit. **3.** A thick, warm quilted bedcover. **4.** *Chiefly Brit.* A woolen neck scarf.

**comfort station** *n.* A public toilet or rest room.

**com·frey** (kŭm′frē) *n., pl.* **-freys.** [ME *conferie* < OFr. *confirie* < Lat. *conferva.*] An Old World, usu. hairy or bristly plant of the genus *Symphytum,* bearing variously colored flower clusters.

**com·fy** (kŭm′fē) *adj.* **-fi·er, -fi·est.** *Informal.* Comfortable.

**com·ic** (kŏm′ĭk) *adj.* [Lat. *comicus* < Gk. *kōmikos* < *kōmos,* revel.] **1.** Of, relating to, or typical of comedy. **2.** Of or relating to comic strips. **3.** Arousing or deserving laughter : HUMOROUS. *—n.* **1. a.** A comedian. **b.** One who is comical. **2. a. comics.** Comic strips. **b.** A comic book. **3.** Something that arouses humor in art or life.

**com·i·cal** (kŏm′ĭ-kəl) *adj.* **1.** Arousing laughter or mirth : AMUSING. **2.** Of or relating to comedy. **—com′i·cal′i·ty** (-kăl′ĭ-tē), **com′i·cal·ness** *n.* **—com′i·cal·ly** *adv.*

**comic book** *n.* A book of comic strips.

**comic opera** *n.* An opera or operetta with an amusing plot, spoken dialogue, and usu. a happy ending.

**com·ic-op·er·a** (kŏm′ĭk-ŏp′rə, -ŏp′ər-ə) *adj.* Not to be taken seriously <*comic-opera* politics>

**comic strip** *n.* A narrative sequence of cartoons.

**com·ing** (kŭm′ĭng) *adj.* **1.** Approaching : following <the *coming* year> **2.** *Informal.* Showing great promise of fame or success. *—n.* Arrival : advent <the *coming* of autumn>

**com·ing-out** (kŭm′ĭng-out′) *n. Informal.* A social debut.

**Com·in·tern** (kŏm′ĭn-tûrn′) *n.* [COM(MUNIST) INTERN(ATIONAL).] The Third International, organized in 1919.

**co·mi·ti·a** (kə-mĭsh′ē-ə, -mĭsh′ə) *n., pl.* **comitia.** [Lat. < pl. of *comitium,* assembly place : *com-,* together + *ire,* to go.] An assembly of citizens in ancient Rome having legislative or electoral powers. **—co·mi′tial** (-mĭsh′əl) *adj.*

**com·i·ty** (kŏm′ĭ-tē) *n., pl.* **-ties.** [Lat. *comitas* < *comis,* friendly.] **1.** Social harmony. **2.** Comity of nations. **3.** Cordial recognition given by the courts of one state or jurisdiction of the laws and judicial decisions of another.

**comity of nations** *n.* **1.** Cordial recognition given by one nation to the laws and institutions of another. **2.** The nations observing international comity.

**com·ix** (kŏm′ĭks) *pl.n.* [Alteration of COMICS.] Comic books and comic strips, esp. of the underground press.

**com·ma** (kŏm′ə) *n.* [LLat. < Gk. *komma,* short clause < *koptein,* to cut.] **1.** A punctuation mark (,) used to indicate a separation of ideas or elements within the structure of a sentence. **2.** A pause or division : CAESURA. **3.** A butterfly of the genus *Polygonia,* having brownish, irregularly notched wings.

**comma bacillus** *n.* [From its commalike shape.] A bacillus, *Vibrio comma,* causing Asiatic cholera.

**comma fault** *n.* Misuse of a comma between independent clauses not joined by a conjunction.

**com·mand** (kə-mănd′) *v.* **-mand·ed, -mand·ing, -mands.** [ME *commaunden* < OFr. *comander* < VLat. **commandare* : *com-* (intensive) + *mandare,* to entrust.] *—vt.* **1.** To give orders to : DIRECT. **2.** To have authoritative control over : RULE. **3.** To have at one's disposal <*commands* seven languages> **4.** To deserve and receive as due <honesty that *commanded* respect> **5.** To dominate by position : OVERLOOK <a promontory *commanding* the harbor> *—vi.* **1.** To give commands. **2.** To be in control. *—n.* **1.** The act of com-

manding. **2.** An authoritative order or direction. **3.** A signal that activates a device, as a computer. **4. a.** The authority to command. **b.** The possession and exercise of the authority to command. **5.** Natural or acquired facility or skill : MASTERY <a *command* of seven languages> **6.** Domination by position. **7. a.** The jurisdiction of a commander. **b.** A military unit, post, district, or territory under the control of one officer. **c.** A unit in the U.S. Air Force having a specified number of wings, gen. three or more, under the authority of an officer. **8.** *Chiefly Brit.* An invitation from a reigning monarch.
☆ ***syns:*** COMMAND, BID, CHARGE, DIRECT, ENJOIN, INSTRUCT, ORDER, REQUIRE, TELL *v. core meaning* : to give orders to <*commanded* the brigade to attack> ***ant:*** comply, obey

**com·man·dant** (kŏm′ən-dănt′, -dänt′) *n.* A military commander.

**com·man·deer** (kŏm′ən-dîr′) *vt.* **-deered, -deer·ing, -deers.** [Afr. *kommanderen* < Fr. *commander,* to command < OFr. *comander.* —see COMMAND.] **1.** To force into military service. **2.** To seize for military use : CONFISCATE. **3.** *Informal.* To take arbitrarily.

**com·mand·er** (kə-măn′dər) *n.* **1.** One who commands : LEADER. **2. a.** An officer in the U.S. Navy who ranks above a lieutenant commander and below a captain. **b.** The chief commissioned officer of a military unit regardless of rank. **3.** A chief or officer in certain knightly or fraternal orders.

**commander in chief** *n.* **1.** *often* **Commander in Chief.** The supreme commander of all the armed forces of a nation. **2.** The officer in command of a major armed force.

**com·mand·er·y** (kə-măn′də-rē) *n., pl.* **-ies. 1.** The territory or office of a commander, esp. of an order of knights. **2.** A lodge or local branch of certain fraternal orders.

**com·mand·ing** (kə-măn′dĭng) *adj.* Dominating, as by size or position <took an early and *commanding* lead in the election> **—com·mand′ing·ly** *adv.*

**commanding officer** *n.* A U.S. Army officer in charge of a unit, as a company or regiment, or of a post or station.

**com·mand·ment** (kə-mănd′mənt) *n.* **1.** An order : command. **2.** One of the Ten Commandments.

**command module** *n.* The part of a spacecraft in which the astronauts live and operate controls during a flight.

**com·man·do** (kə-măn′dō) *n., pl.* **-dos** *or* **-does.** [Afr. *kommando* < Du. *commando,* unit of troops < Sp. *comando* < *comandar,* to command < VLat. **commandare.* —see COMMAND.] **1. a.** A small fighting force specially trained for making quick, destructive raids inside enemy territory. **b.** A member of such a force. **2. a.** A Boer military force. **b.** A raid made by such a force.

**command post** *n.* The field headquarters used by a military commander.

**comma splice** *n.* A comma fault.

**com·meas·ure** (kə-mĕzh′ər) *vt.* **-ured, -ur·ing, -ures.** To coincide or be coextensive with. **—com·meas′ur·a·ble** *adj.*

**com·me·dia dell'ar·te** (kə-mā′dē-ə dĕl-är′tē, -mĕd′ē-ə) *n.* [Ital., comedy of art.] A style of comedy developed in Italy in the 16th cent. and marked by improvisation and the use of stock characters.

**comme il faut** (kŭm′ ēl fō′) *adj.* [Fr., as it should be.] In accordance with accepted standards : PROPER.

**com·mem·o·rate** (kə-mĕm′ə-rāt′) *vt.* **-rat·ed, -rat·ing, -rates.** [Lat. *commemorare, commemorat-,* to remind : *com-* (intensive) + *memorare,* to remind < *memor,* mindful.] **1.** To honor the memory of. **2.** To be a memorial to, as a holiday or monument. **—com·mem′o·ra′tor** *n.*

**com·mem·o·ra·tion** (kə-mĕm′ə-rā′shən) *n.* **1.** The act of commemorating. **2.** Something, as a monument or custom, that commemorates.

**com·mem·o·ra·tive** (kə-mĕm′ər-ə-tĭv, -ə-rā′-) *adj.* Serving to commemorate. *—n.* Something that commemorates.

**com·mem·o·ra·to·ry** (kə-mĕm′ər-ə-tôr′ē, -tōr′ē) *adj.* Commemorative.

**com·mence** (kə-mĕns′) *v.* **-menced, -menc·ing, -menc·es.** [ME *commencen* < OFr. *comencier* < VLat. **cominitiare* : Lat. *com-* (intensive) + LLat. *initiare,* to begin < Lat. *initium,* beginning.] *—vt.* To start : begin. *—vi.* To come into being. **—com·menc′er** *n.*

**com·mence·ment** (kə-mĕns′mənt) *n.* **1.** The act or process of beginning. **2. a.** A graduation ceremony at which academic degrees or diplomas are conferred. **b.** The day on which such a ceremony takes place.

**com·mend** (kə-mĕnd′) *vt.* **-mend·ed, -mend·ing, -mends.** [ME *commenden* < Lat. *commendare* : *com-* (intensive) + *mandare,* to entrust.] **1.** To represent as worthy, qualified, or desirable : RECOMMEND. **2.** To speak highly of : PRAISE. **3.** To put in the care of another : ENTRUST. **—com·mend′a·ble** *adj.* **—com·mend′a·ble·ness** *n.* **—com·mend′a·bly** *adv.*

**com·men·da·tion** (kŏm′ən-dā′shən) *n.* **1.** Recommendation or praise. **2.** An official award or citation.

**com·mend·a·to·ry** (kə-mĕn′də-tôr′ē, -tōr′ē) *adj.* Serving to recommend or praise.

**com·men·sal** (kə-mĕn′səl) *adj.* [ME < Med. Lat. *commensalis* : Lat. *com-,* together + Lat. *mensa,* table.] **1.** Of or relating to those who normally eat at the same table. **2.** *Biol.* Of, relating to, or marked by commensalism. *—n.* **1.** A customary mealtime companion. **2.** *Biol.* An organism participating in a commensal relationship. **—com·men′sal·ly** *adv.*

**com·men·sal·ism** (kə-mĕn′sə-lĭz′əm) *n. Biol.* A relationship in which two or more organisms live in close attachment or partnership and in which one may derive some benefit but neither harms or is parasitic on the other.

**com·men·su·ra·ble** (kə-mĕn′sər-ə-bəl, -shər-) *adj.* [LLat. *commensurabilis* : Lat. *com-,* together + *mensurabilis,* measurable < Lat. *mensura,* measure.] **1.** Capable of being measured by a common standard or unit. **2.** Properly proportioned : SUITABLE. **3.** *Math.* Exactly divisible by the same unit an integral number of times. —Used of two quantities. **—com·men′su·ra·bil′i·ty** *n.* **—com·men′su·ra·bly** *adv.*

**com·men·su·rate** (kə-mĕn′sər-ĭt, -shər-) *adj.* [LLat. *commensuratus* : Lat. *com-,* together + *mensuratus,* p.part. of *mensurare,* to measure < Lat. *mensura,* measure.] **1.** Of the same size, extent, or length of time. **2.** Corresponding in size, amount, or scale : PROPORTIONATE <a salary *commensurate* with performance> **3.** Having a common measure or standard : COMMENSURABLE. **—com·men′su·rate·ly** *adv.* **—com·men′su·ra′tion** *n.*

**com·ment** (kŏm′ĕnt′) *n.* [ME < Lat. *commentum,* interpretation < neuter p.part. of *comminisci,* to devise.] **1.** A written note that explains, illustrates, or criticizes a passage in a book or other writing : ANNOTATION. **2. a.** A brief expression of fact or personal opinion. **b.** An implied conclusion or judgment <a drama that is a *comment* on contemporary society> **3.** Talk : gossip <a scandal causing *comment*> *—v.* **-ment·ed, -ment·ing, -ments.** *—vi.* To make a comment : REMARK. *—vt.* To make comments on : ANNOTATE.

**com·men·tar·y** (kŏm′ən-tĕr′ē) *n., pl.* **-ies. 1.** A series of explanations or interpretations. **2.** *often* **commentaries.** An expository treatise or series of annotations : EXEGESIS. **3.** Something that explains or illustrates. **4.** *often* **commentaries.** A narrative of personal experiences. **—com′men·tar′i·al** (-târ′ē-əl) *adj.*
☆ ***syns:*** COMMENTARY, ANNOTATION, COMMENT, EXEGESIS, INTERPRETATION, NOTE *n. core meaning* : critical comments or analysis <*commentaries* on the Old Testament>

**com·men·tate** (kŏm′ən-tāt′) *vt.* & *vi.* **-tat·ed, -tat·ing, -tates.** To make a commentary on or serve as commentator. ***usage:*** The use of this verb in the sense "to give a commentary" is well established, but in the sense "to provide a commentary for," as in *An announcer commentated the fashion show,* it is unacceptable to a great many.

**com·men·ta·tor** (kŏm′ən-tā′tər) *n.* **1.** An author of commentaries. **2.** One who reports and analyzes events in the news. **3.** A member of the congregation who leads prayers and explains rituals during a service.

**com·merce** (kŏm′ərs) *n.* [OFr. < Lat. *commercium* : *com-,* together + *merx,* merchandise.] **1.** The buying and selling of goods, esp. on a large scale : BUSINESS. **2.** Mutual exchange, as of ideas or social amenities.

**com·mer·cial** (kə-mûr′shəl) *adj.* **1. a.** Of or relating to commerce. **b.** Engaged in commerce. **c.** Involved in work designed or planned for the mass market <a *commercial* artist> **2.** Designating products, often unrefined, made and distributed in large quantities for industrial use. **3.** Having profit as a primary aim <too avant-garde to be a *commercial* film> **4.** Paid for by an advertiser or advertising <*commercial* television> *—n.* An ad presented on radio or television or before or after a film. **—com·mer′cial·ly** *adv.*

**commercial bank** *n.* A bank whose principal functions are to receive demand deposits and make short-term loans.

**com·mer·cial·ism** (kə-mûr′shə-lĭz′əm) *n.* **1.** The practices, systems, aims, and spirit of commerce or business. **2.** An attitude emphasizing tangible profit or success. **—com·mer′cial·ist** *n.* **—com·mer′cial·is′tic** *adj.*

**com·mer·cial·ize** (kə-mûr′shə-līz′) *vt.* **-ized, -iz·ing, -iz·es. 1.** To apply methods of business to for profit. **2. a.** To do, make, or exploit chiefly for financial gain. **b.** To reduce the quality of in order to make a profit. **—com·mer′cial·i·za′tion** *n.*

**commercial paper** *n.* A short-term negotiable paper arising from business transactions.

**commercial traveler** *n.* A traveling sales representative.

**com·mie** *also* **Com·mie** (kŏm′ē) *n. Informal.* A Communist.

**com·mi·na·tion** (kŏm′ə-nā′shən) *n.* [ME *comminacioun* < Lat. *comminatio* < *comminari,* to threaten : *com-* (intensive) + *minari,* to threaten < *minae,* threats.] A formal accusation. **—com·min′a·to′ry** (kə-mĭn′ə-tôr′ē, -tōr′ē, kŏm′ĭ-nə-) *adj.*

**com·min·gle** (kə-mĭng′gəl) *vi.* & *vt.* **-gled, -gling, -gles.** To blend or cause to blend together : MIX.

**com·mi·nute** (kŏm′ə-nōōt′, -nyōōt′) *vt.* **-nut·ed, -nut·ing, -nutes.** [Lat. *comminuere, comminut-* : *com-* (intensive) + *minuere,* to lessen.] To change into a powder : PULVERIZE. **—com′mi·nu′tion** *n.*

**com·mis·er·ate** (kə-mĭz′ə-rāt′) *v.* **-at·ed, -at·ing, -ates.** [Lat. *commiserari, commiserat-* : *com-,* with + *miserari,* to pity < *miser,* wretched.] *—vt.* To experience or express sympathy or pity for. *—vi.*

To experience or express sympathy or pity. **—com·mis′er·a′tion** *n.* **—com·mis′er·a′tive** *adj.* **—com·mis′er·a′tive·ly** *adv.* **—com·mis′er·a′tor** *n.*

**com·mis·sar** (kŏm′ĭ-sär′) *n.* [R. *komissar* < G. *Kommissar,* deputy < Med. Lat. *commissarius,* agent. —see COMMISSARY.] **1.** A Communist Party official in charge of political indoctrination and the enforcement of party loyalty. **2.** A former designation for the head of a commissariat in the Soviet Union.

**com·mis·sar·i·at** (kŏm′ĭ-sâr′ē-ĭt) *n.* [NLat. *commissariatus* < Med. Lat. *commissarius,* agent. —see COMMISSARY.] **1.** An army department that provides food and other supplies for the troops. **2.** A supply of food. **3.** A former major government department in the Soviet Union.

**com·mis·sar·y** (kŏm′ĭ-sĕr′ē) *n., pl.* **-ies.** [ME *commissarie* < Med. Lat. *commissarius,* agent < Lat. *committere,* to entrust. —see COMMIT.] **1. a.** A place where food and equipment are sold, as in a mining camp. **b.** A market for the personnel of a military post. **2.** A cafeteria or lunchroom, esp. one in a motion-picture or television studio. **3.** One to whom a special duty is given by a superior : DEPUTY. **—com′mis·sar′y·ship′** *n.*

**com·mis·sion** (kə-mĭsh′ən) *n.* [ME *commissioun* < Lat. *commissio* < *committere,* to entrust. —see COMMIT.] **1. a.** The act of giving the authority or power to perform a specific task or duty. **b.** The authority so granted. **c.** The task or duty so authorized. **d.** A document conferring such authorization. **2.** A group of people given official authorization to perform certain functions or duties. **3.** The act or process of committing or perpetrating <the *commission* of a felony> **4.** A fee or percentage paid to a salesperson or agent for his or her services. **5. a.** A document issued by a government conferring the rank of a commissioned officer in the armed forces. **b.** The rank and powers so conferred. *—vt.* **-sioned, -sion·ing, -sions. 1.** To give a commission to <was *commissioned* a colonel> **2.** To put in an order for <*commissioned* a self-portrait> **3.** To put (a ship) into active service. **—in commission. 1.** In active service, as a ship. **2.** In use or in working condition. **—out of commission. 1.** Not in active service, as a ship. **2.** Not in use or in working condition. **—com·mis′sion·al** *adj.*

**com·mis·sion·aire** (kə-mĭsh′ə-nâr′) *n.* [Fr. < Med. Lat. *commissionarius* < Lat. *commissio,* commission.] *Chiefly Brit.* A doorman.

**commissioned officer** *n.* An officer who holds a commission and ranks as a second lieutenant or above in the U.S. Army, Air Force, or Marine Corps or as an ensign or above in the U.S. Navy or Coast Guard.

**com·mis·sion·er** (kə-mĭsh′ə-nər) *n.* **1.** One authorized by a commission to perform certain tasks or duties. **2.** A member of a commission. **3.** A governmental official in charge of a department. **4.** An official chosen by an athletic association or league to exercise judicial or regulatory powers <a basketball *commissioner*> **—com·mis′sion·er·ship′** *n.*

**commission merchant** *n.* One who buys and sells goods for others on a commission basis.

**commission plan** *n.* Municipal government in which legislative and administrative functions and powers are exercised by an elected commission rather than by a mayor and city council.

**com·mis·sure** (kŏm′ə-shŏŏr′) *n.* [ME < Lat. *commissura* < *committere,* to join. —see COMMIT.] **1.** A line or place at which two things are joined : JUNCTURE. **2.** *Anat.* **a.** A tract of nerve fibers passing from one side to the other of the spinal cord or brain. **b.** The angle or corner of such structures, as the eyelids or cardiac valves. **3.** *Bot.* A surface by which adhering carpels are joined. **—com′mis·su′ral** *adj.*

**com·mis·sur·ot·o·my** (kŏm′ə-shŏŏ-rŏt′ə-mē) *n., pl.* **-mies.** [COMMISSUR(E) + -TOMY.] Surgical incision of a commissure of the brain.

**com·mit** (kə-mĭt′) *vt.* **-mit·ted, -mit·ting, -mits.** [ME *committen* < Lat. *committere* : *com-,* together + *mittere,* to send.] **1.** To do or perform : PERPETRATE <*commit* a felony> **2.** To give over to another care or use : ENTRUST. **3.** To place officially in confinement or custody. **4.** To put in a certain condition or form, as for future use, reference, or preservation <*commit* to memory> **5.** To put in a place to be disposed of or kept safe. **6. a.** To pledge (oneself) to a position <a politician who refuses to *commit* herself on the issues> **b.** To obligate or bind <was *committed* to attend the dinner> **7.** To refer (e.g., a legislative bill) to a committee. **—com·mit′ta·ble** *adj.*

**com·mit·ment** (kə-mĭt′mənt) *n.* **1.** The act or process of committing, esp.: **a.** The referral of a legislative bill to committee. **b.** Official consignment, as to a prison or mental hospital. **c.** A court order authorizing consignment to a prison : MITTIMUS. **2. a.** A pledge to do something. **b.** Something pledged, esp. a contractual engagement involving financial obligation. **3.** The state of being bound emotionally or intellectually to an ideal or course of action.

**com·mit·tal** (kə-mĭt′l) *n.* **1.** The act of entrusting. **2.** The act of committing to confinement, as to a prison or mental hospital. **3.** The act of pledging oneself to a particular position or course of action.

**com·mit·tee** (kə-mĭt′ē) *n.* [ME *committe,* trustee < AN *comité,* p.part. of *cometre,* to commit < Lat. *committere.* —see COMMIT.] **1.** A group of people delegated to perform a particular function or task. **2.** *Law.* One to whom the care of an estate or incompetent person is committed : TRUSTEE.

**com·mit·tee·man** (kə-mĭt′ē-mən, -măn′) *n.* **1.** A committee member. **2.** A ward or precinct leader of a political party.

**committee of the whole** *n.* A committee composed of all the members present of a legislative house.

**com·mit·tee·wo·man** (kə-mĭt′ē-wŏŏm′ən) *n.* **1.** A woman who is a committee member. **2.** A woman who is a ward or precinct leader of a political party.

**com·mix** (kə-mĭks′, kŏ-) *vi. & vt.* **-mixed, -mix·ing, -mix·es.** [< ME *commixt,* mixed together < Lat. *commixtus,* p.part. of *commiscēre,* to mix together : *com-,* together + *miscēre,* to mix.] To mix or cause to mix together : BLEND.

**com·mix·ture** (kə-mĭks′chər, kŏ-) *n.* **1.** The act or process of mixing. **2.** Something produced by mixing : MIXTURE.

**com·mode** (kə-mōd′) *n.* [Fr. < *commode,* convenient < Lat. *commodus.* —see COMMODIOUS.] **1.** A low cabinet or bureau, often ornately decorated and usu. on legs or short feet. **2. a.** A stand or cupboard containing a washbowl. **b.** A chair enclosing a chamber pot. **c.** A toilet. **3.** An 18th-cent. ornate headdress worn by women.

**com·mo·di·ous** (kə-mō′dē-əs) *adj.* [ME, convenient < Med. Lat. *commodiosus* < Lat. *commodus* : *com-,* with + *modus,* measure.] **1.** Having plenty of room : SPACIOUS. **2.** *Archaic.* Suitable : handy. **—com·mo′di·ous·ly** *adv.* **—com·mo′di·ous·ness** *n.*

**com·mod·i·ty** (kə-mŏd′ĭ-tē) *n., pl.* **-ties.** [ME *commodite* < OFr., convenience < Lat. *commoditas* < *commodus,* convenient. —see COMMODIOUS.] **1.** Something useful or capable of yielding commercial or other advantages. **2.** A commercial article, esp. an agricultural or mining product, that can be transported. **3.** *Obs.* **a.** Convenience : expediency. **b.** A supply of goods.

**com·mo·dore** (kŏm′ə-dôr′, -dōr′) *n.* [Prob. < Du. *komandeur,* commander < Fr. *commandeur* < OFr. *commandeor* < *comander,* to command. —see COMMAND.] **1.** An officer in the U.S. Navy ranking above captain and below rear admiral. **2.** An unofficial designation for a captain in the British Navy temporarily in command of a fleet division or squadron. **3. a.** The senior captain of a naval squadron or merchant fleet. **b.** The highest-ranking officer of a yacht club.

**com·mon** (kŏm′ən) *adj.* **-er, -est.** [ME *commune* < OFr. < Lat. *communis.*] **1. a.** Belonging to, shared by, or applying equally : JOINT <*common* concerns> **b.** Belonging or relating to the community as a whole : PUBLIC. **2.** Widespread : prevalent. **3. a.** Occurring frequently or habitually : USUAL. **b.** Most widely known : ORDINARY <the *common* housefly> **4.** Having no special designation, status, or rank <a *common* soldier> **5. a.** Having no superior characteristics : AVERAGE <the *common* observer> **b.** Having no special quality : STANDARD <*common* procedure> **c.** Of mediocre or inferior quality : SECOND-RATE. **6.** Coarse or vulgar in manner. **7. a.** Either masculine or feminine in grammatical gender. **b.** Representing one or all the members of a grammatical class. *—n.* **1. commons.** COMMONALTY 1. **2. commons.** *(sing. or pl. in number).* **a.** The political class made up of commoners. **b.** The parliamentary representatives of this class. **c.** *often* **Commons.** The House of Commons. **3.** A section of land belonging to or used by a community as a whole. **4.** A legal right to use the lands or waters of another, as for hunting or fishing. **5. commons.** *(sing. in number).* A structure or room for dining. **6.** A church service used for a particular class of festivals. **—in common.** Equally with or by all. **—com′mon·ly** *adv.* **—com′mon·ness** *n.*

☆ ***syns:*** COMMON, COMMUNAL, GENERAL, JOINT, MUTUAL, PUBLIC *adj. core meaning* : belonging to, shared by, or applying to all <world hunger—a *common* concern of all caring peoples>

**com·mon·age** (kŏm′ə-nĭj) *n.* **1.** The right to pasture animals on common land. **2.** The state of being held in common.

**com·mon·al·ty** (kŏm′ə-nəl-tē) *also* **com·mon·al·i·ty** (kŏm′ə-năl′ĭ-tē) *n., pl.* **-ties.** [ME *communalte* < OFr. *comunalte* < Med. Lat. *communalitas* < LLat. *communalis,* of the community. —see COMMUNAL.] **1.** The common people as opposed to the upper classes. **2.** A corporate body : CORPORATION. **3.** A universal group or body.

**common bile duct** *n.* The duct carrying bile from the liver to the duodenum.

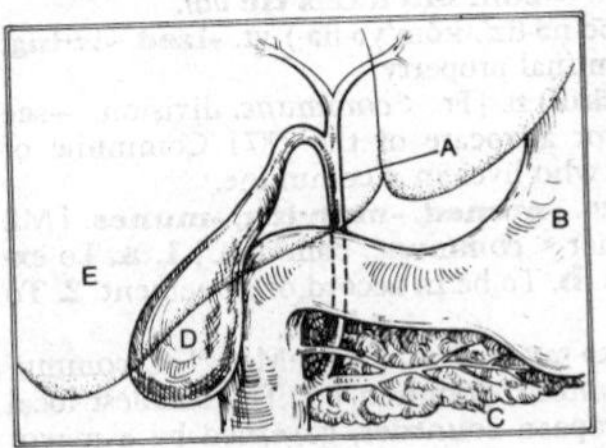


**common bile duct**
*A. common bile duct, B. stomach, C. pancreas, D. gall bladder, E. liver*

**common carrier** *n.* One who transports passsengers or goods for a fee.

**common cold** *n.* Coryza.

**common denominator** *n.* **1.** A number into which all the denominators of a set of fractions may be evenly divided. **2.** A theme or characteristic shared by a number of people or things.

**common divisor** *n.* A number that is a factor of each of two or more quantities.

**com·mon·er** (kŏm′ə-nər) *n.* **1.** One of the common people. **2.** One having no noble rank or title.

**common fraction** *n.* A fraction having an integer as a numerator and an integer as a denominator.

**common gender** *n.* Gender that refers to males or females, as *child* and *person.*

**common law** *n.* The system of laws orig. developed in England and based on court decisions, on the doctrines implicit in those decisions, and on customs and usages rather than on a codified body of written laws or statutes.

**common-law marriage** (kŏm′ən-lô′) *n.* A marriage existing by mutual agreement and cohabitation between a man and a woman without a civil or religious ceremony.

**common logarithm** *n.* A logarithm to the base 10.

**common measure** *n.* **1.** Common time. **2.** A common divisor.

**common multiple** *n.* A number that is a multiple of each of two or more given quantities.

**common noun** *n.* A noun, as *building* or *flower,* that represents one, more, or all of the members of a class and can be preceded by a definite article.

**com·mon·place** (kŏm′ən-plās′) *adj.* [Transl. of Lat. *locus communis,* generally applicable theme, transl. of Gk. *koinos topos.*] Not of any special quality or type : ORDINARY. —*n.* **1. a.** An overused or trite remark : PLATITUDE. **b.** Something customary or regular. **2.** *Archaic.* A passage marked for reference or entered in a commonplace book.

**commonplace book** *n.* A journal or diary in which quotable passages, literary excerpts, and comments are written.

**common pleas** *n.* COURT OF COMMON PLEAS 1.

**common salt** *n.* **1.** Salt. **2.** Sodium chloride. **3.** TABLE SALT 1.

**common school** *n.* An elementary school.

**common sense** *n.* [Transl. of Lat. *sensus communis,* common feelings of humanity.] The ability to make sound judgments.

**common stock** *n.* Ordinary capital shares of a corporation that have claim on the net assets and income of the corporation after all prior or preferred claims have been paid.

**common time** *n. Mus.* A meter having four quarter notes to the measure.

**com·mon·weal** (kŏm′ən-wēl′) *n.* **1.** The public good or welfare. **2.** *Archaic.* A commonwealth.

**com·mon·wealth** (kŏm′ən-wĕlth′) *n.* [ME *commune welthe* : *commune,* common + *welthe,* well-being. —see WEALTH.] **1.** The people of a nation or state. **2.** A nation or state governed by the people : REPUBLIC. **3. Commonwealth.** The official title of some U.S. states, including Kentucky, Virginia, Massachusetts, and Pennsylvania. **4.** The official title of certain political units, as Puerto Rico or Australia, having a self-governing, autonomous, voluntary relationship with a larger political unit. **4.** *Archaic.* COMMONWEAL 1.

**com·mo·tion** (kə-mō′shən) *n.* [ME *commocioun* < Lat. *commotio* < *commovēre,* to disturb : *com-* (intensive) + *movēre,* to move.] **1.** Violent or turbulent motion : AGITATION. **2.** Political or social disturbance or insurrection : DISORDER. **3.** A confused disturbance : HUBBUB. **—com·mo′tion·al** *adj.*

**com·move** (kə-mo͞ov′) *vt.* **-moved, -mov·ing, -moves.** [ME *commeven* < OFr. *commovoir* < Lat. *commovere.* —see COMMOTION.] To agitate : disturb.

**com·mu·nal** (kə-myo͞o′nəl, kŏm′yə-) *adj.* [Fr. < LLat. *communalis* < Lat. *communis,* common.] **1.** Of or relating to a commune. **2.** Of or relating to a community. **3. a.** Of, relating to, or belonging to the people of a community. **b.** Typical of common ownership and control of goods and property. **—com′mu·nal′i·ty** (kŏm′yə-năl′ĭ-tē) *n.* **—com·mu′nal·ly** *adv.*

**com·mu·nal·ism** (kə-myo͞o′nə-lĭz′əm, kŏm′yə-nə-) *n.* **1.** A theory or system of government in which communities are virtually autonomous and loosely bound in a federation. **2.** Belief in or practice of communal ownership, as of assets and property. **3.** Devotion to the concerns of one's own ethnic group rather than those of society as a whole. **—com·mu′nal·ist** *n.* **—com·mu′nal·is′tic** *adj.*

**com·mu·nal·ize** (kə-myo͞o′nə-līz′, kŏm′yə-nə-) *vt.* **-ized, -iz·ing, -iz·es.** To change into communal property.

**Com·mu·nard** (kŏm′yə-närd′) *n.* [Fr. < *commune,* division. —see COMMUNE².] **1.** A member or advocate of the 1871 Commune of Paris. **2. communard.** One who lives in a commune.

**com·mune¹** (kə-myo͞on′) *vi.* **-muned, -mun·ing, -munes.** [ME *communen* < OFr. *communier* < *commune,* common.] **1. a.** To exchange thoughts and feelings. **b.** To be in accord or agreement. **2.** To receive Communion.

**com·mune²** (kŏm′yo͞on′, kə-myo͞on′) *n.* [Fr. < Med. Lat. *communia,* community < Lat. *communis,* common.] **1.** The smallest local political unit of certain European countries, governed by a mayor and municipal council. **2. a.** A local community having a government that promotes local interests. **b.** A medieval municipal corporation. **3. a.** A small group of people, often in a rural area, whose members have common interests and share or own property jointly. **b.** The members of a commune.

**com·mu·ni·ca·ble** (kə-myo͞o′nĭ-kə-bəl) *adj.* **1.** Capable of being transmitted <*communicable* diseases> **2.** Communicative. **—com·mu′ni·ca·bil′i·ty, com·mu′ni·ca·ble·ness** *n.* **—com·mu′ni·ca·bly** *adv.*

**com·mu·ni·cant** (kə-myo͞o′nĭ-kənt) *n.* **1.** One who receives or may receive Communion. **2.** One who communicates. —*adj.* Communicating.

**com·mu·ni·cate** (kə-myo͞o′nĭ-kāt′) *v.* **-cat·ed, -cat·ing, -cates.** [Lat. *communicare, communicat-* < *communis,* common.] —*vt.* **1. a.** To make known : DISCLOSE <*communicate* information> **b.** To manifest : disclose. **2.** To transmit (a disease) to others. —*vi.* **1.** To have an interchange, as of ideas or information. **2.** To express oneself effectively. ***usage:*** In recent years the verb *communicate* has developed the meaning "to express oneself effectively." This new sense, which occurs in contexts such as *a person who lacked the ability to communicate,* is now well established. **3.** To receive Communion. **4.** To be connected <hotel rooms that *communicate*> **—com·mu′ni·ca′tor** *n.*

**com·mu·ni·ca·tion** (kə-myo͞o′nĭ-kā′shən) *n.* **1.** The act or process of communicating : TRANSMISSION. **2.** The exchange of ideas, messages, or information, as by speech, signals, or writing. **3.** Something communicated, as information : MESSAGE. **4. communications.** A means of communicating, esp.: **a.** A system for sending and receiving messages, as by mail, telephone, or television. **b.** A network of routes for sending messages and transporting troops and supplies. **5. communications.** The art and technology of communicating. **—com·mu′ni·ca′tion·al** *adj.*

**communications satellite** *n.* An artificial satellite that aids communications, as by reflecting or relaying a radio signal.

**com·mu·ni·ca·tive** (kə-myo͞o′nĭ-kā′tĭv, -kə-tĭv) *adj.* **1.** Tending to communicate openly and readily : TALKATIVE. **2.** Of or relating to communication. **—com·mu′ni·ca′tive·ly** *adv.* **—com·mu′ni·ca′tive·ness** *n.*

**com·mun·ion** (kə-myo͞on′yən) *n.* [ME *communioun,* Christian fellowship, Eucharist < LLat. *communio,* Eucharist < Lat., mutual participation < *communis,* common.] **1.** An act or instance of sharing or exchanging, as of ideas, opinions, or feelings. **2.** A religious or spiritual fellowship. **3.** A group of Christians with a common religous faith who practice the same rites : DENOMINATION. **4. Communion. a.** The Christian sacrament in which consecrated bread and wine are partaken of in celebration of Christ's Last Supper. **b.** The consecrated elements of this sacrament. **c.** The part of the Mass or other religious ritual in which Communion is received.

**com·mu·ni·qué** (kə-myo͞o′nĭ-kā′, -myo͞o′nĭ-kā′) *n.* [Fr. < *communiquer,* to announce < Lat. *communicare,* to communicate.] An official announcement or bulletin.

**com·mu·nism** (kŏm′yə-nĭz′əm) *n.* [Fr. *communisme* < *commun,* common < OFr. *commune.*] **1.** A social system marked by the common ownership of the means of production and common sharing of labor and products. **2. a.** A system of government in which the state controls the means of production and a single, usu. authoritarian party holds power with the intention of establishing a social order in which all goods are shared equally. **b.** The Marxist-Leninist version of Communist doctrine that is the basis for the system of government in the Soviet Union.

**Com·mu·nist** (kŏm′yə-nĭst) *n.* **1. a.** A member of a Marxist-Leninist party. **b.** A supporter of such a party or movement. **2. communist.** A communalist. **3.** A Communard. **4.** *often* **communist.** One who believes in or advocates communism. —*adj. also* **communist.** Relating to, typical of, or like communism or Communists.

**com·mu·nis·tic** (kŏm′yə-nĭs′tĭk) *adj.* Of, like, or inclined to communism. **—com′mu·nis′ti·cal·ly** *adv.*

**Communist Party** *n.* A Marxist-Leninist party.

**com·mu·ni·tar·i·an** (kə-myo͞o′nĭ-târ′ē-ən) *n.* A member or supporter of a communistic community.

**com·mu·ni·ty** (kə-myo͞o′nĭ-tē) *n., pl.* **-ties.** [ME *communite,* citizenry < OFr. *communite* < Lat. *communitas,* fellowship < *communis,* common.] **1. a.** A group of people residing in the same locality and under the same government. **b.** The area or locality in which such a group resides. **2.** A group or class having common interests <the academic *community*> **3.** Likeness or identity <*a community* of interests> **4.** Society as a whole. **5.** *Ecol.* **a.** A group of plants and animals living in a particular region under more or less similar conditions. **b.** The region in which such a group lives. **6.** Common ownership or participation.

**community antenna television** *n.* Cable television.

**community center** *n.* A meeting place used by members of a community for social, cultural, or recreational purposes.

**community chest** *n.* A fund financed by private contributions for aiding various charitable organizations.

**community college** *n.* A two-year college without residential facilities that is often government-funded.

**community property** *n.* Property owned jointly by a husband and wife.

**com·mu·nize** (kŏm'yə-nīz') *vt.* **-nized, -niz·ing, -niz·es.** [< Lat. *communis,* common.] **1.** To bring under public ownership or control. **2.** To convert to Communist principles or control. **—com'mu·ni·za'tion** *n.*

**com·mut·a·ble** (kə-myōō'tə-bəl) *adj.* Capable of being exchanged or substituted. **—com·mut'a·bil'i·ty** *n.*

**com·mu·tate** (kŏm'yə-tāt') *vt.* **-tat·ed, -tat·ing, -tates.** [Back-formation < COMMUTATION.] To reverse the direction of (an alternating electric current) each half-cycle to yield a unidirectional current.

**com·mu·ta·tion** (kŏm'yə-tā'shən) *n.* [ME *communtacioun* < Lat. *commutatio* < *commutare,* to commute.] **1.** A substitution or exchange. **2. a.** The substitution of one type of payment for another. **b.** The payment substituted. **3.** The travel of a commuter. **4.** *Elect.* **a.** The conversion of alternating to unidirectional current. **b.** The reversing of current direction. **5.** *Law.* A reduction of a penalty to a less severe one.

**commutation ticket** *n.* A ticket sold at a reduced rate by a railroad, bus, or other transportation company for travel over a particular route for a specified number of trips.

**com·mu·ta·tive** (kŏm'yə-tā'tĭv, kə-myōō'tə-tĭv) *adj.* **1.** Relating to, involving, or marked by substitution or exchange. **2.** Independent of order, as a logical or mathematical operation that combines objects two at a time. **—com·mu'ta·tiv'i·ty** (kə-myōō'tə-tĭv'ĭ-tē) *n.*

**commutative group** *n.* A mathematical group in which the result of multiplying one member by another is independent of the order of multiplication.

**com·mu·ta·tor** (kŏm'yə-tā'tər) *n.* A cylindrical arrangement of insulated metal bars connected to the coils of an electric motor or generator to provide a unidirectional current from the generator or a reversal of current into the coils of the motor.

**com·mute** (kə-myōōt') *v.* **-mut·ed, -mut·ing, -mutes.** [ME *commuten,* to change < Lat. *commutare* : *com-,* together + *mutare,* to change.] *—vt.* **1.** To substitute (one thing for another) : EXCHANGE. **2.** To reduce (a penalty, debt, or payment) to a less severe one. *—vi.* **1. a.** To make substitution : EXCHANGE. **b.** To serve as a substitute. **2.** To pay as a lump sum, usu. at a reduced rate, rather than in individual payments. **3.** To travel as a commuter. **4.** *Math. & Logic.* To satisfy or engage in a commutative operation. *—n.* **1.** An act or instance of commuting. **2.** *Informal.* The trip made by a commuter.

**com·mut·er** (kə-myōō'tər) *n.* One who travels regularly from one place to another, as from suburb to city and back.

**co·mose** (kō'mōs') *adj. var. of* COMATE[1]

**comp[1]** (kŏmp) *vi.* **comped, comp·ing, comps.** [Short for ACCOMPANY.] *Mus.* To play a jazz accompaniment, as on a piano or guitar.

**comp[2]** (kŏmp) *n.* [Short for COMPLIMENTARY.] *Informal.* Something given free, as a theater ticket.

**com·pact[1]** (kəm-păkt', kŏm-, kŏm'păkt') *adj.* [ME < Lat. *compactus,* p.part. of *compingere,* to put together : *com-,* together + *pangere,* to fasten.] **1.** Closely and firmly united or packed together : SOLID. **2.** Arranged within a relatively small space. **3.** Terse and meaningful : CONCISE <a *compact* recitation> *—vt.* (kəm-păkt') **-pact·ed, -pact·ing, -pacts. 1.** To put or join firmly together : CONSOLIDATE. **2.** To make by joining firmly together : COMPOSE. *—n.* (kŏm'păkt'). **1.** A small case containing face make-up and a mirror. **2.** A small automobile. **—com·pact'er** *n.* **—com·pact'ly** *adv.* **—com·pact'ness** *n.*

**com·pact[2]** (kŏm'păkt') *n.* [Lat. *compactum* < *compactus,* p.part. of *compacisci,* to make an agreement : *com-,* together + *pacisci,* to agree.] An agreement or covenant : CONTRACT.

**com·pac·tion** (kəm-păk'shən) *n.* The process of compacting or state of being compacted.

**com·pac·tor** (kəm-păk'tər, kŏm'păk'-) *n.* A device that compresses garbage into relatively small packages for easier disposal.

**†com·pa·dre** (kəm-pä'drā) *n.* [Sp., godfather < Med. Lat. *compater* : Lat. *com-,* with + Lat. *pater,* father.] *Southwestern U.S.* A close friend or associate.

**com·pan·ion[1]** (kəm-păn'yən) *n.* [ME *compaignoun* < OFr. *compaignon* < VLat. **companio* : *com-,* together + *panis,* bread.] **1.** One who accompanies or keeps company with another : COMRADE. **2.** One employed to aid, live with, or accompany another. **3.** One of a matched pair or set of things : MATE. *—vt.* **-ioned, -ion·ing, -ions.** To be with or go with another : ACCOMPANY.

**com·pan·ion[2]** (kəm-păn'yən) *n.* A companionway.

**com·pan·ion·a·ble** (kəm-păn'yə-nə-bəl) *adj.* **1.** Suited to be a good companion : SOCIABLE. **2.** Suggestive of companionship. **—com·pan'ion·a·ble·ness** *n.* **—com·pan'ion·a·bly** *adv.*

**companion cell** *n.* A specialized parenchyma cell, located in the phloem of angiosperms, that is closely allied in development and function with a cell that comprises the sieve tube.

**com·pan·ion·ship** (kəm-păn'yən-shĭp') *n.* The relationship of companions : FELLOWSHIP.

**com·pan·ion·way** (kəm-păn'yən-wā') *n.* [Obs. Du. *kompanje,* poop deck.] A staircase from a ship's deck to the cabins or area below.

**com·pa·ny** (kŭm'pə-nē) *n., pl.* **-nies.** [ME *compaignie* < OFr. < *compain,* companion < VLat. **companio.* —see COMPANION.] **1.** A group : assembly. **2. a.** One's friends or associates <moving in fast *company*> **b.** A guest or guests <had *company* for dinner> **c.** A pleasant association : FELLOWSHIP. **3. a.** A commercial enterprise : FIRM. **b.** A partner or partners not named in a firm's title <Smith and *Company*> **4.** A group of dramatic or musical performers. **5.** The lowest administrative subdivision of a regiment or battalion, usu. commanded by a captain. **6.** A ship's crew and officers. *—vt.* **-nied, -ny·ing, -nies.** To be with or go with. **—keep company.** To conduct courtship. **—keep (one) company.** To spend time with : ACCOMPANY. **—part company.** To end an association or friendship.

**com·pa·ra·ble** (kŏm'pər-ə-bəl) *adj.* **1.** Capable of being compared. **2.** Worthy of comparison. **3.** Like or equivalent <artists of *comparable* talent> *—pl.n.* **comparables.** Real properties that can be used to establish the value of a specific property by comparison. **—com'pa·ra·bil'i·ty, com'pa·ra·ble·ness** *n.* **—com'pa·ra·bly** *adv.*

**com·par·a·tist** (kəm-păr'ə-tĭst) *n.* One who employs the comparative method, as in linguistics.

**com·par·a·tive** (kəm-păr'ə-tĭv) *adj.* **1.** Of, based on, or involving comparison. **2.** Estimated by comparison : RELATIVE <*comparative* affluence> **3.** Indicating a degree of comparison of adjectives and adverbs higher than positive and lower than superlative. *—n.* **1.** The comparative degree of an adjective or an adverb. **2.** An adjective or adverb expressing the comparative degree. **—com·par'a·tive·ly** *adv.*

**com·pa·ra·tor** (kŏm'pə-rā'tər, kəm-păr'ə-) *n.* A device used to compare an aspect of an object, as its shape, color, or brightness, with a standard.

**com·pare** (kəm-pâr') *v.* **-pared, -par·ing, -pares.** [ME *comparen* < Lat *comparare* < *compar,* equal : *com-,* together + *par,* equal.] *—vt.* **1.** To represent as similar, equal, or analogous : LIKEN. **2.** To examine so as to note the similarities or differences of <*compared* the prices of new cars> **3.** To form the positive, comparative, or superlative degree of (an adjective or adverb). *—vi.* **1.** To be similar or equal. **2.** To draw comparisons. *—n.* Comparison <an artist beyond *compare*> **—com·par'er** *n.*

**com·par·i·son** (kəm-păr'ĭ-sən) *n.* [ME *comparisoun* < OFr. *comparaison* < Lat. *comparatio* < *comparare,* to compare. —see COMPARE.] **1. a.** The act of comparing or process of being compared. **b.** A statement or estimate of similarities and differences. **2.** The quality of being similar or equivalent : LIKENESS. **3.** The modification or inflection of an adjective or adverb to denote the positive, comparative, and superlative degrees.

**comparison-shop** (kəm-păr'ĭ-sən-shŏp') *vi.* **-shopped, -shop·ping, -shops.** To look for bargains by comparing the prices of competing brands or stores.

**com·part** (kəm-pärt') *vt.* **-part·ed, -part·ing, -parts.** [Ult. < LLat. *compartiri,* to share : Lat. *com-,* with + Lat. *partiri,* to divide < *pars,* a part.] To separate into compartments or parts : PARTITION.

**com·part·ment** (kəm-pärt'mənt) *n.* **1.** One of the sections or spaces into which an area is subdivided. **2.** A separate room, section, or chamber. **—com'part·men'tal** (kŏm'pärt-mĕn'tl) *adj.*

**com·part·men·tal·ize** (kŏm'pärt-mĕn'tl-īz', kəm-pärt'-) *vt.* **-ized, -iz·ing, -iz·es.** To separate or partition into compartments or categories. **—com'part·men'tal·i·za'tion** *n.*

**com·pass** (kŭm'pəs, kŏm'-) *n.* [ME *compas,* compasses < OFr. < *compasser,* to measure < VLat. **compassare* : Lat. *com-,* together + Lat. *passus,* step < *pandere,* to stretch.] **1. a.** A device used to discover geographic direction, usu. having a magnetic needle or needles that is horizontally mounted or suspended and free to pivot until aligned with the magnetic field of the earth. **b.** Any of various other devices used to determine geographic direction, as a gyrocompass. **c.** *often* **compasses.** A V-shaped device for drawing circles or circular arcs, having a pair of rigid, end-hinged, and continuously separable arms, one of which is equipped with a writing implement and the other with a sharp point providing a central anchor or pivot about which the drawing arm is turned. **2. a.** A line around a closed space or area : CIRCUMFERENCE. **b.** An enclosed space or area. **c.** A range or scope : EXTENT <beyond the *compass* of my knowledge> *—vt.* **-passed, -pass·ing, -pass·es. 1.** To go around : CIRCLE <*compassed* the globe> **2.** To shut in on all sides : ENCIRCLE. **3.** To understand the meaning of : COMPREHEND. **4.** To achieve : gain. **5.** To scheme : plot. *—adj.* Circular : round. **—com'pass·a·ble** *adj.*

**compass card** *n.* A freely pivoting disk carrying the magnetic needles of a compass and marked with the 32 points of the compass and the 360° of the circle.

**com·pas·sion** (kəm-păsh'ən) *n.* [ME *compassioun* < LLat. *compassio* < *compati,* to sympathize : Lat. *com-,* with + Lat. *pati,* to suffer.] Sympathetic concern for the suffering of another, together with the inclination to give aid or support or to show mercy.

**com·pas·sion·ate** (kəm-păsh'ə-nĭt) *adj.* Having or displaying compassion : SYMPATHETIC. —*vt.* (-nāt') **-at·ed, -at·ing, -ates.** To have or display compassion for : PITY. **—com·pas'sion·ate·ly** *adv.* **—com·pas'sion·ate·ness** *n.*

**compass plant** *n.* **1.** A tall plant, *Silphium laciniatum* of central North America, bearing yellow flowers and lower leaves that tend to align in a north-south plane. **2.** Any of several similar plants.

**com·pat·i·ble** (kəm-păt'ə-bəl) *adj.* [ME < Med. Lat. *compatibilis* < LLat. *compati,* to sympathize. —see COMPASSION.] **1.** Capable of living or performing in harmonious, agreeable, or friendly association with another or others. **2.** Capable of orderly, efficient integration and operation with other elements in a system <*compatible* software components> **3.** Capable of forming a chemically or biochemically stable system. **4.** Designating a television system in which color broadcasts can be received in black and white by sets incapable of color reception. **—com·pat'i·bil'i·ty, com·pat'i·ble·ness** *n.* **—com·pat'i·bly** *adv.*

**com·pa·tri·ot** (kəm-pā'trē-ət, -ŏt') *n.* [Fr. *compatriote* < LLat. *compatriota* : Lat. *com-,* with + *patriota,* countryman. —see PATRIOT.] **1.** A fellow countryman. **2.** *Informal.* A colleague or associate. **—com·pa'tri·ot'ic** (-ŏt'ĭk) *adj.*

**com·peer** (kŏm'pîr', kəm-pîr') *n.* [ME *comper* < OFr. < Lat. *compar,* equal. —see COMPARE.] **1.** One of equal position or rank : PEER. **2.** A comrade, companion, or associate.

**com·pel** (kəm-pĕl') *vt.* **-pelled, -pel·ling, -pels.** [ME *compellen* < Lat. *compellere* : *com-,* together + *pellere,* to drive.] **1.** To force, drive, or constrain. **2.** To make necessary : EXACT. **—com·pel'la·ble** *adj.* **—com·pel'la·bly** *adv.* **—com·pel'ler** *n.*

**com·pel·la·tion** (kŏm'pə-lā'shən, -pĕl-ā'-) *n.* [Lat. *compellatio* < *compellare,* to address.] **1.** An act of addressing or designating someone by name. **2.** A name : appellation.

**com·pend** (kŏm'pĕnd') *n.* A compendium.

**com·pen·di·ous** (kəm-pĕn'dē-əs) *adj.* [ME < Med. Lat. *compendiosus* < Lat. *compedium,* a shortening. —see COMPENDIUM.] Having or expressing briefly and concisely all the essentials : SUCCINCT. **—com·pen'di·ous·ly** *adv.* **—com·pen'di·ous·ness** *n.*

**com·pen·di·um** (kəm-pĕn'dē-əm) *n., pl.* **-di·ums** *or* **-di·a** (-dē-ə) [Lat., a shortening < *compendere,* to weigh together : *com-,* together + *pendere,* to weigh.] A short detailed summary : ABSTRACT.

**com·pen·sa·ble** (kəm-pĕn'sə-bəl) *adj.* **1.** Entitled to compensation. **2.** Capable of being compensated.

**com·pen·sate** (kŏm'pən-sāt') *v.* **-sat·ed, -sat·ing, -sates.** [Lat. *compensare, compensat-* : *com-,* together + *pensare,* freq. of *pendere,* to weigh.] —*vt.* **1.** To make up for or offset : COUNTERBALANCE. **2.** To make payment or reparation to : REIMBURSE. **3.** To stabilize the purchasing power of (a monetary unit) by changing the gold content in order to counterbalance price variations. —*vi.* To provide with or act as a substitute or counterbalance. **—com'pen·sa'tive** (-sā'tĭv, kəm-pĕn'sə-tĭv) *adj.* **—com'pen·sa'tor** *n.* **—com·pen'sa·to'ry** (kəm-pĕn'sə-tôr'ē, -tōr'ē) *adj.*

**com·pen·sa·tion** (kŏm'pən-sā'shən) *n.* **1.** The act or an instance of compensating. **2.** Something given or received as payment or reparation, as for goods, services, or loss. **3.** *Biol.* The counterbalancing of a defect by the development and activation of another organ or another part of the defective structure. **4.** *Psychol.* Behavior designed to offset real or imagined defects. **—com'pen·sa'tion·al** *adj.*

**com·pete** (kəm-pēt') *vi.* **-pet·ed, -pet·ing, -petes.** [LLat. *competere,* to strive together : Lat. *com-,* together + Lat. *petere,* to strive.] To strive or contend, as for profit or a prize : VIE.

**com·pe·tence** (kŏm'pĭ-təns) *n.* **1.** The quality or state of being competent. **2.** Sufficient resources for a comfortable existence. **3.** *Law.* Legal qualification, eligibility, or admissibility. **4.** *Genetics.* The ability of bacteria to be genetically transformable.

**com·pe·ten·cy** (kŏm'pĭ-tən-se) *n., pl.* **-cies.** Competence.

**com·pe·tent** (kŏm'pĭ-tənt) *adj.* [ME, adequate < Lat. *competens,* pr.part. of *competere,* to be suitable : *com-,* together + *petere,* to seek.] **1.** Properly qualified : CAPABLE. **2.** Adequate for the stipulated purpose : SUFFICIENT <*competent* performance> **3.** *Law.* Legally fit or qualified : ADMISSIBLE. **—com'pe·tent·ly** *adv.*

**com·pe·ti·tion** (kŏm'pĭ-tĭsh'ən) *n.* **1.** The act of competing, as for profit or a prize : RIVALRY. **2.** A test of skill or ability : CONTEST <a gymnastic *competition*> **3.** Rivalry between two or more businesses striving for the same customers or market. **4.** One's competitor or competitors.

**com·pet·i·tive** (kəm-pĕt'ĭ-tĭv) *adj.* **1.** Of, involving, or determined by competition. **2.** Liking or tending to compete. **—com·pet'i·tive·ly** *adv.* **—com·pet'i·tive·ness** *n.*

**com·pet·i·tor** (kəm-pĕt'ĭ-tər) *n.* One that competes against another, as in sports or business : RIVAL.

**com·pet·i·to·ry** (kəm-pĕt'ĭ-tôr'ē, -tōr'ē) *adj.* Competitive.

**com·pi·la·tion** (kŏm'pə-lā'shən) *n.* **1.** The act of compiling. **2.** Matter compiled <a *compilation* of essays>

**com·pile** (kəm-pīl') *vt.* **-piled, -pil·ing, -piles.** [ME *compilen* < OFr. *compiler,* prob. < Lat. *compilare,* to plunder.] **1.** To gather into a single book. **2.** To put together from materials gathered from various sources <*compile* an encyclopedia of music> **3.** *Computer Sci.* To convert to machine language.

**com·pil·er** (kəm-pī'lər) *n.* **1.** One that compiles. **2.** *Computer Sci.* A program that converts a higher level language to machine language.

**com·pla·cen·cy** (kəm-plā'sən-sē) *also* **com·pla·cence** (-səns) *n.* **1.** Satisfaction or contentment : GRATIFICATION. **2.** Smug self-satisfaction.

**com·pla·cent** (kəm-plā'sənt) *adj.* [Lat. *complacens, complacent-,* pr.part. of *complacēre,* to please : *com-* (intensive) + *placēre,* to please.] **1.** Overcontented. **2.** Complaisant. **—com·pla'cent·ly** *adv.*

**com·plain** (kəm-plān') *vi.* **-plained, -plain·ing, -plains.** [ME *compleinen* < OFr. *complaindre* < VLat. **complangere* : Lat. *com-* (intensive) + Lat. *plangere,* to lament.] **1.** To express feelings of dissatisfaction, resentment, or pain. **2.** To make a formal accusation or bring a formal charge. **—com·plain'er** *n.*

☆ ***syns:*** COMPLAIN, BEEF, BELLYACHE, GRIPE, GROUSE, KICK, WHINE *v. core meaning* : to express dissatisfaction or resentment <a staff that was always *complaining* about low salaries>

**com·plain·ant** (kəm-plā'nənt) *n.* One that makes a complaint or files a formal charge, as in a court of law : PLAINTIFF.

**com·plaint** (kəm-plānt') *n.* [ME *compleinte* < OFr. *complainte* < *complaindre,* to complain. —see COMPLAIN.] **1.** Expression of dissatisfaction, resentment, or pain. **2.** A cause for complaining : GRIEVANCE. **3.** A cause of physical pain : ILLNESS. **4.** *Law.* The presentation by the plaintiff in a civil action, setting forth the claim on which relief is sought.

**com·plai·sance** (kəm-plā'səns, -zəns) *n.* Inclination to comply willingly with the wishes of others.

**com·plai·sant** (kəm-plā'sənt, -zənt) *adj.* [Fr. < OFr., pr.part. of *complaire,* to please < Lat. *complacēre.* —see COMPLACENT.] Desiring or willing to please. **—com·plai'sant·ly** *adv.*

**com·pleat** (kəm-plēt') *adj.* [Var. of COMPLETE.] Of or marked by a highly developed, wide-ranging proficiency or skill <was the *compleat* chess player, unparalleled in every facet of strategy>

**com·plect** (kəm-plĕkt') *vt.* **-plect·ed, -plect·ing, -plects.** [Lat. *complecti,* to entwine : *com-,* together + *plectere,* to plait.] To join by weaving or twining together.

**†com·plect·ed** (kəm-plĕk'tĭd) *adj.* [< COMPLEXION.] *Regional.* Marked by or having a particular facial complexion <dark-*complected*>

**com·ple·ment** (kŏm'plə-mənt) *n.* [ME < Lat. *complementum* < *complere,* to fill out. —see COMPLETE.] **1. a.** Something that completes, makes up a whole, or brings to perfection. ***usage:*** *Complement* means "something that completes or brings to perfection": *The wine was a complement to the delicious dinner. Compliment* means "an expression of courtesy or praise": *You have paid me a great compliment.* **b.** The number or quantity required to make up a whole. **c.** Either of two parts that complete the whole or mutually complete each other. **2.** An angle related to another so that the sum of their measures is 90°. **3.** A word or words used after a verb to complete a predicate construction. **4.** *Mus.* An interval that completes an octave when added to a given interval. **5.** The crew needed to operate a ship. **6.** *Biochem.* The thermolabile substance found in normal blood serum that destroys pathogens. —*vt.* (-mĕnt') **-ment·ed, -ment·ing, -ments.** To add or serve as a complement to.

**com·ple·men·tal** (kŏm'plə-mĕn'tl) *adj.* Being complementary. **—com'ple·men'tal·ly** *adv.*

**com·ple·men·ta·ry** (kŏm'plə-mĕn'tə-rē, -trē) *adj.* **1.** Forming or serving as a complement : COMPLETING. **2.** Supplying mutual needs or deficits. **3.** Producing effects in concert different from those produced separately. —Used of genes. **4.** Of or relating to the specific pairing of the purines and pyrimidines between the two strands of the DNA molecule. **—com'ple·men'ta·ri·ly** (-tə-rə-lē, -trə-lē, -mĕn-târ'ə-lē) *adv.* **—com'ple·men'ta·ri·ness** *n.* **—com'ple·men·tar'i·ty** (-tăr'ĭ-tē) *n.*

**complementary angles** *pl.n.* Two angles whose sum is 90°.

**complementary color** *n.* One of two colors that appears white or gray when ideally mixed in proper proportions, as the combination of blue-green with red.

**complement fixation** *n.* The joining of a complement to the antigen-antibody pair for which it is specific.

**com·plete** (kəm-plēt') *adj.* **-plet·er, -plet·est.** [ME < Lat. *completus,* p.part. of *complēre,* to fill out : *com-* (intensive) + *plere,* to fill.] **1.** Having all necessary or normal parts, elements, or steps : WHOLE. **2.** *Bot.* Having all typical floral parts, including sepals, petals, stamens, and a pistil. **3.** Having come to an end. **4.** Thorough <*complete* mastery> **5.** Skilled : accomplished <a *complete* musician> —*vt.* **-plet·ed, -plet·ing, -pletes. 1.** To make complete. **2.** To bring to an end : CONCLUDE. **—com·plete'ly** *adv.* **—com·plete'ness** *n.* **—com·ple'tion** (-plē'shən) *n.* **—com·ple'tive** *adj.*

☆ ***syns:*** COMPLETE, ENTIRE, FULL, INTACT, INTEGRAL, PERFECT, WHOLE *adj. core meaning* : lacking nothing essential <a *complete* marketing strategy> ***ant:*** incomplete

**complete blood count** *n.* Determination of the quantity of each type of blood cell in circulation by extrapolation of the numbers found in a given blood sample.

**com·plex** (kəm-plĕks′, kŏm′plĕks′) *adj.* [Lat. *complexus,* p.part. of *complecti,* to entwine. —see COMPLECT.] **1.** Composed of interconnected or interwoven parts : COMPOSITE. **2.** Involved or intricate, as in structure. **3. a.** Relating to or designating a word, as *slightly,* having at least one bound form. **b.** Relating to or designating a sentence made up of an independent clause and one or more dependent clauses. —*n.* (kŏm′plĕks′). **1.** A whole composed of intricate or interconnected parts <a *complex* of very narrow, twisting streets> **2.** *Psychiat.* A connected group of repressed ideas that compel characteristic or habitual patterns of thought, feeling, and action. **3.** *Informal.* An exaggerated or obsessive concern or fear <had a *complex* about flying> **—com·plex′i·ty** *n.* **—com·plex′ly** *adv.* **—com·plex′ness** *n.*

☆ ***syns:*** COMPLEX, BYZANTINE, COMPLICATED, CONVOLUTED, DAEDAL, ELABORATE, INTRICATE, INVOLUTE, INVOLVED, KNOTTY, LABYRINTHINE *adj. core meaning* : difficult to understand due to intricacy <inflation—a *complex* economic problem> ***ant:*** simple

**complex conjugate** *n.* **1.** A complex number that when multiplied by a given complex number yields a real number. **2.** A complex quantity that when multiplied by a given complex quantity yields a product free of imaginary terms.

**complex fraction** *n.* A fraction in which the numerator or denominator or both contain fractions.

**com·plex·ion** (kəm-plĕk′shən) *n.* [ME *complexioun,* physical constitution < LLat. *complexio* < Lat., combination < *complexus,* p.part. of *complecti,* to entwine. —see COMPLECT.] **1.** The natural color, texture, and appearance of the skin, esp. of the face. **2.** General character, aspect, or appearance <new evidence that alters the *complexion* of the case> **3.** The combination of the four humors of cold, heat, moistness, and dryness in specific proportions believed in medieval physiology to control the temperament and constitution of the body. **—com·plex′ion·al** *adj.*

**com·plex·ioned** (kəm-plĕk′shənd) *adj.* Of or having a specified complexion <light-*complexioned*>

**complex number** *n.* A number of the form $a + bi$, where $a$ and $b$ are real numbers and $i^2 = -1$; a member of the set of ordered pairs $(a,b)$ of real numbers in which a pair equals another pair if and only if corresponding members of the two pairs are identical and in which addition and multiplication are defined by $(a,b)+(c,d)=(a+c, b+d)$ and $(a,b)(c,d)=(ac-bd, ad+bc)$, respectively.

**complex plane** *n.* A plane that has complex numbers as its points.

**complex variable** *n.* An expression of the form $x+iy$, where $x$ and $y$ are real variables and $i^2=-1$.

**com·pli·ance** (kəm-plī′əns) *also* **com·pli·an·cy** (-ən-sē) *n.* **1.** Acquiescence to a wish, request, or demand. **2.** A disposition or tendency to yield to the will of others. **3. a.** Extension or displacement of a loaded structure per unit load. **b.** Flexibility.

**com·pli·ant** (kəm-plī′ənt) *adj.* Inclined or disposed to comply : ACQUIESCENT. **—com·pli′ant·ly** *adv.*

**com·pli·ca·cy** (kŏm′plĭ-kə-sē) *n., pl.* **-cies. 1.** The state of being complicated. **2.** A complication.

**com·pli·cate** (kŏm′plĭ-kāt′) *vt.* & *vi.* **-cat·ed, -cat·ing, -cates.** [Lat. *complicare, complicat-,* to fold up : *com-,* together + *plicare,* to fold.] **1.** To make or become complex, intricate, or bewildering. **2.** To twist or become twisted together. —*adj.* (-kĭt). **1.** Complex : intricate. **2.** *Biol.* Folded longitudinally one or several times, as certain leaves or the wings of some insects.

**com·pli·cat·ed** (kŏm′plĭ-kā′tĭd) *adj.* **1.** Containing intricately combined or involved parts. **2.** Difficult to understand or treat. **—com′pli·cat′ed·ly** *adv.* **—com′pli·cat′ed·ness** *n.*

**com·pli·ca·tion** (kŏm′plĭ-kā′shən) *n.* **1.** The act of complicating. **2.** A confused or intricate relationship. **3.** A complicating factor, condition, or element. **4.** *Med.* A condition occurring during another disease and aggravating it.

**com·plice** (kŏm′plĭs) *n.* [ME < OFr. < LLat. *complex,* closely connected : Lat. *com-,* together + Lat. *plicare,* to fold.] *Obs.* An associate or accomplice.

**com·plic·i·ty** (kəm-plĭs′ĭ-tē) *n., pl.* **-ties.** Involvement as an accomplice in a crime or wrongdoing.

**com·pli·er** (kəm-plī′ər) *n.* One that complies.

**com·pli·ment** (kŏm′plə-mənt) *n.* [Fr. < Sp. *cumplimento* < *cumplir,* to complete < Lat. *complēre,* to complete.] **1.** An expression of praise, admiration, or congratulation. **2.** A formal act of civility, courtesy, or respect. **3. compliments.** Good wishes : REGARDS <Extend my *compliments* to the host and hostess.> —*vt.* **-ment·ed, -ment·ing, -ments. 1.** To pay a compliment to. **2.** To show fondness, regard, or respect for by giving a gift or performing a favor.

**com·pli·men·ta·ry** (kŏm′plə-mĕn′tə-rē, -trē) *adj.* **1.** Expressing, using, or resembling a compliment <*complimentary* reviews of the concert> **2.** Given free to repay a favor or as an act of courtesy <*complimentary* copies of the new novel> **—com′pli·men′ta·ri·ly** *adv.*

**com·plin** (kŏm′plĭn) *also* **com·pline** (-plĭn, -plīn′) *n.* [ME < OFr. *complie* < Med. Lat. (*hora*) *completa,* final hour < Lat. *completus,* p.part. of *complēre,* to complete.] The last of the seven canonical hours.

**com·ply** (kəm-plī′) *vi.* **-plied, -ply·ing, -plies.** [Ital. *complire* < Sp. *cumplir,* to complete < Lat. *complēre.*] **1.** To act in accord with another's request, command, rule, or wish. **2.** *Obs.* To be courteous or obedient.

**com·po** (kŏm′pō) *n., pl.* **-pos.** [Short for COMPOSITION.] A combined substance, as mortar or plaster, formed by mixing ingredients.

**com·po·nent** (kəm-pō′nənt) *n.* [Lat. *componens, component-,* pr.part. of *componere,* to put together : *com-,* together + *ponere,* to put.] **1.** A constituent element, as of a system. **2.** A part of a mechanical or electrical complex. **3.** *Math.* One of a set of vectors having a sum equal to a given vector. **4.** *Chem.* Any of the minimum number of substances required to specify completely the composition of all phases of a chemical system. **—com·po′nent** *adj.* **—com′po·nen′tial** (kŏm′pə-nĕn′shəl) *adj.*

**com·port** (kəm-pôrt′, -pōrt′) *v.* **-port·ed, -port·ing, -ports.** [OFr. *comporter,* to conduct < Lat. *comportare,* to bring together : *com-,* together + *portare,* to carry.] —*vt.* To conduct or behave (oneself) in a given way <*comport* oneself with tact> —*vi.* To be appropriate : HARMONIZE <actions that *comport* with fair play>

**com·port·ment** (kəm-pôrt′mənt, -pōrt′-) *n.* Deportment : bearing.

**com·pose** (kəm-pōz′) *v.* **-posed, -pos·ing, -pos·es.** [ME *composen* < OFr. *composer* < Lat. *componere.* —see COMPONENT.] —*vt.* **1.** To make up the constituent parts of : CONSTITUTE <a collection *composed* of rare antiques> **2.** To create by putting together. **3.** To create or produce (a literary or musical piece). **4.** To calm (one's mind or body) : QUIET. **5.** To reconcile : settle <*composed* their contentions> **6.** To arrange artistically. **7.** To set or arrange (type or matter to be printed). —*vi.* **1.** To create literary or musical pieces. **2.** To set printing type.

☆ ***syns:*** COMPOSE, CREATE, PRODUCE *v. core meaning* : to form by artistic effort <*composed* a piano sonata>

**com·posed** (kəm-pōzd′) *adj.* Self-possessed : tranquil. **—com·pos′ed·ly** (-pō′zĭd-lē) *adv.* **—com·pos′ed·ness** *n.*

**com·pos·er** (kəm-pō′zər) *n.* One who composes, esp. music.

**composing room** *n.* A room for setting type.

**composing stick** *n.* A tray with an adjustable end used by a compositor to hand-set type.

**com·pos·ite** (kəm-pŏz′ĭt) *adj.* [Lat. *compositus,* p.part. of *componere,* to put together. —see COMPONENT.] **1.** Made up of distinct components : COMPOUND. **2.** *Math.* Having factors : FACTORABLE. **3.** *Bot.* Of, belonging to, or typical of the Compositae, a large plant family marked by flower heads having both ray flowers and disk flowers, as in the daisy, of disk flowers only, as in wormwood, or of ray flowers only, as in the dandelion. **4. Composite.** Relating to or designating the Composite order. —*n.* **1.** A compound. **2.** A composite plant. **—com·pos′ite·ly** *adv.* **—com·pos′ite·ness** *n.*

**composite nerve** *n.* A nerve consisting of both sensory and motor fibers.

**composite number** *n.* An integer exactly divisible by at least one number other than itself or 1.

**Composite order** *n.* A Roman capital formed by superimposing Ionic volutes on a Corinthian capital.

**Composite order**

**composite photograph** *n.* A photograph made by combining two or more separate photographs.

**com·po·si·tion** (kŏm′pə-zĭsh′ən) *n.* [ME *composicioun* < Lat. *compositio* < *componere,* to put together. —see COMPONENT.] **1. a.** The act of putting together parts or elements to form a whole. **b.** The way in which such parts are combined or related : CONSTITUTION. **c.** The product of composing : COMPOUND. **2.** The uniform arrangement of artistic parts. **3. a.** The art or act of composing a literary or musical work. **b.** A work of art, literature, or music or its structure or organization. **4.** A short essay, esp. one written as a school exercise. **5.** Mutual agreement or compromised settlement. **6.** The formation of compounds from separate words. **8.** Typesetting. **—com′po·si′tion·al** *adj.* **—com′po·si′tion·al·ly** *adv.*

**composition of forces** *n.* The finding or determination of a vector that is the resultant of a given set of forces.

**com·pos·i·tor** (kəm-pŏz′ĭ-tər) *n.* A typesetter. **—com·pos′i·to′ri·al** (-tôr′ē-əl, -tōr′-) *adj.*

**com·pos men·tis** (kŏm′pəs mĕn′tĭs) *adj.* [Lat., having mastery of the mind.] Of sound mind.

**com·post** (kŏm′pōst′) *n.* [ME, compote < OFr. —see COMPOTE.] **1.** A mixture of decayed organic matter used as fertilizer. **2.** A compound : mixture. —*vt.* **-post·ed, -post·ing, -posts. 1.** To fertilize with compost. **2.** To convert (vegetable matter) to compost.

**com·po·sure** (kəm-pō′zhər) *n.* [< COMPOSE.] Tranquillity of mind.

**com·pote** (kŏm′pōt) *n.* [Fr. < OFr. *composte* < Lat. *compositus*, put together. —see COMPOSITE.] **1.** Fruit cooked or stewed in syrup. **2.** A long-stemmed dish used for holding fruit, nuts, or candy.

**com·pound¹** (kŏm-pound′, kəm-, kŏm′pound′) *v.* **-pound·ed, -pound·ing, -pounds.** [ME *compounen* < Lat. *componere.* —see COMPONENT.] —*vt.* **1.** To combine so as to form a whole : MIX. **2.** To produce or create by combining parts <a druggist *compounding* medicine> **3.** To resolve (e.g., a debt) by settling on an amount less than the claim : ADJUST. **4.** To compute (interest) on the principal and accrued interest. **5.** *Law.* To agree, for a consideration, not to prosecute <*compound* a misdemeanor> **6.** To increase : augment <Fog *compounded* the difficulties of the search party.> —*vi.* **1.** To combine in or form a compound. **2.** To come to terms : AGREE. —*adj.* (kŏm′pound′, kŏm-pound′, kəm-). Comprised of two or more substances or parts. —*n.* (kŏm′pound′). **1.** A combination of two or more elements or parts. **2.** A word, as *loudspeaker* or *baby-sitting*, containing two or more elements having perceptible lexical meaning. **3.** *Chem.* A pure, macroscopically homogeneous substance consisting of atoms or ions of two or more different elements in definite proportions and usu. with properties unlike those of its constituent elements. **—com·pound′a·ble** *adj.* **—com·pound′er** *n.*

**com·pound²** (kŏm′pound′) *n.* [Malay *kampong*, village.] **1.** A group of buildings, esp. residences, enclosed by a fence or wall. **2.** An area for confining prisoners of war.

**com·pound-com·plex sentence** (kŏm′pound-kŏm′plĕks′) *n.* A sentence composed of at least two coordinate independent clauses and one or more dependent clauses.

**compound eye** *n.* The eye of most insects and some crustaceans, composed of many light-sensitive elements, each with its own refractive system and each forming a portion of an image.

**compound flower** *n.* A flower head of a composite plant, consisting of numerous small flowers appearing as a single bloom.

**compound fraction** *n.* A complex fraction.

**compound fracture** *n.* A fracture in which broken bone lacerates soft tissue and usu. protrudes through the skin.

**compound interest** *n.* Interest computed on the accumulated unpaid interest as well as on the original principal.

**compound leaf** *n.* A leaf having two or more separate leaflets borne on a single leafstalk.

**compound lens** *n.* LENS 1b.

**compound microscope** *n.* A microscope with an eyepiece and an objective at opposite ends of an adjustable tube.

**compound number** *n.* A quantity, as 5 pounds 10 ounces or 4 feet 3 inches, involving different units of measure.

**compound sentence** *n.* A sentence of two or more coordinate independent clauses, often joined by a conjunction or conjunctions; e.g., *The climb was arduous, but we finally reached the top.*

**com·pra·dor** *also* **com·pra·dore** (kŏm′prə-dôr′) *n.* [Port. < LLat. *comparator*, buyer < Lat. *comparare*, to buy : *com-*, together + *parare*, to get.] A native agent once employed by a foreign business in China and certain other Asian countries to serve as an intermediary in commercial transactions.

**com·pre·hend** (kŏm′prĭ-hĕnd′) *vt.* **-hend·ed, -hend·ing, -hends.** [ME *comprehenden* < Lat. *comprehendere* : *com-*, together + *prehendere*, to grasp.] **1.** To grasp mentally : UNDERSTAND. **2.** To include or embrace : COMPRISE. **—com′pre·hend′i·ble** *adj.*

**com·pre·hen·si·ble** (kŏm′prĭ-hĕn′sə-bəl) *adj.* [Lat. *comprehensibilis* < *comprehensus*, p.part. of *comprehendere*, to comprehend.] Capable of being understood : INTELLIGIBLE. **—com′pre·hen′si·bil′i·ty, com′pre·hen′si·ble·ness** *n.* **—com′pre·hen′si·bly** *adv.*

**com·pre·hen·sion** (kŏm′prĭ-hĕn′shən) *n.* [ME *comprehensioun* < Lat. *comprehensio* < *comprehensus*, p.part. of *comprehendere*, to comprehend.] **1. a.** The act or fact of comprehending : UNDERSTANDING. **b.** Knowledge acquired by comprehending. **2.** The capacity to include : COMPREHENSIVENESS. **3.** CONNOTATION 3.

**com·pre·hen·sive** (kŏm′prĭ-hĕn′sĭv) *adj.* [Lat. *comprehensivus* < *comprehensus*, p.part. of *comprehendere*, to comprehend.] **1.** Broad in scope or content <a *comprehensive* analysis of the problem> **2.** Marked by or showing extensive understanding <*comprehensive* insight> —*n. often* **comprehensives.** *Informal.* Examinations covering the entire field of major study, given in the final undergraduate or graduate year of college. **—com′pre·hen′sive·ly** *adv.* **—com′pre·hen′sive·ness** *n.*

**com·press** (kəm-prĕs′) *vt.* **-pressed, -press·ing, -press·es.** [ME *compressen* < LLat. *compressare*, freq. of Lat. *comprimere* : *com-*, together + *premere*, to press.] **1.** To squeeze or press together. **2.** To make smaller as if by squeezing. —*n.* (kŏm′prĕs′). **1.** *Med.* A soft pad of material, as gauze, applied to press against a part of the body. **2.** A device for compressing.

**com·pressed** (kəm-prĕst′) *adj.* **1.** Pressed together : COMPACTED. **2.** *Biol.* Flattened lengthwise or laterally, as certain seed pods or the bodies of many fish.

**compressed air** *n.* Air under greater than atmospheric pressure.

**com·press·i·ble** (kəm-prĕs′ə-bəl) *adj.* Capable of being compressed. **—com·press′i·bil′i·ty, com·press′i·ble·ness** *n.*

**com·pres·sion** (kəm-prĕsh′ən) *n.* **1.** The act or process of compressing or state of being compressed. **2. a.** The process by which the vaporized fuel mixture in the cylinder of an internal-combustion engine is compressed. **b.** The engine cycle during which this process occurs. **—com·pres′sion·al** *adj.*

**compression wave** *or* **compressional wave** *n.* A wave, as of sound, propagated by the compression of an elastic medium.

**com·pres·sive** (kəm-prĕs′ĭv) *adj.* Compressing or capable of compressing. **—com·pres′sive·ly** *adv.*

**com·pres·sor** (kəm-prĕs′ər) *n.* One that compresses, esp. a machine used to compress gases.

**com·prise** (kəm-prīz′) *vt.* **-prised, -pris·ing, -pris·es.** [ME *comprisen* < *comprised*, included < OFr. *compris*, p.part. of *comprendre* < Lat. *comprehendere.* —see COMPREHEND.] **1.** To be composed of : consist of. **2.** To include : contain. ***usage:*** The traditional rule states that the whole *comprises* the parts; the parts *compose* the whole. *Comprise*, however, is increasingly used in place of *compose*: *The Union is comprised of 50 states.* This last example is considered unacceptable by many. **—com·pris′a·ble** *adj.*

**com·pro·mise** (kŏm′prə-mīz′) *n.* [ME *compromis* < OFr. < Lat. *compromissum*, mutual promise < *compromittere*, to promise mutually : *com-*, together + *promittere*, to promise. —see PROMISE.] **1. a.** A settlement of differences in which each side makes concessions. **b.** Something resulting from such a settlement. **2.** Something combining qualities of two different things. **3.** A concession to something that is harmful or depreciative <a *compromise* of allegiance> —*v.* **-mised, -mis·ing, -mis·es.** —*vt.* **1.** To settle by concessions. **2.** To expose to suspicion, danger, or disrepute. **3.** *Obs.* To pledge mutually. —*vi.* To make a compromise. **—com′pro·mis′er** *n.*

**Compton effect** (kŏmp′tən) *n.* [After A.H. *Compton* (1892–1962), its discoverer.] The increase in wavelength of electromagnetic radiation, esp. of an x-ray or gamma-ray photon, scattered by an electron.

**comp·trol·ler** (kən-trō′lər) *n. var. of* CONTROLLER 2.

**com·pul·sion** (kəm-pŭl′shən) *n.* [ME < LLat. *compulsio* < Lat. *compellere*, to compel. —see COMPEL.] **1.** The act of compelling or state of being compelled. **2.** *Psychol.* An irresistible impulse to act irrationally.

**com·pul·sive** (kəm-pŭl′sĭv) *adj.* **1.** Having the capacity to compel. **2.** *Psychol.* Caused by obsession or compulsion. **—com·pul′sive·ly** *adv.* **—com·pul′sive·ness** *n.*

**com·pul·so·ry** (kəm-pŭl′sə-rē) *adj.* **1.** Exerting or employing compulsion : COERCIVE. **2.** Required : mandatory. **—com·pul′so·ri·ly** *adv.* **—com·pul′so·ri·ness** *n.*

☆ ***syns:*** COMPULSORY, IMPERATIVE, MANDATORY, NECESSARY, OBLIGATORY, REQUIRED, REQUISITE *adj. core meaning* : imposed on one by authority, command, or convention <*compulsory* attendance><*compulsory* draft registration> ***ant:*** voluntary

**com·punc·tion** (kəm-pŭngk′shən) *n.* [ME *compunccioun* < Med. Lat. *compunctio*, sting of conscience < Lat., puncture < *compungere*, to sting : *com-* (intensive) + *pungere*, to prick.] **1.** A strong uneasiness caused by a sense of guilt <had no *compunctions* about stealing> **2.** Misgiving about the appropriateness of an action : SCRUPLE. **—com·punc′tious** (-shəs) *adj.* **—com·punc′tious·ly** *adv.*

**com·pur·ga·tion** (kŏm′pər-gā′shən) *n.* [LLat. *compurgatio*, complete purification < Lat. *compurgare*, to purify completely : *com-* (intensive) + *purgare*, to purify.] *Law.* The practice of clearing an accused person of a charge by having a number of people swear to belief in his or her innocence.

**com·pu·ta·tion** (kŏm′pyŏŏ-tā′shən) *n.* **1. a.** The act or process of computing. **b.** A method of computing. **2.** The result of computing. **3.** The act of operating a computer. **—com′pu·ta′tion·al** *adj.* **—com′pu·ta′tion·al·ly** *adv.*

**com·pute** (kəm-pyŏŏt′) *v.* **-put·ed, -put·ing, -putes.** [Lat. *computare* : *com-*, together + *putare*, to reckon.] —*vt.* **1.** To determine by mathematics, esp. by numerical methods <*computed* the gallons per mile> **2.** To determine by the use of a computer. —*vi.* **1.** To determine number or amount. **2.** To use a computer. —*n.* Computation. **—com·put′a·bil′i·ty** *n.* **—com·put′a·ble** *adj.*

**com·put·er** (kəm-pyŏŏ′tər) *n.* One that computes, esp. a high-speed electronic device that processes, retrieves, and stores programmed information.

**com·put·er·ese** (kəm-pyŏŏ′tə-rēz′,-rēs′) *n.* The technical language of the computer profession.

**computer graphics** *n. (sing. or pl. in number)* Art obtained by programming a computer to create the desired design or picture.

**com·put·er·ize** (kəm-pyŏŏ′tə-rīz′) *vt.* **-ized, -iz·ing, -iz·es. 1.** To process or store (data) with or in a computer or computer system.

2. To furnish with a computer or computer system. **—com·put′er·iz′a·ble** *adj.* **—com·put′er·i·za′tion** *n.*

**com·put·er·ized** (kəm-pyo͞o′tə-rīzd′) *adj.* Of or relating to a computer or the use of a computer.

**computerized axial tomography** *n.* Tomography in which computer analysis of a series of cross-sectional scans made along a single axis of a structure is used to construct a three-dimensional image.

**computer language** *n.* A code used to provide data and instructions to computers.

**com·rade** (kŏm′răd′, -rəd) *n.* [OFr. *camarade,* roommate < OSp. *camarada* < *camara,* room < LLat. *camera.* —see CHAMBER.] **1.** One who shares another's interests or activities : COMPANION. **2.** *often* **Comrade.** A fellow member of a group, esp. a fellow member of the Communist Party. **—com′rade·ship′** *n.*

**Com·sat** (kŏm′săt′). A trademark for a communications satellite.

**Com·stock·er·y** (kŏm′stŏk′ə-rē, kŭm′-) *n.* [After Anthony *Comstock* (1844–1915).] Overzealous censorship of literature, art, and the theater because of alleged immorality.

**Com·tism** (kŏm′tĭz-əm) *n.* The philosophy of Auguste Comte : POSITIVISM. **—Com′tist** (kŏm′tĭst) *n.*

**Co·mus** (kō′məs) *n.* [Lat. < Gk. *Kōmos* < *kōmos,* revel.] *Myth.* A classical deity of revelry.

**con¹** (kŏn) *adv.* [Short for CONTRA-.] In disagreement with or opposition to : AGAINST <argue pro and *con*> —*n.* **1.** An argument or opinion against something. **2.** One holding an opposing opinion.

**con²** (kŏn) *vt.* **conned, con·ning, cons.** [ME *connen,* to know < OE *cunnan.*] **1.** To peruse, study, or examine carefully. **2.** To memorize. **—con′ner** *n.*

**con³** *also* **conn** (kŏn) [ME *conduen* < OFr. *conduire* < Lat. *conducere,* to lead together. —see CONDUCE.] *Naut.* —*vt.* **conned, con·ning, cons** *also* **conns.** To direct the steering or course of (a vessel). —*n.* **1.** The station or post of the person who cons. **2.** The act or process of conning.

**con⁴** (kŏn) [Short for CONFIDENCE.] *Slang.* —*vt.* **conned, con·ning, cons. 1.** To defraud or swindle (a victim) by soliciting confidence : DUPE. **2.** To win over by benevolent persuasion <*conned* me into buying them ice cream> —*n.* A swindle.

**con⁵** (kŏn) *n. Slang.* A convict.

**con-** *pref. var. of* COM-.

**con a·mo·re** (kŏn′ ə-môr′ē, -mōr′ē, kōn′ ä-mō′rā) *adv.* [Ital., with love.] *Mus.* Tenderly : lovingly. —Used as a direction.

**co·na·tion** (kō-nā′shən) *n.* [Lat. *conatio,* effort < *conatus,* p.part. of *conari,* to try.] *Psychol.* The aspect of mental or behavioral processes directed toward action or change and including impulse, desire, volition, and striving. **—co·na′tion·al** *adj.* **—co′na·tive** (kō′nə-tĭv, kŏn′ə-) *adj.*

**co·na·tus** (kō-nā′təs) *n., pl.* **conatus.** [Lat., effort < p.part. of *conari,* to try.] A natural impulse, tendency, or directed effort.

**con bri·o** (kŏn brē′ō, kōn) *adv.* [Ital., with vigor.] *Mus.* With vigor. —Used as a direction.

**con·cat·e·nate** (kŏn-kăt′n-āt′, kən-) *vt.* **-nat·ed, -nat·ing, -nates.** [LLat. *concatenare, concatenat-* : *com-,* together + *catenare,* to bind < Lat. *catena,* chain.] To connect in a series or chain. —*adj.* (-ĭt, -āt′). Connected or linked in a series. **—con·cat′e·na′tion** *n.*

**con·cave** (kŏn-kāv′, kŏn′kāv′) *adj.* [ME < Lat. *concavus* : *com-* (intensive) + *cavus,* hollow.] Curved like the inner surface of a ball. —*n.* A concave surface, structure, or line. —*vt.* **-caved, -cav·ing, -caves.** To make concave. **—con·cave′ly** *adv.* **—con·cave′ness** *n.* **—con·cav′i·ty** (-kăv′ĭ-tē) *n.*

**con·ca·vo-con·cave** (kŏn-kā′vō-kŏn-kāv′) *adj.* Concave on both surfaces <a *concavo-concave* lens>

**con·ca·vo-con·vex** (kŏn-kā′vō-kŏn-vĕks′) *adj.* **1.** Concave on one side and convex on the other. **2.** Designating a lens with greater concave than convex curvature.

**con·ceal** (kən-sēl′) *vt.* **-cealed, -ceal·ing, -ceals.** [ME *concelen* < OFr. *conceler* < Lat. *concelare* : *com-* (intensive) + *celare,* to hide.] **1.** To keep out of sight : HIDE. **2.** To keep from disclosure : keep secret. **—con·ceal′a·ble** *adj.* **—con·ceal′er** *n.* **—con·ceal′ment** *n.*

**con·cede** (kən-sēd′) *v.* **-ced·ed, -ced·ing, -cedes.** [Fr. *concéder* < Lat. *concedere* : *com-* (intensive) + *cedere,* to yield.] —*vt.* **1.** To acknowledge as true, just, or proper, often unwillingly : ADMIT <*conceding* the point> **2.** To give or grant as a privilege or right. —*vi.* To make a concession : YIELD. **—con·ced′ed·ly** (-sē′dĭd-lē) *adv.* **—con·ced′er** *n.*

**†con·ceit** (kən-sēt′) *n.* [ME *conceite,* mind < *conceiven,* to conceive.] **1.** An overdeveloped opinion of one's abilities, personality, or worth : VANITY. **2.** An imaginative or witty idea. **3. a.** An exaggerated or elaborate metaphor. **b.** The use of such metaphors. **4.** A fancy article or decorative trifle. —*vt.* **-ceit·ed, -ceit·ing, -ceits. 1.** *Obs.* To understand : conceive. **2.** *Regional.* To imagine. **3.** *Chiefly Brit.* To take a fancy to.

**†con·ceit·ed** (kən-sē′tĭd) *adj.* **1.** Holding an exaggerated opinion of oneself : VAIN. **2.** *Regional.* Inclined to be whimsical or fanciful. **—con·ceit′ed·ly** *adv.* **—con·ceit′ed·ness** *n.*

**con·ceive** (kən-sēv′) *v.* **-ceived, -ceiv·ing, -ceives.** [ME *conceiven* < OFr. *concevoir* < Lat. *concipere* : *com-* (intensive) + *capere,* to take.] —*vt.* **1.** To become pregnant with. **2.** To formulate the mind : DEVISE <*conceive* a method to increase production> **3.** To apprehend mentally : UNDERSTAND <couldn't *conceive* what the teacher meant> **4.** To believe or think <couldn't *conceive* that such an event would occur> —*vi.* **1.** To form or hold an idea <Earlier pioneers *conceived* of natural resources as limitless.> **2.** To become pregnant. **—con·ceiv′a·bil′i·ty, con·ceiv′a·ble·ness** *n.* **—con·ceiv′a·ble** *adj.* **—con·ceiv′a·bly** *adv.* **—con·ceiv′er** *n.*

**con·cel·e·brate** (kən-sĕl′ə-brāt′) *vi.* **-brat·ed, -brat·ing, -brates.** To participate in a concelebration. **—con·cel′e·brant** (-brənt) *n.*

**con·cel·e·bra·tion** (kən-sĕl′ə-brā′shən) *n.* Celebration of the Eucharist by two or more members of the clergy.

**con·cen·ter** (kən-sĕn′tər, kŏn-) *vt.* & *vi.* **-tered, -ter·ing, -ters.** [OFr. *concentrer,* to concentrate < VLat. **concentrare.* —see CONCENTRATE.] To direct toward or come together at a common center.

**con·cen·trate** (kŏn′sən-trāt′) *v.* **-trat·ed, -trat·ing, -trates.** [VLat. **concentrare, concentrat-* : Lat. *com-,* together + Lat. *centrum,* center. —see CENTER.] —*vt.* **1. a.** To draw toward a common center : FOCUS. **b.** To gather together in a single main body or power <authority *concentrated* in the Executive branch> **2.** *Chem.* To increase the concentration of (a solution or mixture). —*vi.* **1.** To converge toward or meet in a common center. **2.** To direct one's attention or thoughts <*concentrating* on the business at hand> —*n. Chem.* A product of concentration. —*adj.* Concentrated. **—con·cen·tra′tive** *adj.* **—con′cen·tra′tive·ly** *adv.* **—con′cen·tra′tor** *n.*

**con·cen·tra·tion** (kŏn′sən-trā′shən) *n.* **1. a.** The act or process of concentrating, esp. close, undivided attention. **b.** The state of being concentrated. **2.** Something concentrated. **3.** *Chem.* The amount of a specified substance in a unit amount of another substance.

**concentration camp** *n.* A camp where people, as prisoners of war or political dissidents, are confined.

**concentration gradient** *n.* The difference in concentration of a solute per unit distance of a solution.

**con·cen·tric** (kən-sĕn′trĭk) *also* **con·cen·tri·cal** (-trĭ-kəl) *adj.* [ME *concentrik* < Med. Lat. *concentricus* : Lat. *com-,* same + *centrum,* center. —see CENTER.] Having a common center. **—con·cen′tri·cal·ly** *adv.* **—con′cen·tric′i·ty** (kŏn′sĕn-trĭs′ĭ-tē) *n.*

**con·cept** (kŏn′sĕpt′) *n.* [LLat. *conceptus* < p.part. of *concipere,* to conceive.] **1.** A general idea or understanding, esp. one derived from specific instances or occurrences. **2.** A notion or thought.

**con·cep·ta·cle** (kən-sĕp′tə-kəl) *n.* [Fr. < Lat. *conceptaculum,* receptacle < *concipere,* to conceive.] An external cavity containing reproductive structures in certain algae and fungi.

**con·cep·tion** (kən-sĕp′shən) *n.* [ME *concepcioun* < Lat. *conceptio* < *concipere,* to conceive.] **1. a.** The formation of a zygote capable of survival and maturation in normal conditions. **b.** The entity so formed : EMBRYO. **2. a.** The ability to form or understand mental concepts. **b.** Something conceived in the mind, as a concept, design, or thought. **3.** *Archaic.* A beginning : start. **—con·cep′tion·al** *adj.*

**con·cep·tive** (kən-sĕp′tĭv) *adj.* Capable of or relating to conceiving. **—con·cep′tive·ly** *adv.*

**con·cep·tu·al** (kən-sĕp′cho͞o-əl) *adj.* Of, consisting of, or relating to concepts or conception. **—con·cep′tu·al·ly** *adv.*

**con·cep·tu·al·ism** (kən-sĕp′cho͞o-ə-lĭz′əm) *n. Philos.* The doctrine, intermediate between realism and nominalism, that universals exist only within the mind as abstract concepts and have no external or substantial reality. **—con·cep′tu·al·ist** *n.* **—con·cep′tu·al·is′tic** *adj.* **—con·cep′tu·al·is′ti·cal·ly** *adv.*

**con·cep·tu·al·ize** (kən-sĕp′cho͞o-ə-līz′) *v.* **-ized, -iz·ing, -iz·es.** —*vt.* To form a concept or concepts of. —*vi.* To form concepts. **—con·cep′tu·al·i·za′tion** *n.*

**con·cern** (kən-sûrn′) *v.* **-cerned, -cern·ing, -cerns.** [ME *concernen* < Med. Lat. *concernere* < Lat., to mix : *com-,* together + *cernere,* to sift.] —*vt.* **1.** To pertain or relate to. **2.** To have an effect on : be of interest or importance to <a change that *concerns* everyone> **3.** To engage the attention of : INVOLVE <*concerned* myself with the schedule> **4.** To cause uneasiness or anxiety in : TROUBLE <Management was *concerned* about the union's activities.> —*vi. Obs.* To be of importance. —*n.* **1.** A matter that relates to or affects one. **2.** Regard for or interest in someone or something. **3.** Worry : anxiety. **4.** A business enterprise : COMPANY. **5.** *Informal.* A material article or contrivance.

**con·cerned** (kən-sûrnd′) *adj.* **1.** Interested and involved <*concerned* with equal rights> **2.** Anxious : troubled <*concerned* about nuclear war>

**con·cern·ing** (kən-sûr′nĭng) *prep.* In reference to : REGARDING.

**con·cern·ment** (kən-sûrn′mənt) *n.* **1.** CONCERN 1. **2.** Reference, relation, or importance. **3.** CONCERN 3.

**con·cert** (kŏn′sûrt′, -sərt) *n.* [Fr. < Ital. *concerto* < OItal. *concertare,* to harmonize.] **1.** A musical performance given by one or more artists. **2.** Agreement in purpose, feeling, or action. —*v.* (kən-sûrt′) **-cert·ed, -cert·ing, -certs.** —*vt.* **1.** To plan or arrange by mutual agreement. **2.** To devise or contrive. —*vi.* To act together in harmony. **—in concert.** All together.

**con·cert·ed** (kən-sûr'tĭd) *adj.* **1.** Accomplished or planned together : COMBINED <a *concerted* effort to finish on time> **2.** *Mus.* Arranged in parts for voices or instruments. **—con·cert'ed·ly** *adv.*

**concert grand** *n.* The largest grand piano, approx. nine feet long.

**con·cer·ti·na** (kŏn'sər-tē'nə) *n.* [CONCERT + Ital. *-ina,* fem. dim. suffix.] **1.** A small, hexagonal accordion with bellows and buttons instead of keys. **2.** A coiled, sharp-edged wire used as a barrier.

**con·cer·ti·no** (kŏn'chĕr-tē'nō) *n., pl.* **-nos.** [Ital., dim. of *concerto,* concert. —see CONCERT.] *Mus.* **1.** A short concerto. **2.** The solo instrument group in a concerto grosso.

**con·cer·tize** (kŏn'sər-tīz') *vi.* **-tized, -tiz·ing, -tiz·es.** To give or perform in concerts.

**con·cert·mas·ter** (kŏn'sərt-măs'tər) *n.* The first violinist and assistant conductor in a symphony orchestra.

**con·cer·to** (kən-chĕr'tō) *n., pl.* **-tos** *or* **-ti** (-tē) [Ital., concert.] An orchestral composition for one or more solo instruments, usu. in three movements.

**concerto gros·so** (grō'sō) *n., pl.* **con·cer·ti gros·si** (kən-chĕr'tē grō'sē) [Ital., great concerto.] A composition for a small group of solo instruments and a full orchestra.

**concert pitch** *n.* **1.** A pitch to which orchestral instruments are tuned. **2.** The state of being tensely alert and ready.

**con·ces·sion** (kən-sĕsh'ən) *n.* [ME < Lat. *concessio* < *concedere,* to concede. —see CONCEDE.] **1.** The act of conceding. **2.** Something conceded. **3.** Something, as a land tract, granted esp. by a government to be used for a specific purpose. **4. a.** The privilege of maintaining a subsidiary business within certain premises. **b.** The space allotted for such a business. **—con·ces'sion·al** *adj.* **—con·ces'sion·ar'y** (-ə-nĕr'ē) *adj.*

**con·ces·sion·aire** (kən-sĕsh'ə-nâr') *also* **con·ces·sion·er** (-sĕsh'ə-nər) *n.* [Fr.] The operator or holder of a concession.

**con·ces·sive** (kən-sĕs'ĭv) *adj.* [Lat. *concessivus* < *concessus,* p.part. of *concedere,* to concede. —see CONCEDE.] **1.** Of the nature of or containing a concession. **2.** Expressing concession grammatically, as the conjunction *though.* **—con·ces'sive·ly** *adv.*

**conch** (kŏngk, kŏnch) *n., pl.* **conchs** (kŏngks) *or* **conch·es** (kŏn'chĭz) [ME *conche* < OFr. < Lat. *concha,* mussel < Gk. *konkhē.*] **1.** Any of various tropical marine gastropod mollusks of the genus *Strombus* and other genera, with large, often brightly colored spiral shells and edible flesh. **2.** The shell of a conch. **3.** A concha.

**conch-** *pref. var. of* CONCHO-.

**con·cha** (kŏng'kə) *n., pl.* **-chae** (-kē') [Lat., mussel < Gk. *konkhē.*] **1.** *Anat.* A shell-like structure, as the external ear. **2.** The half dome over an apse. **—con'chal** (-kəl) *adj.*

**conchi-** *pref. var. of* CONCHO-.

**con·chif·er·ous** (kŏng-kĭf'ər-əs) *adj.* Having or forming a shell.

**con·chi·o·lin** (kŏng-kī'ə-lĭn, kŏn-) *n.* [CONCH + -OL + -IN.] A protein substance that is the organic basis of mollusk shells.

**concho-** *or* **conchi-** *or* **conch-** *pref.* [Gk. *konkho-* < *konkhē,* shell.] Shell <conchology>

**con·choi·dal** (kŏng-koid'l) *adj.* [Gk. *konkhoeidēs,* mussellike : *konkhē,* mussel + *-eidēs,* -oid.] Of or relating to rocks, as flint or obsidian, that have shell-like surfaces when fractured. **—con·choi'dal·ly** *adv.*

**con·chol·o·gy** (kŏng-kŏl'ə-jē) *n.* The study of shells and mollusks. **—con'cho·log'i·cal** (-kə-lŏj'ĭ-kəl) *adj.* **—con·chol'o·gist** *n.*

**con·cierge** (kôN-syârzh') *n.* [Fr. < OFr. *cumcerges* < VLat. **conservius* < Lat. *conservus,* fellow slave : *com-* together + *servus,* slave.] **1.** A person, esp. in France, who lives in a building, attends the entrance, and serves as a custodian. **2.** A staff member of a hotel, esp. in Europe, who takes reservations, accommodates guests, and handles luggage and mail.

**con·cil·i·ar** (kən-sĭl'ē-ər) *adj.* [< Lat. *concilium,* council.] Of or relating to a council.

**con·cil·i·ate** (kən-sĭl'ē-āt') *vt.* **-at·ed, -at·ing, -ates.** [Lat. *conciliare, conciliat-* < *concilium,* meeting.] **1.** To overcome the distrust or animosity of : PLACATE. **2.** To regain (e.g., friendship) by pleasant behavior. **3.** To make compatible : RECONCILE. **—con·cil'i·a·ble** (-ə-bəl) *adj.* **—con·cil'i·a'tion** *n.* **—con·cil'i·a'tor** *n.* **—con·cil'i·a·to'ry** (-ə-tôr'ē, -tōr'ē) *adj.*

**con·cin·ni·ty** (kən-sĭn'ĭ-tē) *n., pl.* **-ties.** [< Lat. *concinnitas* < *concinnus,* deftly joined.] Harmonious arrangement of parts, esp. elegance of literary style.

**con·cise** (kən-sīs') *adj.* [Lat. *concisus,* p.part. of *concidere,* to cut up : *com-* (intensive) + *caedere,* to cut.] Short and to the point. **—con·cise'ly** *adv.* **—con·cise'ness** *n.*

**con·clave** (kŏn'klāv', kŏng'-) *n.* [ME, private chamber < Lat. : *com-,* together + *clavis,* key.] **1.** A private or secret meeting. **2. a.** The private rooms in which the cardinals of the Roman Catholic Church meet to elect a pope. **b.** The meeting held to elect a pope.

**con·clude** (kən-klo͞od') *v.* **-clud·ed, -clud·ing, -cludes.** [ME *concluden* < Lat. *concludere* : *com-* (intensive) + *claudere,* to close.] —*vt.* **1.** To bring to an end : CLOSE. **2.** To bring about a final agreement or settlement <*conclude* an arms-limitation treaty> **3.** To reach a decision or form an opinion about : deduce or infer <*concluded* that the evidence warranted an indictment> **4.** To determine : resolve <*concluded* that we had to leave now> **5.** *Obs.* To confine or enclose. —*vi.* **1.** To come to an end : CLOSE. **2.** To come to a decision, agreement, or final judgment. **—con·clud'er** *n.*

**con·clu·sion** (kən-klo͞o'zhən) *n.* [ME *conclusioun* < Lat. *conclusio* < *concludere,* to end. —see CONCLUDE.] **1.** The last part : END. **2.** The outcome or result of an act or process. **3.** A decision reached after deliberation. **4.** A final arrangement or settlement, as of a treaty. **5.** *Law.* The close of a plea or deed. **6.** *Logic.* **a.** The proposition that must follow from the major and minor premises in a syllogism. **b.** The proposition concluded from one or more premises : DEDUCTION.

**con·clu·sive** (kən-klo͞o'sĭv) *adj.* Serving to end doubt or uncertainty : DECISIVE. **—con·clu'sive·ly** *adv.* **—con·clu'sive·ness** *n.*

**con·coct** (kən-kŏkt') *vt.* **-coct·ed, -coct·ing, -cocts.** [Lat. *concoquere, concoct-,* to boil together : *com-,* together + *coquere,* to cook.] **1.** To prepare by combining ingredients, as in cookery. **2.** To invent : contrive <*concoct* an excuse> **—con·coct'er, con·coc'tor** *n.* **—con·coc'tion** *n.* **—con·coc'tive** *adj.*

**con·com·i·tance** (kən-kŏm'ĭ-təns) *n.* **1.** Occurrence together or in connection with another : ACCOMPANIMENT. **2.** A concomitant.

**con·com·i·tant** (kən-kŏm'ĭ-tənt) *adj.* [Lat. *concomitans, concomitant-,* pr.part. of *concomitari,* to accompany : *com-,* together + *comitari,* to accompany < *comes,* companion.] Occurring or existing concurrently : ACCOMPANYING. —*n.* Something that exists or occurs concurrently with something else. **—con·com'i·tant·ly** *adv.*

**con·cord** (kŏn'kôrd', kŏng'-) *n.* [ME *concorde* < OFr. < Lat. *concordia* < *concors,* agreeing : *com-,* same + *cors,* heart.] **1.** A state of harmony : ACCORD. **2.** A treaty establishing peaceful relations. **3.** Grammatical agreement.

**con·cor·dance** (kən-kôr'dns) *n.* **1.** A state of agreement : CONCORD. **2.** An alphabetical index of all the major words in a text or the works of an author, showing every contextual occurrence of a word.

**con·cor·dant** (kən-kôr'dnt) *adj.* [ME *concordaunt* < OFr. *concordant* < Lat. *concordans,* pr.part. of *concordare,* to agree < *concors,* agreeing. —see CONCORD.] Harmonious : accordant. **—con·cor'dant·ly** *adv.*

**con·cor·dat** (kən-kôr'dăt') *n.* [Fr. < Med. Lat. *concordatum* < Lat. *concordare,* to agree. —see CONCORDANT.] **1.** A formal agreement : COMPACT. **2.** An agreement between the pope and a government for the regulation of church affairs.

**Con·cord grape** (kŏng'kərd) *n.* [After *Concord,* Massachusetts.] A grape having purple-black fruit with a bluish bloom, used for making jelly, juice, and wine.

**con·course** (kŏn'kôrs', -kōrs', kŏng'-) *n.* [ME *concours* < OFr. < Lat. *concursus* < p.part. of *concurrere,* to assemble : *com-,* together + *currere,* to run.] **1.** A large crowd : THRONG. **2.** An act of coming or flowing together. **3. a.** A large open space for the gathering or passage of crowds, as in an airport. **b.** A broad thoroughfare.

**con·cres·cence** (kən-krĕs'əns) *n.* [Lat. *concrescentia* < *concrescens,* pr.part. of *concrescere,* to grow together. —see CONCRETE.] The uniting, esp. the growing together, of related parts, as of physical particles or anatomical structures. **—con·cres'cent** *adj.*

**con·crete** (kŏn-krēt', kŏn'krēt') *adj.* [ME *concret* < Lat. *concretus,* p.part. of *concrescere,* to grow together, harden : *com-,* together + *crescere,* to grow.] **1.** Of or relating to an actual, specific thing or instance <*concrete* facts> **2.** Tangible : real **3.** Designating a real thing or group of things <A treaty is *concrete;* peace is abstract.> **4.** Formed by the coalescence of separate particles or parts into one mass : SOLID. **5.** Made of concrete. —*n.* (kŏn'krēt', kŏn-krēt'). **1.** A construction material consisting of conglomerate gravel, pebbles, broken stone, or slag in a mortar or cement matrix. **2.** A mass formed by the coalescence of particles. —*v.* (kŏn'krēt', kŏn-krēt') **-cret·ed, -cret·ing, -cretes.** —*vt.* **1.** To form into a mass by cohesion or coalescence of particles. **2.** To build, treat, or cover with concrete. —*vi.* To harden : solidify. **—con·crete'ly** *adv.* **—con·crete'ness** *n.*

**concrete mixer** *n.* A machine with a revolving drum in which cement, sand, gravel, and water are combined into concrete.

**concrete music** *n.* Musique concrète.

**concrete poetry** *n.* Poetry that visually conveys the poet's meaning through the graphic arrangement of letters, words, or symbols on the page.

**con·cre·tion** (kən-krē'shən) *n.* **1. a.** The act or process of concreting : COALESCENCE. **b.** The state of being concreted. **2.** A solid or concrete mass. **3.** *Geol.* A rounded mass of mineral matter found in sedimentary rock. **4.** *Pathol.* **a.** A solid mass of inorganic material formed in a cavity or tissue of the body : CALCULUS. **b.** Abnormal fusion of otherwise adjacent parts, as of toes or fingers. **—con·cre'tion·ar'y** (-shə-nĕr'ē) *adj.*

**con·cret·ism** (kŏn-krē'tĭz'əm) *n.* The theory or practice of concrete poetry. **—con·cret'ist** *n.*

**con·cre·tize** (kŏn'krĭ-tīz') *vt.* **-tized, -tiz·ing, -tiz·es.** To make real or specific. **—con'cre·ti·za'tion** *n.*

**con·cu·bi·nage** (kŏn-kyo͞o'bə-nĭj, kən-) *n.* **1.** Cohabitation of unmarried people. **2.** The state of being a concubine.

**con·cu·bine** (kŏng'kyə-bīn', kŏn'-) *n.* [ME < OFr. < Lat. *concubina* : *com-* with < *cubare*, to lie down.] **1.** A woman who cohabits with a man without being married to him. **2.** A secondary wife in certain polygamous societies.

**con·cu·pis·cence** (kŏn-kyōō'pĭ-səns) *n.* [LLat. *concupiscentia* < Lat. *concupiscens*, pr.part. of *concupiscere*, inchoative of *concupere*, to desire strongly : *com-* (intensive) + *cupere*, to desire.] A strong, esp. sexual, desire : LUST. **—con·cu'pis·cent** *adj.*

**con·cur** (kən-kûr') *vi.* **-curred, -cur·ring, -curs.** [ME *concurren* < Lat. *concurrere*, to meet : *com-*, together + *currere*, to run.] **1.** To have or express the same opinion : AGREE. **2.** To act together : COOPERATE. **3.** To happen at the same time : COINCIDE. **4.** *Obs.* To come together : CONVERGE.

**con·cur·rence** (kən-kûr'əns) *n.* **1.** Agreement in opinion : ACCORDANCE. **2.** Cooperation, as of agents, circumstances, or events. **3.** Simultaneous occurrence : COINCIDENCE. **4.** *Law.* A claim or power held jointly.

**con·cur·rent** (kən-kûr'ənt) *adj.* [ME < Lat. *concurrens*, pr.part. of *concurrere*, to run together. —see CONCUR.] **1.** Occurring at the same time. **2.** Operating in conjunction. **3.** Meeting or tending to meet at the same point : CONVERGENT. **4.** Being in accordance : HARMONIOUS. **—con·cur'rent** *n.* **—con·cur'rent·ly** *adv.*

**concurrent resolution** *n.* A resolution adopted by both houses of a bicameral legislature that has no force.

**con·cuss** (kən-kŭs') *vt.* **-cussed, -cuss·ing, -cuss·es.** [Lat. *concutere, concuss-*, to strike together : *com-*, together + *quatere*, to strike.] To injure by concussion.

**con·cus·sion** (kən-kŭsh'ən) *n.* **1.** A hard jolt : SHOCK. **2.** An injury of a soft structure, esp. of the brain, resulting from a violent blow. **—con·cus·sive** (-kŭs'ĭv) *adj.* **—con·cus'sive·ly** *adv.*

**con·demn** (kən-dĕm') *vt.* **-demned, -demn·ing, -demns.** [ME *condemnen* < Lat. *condemnare* : *com-* (intensive) + *damnare*, to sentence < *damnum*, penalty.] **1.** To express disapproval of : DENOUNCE. **2. a.** To find guilty : CONVICT. **b.** To pronounce judgment against : SENTENCE. **3.** To declare unfit for use or consumption, usu. by official order <*condemn* an old bridge> **4.** *Law.* To declare appropriated for public use under the right of eminent domain. **—con·dem'na·ble** (-dĕm'nə-bəl) *adj.* **—con·dem'na·to'ry** (-nə-tôr'ē, -tōr'ē) *adj.* **—con·demn'er** *n.*

**con·dem·na·tion** (kŏn'dĕm-nā'shən) *n.* **1. a.** The act of condemning. **b.** The state of being condemned. **2.** Strong reproof or censure. **3.** A cause or occasion for condemning.

**con·den·sate** (kŏn'dən-sāt', -dĕn-, kən-dĕn'sāt') *n.* [Back-formation < CONDENSATION.] A product of condensation.

**con·den·sa·tion** (kŏn'dĕn-sā'shən, -dən-) *n.* **1.** The act of condensing or the state of being condensed. **2.** A product of condensing. **3.** *Physics.* **a.** The physical process by which a liquid is removed from a vapor or vapor mixture. **b.** The liquid so removed : CONDENSATE. **4.** *Chem.* A chemical reaction in which a simple substance, as water, is released by the combination of two or more molecules. **5.** *Psychoanal.* The process by which a single idea or word is invested with the emotional content of a group of ideas or words. **—con'den·sa'tion·al** *adj.*

**con·dense** (kən-dĕns') *v.* **-densed, -dens·ing, -dens·es.** [ME *condensen* < OFr. *condenser* < Lat. *condensare* : *com-* (intensive) + *densare*, to thicken < *densus*, thick.] *—vt.* **1.** To reduce the volume of : COMPRESS. **2.** To make more concise : ABRIDGE <*condense* three articles into one> **3. a.** To form a condensate from (e.g., a vapor). **b.** To subject (e.g., a vapor) to condensation. *—vi.* **1.** To become more compact. **2.** To undergo condensation. **—con·dens'a·bil'i·ty** *n.* **—con·dens'a·ble, con·dens'i·ble** *adj.*

**condensed milk** *n.* Sweetened evaporated milk.

**con·dens·er** (kən-dĕn'sər) *n.* **1.** One that condenses. **2.** *Physics.* An apparatus used to condense vapor. **3.** A capacitor. **4.** A mirror or lens for gathering and concentrating light on an object.

**con·de·scend** (kŏn'dĭ-sĕnd') *vi.* **-scend·ed, -scend·ing, -scends.** [ME *condescenden* < LLat. *condescendere* : Lat. *com-* (intensive) + *descendere*, to descend. —see DESCEND.] **1.** To agree to do something one considers beneath one's rank or dignity : DEIGN. **2.** To deal with or treat people in a superior or patronizing way. **—con'de·scend'er** *n.*

**con·de·scen·dence** (kŏn'dĭ-sĕn'dəns) *n.* Condescension.

**con·de·scend·ing** (kŏn'dĭ-sĕn'dĭng) *adj.* Characterized by condescension : PATRONIZING. **—con'de·scend'ing·ly** *adv.*

**con·de·scen·sion** (kŏn'dĭ-sĕn'shən) *n.* **1.** The act or an instance of condescending. **2.** Patronizing behavior.

**con·dign** (kən-dīn') *adj.* [ME *condigne* < Lat. *condignus* : *com-* (intensive) + *dignus*, worthy.] Adequate : deserved <*condign* criticism> **—con·dign'ly** *adv.*

**con·di·ment** (kŏn'də-mənt) *n.* [ME < Lat. *condimentum* < *condire*, to season < *condere*, to store.] A seasoning for food, as mustard or a spice. **—con'di·men'tal** (-mĕn'tl) *adj.*

**con·di·tion** (kən-dĭsh'ən) *n.* [ME *condicioun* < OFr. *condicion* < Lat. *conditio*, stipulation, prob. < *condicere*, to agree : *com-*, together + *dicere*, to talk.] **1.** Mode or state of being. **2. a.** A state of health. **b.** A state of readiness or physical fitness <getting in *condition* for mountain climbing> **3.** *Informal.* A disease or ailment. **4.** Social position : RANK. **5.** Something indispensable to the appearance or occurrence of something else : PREREQUISITE <Patience is a *condition* of good teaching.> **6.** Something that modifies or restricts : QUALIFICATION. **7.** *often* **conditions.** Existing circumstances <poor boating *conditions*> **8.** The dependent clause of a conditional sentence. **9.** *Logic.* A proposition upon which another proposition depends : the antecedent of a conditional proposition. **10.** *Law.* **a.** A provision making the effect of a legal instrument contingent upon the occurrence of an uncertain future event. **b.** The event itself. **11.** An unsatisfactory grade given a student that may be raised by doing extra work. **12.** *Obs.* Disposition of mind. *—vt.* **-tioned, -tion·ing, -tions. 1.** To make conditional. **2. a.** To render fit. **b.** To air-condition. **3.** To accustom (a person) to : ADAPT. **4.** To give the grade of condition to. **5.** *Psychol.* To cause to respond in a specific manner to a specific stimulus.

☆ ***syns:*** CONDITION, MODE, POSTURE, SITUATION, STATE, STATUS *n. core meaning* : manner of being or form of existence <a patient in a weakened *condition*><an old house restored to its former *condition*>

**con·di·tion·al** (kən-dĭsh'ə-nəl) *adj.* **1.** Imposing, depending on, or containing a condition or conditions <*conditional* acceptance> **2.** Stating, containing, or implying a grammatical condition. **3.** *Psychol.* Brought about by conditioning. *—n.* A mood, tense, clause, or word expressing a condition. **—con·di'tion·al'i·ty** (-dĭsh'ə-năl'ĭ-tē) *n.* **—con·di'tion·al·ly** *adv.*

**conditional probability** *n.* The probability that an event will occur, given that another event has occurred or will occur.

**con·di·tioned** (kən-dĭsh'ənd) *adj.* **1.** Subject to or dependent on a condition or conditions. **2. a.** Physically fit. **b.** Prepared for a specific action or process. **3.** *Psychol.* Exhibiting or trained to exhibit a conditioned response.

**conditioned reflex** *n.* A conditioned response.

**conditioned response** *n. Psychol.* A new or modified response elicited by a stimulus after conditioning.

**conditioned stimulus** *n. Psychol.* A stimulus rendered capable of eliciting a response like that of a specific unconditioned stimulus by conditioning.

**con·di·tion·er** (kən-dĭsh'ə-nər) *n.* **1.** One that conditions. **2.** An additive or application that improves the quality or usability of a substance <a soil *conditioner*><a hair *conditioner*>

**con·di·tion·ing** (kən-dĭsh'ə-nĭng) *n. Psychol.* The process of or complex of organismic processes resulting from presenting an initially inadequate stimulus with an unconditioned stimulus until the former is capable of eliciting a response like that elicited by the latter.

**con·do** (kŏn'dō') *n. Informal.* CONDOMINIUM 2.

**con·dole** (kən-dōl') *vi.* **-doled, -dol·ing, -doles.** [LLat. *condolēre*, to feel another's pain : Lat. *com-*, together + Lat. *dolēre*, to grieve.] To express sorrow or sympathy. **—con·do'la·to'ry** (-dō'lə-tôr'ē, -tōr'ē) *adj.* **—con·dol'er** *n.*

**con·do·lence** (kən-dō'ləns) *n.* **1.** Sympathy with a person in grief. **2.** A formal expression of sympathy. **—con·do'lent** *adj.*

**con·dom** (kŏn'dəm, kŭn'-) *n.* [Orig. unknown.] A sheath, usu. made of thin rubber, designed to cover the penis during sexual intercourse for contraceptive or antivenereal purposes.

**con·do·min·i·um** (kŏn'də-mĭn'ē-əm) *n.* [COM- + Lat. *dominium*, property.] **1. a.** Joint rule, esp. joint sovereignty, of a territory by two or more states. **b.** A territory so governed. **2. a.** An apartment building or housing area in which the living units are owned individually. **b.** A living unit in such a building or housing area. **—con'do·min'i·al** (-ē-əl) *adj.*

**con·do·na·tion** (kŏn'də-nā'shən, -dō-) *n.* The act of condoning, esp. the implied forgiveness of an offense by ignoring it.

**con·done** (kən-dōn') *vt.* **-doned, -don·ing, -dones.** [Lat. *condonare*, to pardon : *com-* (intensive) + *donare*, to give.] To overlook, forgive, or disregard (an offense) without censure or protest. **—con·don'er** *n.*

**con·dor** (kŏn'dôr', -dər) *n.* [Sp. *cóndor* < Quechua *kúntur.*] **1.** A very large New World vulture, *Vultur gryphus* of the Andes or *Gymnogyps californianus* of the California mountains. **2.** A gold coin of some South American countries bearing the figure of a condor.

**con·dot·tie·re** (kŏn'də-tyâr'ē) *n., pl.* **-tie·ri** (-tyâr'ē) [Ital. < *condotta*, troop of mercenaries < fem. p.part. of *condurre*, to conduct < Lat. *conducere*, to lead together. —see CONDUCE.] A 14th–16th-cent. leader of mercenary soldiers.

**con·duce** (kən-dōōs', -dyōōs') *vi.* **-duced, -duc·ing, -duc·es.** [Lat. *conducere*, to lead together : *com-*, together + *ducere*, to lead.] To lead to a specific result : CONTRIBUTE. **—con·duce'ment** *n.* **—con·duc'er** *n.* **—con·duc'ing·ly** *adv.*

**con·du·cive** (kən-dōō'sĭv, -dyōō'-) *adj.* Tending to bring about or cause : CONTRIBUTIVE. **—con·du'cive·ness** *n.*

**con·duct** (kən-dŭkt') *v.* **-duct·ed, -duct·ing, -ducts.** [ME *conducten* < Lat. *conducere*, to lead together. —see CONDUCE.] *—vt.* **1.** To direct the course of : CONTROL <*conduct* an experiment><*conduct* a meeting> **2.** To guide or lead <*conduct* a sightseeing tour> **3.** To lead a musical group, as an orchestra. **4.** To serve as a medium or channel for conveying: TRANSMIT <Some plastics *conduct* light.> **5.** To behave or act in a given way. —*vi.* **1.** To act as a conductor. **2.** To lead. —*n.* (kŏn′dŭkt′). **1.** The way a person acts: BEHAVIOR. **2.** The act of directing or controlling: MANAGEMENT. **3.** *Obs.* An escort. **—con·duct′i·bil′i·ty** *n.* **—con·duct′i·ble** *adj.*

**con·duc·tance** (kən-dŭk′tĕns) *n.* A measure of a material's ability to conduct electric charge, the real part of the complex representation of admittance.

**con·duc·tim·e·try** (kŏn′dŭk-tĭm′ĭ-trē) *n.* Scientific measurement of solution conductance.

**con·duc·tion** (kən-dŭk′shən) *n.* Transmission or conveyance of something through a medium or passage, esp. of electric charge or heat through a conducting medium without perceptible motion of the medium itself.

**con·duc·tive** (kən-dŭk′tĭv) *adj.* Exhibiting conductivity.

**con·duc·tiv·i·ty** (kŏn′dŭk-tĭv′ĭ-tē) *n.* **1.** The ability or power to conduct or transmit. **2.** A measure of the ability of a material to conduct an electric charge, the reciprocal of resistivity.

**con·duc·to·met·ric titration** (kən-dŭk′tə-mĕt′rĭk) *n.* Titration based on changes in the electrical conductance of a solution.

**con·duc·tor** (kən-dŭk′tər) *n.* **1.** One who conducts, esp.: **a.** One who collects fares on a public conveyance, as a railroad train. **b.** The director of a musical ensemble. **2.** *Physics.* A substance or medium that conducts heat, light, sound, or esp. an electric charge. **3.** A lightning rod. **—con′duc·to′ri·al** (kŏn′dŭk-tôr′ē-əl, -tōr′-) *adj.* **—con·duc′tor·ship′** *n.*

**con·duit** (kŏn′dĭt, -dōō-ĭt) *n.* [ME < OFr., conveyance < Med. Lat. *conductus,* transportation < Lat., p.part. of *conducere,* to lead together. —see CONDUCT.] **1.** A channel or pipe for conveying fluids, as water. **2.** A tube or duct for enclosing electric wires or cable. **3.** *Archaic.* A fountain.

**con·du·pli·cate** (kŏn-dōō′plĭ-kĭt, -dyōō′-) *adj.* [Lat. *conduplicatus,* p.part. of *conduplicare,* to double : *com-,* together + *duplicare,* to double < *duplex,* double.] Folded in half lengthwise. —Used of leaves and petals in the bud. **—con′du·pli·ca′tion** *n.*

**con·dyle** (kŏn′dīl′, -dl) *n.* [Lat. *condylus,* knuckle < Gk. *kondulos.*] A rounded articulatory prominence at the end of a bone. **—con′dy·lar** (-də-lər) *adj.* **—con′dy·loid** (-dl-oid′) *adj.*

**con·dy·lo·ma** (kŏn′dl-ō′mə) *n., pl.* **-mas** *or* **-ma·ta** (-mə-tə) [Gk. *kondulōma* < *kondulos,* knuckle.] A wartlike growth near the anus or external genitalia. **—con′dy·lo′ma·tous** (-mə-təs) *adj.*

**cone** (kōn) *n.* [Fr. *cône* < Lat. *conus* < Gk. *kōnos.*] **1. a.** A surface generated by a straight line, passing through a fixed point, and moving along the intersection with a fixed curve. **b.** The surface generated by such a generator passing through a vertex lying on the perpendicular axis of a circular directrix. **2. a.** The figure formed by a cone, bound or regarded as bound by its vertex and a plane section taken anywhere above or below the vertex. **b.** Something having the shape of this figure. **3. a.** A conical, spheroidal, or cylindrical structure borne by certain trees, as the pines, firs, and hemlocks, consisting of clusters of stiff, overlapping, woody scales, between which are the naked ovules. **b.** A similar fruit, as that of the magnolia or hop. **4.** *Physiol.* A photoreceptor in the retina of the eye. **5.** One of various gastropod mollusks of the family Conidae of tropical seas, with a conical, often vividly marked shell. —*vt.* **coned, con·ing, cones.** To shape like a cone or cone segment.

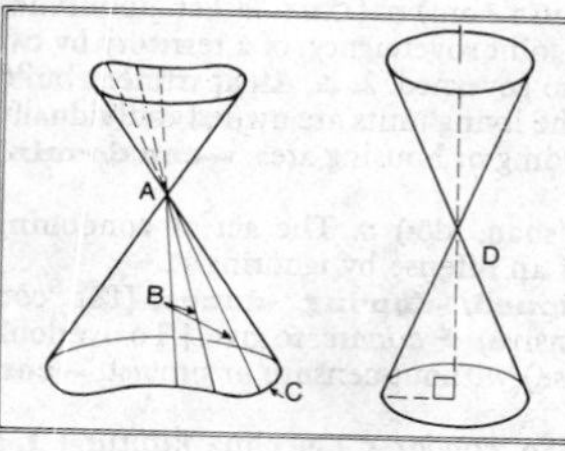


**cone**
*A. vertex, B. elements, C. directrix, D. right circular cone*

**cone·flow·er** (kōn′flou′ər) *n.* A North American plant of the genera *Rudbeckia, Ratibida,* or *Echinacea,* having rayed flowers with a conelike center of tubular florets.

**cone·nose** (kōn′nōz′) *also* **cone-nosed bug** (kōn′nōzd′) *n.* An assassin bug, esp. *Triatoma sanguisuga* of the southern and western United States and Mexico, having sucking mouth parts and capable of inflicting a painful, toxic bite.

**Con·es·to·ga wagon** (kŏn′ĭ-stō′gə) *n.* [After *Conestoga,* Pennsylvania.] A heavy covered wagon with broad wheels, used by American pioneers esp. for prairie travel.

**co·ney** (kō′nē, kŭn′ē) *n. var. of* CONY.

**con·fab** (kŏn′făb′) *Informal.* —*n.* An informal talk: CONFABULATION. —*vi.* (kən-făb′, kŏn′făb′) **-fabbed, -fab·bing, -fabs.** To engage in a confab.

**con·fab·u·late** (kən-făb′yə-lāt′) *vi.* **-lat·ed, -lat·ing, -lates.** [Lat. *confabulari, confabulat-* : *com-,* together + *fabulari,* to talk < *fabula,* conversation < *fari,* to speak.] **1.** To talk informally: CHAT. **2.** *Psychiat.* To replace fact with fantasy in memory. **—con·fab′u·la′tion** *n.* **—con·fab′u·la′tor** *n.* **—con·fab′u·la·to′ry** (-lə-tôr′ē, -tōr′ē) *adj.*

**con·fect** (kən-fĕkt′) *vt.* **-fect·ed, -fect·ing, -fects.** [ME *confecten* < Lat. *conficere,* to prepare : *com-* (intensive) + *facere,* to make.] **1.** To put together by combining elements. **2.** To make into a preserve or confection. —*n.* (kŏn′fĕkt′). CONFECTION 2.

**con·fec·tion** (kən-fĕk′shən) *n.* **1.** The act or process of confecting. **2.** A sweet preparation. **3.** A sweetened medicinal compound. —*vt.* **-tioned, -tion·ing, -tions.** To make into a confection.

**con·fec·tion·ar·y** (kən-fĕk′shə-nĕr′ē) *adj.* Of, relating to, or resembling a confection. —*n., pl.* **-ies.** *var. of* CONFECTIONERY 3.

**con·fec·tion·er** (kən-fĕk′shə-nər) *n.* A maker or seller of confections.

**confectioners' sugar** *n.* Finely pulverized sugar.

**con·fec·tion·er·y** (kən-fĕk′shə-nĕr′ē) *n., pl.* **-ies. 1.** Candies and other confections. **2.** The art or occupation of a confectioner. **3.** *also* **con·fec·tion·ar·y.** A confectioner's shop.

**con·fed·er·a·cy** (kən-fĕd′ər-ə-sē) *n., pl.* **-cies.** [ME *confederacie* < AN < LLat. *confoederatio,* agreement < *confoederare,* to unite. —see CONFEDERATE.] **1. a.** An alliance of persons, parties, or states: LEAGUE. **b.** The persons, parties, or states united in a league. **c. Confederacy.** The 11 Southern states that seceded from the United States (1860–1861). **2.** Two or more people who have united for unlawful practices: CONSPIRACY.

**con·fed·er·ate** (kən-fĕd′ər-ĭt) *n.* [< ME *confederat,* allied < LLat. *confoederatus,* p.part. of *confoederare,* to unite : Lat. *com-,* together + Lat. *foederare,* to unite < *foedus,* league.] **1.** A member of a confederacy: ALLY. **2.** An accomplice. **3. Confederate.** A supporter of the American Confederacy. —*adj.* **1.** United in a confederacy: ALLIED. **2. Confederate.** Of or relating to the American Confederacy. —*vt. & vi.* (-ə-rāt′) **-at·ed, -at·ing, -ates.** To form into or become part of a confederacy. **—con·fed′er·a′tive** *adj.*

**Confederate rose** *n.* COTTON ROSE 1.

**Confederate violet** *n.* A plant, *Viola priceana* of the southeastern United States, with streaked pale-blue flowers.

**con·fed·er·a·tion** (kən-fĕd′ə-rā′shən) *n.* **1.** An act of confederating or a state of being confederated. **2.** A group of confederates, esp. of states or nations, united for a common purpose. **—con·fed′er·a′tion·ism** *n.* **—con·fed′er·a′tion·ist** *n.*

**con·fer** (kən-fûr′) *v.* **-ferred, -fer·ring, -fers.** [Lat. *conferre* : *com-,* together + *ferre,* to bring.] —*vt.* **1.** To bestow (e.g., an honor). **2.** *Obs.* To compare. —*vi.* To hold a conference: consult together. **—con·fer′ment** *n.* **—con·fer′ra·ble** *adj.* **—con·fer′ral** *n.* **—con·fer′rer** *n.*

**con·fer·ee** *also* **con·fer·ree** (kŏn′fə-rē′) *n.* **1.** A participant in a conference. **2.** One on whom something is conferred.

**con·fer·ence** (kŏn′fər-əns, -frəns) *n.* [OFr. < Med. Lat. *conferentia* < Lat. *conferrens,* pr.part. of *conferre,* to bring together. —see CONFER.] **1. a.** A meeting for discussion or consultation. **b.** An exchange of views. **c.** A meeting of committees to settle differences between two legislative bodies. **2.** An assembly of clerical or clerical and lay members of a Protestant church from a particular district. **3.** An association of athletic teams. **4.** The act of conferring, as of a degree. **—con′fer·en′tial** (-fə-rĕn′shəl) *adj.*

**conference call** *n.* A conference by telephone in which several persons participate via a central switching unit.

**con·fer·ree** (kŏn′fə-rē′) *n. var. of* CONFEREE.

**con·fer·va** (kən-fûr′və) *n., pl.* **-vae** (-vē) *or* **-vas.** [Lat., comfrey.] Any of various bright-green, threadlike freshwater algae. **—con·fer′void′** (-void′) *n. & adj.*

**con·fess** (kən-fĕs′) *v.* **-fessed, -fess·ing, -fess·es.** [ME *confessen* < OFr. *confesser* < VLat. **confessare* < Lat. *confiteri,* to acknowledge : *com-* (intensive) + *fateri,* to admit.] —*vt.* **1.** To acknowledge or disclose (something damaging or inconvenient to oneself): ADMIT. **2.** To acknowledge belief or faith in. **3. a.** To make known (one's sins) to God or to a priest. **b.** To hear the confession of (a penitent). —*vi.* **1.** To acknowledge or admit <The child *confessed* to playing hooky.> **2.** To tell one's sins to a priest. **—con·fess′a·ble** *adj.* **—con·fess′ed·ly** (-ĭd-lē) *adv.*

**con·fess·er** (kən-fĕs′ər) *n. var. of* CONFESSOR.

**con·fes·sion** (kən-fĕsh′ən) *n.* **1.** An act of confessing. **2.** Something confessed. **3.** A formal admission of guilt. **4.** Disclosure of sins to a priest for absolution. **5.** An avowal of religious belief: CREED. **6.** A church or group of worshipers adhering to a particular creed.

**con·fes·sion·al** (kən-fĕsh′ə-nəl) *adj.* Of, relating to, or resembling confession. —*n.* A small stall in which a priest hears confessions.

**con·fes·sor** *also* **con·fess·er** (kən-fĕs′ər) *n.* **1.** A priest who hears confessions and gives absolution. **2.** A person who confesses. **3.** A person who confesses faith in Christianity in spite of persecution but does not suffer martyrdom.

**con·fet·ti** (kən-fĕt′ē) *pl.n.* [Ital., pl. of *confetto,* candy < Med. Lat. *confectum* < Lat. *confectus,* p.part. of *conficere,* to prepare. —see

CONFECT.] (*sing. in number*). Small pieces or streamers of colored paper thrown about on festive occasions.

**con·fi·dant** (kŏn′fĭ-dănt′, -dänt′, kŏn′fĭ-dănt′, -dänt′) *n.* [Fr. *confident* < Ital. *confidente* < Lat. *confidens,* p.part. of *confidere,* to rely on. —see CONFIDE.] One to whom private matters are confided.

**con·fi·dante** (kŏn′fĭ-dănt′, -dänt′, kŏn′fĭ-dănt′, -dänt′) *n.* [Fr. *confidente,* fem. of *confident,* confidant.] A woman to whom private matters are confided.

**con·fide** (kən-fīd′) *v.* **-fid·ed, -fid·ing, -fides.** [ME *confiden,* to rely on < OFr. *confider* < Lat. *confidere* : *com-* (intensive) + *fidere,* to trust.] —*vt.* **1.** To tell (something) confidentially. **2.** To place into another's keeping : ENTRUST. —*vi.* To tell private matters in confidence. **—con·fid′er** *n.*

**con·fi·dence** (kŏn′fĭ-dəns) *n.* **1.** Trust or faith. **2.** A trusting relationship. **3.** Something confided : SECRET. **4.** A feeling of assurance, esp. of self-assurance. **5.** The assurance that someone will keep a secret <told them in strict *confidence*>

☆ ***syns:*** **1.** CONFIDENCE, APLOMB, ASSURANCE, SECURITY, SELF-ASSURANCE, SELF-CONFIDENCE, SELF-POSSESSION *n. core meaning* : a firm belief in one's own powers <a person lacking *confidence*> **2.** CONFIDENCE, BELIEF, FAITH, RELIANCE, TRUST *n. core meaning* : absolute certainty in the trustworthiness of another <had *confidence* in the President> ***ant:*** doubt

**confidence game** *n.* A swindle brought about by gaining the victim's confidence.

**confidence interval** *n.* A statistical range with a specified probability that a given parameter lies within the range.

**confidence limit** *n.* One of the two values that specify the range of a confidence interval.

**confidence man** *n.* A person who swindles by using a confidence game.

**con·fi·dent** (kŏn′fĭ-dənt) *adj.* [Lat. *confidens, confident-,* pr.part. of *confidere,* to rely on. —see CONFIDE.] **1.** Marked by assurance, as of success. **2.** Characterized by self-assurance. **3.** Very bold : PRESUMPTUOUS. **4.** *Obs.* Trustful : confiding. **—con′fi·dent·ly** *adv.*

**con·fi·den·tial** (kŏn′fĭ-dĕn′shəl) *adj.* **1.** Communicated or effected secretly. **2.** Entrusted with the confidence of another <a *confidential* secretary> **3.** Denoting intimacy or confidence <a *confidential* nod of the head> **—con′fi·den′ti·al′i·ty** (-shē-ăl′ĭ-tē), **con′fi·den′tial·ness** *n.* **—con′fi·den′tial·ly** *adv.*

**confidential communication** *n. Law.* A statement made to someone, as one's doctor, attorney, or spouse, who cannot be compelled to divulge the information in court.

**con·fid·ing** (kən-fī′dĭng) *adj.* Tending to confide. **—con·fid′ing·ly** *adv.* **—con·fid′ing·ness** *n.*

**con·fig·u·ra·tion** (kən-fĭg′yə-rā′shən) *n.* [LLat. *configuratio* < Lat. *configurare,* to form after : *com-,* with + *figurare,* to form < *figura,* shape.] **1. a.** Arrangement of elements or parts. **b.** The form of a figure determined by the arrangement of its parts : CONTOUR. **2.** *Psychol.* A gestalt. **3.** *Chem.* The structural arrangement of atoms in a chemical compound or molecule. **—con·fig′u·ra′tion·al·ly** *adv.* **—con·fig′u·ra′tive, con·fig′u·ra′tion·al** *adj.*

**con·fine** (kən-fīn′) *v.* **-fined, -fin·ing, -fines.** [OFr. *confiner* < *confin,* boundary < Lat. *confine* < *confinis,* adjoining : *com-,* with + *finis,* border.] —*vt.* **1.** To keep within bounds : RESTRICT. **2.** To keep shut up : IMPRISON. **3.** To restrict in movement <*confined* to a wheelchair> —*vi.* To border : abut. —*n.* (kŏn′fīn′). **1. confines.** The limits of an area : BORDERS <within the *confines* of one's county> **2.** *Obs.* A prison. **—con·fin′a·ble, con·fine′a·ble** *adj.* **—con·fin′er** *n.*

**con·fine·ment** (kən-fīn′mənt) *n.* **1.** The act of confining or state of being confined. **2.** Lying-in.

**con·firm** (kən-fûrm′) *vt.* **-firmed, -firm·ing, -firms.** [ME *confirmen* < OFr. *confermer* < Lat. *confirmare* : *com-* (intensive) + *firmare,* to strengthen < *firmus,* strong.] **1.** To establish or support the certainty or validity of : VERIFY. **2.** To make firmer : STRENGTHEN. **3.** To make binding by formal approval : RATIFY. **4.** To administer the religious rite of confirmation to. **—con·firm′a·bil′i·ty** *n.* **—con·firm′a·ble** *adj.* **—con·firm′a·to′ry** (-fûr′mə-tôr′ē, -tōr′ē) *adj.* **—con·firm′er** *n.*

☆ ***syns:*** CONFIRM, ATTEST, AUTHENTICATE, BACK (up), BEAR OUT, CORROBORATE, SUBSTANTIATE, TESTIFY (to), VALIDATE, VERIFY, WARRANT *v. core meaning* : to assure the certainty or validity of <a suspicion *confirmed* by reliable testimony> ***ant:*** deny

**con·fir·ma·tion** (kŏn′fər-mā′shən) *n.* **1.** An act of confirming. **2.** Something that confirms : VERIFICATION. **3. a.** A Christian rite admitting a baptized person to full church membership. **b.** A ceremony in Judaism that marks the completion of a young person's religious training.

**con·firmed** (kən-fûrmd′) *adj.* **1.** Verified : ratified. **2.** Firmly fixed in habit : INVETERATE <a *confirmed* drunkard> **3.** Having received the rite of confirmation. **—con·firm′ed·ly** (-fûr′mĭd-lē) *adv.*

**con·fis·ca·ble** (kən-fĭs′kə-bəl) *adj.* Subject to confiscation.

**con·fis·cate** (kŏn′fĭ-skāt′) *vt.* **-cat·ed, -cat·ing, -cates.** [Lat. *confiscare, confiscat-* : *com-,* together + *fiscus,* treasury.] **1.** To seize (private property) for the public treasury. **2.** To seize by or as if by authority. —*adj.* (kŏn′fĭ-skāt′, kən-fĭs′kət). **1.** Seized by a government : APPROPRIATED. **2.** Having lost property through confiscation. **—con′fis·ca′tion** *n.* **—con′fis·ca′tor** *n.*

**Con·fi·te·or** (kən-fē′tē-ər, -ôr′) *n.* [Lat., I confess, the first word of the prayer.] *Rom. Cath. Ch.* A prayer in which a confession of sins is made.

**con·fi·ture** (kŏn′fĭ-choŏr′) *n.* [Fr. < OFr. < *confit,* confection. —see COMFIT.] A preserve, confection, or sweetmeat.

**con·fla·grant** (kən-flā′grənt) *adj.* [Lat. *conflagrans, conflagrant-,* pr.part. of *conflagrare,* to burn up : *com-* (intensive) + *flagrare,* to burn.] Burning intensely : BLAZING.

**con·fla·gra·tion** (kŏn′flə-grā′shən) *n.* [Lat. *conflagratio* < *conflagrare,* to burn up. —see CONFLAGRANT.] A large, destructive fire.

**con·flate** (kən-flāt′) *vt.* **-flat·ed, -flat·ing, -flates.** [Lat. *conflare, conflat-,* to fuse : *com-,* together + *flare,* to blow.] To combine (e.g., two variant texts) into one whole. **—con·fla′tion** *n.*

**con·flict** (kŏn′flĭkt′) *n.* [ME < Lat. *conflictus* < p.part. of *confligere,* to strike together : *com-,* together + *fligere,* to strike.] **1.** A state of open, prolonged fighting : WARFARE. **2.** A state of disharmony : CLASH. **3.** *Psychol.* The opposition or simultaneous functioning of mutually exclusive impulses, desires, or tendencies. **4.** A collision. —*vi.* (kən-flĭkt′) **-flict·ed, -flict·ing, -flicts. 1.** To be in or come into opposition : DIFFER. **2.** *Archaic.* To engage in warfare. **—con·flic′tion** *n.* **—con·flic′tive** *adj.*

☆ ***syns:*** **1.** CONFLICT, BELLIGERENCY, HOSTILITIES, STRIFE, WAR, WARFARE *n. core meaning* : a state of open, prolonged fighting <the Vietnam *conflict*> **2.** CONFLICT, CLASH, CONTENTION, DIFFICULTY, DISACCORD, DISCORD, DISSENSION, DISSENT, FRICTION, STRIFE *n. core meaning* : a state of disagreement and disharmony <a family torn by *conflicts*>

**conflict of interest** *n.* A conflict between the public obligations and the private interests of a public official.

**con·flu·ence** (kŏn′flo͞o-əns) *n.* **1. a.** A flowing together of two or more streams. **b.** The point of juncture of such streams. **2.** An assembly : crowd.

**con·flu·ent** (kŏn′flo͞o-ənt) *adj.* [ME < Lat. *confluens,* pr.part. of *confluere,* to flow together : *com-,* together + *fluere,* to flow.] **1.** Flowing together. **2.** *Pathol.* Merging together so as to form a mass, as sores in a rash. **3.** *Anat.* Coalesced, as two orig. separate bones. —*n.* **1.** One of two or more confluent streams. **2.** A tributary.

**con·flux** (kŏn′flŭks′) *n.* [< Lat. *confluxus,* p.part. of *confluere,* to flow together. —see CONFLUENT.] A confluence.

**con·fo·cal** (kŏn-fō′kəl) *adj.* Having the same focus or foci. **—con·fo′cal·ly** *adv.*

**con·form** (kən-fôrm′) *v.* **-formed, -form·ing, -forms.** [ME *conformen* < OFr. *conformer* < Lat. *conformare,* to shape after : *com-,* with + *formare,* to shape < *forma,* shape.] —*vi.* **1.** To be similar in form or character. **2.** To act or be in compliance : COMPLY. **3.** To behave in accordance with prevailing modes or customs. —*vt.* To bring into agreement or correspondence. **—con·form′er** *n.* **—con·form′ist** *n.*

**con·form·a·ble** (kən-fôr′mə-bəl) *adj.* **1.** Similar : corresponding. **2.** Ready to comply : SUBMISSIVE. **3.** *Geol.* Designating strata parallel to each other without interruption. **—con·form′a·bil′i·ty, con·form′a·ble·ness** *n.* **—con·form′a·bly** *adv.*

**con·for·mal** (kən-fôr′məl) *adj.* [LLat. *conformalis,* similar : Lat. *com-,* together + *forma,* shape.] **1.** *Math.* Designating a depiction of a surface or region on another surface so that all angles between intersecting curves remain unchanged. **2.** Of or relating to a map projection in which small areas are rendered with true shape.

**con·for·mance** (kən-fôr′məns) *n.* Conformity.

**con·for·ma·tion** (kŏn′fər-mā′shən) *n.* **1.** The structure or outline of something determined by the arrangement of its parts. **2.** Symmetric arrangement of parts. **3.** An act of conforming or the state of being conformed. **4.** One of the spatial arrangements of atoms in a molecule that can come about through free rotation of the atoms about a single chemical bond. **—con′for·ma′tion·al** *adj.* **—con′for·ma′tion·al·ly** *adv.*

**con·form·i·ty** (kən-fôr′mĭ-tē) *n., pl.* **-ties. 1.** Likeness in form or character : AGREEMENT <behaved in *conformity* with the rules> **2.** Action in correspondence with prevailing customs, rules, or styles <*conformity* to town ordinances>

**con·found** (kən-found′, kŏn-) *vt.* **-found·ed, -found·ing, -founds.** [ME *confounden* < AN *confoundre* < Lat. *confundere,* to mix together : *com-,* together + *fundere,* to pour.] **1.** To cause (a person) to become confused : BEWILDER. **2.** To fail to distinguish : MIX UP <*confound* truth and lies> **3.** To cause to be ashamed : ABASH <an unexpected success that *confounded* our critics> **4.** To damn. **5.** *Archaic.* To defeat : overthrow. **—con·found′er** *n.*

**con·found·ed** (kən-foun′dĭd, kŏn-) *adj.* **1.** Befuddled : confused. **2.** Damned <a *confounded* idiot> **—con·found′ed·ly** *adv.* **—con·found′ed·ness** *n.*

**con·fra·ter·ni·ty** (kŏn′frə-tûr′nĭ-tē) *n., pl.* **-ties.** [ME *confraternite* < OFr. < Med. Lat. *confraternitas* < *confrater,* colleague. —

see CONFRERE.] A group of people united in a common profession or for a common purpose.

**con·frere** (kŏn′frâr′) *n.* [ME < OFr. < Med. Lat. *confrater* : Lat. *com-*, together + Lat. *frater*, brother.] A colleague : associate.

**con·front** (kən-frŭnt′) *vt.* **-front·ed, -front·ing, -fronts.** [OFr. *confronter*, to adjoin < Med. Lat. *confrontare* : Lat. *com-*, together + Lat. *frons*, front.] **1.** To come face to face with, esp. with defiance or hostility. **2.** To bring face to face <*confronted* them with all the evidence> **3.** To come up against : ENCOUNTER <*confronting* new problems everyday> **—con′fron·ta′tion** (kŏn′frŭn-ta′tion), **con·front′ment** *n.* **—con′fron·ta′tion·al** *adj.* **—con·front′er** *n.*

**Con·fu·cian** (kən-fyoo͞′shən) *adj.* Of, relating to, or typical of the Chinese philosopher Confucius, his teachings, or his followers. —*n.* A person who adheres to the teachings of Confucius. **—Con·fu′cian·ism** *n.* **—Con·fu′cian·ist** *n.*

**con·fuse** (kən-fyoo͞z′) *vt.* **-fused, -fus·ing, -fus·es.** [< ME *confused*, perplexed < Lat. *confusus*, p.part. of *confundere*, to mix together. —see CONFOUND.] **1.** To cause to be unclear in mind or intent. **2. a.** To mix up : JUMBLE. **b.** To mistake (one thing for another) <*confused* success with happiness> **c.** To make unclear : BLUR. **—con·fus′ed·ly** (-fyoo͞′zĭd-lē) *adv.* **—con·fus′ed·ness** *n.* **—con·fus′ing·ly** *adv.*

☆ ***syns:*** **1.** CONFUSE, ADDLE, BEFUDDLE, BEWILDER, CONFOUND, DISCOMBOBULATE, FUDDLE, MIX UP, PERPLEX, THROW *v. core meaning* : to cause to be unclear in mind or intent <camouflage that *confused* the enemy><was *confused* by the legal documents> **2.** CONFUSE, DISORDER, JUMBLE, SCRAMBLE, SNARL *v. core meaning*: to put into total disorder <had hopelessly *confused* their finances>

▲ word history: *Confuse* as a verb with a full conjugation did not appear in English until the 19th century. It is derived from *confused*, which first occurs in the 14th century as a past participle of *confound.* This odd situation springs directly from the Latin source of both *confound* and *confuse*, which is *confundere*, literally "to pour together." English *confound* is derived from *confundere. Confused* is derived from *confusus*, the past participle of *confundere*, with the English past participle suffix *-ed* replacing the Latin suffix *-us. Confuse* was formed by deleting the suffix from *confused.* Since its emergence as a full verb in its own right, *confuse* is no longer an exact synonym of *confound.*

**con·fu·sion** (kən-fyoo͞′zhən) *n.* The act of confusing or the state of being confused. **—con·fu′sion·al** *adj.*

**con·fu·ta·tion** (kŏn′fyoo͞-tā′shən) *n.* **1.** The act of confuting. **2.** Something that confutes. **—con·fu′ta·tive** (kən-fyoo͞′tə-tĭv) *adj.*

**con·fute** (kən-fyoo͞t′) *vt.* **-fut·ed, -fut·ing, -futes.** [Lat. *confutare.*] **1.** To refute decisively. **2.** *Archaic.* To confound. **—con·fut′a·ble** *adj.* **—con·fut′er** *n.*

**con·ga** (kŏng′gə) *n.* [Am. Sp. (Caribbean) (*danza*) *Conga*, Congo dance < Sp. *Congo*, of the Congo.] **1.** A dance of Latin-American origin in which the dancers form a long, winding line. **2.** Music for the conga. —*vi.* **-gaed, -ga·ing, -gas.** To dance the conga.

**con game** *n. Slang.* A confidence game.

**con·gé** (kŏn′zhā′, -jā′, kôN-zhā′) *also* **con·gee** (kŏn′jē) *n.* [Fr. < OFr. *congie* < Lat. *commeatus* < *commeare*, to come and go : *com-*, together + *meare*, to go.] **1.** Formal or authoritative permission to depart. **2.** A hasty dismissal. **3. a.** *Archaic.* A formal bow. **b.** A farewell. **4.** A concave molding.

**con·geal** (kən-jēl′) *v.* **-gealed, -geal·ing, -geals.** [ME *congelen* < Lat. *congelare* : *com-*, together + *gelare*, to freeze.] —*vi.* **1.** To solidify, as by freezing. **2.** To jell : coagulate. —*vt.* To cause to solidify. **—con·geal′a·ble** *adj.* **—con·geal′er** *n.* **—con·geal′ment** *n.*

**con·gee** (kŏn′jē) *vi.* **-geed, -gee·ing, -gees.** [ME *congeien* < OFr. *congier* < *congie*, leave. —see CONGÉ.] *Archaic.* **1.** To take ceremonious leave. **2.** To make a formal bow. —*n. var. of* CONGÉ.

**con·ge·la·tion** (kŏn′jə-lā′shən) *n.* **1.** The act or process of congealing. **2.** The state of being congealed. **3.** A coagulation.

**con·ge·ner** (kŏn′jə-nər) *n.* [Lat., of the same race : *com-*, same + *genus*, race.] **1.** A member of the same kind, class, or group. **2.** An organism belonging to the same genus as another or others. **—con′ge·ner′ic** (-nĕr′ĭk), **con·gen′er·ous** (kən-jĕn′ər-əs, kŏn-) *adj.*

**con·ge·net·ic** (kŏn′jə-nĕt′ĭk) *adj.* Similar in origin.

**con·gen·ial** (kən-jēn′yəl) *adj.* **1.** Having the same tastes, habits, or temperament : SYMPATHETIC. **2.** Of a friendly disposition : SOCIABLE. **3.** Agreeably suited to one's needs or nature <*congenial* quarters> **—con·ge′ni·al′i·ty** (-jē′nē-ăl′ĭ-tē), **con·gen′ial·ness** *n.* **—con·gen′ial·ly** *adv.*

**con·gen·i·tal** (kən-jĕn′ĭ-tl) *adj.* [< Lat. *congenitus* : *com-*, with + *genitus*, born, p.part. of *gignere*, to bear.] **1.** Existing at birth but not hereditary <a *congenital* abnormality> **2.** Being such as if by nature <a *congenital* liar> **—con·gen′i·tal·ly** *adv.*

**congenital anomaly** *n.* A physiological or structural abnormality that develops before birth.

**con·ger** or **con·ger eel** (kŏng′gər) *n.* [ME *congre* < OFr. < Lat. *conger* < Gk. *gongros.*] A large, scaleless marine eel of the family Congridae, esp. *Conger oceanicus* of Atlantic waters.

**con·ge·ries** (kən-jîr′ēz′, kŏn′jə-rēz′) *n.* [Lat. < *congerere*, to heap up. —see CONGEST.] (*sing. in number*). A collection : aggregation.

**con·gest** (kən-jĕst′) *v.* **-gest·ed, -gest·ing, -gests.** [Lat. *congerere*, *congest-*, to heap up : *com-*, together + *gerere*, to carry.] —*vt.* **1.** To overfill or overcrowd. **2.** *Pathol.* To cause excessive blood accumulation in (a vessel or organ). —*vi.* To become congested. **—con·ges′tion** *n.* **—con·ges′tive** *adj.*

**con·gi·us** (kŏn′jē-əs) *n., pl.* **-gi·i** (-jē-ī′) [ME, a liquid measure < Lat.] **1.** A gallon. **2.** An ancient Roman measure for liquids, equal to about .84 of the U.S. gallon.

**con·glo·bate** (kŏn-glō′bāt′, kŏng′glō-) *vt.* **-bat·ed, -bat·ing, -bates.** [Lat. *conglobare, conglobat-* : *com-*, together + *globus*, ball.] To form into a ball or globe. **—con·glo′bate′** *adj.* **—con′glo·ba′tion** *n.*

**con·globe** (kən-glōb′) *vt.* **-globed, -glob·ing, -globes.** To conglobate.

**con·glom·er·ate** (kən-glŏm′ə-rāt′) *vi.* & *vt.* **-at·ed, -at·ing, -ates.** [Lat. *conglomerare, conglomerat-* : *com-*, together + *glomerare*, to wind into a ball < *glomus*, ball.] To form or cause to form into an adhering or rounded mass. —*n.* (-ər-ĭt). **1.** A collected heterogeneous mass : CLUSTER. **2.** *Geol.* A rock composed of pebbles and gravel embedded in a loosely cementing material. **3.** A business enterprise consisting of widely diversified companies. —*adj.* (-ər-ĭt). **1.** Gathered into a mass : CLUSTERED. **2.** *Geol.* Composed of loosely cemented heterogeneous material. **—con·glom′er·at′ic** (-ə-răt′ĭk), **con·glom′er·it′ic** (-ə-rĭt′ĭk) *adj.*

**con·glom·er·a·tion** (kən-glŏm′ə-rā′shən) *n.* **1.** The process of conglomerating or the state of being conglomerated. **2.** A collection of miscellaneous things.

**con·glu·ti·nate** (kən-gloo͞t′n-āt′, kŏn-) *vi.* & *vt.* **-nat·ed, -nat·ing, -nates.** [ME *conglutinaten* < Lat. *conglutinare*, to glue together : *com-*, together + *glutinare*, to glue < *gluten*, glue.] **1.** To become or cause to become stuck or glued together. **2.** *Med.* To become or cause to become reunited, as bones or tissues. **—con·glu′ti·na′tion** *n.*

**con·go dye** (kŏng′gō) *n.* [After *Congo*, a region of Africa.] A nitrogen-containing dye usu. derived from benzidine.

**congo eel** *n.* An eellike amphibian, *Amphiuma means* of the southeastern United States, with two pairs of nonfunctioning legs.

**Congo red** *n.* A brownish-red powder, $C_{32}H_{22}N_6O_6S_2Na_2$, used in medicine and as a dye, indicator, and biological stain.

**Congo snake** *n.* Congo eel.

**con·gou** (kŏng′gō, -goo͞) *n.* [Chin. (Amoy) *kong hu* (*tē*), elaborately prepared (tea).] A black tea from China.

**con·grat·u·late** (kən-grăch′ə-lāt′) *vt.* **-lat·ed, -lat·ing, -lates.** [Lat. *congratulari, congratulat-* : *com-*, with + *gratulari*, to rejoice < *gratus*, pleasing.] To express happiness to (a person) because of achievement or good fortune. **—con·grat′u·la′tor** *n.* **—con·grat′u·la·to′ry** (-lə-tôr′ē, -tōr′ē) *adj.*

**con·grat·u·la·tion** (kən-grăch′ə-lā′shən) *n.* **1.** The act of congratulating. **2.** *often* **congratulations.** An expression of happiness for the achievement or good fortune of another.

**con·gre·gate** (kŏng′grĭ-gāt′) *vt.* & *vi.* **-gat·ed, -gat·ing, -gates.** [ME *congregaten* < Lat. *congregare* : *com-*, together + *gregare*, to assemble < *grex*, herd.] To bring or come together in a crowd : ASSEMBLE. —*adj.* (-gĭt). Assembled : gathered. **—con′gre·ga′tive** *adj.* **—con′gre·ga′tive·ness** *n.* **—con′gre·ga′tor** *n.*

**con·gre·ga·tion** (kŏng′grĭ-gā′shən) *n.* **1.** An act of congregating. **2.** A gathering of people : ASSEMBLY. **3. a.** A group of people gathered for religious worship. **b.** The members of a particular religious group who regularly worship at a common church. **4.** *Rom. Cath. Ch.* **a.** A religious institute in which only simple vows are taken. **b.** A division of the Curia.

**con·gre·ga·tion·al** (kŏng′grĭ-gā′shə-nəl) *adj.* **1.** Of or relating to a congregation. **2. Congregational.** Of or relating to Congregationalism or Congregationalists.

**con·gre·ga·tion·al·ism** (kŏng′grĭ-gā′shə-nə-lĭz′əm) *n.* **1.** A type of church government in which each local congregation is self-governing. **2. Congregationalism.** The system of government and religious beliefs of a Protestant denomination in which each member church is autonomous. **—con′gre·ga′tion·al·ist** *n.*

**con·gress** (kŏng′grĭs) *n.* [Lat. *congressus*, meeting < *congredi*, to meet : *com-*, together + *gradi*, to go.] **1.** A formal assembly of representatives, as of various nations, to discuss problems. **2.** The legislature of a nation, esp. of a republic. **3. Congress. a.** The national legislative body of the United States, consisting of the Senate and the House of Representatives. **b.** The two-year session of this legislature between elections of the House of Representatives. **4.** The act of coming together : MEETING. **5.** Sexual intercourse. **—con·gres′sion·al** (kən-grĕsh′ə-nəl) *adj.* **—con·gres′sion·al·ly** *adv.*

**congress boot** *n.* [From its former popularity among members of the U.S. Congress.] An ankle-high shoe with elastic material in the sides.

**congress gaiter** *n.* A congress boot.

**Congressional Medal of Honor** *n.* The Medal of Honor.

**con·gress·man** (kŏng′grĭs-mən) *n.* A member of the U.S. Congress, esp. of the House of Representatives.

**con·gress·wom·an** (kŏng′grĭs-wo͝om′ən) *n.* A woman member of the U.S. Congress, esp. of the House of Representatives.
**con·gru·ence** (kŏng′gro͞o-əns, kən-gro͞o′-) *also* **con·gru·en·cy** (-ən-sē) *n., pl.* **-enc·es** *also* **-en·cies.** **1.** Accord : conformity. **2.** *Math.* **a.** The state of being congruent. **b.** A mathematical statement that two quantities are congruent.
**con·gru·ent** (kŏng′gro͞o-ənt, kən-gro͞o′-) *adj.* [ME < Lat. *congruens,* pr.part. of *congruere,* to agree.] **1.** Corresponding : congruous. **2.** *Math.* **a.** Coinciding exactly when superimposed <*congruent* triangles> **b.** Having a difference divisible by a modulus <*congruent* numbers> **—con′gru·ent·ly** *adv.*
**con·gru·i·ty** (kən-gro͞o′ĭ-tē, kŏn-) *n., pl.* **-ties.** **1.** The quality or fact of being congruous. **2.** The quality or fact of being congruent. **3.** A point of agreement.
**con·gru·ous** (kŏng′gro͞o-əs) *adj.* [Lat. *congruus* < *congruere,* to agree.] **1.** Corresponding in character or kind : HARMONIOUS. **2.** *Math.* Congruent. **—con′gru·ous·ly** *adv.* **—con′gru·ous·ness** *n.*
**con·ic** (kŏn′ĭk) *also* **con·i·cal** (-ĭ-kəl) *adj.* [NLat. *conicus* < Gk. *konikos* < *konos,* cone.] **1.** Shaped like a cone. **2.** Of or relating to a cone. *—n. Math.* A conic section.
**conic projection** *or* **conical projection** *n.* A method of projecting pictures of parts of the earth's spherical surface on a surrounding cone, which is then flattened to a plane surface having concentric circles as parallels of latitude and radiating lines from the apex as meridians.
**conic section** *n.* One of a group of plane curves, including the circle, ellipse, hyperbola, and parabola, generated by: **a.** An intersection of a right circular cone and a plane. **b.** The plane locus of a point that moves so that the ratio of its distance to a fixed point to its distance from a fixed line is a positive constant. **c.** A graph of the general quadratic equation in two variables.
**co·nid·i·o·phore** (kə-nĭd′ē-ə-fôr′, -fōr′) *n.* [CONIDI(UM) + -PHORE.] A specialized hyphal filament in fungi, bearing conidia. **—co·nid′i·o·phor′ous** *adj.*
**co·nid·i·um** (kə-nĭd′ē-əm) *n., pl.* **-i·a** (-ē-ə) [NLat. < Gk. *konis,* dust.] An asexual fungus spore, usu. produced on a specialized sporophore. **—co·nid′i·al** (-əl) *adj.*
**con·i·fer** (kŏn′ə-fər, kō′nə-) *n.* [< NLat. *Coniferae,* family name < Lat. *conifer,* cone-bearing : *conus,* cone (< Gk. *konos*) + *ferre,* to bear.] A predominantly evergreen cone-bearing tree, as a pine, spruce, hemlock, or fir. **—co·nif′er·ous** (kō-nĭf′ər-əs, kə-) *adj.*
**co·ni·ine** (kō′nē-ēn′) *also* **co·nin** (kō′nĭn) *or* **co·nine** (-nēn′) *n.* [G. *Koniin* < LLat. *conium,* conium.] A colorless, poisonous liquid alkaloid, $C_8H_{17}N$, obtained from poison hemlock.
**co·ni·o·sis** (kō′nē-ō′sĭs) *n.* [< Gk. *konia,* dust + -OSIS.] A pathological condition caused by dust inhalation.
**co·ni·um** (kō′nē-əm) *n.* [LLat. *conium,* hemlock < Gk. *kōneion.*] One of several poisonous plants of the genus *Conium,* including the poison hemlock.
**con·i·za·tion** (kō′nĭ-zā′shən, kŏn′ĭ-) *n. Med.* Diagnostic excision of a cone of tissue.
**con·jec·tur·al** (kən-jĕk′chər-əl) *adj.* **1.** Based on or involving conjecture. **2.** Inclined to conjecture. **—con·jec′tur·al·ly** *adv.*
**con·jec·ture** (kən-jĕk′chər) *n.* [ME < Lat. *conjectura* < *conicere,* to infer : *com-,* together + *jacere,* to throw.] **1.** Inference based on incomplete or inconclusive evidence. **2.** A conclusion based on inference. *—v.* **-tured, -tur·ing, -tures.** *—vt.* To infer from incomplete evidence : GUESS. *—vi.* To make a conjecture. **—con·jec′tur·a·ble** *adj.* **—con·jec′tur·a·bly** *adv.* **—con·jec′tur·er** *n.*
**con·join** (kən-join′) *vt.* & *vi.* **-joined, -join·ing, -joins.** [ME *conjoinen* < OFr. *conjoindre* < Lat. *conjungere* : *com-,* together + *jungere,* to join.] To join or become joined together : UNITE. **—con·join′er** *n.*
**con·joint** (kən-joint′) *adj.* [ME < OFr., p.part. of *conjoindre,* to conjoin.] **1.** Joined together : UNITED. **2.** Of, relating to, having, or done by two or more united persons or things. **—con·joint′ly** *adv.*
**con·ju·gal** (kŏn′jə-gəl, kən-jo͞o′-) *adj.* [Lat. *conjugalis* < *conjunx,* spouse < *conjungere,* to join in marriage. —see CONJOIN.] Of or relating to marriage or the marital relationship. **—con′ju·gal′i·ty** (-găl′ĭ-tē) *n.* **—con′ju·gal·ly** *adv.*
**con·ju·gant** (kŏn′jə-gənt) *n.* [< Lat. *conjugans, conjugant-,* pr.part. of *conjugare,* to unite. —see CONJUGATE.] Either of a pair of organisms, cells, or gametes undergoing conjugation.
**con·ju·gate** (kŏn′jə-gāt′) *v.* **-gat·ed, -gat·ing, -gates.** [< ME *conjugat,* joined < Lat. *conjugatus,* p.part. of *conjugare,* to join together : *com-,* together + *jugare,* to join < *jugum,* yoke.] *—vt.* **1.** To give the various inflected forms of (a word, esp. a verb). **2.** To join together. *—vi.* **1.** *Biol.* To undergo conjugation. **2.** To give the various inflected forms of a word, esp. a verb. *—adj.* (-gĭt, -gāt′). **1.** Joined together, esp. in a pair or pairs : COUPLED. **2.** *Math.* & *Physics.* Inversely or oppositely related with respect to one of a group of otherwise identical properties, esp. designating either or both of a pair of complex numbers differing only in the sign of the imaginary term. **3.** Of or relating to words having the same derivation and usu. a related meaning. *—n.* (-gĭt, -gāt′). **1.** One of two or more conjugate words. **2.** *Math.* & *Physics.* Either of a pair of conjugate quantities. **—con′ju·gate′ly** *adv.* **—con′ju·ga′tive** *adj.* **—con′ju·ga′tor** *n.*
**conjugated protein** *n.* A compound consisting of a protein and a nonprotein.
**con·ju·ga·tion** (kŏn′jə-gā′shən) *n.* **1.** The act of conjugating or the state of being conjugated. **2. a.** A verb inflection. **b.** A graphic presentation of the complete set of inflected forms of a verb. **c.** A class of verbs with similar inflected forms. **3. a.** Sexual reproduction in which ciliate protozoans of the same species temporarily couple and exchange genetic material. **b.** Chromosome pairing in the first meiotic division. **c.** Fusion of gamete nuclei : KARYOGAMY. **d.** Union of sex cells : SYNGAMY. **—con′ju·ga′tion·al** *adj.* **—con′ju·ga′tion·al·ly** *adv.*
**conjugation tube** *n.* A slender protoplasmic tube in some algae through which gametes may move to unite sexually with other gametes.
**con·junct** (kən-jŭngkt′, kŏn′jŭngkt′) *adj.* [ME < Lat. *conjunctus,* p.part. of *conjungere,* to join together. —see CONJOIN.] **1.** Joined together : UNITED. **2.** Designating adjacent successive tones of the musical scale. **—con·junct′ly** *adv.*
**con·junc·tion** (kən-jŭngk′shən) *n.* **1.** The act of uniting or the state of being united. **2.** A simultaneous occurrence in space or time : CONCURRENCE. **3.** One of the parts of speech in some languages comprising such words as, in English, *and, but, because,* and *as,* that connect other words, phrases, clauses, or sentences. **4.** *Astron.* The position of two celestial bodies on the celestial sphere when they have the same celestial longitude. **—con·junc′tion·al** *adj.* **—con·junc′tion·al·ly** *adv.*
**con·junc·ti·va** (kŏn′jŭngk-tī′və) *n., pl.* **-vas** *or* **-vae** (-vē) [ME < Med. Lat. *(membrana) conjunctiva,* connective (membrane) < LLat. *conjunctivus,* connective. —see CONJUNCTIVE.] The mucous membrane lining the inner surface of the eyelid and the exposed surface of the eyeball. **—con′junc·ti′val** (-vəl) *adj.*
**con·junc·tive** (kən-jŭngk′tĭv) *adj.* [LLat. *conjunctivus* < Lat. *conjunctus,* p.part. of *conjungere,* to join together. —see CONJOIN.] **1.** Connective : joining. **2.** Joined together : COMBINED. **3. a.** Of or used as a conjunction. **b.** Serving to connect elements of meaning and construction in a sentence, as *and* and *moreover.* *—n.* A connective word, esp. a conjunction. **—con·junc′tive·ly** *adv.*
**con·junc·ti·vi·tis** (kən-jŭngk′tə-vī′tĭs) *n. Pathol.* Inflammation of the conjunctiva.
**con·junc·ture** (kən-jŭngk′chər) *n.* **1.** A combination of circumstances. **2.** A critical set of circumstances : CRISIS.
**con·ju·ra·tion** (kŏn′jə-rā′shən) *n.* **1. a.** The act of conjuring. **b.** A magic spell : INCANTATION. **2.** Magic : legerdemain. **3.** A solemn appeal : INVOCATION.
**con·jure** (kŏn′jər, kən-jo͝or′) *v.* **-jured, -jur·ing, -jures.** [ME *conjuren* < OFr. *conjurer* < Lat. *conjurare,* to swear together : *com-,* together + *jurare,* to swear.] *—vt.* **1.** To entreat or call on solemnly, esp. by an oath. **2. a.** To summon by oath, incantation, or magic spell. **b.** To cause or effect by or as if by magic. **c.** To call to mind : EVOKE <sights and sounds *conjuring* up childhood Christmases> *—vi.* **1.** To practice magic, esp. legerdemain. **2.** To summon a devil by oath, incantation, or magic spell. **3.** *Obs.* To conspire.
**con·jur·er** *also* **con·jur·or** (kŏn′jər-ər, kŭn′-) *n.* **1.** A magician. **2.** A wizard : sorcerer.
**conk**[1] (kŏngk) *n.* [Orig. unknown.] *Slang.* **1.** The head. **2.** The nose. **3.** A blow, esp. on the head. *—v.* **conked, conk·ing, conks.** *—vt.* To hit, esp. on the head. *—vi.* **1.** To fail abruptly <The oven *conked* out.> **2.** To fall asleep <*conked* out after dinner> **3.** To pass out : FAINT.
**conk**[2] (kŏngk) *n.* [Perh. alteration of CONCH.] A hard, shelflike fruiting body of a fungus, esp. of the genera *Polyporus* and *Fomes,* found growing on tree trunks.
**conk**[3] (kŏngk) *n.* [Perh. alteration of *congolene,* a substance for straightening hair.] A hair style in which the hair is straightened, usu. by a chemical process. *—vt.* **conked, conk·ing, conks.** To straighten (hair) usu. by a chemical process.
**con man** *n. Slang.* A confidence man.
**conn** (kŏn) *v.* & *n. var. of* CON[3].
**con·nate** (kŏn′āt′, kŏ-nāt′) *adj.* [LLat. *connatus,* p.part. of *connasci,* to be born with : Lat. *com-,* with + Lat. *nasci,* to be born.] **1.** Inborn : innate. **2.** Coexisting since or associated in birth or origin : COGNATE. **3.** *Biol.* Congenitally or firmly united, as like parts or organs. **—con′nate′ly** *adv.* **—con′nate′ness** *n.*
**con·nat·u·ral** (kə-năch′ər-əl, kŏ-) *adj.* [Med. Lat. *connaturalis* : Lat. *com-,* together + Lat. *naturalis,* by birth. —see NATURAL.] **1.** Innate : inborn. **2.** Similar in nature : COGNATE. **—con·nat′u·ral′i·ty** (-ə-răl′ĭ-tē) *n.* **—con·nat′u·ral·ly** *adv.* **—con·nat′u·ral·ness** *n.*
**con·nect** (kə-nĕkt′) *v.* **-nect·ed, -nect·ing, -nects.** [ME *connecten* < Lat. *conectere* : *com-,* together + *nectere,* to bind.] *—vt.* **1.** To join or fasten together : UNITE. **2.** To associate or relate <reason to *connect* the two burglaries> **3.** To join to a communications circuit. *—vi.* **1.** To become united or joined <two rivers *connecting*> **2.** *Informal.* To hit, throw, or shoot successfully <*connected* for the winning run> **—con·nect′ed·ly** (-nĕk′tĭd-lē) *adv.* **—con·nect′i·ble, con·nect′a·ble** *adj.* **—con·nec′tor, con·nect′er** *n.*

**connecting rod** *n.* A rod linking rotating machine parts in reciprocating motion and connecting the crankshaft of an automobile to a piston.

**con·nec·tion** (kə-nĕk′shən) *n.* **1.** The act of connecting or the state of being connected. **2.** Something that connects : LINK. **3.** An association or relationship <a *connection* between the two incidents> **4.** Logical ordering of words or ideas : COHERENCE. **5.** The relation of a word or idea to the surrounding text : CONTEXT <In this *connection* the wrong word was used.> **6.** A person with whom one is associated, as by kinship, common interests, or marriage <needed *connections* to get the job> **7. a.** The meeting of various means of transportation for the transfer of passengers. **b.** A line of communication between two points. **8.** *Slang* **a.** A narcotics dealer. **b.** A narcotics purchase. **—con·nec′tion·al** *adj.*

**con·nec·tive** (kə-nĕk′tĭv) *adj.* Serving or tending to connect. *—n.* **1.** Something that connects. **2.** A word, as a conjunction, that connects words, phrases, clauses, and sentences. **3.** *Bot.* The tissue of a stamen that forms the division between the two lobes of an anther. **—con·nec′tive·ly** *adv.* **—con′nec·tiv′i·ty** (kŏn′ĕk-tĭv′ĭ-tē) *n.*

**connective tissue** *n.* Tissue arising chiefly from the embryonic mesoderm, including mucous, fibrous, reticular, adipose, cartilage, and bone tissue, marked by a highly vascular matrix structure and forming the supporting and connecting structures of the body.

**connect time** *n. Computer Sci.* The elapsed time during which a user of a remote terminal is connected with a time-sharing system.

**con·nex·ion** (kə-nĕk′shən) *n. Chiefly Brit. var. of* CONNECTION.

**conning tower** *n.* [< CON[3].] **1.** The armored pilothouse of a warship. **2.** An enclosed, raised observation post on a submarine, often used as a means of entrance.

**con·nip·tion** (kə-nĭp′shən) *n.* [Orig. unknown.] *Informal.* A fit of violent emotion.

**con·niv·ance** *also* **con·niv·ence** (kə-nī′vəns) *n.* **1.** The act of conniving. **2.** *Law.* Knowledge of and tacit consent to the commission of a wrongful act.

**con·nive** (kə-nīv′) *vi.* **-nived, -niv·ing, -nives.** [Lat. *conivēre.*] **1.** To feign ignorance of or fail to take measures against a known wrong. **2.** To cooperate secretly. **3.** To plot : conspire. **—con·niv′er** *n.* **—con·niv′er·y** *n.*

**con·niv·ence** (kə-nī′vəns) *n. var. of* CONNIVANCE.

**con·ni·vent** (kə-nī′vənt) *adj.* [Lat. *connivens, connivent-,* pr.part. of *conivere,* to close the eyes.] *Biol.* Converging and touching. —Used esp. of stamens or an insect's wings.

**con·nois·seur** (kŏn′ə-sûr′) *n.* [Obs. Fr. < OFr. *connoisseor* < *connoistre,* to know < Lat. *cognoscere,* to learn. —see COGNITION.] One with knowledgeable and sophisticated discrimination, esp. concerning the arts or matters of taste. **—con′nois·seur′ship′** *n.*

**con·no·ta·tion** (kŏn′ə-tā′shən) *n.* **1.** The act or process of connoting. **2. a.** The configuration of suggestive or associative implications constituting the general sense of an abstract expression beyond its literal, explicit sense. **b.** A secondary meaning suggested by a word in addition to its literal meaning. **3.** *Logic.* The total of the attributes constituting the meaning of a term : INTENSION. **—con′no·ta′tive** *adj.* **—con′no·ta′tive·ly** *adv.*

**con·note** (kə-nōt′) *vt.* **-not·ed, -not·ing, -notes.** [Med. Lat. *connotare,* to mark along with : Lat. *com-,* with + Lat. *notare,* to mark < *nota,* mark.] **1.** To suggest in addition to literal meaning <The word "Orient" often *connotes* mystery.> **2.** To involve as a condition or consequence <Lying often *connotes* guilt.>

**con·nu·bi·al** (kə-nōō′bē-əl, -nyōō′-) *adj.* [Lat. *connubialis* < *connubium,* marriage : *com-,* together + *nubere,* to marry.] Conjugal. **—con·nu′bi·al·ism** *n.* **—con·nu′bi·al′i·ty** (-ăl′ĭ-tē) *n.*

**co·noid** (kō′noid′) *also* **co·noi·dal** (kō-noid′l) *adj.* Shaped like a cone. **—co′noid** *n.*

**con·quer** (kŏng′kər) *v.* **-quered, -quer·ing, -quers.** [ME *conqueren* < OFr. *conquerre* < VLat. **conquaerere* < Lat. *conquirere,* to procure : *com-* (intensive) + *quaerere,* to seek.] *—vt.* **1.** To overcome by force, esp. by force of arms. **2.** To gain control over by surmounting impediments <handicapped people *conquering* mobility problems> **3.** To overcome or surmount by mental or moral force <*conquered* my fear of heights> *—vi.* To be victorious : WIN. **—con′quer·a·ble** *adj.* **—con′quer·or** *n.*

**con·quest** (kŏn′kwĕst′, kŏng′-) *n.* [ME < OFr. < Lat. *conquisitus,* p.part. of *conquirere,* to procure. —see CONQUER.] **1.** The act or process of conquering. **2.** Something, as territory, acquired by conquering. **3.** One whose love or favor has been captivated.

**con·qui·an** (kŏng′kē-ən) *n.* [Mex. Sp. *con quien* < Sp. *con quién,* with whom? : *con,* with (< Lat. *cum*) + *quien,* whom < Lat. *quem,* accusative of *quis,* who.] A card game for two players that resembles rummy.

**con·quis·ta·dor** (kŏn-kwĭs′tə-dôr′, kŏng-kē′stə-) *n., pl.* **-dors** or **-dor·es** (-dôr′ās, -ēz) [Sp. < *conquistar,* to conquer < Med. Lat. *conquestare,* freq. of VLat. **conquaerere.* —see CONQUER.] A conqueror, esp. one of the 16th-cent. Spanish conquerors of Mexico and Peru.

**con·san·guine** (kŏn-săng′gwĭn, kən-) *adj.* [Lat. *consanguineus.* —see CONSANGUINEOUS.] Consanguineous.

**con·san·guin·e·ous** (kŏn′săn-gwĭn′ē-əs, -săng-) *adj.* [Lat. *consanguineus* : *com-,* together + *sanguis,* blood.] Of the same lineage or origin, esp. related by blood. **—con′san·guin′e·ous·ly** *adv.*

**con·san·guin·i·ty** (kŏn′săn-gwĭn′ĭ-tē, -săng-) *n.* **1.** Blood relationship. **2.** A close connection : AFFINITY.

**con·science** (kŏn′shəns) *n.* [ME < OFr. < Lat. *conscientia* < *consciens,* pr.part. of *conscire,* to know wrong : *com-* (intensive) + *scire,* to know.] **1. a.** The faculty of recognizing the difference between right and wrong with regard to one's conduct coupled with a sense that one should act accordingly. **b.** Conformity to one's own sense of proper conduct. **2.** *Obs.* **a.** Consciousness. **b.** Inner thought. **—con′science·less** *adj.*

**conscience clause** *n.* A clause in a law that exempts persons whose conscientious or religious scruples forbid compliance.

**conscience money** *n.* Money paid often anonymously to atone for a dishonest act.

**con·sci·en·tious** (kŏn′shē-ĕn′shəs) *adj.* [Fr. *conscienteux* < Med. Lat. *conscientosus* < Lat. *conscientia,* conscience.] **1.** Governed by or accomplished according to conscience : SCRUPULOUS <a *conscientious* politician> **2.** Careful : painstaking <a *conscientious* editor> **—con′sci·en′tious·ly** *adv.* **—con′sci·en′tious·ness** *n.*

**conscientious objector** *n.* One whose religious and moral principles prohibit participation in military service.

**con·scio·na·ble** (kŏn′shə-nə-bəl) *adj.* [Obs. *conscions,* var. of CONSCIENCE + -ABLE.] Conscientious.

**con·scious** (kŏn′shəs) *adj.* [Lat. *conscius,* knowing with others : *com-,* together + *scire,* to know.] **1. a.** Aware of one's own existence, sensations, and thoughts and of one's environment <received a hard blow but remained *conscious*> **b.** Capable of thought, will, or perception. **2.** Subjectively known or felt <*conscious* regret> **3.** Deliberately conceived or done : INTENTIONAL <a *conscious* rebuke><made a *conscious* effort to keep awake> **4.** Having or showing self-consciousness : AWARE. *—n. Psychoanal.* The component of waking awareness perceptible by an individual at a given instant : CONSCIOUSNESS. **—con′scious·ly** *adv.*

**con·scious·ness** (kŏn′shəs-nĭs) *n.* **1.** The state of being conscious. **2.** The totality of attitudes, opinions, and sensitivities held or thought to be held by an individual or group <moral *consciousness*> **3.** *Psychoanal.* The conscious. **4. a.** A critical awareness of one's own situation and identity. **b.** Awareness : concern.

**consciousness-raising** *n.* **1.** A process of achieving greater awareness of one's needs in order to fulfill one's potential as an individual. **2.** A technique whereby one is made aware of discrimination against a given class of people. **—con′scious·ness-rais′er** *n.*

**con·script** (kŏn′skrĭpt′) *n.* [Lat. *conscriptus,* p.part. of *conscribere,* to enroll : *com-,* together + *scribere,* to write.] A person who is drafted : DRAFTEE. *—adj.* (kŏn′skrĭpt′). Compulsorily enrolled : DRAFTED. *—vt.* (kən-skrĭpt′) **-script·ed, -script·ing, -scripts.** To enroll compulsorily into military service : DRAFT.

**con·scrip·tion** (kən-skrĭp′shən) *n.* **1.** Compulsory enrollment, esp. for the armed forces : DRAFT. **2.** A monetary payment exacted by a government in wartime.

**con·se·crate** (kŏn′sĭ-krāt′) *vt.* **-crat·ed, -crat·ing, -crates.** [ME *consecraten* < Lat. *consecrare* : *com-* (intensive) + *sacrare,* to make sacred < *sacer,* sacred.] **1.** To set apart or declare as holy. **2. a.** To change (the elements of the Eucharist) into the body and blood of Christ. **b.** To initiate (a priest) into an order of bishops. **3.** To dedicate to a given goal or service. **4.** To make venerable <a custom *consecrated* by time> *—adj.* Dedicated to a sacred purpose : SANCTIFIED. **—con′se·cra′tion** *n.* **—con′se·cra′tive** *adj.* **—con′se·cra′tor** *n.* **—con′se·cra·to′ry** (-krə-tôr′ē, -tōr′ē) *adj.*

**con·se·cu·tion** (kŏn′sĭ-kyōō′shən) *n.* [Lat. *consecutio* < *consequi,* to follow closely. —see CONSEQUENT.] **1.** A succession or sequence. **2.** *Logic.* The relation of consequent to antecedent.

**con·sec·u·tive** (kən-sĕk′yə-tĭv) *adj.* [Fr. *consecutif* < Med. Lat. *consecutivus* < Lat. *consequi,* to follow closely. —see CONSEQUENT.] Successively following without interruption : CONTINUOUS. **—con·sec′u·tive·ly** *adv.* **—con·sec′u·tive·ness** *n.*

**con·sen·su·al** (kən-sĕn′shōō-əl) *adj.* [< CONSENSUS.] **1.** *Law.* Existing or brought about by consent. **2.** *Physiol.* Of or relating to an involuntary function or reflex that occurs on the opposite side of the body from the point of stimulation. **—con·sen′su·al·ly** *adv.*

**con·sen·sus** (kən-sĕn′səs) *n.* [Lat. < *consentire,* to agree. —see CONSENT.] **1.** Collective opinion. **2.** General accord : AGREEMENT. ***usage:*** The phrase *consensus of opinion* is considered redundant because *consensus* itself denotes a commonly held opinion.

**con·sent** (kən-sĕnt′) *vi.* **-sent·ed, -sent·ing, -sents.** [ME *consenten* < Lat. *consentire,* to agree : *com-,* together + *sentire,* to feel.] **1.** To give assent : AGREE. **2.** *Archaic.* To be of the same mind or opinion. *—n.* **1.** Voluntary allowance of what is planned or done by another : PERMISSION. **2.** Agreement and acceptance as to opinion or a course of action. **—con·sent′er** *n.*

**con·sen·ta·ne·ous** (kŏn′sĕn-tā′nē-əs) *adj.* [Lat. *consentaneus,* agreeing < *consentire,* to agree. —see CONSENT.] **1.** Showing agreement. **2.** Unanimous. **—con·sen′ta·ne′i·ty** (kən-sĕn′tə-nē′ĭ-tē), **con′sen·ta′ne·ous·ness** *n.* **—con′sen·ta′ne·ous·ly** *adv.*

**con·se·quence** (kŏn′sĭ-kwĕns′, -kwəns) *n.* **1.** That which logically

ă **pat** ā **pay** âr **care** ä **father** ĕ **pet** ē **be** hw **which** ĭ **pit**
ī **tie** îr **pier** ŏ **pot** ō **toe** ô **paw, for** oi **noise** ŏŏ **took**

or naturally follows from an action or condition : EFFECT. **2.** The relation of a result to its cause. **3.** A logical result or inference. **4.** Distinction or importance in rank <a politician of some *consequence*> **5.** Magnitude : importance.

**con·se·quent** (kŏn′sĭ-kwĕnt′, -kwənt) *adj.* [ME < Lat. *consequens*, pr.part. of *consequi*, to follow closely : *com-* (intensive) + *sequi*, to follow.] **1. a.** Following as a natural effect or result. **b.** Following as a logical conclusion. **2.** Logically correct or consistent. **3.** *Geol.* Having a position or direction relating to or resulting from the original slope of the earth's surface. —*n.* **1.** *Logic.* The conclusion, as of a syllogism. **2.** The second term of a ratio.

**con·se·quen·tial** (kŏn′sĭ-kwĕn′shəl) *adj.* **1.** CONSEQUENT 1. **2.** Of consequence : IMPORTANT. **3.** Arrogant : self-important. **—con′se·quen′ti·al′i·ty** (-shē-ăl′ĭ-tē), **con′se·quen′tial·ness** *n.* **—con′se·quen′tial·ly** *adv.*

**con·se·quent·ly** (kŏn′sĭ-kwĕnt′lē, -kwənt-lē) *adv.* As a result.

**con·ser·van·cy** (kən-sûr′vən-sē) *n., pl.* **-cies. 1.** Conservation, esp. of natural resources. **2.** *Chiefly Brit.* A commission supervising fisheries and navigation.

**con·ser·va·tion** (kŏn′sûr-vā′shən) *n.* **1.** The act or process of conserving. **2.** Controlled use and systematic protection of natural resources, as forests and waterways. **—con′ser·va′tion·al** *adj.*

**con·ser·va·tion·ist** (kŏn′sûr-vā′shə-nĭst) *n.* An advocate of conservation of natural resources.

**conservation of charge** *n.* An exact conservation law stating that the total electric charge of an isolated system remains constant regardless of changes within the system.

**conservation of energy** *n.* An exact conservation law stating that the total energy of an isolated system remains constant regardless of changes within the system.

**conservation of mass** *n.* The classical principle that the total mass of an isolated system is unchanged by interaction of its parts.

**conservation of momentum** *n.* An exact conservation law stating that the total linear momentum of an isolated system remains constant regardless of changes within the system.

**con·ser·va·tism** (kən-sûr′və-tĭz′əm) *n.* **1. a.** The disposition in politics to preserve the status quo. **b.** The principles and practices of persons or groups so disposed. **2. Conservatism.** The principles and practices of the Conservative Party in the United Kingdom. **3.** The tendency to resist change in preference to an existing order.

**con·ser·va·tive** (kən-sûr′və-tĭv) *adj.* **1. a.** Tending to oppose change : favoring traditional views and values. **b.** Traditional in style <*conservative* dress> **2.** Moderate : cautious. **3. a.** Belonging to a conservative party or political group. **b. Conservative.** Adhering to or typical of the Conservative Party of the United Kingdom. **4. Conservative.** Of, relating to, or adhering to Conservative Judaism. **5.** Tending to conserve : PRESERVATIVE. —*n.* **1.** One who favors traditional views and values. **2. a.** One who supports political conservatism or a conservative party. **b. Conservative.** A member or supporter of the Conservative Party of the United Kingdom. **3.** A preservative. **—con·ser′va·tive·ly** *adv.* **—con·ser′va·tive·ness** *n.*

☆ ***syns:*** CONSERVATIVE, ORTHODOX, RIGHT, RIGHTIST, RIGHT-WING *adj. core meaning* : strongly favoring the established, traditional order <a *conservative* politician> ***ant:*** liberal

**Conservative Judaism** *n.* The branch of Judaism that holds a modified view of the sanctity of the Torah and is flexible in its submission to the authority of the rabbinical law, accepting some liturgical and ritual changes in the light of the needs of modern life.

**Conservative Party** *n.* A political party of the United Kingdom.

**con·ser·va·tor** (kən-sûr′və-tər, kŏn′sər-vā′tər) *n.* **1.** A person who preserves from injury or infraction : PROTECTOR. **2.** *Law.* A person who is responsible for the welfare and property of a person ruled incompetent. **—con·ser′va·to′ri·al** (-tôr′ē-əl, -tōr′-) *adj.*

**con·ser·va·to·ry** (kən-sûr′və-tôr′ē, -tōr′ē) *n., pl.* **-ries. 1.** A greenhouse for raising and displaying plants. **2.** A school of music or dramatic art.

**con·serve** (kən-sûrv′) *vt.* **-served, -serv·ing, -serves.** [ME *conserven* < OFr. *conserver* < Lat. *conservare* : *com-* (intensive) + *servare*, to preserve.] **1. a.** To protect from loss or depletion <*conserve* a supply of canned goods> **b.** To use carefully, avoiding waste <tried to *conserve* water> **2.** To preserve (fruits) with sugar. —*n.* (kŏn′sûrv′). A jam made of fruits stewed in sugar. **—con·serv′a·ble** *adj.* **—con·serv′er** *n.*

**con·sid·er** (kən-sĭd′ər) *v.* **-ered, -er·ing, -ers.** [ME *consideren* < Lat. *considerare.*] —*vt.* **1.** To think about seriously. **2.** To regard as. **3.** To believe after deliberation : JUDGE. **4.** To take into account : bear in mind. **5.** To show consideration for. **6.** To regard highly : ESTEEM. **7.** To look at thoughtfully. —*vi.* To think carefully : REFLECT. **—con·sid′er·er** *n.*

**con·sid·er·a·ble** (kən-sĭd′ər-ə-bəl) *adj.* **1.** Large in amount, extent, or degree <*considerable* influence> **2.** Worthy of consideration : IMPORTANT. ***usage:*** *Considerable* is not properly used as an adverb; thus correctness requires *You helped me considerably* (not *helped me considerable*). —*n. Informal.* A considerable amount, extent, or degree. **—con·sid′er·a·bly** *adv.*

**con·sid·er·ate** (kən-sĭd′ər-ĭt) *adj.* [Lat. *consideratus*, p.part. of *considerare*, to consider.] **1.** Mindful of the needs or feelings of others : THOUGHTFUL. **2.** Marked by careful thought : DELIBERATE. **—con·sid′er·ate·ly** *adv.* **—con·sid′er·ate·ness** *n.*

**con·sid·er·a·tion** (kən-sĭd′ə-rā′shən) *n.* **1.** Careful thought : DELIBERATION. **2.** Something to be considered in forming a judgment or decision <Safety is the most important *consideration.*> **3.** Mindful concern for others : SOLICITUDE. **4.** A thoughtful opinion. **5.** Payment given for a service rendered : RECOMPENSE. **6.** *Law.* Something promised, given, or done that has the effect of making an agreement a legally enforceable contract. **7.** High regard.

**con·sid·ered** (kən-sĭd′ərd) *adj.* **1.** Reached after deliberation or careful thought <my *considered* decision> **2.** Esteemed : regarded.

**con·sid·er·ing** (kən-sĭd′ər-ĭng) *prep.* In view of. —*adv. Informal.* All things considered <It is not bad, *considering.*>

**con·sign** (kən-sīn′) *v.* **-signed, -sign·ing, -signs.** [OFr. *consigner* < Lat. *consignare*, to attest : *com-* (intensive) + *signare*, to mark < *signum*, mark.] —*vt.* **1.** To entrust to the care of another. **2.** To turn over to another's charge or control : COMMIT. **3.** To deliver (e.g., merchandise) for custody or sale. —*vi. Obs.* To submit : consent. **—con·sign′a·ble** *adj.* **—con′sig·na′tion** (kŏn′sĭ-nā′shən, -sĭg-) *n.* **—con·sig′nor, con·sign′er** *n.*

**con·sign·ee** (kŏn′sī-nē′, kən-sī′nē′) *n.* An agent to whom merchandise is consigned.

**con·sign·ment** (kən-sīn′mənt) *n.* **1.** The act of consigning. **2.** Something consigned. **—on consignment.** Sent to a retailer who is expected to pay following sale.

**con·sist** (kən-sĭst′) *vi.* **-sist·ed, -sist·ing, -sists.** [Lat. *consistere*, to stand still : *com-* (intensive) + *sistere*, to cause to stand.] **1.** To be made up or composed <A triangle *consists* of three sides.> **2.** To have a basis : RESIDE <The beauty of the architecture *consists* in its simplicity.> **3.** To be compatible : ACCORD.

**con·sis·tence** (kən-sĭs′təns) *n.* Consistency.

**con·sis·ten·cy** (kən-sĭs′tən-sē) *n., pl.* **-cies. 1.** Agreement or logical coherence among things or parts. **2.** Compatibility or agreement among successive acts, ideas, or events. **3.** Degree or texture of density, firmness, or viscosity.

**con·sis·tent** (kən-sĭs′tənt) *adj.* [Lat. *consistens, consistent-*, pr.part. of *consistere*, to stand still. —see CONSIST.] **1.** Being in agreement : COMPATIBLE. **2.** Conforming to the same principles or course of action : UNIFORM. **—con·sis′tent·ly** *adv.*

**con·sis·to·ry** (kən-sĭs′tə-rē) *n., pl.* **-ries.** [ME *consistorie* < LLat. *consistorium*, assembly < Lat. *consistere*, to stand together. —see CONSIST.] **1. a.** *Rom. Cath. Ch.* A gathering presided over by the pope for the promulgation of papal acts, as the canonization of a saint. **b.** A governing body of a local congregation in certain Reformed churches. **c.** A court appointed to regulate ecclesiastical affairs in Lutheran state churches. **d.** An Anglican diocesan court presided over by the bishop's chancellor or commissary. **2.** The meeting of a consistory. **3.** A council or tribunal. **—con′sis·to′ri·al** (kŏn′sĭ-stôr′ē-əl, -stōr′ē-əl) *adj.*

**con·so·ci·ate** (kən-sō′shē-āt′) *vt. & vi.* **-at·ed, -at·ing, -ates.** [< ME *consociat*, associated < Lat. *consociatus*, p.part. of *consociare*, to associate : *com-*, together + *sociare*, to associate < *socius*, companion.] To bring into friendly association. —*adj.* (-ĭt). Associated : united. —*n.* (-ĭt). An associate : companion. **—con·so′ci·a′tion** *n.*

**con·sole[1]** (kən-sōl′) *vt.* **-soled, -sol·ing, -soles.** [Fr. *consoler* < OFr. < Lat. *consolari* : *com-* (intensive) + *solari*, to comfort.] To comfort in time of grief or loss : SOLACE. **—con·sol′a·ble** *adj.* **—con′so·la′tion** *n.* **—con·so′la·to′ry** (-sō′lə-tôr′ē, -tōr′ē, -sŏl′ə-) *adj.* **—con·sol′er** *n.* **—con·sol′ing·ly** *adv.*

**con·sole[2]** (kŏn′sōl′) *n.* [Fr., perh. short for *consolider*, to strengthen < Lat. *consolidare*. —see CONSOLIDATE.] **1.** A decorative bracket for supporting an object, as a cornice, shelf, or bust. **2.** A console table. **3.** The desklike part of an organ containing the keyboard, stops, and pedals. **4.** A cabinet for a radio, television set, or phonograph, designed to stand on the floor. **5.** A panel housing the controls for electrical or mechanical equipment. **6.** The part of a computer that houses the apparatus used to operate the machine manually and that provides a means of communication between the operator and the central processing unit.

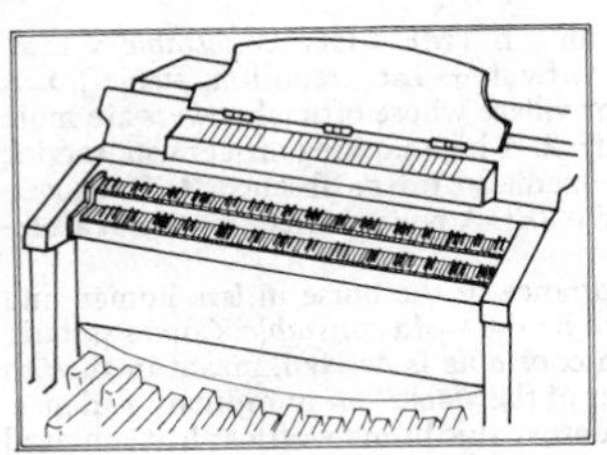

**console[2]**

**con·sole table** (kŏn′sōl′) *n.* **1.** A table fixed to a wall and supported by decorative consoles. **2.** A small table, often with curved legs resembling consoles, designed to be set against a wall.

**con·sol·i·date** (kən-sŏl′ĭ-dāt′) *v.* **-dat·ed, -dat·ing, -dates.** [Lat. *consolidare, consolidat-* : *com-* (intensive) + *solidare,* to make firm < *solidus,* firm.] —*vt.* **1.** To form into a compact mass. **2.** To make stable or strong : STRENGTHEN <*consolidated* their position> **3.** To unite into one system or body : COMBINE <*consolidated* three factions> —*vi.* To become united : MERGE. **—con·sol′i·da′tor** *n.*

**consolidated school** *n.* A usu. rural public school for pupils from several adjacent districts.

**con·sol·i·da·tion** (kən-sŏl′ĭ-dā′shən) *n.* **1. a.** The act or process of consolidating. **b.** The state of being consolidated. **2.** Merger of two or more commercial interests or corporations.

**con·sols** (kŏn′sŏlz′, kən-sŏlz′) *pl.n.* [Short for *consolidated annuities.*] The perpetual governmental securities of Great Britain.

**con·so·lute** (kŏn′sə-lo͞ot′) *adj.* [LLat. *consolutus,* dissolved together: Lat. *con-,* together + Lat. *solutus,* p.part. of *solvere,* to loosen, dissolve.] Of or relating to liquid substances capable of being mixed in all proportions.

**con·som·mé** (kŏn′sə-mā′, kŏn′sə-mā′) *n.* [Fr. < *consommer,* to use up < Lat. *consummare,* to finish. —see CONSUMMATE.] A clear soup of meat or vegetable stock.

**con·so·nance** (kŏn′sə-nəns) *n.* **1.** Harmony : agreement. **2. a.** Correspondence of sounds. **b.** A similarity or repetition of terminal consonants in two or more syllables, words, or lines, as in *sun* and *shine.* **3.** *Mus.* A simultaneous combination of sounds conventionally regarded as pleasing and final in effect.

**con·so·nant** (kŏn′sə-nənt) *adj.* [ME < Lat. *consonans,* pr.part. of *consonare,* to agree : *com-,* together + *sonare,* to sound.] **1.** Being in agreement or accord <statements *consonant* with national policy> **2.** Corresponding in sound. **3.** Harmonious in sound. **4.** Consonantal. —*n.* **1.** A speech sound produced by a partial or complete obstruction of the air stream by any of various constrictions of the speech organs. **2.** A letter or character representing a consonant. **—con′so·nant·ly** *adv.*

**con·so·nan·tal** (kŏn′sə-năn′tl) *adj.* **1.** Of, relating to, or having the nature of a consonant. **2.** Containing a consonant or consonants. **—con′so·nan′tal·ly** *adv.*

**con·sort** (kŏn′sôrt′) *n.* [ME, colleague < OFr. < Lat. *consors,* partner : *com-,* together + *sors,* fate.] **1.** A spouse, esp. the spouse of a monarch. **2.** An associate. **3.** A ship accompanying another in travel. —*v.* (kən-sôrt′) **-sort·ed, -sort·ing, -sorts.** —*vi.* **1.** To keep company : ASSOCIATE <*consorts* with a fast crowd> **2.** To be in accord. —*vt.* **1.** To unite in company : ASSOCIATE. **2.** *Obs.* **a.** To escort : accompany. **b.** To espouse.

**con·sor·ti·um** (kən-sôr′tē-əm, -shē-əm) *n., pl.* **-ti·a** (-tē-ə, -shē-ə) [Lat., fellowship < *consors,* partner. —see CONSORT.] **1.** An association of financial institutions or capitalists esp. in international finance. **2.** An association : partnership. **3.** *Law.* A spouse's legal right to the amenities of marriage, as help and affection.

**con·spe·cif·ic** (kŏn′spĭ-sĭf′ĭk) *adj.* Of the same species.

**con·spec·tus** (kən-spĕk′təs) *n.* [Lat., view < p.part. of *conspicere,* to observe. —see CONSPICUOUS.] **1.** A general survey of a subject. **2.** A synopsis : outline.

**con·spic·u·ous** (kən-spĭk′yo͞o-əs) *adj.* [Lat. *conspicuus* < *conspicere,* to observe : *com-* (intensive) + *specere,* to look.] **1.** Easy to notice : OBVIOUS. **2.** Attracting attention by being unusual or remarkable. **—con·spic′u·ous·ly** *adv.* **—con·spic′u·ous·ness** *n.*

**con·spir·a·cy** (kən-spĭr′ə-sē) *n., pl.* **-cies.** [ME *conspiracie* < AN, prob. var. of OFr. *conspiracious* < Lat. *conspiratio* < *conspirare,* to conspire.] **1.** The act of conspiring. **2. a.** A group of conspirators. **b.** An agreement made by conspirators. **3.** *Law.* An agreement between two or more persons to commit a crime or accomplish a legal purpose through illegal action.

**con·spir·a·tor** (kən-spĭr′ə-tər) *n.* One engaged in a conspiracy.

**con·spir·a·to·ri·al** (kən-spĭr′ə-tôr′ē-əl, -tōr′-) *adj.* Of, relating to, or typical of a conspiracy. **—con·spir′a·to′ri·al·ly** *adv.*

**con·spire** (kən-spīr′) *v.* **-spired, -spir·ing, -spires.** [ME *conspiren,* < Lat. *conspirare* : *com-,* together + *spirare,* to breathe.] —*vi.* **1.** To plan together secretly to commit a crime or wrongful act or accomplish a legal purpose through illegal action. **2.** To act or work together : COMBINE <events that *conspired* to cause the delay> —*vt.* To plan or plot secretly. **—con·spir′er** *n.* **—con·spir′ing·ly** *adv.*

**con spi·ri·to** (kŏn spîr′ĭ-tō′, kōn) *adv.* [Ital.] *Mus.* With spirit and vigor. —Used as a direction.

**con·sta·ble** (kŏn′stə-bəl, kŭn′-) *n.* [ME < OFr. *conestable* < LLat. *comes stabuli* : Lat. *comes,* officer + Lat. *stabulum,* stable.] **1.** A peace officer in a township or village whose official powers are more limited than those of a sheriff. **2.** A high-ranking officer usu. serving as military commander in a medieval ruler's absence. **3.** The governor of a royal castle. **4.** *Chiefly Brit.* A police officer. **—con′sta·ble·ship′** *n.*

▲ word history: The importance of the horse in late Roman and medieval warfare is illustrated by the word *constable. Comes stabuli,* the Latin phrase from which *constable* is derived, meant in the 5th and 6th centuries the "officer of the stable" or, in ordinary parlance, "head groom." By the 13th century this humble officer had achieved the dignity of being the chief household officer of the great lords and kings of France. In the latter case he was called the Constable of France and commanded the army in the king's absence. *Constable* was adopted by the English in the 13th and 14th centuries for a number of officers, including the wardens of royal castles and various officers of the peace. Many of these positions have been abolished or are now ceremonial; in British usage *constable* refers especially to what Americans call a *policeman.*

**con·stab·u·lar** (kən-stăb′yə-lər) *adj.* Constabulary.

**con·stab·u·lar·y** (kən-stăb′yə-lĕr′ē) *n., pl.* **-ies.** **1.** The body of constables of a district or city. **2.** The district under the jurisdiction of a constable. **3.** An armed police force organized like a military unit. —*adj.* Of or relating to constables or constabularies.

**con·stan·cy** (kŏn′stən-sē) *n.* **1.** Steadfastness, as in purpose : FAITHFULNESS. **2.** The quality of being constant : CHANGELESSNESS.

**con·stant** (kŏn′stənt) *adj.* [ME < Lat. *constans,* pr.part. of *constare,* to stand firm : *com-* (intensive) + *stare,* to stand.] **1.** Continually occurring or recurring : PERSISTENT. **2.** Invariable in nature, value, or extent : UNIFORM. **3.** Steadfast in purpose, loyalty, or affection : FAITHFUL. —*n.* **1.** Something unchanging or invariable. **2. a.** *Math.* A quantity taken to have a fixed value in a specified mathematical context. **b.** An experimental or theoretical condition, factor, or quantity that occurs, is held, or is regarded as invariant in given circumstances. **—con′stant·ly** *adv.*

**con·stan·tan** (kŏn′stən-tăn′) *n.* [< CONSTANT.] A copper-nickel alloy used chiefly in electrical instruments.

**constant dollars** *pl.n.* A measure of the cost of goods or services with the effects of inflation removed.

**con·stel·late** (kŏn′stə-lāt′) *vi.* & *vt.* **-lat·ed, -lat·ing, -lates.** [Back-formation < CONSTELLATION.] To form or cause to form a group or cluster.

**con·stel·la·tion** (kŏn′stə-lā′shən) *n.* [ME *constellacioun* < OFr. *constellation* < LLat. *constellatio* : *com-,* together + *stella,* star.] **1.** *Astron.* **a.** One of 88 stellar groups named after and thought to resemble various mythological characters, inanimate objects, and animals. **b.** An area of the celestial sphere occupied by such a group. **2.** The position of the stars, esp. at the time of one's birth. **3.** A gathering or assemblage of similar or related persons or things. **4.** A set or configuration of objects, properties, or individuals, esp. a grouping that is structurally or systematically related. **—con′stel·la′tion·al** *adj.* **—con·stel′la·to′ry** (-stĕl′ə-tôr′ē, -tōr′ē) *adj.*

**con·ster·nate** (kŏn′stər-nāt′) *vt.* **-nat·ed, -nat·ing, -nates.** [Lat. *consternare, consternat-* : *com-* (intensive) + *sternere,* to throw down.] To cause consternation in.

**con·ster·na·tion** (kŏn′stər-nā′shən) *n.* Sudden amazement, confusion, or dismay.

**con·sti·pate** (kŏn′stə-pāt′) *vt.* **-pat·ed, -pat·ing, -pates.** [Lat. *constipare, constipat-,* to crowd together : *com-,* together + *stipare,* to cram.] To cause constipation in.

**con·sti·pa·tion** (kŏn′stə-pā′shən) *n.* Difficult, incomplete, or infrequent evacuation of the bowels.

**con·stit·u·en·cy** (kən-stĭch′o͞o-ən-sē) *n., pl.* **-cies. 1. a.** The body of voters represented by an elected legislator or executive. **b.** The district represented. **2.** A group of supporters.

**con·stit·u·ent** (kən-stĭch′o͞o-ənt) *adj.* [Lat. *constituens, constituent-,* pr.part. of *constituere,* to set up. —see CONSTITUTE.] **1.** Serving as part of a whole : COMPONENT. **2.** Having power to elect or designate. **3.** Authorized to make or amend a constitution. —*n.* **1.** One who authorizes another to represent him or her : CLIENT. **2.** A member of a group represented by an elected official : VOTER. **3.** A constituent part : COMPONENT. **4.** One of two or more elements into which a construction or compound may be divided by analysis, being either immediate, as *He/ works on the railroad,* or ultimate, as *He/ works/ on/ the/ rail/road.* **—con·stit′u·ent·ly** *adv.*

**constituent structure** *n.* An analysis, often in the form of a schematic representation, of the constituents of a grammatical construction, such as a sentence.

**con·sti·tute** (kŏn′stĭ-to͞ot′, -tyo͞ot′) *vt.* **-tut·ed, -tut·ing, -tutes.** [Lat. *constituere, constitut-,* to set up : *com-* (intensive) + *statuere,* to set up.] **1.** To make up : COMPOSE <Four musicians *constitute* a quartet.> **2.** To set up : ENACT (e.g., a law). **3.** To establish formally : FOUND (e.g., an institution). **4.** To appoint to an office, dignity, function, or task : DESIGNATE. **—con′sti·tut′er, con′sti·tu′tor** *n.*

**con·sti·tu·tion** (kŏn′stĭ-to͞o′shən, -tyo͞o′-) *n.* **1.** The act or process of composing, setting up, or establishing. **2. a.** Structure or composition : MAKE-UP. **b.** The physical make-up of a person <a strong *constitution*> **3. a.** The system of fundamental laws and principles that prescribes the nature, functions, and limits of an institution, as a government. **b.** The document on which such a system is recorded.

**con·sti·tu·tion·al** (kŏn′stĭ-to͞o′shə-nəl, -tyo͞o′-) *adj.* **1.** Of or relating to a constitution. **2.** Consistent with or authorized by a constitution. **3.** Established by or operating under a constitution. **4.** Of or proceeding from the basic structure or nature of a person or thing <a *constitutional* reluctance to lie> —*n.* A walk taken usu. daily for one's health. **—con'sti·tu'tion·al'i·ty** (-năl'ĭ-tē) *n.* **—con'sti·tu'tion·al·ly** *adv.*

**con·sti·tu·tion·al·ism** (kŏn'stĭ-to͞o'shə-nə-lĭz'əm, -tyo͞o'-) *n.* **1.** Government in which power is distributed and limited by a system of laws that must be obeyed by the rulers. **2.** Advocacy of constitutionalism. **—con'sti·tu'tion·al·ist** *n.*

**constitutional monarchy** *n.* A monarchy in which the powers of the ruler are restricted to those granted under the constitution and laws of the nation.

**con·sti·tu·tive** (kŏn'stĭ-to͞o'tĭv, -tyo͞o'-) *adj.* **1.** Making a thing what it is : ESSENTIAL. **2.** Having power to establish or enact. **—con'sti·tu'tive·ly** *adv.*

**con·sti·tu·tive enzyme** (kŏn'stĭ-to͞o'tĭv, -tyo͞o'-, kən-stĭch'ə-tĭv) *n.* An enzyme produced by a cell regardless of the presence of its substrate.

**con·strain** (kən-strān') *vt.* **-strained, -strain·ing, -strains.** [ME *constreinen* < Lat. *constringere,* to restrain, compress : *com-,* together + *stringere,* to bind.] **1.** To compel by physical, moral, or circumstantial force : OBLIGE <felt *constrained* to stay> **2.** To keep within close bounds : CONFINE. **3.** To check the movement of : RESTRAIN. **4.** To produce in a strained or artificial manner <a *constrained* smile> **—con·strain'a·ble** *adj.* **—con·strain'ed·ly** (-strā'nĭd-lē) *adv.* **—con·strain'er** *n.*

**con·straint** (kən-strānt') *n.* [ME *constreinte* < OFr. < *constraindre* < Lat. *constringere,* to restrain. —see CONSTRAIN.] **1.** Threat or use of force to prevent, restrict, or dictate the action or thought of others. **2.** The quality, state, or sense of being restricted to a given course of action or inaction. **3.** A restriction, limit, or regulation. **4.** Embarrassed reserve or reticence <"All *constraint* had vanished between the two, and they began to talk" —Edith Wharton>

**con·strict** (kən-strĭkt') *v.* **-strict·ed, -strict·ing, -stricts.** [Lat. *constringere, constrict-,* to compress. —see CONSTRAIN.] *—vt.* **1.** To make narrow, as by shrinking or contracting. **2.** To squeeze or compress by or as if by tightening or narrowing. *—vi.* To become contracted or compressed. **—con·stric'tion** *n.* **—con·stric'tive** *adj.* **—con·stric'tive·ly** *adv.*

**con·stric·tor** (kən-strĭk'tər) *n.* **1.** One that constricts. **2.** *Anat.* A muscle that contracts or compresses a bodily part or organ. **3.** A snake, as a boa, that coils around and crushes its prey.

**con·stringe** (kən-strĭnj') *vt.* **-stringed, -string·ing, -string·es.** [Lat. *constringere,* to compress. —see CONSTRAIN.] To cause to contract : CONSTRICT. **—con·strin'gen·cy** *n.* **—con·strin'gent** *adj.*

**con·struct** (kən-strŭkt') *vt.* **-struct·ed, -struct·ing, -structs.** [Lat. *construere, construct-* : *com-,* together + *struere,* to pile up.] **1.** To put together by assembling parts : BUILD. **2.** To create (e.g., a sentence) by systematically arranging ideas or expressions. **3.** *Math.* To draw (a geometric figure) that meets specific requirements, usu. with instruments limited to a straightedge and compass. *—n.* (kŏn'strŭkt'). Something, esp. a concept, that is synthesized or constructed from simple elements. **—con·struct'i·ble** *adj.* **—con·struc'tor, con·struct'er** *n.*

**con·struc·tion** (kən-strŭk'shən) *n.* **1. a.** The act or process of constructing. **b.** The state of being constructed. **c.** The business of building. **2.** Something constructed. **3.** The way in which something is put together <a cabin of simple *construction*> **4.** The explanation or interpretation given a particular statement. **5. a.** Arrangement of words to form a meaningful phrase, clause, or sentence. **b.** A group of words so arranged. **—con·struc'tion·al** *adj.* **—con·struc'tion·al·ly** *adv.*

**con·struc·tion·ist** (kən-strŭk'shə-nĭst) *n.* One who interprets a legal text or document in a specified way <a strict *constructionist*>

**con·struc·tive** (kən-strŭk'tĭv) *adj.* **1.** Serving to advance a good purpose : HELPFUL. **2.** Of or relating to construction : STRUCTURAL. **3.** *Law.* Based on interpretation : indirectly expressed. **—con·struc'tive·ly** *adv.* **—con·struc'tive·ness** *n.*

**con·struc·tiv·ism** (kən-strŭk'tə-vĭz'əm) *n.* A movement in modern art in which industrial materials, as sheet metal and plastic, are used to create nonrepresentational often geometric objects. **—con·struc'tiv·ist** *n.*

**con·strue** (kən-stro͞o') *v.* **-strued, -stru·ing, -strues.** [ME *construen* < LLat. *construere* < Lat., to build. —see CONSTRUCT.] *—vt.* **1. a.** To analyze the structure of (a clause or sentence). **b.** To use syntactically <The noun "deer" can be *construed* as singular or plural.> **2.** To place a certain meaning on : INTERPRET. **3.** To translate, esp. aloud. *—vi.* To analyze grammatical structure. *—n.* (kŏn'stro͞o'). An interpretation or translation.

**con·sub·stan·tial** (kŏn'səb-stăn'shəl) *adj.* [ME *consubstancial* < LLat. *consubstantialis* : Lat. *com-,* same + LLat. *substantialis,* substantial < Lat. *substantia,* substance.] Having the same substance, essence, or nature.

**con·sub·stan·ti·ate** (kŏn'səb-stăn'shē-āt') *vt.* & *vi.* **-at·ed, -at·ing, -ates.** To unite or become united in one common substance, nature, or essence.

**con·sub·stan·ti·a·tion** (kŏn'səb-stăn'shē-ā'shən) *n.* The theological doctrine that the body and blood of Christ coexist with the elements of bread and wine during the Eucharist.

**con·sue·tude** (kŏn'swĭ-to͞od', -tyo͞od') *n.* [ME < Lat. *consuetudo.* —see CUSTOM.] Usage : custom. **—con'sue·tu'di·nar'y** (-to͞od'n-ĕr'ē, -tyo͞od'-) *adj.*

**con·sul** (kŏn'səl) *n.* [ME *consulat* < Lat. *consulatus* < *consul,* consul.] **1.** An official appointed by a government to reside in a foreign city and represent his or her government's commercial interests and give assistance to its citizens there. **2.** Either of the two chief magistrates of the Roman Republic, elected for a term of one year. **3.** One of the three chief magistrates of the French Republic from 1799 to 1804. **—con'su·lar** (-sə-lər) *adj.*

**con·su·late** (kŏn'sə-lĭt) *n.* [ME *consulat* < Lat. *consulatus* < *consul,* consul.] **1. a.** Government by consuls. **b.** The office or term of office of a consul. **2.** The premises occupied by a consul.

**consul general** *n., pl.* **consuls general.** A consular officer of the highest rank.

**con·sult** (kən-sŭlt') *v.* **-sult·ed, -sult·ing, -sults.** [Lat. *consultare,* freq. of *consulere,* to take counsel.] *—vt.* **1. a.** To seek advice or information <*consult* a specialist> **b.** To refer to <*consult* a dictionary> **2.** To keep in mind : CONSIDER. *—vi.* **1.** To exchange views : CONFER. **2.** To give professional advice. *—n.* (kən-sŭlt', kŏn'sŭlt'). A consultation <a surgical *consult*>

**con·sult·ant** (kən-sŭl'tənt) *n.* **1.** One who gives expert or professional advice. **2.** One who consults another.

**con·sul·ta·tion** (kŏn'səl-tā'shən) *n.* **1.** The act or procedure of consulting. **2.** A conference at which advice is given or views are exchanged.

**con·sul·ta·tive** (kən-sŭl'tə-tĭv) *also* **con·sul·ta·to·ry** (-tôr'ē, -tōr'ē) *adj.* Of or relating to consultation : ADVISORY.

**con·sume** (kən-so͞om') *v.* **-sumed, -sum·ing, -sumes.** [ME *consumen* < Lat. *consumere* : *com-* (intensive) + *sumere,* to take.] *—vt.* **1.** To eat or drink up : INGEST. **2.** To expend (e.g., fuel) : USE UP. **3.** To waste : squander. **4.** To destroy totally, as by fire : LEVEL. **5.** To engross : absorb <*consumed* with curiosity> *—vi.* To be destroyed, expended, or wasted. **—con·sum'a·ble** *adj.* & *n.*

**con·sum·ed·ly** (kən-so͞o'mĭd-lē) *adv.* To an excessive degree.

**con·sum·er** (kən-so͞o'mər) *n.* **1.** One that consumes. **2.** A person who acquires goods or services : BUYER. **3.** A heterotrophic organism in a food chain that ingests other organisms or organic matter. **—con·sum'er·ship'** *n.*

**consumer credit** *n.* Credit granted to permit a consumer to own or use goods while making payments on them.

**consumer goods** *pl.n.* Goods, as food and clothing, that satisfy human needs.

**con·sum·er·ism** (kən-so͞o'mə-rĭz'əm) *n.* **1.** The protection of the rights of the consumer. **2.** The economic theory that a progressively greater consumption of goods is beneficial. **—con·sum'er·ist** *n.*

**consumer price index** *n.* An index of prices used to measure the change in the cost of basic goods and services in comparison with a fixed base period.

**con·sum·mate** (kŏn'sə-māt') *vt.* **-mat·ed, -mat·ing, -mates.** [ME *consummaten* < Lat. *consummare* : *com-,* together + *summa,* sum.] **1.** To bring to fruition : CONCLUDE <*consummate* a business deal> **2.** To fulfill (a marriage) with the first act of sexual intercourse after the ceremony. *—adj.* (kən-sŭm'ĭt). **1.** Perfect in every respect <*consummate* joy> **2.** Highly accomplished or skilled <a *consummate* pianist> **3.** Utter : complete <a *consummate* fool> **—con·sum'mate·ly** *adv.* **—con'sum·ma'tive** (kŏn'sə-mā'tĭv), **con·sum'ma·to'ry** (kən-sŭm'ə-tôr'ē, -tōr'ē) *adj.* **—con'sum·ma'tor** (kŏn'sə-mā'tər) *n.*

**con·sum·ma·tion** (kŏn'sə-mā'shən) *n.* **1.** The act of consummating : FULFILLMENT. **2.** An ultimate goal or end.

**con·sump·tion** (kən-sŭmp'shən) *n.* [Lat. *consumptio,* a consuming < *consumptus,* p.part. of *consumere,* to consume.] **1. a.** The act or process of consuming. **b.** The state of being consumed. **c.** An amount consumed. **2.** The using up of consumer goods and services. **3.** *Pathol.* **a.** A wasting of tissue. **b.** TUBERCULOSIS 2.

**con·sump·tive** (kən-sŭmp'tĭv) *adj.* **1.** Tending to consume. **2.** Of, relating to, or afflicted with consumption. *—n.* One afflicted with consumption. **—con·sump'tive·ly** *adv.*

**con·tact** (kŏn'tăkt') *n.* [Lat. *contactus* < p.part. of *contingere,* to touch : *com-,* together + *tangere,* to touch.] **1.** The touching of two objects or surfaces. **2.** The state of being in communication <in *contact* with their relatives> **3.** One who might be of use : CONNECTION <political *contacts*> **4.** *Elect.* **a.** A connection between two conductors that permits a flow of current. **b.** A part or device that makes or breaks such a connection. **5.** *Med.* One recently exposed to a contagious disease. **6.** *Informal.* A contact lens. *—v.* (kŏn'tăkt', kən-tăkt') **-tact·ed, -tact·ing, -tacts.** *—vt.* **1.** To bring or put in contact. **2.** *Informal.* To communicate with. ***usage:*** Although *contact* occurs frequently as a verb meaning "to get in touch with," some still consider this usage inappropriate to formal style. *—vi.* To be in or come into contact. *—adj.* (kŏn'tăkt'). **1.** Of, sustaining, or making contact. **2.** Caused or transmitted by touching <a *contact* skin inflammation> **—con·tac'tu·al** (kən-tăk'cho͞o-əl) *adj.* **—con·tac'tu·al·ly** *adv.*

**contact dermatitis** *n.* An acute skin inflammation caused by contact with an irritating substance, as a chemical.

**contact flight** *also* **contact flying** *n.* Aircraft navigation by visual reference to the horizon or to landmarks.

**contact inhibition** *n.* Cessation of cellular growth and division due to contact with other cells.

**contact lens** *n.* A thin, usu. corrective lens fitted over the cornea of the eye.

**contact print** *n.* A print made by exposing a photosensitive surface in direct contact with a photographic negative.

**con·ta·gia** (kən-tā′jə) *n. pl. of* CONTAGIUM.

**con·ta·gion** (kən-tā′jən) *n.* [ME *contagioun* < Lat. *contagio* < *contingere*, to touch. —see CONTACT.] **1. a.** Disease transmission by direct or indirect contact. **b.** A disease that is or may be so transmitted. **c.** A contagium. **2.** A corrupting or harmful influence. **3.** The tendency to spread, as of an influence or emotional state.

**con·ta·gious** (kən-tā′jəs) *adj.* **1.** Transmissible by contact. **2.** Carrying or capable of carrying disease. **3.** Spreading or tending to spread from one to another : CATCHING. **—con·ta′gious·ly** *adv.* **—con·ta′gious·ness** *n.*

**contagious abortion** *n.* BRUCELLOSIS 2.

**con·ta·gium** (kən-tā′jəm) *n., pl.* **-gia** (-jə) [Lat., contagion < *contagio.*] The direct cause, as a virus, of an infectious disease.

**con·tain** (kən-tān′) *vt.* **-tained, -tain·ing, -tains.** [ME *conteinen* < OFr. *contenir* < Lat. *continēre* : *com-*, together + *tenēre*, to hold.] **1.** To have within : ENCLOSE. **2.** To have as component parts : COMPRISE. **3.** To have capacity for : HOLD. **4.** *Math.* To be exactly divisible by. **5.** To keep within limits : RESTRAIN <*contain* one's anger> **6.** To restrict the strategic power of (a nation or bloc), as by encircling it with hostile alliances. **—con·tain′a·ble** *adj.* **—con·tain′ment** *n.*

**con·tain·er** (kən-tā′nər) *n.* A receptacle for holding or carrying material.

**con·tain·er·ize** (kən-tā′nə-rīz′) *vt.* **-ized, -iz·ing, -iz·es.** To ship (cargo) in large, standardized containers. **—con·tain′er·i·za′tion** *n.*

**container ship** *n.* A ship for carrying containerized cargo.

**con·tam·i·nant** (kən-tăm′ə-nənt) *n.* A contaminating agent.

**con·tam·i·nate** (kən-tăm′ə-nāt′) *vt.* **-nat·ed, -nat·ing, -nates.** [ME *contaminaten* < Lat. *contaminare.*] To make impure by contact or mixture. **—con·tam′i·na′tion** *n.* **—con·tam′i·na′tive** *adj.* **—con·tam′i·na′tor** *n.*

**conte** (kôNT) *n., pl.* **contes** (kôNT) [Fr. < OFr. *conter*, to relate. —see COUNT[1].] An adventure story.

**con·temn** (kən-tĕm′) *vt.* **-temned, -temn·ing, -temns.** [ME *contempnen*, to slight < Lat. *contemnere* : *com-* (intensive) + *temnere*, to despise.] To view with contempt : ABHOR. **—con·temn′er** (-tĕm′ər, -tĕm′nər) *n.*

**con·tem·plate** (kŏn′təm-plāt′) *v.* **-plat·ed, -plat·ing, -plates.** [Lat. *contemplari, contemplat-* : *com-* (intensive) + *templum*, space for observing auguries.] —*vt.* **1.** To look at pensively. **2.** To ponder or consider thoughtfully. **3.** To have in mind as possible or likely <*contemplate* changing careers> —*vi.* To ponder : meditate. **—con′tem·pla′tor** *n.*

**con·tem·pla·tion** (kŏn′təm-plā′shən) *n.* **1.** Thoughtful observation or meditation. **2.** Expectation or intention.

**con·tem·pla·tive** (kən-tĕm′plə-tĭv, kŏn′təm-plā′-) *adj.* Given to or marked by contemplation. —*n.* **1.** One given to contemplation. **2.** A member of a religious order dedicated to meditation. **—con·tem′pla·tive·ly** *adv.* **—con·tem′pla·tive·ness** *n.*

**con·tem·po·ra·ne·ous** (kən-tĕm′pə-rā′nē-əs) *adj.* [Lat. *contemporaneus* : *com-*, same + *tempus*, time.] Arising, existing, or occurring during the same period of time <*contemporaneous* reigns of two monarchs> **—con·tem′po·ra·ne′i·ty** (-rə-nē′ĭ-tē, -nā′-), **con·tem′po·ra′ne·ous·ness** *n.* **—con·tem′po·ra′ne·ous·ly** *adv.*

**con·tem·po·rar·y** (kən-tĕm′pə-rĕr′ē) *adj.* [Med. Lat. *contemporarius* : Lat. *com-*, same + Lat. *tempus*, time.] **1.** Belonging to the same period of time. **2.** Of about the same age. **3.** Modern : current <*contemporary* trends> —*n., pl.* **-ies. 1.** One of the same time or age. **2.** A person of the present age. **—con·tem′po·rar′i·ly** (-tĕm′pə-rârʹə-lē) *adv.*

**con·tem·po·rize** (kən-tĕm′pə-rīz′) *v.* **-rized, -riz·ing, -riz·es.** [< CONTEMPORARY.] —*vt.* To relate in time : SYNCHRONIZE. —*vi.* To be contemporary.

**con·tempt** (kən-tĕmpt′) *n.* [ME < Lat. *contemptus* < p.part. of *contemnere*, to despise. —see CONTEMN.] **1.** Bitter disdain : SCORN. **2.** The state of being despised or dishonored : DISGRACE. **3.** Open disrespect or willful disobedience of judicial or legislative authority.

**con·tempt·i·ble** (kən-tĕmp′tə-bəl) *adj.* **1.** Deserving contempt : DESPICABLE. **2.** *Obs.* Contemptuous. **—con·tempt′i·bil′i·ty, con·tempt′i·ble·ness** *n.* **—con·tempt′i·bly** *adv.*

**con·temp·tu·ous** (kən-tĕmp′cho͞o-əs) *adj.* Feeling or showing contempt : SCORNFUL. **—con·temp′tu·ous·ly** *adv.* **—con·temp′tu·ous·ness** *n.*

**con·tend** (kən-tĕnd′) *v.* **-tend·ed, -tend·ing, -tends.** [ME *contenden* < Lat. *contendere* : *com-*, with + *tendere*, to strive.] —*vi.* **1.** To struggle, as in battle : FIGHT. **2.** To compete, as in a race : VIE. **3.** To strive in debate : DISPUTE. —*vt.* To maintain or assert. **—con·tend′er** *n.*

**con·tent[1]** (kŏn′tĕnt′) *n.* [ME < Lat. *contentus*, p.part. of *continēre*, to contain.] **1.** *often* **contents.** Something contained in a receptacle <the *contents* of my pockets> **2.** *often* **contents.** Subject matter of a written work. **3.** The meaning or significance of a literary or artistic work. **4.** The proportion of a specified substance <Bonded whiskey has a high alcohol *content.*>

**con·tent[2]** (kən-tĕnt′) *adj.* [ME < OFr. < Lat. *contentus* < p.part. of *continēre*, to restrain. —see CONTAIN.] **1.** Contented : satisfied. **2.** Resigned to circumstances : ASSENTING <had to be *content* with a meager raise> —*vt.* **-tent·ed, -tent·ing, -tents.** To make content or satisfied. —*n.* Contentment : satisfaction.

**content analysis** *n.* The systematic analysis of the content rather than the structure of a communication, esp. the determination for psychological study of the frequency of occurrence of thematic and symbolic elements, including ideas, feelings, assertions, and personal references.

**con·tent·ed** (kən-tĕn′tĭd) *adj.* Satisfied with things as they are <a *contented* attitude> **—con·tent′ed·ly** *adv.* **—con·tent′ed·ness** *n.*

**con·ten·tion** (kən-tĕn′shən) *n.* [ME *contencioun* < OFr. *contention* < Lat. *contentio* < *contendere*, to contend.] **1.** An act of contending. **2.** Competition or rivalry. **3.** An assertion advanced in an argument.

**con·ten·tious** (kən-tĕn′shəs) *adj.* **1.** Given to contention : QUARRELSOME. **2.** Involving contention. **—con·ten′tious·ly** *adv.* **—con·ten′tious·ness** *n.*

**con·tent·ment** (kən-tĕnt′mənt) *n.* The state of being contented : SATISFACTION.

**con·ter·mi·nous** (kən-tûr′mə-nəs) *also* **co·ter·mi·nous** (kō-) *adj.* [Lat. *conterminus* : *com-*, same + *terminus*, boundary.] **1.** CONTIGUOUS 1. **2.** Contained in the same boundaries. **—con·ter′mi·nous·ly** *adv.* **—con·ter′mi·nous·ness** *n.*

**con·test** (kŏn′tĕst′) *n.* [Prob. < OFr. *conteste* < *contester*, to call to witness < Lat. *contestari* : *com-* (intensive) + *testis*, witness.] **1.** A struggle for victory or superiority between rivals. **2.** A competition, esp. one in which entrants perform separately and are rated by judges. —*v.* (kən-tĕst′, kŏn′tĕst′) **-test·ed, -test·ing, -tests.** —*vt.* **1.** To compete or strive for. **2.** To attempt to invalidate : CHALLENGE <*contest* a contract> —*vi.* To compete or struggle : CONTEND. **—con·test′a·ble** *adj.* **—con′tes·ta′tion** (kŏn′tĕ-stā′shən) *n.* **—con·test′er** *n.*

☆ ***syns:*** CONTEST, BUCK, CHALLENGE, COMBAT, DISPUTE, FIGHT, OPPOSE, RESIST *v. core meaning* : to take a stand against <*contested* the court ruling>

**con·tes·tant** (kən-tĕs′tənt, kŏn′tĕs′tənt) *n.* **1.** A person who takes part in a contest : COMPETITOR. **2.** A person who contests something, as an election or a will.

**con·text** (kŏn′tĕkst′) *n.* [ME, composition < Lat. *contextus* < p.part. of *contexere*, to join together : *com-*, together + *texere*, to plait.] **1.** The part of a written or spoken statement that surrounds a word or passage and that often specifies its meaning. **2.** The circumstances in which a particular event occurs : SITUATION. **—con·tex′tu·al** (-tĕks′cho͞o-əl, kŏn-) *adj.* **—con·tex′tu·al·ly** *adv.*

**con·tex·ture** (kən-tĕks′chər, kŏn′tĕks′-) *n.* **1.** The act of weaving or assembling parts into a whole. **2.** An arrangement of interconnected parts : STRUCTURE. **—con·tex′tur·al** *adj.*

**con·tig·u·ous** (kən-tĭg′yo͞o-əs) *adj.* [Lat. *contiguus* < *contingere*, to touch. —see CONTACT.] **1.** Sharing a boundary or edge : TOUCHING. **2.** Nearby : adjacent. **3.** Immediately preceding or following in time. **—con′ti·gu′i·ty** (kŏn′tĭ-gyo͞o′ĭ-tē), **con·tig′u·ous·ness** *n.* **—con·tig′u·ous·ly** *adv.*

**con·ti·nence** (kŏn′tə-nəns) *n.* **1.** Self-restraint : moderation. **2.** Partial or complete abstention from sexual activity. **3.** Voluntary control over bodily discharges.

**con·ti·nent[1]** (kŏn′tə-nənt) *n.* [Lat. *(terra) continens*, continuous (land), pr.part. of *continēre*, to hold together. —see CONTAIN.] **1.** One of the principal land masses of the earth, usu. held to include Africa, Antarctica, Asia, Australia, Europe, North America, and South America. **2. the Continent.** The European mainland.

**con·ti·nent[2]** (kŏn′tə-nənt) *adj.* [ME < Lat. *continens*, pr.part. of *continēre*, to restrain. —see CONTAIN.] Exercising continence. **—con′ti·nent·ly** *adv.*

▲ word history: Although the noun *continent* and the adjective *continent* are derived from the same Latin word, they are treated as separate English words because they were borrowed at different times with different meanings. Both words are derived from Latin *continens*, the present participle of *continēre*, which has the basic meaning "to hold together." *Continēre*, which is also the ancestor of *contain*, was an important Latin word and had a wide range of meanings. From the sense "to restrain, hold back, subdue" the adjective *continent* is derived. The adjective appeared in the 14th century and from its introduction had the meaning "characterized by self-restraint." The Latin participle *continens* meant "continuous," and in the 16th century, during the great age of exploration and discovery, the noun *continent* came into use for land masses of various kinds. The word at first denoted any continuous tract of land but was also applied to a mainland as distinguished from islands or peninsulas, especially to the European continent in contrast to the British Isles. By the 17th century *continent* was used to refer to the large land masses called "continents" today.

**con·ti·nen·tal** (kŏn′tə-nĕn′tl) *adj.* **1.** Of, relating to, or like a continent. **2.** *often* **Continental.** Of or relating to the mainland of Europe : EUROPEAN. **3. Continental.** Of or relating to the American colonies during and immediately after the Revolutionary War. —*n.* **1.** *often* **Continental.** An inhabitant of the mainland of Europe : EUROPEAN. **2. Continental.** A soldier in the Continental Army during the Revolutionary War. **3.** A piece of paper money issued by the Continental Congress during the Revolutionary War. **—con′ti·nen′tal·ism** *n.* **—con′ti·nen′tal·ist** *n.* **—con′ti·nen·tal′i·ty** (-nĕn-tăl′ĭ-tē) *n.* **—con′ti·nen′tal·ly** *adv.*

**continental code** *n.* The international Morse code.

**continental divide** *n.* A divide that separates continental river systems flowing in opposite directions.

**continental drift** *n.* The theoretical slow shifting of continents due to weakness in the suboceanic crust.

**continental shelf** *n.* A gen. shallow, flat submerged portion of a continent, extending to a point of steep descent to the ocean floor.

**continental slope** *n.* The steep descent from the continental shelf to the ocean bottom.

**con·tin·gen·cy** (kən-tĭn′jən-sē) *n.*, *pl.* **-cies. 1. a.** An event that may occur but is not probable : POSSIBILITY. **b.** A possibility that must be prepared against. **2.** Dependency on chance : UNCERTAINTY. **3.** Something incidental to something else.

**contingency table** *n.* A statistical table that shows the observed frequencies of a sample, with the rows indicating one variable and the columns another variable.

**con·tin·gent** (kən-tĭn′jənt) *adj.* [ME < Lat. *contingens*, pr.part. of *contingere*, to touch. —see CONTACT.] **1.** Likely but not certain to occur : POSSIBLE. **2.** Dependent on conditions or events not yet established : CONDITIONAL. **3.** Happening by accident or chance : FORTUITOUS. **4.** *Logic.* Possessing a truth value derived from facts apart from the proposition itself : not necessarily true or false. —*n.* **1.** A contingent event : CONTINGENCY. **2.** A share or quota, as of troops, contributed to a general effort. **3.** A representative group forming part of an assemblage. **—con·tin′gent·ly** *adv.*

**con·tin·u·a** (kən-tĭn′yo͞o-ə) *n. var. pl. of* CONTINUUM.

**con·tin·u·al** (kən-tĭn′yo͞o-əl) *adj.* **1.** Recurring regularly and frequently. **2. a.** Uninterrupted : steady. **b.** Continuous in time : INCESSANT <*continual* worry> **—con·tin′u·al·ly** *adv.*

**con·tin·u·ance** (kən-tĭn′yo͞o-əns) *n.* **1.** The act or fact of continuing. **2.** The time during which something exists or lasts : DURATION. **3.** A continuation : sequel. **4.** *Law.* Postponement of judicial proceedings to a future date.

**con·tin·u·ant** (kən-tĭn′yo͞o-ənt) *n.* A consonant, as *s, z,* or *f,* that can be prolonged as long as the breath lasts without a change in quality.

**con·tin·u·a·tion** (kən-tĭn′yo͞o-ā′shən) *n.* **1. a.** The act or fact of continuing. **b.** The state of being continued. **2.** An extension by which something is carried to a further point. **3.** Resumption after an interruption.

**con·tin·u·a·tive** (kən-tĭn′yo͞o-ā′tĭv, -ə-tĭv) *adj.* Of, relating to, or serving to cause continuation. —*n.* Something that expresses or causes continuation. **—con·tin′u·a′tive·ly** *adv.*

**con·tin·u·a·tor** (kən-tĭn′yo͞o-ā′tər) *n.* One that continues, esp. one who resumes the work of another.

**con·tin·ue** (kən-tĭn′yo͞o) *v.* **-ued, -u·ing, -ues.** [ME *continuen* < Lat. *continuare* < *continuus*, continuous < *continēre*, to hold together. —see CONTAIN.] —*vi.* **1.** To go on with a particular action or in a particular condition : PERSIST. **2.** To exist over an extended period : LAST. **3.** To remain in the same state, capacity, or place. **4.** To go on after an interruption : RESUME. —*vt.* **1.** To carry forward : MAINTAIN <The investigators will *continue* their surveillance.> **2.** To carry further in time, space, or development : EXTEND. **3.** To cause to remain or last : RETAIN. **4.** To carry on after an interruption : RESUME. **5.** *Law.* To postpone or adjourn (a judicial proceeding). **—con·tin′u·a·ble** *adj.* **—con·tin′u·er** *n.*

**continued fraction** *n.* A fraction whose denominator consists of an integer plus a fraction that likewise has a denominator consisting of an integer plus a fraction and so on.

**continuing education** *n.* **1.** An educational program that brings participants up to date in a particular area of knowledge or skills. **2.** Education courses designed esp. for part-time, adult students.

**con·ti·nu·i·ty** (kŏn′tə-no͞o′ĭ-tē, -nyo͞o′-) *n.*, *pl.* **-ties. 1.** The quality or state of being continuous. **2.** An uninterrupted succession or unbroken course. **3. a.** A detailed shooting script consulted to avoid errors and discrepancies from shot to shot in a film. **b.** A script for all the spoken parts of a radio or television program.

**con·tin·u·o** (kən-tĭn′yo͞o-ō′) *n.*, *pl.* **-os.** [Ital. < Lat. *continuus*, continuous.] *Mus.* A usu. keyboard accompaniment for a solo instrument in which numerals indicate the successive chords, the actual notes played being left to the performer.

**con·tin·u·ous** (kən-tĭn′yo͞o-əs) *adj.* [Lat. *continuus.* —see CONTINUE.] **1.** Uninterrupted : unbroken. **2.** *Math.* Designating a function of one or more variables in which the variation of its values can be made arbitrarily small in a sufficiently small neighborhood of every point in a given interval. **—con·tin′u·ous·ly** *adv.* **—con·tin′u·ous·ness** *n.*

☆ ***syns:*** CONTINUOUS, CEASELESS, CONSTANT, CONTINUAL, ENDLESS, ETERNAL, EVERLASTING, INCESSANT, INTERMINABLE, NONSTOP, PERPETUAL, RELENTLESS, ROUND-THE-CLOCK, TIMELESS, UNCEASING, UNREMITTING *adj. core meaning* : existing or occurring without interruption or end <irritated by their *continuous* chatter> ***ant:*** discontinuous

**continuous creation theory** *n.* Steady-state theory.

**continuous spectrum** *n.* A spectrum having no breaks, esp. a spectrum of radiation distributed over an uninterrupted range of wavelengths.

**continuous wave** *adj.* Emitting or capable of emitting continuously : not pulsed. —Used esp. of lasers.

**con·tin·u·um** (kən-tĭn′yo͞o-əm) *n.*, *pl.* **-tin·u·a** (-tĭn′yo͞o-ə) *or* **-tin·u·ums.** [Lat., neuter of *continuus*, continuous.] **1.** A continuous extent, succession, or whole no part of which can be distinguished from neighboring parts except by arbitrary division. **2.** *Math.* A set with the same number of points as all the real numbers in an interval.

**con·tort** (kən-tôrt′) *v.* **-tort·ed, -tort·ing, -torts.** [Lat. *contorquēre, contort-*, to twist : *com-* (intensive) + *torquēre*, to twist.] —*vt.* To twist severely out of shape : WRENCH. —*vi.* To become twisted into a forced expression or shape. **—con·tor′tion** *n.* **—con·tor′tive** *adj.*

**con·tort·ed** (kən-tôr′tĭd) *adj.* **1.** Twisted or wrenched out of shape. **2.** *Bot.* Twisted or bent upon itself. **—con·tort′ed·ly** *adv.* **—con·tort′ed·ness** *n.*

**con·tor·tion·ist** (kən-tôr′shə-nĭst) *n.* An acrobat who exhibits extraordinary bodily positions. **—con·tor′tion·is′tic** *adj.*

**con·tour** (kŏn′to͝or′) *n.* [Fr. < Ital. *contorno* < *contornare*, to draw in outline : Lat. *com-* (intensive) + Lat. *tornare*, to round off < *tornus*, lathe.] **1. a.** The outline of a figure, body, or mass. **b.** A line representing such an outline. **2.** *often* **contours.** A surface, esp. of a curving form. **3.** A contour line. —*vt.* **-toured, -tour·ing, -tours. 1.** To make or shape the outline of. **2.** To build (e.g., a road) to follow the contour of the land. —*adj.* **1.** Following the contour lines of uneven terrain to reduce erosion of topsoil <*contour* plowing> **2.** Shaped to fit the outline or form of something <a *contour* seat>

**contour feather** *n.* One of the outermost feathers of a bird, forming the visible body contour and plumage.

**contour line** *n.* A line, as on a contour map, joining points of equal elevation.

**contour map** *n.* A map showing contour lines.

**contra-** *pref.* [ME < Lat. < *contra*, against.] **1.** Against : opposite : contrasting <*contra*position> **2.** Lower in pitch <*contra*bassoon>

**con·tra·band** (kŏn′trə-bănd′) *n.* [Ital. *contrabbando* : *contra-*, against (< Lat.) + *bando*, proclamation < LLat. *bannus.*] **1.** Goods barred by law or treaty from being imported or exported. **2. a.** Illegal traffic in contraband : SMUGGLING. **b.** Smuggled goods. **3.** Goods that may be seized and confiscated by a belligerent if shipped to another belligerent by a neutral power. **4.** An escaped slave who fled to or was taken behind Union lines during the Civil War. —*adj.* Barred from being imported or exported. **—con′tra·band′age** *n.* **—con′tra·band′ist** *n.*

**con·tra·bass** (kŏn′trə-bās′) [Obs. Ital. *contrabasso* : *contra-*, below (< Lat., against) + *basso*, bass.] *Mus.* —*n.* A double bass. —*adj.* Pitched an octave below the normal bass range. **—con′tra·bass′ist** (-bā′sĭst) *n.*

**con·tra·bas·soon** (kŏn′trə-bə-so͞on′, -bă-) *n.* The largest and lowest-pitched of the double-reed wind musical instruments, sounding an octave below the bassoon.

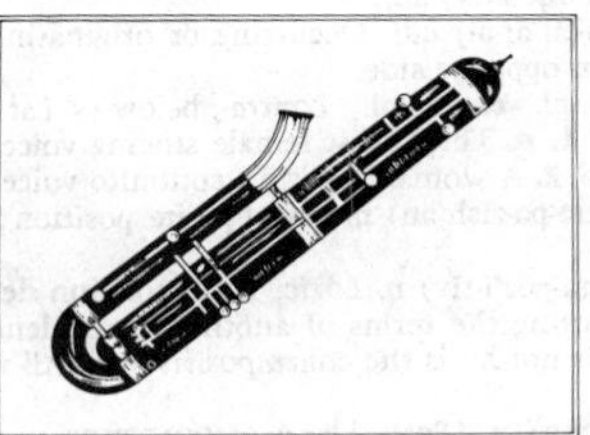

**contrabassoon**

**con·tra·cep·tion** (kŏn′trə-sĕp′shən) *n.* [CONTRA- + (CON)CEPTION.] Prevention of conception.

**con·tra·cep·tive** (kŏn′trə-sĕp′tĭv) *adj.* Capable of preventing conception. **—con′tra·cep′tive** *n.*

**con·tract** (kŏn′trăkt′) *n.* [ME < Lat. *contractus* < p.part. of *contrahere,* to make a contract : *com-,* together + *trahere,* to draw.] **1. a.** A legally enforceable agreement between two or more parties. **b.** The writing or document containing such an agreement. **2.** The branch of law dealing with contracts. **3.** Marriage as a formal agreement : BETROTHAL. **4.** In the game of bridge: **a.** The last and highest bid of one hand. **b.** The number of tricks thus bid. **c.** Contract bridge. **5.** *Slang.* A paid assignment to murder someone. —*v.* (kən-trăkt′, kŏn′trăkt′) **-tract·ed, -tract·ing, -tracts.** —*vt.* **1.** To enter into by contract : settle or establish by formal agreement. **2. a.** To acquire or incur <*contract* obligations> **b.** To become afflicted with (a disease). **3.** To reduce in size by drawing together : SHRINK. **4.** To pull together : WRINKLE. **5.** To shorten (a word or words) by omitting or combining some of the letters or sounds. —*vi.* **1.** To enter into or make a contract <*contract* for lawn care> **2.** To become reduced in size by or as if by being drawn together <The metal spring *contracted.*> **—con·tract′i·bil′i·ty, con·tract′i·ble·ness** *n.* **—con·tract′i·ble** *adj.*

☆ ***syns:*** CONTRACT, CATCH, DEVELOP, GET, SICKEN (with), TAKE *v. core meaning* : to become afflicted with (a disease) <*contracted* diphtheria>

**contract bridge** *n.* Auction bridge in which tricks in excess of the contract may not count toward game bonuses.

**con·trac·tile** (kən-trăk′təl, -tīl′) *adj.* Capable of contracting or causing contraction. **—con′trac·til′i·ty** (kŏn′trăk-tĭl′ĭ-tē) *n.*

**contractile vacuole** *n.* A contracting vesicle in some protozoans that functions in fluid expulsion.

**con·trac·tion** (kən-trăk′shən) *n.* **1.** The act of contracting or state of being contracted. **2. a.** A shortened word or words formed by omitting or combining some of the letters or sounds. **b.** The formation of such a word. **3.** *Physiol.* The shortening and often thickening of functioning muscle. **4.** A period of decreased business activity.

**con·trac·tor** (kŏn′trăk′tər, kən-trăk′-) *n.* **1.** A person who agrees to furnish materials or perform services at a specified price, esp. for construction. **2.** Something that contracts, esp. a muscle.

**con·trac·tu·al** (kən-trăk′cho͞o-əl) *adj.* Of, relating to, or like a contract. **—con·trac′tu·al·ly** *adv.*

**con·trac·ture** (kən-trăk′chər) *n.* **1.** A shrinkage or shortening, as of muscle or scar tissue, resulting in distortion or deformity. **2.** A deformity resulting from a contracture.

**con·tra·cy·cli·cal** (kŏn′trə-sī′klĭ-kəl, -sĭk′lĭ-) *adj.* Acting counter to an economic cycle.

**con·tra·dance** *or* **con·tra·danse** (kŏn′trə-dăns′) *n. vars. of* CONTREDANSE.

**con·tra·dict** (kŏn′trə-dĭkt′) *v.* **-dict·ed, -dict·ing, -dicts.** [Lat. *contradicere, contradict-,* to speak against : *contra-,* against + *dicere,* to speak.] —*vt.* **1.** To express or assert the opposite of (a statement). **2.** To deny the statement of. **3.** To be inconsistent with. —*vi.* To utter a contradictory statement. **—con′tra·dict′a·ble** *adj.* **—con′tra·dict′er, con′tra·dic′tor** *n.*

**con·tra·dic·tion** (kŏn′trə-dĭk′shən) *n.* **1.** An act of contradicting or the state of being contradicted. **2.** A denial. **3.** Inconsistency or discrepancy. **4.** Something containing contradictory elements.

**con·tra·dic·to·ry** (kŏn′trə-dĭk′tə-rē) *adj.* **1.** Involving, like, or being a contradiction. **2.** Tending to contradict. —*n., pl.* **-ries.** *Logic.* Either of two propositions related in such a way that it is impossible for both to be true or both to be false. **—con′tra·dic′to·ri·ly** *adv.* **—con′tra·dic′to·ri·ness** *n.*

**con·tra·dis·tinc·tion** (kŏn′trə-dĭ-stĭngk′shən) *n.* Distinction by contrasting or opposing qualities. **—con′tra·dis·tinc′tive** *adj.* **—con′tra·dis·tinc′tive·ly** *adv.*

**con·tra·dis·tin·guish** (kŏn′trə-dĭ-stĭng′gwĭsh) *vt.* **-guished, -guish·ing, -guish·es.** To distinguish by contrasting qualities.

**con·trail** (kŏn′trāl′) *n.* [CON(DENSATION) + TRAIL.] A visible trail of water droplets or ice crystals sometimes forming in the wake of an aircraft.

**con·tra·in·di·cate** (kŏn′trə-ĭn′dĭ-kāt′) *vt.* **-cat·ed, -cat·ing, -cates.** To indicate the inadvisability of <Adverse reactions *contraindicated* the use of antibiotics.> **—con′tra·in′di·ca′tion** *n.* **—con′tra·in·dic′a·tive** (-ĭn-dĭk′ə-tĭv) *adj.*

**con·tra·lat·er·al** (kŏn′trə-lăt′ər-əl) *adj.* Occurring or originating in a corresponding part on an opposite side.

**con·tral·to** (kən-trăl′tō) *n., pl.* **-tos.** [Ital. : *contra-,* below (< Lat., against) + *alto,* alto.] *Mus.* **1. a.** The lowest female singing voice. **b.** A part sung by a contralto. **2.** A woman having a contralto voice.

**con·tra·po·si·tion** (kŏn′trə-pə-zĭsh′ən) *n.* An opposite position : ANTITHESIS.

**con·tra·pos·i·tive** (kŏn′trə-pŏz′ĭ-tĭv) *n. Logic.* A proposition derived by negating and permuting the terms of another, equivalent proposition; e.g., "All not-Y is not-X" is the *contrapositive* of "All x is y."

**con·trap·tion** (kən-trăp′shən) *n.* [Perh. blend of CONTRIVE and TRAP.] A mechanical contrivance : GADGET.

**con·tra·pun·tal** (kŏn′trə-pŭn′tl) *adj.* [< Ital. *contrapunto,* counterpoint : *contra-,* against (< Lat.) + *punto,* point < Lat. *punctum* < *pungere,* to prick.] *Mus.* Of, relating to, or using counterpoint. **—con′tra·pun′tal·ly** *adv.*

**con·tra·pun·tist** (kŏn′trə-pŭn′tĭst) *n.* A specialist in contrapuntal music.

**con·tra·ri·e·ty** (kŏn′trə-rī′ĭ-tē) *n., pl.* **-ties. 1.** The quality or state of being contrary. **2.** Something that is contrary.

**con·trar·i·ous** (kən-trâr′ē-əs) *adj.* Hostile : inimical. **—con·trar′i·ous·ly** *adv.*

**con·trar·i·wise** (kŏn′trĕr′ē-wīz′, kən-trâr′-) *adv.* **1.** From a contrasting point of view. **2.** In reverse order : VICE VERSA. **3.** Obstinately : contrarily.

**con·trar·y** (kŏn′trĕr′ē) *adj.* [ME *contrarie* < Lat. *contrarius* < *contra,* against.] **1.** Completely different as in character or purpose : OPPOSED. **2.** Opposite in position or direction. **3.** Adverse : unfavorable <a *contrary* wind> **4.** (*also* kən-trâr′ē). Given to recalcitrant behavior : WILLFUL. —*n., pl.* **-ies. 1.** Something contrary : OPPOSITE. **2.** Either of two contrary or opposing things. **3.** *Logic.* A proposition related to another in such a way that if the latter is true, the former must be false, but if the latter is false, the former is not necessarily true. —*adv.* In opposition : CONTRARIWISE. **—con′trar·i·ly** *adv.* **—con′trar·i·ness** *n.*

☆ ***syns:*** CONTRARY, BALKY, DIFFICULT, FROWARD, IMPOSSIBLE, ORNERY, PERVERSE, WAYWARD *adj. core meaning* : obstinately self-willed <a *contrary* person who refused to conform> ***ant:*** complaisant

**con·trast** (kən-trăst′) *v.* **-trast·ed, -trast·ing, -trasts.** [Fr. *contrester* < Ital. *contrastare* < Med. Lat. : Lat. *contra-,* against + Lat. *stare,* to stand.] —*vt.* To set in opposition in order to show or emphasize differences. —*vi.* To show differences when compared. —*n.* (kŏn′trăst′). **1.** An act of contrasting or the state of being contrasted. **2.** A notable dissimilarity between things compared. **3.** Something that shows a marked dissimilarity to something else. **4.** Use of opposing elements, as colors, forms, or lines, in proximity to produce an intensified effect in a work of art. **—con·trast′a·ble** *adj.* **—con·trast′ing·ly** *adv.*

**con·trast·y** (kŏn′trăs′tē) *adj.* Having or producing sharp photographic contrasts between light and dark areas.

**con·tra·vene** (kŏn′trə-vēn′) *vt.* **-vened, -ven·ing, -venes.** [OFr. *contravenir* < LLat. *contravenire,* to oppose : Lat. *contra-,* against + Lat. *venire,* to come.] **1.** To act or be counter to : VIOLATE <*contravene* a legal order> **2.** To oppose in argument. **—con′tra·ven′er** *n.* **—con′tra·ven′tion** (-shən) *n.*

**con·tre·danse** *also* **con·tre·dance** *or* **con·tra·dance** *or* **con·tra·danse** (kŏn′trə-dăns′) *n.* [Fr. < COUNTRY-DANCE.] **1.** A dance performed with the partners facing each other from two lines. **2.** The music for a contredanse.

**con·tre·temps** (kŏn′trə-täN′, kôN′trə-täN′) *n., pl.* **contretemps** (-täNz′, -täN′) [Fr. : *contre-,* against (< Lat. *contra-*) + *temps,* time < Lat. *tempus.*] An inopportune or embarrassing occurrence : MISHAP.

**con·trib·ute** (kən-trĭb′yo͞ot) *v.* **-ut·ed, -ut·ing, -utes.** [Lat. *contribuere, contribut-,* to bring together : *com-* together + *tribuere,* to grant. —see TRIBUTE.] —*vt.* **1.** To give or supply in common with others. **2.** To submit for publication. —*vi.* **1.** To make a contribution. **2.** To act as a determining factor <Poverty *contributes* to crime.> **3.** To submit material for publication. **—con·trib′u·tive** *adj.* **—con·trib′u·tive·ly** *adv.* **—con·trib′u·tive·ness** *n.* **—con·trib′u·tor** *n.*

**con·tri·bu·tion** (kŏn′trĭ-byo͞o′shən) *n.* **1.** The act of contributing. **2.** Something contributed, as an article for publication in a periodical. **3.** A payment, as a levy, for a special purpose.

**con·trib·u·to·ry** (kən-trĭb′yə-tôr′ē, -tōr′ē) *adj.* **1.** Of, relating to, or involving contribution. **2.** Contributing toward a result. **3.** Subject to an impost or levy. —*n., pl.* **-ries.** One that contributes.

**con·trite** (kən-trīt′, kŏn′trīt′) *adj.* [ME *contrit* < Lat. *contritus,* p.part. of *conterere,* to crush : *com-* (intensive) + *terere,* to grind.] **1.** Repentant for one's sins or inadequacies : PENITENT. **2.** Arising from contrition <*contrite* apologies> **—con·trite′ly** *adv.* **—con·trite′ness** *n.*

**con·tri·tion** (kən-trĭsh′ən) *n.* Sincere remorse for wrongdoing.

**con·tri·vance** (kən-trī′vəns) *n.* **1.** The act of contriving or state of being contrived. **2.** Something contrived.

**con·trive** (kən-trīv′) *v.* **-trived, -triv·ing, -trives.** [ME *contreven* < OFr. *controver* < Med. Lat. *contropare,* to compare : Lat. *com-,* with + Lat. *tropus,* trope < Gk. *tropos.*] —*vt.* **1.** To plan or devise. **2.** To bring about by artifice : SCHEME <*contrived* a way to rob the bank> **3.** To invent or fabricate. —*vi.* To plot or scheme. **—con·triv′ed·ly** (-trī′vĭd-lē) *adv.* **—con·triv′er** *n.*

**con·trol** (kən-trōl′) *vt.* **-trolled, -trol·ling, -trols.** [ME *countrollen* < AN < OFr. *contrarotulare* < Med. Lat. *controretulare,* to check by duplicate register < *contrarotulus,* duplicate register : Lat. *contra,* against + Lat. *rotulus,* roll, dim. of *rota,* wheel.] **1.** To exercise authority or influence over : DIRECT. **2.** To hold in restraint : CHECK. **3. a.** To verify or regulate (a scientific experiment) by conducting a parallel experiment or by comparing with another standard. **b.** To verify (e.g., an account) by using a duplicate register for comparison. —*n.* **1.** Authority or ability to regulate, direct, or influence. **2.** A restraining act or influence : CURB <inflation *controls*> **3.** A standard of comparison for checking or verifying the results of an experiment. **4.** *often* **controls.** An instrument or set of instruments for operating, regulating, or guiding a machine or vehicle. **5.** A spirit presumed to act through a spiritualist medium. **6.** An organization for directing a space flight. **—con·trol'la·bil'i·ty** *n.* **—con·trol'la·ble** *adj.*

**control chart** *n.* A graph of a quantitative typical of a manufacturing process, usu. determined from small, periodically repeated samples and evaluated with respect to control limits rendered as parallel horizontal lines above and below a line representing the expected or average value of the characteristic.

**control experiment** *n.* An experiment in which the variable factors are controlled so that the effects of changing one at a time can be observed.

**controlled response** *n.* A response to a military attack by limited military means in an effort to avoid nuclear war.

**con·trol·ler** (kən-trō'lər) *n.* **1.** One that controls. **2.** *also* **comp·trol·ler.** An officer who audits accounts and supervises the finances of a corporation or governmental body. **—con·trol'ler·ship'** *n.*

**control rocket** *n.* A vernier rocket or similar missile used to change the attitude or trajectory of a rocket or spacecraft.

**control rod** *n.* A rod containing a neutron-absorbing substance that is used to regulate fission chain reactions in nuclear reactors.

**control stick** *n.* A lever used in small aircraft to control the angle of the elevators and ailerons.

**control surface** *n.* A movable airfoil for controlling or guiding an aircraft.

**control tower** *n.* A tower at an airport from which air traffic is controlled, usu. by radio.

**con·tro·ver·sial** (kŏn'trə-vûr'shəl, -sē-əl) *adj.* **1.** Of, subject to, or marked by controversy. **2.** Given to controversy : DISPUTATIOUS. **—con'tro·ver'sial·ist** *n.* **—con'tro·ver'sial·ly** *adv.*

**con·tro·ver·sy** (kŏn'trə-vûr'sē) *n.*, *pl.* **-sies.** [ME *controversie* < Lat. *controversia* < *controversus*, disputed < *contro-*, against (var. of *contra-*) + *versus*, p.part. of *vertere*, to turn.] **1.** A dispute characterized esp. by the expression of opposing views. **2.** A quarrel : argument.

**con·tro·vert** (kŏn'trə-vûrt', kŏn'trə-vûrt') *vt.* **-vert·ed, -vert·ing, -verts.** [< CONTROVERSY.] To raise arguments against or voice opposition to. **—con'tro·vert'i·ble** *adj.*

**con·tu·ma·cious** (kŏn'tə-mā'shəs, -tyə-) *adj.* Obstinately disobedient or insubordinate : RECALCITRANT. **—con'tu·ma'cious·ly** *adv.* **—con'tu·ma'cious·ness** *n.*

**con·tu·ma·cy** (kŏn'to͞o-mə-sē, -tyo͞o-) *n.*, *pl.* **-cies.** [ME *contumacie* < Lat. *contumacia* < *contumax*, insolent.] Obstinate or contemptuous resistance, esp. to judicial authority.

**con·tu·me·ly** (kŏn'to͞o-mə-lē, -tyo͞o-, -təm-lē) *n.*, *pl.* **-lies.** [ME *contumelie* < Lat. *contumelia.*] **1.** Rudeness or contempt in behavior or speech : INSOLENCE. **2.** An insulting act or remark. **—con'tu·me'li·ous** (kŏn'tə-mē'lē-əs) *adj.* **—con'tu·me'li·ous·ly** *adv.*

**con·tuse** (kən-to͞oz', -tyo͞oz') *vt.* **-tused, -tus·ing, -tus·es.** [ME *contusen* < Lat. *contundere*, to beat : *com-* (intensive) + *undere*, to beat.] To injure without breaking the skin. **—con·tu'sion** *n.*

**co·nun·drum** (kə-nŭn'drəm) *n.* [Orig. unknown.] **1.** A riddle in which a fanciful question is answered by a pun. **2. a.** A problem with no satisfactory solution. **b.** A complicated problem.

**con·ur·ba·tion** (kŏn'ər-bā'shən) *n.* [CON- + Lat. *urbs*, city + -ATION.] A predominantly urban region including adjacent towns and suburbs : metropolitan area.

**co·nus ar·te·ri·o·sus** (kō'nəs är-tîr'ē-ō'səs) *n.* [NLat., arterial cone.] **1.** A conical extension of the right ventricle in the heart of mammals from which the pulmonary arteries arise. **2.** An extension of the ventricle in the heart of amphibians and fish.

**con·va·lesce** (kŏn'və-lĕs') *vi.* **-lesced, -lesc·ing, -lesc·es.** [Lat. *convalescere* : *com-* (intensive) + *valescere*, to grow strong < *valere*, to be strong.] To return to health after illness : RECUPERATE.

**con·va·les·cence** (kŏn'və-lĕs'əns) *n.* Gradual return to health after illness or the period needed for it. **—con'va·les'cent** *adj.*

**con·vec·tion** (kən-vĕk'shən) *n.* [LLat. *convectio* < *convehere*, to carry together : Lat. *com-*, together + Lat. *vehere*, to carry.] **1.** The act or process of conveying or transmitting. **2.** *Physics.* **a.** Heat transfer by fluid motion between regions of unequal density that result from nonuniform heating. **b.** Fluid motion caused by an external force, as gravity. **3.** *Meteorol.* The transfer of heat or other atmospheric properties by massive motion within the atmosphere, esp. by such motion directed upward. **—con·vec'tion·al** *adj.* **—con·vec'tive** *adj.* **—con·vec'tive·ly** *adv.*

**con·vec·tor** (kən-vĕk'tər) *n.* A partly enclosed, directly heated surface from which warm air circulates by convection.

**con·vene** (kən-vēn') *v.* **-vened, -ven·ing, -venes.** [ME *convenen* < Lat. *convenire* : *com-*, together + *venire*, to come.] —*vi.* To come together, usu. for an official or public purpose : ASSEMBLE. —*vt.* **1.** To cause to assemble : CONVOKE. **2.** To summon to appear, as before a court of law. **—con·ven'a·ble** *adj.* **—con·ven'er** *n.*

**con·ven·ience** (kən-vēn'yəns) *n.* **1.** The quality of being convenient : SUITABILITY. **2.** Personal comfort or material advantage. **3.** Something that increases comfort or makes work less difficult. **4.** *Chiefly Brit.* A lavatory.

**convenience food** *n.* A prepackaged food that can be prepared quickly and easily.

**con·ven·ien·cy** (kən-vēn'yən-sē) *n.*, *pl.* **-cies.** *Archaic.* Convenience.

**con·ven·ient** (kən-vēn'yənt) *adj.* [ME < Lat. *conveniens*, pr.part. of *convenire*, to be suitable. —see CONVENE.] **1.** Appropriate or favorable to one's comfort, purpose, or needs. **2.** Easy to reach : ACCESSIBLE. **—con·ven'ient·ly** *adv.*

☆ ***syns:*** CONVENIENT, APPROPRIATE, FIT, GOOD, PROPER, SUITABLE, TAILOR-MADE *adj. core meaning* : suited to one's end or purpose <found a *convenient* excuse>

**con·vent** (kŏn'vənt, -vĕnt') *n.* [ME *covent* < OFr. < Med. Lat. *conventus* < Lat., assembly < *convenire*, to assemble. —see CONVENE.] **1.** A community, esp. of nuns, bound by vows to a religious life under a superior. **2.** The building or buildings occupied by a convent.

**con·ven·ti·cle** (kən-vĕn'tĭ-kəl) *n.* [ME < Lat. *conventiculum*, meeting, dim. of *conventus*, assembly. —see CONVENT.] **1.** A meeting or an assembly. **2.** An unlawful or questionable assembly. **3.** A religious meeting, esp. a secret or illegal one, as those held by dissenters in England and Scotland in the 16th and 17th cent. **—con·ven'ti·cler** *n.*

**con·ven·tion** (kən-vĕn'shən) *n.* [ME *convencioun* < Lat. *conventio*, meeting < *convenire*, to meet. —see CONVENE.] **1. a.** A formal assembly or meeting of members, representatives, or delegates of a group, as a political party. **b.** The body of persons attending such an assembly. **2.** A compact or agreement, esp. an international agreement dealing with a specific subject <a *convention* on international law> **3.** General agreement on or acceptance of certain practices or attitudes. **4.** A practice or procedure widely observed in a group, esp. to facilitate social interaction : CUSTOM. **5.** A widely accepted device or technique, as in drama, literature, or painting.

**con·ven·tion·al** (kən-vĕn'shə-nəl) *adj.* **1.** Developed, established, or approved by general usage : CUSTOMARY. **2.** Conforming to established practice or accepted standards. **3.** Characterized by or dependent on stereotypical conventions. **4.** Represented, as in a work of art, in simplified or abstract form. **5.** *Law.* Based on consent or agreement : CONTRACTUAL. **6.** Of, relating to, or resembling an assembly. **7.** Using means other than nuclear energy <*conventional* weapons> **—con·ven'tion·al·ism** *n.* **—con·ven'tion·al·ist** *n.* **—con·ven'tion·al·ly** *adv.*

**con·ven·tion·al·i·ty** (kən-vĕn'shə-năl'ĭ-tē) *n.*, *pl.* **-ties.** **1.** The quality, state, or character of being conventional. **2.** A conventional act, principle, or practice. **3. conventionalities.** The rules of conventional social behavior.

**con·ven·tion·al·ize** (kən-vĕn'shə-nə-līz') *vt.* **-ized, -iz·ing, -iz·es.** To make conventional. **—con·ven'tion·al·i·za'tion** *n.*

**con·ven·tion·eer** (kən-vĕn'shə-nîr') *n.* A person who attends a convention.

**con·ven·tu·al** (kən-vĕn'cho͞o-əl) *adj.* Of or relating to a convent. —*n.* **1.** A member of a convent. **2. Conventual.** A member of a branch of the Franciscan order that permits the accumulation and possession of common property.

**con·verge** (kən-vûrj') *v.* **-verged, -verg·ing, -verg·es.** [LLat. *convergere*, to incline together : Lat. *com-*, together + Lat. *vergere*, to incline.] —*vi.* **1.** To approach the same point from different directions : MEET. **2.** To move together toward union or toward a common conclusion or result. **3.** *Math.* To approach a limit. —*vt.* To cause to converge.

**con·ver·gence** (kən-vûr'jəns) *also* **con·ver·gen·cy** (-jən-sē) *n.*, *pl.* **-gen·ces** *also* **-gen·cies.** **1.** The act, quality, state, or fact of converging. **2.** *Math.* The property or manner of approaching a limit such as a point, line, surface, or value. **3.** The point or degree of converging. **4.** *Physiol.* The coordinated turning of the eyes inward to focus on a nearby point. **5.** *Biol.* Adaptive evolution of superficially similar structures, as the wings of birds and insects, in unrelated species living in similar environments. **—con·ver'gent** *adj.*

**convergent evolution** *n.* CONVERGENCE 5.

**con·ver·sance** (kən-vûr'səns, kŏn'vər-) *also* **con·ver·san·cy** (kən-vûr'sən-sē) *n.* The state of being conversant.

**con·ver·sant** (kən-vûr'sənt, kŏn'vər-) *adj.* [ME *conversaunt*, associated with < OFr. *conversant*, pr.part. *converser*, to associate with < Lat. *conversari.*] Familiar, as by study or experience <*conversant* with psycholinguistics> **—con·ver'sant·ly** *adv.*

**con·ver·sa·tion** (kŏn'vər-sā'shən) *n.* **1.** A spoken exchange of opinions, thoughts, and feelings : TALK. **2.** An informal discussion of a matter by representatives of governments, institutions, or organizations. **3.** *Computer Sci.* A real-time interaction with a computer. **—con'ver·sa'tion·al** *adj.* **—con'ver·sa'tion·al·ly** *adv.*

**con·ver·sa·tion·al·ist** (kŏn'vər-sā'shə-nə-lĭst) *also* **con·ver·sa·tion·ist** (-shə-nĭst) *n.* One skilled at or given to conversation.

**conversation piece** *n.* **1.** A genre painting, esp. popular in the 18th cent., depicting a group of fashionable people. **2.** An unusual object that invites comment or arouses interest.

**con·ver·sa·zi·o·ne** (kŏn′vər-sät′sē-ō′nē, kŏn′-) *n., pl.* **-nes** *or* **-ni** (-nē) [Ital. < Lat. *conversatio,* dealings with persons < *conversari,* to associate with. —see CONVERSE[1].] A gathering for conversation, esp. about art or literature.

**con·verse[1]** (kən-vûrs′) *vi.* **-versed, -vers·ing, -vers·es.** [ME *conversen,* to associate with < OFr. *converser* < Lat. *conversari* : *com-,* with + *versari,* to occupy oneself < *vertere,* to depend on.] **1.** To engage in oral exchange of thoughts, opinions, and feelings : TALK. **2.** *Computer Sci.* To interact with a computer on-line. **3.** *Archaic.* To consort : associate. —*n.* (kŏn′vûrs′). **1.** Spoken interchange of thoughts, opinions, and feelings : CONVERSATION. **2.** *Archaic.* Social intercourse.

**con·verse[2]** (kən-vûrs′, kŏn′vûrs′) *adj.* [Lat. *conversus,* p.part. of *convertere,* to turn around. —see CONVERT.] Reversed, as in position, order, or action : CONTRARY. —*n.* (kŏn′vûrs′). **1.** Something reversed : OPPOSITE. **2.** *Logic.* A proposition obtained by conversion. **—con·verse′ly** *adv.*

**con·ver·sion** (kən-vûr′zhən, -shən) *n.* [ME *conversioun,* religious conversion < OFr. *conversion* < Lat. *conversio,* a turning around < *convertere,* to turn around.] **1.** The act of converting or state of being converted. **2.** A change in which one adopts a new religion. **3.** Something converted from one use or purpose to another. **4.** *Law.* **a.** Unlawful appropriation of another's property. **b.** The changing of personal property to real property or vice versa. **5.** Exchange of one type of security or currency for another. **6.** *Logic.* Interchange of the subject and predicate of a proposition. **7.** *Football.* A score made on a try for a point or points after a touchdown. **8.** *Psychiat.* Symbolic manifestation of repressed ideas or impulses in motor or sensory abnormalities, as paralysis. **—con·ver′sion·al, con·ver′sion·ar′y** (-zhə-nĕr′ē, -shə-) *adj.*

**conversion factor** *n.* A numerical factor used to multiply or divide a quantity expressed in one system of units in a conversion to another system.

**conversion reaction** *n.* A neurosis marked by the presence of bodily symptoms with no physical cause.

**con·vert** (kən-vûrt′) *v.* **-vert·ed, -vert·ing, -verts.** [ME *converten,* to convert to a religion < OFr. *convertir* < Lat. *convertere,* to turn around : *com-* (intensive) + *vertere,* to turn.] —*vt.* **1.** To change into another form, substance, state, or product : TRANSFORM <*convert* water into steam> **2.** To persuade or induce to adopt a particular religion, faith, or belief. **3.** To change from one use, function, or purpose to another <*convert* an attic into a study> **4.** To exchange for something of equal value. **5.** To exchange (e.g., a security) by substituting an equivalent of another form. **6.** To express (a quantity) in alternative units. **7.** *Logic.* To transform (a proposition) by conversion. **8.** *Law.* **a.** To appropriate (another's property) without right to one's own use. **b.** To change (property) from real to personal or from joint to separate or vice versa. —*vi.* **1.** To undergo a change. **2.** *Football.* To make a conversion. —*n.* (kŏn′vûrt′). A person who has been converted, esp. from one religion to another.

**con·vert·a·plane** (kən-vûr′tə-plān′) *n. var. of* CONVERTIPLANE.

**con·vert·er** *also* **con·ver·tor** (kən-vûr′tər) *n.* **1.** One that converts. **2.** One employed in converting raw products into finished products. **3.** A furnace for converting pig iron into steel by the Bessemer process. **4. a.** A machine for changing alternating current to direct current or vice versa. **b.** An electronic device for changing the frequency of a radio signal. **c.** A device that transforms information from one code to another.

**con·vert·i·ble** (kən-vûr′tə-bəl) *adj.* **1.** Capable of being converted. **2.** Having a top that can be folded back or removed, as a car. **3.** Capable of being lawfully exchanged for gold or another currency <dollars *convertible* into yen> —*n.* **1.** A convertible automobile. **2.** Something that can be converted. **—con·vert′i·bil′i·ty, con·vert′i·ble·ness** *n.* **—con·vert′i·bly** *adv.*

**con·vert·i·plane** *also* **con·vert·a·plane** (kən-vûr′tə-plān′) *n.* An airplane designed to fly vertically as well as forward.

**con·ver·tor** (kən-vûr′tər) *n. var. of* CONVERTER.

**con·vex** (kŏn′vĕks′, kən-vĕks′) *adj.* [Lat. *convexus* < *convehere,* to bring together.] Having a surface or boundary that curves or bulges outward, as the exterior of a sphere. **—con′vex′ly** *adv.*

**con·vex·i·ty** (kən-vĕk′sĭ-tē) *n., pl.* **-ties. 1.** The state of being convex. **2.** A convex surface or part.

**con·vex·o-con·cave** (kən-vĕk′sō-kən-kāv′) *adj.* **1.** Designating a lens with greater convex than concave curvature. **2.** CONCAVO-CONVEX.

**con·vex·o-con·vex** (kən-vĕk′sō-kən-vĕks′) *adj.* Convex on both sides : BICONVEX.

**con·vey** (kən-vā′) *vt.* **-veyed, -vey·ing, -veys.** [ME *conveien* < OFr. *conveier* < Med. Lat. *conviare,* to escort : Lat. *com-,* with + *via,* way.] **1.** To take or carry from one place to another : TRANSPORT. **2.** To serve as a medium of transmission for : TRANSMIT. **3.** To communicate or make known : IMPART <a look *conveying* jealousy> **4.** *Law.* To transfer ownership of or title to. **5.** *Obs.* To steal. **—con·vey′a·ble** *adj.*

**con·vey·ance** (kən-vā′əns) *n.* **1.** The act of conveying. **2.** A means of conveying, esp. a vehicle like a car or bus. **3.** *Law.* **a.** Transfer of title to property from one person to another. **b.** The document by which this transfer is effected.

**con·vey·anc·ing** (kən-vā′ən-sĭng) *n.* The branch of legal practice dealing with the conveyance of property or real estate. **—con·vey′anc·er** *n.*

**con·vey·er** *also* **con·vey·or** (kən-vā′ər) *n.* **1.** One that conveys. **2.** A mechanical device, as a continuous moving belt, that transports packages or bulk materials from place to place.

**con·vict** (kən-vĭkt′) *vt.* **-vict·ed, -vict·ing, -victs.** [ME *convicten* < Lat. *convincere.*] **1.** To find or prove (someone) guilty of an offense or crime, esp. by the verdict of a court. **2.** To convince of wrongdoing or sinfulness. —*n.* (kŏn′vĭkt′). **1.** One found or declared guilty of an offense or crime. **2.** One serving a sentence of imprisonment. —*adj.* (kən-vĭkt′). *Archaic.* Found guilty : CONVICTED.

**con·vic·tion** (kən-vĭk′shən) *n.* **1.** The act or process of convicting, esp. of an unlawful act. **2. a.** The act or process of convincing. **b.** The state of being convinced. **3.** A fixed or strong belief. **—con·vic′tion·al** *adj.*

**con·vince** (kən-vĭns′) *vt.* **-vinced, -vinc·ing, -vinc·es.** [Lat. *convincere* : *com-,* (intensive) + *vincere,* to conquer.] **1.** To cause to believe with certainty <*convinced* of the need for action> ***usage:*** Traditionalists object to the use of *convince* with a following infinitive, as in *I convinced them to leave;* however, this usage has gained widespread acceptance in current English **2.** *Obs.* To convict. **3.** *Obs.* To conquer. **—con·vince′ment** *n.* **—con·vinc′er** *n.* **—con·vinc′i·ble** *adj.*

**con·vinc·ing** (kən-vĭn′sĭng) *adj.* **1.** Serving to convince <*convincing* statistics> **2.** Plausible : believable <a *convincing* excuse> **—con·vinc′ing·ly** *adv.* **—con·vinc′ing·ness** *n.*

**con·viv·i·al** (kən-vĭv′ē-əl) *adj.* [Lat. *convivialis* < Lat. *convivium,* banquet : *com-,* together + *vivere,* to live.] **1.** Fond of feasting, drinking, and good company : SOCIABLE. **2.** Relating to or like a feast : FESTIVE. **—con·viv′i·al′i·ty** (-ăl′ĭ-tē) *n.* **—con·viv′i·al·ly** *adv.*

**con·vo·ca·tion** (kŏn′və-kā′shən) *n.* **1. a.** The act of convoking. **b.** A group of people convoked. **2.** An Anglican clerical assembly similar to a synod but assembling only when called. **3. a.** An assembly of the clergy and representative laity of a section of a diocese of the Protestant Episcopal Church. **b.** The district represented at such an assembly. **—con′vo·ca′tion·al** *adj.*

**con·voke** (kən-vōk′) *vt.* **-voked, -vok·ing, -vokes.** [OFr. *convoquer* < Lat. *convocare* : *com-,* together + *vocare,* to call.] To cause to assemble in a meeting : CONVENE. **—con·vok′er** *n.*

**con·vo·lute** (kŏn′və-lo͞ot′) *adj.* [Lat. *convolutus,* p.part. of *convolvere,* to convolve.] Rolled or folded together with one part over another : COILED. —*vt.* & *vi.* **-lut·ed, -lut·ing, -lutes.** To coil : twist. **—con′vo·lute′ly** *adv.*

**con·vo·lut·ed** (kŏn′və-lo͞o′tĭd) *adj.* **1.** Coiled : twisted. **2.** Intricate : complicated <*convoluted* legal terminology>

**con·vo·lu·tion** (kŏn′və-lo͞o′shən) *n.* **1.** A convoluted formation or configuration. **2.** One of the convex folds of the surface of the brain. **—con′vo·lu′tion·al** *adj.*

**con·volve** (kən-vŏlv′) *v.* **-volved, -volv·ing, -volves.** [Lat. *convolvere* : *com-,* together + *volvere,* to roll.] —*vt.* To roll or coil together. —*vi.* To form convolutions.

**con·vol·vu·lus** (kən-vŏl′vyə-ləs) *n., pl.* **-lus·es** *or* **-li** (-lī′) [NLat. *Convolvulus,* genus name < Lat. *covolvulus,* bindweed < *convolvere,* to intertwine. —see CONVOLVE.] Any of several trailing or twining plants of the genus *Convolvulus,* including the bindweeds.

**con·voy** (kŏn′voi′, kən-voi′) *vt.* **-voyed, -voy·ing, -voys.** [ME *convoyen,* to escort < OFr. *convoier,* var. of *conveier.* —see CONVEY.] To accompany, esp. for protection : ESCORT. —*n.* (kŏn′voi′). **1.** The act of convoying. **2.** An accompanying and protective force, as of ships. **3.** Something convoyed, as ships or troops. **4.** A group, as of vehicles, traveling together for convenience <a truck *convoy*>

**con·vulse** (kən-vŭls′) *vt.* **-vulsed, -vuls·ing, -vuls·es.** [Lat. *convellere, convuls-,* to pull violently : *com-* (intensive) + *vellere,* to pull.] **1.** To agitate or disturb violently. **2.** To affect with irregular and involuntary muscular contractions.

**con·vul·sion** (kən-vŭl′shən) *n.* **1.** An intense paroxysmal involuntary muscular contraction. **2.** An uncontrolled fit, as of laughter : PAROXYSM. **3.** A violent turmoil.

**con·vul·sion·ar·y** (kən-vŭl′shə-nĕr′ē) *adj.* Of, relating to, affected with, or like convulsions. —*n., pl.* **-ies.** A person affected with convulsions, esp. as a result of religious fanaticism.

**con·vul·sive** (kən-vŭl′sĭv) *adj.* **1.** Characterized by or like convulsions. **2.** Having or producing convulsions. **—con·vul′sive·ly** *adv.* **—con·vul′sive·ness** *n.*

**co·ny** *also* **co·ney** (kō′nē, kŭn′ē) *n., pl.* **-nies** *also* **-neys.** [ME < OFr. *conil* < Lat. *cuniculus.*] **1.** A rabbit, esp. the Old World species *Oryctolagus cuniculus.* **2.** The fur of a rabbit. **3.** A pika. **4.** A hyrax. **5.** *Archaic.* A dupe.

**coo** (ko͞o) *v.* **cooed, coo·ing, coos.** [Imit.] —*vi.* **1.** To utter the typical murmuring sound of a dove or pigeon or a similar sound. **2.** To talk amorously or fondly in murmurs <teenagers billing and *cooing* in the park> —*vt.* To express or utter fondly or amorously, as with a murmuring sound. **—coo′er** *n.*

**cook** (kŏŏk) *v.* **cooked, cook·ing, cooks.** [ME *coken* < *coke,* cook < OE *cōc* < LLat. *cocus* < Lat. *coquus* < *coquere,* to cook.] —*vt.* **1.** To prepare for eating by applying heat. **2.** To prepare or treat by heating. —*vi.* **1.** To prepare food for eating by applying heat. **2.** To undergo cooking. **3.** *Slang.* To happen : occur <wondered what was *cooking* in the meeting> **—cook up.** *Informal.* To fabricate : concoct <*cook up* an alibi> —*n.* One who prepares food for eating.

**cook·book** (kŏŏk'bŏŏk') *n.* A book containing information, as recipes, about food preparation.

**cook·er** (kŏŏk'ər) *n.* **1.** One that cooks, esp. a utensil or an appliance for cooking. **2.** One employed to operate cooking equipment in the commercial preparation of food and drink.

**cook·er·y** (kŏŏk'ə-rē) *n., pl.* **-ies. 1.** The art or practice of preparing food. **2.** A place for cooking.

**cook·ie** *or* **cook·y** (kŏŏk'ē) *n., pl.* **-ies.** [Du. *koekje,* dim. of *koek,* cake < MDu. *koeke.*] **1.** A small, sweetened usu. flat cake. **2. a.** A person : fellow <one mean *cookie*> **b.** An attractive woman.

**cook·out** (kŏŏk'out') *n.* **1.** A meal cooked and served outdoors. **2.** An outing at which such a meal is served.

**cool** (kōōl) *adj.* **-er, -est.** [ME *col* < OE *cōl.*] **1.** Moderately cold <*cool* fall nights> **2.** Affording or allowing relief from heat <a *cool* shower><a *cool* shirt> **3.** Not excited <a *cool* disposition> **4.** Marked by indifference : UNENTHUSIASTIC <a *cool* handshake> **5.** Marked by calm audacity : IMPUDENT. **6.** Designating or typical of colors, as blue and green, that produce the impression of coolness. **7.** *Slang.* Superb : first-rate. **8.** *Informal.* Entire : full <lost a *cool* grand on the stock market> —*v.* **cooled, cool·ing, cools.** —*vt.* **1.** To make less warm. **2.** To make less ardent, intense, or zealous <*cool* one's anger> —*vi.* **1.** To become less warm <took a swim to *cool* off> **2.** To become calm. —*n.* **1.** Something cool or moderately cold <the *cool* of early evening> **2.** The quality or state of being cool. **3.** *Slang.* Composure <lose one's *cool*> **—cool it.** *Slang.* To calm down : RELAX. **—cool (one's) heels.** *Slang.* To wait or be kept waiting. **—cool'ish** *adj.* **—cool'ly** *adv.* **—cool'ness** *n.*

**cool·ant** (kōō'lənt) *n.* A cooling agent, esp. a fluid that draws off heat by circulating through a machine or by bathing a mechanical part.

**cool·er** (kōō'lər) *n.* **1.** A container or device that cools or keeps cool. **2.** A cold drink. **3.** *Slang.* A jail.

**Coo·ley's anemia** (kōō'lēz) *n.* [After Thomas B. *Cooley* (1871–1945).] An inherited form of anemia resulting from a faulty synthesis of hemoglobin.

**cool-head·ed** (kōōl'hĕd'ĭd) *adj.* Unflappable.

**coo·lie** *also* **coo·ly** (kōō'lē) *n., pl.* **-lies.** [Hindi *kulī.*] An unskilled Oriental laborer.

**coon** (kōōn) *n.* [Short for RACCOON.] *Informal.* A raccoon.

**coon·can** (kōōn'kăn') *n.* Conquian.

**coon·hound** (kōōn'hound') *n.* A smooth-coated black and tan hound orig. bred in the southeastern United States to hunt raccoons.

**coon's age** *n. Slang.* A long time.

**coon·skin** (kōōn'skĭn') *n.* **1.** The pelt of a raccoon. **2.** An article, as a hat, made of coonskin.

**coon·tie** (kōōn'tē) *n.* [Of Seminole orig.] An evergreen plant, *Zamia floridana* of southern Florida, with underground stems that yield a starch resembling arrowroot.

**coontie**

**coop** (kōōp) *n.* [ME *coupe.*] **1.** An enclosure or cage, as for poultry or small animals. **2.** *Slang.* A place of confinement. —*vt.* **cooped, coop·ing, coops.** To confine in or as if in a coop <*cooped* up in an office all day>

**co-op** (kō'ŏp', kō-ŏp') *n.* A cooperative.

**coo·per** (kōō'pər) *n.* [ME *couper* < MDu. *cūper* < *cūpe,* cask.] One who makes or repairs wooden tubs and casks. —*v.* **-pered, -per·ing, -pers.** —*vt.* To make or repair (wooden tubs and casks). —*vi.* To work as a cooper. **—coo'per·age** (-ĭj) *n.*

**co·op·er·ate** (kō-ŏp'ə-rāt') *vi.* **-at·ed, -at·ing, -ates.** [LLat. *cooperari, cooperat-* : *co(m)-,* together + *operari,* to work < *opus,* work.] **1.** To work or act together toward a common end or purpose. **2.** To practice economic cooperation. **—co·op'er·a'tor** *n.*

**co·op·er·a·tion** (kō-ŏp'ə-rā'shən) *n.* **1.** An act of cooperating. **2.** An association for mutual benefit. **—co·op'er·a'tion·ist** *n.*

**co·op·er·a·tive** (kō-ŏp'ər-ə-tĭv, -ŏp'rə-, -ə-rā'tĭv) *adj.* **1.** Done in cooperation with others <a *cooperative* venture> **2.** Characterized by willingness to cooperate <a *cooperative* suspect> **3.** Engaged in joint economic activity. —*n.* An enterprise owned jointly by those who use its facilities or services. **—co·op'er·a·tive·ly** *adv.* **—co·op'er·a·tive·ness** *n.*

**co-opt** (kō-ŏpt', kō'ŏpt') *vt.* **-opt·ed, -opt·ing, -opts.** [Lat. *cooptare* : *co(m)-,* together + *optare,* to choose.] **1.** To elect as a fellow member of a group. **2.** To appoint summarily. **3.** To pre-empt : appropriate. **4.** To take or win over (e.g., an independent minority) through assimilation into an established group or culture. **—co'-op·ta'tion** (kō'ŏp-tā'shən) *n.* **—co-op'ta·tive** (-tə-tĭv) *adj.* **—co-op'tion** (-ŏp'shən) *n.* **—co-op'tive** *adj.*

**co·or·di·nate** (kō-ôr'dn-āt', -ĭt) *n.* [Back-formation < COORDINATION.] **1.** One that is equal in rank, importance, or degree. **2.** *Math.* One of a set of numbers that determines the location of a point in a space of a given dimension. **3.** *Math.* Any of a set of two or more magnitudes used to determine the position of a point, line, curve, or plane. —*adj.* (-ĭt, -āt'). **1.** Of equal importance, rank, or degree. **2.** Of or involving coordination. **3.** Of or based on coordinates. —*v.* (-āt') **-nat·ed, -nat·ing, -nates.** —*vt.* **1.** To place in the same order, class, or rank. **2.** To harmonize in a common effort. —*vi.* To work together harmoniously. **—co·or'di·nate·ly** (-ĭt-lē) *adv.* **—co·or'di·nate·ness** (-ĭt-nĭs) *n.* **—co·or'di·na'tive** *adj.* **—co·or'di·na'tor** *n.*

**coordinate bond** *n.* A covalent chemical bond produced when an atom shares a pair of electrons with an atom lacking such a pair.

**coordinate covalent bond** *n.* Coordinate bond.

**coordinating conjunction** *n.* A conjunction connecting two identically constructed grammatical elements, as *or* in *They didn't know whether to laugh or cry.*

**co·or·di·na·tion** (kō-ôr'dn-ā'shən) *n.* [Fr. or < LLat. *coordinatio* : *co(m)-,* same + *ordinatio,* arrangement < *ordinare,* to arrange in order < *ordo,* order.] **1.** The act of coordinating or state of being coordinate. **2.** *Physiol.* Coordinated muscular functioning in executing a complex task.

**coordination complex** *n.* Coordination compound.

**coordination compound** *n.* A chemical compound formed by joining independent molecules or ions to a central metallic atom.

**coot** (kōōt) *n.* [ME *coote.*] **1.** A dark-gray aquatic bird of the genus *Fulica,* esp. *F. americana* of the New World, and *F. atra* of the Old World. **2.** The scoter. **3.** *Informal.* A foolish person.

**coo·tie** (kōō'tē) *n.* [Perh. < Malay *kutu.*] *Slang.* A body louse.

**cop**[1] (kŏp) *n.* [ME, summit < OE.] **1.** A cone-shaped or cylindrical roll of yarn or thread wound on a spindle. **2.** *Archaic.* A summit or crest, as of a hill.

**cop**[2] (kŏp) *n.* [Short for *copper,* prob. < *cop,* to catch.] *Informal.* A police officer. —*vt.* **copped, cop·ping, cops.** *Slang.* **1.** To steal. **2.** To capture : seize. **—cop out.** *Slang.* To back down or out, as from a commitment or task : RENEGE <*copped out* on everything they promised>

**co·pa·cet·ic** *or* **co·pa·set·ic** (kō'pə-sĕt'ĭk) *adj.* [Orig. unknown.] *Slang.* Superb : excellent.

**co·pai·ba** *also* **co·pai·ba balsam** (kō-pī'bə, -pā'-) *n.* [Sp. < Port. *copaíba* < Tupi *copaiba.*] A transparent, yellowish, viscous resin from South American trees of the genus *Copaifera,* used in varnishes and tracing papers and as an expectorant, diuretic, and stimulant.

**co·pal** (kō'pəl, -păl') *n.* [Sp. < Nahuatl *copalli,* resin.] A brittle, aromatic resin of recent or fossil origin, obtained from various tropical trees.

**co·par·ce·nar·y** (kō-pär'sə-nĕr'ē) *n., pl.* **-ies. 1.** *Law.* Joint ownership of inherited property. **2.** Joint ownership.

**co·par·ce·ner** (kō-pär'sə-nər) *n. Law.* One of two persons sharing an undivided inheritance.

**co·part·ner** (kō-pärt'nər) *n.* A partner : associate. **—co·part'ner·ship'** *n.*

**co·pa·set·ic** (kō'pə-sĕt'ĭk) *adj. var. of* COPACETIC.

**cope**[1] (kōp) *vi.* **coped, cop·ing, copes.** [ME *copen,* to strike < OFr. *couper* < *coup,* blow < LLat. *colpus* < Lat. *colaphus* < Gk. *kolaphos.*] **1.** To contend or struggle, esp. on even terms or with success <*coping* with housekeeping and a demanding job> **2.** *Informal.* To contend with difficulties and act to overcome them <A successful teacher must be able to *cope.*>

**cope**[2] (kōp) *n.* [ME *cope* < OE **cāp* < LLat. *capa,* cloak.] **1.** A long ecclesiastical vestment worn by clerics over the alb or surplice. **2.** A covering like a cloak or mantle. **3.** A coping. —*vt.* **coped, cop·ing, copes. 1.** To cover or dress in a cope. **2.** To provide with coping.

**co·peck** (kō'pĕk') *n. var. of* KOPECK.

**co·pen·ha·gen** (kō'pən-hā'gən, -hä'-) *n.* [After *Copenhagen,* Denmark.] A grayish to purplish blue.

**co·pe·pod** (kō'pə-pŏd') *n.* [NLat. *Copepoda,* order name : Gk. *kōpē,* oar + Gk. *pous,* foot.] Any of various small marine and freshwater crustaceans of the order *Copepoda.*

**Co·per·ni·can** (kō-pûr'nĭ-kən) *adj.* Of or relating to the theory of Copernicus that the earth rotates on its axis and revolves around the sun with the other planets in the solar system.

**cope·stone** (kōp'stōn') *n. var. of* CAPSTONE.

**cop·i·er** (kŏp′ē-ər) *n.* One that copies, esp. a machine for making copies of printed matter.
**co·pi·lot** (kō′pī′lət) *n.* The second or relief pilot of an aircraft.
**cop·ing** (kō′pĭng) *n.* [< COPE[2].] The top part of a wall or roof, usu. with a slanting upper surface for drainage.
**coping saw** *n.* [Perh. < COPE[1], to strike (obs.).] A handsaw with a narrow blade set between the ends of a U-shaped frame and used esp. for cutting curves in wood.
**co·pi·ous** (kō′pē-əs) *adj.* [ME < OFr. *copieux* < Lat. *copiosus* < *copia,* abundance.] **1.** Yielding or containing plenty. **2.** Large in quantity : ABUNDANT <a *copious* supply of food> **3.** Abounding in matter, thoughts, or words : WORDY <a *copious* lecture> **—co′pi·ous·ly** *adv.* **—co′pi·ous·ness** *n.*
**co·pla·nar** (kō-plā′nər) *adj.* Lying or occurring in the same plane. **—co′pla·nar′i·ty** (kō′plə-năr′ĭ-tē) *n.*
**co·pol·y·mer** (kō-pŏl′ə-mər) *n.* A polymer of two or more different monomers. **—co·pol′y·mer′ic** (-mĕr′ĭk) *adj.*
**co·pol·y·mer·ize** (kō-pŏl′ə-mə-rīz′, kō′pə-lĭm′ə-) *v.* **-ized, -iz·ing, -iz·es.** —*vt.* To polymerize (different monomers) together. —*vi.* To react to form a copolymer. **—co·pol′y·mer·i·za′tion** *n.*
**cop·per[1]** (kŏp′ər) *n.* [ME *coper* < OE < LLat. *cuprum* < Lat. *Cyprium (aes),* Cyprian (metal).] **1.** *Symbol* **Cu** A ductile, malleable, reddish-brown metallic element that is an excellent conductor of heat and electricity and is used for electrical wiring, water piping, and corrosion-resistant parts; atomic number 29; atomic weight 63.54. **2.** A coin of copper or a copper alloy. **3.** *Chiefly Brit.* A large boiler or pot of copper or iron. **4.** Any of various small butterflies of the subfamily Lycaeninae, with copper-colored wings. —*vt.* **-pered, -per·ing, -pers. 1.** To coat or finish with a layer of copper. **2.** *Slang.* To bet against, as in faro. **—cop′per·y** *adj.*
▲ word history: The ancient Romans called copper the "Cyprian metal" because Cyprus was an important source of copper in the ancient world. The Romans annexed the island in 58 B.C. and mined the copper for their own use.
**cop·per[2]** (kŏp′ər) *n.* [< COP[2].] *Slang.* A police officer.
**cop·per·as** (kŏp′ər-əs) *n.* [ME *coperose,* a metallic sulfate < OFr. < Med. Lat. *cuperosa,* prob. short for *aqua cuperosa,* copper water.] A greenish, crystalline, hydrated ferrous sulfate, $FeSO_4 \cdot 7H_2O$, used in making fertilizers and inks and purifying water.
**cop·per·head** (kŏp′ər-hĕd′) *n.* **1.** A venomous snake, *Agkistrodon contortrix* or *Ancistron contortrix* of the eastern United States, with reddish-brown markings. **2. Copperhead.** A Northerner who sympathized with the South during the American Civil War.
**cop·per·plate** (kŏp′ər-plāt′) *n.* **1.** An engraved or etched copper printing plate. **2.** A print or engraving made by using such a plate.
**copper pyrites** *n.* Chalcopyrite.
**cop·per·smith** (kŏp′ər-smĭth′) *n.* **1.** A person who makes articles of copper. **2.** A brightly colored bird, *Megalaima haemacephala* of southeastern Asia, having a metallic call.
**copper sulfate** *n.* A poisonous blue crystalline copper salt, $CuSO_4 \cdot 5H_2O$, used in agriculture, textile dyeing, leather treatment, electroplating, and the manufacture of germicides.
**cop·per·ware** (kŏp′ər-wâr′) *n.* Articles made of copper.
**cop·pice** (kŏp′ĭs) *n.* [OFr. *copeiz* < *couper,* to cut, strike. —see COPE[1].] *Chiefly Brit.* A thicket : copse.
**co·pra** (kō′prə, kŏp′rə) *n.* [Port. < Malayalam *koppara.*] Dried coconut meat from which coconut oil is extracted.
**copro-** *pref.* [< Gk. *kopros,* dung.] Excrement : dung <*copro*lite>
**cop·ro·an·ti·bod·y** (kŏp′rō-ăn′tē-bŏd′ē) *n.* An immunoglobulin present in the lumen of the intestine.
**cop·ro·lite** (kŏp′rə-līt′) *n.* Fossil excrement. **—cop′ro·lit′ic** (-lĭt′ĭk) *adj.*
**co·prol·o·gy** (kŏ-prŏl′ə-jē) *n.* Pornography or scatology.
**co·proph·a·gous** (kŏ-prŏf′ə-gəs) *adj.* Feeding on excrement. **—co·proph′a·gy** (-ə-jē) *n.*
**cop·ro·phil·i·a** (kŏp′rə-fĭl′ē-ə) *n.* An abnormal interest in fecal matter. **—cop′ro·phil′i·ac′** (-ē-ăk′) *n.*
**co·proph·i·lous** (kŏ-prŏf′ə-ləs) *adj.* Living in excrement.
**copse** (kŏps) *n.* [ME *copys* < OFr. *copeiz,* thicket for cutting < *couper,* to cut. —see COPE[1].] A thicket of small trees or shrubs.
**Copt** (kŏpt) *n.* [Ar. *Qubt,* Copts < Coptic *Gyptias* < Gk. *Aiguptios* < *Aiguptos,* Egypt.] **1.** A native of Egypt descended from ancient Egyptian stock. **2.** A member of the Coptic Church.
**cop·ter** (kŏp′tər) *n.* *Informal.* A helicopter.
**Cop·tic** (kŏp′tĭk) *n.* The Afro-Asiatic language of the Copts, which survives only as a liturgical language of the Coptic Church. —*adj.* Of or relating to the Copts or their language.
**Coptic Church** *n.* The Christian church of Egypt, adhering to the Monophysite doctrine.
**cop·u·la** (kŏp′yə-lə) *n.* [Lat., link.] **1.** A verb, as a form of *be* or *seem,* that identifies the predicate of a sentence with the subject. **2.** *Logic.* The word or group of words serving as a link between the subject and predicate of a proposition. **—cop′u·lar** (-lər) *adj.*
**cop·u·late** (kŏp′yə-lāt′) *vi.* **-lat·ed, -lat·ing, -lates.** [Lat. *copulare, copulat-,* to join together < *copula,* link.] To engage in coitus. **—cop′u·la′tion** *n.* **—cop′u·la·to′ry** (-lə-tôr′ē, -tōr′ē) *adj.*
**cop·u·la·tive** (kŏp′yə-lā′tĭv, -lə-tĭv) *adj.* **1. a.** Serving to connect coordinate words or clauses <a *copulative* conjunction> **b.** Serving as a copula <a *copulative* verb> **2.** Of or relating to copulation. —*n.* A copulative word or group of words. **—cop′u·la′tive·ly** *adv.*
**cop·y** (kŏp′ē) *n., pl.* **-ies.** [ME < OFr. *copie* < Med. Lat. *copia* < Lat., abundance.] **1.** A reproduction or imitation of an original : DUPLICATE. **2.** A specimen or example of a printed text or picture <an autographed *copy* of a book of poems> **3.** Material, as a manuscript, to be set in type. **4.** Suitable source material, as for journalism <Politicians make good *copy.*> —*v.* **-ied, -y·ing, -ies.** —*vt.* **1.** To make a copy of. **2.** To follow as a model or pattern : IMITATE. —*vi.* **1.** To make a copy or reproduction. **2.** To admit of being copied.
☆ ***syns:*** COPY, DUPLICATE, FACSIMILE, REPLICA, REPRODUCTION *n. core meaning* : one closely resembling another <*copies* of Persian miniatures> ***ant:*** original
**cop·y·book** (kŏp′ē-bo͝ok′) *n.* A book of models formerly used to teach penmanship. —*adj.* Unoriginal : trite.
**copy boy** *n.* An employee in a newspaper office who carries copy and runs errands.
**cop·y·cat** (kŏp′ē-kăt′) *n.* An imitator : mimic.
**copy desk** *n.* The desk where newspaper copy is edited.
**cop·y·ed·it** (kŏp′ē-ĕd′ĭt) *vt.* **-it·ed, -it·ing, -its.** To correct and prepare (e.g., a manuscript) for typesetting and printing. **—copy editor** *n.*
**cop·y·hold·er** (kŏp′ē-hōl′dər) *n.* **1.** One who reads manuscript aloud to a proofreader. **2.** A device that holds copy in place, esp. for a typesetter.
**cop·y·ist** (kŏp′ē-ĭst) *n.* **1.** One who makes written copies. **2.** One who imitates.
**cop·y·read·er** (kŏp′ē-rē′dər) *n.* One who edits and corrects copy for publication, as at a newspaper.
**cop·y·right** (kŏp′ē-rīt′) *n.* The legal right granted, as to an author, composer, playwright, or publisher, for exclusive publication, production, sale, or distribution of a literary, musical, dramatic, artistic, or electronically produced work. —*adj.* Protected by copyright. —*vt.* **-right·ed, -right·ing, -rights.** To secure a copyright for. **—cop′y·right′a·ble** *adj.* **—cop′y·right′er** *n.*
**cop·y·writ·er** (kŏp′ē-rī′tər) *n.* A writer esp. of advertising copy.
**coq au vin** (kōk′ ō văN′, kŏk′) *n.* [Fr. : *coq,* cock + *au,* with + *vin,* wine.] Chicken cooked usu. in red wine.
**co·quet** (kō-kĕt′) *vi.* **-quet·ted, -quet·ting, -quets.** [Fr. *coqueter* < *coquet,* flirtatious man, dim. of *coq,* cock < OFr. *coc* < LLat. *coccus* < Lat. *coco,* clucking.] **1.** To engage in coquetry : FLIRT. **2.** To dally : trifle.
**co·quet·ry** (kō′kĭ-trē, kō-kĕt′rē) *n., pl.* **-ries.** [Fr. *coquetterie* < *coquette,* coquette.] Flirtation : dalliance.
**co·quette** (kō-kĕt′) *n.* [Fr., fem. of *coquet,* flirtatious man. —see COQUET.] A woman who flirts with men. **—co·quet′tish** *adj.* **—co·quet′tish·ly** *adv.* **—co·quet′tish·ness** *n.*
**co·quil·la nut** (kō-kēl′yə, -kē′yə) *n.* [Port. *coquilho,* dim. of *côco,* coco.] The nut of a South American palm tree, *Attalea funifera,* with a hard oval shell often used for decorative carving.
**co·quille** (kō-kēl′) *n.* [Fr. < Lat. *conchylia,* pl. of *conchylium,* shellfish < Gk. *konkhulion,* dim. of *konkhē.*] A scallop-shaped dish or a scallop shell in which various seafood dishes are browned and served.
**co·qui·na** (kō-kē′nə) *n.* [Sp., cockle, dim. of *concha,* shell < Lat., mussel. —see CONCH.] **1.** Any of various small bivalve mollusks of warm marine waters, with variously colored, often striped or banded shells. **2.** A soft porous limestone, essentially of shell and coral fragments, used as a construction material.

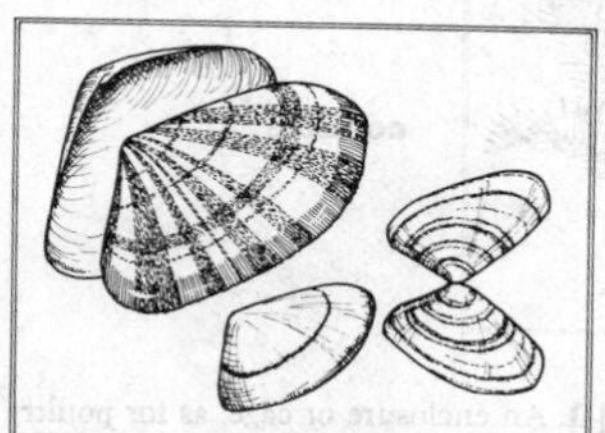

**coquina**

**co·qui·to** (kō-kē′tō) *n., pl.* **-tos.** [Sp., dim. of *coco,* coconut palm < Port. *côco.*] A Chilean palm tree, *Jubaea spectabilis,* from whose sap a sweet edible syrup is obtained.
**co·rac·i·i·form** (kə-răs′ē-ə-fôrm′) *adj.* [< NLat. *Coraciiformes,* order name : Gk. *korakias,* a kind of bird + Lat. *forma,* form.] Of or relating to the order Coraciiformes, which includes the kingfisher.
**cor·a·cle** (kôr′ə-kəl, kŏr′-) *n.* [Welsh *corwgl.*] A small, rounded boat of waterproof material stretched over a wicker or wooden frame.
**cor·a·coid** (kôr′ə-koid′, kŏr′-) *n.* [NLat. *coracoides* < Gk. *korakoiedēs,* like a raven : *korax,* raven + *-oeidēs,* -oid.] A bone or

cartilage projecting from the scapula to the sternum. **—cor'a·coid'** *adj.*

**cor·al** (kôr'əl, kŏr'əl) *n.* [ME < OFr. < Lat. *corallium* < Gk. *korallion*, prob. of Semitic orig.] **1. a.** Any of numerous chiefly colonial marine coelenterates of the class Anthozoa, marked by calcareous skeletons massed in a wide variety of shapes and often forming reefs or islands. **b.** The often hard, rocklike structure formed by such organisms. **c.** The material forming such a structure, esp. the red-orange, pinkish, or white stony substance secreted by corals of the genus *Corallium*, used in making ornaments and jewelry. **2.** An object made of coral. **3.** A deep or strong pink to moderate red or reddish orange. **4.** The unfertilized eggs of a female lobster, which turn a reddish color when cooked. *—adj.* Deep or strong pink to moderate red or reddish orange.

**cor·al-bells** (kôr'əl-bĕlz', kŏr'-) *n. (sing. or pl. in number).* A plant, *Heuchera sanguinea* of the western United States, often cultivated for its small, bell-shaped red flower clusters.

**cor·al·ber·ry** (kôr'əl-bĕr'ē, kŏr'-) *n.* **1.** A North American shrub, *Symphoricarpos orbiculatus*, with red or purplish fruit. **2.** The fruit of the coralberry.

**cor·al·line** (kôr'ə-lĭn, -līn', kŏr'-) *adj.* [Fr. *corallin* < Lat. *corallinus* < *corallium*, coral.] **1.** Of, made of, or producing coral. **2.** Resembling coral, esp. in color. *—n.* **1.** A corallike animal, as certain bryozoans or hydrozoans. **2.** Any of various red algae, esp. of the genus *Corallina*, covered with a calcareous substance and forming stony deposits.

**cor·al·loid** (kôr'ə-loid', kŏr'-) *adj.* [Lat. *corallium*, coral + -OID.] Resembling coral. **—cor'al·loi'dal** (-loid'l) *adj.*

**coral pink** *n.* A moderate to deep yellowish pink.

**coral reef** *n.* An erosion-resistant marine ridge or mound made primarily of compacted coral together with algal material and biochemically deposited magnesium and calcium carbonates.

**cor·al·root** (kôr'əl-rōōt', -rŏŏt', kŏr'-) *n.* An orchid of the genus *Corallorhiza*, with small purplish or yellow-green flowers and branched roots resembling coral.

**coral snake** *n.* A venomous snake of the genus *Micrurus* of tropical America and the southern United States, with brilliant black, red, and yellow banded markings.

**coral vine** *n.* A climbing woody vine, native to Mexico, *Antigonon leptopus*, cultivated for its white or red flowers.

**cor·ban** (kôr'bən, -băn') *n.* [ME < LLat. < Gk. *korban* < Heb. *qurbān*.] An offering to God among the ancient Hebrews.

**cor·beil** *also* **cor·beille** (kôr'bəl, kôr-bā') *n.* [Fr. *corbeille* < LLat. *corbicula*, little basket, dim. of *corbis*, basket.] A sculptured basket of flowers or fruits used as an architectural ornament.

**cor·bel** (kôr'bəl, -bĕl') *n.* [ME < OFr., dim. of *corp*, raven < Lat. *corvus*.] A bracket of building material, as stone or brick, projecting from the face of a wall and gen. used to support a cornice or arch. *—vt.* **-beled, -bel·ing, -bels** *also* **-belled, -bel·ling, -bels.** To provide with or support by a corbel.

**cor·bel·ing** (kôr'bə-lĭng, -bĕl') *n.* **1.** The building of a corbel. **2.** An overlapping arrangement of bricks or stones in which each course extends farther out from the vertical of the wall than the course below.

**cor·bie gable** (kôr'bē) *n.* A gable roof with corbie-steps.

**cor·bie-step** *also* **cor·bie·step** (kôr'bē-stĕp') *n.* [ME *corbie* < OFr. *corbin* < Lat. *corvinus*, ravenlike. —see CORBINA.] One of a series of steps or steplike projections on the top of a gable wall.

**cor·bi·na** (kôr-bē'nə) *also* **cor·vi·na** (-vē'nə) *n.* [Mex. Sp. < Sp. *corvino*, ravenlike (from its color) < *corvus*, raven.] **1.** A food and game fish, *Menticirrhus undulatus* of North American Pacific coastal waters. **2.** A marine fish of the family Sciaenidae related to the corbina.

**cord** (kôrd) *n.* [ME < OFr. *corde* < Lat. *chorda* < Gk. *khordē*.] **1.** A string or small rope of twisted strands or fibers. **2.** A flexible insulated electric wire fitted with a plug. **3.** A hangman's rope. **4.** A binding or restraining influence, feeling, or force. **5.** *also* **chord.** *Anat.* A structure resembling a cord. **6. a.** A raised rib on the surface of cloth. **b.** A fabric or cloth with such ribs. **7. cords.** Trousers made of corduroy. **8.** A unit of quantity for cut fuel wood, equal to 128 cubic feet in a stack measuring 4 by 4 by 8 feet. *—vt.* **cord·ed, cord·ing, cords. 1.** To fasten or bind with a cord. **2.** To furnish with a cord. **3.** To pile (wood) in cords. **—cord'er** *n.*

**cord·age** (kôr'dĭj) *n.* **1.** Ropes, esp. the ropes in the rigging of a ship. **2.** The amount of wood in an area as measured in cords.

**cor·date** (kôr'dāt') *adj.* [NLat. *cordatus* < Lat. *cor*, heart.] Having a heart-shaped outline <a *cordate* leaf> **—cor'date'ly** *adv.*

**cord·ed** (kôr'dĭd) *adj.* **1.** Bound or fastened with cords. **2.** Furnished with or made of cords. **3.** Ribbed or twilled, as corduroy. **4.** Stacked in cords, as firewood.

**cor·dial** (kôr'jəl) *adj.* [ME < Med. Lat. *cordialis* < Lat. *cor*, heart.] **1.** Heartily sincere : WARM <a *cordial* welcome> **2.** Serving to invigorate : STIMULATING. *—n.* **1.** A stimulant. **2.** A liqueur. **—cor·dial'i·ty** (-jăl'ĭ-tē, -jē-ăl'-, -dē-ăl'-), **cor'dial·ness** *n.* **—cor'dial·ly** *adv.*

▲ word history: *Cordial* had its origin as a medical term and in early use the word's derivation from Latin *cor*, "heart," was not forgotten. The first recorded use of *cordial* in English is from Chaucer's *Canterbury Tales*, where it indicates a medicine. Such cordials were supposed to achieve a beneficial effect by stimulating the heart. The heart in medieval physiology was also considered the locus of feelings and affections; from this association the adjective *cordial* meaning "hearty" or "heartfelt" arose.

**cor·di·er·ite** (kôr'dē-ə-rīt') *n.* [Fr., after Pierre L.A. *Cordier* (1777–1861).] A dichroic violet-blue to gray mineral silicate of magnesium, aluminum, and occas. iron.

**cor·di·form** (kôr'də-fôrm') *adj.* [Lat. *cor*, *cord-*, heart + -FORM.] Shaped like a heart.

**cor·dil·le·ra** (kôr'dl-yâr'ə, kôr-dĭl'ər-ə) *n.* [Sp. < *cordilla*, dim. of *cuerda*, cord < Lat. *chorda*.] A chain of mountains, esp. the main mountain system of a large land mass. **—cor'dil·le'ran** (-yâr'ən) *adj.*

**cord·ite** (kôr'dīt') *n.* A smokeless explosive powder that in its processed form resembles brown twine.

**cord·less** (kôrd'lĭs) *adj.* **1.** Having no cord. **2.** Using batteries as a power source <a *cordless* electric drill>

**cord moss** *n.* A moss of the genus *Funaria*, esp. *F. hygrometrica*, usu. growing in burned or waste places.

**cor·do·ba** (kôr'də-bə, -və) *n.* [After Francisco de *Córdoba* (1475–1526).] —See table at CURRENCY.

**cor·don** (kôr'dn) *n.* [Fr. < OFr., dim. of *corde*, cord.] **1.** A line of people, military posts, or ships stationed around an area to guard it. **2.** An ornamental braid or cord worn esp. on costumes. **3.** A ribbon usu. worn diagonally across the breast as a badge of honor or decoration. **4.** A stringcourse. *—vt.* **-doned, -don·ing, -dons.** To form a cordon around (an area) so as to prevent passage <Police *cordoned* off the crime scene.>

**cor·don bleu** (kôr'dôN blœ') *n., pl.* **cor·dons bleus** (kôr'dôN blœ') [Fr. : *cordon*, ribbon + *bleu*, blue.] A person highly distinguished in a particular field, esp. a master chef.

**cor·don sa·ni·taire** (kôr-dôN' sä-nē-târ') *n., pl.* **cor·dons sa·ni·taires** (kôr-doN' sä-nē-târ') [Fr. : *cordon*, line + *sanitaire*, sanitary.] A chain of buffer states organized around a nation regarded as ideologically dangerous or potentially hostile.

**cor·do·van** (kôr'də-vən) *n.* [Sp. *cordobán* < *Córdoba*, Córdoba, Spain.] A fine leather orig. made at Córdoba, Spain, first of goatskin but now more often of split horsehide.

**cor·du·roy** (kôr'də-roi') *n.* [Prob. < CORD + obs. *duroy*, a coarse woolen fabric.] **1.** A durable, usu. cotton cut-pile fabric with vertical ribs or wales. **2. corduroys.** Corduroy trousers. **3.** A road made of logs laid down crosswise. *—adj.* Made of logs laid down crosswise. *—vt.* **-royed, -roy·ing, -roys.** To build (a road) of logs laid down crosswise.

**cord·wood** (kôrd'wŏŏd') *n.* **1.** Wood cut and piled in cords. **2.** Wood sold by the cord.

**core** (kôr, kōr) *n.* [ME.] **1.** The hard or fibrous seed-bearing central part of certain fruits, as the apple or pear. **2.** The innermost or most important part : HEART <the *core* of the discontent> **3.** *Elect.* A soft iron rod in a coil or transformer that intensifies and provides a path for the magnetic field produced by the windings. **4.** *also* **core memory.** *Computer Sci.* A computer memory made up of a series of doughnut-shaped masses of magnetic material. **5.** The central portion of the earth at a depth of approx. 1,800 miles or 2,900 kilometers. **6.** A mass of dry sand placed within a mold to provide openings or shape to a casting. **7.** The base, usu. of soft or inferior wood, to which veneer woods are glued. *—vt.* **cored, cor·ing, cores.** To remove the core of <*core* pears>

**core city** *n.* An inner city.

**core dump** *n.* *Computer Sci.* A listing of the data stored in a core. **—core'-dump'** *v.* **(-dumped, -dump·ing, -dumps).**

**co·re·lig·ion·ist** (kō'rĭ-lĭj'ə-nĭst) *n.* One sharing the same religious beliefs as another.

**cor·e·op·sis** (kôr'ē-ŏp'sĭs, kōr-) *n.* [NLat. *Coreopsis*, genus name : Gk. *koris*, bedbug + Gk. *opsis*, appearance.] A plant of the genus *Coreopsis*, with daisylike variegated or yellow flowers.

**co·re·pres·sor** (kō'rĭ-prĕs'ər) *n.* A substance that combines with a genetic repressor to activate it.

**cor·er** (kôr'ər, kōr'-) *n.* An implement for coring <an apple *corer*>

**co·re·spon·dent** (kō'rĭ-spŏn'dənt) *n.* *Law.* A person charged with having committed adultery with the defendant in a divorce suit. **—co're·spon'den·cy** *n.*

**corf** (kôrf) *n., pl.* **corves** (kôrvz) [ME, basket < MDu. *corf* or < MLG *korf*, both prob. < Lat. *corbis*.] *Chiefly Brit.* A truck, tub, or basket used in a mine.

**cor·gi** (kôr'gē) *n.* [Welsh : *cor*, dwarf + *ci*, dog.] A Welsh corgi.

**co·ri·a·ceous** (kôr'ē-ā'shəs, kōr'-) *adj.* [LLat. *coriaceus* < Lat. *corium*, leather.] Of or like leather.

**co·ri·an·der** (kôr'ē-ăn'dər, kōr'-, kôr'ē-ăn'dər, kōr'-) *n.* [ME *coriandre* < OFr. < Lat. *coriandrum* < Gk. *koriandron*.] **1.** An herb, *Coriandrum sativum*, cultivated for its aromatic seeds. **2.** The dried ripe seeds of the coriander, used as a condiment.

**Co·rin·thi·an** (kə-rĭn'thē-ən) *adj.* **1.** Of or relating to ancient Corinth in Greece. **2.** Given to luxury. **3.** Elaborately or elegantly ornate. **4.** Designating the most ornate of the three classical orders of architecture, marked by a slender fluted column having an ornate bell-shaped capital decorated with acanthus leaves. —*n.* **1.** A native or resident of Corinth, Greece. **2.** A man about town. **3.** A wealthy amateur sportsman, esp. an amateur yachtsman. **4. Corinthians** *(sing. in number).* —See table at BIBLE.

**Co·ri·o·lis force** (kôr'ē-ō'lĭs, kōr'-) *n.* [After Gaspard G. de *Coriolis* (1792–1843).] A fictitious force used mathematically to describe motion, as of aircraft or cloud formations, relative to a noninertial, uniformly rotating frame of reference such as the earth.

**co·ri·um** (kôr'ē-əm, kōr'-) *n., pl.* **-ri·a** (-ē-ə) [Lat., skin.] The layer of the skin beneath the epithelium, having nerve endings, sweat glands, and blood and lymph vessels.

**cork** (kôrk) *n.* [ME < Du. *kurk* or LG *korck,* both < Sp. *alcorque,* cork-soled shoe < dial. Ar. *al-qūrq* < Lat. *quercus,* oak.] **1.** The light, porous, elastic outer bark of the cork oak used esp. for bottle stoppers. **2. a.** Something made of cork, esp. a bottle stopper. **b.** A bottle stopper made of other material, as plastic. **3.** A small angling float. **4.** *Bot.* A tissue of dead cells that forms on the outer side of the cambium in the stems of woody plants. —*vt.* **corked, cork·ing, corks. 1.** To stop or seal with or as if with a cork. **2.** To hold back : RESTRAIN. **3.** To blacken with burnt cork.

**cork·age** (kôr'kĭj) *n.* The amount charged by a restaurant for every bottle of liquor served that was not bought on the premises.

**cork·board** (kôrk'bôrd', -bōrd') *n.* An insulating sheet material made of compressed granulated cork.

**cork cambium** *n. Bot.* Phellogen.

**cork·er** (kôr'kər) *n.* **1.** One that corks. **2.** *Slang.* One that is extraordinary or astounding.

**cork·ing** (kôr'kĭng) *adj. & adv.* [< CORKER.] *Slang.* Excellent : fine.

**cork oak** *n.* An evergreen oak tree, *Quercus suber* of the Mediterranean region, with porous outer bark that is the source of cork.

**cork·screw** (kôrk'skrōō') *n.* A pointed metal spiral attached to a handle, used for drawing corks from bottles. —*adj.* Resembling a corkscrew : SPIRAL. —*vi. & vt.* **-screwed, -screw·ing, -screws.** To move or cause to move in a spiral or winding course.

**cork·wood** (kôrk'wŏŏd') *n.* **1. a.** A small tree or shrub, *Leitneria floridiana* of the southeastern United States, with very light wood. **b.** Any of various other trees with light, porous wood. **2.** The wood of a corkwood.

**cork·y** (kôr'kē) *adj.* **-i·er, -i·est.** Of or like cork. **—cork'i·ness** *n.*

**corm** (kôrm) *n.* [NLat. *cormus* < Gk. *kormos,* a trimmed tree trunk < *keirein,* to cut.] An underground stem, as that of the gladiolus, similar to a bulb but without scales.

**cor·mel** (kôr'məl, kôr-mĕl') *n.* A young corm arising at the base of a fully developed corm.

**cor·mo·rant** (kôr'mər-ənt, -mə-rănt') *n.* [ME *cormoraunt* < OFr. *cormorant* : *corp,* raven (< Lat. *corvus*) + *marenc,* of the sea (< Lat. *marinus*).] **1.** Any of various widely distributed aquatic birds of the genus *Phalacrocorax,* with dark plumage, webbed feet, a hooked bill, and a distensible pouch. **2.** A rapacious or greedy person.

**corn[1]** (kôrn) *n.* [ME, grain < OE.] **1. a.** A variety of a tall, widely cultivated cereal plant, *Zea mays,* bearing seeds or kernels on large ears. **b.** The seeds or kernels of this plant, used for food or fodder and yielding an edible oil. **c.** The ears of the corn plant. **2.** *Chiefly Brit.* **a.** A cereal plant producing edible seed, as wheat, rye, oats, or barley. **b.** The seeds of such a plant or crop : GRAIN. **3. a.** A single seed of a cereal plant : GRAIN. **b.** A seed or fruit of various other plants. **4.** *Informal.* Corn whiskey. **5.** *Slang.* Something, as music or acting, that is corny. —*vt.* **corned, corn·ing, corns. 1.** To granulate or form into small grains. **2. a.** To preserve and season with granulated salt. **b.** To preserve in brine <*corned* beef> **3.** To feed (animals) with corn or grain.

**corn[2]** (kôrn) *n.* [ME *corne* < OFr., horn < Lat. *cornu.*] A hard thickening of the skin, usu. on or near a toe.

**corn·ball** (kôrn'bôl') [< E. *corn ball,* a ball of popcorn and molasses.] *Slang.* —*n.* A mawkish or unsophisticated person : HILLBILLY. **—corn'ball'** *adj.*

**corn borer** *n.* **1.** The larva of an Old World moth, *Pyrausta nubilalis* that feeds on and destroys corn. **2.** Insect larvae similar to the corn borer that infest corn.

**corn bread** *also* **corn·bread** (kôrn'brĕd') *n.* Bread made from cornmeal.

**corn·cake** *also* **corn cake** (kôrn'kāk') *n.* A bread made with white cornmeal : JOHNNYCAKE.

**corn chip** *n. often* **corn chips.** A crisp snack made from cornmeal batter.

**corn·cob** (kôrn'kŏb') *n.* **1.** The woody core of an ear of corn on which the kernels grow. **2.** A corncob pipe.

**corncob pipe** *n.* A pipe with a bowl made from a dried corncob.

**corn cockle** *n.* A plant native to Europe, *Agrostemma githago,* with red flowers that grows in fields and by roadsides.

**corn·crake** (kôrn'krāk') *n.* A common brown-feathered Old World bird, *Crex crex,* that frequents grain fields and meadows.

**corn·crib** (kôrn'krĭb') *n.* A structure for storing and drying corn.

**†corn·dodg·er** (kôrn'dŏj'ər) *n. Southern U.S.* A small, usu. round corncake baked, fried, or broiled.

**cor·ne·a** (kôr'nē-ə) *n.* [Med. Lat. *cornea (tela),* horny (tissue) < Lat. *corneus,* horny < *cornu,* horn.] The transparent anterior portion of the outer fibrous tunic of the eye, a uniformly thick, nearly circular, convex structure covering the lens. **—cor'ne·al** (-əl) *adj.*

**corn earworm** *n.* The large, destructive larva of a moth, *Heliothis armigera,* that feeds esp. on corn.

**cor·ne·i·tis** (kôr'nē-ī'tĭs) *n.* Inflammation of the cornea.

**cor·nel** (kôr'nəl, -nĕl') *n.* [G. *Kornel(baum)* : Lat. *cornus,* cornel tree + G. *Baum,* tree.] A shrub, tree, or plant of the genus *Cornus,* which includes the dogwoods.

**cor·nel·ian** (kôr-nēl'yən) *n. var. of* CARNELIAN.

**cornelian cherry** *n.* [< CORNEL.] A Eurasian shrub or small tree, *Cornus mas,* with small yellow flowers and bright-red fruit.

**cor·ne·ous** (kôr'nē-əs) *adj.* [Lat. *corneus* < *cornu,* horn.] Made of horn or a hornlike substance.

**cor·ner** (kôr'nər) *n.* [ME < OFr. < Lat. *cornu,* horn.] **1. a.** The position at which two lines or surfaces meet. **b.** The immediate interior or exterior region of the angle formed at this position, bounded by the two lines or surfaces. **2.** A vertex, esp. its interior region, formed by the sides of roads or streets that join, meet, or intersect. **3.** An embarrassing or threatening position, esp. one from which escape is difficult or impossible <forced into a *corner* by bragging> **4. a.** A part, quarter, or region <every *corner* of the state> **b.** A remote, secluded, or secret place <a beautiful little *corner* of Venice> **5.** A guard or decoration fitted on a corner, as of a book. **6.** A speculative monopoly of a stock or commodity created by purchasing all or most of the available supply in order to raise its price. —*v.* **-nered, -ner·ing, -ners.** —*vt.* **1.** To furnish with corners. **2.** To place or drive into a corner <*cornered* the outlaws in a blind alley> **3.** To form a corner in (a stock or commodity) <*cornered* the gold market> —*vi.* **1.** To come together or be situated on or at a corner. **2.** To turn, as at a corner <a sports car that *corners* well> **—around the corner.** About to happen : IMMINENT. **—cut corners.** *Informal.* **1.** To take the shortest route around obstacles, often dangerously or illegally. **2.** To reduce expenses : ECONOMIZE.

**cor·ner·back** *also* **corner back** (kôr'nər-băk') *n. Football.* Either of two defensive halfbacks stationed a short distance behind the linebackers and relatively near the sidelines.

**cor·ner·stone** *also* **corner stone** (kôr'nər-stōn') *n.* **1. a.** A stone at the corner of a building uniting two intersecting walls : QUOIN. **b.** Such a stone ceremoniously laid, often inscribed and hollowed to contain historical documents or objects. **2.** The indispensable and fundamental basis <the *cornerstone* of the debate>

**cor·ner·wise** (kôr'nər-wīz') *also* **cor·ner·ways** (-wāz') *adv.* **1.** With a corner toward the front. **2.** So as to form a corner. **3.** From corner to corner : DIAGONALLY.

**cor·net** (kôr-nĕt') *n.* [ME < OFr., dim. of *corn,* horn < Lat. *cornu.*] **1.** A musical wind instrument of the trumpet class, with three valves operated by pistons. **2.** (kôr'nĭt). A piece of paper twisted into a cone for holding small wares such as candy or nuts. **3.** (kôr'nĭt). A headdress, often cone-shaped, worn by women in the 12th and 13th cent.

**cor·net-à-pis·tons** (kôr-nĕt'ə-pĭs'tənz) *n., pl.* **cor·nets-à-pis·tons** (kôr-nĕts'ə-pĭs'tənz) [Fr.] CORNET 1.

**cor·net·ist** *also* **cor·net·tist** (kôr-nĕt'ĭst) *n.* One who plays a cornet.

**corn-fed** (kôrn'fĕd') *adj.* **1.** Fed on corn. **2.** *Slang.* Healthy and strong but provincial and unsophisticated.

**corn flakes** *pl.n.* A crisp, flaky, breakfast cereal made from coarse cornmeal.

**corn·flow·er** (kôrn'flou'ər) *n.* A Eurasian garden plant, *Centaurea cyanus,* with blue, pink, purple, or white flowers.

**corn·husk** (kôrn'hŭsk') *n.* The leafy husk enclosing a corn ear.

**corn·husk·ing** (kôrn'hŭs'kĭng) *n.* **1.** The husking of corn. **2.** A social gathering for husking corn. **—corn'husk'er** *n.*

**cor·nice** (kôr'nĭs) *n.* [OFr. < OItal.] **1. a.** A horizontal molded projection that crowns or completes a wall or building. **b.** The uppermost part of an entablature. **2.** The molding at the top of the walls of a room, between the walls and ceiling. **3.** An ornamental horizontal molding or frame for concealing curtain fixtures. **4.** An overhanging formation of snow, ice, or rock usu. along a ridge. —*vt.* **-niced, -nic·ing, -nic·es.** To supply, decorate, or finish with or as if with a cornice.

**cor·nic·u·late** (kôr-nĭk'yə-lāt', -lĭt) *adj.* [Lat. *corniculatus* < *corniculum,* little horn, dim. of *cornu,* horn.] Having horns or hornlike projections.

**cor·ni·fi·ca·tion** (kôr'nə-fĭ-kā'shən) *n.* [Lat. *cornu,* horn + -FICATION.] Conversion of squamous epithelial cells into a horny material as hair or nails.

**Cor·nish** (kôr'nĭsh) *adj.* Of or relating to Cornwall in England, Cornishmen, or the Brythonic language of Cornwall, which has been extinct since the late 18th cent. —*n.* The Cornish language.

**Cornish hen** *n.* A Rock Cornish hen.

**Cor·nish·man** (kôr'nĭsh-mən) *n.* A native or resident of Cornwall, England.

**corn lily** *n.* A bulbous plant of the genus *Ixia,* native to southern Africa, with variously colored lilylike flowers.

**corn marigold** *n.* A Eurasian plant, *Chrysanthemum segetum,* cultivated for its white or yellow flowers.
**corn·meal** *also* **corn meal** (kôrn′mēl′) *n.* **1.** Meal made from corn. **2.** *Scot.* Oatmeal.
†**corn pone** (pōn) *n. Southern U.S.* Corn bread made without milk or eggs.
**corn poppy** *n.* An Old World plant, *Papaver rhoeas,* with bright-red flowers, frequently a weed in cultivated fields.
**corn rose** *n. Chiefly Brit.* A red-flowered plant growing in grain fields, as the corn poppy.
**corn·row** (kôrn′rō′) *vt.* To arrange or style (hair) by dividing into sections and braiding close to the scalp in rows. **—corn′row′** *n.*
**corn silk** *n.* The silky tuft of styles at the tip of an ear of corn.
**corn snow** *n.* Rough, granular melted and refrozen snow.
**corn·stalk** *also* **corn stalk** (kôrn′stôk′) *n.* A stalk of corn.
**corn·starch** (kôrn′stärch′) *n.* **1.** Starch prepared from corn. **2.** A purified starchy flour used as a thickener in cooking.
**corn sugar** *n.* Dextrose.
**corn syrup** *n.* A syrup prepared from corn and containing glucose combined with dextrin and maltose.
**cor·nu** (kôr′nōō, -nyōō) *n., pl.* **-nu·a** (-nōō-ə, -nyōō-ə) [Lat., horn.] A protuberance of bone similar to a horn. **—cor′nu·al** (-əl) *adj.*
**cor·nu·co·pi·a** (kôr′nə-kō′pē-ə, -nyə-) *n.* [LLat. : Lat. *cornu,* horn + Lat. *copia,* plenty.] **1.** A goat's horn overflowing with fruit, flowers, and corn, signifying prosperity : HORN OF PLENTY. **2.** A cone-shaped receptacle or ornament. **3.** An overflowing store : ABUNDANCE. **—cor′nu·co′pi·an** (-pē-ən) *adj.*
**cor·nute** (kôr-nōōt′, -nyōōt′) *also* **cor·nut·ed** (-nōō′tĭd, -nyōō′-) *adj.* [Lat. *cornutus* < *cornu,* horn.] **1.** Horn-shaped. **2.** Having horns or horn-shaped processes.
**corn whiskey** *n.* Whiskey distilled from corn.
**corn·y** (kôr′nē) *adj.* **-i·er, -i·est.** [< CORN[1].] *Slang.* Trite, melodramatic, or oversentimental. **—corn′i·ly** *adv.* **—corn′i·ness** *n.*
**co·rol·la** (kə-rŏl′ə, -rō′lə) *n.* [NLat. < Lat., small garland, dim. of *corona,* garland. —see CORONA.] *Bot.* The outer envelope of a flower, composed of fused or separate petals. **—co·rol′late′** (-rŏl′āt′) *adj.*
**cor·ol·lar·y** (kôr′ə-lĕr-ē, kŏr′-) *n., pl.* **-ies.** [ME *corolarie* < Lat. *corollarium,* gratuity, money paid for a garland < *corolla,* small garland. —see COROLLA.] **1.** A proposition following with little or no proof required from one already proven. **2.** A deduction or inference. **3.** A natural consequence or effect : RESULT. **—cor′ol·lar·y** *adj.*
**co·ro·na** (kə-rō′nə) *n., pl.* **-nas** *or* **-nae** (-nē) [Lat., crown < Gk. *korōnē.*] **1.** *Astron.* **a.** A faintly colored luminous ring around a celestial body visible through a haze or thin cloud, esp. such a ring around the moon or sun, caused by diffraction of light from suspended matter in the intervening medium. **b.** The luminous irregular envelope of highly ionized gas outside the chromosphere of the sun. **2.** The top projecting part of a cornice. **3.** A long tapering cigar with blunt ends. **4.** *Anat.* A crownlike or upper part or structure, as the top of the head. **5.** *Bot.* A crownlike part of a flower, usu. between the petals and stamens but occas. an appendage of the corolla, as in daffodils. **6.** *Elect.* A faint glow enveloping the high-field electrode in a corona discharge, often accompanied by streamers directed toward the low-field electrode.
**Corona Aus·tra·lis** (ô-strā′lĭs) *n.* [Lat., southern crown.] A constellation in the Southern Hemisphere.
**Corona Bo·re·al·is** (bôr′ē-ăl′ĭs, -ā′lĭs, bōr′ē-) *n.* [Lat., northern crown.] A constellation in the Northern Hemisphere.
**corona discharge** *n.* An electrical discharge marked by a corona and occurring when one of two electrodes in a gas has a shape causing the electric field at its surface to be significantly greater than that between the electrodes.
**co·ro·na·graph** *also* **co·ro·no·graph** (kə-rō′nə-grăf′) *n.* A telescope for examining the sun's corona.
**cor·o·nal** (kôr′ə-nəl, kŏr′-, kə-rō′nəl) *n.* [ME < Lat. *coronalis,* of a crown < *corona,* crown. —see CORONA.] **1.** A garland, wreath, or circlet. **2.** *Anat.* The coronal suture. *—adj.* **1.** Of or relating to a coronal. **2.** *Anat.* Of, designating, or having the direction of the coronal suture.
**coronal suture** *n.* The line of union of the two parietal bones with the frontal bone of the skull.
**cor·o·nar·y** (kôr′ə-nĕr′ē, kŏr′-) *adj.* [Lat. *coronarius,* of a crown < *corona,* crown. —see CORONA.] **1.** Of, relating to, or designating either of two arteries that originate in the aorta and supply blood directly to the heart tissues. **2.** Of or relating to the heart. *—n., pl.* **-ies.** *Informal.* A coronary thrombosis.
**coronary artery** *n.* One of the two arteries that supply blood to the heart.
**coronary occlusion** *n.* Total obstruction of blood flow in a coronary artery.
**coronary thrombosis** *n.* Occlusion of a coronary artery by a blood clot, often causing destruction of heart muscle.
**coronary vein** *n.* One of various blood vessels that drain the blood from the heart.

**cor·o·na·tion** (kôr′ə-nā′shən, kŏr′-) *n.* [ME *coronacioun* < Med. Lat. *coronatio* < Lat. *coronare,* to crown < *corona,* crown. —see CORONA.] The act or ceremony of crowning a sovereign or consort.
**cor·o·ner** (kôr′ə-nər, kŏr′-) *n.* [ME, officer of the Crown < AN *corouner* < Lat. *corona,* crown. —see CORONA.] A public officer whose primary function is to investigate by inquest any death thought to be of other than natural causes. **—cor′o·ner·ship′** *n.*
**cor·o·ner's jury** (kôr′ə-nərz, kŏr-) *n.* A group of people summoned to attend a coroner's inquest and determine the cause of the death under investigation.
**cor·o·net** (kôr′ə-nĕt′, kŏr′-) *n.* [ME *coronette* < OFr., dim. of *corone,* crown < Lat. *corona.* —see CORONA.] **1.** A small crown worn by royalty and nobles below the rank of sovereign. **2.** A chaplet or headband decorated with gold or jewels. **3.** The upper margin of a horse's hoof.
**cor·o·noid** (kôr′ə-noid′, kŏr′-) *adj.* [Lat. *corona,* wreath, crown (< Gk. *korōnē*) + -OID.] Shaped like a crown.
**co·ro·tate** (kō-rō′tāt′) *vi.* **-tat·ed, -tat·ing, -tates.** To rotate along with another body. **—co′ro·ta′tion** *n.* **—co′ro·ta′tion·al** *adj.*
**cor·po·ra** (kôr′pər-ə) *n. pl. of* CORPUS.
**cor·po·ral[1]** (kôr′pər-əl, kôr′prəl) *adj.* [ME < OFr. < Lat. *corporalis* < *corpus,* body.] Of the body : BODILY <*corporal* punishment> **—cor′po·ral′i·ty** (-pə-răl′ĭ-tē) *n.* **—cor′po·ral·ly** *adv.*
**cor·po·ral[2]** (kôr′pər-əl, kôr′prəl) *n.* [OFr., lowest noncommissioned officer, var. of *caporal* < OItal. < *capo,* head < Lat. *caput.*] A noncommissioned officer ranking above private first class and below sergeant in the U.S. Army, Air Force, or Marine Corps.
**cor·po·ral[3]** (kôr′pər-əl, kôr′prəl) *n.* [ME < OFr. and < Med. Lat. *corporale* < Lat. *corporalis,* of the body (from the Eucharistic bread being representative of Christ's body) < *corpus,* body.] A white linen cloth on which the consecrated elements are placed during the celebration of the Eucharist.
**corporal's guard** *n.* **1.** The squad commanded by a corporal. **2.** A small group of people.
**cor·po·rate** (kôr′pər-ĭt, kôr′prĭt) *adj.* [Lat. *corporatus,* p.part. of *corporare,* to make into a body < *corpus,* body.] **1.** Formed into a corporation : INCORPORATED. **2.** Of or relating to a corporation <*corporate* policy> **3.** Combined or united into one body : COLLECTIVE <*corporate* action> **4.** *var. of* CORPORATIVE. **—cor′po·rate·ly** *adv.*
**cor·po·ra·tion** (kôr′pə-rā′shən) *n.* **1. a.** A body of persons granted a charter legally recognizing it as a separate entity having its own rights, privileges, and liabilities distinct from those of its members. **b.** Such a body created for purposes of government. **2.** A combined group of people acting as one body.
**cor·po·ra·tive** (kôr′pər-ə-tĭv, -pə-rā′tĭv) *also* **cor·po·rate** (kôr′pər-ĭt, kôr′prĭt) *adj.* **1.** Of, relating to, or associated with a corporation. **2.** Of a government or political system in which the principal economic functions, as banking, industry, labor, and government, are organized as corporate entities.
**cor·po·ra·tor** (kôr′pə-rā′tər) *n.* A member of a corporation.
**cor·po·re·al** (kôr-pôr′ē-əl, -pōr′-) *adj.* [< Lat. *corporeus* < *corpus,* body.] **1.** Of, relating to, or typical of the body. **2.** Of a material nature : TANGIBLE. **—cor·po′re·al·ly** *adv.* **—cor·po′re·al·ness** *n.*
**cor·po·re·i·ty** (kôr′pə-rē′ĭ-tē, -rā′-) *also* **cor·po·re·al·i·ty** (kôr-pôr′ē-ăl′ĭ-tē, -pōr′-) *n.* Physical existence.
**cor·po·sant** (kôr′pə-zənt) *n.* [Port. and OSp. *corpo santo,* both < Lat. *corpus sanctum,* holy body.] St. Elmo's fire.
**corps** (kôr, kōr) *n., pl.* **corps** (kôrz, kōrz) [Fr. < Lat. *corpus,* body.] **1. a.** A separate branch or department of the armed forces having a specialized function <Signal *Corps*> **b.** A tactical unit of ground combat forces comprising two or more divisions and auxiliary service troops. **2.** A body of persons acting together or associated under common direction <the press *corps*>
**corps de bal·let** (kôr′ də bă-lā′, kōr′) *n.* [Fr.] The dancers in a ballet troupe who perform as a group with no solo parts.
**corpse** (kôrps) *n.* [ME *corps* < Lat. *corpus.*] **1.** A dead body, esp. the body of a deceased human being. **2.** Something lifeless or defunct.
**corps·man** (kôr′mən, kōr′-, kôrz′mən, kōrz′-) *n.* A military enlisted person trained to administer minor medical treatment, as first aid.
**cor·pu·lence** (kôr′pyə-ləns) *n.* [ME, corporality < Lat. *corpulentia,* corpulence < *corpulentus,* corpulent < *corpus,* body.] The state of being too fat : OBESITY. **—cor′pu·lent** *adj.* **—cor′pu·lent·ly** *adv.*
**cor pul·mo·na·le** (kôr′ pōōl′mə-nä′lē, -năl′ē, pŭl′-) *n.* [NLat., pulmonary heart.] Heart disease marked by hypertrophy of the right ventricle that is caused by an obstruction in pulmonary circulation.
**cor·pus** (kôr′pəs) *n., pl.* **-po·ra** (-pər-ə) [ME < Lat.] **1.** A human or animal body, esp. when dead. **2.** *Anat.* A structure constituting the main part of an organ. **3.** The principal or capital, as of a fund or estate. **4.** A large collection of writings of a specific kind or on a specific subject.
**corpus al·bi·cans** (ăl′bĭ-kănz′) *n.* [NLat., white body.] *Anat.* The white fibrous tissue in an ovary that results after the involution and regression of the corpus luteum.
**corpus cal·lo·sum** (kə-lō′səm) *n., pl.* **corpora cal·lo·sa** (kə-lō′sə) [NLat., callous body.] *Anat.* A wide arched band of white matter connecting the cerebral hemispheres at the base of the longitudinal fissure.

**Cor·pus Chris·ti** (kôr′pəs krĭs′tē) *n.* [ME < Med. Lat. : Lat. *corpus*, body + Med. Lat. *Christus*, Christ.] *Rom. Cath. Ch.* A festival celebrating the Eucharist on the first Thursday after Trinity.

**cor·pus·cle** (kôr′pə-səl, -pŭs′əl) *n.* [Lat. *corpusculum*, little particle, dim. of *corpus*, body.] **1.** *Biol.* A cell, as an erythrocyte or leukocyte, that is capable of free movement in a fluid or matrix, as distinguished from a cell fixed in tissue. **2.** A discrete particle, as a photon or electron. **3.** A minute globular particle. **—cor·pus′cu·lar** (kôr-pŭs′kyə-lər) *adj.*

**cor·pus de·lic·ti** (dĭ-lĭk′tī′) *n.* [NLat. : Lat. *corpus*, body + *delictum*, crime.] **1.** *Law.* **a.** The material substance on which a crime has been committed. **b.** The material evidence, as the discovered corpse of a murder victim, of the fact that a crime has been committed. **2.** The victim's corpse in a murder case.

**cor·pus ju·ris** (jo͝or′ĭs) *n.* [Lat., the body of law.] The collective or comprehensive body of all the laws of a nation or state.

**Corpus Juris Ci·vil·is** (sĭ-vĭl′ĭs) *n.* [Lat., the body of civil law.] The body of civil or Roman law assembled and issued during Justinian's reign and forming the basis of most European law.

**corpus lu·te·um** (lo͞o′tē-əm) *n.*, *pl.* **corpora lu·te·a** (lo͞o′tē-ə) [NLat., yellow body.] *Anat.* A yellow mass of endocrine cells in a ruptured mature Graafian follicle of the ovary formed after the release of an ovum.

**corpus stri·a·tum** (strī-ā′təm) *n.*, *pl.* **corpora stri·a·ta** (strī-ā′tə) [NLat., striated body.] *Anat.* Either of two gray-and-white, striated ganglionic masses of the brain stem in the lower lateral wall of each cerebral hemisphere.

**cor·rade** (kə-rād′) *vt.* & *vi.* **-rad·ed, -rad·ing, -rades.** [Lat. *corradere*, to scrape together : *com-*, together + *radere*, to scrape.] To erode or be eroded by abrasion. **—cor·ra′sion** (-rā′zhən) *n.* **—cor·ra′sive** (-sĭv, -zĭv) *adj.*

**cor·ral** (kə-răl′) *n.* [Sp. < Lat. *currus*, cart < *currere*, to run.] **1.** An enclosure for confining livestock. **2.** An enclosure formed by a circle of wagons for defense against attack during an encampment. *—vt.* **-ralled, -ral·ling, -rals. 1.** To drive into and hold in a corral. **2.** To arrange (wagons) in a corral. **3.** *Informal.* To take possession of : SEIZE.

**cor·rect** (kə-rĕkt′) *v.* **-rect·ed, -rect·ing, -rects.** [ME *correcten* < Lat. *corrigere*, to correct : *com-* (intensive) + *regere*, to rule.] *—vt.* **1. a.** To remove the errors from. **b.** To indicate or mark the errors in. **2.** To punish for the purpose of improving. **3.** To remove, remedy, or counteract (e.g., a malfunction). **4.** To adjust so as to meet a specified condition, as a standard <*correct* the sight adjustment> *—vi.* **1.** To make corrections. **2.** To make adjustments : COMPENSATE <*correcting* for the difference in pressure> *—adj.* **1.** Free from error or fault : ACCURATE. **2.** Conforming to standards : PROPER <*correct* manners> **—cor·rect′a·ble, cor·rect′i·ble** *adj.* **—cor·rect′ly** *adv.* **—cor·rect′ness** *n.* **—cor·rec′tor** *n.*

☆ ***syns:*** CORRECT, AMEND, EMEND, MEND, RECTIFY, REMEDY, RIGHT *v. core meaning* : to make right what is wrong <*correct* an error>

**cor·rec·tion** (kə-rĕk′shən) *n.* **1.** The act or process of correcting. **2.** Something substituted for a mistake <made *corrections* in the manuscript> **3.** Punishment intended to rehabilitate or improve. **4.** An amount or quantity added or subtracted to correct. **5.** A decline in stock-market prices or activity following a period of increases. **—cor·rec′tion·al** *adj.*

**cor·rec·ti·tude** (kə-rĕk′tĭ-to͞od′, -tyo͞od′) *n.* The quality or state of being correct, esp. in manners and behavior.

**cor·rec·tive** (kə-rĕk′tĭv) *adj.* Tending or intended to correct. *—n.* Something that corrects. **—cor·rec′tive·ly** *adv.*

**cor·re·late** (kôr′ə-lāt′, kŏr′-) *v.* **-lat·ed, -lat·ing, -lates.** [Back-formation from CORRELATION.] *—vt.* **1.** To put or bring into causal, complementary, parallel, or reciprocal relation. **2.** To establish or demonstrate as having a correlation <*correlated* poverty and crime> *—vi.* To be related by a correlation. *—adj.* (-lĭt, -lāt′). Related by a correlation, esp. having corresponding characteristics. *—n.* (-lĭt, -lāt′). Either of two correlate entities : CORRELATIVE.

**cor·re·la·tion** (kôr′ə-lā′shən, kŏr′-) *n.* [Med. Lat. *correlatio* : Lat. *com-*, together + *relatio*, relation < *referre*, to carry back.] **1.** A causal, complementary, parallel, or reciprocal relationship, esp. a structural, functional, or qualitative correspondence between comparable entities <a *correlation* between drug abuse and crime> **2.** *Statistics.* **a.** Simultaneous increase or decrease in value of two numerically valued random variables <the positive *correlation* between cigarette smoking and the incidence of lung cancer> **b.** Simultaneous increase in the value of one and decrease in the value of the other of two numerically valued random variables <the negative *correlation* between age and normal vision> **3.** The act of correlating or state of being correlated. **—cor′re·la′tion·al** *adj.*

**correlation coefficient** *n.* A measure of the interdependence of two random variables that ranges in value from −1 to +1, indicating perfect negative correlation at −1, absence of correlation at 0, and perfect positive correlation at +1.

**cor·rel·a·tive** (kə-rĕl′ə-tĭv) *adj.* **1.** Corresponding : related. **2.** Indicating a reciprocal or complementary relationship <the *correlative* conjunctions *neither . . . nor*> *—n.* **1.** Either of two correlative entities : CORRELATE. **2.** A correlative word or expression. **—cor·rel′a·tive·ly** *adv.*

**cor·re·spond** (kôr′ĭ-spŏnd′, kŏr′-) *vi.* **-spond·ed, -spond·ing, -sponds.** [OFr. *correspondre* < Med. Lat. *correspondēre* : Lat. *com-*, together + *respondēre*, to respond.] **1.** To be in accord : be consistent or compatible <tastes that *corresponded*> **2. a.** To be similar, analogous, or equal. **b.** To be parallel or closely matched. **3.** To communicate by letter.

**cor·re·spon·dence** (kôr′ĭ-spŏn′dəns, kŏr′-) *n.* **1.** The act, fact, or state of agreeing or conforming. **2.** Similarity or analogy. **3. a.** Communication by exchange of letters. **b.** The letters exchanged.

**correspondence principle** *n.* The principle that predictions of quantum theory approach those of classical physics in the limit of large quantum numbers.

**correspondence school** *n.* A school that offers instruction by mail, sending lessons and examinations to a student who then completes the assigned work and returns the material for grading.

**cor·re·spon·den·cy** (kôr′ĭ-spŏn′dən-sē, kŏr′-) *n.* Correspondence.

**cor·re·spon·dent** (kôr′ĭ-spŏn′dənt, kŏr′-) *n.* **1.** A person who communicates by means of letters. **2.** One employed by a news agency, as a newspaper or television network, to supply news or articles often from remote locations <a foreign *correspondent*> **3.** A person or firm having regular business relations with another, esp. at a distance. **4.** Something that corresponds : CORRELATIVE. *—adj.* Corresponding. **—cor′re·spon′dent·ly** *adv.*

**cor·re·spond·ing** (kôr′ĭ-spŏn′dĭng, kŏr′-) *adj.* **1.** Agreeing or conforming, as in degree or kind : CONSISTENT. **2.** Analogous : equivalent. **—cor′re·spond′ing·ly** *adv.*

**cor·re·spon·sive** (kôr′ĭ-spŏn′sĭv, kŏr′-) *adj.* Jointly responsive. **—cor′re·spon′sive·ly** *adv.*

**cor·ri·da** (kô-rē′də, *-thə*) *n.* [Sp. < *correr*, to run < Lat. *currere*.] A bullfight.

**cor·ri·dor** (kôr′ĭ-dər, -dôr′, kŏr′-) *n.* [OFr. < OItal. *corridore* < *correre*, to run < Lat. *currere*.] **1.** A narrow passageway, often with rooms or apartments opening onto it. **2. a.** A tract of land forming a passageway through another country. **b.** A lane for the passage of aircraft.

**cor·rie** (kôr′ē, kŏr′ē) *n.* [Sc. Gael. *coire*.] *Scot.* A cirque.

**cor·ri·gen·dum** (kôr′ə-jĕn′dəm, kŏr′-) *n.*, *pl.* **-da** (-də) [Lat., neuter gerund. of *corrigere*, to correct.] **1.** An error to be corrected, esp. a typesetting error. **2. corrigenda.** A list of errors in a book with their corrections.

**cor·ri·gi·ble** (kôr′ĭ-jə-bəl, kŏr′-) *adj.* [ME < OFr. < Med. Lat. *corrigibilis* < Lat. *corrigere*, to correct.] Capable of being corrected or improved. **—cor′ri·gi·bil′i·ty** *n.* **—cor′ri·gi·bly** *adv.*

**cor·ri·val** (kə-rī′vəl, kō-) *n.* [OFr. or < Lat. *corrivalis* : *com-* (intensive) + *rivalis*, rival.] A rival or competitor. *—adj.* Rival or competing. **—cor·ri′val·ry** (-rē) *n.*

**cor·rob·o·rant** (kə-rŏb′ər-ənt) *adj. Archaic.* Causing or stimulating physical vigor. —Used of a medicine.

**cor·rob·o·rate** (kə-rŏb′ə-rāt′) *vt.* **-rat·ed, -rat·ing, -rates.** [Lat. *corroborare, corroborat-* : *com-* (intensive) + *roborare*, to strengthen < *robur*, strength.] To attest the truth or accuracy of : CONFIRM. **—cor·rob′o·ra′tion** *n.* **—cor·rob′o·ra′tive** (-ə-rā′tĭv, -ər′ə-tĭv) *adj.* **—cor·rob′o·ra′tor** *n.* **—cor·rob′o·ra·to′ry** (-ər-ə-tôr′ē, -tōr′ē) *adj.*

**cor·rob·o·ree** (kə-rŏb′ə-rē) *n.* [< a native word in Australia.] *Austral.* **1.** An aboriginal dance festival held at night to celebrate victories or other events. **2.** A large or boisterous celebration.

**cor·rode** (kə-rōd′) *v.* **-rod·ed, -rod·ing, -rodes.** [ME *corroden* < Lat. *corrodere*, to gnaw away : *com-* (intensive) + *rodere*, to gnaw.] *—vt.* **1.** To dissolve or eat away gradually, esp. by chemical action. **2.** To injure insidiously : DETERIORATE. *—vi.* To be eaten away or dissolved. **—cor·rod′i·ble, cor·ro′si·ble** (-rō′sə-bəl) *adj.*

**cor·ro·sion** (kə-rō′zhən) *n.* [ME *corosioun*, corrosion of tissue < OFr. *corrosion* < LLat. *corrosio*, the act of gnawing < Lat. *corrodere*, to gnaw away. — see CORRODE.] **1.** The act or process of corroding. **2.** A substance, as rust, produced by corrosion. **3.** The condition resulting from corrosion.

**cor·ro·sive** (kə-rō′sĭv, -zĭv) *adj.* **1. a.** Capable of producing corrosion. **b.** Tending to produce corrosion. **2.** Spiteful or malicious : SARCASTIC <*corrosive* comments about others> *—n.* A corrosive substance. **—cor·ro′sive·ly** *adv.* **—cor·ro′sive·ness** *n.*

**corrosive sublimate** *n.* Mercuric chloride.

**cor·ru·gate** (kôr′ə-gāt′, kŏr′-) *vt.* & *vi.* **-gat·ed, -gat·ing, -gates.** [Lat. *corrugare, corrugat-*, to wrinkle up : *com-* (intensive) + *rugare*, to wrinkle < *ruga*, wrinkle.] To form or become formed into folds or parallel and alternating ridges and grooves. **—cor′ru·gate′, cor′ru·gat′ed** (-gā′tĭd) *adj.*

**corrugated iron** *n.* A structural sheet iron, usu. galvanized, shaped in parallel ridges and grooves for rigidity.

**cor·ru·ga·tion** (kôr′ə-gā′shən, kŏr′-) *n.* **1. a.** The act or process of corrugating. **b.** The state of being corrugated. **2.** A ridge or groove on a corrugated surface.

**cor·rupt** (kə-rŭpt′) *adj.* [ME < Lat. *corruptus,* p.part. of *corrumpere,* to destroy : *com-,* together + *rumpere,* to break.] **1.** Immoral and perverse : DEPRAVED. **2.** Dishonest and venal <a *corrupt* politician> **3.** Rotting : putrid. **4.** Containing mistakes or alterations, as a text <a *corrupt* edition of Shakespeare's plays> —*v.* **-rupt·ed, -rupt·ing, -rupts.** —*vt.* **1.** To ruin or undermine the honesty or integrity of. **2.** To make morally impure : PERVERT. **3.** To taint : contaminate. **4.** To cause to become rotten : SPOIL. **5.** To alter the original form of (e.g., a text). —*vi.* To become corrupt. **—cor·rupt′er, cor·rup′tor** *n.* **—cor·rup′tive** *adj.* **—cor·rupt′ly** *adv.* **—cor·rupt′ness** *n.*

☆ ***syns:*** CORRUPT, CROOKED, DISHONEST, VENAL *adj. core meaning* : lacking in moral restraint, esp. in matters of public trust <*corrupt* tax collectors indicted for bribery>

**cor·rupt·i·ble** (kə-rŭp′tə-bəl) *adj.* Capable of being corrupted, as by bribery. **—cor·rupt′i·bil′i·ty, cor·rupt′i·ble·ness** *n.* **—cor·rupt′i·bly** *adv.*

**cor·rup·tion** (kə-rŭp′shən) *n.* **1. a.** The act or process of corrupting. **b.** The state of being corrupt. **2.** *Archaic.* Something that corrupts. **3.** Decay : rottenness.

**cor·rup·tion·ist** (kə-rŭp′shə-nĭst) *n.* One who defends or engages in corrupt practices.

**cor·sage** (kôr-säzh′, -säj′) *n.* [ME, torso < OFr. < *cors,* body < Lat. *corpus.*] **1.** A small bunch of flowers worn by a woman. **2.** The bodice or waist of a dress.

**cor·sair** (kôr′sâr′) *n.* [OFr. *corsaire* < OProv. *corsari* < OItal. *corsaro* < Med. Lat. *cursarius* < *cursus,* plunder < Lat. *cursus,* course. —see COURSE.] **1.** A privateer, esp. along the Barbary Coast. **2.** A fast-moving pirate ship, often having official sanction. **3.** A pirate.

**corse** (kôrs) *n.* [ME *cors* < OFr. < Lat. *corpus.*] *Archaic.* A corpse.

**corse·let** (kôr′slĭt) *n.* [OFr. *corselet,* dim. of *corsel,* dim. of *cors,* body < Lat. *corpus.*] **1.** *also* **cors·let.** Body armor, esp. a breastplate. **2.** (kôr′sə-lĕt′). A light corset with few or no stays.

**cor·set** (kôr′sĭt) *n.* [ME, bodice < OFr., dim. of *cors,* body < Lat. *corpus.*] **1.** A tight-fitting undergarment, often having stays, worn to support and shape the waistline, hips, and breasts. **2.** A medieval outer garment, esp. a laced jacket or bodice. —*vt.* **-set·ed, -set·ing, -sets.** To enclose in or as if in a corset.

**cor·se·tière** (kôr′sĭ-tyâr′, -tîr′) *n.* [Fr. < *corset,* corset < OFr.] A manufacturer, fitter, or seller of corsets.

**cors·let** (kôr′slĭt) *n. var. of* CORSELET 1.

**cor·tege** *also* **cor·tège** (kôr-tĕzh′) *n.* [Fr. *cortège* < OItal. *corteggio* < *corteggiare,* to pay honor < *corte,* court < Lat. *cohors,* throng.] **1.** A group of attendants, as for royalty : RETINUE. **2. a.** A ceremonial procession. **b.** A funeral procession.

**cor·tex** (kôr′tĕks′) *n., pl.* **-ti·ces** (-tĭ-sēz′) *or* **-tex·es.** [Lat., bark.] **1.** *Anat.* The outer layer of an organ or part, as of the cerebrum or adrenal glands. **2.** *Bot.* **a.** A layer of tissue in roots and stems between the epidermis and the vascular tissue. **b.** An external layer, as bark or rind.

**cortic-** *pref. var. of* CORTICO-.

**cor·ti·cal** (kôr′tĭ-kəl) *adj.* [NLat. *corticalis* < Lat. *cortex,* bark.] **1.** Of, relating to, or made up of cortex. **2.** Of, related to, associated with, or depending on the cerebral cortex. **—cor′ti·cal·ly** *adv.*

**cor·ti·cate** (kôr′tĭ-kĭt, -kāt′) *also* **cor·ti·cat·ed** (-kāt′ĭd) *adj.* [Lat. *corticatus,* having bark < *cortex,* bark.] Having a cortex or similar specialized external layer.

**cor·ti·ces** (kôr′tĭ-sēz′) *n. var. pl. of* CORTEX.

**cortico-** *or* **cortic-** *pref.* [< Lat. *cortex, cortic-,* bark, rind.] Cortex <*cortico*tropin>

**cor·ti·coid** (kôr′tĭ-koid′) *n.* A steroid of the adrenal cortex.

**cor·tic·o·lous** (kôr-tĭk′ə-ləs) *adj. Biol.* Growing or living on tree bark <a *corticolous* fungus>

**cor·ti·co·spi·nal** (kôr′tĭ-kō-spī′nəl) *adj.* Of or relating to the cerebral cortex and the spinal cord.

**cor·ti·co·ste·roid** (kôr′tĭ-kō-stîr′oid′) *n.* A corticoid.

**cor·ti·co·ster·one** (kôr′tĭ-kŏs′tə-rōn′) *n.* [CORTICO- + STER(OL) + -ONE.] A corticoid, $C_{21}H_{30}O_4$, that induces hyperglycemia and deposition of glycogen in the liver.

**cor·ti·co·tro·pin** (kôr′tĭ-kō-trō′pən) *also* **cor·ti·co·tro·phin** (-trō′fĭn) *n.* [CORTICO- + -TROP(IC) + -IN.] ACTH.

**cor·tin** (kôr′tn) *n.* [CORT(EX) + -IN.] An adrenal cortex extract containing several hormones that is used medicinally.

**cor·ti·sol** (kôr′tĭ-sôl′, -zôl′, -sōl′, -zōl′) *n.* [CORTIS(ONE) + -OL.] Hydrocortisone.

**cor·ti·sone** (kôr′tĭ-sōn′, -zōn′) *n.* [Alteration of CORTICOSTERONE.] A corticoid, $C_{21}H_{28}O_5$, active in carbohydrate metabolism and used to treat bursitis, rheumatoid arthritis, adrenal insufficiency, certain allergies, diseases of connective tissue, and gout.

**co·run·dum** (kə-rŭn′dəm) *n.* [Tamil *kuruntam,* prob. of Skt. orig.] A hard mineral, aluminum oxide, sometimes containing iron, magnesia, or silica, that occurs in gem varieties, as ruby and sapphire, and in a common gray, brown, or blue form used esp. in abrasives.

**co·rus·cant** (kə-rŭs′kənt) *adj.* [Lat. *coruscans, coruscant-,* pr.part. of *coruscare,* to flash.] Sparkling.

**cor·us·cate** (kôr′ə-skāt′, kŏr′-) *vi.* **-cat·ed, -cat·ing, -cates.** [Lat. *coruscare, coruscat-,* to flash.] To emit flashes of light : SPARKLE. **—cor′us·ca′tion** *n.*

**cor·vée** (kôr-vā′, kôr′vā′) *n.* [ME *corve* < OFr. *corvée* < LLat. *(opera) corrogata,* work requested < Lat. *corrogare,* to bring together : *com-,* together + *rogare,* to ask.] **1.** A day of unpaid work required of a vassal by his feudal lord. **2.** Labor imposed by a local authority for little or no pay or instead of taxes and used esp. in the maintenance of roads.

**corves** (kôrvz) *n. pl. of* CORF.

**cor·vette** (kôr-vĕt′) *n.* [Fr., a kind of warship.] **1.** A speedy lightly armed warship smaller than a destroyer. **2.** An obsolete sailing warship, smaller than a frigate and usu. armed with one tier of guns.

**cor·vi·na** (kôr-vē′nə) *n. var. of* CORBINA.

**cor·vine** (kôr′vīn′, -vĭn) *adj.* [Lat. *corvinus* < *corvus,* raven.] Of, resembling, or typical of birds such as crows or ravens.

**Cor·vus** (kôr′vəs) *n.* [NLat. < Lat. *corvus,* raven.] A constellation in the Southern Hemisphere.

**co·ry** (kô′rē, kōr′ē) *n., pl.* **cory.** —See table at CURRENCY.

**Cor·y·bant** (kôr′ə-bănt′, kŏr′-) *n., pl.* **Cor·y·bants** *or* **Cor·y·ban·tes** (kôr′ə-băn′tēz′, kŏr′-) [Lat. *Corybas, Corybant-* < Gk. *Korubas.*] *Gk. Myth.* A priest of the Phrygian goddess Cybele, whose rites were celebrated with music and ecstatic dances. **—Cor′y·ban′tic** *adj.*

**co·ryd·a·lis** (kə-rĭd′l-ĭs) *n.* [NLat. *Corydalis,* genus name < Gk. *korudallis,* crested lark (from the shape of the flowers) < *korudos.*] A plant of the genus *Corydalis,* with finely lobed leaves and spurred yellow or pinkish flowers.

**cor·ymb** (kôr′ĭmb, -ĭm, kŏr′-) *n.* [Fr. *corymbe* < Lat. *corymbus,* bunch of flowers < Gk. *korumbos.*] *Bot.* A flat-topped flower cluster in which the stalks grow upward from various points of the main stem to approx. the same height. **—cor′ym·bose′** (-ĭm-bōs′), **co·rym′bous** (kə-rĭm′bəs) *adj.* **—cor′ym·bose′ly** *adv.*

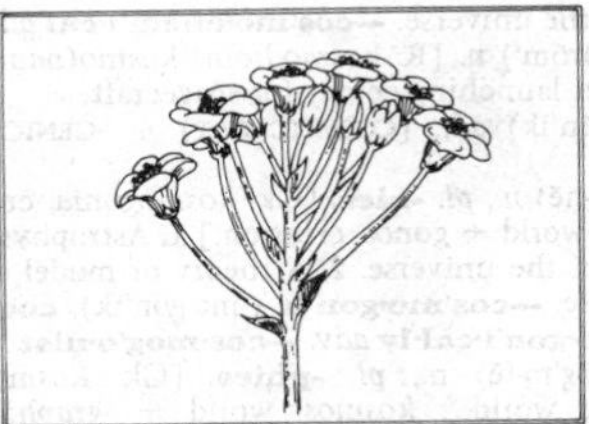
**corymb**

**co·ry·ne·bac·te·ri·um** (kôr′ə-nē-băk-tîr′ē-əm, kə-rĭn′ə-) *n.* [NLat. *Corynebacterium,* genus name : Gk. *korynē,* club + *bacterium,* bacterium.] Any of the genus *Corynebacterium* of Gram-positive rod-shaped bacteria, which includes many animal and plant pathogens, among them the causative agent of diphtheria.

**co·ryn·e·form** (kə-rĭn′ə-fôrm′) *adj.* [CORYNE(BACTERIUM) + -FORM.] Resembling to a corynebacterium.

**cor·y·phae·us** (kôr′ə-fē′əs, kŏr′-) *n., pl.* **-phae·i** (-fē′ī′) [Lat., leader < Gk. *koruphaios* < *koruphē,* head.] **1.** The leader of the chorus in ancient Greek drama. **2.** A leader or spokesperson.

**cor·y·phée** (kôr′ə-fā′) *n.* [Fr. < Lat. *coryphaeus,* leader. —see CORYPHAEUS.] A ballet dancer who ranks above the corps de ballet and below the soloists.

**co·ry·za** (kə-rī′zə) *n.* [Gk. *koruza,* catarrh.] A severe inflammation of the nasal mucous membrane characterized by discharge of mucus, sneezing, and watering of the eyes.

**cos** (kôs, kŏs) *n.* [After *Cos,* the former name for Stanchio, an island in the Aegean.] Romaine.

**Co·sa Nos·tra** (kō′sə nō′strə) *n.* [Ital., our thing.] A crime syndicate active throughout the United States, hierarchic in structure, made up of locally autonomous units known as families, and often thought to have an important relationship with the Sicilian Mafia.

**co·se·cant** (kō-sē′kănt′, -kənt) *n. Math.* The secant of the complement of a directed angle or arc.

**co·seis·mal** (kō-sīz′məl, -sīs′-) *also* **co·seis·mic** (-mĭk) *adj.* Of or designating a line connecting the points on a map indicating the places simultaneously affected by an earthquake shock. —*n.* A coseismal line.

**cosh** (kŏsh) [Prob. < Romany *kosh,* stick.] *Chiefly Brit.* —*n.* **1.** A blackjack. **2.** An assault with a blackjack. —*vt.* **coshed, cosh·ing, cosh·es.** To assault or hit with or as if with a cosh.

**cosh·er** (kŏsh′ər) *vt.* **-ered, -er·ing, -ers.** [Perh. < Ir. *cōisir,* feast.] To pamper.

**co·sign** (kō-sīn′) *vt.* **-signed, -sign·ing, -signs. 1.** To sign (a document) jointly with another or others. **2.** To endorse (a signature), as for a mortgage. **—co·sign′er** *n.*

**co·sig·na·to·ry** (kō-sĭg′nə-tôr′ē, -tōr′ē) *adj.* Signed jointly with another or others. —*n., pl.* **-ries.** One who cosigns.

**co·sine** (kō′sīn′) *n.* **1.** The abscissa of the endpoint of an arc of a unit circle centered at the origin of a Cartesian coordinate system, the arc being of length x and measured counterclockwise from the point (1, 0) if x is positive or clockwise if x is negative. **2.** The function of an acute angle that is the ratio of the adjacent side to the hypotenuse in a right triangle.
**cos lettuce** *n.* Romaine.
**cosm-** *pref. var. of* COSMO-.
**cos·met·ic** (kŏz-mĕt′ĭk) *n.* [< Gk. *kosmētikos,* skilled in arranging < *kosmein,* to arrange < *kosmos,* order.] A preparation, as rouge or lipstick, designed to beautify the body by direct application. —*adj.* **1.** Serving to beautify the body. **2.** Serving to correct physical imperfections <*cosmetic* surgery> **3. a.** Ornamental rather than functional. **b.** Having little or no significance : SUPERFICIAL <*cosmetic* changes in procedures> **—cos·met′i·cal·ly** *adv.*
**cos·me·ti·cian** (kŏz′mĭ-tĭsh′ən) *n.* One whose occupation is manufacturing, selling, or applying cosmetics.
**cos·met·i·cize** (kŏz-mĕt′ĭ-sīz′) *vt.* **-cized, -ciz·ing, -ciz·es.** To make superficially attractive or acceptable.
**cos·me·tol·o·gy** (kŏz′mĭ-tŏl′ə-jē) *n.* [Fr. *cosmétologie* : *cosmétique,* cosmetic + *-logie,* -logy.] The study of cosmetics or the skill of applying them. **—cos′me·tol′o·gist** *n.*
**cos·mic** (kŏz′mĭk) *also* **cos·mi·cal** (-mĭ-kəl) *adj.* [Gk. *kosmikos* < *kosmos,* universe.] **1.** Of or relating to the universe, esp. as distinguished from the earth. **2.** Pervasively or inconceivably extended : VAST. **—cos′mi·cal·ly** *adv.*
**cosmic dust** *n.* Fine particles of matter in interstellar space.
**cosmic noise** *n.* Galactic noise.
**cosmic ray** *n.* A stream of ionizing radiation of extraterrestrial origin, mainly of protons, alpha particles, and other atomic nuclei but also including some high-energy electrons and photons, that enters the atmosphere and produces secondary radiation, principally pions, muons, electrons, and gamma rays.
**cosmo-** *or* **cosm-** *pref.* [< Gk. *kosmos,* order, universe.] Universe : world <*cosmo*logy>
**cos·mo·chem·is·try** (kŏz′mō-kĕm′ĭ-strē) *n.* The scientific study of the chemical make-up of the universe. **—cos′mo·chem′i·cal** *adj.*
**cos·mo·drome** (kŏz′mə-drōm′) *n.* [R. *kosmodrom: kosmo(naut)* + *-drome,* -drome.] A Soviet launching center for spacecraft.
**cos·mo·gen·ic** (kŏz′mə-jĕn′ĭk) *adj.* [COSM(IC RAY) + -GENIC.] Generated by cosmic rays.
**cos·mog·o·ny** (kŏz-mŏg′ə-nē) *n., pl.* **-nies.** [Gk. *kosmogonia,* creation of the world : *cosmos,* world + *gonos,* creation.] **1.** Astrophysical study of the evolution of the universe. **2.** A theory or model of the evolution of the universe. **—cos′mo·gon′ic** (-mə-gŏn′ĭk), **cos′mo·gon′i·cal** *adj.* **—cos′mo·gon′i·cal·ly** *adv.* **—cos·mog′o·nist** *n.*
**cos·mog·ra·phy** (kŏz-mŏg′rə-fē) *n., pl.* **-phies.** [Gk. *kosmographia,* description of the world : *kosmos,* world + *-graphia,* -graphy.] **1.** The scientific study of the general features and constitution of nature. **2.** A description of the world or universe. **—cos·mog′ra·pher** *n.* **—cos′mo·graph′ic** (-mə-grăf′ĭk), **cos′mo·graph′i·cal** *adj.* **—cos′mo·graph′i·cal·ly** *adv.*
**cos·mol·o·gy** (kŏz-mŏl′ə-jē) *n., pl.* **-gies. 1.** A branch of philosophy that treats the genesis, processes, and structure of the universe. **2. a.** The astrophysical study of the structure and dynamics of the universe. **b.** A theory or model of this structure and dynamics. **—cos′mo·log′ic** (-mə-lŏj′ĭk), **cos′mo·log′i·cal** *adj.* **—cos′mo·log′i·cal·ly** *adv.* **—cos·mol′o·gist** *n.*
**cos·mo·naut** (kŏz′mə-nôt′) *n.* [R. *kosmonaut* : Gk. *kosmos,* universe + Gk. *nautēs,* sailor.] A Russian astronaut.
**cos·mo·pol·i·tan** (kŏz′mə-pŏl′ĭ-tn) *adj.* **1.** Common to the entire world. **2.** Of the entire world or from many various parts of the world. **3.** At home throughout the world or in many spheres of interest. **4.** *Biol.* Widely distributed. —*n.* A cosmopolite. **—cos′mo·pol′i·tan·ism** *n.*
**cos·mop·o·lite** (kŏz-mŏp′ə-līt′) *n.* [Gk. *kosmopolitēs* : *kosmos,* world + *politēs,* citizen < *polis,* city.] **1.** A cosmopolitan person. **2.** *Biol.* A cosmopolitan organism. **3.** The painted lady. **—cos·mop′o·lit′ism** (-lĭ′tĭz′əm, -lĭ-tĭz′-) *n.*
**cos·mo·ra·ma** (kŏz′mə-rä′mə, -răm′ə) *n.* [Gk. *kosmos* + *horama,* spectacle.] A display of scenes and pictures from all over the world. **—cos′mo·ram′ic** (-răm′ĭk) *adj.*
**cos·mos** (kŏz′məs, -mōs′, -mŏs′) *n.* [Gk. *kosmos.*] **1.** The universe thought of as a systematically arranged, harmonious whole. **2.** An ordered, harmonious system. **3.** Harmony and order rather than chaos. **4.** A tropical American plant of the genus *Cosmos,* bearing variously colored rayed flowers, esp. *C. bipinnatus,* a tall, widely cultivated garden plant.
**Cos·sack** (kŏs′ăk′) *n.* [R. *kazak* and Ukrainian *kozak,* both < Turk. *kazak,* adventurer.] A member of a people of the southern Soviet Union in Europe and neighboring parts of Asia, noted as cavalrymen. —*adj.* Of or relating to the Cossacks.
**cos·set** (kŏs′ĭt) *vt.* **-set·ed, -set·ing, -sets.** [Orig. unknown.] To pamper : coddle. —*n.* A pet, esp. a pet lamb.
**cost** (kôst) *n.* [ME < OFr. < *coster,* to cost < Lat. *constare.* —see CONSTANT.] **1.** An amount paid or to be paid for a purchase. **2.** A loss, sacrifice, or penalty. **3. costs.** *Law.* The charges fixed for litigation in court, usu. payable by the losing party. —*v.* **cost, cost·ing, costs.** —*vi.* To demand a certain payment, expenditure, effort, o loss. —*vt.* **1.** To have as a price. **2.** To cause to lose, suffer, or sacrifice. **3.** To estimate or determine the cost of. **—cost′less** *adj* **—cost′less·ness** *n.*
☆ ***syns:*** COST, CHARGE, PRICE, TAB *n. core meaning* : the amount paid or to be paid for a purchase <The *cost* of the gown i $1,200.>
**cos·ta** (kŏs′tə) *n., pl.* **-tae** (-tē) [Lat.] *Biol.* A rib or riblike part o segment, as the thickened anterior vein of an insect's wing.
**cost accountant** *n.* An accountant who documents the costs o production and distribution. **—cost accounting** *n.*
**co·star** *also* **co-star** (kō′stär′) —*n.* An actor or actress given equa billing with another or others in a play or motion picture. —*vi.* & *vt* **-starred, -star·ring, -stars.** To act or present as a costar.
**cos·tard** (kŏs′tərd) *n.* [ME < AN, perh. < *coste,* rib (from its ribbe appearance) < Lat. *costa.*] **1.** An English variety of large apple. **2.** *Ar chaic.* The human head.
**cost-ef·fec·tive** (kôst′ĭ-fĕk′tĭv) *adj.* Economical in terms of the goods or services received for the money spent. **—cost′-ef·fec′tive·ness** *n.*
**cos·ter** (kŏs′tər) *n. Chiefly Brit.* A costermonger.
**cos·ter·mon·ger** (kŏs′tər-mŭng′gər, -mŏng′-) *n.* [Obs *costard-monger* : COSTARD + MONGER.] *Chiefly Brit.* One who sell fruit, vegetables, fish, or other goods from a cart, barrow, or stand i the streets.
**cos·tive** (kŏs′tĭv) *adj.* [ME *costif* < OFr. *costeve,* p.part. of *con stever,* to constipate < Lat. *constipare.* —see CONSTIPATE.] **1. a.** Hav ing constipation. **b.** Causing constipation. **2.** Slow-moving : sluggish **3.** Stingy. **—cos′tive·ly** *adv.* **—cos′tive·ness** *n.*
**cost·ly** (kôst′lē) *adj.* **-li·er, -li·est. 1.** Of high price or great value EXPENSIVE <a *costly* necklace> **2.** Involving great loss or sacrifice <a *costly* battle> **—cost′li·ness** *n.*
**cost·mar·y** (kôst′mâr′ē, kŏst′-) *n.* [ME *costmarie* : *cost,* costmar (< OE < Lat. *costum* < Gk. *kostos* < Skt. *kuṣṭhaḥ*) + *marie,* Mary the mother of Jesus.] An herb native to Asia, *Chrysanthemum balsa mita,* with sweet-smelling foliage sometimes used as seasoning.
**cost of living** *n.* **1.** The average cost of the basic necessities of life as food, housing, and clothing. **2.** The cost of basic necessities a defined by an accepted standard.
**cost-of-liv·ing adjustment** (kôst′əv-lĭv′ĭng) *n.* An adjust ment made in wages that reflects a change in the cost of living
**cost-of-living index** *n.* The consumer price index.
**cost-plus** (kôst′plŭs′, kŏst′-) *n.* The cost of production plus a speci fied rate of profit. —*adj.* Paid or negotiated on the basis of cost-plu <a *cost-plus* labor-management agreement>
**cost-push** (kôst′po͝osh′, kŏst′-) *adj.* Specifying a kind of inflatio in which increased production costs, as from higher costs for rav materials, tend to drive prices up.
**cos·trel** (kŏs′trəl) *n.* [ME < OFr. *costerel,* prob. < *costier,* at the sid < *coste,* rib < Lat. *costa.*] A flat, pear-shaped bottle or flask wit loops for attachment to a belt.
**cos·tume** (kŏs′to͞om′, -tyo͞om′) *n.* [Fr. < Ital. < Lat. *consuetudo* custom. —see CUSTOM.] **1.** A prevalent style of dress, includin clothing, accessories, and hairdos. **2.** A style of dress typical of particular time, country, or people, often worn in a play or at festival. **3.** A set of clothes appropriate for a particular occasion o season. —*vt.* (kŏ-sto͞om′, -styo͞om′, kŏs′to͞om′, -tyo͞om′) **-tumed -tum·ing, -tumes. 1.** To put a costume on : DRESS. **2.** To design make, or supply costumes for.

▲ word history: A costume is a custom of dress, a fact that i partly explained by the origin of both *custom* and *costume* in the Latin word *consuetudo,* "custom, habit, usage." English borrowe both words from French, but at different times. *Custom* appeare first; it was used about 1200 in the still current sense of "habitua practice." *Costume* was borrowed much later, in the 18th century, a a term from the technical vocabulary of the fine arts. It meant "th features characteristic of a particular historical period represented i painting or sculpture." This was the meaning of Italian *costume* from which French *costume* and thence English *costume* derive Such characteristic features naturally included styles of dress; as th use of *costume* expanded beyond its original area its meaning wa restricted to clothing styles.

**cos·tum·er** (kŏs′to͞o′mər, -tyo͞o′-, kŏ-sto͞o′mər, -styo͞o′-) *also* **cos tum·i·er** (kŏ-sto͞om′yər, -styo͞om′-) *n.* One who designs, makes, o supplies costumes.
**co·sy** (kō′zē) *adj., v., & n. var. of* COZY.
**cot**[1] (kŏt) *n.* [Hindi *khāṭ,* couch < Skt. *khaṭvā,* of Dravidian orig. **1.** A narrow bed, esp. a collapsible one made of canvas. **2.** *Chiefl Brit.* CRIB 1.
**cot**[2] (kŏt) *n.* [ME < OE.] **1. a.** A small house : COTTAGE. **b.** A smal shelter, esp. one for animals. **2.** A protective covering.
**co·tan·gent** (kō-tăn′jənt) *n. Math.* The tangent of the comple ment of a directed arc or angle. **—co·tan·gen′tial** (-jĕn′shəl) *ad*
**cot death** *n. Chiefly Brit.* Sudden infant death syndrome.

**†cote[1]** (kōt) *n.* [ME < OE.] **1.** A small shed or shelter for sheep or birds. **2.** *Regional.* A cottage : hut.

**cote[2]** (kōt) *vt.* **cot·ed, cot·ing, cotes.** [Orig. unknown.] *Archaic.* To go around the side of : PASS.

**co·ten·ant** (kō-tĕn′ənt) *n.* One of two or more tenants sharing a place or property. **—co·ten′an·cy** *n.*

**co·ter·ie** (kō′tə-rē, kō′tə-rē′) *n.* [Fr. < OFr., peasant association < *cotier,* cottager < Med. Lat. *cotarius* < ME *cot,* cot.] A small group of persons who have similar interests and associate frequently.

**co·ter·mi·nous** (kō-tûr′mə-nəs) *adj. var. of* CONTERMINOUS.

**co·thur·nus** (kō-thûr′nəs) *also* **co·thurn** (kō′thûrn′, kō-thûrn′) *n., pl.* **-ni** (-nī′) *also* **-thurns.** [Lat. < Gk. *kothornos.*] **1.** A buskin worn by actors of classical tragedy. **2.** The style of ancient classic tragedy.

**co·tid·al** (kō-tīd′l) *adj.* **1.** Indicating coincidence of the tides. **2.** Designating lines on a map that indicate where high or low tides occur simultaneously.

**co·til·lion** *also* **co·til·lon** (kō-tĭl′yən, kə-) *n.* [Fr. *cotillon* < OFr., petticoat, dim. of *cote,* coat.] **1. a.** A lively ballroom dance, developed in France in the 18th cent., with varied, intricate patterns and steps. **b.** A quadrille. **c.** Music for these dances. **2.** A formal ball, esp. one at which debutantes are presented to society.

**co·to·ne·as·ter** (kə-tō′nē-ăs′tər) *n.* [NLat. *Cotoneaster,* genus name < Lat. *cotoneum,* quince.] An Old World shrub of the genus *Cotoneaster,* bearing small white or pinkish flowers and often grown for ornament.

**cot·quean** (kŏt′kwēn′) *n. Archaic.* **1.** A vulgar, scolding woman : HUSSY. **2.** A man who does work regarded suitable only for women.

**Cots·wold** (kŏt′swōld′) *n.* A breed of sheep valued for its long wool, orig. developed in the Cotswold Hills of southwestern England.

**cot·ta** (kŏt′ə) *n., pl.* **cot·tae** (kŏt′ē) *or* **cot·tas.** [Med. Lat., of Germanic orig.] A short ecclesiastical surplice.

**cot·tage** (kŏt′ĭj) *n.* [ME *cotage,* prob. < AN < ME *cot,* cot.] **1.** A small, usu. one-story house, esp. in the country. **2.** A small house used during vacations.

**cottage cheese** *n.* A soft, white cheese made of strained and seasoned curds of skim milk.

**cottage industry** *n.* A usu. small-scale industry carried on at home by family members using their own equipment.

**cottage pudding** *n.* Plain cake served with a sweet sauce.

**cot·tag·er** (kŏt′ĭ-jər) *n.* One who lives in a cottage.

**cottage tulip** *n.* A tall-stemmed, mid-season garden tulip, usu. having pointed petals.

**cot·ter** (kŏt′ər) *n.* [Orig. unknown.] **1.** A bolt, wedge, key, or pin inserted through a slot to hold pieces together. **2.** A cotter pin.

**cotter pin** *n.* A split cotter inserted through holes in two or more pieces and bent at the ends for fastening.

**cot·ti·er** (kŏt′ē-ər) *n.* [ME *cotier,* cottager < OFr. < ME *cot,* cot.] A peasant renting land directly from its owner, the rate having been established by public bidding.

**cot·ton** (kŏt′n) *n.* [ME *cotoun* < OFr. *coton* < dial. Ar. *qoṭon* < Ar. *quṭn.*] **1. a.** A plant or shrub of the genus *Gossypium,* grown in warm climates for the fiber surrounding the seeds. **b.** The soft, white, downy fiber attached to the seeds of the cotton plant, used primarily for textiles. **c.** Cotton plants as a whole. **d.** The crop of the cotton plant. **2.** Thread or cloth manufactured from cotton fiber. **3.** A soft, downy substance found in various other plants. *—vi.* **-toned, -ton·ing, -tons.** *Informal.* To become friendly : GET ALONG <doesn't *cotton* to strangers>

**cotton batting** *n.* Batting.

**cotton candy** *n.* Spun sugar.

**cotton flannel** *n.* A soft, warm, napped fabric woven of cotton.

**cotton gin** *n.* A machine that separates the seeds, seed hulls, and unwanted materials from the fibers of cotton.

**cotton grass** *n.* A grasslike bog plant of the genus *Eriophorum,* with densely tufted, cottony flower heads.

**cotton leafworm** *n.* The larva of a New World moth, *Alabama argillacea,* that defoliates cotton plants.

**cot·ton·mouth** (kŏt′n-mouth′) *n.* The water moccasin.

**cotton rose** *n.* **1.** A Chinese shrub, *Hibiscus mutabilis,* grown in warm regions for its white or pink flowers that turn deep red. **2.** CUDWEED 2.

**cotton rust** *n.* A disease of the cotton plant caused by the fungus *Puccinia stakmanii* and characterized by yellowish discolorations on the leaves.

**cot·ton·seed** (kŏt′n-sēd′) *n., pl.* **cottonseed** *or* **-seeds.** The seed of cotton, used as a source of oil and meal.

**cottonseed meal** *n.* Meal made from the residue of cottonseed after the oil has been removed, used as animal feed and fertilizer.

**cottonseed oil** *n.* A yellowish to dark-red oil obtained by pressing cottonseed, used in cooking, as salad oil, and in the manufacture of paints, soaps, and other products.

**cotton stainer** *n.* A small, flat, red insect of the genus *Dysdercus* that punctures cotton bolls and stains the fibers.

**cot·ton·tail** (kŏt′n-tāl′) *n.* A New World rabbit of the genus *Sylvilagus,* with gray or brown fur and a tail having a white underside.

**cotton tree** *n.* A spiny tropical tree, *Bombax malabaricum,* bearing seeds surrounded by a cottony fiber.

**cot·ton·weed** (kŏt′n-wēd′) *n.* Any of various plants covered with cottony down or having cottonlike tufts.

**cot·ton·wood** (kŏt′n-wo͝od′) *n.* A softwood tree of the genus *Populus,* bearing seeds with cottonlike tufts, esp. *P. deltoides* of eastern and central North America.

**cotton wool** *n.* **1.** Cotton in its raw or natural state. **2.** *Chiefly Brit.* Absorbent cotton.

**cot·ton·y** (kŏt′n-ē) *adj.* **1.** Of or like cotton : FLUFFY. **2.** Covered with cottonlike fibers : NAPPY.

**cot·y·le·don** (kŏt′l-ēd′n) *n.* [Lat., navelwort < Gk. *kotulēdōn,* cup-shaped hollow < *kotulē,* anything hollow.] **1.** *Bot.* A leaf of a plant embryo, being the first or one of the first to appear from a sprouting seed. **2.** *Anat.* A lobule of the placenta, esp. of ruminants. **—cot′y·le′don·al** (-ēd′n-əl), **cot′y·le′do·nous** (-ēd′n-əs) *adj.*

**cot·y·loid** (kŏt′l-oid′) *also* **cot·y·loid·al** (kŏt′l-oid′l) *adj.* [Gk. *kotuloeidēs* : *kotulē,* anything hollow + *-oeidēs,* -oid.] Shaped like a cup.

**couch** (kouch) *n.* [ME *couche* < OFr. < *couchier,* to lay down < Lat. *collocare,* to lay. —see COLLOCATE.] **1.** A piece of furniture, usu. upholstered and often having a back, on which one may sit or recline : SOFA. **2. a.** The frame or floor on which grain, esp. barley, is spread in malting. **b.** A layer of grain, esp. barley, spread out to germinate. *—v.* **couched, couch·ing, couch·es.** *—vt.* **1.** To cause (oneself) to lie down, as for rest. **2.** To lower (a lance or spear) to an attack position. **3.** To embroider by placing thread flat on a surface and attaching it by stitches at regular intervals. **4.** To spread (grain) on a couch to germinate, as in malting. **5.** To express in words of a particular form : PHRASE <*couched* the complaint tactfully> *—vi.* **1.** To lie down : RECLINE. **2.** To wait in ambush or concealment : LURK. **3.** To be in a heap or pile, as leaves or grass, for decomposition or fermentation. **—couch′er** *n.*

**couch·ant** (kou′chənt) *adj.* [ME < OFr., pr.part. of *couchier,* to lay down. —see COUCH.] *Heraldry.* Lying down with the head raised.

**couch grass** *n.* [Alteration of QUITCH GRASS.] A grass, *Agropyron repens,* with whitish-yellow rootstocks by means of which it spreads rapidly, becoming a troublesome weed.

**couch·ing** (kou′chĭng) *n.* [< ME *couchen,* to embroider < OFr. *couchier,* to lay down. —see COUCH.] Embroidery in which thread is attached to a material with stitches at regular intervals.

**cou·gar** (ko͞o′gər) *n.* [Fr. *couguar,* ult. < Tupi *suasuarana.*] The mountain lion.

**cougar**
*Approximately 6 feet long*

**cough** (kôf, kŏf) *v.* **coughed, cough·ing, coughs.** [ME *coughen.*] *—vi.* **1.** To eject air from the lungs loudly and suddenly. **2.** To produce a noise similar to that of coughing <The engine *coughed* and died.> *—vt.* To expel by coughing <*coughed* out phlegm> **—cough up.** *Slang.* To relinquish (e.g., money), often reluctantly. *—n.* **1.** An act of coughing. **2.** A condition marked by frequent coughing.

**cough drop** *n.* A small, usu. medicated and flavored lozenge taken to alleviate coughing or soothe a sore throat.

**cough syrup** *n.* A sweetened, usu. medicated liquid that is taken to alleviate coughing.

**could** (ko͝od) *v. p.t. of* CAN[1].

**could·n't** (ko͝od′nt). Could not.

**†cou·lee** (ko͞o′lē) *n.* [Canadian Fr. *coulée* < Fr., flow < *couler,* to flow < Lat. *colare,* to filter < *colum,* sieve.] **1.** *Western U.S.* A deep gulch or ravine formed by rainstorms or melting snow, often dry in summer. **2. a.** A stream of molten lava. **b.** A sheet of solid lava.

**cou·lisse** (ko͞o-lēs′) *n.* [Fr. < *coulis,* sliding < *couler,* to slide. —see COULEE.] **1.** A grooved timber in which something slides. **2. a.** One of the side scenes of a theater stage. **b.** The space between the side scenes.

**cou·loir** (ko͞ol-wär′) *n.* [Fr. < *couler,* to flow. —see COULEE.] A deep gorge or gully in a mountainside, esp. in the Alps.

**cou·lomb[1]** (ko͞o′lŏm′, -lōm′) *n.* [After Charles A. de *Coulomb* (1736–1806).] A meter-kilogram-second unit of electrical charge equal to the quantity of charge transferred in one second by a steady current of one ampere.

**coulomb[2]** *or* **cou·lom·bic** (ko͞o-lŏm′bĭk, -lōm′-) *adj.* Of or pertaining to the Coulomb force.

o͞o **boot** ou **out** th **thin** *th* **this** ŭ **cut** û **urge** y **young**
yo͞o **abuse** zh **vision** ə **about, item, edible, gallop, circus**

**Coulomb field** *n.* An electric field equal to that produced by a point charge so the force at every point is described by Coulomb's law.

**Coulomb force** *n.* An attractive or repulsive electrostatic force described by Coulomb's law.

**cou·lom·bic** (ko͞o-lŏm′bĭk, -lōm′-) *adj. var. of* COULOMB[2].

**Coulomb potential** *n.* The potential at any point in a Coulomb field.

**Coulomb scattering** *n.* The scattering of a charged particle from another charged particle, esp. from an atomic nucleus, chiefly or exclusively as a result of Coulomb forces.

**Coulomb's law** *n.* [After Charles A. de *Coulomb,* its formulator.] The fundamental law of electrostatics stating that the force between two charged particles is directly proportional to the product of their charges and inversely proportional to the square of the distance between them.

**cou·lom·e·try** (ko͞o-lŏm′ĭ-trē) *n.* [COULO(MB) + -METRY.] A method for determining the amount of a substance released during electrolysis in which the number of coulombs used is measured. **—cou′lo·met′ric** (-lə-mĕt′rĭk) *adj.* **—cou′lo·met′ri·cal·ly** *adv.*

**coul·ter** (kōl′tər) *n. var. of* COLTER.

**cou·ma·rin** (ko͞o′mər-ĭn) *n.* [Fr. *coumarine* < *coumarou,* tonka bean tree < Sp. *coumarú* < Tupi.] A sweet-smelling organic compound, $C_9H_6O_2$, derived from tonka beans or synthesized and used in flavorings, perfumes, and soaps. **—cou′ma·ric** (-mər-ĭk) *adj.*

**coun·cil** (koun′səl) *n.* [ME *counceil* < OFr. *concile* < Lat. *concilium.*] **1.** A group of persons gathered together for consultation, deliberation, or discussion. **2.** A body of people elected or appointed to act in an administrative, legislative, or advisory capacity. **3.** The deliberation or discussion that occurs in a council. **4.** A group of church officials and theologians convened for regulating and administering matters of doctrine and discipline. ***usage:*** *Council* and *counsel* are not interchangeable although they are related. A *council* is a deliberative assembly. *Counsel* pertains to advice and guidance in general.

**coun·cil·lor** (koun′sə-lər) *n. var. of* COUNCILOR.

**coun·cil·man** (koun′səl-mən) *n.* A member of a council, esp. of the governing body of a municipality.

**coun·cil-man·ag·er plan** (koun′səl-măn′ə-jər) *n.* A type of municipal government in which the chief executive official is a manager chosen by the city council.

**coun·cil·or** *also* **coun·cil·lor** (koun′sə-lər) *n.* A member of a council. **—coun′cil·or·ship′, coun′cil·lor·ship′** *n.*

**coun·cil·wom·an** (koun′səl-wo͝om′ən) *n.* A woman who is a member of a council, esp. of the governing body of a municipality.

**coun·sel** (koun′səl) *n.* [ME *counseil* < OFr. *conseil* < Lat. *consilium.*] **1.** An exchange of opinions and ideas in order to reach a decision : CONSULTATION. **2.** Advice or guidance, esp. as solicited from a knowledgeable or experienced person. **3.** A deliberate plan of action. **4.** A private purpose or personal opinion <keep one's own *counsel*> **5.** *pl.* **counsel.** A lawyer or group of lawyers, esp. an attorney retained to conduct a case in court. *—v.* **-seled, -sel·ing, -sels** *also* **-selled, -sel·ling, -sels.** *—vt.* **1.** To give counsel to : ADVISE. **2.** To press the adoption of : RECOMMEND. *—vi.* To give or take counsel or advice.

**coun·sel·or** *also* **coun·sel·lor** (koun′sə-lər) *n.* **1.** One who gives counsel : ADVISER. **2.** An attorney, esp. a trial lawyer. **3.** One who supervises children at a camp. **—coun′se·lor·ship′** *n.*

**coun·sel·or-at-law** (koun′sə-lər-ət-lô′) *n., pl.* **coun·sel·ors-at-law.** COUNSELOR 2.

**count**[1] (kount) *v.* **count·ed, count·ing, counts.** [ME *counten* < OFr. *conter* < Lat. *computare,* to compute : *com-,* together + *putare,* to think.] *—vt.* **1. a.** To note (the members of a group or aggregation) one by one in order to determine a total : NUMBER. **b.** To recite numerals in ascending order up to and including <*count* ten before opening your eyes> **c.** To include in a reckoning <eight cats *counting* the kittens> **2.** To believe or think to be : DEEM. **3. a.** To include by or as if by counting <*counted* me in> **b.** To exclude by or as if by counting <*counted* me out> *—vi.* **1.** To recite or list numbers in order or enumerate items by units or groups <*count* by twos> **2. a.** To have significance <You really *count* with me.> **b.** To have a specified significance <theories that *count* for little> **3.** *Mus.* To keep time by counting beats. **—count on. 1.** To rely on <You can *count on* my cooperation.> **2.** To be confident of : EXPECT <*counted on* getting a promotion> *—n.* **1.** The act of enumerating or calculating. **2. a.** A number obtained by counting. **b.** The total number of specific items in a sample <a white blood cell *count*> **3.** *Law.* Any of the distinct charges in an indictment. **4.** The counting from one to ten seconds, during which time a boxer who has been knocked down must get up or be declared the loser. **5.** *Baseball.* The number of balls and strikes made by a player at one turn at bat.

**count**[2] (kount) *n.* [ME *counte* < OFr. *conte* < LLat. *comes,* occupant of any state office < Lat., companion.] A nobleman in some European countries, corresponding in rank to an English earl.

**count·a·ble** (koun′tə-bəl) *adj.* **1.** Capable of being counted. **2.** *Math.* Capable of being put in a one-to-one correspondence with the positive integers. **—count′a·bil′i·ty** *n.* **—count′a·bly** *adv.*

**count·down** (kount′doun′) *n.* **1.** The act or process of counting backward from an arbitrary starting number to show the amount of time remaining before a scheduled event or operation, as the launch of a space vehicle. **2.** Preparations carried out during a countdown.

**coun·te·nance** (koun′tə-nəns) *n.* [ME *contenaunce* < OFr. < *contenir,* to behave. —see CONTAIN.] **1.** Appearance, esp. the expression of the face. **2.** The face or facial features. **3. a.** An apparently encouraging look or expression. **b.** Support or approval in general. **4.** Bearing : demeanor. *—vt.* **-nanced, -nanc·ing, -nanc·es.** To give or express approval to : CONDONE <wouldn't *countenance* lying> **—coun′te·nanc·er** *n.*

**coun·ter**[1] (koun′tər) *adj.* [< COUNTER-.] Diametrically opposed : CONTRARY. *—n.* **1.** One that is diametrically opposed. **2.** A boxing blow given while receiving or parrying another. **3.** A fencing parry in which one foil follows the other in a circular fashion. **4.** A stiffened piece of leather around the heel of a shoe. **5.** The part of a ship's stern from the water line to the extreme outward swell. **6.** The sunken area between the raised lines of a typeface. *—v.* **-tered, -ter·ing, -ters.** *—vt.* **1.** To move or act in opposition to : OPPOSE. **2.** To offer in response. **3.** To meet or return (a blow) by another blow. *—vi.* To move, act, or respond so as to counter. *—adv.* **1.** In an opposing manner or direction. **2.** To or toward an opposite or dissimilar course or result <a theory running *counter* to accepted ideas>

**count·er**[2] (koun′tər) *n.* [ME *countour* < AN < Med. Lat. *computatorium,* counting house < Lat. *computare,* to compute. —see COUNT.] **1.** A flat surface on which money is counted, business is transacted, or food is served. **2.** A piece, as of wood or ivory, used for keeping a tally or a place in games. **3. a.** An imitation coin : TOKEN. **b.** A piece of money.

**count·er**[3] (koun′tər) *n.* **1.** One that counts. **2.** An electronic or mechanical device that automatically counts occurrences or repetitions of phenomena or events.

**counter-** *pref.* [ME *countre-* < OFr. *contre-* < *contre,* counter < Lat. *contra.*] **1.** Contrary : opposite <*counter*claim> **2.** Corresponding : complementary <*counter*foil>

**coun·ter·act** (koun′tər-ăkt′) *vt.* **-act·ed, -act·ing, -acts.** To oppose and render ineffective by contrary action : CHECK. **—coun′ter·ac′tion** *n.* **—coun′ter·ac′tive** *adj.* **—coun′ter·ac′tive·ly** *adv.*

**coun·ter·at·tack** (koun′tər-ə-tăk′) *n.* A return attack. *—vt. & vi.* (koun′tər-ə-tăk′) **-tacked, -tack·ing, -tacks.** To make a counterattack against or deliver a counterattack.

**coun·ter·bal·ance** (koun′tər-băl′əns, koun′tər-băl′əns) *n.* **1.** A force or influence that equally counteracts another. **2.** A weight that balances another : COUNTERPOISE. *—vt.* (koun′tər-băl′əns, koun′tər-băl′əns) **-anced, -anc·ing, -anc·es. 1.** To act as a counterbalance to : COUNTERPOISE. **2.** To oppose with an equal force : OFFSET.

**coun·ter·change** (koun′tər-chānj′) *vt.* **-changed, -chang·ing, -chang·es. 1.** To exchange : transpose. **2.** To make checkered or variegated.

**coun·ter·charge** (koun′tər-chärj′) *n.* A charge in opposition to or answer another charge. *—vt. & vi.* (koun′tər-chärj′) **-charged, -charg·ing, -charg·es.** To bring a countercharge against or make a countercharge.

**coun·ter·check** (koun′tər-chĕk′) *n.* **1.** Something that serves to check something else. **2.** Something that verifies or denies the correctness of a previous check. *—vt.* (koun′ter-chĕk′) **-checked, -check·ing, -checks. 1.** To oppose or check by a counteraction. **2.** To recheck in order to verify.

**coun·ter·claim** (koun′tər-klām′) *n.* A claim filed in opposition to another claim. *—vt. & vi.* (koun′tər-klām′) **-claimed, -claim·ing, -claims.** To make a counterclaim against or plead a counterclaim. **—coun′ter·claim′ant** (-klā′mənt) *n.*

**coun·ter·clock·wise** (koun′tər-klŏk′wīz′) *adv. & adj.* In a direction opposite to the movement of the hands of a clock.

**coun·ter·con·di·tion·ing** (koun′tər-kən-dĭsh′ən-ĭng) *n. Psychol.* Conditioning designed to replace a negative response to a stimulus with a positive response.

**coun·ter·coup** (koun′tər-ko͞o′) *n.* A coup to oust a government that orig. gained power by a coup d'état.

**coun·ter·cul·ture** (koun′tər-kŭl′chər) *n.* A culture, esp. of young people, with values opposed to those of the established culture. **—coun′ter·cul′tur·al** *adj.*

**coun·ter·cur·rent** (koun′tər-kûr′ənt, -kŭr′-) *n.* A current flowing in a direction opposite to the flow of another current. **—coun′ter·cur′rent** *adj.* **—coun′ter·cur′rent·ly** *adv.*

**coun·ter·dem·on·stra·tion** (koun′tər-dĕm′ən-strā′shən) *n.* A demonstration that opposes another demonstration. **—coun′ter·dem′on·stra′tor** *n.*

**coun·ter·es·pi·o·nage** (koun′tər-ĕs′pē-ə-näzh′, -nĭj) *n.* Espionage intended to discover and counteract enemy espionage.

**coun·ter·ex·am·ple** (koun′tər-ĭg-zăm′pəl) *n.* A statement or illustration refuting a hypothesis, proposition, or theorem.

**coun·ter·feit** (koun′tər-fĭt′) *v.* **-feit·ed, -feit·ing, -feits.** [ME *countrefeten* < *countrefet*, made in imitation < OFr. *contrefait* < *contrefaire*, to counterfeit : *contre-*, counter- + *faire*, to make < Lat. *facere*.] —*vt.* **1.** To make a copy of, usu. with the intent to defraud : FORGE <*counterfeiting* dollar bills> **2.** To make a show of : FEIGN <*counterfeited* concern for their suffering> —*vi.* **1.** To carry on a deception : DISSEMBLE. **2.** To make copies. —*adj.* **1.** Made in imitation of what is genuine with the intent to defraud <a *counterfeit* dime> **2.** Pretended : feigned <a *counterfeit* display of tears> —*n.* A fraudulent imitation or facsimile. **—coun′ter·feit′er** *n.*

☆ ***syns:*** COUNTERFEIT, BOGUS, ERSATZ, FAKE, FALSE, FRADULENT, PHONY, SHAM, SPURIOUS *adj. core meaning* : fradulently or deceptively imitative <tried to flood the market with *counterfeit* money> ***ant:*** bona fide, genuine

**coun·ter·foil** (koun′tər-foil′) *n.* The portion of a check or other commercial paper kept by the issuer as a record of a transaction.

**coun·ter·glow** (koun′tər-glō′) *n.* Gegenschein.

**coun·ter·in·sur·gen·cy** (koun′tər-ĭn-sûr′jən-sē) *n.* Political and military action taken to oppose insurgency. **—coun′ter·in·sur′gent** *n.*

**coun·ter·in·tel·li·gence** (koun′tər-ĭn-tĕl′ə-jəns) *n.* The branch of an intelligence agency charged with keeping valuable information from an enemy, preventing subversion and sabotage, and gathering political and military data.

**coun·ter·ir·ri·tant** (koun′tər-ĭr′ĭ-tənt) *n.* An agent that induces local irritation in order to counteract general irritation.

**coun·ter·man** (koun′tər-măn′, -mən) *n.* One who tends a counter, as in a luncheonette.

**coun·ter·mand** (koun′tər-mănd′, koun′tər-mănd′) *vt.* **-mand·ed, -mand·ing, -mands.** [ME *countremaunden* < OFr. *contremander* : *contr-*, counter- + *mander*, to command < Lat. *mandare*.] **1.** To cancel or reverse (e.g., a command). **2.** To recall by an opposing order. —*n.* (koun′tər-mănd′). An order or command reversing another.

**coun·ter·march** (koun′tər-märch′) *n.* **1.** A march back or in a reverse direction. **2.** A total reversal of method or behavior. —*vt.* & *vi.* (koun′tər-märch′) **-marched, -march·ing, -march·es.** To conduct in or execute a countermarch.

**coun·ter·mea·sure** (koun′tər-mĕzh′ər) *n.* A measure taken to oppose another measure.

**coun·ter·mine** (koun′tər-mīn′) *n.* **1. a.** A tunnel dug to intercept and destroy a tunnel dug by besiegers. **b.** An explosive charge placed so as to explode an enemy's mines. **2.** A plot to frustrate or defeat an attack. —*v.* (koun′tər-mīn′) **-mined, -min·ing, -mines.** —*vt.* **1.** To make or use a countermine against. **2.** To frustrate or defeat by secret measures. —*vi.* To make or lay down countermines.

**coun·ter·move** (koun′tər-mōōv′) *n.* A move countering another move. —*vi.* (koun′tər-mōōv′) **-moved, -mov·ing, -moves.** To make a countermove. **—coun′ter·move′ment** *n.*

**coun·ter·of·fen·sive** (koun′tər-ə-fĕn′sĭv) *n.* A massive military attack, designed to stop an enemy offensive.

**coun·ter·of·fer** (koun′tər-ô′fər, -ŏf′ər) *n.* An offer made in return by one who rejects a previous, unsatisfactory offer.

**coun·ter·pane** (koun′tər-pān′) *n.* [Alteration of obs. *counterpoint* < ME *countrepoint*.] A coverlet : bedspread.

**coun·ter·part** (koun′tər-pärt′) *n.* **1. a.** One remarkably similar to another. **b.** One having the same functions and characteristics as another. **2. a.** One of two parts that fit and complete each other. **b.** Something that is a natural complement to another <a rare burgundy that is a perfect *counterpart* to the banquet>

**coun·ter·plea** (koun′tər-plē′) *n. Law.* A plaintiff's answer to a defendant's plea or counterclaim.

**coun·ter·plot** (koun′tər-plŏt′) *n.* A plot intended to frustrate another plot. **—coun′ter·plot′** *v.* **(-plot·ted, -plot·ting, -plots).**

**coun·ter·point** (koun′tər-point′) *n.* **1.** *Mus.* **a.** Melodic material added above or below an existing melody. **b.** The technique of combining two or more melodic lines in such a way that they establish a harmonic relationship while retaining their linear individuality. **c.** Music incorporating or made up of contrapuntal writing. **2.** A contrasting but parallel element, item, or theme.

**coun·ter·poise** (koun′tər-poiz′) *n.* **1.** A counterbalancing weight. **2.** A force or influence that balances or equally counteracts another. **3.** The state of being in equilibrium. —*vt.* **-poised, -pois·ing, -pois·es.** **1.** To oppose with an equal weight : COUNTERBALANCE. **2.** To act against with an equal force or power : OFFSET.

**coun·ter·pro·duc·tive** (koun′tər-prə-dŭk′tĭv) *adj.* Hindering rather than serving one's purpose. **—coun′ter·pro·duc′tive·ly** *adv.*

**coun·ter·pro·pos·al** (koun′tər-prə-pō′zəl) *n.* A proposal made to nullify or take the place of a previous one.

**coun·ter·punch** (koun′tər-pŭnch′) *n.* A countering attack or blow.

**coun·ter·ref·or·ma·tion** (koun′tər-rĕf′ər-mā′shən) *n.* A reformation in opposition to previous reformation.

**Counter Reformation** *n.* A reform movement within the Roman Catholic Church in opposition to the Protestant Reformation.

**coun·ter·rev·o·lu·tion** (koun′tər-rĕv′ə-lōō′shən) *n.* A movement arising in opposition to a revolution and aiming to restore the prerevolutionary government or state. **—coun′ter·rev′o·lu′tion·ar′y** (-shə-nĕr′ē) *adj.* & *n.* **—coun′ter·rev′o·lu′tion·ist** *n.*

**coun·ter·shaft** (koun′tər-shăft′) *n.* An intermediate shaft between the powered and driven shafts in a belt drive.

**coun·ter·sign** (koun′tər-sīn′) *vt.* **-signed, -sign·ing, -signs.** To sign (a previously signed document), as for authentication. —*n.* **1.** A second or verifying signature, as on a previously signed document. **2. a.** A secret sign or signal to be given to a sentry in order to obtain passage : PASSWORD. **b.** A secret sign or signal given in response to another. **—coun′ter·sig′na·ture** (-sĭg′nə-chər) *n.*

**coun·ter·sink** (koun′tər-sĭngk′) *n.* **1.** A hole with the top part enlarged so a screw or bolthead will lie flush with or below the surface. **2.** A tool for making a countersink. —*vt.* **-sunk** (-sŭngk′), **-sink·ing, -sinks.** **1.** To make a countersink on or in. **2.** To drive (a screw or bolt) into a countersink.

**coun·ter·spy** (koun′tər-spī′) *n.* A spy engaged in counterespionage.

**coun·ter·stain** (koun′tər-stān′) *n.* A stain of a contrasting color, used with a principal stain, that colors those microscopic specimen components that are not made visible by the principal stain.

**coun·ter·ten·or** (koun′tər-tĕn′ər) *n.* **1.** An adult male voice with a range above that of tenor. **2.** A singer with a countertenor voice.

**coun·ter·vail** (koun′tər-vāl′, koun′tər-vāl′) *v.* **-vailed, -vail·ing, -vails.** [ME *countrevaillen* < OFr. *contrevaloir* : *contre*, counter- + *valoir*, to be worth < *valere*, to be strong.] —*vt.* **1.** To act against with equal force : COUNTERACT. **2.** To make up for : COMPENSATE. —*vi.* To act against an often detrimental influence or power.

**coun·ter·weigh** (koun′tər-wā′) *vi.* & *vt.* **-weighed, -weigh·ing, -weighs.** To counterbalance or cause to counterbalance.

**coun·ter·weight** (koun′tər-wāt′) *n.* A weight used as a counterbalance. **—coun′ter·weight′ed** (-wā′tĭd) *adj.*

**counter word** *n.* A word, as *nice* or *awful*, frequently used without regard to its precise meaning.

**count·ess** (koun′tĭs) *n.* [ME *countes* < OFr. *contesse*, fem. of *conte*, count. —see COUNT.] **1. a.** The wife or widow of a count in various European countries. **b.** The wife or widow of an earl in Great Britain. **2.** A woman holding the title of count or earl in her own right.

**count·ing·house** (koun′tĭng-hous′) *also* **counting house** *n.* An office in which a company carries on operations such as accounting and correspondence.

**counting room** *n.* A countinghouse.

**count·less** (kount′lĭs) *adj.* Too numerous to be counted : INFINITE. **—count′less·ly** *adv.*

**count noun** *n.* A noun, as *chair* or *pea*, that can form a plural and be used in a noun phrase construction with the indefinite article, with such terms as *many*, or with numerals.

**count palatine** *n.* PALATINE[1] 3.

**coun·tri·fied** *also* **coun·try·fied** (kŭn′trĭ-fīd′) *adj.* **1.** Resembling or typical of country life : RUSTIC. **2.** Lacking in sophistication.

**coun·try** (kŭn′trē) *n., pl.* **-tries.** [ME *countre* < OFr. *contree* < LLat. *contrata* < Lat. *contra*, opposite.] **1.** A large tract of land distinguishable by features of topography, biology, or culture <farming *country*> **2.** An area outside cities and towns. **3. a.** A nation or state. **b.** The territory of a nation or state : LAND. **c.** The people of a nation or state. **4.** The land of a person's birth or citizenship. **5.** *Law.* A jury.

**country and western** *n.* Country music.

**country club** *n.* A club with facilities for golf and other outdoor sports and social activities.

**country cousin** *n.* One whose ingenuousness or rustic ways may embarrass or amuse city dwellers.

**coun·try-dance** (kŭn′trē-dăns′) *n.* A folk dance originating in England in which two lines of dancers face each other.

**coun·try·fied** (kŭn′trĭ-fīd′) *adj. var. of* COUNTRIFIED.

**country gentleman** *n.* **1.** The owner of a country estate. **2.** *often* **Country Gentleman.** A corn with small, sweet white kernels.

**coun·try·man** (kŭn′trē-mən) *n.* **1.** A person from one's own country : COMPATRIOT. **2.** A person from a specific country. **3.** One who lives in the country.

**country music** *n.* A style of popular music based on folk music of the rural United States, esp. of the southern or southwestern United States.

**coun·try·seat** (kŭn′trē-sēt′) *n.* An estate or house in the country.

**coun·try·side** (kŭn′trē-sīd′) *n.* **1.** A rural region. **2.** The residents of a rural region.

**coun·try·wom·an** (kŭn′trē-wōōm′ən) *n.* **1.** A woman from one's own country : COMPATRIOT. **2.** A woman from a specific country. **3.** A woman who lives in the country.

**coun·ty** (koun′tē) *n., pl.* **-ties.** [ME *counte*, territorial division < AN *counte* < OFr. *conte*, the territory of a count < Med. Lat. *comitatus* < LLat., an office of state < Lat., retinue < *comes*, companion.] **1.** An administrative subdivision of a U.S. state. **2.** A British or Irish territorial division having administrative, judicial, and political powers and functions. **3.** The people living in a county. **4.** The territory under the jurisdiction of a count or earl.

**county palatine** *n.* The domain of a count palatine.

**county seat** *n.* A municipality that is the center of government in its county.

**county town** *n. Chiefly Brit.* A county seat.

**coup** (ko͞o) *n., pl.* **coups** (ko͞oz) [Fr., stroke < OFr. < LLat. *colpus* < Lat. *colaphus* < Gk. *kolaphos.*] **1.** A brilliantly conceived and executed stratagem : MASTERSTROKE. **2.** A coup d'état.

**coup de grâce** (ko͞o' də gräs') *n.* [Fr. : *coup,* stroke + *de,* of + *grace,* mercy.] **1.** A deathblow delivered to end the misery of one that is mortally wounded. **2.** A finishing or decisive act or event.

**coup de main** (ko͞o' də mäɴ') *n.* [Fr. : *coup,* stroke + *de,* of + *main,* hand.] A sudden action to surprise an enemy.

**coup d'é·tat** (ko͞o' dā-tä') *n.* [Fr. : *coup,* stroke + *de,* of + *état,* state.] A sudden overthrow of a government in deliberate violation of constitutional forms by a group of persons in or previously in positions of authority.

**coup de thé·â·tre** (ko͞o' də tā-ä'trə) *n.* [Fr. : *coup,* stroke + *de,* of + *théâtre,* theatre.] A sudden, unexpected, and dramatic event, esp. one that reverses a given situation.

**coup d'oeil** (ko͞o dœ'yə) *n.* [Fr. : *coup,* stroke + *de,* of + *oeil,* eye.] A quick survey or glance.

**coupe¹** (ko͞op) *n.* [Fr., cup < LLat. *cuppa.*] **1. a.** A dessert of ice cream or fruit-flavored ice, garnished and served in a special dessert glass. **b.** The tall, narrow, usu. stemmed glass in which a coupe is served. **2.** A shallow, bowl-shaped dessert dish.

**coupe²** (ko͞op) *n. var. of* COUPÉ 2.

**cou·pé** (ko͞o-pā') *n.* [Fr. < p.part. of *couper,* to cut < *coup,* blow. —see COUP.] **1.** A closed four-wheel carriage with two seats inside and one outside. **2.** *also* **coupe** (ko͞op). A closed automobile with two doors.

**cou·ple** (kŭp'əl) *n.* [ME < OFr. < Lat. *copula,* bond.] **1.** Two items of the same kind : PAIR. **2.** Something that unites or connects two things together : LINK. **3.** *(sing. or pl. in number).* **a.** A man and woman united, as by marriage or betrothal. ***usage:*** When referring to a man and woman together, *couple* may be used with either a singular or a plural verb. Whatever the choice, usage should be consistent: *The newlywed couple is* (or *are*) *spending its* (or *their*) *honeymoon in Europe.* **b.** Two people together. **4.** A few : some <a *couple* of hours> **5.** *Physics.* A pair of forces of equal magnitude acting in parallel but opposite directions, capable of causing rotation but not translation. —*v.* **-pled, -pling, -ples.** —*vt.* **1.** To link together : CONNECT <*coupled* my excuse with an apology> **2. a.** To join as man and wife : MARRY. **b.** To join in sexual union. **3.** *Elect.* To link (two circuits or currents) as by magnetic induction. —*vi.* **1.** To form pairs : JOIN. **2.** To copulate. **3.** To unite chemically.

☆ ***syns:*** COUPLE, BRACE, DOUBLET, PAIR *n. core meaning* : two of the same kind together <a *couple* of songs> COUPLE also can mean two closely associated persons <a married *couple*> PAIR stresses the close association and often reciprocal dependence of things <a *pair* of gloves>; sometimes it means a single thing with two interdependent parts <a *pair* of scissors> BRACE and DOUBLET refer to two like things <a *brace* of pistols><a *doublet* of grouse shot on the moors>

**cou·pler** (kŭp'lər) *n.* **1.** One that couples. **2.** A device for coupling two railroad cars. **3.** A device connecting two organ keyboards so they may be played together.

**cou·plet** (kŭp'lĭt) *n.* [OFr., dim. of *couple,* couple.] **1.** A unit of verse made up of two successive lines, usu. rhyming and having the same meter. **2.** Two similar things : PAIR.

**cou·pling** (kŭp'lĭng) *n.* **1.** The act or process of forming couples. **2.** The act of copulating. **3.** Something that unites or connects, as a railroad coupler. **4.** The part of the body connecting the hindquarters and forequarters of a four-footed animal.

**cou·pon** (ko͞o'pŏn', kyo͞o'-) *n.* [Fr. < OFr. *colpon,* piece cut off < *colper,* to cut < *coup,* blow. —see COUP.] **1.** A negotiable certificate attached to a bond that represents a sum of interest due. **2. a.** A detachable part, as of a ticket or advertisement, entitling the bearer to specific benefits, as a gift or cash refund. **b.** A printed form, as in an advertisement, used for ordering merchandise or requesting information. **3.** A detachable slip calling for periodic payments, as for merchandise bought on an installment plan.

**cour·age** (kûr'ĭj, kŭr'-) *n.* [ME *corage* < OFr. < *cuer,* heart < Lat. *cor.*] The quality or state of mind or spirit enabling one to face danger or hardship with confidence and resolution : BRAVERY.

**cou·ra·geous** (kə-rā'jəs) *adj.* Having or marked by courage : BRAVE. **—cou·ra'geous·ly** *adv.* **—cou·ra'geous·ness** *n.*

**cou·rante** (ko͞o-ränt') *n.* [Fr. < fem. pr.part. of *courir,* to run < OFr. *courre* < Lat. *currere.*] **1.** A 17th cent. French dance in which running and gliding steps are performed to an accompaniment in triple time. **2.** The second movement of the classical suite, typically following the allemande.

**cour·gette** (ko͝or-zhĕt') *n.* [Dial. Fr., dim. of *courge,* gourd < OFr. < Lat. *cucurbita.*] *Chiefly Brit.* A zucchini.

**cou·ri·er** (ko͝or'ē-ər, kûr'-, kŭr'-) *n.* [OFr. *courrier* < OItal. *corriere* < *correre,* to run < Lat. *currere.*] **1.** A messenger, esp. one on official diplomatic business. **2.** A personal attendant hired to make arrangements for a journey.

**cour·lan** (ko͝or'lən) *n.* [Fr., alteration of *courliri* < Galibi *kurliri.* The limpkin.

**course** (kôrs, kōrs) *n.* [ME *cours* < OFr. < Lat. *cursus* < *currere,* t run.] **1.** Onward movement in a particular direction : PROGRES **2.** The direction of continuing movement <sailed a westwar *course*> **3.** The route or path taken by something, as a river, tha moves or flows. **4.** A designated section of land or water on which race is held or a sport played <a golf *course*> **5.** Movement in tim : DURATION <in the *course* of a week> **6.** A way of acting or beha ing <chose the wisest *course*> **7.** A typical or normal manner c proceeding : regular development. **8.** A systematic or orderly succe sion : SEQUENCE <a *course* of therapeutic treatments> **9.** A contin ous layer of building material, as brick or tile, on a roof or wall of structure. **10. a.** A body of prescribed studies constituting a curric lum and leading toward an advanced degree. **b.** A unit of such curriculum. **11.** A portion of a meal served as a unit at one tim **12.** The lowest sail on a mast of a square-rigged ship. **13.** A point o the compass, esp. the one toward which a ship is sailing. —v **coursed, cours·ing, cours·es.** —*vt.* **1.** To move rapidly through o over : TRAVERSE <ships *coursing* the open seas> **2. a.** To hur (game) with hounds. **b.** To set (hounds) to follow game. —*vi.* **1.** T follow a direction. **2. a.** To move rapidly : RACE. **b.** To run : flov <tears *coursing* down one's cheeks> **3.** To hunt game with hound **—in due course.** At the right or proper time. **—of course. 1.** In th natural order of things : NATURALLY. **2.** Having no doubt : CERTAINLY

**cours·er¹** (kôr'sər, kōr'-) *n.* A dog trained for coursing.

**cours·er²** (kôr'sər, kōr'-) *n.* A swift horse : CHARGER.

**cours·ing** (kôr'sĭng, kōr'-) *n.* The sport of hunting with dog trained to chase game by sight instead of scent.

**court** (kôrt, kōrt) *n.* [ME < OFr. *cort* < Lat. *cohors.*] **1.** A tract c open ground partly or entirely enclosed by walls or buildings : COUR YARD. **2.** A short street, esp. one having buildings on three sides. **3.** large, open section of a structure, often with a glass roof or skyligh **4.** A mansion or other large building standing in a courtyar **5. a.** The residence of a sovereign or dignitary <Hampton *Court* **b.** The retinue of a sovereign, including the royal family and pe sonal servants, advisers, and ministers. **c.** A sovereign's governin body, including the council of ministers and state advisers. **d.** formal meeting or reception presided over by a sovereign. **6. a.** person or group of persons whose task is to hear and submit a dec sion on cases at law. **b.** The building, hall, or room in which case are heard and decided. **c.** The regular session of a judicial assembl **d.** A similar authorized tribunal with military or ecclesiastical juri diction. **7.** An open, level area, marked with lines, upon which te nis, handball, basketball, etc., is played. **8.** The body of directors of corporation, company, or other organization. **9.** A legislative assen bly. —*v.* **court·ed, court·ing, courts.** —*vt.* **1.** To try to win th favor of by attention or flattery. **2.** To try to win the affections c love of : WOO. **3.** To attempt to gain : SEEK. **4.** To behave so as to invi or incur <*court* disaster> —*vi.* To woo. **—pay court to. 1.** T flatter solicitously in an attempt to obtain something or remove a tagonism. **2.** To woo.

**cour·te·ous** (kûr'tē-əs) *adj.* [ME *corteis* < OFr. < *cort,* court < La *cohors.*] **1.** Marked by graciousness and good manners : POLIT **2.** Marked by consideration toward and respect for others. **—cour te·ous·ly** *adv.* **—cour'te·ous·ness** *n.*

☆ ***syns:*** COURTEOUS, CIVIL, GENTEEL, MANNERLY, POLITE *ad core meaning* : characterized by good manners <a *courteous* esco who opened the door for me> ***ant:*** discourteous, impolite

**cour·te·san** (kôr'tĭ-zən, kōr'-) *n.* [OFr. *courtisane* < OItal. *cortig ana,* female courtier, fem. of *cortigiano,* courtier < *corte,* court < La *cohors.*] A prostitute, esp. one associating with high-ranking c wealthy men.

**cour·te·sy** (kûr'tĭ-sē) *n., pl.* **-sies.** [ME *courtesie* < OFr. < *cortei* courteous.] **1. a.** Polite behavior. **b.** A polite gesture or remar **2.** Agreement or consent in spite of fact : INDULGENCE <was calle "judge" by *courtesy*>

**courtesy title** *n. Chiefly Brit.* A title of nobility with no leg status.

**court·house** (kôrt'hous', kōrt'-) *n.* **1.** A building housing judicia courts. **2.** A building housing county government offices.

**court·i·er** (kôr'tē-ər, kōr'-, -tyər) *n.* [ME *courteour* < AN < OF *courteier,* to be at a royal court < *cort,* court < Lat. *cohors, cohort* courtyard.] **1.** An attendant at the court of a sovereign. **2.** One wh seeks favor, esp. by obsequious behavior or flattery.

**court·ly** (kôrt'lē, kōrt'-) *adj.* **-li·er, -li·est. 1.** Suitable for a roy court : STATELY. **2.** Elegant in manners : REFINED. **3.** Obsequious : fla tering. **—court'li·ness** *n.* **—court'ly** *adv.*

**courtly love** *n.* A code of chivalrous devotion to a married lad that was important in medieval and Renaissance literature.

**court-mar·tial** (kôrt'mär'shəl, kōrt'-) *n., pl.* **courts-mar·tia** **1.** A military or naval court of officers and occas. enlisted personn appointed by a commander to try offenders under military law. **2.**

trial by court-martial. —*vt.* **-tialed, -tial·ing, -tials** *also* **-tialled, -tial·ling, -tials.** To try by court-martial.
**court of appeals** *n.* A court to which appeals are made on points of law arising from the decision of a lower court.
**court of chancery** *n.* CHANCERY 1a.
**court of claims** *n.* A U.S. Federal court that determines claims against the United States.
**court of common pleas** *n.* **1.** A court with general jurisdiction in some U.S. states. **2.** A court in Great Britain that formerly heard civil cases.
**court of law** *n.* A court that hears and decides cases on the basis of statutes or common law.
**Court of St. James's** (sānt jāmz′, jām′zĭz) *n.* The British royal court.
**court plaster** *n.* [From its use by ladies at court to make beauty spots.] Adhesive plaster for covering scratches or cuts on the skin.
**court·room** (kôrt′ro͝om′, kōrt′-, -ro͞om′) *n.* A room in which the proceedings of a court of law are held.
**court·ship** (kôrt′shĭp′, kōrt′-) *n.* The act or time of courting.
**court tennis** *n.* A form of tennis played in a large indoor court with high cement walls off which the ball may be bounced.
**court·yard** (kôrt′yärd′, kōrt′-) *n.* An open space surrounded by buildings or walls, adjoining or within a castle or large building.
**cous·cous** (ko͞os′ko͞os′) *n.* [Fr. < Ar. *kouskous.*] A North African dish of crushed grain steamed and served with meats and vegetables.
**cous·in** (kŭz′ĭn) *n.* [ME *cosin,* a relative < OFr. < Lat. *consobrinus,* cousin : *com-,* together + *sobrinus,* cousin on the mother's side < *soror,* sister.] **1.** A child of one's aunt or uncle. **2.** A relative descended from a common ancestor, as a grandfather, by two or more steps in a diverging line. **3.** A relative by blood or marriage. **4.** A member of a kindred group or country <our Canadian *cousins*> **5.** —Used as a form of address by a sovereign to a nobleman. **—cous′in·ly** *adj.*
**cous·in-ger·man** (kŭz′ĭn-jûr′mən) *n., pl.* **cous·ins-ger·man** (kŭz′inz-). COUSIN 1.
**couth** (ko͞oth) *adj.* [Back-formation < UNCOUTH.] Suave : polished.
**cou·ture** (ko͞o-to͝or′, -tür′) *n.* [Fr., sewing < OFr. *cousture* < VLat. **consutura* < Lat. *consuere,* to sew together : *com-,* together + *suere,* to sew.] **1.** The business of a couturier. **2.** Dressmakers and fashion designers as a group.
**cou·tu·rier** (ko͞o-to͝or′ē-ər, -ē-ā′) *n.* [Fr. < OFr. *cousture,* sewing. —see COUTURE.] **1.** One who designs, makes, and sells fashionable, usu. custom-made women's clothing. **2.** A business establishment engaged in couture.
**cou·tu·rière** (ko͞o-to͝or′ē-ər, -ē-âr′) *n.* [Fr. < OFr. *cousturiere,* fem. of *cousture,* sewing. —see COUTURE.] A woman whose profession is that of a couturier.
**cou·vade** (ko͞o-väd′) *n.* [Fr. < OFr. < *couver,* to incubate < Lat. *cubare,* to lie down.] A practice among some primitive peoples in which the husband of a woman in labor goes to bed as if he were bearing the child.
**co·va·lence** (kō-vā′ləns) *n. Chem.* The number of electron pairs an atom can share with other atoms. **—co·va′len·cy** *n.* **—co·va′lent** *adj.* **—co·va′lent·ly** *adv.*
**covalent bond** *n.* A chemical bond formed by the sharing of one or more electrons, esp. pairs of electrons, between atoms.
**co·var·i·ance** (kō-vâr′ē-əns) *n.* **1.** *Physics.* The principle that the laws of physics have the same form regardless of the system of coordinates in which they are expressed. **2.** *Statistics.* The expected value of the product of the deviations of corresponding values of two variables from their respective means.
**co·var·i·ant** (kō-vâr′ē-ənt) *adj.* **1.** *Physics.* Expressing, displaying, or relating to covariance. **2.** *Math.* Varying with another variable quantity in a manner that leaves a specified relationship unchanged.
**cove**[1] (kōv) *n.* [ME, chamber, cave < OE *cofa.*] **1.** A small, sheltered bay in the shoreline of a sea, river, or lake. **2. a.** A small valley or recess in the side of a mountain. **b.** A cave or cavern. **3.** A narrow pass or gap between woods or hills. **4. a.** A concave molding. **b.** A curved surface forming a junction between a wall and a ceiling. —*vt.* **coved, cov·ing, coves.** To curve inward.
**cove**[2] (kōv) *n.* [Prob. < Romany *kova,* man.] *Chiefly Brit.* A fellow.
**co·vel·lite** (kō-vĕl′īt′, kō′və-līt′) *n.* [After Nicholas *Covelli* (1790–1829), its discoverer.] An indigo-blue mineral, CuS, an important source of copper.
**cov·en** (kŭv′ən, kō′vən) *n.* [Perh. < ME *covent,* assembly < OFr. *convent* < Lat. *conventus.* —see CONVENT.] An assembly of 13 witches.
**cov·e·nant** (kŭv′ə-nənt) *n.* [ME < OFr. < pr.part. of *convenir,* to agree. —see CONVENE.] **1.** A binding agreement made by two or more persons or parties : COMPACT. **2.** *Law.* **a.** A formal sealed contract or agreement. **b.** A suit to recover damages for violation of such a contract. —*v.* **-nant·ed, -nant·ing, -nants.** —*vt.* To promise by a covenant. —*vi.* To enter into a covenant : CONTRACT. **—cov′e·nant′al** (-năn′tl) *adj.* **—cov′e·nant′al·ly** *adv.*
**cov·e·nant·ee** (kŭv′ə-năn-tē′, -nən-) *n.* The participant in a covenant to whom the promise is made.
**cov·e·nant·er** (kŭv′ə-năn′tər) *n.* **1.** One who makes a covenant. **2. Covenanter.** A Scottish Presbyterian who supported either of the agreements intended to defend and extend Presbyterianism.
**cov·e·nan·tor** (kŭv′ə-năn′tər, -nən-, kŭv′ə-năn-tôr′) *n.* The party to a covenant by whom the obligation expressed in it is to be performed.
**cov·er** (kŭv′ər) *v.* **-ered, -er·ing, -ers.** [ME *coveren* < OFr. *covrir* < Lat. *cooperire,* to cover completely : *co(m)-* (intensive) + *operire,* to cover.] —*vt.* **1.** To place something on or over, so as to protect or conceal. **2.** To overlay or spread with something. **3. a.** To put a covering on. **b.** To wrap up : CLOTHE. **4.** To bring upon or invest (oneself) <*covered* themselves with glory> **5.** To serve as a covering for <Sheets *covered* all the furniture.> **6.** To extend over <a park *covering* several square miles> **7.** To copulate with (a female). —Used of animals, esp. horses. **8.** To sit on in order to hatch (eggs). **9.** To hide or screen from view or knowledge : CONCEAL <*covered* up their numerous defalcations> **10.** To protect or shield from harm, loss, or danger. **11.** To protect by insurance. **12.** To make up for or compensate. **13.** To be sufficient to defray, meet, or offset the cost or charge of <*cover* all outstanding checks> **14.** To allow for <ordinances *covering* stray dogs> **15.** To deal with <These books *cover* the subject.> **16.** To travel or pass over : TRAVERSE <*covered* 200 miles in one day> **17.** To have as one's territory <a sales representative who *covers* Georgia> **18.** To hold within the range and aim of a weapon, as a firearm. **19.** To protect, as from enemy attack, by occupying a strategic position. **20.** To be responsible for learning and reporting the details of (an event or situation) <*cover* the nominating convention> **21. a.** To be responsible for guarding the play of (an opponent). **b.** To be responsible for defending (a position) <*cover* left field> **22.** To match (an opponent's stake) in a wager. **23.** To play a higher-ranking card than (the one previously played). —*vi.* **1.** To spread over a surface to conceal or protect something. **2.** *Informal.* To act as a substitute or replacement during another's absence. **3.** To hide something in order to save one from censure or punishment <*cover* up for the absent workers> **4.** To play a higher card than the one previously played. —*n.* **1.** Something that covers or is laid, placed, or spread over or on something else. **2. a.** A position offering shelter or protection from attack. **b.** Strategic protection given by armed units during military action <a *cover* of aerial and naval bombardment> **3. a.** Vegetation covering an area, usu. to provide shade, prevent erosion, or control weeds. **b.** Underbrush or other vegetation serving as shelter for wild animals. **4.** Something that disguises or conceals. **5.** A table setting for one person. **6.** A cover charge. **7.** An envelope or wrapper for mail. **8.** Funds adequate to secure against loss or meet an obligation. **—cover (one's) tracks.** To conceal one's traces so as to elude pursuers. **—cover (the) ground. 1.** To traverse a given distance with satisfying speed. **2.** To accomplish an assignment or task thoroughly and efficiently. **—take cover.** To seek protection or concealment, as from enemy fire. **—cov′er·er** *n.*
**cov·er·age** (kŭv′ər-ĭj) *n.* **1.** The extent or degree to which something is observed, analyzed, and reported. **2.** Extent of protection afforded by an insurance policy. **3.** The amount of funds reserved to meet liabilities.
**cov·er·alls** (kŭv′ər-ôlz′) *pl.n.* A loose-fitting one-piece garment worn by workers to protect their clothes.
**cover charge** *n.* A fixed amount added to the bill, as at a nightclub, for entertainment or services.
**cover crop** *n.* A temporary crop, as winter rye or clover, planted to protect the soil from erosion in winter and provide nitrogen or humus when plowed under in the spring.
**covered bridge** *n.* A roofed bridge.
**covered wagon** *n.* A large wagon covered with an arched canvas top, used by American pioneers for prairie travel.
**cover girl** *n.* A model whose picture appears on magazine covers.
**cover glass** *n.* A small thin piece of glass for covering a specimen on a microscope slide.
**cov·er·ing** (kŭv′ər-ĭng) *n.* Something that covers.
**cov·er·let** (kŭv′ər-lĭt) *n.* A bedspread.
**†cov·er·lid** (kŭv′ər-lĭd′) *n. Regional.* A coverlet.
**co·ver·sine** (kō-vûr′sīn′) *n.* Versed cosine.
**cov·ert** (kŭv′ərt, kō′vərt, kō-vûrt′) *adj.* [ME < OFr. < *covrir,* to cover. —see COVER.] **1.** Covered or covered over : SHELTERED. **2.** Secret : clandestine. **3.** *Law.* Protected by a husband. —*n.* **1.** A covering or cover. **2.** A covered place or shelter. **3.** Thick underbrush or woodland providing cover for game. **4.** Covert cloth. **5.** *Zool.* One of the feathers covering the bases of the longer main feathers of a bird's wings or tail. **—cov′ert·ly** *adv.* **—cov′ert·ness** *n.*
**covert cloth** *n.* Twilled cloth of woolen or worsted yarn with cotton, silk, or rayon, used for garments.
**cov·er·ture** (kŭv′ər-chər, -cho͝or′) *n.* **1. a.** A covering : shelter. **b.** Disguise : concealment. **2.** *Law.* The legal status of a married woman.
**cov·er-up** *also* **cov·er·up** (kŭv′ər-ŭp′) *n.* An effort or strategy intended to conceal something, as a crime or scandal.

**cov·et** (kŭv′ĭt) *vt.* **-et·ed, -et·ing, -ets.** [ME *coveiten* < OFr. *coveitier* < *covitie,* desire < Lat. *cupiditas* < *cupidus,* desirous < *cupire,* to desire.] **1.** To feel envious desire for (that which is another's). **2.** To wish for excessively and longingly : CRAVE. **—cov′et·a·ble** *adj.* **—cov′et·er** *n.*

**cov·et·ous** (kŭv′ĭ-təs) *adj.* **1.** Enviously and culpably desirous of another's possessions. **2.** Marked by extreme desire to acquire <*cov-etous* of learning> **—cov′et·ous·ly** *adv.* **—cov′et·ous·ness** *n.*

**cov·ey** (kŭv′ē) *n., pl.* **-eys.** [ME < OFr. *covee,* brood < *cover,* to incubate < *cubare,* to lie down.] **1.** A small flock of birds, as partridges or grouse. **2.** A small group, as of persons.

**cov·ing** (kō′vĭng) *n.* COVE[1] 4.

**cow**[1] (kou) *n.* [ME *cou* < OE *cū.*] **1.** The mature female of cattle of the genus *Bos.* **2.** The mature female of other animals, as whales, elephants, or moose. **3.** A domesticated bovine. **4.** *Slang.* A fat, slovenly woman.

▲ word history: Everyone knows that *cow* and *beef* refer to the same animal or at least to the male and female of the same species, but it is not generally known that *cow* and *beef* are derived from the same word. This word is the Indo-European form *gwōus.* Indo-European is the hypothetical but unrecorded ancestor of most European and many Asian languages, including English, Latin, and French, the modern descendant of Latin. As individual dialects and languages developed from Indo-European, features preserved in one language were sometimes lost in another. The initial consonant sound *gw–* in *gwōus* is called a *labiovelar* sound because it is produced simultaneously with the lips (Latin *labia*) and the soft palate (*velum*). The English development of *gwōus* preserved the velar element (*g*) and lost the labial (*w*). After other changes peculiar to the Germanic languages, including the change of *g* to the voiceless velar sound *k* (spelled with *c* in Old English), the result in Old English was *cū,* now Modern English *cow.* The Latin development of *gwōus* preserved the labial element as *b* and lost the velar. After other changes peculiar to Latin, the result was *bos,* "ox, bull, cow." *Bos* had a stem form *bov-* that appeared in the oblique cases from which Old French *boef* is derived. The English word is *beef.*

**cow**[2] (kou) *vt.* **cowed, cow·ing, cows.** [Prob. of Scand. orig.] To frighten with threats or a show of force.

**cow·ard** (kou′ərd) *n.* [ME < OFr. *couard* < *coue,* tail < Lat. *cauda.*] One showing ignoble fear in the face of danger or pain.

**cow·ard·ice** (kou′ər-dĭs) *n.* Lack of courage or resoluteness.

**cow·ard·ly** (kou′ərd-lē) *adj.* Ignobly lacking in courage. **—cow′ard·li·ness** *n.* **—cow′ard·ly** *adv.*

☆ ***syns:*** COWARDLY, CHICKEN, CHICKEN-HEARTED, CRAVEN, DASTARDLY, FAINT-HEARTED, GUTLESS, LILY-LIVERED, PUSILLANIMOUS, YELLOW, YELLOW-BELLIED *adj. core meaning* : ignobly lacking in courage <*cowardly* deserters> ***ant:*** brave, courageous

**cow·bane** (kou′bān′) *n.* **1.** A plant, *Oxypolis rigidior* of the southeastern and central United States, with poisonous roots and foliage and small white flower clusters. **2.** A plant related to the cowbane, as the water hemlock.

**cow·bell** (kou′bĕl′) *n.* A bell hung from a collar around a cow's neck to aid in locating it.

**cow·ber·ry** (kou′bĕr′ē) *n.* **1.** A creeping evergreen shrub, *Vaccinium vitis-idaea,* bearing pink or reddish flowers and edible, slightly acid red berries. **2.** A berry of the cowberry.

**cow·bird** (kou′bûrd′) *n.* [From their habit of staying with cattle.] A blackbird of the genus *Molothrus* or related genera that lays its eggs in the nests of other birds, esp. the common North American species *M. ater.*

**cow·boy** (kou′boi′) *n.* A hired hand, esp. in the western United States, who tends cattle and performs many duties on horseback.

**cowboy hat** *n.* A ten-gallon hat.

**cow·catch·er** (kou′kăch′ər, -kĕch′-) *n.* The iron grille or frame projecting from the front of a locomotive or streetcar that serves to clear the track of obstructions.

**cow college** *n. Informal.* **1.** An agricultural college. **2.** A college or university considered unsophisticated and provincial.

**cow·er** (kou′ər) *vi.* **-ered, -er·ing, -ers.** [ME *couren,* of Scand. orig.] To cringe in fear.

**cow·fish** (kou′fĭsh′) *n., pl.* **cowfish** or **-fish·es. 1.** A small whale, porpoise, or a similar aquatic mammal, esp. a whale of the genus *Mesopledon,* having a pointed snout. **2.** A fish, *Lactophrys quadricornis* of warm Atlantic waters, with the body encased in a bony covering and hornlike spines over each eye.

**cow·girl** (kou′gûrl′) *n.* A hired woman, esp. in the western United States, who tends cattle and performs many duties on horseback.

**cow·hand** (kou′hănd′) *n.* A cowboy or cowgirl.

**cow·herb** (kou′ûrb′, -hûrb′) *n.* A European plant, *Saponaria vaccaria,* bearing deep-pink flower clusters.

**cow·herd** (kou′hûrd′) *n.* One who herds cattle.

**cow·hide** (kou′hīd′) *n.* **1. a.** The hide of a cow. **b.** Leather made from cowhide. **2.** A strong, heavy, flexible, usu. braided leather whip. —*vt.* **-hid·ed, -hid·ing, -hides.** To whip with a cowhide.

**cowl** (koul) *n.* [ME *coule* < OE *cugele* < LLat. *cuculla* < Lat. *cucullus,* hood.] **1. a.** A monk's hood or hooded robe. **b.** A draped neckline on a woman's garment. **2.** A hood-shaped covering for increasing the draft of a chimney. **3.** The top portion of the front part of an automotive vehicular body, supporting the dashboard and windshield. **4.** An aircraft engine cowling. —*vt.* **cowled, cowl·ing, cowls.** To cover with or as if with a cowl.

**cowled** (kould) *adj.* **1.** Wearing or supplied with a cowl : HOODED. **2.** Shaped like a cowl.

**cow·lick** (kou′lĭk′) *n.* [From its appearance of having been licked by a cow.] A projecting tuft of hair on the head that grows in a different direction from the rest of the hair and will not lie flat.

**cowl·ing** (kou′lĭng) *n.* A removable metal covering for an aircraft engine.

**cow·man** (kou′mən, -măn′) *n.* **1.** An owner of cattle or a cattle ranch. **2. a.** A cowboy. **b.** *Chiefly Brit.* A cowherd.

**co·work·er** (kō′wûr′kər) *n.* A fellow worker.

**cow parsnip** *n.* A tall, coarse plant of the genus *Heracleum,* esp. *H. lanatum* of North America.

**cow·pea** (kou′pē′) *n.* **1.** A tropical vine, *Vigna sinensis,* bearing long, hanging pods and cultivated in the southern United States for soil improvement and as animal feed. **2.** The edible pealike seed of the cowpea.

**Cow·per's gland** (kou′pərz, ko͞o′-) *n.* [After William *Cowper* (1666–1709).] Either of a pair of small compound racemose glands lying alongside and discharging into the male urethra.

**cow pilot** *n.* Pintano.

**cow·poke** (kou′pōk′) *n. Informal.* A cowboy.

**cow pony** *n.* A small, agile horse used in roundups.

**cow·pox** (kou′pŏks′) *n.* A contagious skin disease of cattle, caused by a virus that is isolated and used to vaccinate humans against smallpox.

**cow·punch·er** (kou′pŭn′chər) *n. Informal.* A cowboy.

**cow·ry** *also* **cow·rie** (kou′rē) *n., pl.* **-ries.** [Hindi *kaurī* < Skt. *kapardaḥ,* shell, of Dravidian orig.] Any of various tropical marine mollusks of the family Cypraeidae, with glossy, often brightly marked shells, some of which are used as money in the South Pacific and Africa.

**cow shark** *n.* Any of several sharks of the family Hexanchidae, indigenous to warm and temperate seas.

**cow·shed** (kou′shed′) *n.* A shed for housing cows.

**cow·slip** (kou′slĭp′) *n.* [ME *cowslyppe* < OE *cūslyppe* : *cū,* cow + *slyppe,* slime.] **1.** An Old World primrose, *Primula veris,* with fragrant yellow flowers. **2.** The marsh marigold.

**cow town** *n.* A small town usu. in a cattle-raising area.

**cox** (kŏks) *Informal.* —*n.* A coxswain. —*vi.* & *vt.* **coxed, cox·ing, cox·es.** To act as coxswain or serve as coxswain for.

**cox·a** (kŏk′sə) *n., pl.* **cox·ae** (kŏk′sē′) [Lat.] **1.** *Anat.* The hip or hip joint. **2.** *Zool.* The first segment of the leg of an insect or other arthropod, adjoining and attached to the body. **—cox′al** *adj.*

**cox·al·gi·a** (kŏk-săl′jē-ə, -jə) *n.* [COX(A) + -ALGIA.] Pain in or disease of the hip. **—cox·al′gic** (-jĭk) *adj.*

**cox·comb** (kŏks′kōm′) *n.* [ME *cokkes comb,* cock's comb.] **1.** *var. of* COCKSCOMB 4. **2.** *Obs.* A cap resembling a cockscomb, worn by a professional jester.

**cox·comb·ry** (kŏk-skōm′rē, -skəm-) *n., pl.* **-ries.** Behavior typical of or appropriate to a coxcomb.

**cox·i·tis** (kŏk-sī′tĭs) *n.* [COX(A) + -ITIS.] Inflammation of the hip joint.

**Cox·sack·ie virus** (ko͝ok-sä′kē, -săk′ē) *n.* [After *Coxsackie,* New York.] Any of a group of enteroviruses that produce a disease resembling poliomyelitis without paralysis.

**cox·swain** (kŏk′sən, -swān′) *n.* [ME *cokswaynne* : *cok,* cockboat + *swain,* servant.] One who steers a racing shell or boat or has charge of its crew. —*vi.* & *vt.* **-swained, -swain·ing, -swains.** To act as coxswain or serve as coxswain for.

**coy** (koi) *adj.* **-er, -est.** [ME < OFr. *coi* < Lat. *quietus.*] **1.** Shy and retiring. **2.** Artfully or affectedly demure. **3.** Annoyingly unwilling to make a commitment. **—coy′ly** *adv.* **—coy′ness** *n.*

**coy·dog** (kī′dôg′, -dŏg′, koi′-) *n.* [COY(OTE) + DOG.] A predatory animal, prob. a cross between a wolf and a western coyote, that now ranges widely throughout the United States.

**coydog**
*Approximately 2½ feet high at shoulder*

ă pat ā pay âr care ä father ĕ pet ē be hw which ĭ pit ī tie îr pier ŏ pot ō toe ô paw, for oi noise o͞o took

**coy·o·te** (kī-ō′tē, kī′ōt′) *n.* [Mex. Sp. < Nahuatl *cóyotl.*] A wolflike carnivorous animal, *Canis latrans,* common in western North America and ranging eastward into New England.

**coy·o·til·lo** (koi′ə-tĭl′ō, -tē′yō, kī′ə-) *n., pl.* **-los.** [Mex. Sp., dim. of *coyote,* coyote.] A poisonous shrub, *Karwinskia humboldtiana* of the southwestern United States and Mexico.

**coy·pu** (koi′po͞o) *n., pl.* **-pus.** [Sp. (South America) *coipú* < Araucanian *kóypu.*] **1.** A large beaverlike South American rodent, *Myocaster coypu.* **2.** NUTRIA 2.

**coz** (kŭz) *n. Informal.* Cousin.

**coz·en** (kŭz′ən) *v.* **-ened, -en·ing, -ens.** [Poss. < Ital. *cozzone,* horse trader < Lat. *coctio,* trader.] *—vt.* **1.** To deceive by means of a fraud or petty trick. **2.** To persuade or induce (someone) to do something by cajoling. **3.** To obtain by cozening. *—vi.* To deceive. **—coz′en·age** (kŭz′ə-nĭj) *n.* **—coz′en·er** *n.*

**co·zy** *also* **co·sy** (kō′zē) [Prob. of Scand. orig.] *—adj.* **-zi·er, -zi·est** *also* **-si·er, -si·est. 1.** Snug and comfortable : WARM. **2.** Characterized by friendly intimacy. **3.** *Informal.* Characterized by close association for devious purposes <a *cozy* little arrangement> *—vi.* **-zied, -zy·ing, -zies** *also* **-sied, -sy·ing, -sies.** *Informal.* To try to get on intimate or friendly terms <workers *cozying* up to the boss> *—n., pl.* **-zies** *also* **-sies.** A knitted or padded covering placed esp. over a teapot to keep the tea hot. **—co′zi·ly** *adv.* **—co′zi·ness** *n.*

**CQ** (sē′kyo͞o′) *n.* [C(ALL TO) Q(UARTERS).] Code letters used at the beginning of radio messages intended for all receivers.

**Cr** *symbol for* CHROMIUM.

**crab**[1] (krăb) *n.* [ME < OE *crabba.*] **1. a.** Any of various predominantly marine crustaceans of the section Brachyura within the order Decapoda, characterized by a broad, flattened cephalothorax covered by a hard carapace and with the small abdomen concealed beneath it and five pairs of legs, of which the anterior pair are large and pincerlike. **b.** A similar related crustacean, as the hermit crab or king crab. **2.** The horseshoe crab. **3. a.** The crab louse. **b. crabs.** Infestation by crab lice. **4. Crab.** CANCER 4. **5.** The maneuvering of an aircraft partially into a crosswind to compensate for drift. **6.** Any of various machines for handling or hoisting heavy weights. *—v.* **crabbed, crab·bing, crabs.** *—vi.* **1.** To catch or hunt crabs. **2.** To move sidewise or diagonally. **3.** To direct an aircraft into a crosswind. *—vt.* **1.** To direct (an aircraft) partially into a crosswind to eliminate drift. **2.** To scurry or move sideways. **—catch a crab.** To strike the water with an oar in recovering a stroke or to miss it in making one. **—crab′ber** *n.*

**crab**[2] (krăb) *n.* [ME.] **1.** The crab apple or its fruit. **2.** A quarrelsome, ill-tempered person. *—v.* **crabbed, crab·bing, crabs.** *—vi. Informal.* To find fault : CRITICIZE. *—vt.* **1.** *Informal.* To interfere with and ruin. **2.** *Informal.* To find fault with. **3.** To make ill-tempered. **—crab′ber** *n.*

**crab apple** *n.* **1.** Any of several trees of the genus *Pyrus,* with white, pink, or red flowers. **2.** The small, tart, edible fruit of the crab apple, used for making jelly.

**crab·bed** (krăb′ĭd) *adj.* [ME.] **1.** Irritable and perverse : ILL-TEMPERED. **2.** Difficult to understand. **3.** Difficult to read <*crabbed* handwriting> **—crab′bed·ly** *adv.* **—crab′bed·ness** *n.*

**crab·by** (krăb′ē) *adj.* **-bi·er, -bi·est.** Grouchy : ill-tempered. **—crab′bi·ly** *adv.* **—crab′bi·ness** *n.*

**crab cactus** *n.* Christmas cactus.

**crab·grass** (krăb′grăs′) *n.* A coarse grass of the genus *Digitaria,* which tends to spread and displace other grasses in lawns.

**crab louse** *n.* A body louse, *Phthirus pubis,* that gen. infests the pubic region and causes severe itching.

**crab·stick** (krăb′stĭk′) *n.* **1.** A stick made of crab-apple wood. **2.** CRAB[2] 2.

**crack** (krăk) *v.* **cracked, crack·ing, cracks.** [ME *craken* < OE *cracian.*] *—vi.* **1.** To break or snap apart. **2.** To make a sharp, snapping sound. **3.** To break without dividing into parts. **4.** To change sharply in timbre or pitch, as from emotion or hoarseness. —Used of the voice. **5.** To break down : GIVE OUT. **6.** To move or go rapidly <was *cracking* along at top speed> **7.** *Informal.* To have a physical or mental breakdown <finally *cracked* under the pressure> **8.** *Chem.* To decompose into simpler compounds. *—vt.* **1.** To cause to make a sharp, snapping sound. **2.** To cause to break or split partially or completely. **3.** To break with a sharp, snapping sound. **4.** To strike with a sudden, sharp sound. **5. a.** To break open or into <*crack* the bank's vault> **b.** To open up for consumption or use. **6.** To discover the solution to, esp. after much effort <*crack* the army's code> **7.** To cause (the voice) to crack. **8.** *Informal.* To tell (a joke). **9.** *Informal.* To cause to have a mental or physical breakdown. **10.** To reduce (petroleum) to simpler compounds by cracking. **—crack down.** To act more forcefully to regulate or restrain < *cracked down* on drunken drivers> **—crack up. 1.** To praise highly <not the great team they were *cracked* up to be> **2.** To damage or wreck <*crack up* a shiny new motorcycle> **3.** To have a physical or mental breakdown <*cracked up* under continual stress> **4.** To experience or cause to experience a great deal of amusement <Their jokes *cracked up* the audience.> *—n.* **1.** A sharp, snapping sound <the *crack* of rifles> **2.** A partial split or break : FISSURE. **3.** A slight, narrow space <a door that was opened a *crack*> **4.** A sharp, resounding blow. **5.** A mental or physical impairment : DEFECT. **6.** A cracking vocal tone or sound, as in hoarseness. **7.** An attempt : chance <get a *crack* at serving on the council> **8.** A witty or sarcastic remark. **9.** A moment : instant <at the *crack* of dawn> *—adj.* Excelling in skill or achievement : FIRST-RATE <a *crack* sharpshooter>

**crack·a·jack** (krăk′ə-jăk′) *adj. & n. Slang. var. of* CRACKERJACK.

**crack·brain** (krăk′brān′) *n.* A foolish or insane person. **—crack′-brained′** *adj.*

**crack·down** (krăk′doun′) *n.* An act or example of cracking down.

**cracked stem** *n.* A disease of the celery plant caused by a deficiency of boron and marked by cracking of the stalks.

**crack·er** (krăk′ər) *n.* **1.** A thin, crisp wafer or biscuit, usu. of unsweetened dough. **2.** A firecracker. **3.** A small cardboard cylinder covered with decorative paper and containing a favor or candy and a weak explosive that makes a sharp popping noise when a paper strip is pulled at one or both ends and torn. **4.** One that cracks.

**crack·er-bar·rel** (krăk′ər-băr′əl) *adj.* [So called because cracker barrels were often features of country stores where such discussions were held.] Resembling or typical of an extended informal discussion <*cracker-barrel* philosophy>

**crack·er·jack** (krăk′ər-jăk′) *also* **crack·a·jack** (krăk′ə-) [< CRACK (first-rate) + JACK.] *Slang. adj.* Of excellent quality or ability : FINE. **—crack′er·jack, crack′a·jack** *n.*

**Cracker Jack.** A trademark for a candied popcorn confection.

**crack·ers** (krăk′ərz) *adj. Chiefly Brit.* Insane.

**crack·ing** (krăk′ĭng) *n. Chem.* Thermal decomposition, occas. with catalysis, of a complex substance, esp. such decomposition of petroleum to extract low-boiling fractions such as gasoline. *—adj.* Extremely good. *—adv. Informal.* Extremely : very. —Used as an intensive.

**crack·le** (krăk′əl) *v.* **-led, -ling, -les.** [Freq. of CRACK.] *—vi.* **1.** To make a succession of slight sharp, snapping noises <a fire *crackling* in the underbrush> **2.** To show liveliness or brilliance <a book that *crackles* with wit> **3.** To become covered with a network of cracks. *—vt.* **1.** To crush (e.g., paper) with sharp, snapping sounds. **2.** To cause (e.g., china) to become covered with a network of fine cracks. *—n.* **1.** The act or sound of crackling. **2.** A network of fine cracks on the surface of glazed pottery, china, or glassware.

**crack·le·ware** (krăk′əl-wâr′) *n.* Ceramic ware made with a surface network of cracks.

**crack·ling** (krăk′lĭng) *n.* **1.** Production of a succession of slight sharp, snapping noises. **2. cracklings.** The crisp bits that remain after rendering fat from meat or roasting or frying the skin, esp. of a pig or a goose.

**crack·ly** (krăk′lē) *adj.* Likely to crackle : CRISP.

**crack·nel** (krăk′nəl) *n.* [ME *craknel.*] **1.** A crisp, hard biscuit. **2. cracknels.** Crisp bits of fried pork fat.

**crack·pot** (krăk′pŏt′) *n.* A bizarre or eccentric person.

**crack·up** (krăk′ŭp′) *n.* **1.** A wreck or collision, as of an aircraft or automotive vehicle. **2.** A physical or mental breakdown.

**-cracy** *suff.* [OFr. *-cratie* < LLat. *-cratia* < Gk. *-kratia* < *kratos,* strength, power.] Government : rule <meritoc*racy*>

**cra·dle** (krād′l) *n.* [ME *cradel* < OE.] **1.** A small low bed for an infant, often furnished with rockers. **2. a.** The earliest period of one's life <from the *cradle* to the grave> **b.** A place of origin : BIRTHPLACE. **3.** A framework of metal or wood used to support something, as a ship undergoing construction or repair. **4.** A framework for protecting an injured limb. **5.** The part of a telephone containing the connecting switch on which the receiver and mouthpiece unit is supported. **6. a.** A frame projecting above a scythe for catching grain as it is cut so that it can be laid flat. **b.** A scythe equipped with such a frame. **7.** A low flat framework on casters, used by a mechanic working beneath a vehicle. **8.** A boxlike device furnished with rockers, used for washing gold-bearing dirt. *—v.* **-dled, -dling, -dles.** *—vt.* **1.** To place or hold in or as if in a cradle. **2.** To care for in infancy. **3.** To reap (grain) with a cradle. **4.** To place or support (a ship) in a cradle. **5.** To wash (gold-bearing dirt) in a cradle. *—vi. Obs.* To lie in or as if in a cradle. **—cra′dler** *n.*

**cradle cap** *n.* Dermatitis occurring in infants and characterized by heavy yellow crusted scalp lesions.

**cra·dle·song** (krād′l-sông′, -sŏng′) *n.* A lullaby.

**craft** (krăft) *n.* [ME < OE *cræft.*] **1.** Skill or ability, esp. in handwork or the arts. **2.** Evasive or deceptive skill : GUILE. **3. a.** An occupation, esp. one requiring manual dexterity. **b.** The membership of such an occupation or trade : GUILD. **4.** *pl.* **craft.** A boat, ship, or aircraft. *—vt.* **craft·ed, craft·ing, crafts.** To make by or as if by hand.

**crafts·man** (krăfts′mən) *n.* A skilled worker who practices a craft. **—crafts′man·ly** *adj.* **—crafts′man·ship′** *n.*

**crafts·wom·an** (krăfts′wo͝om′ən) *n.* A woman who is skilled in or practices a craft.

**craft union** *n.* A labor union limited in membership to workers engaged in the same craft.

**craft·y** (krăf′tē) *adj.* **-i·er, -i·est. 1.** Skillfully underhand and deceptive : SHREWD. **2.** *Archaic.* Ingenious : skillful. **—craft′i·ly** *adv.* **—craft′i·ness** *n.*
**crag** (krăg) *n.* [ME, of Celt. orig.] A steeply projecting mass of rock forming part of a rugged cliff or headland. **—crag′ged** (krăg′ĭd) *adj.*
**crag·gy** (krăg′ē) *adj.* **-gi·er, -gi·est. 1.** Having crags. **2.** Rugged and uneven. **—crag′gi·ly** *adv.* **—crag′gi·ness** *n.*
**crake** (krāk) *n.* [ME, crow, prob. < ON *krāka.*] A bird of the family Rallidae, as the corncrake or a marsh bird of the genus *Porzana.*
**cram** (krăm) *v.* **crammed, cram·ming, crams.** [ME *crammen* < OE *crammian.*] —*vt.* **1.** To force, press, or squeeze into an insufficient space : STUFF. **2.** To fill too tightly. **3.** To gorge with food. **4.** *Informal.* To prepare hastily for an examination. —*vi.* **1.** To gorge oneself with food. **2.** *Informal.* To prepare hastily for an examination. —*n.* **1.** A group crammed together : CRUSH. **2.** *Informal.* Concentrated, hasty study for an examination. **—cram′mer** *n.*
**cram·bo** (krăm′bō) *n., pl.* **-bos.** [Obs. *crambe,* cabbage < Lat. < Gk. *krambē.*] **1.** A word game in which a player or team must find and express a rhyme for a word or line presented by the opposing player or team. **2.** Doggerel.
**cramp**[1] (krămp) *n.* [ME *crampe* < OFr., of Germanic orig.] **1.** Sudden involuntary muscular contraction causing severe pain, often occurring in the shoulder or leg as the result of chill or strain. **2.** Temporary partial paralysis of habitually or excessively used muscles <a swimmer's *cramp*> **3. cramps. a.** Sharp, persistent pains in the abdomen. **b.** DYSMENORRHEA. —*vt.* **cramped, cramp·ing, cramps.** To affect with or as if with a cramp.
**cramp**[2] (krămp) *n.* [MDu. crampe, hook.] **1.** A bar, usu. of iron, with right-angle bends at both ends, used for permanently holding together building materials, as stones or timber. **2.** A frame with an adjustable part to hold pieces together : CLAMP. **3.** Something that restrains or compresses. **4.** A confined position or part. —*vt.* **cramped, cramp·ing, cramps. 1.** To hold together with a cramp. **2.** To shut in so closely as to restrict the physical freedom of <was *cramped* in a narrow dungeon> **3.** To restrict from free action or expression : HAMPER <*cramped* my style> **4. a.** To steer (the wheels of a vehicle) to make a turn. **b.** To jam (a wheel) by a short turn.
**cramp·fish** (krămp′fĭsh′) *n., pl.* **crampfish** *or* **-fish·es.** [< CRAMP[1], from its ability to give electric shocks.] The electric ray.
**cram·pon** (krăm′pŏn′, -pən) *n.* [OFr., of Germanic orig.] **1.** *often* **crampons.** A hinged pair of curved iron bars for raising heavy objects, as timber. **2.** *often* **crampons.** An iron spike attached to the shoe to prevent slipping when climbing or walking on ice.
**cran·ber·ry** (krăn′bĕr′ē) *n.* [Partial transl. of LG *kraanbere* : *kraan,* crane (< MLG *kran*) + *-bere,* berry.] **1. a.** A slender, trailing North American shrub, *Vaccinium macrocarpon,* growing in damp ground and bearing tart red berries. **b.** The edible berry of this plant, often made into sauce or jelly. **2.** A plant similar or related to the cranberry, esp. the European species *V. oxycoccous.*
**cranberry bush** *n.* The high-bush cranberry.
**cranberry tree** *n.* The guelder rose.
**crane** (krān) *n.* [ME < OE *cran.*] **1.** Any of various large wading birds of the family Gruidae, with a long neck, long legs, and a long bill. **2.** A bird similar to a crane, as a heron. **3.** A machine for hoisting and moving heavy objects by cables attached to a movable boom. **4.** A device with a swinging arm, as one in a fireplace for suspending a pot. —*v.* **craned, cran·ing, cranes.** —*vt.* **1.** To hoist or move with or as if with a crane. **2.** To strain and stretch (the neck). —*vi.* To stretch one's neck for a better view.
**crane fly** *n.* Any of numerous long-legged, slender-bodied flies of the family Tipulidae, resembling a large mosquito.
**cranes·bill** (krānz′bĭl′) *n.* GERANIUM 1.
**crani-** *pref. var. of* CRANIO-.
**cra·ni·a** (krā′nē-ə) *n. var. pl. of* CRANIUM.
**cra·ni·al** (krā′nē-əl) *adj.* [< CRANIUM.] Of or relating to the skull. **—cra′ni·al·ly** *adv.*
**cranial index** *n.* The ratio of the maximum width to the maximum length of the cranium, multiplied by 100.
**cranial nerve** *n.* Any of several nerves that arise in pairs from the brainstem and reach the periphery through openings in the skull.
**cra·ni·ate** (krā′nē-ĭt, -āt′) *adj.* Having a skull. —*n.* An animal having a skull : VERTEBRATE.
**cra·ni·ec·to·my** (krā′nē-ĕk′tə-mē) *n., pl.* **-mies.** Surgical removal of part of the cranium.
**cranio-** *or* **crani-** *pref.* [< CRANIUM.] Cranium <*craniometer*>
**cra·ni·o·cer·e·bral** (krā′nē-ō-sĕr′ə-brəl, -sə-rē′brəl) *adj.* Of or relating to the cranium and the brain.
**cra·ni·ol·o·gy** (krā′nē-ŏl′ə-jē) *n.* Scientific study of the characteristics of the skull, esp. in humans. **—cra′ni·o·log′i·cal** (-ə-lŏj′ĭ-kəl) *adj.* **—cra′ni·o·log′i·cal·ly** *adv.* **—cra′ni·ol′o·gist** *n.*
**cra·ni·om·e·ter** (krā′nē-ŏm′ĭ-tər) *n.* An instrument for measuring skulls. **—cra′ni·o·met′ric** (-ə-mĕt′rĭk), **cra′ni·o·met′ri·cal** *adj.* **—cra′ni·om′e·try** *n.*
**cra·ni·o·sac·ral system** (krā′nē-ō-săk′rəl, -sā′krəl) *n.* The parasympathetic nervous system.
**cra·ni·ot·o·my** (krā′nē-ŏt′ə-mē) *n., pl.* **-mies. 1.** Surgical cutting or removal of part of the skull. **2.** Cutting or breaking of the fetal skull to reduce its size when normal delivery is impossible.
**cra·ni·um** (krā′nē-əm) *n., pl.* **-ni·ums** *or* **-ni·a** (-nē-ə) [Med. Lat. < Gk. *kranion.*] **1.** The skull of a vertebrate. **2.** The part of the skull enclosing the brain.
**crank**[1] (krăngk) *n.* [ME < OE *cranc,* as in *crancstæf,* weaving implement.] **1.** A device for transmitting rotary motion, having of a handle or arm attached to a shaft at right angles. **2.** A verbal conceit. **3.** A peculiar idea or act. **4.** *Informal.* **a.** An ill-tempered person. **b.** An eccentric. —*v.* **cranked, crank·ing, cranks.** —*vt.* **1.** To start or operate (e.g., an engine) by turning a crank. **2.** To make into the shape of a crank : BEND. **3.** To provide with a crank. —*vi.* **1.** To turn a crank. **2.** To twist : wind. **—crank out.** To produce, esp. mechanically and rapidly <*cranks out* paperback love stories> **—crank up.** To cause to start or to get started as if by turning a crank <*cranking up* a big ad campaign>
**crank**[2] (krăngk) *adj.* [Orig. unknown.] *Naut.* Liable to capsize : UNSTABLE.
†**crank**[3] (krăngk) *adj.* [ME *cranke.*] *Regional.* **1.** Lively : spirited. **2.** Overconfident.
**crank·case** (krăngk′kās′) *n.* The metal case enclosing the crankshaft and associated parts in a reciprocating engine.
**crank·pin** *also* **crank pin** (krăngk′pĭn′) *n.* A bar or cylinder in the arm of a crank to which a reciprocating member or connecting rod is attached.
**crank·shaft** (krăngk′shăft′) *n.* A shaft that turns or is turned by a crank.
**crank·y**[1] (krăng′kē) *adj.* **-i·er, -i·est. 1.** Peevish : ill-tempered. **2.** Eccentric : odd. **3.** Full of bends and turns : CROOKED. **—crank′i·ly** *adv.* **—crank′i·ness** *n.*
**crank·y**[2] (krăng′kē) *adj.* **-i·er, -i·est.** *Naut.* Liable to capsize.
**cran·ny** (krăn′ē) *n., pl.* **-nies.** [ME *crani* < OFr. *cran,* notch.] A small opening, as in a wall or rock face : CREVICE. **—cran′nied** *adj.*
**crap** (krăp) *n.* [Back-formation < CRAPS.] A losing throw in the game of craps. —*v.* **crapped, crap·ping, craps.** To throw a crap. **—crap out.** To make a losing throw in the game of craps.
**crape** (krāp) *n.* **1.** CREPE 1. **2.** A black band worn on the hat or sleeve as a sign of mourning. —*vt.* **craped, crap·ing, crapes.** To cover or drape with or as if with crape.
**crape·hang·er** (krāp′hăng′gər) *n.* A gloomy or pessimistic person.
**crape jasmine** *n.* [From the crinkled lobes of the corolla.] A fragrant shrub, *Tabernaemontana coronaria* of India, cultivated in warm regions for its white flowers.
**crape myrtle** *also* **crepe myrtle** *n.* An Oriental shrub, *Lagerstroemia indica,* widely cultivated in warm climates for its pink, red, or white flowers.
**crap·pie** (krăp′ē) *n., pl.* **-pies.** [Canadian Fr. *crapet.*] An edible North American freshwater fish, *Pomoxis nigromaculatus,* the black crappie, or *P. annularis,* the white crappie, related to the sunfish.
**craps** (krăps) *pl.n.* [Louisiana Fr. < Fr. *crabs* < E. *crabs,* the lowest throw at hazard.] *(sing. or pl. in number).* A gambling game played with a pair of dice.
**crap·shoot·er** (krăp′sho͞o′tər) *n.* One who plays craps.
**crap·u·lence** (krăp′yə-ləns) *n.* [< LLat. *crapulentus,* very drunk < Lat. *crapula,* intoxication < Gk. *kraipalē.*] **1.** Sickness caused by excessive eating or drinking. **2.** Excessive indulgence : INTEMPERANCE. **—crap′u·lent** *adj.* **—crap′u·lous** *adj.*
**crash**[1] (krăsh) *v.* **crashed, crash·ing, crash·es.** [ME *crasschen.*] —*vi.* **1.** To fall or collide noisily : SMASH. **2.** To undergo sudden damage or destruction on impact. **3.** To make a sudden loud noise. **4.** To move noisily or so as to cause damage. **5.** To fail suddenly <The stock market *crashed* in 1929.> **6.** *Slang.* To return to a normal mental state after an experience caused by the use esp. of a mind-altering drug. **7.** *Slang.* **a.** To find usu. temporary lodging or shelter. **b.** To go to sleep. —*vt.* **1.** To cause to crash. **2.** To dash to pieces : SMASH. **3.** *Informal.* To join or enter without invitation. —*n.* **1.** A sudden loud noise. **2.** A wrecking : collision. **3.** A sudden business or economic failure. —*adj. Informal.* Of or marked by an intensive effort to produce or accomplish something <a *crash* diet> **—crash′er** *n.*
**crash**[2] (krăsh) *n.* [R. *krashenina,* colored linen < *krashenie,* coloring < *krasit′,* to color < *krasa,* beauty.] **1.** Coarse, light, unevenly woven linen or cotton fabric used for towels and curtains. **2.** Starched reinforced fabric for strengthening a book binding or the spine of a bound book.
**crash dive** *n.* A rapid dive by a submarine, esp. in emergency.
**crash helmet** *n.* A padded helmet worn esp. by motorcyclists and aviators to protect the head in case of accident.
**crash·ing** (krăsh′ĭng) *adj. Informal.* Total : absolute <a *crashing* idiot>
**crash-land** (krăsh′lănd′) *v.* **-land·ed, -land·ing, -lands.** —*vt.* To land (an aircraft) in an emergency, usu. causing damage to the aircraft. —*vi.* To crash-land an aircraft in an emergency.
**crash pad** *n.* **1.** Padding inside automobiles, tanks, or other vehicles to protect occupants in the event of an accident, sudden stop, or other quick movement. **2.** *Slang.* A place affording free, usu. temporary lodging.

**crash truck** *also* **crash wagon** *n.* A truck designed and equipped to rescue victims of an air crash.
**crash·wor·thy** (krăsh'wûr'thē) *adj.* Capable of withstanding the effects of a crash <*crashworthy* trucks> **—crash'wor'thi·ness** *n.*
**crass** (krăs) *adj.* **-er, -est.** [Lat. *crassus,* dense.] So crude and unrefined as to be lacking in discrimination and sensibility : COARSE. **—crass'i·tude'** (-ĭ-tōōd', -tyōōd'), **crass'ness** *n.* **—crass'ly** *adv.*
**-crat** *suff.* [Fr. *-crate* < Gk. *-kratēs* < *-kratia,* -cracy.] A participant in or supporter of a specified form of government <technocrat>
**crate** (krāt) *n.* [Lat. *cratis,* wickerwork.] **1.** A container, as a slatted wooden case, for storing or shipping. **2.** *Slang.* A rickety old vehicle, as a car or aircraft. *—vt.* **crat·ed, crat·ing, crates.** To pack into a crate.
**cra·ter** (krā'tər) *n.* [Lat. < Gk. *kratēr,* mixing vessel < *kerannunai,* to mix.] **1.** A bowl-shaped depression at the mouth of a volcano or geyser. **2.** A pit or hole in the ground created as by an explosion or the impact of a meteorite.
**Cra·ter** (krā'tər) *n.* A constellation in the Southern Hemisphere.
**cra·vat** (krə-văt') *n.* [Fr. *cravate,* cravat, necktie worn by Croatian mercenaries in the service of France < *Cravate,* a Croatian < Serbo-Croatian *Hrvat.*] A band or scarf of fabric worn as a necktie.
**crave** (krāv) *v.* **craved, crav·ing, craves.** [ME *craven* < OE *crafian,* to beg.] *—vt.* **1.** To have an intense desire for. **2.** To need urgently : REQUIRE. **3.** To beg earnestly for : IMPLORE. *—vi.* To have an eager or intense desire. **—crav'er** *n.* **—crav'ing·ly** *adv.*
**cra·ven** (krā'vən) *adj.* [ME *cravant.*] Marked by abject fear : COWARDLY. *—n.* A coward. **—cra'ven·ly** *adv.* **—cra'ven·ness** *n.*
**crav·ing** (krā'vĭng) *n.* A consuming desire : YEARNING.
**craw** (krô) *n.* [ME *crawe.*] **1.** A bird's crop. **2.** An animal's stomach.
**craw·fish** (krô'fĭsh') *vi.* **-fished, -fish·ing, -fish·es.** *Informal.* To withdraw from an undertaking. *—n. var. of* CRAYFISH.
**crawl**[1] (krôl) *vi.* **crawled, crawl·ing, crawls.** [ME *craulen* < ON *krafla.*] **1.** To move slowly on the hands and knees or by dragging the body along the ground : CREEP. **2.** To advance slowly, feebly, or laboriously <*crawled* up the hill under heavy enemy fire> **3.** To proceed or act servilely. **4.** To be or feel as if covered with crawling things <skin *crawling* in revulsion> **5.** To swim the crawl. *—n.* **1.** The act of crawling. **2.** A very slow pace. **3.** A rapid swimming stroke consisting of alternating overarm strokes and a flutter kick. **—crawl'er** *n.* **—crawl'ing·ly** *adv.*
☆ ***syns:*** CRAWL, CREEP, SLIDE, SNAKE, WORM *v. core meaning:* to move along in a crouching or prone position <sappers *crawling* through enemy lines at night>
**crawl**[2] (krôl) *n.* [Afr. *kraal,* enclosure for animals. —see KRAAL.] A pen in shallow water, as for confining turtles or fish.
**crawl·space** (krôl'spās) *n.* A low, narrow space, as in the walls of a building, providing workers access to plumbing, wiring, and heating systems.
**crawl·y** (krô'lē) *adj.* **-i·er, -i·est.** *Informal.* **1.** Creepy. **2.** Feeling as if covered with crawling things.
**cray·fish** (krā'fĭsh') *also* **craw·fish** (krô'-) *n., pl.* **crayfish** *or* **-fish·es** *also* **crawfish** *or* **-fish·es.** [By folk etymology < ME *crevis* < OFr. *crevise,* of Germanic orig.] **1.** A freshwater crustacean of the genera *Cambarus* or *Astacus,* resembling a lobster but considerably smaller. **2.** A spiny lobster.
**cray·on** (krā'ŏn', -ən) *n.* [Fr. < *craie,* chalk < Lat. *creta* < *Creta (terra),* Cretan (earth).] **1.** A stick of colored wax, chalk, or charcoal used for drawing. **2.** A drawing made with crayons. *—vt.* **-oned, -on·ing, -ons.** To draw, color, or decorate with crayons. **—cray'on·ist** (-ə-nĭst) *n.*
**craze** (krāz) *v.* **crazed, craz·ing, craz·es.** [ME *crasen,* of Scand. orig.] *—vt.* **1.** To make insane. **2.** To produce a network of fine cracks in the surface or glaze of. *—vi.* **1.** To go insane. **2.** To become covered with fine cracks. *—n.* **1.** A short-lived popular fashion : FAD. **2.** A fine crack.
**cra·zy** (krā'zē) *adj.* **-zi·er, -zi·est.** **1.** Mad : insane. **2.** *Informal.* Departing from proportion or moderation, esp.: **a.** Consumed by enthusiasm or excitement. **b.** Immoderately fond : INFATUATED. **c.** Intensely involved or preoccupied. **d.** Not sensible : IMPRACTICAL. **3.** *Archaic.* Rickety. *—n., pl.* **-zies.** One who is or appears to be crazy. **—like crazy.** *Informal.* To an exceeding degree <dashing in and out of the house *like crazy*> **—cra'zi·ly** *adv.* **—cra'zi·ness** *n.*
**crazy bone** *n. Informal.* The olecranon.
**crazy quilt** *n.* **1.** A patchwork quilt of pieces of cloth of various shapes, colors, and sizes, arranged in random order. **2.** A disorderly mixture : HODGEPODGE.
**cra·zy·weed** (krā'zē-wēd') *n.* [From its toxic effect on some animals.] Locoweed.
**C-re·ac·tive protein** (sē'rē-ăk'tĭv) *n.* [C(ARBOHYDRATE POLYSACCHARIDE) + REACTIVE.] Globulin occurring in the blood in certain acute illnesses, as rheumatic fever.
**creak** (krēk) *vi.* **creaked, creak·ing, creaks.** [ME *creken.*] **1.** To make a squeaking or grating sound. **2.** To move with a creaking sound. **—creak** *n.* **—creak'ing·ly** *adv.*

ōō **boot** ou **out** th **thin** *th* **this** ŭ **cut** ûr **urge** y **young**
yōō ab**use** zh vi**sion** ə **a**bout, it**e**m, ed**i**ble, gall**o**p, circ**u**s

**creak·y** (krē'kē) *adj.* **-i·er, -i·est.** **1.** Tending or liable to creak. **2.** Decrepit : dilapidated. **—creak'i·ly** *adv.* **—creak'i·ness** *n.*
**cream** (krēm) *n.* [ME *creme* < OFr. *craime* < LLat. *cramum.*] **1.** The fatty component of unhomogenized milk that tends to accumulate at the surface. **2.** A pale yellow to yellowish white. **3.** Any of various substances resembling or containing cream, as certain foods or cosmetics. **4.** The choicest part <the *cream* of New Orleans society> *—v.* **creamed, cream·ing, creams.** *—vi.* **1.** To form cream. **2.** To form foam or froth at the top. *—vt.* **1.** To remove the cream from : SKIM. **2.** To select or remove the best part from. **3.** To beat into a creamy consistency. **4.** To prepare or cook in or with a cream sauce. **5.** To add cream to. **6.** *Slang.* To defeat overwhelmingly <The visiting team *creamed* us.> **—cream** *adj.*
**cream cheese** *n.* A soft white cheese made of milk and cream.
**cream·cups** (krēm'kŭps') *n.* *(sing. or pl. in number).* A plant, *Platystemon californicus* of the southwestern United States, bearing long-stalked cream-colored or light-yellow flowers.

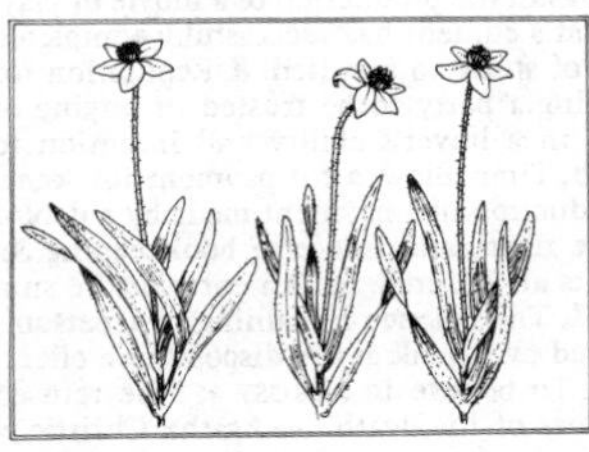
**creamcups**

**cream·er** (krē'mər) *n.* **1.** A small pitcher or jug for cream. **2.** A device for separating cream from milk. **3.** A refrigerator in which milk is placed to form cream.
**cream·er·y** (krē'mə-rē) *n., pl.* **-ies.** A place where dairy products are prepared or sold.
**cream of tartar** *n.* Potassium bitartrate.
**cream puff** *n.* **1.** A shell of light pastry filled with whipped cream, custard, etc. **2.** *Slang.* A sissy.
**cream sauce** *n.* White sauce made by cooking together a flour and butter mixture with cream or milk.
**cream·y** (krē'mē) *adj.* **-i·er, -i·est.** Rich in or like cream. **—cream'i·ly** *adv.* **—cream'i·ness** *n.*
**crease** (krēs) *n.* [Obs. *creast* < ME *crest,* ridge. —see CREST.] **1.** A line made by folding, pressing, or wrinkling. **2.** One of the lines marking off the positions of the bowler and batsman in cricket or the space between two of these lines. **3.** A rectangle marked off in front of the goalkeeper's cage in hockey. *—v.* **creased, creas·ing, creas·es.** *—vt.* **1.** To make a crease in. **2.** To wound superficially with a bullet. *—vi.* To become wrinkled. **—crease'less** *adj.* **—creas'er** *n.* **—creas'y** *adj.*
**cre·ate** (krē-āt') *vt.* **-at·ed, -at·ing, -ates.** [ME *createn* < Lat. *creare.*] **1.** To bring into being. **2.** To give rise to : PRODUCE <The remark *created* great interest.> **3.** To invest with an office or title : APPOINT. **4.** To produce through artistic or imaginative effort <*create* a new ballet> *—adj. Archaic.* Created.
**cre·a·tine** (krē'ə-tēn', -tĭn) *also* **cre·a·tin** (-tĭn) *n.* [Gk. *kreas, kreat-,* flesh + -INE.] A nitrogenous organic acid, $C_4H_9N_3O_2$, found in the muscle tissue of many vertebrates.
**creatine phosphate** *n.* Phosphocreatine.
**cre·at·i·nine** (krē-ăt'n-ēn', -ĭn) *n.* [CREATIN(E) + -INE.] The creatine anhydride $C_4H_7N_3O$, a normal metabolic waste.
**cre·a·tion** (krē-ā'shən) *n.* **1. a.** The act of creating or process of being created. **b.** The fact of having been created. **2.** The act of investing with a title or office. **3. a.** The world and everything in it. **b.** All creatures or a class of creatures. **4.** An original product of human invention or artistic imagination. **5.** A new, usu. specially designed garment. **—cre·a'tion·al** *adj.*
**cre·a·tive** (krē-ā'tĭv) *adj.* **1.** Having the power or ability to create. **2.** Productive. **3.** Marked by originality : IMAGINATIVE. **—cre·a'tive·ly** *adv.* **—cre'a·tiv'i·ty** (-ĭ-tē), **cre·a'tive·ness** *n.*
**cre·a·tor** (krē-ā'tər) *n.* **1.** One that creates. **2. Creator.** GOD 1.
**crea·ture** (krē'chər) *n.* **1.** Something created. **2.** A living being, esp. an animal. **3.** A human being. **4.** One dependent on or subservient to another : TOOL. **—crea'tur·al** *adj.* **—crea'ture·ly** *adj.*
**creature comfort** *n.* Something adding to physical comfort.
**crèche** (krěsh) *n.* [Fr. < OFr. crib, of Germanic orig.] **1.** A usu. three-dimensional representation of the Nativity. **2.** A foundling hospital. **3.** *Chiefly Brit.* A day nursery.
**cre·dence** (krēd'ns) *n.* [ME < OFr. < Med. Lat. *credentia* < Lat. *credere,* to believe.] **1.** Acceptance as true : BELIEF. **2.** Claim to acceptance : TRUSTWORTHINESS. **3.** Credential : recommendation <a letter of *credence*> **4.** A small table used for holding the elements of the Eucharist.
**cre·den·tial** (krĭ-dĕn'shəl) *n.* [< Med. Lat. *credentialis,* giving authority < *credentia,* trust < Lat. *credere,* to believe.] **1.** A factor enti-

tling one to confidence, credit, or authority. **2. credentials.** Evidence attesting one's right to credit, confidence, or authority.

**cre·den·za** (krĭ-dĕn′zə) *n.* [Ital. < Med. Lat. *credentia*, trust, from the practice of placing food and drink on a sideboard to be tasted by a servant before being served to ensure they contained no poison.] A cabinet, buffet, or sideboard, esp. one without legs.

**credibility gap** *n.* **1.** Public skepticism about the truth of official claims and pronouncements. **2.** Lack of trustworthiness. **3.** A disparity or discrepancy.

**cred·i·ble** (krĕd′ə-bəl) *adj.* [ME < Lat. *credibilis* < *credere*, to believe.] **1.** Capable of being believed : PLAUSIBLE. **2.** Deserving confidence. **—cred′i·bil′i·ty** *n.* **—cred′i·ble·ness** *n.* **—cred′i·bly** *adv.*

**cred·it** (krĕd′ĭt) *n.* [OFr. < OItal. *credito* < Lat. *creditum*, loan < *credere*, to entrust.] **1.** Belief or confidence in the truth of something : TRUST. **2.** A reputation for sound character or quality : STANDING. **3.** A source of honor or distinction <a *credit* to one's college> **4.** Approval for an act, ability, or quality : PRAISE. **5.** Influence based on the confidence or good opinion of others. **6.** *often* **credits.** Acknowledgment of work done, as in the production of a movie or play. **7. a.** Official certification that a student has successfully completed a course of study. **b.** A unit of study so certified. **8.** Reputation for solvency and integrity entitling a party to be trusted in buying or borrowing. **9. a.** Confidence in a buyer's ability and intention to fulfill financial obligations. **b.** Time allowed for payment for something sold on trust. **10. a.** Deduction of a payment made by a debtor from an amount due. **b.** The right-hand side of a bookkeeping account on which such amounts are entered. **c.** An entry or the sum of the entries on this side. **11.** The balance remaining in a person's account. **12.** An amount placed by a bank at the disposal of a client. —*vt.* **-it·ed, -it·ing, -its. 1.** To believe in : TRUST <"she refused steadfastly to *credit* the reports of his death" —Agatha Christie> **2.** *Archaic.* To bring distinction or honor to. **3. a.** To give credit to for something <*credited* me with the invention> **b.** To ascribe to a person <*credit* the winning streak to the new manager> **4. a.** To enter as a credit <*credited* the entire sum to my account> **b.** To make a credit entry in <*credit* an account> **5.** To award or give an educational credit to.

**cred·it·a·ble** (krĕd′ĭ-tə-bəl) *adj.* **1.** Deserving commendation : PRAISEWORTHY. **2.** Worthy of belief. **3.** Deserving commercial credit. **4.** Capable of being assigned. **—cred′it·a·bil′i·ty, cred′it·a·ble·ness** *n.* **—cred′it·a·bly** *adv.*

**credit bureau** *n.* An organization from which business firms request credit data on prospective clients.

**credit card** *n.* A card entitling the holder to buy services or goods on credit.

**credit line** *n.* **1.** A line of copy acknowledging the source or origin of a news dispatch, published article, movie, etc. **2.** The maximum amount of credit to be extended to a customer.

**cred·i·tor** (krĕd′ĭ-tər) *n.* One to whom money or its equivalent is owed.

**credit rating** *n.* An estimate of the amount of credit that can be extended to a party without undue risk.

**credit union** *n.* A cooperative organization that makes loans to its members at relatively low interest rates.

**cred·it·wor·thy** (krĕd′ĭt-wûr′*thē*) *adj.* Having an acceptable credit rating. **—cred′it·wor′thi·ness** *n.*

**cre·do** (krē′dō, krā′-) *n., pl.* **-dos.** [Lat., I believe (the first word of the Apostles' Creed) < *credere*, to believe.] A creed.

**cre·du·li·ty** (krĭ-do͞o′lĭ-tē, -dyo͞o′-) *n.* [ME *credulite* < OFr. < Lat. *credulitas* < *credulus*, credulous.] A tendency to believe too readily.

**cred·u·lous** (krĕj′ə-ləs) *adj.* [Lat. *credulus* < *credere*, to believe.] **1.** Tending to believe too readily : GULLIBLE. **2.** Marked by or arising from credulity. **—cred′u·lous·ly** *adv.* **—cred′u·lous·ness** *n.*

**Cree** (krē) *n., pl.* **Cree** or **Crees. 1. a.** A tribe of Indians once living in Ontario, Manitoba, and Saskatchewan. **b.** A member of this tribe. **2.** The Algonquian language of the Cree.

**creed** (krēd) *n.* [ME *crede* < OE *creda* < Lat. *credo*, I believe. —see CREDO.] **1.** A formal statement of religious belief : confession of faith. **2.** A statement or system of belief, principles, or opinions. **—creed′al** (krēd′l) *adj.*

**creek** (krēk, krĭk) *n.* [ME *creke*, prob. < ON *kriki*, bend.] **1.** A small stream, esp. a shallow or intermittent tributary to a river. **2.** *Chiefly Brit.* A small inlet. **—up the creek.** *Informal.* In a difficult or unfortunate position.

**Creek** (krēk) *n., pl.* **Creek** or **Creeks. 1. a.** A confederacy of several Indian tribes once inhabiting parts of Georgia, Alabama, and northern Florida. **b.** A member of any of these tribes. **2.** The Muskhogean language of the Creek.

**creel** (krēl) *n.* [ME *crelle*.] **1.** A wicker basket, esp. one used by anglers for carrying fish. **2.** A frame for holding spools or bobbins in a spinning machine.

**creep** (krēp) *vi.* **crept** (krĕpt), **creep·ing, creeps.** [ME *crepen* < OE *crēopan*.] **1.** To move with the body close to the ground <a baby *creeping* across the rug> **2. a.** To move cautiously or stealthily <a burgler *creeping* from room to room> **b.** To move very slowly <rush hour traffic *creeping* along> **3.** *Bot.* To grow along a surface, rooting at intervals or clinging by means of tendrils or suckers. **4.** To shift gradually. **5.** To have a tingling sensation. —*n.* **1.** The act of creeping. **2.** *Slang.* A repulsive, sinister person. **3.** A slow flow of metal when under high temperature or great pressure. **4.** *Geol.* Slow movement of rock debris and soil down a weathered slope. **5. creeps.** *Informal.* A sensation of repugnance or fear.

**creep·er** (krē′pər) *n.* **1.** One that creeps. **2.** *Bot.* A plant with stems that grow along a surface, either rooting at intervals or clinging for support. **3.** A grappling device for dragging bodies of water, as lakes. **4. creepers.** A metal frame with spikes, attached to a shoe or boot to prevent slipping.

**creep·ing** (krē′pĭng) *adj.* Developing gradually over a period of time <*creeping* inflation>

**creeping Char·lie** (chär′lē) *n.* Moneywort.

**creeping eruption** *n.* A skin disease caused by larvae burrowing beneath the skin and marked by eruptions of reddish lines.

**creeping Jen·nie** *also* **creeping Jen·ny** (jĕn′ē) *n.* The moneywort.

**creeping myrtle** *n.* PERIWINKLE².

**creep·y** (krē′pē) *adj.* **-i·er, -i·est.** *Informal.* Inducing or having a sensation of repugnance or fear <a *creepy* murder mystery> **—creep′i·ness** *n.*

**creese** (krēs) *n. var. of* KRIS.

**cre·mains** (krĭ-mānz′) *pl.n.* [Blend of CREMATED and REMAINS.] Ashes remaining after the cremation of a corpse.

**cre·mate** (krē′māt′, krĭ-māt′) *vt.* **-mat·ed, -mat·ing, -mates.** [Lat. *cremare*, *cremat-*.] To incinerate (a corpse). **—cre·ma′tion** (krĭ-mā′shən) *n.* **—cre′ma′tor** *n.*

**cre·ma·to·ri·um** (krē′mə-tôr′ē-əm, -tōr′-) *n., pl.* **-to·ri·ums** *or* **-to·ri·a** (-tôr′ē-ə, -tōr′-). A crematory.

**cre·ma·to·ry** (krē′mə-tôr′ē, -tōr′ē, krĕm′ə-) *n., pl.* **-ries.** [NLat. *crematorium* < Lat. *cremare*, to cremate.] A furnace or establishment for the cremation of corpses. **—cre′ma·to′ry** *adj.*

**crème de ca·cao** (krĕm′ də kə-kou′, kə-kä′ō) *n.* [Fr. : *crème*, cream + *de*, of + *cacao*, cacao.] A sweet liqueur with a chocolate flavor.

**crème de la crème** (krĕm′ də lä krĕm′) *n.* [Fr. : *crème*, cream + *de*, of + *la*, the + *crème*, cream.] **1.** Something superlative. **2.** The social elite.

**crème de menthe** (krĕm′ də mäɴᴛ′) *n.* [Fr. : *crème*, cream + *de*, of + *menthe*, mint.] A sweet green or white liqueur with a mint flavor.

**cre·nate** (krē′nāt′) *also* **cre·nat·ed** (-nā′tĭd) *adj.* [NLat. *crenatus* < Med. Lat. *crena*, notch.] *Biol.* Having a margin with rounded or scalloped projections <*crenate* leaves> **—cre′nate′ly** *adv.*

**cre·na·tion** (krĭ-nā′shən) *n.* **1.** A rounded projection : CRENATURE. **2.** The fact or condition of being crenate.

**cren·a·ture** (krĕn′ə-chər, krē′nə-) *n.* **1.** CRENATION 1. **2.** A notch between crenations, as on a leaf.

**cren·e·lat·ed** *also* **cren·el·lat·ed** (krĕn′ə-lā′tĭd) *adj.* [< Fr. *crenel*, crenelation < OFr., poss. < Med. Lat. *crena*, notch.] Having battlements. **—cren′e·la′tion** *n.*

**cren·u·late** (krĕn′yə-lĭt, -lāt′) *also* **cren·u·lat·ed** (-lā′tĭd) *adj.* [NLat. *crenulatus* < *crenula*, dim. of Med. Lat. *crena*, notch.] Having minutely notched or scalloped projections. **—cren′u·la′tion** *n.*

**cre·o·dont** (krē′ə-dŏnt′) *n.* [NLat. *Creodonta*, suborder name : Gk. *kreas*, flesh + Gk. *odous*, *odont-*, tooth.] Any of various extinct carnivorous mammals of the suborder Creodonta, of the Paleocene to Pliocene epochs.

**Cre·ole** (krē′ōl′) *n.* [Fr. *créole* < Sp. *criollo* < Port. *crioulo*, prob. < *criar*, to bring up < Lat. *creare*, to beget.] **1.** A person of European descent born in the West Indies or Spanish America. **2. a.** A person descended from or culturally related to the original French settlers of the southern United States, esp. Louisiana. **b.** The French dialect spoken by these people. **3.** A person descended from or culturally related to the Spanish and Portuguese settlers of the Gulf States. **4.** A person of mixed European and Negro ancestry speaking a Creole dialect. **5. creole.** A creolized language. **6.** Haitian Creole. —*adj.* **1.** Of, relating to, or typical of the Creoles. **2. creole.** Cooked with a spicy sauce containing tomatoes, onions, and peppers.

**cre·o·lized language** (krē′ə-līzd′) *n.* A type of mixed language that develops when dominant and subordinate groups who speak different languages have prolonged contact, incorporating the basic vocabulary of the dominant language with the grammar and an admixture of words from the subordinate language and becoming the native tongue of the subordinate group.

**Cre·on** (krē′ŏn′) *n.* *Gk. Myth.* King of Thebes, successor to Oedipus.

**cre·o·sol** (krē′ə-sôl′, -sōl′) *n.* [CREOS(OTE) + -OL¹.] A colorless to yellow aromatic liquid, $C_8H_{10}O_2$, derived from beechwood tar.

**cre·o·sote** (krē′ə-sōt′) *n.* [G. *Kreosot* : Gk. *kreas*, flesh + Gk. *sōtēr*, *preserver* < *sōzein*, to save < *saos*, safe, from its antiseptic properties.] **1.** A colorless to yellowish oily liquid obtained by destructive distillation of wood tar and at one time used to treat tuberculosis and chronic bronchitis. **2.** A yellowish to greenish-brown oily liquid obtained from coal tar and used as a wood preservative and disinfectant. —*vt.* **-sot·ed, -sot·ing, -sotes.** To treat with creosote.

ă **pat** ā **pay** âr **care** ä **father** ĕ **pet** ē **be** hw **which** ĭ **pit** ī **tie** îr **pier** ŏ **pot** ō **toe** ô **paw, for** oi **noise** o͝o **took**

**creosote bush** *n.* A resinous shrub, *Larrea tridentata* of the western United States and Mexico, exuding an odor like that of creosote.
**crepe** *also* **crêpe** (krāp) *n.* [Fr. *crêpe* < OFr. *crespe,* curly < Lat. *crispus.*] **1.** A light, soft, crinkled fabric of silk or other fiber. **2.** CRAPE 2. **3.** Crepe paper. **4.** Crepe rubber. **5.** A thin pancake.
**crêpe de Chine** (krāp′ də shēn′) *n.* [Fr. : *crêpe,* crepe + *de,* of + *Chine,* China.] A silk crepe used for dresses and blouses.
**crepe myrtle** *n. var. of* CRAPE MYRTLE.
**crepe paper** *n.* Crinkled tissue paper used for decorations.
**crepe rubber** *n.* Rubber with a crinkled texture, used esp. for shoe soles.
**crêpe su·zette** (krāp′ sōō-zĕt′) *n., pl.* **crêpe su·zettes** *or* **crêpes su·zettes** (krāp′ sōō-zĕt′) [Fr. : *crêpe,* pancake + *Suzette,* Suzy.] A thin dessert pancake usu. rolled with hot tangerine or orange sauce and often served with a flaming brandy or curaçao sauce.
**crep·i·tate** (krĕp′ĭ-tāt′) *vi.* **-tat·ed, -tat·ing, -tates.** [Lat. *crepitare, crepitat-,* to crackle, freq. of *crepare,* to creak.] To make a creaking or rattling sound : CRACKLE. **—crep′i·tant** (-tənt) *adj.* **—crep′i·ta′tion** *n.*
**crept** (krĕpt) *v. p.t. & p.p. of* CREEP.
**cre·pus·cu·lar** (krĭ-pŭs′kyə-lər) *adj.* **1.** Of or like twilight : DIM. **2.** *Zool.* Becoming active at twilight or before sunrise.
**cre·pus·cule** (krĭ-pŭs′kyōōl) *also* **cre·pus·cle** (-pŭs′əl) *n.* [Lat. *crepusculum* < *creper,* dark.] Twilight.
**cres·cen·do** (krə-shĕn′dō) *n., pl.* **-dos** *or* **-di** (-dē) [Ital. < *crescere,* to increase < Lat.] *Mus.* **1.** A gradual increase, esp. in the volume or intensity of sound in a passage. **2.** A passage played in a crescendo. **—cres·cen′do** *adv. & adj.*
**cres·cent** (krĕs′ənt) *n.* [ME *cressant* < OFr. *creissant* < pr.part. of *creistre,* to grow < Lat. *crescere.*] **1.** The figure of the moon as it appears in its first quarter, with concave and convex edges terminating in points. **2.** Something shaped like a crescent. *—adj.* **1.** Crescent-shaped. **2.** Waxing, as the moon. **—cres·cen′tic** (krə-sĕn′tĭk) *adj.*
**cre·sol** (krē′sôl′, -sōl′) *n.* [Alteration of CREOSOL.] One of three isomeric phenols, $C_7H_8O$, used in resins and as a disinfectant.
**cress** (krĕs) *n.* [ME *cresse* < OE.] A plant, as one of the genera *Cardamine* or *Arabis,* having pungent leaves used in salads and as a garnish.
**cres·set** (krĕs′ĭt) *n.* [ME < OFr. *craisset* < *craisse,* grease < VLat. *crassia* < Lat. *crassus,* fat, thick.] A metal cup, often suspended on a pole, containing burning pitch or oil and used as a torch.
**Cres·si·da** (krĕs′ĭ-də) *n.* A Trojan woman who first returned the love of Troilus but later forsook him for Diomedes.
**crest** (krĕst) *n.* [ME *creste* < OFr. < Lat. *crista.*] **1.** A projection, as a tuft or ridge on the head of a bird or other animal. **2. a.** A plume used as decoration on top of a helmet. **b.** A helmet. **3.** *Heraldry.* **a.** A device placed above the shield on a coat of arms. **b.** A representation of a crest. **4. a.** The top, as of a wave or mountain. **b.** The highest point of a process or action <the *crest* of one's career> **5.** The ridge on a roof. *—v.* **crest·ed, crest·ing, crests.** *—vt.* **1.** To furnish or adorn with a crest. **2.** To reach the crest of. *—vi.* To form into a crest, as a wave.
**crest·fall·en** (krĕst′fô′lən) *adj.* Dispirited : depressed. **—crest′-fall′en·ly** *adv.* **—crest′fall′en·ness** *n.*
**crest·ing** (krĕs′tĭng) *n.* An ornamental ridge, as on the top of a roof or wall.
**cre·syl** (krē′sĭl′) *n.* [CRESO(L) + -YL.] Tolyl.
**cre·syl·ic** (krĭ-sĭl′ĭk) *adj.* [CRES(OL) + -YL + -IC.] *Chem.* Of or relating to creosote or cresol.
**Cre·ta·ceous** (krĭ-tā′shəs) *adj.* [Lat. *cretaceus* : *creta,* chalk < *Creta (terra),* Cretan (earth).] **1.** Of, belonging to, or designating the geologic time, system of rocks, and sedimentary deposits of the third and last period of the Mesozoic era, marked by the development of flowering plants and the disappearance of dinosaurs. **2. cretaceous.** Of, containing, or like chalk. *—n.*The Cretaceous period. **—cre·ta′-ceous·ly** *adv.*
**Cret·an mullein** (krēt′n) *n.* A Mediterranean plant, *Celsia cretica,* having hairy foliage and yellow, purple-splotched flowers.
**cre·tin** (krēt′n) *n.* [Fr. *crétin* < Swiss Fr. *crestin,* Christian, deformed idiot < *Christianus,* Christian. —see CHRISTIAN.] **1.** One afflicted with cretinism. **2.** An idiot. **—cre′tin·oid′** (-oid′) *adj.* **—cre′tin·ous** (-əs) *adj.*
**cre·tin·ism** (krēt′n-ĭz′əm) *n.* Myxedema.
**cre·tonne** (krĭ-tŏn′, krē′tŏn′) *n.* [After *Creton,* France.] A heavy, colorfully printed, unglazed cotton, linen, or rayon fabric used for draperies and slipcovers.
**Cre·ü·sa** (krē-ōō′zə) *n.* [Gk. *Kreousa.*] *Gk. Myth.* **1.** The bride of Jason, killed by Medea. **2.** The daughter of Priam and wife of Aeneas, lost in the flight from Troy.
**cre·val·le** (krĭ-văl′ē) *n.* [Alteration of CAVALLA.] A food and game fish, *Caranx hippos* of warm seas, with a laterally compressed silvery body.
**crevalle jack** *n.* The crevalle.
**cre·vasse** (krĭ-văs′) *n.* [Fr. < OFr. *crevace,* crevice.] **1.** A deep fissure, as in a glacier : CHASM. **2.** A crack in a levee or dike. *—vt.* **-vassed, -vass·ing, -vass·es.** To make crevasses in : FISSURE.
**crev·ice** (krĕv′ĭs) *n.* [ME < OFr. *crevace* < *crever,* to split < Lat. *crepare,* to crack.] A narrow crack : FISSURE. **—crev′iced** *adj.*
**crew**[1] (krōō) *n.* [ME *creue,* military reinforcement < OFr. *creue,* increase < *creistre,* to grow < Lat. *crescere.*] **1. a.** A group of people working together : GANG <*crew* of aircraft mechanics> **b.** A group of people gathered together temporarily : CROWD <fed the whole *crew*> **2. a.** All personnel manning a ship. **b.** All of a ship's personnel except the officers. **c.** All personnel operating an aircraft. **d.** A team of oarsmen. *—vi.* **crewed, crew·ing, crews.** To function as a member of a crew.
**crew**[2] (krōō) *v. var. p.t. & p.p. of* CROW[2] 1.
**crew cut** *n.* [So called because it was worn by oarsmen.] A man's close-cropped haircut.
**crew·el** (krōō′əl) *n.* [ME *crule.*] Loosely twisted worsted yarn used for embroidery and fancywork.
**crew neck** *n.* [From the wearing of similarly-styled sweaters by oarsmen.] A round close-fitting neckline, as on a sweater.
**crew sock** *n.* [From its use by oarsmen.] A heavy usu. ribbed sock.
**crib** (krĭb) *n.* [ME < OE *cribb,* manger.] **1.** A child's bed with high sides. **2.** A small building, usu. with slatted sides, for storing corn. **3.** A rack or trough for fodder : MANGER. **4.** A cattle stall. **5.** A small, crude cottage or room. **6.** A framework to support or strengthen a shaft or mine. **7.** A wicker basket. **8.** *Informal.* **a.** Plagiarism. **b.** A petty theft. **c.** PONY 4. **9.** A set of cards made up from discards by each player in cribbage, used by the dealer. *—v.* **cribbed, crib·bing, cribs.** *—vt.* **1.** To confine in or as if in a crib. **2.** To provide or equip with a crib. **3.** *Informal.* **a.** To plagiarize. **b.** To steal. *—vi. Informal.* To use a crib in examinations : CHEAT. **—crib′ber** *n.*
**crib·bage** (krĭb′ĭj) *n.* [< CRIB.] A card game scored by inserting pegs into holes on a small board.
**crib·bing** (krĭb′ĭng) *n.* A framework support, as of timber lining a shaft.
**crib-bit·ing** (krĭb′bī′tĭng) *n.* An injurious habit of horses of biting at the edge of a feed trough or other object and swallowing air at the same time.
**crib death** *n.* Sudden infant death syndrome.
**crib·ri·form** (krĭb′rə-fôrm′) *adj.* [Lat. *cribrum,* sieve + -FORM.] Having perforations like a sieve.
**crib·work** (krĭb′wûrk′) *n.* A structural framework of logs stacked one above the other, with the logs in each layer at right angles to those in the layer beneath.
**cri·ce·tid** (krĭ-sē′tĭd, -sĕt′ĭd) *n.* [NLat. *Cricetidae,* family name < *Cricetus,* hamster genus, of Slav. orig.] A small rodent of the family Cricetidae, which includes the muskrat and gerbil. **—cri·ce′tid** *adj.*
**crick**[1] (krĭk) *n.* [ME *crike.*] A painful cramp or muscle spasm, as in the neck. *—vt.* **cricked, crick·ing, cricks.** To cause a crick in by turning or wrenching.
**†crick**[2] (krĭk) *n. Regional.* A creek.
**crick·et**[1] (krĭk′ĭt) *n.* [ME *criket* < OFr. *criquet* < *criquer,* to click.] Any of various insects of the family Gryllidae, having long antennae and legs adapted for leaping, and the males of many species producing a shrill, chirping sound by rubbing the front wings together.
**crick·et**[2] (krĭk′ĭt) *n.* [Poss. < OFr. *criquet,* target stick in a bowling game.] **1.** An outdoor game played with bats, a ball, and wickets by 2 teams of 11 players each. **2.** Good sportsmanship <It's not *cricket* to yell at the referee.> *—vi.* **-et·ed, -et·ing, -ets.** To play cricket. **—crick′et·er** *n.*

**cricket**[2]

**cri·coid** (krī′koid′) *n.* [Gk. *krikoeidēs,* ring-shaped : *krikos,* ring + *-eidēs,* -oid.] A ring-shaped cartilage of the lower larynx.
**cri·er** (krī′ər) *n.* **1.** One who cries. **2.** One who shouts out public announcements. **3.** A hawker.
**crime** (krīm) *n.* [ME < OFr. < Lat. *crimen.*] **1.** An act committed or omitted in violation of a law forbidding or commanding it and for which punishment is imposed upon conviction. **2.** Unlawful activity. **3.** A serious moral offense. **4.** An unjust, senseless act or condition. **5.** *Informal.* A regrettable fact : SHAME.
**crim·i·nal** (krĭm′ə-nəl) *adj.* [ME < OFr. *criminel* < LLat. *criminalis* < Lat. *crimen,* accusation.] **1.** Of, involving, or having the nature of crime. **2.** Relating to the administration of penal law. **3.** Guilty of

crime. **4.** Disgraceful : shameful. —*n.* One who has committed or been legally convicted of a crime. **—crim'i·nal·ly** *adv.*

☆ ***syns:*** CRIMINAL, ILLEGAL, ILLEGITIMATE, ILLICIT, LAWLESS, UNLAWFUL, WRONGFUL *adj. core meaning* : constituting a crime <*criminal* acts such as blackmail and extortion>

**criminal conversation** *n. Law.* Adultery.

**crim·i·nal·i·ty** (krĭm'ə-năl'ĭ-tē) *n., pl.* **-ties. 1.** The quality, state, or fact of being criminal. **2.** A criminal act or practice.

**crim·i·nal·ize** (krĭm'ə-nə-līz') *vt.* **-ized, -iz·ing, -iz·es.** To treat as criminal.

**criminal law** *n.* Law dealing with crime and its punishment.

**crim·i·nate** (krĭm'ə-nāt') *vt.* **-nat·ed, -nat·ing, -nates.** [Lat. *criminari, criminat-,* to accuse < *crimen,* accusation.] **1.** To implicate in a crime : INCRIMINATE. **2.** To charge with a crime : ACCUSE. **3.** To condemn as criminal : CENSURE. **—crim'i·na'tion** *n.* **—crim'i·na'tive, crim'i·na·to'ry** (-nə-tôr'ē, -tōr'ē) *adj.* **—crim'i·na'tor** *n.*

**crim·i·nol·o·gy** (krĭm'ə-nŏl'ə-jē) *n.* [Ital. *criminologia* : Lat. *crimen,* accusation + *-logia,* -logy.] The study of crime, criminals, and criminal behavior. **—crim'i·no·log'i·cal** (-nə-lŏj'ĭ-kəl) *adj.* **—crim'i·no·log'i·cal·ly** *adv.* **—crim'i·nol'o·gist** *n.*

**crimp¹** (krĭmp) *vt.* **crimped, crimp·ing, crimps.** [ME *crimpen,* to wrinkle, prob. < LG *krimpen.*] **1.** To press or pinch into small, regular ridges or folds. **2.** To mold or bend (leather) into shape. **3.** To cause (hair) to form tight waves or curls. **4.** To hamper : obstruct. —*n.* **1.** The act of crimping. **2.** Something crimped, esp.: **a.** Hair tightly curled or waved. **b.** A series of curls, as of wool fibers. **3.** An obstruction <My illness put a *crimp* in our summer vacation plans.> **—crimp'er** *n.*

**crimp²** (krĭmp) *n.* [Orig. unknown.] One who procures men to serve as sailors or soldiers by tricking or coercing them. —*vt.* **crimped, crimp·ing, crimps.** To procure (sailors or soldiers) by trickery or coercion.

**crimp·y** (krĭm'pē) *adj.* **-i·er, -i·est.** Full of crimps : WAVY. **—crimp'i·ness** *n.*

**crim·son** (krĭm'zən) *n.* [ME *cremesin* < OSp. < Ar. *qirmizī* < *quirmiz,* kermes insect.] A deep to vivid purplish red to vivid red. —*vt. & vi.* **-soned, -son·ing, -sons.** To make or become crimson.

**cringe** (krĭnj) *vi.* **cringed, cring·ing, cring·es.** [ME *crengen,* prob. ult. < OE *cringan.*] **1.** To recoil, as in fear : COWER. **2.** To behave in a servile manner : FAWN. **—cring'er** *n.*

**crin·gle** (krĭng'gəl) *n.* [LG *kringel,* dim. of *kring,* ring < MLG.] A small ring or grommet of rope or metal fastened to the edge of a sail.

**cri·nite** (krī'nīt') *adj.* [Lat. *crinitus,* p.part. of *crinire,* to cover with hair < *crinis,* hair.] *Biol.* Having hairlike tufts : HAIRY.

**crin·kle** (krĭng'kəl) *v.* **-kled, -kling, -kles.** [ME *crinkelen.*] —*vi.* **1.** To form into wrinkles or ripples. **2.** To make a soft, crackling sound : RUSTLE. —*vt.* To cause to crinkle. —*n.* A wrinkle or ripple : FOLD. **—crin'kly** *adj.*

**crin·kle·root** (krĭng'kəl-ro͞ot', -ro͝ot') *n.* A woodland plant, *Dentaria diphylla* of eastern North America, with fleshy rootstocks and white or pinkish flower clusters.

**cri·noid** (krī'noid') *n.* [NLat. *Crinoidea,* class name : Gk. *krinon,* lily + -OID.] Any of various marine invertebrates of the class Crinoidea, including the sea lilies and feather stars, having feathery, radiating arms and a stalk attaching them to a surface.

**crin·o·line** (krĭn'ə-lĭn) *n.* [Fr. < Ital. *crinolino* : *crino,* horsehair (< Lat. *crinis,* hair) + *lino,* flax (< Lat. *linum*).] **1.** A coarse, stiff fabric of horsehair or cotton used esp. to stiffen and line garments. **2.** A petticoat made of crinoline. **3.** A hoop skirt.

**cri·num** (krī'nəm) *n.* [NLat. *Crinum,* genus name < Gk. *krinon,* lily.] Any of several mostly tropical plants of the genus *Crinum,* with long, strap-shaped leaves and lilylike flower clusters.

**crinum lily** *n.* A crinum.

**cri·o·sphinx** (krī'ə-sfĭngks') *n.* [Gk. *krios,* ram + SPHINX.] A sphinx having the head of a ram.

**crip·ple** (krĭp'əl) *n.* [ME *crepel* < OE *crypel.*] **1.** One who is lame or partially disabled. **2.** Something defective or damaged. —*vt.* **-pled, -pling, -ples. 1.** To make into a cripple. **2.** To disable or damage. **—crip'pler** *n.*

**cri·sis** (krī'sĭs) *n., pl.* **-ses** (-sēz') [Lat. < Gk. < *krinein,* to separate.] **1. a.** A crucial or decisive point or situation : TURNING POINT. **b.** An unstable state of political, international, or economic affairs with an impending abrupt or decisive change. **2.** A sudden change for better or worse during an acute illness. **3.** The point in a story or drama at which hostile forces are in the tensest state of opposition.

**crisis center** *n.* A center staffed esp. by volunteers who counsel individuals experiencing a personal crisis.

**crisis management** *n.* Special measures undertaken to solve problems caused by a crisis.

**crisp** (krĭsp) *adj.* **-er, -est.** [ME, curly < OE < Lat. *crispus.*] **1.** Firm but easily broken or crumbled : BRITTLE <*crisp* crackers> **2.** Firm and fresh <*crisp* lettuce> **3.** Brisk or invigorating : BRACING <*crisp* morning breezes> **4.** Stimulating : animated. **5.** Having small curls, waves, or ripples. —*vt. & vi.* **crisped, crisp·ing, crisps.** To make or become crisp. —*n. Chiefly Brit.* A potato chip. **—crisp'ly** *adv.* **—crisp'ness** *n.*

**cris·pate** (krĭs'pāt') *also* **cris·pat·ed** (-pā'tĭd) *adj.* [Lat. *crispatus* < *crispare,* to curl < *crispus,* curly.] Crimped, curled, or tightly waved.

**cris·pa·tion** (krĭs-pā'shən) *n.* **1.** The act of crisping or curling or the state of being crisped or curled. **2.** A slight involuntary contraction or constriction, as of the skin.

**crisp·er** (krĭs'pər) *n.* One that crisps, esp. a compartment in a refrigerator used to keep vegetables fresh.

**crisp·y** (krĭs'pē) *adj.* **-i·er, -i·est.** Crisp. **—crisp'i·ness** *n.*

**cris·sa** (krĭs'ə) *n. pl. of* CRISSUM.

**criss·cross** (krĭs'krôs', -krŏs') *v.* **-crossed, -cross·ing, -cross·es.** [Alteration of obs. *christcross,* mark of a cross.] —*vt.* **1.** To mark with crossing lines. **2.** To progress crosswise through or over. —*vi.* To move crosswise. —*n.* A mark or pattern made of crossing lines. —*adj.* Crossing one another or marked by crossings. —*adv.* In a crisscross way or direction.

▲ word history: *Crisscross* is a phonetic spelling of *christcross,* which literally means "the cross of Christ." The word originally denoted a mark of two crossed lines resembling the Christian religious symbol. This mark had two main functions: it was a symbol placed at the head of alphabets used for the teaching of children, and it was a figure used in place of a signature by illiterate persons. In the 19th century the religious association was lost and *crisscross* was applied to any pattern of crossed lines.

**cris·sum** (krĭs'əm) *n., pl.* **cris·sa** (krĭs'ə) [NLat. < Lat. *crissare,* to move the buttocks during intercourse.] *Zool.* The feathers or area surrounding a bird's cloacal opening. **—cris'sal** (-əl) *adj.*

**cris·ta** (krĭs'tə) *n., pl.* **-tae** (-tē) [NLat., crest.] *Biol.* **1.** A crest or ridge. **2.** An inward projection of the inner mitochondrial membrane.

**cris·tate** (krĭs'tāt') *also* **cris·tat·ed** (-tā'tĭd) *adj.* [Lat. *cristatus* < *crista,* tuft.] Having or forming a crest.

**cri·te·ri·on** (krī-tîr'ē-ən) *n., pl.* **-te·ri·a** (-tîr'ē-ə) *or* **-te·ri·ons.** [Gk. *kritērion* < *kritēs,* judge < *krinein,* to separate.] A standard on which a judgment is based. ***usage:*** *Criteria* is a plural form and should not be substituted for the singular form *criterion.* **—cri·te'ri·al** (-əl) *adj.*

**crit·ic** (krĭt'ĭk) *n.* [Lat. *criticus* < Gk. *kritikos,* able to discern < *krinein,* to separate.] **1.** One who forms and expresses judgments of the merits, faults, etc., of a matter. **2.** One who judges the quality of literary or artistic works, esp. as a profession. **3.** One who finds fault.

**crit·i·cal** (krĭt'ĭ-kəl) *adj.* **1.** Tending to judge harshly and adversely. **2.** Marked by careful and exact judgment and evaluation <a *critical* study of the poems> **3.** Of, relating to, or typical of critics or criticism. **4.** Forming or having the nature of a crisis : CRUCIAL <a *critical* battle> **5.** Designating materials and products essential to a condition or project but in short supply. **6.** *Med.* Of or relating to a crisis. **7.** *Math.* Of or relating to a point at which a curve has a maximum, minimum, or point of inflection. **8.** *Chem. & Physics.* Of or relating to a condition causing an abrupt change in a quality, property, or phenomenon. **9.** Of sufficient mass to sustain a nuclear chain reaction. **—crit'i·cal·ly** *adv.* **—crit'i·cal·ness** *n.*

☆ ***syns:*** CRITICAL, ACUTE, CLIMACTERIC, CRUCIAL, DESPERATE, DIRE *adj. core meaning* : so serious as to be at the point of crisis <a *critical* shortage of fuel> ***ant:*** noncritical

**critical angle** *n.* **1.** The smallest angle of incidence at which a light ray passing from one medium to another less refractive medium can be totally reflected from the boundary between the two. **2.** The angle of attack of an airfoil at which airflow abruptly changes, causing changes in the lift and drag of an aircraft.

**critical mass** *n.* The smallest mass of a fissionable material that will sustain a nuclear chain reaction.

**critical point** *n.* **1.** *Physics.* The condition in which the liquid and vapor phases of a pure stable substance have the same density. **2.** *Math.* **a.** A maximum, minimum, or point of inflection. **b.** A point at which the derivative of a function is zero or infinite.

**critical pressure** *n.* The least applied pressure required at the critical temperature to liquefy a gas.

**critical state** *n.* CRITICAL POINT 1.

**critical temperature** *n.* The temperature above which a gas cannot be liquefied, regardless of the pressure applied.

**crit·ic·as·ter** (krĭt'ĭ-kăs'tər) *n.* [CRITIC + Lat. *-aster,* pejorative suffix.] A petty critic.

**crit·i·cism** (krĭt'ĭ-sĭz'əm) *n.* **1.** The act of criticizing, esp. adversely. **2.** A critical comment or judgment. **3. a.** The art, skill, or profession of making discriminating judgments, esp. of literary or artistic works. **b.** An article expressing such judgment. **4.** Detailed investigation of the origin and history of literary documents.

**crit·i·cize** (krĭt'ĭ-sīz') *v.* **-cized, -ciz·ing, -ciz·es.** —*vt.* **1.** To analyze and evaluate. **2.** To judge harshly. —*vi.* To act as a critic. **—crit'i·ciz'a·ble** *adj.* **—crit'i·ciz'er** *n.*

**cri·tique** (krĭ-tēk') *n.* [Fr. < Gk. *kritikē,* the art of criticism < *kritikos,* critical < *krinein,* to separate.] **1.** A critical review or commentary, esp. one dealing with a literary or artistic work. **2.** A critical discussion of a specified topic. **3.** The art of criticism. —*vt.* **-tiqued, -tiqu·ing, -tiques.** To criticize.

**†crit·ter** (krĭt′ər) *n.* [Alteration of CREATURE.] *Regional.* **1.** A domestic animal, esp. a steer or horse. **2.** A living creature.

**croak** (krōk) *n.* [< ME *croken,* to croak.] A low, hoarse sound, as that made by frogs. —*v.* **croaked, croak·ing, croaks.** —*vt.* To utter by croaking. —*vi.* **1.** To utter a croak. **2.** To speak with a low, hoarse voice. **3.** To mutter discontentedly : GRUMBLE. **4.** *Slang.* To die. **—croak′i·ly** *adv.* **—croak′y** *adj.*

**croak·er** (krō′kər) *n.* **1. a.** A croaking animal. **b.** A habitual complainer or doomsayer. **2.** Any of various chiefly marine fishes of the family Sciaenidae that make croaking or grunting sounds.

**Croat** (krōt, krō′ăt′) *n.* [NLat. *Croata* < Serbo-Croatian *Hrvat.*] **1.** A Slavic native or resident of Croatia. **2.** Serbo-Croatian as used in Croatia, distinguished from Serbian primarily by its being written in the Latin alphabet. **—Cro·a′tian** (krō-ā′shən) *n.* & *adj.*

**cro·ce·in** (krō′sē-ĭn) *n.* [Lat. *croceus,* saffron-colored (< *crocus,* saffron) + -IN.] A red or orange acid azo dye.

**cro·chet** (krō-shā′) *v.* **-cheted** (-shād′), **-chet·ing** (-shā′ĭng), **-chets** (-shāz′) [< Fr., hook < OFr., dim. of *croc.*] —*vi.* To make needlework by looping thread with a hooked needle. —*vt.* To make by looping thread with a hooked needle. —*n.* Needlework made by crocheting.

**cro·cid·o·lite** (krō-sĭd′l-īt′) *n.* [G. *Krokydolith* : Gk. *krokus,* nap of cloth + *-lith,* -lite.] A fibrous, lavender-blue or greenish mineral, a sodium iron silicate used as a commercial form of asbestos.

**crock**[1] (krŏk) *n.* [ME *crokke* < OE *crocc.*] **1.** An earthenware vessel. **2.** A broken piece of earthenware.

**crock**[2] (krŏk) *n.* [Orig. unknown.] *Chiefly Brit. Regional.* **1.** Soot. **2.** Coloring matter that rubs off from poorly dyed cloth. —*v.* **crocked, crock·ing, crocks.** —*vt.* To soil with or as if with crock. —*vi.* To give off soot or color.

**crock**[3] (krŏk) [ME *crok,* prob. of Scand. orig.] *Chiefly Brit.* —*n.* One that is worn-out, decrepit, or impaired. —*vi.* **crocked, crock·ing, crocks.** To become weak or disabled.

**crocked** (krŏkt) *adj.* [Poss. < CROCK[3].] *Slang.* Intoxicated.

**crock·er·y** (krŏk′ə-rē) *n.* Crocks as a whole : EARTHENWARE.

**crock·et** (krŏk′ĭt) *n.* [ME *croket* < ONFr. *croquet,* var. of OFr. *crochet,* hook. —see CROCHET.] A decorative device, usu. in the form of a curling leaf or cusp, placed along outer angles of gables and pinnacles.

**Crock·pot** (krŏk′pŏt′). A trademark for an electric cooking pot.

**croc·o·dile** (krŏk′ə-dīl′) *n.* [ME *cocodril* < OFr. < Med. Lat. *cocodrillus* < Lat. *crocodillus* < Gk. *krokodilos* : *krokē,* pebble + *drilos,* worm.] **1.** A large tropical aquatic reptile of the genus *Crocodylus* or related genera, with thick armorlike skin and long tapering jaws. **2.** A crocodilian reptile, as an alligator, caiman, or gavial. **3.** Leather made from crocodile skin.

**crocodile bird** *n.* A black and white African bird, *Pluvianus aegyptius,* feeding on insects that infest crocodiles.

**crocodile tears** *pl.n.* [From the belief that crocodiles weep after eating their victims.] An insincere show of grief.

**croc·o·dil·i·an** (krŏk′ə-dĭl′ē-ən, -dĭl′yən) *n.* [< NLat. *Crocodylia,* order name < Lat. *crocodilus,* crocodile.] Any of various reptiles of the order Crocodylia, including the alligators, crocodiles, caimans, and gavials. —*adj.* **1.** Of or relating to a crocodile. **2.** Belonging to the order Crocodylia.

**croc·oi·site** (krŏk′wə-zīt′) *n.* [G. *Krokoisit* < Fr. *crocoise* < Gk. *krokoeis,* saffron-colored < *krokos,* saffron.] Crocoite.

**croc·o·ite** (krŏk′ō-īt′, krō′kō-) *n.* [G. *Krokoit,* alteration of *Krokoisit,* crocoisite.] A rare orange to reddish mineral of lead chromate, $PbCrO_4$, found in oxidized lead deposits.

**cro·cus** (krō′kəs) *n., pl.* **-cus·es** or **-ci** (-sī′) [Lat. < Gk. *krokos,* of Semitic orig.] **1.** A plant of the genus *Crocus,* widely cultivated in gardens and bearing variously colored flowers and grasslike leaves. **2.** A grayish to light reddish purple. **3.** A red variety of iron oxide, $Fe_2O_3$, used in the form of an abrasive powder for polishing.

**croft** (krôft, krŏft) *n.* [ME < OE.] *Chiefly Brit. & Scot.* **1.** A small enclosed field near a house. **2.** A small farm, esp. a tenant farm.

**croft·er** (krôf′tər, krŏf′-) *n. Chiefly Brit. & Scot.* One who rents and cultivates a croft.

**crois·sant** (krwä-säN′) *n.* [Fr., crescent < OFr. *creissant.* —see CRESCENT.] A rich crescent-shaped roll of puff pastry or leavened dough.

**Croix de Guerre** (krwä′ də gâr′) *n.* [Fr. : *croix,* cross + *de,* of + *guerre,* war.] A French military decoration for bravery.

**Cro-Mag·non** (krō-măg′nən, -măn′yən) *n.* An early form of modern human, *Homo sapiens,* marked by a robust physique and known from skeletal parts found in the Cro-Magnon cave in southern France. **—Cro-Mag′non** *adj.*

**crom·lech** (krŏm′lĕk′) *n.* [Welsh : *crom,* fem. of *crwn,* arched + *llech,* stone.] **1.** A prehistoric monument of monoliths surrounding a mound. **2.** A dolmen.

**crone** (krōn) *n.* [ME < ONFr. *carogne,* carrion < VLat. **caronia* < Lat. *caro,* flesh.] A witchlike old woman.

**Cro·nus** (krō′nəs) *n.* [Gk. *Kronos.*] *Gk. Myth.* A Titan who ruled the universe until being dethroned by his son Zeus.

**cro·ny** (krō′nē) *n., pl.* **-nies.** [Poss. < Gk. *khronios,* long-lasting < *khronos,* time.] A close friend.

**cro·ny·ism** (krō′nē-ĭz′əm) *n.* Favoritism shown to cronies regardless of their qualifications, as in filling political positions.

**crook** (krŏŏk) *n.* [ME *crok* < ON *krōkr,* hook.] **1.** A bent or curved object : HOOK. **2.** A tool or implement, as a bishop's crosier or a shepherd's staff, with a curved or bent part. **3.** A curve or bend : TURN. **4.** *Informal.* One who lives by dishonest means. —*vt.* & *vi.* **crooked, crook·ing, crooks.** To curve or become curved : BEND.

**crook·back** (krŏŏk′băk′) *n.* A hunchback. **—crook′backed′** *adj.*

**crook·ed** (krŏŏk′ĭd) *adj.* **1.** Having or marked by bends, curves, or angles. **2.** *Informal.* Dishonest <a *crooked* lawyer> **—crook′ed·ly** *adv.* **—crook′ed·ness** *n.*

**Crookes tube** (krŏŏks) *n.* [After Sir William *Crookes* (1832–1919).] A low-pressure discharge tube for studying the properties of cathode rays.

**crook·neck** (krŏŏk′nĕk′) *n.* A squash with a long, curved neck and yellow flesh.

**croon** (krōōn) *v.* **crooned, croon·ing, croons.** [ME *croynen* < MDu. *kronen,* to lament.] —*vi.* **1.** To sing or hum softly. **2.** To sing popular songs in a soft, sentimental way. **3.** *Chiefly Scot. & Brit. Regional.* To roar : bellow. —*vt.* To sing by crooning. —*n.* A soft singing or humming. **—croon′er** *n.*

**crop** (krŏp) *n.* [ME < OE *cropp,* ear of corn.] **1. a.** Cultivated agricultural plants, as grain, vegetables, or fruit. **b.** The total yield of agricultural produce in a given season or place. **2.** A group, quantity, or supply appearing at one time <a *crop* of presidential candidates> **3.** A short haircut. **4.** An earmark on an animal. **5. a.** A short whip with a loop serving as a lash, used in horseback riding. **b.** The stock of a whip. **6.** *Zool.* **a.** A pouchlike enlargement of a bird's esophagus in which food is stored or partially digested. **b.** A similar organ in earthworms, insects, and other invertebrates. —*v.* **cropped, crop·ping, crops.** —*vt.* **1.** To cut off the stems or top of (a plant). **2.** To cut (e.g., hair) very short. **3.** To clip (e.g., an animal's ears). **4.** To reap : harvest. **5.** To cause to grow or yield a crop or crops. —*vi.* To plant, grow, or yield a crop or crops. **—crop up.** To appear unexpectedly <New facts kept *cropping up.*>

**crop-eared** (krŏp′îrd′) *adj.* **1.** Having the ears cropped. **2.** Having the hair cut so short that the ears show.

**crop·per**[1] (krŏp′ər) *n.* **1.** One that crops. **2.** A sharecropper.

**crop·per**[2] (krŏp′ər) *n.* [Orig. unknown.] **1.** A heavy fall : TUMBLE. **2.** A fiasco.

**crop rotation** *n.* Maintenance and renewal of soil fertility by successive plantings of different crops on the same land.

**cro·quet** (krō-kā′) *n.* [Fr. < ONFr., hook. —see CROCKET.] **1.** An outdoor game in which the players drive wooden balls through a series of wickets using long-handled mallets. **2.** The act of driving away an opponent's croquet ball by hitting one's own ball when the two are in contact. —*vt.* **-queted** (-kād′), **-quet·ing** (-kā′ĭng), **-quets** (-kāz′). To drive away (an opponent's ball) with a croquet.

**cro·quette** (krō-kĕt′) *n.* [Fr. < *croquer,* to crunch.] A small cake of minced food, often coated with bread crumbs and fried in deep fat.

**cro·qui·gnole** (krō′kən-yōl′) *n.* [Fr., a kind of biscuit < *croquer,* to crunch.] A permanent wave in which the hair is wound around metal rods.

**cro·sier** *also* **cro·zier** (krō′zhər) *n.* [ME *croser* < OFr. *crossier,* staff bearer < *crosse,* crosier, of Germanic orig.] **1.** A staff with a cross or crook at the end, carried by or before an abbot, bishop, or archbishop as a symbol of office. **2.** *Bot.* A coiled tip of a plant stalk, as of a young fern frond.

**cross** (krôs, krŏs) *n.* [ME *cros* < OE < ON *kross* < OIr. *cross* < Lat. *crux.*] **1. a.** An upright post with a transverse piece near the top, on which condemned people were executed in ancient times. **b.** *often* **Cross.** The cross on which Jesus was crucified. **c.** A symbolic representation of the cross on which Jesus was crucified. **d.** A crucifix. **2.** A sign of the cross. **3.** An affliction <a heavy *cross* to bear> **4.** A medal, emblem, or insignia shaped like a cross or a modified cross. **5.** A mark formed by the intersection of two lines, esp. such a mark (X) used as a signature. **6.** A pipe fitting with four cross-shaped branches, used as a junction for intersecting pipes. **7.** *Biol.* **a.** A plant or animal produced by crossbreeding : HYBRID. **b.** The process of crossbreeding : HYBRIDIZATION. **8.** One combining the qualities of two things <a movie that is a *cross* between a western and a mystery> **9.** *Slang.* A contest whose outcome has been dishonestly prearranged. —*v.* **crossed, cross·ing, cross·es.** —*vt.* **1.** To pass from one side of to the other <*crossed* the street> **2.** To carry or convey across. **3.** To extend or pass through or over : INTERSECT. **4. a.** To eliminate or delete by or as if by drawing a line through <*crossing* items off a shopping list> **b.** To make or put a line across <*Cross* your *t's.*> **5.** To place crosswise <*cross* our fingers> **6.** To make the sign of the cross upon or over. **7.** To encounter in passing <Their paths *crossed* ours.> **8.** *Informal.* To interfere with : THWART <Don't try to *cross* me.> **9.** *Biol.* To crossbreed or cross-fertilize (plants or animals). —*vi.* **1.** To lie or pass across : INTERSECT. **2.** To move or extend from one side to another. **3.** To encounter in passing

<The spacecrafts' orbits *crossed.*> **4.** *Biol.* To crossbreed or cross-fertilize. *—adj.* **1.** Lying or passing crosswise : INTERSECTING. **2.** Opposing : contrary <working at *cross* purposes> **3.** Showing a bad humor. **4.** Involving interchange : RECIPROCAL. **5.** Hybrid : crossbred. *—adv.* Crosswise. **—cross'er** *n.* **—cross'ly** *adv.* **—cross'ness** *n.*

**cross·bar** (krôs'bär', krŏs'-) *n.* A horizontal bar, stripe, or line.

**cross·bill** (krôs'bĭl', krŏs'-) *n.* A bird of the genus *Loxia,* with curved mandibles having narrow, crossed tips.

**cross·bones** (krôs'bōnz', krŏs'-) *pl.n.* A representation of two bones placed crosswise, usu. beneath a skull, symbolizing death or danger.

**cross·bow** (krôs'bō', krŏs'-) *n.* A powerful weapon composed of a bow fixed crosswise on a wooden stock, with grooves on the stock to direct the projectile. **—cross'bow'man** *n.*

**cross·bred** (krôs'brĕd', krŏs'-) *adj.* Produced by the mating of individuals of different breeds. **—cross'bred'** *n.*

**cross·breed** (krôs'brēd', krŏs'-) *v.* **-bred** (-brĕd'), **-breed·ing, -breeds.** *—vt.* To produce (a hybrid) by the mating of individuals of different varieties or breeds : HYBRIDIZE. *—vi.* To mate so as to produce a hybrid. *—n.* A hybrid produced by crossbreeding.

**cross·check** (krôs'chĕk', krŏs'-) *vt.* **-checked, -check·ing, -checks. 1.** To verify by comparing with parallel or supplementary data. **2.** To check illegally in ice hockey by thrusting with one's stick at an opponent's stick or arms. **—cross'check'** *n.*

**cross-coun·try** (krôs'kŭn'trē, krŏs'-) *adj.* **1.** Moving or directed across open country <a brisk *cross-country* horseback ride> **2.** From one side of a country to the opposite side <a *cross-country* airplane race> **—cross'coun'try** *adv.*

**cross-country skiing** *n.* The sport of skiing over the countryside rather than on downhill runs.

**cross-cul·tur·al** (krôs'kŭl'chər-əl, krŏs'-) *adj.* Comparing or treating two or more different cultures < a *cross-cultural* study>

**cross·cur·rent** (krôs'kûr'ənt, -kŭr'-, krŏs'-) *n.* **1.** A current flowing across another current. **2.** A conflicting movement, tendency, or inclination <*crosscurrents* of unrest and discontent>

**cross·cut** (krôs'kŭt', krŏs'-) *vt. & vi.* **-cut, -cut·ting, -cuts.** To cut or run across or crosswise. *—adj.* **1.** Used or designed for cutting crosswise <a *crosscut* saw> **2.** Cut on the bias or across the grain. *—n.* **1.** A course or cut going crosswise. **2.** A short cut. **3.** A level in a mine driven so as to intersect a vein of ore.

**crosse** (krôs, krŏs) *n.* [Fr. < OFr. *crosse,* staff, of Germanic orig.] A lacrosse stick.

**cross-ex·am·ine** (krôs'ĭg-zăm'ĭn, krŏs'-) *v.* **-ined, -in·ing, -ines.** *—vt.* **1.** To question closely, esp. in order to check the resulting answers against answers given earlier. **2.** *Law.* To question (a witness already examined by the opposing side). *—vi.* To question a person closely. **—cross'-ex·am'i·na'tion** *n.* **—cross'-ex·am'in·er** *n.*

**cross-eye** (krôs'ī', krŏs'ī') *n.* Strabismus in which one or both eyes deviate toward the nose. **—cross'-eyed'** *adj.*

**cross-fer·til·i·za·tion** (krôs'fûr'tl-ĭ-zā'shən, krŏs'-) *n.* **1.** *Biol.* Fertilization by the union of gametes from different individuals, often of different varieties or species. **2.** *Bot.* Fertilization of the ovule of one plant or flower by pollen nuclei from another. **—cross'-fer'tile** *adj.*

**cross-fer·til·ize** (krôs'fûr'tl-īz', krŏs'-) *vt. & vi.* **-ized, -iz·ing, -iz·es.** To fertilize or be fertilized by cross-fertilization.

**cross-file** (krôs'fīl', krŏs'-) *vi. & vt.* **-filed, -fil·ing, -files.** To register as a candidate in the primaries of more than one political party. **—cross'-fil'er** *n.*

**cross·fire** (krôs'fīr', krŏs'-) *n.* **1.** Lines of gunfire from two or more positions crossing each other at a single point. **2.** A situation in which a number of things from different sources come together. **3.** Rapid, often agitated discussion.

**cross-grained** (krôs'grānd', krŏs'-) *adj.* **1.** Having an irregular, transverse, or diagonal grain. **2.** Stubborn : contrary.

**cross hair** *n.* Either of two fine strands of wire crossed in the focus of the eyepiece of an optical instrument and used as a calibration or sighting reference.

**cross·hatch** (krôs'hăch', krŏs'-) *vt.* **-hatched, -hatch·ing, -hatch·es.** To mark with two or more sets of intersecting parallel lines. **—cross'hatch'** *n.*

**cross·head** (krôs'hĕd', krŏs'-) *n.* A beam connecting the piston rod to the connecting rod of a reciprocating engine.

**cross-in·dex** (krôs'ĭn'dĕks', krŏs'-) *v.* **-dexed, -dex·ing, -dex·es.** *—vt.* To furnish (an index) with cross-references. *—vi.* To furnish cross-references. **—cross'-in'dex** *n.*

**cross·ing** (krô'sĭng, krŏs'ĭng) *n.* **1.** A place where roads, lines, or tracks intersect : INTERSECTION. **2.** The place where something, as a river or highway, may be crossed. **3.** The intersection of the nave and transept in a cruciform church.

**crossing over** *n.* Exchange of genetic material between homologous chromosomes.

**cross matching** *n.* The process in which blood compatibilities are tested between a donor and a recipient before transfusion.

**cross-mul·ti·ply** (krôs'mŭl'tə-plī', krŏs'-) *vi.* **-plied, -ply·ing, -plies.** To multiply the numerator of one of a pair of fractions by the denominator of the other. **—cross multiplication** *n.*

**cross·o·ver** (krôs'ō'vər, krŏs'-) *n.* **1.** A place where a crossing is made. **2.** A short connecting track by which a train can be transferred from one line to another. **3.** *Genetics.* **a.** A crossing over. **b.** A character resulting from crossing over. **4.** A registered member of one political party who votes in the primary of another party.

**cross·patch** (krôs'păch', krŏs'-) *n.* [CROSS + obs. *patch,* jester.] *Informal.* A peevish person.

**cross·piece** (krôs'pēs', krŏs'-) *n.* A horizontal bar or beam, as of a structure.

**cross-pol·li·nate** (krôs'pŏl'ə-nāt', krŏs'-) *vt.* **-nat·ed, -nat·ing, -nates.** *Bot.* To cross-fertilize (a plant or flower). **—cross'-pol'li·na'tion** *n.*

**cross product** *n.* Vector product.

**cross-pur·pose** (krôs'pûr'pəs, krŏs'-) *n.* A contrary or conflicting purpose <two factions working at *cross-purposes*>

**cross-ques·tion** (krôs'kwĕs'chən, krŏs'-) *vt.* **-tioned, -tion·ing, -tions.** To cross-examine. *—n.* A question asked during cross-examination.

**cross-re·ac·tion** (krôs'rē-ăk'shən, krŏs'-) *n.* The reaction between an antigen and an antibody generated against a different antigen. **—cross'-re·act'** *v.* **(-act·ed, -act·ing, -acts). —cross'-re·ac'tive** *adj.* **—cross'-re·ac·tiv'i·ty** *n.*

**cross-re·fer** (krôs'rĭ-fûr', krŏs'-) *v.* **-ferred, -fer·ring, -fers.** *—vt.* To refer from one passage or part to another. *—vi.* To make a cross-reference.

**cross-ref·er·ence** (krôs'rĕf'ər-əns, -rĕf'rəns, krŏs'-) *n.* A reference from one part of a book, index, catalogue, etc., to another part containing related information.

**cross·road** (krôs'rōd', krŏs'-) *n.* **1.** A road intersecting another road. **2. crossroads** *(sing. in number).* **a.** A place where two or more roads meet. **b.** A place where different cultures meet. **c.** A crucial point <at the *crossroads* of history>

**cross·ruff** (krôs'rŭf', -rŭf', krŏs'-) *n.* A series of plays in games of the whist family where partnership hands alternately trump suits led by the other partner. **—cross'ruff'** *v.* **(-ruffed, -ruff·ing, -ruffs).**

**cross section** *n.* **1. a.** A section formed by a plane cutting through an object, usu. at right angles to an axis. **b.** A piece so cut or a graphic representation of such a piece. **2.** *Physics.* A measure of the probability of occurrence of a particular atomic or nuclear reaction. **3.** A representative sample intended to be typical of the whole. **—cross'-sec'tion·al** *adj.*

**cross-stitch** (krôs'stĭch', krŏs'-) *n.* **1.** A double needlework stitch forming an X. **2.** Needlework made with cross-stitches. **—cross'-stitch'** *v.* **(-stitched, -stitch·ing, -stitch·es).**

**cross·talk** (krôs'tôk', krŏs'-) *n.* Garbled sounds or noise heard on a telephone or other electronic receiver, caused by interference from another channel.

**cross·tie** (krôs'tī', krŏs'-) *n.* A transverse rod or beam serving as a support, esp. one connecting and supporting the rails of a railroad.

**cross-town** (krôs'toun', krŏs'-) *adj.* Running or extending across a city or town <a *cross-town* subway line> **—cross'-town'** *adv.*

**cross·tree** (krôs'trē', krŏs'-) *n.* *Naut.* One of the two horizontal crosspieces at the upper ends of the lower masts in fore-and-aft-rigged vessels, serving to spread the shrouds.

**cross vault** *n.* Vaulting formed by the intersection of two or more simple vaults.

**cross·walk** (krôs'wôk', krŏs'-) *n.* A street crossing marked for pedestrians.

**cross·way** (krôs'wā', krŏs'-) *n.* A crossroad.

**cross·ways** (krôs'wāz', krŏs'-) *adv. var. of* CROSSWISE.

**cross·wind** (krôs'wĭnd', krŏs'-) *n.* A wind blowing at right angles to a given direction, as to an aircraft's line of flight.

**cross·wise** (krôs'wīz', krŏs'-) *also* **cross·ways** (-wāz') *adv.* Across. *—adj.* Crossing.

**cross·word puzzle** (krôs'wûrd', krŏs'-) *n.* A puzzle in which an arrangement of numbered squares is to be filled with words running both across and down in answer to correspondingly numbered clues.

**crotch** (krŏch) *n.* [Poss. alteration of CRUTCH.] **1.** The angle or region of the angle formed by the junction of two parts or members, as two branches, limbs, or legs. **2.** The fork of a pole or other support. **—crotched** (krŏcht) *adj.*

**crotch·et** (krŏch'ĭt) *n.* [ME *crochet* < OFr. —see CROCHET.] **1.** A small hook or hooklike structure. **2.** An odd, whimsical, or stubborn notion. **3.** *Mus.* A quarter note.

**crotch·et·y** (krŏch'ĭ-tē) *adj.* Capriciously stubborn : PERVERSE. **—crotch'et·i·ness** *n.*

**cro·ton** (krōt'n) *n.* [NLat. *Croton,* genus name < Gk. *kroton,* castor oil plant.] **1.** A chiefly tropical plant, shrub, or tree of the genus *Croton.* **2.** A tropical plant of the genus *Codiaeum,* esp. *C. variegatum pictum,* often grown as a house plant for its showy, varicolored foliage.

**Croton bug** *n.* [After the *Croton* River, New York.] A small light-brown cockroach, *Blatella germanica.*

**Croton bug**
*Approximately one-half inch long*

**cro·ton·ic acid** (krō-tŏn'ĭk) *n.* [< NLat. *Croton,* plant genus. — see CROTON.] An organic acid, $C_4H_6O_2$, used in preparing pharmaceuticals and resins.
**croton oil** *n.* A yellowish-brown, violently cathartic oil obtained from the seeds of a southeast Asian tree, *Croton tiglium.*
**crouch** (krouch) *v.* **crouched, crouch·ing, crouch·es.** [ME *crouchen* < OFr. *crochir,* to be bent < *croc,* hook.] —*vi.* **1.** To stoop with the legs pulled close to the body. **2.** To cringe in a servile way. —*vt.* To cause to bend low, as in humility or fear. **—crouch** *n.* **—crouch'ing·ly** *adv.*
**croup[1]** (kro͞op) *n.* [Orig. unknown.] A condition affecting the larynx in children, marked by respiratory difficulty and a harsh cough. **—croup'ous** (kro͞o'pəs), **croup'y** *adj.*
**croup[2]** *also* **croupe** (kro͞op) *n.* [ME *croupe* < OFr., of Germanic orig.] The rump of an animal, esp. a horse.
**crou·pi·er** (kro͞o'pē-ər, -pē-ā') *n.* [Fr.< *croupe,* rump, croup < OFr.] An attendant at a gaming table who collects and pays bets.
**crou·ton** (kro͞o'tŏn', kro͞o-tŏn') *n.* [Fr. *croûton* < *croûte,* crust < OFr. *crouste* < Lat. *crusta.*] A small crisp piece of toasted or fried bread.
**crow[1]** (krō) *n.* [ME *croue* < OE *crāwe.*] **1.** A large, glossy, black bird of the genus *Corvus,* with a raucous call, esp. *C. brachyrhynchos* of North America. **2.** A crowbar. **—as the crow flies.** In a straight line. **—eat crow.** *Informal.* To be humiliated, as from having been proved mistaken.
**crow[2]** (krō) *vi.* **crowed, crow·ing, crows.** [ME *crouen* < OE *crāwan.*] **1.** *p.t. & p.p.* **crowed** *or* **crew** (kro͞o). To utter the shrill cry characteristic of a cock or rooster. **2. a.** To exult, esp. over another's misfortune. **b.** To boast exultantly. **3.** To make a sound expressing pleasure or well-being, characteristic of an infant. —*n.* **1.** The shrill cry of a cock. **2.** An inarticulate sound expressing delight or pleasure.
**Crow** (krō) *n., pl.* **Crow** *or* **Crows. 1. a.** A tribe of Indians once inhabiting the region between the Platte and Yellowstone rivers and now settled in southeastern Montana. **b.** A member of this tribe. **2.** The Siouan language of the Crow.
**crow·bar** (krō'bär') *n.* [From the resemblance of its forked end to a crow's foot.] A straight bar of steel or iron, with the working end shaped like a forked chisel, used as a lever.
**crow·ber·ry** (krō'bĕr'ē) *n.* **1. a.** A low-growing evergreen shrub, *Empetrum nigrum* of cool regions of the Northern Hemisphere, bearing small purplish flowers and black, berrylike fruit. **b.** A similar or related plant, as the bearberry. **2.** The fruit of a crowberry.
**crow blackbird** *n.* GRACKLE 1.
**crowd[1]** (kroud) *n.* [< ME *crowden,* to crowd < OE *crudan,* to hasten.] **1.** A large number of people gathered together: THRONG. **2.** The common people. **3.** A particular social group : CLIQUE. **4.** A large number of things grouped or considered together. —*v.* **crowd·ed, crowd·ing, crowds.** —*vi.* **1.** To congregate in a close place : THRONG. **2.** To advance by shoving. —*vt.* **1.** To press : shove. **2.** To cram tightly together. **3.** To fill to overflowing. **4.** *Informal.* To put pressure on. **—crowd'er** *n.*
**crowd[2]** (kroud, kro͞od) *n.* [ME *croud* < Welsh *crwth.*] An ancient Celtic stringed musical instrument.
**crow·foot** (krō'fo͝ot') *n.* **1.** *pl.* **-foots. a.** A plant of the genus *Ranunculus,* which includes the buttercups and spearworts, esp. a similar plant, as *R. abortivus* or *R. scleratus,* bearing small inconspicuous yellow flowers. **b.** Any of various other plants bearing leaves or other parts resembling a bird's foot. **2.** *pl.* **-feet.** A caltrop. **3.** *pl.* **-feet.** *Naut.* **a.** A block used in supporting the middle section of an awning. **b.** A set of small lines passed through holes of a batten to help support the backbone of an awning.
**crown** (kroun) *n.* [ME *crowne* < OFr. *corone* < Lat. *corona,* wreath < Gk. *korōnē* < *korōnos,* curved.] **1.** An ornamental circlet or head covering, often made of precious metal set with jewels and worn as an emblem of sovereignty. **2.** *often* **Crown. a.** The power, position, or empire of a monarch. **b.** The monarch as head of state. **3.** A reward or distinction for achievement, esp. a title signifying the championship in a sport. **4.** Something like a crown. **5.** A coin whose reverse side is stamped with a crown or crowned head. **6. a.** A former British coin worth five shillings. **b.** A coin, as the krona or krone, with a name meaning crown. **c.** —See table at CURRENCY. **7. a.** The highest part of the head. **b.** The head itself. **8.** The upper part of a hat. **9.** The highest point : SUMMIT. **10.** The highest or primary attribute, quality, or state. **11. a.** The part of a tooth covered by enamel and projecting beyond the gum line. **b.** An artificial top for a tooth. **12.** The lowest part of an anchor where the arms are joined to it. **13. a.** The upper part of a tree. **b.** The part of a plant, usu. at ground level, between the root and the stem. **c.** CORONA 5. **14.** An animal's crest, esp. that of a bird. **15.** The part of a cut gem above the girdle. —*vt.* **crowned, crown·ing, crowns. 1.** To put a crown or garland on the head of. **2.** To invest with regal power : ENTHRONE. **3.** To confer honor, dignity, or reward on. **4.** To surmount or be the highest part of. **5.** To form the crown, top, or chief ornament of. **6.** To bring to completion or successful conclusion : CLIMAX. **7.** To put a crown on (a tooth). **8.** To make (a checkers piece that has reached the last row) into a king by placing another piece on it. **9.** *Informal.* To hit on the head.
**crown canopy** *n.* The cover formed by the upper branches of trees in a forest.
**crown colony** *n.* A British colony in which the monarch retains some control over legislation, usu. administered by an appointed governor.
**crown cover** *n.* Crown canopy.
**crown glass** *n.* **1.** A clear soda-lime-silica optical glass with low refraction. **2.** A form of window glass made by whirling a glass bubble to make a flat circular disk with a lump in the center formed by the craftsman's rod.
**crown lens** *n.* The crown-glass element in an achromatic lens.
**crown-of-thorns** (kroun'əv-thôrnz') *n.* A spiny, vinelike desert plant, *Euphorbia splendens,* bearing showy scarlet flowers.
**crown prince** *n.* The heir apparent to a throne.
**crown princess** *n.* **1.** The wife of a crown prince. **2.** An heiress presumptive to a throne.
**crown rot** *n.* A disease of plants marked by the localized degeneration of the stem near ground level.
**crown saw** *n.* A cylindrical saw with teeth on the bottom edge of the cylinder.
**crown vetch** *n.* A sprawling plant native to Europe, *Coronilla varia,* with compound leaves and pink flower clusters.
**crow's-foot** (krōz'fo͝ot') *n., pl.* **-feet** (-fēt'). **1.** *often* **crow's-feet.** One of the wrinkles at the outer corner of the eye. **2.** A three-pointed embroidery stitch used as finishing, as at the end of a seam.
**crow's-nest** (krōz'nĕst') *n.* **1.** *Naut.* A small lookout platform with a high protective railing and wind screen near the top of a ship's mast. **2.** A crow's-nest located ashore.
**croze** (krōz) *n.* [Prob. < OFr. *crues,* groove.] The groove at the ends of the staves of a barrel or cask into which the head is set.
**cro·zier** (krō'zhər) *n. var. of* CROSIER.
**cru·ces** (kro͞o'sēz') *n. var. pl. of* CRUX.
**cru·cial** (kro͞o'shəl) *adj.* [OFr., cross-shaped < Lat. *crux,* cross.] **1.** Of supreme importance : CRITICAL <a *crucial* battle> **2.** Difficult : severe. **—cru'cial·ly** *adv.*
**cru·ci·ate** (kro͞o'shē-āt') *adj.* [NLat. *cruciatus* < Lat. *crux,* cross.] **1.** Being in the form of a cross : CRUCIFORM. **2.** Overlapping or crossing, as the wings of some insects when at rest. **—cru'ci·ate'ly** *adv.*
**cru·ci·ble** (kro͞o'sə-bəl) *n.* [ME *crusible* < Med. Lat. *crucibulum,* crucible, night-light, poss. < Lat. *crux,* cross.] **1.** A vessel made of a refractory substance, as porcelain or graphite, and used for melting and calcining materials at high temperatures. **2.** The bottom of an ore furnace, in which the molten metal collects. **3.** A severe trial.
**crucible steel** *n.* High-grade steel made by fusing low-carbon steel with charcoal or cast iron in a graphite crucible, used in tools and dies.
**cru·ci·fer** (kro͞o'sə-fər) *n.* [LLat. : Lat. *crux,* cross + Lat. *-fer,* -fer.] **1.** The crossbearer in a religious procession. **2.** *Bot.* A plant of the family Cruciferae, as a mustard or cress, bearing four-petaled flowers resembling a cross. **—cru·cif'er·ous** (-sĭf'ər-əs) *adj.*
**cru·ci·fix** (kro͞o'sə-fĭks') *n.* [ME < LLat. *crucifixus* < p.part. of *crucifigere,* crucify.] An image of Christ on the cross.
**cru·ci·fix·ion** (kro͞o'sə-fĭk'shən) *n.* **1.** The act of putting to death on a cross. **2.** A representation of Christ on the cross. **3. the Crucifixion.** The crucifying of Christ on Calvary.
**cru·ci·form** (kro͞o'sə-fôrm') *adj.* [Lat. *crux, cruc-,* cross + -FORM.] Forming a cross or arranged in the shape of a cross. **—cru'ci·form'** *n.* **—cru'ci·form'ly** *adv.*
**cru·ci·fy** (kro͞o'sə-fī') *vt.* **-fied, -fy·ing, -fies.** [ME *crucifien* < OFr. *crucifier* < LLat. *crucifigere* : Lat. *crux,* cross + Lat. *figere,* to attach.] **1.** To put (a person) to death by nailing or binding to a cross. **2.** To mortify or subdue (the flesh). **3.** To treat unfairly or cruelly : TORMENT <was *crucified* by the media> **—cru'ci·fi'er** *n.*
**†crud** (krŭd) *n.* [ME *crudde.*] **1.** *Slang.* **a.** A coating or incrustation of filth or refuse. **b.** One that is disgusting or contemptible. **2.** An imaginary or real disease, esp. one affecting the skin. **3.** *Regional.* A milk curd. **—crud'dy** *adj.*
**crude** (kro͞od) *adj.* **crud·er, crud·est.** [ME < Lat. *crudus.*] **1.** Being in an unrefined or natural state : RAW. **2.** *Archaic.* Unripe : immature. **3.** Lacking tact, refinement, or taste. **4.** Not carefully or completely made : ROUGH. **5.** Displaying a lack of knowledge or skill.

**6.** Blunt : undisguised. —*n.* Unrefined petroleum. **—crude'ly** *adv.* **—cru'di·ty** (kro͞o'dĭ-tē), **crude'ness** *n.*

**crude oil** *n.* Petroleum.

**cru·di·tés** (kro͞o'dĭ-tā') *pl.n.* [Fr., pl. of *crudité*, indigestibility < Lat. *cruditas*, indigestion, undigested food < *crudus*, raw.] Raw vegetables, as carrot sticks, radishes, and pepper strips, often served with a dip as an appetizer.

**cru·el** (kro͞o'əl) *adj.* **-el·er, -el·est** *or* **-el·ler, -el·lest.** [ME < OFr. < Lat. *crudelis.*] **1.** Disposed to inflict pain or suffering. **2.** Causing suffering : PAINFUL. **—cru'el·ly** *adv.* **—cru'el·ty** *n.*

**cru·et** (kro͞o'ĭt) *n.* [ME < AN, dim. of OFr. *crue*, flask, of Germanic orig.] A small glass bottle for holding condiments, as vinegar or oil, at the table.

**cruise** (kro͞oz) *v.* **cruised, cruis·ing, cruis·es.** [Du. *kruisen*, to cross < MDu. *crucen* < Lat. *crux*, cross.] —*vi.* **1.** To sail or travel about, as for pleasure or reconnaissance. **2.** To travel at a speed providing maximum operating efficiency for a sustained period. **3. a.** To move leisurely about a place in search of something <taxicabs *cruising* for passengers> **b.** *Slang.* To look for a sexual partner, as in a public place. **4.** To inspect a wooded area to determine its lumber yield. —*vt.* **1.** To cruise or journey over. **2.** To inspect so as to determine lumber yield. —*n.* A sea voyage for pleasure.

**cruise missile** *n.* A long-range, low-flying guided missile that can be launched from air, sea, or land.

**cruis·er** (kro͞o'zər) *n.* **1.** One of a class of fast warships of medium tonnage with a long cruising radius and less firepower and armor than a battleship. **2.** A large motorboat having a cabin with living facilities. **3.** A squad car.

**cruising radius** *n.* The longest distance a ship or aircraft can go and return at cruising speed without refueling.

**crul·ler** (krŭl'ər) *n.* [Du. *krulle* < *krullen*, to curl < *krul*, curly < MDu. *crul.*] A small, usu. ring-shaped or twisted cake of sweet dough fried in deep fat.

**crumb** (krŭm) *n.* [ME *crome* < OE *cruma.*] **1.** A small piece from bread, cake, or other baked goods. **2.** A small scrap or fragment. **3.** The soft inner part of bread. **4.** *Slang.* A contemptible person. —*vt.* **crumbed, crumb·ing, crumbs. 1.** To crumble. **2.** To cover or prepare with bread crumbs : BREAD. **3.** To brush (a table or cloth) clear of crumbs. —*vi.* To break apart in crumbs.

**crum·ble** (krŭm'bəl) *v.* **-bled, -bling, -bles.** [ME *cremelen* < OE *(ge)crymian* < *cruma*, crumb.] —*vt.* To break or cause to break into crumbs. —*vi.* To fall into crumbs : DISINTEGRATE.

**crum·bly** (krŭm'blē) *adj.* **-bli·er, -bli·est.** Easily crumbled. **—crum'bli·ness** *n.*

**crumb·y** (krŭm'ē) *adj. var. of* CRUMMY.

**crum·mie** (krŭm'ē) *n.* [< Sc. *crumb*, crooked < ME < OE.] *Scot.* A cow with crooked horns.

**crum·my** *also* **crumb·y** (krŭm'ē) *adj.* **-mi·er, -mi·est** *also* **-i·er, -i·est.** [< CRUMB.] *Slang.* **1.** Miserable : wretched. **2.** Shabby : cheap.

**crump** (krŭmp) *v.* **crumped, crump·ing, crumps.** [Imit.] —*vt.* **1.** To crunch or crush with the teeth. **2.** To strike heavily with a crunching sound. —*vi.* To make a crunching sound. —*n.* **1.** A heavy crunching sound <the *crump* of heavy artillery> **2.** A heavy blow.

**crum·pet** (krŭm'pĭt) *n.* [Poss. < ME *crompid (cake)*, curled (cake) < *crumpen*, to curl up < *crumb*, crooked < OE.] *Chiefly Brit.* A light, soft bread similar to a muffin, baked on a griddle and often toasted.

**crum·ple** (krŭm'pəl) *v.* **-pled, -pling, -ples.** [Prob. freq. of obs. *crump*, to curl up < ME *crumpen.* —see CRUMPET.] —*vt.* **1.** To crush together or press into wrinkles : RUMPLE. **2.** To cause to fall apart. —*vi.* **1.** To become wrinkled : SHRIVEL. **2.** To fall apart : COLLAPSE. —*n.* An irregular crease, fold, or wrinkle.

**crunch** (krŭnch) *v.* **crunched, crunch·ing, crunch·es.** [Imit.] —*vt.* **1.** To chew with a noisy crackling sound. **2.** To crush, grind, or tread noisily. —*vi.* **1.** To chew noisily with a crackling sound. **2.** To move with a crushing sound. **3.** To produce or emit a crushing sound. —*n.* **1.** The act or sound of crunching. **2.** *Informal.* **a.** A decisive confrontation. **b.** A critical situation. **—crunch'y** *adj.*

**crunch·er** (crŭnch'ər) *n. Slang.* A decisive blow.

**crup·per** (krŭp'ər) *n.* [ME *crouper* < OFr. *cropiere* < *croup*, rump, of Germanic orig.] **1.** A leather strap looped under a horse's tail and attached to a saddle or harness to keep it from slipping forward. **2.** The rump of a horse.

**cru·ra** (kro͝or'ə) *n. pl. of* CRUS.

**cru·ral** (kro͝or'əl) *adj.* [Lat. *cruralis* < *crus*, leg.] Of or relating to the leg, shank, or thigh.

**crus** (kro͞os, krŭs) *n., pl.* **cru·ra** (kro͝or'ə) [NLat. *crus, crur-* < Lat., leg.] **1.** The part of the leg or hind limb between the knee and foot : SHANK. **2.** A leglike part.

**cru·sade** (kro͞o-sād') *n.* [Partly < Fr. *croisade*, and partly < Sp. *cruzada*, both < Med. Lat. *cruciata* < Lat. *crux*, cross.] **1.** *often* **Crusade.** One of the military expeditions undertaken by European Christians in the 11th, 12th, and 13th cent. to win back the Holy Land from the Moslems. **2.** A holy war undertaken with papal sanction. **3.** A vigorous concerted movement against an abuse or for a cause. —*vi.* **-sad·ed, -sad·ing, -sades.** To take part in a crusade. **—cru·sad'er** *n.*

**cruse** (kro͞oz, kro͞os) *n.* [ME *crouse*, perh. < MDu. *cruyse*, pot.] A small pot or jar for holding water, wine, or oil.

**crush** (krŭsh) *v.* **crushed, crush·ing, crush·es.** [ME *crushen* < OFr. *croissir.*] —*vt.* **1.** To press between opposing bodies so as to break or injure. **2.** To extract or obtain by pressing or squeezing <*crush* juice from tomatoes> **3.** To rumple or crumple. **4.** To hug forcibly. **5.** To break, pound, or grind (e.g., stone or ore) into small fragments or powder. **6.** To shove or crowd. **7.** To put down : SUBDUE. **8.** To overwhelm or oppress severely <*crushed* by heavy debts> —*vi.* **1.** To be or become crushed. **2.** To proceed or move by crowding or pressing. —*n.* **1.** The act of crushing or the state of being crushed. **2.** A great crowd : THRONG. **3.** A substance prepared by or as if by crushing. **4.** *Informal.* **a.** An infatuation. **b.** The object of an infatuation. **—crush'a·ble** *adj.* **—crush'er** *n.*

**crust** (krŭst) *n.* [ME *cruste* < OFr. *crouste* < Lat. *crusta*, shell.] **1.** The hard outer portion or surface area of bread. **2.** A piece of bread consisting chiefly of the hard outer portion. **3.** A pastry shell, as of a pie or tart. **4.** A hard, crisp covering or surface. **5.** A hard deposit that maturing wine produces on the interior of bottles. **6. a.** *Geol.* The exterior portion of the earth that lies above the Mohorovičić discontinuity. **b.** The outermost solid layer of a moon or planet. **7.** The hard outer covering or integument of certain plants and animals, as crustaceans and lichens. **8.** *Pathol.* A coating or dry outer layer, as of blood or pus : SCAB. **9.** *Slang.* Insolence : gall. —*v.* **crust·ed, crust·ing, crusts.** —*vt.* **1.** To cover with a crust. **2.** To form (dough) into a crust. —*vi.* **1.** To become covered with a crust. **2.** To harden into a crust.

**crus·ta·cean** (krŭ-stā'shən) *n.* [< NLat. *Crustacea*, class name < Lat. *crusta*, shell.] Any of various predominantly aquatic arthropods of the class Crustacea, including lobsters, crabs, shrimps, and barnacles, having segmented bodies, chitinous exoskeletons, and paired, jointed limbs. —*adj.* Of or belonging to the Crustacea.

**crus·ta·ceous** (krŭ-stā'shəs) *adj.* [Lat. *crusta*, shell + -ACEOUS.] **1.** Having, resembling, or being a hard crust or shell. **2.** Crustacean.

**crust·al** (krŭs'təl) *adj.* Of or relating to a crust, esp. that of the earth or the moon.

**crus·tose** (krŭs'tōs') *adj.* [Lat. *crustosus*, crusted < *crusta*, crust.] Relating to a lichen whose thallus is crusty and thin.

**crust·y** (krŭs'tē) *adj.* **-i·er, -i·est. 1.** Resembling or having a crust. **2.** Brusque. **—crust'i·ly** *adv.* **—crust'i·ness** *n.*

**crutch** (krŭch) *n.* [ME *crucche* < OE *crycc.*] **1.** A staff or support used by the disabled as an aid in walking, usu. designed to fit under the armpit and often used in pairs. **2.** A forked device similar to a crutch, as a forked leg rest on a sidesaddle. **3.** Something depended on for support. **4.** The human crotch. —*vt.* **crutched, crutch·ing, crutch·es.** To support on or as if on crutches : PROP.

**crux** (krŭks, kro͝oks) *n., pl.* **crux·es** *or* **cru·ces** (kro͞o'sēz') [Lat., cross.] **1.** A crucial or vital moment : TURNING POINT. **2.** The basic, essential, or central feature <the *crux* of the difficulties> **3.** A puzzling problem. **4. Crux.** A constellation in the Southern Hemisphere.

**cru·zei·ro** (kro͞o-zâr'ō, -rō͞o) *n., pl.* **-ros.** [Port. < *cruz*, cross (from the figure on the coin) < Lat. *crux.*] —See table at CURRENCY.

**cry** (krī) *v.* **cried, cry·ing, cries.** [ME *crien* < OFr. *crier* < Lat. *quiritare*, to cry for help (from a citizen) < *Quirites*, Roman citizens.] —*vi.* **1.** To make inarticulate sobbing sounds expressing grief, sorrow, or pain : WEEP. **2.** To call loudly : SHOUT. **3.** To utter a characteristic sound or call. —Used of an animal. **4.** To demand or require immediate remedy or action <injustices *crying* out for correction> —*vt.* **1.** To utter loudly. **2.** To proclaim or announce in public <*cried* the good news for all to hear> **3.** To beg for : IMPLORE. **4.** To bring into a particular condition by weeping <*cry* oneself to sleep> **—cry down.** To belittle : disparage. **—cry off.** To break or withdraw from a promise, agreement, or undertaking. **—cry up.** To praise highly : EXTOL. —*n., pl.* **cries. 1.** A loud emotional utterance. **2.** A loud exclamation, as a call or shout. **3.** A fit of weeping. **4.** An urgent entreaty or appeal <a *cry* for clemency> **5.** A general demand or complaint : CLAMOR. **6.** An advertising of wares by calling out. **7.** A rallying call or signal. **8.** A political slogan. **9.** The characteristic call or utterance of a bird or animal. **10.** A pack of hounds. **—cry over spilled milk.** To worry about what cannot be undone or rectified. **—in full cry.** In hot pursuit.

☆ ***syns:*** CRY, BAWL, BOOHOO, HOWL, KEEN, SOB, WAIL, WEEP *v. core meaning* : to make inarticulate sounds of grief or pain, usu. accompanied by tears <*cried* after falling down>

**cry·baby** (krī'bā'bē) *n.* One who cries or complains often with little cause.

**cry·ing** (krī'ĭng) *adj.* Demanding or requiring immediate action or remedy <a *crying* need to feed the hungry children>

**cry·mo·ther·a·py** (krī'mō-thĕr'ə-pē) *n.* [Fr. *crymothérapie* : *crymo-*, cryo- + *thérapie*, therapy.] *var. of* CRYOTHERAPY.

**cryo-** *pref.* [< Gk. *kruos*, icy cold.] Cold : freezing <*cryoscopy*>

**cry·o·bi·ol·o·gy** (krī'ō-bī-ŏl'ə-jē) *n.* Study of the effects of very low temperatures on living organisms. **—cry'o·bi'o·log'i·cal** (-bī'ə-lŏj'ĭ-kəl) *adj.* **—cry'o·bi'o·log'i·cal·ly** *adv.* **—cry'o·bi·ol'o·gist** *n.*

**cry·o·gen** (krī′ə-jən) *n.* A refrigerant for obtaining very low temperatures.

**cry·o·gen·ic** (krī′ə-jĕn′ĭk) *adj.* Of or relating to low temperatures. **—cry′o·gen′i·cal·ly** *adv.*

**cry·o·gen·ics** (krī′ō-jĕn′ĭks) *n. (sing. in number).* The science of low-temperature phenomena.

**cry·og·e·ny** (krī-ŏj′ə-nē) *n.* Cryogenics.

**cry·o·lite** (krī′ə-līt′) *n.* A white, vitreous natural fluoride of aluminum and sodium, $Na_3AlF_6$, used mainly as an electrolyte in aluminum refining and in electrical insulation.

**cry·om·e·ter** (krī-ŏm′ĭ-tər) *n.* A thermometer capable of measuring very low temperatures.

**cry·on·ics** (krī-ŏn′ĭks) *n.* [CRY(O)- + *-onics,* as in *bionics.*] *(sing. in number).* Freezing and storing a corpse to prevent tissue decomposition so that at a future time the individual might be brought back to life when new medical cures have been developed. **—cry·on′ic** (-ĭk) *adj.*

**cry·o·phil·ic** (krī′ə-fĭl′ĭk) *also* **cry·oph·i·lous** (krī-ŏf′ə-ləs) *adj.* Having an affinity for or thriving at low temperatures.

**cry·o·plank·ton** (krī′ə-plăngk′tən) *n.* Minute organisms living in ice, snow, or perpetually icy waters.

**cry·o·probe** (krī′ə-prōb′) *n.* A surgical instrument for applying extreme cold to tissues during cryosurgery.

**cry·o·scope** (krī′ə-skōp′) *n.* [Back-formation < CRYOSCOPY.] An instrument for measureing the freezing point of a substance.

**cry·os·co·py** (krī-ŏs′kə-pē) *n.* Study of the freezing points of solutions. **—cry′o·scop′ic** (-ə-skŏp′ĭk) *adj.*

**cry·o·stat** (krī′ə-stăt′) *n.* A device for maintaining constant low temperature.

**cry·o·sur·ger·y** (krī′ō-sûr′jə-rē) *n.* Selective exposure of tissues to extreme cold to bring about cell destruction. **—cry′o·sur′geon** *n.* **—cry′o·sur′gi·cal** *adj.*

**cry·o·ther·a·py** (krī′ō-thĕr′ə-pē) *also* **cry·mo·ther·a·py** (krī′-mō-) *n.* Use of low temperatures in medical therapy.

**crypt** (krĭpt) *n.* [Lat. *crypta* < Gk. *kruptē* < *kruptos,* hidden < *kruptein,* to hide.] **1.** An underground chamber or vault, esp. one used as a burial place beneath a church. **2.** *Anat.* A small pit, recess, glandular cavity, or follicle in the body.

**crypt-** *pref. var. of* CRYPTO-.

**crypt·aes·the·sia** (krĭp′təs-thē′zhə, -zhē-ə) *n. var. of* CRYPTESTHESIA.

**crypt·a·nal·y·sis** (krĭp′tə-năl′ĭ-sĭs) *n.* [CRYPT(OGRAM) + ANALYSIS.] Analysis and deciphering of cryptograms. **—crypt·an′a·lyst** (-tăn′ə-lĭst) *n.* **—crypt·an′a·lyt′ic** (-tăn′ə-lĭt′ĭk) *adj.*

**crypt·es·the·sia** *or* **crypt·aes·the·sia** (krĭp′təs-thē′zhə, -zhē-ə) *n. Psychol.* One of the modes of paranormal perception, as clairvoyance.

**cryp·tic** (krĭp′tĭk) *also* **cryp·ti·cal** (-tĭ-kəl) *adj.* [LLat. *crypticus* < Gk. *kruptikos* < *kruptos,* hidden < *kruptein,* to hide.] **1.** Having an ambiguous meaning : ENIGMATIC <*cryptic* remarks> **2.** Secret or occult : MYSTIFYING. **3.** *Biol.* Tending to conceal or camouflage <*cryptic* markings> **—cryp′ti·cal·ly** *adv.*

**crypto-** or **crypt-** *pref.* [< Gk. *kruptos,* hidden < *kruptein,* to hide.] Hidden : secret <*crypto*clastic>

**cryp·to·clas·tic** (krĭp′tō-klăs′tĭk) *adj.* Made up of microscopic fragments. —Used of rocks.

**cryp·to·coc·co·sis** (krĭp′tə-kŏ-kō′sĭs) *n.* [NLat. *Cryptoccus,* fungus genus + -OSIS.] A systemic infection caused by the fungus *Cryptococcus neoformans,* which can affect any organ of the body but most often occurs in the central nervous system.

**cryp·to·crys·tal·line** (krĭp′tō-krĭs′tə-lĭn) *adj.* Having a microscopic crystalline structure.

**cryp·to·gam** (krĭp′tə-găm′) *n.* [< NLat. *cryptogamia* : CRYPTO- + *-gamia,* -gamy.] *Bot.* Any of the flowerless and seedless plants that reproduce by spores, as fungi, algae, ferns, and mosses. **—cryp′to·gam′ic, cryp·tog′a·mous** (-tŏg′ə-məs) *adj.*

**cryp·to·gen·ic** (krĭp′tə-jĕn′ĭk) *also* **cryp·tog·e·nous** (krĭp-tŏj′ə-nəs) *adj.* Of unknown or obscure origin. —Used of diseases.

**cryp·to·gram** (krĭp′tə-grăm′) *n.* **1.** Something written in code or cipher. **2.** A figure with a secret or occult significance. **—cryp′to·gram′mic** *adj.*

**cryp·to·graph** (krĭp′tə-grăf′) *n.* [Back-formation < CRYPTOGRAPHY.] **1.** A cryptogram. **2.** A system of secret writing : CIPHER. **3. a.** A device for translating plain text into cipher. **b.** A device for deciphering codes and ciphers.

**cryp·tog·ra·phy** (krĭp-tŏg′rə-fē) *n.* **1.** The art or process of writing in or deciphering secret code. **2.** A system of secret writing. **—cryp·tog′ra·pher, cryp·tog′ra·phist** *n.* **—cryp′to·graph′ic** (-tə-grăf′ĭk) *adj.* **—cryp′to·graph′i·cal·ly** *adv.*

**cryp·to·me·ri·a** (krĭp′tə-mîr′ē-ə) *n.* [NLat. *Cryptomeria,* genus name : CRYPTO- + Gk. *meros,* part.] An evergreen tree native to Japan, *Cryptomeria japonica,* bearing short, inward-curving needles and soft, durable, fragrant wood.

**crypt·or·chism** (krĭp-tôr′kĭz′əm) *also* **crypt·or·chi·dism** (-kĭ-dĭz′əm) *n.* [NLat. *cryptorchidismus* : CRYPT(O)- + *orchis,* testicle < Gk. *orkhis.*] Failure of the testes to descend into the scrotum. **—crypt·or′chid** *n.*

**cryp·to·zo·ite** (krĭp′tə-zō′īt′) *n.* [CRYPTO- + (SPORO)ZOITE.] A malaria parasite as it exists in bodily tissue prior to invasion of the red blood cells.

**crys·tal** (krĭs′təl) *n.* [ME *cristal* < OFr. < Lat. *crystallum* < Gk. *krustallos.*] **1. a.** A three-dimensional atomic, ionic, or molecular structure consisting of periodically repeated, identically constituted, congruent unit cells. **b.** The unit cell of such a structure. **2.** A body, as a piece of quartz, having a crystalline structure, often marked by external planar faces visible without magnification. **3.** An electronic device, as an oscillator or detector, based on crystalline piezoelectricity, magnetism, semiconductivity, or other electric properties. **4. a.** A high-quality clear, colorless glass. **b.** An object, esp. a vessel or ornament, made of such glass. **c.** Such objects as a whole. **5.** A clear glass or plastic protective cover for the face of a watch or clock. *—adj.* Clear or transparent <a *crystal* pond><the *crystal* clarity of the speech>

**crystal ball** *n.* A glass globe for crystal gazing.

**crystal detector** *n.* A rectifying detector used esp. in early radio receivers and consisting of a semiconducting crystal in point contact with a fine metal wire.

**crystal gazing** *n.* Divination by concentrating on a crystal ball. **—crystal gazer** *n.*

**crys·tal·ize** (krĭs′tə-līz′) *v. var. of* CRYSTALLIZE.

**crystall-** *pref. var. of* CRYSTALLO-.

**crys·tal·lif·er·ous** (krĭs′tə-lĭf′ər-əs) *also* **crys·tal·lig·er·ous** (-lĭj′-) *adj.* Generating or containing crystals.

**crys·tal·line** (krĭs′tə-lĭn, -līn′) *adj.* [ME *cristallin* < OFr. < Lat. *crystallinus* < Gk. *krustallinos* < *krustallos,* crystal.] **1.** Relating to or made of crystal. **2.** Clear like crystal : TRANSPARENT. **—crys′tal·lin′i·ty** (-lĭn′ĭ-tē) *n.*

**crystalline lens** *n.* LENS 3.

**crys·tal·lite** (krĭs′tə-līt′) *n.* [G. *Kristallit* < Gk. *krustallos,* crystal.] Any of numerous minute rudimentary, crystalline bodies found in glassy igneous rocks. **—crys′tal·lit′ic** (-lĭt′ĭk) *adj.*

**crys·tal·lize** *also* **crys·tal·ize** (krĭs′tə-līz′) *v.* **-lized, -liz·ing, -liz·es** *also* **-ized, -iz·ing, -iz·es.** *—vt.* **1.** To cause to form crystals or to assume a crystalline structure. **2.** To give a definite and permanent form to <*crystallizing* the overall concept> **3.** To coat with sugar. *—vi.* **1.** To assume a crystalline form. **2.** To take on a definite, permanent form. **—crys′tal·liz′a·ble** *adj.* **—crys′tal·li·za′tion** *n.* **—crys′tal·liz′er** *n.*

**crystallo-** *or* **crystall-** *pref.* [< Gk. *krustallos,* crystal.] Crystal <*crystallize*>

**crys·tal·log·ra·phy** (krĭs′tə-lŏg′rə-fē) *n.* The science of crystal structure. **—crys′tal·log′ra·pher** *n.* **—crys′tal·lo·graph′ic** (-lə-grăf′ĭk), **crys′tal·lo·graph′i·cal** *adj.* **—crys′tal·lo·graph′i·cal·ly** *adv.*

**crys·tal·loid** (krĭs′tə-loid′) *n.* **1.** *Chem.* A water-soluble crystalline substance capable of diffusion through a semipermeable membrane. **2.** *Bot.* Any of various minute crystallike particles consisting of protein, found in certain plant cells, esp. oily seeds. *—adj.* Having properties of a crystal or crystalloid. **—crys′tal·loi′dal** (-loid′l) *adj.*

**crystal pickup** *n.* A phonographic pickup with a piezoelectric crystal to convert stylus vibrations into electric impulses.

**crystal set** *n.* An early radio receiver using a crystal detector.

**crystal violet** *n.* Dye derived from rosaniline and used as a general biological stain.

**Cs** *symbol for* CESIUM.

**cte·nid·i·um** (tĭ-nĭd′ē-əm) *n., pl.* **-i·a** (-ē-ə) [NLat. < Gk. *kteis, kten-,* comb.] *Zool.* A comblike structure, as the respiratory apparatus of a mollusk or a row of spines in some insects.

**cten·oid** (tĕn′oid′, tē′noid′) *adj.* [Gk. *ktenoeidēs,* comblike : *kteis,* comb + *-eidēs,* -oid.] *Biol.* Having narrow segments or spines similar to the teeth of a comb <fish with *ctenoid* scales>

**cten·o·phore** (tĕn′ə-fôr′, -fōr′) *n.* [NLat. *Ctenophora,* phylum name : Gk. *kteis,* comb + Gk. *pherein,* to bear.] Any of various marine animals of the phylum Ctenophora, having transparent, gelatinous bodies bearing eight rows of comblike cilia used for locomotion. **—cte·noph′o·ran** (tĭ-nŏf′ər-ən) *adj.*

**Cu** [Lat. *cuprum.* —see COPPER.] *symbol for* COPPER.

**cub** (kŭb) *n.* [Orig. unknown.] **1.** The young of certain carnivorous animals, as the bear, wolf, or lion. **2.** An awkward or inexperienced youth. **3.** A learner or novice, esp. in newspaper reporting. **4. Cub.** A Cub Scout.

**cub·age** (kyo͞o′bĭj) *n.* Cubic content or volume.

**cu·ba·ture** (kyo͞o′bə-cho͝or′, -chər) *n.* [CUB(E) + (QUADR)ATURE.] **1.** Determination of the cubic contents of a solid. **2.** Cubage.

**cub·by** (kŭb′ē) *n., pl.* **-bies.** [< obs. *cub,* stall, prob. of LG orig.] A cubbyhole.

**cub·by·hole** (kŭb′ē-hōl′) *n.* **1.** A snug or cramped space or room. **2.** A small compartment. **3.** A small cupboard or closet.

**cube** (kyo͞ob) *n.* [OFr. < Lat. *cubus* < Gk. *kubos,* six-sided die.] **1.** *Math.* A regular solid with six congruent square faces. **2.** Something gen. shaped like a cube. **3.** *Math.* The third power of a number or quantity. —*vt.* **cubed, cub·ing, cubes. 1.** To raise (a quantity or number) to the third power. **2.** To ascertain the cubic contents of. **3.** To cut into cubes : DICE. **4.** To tenderize (meat) by breaking the fibers with superficial cuts in a pattern of squares. **—cub'er** *n.*

**cu·bé** *also* **cu·be** (kyōō'bā', kyōō-bā') *n.* [Am. Sp.] **1.** Any of various tropical American shrubs or plants, esp. of the genus *Lonchocarpus,* whose roots yield rotenone. **2.** An extract from the roots of the cubé plants, used as an insecticide and fish poison.

**cu·beb** (kyōō'bĕb') *n.* [ME *cubibe* < OFr. *cubebe* < Med. Lat. *cubeba* < Ar. *kabābah.*] **1.** A treelike woody vine, *Piper cubeba* of southeastern Asia, bearing brownish berries. **2.** The dried, unripe, spicy fruit of the cubeb, used as a diuretic and stimulant and occas. smoked in cigarettes.

**cube root** *n.* A number whose cube is equal to a given number.

**cube steak** *n.* A thin slice of beef made tender by cubing.

**cu·bic** (kyōō'bĭk) *adj.* **1.** Shaped like a cube. **2. a.** Having three dimensions. **b.** Having a volume equal to a cube whose edge is of a stated length <a *cubic* yard> **3.** *Math.* Of the third power, order, or degree. **4.** Isometric. —*n. Math.* A cubic expression, curve, or equation. **—cu'bic·ly** *adv.*

**cu·bi·cal** (kyōō'bĭ-kəl) *adj.* **1.** Cubic. **2.** Of or relating to volume. **—cu'bi·cal·ly** *adv.* **—cu'bi·cal·ness** *n.*

**cu·bi·cle** (kyōō'bĭ-kəl) *n.* [ME < Lat. *cubiculum,* bed chamber < *cubare,* to lie down.] **1.** A small sleeping compartment. **2.** A small partitioned space, as in an office.

**cubic measure** *n.* A unit or a system of units for measuring volume or capacity.

**cu·bi·form** (kyōō'bə-fôrm') *adj.* CUBIC 1.

**cub·ism** (kyōō'bĭz'əm) *n.* [Fr. *cubisme* < *cube,* cube.] A nonrepresentational school of painting and sculpture developed in Paris in the early 20th cent., marked by the reduction and fragmentation of natural forms into abstract, often geometric structures. **—cub'ist** *n.* **—cu·bis'tic** *adj.* **—cu·bis'ti·cal·ly** *adv.*

**cu·bit** (kyōō'bĭt) *n.* [ME *cubite* < Lat. *cubitum,* elbow.] An ancient unit of linear measure, orig. equal to the length of the forearm from the elbow to the tip of the middle finger, or from 17 to 22 inches or 43 to 56 centimeters.

**cu·boid** (kyōō'boid') *adj.* **1.** Having the shape or approximate shape of a cube. **2.** *Anat.* Designating a bone on the side of the tarsus between the calcaneus and the fourth and fifth metatarsal bones of the foot. —*n. Anat.* **1.** The cuboid bone. **2.** *Math.* A rectangular parallelepiped. **—cu·boi'dal** (kyōō-boid'l) *adj.*

**Cub Scout** *n.* A member of the junior division of the Boy Scouts.

**cu·chi·fri·to** (kōō'chĭ-frē'tō) *n.* [Am. Sp. : *cuchi,* pig (< Sp. *cochino*) + Sp. *frito,* p.part. of *freir,* to fry (< Lat. *frigere*).] A small deep-fried cube of pork.

**Cu·chul·ain** (kōō'hōōl-ĭn) *n.* [OIr. *Cú Chulainn* : *Cú,* hound + *Culainn,* of Culann, a legendary Irish smith.] *Myth.* A tribal hero of Ulster who alone defended it against the rest of Ireland.

**cuck·ing stool** (kŭk'ĭng) *n.* [ME *cukking stol* < *cukken,* to defecate, of Scand. orig.] A former instrument of punishment, consisting of a chair in which the offender was tied and exposed to public derision or ducked in water.

**cuck·old** (kŭk'əld, kōōk'-) *n.* [ME *cokewald* < OFr. **cucuald* < *cucu,* the cuckoo.] A man whose wife has committed adultery. —*vt.* **-old·ed, -old·ing, -olds.** To make a cuckold of.

**cuck·oo** (kōō'kōō, kōōk'ōō) *n., pl.* **-oos.** [ME *cuccu.*] **1. a.** An Old World bird, *Cuculus canorus,* with grayish plumage and a two-note call. **b.** Any of various related birds of the family Cuculidae, including several New World species. **2.** The cry or call of a cuckoo. **3.** A fool. —*vt.* **-ooed, -oo·ing, -oos.** To repeat over and over. —*adj.* Foolish.

**cuckoo clock** *n.* A wall clock with a mechanical cuckoo announcing intervals of time.

**cuck·oo·flow·er** (kōō'kōō-flou'ər, kōōk'ōō-) *n.* **1.** A plant, *Cardamine pratensis* of the North Temperate Zone, bearing white or rose-pink flowers. **2.** The ragged robin.

**cuck·oo·pint** (kōō'kōō-pĭnt', kōōk'ōō-) *n.* [Obs. *cuckoopintle* < ME *cokkupyntel* : *cokku,* cuckoo + *pintle,* penis < OE *pintel.*] A European plant, *Arum maculatum,* with arrow-shaped leaves and a spadix enclosed in a purple-spotted spathe.

**cuckoo spit** *n.* A frothy mass of liquid secreted on plant stems as a protective covering by nymphs of the spittlebug.

**cu·cu·li·form** (kyōō'kə-lə-fôrm') *adj.* [NLat. *Cuculiformes,* order name : Lat. *cuculus,* cuckoo + Lat. *forma,* shape.] Of or belonging to the order Cuculiformes, including the cuckoos and related birds.

**cu·cul·late** (kyōō'kə-lāt', kyōō-kŭl'āt') *also* **cu·cul·lat·ed** (kyōō'kə-lā'tĭd) *adj.* [Med. Lat. *cucullatus* < Lat. *cucullus,* hood.] Shaped like a hood or cowl <*cucullate* sepals> **—cu'cul·late'ly** *adv.*

**cu·cum·ber** (kyōō'kŭm'bər) *n.* [ME *cucomer* < OFr. *coucombre* < Lat. *cucumis.*] **1.** A vine, *Cucumis sativus,* cultivated for its edible fruit. **2.** The usu. cylindrical fruit of the cucumber vine, with a hard green rind and succulent white flesh.

**cucumber mosaic** *n.* A viral disease of the cucumber plant that produces a variegated spotting of the fruits and leaves.

**cucumber tree** *n.* A tree, *Magnolia acuminata* of eastern and central North America, bearing cup-shaped greenish-yellow flowers and scarlet or brown cucumber-shaped fruit.

**cu·cur·bit** (kyōō-kûr'bĭt) *n.* [ME *cucurbite,* < OFr. < Lat. *cucurbita,* gourd.] **1.** A gourd-shaped flask forming the body of an alembic, at one time used in distillation. **2.** Any of various vines of the family Cucurbitaceae, including the squash, pumpkin, and cucumber.

**cud** (kŭd) *n.* [ME < OE *cudu.*] **1.** Food regurgitated from the first stomach to the mouth of a ruminant and chewed again. **2.** Something, as a quid of tobacco, suitable to be held in the mouth and chewed. **—chew (one's) cud.** To ponder over : MEDITATE.

**cud·bear** (kŭd'bâr') *n.* [After *Cuthbert* Gordon, 18th-cent. chemist.] A purplish-red coloring substance derived from some lichens.

**cud·dle** (kŭd'l) *v.* **-dled, -dling, -dles.** [Orig. unknown.] —*vt.* To hug tenderly. —*vi.* To snuggle : nestle. **—cud'dle** *n.* **—cud'dle·some** *adj.* **—cud'dly** *adj.*

**cud·dy¹** (kŭd'ē) *n., pl.* **-dies.** [Orig. unknown.] **1.** A small cabin or the cook's galley on a ship. **2.** A small room.

**cud·dy²** (kŭd'ē) *n., pl.* **-dies.** [Perh. < *Cuddy,* nickname for *Cuthbert.*] *Scot.* **1.** A donkey. **2.** A dolt : fool.

**cudg·el** (kŭj'əl) *n.* [ME *cuggel* < OE *cycgel.*] A short, heavy club. —*vt.* **-eled, -el·ing, -els** *also* **-elled, -el·ling, -els.** To strike or beat with a cudgel. **—cudg'el·er** *n.*

**cud·weed** (kŭd'wēd') *n.* **1.** A woolly plant of the genus *Gnaphalium,* bearing whitish or yellow buttonlike flower clusters. **2.** An annual herb, *Filago germanica,* with clusters of woolly heads.

**cue¹** (kyōō) *n.* [Fr. *queue,* tail < OFr. *coue* < Lat. *cauda.*] **1.** A long, tapered rod for propelling the ball in pool and billiards. **2.** A long stick with a concave attachment at one end for shoving disks in shuffleboard. **3.** A queue of hair. —*vt.* **cued, cu·ing, cues. 1.** To strike with a cue. **2.** To braid or twist (hair) into a cue.

**cue²** (kyōō) *n.* [Orig. unknown.] **1.** A bit of stage business or word signaling the start of another action or speech. **2. a.** A reminder or prompting as a signal to do something. **b.** A suggestion or hint. **3.** *Psychol.* A perceived signal for action, esp. one producing an operant response. —*vt.* **cued, cu·ing, cues.** To give (a performer) a cue. **—cue in.** To tell a latecomer what has happened up to now.

**cue³** (kyōō) *n.* The letter *q.*

**cue ball** *n.* The white ball that is propelled with the cue in pool and billiards.

**†cues·ta** (kwĕs'tə) *n.* [Sp., sloping side < Lat. *costa,* side.] *Southwestern U.S.* A land elevation with a gentle slope on one side and a cliff on the other.

**cuff¹** (kŭf) *n.* [ME *cuffe,* mitten.] **1.** A fold or band for trimming the bottom of a sleeve. **2.** The turned-up fold at the bottom of a trouser leg. **3.** The part of a glove covering the wrist. **4.** A handcuff. **—off the cuff.** *Informal.* Extemporaneously.

**cuff²** (kŭf) *vt.* **cuffed, cuff·ing, cuffs.** [Orig. unknown.] To strike with the open hand : SLAP. **—cuff** *n.*

**cuff links** *pl.n.* A pair of linked buttons for fastening shirt cuffs.

**Cu·fic** (kōō'fĭk, kyōō'-) *adj. var. of* KUFIC.

**cui·rass** (kwĭ-răs') *n.* [ME *curace* < OFr. *cuirasse* < Lat. *coriaceus,* of leather < *corium,* hide.] **1. a.** Armor for protecting the back and breast. **b.** An armored breastplate. **2.** *Zool.* A protective covering of bony plates or scales. —*vt.* **-rassed, -rass·ing, -rass·es.** To protect or cover with a cuirass.

**cui·ras·sier** (kwĭr'ə-sîr') *n.* [Fr. < *cuirasse,* cuirasse < OFr.] A mounted trooper in European armies equipped with a cuirass.

**Cui·se·naire** (kwē'zə-nâr'). A trademark for a set of colored rods used to teach arithmetic.

**cuish** (kwĭsh) *n. var. of* CUISSE.

**cui·sine** (kwĭ-zēn') *n.* [Fr. < LLat. *coquina,* cookery, kitchen < *coquere,* to cook.] **1.** A characteristic style or manner of preparing food. **2.** Food prepared in a particular style or way.

**cuisse** (kwĭs) *also* **cuish** (kwĭsh) *n.* [Back-formation < ME *cuisses* (pl.) < OFr. *cuisseaux,* pl. of *cuissel* < *cuisse,* thigh < Lat. *coxa,* hip.] Plate armor for protecting the thighs.

**culch** (kŭlch) *n.* [Perh. ult. < OFr. *culche,* couch.] **1.** A natural bed for oysters, consisting of crushed shells or gravel to which oyster spawn may adhere. **2.** The spawn of the oyster. **3.** Rubbish.

**cul-de-sac** (kŭl'dĭ-săk', kōōl'-) *n., pl.* **cul-de-sacs.** [Fr. : *cul,* bottom + *de,* of the + *sac,* sack.] **1. a.** A dead-end street. **b.** An impasse. **2.** *Anat.* A saclike cavity or tube open only at one end.

**cu·let** (kyōō'lĭt, kŭl'ĭt) *n.* [Fr., dim. of *cul,* rump < Lat. *culus.*] **1.** The flat face of a gem cut as a brilliant. **2.** A series of medieval armor plates covering the lower back.

**cu·lex** (kyōō'lĕks') *n., pl.* **-li·ces** (-lĭ-sēz') [NLat. *Culex,* genus name < Lat. *culex,* gnat.] A mosquito of the genus *Culex,* esp. the common house mosquito, *C. pipiens.*

**cu·li·nar·y** (kyōō'lə-nĕr'ē, kŭl'ə-) *adj.* [Lat. *culinarius* < *culina,* kitchen, alteration of *coquina.* —see CUISINE.] Of or relating to cookery or a kitchen. **—cu'li·nar'i·ly** (-nâr'ə-lē) *adv.*

**cull** (kŭl) *vt.* **culled, cull·ing, culls.** [ME *cullen* < OFr. *cuillir* < Lat. *colligere.* —see COLLECT¹.] **1.** To pick out from others : SELECT.

**2.** To collect : gather. —*n.* Something picked out from others, esp. something rejected due to inferior quality. **—cull'er** *n.*

**cul·len·der** (kŭl'ən-dər) *n. var. of* COLANDER.

**cul·let** (kŭl'ĭt) *n.* [Alteration of *collet,* neck of glass left on the blowing iron < Fr., collar, dim. of *col,* neck < OFr. < Lat. *collum.*] Scraps of broken or waste glass gathered for remelting.

**cul·lis** (kŭl'ĭs) *n.* [ME *colis* < OFr. *coleïs,* channel < *coler,* to pour < Lat. *colare,* to filter < *colum,* sieve.] A groove or gutter in a roof.

**culm**[1] (kŭlm) *n.* [Lat. *culmus,* stalk.] The jointed stem of a grass or sedge.

**culm**[2] (kŭlm) *n.* [ME *colme,* coal dust.] **1.** Waste from anthracite coal mines, composed of fine coal, coal dust, and dirt. **2. a.** Carboniferous shale. **b.** Inferior anthracite coal.

**cul·mi·nant** (kŭl'mə-nənt) *adj.* **1.** At the highest altitude. **2.** Highest : culminating.

**cul·mi·nate** (kŭl'mə-nāt') *vi.* **-nat·ed, -nat·ing, -nates.** [LLat. *culminare, culminat-* < Lat. *culmen,* summit.] **1.** To reach the highest point or degree : CLIMAX. **2.** *Astron.* To reach the highest point above an observer's horizon. —Used of celestial bodies. **—cul'mi·na'tion** *n.*

**cu·lottes** (ko͞o-lŏts', kyo͞o-, ko͞o'lŏts', kyo͞o'-) *pl.n.* [Fr., breeches, dim. of *cul,* rump < Lat. *culus.*] A woman's full trousers cut to resemble a skirt.

**cul·pa** (kŭl'pə, ko͝ol'-) *n.* [Lat.] *Law.* Misconduct : fault.

**cul·pa·ble** (kŭl'pə-bəl) *adj.* [ME *coupable* < OFr. < Lat. *culpabilis* < *culpare,* to blame < *culpa,* fault.] Responsible for wrong or error : BLAMEWORTHY. **—cul'pa·bil'i·ty** *n.* **—cul'pa·bly** *adv.*

**cul·prit** (kŭl'prĭt) *n.* [Prob. < AN *culpable,* culpable < Lat. *culpabilis.*] **1.** One charged with an offense or crime. **2.** One guilty of a fault or crime.

**cult** (kŭlt) *n.* [Fr. *culte* < Lat. *cultus,* worship < p.part. of *colere,* to cultivate.] **1.** A community or system of religious worship and ritual. **2. a.** A religion or religious sect gen. regarded as bogus or extremist. **b.** Followers of such a religion or sect. **3. a.** Obsessive devotion to a person, principle, or ideal. **b.** The object of such devotion. **4.** An exclusive group of persons sharing an esoteric interest. **—cul'tic** (kŭl'tĭk) *adj.* **—cult'ish** *adj.* **—cult'ism** *n.* **—cult'ist** *n.*

**cul·ti·gen** (kŭl'tə-jən) *n.* [CULTI(VATED) + -GEN.] An organism, esp. a cultivated plant such as maize, of a kind not known to have a wild or uncultivated counterpart.

**cul·ti·va·ble** (kŭl'tə-və-bəl) *adj.* That can be cultivated. **—cul'ti·va·bil'i·ty** *n.*

**cul·ti·var** (kŭl'tə-vär', -vâr') *n.* [CULTI(VATED) + VAR(IETY).] A horticulturally or agriculturally derived variety of a plant.

**cul·ti·vate** (kŭl'tə-vāt') *vt.* **-vat·ed, -vat·ing, -vates.** [Med. Lat. *cultivare, cultivat-* < *cultivus,* tilled < Lat. *cultus,* p.part. of *colere,* to till.] **1. a.** To prepare and improve (land), as by fertilizing or plowing, for raising crops : TILL. **b.** To dig or loosen (soil) around growing plants. **2.** To tend or grow (a plant or crop). **3.** To foster the growth of (e.g., a biological culture). **4.** To nurture. **5.** To form and refine, as by education. **6.** To seek the good will or acquaintance of. **—cul'ti·vat'a·ble** *adj.*

**cul·ti·vat·ed** (kŭl'tə-vā'tĭd) *adj.* Refined : cultured.

**cul·ti·va·tion** (kŭl'tə-vā'shən) *n.* **1.** The act of cultivating or state of being cultivated. **2.** Refinement : social polish.

**cul·ti·va·tor** (kŭl'tə-vā'tər) *n.* One that cultivates, esp. an implement or machine for loosening the earth and destroying weeds around growing plants.

**cul·trate** (kŭl'trāt') *also* **cul·trat·ed** (-trā'tĭd) *adj.* [Lat. *cultratus* < *culter,* knife.] Sharp-edged and pointed : KNIFELIKE.

**cul·tur·al** (kŭl'chər-əl) *adj.* **1.** Of or relating to culture. **2.** Obtained by specialized breeding. **—cul'tur·al·ly** *adv.*

**cultural anthropology** *n.* Scientific study of human culture based on archaeological, ethnologic, ethnographic, linguistic, social, and psychological data and methods of analysis.

**cul·ture** (kŭl'chər) *n.* [ME, cultivation < OFr. < Lat. *cultura* < *cultus.* —see CULTIVATE.] **1.** The totality of socially transmitted behavior patterns, arts, beliefs, institutions, and all other products of human work and thought typical of a population or community at a given time. **2.** A style of social and artistic expression peculiar to a class or society. **3.** Intellectual and artistic activity. **4.** The act of developing the social, moral, and intellectual faculties through education. **5.** A high degree of refinement and taste formed by intellectual and aesthetic training. **6.** Development of the body through special training <physical *culture*> **7.** Cultivation of soil : TILLAGE. **8.** The growing of plants or breeding of animals, esp. to produce improved stock. **9.** *Biol.* **a.** The growing of microorganisms in a nutrient medium. **b.** A growth or colony of microorganisms, as bacteria. **c.** MEDIUM 6b. —*vt.* **-tured, -tur·ing, -tures. 1.** CULTIVATE 1. **2.** To develop (e.g., microorganisms or tissues) in a culture medium.

☆ ***syns:*** CULTURE, CIVILIZATION, KULTUR *n. core meaning:* the total product of human creativity and intellect at a particular time <the *culture* of ancient Greece>

**cul·tured** (kŭl'chərd) *adj.* **1.** Cultivated : refined. **2.** Produced under controlled artificial conditions <*cultured* pearls>

**culture shock** *n.* A condition of anxiety and confusion that can affect an individual suddenly exposed to an alien culture or milieu.

**cul·tus** (kŭl'təs) *n., pl.* **-tus·es** *or* **-ti** (-tī') [NLat. < Lat. —see CULT.] A religious cult.

**cul·ver** (kŭl'vər) *n.* [ME < OE *culufre* < VLat. **columbra* < Lat. *columbula,* dim. of *columba,* dove.] A dove : pigeon.

**cul·ver·in** (kŭl'vər-ĭn) *n.* [ME < OFr. *coulevrine* < *couleuvre,* snake < Lat. *colubra,* fem. of *coluber.*] **1.** An early, crudely made musket. **2.** A heavy 16th- and 17th-cent. cannon.

**Cul·ver's root** (kŭl'vərz) *n.* [After Dr. *Culver,* 18th-cent. American physician.] **1.** A North American plant, *Veronicastrum virginicum,* bearing small white or purplish flower spikes. **2.** The root of the Culver's root, once used as a cathartic and emetic.

**cul·vert** (kŭl'vərt) *n.* [Orig. unknown.] A sewer or drain running under a road or embankment.

**cum** (ko͝om, kŭm) *prep.* [Lat.] Together with : PLUS <built a cellar-*cum*-photographic darkroom>

**cum·ber** (kŭm'bər) *vt.* **-bered, -ber·ing, -bers.** [ME *combren,* to annoy, short for *acombren,* perh. of OFr. orig.] **1.** To weigh down : BURDEN. **2.** To hamper : hinder. **3.** To clutter up <Many small structures *cumbered* the lawns.> —*n.* A hindrance : encumbrance. **—cum'ber·er** *n.*

**cum·ber·some** (kŭm'bər-səm) *adj.* **1.** Hard to maneuver due to weight or bulk : UNWIELDY. **2.** Burdensome. **—cum'ber·some·ly** *adv.*

**cum·brous** (kŭm'brəs) *adj.* [ME < *cumbren,* to annoy.] Unwieldy : cumbersome.

**cum gra·no sa·lis** (ko͝om grä'nō sä'lĭs, kŭm grä'nō sā'lĭs) *adv.* [Lat.] With a grain of salt : SKEPTICALLY.

**cum·in** (kŭm'ĭn) *n.* [ME < OFr. < Lat. *cuminum* < Gk. *kuminon,* of Semitic orig.] **1.** An Old World plant, *Cuminum cyminum,* with finely divided leaves and small white or pinkish flowers. **2.** The aromatic seeds of the cumin, used as a condiment.

**cum lau·de** (ko͝om lou'də, lou'dē, kŭm lô'dē) *adv.* & *adj.* [NLat., with praise.] With honor. —Used on diplomas as a mark of high academic standing.

**cum·mer·bund** (kŭm'ər-bŭnd') *n.* [Hindi *kamarband* < Pers., waistband : *kamar,* loins + *band,* band.] **1.** A broad pleated sash worn as an accessory to men's formal dress. **2.** A women's clothing accessory similar to a cummerbund.

**cum·quat** (kŭm'kwŏt') *n. var. of* KUMQUAT.

**cum·shaw** (kŭm'shô') *n.* [Pidgin E. < Chin. (Amoy) *kam sia,* an expression of thanks.] A gratuity.

**cumul-** *pref. var. of* CUMULO-.

**cu·mu·late** (kyo͞om'yə-lāt') *v.* **-lat·ed, -lat·ing, -lates.** [Lat. *cumulare, cumulat-* < *cumulus,* heap.] —*vt.* **1.** To gather in a heap : ACCUMULATE. **2.** To merge into a single unit. **3.** To expand by an increment in new material. —*vi.* To become massed. **—cu'mu·la'tion** *n.*

**cu·mu·la·tive** (kyo͞om'yə-lā'tĭv, -yə-lə-tĭv) *adj.* **1.** Enlarging or increasing by successive addition. **2.** Acquired by or resulting from accumulation. **3.** Of or relating to a dividend or interest that is added to the next payment if not paid when due. **4.** *Law.* Designating additional or supporting evidence. **5.** *Statistics.* **a.** Of or relating to the sum of the frequencies of experimentally determined values of a random variable that are less than or equal to a specified value. **b.** Of or relating to experimental error that increases in magnitude with each successive measurement. **—cu'mu·la'tive·ly** *adv.* **—cu'mu·la'tive·ness** *n.*

**cumulative voting** *n.* A system of voting in proportional representation in which each voter is given as many votes as there are positions to be filled and allowed to cast those votes for one candidate or distributed in any manner among the candidates.

**cu·mu·li** (kyo͞om'yə-lī') *n. pl. of* CUMULUS.

**cumuli-** *pref. var. of* CUMULO-.

**cu·mu·li·form** (kyo͞om'yə-lə-fôrm') *adj.* Shaped like a cumulus.

**cumulo-** *or* **cumuli-** *or* **cumul-** *pref.* [< CUMULUS.] Cumulus <*cumulo*nimbus>

**cu·mu·lo·nim·bus** (kyo͞om'yə-lō-nĭm'bəs) *n., pl.* **-bus·es** *or* **-bi** (-bī'). A very dense, vertically developed cumulus with a rather hazy outline and a glaciated top, usu. producing heavy rains, thunderstorms, or hailstorms.

**cu·mu·lus** (kyo͞om'yə-ləs) *n., pl.* **-li** (-lī') [NLat. < Lat., heap.] **1.** A dense, white, fluffy, flat-based cloud with a multiple rounded top and a well-defined outline, usu. formed by the ascent of thermally unstable air masses. **2.** A pile, mound, or heap. **—cu'mu·lous** *adj.*

**cunc·ta·tion** (kŭngk-tā'shən) *n.* [Lat. *cunctatio, cunctation-* < *cunctus,* p.part. of *cunctari,* to delay.] A delay. **—cunc'ta'tive** (kŭngk'tā'tĭv, -tə-tĭv), **cunc'ta·to'ry** (-tə-tôr'ē, -tōr'ē) *adj.*

**cu·ne·al** (kyo͞o'nē-əl) *adj.* [NLat. *cunealis* < Lat. *cuneus,* wedge.] Wedge-shaped.

**cu·ne·ate** (kyo͞o'nē-ĭt, -āt') *also* **cu·ne·at·ed** (-ā'tĭd) *adj.* [Lat. *cuneatus* < *cuneus,* wedge.] Wedge-shaped <*cuneate* leaves> **—cu'ne·ate'ly** *adv.*

**cu·ne·i·form** (kyo͞o'nē-ə-fôrm', kyo͞o-nē'-) *adj.* [Fr. *cunéiforme* < Lat. *cuneus,* wedge.] **1.** Wedge-shaped. **2. a.** Designating the wedge-shaped characters used in ancient Sumerian, Akkadian, Assyrian, Babylonian, and Persian writing. **b.** Designating documents or inscriptions written in such characters. **3.** *Anat.* Denoting any of the three wedge-shaped bones in the tarsus of the foot. —*n.* **1.** Cuneiform writing. **2.** A cuneiform bone.

**cun·ner** (kŭn′ər) *n.* [Orig. unknown.] A marine fish, *Tautogolabrus adspersus* of North American Atlantic waters.

**†cun·ning** (kŭn′ĭng) *adj.* [ME, pr.part. of *connen,* to know < OE *cunnan.*] **1.** Shrewd : crafty. **2.** Executed with or displaying ingenuity. **3.** *Regional.* Delicately pleasing : PRETTY. —*n.* **1.** Skill in deception : GUILE. **2.** Skill in performance : DEXTERITY. **—cun′ning·ly** *adv.* **—cun′ning·ness** *n.*

**cup** (kŭp) *n.* [ME *cuppe* < OE < LLat. *cuppa,* drinking vessel.] **1.** A small, open container, usu. with a flat bottom and a handle, used for drinking. **2.** The contents of a cup. **3.** A measure of capacity equal to 1/2 pint, 8 ounces, 16 tablespoons, or approx. 237 milliliters. **4.** The bowl of a drinking vessel. **5.** The chalice or the wine used in celebrating the Eucharist. **6.** An ornamented cup-shaped vessel awarded as a prize. **7.** A golf hole or the metal container inside a hole. **8.** Either of the two parts of a brassiere that fit over the breasts. **9.** A rigid athletic supporter. **10.** Any of various beverages, usu. combining wine, fruit, and spices. **11.** Something served in a cup-shaped vessel <fruit *cup*> **12.** Something like a cup. **13.** *Biol.* A cuplike structure or organ. **14.** A lot or portion to be enjoyed or suffered. —*vt.* **cupped, cup·ping, cups. 1.** To place in or as if in a cup. **2.** To shape like a cup <*cup* one's hands> **—cup of tea.** Something one enjoys or is expert in <Political protest is their *cup of tea.*> **—in one's cups.** Intoxicated.

**cup·bear·er** (kŭp′bâr′ər) *n.* One who serves wine.

**cup·board** (kŭb′ərd) *n.* A closet or cabinet, usu. with shelves for storing food, crockery, and utensils.

**cup·cake** (kŭp′kāk′) *n.* A cake baked in a small cup.

**cu·pel** (kyōō′pəl, kyōō-pĕl′) *n.* [Fr. *coupelle,* dim. of *coupe,* cup < LLat. *cuppa,* drinking vessel.] **1.** A small, shallow, porous vessel used in assaying to separate precious metals from less valuable elements such as lead. **2.** The bottom or receptacle in a silver-refining furnace. **—cu′pel** *v.* **(-peled, -pel·ing, -pels** *or* **-pelled, -pel·ling, -pels). —cu′pel·er** *n.*

**cu·pel·la·tion** (kyōō′pə-lā′shən) *n.* A refining process for nonoxidizing metals, as silver and gold, in which the components of a metallic mixture oxidized at high temperatures are separated by absorption into the walls of a cupel.

**cup·ful** (kŭp′fōōl′) *n., pl.* **-fuls. 1.** The amount that a cup will hold. **2.** A measure of capacity equal to 1/2 pint, 8 ounces, 16 tablespoons, or approx. 237 milliliters.

**Cu·pid** (kyōō′pĭd) *n.* [Lat. *Cupido* < *cupido,* desire < *cupere,* to desire.] **1.** *Rom. Myth.* The god of love. **2.** A representation of a winged boy with a bow and arrow, used as a symbol of love.

**cu·pid·i·ty** (kyōō-pĭd′ĭ-tē) *n.* [ME *cupidite* < OFr. < Lat. *cupiditas* < *cupidus,* desiring < *cupere,* to desire.] Excessive desire, esp. for wealth : AVARICE.

**cu·po·la** (kyōō′pə-lə) *n.* [Ital. < LLat. *cupula,* dim. of Lat. *cupa,* tub.] **1. a.** A domed roof or ceiling. **b.** A small, usu. domed structure on a roof. **2.** A cylindrical shaft blast furnace for remelting usu. iron before casting.

**cup·ping** (kŭp′ĭng) *n.* A former therapeutic process in which glass cups, partially evacuated by heating, were applied to the skin to draw blood to the surface.

**cup plant** *n.* [From the cuplike configuration of its leaves.] A coarse North American plant, *Silphium perfoliatum,* bearing yellow-rayed flowers.

**cupr-** *pref. var. of* CUPRO-.

**cu·pre·ous** (kōō′prē-əs, kyōō′-) *adj.* [LLat. *cupreus* < *cuprum,* copper. —see COPPER[1].] Of, relating to, resembling, or containing copper : COPPERY.

**cupri-** *pref. var. of* CUPRO-.

**cu·pric** (kōō′prĭk, kyōō′-) *adj.* Of or containing divalent copper.

**cu·prif·er·ous** (kōō-prĭf′ər-əs, kyōō-) *adj.* Containing copper.

**cu·prite** (kōō′prīt′, kyōō′-) *n.* [G. *Kuprit* < LLat. *cuprum,* copper. —see COPPER[1].] Natural red copper ore, essentially $Cu_2O$.

**cupro-** *or* **cupri-** *or* **cupr-** *pref.* [LLat. *cuprum.* —see COPPER[1].] Copper <*cupri*ferous>

**cu·pro·nick·el** (kōō′prō-nĭk′əl, kyōō′-) *n.* A copper-based alloy containing 10–30% nickel.

**cu·prous** (kōō′prəs, kyōō′-) *adj.* Of, relating to, or containing univalent copper.

**cu·pu·late** (kyōō′pyə-lāt′, -lĭt) *also* **cu·pu·lar** (-lər) *adj.* **1.** Shaped like a cup. **2.** Having or bearing a cupule.

**cu·pule** (kyōō′pyōōl) *n.* [NLat. *cupula* < LLat., little cask, dim. of Lat. *cupa,* tub.] *Biol.* A cup-shaped part, structure, or indentation, esp. the involucre of an acorn.

**cur** (kûr) *n.* [ME *curre.*] **1.** An inferior dog : MONGREL. **2.** A base or cowardly person.

**cur·a·ble** (kyŏŏr′ə-bəl) *adj.* Capable of being healed or cured. **—cur′a·bil′i·ty, cur′a·ble·ness** *n.* **—cur′a·bly** *adv.*

**cu·ra·çao** (kyŏŏr′ə-sō′, -sou′, kōōr′-) *n.* [After *Curaçao,* an island in the Caribbean.] A liqueur flavored with sour orange peel.

**cu·ra·cy** (kyŏŏr′ə-sē) *n., pl.* **-cies.** [CURA(TE) + -CY.] The office, duties, or term of office of a curate.

**cu·ra·re** *or* **cu·ra·ri** (kōō-rä′rē, kyōō-) *n.* [Port. and Sp., both < Cariban *kurari.*] **1.** Any of various resinous extracts of uncertain and variable chemical composition, obtained from several species of South American trees of the genera *Chondodendron* and *Strychnos,* used medicinally as a muscle relaxant and by some South American Indians as an arrow poison. **2.** Any of the trees from which curare is obtained.

**cu·ra·rine** (kōō-rä′rĭn, -rēn′, kyōō-) *n.* [CURAR(E) + -INE[2].] A poisonous alkaloid, $C_{19}H_{26}N_2O$, derived from curare.

**cu·ra·rize** (kōō-rä′rīz′, kyōō-) *vt.* **-rized, -riz·ing, -riz·es. 1.** To poison with curare. **2.** To treat with curare so as to paralyze the motor nerves. **—cu·ra′ri·za′tion** *n.*

**cu·ras·sow** (kŏŏr′ə-sō′, kyŏŏr′-) *n.* [Alteration of *Curaçao,* an island in the Caribbean.] Any of several long-tailed, crested tropical American birds of the family Cracidae, related to the pheasants and domestic fowl.

**cu·rate** (kyŏŏr′ĭt) *n.* [ME *curat* < Med. Lat. *curatus* < *cura,* spiritual charge < Lat., care.] **1.** A member of the clergy in charge of a parish. **2.** A member of the clergy who assists a rector or vicar.

**cu·ra·tive** (kyŏŏr′ə-tĭv) *adj.* **1.** Serving or tending to cure. **2.** Of or relating to the cure of disease. —*n.* A remedy. **—cu′ra·tive·ly** *adv.* **—cu′ra·tive·ness** *n.*

**cu·ra·tor** (kyōō-rā′tər, kyŏŏr′ə-tər) *n.* [ME *curatour,* legal guardian < OFr. *curateur* < Lat. *curator,* overseer < *curare,* to take care of < *cura,* care.] The administrator of an institution, as a museum. **—cu′ra·to′ri·al** (kyŏŏr′ə-tôr′ē-əl, -tōr′-) *adj.* **—cu·ra′tor·ship′** *n.*

**curb** (kûrb) *n.* [OFr. *courbe,* horse bit < *courbe,* curved < Lat. *curvus.*] **1.** A restraint or check. **2.** A concrete border or row of joined stones forming part of a gutter along the edge of a street. **3.** An enclosing framework. **4.** A raised margin along an edge to confine or strengthen. **5.** A strap or chain serving in conjunction with the bit to restrain a horse. —*vt.* **curbed, curb·ing, curbs. 1.** To check, restrain, or control. **2.** To lead (a dog) off the sidewalk into the gutter so that it can excrete waste matter. **3.** To furnish with a curb. **—curb′er** *n.*

**curb·ing** (kûr′bĭng) *n.* **1.** The material for constructing a curb. **2.** CURB 2.

**curb roof** *n.* A roof with two slopes on each side.

**curb·stone** (kûrb′stōn′) *n.* A stone or row of stones forming a curb.

**cur·cu·li·o** (kər-kyōō′lē-ō′) *n., pl.* **-os.** [Lat., a kind of weevil.] Any of several weevils of the family Curculionidae, many of which are destructive to fruit and vegetables.

**cur·cu·ma** (kûr′kyə-mə) *n.* [NLat. *Curcuma,* genus name < Ar. *kurkum,* saffron.] Any of various Old World tropical plants of the genus *Curcuma,* with thick, aromatic rootstocks, including *C. longa,* the source of turmeric.

**curd** (kûrd) *n.* [ME *crud.*] **1.** The coagulated part of milk, used for making cheese. **2.** A coagulation resembling curd. —*vi. & vt.* **curd·ed, curd·ing, curds.** To curdle : coagulate. **—curd′y** *adj.*

**curd cheese** *n. Chiefly Brit.* Cottage cheese.

**cur·dle** (kûr′dl) *vi. & vt.* **-dled, -dling, -dles.** [Freq. of CURD.] To change into or cause to change into curd.

**cure** (kyŏŏr) *n.* [ME, duty, care < OFr. < Lat. *cura.*] **1.** Restoration of health. **2.** A method or course of medical treatment for restoring health. **3.** A restorative agent, as a drug : REMEDY. **4.** Something that relieves or corrects a harmful or disturbing situation. **5.** Spiritual charge or care of souls, as of a priest for a congregation. **6.** The office or duties of a curate. **7.** The act or process of preserving a product, as fish, meat, or tobacco. —*v.* **cured, cur·ing, cures.** —*vt.* **1.** To restore to health. **2.** To get rid of : REMEDY <*cure* an evil> **3.** To preserve (e.g., meat), as by salting, smoking, or aging. **4.** To prepare, preserve, or finish (a substance) by a chemical or physical process. **5.** To vulcanize (rubber). —*vi.* **1.** To effect a cure. **2.** To be prepared, preserved, or finished by a chemical or physical process. **—cure′less** *adj.* **—cur′er** *n.*

☆ **syns:** CURE, ELIXIR, NOSTRUM, REMEDY *n. core meaning* : an agent used to restore health <found no *cure* for cancer>

**cu·ré** (kyōō-rā′, kyŏŏr′ā′) *n.* [Fr.] A parish priest.

**cure-all** (kyŏŏr′ôl′) *n.* A remedy for all diseases or evils : PANACEA.

**cu·ret** (kyōō-rĕt′) *n. var. of* CURETTE.

**cu·ret·tage** (kyŏŏr′ĭ-täzh′) *n.* Surgical cleaning or scraping of a bodily cavity with a curette.

**cu·rette** *also* **cu·ret** (kyōō-rĕt′) *n.* [Fr. < *curer,* to cure < OFr. < Lat. *curare,* to take care of < *cura,* care.] A scoop, spoon, or loop for performing curettage.

**cu·rette·ment** (kyōō-rĕt′mənt) *n.* Curettage.

**cur·few** (kûr′fyōō) *n.* [ME *curfeu* < OFr. *cuevrefeu,* cover the fire : *couvrir,* to cover + *feu,* fire < Lat. *focus,* hearth.] **1.** An order or regulation enjoining specified segments of the population to leave the streets at a prescribed hour. **2. a.** The period during which a curfew regulation is in effect. **b.** The signal announcing curfew.

▲ word history: A *curfew* was originally a medieval regulation requiring that fires be put out or covered at a certain hour at night. The rule was probably instituted as a public safety measure to minimize the risk of a general conflagration. A bell was rung at the prescribed hour, and the word *curfew* has been extended to denote both the signal and the hour in addition to the regulation.

**cu·ri·a** (kŏŏr′ē-ə, kyŏŏr′-) *n., pl.* **cu·ri·ae** (kŏŏr′ē-ē′, kyŏŏr′-) [Lat., council.] **1. a.** One of the ten primitive subdivisions of a tribe in early Rome, consisting of ten gentes. **b.** The curia's place of assembly. **2. a.** The Senate or any of the various buildings in which it met in republican Rome. **b.** The place of assembly of high councils in various Italian cities under Roman administration. **3.** The ensemble of central administrative and governmental services in imperial Rome. **4.** *often* **Curia.** The central administration governing the Roman Catholic Church. **5. a.** A medieval assembly or council. **b.** A royal court of justice. **—cu′ri·al** *adj.*

**cu·rie** (kyŏŏr′ē, kyŏŏ-rē′) *n.* [After Marie *Curie* (1867–1934).] A unit of radioactivity, the amount of any nuclide that undergoes exactly 3.7 x $10^{10}$ radioactive disintegrations per second.

**Curie law** *n.* [After Pierre *Curie* (1859–1906).] The law that the magnetic susceptibility varies inversely with absolute temperature in a paramagnetic substance with negligible interactions among magnetic carriers.

**Curie point** *or* **Curie temperature** *n.* [After Pierre *Curie.*] A transition temperature marking a change in the magnetic properties of a substance, esp. the change from ferromagnetism to paramagnetism.

**Cu·rie-Weiss law** (kyŏŏr′ē-wīs′, -vīs′, kyŏŏ-rē′-) *n.* [After Pierre *Curie* and Pierre-Ernest *Weiss* (1865–1940).] The law that the magnetic susceptibility of a paramagnetic substance above the Curie point varies inversely with the excess of temperature above that point.

**cu·ri·o** (kyŏŏr′ē-ō′) *n., pl.* **-os.** [Short for CURIOSITY.] An unusual object of art.

**cu·ri·o·sa** (kyŏŏr′ē-ō′sə, -zə) *pl.n.* [NLat., neut. pl. of Lat. *curiosus,* inquisitive. —see CURIOUS.] Books or other writings dealing with unusual, esp. pornographic topics.

**cu·ri·os·i·ty** (kyŏŏr′ē-ŏs′ĭ-tē) *n., pl.* **-ties. 1.** A desire to learn or know. **2.** A desire to know about matters of no concern to one : NOSINESS. **3.** Something novel or extraordinary that arouses interest. **4.** A strange aspect. **5.** *Obs.* Fastidiousness.

**cu·ri·ous** (kyŏŏr′ē-əs) *adj.* [ME < OFr. *curios* < Lat. *curiosus,* careful, inquisitive < *cura,* care.] **1.** Eager to acquire information or knowledge. **2.** Unduly inquisitive : NOSY. **3.** Interesting due to novelty or rarity : ODD <a *curious* fact> **4.** *Obs.* **a.** Accomplished with skill or ingenuity. **b.** Very careful or scrupulous. **—cu′ri·ous·ly** *adv.* **—cu′ri·ous·ness** *n.*

**cu·ri·um** (kyŏŏr′ē-əm) *n.* [After Marie *Curie* (1867–1934) and Pierre *Curie* (1859–1906).] *Symbol* **Cm** A silvery, metallic synthetic radioactive element; atomic number 96; longest-lived isotope Cm 247.

**curl** (kûrl) *v.* **curled, curl·ing, curls.** [ME *curlen* < *crulle,* curly, perh. of MLG orig.] *—vt.* **1.** To twist (e.g., the hair) into coils or ringlets. **2.** To form into the spiral shape of a coil or ringlet. **3.** To decorate with curls. *—vi.* **1.** To form coils or ringlets. **2.** To assume a spiral or curved shape. **3.** To move in a curve or spiral. **4.** To play the game of curling. **—curl up.** To assume a position with the legs drawn up. *—n.* **1.** Something shaped like a spiral or coil. **2.** A ringlet of hair. **3.** The act of curling or state of being curled. **4.** Any of various plant diseases in which the leaves roll up. **5.** *Math.* The vector product of the vector differential operator and a vector function.

**curl·er** (kûr′lər) *n.* **1.** One that curls. **2.** A device, as a pin or roller, on which hair is wound for curling. **3.** A player of curling.

**cur·lew** (kûrl′yŏŏ, kûr′lŏŏ) *n.* [ME *curleu* < OFr. *courlieu.*] A brownish, long-legged shore bird of the genus *Numenius,* with a long, slender, downward-curving bill.

**curl·i·cue** *also* **curl·y·cue** (kûr′lĭ-kyŏŏ′) *n.* [CURLY + CUE[1].] A fancy twist or curl.

**curl·ing** (kûr′lĭng) *n.* A game played on ice, in which two four-man teams slide heavy, oblate stones toward a target.

**curling iron** *n.* A rod-shaped metal implement used when heated for curling the hair.

**curl paper** *n.* A piece of soft paper on which a lock of hair is rolled up for curling.

**curl·y** (kûr′lē) *adj.* **-i·er, -i·est. 1.** Having curls <*curly* hair> **2.** Tending to curl. **3.** Having a wavy grain <*curly* maple> **—curl′i·ly** *adv.* **—curl′i·ness** *n.*

**curl·y·cue** (kûr′lĭ-kyŏŏ′) *n. var. of* CURLICUE.

**curly top** *n.* A plant disease caused by a virus, *Ruga verrucosans,* and resulting in severe stunting of growth.

**cur·mudg·eon** (kər-mŭj′ən) *n.* [Orig. unknown.] A cantankerous person. **—cur·mudg′eon·ly** *adj.*

**cur·rach** *also* **cur·ragh** (kûr′əKH, kûr′ə) *n.* [ME *currok* < Ir. Gael. *curach.*] *Scot. & Ir.* A coracle.

**cur·rant** (kûr′ənt, kŭr′-) *n.* [ME *(raysons of) coraunte,* (raisins of) Corinth.] **1.** Any of various usu. prickly shrubs of the genus *Ribes,* bearing clusters of red, black, or greenish fruit. **2.** The small sour fruit of any of the currant plants, used chiefly for making jelly. **3.** A small, dried seedless Mediterranean grape used in cooking.

**cur·ren·cy** (kûr′ən-sē, kŭr′-) *n., pl.* **-cies.** [Med. Lat. *currentia,* a flowing < Lat. *currens,* pr.part. of *currere,* to run.] **1.** Money in use as a medium of exchange. **2.** A passing from hand to hand : CIRCULATION. **3.** General acceptance : PREVALENCE.

**cur·rent** (kûr′ənt, kŭr′-) *adj.* [ME *curraunt* < OFr. *corant,* pr.part. of *courre,* to run < Lat. *currere.*] **1. a.** Belonging to the present time. **b.** Now in progress. **2.** Passing from one to another : CIRCULATING <*current* money> **3.** Being in general or widespread use. **4.** Flowing : running. *—n.* **1.** A smooth and steady onward movement, as of water. **2.** The part of any body of liquid or gas that has a continuous onward movement <river *currents*> **3.** A general tendency, movement, or course. **4.** *Elect.* **a.** A flow of electric charge. **b.** The amount of electric charge flowing past a specified circuit point per unit time. **—cur′rent·ly** *adv.* **—cur′rent·ness** *n.*

**current assets** *pl.n.* Cash or assets readily convertible into cash.

**current density** *n.* **1.** *Elect.* The ratio of the magnitude of current flowing in a conductor to the cross-sectional area perpendicular to the current flow. **2.** *Physics.* The number of subatomic particles per unit time crossing a unit area in a designated plane perpendicular to the direction of motion of the particles.

**current ratio** *n.* The ratio of current assets to liabilities.

**cur·ri·cle** (kûr′ĭ-kəl) *n.* [Lat. *curriculum,* racing chariot, course < *currere,* to run.] A light, open two-wheeled carriage, drawn by two horses.

**cur·ric·u·lum** (kə-rĭk′yə-ləm) *n., pl.* **-la** (-lə) *or* **-lums.** [NLat. < Lat., course < *currere,* to run.] **1.** All the courses of study offered by an educational institution. **2.** A course of study, often in a specialized field. **—cur·ric′u·lar** (-lər) *adj.*

**cur·ric·u·lum vi·tae** (kə-rĭk′yə-ləm vī′tē, kə-rĭk′ə-ləm wē′tī′) *n.* [Lat., course of life.] A résumé of one's career, as for an employer.

**cur·rie** (kûr′ē, kŭr′ē) *n. var. of* CURRY[2].

**cur·ri·er** (kûr′ē-ər, kŭr′-) *n.* [ME *curreiour* < OFr. < Lat. *coriarius,* a tanner < *corium,* leather.] One who curries, esp. leather.

**cur·ri·er·y** (kûr′ē-ə-rē, kŭr′-) *n., pl.* **-ies.** The trade, work, or shop of a leather currier.

**cur·rish** (kûr′ĭsh) *adj.* **1.** Of or like a cur. **2. a.** Snarling : bad-tempered. **b.** Base : cowardly. **—cur′rish·ly** *adv.*

**cur·ry**[1] (kûr′ē, kŭr′ē) *vt.* **-ried, -ry·ing, -ries.** [ME *curreien* < AN *curreier,* to arrange, curry.] **1.** To groom (a horse) with a currycomb. **2.** To prepare (tanned hides) for use by soaking, coloring, or other processes. **—curry favor.** To seek or gain favor by flattery.

**cur·ry**[2] *also* **cur·rie** (kûr′ē, kŭr′ē) *n., pl.* **-ries.** [Tamil *kari,* relish.] **1.** Curry powder. **2.** A heavily spiced relish or sauce made with curry powder and eaten with rice, meat, fish, or other food. **3.** A dish seasoned with curry powder. **—cur′ry** *v.* **(-ried, -ry·ing, -ries).**

**cur·ry·comb** (kûr′ē-kōm′, kŭr′-) *n.* A comb with metal teeth, used for grooming horses. **—cur′ry·comb′** *v.* **(-combed, -comb·ing, -combs).**

**curry powder** *n.* A blended condiment prepared from pungent spices, as cumin, coriander, and turmeric.

**curse** (kûrs) *n.* [ME < OE *curs.*] **1.** An appeal for evil or injury to befall someone or something. **2.** Evil or injury resulting from or as if from an invocation. **3.** One that is accursed. **4.** Something bringing or causing evil : SCOURGE. **5.** A profane oath. **6.** An ecclesiastical censure, ban, or anathema. **7. the curse.** *Slang.* Menstruation. *—v.* **cursed** *or* **curst** (kûrst), **curs·ing, curs·es.** *—vt.* **1.** To invoke evil, calamity, or injury upon : DAMN. **2.** To swear at. **3.** To bring evil upon : AFFLICT. **4.** To put under an ecclesiastical ban or anathema : EXCOMMUNICATE. *—vi.* To utter curses : SWEAR. **—curs′er** *n.*

**curs·ed** (kûr′sĭd, kûrst) *also* **curst** (kûrst) *adj.* That deserves to be cursed : WICKED. **—curs′ed·ly** *adv.* **—curs′ed·ness** *n.*

**cur·sive** (kûr′sĭv) *adj.* [Med. Lat. *(scripta) cursiva,* flowing (script) < Lat. *cursus,* p.part. of *currere,* to run.] Designating writing or printing with letters joined together. *—n.* **1.** A cursive character or letter. **2.** A manuscript written in cursive characters. **3.** Printing type imitative of handwriting. **—cur′sive·ly** *adv.* **—cur′sive·ness** *n.*

**cur·sor** (kûr′sər) *n.* [Lat., runner < *cursus,* p.part. of *currere,* to run.] A visual indicator on a video terminal showing the position of next entry.

**cur·so·ri·al** (kûr-sôr′ē-əl, -sōr′-) *adj.* [< LLat. *cursorius,* of running. —see CURSORY.] Adapted to or specialized for running <*cursorial* birds><*cursorial* legs>

**cur·so·ry** (kûr′sə-rē) *adj.* [LLat. *cursorius,* of running < Lat. *cursor,* runner. —see CURSOR.] Hastily and superficially done. **—cur′so·ri·ly** *adv.* **—cur′so·ri·ness** *n.*

**curst** (kûrst) *adj. var. of* CURSED. *—v. var. p.t. & p.p. of* CURSE.

**curt** (kûrt) *adj.* **-er, -est.** [Lat. *curtus,* cut short.] **1.** Rudely abrupt or brief <a *curt* retort> **2.** Terse or concise. **3.** Shortened. **—curt′ly** *adv.* **—curt′ness** *n.*

**cur·tail** (kər-tāl′) *vt.* **-tailed, -tail·ing, -tails.** [Obs. *curtal,* to

# CURRENCY TABLE

| Country | Basic Unit | Subdivision |
|---|---|---|
| Afghanistan | afghani | pul |
| Albania | lek | quintar |
| Algeria | dinar | centime |
| Andorra | franc | centime |
| | peseta | centimo |
| Angola | kwanza | lwei |
| Argentina | peso | centavo |
| Australia | dollar | cent |
| Austria | schilling | groschen |
| Bahamas | dollar | cent |
| Bahrain | dinar | fil |
| Bangladesh | taka | paisa |
| Barbados | dollar | cent |
| Belgium | franc | centime |
| Belize | dollar | cent |
| Benin | franc | centime |
| Bhutan | ngultrum | chetrum |
| Bolivia | peso | centavo |
| Botswana | pula | thebe |
| Brazil | cruzeiro | centavo |
| Brunei | dollar | cent |
| Bulgaria | lev | stotinki |
| Burma | kyat | pya |
| Burundi | franc | centime |
| Cambodia | riel | sen |
| Cameroon | franc | centime |
| Canada | dollar | cent |
| Cape Verde | escudo | centavo |
| Cayman Islands | dollar | cent |
| Central African Republic | franc | centime |
| Chad | franc | centime |
| Chile | peso | centavo |
| China, People's Republic of | renminbi | chiao |
| China, Republic of (Taiwan) | yuan | cent |
| Colombia | peso | centavo |
| Comoros | franc | centime |
| Congo | franc | centime |
| Costa Rica | colon | centimo |
| Cuba | peso | centavo |
| Cyprus | pound | mil |
| Czechoslovakia | crown | haler |
| Denmark | kroner | öre |
| Djibouti | franc | centime |
| Dominica | dollar | cent |
| Dominican Republic | peso | centavo |
| Ecuador | sucre | centavo |
| Egypt | pound | piaster |
| El Salvador | colon | centavo |
| Equatorial Guinea | ekpwele | cent |
| Ethiopia | birr | cent |
| Fiji | dollar | cent |
| Finland | markka | penni |
| France | franc | centime |
| Gabon | franc | centime |
| Gambia | dalasi | butut |
| Germany, East | mark | pfennig |
| Germany, West | deutsche mark | pfennig |
| Ghana | cedi | pesewa |
| Greece | drachma | lepton |
| Grenada | dollar | cent |
| Guatemala | quetzal | centavo |
| Guinea | syli | cory |
| Guinea-Bissau | peso | centavo |
| Guyana | dollar | cent |
| Haiti | gourde | centime |
| Honduras | lempira | centavo |
| Hong Kong | dollar | cent |
| Hungary | forint | fillér |
| Iceland | krona | eyrir |
| India | rupee | paisa |
| Indonesia | rupiah | sen |
| Iran | rial | dinar |
| Iraq | dinar | fil |
| Ireland, Republic of | pound | penny |
| Israel | shekel | agora |
| Italy | lira | centesimo |
| Ivory Coast | franc | centime |
| Jamaica | dollar | cent |
| Japan | yen | sen |
| Jordan | dinar | fil |
| Kenya | shilling | cent |
| Korea, North | won | jun |
| Korea, South | won | chon |
| Kuwait | dinar | fil |
| Laos | kip | at |
| Lebanon | pound | piaster |
| Lesotho | loti | lisente |
| Liberia | dollar | cent |
| Libya | dinar | dirham |
| Liechtenstein | franc | centime |
| Luxembourg | franc | centime |
| Macao | pataca | avo |
| Madagascar | franc | centime |
| Malawi | kwacha | tambala |
| Malaysia | ringgit | sen |
| Maldives | rupee | laree |
| Mali | franc | centime |
| Malta | pound | cent |
| Mauritania | ouguiya | khoum |
| Mauritius | rupee | cent |
| Mexico | peso | centavo |
| Monaco | franc | centime |
| Mongolia | tugrik | mongo |
| Morocco | dirham | centime |
| Mozambique | metical | centavo |
| Nauru | dollar | cent |
| Nepal | rupee | paisa |
| Netherlands | guilder | cent |
| Netherlands Antilles | guilder | cent |
| New Zealand | dollar | cent |
| Nicaragua | cordoba | centavo |
| Niger | franc | centime |
| Nigeria | naira | kobo |
| Norway | krone | öre |
| Oman | riyal-omani | baiza |
| Pakistan | rupee | paisa |
| Panama | balboa | centesimo |
| Papua New Guinea | kina | toea |
| Paraguay | guarani | centimo |

## CURRENCY *(continued)*

| Country | Basic Unit | Subdivision |
|---|---|---|
| Peru | sol | centavo |
| Philippines | peso | centavo |
| Poland | zloty | grosz |
| Portugal | escudo | centavo |
| Qatar | riyal | dirham |
| Rumania | leu | ban |
| Rwanda | franc | centime |
| Saint Lucia | dollar | cent |
| Saint Vincent | dollar | cent |
| San Marino | lira | centesimo |
| São Tomé and Principe | dobra | centimo |
| Saudi Arabia | riyal | qurush |
| Senegal | franc | centime |
| Seychelles | rupee | cent |
| Sierra Leone | leone | cent |
| Singapore | dollar | cent |
| Solomon Islands | dollar | cent |
| Somalia | schilling | cent |
| South Africa | rand | cent |
| Spain | peseta | centimo |
| Sri Lanka | rupee | cent |
| Sudan | pound | piaster |
| Surinam | guilder | cent |
| Swaziland | lilangeni | cent |
| Sweden | krona | öre |
| Switzerland | franc | centime |
| Syria | pound | piaster |
| Tanzania | shilling | cent |
| Thailand | baht | satang |
| Togo | franc | centime |
| Tonga | pa'anga | seniti |
| Trinidad and Tobago | dollar | cent |
| Tunisia | dinar | millieme |
| Turkey | pound | kuru |
| Uganda | shilling | cent |
| Union of Soviet Socialist Republics | rouble | kopeck |
| United Arab Emirates | dirham | fil |
| United Kingdom of Great Britain and Northern Ireland | pound | penny |
| United States of America | dollar | cent |
| Upper Volta | franc | centime |
| Uruguay | peso | centesimo |
| Vanuatu | franc | centime |
| Vatican City | lira | centesimo |
| Venezuela | bolivar | centimo |
| Vietnam | dong | hao |
| Western Samoa | tala | sene |
| Yemen, People's Democratic Republic of (Southern Yemen) | dinar | fil |
| Yemen Arab Republic (North Yemen) | riyal | fil |
| Yugoslavia | dinar | para |
| Zaire | zaire | likuta |
| Zambia | kwacha | ngwee |
| Zimbabwe | dollar | cent |

dock a horse's tail < CURTAL.] To cut short : ABBREVIATE. **—cur·tail'er** *n.* **—cur·tail'ment** *n.*

**curtail step** *n.* [Orig. unknown.] The widened step or steps at the bottom of a flight of stairs.

**cur·tain** (kûr'tn) *n.* [ME < OFr. *courtine* < LLat. *cortina.*] **1.** Material hanging esp. in a window as a decoration, shade, or screen. **2.** Something that screens, covers, or acts as a barrier. **3. a.** The movable drape or screen between the stage and auditorium in a theater or hall. **b.** The ascent or opening of a theater curtain at the beginning or its descent or closing at the end, as of a play or act. **c.** A line, speech, or situation in a play that occurs at the very end or just before the curtain closes. **d.** The time at which a theatrical performance begins or is scheduled to begin. **4.** The part of a rampart or parapet joining two bastions or gates. **5.** An enclosing wall joining two towers or similar structures. **6. curtains.** *Slang.* **a.** The end. **b.** Ruin. **c.** Death. **—cur'tain** *v.* **(-tained, -tain·ing, -tains).**

**curtain call** *n.* The appearance of a performer or performers at the end of a performance in acknowledgment of applause.

**curtain raiser** *n.* **1.** A short play presented before the principal dramatic production. **2.** A preliminary event.

**curtain speech** *n.* A talk delivered in front of the curtain at the end of a theatrical performance.

**cur·tal** (kûr'tl) [OFr. *courtault,* horse with a cropped tail < *court,* short < Lat. *curtus,* cut short.] *Obs.* —*n.* **1.** An animal with a docked tail. **2.** Something cut short or docked. —*adj.* **1.** Cut short or docked, as an animal's tail. **2.** Wearing a short frock.

**curtal ax** *n.* [By folk ety. < obs. *curtelace, coutelace,* cutlass < OFr. *coutelas.* —see CUTLASS.] *Archaic.* A cutlass.

**cur·tate** (kûr'tāt') *adj.* [Lat. *curtatus,* p.part. of *curtare,* to shorten < *curtus,* cut short.] Abbreviated : shortened.

**cur·te·sy** (kûr'tĭ-sē) *n., pl.* **-sies.** [ME *curtesie.* —see COURTESY.] The life tenure that by common law is held by a man over the property of his deceased wife if children with rights of inheritance were born during the marriage.

**cur·ti·lage** (kûr'tl-ĭj) *n.* [ME < OFr. *courtillage* < *courtil,* dim. of *cort,* court. —see COURT.] *Law.* The enclosed land surrounding a house or dwelling.

**curt·sy** (kûrt'sē) *n., pl.* **-sies.** [Var. of COURTESY.] A gesture of respect or reverence made by bending the knees with one foot forward and lowering the body. —*vi.* **-sied, -sy·ing, -sies.** To make a curtsy.

**cu·rule** (kyo͝or'o͞ol') *adj.* [Lat. *curulis,* of a curule chair < *currus,* chariot < *currere,* to run.] **1.** Of or pertaining to a seat like a campstool that only the highest officials in ancient Rome were permitted to use. **2.** Privileged to sit in a curule chair : of superior rank.

**curule chair** *n.* A backless seat with heavy curved legs, reserved for the use of the highest ancient Roman officials.

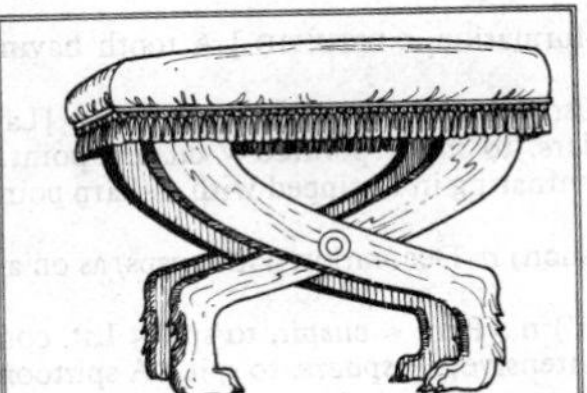

curule chair

**cur·va·ceous** (kûr-vā'shəs) *adj.* Voluptuous in figure. **—cur·va'ceous·ly** *adv.* **—cur·va'ceous·ness** *n.*

**cur·va·ture** (kûr'və-cho͝or', -chər) *n.* [Lat. *curvatura* < *curvatus,* p.part. of *curvare,* to bend < *curvus,* curved.] **1.** The act of curving or state of being curved. **2.** *Math.* **a.** The ratio of the change in tangent inclination over a given arc to the length of the arc. **b.** The limit of this ratio as the length of the arc approaches zero. **3.** *Med.* A curving or bending, esp. an abnormal one <*curvature* of the spine>

**curve** (kûrv) *n.* [ME, curved < Lat. *curvus.*] **1. a.** A line deviating from straightness in a smooth, continuous way. **b.** A surface deviating from planarity in a smooth, continuous way. **2. a.** A rounded part, object, or area. **b.** A rather smooth bend in a road. **3. curves.** *Slang.* A woman's well-proportioned figure. **4. a.** A line representing data on a graph. **b.** A trend derived from or as if from such a graph. **5.** *Math.* **a.** The graph of a function on a coordinate plane. **b.** Intersection of two surfaces in three dimensions. **6.** A graphic representation of the relative performance of individuals as measured against each other, used esp. as a method of grading students with the range of grades based on the proportion of students. **7.** *Baseball.* A curve ball. —*v.* **curved, curv·ing, curves.** —*vi.* To take the shape of or move in a curve. —*vt.* **1.** To cause to curve. **2.** *Baseball.* To pitch a curve ball to. **3.** To grade on a curve. **—curv'ed·ly** (kûr'vĭd-lē) *adv.* **—curv'ed·ness** *n.*

**curve ball** *n.* **1.** *Baseball.* A pitched ball that veers or breaks to the left when thrown with the right hand and to the right when thrown with the left hand. **2.** *Slang.* A trick : deception.

**cur·vet** (kûr-vĕt') *n.* [Ital. *corvetta* < OItal., dim. of *corva,* curve < Lat. *curvus,* curved.] A light leap by a horse, in which both hind legs leave the ground just before the forelegs are set down. —*v.* **-vet·ted, -vet·ting, -vets** or **-vet·ed, -vet·ing, -vets.** —*vi.* **1.** To leap in a curvet. **2.** To prance : frolic. —*vt.* To cause to leap in a curvet.

**cur·vi·lin·e·ar** (kûr'və-lĭn'ē-ər) *also* **cur·vi·lin·e·al** (-əl) *adj.* [Lat. *curvus,* curved + LINEAR.] Formed, bounded, or characterized by curved lines. **—cur'vi·lin'e·ar'i·ty** (-ē-ăr'ĭ-tē) *n.* **—cur'vi·lin'e·ar·ly** *adv.*

**cus·cus** (kŭs'kəs) *n.* [NLat., prob. < a native New Guinean word.] A marsupial of the genus *Phalanger* of New Guinea and adjacent areas, with protruding eyes, a yellow nose, and a prehensile tail.

**cu·sec** (kyo͞o'sĕk') *n.* [CU(BIC) + SEC(OND)[1].] A unit of volumetric flow of liquids, equal to one cubic foot per second.

**cu·shaw** (kə-shô', ko͞o'shô') *n.* [Of Algonquian orig.] A squash, *Cucurbita moschata,* having variably shaped, often crook-necked fruit.

**Cush·ing's disease** *also* **Cushing's syndrome** (ko͝osh'ĭngz) *n.* [After Harvey *Cushing* (1869–1939).] A disease caused by an overgrowth of basophilic cells of the pituitary, marked by obesity and muscular weakness.

**cush·ion** (ko͝osh'ən) *n.* [ME *cushin* < OFr. *cussin* < VLat. **coxinus* < Lat. *coxa,* hip.] **1.** A pad or pillow with a soft filling. **2.** Something resilient used as a rest, support, or shock absorber. **3.** A padlike bodily part. **4.** The rim bordering a billiard table. **5.** A pillow used in lacemaking. **6.** Something that mitigates an adverse effect. —*vt.* **-ioned, -ion·ing, -ions. 1.** To provide with a cushion. **2.** To place or seat on a cushion. **3.** To cover or hide with or as if with a cushion. **4.** To protect against or absorb the shock of. **—cush'ion·y** *adj.*

**Cush·it·ic** (ko͞o-shĭt'ĭk) *n.* A group of Hamitic languages that includes Somali and other languages of Somalia and Ethiopia. **—Cush·it'ic** *adj.*

**cush·y** (ko͝osh'ē) *adj.* **-i·er, -i·est.** [< Hindi *khush* < Pers. *khosh,* pleasant.] *Slang.* Making few demands : COMFORTABLE <a *cushy* position in the firm> **—cush'i·ly** *adv.* **—cush'i·ness** *n.*

**cusk** (kŭsk) *n., pl.* **cusk** or **cusks.** [Prob. alteration of *tusk,* a kind of codfish.] A food fish, *Brosme brosme,* found in North Atlantic coastal waters.

**cusk eel** *n.* Any of various eellike, chiefly marine fishes of the family Ophidiidae.

**cusp** (kŭsp) *n.* [Lat. *cuspis,* point.] **1.** A point or pointed end. **2.** *Anat.* **a.** A prominence or projection on the chewing surface of a tooth. **b.** A fold or flap of a heart valve. **3.** *Math.* A point at which a curve crosses itself and at which the two tangents to the curve coincide. **4.** A pointed projection formed by two intersecting arcs or foils, as on an arch. **5.** *Astron.* Either point of a crescent moon. **6.** The transitional first or last part of an astrological house or sign.

**cus·pate** (kŭs'pāt') *also* **cus·pat·ed** (-pā'tĭd) *adj.* **1.** Having a cusp. **2.** Shaped like a cusp.

**cus·pid** (kŭs'pĭd) *n.* [Back-formation < BICUSPID.] A tooth having one point : canine tooth.

**cus·pi·date** (kŭs'pĭ-dāt') *also* **cus·pi·dat·ed** (-dā'tĭd) *adj.* [Lat. *cuspidatus,* p.part. of *cuspidare,* to make pointed < *cuspis,* point.] **1.** Having a cusp. **2.** *Biol.* Terminating in or tipped with a sharp point <a *cuspidate* leaf>

**cus·pi·da·tion** (kŭs'pĭ-dā'shən) *n.* Decoration with cusps, as on an arch.

**cus·pi·dor** (kŭs'pĭ-dôr', -dōr') *n.* [Port. < *cuspir,* to spit < Lat. *conspuere,* to spit upon : *com-* (intensive) + *spuere,* to spit.] A spittoon.

**cuss** (kŭs) [Var. of CURSE.] *Informal.* —*vi.* & *vt.* **cussed, cuss·ing, cuss·es.** To curse or curse at. —*n.* **1.** A curse. **2.** An odd or perverse creature.

**cuss·ed** (kŭs'ĭd) *adj. Informal.* **1.** Cursed. **2.** Perverse : vexatious. **—cuss'ed·ly** *adv.* **—cuss'ed·ness** *n.*

**cus·tard** (kŭs'tərd) *n.* [ME *crustade,* a pie with a crust < AN < OProv. *croustado* < *crousta,* crust < Lat. *crusta.*] A dessert of milk, sugar, eggs, and flavoring, baked or boiled until set.

**custard apple** *n.* [So called because the pulp of its fruit resembles custard.] **1.** A tropical American tree, *Annona reticulata,* bearing large, heart-shaped fruit. **2.** The fruit of the custard apple, having edible, fleshy pulp. **3.** PAPAW 2.

**cus·to·di·an** (kŭ-stō'dē-ən) *n.* **1.** One in charge of something : CARETAKER. **2.** A janitor. **—cus·to'di·an·ship'** *n.*

**cus·to·dy** (kŭs'tə-dē) *n., pl.* **-dies.** [ME *custodie* < Lat. *custodia* < *custos,* guard.] **1.** The act or right of guarding, esp. such a right granted by a court. **2.** Detention under guard, esp. by the police. **—cus·to'di·al** (-dē-əl) *adj.*

**cus·tom** (kŭs'təm) *n.* [ME *custume* < OFr. *costume* < Lat. *consuetudo* < *consuescere,* to accustom : *com-* (intensive) + *suescere,* to become accustomed.] **1.** A practice followed as a matter of course among a people. **2.** A habitual practice of an individual. **3.** *Law.* A common tradition or usage so long established that it has the force or validity of law. **4.** Habitual patronage, as of a store. **5. customs** *(sing. in number).* **a.** A duty or tax on imported and, less commonly, exported goods. **b.** The governmental agency authorized to collect such duties. **c.** The procedure for inspecting baggage and goods entering a country. **6.** Tribute, service, or rent paid by a feudal tenant to his lord. —*adj.* **1.** Made to order. **2.** Specializing in the making or selling of made-to-order goods.

**cus·tom·a·ble** (kŭs'tə-mə-bəl) *adj.* Subject to tariffs.

**cus·tom·ar·y** (kŭs'tə-mĕr'ē) *adj.* **1.** Commonly practiced : USUAL. **2.** Based on custom or tradition rather than written law or contract. **—cus'tom·ar'i·ly** (-mâr'ə-lē) *adv.* **—cus'tom·ar'i·ness** *n.*

**cus·tom-built** (kŭs'təm-bĭlt') *adj.* Built according to a buyer's specifications <a *custom-built* motorcycle>

**cus·tom·er** (kŭs'tə-mər) *n.* **1.** One who buys goods or services. **2.** *Informal.* A person with whom one must deal <a very shrewd *customer*>

**cus·tom·house** *also* **custom house** (kŭs'təm-hous') *n.* A governmental building or office where customs are collected and ships are cleared for entering or leaving the country.

**cus·tom·ize** (kŭs'tə-mīz') *vt.* **-ized, -iz·ing, -iz·es.** To alter to the tastes of the buyer <*customize* a yacht>

**cus·tom-made** (kŭs'təm-mād') *adj.* Made according to a buyer's specifications <a *custom-made* suit> **—cus'tom-make'** *v.* **(-made, -mak·ing, -makes).**

**customs union** *n.* An international association organized to eliminate customs restrictions on goods exchanged between member nations and to establish a uniform tariff policy toward nonmember nations.

**cut** (kŭt) *v.* **cut, cut·ting, cuts.** [ME *cutten.*] —*vt.* **1.** To penetrate with a sharp edge. **2.** To separate into parts with or as if with a sharp-edged instrument : SEVER <*cut* the material with shears> **3.** To sever the edges or outer extensions of : SHORTEN <*cut* hair> **4.** To reap : harvest <*cut* wheat> **5.** To fell by sawing : HEW. **6.** To have (a new tooth) grow through the gums. **7.** To form or shape by severing or incising <*cut* cookies from dough> **8.** To form by penetrating, probing, or digging. **9.** To separate or dissociate from a main body : DETACH <*cut* slices of cheese> **10.** To pass through or across : CROSS. **11.** To divide (a deck of cards) in two, as before dealing. **12.** To curtail the size, extent, or duration of <*cut* the labor force> **13.** To lessen the strength of : DILUTE <*cut* the punch with water> **14.** To dissolve by breaking down the fat of <Soap *cuts* grease.> **15.** To injure the feelings of. **16.** *Informal.* To fail to attend purposely <*cut* classes> **17.** *Informal.* To stop <*cut* the foolishness><*cut* the motor> **18.** To strike (a ball) so that it spins irregularly or is deflected. **19.** To perform <always *cutting* capers> **20.** To terminate (a scene in a film). **21.** To record a performance on (a phonograph record). **22.** To edit (film or videotape). —*vi.* **1.** To make an incision or separation. **2.** To allow incision or severing <Gingerbread *cuts* easily.> **3.** To use a sharp-edged instrument. **4.** To grow through the gums. —Used of teeth. **5.** To penetrate injuriously. **6.** To change direction abruptly <*cut* to the south> **7.** To go directly and often hastily <*cut* through the alley> **8.** To divide a pack of cards into two parts. **—cut back. 1.** To shorten by cutting : PRUNE. **2.** To reduce <*cut back* production> **—cut down. 1.** To strike down or kill. **2.** To remove additional or extra fittings. **3.** To reduce : curtail. **—cut in. 1.** To move into a line of people or things out of turn. **2.** To interrupt. **3.** To interrupt a dancing couple in order to dance with one of them. **4.** To connect or become connected into an electrical circuit. **5.** To include, esp. to give a share to. **—cut off. 1.** To separate by or as if by cutting : SEVER. **2.** To stop suddenly : DISCONTINUE. **3.** To shut off : BAR. **4.** To interrupt or break off. **5.** To disinherit. **—cut out. 1.** To remove by cutting. **2.** To shape or form by or as if by cutting. **3.** To take the place of : SUPPLANT. **4.** To deprive <heirs *cut out* of the will> **5.** To put an end to : DESIST. **6.** To stop : cease. **7.** *Informal.* To depart suddenly. **—cut up.** *Informal.* **1.** To criticize severely. **2.** To fool around : CLOWN. —*n.* **1.** An act of incising, severing, or separating. **2.** The result of cutting : INCISION. **3.** A part severed from a main body. **4.** A passage resulting from excavating or probing. **5.** An elimination or excision of a part <a *cut* in an article> **6.** A reduction <a budget *cut*> **7.** The style in which a garment is cut. **8.** *Informal.* A share of earnings or profits. **9.** *Informal.* A wounding remark : INSULT. **10.** *Informal.* An unexcused absence, as from a class or school. **11. a.** An engraved block or printing plate. **b.** A print made from such a block. **12.** A stroke that causes a ball to spin irregularly or to deflect. **13.** The act of dividing a deck of cards into two parts, as before dealing. **14.** A sharp transition between scenes or shots in a motion picture. **15.** One of the objects used in drawing lots. **16.** A single selection on a phonograph record. **—cut corners.** To do something in the simplest or least expensive way. **—cut down to size.** To reduce the overblown importance or prestige of. **—cut loose.** To act or speak immoderately. **—cut no**

**ice.** To make no effect or impression. **—cut (one's) losses.** To withdraw from a hopeless situation. **—cut (one's) teeth on.** To learn, do, or use in early years or at the outset of one's career. **—cut short.** To stop suddenly before the end : ABBREVIATE. **—cut the mustard.** To perform up to the standard required for success.

☆ ***syns:*** CUT, GASH, INCISE, PIERCE, SLASH, SLIT *v. core meaning* : to penetrate with a sharp edge <*cut* my finger with a razor blade>

**cut-and-dried** (kŭt′ən-drīd′) *adj.* **1.** Prepared in advance : SETTLED. **2.** Routine : ordinary.

**cu·ta·ne·ous** (kyo͞o-tā′nē-əs) *adj.* [NLat. *cutaneus* < Lat. *cutis,* skin.] Of, relating to, or affecting the skin. **—cu·ta′ne·ous·ly** *adv.*

**cutaneous anaphylaxis** *n.* Anaphylaxis marked by a violent skin reaction upon contact with the sensitizing substance.

**cut·a·way** (kŭt′ə-wā′) *n.* A man's formal daytime coat, with front edges sloping diagonally from the waist and forming tails at the back.

**cut·back** (kŭt′băk′) *n.* **1.** A decrease : curtailment <a *cutback* in imports> **2.** A sharp reversal of direction, as of a ball carrier in football.

**cutch** (kŭch) *n.* [Malay *kachu,* of Dravidian orig.] Catechu.

**cute** (kyo͞ot) *adj.* **cut·er, cut·est.** [Short for ACUTE.] **1.** Delightfully pretty or dainty. **2.** Obviously contrived to charm : PRECIOUS. **3.** Shrewd : clever. **—cute′ly** *adv.* **—cute′ness** *n.*

**cute·sy** (kyo͞ot′sē) *adj.* **-si·er, -si·est.** *Slang.* CUTE 2 <"a *cutesy,* scented-candle haven for the old folks" —Michael W. Robbins>

**cut·ey** (kyo͞o′tē) *n. Slang. var. of* CUTIE.

**cut glass** *n.* Glassware decorated or shaped by cutting instruments or abrasive wheels.

**cut·grass** (kŭt′grăs′) *n.* A swamp grass of the genus *Leersia,* esp. *L. oryzoides,* having leaves with very rough margins.

**cu·ti·cle** (kyo͞o′tĭ-kəl) *n.* [Lat. *cuticula,* dim. of *cutis,* skin.] **1.** The epidermis. **2.** The strip of hardened skin at the base of a fingernail or toenail. **3.** *Zool.* The noncellular, often horny protective outer covering in many invertebrates. **4.** *Bot.* The layer of cutin covering the epidermis of plants. **—cu·tic′u·lar** (-tĭk′yə-lər) *adj.*

**cut·ie** *also* **cut·ey** (kyo͞o′tē) *n., pl.* **-ies** *also* **-eys.** *Slang.* A cute person.

**cu·tin** (kyo͞ot′n) *n.* [Lat. *cutis,* skin + -IN.] *Bot.* Waxlike, water-repellent material in the walls of some plant cells, forming the cuticle which covers the epidermis.

**cut-in** (kŭt′ĭn′) *n.* An inserted shot, often a still close-up, interrupting the continuity of the main action of a motion picture.

**cu·tin·ize** (kyo͞ot′n-īz′) *vt. & vi.* **-ized, -iz·ing, -iz·es.** *Bot.* To impregnate or coat with cutin or become impregnated or coated with cutin. **—cu′tin·i·za′tion** *n.*

**cu·tis** (kyo͞o′tĭs) *n., pl.* **-tes** (-tēz) *or* **-tis·es.** [Lat., skin.] *Anat.* The corium.

**cut·lass** *also* **cut·las** (kŭt′ləs) *n.* [OFr. *coutelas,* aug. of *coutel,* knife < Lat. *cultellus,* dim. of *culter,* knife.] A short, heavy sword with a curved single-edged blade, once used as a weapon by sailors.

**cutlass fish** *n.* A marine fish of the genus *Trichiurus,* with a long, narrow body and a pointed tail.

**cut·ler** (kŭt′lər) *n.* [ME < OFr. *coutelier* < *coutel,* knife. —see CUTLASS.] One that makes, repairs, or sells cutting instruments such as knives.

**cut·ler·y** (kŭt′lə-rē) *n.* **1.** Cutting tools and instruments. **2.** Implements used as tableware. **3.** The occupation of a cutler.

**cut·let** (kŭt′lĭt) *n.* [Fr. *côtelette* < OFr. *costelette,* dim. of *coste,* rib < Lat. *costa.*] **1.** A thin slice of meat, usu. veal or lamb, cut from the ribs or leg of an animal. **2.** A flat croquette of chopped fish or meat.

**cut·off** (kŭt′ôf′, -ŏf′) *n.* **1.** A designated limit or point of termination. **2. a.** A short cut. **b.** A by-pass. **3.** A new channel cut by a river across the neck of an oxbow. **4. a.** A checking or cutting off of a flow of steam, water, or other fluid. **b.** The device that cuts off the flow. **5.** *Mus.* A conductor's signal indicating a stop or break in singing or playing.

**cut-offs** *also* **cut·offs** (kŭt′ôfs′, -ŏfs′) *pl.n.* Pants, as blue jeans, made into shorts by cutting off part of the legs.

**cut·out** (kŭt′out′) *n.* **1.** Something cut out or intended to be cut out. **2.** *Elect.* A device that interrupts, by-passes, or disconnects a circuit or circuit element.

**cut·o·ver** (kŭt′ō′vər) *adj.* Cleared of trees.

**cut·purse** (kŭt′pûrs′) *n. Archaic.* A pickpocket.

**cut-rate** (kŭt′rāt′) *adj.* Sold or on sale at a reduced price.

**cut·ter** (kŭt′ər) *n.* **1.** One that cuts. **2.** *Naut.* **a.** A single-masted fore-and-aft-rigged sailing vessel with a running bowsprit, a mainsail, and two or more headsails that are usu. set flying. **b.** A ship's boat, powered by a motor or oars, and used for transporting passengers or stores. **c.** A small, lightly armed motorboat used by the Coast Guard. **3.** A small sleigh, usu. seating one person and drawn by a single horse.

**cut·throat** (kŭt′thrōt′) *n.* **1.** One who cuts throats : MURDERER. **2.** A ruthless and unprincipled person. —*adj.* **1.** Murderous : cruel. **2.** Relentless or merciless in competition. **3.** Of or designating a form of a game in which each of three players acts and scores for himself or herself.

**cut time** *n.* Alla breve.

**cut·ting** (kŭt′ĭng) *adj.* **1.** Capable of or designed for incising, shearing, or severing. **2.** Sharply penetrating <a *cutting* north wind> **3.** Bitterly sarcastic or insulting. —*n.* **1.** A part cut off from a main body. **2.** An excavation made through high ground in the construction of a road, railway, etc. **3.** *Chiefly Brit.* A clipping, as from a newspaper. **4.** The editing of film or audio tape. **5.** A twig or other plant part removed to form roots and propagate a new plant.

**cut·tle·bone** (kŭt′l-bōn′) *n.* The calcareous internal shell of a cuttlefish, used as a dietary supplement for cage birds or ground into powder for use as a polishing agent.

**cut·tle·fish** (kŭt′l-fĭsh′) *n., pl.* **cuttlefish** *or* **-fish·es.** [ME *codel,* cuttlefish < OE *cudele* + FISH.] A squidlike cephalopod marine mollusk of the genus *Sepia,* having ten arms and a calcareous internal shell and secreting a dark, inky fluid.

**cut·up** (kŭt′ŭp′) *n. Informal.* A mischievous person.

**cut·wa·ter** (kŭt′wô′tər, -wŏt′ər) *n.* **1.** *Naut.* The forward part of a ship's prow. **2.** The wedge-shaped end of a bridge pier, designed for dividing the current and breaking up ice floes.

**cut·work** (kŭt′wûrk′) *n.* Openwork embroidery in which the ground fabric is cut away from the design.

**cut·worm** (kŭt′wûrm′) *n.* [So called because many species eat through stems of plants.] The larva of any of various moths of the family Noctuidae, feeding on many plants.

**cu·vette** (kyo͞o-vĕt′) *n.* [Fr., dim. of *cuve,* tub < Lat. *cupa.*] A small, often tubular laboratory vessel, often made of glass.

**cwm** (ko͞om) *n. Welsh.* A cirque.

**-cy** *suff.* [ME *-cie* < OFr. < Lat. *-cia, -tia* and Gk. *-kia, -tia.*] **1.** State : condition : quality <bankrupt*cy*> **2.** Rank : office <baronet*cy*> **3.** Action : practice <mendican*cy*>

**cy·an** (sī′ăn′, -ən) *n.* [Gk. *kuanos,* blue.] A greenish blue that is one of the subtractive primary colors and a complement of red.

**cyan-** *pref. var. of* CYANO-.

**cy·an·am·ide** *also* **cy·an·am·id** (sī-ăn′ə-mĭd) *n.* **1.** An irritating caustic acidic crystalline compound, $NCNH_2$, prepared by treating calcium cyanamide with sulfuric acid. **2.** Calcium cyanamide. **3.** A salt or ester of cyanamide.

**cy·a·nate** (sī′ə-nāt′, -nət) *n.* A salt or ester of cyanic acid.

**cy·an·ic** (sī-ăn′ĭk) *adj.* **1.** Relating to or containing cyanogen. **2.** Blue or bluish.

**cyanic acid** *n.* A poisonous, unstable, highly volatile organic acid, HOCN, used to prepare certain cyanates.

**cy·a·nide** (sī′ə-nīd′) *also* **cy·a·nid** (-nĭd) *n.* Any of various salts or esters of hydrogen cyanide containing a CN group, esp. the extremely poisonous compounds potassium cyanide and sodium cyanide. —*vt.* **-nid·ed, -nid·ing, -nides. 1.** To treat (a metal surface) with cyanide to produce a hard surface. **2.** To treat (an ore) with cyanide to extract gold or silver.

**cyanide process** *n.* Extraction of gold or silver from ores treated with a solution of sodium or calcium cyanide.

**cy·a·nine** (sī′ə-nēn′, -nĭn) *n.* Any of various blue dyes for extending the range of color sensitivity of photographic emulsions.

**cy·a·nite** (sī′ə-nīt′) *n. var. of* KYANITE.

**cyano-** *or* **cyan-** *pref.* [< Gk. *kuanos,* blue.] **1.** Blue in color <*cyano*type> **2. a.** Cyanogen <*cyan*ic> **b.** Cyanide <*cyano*genesis>

**cy·a·no·ac·ry·late** (sī′ə-nō-ăk′rə-lāt′, sī-ăn′ō-) *n.* An adhesive substance with an acrylate base, used in industry and medicine.

**cy·a·no·co·bal·a·min** (sī′ə-nō′kō-băl′ə-mĭn) *n.* [CYANO- + COBAL(T) + (VIT)AMIN.] Vitamin $B_{12}$.

**cy·an·o·gen** (sī-ăn′ə-jən) *n.* [Fr. *cyanogène* : *cyano-,* cyano- + *-gène,* -gen.] **1.** A colorless, flammable, highly poisonous gas, $C_2N_2$, used in welding and as a rocket propellant, fumigant, and military weapon. **2.** The univalent radical CN found in simple and complex cyanide compounds.

**cy·a·no·gen·e·sis** (sī′ə-nō-jĕn′ĭ-sĭs, sī-ăn′ō-) *n.* Generation of cyanide. **—cy′a·no·ge·net′ic** (-jə-nĕt′ĭk), **cy′a·no·gen′ic** (-jĕn′ĭk) *adj.*

**cy·a·no·hy·drin** (sī′ə-nō-hī′drĭn, sī-ăn′ō-) *n.* [CYANO- + HYDR(O)- + -IN.] A compound containing both the CN and OH radicals.

**cy·a·nosed** (sī′ə-nōzd′) *adj.* [< CYANOSIS.] Afflicted with cyanosis.

**cy·a·no·sis** (sī′ə-nō′sĭs) *n.* Bluish discoloration of the skin, caused by inadequate oxygenation of the blood. **—cy′a·not′ic** (-nŏt′ĭk) *adj.*

**cy·an·o·type** (sī-ăn′ə-tīp′) *n.* A blueprint.

**cy·a·nu·ric acid** (sī′ə-no͝or′ĭk, -nyo͝or′-) *n.* A white crystalline acid, $C_3N_3(OH)_3$, that decomposes with heating to form cyanic acid.

**Cyb·e·le** (sĭb′ə-lē) *n.* [Lat. < Gk. *Kubelē* < *kubelon,* mountain in Phrygia.] *Myth.* The goddess of nature of ancient Asia Minor.

**cy·ber·nate** (sī′bər-nāt′) *v.* **-nat·ed, -nat·ing, -nates.** [CYBERN(ETICS) + -ATE[1].] —*vt.* To control (e.g., an industrial process) automatically by computer. —*vi.* To become so controlled. **—cy′ber·na′tion** *n.*

**cy·ber·net·ics** (sī′bər-nĕt′ĭks) *n.* [< Gk. *kubernētēs,* governor < *kuberna*n, to govern.] *(sing. in number).* Theoretical study of control processes in electronic, mechanical, and biological systems, esp. mathematical analysis of the flow of data in such systems. **—cy′ber·net′ic** *adj.* **—cy′ber·net′i·cal·ly** *adv.* **—cy′ber·net′i·cist** *n.*

**cy·borg** (sīʹbôrg′) *n.* [CYB(ERNETIC) + ORG(ANISM).] A person who has some vital bodily processes controlled by cybernetically operated devices.

**cy·cad** (sīʹkăd′, -kəd) *n.* [NLat. *Cycas,* genus name < Gk. *kukas,* alteration of *koix,* a kind of palm tree.] A seed-bearing plant of the family Cycadaceae, similar to a palm tree but topped by fernlike leaves.

**cycl-** *pref. var. of* CYCLO-.

**cy·cla·mate** (sīʹklə-māt′, sĭkʹlə-) *n.* A salt of cyclamic acid, esp. one of two very sweet crystalline compounds: **a.** Sodium cyclamate. **b.** Calcium cyclamate, $C_{12}H_{24}N_2O_6S_2Ca$.

**cy·cla·men** (sīʹklə-mən, sĭkʹlə-) *n.* [NLat. < Gk. *kuklaminos.*] A plant of the genus *Cyclamen,* having white, pink, or red flowers with reflexed petals.

**cyc·la·mic acid** (sĭkʹlə-mĭk′, sīkʹlə-) *n.* A sour-sweet crystalline acid, $C_6H_{13}NO_3S$.

**cy·clase** (sīʹklās′, -klāz′) *n.* An enzyme that acts as a catalyst in the cyclization of a compound.

**cy·cle** (sīʹkəl) *n.* [Fr. < LLat. *cyclus* < Gk. *kuklos,* circle.] **1.** A time interval in which a characteristic, esp. a regularly repeated event or sequence of events occurs. **2. a.** A single complete execution of a periodically repeated phenomenon. **b.** A periodically repeated sequence of events. **3.** The orbit of a celestial body. **4.** A long period of time : AGE. **5. a.** The aggregate of traditional poems or stories organized around a central theme or hero <the Arthurian *cycle*> **b.** A series of poems or songs on the same theme <Mahler's song *cycles*> **6.** A motorcycle or bicycle. **7.** *Bot.* A circular arrangement of flower parts, as petals or sepals. —*v.* **-cled, -cling, -cles.** —*vi.* **1.** To occur in or pass through a cycle. **2.** To move in or as if in a circle. **3.** To ride a bicycle or motorcycle. —*vt.* To use or employ in a cycle. **—cyʹcler** *n.*

**cy·clic** (sīʹklĭk, sĭkʹlĭk) *also* **cy·cli·cal** (sīʹklĭ-kəl, sĭkʹlĭ-kəl) *adj.* **1. a.** Of, pertaining to, or marked by cycles. **b.** Moving or recurring in cycles. **2.** *Chem.* Of or relating to compounds with atoms arranged in a ring or closed-chain structure. **3.** *Bot.* **a.** Having parts arranged in a whorl. **b.** Forming a whorl. **—cyʹcli·cal·ly** *adv.*

**cyclic AMP** *n.* A cyclic nucleotide that acts as a hormonal mediator on the cellular level in the control of various metabolic processes.

**cyclic GMP** *n.* A cyclic nucleotide of guanosine believed to act as an antagonist to cyclic AMP in cellular processes.

**cy·clist** (sīʹklĭst) *n.* One who rides or races a two-wheeled vehicle, as a bicycle or motorcycle.

**cy·cli·za·tion** (sī′klĭ-zāʹshən, sĭk′lĭ-) *n.* Formation of rings in a hydrocarbon.

**cyclo-** *or* **cycl-** *pref.* [< Gk. *kuklos,* circle.] **1.** Circle : cycle <*cyclorama*> **2.** A cyclic compound <*cyclohexane*>

**cy·clo·hex·ane** (sī′klō-hĕkʹsān′) *n.* A highly flammable, colorless, mobile liquid, $C_6H_{12}$, obtained from petroleum and benzene and used in making nylon and as a solvent, paint and varnish remover.

**cy·clo·hex·i·mide** (sī′klō-hĕkʹsə-mīd′, -mĭd) *n.* A compound, $C_{15}H_{23}NO_4$, used as an agricultural fungicide.

**cy·cloid** (sīʹkloid′) *adj.* [Fr. *cycloïde* < Gk. *kukloeidēs,* circular : *kuklos,* circle + *-eidēs,* -oid.] **1.** Resembling a circle. **2.** *Zool.* Thin, rounded, and smooth-edged. —Used of fish scales. **3.** *Psychiat.* Designating a person afflicted with cyclothymia. —*n.* *Math.* The curve traced by a point on the circumference of a circle rolling on a straight line. **—cy·cloiʹdal** (-kloidʹl) *adj.*

**cy·clom·e·ter** (sī-klŏmʹĭ-tər) *n.* **1.** An instrument for recording the revolutions of a wheel so as to indicate distance traveled. **2.** An instrument for measuring circular arcs. **—cy′clo·metʹric** (-klə-mĕtʹrĭk) *adj.* **—cy·clomʹe·try** *n.*

**cy·clone** (sīʹklōn′) *n.* [Poss. < Gk. *kuklōma,* coil < *kuklos,* circle.] **1.** *Meteorol.* An atmospheric disturbance marked by masses of air rapidly circulating clockwise in the southern and counterclockwise in the northern hemisphere about a low-pressure center, usu. accompanied by stormy, often destructive weather. **2.** A violent, rotating windstorm. **3.** A device using centrifugal force to separate materials. **—cy·clonʹic** (-klŏnʹĭk), **cy·clonʹi·cal** *adj.*

**cyclone cellar** *n.* An underground shelter in or adjacent to a house, used for protection from cyclones or tornadoes.

**cy·clo·pae·di·a** (sī′klə-pēʹdē-ə) *n. var. of* CYCLOPEDIA.

**cy·clo·par·af·fin** (sī′klō-părʹə-fĭn) *n.* Any of a class of hydrocarbons, including cyclopropane, cyclopentane, and cyclohexane, in which at least three carbon atoms per molecule are joined in a ring structure and each such carbon in the ring is bonded to two hydrogen atoms or alkyl groups.

**cy·clo·pe·an** (sī′klə-pēʹən, sī-klōʹpē-) *adj.* **1.** *often* **Cyclopean.** Relating to or like the Cyclopes. **2.** Relating to or designating a primitive masonry style using massive, irregularly shaped stones.

**cy·clo·pe·di·a** *also* **cy·clo·pae·di·a** (sī′klə-pēʹdē-ə) *n.* [Short for ENCYCLOPEDIA.] An encyclopedia. **—cy′clo·peʹdic** (-dĭk) *adj.* **—cy′clo·peʹdist** (-dĭst) *n.*

**cy·clo·pen·tane** (sī′klə-pĕnʹtān′, sĭk′lə-) *n.* A colorless flammable liquid, $C_5H_{10}$, derived from petroleum and used as a solvent and motor fuel.

**Cy·clo·pes** (sī-klōʹpēz) *n. pl. of* CYCLOPS.

**cy·clo·ple·gia** (sī′klə-plēʹjə) *n.* Loss of visual accommodation due to paralysis of the ciliary muscles of the eye.

**cy·clo·pro·pane** (sī′klə-prōʹpān′) *n.* A highly flammable, explosive, colorless gas, $C_3H_6$, used as an anesthetic.

**Cy·clops** (sīʹklŏps′) *n., pl.* **Cy·clo·pes** (sī-klōʹpēz) [Lat. < Gk. *kuklōps* : *kuklos,* circle + *ōps,* eye.] *Gk. Myth.* **1.** Any of the three one-eyed Titans who forged thunderbolts for Zeus. **2.** Any of a race of one-eyed giants, reputedly descended from these Titans, living on the island of Sicily.

**cy·clo·ram·a** (sī′klə-rămʹə, -räʹmə) *n.* [CYCL(O)- + (PAN)ORAMA.] **1.** A large composite picture placed on the interior walls of a cylindrical room so as to appear in natural perspective to a spectator standing in the center. **2.** A large usu. concave curtain or wall placed or hung at the rear of a stage. **—cy′clo·ramʹic** *adj.*

**cy·clo·ser·ine** (sī′klō-sĕrʹēn′) *n.* An antibiotic produced by a species of *Streptomyces,* effective against many Gram-negative bacteria.

**cy·clo·sis** (sī-klōʹsĭs) *n., pl.* **-ses** (-sēz) [NLat. < Gk. *kuklōsis,* a surrounding < *kukloun,* to surround < *kuklos,* circle.] The streaming circulatory motion of protoplasm within certain cells and cell structures.

**cy·clo·stome** (sīʹklə-stōm′) *n.* [NLat. *Cyclostomi* and *Cyclostomata,* class names : CYCLO- + Gk. *stoma,* mouth.] A primitive eellike vertebrate of the class Agnatha, as a lamprey, lacking jaws and true teeth and having a circular, sucking mouth. **—cy·closʹto·mate′** (sĭ-klŏsʹtə-māt′), **cy′clo·stomʹa·tous** (sī′klə-stŏmʹə-təs, -stōʹmə-) *adj.*

**cy·clo·thyme** (sīʹklə-thīm′) *n.* One afflicted with cyclothymia.

**cy·clo·thy·mi·a** (sī′klə-thīʹmē-ə) *n.* A manic-depressive psychosis marked by alternating periods of excitement and activity with periods of depression and inactivity. **—cy′clo·thyʹmic** (-mĭk) *adj.* & *n.*

**cy·clo·tron** (sīʹklə-trŏn′) *n.* A circular accelerator capable of generating particle energies between a few million and several tens of millions of electron volts, in which charged particles generated at a central source are accelerated spirally outward in a plane at right angles to a fixed magnetic field by an alternating electric field.

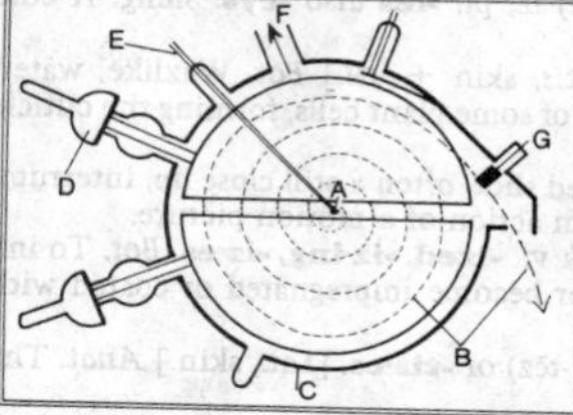


**cyclotron**
*A. source, B. path of ion, C. vacuum chamber, D. high frequency voltage insulator, E. heating wires for ion source, F. pump, G. target*

**cy·der** (sīʹdər) *n. Chiefly Brit. var. of* CIDER.

**cy·e·sis** (sī-ēʹsĭs) *n., pl.* **-ses** (-sēz′) [NLat. < Gk. *kuēsis* < *kuein,* to swell.] Gestation : pregnancy.

**cyg·net** (sĭgʹnĭt) *n.* [ME *sygnett* < OFr. *cygne,* swan < Lat. *cygnus* < Gk. *kuknos.*] A young swan.

**Cyg·nus** (sĭgʹnəs) *n.* [Lat. *cygnus,* swan < Gk. *kuknos.*] A constellation in the Northern Hemisphere.

**cyl·in·der** (sĭlʹən-dər) *n.* [OFr. *cylindre* < Lat. *cylindrus* < Gk. *kulindros* < *kulindein,* to roll.] **1.** *Math.* **a.** A surface generated by a straight line moving parallel to a fixed straight line and intersecting a plane curve. **b.** The part of such a surface bounded by two parallel planes and the regions of the planes bounded by the surface. **c.** A solid bounded by two parallel planes and such a surface having a closed curve, esp. a circle, as a directrix. **2.** A cylindrical container or object. **3.** *Engineer.* **a.** The chamber in which a piston of a reciprocating engine moves. **b.** The chamber of a pump from which fluid is expelled by a piston. **4.** The rotating chamber of a revolver that holds the cartridges. **5.** Any of the rotating cylinders in a printing press that carry the paper or the curved printing plate or receive the ink or impression. **6.** A cylindrical clay or stone object with an engraved design or cuneiform inscription.

**cylinder head** *n.* The closed, often detachable end of a cylinder or cylinders in an internal-combustion engine.

**cy·lin·dri·cal** (sə-lĭnʹdrĭ-kəl) *also* **cy·lin·dric** (-drĭk) *adj.* **1.** Having the shape or properties of a cylinder. **2.** Of or relating to a cylinder. **3.** Of or relating to the coordinate system or to any of three coordinates in it, formed by two polar coordinates in a plane and a rectangular coordinate measured perpendicularly from the plane. **—cy·linʹdri·calʹi·ty** (-kălʹĭ-tē) *n.* **—cy·linʹdri·cal·ly** *adv.*

**cyl·in·droid** (sĭlʹən-droid′) *n.* A cylindrical surface or solid all of whose sections perpendicular to the elements are elliptical. —*adj.* Resembling a cylinder.

**cy·ma** (sīʹmə) *n.* [Gk. *kuma* < *kuein,* to swell.] A molding for a cornice, with a partly concave and partly convex curve in profile, used esp. in classical architecture.

**cy·ma·tium** (sī-māʹshəm, -shē-əm) *n., pl.* **-tia** (-shə, -shē-ə) [Lat. <

ă **pat** ā **pay** âr **care** ä **father** ĕ **pet** ē **be** hw **which** ĭ **pit** ī **tie** îr **pier** ŏ **pot** ō **toe** ô **paw, for** oi **noise** o͝o **took**

Gk. *kumation*, dim. of *kuma*, cyma.] **1.** A cyma. **2.** The topmost molding of a classical cornice.

**cym·bal** (sĭm′bəl) *n.* [ME < OFr. *cymbale* < Lat. *cymbalum* < Gk. *kumbalon* < *kumbē*, bowl.] **1.** One of a pair of concave brass plates struck together as percussion instruments. **2.** A single brass plate, sounded by hitting with a drumstick and often part of a set of drums. **—cym′bal·eer′** (sĭm′bə-lîr′), **cym′bal·er, cym′bal·ist** *n.*

**cym·bid·i·um** (sĭm-bĭd′ē-əm) *n.* [NLat. *Cymbidium*, genus name < Lat. *cymba*, boat < Gk. *kumbē*.] An orchid of the genus *Cymbidium*, with showy flowers often used for decoration.

**cyme** (sīm) *n.* [NLat. *cyma* < Lat., young cabbage sprout < Gk. *kuma*, cyma, sprout.] *Bot.* An often flat-topped flower cluster blooming from the center toward the edges, whose main axis is always terminated by a flower. **—cy·mif′er·ous** (sī-mĭf′ər-əs) *adj.*

**cy·mene** (sī′mēn′) *n.* [Fr. *cymène* < Gk. *kuminon*, cumin, of Semitic orig.] *Chem.* Any of three colorless isomeric liquid hydrocarbons, $C_{10}H_{14}$, obtained chiefly from the essential oils of various plants and used to make synthetic resins.

**cym·ling** (sĭm′lĭng) *also* **cym·lin** (-lĭn) *n.* [Alteration of SIMNEL.] A greenish-white, round, flat squash with a scalloped edge.

**cy·mo·gene** (sī′mə-jēn′) *n.* [CYM(ENE) + -GENE.] A flammable gaseous fraction of petroleum, chiefly butane.

**cy·moid** (sī′moid′) *adj.* Resembling a cyma or cyme.

**cy·mo·phane** (sī′mə-fān′) *n.* [Fr. : Gk. *kuma*, wave, cyma + Fr. *-phane*, -phane.] A chrysoberyl with a shimmering luster.

**cy·mose** (sī′mōs′) *also* **cy·mous** (-məs) *adj.* [CYM(E) + -OSE[1].] **1.** Relating to or resembling a cyme. **2.** Bearing a cyme or cymes. **—cy′mose′ly** *adv.*

**Cym·ric** (kĭm′rĭk, sĭm′-) *adj.* Of or relating to the Cymry. *—n.* **1.** Brythonic. **2.** WELSH 2.

**Cym·ry** (kĭm′rē, sĭm′-) *n.* [Welsh.] The branch of the Celtic people to which the Welsh, the Cornish, and the Bretons belong.

**cyn·ic** (sĭn′ĭk) *n.* [Lat. *cynicus*, Cynic philosopher < Gk. *kunikos*, prob. ult. < *kuōn*, dog.] **1. Cynic.** A member of an ancient Greek philosophical sect who believed virtue to be the only good and practice of self-control to be the only way of achieving virtue. **2.** One who believes all people are motivated by selfishness. *—adj.* **1. Cynic.** Of or relating to the Cynics or their doctrines. **2.** Cynical.

**cyn·i·cal** (sĭn′ĭ-kəl) *adj.* **1.** Scornful of the virtue or motives of others. **2.** Contemptuously and bitterly mocking. **—cyn′i·cal·ly** *adv.* **—cyn′i·cal·ness** *n.*

☆ ***syns:*** CYNICAL, IRONIC, SARDONIC, WRY *adj. core meaning*: marked by or showing contemptuous mockery <*a cynical* attitude toward society>

**cyn·i·cism** (sĭn′ĭ-sĭz′əm) *n.* **1.** A cynical character or attitude. **2.** A cynical act or comment. **3. Cynicism.** The doctrines and beliefs of the Cynics.

**cy·no·sure** (sī′nə-shŏŏr′, sĭn′ə-) *n.* [Fr., Ursa Minor (which contains the guiding star Polaris) < Lat. *cynosura* < Gk. *kunosoura*, dog's tail, Ursa Minor : *kuōn*, dog + *oura*, tail.] A focal point of attention and admiration. **—cy′no·sur′al** *adj.*

▲ word history: A cynosure attracts attention for a compelling reason. The word *cynosure* is derived from Greek *kunosoura*, the Greek name for the constellation now known as Ursa Minor. In ancient times this constellation was always above the horizon in northern latitudes above 18°. It was used in navigation because it is located near the north celestial pole. *Cynosure* was originally borrowed into English as the name of the same constellation but was also used figuratively to mean a guide or center of attention.

**cy·pher** (sī′fər) *n.* & *v. var. of* CIPHER.

**cy·press** (sī′prəs) *n.* [ME *cipres* < OFr. < LLat. *cypressus* < Gk. *kuparissos.*] **1. a.** An evergreen tree of the genus *Cupressus*, growing in warm climates and bearing small, compressed needles. **b.** A similar or related tree, as the bald cypress. **2.** The wood of any of the cypress trees. **3.** Cypress branches used as a symbol of mourning.

**cypress spurge** *n.* A Eurasian plant, *Euphorbia cyparissias*, with densely crowded, narrow leaves and yellow-green flowers.

**cypress vine** *n.* A tropical American vine, *Quamoclit pennata*, with finely divided compound leaves and scarlet flowers.

**Cyp·ri·an** (sĭp′rē-ən) *adj.* **1.** Of or relating to Cyprus, its people, their customs, or their language. **2. a.** Characteristic of or resembling the ancient worship of Aphrodite on Cyprus. **b.** Wanton. *—n.* **1.** A Cypriot. **2.** *Obs.* A wanton person, esp. a prostitute.

**cyp·ri·nid** (sĭp′rə-nĭd) *n.* [NLat. *Cyprinidae*, family name < *Cyprinus*, genus name < Lat. *cyprinus*, carp < Gk. *kuprinos.*] Any of numerous often small freshwater fishes of the family Cyprinidae, including the minnows, carps, and shiners. **—cyp′ri·nid** *adj.*

**cy·prin·o·dont** (sĭ-prĭn′ə-dŏnt′, -prī′nə-) *n.* [Lat. *cyprinus*, carp (< Gk. *kuprinos*) + -ODONT.] Any of various small, soft-finned fishes of the family Cyprinodontidae, including the killifishes, topminnows, and many species popular in home aquariums.

**cyp·ri·noid** (sĭp′rə-noid′, sĭ-prī′-) *adj.* [NLat. *Cyprinoidea*, suborder name < *Cyprinus*, genus name. —see CYPRINID.] Of, relating to, or resembling a carp or related fish. **—cyp′ri·noid′** *n.*

**Cyp·ri·ot** (sĭp′rē-ət, -ŏt′) *also* **Cyp·ri·ote** (-ōt′, -ət) [Fr. *cypriote* < *Cyprus*, Cyprus.] *—n.* **1.** A native or resident of Cyprus. **2.** The ancient Greek dialect of Cyprus. *—adj.* **1.** Of or relating to Cyprus. **2.** Of or relating to the Cypriot language.

**cyp·ri·pe·di·um** (sĭp′rĭ-pē′dē-əm) *n.* [NLat., *Cypripedium*, genus name : Gk. *Kupris*, Aphrodite (< *Kupros*, Cyprus, legendary birthplace of Venus) + Gk. *pedilon*, sandal.] An orchid of the genus *Cypripedium*, including the lady's-slipper.

**cy·prot·er·one** (sī-prŏt′ə-rōn′) *n.* [Prob. Lat. *Cypris*, Venus (< Gk. *Kupris* < *Kupros*, Cyprus) + (TESTOS)TERONE.] A hormone inhibiting secretion of androgens.

**cyp·se·la** (sĭp′sə-lə) *n., pl.* **-lae** (-lē′) [NLat. < Gk. *kupselē*, hollow vessel.] An achene that does not separate from its calyx.

**Cyr·e·na·ic** (sĭr′ə-nā′ĭk, sīr′-) *adj.* **1.** Of or relating to Cyrenaica or Cyrene. **2.** Of or relating to a philosophy emphasizing pleasure as the only good in life. *—n.* **1.** A native or resident of Cyrenaica or Cyrene. **2.** A disciple of the Cyrenaic school of philosophy.

**Cy·ril·lic** (sə-rĭl′ĭk) *adj.* Of or designating the old Slavic alphabet ascribed to Saint Cyril, presently used in modified form for Russian, Bulgarian, and certain other Slavic languages.

**cyst** (sĭst) *n.* [NLat. *cystis* < Gk. *kustis*, bladder.] **1.** *Pathol.* An abnormal membranous sac containing a gaseous, liquid, or semisolid substance. **2.** *Anat.* A sac or vesicle in the body. **3.** *Biol.* A capsulelike membrane of certain organisms in a resting stage. **4.** *Bot.* Any of various cells of nonsexual origin in green algae, which germinate and produce new plants after a resting period.

**cyst-** *pref. var. of* CYSTO-.

**cys·tec·to·my** (sĭ-stĕk′tə-mē) *n., pl.* **-mies. 1.** Surgical removal of a cyst. **2.** Surgical excision of the gallbladder or of part of the urinary bladder.

**cys·te·ine** (sĭs′tə-ēn′, -ĭn) *n.* [CYST(INE) + -EIN.] An amino acid, $C_3H_7NO_2S$, found in most proteins, esp. in keratin.

**cys·tic** (sĭs′tĭk) *adj.* **1.** Of, relating to, or like a cyst. **2.** Having or containing a cyst or cysts. **3.** Enclosed in a cyst. **4.** *Anat.* Relating to the gallbladder or urinary bladder.

**cys·ti·cer·coid** (sĭs′tĭ-sûr′koid) *n.* [CYSTICERC(US) + -OID.] The larval stage of certain tapeworms, like a cysticercus but having the scolex completely filling the enclosing sac.

**cys·ti·cer·co·sis** (sĭs′tĭ-sər-kō′sĭs) *n.* [CYSTICERC(US) + -OSIS.] Infestation with cysticerci.

**cys·ti·cer·cus** (sĭs′tĭ-sûr′kəs) *n., pl.* **-ci** (-sī′) [NLat. : Gk. *kustis*, cyst + Gk. *kerkos*, tail.] The larval stage of many tapeworms, consisting of a scolex enclosed in a fluid-filled sac.

**cystic fibrosis** *n.* A congenital disease of mucous glands throughout the body, usu. developing during childhood and causing pancreatic insufficiency and pulmonary disorders.

**cys·tine** (sĭs′tēn′) *n.* A white crystalline compound, $C_6H_{12}N_2O_4S_2$, the principal sulfur-containing amino acid of protein.

**cys·ti·tis** (sĭ-stī′tĭs) *n.* Inflammation of the urinary bladder.

**cysto-** *or* **cyst-** *pref.* [< Gk. *kustis*, bladder.] Bladder : cyst : sac <*cystocele*>

**cys·to·carp** (sĭs′tə-kärp′) *n.* A structure consisting of fertile filaments and carpospores, developed after fertilization of the carpogonium in red algae.

**cys·to·cele** (sĭs′tə-sēl′) *n.* Hernia of the bladder.

**cys·toid** (sĭs′toid′) *adj.* Formed like or resembling a cyst. **—cys′toid′** *n.*

**cys·to·lith** (sĭs′tə-lĭth′) *n.* **1.** *Bot.* A mineral concretion, usu. calcium carbonate, formed in the cellulose wall of plant cells. **2.** *Pathol.* A urinary calculus.

**cys·to·scope** (sĭs′tə-skōp′) *n.* A tubular instrument equipped with a light for examining the urinary bladder and ureter. **—cys′to·scop′ic** (-skŏp′ĭk) *adj.* **—cys·tos′co·py** (sĭ-stŏs′kə-pē) *n.*

**cys·tos·to·my** (sĭ-stŏs′tə-mē) *n., pl.* **-mies.** Surgical formation of an opening into the bladder.

**cyt-** *pref. var. of* CYTO-.

**-cyte** *suff.* [NLat. *-cyta* < Gk. *kutos*, hollow vessel.] Cell <*leukocyte*>

**Cyth·e·re·a** (sĭth′ə-rē′ə) *n.* [Lat. < Gk. *Kuthereia* < *Kuthēra*, one of the Ionian islands.] *Gk. Myth.* The goddess Aphrodite.

**cyto-** *or* **cyt-** *pref.* [< Gk. *kutos*, hollow vessel.] Cell <*cytoplasm*>

**cy·to·chem·is·try** (sī′tō-kĕm′ĭ-strē) *n.* The chemistry of plant and animal cells. **—cy′to·chem′i·cal** (-kĕm′ĭ-kəl) *adj.*

**cy·to·chrome** (sī′tə-krōm′) *n.* Any of a class of iron-containing proteins important in cell metabolism.

**cytochrome oxidase** *n.* An oxidizing enzyme that functions in cell respiration by reacting with oxygen in the reduced state.

**cy·to·gen·e·sis** (sī′tō-jĕn′ĭ-sĭs) *n.* Formation and development of cells. **—cy′to·ge·net′ic** (sī′tō-jə-nĕt′ĭk) *adj.*

**cy·to·ge·net·ics** (sī′tō-jə-nĕt′ĭks) *n. (sing. in number).* Study of heredity by cytologic and genetic methods. **—cy′to·ge·net′i·cal** (-nĕt′ĭ-kəl) *adj.* **—cy′to·ge·net′i·cal·ly** *adv.* **—cy′to·ge·net′i·cist** (-nĕt′ĭ-sĭst) *n.*

**cy·tog·e·ny** (sī-tŏj′ə-nē) *n.* Cytogenesis.

**cy·to·ki·ne·sis** (sī′tō-kĭ-nē′sĭs, -kī-) *n.* Cleavage of cytoplasm during cell division. **—cy′to·ki·net′ic** (-nĕt′ĭk) *adj.*

**cy·to·ki·nin** (sī′tə-kī′nĭn) *n.* A growth regulator promoting cell division in plants.
**cy·tol·o·gy** (sī-tŏl′ə-jē) *n.* The branch of biology concerned with the formation, structure, pathology, and function of cells. **—cy′to·log′ic** (-tə-lŏj′ĭk), **cy′to·log′i·cal** *adj.* **—cy·tol′o·gist** *n.*
**cy·tol·y·sin** (sī-tŏl′ĭ-sĭn) *n.* [CYTOLYS(IS) + -IN.] An antibody capable of partially or completely destroying an animal cell.
**cy·tol·y·sis** (sī-tŏl′ĭ-sĭs) *n.* Dissolution of a cell. **—cy′to·lyt′ic** (sī′tə-lĭt′ĭk) *adj.*
**cy·to·me·gal·ic** (sī′tō-mĭ-găl′ĭk) *adj.* Relating to or marked by greatly enlarged cells.
**cy·to·meg·a·lo·vi·rus** (sī′tə-mĕg′ə-lō-vī′rəs) *n.* Any of a group of viruses that cause cellular enlargement and a disease of infants marked by microcephaly and circulatory dysfunction.
**cy·to·path·ic** (sī′tə-păth′ĭk) *adj.* Of or relating to pathologic changes in cells.
**cy·toph·a·gy** (sī-tŏf′ə-jē) *n.* Devouring of other cells by the phagocytes. **—cy′to·phag′ic** (sī′tə-făj′ĭk), **cy·toph′a·gous** (-tŏf′ə-gəs) *adj.*
**cy·to·pho·tom·e·try** (sī′tə-fō-tŏm′ĭ-trē) *n.* Photometric study of a cell. **—cy′to·pho′to·met′ric** (-tə-mĕt′rĭk) *adj.*
**cy·to·plasm** (sī′tə-plăz′əm) *n.* Protoplasm outside a cell nucleus. **—cy′to·plas′mic** (-plăz′mĭk) *adj.* **—cy′to·plas′mi·cal·ly** *adv.*
**cy·to·plast** (sī′tə-plăst′) *n.* Cytoplasm within a single cell. **—cy′to·plas′tic** (-plăs′tĭk) *adj.*
**cy·to·sine** (sī′tə-sēn′) *n.* [CYT(O)- + -OS(E)[2] + -INE[2].] A pyrimidine base, $C_4H_5N_3O$, an essential constituent of both ribonucleic and deoxyribonucleic acids.
**cy·to·tax·on·o·my** (sī′tō-tăk-sŏn′ə-mē) *n.* Classification of organisms based on cellular structure, esp. on comparative chromosomal morphology. **—cy′to·tax′o·nom′ic** (-tăk′sə-nŏm′ĭk) *adj.* **—cy′to·tax·on′o·mist** *n.*
**cy·to·tech·nol·o·gist** (sī′tə-tĕk-nŏl′ə-jĭst) *n.* A technician who examines and identifies cellular abnormalities.
**czar** (zär) *n.* [Pol. < R. *tsar′*, ult. < Lat. *Caesar*, emperor. —see CAESAR.] **1.** A former Russian emperor. **2.** An autocrat. **3.** *Informal.* One in authority : LEADER <a *czar* of finance> **—czar′dom** *n.*
**czar·das** (chär′däsh′) *n.* [Hung. *csárdás.*] **1.** An intricate Hungarian dance marked by variations in tempo. **2.** Music for the czardas.
**czar·e·vitch** (zär′ə-vĭch′) *n.* [Pol. < R. *tsarevich* : *tsar′*, czar + *-evich*, masc. patronymic suffix.] A czar's eldest son.
**cza·rev·na** (zä-rĕv′nə) *n.* [Pol. < R. *tsarevna* : *tsar′*, czar + *-evna*, fem. patronymic suffix.] **1.** A czar's daughter. **2.** A czarevitch's wife.
**cza·ri·na** (zä-rē′nə) *also* **cza·rit·za** (-rĭt′sə, -rēt′-) *n.* [Pol. < R. *tsarina* : *tsar′*, czar + *-ina*, fem. suffix.] A czar's wife.
**czar·ism** (zär′ĭz′əm) *n.* The Russian system of government under the czars : AUTOCRACY. **—czar′ist** *adj.*
**cza·rit·za** (zä-rĭt′sə, -rēt′-) *n. var. of* CZARINA.
**Czech** (chĕk) *n.* [Pol. < Czech *Čechy.*] **1.** A native or resident of Czechoslovakia, esp. a Bohemian, Moravian, or Slovak. **2.** The Slavic language of the Czechs. **—Czech** *adj.*

# Dd

**d** *or* **D** (dē) *n., pl.* **d's** *or* **D's. 1.** The fourth letter of the English alphabet. **2.** A speech sound represented by the letter *d.* **3.** The fourth in a series. **4. D** *Mus.* **a.** The second tone in the scale of C major or the fourth tone in the relative minor scale. **b.** The key or a scale in which D is the tonic. **c.** A written or printed note representing this tone. **d.** A string, key, or pipe tuned to the pitch of this tone. **5.** The lowest passing grade given to a student. **6.** Something shaped like the letter D. **7. D** The Roman numeral for 500.
**-′d. 1.** Had <They*′d* already left.> **2.** Would <I*′d* rather walk than jog.> **3.** Did <Who*′d* you ask?>
**dab[1]** (dăb) *v.* **dabbed, dab·bing, dabs.** [ME *dabben.*] *—vt.* **1.** To apply with short, light strokes. **2.** To cover lightly with or as if with a moist substance. **3.** To strike or hit lightly. *—vi.* To tap gently and lightly : PAT. *—n.* **1.** A very small amount. **2.** A quick, light pat.
**dab[2]** (dăb) *n.* [ME *dabbe.*] Any of various flatfishes, chiefly of the genera *Limanda* and *Hippoglossoides,* resembling the flounders.
**dab[3]** (dăb) *n.* [Orig. unknown.] *Chiefly Brit.* An expert.
**dab·ber** (dăb′ər) *n.* **1.** One that dabs. **2.** A cushioned pad used by printers and engravers to apply ink.
**dab·ble** (dăb′əl) *v.* **-bled, -bling, -bles.** [Perh. < Du. *dabbelen,* freq. of *dabben,* to strike, tap.] *—vt.* To spatter or splash, as with a liquid. *—vi.* **1.** To splash liquid gently. **2.** To undertake an activity superficially or casually <*dabbled* in politics> **3.** To bob forward and under in shoal water so as to feed off the bottom. **—dab′bler** *n.*
**dab·chick** (dăb′chĭk′) *n.* A small grebe of the genus *Podiceps.*
**da ca·po** (dä kä′pō, də) *adv.* [Ital.] *Mus.* From the start or beginning. —Used as a direction to repeat a passage.
**dace** (dās) *n., pl.* **dace** *or* **dac·es.** [ME *dars* < OFr.] A small freshwater fish of the family Cyprinidae, resembling the minnow.
**da·cha** (dä′chə) *n.* [R.] A Russian country house.
**dachs·hund** (däks′hŏŏnt′, dăk′sənt) *n.* [G. : *Dachs,* badger + *Hund,* dog.] A small dog orig. bred in Germany for hunting badgers, having a long body, a usu. short-haired brown or black and brown coat, drooping ears, and extremely short legs.
**Da·cron** (dā′krŏn′, dăk′rŏn′). A trademark for a synthetic polyester textile fiber.
**dac·tyl** (dăk′təl) *n.* [ME *dactil* < Lat. *dactylus* < Gk. *daktulos,* finger, dactyl.] **1.** A metrical foot composed of one accented syllable followed by two unaccented or of one long syllable followed by two short. **2.** A bodily part such as a toe : DIGIT. **—dac·tyl′ic** (-tĭl′ĭk) *adj.* & *n.* **—dac·tyl′i·cal·ly** *adv.*
**dactyl-** *pref. var. of* DACTYLO-.
**dac·ty·li** (dăk′tə-lī′) *n. pl. of* DACTYLUS.
**dactylo-** *or* **dactyl-** *pref.* [< Gk. *daktulos,* finger.] Finger : toe : digit <*dactylogram*>
**dac·tyl·o·gram** (dăk-tĭl′ə-grăm′) *n.* A fingerprint.
**dac·ty·log·ra·phy** (dăk′tə-lŏg′rə-fē) *n.* The study of fingerprints as an identification method. **—dac′ty·lo·graph′ic** (-lō-grăf′ĭk) *adj.*
**dac·ty·lol·o·gy** (dăk′tə-lŏl′ə-jē) *n.* The use of the fingers and hands to convey ideas, as in the manual alphabet used by deaf-mutes.
**dac·ty·lus** (dăk′tə-ləs) *n., pl.* **-li** (-lī′) [NLat. < Gk. *daktulos,* finger.] **1.** DACTYL 2. **2.** The tarsus of some insects following the first joint that usu. consists of one or more joints.
**dad** (dăd) *n.* [Prob. of baby-talk orig.] *Informal.* Father.
**Da·da** *or* **da·da** (dä′dä) *also* **Da·da·ism** (-ĭz′əm) *n.* [Fr.] A western European artistic and literary movement (1916–23) that sought the discovery of authentic reality through the abolition of traditional cultural and aesthetic forms by a technique of comic derision in which irrationality, chance, and intuition were the guiding principles. **—Da′da·ist** *n.* **—Da′da·is′tic** (dä′dä-ĭs′tĭk) *adj.*
**dad·dy** (dăd′ē) *n., pl.* **-dies.** *Informal.* Father.
**daddy long·legs** (lông′lĕgz′, lŏng′-) *n., pl.* **daddy longlegs. 1.** Any of various arachnids of the order Phalangida, with a small, rounded body and long, slender legs. **2.** The crane fly.
**da·do** (dā′dō) *n., pl.* **-does.** [Ital. < Lat. *datum,* neuter p.part. of *dare,* to give.] **1.** The section of an architectural pedestal between the base and crown. **2.** The lower part of the wall of a room, decorated differently from the upper section, as with panels. **3. a.** A rectangular groove cut into a board so that a like piece may be fitted into it. **b.** The groove so cut. *—vt.* **-doed, -do·ing, -does. 1.** To furnish with a dado. **2. a.** To cut a dado in. **b.** To fit into a dado.
**dae·dal** (dēd′l) *adj.* [Lat. *daedalus* < Gk. *daidalos.*] **1.** Complex : intricate. **2.** Finely or skillfully made or employed.
**Dae·da·lus** (dĕd′l-əs) *n.* [Lat. < Gk. *Daidalos* < *daidalos,* skillfully made.] *Gk. Myth.* A legendary artist and inventor, builder of the Labyrinth. **—Dae·da′li·an, Dae·da′le·an** (dĭ-dā′lē-ən, -dāl′yən) *adj.*
**dae·mon** (dē′mən) *n. var. of* DEMON 3, 4.
**daf·fo·dil** (dăf′ə-dĭl) *n.* [Alteration of obs. *affodill* < ME *affodylle,* asphodel < Lat. *asphodilus.* —see ASPHODEL.] **1. a.** A bulbous plant, *Narcissus pseudo-narcissus,* having usu. yellow flowers with a trumpet-shaped central crown. **b.** The flower of the daffodil. **2.** A brilliant to vivid yellow.
**daf·fy** (dăf′ē) *adj.* **-fi·er, -fi·est.** [< obs. *daff,* fool < ME *daffe.*] *Informal.* **1.** Silly <a *daffy* idea> **2.** Crazy <a *daffy* person>
**daft** (dăft) *adj.* **-er, -est.** [ME *dafte,* foolish < OE *gedæfte,* meek.] **1.** Crazy : insane. **2.** Foolish : stupid. **3.** *Scot.* Frolicsome. **—daft′ly** *adv.* **—daft′ness** *n.*
**dag** (dăg) *n.* [ME *dagge,* shred.] **1.** A lock of matted or dung-coated wool. **2.** *Archaic.* A hanging end or shred.
**Da·gan** (dä′gän′) *n.* [Akkadian *Dagān.*] *Myth.* The Babylonian god of the earth.
**dag·ger** (dăg′ər) *n.* [ME *daggere.*] **1.** A short pointed weapon with

sharp edges. **2. a.** Something resembling a dagger. **b.** Something that wounds like a dagger. **3. a.** OBELISK 2. **b.** A double dagger.

**Da·gon** (dā'gŏn') *n.* [ME < Lat. < Gk. *Dagōn* < Heb. *Dāgōn*, dim. of *dāg*, fish.] *Myth.* The chief god of the ancient Philistines and later the Phoenicians, represented as half-man and half-fish.

Dagon

**da·guerre·o·type** (də-gâr'ə-tīp') *n.* [After Louis J. M. *Daguerre* (1787–1851), its inventor.] **1.** An early photographic process with the image made on a light-sensitive silver-coated metallic plate. **2.** A picture made by daguerreotype. —*vt.* **-typed, -typ·ing, -types.** To make a daguerreotype of. **—da·guerre'o·typ'er** *n.* **—da·guerre'o·typ'y** *n.*

**dag·wood** *also* **Dag·wood** (dăg'wo͝od') *n.* [After *Dagwood* Bumstead, a character who made such sandwiches in the comic strip *Blondie* by Murat B. Young (1901–1973).] A multilayered sandwich having various fillings.

**dah** (dä) *n.* A dash in Morse code.

**dahl·ia** (dăl'yə, däl'-, dāl'-) *n.* [NLat. *Dahlia*, genus name, after Anders *Dahl* (d. 1789).] **1.** A plant of the genus *Dahlia*, indigenous to Mexico and Central America, with tuberous roots and usu. large, variously colored flowers. **2.** The flower of a dahlia.

**da·hoon** (də-ho͞on') *n.* [Orig. unknown.] An evergreen shrub or small tree, *Ilex cassine* of the southeastern United States, having red fruit.

**dai·ly** (dā'lē) *adj.* [ME *dayly* < OE *dæglic* < *dæg*, day.] **1.** Performed, taking place, or appearing every day or weekday <a *daily* jog> **2.** For each day <a *daily* telephone record> **3.** Day-to-day : everyday <an appliance for *daily* use> —*adv.* **1.** Every day <Take exercise *daily.*> **2.** Once a day <Wind your watch *daily.*> —*n., pl.* **-lies.** A newspaper published every day or every weekday.

**daily double** *n.* A bet won by selecting both winners of two specified races on one day, as in horse racing.

**dai·mi·o** *also* **dai·my·o** (dī'mē-ō', dīm'yō') *n., pl.* **daimio** *or* **-mi·os** *also* **daimyo** *or* **-my·os.** [J. *daimyo* : *dai*, great (< Chin. *da*[4]) + *myo*, name (< Chin. *ming*[2]).] A hereditary nobleman in Japan's feudal period.

**dai·mon** (dī'mōn') *n. var. of* DEMON 3, 4.

**dain·ty** (dān'tē) *adj.* **-ti·er, -ti·est.** [ME *deinte*, excellent < *deinte*, excellence, dignity < OFr. *deintie* < Lat. *dignitas* < *dignus*, worthy.] **1.** Delicately beautiful : EXQUISITE. **2.** Delicious : choice. **3.** Of refined taste : DISCRIMINATING. **4.** Overfastidious. —*n., pl.* **-ties.** A delicacy. **—dain'ti·ly** *adv.* **—dain'ti·ness** *n.*

**dai·qui·ri** (dī'kə-rē, dăk'ə-) *n., pl.* **-ris.** [After *Daiquirí*, Cuba.] An iced cocktail of rum, lime or lemon juice, and sugar.

**dair·y** (dâr'ē) *n., pl.* **-ies.** [ME *daierie* < *daie*, dairymaid < OE *dæge.*] **1.** A commercial establishment that processes or sells milk and milk products. **2.** A place where milk and cream are stored and processed. **3.** A dairy farm. **4.** The dairy business.

**dairy cattle** *pl.n.* Cows bred and raised for milk.

**dairy farm** *n.* A farm for producing milk and milk products.

**dair·y·ing** (dâr'ē-ĭng) *n.* The business of a dairy.

**dair·y·maid** (dâr'ē-mād') *n.* A woman who works in a dairy.

**dair·y·man** (dâr'ē-mən) *n.* **1.** A dairy manager or owner. **2.** A man who works in a dairy.

**da·is** (dā'ĭs, dās) *n.* [ME *deis* < OFr., platform < LLat. *discus*, table. —see DISK.] A raised platform, as in a lecture hall, for honored guests or speakers.

▲ word history: *Dais* is a word that was borrowed into English twice. It first appeared in the 13th century as *deis*, from Old French *deis*, indicating a table raised on a platform at which honored guests were seated. *Deis* was also used of the platform alone. This word died out in England in the 16th century; it survived, however, in Scotland with the meaning "bench." In the late 18th century historical writers revived the word. Although they used the modern French spelling *dais*, the appearance of *dais* probably represents a borrowing from English itself rather than from French, because the modern French word means only "canopy." The French forms *deis* and *dais* are ultimately derived from Latin *discus*, which in medieval times meant "table." *Dais* is thus cognate with *disk*, *dish*, and *desk*.

**dai·sy** (dā'zē) *n., pl.* **-sies.** [ME *daisie* < OE *dægesēage* : *dæg*, day + *ēage*, eye.] **1.** A plant having rayed flowers, esp. a widely naturalized Eurasian plant, *Chrysanthemum leucanthemum*, having flowers with a yellow center and white rays. **2.** A low-growing European plant, *Bellis perennis*, having pink or white rayed flowers. **3.** The flower of a daisy. **4.** *Slang.* Something excellent or notable.

▲ word history: The name *daisy*, a compound word meaning "day's eye," was originally applied to the European plant *Bellis perennis*, which is called in the United States the *English daisy.* The term "day's eye" is especially appropriate to this plant because it folds its petals at night and opens them in the morning with the sun, like an eye that sleeps and wakes.

**Da·kin's solution** (dā'kĭnz) *n.* [After Henry D. *Dakin* (1880–1952).] A dilute sodium hypochlorite solution used as a surgical disinfectant.

**Da·ko·ta** (də-kō'tə) *n., pl.* **Dakota** *or* **-tas. 1. a.** A large group of tribes of Plains Indians, commonly called Sioux, now living on reservations in North and South Dakota, Minnesota, and Montana. **b.** A member of any of these tribes. **2.** The Siouan language of the Dakota. **—Da·ko'tan** *adj.* & *n.*

**Da·lai La·ma** (dä'lī lä'mə) *n.* [Tibetan : Mongolian *dalai*, ocean + Tibetan *bla-ma*, monk.] The traditional governmental ruler and highest priest of the Lamaist religion in Tibet and Mongolia.

**dal·a·pon** (dăl'ə-pŏn') *n.* [Blend of DI-, ALPHA, and PROPIONIC ACID.] An organic acid used as a herbicide.

**da·la·si** (dä-lä'sē) *n., pl.* **dalasi.** [Native word in Gambia.] —See table at CURRENCY.

**dale** (dāl) *n.* [ME < OE *dæl.*] A valley.

**da·leth** (dä'lĕth', -lĕt') *n.* [Heb. *dāleth* < *dālt*, door.] The fourth letter of the Hebrew alphabet. —See table at ALPHABET.

**dalles** (dălz) *pl.n.* [Fr., pl. of *dalle*, gutter < OFr. < ON *dæla.*] The steep precipices forming the sides of a gorge or narrow valley, usu. having rapids at the bottom.

**dal·li·ance** (dăl'ē-əns) *n.* **1.** Frivolous action : dawdling. **2.** Playful flirtation.

**Dal·lis grass** (dăl'ĭs) *n.* [Prob. alteration of *Dallas*, Texas.] A South American grass, *Paspalum dilatatum*, grown in the southern United States for pasturage.

**dal·ly** (dăl'ē) *v.* **-lied, -ly·ing, -lies.** [ME *dalien* < OFr. *dalier.*] —*vi.* **1.** To play amorously : FLIRT. **2.** To trifle. **3.** To waste time : DAWDLE. —*vt.* To waste (time). **—dal'li·er** *n.* **—dal'ly·ing·ly** *adv.*

**Dal·ma·tian** (dăl-mā'shən) *n.* A dog believed to have been bred orig. in Dalmatia, having a short, smooth white coat covered with black or dark-brown spots.

**dal·mat·ic** (dăl-măt'ĭk) *n.* [ME *dalmatik* < Med. Lat. *dalmatica* < Lat. *dalmaticus*, Dalmatian.] **1.** A wide-sleeved garment worn over the alb by a deacon, cardinal, bishop, or abbot at Mass. **2.** A wide-sleeved coronation garment worn by an English monarch.

**dal se·gno** (däl sān'yō) *adv.* [Ital., from the sign.] *Mus.* From a place marked by the sign § to a designated point. —Used as a direction to repeat a passage.

**dal·ton** (dôl'tən) *n.* [After John *Dalton* (1766–1844).] A standard unit of mass equal to one-half the atomic mass of $^{12}_{6}C$ and used to express the masses of atoms, molecules, and nuclear particles.

**dal·ton·ism** *also* **Dal·ton·ism** (dôl'tə-nĭz'əm) *n.* [After John *Dalton* (1766–1844).] Red-green colorblindness. **—dal·to'ni·an** (dôl-tō'nē-ən), **dal·ton'ic** (-tŏn'ĭk) *adj.*

**dam**[1] (dăm) *n.* [ME.] **1.** A barrier built across a waterway to control the flow or raise the level of water. **2.** A body of water controlled by a dam. **3.** An obstruction : hindrance. —*vt.* **dammed, dam·ming, dams. 1.** To build a dam across or hold back by a dam. **2.** To obstruct or restrain : CONFINE.

**dam**[2] (dăm) *n.* [ME *dam, dame.* —see DAME.] **1.** A female parent. —Used of a quadruped. **2.** *Archaic.* A mother.

**dam·age** (dăm'ĭj) *n.* [ME < OFr. < *dam*, loss < Lat. *damnum.*] **1.** Impairment of the usefulness or value of person or property : HARM. **2. damages.** *Law.* Money to be paid as compensation for injury or loss. **3.** *Informal.* Cost : price. —*v.* **-aged, -ag·ing, -ag·es.** —*vt.* To cause injury to : HARM. —*vi.* To suffer or be susceptible to damage. **—dam'age·a·ble** *adj.* **—dam'ag·ing·ly** *adv.*

**dam·ar** (dăm'ər) *n. var. of* DAMMAR.

**dam·as·cene** (dăm'ə-sēn', dăm'ə-sēn') *vt.* **-cened, -cen·ing, -cenes.** [OFr. *damasquiner* < *damasquin*, of Damascus.] To decorate (metal) with wavy inlaid or etched patterns. **—dam'as·cene'** *n.* & *adj.* **—dam'a·scen'er** *n.*

**Da·mas·cus steel** (də-măs'kəs) *n.* An early form of steel with wavy markings, developed in Near Eastern countries, esp. Persia, and used primarily in sword blades.

**dam·ask** (dăm'əsk) *n.* [ME < Med. Lat. *(pannus de) damasco*, (cloth of) Damascus.] **1.** A rich patterned fabric of cotton, linen, silk, or wool. **2.** A fine, twilled table linen. **3.** Damascus steel. **4.** The wavy pattern on Damascus steel. —*vt.* **-asked, -ask·ing, -asks. 1.** To damascene. **2.** To decorate or weave with rich patterns.

**damask rose** *n.* [< obs. *Damask*, Damascan, Damascus.] A rose indigenous to Asia, *Rosa damascena*, with sweet-smelling red or pink flowers used as a source of attar.

**damask steel** *n.* Damascus steel.

**dame** (dām) *n.* [ME < OFr. < Lat. *domina,* fem. of *dominus,* lord, master.] **1.** A title once given to a woman in authority or to the mistress of a household. **2.** A married woman : MATRON. **3.** *Slang.* A woman. **4.** *Chiefly Brit.* **a.** *Archaic.* The legal title of the wife or widow of a knight or baronet. **b.** A title of a woman equivalent to that of a knight.

**dame's rocket** *n.* A plant indigenous to Europe, *Hesperis matronalis,* with fragrant purple or white flower clusters.

**dame's violet** *n.* Dame's rocket.

**dam·mar** *or* **dam·ar** *also* **dam·mer** (dăm′ər) *n.* [Malay *damar,* resin.] Any of various hard resins obtained from Indo-Malayan trees of the genera *Shorea, Balanocarpus,* and *Hopea* and used in varnishes and lacquers.

**damn** (dăm) *v.* **damned, damn·ing, damns.** [ME *dampnen* < OFr. *dampner* < Lat. *damnare,* to condemn, inflict loss upon < *damnum,* loss.] —*vt.* **1.** To pronounce an adverse judgment on. **2.** To bring about the failure of : RUIN. **3.** To condemn as injurious, illegal, or immoral <*damn* drugs and alcohol> **4.** To condemn to eternal punishment : DOOM. **5.** To swear at by using the word "damn." —*vi.* To swear : curse. —*interj.* —Used to express anger, irritation, contempt, or disappointment. —*n.* **1.** The saying of "damn" as a curse. **2.** *Informal.* A bit : jot <a product that isn't worth a *damn*> —*adj.* & *adv.* Damned. **—damn′ing·ly** *adv.*

**dam·na·ble** (dăm′nə-bəl) *adj.* Deserving condemnation : ODIOUS. **—dam′na·ble·ness** *n.* **—dam′na·bly** *adv.*

**dam·na·tion** (dăm-nā′shən) *n.* **1.** The act of damning or state of being damned. **2. a.** Condemnation to everlasting punishment : DOOM. **b.** Everlasting punishment. **3.** Failure or ruination incurred by adverse criticism. —*interj.* —Used to express anger or annoyance.

**dam·na·to·ry** (dăm′nə-tôr′ē, -tōr′ē) *adj.* Threatening with or causing damnation.

**damned** (dămd) *adj.* **-er, -est. 1.** Condemned, esp. to eternal punishment. **2.** *Informal.* Deserving condemnation : DETESTABLE <this *damned* mud> **3.** —Used as an intensive <a *damned* idiot> —*adv. Informal.* Very <a *damned* poor manager> —*n.* Souls doomed to eternal punishment.

**dam·ni·fy** (dăm′nə-fī′) *vt.* **-fied, -fy·ing, -fies.** [OFr. *damnifier* < LLat. *damnificare* < Lat. *damnificus,* harmful : *damnum,* loss, harm + *facere,* to make.] *Law.* To cause loss or damage to : WRONG. **—dam′ni·fi·ca′tion** *n.*

**dam·oi·selle** (dăm′ə-zĕl′) *n. var. of* DAMOSEL.

**Da·mon** (dā′mən) *n.* [Lat. < Gk. *Damōn.*] *Rom. Myth.* A legendary figure who, out of devotion, pledged his life as a hostage for his condemned friend Pythias.

**dam·o·sel** *also* **dam·oi·selle** *or* **dam·o·zel** (dăm′ə-zĕl′) *n.* [ME *damoysele* < OFr. *damoiselle,* damsel.] *Archaic.* A damsel.

**damp** (dămp) *adj.* **-er, -est.** [ME, poison gas, perh. < MLG, vapor.] **1.** Slightly wet : MOIST. **2.** *Archaic.* Dejected. —*n.* **1.** Moisture in the air : HUMIDITY. **2.** Foul or poisonous gas that pollutes the air in mines. **3.** Low spirits : DEPRESSION. **4.** A restraint or check. —*vt.* **damped, damp·ing, damps. 1.** To make damp : MOISTEN. **2.** To extinguish (e.g., a fire) by cutting off air. **3.** To restrain or check : DISCOURAGE. **4.** To provide (the strings of a keyboard instrument) with dampers as a way of reducing the dynamic level. **5.** *Physics.* To decrease the amplitude of (a wave). **—damp off.** *Bot.* To be affected by damping off. **—damp′ish** *adj.* **—damp′ly** *adv.* **—damp′ness** *n.*

**damp·en** (dăm′pən) *v.* **-ened, -en·ing, -ens.** —*vt.* **1.** To make slightly wet : MOISTEN. **2.** To deaden : depress. —*vi.* To become damp. **—damp′en·er** *n.*

**damp·er** (dăm′pər) *n.* **1.** One that damps, restrains, or depresses. **2.** An adjustable plate, as in the flue of a furnace or stove, for controlling draft. **3.** *Mus.* **a.** A device in various keyboard instruments for deadening the vibrations of the strings. **b.** A mute for various brass instruments. **4.** A device that eliminates or progressively diminishes oscillations.

**damp·ing** (dăm′pĭng) *n.* The capacity built into a mechanical or electrical device to prevent excessive correction and the resulting instability or oscillatory conditions.

**damping off** *n.* A disease of planted seeds or very young seedlings caused by fungi and resulting in death of the newly sprouted plants.

**dam·sel** (dăm′zəl) *n.* [ME *damisele* < OFr. *dameisele, damoiselle* < VLat. **dominicella,* dim. of Lat. *domina,* lady. —see DAME.] A young woman or girl : MAIDEN.

**dam·sel·fly** (dăm′zəl-flī′) *n.* Any of various slender-bodied, often brightly colored insects of the order Odonata, related to the dragonflies but differing in having wings that are folded together over the back when at rest.

**dam·son** (dăm′zən) *n.* [ME < Lat. *(prunum) Damascenum,* (plum) of Damascus.] **1.** A Eurasian plum tree, *Prunus institia,* cultivated for its edible fruit. **2.** The oval, bluish-black plum of the damson.

**Dan**[1] (dăn) *n.* [Heb. *Dān.*] One of the 12 tribes of Israel, descended from Dan, the fifth son of Jacob. **—Dan′ite′** (-īt′) *adj.* & *n.*

**Dan**[2] (dăn) *n.* [ME < OFr. < Med. Lat. *Domnus* < Lat. *dominus,* master, lord.] *Obs.* A title of honor equivalent to *master* or *sir.*

**Dan·a·e** *also* **Dan·a·ë** (dăn′ə-ē′) *n.* [Lat. < Gk. *Danaē.*] *Gk. Myth.* The mother of Perseus by Zeus, who visited her in the form of a shower of gold during her imprisonment.

**Dan·a·id** *also* **Dan·a·ïd** (dăn′ə-ĭd) *n.* One of the Danaides.

**Da·na·i·des** *also* **Da·na·ï·des** (də-nā′ĭ-dēz′) *pl.n.* [Gk. < *Danaos,* Danaus.] *Gk. Myth.* The daughters of Danaus, who at their father's command murdered their bridegrooms on their wedding night and were condemned in Hades to pour water eternally into a bottomless vessel.

**Dan·a·us** *also* **Dan·a·üs** (dăn′ē-əs) *n.* [Lat. < Gk. *Danaos.*] *Gk. Myth.* A king of Argos, father of the Danaides.

**dance** (dăns) *v.* **danced, danc·ing, danc·es.** [ME *dauncen* < OFr. *danser.*] —*vi.* **1.** To move rhythmically to music, using improvised or prescribed gestures and steps. **2.** To leap or skip about excitedly : CAPER. **3.** To bob up and down. —*vt.* **1.** To perform (a dance). **2.** To cause to dance. **3.** To bring to a given state by dancing <*danced* me to exhaustion> —*n.* **1.** A series of rhythmical motions and steps, usu. to music. **2.** The art of dancing. **3.** A party or gathering of people for dancing. **4.** A single round of dancing. **5.** A musical or rhythmical accompaniment composed or played for dancing. **6.** An act or instance of dancing. **—danc′er** *n.* **—danc′ing·ly** *adv.*

**dance·a·ble** (dăn′sə-bəl) *adj.* Suitable for dancing <*danceable* tunes>

**dan·de·li·on** (dăn′dl-ī′ən) *n.* [ME *dent-de-lion* < OFr., transl. of Med. Lat. *dens leonis,* lion's tooth (from its sharply indented leaves).] **1.** A Eurasian plant, *Taraxacum officinale,* widely naturalized as a weed in North America and having many-rayed yellow flowers and deeply notched basal leaves occas. used in salads. **2.** A plant related to the dandelion. **3.** A brilliant to vivid yellow.

**dan·der**[1] (dăn′dər) *n.* [Orig. unknown.] *Informal.* Temper : anger.

**dan·der**[2] (dăn′dər) *n.* [Short for DANDRUFF.] Scurf from the coat of various animals, as dogs, cats, or horses, often causing allergies.

**Dan·die Din·mont** (dăn′dē dĭn′mŏnt′) *n.* [After *Dandie Dinmont,* owner of two such dogs in *Guy Mannering,* a novel by Sir Walter Scott (1771–1832).] A small terrier orig. bred in England, having a rough brownish or grayish coat and short legs.

**dan·di·fy** (dăn′də-fī′) *vt.* **-fied, -fy·ing, -fies.** To dress as or cause to look like a dandy. **—dan′di·fi·ca′tion** *n.*

**dan·dle** (dăn′dl) *vt.* **-dled, -dling, -dles.** [Orig. unknown.] **1.** To move (a small child) up and down on the knees or in the arms. **2.** To pamper or pet. **—dan′dler** *n.*

**dan·druff** (dăn′drəf) *n.* [Orig. unknown.] Small white flakes of dead skin shed from the scalp. **—dan′druff·y** *adj.*

**dan·dy** (dăn′dē) *n., pl.* **-dies.** [Perh. short for *jack-a-dandy,* fop.] **1.** A man who affects extreme elegance in manners and clothes : FOP. **2.** *Informal.* Something very good or agreeable. **3.** YAWL 1. —*adj.* **-di·er, -di·est. 1.** Like a dandy : FOPPISH. **2.** *Informal.* Fine : good <had a *dandy* time> **—dan′dy·ish** *adj.*

**dandy roll** *also* **dandy roller** *n.* A cylinder of wire gauze pressed on moist pulp before it starts through the rollers and produces the watermarks in paper.

**Dane** (dān) *n.* [ME *Dan* < ON *Danr.*] **1.** A native or resident of Denmark. **2.** One of Danish ancestry.

**Dane·geld** (dān′gĕld′) *also* **Dane·gelt** (-gĕlt′) *n.* [ME : *Dane,* Danes' + *geld,* tribute < OE *gield.*] A tax levied in England from the 10th to the 12th cent. to finance protection against Danish invasion.

**Dane·law** *also* **Dane·lagh** (dān′lô′) *n.* [ME *Denelage* < OE *Dena lagu,* Danes' law.] **1.** The body of law established by the Danish invaders and settlers in northeastern England in the 9th and 10th cent. **2.** The parts of England under jurisdiction of the Danelaw.

**dan·ger** (dān′jər) *n.* [ME *daunger,* power, dominion, peril < OFr. *dangier* < Lat. *dominium,* sovereignty < *dominus,* lord, master.] **1.** Exposure to possible evil, injury, or harm. **2.** A source or instance of peril or risk. **3.** *Obs.* Power, esp. power to harm.

☆ ***syns:*** DANGER, ENDANGERMENT, HAZARD, IMPERILMENT, JEOPARDY, PERIL, RISK *n. core meaning* : exposure to possible evil, injury, or harm <high flood waters that put the town in *danger*> ***ant:*** safety

**dan·ger·ous** (dān′jər-əs) *adj.* **1.** Perilous. **2.** Able or apt to harm. **—dan′ger·ous·ly** *adv.* **—dan′ger·ous·ness** *n.*

**dan·gle** (dăng′gəl) *v.* **-gled, -gling, -gles.** [Perh. of Scand. orig.] —*vi.* **1.** To hang loosely and swing to and fro. **2.** To be a hanger-on. —*vt.* To cause to dangle. —*n.* **1.** An act of dangling. **2.** Something dangled. **—dan′gler** *n.*

**dangling participle** *n.* A participle usu. in a subordinate clause that lacks a clear grammatical relation with the subject of its sentence; e.g., in the sentence *Approaching Atlanta, the skyline came into view, approaching* is a dangling participle.

**Dan·iel** (dăn′yəl) *n.* [Heb. *Dānī'ēl,* God is my judge.] —See table at BIBLE.

**da·ni·o** (dā′nē-ō′) *n., pl.* **-os.** [NLat. *Danio,* genus name.] A small, often brightly colored freshwater fish of the genera *Danio* or *Brachydanio,* native to Asia and popular as aquarium fish.

**Dan·ish** (dā′nĭsh) *adj.* [ME < OE *Denisc* < *Dene,* the Danes.] Of or relating to Denmark, the Danes, their language, or their culture. —*n.* **1.** The North Germanic language of the Danes. **2.** *Informal.* Danish pastry.

**Dan·ish pastry** *n.* A sweet, buttery pastry made with raised dough.

**dank** (dăngk) *adj.* **-er, -est.** [ME.] Uncomfortably damp and chilly. **—dank'ly** *adv.* **—dank'ness** *n.*

**dan·seur** (däN-sœr') *n., pl.* **-seurs** (-sœr') [Fr. < OFr. < *danser,* to dance.] A man who is a ballet dancer.

**dan·seuse** (däN-sœz') *n., pl.* **-seuses** (-sœz') [Fr., fem. of *danseur,* danseur.] A woman who is a ballet dancer.

**Dan·tesque** (dăn-tĕsk') *adj.* Marked by or having the exalted, visionary style of the Italian poet Dante.

**Da·nu** (*thä'no͞o*) *n.* [Ir.] *Ir. Myth.* The goddess of death and mother of the gods.

**daph·ne** (dăf'nē) *n.* [NLat. *Daphne,* genus name < Lat., laurel < Gk. *daphnē.*] A European shrub of the genus *Daphne,* widely cultivated for its glossy foliage and small, bell-shaped flower clusters.

**Daph·ne** (dăf'nē) *n.* [Lat. < Gk. *Daphnē,* Daphne, laurel.] *Gk. Myth.* A nymph who changed into a laurel tree as a way of escaping Apollo.

**daph·ni·a** (dăf'nē-ə) *n., pl.* **daphnia.** [NLat. *Daphnia,* genus name.] A small freshwater crustacean of the genus *Daphnia,* often used as food for aquarium fish.

**Daph·nis** (daf'nĭs) *n.* [Lat. < Gk.] *Gk. Myth.* A Sicilian shepherd and son of Hermes, famed as a musician and reputed to be the inventor of pastoral poetry.

**dap·per** (dăp'ər) *adj.* [ME *dapyr,* elegant, prob. < MDu. *dapper,* quick, strong.] **1. a.** Neatly dressed : TRIM. **b.** Very stylish. **2.** Small and active. **—dap'per·ly** *adv.* **—dap'per·ness** *n.*

**dap·ple** (dăp'əl) *n.* [Back-formation < DAPPLE-GRAY.] **1. a.** Mottled or spotted marking, as on a horse's skin. **b.** A single spot. **2.** An animal with a mottled or spotted coat. *—vt.* **-pled, -pling, -ples.** To mark or mottle with spots. *—adj. also* **dap·pled** (-əld). Spotted or mottled.

**dap·ple-gray** (dăp'əl-grā') *adj.* [ME *dappel-grai.*] Gray with a mottled pattern of darker gray markings. *—n.* A dapple-gray horse.

▲ word history: *Dapple-gray* is probably an alteration of *apple-gray,* which refers not so much to a color as to the blotchy dark gray markings found on gray horses. The origin of the term is not known with certainty. There are similar words in other Germanic languages, such as Old Norse *apalgrār* and German *apfelgrā,* literally "apple-gray." These words had the same meaning and application as *dapple-gray,* but it is not known if either the Norse or the German term is the direct source of the English word. *Dapple-gray* first appeared in the 14th century in Chaucer's *Canterbury Tales.* In the same work Chaucer also used the synonymous term *pomely grey,* which is apparently a partial translation of French *gris-pommelé,* "apple-gray," from *pomme,* "apple." It is possible that Chaucer coined the term *dapple-gray,* but the origin of the *d* is still unexplained.

**dap·sone** (dăp'sōn', -zōn') *n.* [Contraction of E. *diaminodiphenyl sulfone,* a drug used to treat leprosy.] An antimicrobial agent, $C_{12}H_{12}N_2OS$, used against leprosy.

**Dar·by and Joan** (där'bē) *n.* [Prob. after *Darby and Joan,* a couple in an 18th-cent. English ballad.] An elderly married couple who live a placid, harmonious life together and are seldom seen apart.

**Dard** (därd) *also* **Dar·dic** (där'dĭk) *n.* A group of Indic languages spoken in the upper Indus River valley.

**Dar·dan** (där'dn) *also* **Dar·da·ni·an** (där-dā'nē-ən) *n.* [After *Dardanus,* the legendary founder of Troy.] A Trojan. **—Dar'dan** *adj.*

**Dar·da·nus** (där'dn-əs) *n.* [Lat. < Gk. *Dardanos.*] *Gk. Myth.* The founder of Troy.

**Dar·dic** (där'dĭk) *n. var. of* DARD.

**dare** (dâr) *v.* **dared, dar·ing, dare** *or* **dares.** [ME *daren* < OE *dear,* first and third person pr. indicative of *durran,* to venture, dare.] *—vt.* **1.** To have the courage needed for. **2.** To challenge (another) to do something needing courage. **3.** To confront courageously. *—vi.* To be courageous enough to do or try something. *—n.* An act of daring : CHALLENGE. **—dare say** *also* **dare·say** (dâr-sā'). To consider very likely or almost certain. ***usage:*** The idomatic expression *dare say* (or *daresay*), more common in British than in American English, occurs only in the present tense with *I* as the subject and is never followed by *that,* as in *I daresay they'll regret it.* **—dar'er** *n.*

☆ ***syns:*** DARE, BRAVE, CHALLENGE, DEFY, FACE *v. core meaning* : to confront boldly and courageously <*dared* the steep alpine cliffs in winter>

**dare·dev·il** (dâr'dĕv'əl) *n.* One who is recklessly bold. **—dare'-dev'il** *adj.* **—dare'dev'il·ry, dare'dev'il·try** *n.*

**dar·ing** (dâr'ĭng) *adj.* Recklessly bold. **—dar'ing** *n.* **—dar'ing·ly** *adv.* **—dar'ing·ness** *n.*

**Dar·jee·ling** (där-jē'lĭng) *n.* A fine black tea from Darjeeling, India.

**dark** (därk) *adj.* **-er, -est.** [ME *derk* < OE *deorc.*] **1.** Having little or no light. **2.** Reflecting only a small fraction of the incident light. **3.** Lacking brightness. **4.** Of a shade tending toward black or brown <*dark* meat> **5.** Not fair in complexion : SWARTHY. **6.** Marked by or causing gloom : DISMAL <took a *dark* view of the economic indicators> **7.** Sullen : threatening <a *dark* glare> **8.** Difficult to understand : OBSCURE. **9.** Concealed or secret <a *dark* scheme> **10.** Being without knowledge or enlightenment <a *dark* era in history> **11.** Evil or wicked : SINISTER <a *dark* purpose> **12.** Having richness or depth <a *dark,* melancholy vocal tone> **13.** Not giving performances <a *dark* theater> *—n.* **1.** Absence of light. **2.** A place having little light. **3.** Night : nightfall. **4.** A dark color. **—in the dark. 1.** In secret. **2.** In a state of ignorance. **—dark'ish** *adj.* **—dark'ly** *adv.* **—dark'ness** *n.*

☆ ***syns:*** DARK, DIM, DUSKY, MURKY, OBSCURE, TENEBROUS *adj. core meaning* : lacking brightness and light <a *dark* tunnel> ***ant:*** light

**dark adaptation** *n.* The physical and chemical adjustments of the eye, including dilation of the pupil, that make vision possible in relative darkness. **—dark'-a·dapt'ed** (därk'ə-dăp'tĭd) *adj.*

**Dark Ages** *pl.n.* **1.** The early part of the Middle Ages. **2.** The whole period from the end of classical civilization to the revival of learning in the West.

**dark·en** (där'kən) *v.* **-ened, -en·ing, -ens.** *—vt.* **1.** To make dark or darker. **2.** To give a darker hue to. **3.** To fill with sadness or gloom. **4.** To make vague or uncertain : OBSCURE. **5.** To tarnish : stain <*darken* one's reputation with scandal> *—vi.* To become dark or darker. **—dark'en·er** *n.*

**dark-field microscope** (därk'fēld') *n.* An ultramicroscope.

**dark horse** *n.* **1.** A little-known entrant in a horse race or other contest. **2.** One who receives unexpected support as a candidate for the nomination in a political convention.

**dark lantern** *n.* A lantern whose light can be blocked by a panel.

**dar·kle** (där'kəl) *vi.* **-kled, -kling, -kles.** [Back-formation < DARKLING.] **1.** To appear darkly or indistinctly. **2. a.** To grow dark. **b.** To become gloomy.

**dark·ling** (därk'lĭng) *adv.* [ME *derkeling* < *derk,* dark.] In the dark. *—adj.* **1.** Being or taking place in the dark or the night. **2.** Dim : obscure.

**darkling beetle** *n.* A dark-colored, sluggish, nocturnal plant-eating beetle of the family Tenebrionidae.

**dark·room** (därk'ro͞om', -ro͝om') *n.* A room in which photographic materials are processed, either in complete darkness or with a safelight.

**dark·some** (därk'səm) *adj.* Gloomily dark : SOMBER.

**dark star** *n.* A star normally obscured or too faint for direct visual observation, esp. the component of an eclipsing binary detectable by spectral analysis or in the eclipse of the bright component.

**dar·ling** (där'lĭng) *n.* [ME *dereling* < OE *dēorling* < *dēore,* dear.] **1.** A much-loved person. **2.** One greatly liked or preferred : FAVORITE. *—adj.* **1.** Very dear : BELOVED. **2.** Favorite. **3.** *Informal.* Charming : amusing <a *darling* little jacket>

**darn¹** (därn) *v.* **darned, darn·ing, darns.** [Dial. Fr. *darner.*] *—vt.* To mend by weaving thread or yarn across a gap or hole. *—vi.* To mend a hole or garment by darning. **—darn** *n.* **—darn'er** *n.*

**darn²** (därn) *v. & interj. & adj. & adv.* [Alteration of DAMN.] Damn. **—darned** *adj & adv.*

**dar·nel** (där'nəl) *n.* [ME.] An Old World grass of the genus *Lolium,* esp. *L. tementulum* or *L. perenne.*

**darning egg** *n.* An egg-shaped object for holding the shape of material being darned.

**darning needle** *n.* **1.** A long, large-eyed needle used in darning. **2.** *Informal.* A dragonfly.

**dart** (därt) *n.* [ME < OFr., of Germanic orig.] **1.** A slender, pointed missile, often with tail fins, either thrown by hand or shot from a blowgun. **2.** Something shaped like a dart or having the use or effect of a dart. **3.** An insect's stinger. **4. darts** *(sing. in number).* A game in which darts are thrown at a target. **5.** A rapid, sudden movement. **6.** A tapered tuck adjusting the fit of a garment. *—v.* **dart·ed, dart·ing, darts.** *—vi.* To move suddenly and swiftly. *—vt.* To throw or thrust suddenly or swiftly : SHOOT.

**dart·er** (där'tər) *n.* **1.** One that moves rapidly and suddenly. **2.** A long-necked, long-billed bird of the genus *Anhinga,* as the water turkey. **3.** Any of various small, often brightly colored freshwater fishes of the family Percidae, indigenous to eastern North America.

**Dar·win·ism** (där'wĭ-nĭz'əm) *n.* A theory of biological evolution developed by Charles Darwin and others, stating that species of plants and animals develop through natural selection of variations that increase the organism's ability to survive and reproduce. **—Dar·win'i·an** (-wĭn'ē-ən) *adj.* **—Dar'win·ist** *n.* **—Dar'win·is'tic** *adj.*

**dash¹** (dăsh) *v.* **dashed, dash·ing, dash·es.** [ME *dashen,* prob. of Scand. orig.] *—vt.* **1.** To break by striking violently. **2.** To hurl, knock, or thrust suddenly and violently. **3.** To splash : bespatter. **4.** To perform or finish hastily <*dash* off a paragraph> **5.** To add an altering element to : MIX <delight *dashed* with uneasiness> **6.** To destroy <My dreams were *dashed.*> **7.** To confound : abash. *—vi.* **1.** To strike violently : SMASH. **2.** To move quickly : RUSH <*dashed* down the street> *—n.* **1.** A swift, violent blow or stroke. **2.** A splash. **3.** A small amount of an added ingredient <a *dash* of Tabasco> **4.** A quick stroke, as with a pencil or brush. **5.** A sudden movement : RUSH. **6.** A foot race, usu. less than a quarter-mile long, run at top speed. **7.** Spirited action or style : VERVE. **8.** A punctuation mark (—) used in writing and printing. **9.** The long sound or signal used in combination with the dot and silent intervals to represent letters or numbers in Morse and similar codes. **10.** A dashboard.

**dash²** (dăsh) *v.* & *interj.* & *adj.* & *adv.* Damn. **—dashed** *adj.* & *adv.*

**dash·board** (dăsh′bôrd′, -bōrd′) *n.* A panel under the windshield of a motor vehicle, having indicator dials, storage compartments, and control instruments.

**da·sheen** (dă-shēn′) *n.* [Orig. unknown.] The taro.

**dash·er** (dăsh′ər) *n.* **1.** One that dashes. **2.** The plunger of a churn or ice-cream freezer. **3.** *Informal.* A spirited person.

**da·shi·ki** (də-shē′kē) *n., pl.* **-kis.** [Yoruba *danshiki.*] A loose, brightly colored African tunic, usu. worn by men.

**dash·ing** (dăsh′ĭng) *adj.* **1.** Audacious and gallant. **2.** Marked by showy elegance : STYLISH. **—dash′ing·ly** *adv.*

**dash·pot** (dăsh′pŏt′) *n.* A piston-and-cylinder device used to damp motion.

**das·sie** (dăs′ē) *n.* [Afr., dim. of *das,* badger < MDu.] The hyrax.

**das·tard** (dăs′tərd) *n.* [ME, prob. < ON *dæstr,* p.part. of *dæsa,* to languish, decay.] A base coward.

**das·tard·ly** (dăs′tərd-lē) *adj.* Cowardly and ignoble : BASE. **—das′tard·li·ness** *n.*

**das·y·ure** (dăs′ē-yŏŏr′) *n.* [NLat. *Dasyurus,* genus name : Gk. *dasus,* hairy + *oura,* tail.] Any of various marsupial mammals of the family Dasyuridae of Australia and adjacent regions, ranging widely in size and appearance.

**da·ta** (dā′tə, dăt′ə, dä′tə) *pl.n.* [Lat., pl. of *datum.* —see DATUM.] *(sing. or pl. in number).* **1.** Information, esp. information organized for analysis or used as the basis for decision-making. **2.** Numerical information suitable for computer processing. **3.** *pl. of* DATUM 1. **usage:** *Data,* as the Latin plural of *datum,* in traditional use requires a plural verb, as in *These data are inconclusive.* However, the widespread occurrence of such sentences as *This data is inconclusive* indicates that *data* can now function as a singular form in English.

**data bank** *n.* **1.** A data base. **2.** An organization concerned with building, maintaining, and utilizing a data bank.

**data base** *also* **da·ta·base** (dā′tə-bās′, dăt′ə-) *n.* A collection of data arranged for ease and speed of retrieval, as by a computer.

**data carrier** *n.* The medium, as magnetic tape, selected to transport or communicate data.

**data processing** *n.* **1.** Preparation of information for computer processing. **2.** Storage or processing of raw data by a computer. **—data processor** *n.*

**da·ta·ry** (dā′tə-rē) *n., pl.* **-ries.** [Med. Lat. *dataria* < *data,* date.] *Rom. Cath. Ch.* **1.** The duty, once an official office of the curia, of investigating the fitness of candidates for papal benefices. **2.** A cardinal carrying out the duty of datary.

**data set** *n. Computer Sci.* **1.** An electronic device providing an interface in the transmission of data to a remote station. **2.** A collection of related computer records. **3.** A modem.

**date¹** (dāt) *n.* [ME < OFr. < Med. Lat. *data* < the phrase *data Romae,* issued at Rome (on a certain day) < Lat. *datus,* p.part. of *dare,* to give.] **1. a.** Time stated in terms of the day, month, and year. **b.** A statement of calendar time <the *date* on a letter> **2.** The day of the month. **3.** A particular point or period of time at which something occurred or existed or is to occur. **4.** Duration of something. **5.** The time or historical period to which something belongs <artifacts of a later *date*> **6. a.** An appointment, esp. an engagement to go out socially with a member of the opposite sex. **b.** A person's companion on a date. **7.** An engagement for a performance <have four singing *dates* this month> *—v.* **dat·ed, dat·ing, dates.** *—vt.* **1.** To mark or provide with a date <*date* a letter> **2.** To ascertain the date of <*date* a fossil> **3.** To betray the age of. **4.** To go on a date with. *—vi.* **1.** To have origin in a particular time in the past <This vase *dates* from 400 B.C.> **2.** To become old-fashioned. **3.** To go on dates. **—to date.** Up to the present time. **—dat′a·ble, date′a·ble** *adj.* **—dat′er** *n.*

**date²** (dāt) *n.* [ME < OFr. < OProv. *datil* < Lat. *dactylus* < Gk. *daktulos,* finger (from its shape).] **1.** The sweet, oblong, edible fruit of the date palm, containing a narrow, hard seed. **2.** The date palm.

**dat·ed** (dā′tĭd) *adj.* **1.** Marked with or showing a date. **2.** Old-fashioned. **—dat′ed·ly** *adv.* **—dat′ed·ness** *n.*

**date·less** (dāt′lĭs) *adj.* **1.** Having no date. **2.** Without limits : ENDLESS. **3.** Too ancient to be dated. **4.** Timeless : eternal.

**date·line** (dāt′līn′) *n.* A phrase at the beginning of a printed article giving the date and place of its origin.

**date line** *n.* An imaginary line through the Pacific Ocean roughly corresponding to 180° longitude, to the east of which, by international agreement, the calendar date is one day earlier than to the west.

**date palm** *n.* A tropical tree, *Phoenix dactylifera,* having feather-like leaves and bearing clusters of dates.

**dating bar** *n.* A singles bar.

**da·tive** (dā′tĭv) *also* **da·ti·val** (dā-tī′vəl) *adj.* [ME *datif* < Lat. *(casus) dativus,* (case) of giving < *dare,* to give.] Designating or belonging to a grammatical case in Latin, Russian, and other inflected Indo-European languages that marks the indirect object of a verb and the object of any of certain verbs and prepositions. *—n.* **1.** The dative case. **2.** A word or form in the dative case. **—da′tive·ly** *adv.*

**da·tum** (dā′təm, dăt′əm, dä′təm) *n.* [Lat., something given < p.part. of *dare,* to give.] **1.** *pl.* **-ta** (-tə). An assumed, given, measured, or otherwise determined fact or proposition used to draw a conclusion or make a decision. **2.** *pl.* **-tums.** A point, line, or surface used as a reference, as in surveying, mapping, or geology.

**da·tu·ra** (də-tŏŏr′ə, -tyŏŏr′ə) *n.* [NLat. *Datura,* genus name < Hindi *dhatūrā* < Skt. *dhattūraḥ.*] A plant of the genus *Datura,* with large trumpet-shaped flowers.

**daub** (dôb) *v.* **daubed, daub·ing, daubs.** [ME *dauben* < OFr. *dauber* < Lat. *dealbare,* to whitewash : *de-,* completely + *albus,* white.] *—vt.* **1.** To cover, coat, or smear with an adhesive substance, as plaster. **2.** To cover or smear with a dirty substance. **3.** To apply paint to with hasty or crude strokes. *—vi.* To apply paint or coloring with crude, unskillful strokes. *—n.* **1.** The act or a stroke of daubing. **2.** A soft adhesive coating material, as plaster or mud. **3.** Something daubed on : SMEAR. **4.** A crude painting. **—daub′er** *n.* **—daub′er·y** (dô′bə-rē) *n.*

**daugh·ter** (dô′tər) *n.* [ME *doughter* < OE *dohtor.*] **1.** One's female child. **2.** A female descendant. **3.** A woman considered as if in a relationship of child to parent <a *daughter* of our nation> **4.** Something personified as a female descendant <"Culturally Japan is a *daughter* of Chinese civilization" —Edwin Reischauer> **5.** The immediate product of the radioactive decay of an element. **—daugh′ter·ly** *adj.*

**daughter cell** *n. Biol.* Either of two cells that form from the division of a cell.

**daugh·ter-in-law** (dô′tər-ĭn-lô′) *n., pl.* **daugh·ters-in-law.** The wife of one's son.

**daunt** (dônt, dŏnt) *vt.* **daunt·ed, daunt·ing, daunts.** [ME *daunten* < OFr. *danter* < Lat. *domitare,* freq. of *domare,* to tame.] **1.** To drain the courage of and thus subdue : INTIMIDATE. **2.** To dishearten. **—daunt′er** *n.* **—daunt′ing·ly** *adv.*

**daunt·less** (dônt′lĭs, dŏnt′-) *adj.* Incapable of being intimidated. **—daunt′less·ly** *adv.* **—daunt′less·ness** *n.*

**dau·phin** (dô′fĭn) *n.* [Fr. < OFr. *dalphin,* title of the lords of Dauphiné.] The eldest son of a king of France from 1349 to 1830.

**dau·phin·ess** (dô′fĭ-nĭs) *also* **dau·phine** (dô-fēn′) *n.* A dauphin's wife.

**dav·en·port** (dăv′ən-pôrt′, -pōrt′) *n.* [Orig. unknown.] **1.** A large sofa, often convertible into a bed. **2.** *Chiefly Brit.* A small desk.

**dav·it** (dăv′ĭt, dā′vĭt) *n.* [ME *daviot* < OFr. *daviot,* dim. of *David,* David.] A small crane that projects over the side of a boat or ship and is used to raise anchors and cargo.

**Da·vy Jones** (dā′vē jōnz′) *n.* [*Davy,* nickname for *David.*] The spirit of the sea.

**Davy Jones's locker** *n.* The bottom of the sea, esp. as the grave of all who perish there.

**Davy lamp** *n.* [After Sir Humphrey *Davy* (1778–1829), its inventor.] An early safety oil lamp used by coal miners.

**daw** (dô) *n.* [ME *dawe.*] The jackdaw.

**daw·dle** (dôd′l) *v.* **-dled, -dling, -dles.** [Perh. alteration of dial. *daddle,* to diddle.] *—vi.* **1.** To take more time than needed. **2.** To move aimlessly. *—vt.* To waste (time) <*dawdling* away the day> **—daw′dler** *n.* **—daw′dling·ly** *adv.*

**dawn** (dôn) *n.* [< ME *daunen,* to dawn, prob. back-formation < *dauning,* daybreak, alteration of *dauing* < OE *dagung* < *dagian,* to dawn.] **1.** The time each morning when daylight first appears. **2.** A beginning <the *dawn* of history> *—vi.* **dawned, dawn·ing, dawns. 1.** To begin to become light in the morning. **2.** To begin to appear or develop : EMERGE. **3.** To begin to be perceived.

☆ ***syns:*** DAWN, AURORA, COCKCROW, DAWNING, DAYBREAK, MORN, MORNING, SUNRISE, SUNUP *n. core meaning* : the time of the first appearance of daylight <got up at *dawn*> ***ant:*** dusk

**day** (dā) *n.* [ME < OE *dæg.*] **1.** The period of light between dawn and nightfall. **2. a.** The 24-hour period during which the earth completes one rotation on its axis. **b.** The period during which a celestial body makes a similar rotation. **3.** One of the numbered 24-hour periods into which a week, month, or year is divided. **4.** The part of a day set aside for work. **5.** A day reserved for an activity <a *day* of rest> **6. a.** The period of activity or prominence in one's lifetime. **b.** A period of opportunity. **7.** A period of time : ERA <in Peter the Great's *day*> **8.** The issue at hand <carry the *day*> **—call it a day.** *Informal.* To stop one's work or activity for the day. **—day after day.** For many days : CONTINUOUSLY. **—day in, day out.** Every day without fail.

**Day·ak** (dī′ăk′) *n. var. of* DYAK.

**day bed** *n.* A sofa convertible into a bed.

**day·book** (dā′bŏŏk′) *n.* **1.** A book in which daily financial transactions are recorded. **2.** A diary.

**day·break** (dā′brāk′) *n.* DAWN 1.

**day care** *n.* Provision of daytime supervision, training, and medical services for preschool children or for the elderly.

**day·dream** (dā′drēm′) *n.* A dreamlike fantasy experienced while awake, esp. of the fulfillment of hopes. *—vi.* **-dreamed** *or* **-dreamt**

(-drĕmt'), **-dream·ing, -dreams.** To have daydreams. **—day'-dream'er** *n.*

**day·flow·er** (dā'flou'ər) *n.* A plant of the genus *Commelina,* with blue or purplish flowers that wilt quickly.

**day·fly** (dā'flī') *n.* The mayfly.

**Day-Glo** (dā'glō'). A trademark for fluorescent materials.

**day labor** *n.* Labor hired and paid by the day. **—day laborer** *n.*

**day letter** *n.* A telegram sent during the day, usu. less expensive but slower than a regular telegram.

**day·light** (dā'līt') *n.* **1.** The light of day. **2. a.** DAWN 1. **b.** Daytime. **3.** Exposure to public notice. **4.** Understanding or insight into what was once obscure. **5. daylights.** *Slang.* Mental stability : WITS <scared the *daylights* out of me>

**day·light-sav·ing time** (dā'līt-sā'vĭng) *n.* Time during which clocks are set one hour or more ahead of standard time to provide more daylight at the end of the working day during late spring, summer, and early fall.

**day lily** *n.* **1.** A Eurasian plant of the genus *Hemerocallis,* widely cultivated for its sword-shaped leaves and variously colored funnel-shaped flowers. **2.** The plantain lily.

**day·long** (dā'lông', -lŏng') *adj.* Lasting all day. **—day'long'** *adv.*

**Day of Atonement** *n.* Yom Kippur.

**days** (dāz) *adv.* Regularly or habitually in the daytime <worked *days*>

**day sailer** *n.* A small sailboat for day trips.

**day school** *n.* **1.** A private school for pupils living at home. **2.** A school holding classes during the day.

**day·side** (dā'sīd') *n.* The side of a planet facing the sun.

**days of grace** *pl.n.* [Transl. of Lat. *dies gratiae.*] Extra days, usu. three, allowed for payment of a note or bill after it has fallen due.

**day·spring** (dā'spring') *n.* DAWN 1.

**day·star** (dā'stär') *n.* **1.** The morning star. **2.** The sun.

**day·time** (dā'tīm') *n.* The time between dawn and dark : DAY.

**day-to-day** (dā'tə-dā') *adj.* **1.** Occurring on a routine or daily basis <*day-to-day* tasks> **2.** Marked by subsistence a day at a time with little thought for the future <lived a *day-to-day* existence>

**day-trip·per** (dā'trĭp'ər) *n.* One who takes a trip during the day without staying overnight.

**daze** (dāz) *vt.* **dazed, daz·ing, daz·es.** [ME *dasen,* of Scand. orig.] **1.** To stun, as with a heavy blow or shock : STUPEFY. **2.** To dazzle, as with strong light. **—daze** *n.* **—daz'ed·ly** (dā'zĭd-lē) *adv.*

☆ ***syns:*** DAZE, BEDAZZLE, BLIND, DAZZLE *v. core meaning* : to confuse with bright light <*dazed* by the spotlight>

**daz·zle** (dăz'əl) *v.* **-zled, -zling, -zles.** [Freq. of DAZE.] *—vt.* **1.** To dim the vision of, esp. to blind with intense light. **2.** To bewilder, amaze, or impress with spectacular display. *—vi.* **1.** To become blinded. **2.** To inspire admiration or wonder. **—daz'zle** *n.* **—daz'-zler** *n.* **—daz'zling·ly** *adv.*

**D-day** (dē'dā') *n.* [D (abbr. of DAY) + DAY.] The unnamed day on which an operation or a military offensive is to be launched, esp. Jun. 6, 1944, the day on which the Allied forces invaded France during World War II.

**DDT** (dē'dē-tē') *n.* [D(ICHLORO)D(IPHENYL)T(RICHLOROETHANE).] A colorless contact insecticide, $(ClC_6H_4)_2CHCCl_3$, toxic to humans and animals when swallowed or absorbed through the skin.

**de-** *pref.* [Lat. < *dē,* from.] **1.** Do or make the opposite of : REVERSE <*de*criminalize> **2.** Remove or remove from <*de*louse><*de*throne> **3.** Reduce : degrade <*de*class> **4.** Derived from <*de*verbative>

**de·ac·ces·sion** (dē'ăk-sĕsh'ən) *v.* **-sioned, -sion·ing, -sions.** *—vt.* To remove and sell (a work of art) from a museum's collection, esp. so as to purchase others. *—vi.* To de-accession a work of art.

**de·a·cid·i·fy** (dē'ə-sĭd'ə-fī') *vt.* **-fied, -fy·ing, -fies.** To remove the acid from or reduce the acid content of. **—de'a·cid'i·fi·ca'tion** *n.*

**dea·con** (dē'kən) *n.* [ME *deken* < OE *dīacon* < LLat. *diaconus* < Gk. *diakonos,* attendant.] **1.** An Anglican, Eastern Orthodox, or Roman Catholic clergyman ranking just below a priest. **2.** A layperson who assists the minister in various functions in certain other Christian denominations.

**dea·con·ess** (dē'kə-nĭs) *n.* A woman appointed or elected to serve as an assistant in a church.

**dea·con·ry** (dē'kən-rē) *n., pl.* **-ries. 1.** The office or position of a deacon. **2.** Deacons as a group.

**de·ac·ti·vate** (dē-ăk'tə-vāt') *vt.* **-vat·ed, -vat·ing, -vates. 1.** To render inactive. **2.** To remove from active military status. **—de·ac'ti·va'tion** *n.*

**dead** (dĕd) *adj.* **-er, -est.** [ME *ded* < OE *dēad.*] **1.** No longer alive. **2.** Marked for certain death : DOOMED. **3. a.** Having the physical look of death <a *dead* pallor> **b.** Lacking feeling or sensitivity <*dead* to the pleas for help> **c.** Weary and worn-out. **4. a.** Not having the capacity to live : INANIMATE. **b.** Not having the capacity to produce or sustain life : BARREN <*dead* soil> **5. a.** No longer existing, being in use, or operational <a *dead* language> **b.** Not or no longer active : DORMANT <a *dead* volcano> **6. a.** Not productive <*dead* capital> **b.** Not moving or circulating : STAGNANT <*dead* water> **7. a.** Without activity or traffic : QUIET <a *dead* town> **b.** Without animation : DULL <a *dead* party> **c.** No longer having significance or relevance <a *dead* issue> **8.** Without resonance. —Used of sounds. **9.** Extinguished <a *dead* flame> **10.** Lacking bounce or elasticity <a *dead* soccer ball> **11.** Out of operation due to a fault or breakdown. **12. a.** Sudden : abrupt <a *dead* stop> **b.** Complete <*dead* silence> **c.** Exact : unerring. **13.** Out of play. —Used of a ball. **14. a.** Lacking connection to a source of electric current. **b.** Drained of electric charge, as a battery. **15.** Not transmitting : SILENT <*dead* radio air> *—n.* **1. a.** One that has died. **b.** All those that have died. **2.** The period of greatest intensity, as of cold or darkness <the *dead* of winter> *—adv.* **1.** Absolutely : altogether. **2.** Directly : exactly <*dead* ahead> **—dead'ness** *n.*

**dead-air space** (dĕd'âr') *n.* An unventilated space.

**dead·beat¹** (dĕd'bēt') *adj.* Having an indicator that stops without oscillation.

**dead·beat²** (dĕd'bēt') *n. Slang.* **1.** One who does not pay one's debts. **2.** A lazy person : LOAFER.

**dead center** *n.* Either of two points in the path of a moving crank and connecting rod at the ends of a stroke when the two lie in a straight line.

**dead duck** *n. Slang.* One destined to failure.

**dead·en** (dĕd'n) *v.* **-ened, -en·ing, -ens.** *—vt.* **1.** To render less sensitive, intense, or vigorous. **2.** To make soundproof. **3.** To make less colorful. *—vi.* To become dead or as if dead. **—dead'en·er** *n.*

**dead-end** (dĕd'ĕnd') *adj.* **1.** Lacking an exit. **2.** Posing no opportunity for advancement <a *dead-end* job> **3.** *Informal.* Of or typical of the slums or life therein <a *dead-end* gang>

**dead end** *n.* **1.** An end of a passage, as a street or pipe, that affords no outlet. **2.** A point beyond which no movement or progress can be made : IMPASSE <reached a *dead end* in the negotiations>

**dead·en·ing** (dĕd'n-ĭng) *n.* Material for soundproofing.

**dead·eye** (dĕd'ī') *n.* **1.** *Naut.* A flat hardwood disk with a grooved perimeter, pierced by three holes through which the lanyards are passed, used to fasten the shrouds. **2.** *Slang.* An expert shot.

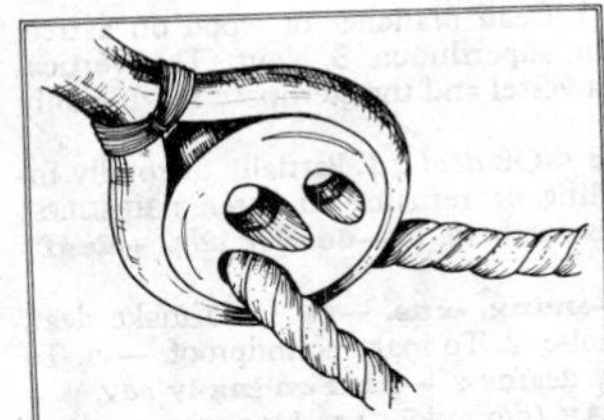

deadeye

**de·aer'ate'** *v.*
**de'·aer·a'tion** *n.*
**de'-A·mer'i·can·i·za'tion** *n.*
**de'-A·mer'i·can·ize'** *v.*
**de·ash'** *v.*
**de'as·sim'i·la'tion** *n.*
**de·cer'ti·fi·ca'tion** *n.*
**de·cer'ti·fy'** *v.*
**de·chlor'i·nate'** *v.*
**de·col'or** *v.*
**de'col'or·a'tion** *n.*
**de'com·mis'sion** *v.*
**de'com'pen·sate'** *v.*
**de'com·pound'** *v.*
**de'con·di'tion** *v.*
**de'con·ges'tion** *n.*
**de'con·ges'tive** *adj.*
**de·con'se·crate'** *v.*
**de·em'pha·size'** *v.*
**de·en'er·gize'** *v.*
**de·hull'** *v.*
**de·lead'** *v.*
**de·lime'** *v.*
**de·lint'** *v.*
**de·lus'ter** *v.*
**de·mast'** *v.*
**de'ma·te'ri·al·i·za'tion** *n.*
**de'ma·te'ri·al·ize'** *v.*
**de·mes'mer·ize'** *v.*
**de·mount'** *v.*
**de·ni'trate'** *v.*
**de·pas'ture** *v.*
**de·pig'ment** *v.*
**de·pol'ish** *v.*
**de·ra'tion** *v.*
**de·req'ui·si'tion** *v.*
**de're·strict'** *v.*
**de·sanc'ti·fy'** *v.*
**de·sat'u·rate'** *v.*
**de·scale'** *v.*
**de·sil'ver** *v.*
**de·size'** *v.*
**de·so'cial·ize'** *v.*
**de·soil'** *v.*
**de·sug'ar** *v.*
**de·vow'** *v.*
**de·wool'** *v.*
**de·worm'** *v.*

**dead·fall** (dĕd'fôl') *n.* **1.** A trap for large animals in which a heavy weight falls on and kills or disables the prey. **2.** A mass of fallen timber and tangled brush.

**dead hand** *n.* [ME *dede hond,* transl. of OFr. *mortemain,* mortmain.] *Law.* Mortmain.

**dead·head** (dĕd'hĕd') *n. Informal.* **1.** One who uses a free ticket for admittance, accommodation, or entertainment. **2.** A vehicle, as a railroad car or an aircraft, carrying no passengers or freight. **3.** A slow or dull-witted person. —*vt.* **-head·ed, -head·ing, -heads. 1.** *Informal.* To drive or pilot (a vehicle) carrying no passengers or freight. **2.** To pull dead or dying blossoms off (a flower). —*adv. Informal.* Without passengers or freight : EMPTY.

**dead heat** *n.* A race in which two or more contestants finish at the same time.

**dead letter** *n.* **1.** An undeliverable or unclaimed letter that after a certain time is destroyed or returned to the sender by the post office. **2.** A law, directive, or factor still in effect but no longer valid or enforced.

**dead·light** (dĕd'līt') *n.* **1.** *Naut.* **a.** A strong shutter or plate placed over a ship's porthole or cabin window in stormy weather. **b.** A thick window set in a ship's side or deck. **2.** An unopenable skylight.

**dead·line** (dĕd'līn') *n.* **1.** A time limit, as for payment of a debt or completion of an assignment. **2.** A boundary line in a prison that prisoners can cross only at the risk of being shot.

**dead load** *n.* The fixed weight of a structure or piece of equipment, as a bridge on its supports.

**dead·lock** (dĕd'lŏk') *n.* A stoppage or standstill caused by the opposition of two equally strong, unrelenting forces. —*vt. & vi.* **-locked, -lock·ing, -locks.** To bring or come to a deadlock.

**dead·ly** (dĕd'lē) *adj.* **-li·er, -li·est. 1.** Causing or tending to cause death : LETHAL. **2.** Like death <a *deadly* white> **3.** Implacable <*deadly* enemies> **4.** Destructive in effect <gave the book a *deadly* review> **5.** Utter <*deadly* earnestness> **6.** Extreme <under *deadly* strain> **7.** Very accurate <a *deadly* marksman> **8.** *Informal.* Dull and boring. —*adv.* **1.** So as to suggest death. **2.** To an extreme <I'm *deadly* serious.> **—dead'li·ness** *n.*

☆ ***syns:*** DEADLY, FATAL, LETHAL, MORTAL *adj. core meaning* : causing or tending to cause death <a *deadly* poison>

**deadly nightshade** *n.* BELLADONNA 1.

**dead march** *n.* A slow, solemn march played for a funeral.

**dead nettle** *n.* An Old World weedy plant of the genus *Lamium,* with small purplish, white, or yellow flower clusters.

**dead·pan** (dĕd'păn') *n.* **1.** A blank, expressionless face. **2.** A person, esp. a theatrical performer, who has or assumes a deadpan. —*v.* **-panned, -pan·ning, -pans.** —*vt.* To express in a deadpan. —*vi.* To express oneself in a deadpan way. **—dead'pan'** *adj. & adv.*

**dead point** *n.* Dead center.

**dead reckoning** *n.* **1.** A method of estimating the position of an aircraft or ship without astronomical observations, as by applying to a previously determined position the course and distance traveled since. **2.** Calculation based on inference or guesswork.

**dead spot** *n.* A region where the reception of radio transmissions over a given frequency range is extremely weak.

**dead weight** *n.* **1.** The unrelieved weight of a heavy, motionless mass. **2.** An oppressive burden or difficulty affording no advantage whatever. **3.** A dead load.

**dead·wood** (dĕd'wo͝od') *n.* **1.** Dead branches or wood on a tree. **2.** One that is burdensome or superfluous. **3.** *Naut.* The vertical planking between the keel of a vessel and the sternpost, serving only as reinforcement.

**deaf** (dĕf) *adj.* **-er, -est.** [ME < OE *dēaf.*] **1.** Partially or totally incapable of hearing. **2.** Unwilling or refusing to listen : HEEDLESS <They were *deaf* to our cries for help.> **—deaf'ly** *adv.* **—deaf'ness** *n.*

**deaf·en** (dĕf'ən) *v.* **-ened, -en·ing, -ens.** —*vt.* **1.** To make deaf, esp. momentarily, by a loud noise. **2.** To make soundproof. —*vi.* To cause permanent or temporary deafness. **—deaf'en·ing·ly** *adv.*

**deaf-mute** *also* **deaf mute** (dĕf'myo͞ot') *n.* One who can neither speak nor hear. —*adj.* (dĕf-myo͞ot'). Unable to speak or hear.

**deal[1]** (dēl) *v.* **dealt** (dĕlt), **deal·ing, deals.** [ME *delen* < OE *dǣlan,* to divide, share.] —*vt.* **1.** To give to someone as a share : APPORTION. **2.** To distribute or pass out among several people. **3.** To administer : deliver <*dealt* a blow to the chest> **4. a.** To distribute (cards) among players. **b.** To give (a specific card) to a player while so distributing. —*vi.* **1.** To be concerned <a book *dealing* with the Renaissance> **2.** To behave in a particular way toward another or others <*deal* honestly with our competitors> **3.** To take action <The management will *deal* with this complaint.> **4.** To do business : TRADE <*dealing* in precious gems> **5.** To distribute playing cards. —*n.* **1.** The act or a round of apportioning or distributing. **2. a.** Distribution of playing cards. **b.** The cards distributed : HAND. **c.** The right or turn of a player to distribute the cards. **d.** The playing of one hand. **3.** *Informal.* An indefinite quantity, extent, or degree <a great *deal* of technical experience> **4.** A secretly arranged agreement, as in politics or business. **5.** *Informal.* A business transaction. **6.** *Informal.* A favorable bargain or sale <got a terrific *deal* on that car> **7.** *Informal.* Treatment received <a raw *deal*> **8.** *Slang.* An important issue <make a big *deal* out of nothing>

**deal[2]** (dēl) *n.* [ME *dele* < MLG *dele,* plank.] **1. a.** A fir or pine board cut to standard dimensions. **b.** Such boards or planks as a whole. **2.** Fir or pine wood.

**de·a·late** (dē-ā'lāt') *or* **de·a·lat·ed** (-lā'tĭd) *adj.* Having lost the wings. —Used of certain insects, as ants, that shed their wings after a mating flight.

**deal·er** (dē'lər) *n.* **1.** One engaged in buying and selling. **2.** One who distributes the cards in a card game.

**deal·er·ship** (dē'lər-shĭp') *n.* A franchise to sell an item in a particular area.

**deal·fish** (dēl'fĭsh') *n., pl.* **dealfish** or **-fish·es.** [< DEAL[2].] A marine fish, *Trachipterus arcticus,* indigenous to Atlantic waters, resembling the ribbonfish.

**deal·ing** (dē'lĭng) *n.* **1. dealings.** Transactions with others, usu. in business. **2.** Method or manner of conduct in relation to others : TREATMENT <honest *dealing*>

**dealt** (dĕlt) *v. p.t. & p.p. of* DEAL[1].

**de·am·i·nase** (dē-ăm'ə-nās', -nāz') *n.* An enzyme that catalyzes the hydrolysis of amino compounds, as amino acids.

**de·am·i·nate** (dē-ăm'ə-nāt') *vt.* **-nat·ed, -nat·ing, -nates.** To remove an amino group from (an organic compound). **—de·am'i·na'tion** *n.*

**de·am·i·nize** (dē-ăm'ə-nīz') *vt.* **-nized, -niz·ing, -niz·es.** To deaminate. **—de·am'i·ni·za'tion** *n.*

**dean** (dēn) *n.* [ME *deen* < OFr. *deien* < LLat. *decanus,* chief of ten < *decem,* ten.] **1. a.** An administrative officer in charge of a college, faculty, or university division. **b.** An officer of a college or high school who counsels students and enforces rules. **2.** The head of the chapter of canons governing a cathedral or collegiate church. **3.** *Chiefly Brit.* A priest appointed to oversee a group of parishes within a diocese. **4.** The senior member of a body or group : DOYEN. **—dean'ship'** *n.*

**dean·er·y** (dē'nə-rē) *n., pl.* **-ies. 1.** The office, jurisdiction, or authority of a dean. **2.** A dean's official residence.

**dean's list** *n.* A periodically issued list of students in a college or university who have attained high academic rank.

**dear[1]** (dîr) *adj.* **-er, -est.** [ME *dere* < OE *dēore.*] **1. a.** Loved and cherished <my *dear* sister> **b.** Greatly valued : PRECIOUS <left everything *dear* to me> **2.** Highly esteemed or regarded. —Used in direct address, esp. in salutations. **3. a.** High-priced. **b.** Charging high prices. **4.** *Obs.* Noble : worthy. —*n.* A greatly loved person. —*adv.* **1.** Fondly or affectionately. **2.** At a high cost. —*interj.* —Used as a polite exclamation, primarily of surprise or distress. **—dear'ly** *adv.* **—dear'ness** *n.*

**dear[2]** (dîr) *adj.* [ME *dere* < OE *dēor.*] *Obs.* Severe : grievous.

**Dear John** (jŏn) *n.* A letter from one's wife or girlfriend requesting termination of the relationship <The soldier got a *Dear John.*>

**dearth** (dûrth) *n.* [ME < *dere,* dear.] **1.** Scarcity : lack <a *dearth* of good will> **2.** Shortage of food : FAMINE.

**death** (dĕth) *n.* [ME *deeth* < OE *dēað.*] **1.** The act of dying : cessation of life. **2.** The state of being dead. **3.** *often* **Death.** A personification of the destroyer of life, usu. represented as a skeleton holding a scythe. **4.** Termination or extinction <the *death* of slavery> **5.** The cause of dying. **6.** A manner of dying <a martyr's *death*> **7. a.** Bloodshed or murder. **b.** Execution. **8.** Civil death. **9.** *Christian Science.* The product of human belief of life in matter. **—to death.** To an intolerable degree <scared *to death*>

**death·bed** (dĕth'bĕd') *n.* **1.** The bed on which a person dies. **2.** The last hours before death.

**death·blow** (dĕth'blō') *n.* **1.** A blow that causes death. **2.** A fatal occurrence or event.

**death camas** *also* **death camass** *n.* A plant of the genus *Zygadenus* of western North America, bearing grasslike leaves and greenish-white flower clusters poisonous to livestock.

**death cup** *n.* A poisonous, usu. white mushroom, *Amanita phalloides,* with a prominent bulbous base.

**death duty** *n. Chiefly Brit.* Inheritance tax.

**death house** *n.* A prison cell block where condemned prisoners await execution.

**death·less** (dĕth'lĭs) *adj.* Not subject to death : IMMORTAL. **—death'less·ly** *adv.* **—death'less·ness** *n.*

**death·ly** (dĕth'lē) *adj.* **1.** Of, resembling, or typical of death. **2.** Causing death : FATAL. —*adv.* **1.** In the manner of death. **2.** Very : extremely <*deathly* pale>

**death mask** *n.* A cast of a person's face taken after death.

**death point** *n.* An environmental limit, as of temperature or radiation, beyond which a specified life form cannot survive.

**death rate** *n.* The ratio of total deaths to total population in a specified community.

**death rattle** *n.* A rare respiratory gurgling or rattling in the throat of a dying person, caused by loss of the cough reflex and passage of breath through accumulating mucus in the throat.

**death's-head** (dĕths'hĕd') *n.* The human skull as a symbol of mortality or death.

**deaths·man** (dĕths′mən) *n. Archaic.* An executioner.

**death tax** *n.* Inheritance tax.

**death·trap** (dĕth′trăp′) *n.* **1.** An unsafe building or structure. **2.** A perilous situation or circumstance.

**death warrant** *n.* **1.** *Law.* An official order authorizing a person's execution. **2.** DEATHBLOW 2.

**death·watch** (dĕth′wŏch′) *n.* **1.** A vigil kept beside a dying or dead person. **2.** One who guards a condemned person before his or her execution. **3. a.** Any of several beetles of the family Anobiidae that strike their heads against the wood into which they burrow with a hollow, clicking sound. **b.** A booklouse that makes a similar sound.

**death wish** *n. Psychiat.* A conscious or unconscious desire for one's own death or the death of another.

**de·ba·cle** (dĭ-bä′kəl, -băk′əl) *n.* [Fr. *débâcle* < *débâcler,* to unbar < OFr. *desbacler* : *des-,* away (< Lat. *de-*) + *bacler,* to bar < OProv. *baclar* < VLat. **bacclare* < Lat. *baculum,* rod.] **1.** A sudden, disastrous collapse, downfall, or defeat. **2.** A total, often ludicrous failure. **3.** The breakup of ice in a river. **4.** A violent flood.

**de·bar** (dē-bär′) *vt.* **-barred, -bar·ring, -bars.** [ME *debarren* < OFr. *desbarrer,* to unbar : *des-,* away (< Lat. *de-*) + *barrer,* to bar < *barre,* bar.] **1.** To bar or exclude : shut out. **2.** To forbid, hinder, or prevent. **—de·bar′ment** *n.*

**de·bark** (dĭ-bärk′) *v.* **-barked, -bark·ing, -barks.** [Fr. *débarquer* < OFr. *debarquer* : *de-,* from (< Lat. *dē-*) + *barque,* ship. —see BARQUE.] —*vt.* To unload, as from a ship. —*vi.* To disembark. **—de′bar·ka·tion** (dē′bär-kā′shən) *n.*

**de·base** (dĭ-bās′) *vt.* **-based, -bas·ing, -bas·es.** To lower in character, quality, or value : DEMEAN. **—de·base′ment** *n.* **—de·bas′er** *n.*

**de·bat·a·ble** (dĭ-bā′tə-bəl) *adj.* **1.** Capable of being formally argued or discussed. **2.** Open to debate : QUESTIONABLE. **3.** In dispute, as land. **—de·bat′a·bly** *adv.*

**de·bate** (dĭ-bāt′) *v.* **-bat·ed, -bat·ing, -bates.** [ME *debaten* < OFr. *debatre* : *de-,* apart (< Lat. *de-*) + *battre,* to fight < Lat. *battuere,* to batter.] —*vi.* **1.** To consider or deliberate <*debating* whether to go> **2.** To engage in argument by discussing opposing points. **3.** To engage in a formal discussion or argument. **4.** *Obs.* To fight : quarrel. —*vt.* **1.** To deliberate on : CONSIDER. **2.** To dispute or argue about. **3.** To discuss or argue (e.g., a question) formally. **4.** *Obs.* To fight or argue for or over. —*n.* **1.** A discussion involving opposing points : ARGUMENT. **2.** Consideration : deliberation. **3.** A formal contest of argumentation in which two opposing teams defend and attack a given proposition. **4.** *Obs.* Conflict : strife. **—de·bat′er** *n.*

**de·bauch** (dĭ-bôch′) *v.* **-bauched, -bauch·ing, -bauch·es.** [Fr. *débaucher* < OFr. *desbaucher,* to lead astray, roughhew timber : *des-,* apart (< Lat. *de-*) + *bauch,* beam, of Germanic orig.] —*vt.* **1. a.** To corrupt morally : SEDUCE. **b.** To lead away from virtue or excellence. **2.** *Obs.* To cause to abandon allegiance. —*vi.* To indulge in dissipation. —*n.* **1.** An act or time of debauchery. **2.** An orgy. **—de·bauch′ed·ly** (-bô′chĭd-lē) *adv.* **—de·bauch′er** *n.*

**de·bauch·ee** (dĭ-bô′chē′, dĕb′ə-shē′, -shā′) *n.* One who habitually indulges in debauchery : LIBERTINE.

**de·bauch·er·y** (dĭ-bô′chə-rē) *n., pl.* **-ies. 1.** Excessive indulgence in sensual pleasures : DISSIPATION. **2.** *Archaic.* Seduction from morality, allegiance, or duty.

**de·ben·ture** (dĭ-bĕn′chər) *n.* [ME *debentur* < Lat., they are due.] **1.** A certificate or voucher acknowledging a debt. **2.** An unsecured bond issued by a civil or governmental corporation or agency and backed only by the credit standing of the issuer. **3.** A customhouse certificate providing for the payment of a drawback.

**de·bil·i·tate** (dĭ-bĭl′ĭ-tāt′) *vt.* **-tat·ed, -tat·ing, -tates.** [Lat. *debilitare, debilitat-* < *debilis,* weak.] To make weak or feeble : ENERVATE. **—de·bil′i·ta′tion** *n.* **—de·bil′i·ta′tive** *adj.*

**de·bil·i·ty** (dĭ-bĭl′ĭ-tē) *n.* [ME *debilite* < OFr. < Lat. *debilitas* < *debilis,* weak.] Abnormal bodily weakness : ENERVATION.

**deb·it** (dĕb′ĭt) *n.* [ME *debite* < Lat. *debitum.* —see DEBT.] **1.** An item of debt charged to and recorded in an account. **2. a.** An entry of a sum in the debit side of an account. **b.** The sum of such entries. **3.** The left-hand side of an account or an accounting ledger where bookkeeping entries are made. **4.** A disadvantage. —*vt.* **-it·ed, -it·ing, -its. 1.** To enter (a sum) on the left-hand side of an account or accounting ledger. **2.** To charge with a debt.

**deb·o·nair** *also* **deb·o·naire** (dĕb′ə-nâr′) *adj.* [ME *debonaire,* gracious, kindly < OFr. < *de bonne aire,* of good disposition.] **1.** Suave : urbane. **2.** Affable : genial. **3.** Jaunty : nonchalant. **—deb′o·nair′ly** *adv.* **—deb′o·nair′ness** *n.*

**de·bouch** (dĭ-bouch′, -bo͞osh′) *v.* **-bouched, -bouch·ing, -bouch·es.** [Fr. *déboucher* : *de-,* out of (< Lat. *de-*) + *bouche,* mouth < Lat. *bucca.*] —*vi.* **1.** To march from a narrow or confined area into the open. **2.** To emerge or issue. —*vt.* To cause to emerge or issue.

**dé·bou·ché** (dā′bo͞o-shā′) *n.* [Fr. < *déboucher,* to debouch.] **1.** An outlet in military works for the passage of troops. **2.** An outlet, as for goods.

**de·bouch·ment** (dĭ-bouch′mənt, -bo͞osh′-) *n.* **1.** The act or an instance of debouching. **2.** A debouchure.

**de·bou·chure** (dĭ-bo͞o′sho͝or′) *n.* A mouth or outlet, esp. of a river.

**dé·bride·ment** (dā′brēd-mäɴ′, dĭ-brēd′mənt) *n.* [Fr. < *débrider* < OFr. *desbrider,* to unbridle : *des-,* away (< Lat. *de-*) + *bride,* bridle, of Germanic orig.] Surgical excision of dead, devitalized, or contaminated tissue from a wound. **—de·bride′** (dĭ-brēd′, dā-) *v.* **(-brid·ed, -brid·ing, -brides).**

**de·brief** (dē-brēf′) *vt.* **-briefed, -brief·ing, -briefs. 1.** To question or interrogate to obtain intelligence gathered esp. on a military mission. **2.** To instruct (e.g., a government agent) not to reveal classified information after his or her employment has terminated.

**de·brief·ing** (dē-brē′fĭng) *n.* **1.** The act or process of debriefing or being debriefed. **2.** Information conveyed during debriefing.

**de·bris** *also* **dé·bris** (də-brē′, dā-, dā′brē′) *n.* [Fr. *débris* < OFr. *desbrisier,* to break to pieces : *des-* (intensive < Lat. *dē-*) + *brisier,* to break.] **1.** Scattered remains : RUINS. **2.** Discarded waste : LITTER. **3.** *Geol.* An accumulation of rather large rock fragments.

**debt** (dĕt) *n.* [ME *dette* < OFr. < VLat. **debita,* fem. of Lat. *debitum,* debt < *debēre,* to owe.] **1.** Something owed, as money, goods, or services. **2. a.** An obligation or liability to pay or render something to another. **b.** The condition of having such an obligation. **3.** An offense requiring forgiveness or reparation : TRESPASS.

▲ word history: The pronunciation of *debt* represents the original pronunciation of the word, which was borrowed from Old French *dette.* Medieval writers knew that *dette* was derived from Latin *debitum* and in the 15th century the spelling *debt* first appeared. This form was promoted as more correct by the language reformers of the 16th century and is now the only acceptable spelling of the word.

**debt·or** (dĕt′ər) *n.* [ME *dettour* < OFr. *dettor* < Lat. *debitor* < *debēre,* to owe.] **1.** One who owes something to another. **2.** One guilty of a trespass or sin : SINNER.

**de·bug** (dē-bŭg′) *vt.* **-bugged, -bug·ging, -bugs. 1.** To remove insects from. **2. a.** To remove a hidden electronic device, as a microphone, from (e.g., a conference room). **b.** To make (e.g., a hidden microphone) ineffective. **3.** To search for and eliminate malfunctions or errors in (e.g., a spacecraft or a computer program).

**de·bunk** (dē-bŭngk′) *vt.* **-bunked, -bunk·ing, -bunks.** *Informal.* To expose or ridicule the fallacy or fraudulence of <*debunk* an old wives' tale> **—de·bunk′er** *n.*

**de·but** *also* **dé·but** (dā-byo͞o′, dā′byo͞o′) *n.* [Fr. *début* < *débuter,* to lead off in a game, debut : *de-,* away (< Lat. *de-*) + *but,* target < OFr. *butte.*] **1.** A first public appearance, as of a performer. **2.** The formal presentation of a girl to society. **3.** The beginning of a course of action, as a career. —*vt.* & *vi.* **-buted** (-byo͞od′), **-but·ing** (-byo͞o′ĭng), **-buts** (-byo͞oz′). *Informal.* To present in or make a debut.

**deb·u·tante** *also* **dé·bu·tante** (dĕb′yo͞o-tänt′, dā′byo͞o-) *n.* [Fr. *débutante* < fem. pr.part. of *débuter,* to debut.] A young woman making a debut into society.

**deca-** or **dec-** *also* **deka-** or **dek-** *pref.* [Gk. *deka-* < *deka,* ten.] Ten <*decane*>

**dec·ade** (dĕk′ād′, dĕ-kād′) *n.* [ME, a group of ten < OFr. < LLat. *decas* < Gk. *dekas* < *deka,* ten.] **1.** A period of ten years. **2.** A group or series of ten.

**de·ca·dence** (dĭ-kād′ns, dĕk′ə-dəns) *n.* [OFr. < Med. Lat. *decadentia,* a falling : Lat. *de-,* down + *cadere,* to fall.] A process, state, or period of decline or deterioration, as in art or morals : DECAY.

**de·ca·den·cy** (dĭ-kād′n-sē, dĕk′ə-dən-) *n.* Decadence.

**de·ca·dent** (dĭ-kād′nt, dĕk′ə-dənt) *adj.* **1.** Being in a state or condition of decline or decay. **2.** Of or relating to the decadents. —*n.* **1.** A person in a condition or process of mental or moral decay. **2.** A member of a group of French and English writers of the 19th cent. who often sought inspiration in the morbid, neurotic, or macabre and tended toward stylistic overrefinement. **—de·ca′dent·ly** *adv.*

**de·caf·fein·at·ed** (dē-căf′ə-nāt′əd) *adj.* Having the caffeine removed <*decaffeinated* coffee>

**dec·a·gon** (dĕk′ə-gŏn′) *n.* [NLat. *decagonum* < Gk. *dekagōnon* : *deka,* ten + *-gōnon,* -gon.] A polygon having ten angles and ten sides. **—de·cag′o·nal** (dĭ-kăg′ə-nəl) *adj.* **—de·cag′o·nal·ly** *adv.*

**dec·a·gram** or **dek·a·gram** (dĕk′ə-grăm′) *n.* Ten grams.

**dec·a·he·dron** (dĕk′ə-hē′drən) *n., pl.* **-drons** or **-dra** (-drə). A polyhedron having ten faces. **—dec′a·he′dral** *adj.*

**de·cal** (dē′kăl′, dĭ-kăl′) *n.* A picture or design transferred by decalcomania.

**de·cal·ci·fy** (dē-kăl′sə-fī′) *vt.* **-fied, -fy·ing, -fies.** To remove calcium or calcareous matter from (e.g., bones). **—de·cal′ci·fi·ca′tion** *n.* **—de·cal′ci·fi′er** *n.*

**de·cal·co·ma·ni·a** (dē-kăl′kə-mā′nē-ə, -mān′yə) *n.* [Fr. *décalcomanie* : *decalquer,* to transfer by tracing (*de-,* from + *calquer,* to trace < Ital. *calcare,* to trace, trample < Lat., to tread < *calx,* heel) + *manie,* madness < Lat. *mania* (from its popularity in the 19th cent.).] **1.** The process of transferring pictures or designs printed on specially prepared paper to material such as glass or metal. **2.** Decal.

**de·ca·les·cence** (dē′kə-lĕs′əns) *n.* [DE- + Lat. *calescens,* pr.part. of *calescere,* to become warm < *calēre,* to be warm.] A sudden slowing in the rate of temperature increase in a metal being heated, due to endothermic structural changes. **—de′ca·les′cent** *adj.*

**dec·a·li·ter** or **dek·a·li·ter** (dĕk′ə-lē′tər) *n.* Ten liters.

**Dec·a·logue** or **Dec·a·log** (dĕk′ə-lôg′, -lŏg′) *n.* [ME *decalog* <

OFr. *decalogue* < LLat. *decalogus* < Gk. *dekalogos* : *deka,* ten + *logos,* speech, word.] The Ten Commandments.

**dec·a·me·ter** or **dek·a·me·ter** (dĕk′ə-mē′tər) *n.* Ten meters.

**dec·a·met·ric** (dĕk′ə-mĕt′rĭk) *adj.* Of, pertaining to, or being a radio wave between one and ten decameters of wavelength.

**de·camp** (dĭ-kămp′) *vi.* **-camped, -camp·ing, -camps.** [Fr. *décamper* < OFr. *descamper* : *des-,* away (< Lat. *dē-*) + *camper,* to camp < *camp,* camp < Lat. *campus.*] **1.** To leave a camping ground. **2.** To depart suddenly or furtively. **—de·camp′ment** *n.*

**dec·a·nal** (dĕk′ə-nəl, dĭ-kā′nəl) *adj.* [< LLat. *decanus,* dean. —see DEAN.] Of or relating to a dean or deanery.

**dec·ane** (dĕk′ān′) *n.* Any of various liquid isomers, $C_{10}H_{22}$, of the methane series.

**dec·a·no·ic acid** (dĕk′ə-nō′ĭk) *n.* Capric acid.

**de·cant** (dĭ-kănt′) *vt.* **-cant·ed, -cant·ing, -cants.** [Fr. *décanter* < Med. Lat. *decanthare* : Lat. *de-,* from + Lat. *canthus,* rim of a wheel, of Celtic orig.] **1.** To pour off (e.g., wine) without disturbing the sediment. **2.** To pour (a liquid) from one receptacle into another. **—de′can·ta′tion** (dē′kăn-tā′shən) *n.*

**de·cant·er** (dĭ-kăn′tər) *n.* **1.** A decorative bottle for serving liquids, as wine. **2.** A container for decanting.

**de·cap·i·tate** (dĭ-kăp′ĭ-tāt′) *vt.* **-tat·ed, -tat·ing, -tates.** [LLat. *decapitare, decapitat-* : Lat. *de-,* off + *caput,* head.] To cut off the head of : BEHEAD. **—de·cap′i·ta′tion** *n.* **—de·cap′i·ta′tor** *n.*

**dec·a·pod** (dĕk′ə-pŏd′) *n.* [< NLat. *Decapoda,* order name : DECA- + Gk. *pous,* foot.] **1.** A crustacean of the order Decapoda, as a crab, lobster, or shrimp, bearing five pairs of locomotor appendages, each joined to a segment of the thorax. **2.** A cephalopod mollusk, as a squid or cuttlefish, bearing ten armlike tentacles. *—adj.* Of or relating to the Decapoda or a decapod. **—de·cap′o·dal** (dĭ-kăp′ə-dəl), **de·cap′o·dan** (-dən), **de·cap′o·dous** (-dəs) *adj.*

**de·car·bon·ate** (dē-kär′bə-nāt′) *vt.* **-at·ed, -at·ing, -ates.** To remove carbon dioxide or carbonic acid from. **—de·car′bon·a′tion** *n.*

**de·car·bon·ize** (dē-kär′bə-nīz′) *vt.* **-ized, -iz·ing, -iz·es.** To remove the element carbon from. **—de·car′bon·i·za′tion** *n.* **—de·car′bon·iz′er** *n.*

**de·car·box·yl·ase** (dē′kär-bŏk′sə-lās′, -lāz′) *n.* Any of various enzymes that hydrolize the carboxyl radical.

**de·car·box·yl·a·tion** (dē′kär-bŏk′sə-lā′shən) *n.* Removal of a carboxyl group from a chemical compound.

**de·car·bu·rize** (dē-kär′bə-rīz′, -byə-) *vt.* **-rized, -riz·ing, -riz·es.** To decarbonize. **—de·car′bu·ri·za′tion** *n.*

**dec·are** (dĕk′âr′, -är′) *n.* A metric unit of area measure equal to 10 ares or 0.2471 acre.

**dec·a·stere** or **dek·a·stere** (dĕk′ə-stîr′) *n.* Ten steres.

**dec·a·syl·la·ble** (dĕk′ə-sĭl′ə-bəl) *n.* A line of verse having ten syllables. **—dec′a·syl·lab′ic** (-sə-lăb′ĭk) *adj.*

**de·cath·lon** (dĭ-kăth′lən, -lŏn′) *n.* [Fr. *décathlon* : *déca-,* deca- + Gk. *athlon,* contest.] An athletic contest in which each contestant participates in ten different track and field events.

**de·cay** (dĭ-kā′) *v.* **-cayed, -cay·ing, -cays.** [ME *decayen* < AN *decair* < VLat. **decadere* : Lat. *de-,* down + Lat. *cadere,* to fall.] *—vi.* **1.** *Biol.* To break down into component parts : ROT. **2.** *Physics.* To disintegrate or diminish by radioactive decay. **3.** *Aerospace.* To decrease in orbit, as an artificial satellite. **4.** To fall into ruin. **5.** *Pathol.* To decline in health or vigor : become enervated. **6.** To decline from a normal or prosperous state. *—vt.* To cause to decay. *—n.* **1.** Decomposition or destruction of organic matter as a result of bacterial or fungal action. **2.** Radioactive decay. **3.** A decrease in orbital altitude of an artificial satellite due to conditions such as atmospheric drag. **4.** Gradual deterioration to an inferior state, as of health or mental capability.

☆ ***syns:*** DECAY, BREAK DOWN, DECOMPOSE, DETERIORATE, MOLDER, PUTREFY, ROT, SPOIL, TURN *v. core meaning* : to become or cause to become rotten or unsound <*decaying* vegetation>

**de·cease** (dĭ-sēs′) *vi.* **-ceased, -ceas·ing, -ceas·es.** [ME *decesen* < *deces,* death < OFr. < Lat. *decessus* < p.part. of *decedere,* to depart : *de-,* away + *cedere,* to go.] To die. **—de·cease′** *n.*

**de·ceased** (dĭ-sēst′) *adj.* No longer living. *—n.* A dead person.

**de·ce·dent** (dĭ-sēd′nt) *n.* [Lat. *decedens, decedent-,* pr.part. of *decedere,* to die. —see DECEASE.] *Law.* A deceased person.

**de·ceit** (dĭ-sēt′) *n.* [ME < AN < Lat. *decepta,* fem. p.part. of *decipere,* to deceive.] **1.** The act or practice of deceiving : DECEPTION. **2.** A stratagem or trick. **3.** The quality of being deceitful.

**de·ceit·ful** (dĭ-sēt′fəl) *adj.* **1.** Inclined to cheat or deceive. **2.** Deliberately misleading : DECEPTIVE. **—de·ceit′ful·ly** *adv.* **—de·ceit′ful·ness** *n.*

**de·ceive** (dĭ-sēv′) *v.* **-ceived, -ceiv·ing, -ceives.** [ME *deceiven* < AN *deceiver* < Lat. *decipere* : *de-* (pejorative) + *capere,* to seize.] *—vt.* **1.** To cause to believe what is not true : MISLEAD. **2.** *Archaic.* To catch by guile : ENSNARE. *—vi.* To practice deceit. **—de·ceiv′a·ble** *adj.* **—de·ceiv′er** *n.* **—de·ceiv′ing·ly** *adv.*

☆ ***syns:*** DECEIVE, DELUDE, DUPE, FOOL, HOODWINK, MISLEAD, TAKE IN *v. core meaning* : to cause to accept what is false, esp. by misrepresentation or trickery <was *deceived* by their show of friendship>

**de·cel·er·ate** (dē-sĕl′ə-rāt′) *vt. & vi.* **-at·ed, -at·ing, -ates.** [DE- + (AC)CELERATE.] To decrease the velocity of or decrease in velocity. **—de·cel′er·a′tion** *n.* **—de·cel′er·a′tor** *n.*

**De·cem·ber** (dĭ-sĕm′bər) *n.* [ME *decembre* < OFr. < Lat. *December,* the tenth month < *decem,* ten.] The 12th month of the year according to the Gregorian calendar. —See table at CALENDAR.

**De·cem·brist** (dĭ-sem′brĭst) *n.* A participant in an unsuccessful plot to overthrow Czar Nicholas I of Russia in Dec., 1825.

**de·cem·vir** (dĭ-sĕm′vər) *n., pl.* **-virs** or **-vi·ri** (-və-rī′) [ME < Lat., back-formation < *decemviri,* commission of ten < *decem viri,* ten men.] One of a body of ten Roman magistrates, esp. a member of one of two such bodies appointed in 451 and 450 B.C. to draw up a code of laws. **—de·cem′vi·ral** *adj.* **—de·cem′vi·rate** *n.*

**de·cen·cy** (dē′sən-sē) *n., pl.* **-cies. 1.** The quality or state of being decent : PROPRIETY. **2.** Conformity to accepted standards of propriety or modesty. **3. decencies.** Social or moral proprieties.

**de·cen·na·ry** (dĭ-sĕn′ə-rē) *adj.* [< Lat. *decennis,* of ten years. —see DECENNIUM.] Of or relating to a ten-year period. *—n., pl.* **-ries.** A decennium.

**de·cen·ni·a** (dĭ-sĕn′ē-ə) *n. var. pl. of* DECENNIUM.

**de·cen·ni·al** (dĭ-sĕn′ē-əl) *adj.* [< Lat. *decennium,* decennium.] **1.** Relating to or lasting for ten years. **2.** Occurring once every ten years. *—n.* A tenth anniversary. **—de·cen′ni·al·ly** *adv.*

**de·cen·ni·um** (dĭ-sĕn′ē-əm) *n., pl.* **-cen·ni·ums** or **-cen·ni·a** (-sĕn′ē-ə) [Lat. < *decennis,* of ten years : *decem,* ten + *annus,* year.] DECADE 1.

**de·cent** (dē′sənt) *adj.* [Lat. *decens, decent-,* pr.part. of *decēre,* to be fitting.] **1.** Marked by conformity to traditional standards of propriety or morality. **2.** Free from vulgarity or immodesty. **3.** Meeting accepted standards : ADEQUATE <a *decent* income> **4.** Kind or obliging <very *decent* of you to help> **5.** *Informal.* Properly or modestly dressed. **—de′cent·ly** *adv.* **—de′cent·ness** *n.*

**de·cen·tral·ize** (dē-sĕn′trə-līz′) *vt.* **-ized, -iz·ing, -iz·es. 1.** To distribute the administrative functions or powers of (a central authority) among several local authorities. **2.** To redistribute a concentration of (e.g., population or industry) over a wider area. **—de·cen′tral·i·za′tion** *n.*

**de·cep·tion** (dĭ-sĕp′shən) *n.* [ME *decepcioun* < OFr. *deception* < LLat. *deceptio* < Lat. *deceptus,* p.part. of *decipere,* to deceive.] **1.** Use of deceit. **2.** The fact or state of being deceived. **3.** A ruse.

**de·cep·tive** (dĭ-sĕp′tĭv) *adj.* Intended or tending to deceive : MISLEADING. **—de·cep′tive·ly** *adv.* **—de·cep′tive·ness** *n.*

**deci-** *pref.* [Fr. *déci-* < Lat. *decimus,* tenth < *decem,* ten.] One tenth ($10^{-1}$) <deciliter>

**dec·i·are** (dĕs′ē-âr′, -ar′) *n.* One tenth ($10^{-1}$) of an are.

**dec·i·bel** (dĕs′ə-bəl, -bĕl′) *n.* A unit for expressing relative difference in power, usu. between acoustic or electric signals, equal to ten times the common logarithm of the ratio of the two levels.

**de·cide** (dĭ-sīd′) *v.* **-cid·ed, -cid·ing, -cides.** [ME *deciden* < AN *decider* < Lat. *decidere* : *de-,* off + *caedere,* to cut.] *—vt.* **1.** To settle or conclude. **2.** To influence or determine the conclusion of <One more lap will *decide* the race.> **3.** To cause to make or reach a decision. *—vi.* **1.** To pronounce a judgment. **2.** To make up one's mind. **—de·cid′a·ble** *adj.* **—de·cid′er** *n.*

☆ ***syns:*** DECIDE, CONCLUDE, DETERMINE, RESOLVE, SETTLE *v. core meaning* : to make up or cause to make up one's mind <*decided* to buy the house>

**de·cid·ed** (dĭ-sī′dĭd) *adj.* **1.** Definite : unquestionable <a *decided* victory> **2.** Free from hesitation or vacillation : RESOLUTE. **—de·cid′ed·ly** *adv.* **—de·cid′ed·ness** *n.*

**de·cid·ing** (dĭ-sī′dĭng) *adj.* **1.** Having the quality or power that decides <The *deciding* factor was lack of time.> **2.** Effecting a decision <made the *deciding* knockout>

**de·cid·u·a** (dĭ-sĭj′o͞o-ə) *n.* [NLat. *(membrana) decidua,* (membrane) that falls off < Lat. *deciduus,* deciduous.] A uterine mucous membrane that is modified during pregnancy and cast off during menstruation or at parturition. **—de·cid′u·al** *adj.*

**de·cid·u·ate** (dĭ-sĭj′o͞o-ĭt) *adj.* **1.** Marked by or having a decidua. **2.** Marked by shedding.

**de·cid·u·ous** (dĭ-sĭj′o͞o-əs) *adj.* [Lat. *deciduus* < *decidere,* to fall off : *de-,* off + *caedere,* to cut.] **1.** Falling off or shed at a specific season or stage of growth, as leaves or antlers. **2.** Shedding foliage at the end of the growing season. **3.** Temporary : impermanent. **—de·cid′u·ous·ly** *adv.* **—de·cid′u·ous·ness** *n.*

**dec·i·gram** (dĕs′ĭ-grăm′) *n.* One tenth ($10^{-1}$) of a gram.

**dec·ile** (dĕs′īl′, -əl) *n.* [< Lat. *decem,* ten.] *Statistics.* **1.** One of the numbers or values in a series dividing the distribution of the individuals in the series into ten groups of equal frequency. **2.** Any of the ten groups.

**dec·i·li·ter** (dĕs′ə-lē′tər) *n.* One tenth ($10^{-1}$) of a liter.

**de·cil·lion** (dĭ-sĭl′yən) *n.* [Lat. *decem,* ten + (M)ILLION.] **1.** The cardinal number equal to $10^{33}$. **2.** *Chiefly Brit.* The cardinal number equal to $10^{60}$. **—de·cil′lionth** *adj & adv.*

**dec·i·mal** (dĕs′ə-məl) *n.* [Med. Lat. *decimalis,* of tithes < Lat. *decimus,* tenth < *decem,* ten.] **1.** A linear array of integers that represents a fraction, every decimal place indicating a multiple of a positive or negative power of 10; e.g., the decimal .1 = 1/10, .12 = 12/100, .003 = 3/1000. **2.** A number written using base 10 : a number containing a decimal point. —*adj.* **1.** Expressed or expressible as a decimal. **2. a.** Based on ten. **b.** Numbered or ordered by tens. **—dec'i·mal·ly** *adv.*

**decimal fraction** *n.* DECIMAL 1.

**dec·i·mal·ize** (dĕs'ə-mə-līz') *vt.* **-ized, -iz·ing, -iz·es.** To change to a decimal system. **—dec'i·mal·i·za'tion** *n.*

**decimal place** *n.* The position of a digit to the right of a decimal point, usu. identified by successive ascending ordinal numbers with the digit immediately to the right of the decimal point being first.

**decimal point** *n.* A period placed to the left of a decimal.

**decimal system** *n.* **1.** A number system using the base ten. **2.** A system of measurement in which all derived units are multiples of ten of fundamental units.

**dec·i·mate** (dĕs'ə-māt') *vt.* **-mat·ed, -mat·ing, -mates.** [Lat. *decimare, decimat-* < *decimus,* tenth < *decem,* ten.] **1.** To destroy or kill a large proportion of. **2.** To select by lot and kill one in every ten of. ***usage:*** *Decimate* orig. meant "to kill every tenth person," but its English meaning has been extended to include the destruction of any large proportion of a group, as in *Fire, famine, and sword decimated the population.* Many, however, still avoid the use of *decimate* in describing the destruction of a single person, an entire group, or a specified percentage other than 10%. **—dec'i·ma'tion** *n.*

**dec·i·me·ter** (dĕs'ə-mē'tər) *n.* One tenth ($10^{-1}$) of a meter.

**de·ci·pher** (dĭ-sī'fər) *vt.* **-phered, -pher·ing, -phers. 1.** To read or interpret (something hard to understand or illegible) <couldn't *decipher* the handwriting> **2.** To convert from a cipher or code to plain text : DECODE. **—de·ci'pher·a·ble** *adj.* **—de·ci'pher·er** *n.* **—de·ci'pher·ment** *n.*

**de·ci·sion** (dĭ-sĭzh'ən) *n.* [ME *decisioun* < OFr. *decision* < Lat. *decisio* < *decidere,* to decide.] **1.** Judgment on an issue under consideration. **2.** The act of making up one's mind or reaching a conclusion. **3.** A verdict reached or judgment pronounced. **4.** Firmness of character or action : RESOLUTENESS. **5.** A boxing victory won on points when no knockout has occurred.

**de·ci·sive** (dĭ-sī'sĭv) *adj.* **1.** Having the power to decide : CONCLUSIVE. **2.** Marked by firm determination : RESOLUTE. **3.** Beyond doubt <a *decisive* win> **—de·ci'sive·ly** *adv.* **—de·ci'sive·ness** *n.*

**deck**[1] (dĕk) *n.* [ME *dekke* < MDu. *dec,* covering.] **1.** *Naut.* A platform extending horizontally from one side of a ship to the other. **2.** A platform or surface similar to a ship's deck. **3. a.** A pack of playing cards. **b.** *Computer Sci.* A group of data processing cards. **4.** A tape deck. **5.** *Slang.* A packet of narcotics. —*vt.* **decked, deck·ing, decks. 1.** To furnish with a deck. **2.** To knock to the ground, as by punching. **—clear the deck.** To prepare for action. **—hit the deck.** *Slang.* **1.** To get out of bed. **2.** To clear the deck. **3.** To fall or drop to a prone position. **—on deck.** *Slang.* **1.** On hand : PRESENT. **2.** Waiting to take one's turn, esp. at bat.

**deck**[2] (dĕk) *vt.* **decked, deck·ing, decks.** [MDu. *dekken,* to cover.] To clothe with finery : ADORN <all *decked* out for a dance>

**deck chair** *n.* A folding chair usu. with arms and a leg rest.

**deck hand** *n.* A ship's crew member who works on deck.

**deck·house** (dĕk'hous') *n.* A superstructure on a ship's upper deck.

**deck·le** (dĕk'əl) *n.* [G. *Deckel,* dim. of *Decke,* cover.] **1.** A frame used to form paper pulp into sheets of a desired size in making paper by hand. **2.** A deckle edge.

**deckle edge** *n.* The rough edge of handmade paper formed in a deckle. **—deck'le-edged'** *adj.*

**deck tennis** *n.* A game in which a small ring or quoit is tossed back and forth over a net.

**de·claim** (dĭ-klām') *v.* **-claimed, -claim·ing, -claims.** [ME *declamen* < Lat. *declamare* : *de-* (intensive) + *clamare,* to cry out.] —*vi.* **1.** To deliver an elocutionary recitation. **2.** To speak loudly and vehemently, esp. in opposition or censure : RANT. —*vt.* To utter or recite with rhetorical effect. **—de·claim'er** *n.*

**dec·la·ma·tion** (dĕk'lə-mā'shən) *n.* [ME *declamacioun* < Lat. *declamatio* < *declamare,* to declaim.] **1.** An elocutionary recitation. **2. a.** Vehement oratory. **b.** A tirade : harangue.

**de·clam·a·to·ry** (dĭ-klăm'ə-tôr'ē, -tōr'ē) *adj.* **1.** Having the quality of a declamation. **2.** Pretentiously rhetorical : BOMBASTIC.

**de·clar·ant** (dĭ-klâr'ənt) *n.* One who has signed a declaration of intent to become a U.S. citizen.

**dec·la·ra·tion** (dĕk'lə-rā'shən) *n.* **1. a.** An explicit or formal announcement or statement. **b.** Such a statement in written form. **2.** The act or process of declaring. **3.** A statement of taxable goods or of properties subject to duty. **4.** *Law.* **a.** A formal statement by a plaintiff specifying the facts and circumstances of his or her cause of action. **b.** An unsworn statement of facts admissible as evidence. **5.** A bid, esp. the final bid of a hand in certain card games.

**de·clar·a·tive** (dĭ-klâr'ə-tĭv, -klăr'-) *also* **de·clar·a·to·ry** (-tôr'ē, -tōr'ē) *adj.* Serving to declare. **—de·clar'a·tive·ly** *adv.*

**de·clare** (dĭ-klâr') *v.* **-clared, -clar·ing, -clares.** [ME *declaren* < OFr. *declarer* < Lat. *declarare* : *de-* (intensive) + *clarare,* to make clear < *clarus,* clear.] —*vt.* **1.** To state formally or officially. **2.** To state authoritatively or emphatically : AFFIRM. **3.** To reveal or manifest : SHOW. **4.** To make a full statement of (e.g., dutiable goods). **5.** To designate (a trump suit or no-trump) in bridge with the final bid of a hand. —*vi.* **1.** To make a declaration. **2.** To proclaim one's choice, opinion, or resolution <*declared* against adoption of the resolution> **—de·clar'a·ble** *adj.* **—de·clar'er** *n.*

**de·class** (dē-klăs') *vt.* **-classed, -class·ing, -class·es.** To lower in class or status.

**dé·clas·sé** (dā'klä-sā') *adj.* [Fr., p.part. of *déclasser,* to lower in class : *de-,* down (< Lat. *dē-*) + *classe,* class < LLat. *classis.*] **1.** Lowered in rank or social position. **2.** Lacking high rank or birth.

**de·clas·si·fy** (dē-klăs'ə-fī') *vt.* **-fied, -fy·ing, -fies.** To remove official security classification from (a document). **—de·clas'si·fi'a·ble** *adj.* **—de·clas'si·fi·ca'tion** *n.*

**de·clen·sion** (dĭ-klĕn'shən) *n.* [ME *declenson* < OFr. *declinaison* < LLat. *declinatio,* grammatical declension < Lat., declination.] **1. a.** Inflection of nouns, pronouns, and adjectives in categories such as case, number, and gender in certain languages. **b.** A class of words of one language with the same or a similar system of inflections, as the first declension in Latin. **2.** A descending slope : DESCENT. **3.** A decrease or decline : DETERIORATION. **4.** A deviation, as from a standard or practice. **—de·clen'sion·al** *adj.*

**dec·li·na·tion** (dĕk'lə-nā'shən) *n.* [ME *declinacioun* < OFr. *declination* < Lat. *declinatio* < *declinare,* to turn aside. —see DECLINE.] **1.** A sloping or bending downward. **2.** A decline esp. from prosperity or vigor. **3.** A deviation, as from a specific direction or standard. **4.** A refusal to accept. **5.** Magnetic declination. **6.** *Astron.* The angular distance to a point on the celestial sphere, measured north or south from the celestial equator along the hour circle to the point. **—dec'li·na'tion·al** *adj.*

**de·cline** (dĭ-klīn') *v.* **-clined, -clin·ing, -clines.** [ME *declinen* < OFr. *decliner* < Lat. *declinare,* to turn aside : *de-,* away + *clinare,* to incline, bend.] —*vi.* **1.** To refuse to do, consider, or accept something. **2. a.** To slope downward : DESCEND. **b.** To bend downward : DROOP. **3.** To degrade or lower oneself. **4.** To deteriorate gradually, as from disease. **5. a.** To sink, as the setting sun. **b.** To decrease gradually : WANE. —*vt.* **1.** To refuse <*decline* an offer> **2.** To cause to slope or bend downward. **3.** To give the inflected forms of (a noun, pronoun, or adjective) in certain languages. —*n.* **1.** The process or result of declining, esp. gradual deterioration. **2.** A downward movement. **3.** The period when something is nearing an end. **4.** A downward slope : DECLIVITY. **5.** A disease, as tuberculosis, that causes gradual bodily deterioration. **—de·clin'a·ble** *adj.* **—de·clin'er** *n.*

**de·cliv·i·tous** (dĭ-klĭv'ĭ-təs) *adj.* Rather steep.

**de·cliv·i·ty** (dĭ-klĭv'ĭ-tē) *n., pl.* **-ties.** [Lat. *declivitas* < *declivis,* sloping down : *de-,* down + *clivus,* slope.] A descending slope, as of a hill.

**de·coct** (dĭ-kŏkt') *vt.* **-coct·ed, -coct·ing, -cocts.** [ME *decocten,* to boil < *decoct,* boiled < Lat. *decoctus* < *decoquere,* to boil away : *de-* (intensive) + *coquere,* to cook.] **1.** To extract the flavor of by boiling. **2.** To concentrate by boiling down. **—de·coc'tion** *n.*

**de·code** (dē-kōd') *vt.* **-cod·ed, -cod·ing, -codes.** To convert from code into plain text. **—de·cod'er** *n.*

**de·col·late**[1] (dĭ-kŏl'āt') *vt.* **-lat·ed, -lat·ing, -lates.** [Lat. *decollare, decollat-* : *de-,* off + *collum,* neck.] To behead. **—de'col·la'tion** *n.*

**de·col·late**[2] (dĕk'ə-lāt', dē-kŏ'-) *vt.* **-lat·ed, -lat·ing, -lates.** To separate the copies of. **—de'col·la'tor** *n.*

**dé·colle·tage** (dā'kôl-täzh') *n.* [Fr. < *décolleté,* décolleté.] **1.** A low neckline on a garment. **2.** A décolleté garment.

**dé·colle·té** (dā'kôl-tā') *adj.* [Fr., p.part. of *décolleter,* to cut a low neckline : *dé-,* off (< Lat. *de-*) + *collet,* collar, dim. of *col,* neck < OFr. < Lat. *collum.*] **1.** Having a low neckline <a *décolleté* dress> **2.** Wearing a garment with a low neckline.

**de·col·o·nize** (dē-kŏl'ə-nīz') *vt.* **-nized, -niz·ing, -niz·es.** To liberate (e.g., a colony) from dependency. **—de·col'o·ni·za'tion** *n.*

**de·col·or·ant** (dē-kŭl'ər-ənt) *n.* A bleaching agent.

**de·col·or·ize** (dē-kŭl'ə-rīz') *vt.* **-ized, -iz·ing, -iz·es.** To remove the color from. **—de·col'or·i·za'tion** *n.* **—de·col'or·iz'er** *n.*

**de·com·pose** (dē'kəm-pōz') *v.* **-posed, -pos·ing, -pos·es.** [Fr. *décomposer* : *dé-,* de- (< Lat.) + *composer,* to compose < OFr. —see COMPOSE.] —*vt.* **1.** To separate into component parts. **2.** To cause to decay. —*vi.* **1.** To break down into component parts : DISINTEGRATE. **2.** To putrefy : decay. **—de'com·pos'a·ble** *adj.* **—de'com·pos'er** *n.*

**de·com·po·si·tion** (dē-kŏm'pə-zĭsh'ən) *n.* **1.** The act or result of decomposing. **2. a.** *Chem.* Separation into constituents by chemical reaction. **b.** *Biol.* Organic decay. **—de·com'po·si'tion·al** *adj.*

**de·com·pound** (dē-kŏm'pound', dē'kəm-pound') *adj.* **1.** Compounded or made up of things or parts that are already compound. **2.** *Bot.* Having or composed of subdivided or compound leaflets.

**de·com·press** (dē'kəm-prĕs') *vt.* **-pressed, -press·ing, -press·es. 1.** To relieve of pressure. **2.** To bring (a person working in compressed air) back to normal air pressure.

**de·com·pres·sion** (dē'kəm-prĕsh'ən) *n.* **1.** The act or process of decompressing. **2.** A surgical procedure used to relieve pressure on an organ or part.

**decompression sickness** *n.* Caisson disease.

**de·con·ges·tant** (dē'kən-jĕs'tənt) *n.* A medication or treatment that alleviates congestion, as of the sinuses.

**de·con·tam·i·nate** (dē'kən-tăm'ə-nāt') *vt.* **-nat·ed, -nat·ing, -nates. 1.** To free of contamination. **2.** To make safe by eliminating harmful substances, as noxious chemicals or radioactive material. **—de'con·tam'i·nant** *n.* **—de'con·tam'i·na'tion** *n.*

**de·con·trol** (dē'kən-trōl') *vt.* **-trolled, -trol·ling, -trols.** To free from control, esp. from governmental control. **—de'con·trol'** *n.*

**dé·cor** *also* **de·cor** (dā'kôr', dā-kôr') *n.* [Fr. < *décorer*, to decorate < Lat. *decorare.*] **1.** A decorative style or scheme, as of a room. **2.** A stage setting : SCENERY.

**dec·o·rate** (dĕk'ə-rāt') *vt.* **-rat·ed, -rat·ing, -rates.** [Lat. *decorare, decorat-* < *decus*, ornament.] **1.** To furnish or adorn with fashionable or beautiful things. **2.** To confer an emblem of honor upon <was *decorated* for bravery>

**dec·o·ra·tion** (dĕk'ə-rā'shən) *n.* **1.** The act, process, technique, or art of decorating. **2.** Objects or material used to decorate. **3.** An emblem of honor, as a medal or badge.

**Decoration Day** *n.* Memorial Day.

**dec·o·ra·tive** (dĕk'ər-ə-tĭv, -ə-rā'-) *adj.* Serving to decorate : ORNAMENTAL. **—dec'o·ra·tive·ly** *adv.* **—dec'o·ra·tive·ness** *n.*

**dec·o·ra·tor** (dĕk'ə-rā'tər) *n.* One that decorates, esp. an interior decorator.

**dec·o·rous** (dĕk'ər-əs, dĭ-kôr'əs, -kōr'-) *adj.* [Lat. *decorus* < *decor*, seemliness, beauty.] Marked by or displaying decorum : SEEMLY. **—dec'o·rous·ly** *adv.* **—dec'o·rous·ness** *n.*

**de·cor·ti·cate** (dē-kôr'tĭ-kāt') *vt.* **-cat·ed, -cat·ing, -cates.** [Lat. *decorticare, decorticat-* : *de-*, off + *cortex*, bark, rind.] **1.** To remove the cortex from (an organ or structure), esp. in surgery. **2.** To remove the bark, husk, or outer layer from : PEEL. **—de·cor'ti·ca'tion** *n.* **—de·cor'ti·ca'tor** *n.*

**de·co·rum** (dĭ-kôr'əm, -kōr'-) *n.* [Lat. *decorum* < *decorus*, decorous.] **1.** Suitability of behavior or conduct : PROPRIETY. **2.** The conventions of polite behavior. **3.** Something proper to the harmony, essence, or unity of a literary or artistic composition.

**de·cou·page** *also* **dé·cou·page** (dā'ko͞o-päzh') *n.* [Fr. < OFr. *decouper*, to cut out : *de-*, away (< Lat.) + *couper*, to cut < *coup*, stroke. —see COUP.] **1.** The technique of decorating a surface with cutouts, as of paper. **2.** A product of decoupage.

**de·cou·ple** (dē-kŭp'əl) *vt.* **-pled, -pling, -ples. 1.** *Electron.* To reduce or eliminate the coupling of circuits or mechanical parts. **2.** *Physics.* To decrease the seismic effect of (an explosion) by carrying out in an underground cavity. **—de·cou'pler** *n.*

**de·coy** (dē'koi', dĭ-koi') *n.* [Poss. < Du. *de kooi*, the cage : *de*, the + *kooi*, cage < MDu. *cōie* < Lat. *cavea* < *cavus*, hollow.] **1.** An enclosed place, as a pond, into which wildfowl are lured for capture. **2.** A living or artificial bird or other animal used to entice game into a trap or within shooting range. **3.** A means used to trap, mislead, or lure into danger. *—vt.* (dĭ-koi') **-coyed, -coy·ing, -coys.** To lure into danger or a trap by or as if by a decoy. **—de·coy'er** *n.*

**de·crease** (dĭ-krēs') *vi.* & *vt.* **-creased, -creas·ing, -creas·es.** [ME *decresen* < OFr. *decreistre, decreiss-* < Lat. *decrescere* : *de-*, from, away + *crescere*, to grow.] To grow or cause to grow gradually smaller or less, as in number, amount, or intensity. *—n.* (dē'krēs'). **1.** The act or process of decreasing. **2.** An amount of decreasing. **—de·creas'ing·ly** *adv.*

☆ ***syns:*** DECREASE, ABATE, DIMINISH, DWINDLE, EBB, LESSEN, REDUCE, TAPER (off) *v. core meaning* : to grow or cause to grow gradually less <an appetite that *decreased*><pain slowly *decreasing*> ***ant:*** increase

**de·cree** (dĭ-krē') *n.* [ME < OFr. *decret* < Lat. *decretum* < *decernere*, to decide : *de-*, away + *cernere*, to sift.] **1.** An authoritative order having legal force. **2.** The judgment of a court of equity, admiralty, probate, or divorce. **3.** *Rom. Cath. Ch.* **a.** A disciplinary or doctrinal act of an ecumenical council. **b.** An administrative act applying or interpreting articles of canon law. *—v.* **-creed, -cree·ing, -crees.** *—vt.* To ordain, establish, or decide by decree. *—vi.* To issue a decree. **—de·cree'a·ble** *adj.* **—de·cre'er** *n.*

**de·cree-law** (dĭ-krē'lô') *n.* A decree having the force of a law enacted by a legislature but usu. issued on the sole authority of an absolute ruler or the executive branch of a government.

**dec·re·ment** (dĕk'rə-mənt) *n.* [Lat. *decrementum* < *decrescere*, to decrease.] **1.** The act or process of becoming gradually less : DECREASE. **2.** The amount lost by gradual waste or diminution. **3.** *Math.* The amount by which a variable is decreased : negative increment. **4.** *Computer Sci.* A specific part of an instruction word. **—dec're·ment'al** (-mĕn'tl) *adj.*

**de·crep·it** (dĭ-krĕp'ĭt) *adj.* [ME < OFr. < Lat. *decrepitus*, very old : *de-*, from, without + *crepitus*, p.part. of *crepare*, to rattle.] Broken-down or worn-out, as from long use or old age. **—de·crep'it·ly** *adv.* **—de·crep'i·tude'** (-to͞od', -tyo͞od') *n.*

**de·crep·i·tate** (dĭ-krĕp'ĭ-tāt') *v.* **-tat·ed, -tat·ing, -tates.** [Med. Lat. *decrepitare, decrepitat-* : Lat. *de-* (intensive) + *crepitare*, freq. of *crepare*, to crack.] *—vt.* To roast or calcine (crystals or salts) until a crackling sound is produced or until crackling stops. *—vi.* To make a crackling sound when roasted. **—de·crep'i·ta'tion** *n.*

**de·cre·scen·do** (dā'krə-shĕn'dō, dē'-) [Ital., decreasing < Lat. *decrescendum*, gerund of *decrescere*, to decrease.] *Mus.* *—n., pl.* **-dos. 1.** A gradual decrease in force or loudness. **2.** A passage marked or performed in a decrescendo. *—adj.* Gradually diminishing in force or loudness. **—de'cre·scen'do** *adv.*

**de·cres·cent** (dĭ-krĕs'ənt) *adj.* [Lat. *decrescens, decrescent-*, pr.part. of *decrescere*, to decrease.] Diminishing or becoming gradually less : WANING.

**de·cre·tal** (dĭ-krēt'l) *n.* [ME < OFr. < Med. Lat. (*epistola*) *decretalis*, (letter) of decree < Lat. *decretum*, decree.] *Rom. Cath. Ch.* **1.** A decree, esp. a papal letter giving a decision on a point or question of canon law. **2. Decretals.** The body of papal laws and decrees forming a part of canon law.

**de·cre·tive** (dĭ-krē'tĭv) *adj.* Having the force of a decree.

**dec·re·to·ry** (dĕk'rĭ-tôr'ē, -tōr'ē, dĭ-krē'tə-rē) *adj.* Of or resulting from a decree.

**de·crim·i·nal·ize** (dē-krĭm'ə-nə-līz') *vt.* **-ized, -iz·ing, -iz·es.** To make no longer criminal or illegal. **—de·crim'i·nal·i·za'tion** *n.*

**de·cry** (dĭ-krī') *vt.* **-cried, -cry·ing, -cries.** [Fr. *décrier* < OFr. *descrier* : *des-*, down (< Lat. *de-*) + *crier*, to cry. —see CRY.] **1.** To belittle or disparage openly : DENOUNCE. **2.** To depreciate (e.g., currency) by official proclamation or by rumor. **—de·cri'er** *n.*

**de·cum·bent** (dĭ-kŭm'bənt) *adj.* [Lat. *decumbens, decumbent-*, pr.part. of *decumbere*, to lie down.] **1.** Reclining : prostrate. **2.** *Bot.* Lying or growing along the ground but erect at or near the apex, as some stems. **—de·cum'bence** (-bəns), **de·cum'ben·cy** (-bən-se) *n.*

**dec·u·ple** (dĕk'yə-pəl) *adj.* [ME < OFr. < LLat. *decuplus* < Lat. *decem*, ten.] **1.** Ten times as great : TENFOLD. **2.** In groups of ten.

**de·cur·rent** (dĭ-kûr'ənt, -kŭr'-) *adj.* [Lat. *decurrens, decurrent-*, pr.part. of *decurrere*, to run down : *de-*, down + *currere*, to run.] *Bot.* Extending downward from the base along a stem, as some leaves. **—de·cur'rent·ly** *adv.*

**de·cus·sate** (dĭ-kŭs'āt', dĕk'ə-sāt') *vt.* & *vi.* **-sat·ed, -sat·ing, -sates.** [Lat. *decussare, decussat-* < *decussis*, the number 10, intersection (from the Romans' use of X for the numeral 10) : *decem*, ten + *as*, unit.] To intersect so as to form an X : CROSS. *—adj.* **1.** Intersected or crossed in the form of an X. **2.** *Bot.* Arranged on a stem in opposite pairs at right angles to those above or below. **—de·cus'sate'ly** *adv.*

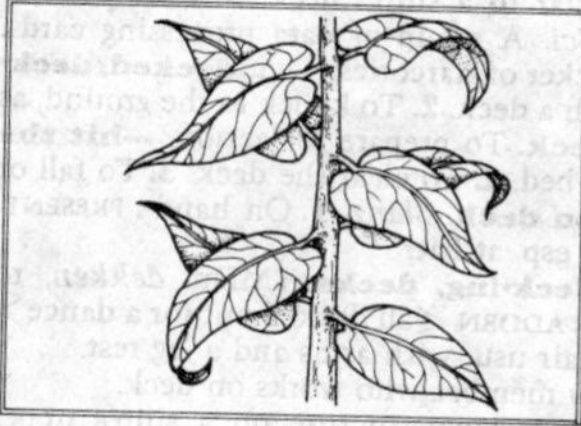

**decussate**
*Decussate leaves*

**dec·us·sa·tion** (dĕk'ə-sā'shən, dē'kə-) *n.* **1.** Intersection in the form of an X. **2.** An X-shaped crossing of nerve fibers connecting dissimilar parts on the two sides of the spinal cord or brain.

**de·dans** (də-däN') *n., pl.* **dedans** (-däN', -däNz'). [Fr. < *dedans*, inside : *de*, from (< Lat.) + *dans*, within < LLat. *deintus* (*de*, from + *intus*, within).] **1.** A screened gallery for spectators at the service end of a court-tennis court. **2.** The spectators at a court-tennis match.

**ded·i·cate** (dĕd'ĭ-kāt') *vt.* **-cat·ed, -cat·ing, -cates.** [ME *dedicaten* < Lat. *dedicare* : *de-*, apart + *dicare*, to say.] **1.** To set apart for a deity or for religious purposes : CONSECRATE. **2.** To set apart for a special use <*dedicated* the funds to cancer research> **3.** To commit (oneself) to a particular course of thought or action <*dedicated* ourselves to teaching> **4.** To address or inscribe (e.g., a literary work) to someone as a mark of respect or affection. **5. a.** To open (e.g., a building) to public use. **b.** To unveil (a monument). **—ded'i·ca·tee'** (-kə-tē') *n.* **—ded'i·ca'tive, ded'i·ca·to'ry** (-kə-tôr'ē, -tōr'ē) *adj.* **—ded'i·ca'tor** *n.*

**ded·i·cat·ed** (dĕd'ĭ-kā'tĭd) *adj.* Used for a single, special electronic business application <a *dedicated* word processor>

**ded·i·ca·tion** (dĕd'ĭ-kā'shən) *n.* **1.** The act of dedicating or state of being dedicated. **2.** A name often accompanied by a message of dedication, prefixed to a literary, artistic, or musical composition. **3.** A rite or ceremony of dedicating.

**de·dif·fer·en·ti·a·tion** (dē'dĭf-ə-rĕn'shē-ā'shən) *n. Biol.* Loss of specialized cellular form, esp. prior to redifferentiation.

**de·duce** (dĭ-do͞os', -dyo͞os') *vt.* **-duced, -duc·ing, -duc·es.** [ME *deducen* < Lat. *deducere*, to lead away : *de-*, away + *ducere*, to lead.] **1.** To reach (a conclusion) by reasoning. **2.** To infer from a general principle : reason deductively. **3.** To trace the origin or derivation of. **—de·duc′i·ble** *adj.*

▲ word history: *Deduce* and *deduct* were both borrowed in the early 16th century from Latin *deducere,* "to lead or bring away." *Deducere* also meant "to subtract" in classical Latin and developed the sense "to derive a logical conclusion from" in Medieval Latin. In English *deduce* and *deduct* were at first used interchangeably with both meanings of the Latin word but gradually *deduce* was restricted to "derive" and *deduct* was used primarily to mean "subtract." *Deduction,* however, still serves as the noun for both.

**de·duct** (dĭ-dŭkt′) *v.* **-duct·ed, -duct·ing, -ducts.** [Lat. *deducere, deduct-,* to lead away. —see DEDUCE.] —*vt.* **1.** To take away (a quantity) from another : SUBTRACT. **2.** To derive by deduction : DEDUCE. —*vi.* To diminish : detract <Faulty brakes will *deduct* from the value of a used car.>

**de·duct·i·ble** (dĭ-dŭk′tə-bəl) *adj.* **1.** Capable of being deducted. **2.** Allowable as a tax deduction. **—de·duct′i·bil′i·ty** *n.*

**de·duc·tion** (dĭ-dŭk′shən) *n.* **1.** The act of deducting : SUBTRACTION. **2.** An amount that is or may be deducted. **3. a.** The drawing of a conclusion by reasoning. **b.** *Logic.* The process of reasoning in which a conclusion follows necessarily from the stated premises : inference by reasoning from the general to the specific. **c.** *Logic.* A conclusion reached by this process.

**de·duc·tive** (dĭ-dŭk′tĭv) *adj.* **1.** Of or based on deduction. **2.** Involving deduction in reasoning. **—de·duc′tive·ly** *adv.*

**dee** (dē) *n.* The letter *d.*

**deed** (dēd) *n.* [ME *dede* < OE *dǣd.*] **1.** An act. **2.** An exploit : feat. **3.** Action or performance in general. **4.** *Law.* A document sealed as an instrument of bond, contract, or conveyance, esp. pertaining to property. —*vt.* **deed·ed, deed·ing, deeds.** To transfer by means of a deed.

**deem** (dēm) *v.* **deemed, deem·ing, deems.** [ME *demen* < OE *dēman.*] —*vt.* To consider or judge <I *deem* it necessary to stay.> —*vi.* To have an opinion : SUPPOSE.

**deep** (dēp) *adj.* **-er, -est.** [ME *dep* < OE *dēop.*] **1. a.** Extending far downward below a surface <a *deep* well> **b.** Extending far backward from front to rear <a *deep* walk-in refrigerator> **c.** Extending far inward from an outer surface <a *deep* wound> **d.** Extending far from side to side from a center <*deep* ruffles on the collar> **e.** Far distant down or in <*deep* in the forest> **f.** Coming from or penetrating to a depth <take a *deep* breath> **2.** Extending a specific distance in a given direction <water six feet *deep*> **3.** Far distant in time or space <in the *deep* past> **4. a.** Difficult to penetrate or understand : RECONDITE <a *deep* film> **b.** Mysterious or obscure <*deep* rituals> **c.** Very learned or intellectual : WISE. **d.** Very crafty or cunning <*deep* machinations> **5. a.** Extremely serious <*deep* scandal> **b.** Very absorbed or involved <in *deep* meditation> **c.** Showing strong feelings : PROFOUND <a *deep* devotion> **6.** Rich and vivid in shade <a *deep* green> **7.** Low in pitch : RESONANT. —*adv.* **1.** To a great depth : DEEPLY <delve *deep*><hostility that ran *deep*> **2.** Well on in time : LATE. —*n.* **1.** A deep place in land or in a body of water, esp. in the ocean. **2. a.** The extent of encompassing time or space. **b.** A vast, immeasurable extent : ABYSS. **3.** The most intense or extreme part. **4.** The ocean. **5.** A distance estimated in fathoms between successive marks on a sounding line. **—in deep water.** In trouble. **—deep′ly** *adv.* **—deep′ness** *n.*

☆ ***syns:*** DEEP, ABSTRUSE, ESOTERIC, HEAVY, PROFOUND, RECONDITE *adj. core meaning* : beyond an average person's understanding <a *deep* book> ***ant:*** shallow

**deep·en** (dē′pən) *vt.* & *vi.* **-ened, -en·ing, -ens.** To make or become deep or deeper.

**Deep·freeze** (dēp′frēz′). A trademark for a refrigerator designed to freeze food for long periods.

**deep-fry** (dēp′frī′) *vt.* **-fried, -fry·ing, -fries.** To fry by immersing in a deep pan of fat or oil.

**deep-root·ed** (dēp′rōō′tĭd, -rŏŏt′ĭd) *adj.* Firmly implanted.

**deep-sea** (dēp′sē′) *adj.* Relating to deep parts of the sea.

**deep-seat·ed** (dēp′sē′tĭd) *adj.* Deeply entrenched : INGRAINED.

**deep-six** (dēp′sĭks′) *vt.* **-sixed, -six·ing, -six·es.** *Slang.* **1.** To toss overboard. **2.** To toss out : get rid of <*deep-sixed* the leftovers>

**deep space** *n.* The regions beyond the moon, encompassing interplanetary, interstellar, and intergalactic space.

**deep structure** *n.* An underlying structure determining the semantic interpretation of a sentence.

**deer** (dîr) *n., pl.* **deer.** [ME *der,* wild animal < OE *dēor.*] **1.** Any of various hoofed ruminant mammals of the family Cervidae, with deciduous antlers borne only by the males. **2.** Any of various smaller deerlike mammals, as the mouse deer.

**deer fly** *n.* A blood-sucking fly of the genus *Chrysops,* bearing dark bars or spots on the wings.

**deer grass** *n.* Meadow beauty.

**deer·hound** (dîr′hound′) *n.* A wiry-coated dog of a breed developed in Scotland, resembling but larger than a greyhound.

**deer mouse** *n.* [From its deerlike agility.] A large-eared, long-tailed New World mouse of the genus *Peromyscus,* with white feet and underparts.

**deer·skin** (dîr′skĭn′) *n.* **1.** Leather made from the hide of a deer. **2.** A garment made from deerskin.

**deer·stalk·er** (dîr-stô′kər) *n.* A tight-fitting hat with visors in the front and back, orig. worn by hunters.

**de-es·ca·late** (dē-ĕs′kə-lāt′) *v.* **-lat·ed, -lat·ing, -lates.** —*vt.* To decrease the size, scope, or intensity of (e.g., a war) —*vi.* To decrease or diminish in size, scope, or intensity. **—de-es′ca·la′tion** *n.* **—de-es′ca·la·tor′y** (-lə-tôr′ē, -tōr′ē) *adj.*

**deet** (dēt) *n.* [Pronunciation of *d.t.,* abbr. of DIETHYL TOLUAMIDE.] A colorless, oily, mild-smelling liquid, $C_{12}H_{17}NO$, used as an insect repellent.

**de·face** (dĭ-fās′) *vt.* **-faced, -fac·ing, -fac·es.** [ME *defacen* < OFr. *desfacier* : *des-,* de- (< Lat *de-*) + *face,* face. —see FACE.] **1.** To spoil or mar the surface or appearance of : DISFIGURE. **2.** To impair the usefulness, value, or influence of. **3.** To wipe out : OBLITERATE. **—de·face′a·ble** *adj.* **—de·face′ment** *n.* **—de·fac′er** *n.*

**de fac·to** (dĭ făk′tō, dā) *adv.* [Lat., according to the fact.] In reality or fact : ACTUALLY. —*adj.* **1.** Actual. **2.** Actually exercising power.

**de·fal·cate** (dĭ-făl′kāt′, -fôl′-, dĕf′əl-) *vi.* **-cat·ed, -cat·ing, -cates.** [Med. Lat. *defalcare, defalcat-* : Lat. *de-,* off + Lat. *falx,* sickle.] To misuse funds : EMBEZZLE. **—de′fal·ca′tion** (dē′făl-kā′shən, -fôl-, dĕf′-əl-) *n.* **—de·fal′ca′tor** *n.*

**def·a·ma·tion** (dĕf′ə-mā′shən) *n.* Libel or slander : CALUMNY. **—de·fam′a·to′ry** (dĭ-făm′ə-tôr′ē, -tōr′ē) *adj.*

**de·fame** (dĭ-fām′) *vt.* **-famed, -fam·ing, -fames.** [ME *defamen* < OFr. *defamer* < Lat. *diffamare* : *dis-,* apart + *fama,* reputation.] **1.** To attack the good name of by libel or slander. **2.** *Archaic.* To disgrace. **—de·fam′er** *n.*

**de·fault** (dĭ-fôlt′) *n.* [ME *defaute* < OFr. < VLat. ***defallita** : Lat. *de-* (intensive) + *fallere,* to fail.] **1.** Failure to perform a task or fulfill an obligation, esp. failure to meet a financial obligation. **2.** Failure to make a required appearance in court. **3.** Failure to participate in or complete a competition. —*v.* **-fault·ed, -fault·ing, -faults.** —*vi.* **1.** To fail to do what is required. **2.** To fail to pay money when it is due. **3.** *Law.* **a.** To fail to appear in court when summoned. **b.** To lose a case by not appearing. **4.** To fail to compete in or complete a competition. —*vt.* **1.** To fail to perform or pay. **2.** *Law.* To lose (a case) by failing to appear in court. **3.** To fail to participate in or complete (e.g., a competition). **—de·fault′er** *n.*

**de·fea·sance** (dĭ-fē′zəns) *n.* [ME *defesaunce* < AN < OFr. *defesance* < *defesant,* pr.part. of *desfaire,* to destroy. —see DEFEAT.] **1.** An annulment or invalidation. **2.** The nullification of a contract or deed. **3.** A clause within a contract or deed providing for annulment.

**de·fea·si·ble** (dĭ-fē′zə-bəl) *adj.* Capable of being annulled or terminated. **—de·fea′si·bil′i·ty, de·fea′si·ble·ness** *n.*

**de·feat** (dĭ-fēt′) *vt.* **-feat·ed, -feat·ing, -feats.** [ME *defeten* < *defet,* disfigured < OFr. *desfait,* p.part. of *desfaire,* to destroy < Med. Lat. *disfacere* : Lat. *dis-,* asunder + *facere,* to do.] **1.** To win victory over : BEAT. **2.** To prevent the success of : THWART <*defeated* my plan> **3.** *Law.* To annul or make void. —*n.* **1.** The act of defeating or state of being defeated. **2.** Failure to win. **3.** Frustration. **4.** *Law.* The act of making null and void. **—de·feat′er** *n.*

☆ ***syns:*** DEFEAT, BEAT, BEST, CLOBBER, CONQUER, DRUB, LICK, OVERCOME, ROUT, SHELLAC, SUBDUE, THRASH, TRIM, TRIUMPH (over), TROUNCE, VANQUISH, WHIP, WORST *v. core meaning* : to win a victory over <a team that *defeated* its rival>

**de·feat·ism** (dĭ-fē′tĭz′əm) *n.* Expectation or unresisting acceptance of the prospect of defeat. **—de·feat′ist** *n.*

**def·e·cate** (dĕf′ĭ-kāt′) *v.* **-cat·ed, -cat·ing, -cates.** [Lat. *defaecare, defaecat-* : *de-,* away + *faex,* dregs.] —*vi.* To void feces from the bowels. —*vt.* To clarify (a chemical solution). **—def′e·ca′tion** *n.* **—def′e·ca′tor** *n.*

**de·fect** (dē′fĕkt′, dĭ-fĕkt′) *n.* [ME < Lat. *defectus* < p.part. of *deficere,* to depart, fail : *de-,* from + *facere,* to do.] **1.** Insufficiency of something required or desirable for completion or perfection : DEFICIENCY. **2.** A fault or imperfection. —*vi.* (dĭ-fĕkt′) **-fect·ed, -fect·ing, -fects.** To desert one's country or party or a previously espoused cause. **—de·fec′tion** *n.* **—de·fec′tor** *n.*

☆ ***syns:*** DEFECT, BLEMISH, FAULT, FLAW, IMPERFECTION *n. core meaning* : something that mars the appearance of <a *defect* in the crystal>

**de·fec·tive** (dĭ-fĕk′tĭv) *adj.* **1.** Having a defect : FAULTY <*defective* wiring> **2.** Lacking one or more of the inflected forms normal for a particular category of word, as the verb *may* in English. **3.** Of subnormal intelligence. —*n.* Someone physically or mentally incapacitated. **—de·fec′tive·ly** *adv.* **—de·fec′tive·ness** *n.*

**de·fence** (dĭ-fĕns′) *n. Chiefly Brit. var. of* DEFENSE.

**de·fend** (dĭ-fĕnd′) *v.* **-fend·ed, -fend·ing, -fends.** [ME *defenden* < OFr. *defendre* < Lat. *defendere,* to ward off.] —*vt.* **1.** To protect from danger, attack, or harm : GUARD. **2.** To support or maintain, as by argument or action : JUSTIFY. **3.** *Law.* **a.** To represent (a defendant) in a court of law. **b.** To contest (a legal action or claim). —*vi.* To make a defense. **—de·fend′a·ble** *adj.* **—de·fend′er** *n.*

☆ ***syns:*** DEFEND, GUARD, PROTECT, SAFEGUARD, SECURE, SHIELD *v. core meaning* : to keep safe from danger, attack, or harm <*defend* a castle><*defend* the right of free speech> ***ant:*** attack

**de·fen·dant** (dĭ-fĕn′dənt) *n. Law.* One against whom an action is brought.

**de·fen·es·tra·tion** (dē-fĕn′ĭ-strā′shən) *n.* [DE- + Lat. *fenestra,* window.] An act of throwing something or someone out of a window. **—de·fen′es·trate′** *v.* **(-trat·ed, -trat·ing, -trates).**

**de·fense** (dĭ-fĕns′) *n.* [ME < OFr. < Lat. *defensa* < fem. p.part. of *defendere,* to ward off.] **1.** The act of defending against attack, danger, or injury : PROTECTION. **2.** Something that defends or protects. **3.** *Psychoanal.* A defense mechanism. **4.** An argument in support or justification. **5.** *Law.* **a.** The action of the defendant in opposition to complaints against him or her. **b.** The defendant and his or her legal counsel. **6.** The science or art of defending oneself : SELF-DEFENSE. **7.** The sports team or those players on the team attempting to keep the opposition from scoring. *—vt.* **-fensed, -fens·ing, -fens·es.** To attempt to keep (the opposition) from scoring in a sport. **—de·fense′less** *adj.* **—de·fense′less·ly** *adv.* **—de·fense′less·ness** *n.*

**defense mechanism** *n.* **1.** *Biol.* A reaction of an organism used in self-defense, as against germs. **2.** *Psychoanal.* A usu. involuntary mental mechanism, as repression or projection, that protects an individual from shame, anxiety, or loss of self-esteem.

**de·fen·si·ble** (dĭ-fĕn′sə-bəl) *adj.* Capable of being defended, protected, or justified. **—de·fen′si·bil′i·ty, de·fen′si·ble·ness** *n.* **—de·fen′si·bly** *adv.*

**de·fen·sive** (dĭ-fĕn′sĭv) *adj.* **1.** Appropriate or meant for defense. **2.** Done for defense. **3.** Of or relating to defense. *—n.* **1.** A means of defense. **2.** An attitude of defense <on the *defensive*> **—de·fen′sive·ly** *adv.* **—de·fen′sive·ness** *n.*

**de·fer[1]** (dĭ-fûr′) *v.* **-ferred, -fer·ring, -fers.** [ME *differren* < OFr. *diferer* < Lat. *differre.* —see DIFFER.] *—vt.* **1.** To put off until a future time : POSTPONE. **2.** To postpone the induction of (one eligible for the military draft). *—vi.* To delay : procrastinate. **—de·fer′ra·ble** *adj.* **—de·fer′rer** *n.*

**de·fer[2]** (dĭ-fûr′) *vi.* **-ferred, -fer·ring, -fers.** [ME *deferen* < OFr. *defere* < Lat. *deferre,* to carry away, bring to : *de-,* away + *ferre,* to carry.] To comply with or submit to the wishes, opinion, or decision of another. **—de·fer′rer** *n.*

**def·er·ence** (dĕf′ər-əns, dĕf′rəns) *n.* Courteous respect for or submission to another's opinion, wishes, or judgment.

**def·er·ent[1]** (dĕf′ər-ənt, dĕf′rənt) *adj.* Deferential.

**def·er·ent[2]** (dĕf′ər-ənt, dĕf′rənt) *adj.* [Lat. *deferens, deferent-,* pr.part. of *deferre,* to carry away. —see DEFER².] **1.** Carrying down or away. **2.** Adapted to carry or transport.

**def·er·en·tial** (dĕf′ə-rĕn′shəl) *adj.* Marked by deference <a *deferential* attitude toward their parents> **—def′er·en′tial·ly** *adv.*

**de·fer·ment** (dĭ-fûr′mənt) *n. also* **de·fer·ral** (-fûr′əl) *n.* The act of delaying : POSTPONEMENT.

**de·ferred** (dĭ-fûrd′) *adj.* **1.** Delayed or postponed. **2.** With benefits or payments withheld until a future date. **3.** Having had one's compulsory military service postponed.

**de·fer·ves·cence** (dē′fər-vĕs′əns) *n.* [< Lat. *defervescens, defervescent-,* pr.part. of *defervescere,* to stop boiling : *de-,* off, away + *fervescere,* to begin to boil < *fervēre,* to boil.] The abatement of a fever.

**de·fi·ance** (dĭ-fī′əns) *n.* [ME *defiaunce* < OFr. *desfiance* < *desfier,* to defy.] **1.** Bold resistance to an opposing force or authority. **2.** Deliberately provocative behavior or attitude.

**de·fi·ant** (dĭ-fī′ənt) *adj.* Marked by defiance. **—de·fi′ant·ly** *adv.*

**de·fib·ril·late** (dē-fĭb′rə-lāt′) *vt.* **-lat·ed, -lat·ing, -lates.** To stop the fibrillating or restore the normal rhythm of (a heart). **—de·fib′ril·la′tion** *n.* **—de·fib′ril·la′tive** *adj.* **—de·fib′ril·la′tor** *n.* **—de·fib′ril·la·to′ry** (-lə-tôr′ē, -tōr′ē) *adj.*

**de·fi·cien·cy** (dĭ-fĭsh′ən-sē) *n., pl.* **-cies.** **1.** The quality or condition of being deficient. **2.** A shortage or lack : INSUFFICIENCY.

☆ ***syns:*** DEFICIENCY, INSUFFICIENCY, LACK, SCARCITY, SHORTAGE *n. core meaning* : inadequacy in amount or degree <a *deficiency* of fuel oil> ***ant:*** sufficiency

**deficiency disease** *n.* A disease, as scurvy or rickets, resulting from a dietary deficiency of specific vitamins and minerals.

**de·fi·cient** (dĭ-fĭsh′ənt) *adj.* [Lat. *deficiens, deficient-,* pr.part. of *deficere,* to fail. —see DEFECT.] **1.** Lacking an essential quality or element <*deficient* in tact> **2.** Inadequate in amount or degree : INSUFFICIENT <*deficient* rainfall> **—de·fi′cient·ly** *adv.*

**def·i·cit** (dĕf′ĭ-sĭt) *n.* [Fr. *déficit* < Lat. *deficit,* it is lacking.] **1.** The amount by which something, as a sum of money, falls short of the required or expected amount : SHORTAGE. **2.** A lack or impairment <a visual *deficit*>

**deficit spending** *n.* The spending of public funds obtained by borrowing.

**def·i·lade** (dĕf′ə-lād′, -läd′) *vt.* **-lad·ed, -lad·ing, -lades.** [DE- + (EN)FILADE.] To arrange (fortifications) so as to give protection from enfilading and other fire. *—n.* The act or procedure of defilading.

**de·file[1]** (dĭ-fīl′) *vt.* **-filed, -fil·ing, -files.** [ME *defilen,* blend of *filen,* to defile (< OE *fȳlan*) and *defoulen,* to injure < OFr. *defouler* : *de-,* down (< Lat.) + *fouler,* to trample. —see FULL².] **1.** To make dirty : POLLUTE. **2.** To render impure : CORRUPT. **3.** To profane or sully (e.g., a good reputation). **4.** To make unclean or unfit for ceremonial use : DESECRATE. **5.** To violate the chastity of. **—de·file′ment** *n.* **—de·fil′er** *n.* **—de·fil′ing·ly** *adv.*

**de·file[2]** (dĭ-fīl′) *vi.* **-filed, -fil·ing, -files.** [Fr. *défiler* : *dé-,* off (< Lat. *de-*) + *filer,* to march in files < OFr., to spin < LLat. *filare* < Lat. *filum,* thread.] To march in single file or in files or columns. *—n.* **1.** A narrow gorge or pass that prevents easy passage, as of troops. **2.** A march in single file.

**de·fine** (dĭ-fīn′) *v.* **-fined, -fin·ing, -fines.** [ME *diffinen* < OFr. *definer* < Lat. *definire,* to limit : *de-,* off + *finis,* end.] *—vt.* **1.** To state the precise meaning of (e.g., a word or sense of a word). **2.** To describe the nature or basic qualities of : EXPLAIN <*define* the purpose of the new law> **3.** To delineate the outline or form of. **4.** To specify or fix distinctly <*define* the materials required for the project> **5.** To serve to distinguish : CHARACTERIZE <*define* oneself by one's beliefs> *—vi.* To make a definition. **—de·fin′a·bil′i·ty** *n.* **—de·fin′a·ble** *adj.* **—de·fin′a·bly** *adv.* **—de·fine′ment** *n.* **—de·fin′er** *n.*

**de·fin·i·en·dum** (dĭ-fĭn′ē-ĕn′dəm) *n., pl.* **-da** (-də) [Lat., neuter gerund. of *definire,* to define.] A word or expression defined by a definiens.

**de·fin·i·ens** (dĭ-fĭn′ē-ĕnz′) *n., pl.* **-fin·i·en·ti·a** (-fĭn′ē-ĕn′shē-ə, -shə) [Lat., pr.part. of *definire,* to define.] The word or words serving to define another word or expression, as in a dictionary entry.

**def·i·nite** (dĕf′ə-nĭt) *adj.* [ME *diffinite* < Lat. *definitus,* p.part. of *definire,* to define.] **1.** Having distinct limits <*definite* constraints on their power> **2.** Certain : sure <a *definite* success> **3.** Clearly defined : PRECISE. **4.** Limiting or particularizing <the *definite* article> **5.** *Bot.* **a.** Of a specified number not exceeding 20, as certain floral organs, esp. stamens. **b.** Cymose : determinate. **—def′i·nite·ly** *adv.* **—def′i·nite·ness** *n.*

☆ ***syns:*** DEFINITE, CATEGORICAL, CLEAR-CUT, DECIDED, EXPLICIT, EXPRESS, POSITIVE, PRECISE, SPECIFIC, UNAMBIGUOUS, UNEQUIVOCAL *adj. core meaning* : clearly, fully, and sometimes emphatically expressed <a *definite* denial of the rumor>

**definite article** *n.* An article, as *the,* that restricts or particularizes the noun or noun phrase following it.

**definite integral** *n.* The limit of sums with terms of the form $f(x_i)\Delta x_i$, where $f$ is a function defined in the interval between two numbers $a$ and $b$, $\Delta x_i$ is the length of one of several intervals into which the interval from $a$ to $b$ is divided, $x_i$ is a number in that interval, and the limit is taken as the lengths of the subintervals become smaller.

**def·i·ni·tion** (dĕf′ə-nĭsh′ən) *n.* [ME *diffinicioun* < OFr. *definition* < Lat. *definitio* < *definire,* to define.] **1.** The act of stating a precise meaning or significance. **2.** The statement of the meaning of a word, phrase, or expression. **3.** The act of making clear and distinct <a *definition* of my objectives> **4.** The state of being closely outlined or determined <The moon gave the shadows *definition.*> **5.** A determination of outline, extent, or limits <the *definition* of a leader's jurisdiction> **6.** Degree of clarity with which a televised image or radio broadcast is received. **7.** Clarity of detail in an optically produced image, as a photograph, effected by a combination of contrast and resolution. **—def′i·ni′tion·al** *adj.*

**de·fin·i·tive** (dĭ-fĭn′ĭ-tĭv) *adj.* **1.** Precisely outlining or defining : EXPLICIT. **2.** Determining finally : DECISIVE <a *definitive* maneuver> **3.** Complete and authoritative, as a biographical work. *—n.* A word that defines or limits, as the definite article or a demonstrative pronoun. **—de·fin′i·tive·ly** *adv.* **—de·fin′i·tive·ness** *n.*

**def·la·grate** (dĕf′lə-grāt′) *vi.* & *vt.* **-grat·ed, -grat·ing, -grates.** [Lat. *deflagrare, deflagrat-* : *de-* (intensive) + *flagrare,* to burn.] To burn or cause to burn with great heat and intense light. **—def′la·gra′tion** *n.*

**de·flate** (dĭ-flāt′) *v.* **-flat·ed, -flat·ing, -flates.** [DE- + (IN)FLATE.] *—vt.* **1. a.** To release contained air or gas from. **b.** To collapse by releasing contained air or gas. **2.** To reduce or lessen in importance or magnitude <*deflate* someone's pride> **3.** To reduce (currency) in value or amount, causing a decline in prices. *—vi.* To be or become deflated. **—de·fla′tor** *n.*

**de·fla·tion** (dĭ-flā′shən) *n.* **1.** The act of deflating or state of being deflated. **2.** A reduction in available currency and credit that causes a decrease in the general price level. **—de·fla′tion·ar′y** (-shə-nĕr′ē) *adj.* **—de·fla′tion·ist** *n.*

**de·flect** (dĭ-flĕkt′) *vi.* & *vt.* **-flect·ed, -flect·ing, -flects.** [Lat. *deflectere* : *de-,* away + *flectere,* to bend.] To turn aside or cause to turn aside : SWERVE. **—de·flect′a·ble** *adj.* **—de·flec′tive** *adj.* **—de·flec′tor** *n.*

**de·flec·tion** (dĭ-flĕk′shən) *n.* **1.** The act of deflecting or state of being deflected. **2.** Deviation or an amount of deviation. **3.** Deviation from zero shown by the indicator of a measuring instrument. **4.** Movement of a structure or structural part due to stress.

**de·flexed** (dĭ-flĕkst′, dē′flĕkst′) *adj.* [< Lat. *deflexus,* p.part. of *deflectere,* to deflect.] *Bot.* Bent or turned downward at a sharp angle <*deflexed* branches>

**de·flex·ion** (dĭ-flĕk′shən) *n. Chiefly Brit. var. of* DEFLECTION.

**def·lo·ra·tion** (dĕf′lə-rā′shən) *n.* [ME *defloracioun* < LLat. *defloratio* < *deflorare,* to deflower.] The act of deflowering.

**de·flow·er** (dē-flou′ər) *vt.* **-ered, -er·ing, -ers.** [ME *deflouren* < OFr. *deflorer* < LLat. *deflorare* : Lat. *de-*, away + Lat. *flos*, flower.] **1.** To rupture the hymen of (a virgin) by sexual intercourse. **2.** To destroy the innocence of : VIOLATE. **3.** To spoil the nature or appearance of : MAR. **—de·flow′er·er** *n.*

**de·fo·cus** (dē-fō′kəs) *vt.* **-cused, -cus·ing, -cus·es** *also* **-cussed, -cus·sing, -cus·ses.** To cause (a beam or a lens) to deviate from an accurate focus. —*n.* The result of defocusing.

**de·fog** (dē-fŏg′, -fôg′) *vt.* **-fogged, -fog·ging, -fogs.** To remove fog from. **—de·fog′ger** *n.*

**de·fo·li·ant** (dē-fō′lē-ənt) *n.* A chemical sprayed or dusted on plants or trees to cause the leaves to fall off.

**de·fo·li·ate** (dē-fō′lē-āt′) *v.* **-at·ed, -at·ing, -ates.** [LLat. *defoliare, defoliat-* : Lat. *de-*, off + Lat. *folium*, leaf.] —*vt.* **1.** To deprive (a tree or plant) of leaves. **2.** To cause the leaves of (a tree or plant) to fall off, esp. by the use of a defoliant. —*vi.* To lose foliage. **—de·fo′li·ate** (-ĭt) *adj.* **—de·fo′li·a′tion** *n.* **—de·fo′li·a′tor** *n.*

**de·force** (dē-fôrs′, -fōrs′) *vt.* **-forced, -forc·ing, -forc·es.** [ME *deforcen* < AN *deforcier* : *de-*, away (< Lat.) + *forcier*, to force < VLat. **fortiare* < Lat. *fortis*, strong.] *Law.* To withhold (e.g., property) by force from the rightful owner. **—de·force′ment** *n.*

**de·for·ciant** (dĭ-fôr′shənt, -fōr′-) *n.* *Law.* One that deforces a rightful owner.

**de·for·est** (dē-fôr′ĭst, -fŏr′-) *vt.* **-est·ed, -est·ing, -ests.** To clear away the trees or forests from. **—de·for′es·ta′tion** (-ĭ-stā′shən) *n.* **—de·for′est·er** *n.*

**de·form** (dĭ-fôrm′) *v.* **-formed, -form·ing, -forms.** [ME *deformen* < OFr. *deformer* < Lat. *deformare* : *de-*, off + *forma*, form.] —*vt.* **1.** To spoil the natural form of : MISSHAPE <trees *deformed* by the north wind> **2.** To mar the appearance of : DISFIGURE <hands *deformed* by arthritis> **3.** *Physics.* To alter the shape of by pressure or stress. —*vi.* To become deformed. **—de·form′a·bil′i·ty** *n.* **—de·form′a·ble** *adj.*

**de·for·ma·tion** (dē′fôr-mā′shən, dĕf′ər-) *n.* **1.** The act of deforming or state of being deformed. **2.** A change for the worse. **3.** *Physics.* **a.** Alteration of shape by pressure or stress. **b.** The altered shape resulting from deformation.

**de·formed** (dĭ-fôrmd′) *adj.* Misshapen or distorted.

**de·for·mi·ty** (dĭ-fôr′mĭ-tē) *n., pl.* **-ties. 1.** The state of being deformed. **2.** A bodily malformation. **3.** One that is deformed. **4.** Gross ugliness or distortion, esp. in art or morals.

**de·fraud** (dĭ-frôd′) *vt.* **-fraud·ed, -fraud·ing, -frauds.** [ME *defrauden* < OFr. *defrauder* < Lat. *defraudare* : *de-* (intensive) + *fraudare*, to cheat < *fraus*, fraud.] To take from by fraud : SWINDLE. **—de′fraud·a′tion** (dē′frô-dā′shən) *n.* **—de·fraud′er** *n.*

**de·fray** (dĭ-frā′) *vt.* **-frayed, -fray·ing, -frays.** [Fr. *défrayer* < OFr. *desfrayer* : *des-*, away (< Lat. *de-*) + **frai*, expense < Lat. *fractum*, neuter p.part. of *frangere*, to break.] To pay or provide for payment of <*defray* expenses> **—de·fray′a·ble** *adj.* **—de·fray′al** *n.*

**de·frock** (dē-frŏk′) *vt.* **-frocked, -frock·ing, -frocks.** To unfrock, as a priest.

**de·frost** (dē-frôst′, -frŏst′) *v.* **-frost·ed, -frost·ing, -frosts.** —*vt.* **1.** To remove ice or frost from. **2.** To cause to thaw. —*vi.* **1.** To become free of ice or frost. **2.** To thaw out.

**de·frost·er** (dē-frô′stər, -frŏs′tər) *n.* A heating device designed to remove ice or frost or prevent its formation.

**deft** (dĕft) *adj.* **-er, -est.** [ME, gentle, humble, var. of *dafte*, foolish. —see DAFT.] Dexterous : skillful. **—deft′ly** *adv.* **—deft′ness** *n.*

**de·funct** (dĭ-fŭngkt′) *adj.* [Lat. *defunctus*, p.part. of *defungi*, to finish : *de-* (intensive) + *fungi*, to discharge.] No longer in existence or use. **—de·func′tive** *adj.* **—de·funct′ness** *n.*

**de·fuse** (dē-fyo͞oz′) *vt.* **-fused, -fus·ing, -fus·es. 1.** To remove the fuse from (an explosive device). **2.** To make less dangerous, tense, or hostile <*defuse* angry feelings>

**de·fy** (dĭ-fī′) *vt.* **-fied, -fy·ing, -fies.** [ME *defien* < OFr. *desfier* < VLat. **disfidare* : Lat. *dis-*, away + *fidere*, to trust.] **1.** To confront boldly or stand up to : CHALLENGE. **2.** To resist successfully : WITHSTAND <a disease that *defied* all remedies> **3.** To challenge or dare to perform something deemed impossible. **—de·fi′er** *n.*

**dé·ga·gé** (dā′gä-zhā′) *adj.* [Fr., p.part. of *dégager*, to disengage < OFr. *desgagier* : *des-*, from (< Lat.) + *gage*, pledge, of Germanic orig.] **1.** Nonchalant. **2.** Casual <a *dégagé* atmosphere>

**de·gas** (dē-găs′) *vt.* **-gassed, -gas·sing, -gas·ses** *or* **-gas·es.** To remove gas from.

**de·gauss** (dē-gous′) *vt.* **-gaussed, -gauss·ing, -gauss·es.** To neutralize the magnetic field of (e.g., a ship). **—de·gauss′er** *n.*

**de·gen·er·a·cy** (dĭ-jĕn′ər-ə-sē) *n., pl.* **-cies. 1.** The state of being degenerate. **2.** The process of degenerating. **3.** Degenerate behavior.

**de·gen·er·ate** (dĭ-jĕn′ər-ĭt) *adj.* [Lat. *degenerare, degenerat-* : *de-*, from + *genus*, race.] **1.** Having declined, as in nature or function, from a former or original state. **2.** Having fallen to a state below what is held to be normal or desirable, esp. in mental or moral qualities. **3.** Marked by or displaying degeneracy. **4.** *Physics.* Taking on several discrete values or states. —*n.* **1.** A morally degraded person. **2. a.** A person lacking or having progressively lost normative biological or psychological characteristics. **b.** A person displaying antisocial, esp. sexually deviant, behavior. —*vi.* (-ə-rāt′) **-at·ed, -at·ing, -ates. 1.** To decline from a former or original state : DETERIORATE <a neglected house *degenerating* with age> **2.** To fall below a normal or desirable state, esp. mentally or morally. **3.** To decline or decrease in quality. **4.** *Physics.* To undergo degeneration. **—de·gen′er·ate·ly** *adv.* **—de·gen′er·ate·ness** *n.* **—de·gen′er·a·tive** (-ə-tĭv) *adj.*

**de·gen·er·a·tion** (dĭ-jĕn′ə-rā′shən) *n.* **1.** The process of degenerating or state of being degenerate. **2.** *Biol.* The usu. irreversible deterioration of specific cells or organs with corresponding functional impairment, caused by injury or disease and often resulting in necrosis or death. **3.** *Electron.* Negative feedback of output power to an input signal in an amplifying circuit.

**de·glu·ti·nate** (dĭ-glo͞ot′n-āt′) *vt.* **-nat·ed, -nat·ing, -nates.** [Lat. *deglutinare, deglutinat-* : *de-*, away + *gluten*, glue.] To extract the gluten from <*deglutinate* wheat flour> **—de·glu′ti·na′tion** *n.*

**de·glu·ti·tion** (dē′glo͞o-tĭsh′ən) *n.* [Fr. < Lat. *deglutire*, to swallow down : *de-*, down + *glutire*, to gulp.] The act or process of swallowing. **—de·glu′ti·to′ry** (-tĭ-tôr′ē, -tōr′ē) *adj.*

**de·grad·a·ble** (dĭ-grā′də-bəl) *adj.* Capable of being chemically degraded.

**deg·ra·da·tion** (dĕg′rə-dā′shən) *n.* **1.** The act or process of degrading. **2.** Transition from a higher to a lower level or quality. **3.** The state of being degraded : DEGENERATION. **4.** *Geol.* A general lowering of the earth's surface by erosion or transportation in running water. **5.** *Chem.* Decomposition of a compound by stages, exhibiting well-defined intermediate products.

**de·grade** (dĭ-grād′) *vt.* **-grad·ed, -grad·ing, -grades.** [ME *degraden* < OFr. *degrader* < LLat. *degradare* : *de-*, down + *gradus*, step.] **1.** To reduce in grade, rank, or status, esp. to deprive of an office or dignity : DEMOTE. **2.** To lower in moral or intellectual character : DEBASE. **3.** To reduce in worth or value. **4.** To expose to contempt, dishonor, or disgrace. **5.** *Geol.* To lower or wear by erosion. **6.** *Chem.* To decompose (a compound) by stages. **—de·grad′er** *n.*

**de·grad·ed** (dĭ-grā′dĭd) *adj.* **1.** Reduced in rank, honor, or position. **2.** Reduced in quality or value. **3.** Having declined in moral qualities : DEPRAVED. **4.** Regarded as below normal standards of civilized life. **—de·grad′ed·ly** *adv.* **—de·grad′ed·ness** *n.*

**de·grad·ing** (dĭ-grā′dĭng) *adj.* Tending or meant to degrade : demeaning or debasing. **—de·grad′ing·ly** *adv.*

**de·gran·u·la·tion** (dē-grăn′yə-lā′shən) *n.* The process of losing granules <the *degranulation* of leukocytes>

**de·gree** (dĭ-grē′) *n.* [ME *degre* < OFr. < VLat. **degradus* : Lat. *de-*, down + Lat. *gradus*, step.] **1.** One of a series of steps or stages in a process, course of action, progression, or retrogression. **2.** The relative distance or a step in a direct hereditary line of descent or ascent. **3.** Relative social or official status, dignity, or position. **4.** Relative amount or intensity, as of a quality or attribute <a high *degree* of sensitivity> **5.** The measure or extent of a state of being, action, or relation <improved my playing to a great *degree*> **6.** A unit division of a temperature scale. **7.** *Math.* A unit of angular measure equal in magnitude to the central angle subtended by 1/360 of the circumference of a circle. **8.** A unit of latitude or longitude, 1/360 of a great circle. **9. a.** The greatest sum of the exponents of the variables in a term of a polynomial or polynomial equation. **b.** The exponent of the derivative of highest order in a differential equation in standard form. **10. a.** An academic title granted by a university or college to a student who has completed a given course of study. **b.** A similar title conferred as an honorary distinction. **11.** *Law.* A division or classification of a specific crime according to its seriousness. **12.** One of the grammatical forms used in the comparison of adjectives and adverbs. **13.** *Mus.* **a.** One of the seven notes of a diatonic scale. **b.** A space or line of the staff.

☆ ***syns:*** DEGREE, EXTENT, MAGNITUDE, MEASURE, PROPORTION *n. core meaning* : relative intensity or amount <various *degrees* of intelligence>

**de·gree-day** (dĭ-grē′dā′) *n.* **1.** An indication of the extent of departure from a standard of mean daily temperature. **2.** A unit used in estimating quantities of fuel and power consumption, based on a daily ratio of consumption and the mean temperature below 65°F.

**degree of freedom** *n.* **1.** *Statistics.* Any of the unrestricted, independent random variables that make up a statistic. **2.** *Physics.* **a.** Any of the minimum number of coordinates needed to specify completely the motion of a mechanical system. **b.** Any of the independent thermodynamic variables, as pressure, temperature, or composition, necessary to specify a system with a given number of phases and components.

**de·gres·sion** (dĭ-grĕsh′ən, dē-) *n.* [Med. Lat. *degressio*, descent < Lat. *degressus*, p.part. of *degredi*, to descend : *de-*, down + *gradi*, to step.] A decrease by steps or degrees : DESCENT. **—de·gres′sive** *adj.*

**de·gust** (dĭ-gŭst′, dē-) *vt.* **-gust·ed, -gust·ing, -gusts.** [Lat. *degustare* : *de-* (intensive) + *gustare*, to taste < *gustus*, taste.] To taste esp. with enjoyment : SAVOR. **—de′gus·ta′tion** (dē′gŭ-stā′shən) *n.*

**de·hisce** (dĭ-hĭs′) *vi.* **-hisced, -hisc·ing, -hisc·es.** [Lat. *dehiscere* : *de-*, off + *hiscere*, to split, inchoative of *hiare*, to be open.] To burst or split open along a line or slit, as do the ripe capsules or pods of some plants.

**de·his·cent** (dĭ-hĭs′ənt) *adj.* Opening at pores or by splitting to release seeds within a fruit or pollen from an anther. **—de·his′cence** *n.*

**de·horn** (dē-hôrn′) *vt.* **-horned, -horn·ing, -horns. 1.** To remove the horns from (an animal). **2.** To prevent growth in the horns of, as by cauterization.

**de·hu·man·ize** (dē-hyōō′mə-nīz′) *vt.* **-ized, -iz·ing, -iz·es. 1.** To deprive of human qualities or attributes. **2.** To make routine and mechanical <The assembly line *dehumanized* their work.> **—de·hu′man·i·za′tion** *n.*

**de·hu·mid·i·fy** (dē′hyōō-mĭd′ə-fī′) *vt.* **-fied, -fy·ing, -fies.** To remove atmospheric moisture from <*dehumidify* a basement> **—de′hu·mid′i·fi·ca′tion** *n.* **—de′hu·mid′i·fi′er** *n.*

**de·hy·drase** (dē-hī′drās′, -drāz′) *n. Biochem.* **1.** Dehydratase. **2.** Dehydrogenase.

**de·hy·dra·tase** (dē-hī′drə-tās′, -tāz′) *n. Biochem.* An enzyme that catalyzes the removal of oxygen and hydrogen from a metabolite in the ratio in which they occur in water.

**de·hy·drate** (dē-hī′drāt′) *v.* **-drat·ed, -drat·ing, -drates.** —*vt.* **1.** *Chem.* To eliminate water from or make anhydrous. **2.** To remove water from (e.g., fruit or vegetables) for preservation. —*vi.* To lose water or moisture. **—de·hy′dra′tor** *n.*

**de·hy·dra·tion** (dē′hī-drā′shən) *n.* **1.** Removal of water from a substance or compound. **2.** *Pathol.* Excessive water loss from the body or from an organ or bodily part.

**de·hy·dro·chlo·rin·ase** (dē-hī′drə-klôr′ə-nās′, -nāz′, -klōr′-) *n. Biochem.* An enzyme that catalyzes the removal of hydrogen and chlorine from a chlorinated hydrocarbon.

**de·hy·dro·chlo·rin·ate** (dē-hī′drə-klôr′ə-nāt′, -klōr′-) *vt.* **-at·ed, -at·ing, -ates.** *Biochem.* To remove hydrogen and chlorine or hydrogen chloride from. **—de·hy′dro·chlo′ri·na′tion** *n.*

**de·hy·dro·gen·ase** (dē′hī-drŏj′ə-nās′, -nāz′, dē-hī′drə-jə-) *n.* An enzyme that removes hydrogen from metabolites.

**de·hy·dro·ge·nate** (dē′hī-drŏj′ə-nāt′, dē-hī′drə-jə-) *vt.* **-nat·ed, -nat·ing, -nates.** To remove hydrogen from. **—de·hy′dro·ge·na′tion** *n.*

**de·hy·dro·ge·nize** (dē′hī-drŏj′ə-nīz′, dē-hī′drə-jə-) *vt.* **-nized, -niz·ing, -niz·es.** To dehydrogenate. **—de·hy′dro·ge′ni·za′tion** *n.*

**de·hyp·no·tize** (dē-hĭp′nə-tīz′) *vt.* **-tized, -tiz·ing, -tiz·es.** To arouse from a hypnotic state.

**de·ice** (dē-īs′) *vt.* **-iced, -ic·ing, -ic·es.** To rid or keep free of ice.

**de·ic·er** (dē-ī′sər) *n.* **1.** A device used on an aircraft in flight to remove surface ice or keep it from forming. **2.** A compound used to prevent the formation of ice, as on windshields.

**de·i·cide** (dē′ĭ-sīd′) *n.* [Lat. *deus,* god + -CIDE.] **1.** The murder of a god. **2.** One who murders a god.

**deic·tic** (dīk′tĭk) *adj.* [Gk. *deiktikos* < *deiktos,* able to show directly < *deiknunai,* to show.] **1.** *Logic.* Directly proving by argument. **2.** Specifying or pointing out, as a demonstrative pronoun. **—deic′ti·cal·ly** *adv.*

**de·if·ic** (dē-ĭf′ĭk) *adj.* [OFr. *deifique* < LLat. *deificus* : Lat. *deus,* god + *-ficus,* -fic.] **1.** Making or tending to make divine. **2.** Divine or godlike in nature.

**de·i·fi·ca·tion** (dē′ə-fĭ-kā′shən) *n.* **1. a.** The act or process of deifying. **b.** The state of having been deified. **2.** One embodying the qualities of a god.

**de·i·fy** (dē′ə-fī′) *vt.* **-fied, -fy·ing, -fies.** [ME *deifien* < OFr. *deifier* < LLat. *deificare* < *deificus,* deific.] **1.** To raise to divine rank. **2.** To worship or revere as a god. **3.** To exalt or idealize. **—de′i·fi′er** *n.*

**deign** (dān) *v.* **deigned, deign·ing, deigns.** [ME *deinen* < OFr. *deignier,* to regard as worthy < Lat. *dignari* < *dignus,* worthy.] —*vi.* To deem something barely worthy of oneself : CONDESCEND <wouldn't *deign* to wash the floor> —*vt.* To condescend to give or grant <*deigned* an interview>

**deil** (dēl) *n.* [Sc. < ME *dele,* alteration of *devel.* —see DEVIL.] *Scot.* **1.** DEVIL 1. **2.** A mischievous person : IMP.

**de·in·sti·tu·tion·al·ize** (dē-ĭn′stĭ-tōō′shə-nə-līz′, -tyōō′-) *vt.* **-ized, -iz·ing, -iz·es. 1.** To remove the status of an institution from. **2.** To enable (e.g., one who is disabled or mentally ill) to live away from an institution. **—de·in′sti·tu′tion·al·i·za′tion** *n.*

**Deir·dre** (dîr′drə, -drē) *n.* [OIr. *Deidru.*] *Ir. Myth.* A princess of Ulster who killed herself after her husband, Naoise, was murdered.

**de·ism** (dē′ĭz′əm) *n.* [Fr. *déisme* < Lat. *deus,* god.] Belief in the existence of God as the creator of the universe who after setting it in motion abandoned it, assumed no control over life, exerted no influence on natural phenomena, and gave no supernatural revelation. **—de′ist** *n.* **—de·is′tic** *adj.* **—de·is′ti·cal·ly** *adv.*

**de·i·ty** (dē′ĭ-tē) *n., pl.* **-ties.** [ME *deite* < OFr. < LLat. *deitas,* divine nature < Lat. *deus,* god.] **1.** A god or goddess. **2. a.** The essential nature or condition of being a god : DIVINITY. **b. Deity.** GOD 1a, b.

**dé·jà vu** (dā′zhä vü′) *n.* [Fr., already seen.] An illusion of having already experienced something actually being experienced for the first time.

**de·ject** (dĭ-jĕkt′) *vt.* **-ject·ed, -ject·ing, -jects.** [ME *dejecten* < Lat. *dejectus,* p.part. of *deicere,* to cast down : *de-,* down + *jacere,* to throw.] To lower the spirits of : DISHEARTEN.

**de·jec·ta** (dĭ-jĕk′tə) *pl.n.* [NLat. < Lat. *dejectus,* p.part. of *deicere,* to cast down. —see DEJECT.] Feces.

**de·ject·ed** (dĭ-jĕk′tĭd) *adj.* Marked by low spirits : DEPRESSED. **—de·ject′ed·ly** *adv.* **—de·ject′ed·ness** *n.*

**de·jec·tion** (dĭ-jĕk′shən) *n.* **1.** The state of being depressed : MELANCHOLY. **2.** *Med.* **a.** Evacuation of the bowels. **b.** Fecal matter.

**de ju·re** (dē jŏŏr′ē, dā yŏŏr′ā) *adv. & adj.* [Lat.] By law : by right.

**deka-** or **dek-** *pref. vars. of* DECA-.

**dek·a·gram** (dĕk′ə-grăm′) *n. var. of* DECAGRAM.

**dek·a·li·ter** (dĕk′ə-lē′tər) *n. var. of* DECALITER.

**dek·a·me·ter** (dĕk′ə-mē′tər) *n. var. of* DECAMETER.

**dek·a·stere** (dĕk′ə-stîr′) *n. var. of* DECASTERE.

**de·laine** (də-lān′) *n.* [Short for Fr. *mousseline de laine,* muslin of wool.] A light dress fabric of wool or cotton and wool.

**de·lam·i·nate** (dē-lăm′ə-nāt′) *vi.* **-nat·ed, -nat·ing, -nates.** To split into thin layers. **—de·lam′i·na′tion** *n.*

**Del·a·ware¹** (dĕl′ə-wâr′) *n., pl.* **Delaware** or **-wares. 1. a.** A group of Indian tribes once inhabiting the Delaware River valley. **b.** A member of any of these tribes. **2.** The Algonquian language of the Delaware. **—Del′a·war′e·an** *adj.*

**Del·a·ware²** (dĕl′ə-wâr′) *n.* [After the state of *Delaware.*] A grape with sweet light-red fruit.

**de·lay** (dĭ-lā′) *v.* **-layed, -lay·ing, -lays.** [ME *delaien* < OFr. *deslaier* : *des-,* off (< Lat. *de-*) + *laier,* alteration of *laissier,* to leave < Lat. *laxare,* to slacken < *laxus,* loose.] —*vt.* **1.** To postpone until a later time : DEFER. **2.** To cause to be detained or late. —*vi.* To procrastinate. —*n.* **1.** The act of delaying or state of being delayed. **2.** The period of time during which one is delayed. **3.** The time interval between two events. **—de·lay′er** *n.*

☆ ***syns:*** DELAY, DETAIN, HANG UP, HOLD UP, RETARD, SLOW *v. core meaning* : to cause to be later or slower than expected or desired <was *delayed* by heavy traffic>

**de·le** (dē′lē) *n.* [Lat., imper. of *delēre,* to delete.] A sign indicating that something is to be removed from typeset matter. —*vt.* **-led, -le·ing, -les. 1.** To take out or delete. **2.** To mark with a dele.

**de·leave** (dē-lēv′) *vt.* **-leaved, -leav·ing, -leaves.** To decollate.

**de·lec·ta·ble** (dĭ-lĕk′tə-bəl) *adj.* [ME < OFr. < Lat. *delectabilis* < *delectare,* to please. —see DELIGHT.] **1.** Greatly pleasing : DELIGHTFUL. **2.** Savory : delicious. **—de·lec′ta·bil′i·ty, de·lec′ta·ble·ness** *n.* **—de·lec′ta·bly** *adv.*

**de·lec·ta·tion** (dē′lĕk-tā′shən) *n.* Pleasure : delight.

**del·e·ga·cy** (dĕl′ĭ-gə-sē) *n., pl.* **-cies. 1.** The authority, office, or position of a delegate. **2.** The act of delegating or state of being delegated. **3.** DELEGATION 2.

**del·e·gate** (dĕl′ĭ-gāt′, -gĭt) *n.* [ME *delegat* < Med. Lat. *delegatus* < Lat., p.part. of *delegare,* to dispatch : *de-,* away + *legare,* to send < *lex,* law.] **1.** One authorized to act as representative or agent for another : DEPUTY. **2. a.** An elected or appointed representative of a U.S. territory in the House of Representatives who is entitled to speak but not vote. **b.** A member of the House of Delegates, the lower house of the Maryland, Virginia, and West Virginia legislatures. **c.** A representative to a convention or conference. —*vt.* (-gāt′) **-gat·ed, -gat·ing, -gates. 1.** To authorize and send as one's representative. **2.** To commit or entrust to another. **3.** *Law.* To appoint (one's debtor) as a debtor to one's creditor to replace oneself in satisfying a claim.

**del·e·ga·tion** (dĕl′ĭ-gā′shən) *n.* **1.** The act of delegating or state of being delegated. **2.** A person or group of persons officially elected or appointed to represent another or others.

**de·lete** (dĭ-lēt′) *vt.* **-let·ed, -let·ing, -letes.** [Lat. *delēre, delet-,* to wipe out.] To strike out or cancel, as from a text : OMIT.

**del·e·te·ri·ous** (dĕl′ĭ-tîr′ē-əs) *adj.* [Med. Lat. < Gk. *dēlētērios* < *dēleisthai,* to harm.] Having an injurious effect : HARMFUL. **—del′e·te′ri·ous·ly** *adv.* **—del′e·te′ri·ous·ness** *n.*

**de·le·tion** (dĭ-lē′shən) *n.* **1.** An act of deleting. **2.** Something, as a word, deleted from written or printed matter.

**delft** (dĕlft) *n.* **1. a.** A style of glazed, usu. blue and white earthenware orig. made in Delft, Netherlands. **b.** A piece of pottery of this style. **2.** Pottery made in imitation of delft.

**delft·ware** (dĕlft′wâr′) *n.* Delft.

**del·i** (dĕl′ē) *n., pl.* **-is.** *Informal.* A delicatessen.

**de·lib·er·ate** (dĭ-lĭb′ər-ĭt) *adj.* [Lat. *deliberatus,* resolved, p.part. of *deliberare,* to consider : *de-,* thoroughly + *librare,* to balance < *libra,* a balance, scales.] **1. a.** Thought out or planned in advance : PREMEDITATED <a *deliberate* delay> **b.** Said or done intentionally <a *deliberate* snub> **2.** Careful and thorough in deciding or determining. **3.** Leisurely or slow in motion or manner <*deliberate* speech> —*v.* (-ə-rāt′) **-at·ed, -at·ing, -ates.** —*vi.* **1.** To take careful thought : REFLECT. **2.** To consult with others as a process in reaching a decision. —*vt.* To consider carefully, as by weighing alternatives. **—de·lib′er·ate·ly** *adv.* **—de·lib′er·ate·ness** *n.*

**de·lib·er·a·tion** (dĭ-lĭb′ə-rā′shən) *n.* **1.** The act or process of deliberating. **2.** *often* **deliberations.** Formal discussion and debate of all sides of an issue. **3.** Thoughtfulness in action or decision.

**de·lib·er·a·tive** (dĭ-lĭb′ə-rā′tĭv, -ər-ə-tĭv) *adj.* **1.** Assembled or organized for deliberation or debate <a *deliberative* council> **2.** Of, marked by, or for use in deliberation or debate. **—de·lib′er·a′tive·ly** *adv.* **—de·lib′er·a′tive·ness** *n.*

**del·i·ca·cy** (dĕl′ĭ-kə-sē) *n., pl.* **-cies.** [ME *delicacie* < *delicat,* delicate.] **1.** The quality of being delicate. **2.** Something pleasing and appealing, esp. a choice food. **3.** Exquisite fineness or daintiness of structure or appearance. **4.** Frailty of bodily constitution or health. **5.** Sensitivity of perception, discrimination, or taste : REFINEMENT. **6. a.** Sensitivity to the feelings of others : TACT. **b.** Sensitivity to what is proper : PROPRIETY. **c.** Undue sensitivity to the offensive or improper : FASTIDIOUSNESS. **7.** The need for tact in treatment or handling <an issue of some *delicacy*> **8.** Fineness of touch <the *delicacy* of a surgeon's work> **9.** Sensitivity or keenness of response or reaction <the *delicacy* of a precision instrument>

**del·i·cate** (dĕl′ĭ-kĭt) *adj.* [ME *delicat* < Lat. *delicatus,* pleasing.] **1.** Pleasing to the senses, esp. in a subtle way <a *delicate* fragrance><a *delicate* flute solo> **2.** Exquisitely fine or dainty <*delicate* embroidery> **3.** Frail in constitution or health. **4.** Easily damaged or broken. **5.** Marked by sensitivity of discrimination. **6. a.** Considerate of the feelings of others. **b.** Concerned with propriety. **c.** Squeamish or fastidious. **7.** Requiring tactful treatment. **8.** Fine or soft in touch or skill <an artist's *delicate* brushstrokes> **9.** Keenly accurate in response or reaction. **10.** Subtle in difference or distinction. **—del′i·cate·ly** *adv.* **—del′i·cate·ness** *n.*

☆ ***syns:*** DELICATE, CHOICE, DAINTY, ELEGANT, EXQUISITE, FINE *adj. core meaning* : appealing to refined tastes <a *delicate* dessert>

**del·i·ca·tes·sen** (dĕl′ĭ-kə-tĕs′ən) *n.* [G. *Delikatessen,* pl. of *Delikatesse,* delicacy < Fr. *délicatesse* < Ital. *delicatezza* < *delicato,* delicate < Lat. *delicatus,* pleasing.] A shop that sells cooked or prepared foods ready for serving.

**de·li·cious**[1] (dĭ-lĭsh′əs) *adj.* [ME < OFr. < LLat. *deliciosus,* pleasing < Lat. *delicia,* pleasure < *delicere,* to allure. —see DELIGHT.] **1.** Highly pleasing or enjoyable, esp. to taste or smell. **2.** Very pleasant : DELIGHTFUL. **—de·li′cious·ly** *adv.* **—de·li′cious·ness** *n.*

☆ ***syns:*** DELICIOUS, DELECTABLE, HEAVENLY, LUSCIOUS, SAVORY, SCRUMPTIOUS, TASTY, TOOTHSOME, YUMMY *adj. core meaning* : highly pleasing to the taste <a *delicious* cake>

**de·li·cious**[2] (dĭ-lĭsh′əs) *n. often* **Delicious.** A variety of apple with sweet fruit often streaked with yellow and red.

**de·lict** (dĭ-lĭkt′) *n.* [Lat. *delictum* < neuter p.part. of *delinquere,* to offend. —see DELINQUENT.] *Law.* A misdemeanor.

**de·light** (dĭ-līt′) *n.* [ME *delit* < OFr. < *delitier,* to please < Lat. *delectare,* freq. of *delicere,* to allure : *de-,* away + *lacere,* to entice.] **1.** Great pleasure or gratification : JOY. **2.** Something that affords great pleasure or enjoyment. —*v.* **-light·ed, -light·ing, -lights.** —*vi.* To take or give great pleasure or joy. —*vt.* To please greatly.

▲ word history: The word *delight* is unrelated to either *light*[1] or *light*[2]. The Old French and Middle English form of the noun *delight* was *delit,* a regular derivation from the Latin verb *delectare.* The spelling *delight* appeared in the 16th century, a time when the orthography of English was unsettled. Several spellings could represent the same sound, and *delit* was respelled to resemble words that rhymed with it, like *light* and *night.*

**de·light·ed** (dĭ-lī′tĭd) *adj.* **1.** Filled with delight : greatly pleased. **2.** *Obs.* Delightful. **—de·light′ed·ly** *adv.* **—de·light′ed·ness** *n.*

**de·light·ful** (dĭ-līt′fəl) *adj.* Greatly pleasing. **—de·light′ful·ly** *adv.* **—de·light′ful·ness** *n.*

**de·light·some** (dĭ-līt′səm) *adj.* Delightful. **—de·light′some·ly** *adv.* **—de·light′some·ness** *n.*

**de·lim·it** (dĭ-lĭm′ĭt) *also* **de·lim·i·tate** (-ĭ-tāt′) *vt.* **-it·ed, -it·ing, -its** *also* **-tat·ed, -tat·ing, -tates.** [Fr. *délimiter* < Lat. *delimitare* : *de-,* thoroughly + *limitare,* to limit < *limes,* limit.] To establish the limit or boundaries of : DEMARCATE. **—de·lim′i·ta′tion** *n.* **—de·lim′i·ta′tive** *adj.*

**de·lim·it·er** (dĭ-lĭm′ĭ-tər) *n. Computer Sci.* A character marking the beginning or end of a unit of data.

**de·lin·e·ate** (dĭ-lĭn′ē-āt′) *vt.* **-at·ed, -at·ing, -ates.** [Lat. *delineare, delineat-* : *de-,* thoroughly + *linea,* line < *linum,* flax.] **1.** To draw or trace the outline of : SKETCH. **2.** To represent pictorially : DEPICT. **3.** To represent in words or gestures : PORTRAY. **—de·lin′e·a′tion** *n.* **—de·lin′e·a′tive** *adj.* **—de·lin′e·a′tor** *n.*

**de·lin·quen·cy** (dĭ-lĭng′kwən-sē, -lĭn′-) *n., pl.* **-cies. 1.** Negligence or failure in doing what is required. **2.** A delinquent act : MISDEED. **3.** Juvenile delinquency.

**de·lin·quent** (dĭ-lĭng′kwənt, -lĭn′-) *adj.* [Lat. *delinquens, delinquent-,* pr.part. of *delinquere,* to offend : *de-* (intensive) + *linquere,* to leave, abandon.] **1.** Failing to do what is required by law or obligation. **2.** Overdue in payment, as an account. —*n.* **1.** One who neglects or fails to do what law or obligation requires. **2.** A juvenile delinquent. **—de·lin′quent·ly** *adv.*

**del·i·quesce** (dĕl′ĭ-kwĕs′) *vi.* **-quesced, -quesc·ing, -quesc·es.** [Lat. *deliquescere* : *de-,* thoroughly +, *liquescere,* to melt, inchoative of *liquēre,* to be liquid.] **1.** To melt away or disappear as if by melting. **2.** *Chem.* To dissolve and become liquid by absorbing moisture from the air. **3.** *Bot.* **a.** To branch out into numerous subdivisions that lack a main axis. **b.** To become soft or fluid on maturing, as do certain fungi. **—del′i·ques′cence** *n.* **—del′i·ques′cent** *adj.*

**de·lir·i·a** (dĭ-lîr′ē-ə) *n. var. pl. of* DELIRIUM.

**de·lir·i·ous** (dĭ-lîr′ē-əs) *adj.* **1.** Suffering from delirium. **2.** Typical of or relating to delirium <*delirious* raving> **—de·lir′i·ous·ly** *adv.* **—de·lir′i·ous·ness** *n.*

**de·lir·i·um** (dĭ-lîr′ē-əm) *n., pl.* **-i·ums** *or* **-i·a** (-ē-ə) [Lat. < *delirare,* to be deranged : *de-,* from + *lira,* furrow.] **1.** A state of temporary mental confusion and clouded consciousness resulting from high fever, intoxication, or shock and marked by anxiety, tremors, hallucinations, delusions, and incoherence. **2.** A state of uncontrolled emotion, esp. excitement. **—de·lir′i·ant** *adj.*

**delirium tre·mens** (trē′mənz) *n.* [NLat., trembling delirium.] An acute delirium caused by alcohol poisoning.

**de·liv·er** (dĭ-lĭv′ər) *v.* **-ered, -er·ing, -ers.** [ME *deliveren* < OFr. *delivrer* < LLat. *deliberare* : Lat. *de-,* thoroughly + *liberare,* to free < *liber,* free.] —*vt.* **1.** To set free : LIBERATE. **2. a.** To assist in giving birth. **b.** To assist or aid in the birth of. **3.** To surrender to another : HAND OVER <*delivered* the suspect to the FBI> **4.** To take to the intended recipient <*deliver* mail> **5. a.** To send (e.g., a blow) to an intended goal or target. **b.** To throw or hurl (e.g., a ball). **6.** To give or utter <*deliver* a speech> **7.** To secure (something promised or desired), as for a candidate or political party. —*vi.* To do what is desired or expected <*deliver* on a pledge> **—de·liv′er·a·bil′i·ty** *n.* **—de·liv′er·a·ble** *adj.* **—de·liv′er·er** *n.*

**de·liv·er·ance** (dĭ-lĭv′ər-əns, -lĭv′rəns) *n.* **1.** The act of delivering or state of being delivered, as from bondage or danger. **2.** A publicly expressed opinion or judgment, as a jury's verdict.

**de·liv·er·y** (dĭ-lĭv′ə-rē, -lĭv′rē) *n., pl.* **-ies. 1.** The act of delivering or conveying. **2.** Something delivered. **3.** The act of releasing or rescuing. **4.** The act of giving birth : PARTURITION. **5.** The act of transferring to another. **6.** The act of surrendering. **7. a.** Utterance. **b.** Manner of speaking or singing. **8.** The act or manner of throwing or discharging <a pitcher's *delivery*>

**dell** (dĕl) *n.* [ME *del* < OE *dell.*] A small, secluded wooded valley.

**dells** (dĕlz) *pl.n.* [By folk ety. < DALLES.] The rapids of a river.

**Del·mon·i·co steak** (dĕl-mŏn′ĭ-kō′) *n.* [After the *Delmonico* Restaurant in New York City.] A small, often boned steak from the front section of the short loin of beef.

**de·lo·cal·ize** (dē-lō′kə-līz′) *vt.* **-ized, -iz·ing, -iz·es. 1.** To remove (something) from its native or usual locality. **2.** To broaden the range or scope of. **—de·lo′cal·i·za′tion** *n.*

**de·louse** (dē-lous′) *vt.* **-loused, -lous·ing, -lous·es.** To rid of parasitic infestation by physical or chemical means.

**Del·phic** (dĕl′fĭk) *also* **Del·phi·an** (-fē-ən) *adj.* **1.** Of or pertaining to Delphi or to the oracle of Apollo at Delphi. **2.** Ambiguous in meaning : OBSCURE.

**del·phin·i·um** (dĕl-fĭn′ē-əm) *n.* [NLat. *Delphinium,* genus name < Gk. *delphinion,* larkspur, dim. of *delphis,* dolphin (from the shape of the nectary).] A plant of the genus *Delphinium,* esp. any of several tall cultivated varieties bearing spikes of variously colored spurred flowers.

**Del·phi·nus** (dĕl-fī′nəs) *n.* [Lat. < *delphinus,* dolphin. —see DOLPHIN.] A constellation in the Northern Hemisphere.

**del·ta** (dĕl′tə) *n.* [Gk., of Phoenician orig.; akin to Heb. *dāleth,* daleth.] **1.** The fourth letter of the Greek alphabet. —See table at ALPHABET. **2. a.** A usu. triangular alluvial deposit at the mouth of a river. **b.** A similar deposit at the mouth of a tidal inlet, caused by tidal currents. **3.** An object shaped like a triangle. **4.** *Math.* A finite increment in a variable. **—del·ta′ic** (-tā′ĭk), **del′tic** (-tĭk) *adj.*

**delta ray** *n.* An electron ejected from matter by ionizing radiation.

**delta wave** *or* **delta rhythm** *n.* A low-frequency brain wave that emanates from the forward portion of the brain during deep sleep in normal adults.

**delta wing** *n.* An aircraft with swept-back wings that give it the appearance of an isosceles triangle.

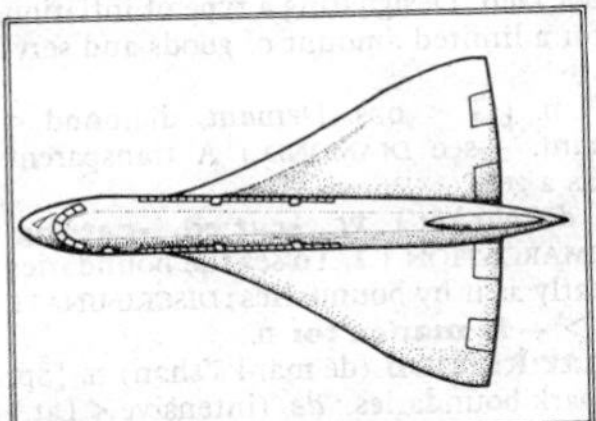

**delta wing**

**del·toid** (dĕl′toid′) *n.* [NLat. *deltoides* < Gk. *deltoeidēs,* triangular : *delta,* delta + *-eides,* -oid.] A thick, triangular muscle covering the shoulder joint, used to raise the arm from the side. —*adj.* **1.** Triangular. **2.** Relating to the deltoid.

**de·lude** (dĭ-lo͞od′) *vt.* **-lud·ed, -lud·ing, -ludes.** [ME *deluden* < Lat. *deludere* : *de-* (pejorative) + *ludere,* to play < *ludus,* game.] **1.** To cause to be deceived <*deluded* by a clever scheme> **2.** *Obs.*

To elude or evade. **3.** *Obs.* To frustrate the hopes or plans of. **—de·lud′a·ble** *adj.* **—de·lud′er** *n.* **—de·lud′ing·ly** *adv.*

**del·uge** (dĕl′yo͞oj) *vt.* **-uged, -ug·ing, -ug·es.** [< ME *deluge*, flood < OFr. < Lat. *diluvium* < *diluere*, to wash away : *dis-*, apart + *lavere*, to wash.] **1.** To overflow with water. **2.** To inundate in overwhelming numbers <was *deluged* with fan mail> —*n.* **1. a.** A great flood. **b.** A heavy downpour. **2.** Something overwhelming <a *deluge* of praise> **3. Deluge.** The great flood that occurred in the time of Noah, as described in the Old Testament.

**de·lu·sion** (dĭ-lo͞o′zhən) *n.* [ME *delusioun* < Lat. *delusio* < *deludere*, to delude.] **1. a.** The act or process of deluding. **b.** The state of being deluded. **2.** Something falsely disseminated or believed : DECEPTION. **3.** A false belief held in spite of invalidating evidence, esp. as a symptom of certain forms of mental illness. **—de·lu′sion·al** *adj.*

**de·lu·sive** (dĭ-lo͞o′sĭv) *also* **de·lu·so·ry** (-sə-rē) *adj.* **1.** Tending to mislead or deceive : DECEPTIVE. **2.** Having the nature of a delusion. **—de·lu′sive·ly** *adv.* **—de·lu′sive·ness** *n.*

**de luxe** *also* **de·luxe** (dĭ-lo͝oks′, -lŭks′) [Fr., of luxury.] *adj.* Elegant and luxurious : SUMPTUOUS <a *de luxe* hotel suite> **—de luxe, de·luxe′** *adv.*

**delve** (dĕlv) *v.* **delved, delv·ing, delves.** [ME *delven*, to dig < OE *delfan.*] —*vi.* **1.** To search carefully and laboriously. **2.** *Archaic.* To dig the ground, as with a spade. —*vt. Archaic.* To dig (ground) with a spade. **—delv′er** *n.*

**de·mag·net·ize** (dē-măg′nĭ-tīz′) *vt.* **-ized, -iz·ing, -iz·es.** To remove magnetic properties from. **—de·mag′net·i·za′tion** *n.*

**dem·a·gog** (dĕm′ə-gôg′, -gŏg′) *n. var. of* DEMAGOGUE.

**dem·a·gog·ic** (dĕm′ə-gŏj′ĭk, -gŏg′-, -gō′jĭk) *also* **dem·a·gog·i·cal** (-gŏj′ĭ-kəl, -gŏg′-, -gō′jĭ-kəl) *adj.* Of, relating to, or typical of a demagogue. **—dem′a·gog′i·cal·ly** *adv.*

**dem·a·gog·ism** (dĕm′ə-gô′gĭz-əm, -gŏg′ĭz-) *n.* Demagoguery.

**dem·a·gogue** *also* **dem·a·gog** (dĕm′ə-gôg′, -gŏg′) *n.* [Gk. *dēmagōgos*, popular leader : *dēmos*, common people + *agōgos*, leading < *agein*, to lead.] **1.** A leader who obtains power by means of impassioned appeals to the emotions and prejudices of the populace. **2.** A leader of the common people in ancient times. **—dem′a·gog′y** (dĕm′ə-gŏj′ē, -gô′jē, -gŏg′ē, -gō′jē) *n.*

**dem·a·gogu·er·y** (dĕm′ə-gô′gə-rē, -gŏg′ə-) *n.* The rhetoric or practices of a demagogue.

**de·mand** (dĭ-mănd′) *v.* **-mand·ed, -mand·ing, -mands.** [ME *demanden* < OFr. *demander*, to charge with doing < Lat. *demandare*, to entrust : *de-* (intensive) + *mandare*, to entrust.] —*vt.* **1.** To ask for urgently or firmly : COMMAND. **2.** To claim as just or due <*demand* immediate payment> **3.** To ask to be informed of <*demand* the reason for the roadblock> **4.** To need or require as useful, just, proper, or necessary <The affair *demands* a tuxedo.> **5.** *Law.* **a.** To summon to court. **b.** To claim legally and formally. —*vi.* To make a demand. —*n.* **1.** The act of demanding. **2.** Something demanded. **3. a.** The state of being sought after <in great *demand*> **b.** An urgent requirement, need, or claim <an increased *demand* for oil> **4.** *Archaic.* An emphatic question or inquiry. **5.** *Law.* A formal claim. **6.** *Computer Sci.* A coding technique in which a read or write order is initiated as the need for a new block of data occurs. **7. a.** The desire to possess something combined with the ability to purchase it. **b.** The amount of a commodity that people are ready and able to buy at a given time for a given price. **—on demand.** On presentation <pay *on demand*> **—de·mand′a·ble** *adj.* **—de·mand′er** *n.*

**de·mand·ant** (dĭ-măn′dənt) *n. Law.* A plaintiff.

**demand deposit** *n.* A bank deposit that can be withdrawn without advance notice.

**de·mand·ing** (dĭ-măn′dĭng) *adj.* **1.** Making rigorous demands. **2.** Requiring constant effort or attention. **—de·mand′ing·ly** *adv.*

**demand loan** *n.* A call loan.

**demand note** *n.* A bill or draft payable in lawful money on presentation or demand.

**de·mand-pull** (dĭ-mănd′po͝ol′) *adj.* Designating a type of inflation in which increased demand for a limited amount of goods and services tends to drive up prices.

**de·man·toid** (dĭ-măn′toid′) *n.* [G. < obs. *Demant*, diamond < MHG *diemant* < OFr. *diamant.* —see DIAMOND.] A transparent green variety of garnet, used as a gem.

**de·mar·cate** (dĭ-mär′kāt′, dē′mär-kāt′) *vt.* **-cat·ed, -cat·ing, -cates.** [Back-formation < DEMARCATION.] **1.** To set the boundaries of : DELIMIT. **2.** To separate clearly as if by boundaries : DISCRIMINATE <*demarcate* right and wrong> **—de·mar′ca′tor** *n.*

**de·mar·ca·tion** *also* **de·mar·ka·tion** (dē′mär-kā′shən) *n.* [Sp. *demarcacion* < *demarcar*, to mark boundaries : *de-* (intensive < Lat.) + *marcar*, to mark < Ital. *marcare*, of Germanic orig.] **1.** The setting or marking of boundaries or limits. **2.** A distinction : separation.

**dé·marche** (dā-märsh′) *n.* [Fr. < OFr. *demarche*, gait < *demarchier*, to march : *de-*, from (< Lat.) + *marchier*, to march, prob. of Germanic orig.] **1.** A course of action : MANEUVER. **2.** A diplomatic representation or protest. **3.** A civil statement or protest addressed to public authorities.

**de·mar·ka·tion** (dē′mär-kā′shən) *n. var. of* DEMARCATION.

**deme** (dēm) *n.* [Gk. *dēmos.*] **1.** One of the townships of ancient Attica. **2.** *Ecol.* A local, usu. stable population of organisms of the same kind or species.

**de·mean¹** (dĭ-mēn′) *vt.* **-meaned, -mean·ing, -means.** [ME *demeinen*, to govern < OFr. *demener* : *de-*, thoroughly (< Lat.) + *mener*, to conduct < Lat. *minare*, to drive < *minari*, to threaten < *minae*, threats.] To conduct or behave (oneself) in a particular way.

**de·mean²** (dĭ-mēn′) *vt.* **-meaned, -mean·ing, -means. 1.** To debase or degrade. **2.** To humble (oneself).

**de·mean·or** (dĭ-mē′nər) *n.* The way in which one behaves or conducts oneself : DEPORTMENT.

**de·mean·our** (dĭ-mē′nər) *n. Chiefly Brit. var. of* DEMEANOR.

**de·ment** (dĭ-mĕnt′) *vt.* **-ment·ed, -ment·ing, -ments.** [LLat. *dementare* < Lat. *demens*, mad : *de-*, from + *mens*, mind.] To make insane.

**de·ment·ed** (dĭ-mĕn′tĭd) *adj.* **1.** Insane. **2.** Afflicted with dementia. **—de·ment′ed·ly** *adv.* **—de·ment′ed·ness** *n.*

**de·men·tia** (dĭ-mĕn′shə) *n.* [Lat., madness < *demens*, mad. —see DEMENT.] **1.** Irreversible deterioration of intellectual faculties with accompanying emotional disturbance resulting from organic brain disorder. **2.** Insanity. **—de·men′tial** *adj.*

**dementia prae·cox** (prē′kŏks′) *n.* [NLat., premature dementia.] Schizophrenia.

**de·mer·it** (dĭ-mĕr′ĭt) *n.* [ME *demerite*, offense < OFr. *desmerite* < Lat. *demeritum*, neuter p.part. of *demerēre*, to deserve : *de-* (intensive) + *merēre*, to earn.] **1. a.** A quality deserving blame : FAULT. **b.** Absence of merit. **2.** A mark made against one's record and usu. involving loss of privileges for bad conduct or failure. **—de·mer′i·to′ri·ous** (-tôr′ē-əs, -tōr′-) *adj.* **—de·mer′i·to′ri·ous·ly** *adv.*

**Dem·er·ol** (dĕm′ə-rôl′, -rōl′). A trademark for a synthetically produced morphine.

**de·mesne** (dĭ-mān′, -mēn′) *n.* [ME *demeine* < OFr. *demaine.* —see DOMAIN.] **1.** *Law.* Possession and use of one's own land. **2.** Lands retained by a feudal lord for his own use. **3.** The grounds belonging to a mansion or country house. **4.** An extensive piece of landed property : ESTATE. **5.** A region or territory. **6.** A domain : realm.

**De·me·ter** (dĭ-mē′tər) *n.* [Gk. *Dēmētēr.*] *Gk. Myth.* The goddess of agriculture, fertility, and marriage.

**dem·e·ton** (dĕm′ĭ-tŏn′) *n.* [Blend of DIETHYL, MERCAPTAN, and THIONATE.] A highly toxic, pale-yellow organophosphorous liquid used as a systemic insecticide.

**demi-** *pref.* [Fr. < *demi*, half < Med. Lat. *dimedius* < Lat. *dimidius*, divided in half : *dis-*, apart + *medius*, half.] **1.** Half <*demi*relief> **2.** Part : partly : to some degree <*demi*god>

**dem·i·god** (dĕm′ē-gŏd′) *n.* **1. a.** A mythological semidivine being, esp. the offspring of a god and a mortal. **b.** A minor deity. **2.** One with godlike attributes.

**dem·i·god·dess** (dĕm′ē-gŏd′ĭs) *n.* A woman who is thought to be a demigod.

**dem·i·john** (dĕm′ē-jŏn′) *n.* [Prob. alteration of Fr. *dame-Jeanne*, lady Jane.] A large, narrow-necked glass or earthenware bottle, usu. encased in wickerwork.

**de·mil·i·ta·rize** (dē-mĭl′ĭ-tə-rīz′) *vt.* **-rized, -riz·ing, -riz·es. 1.** To eliminate the military character of. **2.** To prohibit military forces or installations in. **3.** To replace military control of with civilian control. **—de·mil′i·ta·ri·za′tion** *n.*

**dem·i·mon·daine** (dĕm′ē-mŏn-dān′, -mŏn′dān′) *n.* [Fr. < *demimonde*, demimonde.] A woman belonging to the demimonde.

**dem·i·monde** (dĕm′ē-mŏnd′) *n.* [Fr. : *demi-*, demi- + *monde*, world < Lat. *mundus.*] **1. a.** The social class comprising those who are supported by wealthy lovers or protectors. **b.** Prostitutes. **2.** A group that exists on the fringes of respectability.

**dem·i·rep** (dĕm′ē-rĕp′) *n.* [DEMI- + REP(UTATION).] A demimondaine.

**de·mise** (dĭ-mīz′) *n.* [ME, transfer of property < OFr. *dimis*, p.part. of *demettre*, to release. —see DEMIT.] **1.** Death. **2.** The transfer of an estate by will or lease. **3.** The transfer of a ruler's authority by abdication or death. —*v.* **-mised, -mis·ing, -mis·es.** —*vt.* **1.** To transfer (an estate) by will or lease. **2.** To transfer (sovereignty) by abdication or will. —*vi.* **1.** To be transferred by will or descent. **2.** To die. **—de·mis′a·ble** *adj.*

**dem·i·sem·i·qua·ver** (dĕm′ē-sĕm′ē-kwā′vər) *n. Chiefly Brit.* A thirty-second note.

**de·mis·sion** (dĭ-mĭsh′ən) *n.* [ME *dimissioun* < AN < Lat. *dimissio*, dismissal < *dimissus*, p.part. of *dimittere*, to release. —see DEMIT.] Relinquishment of an office or function : RESIGNATION.

**de·mit** (dĭ-mĭt′) *v.* **-mit·ted, -mit·ting, -mits.** [ME *dimitten*, to release < OFr. *demettre* < Lat. *dimittere* : *dis-*, away + *mittere*, to send.] —*vt.* **1.** To relinquish (e.g., an office or function). **2.** *Obs.* To dismiss. —*vi.* To resign.

**dem·i·tasse** (dĕm′ē-tăs′, -täs′) *n.* [Fr. : *demi-*, half + *tasse*, cup < OFr. < Ar. *ṭašt*, basin < Pers.] **1.** A small cup of strong black coffee. **2.** The cup used to serve demitasse.

**dem·i·urge** (dĕm′ē-ûrj′) *n.* [LLat. *demiurgus* < Gk. *dēmiourgos*, artisan : *dēmios*, public (< *dēmos*, people) + *ergon*, work.] **1. Demi-**

**urge.** The name used by Plato to designate the deity who fashions the material world. **2. Demiurge.** The Gnostic creator of the material world. **3.** A public magistrate in some ancient Greek states. **4. a.** A creative force. **b.** An authoritative power. **—dem'i·ur'geous** (-ûr'jəs), **dem'i·ur'gic** (-jĭk), **dem'i·ur'gi·cal** (-jĭ-kəl) *adj.* **—dem'i·ur'gi·cal·ly** *adv.*

**dem·i·world** (dĕm'ē-wûrld') *n.* The demimonde.

**dem·o** (dĕm'ō) *n., pl.* **-os.** *Informal.* **1. a.** A demonstration, as of a service or product. **b.** A brief tape or recording used to demonstrate the qualities of a performer, as a musician. **2.** A product, as a stereo, used for demonstration and often sold later at a discount.

**de·mob** (dē-mŏb') *Chiefly Brit.* —*vt.* **-mobbed, -mob·bing, -mobs.** To demobilize. —*n.* Demobilization.

**de·mo·bil·ize** (dē-mō'bə-līz') *vt.* **-ized, -iz·ing, -iz·es. 1.** To dismiss from military use or service. **2.** To disband. **—de·mo'bil·i·za'tion** *n.*

**de·moc·ra·cy** (dĭ-mŏk'rə-sē) *n., pl.* **-cies.** [OFr. *democratie* < LLat. *democratia* < Gk. *dēmokratia* : *dēmos,* people + *-kratia,* -cracy.] **1.** Government exercised either directly by the people or through elected representatives. **2.** A political or social unit based on democratic rule. **3.** The populace, esp. as the primary source of political power. **4.** Rule by the majority. **5.** The principles of social equality and respect for the individual within a community. **6. Democracy.** The ideology and policies of the U.S. Democratic Party.

**dem·o·crat** (dĕm'ə-krăt') *n.* [Fr. *démocrate,* back-formation < *democratie,* democracy < OFr.] **1.** An advocate of democracy. **2. Democrat.** A member of the Democratic Party.

**dem·o·crat·ic** (dĕm'ə-krăt'ĭk) *adj.* **1.** Of, marked by, or advocating democracy. **2. a.** Relating to, encompassing, or promoting the interests of the people. **b.** Carried on by the general populace. **3.** Believing in or practicing social equality <"a proper *democratic* scorn for bloated dukes and lords" —George Du Maurier> **4. Democratic.** Of, relating to, or typical of the Democratic Party. **—dem'o·crat'i·cal·ly** *adv.*

**Democratic Party** *n.* One of the two major political parties in the United States, originating from a split in the Democratic-Republican Party under Andrew Jackson in 1828.

**Dem·o·crat·ic-Re·pub·li·can Party** (dĕm'ə-krăt'ĭk-rĭ-pŭb'lĭ-kən) *n.* A U.S. political party opposed to the Federalist Party, founded by Thomas Jefferson in 1792 and dissolved in 1828.

**de·moc·ra·tize** (dĭ-mŏk'rə-tīz') *vt.* **-tized, -tiz·ing, -tiz·es.** To make democratic. **—de·moc'ra·ti·za'tion** *n.*

**dé·mo·dé** (dā'mō-dā') *adj.* [Fr.] No longer fashionable.

**de·mod·u·late** (dē-mŏj'ə-lāt', -mŏd'yə-) *vt.* **-lat·ed, -lat·ing, -lates.** To extract (information) from a modulated carrier wave. **—de·mod'u·la'tion** *n.* **—de·mod'u·la'tor** *n.*

**De·mo·gor·gon** (dē'mə-gôr'gən, dē'mə-gôr'-) *n.* [LLat.] *Myth.* A terrifying deity or spirit, whose very name if mentioned was believed to bring disaster or death.

**dem·o·graph·ics** (dĕm'ə-grăf'ĭks, dē'mə-) *n. (pl. in number).* Demographic data used esp. to identify consumer markets.

**de·mog·ra·phy** (dĭ-mŏg'rə-fē) *n.* [Fr. *démographie* : Gk. *dēmos,* people + Fr. *-graphie,* -graphy.] Study of the vital statistics of human populations, as size, growth, density, and distribution. **—de·mog'ra·pher** *n.* **—dem'o·graph'ic** (dĕm'ə-grăf'ĭk, dē'mə-), **dem'o·graph'i·cal** (-ĭ-kəl) *adj.* **—dem'o·graph'i·cal·ly** *adv.*

**dem·oi·selle** (dĕm'wə-zĕl') *n.* [Fr. < OFr. *dameisele.* —see DAMSEL.] **1.** A young lady. **2.** An Old World crane, *Anthropoides virgo,* bearing gray and black plumage and white plumes at the sides of the head. **3.** A damselfly.

**demoiselle**
*Approximately 3 feet long*

**de·mol·ish** (dĭ-mŏl'ĭsh) *vt.* **-ished, -ish·ing, -ish·es.** [OFr. *demolir, demoliss-* < Lat. *demoliri* : *de-,* apart + *moliri,* to build < *moles,* mass.] **1.** To tear down completely : RAZE. **2.** To do away with completely. **3.** To damage (e.g., a good name) severely.

**dem·o·li·tion** (dĕm'ə-lĭsh'ən, dē'mə-) *n.* [OFr. < Lat. *demolitio* < *demoliri,* to demolish.] **1.** The act or process of wrecking or destroying, esp. destruction by explosives. **2. demolitions.** Explosives, esp. when designed or used as weapons. **—dem'o·li'tion·ist** *n.*

**demolition derby** *n.* A sports contest in which drivers crash old cars into each other until only one remains running.

**de·mon** (dē'mən) *n.* [ME < LLat. *daemon* < Lat., spirit < Gk. *daimōn,* divine power.] **1.** An evil being : devil or fiend. **2.** A persistently tormenting person, force, or passion. **3.** *also* **dae·mon** or **dai·mon** (dī'mōn'). *Gk. Myth.* An inferior divinity, as a deified hero. **4.** *also* **dae·mon** or **dai·mon** (dī'mōn'). An attendant spirit : GENIUS. **5.** One who is extremely zealous or skillful in a given activity.

**de·mon·e·tize** (dē-mŏn'ĭ-tīz', -mŭn'-) *vt.* **-tized, -tiz·ing, -tiz·es.** [Fr. *démonétiser* : *dé-,* de- + Lat. *moneta,* coin. —see MONEY.] **1.** To divest (e.g., a coin) of monetary value. **2.** To stop using (a metal) as a monetary standard. **—de·mon'e·ti·za'tion** *n.*

**de·mo·ni·ac** (dĭ-mō'nē-ăk') *also* **de·mo·ni·a·cal** (dē'mə-nī'ə-kəl) *adj.* [ME *demoniak* < LLat. *daemoniacus* < Gk. *daimoniakos* < *daimonios,* of a spirit < *daimōn,* divine power.] **1.** Arising or appearing to arise from possession by a demon. **2.** Of, resembling, or suggestive of a devil : FIENDISH. —*n.* **demoniac.** One who is or appears to be possessed by a demon. **—de'mo·ni'a·cal·ly** *adv.*

**de·mon·ic** (dĭ-mŏn'ĭk) *adj.* **1.** Befitting a demon : FIENDISH. **2.** Motivated by a spiritual force : INSPIRED. **—de·mon'i·cal·ly** *adv.*

**de·mon·ize** (dē'mə-nīz') *vt.* **-ized, -iz·ing, -iz·es. 1.** To transform into or as if into a demon. **2.** To possess by a demon.

**de·mon·ol·o·gy** (dē'mə-nŏl'ə-jē) *n.* **1.** Study of demons. **2.** A treatise on demons or demon worship. **3.** Belief in demons. **—de'mon·ol'o·gist** *n.*

**de·mon·stra·ble** (dĭ-mŏn'strə-bəl) *adj.* **1.** Capable of being demonstrated. **2.** Obvious or apparent <a *demonstrable* falsehood> **—de·mon'stra·bil'i·ty, de·mon'stra·ble·ness** *n.* **—de·mon'stra·bly** *adv.*

**dem·on·strate** (dĕm'ən-strāt') *v.* **-strat·ed, -strat·ing, -strates.** [Lat. *demonstrare, demonstrat-* : *de-,* completely + *monstrare,* to show < *monstrum,* divine portent < *monēre,* to warn.] —*vt.* **1.** To prove or make evident by reasoning or adducing evidence. **2.** To describe or illustrate by experiment or practical application <*demonstrate* the effect of marijuana> **3.** To show or reveal <*demonstrates* talent> **4.** To display, operate, and explain (a product). —*vi.* To present or participate in a demonstration.

**dem·on·stra·tion** (dĕm'ən-strā'shən) *n.* **1.** The act of making evident or proving. **2.** Conclusive evidence : PROOF. **3.** An illustration or explanation, as of a theory or product, by exemplification or practical application. **4.** A manifestation, as of one's feelings. **5.** A public display of group opinion, as by a march or rally. **6.** A show of military strength.

**de·mon·stra·tive** (dĭ-mŏn'strə-tĭv) *adj.* **1.** Serving to manifest or prove. **2.** Involving or marked by demonstration. **3.** Given to or marked by the open expression of emotion, esp. affection. **4.** Specifying or singling out the person or thing referred to in a phrase or sentence. —*n.* A demonstrative adjective or pronoun. **—de·mon'stra·tive·ly** *adv.* **—de·mon'stra·tive·ness** *n.*

**dem·on·stra·tor** (dĕm'ən-strā'tər) *n.* **1.** One that demonstrates. **2.** A sample used in a demonstration.

**dem·o·pho·bi·a** (dĕm'ə-fō'bē-ə, dē'mə-) *n.* [Gk. *dēmos,* people + -PHOBIA.] Abnormal fear of crowds. **—dem'o·pho'bic** *adj.*

**de·mor·al·ize** (dĭ-môr'ə-līz', -mŏr'-) *vt.* **-ized, -iz·ing, -iz·es. 1.** To corrupt the morals of : DEGRADE. **2.** To undermine the confidence or morale of : DISHEARTEN. **3.** To throw into disorder or confusion. **—de·mor'al·i·za'tion** *n.* **—de·mor'al·iz'er** *n.*

**de·mos** (dē'mŏs') *n.* [Gk. *dēmos,* district, people.] **1.** The common people of an ancient Greek state. **2.** The populace.

**de·mote** (dĭ-mōt') *vt.* **-mot·ed, -mot·ing, -motes.** [DE- + (PRO)-MOTE.] To lower in grade or rank. **—de·mo'tion** *n.*

**de·mot·ic** (dĭ-mŏt'ĭk) *adj.* [Gk. *dēmotikos* < *dēmotēs,* a commoner < *dēmos,* people.] **1.** Of or relating to the common people. **2.** Of, relating to, or written in the simplified form of ancient Egyptian hieratic writing. **3. Demotic.** Of or relating to a form of modern Greek based on colloquial use. —*n.* **Demotic.** Demotic Greek.

**de·mount** (dē-mount') *vt.* **-mount·ed, -mount·ing, -mounts.** To remove (e.g., a motor) from a mounted position or other support. **—de·mount'a·ble** *adj.*

**de·mul·cent** (dĭ-mŭl'sənt) *adj.* [Lat. *demulcens, demulcent-,* pr.part. of *demulcēre,* to soften : *de-,* down + *mulcēre,* to stroke.] Soothing. —*n.* A soothing, usu. mucilaginous or oily substance used esp. for relieving pain in irritated mucous surfaces.

**de·mur** (dĭ-mûr') *vi.* **-murred, -mur·ring, -murs.** [ME *demuren,* to delay < OFr. *demorer* < Lat. *demorari* : *de-* (intensive) + *morari,* to delay < *mora,* delay.] **1.** To take exception : OBJECT. **2.** *Law.* To enter a demurrer. **3.** To delay. —*n.* **1.** The act of demurring. **2.** An objection. **3.** A delay. **—de·mur'ra·ble** *adj.* **—de·mur'ral** *n.*

**de·mure** (dĭ-myŏŏr') *adj.* **-mur·er, -mur·est.** [ME, prob. < AN.] **1.** Modest and reserved. **2.** Feigning modesty or shyness : COY. **—de·mure'ly** *adv.* **—de·mure'ness** *n.*

**de·mur·rage** (dĭ-mûr'ĭj) *n.* **1.** Detention of a cargo conveyance, as a ship or freight car, during loading or unloading beyond the scheduled time of departure. **2.** Compensation paid for demurrage.

**de·mur·rer** (dĭ-mûr'ər) *n.* **1.** One who demurs : OBJECTOR. **2.** *Law.* A plea to dismiss a lawsuit on the grounds that although the opposition's statements may be true, they are insufficient to sustain the claim. **3.** An objection.

**de·my** (dĭ-mī′) *n., pl.* **-mies.** [Alteration of DEMI-.] Any of several standard sizes of paper, esp. paper measuring 16 by 21 inches.

**de·my·e·lin·ate** (dē-mī′ə-lə-nāt′) *vt.* **-at·ed, -at·ing, -ates.** To destroy or remove the myelin of (a nerve). **—de·my′e·lin·a′tion** *n.*

**de·mys·ti·fy** (dē-mĭs′tə-fī′) *vt.* **-fied, -fy·ing, -fies.** To make less difficult to understand. **—de·mys′ti·fi·ca′tion** *n.*

**de·my·thol·o·gize** (dē′mĭ-thŏl′ə-jīz′) *vt.* **-gized, -giz·ing, -giz·es. 1.** To rid of mythological elements in order to discover the underlying meaning <*demythologize* Indian legends> **2.** To remove mythical aspects from <"providing an antiheroic age with heroes suitably *demythologized,* yet also grand" —John Simon> **—de′my·thol′o·gi·za′tion** *n.* **—de′my·thol′o·giz′er** *n.*

**den** (dĕn) *n.* [ME < OE *denn.*] **1.** A wild animal's shelter or retreat : LAIR. **2.** A cave used esp. as a refuge or hiding place. **3.** A small, usu. hidden place, esp. one used as a meeting place for wrongdoers <a *den* of thieves> **4.** A small secluded room for study or relaxation. **5.** A unit of approx. ten Cub Scouts. *—vi.* **denned, den·ning, dens.** To inhabit or hide in a den.

**de·nar·i·us** (dĭ-nâr′ē-əs) *n., pl.* **-i·i** (-ē-ī′) [ME < Lat. —see DENARY.] **1.** An ancient Roman silver coin. **2.** An ancient Roman gold coin valued at 25 silver denarii.

**den·a·ry** (dĕn′ə-rē) *adj.* [Lat. *denarius* < *deni,* by tens.] **1.** Tenfold. **2.** DECIMAL 2.

**de·na·tion·al·ize** (dē-năsh′ə-nə-līz′) *vt.* **-ized, -iz·ing, -iz·es. 1.** To deprive of national characteristics or rights. **2.** To return to private ownership <*denationalize* an industry> **—de·na′tion·al·i·za′tion** *n.*

**de·nat·u·ral·ize** (dē-năch′ər-ə-līz′) *vt.* **-ized, -iz·ing, -iz·es. 1.** To make unnatural. **2.** To deprive of the rights of naturalization or citizenship. **—de·nat′u·ral·i·za′tion** *n.*

**de·na·ture** (dē-nā′chər) *vt.* **-tured, -tur·ing, -tures. 1.** To change the nature or natural qualities of. **2.** To render unfit for consumption without destroying usefulness in other applications, esp. to add methanol to ethyl alcohol. **3.** To change the structure of (a protein), as with heat, alkali, or acid, so that some of the original properties are diminished or eliminated. **4.** To add nonfissionable matter to (fissionable material) to prevent use in an atomic weapon. **—de·na′tur·ant** *n.* **—de·na′tur·a′tion** *n.*

**dendri-** *or* **dendr-** *pref. vars. of* DENDRO-.

**den·dri·form** (dĕn′drə-fôrm′) *adj.* Treelike in form or structure.

**den·drite** (dĕn′drīt′) *n.* **1. a.** A mineral crystallizing in another mineral in the form of a branching or treelike mark. **b.** A rock or mineral bearing such a mark. **2.** A branched part of a nerve cell that transmits impulses toward the cell body.

**den·drit·ic** (dĕn-drĭt′ĭk) *also* **den·drit·i·cal** (-ĭ-kəl) *adj.* **1.** Of, relating to, or resembling a dendrite. **2.** Dendriform. **—den·drit′i·cal·ly** *adv.*

**dendro-** *or* **dendri-** *or* **dendr-** *pref.* [< Gk. *dendron,* tree.] Tree : treelike <*dendro*chronology>

**den·dro·chro·nol·o·gy** (dĕn′drō-krə-nŏl′ə-jē) *n.* Study of growth rings in trees to date past events. **—den′dro·chron′o·log′i·cal** (-krŏn′ə-lŏj′ĭ-kəl) *adj.* **—den′dro·chron′o·log′i·cal·ly** *adv.*

**den·droid** (dĕn′droid′) *also* **den·droid·al** (dĕn-droid′l) *adj.* Shaped like a tree : ARBORESCENT.

**den·drol·o·gy** (dĕn-drŏl′ə-jē) *n.* Botanical study of trees. **—den′dro·log′ic** (-drə-lŏj′ĭk), **den′dro·log′i·cal** *adj.* **—den·drol′o·gist** *n.*

**den·dron** (dĕn′drŏn′) *n.* [Gk., tree.] DENDRITE 2.

**dene** (dēn) *n.* [ME *den,* sandy seashore.] *Chiefly Brit.* A sandy dune or tract by the seashore.

**De·neb** (dĕn′ĕb′) *n.* [Ar. *dhanab,* tail.] The brightest star in the constellation Cygnus.

**den·e·ga·tion** (dĕn′ĭ-gā′shən) *n.* [Fr. < Lat. *denegatio* < *denegatus,* p.part. of *denegare,* to deny.] A denial.

**den·gue** (dĕng′gē, -gā′) *n.* [Sp., of African orig.] An infectious tropical disease transmitted by mosquitoes and marked by fever, rash, and severe joint pains.

**de·ni·a·ble** (dĭ-nī′ə-bəl) *adj.* Able to be denied. **—de·ni′a·bly** *adv.*

**de·ni·al** (dĭ-nī′əl) *n.* [< DENY.] **1.** A refusal to comply with or satisfy a request. **2.** A refusal to acknowledge the truth of a statement or allegation : CONTRADICTION. **3.** A rejection, as of a doctrine or belief. **4.** The act of disowning or disavowing : REPUDIATION. **5.** *Law.* The opposing by a defendant of the allegation against him or her by the plaintiff. **6.** Self-denial : abstinence.

**de·ni·er[1]** (dĭ-nī′ər) *n.* One who denies.

**den·ier[2]** (dən-yā′) *n.* [ME *denere,* a coin < OFr. *dener* < Lat. *denarius.* —see DENARIUS.] **1.** (*also* dĕn′yər). A unit of fineness for rayon, nylon, and silk yarns, based on a standard of 50 milligrams per 450 meters of yarn. **2.** (*also* də-nîr′). **a.** A small coin current in France and western Europe from the 8th cent. until the French Revolution. **b.** *Archaic.* A small sum.

**den·i·grate** (dĕn′ĭ-grāt′) *vt.* **-grat·ed, -grat·ing, -grates.** [Lat. *denigrare, denigrat-,* to blacken, defame : *de-* (intensive) + *niger,* black.] **1.** To deny the validity or importance of : DISPARAGE. **2.** To slander the character or reputation of. **—den′i·gra′tion** *n.* **—den′i·gra′tor** *n.*

**den·im** (dĕn′ĭm) *n.* [Fr. *(serge) de Nîmes,* (serge) of Nîmes, a city in southern France.] **1. a.** A coarse twilled cloth used for jeans, overalls, and work uniforms. **b. denims.** Garments made of coarse denim. **2.** A finer grade of denim material used in draperies and upholstery.

**de·ni·tri·fy** (dē-nī′trə-fī′) *vt.* **-fied, -fy·ing, -fies.** To remove nitrogen from (a material or chemical compound), as by bacterial action on soil. **—de·ni′tri·fi·ca′tion** *n.*

**den·i·zen** (dĕn′ĭ-zən) *n.* [ME *denisein* < AN *deinzein* < OFr. *deinz,* within < LLat. *deintus,* from within : Lat. *de-,* from + Lat. *intus,* within.] **1.** An inhabitant. **2.** *Chiefly Brit.* A foreigner permitted certain rights and privileges of citizenship. **3.** One that frequents a certain place : HABITUÉ. **4.** *Ecol.* An animal or plant naturalized in a region to which it is not indigenous. *—vt.* **-zened, -zen·ing, -zens.** *Chiefly Brit.* To naturalize. **—den′i·zen·a′tion** *n.*

**den mother** *n.* A woman who supervises a den of Cub Scouts.

**de·nom·i·nate** (dĭ-nŏm′ə-nāt′) *vt.* **-nat·ed, -nat·ing, -nates.** [Lat. *denominare, denominat-* : *de-* (intensive) + *nominare,* to name < *nomen,* name.] To give a name to : DESIGNATE. **—de·nom′i·na·ble** (-nə-bəl) *adj.*

**de·nom·i·nate number** (dĭ-nŏm′ə-nĭt) *n.* A number that designates a quantity as a multiple of a unit; e.g., *20* in the expression *20 pounds.*

**de·nom·i·na·tion** (dĭ-nŏm′ə-nā′shən) *n.* **1.** The act of naming. **2.** A name or title. **3.** The name of a class or group : CLASSIFICATION. **4.** A class of units having given values, as in a system of weights or currency. **5.** An organized group of religious congregations. **—de·nom′i·na′tion·al** *adj.* **—de·nom′i·na′tion·al·ly** *adv.*

**de·nom·i·na·tion·al·ism** (dĭ-nŏm′ə-nā′shə-nə-lĭz′əm) *n.* **1.** The tendency to separate into religious sects or denominations. **2.** Advocacy of the principles of denominationalism. **3.** Strict adherence to a denomination : SECTARIANISM. **—de·nom′i·na′tion·al·ist** *n.*

**de·nom·i·na·tive** (dĭ-nŏm′ə-nā′tĭv, -nə-tĭv) *adj.* **1.** Giving or being a name : DESIGNATIVE. **2.** Formed from a noun or adjective. *—n.* A word, esp. a verb, derived from a noun or adjective.

**de·nom·i·na·tor** (dĭ-nŏm′ə-nā′tər) *n.* **1.** *Math.* The quantity below the line indicating the number of units into which a whole is divided. **2.** A common trait or characteristic. **3.** An average level or standard <the *denominator* of public taste>

**de·no·ta·tion** (dē′nō-tā′shən) *n.* **1.** The act of denoting : INDICATION. **2.** A sign, symbol, or reference that denotes : INDICATOR. **3.** Something signified or referred to. **4.** The explicit meaning of a word as opposed to its connotation.

**de·no·ta·tive** (dĭ-nō′tə-tĭv, dē′nō-tā′-) *adj.* **1.** Able to denote : DESIGNATIVE. **2.** Explicit. **—de·no′ta·tive·ly** *adv.*

**de·note** (dĭ-nōt′) *vt.* **-not·ed, -not·ing, -notes.** [OFr. *denoter* < Lat. *denotare* : *de-* (intensive) + *notare,* to mark < *nota,* mark.] **1.** To indicate or reveal : MARK. **2.** To serve as a name or symbol for : SIGNIFY. **3.** To refer to specifically : mean explicitly. ***usage:*** In speaking of words, *denote* is used to indicate the thing a word names whereas *connote* is used to indicate our associations with that thing, as in *The word "bachelor" denotes an unmarried man and connotes a life of parties and carefree amusements.* **—de·not′a·ble** *adj.* **—de·no′tive** *adj.*

**dé·noue·ment** *also* **de·noue·ment** (dā′no͞o-mäN′) *n.* [Fr. < OFr. *desnouement,* an untying < *desnouer,* to undo : *des-,* de- + *nouer,* to tie < Lat. *nodare* < *nodus,* knot.] **1.** The outcome of the plot of a play or novel. **2.** The outcome or final solution of a sequence of events.

**de·nounce** (dĭ-nouns′) *vt.* **-nounced, -nounc·ing, -nounc·es.** [ME *denouncen,* to proclaim < OFr. *denoncier* < Lat. *denuntiare* : *de-* (intensive) + *nuntiare,* to announce < *nuntius,* messenger.] **1.** To condemn openly, esp. as evil : CENSURE. **2.** To accuse formally. **3.** To announce formally the ending of (a treaty). **—de·nounce′ment** *n.* **—de·nounc′er** *n.*

**dense** (dĕns) *adj.* **dens·er, dens·est.** [Lat. *densus.*] **1. a.** Having relatively high density. **b.** Crowded closely together : COMPACT. **2.** Thick or impenetrable <*dense* smoke> **3.** Requiring effort to understand because of complexity in structure or content <a *dense* book> **4.** Slow to understand : STUPID. **5.** Opaque, with good contrast between light and dark areas. —Used of a developed photographic negative. **—dense′ly** *adv.* **—dense′ness** *n.*

**den·sim·e·ter** (dĕn-sĭm′ĭ-tər) *n.* [Lat. *densus,* dense + -METER.] A device for determining density. **—den′si·met′ric** (-sə-mĕt′rĭk) *adj.*

**den·si·tom·e·ter** (dĕn′sĭ-tŏm′ĭ-tər) *n.* [DENSIT(Y) + -METER.] A device for measuring the optical density of a material, as a negative.

**den·si·ty** (dĕn′sĭ-tē) *n., pl.* **-ties. 1.** The quality or state of being dense. **2.** *Physics.* **a.** The amount of something per unit measure, esp. per unit length, area, or volume. **b.** The mass per unit volume of a substance under specified or standard conditions of temperature and pressure. **3.** *Computer Sci.* The number of units of useful information contained within a linear dimension. **4.** The number of inhabitants per unit geographic region. **5.** The degree of optical opacity of a medium or material, as of a photographic negative. **6.** Thickness of consistency. **7.** Complexity of structure or content. **8.** Stupidity.

**dent[1]** (dĕnt) *n.* [ME *dent,* alteration of *dint,* blow < OE *dynt.*] **1.** A depression in a surface made by pressure or a blow. **2.** A usu. weakening effect or impression made by or as if by lessening. **3.** Meaningful progress : HEADWAY <couldn't make a *dent* in the workload> —*vt. & vi.* **dent·ed, dent·ing, dents.** To make a dent in or to become dented.

**dent²** (dĕnt) *n.* [Fr. < Lat. *dens.*] TOOTH 3.

**dent-** *pref. var. of* DENTI-.

**den·tal** (dĕn'tl) *adj.* [NLat. *dentalis* < Lat. *dens,* tooth.] **1.** Of, relating to, or for the teeth. **2.** Of, relating to, or intended for dentistry. **3.** Articulated with the tip of the tongue near or against the upper front teeth. —Used of a speech sound. —*n.* A dental consonant.

**dental floss** *n.* A strong waxed or unwaxed thread used to clean between the teeth.

**dental hygienist** *n.* A licensed professional who provides preventive dental services, as cleaning and taking x-rays.

**dental plate** *n.* A denture.

**dental technician** *n.* One who makes dental appliances, as bridges or dentures.

**den·tate** (dĕn'tāt') *adj.* [Lat. *dentatus* < *dens,* tooth.] Edged with toothlike projections <a *dentate* leaf> **—den'tate'ly** *adv.*

**den·ta·tion** (dĕn-tā'shən) *n.* **1.** The condition of being dentate. **2.** A toothlike projection or part.

**dent corn** *n.* A tall-growing variety of corn, *Zea mays indentata,* with yellow or white kernels that are indented at the tip.

**denti-** or **dent-** *pref.* [< Lat. *dens, dent-,* tooth.] **1.** Tooth <*dentoid*> **2.** Dental <*denti*labial>

**den·ti·cle** (dĕn'tĭ-kəl) *n.* [ME < Lat. *denticulus,* dim. of *dens,* tooth.] A small tooth or toothlike projection.

**den·tic·u·late** (dĕn-tĭk'yə-lĭt) *also* **den·tic·u·lat·ed** (-lā'tĭd) *adj.* [Lat. *denticulatus* < *denticulus,* denticle.] **1.** Finely dentate. **2.** Having dentils. **—den·tic'u·late·ly** *adv.* **—den·tic'u·la'tion** *n.*

**den·ti·form** (dĕn'tə-fôrm') *adj.* Tooth-shaped.

**den·ti·frice** (dĕn'tə-frĭs) *n.* [OFr. < Lat. *dentifricium* : *dens,* tooth + *fricare,* to rub.] A paste or powder for cleaning the teeth.

**den·tig·er·ous** (dĕn-tĭj'ər-əs) *adj.* [DENTI- + Lat. *gerere,* to bear.] Having teeth.

**den·til** (dĕn'tĭl) *n.* [Obs. Fr. *dentille* < OFr., dim. of *dent,* tooth < Lat. *dens.*] One of a series of small rectangular blocks forming an architectural molding or projecting beneath a cornice.

**den·tine** (dĕn'tēn') *also* **den·tin** (-tĭn) *n.* The calcareous part of a tooth beneath the enamel, containing the pulp chamber and root canals. **—den·tin'al** (dĕn-tē'nəl, dĕn'tə-) *adj.*

**den·tist** (dĕn'tĭst) *n.* [Fr. *dentiste* < *dent,* tooth < Lat. *dens.*] One whose profession is dentistry.

**den·tist·ry** (dĕn'tĭ-strē) *n.* The diagnosis, prevention, and treatment of diseases of the teeth, gums, and related structures.

**den·ti·tion** (dĕn-tĭsh'ən) *n.* [Lat. *dentitio* < *dentitus,* p.part. of *dentire,* to teethe < *dens,* tooth.] **1.** *Biol.* The type, number, and arrangement of teeth, esp. in animals. **2.** The development and cutting of teeth : TEETHING.

**den·toid** (dĕn'toid') *adj.* Toothlike.

**den·tu·lous** (dĕn'chə-ləs) *adj.* [Back-formation < E. *edentulous,* toothless < Lat. *edentulus* : *ex-,* out + *dens,* tooth.] Having teeth.

**den·ture** (dĕn'chər) *n.* [Fr. < OFr. < *dent,* tooth < Lat. *dens.*] *often* **dentures.** A set of artificial teeth.

**de·nu·cle·ar·ize** (dē-nōō'klē-ə-rīz', -nyōō'-) *vt.* **-ized, -iz·ing, -iz·es.** To prohibit or remove nuclear arms from. **—de·nu'cle·ar·i·za'·tion** *n.*

**de·nu·date** (dĭ-nōō'dāt', -nyōō'-) *vt.* **-dat·ed, -dat·ing, -dates.** [Lat. *denudare, denudat-,* to denude.] To denude. —*adj.* Bare.

**de·nude** (dĭ-nōōd', -nyōōd') *vt.* **-nud·ed, -nud·ing, -nudes.** [Lat. *denudare* : *de-,* completely + *nudare,* to make bare < *nudus,* nude.] **1.** To divest of covering : make bare. **2.** *Geol.* To expose (rock strata) by erosion. **—de'nu·da'tion** (dē'nōō-dā'shən, -nyōō-, dĕn'yōō-) *n.*

**de·nu·mer·a·ble** (dĭ-nōō'mər-ə-bəl, -nyōō'-) *adj.* Capable of being put into one-to-one correspondence with the positive integers : COUNTABLE. **—de·nu'mer·a·bil'i·ty** *n.* **—de·nu'mer·a·bly** *adv.*

**de·nun·ci·a·tion** (dĭ-nŭn'sē-ā'shən, -shē-) *n.* **1.** Open censure or condemnation. **2.** The act of accusing another of a crime before a public prosecutor. **—de·nun'ci·a'tive, de·nun'ci·a·to'ry** (-ə-tôr'ē, -tōr'ē) *adj.*

**de·ny** (dĭ-nī') *vt.* **-nied, -ny·ing, -nies.** [ME *denien* < OFr. *denier* < Lat. *denegare* : *de-* (intensive) + *negare,* to say no.] **1.** To declare untrue : CONTRADICT. **2.** To refuse to believe : REJECT. **3.** To refuse to recognize or acknowledge : DISAVOW. **4. a.** To refuse to grant : WITHHOLD. **b.** To abstain from indulging (oneself) in <*denied* myself dessert>

☆ ***syns:*** DENY, CONTRADICT, REFUTE *v. core meaning* : to dispute the truth, reality, or worth of <couldn't *deny* that they had witnessed the crime> DENY is the most general and usu. implies an open declaration that something is untrue <*denied* the accusation> To CONTRADICT is to assert that the opposite of a given statement is true <You just *contradicted* my explanation.> REFUTE implies the use of evidence to disprove a claim <voting statistics that *refute* your claim to victory> ***ant:*** confirm

**de·o·dar** (dē'ə-där') *or* **de·o·dar·a** (dē'ə-där'ə) *n.* [Hindi *dē'odār* < Skt. *devadāru* : *deva-,* divine + *dāru,* wood.] A very tall cedar native to the Himalayas, *Cedrus deodara,* with drooping branches and wood valued as timber.

**de·o·dor·ant** (dē-ō'dər-ənt) *n.* **1.** A substance applied to counteract body odors. **2.** A chemical exposed to or sprayed into the air to counteract staleness. **—de·o'dor·ant** *adj.*

**de·o·dor·ize** (dē-ō'də-rīz') *vt.* **-ized, -iz·ing, -iz·es.** To disguise or absorb the odor of. **—de·o'dor·i·za'tion** *n.* **—de·o'dor·iz'er** *n.*

**de·on·tol·o·gy** (dē'ŏn-tŏl'ə-jē) *n.* [Gk. *deon, deont-,* obligation, necessity < neuter p.part. of *dein,* to need, lack + -LOGY.] The theory or study of moral obligation or commitment : ETHICS. **—de·on'to·log'i·cal** (-tə-lŏj'ĭ-kəl) *adj.* **—de'on·tol'o·gist** *n.*

**de·or·bit** (dē-ôr'bĭt) *vi. & vt.* **-bit·ed, -bit·ing, -bits.** To go or cause to go out of orbit. —*n.* The process of deorbiting.

**De·o vo·len·te** (dē'ō və-lĕn'tē, dā'ō) *adv.* [Lat.] God being willing.

**de·ox·i·dize** (dē-ŏk'sĭ-dīz') *vt.* **-dized, -diz·ing, -diz·es.** To remove oxygen, esp. oxygen in chemical combination, from. **—de·ox'·i·di·za'tion** *n.* **—de·ox'i·diz'er** *n.*

**deoxy-** *pref.* [DE- + OXY-.] A molecule having less oxygen than another to which it is closely related <*deoxy*corticosterone>

**de·ox·y·cor·ti·cos·ter·one** (dē-ŏk'sē-kôr'tĭ-kŏs'tə-rōn') *n.* A steroid hormone, $C_{21}H_{30}O_3$, derived from the adrenal cortex and used in the treatment of adrenal insufficiency.

**de·ox·y·gen·ate** (dē-ŏk'sə-jə-nāt') *vt.* **-at·ed, -at·ing, -ates.** To remove oxygen from. **—de·ox'y·gen·a'tion** *n.*

**de·ox·y·ri·bo·nu·cle·ic acid** (dē-ŏk'sē-rī'bō-nōō-klē'ĭk, -nyōō-) *n.* DNA.

**de·ox·y·ri·bo·nu·cle·o·tide** (dē-ŏk'sē-rī'bō-nōō'klē-ə-tīd', -nyōō'-) *n.* [DEOXYRIBO(SE) + NUCLEOTIDE.] A nucleotide containing deoxyribose, a constituent of DNA.

**de·ox·y·ri·bose** (dē-ŏk'sē-rī'bōs') *n.* A sugar, $C_5H_{10}O_4$, a constituent of deoxyribonucleic acid.

**de·part** (dĭ-pärt') *v.* **-part·ed, -part·ing, -parts.** [ME *departen* < OFr. *departir,* to divide : *de-,* away (< Lat.) + *partir,* to divide < Lat. *partire* < *pars,* part.] —*vi.* **1.** To go away : LEAVE. **2.** To die. **3.** To vary, as from a regular course : DEVIATE. —*vt.* To go away from.

**de·part·ed** (dĭ-pär'tĭd) *adj.* **1.** Gone by : PAST <*departed* glory> **2.** Dead.

**de·part·ment** (dĭ-pärt'mənt) *n.* [Fr. *département* < OFr., separation < *departir,* to divide. —see DEPART.] **1.** A distinct, usu. specialized division of a large organization, as a government or business. **2. Department.** One of the principal executive divisions of the U.S. federal government, headed by a cabinet officer. **3.** A French governmental administrative district. **4.** A division of a school or college dealing with a particular field of knowledge. **5.** *Informal.* A sphere of special knowledge or activity. **—de'part·men'tal** (dē'pärt-mĕn'tl) *adj.* **—de'part·men'tal·ly** *adv.*

**de·part·men·tal·ize** (dē'pärt-mĕn'tl-īz') *vt.* **-ized, -iz·ing, -iz·es.** To organize into departments. **—de'part·men'tal·i·za'tion** *n.*

**department store** *n.* A retail establishment organized in departments and offering a wide variety of merchandise and services.

**de·par·ture** (dĭ-pär'chər) *n.* **1.** The act of leaving. **2.** A starting out, as on a trip or a new course of action. **3.** A deviation or divergence, as from an established rule, plan, or procedure. **4.** The distance sailed due east or west by a ship on its course.

**de·pend** (dĭ-pĕnd') *vi.* **-pend·ed, -pend·ing, -pends.** [ME *dependen,* to hang down < OFr. *dependre* < Lat. *dependēre* : *de-,* down + *pendēre,* to hang.] **1.** To place trust in or rely on, as for support or aid <I *depend* on your skill.> **2.** To be determined, conditioned, or dependent <Everything *depends* on the weather.> **3.** To hang down. **4.** *Informal.* To be pending or undecided. ***usage:*** *Depend,* when it indicates condition or contingency, is followed by *on* or *upon,* as in *The success of the mission depends on (*or *upon) careful planning.*

**de·pend·a·ble** (dĭ-pĕn'də-bəl) *adj.* Capable of being depended on : RELIABLE. **—de·pend'a·bil'i·ty, de·pend'a·ble·ness** *n.* **—de·pend'a·bly** *adv.*

**de·pend·ance** (dĭ-pĕn'dəns) *n. var. of* DEPENDENCE.

**de·pend·an·cy** (dĭ-pĕn'dən-sē) *n. var. of* DEPENDENCY.

**de·pend·ant** (dĭ-pĕn'dənt) *adj. & n. var. of* DEPENDENT.

**de·pend·ence** *also* **de·pend·ance** (dĭ-pĕn'dəns) *n.* **1.** The state of being dependent. **2.** Subordination to one needed or greatly desired. **3.** The state of being determined, influenced, or controlled by something else. **4.** Reliance : trust.

**de·pend·en·cy** *also* **de·pend·an·cy** (dĭ-pĕn'dən-sē) *n., pl.* **-cies. 1.** Dependence. **2.** Something dependent or subordinate. **3.** A territory or state under the jurisdiction of another country from which it is separated geographically.

**de·pend·ent** *also* **de·pend·ant** (dĭ-pĕn'dənt) —*adj.* **1.** Contingent on something or someone else. **2.** Subordinate. **3.** Relying on the aid of another for support. **4.** Hanging down. —*n.* One, as a child, who relies on another for support. **—de·pend'ent·ly** *adv.*

**dependent clause** *n.* A clause that cannot stand alone as a sentence and acts as a noun, adjective, or adverb within a sentence.

**dependent variable** *n.* A mathematical variable whose value is determined by the value assumed by an independent variable.

**de·per·son·al·ize** (dē-pûr′sə-nə-līz′) *vt.* **-ized, -iz·ing, -iz·es.** **1.** To deprive of personal or individual character. **2.** To make impersonal <*depersonalize* an interview> **—de·per′son·al·i·za′tion** *n.*

**de·pict** (dĭ-pĭkt′) *vt.* **-pict·ed, -pict·ing, -picts.** [Lat. *depingere, depict-* : *de-*, completely + *pingere*, to picture.] **1.** To represent, as in a picture or sculpture. **2.** To represent in words : DESCRIBE. **—de·pic′tion** *n.*

**de·pig·men·ta·tion** (dē-pĭg′mən-tā′shən, -mĕn-) *n.* Loss of normal pigmentation.

**dep·i·late** (dĕp′ə-lāt′) *vt.* **-lat·ed, -lat·ing, -lates.** [Lat. *depilare, depilat-* : *de-*, completely + *pilare*, to deprive of hair < *pilus*, hair.] To remove hair from. **—dep′i·la′tion** *n.* **—dep′i·la′tor** *n.*

**de·pil·a·to·ry** (dĭ-pĭl′ə-tôr′ē, -tōr′ē) *adj.* Capable of removing hair. —*n., pl.* **-ries.** A cream or liquid used to remove unwanted body hair.

**de·plane** (dē-plān′) *vi.* **-planed, -plan·ing, -planes.** To disembark from an aircraft.

**de·plete** (dĭ-plēt′) *vt.* **-plet·ed, -plet·ing, -pletes.** [Lat. *deplere, deplet-*, to empty : *de-* (reversal) + *plere*, to fill.] **1.** To lessen or reduce in quantity, value, or effectiveness : EXHAUST <*depleted* by the exam> **2.** To empty <*deplete* a barrel of oil> **—de·plet′a·ble** *adj.* **—de·ple′tion** *n.*

**de·plor·a·ble** (dĭ-plôr′ə-bəl, -plōr′-) *adj.* **1.** Deserving severe reproach. **2.** Grievous : lamentable. **3.** Very bad : WRETCHED. **—de·plor′a·ble·ness, de·plor′a·bil′i·ty** *n.* **—de·plor′a·bly** *adv.*

**de·plore** (dĭ-plôr′, -plōr′) *vt.* **-plored, -plor·ing, -plores.** [OFr. *deplorer* < Lat. *deplorare* : *de-* (intensive) + *plorare*, to wail.] **1.** To feel or express sorrow over. **2.** To feel or express regret about. **3.** To feel or express strong disapproval of : CENSURE.

**de·ploy** (dĭ-ploi′) *v.* **-ployed, -ploy·ing, -ploys.** [Fr. *déployer* < OFr. *despleier* < Lat. *displicare*, to scatter : *dis-* (reversal) + *plicare*, to fold.] —*vt.* **1.** To station (persons or forces) systematically over an area. **2.** To spread out (troops) to form an extended front. —*vi.* To be or become deployed. **—de·ploy′ment** *n.*

**de·plume** (dē-plo͞om′) *vt.* **-plumed, -plum·ing, -plumes.** [ME *deplumen* < OFr. *deplumer* < Med. Lat. *deplumare* : *de-*, off + *pluma*, feather.] **1.** To pluck the feathers from. **2.** To deprive of pride or honor. **—de′plu·ma′tion** *n.*

**de·po·lar·ize** (dē-pō′lə-rīz′) *vt.* **-ized, -iz·ing, -iz·es.** To counteract or eliminate the polarization of. **—de·po′lar·i·za′tion** *n.*

**de·po·lit·i·cize** (dē′pə-lĭt′ĭ-sīz′) *vt.* **-cized, -ciz·ing, -ciz·es.** To remove from the political sphere <*depoliticize* a world hunger program> **—de′po·lit′i·ci·za′tion** *n.*

**de·pol·lute** (dē′pə-lo͞ot′) *vt.* **-lut·ed, -lut·ing, -lutes.** To remove the pollution from <*depollute* a lake>

**de·pone** (dĭ-pōn′) *v.* **-poned, -pon·ing, -pones.** [ME *deponen* < Med. Lat. *deponere* < Lat., to put down : *de-*, down + *ponere*, to put.] —*vt.* To declare or testify under oath. —*vi.* To give testimony.

**de·po·nent** (dĭ-pō′nənt) *adj.* [LLat. *deponens, deponent-* < Lat., pr.part. of *deponere*, to put down. —see DEPONE.] Denoting a verb of active meaning but passive form, as certain Latin and Greek verbs. —*n.* **1.** A deponent verb. **2.** *Law.* A person who testifies under oath, esp. in writing.

**de·pop·u·late** (dē-pŏp′yə-lāt′) *vt.* **-lat·ed, -lat·ing, -lates.** [Lat. *depopulari, depopulat-*, to lay waste : *de-* (intensive) + *populari*, to ravage < *populus*, people, throng.] To reduce greatly the population of, as by expulsion, disease, or massacre. **—de·pop′u·la′tion** *n.* **—de·pop′u·la′tor** *n.*

**de·port** (dĭ-pôrt′, -pōrt′) *vt.* **-port·ed, -port·ing, -ports.** [Partly < Fr. *déporter*, to banish, and partly < OFr. *deporter*, to behave, both < Lat. *deportare*, to carry away : *de-*, away + *portare*, to carry.] **1.** To expel from a country. **2.** To behave or conduct (oneself) in a specified manner.

**de·port·a·ble** (dĭ-pôr′tə-bəl, -pōr′-) *adj.* Subject to or punishable by deportation.

**de·por·ta·tion** (dē′pôr-tā′shən, -pōr-) *n.* **1.** An act or instance of deporting. **2.** Expulsion of an undesirable alien from a country.

**de·port·ee** (dē′pôr-tē′, -pōr-) *n.* A deported individual.

**de·port·ment** (dĭ-pôrt′mənt, -pōrt′-) *n.* Behavior : demeanor.

**de·pos·al** (dĭ-pō′zəl) *n.* The act of deposing from office.

**de·pose** (dĭ-pōz′) *v.* **-posed, -pos·ing, -pos·es.** [ME *deposen* < OFr. *deposer* : *de-*, away (< Lat.) + *poser*, to put. —see POSE[1].] —*vt.* **1.** To remove from office or a powerful position. **2.** *Archaic.* To put or lay down : DEPOSIT. **3.** *Law.* To declare under oath, esp. in writing. —*vi.* *Law.* To testify, esp. in writing. **—de·pos′a·ble** *adj.*

**de·pos·it** (dĭ-pŏz′ĭt) *v.* **-it·ed, -it·ing, -its.** [Lat. *deponere, deposit-* : *de-*, aside + *ponere*, to put.] —*vt.* **1.** To lay or set down : PLACE. **2.** To put down (e.g., layers of sediment) by a natural process. **3.** To give as partial payment or security. **4.** To entrust (money) to a bank. —*vi.* To become deposited : SETTLE. —*n.* **1.** Something entrusted for safekeeping, as money in a bank. **2.** The state of being deposited. **3.** A partial or initial payment of a cost or debt. **4.** A sum of money given as security for an item acquired for temporary use. **5.** A depository. **6.** Something deposited esp. by a natural process, as mineral or sandy material settled out of water. **—de·pos′i·tor** *n.*

**de·pos·i·tar·y** (dĭ-pŏz′ĭ-tĕr′ē) *n., pl.* **-ies.** **1.** A person entrusted with something. **2.** DEPOSITORY 1.

**dep·o·si·tion** (dĕp′ə-zĭsh′ən) *n.* **1.** The act of deposing, as from office. **2.** The act of depositing. **3.** Something deposited : DEPOSIT. **4.** *Law.* Testimony under oath, esp. a written statement by a witness for use in court in his or her absence. **—dep′o·si′tion·al** *adj.*

**de·pos·i·to·ry** (dĭ-pŏz′ĭ-tôr′ē, -tōr′ē) *n., pl.* **-ries.** **1.** A place where something is deposited for safekeeping. **2.** DEPOSITARY 1.

**de·pot** (dē′pō, dĕp′ō) *n.* [Fr. *dépôt* < OFr. *depost* < Lat. *depositum*, deposit < neuter p.part. of *deponere*, to deposit.] **1.** A bus or railroad station. **2.** A warehouse. **3. a.** A storage installation for military materials. **b.** A station for receiving, classifying, and assembling military personnel.

**de·prave** (dĭ-prāv′) *vt.* **-praved, -prav·ing, -praves.** [ME *depraven*, to corrupt < OFr. *depraver* < Lat. *depravare* : *de-* (intensive) + *pravus*, crooked.] To debase morally : CORRUPT. **—dep′ra·va′tion** (dĕp′rə-vā′shən) *n.* **—de·prav′er** *n.*

**de·praved** (dĭ-prāvd′) *adj.* Morally debased and corrupt : PERVERTED. **—de·prav′ed·ly** (-prā′vĭd-lē, -prāvd′lē) *adv.*

**de·prav·i·ty** (dĭ-prăv′ĭ-tē) *n., pl.* **-ties.** **1.** Moral corruption. **2.** A wicked or perverse act.

**dep·re·cate** (dĕp′rĭ-kāt′) *vt.* **-cat·ed, -cat·ing, -cates.** [Lat. *deprecari, deprecat-*, to ward off by prayer : *de-*, against + *precari*, to pray.] **1.** To express disapproval of. **2.** To belittle : depreciate. ***usage:*** In recent times, *deprecate* has encroached upon the meaning of *depreciate*, coming into use almost to the exclusion of the latter in the sense of "to belittle." **—dep′re·ca′tion** *n.* **—dep′re·ca′tor** *n.*

**dep·re·ca·to·ry** (dĕp′rĭ-kə-tôr′ē, -tōr′ē) *also* **dep·re·ca·tive** (-kā′tĭv) *adj.* Expressing deprecation : DISAPPROVING. **—dep′re·ca·to′ri·ly** *adv.*

**de·pre·cia·ble** (dĭ-prē′shə-bəl) *adj.* Capable of being depreciated in value.

**de·pre·ci·ate** (dĭ-prē′shē-āt′) *v.* **-at·ed, -at·ing, -ates.** [Med. Lat. *depreciare, depreciat-*, alteration of LLat. *depretiare* : Lat. *de-*, down + Lat. *pretium*, price.] —*vt.* **1.** To lessen the value or price of. **2.** To cause to seem less valuable or important : DISPARAGE. —*vi.* To diminish in value. **—de·pre′ci·a′tor** *n.*

☆ ***syns:*** DEPRECIATE, CHEAPEN, DEVALUATE, DEVALUE, DOWNGRADE, LOWER *v. core meaning* : to make or become less in price or value <The value of the dollar has *depreciated*.><used cars that steadily *depreciate*>

**de·pre·ci·a·tion** (dĭ-prē′shē-ā′shən) *n.* **1.** A decrease or loss in value esp. because of wear or age. **2.** An allowance made for a loss in value of property. **3.** A reduction in the purchasing value of money. **4.** An instance of disparaging.

**de·pre·cia·to·ry** (dĭ-prē′shə-tôr′ē, -tōr′ē) *also* **de·pre·cia·tive** (-shə-tĭv, -shē-ā′tĭv) *adj.* **1.** Diminishing in value. **2.** Disparaging.

**dep·re·date** (dĕp′rĭ-dāt′) *v.* **-dat·ed, -dat·ing, -dates.** [LLat. *depraedari, depraedat-* : Lat. *de-* (intensive) + Lat. *praedari*, to plunder < *praeda*, booty.] —*vt.* To prey on : PLUNDER. —*vi.* To engage in plundering. **—dep′re·da′tion** *n.* **—dep′re·da′tor** *n.* **—de·pred′a·to′ry** (dĭ-prĕd′ə-tôr′ē, -tōr′ē, dĕp′rĭ-də-) *adj.*

**de·press** (dĭ-prĕs′) *vt.* **-pressed, -press·ing, -press·es.** [ME *depressen*, to push down < OFr. *depresser* < Lat. *depressus*, p.part. of *deprimere* : *de-*, down + *premere*, to press.] **1.** To lower in spirits : SADDEN. **2.** To press down : LOWER <*depress* a pedal on a piano> **3.** To lessen the activity or force of : WEAKEN. **4.** To lower prices in (a stock market). **—de·press′i·ble** *adj.*

**de·pres·sant** (dĭ-prĕs′ənt) *adj.* Serving to lower the rate of vital activities. —*n.* A depressant drug.

**de·pressed** (dĭ-prĕst′) *adj.* **1.** Low in spirits : DEJECTED. **2.** *Bot.* Flattened downward, as if pressed from above. **3.** *Zool.* Flattened along the dorsal and ventral surfaces. **4.** Sunk below the surrounding region : HOLLOW. **5.** Suffering from social and economic hardship.

☆ ***syns:*** DEPRESSED, BACKWARD, DEPRIVED, DISADVANTAGED, IMPOVERISHED, UNDERPRIVILEGED *adj. core meaning* : economically and socially below standard <aid to *depressed* urban areas>

**de·press·ing** (dĭ-prĕs′ĭng) *adj.* Causing esp. emotional depression <a sad, *depressing* movie> **—de·press′ing·ly** *adv.*

**de·pres·sion** (dĭ-prĕsh′ən) *n.* **1.** The act of depressing or state of being depressed. **2.** An area sunk below its surroundings : HOLLOW. **3.** *Meteorol.* A region of low barometric pressure. **4.** The angular distance below the horizontal plane through the point of observation. **5.** *Astron.* The angular distance of a celestial body below the horizon. **6.** A reduction in force or activity. **7.** Melancholy : sadness. **8.** *Psychiat.* A neurotic or psychotic condition marked by an inability to concentrate, insomnia, and feelings of dejection and guilt. **9.** A period of severe economic decline, marked by unemployment, decreasing business activity, and falling prices.

**Depression glass** *n.* [After the Great *Depression*, a period of severe economic hardship during the 1930's.] Variously colored glassware produced in large quantities during the 1920's and 1930's.

**de·pres·sive** (dĭ-prĕs′ĭv) *adj.* **1.** Causing depression. **2.** Of or relating to psychological depression. **—de·pres′sive·ly** *adv.* **—de·pres′sive·ness** *n.*

**de·pres·sor** (dĭ-prĕs′ər) *n.* **1.** Something that depresses or is used to depress. **2.** A depressor nerve. **3.** A muscle that causes depression or contraction of a part. **4.** An instrument for depressing a part.

**depressor nerve** *n.* A nerve that lowers arterial blood pressure.

**de·pres·sur·ize** (dē-prĕsh'ə-rīz') *vt.* **-ized, -iz·ing, -iz·es.** To release from pressure. **—de·pres'sur·i·za'tion** *n.*

**de·priv·al** (dĭ-prī'vəl) *n.* Deprivation.

**dep·ri·va·tion** (dĕp'rə-vā'shən) *n.* **1.** The act of depriving or state of being deprived. **2.** A removal of rank or office.

**de·prive** (dĭ-prīv') *vt.* **-prived, -priv·ing, -prives.** [ME *depriven* < OFr. *depriver* < Med. Lat. *deprivare* : Lat. *de-*, completely + Lat. *privare,* to rob < *privus,* without.] **1.** To take something away from. **2.** To keep from having or enjoying : DENY <was *deprived* of the gold medal> **3.** To remove from office. **—de·priv'a·ble** *adj.*

**de·prived** (dĭ-prīvd') *adj.* Marked by deprivation, esp. of economic or social necessities.

**de·pro·gram** (dē-prō'grăm', -grəm) *vt.* **-grammed, -gram·ming, -grams** or **-gramed, -gram·ing, -grams.** To try to counteract the effect of an indoctrination, esp. a religious indoctrination. **—de·pro'gram'mer** *n.*

**depth** (dĕpth) *n.* [ME *depthe* < *dep,* deep.] **1.** The quality or condition of being deep. **2. a.** The extent, measurement, or dimension downward, backward, or inward. **b.** The linear measurement or sense of distance from an observation point, as perspective in painting. **3.** *often* **depths.** A deep part of or place <in the *depths* of the jungle> **4.** The most profound or intense part or stage <the *depth* of misery><the *depth* of meditation> **5.** The severest or worst part, as of winter. **6.** A bad or deteriorated condition. **7.** Intellectual complexity or penetration : PROFUNDITY. **8.** The range of one's understanding or competence <out of my *depth*> **9.** The degree of richness or intensity, as of color. **10.** Lowness in pitch. **—in depth.** With thoroughness <a report *in depth*>

**depth charge** *n.* A charge designed for explosion under water, used esp. against submarines.

**depth perception** *n.* Perception of spatial relationships, esp. of distances between objects, in three dimensions.

**depth psychology** *n.* **1.** Psychology of the unconscious. **2.** Psychoanalysis.

**dep·u·rate** (dĕp'yə-rāt') *vt.* & *vi.* **-rat·ed, -rat·ing, -rates.** [Med. Lat. *depurare, depurat-* : Lat. *de-*, away + *purus,* pure.] To cleanse or purify or become cleansed or purified. **—dep'u·ra'tion** *n.* **—dep'u·ra'tor** *n.*

**dep·u·ta·tion** (dĕp'yə-tā'shən) *n.* **1.** The act of deputing or state of being deputed. **2.** DELEGATION 2.

**de·pute** (dĭ-pyo͞ot') *vt.* **-put·ed, -put·ing, -putes.** [ME *deputen* < OFr. *deputer* < LLat. *deputare,* to allot < Lat., to consider : *de-*, apart, away + *putare,* to trim, arrange.] **1.** To authorize or appoint as an agent or representative. **2.** To assign (authority or duties) to another.

**dep·u·tize** (dĕp'yə-tīz') *vt.* & *vi.* **-tized, -tiz·ing, -tiz·es.** To appoint as or serve as a deputy. **—dep'u·ti·za'tion** *n.*

**dep·u·ty** (dĕp'yə-tē) *n., pl.* **-ties.** [ME *depute* < OFr. < p.part. of *deputer,* to depute.] **1.** One designated or empowered to act for another. **2.** An assistant exercising full authority in the absence of his or her superior and equal authority in emergencies. **3.** A representative in a legislative body in certain countries.

**de·rac·i·nate** (də-răs'ə-nāt') *vt.* **-nat·ed, -nat·ing, -nates.** [< Fr. *déraciner* < OFr. *desraciner* : *des-*, apart (< Lat. *dis-*) + *racine,* root < LLat. *radicina* < Lat. *radix.*] To pull out by or as if by the roots : UPROOT. **—de·rac'i·na'tion** *n.*

**de·rail** (dē-rāl') *vi.* & *vt.* **-railed, -rail·ing, -rails.** [Fr. *dérailler* : *dé-*, off (< Lat. *de-*) + *rail,* rail < E.] **1.** To run or cause to run off the rails. **2.** To be thrown or throw off course. **—de·rail'ment** *n.*

**de·rail·leur** (dĭ-rā'lər) *n.* [Fr. *dérailler,* to become derailed : *dé-*, off (< Lat. *de-*)+ *rail,* rail < E.] A gear mechanism on a bicycle that changes the gear ratio by moving the chain from one sprocket to another.

**derailleur**

**de·range** (dĭ-rānj') *vt.* **-ranged, -rang·ing, -rang·es.** [Fr. *déranger* < OFr. *desrengier* : *des-*, apart (< Lat. *dis-*) + *reng,* line, of Germanic orig.] **1.** To disturb the order of. **2.** To disturb the normal functioning or condition of. **3.** To make insane. **—de·range'ment** *n.*

**der·by** (dûr'bē; *Brit.* där'bē) *n., pl.* **-bies.** [After Edward Smith Stanley (1752–1834), 12th Earl of *Derby,* founder of the English Derby.] **1.** Any of various annual horse races, esp. for three-year-olds. **2.** A formal race either open to all or to a given category of contestants. **3.** A stiff felt hat with a round crown and a narrow, curved brim.

**de·reg·u·late** (dē-rĕg'yə-lāt') *vt.* **-lat·ed, -lat·ing, -lates.** To decontrol. **—de·reg'u·la'tion** *n.*

**der·e·lict** (dĕr'ə-lĭkt') *adj.* [Lat. *derelictus,* p.part. of *derelinquere,* to abandon : *de-*, completely + *relinquere,* to leave behind (*re-*, behind + *linquere,* to leave).] **1.** Neglectful of duty or obligation : REMISS. **2.** Deserted by an owner or guardian : ABANDONED. —*n.* **1.** Abandoned property, esp. a ship abandoned at sea. **2.** A homeless person lacking means of support : VAGRANT. **3.** *Law.* Land left dry by a permanent recession of the water line.

**der·e·lic·tion** (dĕr'ə-lĭk'shən) *n.* **1.** Willful neglect, as of duty. **2.** Abandonment. **3.** *Law.* **a.** Accession of land by the permanent recession of the water line. **b.** The land so gained.

**de·ride** (dĭ-rīd') *vt.* **-rid·ed, -rid·ing, -rides.** [Lat. *deridēre* : *de-* (pejorative) + *ridēre,* to laugh at.] To laugh at or ridicule contemptuously. **—de·rid'er** *n.* **—de·rid'ing·ly** *adv.*

**de ri·gueur** (də rē-gœr') *adj.* [Fr.] Required by the current custom or fashion.

**de·ri·sion** (dĭ-rĭzh'ən) *n.* [ME *derisioun* < OFr. *derision* < LLat. *derisio* < Lat. *derisus,* p.part. of *deridēre,* to deride.] **1.** The act of deriding or state of being derided. **2.** An object of ridicule : LAUGHINGSTOCK.

**de·ri·sive** (dĭ-rī'sĭv, -zĭv, -rĭs'ĭv, -rĭz'-) *adj.* Expressing or provoking derision. **—de·ri'sive·ly** *adv.* **—de·ri'sive·ness** *n.*

**de·ri·so·ry** (dĭ-rī'sə-rē, -zə-) *adj.* Derisive.

**der·i·vate** (dĕr'ə-vāt') *adj.* Derivative.

**der·i·va·tion** (dĕr'ə-vā'shən) *n.* **1. a.** The act or process of deriving. **b.** The state of being derived. **2.** Something derived. **3.** The form or source from which something is derived : ORIGIN. **4.** The historical origin and development of a word : ETYMOLOGY. **5.** The process by which new words are formed from existing words, chiefly by the addition of affixes to roots, stems, or words. **6.** *Math.* A logical or mathematical process indicating through a sequence of statements that a result such as a theorem or a formula necessarily follows from the initial assumptions. **—der'i·va'tion·al** *adj.*

**de·riv·a·tive** (dĭ-rĭv'ə-tĭv) *adj.* **1.** Resulting from derivation. **2.** Copied or adapted from others. —*n.* **1.** Something derived. **2.** A word formed from another by derivation. **3.** *Math.* The limit, as the increment in the argument of a function approaches zero, of the ratio of the increment in its value to the corresponding increment in the argument : the instantaneous rate of change of a function with respect to a variable. **4.** *Chem.* A compound derived or obtained from known or hypothetical substances and containing essential elements of the parent substance. **—de·riv'a·tive·ly** *adv.* **—de·riv'a·tive·ness** *n.*

**de·rive** (dĭ-rīv') *v.* **-rived, -riv·ing, -rives.** [ME *deriven,* to be derived from < OFr. *deriver* < Lat. *derivare,* to derive : *de-*, away + *rivus,* stream.] —*vt.* **1.** To receive or obtain from a source. **2.** To arrive at by reasoning : deduce or infer. **3.** To trace the origin or development of (e.g., a word). **4.** *Chem.* To produce or obtain (a compound) from another substance by chemical reaction. —*vi.* To issue from a source : ORIGINATE. **—de·riv'a·ble** *adj.* **—de·riv'er** *n.*

**derm** (dûrm) *n. var. of* DERMA[1].

**derm-** *pref. var. of* DERMA-.

**-derm** *suff.* [< Gk. *derma,* skin.] Skin : covering <blasto*derm*>

**der·ma[1]** (dûr'mə) *also* **derm** (dûrm) *or* **der·mis** (dûr'mĭs) *n.* [NLat. < Gk., skin.] *Anat.* The corium.

**der·ma[2]** (dûr'mə) *n.* [Yiddish *derme,* pl. of *darm,* intestine < MHG < OHG.] Beef casing stuffed with a seasoned mixture of matzo meal or flour, onion, and suet, prepared by boiling, then roasting.

**derma-** or **derm-** or **dermo-** *pref.* [< Gk. *derma,* skin.] Skin <*derma*l>

**-derma** *suff.* [NLat. < Gk. *derma,* skin.] Skin : skin disease <sclero*derma*>

**der·ma·bra·sion** (dûr'mə-brā'zhən) *n.* A surgical procedure for removing skin imperfections, as wrinkles or scars, by abrasion of the frozen epidermis.

**der·mal** (dûr'məl) *also* **der·mic** (-mĭk) *adj.* Of or relating to the skin.

**dermat-** *pref. var. of* DERMATO-.

**der·ma·ti·tis** (dûr'mə-tī'tĭs) *n.* Inflammation of the skin.

**dermato-** or **dermat-** *pref.* [< Gk. *derma, dermat-*, skin.] Skin <*dermato*me>

**der·mat·o·gen** (dûr-măt'ə-jən) *n. Bot.* The outer layer of meristem, from which the epidermis is formed.

**der·ma·toid** (dûr'mə-toid') *n.* Resembling skin.

**der·ma·tol·o·gy** (dûr'mə-tŏl'ə-jē) *n.* Medical study of skin physiology and pathology. **—der'ma·to·log'i·cal** (-tə-lŏj'ĭ-kəl) *adj.* **—der'ma·tol'o·gist** *n.*

**der·ma·tome** (dûr'mə-tōm') *n.* The lateral wall of a somite, from which the corium is formed.

**der·mat·o·phyte** (dûr-măt'ə-fīt', dûr'mə-tə-) *n.* Any of various fungi causing skin disease. **—der·mat'o·phyt'ic** (-fĭt'ĭk) *adj.*

**der·ma·to·phy·to·sis** (dûr'mə-tō'fī-tō'sĭs) *n.* Athlete's foot.

**der·ma·to·plas·ty** (dûr′mə-tō-plăs′tē) *n.* Use of skin grafts in plastic surgery to replace skin loss or correct defects.
**der·ma·to·sis** (dûr′mə-tō′sĭs) *n., pl.* **-ses** (-sēz′). A skin disease.
**-dermatous** *suff.* [< Gk. *derma, dermat-,* skin.] Having a specified type of skin <sclero*dermatous*>
**der·mic** (dûr′mĭk) *adj. var. of* DERMAL.
**der·mis** (dûr′mĭs) *n. var. of* DERMA[1].
**dermo-** *pref. var. of* DERMA-.
**der·nier cri** (dĕr′nyā krē′) *n.* [Fr. : *dernier,* last + *cri,* cry.] The newest or latest fashion.
**der·o·gate** (dĕr′ə-gāt′) *v.* **-gat·ed, -gat·ing, -gates.** [Lat. *derogare, derogat-,* to take away : *de-,* away + *rogare,* to ask.] —*vi.* **1.** To take away : DETRACT. **2.** To stray from a standard or expectation : DEVIATE. —*vt.* To disparage. **—der′o·ga′tion** *n.* **—de·rog′a·tive** (dĭ-rŏg′ə-tĭv, dĕr′ə-gā′-) *adj.*
**de·rog·a·to·ry** (dĭ-rŏg′ə-tôr′ē, -tōr′ē) *adj.* **1.** Disparaging. **2.** Detracting. **—de·rog′a·to′ri·ly** *adv.* **—de·rog′a·to′ri·ness** *n.*
**der·rick** (dĕr′ĭk) *n.* [Obs. *derick,* gallows, hangman, after *Derick,* 16th-cent. English hangman.] **1.** A large crane for hoisting and moving heavy objects, connected to the base of an upright stationary beam and consisting of a movable boom equipped with cables and pulleys. **2.** A tall framework over the opening of a drilled hole, as an oil well, used to support boring equipment or to hoist and lower pipe lengths.
**der·ri·ère** *also* **der·ri·ere** (dĕr′ē-âr′) *n.* [Fr. < *derrière,* behind < OFr. *deriere,* in back of < Lat. *de retro.*] The buttocks.
**der·ring-do** (dĕr′ĭng-dōō′) *n.* [ME < *dorring don,* daring to do.] Daring spirit and action : VALOR.
**der·rin·ger** (dĕr′ĭn-jər) *n.* [After Henry *Deringer,* 19th-cent. American gunsmith.] A short-barreled pistol having a large bore.
**der·ris** (dĕr′ĭs) *n.* [NLat. < Gk., covering.] A tropical Asian woody vine of the genus *Derris,* whose roots yield rotenone.
**der·vish** (dûr′vĭsh) *n.* [Turk. *derviş,* mendicant < Pers. *dārvīsh.*] A member of any of various Moslem orders of ascetics, some of which employ whirling dances and the chanting of religious formulas to produce a collective ecstasy.
**DES** (dē′ē-ĕs′) *n.* Diethylstilbestrol.
**de·sal·i·nate** (dē-săl′ə-nāt′) *vt.* **-nat·ed, -nat·ing, -nates.** To desalinize. **—de·sal′i·na′tion** *n.* **—de·sal′i·na′tor** *n.*
**de·sal·i·nize** (dē-săl′ə-nīz′) *vt.* **-nized, -niz·ing, -niz·es.** To remove (salts and other chemicals) from saline water, esp. sea water. **—de·sal′i·ni·za′tion** *n.*
**de·salt** (dē-sôlt′) *vt.* **-salt·ed, -salt·ing, -salts.** To desalinize.
**des·cant** (dĕs′kănt′) *n.* [ME < AN *descaunt* <Med. Lat. *discantus,* a refrain : Lat. *dis-,* apart + Lat. *cantus,* song < p.part. of *canere,* to sing.] **1.** *also* **dis·cant** (dĭs′-). *Mus.* **a.** An ornamental melody or counterpoint sung or played above a musical theme. **b.** The highest part sung in part music. **2.** A discourse or discussion on a theme. —*vi.* (dĕs′kănt′, dĕ-skănt′) **-cant·ed, -cant·ing, -cants. 1.** To comment at length : DISCOURSE. **2.** *also* **dis·cant** (dĭ-skănt′). *Mus.* **a.** To sing or play a descant. **b.** To sing melodiously. **—des′cant′er** *n.*
**de·scend** (dĭ-sĕnd′) *v.* **-scend·ed, -scend·ing, -scends.** [ME *descenden* < OFr. *descendre* < Lat. *descendere* : *de-,* down + *scandere,* to climb.] —*vi.* **1.** To move from a higher to a lower place. **2.** To slope, extend, or incline downward <"A rough path *descended* like a steep stair into the plain" —J.R.R. Tolkien> **3. a.** To come down from a source : DERIVE <*descended* from the Pilgrims> **b.** To pass through inheritance <The estate has *descended* through several generations.> **4.** To sink : stoop <*descended* to whining and wheedling> **5.** To arrive or attack suddenly or overwhelmingly <The guests *descended* on the food.> —*vt.* To move from a higher to a lower part of. **—de·scend′i·ble, de·scend′a·ble** *adj.*
**de·scen·dant** (dĭ-sĕn′dənt) *n.* **1.** An individual descended from another. **2.** Something derived from an earlier form or prototype. —*adj. var. of* DESCENDENT.

☆ ***syns:*** DESCENDANT, CHILD, OFFSPRING, SCION *n. core meaning* : one descended directly from the same parents or ancestors <Elizabeth II—a *descendant* of Queen Victoria and Prince Albert> ***ant:*** ancestor, ascendant

**de·scen·dent** *also* **de·scen·dant** (dĭ-sĕn′dənt) *adj.* **1.** Moving downward : DESCENDING. **2.** Proceeding by descent from an ancestor.
**de·scend·er** (dĭ-sĕn′dər) *n.* The part of certain letters, as *j, p,* or *y,* that extends below the bottom of most lower-case letters.
**de·scent** (dĭ-sĕnt′) *n.* [ME < OFr. < *descendre,* to descend.] **1.** The act or an instance of descending. **2.** A way down. **3.** A downward incline or passage : SLOPE. **4. a.** Hereditary derivation : LINEAGE. **b.** The fact or process of coming down or being derived from a source <traced the *descent* of the folktale> **c.** Development in form or structure during transmission from an original source. **5.** One generation of a specific lineage. **6.** *Law.* Transference of property by inheritance. **7.** A lowering or decline, as in status or level. **8.** A sudden attack : ONSLAUGHT.
**de·scram·ble** (dē-skrăm′bəl) *vt.* **-bled, -bling, -bles.** To decode.
**de·scribe** (dĭ-skrīb′) *vt.* **-scribed, -scrib·ing, -scribes.** [Lat. *describere,* to delineate : *de-,* down + *scribere,* to write.] **1.** To give a verbal account of. **2. a.** To transmit a mental image or impression of with words. **b.** To present a lifelike image of. **3.** To trace or draw the outline of, as a circle. **—de·scrib′a·ble** *adj.* **—de·scrib′er** *n.*

☆ ***syns:*** DESCRIBE, NARRATE, RECITE, RECOUNT, RELATE, REPORT *v. core meaning* : to give a verbal account of <*described* the incident to the police>

**de·scrip·tion** (dĭ-skrĭp′shən) *n.* [ME *descripcioun* < OFr. *description* < Lat. *descriptio* < *descriptus,* p.part. of *describere,* to delineate. —see DESCRIBE.] **1.** The act, process, or technique of describing. **2.** A statement or account that describes. **3.** The act of drawing or tracing a figure. **4.** A kind or type <flowers of every *description*>
**de·scrip·tive** (dĭ-skrĭp′tĭv) *adj.* **1.** Serving to describe. **2.** Concerned with description or classification rather than explanation <*descriptive* science> **3.** Expressing an attribute of the modified noun, e.g., *blue* in *blue sky.* —Used of an adjective or adjectival clause. **—de·scrip′tive·ly** *adv.* **—de·scrip′tive·ness** *n.*
**descriptive geometry** *n.* The collection of mathematical techniques used to describe geometric relationships among three-dimensional structures on a plane surface.
**descriptive linguistics** *n. (sing. in number).* The study of a language or languages at a specific stage of development, with emphasis on constructing a complete grammar rather than on historical development or comparison with other languages.
**de·scrip·tor** (dĭ-skrĭp′tər) *n.* [LLat., describer < Lat. *descriptus,* p.part. of *describere,* to describe.] *Computer Sci.* A word, phrase, or alphanumeric character used to identify an item in an information retrieval system.
**de·scry** (dĭ-skrī′) *vt.* **-scried, -scry·ing, -scries.** [ME *descrien,* to proclaim < OFr. *descrier.* —see DECRY.] **1.** To catch sight of : DISCERN. **2.** To discover by careful observation or investigation. **—de·scri′er** *n.*
**des·e·crate** (dĕs′ĭ-krāt′) *vt.* **-crat·ed, -crat·ing, -crates.** [DE- + (CON)SECRATE.] To violate the sacredness of : PROFANE. **—des′e·crat′er, des′e·cra′tor** *n.* **—des′e·cra′tion** *n.*
**de·seg·re·gate** (dē-sĕg′rĭ-gāt′) *v.* **-gat·ed, -gat·ing, -gates.** —*vt.* To abolish segregation, esp. racial segregation, in. —*vi.* To become desegregated. **—de·seg′re·ga′tion** *n.* **—de·seg′re·ga′tion·ist** *n.*
**de·sen·si·tize** (dē-sĕn′sĭ-tīz′) *vt.* **-tized, -tiz·ing, -tiz·es.** To make insensitive or less sensitive, as to pain or light. **—de·sen′si·ti·za′tion** *n.* **—de·sen′si·tiz′er** *n.*
**des·ert**[1] (dĕz′ərt) *n.* [ME < OFr. < LLat. *desertum* < p.part. of *deserere,* to desert.] **1.** A dry, barren, often sandy region that can naturally support little or no vegetation. **2.** *Archaic.* A wild uncultivated and uninhabited region. **3.** An area devoid of positive character or quality : WASTELAND.
**de·sert**[2] (dĭ-zûrt′) *n.* [ME *deserte* < OFr. < fem. p.part. of *deservir,* to deserve.] **1.** *often* **deserts.** Something deserved or merited, esp. a punishment <just *deserts*> **2.** The state or fact of deserving reward or punishment.
**de·sert**[3] (dĭ-zûrt′) *v.* **-sert·ed, -sert·ing, -serts.** [Fr. *déserter* < LLat. *desertare* < Lat. *desertus,* p.part. of *deserere,* to abandon : *de-,* apart + *serere,* to join.] —*vt.* **1.** To forsake or leave, esp. when most needed : ABANDON. **2.** To abandon (e.g., a military post) in violation of oath or orders. —*vi.* To forsake one's duty or post, esp. to be absent without leave from the armed forces with no intention of returning. **—de·sert′er** *n.*
**de·ser·tion** (dĭ-zûr′shən) *n.* **1.** The act of deserting or state of being deserted. **2.** *Law.* Willful abandonment of one's spouse or children or both without their consent and with the intention of forsaking all legal obligation.
**de·serve** (dĭ-zûrv′) *v.* **-served, -serv·ing, -serves.** [ME *deserven* < OFr. *deservir* < Lat. *deservire,* to serve zealously : *de-* (intensive) + *servire,* to serve < *servus,* slave.] —*vt.* To be worthy of : MERIT <*deserves* much praise> —*vi.* To be worthy.
**de·served** (dĭ-zûrvd′) *adj.* Merited : earned <a *deserved* reward> **—de·serv′ed·ly** (-zûr′vĭd-lē) *adv.* **—de·serv′ed·ness** *n.*
**de·serv·ing** (dĭ-zûr′vĭng) *adj.* Worthy of reward, praise, or aid : MERITORIOUS. —*n.* Merit. **—de·serv′ing·ly** *adv.*
**de·sex** (dē-sĕks′) *vt.* **-sexed, -sex·ing, -sex·es.** To remove part or all of the reproductive organs of (an animal) : spay or castrate.
**de·sex·u·al·ize** (dē-sĕk′shōō-ə-līz′) *vt.* **-ized, -iz·ing, -iz·es. 1.** To desex. **2.** To take away the sexual quality of. **—de·sex′u·al·i·za′tion** *n.*
**des·ha·bille** (dĕs′ə-bēl′, -bē′) *n. var. of* DISHABILLE.
**des·ic·cant** (dĕs′ĭ-kənt) *n.* [Lat. *desiccans, desiccant-,* pr.part. of *desiccare,* to desiccate.] A substance, as calcium oxide, used as a drying agent. **—des′ic·cant** *adj.*
**des·ic·cate** (dĕs′ĭ-kāt′) *v.* **-cat·ed, -cat·ing, -cates.** [Lat. *desiccare, desiccat-* : *de-,* completely + *siccare,* to dry up < *siccus,* dry.] —*vt.* **1.** To dry out thoroughly. **2.** To preserve (foods) by removing the moisture. **3.** To divest of spirit, spontaneity, or animation. —*vi.* To become dry. —*adj.* Lacking spirit, spontaneity, or animation : ARID <"there was only the sun-bruised and *desiccate* feeling in his mind" —J.R. Salamanca> **—des′ic·ca′tion** *n.* **—des′ic·ca′tive** *adj.* **—des′ic·ca′tor** *n.*
**de·sid·er·a·ta** (dĭ-sĭd′ə-rā′tə, -rä′-) *n. pl. of* DESIDERATUM.

**de·sid·er·ate** (dĭ-sĭd′ə-rāt′) *vt.* **-at·ed, -at·ing, -ates.** [Lat. *desiderare, desiderat-*, to desire.] To long or yearn for. **—de·sid′er·a′tion** *n.* **—de·sid′er·a′tive** *adj.*

**de·sid·er·a·tum** (dĭ-sĭd′ə-rā′təm, -rä′-) *n., pl.* **-ta** (-tə) [Lat., neuter p.part. of *desiderare*, to desire.] Something needed and desired <"A journalist of spirit is a *desideratum* in a revolution" —Hugh H. Brackenridge>

**de·sign** (dĭ-zīn′) *v.* **-signed, -sign·ing, -signs.** [OFr. *designer* < Lat. *designare*, to designate.] *—vt.* **1.** To conceive in the mind : INVENT. **2.** To form a plan for <*designed* a marketing strategy> **3.** To have as a goal or purpose : INTEND. **4.** To plan by making a preliminary sketch, outline, or drawing. **5.** To create or execute in an artistic or highly skilled manner. *—vi.* **1.** To make or execute plans. **2.** To create designs. *—n.* **1.** A sketch or drawing. **2.** Invention and disposition of the forms, parts, or details of something according to a plan. **3.** A decorative or artistic work. **4.** A visual composition : PATTERN. **5.** The art of creating designs. **6.** A project or plan. **7.** A reasoned purpose : INTENTION. **8.** *often* **designs.** A sinister or hostile scheme <had *designs* on the jewels> **—de·sign′a·ble** *adj.* **—de·sign′er** *n.*

☆ ***syns:*** DESIGN, BLUEPRINT, IDEA, LAYOUT, PLAN, SCHEMA, SCHEME, STRATEGY *n. core meaning* : a method for making, doing, or accomplishing <a *design* for a building><a grand *design* for energy conservation>

**des·ig·nate** (dĕz′ĭg-nāt′) *vt.* **-nat·ed, -nat·ing, -nates.** [Lat. *designare, designat-* : *de-*, out + *signare*, to mark < *signum*, sign.] **1.** To indicate or specify : point out. **2.** To give a name or title to : CHARACTERIZE. **3.** To select for a duty, office, or purpose : APPOINT. *—adj.* (-nĭt). Appointed but not yet installed in office. **—des′ig·na′tive, des′ig·na·to′ry** (-nə-tôr′ē, -tōr′ē) *adj.* **—des′ig·na′tor** *n.*

**designated hitter** *n. Baseball.* A player designated at the start of a game to bat instead of the pitcher in the lineup.

**des·ig·na·tion** (dĕz′ĭg-nā′shən) *n.* **1.** The act of designating. **2.** Appointment or nomination. **3.** A distinguishing name or mark : TITLE.

**de·sign·ed·ly** (dĭ-zī′nĭd-lē) *adv.* On purpose : DELIBERATELY.

**des·ig·nee** (dĕz′ĭg-nē′) *n.* One designated.

**de·sign·ing** (dĭ-zī′nĭng) *adj.* **1.** Crafty : conniving. **2.** Showing or exercising forethought. **—de·sign′ing·ly** *adv.*

**de·sir·a·ble** (dĭ-zīr′ə-bəl) *adj.* **1.** Of such quality as to be worth seeking : pleasing or valuable. **2.** Worth wanting or doing : ADVANTAGEOUS. *—n.* One that is desirable. **—de·sir′a·bil′i·ty, de·sir′a·ble·ness** *n.* **—de·sir′a·bly** *adv.*

**de·sire** (dĭ-zīr′) *vt.* **-sired, -sir·ing, -sires.** [ME *desiren* < OFr. *desirer* < Lat. *desiderare.*] **1.** To wish or long for : CRAVE. **2.** To express a wish for : REQUEST. *—n.* **1.** A wish, longing, or craving. **2.** A request or petition. **3.** One that is longed for. **4.** Sexual appetite : PASSION. **—de·sir′er** *n.*

☆ ***syns:*** DESIRE, ACHE, CRAVE, HANKER, LONG, WISH, YEARN (for or after) *v. core meaning* : to have a strong longing for <We *desire* continuing world peace.>

**de·sir·ous** (dĭ-zīr′əs) *adj.* Having, displaying, or marked by desire. **—de·sir′ous·ly** *adv.* **—de·sir′ous·ness** *n.*

**de·sist** (dĭ-zĭst′, -sĭst′) *vi.* **-sist·ed, -sist·ing, -sists.** [OFr. *desister* < Lat. *desistere* : *de-*, from + *sistere*, to stop.] To cease doing something : STOP.

**desk** (dĕsk) *n.* [ME < Med. Lat. *desca* < OItal. *desco*, table < Lat. *discus*, quoit. —see DISK.] **1.** A piece of furniture usu. with drawers or compartments and a flat top for writing. **2.** A table, counter, or booth at which specified, usu. public services or functions are performed. **3.** A lectern. **4.** A department of a large organization, as a newspaper publisher, in charge of a specified operation. **5.** A music stand in an orchestra.

**desk·man** (dĕsk′măn′, -mən) *n.* A person, esp. a newspaper writer, who works at a desk.

**des·man** (dĕs′mən) *n., pl.* **-mans.** [Short for Swedish *desmansråtta*, muskrat : *desman*, musk (< MLG *desem* < Med. Lat. *bisamum*, of Semitic orig.) + *råtta*, rat.] An aquatic, insectivorous, molelike mammal, *Desmana moschata* of eastern Europe and western Asia or *Galemys pyrenaicus* of southwestern Europe, with dense, brownish fur, a long snout, and a flattened, scaly tail.

**des·mid** (dĕs′mĭd) *n.* [< NLat. *Desmidiaceae*, family name < *Desmidium*, genus name < Gk. *desmos*, bond < *dein*, to bind.] Any of various green, unicellular freshwater algae of the family Desmidiaceae, often forming chainlike colonies.

**des·o·late** (dĕs′ə-lĭt, dĕz′-) *adj.* [ME *desolat* < Lat. *desolatus*, p.part. of *desolare*, to abandon : *de-* (intensive) + *solus*, alone.] **1.** Devoid of inhabitants : DESERTED. **2.** Made unfit for habitation or use. **3.** Dismal : dreary. **4.** Lacking friends or hope : FORLORN. *—vt.* (-lāt′) **-lat·ed, -lat·ing, -lates. 1.** To rid or deprive of inhabitants. **2.** To lay waste : DEVASTATE. **3.** To abandon or forsake. **4.** To make lonely, forlorn, or wretched. **—des′o·late·ly** *adv.* **—des′o·late·ness** *n.* **—des′o·lat′er, des′o·la′tor** *n.*

**des·o·la·tion** (dĕs′ə-lā′shən, dĕz′-) *n.* **1.** The act of making desolate or state of being desolate. **2.** A wasteland. **3.** Loneliness or misery : WRETCHEDNESS <"an air of tranquil and unwitting *desolation* . . . as if she had never lived at all" —Faulkner>

**de·sorb** (dē-sôrb′, -zôrb′) *vt.* **-sorbed, -sorb·ing, -sorbs.** To remove (an absorbed or adsorbed substance) from. **—de·sorp′tion** *n.*

**de·spair** (dĭ-spâr′) *vi.* **-spaired, -spair·ing, -spairs.** [ME *despeiren* < OFr. *desperer* < Lat. *desperare* : *de-* (reversal) + *sperare*, to hope.] To be overcome by a sense of futility or defeat. *—n.* **1.** Utter lack of hope. **2.** Something destroying all hope. **—de·spair′ing·ly** *adv.*

**des·patch** (dĭ-spăch′) *v. & n. var. of* DISPATCH.

**des·per·a·do** (dĕs′pə-rä′dō, -rā′-) *n., pl.* **-does** *or* **-dos.** [Alteration of DESPERATE.] A desperate, dangerous criminal.

▲ word history: In the 16th century, Spain was the most powerful nation in Europe. It had close ties with England and had a strong influence on English life. A small instance of this influence was the vogue for the Spanish noun suffix *-ado.* This suffix was added to words that were not borrowed from Spanish to form words that did not exist in Spanish. *Desperado* is simply a refashioning of the English word *desperate.* Of the dozen or so words ending with *-ado* that were coined in the 16th century, most had died out by 1800. Two others besides *desperado* that survive in the living language are *bravado* and *tornado.*

**des·per·ate** (dĕs′pər-ĭt) *adj.* [Lat. *desperatus*, p.part. of *desperare*, to despair.] **1.** Reckless or violent because of despair. **2.** Undertaken as a last resort <*desperate* measures> **3.** Nearly hopeless : GRAVE <a *desperate* illness> **4.** Marked by, arising from, or showing despair <a *desperate* plea for help> **5.** Suffering unbearable need or anxiety <*desperate* for companionship> **6.** Overpowering : intense <*desperate* privation> **—des′per·ate·ly** *adv.* **—des′per·ate·ness** *n.*

**des·per·a·tion** (dĕs′pə-rā′shən) *n.* **1.** The condition of being desperate. **2.** Recklessness arising from despair.

**des·pi·ca·ble** (dĕs′pĭ-kə-bəl, dĭ-spĭk′ə-) *adj.* [LLat. *despicabilis* < Lat. *despicari*, to despise < *despicere.*] Deserving scorn or contempt : VILE. **—des′pi·ca·ble·ness** *n.* **—des′pi·ca·bly** *adv.*

**de·spise** (dĭ-spīz′) *vt.* **-spised, -spis·ing, -spis·es.** [ME *despisen* < OFr. *despire, despis-* < Lat. *despicere* : *de-*, down + *specere*, to look.] **1.** To regard with scorn or contempt. **2.** To regard with extreme hostility or dislike. **3.** To regard as trivial or worthless. **—de·spis′er** *n.*

☆ ***syns:*** DESPISE, ABHOR, CONTEMN, DISDAIN, SCORN *v. core meaning* : to regard with utter contempt and disdain <*despised* the idle rich> ***ant:*** esteem

**de·spite** (dĭ-spīt′) *prep.* [ME *despit*, spite < OFr. < Lat. *despectus*, p.part. of *despicere*, to despise.] In spite of : NOTWITHSTANDING <succeeded *despite* adversity> *—n.* **1.** Contemptuous defiance or disregard. **2.** Spite or malice. **3.** An act of contemptuous defiance : INDIGNITY. **—in despite of.** In spite of.

**de·spite·ful** (dĭ-spīt′fəl) *adj.* Full of malice : SPITEFUL. **—de·spite′ful·ly** *adv.* **—de·spite′ful·ness** *n.*

**de·spit·e·ous** (dĭ-spĭt′ē-əs) *adj. Obs.* Despiteful : malicious. **—de·spit′e·ous·ly** *adv.*

**de·spoil** (dĭ-spoil′) *vt.* **-spoiled, -spoil·ing, -spoils.** [ME *despoilen* < OFr. *despoiller* < Lat. *despoliare* : *de-*, away + *spoliare*, to plunder < *spolium*, booty.] To deprive of possessions or property by force : PLUNDER. **—de·spoil′er** *n.* **—de·spoil′ment** *n.*

**de·spo·li·a·tion** (dĭ-spō′lē-ā′shən) *n.* [LLat. *despoliatio* < Lat. *despoliatus*, p.part. of *despoliare*, to despoil.] The act of despoiling or state of being despoiled.

**de·spond** (dĭ-spŏnd′) *vi.* **-spond·ed, -spond·ing, -sponds.** [Lat. *despondere*, to give up : *de-*, away + *spondere*, to promise.] To become discouraged. **—de·spond′ing·ly** *adv.*

**de·spon·dence** (dĭ-spŏn′dəns) *n.* Despondency.

**de·spon·den·cy** (dĭ-spŏn′dən-sē) *n.* Depression of spirits from loss of hope, courage, or confidence : DEJECTION.

**de·spon·dent** (dĭ-spŏn′dənt) *adj.* In low spirits : DISHEARTENED. **—de·spon′dent·ly** *adv.*

**des·pot** (dĕs′pət) *n.* [OFr. < Gk. *despotēs.*] **1.** An absolute ruler. **2.** One who wields power oppressively. **3. a.** A Byzantine emperor or prince. **b.** An Eastern Orthodox bishop or patriarch. **—des·pot′ic** (dĭ-spŏt′ĭk) *adj.* **—des·pot′i·cal·ly** *adv.*

**des·pot·ism** (dĕs′pə-tĭz′əm) *n.* **1.** Rule by or as if by absolute power or authority. **2.** The actions of a despot : TYRANNY. **3. a.** A government or political system in which the ruler exercises absolute power : ABSOLUTISM. **b.** A state so ruled.

**des·qua·mate** (dĕs′kwə-māt′) *vi.* **-mat·ed, -mat·ing, -mates.** [Lat. *desquamare, desquamat-* : *de-*, off + *squama*, scale.] To shed, peel, or scale off. —Used of skin. **—des′qua·ma′tion** *n.*

**des·sert** (dĭ-zûrt′) *n.* [OFr. < *desservir*, to clear the table : *des-*, away (< Lat. *de-*) + *servir*, to serve. —see SERVE.] **1.** A usu. sweet food, as fruit, ice cream, or pastry, served as the last course of an afternoon or evening meal. **2.** *Chiefly Brit.* Fresh fruit, nuts, or sweetmeats served after the sweet course of a dinner.

**des·sert·spoon** (dĭ-zûrt′spōōn′) *n.* A spoon intermediate in size between a tablespoon and a teaspoon, used for eating dessert. **—des·sert′spoon′ful′** (-fōōl′) *n.*

**de·sta·bi·lize** (dē-stā′bə-līz′) *vt.* **-lized, -liz·ing, -liz·es.** To disturb the stability or smooth functioning of. **—de·sta′bi·li·za′tion** *n.*

**de·stain** (dē-stān′) *vt.* **-stained, -stain·ing, -stains.** To remove stain from (a specimen) to aid in microscopic study.

**de·stal·in·i·za·tion** (dē-stä′lĭ-nĭ-zā′shən) *n.* The process of discrediting and eliminating the political policies, methods, and personal image of Joseph Stalin.

**de·ster·i·lize** (dē-stĕr′ə-līz′) *vt.* **-lized, -liz·ing, -liz·es.** To release (gold) from an inactive status and return it to use as a backing for credit and new currency.

**de Stijl** (də stīl′, stāl′) *n.* [Du., the style.] A school of art originated in the Netherlands in 1917 and marked by the use of rectangular shapes and primary colors.

**des·ti·na·tion** (dĕs′tə-nā′shən) *n.* **1.** The place or point to which one is going or something is directed. **2.** The purpose for which something is created or intended. **3.** *Archaic.* An act of appointing or setting aside for a specific purpose.

**des·tine** (dĕs′tĭn) *vt.* **-tined, -tin·ing, -tines.** [ME *destinen* < OFr. *destiner* < Lat. *destinare,* to determine.] **1.** To determine beforehand : PREORDAIN. **2.** To assign for a specific end, use, or purpose <clothes *destined* for a thrift shop> **3.** To direct toward a given destination <a streamliner *destined* for Chicago>

**des·ti·ny** (dĕs′tə-nē) *n., pl.* **-nies.** [ME *destine* < OFr. *destinee* < fem. p.part. of *destiner,* to destine.] **1.** The inevitable or necessary lot to which a particular person or thing is destined : FORTUNE. **2.** The predetermined or inevitable course of events considered beyond the power or control of people. **3.** The power or agency held to predetermine events : FATE.

**des·ti·tute** (dĕs′tĭ-to͞ot′, -tyo͞ot′) *adj.* [ME *destitut* < Lat. *destitutus,* p.part. of *destituere,* to abandon : *de-,* away + *statuere,* to place.] **1.** Utterly devoid <*destitute* of talent> **2.** Utterly impoverished. **—des′ti·tute′ness** *n.*

**des·ti·tu·tion** (dĕs′tĭ-to͞o′shən, -tyo͞o′-) *n.* **1.** Extreme lack of resources or the means of subsistence : utter poverty. **2.** Deprivation : deficiency.

**des·tri·er** (dĕs′trē-ər, dĭ-strîr′) *n.* [ME < OFr. < *destre,* right hand < Lat. *dexter,* right.] *Archaic.* A war horse.

**de·stroy** (dĭ-stroi′) *v.* **-stroyed, -stroy·ing, -stroys.** [ME *destruyen* < OFr. *destruire* < VLat. **destrugere* < Lat. *destruere* : *de-,* away + *struere,* to pile up.] —*vt.* **1.** To ruin completely : SPOIL <valuable books *destroyed* by the flood> **2.** To tear down or break up : DEMOLISH. **3.** To put an end to <*destroy* a crime syndicate> **4.** To kill <*destroy* a diseased animal> **5.** To render useless or ineffective <*destroyed* the witness for the prosecution> **6.** To subdue or defeat completely : CRUSH. —*vi.* To be harmful or destructive.

☆ ***syns:*** **1.** DESTROY, DEMOLISH, LEVEL, RAZE, TEAR DOWN *v. core meaning* : to break up so that rebuilding is impossible <*destroy* a condemned building> **2.** DESTROY, DYNAMITE, FINISH, RUIN, SHATTER, SMASH, TORPEDO, TOTAL, WRECK *v. core meaning* : to cause the complete ruin of <drugs that *destroyed* their health><news that *destroyed* our hopes>

**de·stroy·er** (dĭ-stroi′ər) *n.* **1.** One that destroys. **2.** A small, fast warship armed with guns, torpedoes, and depth charges and noted for its high maneuverability.

**destroyer escort** *n.* A warship, usu. smaller than a destroyer, used to convoy merchant vessels.

**destroying angel** *n.* Any of several poisonous mushrooms of the genus *Amanita.*

**de·struct** (dĭ-strŭkt′, dē′strŭkt′) *n.* [Back-formation < DESTRUCTION.] The deliberate destruction of a space vehicle, rocket, or missile after launching.

**de·struc·ti·ble** (dĭ-strŭk′tə-bəl) *adj.* Capable of being destroyed. **—de·struc′ti·bil′i·ty, de·struc′ti·ble·ness** *n.*

**de·struc·tion** (dĭ-strŭk′shən) *n.* [ME *destruccioun* < OFr. *destruction* < Lat. *destructio* < *destructus,* p.part. of *destruere,* to destroy.] **1. a.** The act of destroying. **b.** The state or fact of being destroyed. **2.** The cause or means of destroying.

**de·struc·tion·ist** (dĭ-strŭk′shə-nĭst) *n.* One who favors or advocates destruction, esp. of existing social institutions.

**de·struc·tive** (dĭ-strŭk′tĭv) *adj.* **1.** Causing or bringing destruction : RUINOUS. **2.** Designed or tending to disprove or discredit <*destructive* comments about the article> **—de·struc′tive·ly** *adv.* **—de·struc′tive·ness, de′struc·tiv′i·ty** (dē′strŭk-tĭv′ĭ-tē) *n.*

**destructive distillation** *n.* Simultaneous decomposition by heat and distillation of substances, as wood, coal, and oil shale, to produce useful by-products, as coke, charcoal, oils, and gases.

**de·struc·tor** (dĭ-strŭk′tər) *n.* **1.** An incinerator for refuse. **2.** An explosive device for effecting a destruct.

**des·ue·tude** (dĕs′wĭ-to͞od′, -tyo͞od′) *n.* [Fr. *désuétude* < Lat. *desuetudo* < *desuescere,* to put out of use : *de-* (reversal) + *suescere,* to become accustomed.] A state of disuse <traditions now fallen into *desuetude*>

**de·sul·fur·ize** (dē-sŭl′fə-rīz′) *vt.* **-ized, -iz·ing, -iz·es.** To eliminate sulfur from. **—de·sul′fur·i·za′tion** *n.*

**des·ul·to·ry** (dĕs′əl-tôr′ē, -tōr′ē, dĕz′-) *adj.* [Lat. *desultorius* < *desultor,* a leaper < *desultus,* p.part. of *desilire,* to leap down : *de-,* down + *salire,* to jump.] **1.** Marked by lack of order or planning : DISCONNECTED <a *desultory* discussion> **2.** Occurring haphazardly : RANDOM. **—des′ul·to′ri·ly** *adv.* **—des′ul·to′ri·ness** *n.*

**de·tach** (dĭ-tăch′) *vt.* **-tached, -tach·ing, -tach·es.** [Fr. *détacher* < OFr. *destachier* : *des-,* apart (< Lat. *de-*) + *attachier,* to attach, of Germanic orig.] **1.** To disconnect : separate. **2.** To cut off from association with. **3.** To send (e.g., troops) on a special mission. **—de·tach′a·bil′i·ty** *n.* **—de·tach′a·ble** *adj.* **—de·tach′a·bly** *adv.*

☆ ***syns:*** DETACH, DISCONNECT, DISENGAGE, UNCOUPLE, UNFASTEN *v. core meaning* : to separate one thing from another <*detached* the side panels from the truck> ***ant:*** attach

**de·tached** (dĭ-tăcht′) *adj.* **1.** Standing apart : SEPARATE <a *detached* dwelling> **2. a.** Free from emotional, intellectual, or social involvement : DISINTERESTED. **b.** Indifferent : aloof. **—de·tach′ed·ly** (-tăch′ĭd-lē, -tăcht′lē) *adv.* **—de·tach′ed·ness** *n.*

**de·tach·ment** (dĭ-tăch′mənt) *n.* **1. a.** The act or process of disconnecting or detaching : SEPARATION. **b.** The state of being separate. **2.** Indifference to worldly affairs or the concerns of others : ALOOFNESS. **3.** Absence of bias or prejudice : DISINTEREST. **4. a.** Dispatch of troops or ships selected from a larger unit for a special duty or mission. **b.** The unit of troops or ships so dispatched. **c.** A permanent unit, usu. smaller than a platoon, organized for special duties.

**de·tail** (dĭ-tāl′, dē′tāl′) *n.* [Fr. *détail* < OFr. *detail,* a piece cut off < *detailler,* to cut up : *de-,* completely (< Lat. *de-*) + *tailler,* to cut. —see TAILOR.] **1.** An individual part or item : PARTICULAR. **2.** Particulars considered individually and in relation to a whole <attentiveness to *detail*> **3.** The act of dealing with things item by item. **4.** A small or secondary part of a work of art, as a painting, statue, or building, esp. when considered or represented in isolation. **5. a.** The selection of military personnel for a particular duty. **b.** The personnel so selected. **c.** The duty assigned. —*vt.* (dĭ-tāl′) **-tailed, -tail·ing, -tails.** **1.** To report or relate in detail. **2.** To name or state explicitly. **3.** To select and dispatch for a particular duty.

**detailed** (dĭ-tāld′, dē′tāld′) *adj.* Marked by abundant use of or careful attention to detail.

**detail man** *n.* A drug or medical supplies sales representative.

**de·tain** (dĭ-tān′) *vt.* **-tained, -tain·ing, -tains.** [ME *deteynen* < OFr. *detenir* < Lat. *detinēre* : *de-,* away + *tenēre,* to hold.] **1.** To keep from proceeding : DELAY. **2.** To keep in custody : CONFINE. **3.** *Obs.* To retain or withhold. **—de·tain′ment** *n.*

**de·tain·ee** (dē′tā-nē′, dĭ-tā′-) *n.* One who is held in custody.

**de·tain·er** (dĭ-tā′nər) *n. Law.* **1. a.** The unlawful withholding of the property of another. **b.** The detention of a person, esp. in custody. **2.** A writ authorizing the further detention of a person in custody pending action.

**de·tect** (dĭ-tĕkt′) *vt.* **-tect·ed, -tect·ing, -tects.** [ME *detecten* < Lat. *detectus,* p.part. of *detegere,* to uncover : *de-,* off + *tegere,* to cover.] **1.** To discover or discern the existence, presence, or fact of <*detect* a loophole in the contract> **2.** To find out the true nature of. **3.** *Electron.* To demodulate. **—de·tect′a·ble, de·tect′i·ble** *adj.* **—de·tect′er** *n.*

**de·tect·a·phone** (dĭ-tĕk′tə-fōn′) *n.* A device used for secretly listening to another's telephone conversations.

**de·tec·tion** (dĭ-tĕk′shən) *n.* **1.** The act of detecting or fact of being detected. **2.** *Electron.* Demodulation.

**de·tec·tive** (dĭ-tĕk′tĭv) *n.* One whose work is investigating crimes and obtaining evidence.

**de·tec·tor** (dĭ-tĕk′tər) *n.* One that detects, esp. a mechanical, electrical, or chemical device that automatically identifies and records a stimulus, as an environmental change in pressure or temperature, an electric signal, or radiation from a radioactive material.

**de·tent** (dĭ-tĕnt′) *n.* [Fr. *détente,* a loosening < OFr. *destente* < *destendre,* to release : *des-,* apart (< Lat. *de-*) + *tendre,* to stretch < Lat. *tendere.*] A pawl.

**dé·tente** (dā-tänt′, -tänt′) *n.* [Fr. —see DETENT.] A relaxation or reduction, as of tension between nations.

**de·ten·tion** (dĭ-tĕn′shən) *n.* [ME *detencioun,* act of withholding < OFr. *detention* < LLat. *detentio* < Lat. *detentus,* p.part. of *detinēre,* to detain.] **1. a.** The act of detaining. **b.** The state of being detained, esp. a period of temporary custody while awaiting trial. **2.** A forced or punitive delay.

**detention home** *n.* A place where juvenile delinquents or offenders are held in custody, esp. while awaiting legal action.

**de·ter** (dĭ-tûr′) *vt.* **-terred, -ter·ring, -ters.** [Lat. *deterrēre* : *de-,* away + *terrēre,* to frighten.] To prevent or discourage from acting, esp. by means of doubt or fear. **—de·ter′ment** *n.* **—de·ter′rer** *n.*

**de·terge** (dĭ-tûrj′) *vt.* **-terged, -terg·ing, -terg·es.** [Fr. *déterger* < Lat. *detergēre* : *de-,* off + *tergēre,* to wipe.] To cleanse or wipe off.

**de·ter·gen·cy** (dĭ-tûr′jən-sē) *also* **de·ter·gence** (-jəns) *n.* Cleansing power or quality.

**de·ter·gent** (dĭ-tûr′jənt) *n.* A cleansing substance, esp. one made synthetically from chemical compounds rather than from fats and lye and used as a wetting agent and emulsifier. **—de·ter′gent** *adj.*

**de·te·ri·o·rate** (dĭ-tîr′ē-ə-rāt′) *v.* **-rat·ed, -rat·ing, -rates.** [LLat. *deteriorare, deteriorat-* < Lat. *deterior,* worse.] —*vt.* To lower or impair in quality, character, or value. —*vi.* To degenerate. **—de·te′ri·o·ra′tion** *n.* **—de·te′ri·o·ra′tive** *adj.*

**de·ter·min·a·ble** (dĭ-tûr′mə-nə-bəl) *adj.* **1.** Capable of being settled, fixed, or determined. **2.** *Law.* Liable to be terminated. **—de·ter′min·a·ble·ness** *n.* **—de·ter′min·a·bly** *adv.*

**de·ter·mi·na·cy** (dĭ-tûr′mə-nə-sē) *n.* **1.** The quality or condition of being determinate. **2.** The condition of being determined.

**de·ter·mi·nant** (dĭ-tûr′mə-nənt) *adj.* Determinative. —*n.* **1.** An influencing or determining factor. **2.** *Math.* A square array of quantities or elements having a value determined by a rule of combination for the elements and used esp. in solving certain classes of simultaneous equations.

**de·ter·mi·nate** (dĭ-tûr′mə-nĭt) *adj.* [ME *determinat* < Lat. *determinatus*, p.part. of *determinare*, to determine.] **1.** Precisely defined or limited. **2.** Conclusively settled. **3.** Firm in purpose : RESOLUTE. **4.** *Bot.* **a.** Terminating in a flower and blooming in a sequence beginning with the topmost or central flower. **b.** Not continuing indefinitely at the tip of an axis. **—de·ter′mi·nate·ly** *adv.* **—de·ter′mi·nate·ness** *n.*

**de·ter·mi·nat·er** (dĭ-tûr′mə-nā-tər) *n.* A determiner.

**de·ter·mi·na·tion** (dĭ-tûr′mə-nā′shən) *n.* **1. a.** The act of making or arriving at a decision. **b.** The decision reached. **2.** The quality of being resolute or firm in purpose. **3. a.** The act of settling a dispute, suit, or other question by an authoritative decision or pronouncement, esp. by a judicial body. **b.** The decision or pronouncement made. **4. a.** The ascertainment or establishment of the extent, quality, position, or character of something. **b.** The result of such ascertainment. **5.** A fixed movement or tendency toward an object or end. **6.** *Logic.* **a.** More definite rendition of a concept or proposition by further qualification. **b.** Definition of a concept through its constituent elements.

**de·ter·mi·na·tive** (dĭ-tûr′mə-nā′tĭv, -nə-) *adj.* Able, tending, or serving to determine. —*n.* Something that determines. **—de·ter′mi·na′tive·ly** *adv.* **—de·ter′mi·na′tive·ness** *n.*

**de·ter·mine** (dĭ-tûr′mĭn) *v.* **-mined, -min·ing, -mines.** [ME *determinen* < OFr. *determiner* < Lat. *determinare*, to limit : *de-*, off + *terminus*, boundary.] —*vt.* **1. a.** To decide or settle (e.g., a dispute) authoritatively and conclusively. **b.** To end or decide by final, esp. judicial action. **2.** To establish or ascertain definitely, as after consideration, investigation, or calculation. **3.** To cause to come to a conclusion or resolution. **4.** To be the cause of : REGULATE <Need should *determine* expenditures.> **5.** To give direction to <Their parents *determined* their religious beliefs.> **6.** To limit in scope or extent : fix the limits of. **7.** *Math.* To fix or define the position, form, or configuration of. **8.** *Logic.* To explain or limit by adding differences. **9.** *Law.* To put an end to : TERMINATE. —*vi.* **1.** To reach a decision : RESOLVE. **2.** *Law.* To come to an end.
☆ ***syns:*** DETERMINE, BOUND, DELIMIT, DEMARCATE, LIMIT *v. core meaning* : to fix the limits of <Surveyors *determined* the property lines.>

**de·ter·mined** (dĭ-tûr′mĭnd) *adj.* **1.** Marked by or showing determination : RESOLUTE. **2.** Decided or resolved <Has the cause been *determined*?> **—de·ter′mined·ly** *adv.* **—de·ter′mined·ness** *n.*

**de·ter·min·er** (dĭ-tûr′mə-nər) *n.* **1.** One that determines. **2.** A word belonging to a group of noun modifiers gen. regarded as including articles, demonstratives, possessive adjectives, and a few other words such as *any, both, several,* and *whose* and that occupies the first position in a noun phrase or the second or third position after another determiner.

**de·ter·min·ism** (dĭ-tûr′mə-nĭz′əm) *n.* *Philos.* The doctrine that every event, act, and decision is the inevitable consequence of antecedents that are independent of the human will.

**de·ter·rence** (dĭ-tûr′əns, -tŭr′-) *n.* **1.** The act or a means of deterring. **2.** Measures taken by a state or an alliance of states to prevent hostile action by another state.

**de·ter·rent** (dĭ-tûr′ənt, -tŭr′-) *adj.* Tending to deter. —*n.* **1.** Something that deters. **2.** A retaliatory means to deter enemy attack.

**de·ter·sive** (dĭ-tûr′sĭv, -zĭv) *adj.* [OFr. *detersif* < Lat. *detersus*, p.part. of *detergēre*, to deterge.] Detergent. **—de·ter′sive** *n.*

**de·test** (dĭ-tĕst′) *vt.* **-test·ed, -test·ing, -tests.** [Lat. *detestari*, to curse : *de-* (pejorative) + *testari*, to invoke < *testis*, witness.] To dislike intensely : ABHOR. **—de·test′er** *n.*

**de·test·a·ble** (dĭ-tĕs′tə-bəl) *adj.* Deserving abhorrence. **—de·test′a·bil′i·ty, de·test′a·ble·ness** *n.* **—de·test′a·bly** *adv.*

**de·tes·ta·tion** (dē′tĕ-stā′shən) *n.* **1.** Strong dislike or hatred : ABHORRENCE. **2.** One that is detested.

**de·throne** (dē-thrōn′) *vt.* **-throned, -thron·ing, -thrones. 1.** To remove from a throne : DEPOSE. **2.** To remove from a powerful or prominent position. **—de·throne′ment** *n.*

**det·i·nue** (dĕt′n-o͞o′, -yo͞o′) *n.* [ME *detenue* < OFr., detention < p.part. of *detenir*, to detain.] *Law.* **1. a.** An action to recover possession or the value of property wrongfully detained. **b.** A writ authorizing detinue. **2.** *Obs.* The unlawful detention of personal property.

**det·o·na·ble** (dĕt′n-ə-bəl) *adj.* Capable of being detonated.

**det·o·nate** (dĕt′n-āt′) *vi.* & *vt.* **-nat·ed, -nat·ing, -nates.** [Lat. *detonare, detonat-*, to thunder down : *de-*, down + *tonare*, to thunder.] To explode or cause to explode. **—det′o·nat′a·ble** *adj.* **—det′o·na′tion** *n.*

**det·o·na·tor** (dĕt′n-ā′tər) *n.* **1.** A device, as a fuse or percussion cap, used to set off explosives. **2.** An explosive.

**de·tour** (dē′to͝or′, dĭ-to͝or′) *n.* [Fr. *détour* < OFr. *destor* < *destorner*, to turn away : *des-*, away (< Lat. *de-*) + *tourner*, to turn. —see TURN.] **1.** A roundabout way, esp. a road used temporarily instead of a main route. **2.** A deviation from a direct course of action. —*vi.* & *vt.* **-toured, -tour·ing, -tours.** To go or cause to go by a detour.

**de·tox** (dē-tŏks′) *Informal.* —*vt.* **-toxed, -tox·ing, -tox·es.** To subject to detoxification. —*n.* (dē′tŏks′). A section of a hospital or clinic where patients are detoxified.

**de·tox·i·fy** (dē-tŏk′sə-fī′) *also* **de·tox·i·cate** (-sĭ-kāt′) *vt.* **-fied, -fy·ing, -fies** *also* **-cat·ed, -cat·ing, -cates.** [DE- + TOXI(C) + -FY.] **1.** To counteract or destroy the toxic properties of. **2.** To remove the effects of a toxic substance from (e.g., one who abuses alcohol or drugs) or to free from dependence on (e.g., alcohol or drugs). **—de·tox′i·fi·ca′tion** *n.*

**de·tract** (dĭ-trăkt′) *v.* **-tract·ed, -tract·ing, -tracts.** [ME *detracten* < Lat. *detractus*, p.part. of *detrahere*, to remove : *de-*, away + *trahere*, to pull.] —*vi.* To take away something desirable : DIMINISH <Poor grooming *detracts* from one's appearance.> —*vt.* **1.** To distract. **2.** *Archaic.* To speak ill of : BELITTLE. **—de·trac′tive** *adj.* **—de·trac′tor** *n.*

**de·trac·tion** (dĭ-trăk′shən) *n.* **1.** Disparagement. **2.** The act of taking away.

**de·train** (dē-trān′) *vi.* & *vt.* **-trained, -train·ing, -trains.** To leave or cause to leave a railroad train. **—de·train′ment** *n.*

**de·trib·al·ize** (dē-trī′bə-līz′) *vt.* **-ized, -iz·ing, -iz·es.** To cause to lose tribal customs by means of acculturation. **—de·trib′al·i·za′tion** *n.*

**det·ri·ment** (dĕt′rə-mənt) *n.* [ME < OFr. < Lat. *detrimentum* < *deterere*, to lessen : *de-*, away + *terere*, to rub.] **1.** Damage, harm, or loss. **2.** Something that causes damage, harm, or loss.

**det·ri·men·tal** (dĕt′rə-mĕn′tl) *adj.* Causing harm or damage : INJURIOUS. **—det′ri·men′tal·ly** *adv.*

**de·tri·tion** (dĭ-trĭsh′ən) *n.* [Med. Lat. *detritio* < Lat. *detritus*, p.part. of *deterere*, to lessen. —see DETRIMENT.] The act of wearing away by rubbing or friction.

**de·tri·tus** (dĭ-trī′təs) *n., pl.* **detritus.** [Fr. *détritus* < Lat. *detritus*, p.part. of *deterere*, to lessen. —see DETRIMENT.] **1.** Loose fragments, particles, or grains that have been formed by the disintegration of rocks. **2.** Disintegrated matter : DEBRIS.

**de trop** (də trō′) *adj.* [Fr.] Too much : SUPERFLUOUS.

**de·tu·mes·cence** (dē′to͞o-mĕs′əns, -tyo͞o-) *n.* [< Lat. *detumescere*, to subside : *de-* (reversal) + *tumescere*, to swell up < *tumere*, to subside.] Contraction following expansion, esp. return of a swollen organ or part to normal size. **—de′tu·mes′cent** *adj.*

**Deu·ca·li·on** (do͞o-kā′lē-ən, dyo͞o-) *n.* [Lat. < Gk. *Deukaliōn.*] *Gk. Myth.* A son of Prometheus who with his wife, Pyrrha, survived a deluge sent by Zeus and became the ancestor of the renewed human race.

**deuce**[1] (do͞os, dyo͞os) *n.* [OFr. *deus*, two < Lat. *duos*, accusative of *duo.*] **1. a.** A playing card or side of a die bearing two spots. **b.** A cast of the dice totaling two. **2.** A tennis score in which each player or side has 40 points or 5 or more games each and either player or side must win 2 successive points or games to win the game or set.

**deuce**[2] (do͞os, dyo͞os) *n.* [Prob. < LG *duus*, a throw of two in dice games, bad luck, ult. < Lat. *duo*, two.] *Informal.* The devil. —Used as a mild oath or exclamation of annoyance, impatience, or surprise.

**deuc·ed** (do͞o′sĭd, dyo͞o′-) *adj.* [< DEUCE[2].] *Informal.* Confounded <a *deuced* nuisance> **—deuc′ed, deuc′ed·ly** *adv.*

**de·us ex ma·chi·na** (dā′əs ĕks mä′kə-nə, -nä′, măk′ə-nə) *n.* [NLat., god from a machine, transl. of Gk. *theos ek mēkhanēs.*] **1.** A deity in Greek and Roman drama who was brought in by stage machinery to intervene in a difficult situation. **2.** An improbable character or a contrived device or event suddenly introduced to untangle a plot or resolve a situation.

**deut-** *pref. var. of* DEUTO-.

**deuter-** *pref. var. of* DEUTERO-.

**deu·ter·ag·o·nist** (do͞o′tə-răg′ə-nĭst) *n.* [Gk. *deuteragōnistēs* : *deuteros*, second + *agōnistēs*, actor. —see PROTAGONIST.] The character second in importance to the protagonist in classical Greek drama.

**deu·ter·a·no·pi·a** (do͞o′tər-ə-nō′pē-ə, dyo͞o′-) *n.* [DEUTER(O)- + AN- + -OPIA (so called from the blindness to green, which is considered the second of the primary colors).] Colorblindness marked by confusion of green, bluish red, and neutral. **—deu′ter·a·nope′** (-nōp′) *n.* **—deu′ter·a·nop′ic** (-nŏp′ĭk, -nō′pĭk) *adj.*

**deu·ter·ate** (do͞o′tə-rāt′, dyo͞o′-) *vt.* **-at·ed, -at·ing, -ates.** [DEUTER(IUM) + -ATE.] To introduce deuterium into. **—deu′ter·a′tion** *n.*

**deu·te·ri·um** (do͞o-tîr′ē-əm, dyo͞o-) *n.* An isotope of hydrogen having an atomic weight of 2.0141.

**deuterium oxide** *n.* An isotopic form of water with composition $D_2O$, present in natural water as approx. 1 part in 6,500 and isolated for use as a moderator in certain nuclear reactors.

**deutero-** *or* **deuter-** *pref.* [Gk. *deuteros*, second.] Second : secondary <*deutero*canonical>

**deu·ter·o·ca·non·i·cal** (do͞o′tə-rō′kə-nŏn′ĭ-kəl, dyo͞o′-) *adj.* Of or relating to books or sections of books in the Old Testament held to be canonical by the Eastern Orthodox and Roman Catholics and apocryphal by many Protestants.

**deu·ter·og·a·my** (do͞o'tə-rŏg'ə-mē, dyo͞o'-) *n.* A second legal marriage after the death or divorce of a first spouse.

**deu·ter·on** (do͞o'tə-rŏn', dyo͞o'-) *n.* [DEUTER(IUM) + -ON.] The nucleus of a deuterium atom, a composite of a proton and a neutron, regarded as a subatomic particle with unit positive charge.

**Deu·ter·on·o·my** (do͞o'tə-rŏn'ə-mē, dyo͞o'-) *n.* [LLat. *deuteronomium* < Gk. *deuteronomion* : *deuteros*, second + *nomos*, law.] —See table at BIBLE.

**deuto-** or **deut-** *pref.* [< DEUTERO-.] Second : secondary <*deutoplasm*>

**deu·to·plasm** (do͞o'tə-plăz'əm, dyo͞o'-) *also* **deu·ter·o·plasm** (-tə-rō-plăz'əm) *n.* Food substance or yolk in the cytoplasm of an ovum or other cell.

**deut·sche mark** *also* **deut·sche·mark** (doi'chə-märk') *n.* [G., German mark.] —See table at CURRENCY.

**deut·zi·a** (do͞ot'sē-ə, dyo͞ot'-) *n.* [NLat. *Deutzia*, genus name, after Jean Deutz (d. 1784?).] A shrub of the genus *Deutzia*, cultivated for its white or pinkish flower clusters.

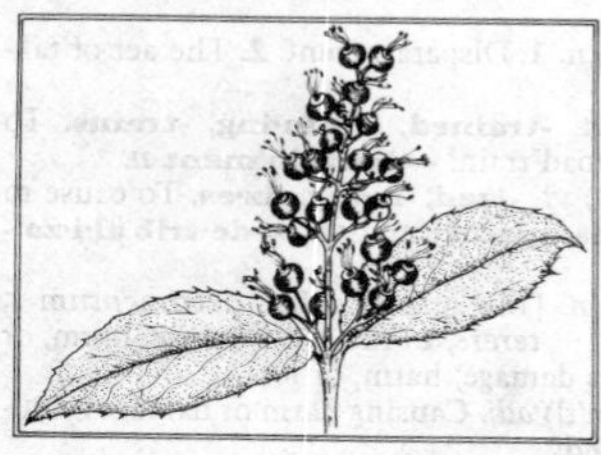

**deutzia**

**de·val·u·ate** (dē-văl'yo͞o-āt') *also* **de·val·ue** (-văl'yo͞o) *vt.* **-at·ed, -at·ing, -ates** *also* **-ued, -u·ing, -ues. 1.** To lessen or annul the value of. **2.** To lower the exchange value of (currency) by lowering its gold equivalency. **—de·val'u·a'tion** *n.*

**De·va·na·ga·ri** (dā'və-nä'gə-rē) *n.* [Skt. *devanāgarī* : *deva-*, divine + *nagaram*, city.] The alphabet in which Sanskrit and many modern Indian languages are written.

**dev·as·tate** (dĕv'ə-stāt') *vt.* **-tat·ed, -tat·ing, -tates.** [Lat. *devastare, devastat-* : *de-* (intensive) + *vastare*, to lay waste < *vastus*, waste.] **1.** To lay waste : RUIN. **2.** To cause to be overwhelmed <*devastated* by guilt> **—dev'as·tat'ing·ly** *adv.* **—dev'as·ta'tion** *n.* **—dev'as·ta'tor** *n.*

**de·vel·op** (dĭ-vĕl'əp) *v.* **-oped, -op·ing, -ops.** [Fr. *développer* < OFr. *desveloper* : *des-*, apart (< Lat. *dis-*) + *voloper*, to wrap.] —*vt.* **1. a.** To realize the potentialities of. **b.** To aid in the growth of : STRENGTHEN <*develop* the biceps> **2. a.** To cause to unfold gradually <an author *developing* a story> **b.** To cause to expand or grow gradually <a plant *developing* buds> **3. a.** To bring into being : make active <*develop* a business> **b.** To make more available or effective <*develop* the ocean's resources> **4. a.** To set forth or clarify by degrees <*develop* a strategy> **b.** To elaborate or enlarge <*develop* a basic theory> **5.** *Mus.* To unfold (a theme) with rhythmic and harmonic variations. **6.** To convert (a tract of land) to a specific purpose, as by building extensively. **7.** To acquire gradually <*develop* driving skill> **8.** To become affected with : CONTRACT <*develop* pneumonia> **9.** To process (a photosensitive material), esp. with chemicals, in order to render a recorded image visible. —*vi.* **1.** To grow : expand. **2.** To come gradually into existence or activity. **3.** To be disclosed <It *developed* that we had been cheated.> **4.** *Biol.* **a.** To progress from earlier to later stages of individual maturation. **b.** To progress from earlier to later or from simpler to more complex stages of evolution. **—de·vel'op·a·ble** *adj.*

**de·vel·op·er** (dĭ-vĕl'ə-pər) *n.* **1.** One that develops. **2.** A person who develops real estate. **3.** A chemical used to render visible the image recorded on a photosensitive surface.

**de·vel·op·ing** (dĭ-vĕl'ə-pĭng) *adj.* Underdeveloped <aid to *developing* countries>

**de·vel·op·ment** (dĭ-vĕl'əp-mənt) *n.* **1.** The act of developing or state of being developed. **2.** A product or result of developing. **3.** A significant event. **4.** A group of buildings, as dwellings, usu. built by the same contractor. **—de·vel'op·men'tal** (-mĕn'tl) *adj.* **—de·vel'op·men'tal·ly** *adv.*

**de·verb·a·tive** (dē-vûr'bə-tĭv) *adj.* **1.** Designating a word derived from a verb; e.g., *dancer* is a deverbative noun derived from the verb *dance.* **2.** Designating an element used in derivation from a verb; e.g., the suffix *-er* in *worker* is deverbative. **—de·verb'a·tive** *n.*

**de·vest** (dĭ-vĕst') *vt.* **-vest·ed, -vest·ing, -vests.** [OFr. *desvestir*, to undress < VLat. *disvestire : Lat. *dis-* apart + Lat. *vestis*, garment.] *Law.* To take (e.g., a right) away.

**De·vi** (dā'vē) *n.* [Skt. *devī*, fem. of *devaḥ*, god.] A general appellation for all feminine Hindu deities, used esp. for the wife of Shiva.

**de·vi·ant** (dē'vē-ənt) *adj.* [ME *deviaunt* < LLat. *devians*, pr.part. of *deviare*, to deviate.] Differing from a norm or from accepted moral or societal standards. —*n.* One whose behavior and attitudes differ from the norm or from accepted moral and social standards. **—de'vi·ance, de'vi·an·cy** *n.*

**de·vi·ate** (dē'vē-āt') *v.* **-at·ed, -at·ing, -ates.** [LLat. *deviare, deviat-* : Lat. *de-*, away + Lat. *via*, road.] —*vi.* To turn or move increasingly away from a specified course or prescribed mode of behavior. —*vt.* To cause to turn aside or differ. —*n.* (-ĭt). A deviant. **—de'vi·a'tor** *n.*

**de·vi·a·tion** (dē'vē-ā'shən) *n.* **1.** The act of deviating or turning aside. **2.** A departure from normality. **3.** Deviant behavior or attitudes. **4.** Divergence from an accepted political policy or party line. **5.** Deflection of a compass needle caused by a magnetic influence, as in a ship. **6.** *Statistics.* The difference, esp. the absolute difference, between one of a set of numbers and their mean. **—de'vi·a'tion·ism** *n.* **—de'vi·a'tion·ist** *n.*

**de·vice** (dĭ-vīs') *n.* [ME < OFr. *devis*, division, and *devise*, design, both < *deviser*, to devise.] **1.** Something constructed or devised for a particular purpose, esp. a machine used to perform one or more relatively simple tasks. **2.** An artistic contrivance in a literary work used to achieve a particular effect. **3.** A plan or scheme, esp. a malign one. **4.** A decorative design, figure, or pattern, as one used in embroidery. **5.** A graphic motto or symbol, esp. in heraldry. **6.** *Archaic.* The act, state, or power of devising.

**dev·il** (dĕv'əl) *n.* [ME *devel* < OE *dēofol* < LLat. *diabolus* < LGk. *diabolos* < Gk., slanderer < *diaballein*, to slander : *dia-*, across + *ballein*, to throw.] **1.** *often* **Devil.** The major spirit of evil, ruler of Hell, and foe of God : SATAN. **2.** A subordinate evil spirit : DEMON. **3.** A wicked or malevolent person. **4.** A person : individual <you lucky *devil*> <poor *devil*> **5.** An energetic, mischievous, daring, or clever person. **6.** A printer's devil. **7.** A toothed machine, as for tearing up rags. **8.** *Informal.* Something annoying or hard to manage <The car gave me the very *devil* of a time.> **9.** *Christian Science.* The opposite of Truth : ERROR. —*v.* **-iled, -il·ing, -ils** or **-illed, -il·ling, -ils.** —*vt.* **1.** To season (food) heavily. **2.** To tear up (cloth or rags) in a toothed machine. **3.** To annoy, torment, or harass. —*vi.* To serve as a printer's devil. **—give the devil his due.** To acknowledge the ability or success of an evil or disliked person. **—the devil.** *Informal.* —Used as an exclamation indicating surprise, anger, disgust, or vexation. **—the devil to pay.** Unpleasant consequences.

**dev·il·fish** (dĕv'əl-fĭsh') *n., pl.* **devilfish** or **-fish·es. 1.** MANTA 2. **2.** OCTOPUS 1.

**dev·il·ish** (dĕv'ə-lĭsh) *adj.* **1.** Of, resembling, or typical of a devil : FIENDISH. **2.** *Informal.* Extreme : excessive <*devilish* thirst> —*adv.* *Informal.* Extremely. **—dev'il·ish·ly** *adv.* **—dev'il·ish·ness** *n.*

**dev·il·kin** (dĕv'əl-kĭn) *n.* A little devil : IMP.

**dev·il-may-care** (dĕv'əl-mā-kâr') *adj.* **1.** Incautious : reckless. **2.** Rakish and jovial.

**dev·il·ment** (dĕv'əl-mənt) *n.* Devilish mischief.

**dev·il·ry** (dĕv'əl-rē) *n. var. of* DEVILTRY.

**devil's advocate** *n.* **1.** *Rom. Cath. Ch.* An official appointed to present arguments against a proposed canonization or beatification. **2.** A person who opposes an argument with which he or she does not necessarily disagree, as to determine its validity. **3.** An adverse critic, esp. of a good cause.

**devil's bit** *n.* The blazing star.

**devil's club** *n.* A spiny shrub, *Oplopanax horridus* of western North America, bearing greenish-white flowers and scarlet fruit.

**devil's darning needle** *n.* **1.** A dragonfly. **2.** A damselfly.

**dev·il's-food cake** (dĕv'əlz-fo͞od') *n.* A rich chocolate cake.

**devil's paintbrush** *n.* The orange hawkweed.

**devil's walking stick** *n.* The Hercules'-club.

**dev·il·try** (dĕv'əl-trē) *also* **dev·il·ry** (-əl-rē) *n., pl.* **-tries** *also* **-ries. 1.** Wanton or reckless mischief. **2.** Extreme cruelty : WICKEDNESS. **3.** Evil magic : WITCHCRAFT.

**dev·il·wood** (dĕv'əl-wo͝od') *n.* A tree, *Osmanthus americanus* of the southeastern United States, with fragrant greenish flowers and hard wood.

**de·vi·ous** (dē'vē-əs) *adj.* [Lat. *devius*, out-of-the-way : *de-*, away from + *via*, road.] **1.** Deviating from the straight or direct course. **2.** Departing from the correct or proper way : ERRING. **3.** Not straightforward : SHIFTY. **—de'vi·ous·ly** *adv.* **—de'vi·ous·ness** *n.*

**de·vise** (dĭ-vīz') *vt.* **-vised, -vis·ing, -vis·es.** [ME *devisen* < OFr. *deviser* < VLat. *divisare, freq. of Lat. *dividere*, to divide.] **1.** To form or arrange in the mind : CONTRIVE <*devise* a plan> **2.** *Law.* To transmit or give (real property) by will. **3.** *Obs.* To suppose : imagine. —*n.* *Law.* **1.** The act of transmitting or giving real property by will. **2.** The property or lands transmitted by will. **3.** A will or clause in a will devising real property. **—de·vis'a·ble** *adj.* **—de·vis'er** *n.*

**de·vi·see** (dĭ-vī'zē', dĕv'ĭ-zē') *n.* *Law.* One to whom a devise is made.

**de·vi·sor** (dĭ-vī'zər, dĕv'ĭ-zôr') *n.* *Law.* One who makes a devise.

**de·vi·tal·ize** (dē-vīt'l-īz') *vt.* **-ized, -iz·ing, -iz·es.** To reduce or destroy the vitality of. **—de·vit'al·i·za'tion** *n.*

**de·vit·ri·fy** (dē-vĭt′rə-fī′) *vt.* **-fied, -fy·ing, -fies.** [Fr. *dévitrifier* : *dé-*, de- + *vitrifier*, to vitrify. —see VITRIFY.] **1.** To remove or destroy the glassy quality of. **2.** To treat (e.g., glass) so as to cause crystallization, brittleness, and loss of transparency. **—de·vit′ri·fi′a·ble** *adj.* **—de·vit′ri·fi·ca′tion** *n.*

**de·vo·cal·ize** (dē-vō′kə-līz′) *vt.* **-ized, -iz·ing, -iz·es.** To devoice. **—de·vo′cal·i·za′tion** *n.*

**de·voice** (dē-vois′) *vt.* **-voiced, -voic·ing, -voic·es.** To unvoice (a speech sound).

**de·void** (dĭ-void′) *adj.* [ME, p.part. of *devoiden*, to remove < OFr. *desvoidier* : *des-*, completely (< Lat. *de-*) + *voidier*, to empty < *voide*, empty. —see VOID.] Completely lacking : DESTITUTE <*devoid* of wit>

**de·voir** (dəv-wär′, dĕv′wär′) *n.* [ME, duty < OFr. < *devoir*, to owe < Lat. *debēre*.] **1.** *often* **devoirs.** An act or expression of respect or courtesy : CIVILITY. **2.** Responsibility or duty.

**de·vol·a·til·ize** (dē-vŏl′ə-tl-īz′) *vt.* **-ized, -iz·ing, -iz·es.** To remove volatile material from. **—de·vol′a·til·i·za′tion** *n.*

**dev·o·lu·tion** (dĕv′ə-lo͞o′shən, dē′və-) *n.* [Med. Lat. *devolutio* < Lat. *devolutus*, p.part. of *devolvere*, to roll down. —see DEVOLVE.] **1.** A passing down through successive stages. **2.** The transference of something, as properties, rights, and qualities, to a successor. **3.** Delegation of duties or authority to a subordinate or substitute. **4.** Transfer of powers from a central government to local units. **5.** *Biol.* DEGENERATION 2. **—dev′o·lu′tion·ar′y** (-shə-nĕr′ē) *adj.* **—dev′o·lu′tion·ist** *n.*

**de·volve** (dĭ-vŏlv′) *v.* **-volved, -volv·ing, -volves.** [ME *devolven*, to transfer < Lat. *devolvere*, to roll down : *de-*, down + *volvere*, to roll.] —*vt.* **1.** *Archaic.* To cause to roll onward or downward. **2.** To hand down or delegate to another. —*vi.* **1.** *Archaic.* To roll onward or downward. **2.** To be passed on or transferred to another. **—de·volve′ment** *n.*

**Dev·on** (dĕv′ən) *n.* Any of a breed of reddish cattle developed in Devonshire, England, and raised primarily for beef.

**De·vo·ni·an** (dĭ-vō′nē-ən) *adj.* [After *Devon*, a county in England.] Of, belonging to, or designating the geologic time, system of rocks, or sedimentary deposits of the fourth period of the Paleozoic era, preceded by the Silurian and followed by the Mississippian or Carboniferous period and marked by the appearance of forests and amphibians. —*n.* The Devonian period or system of deposits.

**de·vote** (dĭ-vōt′) *vt.* **-vot·ed, -vot·ing, -votes.** [Lat. *devovēre*, *devot-*, to vow : *de-* (intensive) + *vovēre*, to vow.] **1.** To give or apply (one's time, attention, or self) entirely to a particular activity or cause. **2.** To set apart for a specific purpose or use <money *devoted* to research> **3.** To set apart by or as if by a vow or solemn act : CONSECRATE. **—de·vote′ment** *n.*

☆ ***syns:*** DEVOTE, CONSECRATE, DEDICATE, HALLOW *v. core meaning* : to give over by or as if by a vow to a higher purpose <doctors who *devote* their lives to healing the sick>

**de·vot·ed** (dĭ-vō′tĭd) *adj.* **1.** Feeling or exhibiting strong attachment or loyalty : ARDENT. **2.** Dedicated : consecrated. **—de·vot′ed·ly** *adv.* **—de·vot′ed·ness** *n.*

**dev·o·tee** (dĕv′ə-tē′, -tā′) *n.* **1.** A zealous follower or enthusiast. **2.** An ardent or fanatical adherent of a religion.

**de·vo·tion** (dĭ-vō′shən) *n.* **1.** Ardent attachment or loyalty. **2.** Religious ardor or zeal : PIETY. **3.** *often* **devotions.** An act of religious observance or prayer, esp. when private. **4.** The act of devoting or state of being devoted.

**de·vo·tion·al** (dĭ-vō′shə-nəl) *adj.* **1.** Of or relating to devotion. **2.** Used in worship. —*n.* A short religious or prayer service. **—de·vo′tion·al·ly** *adv.*

**de·vour** (dĭ-vour′) *vt.* **-voured, -vour·ing, -vours.** [ME *devouren* < OFr. *devourer* < Lat. *devorare* : *de-*, completely + *vorare*, to swallow.] **1.** To eat up greedily. **2.** To destroy, consume, or waste <An earthquake *devoured* the countryside.> **3.** To take in eagerly <*devour* a mystery novel> **4.** To swallow up : ENGULF <*devoured* by the cheering fans> **—de·vour′er** *n.* **—de·vour′ing·ly** *adv.*

**de·vout** (dĭ-vout′) *adj.* **-er, -est.** [ME < OFr. < LLat. *devotus* < Lat., p.part. of *devovere*, to vow. —see DEVOTE.] **1.** Deeply religious : PIOUS. **2.** Displaying piety or reverence. **3.** Earnest : sincere <*devout* wishes for their happiness> **—de·vout′ly** *adv.* **—de·vout′ness** *n.*

**dew** (do͞o, dyo͞o) *n.* [ME *deu* < OE *dēaw*.] **1.** Water droplets condensed from the air onto cool surfaces, usu. at night. **2.** Something moist, refreshing, or pure. **3.** Moisture appearing in small drops, as tears. —*vt.* **dewed, dew·ing, dews.** To wet with or as if with dew.

**de·wan** (dĭ-wän′) *n.* [Hindi *dīwān* < Pers. *divan*, account book.] A governmental official in India, esp. a prime minister.

**Dew·ar flask** (do͞o′ər, dyo͞o′-) *n.* [After Sir James *Dewar* (1842–1923), its inventor.] An insulated container having a double wall with evacuated space between the walls and silvered surfaces, used esp. for storing liquefied gases.

**dew·ber·ry** (do͞o′bĕr′ē, dyo͞o′-) *n.* **1.** A trailing form of the blackberry, as *Rubus hispidus* of North America or *R. caesius* of Europe. **2.** The fruit of a dewberry.

**dew·claw** (do͞o′klô′, dyo͞o′-) *n.* A vestigial digit, claw, or hoof on the foot of certain mammals.

**dew·drop** (do͞o′drŏp′, dyo͞o′-) *n.* A drop of dew.

**Dew·ey decimal system** (do͞o′ē, dyo͞o′ē) *n.* [After Melvil *Dewey* (1851–1931), its inventor.] A system used for the classification of library books and other publications into ten major categories, each category being further subdivided by number.

**dew·lap** (do͞o′lăp′, dyo͞o′-) *n.* [ME *dewlappe*.] **1.** A fold of loose skin hanging from the neck of certain animals, as cattle. **2.** A pendulous part similar to a dewlap, as the wattle of a bird.

**DEW line** (do͞o, dyo͞o) *n.* [*D(istant) E(arly) W(arning)*.] A line of radar stations at about the 70th parallel across the North American continent, designed to give advance warning of approaching enemy aircraft and missiles.

**dew point** *n.* The temperature at which air becomes saturated and produces dew.

**dew-worm** (do͞o′wûrm′, dyo͞o′-) *n.* An earthworm found on or near the surface of the ground and used as fishing bait : NIGHT CRAWLER.

**dew·y** (do͞o′ē, dyo͞o′ē) *adj.* **-i·er, -i·est.** **1.** Moist with dew. **2.** Of or resembling dew. **3.** Like the purity or freshness of dew. **—dew′i·ly** *adv.* **—dew′i·ness** *n.*

**dew·y-eyed** (do͞o′ē-īd′, dyo͞o′-) *adj.* Innocent, credulous, or naive.

**dex** (dĕks) *n.* Dextroamphetamine sulfate.

**Dex·e·drine** (dĕk′sĭ-drĭn, -drēn′). A trademark for a preparation of dextroamphetamine sulfate.

**dex·ter** (dĕk′stər) *adj.* [Lat.] **1.** Of or situated on the right side. **2.** *Heraldry.* Located on the wearer's right and the observer's left. **3.** *Obs.* Auspicious : favorable.

**dex·ter·i·ty** (dĕk-stĕr′ĭ-tē) *n.* [OFr. *dexterite* < Lat. *dexteritas* < *dexter*, skillful, on the right side.] **1.** Skill in the use of the hands or body : ADROITNESS. **2.** Mental skill or facility : CLEVERNESS.

**dex·ter·ous** (dĕk′stər-əs, -strəs) *also* **dex·trous** (-strəs) *adj.* [< Lat. *dexter*, skillful, on the right side.] **1.** Skillful or adroit in the use of the hands, body, or mind. **2.** Performed with dexterity. **—dex′ter·ous·ly** *adv.* **—dex′ter·ous·ness** *n.*

☆ ***syns:*** DEXTEROUS, ADROIT, DEFT, FACILE, NIMBLE, SLICK *adj. core meaning* : exhibiting or possessing skill and ease in performance <*dexterous* handling of the sports car>

**dextr-** *pref. var. of* DEXTRO-.

**dex·tral** (dĕk′strəl) *adj.* **1.** Of, relating to, or situated on the right side : RIGHT. **2.** Right-handed. **3.** *Zool.* Designating or relating to a gastropod shell that has its aperture to the right when facing the observer with the apex upward. **—dex·tral′i·ty** (-străl′ĭ-tē) *n.* **—dex′tral·ly** *adv.*

**dex·tran** (dĕk′străn′, -strən) *n.* Any of various heavy long-chain glucose polymers that are used, depending on molecular weight, as a blood-plasma substitute and in candy, lacquers, and food additives.

**dex·trin** (dĕk′strĭn) *also* **dex·trine** (dĕk′strĭn, -strēn′) *n.* A white or yellow powder formed by the hydrolysis of starch, having colloidal properties and used chiefly as an adhesive and thickening agent.

**dex·tro** (dĕk′strō) *adj.* Dextrorotatory.

**dextro-** *or* **dextr-** *pref.* [Lat. < *dexter*, on the right side.] **1.** On or to the right : RIGHT <*dextrorotation*> **2.** Dextrorotatory <*dextrose*>

**dex·tro·am·phet·a·mine** (dĕk′strō-ăm-fĕt′ə-mēn′, -mĭn) *n.* AMPHETAMINE 2.

**dex·tro·glu·cose** (dĕk′strə-glo͞o′kōs′, -kōz′) *n.* Dextrose.

**dex·tro·ro·ta·ry** (dĕk′strə-rō′tə-rē) *adj. var. of* DEXTROROTATORY.

**dex·tro·ro·ta·tion** (dĕk′strə-rō-tā′shən) *n.* A rotation to the right. —Used esp. of the plane of polarization of light.

**dex·tro·ro·ta·to·ry** (dĕk′strə-rō′tə-tôr′ē, -tōr′ē) *also* **dex·tro·ro·ta·ry** (-rō′tə-rē) *adj.* **1.** Turning or rotating the plane of polarization of light to the right or clockwise <*dextrorotatory* crystals> **2.** *Chem.* Of or relating to a solution that rotates the plane of polarized light to the right or clockwise.

**dex·trorse** (dĕk′strôrs′) *adj.* [NLat. *dextrorsus* < Lat., turned toward the right : *dexter*, right + *versus*, p.part. of *vertere*, to turn.] Growing upward in a spiral that turns from left to right <a *dextrorse* vine> **—dex′trorse′ly** *adv.*

**dex·trose** (dĕk′strōs′, -strōz′) *n.* A dextrorotatory sugar, $C_6H_{12}O_6 \cdot H_2O$, found in plant and animal tissue and derived synthetically from starch.

**dex·trous** (dĕk′strəs) *adj. var. of* DEXTEROUS.

**dey** (dā) *n.* [Fr. < Turk. *dayi*, maternal uncle.] **1.** The title of the governor of Algiers before the French conquest in 1830. **2.** A title held by a ruler of the former states of Tunis or Tripoli.

**dhar·ma** (dûr′mə, där′-) *n.* [Skt. *darmaḥ*, law.] **1.** The ultimate law of all things in Hinduism and Buddhism. **2.** Individual right conduct in conformity to dharma.

**Dhe·gi·ha** (dā′jē-hä′) *n., pl.* **Dhegiha** *or* **-has.** **1.** A Siouan language of the Osage, Omaha, and other neighboring tribes. **2. a.** The tribes speaking Dhegiha. **b.** A member of any of these tribes.

**dhole** (dōl) *n.* [Perh. < Kanarese *tōla*, wolf.] A doglike carnivorous mammal, *Cuon alpinus* of Asia, with yellowish fur and often hunting in packs.

**dho·ti** (dō′tē) *also* **dhoo·ti** (do͞o′-) *n., pl.* **-tis.** [Hindi *dhōtī*.] **1.** A loincloth worn by Hindu men in India. **2.** The cloth used in dhotis.

**dhow** (dou) *n.* [Ar. *dāw.*] A lateen-rigged Arabian vessel.

**Dhul-Hij·ja** (dŏŏl-hĭj′ä) *n.* [Ar. *dhū'l-ḥijja,* the one of the pilgrimage.] The 12th month of the Moslem year. —See table at CALENDAR.

**Dhul-Qa·dah** (dŏŏl-kä′dä) *n.* [Ar. *dhu'l-ga'dah,* the one of the sitting.] The 11th month of the Moslem year. —See table at CALENDAR.

**di-**[1] *pref.* [Gk.] **1.** Two : twice : double <*dichromatic*> **2.** Containing two atoms, radicals, or groups <*dichloride*>

**di-**[2] *pref. var. of* DIA-.

**dia-** *or* **di-** *pref.* [Gk. < *dia,* through.] **1.** Through <*diachronic*> **2.** Across <*diactinic*>

**di·a·base** (dī′ə-bās′) *n.* [Fr. < Gk. *diabasis,* a crossing over < *diabainein,* to cross over : *dia-,* across + *bainein,* to go.] A dark-gray to black, fine-textured igneous rock consisting mainly of feldspar and pyroxene and used for monuments and as crushed stone.

**di·a·be·tes** (dī′ə-bē′tĭs, -tēz) *n.* [ME *diabete* < Med. Lat. *diabetes* < Gk. *diabētēs,* a passing through < *diabainein,* to cross over. —see DIABASE.] Any of several metabolic disorders marked by persistent thirst and excessive discharge of urine.

**diabetes in·sip·i·dus** (ĭn-sĭp′ĭ-dəs) *n.* [NLat., insipid diabetes.] A disease caused by a disorder of the pituitary gland and marked by intense thirst and excessive urination.

**diabetes mel·li·tus** (mə-lī′təs, mĕl′ĭ-) *n.* [NLat., honey-sweet diabetes.] A chronic disease of pancreatic etiology, marked by insulin deficiency, subsequent inability to utilize carbohydrates, excess sugar in the blood and urine, excessive thirst, hunger, and urination, weakness, emaciation, imperfect combustion of fats resulting in acidosis, and, without insulin injection, eventual coma and death.

**di·a·bet·ic** (dī′ə-bĕt′ĭk) *adj.* Of, relating to, or having diabetes. —*n.* One afflicted with diabetes mellitus.

**di·a·ble·rie** (dē-ä′blə-rē, -äb′lə-) *n.* [Fr. < *diable,* devil < LLat. *diabolus.* —see DEVIL.] **1.** Witchcraft or sorcery. **2.** The representation of devils or demons, as in art or fiction. **3.** Devilish conduct.

**di·a·bol·ic** (dī′ə-bŏl′ĭk) *also* **di·a·bol·i·cal** (-ĭ-kəl) *adj.* [ME *deabolik* < OFr. *diabolique* < LLat. *diabolicus* < *diabolus,* devil. —see DEVIL.] **1.** Of, concerning, or typical of the devil : SATANIC. **2.** Wicked or cruel. **—di′a·bol′i·cal·ly** *adv.* **—di′a·bol′i·cal·ness** *n.*

**di·ab·o·lism** (dī-ăb′ə-lĭz′əm) *n.* **1.** Worship of or dealings with the devil or demons. **2.** Devilish character or conduct. **—di·ab′o·list** *n.*

**di·ab·o·lize** (dī-ăb′ə-līz′) *vt.* **-lized, -liz·ing, -liz·es. 1.** To cause to be diabolic or devilish. **2.** To portray as diabolic.

**di·a·ce·tyl·mor·phine** (dī′ə-sēt′l-môr′fēn′, dī-ăs′ĭ-tl-) *n.* Heroin.

**di·a·chron·ic** (dī′ə-krŏn′ĭk) *adj.* [DIA- + Gk. *khronos,* time.] Of or concerned with phenomena, esp. of language, as they occur or change through time. **—di′a·chron′i·cal·ly** *adv.*

**di·ac·id** (dī-ăs′ĭd) *also* **di·a·cid·ic** (dī′ə-sĭd′ĭk) *adj.* **1.** Capable of combining with two monoprotic acid molecules or one diprotic acid molecule to form a salt or ester. —Used esp. of bases. **2.** Having two hydrogen atoms replaceable by metal atoms. —Used of a salt. —*n.* **diacid.** An acid having two readily replaceable hydrogen atoms.

**di·ac·o·nal** (dī-ăk′ə-nəl) *adj.* [LLat. *diaconalis* < *diaconus,* deacon.] Of or concerning a deacon or the diaconate.

**di·ac·o·nate** (dī-ăk′ə-nĭt, -nāt′) *n.* [LLat. *diaconatus* < *diaconus,* deacon.] **1.** The rank or office of a deacon. **2.** Deacons as a group.

**di·a·crit·ic** (dī′ə-krĭt′ĭk) *adj.* **1.** Diacritical. **2.** *Med.* Serving to identify : diagnostic or distinctive. —*n.* A diacritical mark.

**di·a·crit·i·cal** (dī′ə-krĭt′ĭ-kəl) *adj.* [< Gk. *diakritikos,* distinguishing < *diakrinein,* to distinguish : *dia-,* apart + *krinein,* to separate.] Marking a distinction : DISTINGUISHING. **—di′a·crit′i·cal·ly** *adv.*

**diacritical mark** *n.* A mark, as a circumflex, added to a letter to indicate a special phonetic value or to distinguish words otherwise graphically identical.

**di·ac·tin·ic** (dī′ăk-tĭn′ĭk) *adj.* Capable of transmitting chemically active or actinic radiation. **—di·ac′tin·ism** (-ăk′tə-nĭz′əm) *n.*

**di·a·del·phous** (dī′ə-dĕl′fəs) *adj. Bot.* Having the filaments united so as to form two groups. —Used of stamens.

**di·a·dem** (dī′ə-dĕm′, -dəm) *n.* [ME *diademe* < OFr. < Lat. *diadema* < Gk. *diadēma* < *diadein,* to bind on either side : *dia-,* across + *dein,* to bind.] **1.** A headband or crown worn as a sign of royalty. **2.** Royal power or dignity. —*vt.* **-demed, -dem·ing, -dems.** To adorn with or as if with a diadem.

**di·aer·e·sis** (dī-ĕr′ĭ-sĭs) *n. var. of* DIERESIS.

**di·a·gen·e·sis** (dī′ə-jĕn′ĭ-sĭs) *n.* The process of physical and chemical change in deposited sediment during its conversion to rock. **—di′a·ge·net′ic** (-jə-nĕt′ĭk) *adj.*

**di·a·ge·ot·ro·pism** (dī′ə-jē-ŏt′rə-pĭz′əm) *n. Bot.* The tendency of growing parts, as roots, to become oriented at right angles to the direction of gravitational force. **—di′a·ge′o·trop′ic** (-ə-trŏp′ĭk, -trō′pĭk) *adj.*

**di·ag·nose** (dī′əg-nōs′, -nōz′) *v.* **-nosed, -nos·ing, -nos·es.** [Back-formation < DIAGNOSIS.] —*vt.* To distinguish or identify (e.g., a disease) by diagnosis. —*vi.* To make a diagnosis.

**di·ag·no·sis** (dī′əg-nō′sĭs) *n., pl.* **-ses** (-sēz′) [NLat. < Gk. *diagnōsis,* discernment < *diagignoskein,* to distinguish : *dia-,* apart + *gignoskein,* to know.] **1.** *Med.* **a.** The act or process of identifying or determining the nature of a disease by way of examination. **b.** The opinion derived from such an examination. **2. a.** A critical analysis. **b.** The conclusion reached by such analysis. **3.** *Biol.* A precise, detailed description of an organism's characteristics for taxonomic classification.

**di·ag·nos·tic** (dī′əg-nŏs′tĭk) *adj.* [Gk. *diagnostikos* < *diagnōstos,* to be distinguished < *diagignoskein,* to distinguish. —see DIAGNOSIS.] **1.** Of, relating to, or used in a diagnosis. **2.** Serving to identify a disease : CHARACTERISTIC. —*n.* **1.** *often* **diagnostics.** The art or practice of medical diagnosis. **2.** A symptom serving as supporting evidence in a diagnosis. **—di′ag·nos′ti·cal·ly** *adv.*

**di·ag·nos·ti·cian** (dī′əg-nŏ-stĭsh′ən) *n.* One who diagnoses, esp. a physician specializing in medical diagnostics.

**di·ag·o·nal** (dī-ăg′ə-nəl) *adj.* [Lat. *diagonalis* < Gk. *diagonios,* from angle to angle : *dia-,* across + *gonia,* angle.] **1.** *Math.* **a.** Joining two nonadjacent vertices of a polygon. **b.** Joining two vertices of a polyhedron not in the same face. **2.** Oblique or slanted. **3.** Having oblique lines or markings. —*n.* **1.** *Math.* A diagonal line or plane. **2.** Something arranged obliquely, as a row, course, or part. **3.** A fabric woven with diagonal lines. **—di·ag′o·nal·ly** *adv.*

**di·ag·o·nal·ize** (dī-ăg′ə-nə-līz′) *vt.* **-ized, -iz·ing, -iz·es.** To order a matrix so that all the nonzero elements occur on the diagonal from upper left to lower right. **—di·ag′o·nal·iz′a·ble** *adj.* **—di·ag′o·nal·i·za′tion** *n.*

**diagonal matrix** *n.* A matrix that has been diagonalized.

**di·a·gram** (dī′ə-grăm′) *n.* [Lat. *diagramma* < Gk. < *diagraphein,* to mark out : *dia-,* apart + *graphein,* to write.] **1.** A plan, sketch, drawing, or outline designed to explain or demonstrate how something works or to clarify the relationship between the parts of a whole. **2.** *Math.* A graphic representation of an algebraic or geometric relationship. **3.** A graph or chart. —*vt.* **-grammed, -gram·ming, -grams** *or* **-gramed, -gram·ing, -grams.** To represent or indicate by or as if by a diagram. **—di′a·gram′ma·ble** *adj.* **—di′a·gram·mat′ic** (-grə-măt′ĭk), **di′a·gram·mat′i·cal** *adj.* **—di′a·gram·mat′i·cal·ly** *adv.*

**di·a·ki·ne·sis** (dī′ə-kə-nē′sĭs, -kī-) *n., pl.* **-ses** (-sēz′). *Genetics.* The final stage of the prophase in meiosis, during which the shortening, thickening, and dispersion of the chromosomes and the disappearance of the nucleolus occur. **—di′a·ki·net′ic** (-nĕt′ĭk) *adj.*

**di·al** (dī′əl) *n.* [ME *diall* < Med. Lat. *diale* < *dialis,* daily < Lat. *dies,* day.] **1.** A graduated, usu. circular face on which a measurement, as speed, is indicated by a moving pointer or needle. **2. a.** A clock face. **b.** A sundial. **3. a.** The face or panel on a radio or television receiver on which the frequencies or channels are indicated. **b.** A device, as a movable control knob, on a radio or television receiver used to change the frequency or channel. **4.** A rotatable disk on a telephone with numbers and letters, used to signal the number to which a call is made. —*v.* **-aled, -al·ing, -als** *or* **-alled, -al·ling, -als.** —*vt.* **1.** To measure with or as if with a dial. **2.** To point to, indicate, or register by a dial. **3.** To select or control by a dial. —*vi.* To use a dial, as on a telephone. **—di′al·er** *n.*

**di·a·lect** (dī′ə-lĕkt′) *n.* [OFr. *dialecte* < Lat. *dialectus* < Gk. *dialektos,* speech < *dialegesthai,* to discuss : *dia-,* between + *legesthai,* to speak < *legein,* to tell.] **1. a.** A regional variety of a language distinguished by pronunciation, grammar, or vocabulary, esp. a variety of speech differing from the standard literary language or speech pattern of the culture in which it exists <West Saxon was a *dialect* of Old English.> **b.** A variety of language that with other varieties constitutes a single language of which no single variety is standard <Ancient Greek had many *dialects.*> **2.** The language peculiar to an occupational group or a particular social class <the *dialect* of law> **3.** The manner or style of expressing oneself in language or the arts. **4.** A language considered as part of a larger family of languages or a linguistic branch <the Indic and Italic *dialects* of Indo-European> **—di′a·lec′tal** *adj.* **—di′a·lec′tal·ly** *adv.*

**dialect atlas** *n.* A linguistic atlas.

**dialect geography** *n.* Linguistic geography.

**di·a·lec·tic** (dī′ə-lĕk′tĭk) *n.* [ME *dialetik* < OFr. *dialetique* < Lat. *dialectica* < Gk. *dialektikē (tekhnē),* (art of) debate < *dialektos,* speech. —see DIALECT.] **1.** The art or practice of arriving at the truth by disclosing the contradictions in an opponent's argument and overcoming them. **2. a.** The Hegelian process of change whereby a thesis is transformed into an antithesis, and preserved and fulfilled by it, the combination of the two being resolved in a synthesis. **b.** Hegel's critical method for the investigation of this process. **3. a.** *often* **dialectics** *(sing. in number).* The Marxian process of change through the conflict of opposing forces, whereby a given contradiction is marked by a primary and a secondary aspect, the secondary succumbing to the primary, which is then transformed into an aspect of a new contradiction. **b.** The Marxian critique of this process. **4. dialectics** *(sing. in number).* A method of argument or exposition that systematically weighs contradictory facts or ideas with a view to the resolution of their real or apparent contradictions. **5.** The contradiction between two conflicting forces viewed as the determining factor in their continuing interaction. **—di′a·lec′ti·cal, di′a·lec′tic** *adj.* **—di′a·lec′ti·cal·ly** *adv.*

**dialectical materialism** *n.* Marxian interpretation of reality that views matter as the sole subject of change and all change as the product of a constant conflict between opposites arising from the internal contradictions inherent in all events, ideas, and movements.

**di·a·lec·ti·cian** (dī′ə-lĕk-tĭsh′ən) *n.* **1.** A specialist in the study of dialects. **2.** One who is skilled in or practices dialectic.

**di·a·lec·tol·o·gy** (dī′ə-lĕk-tŏl′ə-jē) *n.* Study of dialects. **—di′a·lec′to·log′i·cal** (-tə-lŏj′ĭ-kəl) *adj.* **—di′a·lec′to·log′i·cal·ly** *adv.* **—di′a·lec·tol′o·gist** *n.*

**di·a·log** (dī′ə-lôg′, -lŏg′) *n.* & *v. var. of* DIALOGUE.

**di·a·log·ic** (dī′ə-lŏj′ĭk) *also* **di·a·log·i·cal** (-ĭ-kəl) *adj.* Of, relating to, or written in dialogue. **—di′a·log′i·cal·ly** *adv.*

**di·al·o·gist** (dī-ăl′ə-jĭst, dī′ə-lô′gĭst, -lŏg′ĭst) *n.* **1.** A writer of dialogue. **2.** One who speaks in a dialogue. **—di′a·lo·gis′tic** (dī′ə-lə-jĭs′tĭk), **di′a·lo·gis′ti·cal** *adj.*

**di·a·logue** *also* **di·a·log** (dī′ə-lôg′, -lŏg′) [ME < OFr. < Lat. *dialogus* < Gk. *dialogos* < *dialegesthai,* to discuss. —see DIALECT.] —*n.* **1.** A conversation between two or more people. **2.** A conversational passage in a narrative or play. **3.** A literary work written in the form of a conversation <the *dialogues* of Plato> **4.** A musical composition or passage for two or more parts that is suggestive of conversational interplay. **5.** An exchange of opinions or ideas. —*v.* **-logued, -logu·ing, -logues** *also* **-loged, -log·ing, -logs.** —*vt.* To express as or in a dialogue. —*vi.* To converse in a dialogue. **—di′a·log′uer** *n.*

**dial tone** *n.* A low, steady tone in a telephone receiver indicating that the line is open and a number may be dialed.

**di·al·y·sis** (dī-ăl′ĭ-sĭs) *n., pl.* **-ses** (-sēz′) [NLat. < Gk. *dialusis,* separating < *dialuein,* to tear apart : *dia-,* apart + *luein,* to loosen.] Separation of smaller molecules from larger molecules or of crystalloid particles from colloidal particles in a solution by selective diffusion through a semipermeable membrane. **—di′a·lyt′ic** (-ə-lĭt′ĭk) *adj.* **—di′a·lyt′i·cal·ly** *adv.*

**di·a·lyze** (dī′ə-līz′) *vt.* & *vi.* **-lyzed, -lyz·ing, -lyz·es.** [Back-formation < DIALYSIS.] To subject to or undergo dialysis. **—di′a·lyz′a·bil′i·ty** *n.* **—di′a·lyz′a·ble** *adj.* **—di′a·lyz′er** *n.*

**di·a·mag·net** (dī′ə-măg′nĭt) *n.* [< DIAMAGNETIC.] A diamagnetic substance.

**di·a·mag·net·ic** (dī′ə-măg-nĕt′ĭk) *adj.* Of or relating to a substance in which an induced magnetic field is in the opposite direction to and much weaker than the magnetizing field. **—di′a·mag′ne·tism** (-nĭ-tĭz′əm) *n.*

**di·am·e·ter** (dī-ăm′ĭ-tər) *n.* [ME *diametre* < OFr. < Lat. *diametros* < Gk. *diametros (grammē),* diagonal (line) : *dia-,* through + *metron,* measure.] **1.** *Math.* **a.** A straight line segment passing through the center of a figure, esp. of a circle or sphere, and terminating at the periphery. **b.** The length of such a segment. **2.** Width or thickness. **—di·am′e·tral** (-trəl) *adj.*

**di·a·met·ri·cal** (dī′ə-mĕt′rĭ-kəl) *also* **di·a·met·ric** (-rĭk) *adj.* **1.** Of, relating to, or along a diameter. **2.** Exactly opposite : CONTRARY <thinking *diametrical* to mine> **—di′a·met′ri·cal·ly** *adv.*

**di·am·ine** (dī-ăm′ēn′, -ĭn, dī′ə-mēn′, -mĭn) *n.* Any of various chemical compounds having two amino groups, esp. hydrazine.

**di·a·mond** (dī′ə-mənd, dī′mənd) *n.* [ME *diamaunt* < OFr. *diamant* < LLat. *diamas* < Lat. *adamas* < Gk.] **1.** A very hard, highly refractive colorless or white crystalline allotrope of carbon, used when pure as a gemstone and otherwise chiefly in abrasives. **2.** A figure with four equal sides forming two inner obtuse angles and two inner acute angles : rhombus or lozenge. **3. a.** A red, lozenge-shaped figure on certain playing cards. **b.** A playing card with this figure. **c. diamonds** *(sing. or pl. in number).* The suit of cards represented by this figure. **4.** *Baseball.* **a.** An infield. **b.** The entire playing field. —*vt.* **-mond·ed, -mond·ing, -monds.** To adorn with or as if with diamonds.

**di·a·mond·back** (dī′ə-mənd-băk′, dī′mənd-) *n.* **1.** A large venomous rattlesnake of the genus *Crotalus* of the southern and western United States and Mexico, bearing diamond-shaped markings. **2.** A turtle of the genus *Malaclemys* of the southern Atlantic and Gulf coasts of the United States, with edible flesh and a carapace bearing diamond-shaped ridged or knobbed markings.

**diamondback terrapin** *n.* DIAMONDBACK 2.

**di·a·mond·if·er·ous** (dī′ə-mən-dĭf′ər-əs, dī′mən-) *adj.* Bearing or yielding diamonds.

**Di·an·a** (dī-ăn′ə) *n.* [ME < Lat.] *Rom. Myth.* **1.** The goddess of chastity, hunting, and the moon. **2.** The moon.

**di·an·drous** (dī-ăn′drəs) *adj. Bot.* Having two stamens.

**di·an·thus** (dī-ăn′thəs) *n.* [NLat. *Dianthus,* genus name : DI- + Gk. *anthos,* flower.] A plant of the genus *Dianthus,* which includes carnations and pinks.

**di·a·pa·son** (dī′ə-pā′zən, -sən) *n.* [ME *diapasoun* < Lat. *diapason* < Gk. *(hē) dia pasōn (khordōn sumphonia),* (concord) through all (the notes).] **1.** A full, rich outpouring of harmonious sound. **2. a.** The entire range of an instrument or voice. **b.** Entire range or scope. **3.** Either of the two principal stops on a pipe organ that form the tonal basis for the entire scale of the instrument. **4.** The musical interval and the consonance of an octave. **5.** A standard indication of musical pitch. **6.** A tuning fork.

**di·a·pause** (dī′ə-pôz′) *n.* [Gk. *diapausis,* pause < *diapauein,* to pause : *dia-,* between + *pauein,* to stop.] A period during which growth or development is suspended, as in certain insects.

**di·a·pe·de·sis** (dī′ə-pĭ-dē′sĭs) *n., pl.* **-ses** (-sēz′) [NLat. < Gk. *diapēdēsis,* transudation < *diapēdan,* to ooze : *dia-,* through + *pēdan,* to leap.] The passage of blood or any constituents, esp. erythrocytes, through intact blood-vessel walls. **—di′a·pe·det′ic** (-dĕt′ĭk) *adj.*

**di·a·per** (dī′ə-pər, dī′pər) *n.* [ME, a patterned fabric < OFr. < Med. Lat. *diasprum* < Med. Gk. *diaspros,* pure white : *dia-,* thoroughly + *aspros,* white < Lat. *asper,* rough.] **1.** A folded piece of absorbent material placed between a baby's legs and fastened at the waist. **2. a.** A white cotton or linen fabric patterned with small diamond-shaped figures. **b.** A piece of such cloth. **c.** Such a pattern. —*vt.* **-pered, -per·ing, -pers. 1.** To put a diaper on <*diaper* a baby> **2.** To weave or decorate in a diamond-shaped pattern.

**di·aph·a·nous** (dī-ăf′ə-nəs) *adj.* [Med. Lat. *diaphanus* Gk. *diaphanēs* < *diaphanein,* to be transparent : *dia-,* through + *phainein,* to show.] **1.** Of such fine texture as to be transparent or translucent. **2.** Delicate in form. **3.** Vague or insubstantial <*diaphanous* fantasies> **—di′a·pha·ne′i·ty** (dī′ə-fə-nē′ĭ-tē), **di·aph′a·nous·ness** *n.* **—di·aph′a·nous·ly** *adv.*

**di·a·pho·re·sis** (dī′ə-fə-rē′sĭs, dī-ăf′ə-) *n.* [LLat. < Gk. *diaphorēsis* < *diaphorein,* to disperse : *dia-,* apart + *phorein,* to convey, freq. of *pherein,* to bear.] Perspiration, esp. when copious and induced by a diaphoretic agent.

**di·a·pho·ret·ic** (dī′ə-fə-rĕt′ĭk, dī-ăf′ə-) *adj.* Inducing perspiration. —*n.* A diaphoretic medicine or agent.

**di·a·phragm** (dī′ə-frăm′) *n.* [ME *diafragma* < LLat. *diaphragma* < Gk. < *diaphrassein,* to barricade : *dia-,* completely + *phrassein,* to enclose.] **1.** *Anat.* A muscular membranous partition separating the abdominal and thoracic cavities and functioning in respiration. **2.** A membranous part that separates or divides. **3.** A thin disk, esp. in a microphone or telephone receiver, whose vibrations convert electric signals to sound waves or sound waves to electric signals. **4.** A contraceptive device consisting of a flexible disk that covers the uterine cervix. **5.** A disk having a fixed or variable opening for restricting the amount of light traversing a lens or optical system. **—di′a·phrag·mat′ic** (-frăg-măt′ĭk) *adj.* **—di′a·phrag·mat′i·cal·ly** *adv.*

**di·aph·y·sis** (dī-ăf′ĭ-sĭs) *n., pl.* **-ses** (-sēz′) [NLat. < Gk. *diaphusis,* spinous process of the tibia < *diaphuesthai,* to grow between : *dia-,* between + *phuesthai,* to grow < *phuein,* to produce.] The shaft of a long bone. **—di′a·phys′i·al, di′a·phys′e·al** (dī′ə-fĭz′ē-əl) *adj.*

**di·a·poph·y·sis** (dī′ə-pŏf′ĭ-sĭs) *n., pl.* **-ses** (-sēz′). The superior or articular surface of a transverse vertebral process. **—di·ap′o·phys′i·al** (-ăp′ə-fĭz′ē-əl) *adj.*

**di·ar·chy** *also* **dy·ar·chy** (dī′är′kē) *n., pl.* **-chies.** Government by two joint rulers.

**di·a·rist** (dī′ə-rĭst) *n.* One who keeps a diary.

**di·ar·rhe·a** *also* **di·ar·rhoe·a** (dī′ə-rē′ə) *n.* [ME *diaria* < Med. Lat. < LLat. *diarrhoea* < Gk. *diarrhoia* < *diarrhein,* to flow through : *dia-,* through + *rhein,* to flow.] *Pathol.* Excessive evacuation of watery feces. **—di′ar·rhe′al** (-əl), **di′ar·rhe′ic** (-ĭk), **di′ar·rhet′ic** (-rĕt′ĭk) *adj.*

**di·ar·thro·sis** (dī′är-thrō′sĭs) *n., pl.* **-ses** (-sēz′) [NLat. < Gk. *diarthrōsis* < *diarthroun,* to articulate : *dia-,* between + *arthroun,* to fasten < *arthron,* joint.] A type of bone articulation allowing free motion in a joint. **—di′ar·thro′di·al** (-dē-əl) *adj.*

**di·a·ry** (dī′ə-rē) *n., pl.* **-ries.** [Lat. *diarium,* daily allowance, diary < *dies,* day.] **1.** A daily record, esp. a personal record of events, experiences, and observations : JOURNAL. **2.** A book for keeping a diary.

**Di·as·po·ra** (dī-ăs′pər-ə) *n.* [Gk., dispersion < *diaspeirein,* to disperse : *dia-,* apart + *speirein,* to scatter.] **1.** *often* **diaspora.** The body of Jews or Jewish communities settled outside Palestine or modern Israel. **2.** The body of Jews living dispersed among the Gentiles after the Babylonian captivity. **3. diaspora. a.** A dispersion of an orig. homogeneous people. **b.** A migration.

**di·a·spore** (dī′ə-spôr′, -spōr′) *n.* [Gk. *diaspora,* dispersion. —see DIASPORA.] A white to greenish, pearly hydrous aluminum oxide, $Al_2O_3 \cdot H_2O$, found in bauxite, corundum, and dolomite and used as a refractory and abrasive.

**di·a·stal·sis** (dī′ə-stôl′sĭs, -stăl′-) *n., pl.* **-ses** (-sēz′) [DIA- + (PERI)STALSIS.] Peristaltic contraction of the small intestine in digestion. **—di′a·stal′tic** (-tĭk) *adj.*

**di·a·stase** (dī′ə-stās′, -stāz′) *n.* [Fr. < Gk. *diastasis,* separation. —see DIASTASIS.] An amylase or a mixture of amylases that converts starch to maltose, found in germinating grains such as malt. **—di′a·sta′sic** (-stā′sĭk, -zĭk) *adj.*

**di·as·ta·sis** (dī-ăs′tə-sĭs) *n., pl.* **-ses** (-sēz′) [NLat. < Gk., separation < *diistanai,* to separate : *dia-,* apart + *histanai,* to cause to stand.] **1.** *Pathol.* Separation of normally adjacent, unjoined bones without fracture or of certain muscles during pregnancy. **2.** *Physiol.* The final stage of diastole in the heart, which occurs prior to contraction and during which little blood enters the filled ventricle. **—di′a·stat′ic** (dī′ə-stăt′ĭk) *adj.*

**di·a·ste·ma** (dī′ə-stē′mə) *n., pl.* **-ma·ta** (-mə-tə) [NLat. < LLat., interval < Gk. *diastēma* < *diistanai,* to separate. —see DIASTASIS.]

1. *Pathol.* A bodily cleft or fissure, esp. if congenital. **2.** An abnormally large space between teeth. **—di′a·ste·mat′ic** (-stə-măt′ĭk) *adj.*

**di·as·to·le** (dī-ăs′tə-lē) *n.* [Gk. *diastolē*, dilation < *diastellein*, to expand : *dia-*, apart + *stellein*, to put.] **1.** *Physiol.* The normal rhythmically occurring relaxation and dilatation of the heart cavities during which the cavities are filled with blood. **2.** The lengthening of a normally short syllable in Greek and Latin verse. **—di′a·stol′ic** (dī′ə-stŏl′ĭk) *adj.*

**di·as·tro·phism** (dī-ăs′trə-fĭz′əm) *n.* [< Gk. *diastrophē*, distortion < *diastrephein*, to distort : *dia-*, apart + *strephein*, to twist.] The process by which the major features of the earth's crust, including continents, mountains, ocean beds, folds, and faults, are formed. **—di′a·stroph′ic** (dī′ə-strŏf′ĭk) *adj.*

**di·a·tes·sa·ron** (dī′ə-tĕs′ər-ən) *n.* [Gk. *(evangelion) dia tessarōn*, (gospel) consisting of four parts.] The four Gospels combined into a single narrative.

**di·a·ther·my** (dī′ə-thûr′mē) *n.* [DIA- + Gk. *thermē*, heat.] Therapeutic generation of local heat in body tissues by high-frequency electromagnetic waves. **—di′a·ther′mic** (-mĭk) *adj.*

**di·ath·e·sis** (dī-ăth′ĭ-sĭs) *n., pl.* **-ses** (-sēz′) [NLat. < Gk., condition < *diatithenai*, to distribute : *dia-*, apart + *tithenai*, to put.] An often hereditary, congenital bodily predisposition to a disease or metabolic or structural abnormality. **—di′a·thet′ic** (dī′ə-thĕt′ĭk) *adj.*

**di·a·tom** (dī′ə-tŏm′) *n.* [NLat. *diatoma* < Gk. *diatomos*, cut in half < *diatemnein*, to cut in half : *dia-*, through + *temnein*, to cut.] Any of various minute unicellular or colonial algae of the class Bacillariophyceae, with siliceous cell walls consisting of two overlapping symmetric parts.

**di·a·to·ma·ceous** (dī′ə-tə-mā′shəs, dī-ăt′ə-) *adj.* Made up of diatoms or their siliceous skeletons.

**diatomaceous earth** *n.* A white or cream-colored siliceous earth composed of the shells of diatoms.

**di·a·tom·ic** (dī′ə-tŏm′ĭk) *adj.* **1.** Having two atoms in the molecule. **2.** Having two replaceable atoms or radicals.

**di·at·o·mite** (dī-ăt′ə-mīt′) *n.* A fine, powdered diatomaceous earth used industrially as a filler, filtering agent, absorbent, clarifier, and insulator.

**di·a·ton·ic** (dī′ə-tŏn′ĭk) *adj.* [OFr. *diatonique* < LLat. *diatonicus* < Gk. *diatonikos* : *dia-*, through + *tonos*, tone.] *Mus.* Of or using only the eight tones of a standard major or minor scale without chromatic deviations. **—di′a·ton′i·cal·ly** *adv.* **—di′a·ton′i·cism** (-ĭ-sĭz′əm) *n.*

**di·a·tribe** (dī′ə-trīb′) *n.* [Lat. *diatriba*, learned discourse < Gk. *diatribē*, lecture, pastime < *diatribein*, to consume : *dia-*, completely + *tribein*, to rub.] Bitterly abusive criticism or denunciation.

**di·a·tron** (dī′ə-trŏn′) *n.* A circuitry design using diodes.

**di·at·ro·pism** (dī-ăt′rə-pĭz′əm) *n.* The tendency of certain plants or their parts to become oriented at right angles to the line of force of a stimulus. **—di′a·trop′ic** (dī′ə-trŏp′ĭk, -trō′pĭk) *adj.*

**di·az·e·pam** (dī-ăz′ə-păm′) *n.* [DIAZ(O) + EP(OXIDE) + AM(MONIA).] An antianxiety drug, $C_{16}H_{13}ClN_2O$.

**di·a·zine** (dī′ə-zēn′, dī-ăz′ĭn) *n.* [DI- + AZ(O)- + -INE.] A compound containing a benzene ring in which two of the carbon atoms have been replaced by nitrogen atoms, esp. any of three compounds so structured and having the composition $C_4H_4N_2$.

**di·az·o** (dī-ăz′ō) *adj.* Of or relating to a pair of nitrogen atoms bonded together and to an organic compound.

**di·a·zo·ni·um** (dī′ə-zō′nē-əm) *n.* [DIAZ(O) + (AMM)ONIUM.] The univalent cation $RN_2$, where R is an aromatic hydrocarbon.

**di·ba·sic** (dī-bā′sĭk) *adj.* **1.** Having two replaceable hydrogen atoms. **2.** Designating salts or acids forming salts with two atoms of a univalent metal.

**dib·ble** (dĭb′əl) *n.* [ME *debylle*.] A pointed gardening implement used to make holes in soil, esp. for planting bulbs or seedlings. *—vt.* **-bled, -bling, -bles. 1.** To make holes in (soil) with a dibble. **2.** To plant by means of a dibble. **—dib′bler** *n.*

**di·bran·chi·ate** (dī-brăng′kē-ĭt) *n.* [NLat. *Dibranchiata*, order name : DI- + Gk. *brankhia*, gills.] Any of various cephalopod mollusks of the order Dibranchiata, including the octopuses, cuttlefish, and squids. *—adj.* Of or belonging to the Dibranchiata.

**di·bro·mide** (dī-brō′mīd′, -mĭd) *n.* A binary chemical compound containing two bromine atoms per molecule.

**dibs** (dĭbz) *pl.n.* [Short for *dibstones*, counters used in a game, prob. < obs. *dib*, to tap.] *Slang.* **1.** Money, esp. in small amounts. **2.** A claim <thought I had *dibs* on the car tonight>

**di·car·box·yl·ic** (dī-kär′bŏk-sĭl′ĭk) *adj.* Containing two carboxyl groups per molecule.

**di·cast** (dī′kăst′, dĭk′ăst′) *n.* [Gk. *dikastēs*, judge < *dikazein*, to judge < *dikē*, right.] One of the 6,000 ancient Athenian citizens chosen annually to sit in the law courts, acting as both judge and juror. **—di·cas′tic** *adj.*

**dice** (dīs) *n., pl.* **dice.** [Pl. of DIE².] **1. a.** *pl. of* DIE² 3, 4. **b.** A game of chance using dice. **2.** *pl. also* **dic·es.** A small cube, as of food. *—v.* **diced, dic·ing, dic·es.** *—vi.* To play or gamble with dice. *—vt.* **1.** To win or lose (money) by gambling with dice. **2.** To cut (food) into small cubes. **3.** To decorate with dice-shaped figures. **—no dice.** *Slang.* **1.** Of no use : FUTILE. **2.** No. —Used as a refusal to a request.

**di·cen·tra** (dī-sĕn′trə) *n.* [NLat. *Dicentra*, genus name : DI- + Gk. *kentron*, center < *kentein*, to prick.] A plant of the genus *Dicentra*, including the bleeding-heart and Dutchman's-breeches.

**di·ceph·a·lous** (dī-sĕf′ə-ləs) *adj.* Having two heads, as a monster.

**dic·er** (dī′sər) *n.* **1.** A device used for dicing food. **2.** One who uses dice in gambling.

**dic·ey** (dī′sē) *adj.* **-i·er, -i·est.** [< DICE.] Risky : chancy.

**dich-** *pref. var. of* DICHO-.

**di·cha·si·um** (dī-kā′zē-əm, -zhē-əm, -zhəm) *n., pl.* **-si·a** (-zē-ə, -zhē-ə, -zhə) [NLat. < Gk. *dikhasis*, division < *dikhazein*, to divide in two < *dikha*, in two.] *Bot.* A cyme with two lateral stems branching from a main axis. **—di·cha′si·al** (-zē-əl, -zhē-əl, -zhəl) *adj.* **—di·cha′si·al·ly** *adv.*

**di·chlo·ride** (dī-klôr′īd′, -klōr′-) *n.* A binary chemical compound having two chloride atoms per molecule.

**di·chlo·ro·di·phen·yl·tri·chlo·ro·eth·ane** (dī-klôr′ō-dī-fĕn′əl-trī-klôr′ō-ĕth′ān′, -klōr′-, -fē′nəl-, dī-klōr′-) *n.* DDT.

**di·chlor·vos** (dī-klôr′vŏs′, -vəs, -klōr′-) *n.* [DI- + CHLOR(O) + V(INYL) + (PH)OS(PHATE).] A nonpersistent organophosphorous pesticide, $C_4H_7Cl_2O_4P$, of low toxicity to humans.

**dicho-** or **dich-** *pref.* [LLat. < Gk. *dikho-* < *dikha*, in two.] In two : into two parts <*dichogamous*>

**di·chog·a·mous** (dī-kŏg′ə-məs) *adj. Bot.* Having pistils and stamens that mature at different times, thus ensuring cross-fertilization rather than self-pollination. **—di·chog′a·my** (-mē) *n.*

**di·chot·o·mize** (dī-kŏt′ə-mīz′) *v.* **-mized, -miz·ing, -miz·es.** *—vt.* To divide into two parts or classifications. *—vi.* To be or become divided into parts or branches : FORK. **—di·chot′o·mist** *n.* **—di·chot′o·mi·za′tion** *n.*

**di·chot·o·mous** (dī-kŏt′ə-məs) *adj.* **1.** Divided or dividing into two parts or classifications. **2.** Marked by dichotomy. **—di·chot′o·mous·ly** *adv.* **—di·chot′o·mous·ness** *n.*

**di·chot·o·my** (dī-kŏt′ə-mē) *n., pl.* **-mies.** [Gk. *dikhotomia* < *dikhotomos*, divided : *dikha*, in two + *temnein*, to cut.] **1.** Division into two usu. contradictory parts or opinions. **2.** *Astron.* The phase of the moon, Mercury, or Venus when half of the disk is illuminated. **3.** *Bot.* Branching in which successive forking into two approx. equal divisions occurs.

**di·chro·ic** (dī-krō′ĭk) *also* **di·chro·it·ic** (dī′krō-ĭt′ĭk) *adj.* [< Gk. *dikhroos*, bicolored : *di-*, two + *khrōs*, color.] **1.** Manifesting dichroism. **2.** Dichromatic.

**di·chro·ism** (dī′krō-ĭz′əm) *n. Chem.* **1.** The property of showing different colors depending on the thickness of the medium or the relative concentration of coloring matter in it. **2.** The property possessed by some crystals of exhibiting different colors, esp. two different colors, when viewed along different axes.

**di·chro·ite** (dī-krō′īt′) *n.* [DICHRO(IC) + -ITE.] Cordierite.

**di·chro·mate** (dī-krō′māt′, dī′krō-) *n.* A usu. orange-red chemical compound with two chromium atoms per anion.

**di·chro·mat·ic** (dī′krō-măt′ĭk) *adj.* **1.** Having or exhibiting two colors. **2.** *Zool.* Having two distinct color phases in the adult, as certain species of birds do. **3.** *Pathol.* Capable of distinguishing only two colors.

**di·chro·ma·tism** (dī-krō′mə-tĭz′əm) *also* **di·chro·mism** (-mĭz′əm) *n.* The quality or state of being dichromatic.

**di·chro·mic** (dī-krō′mĭk) *adj.* **1.** Dichromatic. **2.** *Chem.* Having two chromium atoms per molecule.

**dichromic acid** *n.* An acid, $H_2Cr_2O_7$, known only in solution.

**di·chro·mism** (dī-krō′mĭz′əm) *n. var. of* DICHROMATISM.

**dick¹** (dĭk) *n.* [Shortening and alteration of DETECTIVE.] *Slang.* A detective.

**dick²** (dĭk) *n.* [< *Dick*, nickname for *Richard*.] *Chiefly Brit.* A fellow : chap.

**dick·cis·sel** (dĭk-sĭs′əl, dĭk′sĭs′-) *n.* [Imit.] A sparrowlike bird, *Spiza americana* of central North America, of which the male has a yellow breast marked with black.

**dick·ens** (dĭk′ənz) *n.* [Perh. alteration of OLD NICK.] Deuce : devil.

**dick·er** (dĭk′ər) *v.* **-ered, -er·ing, -ers.** [Perh. < *dicker*, a quantity of ten, ten hides < ME *diker*, ult. < Lat. *decuria*, set of ten < *decem*, ten.] *—vi.* To barter : bargain. *—vt.* To exchange or trade. *—n.* The act or process of dickering.

**dick·ey** *also* **dick·ie** *or* **dick·y** (dĭk′ē) *n., pl.* **-eys** *also* **-ies.** [< *Dick*, nickname for *Richard*.] **1.** A woman's blouse front worn under a suit jacket or low-necked garment. **2.** A man's detachable shirt front. **3.** A collar for a shirt. **4.** A child's pinafore or bib. **5.** A donkey. **6.** A small bird. **7. a.** The forward outside driver's seat on a carriage. **b.** A rear seat for servants on a carriage.

**Dick test** (dĭk) *n.* [After George *Dick* (1881–1967) and Gladys Henry *Dick* (1881–1963), its devisers.] A test of susceptibility to scarlet fever.

**dick·y** (dĭk′ē) *n. var. of* DICKEY.

**di·cli·nous** (dī-klī′nəs) *adj.* [DI- + Gk. *klinē*, bed + -OUS.] *Bot.* **1.** Having stamens and pistils in separate flowers, as do some plants. **2.** Having pistils but not stamens or stamens but not pistils, as some flowers do. **—di′cli′ny** (dī′klī′nē) *n.*

**di·cot·y·le·don** (dī-kŏt′l-ēd′n) *also* **di·cot** (dī′kŏt′) *n.* A plant of the subclass Dicotyledonae, characterized by a pair of embryonic seed leaves that appear at germination. **—di′cot′y·le′don·ous** (-n-əs) *adj.*

**di·cro·tism** (dī′krə-tĭz′əm) *n.* [< Gk. *dikrotos*, double-beating : *di-*, two + *krotein*, to strike.] A pathological doubling of the pulse with each heartbeat. **—di·crot′ic** (-krŏt′ĭk) *adj.*

**dic·ta** (dĭk′tə) *n. var. pl. of* DICTUM.

**Dic·ta·belt** (dĭk′tə-bĕlt′). A trademark for a plastic belt on which dictation is recorded in a dictating machine.

**Dic·ta·phone** (dĭk′tə-fōn′). A trademark for a phonographic apparatus that records and reproduces dictation for transcription.

**dic·tate** (dĭk′tāt′, dĭk-tāt′) *v.* **-tat·ed, -tat·ing, -tates.** [Lat. *dictare, dictat-*, freq. of *dicere*, to say.] —*vt.* **1.** To say or read aloud to be recorded or written by another. **2.** To issue authoritatively <*dictate* an order> —*vi.* **1.** To say or read aloud material to be recorded or written by another. **2.** To issue orders or commands. —*n.* (dĭk′tāt′). **1.** A directive or command. **2.** A guiding principle, as of one's conscience.

**dic·ta·tion** (dĭk-tā′shən) *n.* **1. a.** The process of dictating material to another for transcription. **b.** The material dictated. **2.** An authoritative order or command.

**dic·ta·tor** (dĭk′tā′tər, dĭk-tā′-) *n.* **1. a.** A ruler having absolute authority and supreme jurisdiction over the government of a state. **b.** A tyrant. **2.** One who dictates. **3.** A magistrate in ancient Rome appointed temporarily to deal with a crisis or emergency.

**dic·ta·to·ri·al** (dĭk′tə-tôr′ē-əl, -tōr′-) *adj.* **1.** Tending to dictate : DOMINEERING. **2.** Of or typical of a dictator or dictatorship : AUTOCRATIC. **—dic′ta·to′ri·al·ly** *adv.* **—dic′ta·to′ri·al·ness** *n.*

☆ ***syns:*** DICTATORIAL, BOSSY, DOGMATIC, DOMINEERING, IMPERIOUS, MASTERFUL, OVERBEARING, PEREMPTORY *adj. core meaning* : tending to assert authority and control over others <a *dictatorial* executive>

**dic·ta·tor·ship** (dĭk-tā′tər-shĭp′, dĭk′tā′-) *n.* **1.** The office or tenure of a dictator. **2.** A state or government under dictatorial rule. **3.** Absolute or despotic power or control.

**dic·tion** (dĭk′shən) *n.* [Lat. *dictio* < *dictus*, p.part. of *dicere*, to say.] **1.** Choice and use of words in speech or writing. **2.** Degree of clarity and distinctness of pronunciation in speech or singing : ENUNCIATION. **—dic′tion·al** *adj.* **—dic′tion·al·ly** *adv.*

**dic·tion·ar·y** (dĭk′shə-nĕr′ē) *n., pl.* **-ies.** [Med. Lat. *dictionarium* < Lat. *dictio*, diction.] **1.** A reference book having an explanatory alphabetical list of words, with information given for each word, including meaning, pronunciation, etymology, and often usage guidance. **2.** A book listing the words of a language with translations into another language. **3.** A book listing linguistic items, as words, with specialized information about them <a legal *dictionary*> <a biographical *dictionary*> **4.** *Computer Sci.* A list stored in machine-readable form for reference by an automatic system.

**Dic·to·graph** (dĭk′tə-grăf′). A trademark for a telephonic instrument that reproduces or records sounds from a transmitter by means of a small, often concealed microphone.

**dic·tum** (dĭk′təm) *n., pl.* **-ta** (-tə) *or* **-tums.** [Lat. < neuter p.part. of *dicere*, to say.] **1.** A dogmatic and authoritative pronouncement. **2.** *Law.* An obiter dictum. **3.** A popular saying : MAXIM.

**did** (dĭd) *v. p.t. of* DO.

**di·dact** (dī′dăkt′) *n.* [Back-formation < DIDACTIC.] One who is didactic in manner.

**di·dac·tic** (dī-dăk′tĭk) *also* **di·dac·ti·cal** (-tĭ-kəl) *adj.* [Gk. *didaktikos*, skillful in teaching < *didaktos*, taught < *didaskein*, to teach.] **1.** Intended to instruct. **2.** Morally instructive. **3.** Inclined to teach or moralize excessively. **—di·dac′ti·cal·ly** *adv.* **—di·dac′ti·cism** (-tĭ-sĭz′əm) *n.*

**di·dac·tics** (dī-dăk′tĭks) *n. (sing. in number).* The art or science of teaching or instruction : PEDAGOGY.

**di·dap·per** (dī′dăp′ər) *n.* [ME *didopper*, alteration of *divedap* < OE *dūfeedoppa* : *dūfan*, to dive + *-doppa*, a kind of bird.] A small grebe, as the dabchick.

**did·dle**[1] (dĭd′l) *v.* **-dled, -dling, -dles.** [Orig. unknown.] —*vt.* To swindle : cheat. —*vi.* To waste time. **—did′dler** *n.*

**did·dle**[2] (dĭd′l) *v.* **-dled, -dling, -dles.** [Alteration of dial. *didder*, to quiver < ME *dideren*.] —*vt.* To jerk up and down or back and forth. —*vi.* To shake rapidly : JIGGLE.

**did·n't** (dĭd′nt). Did not.

**di·do** (dī′dō) *n., pl.* **-dos** *or* **-does.** [Orig. unknown.] *Informal.* A mischievous antic or prank : CAPER. —Used esp. in the phrase *cut didos*.

**Di·do** (dī′dō) *n.* [Lat. < Gk. *Didō*.] *Rom. Myth.* A Tyrian princess, founder and queen of Carthage.

**didst** (dĭdst) *v. Archaic. 2nd person sing. p.t. of* DO.

**di·dym·i·um** (dī-dĭm′ē-əm) *n.* [NLat. < Gk. *didumos*, twin.] **1.** A metallic mixture, once considered an element, composed of neodymium and praseodymium. **2.** A mixture of rare-earth elements and oxides used primarily in manufacturing and coloring glass esp. for optical filters.

**did·y·mous** (dĭd′ə-məs) *adj.* [Gk. *didumos*, twin.] Occurring or arranged in pairs : TWIN.

**di·dyn·a·mous** (dī-dĭn′ə-məs) *adj.* [< NLat. *Didynamia*, former class name : DI- + Gk. *dunamis*, power < *dunasthai*, to be able.] *Bot.* Having four stamens arranged in pairs that differ from one another, esp. in length.

**die**[1] (dī) *vi.* **died, dy·ing, dies.** [ME *deien* < ON *deyja*.] **1.** To cease living : EXPIRE. **2.** To cease existing, esp. by degrees : FADE <The daylight slowly *died.*> **3.** To lose vitality, activity, or force : SUBSIDE <The storm began to *die.*> **4.** To cease existing completely <a barbaric custom that *died* with the Middle Ages> **5.** To experience an agony or suffering suggestive of that of death <nearly *died* of despair> **6.** *Informal.* To desire greatly <*dying* to buy a house> **7.** To cease operation : STOP <The engine suddenly *died.*> **—die back.** To be affected by dieback. **—die off.** To undergo a sudden, sharp decline in population.

☆ ***syns:*** DIE, CROAK, DECEASE, DEMISE, DEPART, EXPIRE, GO, PASS AWAY, PASS (on), PERISH, SUCCUMB *v. core meaning* : to cease living <*died* young> ***ant:*** live

**die**[2] (dī) *n., pl.* **dies.** [ME *de*, gaming dice < OFr. < Lat. *datum*, neuter p.part. of *dare*, to give.] **1.** A device for cutting out, forming, or stamping material. **2. a.** An engraved metal piece for impressing a design on a softer metal, as in coining money. **b.** Any of several components fitted into a diestock to cut threads on screws or bolts. **c.** A machine part that punches shaped holes in, cuts, or forms sheet metal, cardboard, or other stock. **d.** A metal block containing small conical holes through which plastic, metal, or other ductile stock is extruded or drawn. **3.** *pl.* **dice** (dīs). The dado of an architectural pedestal, esp. when cube-shaped. **4.** *pl.* **dice.** A small cube marked on each side with from one to six dots, usu. used in pairs in games and gambling. —*vt.* **died, die·ing, dies.** To cut, form, or stamp with or as if with a die.

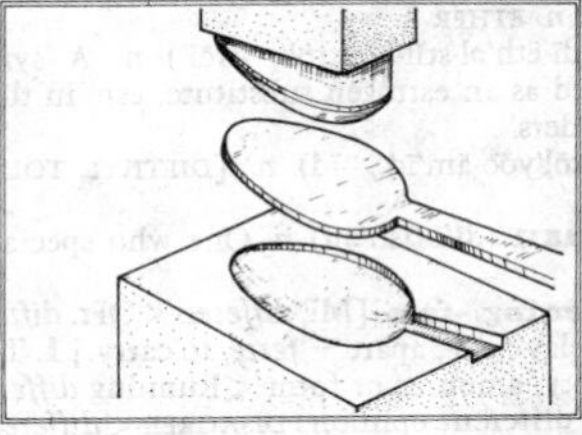
**die**[2]

**die·back** (dī′băk′) *n.* The gradual dying of plant shoots, starting at the tips, as a result of various diseases or climatic conditions.

**di·e·cious** (dī-ē′shəs) *adj. var. of* DIOECIOUS.

**die-hard** *also* **die·hard** (dī′härd′) *n.* One who stubbornly resists change or refuses to abandon a position. **—die′-hard′** *adj.* **—die′-hard′ism** *n.*

**di·e·lec·tric** (dī′ĭ-lĕk′trĭk) *n.* [DI(A)- + ELECTRIC.] A nonconductor of electricity, esp. a substance with electrical conductivity less than a millionth ($10^{-6}$) of a mho. **—di′e·lec′tric** *adj.* **—di′e·lec′tri·cal·ly** *adv.*

**dielectric constant** *n.* Permittivity.

**dielectric heating** *n.* The heating of electrically nonconducting materials by a rapidly varying electromagnetic field.

**di·en·ceph·a·lon** (dī′ĕn-sĕf′ə-lŏn′, -lən) *n.* [DI(A)- + ENCEPHALON.] The posterior part of the forebrain that connects the midbrain with the cerebral hemispheres, encloses the third ventricle, and contains the pituitary gland.

**di·er·e·sis** *also* **di·aer·e·sis** (dī-ĕr′ĭ-sĭs) *n., pl.* **-ses** (-sēz′) [LLat. *diaeresis* < Gk. *diairesis*, separation < *diairein*, to divide : *dia-*, apart + *hairein*, to take.] **1.** A mark (¨) over the second of two adjacent vowels indicating that two separate sounds are to be pronounced. **2.** A slight pause at the end of a line of verse that occurs when the end of a word and the end of a metrical foot coincide.

**die·sel** (dē′zəl, -səl) *n.* **1.** A diesel engine. **2.** A vehicle driven by a diesel engine.

**diesel engine** *n.* [After Rudolf *Diesel* (1858–1913), its inventor.] An internal-combustion engine that uses the heat of highly compressed air to ignite a spray of fuel introduced after the start of the compression stroke.

**die·sink·er** (dī′sĭng′kər) *n.* A maker of stamping and shaping dies. **—die′sink′ing** *n.*

**Di·es I·rae** (dē′ās ĭr′ā′) *n.* [Med. Lat., day of wrath, the first words of the hymn.] A medieval Latin hymn describing the Day of Judgment, used in some requiem Masses.

**di·e·sis** (dī′ĭ-sĭs) *n., pl.* **-ses** (-sēz′) [ME, semitone (which was indicated by a double dagger) < Lat., quarter tone < Gk., a letting through < *diienai*, to send through : *dia-*, through + *hienai*, to send.] The double dagger.

**die·stock** (dī'stŏk') *n.* An apparatus for holding dies that cut threads on screws, bolts, pipes, or rods.

**di·et**[1] (dī'ĭt) *n.* [ME *diete* < OFr. < Lat. *diaeta* < Gk. *diaita* < *diaitan,* to lead one's life.] **1.** An organism's usual food and drink. **2.** A regulated selection of foods, esp. as medically prescribed. **3.** Something taken or provided regularly <a *diet* of horror movies> —*v.* **-et·ed, -et·ing, -ets.** —*vt.* To regulate or prescribe food and drink for. —*vi.* **1.** To eat and drink according to a prescribed regimen. **2.** To eat or feed. **—di'et·er** *n.*

**di·et**[2] (dī'ĭt) *n.* [ME *diete,* day's journey, day for meeting < Med. Lat. *dieta* < Lat. *dies,* day.] **1.** A national or local legislative assembly in certain countries. **2.** A formal deliberative assembly of high personages, esp. of princes or electors.

▲ word history: *Diet*[1] meaning "one's usual food" and *diet*[2] meaning "assembly" are not related words. *Diet*[1] is ultimately a Greek word, *diaita,* meaning "mode of life," especially one prescribed by a physician that includes the regulation of eating habits. In the latter sense *diaita* was borrowed into Latin, whence it passed into French and then into English. The earliest recorded English sense of *diet*[1] is "food, daily provisions," which is still current. *Diet*[2] comes from Medieval Latin *dieta,* which is derived from Latin *dies,* "day." The Medieval Latin word had a range of meanings based on the sense "day," such as "a day's journey," "a day's work," and "an assembly (held on an appointed day)." All of these senses were current for Middle English *diete* but only the sense "assembly" has survived to modern times.

**di·e·tar·y** (dī'ĭ-tĕr'ē) *adj.* Of or relating to diet. —*n., pl.* **-ies. 1.** A system or regimen of dieting. **2.** A regulated daily food allowance. **—di'e·tar'i·ly** (-târ'ə-lē) *adv.*

**dietary law** *n.* One of the regulations prescribing the kinds and combinations of food that may be eaten by Jews.

**di·e·tet·ic** (dī'ĭ-tĕt'ĭk) *adj.* [LLat. *diaeteticus* < Gk. *diaitētikos* < *diaita,* diet.] **1.** Of or relating to diet or its regulation. **2.** Specially prepared for restrictive diets. **—di'e·tet'i·cal·ly** *adv.*

**di·e·tet·ics** (dī'ĭ-tĕt'ĭks) *n. (sing. in number).* The study of diet and dieting as it relates to health and hygiene.

**di·eth·yl ether** (dī-ĕth'əl) *n.* ETHER 2.

**di·eth·yl·stil·bes·trol** (dī-ĕth'əl-stĭl-bĕs'trōl', -trŏl') *n.* A synthetic estrogen, $C_{18}H_{20}O_2$, used as an estrogen substitute, esp. in the treatment of menstrual disorders.

**diethyl tol·u·am·ide** (tŏl'yo͞o-ăm'īd', -ĭd) *n.* [DIETHYL TOLU(ENE) + AMIDE.] Deet.

**di·e·ti·tian** *also* **di·e·ti·cian** (dī'ĭ-tĭsh'ən) *n.* One who specializes in dietetics.

**dif·fer** (dĭf'ər) *vi.* **-fered, -fer·ing, -fers.** [ME *differen* < OFr. *differer* < Lat. *differre,* to differ, delay : *dis-,* apart + *ferre,* to carry.] **1.** To be dissimilar in nature, quality, amount, or form <Running *differs* from jogging.> **2.** To be of a different opinion : DISAGREE <*differed* with the findings> **3.** To dispute : quarrel.

**dif·fer·ence** (dĭf'ər-əns, dĭf'rəns) *n.* **1.** The quality or condition of being different. **2. a.** An instance of disparity or unlikeness. **b.** A degree or amount of difference. **c.** A specific point or element constituting a difference. **d.** *Archaic.* A distinct mark or peculiarity. **3.** A disagreement or controversy or the cause of one. **4.** Discrimination in taste or choice : DISTINCTION. **5.** *Math.* **a.** The amount by which one quantity is greater or less than another. **b.** The amount that remains after one quantity is subtracted from another. —*vt.* **-enced, -enc·ing, -enc·es.** To differentiate or distinguish.

**dif·fer·ent** (dĭf'ər-ənt, dĭf'rənt) *adj.* [ME < OFr. < Lat. *differens,* pr.part. of *differe,* to differ.] **1.** Dissimilar in form, quality, amount, or nature : UNLIKE. **2. a.** Separate or distinct <We're talking about two *different* problems.> **b.** Various or assorted <saw several *different* cars we liked> **3.** Differing from all others : UNUSUAL <a *different* kind of meal> **—dif'fer·ent·ly** *adv.* **—dif'fer·ent·ness** *n.*

☆ ***syns:*** DIFFERENT, DISPARATE, DISSIMILAR, UNLIKE *adj. core meaning* : not like another <two *different* tasks><siblings who were very *different*> ***ant:*** alike, identical, same

**dif·fer·en·ti·a** (dĭf'ə-rĕn'shē-ə, -shə) *n., pl.* **-ti·ae** (-shē-ē') [Lat., difference < *differens,* different.] An attribute that characterizes and distinguishes, esp. one that differentiates a species from others of the same genus.

**dif·fer·en·tia·ble** (dĭf'ə-rĕn'shə-bəl, -shē-ə-) *adj.* **1.** Capable of being differentiated. **2.** *Math.* Having a derivative. **—dif'fer·en'tia·bil'i·ty** *n.*

**dif·fer·en·ti·ae** (dĭf'ə-rĕn'shē-ē') *n. pl. of* DIFFERENTIA.

**dif·fer·en·tial** (dĭf'ə-rĕn'shəl) *adj.* **1.** Of, relating to, or showing a difference. **2.** Constituting a difference : DISTINCTIVE. **3.** Dependent on or making use of a difference or distinction. **4.** *Math.* Of or relating to differentiation. **5.** Involving differences in speed or direction of motion. —*n.* **1.** *Math.* **a.** An infinitesimal increment in a variable. **b.** The product of the derivative of a function of one variable multiplied by the independent variable increment. **2.** A differential gear. **3.** A differential rate **—dif'fer·en'tial·ly** *adv.*

**differential analyzer** *n.* A mechanical or electronic analog computer used to solve esp. complicated differential equations.

**differential calculus** *n.* The mathematics of the variation of a function with respect to changes in independent variables : the study of slopes of curves, accelerations, maxima, and minima by means of derivatives and differentials.

**differential coefficient** *n.* DERIVATIVE 3.

**differential equation** *n.* An equation having derivatives or differentials of an unknown function.

**differential gear** *also* **differential gearing** *n.* An arrangement of gears in an epicycle train permitting the rotation of two shafts at different speeds, used on the rear axle of automotive vehicles to allow different rates of wheel rotation on curves.

**differential rate** *n.* **1.** A difference in wage rate paid for the same work because of differing conditions. **2. a.** A difference in transportation rates to the same destination over different routes, to equalize traffic. **b.** A rate difference over the same route owing to differences in the commodities being shipped.

**differential windlass** *n.* A hoisting device that has two drums of different sizes on the same axis.

**dif·fer·en·ti·ate** (dĭf'ə-rĕn'shē-āt') *v.* **-at·ed, -at·ing, -ates.** —*vt.* **1.** To constitute a distinctive difference in or between : DISTINGUISH. **2.** To perceive or show the difference in or between : DISCRIMINATE. **3.** To develop differences in by alteration or modification. **4.** *Math.* To calculate the derivative or differential of. —*vi.* **1.** To become distinct or specialized. **2.** To make distinctions. **3.** To develop into specialized organs. —Used esp. of embryonic cells or tissues. **—dif'fer·en'ti·a'tion** *n.*

**dif·fi·cult** (dĭf'ĭ-kŭlt', -kəlt) *adj.* [ME, back-formation < *difficulte,* difficulty.] **1. a.** Hard to do, achieve, or perform. **b.** Imposing a severe test of physical or spiritual strength : ARDUOUS. **c.** Causing difficulty or trouble. **2.** Hard to comprehend or solve. **3.** Hard to please, satisfy, or manage <a *difficult* employer> **4.** Hard to convince or persuade. **—dif'fi·cult'ly** *adv.*

☆ ***syns:*** DIFFICULT, HARD, KNOTTY, TOUGH *adj. core meaning* : not easy to do, achieve, or master <*difficult* mathematical problems><a *difficult* flight maneuver> ***ant:*** easy

**dif·fi·cul·ty** (dĭf'ĭ-kŭl'tē, -kəl-) *n., pl.* **-ties.** [ME *difficulte* < OFr. *dificulte* < Lat. *difficultas* < *difficilis,* difficult : *dis-,* not + *facilis,* easy.] **1.** The quality or condition of being difficult. **2.** Something not easily done, accomplished, comprehended, or solved. **3.** *often* **difficulties.** A troublesome or embarrassing state of affairs. **4.** Something that causes trouble or worry. **5.** A disagreement. **6.** An objection or reluctance.

**dif·fi·dent** (dĭf'ĭ-dənt, -dĕnt') *adj.* [ME < Lat. *diffidens,* pr.part. of *diffidere,* to mistrust : *dis-,* not + *fidere,* to trust.] Hesitant to assert oneself from a lack of self-confidence : TIMID. **—dif'fi·dence** (-dəns, -dĕns') *n.* **—dif'fi·dent·ly** *adv.*

**dif·fract** (dĭ-frăkt') *vi.* & *vt.* **-fract·ed, -fract·ing, -fracts.** [Back-formation < DIFFRACTION.] To undergo or cause to undergo diffraction. **—dif·frac'tive** *adj.* **—dif·frac'tive·ly** *adv.* **—dif·frac'tive·ness** *n.*

**dif·frac·tion** (dĭ-frăk'shən) *n.* [NLat. *diffractio* < Lat. *diffractus,* p.part. of *diffringere,* to shatter : *dis-,* apart + *frangere,* to break.] Modification of the behavior of light or of other waves resulting from limitation of their lateral extent, as by an obstacle or aperture.

**diffraction grating** *n.* A usu. glass or polished metal surface having a large number of very fine parallel grooves or slits cut in the surface and used to produce optical spectra by diffraction of transmitted or reflected light.

**dif·fuse** (dĭ-fyo͞oz') *v.* **-fused, -fus·ing, -fus·es.** [ME, dispersed < OFr. *diffus* < Lat. *diffusus,* p.part. of *diffundere,* to spread : *dis-,* apart + *fundere,* to pour.] —*vt.* **1.** To pour out and cause to spread freely. **2.** To scatter or spread about : DISSEMINATE. **3.** To make less brilliant : SOFTEN <*diffuse* bright light> —*vi.* **1.** To spread out or soften. **2.** *Physics.* To undergo diffusion. —*adj.* (dĭ-fyo͞os'). **1.** Widely spread or scattered. **2.** Marked by wordiness : VERBOSE. **—dif·fuse'ly** (-fyo͞os'lē) *adv.* **—dif·fuse'ness** (-fyo͞os'nĭs) *n.* **—dif·fus'i·ble** *adj.*

**dif·fus·er** (dĭ-fyo͞o'zər) *n.* **1.** One that diffuses. **2.** A lighting fixture, as a frosted globe, that spreads light evenly. **3.** A flow passage in a wind tunnel that decelerates a stream of gas or liquid from a high to a low velocity.

**dif·fu·sion** (dĭ-fyo͞o'zhən) *n.* **1.** The process of diffusing or state of being diffused. **2.** *Physics.* Angular redistribution of radiation by a scattering, reflecting, or refracting system, ideally producing an isotropic distribution of intensity. **3.** *Physics.* Gradual mixing of the molecules of two or more substances due to random thermal motion. **4.** Needless profusion of words : VERBOSITY. **—dif·fu'sion·al** *adj.*

**dif·fu·sive** (dĭ-fyo͞o'sĭv, -zĭv) *adj.* Marked by diffusion <*diffusive* prose> **—dif·fu'sive·ly** *adv.* **—dif·fu'sive·ness** *n.*

**dig** (dĭg) *v.* **dug** (dŭg), **dig·ging, digs.** [ME *diggen.*] —*vt.* **1.** To break up, turn over, or remove (e.g., earth or sand) with a tool or the hands. **2.** To make (an excavation) by or as if by digging. **3.** To obtain by digging. **4.** To learn or discover by careful research or investigation. **5.** To force down and into, as for support <*dug* my heels into the sand> **6.** To force or prod against <always *digging* me with your elbow> **7.** *Slang.* **a.** To understand and appreciate <Can you *dig* what I said?> **b.** To like or enjoy <doesn't *dig* rock music> **c.** To notice, esp. in amusement or disbelief <Did you *dig* that bill-board?> —*vi.* **1.** To loosen or turn over the earth. **2.** To proceed along one's way by or as if by digging. **3.** *Informal.* To work industriously. **—dig in. 1.** To dig holes or trenches. **2.** To entrench oneself. **3.** *Informal.* **a.** To begin to work intensively. **b.** To begin to eat. —*n.* **1.** A poke or punch. **2.** A sarcastic, taunting remark : GIBE. **3.** An archaeological excavation. **4. digs.** *Chiefly Brit.* Lodgings : diggings.

**di·ga·met·ic** (dī'gə-mĕt'ĭk) *adj. Biol.* Having two types of gametes, one producing males and the other producing females.

**di·gam·ma** (dī-găm'ə) *n.* [Lat. < Gk. : *di-* two + *gamma,* gamma (from its shape).] A letter occurring in certain early forms of Greek, transliterated in English as *w.*

**dig·a·my** (dĭg'ə-mē) *n.* [LLat. *digamia* < Gk. : *di-,* two + *gamos,* marriage.] Remarriage after the death or divorce of one's first spouse. **—dig'a·mous** (-məs) *adj.*

**di·gas·tric** (dī-găs'trĭk) *adj.* Having two fleshy ends connected by a thinner tendinous portion. —Used of certain muscles. —*n.* A lower jaw muscle that assists in lowering the jaw.

**di·gen·e·sis** (dī-jĕn'ĭ-sĭs) *n.* Metagenesis.

**di·gest** (dī-jĕst', dĭ-) *v.* **-gest·ed, -gest·ing, -gests.** [ME *digesten* < Lat. *digestus,* p.part. of *digerere,* to separate, arrange : *dis-,* apart + *gerere,* to carry.] —*vt.* **1.** To transform (food) into an assimilable condition, as by chemical and muscular action in the alimentary canal. **2.** To assimilate mentally. **3.** To organize into a systematic arrangement, usu. by classifying or summarizing. **4.** To endure patiently. **5.** *Chem.* To soften or disintegrate by chemical action, heat, or moisture. —*vi.* **1.** To become assimilated into the body. **2.** To assimilate food substances. **3.** *Chem.* To undergo exposure to heat, liquids, or chemical agents. —*n.* (dī'jĕst'). **1.** A systematic arrangement of condensed data, esp. of literary or scientific material. **2.** *Law.* A systematic arrangement of statutes or court decisions. **3. Digest.** PANDECT 3.

**di·gest·er** (dī-jĕs'tər, dĭ-) *n.* **1.** One that organizes a digest. **2.** *Chem.* A vessel in which substances are softened or decomposed, usu. for further processing.

**di·gest·i·ble** (dī-jĕs'tə-bəl, dĭ-) *adj.* Capable of being digested. **—di·gest'i·bil'i·ty, di·gest'i·ble·ness** *n.* **—di·gest'i·bly** *adv.*

**di·ges·tion** (dī-jĕs'chən, dĭ-) *n.* **1.** *Physiol.* **a.** The primarily enzymatic bodily process by which foodstuffs are decomposed into simple, assimilable substances. **b.** The ability to digest food. **c.** The result of this process. **2.** The process of bacterial decomposition of organic matter in sewage. **3.** The assimilation of ideas.

**di·ges·tive** (dī-jĕs'tĭv, dĭ-) *adj.* **1.** Relating to, involving, or aiding digestion. **2.** Functioning to digest food. —*n.* A digestive substance. **—di·ges'tive·ly** *adv.* **—di·ges'tive·ness** *n.*

**digestive gland** *n.* Any of various endocrine and exocrine glands that secrete enzymes necessary for digestion.

**digestive system** *n.* The alimentary canal together with accessory glands including the salivary glands, liver, and pancreas, regarded as an integrated system responsible for digestion.

**dig·ger** (dĭg'ər) *n.* **1.** One that digs, esp. a tool or machine for digging or excavating. **2.** *Informal.* A soldier from Australia or New Zealand.

**digger wasp** *n.* Any of various burrowing wasps of the family Sphecidae that build their nests in the ground.

**dig·gings** (dĭg'ĭngz) *pl.n.* **1.** An excavation site. **2.** Materials dug out. **3.** *Chiefly Brit.* Living quarters : LODGINGS.

**dight** (dīt) *vt.* **dight** *or* **dight·ed, dight·ing, dights.** [ME *dighten* < OE *dihtan,* to arrange < Lat. *dictare,* to dictate.] *Archaic.* To dress : adorn.

**dig·it** (dĭj'ĭt) *n.* [ME < Lat. *digitus,* finger.] **1.** A finger or toe. **2.** The breadth of a finger, used as a unit of length, equal to approx. 3/4 inch. **3. a.** Any of the ten Arabic number symbols, 0 through 9. **b.** Such a symbol used in a system of numeration.

**dig·i·tal** (dĭj'ĭ-tl) *adj.* **1.** Of, like, or involving a digit, esp. a finger. **2.** Having digits. **3.** Expressed in digits, esp. for use by a computer. **4.** Giving or using a reading in digits. —*n.* A key played with the finger, as on a piano. **—dig'i·tal·ly** *adv.*

**digital computer** *n.* A computer that performs operations on data represented as a series of digits.

**dig·i·tal·in** (dĭj'ĭ-tăl'ĭn) *n.* [DIGITAL(IS) + -IN.] A poisonous white powder, $C_{36}H_{56}O_{14}$, used in treating heart disease.

**dig·i·tal·is** (dĭj'ĭ-tăl'ĭs) *n.* [NLat. *Digitalis,* genus name < Lat. *digitalis,* digital (from the finger-shaped corollas of foxglove) < *digitus,* finger.] **1.** A plant of the genus *Digitalis,* which includes the foxgloves. **2.** A drug prepared from the seeds and dried leaves of digitalis, used as a cardiac stimulant.

**dig·i·tal·ize** (dĭj'ĭ-tl-īz') *vt.* **-ized, -iz·ing, -iz·es.** To treat with digitalis until the desired physiological or medical effect has been achieved. **—dig'i·tal·i·za'tion** *n.*

**dig·i·tate** (dĭj'ĭ-tāt') *also* **dig·i·tat·ed** (-tā'tĭd) *adj.* **1.** Having digits or fingerlike parts. **2.** *Bot.* Having radiating fingerlike leaflets or lobes. **—dig'i·tate'ly** *adv.*

**dig·i·ta·tion** (dĭj'ĭ-tā'shən) *n.* **1. a.** The condition of being digitate. **b.** Division into fingerlike parts. **2.** A fingerlike process or part.

**dig·i·ti·grade** (dĭj'ĭ-tĭ-grād') *adj.* [Fr. : Lat. *digitus,* digit + Lat. *gradus,* step.] Walking so that only the toes touch the ground, as do horses, cats, and dogs. —*n.* A digitigrade animal.

**dig·it·ize** (dĭj'ĭ-tīz') *vt.* **-tized, -tiz·ing, -tiz·es.** To change (e.g., data) into digital form. **—dig'i·ti·za'tion** *n.* **—dig'i·tiz'er** *n.*

**dig·i·tox·in** (dĭj'ĭ-tŏk'sĭn) *n.* [DIGI(TALIS) + TOXIN.] A highly active glycoside, $C_{41}H_{64}O_{13}$, obtained from digitalis.

**dig·ni·fied** (dĭg'nə-fīd') *adj.* Having or expressing dignity : POISED. **—dig'ni·fied'ly** (-fīd'lē, -fī'ĭd-lē) *adv.*

**dig·ni·fy** (dĭg'nə-fī') *vt.* **-fied, -fy·ing, -fies.** [ME *dignifien* < OFr. *dignifier* < LLat. *dignificare* : Lat. *dignus,* worthy + Lat. *facere,* to do.] **1.** To give dignity or honor to. **2.** To add to the status, importance, or prestige of.

**dig·ni·tar·y** (dĭg'nĭ-tĕr'ē) *n., pl.* **-ies.** A high-ranking, influential person : NOTABLE.

☆ ***syns:*** DIGNITARY, BIG SHOT, BIG WHEEL, BIGWIG, EMINENCE, HEAVYWEIGHT, LEADER, LION, LUMINARY, MUCKAMUCK, NABOB, NOTABILITY, NOTABLE, PERSONAGE, PERSONALITY, SOMEBODY, SOMEONE, VIP *n. core meaning* : an important, influential person <government *dignitaries*>

**dig·ni·ty** (dĭg'nĭ-tē) *n., pl.* **-ties.** [ME *dignite* < OFr. < Lat. *dignitas* < *dignus,* worthy.] **1.** The quality or condition of being esteemed, honored, or worthy. **2. a.** Self-esteem and poise. **b.** Stately reserve in appearance and demeanor. **3.** The respect and honor associated with an important position. **4.** A high rank or office. **5. dignities.** The ceremonial symbols and observances attached to high office <received the *dignities* of the Presidency>

**di·graph** (dī'grăf') *n.* **1.** A pair of letters representing a single speech sound, as the *ph* in *phone* or the *ea* in *seat.* **2.** Two letters run together to represent a special sound, as the Old English *æ.* **—di·graph'ic** (dī-grăf'ĭk) *adj.*

**di·gress** (dī-grĕs', dĭ-) *vi.* **-gressed, -gress·ing, -gress·es.** [Lat. *digredi, digress-* : *dis-,* apart + *gradi,* to go.] **1.** To stray from the main subject in speaking or writing. **2.** To turn aside from a prescribed course of action or conduct.

**di·gres·sion** (dī-grĕsh'ən, dĭ-) *n.* An act or instance of digressing. **—di·gres'sion·al** *adj.*

**di·gres·sive** (dī-grĕs'ĭv, dĭ-) *adj.* Marked by digression. **—di·gres'sive·ly** *adv.* **—di·gres'sive·ness** *n.*

**di·he·dral** (dī-hē'drəl) *adj.* **1.** Formed by or having two plane faces : TWO-SIDED. **2.** Forming, having, or relating to a dihedral angle. —*n.* **1.** *Math.* A dihedral angle. **2.** The upward or downward inclination of an aircraft wing from true horizontal.

**dihedral angle** *n.* **1.** The angle formed by two intersecting planes. **2.** The dihedral of an aircraft wing.

**di·hy·brid** (dī-hī'brĭd) *n. Genetics.* An individual heterozygous for two pairs of genes.

**di·hy·dric** (dī-hī'drĭk) *adj.* Having two hydroxyl radicals.

**dik-dik** (dĭk'dĭk') *n.* [Native word in East Africa.] A very small African antelope of the genus *Madoqua.*

**dike** *also* **dyke** (dīk) [ME, partly < OE *dīc,* trench, and partly < ON *dīki,* ditch.] —*n.* **1.** An embankment built to prevent floods : LEVEE. **2.** *Chiefly Brit.* A low wall, often of sod, dividing or enclosing lands. **3.** A protective barrier blocking a passage. **4.** A raised causeway. **5.** A channel or ditch. **6.** *Geol.* A long mass of igneous rock that cuts across the structure of adjacent rock. —*vt.* **diked, dik·ing, dikes** *also* **dyked, dyk·ing, dykes. 1.** To enclose, protect, or provide with a dike. **2.** To drain with ditches. **—dik'er** *n.*

**dik·tat** (dĭk-tät') *n.* [G. < Lat. *dictatum,* neuter p.part. of *dictare,* to dictate.] **1.** A unilaterally imposed settlement that deals harshly with a defeated party or government. **2.** An authoritarian decree.

**Di·lan·tin** (dī-lăn'tĭn). A trademark for diphenylhydantoin sodium, an anticonvulsant drug used to treat epilepsy.

**di·lap·i·date** (dĭ-lăp'ĭ-dāt') *vt. & vi.* **-dat·ed, -dat·ing, -dates.** [Lat. *dilapidare, dilapidat-,* to throw away, destroy : *dis-,* apart + *lapidare,* to throw stones < *lapis,* stone.] To bring or fall into a state of disrepair, decay, or ruin. **—di·lap'i·da'tion** *n.*

**di·lap·i·dat·ed** (dĭ-lăp'ĭ-dā'tĭd) *adj.* In a state of disrepair or decay.

**di·la·tan·cy** (dī-lāt'n-sē, dĭ-) *n.* **1.** Increase in volume of a fixed amount of certain materials, as of wet sand, subjected to a deformation that changes the interparticle distances of its constituents from their minimum-value configuration. **2.** Any of various phenomena, as increase in solidification or viscosity, that result from dilatancy deformation.

**di·la·tant** (dī-lāt'nt, dĭ-) *adj.* **1.** Tending to dilate. **2.** Manifesting dilatancy. —*n.* A dilator.

**dil·a·ta·tion** (dĭl'ə-tā'shən, dī'lə-) *n.* **1.** The act or process of dilating. **2.** The condition of being dilated. **3.** *Med.* The condition of being abnormally dilated or enlarged. **4.** Expatiation in speech or writing. **—dil'a·ta'tion·al** *adj.*

**dil·a·ta·tor** (dĭl'ə-tā'tər, dī'lə-) *n. var. of* DILATOR.

**di·late** (dī-lāt', dī'lāt') *v.* **-lat·ed, -lat·ing, -lates.** [ME *dilaten* < OFr. *dilater* < Lat. *dilatare,* to enlarge : *dis-,* apart + *latus,* wide.] —*vt.* To enlarge or expand : DISTEND. —*vi.* **1.** To become larger or wider. **2.** To express lengthily : EXPATIATE. **—di·lat'a·bil'i·ty** *n.* **—di·lat'a·ble** *adj.* **—di·lat'a·bly** *adv.* **—di·la'tion** *n.*

**di·lat·ed** (dī-lā'tĭd, dī'lā'-) *adj.* **1.** Expanded : widened. **2.** Distended : stretched. **—di·lat'ed·ly** *adv.* **—di·lat'ed·ness** *n.*

**di·lat·er** (dī-lā′tər, dĭ′lā′tər, dī-lā′tər) *n. var. of* DILATOR.

**di·la·tive** (dī-lā′tĭv) *n.* **1.** Causing or producing dilation. **2.** Tending to dilate.

**dil·a·tom·e·ter** (dĭl′ə-tŏm′ĭ-tər, dĭ′lə-) *n.* [DILATE + -METER.] An instrument for measuring thermal expansion in solids, liquids, and gases. **—dil′a·to·met′ric** (-tə-mĕt′rĭk) *adj.* **—dil′a·tom′e·try** *n.*

**di·la·tor** *also* **di·lat·er** (dī-lā′tər, dĭ′lā′tər, dī-lā′tər) *or* **dil·a·ta·tor** (dĭl′ə-tā′tər, dĭ′lə-) *n.* Something that dilates an object or organ, esp. a muscle, surgical instrument, or drug that produces dilation.

**dil·a·to·ry** (dĭl′ə-tôr′ē, -tōr′ē) *adj.* [ME *dilatorie* < Lat. *dilatorius* < *dilator*, delayer < *dilatus*, p.part. of *differre*, to delay. —see DIFFER.] **1.** Tending or meant to delay. **2.** Marked by procrastination <*dilatory* study habits> **3.** Proceeding at an undesirably slow rate. **—dil′a·to′ri·ly** *adv.* **—dil′a·to′ri·ness** *n.*

**di·lem·ma** (dĭ-lĕm′ə, dī-) *n.* [Lat. < Gk. *dilēmma*, ambiguous proposition : *di-*, two + *lēmma*, proposition.] **1. a.** A situation that requires a choice between two evenly balanced alternatives. **b.** A predicament that apparently defies a satisfactory solution. ***usage:*** Traditionally *dilemma* has been applied to a situation in which there is a choice between evenly balanced alternatives, both of which are usu. unpleasant. The use of *dilemma* as a synonym for *problem* or *predicament*, although still unacceptable to some, is a practice currently widespread in all contexts. **2.** *Logic.* An argument in which a choice of two or more alternatives, each being conclusive and fatal, is presented to an antagonist. **—dil′em·mat′ic** (dĭl′ə-măt′ĭk) *adj.*

**dil·et·tante** (dĭl′ĭ-tänt′, -tän′tē, -tănt′, -tăn′tē, dĭl′ĭ-tänt′) *n., pl.* **-tantes** *or* **-tan·ti** (-tän′tē, -tăn′-) [Ital., lover of the arts, pr.part. of *dilettare*, to delight < Lat. *delectare*. —see DELIGHT.] **1.** One superficially interested in the arts or in a branch of knowledge : AMATEUR. **2.** A lover of the fine arts : CONNOISSEUR. —*adj.* Superficial or amateurish. **—dil′et·tan′tish** *adj.* **—dil′et·tan′tism** *n.*

**dil·i·gence**[1] (dĭl′ə-jəns) *n.* [ME < OFr. < Lat. *diligentia* < pr.part. of *diligere*, to esteem. —see DILIGENT.] **1.** Persistent application to one's work : assiduous effort. **2.** Careful attention.

**dil·i·gence**[2] (dĭl′ə-jəns, dē′lē-zhäns′) *n.* [Fr. < *diligence*, speed < *diligent*, diligent.] A large public stagecoach.

**dil·i·gent** (dĭl′ə-jənt) *adj.* [ME < OFr. < Lat. *diligens*, pr.part. of *diligere*, to esteem, love : *dis-*, apart + *legere*, to choose.] Marked by persevering, painstaking effort : ASSIDUOUS. **—dil′i·gent·ly** *adv.*

**dill** (dĭl) *n.* [ME *dile* < OE.] **1.** An aromatic herb indigenous to the Old World, *Anethum graveolens*, bearing finely dissected leaves and small yellow flowers. **2.** The leaves or seeds of the dill plant, used as seasoning.

**dill pickle** *n.* A pickled cucumber flavored with dill.

**dil·ly** (dĭl′ē) *n., pl.* **-lies.** [Obs. *dilly*, delightful < DELIGHTFUL.] *Slang.* One that is remarkable or extraordinary <a *dilly* of a show>

**dilly bag** *n.* [< *dilli*, native word in Australia.] A bag or basket woven of rushes or bark, used in Australia.

**dil·ly-dal·ly** (dĭl′ē-dăl′ē) *vi.* **-lied, -ly·ing, -lies.** [Redup. of DALLY.] **1.** To waste time. **2.** To vacillate. **—dil′ly-dal′li·er** *n.*

**dil·u·ent** (dĭl′yo͞o-ənt) *adj.* [Lat. *diluens*, *diluent-*, pr.part. of *diluere*, to dilute.] Capable of diluting. —*n.* A substance used to dilute.

**di·lute** (dī-lo͞ot′, dĭ-) *vt.* **-lut·ed, -lut·ing, -lutes.** [Lat. *diluere*, *dilut-* : *dis-*, apart + *lavere*, to wash.] **1.** To thin or reduce the concentration of <*dilute* a solution> **2.** To reduce the potency, strength, purity, or brilliance of by or as if by admixture. —*adj.* Reduced in strength : DILUTED. **—di·lut′er** *n.*

☆ ***syns:*** DILUTE, CUT, THIN, WATER DOWN, WEAKEN *v. core meaning* : to lessen the strength of by or as if by admixture <*dilute* whiskey with a splash of soda><*dilute* a report with trivia>

**di·lu·tion** (dī-lo͞o′shən, dĭ-) *n.* **1. a.** The process of diluting. **b.** A dilute or weakened condition. **2.** A diluted substance.

**di·lu·vi·al** (dĭ-lo͞o′vē-əl) *also* **di·lu·vi·an** (-ən) *adj.* [LLat. *diluvialis* < Lat. *diluvium*, flood < *diluere*, to dilute.] Of or produced by a flood.

**dim** (dĭm) *adj.* **dim·mer, dim·mest.** [ME < OE *dimm.*] **1. a.** Lacking brightness. **b.** Shedding minimal light : FAINT. **c.** Negative or pessimistic <taking a *dim* view of the future> **2.** Lacking brightness or luster : DULL. **3.** Lacking distinctness : OBSCURE. **4.** Lacking sharpness or clarity of understanding : STUPID. **5.** Lacking keenness or vigor. —*v.* **dimmed, dim·ming, dims.** —*vt.* **1.** To make dim. **2.** To put (headlights) on low beam. —*vi.* To become dim. —*pl.n.* **dims.** The parking lights on an automobile. **—dim′ly** *adv.* **—dim′ness** *n.*

**dime** (dīm) *n.* [ME, tenth part < OFr. < Lat. *decima (pars)*, tenth (part) < *decem*, ten.] A U.S. coin worth ten cents.

**di·men·hy·dri·nate** (dī′mĕn-hī′drə-nāt′) *n.* [DIME(THYL) + (AMI)N(E) + HYDR(O)- + -IN + -ATE.] An antihistamine, $C_{24}H_{28}ClN_5O_3$, used to treat motion sickness and allergies.

**dime novel** *n.* A usu. paperbound melodramatic novel. **—dime novelist** *n.*

**di·men·sion** (dĭ-mĕn′shən, dī-) *n.* [ME *dimensioun* < OFr. *dimension* < Lat. *dimensio*, extent < *dimensus*, p.part. of *dimetiri*, to measure : *dis-* (intensive) + *metiri*, to measure.] **1.** A measure of spatial extent, esp. width, height, or length. **2.** *often* **dimensions.** Scope or magnitude : EXTENT. **3.** *Math.* **a.** Any of the least number of independent coordinates required to specify a point in space uniquely. **b.** The range of any of these coordinates. **4.** *Physics.* A physical property, often mass, length, time, or a combination thereof, regarded as a fundamental measure or as one of a set of fundamental measures of a physical quantity. —*vt.* **-sioned, -sion·ing, -sions.** To shape or cut to specified dimensions. **—di·men′sion·al** *adj.* **—di·men′sion·al′i·ty** (-shə-năl′ĭ-tē) *n.* **—di·men′sion·al·ly** *adv.* **—di·men′sion·less** *adj.*

**di·mer** (dī′mər) *n.* **1.** A molecule having two identical simpler molecules. **2.** A chemical compound made up of dimers.

**di·mer·ic** (dī-mĕr′ĭk) *adj. Biol.* Made up of two divisions or parts.

**dim·er·ous** (dĭm′ər-əs) *adj.* **1.** Having two parts or segments, as the tarsus in certain insects. **2.** *Bot.* Having flower parts, as petals, sepals, and stamens, in sets of two. **—dim′er·ism** *n.*

**dime store** *n.* A five-and-ten.

**dim·e·ter** (dĭm′ĭ-tər) *n.* [LLat. < Gk. *dimetros*, having two meters : *di-*, two + *metron*, meter.] A line of verse having two metrical feet or two dipodies.

**di·meth·yl** (dī-mĕth′əl) *n.* Ethane.

**di·meth·yl·sulf·ox·ide** (dī-mĕth′əl-sŭl-fŏk′sīd′) *n.* A colorless hygroscopic liquid, $(CH_3)_2SO$, obtained from lignin and used as a solvent and skin penetrant for conveying medications into tissues.

**di·min·ish** (dĭ-mĭn′ĭsh) *v.* **-ished, -ish·ing, -ish·es.** [ME *deminishen*, blend of *diminuen*, to lessen (< OFr. *diminuer* < Lat. *deminuere* : *de-*, from + *minuere*, to lessen) and *minishen*, to reduce (< OFr. *minuiser* < VLat. **minutiare* < Lat. *minutia*, smallness < *minutus*, small, p.part. of *minuere*, to lessen).] —*vt.* **1. a.** To make or cause to seem smaller or less. **b.** To detract from the authority, rank, or prestige of. **2.** To cause to taper. **3.** *Mus.* To reduce (a perfect or minor interval) by a semitone. —*vi.* **1.** To become smaller or less. **2.** To taper. **—di·min′ish·a·ble** *adj.* **—di·min′ish·ment** *n.*

**diminishing returns** *pl.n.* The rate at which profits diminish in proportion to the amount of further investment after a certain point.

**di·min·u·en·do** (dĭ-mĭn′yo͞o-ĕn′dō) *n.* & *adj.* & *adv.* [Ital., diminishing < Lat. *deminuendum*, gerund of *deminuere*, to diminish.] *Mus.* Decrescendo.

**dim·i·nu·tion** (dĭm′ə-no͞o′shən, -nyo͞o′-) *n.* [ME *diminucioun* < OFr. *diminution* < Lat. *deminutio* < *deminuere*, to diminish.] **1. a.** The act or process of diminishing. **b.** The resulting decrease. **2.** *Mus.* The repetition of a theme in notes one-quarter or one-half as long as the original. **—dim′i·nu′tion·al** *adj.*

**di·min·u·tive** (dĭ-mĭn′yə-tĭv) *adj.* [ME *diminutif* < OFr. < Lat. *deminutivus* < *deminutus*, p.part. of *deminuere*, to diminish.] **1.** Very small : TINY. **2.** Designating certain suffixes that denote smallness, youth, familiarity, or affection, as *-let* in *playlet* or *-ling* in *duckling*. —*n.* A diminutive suffix, word, or name. **—di·min′u·tive·ly** *adv.* **—di·min′u·tive·ness** *n.*

**dim·i·ty** (dĭm′ĭ-tē) *n., pl.* **-ties.** [ME *demyt* < Med. Lat. *dimitum* < Med. Gk. *dimitos*, double-threaded : Gk. *di-*, two + *mitos*, thread.] A thin, crisp, usu. corded or checked cotton fabric.

**dim·mer** (dĭm′ər) *n.* **1.** A device for reducing the intensity of an electric light or lighting system. **2.** **dimmers. a.** Automobile parking lights. **b.** Low-beam automobile headlights.

**di·morph** (dī′môrf′) *n.* [Back-formation < DIMORPHISM.] Either of two dimorphic forms.

**di·mor·phic** (dī-môr′fĭk) *also* **di·mor·phous** (-fəs) *adj.* Having or occurring in two distinct forms.

**di·mor·phism** (dī-môr′fĭz′əm) *n.* [< Gk. *dimorphos*, having two forms : *di-*, two + *morphē*, shape.] **1.** *Bot.* The occurrence of two distinct forms of the same parts, as leaves, flowers, or stamens, in a single plant or in plants of the same kind. **2.** *Chem. & Physics.* Dimorphic crystallization. **3.** *Zool.* The state of having two distinct forms in the same species when the sexes differ in secondary as well as primary sexual characteristics.

**di·mor·phous** (dī-môr′fəs) *adj. var. of* DIMORPHIC.

**dim-out** (dĭm′out′) *n.* **1.** The restricted use or exposure of lights at night, esp. for protection against air raids. **2.** The semidarkness produced by a dim-out.

**dim·ple** (dĭm′pəl) *n.* [ME *dimpel.*] **1.** A small natural indentation in the flesh on a part of the human body <a *dimple* in one's chin> **2.** A slight surface depression. —*v.* **-pled, -pling, -ples.** —*vt.* To produce dimples in <The rain *dimpled* the lake.> —*vi.* To form dimples by smiling. **—dim′ply** *adj.*

**dim sum** (dĭm′ so͞om′, sŭm′) *pl.n.* [Cantonese.] A variety of Chinese delicacies, esp. small steamed or fried dumplings, served as a light meal.

**dim·wit** (dĭm′wĭt′) *n. Slang.* A stupid person. **—dim′wit′ted** *adj.* **—dim′wit′ted·ly** *adv.* **—dim′wit′ted·ness** *n.*

**din** (dĭn) *n.* [ME *dine* < OE *dyne.*] A mixture of loud, confused, and disagreeable noises. —*v.* **dinned, din·ning, dins.** —*vt.* **1.** To stun with deafening noise. **2.** To impress by wearying repetition <tried to *din* the lesson into our heads> —*vi.* To make a din.

**di·nar** (dĭ-när′, dē′när′) *n.* [Ar. *dīnār* < LGk. *dēnarion* < Lat. *denarius*, denarius.] **1.** —See table at CURRENCY. **2.** Any of several units of gold and silver currency used in the Middle East from the 8th to the 19th cent.

**dine** (dīn) *vi. & vt.* **dined, din·ing, dines.** [ME *dinen* < OFr. *diner* < VLat. **disjejunare* : Lat. *dis-* (reversal) + Lat. *jejunus,* fast, hunger.] To eat dinner or give dinner to.

**din·er** (dī′nər) *n.* **1.** One that dines. **2.** A railroad dining car. **3.** A restaurant with a long counter, orig. shaped like a railroad car.

**di·nette** (dī-nĕt′) *n.* [< DINE.] **1.** A nook or alcove for informal meals. **2.** The table and chairs used in a dinette.

**ding** (dĭng) *v.* **dinged, ding·ing, dings.** [Prob. imit.] —*vi.* **1.** To ring or clang. **2.** To speak persistently and repetitiously. —*vt.* **1.** To cause to clang, as by striking. **2.** To hammer into or at with repetitious talk. —*n.* A ringing sound.

**ding-a-ling** (dĭng′ə-lĭng′) *n. Informal.* A silly person.

**ding·bat** (dĭng′băt′) *n.* [Orig. unknown.] **1.** A small object, as a stick or stone, suitable for hurling. **2.** A typographical ornament or symbol. **3.** *Informal.* A ding-a-ling.

**ding-dong** (dĭng′dông′, -dŏng′) *n.* [Imit.] The peal of a bell. —*vi.* **-donged, -dong·ing, -dongs.** To ring. —*adj.* Marked by a rapid, hammering exchange, as of blows.

**din·ghy** (dĭng′ē) *n., pl.* **-ghies.** [Hindi *ḍĩgī,* dim. of *ḍẽgā,* long.] **1.** A small boat powered by sails, oars, or a motor carried as a lifeboat or pleasure craft on a larger boat. **2.** A small rowboat. **3.** An inflatable rubber life raft.

**din·gle** (dĭng′gəl) *n.* [ME, dell, hollow.] A small, wooded valley : DELL.

**din·go** (dĭng′gō) *n., pl.* **-goes.** [Native word in Australia.] A wild dog, *Canis dingo* of Australia, with a yellowish-brown coat.

**din·gus** (dĭng′əs) *n.* [Du. *dinges,* prob. < G., genitive of *Ding,* thing.] *Slang.* An object whose name is forgotten or unknown.

**din·gy** (dĭn′jē) *adj.* **-gi·er, -gi·est.** [Orig. unknown.] **1.** Dirtied with smoke and grime. **2.** Squalid or dreary. **3.** Shabby : worn. **—din′gi·ly** *adv.* **—din′gi·ness** *n.*

**dining car** *n.* A railroad car in which meals are served.

**dining room** *n.* A room in which meals are served.

**di·ni·tro·ben·zene** (dī-nī′trō-bĕn′zēn′, -bĕn-zēn′) *n.* Any of three isomeric compounds, $C_6H_4(NO_2)_2$, made from a blend of nitric acid, sulfuric acid, and heated benzene and used in organic syntheses, dyes, and celluloid.

**din·key** *also* **din·ky** (dĭng′kē) *n., pl.* **-keys** *also* **-kies.** [< DINKY.] *Informal.* A small locomotive used in a railroad yard.

**din·kum** (dĭn′kəm) [Orig. unknown.] *Austral.* —*adj.* Real : genuine. —*adv.* Truly : honestly.

**din·ky** (dĭng′kē) [Prob. < Sc. *dink,* neat.] *Informal.* —*adj.* **-ki·er, -ki·est.** Small or insignificant. —*n. var. of* DINKEY.

**din·ner** (dĭn′ər) *n.* [ME *diner* < OFr. < *diner,* to dine.] **1.** The main meal of the day. **2.** A formal meal or banquet honoring a person or commemorating an occasion. **3.** The food prepared for a dinner. **4.** TABLE D'HÔTE 2.

**dinner jacket** *n.* TUXEDO 1.

**dinner theater** *n.* A restaurant that presents a play during or after dinner.

**din·ner·ware** (dĭn′ər-wâr′) *n.* **1.** Tableware, as dishes, serving bowls, and platters, used in serving a meal. **2.** A set of dishes.

**di·no·flag·el·late** (dī′nō-flăj′ə-lĭt, -lāt′, -flə-jĕl′ĭt) *n.* [NLat. *Dinoflagellata,* class name : Gk. *dinos,* eddy (< *dinein,* to whirl) + NLat. *flagellum,* flagellum.] Any of numerous minute, chiefly marine protozoans of the class Dinoflagellata, having two flagella and a cellulose outer envelope and forming a primary constituent of plankton.

**di·no·saur** (dī′nə-sôr′) *n.* [Gk. *deinos,* monstrous + Gk. *sauros,* lizard.] Any of various extinct, often gigantic reptiles of the orders Saurischia and Ornithischia, existing during the Mesozoic era. **—di′no·sau′ri·an** (-sôr′ē-ən) *adj. & n.* **—di′no·sau′ric** (-sôr′ĭk) *adj.*

**di·no·there** (dī′nə-thîr′) *n.* [NLat. *Dinotherium,* genus name : Gk. *deinos,* monstrous + Gk *thērion,* dim. of *thēr,* beast.] Any of various extinct elephantlike mammals of the genus *Dinotherium,* existing during the Miocene, Pliocene, and Pleistocene epochs.

**dint** (dĭnt) *n.* [ME < OE *dynt.*] **1.** Force or effort <won by *dint* of sheer perseverance> **2.** A dent. —*vt.* **dint·ed, dint·ing, dints.** **1.** To put a dent in. **2.** To drive in or impress forcibly.

**di·oc·e·san** (dī-ŏs′ĭ-sən) *adj.* Of or relating to a diocese. —*n.* A bishop of a diocese.

**di·o·cese** (dī′ə-sĭs, -sēs′, -sēz′) *n.* [ME *diocise* < OFr. < LLat. *diocesis* < Lat. *dioecesis,* jurisdiction < Gk. *dioikēsis,* administration < *dioikein,* to keep house, administer : *dia-* (intensive) + *oikein,* to inhabit < *oikos,* house.] The territory under the jurisdiction of a bishop : BISHOPRIC.

**di·ode** (dī′ōd′) *n.* **1.** An electronic device that restricts current flow chiefly to one direction. **2.** A vacuum tube having two electrodes, a cathode, and an anode. **3.** A two-terminal semiconductor device used mainly as a rectifier.

**di·oe·cious** *also* **di·e·cious** (dī-ē′shəs) *adj.* [DI- + Gk. *oikia,* a dwelling < *oikos,* house.] *Bot.* Having male and female flowers borne on separate plants. **—di·oe′cious·ly** *adv.*

**di·oi·cous** (dī-oi′kəs) *adj.* [NLat. *dioecus* : DI- + Gk. *oikos,* house.] *Bot.* Having antheridia and archegonia on separate plants : UNISEXUAL. —Used of mosses and related plants.

**Di·o·me·des** (dī′ə-mē′dēz) *n.* [Lat. < Gk. *Diomēdēs.*] *Gk. Myth.* A prince of Argos and one of the chief heroes of the Trojan War.

**Di·o·ne** (dī-ō′nē) *n.* [Lat. < Gk. *Diōnē.*] *Gk. Myth.* The mother of Aphrodite by Zeus.

**Di·o·nys·i·a** (dī′ə-nĭz′ē-ə, -nĭzh′ē-ə, -nĭs′ē-ə) *pl.n.* Any of various seasonal festivals of ancient Attica in honor of the god Dionysus, esp. the autumnal one from which Greek tragedy is thought to have originated.

**Di·o·nys·i·ac** (dī′ə-nĭs′ē-ăk′) *adj.* **1.** Of or relating to Dionysus or the Dionysia. **2.** *often* **dionysiac.** DIONYSIAN 2b.

**Di·o·nys·i·an** (dī′ə-nĭsh′ən, -nĭzh′ən, -nĭs′ē-ən) *adj.* **1. a.** Of or relating to Dionysus or the Dionysia. **b.** Of or relating to any of several historical persons named Dionysus. **2. a.** Of or devoted to the worship of Dionysus. **b.** *often* **dionysian.** Of an ecstatic, frenzied, or irrational nature : ORGIASTIC. **3.** *often* **dionysian.** Of or typical of creative-intuitive power, as opposed to critical-rational power, in the philosophy of Nietzsche.

**Di·o·ny·sus** (dī′ə-nī′səs, -nē′-) *n.* [Lat. < Gk. *Dionusios.*] *Gk. Myth.* The god of wine and of an orgiastic religion celebrating the power and fertility of nature.

**di·o·phan·tine analysis** (dī′ə-făn′tīn′, -tīn) *n.* [After *Diophantus,* Greek mathematician of the 3rd cent. B.C.] A method for determining integral solutions of certain algebraic equations.

**di·op·side** (dī-ŏp′sīd′) *n.* [Fr. : *di-,* two + Gk. *opsis,* appearance.] A monoclinic pyroxene mineral, $CaMgSi_2O_6$, used as a refractory and a gemstone.

**di·op·ter** (dī-ŏp′tər) *n.* [Obs. *diopter,* an instrument for measuring angles < Lat. *dioptra* < Gk. : *dia-,* through + *optos,* visible.] A unit of curvature and of the power of lenses, refracting surfaces, and other optical systems, equal to a reciprocal meter. **—di·op′tral** (-trəl) *adj.*

**di·op·tom·e·ter** (dī′ŏp-tŏm′ĭ-tər) *n.* [DIOPT(ER) + -METER.] An instrument for measuring ocular refraction. **—di′op·tom′e·try** *n.*

**di·op·tric** (dī-ŏp′trĭk) *also* **di·op·tri·cal** (-trĭ-kəl) *adj.* **1.** Of or relating to dioptrics. **2.** Relating to optical refraction : REFRACTIVE.

**di·op·trics** (dī-ŏp′trĭks) *n.* [Gk. *dioptrikas* < *dioptra,* an instrument for measuring angles. —see DIOPTER.] *(sing. in number).* The study of the refraction of light.

**di·o·ram·a** (dī′ə-răm′ə, -rä′mə) *n.* [Fr. : *dia-,* dia- + *(pan)orama,* panorama < E.] **1.** A three-dimensional miniature scene with modeled figures and a painted, realistic background. **2.** A scene reproduced on cloth transparencies with various lights shining through the cloths to produce changes in effect and viewed through a small opening. **—di′o·ram′ic** (-răm′ĭk) *adj.*

**di·o·rite** (dī′ə-rīt′) *n.* [Fr. < Gk. *diorizein,* to distinguish : *dia-,* apart + *horizein,* to divide < *horos,* boundary.] A crystalline, granite-textured dark rock rich in plagioclase. **—di′o·rit′ic** (-rĭt′ĭk) *adj.*

**Di·os·cu·ri** (dī-ŏs′kyə-rī′, dī′ə-skyŏŏr′ī′) *pl.n.* [Gk. *Dioskouroi* : *Dios,* genitive of *Zeus,* Zeus + *kouroi,* pl. of *kouros,* boy.] *Gk. Myth.* Castor and Pollux, the twin sons of Leda and brothers of Helen and Clytemnestra, who were transformed by Zeus into the constellation Gemini.

**di·ox·ane** (dī-ŏk′sān′) *n.* A flammable, potentially explosive, colorless liquid, $C_4H_8O_2$, used as a solvent for fats, greases, and resins and in paints, lacquers, glues, cosmetics, and fumigants.

**di·ox·ide** (dī-ŏk′sīd′) *n.* An oxide having two oxygen atoms per molecule.

**di·ox·in** (dī-ŏk′sĭn) *n.* [DI- + OX(A)- + -IN.] Any of several carcinogenic or teratogenic heterocyclic hydrocarbons that occur as impurities in petroleum-derived herbicides.

**dip** (dĭp) *v.* **dipped, dip·ping, dips.** [ME *dippen* < OE *dyppan.*] —*vt.* **1.** To plunge briefly into a liquid, usu. in order to wet, coat, or saturate. **2.** To color or dye (e.g., eggs) by immersing. **3.** To immerse (an animal) in a disinfectant solution. **4.** To make (a candle) by repeatedly immersing a wick in melted wax or tallow. **5.** To galvanize or plate (metal) by immersion. **6.** To scoop up by plunging something into and out of a liquid : LADLE. **7.** To lower and raise (a flag) in salute. **8.** To lower or drop suddenly : DUCK. —*vi.* **1.** To plunge into liquid and come out quickly. **2. a.** To plunge the hand or a container into a liquid, esp. to take something up or out. **b.** To make inroads for money <had to *dip* into our nest egg> **3.** To drop or sink out of sight, esp. suddenly. **4.** To appear to sink <The sun *dipped* below the horizon.> **5.** To drop suddenly before climbing. —Used of an airplane. **6.** To slope downward : DECLINE. **7.** To decline slightly and usu. temporarily, as sales or prices. **8.** *Geol.* To lie at an angle to the horizontal plane, as a rock stratum or vein. **9. a.** To read here and there in a book or magazine : BROWSE. **b.** To investigate a subject superficially : DABBLE. —*n.* **1.** A brief plunge or immersion. **2.** A liquid into which something is dipped. **3.** A preparation into which food, as crackers or chips, may be dipped. **4.** An amount taken up by dipping. **5.** A container for dipping. **6.** A candle made by repeated dipping in wax or tallow. **7.** A downward slope. **8.** A downward course : DROP <a *dip* in sales> **9.** *Geol.* The downward inclination of a rock stratum or vein in reference to the plane of the horizon.

**10.** Magnetic dip. **11.** A hollow. **12.** A gymnastic exercise on parallel bars in which the body is lowered by bending the elbows until the chin reaches the level of the bars and then is raised by straightening the arms. **13.** *Slang.* A foolish or gullible person.

**di·pet·al·ous** (dī-pĕt′l-əs) *adj.* Having two petals.

**di·phase** (dī′fāz′) *also* **di·pha·sic** (dī-fā′zĭk) *adj.* Having two phases.

**di·phen·yl** (dī-fĕn′əl, -fē′nəl) *n.* Biphenyl.

**di·phen·yl·a·mine** (dī-fĕn′əl-ə-mēn′, -ăm′ĭn, -fē′nəl-) *n.* A colorless crystalline compound, $(C_6H_5)_2NH$, used to make dyes, explosives, pesticides, and pharmaceuticals and as a stabilizer for plastics.

**di·phen·yl·a·mine·chlo·ro·ar·sine** (dī-fĕn′əl-ə-mēn′klôr′ō-är-sēn′, -är′sēn′, -fē′nəl-, -klōr′-) *n.* Phenarsazine chloride.

**di·phen·yl·hy·dan·to·in sodium** (dī-fĕn′əl-hī-dăn′tō-ĭn, -fē′nəl-) *n.* A white powder, $C_{15}H_{11}N_2O_2Na$, used as an anticonvulsant.

**di·phen·yl·ke·tone** (dī-fĕn′əl-kē′tōn′, -fē′nəl-) *n.* Benzophenone.

**di·phos·gene** (dī-fŏz′jēn′) *n.* A colorless mobile liquid, $ClCOOCCl_3$, used in vapor form as a poison gas.

**diph·the·ri·a** (dĭf-thîr′ē-ə, dĭp-) *n.* [NLat. < Fr. *diphthérie* < Gk. *diphthera,* piece of leather.] An acute febrile contagious disease caused by infection with the bacillus *Corynebacterium diphtheriae* and characterized by weakness and the formation of false membranes in the throat and other air passages, causing respiratory difficulty. **—diph′the·rit′ic** (-thə-rĭt′ĭk), **diph·ther′ic** (-thĕr′ĭk), **diph·the′ri·al** *adj.*

**diph·thong** (dĭf′thông′, -thŏng′, dĭp′-) *n.* [ME *diptonge* < OFr. *diptongue* < LLat. *dipthongus* < Gk. *diphthongos* : *di-,* two + *phthongos,* sound.] **1.** A complex speech sound beginning with one vowel sound and moving to another vowel or semivowel position within the same syllable; e.g., *oi* in the word *oil* is a diphthong. **2.** Either of the two ligatures æ or œ, orig. pronounced as diphthongs in Classical Latin but now pronounced as single vowels. **—diph·thon′gal** *adj.*

**diph·thong·ize** (dĭf′thông-īz′, -thŏng-, dĭp′-) *v.* **-ized, -iz·ing, -iz·es.** *—vt.* To pronounce as a diphthong. *—vi.* To become a diphthong. **—diph′thong·i·za′tion** *n.*

**diph·y·cer·cal** (dĭf′ĭ-sûr′kəl) *adj.* [< Gk. *diphuēs,* double (*di-,* two + *phuein,* to grow) + *kerkos,* tail.] Having or designating a tail fin in which the vertebral column extends to the tip, with symmetric upper and lower parts. **—diph·y·cer′cy** (-sûr′sē) *n.*

**di·phy·let·ic** (dī′fī-lĕt′ĭk) *adj.* Descended from two ancestral lines or individuals.

**di·phyl·lous** (dī-fĭl′əs) *adj.* Having two leaves.

**di·phy·o·dont** (dī-fī′ə-dŏnt′) *adj.* [Gk. *diphuēs,* double (*di-,* two + *phuein,* to grow) + -ODONT.] Having two successive sets of teeth, as do most mammals.

**dipl-** *pref. var. of* DIPLO-.

**di·ple·gia** (dī-plē′jə, -jē-ə) *n.* Paralysis of corresponding parts on both sides of the body.

**di·plex** (dī′plĕks′) *adj.* [DI- + (DU)PLEX.] Capable of simultaneous transmission or reception of two messages in the same radio channel.

**di·plex·er** (dī′plĕk-sər) *n.* A coupling device that permits two radio transmitters to share the same antenna.

**diplo-** *or* **dipl-** *pref.* [Gk. < *diploos,* double.] **1.** Double <*diploblastic*> **2.** Having double the basic number of chromosomes : DIPLOID <*diplont*>

**dip·lo·blas·tic** (dĭp′lō-blăs′tĭk) *adj.* Having two distinct cellular layers. —Used of embryos and lower invertebrate animals.

**dip·lo·car·di·ac** (dĭp′lō-kär′dē-ăk′) *adj.* [DIPLO- + Gk. *kardia,* heart.] Having or describing a heart in which the two sides are distinctly separated, as in birds and mammals.

**dip·lo·coc·cus** (dĭp′lō-kŏk′əs) *n., pl.* **-coc·ci** (-kŏk′sī′, -kŏk′ī′) [NLat. *Diplococcus,* genus name : DIPLO- + *coccus,* coccus.] Any of various paired spherical bacteria of the genus *Diplococcus,* some of which are pathogenic. **—dip′lo·coc′cal** (-kŏk′əl), **dip′lo·coc′cic** (-kŏk′sĭk, -kŏk′ĭk) *adj.*

**di·plod·o·cus** (dĭ-plŏd′ə-kəs, dī-) *n.* [NLat. *Diplodocus,* genus name : DIPLO- + Gk. *dokos,* beam.] A very large, extinct, herbivorous dinosaur of the genus *Diplodocus,* existing during the Jurassic period.

**dip·lo·e** (dĭp′lō-ē′) *n.* [NLat. < Gk. *diploē* < *diploos,* double.] The spongy, bony tissue between the inner and outer layers of the cranial bones.

**dip·loid** (dĭp′loid′) *adj.* **1.** Twofold or double. **2.** *Genetics.* Having a homologous pair of chromosomes for each characteristic except sex, the total number of chromosomes being twice that of a gamete. *—n. Genetics.* **1.** A diploid cell. **2.** An individual characterized by a diploid chromosome number. **—dip′loid′ly** *adv.* **—dip′loi′dy** *n.*

**di·plo·ma** (dĭ-plō′mə) *n.* [Lat. < Gk. *diplōma,* document, folded paper < *diploos,* double.] **1.** A document issued by an institution of learning testifying that a student has earned a degree or finished a course of study. **2.** A certificate conferring a privilege or honor. **3.** An official document or charter.

**di·plo·ma·cy** (dĭ-plō′mə-sē) *n., pl.* **-cies. 1.** The art or practice of conducting international relations, as in negotiating alliances, treaties, and agreements. **2.** Tact and skill in dealing with people.

**dip·lo·mat** (dĭp′lə-măt′) *n.* [Fr. *diplomate,* back-formation < *diplomatique,* diplomatic.] One skilled or working in diplomacy.

**dip·lo·mate** (dĭp′lə-māt′) *n.* A physician certified as a specialist by a board of examiners.

**dip·lo·mat·ic** (dĭp′lə-măt′ĭk) *adj.* [Fr. *diplomatique* < NLat. *diplomaticus,* relating to documents < Lat. *diploma,* diploma.] **1.** Of, relating to, or involving diplomacy. **2.** Marked by sensitivity or tact in dealing with people : smoothly adroit. **—dip′lo·mat′i·cal·ly** *adv.*

**diplomatic corps** *n.* The body of diplomatic personnel in residence at the capital of a nation.

**diplomatic immunity** *n.* Exemption from ordinary legal procedures granted to diplomatic personnel in a foreign country.

**dip·lo·mat·ics** (dĭp′lə-măt′ĭks) *n.* (*sing. in number*). **1.** Diplomacy. **2.** The branch of paleography devoted to the study of ancient documents and the determination of their age and authenticity.

**di·plo·ma·tist** (dĭ-plō′mə-tĭst) *n.* A diplomat.

**dip·lont** (dĭp′lŏnt′) *n.* An organism having somatic cells with diploid chromosomes. **—dip·lont′ic** (-lŏn′tĭk) *adj.*

**dip·lo·pi·a** (dĭ-plō′pē-ə) *n. Pathol.* A visual disorder that causes objects to appear double. **—dip·lo′pic** (-plō′pĭk, -plŏp′ĭk) *adj.*

**dip·lo·pod** (dĭp′lə-pŏd′) *n.* [NLat. *Diplopoda,* class name : DIPLO- + Gk. *pous,* foot.] Any of various cylindrical, segmented arthropods of the class Diplopoda, which includes the millipedes. **—dip·lop′o·dous** (-lŏp′ə-dəs) *adj.*

**dip·lo·sis** (dĭ-plō′sĭs) *n.* [NLat. < Gk. *diplōsis,* a doubling < *diploun,* to double < *diploos,* double.] The formation of the full number of chromosomes found in a somatic cell by the fusion of gamete nuclei containing haploid sets in fertilization.

**dip needle** *n.* **1.** *Physics.* A magnetic needle vertically balanced and pivoted to rotate freely in order to indicate the local inclination of the earth's magnetic field. **2.** INCLINOMETER 2.

**dip·no·an** (dĭp′nō-ən) *n.* [< NLat. *Dipnoi,* group name < Gk. *dipnoos,* having two apertures for breathing : *di-,* two + *pnoē,* breath < *pnein,* to breathe.] Any of various fishes of the group Dipnoi, as the lungfishes, having the ability to breathe through modified lungs as well as gills. *—adj.* Of or belonging to the Dipnoi.

**dip·o·dy** (dĭp′ə-dē) *n., pl.* **-dies.** [LLat. *dipodia* < Gk. < *dipous,* two-footed : *di-,* two + *pous,* foot.] A prosodic unit made up of two metrical feet.

**di·po·lar** (dī′pō′lər, dī-pō′-) *adj.* Of, relating to, or having a dipole.

**di·pole** (dī′pōl′) *n.* **1.** *Physics.* A pair of magnetic poles or electric charges of equal magnitude but of opposite polarity or sign, separated by a small distance. **2.** *Electron.* An antenna, usu. fed from the center, having two equal rods extending outward in a straight line.

**dipole moment** *n.* **1.** The product of either charge in an electric dipole with the distance separating them. **2.** The product of the strength of either pole in a magnetic dipole with the distance separating them.

**dip·per** (dĭp′ər) *n.* **1.** One that dips. **2.** A container for dipping, esp. a long-handled cup for dipping water. **3. Dipper. a.** The Big Dipper in Ursa Major. **b.** The Little Dipper in Ursa Minor. **4.** The water ouzel.

**dip·py** (dĭp′ē) *adj.* **-pi·er, -pi·est.** [Orig. unknown.] *Slang.* Silly.

**di·pro·pel·lant** (dī′prə-pĕl′ənt) *n.* A bipropellant.

**dip·so·ma·ni·a** (dĭp′sə-mā′nē-ə, -mān′yə) *n.* [Gk. *dipsa,* thirst + -MANIA.] An often periodic, insatiable craving for alcohol. **—dip′so·ma′ni·ac′** (-ăk′) *adj.* & *n.* **—dip′so·ma·ni′a·cal** (-mə-nī′ə-kəl) *adj.*

**dip·stick** (dĭp′stĭk′) *n.* A graduated rod for measuring the depth or amount of liquid in a container, as of oil in a crankcase.

**dip·ter·an** (dĭp′tər-ən) *n.* A dipterous insect. *—adj.* Of or belonging to the order Diptera : DIPTEROUS.

**dip·ter·on** (dĭp′tə-rŏn′) *n.* [Gk., neuter of *dipteros,* dipterous.] A dipteran.

**dip·ter·ous** (dĭp′tər-əs) *adj.* [< NLat. *Diptera,* order name < Gk. *dipteros,* having two wings : *di-,* two + *pteron,* wing.] **1.** Of, relating to, or belonging to the Diptera, a large order of insects that includes the true flies and mosquitoes, marked by a single pair of membranous wings and a pair of halteres. **2.** Having two winglike parts.

**dip·tych** (dĭp′tĭk) *n.* [LLat. *diptycha* < Gk. *diptukha* < *diptukhos,* folded double : *di-,* two + *ptukhē,* fold < *ptussein,* to fold.] **1.** An ancient writing tablet having two hinged leaves. **2.** A pair of carved or painted panels hinged together.

**diptych**

**di·quat** (dī′kwät′) *n.* [DI- + QUAT(ERNARY).] A strong, nonpersistent herbicide, $C_{12}H_{12}Br_2N_2$, used to control water weeds.

**dire** (dīr) *adj.* **dir·er, dir·est.** [Lat. *dirus,* ill-omened.] **1.** Having or warning of terrible consequences : DISASTROUS. **2.** Requiring urgent remedial action or treatment <*dire* hunger> **—dire′ly** *adv.* **—dire′ness** *n.*

**di·rect** (dĭ-rĕkt′, dī-) *v.* **-rect·ed, -rect·ing, -rects.** [ME *directen* < Lat. *directus,* p.part. of *dirigere,* to give direction to : *dis-,* apart + *regere,* to guide.] *—vt.* **1.** To regulate or conduct the affairs of : MANAGE. **2.** To take authoritative charge of : CONTROL. **3.** To order or command <*directed* the jury to reach a decision> **4. a.** To move or guide (someone) toward a goal. **b.** To show or indicate the way to. **5.** To cause to move in or follow a direct or straight course <*directed* the arrow at the bull's-eye> **6.** To address (e.g., a letter) to a destination. **7.** To bestow orally <Let me *direct* a few remarks.> **8. a.** To give guidance and instruction to in the rehearsal and performance of a work. **b.** To supervise the performance of. *—vi.* **1.** To give directions or commands. **2.** To conduct a rehearsal or performance. *—adj.* **1.** Proceeding or lying in a straight course or line. **2.** Straightforward and candid <a *direct* question> **3.** With nothing intervening : IMMEDIATE <*direct* contact> **4.** By action of the voters, rather than through elected representatives or delegates. **5.** Of unbroken descent : LINEAL. **6.** Consisting of the exact words of the writer or speaker <a *direct* quotation> **7.** Lacking compromising or mitigating elements : ABSOLUTE. **8.** *Math.* Varying in the same manner as another quantity, esp. increasing if another quantity increases or decreasing if it decreases. **9.** *Astron.* Designating a west-to-east motion of a planet in the same direction as the sun's movement among the stars. *—adv.* Directly.

☆ ***syns:*** DIRECT, STRAIGHT, THROUGH *adj. core meaning* : in an uninterrupted line or course <a *direct* flight> ***ant:*** indirect

**direct action** *n.* The strategic use of immediately effective acts, as strikes or sabotage, to achieve a social or political objective.

**di·rect-ac·tion** (dĭ-rĕkt′ăk′shən, dī-) *adj.* Operating without intermediate processes, stages, components, or ingredients.

**direct current** *n.* An electric current flowing in one direction.

**directed angle** *n.* An angle having an indicated positive sense.

**directed distance** *n.* A segment of a line having an indicated positive sense.

**di·rec·tion** (dĭ-rĕk′shən, dī-) *n.* [ME, arrangement < OFr. < Lat. *directio* < *directus.* —see DIRECT.] **1.** The act or function of directing. **2.** Guidance, supervision, or management of an action or operation. **3.** The art or action of musical or theatrical directing. **4.** A word or phrase in a musical score indicating how a particular passage is to be performed. **5.** *often* **directions.** An instruction or series of instructions. **6.** An authoritative indication : ORDER. **7. a.** The relationship by which the alignment or orientation of any position with respect to any other position is established. **b.** A position to which motion or another position is referred. **c.** A line leading to a place or point. **d.** Line or course of movement. **8.** The statement, in degrees, of the angle measured between due north and a given line or course on a compass. **9.** A course or area of development <gave new *direction* to my life by going back to school> **—di·rec′tion·less** *adj.*

**di·rec·tion·al** (dĭ-rĕk′shə-nəl, dī-) *adj.* **1.** Of or indicating direction. **2.** *Electron.* Capable of receiving or transmitting signals in one direction only. **3.** Of or relating to guidance in effort or behavior. *—n.* A directional signal. **—di·rec′tion·al′i·ty** (-năl′ĭ-tē) *n.*

**directional antenna** *n.* An antenna adapted for receiving signals from or transmitting signals in a particular direction.

**direction finder** *n.* A device for determining the source of a transmitted signal, consisting chiefly of a radio receiver and a coiled rotating antenna.

**di·rec·tive** (dĭ-rĕk′tĭv, dī-) *n.* An order or instruction, esp. one issued by a government or military unit. *—adj.* Serving to direct, indicate, or point out.

**di·rect·ly** (dĭ-rĕkt′lē, dī-) *adv.* **1.** In a direct line or way : STRAIGHT. **2.** Without intervention : IMMEDIATELY. **3.** Exactly or completely. **4. a.** At once. **b.** In a little while. *—conj. Chiefly Brit.* As soon as <They went *directly* the taxi arrived.>

**direct object** *n.* The word or words in a sentence designating the person or thing receiving the action of a transitive verb; e.g., in *The dog buried the bone,* the direct object is *the bone.*

**di·rec·tor** (dĭ-rĕk′tər, dī-) *n.* **1.** One who controls, manages, or supervises. **2.** A member of a board of persons who control or govern the affairs of an institution or corporation. **3. a.** One whose profession is the supervision and guidance of the actors in a dramatic production. **b.** The conductor of an orchestra or chorus. **—di·rec′tor·ship′** *n.*

**di·rec·tor·ate** (dĭ-rĕk′tər-ĭt, dī-) *n.* **1.** The office or position of a director. **2.** A board of directors.

**di·rec·to·ri·al** (dĭ-rĕk′tôr′ē-əl, -tōr′-, dī-) *adj.* **1.** Of or relating to a director or directorate. **2.** Directive. **—di·rec′to′ri·al·ly** *adv.*

**director's chair** *n.* [From its use by motion-picture directors on the set.] A collapsible armchair having a back and seat usu. made of canvas.

**di·rec·to·ry** (dĭ-rĕk′tə-rē, dī-) *n., pl.* **-ries. 1.** One that directs. **2.** A book listing names, addresses, and other data about a specific group of persons or organizations. **3.** A book of directions or rules. **4.** A body or group of directors. *—adj.* Directive.

**direct primary** *n.* A preliminary election in which a party's candidates for public office are nominated by popular vote.

**direct tax** *n.* A tax, as an income or property tax, levied directly on the taxpayer.

**di·rec·trix** (dĭ-rĕk′trĭks, dī-) *n., pl.* **di·rec·trix·es** or **di·rec·tri·ces** (dĭ′rĕk-trī′sēz). **1.** The fixed curve traversed by a generatrix in generating a conic or a cylinder. **2.** The median line in the trajectory of fire of an artillery piece.

**dire·ful** (dīr′fəl) *adj.* **1.** Dreadful or frightful. **2.** Ominous : threatening. **—dire′ful·ly** *adv.* **—dire′ful·ness** *n.*

**dirge** (dûrj) *n.* [ME *dirige,* an antiphon in the Office for the Dead < Med. Lat. *dirige Domine Deus meus in conspectu tuo viam meam,* direct, O Lord my God, my way in your sight (the opening words of the antiphon).] **1.** A funeral hymn or lament. **2. a.** A slow, mournful musical composition. **b.** Something, as a poem, resembling a dirge. **—dirge′ful** *adj.*

**dir·ham** (də-răm′) *n.* [Ar. < Gk. *drakhmē,* drachma.] —See table at CURRENCY.

**dir·i·gi·ble** (dĭr′ə-jə-bəl, də-rĭj′ə-bəl) *n.* [< Lat. *dirigere,* to direct.] A steerable lighter-than-air craft with a rigid body : AIRSHIP. *—adj.* Capable of being steered. **—dir′i·gi·bil′i·ty** *n.*

**dirk** (dûrk) *n.* [Sc. *durk.*] A straight-bladed dagger. *—vt.* **dirked, dirk·ing, dirks.** To stab with a dirk.

**dirn·dl** (dûrn′dl) *n.* [G., short for *Dirndlkleid* : *Dirndl,* dim. of *Dirne,* girl + *Kleid,* dress.] **1.** A full-skirted dress with a tight bodice, patterned after Tyrolean peasant wear. **2.** A full skirt similar to a dirndl.

**dirt** (dûrt) *n.* [ME *drit* < ON.] **1.** Earth or soil. **2. a.** A soiling substance, as mud or grime. **b.** Excrement. **3.** Something mean, contemptible, or vile. **4. a.** Obscene language. **b.** Scandalous or malicious gossip. **5.** A squalid or filthy condition. **6.** Unethical behavior or practice : CORRUPTION. **7.** Material, as gravel or slag, from which metal is extracted in mining.

**dirt bike** *n.* A lightweight motorbike for use on rough surfaces, as dirt roads or trails.

**dirt-cheap** (dûrt′chēp′) *adj. & adv.* Extremely cheap.

**dirt farmer** *n. Informal.* A farmer who does all his own work.

**dirt·y** (dûr′tē) *adj.* **-i·er, -i·est. 1. a.** Soiled, as with dirt : not clean. **b.** Causing to become soiled <Gardening can be *dirty* work.> **2. a.** Indecent or obscene <*dirty* stories> **b.** Malicious or scandalous. **3.** Squalid or filthy. **4. a.** Unethical or corrupt : SORDID <*dirty* business deals> **b.** Contrary to honor or rules : UNSPORTSMANLIKE. **5. a.** Unavoidably distasteful and offensive <has to do the *dirty* deed of firing people> **b.** Demanding, unpleasant, and thankless <the *dirty* work> **c.** Unfortunate or regrettable <It's a *dirty* shame.> **6.** Of a clouded or muddy appearance <a *dirty* blue> **7.** Designating a nuclear weapon that produces an excessive amount of radioactive fallout. **8.** Stormy or rough, as weather. **9.** Expressing hostility or disapproval <gave me a *dirty* look> *—v.* **-ied, -y·ing, -ies.** *—vt.* **1.** To make dirty. **2.** To stain or tarnish with dishonor. *—vi.* To become dirty. **—dirt′i·ly** *adv.* **—dirt′i·ness** *n.*

☆ ***syns:*** DIRTY, BLACK, FILTHY, GRIMY, GRUBBY, SOILED, UNCLEAN *adj. core meaning* : covered with dirt <*dirty* clothes> ***ant:*** clean

**dirty old man** *n. Informal.* A lecherous middle-aged or older man.

**dirty pool** *n.* [< POOL[2].] *Slang.* Unfair or dishonest conduct.

**dirty tricks** *pl.n. Informal.* **1.** Covert intelligence operations. **2.** Unethical behavior, esp. in politics.

**dirty word** *n.* An offensive or inappropriate word or expression.

**dis-** *pref.* [Lat. < *dis,* apart, asunder.] **1.** Not <*dis*similar> **2. a.** Absence of <*dis*interest> **b.** Opposite of <*dis*favor> **3.** Do the opposite of : UNDO <*dis*arrange> **4. a.** Deprive of <*dis*franchise> **b.** Remove <*dis*bud> **c.** Free from <*dis*intoxicate> **5.** —Used as an intensive <*dis*annul>

**dis·a·bil·i·ty** (dĭs′ə-bĭl′ĭ-tē) *n., pl.* **-ties. 1.** A disabled state : INCAPACITY. **2.** Something that disables : HANDICAP. **3.** A legal incapacity or disqualification.

**dis·a·ble** (dĭs-ā′bəl) *vt.* **-bled, -bling, -bles. 1. a.** To make motionless or powerless by damage or injury. **b.** To weaken or destroy the normal physical or mental abilities of. **2.** To render legally disqualified. **3.** *Computer Sci.* To suppress an interrupt feature.

☆ ***syns:*** DISABLE, CRIPPLE, IMMOBILIZE, INCAPACITATE, KNOCK OUT, PARALYZE *v. core meaning* : to render powerless or motionless by infliction of severe injury or damage <veterans *disabled* in the war><*disabled* the tank with a rocket>

**dis·a·buse** (dĭs′ə-byo͞oz′) *vt.* **-bused, -bus·ing, -bus·es.** [Fr. *désabuser* : *dés-,* dis- + *abuser,* to delude < OFr., to abuse. —see ABUSE.] To free from a misconception or delusion.

**di·sac·cha·ride** (dī-săk′ə-rīd′) *n.* Any of a class of carbohydrates, including lactose and sucrose, that yield two monosaccharides on hydrolysis.

o͞o **boot** ou **out** th **thin** *th* **this** ŭ **cut** û r **urge** y **young**
yo͞o abuse zh vision ə **a**bout, it**e**m, edib**i**le, gall**o**p, circ**u**s

**dis·ac·cord** (dĭs′ə-kôrd′) *n.* [ME *disaccorden,* to disagree < OFr. *desacorder* : *des-,* not (Lat. *dis-*) + *acorder,* to agree. —see ACCORD.] Lack of agreement. —*vi.* **-cord·ed, -cord·ing, -cords.** To disagree.
**dis·ac·cus·tom** (dĭs′ə-kŭs′təm) *vt.* **-tomed, -tom·ing, -toms.** [OFr. *desacostumer* : *dis-,* not (< Lat. *dis-*) + *acostumer,* to accustom. —see ACCUSTOM.] To cause to become unaccustomed.
**dis·ad·van·tage** (dĭs′əd-văn′tĭj) *n.* [ME *disavauntage* < OFr. *desavantage* : *des-,* not (< Lat. *dis-*) + *avantage,* advantage.] **1.** An unfavorable condition or circumstance : HANDICAP. **2.** Damage or loss, esp. to reputation : DETRIMENT. —*vt.* **-taged, -tag·ing, -tag·es.** To put at a disadvantage.
**dis·ad·van·taged** (dĭs′əd-văn′tĭjd) *adj.* Suffering under severe economic and social disadvantage. —*n.* A group that suffers under severe economic and social disadvantage.
**dis·ad·van·ta·geous** (dĭs-ăd′vən-tā′jəs, dĭs′ăd-vən-) *adj.* Detrimental. **—dis·ad′van·ta′geous·ly** *adv.* **—dis·ad′van·ta′geous·ness** *n.*
**dis·af·fect** (dĭs′ə-fĕkt′) *vt.* **-fect·ed, -fect·ing, -fects.** To cause to lose affection or loyalty : ESTRANGE. **—dis′af·fec′tion** *n.*
**dis·af·fect·ed** (dĭs′ə-fĕk′tĭd) *adj.* No longer contented and loyal : RESENTFUL. **—dis′af·fect′ed·ly** *adv.*
**dis·af·fil·i·ate** (dĭs′ə-fĭl′ē-āt′) *vt.* **-at·ed, -at·ing, -ates.** To disassociate from or sever an affiliation or alliance with. **—dis′af·fil′i·a′tion** *n.*
**dis·af·firm** (dĭs′ə-fûrm′) *vt.* **-firmed, -firm·ing, -firms. 1.** To contradict or deny. **2.** *Law.* To repudiate. **—dis′af·fir′mance** (dĭs′ə-fûr′məns), **dis·af′fir·ma′tion** (dĭs-ăf′ər-mā′shən) *n.*
**dis·ag·gre·gate** (dĭs-ăg′rə-gĭt′, -gāt′) *vi.* **-gat·ed, -gat·ing, -gates.** To break up or apart. **—dis·ag′gre·ga′tive** *adj.*
**dis·a·gree** (dĭs′ə-grē′) *vi.* **-greed, -gree·ing, -grees.** [ME *disagreen* < OFr. *desagreer* : *des-,* not (< Lat. *dis-*) + *agreer,* to agree. —see AGREE.] **1.** To fail to correspond <Our findings *disagree.*> **2. a.** To have a different opinion. **b.** To quarrel. **3.** To cause adverse effects <Chocolate *disagrees* with me.>
☆ ***syns:*** DISAGREE, DIFFER, DISACCORD, DISSENT, VARY *v. core meaning* : to be of different opinion <The prosecutor *disagreed* with the court ruling.> ***ant:*** agree
**dis·a·gree·a·ble** (dĭs′ə-grē′ə-bəl) *adj.* **1.** Unpleasant or offensive <a *disagreeable* chore> **2.** Quarrelsome : bad-tempered. **—dis′a·gree′a·ble·ness** *n.* **—dis′a·gree′a·bly** *adv.*
**dis·a·gree·ment** (dĭs′ə-grē′mənt) *n.* **1.** A failure or refusal to agree. **2.** A disparity or inconsistency. **3.** A difference of opinion.
**dis·al·low** (dĭs′ə-lou′) *vt.* **-lowed, -low·ing, -lows.** [ME *disallowen* < OFr. *desalouer,* to reprimand : *des-,* not (< Lat. *dis-*) + *alouer,* to approve. —see ALLOW.] **1.** To refuse to allow. **2.** To reject as improper, invalid, or untrue. **—dis′al·low′a·ble** *adj.* **—dis′al·low′ance** *n.*
**dis·am·big·u·ate** (dĭs′ăm-bĭg′yo͞o-āt′) *vt.* **-at·ed, -at·ing, -ates.** To establish a single grammatical or semantic interpretation for. **—dis′am·big′u·a′tion** *n.*
**dis·an·nul** (dĭs′ə-nŭl′) *vt.* **-nulled, -null·ing, -nuls.** To annul : nullify. **—dis′an·nul′ment** *n.*
**dis·ap·pear** (dĭs′ə-pîr′) *vi.* **-peared, -pear·ing, -pears. 1.** To pass out of sight : VANISH. **2.** To cease to exist. **—dis′ap·pear′ance** *n.*
**dis·ap·point** (dĭs′ə-point′) *v.* **-point·ed, -point·ing, -points.** [ME *disappointen,* to remove from office < OFr. *desapointier* : *des-* (reversal < Lat. *dis-*) + *apointer,* to appoint.] —*vt.* **1.** To fail to satisfy the hope, desire, or expectation of. **2.** To frustrate <Rain *disappointed* our picnic plans.> —*vi.* To cause disappointment **—dis′ap·point′ing·ly** *adv.*
**dis·ap·point·ment** (dĭs′ə-point′mənt) *n.* **1.** The act of disappointing or state of being disappointed. **2.** One that disappoints.
**dis·ap·pro·ba·tion** (dĭs-ăp′rə-bā′shən) *n.* Moral disapproval : CONDEMNATION.
**dis·ap·prov·al** (dĭs′ə-pro͞o′vəl) *n.* The act of disapproving.
**dis·ap·prove** (dĭs′ə-pro͞ov′) *v.* **-proved, -prov·ing, -proves.** —*vt* **1.** To have an unfavorable opinion of : CONDEMN. **2.** To refuse to approve : REJECT. —*vi.* To have an unfavorable opinion. **—dis′ap·prov′er** *n.* **—dis′ap·prov′ing·ly** *adv.*
**dis·arm** (dĭs-ärm′) *v.* **-armed, -arm·ing, -arms.** [ME *disarmen* < OFr. *desarmer* : *des-* (reversal < Lat. *dis-*) + *armer,* to arm < Lat. *armare* < *arma,* weapons.] —*vt.* **1.** To divest of weapons. **2. a.** To deprive of the means of attack or defense. **b.** To make harmless. **3. a.** To allay the suspicion, hostility, or antagonism of. **b.** To win the confidence of. —*vi.* **1.** To lay down arms. **2.** To reduce or abolish armed forces.
**dis·ar·ma·ment** (dĭs-är′mə-mənt) *n.* **1.** The act of laying down arms, esp. the reduction or abolition of a nation's armed forces. **2.** The state of being disarmed.
**dis·arm·ing** (dĭs-är′mĭng) *adj.* Tending to remove suspicion or hostility : ENDEARING <a *disarming* smile> **—dis·arm′ing·ly** *adv.*
**dis·ar·range** (dĭs′ə-rānj′) *vt.* **-ranged, -rang·ing, -rang·es.** To disturb the proper arrangement or order of. **—dis′ar·range′ment** *n.*
**dis·ar·ray** (dĭs′ə-rā′) *n.* [ME *disaraien* < OFr. *desareer* : *des-* (reversal < Lat. *dis-*) + *areer,* to array.] **1.** A state of disorder or confusion. **2.** Disorderly dress. —*vt.* **-rayed, -ray·ing, -rays.** To throw into confusion : UPSET.
**dis·ar·tic·u·late** (dĭs′är-tĭk′yə-lāt′) *v.* **-lat·ed, -lat·ing, -lates.** —*vt.* To separate at the joints : DISJOINT. —*vi.* To become disjointed. **—dis′ar·tic′u·la′tion** *n.* **—dis′ar·tic′u·la′tor** *n.*
**dis·as·sem·ble** (dĭs′ə-sĕm′bəl) *v.* **-bled, -bling, -bles.** —*vt.* To take apart. —*vi.* To come apart <The stereo doesn't *disassemble* easily.> **—dis′as·sem′bly** *n.*
**dis·as·so·ci·ate** (dĭs′ə-sō′shē-āt′, -sē-) *vt.* **-at·ed, -at·ing, -ates.** To dissociate. **—dis′as·so′ci·a′tion** *n.*
**dis·as·ter** (dĭ-zăs′tər, -săs′-) *n.* [Fr. *désastre* < Ital. *disastro* : *dis-* (pejorative < Lat. *dis-*) + *astro,* star < Lat. *astrum* < Gk. *astron.*] **1. a.** An occurrence causing widespread destruction and distress. **b.** A grave misfortune. **2.** A total failure.
☆ ***syns:*** DISASTER, CALAMITY, CATACLYSM, CATASTROPHE, TRAGEDY *n. core meaning* : a grave occurrence having ruinous results <floods, air crashes, and other *disasters*>
▲ word history: A *disaster,* etymologically speaking, is a calamity brought about by the evil influence of a star or planet. *Disaster* is ultimately derived from Latin *dis–,* a pejorative prefix, and Greek *astron,* "star." Although astrology was very important in ancient and medieval times, the word *disaster* is not a Latin word, and it does not appear in either French or English until the 16th century. The meaning "evil celestial influence" did not survive the 17th century and

**dis′ac·com′mo·date′** *v.*
**dis′ac·cred′it** *v.*
**dis′a·cid′i·fy′** *v.*
**dis′ac·knowl′edge** *v.*
**dis′ac·knowl′edg·ment** *n.*
**dis′ac·quain′tance** *n.*
**dis′ac·quaint′ed** *adj.*
**dis′ad·vise′** *v.*
**dis′al·ly′** *v.*
**dis′a·men′i·ty** *n.*
**dis·an′chor** *v.*
**dis′an·nex′** *v.*
**dis′a·noint′** *v.*
**dis·bal′ance** *n.*
**dis′com·mend′** *v.*
**dis′com·mend′a·ble** *adj.*
**dis′com·mod′i·ty** *n.*
**dis′con·firm′** *v.*
**dis′con′fir·ma′tion** *n.*
**dis′con·gru′i·ty** *n.*
**dis′con′gru·ous** *adj.*
**dis′con·sid′er** *v.*
**dis·con′so·nant** *adj.*
**dis·crown′** *v.*
**dis′e·con′o·my** *n.*
**dis·edge′** *v.*
**dis·ed′i·fy′** *v.*
**dis′em·balm′** *v.*
**dis′em·bel′lish** *v.*
**dis′em·bel′lish·ment** *n.*
**dis′em·broil′** *v.*
**dis′em·broil′ment** *n.*
**dis′em·ploy′** *v.*
**dis′em·ploy′ment** *n.*
**dis′en·a′ble** *v.*
**dis′en·cour′age** *v.*
**dis′en·cour′age·ment** *n.*
**dis′en·dow′** *v.*
**dis′en·dow′ment** *n.*
**dis′en·joy′** *v.*
**dis′en·joy′ment** *n.*
**dis′en·roll′** *v.*
**dis′en·roll′ment** *n.*
**dis′en·thrall′** *v.*
**dis′en·thrall′ment** *n.*
**dis′en·ti′tle** *v.*
**dis′en·tomb′** *v.*
**dis′en·tomb′ment** *n.*
**dis′en·trance′** *v.*
**dis′en·twine′** *v.*
**dis·e′qui·lib′ri·um** *n.*
**dis·fel′low·ship′** *n.*
**dis·for′est** *v.*
**dis·gar′ri·son** *v.*
**dis·hal′low** *v.*
**dis·horn′** *v.*
**dis′im·ag′ine** *v.*
**dis′im·pas′sioned** *adj.*
**dis′im·pris′on** *v.*
**dis′im·pris′on·ment** *n.*
**dis′im·prove′** *v.*
**dis′im·prove′ment** *n.*
**dis′in·car′nate** *adj.*
**dis′in·car·na′tion** *n.*
**dis′in·cor′po·rate′** *v.*
**dis′in·cor′po·ra′tion** *n.*
**dis′in·flate′** *v.*
**dis′in·hib′i·to′ry** *adj.*
**dis′in·vest′** *v.*
**dis′in·vite′** *v.*
**dis′in·volve′** *v.*
**dis·lus′ter** *v.*
**dis′o·blig′ing·ly** *adv.*
**dis·oc′cu·pa′tion** *n.*
**dis·pau′per·ize′** *v.*
**dis·peace′** *v.*
**dis′per·son′i·fy′** *v.*
**dis·pet′al** *v.*
**dis·pope′** *v.*
**dis·priv′i·lege** *v.*
**dis′re·late′** *v.*
**dis′re·la′tion** *n.*
**dis′re·mem′ber** *v.*
**dis′re·spect′a·ble** *adj.*
**dis·roof′** *v.*
**dis·root′** *v.*
**dis·throne′** *v.*
**dis·un′i·fi·ca′tion** *n.*
**dis·u′ni·form′** *adj.*
**dis·val′ue** *v.* & *n.*
**dis·weap′on** *v.*
**dis·yoke′** *v.*

the more general sense has been current since the introduction of *disaster* as an English word.

**disaster area** *n.* An area that officially qualifies for emergency governmental aid as a result of a disaster such as an earthquake or flood.

**disaster dump** *n. Computer Sci.* A printout that occurs as a result of a nonrecoverable program error.

**dis·as·trous** (dĭ-zăs′trəs, -săs′-) *adj.* **1.** Causing disaster : CALAMITOUS. **2.** Awful : terrible <a *disastrous* double date> **—dis·as′trous·ly** *adv.* **—dis·as′trous·ness** *n.*

**dis·a·vow** (dĭs′ə-vou′) *vt.* **-vowed, -vow·ing, -vows.** [ME *disavowen* < OFr. *desavouer* : *des-*, not (< Lat. *dis-*) + *avouer*, to avow.] To disclaim knowledge of, responsibility for, or association with : DENY. **—dis′a·vow′al** *n.*

**dis·band** (dĭs-bănd′) *v.* **-band·ed, -band·ing, -bands.** —*vt.* To break up : DISSOLVE <*disband* a government commission> —*vi.* To become disbanded. **—dis·band′ment** *n.*

**dis·bar** (dĭs-bär′) *vt.* **-barred, -bar·ring, -bars.** To expel (an attorney) from the legal profession by official action or procedure. **—dis·bar′ment** *n.*

**dis·be·lief** (dĭs′bĭ-lēf′) *n.* Reluctance or refusal to believe.

**dis·be·lieve** (dĭs′bĭ-lēv′) *v.* **-lieved, -liev·ing, -lieves.** —*vt.* To refuse to believe in : reject or deny. —*vi.* To withhold belief. **—dis′be·liev′er** *n.* **—dis′be·liev′ing·ly** *adv.*

**dis·bound** (dĭs-bound′) *adj.* No longer having a binding or having one in poor condition <a *disbound* novel>

**dis·branch** (dĭs-brănch′) *vt.* **-branched, -branch·ing, -branch·es. 1.** To cut or break a branch from (a tree). **2.** To remove (a branch).

**dis·bud** (dĭs-bŭd′) *vt.* **-bud·ded, -bud·ding, -buds. 1.** To prune buds from to control the shape of or promote a better quality of bloom <*disbud* a plant> **2.** To remove newly developing horns from (livestock).

**dis·bur·den** (dĭs-bûr′dn) *v.* **-dened, -den·ing, -dens.** —*vt.* **1.** To relieve of a burden. **2.** To remove or unload (a burden). —*vi.* To remove or unload a burden. **—dis·bur′den·ment** *n.*

**dis·bur·sal** (dĭs-bûr′səl) *n.* Disbursement.

**dis·burse** (dĭs-bûrs′) *vt.* **-bursed, -burs·ing, -burs·es.** [OFr. *desbourser* : *des-* (reversal < Lat. *dis-*) + *bourse*, purse < Med. Lat. *bursa* < Gk.] To pay out, as from a fund. **—dis·burs′a·ble** *adj.* **—dis·burs′er** *n.*

**dis·burse·ment** (dĭs-bûrs′mənt) *n.* **1.** The act of disbursing. **2.** Money paid out : EXPENDITURE.

**disc** (dĭsk) *n.* **1.** *also* **disk.** A phonograph record. **2.** *var. of* DISK.

**disc-** *pref. var. of* DISCO-.

**dis·calced** (dĭ-skălst′) *adj.* [< Lat. *discalceatus* : *dis-*, not + *calceatus*, shod < *calceus*, shoe < *calx*, heel.] Barefoot or wearing sandals. —Used of certain religious orders.

**dis·cant** (dĭs′kănt′) *n. var. of* DESCANT 1. —*v. var. of* DESCANT 2.

**dis·card** (dĭ-skärd′) *v.* **-card·ed, -card·ing, -cards.** —*vt.* **1.** To throw away : get rid of. **2. a.** To throw out (a playing card) from one's hand. **b.** To play (a card other than a trump) from a suit unlike the card led. —*vi.* To throw out a playing card. —*n.* (dĭs′kärd′). **1.** The act or an instance of discarding. **2.** One that is discarded, esp. a playing card. **—dis·card′er** *n.*

**disc brake** *also* **disk brake** *n.* A brake in which the retarding friction is generated between a set of stationary pads and a rotating disk.

**dis·cern** (dĭ-sûrn′, -zûrn′) *v.* **-cerned, -cern·ing, -cerns.** [ME *discernen* < OFr. *discerner* < Lat. *discernere* : *dis-*, apart + *cernere*, to perceive.] —*vt.* **1.** To perceive (something hidden or obscure) : DETECT. **2.** To comprehend mentally. **3.** To perceive as separate and distinct : DISCRIMINATE. —*vi.* To perceive differences. **—dis·cern′er** *n.* **—dis·cern′i·ble** *adj.* **—dis·cern′i·bly** *adv.*

**dis·cern·ing** (dĭ-sûr′nĭng, -zûr′-) *adj.* Showing insight and judgment : perceptive and discriminating. **—dis·cern′ing·ly** *adv.*

**dis·cern·ment** (dĭ-sûrn′mənt, -zûrn′-) *n.* **1.** The act or process of discerning. **2.** Acuteness of discrimination : PERSPICACITY.

**dis·charge** (dĭs-chärj′) *v.* **-charged, -charg·ing, -charg·es.** [ME *dischargen* < OFr. *deschargier* < VLat. **discarricare* : Lat. *dis-*, apart + LLat. *carricare*, to load < *carrus*, cart, of Celtic orig.] —*vt.* **1.** To relieve of a burden or of contents : UNLOAD. **2.** To unload or empty (contents). **3.** To release, as from duty or service. **4.** To dismiss from employment. **5.** To send or pour forth : EMIT. **6.** To shoot or fire <*discharge* a weapon> **7.** To perform the obligations or demands of (an office, duty, or task). **8.** To comply with the terms of (e.g., a promise or debt). **9.** *Law.* **a.** To release (e.g., a defendant). **b.** To set aside : ANNUL. **10.** To remove (color) from cloth, as by chemical bleaching. **11.** *Elect.* To cause electrical discharge in (e.g., a battery). **12. a.** To apportion (weight) evenly, as over a door. **b.** To relieve (a structural part) of excess weight by distribution of pressure. —*vi.* **1.** To get rid of a burden, load, or weight. **2.** To go off : FIRE <The gun *discharged.*> **3.** To pour forth contents. **4.** To run, as a dye. **5.** To undergo electrical discharge. —*n.* (dĭs′chärj′, dĭs-chärj′). **1.** The act of removing a load or burden. **2.** The act of shooting or firing a projectile or weapon. **3. a.** An instance of pouring forth : EMISSION <a *discharge* of blood> **b.** The amount or rate of emission or ejection. **4.** Something released or emitted. **5.** An instance of relieving from or elimination of an obligation, burden, or responsibility. **6.** Fulfillment or performance <*discharge* of duties> **7. a.** Dismissal or release from employment, service, or confinement. **b.** A document certifying such release, esp. from military service. **8.** *Law.* An annulment or acquittal. **9.** *Elect.* **a.** The release of stored energy in a capacitor by the flow of electric current between its terminals. **b.** The conversion of chemical energy to electric energy in a storage battery. **c.** A flow of electricity in a dielectric, esp. in a rarefied gas. **d.** The elimination of net electric charge from a charged body. **—dis·charge′a·ble** *adj.* **—dis·charg′er** *n.*

**discharge lamp** *n.* A lamp that produces light by means of an internal electrical discharge.

**discharge tube** *n.* A closed insulating vessel fitted with electrodes and containing a gas in which an electrical discharge is induced by high applied potentials.

**dis·ci·ple** (dĭ-sī′pəl) *n.* [ME, partly < OE *discipul* and partly < OFr. *desciple*, both < Lat. *discipulus*, pupil < *discere*, to learn.] **1. a.** One who believes in and helps disseminate the teachings of a master. **b.** An active adherent, as of a movement or philosophy. **2.** *often* **Disciple.** One of the companions of Christ. **3. Disciple.** A member of the Disciples of Christ. **—dis·ci′ple·ship′** *n.*

**Disciples of Christ** *n.* A Christian denomination, founded in 1809, that accepts the Bible as the only rule of Christian faith and practice, rejects denominational creeds, and practices baptism by immersion.

**dis·ci·plin·a·ble** (dĭs′ə-plĭn′ə-bəl, dĭs′ə-plĭn′-) *adj.* **1.** Subject to or deserving of discipline. **2.** Responsive to training.

**dis·ci·pli·nar·i·an** (dĭs′ə-plə-nâr′ē-ən) *n.* One who believes in or enforces strict discipline. —*adj.* Disciplinary.

**dis·ci·pli·nar·y** (dĭs′ə-plə-nĕr′ē) *adj.* **1.** Of, relating to, or used for discipline. **2.** Of or relating to a particular field of academic study. **—dis′ci·pli·nar′i·ly** (-nâr′ə-lē) *adv.*

**dis·ci·pline** (dĭs′ə-plĭn) *n.* [ME < OFr. *descepline* < Lat. *disciplina* < *discipulus*, pupil < *discere*, to learn.] **1.** Training expected to produce a specific type or pattern of behavior, esp. training that produces moral or mental improvement. **2.** Controlled behavior resulting from disciplinary training. **3.** A systematic method to obtain obedience. **4.** A state of order based on submission to rules and authority. **5.** Punishment intended to train or correct. **6.** A set of methods or rules, as those regulating the practice of a church or monastic order. **7.** A branch of knowledge or of teaching. —*vt.* **-plined, -plin·ing, -plines. 1.** To train by instruction and control. **2.** To punish or penalize. **3.** To impose order on <*discipline* one's work habits> **—dis′ci·pli·nal** (-plə-nəl) *adj.* **—dis′ci·plin′er** *n.*

**disc jockey** *also* **disk jockey** *n.* A radio announcer who presents and comments on phonograph records.

**dis·claim** (dĭs-klām′) *v.* **-claimed, -claim·ing, -claims.** [ME *disclaimen* < AN *desclaimer* : *des-* (reversal < Lat. *dis-*) + *claimer*, to claim < OFr. *clamer*.] —*vt.* **1.** To deny or renounce any claim to or connection with : DISOWN. **2.** To deny the validity of : REPUDIATE. **3.** *Law.* To renounce one's right or claim to. —*vi. Law.* To renounce a right or claim.

**dis·claim·er** (dĭs-klā′mər) *n.* **1.** A repudiation or denial of responsibility or connection. **2.** *Law.* A renunciation of one's right or claim.

**dis·cla·ma·tion** (dĭs′klə-mā′shən) *n.* Renunciation or disavowal.

**dis·cli·max** (dĭs-klī′măks′) *n. Ecol.* A normally stable climax community that has been altered by human beings or other influences.

**dis·close** (dĭ-sklōz′) *vt.* **-closed, -clos·ing, -clos·es.** [ME *disclosen* < OFr. *desclore, desclos-* : *des-* (reversal < Lat. *dis-*) + *clore*, to close.] **1.** To expose to view : REVEAL. **2.** To make known : DIVULGE. **—dis·clos′er** *n.*

**dis·clo·sure** (dĭ-sklō′zhər) *n.* **1.** The act or process of disclosing. **2.** Something disclosed : REVELATION.

**dis·co** (dĭs′kō′) *n., pl.* **-cos.** [Short for DISCOTHEQUE.] **1.** A nightclub with showy decor and usu. special lighting effects and featuring electronically amplified music for dancing. **2. a.** Popular dance music marked by strong repetitive bass rhythms. **b.** A style of dancing done esp. to disco music. —*vi.* **-coed, -co·ing, -cos.** To dance to disco music.

**disco-** *or* **disc-** *pref.* [Lat. *disco-* < Gk. *disko-* < *diskos*, disk.] **1.** Disk <*discoid*> **2.** Phonograph record <*discophile*>

**dis·cog·ra·phy** (dĭ-skŏg′rə-fē) *n., pl.* **-phies.** [Fr. *discographie* : *disco-*, disco- + *-graphie*, -graphy.] **1.** The study and cataloguing of phonograph records. **2.** A comprehensive list of a particular composer's works or of the recordings made by a particular performer. **—dis·cog′ra·pher** *n.*

**dis·coid** (dĭs′koid′) *also* **dis·coi·dal** (dĭ-skoid′l) *adj.* **1.** Shaped like a disk. **2.** *Bot.* Having disk flowers but no ray flowers, as a composite flower head. **—dis′coid′** *n.*

**dis·col·or** (dĭs-kŭl′ər) *v.* **-ored, -or·ing, -ors.** [ME *discolouren* < OFr. *discolerer* < LLat. *discolorare* : *dis-* (reversal) + *colorare*, to color < *color*, color.] —*vt.* To spoil or alter the proper color of : STAIN. —*vi.* To become faded or stained. **—dis·col′or·a′tion** (-kŭl′ə-rā′shən) *n.*

**dis·com·bob·u·late** (dĭs'kəm-bŏb'yə-lāt') *vt.* **-lat·ed, -lat·ing, -lates.** [Perh. alteration of DISCOMPOSE.] *Slang.* To confuse : upset. **—dis'com·bob'u·la'tion** *n.*

**dis·com·fit** (dĭs-kŭm'fĭt) *vt.* **-fit·ed, -fit·ing, -fits.** [ME *discomfiten* < OFr. *disconfit,* p.part. of *desconfire,* to defeat < VLat. **disconficere* : Lat. *dis-* (reversal) + Lat. *conficere,* to prepare. —see COMFIT.] **1.** To frustrate the plans of : THWART. **2.** To defeat in battle : VANQUISH. **3.** To make uneasy : DISCONCERT. ***usage:*** Although *discomfit* was once used strictly in the sense of "to defeat or frustrate," it has also acquired the sense of "to disconcert or make uncomfortable" through confusion with the unrelated word *discomfort.* —*n.* Discomfiture.

**dis·com·fi·ture** (dĭs-kŭm'fĭ-cho͝or', -chər) *n.* **1.** Disappointment or frustration. **2.** Defeat. **3.** Lack of ease : DISCOMFORT.

**dis·com·fort** (dĭs-kŭm'fərt) *n.* [ME, distress < OFr. *desconfort* < *desconforter,* to discourage : *des-* (reversal < Lat. *dis-*) + *conforter,* to strengthen. —see COMFORT.] **1.** Physical or mental distress. **2.** Something that disturbs one's comfort : annoyance or inconvenience. —*vt.* **-fort·ed, -fort·ing, -forts.** To make uncomfortable. **—dis·com'fort·a·ble** (-kŭm'fər-tə-bəl, -kŭmf'tə-bəl) *adj.*

**dis·com·mend** (dĭs'kə-mĕnd') *vt.* **-mend·ed, -mend·ing, -mends.** **1.** To express disapproval of. **2.** To cause to be regarded with disfavor. **—dis'com·mend'a·ble** *adj.*

**dis·com·mode** (dĭs'kə-mōd') *vt.* **-mod·ed, -mod·ing, -modes.** [Fr. *discommoder* : Lat. *dis-* (reversal) + Fr. *commode,* convenient. —see COMMODE.] To inconvenience.

**dis·com·pose** (dĭs'kəm-pōz') *vt.* **-posed, -pos·ing, -pos·es. 1.** To disturb the composure or tranquillity of. **2.** To disorder. **—dis'com·pos'ed·ly** (-pō'zĭd-lē) *adv.* **—dis'com·pos'ing·ly** *adv.*

**dis·com·po·sure** (dĭs'kəm-pō'zhər) *n.* Lack of composure.

**dis·con·cert** (dĭs'kən-sûrt') *vt.* **-cert·ed, -cert·ing, -certs.** [Obs. Fr. *disconcerter* < OFr. *desconcerter* : *des-* (reversal < Lat. *dis-*) + *concerter,* to bring into agreement < OItal. *concertare.*] **1. a.** To upset the self-possession of : DISCOMPOSE. **b.** To embarrass. **2.** To throw into disorder. **—dis'con·cert'ing·ly** *adv.*

**dis·con·form·i·ty** (dĭs'kən-fôr'mĭ-tē) *n., pl.* **-ties.** *Geol.* An interruption of sedimentation caused by erosion resulting in the formation of two parallel strata.

**dis·con·nect** (dĭs'kə-nĕkt') *vt.* **-nect·ed, -nect·ing, -nects. 1.** To interrupt or break the connection of or between. **2.** *Elect.* To shut off the current in (an appliance) by removing its connection with the power source. **—dis'con·nec'tion** *n.*

**dis·con·nect·ed** (dĭs'kə-nĕk'tĭd) *adj.* **1.** Not connected. **2.** Incoherent <*disconnected* speech> **—dis'con·nect'ed·ly** *adv.* **—dis'con·nect'ed·ness** *n.*

**dis·con·so·late** (dĭs-kŏn'sə-lĭt) *adj.* [ME < Med. Lat. *disconsolatus* : Lat. *dis-,* not + *consolatus,* p.part. of *consolari,* to console. —see CONSOLE.] **1.** Beyond consolation : utterly dejected. **2.** Cheerless or gloomy <a *disconsolate* landscape> **—dis·con'so·late·ly** *adv.* **—dis·con'so·late·ness, dis·con'so·la'tion** (-kŏn'sə-lā'shən) *n.*

**dis·con·tent** (dĭs'kən-tĕnt') *n.* **1.** Lack of contentment : DISSATISFACTION. **2.** A sense of resentment and grievance. —*adj.* Discontented. —*vt.* **-tent·ed, -tent·ing, -tents.** To make discontented. **—dis'con·tent'ment** *n.*

**dis·con·tent·ed** (dĭs'kən-tĕn'tĭd) *adj.* Not content or satisfied : UNHAPPY. **—dis'con·tent'ed·ly** *adv.* **—dis'con·tent'ed·ness** *n.*

**dis·con·tin·u·ance** (dĭs'kən-tĭn'yo͞o-əns) *n.* **1.** The act of discontinuing or the condition of being discontinued : CESSATION. **2.** *Law.* The termination of an action by the plaintiff.

**dis·con·tin·u·a·tion** (dĭs'kən-tĭn'yo͞o-ā'shən) *n.* DISCONTINUANCE 1.

**dis·con·tin·ue** (dĭs'kən-tĭn'yo͞o) *v.* **-ued, -u·ing, -ues.** [ME *discontinuen* < OFr. *descontinuer* < Med. Lat. *discontinuare* : Lat. *dis-,* not + Lat. *continuare,* to continue. —see CONTINUE.] —*vt.* **1.** To put an end to : TERMINATE. **2.** To cease trying to continue or accomplish : ABANDON <*discontinue* a search> **3.** *Law.* To terminate (an action) by discontinuance. —*vi.* To come to an end.

**dis·con·ti·nu·i·ty** (dĭs-kŏn'tə-no͞o'ĭ-tē, -nyo͞o'-) *n., pl.* **-ties. 1.** A lack of continuity, cohesion, or logical sequence. **2.** A break or interruption. **3.** *Math.* **a.** The property of being discontinuous. **b.** A point at which a function is defined but is not continuous. **c.** A point at which a function is undefined.

**dis·con·tin·u·ous** (dĭs'kən-tĭn'yo͞o-əs) *adj.* **1.** Marked by breaks or interruptions. **2.** *Math.* Having one or more discontinuities. **—dis'con·tin'u·ous·ly** *adv.* **—dis'con·tin'u·ous·ness** *n.*

**dis·co·phile** (dĭs'kə-fīl') *n.* One who collects or is knowledgeable about phonograph records.

**dis·cord** (dĭs'kôrd') *n.* [ME < OFr. *descorde* < Lat. *discordia,* strife < *discors,* disagreeing : *dis-,* apart + *cors,* heart.] **1. a.** Lack of agreement, as among persons or things. **b.** Tension and strife resulting from discord : DISSENSION. **2.** A harsh or confused mingling of sounds. **3.** *Mus.* Inharmonious combination of simultaneously sounded tones : DISSONANCE. —*vi.* (dĭ-skôrd', dĭs'kôrd') **-cord·ed, -cord·ing, -cords.** To fail to agree or harmonize : CLASH.

**dis·cor·dant** (dĭ-skôr'dnt) *adj.* **1.** Not in accord : CONFLICTING. **2.** Not agreeable in sound : harsh or dissonant. **—dis·cor'dance, dis·cor'dan·cy** *n.* **—dis·cor'dant·ly** *adv.*

**Dis·cor·di·a** (dĭ-skôr'dē-ə) *n.* [Lat. < *discordia,* discord.] *Rom. Myth.* The goddess of strife.

**dis·co·theque** *also* **dis·co·thèque** (dĭs'kə-tĕk', dĭs'kə-tĕk') *n.* [Fr. : *disco-,* disco- + *(biblio)thèque,* library < Lat. *bibliotheca.* —see BIBLIOTHECA.] **1.** A small nightclub featuring dancing to recorded or sometimes live music. **2.** DISCO 1.

**dis·count** (dĭs'kount', dĭs-kount') *v.* **-count·ed, -count·ing, -counts.** [OFr. *desconter* < Med. Lat. *discomputare* : Lat. *dis-* (reversal) + Lat. *computare,* to compute. —see COMPUTE.] —*vt.* **1.** To deduct or subtract from a cost or price. **2. a.** To purchase or sell (a commercial paper, as a bill or note) after deducting the amount of interest that will accumulate before it matures. **b.** To loan money on (a commercial paper not immediately payable) after deducting the interest. **3.** To reduce in cost, quantity, or value. **4.** To regard (e.g., a rumor) as being untrustworthy or exaggerated. **5.** To underestimate the effectiveness or significance of. **6.** To anticipate and make allowance for. —*vi.* To lend money after deduction of interest. —*n.* (dĭs'kount'). **1.** A reduction from the full or standard amount of a price or debt. **2.** Discount rate. **3.** The act or an instance of discounting a commercial paper, as a note or bill of exchange. **—dis'count'a·ble** *adj.* **—dis'count'er** *n.*

**dis·coun·te·nance** (dĭs-koun'tə-nəns) *vt.* **-nanced, -nanc·ing, -nanc·es. 1.** To treat or regard with disfavor. **2.** To make uneasy : abash or disconcert. —*n.* Disfavor or disapproval.

**discount rate** *n.* **1.** The interest deducted in advance in purchasing, selling, or lending a bill, note, or other commercial paper. **2.** The rate of interest deducted in a discount rate transaction.

**discount store** *n.* A store that sells merchandise at a discount from the manufacturer's suggested retail price.

**dis·cour·age** (dĭ-skûr'ĭj, -skŭr'-) *vt.* **-aged, -ag·ing, -ag·es.** [ME *discoragen* < OFr. *descoragier* : *des-* (reversal < Lat. *dis-*) + *corage,* courage. —see COURAGE.] **1.** To deprive of confidence, hope, or spirit. **2. a.** To deter or dissuade <refused to be *discouraged* by the setback> **b.** To try to prevent by expressing disapproval or raising objections <They *discouraged* me from applying for the job.> **—dis·cour'ag·er** *n.* **—dis·cour'ag·ing·ly** *adv.*

☆ ***syns:*** DISCOURAGE, DISHEARTEN, DISPIRIT *v. core meaning* : to make less hopeful or enthusiastic <problems that *discouraged* me> ***ant:*** encourage

**dis·cour·age·ment** (dĭ-skûr'ĭj-mənt, -skŭr'-) *n.* **1.** The act of discouraging or state of being discouraged. **2.** One that discourages.

**dis·course** (dĭs'kôrs', -kōrs') *n.* [ME *discours* < LLat. *discursus,* discussion < p.part. of Lat. *discurrere,* to run about, to speak at length : *dis-,* apart + *currere,* to run.] **1.** Spoken or written expression. **2.** A conversation. **3.** A formal and lengthy, written or spoken discussion of a subject. **4.** *Archaic.* The process or power of reasoning. —*v.* (dĭ-skôrs', -skōrs') **-coursed, -cours·ing, -cours·es.** —*vi.* **1.** To speak or write formally and at length. **2.** To engage in conversation or discussion : CONVERSE. —*vt. Archaic.* UTTER¹ 2. **—dis·cours'er** *n.*

**dis·cour·te·ous** (dĭs-kûr'tē-əs) *adj.* Lacking courteous manners : RUDE. **—dis·cour'te·ous·ly** *adv.* **—dis·cour'te·ous·ness** *n.*

**dis·cour·te·sy** (dĭs-kûr'tĭ-sē) *n., pl.* **-sies. 1.** Lack of courtesy : RUDENESS. **2.** A rude statement or act.

**dis·cov·er** (dĭ-skŭv'ər) *vt.* **-ered, -er·ing, -ers.** [ME *discoveren,* to reveal < OFr. *descovrir* < LLat. *discooperire* : Lat. *dis-* (reversal) + *cooperire,* to cover. —see COVER.] **1.** To gain knowledge of through observation, study, or search. **2.** To be the first to find, learn of, or observe. **3.** *Archaic.* To reveal : expose. **—dis·cov'er·a·ble** *adj.* **—dis·cov'er·er** *n.*

**dis·cov·er·y** (dĭ-skŭv'ə-rē) *n., pl.* **-ies. 1.** The act or an instance of discovering. **2.** Something discovered. **3.** *Law.* Data or documents that a party to a legal action is compelled to disclose to another party either prior to or during a proceeding.

**dis·cred·it** (dĭs-krĕd'ĭt) *vt.* **-it·ed, -it·ing, -its. 1.** To damage the reputation of : DISGRACE. **2.** To cause to be distrusted or doubted. **3.** To refuse to believe <*discredit* a rumor> —*n.* **1.** Loss of or damage to one's reputation. **2.** Lack or loss of trust or belief : DOUBT. **3.** Something damaging to one's reputation or position.

**dis·cred·it·a·ble** (dĭs-krĕd'ĭ-tə-bəl) *adj.* Deserving of or resulting in discredit : BLAMEWORTHY. **—dis·cred'it·a·bly** *adv.*

**dis·creet** (dĭ-skrēt') *adj.* [ME < OFr. *discret* < Med. Lat. *discretus* < Lat., p.part. of *discernere,* to discern, separate. —see DISCERN.] **1.** Having or displaying a judicious reserve in one's speech or conduct : PRUDENT. **2.** Lacking pretension or ostentation : MODEST. **—dis·creet'ly** *adv.* **—dis·creet'ness** *n.*

**dis·crep·ance** (dĭ-skrĕp'əns) *n.* Discrepancy.

**dis·crep·an·cy** (dĭ-skrĕp'ən-sē) *n., pl.* **-cies. 1.** Disagreement or divergence, as between facts or claims : INCONSISTENCY. **2.** An instance of discrepancy.

**dis·crep·ant** (dĭ-skrĕp'ənt) *adj.* [ME *discrepaunt* < Lat. *discrepans,* pr.part. of *discrepare,* to disagree : *dis-,* apart + *crepare,* to rattle.] Showing discrepancy. **—dis·crep'ant·ly** *adv.*

**dis·crete** (dĭ-skrēt') *adj.* [ME < Lat. *discretus.* —see DISCREET.]

1. Constituting a separate thing : DISTINCT. 2. Made up of unconnected distinct parts. **—dis·crete'ly** *adv.* **—dis·crete'ness** *n.*

**discrete variable** *n.* A mathematical variable that assumes only whole number values.

**dis·cre·tion** (dĭ-skrĕsh'ən) *n.* **1.** The quality of being discreet : CIRCUMSPECTION. **2.** Freedom or power to act or judge on one's own <used our own *discretion* in dealing with the crisis> **—dis·cre'tion·al** *adj.* **—dis·cre'tion·al·ly** *adv.*

**dis·cre·tion·ar·y** (dĭ-skrĕsh'ə-nĕr'ē) *adj.* **1.** Left to or regulated by one's own judgment or discretion. **2.** Something, as a fund of money, that is to be used responsibly as needed. **—dis·cre'tion·ar'i·ly** (-nâr'ə-lē) *adv.*

**discretionary account** *n.* A stock or commodity account in which an agent is free to trade for the customer at his or her own discretion.

**dis·cret·i·za·tion** (dĭ-skrē'tə-zā'shən) *n.* The act of making mathematically discrete.

**dis·crim·i·nate** (dĭ-skrĭm'ə-nāt') *v.* **-nat·ed, -nat·ing, -nates.** [Lat. *discriminare, discriminat-* < *discrimen,* distinction.] *—vi.* **1.** To make a clear distinction : DIFFERENTIATE. **2.** To act on the basis of prejudice <accused of *discriminating* against the elderly> *—vt.* **1.** To perceive the distinguishing features of. **2.** To serve to mark : DIFFERENTIATE. **—dis·crim'i·nate·ly** *adv.*

**dis·crim·i·nat·ing** (dĭ-skrĭm'ə-nā'tĭng) *adj.* **1.** Capable of recognizing or making fine distinctions : PERCEPTIVE. **2.** Fastidiously selective. **3.** Serving to distinguish : DISTINCTIVE. **4.** Discriminatory : prejudiced. **—dis·crim'i·nat'ing·ly** *adv.*

**dis·crim·i·na·tion** (dĭ-skrĭm'ə-nā'shən) *n.* **1.** The act of discriminating. **2.** The ability or power to discern. **3.** A prejudiced act.

**dis·crim·i·na·tive** (dĭ-skrĭm'ə-nā'tĭv, -nə-tĭv) *adj.* **1.** Making distinctions. **2.** Discriminatory. **—dis·crim'i·na'tive·ly** *adv.*

**dis·crim·i·na·tor** (dĭ-skrĭm'ə-nā'tər) *n.* **1.** One that discriminates. **2.** *Electron.* A device that alters a property of a signal, as frequency or phase, into an amplitude variation.

**dis·crim·i·na·to·ry** (dĭ-skrĭm'ə-nə-tôr'ē, -tōr'ē) *adj.* **1.** Displaying or marked by prejudice : BIASED. **2.** Discriminating. **—dis·crim'i·na·to'ri·ly** *adv.*

**dis·cur·sive** (dĭ-skûr'sĭv) *adj.* [Med. Lat. *discursivus* < Lat. *discursus,* discussion. —see DISCOURSE.] **1.** Covering a wide field of subjects : DIGRESSIVE. **2.** Proceeding to a conclusion through reason. **—dis·cur'sive·ly** *adv.* **—dis·cur'sive·ness** *n.*

**dis·cus** (dĭs'kəs) *n., pl.* **-cus·es.** [Lat. —see DISK.] **1.** A disk, typically wooden with a metal rim and weighing approx. 4½ pounds or 2 kilograms, thrown for distance in athletic competitions. **2.** A field event in which a discus is thrown. **3.** A small, brilliantly colored disk-shaped South American freshwater fish, *Symphysodon discus,* that is popular in home aquariums.

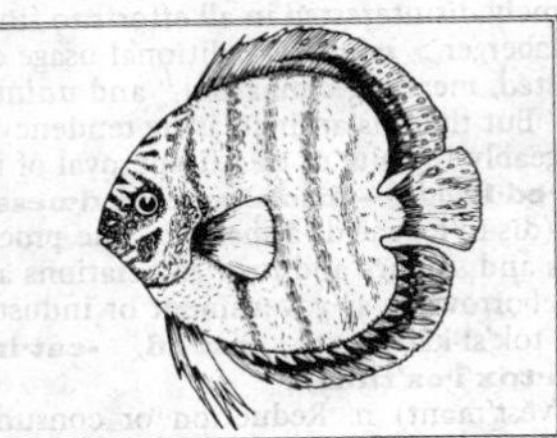

**discus**
*Up to 6 inches long*

**dis·cuss** (dĭ-skŭs') *vt.* **-cussed, -cuss·ing, -cuss·es.** [ME *discussen* < LLat. *discussus,* p.part. of *discutere,* to discuss < Lat., to break up : *dis-,* apart + *quatere,* to shake.] **1.** To speak together about : talk over. **2.** To examine (a subject) in speech or writing. **—dis·cuss'er** *n.* **—dis·cuss'i·ble** *adj.*

**dis·cuss·ant** (dĭ-skŭs'ənt) *n.* A participant in a discussion.

**dis·cus·sion** (dĭ-skŭsh'ən) *n.* **1.** Informal group consideration of a topic. **2.** A formal discourse upon a topic : EXPOSITION.

**dis·dain** (dĭs-dān') *vt.* **-dained, -dain·ing, -dains.** [ME *disdeinen* < OFr. *desdeignier* < Lat. *dedignari* : *de-,* not + *dignari,* to deem worthy < *dignus,* worthy.] **1.** To treat or regard with contempt : DESPISE. **2.** To regard or reject as unworthy of oneself. *—n.* A feeling, attitude, or display of contempt and scornful aloofness.

**dis·dain·ful** (dĭs-dān'fəl) *adj.* Feeling or displaying disdain. **—dis·dain'ful·ly** *adv.* **—dis·dain'ful·ness** *n.*

**dis·ease** (dĭ-zēz') *n.* [ME *disese,* misery < OFr. : *des-,* not (< Lat. *dis-*) + *aise,* ease. —see EASE.] **1.** An abnormal condition of an organism or part that impairs normal physiological functioning, esp. as a result of infection, inherent weakness, or environmental stress. **2.** A condition or tendency, as of society, regarded as abnormal and harmful. **3.** *Obs.* Lack of ease. **—dis·eased'** (dĭ-zēzd') *adj*

**dis·em·bark** (dĭs'ĕm-bärk') *v.* **-barked, -bark·ing, -barks.** *—vt.* To cause to go ashore from a ship. *—vi.* **1.** To go ashore from a ship. **2.** To leave a means of transportation <*disembark* from an airplane> **—dis·em'bar·ka'tion** *n.*

**dis·em·bar·rass** (dĭs'ĕm-băr'əs) *vt.* **-rassed, -rass·ing, -rass·es.** To free from something bothersome or encumbering : RELIEVE. **—dis'em·bar'rass·ment** *n.*

**dis·em·bod·y** (dĭs'ĕm-bŏd'ē) *vt.* **-ied, -y·ing, -ies.** To free (the soul or spirit) from the body. **—dis'em·bod'i·ment** *n.*

**dis·em·bogue** (dĭs'ĕm-bōg') *v.* **-bogued, -bogu·ing, -bogues.** [< Sp. *desembogue,* river mouth < *desembocar,* to flow out : des- (reversal < Lat. *dis-*) + *embocar,* to put into the mouth (*em-,* in + *boca,* mouth < Lat. *bucca,* cheek).] *—vi.* To empty its water at the mouth. —Used of a river. *—vt.* To discharge (waters) at the mouth. —Used of a river. **—dis'em·bogue'ment** *n.*

**dis·em·bow·el** (dĭs'ĕm-bou'əl) *vt.* **-eled, -el·ing, -els** *also* **-elled, -el·ling, -els. 1.** To remove the entrails from. **2.** To deprive of meaning or substance by cutting or altering. **—dis'em·bow'el·ment** *n.*

**dis·en·chant** (dĭs'ĕn-chănt') *vt.* **-chant·ed, -chant·ing, -chants.** To free from false belief : DISILLUSION. **—dis'en·chant'er** *n.* **—dis'en·chant'ment** *n.*

**dis·en·cum·ber** (dĭs'ĕn-kŭm'bər) *vt.* **-bered, -ber·ing, -bers.** To relieve of burdens or hardships. **—dis'en·cum'ber·ment** *n.*

**dis·en·fran·chise** (dĭs'ĕn-frăn'chīz') *vt.* **-chised, -chis·ing, -chis·es.** To disfranchise. **—dis'en·fran'chise'ment** (-chīz'mənt, -chĭz'-) *n.*

**dis·en·gage** (dĭs'ĕn-gāj') *v.* **-gaged, -gag·ing, -gag·es.** *—vt.* **1.** To release from something that holds fast, joins, or entangles. **2.** To release (oneself) from an engagement, pledge, or obligation. *—vi.* To free or detach oneself : WITHDRAW. **—dis'en·gage'ment** *n.*

**dis·en·tail** (dĭs'ĕn-tāl') *vt.* **-tailed, -tail·ing, -tails.** *Law.* To release (an estate) from entail. **—dis'en·tail'ment** *n.*

**dis·en·tan·gle** (dĭs'ĕn-tăng'gəl) *v.* **-gled, -gling, -gles.** *—vt.* **1.** To free from entanglement or involvement : EXTRICATE. **2.** To resolve (e.g., a plot) : UNRAVEL. *—vi.* To become disentangled. **—dis'en·tan'gle·ment** *n.*

**dis·en·tomb** (dĭs'ĕn-tōōm') *vt.* **-tombed, -tomb·ing, -tombs.** To remove from or as if from a tomb.

**dis·en·twine** (dĭs'ĕn-twīn') *vt.* & *vi.* **-twined, -twin·ing, -twines.** To untwine or become untwined.

**dis·e·qui·lib·ri·um** (dĭs-ē'kwə-lĭb'rē-əm, -ĕk'wə-) *n.* Lack or loss of equilibrium or stability.

**dis·es·tab·lish** (dĭs'ĭ-stăb'lĭsh) *vt.* **-lished, -lish·ing, -lish·es. 1.** To alter the status of (something established by authority or general acceptance). **2.** To deprive (a church) of official governmental support. **—dis'es·tab'lish·ment** *n.*

**dis·es·teem** (dĭs'ĭ-stēm') *vt.* **-teemed, -teem·ing, -teems.** To regard with disfavor. *—n.* Lack of esteem.

**dis·fa·vor** (dĭs-fā'vər) *n.* **1.** Unfavorable opinion or regard : DISAPPROVAL. **2.** The condition of being regarded with disapproval. *—vt.* **-vored, -vor·ing, -vors.** To treat or view with disapproval or dislike.

**dis·fea·ture** (dĭs-fē'chər) *vt.* **-tured, -tur·ing, -tures.** To disfigure. **—dis·fea'ture·ment** *n.*

**dis·fig·ure** (dĭs-fĭg'yər) *vt.* **-ured, -ur·ing, -ures.** [ME *disfiguren* < OFr. *desfigurer* < VLat. **disfigurare* : Lat. *dis-,* apart + Lat. *figura,* figure < *fingere,* to form.] To mar or spoil the appearance or shape of. **—dis·fig'u·ra'tion, dis·fig'ure·ment** *n.* **—dis·fig'ur·er** *n.*

**dis·fran·chise** (dĭs-frăn'chīz') *vt.* **-chised, -chis·ing, -chis·es. 1.** To deprive (an individual) of a right of citizenship, esp. of the right to vote. **2.** To deprive (e.g., a corporation) of a franchise or privilege. **—dis·fran'chise'ment** (-chīz'mənt, -chĭz'-) *n.* **—dis·fran'chis'er** *n.*

**dis·frock** (dĭs-frŏk') *vt.* **-frocked, -frock·ing, -frocks.** To unfrock.

**dis·gorge** (dĭs-gôrj') *v.* **-gorged, -gorg·ing, -gorg·es.** *—vt.* **1.** To bring up and expel from the throat or stomach : VOMIT. **2.** To discharge violently : SPEW <The volcano *disgorged* lava.> *—vi.* To pour forth contents. **—dis·gorge'ment** *n.*

**dis·grace** (dĭs-grās') *n.* [Fr. *disgrâce* < Ital. *disgrazia* : *dis-,* not (< Lat. *dis-*) + *grazia,* favor < Lat. *gratia* < *gratus,* pleasing.] **1.** Loss of honor, respect, or reputation : SHAME. **2.** The condition of being regarded with disapproval. **3.** Something that brings disfavor <Your manners are a *disgrace.*> *—vt.* **-graced, -grac·ing, -grac·es. 1.** To bring shame or dishonor upon. **2.** To put (someone) out of grace or favor. **—dis·grac'er** *n.*

☆ ***syns:*** DISGRACE, DISCREDIT, DISHONOR, DISREPUTE, IGNOMINY, OBLOQUY, OPPROBRIUM, SHAME *n. core meaning* : loss of or severe damage to one's reputation <a President who resigned in *disgrace*> ***ant:*** esteem, respect

**dis·grace·ful** (dĭs-grās'fəl) *adj.* Causing or deserving disgrace. **—dis·grace'ful·ly** *adv.* **—dis·grace'ful·ness** *n.*

**dis·grun·tle** (dĭs-grŭn'tl) *vt.* **-tled, -tling, -tles.** [DIS- + dial. *gruntle,* to grumble < ME *gruntlen,* freq. of *grunten,* to grunt.] To make discontented or irritable. **—dis·grun'tle·ment** *n.*

**dis·guise** (dĭs-gīz') *vt.* **-guised, -guis·ing, -guis·es.** [ME *disguisen* < OFr. *disguiser* : *des-* (reversal < Lat. *dis-*) + *guise,* manner, of Germanic orig.] **1. a.** To change the manner or appearance of in order to prevent recognition. **b.** To furnish with a disguise. **2.** To hide or obscure by dissemblance : MISREPRESENT <*disguise* one's real

feelings> —*n.* **1.** The act of disguising or state of being disguised. **2.** Clothes or accessories worn to hide one's true identity. **3.** A pretense that conceals the truth. **—dis·guis′er** *n.*
**dis·gust** (dĭs-gŭst′) *vt.* **-gust·ed, -gust·ing, -gusts.** [OFr. *desgouster* : *des-*, not (< Lat. *dis-*) + *goust,* taste < Lat. *gustus.*] **1.** To arouse nausea or loathing in : SICKEN. **2.** To offend the taste or moral sense of : REPEL. **3.** To cause to become impatient or annoyed <*disgusted* by the long lines at the check-out counter> —*n.* Profound aversion or repugnance provoked by something offensive.
**dis·gust·ed** (dĭs-gŭs′tĭd) *adj.* Filled with disgust or irritated impatience. **—dis·gust′ed·ly** *adv.*
**dis·gust·ful** (dĭs-gŭst′fəl) *adj.* **1.** Provoking disgust : REPUGNANT. **2.** Full of or marked by disgust. **—dis·gust′ful·ly** *adv.*
**dis·gust·ing** (dĭs-gŭs′tĭng) *adj.* Causing disgust : REPELLENT. **—dis·gust′ing·ly** *adv.*
**dish** (dĭsh) *n.* [ME < OE *disc,* plate < Lat. *discus,* quoit. —see DISK.] **1. a.** An open, gen. shallow and concave container for holding or serving food. **b.** The amount that a dish holds. **2. a.** The food served or contained in a dish. **b.** A particular variety or preparation of food. **3. a.** A concavity like that in a dish. **b.** The degree of such a concavity. **4.** A microwave transmitter or receiver consisting of a concave parabolic reflector. **5.** *Informal.* Something one particularly likes or excels in <a game that is just my *dish*> **6.** *Slang.* A good-looking person, esp. a woman. —*vt.* **dished, dish·ing, dish·es. 1.** To serve (food) in or as if in a dish. **2.** To hollow out : make concave. **3.** *Chiefly Brit.* To foil or cheat. **—dish out.** *Informal.* To give out : DISPENSE <*dishing out* homespun philosophy>

▲ word history: The Latin word *discus,* a borrowing of the Greek word *diskos,* meant primarily the "discus," a flat round object of metal or stone used in athletic contests. The word *discus* appears in English in several guises. In Late Latin *discus* also meant "dish," and with that sense *discus* was borrowed into Old English as *disc,* the ancestor of Modern English *dish.* A thousand years later, in the late 17th and 18th centuries, the word *discus* was borrowed into English with the shortened form *disk* or *disc. Disk* was used especially in scientific writings to indicate various flat round objects. During the medieval period Latin *discus* acquired the meaning "table." The Italian descendent of *discus, desco,* preserves this later meaning and is the ancestor of English *desk.*

**dis·ha·bille** (dĭs′ə-bēl′, -bē′) *also* **des·ha·bille** (dĕs′-) *n.* [Fr. *déshabillé* < p.part. of *déshabiller,* to undress : *des-* (reversal < Lat. *dis-*) + *habiller,* to clothe. —see HABILIMENT.] **1. a.** The state of being casually dressed. **b.** A state of undress. **2.** Casual attire.
**dis·har·mo·ny** (dĭs-här′mə-nē) *n.* Lack of harmony : DISCORD. **—dis′har·mo′ni·ous** (-mō′nē-əs) *adj.*
**dish·cloth** (dĭsh′klôth′, -klŏth′) *n.* A cloth for washing dishes.
**dishcloth gourd** *n.* LOOFA 1.
**dis·heart·en** (dĭs-här′tn) *vt.* **-ened, -en·ing, -ens.** To cause to become discouraged or demoralized : DISPIRIT. **—dis·heart′en·ing·ly** *adv.* **—dis·heart′en·ment** *n.*
**dished** (dĭsht) *adj.* Slanting toward one another at the bottom. —Used of a pair of wheels.
**di·shev·el** (dĭ-shĕv′əl) *vt.* **-eled, -el·ing, -els** *also* **-elled, -el·ling, -els.** [Back-formation < DISHEVELED.] To disorder or disarrange (e.g., hair or clothing). **—di·shev′el·ment** *n.*
**di·shev·eled** *also* **di·shev·elled** (dĭ-shĕv′əld) *adj.* [ME *discheveled* < OFr. *deschevele,* p.part. of *descheveler,* to disarrange the hair : *des-,* apart (< Lat. *dis-*) + *chevel,* hair < Lat. *capillus.*] **1.** Hanging in loose disarray, as hair : UNKEMPT. **2.** In a disordered state : UNTIDY.
**dis·hon·est** (dĭs-ŏn′ĭst) *adj.* [ME *dishoneste* < OFr. *deshoneste* : *des-,* not (< Lat. *dis-*) + *honeste,* honest. —see HONEST.] **1.** Tending to lie, cheat, or deceive. **2.** Arising from, gained by, or showing falseness or improbity. **—dis·hon′est·ly** *adv.*
**dis·hon·es·ty** (dĭs-ŏn′ĭ-stē) *n., pl.* **-ties. 1.** Lack of integrity. **2.** A dishonest statement or act.

☆ ***syns:*** DISHONESTY, CORRUPTION, IMPROBITY *n. core meaning* : lack of integrity <cheating—a form of *dishonesty*> ***ant:*** honesty

**dis·hon·or** (dĭs-ŏn′ər) *n.* [ME *dishonour* < OFr. *deshonor* < VLat. **dishonor* : Lat. *dis-,* not + Lat. *honor,* honor.] **1.** Loss of honor, respect, or reputation : DISGRACE. **2.** The condition of being in disgrace. **3.** Something that causes loss of honor. **4.** Failure to pay a commercial obligation, as a note or bill. —*vt.* **-ored, -or·ing, -ors. 1.** To bring shame or disgrace on. **2.** To treat in a disrespectful or demeaning way. **3.** To fail to pay (e.g., a note). **—dis·hon′or·er** *n.*
**dis·hon·or·a·ble** (dĭs-ŏn′ər-ə-bəl) *adj.* **1.** Marked by or causing dishonor or discredit. **2.** Lacking probity. **—dis·hon′or·a·ble·ness** *n.* **—dis·hon′or·a·bly** *adv.*
**dish·pan** (dĭsh′păn′) *n.* A container in which to wash dishes.
**dish·rag** (dĭsh′răg′) *n.* A dishcloth.
**dish·tow·el** (dĭsh′tou′əl) *n.* A towel for drying dishes.
**dish·ware** (dĭsh′wâr′) *n.* Dishes in which food is served.
**dish·wash·er** (dĭsh′wŏsh′ər, -wô′shər) *n.* **1.** One who washes dishes, esp. one hired to wash dishes in a restaurant. **2.** A machine that washes dishes.
**dish·wa·ter** (dĭsh′wô′tər, -wŏt′ər) *n.* Water for washing dishes.
**dish·y** (dĭsh′ē) *adj.* **-i·er, -i·est.** *Chiefly Brit.* Good-looking.
**dis·il·lu·sion** (dĭs′ĭ-lo͞o′zhən) *vt.* **-sioned, -sion·ing, -sions.** To disenchant, esp. to disappoint or embitter by leaving without illusion. —*n.* **1.** The act of disenchanting, esp. to disappoint or embitter by leaving without illusion. **2.** The condition of being disenchanted. **—dis′il·lu′sion·ment** *n.* **—dis′il·lu′sive** (-sĭv, -zĭv) *adj.*
**dis·in·cen·tive** (dĭs′ĭn-sĕn′tĭv) *n.* A deterrent.
**dis·in·cli·na·tion** (dĭs-ĭn′klə-nā′shən) *n.* Lack of willingness or inclination : RELUCTANCE.
**dis·in·cline** (dĭs′ĭn-klīn′) *vt. & vi.* **-clined, -clin·ing, -clines.** To make or be reluctant or unwilling.
**dis·in·clined** (dĭs′ĭn-klīnd′) *adj.* Reluctant <*disinclined* to go>
**dis·in·fect** (dĭs′ĭn-fĕkt′) *vt.* **-fect·ed, -fect·ing, -fects.** To cleanse of harmful microorganisms. **—dis′in·fec′tion** *n.*
**dis·in·fec·tant** (dĭs′ĭn-fĕk′tənt) *n.* An agent that disinfects by inhibiting, neutralizing, or destroying the growth of harmful microorganisms. —*adj.* Serving to disinfect.
**dis·in·fest** (dĭs′ĭn-fĕst′) *vt.* **-fest·ed, -fest·ing, -fests.** To rid of vermin. **—dis·in′fes·ta′tion** (-fĕ-stā′shən) *n.*
**dis·in·fla·tion** (dĭs′ĭn-flā′shən) *n.* Downward movement of inflated prices to a more normal level. **—dis′in·fla′tion·ar′y** (-shə-nĕr′ē) *adj.*
**dis·in·for·ma·tion** (dĭs-ĭn′fər-mā′shən) *n.* Deliberately incorrect and misleading information leaked esp. by an intelligence agency as a means of counteracting and discrediting authentic information that an enemy has obtained.
**dis·in·gen·u·ous** (dĭs′ĭn-jĕn′yo͞o-əs) *adj.* Not straightforward : CRAFTY. **—dis′in·gen′u·ous·ly** *adv.* **—dis′in·gen′u·ous·ness** *n.*
**dis·in·her·it** (dĭs′ĭn-hĕr′ĭt) *vt.* **-it·ed, -it·ing, -its. 1.** To exclude from inheritance or the right of inheriting. **2.** To deprive of a natural or established right or privilege. **—dis′in·her′i·tance** *n.*
**dis·in·te·grate** (dĭs-ĭn′tĭ-grāt′) *v.* **-grat·ed, -grat·ing, -grates.** —*vi.* **1.** To separate into components : FRAGMENT. **2.** To decay or undergo a transformation, as an atomic nucleus. —*vt.* **1.** To cause to separate into components. **2.** To destroy the wholeness or unity of. **—dis·in′te·gra′tive** *adj.* **—dis·in′te·gra′tor** *n.*
**dis·in·te·gra·tion** (dĭs-ĭn′tĭ-grā′shən) *n.* **1.** The process of disintegrating or state of being disintegrated. **2.** *Physics.* The natural or induced transformation of an atomic nucleus from a more massive to a less massive configuration by the emission of radiation, an electron, or a nuclear fragment.
**dis·in·ter** (dĭs′ĭn-tûr′) *vt.* **-terred, -ter·ring, -ters. 1.** To dig up or remove (a corpse) from a grave or tomb : EXHUME. **2.** To disclose or expose. **—dis′in·ter′ment** *n.*
**dis·in·ter·est** (dĭs-ĭn′tər-ĭst, -ĭn′trĭst) *n.* **1.** Freedom from bias or self-interest : IMPARTIALITY. **2.** Lack of interest.

☆ ***syns:*** DISINTEREST, APATHY, INDIFFERENCE, UNCONCERN *n. core meaning* : lack of interest <viewed the film with utter *disinterest*> ***ant:*** interest

**dis·in·ter·est·ed** (dĭs-ĭn′trĭ-stĭd, -ĭn′tə-rĕs′tĭd) *adj.* **1.** Free of bias and self-interest : IMPARTIAL <*disinterested* concern> **2.** Not interested : INDIFFERENT <"supremely *disinterested* in all efforts to find a peaceful solution" —C.L. Sulzberger> ***usage:*** Traditional usage distinguishes between *disinterested,* meaning "impartial," and *uninterested,* meaning "indifferent." But there is an increasing tendency to use the two terms interchangeably in spite of the disapproval of traditionalists. **—dis·in′ter·est·ed·ly** *adv.* **—dis·in′ter·est·ed·ness** *n.*
**dis·in·ter·me·di·a·tion** (dĭs-ĭn′tər-mē′dē-ā′shən) *n.* The process whereby savers by-pass banks and savings and loan associations and lend their money directly to borrowers, as government or industry.
**dis·in·tox·i·cate** (dĭs′ĭn-tŏk′sĭ-kāt′) *vt.* **-cat·ed, -cat·ing, -cates.** DETOXIFY 2. **—dis′in·tox′i·ca′tion** *n.*
**dis·in·vest·ment** (dĭs′ĭn-vĕst′mənt) *n.* Reduction or consumption of capital investment.
**dis·join** (dĭs-join′) *v.* **-joined, -join·ing, -joins.** [ME *disjoinen* < OFr. *desjoindre* < Lat. *disjungere* : *dis-* (reversal) + *jungere,* to join.] —*vt.* To disconnect (one thing) from another : SEPARATE. —*vi.* To become disconnected.
**dis·joint** (dĭs-joint′) *v.* **-joint·ed, -joint·ing, -joints.** [ME *disjointen,* to destroy < OFr. *desjoint,* p.part. of *desjoindre,* to disjoin.] —*vt.* **1.** To put out of joint : DISLOCATE. **2.** To take apart at the joints. **3.** To destroy the coherence or orderly arrangement of. **4.** To disjoin. —*vi.* **1.** To come apart at the joints. **2.** To become dislocated. —*adj. Math.* Having no elements in common.
**dis·joint·ed** (dĭs-join′tĭd) *adj.* **1.** Separated at the joints. **2.** Out of joint : DISLOCATED. **3.** Lacking order or coherence <a *disjointed* lecture> **—dis·joint′ed·ly** *adv.* **—dis·joint′ed·ness** *n.*
**dis·junct** (dĭs-jŭngkt′) *adj.* [ME *disjuncte* < Lat. *disjunctus,* p.part. of *disjungere,* to disjoin.] **1.** Marked by separation. **2.** *Mus.* Relating to progression by intervals larger than major seconds. **3.** *Zool.* Having the head, thorax, and abdomen separated by deep constrictions, as certain insects.
**dis·junc·tion** (dĭs-jŭngk′shən) *n.* **1.** The act of disjoining or state of being disjointed. **2.** *Logic.* A proposition that presents two or more alternative terms, with the assertion that only one is true.
**dis·junc·tive** (dĭs-jŭngk′tĭv) *adj.* **1.** Serving to divide or separate. **2.** Serving to establish a relationship of contrast or opposition; e.g.,

the conjunction *but* in the phrase *strange but true* is disjunctive. **3.** *Logic.* **a.** Of a proposition that presents two or more alternative terms. **b.** Of a syllogism that has a disjunction as one premise. —*n.* A disjunctive conjunction. **—dis·junc'tive·ly** *adv.*

**dis·junc·ture** (dĭs-jŭngk'chər) *n.* Disjunction.

**disk** *also* **disc** (dĭsk) *n.* [Lat. *discus,* quoit < Gk. *diskos* < *dikein,* to throw.] **1.** A thin, flat, circular plate. **2.** Something resembling a disk, as an anatomical structure or an astronomical body. **3.** *Bot.* The enlarged receptacle containing numerous tiny flowers in the flower head of many composite plants, as the daisy and the coneflower. **4.** *var. of* DISC 1. **5.** *Computer Sci.* A round, flat plate coated with a magnetic substance on which data may be stored. **6.** A circular grid in a phototypesetting machine. —*vt.* **disked, disk·ing, disks** *also* **disced, disc·ing, discs.** To work (soil) with a disk harrow.

**disk brake** *n. var. of* DISC BRAKE.

**disk·ette** (dĭ-skĕt') *n.* A floppy disk.

**disk flower** *n.* A segment of tiny tubular flowers forming the center of the flower head of certain composite plants, as the daisy.

**disk harrow** *n.* A harrow with a series of disks set on edge or at an angle on one or more axles.

**disk jockey** *n. var. of* DISC JOCKEY.

**disk pack** *n.* A computer storage device consisting of several magnetic disks that can be used and stored as a unit.

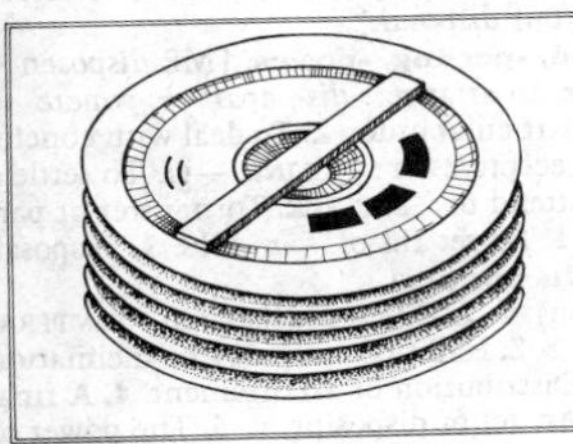

disk pack

**dis·like** (dĭs-līk') *vt.* **-liked, -lik·ing, -likes.** To regard with aversion or distaste. —*n.* An attitude or feeling of aversion or distaste.

**dis·lo·cate** (dĭs'lō-kāt', dĭs-lō'kāt') *vt.* **-cat·ed, -cat·ing, -cates.** [ME *dislocare, dislocat-* : Lat. *dis-* (reversal) + Lat. *locare,* to place < *locus,* place.] **1.** To displace from the proper or usual relationship with adjoining parts. **2.** *Pathol.* To displace (a limb or organ) from the normal position, esp. to displace (a bone) from the socket or joint. **3.** To disorder or disrupt. **—dis'lo·ca'tion** *n.*

**dis·lodge** (dĭs-lŏj') *v.* **-lodged, -lodg·ing, -lodg·es.** [ME *disloggen* < OFr. *deslogier* : *des-* (reversal < Lat. *dis-*) + *logier,* to lodge < *loge,* shed, of Germanic orig.] —*vt.* To remove or force out from a dwelling or position. —*vi.* To move or go from a dwelling or former position. **—dis·lodge'ment, dis·lodg'ment** *n.*

**dis·loy·al** (dĭs-loi'əl) *adj.* Being untrue to duty or obligation : FAITHLESS. **—dis·loy'al·ly** *adv.*

**dis·loy·al·ty** (dĭs-loi'əl-tē) *n., pl.* **-ties. 1.** The quality of being disloyal. **2.** A disloyal act.

**dis·mal** (dĭz'məl) *adj.* [< ME, unlucky days < OFr. *dis mal* < Med. Lat. *dies mali* : Lat. *dies,* day + Lat. *malus,* evil.] **1.** Causing gloom or depression : DREARY <*dismal* rain and fog> **2.** Marked by a lack of hope : DEJECTED. **3.** Causing dread or dismay : DIRE <"We beheld the *dismal* spectacle, the whole city in dreadful flames" —John Evelyn> **4.** Without substance or interest <a tedious, *dismal* novel> **—dis'mal·ly** *adv.* **—dis'mal·ness** *n.*

▲ word history: The word *dismal* was originally a noun phrase, *dis mal,* meaning "evil days," from Latin *dies mali.* In the medieval calendar two days each month were considered unlucky. A *dismal day* was one of these unlucky days. By the 16th century the phrase *dismal day* had become so common that *dismal* was interpreted as an adjective and came to mean "unlucky" and "causing dread" in general. In more recent times these senses have been weakened to "depressing" and "gloomy," and the word does not connote disaster so much as boredom.

**dis·man·tle** (dĭs-măn'tl) *vt.* **-tled, -tling, -tles.** [OFr. *desmanteler* : *des-,* apart (< Lat. *dis-*) + *mantel,* cloak. —see MANTLE.] **1.** To strip of equipment or furnishings. **2. a.** To take apart. **b.** To put an end to in a gradual systematic way. **3.** To divest of clothing or covering. **—dis·man'tle·ment** *n.*

**dis·mast** (dĭs-măst') *vt.* **-mast·ed, -mast·ing, -masts.** *Naut.* To remove or break off the mast of.

**dis·may** (dĭs-mā') *vt.* **-mayed, -may·ing, -mays.** [ME *dismaien* : *dis-,* dis- + OFr. *esmaier,* to frighten, of Germanic orig.] **1.** To fill with apprehension or dread. **2.** To discourage, trouble, or perplex greatly. —*n.* **1.** A sudden or complete loss of courage or confidence. **2. a.** Sudden disappointment. **b.** Anxiety or perplexity. **—dis·may'ing·ly** *adv.*

**dis·mem·ber** (dĭs-mĕm'bər) *vt.* **-bered, -ber·ing, -bers.** [ME *dismembren* < OFr. *desmembrer* < VLat. **dismembrare* : Lat. *dis-,* apart + Lat. *membrum,* limb.] **1.** To cut, tear, or pull off the limbs of. **2.** To separate into pieces. **—dis·mem'ber·ment** *n.*

**dis·miss** (dĭs-mĭs') *vt.* **-missed, -miss·ing, -miss·es.** [ME *dismissen* < Med. Lat. *dismissus,* p.part. of *dismittere,* alteration of Lat. *dimittere* : *dis-,* apart + *mittere,* to send.] **1.** To discharge, as from employment. **2.** To permit or direct to leave <*dismiss* students> **3.** To rid one's mind of : DISPEL <*dismiss* one's doubts> **4.** To refuse to accept or recognize <"unanimous in *dismissing* the claim as highly improbable" —Richard Aldington> **5.** *Law.* To put (a claim or action) out of court without further hearing. **—dis·miss'i·ble** *adj.*

**dis·miss·al** (dĭs-mĭs'əl) *n.* **1.** The act of dismissing or state of being dismissed. **2.** An order or notice of discharge.

**dis·mis·sion** (dĭs-mĭsh'ən) *n.* Dismissal.

**dis·mount** (dĭs-mount') *v.* **-mount·ed, -mount·ing, -mounts.** —*vi.* To get off or down, as from a horse or bicycle : ALIGHT. —*vt.* **1.** To remove from a mounting, setting, or support. **2.** To unseat, as from a horse. **3.** To disassemble (a mechanism). —*n.* (dĭs'mount'). The act or manner of dismounting esp. from a horse. **—dis·mount'a·ble** *adj.*

**dis·o·be·di·ence** (dĭs'ə-bē'dē-əns) *n.* Failure or refusal to obey. **—dis'o·be'di·ent** *adj.* **—dis'o·be'di·ent·ly** *adv.*

**dis·o·bey** (dĭs'ə-bā') *v.* **-beyed, -bey·ing, -beys.** [ME *disobeien* < OFr. *desobeir* < VLat. **disobedire* : Lat. *dis-,* not + Lat. *oboedire,* to obey. —see OBEY.] —*vi.* To refuse or fail to obey an order or rule. —*vt.* To refuse or fail to obey. **—dis'o·bey'er** *n.*

**dis·o·blige** (dĭs'ə-blīj') *vt.* **-bliged, -blig·ing, -blig·es. 1.** To neglect or refuse to act in accord with the wishes of. **2.** To offend, as by slighting. **3.** To inconvenience. **—dis'o·blig'ing·ly** *adv.*

**dis·or·der** (dĭs-ôr'dər) *n.* **1.** Lack of order or regular arrangement : CONFUSION. **2.** A breach of civic order or peace : public disturbance. **3.** An ailment that affects normal, healthy functioning. —*vt.* **-dered, -der·ing, -ders. 1.** To throw into disorder or confusion. **2.** To disturb the normal physical or mental health of : DERANGE.

**dis·or·dered** (dĭs-ôr'dərd) *adj.* **1.** In a condition of disorder : DISARRANGED. **2.** Physically or mentally ill : DERANGED. **—dis·or'dered·ly** *adv.* **—dis·or'dered·ness** *n.*

**dis·or·der·ly** (dĭs-ôr'dər-lē) *adj.* **1.** Lacking regular or logical order or arrangement. **2.** Not disciplined : UNRULY. **3.** Disturbing the public peace or decorum. **—dis·or'der·li·ness** *n.*

☆ ***syns:*** **1.** DISORDERLY, MESSY, UNSYSTEMATIC *adj. core meaning* : lacking regular, logical order <a *disorderly* room full of junk> ***ant:*** orderly **2.** DISORDERLY, RIOTOUS, ROWDY, UNRULY *adj. core meaning* : upsetting civil order or the peace <a *disorderly* mob>

**disorderly conduct** *n. Law.* A petty offense involving a disturbance of public peace or decorum.

**dis·or·gan·ize** (dĭs-ôr'gə-nīz') *vt.* **-ized, -iz·ing, -iz·es.** To destroy the organization, structure, or unity of. **—dis·or'gan·i·za'tion** *n.* **—dis·or'gan·ized** *adj.*

**dis·o·ri·ent** (dĭs-ôr'ē-ĕnt', -ōr'-) *vt.* **-ent·ed, -ent·ing, -ents.** To cause to lose one's sense of direction, position, or relationship with one's surroundings. **—dis·o'ri·en·ta'tion** *n.*

**dis·own** (dĭs-ōn') *vt.* **-owned, -own·ing, -owns.** To refuse to claim or accept as one's own : REPUDIATE.

**dis·par·age** (dĭ-spăr'ĭj) *vt.* **-aged, -ag·ing, -ag·es.** [ME *disparagen,* to degrade < OFr. *desparager* : *des-,* apart (< Lat. *dis-*) + *parage,* rank < *per,* peer. —see PEER².] **1.** To speak of in a belittling way : DECRY. **2.** To reduce in rank or esteem. **—dis·par'age·ment** *n.* **—dis·par'ag·er** *n.* **—dis·par'ag·ing·ly** *adv.*

**dis·pa·rate** (dĭs'pər-ĭt, dĭ-spăr'ĭt) *adj.* [Lat. *disparatus,* p.part. of *disparare,* to separate : *dis-,* apart + *parare,* to prepare.] Completely dissimilar. **—dis'pa·rate·ly** *adv.* **—dis'pa·rate·ness** *n.*

**dis·par·i·ty** (dĭ-spăr'ĭ-tē) *n., pl.* **-ties.** [OFr. *disparite* < LLat. *disparitas* : Lat. *dis-,* not + Lat. *paritas,* equality < *par,* equal.] **1.** The condition or fact of being unequal in age, rank, or degree : DIFFERENCE. **2.** Incongruity or unlikeness.

**dis·pas·sion** (dĭs-păsh'ən) *n.* Freedom from passion, bias, or emotion : OBJECTIVITY.

**dis·pas·sion·ate** (dĭs-păsh'ə-nĭt) *adj.* Devoid of or unaffected by passion, bias, or emotion : OBJECTIVE <a *dispassionate* ruling> **—dis·pas'sion·ate·ly** *adv.* **—dis·pas'sion·ate·ness** *n.*

**dis·patch** *also* **des·patch** (dĭ-spăch') [Sp. *despachar* or Ital. *dispacciare,* both < OFr. *despeechier,* to set free : *des-,* apart (< Lat. *dis-*) + *(em)peechier,* to hinder < LLat. *impedicare,* to entangle (Lat. *in-,* in + Lat. *pedica,* shackle).] —*vt.* **-patched, -patch·ing, -patch·es. 1.** To send off to a particular destination or on specific business. **2.** To finish or deal with promptly. **3.** To put to death summarily. —*n.* **1.** The act of sending off. **2.** The act of putting to death. **3.** Speed in movement or performance : HASTE. **4.** A written message, esp. an official communication, sent with speed. **5.** An item sent to a news organization, as by a correspondent.

**dis·patch·er** (dĭ-spăch'ər) *n.* **1.** One that dispatches. **2.** One who sends out trains, buses, trucks, or taxis according to a schedule. **3.** *Computer Sci.* A routine that controls the order in which input and output devices obtain access to the processing system.

**dis·pel** (dĭ-spĕl′) *v.* **-pelled, -pel·ling, -pels.** [ME *dispellen* < Lat. *dispellere* : *dis-*, apart + *pellere*, to drive.] —*vt.* **1.** To rid one's mind of <*dispel* all fears> **2.** To cause to go in various directions : SCATTER. —*vi.* To disappear by or as if by rising : DISSIPATE.

**dis·pen·sa·ble** (dĭ-spĕn′sə-bəl) *adj.* **1.** Capable of being dispensed with. **2.** That can be administered or distributed. **—dis·pen′sa·bil′i·ty, dis·pen′sa·ble·ness** *n.*

**dis·pen·sa·ry** (dĭ-spĕn′sə-rē) *n., pl.* **-ries. 1.** An office in an institution, as a hospital or school, from which medical supplies and preparations are dispensed. **2.** A public institution that dispenses medicines or medical care.

**dis·pen·sa·tion** (dĭs′pən-sā′shən, -pĕn-) *n.* **1.** The act of dispensing. **2.** Something dispensed. **3.** A specific system or arrangement by which something is dispensed or administered. **4.** An exemption or release from an obligation or rule, granted by or as if by an authority. **5. a.** An exemption from an obligation, as a church law or vow, granted in a particular case by an ecclesiastical authority. **b.** The document containing this exemption. **—dis′pen·sa′tion·al** *adj.*

**dis·pen·sa·to·ry** (dĭ-spĕn′sə-tôr′ē, -tōr′ē) *n., pl.* **-ries. 1.** A book in which the preparation, uses, and contents of medicines are described : PHARMACOPOEIA. **2.** *Archaic.* A dispensary.

**dis·pense** (dĭ-spĕns′) *v.* **-pensed, -pens·ing, -pens·es.** [ME *dispensen* < Med. Lat. *dispensare*, to exempt < Lat., to distribute, freq. of *dispendere*, to weigh out : *dis-*, apart + *pendere*, to weigh.] —*vt.* **1.** To distribute in portions or parts. **2.** To prepare and give out (medicines). **3.** To administer (e.g., laws). **4.** To release or exempt, as from a duty or religious obligation. —*vi.* To grant dispensation or exemption. **—dispense with. 1.** To forgo. **2.** To dispose of.

**dis·pens·er** (dĭ-spĕn′sər) *n.* One that dispenses, esp. a machine or container that allows the contents to be taken out and used in convenient or prescribed amounts <a cup *dispenser*>

**dis·peo·ple** (dĭs-pē′pəl) *vt.* **-pled, -pling, -ples.** To depopulate.

**dis·per·sal** (dĭ-spûr′səl) *n.* The act or process of dispersing or the condition of being dispersed : DISTRIBUTION.

**dis·perse** (dĭ-spûrs′) *v.* **-persed, -pers·ing, -pers·es.** [ME *dispersen* < OFr. *disperser* < Lat. *dispersus*, p.part. of *dispergere*, to disperse : *dis-*, apart + *spargere*, to scatter.] —*vt.* **1.** To break up and scatter in various directions. **2.** To cause to vanish or disappear : DISPEL. **3.** To disseminate (e.g., knowledge). **4.** To separate (light) into spectral rays. —*vi.* **1.** To move or scatter in different directions. **2.** To vanish : disappear <The fog *dispersed* by nine.> **—dis·per′sant** *n.* **—dis·pers′ed·ly** (-spûr′sĭd-lē) *adv.* **—dis·pers′i·ble** *adj.*

**disperse system** *n.* A continuous medium containing dispersed entities of any size or state.

**dis·per·sion** (dĭ-spûr′zhən, -shən) *n.* **1. a.** The act or process of dispersing. **b.** The state of being dispersed. **2.** *Statistics.* The degree of scatter of data, usu. about some mean or median value. **3.** *Physics.* **a.** The separation of a complex wave into component parts according to a given characteristic, as wavelength or frequency. **b.** The separation of visible light into its color components by diffraction or refraction. **4.** *Chem.* A suspension, as smog or homogenized milk, of solid, liquid, or gaseous particles, of colloidal size or larger, in a liquid, solid, or gaseous medium.

**dis·per·sive** (dĭ-spûr′sĭv, -zĭv) *adj.* Tending to disperse or become dispersed. **—dis·per′sive·ly** *adv.* **—dis·per′sive·ness** *n.*

**dis·pir·it** (dĭ-spĭr′ĭt) *vt.* **-it·ed, -it·ing, -its.** [DI(S)- + SPIRIT.] To lower in or deprive of spirit : DISHEARTEN.

**dis·pir·it·ed** (dĭ-spĭr′ĭ-tĭd) *adj.* Marked by low spirits : disheartened or dejected. **—dis·pir′it·ed·ly** *adv.*

**dis·place** (dĭs-plās′) *vt.* **-placed, -plac·ing, -plac·es. 1.** To change the place or position of : REMOVE. **2.** To take the place of : SUPPLANT. **3.** To discharge from an office or position. **4.** To cause a physical displacement of <water *displaced* by a ship> **—dis·place′a·ble** *adj.* **—dis·plac′er** *n.*

**displaced person** *n.* One who has been driven from his or her homeland by war.

**dis·place·ment** (dĭs-plās′mənt) *n.* **1.** The act of displacing or state of being displaced. **2.** *Chem.* A reaction in which one kind of atom, molecule, or radical is removed from combination and replaced by another. **3.** *Physics.* **a.** The weight or volume of a fluid displaced by a floating body, used esp. as a measurement of the weight or bulk of ships. **b.** A vector or the magnitude of a vector from the initial position to a subsequent position assumed by a body. **4.** *Psychoanal.* The shifting of an emotional affect, as of anger, from an appropriate to a more acceptable object.

**displacement ton** *n. Naut.* A unit for measuring the displacement of a ship afloat, equivalent to one long ton or approx. one cubic meter of salt water.

**dis·play** (dĭ-splā′) *vt.* **-played, -play·ing, -plays.** [ME *displayen*, *displaien* < AN *despleier* < Med. Lat. *displicare* < Lat., to scatter : *dis-*, apart + *plicare*, to fold.] **1.** To put forth for viewing : EXHIBIT. **2.** To make noticeable : MANIFEST <*displayed* their know-how> **3.** To exhibit ostentatiously : FLAUNT <*display* one's riches> **4.** To be endowed with an identifiable form or character. **5.** To express, as by gestures or bodily posture. **6.** To spread out : UNFURL. —*n.* **1.** The act of displaying, esp. a public exhibition. **2.** A demonstration or manifestation <a *display* of genius> **3.** Vulgar ostentation <made a *display* of their wealth> **4.** An advertisement designed to catch the eye. **5.** *Computer Sci.* A device that gives information in a visual form, as on a cathode-ray tube.

**dis·please** (dĭs-plēz′) *v.* **-pleased, -pleas·ing, -pleas·es.** [ME *displesen* < OFr. *desplaisir* < VLat. **displacare* : Lat. *dis-* (reversal) + Lat. *placare*, to calm.] —*vt.* To cause annoyance or dissatisfaction to. —*vi.* To cause displeasure. **—dis·pleas′ing·ly** *adv.*

**dis·pleas·ure** (dĭs-plĕzh′ər) *n.* **1.** The condition or fact of being displeased or dissatisfied. **2.** *Archaic.* Discomfort : uneasiness. **3.** *Archaic.* An annoying or injurious offense.

**dis·plode** (dĭ-splōd′) *vt.* & *vi.* **-plod·ed, -plod·ing, -plodes.** [Lat. *displodere*, to extend : *dis-*, apart + *plaudere*, to clap, beat.] *Archaic.* To explode.

**dis·port** (dĭ-spôrt′, -spōrt′) *v.* **-port·ed, -port·ing, -ports.** [ME *disporten* < OFr. *desporter*, to divert : *des*, apart (< Lat. *dis-*) + *porter*, to carry < Lat. *portare*.] —*vi.* To play. —*vt.* To divert or amuse (oneself). —*n.* Diversion : play.

**dis·pos·a·ble** (dĭ-spō′zə-bəl) *adj.* **1.** Designed to be disposed of after use <a *disposable* syringe> **2.** Available for use (e.g., assets). —*n.* Something that can be disposed of after use. **—dis·pos′a·bil′i·ty** *n.*

**dis·pos·al** (dĭ-spō′zəl) *n.* **1.** Particular order, distribution, or placement : ARRANGEMENT. **2.** A particular method of dealing with or settling matters. **3.** The transference of something by sale or gift. **4.** An act of throwing out or away. **5.** An apparatus for disposing of something, as garbage. **6.** The freedom or power to dispose of or use <The swimming pool was at our *disposal*.>

**dis·pose** (dĭ-spōz′) *v.* **-posed, -pos·ing, -pos·es.** [ME *disposen* < OFr. *disposer* < Lat. *disponere*, to arrange : *dis-*, apart + *ponere*, to put.] —*vt.* **1.** To arrange in a particular order. **2.** To deal with conclusively. **3.** To make willing or receptive for : INCLINE. —*vi.* To settle a matter. **—dispose of. 1.** To attend to : SETTLE. **2.** To transfer or part with, as by giving or selling. **3.** To get rid of. —*n. Obs.* **1.** Disposal. **2.** Disposition : demeanor. **—dis·pos′er** *n.*

**dis·po·si·tion** (dĭs′pə-zĭsh′ən) *n.* **1.** One's usual mood : TEMPERAMENT <a grumpy *disposition*> **2.** Habitual tendency or inclination <a *disposition* to argue> **3.** Distribution or arrangement. **4.** A final settlement, as of property. **5.** An act of disposing of. **6.** The power or freedom to dispose, direct, or control.

☆ ***syns:*** DISPOSITION, HUMOR, NATURE, TEMPERAMENT *n. core meaning* : one's usual manner of emotional response <an affectionate *disposition*>

**dis·pos·sess** (dĭs′pə-zĕs′) *vt.* **-sessed, -sess·ing, -sess·es.** To deprive of possession or occupancy of, as land or property. **—dis′pos·ses′sion** *n.* **—dis′pos·ses′sor** *n.* **—dis′pos·ses′so·ry** (-zĕs′ə-rē) *adj.*

**dis·praise** (dĭs-prāz′) *vt.* **-praised, -prais·ing, -prais·es.** [ME *dispreisen* < OFr. *despreiser* < Lat. *depretiare*, to depreciate. —see DEPRECIATE.] To express disapproval of : DISPARAGE. —*n.* Disapproval : censure. **—dis·prais′er** *n.* **—dis·prais′ing·ly** *adv.*

**dis·prize** (dĭs-prīz′) *vt.* **-prized, -priz·ing, -priz·es.** [ME *dispreisen*, to dispraise.] *Archaic.* To disdain.

**dis·proof** (dĭs-pro͞of′) *n.* **1.** The act of disproving or refuting. **2.** Evidence that disproves or refutes.

**dis·pro·por·tion** (dĭs′prə-pôr′shən, -pōr-) *n.* **1.** Absence of due proportion : DISPARITY. **2.** An instance of a disproportionate relation, as in size. —*vt.* **-tioned, -tion·ing, -tions.** To make disproportionate.

**dis·pro·por·tion·al** (dĭs′prə-pôr′shə-nəl, -pōr-) *adj.* Disproportionate. **—dis′pro·por′tion·al·ly** *adv.*

**dis·pro·por·tion·ate** (dĭs′prə-pôr′shə-nĭt, -pōr-) *adj.* Being out of proportion, as in relative size, shape, or amount. **—dis′pro·por′tion·ate·ly** *adv.* **—dis′pro·por′tion·ate·ness** *n.*

**dis·prove** (dĭs-pro͞ov′) *vt.* **-proved, -prov·ing, -proves.** [ME *disproven* < OFr. *desprover* : *des-* (reversal < Lat. *dis-*) + *prover*, to prove. —see PROVE.] To prove to be false or erroneous : REFUTE. **—dis·prov′a·ble** *adj.* **—dis·prov′al** *n.*

☆ ***syns:*** DISPROVE, BELIE, CONFUTE, DISCREDIT, REBUT, REFUTE, *v. core meaning* : to show to be false <evidence that *disproved* previous testimony> ***ant:*** prove

**dis·put·a·ble** (dĭ-spyo͞o′tə-bəl, dĭs′pyə-) *adj.* Capable of being disputed : ARGUABLE. **—dis·put′a·bil′i·ty** *n.* **—dis·put′a·bly** *adv.*

**dis·pu·tant** (dĭ-spyo͞ot′nt, dĭs′pyə-tənt) *adj.* Engaged in dispute or argument. —*n.* One who disputes : DEBATER.

**dis·pu·ta·tion** (dĭs′pyə-tā′shən) *n.* **1.** The act of disputing : argument or debate. **2.** An academic exercise including a formal debate or an oral defense of a thesis.

**dis·pu·ta·tious** (dĭs′pyə-tā′shəs) *adj.* Tending to dispute. **—dis′pu·ta′tious·ly** *adv.* **—dis′pu·ta′tious·ness** *n.*

**dis·pute** (dĭ-spyo͞ot′) *v.* **-put·ed, -put·ing, -putes.** [ME *disputen* < OFr. *desputer* < LLat. *disputare* < Lat., to examine : *dis-*, apart + *putare*, to reckon.] —*vt.* **1.** To argue about : DEBATE. **2.** To question the truth or validity of : DOUBT <*disputed* my claim> **3.** To strive to win (e.g., a prize). **4.** To strive against : RESIST. —*vi.* **1.** To discuss or debate. **2.** To quarrel vehemently. —*n.* **1.** A verbal controversy : DEBATE. **2.** A quarrel. **—dis·put′er** *n.*

**dis·qual·i·fy** (dĭs-kwŏl′ə-fī′) *vt.* **-fied, -fy·ing, -fies. 1.** To render unqualified or unfit. **2.** To declare unqualified or ineligible. **3.** To deprive of legal rights or privileges. **—dis·qual'i·fi·ca·tion** *n.*

**dis·qui·et** (dĭs-kwī'ĭt) *vt.* **-et·ed, -et·ing, -ets.** To deprive of peace or tranquillity : TROUBLE. —*n.* Absence of mental peace or rest : ANXIETY. —*adj. Archaic.* Uneasy : restless. **—dis·qui'et·ing·ly** *adv.* **—dis·qui'et·ly** *adv.* **—dis·qui'et·ness** *n.*

**dis·qui·e·tude** (dĭs-kwī'ĭ-to͞od', -tyo͞od') *n.* A state of uneasiness.

**dis·qui·si·tion** (dĭs'kwĭ-zĭsh'ən) *n.* [Lat. *disquisitio,* investigation < *disquirere,* to investigate : *dis-* (intensive) + *quaerere,* to search for.] A formal, often written discourse on a subject.

**dis·rate** (dĭs-rāt') *vt.* **-rat·ed, -rat·ing, -rates.** To reduce in rating or rank : DEMOTE.

**dis·re·gard** (dĭs'rĭ-gärd') *vt.* **-gard·ed, -gard·ing, -gards. 1.** To pay no attention to : IGNORE. **2.** To treat without proper respect or attentiveness : NEGLECT. —*n.* Lack of thoughtful attention or due regard, esp. when willful. **—dis're·gard'er** *n.* **—dis're·gard'ful** *adj.*

**dis·rel·ish** (dĭs-rĕl'ĭsh) *vt.* **-ished, -ish·ing, -ish·es.** To have distaste for : DISLIKE. —*n.* Aversion or distaste.

**†dis·re·mem·ber** (dĭs'rĭ-mĕm'bər) *v.* **-bered, -ber·ing, -bers.** *Regional.* —*vt.* To fail to remember. —*vi.* To forget.

**dis·re·pair** (dĭs'rĭ-pâr') *n.* The condition of being neglected or in need of repairs.

**dis·rep·u·ta·ble** (dĭs-rĕp'yə-tə-bəl) *adj.* Not respectable. **—dis·rep'u·ta·bil'i·ty, dis·rep'u·ta·ble·ness** *n.* **—dis·rep'u·ta·bly** *adv.*

**dis·re·pute** (dĭs'rĭ-pyo͞ot') *n.* Loss of reputation : DISGRACE.

**dis·re·spect** (dĭs'rĭ-spĕkt') *n.* Lack of respect or courteous regard. —*vt.* **-spect·ed, -spect·ing, -spects.** To show a lack of respect for.

**dis·re·spect·a·ble** (dĭs'rĭ-spĕk'tə-bəl) *adj.* Not deserving of respect. **—dis're·spect'a·bil'i·ty** *n.*

**dis·re·spect·ful** (dĭs'rĭ-spĕkt'fəl) *adj.* Having or exhibiting a lack of respect. **—dis're·spect'ful·ly** *adv.* **—dis're·spect'ful·ness** *n.*

**dis·robe** (dĭs-rōb') *v.* **-robed, -rob·ing, -robes.** —*vt.* To divest of clothing or covering. —*vi.* To undress oneself. **—dis·robe'ment** *n.* **—dis·rob'er** *n.*

**dis·rupt** (dĭs-rŭpt') *vt.* **-rupt·ed, -rupt·ing, -rupts.** [Lat. *disrumpere, disrupt-,* to break apart : *dis-,* apart + *rumpere,* to break.] **1.** To throw into disorder or confusion. **2.** To interrupt or impede the usual course or harmony of. **3.** To break or burst : RUPTURE. **—dis·rupt'er, dis·rup'tor** *n.* **—dis·rup'tion** *n.*

**dis·rup·tive** (dĭs-rŭp'tĭv) *adj.* Tending to disrupt. **—dis·rup'tive·ly** *adv.*

**dis·sat·is·fac·tion** (dĭs-săt'ĭs-făk'shən) *n.* **1.** The condition or feeling of being displeased : DISCONTENT. **2.** A cause of discontent.

**dis·sat·is·fac·to·ry** (dĭs-săt'ĭs-făk'tə-rē) *adj.* Unsatisfactory.

**dis·sat·is·fied** (dĭs-săt'ĭs-fīd') *adj.* Feeling or displaying a lack of satisfaction or contentment. **—dis·sat'is·fied'ly** *adv.*

**dis·sat·is·fy** (dĭs-săt'ĭs-fī') *vt.* **-fied, -fy·ing, -fies.** To disappoint.

**dis·seat** (dĭs-sēt') *vt.* **-seat·ed, -seat·ing, -seats.** *Archaic.* To unseat.

**dis·sect** (dĭ-sĕkt', dī-, dī'sĕkt') *vt.* **-sect·ed, -sect·ing, -sects.** [Lat. *dissecare, dissect-,* to cut apart : *dis-,* apart + *secare,* to cut.] **1.** To cut apart or separate (tissue), esp. surgically or for anatomical study. **2.** To examine, analyze, or criticize in minute detail <*dissected* the theory> **—dis·sec'ti·ble** *adj.* **—dis·sec'tor** *n.*

**dis·sect·ed** (dĭ-sĕk'tĭd, dī-) *adj. Bot.* Divided into numerous narrow segments or lobes, as certain leaves.

**dissected**
*Dissected leaves*

**dis·sec·tion** (dĭ-sĕk'shən, dī-) *n.* **1.** The act of dissecting. **2.** Something dissected, as tissue under study. **3.** A minutely detailed examination, analysis, or critique.

**dis·seise** (dĭs-sēz') *v. var. of* DISSEIZE.

**dis·sei·sin** (dĭs-sē'zĭn) *n. var. of* DISSEIZIN.

**dis·seize** *also* **dis·seise** (dĭs-sēz') *vt.* **-seized, -seiz·ing, -seiz·es** *also* **-seised, -seis·ing, -seis·es.** [ME *disseisen* < AN *disseisir* < OFr. *desseisir* : *des-* (reversal < Lat. *dis-*) + *seisir,* to seize, of Germanic orig.] *Law.* To dispossess unlawfully of real property.

**dis·sei·zin** *also* **dis·sei·sin** (dĭs-sē'zĭn) *n.* [ME *disseisine* < AN < OFr. *desseisine* : *des-* (reversal < Lat. *dis-*) + *seisine,* seisin. —see SEISIN.] *Law.* Unlawful dispossession from real property.

**dis·sem·ble** (dĭ-sĕm'bəl) *v.* **-bled, -bling, -bles.** [ME *dissemblen* < OFr. *dessembler,* to be different : *des-* (reversal < Lat. *dis-*) + *sembler,* to appear, seem. —see SEMBLABLE.] —*vt.* **1.** To hide or disguise the real nature of, as feelings or motives. **2.** To make a pretense of : FEIGN. —*vi.* To hide one's real motives, nature, or feelings under a pretense. **—dis·sem'blance** *n.* **—dis·sem'bler** *n.* **—dis·sem'bling·ly** *adv.*

**dis·sem·i·nate** (dĭ-sĕm'ə-nāt') *v.* **-nat·ed, -nat·ing, -nates.** [Lat. *disseminare, disseminat-* : *dis-,* apart + *seminare,* to sow < *semen,* seed.] —*vt.* **1.** To scatter widely, as in sowing seed. **2.** To spread abroad : PROMULGATE <*disseminate* propaganda> —*vi.* To become spread out : DIFFUSE. **—dis·sem'i·na'tion** *n.* **—dis·sem'i·na'tor** *n.*

**dis·sem·i·nule** (dĭ-sĕm'ə-nyo͞ol') *n.* [DISSEMIN(ATE) + -ULE.] A reproductive plant part, as a seed, fruit, or spore, that is modified for dispersal.

**dis·sen·sion** (dĭ-sĕn'shən) *n.* [ME *dissencioun* < OFr. *dissension* < Lat. *dissensio* < *dissentire,* to dissent.] A difference of opinion, esp. one that leads to argument or strife.

**dis·sent** (dĭ-sĕnt') *vi.* **-sent·ed, -sent·ing, -sents.** [ME *dissenten* < Lat. *dissentire* : *dis-,* apart + *sentire,* to feel.] **1.** To differ in opinion or feeling : DISAGREE. **2.** To withhold approval or assent. —*n.* **1.** Difference of opinion or feeling : DISAGREEMENT. **2.** The refusal to conform to the authority, doctrine, or usages of an established church. **—dis·sent'ing·ly** *adv.*

**dis·sent·er** (dĭ-sĕn'tər) *n.* **1.** One who dissents. **2.** *often* **Dissenter.** One who refuses to accept the authority, doctrine, or usages of an established church, esp. a Protestant who dissents from the Church of England.

**dis·sen·tient** (dĭ-sĕn'shənt) *adj.* Dissenting, esp. from the policies or sentiment of a majority. —*n.* DISSENTER 1. **—dis·sen'tience** *n.*

**dis·sen·tious** (dĭ-sĕn'shəs) *adj.* Given to dissension.

**dis·sep·i·ment** (dĭ-sĕp'ə-mənt) *n.* [Lat. *dissaepimentum,* partition < *dissaepire,* to divide : *dis-,* apart + *saepire,* to enclose < *saepes,* fence.] A membranous or calcareous partition between organs or parts : SEPTUM. **—dis·sep'i·men'tal** (-mĕn'tl) *adj.*

**dis·ser·tate** (dĭs'ər-tāt') *also* **dis·sert** (dĭ-sûrt') *vi.* **-tat·ed, -tat·ing, -tates** *also* **-sert·ed, -sert·ing, -serts.** [Lat. *dissertare, dissertat-,* freq. of *disserere,* to discuss : *dis-,* apart + *serere,* to connect.] To discourse formally. **—dis'ser·ta'tor** *n.*

**dis·ser·ta·tion** (dĭs'ər-tā'shən) *n.* A formal, often lengthy treatise or discourse, esp. one written by a candidate for a doctoral degree.

**dis·serve** (dĭs-sûrv') *vt.* **-served, -serv·ing, -serves.** To mistreat.

**dis·serv·ice** (dĭs-sûr'vĭs) *n.* A harmful action : INJURY.

**dis·sev·er** (dĭ-sĕv'ər) *v.* **-ered, -er·ing, -ers.** [ME *disseveren* < OFr. *dessevrer* < LLat. *disseparare* : Lat. *dis-,* apart + Lat. *separare,* to separate. —see SEPARATE.] —*vt.* **1.** To cut apart : SEPARATE. **2.** To divide into parts. —*vi.* To become separated or disunited. **—dis·sev'er·ance, dis·sev'er·ment** *n.*

**dis·si·dence** (dĭs'ĭ-dəns) *n.* Disagreement : dissent.

**dis·si·dent** (dĭs'ĭ-dənt) *adj.* [Lat. *dissidens, dissident-,* pr.part. of *dissidēre,* to disagree : *dis-,* apart + *sedēre,* to sit.] Disagreeing, as in opinion or belief. —*n.* DISSENTER 1.

**dis·sil·i·ent** (dĭ-sĭl'ē-ənt) *adj.* [Lat. *dissiliens, dissilient-,* pr.part. of *dissilire,* to burst apart : *dis-,* apart + *salire,* to leap.] Bursting apart, as do some ripe seed pods.

**dis·sim·i·lar** (dĭ-sĭm'ə-lər) *adj.* Different. **—dis·sim'i·lar·ly** *adv.*

**dis·sim·i·lar·i·ty** (dĭ-sĭm'ə-lăr'ĭ-tē) *n., pl.* **-ties. 1.** The quality of being unlike : DIFFERENCE. **2.** A point of difference.

**dis·sim·i·late** (dĭ-sĭm'ə-lāt') *vt. & vi.* **-lat·ed, -lat·ing, -lates.** [DIS- + (AS)SIMILATE.] To make or become dissimilar.

**dis·sim·i·la·tion** (dĭ-sĭm'ə-lā'shən) *n.* **1.** The act or process of making or becoming dissimilar. **2.** The process by which one of two similar phonemes is displaced or changed by the other, e.g., the English word *pilgrim* is derived from Latin *pelegrinus,* a dissimilated form of *peregrinus.*

**dis·si·mil·i·tude** (dĭs'ə-mĭl'ĭ-to͞od', -tyo͞od') *n.* [ME < Lat. *dissimilitudo* < *dissimilis,* different : *dis-,* not + *similis,* like.] Lack of resemblance : DISSIMILARITY.

**dis·sim·u·late** (dĭ-sĭm'yə-lāt') *v.* **-lat·ed, -lat·ing, -lates.** [ME *dissimulaten* < Lat. *dissimulare* : *dis-* (reversal) + *simulare,* to simulate.] —*vt.* To conceal (e.g., one's intentions) under a false appearance. —*vi.* To disguise one's true feelings or intentions. **—dis·sim'u·la'tion** *n.* **—dis·sim'u·la'tive** *adj.* **—dis·sim'u·la'tor** *n.*

**dis·si·pate** (dĭs'ə-pāt') *v.* **-pat·ed, -pat·ing, -pates.** [ME *dissipaten* < Lat. *dissipare,* to disperse : *dis-,* apart + *supare,* to throw.] —*vt.* **1.** To drive away or dispel by or as if by dispersing : SCATTER. **2.** To waste or squander <*dissipated* their inheritance> **3.** To cause to lose (e.g., heat) irreversibly. —*vi.* **1.** To vanish by dispersion. **2.** To disappear by or as if by rising. **3.** To indulge in extravagant pursuit of pleasure : CAROUSE. **—dis'si·pat'er, dis'si·pa'tor** *n.* **—dis'si·pa'tive** *adj.*

**dis·si·pat·ed** (dĭs'ə-pā'tĭd) *adj.* **1.** Extravagant in the pursuit of pleasure : DISSOLUTE. **2.** Wasted or squandered <a *dissipated* fortune> **—dis'si·pat'ed·ly** *adv.* **—dis'si·pat'ed·ness** *n.*

**dis·si·pa·tion** (dĭs'ə-pā'shən) *n.* **1.** The act of dissipating or state of being dissipated. **2.** Wasteful expenditure or use. **3.** Dissolute indulgence in pleasure : INTEMPERANCE. **4.** A diversion.

**dis·so·cia·ble** (dĭ-sō′shə-bəl, -shē-ə-bəl) *adj.* Capable of being dissociated : SEPARABLE. **—dis·so′cia·bil′i·ty, dis·so′cia·ble·ness** *n.* **—dis·so′cia·bly** *adv.*

**dis·so·ci·ate** (dĭ-sō′shē-āt′, -sē-) *v.* **-at·ed, -at·ing, -ates.** [Lat. *dissociare, dissociat-* : *dis-* (reversal) + *sociare,* to unite < *socius,* companion.] —*vt.* **1.** To remove from association : SEPARATE <"Marx never *dissociated* man from his social environment" —Sidney Hook> **2.** *Chem.* To cause to undergo dissociation. —*vi.* **1.** To cease associating : PART. **2.** *Chem.* To undergo dissociation. **—dis·so′ci·a′tive** *adj.*

**dis·so·ci·a·tion** (dĭ-sō′sē-ā′shən, -shē-) *n.* **1.** The act of dissociating or state of being dissociated. **2.** *Chem.* **a.** The process by which a change in physical condition, as in temperature or pressure, or the action of a solvent causes a molecule to split into less complex groups of atoms, single atoms, or ions. **b.** The separation of an electrolyte into ions of opposite sign. **3.** *Psychiat.* The separation of a group of related psychological activities into autonomously functioning units, as in the generation of multiple personalities.

**dis·sol·u·ble** (dĭ-sŏl′yə-bəl) *adj.* [Lat. *dissolubilis* < *dissolvere,* to dissolve.] Capable of being dissolved. **—dis′sol·u·bil′i·ty, dis·sol′u·ble·ness** *n.*

**dis·so·lute** (dĭs′ə-lo͞ot′) *adj.* [ME < Lat. *dissolutus,* p.part. of *dissolvere,* to dissolve.] Lacking moral restraint : PROFLIGATE. **—dis′so·lute′ly** *adv.* **—dis′so·lute′ness** *n.*

**dis·so·lu·tion** (dĭs′ə-lo͞o′shən) *n.* **1.** Disintegration into component parts : DECOMPOSITION. **2.** Lack of moral restraint. **3.** Termination or extinction by deconcentration or dispersion. **4.** Death. **5.** Annulment or termination of a formal or legal bond, tie, or contract. **6.** Formal adjournment or dismissal of an assembly or legislature. **7.** Reduction to a liquid form. **—dis′so·lu′tive** *adj.*

**dis·solve** (dĭ-zŏlv′) *v.* **-solved, -solv·ing, -solves.** [ME *dissolven* < Lat. *dissolvere* : *dis-,* apart + *solvere,* to release.] —*vt.* **1.** To cause to pass into solution <*dissolve* instant coffee in water> **2.** To reduce to liquid form : MELT. **3.** To cause to disappear : DISPEL. **4.** To separate into component parts : DISINTEGRATE. **5.** To bring to an end by or as if by breaking up : TERMINATE. **6.** To dismiss (e.g., an assembly or legislature). **7.** To affect emotionally. **8.** To cause to lose definition : BLUR <"Morality has finally been *dissolved* in pity" —Leslie Fiedler> **9.** *Law.* To render null : ABROGATE. —*vi.* **1.** To pass into solution. **2.** To melt. **3.** To disperse or break up. **4.** To become disintegrated. **5.** To be moved emotionally <*dissolved* in tears> **6.** To lose definition or clarity : fade away. **7.** To shift scenes in a motion-picture film or videotape by having one scene fade out while the next appears behind it and grows clearer as the first dims. —*n.* A scene transition in a motion-picture film or videotape made by dissolving. **—dis·solv′a·ble** *adj.* **—dis·solv′er** *n.*

**dis·sol·vent** (dĭ-zŏl′vənt) *n.* A solvent. **—dis·sol′vent** *adj.*

**dis·so·nance** (dĭs′ə-nəns) *also* **dis·so·nan·cy** (-nən-sē) *n.* **1.** A harsh or unpleasant combination of sounds : DISCORD. **2.** Lack of agreement : CONFLICT. **3.** *Mus.* A combination of tones conventionally held to suggest unrelieved tension and to require resolution.

**dis·so·nant** (dĭs′ə-nənt) *adj.* [ME *dissonaunt* < OFr. *dissonant* < Lat. *dissonans,* pr.part. of *dissonare,* to be dissonant : *dis-,* apart + *sonare,* to sound.] **1.** Harsh or unpleasant in sound : DISCORDANT. **2.** Disagreeing : conflicting. **3.** *Mus.* Constituting or producing a dissonance. **—dis′so·nant·ly** *adv.*

**dis·suade** (dĭ-swād′) *vt.* **-suad·ed, -suad·ing, -suades.** [Lat. *dissuadēre* : *dis-* (reversal) + *suadēre,* to advise.] To discourage or deter from a course of action or intention by exhortation or persuasion. **—dis·suad′er** *n.*

**dis·sua·sion** (dĭ-swā′zhən) *n.* [Lat. *dissuasio* < *dissuadēre,* to dissuade.] The act or an instance of dissuading. **—dis·sua′sive** *adj.* **—dis·sua′sive·ly** *adv.* **—dis·sua′sive·ness** *n.*

**dis·syl·la·ble** (dĭ-sĭl′ə-bəl, dĭs′sĭl-, dī′sĭl′-) *n. var. of* DISYLLABLE.

**dis·sym·me·try** (dĭs-sĭm′ĭ-trē) *n., pl.* **-tries.** Lack of symmetry. **—dis′sym·met′ric** (dĭ′sĭ-mĕt′rĭk), **dis′sym·met′ri·cal** *adj.* **—dis′sym·met′ri·cal·ly** *adv.*

**dis·taff** (dĭs′tăf′) *n.* [ME *distaf* < OE *distæf* : *dis-,* bunch of flax + *stæf,* staff.] **1.** A staff having a cleft end that holds the unspun flax, wool, or tow from which thread is drawn in spinning by hand. **2.** A woman's work and domain. **3.** Women as a group.

**distaff side** *n.* The maternal branch or female side of a family.

**dis·tal** (dĭs′təl) *adj.* [DIST(ANT) + -AL.] *Anat.* Located far from the origin or line of attachment, as a bone. **—dis′tal·ly** *adv.*

**dis·tance** (dĭs′təns) *n.* **1.** The fact or condition of being apart in space or time. **2. a.** A nonnegative number designating the magnitude of a path along a straight line or curve. **b.** The length of a line segment joining two points. **c.** The length of the perpendicular from a given point to a given line. **3.** The interval separating two specified instants in time. **4.** The extent of space between points on a linearly measured course. **5. a.** The degree of deviation or difference that separates two things in relationship <the *distance* between liberal and conservative> **b.** The degree of progress between two points in a course or trend. **6.** A stretch of linear space without definite limits. **7.** A point removed in space or time. **8.** Aloofness of manner : RESERVE. —*vt.* **-tanced, -tanc·ing, -tanc·es. 1.** To place or keep at a distance. **2.** To cause to appear at a distance. **3.** To leave far behind : OUTSTRIP.

**dis·tant** (dĭs′tənt) *adj.* [ME *distaunt* < OFr. < Lat. *distans,* pr.part. of *distare,* to be remote : *dis-,* apart + *stare,* to stand.] **1.** Apart or separate in space or time. **2.** Far removed in space or time <the *distant* future> **3.** Located at, coming from, or going to a distance <*distant* travels> **4.** Remotely related <a *distant* cousin> **5.** Of or relating to mental distance or absent-mindedness <a *distant* reverie> **6.** Reserved in manner : ALOOF. **—dis′tant·ly** *adv.*

**dis·taste** (dĭs-tāst′) *n.* Dislike or aversion. —*vt.* **-tast·ed, -tast·ing, -tastes.** *Archaic.* **1.** To feel repugnance for. **2.** To offend.

**dis·taste·ful** (dĭs-tāst′fəl) *adj.* **1. a.** Disagreeable or unpleasant <the *distasteful* job of laying off workers> **b.** Objectionable or offensive <*distasteful* magazines> **2.** Expressing distaste <a *distasteful* glare> **—dis·taste′ful·ly** *adv.* **—dis·taste′ful·ness** *n.*

**dis·tem·per**[1] (dĭs-tĕm′pər) *n.* [ME *distemperen,* to upset the balance of the humors < OFr. *destemprer* < Med. Lat. *distemperare* : Lat. *dis-* (reversal) + Lat. *temperare,* to temper.] **1. a.** An infectious virus disease occurring in certain mammals, esp. dogs, marked by loss of appetite, a catarrhal discharge from the eyes and nose, and often partial paralysis and death. **b.** Any of various similar mammalian diseases. **2.** Bad temper : PEEVISHNESS. **3.** Social or political disorder. —*vt.* **-pered, -per·ing, -pers.** To upset.

**dis·tem·per**[2] (dĭs-tĕm′pər) *n.* [ME *distemperen,* to dilute < Med. Lat. *distemperare.* —see DISTEMPER[1].] **1. a.** A process of painting in which pigments are mixed with water and a glue-size or casein binder, used for flat wall decoration or for scenic and poster painting. **b.** The paint used in distemper. **2.** A painting done in distemper. —*vt.* **-pered, -per·ing, -pers. 1.** To mix (powdered pigments or colors) with water and size. **2.** To paint in distemper.

**dis·tend** (dĭ-stĕnd′) *v.* **-tend·ed, -tend·ing, -tends.** [ME *distenden* < Lat. *distendere* : *dis-,* apart + *tendere,* to stretch.] —*vi.* To swell out or expand from or as if from internal pressure. —*vt.* **1.** To cause to expand by or as if by internal pressure : DILATE. **2.** To stretch out in all directions : EXTEND.

**dis·ten·si·ble** (dĭ-stĕn′sə-bəl) *adj.* Capable of being distended. **—dis·ten′si·bil′i·ty** *n.*

**dis·ten·tion** *also* **dis·ten·sion** (dĭ-stĕn′shən) *n.* [ME *distensioun* < Lat. *distentio* < *distentus,* p.part. of *distendere,* to distend.] The act of distending or state of being distended.

**dis·tich** (dĭs′tĭk) *n., pl.* **-tichs.** [Lat. *distichon* < Gk. *distikhon* < *distikhos,* having two rows or verses : *di-,* two + *stikhos,* line of verse.] A verse couplet, esp. one used in a Latin or Greek elegiac.

**dis·ti·chous** (dĭs′tĭ-kəs) *adj.* [LLat. *distichus,* having two rows < Gk. *distikhos.* —see DISTICH.] Arranged in two vertical rows or ranks on opposite sides of an axis. —Used of leaves. **—dis′ti·chous·ly** *adv.*

**dis·til** (dĭ-stĭl′) *v. Chiefly Brit. var. of* DISTILL.

**dis·till** (dĭ-stĭl′) *v.* **-tilled, -till·ing, -tills.** [ME *distillen* < OFr. *distiller* < Lat. *destillare,* to trickle : *de-,* down + *stillare,* to drip < *stilla,* drop.] —*vt.* **1.** To subject (a substance) to distillation. **2.** To extract (a distillate) by distillation. **3.** To refine or purify by or as if by distillation. **4.** To separate or extract the essence of <*distill* the main ideas of a film> **5.** To exude or give off in drops. —*vi.* **1.** To undergo or be produced by distillation. **2.** To fall or exude in drops. **—dis·till′a·ble** *adj.*

**dis·til·late** (dĭs′tə-lāt′, -lĭt, dĭ-stĭl′ĭt) *n.* **1.** The liquid condensed from vapor in distillation. **2.** An essence or purified form.

**dis·til·la·tion** (dĭs′tə-lā′shən) *n.* **1.** Any of various heat-dependent processes used to purify or separate a fraction of a relatively complex mixture or substance, esp. the vaporization of a liquid mixture with subsequent collection of components by differential cooling to condensation. **2.** A distillate.

**distillation column** *n.* A tall cylindrical metal shell internally fitted with perforated horizontal plates used to promote separation of miscible liquids ascending in the shell as vapor.

**dis·till·er** (dĭ-stĭl′ər) *n.* **1.** One that distills, as a condenser. **2.** A maker of alcoholic liquors by distillation.

**dis·till·er·y** (dĭ-stĭl′ə-rē) *n., pl.* **-ies.** A plant or establishment for distilling, esp. alcoholic liquors.

**dis·tinct** (dĭ-stĭngkt′) *adj.* [ME < OFr. < Lat. *distinctus,* p.part. of *distinguere,* to distinguish.] **1.** Distinguished from all others : INDIVIDUAL <met us on three *distinct* days> ***usage:*** Something is *distinct* if it is sharply distinguished or set apart from other things; a characteristic or property is *distinctive* if it enables us to distinguish one thing from another. **2.** Easily perceived : CLEAR <a *distinct* fragrance> **3.** Unquestionable : decided <a *distinct* drawback> **4.** Very likely : PROBABLE <a *distinct* chance of rain> **5.** Marked by excellence : NOTABLE <a *distinct* achievement> **—dis·tinct′ly** *adv.* **—dis·tinct′ness** *n.*

**dis·tinc·tion** (dĭ-stĭngk′shən) *n.* **1.** The act of distinguishing : DIFFERENTIATION. **2.** The condition or fact of being dissimilar : DIFFERENCE. **3.** A distinguishing factor, attribute, or characteristic. **4. a.** Excellence or eminence, as of performance, character, or reputation. **b.** A special feature or quality conferring superiority : VIRTUE. **5.** Recognition of achievement or superiority : HONOR <served their country with *distinction*>

**dis·tinc·tive** (dĭ-stĭngk′tĭv) *adj.* **1.** Serving to identify or set apart : DISTINGUISHING <*distinctive* cattle brands> **2.** Characteristic or typical <*distinctive* regional cuisine> **3.** Phonemically relevant. **—dis·tinc′tive·ly** *adv.* **—dis·tinc′tive·ness** *n.*

**dis·tin·gué** (dēs′tăng-gā′, dĭs′-, dĭ-stăng′gā) *adj.* [Fr., p.part. of *distinguer*, to distinguish < OFr.] Distinguished in appearance, manner, or demeanor.

**dis·tin·guish** (dĭ-stĭng′gwĭsh) *v.* **-guished, -guish·ing, -guish·es.** [< ME *distinguen* < OFr. *distinguer* < Lat. *distinguere*, to separate.] —*vt.* **1.** To recognize as being different or distinct. **2. a.** To perceive distinctly <*distinguish* a light in the window> **b.** To pick out : DISCERN <*distinguished* my child's voice in the chorus> **3. a.** To separate into different categories. **b.** To make noticeable : SET APART. **c.** To indicate as separate or different. **4.** To cause (oneself) to be eminent or recognized <*distinguish* oneself as a scholar> —*vi.* To perceive or indicate differences : DISCRIMINATE. **—dis·tin′guish·a·ble** *adj.* **—dis·tin′guish·a·bly** *adv.*

**dis·tin·guished** (dĭ-stĭng′gwĭsht) *adj.* **1.** Marked by excellence : EMINENT. **2.** Dignified in appearance or deportment.

**Distinguished Conduct Medal** *n.* A British military decoration awarded for distinguished conduct in the field.

**Distinguished Flying Cross** *n.* **1.** A U.S. military decoration awarded for heroism or extraordinary achievement in aerial combat. **2.** A British decoration awarded to officers of the Royal Air Force for extraordinary achievement.

**Distinguished Service Cross** *n.* **1.** A U.S. Army decoration awarded for exceptional heroism in combat. **2.** A British decoration awarded to officers of the Royal Navy for bravery in action.

**Distinguished Service Medal** *n.* **1.** A U.S. military decoration awarded for distinguished performance of duty. **2.** A British decoration awarded to noncommissioned officers and enlisted personnel in the Royal Navy and Royal Marines for distinguished conduct in war.

**Distinguished Service Order** *n.* A British military decoration awarded for bravery in action.

**dis·tort** (dĭ-stôrt′) *vt.* **-tort·ed, -tort·ing, -torts.** [Lat. *distorquēre*, *distort-* : *dis-*, apart + *torquēre*, to twist.] **1.** To twist out of proper shape or relation : CONTORT. **2.** To give a false or misleading account of : MISREPRESENT. **3.** To cause to function in a twisted or disorderly way : PERVERT. **—dis·tort′er** *n.*

**dis·tor·tion** (dĭ-stôr′shən) *n.* **1.** The act or an instance of distorting. **2.** The condition of being distorted. **3.** A factual misrepresentation. **4.** A distorted image caused by imperfections in an optical system, as a lens. **5.** *Electron.* **a.** An undesired change in the waveform of a signal. **b.** A consequence of such a change, esp. diminished clarity in reproduction or reception. **6.** *Psychoanal.* The modification of unconscious impulses into forms acceptable by conscious or dreaming perception. **—dis·tor′tion·al** *adj.*

**dis·tract** (dĭ-străkt′) *vt.* **-tract·ed, -tract·ing, -tracts.** [ME *distracten* < Lat. *distractus*, p.part. of *distrahere*, to pull away : *dis-*, apart + *trahere*, to draw.] **1.** To cause to turn away from a focus of attention : DIVERT. **2.** To pull in conflicting emotional directions : DISTURB. **—dis·tract′ing·ly** *adv.* **—dis·trac′tive** *adj.*

**dis·tract·ed** (dĭ-străk′tĭd) *adj.* **1.** Having the attention diverted. **2.** Distraught. **—dis·tract′ed·ly** *adv.*

**dis·tract·er** *also* **dis·trac·tor** (dĭ-străk′tər) *n.* **1.** One that distracts. **2.** An incorrect answer presented as a choice in a multiple-choice test.

**dis·trac·tion** (dĭ-străk′shən) *n.* **1.** The act of distracting or state of being distracted. **2.** Something that distracts, esp. an amusement : DIVERSION. **3.** Mental or emotional confusion or disturbance.

**dis·trac·tor** (dĭ-străk′tər) *n.* *var. of* DISTRACTER.

**dis·train** (dĭ-strān′) *v.* **-trained, -train·ing, -trains.** [ME *distreinen* < OFr. *destreindre*, to seize, compel < Med. Lat. *distringere* < Lat., to hinder : *dis-*, apart + *stringere*, to draw tight.] *Law.* —*vt.* **1.** To seize and hold (property) to compel reparation or payment of debts. **2.** To seize the property of in order to compel payment of debts : DISTRESS. —*vi.* To levy a distress. **—dis·train′a·ble** *adj.* **—dis·train′er, dis·trai′nor** *n.* **—dis·train′ment** *n.*

**dis·train·ee** (dĭs′trā-nē′) *n.* *Law.* One who has been distrained.

**dis·traint** (dĭ-strānt′) *n.* [< DISTRAIN.] *Law.* The act or process of distraining : DISTRESS.

**dis·trait** (dĭ-strā′) *adj.* [Fr. < Lat. *distractus.* —see DISTRACT.] Inattentive or absent-minded, esp. due to anxiety.

**dis·traught** (dĭ-strôt′) *adj.* [ME alteration of *distract*, p.part. of *distracten*, to distract.] **1.** Anxious : worried. **2.** Crazed : insane.

**dis·tress** (dĭ-strĕs′) *vt.* **-tressed, -tress·ing, -tress·es.** [ME *distressen* < OFr. *destresser* < *destresse*, constraint < VLat. **districtia* < Lat. *districtus*, p.part. of *distringere*, to hinder. —see DISTRAIN.] **1.** To cause anxiety or suffering to : WORRY. **2.** *Archaic.* To constrain by harassment. **3.** *Law.* To hold the property of against the payment of debts : DISTRAIN. **4.** To antique the appearance of (e.g., furniture). —*n.* **1.** Anxiety or suffering. **2. a.** Severe mental strain resulting from exhaustion or an accident. **b.** Acute physical discomfort. **3.** The condition of being in need of immediate assistance <a swimmer in *distress*> **4.** *Law.* **a.** The act of seizing to compel payment. **b.** The goods thus seized. —*adj.* **1.** Offered for sale at a loss <*distress* merchandise> **2.** Pertaining to distress goods <a *distress* sale> **—dis·tress′ing·ly** *adv.*

☆ ***syns:*** DISTRESS, AFFLICTION, AGONY, ANGUISH, HURT, MISERY, PAIN, WOE *n. core meaning* : a state of physical or mental suffering <felt great *distress* over the death in the family>

**dis·tress·ful** (dĭ-strĕs′fəl) *adj.* **1.** Causing distress. **2.** Experiencing or showing distress. **—dis·tress′ful·ly** *adv.* **—dis·tress′ful·ness** *n.*

**dis·trib·u·tar·y** (dĭ-strĭb′yə-tĕr′ē) *n.*, *pl.* **-ies.** A branch of a river that flows away from the main stream.

**dis·trib·ute** (dĭ-strĭb′yo͞ot) *v.* **-ut·ed, -ut·ing, -utes.** [ME *distributen* < Lat. *distribuere* : *dis-*, apart + *tribuere*, to give. —see TRIBUTE.] —*vt.* **1.** To divide and dispense in portions. **2. a.** To supply (goods) to retailers. **b.** To deliver or give out <*distributing* campaign fliers> **3.** To diffuse or spread over an area. **4.** To divide into categories : CLASSIFY. **5.** *Logic.* To use (a term) so as to include all members of a given class. **6.** To separate (printing type) and replace in the proper boxes. —*vi.* To be mathematically distributive.

**dis·tri·bu·tion** (dĭs′trə-byo͞o′shən) *n.* **1.** The act of distributing or condition of being distributed. **2.** Something distributed. **3.** The act of diffusing or condition of being diffused : DISPERSAL. **4. a.** The geographic occurrence or range of an organism. **b.** The geographic occurrence or range of a custom or usage. **5.** Separation into categories : CLASSIFICATION. **6.** *Law.* The division of an estate or property among rightful heirs. **7.** The process of marketing and supplying goods, esp. to retailers. **8.** An array of objects or events in space or time <the *distribution* of fast-food restaurants on the highway> **9.** *Statistics.* A set of numbers collected from a well-defined universe of possible measurements arising from a property or relationship under study. **—dis′tri·bu′tion·al** *adj.*

**dis·trib·u·tive** (dĭ-strĭb′yə-tĭv) *adj.* **1.** Of or relating to distribution. **2.** Serving to distribute. **3.** Of or relating to the distributive property. **4.** Referring to each individual or entity of a group separately rather than collectively; e.g., *every* in the sentence *Every animal in the zoo was examined.* —*n.* A distributive word or term. **—dis·trib′u·tive·ly** *adv.* **—dis·trib′u·tive·ness** *n.*

**distributive education** *n.* An educational program in which students receive both classroom instruction and on-the-job training.

**dis·trib·u·tor** (dĭ-strĭb′yə-tər) *n.* **1.** One that distributes. **2.** One that markets or sells merchandise, esp. a wholesaler. **3.** A device that applies electric current in proper sequence to the spark plugs of an engine. **4.** *Computer Sci.* Electronic circuitry that acts as an intermediate link between a computer's accumulator and drum storage.

**dis·trict** (dĭs′trĭkt) *n.* [Fr. < Med. Lat. *districtus*, jurisdiction < Lat., p.part. of *distringere*, to hinder. —see DISTRAIN.] **1.** An administrative or political division of a territory. **2.** A distinctive area <the theater *district*> —*vt.* **-trict·ed, -trict·ing, -tricts.** To divide into districts.

**district attorney** *n.* The prosecuting officer of a designated judicial district.

**district court** *n.* **1.** A U.S. Federal trial court serving a judicial district. **2.** A court of general jurisdiction in some states.

**dis·trust** (dĭs-trŭst′) *n.* Lack of trust. —*vt.* **-trust·ed, -trust·ing, -trusts.** To lack confidence in : doubt or suspect.

☆ ***syns:*** DISTRUST, DOUBT, DOUBTFULNESS, MISTRUST, SUSPICION *n. core meaning* : lack of trust <regarded the stranger with *distrust*> ***ant:*** trust

**dis·trust·ful** (dĭs-trŭst′fəl) *adj.* Feeling or exhibiting distrust or doubt. **—dis·trust′ful·ly** *adv.* **—dis·trust′ful·ness** *n.*

**dis·turb** (dĭ-stûrb′) *vt.* **-turbed, -turb·ing, -turbs.** [ME *destourben* < OFr. *destorber* < Lat. *disturbare* : *dis-* (intensive) + *turbare*, to agitate < *turba*, confusion < Gk. *turbē.*] **1.** To trouble or destroy the tranquillity or serenity of. **2.** To trouble emotionally or mentally : UPSET. **3. a.** To interfere with : INTERRUPT <*disturb* one's concentration> **b.** To intrude on : INCONVENIENCE <The incessant chatter *disturbed* my studies.> **4.** To put into disorder : DISARRANGE. **—dis·turb′er** *n.* **—dis·turb′ing·ly** *adv.*

**dis·tur·bance** (dĭ-stûr′bəns) *n.* **1.** The act of disturbing or state of being disturbed. **2.** Something that disturbs. **3.** A commotion or scuffle, esp. a public tumult. **4.** Mental disorder or unbalance. **5.** A variation in a usual course or condition.

**di·sul·fide** (dī-sŭl′fīd′) *n.* A chemical compound having two sulfur atoms combined with other elements or radicals.

**di·sul·fo·ton** (dī-sŭl′fə-tŏn′) *n.* [DI(ETHYL) + SULFO- + T(HI)O-N(ATE).] A highly toxic, pale yellow, organophosphorous systemic insecticide, $C_8H_{19}O_2PS_3$.

**dis·un·ion** (dĭs-yo͞on′yən) *n.* **1.** The state of being disunited : SEPARATION. **2.** Lack of unity : DISSENSION.

**dis·un·ion·ist** (dĭs-yo͞on′yə-nĭst) *n.* One who advocates disunion, esp. a secessionist during the American Civil War.

**dis·u·nite** (dĭs′yo͞o-nīt′) *v.* **-nit·ed, -nit·ing, -nites.** —*vt.* **1.** To separate or sever. **2.** To cause dissension among. —*vi.* To become separate.

**dis·u·ni·ty** (dĭs-yo͞o′nĭ-tē) *n.*, *pl.* **-ties.** Lack of unity : DISCORD.

**dis·use** (dĭs-yo͞os′) *n.* The state of not being used or of cessation of use : DESUETUDE.

**dis·u·til·i·ty** (dĭs'yo͞o-tĭl'ĭ-tē) *n.* **1.** Lack of usefulness. **2.** The quality of causing harm, inconvenience, or tiredness.

**dis·val·ue** (dĭs-văl'yo͞o) *vt.* **-ued, -u·ing, -ues. 1.** To consider as of little or no value. **2.** *Archaic.* To disparage. —*n.* Disesteem.

**di·syl·la·ble** (dī'sĭl'ə-bəl, dī-sĭl'-) *also* **dis·syl·la·ble** (dĭ-sĭl'-, dĭs'-sĭl-, dī'sĭl'-) *n.* A word composed of two syllables. **—di'syl·lab'ic** (dī'sĭ-lăb'ĭk, dĭs'ĭ-) *adj.*

**dit** (dĭt) *n.* [Imit.] The oral representation of the dot in radio and telegraphic code.

**ditch** (dĭch) *n.* [ME *dich* < OE *dīc.*] A long narrow trench or furrow dug in the ground, as for irrigation, drainage, or a boundary line. —*vt.* **ditched, ditch·ing, ditch·es. 1.** To dig or make a ditch in. **2.** To surround with a ditch. **3. a.** To drive (a vehicle) into a ditch. **b.** To derail (a train). **4.** *Slang.* To throw away : DISCARD. **5.** *Slang.* To avoid or escape from. **6.** To bring (a disabled aircraft) down on water.

**dith·er** (dĭth'ər) *n.* [< ME *didderen,* to tremble.] A state of nervous agitation and indecision. —*vi.* **-ered, -er·ing, -ers.** To be nervously irresolute.

**dith·y·ramb** (dĭth'ĭ-răm', -rămb') *n.* [Lat. *dithyrambus* < Gk. *dithurambos.*] **1.** A frenzied and impassioned choric hymn and dance of ancient Greece in honor of Dionysus. **2.** An irregular poetic expression suggestive of the ancient Greek dithyramb. **—dith'y·ramb'ic** *adj.*

**dit·ta·ny** (dĭt'n-ē) *n., pl.* **-nies.** [ME *ditaine* < OFr. *ditain* < Med. Lat. *diptamnus* < Lat. *dictamnus* < Gk. *diktamnon.*] **1.** An aromatic Old World plant, *Origanum dictamnus,* once thought to have magical powers. **2.** The stone mint. **3.** The gas plant.

**dit·to** (dĭt'ō) *n., pl.* **-tos.** [Dial. Ital. p.part. of Ital. *dire,* to say < Lat. *dicere.*] **1.** The same as stated above or before. —Used to avoid repetition and indicated by a pair of small marks (") placed under the original printed material. **2.** A copy or duplicate. —*adv.* As before. —*vt.* **-toed, -to·ing, -tos.** To duplicate or repeat. —*interj.* —Used to express similarity or accord.

**dit·ty** (dĭt'ē) *n., pl.* **-ties.** [ME *dite,* a literary composition < OFr. *dite* < Lat. *dictatum,* p.part. of *dictare,* to dictate, freq. of *dicere,* to say.] A short, simple song.

**ditty bag** *n.* [Orig. unknown.] A bag used by soldiers and sailors to carry small items, as sewing implements or toilet articles.

**di·u·re·sis** (dī'ə-rē'sĭs) *n.* [NLat. < LLat. *diureticus,* diuretic.] Excessive discharge of urine.

**di·u·ret·ic** (dī'ə-rĕt'ĭk) *adj.* [ME *diuretik* < LLat. *diureticus* < Gk. *diouretikos* < *diourein,* to pass urine : *dia-,* through + *ourein,* to urinate < *ouron,* urine.] Tending to increase discharge of urine. —*n.* A diuretic substance or drug. **—di'u·ret'ic·al·ly** *adv.*

**di·ur·nal** (dī-ûr'nəl) *adj.* [ME < LLat. *diurnalis* < Lat. *diurnus* < *dies,* day.] **1.** Relating to or happening in a day or each day : DAILY. **2.** Active or occurring during the daytime rather than at night, as certain insects and animals. **3.** *Bot.* Opening during daylight hours and closing at night. —*n. Archaic.* **1.** A diary. **2.** A daily newspaper. **—di·ur'nal·ly** *adv.*

**di·u·ron** (dĭ'ə-rŏn') *n.* [DI- + UR(EA) + -ON.] A persistent, white, crystalline solid, $C_9H_{10}Cl_2N_2O_3$, used as a pre-emergence herbicide.

**di·va** (dē'və) *n., pl.* **-vas** or **-ve** (-vā) [Ital. < Lat., goddess, fem. of *divus,* god.] An operatic prima donna.

**di·va·gate** (dī'və-gāt', dĭv'ə-) *vi.* **-gat·ed, -gat·ing, -gates.** [LLat. *divagari, divagat-* : Lat. *dis-,* apart + Lat. *vagari,* to wander < *vagus,* wandering.] **1.** To wander or drift about. **2.** To digress from a subject. **—di'va·ga'tion** *n.*

**di·va·lent** (dī-vā'lənt) *adj.* Having a valence of 2 : BIVALENT.

**di·van** (dĭ-văn') *n.* [Fr. < Turk. *dīvān* < Pers., register, office of accounts.] **1.** (*also* dī'văn'). A long backless and armless couch. **2. a.** (*also* dī-văn', dī-văn'). A counting room, tribunal, or public audience room in Moslem countries. **b.** The seat used by an administrator when holding audience. **c.** A government bureau or council chamber. **3.** (*also* dī-văn', dī-văn'). A coffeehouse or smoking lounge with divans. **4.** (*also* dī-văn', dī-văn'). A book of poems by usu. one Middle Eastern author.

**di·var·i·cate** (dī-văr'ĭ-kāt', dĭ-) *vi.* **-cat·ed, -cat·ing, -cates.** [Lat. *divaricare, divaricat-,* to be spread out : *dis-,* apart + *varicare,* to straddle < *varus,* bent.] To diverge at a wide angle : spread apart. —*adj.* (dī-văr'ə-kĭt, -kāt', dĭ-). *Biol.* Spreading or branching widely from a point or axis : DIVERGING. **—di·var'i·cate·ly** *adv.*

**di·var·i·ca·tion** (dī-văr'ĭ-kā'shən, dĭ-) *n.* **1.** The act of divaricating. **2.** A divergence of opinion. **3.** The point at which branching occurs.

**dive**[1] (dīv) *v.* **dived** or **dove** (dōv), **dived, div·ing, dives.** [ME *diven* < OE *dȳfan,* to dip and *dūfan,* to sink.] —*vi.* **1. a.** To plunge headfirst into water, esp. as a sport. **b.** To descend toward the bottom of a body of water : SUBMERGE. **c.** To submerge under power. —Used of a submarine. **d.** To fall head down through the air. **e.** To descend nose down at an acceleration usu. exceeding that of free fall. —Used of an aircraft. **f.** To engage in the sport of skydiving. **g.** To drop sharply and rapidly : PLUMMET. **2. a.** To rush headlong and disappear <*dive* into a crowded street> **b.** To plunge one's hand into something. **3.** To make a lunge <*dove* for the scattered money> **4.** To plunge vigorously into an activity or enterprise. —*vt.* To cause (e.g., an aircraft or a submarine) to dive. ***usage:*** As a past tense, *dove* is actually a more recent form than the historically correct *dived,* but is equally acceptable at all levels. —*n.* **1. a.** A headlong plunge into water, esp. one executed with athletic skill and form. **b.** A nearly vertical descent at an accelerated speed through water or space. **c.** A quick, sharp drop <prices took a *dive*> **2.** *Slang.* A run-down or disreputable bar or nightclub. **3.** *Slang.* A prearranged knockout feigned between prizefighters.

**di·ve**[2] (dē'vā) *n.* [Ital.] *var. pl. of* DIVA.

**dive-bomb** (dīv'bŏm') *vt.* **-bombed, -bomb·ing, -bombs.** To bomb from an airplane at the end of a steep dive toward the target. **—dive'-bomb'er** *n.*

**div·er** (dī'vər) *n.* **1.** One that dives. **2.** One who works under water, esp. one equipped with breathing apparatus. **3.** Any of several diving birds, esp. the loon.

**di·verge** (dĭ-vûrj', dī-) *v.* **-verged, -verg·ing, -verg·es.** [LLat. *divergere,* to turn aside : Lat. *dis-,* apart + Lat. *vergere,* to bend.] —*vi.* **1.** To move or extend in different directions from a common point : branch out. **2.** To differ, as in opinion. **3.** To turn aside from a set course or norm : DEVIATE. **4.** *Math.* To fail to approach a limit. —*vt.* To cause to turn aside : DEFLECT.

**di·ver·gence** (dĭ-vûr'jəns, dī-) *also* **di·ver·gen·cy** (-jən-sē) *n., pl.* **-genc·es** *also* **-gen·cies. 1. a.** The act of diverging or state of being divergent. **b.** Degree of divergence between things. **2.** Departure from a course or convention : DEVIATION. **3.** Difference, as of opinion. **4.** *Math.* **a.** Failure to approach a limit. **b.** The scalar product of the del operator and a vector function. **5.** *Meteorol.* A condition marked by the uniform expansion in volume of a mass of air over a region, usu. accompanied by fine dry weather.

**di·ver·gent** (dĭ-vûr'jənt, dī-) *adj.* **1. a.** Moving apart from a common point : DIVERGING. **b.** Causing divergence of radiation. **2.** Departing from a norm : DEVIANT. **3.** Differing from another <a *divergent* belief> **4.** *Math.* Failing to approach a limit : not convergent. **—di·ver'gent·ly** *adv.*

**di·vers** (dī'vərz) *adj.* [ME. —see DIVERSE.] Various : sundry.

**di·verse** (dĭ-vûrs', dī-, dī'vûrs') *adj.* [ME *divers, diverse* < OFr. *divers* < Lat. *diversus* < p.part. of *divertere,* to divert.] **1.** Unlike in kind : DISTINCT. **2.** Having diversity in form : VARIED. **—di·verse'ly** *adv.* **—di·verse'ness** *n.*

**di·ver·si·form** (dĭ-vûr'sə-fôrm', dī-) *adj.* Variform.

**di·ver·si·fy** (dĭ-vûr'sə-fī', dī-) *v.* **-fied, -fy·ing, -fies.** [ME *diversifien* < OFr. *diversifier* < Med. Lat. *diversificare* : Lat. *diversus,* diverse + Lat. *facere,* to make.] —*vt.* **1. a.** To give variety to : VARY. **b.** To extend (e.g., business activities) into disparate fields. **2.** To distribute (investments) among several companies in order to average the risk of loss. —*vi.* To extend or distribute activities or investments, esp. in business. **—di·ver'si·fi·ca'tion** (-fĭ-kā'shən) *n.*

**di·ver·sion** (dĭ-vûr'zhən, -shən, dī-) *n.* **1.** An act or instance of diverting : DEVIATION. **2.** Something that distracts the mind and relaxes or entertains : amusement or pastime. **3.** A maneuver that draws the attention of an opponent away from the planned point of action, esp. as part of military strategy.

**di·ver·sion·ar·y** (dĭ-vûr'zhən-ər-ē, -shən-, dī-) *adj.* Serving to distract attention, esp. in military strategy.

**di·ver·sion·ist** (dĭ-vûr'zhən-ĭst, -shən-, dī-) *n.* One engaged in diversionary, disruptive, or subversive activities.

**di·ver·si·ty** (dĭ-vûr'sĭ-tē, dī-) *n., pl.* **-ties. 1. a.** The fact or quality of being diverse : DIFFERENCE. **b.** A point or respect in which things differ. **2.** Variety or multiformity <a *diversity* of recreation>

**di·vert** (dĭ-vûrt', dī-) *v.* **-vert·ed, -vert·ing, -verts.** [ME *diverten* < OFr. *divertir* < Lat. *divertere* : *dis-,* aside + *vertere,* to turn.] —*vt.* **1.** To turn aside from a direction or course : DEFLECT. **2.** To distract. **3.** To entertain : amuse. —*vi.* To turn aside. **—di·vert'er** *n.* **—di·vert'ing·ly** *adv.*

**di·ver·tic·u·la** (dī'vûr-tĭk'yə-lə) *n. pl. of* DIVERTICULUM.

**di·ver·tic·u·li·tis** (dī'vûr-tĭk'yə-lī'tĭs) *n.* Inflammation of a diverticulum.

**di·ver·tic·u·lo·sis** (dī'vûr-tĭk'yə-lō'sĭs) *n.* [DIVERTICUL(UM) + -OSIS.] A condition characterized by the presence of numerous diverticula in the colon.

**di·ver·tic·u·lum** (dī'vûr-tĭk'yə-ləm) *n., pl.* **-la** (-lə) [NLat. < Lat. *deverticulum,* by-path < *devertere,* to turn aside : *de-,* away + *vertere,* to turn.] A pouch or sac branching out from a hollow organ or structure, as the intestine. **—di'ver·tic'u·lar** *adj.*

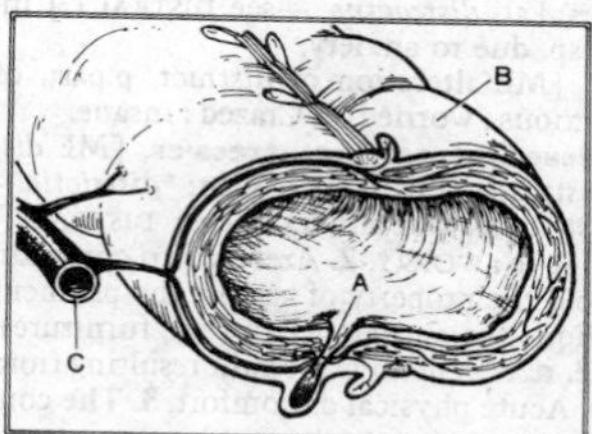


**diverticulum**
*A diagram of an intestinal diverticulum:*
*A. diverticulum, B. large intestine, C. blood vessel*

**di·ver·ti·men·to** (dĭ-vĕr'tə-mĕn'tō) *n., pl.* **-tos** *or* **-ti** (-tē) [Ital., diversion < *divertire,* to divert < OFr. *divertir.*] *Mus.* A chiefly 18th-cent. form of chamber music having several short movements.

**di·ver·tisse·ment** (də-vûr'tĭs-mənt, dē-vĕr-tēs-mäN') *n.* [Fr. < *divertir,* to divert < OFr.] **1.** A short performance, esp. a ballet, given as an interlude in the opera or theater. **2.** *Mus.* A divertimento. **3.** A diversion : amusement.

**Di·ves** (dī'vēz') *n.* [ME < Lat. < *dives,* rich.] A wealthy man.

**di·vest** (dĭ-vĕst', dī-) *vt.* **-vest·ed, -vest·ing, -vests.** [Alteration of DEVEST.] **1.** To strip, as of clothes. **2.** To deprive, as of property or rights : DISPOSSESS. **3.** *Law.* To devest. **4.** To rid (oneself), as of financial holdings.

**di·vide** (dĭ-vīd') *v.* **-vid·ed, -vid·ing, -vides.** [ME *dividen* < Lat. *dividere.*] —*vt.* **1. a.** To separate into parts. **b.** To sector into units of measurement : GRADUATE. **c.** To separate and group according to kind : CLASSIFY. **2. a.** To separate into opposing factions : DISUNITE. **b.** *Chiefly Brit.* To cause (members of Parliament) to vote by separating into groups, as pro and con. **3.** To cause to be separate or cut off from. **4.** To apportion among a number : DISTRIBUTE. **5.** *Math.* **a.** To subject to the process of division. **b.** To be an exact divisor of. —*vi.* **1. a.** To become separated into parts. **b.** To branch out, as a river. **c.** To take sides : form into factions. **d.** *Chiefly Brit.* To vote by dividing. **2.** To perform the mathematical operation of division. —*n.* **1.** A dividing point or line. **2.** A ridge of land : WATERSHED. **—di·vid'a·ble** *adj.*

**di·vid·ed** (dĭ-vī'dĭd) *adj.* **1.** Separated into parts or pieces. **2.** Disunited by disagreement. **3.** Pulled by conflicting interests or activities. **4.** Separated by distance <*divided* from our loved ones> **5.** Having the lanes for opposing traffic separated, as a highway. **6.** *Bot.* Having indentations extending to the midrib or base and forming distinct divisions. —Used of leaves.

**div·i·dend** (dĭv'ĭ-dĕnd') *n.* [Lat. *dividendum,* neuter gerund. of *dividere,* to divide.] **1.** *Math.* A quantity to be divided. **2. a.** A share of profits received by a stockholder or by a policyholder in a mutual insurance society. **b.** A payment pro rata to a creditor of one adjudged bankrupt. **3.** *Informal.* **a.** A share of a surplus. **b.** An unexpected gain, benefit, or advantage.

**di·vid·er** (dĭ-vī'dər) *n.* **1.** One that divides, as a screen or partition. **2. dividers.** A device resembling a compass, used for dividing lines and transferring measurements.

**div·i-div·i** (dĭv'ē-dĭv'ē) *n., pl.* **-is.** [Sp. *dividivi,* perh. of Cariban orig.] **1.** A tropical American tree, *Caesalpina coriaria,* bearing compound leaves and long pods. **2.** The dried pods of the divi-divi, yielding an extract used in tanning leather.

**div·i·na·tion** (dĭv'ə-nā'shən) *n.* **1.** The art or act of foretelling future events or revealing occult knowledge by means of augury or alleged supernatural agency. **2.** A presentiment or inspired guess. **3.** Something that has been divined. **—di·vin'a·to'ry** (dĭ-vĭn'ə-tôr'ē, -tōr'ē) *adj.*

**di·vine**[1] (dĭ-vīn') *adj.* **-vin·er, -vin·est.** [ME < OFr. *devine* < Lat. *divinus* < *divus,* god.] **1. a.** Being or having the nature of a deity. **b.** Of, relating to, emanating from, or being the expression of a deity <*divine* guidance> **c.** Inspired by or devoted to a deity <*divine* worship> **2.** *Informal.* **a.** Godlike : perfect. **b.** Supremely pleasing <a *divine* little house in the country> —*n.* **1.** A member of the clergy. **2.** A theologian. **—di·vine'ly** *adv.* **—di·vine'ness** *n.*

**di·vine**[2] (dĭ-vīn') *v.* **-vined, -vin·ing, -vines.** [ME *divinen* < OFr. *deviner* < Lat. *divinare* < *divinus,* prophet < *divinus,* divine.] —*vt.* **1.** To reveal or foretell through the art of divination. **2. a.** To know by inspiration, intuition, or reflection. **b.** To guess. **3.** To locate (water) with a divining rod. —*vi.* **1.** To practice divination. **2.** To guess. **—di·vin'er** *n.*

**Divine Liturgy** *n.* The Eastern Orthodox Eucharistic rite.

**Divine Office** *n. Rom. Cath. Ch.* The canonical hours.

**divine right** *n.* The doctrine that monarchs derive their right to rule directly from God and are accountable only to Him.

**diving beetle** *n.* Any of various predatory aquatic beetles of the family Dytiscidae.

**diving bell** *n.* A large, open-bottomed vessel for underwater work that is supplied with air under pressure.

**diving board** *n.* A flexible board secured at one end and projecting over water at the other, from which a dive may be executed.

**diving suit** *n.* A heavy waterproof garment with a detachable air-fed helmet, used for underwater work.

**divining rod** *n.* A forked branch or stick that allegedly indicates subterranean water or minerals by bending downward when held over a source.

**di·vin·i·ty** (dĭ-vĭn'ĭ-tē) *n., pl.* **-ties. 1.** The quality or state of being divine. **2. a. the Divinity.** GOD 1a, b. **b.** A deity. **3.** Godlike character. **4.** Theology. **5.** A soft, white fudge usu. having nuts.

**di·vis·i·ble** (dĭ-vĭz'ə-bəl) *adj.* Capable of being divided, esp. of being divided evenly with no remainder. **—di·vis'i·bil'i·ty, di·vis'i·ble·ness** *n.* **—di·vis'i·bly** *adv.*

**di·vi·sion** (dĭ-vĭzh'ən) *n.* [ME *divisioun* < OFr. *division* < Lat. *divisio* < *divisus,* p.part. of *dividere,* to divide.] **1. a.** The act or process of dividing. **b.** The state of being divided. **2.** Proportional distribution : APPORTIONMENT. **3.** Something, as a boundary or partition, that serves to divide. **4.** One of the parts, sections, or groups into which something is divided. **5. a.** An administrative or functional unit of a governmental, educational, or business organization. **b.** A territorial section marked off for political or governmental purposes. **6. a.** A self-contained administrative and tactical military unit capable of independent combat operations. **b.** A group of several ships of similar type forming a tactical unit under a single command in the U.S. Navy. **c.** An air combat group of two or more combat wings and required service units in the U.S. Air Force. **7.** A major taxonomic category corresponding approx. to a phylum, used esp. in botany. **8.** A category created for purposes of competition, as in wrestling or boxing. **9. a.** Difference of opinion : DISAGREEMENT. **b.** A splitting into factions : DISUNITY. **10.** *Chiefly Brit.* The physical separation and regrouping of members of Parliament according to their stand on an issue put to vote. **11.** *Math.* The operation of determining how many times one quantity is contained in another. **12.** A propagation typical of plants that spread by means of newly formed parts, as bulbs, suckers, or rhizomes. **—di·vi'sion·al** *adj.*

**di·vi·sion·ism** (də-vĭzh'ə-nĭz'əm) *n.* A branch of neo-impressionist painting in which colors are divided into their components and arranged so that the eye organizes the shape. **—di·vi'sion·ist** *n.*

**division sign** *n.* **1.** The symbol (÷) placed between two quantities to indicate the division of the first by the second. **2.** The symbol (/ *or* –) placed between two quantities to indicate a fraction.

**di·vi·sive** (dĭ-vī'sĭv) *adj.* Creating dissension or disunity. **—di·vi'sive·ly** *adv.* **—di·vi'sive·ness** *n.*

**di·vi·sor** (dĭ-vī'zər) *n. Math.* The quantity by which another quantity, the dividend, is to be divided.

**di·vorce** (dĭ-vôrs', -vōrs') *n.* [ME < OFr. < Lat. *divortium,* separation < *divertere,* to divert.] **1.** The legal dissolution of a marriage. **2.** A complete or radical separation of closely connected things. —*v.* **-vorced, -vorc·ing, -vorc·es.** —*vt.* **1.** To dissolve the marriage bond between. **2.** To terminate marriage with (one's spouse) by legal divorce. **3.** To remove or separate : DISUNITE. —*vi.* To obtain a legal divorce.

**di·vor·cé** (dĭ-vôr-sā', -sē', -vōr-, -vôr'sā', -sē', -vōr'-) *n.* [Fr., masc. p.part. of *divorcer,* to divorce < OFr.] A divorced man.

**di·vor·cée** (dĭ-vôr-sā', -sē', -vōr-, -vôr'sā', -sē', -vōr'-) *n.* [Fr., fem. p.part. of *divorcer,* to divorce < OFr.] A divorced woman.

**di·vorce·ment** (dĭ-vôrs'mənt, -vōrs'-) *n.* A complete separation.

**div·ot** (dĭv'ət) *n.* [Sc.] **1.** A piece of turf torn up by a golf club in striking a ball. **2.** *Scot.* A thin square of sod or turf used for roofing.

**di·vulge** (dĭ-vŭlj') *vt.* **-vulged, -vulg·ing, -vulg·es.** [ME *divulgen* < Lat. *divulgare,* to publish : *dis-,* among + *vulgus,* common people.] **1.** To reveal or make known (a secret) : DISCLOSE. **2.** *Archaic.* To proclaim publicly. **—di·vul'gence** *n.* **—di·vulg'er** *n.*

**di·vul·sion** (dĭ-vŭl'shən) *n.* [Lat. *divulsio* < *divulsus,* p.part. of *divellere,* to tear apart : *dis-,* apart + *vellere,* to pluck.] An act of tearing apart. **—di·vul'sive** *adj.*

**div·vy** (dĭv'ē) [Shortening and alteration of DIVIDEND.] *Slang.* —*vt.* **-vied, -vy·ing, -vies.** To divide <*divvy* up the goods> —*n., pl.* **-vies.** A portion or share.

**Dix·ie** (dĭk'sē) *n.* [After *Dixie,* a name for the Southern states in the song *Dixie* by Daniel D. Emmett (1815–1904).] The Southern states, esp. those that joined the Confederacy during the Civil War.

**Dix·ie·crat** (dĭk'sē-krăt') *n.* [DIXIE + (DEMO)CRAT.] A member of a dissenting group of Southern Democrats who formed the States' Rights Party in 1948. **—Dix'ie·crat'ic** *adj.*

**Dix·ie·land** (dĭk'sē-lănd') *n.* A style of instrumental jazz associated with New Orleans and marked by a relatively fast two-beat rhythm and by group and solo improvisations.

**di·zen** (dī'zən, dĭz'ən) *vt.* **-zened, -zen·ing, -zens.** [Obs. *disen,* to prepare a distaff with flax for spinning.] *Archaic.* To array in finery : BEDIZEN. **—di'zen·ment** *n.*

**di·zy·got·ic** (dī'zī-gŏt'ĭk) *adj.* [DI- + ZYGOT(E) + -IC.] Derived from two separate and separately fertilized ova. —Used esp. of fraternal twins.

**diz·zy** (dĭz'ē) *adj.* **-zi·er, -zi·est.** [ME *dusie,* foolish < OE *dysig.*] **1.** Experiencing a whirling sensation or feeling a tendency to fall. **2.** Confused or bewildered. **3. a.** Producing or tending to produce giddiness or vertigo <climbed to a *dizzy* height> **b.** Caused by giddiness : REELING. **4.** Marked by impulse and haste : RAPID. **5.** *Informal.* Silly or scatterbrained. —*vt.* **-zied, -zy·ing, -zies. 1.** To make dizzy. **2.** To bewilder or confuse. **—diz'zi·ly** *adv.* **—diz'zi·ness** *n.*

**djel·la·ba** *also* **djel·la·bah** (jə-lä'bə) *n.* [Ar. *jallabah.*] A long hooded garment with full sleeves, worn esp. in Moslem countries.

**djin·ni** *or* **djin·ny** (jĭn'ē, jĭ-nē') *n. vars. of* JINNI.

**D layer** *n.* The lowest area of the ionosphere, existing only during the daytime as a layer in the D region.

**DNA** (de'ĕn-ā') *n.* [D(EOXYRIBO)N(UCLEIC) A(CID).] A polymeric chromosomal constituent of living cell nuclei, having two long chains of alternating phosphate and deoxyribose units twisted into a double helix and joined by hydrogen bonds between the complementary bases adenine and thymine or cytosine and guanine, each of which projects toward the axis of the helix from one of the strands where it is bonded in a sequence that determines individual hereditary characteristics.

**DNase** (dē-ĕn′ās) *also* **DNAase** (dē-ĕn-ā′ās) *n.* [DN(A) + -ASE.] An enzyme that hydrolyzes DNA to its component nucleotides.

**do**[1] (dōō) *v.* **did** (dĭd), **done** (dŭn), **do·ing, does** (dŭz) [ME *don* < OE *dōn.*] —*vt.* **1.** To perform or execute. **2.** To carry out the requirements of : COMPLETE. **3.** To produce, esp. by creative effort. **4.** To bring about : EFFECT <The remedy might *do* some good.> **5.** To bring or put forth : EXERT <*did* our best to help> **6.** To attend to : put in order <have to *do* the bathroom> **7.** To deal with in order to prepare for use <*did* the laundry> **8.** To render or give <*do* honor to the war dead> **9.** To have as an occupation or profession. **10.** To work out by studying <*do* an assignment for school> **11.** To play the role of. **12. a.** To travel (a specified distance) <*do* a mile in six minutes> **b.** To travel at a speed of <*doing* 70 m.p.h.> **13.** To travel about in : TOUR. **14.** To meet the needs of sufficiently <Fifty dollars will *do* me until payday.> **15. a.** To set or style (the hair). **b.** To apply cosmetics to. **16.** To decorate <*did* the den in pine> **17.** *Informal.* To serve a prison term. **18.** To cheat or swindle <*did* them out of their savings> **19.** *Slang.* To partake of : USE <I don't *do* drugs.> —*vi.* **1.** To behave or conduct oneself : ACT. **2.** To act effectively or energetically : STRIVE. **3.** To get along : FARE <not *doing* very well> **4.** To serve the purpose <This car will *do* for another year.> **5.** To be proper or fitting <A poor attitude toward school just won't *do.*> **6.** —Used as a substitute for an antecedent verb <earned as much as you *did*> **7.** —Used after a verb for emphasis <Eat slowly, *do!*> —*aux.* **1.** —Used with a simple infinitive to indicate the tense in questions, negative statements, and inverted phrases <*Do* you hear it?><I *did* not play well.><Little *did* they realize.> **2.** —Used as a means of emphasis <I *do* want you to come.><*Do* be quick!> **—do away with. 1.** To make an end of : ELIMINATE. **2. a.** To destroy. **b.** To kill. **—do by.** To act with respect to : deal with. **—do for.** To take care of or provide for. **—do in.** *Slang.* **1.** To tire completely : EXHAUST. **2.** To kill. **—do up. 1.** To adorn or dress lavishly. **2.** To wrap and tie (a package). —*n., pl.* **do's** *or* **dos. 1.** Commotion : ado. **2.** An entertainment : PARTY. **3.** A statement of what should be done <*do's* and don'ts> **4.** *Chiefly Brit.* A swindle. **5.** *Archaic.* Duty. **—do a number on.** *Slang.* **1.** To confuse or defeat completely, esp. in a deceitful way. **2.** To ridicule. **—do (oneself) proud.** To act or perform in a way that gives cause for pride. **—do (one's) thing.** To do what is personally pleasing.

**do**[2] (dō) *n.* [Ital. —see GAMUT.] *Mus.* The first tone of the diatonic scale in solfeggio.

**do·a·ble** (dōō′ə-bəl) *adj.* Capable of being done.

**do-all** (dōō′ôl′) *n.* One employed to do all kinds of work.

**dob·bin** (dŏb′ĭn) *n.* [< *Dobbin,* nickname for *Robert.*] A horse, esp. a workhorse.

**Do·bell's solution** (dō′bĕlz′) *n.* [After Horace B. *Dobell* (1828–1917).] An aqueous solution of sodium borate, sodium bicarbonate, glycerol, and phenol, used as an antiseptic and astringent for the mucous membranes, esp. of the nose and throat.

**Do·ber·man pin·scher** (dō′bər-mən pĭn′shər) *n.* [G. *Dobermann* (after Ludwig *Dobermann,* 19th-cent. German dog-breeder) + G. *Pinscher,* terrier.] A medium-sized, short-haired dog orig. bred in Germany.

**do·bra** (dō′brə) *n.* [Port., ult. < Lat. *duplus,* double.] —See table at CURRENCY.

**dob·son** (dŏb′sən) *n.* [Prob. < the name *Dobson.*] Hellgrammite.

**dobson fly** *n.* An insect, *Corydalus cornutus,* with four large, many-veined wings and long, pincerlike mandibles.

**do·cent** (dō′sənt, dō-sĕnt′) *n.* [Obs. G. < Lat. *Docens,* pr.part. of *docēre,* to teach.] **1.** A visiting teacher or lecturer at certain universities. **2.** A museum lecturer or tour guide.

**Do·ce·tism** (dō-sē′tĭz′əm, dō′sə-tĭz′əm) *n.* [LLat. *Docetae,* espousers of Docetism < LGk. *Dokētai* < Gk. *dokein,* to seem.] The doctrine, advocated by a sect regarded as heretical in the early Christian Church, that Christ had no human body and only appeared to have died on the cross. **—Do·ce′tist** *n.*

**doc·ile** (dŏs′əl, -sīl′) *adj.* [Lat. *docilis* < *docēre,* to teach.] **1.** Easily instructed : TEACHABLE. **2.** Submissive to training or control : TRACTABLE. **—doc′ile·ly** *adv.* **—do·cil′i·ty** (dŏ-sĭl′ə-tē, dō-) *n.*

**dock**[1] (dŏk) *n.* [MDu. *docke,* prob. < Lat. *ductia,* act of leading < *ducere,* to lead.] **1.** The area of water between two piers or alongside a pier that receives a ship for loading, unloading, or repairs. **2.** A wharf or pier. **3.** *often* **docks.** A group of piers on a protected basin or other waterway serving as a general landing area for watercraft. **4.** A loading platform for trucks or trains. —*v.* **docked, dock·ing, docks.** —*vt.* **1.** To maneuver (a vessel or other vehicle) into or next to a dock. **2.** *Aerospace.* To couple (e.g., two or more spacecraft) in space. —*vi.* To move or come into a dock.

**dock**[2] (dŏk) *n.* [ME *dok.*] **1.** The fleshy part of an animal's tail. **2.** The tail of an animal after it has been clipped or bobbed. —*vt.* **docked, dock·ing, docks. 1.** To clip or bob (e.g., an animal's ears or tail). **2.** To withhold or deprive of a benefit or a part of the wages, esp. as a punitive measure. **3.** To subject (e.g., wages) to a deduction.

**dock**[3] (dŏk) *n.* [Flem. *docke,* cage.] An enclosure where the defendant stands or sits in a criminal court.

**dock**[4] (dŏk) *n.* [ME < OE *docce.*] A weedy plant of the genus *Rumex,* with small greenish or reddish flower clusters.

**dock·age** (dŏk′ĭj) *n.* **1.** A charge for docking privileges. **2.** Facilities for docking vessels. **3.** The docking of ships.

**dock·er**[1] (dŏk′ər) *n.* One that docks something, as the ears or tail of an animal.

**dock·er**[2] (dŏk′ər) *n.* A longshoreman.

**dock·et** (dŏk′ĭt) *n.* [ME *doggett.*] **1.** A summary of the contents of a document : ABSTRACT. **2.** *Law.* **a.** A brief entry of the proceedings in a court of justice. **b.** The book containing such entries. **c.** A calendar of the cases awaiting action in a court. **3.** A list of things to be done : AGENDA. **4.** A label or ticket affixed to a package listing the contents or directions for assembling or operating. —*vt.* **-et·ed, -et·ing, -ets. 1.** To provide with a summary. **2.** To enter in a docket. **3.** To label or ticket (a parcel).

**dock·hand** (dŏk′hănd′) *n.* A longshoreman.

**dock·mack·ie** (dŏk′măk′ē) *n.* [Prob. < Du. < Delaware *dogekumak.*] A shrub, *Viburnum acerifolium* of eastern North America, bearing white flower clusters.

**dock·work·er** (dŏk′wûr′kər) *n.* A longshoreman.

**dock·yard** (dŏk′yärd′) *n.* **1.** An area with facilities for building, repairing, or dry-docking ships. **2.** *Chiefly Brit.* A navy yard.

**doc·tor** (dŏk′tər) *n.* [ME, an expert, authority < OFr. *docteur* < Lat. *doctor,* teacher < *docēre,* to teach.] **1. a.** A person who has earned the highest academic degree awarded by a college or university in a specified discipline. **b.** A person awarded an honorary degree by a college or university. **2.** A person trained in the healing arts and licensed to practice, esp. a physician, surgeon, dentist, or veterinarian. **3.** The title used in addressing a person who holds the degree of doctor. **4.** *Obs.* A learned person : TEACHER. **5.** A contrivance for emergency or special use. **6.** Any of several brightly colored artificial flies used in fly fishing. —*v.* **-tored, -tor·ing, -tors.** *Informal.* —*vt.* **1.** To give medical treatment to. **2.** To repair, esp. in a makeshift way. **3.** To tamper with so as to deceive <*doctored* the coroner's report> **4.** To add ingredients to (food or drink) in order to improve taste or appearance. **5.** To modify or alter for a specific end <*doctor* a puppet show for an adult audience> —*vi.* **1.** To practice medicine. **2.** To receive medical treatment. **—doc′tor·al** *adj.*

**doc·tor·ate** (dŏk′tər-ĭt) *n.* The degree or status of a doctor.

**doc·tri·naire** (dŏk′trə-nâr′) *n.* [Fr. < *doctrine,* doctrine < OFr.] A person inflexibly attached to a practice or theory without regard to its practicality. —*adj.* Of, relating to, or typical of a doctrinaire : DOGMATIC. **—doc′tri·nair′ism** *n.*

**doc·tri·nal** (dŏk′trə-nəl) *adj.* Belonging to, marked by, or concerning doctrine. **—doc′tri·nal·ly** *adv.*

**doc·trine** (dŏk′trĭn) *n.* [ME < OFr. < Lat. *doctrina,* teaching, learning < *doctor,* teacher < *docēre,* to teach.] **1.** Something taught. **2.** A principle or body of principles presented by a specific field, system, or organization for acceptance or belief : DOGMA. **3.** A rule or principle of law, esp. when established by precedent. **4.** A statement of official government policy, esp. in foreign affairs.

**doc·u·dra·ma** (dŏk′yə-drä′mə, -drăm′ə) *n.* [DOCU(MENTARY) + DRAMA.] A television or motion-picture dramatization based on factual events.

**doc·u·ment** (dŏk′yə-mənt) *n.* [ME, precept < OFr. < Lat. *documentum,* lesson < *docēre,* to teach.] **1.** A written or printed paper bearing the original, official, or legal form of something that can be used to provide decisive information or proof. **2.** Something serving as proof or evidence, esp. an object, as a coin, bearing a revealing mark or symbol. —*vt.* **-ment·ed, -ment·ing, -ments. 1.** To provide with a document. **2.** To support (e.g., an assertion or claim) with decisive information or proof. **3.** To support (e.g., statements in a book) with written references or citations : ANNOTATE. **—doc′u·ment′al** *adj.*

**doc·u·ment·al·ist** (dŏk′yə-mĕn′tl-ĭst′) *n.* One who specializes in documentation.

**doc·u·men·ta·ry** (dŏk′yə-mĕn′tə-rē) *adj.* **1.** Consisting of, concerning, or based on documents. **2.** Presenting facts objectively without editorializing or inserting fictional matter, as in a book or motion picture. —*n., pl.* **-ries.** A television or motion-picture presentation of factual political, social, or historical events or circumstances, often consisting of actual news films accompanied by narration.

**doc·u·men·ta·tion** (dŏk′yə-mĕn-tā′shən) *n.* **1. a.** The act or an instance of the provision of documents or supporting references or records. **b.** The documents or references provided. **2.** The gathering, collation, synopsizing, and coding of printed material for future reference. **3.** *Computer Sci.* The orderly presentation, organization, and communication of recorded special knowledge to produce a historical record of changes in variables.

**dod·der**[1] (dŏd′ər) *vi.* **-dered, -der·ing, -ders.** [ME *dadеren.*] **1.** To tremble or shake, as from old age. **2.** To progress in a feeble, unsteady way : TOTTER. **—dod′der·er** *n.*

**dod·der**[2] (dŏd′ər) *n.* [ME *doder* < MLG, yolk of an egg.] A parasitic

vine of the genus *Cuscuta*, bearing slender, twining reddish or yellow stems with a few minute, scalelike leaves and small whitish flowers.
**dod·dered** (dŏd′ərd) *adj.* [Alteration of *dodded*, p.part.of dial. *dod*, to lop off < ME *dodden*.] **1.** Lacking the top branches as a result of age or decay. **2.** Feeble or infirm.
**dod·der·ing** (dŏd′ər-ĭng) *adj.* Feeble or feeble-minded, as from old age.
**do·dec·a·gon** (dō-dĕk′ə-gŏn′) *n.* [Gk. *dōdekagōnon* : *dōdeka*, twelve (*duo*, two + *deka*, ten) + *-gōnon*, -gon.] A polygon with 12 sides. **—do′de·cag′o·nal** (dō′dĕ-kăg′ə-nəl) *adj.*
**do·dec·a·he·dron** (dō′dĕk-ə-hē′drən) *n., pl.* **-drons** or **-dra** (-drə) [Gk. *dōdekaedron* : *dōdeka*, twelve (*duo*, two +*deka*, ten) + *-edron*, -hedron.] A polyhedron with 12 faces. **—do′dec·a·he′dral** *adj.*
**do·dec·a·phon·ic** (dō′dĕk-ə-fŏn′ĭk) *adj.* [Gk. *dōdeka*, twelve (*duo*, two + *deka*, ten) + PHONIC.] Relating to, composed in, or consisting of 12-tone music. **—do·dec′a·phon·ist** (-fə-nĭst, dō′-də-kăf′ə-) *n.* **—do·dec′a·phon′y** (-fō′nē, dō′də-kăf′ə-), **do·dec′a·phon·ism** *n.*
**dodge** (dŏj) *v.* **dodged, dodg·ing, dodg·es.** [Orig. unknown.] *—vt.* **1.** To avoid (e.g., a blow) by moving quickly aside. **2.** To evade esp. by tricky or deceitful means <kept *dodging* the real issue> *—vi.* **1.** To move aside or in a given direction by shifting or twisting suddenly. **2.** To practice trickery or cunning : EQUIVOCATE. **3.** To shade (a section of a photograph) during the printing process to blunt or reduce intensity. *—n.* **1.** An act of dodging. **2.** A clever or evasive plan or device : STRATAGEM.
**†dodg·er** (dŏj′ər) *n.* **1.** One who dodges or evades. **2.** A dishonest or shifty person : TRICKSTER. **3.** A small printed handbill. **4.** *Southern U.S.* A corndodger.
**dodg·y** (dŏj′ē) *adj.* **-i·er, -i·est.** *Chiefly Brit.* Shifty : evasive.
**do·do** (dō′dō) *n., pl.* **-does** or **-dos.** [Port. *doudo* < *doudo*, stupid.] **1.** A large flightless bird, *Raphus cucullatus* of the island of Mauritius in the Indian Ocean, that has been extinct since the late 17th cent. **2.** *Informal.* One who is entirely out-of-date : FOGY. **3.** *Informal.* A stupid person.
**doe** (dō) *n., pl.* **doe** or **does.** [ME *do* < OE *dā.*] **1.** The female of a deer or related animal. **2.** The female of various mammals, as the hare or kangaroo.
**do·er** (do͞o′ər) *n.* **1.** A person who does something, as an agent. **2.** A vigorously active person.
**does** (dŭz) *v. 3rd person sing. present tense of* DO[1].
**doe·skin** (dō′skĭn′) *n.* **1. a.** The skin of a doe, deer, or goat. **b.** Leather made from this skin, used esp. for gloves. **2.** A fine, soft, smooth woolen fabric. **3.** A densely napped finish for certain woolen fabrics, as flannel.
**does·n't** (dŭz′ənt). Does not.
**do·est** (do͞o′ĭst) *v. Archaic. 2nd person sing. present tense of* DO[1].
**do·eth** (do͞o′əth) *v. Archaic. 3rd person sing. present tense of* DO[1].
**doff** (dôf, dŏf) *vt.* **doffed, doff·ing, doffs.** [ME *doffen* < *don off*, to do off.] **1.** To take off or remove <*doff* one's shirt> **2.** To tip or remove (one's hat) in greeting. **3.** To throw out or away : DISCARD.
**dog** (dôg, dŏg) *n.* [ME *dogge* < OE *docga.*] **1.** A domesticated carnivorous mammal, *Canis familiaris*, raised in a wide variety of breeds and held to have orig. derived from several wild species. **2.** Any of various animals of the family Canidae, as the dingo. **3.** A male canine animal, esp. of a domesticated breed or of the fox. **4.** Any of various mammals, as the prairie dog. **5.** *Informal.* A fellow <you witty *dog*> **6.** *Slang.* **a.** An uninteresting or unattractive person. **b.** A hopelessly inferior product or creation. **c.** A contemptible person. **7. dogs.** *Slang.* The feet. **8.** An andiron : firedog. **9.** *Slang.* A hot dog. **10.** Any of various hooked or U-shaped metallic devices used for gripping or holding heavy objects. **11. a.** *Astron.* A sun dog. **b.** A fogbow. *—adv.* Completely <*dog*-weary> *—vt.* **dogged, dog·ging, dogs. 1.** To track or trail persistently <"A stranger then is still *dogging* us" —Arthur Conan Doyle> **2.** To hold or fasten with a mechanical dog. **—go to the dogs.** *Informal.* To go to ruin : DETERIORATE. **—put on the dog.** *Informal.* To make an affected display of luxury, style, or culture.
**dog·bane** (dôg′bān′, dŏg′-) *n.* A plant of the genus *Apocynum*, with bell-shaped white or pink flowers and milky juice.
**dog·ber·ry** (dôg′bĕr′ē, dŏg′-) *n.* **1.** A wild gooseberry, *Ribes cynosbati* of eastern North America, yielding large, prickly berries. **2.** Any of several plants or shrubs with berrylike fruit. **3.** The fruit of any of the dogberries.
**dog biscuit** *n.* **1.** A hard cracker for dogs. **2.** *Slang.* A hard biscuit used as an army field ration.
**dog·cart** (dôg′kärt′, dŏg′-) *n.* **1.** A vehicle drawn by one horse and accommodating two persons seated back to back. **2.** A small cart pulled by dogs.
**dog·catch·er** (dôg′kăch′ər, dŏg′-) *n.* A dog officer.
**dog collar** *n.* **1.** A collar for a dog. **2.** *Slang.* A collar worn by members of the clergy. **3.** CHOKER 2a.
**dog days** *pl.n.* [Transl. of LLat. *dies caniculares*, Dog Star days (so called because the Dog Star (Sirius) rises and sets with the sun during this time).] **1.** The hot, sultry period between mid-July and Sept. **2.** A time of stagnation or lessened activity.
**doge** (dōj) *n.* [Fr. < dial. Ital. < Lat. *dux*, leader < *ducere*, to lead.] The elected chief magistrate of the former republics of Venice and Genoa.
**dog-ear** (dôg′ìr′, dŏg′-) *n.* A turned-down corner of the page of a book. *—vt.* **-eared, -ear·ing, -ears. 1.** To turn down the corner of (a page). **2.** To make shabby or tattered, as from overuse. **—dog′-eared′** *adj.*
**dog-eat-dog** (dôg′ēt-dôg′, dŏg′ēt-dŏg′) *adj.* Ruthlessly acquisitive or competitive <a *dog-eat-dog* business world>
**dog·face** (dôg′fās′, dŏg′-) *n. Slang.* An infantryman in the U.S. Army in World War II.
**dog fennel** *n.* **1.** Any of various strong-smelling plants of the genus *Anthemis*, as the mayweed. **2.** A weedy plant, *Eupatorium capillifolium* of the southeastern United States, bearing divided leaves and long greenish flower clusters.
**dog·fight** (dôg′fīt′, dŏg′-) *n.* **1.** A violent fight between or as if between dogs : BRAWL. **2.** An aerial battle between fighter planes.
**dog·fish** (dôg′fĭsh′, dŏg′-) *n., pl.* **dogfish** or **-fish·es. 1.** Any of various small sharks of Atlantic and Pacific coastal waters, chiefly of the family Squalidae. **2.** The bowfin.
**dog·ged** (dô′gĭd, dŏg′ĭd) *adj.* Stubbornly unyielding : OBSTINATE. **—dog′ged·ly** *adv.* **—dog′ged·ness** *n.*
**dog·ger·el** (dô′gər-əl, dŏg′ər-) *also* **dog·grel** (dôg′rəl, dŏg′-) [ME, poor, worthless < *dogge*, dog.] *—n.* Loose, irregular verse, esp. of an inferior or trivial nature. *—adj.* Relating to or written in doggerel.
**dog·ger·y** (dô′gə-rē, dŏg′ə-) *n., pl.* **-ies. 1.** Ill-tempered behavior. **2.** Rabble or riffraff. **3.** A cheap bar or saloon : DIVE.
**dog·gie** (dô′gē, dŏg′ē) *n. var. of* DOGGY.
**doggie bag** *n. var. of* DOGGY BAG.
**dog·gish** (dô′gĭsh, dŏg′ĭsh) *adj.* **1.** Relating to or like a dog. **2.** Ill-tempered or gruff. **3.** *Informal.* Showily stylish.
**dog·go** (dô′gō, dŏg′ō) *adv.* [< DOG.] *Slang.* Out of sight : in hiding <lying *doggo* until the scandal blows over>
**dog·gone** (dôg′gôn′, -gŏn′, dŏg′-) *v. & interj. & n.* Damn.
**dog·grel** (dôg′rəl, dŏg′-) *n. & adj. var. of* DOGGEREL.
**dog·gy** or **dog·gie** (dô′gē, dŏg′ē) *n., pl.* **-gies.** A dog, esp. a small one. *—adj.* **-gi·er, -gi·est.** Of or like a dog.
**doggy bag** or **doggie bag** *n.* [From the assumption that such food would be given to the customer's dog.] A bag for leftover food that a restaurant patron may take home.
**dog·house** (dôg′hous′, dŏg′-) *n.* A shelter for a dog. **—in the doghouse.** *Slang.* Out of favor.
**†do·gie** *also* **do·gy** (dō′gē) *n., pl.* **-gies.** [Orig. unknown.] *Western U.S.* A stray or motherless calf.
**dog in the manger** *n.* [From a fable in which a dog prevented an ox from eating hay he did not want himself.] One who prevents others from enjoying what he or she has no use or liking for.
**dog·leg** (dôg′lĕg′, dŏg′-) *n.* **1.** Something with a sharp bend, esp. a road that bends or curves abruptly. **2.** A golf hole in which the fairway is sharply angled *—vi.* **-legged, -leg·ging, -legs.** To move along a dogleg course. **—dog′leg′ged** (-lĕg′ĭd, -lĕgd′) *adj.*
**dog·ma** (dôg′mə, dŏg′-) *n., pl.* **-mas** or **-ma·ta** (-mə-tə) [Lat. < Gk., opinion, belief < *dokein*, to seem, think.] **1.** A system of doctrines proclaimed true by a religious sect. **2.** A principle, belief, idea, or opinion, esp. one authoritatively considered to be absolute truth : TENET. **3.** A system of principles or beliefs.
**dog·mat·ic** (dôg-măt′ĭk, dŏg-) *also* **dog·mat·i·cal** (-ĭ-kəl). *adj.* [LLat. *dogmaticus* < Gk. *dogmatikos* < *dogma*, belief. —see DOGMA.] **1.** Relating to or typical of dogma. **2.** Marked by an authoritative, arrogant assertion of unproved or unprovable principles. **—dog·mat′i·cal·ly** *adv.*
**dog·mat·ics** (dôg-măt′ĭks, dŏg-) *n. (sing. in number).* The study of religious dogmas, esp. those of the Christian church.
**dog·ma·tism** (dôg′mə-tĭz′əm, dŏg′-) *n.* Dogmatic assertion of opinion or belief.
**dog·ma·tist** (dôg′mə-tĭst, dŏg′-) *n.* **1.** An arrogantly assertive person. **2.** One who proclaims or sets forth dogma.
**dog·ma·tize** (dôg′mə-tīz′, dŏg′-) *v.* **-tized, -tiz·ing, -tiz·es.** *—vi.* To speak or write dogmatically. *—vt.* To proclaim as dogma. **—dog′ma·ti·za′tion** *n.*
**dog·nap** (dôg′năp′) *vt.* **-naped, -nap·ing, -naps** or **-napped, -nap·ping, -naps.** [DOG + (KID)NAP.] To steal (a dog), esp. for sale to a research laboratory. **—dog′nap′er, dog′nap′per** *n.*
**dog officer** *n.* One whose duty it is to impound stray dogs.
**do-good·er** (do͞o′go͝od′ər) *n. Informal.* An earnest but naive supporter of philanthropic or humanitarian reforms. **—do′-good′ism** *n.*
**dog paddle** *n.* A prone swimming stroke in which the arms paddle and the legs kick up and down.
**dog rose** *n.* A prickly European wild rose, *Rosa canina*, with fragrant pink or white flowers.
**dog's age** *n. Informal.* A long period of time.
**dogs·bod·y** (dôgz′bŏd′ē, dŏgz′-) *n.* [Brit. slang, midshipman.] *Chiefly Brit.* One who does menial work : DRUDGE.
**dog's chance** *n.* A slim possibility, as of success.
**dog sled** *also* **dog sledge** *n.* A sled pulled by one or more dogs.
**dog's life** *n. Informal.* A very unhappy existence.

**dog's mercury** *n.* A creeping, malodorous Old World weed, *Mercurialis perennis,* with small greenish flowers.
**Dog Star** *n.* **1.** Sirius. **2.** Procyon.
**dog tag** *n.* **1.** A metal identification disk attached to a dog's collar. **2.** A military identification tag worn on a chain around one's neck.
**dog·tired** (dôg′tīrd′, dŏg′-) *adj.* Extremely tired.
**dog·tooth** (dôg′to͞oth′, dŏg′-) *n.* **1.** A canine tooth : EYETOOTH. **2.** A medieval architectural ornament having four leaflike projections radiating from a raised center.
**dogtooth violet** *n.* A plant of the genus *Erythronium,* esp. *E. americanum* of North America, bearing leaves with reddish blotches and lilylike yellow flowers.
**†dog·trot** (dôg′trŏt′, dŏg′-) *n.* **1.** A steady trot like that of a dog. **2.** *Regional.* A roofed passage between two parts of a structure.
**dog·watch** (dôg′wŏch′, dŏg′-) *n. Naut.* Either of two short periods of watch duty, from 4 to 6 P.M. or from 6 to 8 P.M.
**dog·wood** (dôg′wo͝od′, dŏg′-) *n.* **1.** A tree, *Cornus florida* of eastern North America, bearing small greenish flowers surrounded by petallike white or pink bracts. **2.** Any of various trees or shrubs of the genus *Cornus.*
**do·gy** (dō′gē) *n. var. of* DOGIE.
**doi·ly** *also* **doy·ly** *or* **doy·ley** (doi′lē) *n., pl.* **-lies** *also* **-leys.** [After *Doily* or *Doyly,* 18th-cent. London draper.] **1.** A small, usu. lace or linen ornamental mat. **2.** A small table napkin.
**do·ing** (do͞o′ĭng) *n.* **1.** The act of performing something <Is it worth the *doing?*> **2. doings.** Events or activities, esp. social activities.
**doit** (doit) *n.* [Du. *duit* < MDu.] **1.** A former Dutch coin worth about 1/4 cent. **2.** A small part : BIT.
**do-it-your·self** (do͞o′ĭt-yər-sĕlf′) *adj. Informal.* Of or designed to be done by an amateur or as a hobby <*do-it-yourself* auto repair> <*do-it-yourself* wine-making> **—do′-it-your·self′er** *n.*
**do·jo** (dō′jō) *n.* [J. : *do,* art + *-jo,* ground.] A school for training in Japanese arts of self-defense, as judo and karate.
**dol** (dŏl) *n.* [< Lat. *dolor,* pain. —see DOLOR.] A unit used to measure pain, or by inference analgesia, based on application of heat to the skin.
**do·lab·ri·form** (dō-lăb′rə-fôrm′) *also* **do·lab·rate** (-rāt′) *adj.* [Lat. *dolabra,* pickax < *dolare,* to hew + -FORM.] *Biol.* Shaped like the head of an ax.
**Dol·by System** (dŏl′bē). A trademark for an electronic device that eliminates noise from recorded sound.
**dol·ce** (dōl′chā′) *adv.* [Ital. < Lat. *dulcis,* sweet.] *Mus.* Sweetly and gently. —Used as a direction. **—dol′ce** *adj.*
**dolce vi·ta** (vē′tä) *n.* [Ital. : *dolce,* sweet + *vita,* life.] A sensual and self-indulgent life.
**dol·drums** (dōl′drəmz′, dôl′-, dŏl′-) *n.* [< Obs. *doldrum,* dullard < *dol,* var. of DULL.] *(sing. in number).* **1. a.** Ocean regions near the equator, marked by calms or light winds. **b.** The calms typical of these areas. **2. a.** A period of recession or inactivity. **b.** A period of apathy or depression.
**dole[1]** (dōl) *n.* [ME *dol,* part, share < OE *dāl.*] **1.** The charitable distribution or dispensing of money, food, or clothing. **2.** A gift or share of money, food, or clothing distributed as charity. **3.** *Chiefly Brit.* The governmental distribution of relief payments to the unemployed. **4.** *Archaic.* One's fate. —*vt.* **doled, dol·ing, doles. 1.** To distribute or dispense as charity. **2.** To distribute in small portions <*dole* out soup> **—on the dole.** Receiving regular relief payments from or as if from the government.
**dole[2]** (dōl) *n.* [ME *dol* < OFr. < LLat. *dolus* < Lat. *dolēre,* to feel pain, grieve.] *Archaic.* Grief : sorrow.
**dole·ful** (dōl′fəl) *adj.* **1.** Filled with or expressing grief <a *doleful* tune> **2.** Causing grief. **—dole′ful·ly** *adv.* **—dole′ful·ness** *n.*
**dol·er·ite** (dŏl′ə-rīt′) *n.* [Fr. *dolérite* < Gk. *doleros,* deceitful < *dolos,* trick (from its easily being mistaken for diorite).] **1.** *Chiefly Brit.* A coarse variety of basalt : DIABASE. **2.** A dark igneous rock with macroscopically indeterminate composition. **—dol′er·it′ic** *adj.*
**dol·i·cho·ce·phal·ic** (dŏl′ĭ-kō-sə-făl′ĭk) *also* **dol·i·cho·ceph·a·lous** (-sĕf′ə-ləs) *adj.* [Gk. *dolikhas,* long + -CEPHALIC.] **1.** Having a relatively long head. **2.** Designating a skull that is longer than it is broad, with a cephalic index of 75.9 or less. **—dol′i·cho·ceph′a·lism** (-sĕf′ə-lĭz′əm), **dol′i·cho·ceph′a·ly** (-sĕf′ə-lē) *n.*
**dol·i·cho·cran·i·al** (dŏl′ə-kō-krā′nē-əl) *also* **dol·i·cho·cran·ic** (-nĭk) *adj.* [Gk. *dolikhos,* long + CRANIAL.] Dolichocephalic. **—dol′i·cho·cran′y** *n.*
**do-lit·tle** (do͞o′lĭt′l) *n.* A lazy, self-indulgent person.
**doll** (dŏl) *n.* [< *Doll,* nickname for *Dorothy.*] **1.** A child's toy representing a human being. **2.** A pretty child. **3.** *Slang.* **a.** An attractive person. **b.** A woman. **4. a.** A sweetheart. **b.** A person regarded with fond familiarity <My roommate is a living *doll.*> —*v.* **dolled, doll·ing, dolls.** *Slang.* —*vi.* To array oneself elegantly, as for a special occasion. —*vt.* To dress (oneself) up elegantly.
**dol·lar** (dŏl′ər) *n.* [LG *daler,* taler < G. *Taler,* short for *Joachimstaler,* after *Joachimstal,* a mining community in Bohemia where the coin was first minted.] —See table at CURRENCY.

▲ word history: The word *dollar* has been used to indicate several different coins. The German form of *dollar* is *taler,* which is short for *Joachimstaler,* a silver coin minted in Joachimstal in the 16th century. The north German and Dutch form of *taler* was *daler,* the form borrowed into English as *dollar.* From the 16th to the 18th century the English used *dollar* to refer to the Spanish coin also known as a *piece of eight* or *peso* that was current in Spain and the Spanish-American colonies. Because the Spanish dollar was also familiar to North American colonists, Thomas Jefferson proposed that the monetary unit of the newly independent United States be called a *dollar,* and his proposal was adopted in 1785.

**dol·lar-a-year** (dŏl′ər-ə-yîr′) *adj.* Designating U.S. Federal employees, as consultants, who receive token payment for patriotic service.
**dollar cost averaging** *n.* The periodic investment in the stock market of a fixed dollar amount regardless of prevailing prices.
**dollar diplomacy** *n.* **1.** A policy aimed at furthering the interests of the United States abroad by encouraging the investment of U.S. capital in foreign countries. **2.** A policy designed to protect a nation's foreign investments.
**dol·lar·fish** (dŏl′ər-fĭsh′) *n., pl.* **dollarfish** *or* **-fish·es.** A fish of the family Carangidae, as the moonfish.
**dollar sign** *n.* The symbol ($) for a dollar when placed before a numeral.
**dol·lop** (dŏl′əp) *n.* [Orig. unknown.] **1.** A large lump or portion, as of sour cream. **2.** A small amount of liquid <a *dollop* of whiskey> **3.** A small amount : BIT.
**dol·ly** (dŏl′ē) *n., pl.* **-lies. 1.** DOLL 1. **2. a.** A low mobile platform on casters, used for moving heavy loads. **b.** Such a platform used by one working underneath a vehicle, as an automobile. **3.** A wheeled apparatus for moving a motion-picture or television camera about a set. **4.** A small locomotive for use esp. in a railroad yard or construction site. **5.** A wooden implement for stirring clothes in a washtub. **6.** A tool for holding one end of a rivet while the opposite end is being hammered to form a head. **7.** A small piece of metal or wood placed on the head of a pile to prevent damage while the pile is being driven. —*vi.* **-lied, -ly·ing, -lies.** To move a camera dolly toward or away from a scene of action.
**Dol·ly Var·den** (dŏl′ē vär′dn) *n.* [After *Dolly Varden,* a character in the novel *Barnaby Rudge* by Charles Dickens (1812–1870).] **1.** A 19th-cent. costume consisting of a dress with a tight bodice and a flowered skirt draped over a colored petticoat. **2.** A colorfully spotted trout, *Salvelinus malma* of northwestern North America.
**dol·man** (dōl′mən) *n.* [Fr. < G. < Turk. *dolaman,* robe < *dolamak,* to wind.] **1.** A long Turkish outer robe. **2.** A woman's cloak or coat made with dolman sleeves. **3.** A decorated jacket often worn like a cape as part of a hussar's uniform.
**dolman sleeve** *n.* A sleeve that is very wide at the armhole and narrow at the wrist.
**dol·men** (dōl′mən) *n.* [Fr. : Breton *tol,* table (< Lat. *tabula*) + Breton *men,* stone.] A prehistoric megalithic structure consisting of two or more upright stones supporting a horizontal stone.
**dol·o·mite** (dŏl′ə-mīt′) *n.* [Fr., after Déodat de *Dolomieu* (1705–1801).] **1.** A usu. gray, pink, or white mineral, essentially $CaMg(CO_3)_2$, used as a construction and ceramic material, a furnace refractory, and in fertilizer. **2.** A magnesia-rich sedimentary rock resembling limestone. **—dol′o·mit′ic** (-mĭt′ĭk) *adj.* **—dol′o·mit′i·za′tion** (-mĭt′ĭ-zā′shən) *n.* **—dol′o·mit·ize′** (-mĭt-īz′) *v.* **(-ized, -iz·ing, -iz·es).**
**do·lor** (dō′lər) *n.* [ME *dolour* < OFr. < Lat. *dolor,* pain < *dolēre,* to feel pain.] *Archaic.* Sorrow : grief.
**do·lo·ro·so** (dō′lə-rō′sō) *adv.* [Ital. < Lat. *dolorosus,* dolorous.] *Mus.* With a mournful or plaintive tempo or quality. —Used as a direction. **—do′lo·ro′so** *adj.*
**do·lor·ous** (dō′lə-rəs, dŏl′-) *adj.* [ME < Lat. *dolorus* < *dolor,* dolor.] Marked by or expressive of sorrow or pain. **—do′lor·ous·ly** *adv.* **—do′lor·ous·ness** *n.*
**do·lour** (dō′lər) *n. Chiefly Brit. var. of* DOLOR.
**dol·phin** (dŏl′fĭn, dôl′-) *n.* [ME < OFr. *dalfin* < Med. Lat. *dalfinus* < Lat. *delphinus* < Gk. *delphis.*] **1.** Any of various marine mammals related to the whales but gen. smaller and having a beaklike snout, chiefly of the family Delphinidae, esp. the common, widely distributed species *Delphinus delphis.* **2.** Either of two iridescently colored marine fishes, *Coryphaena hippurus* or *C. equisetis.*
**dolphin striker** *n. Naut.* A small vertical spar under the bowsprit of a sailboat that extends and helps support the jib boom.
**dolt** (dōlt) *n.* [Perh. < ME *dol,* dull.] A stupid person. **—dolt′ish** *adj.* **—dolt′ish·ly** *adv.* **—dolt′ish·ness** *n.*
**Dom** (dŏm) *n.* [Port. < Lat. *dominus,* lord.] **1.** *(also* dōɴ*).* A title once bestowed on men of rank in Portugal and Brazil. **2.** *Rom. Cath. Ch.* A title used for monks of certain orders.
**-dom** *suff.* [ME < OE *-dōm.*] **1.** State : condition <star*dom*> **2. a.** Domain : position : rank <duke*dom*> **b.** Those that have a specified position, office, or character as a group <official*dom*>
**do·main** (dō-mān′) *n.* [Fr. *domaine* < OFr. *demaine* < Lat. *dominium,* property < *dominus,* lord.] **1.** A territory over which rule or control is exercised. **2.** A sphere of activity, interest, or function : FIELD <the *domain* of science> **3.** *Physics.* Any of numerous contiguous regions in a ferromagnetic material in which the direction of

spontaneous magnetization is uniform and different from that in neighboring regions. **4.** *Law.* **a.** The ownership and right of disposal of property. **b.** The right of eminent domain. **5.** *Math.* **a.** The set of possible values of an independent variable of a function. **b.** An open connected set that contains at least one point.

**dome** (dōm) *n.* [Fr. *dôme*, dome, cathedral < Ital. *duomo*, cathedral < Lat. *domus*, house.] **1.** A hemispherical vault or roof. **2.** An object or structure suggesting a dome. **3.** A large, stately building. **4.** *Slang.* HEAD 1a. **5.** A form of crystal in which two similarly inclined faces intersect in a line parallel to the horizontal axis. —*v.* **domed, dom·ing, domes.** —*vt.* **1.** To cover with or as if with a dome. **2.** To shape like a dome. —*vi.* To take on the shape of a dome, as by swelling.

**dome car** *n.* A railroad passenger car with an elevated glassed-in section for scenic viewing.

**Domes·day Book** (do͞omz'dā', dōmz'-) *also* **Dooms·day Book** (do͞omz'-) *n.* [< ME *domesday*, doomsday.] The written record of a census and survey of English landowners and their property made by order of William the Conqueror in 1085–86.

**do·mes·tic** (də-mĕs'tĭk) *adj.* [OFr. *domestique* < Lat. *domesticus* < *domus*, house.] **1.** Of or pertaining to the family or household. **2.** Fond of home life and household affairs. **3.** Domesticated or tame. —Used of animals. **4.** Of or relating to a country's internal affairs. **5.** Indigenous to or produced in a particular country <*domestic* grapes> —*n.* **1.** A household servant. **2.** Cotton cloth as distinguished from linen. **3.** A product of domestic origin. **4. domestics.** Household linens. **—do·mes'ti·cal·ly** *adv.*

**do·mes·ti·cate** (də-mĕs'tĭ-kāt') *vt.* **-cat·ed, -cat·ing, -cates. 1.** To accustom to domestic life. **2.** To adopt for domestic use. **3. a.** To train or adapt (an animal or plant) to live in a human environment and be of use to human beings. **b.** To introduce and accustom (an animal or plant) to another region : NATURALIZE. **4.** To make familiar to the ordinary person <*domesticate* computer terminology> **—do·mes'ti·ca'tion** *n.*

**do·mes·tic·i·ty** (dō'mĕ-stĭs'ĭ-tē) *n., pl.* **-ties. 1.** The quality or condition of being domestic. **2.** Domestic life or devotion to it. **3. domesticities.** Household affairs.

**do·mes·ti·cize** (də-mĕs'tĭ-sīz') *vt.* **-cized, -ciz·ing, -ciz·es.** To domesticate, as an animal.

**domestic prelate** *n.* *Rom. Cath. Ch.* A priest who is an honorary member of the papal household.

**domestic relations court** *n.* A court in certain U.S. states with jurisdiction over family disputes, esp. those involving child custody, support, and welfare.

**do·mi·cal** (dō'mĭ-kəl, dŏm'ĭ-) *also* **do·mic** (dō'mĭk, dŏm'ĭk) *adj.* Relating to, having, or shaped like a dome. **—do'mi·cal·ly** *adv.*

**dom·i·cile** (dŏm'ĭ-sīl', -səl, dō'mĭ-) *also* **dom·i·cil** (-səl) *n.* [OFr. < Lat. *domicilium* < *domus*, house.] **1.** A residence : home. **2.** One's legal residence. —*v.* **-ciled, -cil·ing, -ciles.** —*vt.* **1.** To establish (oneself or someone) in a residence. **2.** To provide with often temporary lodging. —*vi.* To dwell or reside. **—dom'i·cil'i·ar'y** (dŏm'ĭ-sĭl'ē-ĕr'ē, dō'mĭ-) *adj.*

**dom·i·nance** (dŏm'ə-nəns) *also* **dom·i·nan·cy** (-nən-sē) *n.* The state or fact of being dominant.

**dom·i·nant** (dŏm'ə-nənt) *adj.* [OFr. < Lat. *dominans*, pr.part. of *dominare*, to dominate.] **1.** Exercising the most control or influence : RULING. **2.** Most prominent in position or prevalence : ASCENDANT. **3.** *Genetics.* Producing the same phenotypic effect whether paired with an identical or a dissimilar gene. **4.** *Ecol.* Designating or relating to the species that is most typical of a habitat and that may determine the presence and type of other species. **5.** *Mus.* Pertaining to or based on the fifth tone of a diatonic scale. —*n.* **1.** *Genetics.* A dominant character. **2.** *Ecol.* A dominant species. **3.** *Mus.* The fifth tone of a diatonic scale. **—dom'i·nant·ly** *adv.*

**dominant wavelength** *n.* The wavelength of the light that when combined in specific proportions with an achromatic standard light matches a given color.

**dom·i·nate** (dŏm'ə-nāt') *v.* **-nat·ed, -nat·ing, -nates.** [Lat. *dominari*, to rule < *dominus*, lord.] —*vt.* **1.** To influence, control, or rule by superior power or authority. **2.** To occupy the pre-eminent position in or over. **3.** To order or command imperiously. **4.** To overlook from a height. —*vi.* To be dominant in authority or position. **—dom'i·na'tive** *adj.* **—dom'i·na'tor** *n.*

**dom·i·na·tion** (dŏm'ə-nā'shən) *n.* **1.** The act of dominating or condition of being dominated. **2. dominations.** The fourth of the nine orders of angels.

**dom·i·neer** (dŏm'ə-nîr') *v.* **-neered, -neer·ing, -neers.** [Du. *domineren* < Fr. *dominer* < Lat. *dominari*, to dominate.] —*vt.* To rule over arbitrarily : TYRANNIZE. —*vi.* To rule tyrannically.

**dom·i·neer·ing** (dŏm'ə-nîr'ĭng) *adj.* Tending to domineer : IMPERIOUS. **—dom'i·neer'ing·ly** *adv.*

**do·min·i·cal** (də-mĭn'ĭ-kəl) *adj.* [Med. Lat. *dominicalis* < Lat. *dominicus*, of a lord < *dominus*, lord.] **1.** Of or associated with the Christ as the Lord. **2.** Relating to Sunday as the Lord's day.

**Do·min·i·can** (də-mĭn'ĭ-kən) *adj.* Of or relating to the mendicant order of preaching friars established in 1215 by Saint Dominic. —*n.* A friar of the order of Saint Dominic.

**Dom·i·nick** (dŏm'ə-nĭk) *n. var. of* DOMINQUE.

**dom·i·nie** (dŏm'ə-nē', dō'mə-) *n.* [Obs. *domine*, clergyman < Lat., vocative of *dominus*, lord.] **1.** A minister of the Dutch Reformed Church. **2.** *Informal.* A minister. **3.** *Scot.* A schoolmaster.

**do·min·ion** (də-mĭn'yən) *n.* [ME *dominioun* < OFr. *dominion* < Med. Lat. *dominio* < Lat. *dominium*, property < *dominus*, lord.] **1.** Supreme authority or control : SOVEREIGNTY. **2.** A territory or sphere of control or influence : REALM. **3.** *often* **Dominion.** One of the self-governing nations within the British Commonwealth. **4.** *Law.* Dominium. **5. dominions.** DOMINATION 2.

**Dominion Day** *n.* Jul. 1, a legal holiday in Canada, the anniversary of the Dominion's formation in 1867.

**Dom·i·nique** (dŏm'ə-nēk', dŏm'ə-nĭk) *also* **Dom·i·nick** (dŏm'ə-nĭk) *n.* [After *Dominica*, an island in the West Indies.] An American breed of domestic fowl with a rose-colored comb, yellow legs, and gray, barred plumage.

**do·min·i·um** (də-mĭn'ē-əm) *n.* [Lat., property < *dominus*, lord.] *Law.* Ownership of property and the right to its disposition.

**dom·i·no[1]** (dŏm'ə-nō') *n., pl.* **-noes** *or* **-nos.** [Fr. < Lat. *benedicamus domino*, let us bless the Lord.] **1.** A hooded cape worn by the clergy. **2. a.** A hooded robe worn with an eye mask at a masquerade. **b.** The mask itself. **3.** One wearing a domino.

**dom·i·no[2]** (dŏm'ə-nō') *n., pl.* **-noes** *or* **-nos.** [Fr. < Lat.] **1.** A small, rectangular block whose face is divided into halves, each half being blank or marked by one to six dots. **2. dominoes** *or* **dominos** *(sing. in number).* A game played with a set of gen. 28 dominoes.

**domino effect** *n.* [So-called from the fact that a row of dominoes stood on end will fall in succession if the first one is pushed.] A cumulative effect brought about when one event sets off a chain of comparable events.

**domino theory** *n.* **1.** A theory that if one nation comes under Communist control, then neighboring nations will also come under Communist control. **2.** A theory that one event will set off a train of comparable events.

**don[1]** (dŏn) *n.* [Sp. < Lat. *dominus*, lord.] **1. Don.** Sir. —Used gen. as a courtesy title with a man's given name in Spanish-speaking countries. **2.** A Spanish gentleman. **3.** *Chiefly Brit.* A head, tutor, or fellow at an Oxford or Cambridge college. **4.** The leader of an organized-crime family.

**don[2]** (dŏn) *vt.* **donned, don·ning, dons.** [Contraction of *do on.*] **1.** To put on (clothing). **2.** To take on, as a role or posture.

**do·ña** (dō'nyä) *n.* [Sp. < Lat. *domina*, fem. of *dominus*, lord.] **1. Doña.** Lady. —Used gen. as a courtesy title with a woman's given name in Spanish-speaking countries. **2.** A Spanish gentlewoman.

**do·nate** (dō'nāt', dō-nāt') *vt.* **-nat·ed, -nat·ing, -nates.** [Back-formation < DONATION.] To contribute as a gift to a fund or cause. **—do'na·tor** *n.*

**do·na·tion** (dō-nā'shən) *n.* [ME, gift, benefice < OFr. < Lat. *donatio* < *donatus*, p.part. of *donare*, to give < *donum*, gift.] **1.** The act of giving something to a charity or cause. **2.** An offering or gift : CONTRIBUTION.

**Don·a·tist** (dŏn'ə-tĭst, dō'nə-) *n.* [Med. Lat. *Donatista*, after *Donatus*, 4th-cent. Bishop of Carthage.] A member of a schismatic Christian sect that arose in North Africa in the 4th cent. A.D. **—Don'a·tism** *n.*

**don·a·tive** (dō'nə-tĭv, dŏn'ə-) *n.* [Lat. *donativum* < *donativus*, of a donation < *donatus*, p.part. of *donare*, to give < *donum*, gift.] **1.** A largess or bounty. **2.** A benefice. —*adj.* Constituting a benefice.

**done** (dŭn) *adj.* [< p.part. of DO[1].] **1.** Completely finished : ACCOMPLISHED. *usage:* *Done* in this sense is acceptable in all contexts, although its use can occas. lead to ambiguity in sentences such as *The work will be done next month.* Rephrasing (*The work will get done next month* or *The work will be done by next month*) will usu. eliminate the unclarity. **2.** Cooked adequately. **3.** Socially acceptable <That just isn't *done* in our town.> **—done for. 1.** Defeated. **2.** Doomed : dying. **3.** Ruined. **—done'ness** *n.*

**do·nee** (dō-nē') *n.* [DON(OR) + -EE.] A recipient of a gift.

**dong** (dŏng) *n.* [Vietnamese < Chin. *tong[2]*, copper coin.] —See table at CURRENCY.

**don·jon** (dŏn'jən, dŭn'-) *n.* [Variant of DUNGEON.] The fortified main tower of a castle.

**Don Juan** (dŏn wŏn', jo͞o'ən) *n.* [After *Don Juan*, legendary Spanish nobleman and libertine.] **1.** A libertine : profligate. **2.** A man obsessed with seducing women.

**don·key** (dŏng'kē, dŭng'-, dông'-) *n., pl.* **-keys.** [Perh. DUN (dark) + *-key* as in *monkey.*] **1.** The domesticated ass, prob. descended from the wild ass *Equus asinus.* **2.** *Informal.* **a.** A stubborn person. **b.** A stupid person.

**donkey engine** *n.* **1.** A small auxiliary steam engine used for hoisting or pumping, esp. aboard ship. **2.** A small locomotive.

**don·na** (dŏn'ə, dôn'nä) *n.* [Ital. < Lat. *domina*, fem. of *dominus*, lord.] **1. Donna.** Lady. —Used gen. as a courtesy title with a woman's given name in Italian-speaking countries. **2.** An Italian gentlewoman.

**don·nish** (dŏn′ĭsh) *adj.* Of, resembling, or typical of a university don : BOOKISH.

**don·ny·brook** (dŏn′ē-brŏŏk′) *n.* [After *Donnybrook* fair, held annually in Donnybrook, Ireland, and noted for its brawls.] A brawl.

**do·nor** (dō′nər) *n.* [AN *donour* < Lat. *donator* < *donare,* to give < *donum,* gift.] **1.** One who contributes something, as money, to a fund or cause. **2.** One from whom blood, tissue, or an organ is taken for use in a transfusion or transplant. **3.** *Electron.* An element introduced into a semiconductor with a negative valence greater than that of the pure semiconductor.

**do-noth·ing** (dōō′nŭth′ĭng) *adj.* Offering no initiative for change, esp. in politics. —*n.* An idle, lazy person. **—do′-no′thing·ism** *n.*

**Don Qui·xo·te** (dŏn′ kē-hō′tē, kwĭk′sət) *n.* [After *Don Quixote,* hero of a satirical chivalric romance by Miguel de Cervantes (1547–1616).] An impractical idealist.

**don't¹** (dōnt). Do not.

**don't²** (dōnt) *n.* A prohibition <do's and *don'ts*>

**do·nut** (dō′nŭt′, -nət) *n. var. of* DOUGHNUT.

**doo·dad** (dōō′dăd′) *n. Informal.* An unnamed or nameless gadget.

**doo·dle** (dōōd′l) *v.* **-dled, -dling, -dles.** [Dial. E., to fritter away time.] *Informal.* —*vi.* **1.** To scribble aimlessly, esp. when preoccupied. **2.** To spend time idly : DAWDLE. —*vt.* To make by doodling. —*n. Informal.* A figure, design, or scribble produced absentmindedly.

**doo·dle·bug** (dōōd′l-bŭg′) *n.* [Perh. dial. E. *doodle,* fool + BUG.] **1.** The larva of the ant lion. **2.** A divining rod.

**doo·hick·ey** (dōō′hĭk′ē) *n., pl.* **-eys.** [Perh. DOO(DAD) + HICKEY.] *Informal.* A doodad.

**doom** (dōōm) *n.* [ME *dom* < OE *dōm.*] **1.** A decision or judgment, esp. an official condemnation to a severe penalty. **2.** Destiny or fate, esp. a ruinous or tragic fate. **3.** Inevitable destruction or ruin : ANNIHILATION. **4.** Judgment Day. **5.** *Archaic.* A statute or ordinance. **6.** *Informal.* A crucial, esp. negative, judgment or reckoning. —*vt.* **doomed, doom·ing, dooms. 1.** To pronounce judgment against : CONDEMN. **2.** To set the destiny of, esp. to an unhappy end.

▲ word history: The semantics of *doom* illustrates the process of *pejoration,* by which a word of a good or neutral connotation acquires an evil one. *Doom* is related to the verb *do* and at first basically meant "something set up," specifically "a law, statute." Even by Old English times, however, *doom* meant "judgment," especially an adverse judgment, condemnation, or punishment. *Doomsday,* the day of the last judgment in Christian theology, will doubtless be a day of great rejoicing for the just, but the fate of the damned has cast a pall over the connotations of the word.

**doom·say·er** (dōōm′sā′ər) *n.* One inclined to predict future misfortune or disaster.

**dooms·day** (dōōmz′dā′) *n.* [ME *domesday* < OE *dōmes dæg* : *dōm,* judgment + *dæg,* day.] **1.** The day of the Last Judgment. **2.** A dreaded day of judgment or reckoning.

**Doomsday Book** (dōōmz′dā′) *n. var. of* DOMESDAY BOOK.

**door** (dôr, dōr) *n.* [ME *dor* < OE *duru.*] **1. a.** A movable structure for closing off an entrance, typically consisting of a panel that swings on hinges, slides, or rotates. **b.** A similar part on a vehicle or a piece of furniture. **2.** The entranceway to a passage, room, or building. **3.** A means of access or approach. **4.** The room or building to which a door belongs <two *doors* down the corridor>

**door·bell** (dôr′bĕl′, dōr′-) *n.* A bell outside a door, used as a signal for admission.

**door·jamb** (dôr′jăm′, dōr′-) *n.* Either of the two vertical pieces framing a doorway and supporting the lintel.

**door·keep·er** (dôr′kē′pər, dōr′-) *n.* A person employed to guard an entrance or gateway.

**door·knob** (dôr′nŏb′, dōr′-) *n.* A knob for opening and closing a door.

**door·man** (dôr′măn′, -mən, dōr′-) *n.* A person employed to attend the entrance of a hotel, apartment house, or building.

**door·mat** (dôr′măt′, dōr′-) *n.* **1.** A mat placed before or inside a doorway for wiping the shoes. **2.** *Slang.* One who unprotestingly allows mistreatment by others.

**door·nail** (dôr′nāl′, dōr′-) *n.* A large-headed nail once used as a stud on doors. **—dead as a doornail.** Undoubtedly dead.

**door·post** (dôr′pōst′, dōr′-) *n.* A doorjamb.

**door prize** *n.* A lottery prize awarded to the holder of a ticket purchased at or before a function.

**door·sill** (dôr′sĭl′, dōr′-) *n.* The threshold of a doorway.

**door·step** (dôr′stĕp′, dōr′-) *n.* A step leading to a door.

**door·stop** (dôr′stŏp′, dōr′-) *n.* **1.** A wedge inserted beneath a door to hold it open at a desired position. **2.** A weight or spring that prevents a door from slamming. **3.** A rubber-tipped projection attached to a wall to protect it from the impact of an opening door.

**door-to-door** (dôr′tə-dôr′, dōr′tə-dōr′) *adj.* Making or being unsolicited calls from one house to another in an area <a *door-to-door* salesperson> **—door′-to-door′** *adv.*

**door·way** (dôr′wā′, dōr′-) *n.* The entrance to a room or building.

**door·yard** (dôr′yärd′, dōr′-) *n.* A yard in front of the door of a house.

**do·pa** (dō′pə) *n.* [Contraction of E. *dihydroxyphenylalanine.*] An amino acid, $C_9H_{11}NO_4$, that is converted to dopamine in the bloodstream and used to treat Parkinson's disease.

**do·pa·mine** (dō′pə-mēn′) *n.* [DOP(A) + AMINE.] A monoamine neurotransmitter that is a carboxylated form of dopa and is essential to normal nerve activity.

**dop·ant** (dō′pənt) *n.* [DOP(E) + -ANT.] A small quantity of a substance, as phosphorus, that is added to another substance, as a semiconductor, to alter the latter's properties.

**dope** (dōp) *n.* [Du. *doop,* sauce < *doopen,* to dip.] **1.** A viscid substance or liquid, esp. a lubricant such as axle grease or an absorbent material such as the nitroglycerin used to make dynamite. **2.** A preparation resembling varnish formerly used to protect, waterproof, and tauten the cloth surfaces of airplane wings. **3.** *Informal.* A narcotic, esp. one that is addictive. **4.** A narcotic used to alter the performance of a racehorse. **5.** *Slang.* A very stupid person. **6.** *Slang.* Factual, esp. private, information. —*v.* **doped, dop·ing, dopes.** —*vt.* **1.** To add or apply dope to. **2.** *Informal.* To administer a narcotic to. **3.** *Informal.* To figure out (an outcome or puzzle). **4.** *Informal.* To make a rough plan of. —*vi. Informal.* To take drugs. **—dop′er** *n.*

**dope sheet** *n. Slang.* A publication giving information on the horses running in the day's races.

**dope·ster** (dōp′stər) *n.* One who analyzes and predicts future events, as in sports or politics.

**dop·ey** *also* **dop·y** (dō′pē) *adj.* **dop·i·er, dop·i·est** *Slang.* **1.** Dazed or lethargic from or as if from drugs. **2.** Stupid. **3.** Foolish.

**dop·pel·gäng·er** *or* **dop·pel·gang·er** (dŏp′əl-găng′ər, dôp′əl-gĕng′ər) *n.* [G. : *doppel,* double + *gänger,* goer < *gehen,* to go.] A ghostly double of a living person, esp. one that haunts its own counterpart.

**Dop·pler** (dŏp′lər) *adj.* Of, relating to, or utilizing the Doppler effect or Doppler radar.

**Dop·pler effect** (dŏp′lər) *n.* [After Christian *Doppler* (1803–1853).] An apparent change in the frequency of waves, as of sound or light, occurring when the source and observer are in motion relative to one another, the frequency increasing when the source and observer approach one another and decreasing when they move apart.

**Doppler radar** *n.* Radar that uses the Doppler effect to measure velocity.

**do·py** (dō′pē) *adj. var. of* DOPEY.

**Do·ra·do** (də-rä′dō) *n.* A constellation of the Southern Hemisphere.

**dor·bee·tle** (dôr′bēt′l) *n.* [Obs. *dor,* a buzzing bee or beetle (< ME *dorre* < OE *dora*) + BEETLE.] An Old World dung beetle, *Geotrupes stercorarius,* that makes a droning sound in flight.

**Dor·cas society** (dôr′kəs) *n.* [After *Dorcas,* a Christian woman of the 1st or 2nd cent. A.D.] A women's auxiliary group, often sponsored by a church, that provides clothes for the poor.

**Do·ri·an** (dôr′ē-ən, dōr′-) *n.* One of a Hellenic people that invaded Greece around 1100 B.C. and remained culturally and linguistically distinct within the Greek world, esp. in Sparta, Corinth, and Argos. **—Do′ri·an** *adj.*

**Dor·ic** (dôr′ĭk, dŏr′-) *n.* [Lat. *Doricus* < Gk. *Dorikos* < *Doris,* a region of ancient Greece.] A dialect of ancient Greek spoken in the Peloponnesus, Crete, certain of the Aegean islands, Sicily, and in southern Italy. —*adj.* **1.** Of, pertaining to, typical of, or designating Doric. **2.** Designating or in the style of the Doric order.

**Doric order** *n.* The oldest and simplest of the three orders of classical Greek architecture, marked by heavy, fluted columns having no base and plain saucer-shaped capitals.

**Dor·king** (dôr′kĭng) *n.* [After *Dorking,* a town in England.] A domestic fowl of a breed with a heavy body, raised mainly for the table.

**dorm** (dôrm) *n. Informal.* A dormitory.

**dor·mant** (dôr′mənt) *adj.* [ME *dormaunt* < OFr. *dormant* < pr.part. of *dormir,* to sleep < Lat. *dormire.*] **1.** Asleep or inactive. **2.** Latent but capable of being activated <"a harrowing experience which . . . lay *dormant* but still menacing" —Charles Jackson> **3.** Temporarily quiescent, as a volcano. **4.** *Biol.* Being in a relatively inactive or resting condition in which some processes are slowed down or suspended. **—dor′man·cy** *n.*

**dor·mer** (dôr′mər) *n.* [OFr. *dormeor,* bedroom < *dormir,* to sleep < Lat. *dormire.*] **1.** A window set vertically in a small gable projecting from a sloping roof. **2.** The gable holding a dormer.

**dor·mie** (dôr′mē) *adj. var. of* DORMY.

**dor·min** (dôr′mĭn) *n.* [DORM(ANCY) + -IN.] Abscisic acid.

**dor·mi·to·ry** (dôr′mĭ-tôr′ē, -tōr′ē) *n., pl.* **-ries.** [Lat. *dormitorium* < *dormitorius,* of sleep < *dormire,* to sleep.] **1.** A room furnished with beds for a number of persons. **2.** A structure for housing a number of persons, as at a school. **3.** A suburban community whose residents commute to a nearby metropolis for employment and recreation.

**dor·mouse** (dôr′mous′) *n.* [ME *dormowse.*] Any of various small, squirrellike Old World rodents of the family Gliridae.

**dor·my** *also* **dor·mie** (dôr′mē) *adj.* [Orig. unknown.] Ahead of an opponent by as many holes in a golf match as remain to be played.

ă **pat** ā **pay** âr **care** ä **father** ĕ **pet** ē **be** hw **which** ĭ **pit**
ī **tie** îr **pier** ŏ **pot** ō **toe** ô **paw, for** oi **noise** ŏŏ **took**

**dor·nick¹** (dôr′nĭk) *n.* [ME *dornick*, after *Doornik* (Tournai), Belgium.] A coarse damask cloth.

**†dor·nick²** (dôr′nĭk) *n.* [Perh. of Celtic orig.] *Regional.* A small chunk of rock : STONE.

**do·ron·i·cum** (də-rŏn′ĭ-kəm) *n.* [NLat. < Ar. *doronaj.*] A plant of the genus *Doronicum*, which includes the leopard's-bane.

**dors-** *pref. var. of* DORSO-.

**dor·sad** (dôr′săd′) *adv. Anat.* In the direction of the back.

**dor·sal** (dôr′səl) *adj.* [LLat. *dorsalis* < Lat. *dorsualis* < *dorsum*, back.] **1.** *Anat.* Of, toward, on, in, or near the back. **2.** *Bot.* Of or on the outer surface, underside, or back of an organ. **—dor′sal·ly** *adv.*

**dorsal fin** *n.* The main fin on the dorsal surface of fishes or certain marine mammals.

**Dor·set Horn** (dôr′sĭt) *n.* [After *Dorset*, a county in England.] A long-horned domestic sheep of a breed with fine-textured wool.

**dorsi-** *pref. var. of* DORSO-.

**dor·si·ven·tral** (dôr′sĭ-vĕn′trəl) *adj.* Having distinct upper and lower surfaces.

**dorso-** *or* **dorsi-** *or* **dors-** *pref.* [< Lat. *dorsum*, back.] **1.** Back <*dorsad*> **2.** Dorsal <*dorsoventral*>

**dor·so·ven·tral** (dôr′sō-vĕn′trəl) *adj.* [DORSO- + VENTRAL.] Extending from a dorsal to a ventral surface.

**dor·sum** (dôr′səm) *n., pl.* **-sa** (-sə) [Lat., back.] *Anat.* **1.** The back. **2.** A part of an organ or appendage analogous to the back.

**do·ry¹** (dôr′ē, dōr′ē) *n., pl.* **-ries.** [Mosquito *dōri*, dugout.] A small, narrow, flat-bottomed fishing boat with high sides and a sharp prow.

**do·ry²** (dôr′ē, dōr′ē) *n., pl.* **-ries.** [ME *dorre* < OFr. *doree*, gilded, fem. p.part. of *dorer*, to gild < LLat. *deaurare* : Lat. *de-*, (intensive) + Lat. *aurum*, gold.] **1.** The John Dory. **2.** WALLEYE 3.

**dos-à-dos** (dō′zä-dō′) *n., pl.* **dos-à-dos** (-dōz′, -dō′) [Fr. : *dos*, back + *à*, to + *dos*, back.] A sofa or carriage accommodating two people seated back to back.

dos-à-dos

**dos·age** (dō′sĭj) *n.* **1. a.** Administration of a therapeutic agent in prescribed amounts. **b.** Determination of the amount to be administered. **c.** DOSE 1. **2.** Addition of an ingredient to a substance, esp. to wine, in a specific dose.

**dose** (dōs) *n.* [Fr. < LLat. *dosis* < Gk. < *didonai*, to give.] **1.** A specified amount of a therapeutic agent prescribed to be taken at one time or at stated intervals. **2.** *Med.* The amount of radiation administered to a certain bodily part. **3.** *Informal.* A portion of an experience, esp. of something unpleasant, to which one is subjected <a *dose* of misfortune> **4.** An ingredient added, esp. to wine, to impart flavor or strength. **5.** *Slang.* A venereal infection. *—vt.* **dosed, dos·ing, dos·es. 1.** To give a dose to, as of medicine. **2.** To give or prescribe (medicine) in doses. **—dos′er** *n.*

**do-si-do** (dō′sē-dō′) *n., pl.* **-dos.** [Var. of DOS-À-DOS.] **1.** A square-dance movement in which two dancers approach each other and circle back to back, then return to their original positions. **2.** The call given for a do-si-do.

**do·sim·e·ter** (dō-sĭm′ĭ-tər) *n.* [DOS(E) + -METER.] A device that measures and indicates the amount of x-rays or radioactivity absorbed. **—do·si·met′ric** *adj.* **—do·sim′e·try** *n.*

**doss** (dŏs) [Perh. alteration of *dorse*, back < Lat. *dorsum.*] *Chiefly Brit.* *—n.* **1.** A crude or makeshift bed. **2.** A cheap lodging house. *—vi.* **dossed, doss·ing, doss·es.** To bed down or sleep, esp. in a doss.

**dos·sal** *also* **dos·sel** (dŏs′əl) *n.* [Med. Lat. *dossale*, neuter of LLat. *dorsalis*, dorsal.] **1.** An ornamental hanging of rich fabric, as behind an altar. **2.** An ornamental covering for the back of a throne or chair.

**dos·ser** (dŏs′ər) *n.* [ME *doser* < OFr. *dossier* < Med. Lat. *dorsarium* < Lat. *dorsum*, back.] **1.** PANNIER 1. **2.** DOSSAL 2.

**dos·si·er** (dŏs′ē-ā′, dô′sē-ā′) *n.* [Fr. < OFr., bundle of papers labeled on the back < *dos*, back < Lat. *dorsum.*] A collection of documents or papers giving detailed information about a person or subject.

**dost** (dŭst) *v. Archaic. 2nd person sing. present tense of* DO¹.

**dot¹** (dŏt) *n.* [ME **dot* < OE *dott*, head of a boil.] **1. a.** A tiny round mark made by or as if by a pointed instrument : SPOT. **b.** Such a mark used in orthography, as above an *i.* **2.** A very small amount. **3.** A short sound or signal used in combination with the dash and written as a dot to represent letters, numbers, or punctuation in Morse and similar codes. **4.** *Math.* **a.** A decimal point. **b.** A symbol of multiplication. **5.** *Mus.* A mark after a note indicating an increase in time value by half. *—v.* **dot·ted, dot·ting, dots.** *—vt.* **1.** To mark with a dot. **2.** To form or make with dots. **3.** To cover with or as if with dots. *—vi.* To make a dot. **—on (***or* **at) the dot.** *Informal.* On time. **—dot′ter** *n.*

**dot²** (dŏt, dō) *n.* [Fr. < Lat. *dos*, dowry.] A woman's marriage portion : DOWRY. **—do′tal** (dōt′l) *adj.*

**do·tage** (dō′tĭj) *n.* [ME < *doten*, to dote.] **1.** A condition of mental deterioration : SENILITY. **2.** Foolish or excessive fondness.

**do·tard** (dō′tərd) *n.* [ME < *doten*, to dote.] A senile person.

**dote** (dōt) *vi.* **dot·ed, dot·ing, dotes.** [ME *doten.*] **1.** To show excessive fondness or love. **2.** To show mental deterioration, esp. as a result of senility. **—dot′er** *n.*

**doth** (dŭth) *v. Archaic. 3rd person sing. present tense of* DO¹.

**dot product** *n.* [From the use of a dot to indicate the function, as in *x·y.*] *Math.* Scalar product.

**dot-se·quen·tial** (dŏt′sĭ-kwĕn′shəl) *adj.* Relating to a color-television system in which the primary colors red, green, and blue are transmitted as dots in sequence and displayed in the same sequence to produce a complete color image.

**dotted swiss** *n.* A crisp cotton fabric with woven, flocked, or embroidered dots.

**dot·tle** *also* **dot·tel** (dŏt′l) *n.* [< DOT¹, lump (obs.).] The plug of smoked tobacco ash left in the bowl of a pipe.

**dot·ty** (dŏt′ē) *adj.* **-ti·er, -ti·est.** [Alteration of Sc. *dottle*, silly < ME *doten*, to dote.] **1.** Having a feeble or shaky gait : UNSTEADY. **2. a.** Mentally unbalanced : CRAZY. **b.** Eccentric. **c.** Ridiculous or absurd. **3.** Obsessively infatuated.

**Dou·ay Bible** (dōō-ā′, dōō′ā) *n.* [After *Douai*, France.] An English translation of the Latin Vulgate Bible by Roman Catholic scholars.

**Douay Version** *n.* The Douay Bible.

**dou·ble** (dŭb′əl) *adj.* [ME < OFr. < Lat. *duplus.*] **1.** Twice as much in size, strength, number, or amount <*double* trouble> **2.** Having two like parts <*double* windows> **3.** Having two unlike parts : DUAL <a *double* standard> **4.** Designed for or accommodating two <a *double* hotel room> **5. a.** Acting two parts. **b.** Marked by duplicity : DECEITFUL. **6.** *Bot.* Having many more than the usual number of petals, usu. in a crowded or overlapping arrangement. *—n.* **1.** Something increased twofold. **2. a.** A duplicate of another : COUNTERPART. **b.** An apparition. **3.** An understudy. **4. a.** A sharp turn in running : REVERSAL. **b.** An evasive shift in argument. **5. doubles.** A game, as tennis or handball, having two players on each side. **6.** *Baseball.* A two-base hit. **7. a.** A request for a bid in bridge indicating strength to one's partner. **b.** A bid doubling one's opponent's bid in bridge, thus increasing the penalty for failure to fulfill the contract. **c.** A hand justifying such a bid. *—v.* **-bled, -bling, -bles.** *—vt.* **1.** To make twice as great. **2.** To be twice as much as. **3.** To fold in two. **4.** To repeat : duplicate. **5.** *Baseball.* **a.** To cause the scoring of (a run) by hitting a double. **b.** To advance or score (a runner) by hitting a double. **6.** *Baseball.* To put out (a runner) as the second part of a double play. **7.** To challenge (an opponent's bid) with a double in bridge. **8.** *Mus.* To duplicate (another part or voice) an octave higher or lower or in unison. **9.** *Naut.* To sail around <*doubled* the cape> *—vi.* **1.** To be increased twofold. **2.** To turn sharply backward : REVERSE. **3.** To serve in an additional capacity <a math teacher who *doubles* as coach> **4.** To replace an actor in the performance of a given action or in the actor's absence. **5.** *Baseball.* To hit a double. **6.** To announce a double in bridge. **—double up. 1.** To bend suddenly, as in laughter or pain. **2.** To share accommodations meant for one person. *—adv.* **1.** To twice the extent or amount : DOUBLY. **2.** Two together <riding *double*> **—on (***or* **at) the double.** *Informal.* **1.** In double time. **2.** Immediately. **—see double.** To see two images of a single object, usu. as a result of visual impairment. **—dou′ble·ness** *n.*

**double agent** *n.* A spy who infiltrates a government and pretends to work for it while actually working for another.

**double bar** *n. Mus.* A double vertical or heavy black line drawn through a staff to indicate the end of any of the main sections of a musical composition.

**dou·ble-bar·reled** (dŭb′əl-băr′əld) *adj.* **1.** Having two barrels side by side. **2.** Serving a double purpose : TWOFOLD.

**double bass** *n.* The cello-shaped, largest member of the violin family, having a deep range of about three octaves and played usu. with a bow.

**double bassoon** *n.* The contrabassoon.

**double bed** *n.* A bed sleeping two people.

**dou·ble-blind** (dŭb′əl-blīnd′) *adj.* Of or designating a research procedure in which neither the researcher nor the subjects knows who is receiving the experimental substance or treatment.

**double boiler** *n.* A cooking utensil with two nested pans, designed for slow, even cooking or heating of food in the upper pan by the action of the water boiling in the lower.

**dou·ble-breast·ed** (dŭb′əl-brĕs′tĭd) *adj.* **1.** Fastened by lapping one half over the other and usu. having a double row of buttons with a single row of buttonholes. **2.** Having a double-breasted coat.

**double check** *n.* A careful re-examination to assure efficiency or accuracy : VERIFICATION.

**dou·ble-check** (dŭb′əl-chĕk′) *v.* **-checked, -check·ing, -checks.** —*vt.* To check again : VERIFY. —*vi.* To make a double check.

**double chin** *n.* A fold of fatty flesh beneath the chin.

**dou·ble-cross** (dŭb′əl-krôs′, -krŏs′) *Slang.* —*vt.* **-crossed, -cross·ing, -cross·es.** To betray by acting in contradiction to an agreed course of action. —*n.* A betrayal. **—dou′ble-cross′er** *n.*

**double dagger** *n.* A reference mark (‡) in writing and printing.

**double date** *n.* A date in which two couples participate. **—dou′ble-date′** (dŭb′əl-dāt′) *v.* **(-dat·ed, -dat·ing, -dates).**

**dou·ble-deal·ing** (dŭb′əl-dē′lĭng) *adj.* Marked by duplicity : DECEITFUL. —*n.* Treachery : duplicity. **—dou′ble-deal′er** *n.*

**dou·ble-deck·er** (dŭb′əl-dĕk′ər) *n.* **1.** A vehicle with two decks or tiers for passengers. **2.** Two beds, one built above the other. **3.** *Informal.* A sandwich with three slices of bread and two layers of filling.

**double decomposition** *n.* A chemical reaction between two compounds in which the first and second parts of one reactant are united, respectively, with the second and first parts of the other reactant.

**dou·ble-dig·it** (dŭb′əl-dĭj′ĭt) *adj.* Pertaining to percentage rates between 10 and 99% <*double-digit* price hikes>

**double dipping** *n.* The practice of drawing two incomes from the government, usu. by holding a U.S. government position while receiving a pension. **—double dipper** *n.*

**dou·ble-dome** (dŭb′əl-dōm′) *n. Slang.* An intellectual.

**double dribble** *n. Basketball.* An illegal dribble in which a player uses both hands simultaneously to dribble the ball or begins to dribble the ball a second time after having come to a complete stop.

**double eagle** *n.* A former U.S. gold coin having a face value of 20 dollars.

**dou·ble-edged** (dŭb′əl-ĕjd′) *adj.* **1.** Having two cutting edges, as a sword or razor blade. **2.** Capable of being effective or interpreted in two ways <*double-edged* criticism>

**dou·ble-en·ten·dre** (dŭb′əl-än-tän′drə, dōō-blän-täN′dr′) *n.* [Fr.] **1.** A word or phrase having a double meaning, esp. when the second meaning is risqué. **2.** The use of double-entendres : AMBIGUITY.

**double entry** *n.* A method of bookkeeping in which a transaction is entered both as a debit to one account and a credit to another account, so that the totals of debits and credits are equal.

**dou·ble-faced** (dŭb′əl-fāst′) *adj.* **1.** Having two faces, aspects, or sides. **2.** Marked by hypocrisy : DUPLICITOUS.

**double feature** *n.* A motion-picture program consisting of two full-length feature films.

**dou·ble-head·er** (dŭb′əl-hĕd′ər) *n.* **1.** Two games or events held in succession on the same program, esp. in baseball. **2.** A train pulled by two locomotives.

**double indemnity** *n.* A clause in an insurance policy providing for payment of double the face value of the contract in case of accidental death.

**double jeopardy** *n. Law.* The act of trying a person a second time for an offense for which he or she has already been prosecuted.

**dou·ble-joint·ed** (dŭb′əl-join′tĭd) *adj.* Having exceptionally flexible joints permitting connected parts to be bent at unusual angles.

**double knit** *n.* **1.** A jerseylike fabric knitted on a machine equipped with two sets of needles so that a double thickness of fabric with interlocking stitches is produced. **2.** An article of clothing made of double knit.

**double negative** *n.* A syntactic construction having two negatives, esp. to express a single negation. ***usage:*** A double negative (*We aren't happy neither*) is gen. unacceptable in standard usage. Such constructions once were employed for the purpose of emphasis or intensification and were considered part of the standard language; an example is Hamlet's advice to the players: *"Be not too tame neither, but let your discretion be your tutor."*

**dou·ble-park** (dŭb′əl-pärk′) *v.* **-parked, -park·ing, -parks.** —*vt.* To park (a vehicle) alongside a vehicle parked parallel to the curb. —*vi.* To double-park a vehicle.

**double play** *n. Baseball.* A play in which two players are put out.

**double pneumonia** *n.* Pneumonia afflicting both lungs.

**dou·ble-quick** (dŭb′əl-kwĭk′) *adj.* Very quick. —*n.* DOUBLE TIME 1. —*vi.* **-quicked, -quick·ing, -quicks.** To move at double time.

**dou·bler** (dŭb′lər) *n.* A device that doubles the frequency of an input signal.

**dou·ble-reed** (dŭb′əl-rēd′) *n.* Any of a group of wind instruments that have a mouthpiece formed of two joined reeds that vibrate against each other.

**double refraction** *n.* Birefringence.

**double salt** *n. Chem.* A salt consisting or regarded as consisting of a molecular combination of two simple salts.

**dou·ble-space** (dŭb′əl-spās′) *v.* **-spaced, -spac·ing, -spac·es.** —*vi.* To type so that there is a full space between lines. —*vt.* To type (copy) on every other line.

**dou·ble·speak** (dŭb′əl-spēk′) *n.* DOUBLE TALK 2.

**double standard** *n.* A set of principles allowing greater opportunity or liberty to one than to another, esp. the allowance of greater sexual freedom to men than to women.

**double star** *n.* A binary star.

**dou·blet** (dŭb′lĭt) *n.* [ME < OFr. < *double,* double.] **1.** A close-fitting jacket worn by men between the 15th and 17th cent. **2. a.** A pair of similar things. **b.** One of a pair. **c.** *Physics.* A multiplet with two members. **3.** One of two words derived from the same source by different routes of transmission. **4. doublets.** A throw of two dice in which the same number of dots appears on the upper face of each.

**double take** *n.* A delayed reaction to an unusual circumstance or remark, often used as a comic device.

**double talk** *n.* **1.** Meaningless speech consisting of nonsense syllables mixed with intelligible words : GIBBERISH. **2.** Evasive or ambiguous language.

**dou·ble-team** (dŭb′əl-tēm′) *vt.* **-teamed, -team·ing, -teams.** To guard or cover an offensive player with two defensive players.

**dou·ble·think** (dŭb′əl-thĭngk′) *n.* Simultaneous belief in two contradictory ideas or points of view.

▲ word history: *Doublethink* was coined by George Orwell in his novel *Nineteen Eighty-Four,* which was published in 1949. It is rare that an invented word gains sufficient currency to be included in a dictionary.

**double time** *n.* **1.** A marching pace of 180 three-foot steps per minute. **2.** *Mus.* Duple time. **—dou′ble-time** *vi.* **(-timed, -tim·ing, -times).**

**dou·ble-tongue** (dŭb′əl-tŭng′) *vi.* **-tongued, -tongu·ing, -tongues.** *Mus.* To play a rapidly repeated series of notes on a wind instrument by placing the tongue alternately between the positions for *t* and *k.*

**dou·ble·tree** (dŭb′əl-trē′) *n.* A crossbar on a wagon or coach to which two whiffletrees are attached for harnessing two animals abreast.

**dou·ble-u** (dŭb′əl-yōō′) *n.* The letter *w.*

▲ word history: The name *double-u* for the letter W is recorded from the 15th century, but the letter itself was invented in England in the 7th century. Although the Anglo-Saxons used the Latin alphabet to write their language, Old English contained sounds for which letters did not exist in early Medieval Latin. The letter V, which represented the sound of both *u* and *w* in classical Latin, was pronounced *v* in the 7th century. The sound *w* did exist in Old English; the letter W representing this sound was formed by writing together two rounded V's to form a single character. The Anglo-Saxons introduced the letter W to the Continent, although they replaced it in English writing with the runic character *wyn.* W was reintroduced into England by the Normans after the Norman Conquest in 1066.

**double vision** *n.* Diplopia.

**dou·ble·word** (dŭb′əl-wûrd′) *n. Computer Sci.* Two computer words considered as a single 64-bit quantity.

**dou·bloon** (dŭ-blōōn′) *n.* [Sp. *doblón,* aug. of *dobla,* Sp. coin < Lat. *dupla,* fem. of *duplus,* double.] An obsolete Spanish gold coin.

**dou·blure** (dōō-blŏŏr′) *n.* [Fr., lining < OFr. *doubler,* to double, line < Lat. *duplare,* to double < *duplus,* double.] An ornamental lining, as of leather or vellum, on the inside face of a book cover.

**dou·bly** (dŭb′lē) *adv.* **1.** To a double degree. **2.** In a twofold way.

**doubt** (dout) *v.* **doubt·ed, doubt·ing, doubts.** [ME *douten* < OFr. *douter* < Lat. *dubitare,* to waver.] —*vt.* **1.** To be skeptical or undecided about. **2.** To tend to distrust : DISBELIEVE <Why do you *doubt* my word?> **3.** *Archaic.* To suspect : fear. —*vi.* To be undecided. ***usage:*** In informal speech, *doubt* is often followed by *but,* as in *I don't doubt but they'll go.* Such a usage should be avoided in formal style by substituting *that* or *whether* for *but* as the case requires. —*n.* **1.** A lack of certainty or conviction. **2.** A lack of trust. **3.** A point about which one is undecided or skeptical. **4.** An uncertain state of affairs <Election returns are still in *doubt.*> **—beyond (or without) doubt.** Without question : CERTAINLY. **—no doubt. 1.** Certainly. **2.** Probably. **—doubt′er** *n.*

☆ ***syns:*** DOUBT, DOUBTFULNESS, DUBIETY, QUESTION, SKEPTICISM, UNCERTAINTY *n. core meaning* : lack of certainty or conviction <had *doubts* about their technical qualifications> ***ant:*** certitude

**doubt·ful** (dout′fəl) *adj.* **1.** Subject to or tending to cause doubt : UNCERTAIN. **2.** Experiencing or exhibiting doubt <*doubtful* about the future> **3.** Of uncertain outcome : UNDECIDED. **4.** Questionable, as to honesty or value : SUSPICIOUS <a *doubtful* past> **—doubt′ful·ly** *adv.* **—doubt′ful·ness** *n.*

**doubting Thom·as** (tŏm′əs) *n.* [After St. *Thomas,* an apostle who doubted Jesus' resurrection until he had proof of it.] One who is habitually doubtful.

**doubt·less** (dout′lĭs) *adj.* Assured : certain. —*adv.* **1.** Certainly. **2.** Probably. **—doubt′less·ly** *adv.*

**dou·ceur** (dōō-sûr′) *n.* [Fr. < LLat. *dulcor,* sweetness < Lat. *dulcis,* sweet.] Money given as a gratuity or bribe.

**douche** (dōōsh) *n.* [Fr., shower < Ital. *doccia,* conduit, douche < *doccione,* pipe < Lat. *ductio,* act of leading < *ductus,* p.part. of *ducere,* to lead.] **1. a.** A stream of water or air applied to a bodily part or cavity for cleansing or medicinal purposes. **b.** The application of a douche. **2.** An instrument for applying a douche. —*vt.* & *vi.*

**douched, douch·ing, douch·es.** To cleanse or treat or be cleansed or treated by a douche.

**dough** (dō) *n.* [ME *dogh* < OE *dāg.*] **1.** A thick, soft mixture of flour or meal, liquids, and various dry ingredients that is baked, esp. bread or pastry. **2.** A pasty mass similar to dough. **3.** *Slang.* Money.

**dough·boy** (dō′boi′) *n.* **1.** Bread dough rolled thin and cut into various shapes and deep-fried. **2.** An infantryman in World War I.

**dough·face** (dō′fās) *n.* A Northerner sympathetic to the South in the American Civil War, esp. a congressman who supported slavery.

**dough·nut** *also* **do·nut** (dō′nŭt′, -nət) *n.* A small, usu. ring-shaped cake made of deep-fried rich, light dough.

**dough·ty** (dou′tē) *adj.* **-ti·er, -ti·est.** [ME < OE *dohtig.*] Marked by valor : COURAGEOUS. **—dough′ti·ly** *adv.* **—dough′ti·ness** *n.*

**dough·y** (dō′ē) *adj.* **-i·er, -i·est.** Having the appearance or consistency of dough.

**Doug·las fir** (dŭg′ləs) *n.* [After David *Douglas* (1798–1834).] A tall, short-needled evergreen timber tree, *Pseudotsuga taxifolia* or *P. menziesii* of northwestern North America, with egg-shaped cones.

**Dou·kho·bor** (dōō′kə-bôr′) *n. var. of* DUKHOBOR.

**dour** (dōōr, dour) *adj.* [ME < Lat. *durus,* hard.] **1.** Characterized by intractable sternness : FORBIDDING. **2.** Morose and gloomy : ILL-HUMORED. **3.** Sternly unyielding : OBSTINATE.

**dou·rine** (dōō-rēn′) *n.* [Fr. < Ar. *darina,* to be dirty.] A contagious equine venereal disease, caused by a microorganism.

**douse**[1] *also* **dowse** (dous) [Perh. < obs. *douse,* to strike.] *—v.* **doused, dous·ing, dous·es** *also* **dowsed, dows·ing, dows·es.** *—vt.* **1.** To plunge into liquid : IMMERSE. **2.** To drench thoroughly. **3.** To put out (a light or fire) : EXTINGUISH. *—vi.* To become thoroughly wet. *—n.* A thorough drenching. **—dous′er** *n.*

**douse**[2] (douz) *v. var. of* DOWSE[1].

**dove**[1] (dŭv) *n.* [ME *douve* < OE *dūfe.*] **1.** Any of various birds of the family Columbidae, which includes the pigeons. **2.** A gentle or innocent child or woman. **3.** A messenger of peace or deliverance from care in the Old Testament. **4.** One who advocates peace and conciliatory measures as opposed to armed conflict. **5. Dove.** The constellation Columba. **—dov′ish** *adj.* **—dov′ish·ness** *n.*

**dove**[2] (dōv) *v. var. p.t. of* DIVE[1].

**dove·cote** (dŭv′kōt′, -kŏt′) *also* **dove·cot** (-kŏt′) *n.* A roost for domesticated pigeons.

**dove·kie** *also* **dove·key** (dŭv′kē) *n.* [Dim. of DOVE[1].] A small black-and-white bird, *Plautus alle* of arctic and northern Atlantic regions.

**Do·ver's powder** (dō′vərz) *n.* [After Thomas *Dover* (1660–1742).] A powdered drug of ipecac and opium, formerly used as an analgesic and a sudorific.

**dove·tail** (dŭv′tāl′) *n.* **1.** A fan-shaped tenon that forms a tight interlocking joint when fitted into a corresponding mortise. **2.** A joint formed by interlocking dovetails and mortises. *—v.* **-tailed, -tail·ing, -tails.** *—vt.* **1.** To cut into or join by means of dovetails. **2.** To combine or connect harmoniously or precisely. *—vi.* To interlock or combine into a unified whole.

**dow·a·ger** (dou′ə-jər) *n.* [OFr. *douagiere* < *douage,* dower < *douer,* to endow < Lat. *dotare* < *dos,* dowry.] **1.** A widow who holds property or a title from her deceased husband. **2.** An elderly woman of high rank.

**dow·dy** (dou′dē) *adj.* **-di·er, -di·est.** [< ME *doude,* unattractive woman.] **1.** Not stylish or neat : SHABBY. **2.** Old-fashioned in manner or appearance. *—n., pl.* **-dies.** A dowdy woman : FRUMP. **—dow′di·ly** *adv.* **—dow′di·ness** *n.* **—dow′dy·ish** *adj.*

**dow·el** (dou′əl) *n.* [ME *doule,* part of a wheel.] **1.** A usu. round pin that fits tightly into a corresponding hole to join or align two adjacent pieces. **2.** A piece of wood driven into a wall to act as an anchor for nails. *—vt.* **-eled, -el·ing, -els. 1.** To join or align with dowels. **2.** To provide with dowels.

**dow·er** (dou′ər) *n.* [ME *douere* < OFr. *douaire* < Med. Lat. *dotarium* < Lat. *dos,* dowry.] **1.** The part or interest of a deceased man's real estate allotted by law to his widow for her lifetime. **2.** DOWRY 1. *—vt.* **-ered, -er·ing, -ers.** To assign a dower to : ENDOW.

**dow·itch·er** (dou′ĭ-chər) *n.* [Of Iroquoian orig.] Either of two shore birds, *Limnodromus griseus* or *L. scolopaeus* of northern regions, with brownish plumage and a long, straight bill.

**Dow Jones Averages** (dou′jōnz′). A trademark used for an index of the relative price of selected industrial, transportation, and utility stocks based on a formula developed and periodically revised by Dow Jones & Company, Inc.

**down**[1] (doun) *adv.* [ME *doun* < OE *dūne* < *adūne* : *a-,* from (< *of*) + *dūn,* hill.] **1. a.** From a higher to a lower position or place. **b.** Toward, to, or on the bottom, ground, or floor. **2. a.** Into a lower posture. **b.** In or into a prostrate position. **3.** Toward or in the south or in a southerly direction. **4. a.** Toward or in a center of activity <went *down* to the bank> **b.** Away from the speaker's point of reference <*down* on the ranch> **5.** To the source <tracked the gossip *down*> **6.** Toward or at a low or lower point on a scale. **7.** To or in a subdued or quiescent state. **8.** To or in a low status, as of subjection or disgrace. **9.** To an extreme degree. **10.** Seriously or vigorously <got *down* to business> **11.** From earlier times or people. **12.** To a reduced or concentrated form. **13.** In writing : on paper <took *down* my remarks> **14.** In partial payment at the time of purchase <ten dollars *down*> *—adj.* **1. a.** Moving or directed downward <a *down* staircase> **b.** In a low position. **c.** At a reduced level. **2. a.** Sick <*down* with the flu> **b.** Low in spirits : DEJECTED. **3. a.** Trailing an opponent by a given number of points, goals, or strokes <was *down* by two points> **b.** *Football.* Not in play. —Used of the ball. **c.** *Baseball.* Having been put out. **4.** Being the first installment <a *down* payment> *—prep.* **1.** In a descending direction along, upon, into, or through. **2.** Along the course of. **3.** Toward the mouth of a river. *—n.* **1.** A downward movement : DESCENT. **2.** *Football.* Any of a series of four plays during which a team must advance at least ten yards to retain possession of the ball. *—v.* **downed, down·ing, downs.** *—vt.* **1.** To bring, put, strike, or throw down. **2.** To swallow hastily : GULP. **3.** *Football.* To cause (the ball) to be out of play. *—vi.* To go or come down : DESCEND. **—down and out. 1.** Physically weakened. **2.** Impoverished : destitute. **—down in the mouth.** Discouraged : depressed. **—down on.** *Informal.* Negative or hostile toward.

**down**[2] (doun) *n.* [ME *doun* < ON *dūnn.*] **1.** Fine, soft, fluffy feathers on a young bird or underlying the contour feathers in adult birds. **2.** *Bot.* A covering of soft, short fibers, as on some leaves. **3.** A soft, silky, or feathery substance, as the first growth of human beard.

**down**[3] (doun) *n.* [ME *doune* < OE *dūn.*] **1.** *often* **downs.** A rolling, grassy upland used for grazing. **2.** *often* **Down.** Any of several breeds of sheep with short wool, developed in the downs of England.

**down-at-heel** (doun′ət-hēl′) *or* **down-at-the-heel** (-ət-thə-hēl′) *adj.* Showing signs of wear and tear : SHABBY.

**down·beat** (doun′bēt′) *n.* **1.** *Mus.* The downward stroke made by a conductor to indicate the first beat of a measure. **2.** *Informal.* A period of inactivity or stagnation. *—adj.* Gloomy : pessimistic.

**down-bow** (doun′bō′) *n. Mus.* A stroke made by drawing a bow from handle to tip across the strings of a bowed instrument, as a violin.

**down·cast** (doun′kăst′) *adj.* **1.** Directed downward <*downcast* eyes> **2.** Low in spirits : DEPRESSED.

**down·court** (doun-kôrt′, -kōrt′) *adv. & adj.* To, into, or in the far end of the court, esp. in basketball.

**Down East** *also* **down East** *n.* New England, esp. Maine.

**down·er** (dou′nər) *n. Slang.* **1.** A depressant or sedative drug, as a barbiturate or tranquilizer. **2.** One that is depressing.

**down·fall** (doun′fôl′) *n.* **1. a.** A sudden loss of wealth, position, reputation, or happiness : RUIN. **b.** Something causing a downfall. **2.** A fall of rain or snow, esp. when heavy or unexpected.

**down·fall·en** (doun′fô′lən) *adj.* Having experienced loss or ruin.

**down·field** (doun-fēld′) *adv. & adj.* To, into, or in a defensive team's end of the field.

**down·grade** (doun′grād′) *n.* **1.** A descending slope, as in a road. **2.** A downward turn or trend. *—vt.* **-grad·ed, -grad·ing, -grades. 1.** To lower the status or salary of. **2.** To lower or minimize the importance or reputation of. **—on the downgrade.** Declining, as in wealth, position, or influence.

**down·haul** (doun′hôl′) *n.* A rope or set of ropes for hauling down or securing a sail or spar.

**down·heart·ed** (doun′här′tĭd) *adj.* Low in spirits : DOWNCAST. **—down′heart′ed·ly** *adv.* **—down′heart′ed·ness** *n.*

**down·hill** (doun′hĭl′) *adv.* Down the slope of a hill. *—adj.* (doun′-hĭl′). Sloping downward : DESCENDING. **—go downhill.** To decline, as in status or health.

**down-home** (doun′hōm′) *adj.* Of, pertaining to, or typical of the rural southern United States or its people, as in naturalness.

**Down·ing Street** (dou′nĭng) *n.* [From the location of the prime minister's residence at No. 10 Downing Street, London.] The British government.

**down·play** (doun′plā′) *vt.* **-played, -play·ing, -plays.** To minimize the significance of.

**down·pour** (doun′pôr′, -pōr′) *n.* A heavy fall of rain.

**down·range** (doun′rānj′) *adv.* In a direction away from the launch site and along the flight line of a missile test range. *—adj.* (doun′-rānj′). Designating the area and airspace along the flight line of a missile test range.

**down·right** (doun′rīt′) *adj.* **1.** Thoroughgoing or unambiguous <*downright* slander> **2.** Candid : forthright. *—adv.* Thoroughly : absolutely.

**down·size** (doun′sīz′) *vt.* **-sized, -siz·ing, -siz·es.** To make in a smaller size <will *downsize* next year's automobiles>

**Down's syndrome** (dounz) *n.* [After John L.H. *Down* (1828–1896).] A congenital disorder marked by moderate to severe mental retardation, a short flattened skull, and slanting eyes.

**down·stage** (doun′stāj′) *adv.* Toward or at the front part of a stage. **—down′stage′** *adj. & n.*

**down·stairs** (doun′stârz′) *adv.* **1.** Down the stairs. **2.** To or on a lower floor. *—adj.* (doun′stârz′) *also* **down·stair** (-stâr′). Located on

a lower or main floor. —*n.* (doun'stârz') *(sing. in number).* The lower or main floor.
**down·state** (doun'stāt') *n.* The southerly section of a U.S. state. —*adv. & adj.* To, from, or in the southerly section of a state. **—down'stat'er** *n.*
**down·stream** (doun'strēm') *adj.* In the direction of a stream's current. —*adv.* (doun'strēm'). Down a stream.
**down·swing** (doun'swĭng') *n.* **1.** A swing downward. **2.** A trend downward in business.
**down·tick** (doun'tĭk') *n.* A transaction in a stock market security below the price of the previous transaction.
**down·time** (doun'tīm') *n.* The period of time when something, as a machine or a factory, is inactive.
**down-to-earth** (doun'tŏŏ-ûrth', -tə-) *adj.* Sensible : realistic.
**down·town** (doun'toun') *n.* The lower part or the business center of a city or town. —*adv.* (doun'toun'). To, toward, or in the lower part or the business center of a city or town. —*adj.* (doun'toun'). Of or located downtown.
**down·trend** (doun'trĕnd') *n.* A downturn.
**down·trod·den** (doun'trŏd'n) *adj.* Subjected to oppression.
**down·turn** (doun'tûrn') *n.* A tendency downward, esp. in economic or business activity.
**down under** *n. Informal.* Australia or New Zealand.
**down·ward** (doun'wərd) *adv.* **1.** From a higher to a lower place, point, level, or condition. **2.** From an earlier to a more recent time. —*adj.* **1.** Descending downward. **2.** Descending from an origin or source. **—down'wards, down'ward·ly** *adv.*
**down·wind** (doun'wĭnd') *adv.* In the direction in which the wind blows : LEEWARD. **—down'wind'** *adj.*
**down·y** (dou'nē) *adj.* **-i·er, -i·est. 1.** Made of or covered with down. **2. a.** Resembling down. **b.** Soothing.
**downy mildew** *n.* A plant disease caused by fungi of the order Peronosporales and marked by gray, velvety patches of spores on the lower surfaces of leaves.
**dow·ry** (dou'rē) *n., pl.* **-ries.** [ME *douerie* < AN *dowarie,* ult. < Med. Lat. *dotarium,* dower.] **1.** Money or property brought by a bride to her husband at marriage. **2.** *Archaic.* DOWER 1. **3.** A sum of money required of a postulant at a convent. **4.** A natural gift or endowment.
**dowse**[1] *also* **douse** (douz) *vi.* **dowsed, dows·ing, dows·es** *also* **doused, dous·ing, dous·es.** [Orig. unknown.] To use a divining rod to find underground water or minerals. **—dows'er** *n.*
**dowse**[2] (dous) *v. & n. var. of* DOUSE[1].
**Dow theory** (dou) *n.* [After Charles H. *Dow* (1851–1902).] A theory of stock market forecasting based on the activity of the market itself.
**dox·ol·o·gy** (dŏk-sŏl'ə-jē) *n., pl.* **-gies.** [Med. Lat. *doxologia* < Gk., praise : *doxa,* glory, honor (< *dokein,* to seem) + *logos,* speech.] A liturgical prayer or hymn of praise to God. **—dox'o·log'i·cal** (-sə-lŏj'ĭ-kəl) *adj.* **—dox'o·log'i·cal·ly** *adv.*
**dox·y** (dŏk'sē) *n., pl.* **-ies.** [Perh. < obs. Du. *docke,* doll.] *Slang.* **1.** A prostitute. **2.** MISTRESS 6.
**dox·y·cy·cline** (dŏk'sĭ-sī'klēn') *n.* [D(E)- + OX(Y)- + (TETRA)CYCLINE.] A broad-spectrum antibiotic, $C_{22}H_{24}N_2O_8$, obtained from tetracycline.
**doy·en** (doi-ĕn', doi'ən, dwä-yăN') *n.* [Fr. < LLat. *decanus,* chief of ten. —see DEAN.] The eldest or senior male member of a group.
**doy·enne** (doi-ĕn', dwä-yĕn') *n.* The eldest or senior female member of a group.
**doy·ly** or **doy·ley** (doi'lē) *n. vars. of* DOILY.
**doze** (dōz) *v.* **dozed, doz·ing, doz·es.** [Prob. of Scand. orig.] —*vi.* To sleep lightly. —*vt.* To spend (time) dozing or as if dozing <*dozed* the afternoon away> **—doze off.** To fall into a light sleep. —*n.* A brief, light sleep. **—doz'er** *n.*
**doz·en** (dŭz'ən) *n.* [ME *dozeine* < OFr. *dozaine* < *doze,* twelve < Lat. *duodecim* : *duo,* two + *decem,* ten.] **1.** *pl.* **dozen.** A set of 12. **2.** *pl.* **-ens.** An indefinite number <*dozens* of things to do> —*adj.* Twelve. **—doz'enth** *adj.*
**do·zy** (dō'zē) *adj.* **-zi·er, -zi·est.** Half asleep : DROWSY. **—doz'i·ly** *adv.* **—do'zi·ness** *n.*
**drab**[1] (drăb) *adj.* **drab·ber, drab·best.** [Obs. *drap,* cloth < OFr. —see DRAPE.] **1. a.** Of a light dull brown color. **b.** Of a light olive brown or khaki color. **2.** Faded in appearance. **3.** Of a commonplace or dreary character. —*n.* **1.** Cloth of a light dull brown or grayish brown or unbleached natural color, esp. a heavy woolen or cotton fabric. **2.** A light yellowish or olive brown. **—drab'ly** *adv.* **—drab'ness** *n.*
**drab**[2] (drăb) *n.* [Of Celtic orig.] **1.** A slattern. **2.** A whore. —*vi.* **drabbed, drab·bing, drabs.** To consort with whores.
**drab·bet** (drăb'ĭt) *n.* [< DRAB[1].] A coarse, unbleached linen.
**drab·ble** (drăb'əl) *v.* **-bled, -bling, -bles.** [ME *drabelen.*] —*vi.* To become muddy and wet : DRAGGLE. —*vt.* To bedraggle.
**dra·cae·na** (drə-sē'nə) *n.* [NLat. *Dracaena,* genus name < LLat. *dracaena,* female dragon < Gk. *drakaina,* fem. of *drakōn,* serpent.] Any of several tropical plants of the genera *Dracaena* and *Cordyline,* some species of which are widely grown as house plants for their decorative foliage.

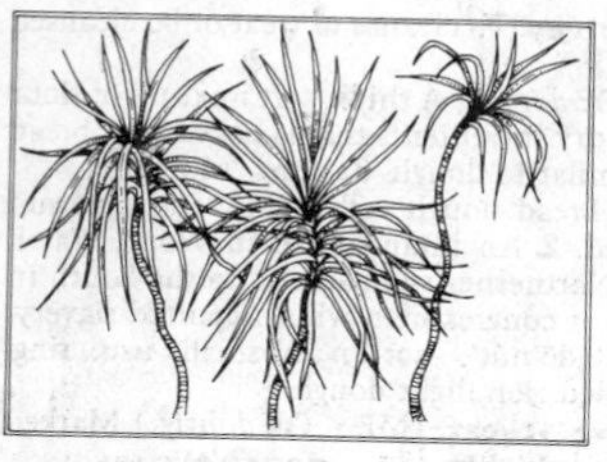
**dracaena**

**drachm** (drăm) *n. Chiefly Brit.* **1.** A dram. **2.** A drachma.
**drach·ma** (drăk'mə) *n., pl.* **-mas** or **-mae** (-mē) [Lat. < Gk. *drakhmē.*] **1.** —See table at CURRENCY. **2.** A silver coin of ancient Greece. **3.** One of several modern units of weight, esp. the dram.
**Dra·co** (drā'kō) *n.* [Lat. —see DRAGON.] A constellation in the Northern Hemisphere.
**dra·co·ni·an** (drā-kō'nē-ən, drə-) *also* **dra·con·ic** (-kŏn'ĭk) *adj.* [After *Draco,* Athenian lawgiver of the 7th cent. B.C., whose laws were proverbially harsh.] **1.** Of or designating a law or code of extreme severity. **2.** Exceedingly rigorous and harsh <a *draconian* punishment> **—dra·con'i·cal·ly** *adv.*
**dra·con·ic**[1] (drā-kŏn'ĭk) *adj.* [< Lat. *draco, dracon-,* dragon < Gk. *drakōn,* serpent.] Of or relating to a dragon.
**dra·con·ic**[2] (drā-kŏn'ĭk, drə-) *adj. var. of* DRACONIAN.
**draft** (drăft, dräft) *n.* [ME *draught,* act of drawing or pulling.] **1. a.** A current of air in an enclosed space. **b.** A device in a flue controlling the circulation of air. **2. a.** A pull or traction of a load. **b.** Something pulled or drawn. **c.** A team of animals used to pull or draw a load. **3.** The depth of a vessel's keel below the water line, esp. when loaded. **4.** A heavy demand on resources. **5.** A documentary instrument for transferring money. **6. a.** A gulp, swallow, or inhalation. **b.** The amount taken in by an act of drinking or inhaling. **c.** A measured portion : DOSE. **7. a.** The withdrawal of a liquid, as from a keg. **b.** The amount withdrawn. **8. a.** A purposeful selection of one or more individuals from a group <the committee *drafted* a candidate> **b.** Conscription for military service. **c.** The group selected or conscripted. **d.** A system in which professional sports teams get the exclusive rights to new players. **9. a.** The act of drawing in a fishnet. **b.** The catch drawn in. **10. a.** A preliminary outline, plan, or version <a rough *draft* of the memo> **b.** A representation of something to be constructed. **11.** A narrow line chiseled on a stone to guide the stonecutter in leveling its surface. **12.** A slight taper given a die to ease the removal of a casting. **13.** An allowance made to a buyer for loss in weight of merchandise. —*v.* **draft·ed, draft·ing, drafts.** —*vt.* **1.** To select and draw from a group for some usu. compulsory assignment, as military service. **2. a.** To draw up a preliminary version of or plan for. **b.** To compose <*draft* a lecture> —*vi.* To drive or ride close behind another vehicle in a race to take advantage of the reduced air pressure in its wake. —*adj.* **1.** Suited or used for drawing heavy loads <a *draft* animal> **2.** Drawn from a keg or tap <*draft* beer> **—on draft.** Tapped from the keg.
**draft board** *n.* A local board of civilians in charge of the selection of individuals for compulsory military service.
**draft·ee** (drăf-tē', dräf-) *n.* One drafted, esp. for military service.
**draft·er** (drăf'tər, dräf'-) *n.* One who drafts, esp.: **a.** One who draws plans or designs. **b.** One who draws up documents.
**draft·ing** (drăf'tĭng, dräf'-) *n.* The systematic representation and dimensional specification of mechanical or architectural structures.
**drafts·man** (drăfts'mən, dräfts'-) *n.* **1.** A man who functions as a drafter. **2.** One who excels in drawing. **—drafts'man·ship'** *n.*
**drafts·per·son** (drăfts'pûr'sən, dräfts'-) *n.* A drafter.
**drafts·wom·an** (drăfts'wŏŏm'ən, dräfts'-) *n.* A woman who functions as a drafter.
**draft·y** (drăf'tē, dräf'-) *adj.* **-i·er, -i·est.** Having or exposed to drafts of air. **—draft'i·ly** *adv.* **—draft'i·ness** *n.*
**drag** (drăg) *v.* **dragged, drag·ging, drags.** [ME *draggen* < ON *draga.*] —*vt.* **1. a.** To pull or draw forcefully along the ground. **b.** To cause to trail along the ground. **2. a.** To search or dredge the bottom of (a body of water), as with a grappling hook or dragnet. **b.** To bring up or catch by such means. **3.** To bring forcibly to or into <always had to *drag* us to the doctor> **4.** To cause to move with great reluctance, weariness, or difficulty <*dragged* myself to work> **5.** To prolong tediously, as a story. **6.** To introduce gratuitously into a discussion. **7.** *Baseball.* To hit (a bunt) down either foul line by pushing or pulling the ball with the bat while taking the first step to first base. —*vi.* **1.** To trail along the ground. **2.** To move slowly or with effort. **3.** To lag behind. **4.** To advance slowly, tediously, or laboriously <The time just *dragged.*> **5.** To search or dredge the bottom of a body of water. **6.** *Baseball.* To hit a drag bunt. **7.** To participate in or as if in a drag race. **8.** *Slang.* To draw on a cigarette, pipe, or cigar. —*n.* **1.** The act of dragging. **2.** Something dragged along the

ground, as a harrow. **3.** A device for dragging under water, as a grappling hook. **4.** A heavy cart or sledge for conveying loads. **5.** A large four-horse coach with seats inside and on top. **6.** Something that stops or slows motion, as a brake on a fishing reel. **7.** One that hinders or prohibits progress **:** DRAWBACK. **8.** Degree of resistance involved in dragging or hauling. **9.** The retarding force exerted on a moving body by a fluid medium in aviation. **10.** Slow, laborious movement or motion. **11. a.** The scent or trail of an animal, as a fox. **b.** Something that provides an artificial scent. **12.** *Slang.* An obnoxious or tiresome bore. **13.** *Slang.* A puff on a cigarette, pipe, or cigar. **14.** *Slang.* A street or road <turned off the main *drag*>

**drag bunt** *n. Baseball.* A bunt used esp. by left-handed batters in which the batter tries to push or pull the ball down either foul line by hitting it with the bat trailing behind after already taking the first step toward first base.

**dra·gée** (drä-zhā′) *n.* [Fr. < OFr. *dragie.* —see DREDGE[2].] **1.** A small, often medicated candy. **2.** A small, silver-colored ball used in cake decorating.

**drag·ger** (drăg′ər) *n.* **1.** One that drags. **2.** A fishing vessel that makes its catch in nets dragged along the bottom.

**drag·gle** (drăg′əl) *v.* **-gled, -gling, -gles.** [< DRAG.] —*vt.* To make wet and dirty by dragging in mud. —*vi.* **1.** To become muddy by being dragged along the ground. **2.** To follow slowly **:** STRAGGLE.

**drag·gy** (drăg′ē) *adj.* **-gi·er, -gi·est. 1.** Listless and dull. **2.** *Slang.* Obnoxiously tiresome.

**drag·line** (drăg′līn′) *n.* **1.** A line used for dragging. **2.** An excavating machine.

**drag link** *n.* A link for transmitting rotary motion between cranks on two parallel but slightly offset shafts, as the rod connecting the lever of the steering gear to the steering arm in an automobile.

**drag·net** (drăg′nĕt′) *n.* **1. a.** A trawling net. **b.** A net for catching small game. **2.** A system of procedures used in apprehension, esp. of criminal suspects.

**drag·o·man** (drăg′ə-mən) *n., pl.* **-mans** *or* **-men.** [ME *drugeman* < OFr. < Med. Lat. *dragumannus* < Gk. *dragoumanos* < Ar. *targumān* < Aram. *tūrgemānā* < Akkadian *targumānu,* interpreter < *ragāmu,* to call.] An interpreter or guide in countries where Arabic, Turkish, or Persian is spoken.

**drag·on** (drăg′ən) *n.* [ME < OFr. < Lat. *draco* < Gk. *drakōn,* serpent.] **1.** A monster represented as a gigantic reptile having the tail of a serpent, a lion's claws, wings, and a scaly skin. **2.** *Archaic.* A large snake. **3.** A fierce, violent, or strict person. **4.** Any of various lizards, as one of the genus *Draco.* **5. Dragon.** Draco.

**drag·on·et** (drăg′ə-nĭt) *n.* [ME < *dragon,* dragon.] Any of various small, slender-bodied, often brightly colored marine fishes of the family Callionymidae, with a flattened head.

**drag·on·fly** (drăg′ən-flī′) *n., pl.* **-flies.** Any of various large insects of the order Odonata, with two pairs of narrow, net-veined wings and a long, slender body.

**drag·on·head** (drăg′ən-hĕd′) *n.* A plant of the genera *Dracocephalum* or *Physostegia,* with rose-pink or purplish flower spikes.

**drag·on·root** (drăg′ən-ro͞ot′, -ro͝ot′) *n.* The green dragon.

**dragon's blood** (drăg′ənz) *n.* **1.** A red, resinous substance obtained from the fruit of a tropical Asian tree, *Daemonorops draco,* once used in making varnishes and lacquers. **2.** Any of several resins similar to dragon's blood.

**dragon tree** *n.* A tree, *Dracaena draco* of the Canary Islands, with a thick trunk, sword-shaped leaf clusters, and orange fruit.

**dra·goon** (drə-go͞on′, dră-) *n.* [Fr. *dragon,* dragoon, carbine < OFr., dragon. —see DRAGON.] A heavily armed trooper in some European armies of the 17th and 18th cent. —*vt.* **-gooned, -goon·ing, -goons. 1.** To persecute by the use of troops. **2.** To force by violent measures **:** HARASS.

**drag race** *n.* A race between cars to determine which can accelerate faster from a standstill. **—drag racing** *n.*

**drag strip** *n.* A straight strip of dirt or pavement at least one-fourth mile long, used for drag racing.

**drain** (drān) *v.* **drained, drain·ing, drains.** [ME *dreinen* < OE *drēahnian.*] —*vt.* **1.** To draw off (a liquid) by a gradual process <*drained* water from the pool> **2. a.** To cause liquid to go out from **:** EMPTY <*drained* the pool> **b.** To draw off the surface water of. **3.** To drink all the contents of. **4. a.** To use up completely **:** EXHAUST <medical bills that *drained* our earnings> **b.** To fatigue or spend physically or emotionally. —*vi.* **1.** To flow off or go out of something. **2.** To become empty or dry by the drawing off of liquid. **3.** To discharge surface waters through natural drainage channels in a given region or tract of land. —*n.* **1.** A pipe or channel by which liquid is drawn off. **2.** A device, as a tube, inserted into a cavity or wound to facilitate discharge of fluid. **3.** The act or process of draining. **4. a.** A gradual outflow or depletion <a *drain* on the economy> **b.** Something that causes a depletion. **—down the drain. 1.** Used wastefully. **2.** Proving fruitless. **—drain′a·ble** *adj.* **—drain′er** *n.*

**drain·age** (drā′nĭj) *n.* **1.** The action or a method of draining. **2.** A natural or artificial system of drains. **3.** Something drained off.

**drainage basin** *n.* The region drained by a river system.

**drain·pipe** (drān′pīp′) *n.* A pipe for removing rainwater or sewage.

**drake**[1] (drāk) *n.* [ME.] A male duck.

**drake**[2] (drāk) *n.* [ME, dragon < OE *draca* < Lat. *draco.* —see DRAGON.] A mayfly used as fishing bait.

**dram** (drăm) *n.* [ME *dragme* < OFr. < Med. Lat. *dragma* < Lat. *drachma.* —see DRACHMA.] **1. a.** A unit of weight in the U.S. Customary System, an avoirdupois unit equal to 1.771 grams or 0.0625 ounce. **b.** A unit of apothecary weight, equal to 3.889 grams or 0.125 ounce. **2. a.** A small drink, as of a liqueur. **b.** A small amount **:** BIT <not a *dram* of sympathy>

**dra·ma** (drä′mə, drăm′ə) *n.* [LLat. < Gk. < *dran,* to do.] **1.** A play in prose or verse, esp. one recounting a serious story. **2.** Dramatic art of a particular kind or period <Shakespearean *drama*> **3.** The art or practice of writing or producing plays. **4.** A real-life situation or succession of events having the dramatic progression or emotional content typical of a play. **5.** The quality or condition of being dramatic.

**Dram·a·mine** (drăm′ə-mēn′). A trademark for dimenhydrinate, a drug used to treat motion sickness.

**dra·mat·ic** (drə-măt′ĭk) *adj.* [LLat. *dramaticus* < Gk. *dramatikos* < *drama,* drama.] **1.** Of or relating to drama or the theater. **2.** Like a drama in emotional content or sequential progression. **3.** Striking, as in appearance or effect. **—dra·mat′i·cal·ly** *adv.*

**dramatic monologue** *n.* A literary work, esp. in verse, in which a figure reveals his or her character in a monologue addressed directly to the reader or to another person.

**dra·mat·ics** (drə-măt′ĭks) *n. (sing. or pl. in number).* **1.** The art and practice of acting and stagecraft. **2.** Exaggerated behavior.

**dram·a·tis per·so·nae** (drăm′ə-tĭs pər-sō′nē, drä′mə-tĭs pər-sō′nī′) *pl.n.* [NLat., persons of the drama.] **1.** The characters in a play or story. **2.** A list of the characters in a play or story.

**dram·a·tist** (drăm′ə-tĭst, drä′mə-) *n.* A playwright.

**dram·a·ti·za·tion** (drăm′ə-tĭ-zā′shən, drä′mə-) *n.* **1.** The art or act of transforming into a drama or play. **2.** A dramatic version.

**dram·a·tize** (drăm′ə-tīz′, drä′mə-) *v.* **-tized, -tiz·ing, -tiz·es.** —*vt.* **1.** To adapt for presentation as a drama. **2.** To present or regard in a dramatic or melodramatic way. —*vi.* **1.** To be adaptable to dramatic form. **2.** To indulge in dramatic behavior.

**dram·a·turge** (drăm′ə-tûrj′, drä′mə-) *n.* [Fr. < Gk. *dramatourgos* : *drama,* drama + *ergon,* work.] A playwright.

**dram·a·tur·gy** (drăm′ə-tûr′jē, drä′mə-) *n.* The art of the theater. **—dram′a·tur′gic, dram′a·tur′gi·cal** *adj.*

**drank** (drăngk) *v. p.t. of* DRINK.

**drape** (drāp) *v.* **draped, drap·ing, drapes.** [ME *drapen,* to weave < OFr. < *drap,* cloth < LLat. *drappus,* of Celtic orig.] —*vt.* **1.** To cover or hang with or as if with loose folds of cloth. **2.** To let fall or arrange in loose folds, as a garment. **3.** To rest or hang limply <*draped* my arms over the counter> —*vi.* To hang or fall in loose folds. —*n.* **1.** *often* **drapes. a.** A drapery. **b.** A cloth used in a hospital operating room to reduce the site of maximum sterility. **2.** The manner in which cloth hangs or falls.

**drap·er** (drā′pər) *n.* [ME, weaver < OFr. *drapier* < *drap,* cloth. —see DRAPE.] *Chiefly Brit.* A dealer in cloth or clothing and dry goods.

**drap·er·y** (drā′pə-rē) *n., pl.* **-ies. 1.** Cloth or clothing gracefully arranged in loose folds. **2.** *often* **draperies.** Curtains, usu. of heavy fabric, that hang straight in loose folds. **3.** Fabric **:** cloth. **4.** *Chiefly Brit.* The business of a draper.

**dras·tic** (drăs′tĭk) *adj.* [Gk. *drastikos,* active < *dran,* to do.] **1.** Taking effect rapidly or violently. **2.** Extremely radical or severe <*drastic* measures to economize> **—dras′ti·cal·ly** *adv.*

**drat** (drăt) *interj.* [Short for *God rot.*] —Used to express annoyance.

**draught** (drăft) *n. & v. & adj. Chiefly Brit. var. of* DRAFT.

**draughts** (drăfts, drăfts) *n.* [ME *draughtes,* pl. of *draught,* move at chess, act of pulling.] *(sing. in number). Chiefly Brit.* The game of checkers.

**Dra·vid·i·an** (drə-vĭd′ē-ən) *n.* [< Skt. *drāviḍaḥ,* a Dravidian.] **1.** A large family of languages spoken esp. in southern India and northern Sri Lanka that includes Tamil, Telegu, Malayalam, and Kanarese. **2.** A member of any of the peoples that speak one of the Dravidian languages, esp. a member of the aboriginal population of southern India. **—Dra·vid′i·an, Dra·vid′ic** (-vĭd′ĭk) *adj.*

**draw** (drô) *v.* **drew** (dro͞o), **drawn** (drôn), **draw·ing, draws.** [ME *drawen* < OE *dragan.*] —*vt.* **1. a.** To cause to move after or toward one by exerting continuous force. **b.** To cause to move in a specified direction or to a specified position, as by leading <The lawyer *drew* us into the office.> **c.** To move or pull so as to cover or uncover <*draw* the curtains> **2.** To cause to flow forth <*draw* water from a faucet> **3.** To suck or take in (air) **:** INHALE. **4.** To displace (a specified depth of water) in floating <a vessel *drawing* 20 inches> **5. a.** To take or pull out for use, as a weapon. **b.** To eviscerate. **c.** To derive for one's own use or benefit <*drew* comfort from my family's support> **6. a.** To attract <The accident *drew* a crowd.> **b.** To select or take in from a particular group, type, or region <*draw* students from ethnic backgrounds> **7. a.** To induce to act. **b.** To provoke <*drew* fire from the enemy> **c.** To elicit in response **:** EVOKE <*drew* cheers from the fans> **8. a.** To earn or bring in <*draw* interest> **b.** To withdraw (money). **c.** To use (e.g., a check) when paying. **d.** To receive on a regular basis or at a specified time

<*draw* weekly wages> **9.** To take or receive by chance or in a chance drawing. **10. a.** To take (cards) from a dealer or central stack. **b.** To force (a card) to be played. **11.** To end or leave (a contest) undecided or tied. **12.** To hit or strike (a ball) so as to impart backspin. **13.** To pull back the string of (a bow). **14.** To distort the shape of <*draw* one's mouth into a sneer> **15.** To stretch tight. **16. a.** To flatten, stretch, or mold (metal) by hammering or die stamping. **b.** To shape or elongate (e.g., a wire) by drawing through dies. **17. a.** To describe (a line or figure) with a drafting implement. **b.** To draft or sketch (a picture). **18. a.** To portray, as by speech or writing <*draws* believable characters> **b.** To devise or formulate from evidence or information at hand <*draw* similes> **c.** To compose or write in a set form <*draw* a legal document> —*vi.* **1. a.** To advance steadily <The bus *drew* near.> **b.** To wield an attracting force <The new exhibit is *drawing* well.> **2. a.** To pour forth liquid <veins that *draw* easily> **b.** To take in a draft of air <*draw* on a pipe> **3. a.** To cause suppuration. **b.** To steep in the manner of tea. **4.** To pull out a weapon for use. **5.** To use or call upon part of a fund, as of a bank account, or a supply, as of experience. **6.** To contract or tighten (e.g., material). **7.** To tie in a contest. **8.** To sketch forms and figures. **—draw away.** To move ahead of (e.g., a competitor). **—draw back.** To pull back in order to avoid : RETREAT. **—draw down.** To deplete by using or spending. **—draw in.** To sketch roughly. **—draw on.** To approach <Night *draws on.*> **—draw out. 1.** To cause to converse easily. **2.** To drag out : PROLONG. **—draw up. 1.** To bring into order (as troops). **2.** To write in established or proper form <*draw up* a will> **3.** To bring or come to a halt. **4.** To bring (oneself) into an erect posture, often as a result of indignation or contempt. —*n.* **1.** An act of drawing. **2.** The result of drawing. **3.** Something drawn. **4.** The cards received as replacements in draw poker. **5.** Something that attracts interest or attendance. **6.** The movable part of a drawbridge. **7.** An advantage : edge. **8.** A tied contest. **9.** A natural drainage basin : GULLY. **—draw a blank.** To fail to find or recollect something. **—draw and quarter. 1.** To execute (a prisoner) by tying each limb to a horse and driving the horses in different directions. **2.** To eviscerate and dismember after hanging. **—draw straws.** To decide by a lottery with straws of unequal lengths. **—draw the line.** To set a limit, as of acceptable activity or behavior.

**draw·back** (drô′băk′) *n.* **1.** An undesirable feature or circumstance : DISADVANTAGE. **2.** A refund or remittance, as a discount on duties or taxes for goods destined for favored uses.

**draw·bar** (drô′bär′) *n.* **1.** A bar across the rear of a tractor for hitching machinery. **2.** A railroad coupler.

**draw·bridge** (drô′brĭj′) *n.* A bridge that can be raised or turned aside either to prevent access or to allow passage beneath it.

**draw·down** (drô′doun′) *n.* **1.** A reduction of the water level in a reservoir. **2.** The act, process, or result of depleting.

**draw·ee** (drô′ē′) *n.* One on whom an order for the payment of money is drawn.

**draw·er** (drô′ər) *n.* **1.** One who draws, esp. one who draws an order for the payment of money. **2.** (*also* drôr). A sliding boxlike compartment in furniture. **3. drawers** (drôrz). Underpants.

**draw·ing** (drô′ĭng) *n.* **1.** The act or an instance of drawing. **2.** The art of depicting forms or figures on a surface by means of lines. **3.** A representation or portrayal by drawing.

**drawing account** *n.* An account recording cash payments to an employee to cover expenses or as advances on commissions.

**drawing card** *n.* One that attracts interest or attendance.

**drawing pin** *n.* *Chiefly Brit.* A thumbtack.

**drawing room** *n.* [Short for *withdrawing room.*] **1.** A formal reception room. **2.** A ceremonial reception. **3.** A private room on a railroad sleeping car.

**draw·knife** (drô′nīf′) *n.* A woodworking tool with a handle at each end of the blade, used to shave a surface.

**drawl** (drôl) *v.* **drawled, drawl·ing, drawls.** [Perh. < DRAW.] —*vi.* To speak with drawn-out vowels. —*vt.* To utter with a drawl. —*n.* A drawling manner of speech. **—drawl′er** *n.*

**drawn** (drôn) *adj.* [< p.part. of DRAW.] Haggard, as from weariness.

**drawn butter** *n.* Melted butter, often seasoned and used as a sauce.

**draw poker** *n.* A kind of poker in which each player is dealt five cards face down and may then discard and get replacements for a specified number of cards after the first round of betting.

**draw·shave** (drô′shāv′) *n.* A drawknife.

**draw·string** (drô′strĭng′) *n.* A cord or ribbon run through a casing or hem and pulled to close or tighten an opening.

**draw·tube** (drô′to͞ob′, -tyo͞ob′) *n.* A tube that slides within another.

**dray** (drā) *n.* [ME *draye,* sled < OE *dræge,* dragnet.] A low, heavy cart without sides, used for haulage. —*vt.* **drayed, dray·ing, drays.** To transport by dray.

**dray·age** (drā′ĭj) *n.* **1.** Transport by dray. **2.** A charge for drayage.

**dray·man** (drā′mən) *n.* A driver of a dray.

**dread** (drĕd) *v.* **dread·ed, dread·ing, dreads.** [ME *dreden* < OE *drǣdan.*] —*vt.* **1.** To be in great fear of. **2.** To hold in reverence or awe. **3.** To anticipate with alarm, anxiety, or reluctance <*dreaded* surgery> —*vi.* To be very afraid. —*n.* **1.** Profound fear : TERROR. **2.** Reverence : awe. **3.** Fearful or anxious anticipation. **4.** The object of fear, awe, or reverence. —*adj.* **1.** Causing fear : TERRIFYING. **2.** Inspiring awe.

**dread·ful** (drĕd′fəl) *adj.* **1. a.** Causing fear or dread. **b.** Inspiring awe. **2.** Extremely disagreeable or distasteful. **—dread′ful·ly** *adv.* **—dread′ful·ness** *n.*

**dread·locks** (drĕd′lŏks′) *pl.n.* Hair that has formed long matted clumps due to lack of washing and combing.

dreadlocks

**dread·nought** (drĕd′nôt′) *n.* A heavily armed battleship.

**dream** (drēm) *n.* [ME *drem* < OE *drēam,* joy and ON *draumr,* dream.] **1.** A series of mental images, ideas, and emotions occurring in certain stages of sleep. **2.** A reverie : daydream. **3.** A state of abstraction : TRANCE. **4.** A wild hope or fancy. **5.** An ambition. **6.** One that is extremely pleasant, beautiful, or fine. —*v.* **dreamed** or **dreamt** (drĕmt), **dream·ing, dreams.** —*vi.* **1.** To experience a dream in sleep. **2.** To daydream. **3.** To have an ambition <*dream* of wealth and power> —*vt.* **1.** To experience an image sequence of in sleep. **2.** To conceive of : IMAGINE. **3.** To pass (time) idly or in reverie. **—dream of.** To consider something practical or feasible <wouldn't *dream of* selling> **—dream up.** To invent <*dream up* an excuse>

**dream·er** (drē′mər) *n.* **1.** One who dreams. **2. a.** A visionary. **b.** An idealist. **3.** An impractical person.

**dream·land** (drēm′lănd′) *n.* **1.** An ideal or imaginary land. **2.** A state of sleep.

**dream·scape** (drēm′skāp′) *n.* A scene or picture having surreal qualities.

**dreamt** (drĕmt) *v. var. p.t. & p.p. of* DREAM.

**dream·y** (drē′mē) *adj.* **-i·er, -i·est. 1.** Like a dream : VAGUE. **2.** Given to daydreams or reverie. **3.** Soothing and peaceful. **4.** *Informal.* Inspiring pleasure or delight : MARVELOUS. **—dream′i·ly** *adv.* **—dream′i·ness** *n.*

**drear** (drîr) *adj.* Dreary.

**drea·ry** (drîr′ē) *adj.* **-ri·er, -ri·est.** [ME *dreri* < OE *drēorig,* sad, bloody < *drēor,* gore.] **1.** Dark and depressing : GLOOMY <*dreary* weather> **2.** Dull or boring <*dreary* chores> **—drea′ri·ly** *adv.* **—drea′ri·ness** *n.*

**dreck** (drĕk) *n.* [Yiddish *drek* and G. *Dreck,* dung < MHG *drec.*] *Slang.* **1.** Excrement. **2. a.** Trash. **b.** Shoddy or cheap merchandise : JUNK. **—dreck′y** *adj.*

**dredge**[1] (drĕj) *n.* [Sc. *dreg.*] **1.** Any of various machines equipped with scooping or suction devices used in deepening harbors and waterways and in underwater mining. **2.** A boat or barge equipped with a dredge. **3.** An implement consisting of a net on a frame, used for gathering shellfish. —*v.* **dredged, dredg·ing, dredg·es.** —*vt.* **1.** To clean, deepen, or widen with a dredge. **2.** To bring up with a dredge. **3.** To come up with : UNEARTH <*dredged* up old regrets> —*vi.* To use a dredge. **—dredg′er** *n.*

**dredge**[2] (drĕj) *vt.* **dredged, dredg·ing, dredg·es.** [< obs. *dredge,* a sweetmeat < ME *dragge* < OFr. *dragie* < Med. Lat. *dragia* < Lat. *tragemata,* confectionery < Gk. *tragēmata,* pl. of *tragēma,* sweetmeat < *trōgein,* to gnaw.] To coat (food) by sprinkling with a powder, as flour or sugar. **—dredg′er** *n.*

**D region** *n.* The region of the ionosphere approx. 25–40 miles or 40–65 kilometers above the earth.

**dregs** (drĕgz) *pl.n.* [ME *dregges* < ON *dregg.*] **1.** The sediment of a liquid : LEES. **2.** The least desirable part, as of society. **3.** A small amount : RESIDUE.

**drei·del** *also* **drei·dl** (drād′l) *n.* [Yiddish *dreydl* < *dreyen,* to turn < MHG *dræjen* < OHG *drǣen.*] **1.** A toy similar to a top with four sides marked with Hebrew letters. **2.** A game of chance played by children at Chanukah.

**drench** (drĕnch) *vt.* **drenched, drench·ing, drench·es.** [ME *drenchen,* to drown < OE *drencan,* to cause to drink.] **1.** To wet thoroughly : SOAK. **2.** To administer a dose of liquid medicine to (an animal). **3.** To soak or cover completely as if by drenching. —*n.* **1.** An act of drenching. **2.** A large dose of liquid medicine. **—drench′er** *n.*

**Dres·den china** (drĕz′dən) *n.* [After *Dresden,* a city in East Germany.] Meissen.

**dress** (drĕs) *v.* **dressed, dress·ing, dress·es.** [ME *dressen,* to prepare < OFr. *dresser,* to arrange < VLat. **directiare* < Lat. *directus,* p.part. of *dirigere,* to direct.] —*vt.* **1. a.** To put on (clothes). **b.** To furnish with clothing. **2.** To decorate or trim : ADORN. **3.** To arrange a display in <*dress* a showcase> **4.** To arrange (troops) in ranks : ALIGN. **5.** To apply therapeutic materials to (a wound). **6.** To groom (the hair). **7.** To curry (an animal). **8.** To cultivate (land or plants). **9.** To clean (fish or fowl) for cooking or sale. **10.** To put a finish on, as wood or stone. —*vi.* **1.** To put on clothes. **2.** To wear clothes. **3.** To wear formal clothes. **4.** To get into proper alignment. **—dress down.** To reprimand. **—dress up. 1.** To wear formal or fancy clothes. **2.** To arrange in ranks. —*n.* **1.** Clothing. **2.** A one-piece outer garment for women and girls. **3.** Outer covering or appearance. **4.** A style of clothing <informal *dress*>

**dres·sage** (drə-säzh′, drĕ-) *n.* [Fr., preparation < *dresser,* to arrange < OFr. —see DRESS.] Guidance of a horse through a series of complex maneuvers by slight movements of the hands, legs, and weight.

**dress circle** *n.* A section of seats in a theater or opera house, usu. the first tier above the orchestra.

**dress code** *n.* A set of rules, as in a school, indicating the approved manner of dress.

**dress·er[1]** (drĕs′ər) *n.* **1.** One that dresses. **2.** A wardrobe assistant, as for a performer. **3.** One who dresses well or in a specified way.

**dress·er[2]** (drĕs′ər) *n.* [ME *dressour,* table for preparing food < OFr. *dreceur* < *dresser,* to arrange. —see DRESS.] **1.** A chest of drawers with a mirror. **2.** A set of shelves or a cupboard for dishes or kitchen utensils.

**dress·ing** (drĕs′ĭng) *n.* **1.** The act of one that dresses. **2.** Medicine or bandages applied to a wound. **3.** A sauce, as for a salad. **4.** A stuffing, as for fish or fowl. **5.** Manure for dressing soil.

**dressing gown** *n.* A robe worn before dressing or for lounging.

**dressing room** *n.* A room esp. in a theater for changing costumes or clothes and applying make-up.

**dressing table** *n.* A low table with a mirror at which one sits while applying make-up.

**dress·mak·er** (drĕs′mā′kər) *n.* A maker of women's dresses. **—dress′mak′ing** *n.*

**dress parade** *n.* A military parade in dress uniform.

**dress rehearsal** *n.* A final rehearsal, as of a play, with costumes and stage properties.

**dress suit** *n.* A man's formal suit.

**dress uniform** *n.* A military uniform for formal occasions.

**dress·y** (drĕs′ē) *adj.* **-i·er, -i·est. 1.** Elegant or showy in dress or appearance. **2.** Smart or stylish. **—dress′i·ness** *n.*

**drew** (drōō) *v. p.t. of* DRAW.

**drib·ble** (drĭb′əl) *v.* **-bled, -bling, -bles.** [Freq. of *drib,* var. of DRIP.] —*vi.* **1.** To flow or fall in drops or an unsteady stream : TRICKLE. **2.** To let saliva drip from the mouth : DROOL. **3. a.** To dribble a ball. **b.** To proceed by dribbling. —*vt.* **1.** To let trickle or drip. **2. a.** To move (a ball or puck) by repeated light bounces, kicks, or light taps, as in basketball, soccer, or hockey. **b.** To hit (a ball) so as to cause slow bouncing. —*n.* **1.** A trickle or drip. **2.** A small quantity : BIT. **3.** The act of dribbling a ball. **—drib′bler** *n.*

**drib·let** (drĭb′lĭt) *n.* [< obs. *drib,* drop < *drib,* var. of DRIP.] **1.** A tiny falling drop of liquid. **2.** A small portion or amount.

**dribs and drabs** (drĭbz′n drăbz′) *pl.n.* [Redup. of obs. *drib,* drop. —see DRIBLET.] Small, sporadic amounts.

**dri·er[1]** *also* **dry·er** (drī′ər) *n.* **1.** One that dries. **2.** A substance added to paint, varnish, or ink to expedite drying.

**dri·er[2]** (drī′ər) *adj. compar. of* DRY.

**dri·est** (drī′ĭst) *adj. superl. of* DRY.

**†drift** (drĭft) *v.* **drift·ed, drift·ing, drifts.** [< ME, drove, herd, act of driving.] —*vi.* **1.** To be borne along by or as if by currents of air or water. **2.** To proceed or move smoothly and unhurriedly. **3.** To wander leisurely or sporadically from place to place, esp. with no particular goal. **4. a.** To wander from a set course or point of attention : STRAY. **b.** To vary from or oscillate randomly about a fixed setting, position, or mode of operation. **5.** To be piled up in banks or heaps, as of snow or sand, by the force of a current. —*vt.* **1.** To cause to drift. **2.** To cover with drifts. **3.** *Western U.S.* To drive (livestock) slowly or far afield, esp. for grazing. —*n.* **1.** The act or condition of drifting. **2.** Something moving along on a current of air or water. **3.** A bank or pile, as of sand or snow, heaped up by air or water currents. **4.** *Geol.* Rock debris carried and deposited by or from ice, esp. by or from a glacier. **5. a.** A trend or general direction <a *drift* toward conservatism> **b.** General meaning or purport : TENOR <got the *drift* of the speech> **6.** A variation from an original model, method, or intention. **7. a.** Lateral displacement or deviation of an object or vehicle from a planned course, esp. as a result of wind, ocean current, or other disturbance in the medium of travel. **b.** Variation or random oscillation about a fixed setting, position, or mode of operation. **8.** A change in the output of a circuit or amplifier that occurs slowly. **9.** The rate of flow of a water current. **10. a.** A tool for driving or ramming something down. **b.** A tapered steel pin for enlarging and aligning holes. **11. a.** A horizontal or nearly horizontal passageway in a mine running through or parallel to a vein. **b.** A secondary mine passageway between two main tunnels or shafts. **12.** A drove or herd, esp. of swine. **—drift′er** *n.* **—drift′y** *adj.*

**drift·age** (drĭf′tĭj) *n.* **1.** Deviation from a set course caused by drifting. **2.** Something that has been borne along or deposited by air or water currents.

**drift·wood** (drĭft′wŏŏd′) *n.* **1.** Wood floating in or washed up by water. **2.** A collection of unimportant or worthless elements.

**drill[1]** (drĭl) *n.* [Du. *dril* < *drillen,* to drill.] **1. a.** An implement with cutting edges or a pointed end for boring holes in hard materials, usu. by a rotating abrasion or repeated blows. **b.** The hand-operated or hand-powered holder for a drill. **2.** Disciplined, repetitious exercise as a means of teaching and perfecting a skill or procedure, esp. as part of military training. **3.** Any of several marine gastropod mollusks, primarily of the genus *Urosalpinx,* that drill holes into the shells of bivalve mollusks. —*v.* **drilled, drill·ing, drills.** —*vt.* **1.** To make a hole in with a drill <*drill* rock><*drill* a tooth> **2.** To strike with a hard jabbing blow. **3. a.** To instruct thoroughly in a skill or procedure by repetition. **b.** To infuse knowledge of or skill in by repetitious instruction. —*vi.* **1.** To make a hole with a drill. **2.** To perform an instructive exercise.

**drill[2]** (drĭl) *n.* [Perh. < DRILL[1], rill (obs.).] **1.** A furrow or trench in which seeds are planted. **2.** A row of planted seeds. **3.** An implement or machine for planting seeds in holes or furrows. —*vt.* **drilled, drill·ing, drills. 1.** To sow (seeds) in rows. **2.** To plant (a field) in drills.

**drill[3]** (drĭl) *n.* [Short for *drilling* < G. *Drillich* < OHG *drilīch* < Lat. *trilix,* triple-twilled : *tri-,* three + *licum,* thread.] Long-wearing cotton or linen twill, gen. used for work clothes.

**drill[4]** (drĭl) *n.* [Native word in West Africa.] A monkey, *Mandrillus leucophaeus* of western Africa, resembling the mandrill.

**drill instructor** *n.* A noncommissioned officer who instructs recruits in military drill and discipline.

**drill·mas·ter** (drĭl′măs′tər) *n.* **1.** A drill instructor. **2.** An instructor given to severely rigorous training and strict discipline.

**drill press** *n.* A powered vertical drilling machine in which the drill is pressed to the work automatically or by a hand lever.

**drill·ship** (drĭl′shĭp′) *n.* A ship equipped for ocean floor drilling.

**drill·stock** (drĭl′stŏk′) *n.* The part of a drilling tool or machine that holds the shank of a drill or bit.

**drink** (drĭngk) *v.* **drank** (drăngk), **drunk, drink·ing, drinks.** [ME *drinken* < OE *drincan.*] —*vt.* **1.** To take into the mouth and swallow (a liquid). **2.** To soak up (liquid or moisture) <*drank* the country air> **3.** To take in avidly through the senses or intellect <*drank* in every phrase of the sermon> **4.** To swallow the liquid contents of (e.g., a glass). **5.** To give or make a toast to (e.g., a person or occasion). **6.** To bring to a specific state by drinking alcoholic liquors <*drank* our sorrows away> —*vi.* **1.** To swallow liquid. **2.** To imbibe alcoholic liquors. **3.** To salute a person or occasion with a toast. —*n.* **1.** A liquid suitable for drinking : BEVERAGE. **2.** An alcoholic beverage. **3.** An amount of liquid swallowed. **4.** Excessive or habitual indulgence in alcoholic liquor. **5.** *Slang.* A fairly large body of water <fell off the boat and into the *drink*>

**drink·a·ble** (drĭng′kə-bəl) *adj.* Fit for drinking : POTABLE. —*n.* A beverage.

**drink·er** (drĭng′kər) *n.* **1.** One who drinks. **2.** One who drinks alcoholic liquors habitually or excessively.

**drip** (drĭp) *v.* **dripped, drip·ping, drips.** [ME *drippen* < OE *dryppan.*] —*vi.* **1.** To fall in drops. **2.** To shed drops. **3.** To overflow or ooze with or as if with liquid <a gown *dripping* with sequins> —*vt.* To let fall in or as if in drops <eaves *dripping* rainwater><a play that *drips* cynicism> —*n.* **1.** The process of forming and falling in drops. **2.** Liquid or moisture that falls in drops. **3.** The sound made by dripping liquid. **4.** A projection on a cornice or sill that protects the area below from rainwater. **5.** *Slang.* A very boring or dull person.

**drip-dry** (drĭp′drī′) *adj.* Made of a fabric that will not wrinkle when hung dripping wet. —*vi.* **-dried, -dry·ing, -dries.** To dry with no wrinkles when hung dripping wet. —*n., pl.* **-dries.** A drip-dry garment.

**drip pan** *n.* A pan for catching the drippings from roasting meat.

**drip·ping** (drĭp′ĭng) *n.* **1.** The act or sound of something that drips. **2.** *often* **drippings.** The fat and juice exuded from roasting meat. —*adv.* Thoroughly <*dripping* wet>

**drip·py** (drĭp′ē) *adj.* **-pi·er, -pi·est. 1.** Very wet : DRIZZLY <*drippy* weather> **2.** *Slang.* Mawkishly sentimental <a *drippy* love letter> **—drip′pi·ness** *n.*

**drip·stone** (drĭp′stōn′) *n.* **1.** A drip made of stone, as on a cornice over a door or window. **2.** Calcium carbonate in the form of stalactites or stalagmites.

**drive** (drīv) *v.* **drove** (drōv), **driv·en** (drĭv′ən), **driv·ing, drives.** [ME *driven* < OE *drīfan.*] —*vt.* **1.** To push, propel, or press onward forcibly. **2.** To repulse by authority or force <*drove* the mugger away> **3.** To force to work, usu. excessively. **4.** To force into or from a particular act or state <Waiting *drives* me crazy.> **5. a.** To throw, strike, or cast (e.g., a ball) hard or rapidly in a sport. **b.** *Basketball.* To move with the ball directly through. **6.** *Baseball.* To cause (a base runner or run) to be scored. **7.** To force to penetrate. **8.** To produce or create by penetrating forcibly. **9.** To guide, control, or direct (a vehicle). **10.** To transport or convey in a vehicle. **11.** To supply the motive force to and cause to function. **12.** To carry through vigorously to a conclusion <*drove* the message home> **13. a.** To chase (game) into the open or into nets or traps. **b.** To search (an area) for game in this way. —*vi.* **1.** To move along or advance quickly as if impelled. **2.** To rush, dash, or advance violently against something <The icy blast *drove* into our faces.> **3.** To hit, throw, or impel a ball or other missile forcibly. **4.** *Basketball.* To move with the ball directly to the basket. **5.** To operate a vehicle. **6.** To go or be transported in a vehicle. **7.** To attempt to achieve an objective : AIM. **—drive at.** To intend to do or say <What are you *driving at?*> —*n.* **1.** The act of driving. **2.** A road for vehicles. **3.** A trip in a vehicle. **4. a.** The means or apparatus for transmitting motion to a machine or machine part. **b.** The means by which automotive power is applied to a roadway <four-wheel *drive*> **c.** The means or apparatus for controlling and directing an automobile <right-hand *drive*> **5.** An organized effort to accomplish a purpose, as raising money. **6.** Initiative, energy, or aggressiveness. **7.** *Psychoanal.* A strong motivating tendency or instinct, esp. of sexual or aggressive origin, that prompts activity toward a particular end. **8.** A massive and sustained military offensive. **9. a.** The act of hitting, stroking, or thrusting the ball swiftly. **b.** The stroke or thrust by which the ball is driven. **c.** *Basketball.* The act of moving with the ball directly to the basket. **10. a.** A collection and driving of cattle to new pastures or to market. **b.** An amassment and driving of logs down a river. **c.** The cattle or logs thus driven.

**drive-in** (drīv′ĭn′) *n.* A retail establishment, as a restaurant or movie theater, designed to permit customers to remain in their vehicles while being served or entertained.

**driv·el** (drĭv′əl) *v.* **-eled, -el·ing, -els** *or* **-elled, -el·ling, -els.** [ME *drivelen* < OE *dreflian.*] —*vi.* **1.** To drool or slobber. **2.** To flow like saliva. **3.** To talk nonsensically. —*vt.* **1.** To allow to flow from the mouth. **2.** To say stupidly <*drivel* a reply> —*n.* **1.** Saliva flowing from the mouth. **2.** Senseless or trivial talk. **—driv′el·er** *n.*

**drive·line** (drīv′līn′) *n.* The components of an automotive vehicle that connect the transmission with the driving axles and include the universal joint and drive shaft.

**driv·en** (drĭv′ən) *adj.* [< p.part. of DRIVE.] **1.** Carried along or piled up by a current. **2.** Having or motivated by a compulsive quality or need <a *driven* executive>

**driv·er** (drī′vər) *n.* **1.** One that drives, esp. a chauffeur. **2.** A tool or device used for driving, as a hammer. **3.** A machine part that transmits force or motion to another part. **4.** A wooden-headed golf club with a long shaft, used for making long shots from the tee.

**driver ant** *n.* Any of various rapacious tropical Old World ants of the subfamily Dorylinae that move in huge colonies.

**driver's seat** *n.* A position of authority or control.

**drive shaft** *n.* A rotating shaft that transmits mechanical power to a point or area of application.

**drive-up** (drīv′ŭp′) *adj.* Enabling customers to remain in their vehicles while being served, as at a bank.

**drive·way** (drīv′wā′) *n.* A private road connecting a building, as a house or garage, with the street.

**driv·ing** (drī′vĭng) *adj.* **1.** Transmitting motion or power. **2.** Violent, intense, or forceful <a *driving* snowstorm> **3.** Highly energetic <a hard-*driving* politician>

**driz·zle** (drĭz′əl) *v.* **-zled, -zling, -zles.** [Perh. < ME *dresen,* to fall < OE *drēosan.*] —*vi.* To rain gently in fine, mistlike drops. —*vt.* **1.** To let fall in fine drops or particles. **2.** To moisten with fine drops. —*n.* A fine, gentle rain. **—driz′zly** *adj.*

**drogue** (drōg) *n.* [Perh. alteration of DRAG.] **1.** A sea anchor. **2.** A drogue parachute. **3.** A funnel- or cone-shaped device towed behind an aircraft as a target. **4.** A funnel-shaped device at the end of the hose of a tanker aircraft, used as a stabilizer and receptacle for the probe of a receiving aircraft.

**drogue parachute** *n.* **1.** A parachute used in decelerating a fast-moving object, esp. a small parachute used to slow down a re-entering spacecraft or satellite prior to deployment of the main parachute. **2.** A small parachute that is used to pull a main parachute from its storage pack.

**droit** (droit, drə-wä′) *n.* [ME, a fee allowed by law < OFr., right < LLat. *directum* < Lat. *directus,* right, p.part. of *dirigere,* to direct.] **1.** A legal right. **2.** Something to which one has legal right.

**droll** (drōl) *adj.* **-er, -est.** [Fr. *drôle* < *drôle,* buffoon < MDu. *drol,* little man.] Whimsically comical or amusingly odd. —*n.* A buffoon. **—droll′ness** *n.* **—drol′ly** *adv.*

**droll·er·y** (drō′lə-rē) *n., pl.* **-ies. 1.** A droll quality. **2.** A droll manner of talking or acting. **3. a.** The act of joking or clowning. **b.** Something droll, as an amusing story.

**-drome** *suff.* [Lat. *-dromos* < Gk. *dromos,* racecourse.] **1. a.** Racetrack <motor*drome*> **b.** Field : arena <air*drome*> **2.** Running <acro*drome*>

**drom·e·dar·y** (drŏm′ĭ-dĕr′ē, drŭm′-) *n., pl.* **-ies.** [ME *dromedarie* < OFr. *dromedaire* < LLat. *dromedarius* < Lat. *dromas* < Gk., running.] The one-humped domesticated camel, *Camelus dromedarius,* widely used in northern Africa and western Asia as a beast of burden.

**drom·ond** (drŏm′ənd, drŭm′-) *n.* [ME < OFr. *dromont* < LLat. *dromo,* a kind of ship < LGk. *dromōn* < Gk. *dromos,* race.] A large medieval sailing galley.

**-dromous** *suff.* [NLat. *-dromus* < Gk. *-dromos* < *dromos,* act of running.] Running : moving <cata*dromous*>

**drone[1]** (drōn) *n.* [ME < OE *drān.*] **1.** A male bee, esp. a honeybee, that is usu. stingless, performs no work, and produces no honey. **2.** A person who lives off others : SPONGE. **3.** A pilotless aircraft operated by remote control. **4.** One who performs dull and perfunctory tasks.

**drone[2]** (drōn) *v.* **droned, dron·ing, drones.** [< DRONE[1] (from the bee's humming sound).] —*vi.* **1.** To make a continuous low dull humming sound. **2.** To speak in a monotonous tone. **3.** To pass or act in a monotonous way <a speech that *droned* on and on> —*vt.* To utter in a monotonous low tone. —*n.* **1.** A continuous low humming or buzzing sound. **2.** Any of the pipes of a bagpipe tuned to produce a single tone. **3.** *Mus.* A single sustained tone.

**drool** (drōōl) *v.* **drooled, drool·ing, drools.** [Perh. alteration of DRIVEL.] —*vi.* **1.** To let saliva flow from the mouth : DRIVEL. **2.** *Informal.* To make an exaggerated show of desire or appreciation. **3.** *Informal.* To talk nonsense. —*vt.* To let flow from the mouth. —*n.* **1.** Saliva. **2.** *Informal.* Nonsense.

**droop** (drōōp) *v.* **drooped, droop·ing, droops.** [ME *droupen* < ON *drūpa.*] —*vi.* **1.** To hang or bend downward. **2.** To bend or sag gradually. **3.** To sag in dejection, exhaustion, or lifelessness. —*vt.* To let hang or bend down. —*n.* The act or condition of drooping. **—droop′i·ly, droop′ing·ly** *adv.* **—droop′y** *adj.*

**drop** (drŏp) *n.* [ME < OE *dropa.*] **1.** The smallest quantity of liquid heavy enough to fall in a spherical mass. **2.** A minute quantity of a substance. **3. drops.** Liquid medicine administered in drops. **4.** A small amount : DRAM <not a *drop* of compassion> **5. a.** Something, as an earring, shaped or hanging like a drop. **b.** A small globular piece of hard candy. **6.** The act of falling : DESCENT. **7.** A swift decrease or decline, as in quality, quantity, or intensity. **8. a.** The vertical distance from a higher to a lower level. **b.** The distance through which something drops or falls. **9.** A sheer incline, as the face of a cliff. **10. a.** A descent by parachute. **b.** Military personnel and equipment landed by parachute. **11.** Something arranged to fall or be lowered, as a trap door on a gallows. **12.** A drop curtain. **13.** A slot through which something is deposited in a receptacle <a garbage *drop*> **14.** *Electron.* A connection made available for a terminal unit on a transmission line. **15.** A central place where something, as mail, is brought and subsequently distributed. —*v.* **dropped, drop·ping, drops.** —*vi.* **1.** To fall in drops. **2.** To fall from a higher to a lower position or place. **3.** To become less, as in number, intensity, or volume. **4.** To descend from one level to another. **5.** To fall or sink into a state of exhaustion or death. **6.** To pass into a specified state or condition <*dropped* into a light sleep> **7.** To fall or roll into a basket or hole. —Used of a ball. —*vt.* **1.** To let or cause to fall. **2.** To let fall in drops. **3.** To cause to become less : REDUCE <*dropped* the price ten dollars> **4.** To cause to fall, as by hitting or shooting. **5.** To hurl or strike (a ball) into a basket or hole. **6.** To give birth to. —Used of animals. **7.** To say or offer casually, as a hint. **8.** To write (e.g., a note) at one's leisure. **9.** To cease consideration or treatment of <*dropped* the charges for lack of evidence> **10.** To terminate an association or relationship. **11.** To leave unfinished <*dropped* what we were doing to help them> **12.** To leave out (e.g., a letter) in speaking or writing. **13.** To leave at a particular place : UNLOAD. **14.** To parachute. **15.** To lower the level of (the voice). **16.** To lose (e.g., a game or contest). **17.** *Slang.* To take, as a drug, by mouth <*drop* acid> **—drop behind.** To fall behind. **—drop by** (*or* **in**). To stop in for a short visit. **—drop off. 1.** To fall asleep. **2.** To decrease. **—drop out. 1.** To withdraw from participation or membership, as in a game, club, or school. **2.** To withdraw from established society, esp. because of disillusion with conventional values.

**drop cloth** *n.* A sheet, as of cloth or plastic, for protection against dripping or spills, used esp. by painters.

**drop curtain** *n.* **1.** An unframed curtain that forms part of the scenery on a stage. **2.** A theater curtain that can be raised or lowered.

**drop·forge** (drŏp′fôrj′, -fōrj′) *vt.* **-forged, -forg·ing, -forg·es.** To forge (a metal) between dies by the force of a drop hammer.

**drop hammer** *n.* A machine used to forge or stamp metal, having an anvil or base aligned with a hammer that is forced down upon the molten metal.

**drop-in** (drŏp′ĭn′) *n.* **1.** One who drops in casually, as to visit. **2.** An informal social event.

**drop kick** *n.* *Football.* A kick made by dropping the ball to the ground and kicking it when it starts to rebound.

**drop-kick** (drŏp′kĭk′) *vi.* & *vt.* **-kicked, -kick·ing, -kicks.** *Football.* To make or kick with a drop kick. **—drop′-kick′er** *n.*

**drop leaf** *n.* A hinged wing on a table that can be folded down.

**drop·let** (drŏp′lĭt) *n.* A small drop.

**drop letter** *n.* A letter that is mailed and delivered from the same post office.

**drop·light** (drŏp′līt′) *n.* A hanging lamp that can be raised and lowered on its cord.
**drop-off** (drŏp′ôf′, -ŏf′) *n.* **1.** A steep slope. **2.** A noticeable decrease, as in attendance or sales.
**drop·out** (drŏp′out′) *n.* **1.** One who quits school. **2.** One who has withdrawn from a given social group, environment, or pursuit or from established society. **3.** *Computer Sci.* A segment of magnetic tape on which expected information is absent.
**dropped egg** *n.* A poached egg.
**drop·per** (drŏp′ər) *n.* **1.** One that drops. **2.** A small tube with a suction bulb at one end for drawing in a liquid and releasing it in drops.
**drop·ping** (drŏp′ĭng) *n.* **1.** Something dropped. **2. droppings.** Dung left by animals.
**drop·shot** (drŏp′shŏt′) *n.* A shot in a racquet game in which a ball or shuttlecock drops quickly after crossing the net or hitting the wall.
**drop·sy** (drŏp′sē) *n.* [ME *dropesie,* short for *ydropesie* < OFr. < Lat. *hydropsis* < Gk. *hudrōpsis* < *hudrōps* < *hudōr,* water.] Pathological accumulation of diluted lymph in body tissues and cavities. **—drop′si·cal** (-sĭ-kəl) *adj.* **—drop′si·cal·ly** *adv.*
**drop·wort** (drŏp′wôrt′, -wûrt′) *n.* A Eurasian plant, *Filipendula hexapetala,* bearing finely divided leaflets and white flower clusters.
**dros·er·a** (drŏs′ə-rə) *n.* [NLat. < Gk., fem. of *droseros,* dewy < *drosos,* dew.] Sundew.
**drosh·ky** (drŏsh′kē) *also* **dros·ky** (drŏs′kē) *n., pl.* **-kies.** [R. *drozhki,* dim. of *drogi,* wagon.] An open, four-wheeled, horse-drawn carriage once common in Russia.

**droshky**

**dro·soph·i·la** (drō-sŏf′ə-lə) *n.* [NLat. *Drosophila,* genus name : Gk. *drosos,* dew + NLat. *-philus,* -phile.] A small fly of the genus *Drosophila,* esp. the fruit fly *D. melanogaster,* used extensively in genetic studies.
**dross** (drŏs, drôs) *n.* [ME *dros* < OE *drōs,* dregs.] **1.** Waste product or impurities formed on the surface of molten metal. **2.** Inferior, trivial, or worthless matter. **—dross′y** *adj.*
**drought** (drout) *also* **drouth** (drouth) *n.* [ME < OE *drūgoð.*] **1.** A long period with little or no rain. **2.** A shortage or dearth. **—drought′y** *adj.*
**drove**[1] (drōv) *v. p.t. of* DRIVE.
**drove**[2] (drōv) *n.* [ME < OE *drāf* < *drīfan,* to drive.] **1.** A flock or herd being driven in a body. **2. a.** A large mass of people moving or acting together. **b.** A sizable group of similar things. **3. a.** A stonemason's broad-edged chisel used for roughhewing. **b.** A stone surface dressed with a drove.
**drov·er** (drō′vər) *n.* A driver of sheep or cattle.
**drown** (droun) *v.* **drowned, drown·ing, drowns.** [ME *drounen,* of Scand. orig.] *—vt.* **1.** To kill by submerging and suffocating in a liquid. **2.** To drench completely or cover with or as if with a liquid. **3.** To deaden one's awareness of, as if by immersion <*drowned* our problems in drink> **4.** To overwhelm and blur (a sound) by a louder sound. *—vi.* To die by suffocating in a liquid.
**drowse** (drouz) *v.* **drowsed, drows·ing, drows·es.** [Perh. ult. < OE *drusian,* to be sluggish.] *—vi.* To be half asleep. *—vt.* **1.** To make drowsy. **2.** To pass (time) by drowsing. *—n.* Sleepiness.
**drows·y** (drou′zē) *adj.* **-i·er, -i·est. 1.** Sluggish because of sleepiness. **2.** Produced or marked by sleepiness. **3.** Inducing sleepiness : SOPORIFIC. **—drows′i·ly** *adv.* **—drows′i·ness** *n.*
**drub** (drŭb) *v.* **drubbed, drub·bing, drubs.** [Ar. *ḍaraba,* to beat.] *—vt.* **1.** To beat with a stick. **2.** To instill (e.g., an idea or lesson) forcefully. **3. a.** To defeat completely. **b.** To berate or criticize harshly. **4.** To stamp (the feet). *—vi.* **1.** To stamp the ground with the feet. **2.** To pound : throb. *—n.* A blow with a stick. **—drub′ber** *n.*
**drub·bing** (drŭb′ĭng) *n.* **1.** A severe beating. **2.** A total defeat.
**drudge** (drŭj) *n.* [< ME *druggen,* to labor.] One who does unpleasant, tedious, or menial work. *—vi.* **drudged, drudg·ing, drudg·es.** To do the work of a drudge. **—drudg′er** *n.* **—drudg′ing·ly** *adv.*
**drudg·er·y** (drŭj′ə-rē) *n., pl.* **-ies.** Work that is unpleasant, tedious, or menial.
**drug** (drŭg) *n.* [ME *drogge* < OFr. *drogue,* a chemical substance.] **1.** A substance used as medicine in the treatment of illness or disease. **2.** A narcotic, esp. one that is addictive. **3.** *Obs.* A chemical or dye. *—vt.* **drugged, drug·ging, drugs. 1.** To administer a drug to. **2.** To mix or poison (food or drink) with drugs. **3.** To stupefy with or as if with a drug.
**drug·get** (drŭg′ĭt) *n.* [OFr. *droguet,* dim. of *drogue,* stuff.] **1.** A heavy felted fabric of wool or wool and cotton, used for floor covering. **2.** A coarse rug of drugget, made in India. **3.** A wholly or partly woolen fabric formerly used for clothing.
**drug·gist** (drŭg′ĭst) *n.* **1.** A pharmacist. **2.** One who manages or owns a drugstore.
**drug·store** *also* **drug store** (drŭg′stôr′, -stōr′) *n.* A store where prescriptions are filled and drugs and sundries are sold.
**dru·id** *also* **Dru·id** (drōō′ĭd) *n.* [Lat. *druides,* druids, prob. of Celtic orig.] A member of an order of priests in ancient Britain and Gaul who appear in Welsh and Irish legend as prophets and sorcerers. **—dru·id′ic** (drōō-ĭd′ĭk), **dru·id′i·cal** *adj.* **—dru′id·ism** *n.*
**drum** (drŭm) *n.* [Prob. < Du. *trom.*] **1.** A percussion instrument having a hollow cylinder or hemisphere with a membrane stretched tightly over one or both ends, played by beating with the hands or with sticks. **2.** A sound produced by a drum. **3.** Something resembling a drum in structure or shape, esp. a metal cylinder wound with cable, wire, or heavy rope or a barrellike metal container. **4.** Any of various marine and freshwater fishes of the family Sciaenidae that make a drumming sound. **5.** The tympanic membrane. *—v.* **drummed, drum·ming, drums.** *—vi.* **1.** To play drums. **2.** To thump or tap rhythmically or incessantly. **3.** To produce a booming, reverberating sound by beating the wings, as do certain birds. *—vt.* **1.** To produce (rhythmic sounds) by tapping repeatedly on or as if on a drum. **2.** To summon by or as if by beating a drum. **3.** To instill by constant repetition <*drummed* the lesson into their heads> **4.** To dismiss or expel in disgrace <*drummed* out of the service> **—drum up. 1.** To bring about by persistent effort <*drum up* sales> **2.** To invent : devise <*drum up* an excuse>
**drum·beat** (drŭm′bēt′) *n.* **1.** The sound produced by beating a drum. **2.** A cause supported ardently and esp. vehemently.
**drum·beat·er** (drŭm′bē′tər) *n.* A vehement advocate of a cause. **—drum′beat′ing** *n.*
**drum·fire** (drŭm′fīr′) *n.* Heavy, continuous gunfire.
**drum·head** (drŭm′hĕd′) *n.* **1.** The membrane stretched over the open end of a drum. **2.** *Naut.* The circular top part of a capstan, used to hold bars for turning.
**drumhead court-martial** *n.* [So called because it was sometimes held around a drumhead.] A court-martial held for the summary trial of an offense committed during military operations.
**drum·lin** (drŭm′lĭn) *n.* [Ir. Gael. *druim,* ridge < OIr. + -LIN(G)[1].] A streamlined hill or ridge of glacial drift.
**drum major** *n.* A man who leads a marching band or drum corps, often twirling a baton.
**drum majorette** *n.* A woman who leads a marching band or drum corps, often twirling a baton.
**drum memory** *n. Computer Sci.* A memory device with a magnetizable coating on the outer surface of a rotating cylinder.
**drum·mer** (drŭm′ər) *n.* **1.** One who plays a drum, as in a band. **2.** A traveling sales representative.
**drum printer** *n.* A line printer in which a revolving drum acts as the printing element.
**drum·stick** (drŭm′stĭk′) *n.* **1.** A stick for beating a drum. **2.** The lower part of the leg of a fowl.
**drunk** (drŭngk) *adj.* [< p.part. of DRINK.] **1.** Intoxicated with alcohol to the point of impairment of physical and mental faculties. **2.** Caused or influenced by intoxication : DRUNKEN. **3.** Overcome by strong feeling or emotion <*drunk* with rage> *—n.* **1.** A drunkard. **2.** A bout of drinking.
**drunk·ard** (drŭng′kərd) *n.* One who is habitually drunk.
**drunk·en** (drŭng′kən) *adj.* **1.** Intoxicated with or as if with alcohol. **2.** Habitually drunk. **3.** Of or occurring during intoxication <a *drunken* stupor> **—drunk′en·ly** *adv.* **—drunk′en·ness** *n.*
**drunk·o·me·ter** (drŭng′kə-mē′tər, drŭng-kŏm′ĭ-tər) *n.* A device for determining the alcoholic content of the blood by analysis of the breath.
**dru·pa·ceous** (drōō-pā′shəs) *adj.* **1.** Relating to or consisting of a drupe <*drupaceous* fruit> **2.** Producing drupes, as a tree.
**drupe** (drōōp) *n.* [NLat. *drupa* < Lat. *druppa,* overripe olive < Gk. < *drupepēs,* overripe : *drus,* tree + *peptein,* to cook, ripen.] A fleshy fruit, as the peach, plum, or cherry, usu. having a single hard stone that encloses a seed.
**drupe·let** (drōōp′lĭt) *n.* A small drupe, as one of the numerous subdivisions of the raspberry or blackberry.
**druse** (drōōz) *n.* [G. < OHG *druos,* bump.] **1.** A crust of tiny crystals lining a rock cavity. **2.** A cavity lined by a druse.
**Druse** *also* **Druze** (drōōz) *n.* [Ar. *durūz,* pl. of *darazi,* a Druse, after Ismail al-*Darazi* (d. 1019).] A member of a sect in Syria and Lebanon whose primarily Moslem religion contains some elements of Christianity. **—Dru′se·an, Dru′si·an** *adj.*
**druth·ers** (drŭth′ərz) *pl.n.* [Alteration of the phrase *I would rather.*] *Informal.* A choice or preference. —Used chiefly in the phrase *if one had one's druthers.*
**Druze** (drōōz) *n. var. of* DRUSE.

**dry** (drī) *adj.* **dri·er, dri·est** *also* **dry·er, dry·est.** [ME *drie* < OE *drȳge.*] **1.** Free from moisture or liquid. **2.** Having or marked by little or no rain <a *dry* desert region> **3.** Marked by the absence of natural or normal moisture <a *dry* mouth> **4.** Not under water. **5.** Having all the water or liquid drained away, evaporated, or exhausted <a *dry* stream> **6.** No longer yielding liquid <a *dry* well><a *dry* cow> **7.** Lacking a mucous or watery discharge, as a cough. **8.** Not accompanied by tears <racked by *dry* sobs> **9.** Needing or desiring drink : THIRSTY. **10.** No longer wet. **11.** Of or pertaining to solid rather than liquid substances or commodities. **12.** Not sweet as a result of the decomposition of sugar during fermentation. —Used of wines. **13.** Having a large proportion of strong liquor to other ingredients <a *dry* cocktail> **14.** Eaten without butter, gravy, or other garnish. **15.** Without adornment : PLAIN <gave us just the *dry* facts> **16.** Without bias or personal concern <a *dry* review> **17.** Without tenderness, warmth, or involvement : SEVERE <a *dry* style of sculpture> **18.** Marked by matter-of-factness or indifference. **19.** Wearisome and dull. **20.** Humorous or sarcastic in a shrewd, impersonal way <has a very *dry* sense of humor> **21.** *Informal.* Prohibiting or opposed to the sale or consumption of alcoholic beverages. **22.** Not yielding the expected results : UNPRODUCTIVE <a mind *dry* of creative thoughts> —*v.* **dried, dry·ing, dries.** —*vt.* **1.** To make dry. **2.** To preserve (e.g., meat or fruit) by extracting the moisture. —*vi.* To become dry. **—dry out.** *Informal.* To undergo a cure for alcoholism. **—dry up. 1.** To make or become gradually unproductive. **2.** *Slang.* To stop talking. —*n., pl.* **drys.** *Informal.* A prohibitionist. **—dry'ly, dri'ly** *adv.* **—dry'ness** *n.*

**dry·ad** (drī'əd, -ăd') *n.* [Lat. *dryas, dryad-* < Gk. *druas* < *drus,* tree.] *Gk. Myth.* A divinity presiding over forests and trees. **—dry·ad'ic** *adj.*

**dry·as·dust** (drī'əz-dŭst') *n.* [After Dr. Jonas *Dryasdust,* a fictitious character to whom Sir Walter Scott (1771–1832) dedicated some of his novels.] A dull, pedantic writer or speaker.

**dry battery** *n.* An electric battery having two or more dry cells.

**dry cell** *n.* [So called because its contents are not spillable.] A primary cell having an electrolyte in the form of moist paste.

**dry-clean** (drī'klēn') *v.* **-cleaned, -clean·ing, -cleans.** —*vt.* To cleanse (clothing or fabrics) with chemical solvents rather than water. —*vi.* To undergo dry-cleaning. **—dry cleaner** *n.* **—dry cleaning** *n.*

**dry dock** *n.* A basin-shaped dock from which water can be emptied, used for building or repairing ships.

**dry-dock** (drī'dŏk') *vt.* & *vi.* **-docked, -dock·ing, -docks.** To place in or go into a dry dock.

**dry·er** (drī'ər) *n.* **1.** An appliance that removes moisture, as by heating <a hair *dryer*> <a clothes *dryer*> **2.** *var. of* DRIER[1].

**dry farming** *n.* Farming practiced in arid areas without irrigation by maintaining a fine surface mulch or tilth that protects the natural moisture of the soil from evaporation. **—dry-farm** (drī'färm') *v.* **(-farmed, -farm·ing, -farms). —dry farm** *n.* **—dry farmer** *n.*

**dry fly** *n.* An artificial fishing fly that floats on the water.

**dry gangrene** *n.* Gangrene resulting from arterial obstruction and marked by mummification of the dead tissue and asepsis.

**dry goods** *pl.n.* Textiles, clothing, and related articles of trade.

**dry ice** *n.* [Orig. a trademark.] Solid carbon dioxide that evaporates to gas at −78.5°C (−110°F) and is used mainly as a refrigerant.

**drying oil** *n.* An oily organic liquid, as linseed oil, that is used as a binder in paints and varnishes and forms a tough plastic layer when exposed to air in a thin film.

**dry kiln** *n.* A heated chamber in which cut lumber is dried and seasoned.

**dry measure** *n.* A system of units for measuring dry quantities, as grains, fruits, and vegetables.

**dry mop** *n.* A mop for dusting floors. **—dry'-mop'** (drī'mŏp') *v.* **(-mopped, -mop·ping, -mops).**

**dry nurse** *n.* A nurse employed to care for an infant without breast-feeding it.

**dry·o·pith·e·cine** (drī'ō-pĭth'ĭ-sēn') *n.* [< NLat. *Dryopithecus,* genus name : Gk. *drus,* tree + Gk. *pithēkos,* ape.] An extinct ape of the genus *Dryopithecus,* known from Old World fossil remains of the Miocene and Pliocene epochs and held to be an ancestor of the chimpanzees, gorillas, and humans. —*adj.* Of or belonging to the genus *Dryopithecus.*

**dry point** *n.* **1.** A technique of intaglio engraving without the use of acid, in which a hard steel needle is used to incise lines in the metal plate. **2.** An engraving or print made with dry point.

**dry rot** *n.* **1.** A fungous disease of timber, causing it to become brittle and crumble into powder. **2.** A plant disease in which the plant tissue remains relatively dry while fungi invade and ultimately decay bulbs, fruit, or woody tissue. **—dry'-rot'** (drī'rŏt') *v.* **(-rot·ted, -rot·ting, -rots).**

**dry run** *n.* **1.** A test exercise in combat skills without the use of live ammunition. **2.** A trial exercise or practice : REHEARSAL.

**dry·salt·er** (drī'sôlt'ər) *n. Chiefly Brit.* A dealer in chemical products and dyes. **—dry'salt'er·y** *n.*

**dry socket** *n.* An inflamed, painful condition of a tooth socket after the tooth has been extracted.

**dry wall** *n.* A wall made of prefabricated material, as wallboard.

**dry wash[1]** *n.* Washed and dried laundry that has not been ironed.

**dry wash[2]** *n.* WASH 11.

**D.T.'s** (dē'tēz') *pl.n.* Delirium tremens.

**du·ad** (do͞o'ăd', dyo͞o'-) *n.* [Gk. *duas, duad-,* two < *duo.*] A unit of two objects : PAIR.

**du·al** (do͞o'əl, dyo͞o'-) *adj.* [Lat. *dualis* < *duo,* two.] **1.** Made up of two parts : DOUBLE. **2.** Having a double nature, character, or purpose. **3.** Designating or relating to a number category that indicates two persons or things, as in Greek, Sanskrit, and Old English. —*n.* **1.** The dual number, as in Greek, Sanskrit, and Old English. **2.** A word or expression in the dual number. **—du'al·ly** *adv.*

**du·al·ism** (do͞o'ə-lĭz'əm, dyo͞o'-) *n.* **1.** The state of being twofold : DUALITY. **2.** *Philos.* The view that the world consists of or is explicable as two fundamental entities, as mind and matter. **3.** *Psychol.* The view that there is a phenomenal distinction between mental and physical processes. **4. a.** The theological concept that the world is ruled by the antagonistic forces of good and evil. **b.** The theological concept that a person has two basic natures, the physical and the spiritual. **—du'al·ist** *n.*

**du·al·is·tic** (do͞o'ə-lĭs'tĭk, dyo͞o'-) *adj.* **1.** Relating to or like dualism. **2.** Dual. **—du'al·is'ti·cal·ly** *adv.*

**du·al·i·ty** (do͞o-ăl'ĭ-tē, dyo͞o-) *n.* The quality or state of being twofold.

**du·al-pur·pose** (do͞o'əl-pûr'pəs, dyo͞o'-) *adj.* Serving or designed for two purposes.

**dub[1]** (dŭb) *vt.* **dubbed, dub·bing, dubs.** [ME *dubben* < OE *dubbian.*] **1.** To confer knighthood on. **2.** To honor with a new title or description. **3.** To name playfully or facetiously : NICKNAME. **4.** To strike, cut, or rub (e.g., timber or leather) so as to make even or smooth. **5.** To dress (a fowl). **6.** *Slang.* To execute (e.g., a golf stroke) poorly. —*n. Slang.* A clumsy person or player.

**dub[2]** (dŭb) *v.* **dubbed, dub·bing, dubs.** [Perh. < LG *dubben,* to hit, strike.] —*vt.* **1.** To thrust at : POKE. **2.** To beat (a drum). —*vi.* **1.** To make a thrust. **2.** To beat on a drum. —*n.* **1.** The act of dubbing. **2.** A drumbeat.

**dub[3]** (dŭb) *vt.* **dubbed, dub·bing, dubs.** [Short for DOUBLE.] **1.** To make a new recording from the original of (a record or tape). **2.** To insert a new sound track, often a synchronized translation of the original dialogue, into (a film). **3.** To insert (sound) into a film or tape. —*n.* The new sounds added by dubbing. **—dub'ber** *n.*

**dub[4]** (dŭb) *n.* [ME *dubbe.*] *Scot.* A small pool or puddle.

**dub·bin** (dŭb'ĭn) *also* **dub·bing** (-ĭng) *n.* [< DUB[1].] An application of oil and tallow for treating leather.

**du·bi·e·ty** (do͞o-bī'ĭ-tē, dyo͞o-) *also* **du·bi·os·i·ty** (do͞o'bē-ŏs'ĭ-tē, dyo͞o'-) *n.* [LLat. *dubietas* < Lat. *dubius,* dubious.] **1.** The quality or condition of being dubious. **2.** An uncertainty.

**du·bi·ous** (do͞o'bē-əs, dyo͞o'-) *adj.* [Lat. *dubius.*] **1.** Causing doubt or uncertainty : EQUIVOCAL. **2.** Reluctant to agree : SKEPTICAL. **3. a.** Questionable as to quality or validity <a person of *dubious* character> **b.** Verging on impropriety. **4.** Not yet decided <still *dubious* as to what to do> **—du'bi·ous·ly** *adv.* **—du'bi·ous·ness** *n.*

**du·bi·ta·ble** (do͞o'bĭ-tə-bəl, dyo͞o'-) *adj.* [Lat. *dubitabilis* < *dubitare,* to doubt.] Subject to uncertainty : DOUBTFUL. **—du'bi·ta·bly** *adv.*

**Du·bon·net** (do͞o'bə-nā', dyo͞o'-). A trademark for a fortified sweet wine of French origin.

**du·cal** (do͞o'kəl, dyo͞o'-) *adj.* [OFr. < LLat. *ducalis* < Lat. *dux,* leader. —see DUKE.] Of or relating to a duke or dukedom. **—du'cal·ly** *adv.*

**duc·at** (dŭk'ət) *n.* [ME < OFr. < OItal. *ducato* < Med. Lat. *ducatus,* duchy (a word used on one of the early ducats).] **1.** Any of various gold coins once used in Europe. **2.** *Slang.* An admission ticket.

**du·ce** (do͞o'chā) *n.* [Ital. < Lat. *dux* < *ducere,* to lead.] A leader or commander : CHIEF.

**duch·ess** (dŭch'ĭs) *n.* [ME *duchesse* < OFr. < Med. Lat. *ducissa* < Lat. *dux,* leader. —see DUKE.] **1.** The wife or widow of a duke. **2.** A woman holding title to a duchy in her own right.

**duch·y** (dŭch'ē) *n., pl.* **-ies.** [ME *duchie* < OFr. *duche* < Med. Lat. *ducatus* < Lat. *dux,* leader. —see DUKE.] The territory ruled by a duke or duchess : DUKEDOM.

**duck[1]** (dŭk) *n.* [ME *doke* < OE *duce.*] **1.** Any of various wild or domesticated aquatic birds of the family Anatidae, with a broad, flat bill, short legs, and webbed feet. **2.** A female duck, as distinguished from a drake. **3.** The flesh of a duck used as food. **4.** *Slang.* A person, esp. an eccentric one. **5.** *often* **ducks** *(sing. in number). Chiefly Brit.* One regarded fondly : DEAR.

**duck[2]** (dŭk) *v.* **ducked, duck·ing, ducks.** [ME *douken.*] —*vt.* **1.** To lower (the head or body) quickly, esp. so as to avoid something. **2.** To dodge or evade <*duck* one's obligations> **3.** To push suddenly under water, as someone's head. —*vi.* **1.** To lower the head or body. **2.** To move swiftly, esp. so as to avoid being seen. **3.** To submerge the head or body briefly in water. —*n.* **1.** A quick lowering of the head or body. **2.** A plunge into water. **—duck'er** *n.*

**duck[3]** (dŭk) *n.* [Du. *doek,* cloth.] **1.** A durable, closely woven heavy cotton or linen fabric. **2. ducks.** Clothing made of duck, esp. white trousers.

**duck**[4] (dŭk) *n.* [< *DUKW,* its code designation.] An amphibious military truck used during World War II.

**duck·bill** (dŭk′bĭl′) *n.* The platypus.

**duck blind** *n.* A wooden or canvas structure, often camouflaged with reeds and grasses, behind which a hunter can hide while awaiting a flight of ducks.

**duck·board** (dŭk′bôrd′, -bōrd′) *n.* A board or boardwalk laid across wet or muddy ground or flooring.

**duck hawk** *n.* The peregrine falcon.

**ducking stool** *n.* A punishment device once used in Europe and New England, having a chair in which an offender was tied and ducked into water.

**duck·ling** (dŭk′lĭng) *n.* A young duck.

**duck·pin** (dŭk′pĭn′) *n.* [From its squat appearance.] **1.** A bowling pin, shorter and squatter than a tenpin. **2. duckpins** *(sing. in number).* A bowling game played with duckpins and small balls.

**ducks and drakes** *n.* The game of skipping flat stones along the surface of water. **—make ducks and drakes of (***or* **play ducks and drakes with).** To waste : squander.

**duck soup** *n. Slang.* Something easy to accomplish.

**duck·weed** (dŭk′wēd′) *n.* A small, free-floating, stemless aquatic plant of the genus *Lemna.*

**duck·y** (dŭk′ē) *adj.* **-i·er, -i·est.** *Slang.* Causing satisfaction : FINE.

**duct** (dŭkt) *n.* [Lat. *ductus,* act of leading < p.part. of *ducere,* to lead.] **1.** A tubular passage through which a substance, esp. a fluid, is conveyed. **2.** A bodily passage, esp. one for a secretion. **3.** *Elect.* A pipe or tube for carrying wires or cables.

**duc·tile** (dŭk′təl, -tīl′) *adj.* [OFr. < Lat. *ductilis* < *ductus,* p.part. of *ducere,* to lead.] **1.** Capable of being drawn into wire or hammered thin. **2.** Capable of being easily shaped or molded : PLASTIC. **3.** Readily influenced or persuaded : AMENABLE. **—duc·til′i·ty** (-tĭl′ĭ-tē), **duc′ti·li·bil′i·ty** (-lə-bĭl′ĭ-tē) *n.*

**duct·less gland** (dŭkt′lĭs) *n.* An endocrine gland.

**duct·ule** (dŭk′to͞ol′) *n.* A small duct.

**dud** (dŭd) [ME *dudde,* an article of clothing.] *Informal.* —*n.* **1.** A bomb, shell, or explosive round that fails to detonate. **2.** One that is dissatisfyingly unsuccessful or ineffective. **3. duds. a.** Clothing. **b.** Personal belongings.

**dude** (do͞od, dyo͞od) *n.* [Orig. unknown.] **1.** *Informal.* An easterner or city person who vacations on a Western ranch. **2.** *Informal.* A man who is a very stylish dresser. **3.** *Slang.* A guy : fellow.

**du·deen** (do͞o-dēn′) *n.* [Ir. Gael. *dūidīn,* dim. of *dūd,* pipe.] A short-stemmed clay pipe.

**dude ranch** *n.* A resort patterned after a Western ranch, featuring outdoor activities such as horseback riding and camping.

**dudg·eon**[1] (dŭj′ən) *n.* [Orig. unknown.] A sullen, angry, or indignant mood <"Slamming the door in Meg's face, Aunt March drove off in high *dudgeon*" —Louisa May Alcott>

**dudg·eon**[2] (dŭj′ən) *n.* [ME *dogeon* < AN.] **1.** *Obs.* A wood for making knife handles. **2.** *Archaic.* **a.** A dagger with a hilt made of dudgeon. **b.** The hilt of a dagger.

**due** (do͞o, dyo͞o) *adj.* [ME < OFr. *deu* < VLat. **debutus* < Lat. *debitus,* p.part. of *debēre,* to owe.] **1.** Payable immediately or on demand. **2.** Owed as a debt : OWING <40 dollars still *due*> **3.** Fitting or appropriate <*due* respect> **4.** Meeting special requirements : SUFFICIENT <We have *due* reason to suspect them.> **5.** Expected or scheduled, esp. appointed to arrive. **6.** Capable of being attributed <death *due* to cancer> —*n.* **1.** Something owed or deserved. **2. dues.** A fee or charge for membership, as in a club. —*adv.* **1.** Directly : straight <*due* north> **2.** *Archaic.* Duly.

**due bill** *n.* A written acknowledgment of indebtedness to a party, but not payable to his or her order or transferable by endorsement.

**du·el** (do͞o′əl, dyo͞o′-) *n.* [Med. Lat. *duellum* < Lat., war.] **1.** A formal prearranged combat between two persons, usu. fought to settle a point of honor. **2.** A struggle for ascendancy between two contending persons, groups, or ideas. —*v.* **-eled, -el·ing, -els** *or* **-elled, -el·ling, -els.** —*vt.* **1.** To fight with in a duel. **2.** To oppose vigorously. —*vi.* To fight a duel. **—du′el·er, du′el·ist** *n.*

**du·el·lo** (do͞o-ĕl′ō, dyo͞o-) *n.* [Ital. < Lat. *duellum,* war.] The art or rules of the duel.

**du·en·de** (do͞o-ĕn′dā′) *n.* [Dial. Sp., charm < Sp., ghost.] Unusual power to attract or charm.

**du·en·na** (do͞o-ĕn′ə, dyo͞o-) *n.* [Sp. *dueña* < Lat. *domina,* lady, fem. of *dominus,* lord.] **1.** An elderly woman retained by a Spanish or Portuguese family to act as companion and governess to the daughters. **2.** A chaperon.

**due process** *n.* An established course esp. for judicial proceedings designed to protect the legal rights of the individual.

**du·et** (do͞o-ĕt′, dyo͞o-) *n.* [Ital. *duetto,* dim. *duo* < Lat., two.] **1.** A musical composition for two voices or instruments. **2.** The two performers of a duet. **3.** A pair.

**due to** *prep.* **1.** Caused by : attributable to. **2.** Because of.

**duff**[1] (dŭf) *n.* [Dial. var. of DOUGH.] A stiff flour pudding boiled in a cloth bag or steamed.

**duff**[2] (dŭf) *n.* [Back-formation < DUFFER.] **1.** Decaying leaves and branches on a forest floor. **2.** Fine coal : SLACK.

**duff**[3] (dŭf) *n.* [Orig. unknown.] *Slang.* The buttocks.

**duf·fel** *also* **duf·fle** (dŭf′əl) *n.* [Du., after *Duffel,* a town in Belgium.] **1.** A blanket fabric made of low-grade woolen cloth with a two-sided nap. **2.** Personal gear and clothing carried by a camper.

**duffel bag** *n.* A large cloth bag of canvas or duck for carrying personal belongings.

**duf·fer** (dŭf′ər) *n.* [Orig. unknown.] **1.** *Informal.* An inept, incompetent, or stupid person. **2.** *Slang.* A peddler of cheap merchandise. **3.** *Slang.* Something useless or worthless.

**duf·fle** (dŭf′əl) *n. var. of* DUFFEL.

**dug**[1] (dŭg) *n.* [Orig. unknown.] An udder, breast, or teat of a female mammal.

**dug**[2] (dŭg) *v. p.t. & p.p. of* DIG.

**du·gong** (do͞o′gŏng′, -gông′) *n.* [NLat. *Dugong,* genus name < Malay *duyong.*] A herbivorous marine mammal, *Dugong dugon* of tropical coastal waters of the Old World, with a deeply notched tail fin and flipperlike forelimbs.

**dug·out** (dŭg′out′) *n.* **1.** A boat or canoe made by hollowing out a log. **2.** A shelter dug into the ground or on a hillside. **3.** *Baseball.* Either of two usu. sunken shelters at the side of a field where the players stay while not on the field.

**dui·ker** (dī′kər) *n.* [Afr. < Du. *duiken,* to dive.] Any of various small African antelopes chiefly of the genus *Cephalopus,* with short, backward-pointing horns.

**duiker**
*2½–4 feet long*

**duke** (do͞ok, dyo͞ok) *n.* [ME < OFr. *duc* < Lat. *dux,* leader < *ducere,* to lead.] **1.** A nobleman with the highest hereditary rank, esp. in Great Britain. **2.** A prince who rules an independent duchy. **3.** A type of cherry intermediate between a sweet and a sour cherry.

**duke·dom** (do͞ok′dəm, dyo͞ok′-) *n.* **1.** A duchy. **2.** The office, rank, or title of a duke.

**dukes** (do͞oks, dyo͞oks) *pl.n.* [Short for *Duke of Yorks,* rhyming slang for *forks,* fingers.] *Slang.* The fists.

**Du·kho·bor** *also* **Dou·kho·bor** (do͞o′kə-bôr′) *n.* [R. *dukhoborets* : *dukh,* spirit + *borets,* wrestler < *borot′,* to overcome.] A member of an 18th-cent. Russian Christian sect, many of whom migrated to Canada in the 1890's to escape persecution.

**dul·cet** (dŭl′sĭt) *adj.* [ME *doucet* < OFr. < *doux,* sweet < Lat. *dulcis.*] **1. a.** Pleasing to the ear : MELODIOUS. **b.** Having an agreeably soothing quality. **2.** *Archaic.* Sweet to the taste.

**dul·ci·fy** (dŭl′sə-fī′) *vt.* **-fied, -fy·ing, -fies.** [LLat. *dulcificare,* to sweeten : Lat. *dulcis,* sweet + *facere,* to do.] **1.** To make gentle or agreeable : MOLLIFY. **2.** To sweeten. **—dul′ci·fi·ca′tion** *n.*

**dul·ci·mer** (dŭl′sə-mər) *n.* [ME *doucemer* < OFr. *doulcemer.*] A musical instrument with wire strings of graduated lengths stretched over a sound box, played with two padded hammers or by plucking.

**dull** (dŭl) *adj.* **-er, -est.** [ME *dul* < OE *dol.*] **1.** Lacking mental agility : STUPID. **2.** Lacking alertness or responsiveness : INSENSITIVE. **3.** Dispirited : dejected. **4.** Not brisk or rapid : SLUGGISH. **5.** Having a blunt edge or point : not sharp. **6.** Not intensely or keenly felt <a *dull* pain> **7.** Arousing no interest or curiosity : UNEXCITING. **8.** Not bright or vivid : dim or drab, as a color. **9.** Cloudy or overcast. **10.** Not clear or resonant <a *dull* thump> —*v.* **dulled, dull·ing, dulls.** —*vt.* **1.** To make less sharp. **2.** To make less bright or clear. **3.** To make (e.g., the senses) less keen or perceptive. —*vi.* To become dull. **—dull′ish** *adj.* **—dull′ness, dul′ness** *n.* **—dul′ly** *adv.*

☆ ***syns:*** DULL, COLORLESS, DRAB, DRY, FLAT, LACKLUSTER, LIFELESS, PEDESTRIAN, PROSAIC, UNINSPIRED *adj. core meaning* : lacking liveliness, charm, or surprise <a *dull* conversation>

**dull·ard** (dŭl′ərd) *n.* One who is mentally dull or slow.

**du·ly** (do͞o′lē, dyo͞o′-) *adv.* [ME *duely* < *due,* due.] **1.** In a proper manner. **2.** At the expected time.

**Du·ma** (do͞o′mə) *n.* [R., of Germanic orig.] A Russian national parliament, convened and dissolved four times between 1905 and 1917.

**dumb** (dŭm) *adj.* **-er, -est.** [ME < OE.] **1.** Lacking the power of speech : MUTE. **2.** Temporarily speechless, as with fear or shock. **3.** Reluctant to speak : TACITURN. **4.** Not producing or accompanied by speech or sound. **5.** *Naut.* Not self-propelling. **6.** *Informal.* Stupid. **—dumb′ly** *adv.* **—dumb′ness** *n.*

**dumb·bell** (dŭm′bĕl′) *n.* **1.** A weight consisting of a short bar with a metal ball or disk at each end that is lifted for exercise and muscular development. **2.** *Slang.* DUMMY 3.

**dumb·found** (dŭm′found′) *v. var. of* DUMFOUND.

**dumb show** *n.* **1.** A part of a dramatic performance unaccompanied by speech **:** PANTOMIME. **2.** Communication by means of gestures.

**dumb·struck** (dŭm′strŭk′) *adj.* Astounded or shocked into speechlessness.

**dumb·wait·er** (dŭm′wā′tər) *n.* **1.** A small elevator for conveying goods, as food, from one floor to another. **2.** A portable serving table.

**dum-dum** (dŭm′dŭm) *n.* [Redup. and alteration of DUMB.] *Slang.* DUMMY 3.

**dum·dum bullet** (dum′dum′) *n.* [After *Dum-Dum,* town in India.] A small-arms bullet with a soft nose designed to expand upon contact with the target.

**dum·found** *or* **dumb·found** (dŭm′found′) *vt.* **-found·ed, -found·ing, -founds.** [DUM(B) + (CON)FOUND.] To confound with amazement or astonishment.

**dum·my** (dŭm′ē) *n., pl.* **-mies.** [< DUMB.] **1.** An imitation of a real object, intended to be used as a practical substitute, as in testing automobile safety devices. **2. a.** A mannequin for displaying clothes. **b.** A stuffed or pasteboard figure used as a target. **c.** A figure of a person or animal manipulated by a ventriloquist. **3.** A stupid person **:** DOLT. **4.** A person or agency secretly in the service of another. **5. a.** A model of a work being published, indicating its general dimensions and appearance. **b.** A model page with text and illustrations pasted into place to direct the printer. **6. a.** The partner who exposes his or her hand to be played by the declarer in bridge. **b.** The hand thus exposed. —*v.* **-mied, -my·ing, -mies.** —*vt.* To make a dummy of (a page or publication) in printing. —*vi.* To keep silent <told me to *dummy* up>

**dummy variable** *n.* A mathematical variable that can be replaced arbitrarily by another.

**du·mor·ti·er·ite** (do͞o-môr′tē-ə-rīt′, dyo͞o-) *n.* [Fr., after Eugène *Dumortier* (d.1876).] A greenish-blue aluminum borosilicate mineral, used in special refractories and spark-plug porcelain.

**dump** (dŭmp) *v.* **dumped, dump·ing, dumps.** [ME *dumpen,* of Scand. orig.] —*vt.* **1.** To throw down or release in a large mass. **2. a.** To empty (material) out of a container or vehicle. **b.** To empty out (a container or vehicle). **3. a.** To get rid of **:** JETTISON. **b.** To reject or discard unceremoniously <*dumped* me for a football player> **4.** To place (goods) on the market in large quantities and at a low price, esp. in a foreign country. **5.** *Computer Sci.* To reproduce (data stored internally) onto an external storage medium, as a printout. **6.** *Slang.* To knock down **:** beat up. —*vi.* **1.** To drop or fall abruptly. **2.** To discharge contents or cargo **:** UNLOAD. —*n.* **1.** A place where refuse is dumped. **2.** A storage place for goods or supplies **:** DEPOT. **3.** A disorderly accumulation or pile. **4.** *Computer Sci.* An instance or the result of dumping stored data. **5.** *Slang.* A poorly maintained or disreputable place. **—dump′er** *n.*

**dump·ling** (dŭmp′lĭng) *n.* [Orig. unknown.] **1.** A small ball of dough cooked with soup or stew. **2.** Fruit, as an apple, wrapped and baked in sweetened dough. **3.** *Informal.* A short, chubby creature.

**dumps** (dŭmps) *pl.n.* [< Du. *domp,* haze < MDu. *damp,* vapor.] A gloomy, melancholy state of mind **:** DEPRESSION.

**dump truck** *n.* A heavy-duty truck with a bed that tilts backward to dump loose material, as sand or rock.

**dump·y**[1] (dŭm′pē) *adj.* **-i·er, -i·est.** [< dial. *dump,* lump.] Short and stout **:** SQUAT. **—dump′i·ly** *adv.* **—dump′i·ness** *n.*

**dump·y**[2] (dŭm′pē) *adj.* **-i·er, -i·est.** Dejected or discontent.

**dumpy level** *n.* A surveyor's instrument having a short telescope fixed rigidly to a horizontally rotating table.

**dun**[1] (dŭn) *vt.* **dunned, dun·ning, duns.** [Orig. unknown.] To importune (a debtor) persistently for payment. —*n.* **1.** One that duns. **2.** An urgent demand esp. for payment.

**dun**[2] (dŭn) *n.* [ME < OE *dunn.*] **1.** A brownish gray to dull grayish brown. **2.** A dun-colored fishing fly. **3.** A dun-colored horse.

**dunce** (dŭns) *n.* [After John *Duns* Scotus (1265?–1308), whose writings and philosophy were ridiculed in the 16th cent.] DUMMY 3.

▲ word history: The word *dunce* comes from the name of John Duns Scotus, an eminent 13th-century Scholastic theologian. In the early 16th century the humanist scholars of classical Greek and Latin and the religious reformers criticized the *Dunses,* or followers of Scotus, for their resistance to the new learning and theology. By the end of the 16th century *dunse* or *dunce* had acquired its current meaning of "a stupid person."

**dunce cap** *also* **dunce's cap** *n.* A cone-shaped paper cap, formerly placed upon the head of a mentally slow or lazy pupil.

**dun·der·head** (dŭn′dər-hĕd′) *n.* [Perh. Du. *donder,* thunder + HEAD.] A numbskull.

**dun·drear·ies** (dŭn-drîr′ēz) *pl.n.* [After Lord *Dundreary,* a character in the play *Our American Cousin* by Tom Taylor (1817–1880).] Long sideburns with a clean-shaven chin.

**dune** (do͞on, dyo͞on) *n.* [Fr. < OFr. < MDu. *dūne.*] A ridge or hill of wind-blown sand.

**dune buggy** *n.* A small, light vehicle, often without doors and roof and usu. equipped with a high-powered engine and oversize tires for driving on sand dunes.

**dung** (dŭng) *n.* [ME < OE.] **1.** The excrement of animals **:** MANURE. **2.** Something foul or repugnant. —*vt.* **dunged, dung·ing, dungs.** To fertilize with manure. **—dung′y** *adj.*

**dun·ga·ree** (dŭng′gə-rē′) *n.* [Hindi *dungrī.*] **1.** A sturdy, usu. blue denim fabric. **2. dungarees.** Pants or overalls made of dungaree.

**dung beetle** *n.* Any of various beetles of the family Scarabaeidae that roll dung into balls on which they feed and in which they lay their eggs.

**dun·geon** (dŭn′jən) *n.* [ME *donjon,* castle keep, dungeon < OFr., keep < Med. Lat. *dominio,* lordship < Lat. *dominus,* master.] **1.** A dark, often underground cell or chamber for confining prisoners. **2.** A donjon.

**dung·hill** (dŭng′hĭl′) *n.* **1.** A heap of animal dung. **2.** A foul, degraded place or condition.

**du·nite** (do͞o′nīt′, dŭn′īt′) *n.* [After Mt. *Dun* in New Zealand, where it is found.] A dense igneous rock consisting primarily of olivine. **—du·nit′ic** (do͞o-nĭt′ĭk, də-) *adj.*

**dunk** (dŭngk) *v.* **dunked, dunk·ing, dunks.** [Pennsylvania Dutch *dunke* < MHG *dunken* < OHG *dunkōn.*] —*vt.* **1.** To immerse in liquid. **2.** To dip (food) into a liquid before eating. **3.** *Basketball.* To slam (a ball) through the basket from above. —*vi.* **1.** To submerge oneself briefly in water. **2.** *Basketball.* To dunk the ball. —*n.* **1.** The act or an instance of dunking. **2.** *Basketball.* A shot made by jumping and slamming the ball down through the basket. **—dunk′er** *n.*

**Dun·ker** (dŭng′kər) *also* **Dun·kard** (-kərd) *n.* [Pennsylvania Dutch < *dunke,* to dunk (from the practice of baptism by immersion).] A member of the German Baptist Brethren, a sect of German-American Baptists opposed to military service and the taking of legal oaths.

**dunk shot** *n. Basketball.* DUNK 2.

**dun·lin** (dŭn′lĭn) *n.* [DUN[2] + -LIN(G)[1].] A brown and white sandpiper of northern regions, *Erolia* or *Calidris alpina.*

**dun·nage** (dŭn′ĭj) *n.* [ME *dennage.*] **1.** Loose packing material protecting a ship's cargo from damage during transport. **2.** Baggage.

**du·o** (do͞o′ō, dyo͞o′ō) *n., pl.* **-os.** [Ital. < Lat., two.] **1.** *Mus.* **a.** A duet. **b.** Two performers singing or playing together. **2.** Two people or objects in close association **:** PAIR.

**duo-** *pref.* [Lat. < *duo,* two.] Two <*duopsony*>

**du·o·dec·i·mal** (do͞o′ə-dĕs′ə-məl, dyo͞o′-) *adj.* [< Lat. *duodecimus,* twelfth < *duodecim,* twelve : *duo,* two + *decem,* ten.] **1.** Of, relating to, or based on the number 12. **2.** Of or relating to twelfths. —*n.* A twelfth.

**du·o·dec·i·mo** (do͞o′ə-dĕs′ə-mō′, dyo͞o′-) *n., pl.* **-mos.** [Lat., ablative of *duodecimus,* twelfth. —see DUODECIMAL.] **1.** The page size (5 by 7¾ inches) of a book, formed by folding a single printer's sheet into 12 leaves. **2.** A book made up of duodecimos.

**du·o·de·num** (do͞o′ə-dē′nəm, dyo͞o′-, do͞o-ŏd′n-əm, dyo͞o-) *n., pl.* **du·od·e·na** (do͞o-ŏd′n-ə, dyo͞o-) *or* **du·o·de·nums.** [ME < Med. Lat., short for *duodenum digitorum,* intestine of twelve fingers (in length) < Lat. *duodeni,* twelve each < *duodecim,* twelve. —see DUODECIMAL.] The beginning portion of the small intestine, extending from the lower end of the stomach to the jejunum. **—du′o·de′nal** (do͞o′ə-dē′nəl, dyo͞o′-, do͞o-ŏd′n-əl, dyo͞o-) *adj.*

**du·o·logue** (do͞o′ə-lôg′, -lŏg′) *n.* A conversation between two people, esp. in a play.

**du·op·so·ny** (do͞o-ŏp′sə-nē, dyo͞o-) *n., pl.* **-nies.** [DUO- + Gk. *opsōnia,* purchasing of provisions : *opson,* food + *ōneisthai,* to buy.] A stock-market condition wherein two rival buyers exert a controlling influence on numerous sellers.

**dupe** (do͞op, dyo͞op) *n.* [Fr. < OFr., prob. alteration of *huppe,* hoopoe.] **1.** One who is easily deceived or taken in. **2.** One who is the tool of another person or a power. —*vt.* **duped, dup·ing, dupes.** To make a dupe of. **—dup′a·bil′i·ty** *n.* **—dup′a·ble** *adj.* **—dup′er** *n.*

**du·ple** (do͞o′pəl, dyo͞o′-) *adj.* [Lat. *duplus.*] **1.** Consisting of two **:** DOUBLE. **2.** *Mus.* Consisting of two or a multiple of two beats to the measure.

**du·plex** (do͞o′plĕks′, dyo͞o′-) *adj.* [Lat.] **1.** Double **:** twofold. **2.** Designating machinery having two identical units that operate in a single frame, each capable of operating independently. **3.** *Electron.* Able to transmit two messages simultaneously in the same or opposite directions over a single wire. —*n.* A duplex apartment or house. **—du·plex′i·ty** (do͞o-plĕk′sĭ-tē, dyo͞o-) *n.*

**duplex apartment** *n.* An apartment with rooms on two adjoining floors connected by an inner staircase.

**duplex house** *n.* A house divided into two living units.

**du·pli·cate** (do͞o′plĭ-kĭt, dyo͞o′-) *adj.* [ME < Lat. *duplicatus,* p.part. of *duplicare,* to double < *duplex,* twofold.] **1.** Copied identically from an original. **2.** Existing or growing in double corresponding parts. **3.** Designating a manner of play in certain card games in which all partnerships play the same hands and compare scores at the end. —*n.* **1.** An identical copy **:** FACSIMILE. **2.** Something that corresponds exactly to something else, esp. an original. **3.** A duplicate card game. —*vt.* (-kāt') **-cat·ed, -cat·ing, -cates. 1.** To make an identical copy of. **2.** To make twofold : DOUBLE. **3.** To make or execute again : REPEAT <couldn't *duplicate* such a performance> —*vi.* To become duplicate. **—du'pli·cate·ly** *adv.* **—du'pli·ca'tive** *adj.*

**du·pli·ca·tion** (dōō'plĭ-kā'shən, dyōō'-) *n.* **1. a.** The act or procedure of duplicating. **b.** The state of being duplicated. **2.** A duplicate.

**du·pli·ca·tor** (dōō'plĭ-kā'tər, dyōō'-) *n.* A machine that reproduces written or printed material.

**du·plic·i·tous** (dōō-plĭs'ĭ-təs, dyōō-) *adj.* Marked by duplicity : DECEITFUL. **—du·plic'i·tous·ly** *adv.* **—du·plic'i·tous·ness** *n.*

**du·plic·i·ty** (dōō-plĭs'ĭ-tē, dyōō-) *n., pl.* **-ties.** [ME *duplicite* < OFr. < LLat. *duplicitas* < Lat. *duplex*, twofold.] **1.** Deliberate deception : DOUBLE-DEALING. **2.** The quality or state of being double.

**du·ra·ble** (dŏŏr'ə-bəl, dyŏŏr'-) *adj.* [ME < OFr. < Lat. *durabilis* < *durare*, to last.] Capable of withstanding wear or decay <a *durable* textile> **—du'ra·bil'i·ty, du'ra·ble·ness** *n.* **—du'ra·bly** *adv.*

**durable goods** *pl.n.* Manufactured products designed to be durable, as appliances and automobiles.

**durable press** *n.* **1.** A chemical process in which fabrics are permanently shaped and treated for wrinkle resistance. **2.** A fabric treated by durable press.

**du·ral** (dŏŏr'əl, dyŏŏr'-) *adj.* Relating to the dura mater.

**du·ral·u·min** (dōō-răl'yə-mĭn, dyōō-) *n.* [Orig. a trademark.] A corrosion-resistant alloy of aluminum containing copper, manganese, magnesium, iron, and silicon.

**du·ra ma·ter** (dŏŏr'ə mā'tər, mä-, dyŏŏr'ə) *n.* [ME < Med. Lat. *dura mater* (*cerebri*), hard mother (of the brain).] A tough fibrous membrane, lying over the arachnoid and the pia mater, that covers the brain and the spinal cord.

**du·ra·men** (dōō-rā'mən, dyōō-) *n.* [NLat. < Lat., hardness < *durare*, to harden < *durus*, hard.] Heartwood.

**du·rance** (dŏŏr'əns, dyŏŏr'-) *n.* [ME *duraunce*, duration < OFr. *durance* < *durer*, to last < Lat. *durare*.] Imprisonment, esp. for a long time. —Often used in the phrase *in durance vile*.

**du·ra·tion** (dōō-rā'shən, dyōō-) *n.* [Med. Lat. *duratio, duration-* < Lat. *durare*, to last.] **1.** Continuance in time. **2.** The period of time during which something exists or lasts.

**dur·bar** (dûr'bär') *n.* [Hindi *darbār* < Pers., court : *dar*, door + *bār*, admission, audience.] **1.** A state reception formerly given by an Indian prince or a British governor in India. **2.** The court of an Indian prince.

**du·ress** (dōō-rĕs', dyōō-) *n.* [ME *duresse*, harshness, compulsion < OFr. *durece*, hardness < Lat. *duritia* < *durus*, hard.] **1.** Constraint by threat : COERCION. **2.** *Law.* **a.** Coercion illegally applied. **b.** Forced restraint : DURANCE.

**Dur·ham** (dûr'əm) *n.* [After *Durham*, a county in England.] A shorthorn.

**du·ri·an** (dŏŏr'ē-ən, dyŏŏr'-, -än') *n.* [Malay.] **1.** A tree, *Durio zibethinus* of southeastern Asia, bearing edible fruit. **2.** The fruit of the durian, with a hard, prickly rind and soft, malodorous yet pleasant-tasting pulp.

**dur·ing** (dŏŏr'ĭng, dyŏŏr'-) *prep.* [ME < *duren*, to last < OFr. *durer* < Lat. *durare*.] **1.** Throughout the course or duration of <talked all *during* the movie> **2.** At some time in <died *during* the night>

**dur·mast** (dûr'măst') *n.* [Perh. alteration of *dun mast* : DUN² + MAST².] A European oak, *Quercus petraea*, yielding tough, elastic wood and bark rich in tannin.

**durn** (dûrn) *v.* & *interj.* DARN². **—durn, durned** *adj.* & *adv.*

**du·roc** *also* **Du·roc** (dŏŏ'rŏk', dyŏŏ'-) *n.* [After *Duroc*, a horse owned by the developer of the breed.] A large red hog of a breed developed in the United States during the 19th cent.

**dur·ra** (dŏŏr'ə) *n.* [Ar. *dhurah*.] A cereal grain, *Sorghum vulgare durra* of Asia and northern Africa, widely cultivated in warm dry regions.

**durst** (dûrst) *v. Archaic. var. p.t.* & *p.p. of* DARE.

**du·rum** (dŏŏr'əm, dyŏŏr'-, dûr'-, dŭr'-) *n.* [NLat. < Lat., neuter of *durus*, hard.] A hardy wheat, *Triticum aestivum durum*, used primarily in pasta-making.

**dusk** (dŭsk) *n.* [ME, dark < OE *dox*.] The darker stage of twilight, esp. in the evening. —*adj.* Tending to darkness : DUSKY. —*vt.* & *vi.* **dusked, dusk·ing, dusks.** To darken or become dark.

**dusk·y** (dŭs'kē) *adj.* **-i·er, -i·est. 1.** Marked by dim or inadequate light : SHADOWY. **2.** Somewhat dark in color, as skin. **—dusk'i·ly** *adv.* **—dusk'i·ness** *n.*

**dusky grouse** *n.* The blue grouse.

**dust** (dŭst) *n.* [ME < OE *dūst*.] **1.** Fine, dry particles of matter. **2.** A cloud of dust. **3.** Dust viewed as the result of disintegration. **4. a.** Earth, esp. when considered as the substance of the grave. **b.** The surface of the ground. **5.** A debased or scorned condition. **6.** Something worthless. **7.** *Chiefly Brit.* Household refuse readied for disposal. **8.** Confusion or agitation resulting from a commotion. —*v.* **dust·ed, dust·ing, dusts.** —*vt.* **1.** To remove dust from by wiping, brushing, or beating. **2.** To sprinkle or cover with a powdery substance. **3.** To strew like dust <Freckles *dusted* the child's face.> **4.** *Baseball.* To deliver a pitch so close to (the batter) as to cause backing away. **5.** To restore to use <*dust* off one's tuxedo> **6.** *Archaic.* To cover with dust. —*vi.* **1.** To clean by removing dust. **2.** To cover itself with dust.—Used of a bird.

**dust·bin** (dŭst'bĭn') *n. Chiefly Brit.* A can for garbage or trash.

**dust bowl** *n.* A region reduced to aridity by dust storms and drought.

**dust cover** *n.* **1.** A removable or hinged plastic cover for protecting a turntable. **2.** DUST JACKET 1.

**dust devil** *n.* A small whirlwind that swirls dust, debris, and sand to great heights.

**dust·er** (dŭs'tər) *n.* **1.** One that dusts. **2.** A brush or cloth for removing dust. **3.** A device for sifting or scattering a powdered substance. **4.** A smock worn to protect one's clothing from dust. **5.** A woman's loose dress-length housecoat.

**dust jacket** *n.* **1.** A removable paper cover for protecting the binding of a book. **2.** A cardboard sleeve for packaging and protecting a phonograph record.

**dust·man** (dŭst'mən) *n. Chiefly Brit.* One employed to remove trash.

**dust·pan** (dŭst'păn') *n.* A short-handled pan for swept-up refuse.

**dust storm** *n.* A severe windstorm that sweeps clouds of dust across an arid region.

**dust·up** (dŭst'ŭp') *n.* An argument.

**dust·y** (dŭs'tē) *adj.* **-i·er, -i·est. 1.** Filled or covered with dust. **2.** Consisting of or like dust : POWDERY. **3.** Tinged with gray. **—dust'i·ly** *adv.* **—dust'i·ness** *n.*

**dusty miller** *n.* Any of various plants bearing leaves and stems covered with dustlike down.

**Dutch** (dŭch) *adj.* [ME *Duch*, German, Dutch < MDu. *Duutsch*.] **1.** Of or relating to the Netherlands, its inhabitants, or their language. **2.** *Archaic.* German. —*n.* **1. a.** The people of the Netherlands. **b.** *Archaic.* The Germans. **2. a.** The West Germanic language of the Netherlands. **b.** *Archaic.* GERMAN 2. **c.** Pennsylvania Dutch. **3.** *Slang.* Temper or anger. **—go Dutch.** To participate in a Dutch treat. **—in Dutch.** *Informal.* In difficulty or disfavor.

**Dutch cheese** *n.* Cottage cheese.

**Dutch clover** *n.* The white clover.

**Dutch courage** *n. Informal.* Courage gained from drinking alcoholic beverages.

**Dutch door** *n.* A door divided in half horizontally so that either part can be opened or closed.

**Dutch elm disease** *n.* A disease of elm trees caused by a fungus, *Ceratocystis ulmi*, and resulting in brown streaks in the wood and eventual death of the tree.

**Dutch·man** (dŭch'mən) *n.* **1. a.** A native or resident of the Netherlands. **b.** A person of Dutch descent. **2.** *Archaic.* A German. **3. dutchman.** Something used to conceal structural faults.

**Dutch·man's-breech·es** (dŭch'mənz-brĭch'ĭz) *n.* A woodland plant, *Dicentra cucullaria* of eastern North America, bearing finely divided leaves and yellowish-white two-spurred flowers.

**Dutchman's-breeches**

**Dutch·man's-pipe** (dŭch'mənz-pīp') *n.* The pipe vine.

**Dutch metal** *n.* An alloy of copper and zinc used in thin sheets as an imitation of gold leaf.

**Dutch oven** *n.* **1.** A large, heavy pot or kettle, usu. of cast iron and with a tight lid, used for slow cooking. **2.** A metal utensil open on one side and equipped with shelves, placed before an open fire for baking or roasting food. **3.** A brick wall oven in which food is baked by the preheated bricks.

**Dutch treat** *n.* An outing, as for dinner or a movie, in which each person pays his or her own expenses.

**Dutch uncle** *n.* A stern and candid critic or adviser.

**du·te·ous** (dōō'tē-əs, dyōō'-) *adj.* [< DUTY.] Dutiful or obedient. **—du'te·ous·ly** *adv.*

**du·ti·a·ble** (dōō'tē-ə-bəl, dyōō'-) *adj.* Subject to import tax.

**du·ti·ful** (dōō'tĭ-fəl, dyōō'-) *adj.* **1.** Careful to perform duties. **2.** Imbued with or expressing a sense of duty. **—du'ti·ful·ly** *adv.* **—du'ti·ful·ness** *n.*

**du·ty** (dōō'tē, dyōō'-) *n., pl.* **-ties.** [ME *duete* < AN < OFr. *deu*, due. —see DUE.] **1.** An act or a course of action required of one by position, custom, law, or religion. **2. a.** Moral obligation. **b.** The compulsion felt to meet such obligation. **3.** A service, action, or task assigned to one, esp. in the armed forces. **4.** Function or work : SERVICE. **5.** A government tax, esp. on imports. **6. a.** The work capability of a machine under specified conditions. **b.** A measure of efficiency expressed as work per unit energy input. **7.** The amount of water needed to irrigate a specified area for the cultivation of a crop.

**du·ty-free** (do͞o′tē-frē′, dyo͞o′-) *adj.* Exempt from customs duties. **—du′ty-free′** *adv.*

**du·um·vir** (do͞o-ŭm′vər, dyo͞o-) *n.* [Lat., alteration of *duovir* : *duo*, two + *vir*, man.] A member of a duumvirate.

**du·um·vi·rate** (do͞o-ŭm′vər-ĭt, dyo͞o-) *n.* **1.** Any of various two-man administrative or judicial positions in the Roman Republic. **2.** A regime or partnership of two persons.

**du·ve·tyn** *also* **du·ve·tyne** (do͞o′və-tēn′, dyo͞o′-, do͞o′və-tēn′, dyo͞o′-) *n.* [Fr. *duvetine* < *duvet*, down < OFr. < *dum, dun* < ON *dūnn.*] A soft, short-napped wool, cotton, rayon, or silk fabric with a twill weave.

**dwarf** (dwôrf) *n., pl.* **dwarfs** *or* **dwarves** (dwôrvz) [ME *dwerf* < OE *dweorh.*] **1. a.** A very small person, esp. one afflicted with dwarfism. **b.** An unusually small animal or plant. **2.** A diminutive, often ugly or deformed humanlike creature of fairy tales and legend. **3.** A dwarf star. *—v.* **dwarfed, dwarf·ing, dwarfs.** *—vt.* **1.** To stunt the natural growth or development of <"The oaks were *dwarfed* from lack of moisture" —John Steinbeck> **2.** To cause to appear small by comparison <"Together these two big men *dwarfed* the tiny Broadway office" —Saul Bellow> *—vi.* To grow smaller or become stunted. **—dwarf** *adj.* **—dwarf′ish** *adj.* **—dwarf′ish·ness** *n.*

**dwarf cornel** *n.* A woody plant, *Cornus canadensis* of northern North America, bearing inconspicuous greenish flowers surrounded by white, petallike bracts and scarlet fruit.

**dwarf·ism** (dwôr′fĭz′əm) *n.* A pathological condition of arrested growth.

**dwarf star** *n.* A star, as the sun, having relatively low mass and average or below average luminosity.

**dwarves** (dwôrvz) *n. var. pl. of* DWARF.

**dwell** (dwĕl) *vi.* **dwelt** (dwĕlt) *or* **dwelled, dwell·ing, dwells.** [ME *dwellen* < OE *dwellan*, to delay.] **1.** To live as an inhabitant : RESIDE. **2.** To exist in a given place or condition. **3. a.** To focus one's attention <kept *dwelling* on past mistakes> **b.** To treat at length <*dwelt* on the need to economize> **4.** *Computer Sci.* A programmed time delay of variable duration. **—dwell′er** *n.*

**dwell·ing** (dwĕl′ĭng) *n.* An abode : residence.

**dwelt** (dwĕlt) *v. var. p.t. & p.p. of* DWELL.

**dwin·dle** (dwĭn′dl) *v.* **-dled, -dling, -dles.** [Freq. of obs. *dwine*, to diminish < ME *dwinen*, to shrink < OE *dwīnan.*] *—vi.* To become gradually less : DIMINISH. *—vt.* To make continuously smaller or less.

**Dy** *symbol for* DYSPROSIUM.

**dy·ad** (dī′ăd′, -əd) *n.* [< Gk. *duas, duad-*, pair < *duo*, two.] **1.** Two units considered as a pair. **2.** *Biol.* One pair of chromosomes separated from a tetrad in meiosis. **3.** *Chem.* A divalent atom or radical. **4.** A mathematical operator represented as a pair of vectors juxtaposed without multiplication. *—adj.* Composed of two units.

**dy·ad·ic** (dī-ăd′ĭk) *adj.* **1.** Twofold. **2.** Of or relating to a dyad. *—n.* *Math.* The direct product *(B·C) AD* of two dyads *AB* and *CD.*

**Dy·ak** *also* **Day·ak** (dī′ăk′) *n.* [Malay *Dayak* < *darat*, land.] **1.** A member of any of various Indonesian peoples of Borneo and the Sulu Sea islands. **2.** The language of the Dyaks.

**dy·ar·chy** (dī′är′kē) *n. var. of* DIARCHY.

**dyb·buk** (dĭb′ək) *n.* [Yiddish *dibek* < Heb. *dibbūq.*] The wandering soul of a deceased person that enters the body of a living person and controls his or her behavior in Jewish folklore.

**dye** (dī) *n.* [ME *deie* < OE *dēah*, hue.] **1.** A substance used to color materials. **2.** A color imparted by dyeing. *—v.* **dyed, dye·ing, dyes.** *—vt.* To color (a material) with or as if with a dye, esp. by soaking in a coloring solution. *—vi.* To impart or take on color. **—dye′a·ble** *adj.* **—dy′er** *n.*

**dyed-in-the-wool** (dīd′ĭn-*th*ə-wo͝ol′) *adj.* **1.** Dyed before being woven into cloth. **2.** Thoroughgoing <a *dyed-in-the-wool* rebel>

**dy·er's broom** (dī′ərz) *n.* The dyer's greenweed.

**dyer's greenweed** *n.* A small Eurasian shrub, *Genista tinctoria*, with yellow flower clusters.

**dyer's rocket** *n.* A European plant, *Reseda luteola*, bearing spikes of small, yellowish-green flowers and yielding a yellow dye.

**dy·er's-weed** (dī′ərz-wēd′) *n.* Any of various plants yielding coloring matter used as dye.

**dye·stuff** (dī′stŭf′) *n.* A material used as or yielding a dye.

**dye·wood** (dī′wo͝od′) *n.* A wood used as a dyestuff.

**dy·ing** (dī′ĭng) *adj.* **1.** About to die. **2.** Drawing to a close : DECLINING <the *dying* year> **3.** Done or uttered just before death.

**dyke** (dīk) *n. & v. var. of* DIKE.

**dy·nam·ic** (dī-năm′ĭk) *also* **dy·nam·i·cal** (-ĭ-kəl) *adj.* [Fr. *dynamique* < Gk. *dunamikos*, powerful < *dunamis*, power < *dunasthai*, to be able.] **1. a.** Of or relating to energy, force, or motion in relation to force. **b.** Of or relating to dynamics. **2.** Marked by continuous change, activity, or progress. **3.** Marked by vigor and energy. **4.** Of or relating to variation of intensity, as in musical sound. **—dy·nam′i·cal·ly** *adv.*

**dy·nam·ics** (dī-năm′ĭks) *n. (sing. or pl. in number).* **1. a.** The study of the relationship between motion and the forces affecting motion. **b.** The combined study of kinetics and kinematics. **2.** The physical, intellectual, or moral forces that produce motion, activity, and change in a given sphere. **3.** Variation in force or intensity, esp. in musical sound. **4.** *Psychoanal.* **a.** The action of psychic forces or mechanisms. **b.** The psychological aspect or conduct of an interpersonal relationship.

**dy·na·mism** (dī′nə-mĭz′əm) *n.* [Fr. *dynamisme* < Gk. *dunamis*, power < *dunasthai*, to be able.] **1.** Any of various philosophical systems or theories that explain the universe in terms of force or energy. **2.** A process or mechanism responsible for the development or motion of a system. **3.** The quality of being dynamic. **—dy′na·mist** *n.* **—dy′na·mis′tic** *adj.*

**dy·na·mite** (dī′nə-mīt′) *n.* [Swed. *dynamit* < Gk. *dunamis*, power < *dunasthai*, to be able.] **1.** A powerful explosive composed of nitroglycerin or ammonium nitrate dispersed in an absorbent medium with a combustible dope, as wood pulp, and an antacid, as calcium carbonate. **2.** *Slang.* Something highly effective, outstanding, or wonderful. *—vt.* **-mit·ed, -mit·ing, -mites. 1.** To blow up, shatter, or destroy with or as if with dynamite. **2.** To charge with dynamite. **—dy′na·mit′er** *n.*

**dy·na·mo** (dī′nə-mō′) *n., pl.* **-mos.** [Short for *dynamoelectric machine.*] **1.** A generator, esp. one for producing direct current. **2.** An extremely forceful and energetic person.

**dy·na·mo·e·lec·tric** (dī′nə-mō′ĭ-lĕk′trĭk) *also* **dy·na·mo·e·lec·tri·cal** (-trĭ-kəl) *adj.* [Gk. *dunamis*, power + ELECTRIC.] Of or relating to the interconversion of mechanical and electrical energy.

**dy·na·mom·e·ter** (dī′nə-mŏm′ĭ-tər) *n.* [Fr. *dynamomètre* : Gk. *dunamis*, power + *-mètre*, meter.] Any of several instruments for measuring power or force. **—dy′na·mo·met′ric** (-mō-mĕt′rĭk), **dy′na·mo·met′ri·cal** *adj.* **—dy′na·mom′e·try** *n.*

**dy·na·mo·tor** (dī′nə-mō′tər) *n.* [Gk. *dunamis*, power + MOTOR.] A rotating electric machine with two armatures, used to convert alternating to direct current.

**dy·nast** (dī′năst′, -nəst) *n.* [Lat. *dynastes* < Gk. *dunastēs*, lord < *dunasthai*, to be able.] A ruler, esp. a hereditary ruler.

**dy·nas·ty** (dī′nə-stē) *n., pl.* **-ties.** [Fr. *dynastie* < Gk. *dunasteia*, lordship < *dunastēs*, dynast.] **1.** A succession of rulers from the same line or family. **2.** A family or group that maintains power for several generations. **—dy·nas′tic** (dī-năs′tĭk) *adj.* **—dy·nas′ti·cal·ly** *adv.*

**dy·na·tron** (dī′nə-trŏn′) *n.* [Gk. *dunamis*, power + -TRON.] *Electron.* A tetrode with grid and plate potentials so arranged that plate current decreases when plate potential increases.

**dyne** (dīn) *n.* [Fr. < Gk. *dunamis*, power. —see DYNAMIC.] A centimeter-gram-second unit of force, equal to the force needed to impart an acceleration of one centimeter per second per second to a mass of one gram.

**Dy·nel** (dī-nĕl′). A trademark for a copolymer of vinyl chloride and acrylonitrile, used to make fire-resistant, insect-resistant, and easily dyed textile fiber.

**dy·node** (dī′nōd′) *n.* [Gk. *dunamis*, power + -ODE.] An electrode used in certain electron tubes to provide secondary emission.

**dys-** *pref.* [ME *dis-*, bad < OFr. < Lat. *dys-* < Gk. *dus-*.] **1.** Abnormal <*dys*plasia> **2. a.** Impaired <*dys*graphia> **b.** Difficult <*dys*phonia> **3.** Bad <*dys*logistic>

**dys·cra·sia** (dĭs-krā′zhə, -zhē-ə) *n.* [NLat. < Med. Lat., disease < Gk. *duskrasia* : *dus-*, bad + *krasis*, mixing.] An abnormal bodily condition caused by poisons in the blood.

**dys·en·ter·y** (dĭs′ən-tĕr′ē) *n.* [ME *dissenterie* < Lat. *dysenteria* < Gk. *dusenteria* : *dus-*, bad + *enteron*, intestine.] An infection of the lower intestinal tract producing pain, fever, and severe diarrhea, often with the passage of blood and mucus. **—dys′en·ter′ic** *adj.*

**dys·func·tion** (dĭs-fŭngk′shən) *n.* Impaired or disordered functioning of a bodily system or organ. **—dys·func′tion·al** *adj.*

**dys·gen·ic** (dĭs-jĕn′ĭk) *adj.* Causing or relating to the deterioration of hereditary qualities.

**dys·gen·ics** (dĭs-jĕn′ĭks) *n. (sing. in number).* The biological study of the factors producing degeneration in offspring.

**dys·graph·ia** (dĭs-grăf′ē-ə) *n.* [NLat. : DYS- + *-graphia*, -graphy.] Impairment of the ability to write.

**dys·lex·i·a** (dĭs-lĕk′sē-ə) *n.* [NLat. DYS- + Gk. *lexis*, speech < *legein*, to speak.] Impairment of the ability to read. **—dys·lec′tic** (-lĕk′tĭk) *n.* **—dys·lex′ic** *adj.*

**dys·lo·gis·tic** (dĭs′lə-jĭs′tĭk) *adj.* [DYS- + (EU)LOGISTIC.] Conveying disapproval. **—dys′lo·gis′ti·cal·ly** *adv.*

**dys·men·or·rhe·a** *also* **dys·men·or·rhoe·a** (dĭs-mĕn′ə-rē′ə) *n.* [DYS- + Gk. *mēn*, month + -RRHEA.] Painful or difficult menstruation. **—dys·men′or·rhe′al, dys·men′or·rhe′ic** *adj.*

**dys·pep·sia** (dĭs-pĕp′shə, -sē-ə) *n.* [Lat. < Gk. *duspepsia* : *dus-*, bad + *pepsis*, digestion < *peptein*, to cook.] Indigestion.

**dys·pep·tic** (dĭs-pĕp′tĭk) *adj.* **1.** Relating to or having dyspepsia. **2.** Of or displaying a sullen disposition. *—n.* One who suffers from dyspepsia. **—dys·pep′ti·cal·ly** *adv.*

**dys·pha·gia** (dĭs-fā′jə, -jē-ə) *n.* Difficulty in swallowing. **—dys·phag′ic** (-făj′ĭk) *adj.*

**dys·pha·sia** (dĭs-fā′zhə, -zhē-ə) *n.* Impairment of speech and verbal comprehension, esp. when associated with brain damage. **—dys·pha′sic** (-zĭk) *adj.* & *n.*
**dys·pho·ni·a** (dĭs-fō′nē-ə) *n.* [NLat. DYS- + Gk. *phonē*, sound.] Difficulty in speaking, usu. indicated by hoarseness. **—dys·phon′ic** (-fŏn′ĭk) *adj.*
**dys·pho·ri·a** (dĭs-fôr′ē-ə, -fōr′-) *n.* [NLat. < Gk. *dusphoria*, distress < *dusphoros*, hard to bear : *dus-*, bad + *pherein*, to bear.] An emotional state marked by anxiety, depression, and restlessness. **—dys·phor′ic** (-fôr′ĭk, -fŏr′ĭk) *adj.*
**dys·pla·sia** (dĭs-plā′zhə, -zhē-ə) *n.* [NLat. DYS- + Gk. *plasis*, formation < *plassein*, to mold.] Abnormal development of tissues, organs, or cells. **—dys·plas′tic** (-plăs′tĭk) *adj.*
**dysp·ne·a** (dĭsp-nē′ə) *n.* [NLat. < Gk. *duspnoia* < *duspnoos*, short of breath : *dus-*, bad + *pnoē*, breathing < *pnein*, to breathe.] Difficulty in breathing, often associated with lung or heart disease. **—dysp·ne′ic** (-nē′ĭk) *adj.*
**dys·pro·si·um** (dĭs-prō′zē-əm, -zhē-əm) *n.* [NLat. < Gk. *dusprositos*, difficult to approach : *dus-*, bad + *prositos*, approachable < *prosienai*, to approach (*pros-*, toward + *ienai*, to go).] *Symbol* **Dy** A soft, silvery metal used in nuclear research; atomic number 66; atomic weight 162.50.
**dys·tel·e·ol·o·gy** (dĭs-tĕl′ē-ŏl′ə-jē, -tē′lē-) *n.* [G. *Dysteleologie* : *dys-*, dys- + *Teleologie*, teleology.] **1.** The doctrine of purposelessness in nature. **2.** Purposelessness in natural structures, as evidenced by the existence of nonfunctional or vestigial organs or parts. **—dys·tel′e·o·log′i·cal** (-ə-lŏj′ĭ-kəl) *adj.* **—dys·tel′e·ol′o·gist** *n.*
**dys·to·pi·a** (dĭs-tō′pē-ə) *n.* [DYS- + (U)TOPIA.] An imaginary place of total misery and wretchedness. **—dys·to′pi·an** *adj.*
**dys·tro·phy** (dĭs′trə-fē) *also* **dys·tro·phi·a** (dĭ-strō′fē-ə) *n.* **1. a.** Defective nutrition. **b.** A disorder caused by defective nutrition. **2.** Atrophy of muscle tissue, esp. muscular dystrophy. **—dys·troph′ic** (-trŏf′ĭk, -trō′fĭk) *adj.*
**dys·u·ri·a** (dĭs-yŏŏr′ē-ə) *n.* Difficult or painful urination. **—dys·u′ric** (-yŏŏr′ĭk) *adj.*

# Ee

**e** *or* **E** (ē) *n.*, *pl.* **e's** *or* **E's. 1.** The fifth letter of the English alphabet. **2.** A speech sound represented by the letter *e*. **3.** The fifth in a series. **4. e** *Math.* The base of the natural system of logarithms, having a numerical value of approx. 2.718 . . . . **5. E** *Mus.* The third tone in the scale of C major. **6. E** A failing grade. **7.** Something shaped like the letter E.
**each** (ēch) *adj.* [ME *ech* < OE *ælc*.] Being one of two or more regarded individually : EVERY. —*pron.* Every one of a group regarded individually : each one. ***usage:*** As the pronoun subject of a verb, *each* is traditionally considered to be singular, as in *Each of the students has a particular job to do.* But when *each* follows a plural subject, the verb then is in agreement with the plural form, as in *The students each have a particular job to do.* —*adv.* For or to each one : APIECE <five dollars *each*>
**each other** *pron.* Each the other. —Used to indicate a reciprocal relationship or action. ***usage:*** Although some traditional grammarians state that *each other* is used of two and *one another* of more than two, this distinction has been largely ignored by the best writers, and few people question the acceptability of sentences such as *The four players regarded each other with suspicion* or *A husband and wife should confide in one another.* When speaking of the relations between the members of a series or set, however, only *one another* is appropriate, as in *The Caesars exceeded one another* (not *each other*) *in cruelty.*
**ea·ger¹** (ē′gər) *adj.* **-er, -est.** [ME *eger* < OFr. *aigre*, sharp < Lat. *acer*.] **1.** Intensely desirous or impatiently expectant <*eager* to win> **2.** *Obs.* Tart : sharp. **—ea′ger·ly** *adv.* **—ea′ger·ness** *n.*
☆ ***syns:*** EAGER, AGOG, ANXIOUS, ARDENT, ATHIRST, AVID, IMPATIENT, KEEN, RARING, THIRSTY *adj. core meaning* : intensely desirous or interested <*eager* for success><*eager* football fans>
**ea·ger²** (ē′gər, ā′-) *n. var. of* EAGRE.
**eager beaver** *n. Informal.* An industrious or overzealous person.
**ea·gle** (ē′gəl) *n.* [ME *egle* < OFr. *aigle* < Lat. *aquila*.] **1.** Any of various large birds of prey of the family Accipitridae, including members of the genera *Aquila* and *Haliaeetus*, marked by a powerful hooked bill, long broad wings, and strong, soaring flight. **2.** A representation of an eagle used as an insignia or emblem. **3.** A former gold U.S. coin with a face value of ten dollars. **4.** A golf score of two under par on a hole.
**ea·gle-eyed** (ē′gəl-īd′) *adj.* Having keen eyesight.
**eagle owl** *n.* A large Eurasian owl, *Bubo bubo*, with brownish plumage and prominent ear tufts.
**Eagle Scout** *n.* A Boy Scout of the highest rank.
**ea·glet** (ē′glĭt) *n.* A young eagle.
**ea·gre** *also* **ea·ger** (ē′gər, ā′-) *n.* [Orig. unknown.] BORE³.
**eal·dor·man** (ôl′dər-mən) *n.* [OE. —see ALDERMAN.] The chief magistrate of a district in Anglo-Saxon England.
**-ean** *suff. var. of* -IAN.
**ear¹** (îr) *n.* [ME *ere* < OE *ēare*.] **1.** *Anat.* The organ of hearing in vertebrates, responsible for maintaining equilibrium as well as sensing sound. **2.** An organ analogous to the mammalian ear in invertebrates. **3.** The sense of hearing. **4.** Keenness of hearing. **5.** Favorable or sympathetic attention. **6.** Something resembling the external ear in shape or position, esp.: **a.** A tuft of feathers on the head of a bird. **b.** A projecting handle, as on a vase. **7.** A small box in the upper corner of the page in a periodical, as a newspaper, often containing an advertisement or weather information. **—all ears.** Acutely attentive. **—by ear.** Without reference to a musical score. **—give (***or* **lend) an ear.** To pay close attention to. **—have (***or* **keep) an ear to the ground.** To give attention to or watch for coming events or trends. **—in one ear and out the other.** Heard but without influence or effect. **—play it by ear.** To improvise. **—up to one's (***or* **the) ears.** Deeply involved or committed. **—ear′less** *adj.*
**ear²** (îr) *n.* [ME *ere* < OE *ær*.] The seed-bearing spike of a cereal plant, as corn. —*vi.* **eared, ear·ing, ears.** To form or grow ears.
**ear·ache** (îr′āk′) *n.* Ache in the ear.
**ear·drop** (îr′drŏp′) *n.* An earring, esp. one with a pendant.
**ear·drum** (îr′drŭm′) *n. Anat.* The tympanic membrane.
**eared** (îrd) *adj.* **1.** Having ears or earlike projections. **2.** Having a specified kind or number of ears <floppy-*eared*>
**eared seal** *n.* Any of various seals of the family Otariidae, including the fur seals and sea lions, having external ears and hind flippers that can be turned forward for walking on land.

**eared seal**
(Right) *sea lion,*
*5½–8 feet long*

**earless seal**
(Left) *elephant seal,*
*9–16 feet long*

**ear·flap** (îr′flăp′) *n.* A flap attached to a cap that may be turned down to cover the ears.
**ear·ful** (îr′fŏŏl′) *n.* **1.** A flow of gossip or information. **2.** A scolding.
**ear·ing** (îr′ĭng) *n.* [Perh. < EAR¹.] *Naut.* A short line attaching an upper corner of a sail to the yard.
**earl** (ûrl) *n.* [ME *erl*, nobleman of high rank < OE *eorl*.] A British peer ranking above a viscount and below a marquis.
**ear·lap** (îr′lăp′) *n.* An earflap.
**earl·dom** (ûrl′dəm) *n.* [ME *erldom* < OE *eorldōm*.] **1.** The title or rank of an earl. **2.** The territory of an earl.
**earless seal** *n.* Any of various seals of the family Phocidae, including the hair seals, marked by the absence of external ears, short fore flippers, and reduced hind flippers specialized for swimming.
**ear lobe** *n.* The soft, fleshy tissue at the lowest portion of the external ear.
**ear·ly** (ûr′lē) *adj.* **-li·er, -li·est.** [ME *erli* < *ēr*, before, soon < OE *ær* and ON *ār*.] **1.** Near the beginning of a given series, time period, or

course of events. **2.** Belonging to a remote time period : PRIMITIVE. **3.** Taking place, developing, or appearing before the usual or expected time. **4.** Taking place in the near future. *—adv.* **1.** Near the beginning of a series, time period, or course of events. **2.** Far back in time. **3.** Before the usual or expected time. **—ear′li·ness** *n.*

☆ ***syns:*** **1.** EARLY, BEGINNING, FIRST, INITIAL *adj. core meaning* : at or near the start of a period, development, or series <*early* stages of cancer><*early* Impressionism> **2.** EARLY, ANCIENT, PRIMITIVE *adj. core meaning* : of, existing, or occurring in a distant period <*early* fossils><*early* humans> **3.** EARLY, PREMATURE, UNTIMELY *adj. core meaning* : developing or appearing before the expected time <an *early* death><an *early* frost> ***ant:*** late

**early bird** *n.* **1.** An early riser. **2.** One who arrives early.

**early on** *adv.* At or during an early stage <The weakness of the story was evident *early on.*>

**ear·mark** (îr′märk′) *n.* **1.** An identifying mark on a domestic animal's ear. **2.** An identifying characteristic. *—vt.* **-marked, -mark·ing, -marks. 1.** To mark the ear of (a domestic animal) for identification. **2.** To place an identifying mark on. **3.** To reserve or set aside for a specific purpose <*earmarked* funds for research>

**ear·muff** (îr′mŭf′) *n.* Either of a pair of ear coverings often attached to an adjustable headband and worn to protect esp. against the cold.

**earn**[1] (ûrn) *vt.* **earned, earn·ing, earns.** [ME *ernen* < OE *earnian.*] **1.** To gain esp. for the performance of service or work. **2. a.** To acquire as a result of effort or action. **b.** To make worthy of. **3.** To produce as profit or return. **—earn′er** *n.*

☆ ***syns:*** **1.** EARN, GAIN, GET, MAKE *v. core meaning* : to receive, as wages, for one's labor <*earns* $40,000 a year> **2.** EARN, DESERVE, GAIN, GET, MERIT, RATE, WIN *v. core meaning* : to acquire as the result of one's efforts or behavior <They *earned* the respect of their peers.>

**earn**[2] (ûrn) *vi.* **earned, earn·ing, earns.** [Var. of YEARN.] *Obs.* To yearn.

**ear·nest**[1] (ûr′nĭst) *adj.* [ME *ernest* < OE *eornoste.*] **1.** Characterized by or showing deep sincerity or seriousness <an *earnest* expression of regret> **2.** Important : grave. **—in earnest.** With a serious or purposeful intent. **—ear′nest·ly** *adv.* **—ear′nest·ness** *n.*

**ear·nest**[2] (ûr′nĭst) *n.* [ME *ernest* < OFr. *erres*, pl. of *erre*, pledge < Lat. *arra*, short for *arrabo* < Gk. *arrabōn* < Heb. *'ērābhōn* < *'ārabh*, he pledged.] **1.** Money paid in advance as part payment to bind a contract or bargain. **2.** A token of something to come : PROMISE.

**earn·ings** (ûr′nĭngz) *pl.n.* Something earned, esp.: **a.** Salary or wages. **b.** Business profits. **c.** Investment gains.

**ear·phone** (îr′fōn′) *n.* A device fitting over or in the ear that converts electric signals, as from a radio or telephone receiver, to audible sound.

**ear·ring** (îr′rĭng, îr′ĭng) *n.* An ornament worn on the ear.

**ear rot** *n.* Any of various fungal diseases of corn marked by molding and decay of the ears.

**ear shell** *n.* The shell of the abalone.

**ear·shot** (îr′shŏt′) *n.* The range within which sound can be heard.

**ear·split·ting** (îr′splĭt′ĭng) *adj.* Loud and shrill : DEAFENING.

**earth** (ûrth) *n.* [ME *erthe* < OE *eorðe.*] **1. a.** The land surface of the world. **b.** The softer, friable part of land : SOIL. **2. Earth.** The third planet from the sun, having a sidereal period of revolution about the sun of 365.26 days at a mean distance of 92.96 million miles or approx. 149 million kilometers, an axial rotation period of 23 hours 56.07 minutes, an average radius of 3,959 miles or approx. 6,370 kilometers, and a mass of 13.17 x $10^{24}$ pounds. **3.** The dwelling place of mortals. **4.** All of the human inhabitants of the world <The *earth* sorrowed over the deaths.> **5.** Worldly pursuits and affairs. **6.** The material body of a human being. **7.** The den of a burrowing animal. **8.** *Chiefly Brit.* The ground of an electrical circuit. **9.** *Chem.* A metallic oxide, as alumina or zirconia, that is difficult to reduce and was once held to be an element. *—v.* **earthed, earth·ing, earths.** *—vt.* **1.** To cover (plants) with soil for protection. **2.** To chase into an underground hiding place. *—vi.* To hide or burrow in the ground, as a hunted animal. **—down to earth.** Realistic : sensible. **—run to earth.** To hunt and run down.

**earth·born** (ûrth′bôrn′) *adj.* **1. a.** Born on or springing from the earth. **b.** Mortal : human. **2.** Of, relating to, or connected with earthly life.

**earth·bound** *also* **earth-bound** (ûrth′bound′) *adj.* **1.** Fastened in or to the soil. **2. a.** Attached or confined to earthly interests. **b.** Unimaginative or ordinary : DULL. **3.** Headed for the earth <an *earthbound* satellite>

**earth·en** (ûr′thən, -thən) *adj.* **1.** Made of earth. **2.** Worldly : earthly.

**earth·en·ware** (ûr′thən-wâr′, -thən-) *n.* Ware, as dishes or pots, made from a porous baked clay.

**earth·light** (ûrth′līt′) *n.* Earthshine.

**earth·ling** (ûrth′lĭng) *n.* **1.** An inhabitant of the earth : HUMAN BEING. **2.** One devoted to the world : WORLDLING.

**earth·ly** (ûrth′lē) *adj.* **1.** Of or pertaining to the earth. **2. a.** Not heavenly or divine : SECULAR. **b.** Terrestrial. **3.** Possible : conceivable <no *earthly* excuse for such conduct> **—earth′li·ness** *n.*

☆ ***syns:*** EARTHLY, EARTHBOUND, MUNDANE, SECULAR, TELLURIAN, TELLURIC, TEMPORAL, TERRENE, TERRESTRIAL, WORLDLY *adj. core meaning* : pertaining to or characteristic of the earth or of human life thereon <*earthly* pursuits>

**earth·nut** (ûrth′nŭt′) *n.* **1. a.** An Old World plant, *Conopodium denudatum*, with edible, nutlike tubers. **b.** The tuber of the earthnut. **2.** A plant, as the peanut, similar to the earthnut.

**earth·quake** (ûrth′kwāk′) *n.* A series of elastic waves in the crust of the earth, caused by abrupt easing of strains built up along geologic faults and by volcanic action, and resulting in movements in the earth's surface.

**earth·rise** (ûrth′rīz′) *n.* The rising of the earth from the horizon as viewed from the moon.

**earth satellite** *n.* A satellite that orbits the earth.

**earth science** *n.* Any of several geologic sciences concerned with the origin, structure, and physical phenomena of the earth.

**earth·shak·ing** (ûrth′shā′kĭng) *adj.* Enormously or fundamentally important. **—earth′shak′ing·ly** *adv.*

**earth·shine** (ûrth′shīn′) *n.* The sunlight reflected from the earth's surface that illuminates part of the moon that is not directly lighted by the sun.

**earth·star** (ûrth′stär′) *n.* A fungus of the genus *Geastrum*, resembling the puffball and having an outer covering that splits open in a starlike form.

**earth station** *n.* An on-ground terminal linked to a spacecraft or satellite by an antenna and associated electronic equipment, for the purpose of transmitting or receiving messages, tracking, or control.

**earth·ward** (ûrth′wərd) *adj.* & *adv.* To or toward the earth. **—earth′wards** (-wərdz) *adv.*

**earth·work** (ûrth′wûrk′) *n.* **1.** An earthen embankment, esp. when used as a military fortification. **2.** *Engineer.* Excavation and embankment of earth.

**earth·worm** (ûrth′wûrm′) *n.* Any of various terrestrial annelid worms of the class Oligochaeta and esp. of the family Lumbricidae, that burrow into and help enrich and aerate soil.

**earth·y** (ûr′thē) *adj.* **-i·er, -i·est. 1.** Consisting of or like earth or soil. **2.** Relating to or typical of this world : WORLDLY. **3.** Crude : unrefined <*earthy* jokes> **—earth′i·ness** *n.*

**ear trumpet** *n.* A horn-shaped instrument once used to direct sound into the ear of a partially deaf person.

**ear·wax** (îr′wăks′) *n.* The waxlike secretion of certain glands lining the canal of the outer ear : CERUMEN.

**ear·wig** (îr′wĭg′) *n.* [ME *erwig* < OE *ēarwicga* : *eare*, ear + *wicga*, insect.] Any of various insects of the order Dermaptera, with pincerlike appendages protruding from the rear of the abdomen. *—vt.* **-wigged, -wig·ging, -wigs. 1.** To annoy by persistent and confidential solicitation. **2.** To try to influence by private communications.

**ear·worm** (îr′wûrm′) *n.* The corn earworm.

**ease** (ēz) *n.* [ME *ese* < OFr. *aise* < Lat. *adjacens*, pr.part. of *adjacēre*, to lie near. —see ADJACENT.] **1. a.** A state of comfort. **b.** Freedom from worry, pain, or agitation. **2.** Freedom from constraint or embarrassment : NATURALNESS. **3. a.** Freedom from difficulty, great effort, or hard work. **b.** Readiness in performance : FACILITY. **4.** Freedom from financial difficulty : AFFLUENCE. *—v.* **eased, eas·ing, eas·es.** *—vt.* **1.** To free from pain, worry, or trouble. **2.** To alleviate the pain or discomfort of. **3.** To slacken the strain, pressure, or tension of : LOOSEN. **4.** To move into place slowly and carefully <*eased* the boat into the slip> *—vi.* To lessen, as in stress, discomfort, or pressure.

☆ ***syns:*** EASE, EASINESS, EFFORTLESSNESS, FACILITY, READINESS *n. core meaning* : the ability to perform without apparent effort <translated the document with *ease*> ***ant:*** effort

**ease·ful** (ēz′fəl) *adj.* Affording or marked by comfort and peace : RESTFUL. **—ease′ful·ly** *adv.* **—ease′ful·ness** *n.*

**ea·sel** (ē′zəl) *n.* [Du. *ezel*, ass < MDu. *esel* < Lat. *asinus.*] An upright frame for holding or displaying something, as an artist's canvas.

**ease·ment** (ēz′mənt) *n.* **1.** The act of easing or state of being eased. **2.** *Law.* A right, as a right of way, afforded a person to make limited use of another's real property.

**eas·i·ly** (ē′zə-lē) *adv.* **1.** Without stress or difficulty. **2.** Without question or doubt : CERTAINLY <*easily* the best player>

**east** (ēst) *n.* [ME < OE *ēast.*] **1. a.** The direction of the earth's axial rotation. **b.** The cardinal point on the mariner's compass 90° clockwise from north and directly opposite west. **2. East.** The eastern part of a region or country, esp. the Orient. *—adj.* **1.** To, toward, of, facing, or in the east. **2.** Originating in or coming from the east. *—adv.* In, from, or toward the east.

**east·bound** (ēst′bound′) *adj.* Going toward the east.

**east by north** *n.* The direction or point on the mariner's compass halfway between due east and east-northeast that is 78°45′ east of due north. *—adv.* & *adj.* Toward or from east by north.

**east by south** *n.* The direction or point on the mariner's compass halfway between due east and east-southeast that is 101°15′ east of due north. *—adv.* & *adj.* Toward or from east by south.

**Eas·ter** (ē′stər) *n.* [ME *ester* < OE *ēastre.*] **1.** A festival in the Christian Church commemorating the Resurrection of Christ, celebrated on the first Sunday following the full moon that occurs on or next after Mar. 21. **2.** The Sunday on which the festival of Easter is held.

▲ word history: The word *Easter*, although the name of a Christian festival, had its origins in pagan times. *Eastre* or *Eostre*, the Old English spelling of *Easter*, was originally the name of a Germanic goddess who was worshiped at a festival at the spring equinox. Her name is closely related to Latin *aurora* and Greek *ēōs*, both of which mean "dawn." *Easter* is also derived from the same root word as *east*, the direction of sunrise.

**Easter egg** *n.* A dyed or painted egg, associated with Easter.

**Easter lily** *n.* Any of various white-colored lilies that bloom in the Easter season.

**east·er·ly** (ē′stər-lē) *adj.* **1.** Situated toward the east. **2.** Coming or being from the east. —*n., pl.* **-lies.** A wind or storm from the east. **—east′er·ly** *adv.*

**Easter Monday** *n.* The Monday after Easter.

**east·ern** (ē′stərn) *adj.* [ME *estern* < OE *ēasterne.*] **1.** Situated in, toward, or facing the east. **2.** Coming from the east. **3.** Growing in the east. **4.** *often* **Eastern.** Of, relating to, or typical of eastern regions or the East. **5. Eastern.** Of, relating to, or typical of the Orient. **6. Eastern. a.** Of or relating to the Eastern Church. **b.** Of or relating to the Eastern Orthodox Church.

**Eastern Church** *n.* **1.** The church of the Eastern Roman Empire, including the patriarchates of Constantinople, Antioch, Alexandria, and Jerusalem. **2.** The Eastern Orthodox Church. **3.** The Uniat Church.

**east·ern·er** *also* **East·ern·er** (ē′stər-nər) *n.* A native or resident of the east, esp. of the eastern United States.

**Eastern Hemisphere** *n.* The part of the earth including the continents of Europe, Africa, Asia, and Australia.

**east·ern·most** (ē′stərn-mōst′) *adj.* Farthest east.

**Eastern Orthodox Church** *n.* The body of modern churches, including the Greek and Russian Orthodox, that is derived from the church of the Byzantine Empire and adheres to the Byzantine rite and primacy of the patriarch of Constantinople.

**Eastern Standard Time** *n.* The time in the zone that includes the eastern part of North America and is in the fifth time zone west of Greenwich, England.

**Eas·ter·tide** (ē′stər-tīd′) *n.* The Easter season.

**East Germanic** *n.* The subdivision of the Germanic languages including Gothic.

**east·ing** (ē′stĭng) *n.* **1.** *Naut.* **a.** The distance sailed by a ship on an easterly course. **b.** The longitudinal distance from a given meridian on an easterly course. **2.** An easterly direction.

**east-north·east** (ēst′nôrth-ēst′) *n.* The direction or point on the mariner's compass halfway between due east and northeast that is 67°30′ east of due north. —*adj.* & *adv.* In, from, or toward east-northeast.

**east-south·east** (ēst′south′ēst′) *n.* The direction or point on the mariner's compass halfway between due east and southeast that is 112°30′ east of due north. —*adj.* & *adv.* In, from, or toward east-southeast.

**east·ward** (ēst′wərd) *adj.* & *adv.* Toward, to, or in the east. —*n.* An eastward direction, point, or region. **—east′ward·ly** *adv.* **—east′wards** (-wərdz) *adv.*

**eas·y** (ē′zē) *adj.* **-i·er, -i·est.** [ME *esi* < OFr. *aaisie*, p.part. of *aaisier*, to put at ease : *a-*, to (< Lat. *ad-*) + < *aise*, ease.] **1.** Capable of being accomplished or obtained with ease : not difficult. **2.** Free from worry, anxiety, or pain. **3.** Conducive to comfort or rest. **4.** Socially at ease. **5. a.** Relaxed in attitude : EASYGOING. **b.** Not strict or severe : LENIENT <an *easy* singing coach> **6.** Readily persuaded or influenced. **7.** Not hurried or forced : MODERATE. **8. a.** Less in demand and therefore readily obtainable <Fresh citrus fruit is *easier.*> **b.** Plentiful and therefore at low interest rates <*easy* loans> —*adv.* **1.** Without haste or agitation <Go very *easy.*> **2.** Easily <flattery spoken too *easy*> **—eas′i·ness** *n.*

☆ ***syns:*** EASY, EFFORTLESS, FACILE, READY, SIMPLE, SMOOTH *adj. core meaning* : posing no difficulty <an *easy* solution to the problem> ***ant:*** difficult, hard

**easy chair** *n.* A large, comfortable, well-upholstered chair.

**eas·y·go·ing** *also* **eas·y-go·ing** (ē′zē-gō′ĭng) *adj.* **1. a.** Living without concern or worry : PLACID. **b.** Lazy and careless. **c.** Morally lax. **d.** Undemanding <an *easygoing* routine> **2.** Having or moving at a comfortable, unhurried pace.

**easy street** *n. Slang.* Financial security or independence.

**eat** (ēt) *v.* **ate** (āt; *nonstandard* ĕt), **eat·en** (ēt′n), **eat·ing, eats.** [ME *eten* < OE *etan.*] —*vt.* **1.** To take into the mouth, chew, and swallow as food. **2.** To consume, ravage, or destroy by or as if by eating. **3.** To erode or corrode. **4.** *Slang.* To bother or annoy <Something's *eating* them.> —*vi.* **1. a.** To consume food. **b.** To take or have a meal. **2.** To wear away by or as if by eating. **—eat crow.** *Informal.* To be forced to accept defeat. **—eat one's heart out.** To feel bitter, hopeless anguish or longing. **—eat one's words.** To take back something that one has said. **—eat′er** *n.*

☆ ***syns:*** EAT, CONSUME, DEVOUR, INGEST *v. core meaning* : to take (food) into the body as nourishment <*ate* a hearty meal>

**eat·a·ble** (ē′tə-bəl) *adj.* Fit to be eaten : EDIBLE. —*n.* **1.** Something fit to be eaten. **2. eatables.** Food.

**eat·er·y** (ē′tə-rē) *n., pl.* **-ies.** *Informal.* A restaurant.

**eat·ing** (ē′tĭng) *adj.* Suitable for eating raw <*eating* apples>

**eats** (ēts) *pl.n. Slang.* Food.

**eau de co·logne** (ō′ də kə-lōn′) *n., pl.* **eaux de cologne** (ō′, ōz′) [Fr., water of Cologne < *Cologne*, Köln, West Germany.] Cologne.

**eau de vie** (ō′ də vē′) *n., pl.* **eaux de vie** (ō′, ōz′) [Fr. *eau-de-vie* : *eau*, water + *de*, of + *vie*, life.] Brandy.

**eaves** (ēvz) *n.* [ME *eves* < OE *efes.*] (*pl. in number*). The projecting overhang at the lower border of a roof.

**eaves·drop** (ēvz′drŏp′) *vi.* **-dropped, -drop·ping, -drops.** [Prob. back-formation < *eavesdropper*, one who eavesdrops < ME *evesdropper* < *evesdrop*, place where water falls from eaves < OE *yfæs drypæ.*] To listen secretly to a private conversation. **—eaves′drop′per** *n.*

▲ word history: The "eavesdrop" of a building is the space of ground where water falls from the eaves. An *eavesdropper* was originally someone who stood in the eavesdrop in order to overhear private conversations taking place inside. The noun *eavesdrop* did not survive Old English times, but *eavesdropper* and *eavesdropping*, which was a legal offense, were preserved in legal texts. The modern verb *eavesdrop* is most probably derived from *eavesdropper*.

**ebb** (ĕb) *n.* [ME *ebbe* < OE *ebba.*] **1.** Ebb tide. **2.** A time of decline or diminution <"insistence upon rules of conduct marks the *ebb* of religious fervor" —A.N. Whitehead> —*vi.* **ebbed, ebb·ing, ebbs. 1.** To fall back from the flood. **2.** To fall away or decline.

**ebb tide** *n.* The period of a tide between high water and a succeeding low water.

**EBCDIC** (ĕb′sə-dĭk′) *n.* [E(XTENDED) + B(INARY) + C(ODED) + D(ECIMAL) + I(NTERCHANGE) + C(ODE).] A code for representing alphanumeric information.

**Eb·lis** (ĕb′lĭs) *n.* [Ar. *Iblīs* < LGk. *diabolos*, devil. —see DEVIL.] *Myth.* The principal Islamic evil spirit or devil.

**eb·on** (ĕb′ən) *adj.* [ME *eban*, ebony wood < Lat. *ebenus*, ebony tree < Gk. *ebenos*, of Egypt. orig.] **1.** Made of ebony. **2.** Black. —*n.* Ebony.

**eb·on·ite** (ĕb′ə-nīt′) *n.* A hard rubber.

**eb·on·ize** (ĕb′ə-nīz′) *vt.* **-ized, -iz·ing, -iz·es.** To stain black.

**eb·on·y** (ĕb′ə-nē) *n., pl.* **-ies.** [ME *hebenyf*, ebony wood < LLat. *ebeninus*, of ebony < Gk. *ebeninos* < *ebenos*, ebony tree, of Egypt. orig.] **1.** A chiefly tropical tree of the genus *Diospyros*, esp. *D. ebenum* of southern Asia, with hard dark-colored heartwood. **2.** The wood of the ebony, used in cabinetwork and for piano keys. —*adj.* **1.** Made of or suggesting ebony. **2.** Black.

**e·brac·te·ate** (ē-brăk′tē-āt′) *adj.* [NLat. *ebracteatus* : Lat. *ex-*, without + Lat. *bractea*, bract.] *Bot.* Without bracts.

**e·bul·lience** (ĭ-bo͝ol′yəns, ĭ-bŭl′-) *n.* [< Lat. *ebulliens*, pr.part. of *ebullire*, to boil up : *ex-*, up + *bullire*, to boil.] The quality of expressing feelings or ideas in an enthusiastic, lively way.

**e·bul·lien·cy** (ĭ-bo͝ol′yən-sē, ĭ-bŭl′-) *n.* Ebullience.

**e·bul·lient** (ĭ-bo͝ol′yənt, ĭ-bŭl′-) *adj.* **1.** Boiling : bubbling. **2.** Marked by ebullience. **—e·bul′lient·ly** *adv.*

**eb·ul·li·tion** (ĕb′ə-lĭsh′ən) *n.* [LLat. *ebullitio* < Lat. *ebullire*, to boil up. —see EBULLIENT.] **1.** The process or state of boiling or bubbling. **2.** A sudden, violent outpouring, as of emotion.

**ec·ce ho·mo** (ĕk′sē hō′mō, ĕk′ĕ) *n.* [LLat., behold the man.] A statue or picture of Christ wearing the crown of thorns.

**ec·cen·tric** (ĭk-sĕn′trĭk) *adj.* [ME *eccentrik*, planetary orbit of which the earth is not center < Med. Lat. *eccentricus* < *eccentricus*, not having the same center < Gk. *ekkentros* : *ek-*, out + *kentron*, center.] **1. a.** Departing from the established or traditional norm. **b.** Departing from a direct or charted course : ERRATIC. **2.** Deviating from a circular path or form. **3.** Not situated at or in the geometric center. **4.** With the axis located somewhere other than at the geometric center. **5.** Not having the same center. —*n.* **1.** A bizarre person. **2.** A disk or wheel with its axis of revolution displaced from its center so that it is able to impart reciprocating motion. **—ec·cen′tri·cal·ly** *adv.*

**ec·cen·tric·i·ty** (ĕk′sĕn-trĭs′ĭ-tē) *n., pl.* **-ties. 1. a.** The quality of being eccentric. **b.** Deviation from the normal, expected, or traditional. **2.** An example or instance of eccentric behavior. **3.** The distance between the center of an eccentric and its axis. **4.** *Math.* The ratio of the distance of any point on a conic section from a focus to its distance from the corresponding directrix.

☆ ***syns:*** ECCENTRICITY, IDIOSYNCRASY, PECULIARITY, QUIRK, SINGULARITY *n. core meaning* : peculiar behavior <a hermit with many *eccentricities*>

**ec·chy·mo·sis** (ĕk′ĭ-mō′sĭs) *n.* [NLat. < Gk. *ekkhumōsis* < *ekkhumousthai*, to extravasate blood : *ek-*, out + *khumos*, juice < *khein*, to pour.] Passage of blood from ruptured blood vessels into subcutaneous tissue, marked by a purple discoloration of the skin. **—ec′chy·mot′ic** (-mŏt′ĭk) *adj.*

**ec·cle·si·a** (ĭ-klē′zhē-ə, -zē-ə) *n., pl.* **-si·ae** (-zhē-ē′, -zē-ē′) [Lat. < Gk. *ekklēsia* < *ekkalein*, to summon : *ek-*, out + *kalein*, to call.] **1.** The political assembly of citizens of an ancient Greek state. **2.** A church or congregation.

**Ec·cle·si·as·tes** (ĭ-klē'zē-ăs'tēz') *pl.n.* [LLat. < Gk. *Ekklēsiastēs,* preacher (transl. of Heb. *Qoholeth*) < *ekklēsiastēs,* a member of the ecclesia.] *(sing in number).* —See table at BIBLE.

**ec·cle·si·as·tic** (ĭ-klē'zē-ăs'tĭk) *adj.* Ecclesiastical. —*n.* A cleric.

**ec·cle·si·as·ti·cal** (ĭ-klē'zē-ăs'tĭ-kəl) *adj.* **1.** Of or relating to a church, esp. as an organized institution. **2.** Suitable for use in a church. **—ec·cle'si·as'ti·cal·ly** *adv.*

**ec·cle·si·as·ti·cism** (ĭ-klē'zē-ăs'tə-sĭz'əm) *n.* **1.** Ecclesiastical principles and activities. **2.** Excessive adherence to ecclesiastical principles and forms.

**Ec·cle·si·as·ti·cus** (ĭ-klē'zē-ăs'tĭ-kəs) *n.* [LLat. < *ecclesiasticus,* ecclesiastical < Gk. *ekklēsiastikos,* ecclesiastical, of an ecclesia.] —See table at BIBLE.

**ec·cle·si·ol·o·gy** (ĭ-klē'zē-ŏl'ə-jē) *n.* [ECCLESI(A) + -LOGY.] Study of ecclesiastical architecture and ornamentation. **—ec·cle'si·o·log'i·cal** (-ə-lŏj'ĭ-kəl) *adj.*

**ec·crine** (ĕk'rĭn, -rīn', -rēn') *adj.* [< Gk. *ekkrinein,* to secrete : *ek-,* out + *krinein,* to separate.] **1.** Secreting externally. —Used esp. of an eccrine gland or its secretion. **2.** Exocrine.

**eccrine gland** *n.* Any of the small sweat glands distributed over the body's surface.

**ec·crin·ol·o·gy** (ĕk'rə-nŏl'ə-jē) *n.* Study of eccrine secretions and secretory organs.

**ec·dys·i·ast** (ĕk-dĭz'ē-ăst', -əst) *n.* [< ECDYSIS; coined by Henry L. Mencken (1880–1956).] A stripteaser.

**ec·dy·sis** (ĕk'dĭ-sĭs) *n., pl.* **-ses** (-sēz') [NLat. < Gk. *ekdusis,* a stripping < *ekduein,* to take off : *ek-,* out + *duein,* to put on.] The shedding of an outer integument or layer of skin, as in insects, crustaceans, and snakes.

**ec·dy·sone** (ĕk'də-sōn') *n.* [ECDYS(IS) + (HORM)ONE.] Any of various hormones of insects that control molting.

**e·ce·sis** (ĭ-sē'sĭs) *n.* [Gk. *oikēsis,* inhabitation < *oikein,* to dwell < *oikos,* house.] Successful establishment of an organism in a new environment.

**ec·hard** (ĕk'härd') *n.* [Gk. *ekhein,* to hold back + *ardein,* to irrigate.] *Ecol.* Soil water unavailable for absorption by plants.

**ech·e·lon** (ĕsh'ə-lŏn') *n.* [Fr. *échelon* < OFr. *eschelon,* rung of a ladder < *eschile,* ladder < Lat. *scala.*] **1. a.** A troop formation with parallel units arranged to the left or right of the rear unit in a steplike manner. **b.** A similar formation of groups, units, or individuals. **c.** A flight formation or arrangement of vessels in an echelon. **2.** A level of command or authority, as in a military force or in an organization. **3.** A diffraction grating of parallel glass plates of successively varying sizes, for examining fine structures. —*vt.* & *vi.* **-loned, -lon·ing, -lons.** To arrange or take place in an echelon.

**ech·e·ve·ri·a** (ĕch'ə-və-rē'ə) *n.* [NLat. *Echeveria,* genus name, after *Echeveria,* a 19th-cent. Mexican botanical illustrator.] A tropical American plant of the genus *Echeveria,* bearing thick, succulent leaves often clustered in a rosette.

**e·chid·na** (ĭ-kĭd'nə) *n.* [NLat. < Lat., viper < Gk. *ekhidna.*] Any of several burrowing, egg-laying mammals of the genera *Tachyglossus* and *Zaglossus* of Australia, Tasmania, and New Guinea, with a spiny coat, slender snout, and sticky tongue for catching insects.

**echin-** *pref. var. of* ECHINO-.

**ech·i·nate** (ĕk'ə-nāt') *adj.* Bearing or covered with spines : PRICKLY.

**echino-** *or* **echin-** *pref.* [NLat. < Lat. *echinus,* sea urchin < Gk. *ekhinos.*] **1.** Spiny : prickly <*echinate*> **2.** Echinoderm <*echinoid*>

**e·chi·no·coc·co·sis** (ĭ-kī'nə-kə-kō'sĭs) *n., pl.* **-ses** (-sēz') [ECHINOCOCC(US) + -OSIS.] Infestation with echinococci.

**e·chi·no·coc·cus** (ĭ-kī'nə-kŏk'əs) *n., pl.* **-coc·ci** (-kŏk'sī') [NLat. *Echinococcus,* genus name : ECHINO- + COCCUS.] A parasitic tapeworm of the genus *Echinococcus,* the larvae of which infest mammals and form spherical cysts, causing serious or fatal disease.

**e·chi·no·derm** (ĭ-kī'nə-dûrm') *n.* Any of numerous radially symmetric marine invertebrates of the phylum Echinodermata, including the starfishes, sea urchins, and sea cucumbers, having a spiny body. **—e·chi'no·der'mal, e·chi'no·der'ma·tous** *adj.*

**e·chi·noid** (ĭ-kī'noid') *n.* An echinoderm of the class Echinoidea, including the sand dollars and sea urchins.

**e·chi·nus** (ĭ-kī'nəs) *n., pl.* **-ni** (-nī') [Lat. < Gk. *ekhinos,* echinus, sea urchin, hedgehog.] A curved architectural molding just below the abacus of a Doric capital.

**ech·o** (ĕk'ō) *n., pl.* **-oes.** [ME < OFr. < Lat. < Gk. *ēkhō.*] **1. a.** Repetition of a sound by sound waves reflecting from a surface. **b.** The sound produced in this way. **2.** A repetition or imitation <a literary style that is an *echo* of Walt Whitman's poetry> **3.** One who imitates another, as in opinions, speech, or dress. **4.** A sympathetic response. **5.** Repetition of certain sounds or syllables in poetry. **6.** Soft repetition of a musical phrase or note. **7.** *Electron.* A reflected wave received by a radio or radar. **8. Echo.** *Aerospace.* One of a series of U.S. passive satellites that are often visible at night. **9. Echo.** *Gk. Myth.* A nymph whose unrequited love for Narcissus caused her to pine away until nothing but her voice remained. —*v.* **-oed, -o·ing, -oes.** —*vt.* **1.** To repeat by or as if by an echo. **2.** To repeat or imitate <a comment that *echoed* my own thoughts> —*vi.* **1.** To produce an echo. **2.** To resound with an echo : REVERBERATE. **—ech'o·er** *n.*

☆ ***syns:*** ECHO, REPERCUSSION, RESOUNDING, REVERBERATION *n. core meaning* : repetition of sound via reflection from a surface <an *echo* of the cry across the canyon>

**ech·o·car·di·og·ra·phy** (ĕk'ō-kär'dē-ŏg'rə-fē) *n., pl.* **-phies.** *Med.* A diagnostic technique utilizing ultrasound to visualize the internal structure of the heart. **—ech'o·car'di·o·graph'** (-kär'dē-ə-grăf') *n.* **—ech'o·car'di·o·graph'ic** *adj.*

**echo check** *n. Computer Sci.* An error control technique in which the receiving terminal or computer returns the original message to verify its correct reception.

**e·cho·ic** (ĭ-kō'ĭk) *adj.* **1.** Resembling an echo. **2.** Imitative of sounds : ONOMATOPOEIC.

**ech·o·la·li·a** (ĕk'ō-lā'lē-ə) *n.* [ECHO + Gk. *lalia,* talk < *lalos,* talkative.] Involuntary repetition of phrases or words just uttered by others. **—ech'o·la'lic** (-lĭk) *adj.*

**ech·o·lo·ca·tion** (ĕk'ō-lō-kā'shən) *n.* **1.** The ability of an animal, as a bat or dolphin, to orient itself by the reflection of the sound that it has produced. **2.** *Electron.* Ranging by acoustical echo analysis. **—ech'o·lo·cate'** *v.* **(-cat·ed, -cat·ing, -cates).**

**echo sounder** *n.* A device for measuring sea depth by sending pressure waves down from the surface and recording the time until the echo returns from the sea floor.

**é·clair** (ā-klâr', ā'klâr') *n.* [Fr. < OFr. *esclair,* lightning < *esclairier,* to flash < VLat. ***exclariare** : Lat. *ex-,* completely + Lat. *clarus,* clear.] A tube-shaped cream puff with custard or cream filling, usu. iced with chocolate.

**é·clair·cisse·ment** (ā-klâr-sēs-mäN') *n.* [Fr. < OFr. *ésclarcir,* to clarify < VLat. **exclaricire* : Lat. *ex-,* completely + Lat. *clarus,* clear.] A clarification : enlightenment.

**e·clamp·si·a** (ĭ-klămp'sē-ə) *n.* [NLat. < Gk. *eklampsis,* a shining forth < *eklampein,* to shine forth : *ek-,* out + *lampein,* to shine.] Convulsions followed by coma, arising from conditions during or immediately after pregnancy. **—e·clamp'tic** (-tĭk) *adj.*

**é·clat** (ā-klä', ā'klä') *n.* [Fr. < OFr. *esclat* < *esclater,* to burst, prob. of Germanic orig.] **1.** Great brilliance, as of achievement or performance. **2.** Conspicuous success. **3.** Great acclamation or applause. **4.** *Archaic.* Notoriety : scandal.

**e·clec·tic** (ĭ-klĕk'tĭk) *adj.* [Gk. *eklektikos,* selective < *eklegein,* to select : *ek-,* out + *legein,* to choose.] **1.** Selecting what appears to be the best from diverse sources, systems, or styles. **2.** Made up of components selected from diverse sources. —*n.* One that follows an eclectic method. **—e·clec'ti·cal·ly** *adv.*

**e·clec·ti·cism** (ĭ-klĕk'tə-sĭz'əm) *n.* An eclectic method or system.

**e·clipse** (ĭ-klĭps') *n.* [ME < OFr. < Lat. *eclipsis* < Gk. *ekleipsis* < *ekleipein,* to fail to appear : *ek-,* out + *leipein,* to leave.] **1. a.** The complete or partial obscuring, relative to a designated observer, of one celestial body by another. **b.** The time period during which an eclipse occurs. **2.** A permanent or temporary dimming or cutting off of light. **3. a.** A fall into obscurity : DECLINE. **b.** Disgrace : downfall. —*vt.* **e·clipsed, e·clips·ing, e·clips·es. 1. a.** To cause an eclipse of. **b.** To darken or make dim : OBSCURE. **2. a.** To obscure or diminish in importance or reputation. **b.** To surpass : outshine.

**eclipsing binary** *n.* A binary star, the components of which pass in front of each other as viewed from the earth.

**e·clip·tic** (ĭ-klĭp'tĭk) *n.* [ME < LLat. *ecliptica (linea),* (line) of eclipses < Lat. *eclipticus,* of an eclipse < Gk. *ekleiptikos* < *ekleipein,* to fail to appear. —see ECLIPSE.] **1.** The apparent path of the sun among the stars : the intersection plane of the earth's solar orbit with the celestial sphere. **2.** A great circle on a terrestrial globe inclined at an angle of approx. 23°27' to the equator.

**ec·logue** (ĕk'lôg', -lŏg') *n.* [ME *eclog* < Lat. *ecloga* < Gk. *eklogē,* selection < *eklegein,* to select. —see ECLECTIC.] A poem usu. in the form of a pastoral dialogue.

**e·clo·sion** (ĭ-klō'zhən) *n.* [Fr. *éclosion* < *éclore,* to open < Lat. *excludere,* to shut out : *ex-,* out + *claudere,* to shut.] Emergence of an adult insect from a pupal case or of an insect larva from an egg.

**eco-** *pref.* [< ECOLOGY.] Ecology : ecological <*ecosystem*>

**ec·o·cide** (ĕk'ō-sīd', ē'kō-) *n.* Destruction of the natural environment, as by pollutants.

**ec·o·ge·o·graph·ic** (ĕk'ō-jē'ə-grăf'ĭk) *or* **ec·o·ge·o·graph·i·cal** (-ĭ-kəl) *adj.* Of or pertaining to the ecology and geography of the environment. **—ec'o·ge'o·graph'i·cal·ly** *adv.*

**e·col·o·gy** (ĭ-kŏl'ə-jē) *n.* [G. *Ökologie* : Gk. *oikos,* house + G. *-logie,* -logy.] **1.** The science of the relationships between organisms and their environments. **2.** The relationship between organisms and their environment. **—ec'o·log'i·cal** (ĕk'ə-lŏj'ĭ-kəl, ē'kə-), **ec'o·log'ic** (-ĭk) *adj.* **—ec'o·log'i·cal·ly** *adv.* **—e·col'o·gist** *n.*

**e·con·o·met·rics** (ĭ-kŏn'ə-mĕt'rĭks) *n.* [ECONO(MICS) + -METRICS.] *(sing. in number).* Application of statistical techniques to economics in the study of problems, analysis of data, and development of theory. **—e·con'o·met'ric** *adj.*

**ec·o·nom·ic** (ĕk'ə-nŏm'ĭk, ē'kə-) *adj.* **1.** Of or relating to the development, production, and management of material wealth, as of a country, household, or business enterprise. **2.** Of or relating to economics. **3.** Of or relating to matters of finance. **4.** Of or relating to the necessities of life : UTILITARIAN.

**ec·o·nom·i·cal** (ĕk'ə-nŏm'ĭ-kəl, ē'kə-) *adj.* **1.** Not wasteful or extravagant : PRUDENT. **2.** Economic. **—ec'o·nom'i·cal·ly** *adv.*

**economic rent** *n.* RENT[1] 2.

**ec·o·nom·ics** (ĕk'ə-nŏm'ĭks, ē'kə-) *n. (sing. in number).* The science that deals with the production, distribution, and consumption of commodities.

**e·con·o·mist** (ĭ-kŏn'ə-mĭst) *n.* **1.** A specialist in economics. **2.** *Archaic.* One who is economical.

**e·con·o·mize** (ĭ-kŏn'ə-mīz') *v.* **-mized, -miz·ing, -miz·es.** *—vi.* To be thrifty : practice economy. *—vt.* To manage or use with thrift. **—e·con'o·miz'er** *n.*

**e·con·o·my** (ĭ-kŏn'ə-mē) *n., pl.* **-mies.** [OFr. *economie,* management of household < Lat. *oeconomia* < Gk. *oikonomia* < *oikonomos,* household manager : *oikos,* house + *nemein,* to manage.] **1. a.** Careful or thrifty management or use of resources, as income, materials, or labor. **b.** An example of economy. **2.** Management of the resources of a country, community, or business. **3. a.** A system for the management and development of resources. **b.** The economic system of a country or area. **4.** The functional arrangement of elements within a structure or system <the *economy* of the community>

**ec·o·spe·cies** (ĕk'ə-spē'shēz, -sēz, ē'kə-) *n.* A taxonomic species regarded in terms of its ecological characteristics and usu. including several ecotypes.

**ec·o·sphere** (ĕk'ō-sfîr') *n.* The regions of the universe capable of supporting living organisms.

**ec·o·sys·tem** (ĕk'ə-sĭs'təm, ē'kə-) *n.* An ecological community with its physical environment, regarded as a unit.

**ec·o·tone** (ĕk'ə-tōn', ē'kə-) *n.* [ECO- + Gk. *tonos,* tension.] An ecological community of mixed vegetation formed by the overlapping of adjoining communities.

**ec·o·type** (ĕk'ə-tīp', ē'kə-) *n.* The smallest taxonomic subdivision of an ecospecies, made up of subspecies or varieties adapted to a particular set of environmental conditions. **—ec'o·typ'ic** (-tĭp'ĭk) *adj.*

**ec·ru** (ĕk'rōō, ā'krōō) *n.* [Fr. *écru* : *é-,* completely (< Lat. *ex-*) + *cru,* crude (< Lat. *crudus*).] A grayish to pale yellow or a light grayish yellowish brown.

**ec·sta·sy** (ĕk'stə-sē) *n., pl.* **-sies.** [ME *extasie* < OFr. < LLat. *extasis,* terror < Gk. *ekstasis* < *existanai,* to drive out of one's senses : *ek-,* out + *histanai,* to place.] **1.** Intense joy or delight. **2.** A state of emotion so intense that one is carried beyond thought and self-control. **3.** The trance or rapture of mystic or prophetic exaltation.

**ec·stat·ic** (ĕk-stăt'ĭk) *adj.* **1.** Of or pertaining to ecstasy. **2.** In a state of ecstasy : ENRAPTURED <*ecstatic* over the gorgeous sunrise> **—ec·stat'i·cal·ly** *adv.*

**ecto-** *pref.* [Gk. *ekto-* < *ektos,* outside < *ek,* out.] Outer : external <*ecto*parasite>

**ec·to·com·men·sal** (ĕk'tə-kə-mĕn'səl) *n.* A commensal organism living on the outer body surface of another organism.

**ec·to·derm** (ĕk'tə-dûrm') *n.* The outermost of the three primary germ layers of an embryo, developing the epidermis, nervous tissue, and, in vertebrates, sense organs. **—ec'to·derm'al** (-dûr'məl), **ec'to·der'mic** (-mĭk) *adj.*

**ec·tog·e·nous** (ĕk-tŏj'ə-nəs) *adj.* Capable of living and developing outside a host. —Used of certain pathogenic microorganisms.

**ec·to·mere** (ĕk'tə-mîr') *n.* A blastomere that develops into ectoderm. **—ec·to·mer'ic** (-mîr'ĭk, -mĕr'-) *adj.*

**ec·to·morph** (ĕk'tə-môrf') *n.* An ectomorphic individual.

**ec·to·mor·phic** (ĕk'tə-môr'fĭk) *adj.* [ECTO(DERM) + -MORPHIC.] Lean and slightly muscular. **—ec'to·mor'phy** *n.*

**-ectomy** *suff.* [NLat. *-ectomia* : Gk. *ek-,* out + NLat. *-tomia,* -tomy.] Surgical removal <aden*ectomy*>

**ec·to·par·a·site** (ĕk'tə-păr'ə-sīt') *n.* A parasite, as a flea, living on the exterior of another organism. **—ec'to·par'a·sit'ic** (-sĭt'ĭk) *adj.*

**ec·to·pi·a** (ĕk-tō'pē-ə) *n.* [NLat. < Gk. *ektopos,* away from a place : *ek-,* out + *topos,* place.] Congenital positional abnormality of an organ or part. **—ec·top'ic** (-tŏp'ĭk) *adj.*

**ectopic pregnancy** *n.* Gestation outside the uterus, often in a Fallopian tube.

**ec·to·plasm** (ĕk'tə-plăz'əm) *n.* **1.** *Biol.* A portion of the phase of cytoplasm distinguishable in some cells as a relatively rigid cortex limited on the outside by the cell membrane. **2.** The luminous substance believed to emanate from a spiritualistic medium.

**ec·to·sarc** (ĕk'tə-särk') *n.* [ECTO- + Gk. *sarx,* flesh.] The rather clear outermost layer of protoplasm of certain protozoans, as the amoeba.

**ec·to·therm** (ĕk'tə-thûrm') *n.* Poikilotherm. **—ec'to·therm'ic** *adj.*

**ec·type** (ĕk'tīp') *n.* [Gk. *ektupos,* worked in relief : *ek-,* out of + *tupos,* mold.] A reproduction of an original : COPY. **—ec'ty·pal** (-tə-pəl) *adj.*

**é·cu** (ā-kyōō') *n., pl.* **é·cus** (ā-kyōō') [Fr. < OFr. *escu* < Lat. *scutum,* shield (from the shield stamped on the coin).] Any of various old French coins, esp. a silver five-franc piece.

**ec·u·men·i·cal** (ĕk'yə-mĕn'ĭ-kəl) *or* **ec·u·men·ic** (-mĕn'ĭk) *adj.* [LLat. *oecumenicus* < Gk. *oikoumenikos* < *oikoumenē,* the inhabited world < *oikein,* to inhabit < *oikos,* house.] **1.** Worldwide in range or applicability : UNIVERSAL. **2.** Of or relating to the worldwide Christian church, esp. in regard to unity. **—ec'u·men'i·cal·ism** *n.* **—ec'u·men'i·cal·ly** *adv.*

**ecumenical patriarch** *n.* The patriarch of Constantinople, the highest ecclesiastical official of the Eastern Orthodox Church.

**ec·u·me·nism** (ĕk'yə-mə-nĭz'əm, ĭ-kyōō'-) *n.* A movement seeking worldwide unity among religions through greater cooperation and improved understanding. **—ec'u·men'ist** *n.*

**ec·ze·ma** (ĕk'sə-mə, ĭg-zē'-) *n.* [NLat. < Gk. *ekzema* < *ekzein,* to break out : *ek-,* out + *zeein,* to boil.] A noncontagious inflammation of the skin, characterized mainly by redness, itching, and the outbreak of lesions that discharge serous matter and become encrusted and scaly. **—ec·zem'a·tous** (ĕg-zĕm'ə-təs, -zē'mə-təs, ĭg-) *adj.*

**-ed[1]** *suff.* [ME *-ed* < OE *-ad, -ed, -od.*] **1.** —Used to form the past participle of regular verbs <absorb*ed*> **2.** Having : characterized by : resembling <blackheart*ed*>

**-ed[2]** *suff.* [ME *-ede* < OE *-ade, -ede, -ode.*] —Used to form the past tense of regular verbs <inhabit*ed*>

**e·da·cious** (ĭ-dā'shəs) *adj.* [< Lat. *edax, edāc-* < *edere,* to eat.] Marked by voracity : DEVOURING. **—e·dac'i·ty** (ĭ-dăs'ə-tē) *n.*

**E·dam cheese** (ē'dəm, ē'dăm') *n.* [After *Edam,* Netherlands.] A mild, yellow Dutch cheese, pressed into balls and usu. covered with red wax.

**e·daph·ic** (ĭ-dăf'ĭk) *adj.* [< Gk. *edaphos,* soil.] Of or relating to soil, esp. as it affects living organisms.

**Ed·da** (ĕd'ə) *n.* [ON.] **1.** A collection of Old Norse poems called the Elder or Poetic Edda, assembled in the early 13th cent. **2.** A manual of Icelandic poetry called the Younger or Prose Edda. **—Ed'dic** *adj.*

**ed·do** (ĕd'ō) *n., pl.* **-does.** [Of African orig.] The taro.

**ed·dy** (ĕd'ē) *n., pl.* **-dies.** [ME *ydy,* prob. < ON *ðā.*] **1.** A current, as of air or water, moving contrary to the direction of the main current, esp. in a circular motion. **2.** A current that runs contrary to the main current or tradition. *—vi. & vt.* **-died, -dy·ing, -dies.** To move or cause to move against the main current, as in an eddy.

**e·del·weiss** (ā'dəl-vīs', -wīs') *n.* [G. : *edel,* noble (< OHG *edili*) + *weiss,* white (< OHG *wīz*).] A plant indigenous to the high Alps, *Leontopodium alpinum,* bearing leaves covered with whitish down and small flowers surrounded by conspicuous whitish bracts.

**e·de·ma** (ĭ-dē'mə) *n., pl.* **-mas** *or* **-ma·ta** (-mə-tə) [NLat. < Gk. *oidēma,* a swelling < *oidein,* to swell.] **1.** *Pathol.* Excessive accumulation of serous fluid in the tissues. **2.** *Bot.* Extended swellings in plant organs caused mainly by excessive accumulation of water. **—e·dem'a·tous** (-dĕm'ə-təs) *adj.*

**E·den** (ēd'n) *n.* [ME < LLat. < Gk. *Ēdēn* < Heb. *'Ēdhen* < *'ēdhen,* delight.] **1.** The first home of Adam and Eve. **2.** A delightful place or dwelling : PARADISE. **3.** Ultimate happiness : BLISS.

**e·den·tate** (ē-dĕn'tāt') *adj.* [Lat. *edentatus,* p.part. of *edentare,* to knock out the teeth : *ex-,* out + *dens,* tooth.] **1.** *Biol.* Lacking teeth. **2.** Of or belonging to the order Edentata, including mammals, as anteaters, armadillos, and sloths, having few or no teeth. *—n.* A member of the Edentata.

**e·den·tu·lous** (ē-dĕn'chə-ləs) *adj.* [Lat. *edentulus* : *ex-,* without + *dens,* tooth.] Having no teeth : TOOTHLESS.

**edge** (ĕj) *n.* [ME *egge* < OE *ecg.*] **1. a.** The usu. thin, sharpened side of the blade of a cutting instrument. **b.** Degree of sharpness of a cutting blade. **2.** Keenness, as of desire or enjoyment : ZEST. **3.** A rim, brink, or crest, as of a cliff or ridge of hills. **4.** A dividing line or point of transition <the *edge* of defeat> **5.** The line of intersection of two surfaces of a solid <the *edge* of a building block> **6.** A margin of superiority : ADVANTAGE <had an *edge* on their opponents> *—v.* **edged, edg·ing, edg·es.** *—vt.* **1.** To give an edge to : SHARPEN. **2.** To put a border on. **3.** To advance or push slowly. *—vi.* To move slowly or hesitantly <*edged* away from the snarling dog> **—on edge.** Highly tense or nervous. **—edge'less** *adj.*

**edg·er** (ĕj'ər) *n.* One that edges, as a tool for trimming the edges of a lawn.

**edge tool** *n.* A tool, as a chisel, with a cutting edge.

**edge·wise** (ĕj'wīz') *also* **edge·ways** (-wāz') *adv.* **1.** With the edge foremost. **2.** On, by, with, or toward the edge.

**edg·ing** (ĕj'ĭng) *n.* Something that serves as or forms an edge or border around something else.

**edg·y** (ĕj'ē) *adj.* **-i·er, -i·est.** **1.** Feeling or exhibiting nervous tension. **2.** Having a sharp edge. **3.** Having excessively sharp definition, as in painting. **—edg'i·ness** *n.*

☆ ***syns:*** EDGY, FIDGETY, JITTERY, JUMPY, NERVOUS, RESTIVE, RESTLESS, SKITTISH, TENSE, UNEASY, UPTIGHT *adj. core meaning* : feeling or showing nervous tension <pilots *edgy* before combat>

**edh** *also* **eth** (ĕth) *n.* [Icel. *eð.*] **1.** A letter (ð) in Old English, Old Saxon, Old Norse, and modern Icelandic that represents an interdental fricative. **2.** The symbol in the International Phonetic Alphabet representing the interdental voiced fricative, as in *the* or *with.*

**ed·i·ble** (ĕd'ə-bəl) *adj.* [LLat. *edibilis* < Lat. *edere,* to eat.] Fit to eat. *—n.* Something fit to be eaten. **—ed'i·bil'i·ty, ed'i·ble·ness** *n.*

**e·dict** (ē′dĭkt′) *n.* [Lat. *edictum* < neuter p.part of *edicere*, to declare : *ex-*, out + *dicere*, to speak.] **1.** An authoritative proclamation or decree having the force of law. **2.** A formal proclamation, command, or decree.

**ed·i·fi·ca·tion** (ĕd′ə-fə-kā′shən) *n.* Intellectual, moral, or spiritual improvement : ENLIGHTENMENT.

**ed·i·fice** (ĕd′ə-fĭs) *n.* [ME < OFr. < Lat. *aedificium* < *aedificare*, to build : *aedis*, a building + *facere*, to make.] A building, esp. one of imposing size or appearance.

**ed·i·fy** (ĕd′ə-fī′) *vt.* **-fied, -fy·ing, -fies.** [ME *edifien* < OFr. *edifier* < LLat. *aedificare* < *Lat.*, to build. —see EDIFICE.] To enlighten or instruct so as to encourage intellectual, moral, or spiritual improvement. **—ed′i·fi′er** *n.*

**ed·it** (ĕd′ĭt) *vt.* **-it·ed, -it·ing, -its.** [Back-formation < EDITOR.] **1. a.** To prepare for publication or presentation, as by adapting or correcting. **b.** To prepare an edition of for publication <*edit* a collection of Jefferson's letters> **2.** To supervise the publication of. **3.** To delete <*edited* the longest chase sequence out> **4.** To put together the component parts of (e.g., a film) by cutting, combining, and splicing. **—ed′it** *n.*

▲ word history: The verb *edit* was derived in English from the noun *editor*, and not vice versa as one would expect. The process by which *edit* was formed is called back-formation. By this process a syllable thought to be an affix is removed from what is incorrectly considered to be a base word, and the base becomes a new word in its own right. *Editor* can thus be analyzed as a verb *edit* plus the noun suffix *-or*. The English word *editor*, however, was borrowed from Latin *editor* and is recorded 150 years earlier than the verb *edit*.

**e·di·tion** (ĭ-dĭsh′ən) *n.* [OFr. < Lat. *editio*, publication < *edere*, to publish : *ex-*, out + *dare*, to give.] **1. a.** The entire number of copies of a publication printed from a single typesetting or other form of reproduction. **b.** A single copy from this group. **c.** A facsimile of an earlier publication having substantial changes or additions. **2. a.** Any of the various forms in which something, as a publication, is issued or produced <a variorum *edition* of Melville> **b.** One closely similar to an original : VERSION. **3.** All the copies of a single press run of a newspaper <the late city *edition*>

**ed·i·tor** (ĕd′ĭ-tər) *n.* [LLat., publisher < Lat. *edere*, to publish. —see EDITION.] **1.** One who edits, esp. as an occupation. **2.** One who writes editorials. **3.** A device consisting of a splicer and viewer and used to edit film. **4.** *Computer Sci.* A routine that performs editing functions. **—ed′i·tor·ship′** *n.*

**ed·i·to·ri·al** (ĕd′ĭ-tôr′ē-əl, -tōr′-) *n.* **1.** An article in a publication expressing the opinion of its publishers or editors. **2.** A commentary on radio or television expressing the opinion of the station or network. *—adj.* **1.** Of, concerning, or pertaining to an editor. **2.** Of the nature of an editorial. **—ed′i·to′ri·al·ly** *adv.*

**ed·i·to·ri·al·ize** (ĕd′ĭ-tôr′ē-ə-līz′, -tōr′-) *vi.* **-ized, -iz·ing, -iz·es.** **1.** To express an opinion in or as if in an editorial. **2.** To present an opinion in the guise of an objective report. **—ed′i·to′ri·al·i·za′tion** *n.* **—ed′i·to′ri·al·i′zer** *n.*

**editor in chief** *n., pl.* **editors in chief.** The editor bearing final responsibility for the policies and operations of a publication.

**E·dom·ite** (ē′də-mīt′) *n.* [After *Edom*, a region of ancient Palestine.] A member of a Semitic people living in an area southeast of the Dead Sea in ancient times. **—E′dom·it′ish** *adj.*

**EDTA** (ē′dē-tē-ā′) *n.* [Abbr. of *ethylenediaminetetraacetic acid.*] A crystalline acid, $C_{10}H_{16}N_2O_8$, used as a chelating agent.

**ed·u·ca·ble** (ĕj′ə-kə-bəl) *adj.* [EDUC(ATE) + -ABLE.] Capable of being educated.

**ed·u·cate** (ĕj′ə-kāt′) *v.* **-cat·ed, -cat·ing, -cates.** [ME *educaten* < Lat. *educare.*] *—vt.* **1. a.** To provide with training or knowledge, esp. via formal schooling : TEACH. **b.** To provide with training for a specific purpose, as a vocation. **2.** To provide with information : INFORM. **3.** To stimulate or develop the mental or moral growth of. *—vi.* To teach or instruct another or others.

▲ word history: It has often been said that *educate* means "to draw out" a person's talents as opposed to putting in knowledge or instruction. This is an interesting idea, but it is not quite true in terms of the etymology of the word. *Educate* comes from Latin *educare*, "to educate," which is derived from a specialized use of Latin *educere* (from *e-*, "out," and *ducere*, "to lead") meaning "to assist at the birth of a child."

**ed·u·cat·ed** (ĕj′ə-kā′tĭd) *adj.* **1.** Having an education. **2. a.** Exhibiting evidence of schooling or instruction. **b.** Cultured : cultivated. **3.** Based on experience and some factual knowledge <an *educated* opinion>

**ed·u·ca·tion** (ĕj′ə-kā′shən) *n.* **1.** The act or process of educating or of being educated. **2.** The knowledge or skill obtained or developed by such a process : LEARNING. **3.** The field of study concerned with teaching and learning pedagogy.

**ed·u·ca·tion·al** (ĕj′ə-kā′shən-əl) *adj.* **1.** Of or pertaining to education. **2.** That educates : INSTRUCTIVE <an *educational* tour> **—ed′u·ca′tion·al·ly** *adv.*

**educational television** *n.* **1.** Public television. **2.** A video system providing instructional material.

**ed·u·ca·tive** (ĕj′ə-kā′tĭv) *adj.* Educational.

**ed·u·ca·tor** (ĕj′ə-kā′tər) *n.* **1.** One trained in teaching : TEACHER. **2.** A specialist in the theory and practice of education.

**e·duce** (ĭ-do͞os′, ĭ-dyo͞os′) *vt.* **e·duced, e·duc·ing, e·duc·es.** [Lat. *educere* : *ex-*, out + *ducere*, to lead.] **1.** To draw or bring out : ELICIT. **2.** To assume or work out from given facts : DEDUCE. **—e·duc′i·ble** *adj.* **—e·duc′tion** (ĭ-dŭk′shən) *n.*

**Ed·ward·i·an** (ĕd-wôr′dē-ən, -wär′-) *adj.* Of, pertaining to, or typical of the reign of Edward VII of England.

**-ee**[1] *suff.* [ME *-e* < OFr. < Lat. *-atus*, -ate.] **1. a.** One that receives or benefits from a specified action <address*ee*> **b.** One that possesses a specified thing <mortgag*ee*> **2.** One that performs a specified action <absent*ee*>

**-ee**[2] *suff.* [Alteration of -Y¹.] **1. a.** One resembling <goat*ee*> **b.** A particular kind of <boot*ee*> —Used often as a diminutive. **2.** One connected with <barg*ee*>

**eel** (ēl) *n., pl.* **eel** *or* **eels.** [ME *ele* < OE *ǣl.*] **1.** A long snakelike marine or freshwater fish of the order Anguilliformes or Apodes, esp. *Anguilla rostrata* of eastern North America or *A. anguilla* of Europe, migrating from fresh water to the Sargasso Sea to spawn. **2.** A fish related to or resembling the eel.

**eel·grass** (ēl′grăs′) *n.* **1.** A submerged aquatic plant of the genus *Zostera*, growing along the North Atlantic coast and having narrow, grasslike leaves. **2.** A similar or related plant, as tape grass.

**eel·pout** (ēl′pout′) *n., pl.* **eelpout** *or* **-pouts.** [ME **elepout* < OE *ǣle-puta.*] Any of various marine fishes of the family Zoarcidae, with an elongated body and a large head.

**eel·worm** (ēl′wûrm′) *n.* An often parasitic nematode worm, as the vinegar eel.

**e'en**[1] (ēn) *n.* EVENING 1, 2a.

**e'en**[2] (ēn) *adv.* Even.

**-eer** *suff.* [OFr. *-ier* < Lat. *-arius*, -ary.] One associated with, concerned with, or engaged in <ballad*eer*>

**e'er** (âr) *adv.* Ever.

**ee·rie** *or* **ee·ry** (îr′ē) *adj.* **-ri·er, -ri·est.** [ME *eri*, fearful < OE *earg*, cowardly.] **1. a.** Arousing fear or dread : WEIRD. **b.** Supernatural in aspect or character : MYSTERIOUS <an *errie* presence in the castle> **2.** *Scot.* Frightened or intimidated by superstition. **—ee′ri·ness** *n.*

**ef** (ĕf) *n.* The letter *f.*

**ef·face** (ĭ-fās′) *vt.* **-faced, -fac·ing, -fac·es.** [Fr. *effacer* < OFr. : Lat. *ex-*, out + Lat. *facies*, face.] **1.** To make indistinct or erase by or as if by rubbing. **2.** To conduct (oneself) inconspicuously or humbly. **—ef·face′a·ble** *adj.* **—ef·face′ment** *n.* **—ef·fac′er** *n.*

**ef·fect** (ĭ-fĕkt′) *n.* [ME < OFr. < Lat. *effectus*, p.part. of *efficere*, to accomplish : *ex-*, out + *facere*, to make.] **1.** Something brought about by an agent or cause : RESULT. **2.** The way in which something acts on or influences an object <the *effect* of music on the mentally ill> **3.** The capacity or power to achieve the desired result : INFLUENCE. **4.** The condition of being in full force or execution <The law had an immediate *effect* on traffic violations.> **5. a.** Something that produces a specific impression or supports a general design or intention <vivid lighting *effects*> **b.** A particular impression <an *effect* of graciousness> **c.** The production of a particular impression <screams and carries on just for *effect*> **6. a.** The basic meaning. **b.** Purport : intention. **7. effects.** Movable goods : PROPERTY. *—vt.* **-fect·ed, -fect·ing, -fects.** **1.** To bring into existence. **2.** To bring about. **—in effect. 1.** In fact : ACTUALLY. **2.** In essence : VIRTUALLY. **3.** EFFECTIVE 3. **—take effect.** To become operative. **—ef·fect′er** *n.* **—ef·fect′i·ble** *adj.*

☆ **syns:** EFFECT, CONSEQUENCE, FRUIT, ISSUE, OUTCOME, RESULT, SEQUEL, SEQUENT, UPSHOT *n. core meaning* : something brought about by a cause <The *effect* of good advertising is increased sales.> ***ant:*** cause

**ef·fec·tive** (ĭ-fĕk′tĭv) *adj.* **1.** Having an expected or intended effect. **2.** Producing or designed to produce a desired effect. **3.** In effect : OPERATIVE <The new requirements are *effective* immediately.> **4.** Existing in fact : ACTUAL <an abrupt increase in the *effective* demand> **5.** Prepared for use or action, esp. in warfare. *—n.* A combat-ready soldier or piece of military equipment. **—ef·fec′tive·ly** *adv.* **—ef·fec′tive·ness, ef′fec·tiv′i·ty** *n.*

☆ **syns:** EFFECTIVE, EFFECTUAL, EFFICACIOUS, EFFICIENT, PRODUCTIVE *adj. core meaning* : producing or able to produce a desired effect <an *effective* cold remedy> ***ant:*** ineffective

**ef·fec·tor** (ĭ-fĕk′tər) *n.* **1.** An organ at the end of a nerve that activates either gland secretion or muscular contraction. **2.** *Computer Sci.* A device for producing a desired change in an object in response to input.

**ef·fec·tu·al** (ĭ-fĕk′cho͞o-əl) *adj.* EFFECTIVE 2. **—ef·fec′tu·al′i·ty, ef·fec′tu·al·ness** *n.* **—ef·fec′tu·al·ly** *adv.*

**ef·fec·tu·ate** (ĭ-fĕk′cho͞o-āt′) *vt.* **-at·ed, -at·ing, -ates.** [Med. Lat. *effectuare* < Lat. *efficere.* —see EFFECT.] To bring about : EFFECT. **—ef·fec′tu·a′tion** *n.*

**ef·fem·i·nate** (ĭ-fĕm′ə-nĭt) *adj.* [ME *effeminat* < Lat. *effeminatus*, p.part. of *effeminare*, to make feminine : *ex*, from + *femina*, woman.] **1.** Having qualities or characteristics more often associated with women than men : UNMANLY. **2.** Marked by weakness and excessive refinement. **—ef·fem'i·na·cy** (ĭ-fĕm'ə-nə-sē) *n.* **—ef·fem'i·nate** *n.* **—ef·fem'i·nate·ly** *adv.* **—ef·fem'i·nate·ness** *n.*

**ef·fen·di** (ĭ-fĕn'dē) *n.* [Turk. *efendi* < Mod. Gk. *aphentēs*, master, alteration of Gk. *authentēs*, master.] Sir. —Used as a title of respect in Turkey.

**ef·fer·ent** (ĕf'ər-ənt) *adj.* [Fr. *efférent* < Lat. *efferens*, pr.part. of *efferre*, to carry off : *ex*, away + *ferre*, to carry.] **1.** Directed away from a central organ or section. **2.** Carrying impulses from the central nervous system to an effector. **—ef'fer·ent** *n.* **—ef'fer·ent·ly** *adv.*

**ef·fer·vesce** (ĕf'ər-vĕs') *vi.* **-vesced, -vesc·ing, -vesc·es.** [Lat. *effervescere* : *ex-* (intensive) + *fervescere*, to start boiling < *fervēre*, to boil.] **1.** To emit small bubbles as gas comes out of a liquid. **2.** To show high spirits or vivacity. **—ef'fer·ves'cence, ef'fer·ves'cen·cy** *n.* **—ef'fer·ves'cent** *adj.* **—ef'fer·ves'cent·ly** *adv.*

**ef·fete** (ĭ-fēt') *adj.* [Lat. *effetus*, worn out by childbearing : *ex-*, out + *fetus*, childbearing.] **1.** No longer productive : INFERTILE. **2.** Exhausted of vitality, force, or effectiveness : WORN-OUT. **3.** Marked by self-indulgence, decadence, or weakness. **4.** Effeminate. : overrefined. **—ef·fete'ly** *adv.* **—ef·fete'ness** *n.*

**ef·fi·ca·cious** (ĕf'ĭ-kā'shəs) *adj.* [Lat. *efficax*, *efficac-* < *efficere*, to effect.] Capable of producing a desirable effect. **—ef'fi·ca'cious·ly** *adv.* **—ef'fi·ca'cious·ness** *n.*

**ef·fi·ca·cy** (ĕf'ĭ-kə-sē) *n.* [Lat. *efficacia* < *efficax*, efficacious.] Power or capacity to produce a desired effect : EFFECTIVENESS.

**ef·fi·cien·cy** (ĭ-fĭsh'ən-sē) *n.*, *pl.* **-cies. 1. a.** The quality of being efficient. **b.** The degree to which this quality is exercised. **2. a.** The ratio of the effective or useful output to the total input in a system. **b.** The ratio of the energy delivered by a machine to the energy supplied for its operation. **3.** *Informal.* An efficiency apartment.

**efficiency apartment** *n.* A small, usu. furnished apartment having a private bathroom and kitchenette.

**efficiency expert** *n.* An expert who analyzes esp. industrial operations in order to improve efficiency and productivity.

**ef·fi·cient** (ĭ-fĭsh'ənt) *adj.* [ME < OFr. < Lat. *efficiens*, pr.part. of *efficere*, to effect.] **1.** Acting directly to produce an effect <an *efficient* reason> **2. a.** Acting or producing effectively with a minimum of waste or unnecessary effort. **b.** Exhibiting a high ratio of output to input. **—ef·fi'cient·ly** *adv.*

☆ ***syns:*** EFFICIENT, HIGH-PERFORMANCE, PRODUCTIVE, PROFICIENT *adj. core meaning* : acting effectively with minimal waste <an *efficient* production line> ***ant:*** inefficient

**ef·fi·gy** (ĕf'ə-jē) *n.*, *pl.* **-gies.** [ME *effigie* < Lat. *effigies* < *effingere*, to portray : *ex-*, from + *fingere*, to shape.] **1.** A painted or sculptured representation, esp. of a person. **2.** A crude image of a hated person.

**ef·flo·resce** (ĕf'lə-rĕs') *vi.* **-resced, -resc·ing, -resc·es.** [Lat. *efflorescere* : *ex-*, out + *florescere*, inchoative of *florēre*, to blossom < *flos*, flower.] **1.** To bloom or flower : BLOSSOM. **2.** *Chem.* **a.** To become a powder by losing water of crystallization. **b.** To become covered with a powdery deposit.

**ef·flo·res·cence** (ĕf'lə-rĕs'əns) *n.* **1.** A time or state of flowering. **2. a.** A gradual process of unfolding or developing. **b.** The highest point : ACME <the *efflorescence* of French Impressionism> **3.** *Chem.* **a.** The process of efflorescing. **b.** The deposit that results from this process. **c.** A growth of salt crystals on a surface due to evaporation of salt-laden water. **—ef'flo·res'cent** *adj.*

**ef·flu·ence** (ĕf'lo͞o-əns) *n.* **1.** An act or instance of flowing out. **2.** Something that flows out or forth : EMANATION.

**ef·flu·ent** (ĕf'lo͞o-ənt) *adj.* [ME < Lat. *effluens*, pr.part. of *effluere*, to flow out : *ex-*, out + *fluere*, to flow.] Flowing out or forth. —*n.* Something that flows out, esp.: **a.** A stream flowing out of a lake or other body of water. **b.** An outflow of waste, as from a sewer.

**ef·flu·vi·um** (ĭ-flo͞o'vē-əm) *n.*, *pl.* **-vi·a** (-vē-ə) or **-ums.** [Lat. < *effluere*, to flow out. —see EFFLUENT.] **1.** A usu. foul-smelling outflow or rising vapor. **2.** An invisible outflow of radiation or vapor : AURA. **—ef·flu'vi·al** *adj.*

**ef·flux** (ĕf'lŭks') *n.* [< Lat. *effluxus*, p.part. of *effluere*, to flow out. —see EFFLUENT.] **1.** An outward flow. **2.** EFFLUENCE 2. **3.** An expiration : ending. **—ef·flux'ion** *n.*

**ef·fort** (ĕf'ərt) *n.* [OFr. < *efforcier*, to force < Med. Lat. *exfortiare* : Lat. *ex-*, out + Lat. *fortis*, strong.] **1.** Exertion of physical or mental energy to do something. **2.** A difficult or tiring exertion of the strength or will <an *effort* to climb the mountain> **3.** A usu. earnest attempt <an *effort* to learn the lesson> **4.** Something done through exertion : ACHIEVEMENT. **5.** *Physics.* Force applied against inertia. **—ef'fort·ful** *adj.* **—ef'fort·ful·ly** *adv.*

☆ ***syns:*** EFFORT, EXERTION, PAINS, STRAIN, STRUGGLE, TROUBLE *n. core meaning* : the use of energy to do something <a job that isn't worth the *effort*>

**ef·fort·less** (ĕf'ərt-lĭs) *adj.* Requiring or displaying little or no effort. **—ef'fort·less·ly** *adv.* **—ef'fort·less·ness** *n.*

**ef·front·er·y** (ĭ-frŭn'tə-rē) *n.*, *pl.* **-ies.** [Fr. *effronterie* < LLat. *effrons*, shameless : *ex-*, from + *frons*, forehead.] Impudent boldness.

**ef·ful·gence** (ĭ-fo͝ol'jəns, ĭ-fŭl'-) *n.* A brilliant radiance.

**ef·ful·gent** (ĭ-fo͝ol'jənt, ĭ-fŭl'-) *adj.* [Lat. *effulgens*, *effulgent-*, pr.part. of *effulgere*, to shine out : *ex-*, out + *fulgēre*, to shine.] Shining brilliantly : RADIANT.

**ef·fuse** (ĭ-fyo͞os') *adj.* [< Lat. *effusus*, p.part. of *effundere*, to pour out : *ex-*, out + *fundere*, to pour.] *Bot.* Spreading out loosely on a surface. —*v.* (ĭ-fyo͞oz') **-fused, -fus·ing, -fus·es.** —*vt.* To pour or spread out : DISSEMINATE. —*vi.* **1.** To spread or flow out. **2.** To exude.

**ef·fu·sion** (ĭ-fyo͞o'zhən) *n.* **1. a.** An act or instance of effusing. **b.** Something poured out. **2.** An unrestrained outpouring of feeling, as in writing or speech. **3.** *Pathol.* **a.** The seeping of serous, purulent, or bloody fluid into a cavity. **b.** The effused fluid.

**ef·fu·sive** (ĭ-fyo͞o'sĭv) *adj.* Excessive or unrestrained in emotional expression : GUSHY. **—ef·fu'sive·ly** *adv.* **—ef·fu'sive·ness** *n.*

**Ef·ik** (ĕf'ĭk) *n.* **1.** One of a people of southern Nigeria. **2.** The Ibibio language of the Efik people. **—Efik** *adj.*

**eft** (ĕft) *n.* [ME *evete* < OE *efeta*.] A newt.

**eft·soons** (ĕft-so͞onz') *adv.* [ME *eftsone* < OE *eftsōna* : *eft*, again + *sōna*, soon.] *Archaic.* **1.** Soon afterward : PRESENTLY. **2.** Once again.

**e·gad** (ĭ-găd') *interj.* [Alteration of *ah God.*] —Used as a mild oath expressing surprise or enthusiasm.

**e·gal·i·tar·i·an** (ĭ-găl'ĭ-târ'ē-ən) *n.* [Fr. *égalitaire*, an egalitarian < *égalité*, equality < Lat. *aequalitas* < *aequalis*, equal.] An adherent of the doctrine of equal political, economic, and legal rights for all human beings. **—e·gal'i·tar'i·an** *adj.* **—e·gal'i·tar'i·an·ism** *n.*

**E·ge·ri·a** (ĭ-jîr'ē-ə) *n.* [After *Egeria*, an adviser to Numa Pompilius, legendary Roman king.] A woman counselor or adviser.

**e·gest** (ē-jĕst') *vt.* **e·gest·ed, e·gest·ing, e·gests.** [Lat. *egerere*, *egest-*, to carry out : *ex-*, out + *gerere*, to carry.] To excrete or discharge from the body. **—e·ges'tion** *n.* **—e·ges'tive** *adj.*

**e·ges·ta** (ē-jĕs'tə) *pl.n.* [Lat. < *egerere*, to carry out. —see EGEST.] Egested matter, esp. excrement.

**egg¹** (ĕg) *n.* [ME *egge*, bird's egg < ON *egg*.] **1.** A female gamete : OVUM. **2.** One of the female reproductive cells of various animals, consisting usu. of an embryo surrounded by nutrient material with a protective covering. **3.** The oval, thin-shelled ovum of a bird, esp. of a domestic fowl, used as food. **4.** Something having the typically ovoid shape of a hen's egg. **5.** *Slang.* A person <not a bad *egg*> —*vt.* **egged, egg·ing, eggs. 1.** To cover with beaten egg, as in cooking. **2.** *Informal.* To throw eggs at. **—put (**or **have) all one's eggs in one basket.** To risk everything on a single venture. **—with egg on one's face.** *Informal.* Embarrassed.

**egg²** (ĕg) *vt.* **egged, egg·ing, eggs.** [ME *eggen* < ON *eggja*.] To encourage or incite to action <The crowd *egged* the fighters on.>

**egg-and-dart** (ĕg'ən-därt') *n.* A decorative molding alternating egg-shaped figures with anchor, dart, or tongue-shaped figures.

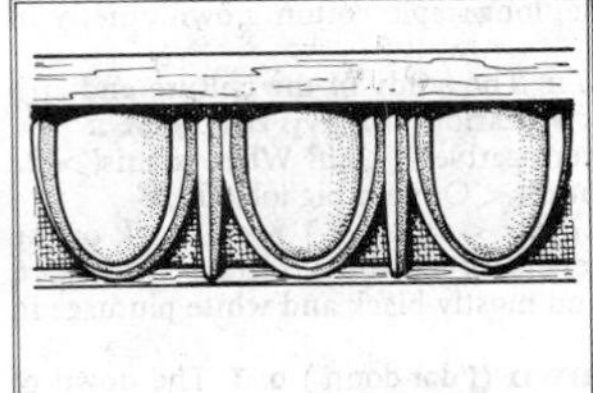

**egg-and-dart**

**eg·gar** (ĕg'ər) *n. var. of* EGGER.

**egg·beat·er** (ĕg'bē'tər) *n.* A kitchen utensil with rotating blades for whipping, beating, or mixing.

**egg case** *n.* A protective capsule in some animals that contains eggs.

**egg·cup** (ĕg'kŭp') *n.* A cup for a usu. soft-boiled egg.

**eg·ger** *also* **eg·gar** (ĕg'ər) *n.* Any of various moths of the family Lasiocampidae, whose larvae often construct tentlike webs.

**egg·head** (ĕg'hĕd') *n. Slang.* An intellectual.

**egg·nog** (ĕg'nŏg') *n.* [EGG¹ + obs. *nog*, ale.] A drink of beaten eggs and milk, often mixed with rum, brandy, or wine.

**egg·plant** (ĕg'plănt') *n.* **1. a.** A tropical Old World plant, *Solanum melongena*, cultivated for its edible fruit. **b.** The glossy, ovoid fruit of the eggplant. **2.** A blackish purple.

**egg roll** *n.* A cylindrical case made of an egg dough, filled with minced vegetables and sometimes meat or seafood and fried.

**eggs Ben·e·dict** (bĕn'ĭ-dĭkt') *n.* [Prob. < the name *Benedict*.] A dish of poached eggs on slices of toast and ham, covered with hollandaise sauce.

**egg·shell** (ĕg'shĕl') *n.* **1.** The thin, brittle exterior covering of a bird's egg. **2.** A pale yellow to yellowish white. **—egg'shell'** *adj.*

**egg white** *n.* Egg albumen.

**e·gis** (ē'jĭs) *n. var. of* AEGIS.

**eg·lan·tine** (ĕg'lən-tīn', -tēn') *n.* [ME *eglentin* < OFr. *aiglent*, prob. ult. < Lat. *acus*, needle.] The sweetbrier.

**e·go** (ē'gō, ĕg'ō) *n.* [NLat. < Lat., I.] **1.** The self, esp. as distinct from the world and other selves. **2.** *Psychoanal.* The personality component that is conscious, most directly controls behavior, and is most in touch with external reality. **3. a.** Self-love : egotism. **b.** Self-confidence : self-esteem.

**e·go·cen·tric** (ē'gō-sĕn'trĭk, ĕg'ō-) *adj.* **1.** Having the view that the ego is the center, object, and norm of all experience. **2.** Self-centered : selfish. **3.** *Philos.* Seen or perceived from one's own mind as a center. **—e'go·cen'tric** *n.* **—e'go·cen·tric'i·ty** (-trĭs'ĭ-tē) *n.*

**ego ideal** *n. Psychoanal.* The entirety of an individual's positive identifications with loving, reassuring parents or parental substitutes, regarded as a differentiated component of the mature ego.

**e·go·ism** (ē'gō-ĭz'əm, ĕg'ō-) *n.* **1. a.** The ethical doctrine that morality is based on self-interest. **b.** The ethical belief that self-interest is the proper motive for all human conduct. **2.** Egotism : conceit.

**e·go·ist** (ē'gō-ĭst, ĕg'ō-) *n.* [Fr. *égoiste* < *égo*, ego < Lat. *ego*, I.] **1.** One devoted to one's own interests and advancement. **2.** An adherent of egoism. **3.** EGOTIST 2. **—e'go·is'tic, e'go·is'ti·cal** *adj.* **—e'go·is'ti·cal·ly** *adv.*

**e·go·ma·ni·a** (ē'gō-mā'nē-ə, -mān'yə, ĕg'ō-) *n.* Obsession with the self. **—e'go·ma'ni·ac'** (-nē-ăk') *n.* **—e'go·ma·ni'a·cal** (-mə-nī'ə-kəl) *adj.* **—e'go·ma·ni'a·cal·ly** *adv.*

**e·go·tism** (ē'gə-tĭz'əm, ĕg'ə-) *n.* [EGO + *-tism*, as in *nepotism.*] **1.** The tendency to refer to oneself in a boastful and excessive way. **2.** An extreme sense of self-importance : CONCEIT.

**e·go·tist** (ē'gə-tĭst, ĕg'ə-) *n.* **1.** A conceited, boastful, self-absorbed person. **2.** One who acts selfishly : EGOIST. **—e'go·tis'tic, e'go·tis'ti·cal** *adj.* **—e'go·tis'ti·cal·ly** *adv.*

☆ ***syns:*** EGOTIST, EGOCENTRIC, EGOIST, EGOMANIAC, NARCISSIST *n. core meaning* : a conceited, self-absorbed person <Powerful people are often *egotists.*>

**ego trip** *n. Slang.* Something that gratifies the ego.

**e·go-trip** (ē'gō-trĭp', ĕg'ō-) *vi.* **-tripped, -trip·ping, -trips.** *Slang.* To behave in an egotistic or self-seeking way. **—e'go-trip'per** *n.*

**e·gre·gious** (ĭ-grē'jəs, -jē-əs) *adj.* [Lat. *egregius*, distinguished : *ex-*, out from + *grex*, herd.] Outstandingly bad : BLATANT. **—e·gre'gious·ly** *adv.* **—e·gre'gious·ness** *n.*

**e·gress** (ē'grĕs') *n.* [Lat. *egressus* < *egredi*, to go out : *ex-*, out + *gradi*, to go.] **1.** The act of going out : EMERGENCE. **2.** The right of going out. **3.** An opening or means of going out : EXIT. **4.** *Astron.* Emergence of a celestial body from eclipse or occultation.

**e·gres·sion** (ĭ-grĕsh'ən) *n.* Egress.

**e·gret** (ē'grĭt, ĕg'rĭt) *n.* [ME < OFr. *aigrette* < Prov. *aigreta* < *aigron*, heron, of Germanic orig.] Any of several usu. white wading birds of the genera *Bubulcus, Casmerodius, Leucophoyx*, and related genera, with long, drooping plumes during the breeding season.

**E·gyp·tian** (ĭ-jĭp'shən) *n.* **1.** A native or resident of Egypt. **2.** The Afro-Asiatic language of the ancient Egyptians. **—E·gyp'tian** *adj.*

**Egyptian clover** *n.* Berseem.

**Egyptian corn** *n.* Durra.

**Egyptian cotton** *n.* A fine, long-staple cotton grown chiefly in northern Africa.

**E·gyp·tol·o·gy** (ē'jĭp-tŏl'ə-jē) *n.* The study of the culture and artifacts of the ancient Egyptian civilization. **—E'gyp·tol'o·gist** *n.*

**eh** (ā, ĕ) *interj.* **1.** —Used interrogatively <*Eh?* What is this?> **2.** —Used in asking for confirmation <Quite a big job, *eh?*>

**ei·der** (ī'dər) *n.* [Ult. < Icel. *æður* < ON *æðr.*] A sea duck of the genus *Somateria* or related genera of northern regions, with soft, commercially valuable down and mostly black and white plumage in the males.

**ei·der·down** *also* **eider down** (ī'dər-doun') *n.* **1.** The down of the eider duck, used to stuff quilts and pillows. **2.** A quilt stuffed with eiderdown. **3.** A warm, napped fabric.

**ei·det·ic** (ī-dĕt'ĭk) *adj.* [G. *eidetisch* < Gk. *eidētikos*, relating to images < *eidēsis*, knowledge < *eidos*, form.] Of, pertaining to, or characterized by extremely detailed and vivid recall of visual images. **—ei·det'i·cal·ly** *adv.*

**ei·do·lon** (ī-dō'lən) *n., pl.* **-lons** *or* **-la** (-lə) [Gk. *eidōlon* < *eidos*, form.] **1.** A phantom : apparition. **2.** An image of an ideal.

**eight** (āt) *n.* [ME *eighte* < OE *eahta* : akin to G. *acht*, Lat. *octo*, Gk. *oktō*, Skt. *aṣṭā.*] **1.** The cardinal number equal to 7 + 1. **2.** The eighth in a set or sequence. **3.** Something having eight parts, units, or members, esp.: **a.** An eight-oared racing shell. **b.** An eight-cylinder automobile or engine. **—eight** *adj. & pron.*

**eight ball** *n.* A black pool ball bearing the number eight that may place a player at a disadvantage. **—behind the eight ball.** *Slang.* In a disadvantageous or uncomfortable position.

**eight·een** (ā-tēn') *n.* [ME *eightetene* < OE *eahtatīene.*] **1.** The cardinal number equal to 17 + 1. **2.** The 18th in a set or sequence. **—eight·een'** *adj. & pron.*

**eight·een·mo** (ā-tēn'mō) *n., pl.* **-mos.** An octodecimo.

**eight·eenth** (ā-tēnth') *n.* **1.** The ordinal number matching the number 18 in a series. **2.** One of 18 equal parts. **—eight·eenth'** *adj. & adv.*

**eighth** (ātth, āth) *n.* [ME < OE < *eahta*, eight.] **1.** The ordinal number matching the number eight in a series. **2.** One of eight equal parts. **—eighth** *adj. & adv.*

**eighth note** *n. Mus.* A note with one-eighth the time value of a whole note.

**eight·i·eth** (ā'tē-ĭth) *n.* **1.** The ordinal number matching the number 80 in a series. **2.** One of 80 equal parts. **—eight'i·eth** *adj. & adv.*

**eight·vo** (āt'vō') *n., pl.* **-vos.** [EIGHT + (OCTA)VO.] An octavo.

**eight·y** (ā'tē) *n., pl.* **-ies.** [ME *eighti* < OE *hundeahtatig* : *hund*, hundred + *eahta*, eight + *-tig*, -ty.] The cardinal number equal to 8 × 10. **—eight'y** *adj. & pron.*

**-ein** *suff.* [Alteration of -IN.] A chemical compound related to a compound with a similar name ending in *-in* <phthal*ein*>

**ein·korn** (īn'kôrn') *n.* [G. *Einkorn* : *ein*, one (< OHG) + *Korn*, grain (< OHG).] A one-seeded wheat, *Triticum manococcum*, grown in arid regions.

**ein·stein·i·um** (īn-stī'nē-əm) *n.* [After Albert *Einstein* (1879–1955).] *Symbol* **Es** A synthetic element first produced by neutron irradiation of uranium in a thermonuclear explosion; atomic number 99; longest-lived isotope Es 254.

**eis·tedd·fod** (ā-stĕth'vŏd, ĕs-tĕth'-) *n.* [Welsh : *eistedd*, to sit (< *sedd*, seat) + *bod*, to be.] An annual assembly of Welsh musicians and poets.

**ei·ther** (ē'thər, ī'thər) *pron.* [ME < OE *ǣgðer.*] One or the other. —*conj.* —Used before the first of two or more coordinates or clauses linked by *or* <*Either* we go now, or stay here all day.> —*adj.* **1.** One or the other <Use *either* tool.> **2.** One and the other : EACH <repairs on *either* side of the highway> —*adv.* Also : likewise. —Used as an intensifier following negative statements <If you don't diet, I won't *either.*> ***usage:*** In a sentence whose subject is one of the elements of an *either . . . or* construction, the verb is singular when both elements of the construction are singular (*Either he or she knows the answer*) and plural when both elements are plural (*Either the lords or the hunters are liars*). A problem arises when one of the linked elements is singular and the other plural; traditionalists insist that rephrasing is the only solution. It has also been suggested that the verb should agree with the closest element of the construction, as in *Either she or the boys are going.*

**either-or** (ē'thər-ôr, ī'thər-) *adj.* Characterized by a choice between only two possibilities <caught in an *either-or* situation>

**e·jac·u·late** (ĭ-jăk'yə-lāt') *v.* **-lat·ed, -lat·ing, -lates.** [Lat. *ejaculari, ejaculat-* : *ex-*, out + *jaculari*, to throw < *jaculum*, dart.] —*vt.* **1.** To discharge abruptly, esp. to discharge (semen) in orgasm. **2.** To utter suddenly and passionately : EXCLAIM. —*vi.* To eject semen. —*n.* (ĭ-jăk'yə-lĭt). Semen ejaculated in orgasm. **—e·jac'u·la'tor** *n.*

**e·jac·u·la·tion** (ĭ-jăk'yə-lā'shən) *n.* **1.** The act of ejaculating. **2.** An abrupt discharge of fluid, esp. of seminal fluid. **3. a.** A sudden, short utterance : EXCLAMATION. **b.** A brief, pious utterance or prayer.

**e·jac·u·la·to·ry** (ĭ-jăk'yə-lə-tôr'ē, -tōr'ē) *adj.* **1.** Of or relating to ejaculation. **2.** Relating to or constituting a sudden, brief utterance.

**e·ject** (ĭ-jĕkt') *v.* **e·ject·ed, e·ject·ing, e·jects.** [ME *ejecten* < Lat. *eicere* : *ex-*, out + *jacere*, to throw.] —*vt.* **1.** To drive out forcefully : EXPEL. **2. a.** To force to leave. **b.** To evict. —*vi.* To make an emergency exit from a fighter aircraft by ejection seat or capsule.

☆ ***syns:*** EJECT, BOOT (out), BOUNCE, CHUCK, DISMISS, EVICT, EXPEL, KICK OUT, OUST, THROW OUT *v. core meaning* : to put out by force <*eject* a tenant from an apartment>

**e·jec·ta** (ĭ-jĕk'tə) *pl.n.* [NLat. < Lat. *ejectus*, p.part. of *eicere*, to throw out. —see EJECT.] Ejected matter, as that from an erupting volcano.

**e·jec·tion** (ĭ-jĕk'shən) *n.* **1.** The act of ejecting or state of being ejected. **2.** Ejected matter.

**ejection seat** *n.* A seat made to eject clear of an aircraft and parachute to the ground in an emergency.

**e·ject·ment** (ĭ-jĕkt'mənt) *n.* **1.** The act of ejecting : DISPOSSESSION. **2.** An action to regain possession of real estate held by another.

**e·jec·tor** (ĭ-jĕk'tər) *n.* **1.** One that ejects. **2.** A device in a gun that ejects the empty shell after each firing.

**eke**[1] (ēk) *vt.* **eked, ek·ing, ekes.** [ME *eken*, to increase < OE *ēcan.*] **1.** To supplement with great effort <*eked* out my income with extra night work> **2.** To earn with great effort or strain <*eke* out a bare existence> **3.** To make (a supply) last by economy.

**eke**[2] (ēk) *adv. Archaic.* [ME < OE *ēc.*] Also.

**e·kis·tics** (ĭ-kĭs'tĭks) *n.* [< Gk. *oikistikos*, of settlements < *oikizein*, to settle < *oikos*, house.] *(sing. in number).* The science of human settlements, including city or community planning and design. **—e·kis'tic, e·kis'ti·cal** *adj.* **—ek'is·ti'cian** (ĕk'ĭ-stĭsh'ən) *n.*

**ek·pwe·le** (ĕk-pwē'lē) *n.* —See table at CURRENCY.

**el**[1] *also* **ell** (ĕl) *n.* The letter *l.*

**el**[2] (ĕl) *n. Informal.* An elevated railway.

**e·lab·o·rate** (ĭ-lăb'ər-ĭt) *adj.* [Lat. *elaborare, elaborat-* : *ex-*, out + *laborare*, to work < *labor*, work.] **1.** Planned or done with careful attention to numerous details or parts. **2.** Intricate and rich in detail. —*v.* (ĭ-lăb'ə-rāt') **-rat·ed, -rat·ing, -rates.** —*vt.* **1.** To work out with detail and care. **2.** To produce by effort : CREATE. —*vi.* **1.** To become elaborate. **2.** To express at greater length or in greater detail. **—e·lab'o·rate·ly** *adv.* **—e·lab'o·rate·ness** *n.* **—e·lab'o·ra'tion** (ĭ-lăb'ə-rā'shən) *n.* **—e·lab'o·ra'tor** *n.*

**Elaine** (ĭ-lān') *n.* One of two women in Arthurian legend who loved Lancelot: **a.** One who died of unrequited love for him. **b.** One who was the mother of Galahad by Lancelot.

**é·lan** (ā-län′, ā-lăn′) *n.* [Fr. < OFr. *eslan,* rush < *eslancer,* to throw out : *es-,* out (< Lat. *ex-*) + *lancer,* to throw (< LLat. *lanceare,* to throw a lance < Lat. *lancea,* lance).] **1.** Enthusiastic liveliness and vigor : ZEST. **2.** Flair : style.
**e·land** (ē′lənd) *n.* [Afr. < Du., elk < obs. G. *elen,* prob. < Lith. *élnis.*] A large African antelope, *Taurotragus oryx* or *T. derbianus,* with a light-brown or grayish coat and spirally twisted horns.
**élan vi·tal** (vē-tăl′) *n.* [Fr.] The vital force hypothesized by Henri Bergson as a source of efficient causation and evolution in nature.
**el·a·pid** (ĕl′ə-pĭd) *n.* [NLat. *Elapidae,* family name < Gk. *elaps,* fish, var. of *ellops.*] Any of various venomous snakes of the family Elapidae, including the cobras and coral snakes.
**e·lapse** (ĭ-lăps′) *vi.* **e·lapsed, e·laps·ing, e·laps·es.** [Lat. *elabi, elaps-* : *ex-,* away + *labi,* to slip.] To pass or slip by <Time *elapses* between the acts.> *—n.* Passage : lapse.
**elapsed time** *n.* The measured duration of an event, esp. the actual time spent in transit, as in flight, by a moving body.
**e·las·mo·branch** (ĭ-lăz′mə-brăngk′) *n.* [NLat. *Elasmobranchii,* subclass name : Gk. *elasmos,* metal plate (< *elaunein,* to beat) + Lat. *branchia,* gill < Gk., gills.] Any of numerous fishes of the subclass Elasmobranchii within the class Chondricthyes, including the sharks, rays, and skates, marked by a cartilaginous skeleton. **—e·las′-mo·branch′** *adj.*
**e·las·tic** (ĭ-lăs′tĭk) *adj.* [NLat. *elasticus* < Gk. *elastos,* impulsive < *elaunein,* to drive.] **1.** *Physics.* Returning or capable of returning to an initial state or form after deformation. **2.** Capable of adapting to change or a variety of circumstances : FLEXIBLE <an *elastic* interpretation of the law> **3.** Quick to recover or revive <an *elastic* attitude> **4.** Springy : firm. *—n.* **1. a.** A stretchable fabric made with interwoven strands of real or synthetic rubber. **b.** Something made of elastic. **2.** A rubber band. **—e·las′ti·cal·ly** *adv.*
**elastic collision** *n.* A collision of particles that conserves the total kinetic energy of translation.
**e·las·tic·i·ty** (ĭ-lă-stĭs′ĭ-tē, ē′lă-) *n.* **1.** The condition or property of being elastic : FLEXIBILITY. **2.** *Physics.* **a.** The property of returning to an initial state or form following deformation. **b.** The degree to which this property is exhibited.
**e·las·ti·ci·zer** (ĭ-lăs′tə-sī′zər) *n.* An additive that increases the elasticity of a solid propellant to prevent cracking of the propellant grain in the combustion chamber.
**e·las·tin** (ĭ-lăs′tĭn) *n.* [ELAST(IC) + -IN.] The albuminoid base of elastic tissue, as cartilage, tendons, and connective tissue.
**e·las·to·mer** (ĭ-lăs′tə-mər) *n.* [ELAST(IC) + -O- + -MER(E).] Any of various polymers with elastic properties resembling those of natural rubber. **—e·las′to·mer′ic** (-mĕr′ĭk) *adj.*
**e·late** (ĭ-lāt′) *vt.* **e·lat·ed, e·lat·ing, e·lates.** [Lat. *effere, elat-,* to carry out : *ex-,* out + *ferre,* to carry.] To raise the spirits of. *—adj.* Joyful : elated. **—e·la′tion** *n.*
☆ ***syns:*** ELATE, ANIMATE, BUOY (up), ELEVATE, EXALT, EXHILARATE, INSPIRE, INSPIRIT, LIFT, UPLIFT *v. core meaning* : to raise the spirits of <Winning the gold medal *elated* the team.>
**e·lat·ed** (ĭ-lā′tĭd) *adj.* Lively or joyful : EXULTANT. **—e·lat′ed·ly** *adv.* **—e·lat′ed·ness** *n.*
**el·a·ter** (ĕl′ə-tər) *n.* [NLat. < Gk. *elatēr,* driver < *elaunein,* to drive.] **1.** An elaterid beetle. **2.** *Bot.* An elongated, often spirally thickened filament occurring among the spores of liverworts.
**e·lat·er·id** (ĭ-lăt′ər-ĭd) *n.* [NLat. *Elateridae,* family name < Gk. *elatēr,* driver. —see ELATER.] Any of numerous beetles of the family Elateridae, including the click beetles.
**e·lat·er·ite** (ĭ-lăt′ə-rīt′) *n.* [G. *Elaterit* < Gk. *elatēr,* driver < *elaunein,* to drive.] A brown, elastic hydrocarbon resin.
**E layer** *n.* A region of the ionosphere occurring between approx. 55 and 95 miles or 90 and 150 kilometers above the earth and influencing long-distance communications by strongly reflecting radio waves in the range from one to three megahertz.
**el·bow** (ĕl′bō′) *n.* [ME *elbowe* < OE *elnboga.*] **1. a.** The joint or bend of the arm between the upper arm and the forearm. **b.** The bony outer projection of this joint. **2.** A joint, as of a quadruped or bird, corresponding to the human elbow. **3.** Something having an angle or bend similar to an elbow, esp.: **a.** A length of pipe with a sharp bend in it. **b.** A sharp bend in a river or a road. *—v.* **-bowed, -bow·ing, -bows.** *—vt.* **1. a.** To jostle or push with the elbow. **b.** To shove aside with the elbow. **2.** To make (one's way) by jostling, pushing, or shoving with one's elbow. *—vi.* To advance by elbowing.
**elbow grease** *n. Informal.* Strenuous physical effort.
**el·bow·room** (ĕl′bō-ro͞om′, -ro͝om′) *n.* **1.** Room enough to move about or function in. **2.** Ample scope.
**eld·er[1]** (ĕl′dər) *n.* [ME *eldre* < OE *eldra.*] **1.** An older person. **2.** An older, influential man of a family, tribe, or community. **3.** One of the governing officers of the church, often with pastoral or teaching functions. **4.** *Mormon Ch.* A member of the higher order of priesthood. *—adj.* **1.** Greater in age or seniority. **2.** Earlier or higher in order, rank, or office. **—el′der·ship′** *n.*
**el·der[2]** (ĕl′dər) *n.* [ME < OE *ellen.*] A shrub or small tree of the genus *Sambucus,* bearing white flower clusters and red or blackish berrylike fruit.
**el·der·ber·ry** (ĕl′dər-bĕr′ē) *n.* **1.** The small, edible fruit of an elder, used for making wine or preserves. **2.** A shrub or tree that bears elderberries.
**eld·er·ly** (ĕl′dər-lē) *adj.* Approaching old age. **—el′der·li·ness** *n.*
**elder statesman** *n.* A prominent, experienced older person, esp. a statesman acting as an unofficial adviser.
**eld·est** (ĕl′dĭst) *adj.* [ME < OE *eldesta.*] Greatest in age or seniority.
**El Do·ra·do** (ĕl də-rä′dō) *n.* [Sp., the gilded (land).] **1.** A legendary kingdom or city in Spanish America rich in jewels and precious metals, sought after by 16th-cent. explorers. **2.** A place of fabulous wealth or opportunity.
**el·dritch** (ĕl′drĭch) *adj.* [Of Sc. orig.] Weird or unearthly.
**El·e·at·ic** (ĕl′ē-ăt′ĭk) *adj.* [Lat. *Eleaticus* < Gk. *Eleatikos* < *Elea,* Velia, an ancient town in south Italy.] Of or typical of the school of philosophy founded by Xenophanes and Parmenides and believing that immutable being is the only knowable reality. **—El′e·a′tic** *n.* **—El′e·at′i·cism** *n.*
**el·e·cam·pane** (ĕl′ĭ-kăm-pān′) *n.* [ME *elecampana* < Med. Lat. *enula campana* : Lat. *inula,* elecampane (< Gk. *helenion*) + Lat. *campanea,* fem. of *campaneus,* of the field < *campus,* field.] A tall Eurasian plant, *Inula helenium,* bearing rayed yellow flowers.
**e·lect** (ĭ-lĕkt′) *v.* **e·lect·ed, e·lect·ing, e·lects.** [ME *electen* < Lat. *eligere,* to select : *ex-,* out + *legere,* to choose.] *—vt.* **1.** To select by vote for an office or for membership. **2.** To choose : PICK OUT <*elect* a course in geology> **3.** To decide on <*elected* to make the trip> *—vi.* To make a selection or choice. *—adj.* **1.** Singled out on purpose. **2.** Elected but not yet installed <the mayor-*elect*> **3.** Chosen for marriage <the bride-*elect*><the groom-*elect*> **4.** Selected by the divine will for salvation. *—n.* **1.** One selected or chosen. **2.** One selected by the divine will for salvation. **3.** *(pl. in number).* An exclusive group <one of the *elect* at the party> **—e·lect′a·bil′i·ty** *n.* **—e·lect′a·ble** *adj.*
**e·lec·tion** (ĭ-lĕk′shən) *n.* **1. a.** The act or power of electing. **b.** The fact of being elected. **2.** The right or ability to choose. **3.** Predestined salvation, esp. as conceived by Calvinists.
**Election Day** *n.* A day set by law for electing public officials.
**e·lec·tion·eer** (ĭ-lĕk′shə-nîr′) *vi.* **-eered, -eer·ing, -eers.** To work actively for a candidate or political party.
**e·lec·tive** (ĭ-lĕk′tĭv) *adj.* **1.** Of or relating to a selection by vote. **2.** Filled or obtained by election <*elective* positions> **3.** Having the power or authority to elect : ELECTORAL. **4.** Permitting or involving a choice : OPTIONAL <*elective* courses> *—n.* An optional academic course or subject. **—e·lec′tive·ly** *adv.* **—e·lec′tive·ness** *n.*
**e·lec·tor** (ĭ-lĕk′tər) *n.* **1.** A qualified voter. **2.** A member of the U.S. Electoral College. **3.** Any of the German princes in the Holy Roman Empire entitled to elect the emperor.
**e·lec·tor·al** (ĭ-lĕk′tər-əl) *adj.* **1.** Of, relating to, or made up of electors. **2.** Of or relating to election.
**Electoral College** *n.* A body of electors chosen to elect the U.S. President and Vice President.
**e·lec·tor·ate** (ĭ-lĕk′tər-ĭt) *n.* **1.** A body of qualified voters. **2.** The dignity or territory of a Holy Roman Empire elector.
**electr-** *pref. var. of* ELECTRO-.
**E·lec·tra** (ĭ-lĕk′trə) *n.* [Lat. < Gk. *Ēlektra.*] *Gk. Myth.* A daughter of Clytemnestra and Agamemnon who helped her brother Orestes avenge the murder of Agamemnon by killing their mother and her lover.
**Electra complex** *n. Psychoanal.* Unconscious libidinal feeling of a daughter toward her father.
**e·lec·tret** (ĭ-lĕk′trĭt) *n.* [ELECTR(ICITY) + (MAGN)ET.] A solid dielectric exhibiting persistent dielectric polarization.
**e·lec·tric** (ĭ-lĕk′trĭk) or **e·lec·tri·cal** (-trĭ-kəl) *adj.* [NLat. *electricus* < Lat. *electrum,* amber < Gk. *ēlektron.*] **1.** Of, relating to, or operated by electricity. **2. a.** Emotionally exciting : THRILLING. **b.** Exceptionally tense and highly charged with emotion <an atmosphere *electric* with hate> *—n.* An electrically powered machine, esp. a vehicle. **—e·lec′tri·cal·ly** *adv.*
**electrical engineering** *n.* The scientific technology of electricity, esp. the design and application of circuitry and equipment for power generation and distribution, machine control, and communications. **—electrical engineer** *n.*
**electrical storm** *n.* Thunderstorm.
**electric chair** *n.* **1.** A chair for electrocuting a prisoner sentenced to death by law. **2.** Execution by electrocution. **3.** The sentence of death by electrocution.
**electric displacement** *n.* The product of electric intensity and permittivity.
**electric eel** *n.* A long, eellike freshwater fish, *Electrophorus electricus* of northern South America, with organs capable of producing a powerful electric discharge.
**electric eye** *n.* A photoelectric cell.
**electric field** *n.* A region of space marked by the presence of a detectable electric intensity at every point.
**electric flux** *n.* The integral over a designated surface of the component of electric displacement normal to the surface.

**electric flux density** *n.* Electric displacement.

**electric furnace** *n.* A laboratory or industrial furnace heated by an electric arc, electric induction, or electric resistance.

**electric guitar** *n.* A guitar that transmits tones to an amplifier by an electronic pickup under the strings.

**e·lec·tri·cian** (ĭ-lĕk-trĭsh′ən, ē′lĕk-) *n.* One whose occupation is the installation, maintenance, repair, or operation of electric equipment and circuitry.

**electric intensity** *n.* The ratio of the electrostatic force exerted on a body to the charge on the body.

**e·lec·tric·i·ty** (ĭ-lĕk-trĭs′ə-tē, ē′lĕk-) *n.* **1. a.** The class of physical phenomena arising from the existence and interactions of electric charge. **b.** The physical science of such phenomena. **2.** Electric current used or regarded as a source of power. **3.** Intense emotional excitement.

▲ word history: The effects of electricity have been observed since ancient times, when it was noticed that amber, when rubbed, attracts small bits of straw, wood, and other light materials. When this electromagnetic effect was first scientifically studied in the early 17th century, the words *electric* and *electricity* were coined from Greek *ēlektron,* "amber." The English word *electron,* however, does not come directly from the Greek word; it is derived from *electric* and the suffix *-on,* meaning "subatomic particle."

**electric lamp** *n.* A lamp using electricity to produce light.

**electric light** *n.* **1.** An electric lamp. **2.** Light produced electrically.

**electric moment** *n.* The dipole moment of an electric dipole.

**electric motor** *n.* A motor powered by electricity.

**electric ray** *n.* Any of various fishes of the family Torpedinidae, with a rounded body and a pair of organs that produce a relatively strong electric discharge.

**e·lec·tri·fy** (ĭ-lĕk′trə-fī′) *vt.* **-fied, -fy·ing, -fies.** [ELECTRI(C) + -FY.] **1.** To produce electric charge on or in (a conductor). **2. a.** To wire or equip (e.g., a building) for the use of electric power. **b.** To provide with electric power. **3.** To thrill, shock, or startle greatly. **—e·lec′tri·fi′a·ble** *adj.* **—e·lec′tri·fi·ca′tion** *n.* **—e·lec′tri·fi′er** *n.*

**electro-** *or* **electr-** *pref.* [NLat. < Lat. *electrum,* amber < Gk. *ēlektron.*] **1. a.** Electricity <*electro*magnet> **b.** Electric : electrically <*electro*cute> **2.** Electrolysis <*electro*deposit> **3.** Electron <*electro*negative>

**e·lec·tro·a·cous·tics** (ĭ-lĕk′trō-ə-ko͞o′stĭks) *n.* (*sing. in number*). The science of the interaction or interconversion of electric and acoustic phenomena. **—e·lec′tro·a·cous′tic** *adj.* **—e·lec′tro·a·cous′tic·al·ly** *adv.*

**e·lec·tro·a·nal·y·sis** (ĭ-lĕk′trō-ə-năl′ĭ-sĭs) *n., pl.* **-ses** (-sēz′). Chemical analysis by means of electrolytic techniques. **—e·lec′tro·an′a·lyt′ic** (-trō-ăn′ə-lĭt′ĭk), **e·lec′tro·an·a·lyt′i·cal** *adj.*

**e·lec·tro·car·di·o·gram** (ĭ-lĕk′trō-kär′dē-ə-grăm′) *n.* The tracing produced by an electrocardiograph, used for diagnosing heart disease.

**e·lec·tro·car·di·o·graph** (ĭ-lĕk′trō-kär′dē-ə-grăf′) *n.* An instrument for recording electric potentials associated with the electric currents traversing the heart. **—e·lec′tro·car′di·o·graph′ic** (-grăf′ĭk) *adj.* **—e·lec′tro·car′di·o·graph′i·cal·ly** *adv.* **—e·lec′tro·car′di·og′ra·phy** (-kär′dē-ŏg′rə-fē) *n.*

**e·lec·tro·chem·is·try** (ĭ-lĕk′trō-kĕm′ĭs-trē) *n.* The science of the interaction or interconversion of electric and chemical phenomena. **—e·lec′tro·chem′i·cal** (-kĕm′ĭ-kəl) *adj.* **—e·lec′tro·chem′i·cal·ly** *adv.* **—e·lec′tro·chem′ist** *n.*

**e·lec·tro·co·ag·u·la·tion** (ĭ-lĕk′trō-kō-ăg′yə-lā′shən) *n. Med.* Use of a high frequency electric current to coagulate and destroy tissue.

**e·lec·tro·cute** (ĭ-lĕk′trə-kyo͞ot′) *vt.* **-cut·ed, -cut·ing, -cutes.** [ELECTRO- + (EXE)CUTE.] **1.** To kill with electricity. **2.** To execute (a condemned prisoner) by electricity. **—e·lec′tro·cu′tion** (-kyo͞o′shən) *n.*

**e·lec·trode** (ĭ-lĕk′trōd′) *n.* **1.** A solid electric conductor through which an electric current enters or leaves a medium, as an electrolyte, a nonmetallic solid, a molten metal, a gas, or a vacuum. **2.** A collector or emitter of electric charge or electric-charge carriers, as in a semiconducting device.

**e·lec·tro·de·pos·it** (ĭ-lĕk′trō-dĭ-pŏz′ĭt) *vt.* **-it·ed, -it·ing, -its.** To deposit (a suspended or dissolved substance) on an electrode by electrolysis. —*n.* The substance deposited on an electrode by electrolysis. **—e·lec′tro·dep′o·si′tion** (-dĕp′ə-zĭsh′ən, -dē′pə-zĭsh′ən) *n.*

**e·lec·tro·di·al·y·sis** (ĭ-lĕk′trō-dī-ăl′ĭ-sĭs) *n., pl.* **-ses** (-sēz′). Dialysis at a rate speeded by applying an electric potential across the dialysis membrane, used esp. for removing electrolytes from a colloid.

**e·lec·tro·dy·nam·ics** (ĭ-lĕk′trō-dī-năm′ĭks) *n.* (*sing. in number*). The physics of the relationship between electric, magnetic, and mechanical phenomena. **—e·lec′tro·dy·nam′ic** *adj.*

**e·lec·tro·dy·na·mom·e·ter** (ĭ-lĕk′trō-dī′nə-mŏm′ə-tər) *n.* An instrument using the interaction of the magnetic fields of fixed and moving sets of coils for measuring current, voltage, or power.

**e·lec·tro·en·ceph·a·lo·gram** (ĭ-lĕk′trō-ĕn-sĕf′ə-lə-grăm′) *n.* A graphic tracing of the electrical activity of the brain as recorded by an electroencephalograph.

**e·lec·tro·en·ceph·a·lo·graph** (ĭ-lĕk′trō-ĕn-sĕf′ə-lə-grăf′) *n.* A device that detects and records the electrical activity of the brain. **—e·lec′tro·en·ceph′a·lo·graph′ic** *adj.* **—e·lec′tro·en·ceph′a·log′ra·phy** (-lŏg′rə-fē) *n.*

**e·lec·tro·form** (ĭ-lĕk′trə-fôrm′) *vt.* **-formed, -form·ing, -forms.** To produce or reproduce by electrodeposition in a mold.

**e·lec·tro·gas·dy·nam·ics** (ĭ-lĕk′trə-găs′dī-năm′ĭks) *n.* (*sing. in number*). Generation of electrical energy based on the conversion of the kinetic energy contained in a high-pressure, ionized, moving combustion gas. **—e·lec′tro·gas′dy·nam′ic** *adj.*

**e·lec·tro·graph** (ĭ-lĕk′trə-grăf′) *n.* **1.** An electrically produced graph or tracing. **2.** Equipment for producing electrographs in facsimile transmission.

**e·lec·tro·hy·drau·lic** (ĭ-lĕk′trō-hī-drô′lĭk) *adj.* Of, pertaining to, or involving a combination of electric and hydraulic mechanisms. **—e·lec′tro·hy·drau′lic·al·ly** *adv.*

**e·lec·tro·kin·et·ics** (ĭ-lĕk′trō-kĭ-nĕt′ĭks) *n.* (*sing. in number*). The electrodynamics of heating effects and of current distribution in electric networks.

**e·lec·trol·o·gist** (ĭ-lĕk-trŏl′ə-jĭst, ē′lĕk-) *n.* One who removes esp. body hair by means of an electric current.

**e·lec·tro·lu·mi·nes·cence** (ĭ-lĕk′trō-lo͞o′mə-nĕs′əns) *n.* **1.** Direct conversion of electric energy to light by a solid phosphor subjected to an alternating electric field. **2.** Emission of light caused by electric discharge in a gas. **—e·lec′tro·lu′mi·nes′cent** *adj.*

**e·lec·trol·y·sis** (ĭ-lĕk-trŏl′ə-sĭs, ē′lĕk-) *n.* **1.** Chemical change, esp. decomposition, produced in an electrolyte by an electric current. **2.** Destruction of living tissue, esp. hair roots, by an electric current.

**e·lec·tro·lyte** (ĭ-lĕk′trə-līt′) *n.* A substance that dissociates into ions in solution or when fused, thereby becoming an electrical conductor.

**e·lec·tro·lyt·ic** (ĭ-lĕk′trə-lĭt′ĭk) *adj.* **1. a.** Of or relating to electrolysis. **b.** Produced by electrolysis. **2.** Of or relating to an electrolyte. **—e·lec′tro·lyt′i·cal·ly** *adv.*

**electrolytic cell** *n.* **1.** A cell containing an electrolyte through which a system of electrodes passes an externally generated electric current to create an electrochemical reaction. **2.** A cell containing an electrolyte in which an electrochemical reaction creates an electromotive force.

**e·lec·tro·lyze** (ĭ-lĕk′trə-līz′) *vt.* **-lyzed, -lyz·ing, -lyz·es.** To decompose by electrolysis.

**e·lec·tro·mag·net** (ĭ-lĕk′trō-măg′nĭt) *n.* A magnet made of a soft-iron core wound with a current-carrying coil of insulated wire, the current in which magnetizes the core.

**e·lec·tro·mag·net·ic** (ĭ-lĕk′trō-măg-nĕt′ĭk) *adj.* Of or caused by electromagnetism. **—e·lec′tro·mag·net′i·cal·ly** *adv.*

**electromagnetic field** *n.* The field of force associated with electric charge in motion, with both electric and magnetic components and containing a specific amount of electromagnetic energy.

**electromagnetic spectrum** *n.* The total range of radiation extending in frequency approx. from $10^{23}$ cycles per second to 0 cycles per second or, in corresponding wavelengths, from $10^{-13}$ centimeter to infinity and including cosmic-ray photons, gamma rays, x-rays, ultraviolet radiation, visible light, infrared radiation, microwaves, radio waves, heat, and electric currents.

**electromagnetic unit** *n.* Any of a system of units for electricity and magnetism based on a system of equations in which the permeability of free space is taken as unity and by means of which the abampere is defined as the fundamental unit of current.

**electromagnetic wave** *n.* A wave propagating as a periodic disturbance of the electromagnetic field and with a frequency in the electromagnetic spectrum.

**e·lec·tro·mag·net·ism** (ĭ-lĕk′trō-măg′nə-tĭz′əm) *n.* **1.** Magnetism arising from electric charge in motion. **2.** The physics of electricity and magnetism.

**e·lec·tro·met·al·lur·gy** (ĭ-lĕk′trō-mĕt′l-ûr′jē) *n.* Use of electricity for purifying metals or reducing metallic compounds to metals. **—e·lec′tro·met′al·lur′gi·cal** *adj.*

**e·lec·trom·e·ter** (ĭ-lĕk′trŏm′ĭ-tər, ē′lĕk-) *n.* An instrument for detecting or measuring potential differences, electric charge, or, indirectly, electric current by means of mechanical forces exerted between electrically charged bodies.

**e·lec·tro·mo·tive** (ĭ-lĕk′trō-mō′tĭv) *adj.* Of, relating to, or producing electric current.

**electromotive force** *n.* The energy per unit charge that is converted reversibly from chemical, mechanical, or other forms of energy into electrical energy in a conversion device such as a dynamo or battery.

**e·lec·tro·my·o·gram** (ē-lĕk′trō-mī′ō-grăm′) *n.* A graphic recording of a muscle response as a result of electrical stimulation.

**e·lec·tro·my·og·ra·phy** (ĭ-lĕk′trō-mī-ŏg′rə-fē) *n.* Preparation and study of electromyograms. **—e·lec′tro·my′o·graph′ic** (-mī′ə-grăf′ĭk) *adj.* **—e·lec′tro·my′o·graph′ic·al·ly** *adv.*

**e·lec·tron** (ĭ-lĕk′trŏn′) *n.* A subatomic particle in the lepton family having a rest mass of $9.1066 \times 10^{-28}$ gram and a unit negative electric charge of approx. $1.602 \times 10^{-19}$ coulomb.

**e·lec·tro·neg·a·tive** (ĭ-lĕk′trō-nĕg′ə-tĭv) *adj.* **1.** Having a negative electric charge. **2.** Tending to attract electrons to form a chemical bond.

**electron gun** *n.* An electron-emitting electrode and associated elements, esp. in a cathode-ray tube, that produce a beam of accelerated electrons.

**e·lec·tron·ic** (ĭ-lĕk-trŏn′ĭk, ē′lĕk-) *adj.* **1.** Of or relating to electrons. **2.** Of, relating to, based on, operated by, or otherwise involving the controlled conduction of electrons or other charge carriers, esp. in a vacuum, gas, or semiconducting material. **3.** Of or relating to electronics. **—e·lec′tron′i·cal·ly** *adv.*

**electronic flash** *n.* A strobe light.

**electronic music** *n.* Music produced by electronic devices.

**e·lec·tron·ics** (ĭ-lĕk′trŏn′ĭks, ē′lĕk-) *n.* *(sing. in number).* **1.** The science and technology of electronic phenomena. **2.** The commercial industry of electronic devices and systems.

**electronic stylus** *n.* A penlike input device that signals the computer by means of an electronic pulse and is commonly used in conjunction with a cathode-ray tube.

**electron lens** *n.* Any of various devices using an electric or a magnetic field to focus a beam of electrons.

**electron micrograph** *n.* A micrograph made by an electron microscope.

**electron microscope** *n.* Any of a class of microscopes using electrons rather than visible light to produce magnified images, esp. of objects with dimensions smaller than the wavelengths of visible light, with linear magnification up to or exceeding a million ($10^6$).

**electron multiplier** *n.* A vacuum tube in which a single electron produces a large number of secondary electrons by collision with an anode, the process gen. being repeated through a number of stages to achieve great amplification.

**electron optics** *n.* *(sing. in number).* The science of the control of electron motion by electron lenses in systems or under conditions analogous to those involving or affecting visible light.

**electron pair** *n.* **1.** Two electrons functioning or considered as functioning in concert, esp. two electrons shared by two atoms joined by a covalent chemical bond. **2.** The combination of an electron and a positron as produced by a high-energy photon.

**electron tube** *n.* A sealed enclosure, either highly evacuated or containing a controlled quantity of gas, in which electrons can be made sufficiently mobile to act as the principal carriers of current between at least one pair of electrodes, often under the control of one or more additional electrodes.

**electron volt** *n.* A unit of energy equal to the energy acquired by an electron falling through a potential difference of one volt, approx. $1.602 \times 10^{-19}$ joule.

**e·lec·tro·pho·rese** (ĭ-lĕk′trō-fə-rēs′) *vt.* **-resed, -res·ing, -res·es.** [Back-formation < ELECTROPHORESIS.] To subject to electrophoresis.

**e·lec·tro·pho·re·sis** (ĭ-lĕk′trō-fə-rē′sĭs) *n.* The motion of charged particles, esp. colloidal particles, through a relatively stationary liquid under the influence of an applied electric field provided, in general, by immersed electrodes.

**e·lec·tro·pho·ret·o·gram** (ĭ-lĕk′trō-fə-rē′tə-grăm′) *n.* [ELECTROPHORET(IC) + -GRAM.] A record of the separated components of a mixture produced by electrophoresis.

**e·lec·troph·o·rus** (ĭ-lĕk′trŏf′ər-əs, ē′lĕk-) *n., pl.* **-o·ri** (-ə-rī′) [NLat. : ELECTRO- + Gk. *phoros,* bearer < *pherein,* to bear.] A device for generating static electricity, having a disk given a negative charge by friction and a metal plate charged by induction when in contact with the disk.

**e·lec·tro·plate** (ĭ-lĕk′trə-plāt′) *vt.* **-plat·ed, -plat·ing, -plates.** To cover or coat with a thin layer of metal by electrodeposition.

**e·lec·tro·pos·i·tive** (ĭ-lĕk′trō-pŏz′ĭ-tĭv) *adj.* **1.** Having a positive electric charge. **2.** Tending to release electrons to form a chemical bond.

**e·lec·tro·scope** (ĭ-lĕk′trə-skōp′) *n.* An instrument for detecting the presence, sign, and in some configurations the magnitude of an electric charge by the mutual attraction or repulsion of metal foils or pith balls. **—e·lec′tro·scop′ic** (-skŏp′ĭk) *adj.*

**e·lec·tro·shock** (ĭ-lĕk′trō-shŏk′) *n.* Shock therapy in which an electric current is passed through the brain.

**e·lec·tro·stat·ic** (ĭ-lĕk′trō-stăt′ĭk) *adj.* **1. a.** Of or relating to stationary electric charges. **b.** Produced or caused by such charges. **2.** Of or relating to electrostatics. **—e·lec′tro·stat′i·cal·ly** *adv.*

**electrostatic generator** *n.* Any of various devices, esp. the Van de Graaff generator, that generate high voltages by accumulating large quantities of electric charge.

**electrostatic precipitation** *n.* Removal of particles suspended in a gas by electrostatic charging and subsequent precipitation onto a collector in a strong electric field.

**electrostatic printing** *n.* A printing or copying process that uses electrostatic forces to form the image in powder or ink directly on the surface to be printed.

**e·lec·tro·stat·ics** (ĭ-lĕk′trō-stăt′ĭks) *n.* *(sing. in number).* The physics of electrostatic phenomena.

**electrostatic unit** *n.* Any of a system of units for electricity and magnetism based on a system of equations in which the permittivity of empty space is defined as unity and by means of which a fundamental unit of charge is defined.

**e·lec·tro·ther·a·peu·tics** (ĭ-lĕk′trō-thĕr′ə-pyōō′tĭks) *n.* *(sing. in number).* Electrotherapy.

**e·lec·tro·ther·a·py** (ĭ-lĕk′trō-thĕr′ə-pē) *n.* Medical therapy, as diathermy, using electric currents.

**e·lec·tro·ther·mal** (ĭ-lĕk′trō-thûr′məl) *adj.* **1.** Of, relating to, or involving both electricity and heat. **2.** Of or relating to the production of heat by electricity. **—e·lec′tro·ther′mal·ly** *adv.*

**e·lec·trot·o·nus** (ĭ-lĕk′trŏt′ə-nəs, ē′lĕk-) *n.* Alteration in sensitivity of a nerve caused by the passage of an electric current through any part of it. **—e·lec′tro·ton′ic** (-trə-tŏn′ĭk) *adj.*

**e·lec·tro·type** (ĭ-lĕk′trə-tīp′) *n.* **1.** A duplicate metal plate used in letterpress printing, made by electroplating a lead or plastic mold of the original plate. **2.** The process of making an electrotype. **—e·lec′tro·type′** *v.* **(-typed, -typ·ing, -types). —e·lec′tro·typ′er** *n.* **—e·lec′tro·typ′ic** (-trō-tĭp′ĭk) *adj.*

**e·lec·tro·va·lence** (ĭ-lĕk′trō-vā′ləns) *also* **e·lec·tro·va·len·cy** (-lən-sē) *n.* **1.** Valence marked by the transfer of electrons from atoms of one element to atoms of another. **2.** The number of electric charges lost or gained by an atom in such a transfer. **—e·lec′tro·va′lent** *adj.*

**electrovalent bond** *n. Chem.* An ionic bond.

**e·lec·trum** (ĭ-lĕk′trəm) *n.* [ME *electrum* < Lat., amber < Gk. *ēlektron.*] An alloy of silver and gold.

**e·lec·tu·ar·y** (ĭ-lĕk′chōō-ĕr′ē) *n., pl.* **-ies.** [ME *electuarie* < LLat. *electuarium,* prob. alteration of Gk. *ekleikton* < *ekleikhein,* to lick up : *ek-,* out + *leikhein,* to lick.] A drug mixed with sugar and water or honey into a pasty mass suitable for oral administration.

**el·ee·mos·y·nar·y** (ĕl′ə-mŏs′ə-nĕr′ē, ĕl′ē-ə-) *adj.* [Med. Lat. *eleemosynarius* < LLat. *eleemosyna,* alms. —see ALMS.] **1.** Of or relating to alms or the giving of alms. **2.** Depending on or supported by alms. **3.** Contributed as an act of charity : GRATUITOUS.

**el·e·gance** (ĕl′ĭ-gəns) *also* **el·e·gan·cy** (-gən-sē) *n.* **1. a.** Grace and refinement in appearance, movement, or manners. **b.** Tasteful opulence in form, decoration, or presentation. **2 a.** Grace and restraint of style. **b.** Scientific precision and exactness. **3.** Something elegant or luxurious.

**el·e·gant** (ĕl′ĭ-gənt) *adj.* [OFr. < Lat. *elegans.*] Marked by or displaying elegance. **—el′e·gant·ly** *adv.*

**el·e·gi·ac** (ĕl′ə-jī′ək, ĭ-lē′jē-ăk′) *adj.* [LLat. *elegiacus* < Gk. *elegeiakos* < *elegeia,* elegy.] **1. a.** Relating to an elegy. **b.** Expressing sorrow : MOURNFUL. **2.** Of, relating to, or written in couplets whose first line is a dactylic hexameter and second a pentameter. **—el′e·gi′ac** *n.* **—el′e·gi′a·cal** *adj.* **—el′e·gi′ac·al·ly** *adv.*

**e·le·git** (ĭ-lē′jĭt) *n.* [Lat. *elegit,* he has chosen, the first word of a phrase frequently used in the writ.] *Law.* A writ of execution against a debtor by which the debtor's goods or property are delivered to the plaintiff until the debtor can settle the debt.

**el·e·gize** (ĕl′ə-jīz′) *v.* **-gized, -giz·ing, -giz·es.** —*vi.* To compose an elegy. —*vt.* To compose an elegy on or for.

**el·e·gy** (ĕl′ə-jē) *n., pl.* **-gies.** [Fr. *élégie* < Lat. *elegia* < Gk. *elegeia* < *elegos,* mournful song.] **1.** A poem in elegiac couplets. **2.** A poem or song written esp. as a lament for one who is dead. **3.** A melancholy or pensive musical composition or poem.

**el·e·ment** (ĕl′ə-mənt) *n.* [ME < OFr. < Lat. *elementum.*] **1.** A fundamental, irreducible constituent. **2. elements.** The basic principles or assumptions of a subject. **3.** *Math.* **a.** A member of a set. **b.** A point, line, or plane. **c.** A part of a geometric configuration, as an angle in a triangle. **d.** The generatrix of a geometric figure. **e.** Any of the terms in the rectangular array of terms that constitute a matrix or determinant. **4.** *Chem. & Physics.* A substance made up of atoms bearing an identical number of protons in each nucleus. **5.** One of four substances, earth, air, fire, or water, once held to be a basic constituent of the universe. **6. elements.** The forces that constitute the weather, esp. severe or inclement weather. **7.** An environment naturally appropriate to or associated with an individual. **8. elements.** The bread and wine of the Eucharist.

☆ ***syns:*** **1.** ELEMENT, BASIC, ESSENTIAL, FUNDAMENTAL, RUDIMENT *n. core meaning* : an irreducible constituent of a whole <the *elements* of physics> **2.** ELEMENT, COMPONENT, CONSTITUENT, FACTOR, INGREDIENT, PART *n. core meaning* : one of the individual entities contributing to a whole <intelligence and ambition—two key *elements* to success>

**el·e·men·tal** (ĕl′ə-mĕn′tl) *adj.* **1.** Of, relating to, or constituting an element. **2. a.** Fundamental or essential : INDISPENSABLE. **b.** Of or relating to the fundamentals : ELEMENTARY. **c.** Belonging to inherently. **3.** Being like a force of nature in power or effect. **—el′e·men′tal** *n.* **—el′e·men′tal·ly** *adv.*

**el·e·men·ta·ry** (ĕl′ə-mĕn′tə-rē, -trē) *adj.* **1.** Fundamental, essential, or irreducible. **2.** Of, involving, or introducing the fundamental or simplest aspects of a subject <an *elementary* arithmetic textbook> **3.** Of or pertaining to an elementary school <the *elementary* class

# PERIODIC TABLE: ELEMENTS

KEY

Atomic Number — 1; Symbol — H; Hydrogen 1.00797 — Atomic Weight (or Mass Number of most stable isotope if in parentheses)

| 1a | 2a | 3b | 4b | 5b | 6b | 7b | 8 | 8 | 8 | 1b | 2b | 3a | 4a | 5a | 6a | 7a | 0 |
|---|---|---|---|---|---|---|---|---|---|---|---|---|---|---|---|---|---|
| 1<br>**H**<br>Hydrogen<br>1.00797 | | | | | | | | | | | | | | | | | 2<br>**He**<br>Helium<br>4.0026 |
| 3<br>**Li**<br>Lithium<br>6.939 | 4<br>**Be**<br>Beryllium<br>9.0122 | | | | | | | | | | | 5<br>**B**<br>Boron<br>10.811 | 6<br>**C**<br>Carbon<br>12.01115 | 7<br>**N**<br>Nitrogen<br>14.0067 | 8<br>**O**<br>Oxygen<br>15.9994 | 9<br>**F**<br>Fluorine<br>18.9984 | 10<br>**Ne**<br>Neon<br>20.183 |
| 11<br>**Na**<br>Sodium<br>22.9898 | 12<br>**Mg**<br>Magnesium<br>24.312 | | | | | | | | | | | 13<br>**Al**<br>Aluminum<br>26.9815 | 14<br>**Si**<br>Silicon<br>28.086 | 15<br>**P**<br>Phosphorus<br>30.9738 | 16<br>**S**<br>Sulfur<br>32.064 | 17<br>**Cl**<br>Chlorine<br>35.453 | 18<br>**Ar**<br>Argon<br>39.948 |
| 19<br>**K**<br>Potassium<br>39.102 | 20<br>**Ca**<br>Calcium<br>40.08 | 21<br>**Sc**<br>Scandium<br>44.956 | 22<br>**Ti**<br>Titanium<br>47.90 | 23<br>**V**<br>Vanadium<br>50.942 | 24<br>**Cr**<br>Chromium<br>51.996 | 25<br>**Mn**<br>Mangenese<br>54.9380 | 26<br>**Fe**<br>Iron<br>55.847 | 27<br>**Co**<br>Cobalt<br>58.9332 | 28<br>**Ni**<br>Nickel<br>58.71 | 29<br>**Cu**<br>Copper<br>63.546 | 30<br>**Zn**<br>Zinc<br>65.37 | 31<br>**Ga**<br>Gallium<br>69.72 | 32<br>**Ge**<br>Germanium<br>72.59 | 33<br>**As**<br>Arsenic<br>74.9216 | 34<br>**Se**<br>Selenium<br>78.96 | 35<br>**Br**<br>Bromine<br>79.904 | 36<br>**Kr**<br>Krypton<br>83.80 |
| 37<br>**Rb**<br>Rubidium<br>85.47 | 38<br>**Sr**<br>Strontium<br>87.62 | 39<br>**Y**<br>Yttrium<br>88.905 | 40<br>**Zr**<br>Zirconium<br>91.22 | 41<br>**Nb**<br>Niobium<br>92.906 | 42<br>**Mo**<br>Molybdenum<br>95.94 | 43<br>**Tc**<br>Technetium<br>(97) | 44<br>**Ru**<br>Ruthenium<br>101.07 | 45<br>**Rh**<br>Rhodium<br>102.905 | 46<br>**Pd**<br>Palladium<br>106.4 | 47<br>**Ag**<br>Silver<br>107.868 | 48<br>**Cd**<br>Cadmium<br>112.40 | 49<br>**In**<br>Indium<br>114.82 | 50<br>**Sn**<br>Tin<br>118.69 | 51<br>**Sb**<br>Antimony<br>121.75 | 52<br>**Te**<br>Tellurium<br>127.60 | 53<br>**I**<br>Iodine<br>126.9044 | 54<br>**Xe**<br>Xenon<br>131.30 |
| 55<br>**Cs**<br>Cesium<br>132.905 | 56<br>**Ba**<br>Barium<br>137.34 | 51-71*<br>Lanthanides | 72<br>**Hf**<br>Hafnium<br>178.49 | 73<br>**Ta**<br>Tantalum<br>180.948 | 74<br>**W**<br>Tungsten<br>183.85 | 75<br>**Re**<br>Rhenium<br>186.2 | 76<br>**Os**<br>Osmium<br>190.2 | 77<br>**Ir**<br>Iridium<br>192.2 | 78<br>**Pt**<br>Platinum<br>195.09 | 79<br>**Au**<br>Gold<br>196.967 | 80<br>**Hg**<br>Mercury<br>200.59 | 81<br>**Tl**<br>Thallium<br>204.37 | 82<br>**Pb**<br>Lead<br>207.19 | 83<br>**Bi**<br>Bismuth<br>208.980 | 84<br>**Po**<br>Polonium<br>(210) | 85<br>**At**<br>Astatine<br>(210) | 86<br>**Rn**<br>Radon<br>(222) |
| 87<br>**Fr**<br>Francium<br>(223) | 88<br>**Ra**<br>Radium<br>(226) | 89-103**<br>**Actinides | | | | | | | | | | | | | | | |

| | | | | | | | | | | | | | | | |
|---|---|---|---|---|---|---|---|---|---|---|---|---|---|---|---|
| *Lanthanides | 57<br>**La**<br>Lanthanum<br>138.91 | 58<br>**Ce**<br>Cerium<br>140.12 | 59<br>**Pr**<br>Praseodymium<br>140.907 | 60<br>**Nd**<br>Neodymium<br>144.24 | 61<br>**Pm**<br>Promethium<br>(145) | 62<br>**Sm**<br>Samarium<br>150.35 | 63<br>**Eu**<br>Europium<br>151.96 | 64<br>**Gd**<br>Gadolinium<br>157.25 | 65<br>**Tb**<br>Terbium<br>158.924 | 66<br>**Dy**<br>Dysprosium<br>162.50 | 67<br>**Ho**<br>Holmium<br>164.930 | 68<br>**Er**<br>Erbium<br>167.26 | 69<br>**Tm**<br>Thulium<br>168.934 | 70<br>**Yb**<br>Ytterbium<br>173.04 | 71<br>**Lu**<br>Lutetium<br>174.97 |
| **Actinides | 89<br>**Ac**<br>Actinium<br>(227) | 90<br>**Th**<br>Thorium<br>232.038 | 91<br>**Pa**<br>Protactinium<br>(231) | 92<br>**U**<br>Uranium<br>238.03 | 93<br>**Np**<br>Neptunium<br>(237) | 94<br>**Pu**<br>Plutonium<br>(244) | 95<br>**Am**<br>Americium<br>(243) | 96<br>**Cm**<br>Curium<br>(247) | 97<br>**Bk**<br>Berkelium<br>(247) | 98<br>**Cf**<br>Californium<br>(251) | 99<br>**Es**<br>Einsteinium<br>(254) | 100<br>**Fm**<br>Fermium<br>(257) | 101<br>**Md**<br>Mendelevium<br>(256) | 102<br>**No**<br>Nobelium<br>(254) | 103<br>**Lw**<br>Lawrencium<br>(257) |

schedules> **—el′e·men·ta′ri·ly** (-tĕr′ə-lē) *adv.* **—el′e·men′ta·ri·ness** *n.*

**elementary particle** *n.* A subatomic particle hypothesized or regarded as an irreducible constituent of matter.

**elementary school** *n.* **1.** The first six to eight years of formal education. **2.** A school usu. for the first six or eight grades.

**el·e·mi** (el′ə-mē) *n., pl.* **-mis.** [Sp. *elemí* < Ar. *elemī,* var. of *allāmi,* the elemi.] Any of various oily resins derived from certain tropical trees, esp. *Canarium luzonicum* of the Philippines, used for making inks and varnishes.

**el·e·phant** (ĕl′ə-fənt) *n.* [ME *elefaunt* < OFr. *olifant* < VLat. **olifantus* < Lat. *elephantus* < Gk. *elephas.*] Either of two very large herbivorous mammals, *Elephas maximus* of south-central Asia or *Loxodonta africana* of Africa, with thick, nearly hairless skin, a long, flexible, prehensile trunk, upper incisors forming long, curved tusks, and, in the African species, large fan-shaped ears.

**el·e·phan·ti·a·sis** (ĕl′ə-fən-tī′ə-sĭs) *n.* [Lat. < Gk.: *elephas,* elephant + *-iasis,* -iasis.] A chronic, often extreme enlargement and hardening of the cutaneous and subcutaneous tissue, esp. of the legs and the scrotum, resulting from lymphatic obstruction and usu. caused by a nematode worm, *Wuchereria bancrofti.*

**el·e·phan·tine** (ĕl′ə-făn′tēn, -tīn′, ĕl′ə-fən-) *adj.* **1.** Of or relating to an elephant. **2 a.** Enormous in size or strength : GIGANTIC. **b.** Heavy-footed : ponderous.

**elephant seal** *n.* Either of two large seals, *Mirounga angustirostris* or *M. leonina* of Pacific coastal waters of North and South America, with a trunklike proboscis.

**el·e·vate** (ĕl′ə-vāt′) *vt.* **-vat·ed, -vat·ing, -vates.** [ME *elevaten* < Lat. *elevare* : *ex-,* up + *levare,* to raise.] **1.** To raise to a higher position or place. **2.** To increase the amplitude, intensity, or volume of. **3.** To promote to a higher rank. **4.** To raise to a higher moral, cultural, or intellectual level. **5.** To lift the spirits of : ELATE.

☆ ***syns:*** ELEVATE, BOOST, HOIST, LIFT, PICK UP, RAISE, UPLIFT *v. core meaning* : to move (something) to a higher position <Cranes *elevated* the crates onto the ship.> ***ant*** : lower

**el·e·vat·ed** (ĕl′ə-vā′tĭd) *adj.* **1.** Raised above a given level. **2. a.** Morally or intellectually superior. **b.** Lofty : exalted <*elevated* speech> **3.** Elated : high-spirited. *—n. Informal.* An elevated railway.

**elevated railway** *n.* A railway operating on a raised structure in order to allow vehicles or pedestrians to pass beneath it.

**el·e·va·tion** (ĕl′ə-vā′shən) *n.* **1. a.** The act or an instance of elevating. **b.** The state of being elevated. **2.** An elevated place or position. **3.** The height to which something is elevated above a point of reference, as the ground. **4.** Loftiness of feeling or thought. **5.** A scale drawing of the side, front, or rear of a particular structure. **6.** Altitude. **7. a.** A leap, as by a dancer, in which the performer appears to be suspended. **b.** A performer's ability to execute an elevation.

**el·e·va·tor** (ĕl′ə-vā′tər) *n.* **1. a.** A platform or enclosure raised and lowered in a vertical shaft to carry people or freight. **b.** The enclosure or platform with its operating equipment, motor, cables, and accessories. **2.** A mechanism, often with scoops or buckets attached to a conveyor, for hoisting materials. **3.** A granary equipped with devices for hoisting and discharging grain. **4.** A movable control surface, usu. attached to the horizontal stabilizer of an aircraft, for producing up or down motion.

**e·lev·en** (ĭ-lĕv′ən) *n.* [ME *elleven* < OE *endleofan.*] **1.** The cardinal number equal to 10 + 1. **2.** The 11th in a set or sequence. **3.** Something having 11 parts, units, or members, esp. a football team. **—e·lev′en** *adj.* & *pron.*

▲ word history: The system of counting used by the prehistoric Germanic peoples was apparently based on ten. The words *eleven* and *twelve* attest to this. The Germanic ancestor of *eleven* can be reconstructed as *ainlif-,* a compound of *ainaz,* "one," and *lif-,* a form derived from an Indo-European root meaning "to leave." *Eleven* therefore ultimately means "one left over after counting to ten."

**e·lev·ens·es** (ĭ-lĕv′ən-zəs) *pl.n. Chiefly Brit.* Midmorning tea or coffee often accompanied by a snack.

**e·lev·enth** (ĭ-lĕv′ənth) *n.* **1.** The ordinal number matching the number 11 in a series. **2.** One of 11 equal parts. **—e·lev′enth** *adj.* & *adv.*

**eleventh hour** *n.* The latest possible time.

**el·e·von** (ĕl′ə-vŏn′) *n.* [ELEV(ATOR) + (AILER)ON.] An airplane control surface combining elevator and aileron functions.

**elf** (ĕlf) *n., pl.* **elves** (ĕlvz) [ME < OE *ælf.*] **1.** A small, often mischievous, imaginary creature believed to possess magical powers. **2.** A mischievous child. **3.** A dwarf.

**elf·in** (ĕl′fĭn) *adj.* [Prob. < ME *elvene,* genitive pl. of *elf,* elf.] **1. a.** Of, relating to, or like an elf. **b.** Made, done, or produced by an elf. **2. a.** Small and sprightly : MISCHIEVOUS. **b.** Magical : fairylike.

**elf·ish** (ĕl′fĭsh) *also* **elv·ish** (ĕl′vĭsh) *adj.* **1.** Of or relating to elves. **2.** Mischievous : prankish. **—elf′ish·ly** *adv.* **—elf′ish·ness** *n.*

**elf·lock** (ĕlf′lŏk′) *n.* A tangled lock of hair.

**el·hi** (ĕl′hī′) *adj.* [EL(EMENTARY) + HI(GH SCHOOL).] Of, relating to, involving, or intended for use in elementary school and high school.

**e·lic·it** (ĭ-lĭs′ĭt) *vt.* **-it·ed, -it·ing, -its.** [Lat. *elicere, elicit-* : *ex-,* out

\+ *lacere,* to entice.] **1.** To bring out : EVOKE. **2.** To call forth (e.g., a reaction). **3.** To educe. **—e·lic'i·ta'tion** *n.* **—e·lic'i·tor** *n.*

**e·lide** (ĭ-līd') *vt.* **e·lid·ed, e·lid·ing, e·lides.** [Lat. *elidere,* to strike out : *ex-,* out + *laedere,* to strike.] **1.** To omit or slur over (e.g., a syllable) in pronunciation. **2.** To leave out or suppress.

**el·i·gi·ble** (ĕl'ĭ-jə-bəl) *adj.* [ME < OFr. < VLat. **eligibilis* < Lat. *eligere,* to select. —see ELECT.] **1.** Qualified, as for an office or position. **2.** Desirable and worthy of choice, esp. for marriage. **—el'i·gi·bil'i·ty** *n.* **—el'i·gi·ble** *n.* **—el'i·gi·bly** *adv.*

**e·lim·i·nate** (ĭ-lĭm'ə-nāt') *vt.* **-nat·ed, -nat·ing, -nates.** [Lat. *eliminare, eliminat-,* to banish : *ex-,* out + *limen,* threshold.] **1. a.** To get rid of : REMOVE <*eliminated* overtime for all workers> **b.** To get rid of by banishment or execution. **2. a.** To omit from consideration : REJECT. **b.** To remove from consideration by defeating, as in a contest. **3.** To remove by combining equations. **4.** *Physiol.* To excrete as waste. **—e·lim'i·na'tion** *n.* **—e·lim'i·na'tive, e·lim'i·na·to'ry** (-nə-tôr'ē, -tōr'ē) *adj.* **—e·lim'i·na'tor** *n.*

☆ ***syns:*** ELIMINATE, ERADICATE, LIQUIDATE, PURGE, REMOVE, WIPE OUT *v. core meaning* : to get rid of, esp. by banishment or execution <a dictator who had *eliminated* all dissidents>

**e·li·sion** (ĭ-lĭzh'ən) *n.* [Lat. *elisio, elision-* < *elidere,* to strike out. —see ELIDE.] **1.** The act of eliding. **2.** Omission of an unstressed syllable or vowel, as in scanning a verse. **3.** An omission.

**e·lite** or **é·lite** (ĭ-lēt', ā-lēt') *n.* [Fr. *élite* < OFr. *eslite,* < fem. p.part. of *eslire,* to choose < Lat. *eligere.* —see ELECT.] **1. a.** The best or superior members of a society or group. **b.** A small, privileged, and often powerful group. **2.** A size of type on a typewriter, equal to 12 characters per inch. **—e·lite'** *adj.*

**e·lit·ism** or **é·lit·ism** (ĭ-lē'tĭz'əm, ā-lē'-) *n.* **1. a.** Belief in rule by an elite. **b.** Domination or rule by an elite. **2.** A sense of being part of an elite. **—e·lit'ist** *adj.* & *n.*

**e·lix·ir** (ĭ-lĭk'sər) *n.* [ME, a substance of transmutative properties < Med. Lat. < Ar. *al-iksīr* : *al,* the + *iksīr,* elixir, prob. < Gk. *xērion,* desiccative powder < *xēros,* dry.] **1.** A sweetened aromatic solution of alcohol and water, used as a vehicle for medicine. **2.** A medicine regarded as a cure for all ills. **3.** The philosophers' stone. **4.** The quintessence or underlying principle.

**E·liz·a·be·than** (ĭ-lĭz'ə-bē'thən, -bĕth'ən) *adj.* Of, relating to, or typical of the reign of Elizabeth I of England.

**Elizabethan sonnet** *n.* A Shakespearean sonnet.

**elk** (ĕlk) *n., pl.* **elks** *or* **elk.** [ME, prob. < OE *eolh.*] **1.** The wapiti. **2.** A large deer, *Alces alces* of northern regions, with palmate antlers. **3.** A light, pliant leather of calfskin or horsehide, tanned and finished to resemble elk hide.

**elk·hound** (ĕlk'hound') *n.* A hunting dog orig. bred in Scandinavia, with a grayish coat and a tail curled up over the back.

**ell¹** (ĕl) *n.* [From its resemblance to the shape of the capital letter L.] A wing of a building at right angles to the main structure.

**ell²** (ĕl) *n.* [ME < OE *eln,* the length from elbow to the middle finger's tip.] An English linear measure equal to 114 centimeters or 45 inches.

**ell³** (ĕl) *n. var. of* EL¹.

**el·lag·ic acid** (ĭ-lăj'ĭk) *n.* [Fr. *ellagique* < *ellag,* backward spelling of *galle,* plant gall < Lat. *galla.*] A yellow crystalline compound, $C_{14}H_6O_8$, derived from tannins.

**el·lipse** (ĭ-lĭps') *n.* [Gk. *elleipsis,* ellipse, ellipsis.] **1.** A plane curve, esp.: **a.** A conic section taken neither parallel to an element nor parallel to the axis of the intersected cone. **b.** The locus of points the sum of the distances of each of which from two fixed points is the same constant. **2.** Ellipsis.

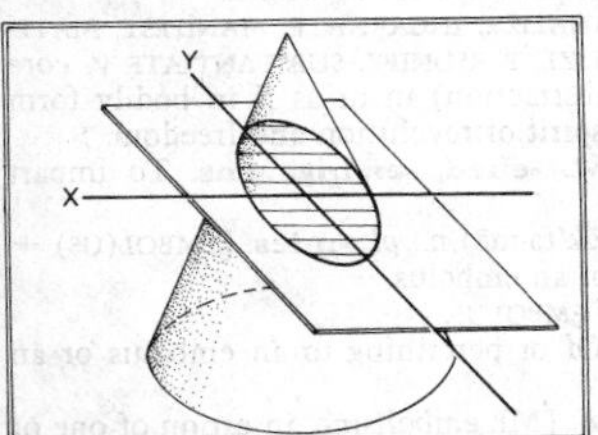

ellipse

**el·lip·sis** (ĭ-lĭp'sĭs) *n., pl.* **-ses** (-sēz') [Lat. *ellipsis* < Gk. *elleipsis* < *elleipein,* to fall short : *en-,* in + *leipein,* to leave.] **1. a.** The omission of a word or phrase required for a complete syntactical construction but not necessary for understanding. **b.** An example of ellipsis. **2.** A mark or series of marks (e.g., . . . or ***) used in writing or printing for indicating an omission, esp. of letters or words.

**el·lip·soid** (ĭ-lĭp'soid') *n.* A geometric surface whose plane sections are all ellipses or circles. **—el·lip'soid', el'lip·soid'al** (-soid'l) *adj.*

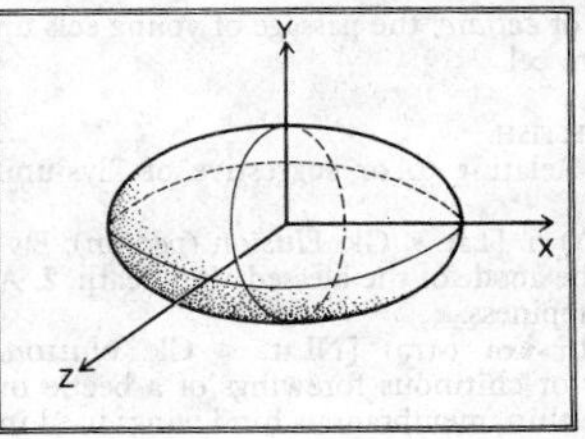

ellipsoid

**el·lip·tic** (ĭ-lĭp'tĭk) or **el·lip·ti·cal** (-tĭ-kəl) *adj.* [Gk. *elleiptikos,* defective < *elleipein,* to fall short. —see ELLIPSIS.] **1.** Of, relating to, or shaped like an ellipse. **2.** Containing or marked by ellipsis. **3. a.** Of or relating to economy of expression. **b.** Characterized by an obscurity of expression or style. **—el·lip'ti·cal·ly** *adv.*

**elliptic geometry** *n.* Riemannian geometry.

**el·lip·tic·i·ty** (ĭ-lĭp'-tĭs'ĭ-tē) *n.* **1.** Deviation from perfect circular or spherical form toward elliptic or ellipsoidal form. **2.** Degree of ellipticity.

**elm** (ĕlm) *n.* [ME < OE.] **1.** A deciduous tree of the genus *Ulmus,* with arching or curving branches, widely planted as shade trees. **2.** The wood of an elm.

**el·o·cu·tion** (ĕl'ə-kyōō'shən) *n.* [ME *elocucion* < Lat. *elocutio, elocution-* < *eloqui,* to speak out : *ex-,* out + *loqui,* to speak.] **1.** The art of public speaking, emphasizing gesture, vocal production, and delivery. **2.** Public speaking style. **—el'o·cu'tion·ar'y** (-shə-nĕr'ē) *adj.* **—el'o·cu'tion·ist** *n.*

**e·lo·de·a** (ĭ-lō'dē-ə) *n.* [NLat. *Elodea,* genus name < Gk. *helōdēs,* marshy < *helos,* marsh.] A small aquatic herb of the genus *Elodea.*

**E·lo·him** (ĕ-lō'hĭm, ĕl'ō-hēm') *n.* [Heb. *'Elōhim,* pl. of *'Elōah,* God.] A Hebrew name for God in the Old Testament. **—E·lo'hism** *n.*

**e·loign** (ĭ-loin') *vt.* **e·loigned, e·loign·ing, e·loigns.** [ME *elongen* < OFr. *esloigner* < LLat. *elongare* : *ex-,* away + *longe,* distant < *longus,* long.] *Archaic.* **1.** To remove or carry away to a distance, esp. to conceal. **2.** To remove (oneself) : ABSCOND.

**e·lon·gate** (ĭ-lông'gāt', ĭ-lŏng'-) *vt.* & *vi.* **-gat·ed, -gat·ing, -gates.** [LLat. *elongare, elongat-* : *ex-,* out + *longus,* long.] To make or grow longer. —*adj.* **1.** Extended : lengthened. **2.** Slender. **—e·lon'ga'tion** (ĭ-lông'gā'shən, ĭ-lŏng'-, ē'lông-, ē'lŏng-) *n.*

**e·lope** (ĭ-lōp') *vi.* **e·loped, e·lop·ing, e·lopes.** [AN *aloper,* to run away from one's husband with a lover.] **1.** To run away with a lover, esp. with the intention of getting married. **2.** To run away : ABSCOND. **—e·lope'ment** *n.* **—e·lop'er** *n.*

**el·o·quence** (ĕl'ə-kwəns) *n.* **1.** Fluent and persuasive discourse. **2.** The ability to persuade with discourse.

**el·o·quent** (ĕl'ə-kwənt) *adj.* [ME < OFr. < Lat. *eloquens,* pr.part. of *eloqui,* to speak out. —see ELOCUTION.] **1.** Fluent and persuasive in discourse. **2.** Vividly or movingly expressive <a gesture *eloquent* with distaste> **—el'o·quent·ly** *adv.* **—el'o·quent·ness** *n.*

☆ ***syns:*** ELOQUENT, ARTICULATE, FLUENT, SILVER-TONGUED *adj. core meaning* : fluently persuasive and forceful in discourse <an *eloquent* rebuttal>

**else** (ĕls) *adj.* [ME *elles* < OE.] **1.** Different : other <anybody *else*> **2.** More : additional <Do they require anything *else?*> —*adv.* **1.** In a different time, place, or manner : DIFFERENTLY <How *else* could it be cooked?> **2.** If not : OTHERWISE <Watch out, or *else* you will make a big mistake.>

**else·where** (ĕls'hwâr', -wâr') *adv.* To or in another or different place.

**e·lu·ci·date** (ĭ-lōō'sĭ-dāt') *v.* **-dat·ed, -dat·ing, -dates.** [LLat. *elucidare, elucidat-,* : *ex-* (intensive) + *lucidus,* bright.] —*vt.* To make clear : CLARIFY. —*vi.* To give a clarification. **—e·lu'ci·da'tion** *n.* **—e·lu'ci·da'tive** *adj.* **—e·lu'ci·da'tor** *n.*

**e·lude** (ĭ-lōōd') *vt.* **e·lud·ed, e·lud·ing, e·ludes.** [Lat. *eludere* : *ex-,* away + *ludere,* to play < *ludus,* play.] **1.** To escape from or evade, as by daring or artifice <*eluded* their pursuers> **2.** To escape the understanding or grasp of <The meaning of the poem *eluded* me.>

**E·lul** (ĕ-lōōl', ĕl'ōōl) *n.* [Heb. *'Elul* < Akkadian *elūlu,* harvest time.] The 12th month of the Hebrew year. —See table at CALENDAR.

**e·lu·sive** (ĭ-lōō'sĭv, -zĭv) *adj.* [< Lat. *eludere, elus-,* to elude.] **1.** Tending to elude perception or comprehension <an *elusive* set of ideas> **2.** Difficult to describe <an *elusive* attraction> **—e·lu'sive·ly** *adv.* **—e·lu'sive·ness** *n.*

**e·lute** (ĭ-lōōt) *vt.* **e·lut·ed, e·lut·ing, e·lutes.** [< Lat. *elutus,* p.part. of *eluere,* to wash out : *ex-,* out + *lavere,* to wash.] To extract one material from another, usu. by use of a solvent. **—e·lu'tion** *n.*

**e·lu·tri·ate** (ĭ-lōō'trē-āt') *vt.* **-at·ed, -at·ing, -ates.** [Lat. *elutriare, elutriat-* < *eluere,* to wash out. —see ELUVIUM.] To purify, separate, or remove by washing, decanting, and settling. **—e·lu'tri·a'tion** *n.*

**e·lu·vi·a·tion** (ĭ-lōō'vē-ā'shən) *n.* [ELUVI(UM) + -ATION.] Internal movement of soil when rainfall is greater than evaporation.

**e·lu·vi·um** (ĭ-lōō'vē-əm) *n.* [NLat. < Lat. *eluere,* to wash out : *ex-,* out + *lavere,* to wash.] Residual deposits of soil, dust, and rock particles produced by the action of the wind. **—e·lu'vi·al** *adj.*

**el·ver** (ĕl′vər) *n.* [Alteration of *eelfare,* the passage of young eels up a river.] A young or immature eel.
**elves** (ĕlvz) *n. pl. of* ELF.
**elv·ish** (ĕl′vĭsh) *adj. var. of* ELFISH.
**E·ly·sian** (ĭ-lĭzh′ən) *adj.* **1.** Relating to or suggestive of Elysium. **2.** Blissful : delightful.
**E·ly·si·um** (ĭ-lĭz′ē-əm, ĭ-lĭzh′-) *n.* [Lat. < Gk. *Ēlusion (pedion),* Elysian (fields).] **1.** *Gk. Myth.* The abode of the blessed after death. **2.** A place or condition of ideal happiness.
**el·y·tron** (ĕl′ə-trŏn′) *n., pl.* **-tra** (-trə) [NLat. < Gk. *elutron,* sheath.] One of the leathery or chitinous forewings of a beetle or related insect that encases the thin, membranous hind wings used in flight. **—el′y·troid′** *adj.*
**em** (ĕm) *n.* **1.** The letter *m.* **2.** The square of the body size of any typeface, used as a unit of measure, esp. of a pica M.
**'em** (əm) *pron.* [ME *hem* < OE *heom,* dative and accusative pl. of *hē,* he.] *Informal.* Them.
**em-**[1] *pref. var. of* EN-[1]. —Used before *b, m,* and *p.*
**em-**[2] *pref. var. of* EN-[2]. —Used before *b, m,* and *p.*
**e·ma·ci·ate** (ĭ-mā′shē-āt′) *vi.* & *vt.* **-at·ed, -at·ing, -ates.** [Lat. *emaciare, emaciat-* : *ex-* (intensive) + *maciare,* to make thin < *macer,* thin.] To become or cause to become very thin, esp. due to starvation. **—e·ma′ci·a′tion** *n.*
**em·a·nate** (ĕm′ə-nāt′) *vi.* & *vt.* **-nat·ed, -nat·ing, -nates.** [Lat. *emanare, emanat-,* flow out : *ex-,* out + *manare,* to flow.] To come or send forth, as from a source. **—em′a·na′tive** *adj.*
**em·a·na·tion** (ĕm′ə-nā′shən) *n.* **1.** An act or instance of emanating. **2. a.** Something emanating from a source : EFFLUENCE. **b.** *Chem.* A gaseous product of radioactive disintegration.
**e·man·ci·pate** (ĭ-măn′sə-pāt′) *vt.* **-pat·ed, -pat·ing, -pates.** [Lat. *emancipare, emancipat-* : *ex-,* out of, + *mancipium,* ownership < *manceps,* purchaser.] **1.** To free from bondage, oppression, or restraint : LIBERATE. **2.** *Law.* To release (a child) from parental or guardian control. **—e·man′ci·pa′tion** (-pā′shən) *n.* **—e·man′ci·pa′tive** *adj.* **—e·man′ci·pa′tor** *n.*
**Emancipation Proclamation** *n.* A proclamation issued by President Abraham Lincoln, effective Jan. 1, 1863, freeing all slaves in territory still at war with the Union.
**e·mar·gi·nate** (ĭ-mär′jə-nĭt, -nāt′) *adj.* [Lat. *emarginare, emarginat-,* to take the edge away : *ex-,* away + *margo,* margin.] Having a notched tip, as a leaf. **—e·mar′gi·na′tion** (-nā′shən) *n.*
**e·mas·cu·late** (ĭ-măs′kyə-lāt′) *vt.* **-lat·ed, -lat·ing, -lates.** [Lat. *emasculare, emasculat-* : *ex-,* away + *masculus,* manly.] **1.** To castrate. **2.** To deprive of vigor : WEAKEN. **—e·mas′cu·late′** (-lĭt, -lāt′) *adj.* **—e·mas′cu·la′tion** *n.* **—e·mas′cu·la′tive, e·mas′cu·la·to′ry** (-lə-tôr′ē, -tōr′ē) *adj.* **—e·mas′cu·la′tor** *n.*
**em·balm** (ĕm-bäm′) *vt.* **-balmed, -balm·ing, -balms.** [ME *embaumen* < OFr. *embasmer* : *en-,* in (< Lat. *in-*) + *basme,* balm. —see BALM.] **1.** To prevent the decay of (a corpse) by treating it with preservatives. **2.** To preserve the memory of. **3.** To give fragrance to : PERFUME. **—em·balm′er** *n.* **—em·balm′ment** *n.*
**em·bank** (ĕm-băngk′) *vt.* **-banked, -bank·ing, -banks.** To confine, protect, or support with an embankment.
**em·bank·ment** (ĕm-băngk′mənt) *n.* **1.** A mound of earth or stone to hold back water or support a road. **2.** The act of embanking.
**em·bar·go** (ĕm-bär′gō) *n., pl.* **-goes.** [Sp. < *embargar,* to impede, of Lat. orig.] **1.** A government order prohibiting passage of merchant ships into or out of its ports. **2.** A suspension of trade, esp. with respect to a particular commodity. **3.** A prohibition. —*vt.* **-goed, -go·ing, -goes.** To impose an embargo on.
**em·bark** (ĕm-bärk′) *v.* **-barked, -bark·ing, -barks.** [OFr. *embarquer* < LLat. *imbarcare* : Lat. *in-,* in + *barca,* boat.] —*vt.* **1.** To cause to board a vessel or aircraft. **2.** To enlist or invest in an enterprise. —*vi.* **1.** To go aboard a vessel. **2.** To set out on a venture : COMMENCE. **—em′bar·ka′tion, em·bark′ment** *n.*
**em·bar·rass** (ĕm-băr′əs) *vt.* **-rassed, -rass·ing, -rass·es.** [Fr. *embarrasser,* to impede < Sp. *embarazar* < Ital. *imbarazzare,* of Lat. orig.] **1.** To cause to feel self-consciously distressed : DISCONCERT. **2.** To involve in or hamper with financial difficulties. **3. a.** To beset with difficulties. **b.** To impede : hinder. **4.** To complicate. **—em·bar′rass·ing·ly** *adv.*
☆ ***syns:*** EMBARRASS, ABASH, CHAGRIN, CONFOUND, CONFUSE, DISCOMFORT, DISCONCERT, FAZE, MORTIFY, RATTLE *v. core meaning* : to cause (a person) to be self-consciously distressed <Personal questions *embarrass* me.>
**em·bar·rass·ment** (ĕm-băr′əs-mənt) *n.* **1. a.** An act or instance of embarrassing. **b.** The state of being embarrassed. **2.** Something that embarrasses. **3.** An overabundance : excess <an *embarrassment* of treasures>
**em·bas·sage** (ĕm′bə-sĭj) *n.* [ME *ambassage,* the function of a messenger, perh. < OFr. *ambassee,* vassal < Lat. *ambactia,* ult. of Celt. orig.] *Archaic.* Embassy.
**em·bas·sy** (ĕm′bə-sē) *n., pl.* **-sies.** [OFr. *ambassee.* —see EMBASSAGE.] **1.** The position, function, or assignment of an ambassador. **2.** A mission to a foreign government led by an ambassador. **3.** An ambassador and his or her staff. **4.** The official headquarters of an ambassador and his or her staff.
**em·bat·tle** (ĕm-băt′l) *vt.* **-tled, -tling, -tles.** [ME *embatailen* < OFr. *embataillier* : *en-,* in < Lat. *in-,* + *batailler,* to battle < *bataille,* battle, battlement. —see BATTLE.] **1.** To prepare for battle. **2.** To prepare to resist or struggle. **3.** To equip with battlements for defense.
**em·bat·tle·ment** (ĕm-băt′l-mənt, ĭm-) *n.* A battlement.
**em·bay** (ĕm-bā′) *vt.* **-bayed, -bay·ing, -bays. 1.** To put, shelter, or detain in a bay. **2.** To enclose in or as if in a bay.
**em·bed** (ĕm-bĕd′) *v.* **-bed·ded, -bed·ding, -beds.** —*vt.* **1. a.** To fix securely in a surrounding mass. **b.** To enclose in a matrix. **2.** To enclose firmly. **3.** To make an integral part of. —*vi.* To become embedded. **—em·bed′ment** *n.*
**em·bel·lish** (ĕm-bĕl′ĭsh) *vt.* **-lished, -lish·ing, -lish·es.** [ME *embelishen* < OFr. *embellir, embelliss-* : *en* (causative < Lat. *in-,* in) + *bel,* beautiful (< Lat. *bellus*).] **1.** To make beautiful, as by ornamentation : ADORN. **2.** To add fanciful or ornamental details to <*embellish* the account with exaggerated details> **—em·bel′lish·er** *n.*
**em·bel·lish·ment** (ĕm-bĕl′ĭsh-mənt) *n.* **1.** The act of embellishing or state of being embellished. **2.** Something that embellishes : ADORNMENT. **3.** *Mus.* A note that embellishes a melody.
**em·ber** (ĕm′bər) *n.* [ME *embre* < OE *ǣmerge.*] **1.** A small piece of live wood or coal, as in a dying fire. **2. embers.** The smoldering ash or coal of a dying fire.
**Ember day** *n.* [ME *Ymber Dayes* < OE *Ymbrendagas* : *ymbryne,* revolution of time (*ymbe,* around + *ryne,* running) + *dæg,* day.] A day for prayer and fasting in some Christian churches, observed on the Wednesday, Friday, and Saturday after the first Sunday of Lent, after Whitsunday, after Sept. 14, and after Dec. 13.
**em·bez·zle** (ĕm-bĕz′əl) *vt.* **-zled, -zling, -zles.** [ME *embesilen* < AN *enbesiler* : OFr. *en-* (intensive) + OFr. *besillier,* to ravage.] To take (e.g., money) for one's own use in violation of a trust and by fraudulent means. **—em·bez′zle·ment** *n.* **—em·bez′zler** *n.*
**em·bit·ter** (ĕm-bĭt′ər) *vt.* **-tered, -ter·ing, -ters. 1.** To make bitter in flavor. **2.** To arouse bitter feelings in. **—em·bit′ter·ment** *n.*
**em·blaze**[1] (ĕm-blāz′) *vt.* **-blazed, -blaz·ing, -blaz·es. 1.** To set on fire. **2.** To cause to glow : light up.
**em·blaze**[2] (ĕm-blāz′) *vt.* **-blazed, -blaz·ing, -blaz·es.** *Archaic.* To emblazon.
**em·bla·zon** (ĕm-blā′zən) *vt.* **-zoned, -zon·ing, -zons. 1.** To ornament, esp. with heraldic devices. **2.** To make resplendent with brilliant colors. **3.** To make illustrious : CELEBRATE. **—em·bla′zon·er** *n.* **—em·bla′zon·ment** *n.* **—em·bla′zon·ry** *n.*
**em·blem** (ĕm′bləm) *n.* [ME, pictorial fable < Lat. *emblema,* raised ornament < Gk. *emblēma* < *emballein,* to insert : *en-,* in + *ballein,* to throw.] **1.** An object or representation serving as a symbol. **2.** A distinctive badge, design, or device. **3.** An allegorical picture usu. inscribed with a motto or verse presenting a moral lesson.
**em·blem·at·ic** (ĕm′blə-măt′ĭk) *or* **em·blem·at·i·cal** (-ĭ-kəl) *adj.* Of, pertaining to, or serving as an emblem : SYMBOLIC. **—em′blem·at′i·cal·ly** *adv.*
**em·blem·a·tize** (ĕm-blĕm′ə-tīz′) *also* **em·blem·ize** (ĕm′blə-mīz′) *vt.* **-tized, -tiz·ing, -tiz·es** *also* **-ized, -iz·ing, -iz·es.** To represent with or as if with an emblem : SYMBOLIZE.
**em·ble·ments** (ĕm′blə-mənts) *pl.n.* [ME *emblaiment* < OFr. *emblaement* < *emblaer,* to sow with grain < Med. Lat. *imbladare* : Lat. *in-,* in + *bladum,* grain.] *Law.* The crops or products of the land legally belonging to a tenant.
**em·bod·i·ment** (ĕm-bŏd′ē-mənt, ĭm-) *n.* **1.** The act of embodying or state of being embodied. **2.** One that embodies something <regarded the leader as the *embodiment* of honor>
**em·bod·y** (ĕm-bŏd′ē) *vt.* **-bod·ied, -bod·y·ing, -bod·ies. 1.** To invest with or as if with bodily form. **2. a.** To represent in concrete form. **b.** To personify <young people who *embodied* the aspirations of a new age> **3.** To make part of a system or whole.
☆ ***syns:*** EMBODY, EXTERNALIZE, INCARNATE, MANIFEST, MATERIALIZE, OBJECTIFY, PERSONALIZE, PERSONIFY, SUBSTANTIATE *v. core meaning* : to represent (an abstraction) in or as if in bodily form <The general *embodies* the spirit of revolution and freedom.>
**em·bold·en** (ĕm-bōl′dən) *vt.* **-ened, -en·ing, -ens.** To impart courage to : ENCOURAGE.
**em·bo·lec·to·my** (ĕm′bə-lĕk′tə-mē) *n., pl.* **-mies.** [EMBOL(US) + -ECTOMY.] Medical removal of an embolus.
**em·bo·li** (ĕm′bə-lī′) *n. pl. of* EMBOLUS.
**em·bol·ic** (ĕm-bŏl′ĭk) *adj.* Of or pertaining to an embolus or an embolism.
**em·bo·lism** (ĕm′bə-lĭz′əm) *n.* [ME *embolisme,* insertion of one or more days in a calendar < LLat. *embolismus* < Gk. *embolismos* < *emballein,* to insert. —see EMBLEM.] Occlusion or obstruction of a blood vessel by an embolus. **—em′bo·lis′mic** *adj.*
**em·bo·lus** (ĕm′bə-ləs) *n., pl.* **-li** (-lī′) [NLat. < Lat., pump's piston < Gk. *embolos,* stopper < *emballein,* to insert. —see EMBLEM.] An air bubble, detached clot, mass of bacteria, or other foreign body occluding a blood vessel.
**em·bo·ly** (ĕm′bə-lē) *n.* [Gk. *embolē,* insertion < *emballein,* to insert. —see EMBLEM.] Development of a gastrula from a blastula by invagination.

**em·bon·point** (äɴ′bôɴ-pwăɴ′) *n.* [Fr. < OFr. *en bon point,* in good condition.] Plumpness : corpulence.

**em·bos·om** (ĕm-bo͝oz′əm, -bo͞o′zəm) *vt.* **-omed, -om·ing, -oms. 1.** *Archaic.* To hold in the bosom. **2.** To envelop or enclose protectively : SHELTER.

**em·boss** (ĕm-bôs′, -bŏs′) *vt.* **-bossed, -boss·ing, -boss·es.** [ME *embosen* < OFr. *embocer* : *en-,* in (< Lat. *in-*) + *boce,* knob.] **1.** To mold or carve in relief. **2. a.** To decorate with or as if with a raised design. **b.** To raise the surface of in relief. **3.** To adorn lavishly : EMBELLISH. **—em·boss′er** *n.*

**em·boss·ment** (ĕm-bôs′mənt, -bŏs′-) *n.* The distance between the nondeformed part of a document surface and a specified point on a printed character in optical character recognition.

**em·bou·chure** (äm′bo͞o-sho͝or′) *n.* [Fr. < OFr. *emboucher,* to occlude : *en-,* in (< Lat. *in-*) + *bouche,* mouth < Lat. *bucca,* cheek.] **1.** The mouth of a river. **2. a.** The mouthpiece of a wind instrument. **b.** The way in which the lips and tongue are applied to a mouthpiece in order to produce a musical tone.

**em·bowed** (ĕm-bōd′) *adj.* **1.** Curved like a bow. **2. a.** Arched <an *embowed* ceiling> **b.** Protruding in an outward curve so as to form a recess.

**em·bow·el** (ĕm-bou′əl) *vt.* **-eled, -el·ing, -els** or **-elled, -el·ling, -els. 1.** To disembowel. **2.** *Obs.* To bury deeply.

**em·bow·er** (ĕm-bou′ər) *vt.* **-ered, -er·ing, -ers.** To enclose in or as if in a bower.

**em·brace** (ĕm-brās′) *v.* **-braced, -brac·ing, -brac·es.** [ME *embracen* < OFr. *embracer* : *en-,* in (< Lat. *in-*) + *brace,* arms < Lat. *bracchium,* arm < Gk. *brakhiōn.*] —*vt.* **1.** To clasp or hold with the arms, usu. as a show of affection : HUG. **2. a.** To encircle. **b.** To twine around. **3.** To include, comprise, or contain : ENCOMPASS. **4.** To take up willingly or eagerly <*embrace* the fight for justice> **5.** To avail oneself of <*embrace* the chance> —*vi.* To join in an embrace. —*n.* **1.** An act of embracing : HUG. **2.** An enclosure. **3.** Eager acceptance. **—em·brace′ment** *n.* **—em·brac′er** *n.*

**em·brac·er** (ĕm-brā′sər) *n.* [ME *embracer* < *embracen,* to influence a jury by illegal means. —see EMBRACE.] *Law.* One guilty of attempting to influence a court illegally.

**em·brac·er·y** (ĕm-brā′sə-rē) *n., pl.* **-ies.** [ME *embracerie* < *embracen,* to influence a jury by illegal means, to embrace.] *Law.* An attempt to corrupt a jury, as with bribery.

**em·branch·ment** (ĕm-brănch′mənt) *n.* A branching out, as of a river or mountain range.

**em·bran·gle** (ĕm-brăng′gəl) *vt.* **-gled, -gling, -gles.** [EN-[1] + dial. *brangle,* to wrangle.] To embroil : entangle. **—em·bran′gle·ment** *n.*

**em·bra·sure** (ĕm-brā′zhər) *n.* [Fr. < *embraser,* to widen an opening.] **1.** An opening in a wall for a door or window. **2.** A flared opening for a gun in a wall or parapet.

**em·bro·cate** (ĕm′brə-kāt′) *vt.* **-cat·ed, -cat·ing, -cates.** [Med. Lat. *embrocare, embrocat-* < LLat. *embrocha,* lotion < Gk. *embrokhē* < *embrekhein,* to foment : *en-,* in + *brekhein,* to wet.] To moisten and rub with a lotion or liniment.

**em·broi·der** (ĕm-broi′dər) *v.* **-dered, -der·ing, -ders.** [ME *embrouderen* < OFr. *embroder* : *en-,* in (< Lat. *in-*) + *broder,* to embroider, of Germanic orig.] —*vt.* **1.** To decorate with needlework. **2.** To fashion by needlework. **3.** To add ornamentation to. —*vi.* **1.** To make embroidery. **2.** To supply ornamentation. **—em·broi′der·er** *n.*

**em·broi·der·y** (ĕm-broi′də-rē) *n., pl.* **-ies. 1.** The art or act of embroidering. **2.** Ornamentation of fabric with needlework. **3.** A piece of embroidered fabric. **4.** Embellishment with fanciful details.

**em·broil** (ĕm-broil′) *vt.* **-broiled, -broil·ing, -broils.** [Fr. *embrouiller* : *en-,* in (< Lat. *in-*) + *brouiller,* to confuse, of Germanic orig.] **1.** To involve in argument, contention, or hostile actions <*embroiled* me in a family dispute> **2.** To throw into confusion : ENTANGLE. **—em·broil′ment** *n.*

**em·brown** (ĕm-broun′) *vt.* **-browned, -brown·ing, -browns. 1.** To make brown or dusky. **2.** To darken.

**em·brue** (ĕm-bro͞o′) *v. var. of* IMBRUE.

**em·bry·ec·to·my** (ĕm′brē-ĕk′tə-mē) *n., pl.* **-mies.** Surgical removal of an extrauterine embryo.

**em·bry·o** (ĕm′brē-ō′) *n., pl.* **-os.** [Med. Lat. < Gk. *embruon* : *en-,* in + *bruein,* to grow.] **1.** *Biol.* **a.** An organism in its early developmental stages, esp. before it has reached a distinctively recognizable form. **b.** Such an organism at any time before full development, birth, or hatching. **2. a.** The fertilized egg of a vertebrate animal. **b.** The prefetal product of human conception up to the beginning of the third month of pregnancy. **3.** *Bot.* The rudimentary plant contained within a seed or archegonium. **4.** A beginning or rudimentary stage.

**em·bry·o·gen·e·sis** (ĕm′brē-ō-jĕn′ə-sĭs) *also* **em·bry·og·e·ny** (-ŏj′ə-nē) *n.* The growth and development of an embryo. **—em′bry·o·ge·net′ic** (-ō-jə-nĕt′ĭk) *adj.*

**em·bry·ol·o·gy** (ĕm′brē-ŏl′ə-jē) *n.* The science concerned with the formation, early growth, and development of living organisms. **—em′bry·o·log′ic** (-ə-lŏj′ĭk), **em′bry·o·log′i·cal** *adj.* **—em′bry·o·log′i·cal·ly** *adv.* **—em′bry·ol′o·gist** *n.*

**em·bry·on·ic** (ĕm′brē-ŏn′ĭk) *also* **em·bry·on·al** (ĕm′brē-ə-nəl) *adj.* **1. a.** Of or relating to an embryo. **b.** Being an embryo. **2.** Rudimentary : incipient. **—em′bry·on′ic·al·ly** *adv.*

**embryonic layer** *n.* The germ layer.

**em·bry·op·a·thy** (ĕm′brē-ŏp′ə-thē) *n., pl.* **-thies.** Abnormal development of an embryo.

**embryo sac** *n.* A structure formed by the female gametophyte of a seed plant, in which the embryo develops.

**em·cee** (ĕm′sē′) [Pronunciation of *M.C.,* abbr. of *master of ceremonies.*] *Informal.* —*n.* A master of ceremonies. —*v.* **-ceed, -cee·ing, -cees.** —*vt.* To serve as master of ceremonies of. —*vi.* To act as master of ceremonies.

**-eme** *suff.* [Fr. *-ème* < *phonème,* phoneme.] A distinctive unit of linguistic structure <semant*eme*>

**e·meer** (ĭ-mîr′, ā-mîr′) *n. var. of* EMIR.

**e·mend** (ĭ-mĕnd′) *vt.* **e·mend·ed, e·mend·ing, e·mends.** [ME *emenden* < Lat. *emendare* : *ex-,* away + *mendum,* fault.] **1.** To improve by scholarly editing. **2.** *Archaic.* To correct. **—e·mend′er** *n.*

**e·men·date** (ē′mĕn-dāt′, ĭ-mĕn′-) *vt.* **-dat·ed, -dat·ing, -dates.** [Lat. *emendare, emendat-,* to emend.] To make corrections in (a printed text). **—e′men·da′tor** (-dā′tər) *n.* **—e·men′da·to′ry** (ĭ-mĕn′də-tôr′ē, -tōr′ē) *adj.*

**e·men·da·tion** (ĭ-mĕn-dā′shən, ē′mĕn-) *n.* **1.** The act of emending. **2.** An alteration meant to improve or correct.

**em·er·ald** (ĕm′ər-əld, ĕm′rəld) *n.* [ME *emeraude* < OFr. < Lat. *smaragdus* < Gk. *smaragdos.*] **1.** A brilliant, transparent green beryl used as a gemstone. **2.** A strong yellowish green.

**e·merge** (ĭ-mûrj′) *vi.* **e·merged, e·merg·ing, e·merg·es.** [Lat. *emergere* : *ex-,* out + *mergere,* to immerse.] **1.** To come forth or rise up from or as if from immersion. **2.** To become obvious or evident. **3.** To issue, as from obscurity. **4.** To come into existence.

**e·mer·gence** (ĭ-mûr′jəns) *n.* **1.** The act or process of emerging. **2.** *Bot.* A superficial outgrowth of plant tissue, as a thorn.

**e·mer·gen·cy** (ĭ-mûr′jən-sē) *n., pl.* **-cies.** An unexpected, serious occurrence or situation urgently requiring prompt action. **—e·mer′gen·cy** *adj.*

▲ word history: The word *emergency* is derived from *emerge,* which is from Latin *ex-,* "out," and *mergere,* "to sink, dip." An *emergency* is something that arises, especially unexpectedly, and requires immediate attention.

**emergency brake** *n.* A separate brake system in a vehicle for use in case the regular brakes fail and commonly used as a parking brake.

**emergency room** *n.* The area in a hospital where emergency cases are treated.

**e·mer·gent** (ĭ-mûr′jənt) *adj.* **1.** Coming unexpectedly into being. **2.** Requiring immediate action. **3.** Coming into being as a logical result. **4.** Newly created <*emergent* nations>

**emergent evolution** *n.* A theory that entirely new types of organisms, modes of behavior, and consciousness appear at certain stages of the evolutionary process, usu. because of an unpredictable rearrangement of the pre-existing elements.

**e·mer·i·ta** (ĭ-mĕr′ĭ-tə) *adj.* [Lat., fem. of *emeritus.*] Emeritus. —Used of a woman. **—e·mer′i·ta** *n.*

**e·mer·i·tus** (ĭ-mĕr′ĭ-təs) *adj.* [Lat. *emeritus,* p.part. of *emereri,* to earn by service : *ex-* (intensive) + *mereri,* to earn.] Retired but retaining an honorary title corresponding to that held immediately before retirement <a professor *emeritus*> —*n., pl.* **-ti** (-tī′). One who is emeritus.

**e·mersed** (ĭ-mûrst′) *adj. Bot.* Rising above the surface of a fluid, as certain aquatic plants.

**e·mer·sion** (ĭ-mûr′zhən, -shən) *n.* [< Lat. *emersus* < *emergere,* to emerge.] EMERGENCE 1.

**em·er·y** (ĕm′ə-rē, ĕm′rē) *n.* [ME < OFr. *emeri* < LLat. *smericulum* < Gk. *smeris.*] A fine-grained impure corundum used for polishing and grinding.

**emery board** *n.* A nail file consisting of a strip of cardboard with a coating of powdered emery.

**emery cloth** *n.* An abrasive cloth coated with powdered emery.

**em·e·sis** (ĕm′ĭ-sĭs) *n.* [NLat. < Gk. < *emein,* to vomit.] Vomiting.

**e·met·ic** (ĭ-mĕt′ĭk) *n.* [Lat. *emetica* < Gk. *emetikos,* provoking vomiting < *emetos,* vomiting < *emein,* to vomit.] An agent that induces vomiting. **—e·met′ic** *adj.* **—e·met′i·cal·ly** *adv.*

**em·e·tine** (ĕm′ĭ-tēn′) *n.* [Fr. *émétine* : *émétique,* emetic + *-ine,* -ine.] A bitter-tasting crystalline alkaloid, $C_{29}H_{40}O_4N_2$, derived from ipecac root and used as an emetic.

**e·meu** (ē′myo͞o) *n. var. of* EMU.

**-emia** *suff.* [NLat. < Gk. *-aimia* < *haima,* blood.] Blood <leuk*emia*>

**em·i·grant** (ĕm′ĭ-grənt) *n.* One that emigrates. **—em′i·grant** *adj.*

**em·i·grate** (ĕm′ĭ-grāt′) *vi.* **-grat·ed, -grat·ing, -grates.** [Lat. *emigrare, emigrat-* : *ex-,* away + *migrare,* to move.] To leave one country or area to settle elsewhere. **—em′i·gra′tion** (-grā′shən) *n.*

**é·mi·gré** (ĕm′ĭ-grā′) *n.* [Fr. < p.part. of *émigrer,* to emigrate < Lat. *emigrare.*] An emigrant, esp. one compelled to emigrate for political reasons.

**em·i·nence** (ĕm′ə-nəns) *also* **em·i·nen·cy** (-nən-sē) *n., pl.* **-cies. 1.** A position of great distinction or superiority. **2.** A rise of ground : HILL. **3. a.** A person of high rank or great accomplishments.

**b. Eminence.** —Used as a title of honor, esp. for a Roman Catholic cardinal.

**em·i·nent** (ĕm′ə-nənt) *adj.* [ME < OFr. or < Lat. *eminens*, pr.part. of *eminēre*, to project.] **1.** Standing out above others : PROMINENT. **2.** Outstanding in performance, rank, or attainments : DISTINGUISHED <an *eminent* author> **3.** Having or exhibiting eminence : NOTEWORTHY. **—em′i·nent·ly** *adv.*

**eminent domain** *n. Law.* A government's right to take private property for public use, usu. with compensation to the owner.

**e·mir** *or* **e·meer** (ĭ-mîr′, ā-mîr′) *n.* [Fr. *émir* < Sp. *emir* < Ar. *'amīr*, commander < *amara*, he commanded.] A prince, chieftain, or governor in the Middle East and some parts of Africa.

**e·mir·ate** (ĭ-mîr′ĭt, -āt′) *n.* **1.** The office of an emir. **2.** The territory or nation ruled by an emir.

**em·is·sar·y** (ĕm′ĭ-sĕr′ē) *n., pl.* **-ies.** [Lat. *emissarius* < *emissus*, p.part of *emittere*, to send out. —see EMIT.] An agent, esp. a secret agent, sent to represent or promote the interests of another.

**e·mis·sion** (ĭ-mĭsh′ən) *n.* [< Lat. *emittere, emiss-*, to send out. —see EMIT.] **1.** An act or instance of emitting. **2.** Something emitted. **3.** The substance discharged into the air, esp. by an internal-combustion engine.

**emission nebula** *n.* A nebula that absorbs ultraviolet radiation from stars and re-emits it as visible light.

**emission spectrum** *n.* The spectrum of bright lines, bands, or continuous radiation typical of and determined by a specific emitting substance subjected to a specific kind of excitation.

**e·mis·sive** (ĭ-mĭs′ĭv) *adj.* Having the power or tendency to emit.

**em·is·siv·i·ty** (ĕm′ĭ-sĭv′ĭ-tē) *n.* The ratio of radiation intensity from a surface to the radiation intensity at the same wavelength from a blackbody at the same temperature.

**e·mit** (ĭ-mĭt′) *vt.* **e·mit·ted, e·mit·ting, e·mits.** [Lat. *emittere*, to send out : *ex-*, out + *mittere*, to send.] **1.** To send or give out (e.g., radiation). **2. a.** To utter : express. **b.** To give out as sound <*emitted* a shriek> **3.** To issue with authority, esp. to put into circulation (e.g., currency). **—e·mit′ter** *n.*

**em·men·a·gogue** (ĭ-mĕn′ĭ-gôg′, -gŏg′) *n.* [Gk. *emmēna*, the menses (*en-*, in + *mēn*, month) + -AGOGUE.] A medicine that induces or hastens menstrual flow.

**em·mer** (ĕm′ər) *n.* [G. < OHG *amaro.*] A Eurasian wheat, *Triticum dicoccum*, cultivated as a cereal grain and livestock feed.

**em·met** (ĕm′ĭt) *n.* [ME *emete* < OE *ǣmete.*] *Archaic.* An ant.
▲ word history: The words *emmet* and *ant* are descended from the same Old English word, *ǣmete*, "ant." The Modern English words result from the different development of *ǣmete* in two different dialects of Old and Middle English. In one dialect *ǣmete* became *amte* and finally *ant*. In another, *ǣmete* became *emete* and then *emmet*. In Modern English *ant* is the standard word and *emmet* is archaic.

**em·me·tro·pi·a** (ĕm′ĭ-trō′pē-ə) *n.* [Gk. *emmetros*, in measure (*en*, in + *metron*, measure) + -OPIA.] The state of the normal eye when parallel rays are focused precisely on the retina and vision is perfect. **—em′me·trop′ic** (-trŏp′ĭk) *adj.*

**Em·my** (ĕm′ē) *n., pl.* **-mys** *or* **-mies.** [Alteration of *Immy*, nickname for *image orthicon tube.*] A statuette awarded annually by the Academy of Television Arts and Sciences for outstanding achievement in television.

**em·o·din** (ĕm′ə-dĭn) *n.* [NLat. *emodi*, specific epithet of a species of rhubarb + -IN.] An orange crystalline compound, $C_{15}H_{10}O_5$, derived from plants and used as a laxative.

**e·mol·lient** (ĭ-mŏl′yənt) *adj.* [Lat. *emolliens, emollient-*, pr.part. of *emollire*, to soften : *ex-* (intensive) + *mollire*, to soften < *mollis*, soft.] **1.** Soothing and softening, esp. to the skin. **2.** Making less harsh or abrasive : MOLLIFYING. —*n.* **1.** An agent that soothes or softens the skin. **2.** Something that assuages or mollifies.

**e·mol·u·ment** (ĭ-mŏl′yə-mənt) *n.* [ME < Lat. *emolumentum*, gain, perh. < *emoliri*, to bring about by effort.] Compensation or payment from an office or employment.

**e·mote** (ĭ-mōt′) *vi.* **e·mot·ed, e·mot·ing, e·motes.** [Back-formation < EMOTION.] *Informal.* To express emotion excessively and theatrically.

**e·mo·tion** (ĭ-mō′shən) *n.* [Fr. *émotion* < OFr. *esmovoir*, to excite < VLat. **exmovēre* < Lat. *emovēre*, to move out : *ex-*, out + *movēre*, to move.] **1. a.** A complex, usu. strong subjective response, as love or fear. **b.** Such a response involving physiological changes as a preparation for action. **2.** A state of agitation or disturbance <a struggle to control my *emotions*> **3.** The part of the consciousness that involves feeling or sensibility <*Emotions* won out over good sense.>

**e·mo·tion·al** (ĭ-mō′shə-nəl) *adj.* **1.** Of or relating to emotion. **2.** Easily affected with or stirred by emotion. **3.** Capable of stirring the emotions <an *emotional* speech> **4.** Marked by or displaying emotion : AGITATED. **—e·mo′tion·al′i·ty** (-năl′ĭ-tē) *n.* **—e·mo′tion·al·ly** *adv.*

**e·mo·tion·al·ism** (ĭ-mō′shə-nə-lĭz′əm) *n.* **1.** A tendency to rely on or place too much value on emotion. **2.** Excessive display of emotion.

**e·mo·tion·al·ist** (ĭ-mō′shə-nə-lĭst) *n.* **1.** One whose conduct, thought, or rhetoric is ruled by emotion as opposed to reason. **2.** An excessively emotional person. **—e·mo′tion·al·is′tic** *adj.*

**e·mo·tion·al·ize** (ĭ-mō′shə-nə-līz′) *vt.* **-ized, -iz·ing, -iz·es.** To impart an emotional character to.

**e·mo·tion·less** (ĭ-mō′shən-lĭs) *adj.* Lacking emotion. **—e·mo′tion·less·ness** *n.*

**e·mo·tive** (ĭ-mō′tĭv) *adj.* **1.** Of or relating to emotion. **2.** Expressing or arousing emotion. **—e·mo′tive·ly** *adv.* **—e·mo′tive·ness, e′mo·tiv′i·ty** *n.*

**em·pale** (ĕm-pāl′) *v. var. of* IMPALE.

**em·pan·el** (ĕm-păn′əl) *v. var. of* IMPANEL.

**em·pa·thet·ic** (ĕm′pə-thĕt′ĭk) *adj.* Empathic. **—em′pa·thet′i·cal·ly** *adv.*

**em·path·ic** (ĕm-păth′ĭk) *adj.* Of or marked by empathy.

**em·pa·thize** (ĕm′pə-thīz′) *vi.* **-thized, -thiz·ing, -thiz·es.** To feel or experience empathy.

**em·pa·thy** (ĕm′pə-thē) *n.* [Transl. of G. *Einfühlung*, transl. of Gk. *empatheia*, passion.] **1.** Identification with and understanding of another's feelings, situation, and motives. **2.** Attribution of one's own feelings to an object.

**em·pen·nage** (ĕm′pĭ-nĭj) *n.* [Fr. < *empenner*, to feather an arrow : *en-*, in (< Lat. *in-*) + *penne*, feather < Lat. *penna.*] The tail of an aircraft.

**em·per·or** (ĕm′pər-ər) *n.* [ME *emperour* < OFr. *empereor* < Lat. *imperator* < *imperare*, to command : *in-*, in + *parare*, to prepare.] **1.** A man who rules an empire. **2. a.** A brightly colored butterfly of the family Nymphalidae, as *Asterocampa clyton*, with orange-tawny wings bearing dark markings. **b.** A moth of the family Saturniidae, esp. an Old World species, *Saturnia pavonia*, having distinctively patterned wings. **—em′per·or·ship′** *n.*

**emperor butterfly** *n.* EMPEROR 2a.

**emperor moth** *n.* EMPEROR 2b.

**emperor penguin** *n.* A large penguin, *Aptenodytes forsteri* of Antarctic regions.

**em·per·y** (ĕm′pə-rē) *n., pl.* **-ies.** [ME *emperie* < OFr. < Lat. *imperium.*] Absolute dominion or jurisdiction : SOVEREIGNTY.

**em·pha·sis** (ĕm′fə-sĭs) *n., pl.* **-ses** (-sēz′) [Lat. < Gk. < *emphainein*, to indicate : *en-*, in + *phainein*, to show.] **1.** Special weight or significance <an *emphasis* on promptness> **2.** Stress applied, as to a word or syllable. **3.** Force of expression.

**em·pha·size** (ĕm′fə-sīz′) *vt.* **-sized, -siz·ing, -siz·es.** [< EMPHASIS.] To place emphasis on : STRESS.

**em·phat·ic** (ĕm-făt′ĭk) *adj.* [Med. Lat. *emphaticus* < Gk. *emphatikos* < *emphainein*, to indicate. —see EMPHASIS.] **1.** Expressed or carried out with emphasis. **2.** Forceful in action or expression. **3.** Striking and distinctly defined. **—em·phat′i·cal·ly** *adv.*

**em·phy·se·ma** (ĕm′fĭ-sē′mə) *n.* [NLat. < Gk. *emphusēma*, inflation < *emphusan*, to blow in : *en-*, in + *phusan*, to blow.] **1.** A pulmonary condition characterized by dilation of the air vesicles in the lungs following atrophy of the septa, resulting in labored breathing and greater susceptibility to infection. **2.** A distention of connective tissues because of retention of air. **—em′phy·sem′a·tous** (-sĕm′ə-təs) *adj.*

**em·pire** (ĕm′pīr′) *n.* [ME < OFr. < Lat. *imperium* < *imperare*, to command. —see EMPEROR.] **1. a.** A political unit, often made up of a number of territories or nations, ruled by a single supreme authority. **b.** The territory of such a unit. **2.** An extensive enterprise under a unified authority <a cattle *empire*>

**Em·pire** (ŏm-pîr′, ĕm′pīr′) *adj.* [After the 1st *Empire* of France (1804–1815).] Of, pertaining to, or typical of a neoclassic style common in France during the first part of the 19th cent.

**em·pir·ic** (ĕm-pîr′ĭk) *n.* [Lat. *empiricus* < Gk. *empeirikos* < *empeirikos*, experienced < *empeiros* : *en-*, in + *peiran*, to try.] **1.** One who believes that practical experience is the only way to gain knowledge. **2.** *Archaic.* A charlatan.

**em·pir·i·cal** (ĕm-pîr′ĭ-kəl) *adj.* **1. a.** Relying upon or gained from experiment or observation <*empirical* techniques> **b.** Capable of proof or verification by means of experiment or observation <*empirical* knowledge **2.** Relying solely on practical experience and without regard for theory or system. **—em·pir′i·cal·ly** *adv.*

**empirical formula** *n.* A chemical formula indicating ratio of the elements rather than the total number of atoms in a molecule.

**em·pir·i·cism** (ĕm-pîr′ĭ-sĭz′əm) *n.* **1.** The view that experience, esp. of the senses, is the single source of knowledge. **2. a.** Employment of empirical methods, as in science. **b.** An empirical conclusion. **3.** Medical practice based on practical experience rather than scientific theory. **—em·pir′i·cist** *n.*

**em·place** (ĕm-plās′) *vt.* **-placed, -plac·ing, -plac·es.** To put in place or position.

**em·place·ment** (ĕm-plās′mənt) *n.* [Fr. < obs. *emplacer*, to place in position : *en-*, in (< Lat. *in-*) + OFr. *place*, open space. —see PLACE.] **1.** A prepared position, as a platform or mounting, for guns or other military equipment. **2.** The action of putting in a certain position : PLACEMENT. **3.** Location : position.

**em·plane** (ĕm-plān′) *v. var. of* ENPLANE.

**em·ploy** (ĕm-ploi′) *vt.* **-ployed, -ploy·ing, -ploys.** [ME *emploien* < OFr. *emploier* < Lat. *implicare,* to involve : *in-,* in + *plicare,* to fold.] **1.** To put to service or use. **2.** To apply or devote (e.g., time) to an activity. **3. a.** To put to work. **b.** To provide with gainful work. —*n.* **1.** The state of being employed. **2.** *Archaic.* Occupation. **—em·ploy′a·bil′i·ty** *n.* **—em·ploy′a·ble** *adj.* **—em·ploy′er** *n.*

**em·ploy·ee** *also* **em·ploy·e** *or* **em·ploy·é** (ĕm-ploi′ē, ĕm′-ploi-ē′) *n.* A person who works for another in exchange for financial compensation.

**em·ploy·ment** (ĕm-ploi′mənt) *n.* **1.** The act of employing or state of being employed. **2.** The work in which one is engaged : BUSINESS. **3.** An activity to which one devotes time.

**employment agency** *n.* An agency whose business it is to find jobs for people and find people qualified to fill jobs.

**em·poi·son** (ĕm-poi′zən) *vt.* **-soned, -son·ing, -sons.** [ME *empoisounen* < OFr. *empoisoner* : *en-,* in + *poison,* poison.] **1.** To fill with venom : EMBITTER. **2.** *Archaic.* To poison.

**em·po·ri·um** (ĕm-pôr′ē-əm, -pōr′-) *n., pl.* **-ri·ums** *or* **-ri·a** (-ē-ə) [Lat. < Gk. *emporion* < *emporos,* merchant : *en-,* in + *poros,* journey.] **1.** An important trade center : MARKETPLACE. **2.** A large retail store stocked with a great variety of merchandise.

**em·pow·er** (ĕm-pou′ər) *vt.* **-ered, -er·ing, -ers.** To invest with legal power : AUTHORIZE.

**em·press** (ĕm′prĭs) *n.* [ME *emperesse* < OFr., fem. of *empereor,* emperor.] **1.** A woman who rules an empire. **2.** The wife or widow of an emperor.

**em·presse·ment** (äN′prĕs-mäN′) *n.* [Fr. < *s'empresser,* to be eager.] Excessive cordiality.

**empress tree** *n.* A tree indigenous to China, *Paulownia tomentosa,* with large, hairy leaves and lavender flowers.

**em·prise** *also* **em·prize** (ĕm-prīz′) *n.* [ME < OFr., fem. p.part. of *emprendre,* to undertake < VLat. **imprendere* : Lat. *in-,* in + Lat. *prendere,* to take.] **1.** An undertaking, esp. a chivalrous or adventurous one. **2.** Chivalrous prowess or daring.

**emp·ty** (ĕmp′tē) *adj.* **-ti·er, -ti·est.** [ME < OE *ǣmtig* < *ǣmetta,* leisure.] **1. a.** Holding or containing nothing. **b.** Having no elements or members : NULL. **2.** Having no occupants or inhabitants : VACANT. **3.** Lacking force or power <*empty* guarantees> **4.** Lacking purpose or substance : MEANINGLESS <an *empty* existence> **5.** Idle <*empty* days> **6.** Vacuous : inane <an *empty* brain> **7.** Needing nourishment : HUNGRY. **8.** Devoid : destitute <*empty* of human warmth> —*v.* **-tied, -ty·ing, -ties.** —*vt.* **1.** To remove the contents of. **2.** To transfer or pour off <*empty* the trash into a bin> **3.** To unburden : RELIEVE <*empty* one's mind of fears> —*vi.* **1.** To become empty. **2.** To discharge or flow. —*n., pl.* **-ties.** An empty container. **—emp′ti·ly** *adv.* **—emp′ti·ness** *n.*

☆ ***syns:*** EMPTY, BARE, CLEAR, DEVOID (of), VACUOUS, VOID *adj. core meaning* : not containing anything <an *empty* closet> <streets *empty* of people> ***ant:*** full

**emp·ty-hand·ed** (ĕmp′tē-hăn′dĭd) *adj.* Bringing or taking away nothing.

**emp·ty-head·ed** (ĕmp′tē-hĕd′ĭd) *adj.* Scatterbrained.

**em·pur·ple** (ĕm-pûr′pəl) *vt. & vi.* **-pled, -pling, -ples.** To become or make purple.

**em·py·e·ma** (ĕm′pī-ē′mə) *n., pl.* **-ma·ta** (-mə-tə) [Med. Lat. < Gk. *empuēma* < *empuein,* to suppurate.] Pus in a bodily cavity, as the gallbladder or pleural cavity. **—em′py·e′mic** *adj.*

**em·pyr·e·al** (ĕm′pī-rē′əl, ĕm-pîr′ē-əl) *adj.* [ME *imperyale* < Med. Lat. *empyreus* < Gk. *empurios,* fiery : *en-,* in + *pur,* fire.] **1.** Empyrean. **2.** Of or relating to the sky : CELESTIAL. **3.** Sublime : elevated.

**em·py·re·an** (ĕm′pī-rē′ən, ĕm-pîr′ē-ən) *n.* [< Med. Lat. *empyreum* < *empyreus,* empyreal.] **1. a.** The highest part of heaven, believed by ancient people to be a realm of pure fire or light. **b.** The abode of God and the angels : PARADISE. **2.** The sky. **—em′py·re′an** *adj.*

**e·mu** *also* **e·meu** (ē′myōō) *n.* [Port. *ema,* rhea.] A large, fast-running, flightless Australian bird, *Dromiceius novaehollandiae,* related to the ostrich.

**em·u·late** (ĕm′yə-lāt′) *vt.* **-lat·ed, -lat·ing, -lates.** [Lat. *aemulari, aemulat-* < *aemulus,* emulous.] **1.** To strive to equal or excel, esp. by imitating. **2.** To vie with successfully. **3.** *Computer Sci.* To cause one system to accept the same data, execute the same programs, and achieve identical results as another system. **—em′u·la′tive** *adj.* **—em′u·la′tive·ly** *adv.* **—em′u·la′tor** *n.*

**em·u·la·tion** (ĕm′yə-lā′shən) *n.* **1.** Ambition or effort to equal or surpass another. **2.** Imitation of another. **3.** *Obs.* Jealous rivalry.

**em·u·lous** (ĕm′yə-ləs) *adj.* [Lat. *aemulus.*] **1.** Ambitious or eager to surpass or equal another. **2.** Marked by or resulting from rivalry. **3.** *Obs.* Coveting power or honor : ENVIOUS. **—em′u·lous·ly** *adv.* **—em′u·lous·ness** *n.*

**e·muls·i·ble** (ĭ-mŭl′sə-bəl) *adj.* Able to undergo emulsification.

**e·mul·si·fy** (ĭ-mŭl′sə-fī′) *vt.* **-fied, -fy·ing, -fies.** [EMULSI(ON) + -FY.] To make into an emulsion. **—e·mul′si·fi·ca′tion** *n.* **—e·mul′si·fi′er** *n.*

**e·mul·sion** (ĭ-mŭl′shən) *n.* [NLat. < Lat. *emulgēre, emuls-,* to milk out : *ex,* out + *mulgēre,* to milk.] **1.** *Chem.* A suspension of small globules of one liquid in a second liquid with which the first will not mix, as milk fats in milk. **2.** A light-sensitive coating, usu. of silver halide grains in a thin gelatin layer, on photographic film, paper, or glass. **—e·mul′sive** *adj.*

**e·munc·to·ry** (ĭ-mŭngk′tə-rē) *adj.* [ME *emunctorie* < Med. Lat. *emunctorius* < Lat. *emungere,* to blow the nose : *ex-* (intensive) + *mungere,* to blow the nose.] Functioning to carry waste matter out of the body. —*n., pl.* **-ries.** An emunctory passage or organ.

**en** (ĕn) *n.* **1.** The letter *n.* **2.** A space equal to half the width of an em in printed matter.

**en-[1]** *or* **em-** *pref.* [ME < OFr. < Lat. *in-,* in.] **1. a.** To put into or onto <*en*capsulate> **b.** To go into or onto <*en*train> **2.** To cover or provide with <*en*robe> **3.** To cause to be <*en*dear> **4.** Thoroughly. —Used often as an intensive <*en*tangle>

**en-[2]** *or* **em-** *pref.* [ME < Lat. < Gk.] In : into <*en*zootic>

**-en[1]** *suff.* [ME *-nen* < OE *-nian.*] **1. a.** To cause to be <cheap*en*> **b.** To become <redd*en*> **2. a.** To cause to have <heart*en*> **b.** To come to have <length*en*>

**-en[2]** *suff.* [ME < OE.] Made of : RESEMBLING <earth*en*>

**en·a·ble** (ĕn-ā′bəl) *vt.* **-bled, -bling, -bles. 1. a.** To supply with the means, knowledge, or chance to be or do something. **b.** To make possible. **2.** To give legal power, capacity, or sanction to.

**en·act** (ĕn-ăkt′) *vt.* **-act·ed, -act·ing, -acts. 1.** To make (e.g., a legislative bill) into law. **2.** To act out, as on a stage : REPRESENT. **—en·act′a·ble** *adj.* **—en·ac′tor** *n.*

**en·act·ment** (ĕn-ăkt′mənt) *n.* **1. a.** The act or process of enacting. **b.** The state of being enacted. **2.** Something, as a law or statute, that has been enacted.

**en·am·el** (ĭ-năm′əl) *n.* [< ME *enamelen,* to put on enamel < AN *enamailler* : *en-,* en- + *amail,* enamel < OFr. *esmail,* of Germanic orig.] **1.** A vitreous, usu. opaque, protective or decorative coating baked on metal, glass, or ceramic ware. **2.** An object, as a piece of cloisonné, with an enameled surface. **3.** A paint that dries to a hard, glossy surface. **4.** A glossy, hard surface coating resembling enamel. **5.** *Anat.* The hard, calcareous covering on the exposed portion of a tooth. —*vt.* **-eled, -el·ing, -els** *or* **-elled, -el·ling, -els. 1.** To coat, inlay, or decorate with enamel. **2.** To give a glossy surface to. **3.** To adorn, as with bright colors. **—en·am′el·er, en·am′el·ist** *n.*

**en·am·el·ware** (ĭ-năm′əl-wâr′) *n.* Ware coated with enamel.

**en·a·mine** (ĕn′ə-mēn′, ĭ-năm′ēn) *n.* [E. *en-,* chemically unsaturated + -AMINE.] An unsaturated amine having the double bond linkage C=C–N.

**en·am·or** (ĭ-năm′ər) *vt.* **-ored, -or·ing, -ors.** [ME *enamouren* < OFr. *enamourer* : *en-,* in (< Lat. *in-*) + *amour,* love < Lat. *amor* < *amare,* to love.] To inspire with love : CAPTIVATE <*enamored* of the area's natural beauty>

**en·am·our** (ĭ-năm′ər) *v. Chiefly Brit. var. of* ENAMOR.

**en·an·ti·o·mer** (ĭ-năn′tē-ə-mər) *n.* [Gk. *enantios,* opposite + -MER.] Enantiomorph. **—en·an′ti·o·mer′ic** (-mĕr′ĭk) *adj.*

**en·an·ti·o·morph** (ĕn-ăn′tē-ə-môrf′) *n.* [Gk. *enantios,* opposite + -MORPH.] Either of a pair of crystals that are similar in form but cannot be superimposed, one crystal being the mirror image of the other. **—en·an′ti·o·mor′phic, en·an′ti·o·mor′phous** *adj.* **—en·an′ti·o·morph′ism** *n.*

**en·ar·thro·sis** (ĕn′är-thrō′sĭs) *n., pl.* **-ses** (-sēz′) [NLat. < Gk. *enarthrōsis* < *enarthros,* jointed : *en-,* in + *arthron,* joint.] BALL-AND-SOCKET JOINT 2.

**e·nate** (ĭ-nāt′, ē′nāt′) *adj.* [Lat. *enatus,* p.part. of *enasci,* to issue forth : *ex-,* out + *nasci,* to be born.] **1.** Growing outward. **2.** *also* **e·nat·ic** (ĭ-năt′ĭk). Related on the mother's side. —*n.* A relative on one's mother's side.

**en bro·chette** (äN′ brô-shĕt′) *adv.* [Fr.] On a skewer.

**en·cage** (ĕn-kāj′) *vt.* **-caged, -cag·ing, -cag·es.** To confine in or as if in a cage.

**en·camp** (ĕn-kămp′) *v.* **-camped, -camp·ing, -camps.** —*vi.* To set up or live in a camp. —*vt.* To provide quarters for in a camp.

**en·camp·ment** (ĕn-kămp′mənt) *n.* **1.** The act of encamping or state of being encamped. **2.** A camp or campsite.

**en·cap·su·lant** (ĕn-kăp′sə-lənt) *n.* A material for encapsulating.

**en·cap·su·late** (ĕn-kăp′sə-lāt′) *vt. & vi.* **-lat·ed, -lat·ing, -lates.** To encase in or as if in a capsule or to become encapsulated. **—en·cap′su·la′tion** *n.*

**en·cap·su·lat·ed** (ĕn-kăp′sə-lā′təd) *adj.* Enclosed by a protective coating or membrane, as in certain bacteria.

**en·case** (ĕn-kās′) *vt.* **-cased, -cas·ing, -cas·es.** To enclose in or as if in a case. **—en·case′ment** *n.*

**en·caus·tic** (ĕn-kô′stĭk) *n.* [Lat. *encausticus* < Gk. *enkaustikos* < *enkaiein,* to paint in encaustic : *en-,* in + *kaiein,* to burn.] **1.** A paint made of pigment mixed with beeswax and fixed with heat after application. **2. a.** The art of painting with encaustic. **b.** A painting done with encaustic.

**-ence** *suff.* [ME < OFr. < Lat. *-entia* < *-ens,* -ent.] **1.** State or condition <depend*ence*> **2.** Action <emerg*ence*>

**en·ceinte[1]** (ĕn-sānt′) *adj.* [Fr. < OFr. < VLat. **incienta* < Lat. *inciens.*] With child : PREGNANT.

**en·ceinte[2]** (ĕn-sānt′, äN-săNt′) *n.* [Fr. < LLat. *incincta,* fem. p.part. of *incingere,* to surround closely : *in-,* in + *cingere,* to gird.] **1.** A fortification encircling a fort, castle, or town. **2.** The structures or area protected by an encircling fortification.

**encephal-** *pref. var. of* ENCEPHALO-.

**en·ceph·a·la** (ĕn-sĕf′ə-lə) *n. pl. of* ENCEPHALON.

**en·ce·phal·ic** (ĕn′sə-făl′ĭk) *adj.* **1.** Of or relating to the brain. **2.** Situated inside the cranial cavity.

**en·ceph·a·li·tis** (ĕn-sĕf′ə-lī′tĭs) *n.* Brain inflammation. **—en·ceph′a·lit′ic** (-lĭt′ĭk) *adj.*

**encephalitis le·thar·gi·ca** (lə-thär′jĭ-kə) *n.* [NLat., lethargic encephalitis.] A viral epidemic encephalitis often associated with influenza and characterized by apathy, double vision, and extreme muscular weakness.

**encephalo-** *or* **encephal-** *pref.* [NLat. < Gk. *(muelos) enkephalos,* (marrow) in the head: *en-,* in + *kephalē,* head.] Brain <*encephalitis*><*encephalogram*>

**en·ceph·a·lo·gram** (ĕn-sĕf′ə-lō-grăm′) *n.* **1.** An x-ray picture of the brain taken by encephalography. **2.** An electroencephalogram.

**en·ceph·a·log·ra·phy** (ĕn-sĕf′ə-lŏg′rə-fē) *n., pl.* **-phies.** Roentgenography of the brain. **—en·ceph′a·lo·graph′** (-lō-grăf′) *n.* **—en·ceph′a·lo·graph′ic** *adj.* **—en·ceph′a·lo·graph′i·cal·ly** *adv.*

**en·ceph·a·lo·ma** (ĕn-sĕf′ə-lō′mə) *n., pl.* **-mas** *or* **-ma·ta** (-mə-tə). A brain tumor.

**en·ceph·a·lo·my·e·li·tis** (ĕn-sĕf′ə-lō-mī′ə-lī′tĭs) *n.* A viral disease causing inflammation of the brain and spinal cord.

**en·ceph·a·lon** (ĕn-sĕf′ə-lŏn′) *n., pl.* **-la** (-lə) [NLat. < Gk. *enkephalon,* neuter of *enkephalos,* in the head. —see ENCEPHALO-.] The brain of a vertebrate. **—en·ceph′a·lous** *adj.*

**en·ceph·a·lop·a·thy** (ĕn-sĕf′ə-lŏp′ə-thē) *n., pl.* **-thies.** A disease of the brain. **—en·ceph′a·lo·path′ic** (-lə-păth′ĭk) *adj.*

**en·chain** (ĕn-chān′) *vt.* **-chained, -chain·ing, -chains.** [ME *encheinen* < OFr. *enchaener* : *en-,* in (< Lat. *in-*) + *chaeine,* chain < Lat. *catena.*] To fetter or confine with or as if with chains. **—en·chain′ment** *n.*

**en·chant** (ĕn-chănt′) *vt.* **-chant·ed, -chant·ing, -chants.** [ME *enchanten* < OFr. *enchanter* < Lat. *incantare* : *in-* against + *cantare,* to sing.] **1.** To cast a spell over : BEWITCH. **2.** To charm and delight completely. **—en·chant′ment** *n.*

**en·chant·er** (ĕn-chăn′tər) *n.* **1.** One that enchants. **2. a.** A sorcerer. **b.** A magician.

**en·chant·ing** (ĕn-chăn′tĭng) *adj.* Capable of enchanting : CHARMING. **—en·chant′ing·ly** *adv.*

**en·chant·ress** (ĕn-chăn′trĭs) *n.* **1. a.** A woman who practices magic. **b.** A sorceress. **2.** A fascinating woman.

**en·chase** (ĕn-chās′) *vt.* **-chased, -chas·ing, -chases.** [ME, to engrave < OFr. *enchasser,* to set gems : *en-,* in (< Lat. *in-*) + *chasse,* case < Lat. *capsa,* box.] **1.** To set (e.g., a gem). **2.** To set with or as if with gems. **3.** To adorn or decorate by inlaying or engraving.

**en·chi·la·da** (ĕn′chə-lä′də) *n.* [Mex. Sp. : *en-,* in (< Lat. *in-*) + *chile,* chili pepper < Nahuatl *chilli.*] A tortilla rolled and stuffed usu. with a mixture of meat or cheese and served with a sauce spiced with chili.

**en·chi·rid·i·on** (ĕn′kī-rĭd′ē-ən) *n., pl.* **-i·ons** *or* **-i·a** (-ē-ə) [Gk. *enkheiridion* : *en-,* in + *kheir,* hand + *-idion,* dim. suffix.] A handbook or manual.

**-enchyma** *suff.* [< PARENCHYMA.] Cellular tissue <*chlorenchyma*>

**en·ci·na** (ĕn-sē′nə) *n.* [Sp., holm oak < LLat. *ilicina* < Lat. *ilex.*] An evergreen oak, *Quercus agrifolia* of southwestern North America, grown as a shade tree.

**en·ci·pher** (ĕn-sī′fər) *vt.* **-phered, -pher·ing, -phers.** ENCODE 1. **—en·ci′pher·er** *n.* **—en·ci′pher·ment** *n.*

**en·cir·cle** (ĕn-sûr′kəl) *vt.* **-cled, -cling, -cles. 1.** To form a circle around : SURROUND. **2.** To move or go around completely : make a circuit of. **—en·cir′cle·ment** *n.*

**en·clasp** (ĕn-klăsp′) *vt.* **-clasped, -clasp·ing, -clasps.** To embrace closely.

**en·clave** (ĕn′klāv′, ŏn′-) *n.* [Fr. < OFr. *enclaver,* to enclose < VLat. **inclavare* : Lat. *in-,* in + Lat. *clavis,* key.] **1.** A country or part of a country within the boundaries of another country. **2.** A minority group preserving its own distinct culture while living within a larger group <Chinese *enclaves* in California cities>

**en·clit·ic** (ĕn-klĭt′ĭk) *n.* [LLat. *encliticus,* being an enclitic < Gk. *enklitikos* < *enklinein,* to lean on : *en-,* in + *klinein,* to lean.] A word or particle that has no independent accent and forms an accentual and sometimes also graphemic unit with the preceding word; e.g., in *"Give 'em the business,"* the particle *'em* is an enclitic.

**en·close** (ĕn-klōz′) *vt.* **-closed, -clos·ing, -clos·es.** [ME *enclosen* < OFr. *enclos,* p.part. of *enclore* < Lat. *includere,* to include : *in-,* in + *claudere,* to close.] **1.** To surround on all sides. **2. a.** To place inside a container. **b.** To insert in the same envelope or package. **3.** To place an enclosure around so as to prevent common use.

**en·clo·sure** (ĕn-klō′zhər) *n.* **1.** The act of enclosing or state of being enclosed. **2.** Something enclosed. **3.** Something that encloses.

**en·code** (ĕn-kōd′) *vt.* **-cod·ed, -cod·ing, -codes. 1.** To put (a message) into code. **2.** *Computer Sci.* To convert (a character) into an equivalent combination of bits. **—en·cod′er** *n.*

**en·co·mi·ast** (ĕn-kō′mē-ăst′, -əst) *n.* [Gk. *enkōmiastēs* < *enkōmiazein,* to praise < *enkōmion,* encomium.] One who delivers or writes an encomium : EULOGIST. **—en·co′mi·as′tic** (ĕn-kō′mē-ăs′tĭk), **en·co′mi·as′ti·cal** (-tĭ-kəl) *adj.*

**en·co·mi·um** (ĕn-kō′mē-əm) *n., pl.* **-mi·ums** *or* **-mi·a** (-mē-ə) [Lat. *encomium* < Gk. *enkōmion (epos),* (speech) praising a victor < *enkōmios,* of the victory procession : *en-,* in + *kōmos,* celebration.] **1.** Warm praise. **2.** A formal expression of praise : TRIBUTE.

**en·com·pass** (ĕn-kŭm′pəs) *vt.* **-passed, -pass·ing, -pass·es. 1.** To form a circle about : SURROUND. **2.** To envelop : enclose. **3.** To have as an integral part : INCLUDE. **4.** To achieve : accomplish. **—en·com′pass·ment** *n.*

**en·core** (ŏn′kôr′, -kōr′) *n.* [Fr., again.] **1.** An audience's demand for an additional performance. **2.** An additional performance in response to audience demand. *—vt.* **-cored, -cor·ing, -cores.** To demand an encore of. *—interj.* —Used to demand an additional performance.

**en·coun·ter** (ĕn-koun′tər) *n.* [ME *encountre* < OFr. *encontrer,* to meet < LLat. *incontrare* : Lat. *in-,* in + Lat. *contra,* against.] **1.** An unexpected or unplanned meeting. **2.** A usu. brief meeting. **3. a.** A hostile confrontation. **b.** An often violent meeting : CLASH. *—v.* **-tered, -ter·ing, -ters.** *—vt.* **1.** To meet or come upon (someone or something), esp. unexpectedly. **2.** To confront in a hostile situation. **3.** To be faced with <*encounter* strong enemy forces> *—vi.* To meet, esp. unexpectedly.

**encounter group** *n.* A usu. unstructured therapy group in which individuals try to increase their sensitivity and responsiveness, reveal their feelings, and relate to others openly and intimately, as by touching or speaking freely.

**en·cour·age** (ĕn-kûr′ĭj, -kŭr′-) *vt.* **-aged, -ag·ing, -ag·es.** [ME *encouragen* < OFr. *encoragier* : *en-* (causative) < Lat. *in-,* in + *corage,* courage < Lat. *cor,* heart.] **1.** To inspire with hope, courage, or confidence : HEARTEN. **2.** To give support to : FOSTER. **3.** To stimulate. **—en·cour′age·ment** (-mənt) *n.* **—en·cour′ag·er** *n.*

☆ ***syns:*** ENCOURAGE, CHEER (up), HEARTEN, NERVE, PERK UP *v. core meaning* : to impart strength and confidence to <were *encouraged* by the doctors' findings> ***ant:*** discourage

**en·cour·ag·ing** (ĕn-kûr′ə-jĭng, -kŭr′-) *adj.* Imparting hope or courage. **—en·cour′ag·ing·ly** *adv.*

**en·croach** (ĕn-krōch′) *vi.* **-croached, -croach·ing, -croach·es.** [ME *encrochen,* to seize illegally < OFr. *encrochier,* to seize : *en-,* in (< Lat. *in-* ) + *croc,* hook, of Scand. orig.] **1.** To intrude gradually on the rights or possessions of another <road construction *encroaching* on the city park> **2.** To advance beyond proper or prescribed limits. **—en·croach′er** *n.* **—en·croach′ment** *n.*

**en·crust** (ĕn-krŭst′) *vt.* **-crust·ed, -crust·ing, -crusts.** [Prob. < Fr. *incruster* < Lat. *incrustare* : *in-* upon + *crusta,* crust.] To cover or surmount with or as if with a crust <a coffee cake *encrusted* with icing> **—en′crust·a′tion** *n.*

**en·crypt** (ĕn-krĭpt′) *vt.* **-crypted, -crypt·ing, -crypts. 1.** ENCODE 1. **2.** *Computer Sci.* To scramble (access codes) in order to prevent illicit entry into a system.

**en·cryp·tion** (ĕn-krĭp′shən) *n. Computer Sci.* A process for scrambling access codes to prevent illicit entry into a system.

**en·cum·ber** (ĕn-kŭm′bər) *vt.* **-bered, -ber·ing, -bers.** [ME *encombren* < OFr. *encombrer,* to block up : *en-,* in (< Lat. *in-*) + *combre,* hindrance.] **1.** To weigh down unduly. **2.** To impede or hinder the action or performance of. **3.** To burden, as with legal claims.

**en·cum·brance** (ĕn-kŭm′brəns) *n.* **1.** One that encumbers : IMPEDIMENT. **2.** *Law.* A lien or claim on property.

**en·cum·branc·er** (ĕn-kŭm′brən-sər) *n. Law.* One holding an encumbrance.

**-ency** *suff.* [Lat. *-entia.* —see -ENCE.] -ENCE.

**en·cyc·li·cal** (ĕn-sĭk′lĭ-kəl) *adj.* [LLat. *encyclicus,* circular < Gk. *enkuklios* : *en-,* in + *kuklos,* circle.] Intended for general circulation. *—n. Rom. Cath. Ch.* A papal letter addressed to bishops or the hierarchy of a specific country.

**en·cy·clo·pe·di·a** *or* **en·cy·clo·pae·di·a** (ĕn-sī′klə-pē′dē-ə) *n.* [Med. Lat. *encyclopaedia,* general education course < Gk. *enkuklopaideiā* < *enkuklios paideia,* general education.] A comprehensive reference work having articles on a broad range of subjects or on numerous aspects of a given field, usu. arranged alphabetically.

**en·cy·clo·pe·dic** *or* **en·cy·clo·pae·dic** (ĕn-sī′klə-pē′dĭk) *adj.* **1.** Of, relating to, or typical of an encyclopedia. **2.** Embracing many subjects : COMPREHENSIVE <*encyclopedic* learning> **—en·cy′clo·pe′di·cal·ly** *adv.*

**en·cy·clo·pe·dism** *or* **en·cy·clo·pae·dism** (ĕn-sī′klə-pē′dĭz′əm) *n.* Encyclopedic knowledge : ERUDITION.

**en·cy·clo·pe·dist** *or* **en·cy·clo·pae·dist** (ĕn-sī′klə-pē′dĭst) *n.* A writer for or compiler of an encyclopedia.

**en·cyst** (ĕn-sĭst′) *v.* **-cyst·ed, -cyst·ing, -cysts.** *—vt.* To enclose in or as if in a cyst. *—vi.* To form or become enclosed in a cyst. **—en·cyst′ment, en′cys·ta′tion** *n.*

**end** (ĕnd) *n.* [ME *ende* < OE.] **1.** Either extremity of an object having length. **2.** The outside or extreme edge : BOUNDARY. **3.** The point at which an act, event, or phenomenon ceases or is completed :

CONCLUSION <the *end* of the campaign> **4.** A result : outcome. **5.** Termination of life : DEATH. **6.** An ultimate extent : LIMIT <the *end* of one's patience> **7.** A goal toward which one strives. **8.** A remnant. **9. a.** A share of a responsibility or obligation <your *end* of the project> **b.** A particular area of responsibility <in charge of the publicity *end* of the campaign> **10.** *Football.* **a.** Either of the players in the outermost position at the line of scrimmage. **b.** The position played by such a player. —*v.* **end·ed, end·ing, ends.** —*vt.* **1.** To bring to an end : FINISH. **2.** To form the end or concluding part of. **3.** To destroy. —*vi.* **1.** To come to an end : CEASE. **2.** To die. **—make (both) ends meet.** To live within one's means. **—no end.** *Informal.* A great deal <*no end* of odd jobs to finish>

☆ ***syns:*** END, CLOSE, COMPLETE, CONCLUDE, FINISH, TERMINATE, WIND UP *v. core meaning* : to bring to a natural or proper end <*ended* the letter with thanks><a performance that *ended* at 10 P.M.> ***ant:*** begin, start

**end-** *pref. var. of* ENDO-.

**en·da·moe·ba** *or* **en·da·me·ba** (ĕn'də-mē'bə) *n. vars. of* ENTAMOEBA.

**en·dan·ger** (ĕn-dān'jər) *vt.* **-gered, -ger·ing, -gers.** To expose to danger or harm : IMPERIL. **—en·dan'ger·ment** *n.*

**endangered** *adj.* Threatened with extinction <an *endangered* species>

**end·ar·te·rec·to·my** (ĕn'dăr-tə-rĕk'tə-mē) *n., pl.* **-mies.** [NLat. *endarterium*, inner lining of an artery + -ECTOMY.] Surgical excision of the inner lining of an artery clogged with atherosclerotic buildup.

**end·ar·te·ri·tis** (ĕn'dăr-tə-rī'tĭs) *n.* [NLat. *endarterium*, inner lining of an artery + -ITIS.] Inflammation of the inner lining of an artery.

**end·brain** (ĕnd'brān') *n.* The telencephalon.

**en·dear** (ĕn-dîr') *vt.* **-deared, -dear·ing, -dears. 1.** To make beloved. **2.** *Obs.* To increase the value of.

**en·dear·ing** (ĕn-dîr'ĭng) *adj.* Arousing affection or warm sympathy <an *endearing* child> **—en·dear'ing·ly** *adv.*

**en·dear·ment** (ĕn-dîr'mənt) *n.* **1.** The act of endearing. **2.** An expression of affection, as a caress.

**en·deav·or** (ĕn-dĕv'ər) *n.* [ME *endevour* < *endeveren*, to make an effort : *en-*, en- + *dever*, duty < OFr. *devoir*, duty. —see DEVOIR.] A concerted or conscientious effort toward a given end. —*vi.* **-ored, -or·ing, -ors.** To make an earnest attempt. **—en·deav'or·er** *n.*

**en·deav·our** (ĕn-dĕv'ər) *n.* & *v. Chiefly Brit. var. of* ENDEAVOR.

**en·dem·ic** (ĕn-dĕm'ĭk) *adj.* [Gk. *endēmos* : *en-*, in + *dēmos*, people.] **1.** Prevalent in or peculiar to a particular locality or people. **2.** *Ecol.* Native or limited to a certain region. **3.** *Med.* Peculiar to a particular locality <an *endemic* disease> —*n. Ecol.* An endemic plant or animal. **—en·dem'i·cal·ly** *adv.* **—en·dem'ism** *n.*

**end·er·gon·ic** (ĕn'dər-gŏn'ĭk) *adj.* [END(O)- + Gk. *ergon*, work + -IC.] Needing energy.

**en·der·mic** (ĕn-dûr'mĭk) *adj.* Acting medicinally by absorption through the skin. **—en·der'mi·cal·ly** *adv.*

**end game** *n.* The last stage of a game, esp. the stage of a game of chess following major reduction of forces.

**end·ing** (ĕn'dĭng) *n.* **1.** A conclusion : termination. **2.** The concluding part : FINALE. **3.** An element added to a word base, esp. to create an inflected form.

**en·dive** (ĕn'dīv', ŏn'dēv') *n.* [ME < OFr. < Lat. *intibum.*] **1.** A plant, *Cichorium endivia*, cultivated for its crown of crisp, succulent leaves used in salads. **2.** A variety of the common chicory, *Cichorium intybus*, cultivated to produce a narrow, pointed cluster of whitish leaves used in salads.

**end leaf** *n.* An endpaper.

**end·less** (ĕnd'lĭs) *adj.* **1.** Being or appearing to be without an end : BOUNDLESS. **2.** Interminable : incessant <an *endless* sermon> **3.** Formed with the ends joined : CONTINUOUS <*endless* coils of rope> **—end'less·ly** *adv.* **—end'less·ness** *n.*

**end·long** (ĕnd'lông', -lŏng') *adv. Archaic.* Lengthwise.

**end man** *n.* **1.** The person at the end of a line or row. **2.** The man in a minstrel show who sits at one end of the company and banters with the interlocutor.

**end matter** *n.* Back matter.

**end·most** (ĕnd'mōst') *adj.* Being at or nearest to the end : LAST.

**endo-** *or* **end-** *pref.* [Gk. < *endon*, within.] Inside : within <*endometrium*>

**en·do·bi·ot·ic** (ĕn'də-bī-ŏt'ĭk) *adj.* Living within the tissues of a host.

**en·do·blast** (ĕn'də-blăst') *also* **en·to·blast** (ĕn'tə-) *n.* The inner layer of the blastoderm. **—en'do·blas'tic** *adj.*

**en·do·car·di·tis** (ĕn'dō-kär-dī'tĭs) *n.* [ENDOCARD(IUM) + -ITIS.] Inflammation of the endocardium. **—en'do·car·dit'ic** (-dĭt'ĭk) *adj.*

**en·do·car·di·um** (ĕn'dō-kär'dē-əm) *n., pl.* **-di·a** (-dē-ə) [NLat. : ENDO- + Gk. *kardia*, heart.] The thin, endothelial, serous membrane lining the interior of the heart. **—en'do·car'di·al** *adj.*

**en·do·carp** (ĕn'də-kärp') *n.* The often hard or leathery inner layer of the pericarp of many fruits. **—en'do·car'pal** *adj.*

**en·do·cra·ni·um** (ĕn'dō-krā'nē-əm) *n., pl.* **-ni·a** (-nē-ə). The dura mater.

**en·do·crine** (ĕn'də-krĭn, -krēn', -krīn') *also* **en·do·crin·ic** (-krĭn'ĭk) *or* **en·doc·ri·nous** (-dŏk'rə-nəs) *adj.* [ENDO- + Gk. *krinein*, to separate.] **1.** Secreting internally. **2.** Of or relating to any of the ductless or endocrine glands. —*n.* **1.** The internal secretion of a gland. **2.** An endocrine gland.

**endocrine gland** *n.* One of the ductless glands, as the thyroid or adrenal, the secretions of which pass directly into the bloodstream from the cells of the gland.

**en·do·cri·nol·o·gy** (ĕn'də-krə-nŏl'ə-jē) *n.* Physiology of the endocrine glands. **—en'do·cri'no·log'ic** (-krĭn'ə-lŏj'ĭk) *adj.* **—en'do·cri·nol'o·gist** *n.*

**en·do·derm** (ĕn'də-dûrm') *also* **en·to·derm** (ĕn'tə-) *n.* The innermost of the three primary embryonic germ layers, developing into the intestinal tract and allied structures. **—en'do·der'mal** *adj.*

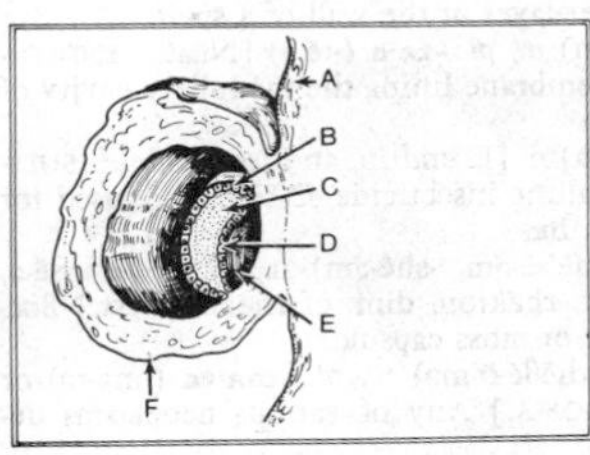


**endoderm**
*A. uterine wall, B. ectoderm, C. mesoderm, D. endoderm, E. amniotic cavity, F. trophoblast*

**en·do·der·mis** (ĕn'də-dûr'mĭs) *n.* [NLat. : ENDO- + Gk. *derma*, skin.] *Bot.* The innermost layer of the cortex in many plants.

**en·do·don·tics** (ĕn'dō-dŏn'tĭks) *also* **en·do·don·tia** (-shə, -shē-ə) *n.* (*sing. in number*). The branch of dentistry concerned with tooth pulp diseases. **—en'do·don'tic** *adj.* **—en'do·don'tist** *n.*

**en·do·en·zyme** (ĕn'dō-ĕn'zīm') *n.* An enzyme that acts inside the cell that produces it.

**en·do·er·gic** (ĕn'dō-ûr'jĭk) *adj.* [ENDO- + Gk. *ergon*, work + -IC.] Endothermic.

**en·dog·a·my** (ĕn-dŏg'ə-mē) *n.* **1.** Marriage within a particular group, caste, class, or tribe consistent with custom or law. **2.** *Bot.* Fertilization in which pollen is transferred to another flower of the same plant. **—en·dog'a·mous** *adj.*

**en·dog·e·nous** (ĕn-dŏj'ə-nəs) *adj.* **1.** Produced from within. **2.** *Biol.* Originating within an organ or part. **—en·dog'e·nous·ly** *adv.* **—en·dog'e·ny** *n.*

**en·do·lymph** (ĕn'də-lĭmf') *n.* The fluid in the cochlear duct of the labyrinth of the ear. **—en'do·lym·phat'ic** *adj.*

**en·do·me·tri·um** (ĕn'dō-mē'trē-əm) *n., pl.* **-tri·a** (-trē-ə). The mucous membrane lining the uterus. **—en'do·me'tri·al** *adj.*

**en·do·morph** (ĕn'də-môrf') *n.* **1.** A mineral found as an inclusion in another, as rutile or tourmaline in quartz. **2.** *Physiol.* An endomorphic person.

**en·do·mor·phic** (ĕn'də-môr'fĭk) *adj.* **1. a.** Of or relating to an endomorph. **b.** Created through endomorphism. **2.** *Physiol.* Marked by prominence of the abdomen and other soft body parts developed from the embryonic endodermal layer. **—en'do·mor'phy** *n.*

**en·do·morph·ism** (ĕn'də-môr'fĭz'əm) *n.* **1.** Metamorphism of igneous rock as it cools, resulting from contact with and the assimilation of the wall rock. **2.** *Math.* A homomorphism that maps a mathematical set into itself.

**en·do·par·a·site** (ĕn'dō-păr'ə-sīt') *n.* An organism, as a tapeworm, living parasitically within another organism.

**en·do·phyte** (ĕn'də-fīt') *n.* A plant, as a fungus, growing within another plant. **—en'do·phyt'ic** (-dō-fĭt'ĭk) *adj.*

**en·do·plasm** (ĕn'də-plăz'əm) *n.* A low-viscosity portion of the continuous phase of cytoplasm distinguishable within some cells.

**en·do·ra·di·o·sonde** (ĕn'də-rā'dē-ō-sŏnd') *n.* A microelectronic device introduced into the body by swallowing so as to record physiological data.

**end organ** *n.* The expanded functional termination of a sensory nerve or of a motor nerve in tissue.

**en·dor·phin** (ĕn-dôr'fĭn) *n.* [END(O)- + (M)ORPHIN(E).] Any of a group of hormones with tranquilizing and pain-killing capabilities that are secreted by the brain.

**en·dorse** (ĕn-dôrs') *vt.* **-dorsed, -dors·ing, -dors·es.** [ME *endosen* < AN *endosser* < Med. Lat. *indorsare* : Lat. *in-*, upon + Lat. *dorsum*, back.] **1.** To write one's signature on the back of (e.g., a check) as evidence that ownership has been legally transferred, esp. in return for the cash or credit indicated on the front. **2.** To place (one's signature), as on a contract, to show approval of its contents or terms. **3.** To acknowledge (receipt of payment) by signing a bill, draft, or other instrument. **4.** To give approval of or support to : SANCTION <*endorse* a political candidate> **—en·dors'a·ble** *adj.* **—en·dors'er, en·dor'sor** *n.*

**en·dor·see** (ĕn'dôr-sē') *n.* One to whom ownership of a negotiable document is transferred by endorsement.

**en·dorse·ment** (ĕn-dôrs'mənt) *n.* **1.** An act of endorsing. **2.** Something, as a signature or voucher, that endorses or validates. **3.** Sanction : approbation. **4.** An amendment to a contract, as an insurance policy, allowing a change in the original terms.

**endorsement in blank** *n.* A blank endorsement.

**en·do·scope** (ĕn'də-skōp') *n.* An instrument for examining the interior of a bodily canal or hollow organ. **—en'do·scop'ic** (ĕn'də-skŏp'ĭk) *adj.* **—en'do·scop'ic·al·ly** *adv.*

**en·do·skel·e·ton** (ĕn'dō-skĕl'ĭ-tən) *n.* An internal supporting skeleton typical of vertebrates. **—en'do·skel'e·tal** *adj.*

**en·dos·mo·sis** (ĕn'dŏz-mō'sĭs, -dŏs-) *n.* Osmosis toward the interior of a cell or cavity. **—en'dos·mot'ic** (-mŏt'ĭk) *adj.*

**en·do·some** (ĕn'də-sōm') *n.* A discrete cellular particle of chromatin near the center of a nucleus.

**en·do·sperm** (ĕn'də-spûrm') *n.* The nutritive tissue of a plant seed, surrounding and absorbed by the embryo.

**en·do·spore** (ĕn'də-spôr', -spōr') *n.* **1.** A small asexual spore formed by some bacteria. **2.** The inner layer of the wall of a spore.

**en·dos·te·um** (ĕn-dŏs'tē-əm) *n., pl.* **-te·a** (-tē-ə) [NLat. : END(O)- + Gk. *osteon,* bone.] The membrane lining the medullary cavity of a bone. **—en·dos'te·al** *adj.*

**en·do·sul·fan** (ĕn'də-sŭl'fən) *n.* [E. *endrin,* an insecticide + SULF- + -AN.] A highly toxic crystalline insecticide, $C_9H_6Cl_6O_3S$, used for controlling crop insects and mites.

**en·do·the·ci·um** (ĕn'dō-thē'sē-əm, -shē-əm) *n., pl.* **-ci·a** (-sē-ə, -shē-ə) [NLat. : ENDO- + Gk. *thēkion,* dim. of *thēkē,* chest.] *Bot.* The inner tissue of an anther or moss capsule.

**en·do·the·li·o·ma** (ĕn'dō-thē'lē-ō'mə) *n., pl.* **-ma·ta** (-mə-tə) *or* **-mas.** [ENDOTHELI(UM) + -OMA.] Any of various neoplasms derived from endothelial tissue.

**en·do·the·li·um** (ĕn'dō-thē'lē-əm) *n., pl.* **-li·a** (-lē-ə) [NLat. : ENDO- + Gk. *thēlē,* nipple.] A thin layer of flat cells that lines serous cavities, lymph vessels, and blood vessels. **—en'do·the'li·al, en'do·the'loid'** (-thē'loid') *adj.*

**en·do·therm** (ĕn'də-thûrm') *n.* Homoiotherm.

**en·do·ther·mic** (ĕn'dō-thûr'mĭk) *also* **en·do·ther·mal** (-məl) *adj.* Marked by or causing the absorption of heat.

**en·do·tox·in** (ĕn'dō-tŏk'sən) *n.* A toxin produced within a microorganism and released upon destruction of the cell in which it is produced.

**en·do·tra·che·al** (ĕn'də-trā'kē-əl) *adj.* Within the trachea.

**en·dow** (ĕn-dou') *vt.* **-dowed, -dow·ing, -dows.** [ME *endowen* < AN *endouer* : OFr. *en-* (intensive < Lat. *in-*) + *douer,* to provide with a dowry (< Lat. *dotare* < *dos,* dowry).] **1.** To supply with property, income, or a source of income. **2.** To equip with a quality or talent. **3.** *Obs.* To provide with a dower.

**en·dow·ment** (ĕn-dou'mənt) *n.* **1.** An act of endowing. **2.** Funds or property donated to an institution, individual, or group to produce income. **3.** A natural ability, gift, or quality.

**end·pa·per** *also* **end paper** (ĕnd'pā'pər) *n.* Either of two folded sheets of heavy paper with one half pasted to the inside front or back cover of a book and the other half pasted to the base of the first or last page to form a flyleaf.

**end·plate** (ĕnd'plāt') *n.* A motor nerve terminal that transmits nerve impulses to muscle.

**end·point** (ĕnd'point') *n.* **1.** Either of two points marking the end of a line segment. **2.** *Chem.* The point in a volumetric titration at which the amount of added reagent is chemically equivalent to the solution titrated.

**end table** *n.* A small table, usu. placed at either end of a sofa or next to a chair.

**en·due** (ĕn-do͞o', -dyo͞o') *vt.* **-dued, -du·ing, -dues.** [Partly < ME *enduen* < OFr. *enduire,* to lead in < Lat. *inducere* and partly < ME *induen,* to clothe (< Lat. *induere*).] To provide with a quality, trait, or power.

**en·dur·a·ble** (ĕn-do͝or'ə-bəl, -dyo͝or'-) *adj.* Capable of being endured : BEARABLE. **—en·dur'a·bly** *adv.*

**en·dur·ance** (ĕn-do͝or'əns, -dyo͝or'-) *n.* **1.** The quality, act, or power of withstanding stress. **2.** The state or fact of persevering. **3.** Continuing existence : DURATION.

**en·dure** (ĕn-do͝or', -dyo͝or') *v.* **-dured, -dur·ing, -dures.** [ME *enduren* < OFr. *endurer* < Lat. *indurare,* to make hard : *in-* (intensive) + *durare,* to harden < *durus,* hard.] *—vt.* **1.** To carry on through despite hardships : UNDERGO <*endure* weeks of travel in the desert> **2.** To bear with tolerance <could not *endure* the bitter cold> *—vi.* **1.** To continue in existence : LAST <fine paper that *endures* for many years> **2.** To suffer patiently without yielding.

**en·dur·ing** (ĕn-do͝or'ĭng, -dyo͝or'-) *adj.* **1.** Lasting : durable. **2.** Unresolved : chronic <an *enduring* shortage> **3.** Long-suffering. **—en·dur'ing·ly** *adv.* **—en·dur'ing·ness** *n.*

**en·dur·o** (ĕn-do͝or'ō, -dyo͝or'ō) *n., pl.* **-os.** [Shortening and alteration of ENDURANCE.] A long race, as of cars or runners, emphasizing endurance instead of speed.

**end·wise** (ĕnd'wīz') *also* **end·ways** (-wāz') *adv.* **1.** On end. **2.** With the end foremost. **3.** Lengthwise. **4.** End to end.

**En·dym·i·on** (ĕn-dĭm'ē-ən) *n.* [Lat. < Gk. *Endumiōn.*] *Gk. Myth.* A handsome young man who was loved by a moon goddess and whose youth was preserved by eternal sleep.

**-ene** *suff.* [< Gk. *-ēnē,* fem. adj. suffix.] An unsaturated organic compound, esp. one having a double bond <acetyl*ene*>

**en·e·ma** (ĕn'ə-mə) *n., pl.* **en·e·mas** *or* **en·e·ma·ta** (ĕn'ə-mä'tə) [LLat. < Gk. < *enienai,* to inject : *en-,* in + *hienai,* to send.] **1.** Injection of liquid into the rectum through the anus for cleansing or for other therapy. **2.** The fluid injected as an enema.

**en·e·my** (ĕn'ə-mē) *n., pl.* **-mies.** [ME *enemi* < OFr. < Lat. *inimicus* : *in-,* not + *amicus,* friend.] **1.** One feeling or displaying hostility or malice toward another : FOE. **2. a.** A hostile force or power. **b.** A member or unit of such a force. **3.** Something having destructive effects <Greed is the *enemy* of loyalty.> *—adj.* Of or relating to a hostile power or force.

**en·er·get·ic** (ĕn'ər-jĕt'ĭk) *adj.* [Gk. *energētikos* < *energein,* to be active < *energos,* active. —see ENERGY.] Having, exerting, or displaying energy. **—en'er·get'i·cal·ly** *adv.*

☆ ***syns:*** ENERGETIC, DYNAMIC, FORCEFUL, KINETIC, LIVELY, PEPPY, SPRIGHTLY, VIGOROUS *adj. core meaning* : having, exerting, or displaying energy <an *energetic* worker><an *energetic* attempt to win the race> ***ant:*** phlegmatic, unenergetic

**en·er·get·ics** (ĕn'ər-jĕt'ĭks) *n. (sing. in number).* The physics of energy and its transformations.

**en·er·gid** (ĕn'ər-jĭd') *n.* [ENERG(Y) + -ID.] A unit made up of a nucleus surrounded by cytoplasm, with or without a cell wall, that does not constitute a cell.

**en·er·gize** (ĕn'ər-jīz') *v.* **-gized, -giz·ing, -giz·es.** *—vt.* To impart energy to. *—vi.* To release or put out energy. **—en'er·giz'er** *n.*

**en·er·gy** (ĕn'ər-jē) *n., pl.* **-gies.** [LLat. *energia* < Gk. *energeia* < *energos,* active : *en-,* at + *ergon,* work.] **1. a.** Vigor in performance of an action. **b.** Vitality in expression. **2.** The capacity for action or accomplishment <enough *energy* to finish the job ahead of time> **3.** *Physics.* The work a physical system is capable of doing in changing from its actual state to a specified reference state, with the total gen. including contributions of potential energy, kinetic energy, and rest energy. **4.** Usable heat or electric power.

**energy density** *n.* The energy per unit volume of a region of space.

**energy level** *n.* **1.** The energy typical of a stationary state of a quantum mechanical system. **2.** The stationary state of a quantum mechanical system.

**en·er·vate** (ĕn'ər-vāt') *vt.* **-vat·ed, -vat·ing, -vates.** [Lat. *enervare, enervat-* : *ex-,* out + *nervus,* sinew.] To deprive of vitality : WEAKEN. *—adj.* (ĭ-nûr'vĭt). Deprived of vitality. **—en'er·va'tion** *n.* **—en'er·va'tor** *n.*

**en·face** (ĕn-fās') *vt.* **-faced, -fac·ing, -fac·es.** To write on the face of (e.g., a check or document). **—en·face'ment** *n.*

**en·fant ter·ri·ble** (äɴ-fäɴ' tĕ-rē'blə) *n., pl.* **en·fants ter·ri·bles** (äɴ-fäɴ' tĕ-rē'blə) [Fr. : *enfant,* child + *terrible,* terrible.] One whose unconventional behavior and ideas cause embarrassment or dismay.

**en·fee·ble** (ĕn-fē'bəl) *vt.* **-bled, -bling, -bles.** [ME *enfeblen* < OFr. *enfeblir* : *en-* (causative < Lat. *in-,* in) + *feble,* feeble. —see FEEBLE.] To deprive of strength. **—en·fee'ble·ment** *n.* **—en·fee'bler** *n.*

**en·feoff** (ĕn-fĕf', -fēf') *vt.* **-feoffed, -feoff·ing, -feoffs.** [ME *enfeffen* < OFr. *enfeffer* : *en-,* in (< Lat. *in-*) + *fief,* fee.] To invest with an estate or fee. **—en·feoff'ment** *n.*

**en·fet·ter** (ĕn-fĕt'ər, ĭn-) *vt.* **-tered, -ter·ing, -ters.** To enchain.

**En·field rifle** (ĕn'fēld') *n.* [After *Enfield,* England.] Any of several rifles of varying calibers used at one time by British and American troops, esp. the .30 or .303 breechloading, belt-action model.

**en·fi·lade** (ĕn'fə-lād', -läd') *n.* [Fr. < *enfiler,* to rake with gunfire < OFr. to thread : *en-,* in (< Lat. *in-*) + *fil,* thread < Lat. *filum.*] The firing of a gun or guns so as to sweep the length of a target. *—vt.* **-lad·ed, -lad·ing, -lades.** To rake with gunfire lengthwise.

**en·fleu·rage** (ŏɴ'flə-räzh', -räj') *n.* [Fr. < *enfleurer,* to saturate with the perfume of flowers : *en-* (< Lat. *in-*) + *fleur,* flower < OFr. *flour* < Lat. *flos.*] A process in perfumery by which odorless fats or oils are exposed to the exhaled fragrance of fresh flowers.

**en·flur·ane** (ĕn-flo͝or'ān') *n.* [EN- + (TRI)FLU(O)R(OETH)ANE.] A nonexplosive anesthetic, $C_3H_2ClF_5O$.

**en·fold** (ĕn-fōld') *vt.* **-fold·ed, -fold·ing, -folds. 1.** To cover with or as if with folds : ENVELOP. **2.** To hold within limits : ENCLOSE. **3.** EMBRACE 1. **—en·fold'er** *n.*

**en·force** (ĕn-fôrs', -fōrs') *vt.* **-forced, -forc·ing, -forc·es.** [ME *enforcen* < OFr. *enforcier,* to make strong < VLat. **infortiare* : *in-,* in + *fortis,* strong.] **1.** To compel observance of or obedience to. **2.** To bring about by force. **3.** To give force to : REINFORCE. **—en·force'a·ble** *adj.* **—en·force'ment** *n.* **—en·forc'er** *n.*

☆ ***syns:*** ENFORCE, EXECUTE, IMPLEMENT, INVOKE *v. core meaning* : to compel observance of (a law or regulation) <strictly *enforced* the curfew>

**en·fran·chise** (ĕn-frăn'chīz') *vt.* **-chised, -chis·ing, -chis·es.** [ME *enfraunchisen* < OFr. *enfranchir, enfranchiss-* : *en-* (causative < Lat. *in-,* in) + *franc,* free. —see FRANK[1].] **1.** To bestow a franchise on. 2. To endow with the rights of citizenship, esp. the right to vote. 3. To free, as from bondage.

**en·gage** (ĕn-gāj′) *v.* **-gaged, -gag·ing, -gag·es.** [ME *engagen,* to pledge something as security for repayment of debt < OFr. *engager* : *en-,* in (< Lat. *in-*) + *gage,* pledge, of Germanic orig.] —*vt.* 1. To obtain or contract for the services of : EMPLOY <*engage* an architect> 2. To contract for the use of : RESERVE <*engage* a car> 3. To obtain and hold the attention of : ENGROSS <*engaged* the children's full support> 4. To require the use of : OCCUPY <Training *engages* most of an athlete's time.> 5. To promise, esp. to marry. 6. To enter or bring into conflict with <We have *engaged* the enemy.> 7. To interlock or cause to interlock : MESH. 8. To win : attract. 9. To involve : entangle <*engaged* me in their silly quarrel> 10. *Archaic.* To give or take as security. —*vi.* 1. To involve oneself or become occupied : PARTICIPATE <*engage* in a discussion> 2. To assume an obligation : AGREE. 3. To enter into conflict or battle. 4. To become meshed or interlocked. **—en·gag′er** *n.*

**en·ga·gé** (ĕn′gä-zhā′) *adj.* [Fr., p.part. of *engager,* to engage < OFr.] Actively committed, as to a political cause.

**en·gaged** (ĕn-gājd′) *adj.* 1. Employed or occupied : BUSY. 2. Committed to. 3. Pledged to marry. 4. Involved in conflict or battle. 5. Being in gear : MESHED. 6. Partly sunk, built into, or attached to another part, as columns on a wall.

**en·gage·ment** (ĕn-gāj′mənt) *n.* 1. The act of engaging or state of being engaged. 2. Betrothal. 3. One that engages. 4. A promise to appear at a certain time : APPOINTMENT. 5. a. Employment, esp. for a given period of time. b. A period of employment. 6. A battle. 7. The state of being in gear.

**en·gag·ing** (ĕn-gā′jĭng) *adj.* Charming. **—en·gag′ing·ly** *adv.*

**en garde** (äN gärd′) *interj.* [Fr. : *en,* on + *garde,* guard.] —Used to warn a fencer to assume the first position preparatory to a match.

**en·gar·land** (ĕn-gär′lənd) *vt.* **-land·ed -land·ing, -lands.** To deck or encircle with or as if with a garland.

**en·gen·der** (ĕn-jĕn′dər) *v.* **-dered, -der·ing, -ders.** [ME *engendren* < OFr. *engendrer* < Lat. *ingenerare* : *in-,* in + *generare,* to produce < *genus,* birth.] —*vt.* 1. To give rise to. 2. To procreate : propagate. —*vi.* To come into existence.

**en·gine** (ĕn′jĭn) *n.* [ME *engin* < OFr., skill < Lat. *ingenium.*] 1. a. A machine that converts energy into mechanical motion. b. A mechanical appliance, instrument, or tool <*engines* of destruction> 2. A locomotive. 3. *Archaic.* An agent, instrument, or means of accomplishment.

**engine block** *n.* The cast metal block housing the cylinders of an internal-combustion engine.

**en·gi·neer** (ĕn′jə-nîr′) *n.* [ME *enginer* < OFr. *engigneor* < Med. Lat. *ingeniator,* contriver < *ingeniare,* to contrive < *ingenium,* skill.] 1. One trained or professionally engaged in a branch of engineering. 2. One who manages an enterprise in a shrewd or skillful way. 3. One who operates an engine. —*vt.* **-neered, -neer·ing, -neers.** 1. To plan, construct, and manage as an engineer. 2. To plan, manage, and accomplish by skillful acts or contrivance : MANEUVER <*engineered* a meeting for the couple>

**en·gi·neer·ing** (ĕn′jə-nîr′ĭng) *n.* 1. The application of mathematical and scientific principles to practical ends, as the design, construction, and operation of economical and efficient structures, equipment, and systems. 2. An engineer's profession or work.

**en·gird** (ĕn-gûrd′) *vt.* **-girt** (-gûrt′) *or* **-gird·ed, -gird·ing, -girds.** To encircle : surround.

**en·gir·dle** (ĕn-gûr′dl) *vt.* **-dled, -dling, -dles.** To encircle or surround with or as if with a girdle.

**en·girt** (ĕn-gûrt′) *v. var. p.t. & p.p. of* ENGIRD.

**en·gla·cial** (ĕn-glā′shəl) *adj.* Located or occurring within a glacier.

**Eng·lish** (ĭng′glĭsh) *adj.* [ME *Englisch* < OE *Englisc* < *Engle,* the Angles.] Of, relating to, derived from, or typical of England, its people, or its culture. —*n.* 1. The people of England. 2. a. The West Germanic language of England, the United States, and other countries that are or have been under English control or influence. b. The English language of a specific time, area, person, or group of people <Australian *English*> 3. A translation into or an equivalent in the English language. 4. A course or class in the study of English literature, language, or composition. 5. Fourteen-point printing type. 6. *often* **english.** The spin given to a ball by striking it on one side, as in pool, or releasing it with a sharp twist, as in bowling. —*vt.* **-lished, -lish·ing, -lish·es.** 1. To translate into English. 2. To adapt into English : ANGLICIZE.

**English daisy** *n.* DAISY 2.

**English finish** *n.* A smooth, nonglossy paper finish.

**English horn** *n.* A double-reed woodwind musical instrument similar to but larger than the oboe and pitched lower by a fifth.

**Eng·lish·man** (ĭng′glĭsh-mən) *n.* A man who is English by birth, descent, or naturalization.

**English muffin** *n.* A flat round of yeast dough baked on a griddle and usu. split and toasted before eating.

**English plantain** *n.* Ribgrass.

**English setter** *n.* A dog orig. bred in England, having a silky white coat usu. with black or brownish markings.

**English sheepdog** *n.* The Old English sheepdog.

**English sonnet** *n.* A Shakespearean sonnet.

**English sparrow** *n.* The house sparrow.

**English walnut** *n.* 1. A Eurasian tree, *Juglans regia,* cultivated in southern Europe and California. 2. The large edible nut of the English walnut tree.

**Eng·lish·wom·an** (ĭng′glĭsh-wo͝om′ən) *n.* A woman who is English by birth, descent, or naturalization.

**en·glut** (ĕn-glŭt′) *vt.* **-glut·ted, -glut·ting, -gluts.** [OFr. *englotir* < LLat. *inglutire* : Lat. *in-,* in + Lat. *gluttire,* to swallow.] To swallow greedily : GULP.

**en·gorge** (ĕn-gôrj′) *v.* **-gorged, -gorg·ing, -gorg·es.** [OFr. *engorgier* : *en-,* in (< Lat. *in-*) + *gorge,* throat < Lat. *gurges,* gulf.] —*vt.* 1. To devour greedily. 2. To gorge : glut. 3. To fill to excess, as with blood or another bodily fluid. —*vi.* To feed ravenously. **—en·gorge′ment** *n.*

**en·graft** (ĕn-grăft′) *vt.* **-graft·ed, -graft·ing, -grafts.** 1. To graft (a scion) onto or into another plant. 2. To plant firmly : ESTABLISH. **—en·graft′ment** *n.*

**en·grailed** (ĕn-grāld′) *adj.* [ME *engrelen* < OFr. *engresler* : *en-,* in (< Lat. *in-*) + *gresle,* slender < Lat. *gracilis.*] 1. Indented along the edge with small curves. 2. Having an edge formed by a ring of dots.

**en·grain** (ĕn-grān′) *vt.* **-grained, -grain·ing, -grains.** [ME *engreinen,* to dye red < OFr. *engrainer,* to dye < *en graine,* in cochineal dye.] To ingrain.

**en·gram** *also* **en·gramme** (ĕn′grăm′) *n.* A hypothetical alteration of living neural tissue, posited as an explanation for memory.

**en·grave** (ĕn-grāv′) *vt.* **-graved, -grav·ing, -graves.** 1. To carve, cut, or etch into a material. 2. a. To carve, cut, or etch into a block or printing surface. b. To print from a block or plate made by such a process. 3. To impress deeply <terrible events that were forever *engraved* in their memory> **—en·grav′er** *n.*

**en·grav·ing** (ĕn-grā′vĭng) *n.* 1. The art or technique of one that engraves. 2. An engraved printing surface. 3. A print made from an engraved plate or block.

**en·gross** (ĕn-grōs′) *vt.* **-grossed, -gross·ing, -gross·es.** [Partly < ME *engrossen,* to collect in large quantity < OFr. *engrossier* < *en gros,* in large quantity, and partly < ME *engrossen,* to write in a large hand < AN *engrosser,* prob. < Med. Lat. *ingrossare* : Lat. *in-,* in + *grossa,* a copy in a large hand < Lat. *grossus,* thick.] 1. To occupy the attention of : ABSORB. 2. To acquire most or all of a commodity so as to monopolize a market. 3. a. To write or transcribe in a large, legible hand. b. To prepare the text of (an official document) by writing or printing. **—en·gross′er** *n.* **—en·gross′ment** *n.*

**en·gross·ing** (ĕn-grō′sĭng) *adj.* Occupying one's complete attention <an *engrossing* tale> **—en·gross′ing·ly** *adv.*

**en·gulf** (ĕn-gŭlf′) *vt.* **-gulfed, -gulf·ing, -gulfs.** 1. To enclose and surround completely. 2. To swallow up or overwhelm by or as if by overflowing and enclosing. **—en·gulf′ment** *n.*

**en·hance** (ĕn-hăns′) *vt.* **-hanced, -hanc·ing, -hanc·es.** [ME *enhauncen* < AN *enhauncer* < OFr. *enhaucier* < VLat. **inaltiare* : Lat. *in-* (intensive) + *altus,* high.] To increase or make greater, as in value, beauty, or reputation : AUGMENT. **—en·hance′ment** *n.* **—en·hanc′er** *n.* **—en·hanc′ive** *adj.*

**en·har·mon·ic** (ĕn′här-mŏn′ĭk) *adj.* [Fr. *enharmonique* < OFr., of a scale employing quarter tones < Gk. *enarmonios* : *en-,* in + *harmonia,* harmony.] *Mus.* Of, pertaining to, or involving the use of two different written representations, as C# and D♭, for the same tone. **—en′har·mon′i·cal·ly** *adv.*

**e·nig·ma** (ĭ-nĭg′mə) *n.* [Lat. *aenigma* < Gk. *ainigma* < *ainissesthai,* to speak in riddles < *ainos,* fable.] 1. One that is puzzling, ambiguous, or inexplicable. 2. Cryptic writing or speech.

**en·ig·mat·ic** (ĕn′ĭg-măt′ĭk) *or* **en·ig·mat·i·cal** (-ĭ-kəl) *adj.* Of or like an enigma : PUZZLING. **—en′ig·mat′i·cal·ly** *adv.*

**en·isle** (ĕn-īl′) *vt.* **-isled, -isl·ing, -isles.** 1. To make into an island. 2. To set apart from others : ISOLATE.

**en·jamb·ment** *or* **en·jambe·ment** (ĕn-jăm′mənt, -jămb′mənt) *n.* [Fr. *enjambement* < OFr. *enjamber,* to straddle : *en-,* in (< Lat. *in-*) + *jambe,* leg < Lat. *gamba,* hoof, perh. < Gk. *kampē,* bend.] Continuation of a sentence from one line or couplet of a poem to the next so that closely related words fall on different lines.

**en·join** (ĕn-join′) *vt.* **-joined, -join·ing, -joins.** [ME *enjoinen* < OFr. *enjoindre* < Lat. *injungere* : *in-,* in + *jungere,* to join.] 1. To direct with emphasis and authority : COMMAND. 2. To forbid. **—en·join′er** *n.* **—en·join′ment** *n.*

**en·joy** (ĕn-joi′) *vt.* **-joyed, -joy·ing, -joys.** [ME *enjoien* < OFr. *enjoir* : *en-,* in (< Lat. *in-*) + *joir,* to rejoice < Lat. *gaudere.*] 1. To derive pleasure from : RELISH. 2. To have the benefit or use of <*enjoys* good living> 3. To make happy <*enjoying* themselves with a new game> **—en·joy′a·ble** *adj.* **—en·joy′a·bly** *adv.* **—en·joy′er** *n.*

**en·joy·ment** (ĕn-joi′mənt) *n.* 1. The act or state of enjoying. 2. Use or possession of something pleasurable or beneficial. 3. Something giving pleasure.

**en·keph·a·lin** (ĕn-kĕf′ə-lĭn′) *n.* [Gk. *enkephalos,* in the head (*en-,* in + *kephalē,* head) + -IN.] One of two closely related proteins occurring in the brain and having opiate qualities.

**en·kin·dle** (ĕn-kĭn′dl) *v.* **-dled, -dling, -dles.** —*vt.* **1.** To set on fire : LIGHT. **2.** To arouse : incite. **3.** To make luminous and glowing. —*vi.* To catch fire. **—en·kin′dler** *n.*

**en·lace** (ĕn-lās′) *vt.* **-laced, -lac·ing, -lac·es. 1.** To wrap with or as if with a lace or laces : ENCIRCLE. **2.** To entwine : interlace. **—en·lace′ment** *n.*

**en·large** (ĕn-lärj′) *v.* **-larged, -larg·ing, -larg·es.** [ME *enlargen* < OFr. *enlargier* : *en-*, in (< Lat. *in-*) + *large*, large < Lat. *largus.*] —*vt.* **1.** To make larger. **2.** To give greater scope to : EXPAND. —*vi.* **1.** To become larger : GROW. **2.** To speak or write in greater detail or at greater length <*enlarged* upon the proposal> **—en·larg′er** *n.*

**en·large·ment** (ĕn-lärj′mənt) *n.* **1.** The act of enlarging or state of being enlarged. **2.** Something that enlarges. **3.** A photographic reproduction or copy larger than an original.

**en·light·en** (ĕn-līt′n) *vt.* **-ened, -en·ing, -ens. 1.** To provide with spiritual wisdom. **2.** To give information to. **—en·light′en·er** *n.*

**en·light·en·ment** (ĕn-līt′n-mənt) *n.* **1. a.** An act or means of enlightening. **b.** The state of being enlightened. **2. Enlightenment.** An 18th-cent. philosophical movement devoted to critical examination of previously accepted doctrines and institutions from the viewpoint of rationalism.

**en·list** (ĕn-lĭst′) *v.* **-list·ed, -list·ing, -lists.** [EN- + LIST[1].] —*vt.* **1.** To engage or obtain for service in the armed forces. **2.** To engage the support or cooperation of. —*vi.* **1.** To enter the armed forces voluntarily. **2.** To participate actively in an enterprise. **—en·list′ment** *n.*

**enlisted man** *n.* One who has enlisted in the armed forces without an officer's commission or warrant.

**enlisted woman** *n.* A woman who has enlisted in the armed forces without an officer's commission or warrant.

**en·liv·en** (ĕn-lī′vən) *vt.* **-ened, -en·ing, -ens.** To make lively or spirited : ANIMATE. **—en·liv′en·er** *n.* **—en·liv′en·ment** *n.*

**en masse** (ŏn măs′) *adv.* [Fr. : *en*, on + *masse*, crowd.] In one body or group : all together.

**en·mesh** (ĕn-mĕsh′) *vt.* **-meshed, -mesh·ing, -mesh·es.** To entangle in or as if in a mesh.

**en·mi·ty** (ĕn′mĭ-tē) *n., pl.* **-ties.** [ME *enemite* < OFr. *enemitie* < VLat. **inimicitas* < Lat. *inimicus*, enemy. —see ENEMY.] Deep-rooted mutual hatred.

**en·ne·ad** (ĕn′ē-ăd′) *n.* [Gk. *enneas*, *ennead-* < *ennea*, nine.] A group or set of nine.

**en·no·ble** (ĕn-nō′bəl) *vt.* **-bled, -bling, -bles.** [ME *ennoblen* < OFr. *ennoblir* : *en-*, in (< Lat. *in-*) + *noble*, noble < Lat. *nobilis.*] **1.** To make noble in quality. **2.** To confer a rank of nobility on. **—en·no′ble·ment** *n.* **—en·no′bler** *n.*

**en·nui** (ŏn-wē′, ŏn′wē) *n.* [Fr. < OFr. *enui* < Lat. *in odio*, odious : *in*, in + *odium*, hate.] Listlessness and dissatisfaction : BOREDOM.

**e·nol** (ē′nôl′, ē′nōl′) *n.* [< -EN(E) + -OL.] An organic compound having a hydroxyl group bonded to a carbon atom that in turn is doubly bonded to another carbon atom. **—e·nol′ic** (ē-nŏl′ĭk) *adj.*

**e·no·lase** (ē′nə-lās′) *n.* An enzyme present in muscle tissue that acts in carbohydrate metabolism.

**e·nol·o·gy** (ē-nŏl′ə-jē) *n. var. of* OENOLOGY.

**e·nor·mi·ty** (ĭ-nôr′mĭ-tē) *n., pl.* **-ties. 1.** Excessive wickedness or outrageousness. **2.** A monstrous offense : OUTRAGE.

**e·nor·mous** (ĭ-nôr′məs) *adj.* [ME *enorme* < Lat. *enormis* : *ex-*, out of + *norma*, norm.] **1.** Very great in size, extent, number, or degree : IMMENSE. **2.** *Archaic.* Extremely wicked : HEINOUS. **—e·nor′mous·ly** *adv.* **—e·nor′mous·ness** *n.*

**e·nough** (ĭ-nŭf′) *adj.* [ME *enogh* < OE *genōg.*] Sufficient to meet a need or satisfy a desire : ADEQUATE. —*pron.* An adequate quantity <*enough* for everyone> —*adv.* **1.** To a satisfactory amount or degree : SUFFICIENTLY. **2.** Fully : quite <Everybody was happy *enough* to go.> **3.** Tolerably : rather <The photography was good *enough*, but the movie itself was dull.>

**e·nounce** (ĭ-nouns′) *vt.* **e·nounced, e·nounc·ing, e·nounc·es.** [Fr. *énoncer* < Lat. *enuntiare*, to speak out : *ex-*, out + *nuntiare*, to declare.] **1.** To declare formally : STATE. **2.** To pronounce clearly : ENUNCIATE. **—e·nounce′ment** *n.*

**e·now** (ĭ-nou′) *adj. & adv.* [ME *inow* < OE *genōg.*] *Archaic.* Enough.

**en pas·sant** (äN′ pä-säN′) *adv.* [Fr.] In passing : by the way. —*n.* Capture of a chess pawn after an initial move of two squares by an enemy pawn in a position to make a capture on the first of the two squares so crossed.

**en·phy·tot·ic** (ĕn′fī-tŏt′ĭk) *adj.* [EN- + -PHYT(E) + -OTIC.] Designating or characterizing a plant disease causing a relatively constant amount of damage each year. **—en′phy·tot′ic** *n.*

**en·plane** (ĕn-plān′) *also* **em·plane** (ĕm-) *vi.* **-planed, -plan·ing, -planes.** To board an aircraft.

**en·quire** (ĕn-kwīr′) *v. var. of* INQUIRE.

**en·rage** (ĕn-rāj′) *vt.* **-raged, -rag·ing, -rag·es.** To put into a rage.

**en·rapt** (ĕn-răpt′) *adj.* **1.** Enraptured. **2.** Enthralled.

**en·rap·ture** (ĕn-răp′chər) *vt.* **-tured, -tur·ing, -tures.** To fill with rapture. **—en·rap′ture·ment** *n.*

**en·rich** (ĕn-rĭch′) *vt.* **-riched, -rich·ing, -rich·es.** [ME *enrichen* < OFr. *enricher* : *en-* (causative < Lat. *in-*, in) + *riche*, rich, of Germanic orig.] **1.** To make rich or richer. **2.** To make fuller, more meaningful, or more rewarding <Field trips *enriched* the ecology course.> **3.** To add fertilizer to. **4.** To add nutrients to. **5.** To add to the beauty or character of : ADORN <mosaics that *enriched* the floors> **6.** *Physics.* To increase the ratio of radioactive isotopes in. **—en·rich′er** *n.* **—en·rich′ment** (-mənt) *n.*

**en·robe** (ĕn-rōb′) *vt.* **-robed, -rob·ing, -robes.** To dress in or as if in a robe <*enrobed* the monarch for the coronation>

**en·roll** *also* **en·rol** (ĕn-rōl′) *v.* **-rolled, -roll·ing, -rolls** *also* **-rolled, -rol·ling, -rols.** [ME *enrollen* < OFr. *enroller* : *en-*, in (< Lat. *in-*) + *rolle*, roll < Lat. *rotulus*, dim. of *rota*, wheel.] —*vt.* **1.** To enter the name of in a record, register, or roll. **2.** To roll or wrap up. —*vi.* To place one's name on a roll or register. **—en·roll·ee′** *n.*

**en·roll·ment** *or* **en·rol·ment** (ĕn-rōl′mənt) *n.* **1.** The act of enrolling or process of being enrolled. **2.** A record or entry. **3.** A number enrolled.

**en·root** (ĕn-ro͞ot′, -ro͝ot′) *vt.* **-root·ed, -root·ing, -roots.** To establish firmly by or as if by roots : IMPLANT.

**en route** (ŏn ro͞ot′, ĕn) *adv. & adj.* [Fr.] On the way.

**en·sam·ple** (ĕn-săm′pəl) *n.* [ME < OFr. *example.* —see EXAMPLE.] *Archaic.* An example.

**en·san·guine** (ĕn-săng′gwĭn) *vt.* **-guined, -guin·ing, -guines. 1.** To stain or cover with blood. **2.** To make crimson.

**en·sconce** (ĕn-skŏns′) *vt.* **-sconced, -sconc·ing, -sconc·es. 1.** To settle comfortably or securely <*ensconced* ourselves in a corner of the library> **2.** To conceal or place in a secure place.

**en·sem·ble** (ŏn-sŏm′bəl) *n.* [Fr. < LLat. *insimul*, at the same time : *in*, in + *simul*, at the same time.] A unit or group of complementary parts contributing to a single effect, esp.: **a.** A coordinated outfit or costume. **b.** A group of supporting musicians, singers, dancers, or actors who perform together. **c.** Music for two or more vocalists or instrumentalists. **d.** Musicians performing in a musical ensemble.

**en·shrine** (ĕn-shrīn′) *vt.* **-shrined, -shrin·ing, -shrines. 1.** To enclose in or as if in a shrine. **2.** To cherish as sacred. **—en·shrine′ment** *n.*

**en·shroud** (ĕn-shroud′) *vt.* **-shroud·ed, -shroud·ing, -shrouds.** To cover with or as if with a shroud.

**en·si·form** (ĕn′sə-fôrm′) *adj.* [Lat. *ensis*, sword + -FORM.] Sword-shaped, as the leaf of an iris.

**en·sign** (ĕn′sən) *n.* [ME *ensigne* < OFr. *enseigne* < Lat. *insignia*, insignia. —see INSIGNIA.] **1.** (*also* ĕn′sīn′). A national flag displayed on ships and aircraft, often with the special insignia of a branch or unit of the armed forces <the naval *ensign*> **2.** (*also* ĕn′sīn′). A standard or banner, as of a military unit. **3.** (*also* ĕn′sīn′). *Archaic.* A standard-bearer. **4.** A commissioned officer of the lowest rank in the U.S. Navy or Coast Guard. **5.** (*also* ĕn′sīn′). **a.** A badge : emblem. **b.** A sign : token.

**en·si·lage** (ĕn′sə-lĭj) *n.* [Fr. < *ensiler*, to ensile.] **1.** The process of storing and fermenting fodder in a silo. **2.** Fodder preserved in a silo : SILAGE. —*vt.* **-laged, -lag·ing, -lag·es.** To ensile.

**en·sile** (ĕn-sīl′) *vt.* **-siled, -sil·ing, -siles.** [Fr. *ensiler* < Sp. *ensilar* : *en-*, in (< Lat. *in-*) + *silo*, silo. —see SILO.] To store (fodder) in a silo for preservation.

**en·slave** (ĕn-slāv′) *vt.* **-slaved, -slav·ing, -slaves.** To make a slave of : SUBJUGATE. **—en·slave′ment** *n.* **—en·slav′er** *n.*

**en·snare** (ĕn-snâr′) *vt.* **-snared, -snar·ing, -snares.** To catch in or as if in a snare. **—en·snare′ment** *n.* **—en·snar′er** *n.*

**en·soul** (ĕn-sōl′) *vt.* **-souled, -soul·ing, -souls. 1.** To endow with a soul. **2.** To unite with the soul.

**en·sphere** (ĕn-sfîr′) *vt.* **-sphered, -spher·ing, -spheres. 1.** To enclose in or as if in a sphere. **2.** To give spherical form to.

**en·sta·tite** (ĕn′stə-tīt′) *n.* [Gk. *enstatēs*, adversary + -ITE.] A variety of orthorhombic pyroxene with a magnesium silicate base, essentially $Mg_2Si_2O_6$, usu. found embedded in igneous rocks.

**en·sue** (ĕn-so͞o′) *vi.* **-sued, -su·ing, -sues.** [ME *ensuen* < OFr. *ensuir* < Lat. *insequi*, to pursue : *in-*, in + *sequi*, to follow.] To follow as a result or consequence.

**en·sure** (ĕn-sho͝or′) *vt.* **-sured, -sur·ing, -sures.** To make certain or sure of : INSURE.

**ent-** *pref. var. of* ENTO-.

**-ent** *suff.* [ME < OFr. < Lat *-ens*, pr.part. suffix.] **1. a.** Performing, promoting, or causing a specified action <absorb*ent*> **b.** Being in a specified state or condition <bival*ent*> **2.** One that performs, promotes, or brings about a specified action <refer*ent*>

**en·tab·la·ture** (ĕn-tăb′lə-cho͝or′) *n.* [Obs. Fr. < Ital. *intavolatura* < *intavolare*, to put on a table : *in-*, in (< Lat.) + *tavola*, table < Lat. *tabula*, board.] **1.** The upper part of a classical architectural order, resting on the capital and including the architrave, frieze, and cornice. **2.** A raised horizontal structure, esp. a support for machinery.

**en·ta·ble·ment** (ĕn-tā′bəl-mənt) *n.* [Fr. < OFr. : *en-*, in (< Lat. *in-*) + *table*, table (< Lat. *tabula*, board) + *-ment*, -ment.] A platform supporting a statue and situated above the base and the dado.

**en·tail** (ĕn-tāl′, ĭn-) *vt.* **-tailed, -tail·ing, -tails.** [ME *entaillen*, to limit inheritance to specific heirs : *en-*, in (< Lat. *in-*) + *taille*, tail. —see TAIL[2].] **1.** To have, impose, or require as a necessary accompaniment or consequence <an expansion plan that *entailed* heavy ex-

penses> **2.** To limit the inheritance of (property) to a specified succession of heirs. —*n.* **1.** The act of entailing or state of being entailed. **2.** An entailed estate. **3.** A predetermined order of succession, as to an estate or to an office. **4.** Something transmitted as if by unalterable inheritance. **—en·tail′ment** *n.*

**en·ta·moe·ba** *or* **en·ta·me·ba** (ĕn′tə-mē′bə) *also* **en·da·moe·ba** *or* **en·da·me·ba** (ĕn′də-) *n., pl.* **-bas** *or* **-bae** (-bē). A parasitic amoeba of the genus *Entamoeba*, esp. *E. histolytica*, causing dysentery and ulceration of the colon and liver.

**en·tan·gle** (ĕn-tăng′gəl) *vt.* **-gled, -gling, -gles. 1.** To make tangled : SNARL. **2.** To complicate : confuse. **3.** To involve in or as if in a tangle. **—en·tan′gle·ment** *n.* **—en·tan′gler** *n.*

**en·tel·e·chy** (ĕn-tĕl′ĭ-kē) *n., pl.* **-chies.** [LLat. *entelechia* < Gk. *entelekheia* : *entelēs*, complete (*en-*, in + *telos*, perfection) + *ekhein*, to have.] **1.** Actuality as distinguished from potentiality in the philosophy of Aristotle. **2.** A vital force urging an organism toward self-fulfillment in some philosophical systems.

**en·tente** (ŏn-tŏnt′) *n.* [Fr. < OFr. *entendre*, to understand. —see INTEND.] **1.** An agreement between two or more governments or powers for cooperative policy or action. **2.** The parties to an entente.

**en·ter** (ĕn′tər) *v.* **-tered, -ter·ing, -ters.** [ME *entren* < OFr. *entrer* < Lat. *intrare* < *intra*, within.] —*vt.* **1.** To come or go into. **2.** To penetrate : pierce. **3.** To introduce : insert. **4.** To become a part of or an element in. **5.** To embark on : BEGIN. **6. a.** To obtain admission to. **b.** To gain admission for. **c.** To enroll. **7. a.** To register as an entry in a competition or exhibition <runners *enter* a track meet> **b.** To become a participant in <*enter* an art show> **c.** To take part as a contestant in <*enter* the marathon> **8.** To make a beginning in <*enter* the legal profession> **9.** *Law.* **a.** To place formally upon the records <*enter* a plea> **b.** To go upon in order to take possession of (land). **10.** To report (e.g., a ship) to customs. —*vi.* **1.** To come or go in. **2.** To gain entry. **3.** To become a member of a group. **4. a.** To participate <*enter* into a controversy> **b.** To be a part or component of. **5.** To consider : investigate <an argument that *entered* into the effect of sulfuric emissions> **6.** To become party to a contract. **7.** To set out : EMBARK. **8.** To go upon in order to take legal possession of land. **9.** To begin to deal with or consider a subject.

**enter-** *pref. var. of* ENTERO-.

**en·ter·al** (ĕn′tər-əl) *adj.* Enteric. **—en′ter·al·ly** *adv.*

**en·ter·ic** (ĕn-tĕr′ĭk) *adj.* [Gk. *enterikos* < *enteron*, intestine.] Of or within the intestine.

**enteric fever** *n.* Typhoid fever.

**en·ter·i·tis** (ĕn′tə-rī′tĭs) *n.* Intestinal inflammation.

**entero-** *or* **enter-** *pref.* [NLat. < Gk. *enteron*, intestine.] Intestine <enteritis>

**en·ter·o·bac·te·ri·um** (ĕn′tə-rō-băk-tîr′ē-əm) *n., pl.* **-i·a** (-ē-ə). Any of various Gram-negative rod-shaped bacteria of the family Enterobacteriaceae, including some pathogens of animals and plants.

**en·ter·o·bi·a·sis** (ĕn′tə-rō-bī′ə-sĭs) *n.* [NLat. *Enterobius*, pinworm genus + -IASIS.] Infestation of the intestine with pinworms.

**en·ter·o·coc·cus** (ĕn′tə-rō-kŏk′əs) *n., pl.* **-coc·ci** (-kŏk′sī′, -kŏk′ī′). A streptococcus in the intestine. **—en′ter·o·coc′cal** *adj.*

**en·ter·o·gas·trone** (ĕn′tə-rō-găs′trōn′) *n.* [ENTERO- + GASTR(O)- + (HORM)ONE.] A hormone of the upper intestinal mucosa that inhibits gastric secretion and motility.

**en·ter·o·ki·nase** (ĕn′tə-rō-kī′nās′, -kĭn′ās′) *n.* An enzyme found in intestinal juice that converts trypsinogen to trypsin.

**en·ter·on** (ĕn′tə-rŏn′) *n.* [NLat. < Gk.] The intestine.

**en·ter·o·path·o·gen·ic** (ĕn′tə-rō-păth′ə-jĕn′ĭk) *adj.* Capable of causing intestinal disease.

**en·ter·op·a·thy** (ĕn′tə-rŏp′ə-thē) *n.* An intestinal disease.

**en·ter·os·to·my** (ĕn′tə-rŏs′tə-mē) *n., pl.* **-mies.** Surgical creation of an opening into the intestine through the abdominal wall.

**en·ter·ot·o·my** (ĕn′tə-rŏt′ə-mē) *n., pl.* **-mies.** Surgical incision into the intestine.

**en·ter·o·tox·in** (ĕn′tə-rō-tŏk′sĭn) *n.* A toxin produced by bacteria specific for intestinal cells and causing symptoms of food poisoning.

**en·ter·prise** (ĕn′tər-prīz′) *n.* [ME < OFr. *entreprise* < *entreprendre*, to undertake : *entre-*, between (< Lat. *inter-*) + *prendre*, to take < Lat. *prendere*.] **1.** An undertaking, esp. one of great scope, complication, or risk. **2.** A business organization. **3.** Systematic and industrious activity. **4.** Eagerness to venture : INITIATIVE. **—en′ter·pris′er** *n.*

**en·ter·pris·ing** (ĕn′tər-prī′zĭng) *adj.* Showing imagination, initiative, and eagerness to undertake challenging or risky projects. **—en′ter·pris′ing·ly** *adv.*

**en·ter·tain** (ĕn′tər-tān′) *v.* **-tained, -tain·ing, -tains.** [ME *entertinen*, to maintain < OFr. *entretenir* < VLat. *intertenēre : Lat. *inter*, among + Lat. *tenēre*, to hold.] —*vt.* **1.** To hold the attention of : AMUSE. **2.** To extend hospitality toward <*entertain* business associates at lunch> **3.** To consider : contemplate <*entertain* a notion> **4.** To hold in mind : HARBOR <*entertained* no false hopes> **5.** *Archaic.* To continue with : MAINTAIN. —*vi.* **1.** To show hospitality to guests. **2.** To provide entertainment. **—en′ter·tain′er** *n.*

**en·ter·tain·ing** (ĕn′tər-tā′nĭng) *adj.* Pleasingly diverting : AMUSING. **—en′ter·tain′ing·ly** *adv.*

**en·ter·tain·ment** (ĕn′tər-tān′mənt) *n.* **1.** An act of entertaining. **2.** The art or field of entertaining. **3.** Something that entertains. **4.** Pleasure afforded by being entertained. **5.** *Obs.* **a.** Maintenance : support. **b.** Employment.

**en·thal·py** (ĕn′thăl′pē, ĕn-thăl′-) *n., pl.* **-pies.** [< Gk. *enthalpein*, to heat in : *en-*, in + *thalpein*, to heat.] A thermodynamic function of a system, equivalent to the internal energy plus the product of the pressure and the volume.

**en·thrall** (ĕn-thrôl′) *vt.* **-thralled, -thrall·ing, -thralls.** [ME, to put in bondage : *en-*, en- + *thral*, slave. —see THRALL.] **1.** To hold spellbound : CAPTIVATE. **2.** To enslave. **—en·thrall′ment** *n.*

**en·throne** (ĕn-thrōn′) *vt.* **-throned, -thron·ing, -thrones. 1. a.** To seat on a throne. **b.** To invest with sovereign power or with the authority of high office. **2.** To raise to a high position. **—en·throne′ment** *n.*

**en·thuse** (ĕn-tho͞oz′) *vt.* & *vi.* **-thused, -thus·ing, -thus·es.** [Back-formation < ENTHUSIASM.] *Informal.* To stimulate enthusiasm in or to show enthusiasm.

**en·thu·si·asm** (ĕn-tho͞o′zē-ăz′əm) *n.* [LLat. *enthusiasmus* < Gk. *enthousiasmos* < *enthousiazein*, to be inspired by a god < *entheos*, possessed : *en-*, in + *theos*, god.] **1. a.** Intense feeling for a subject or cause. **b.** Eagerness : zeal. **2.** Something inspiring enthusiasm. **3.** *Archaic.* Excessive religious fervor.

**en·thu·si·ast** (ĕn-tho͞o′zē-ăst′) *n.* [Gk. *enthousiastēs* < *enthousiazein*, to be inspired. —see ENTHUSIASM.] **1.** One intensely involved or preoccupied with a particular subject <a chess *enthusiast*> **2.** A zealot : fanatic.

**en·thu·si·as·tic** (ĕn-tho͞o′zē-ăs′tĭk) *adj.* Having or showing enthusiasm. **—en·thu′si·as′ti·cal·ly** *adv.*

**en·thy·meme** (ĕn′thə-mēm′) *n.* [Lat. *enthymema* < Gk. *enthumēma*, a rhetorical argument < *enthumeisthai*, to consider : *en-*, in + *thumos*, mind.] *Logic.* A syllogism with an implicit premise.

**en·tice** (ĕn-tīs′) *vt.* **-ticed, -tic·ing, -tic·es.** [ME *enticen* < OFr. *enticier*, to instigate < VLat. *intitiare : Lat. *in-*, in + Lat. *titio*, firebrand.] To beguile by arousing hope or desire : LURE. **—en·tice′ment** *n.* **—en·tic′er** *n.* **—en·tic′ing·ly** *adv.*

**en·tire** (ĕn-tīr′) *adj.* [ME < OFr. *entir* < Lat. *integer* : *in-*, not + *tangere*, to touch.] **1.** Having no part left out : WHOLE. **2.** Being without limitation or reservation : COMPLETE <gave the speaker our *entire* attention> **3.** In one piece : INTACT. **4.** Of one piece : CONTINUOUS. **5.** Not castrated. **6.** *Bot.* Not having an indented margin. —*n.* **1.** The whole : entirety. **2.** An uncastrated horse. **—en·tire′ness** *n.*

**en·tire·ly** (ĕn-tīr′lē) *adv.* **1.** Completely : wholly. **2.** Solely or exclusively <We were *entirely* to blame.>

**en·tire·ty** (ĕn-tī′rə-tē) *n., pl.* **-ties. 1.** The state of being entire. **2.** The whole amount or extent : TOTAL.

**en·ti·tle** (ĕn-tīt′l) *vt.* **-tled, -tling, -tles.** [ME *entitlen* < OFr. *entiteler* < Med. Lat. *intitulare* : Lat. *in-*, in + Lat. *titulus*, title.] **1.** To give a name or title to. **2.** To furnish with a right <a ticket *entitling* me to free admission> **—en·ti′tle·ment** *n.*

**en·ti·ty** (ĕn′tə-tē) *n., pl.* **-ties.** [Med. Lat. *entitas* < Lat. *ens*, pr.part. of *esse*, to be.] **1.** The fact of existence : BEING. **2.** The existence of something considered apart from its properties. **3.** Something that exists as a particular and discrete unit <Individuals and corporations are equivalent *entities* under the law.>

**ento-** or **ent-** *pref.* [NLat. < Gk. *entos*, within.] Inside : within <entozoan>

**en·to·blast** (ĕn′tə-blăst′) *n. var. of* ENDOBLAST.

**en·to·derm** (ĕn′tə-durm′) *n. var. of* ENDODERM.

**en·toil** (ĕn-toil′) *vt.* **-toiled, -toil·ing, -toils.** *Archaic.* To ensnare. **—en·toil′ment** *n.*

**en·tomb** (ĕn-to͞om′) *vt.* **-tombed, -tomb·ing, -tombs.** [OFr. *entoumber* : *en-*, in (< Lat. *in-*) + *tombe*, tomb. —see TOMB.] **1.** To place in or as if in a tomb or grave. **2.** To serve as a tomb for. **—en·tomb′ment** *n.*

**entomo-** *pref.* [Fr. < Gk. *entomon*, insect : *en-*, in + *temnein*, to cut (from the insect's segmented body).] Insect <entomology>

**en·to·mol·o·gy** (ĕn′tə-mŏl′ə-jē) *n.* Scientific study of insects. **—en′to·mo·log′ic** (-mə-lŏj′ĭk), **en′to·mo·log′i·cal** *adj.* **—en′to·mo·log′i·cal·ly** *adv.* **—en′to·mol′o·gist** *n.*

**en·to·moph·a·gous** (ĕn′tə-mŏf′ə-gəs) *adj.* Feeding on insects.

**en·to·moph·i·lous** (ĕn′tə-mŏf′ə-ləs) *adj.* Pollinated by insects. **—en′to·moph′i·ly** *n.*

**en·tou·rage** (ŏn′to͞o-räzh′) *n.* [Fr. < *entourer*, to surround < OFr. *entour*, surroundings : *en-*, in + *tour*, circuit.] **1.** A group of attendants or associates : RETINUE. **2.** One's surroundings or environment.

**en·to·zo·an** (ĕn′tə-zō′ən) *n., pl.* **-zo·a** (-zō′ə). An animal, as a tapeworm, that lives within another animal, usu. as a parasite. **—en′to·zo′ic** *adj.*

**en·tr'acte** (ŏn′trăkt′, äN-trăkt′) *n.* [Fr. : OFr. *entre*, between + *acte*, act.] **1.** The interval between acts of a theatrical performance. **2.** An entertainment, as a musical performance, given between acts of a play.

**en·trails** (ĕn′trālz′, -trəlz) *pl.n.* [ME *entraille* < OFr. < Med. Lat. *intralia* < Lat. *inter*, within.] The internal organs, esp. the intestines.

**en·train¹** (ĕn-trān′) *vt.* **-trained, -train·ing, -trains.** [OFr. *entrainer* : *en-*, away (< Lat. *inde*, thence) + *trainer*, to drag < Lat. *trahere.*] To pull along after itself.

**en·train²** (ĕn-trān′) *vt.* & *vi.* **-trained, -train·ing, -trains.** To put on or board a train. **—en·train′ment** *n.*

**en·trance¹** (ĕn′trəns) *n.* [ME *entraunce*, right to enter < OFr. < *entrer*, to enter. —see ENTER.] **1.** The act or an instance of entering. **2.** Something allowing entry. **3.** Power or permission to enter : ADMISSION. **4.** The point, as in a musical score, at which a performer is to begin. **5.** The first entry of an actor into a scene.

**en·trance²** (ĕn-trăns′) *vt.* **-tranced, -tranc·ing, -tranc·es. 1.** To put into a trance. **2.** To fill with delight, wonder, or enchantment <The sunset *entranced* the visitors.> **—en·trance′ment** *n.* **—en·tranc′ing·ly** *adv.*

**en·trant** (ĕn′trənt) *n.* [Fr. < pr.part. of *entrer*, to enter. —see ENTER.] One who enters, esp. one who enters a contest or competition.

**en·trap** (ĕn-trăp′) *vt.* **-trapped, -trap·ping, -traps.** [OFr. *entraper* : *en-*, in + *trappe*, trap.] **1.** To catch in or as if in a trap. **2.** To lure into difficulty or danger. **—en·trap′ment** *n.*

**en·treat** (ĕn-trēt′) *v.* **-treat·ed, -treat·ing, -treats.** [ME *entreten* < OFr. *entraitier* : *en-*, in (< Lat. *in-*) + *traiter*, to treat < Lat. *trahere*, to drag.] —*vt.* **1.** To make an earnest request of. **2.** To ask for earnestly. **3.** *Archaic.* To treat. —*vi.* To make an earnest request or petition : PLEAD. **—en·treat′ing·ly** *adv.* **—en·treat′ment** *n.*

**en·treat·y** (ĕn-trē′tē) *n.*, *pl.* **-ies.** An earnest request : PLEA.

**en·tre·chat** (ŏn′trə-shä′) *n.* [Fr. < Ital. (*capriola*) *intrecciata*, intricate (caper) < *intrecciare*, to complicate : *in-*, in (< Lat.) + *treccia*, tress.] A balletic leap during which the dancer crosses the feet several times, often beating them together.

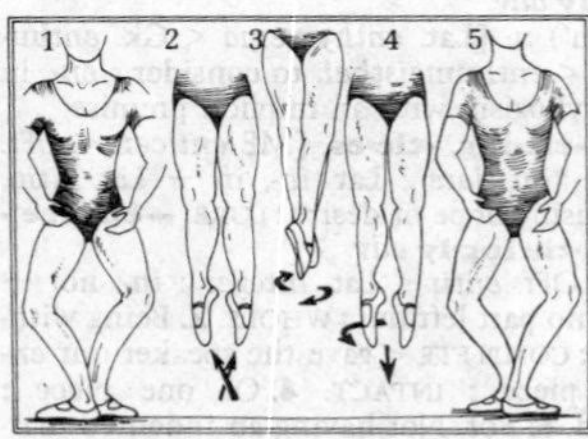

entrechat

**en·trée** *or* **en·tree** (ŏn′trā, ŏn-trā′) *n.* [Fr. < fem. p.part. of *entrer*, to enter < OFr. —see ENTER.] **1. a.** An act of entering. **b.** Power, permission, or liberty to enter. **2.** The main course of a meal.

**en·tre·mets** (ŏn′trə-mā′) *n.*, *pl.* **-mets** (-māz′) [Fr. < OFr. *entremes* < Lat. *intermissus*, p.part. of *intermittere*, to intermit. —see INTERMIT.] A side dish in addition to the principal course.

**en·trench** (ĕn-trĕnch′) *v.* **-trenched, -trench·ing, -trench·es.** —*vt.* **1.** To provide with a trench, esp. in order to fortify or defend. **2.** To fix firmly or securely <statements that only *entrench* them in awkward positions> —*vi.* **1.** To dig a trench. **2.** To encroach, infringe, or trespass. **—en·trench′ment** *n.*

**en·tre·pôt** (ŏn′trə-pō′) *n.* [Fr. < *entreposer*, to store : *entre-*, among (< Lat. *inter-*) + *poser*, to place < OFr. —see POSE¹.] **1.** A place where goods are stored and from which they are distributed. **2.** A trading or market center.

**en·tre·pre·neur** (ŏn′trə-prə-nûr′) *n.* [Fr. < OFr. < *entreprendre*, to undertake. —see ENTERPRISE.] One who organizes, operates, and assumes the risk in a business venture in expectation of gaining the profit. **—en′tre·pre·neu′ri·al** *adj.*

**en·tre·sol** (ĕn′tər-sŏl′, ĕn′trə-) *n.* [Fr. : *entre-*, between (< Lat. *inter-*) + *sol*, floor < Lat. *solum.*] The floor just above a ground floor.

**en·tro·py** (ĕn′trə-pē) *n.*, *pl.* **-pies.** [G. *Entropie* : Gk. *en-*, in + Gk. *tropē*, transformation.] **1.** A measure of a system's capacity to undergo spontaneous change, thermodynamically specified by the relationship $dS = dQ/T$, where $dS$ is an infinitesimal change in the measure for a system absorbing an infinitesimal quantity of heat $dQ$ at absolute temperature $T$. **2.** A measure of the disorder in a system specified in statistical mechanics by the relationship $S = k\ln P + c$, where $S$ is the value of the measure for a system in a given state, $P$ is the probability of occurrence of that state, and $k$ is a fixed and $c$ an arbitrary constant.

**en·trust** (ĕn-trŭst′) *vt.* **-trust·ed, -trust·ing, -trusts. 1.** To give over (something) to another for care, protection, or performance <*entrusted* the operation to mercenaries> **2.** To give as a trust to (someone) <*entrusted* management of the funds to a lawyer>

**en·try** (ĕn′trē) *n.*, *pl.* **-tries.** [ME < OFr. *entree*, fem. p.part. of *entrer*, to enter. —see ENTER.] **1. a.** An act or instance of entering. **b.** The right or privilege of entering. **2. a.** A place of entrance. **b.** A means of entrance. **3. a.** The act of entering an item, as in a record. **b.** An item entered in a record. **4. a.** A word or expression entered in a dictionary. **b.** A lexical item along with its related text. **5.** One entered in a competition.

**en·try·way** (ĕn′trē-wā′) *n.* ENTRY 2.

**en·twine** (ĕn-twīn′) *v.* **-twined, -twin·ing, -twines.** —*vt.* To twine around or together. —*vi.* To twine or twist together.

**en·twist** (ĕn-twĭst′) *vt.* **-twist·ed, -twist·ing, -twists.** To entwine.

**e·nu·cle·ate** (ĭ-nōō′klē-āt′, ĭ-nyōō′-) *vt.* **-at·ed, -at·ing, -ates.** [Lat. *enucleare, enucleat-*, to take out the kernel : *ex-*, out + *nucleus*, kernel.] **1.** *Archaic.* To explain. **2.** *Med.* To remove from an enveloping cover or sac. **3.** *Biol.* To remove the nucleus of. —*adj.* (-ĭt, -āt′). Lacking a nucleus. **—e·nu′cle·a′tion** *n.* **—e·nu′cle·a′tor** *n.*

**e·nu·mer·ate** (ĭ-nōō′mə-rāt′, ĭ-nyōō′-) *vt.* **-at·ed, -at·ing, -ates.** [Lat. *enumerare, enumerat-*, to count out : *ex-*, out + *numerus*, number.] **1.** To name or count off one by one : LIST. **2.** To determine the number of : COUNT. **—e·nu′mer·a′tion** *n.* **—e·nu′mer·a′tive** *adj.* **—e·nu′mer·a′tor** *n.*

**e·nun·ci·ate** (ĭ-nŭn′sē-āt′) *v.* **-at·ed, -at·ing, -ates.** [Lat. *enuntiare, enuntiat-* : *ex-*, out + *nuntiāre*, to announce < *nuntius*, messenger.] —*vt.* **1.** To pronounce clearly : ARTICULATE. **2.** To set forth precisely <*enunciate* acceptable treaty terms> **3.** To proclaim : announce. —*vi.* To make articulate sounds. **—e·nun′ci·a·ble** (-ə-bəl) *adj.* **—e·nun′ci·a′tion** *n.* **—e·nun′ci·a′tive** (-sē-ā′tĭv, -sē-ə-tĭv) *adj.* **—e·nun′ci·a′tive·ly** *adv.* **—e·nun′ci·a′tor** *n.*

**en·ure** (ĭn-yōōr′) *v. var. of* INURE.

**en·u·re·sis** (ĕn′yə-rē′sĭs) *n.* [NLat. < Gk. *enourein*, to urinate in : *en-*, in + *ourein*, to urinate < *ouron*, urine.] Involuntary urination. **—en′u·ret′ic** (-rĕt′ĭk) *adj.*

**en·vel·op** (ĕn-vĕl′əp) *vt.* **-oped, -op·ing, -ops.** [ME *envolupen*, to be involved in < OFr. *envoloper* : *en-*, in (< Lat. *in-*) + *voloper*, to wrap up.] **1.** To enclose or encase entirely with or as if with a covering. **2.** To attack (an enemy's flank). **—en·vel′op·er** *n.* **—en·vel′op·ment** *n.*

**en·ve·lope** (ĕn′və-lōp′, ŏn′-) *n.* [Fr. *enveloppe* < *envelopper*, to envelop < OFr. *envoloper.*] **1.** Something that envelops. **2.** A flat, folded paper container esp. for a letter. **3.** *Biol.* An enclosing covering, as a membrane or shell. **4.** The gasbag in a balloon. **5.** *Math.* A curve or surface tangent to all curves or surfaces of a family of curves or surfaces.

**en·ven·om** (ĕn-vĕn′əm) *vt.* **-omed, -om·ing, -oms.** [ME *envenimen* < OFr. *envenimer* : *en-*, in + *venim*, venom. —see VENOM.] **1.** To make poisonous or noxious. **2.** To embitter.

**en·vi·a·ble** (ĕn′vē-ə-bəl) *adj.* Arousing envy : DESIRABLE. **—en′vi·a·bly** *adv.*

**en·vi·ous** (ĕn′vē-əs) *adj.* **1.** Feeling, expressing, or marked by envy. **2.** *Obs.* Eager to emulate : EMULOUS. **—en′vi·ous·ly** *adv.* **—en′vi·ous·ness** *n.*

**en·vi·ron** (ĕn-vī′rən) *vt.* [ME *envirounen* < OFr. *environner* < *environ*, round about : *en-*, in (< Lat. *in-*) + *viron*, circle < *virer*, to turn, poss. of Celt. orig.] To encircle : surround.

**en·vi·ron·ment** (ĕn-vī′rən-mənt) *n.* **1.** The circumstances or conditions surrounding one : SURROUNDINGS. **2.** The total of circumstances surrounding an organism or group of organisms, esp.: **a.** The combination of external or extrinsic physical conditions affecting and influencing the growth and development of organisms. **b.** The complex of social and cultural conditions affecting the nature of an individual or community. **3.** An artistic or theatrical work surrounding or involving the audience. **—en·vi′ron·men′tal** (-mĕn′tl) *adj.* **—en·vi′ron·men′tal·ly** *adv.*

**en·vi·ron·men·tal·ism** (ĕn-vī′rən-mĕn′tl-ĭz′əm) *n.* The theory that environment rather than heredity is the primary influence on intellectual growth and cultural development.

**en·vi·ron·men·tal·ist** (ĕn-vī′rən-mĕn′tl-ĭst) *n.* **1.** One who wants to protect the natural environment. **2.** A supporter of environmentalism.

**en·vi·rons** (ĕn-vī′rənz) *pl.n.* [Fr. < OFr. *environ*, about. —see ENVIRON.] **1.** A surrounding area, esp. of a city. **2.** Surroundings.

**en·vis·age** (ĕn-vĭz′ĭj) *vt.* **-aged, -ag·ing, -ag·es.** [Fr. *envisager* : OFr. *en-*, in (< Lat. *in-*) + *visage*, face. —see VISAGE.] **1.** To conceive an image of, esp. as a future possibility. **2.** To consider or regard in a particular way.

**en·vi·sion** (ĕn-vĭzh′ən) *vt.* **-sioned, -sion·ing, -sions.** To picture in the mind.

**en·voi** *also* **en·voy** (ĕn′voi′, ŏn′-) *n.* [ME *envoie* < OFr. *envoier*, to send. —see ENVOY.] The closing stanza of some verse forms, as the ballade, either dedicating the poem to a patron or summarizing it.

**en·voy¹** (ĕn′voi′, ŏn′-) *n.* [Fr. *envoyé* < *envoyer*, to send < OFr. *envoier* < LLat. *inviare*, to put on the way : Lat. *in-*, on + Lat. *via*, way.] **1.** A messenger : agent. **2.** A governmental representative dispatched on a special diplomatic mission. **3.** A minister plenipotentiary to a foreign embassy, ranking below an ambassador.

**en·voy²** (ĕn′voi′, ŏn′-) *n. var. of* ENVOI.

**en·vy** (ĕn′vē) *n.*, *pl.* **-vies.** [ME *envie* < OFr. < Lat. *invidia* < *invidus*, envious < *invidēre* to envy : *in-*, in + *vidēre*, to see.] **1.** Resentful desire for another's possessions or advantages. **2.** The object of envy <the *envy* of all the neighbors> **3.** *Obs.* Malevolence. —*vt.*

**-vied, -vy·ing, -vies. 1.** To feel envy toward. **2.** To feel envy because of. **—en'vi·er** *n.* **—en'vy·ing·ly** *adv.*

☆ **syns:** ENVY, COVETOUSNESS, ENVIOUSNESS, JEALOUSY *n. core meaning* : resentful desire for another's possessions or advantages <Their riches provoked *envy* among their poorer relatives.>

**en·wind** (ĕn-wīnd') *vt.* **-wound** (-wound'), **-wind·ing, -winds.** To wind around.

**en·wrap** (ĕn-răp') *vt.* **-wrapped, -wrap·ping, -wraps. 1.** To wrap up : ENCLOSE. **2.** To envelop. **3.** To absorb completely : ENGROSS.

**en·wreathe** (ĕn-rēth') *vt.* **-wreathed, -wreath·ing, -wreathes.** To surround with or as if with a wreath.

**en·zo·ot·ic** (ĕn'zō-ŏt'ĭk) *adj.* Affecting or peculiar to animals of a particular area or limited district. —Used of a disease. —*n.* An enzootic disease.

**en·zyme** (ĕn'zīm') *n.* [G. *Enzym* < Med. Gk. *enzumos*, leavened : Gk. *en-*, in + *zumē*, leaven.] Any of numerous proteins or conjugated proteins produced by living organisms and functioning as biochemical catalysts in living organisms. **—en'zy·mat'ic** (-zə-măt'ĭk) *adj.* **—en'zy·mat'i·cal·ly** *adv.*

**en·zy·mol·o·gy** (ĕn'zə-mŏl'ə-jē) *n.* The biochemistry of enzymes. **—en'zy·mol'o·gist** *n.*

**eo-** *pref.* [Gk. *ēō-* < *ēōs*, dawn.] Most primitive : EARLIEST <*eo*hippus><*eo*lith>

**E·o·cene** (ē'ə-sēn') *adj.* Of, relating to, or designating the geologic time, rock series, sedimentary deposits, and fossils of the second oldest of the five major epochs of the Tertiary period of the Cenozoic era, extending from the end of the Paleocene to the beginning of the Oligocene and marked by the rise of mammals. —*n.* The Eocene epoch.

**e·o·hip·pus** (ē'ō-hĭp'əs) *n.* [EO- + Gk. *hippos*, horse.] An extinct, small, herbivorous mammal of the genus *Hyracotherium* or *Eohippus* of the Eocene epoch, with four-toed front feet and three-toed hind feet and ancestrally related to the horse.

**e·o·li·an** (ē-ō'lē-ən) *adj.* [After AEOLUS.] Relating to, caused by, or transmitted by the wind.

**e·o·lith** (ē'ə-lĭth') *n.* A crude stone artifact.

**E·o·lith·ic** (ē'ə-lĭth'ĭk) *adj.* Of or pertaining to the postulated earliest period of human culture preceding the Lower Paleolithic.

**e·on** (ē'ŏn', ē'ən) *n.* [LLat. *aeon* < Gk. *aiōn*.] **1.** An indefinitely long time period : AGE. **2.** *Geol.* The longest division of geologic time, having two or more eras. **—e·o'ni·an** (ē-ō'nē-ən) *adj.*

**E·os** (ē'ŏs') *n.* [Gk. *Ēos* < *ēōs*, dawn.] *Gk. Myth.* The goddess of the dawn.

**e·o·sin** (ē'ə-sən) *n.* [Gk. *ēōs*, dawn + -IN.] A red crystalline powder, $C_{20}H_8Br_4O_5$, used to color gasoline and in textile dyeing and ink manufacturing.

**e·o·sin·o·phil** (ē'ə-sĭn'ə-fĭl') *also* **e·o·sin·o·phile** (-fīl') *n.* **1.** *Physiol.* A leukocyte in vertebrate blood that accepts an eosin stain. **2.** *Biochem.* A microorganism, cell, or histological element easily stained by eosin dye. **—e'o·sin'o·phil', e'o·sin'o·phil'ic, e'o·si·noph'i·lous** (ē'ō-sĭ-nŏf'ə-ləs) *adj.*

**e·o·sin·o·phil·i·a** (ē'ə-sĭn'ə-fĭl'ē-ə) *n.* An increase in the number of eosinophils in the blood.

**-eous** *suff.* [Lat. *-eus*.] Having the nature of : RESEMBLING <gas*eous*>

**ep-** *pref. var. of* EPI-.

**e·pact** (ē'păkt') *n.* [OFr. *epacte* < Lat. *epacta* < Gk. *epaktai* < *epagein*, to intercalate : *epi-*, on + *agein*, to lead.] The period required to harmonize the solar calendar with the lunar calendar.

**ep·arch** (ĕp'ärk') *n.* [Gk. *eparkhos*, commander : *epi-*, over + *arkhein*, to rule.] **1.** The administrator of an eparchy. **2.** An Eastern Orthodox bishop or metropolitan. **—e·par'chi·al** (ĭ-pär'kē-əl) *adj.*

**ep·ar·chy** (ĕp'är'kē) *n.*, *pl.* **-chies. 1.** An administrative subdivision of Greece. **2.** An Eastern Orthodox diocese.

**ep·au·let** *also* **ep·au·lette** (ĕp'ə-lĕt', ĕp'ə-lĕt') *n.* [Fr. *épaulette*, dim. of *épaule*, shoulder < OFr. *espaule* < Med. Lat. *spatula*. —see ESPALIER.] A military uniform shoulder ornament.

**é·pée** *also* **e·pee** (ā-pā') *n.* [Fr. < Lat. *spatha*, sword.] **1.** A fencing sword with a bowl-shaped guard and a long, narrow, fluted blade lacking a cutting edge and tapering to a blunted point. **2.** The art or sport of fencing with the épée. **—é·pée'ist** *n.*

**ep·ei·rog·e·ny** (ĕp'ī-rŏj'ə-nē) *n.*, *pl.* **-nies.** [Gk. *ēpeiros*, continent + -GENY.] Deformation of the earth's crust that forms continents and oceanic basins or parts of these. **—e·pei'ro·gen'ic** (ĭ-pī'rō-jĕn'ĭk) *adj.* **—e·pei'ro·gen'i·cal·ly** *adv.*

**e·pen·the·sis** (ĭ-pĕn'thə-sĭs) *n.*, *pl.* **-ses** (-sēz') [LLat. < Gk. < *epentithenai*, to insert : *epi-*, in addition to + *en-*, in + *tithenai*, to place.] Insertion of a sound or letter into a word. **—ep'en·thet'ic** (ĕp'ĭn-thĕt'ĭk) *adj.*

**e·pergne** (ĭ-pûrn', ā-pârn') *n.* [Perh. alteration of Fr. *épargne*, saving < *épargner*, to save.] A large table centerpiece having a frame with extended arms or branches supporting holders, as for flowers, fruit, or sweetmeats.

**ep·ex·e·ge·sis** (ĕp-ĕk'sə-jē'sĭs) *n.* [Gk. *epexēgēsis* < *epexēgeisthai*, to explain in detail : *epi-*, in addition to + *exēgeisthai*, to explain. —see EXEGESIS.] Additional explanation or explanatory material. **—ep·ex'e·get'ic** (-jĕt'ĭk), **ep·ex'e·get'i·cal** *adj.*

**e·phah** *also* **e·pha** (ē'fə) *n.* [Heb. *'ēphāh*, prob. of Egypt. orig.] An ancient Hebrew unit of dry measure equal to slightly more than a bushel.

**e·phebe** (ĕf'ēb', ĭ-fēb') *n.* [Lat. *ephebus* < Gk. *ephēbos* : *epi-*, upon + *hēbē*, early manhood.] A youth aged 18 to 20 years in ancient Greece. **—e·phe'bic** *adj.*

**e·phe·bus** (ĭ-fē'bəs) *n.*, *pl.* **-bi** (-bī') [Lat.] An ephebe.

**e·phed·rine** (ĭ-fĕd'rĭn, ĕf'ĭ-drēn') *n.* [< NLat. *Ephedra*, genus name < Lat. *ephedra*, horsetail < Gk. *ephedros*, sitting upon : *epi-*, upon + *hedra*, seat).] A white, odorless, crystalline alkaloid, $C_{10}H_{15}NO$, isolated from the mahuang shrub or made synthetically and used to treat allergies and asthma and as a vasoconstrictor.

**e·phem·er·al** (ĭ-fĕm'ər-əl) *adj.* [Gk. *ephēmeros* : *epi-*, on + *hēmera*, day.] **1.** Lasting a short time : TRANSITORY. **2.** Lasting or living but a day. —*n.* An ephemeral thing or organism. **—e·phem'er·al'i·ty** *n.* **—e·phem'er·al·ly** *adv.*

**e·phem·er·id** (ĭ-fĕm'ər-ĭd) *n.* [NLat. *Ephemeridae*, former order name < Gk. *ephēmeron*, mayfly < *ephēmeros*, ephemeral.] An insect of the order Ephemeroptera, including the mayflies.

**e·phem·er·is** (ĭ-fĕm'ər-ĭs) *n.*, *pl.* **eph·e·mer·i·des** (ĕf'ə-mĕr'ə-dēz') [LLat., diary < Gk. < *ephēmeros*, daily, ephemeral.] A table giving the coordinates of a celestial body at a number of specific times within a specific period.

**ephemeris time** *n.* A highly accurate astronomical system for the measurement of time based on the period of the earth's orbit, but in practice depending on lunar observations and an accurate lunar ephemeris to calculate corrections to be applied to clocks.

**e·phem·er·on** (ĭ-fĕm'ə-rŏn') *n.*, *pl.* **-er·a** (-ər-ə) *or* **-er·ons.** [Gk. *ephemēron*, mayfly. —see EPHEMERIS.] **1.** A short-lived thing or organism. **2.** *pl.* **ephemera.** Printed matter of passing interest.

**E·phe·sians** (ĭ-fē'zhənz) *pl.n.* *(sing. in number).* —See table at BIBLE.

**eph·od** (ĕf'ŏd', ē'fŏd') *n.* [ME < Med. Lat. < Heb. *ēphōdh*.] A vestment once worn by Hebrew priests.

**eph·or** (ĕf'ôr', -ər) *n.*, *pl.* **-ors** *or* **-o·ri** (-ə-rī') [Lat. *ephorus* < Gk. *ephoros* < *ephoran*, to oversee : *epi-*, over + *horan*, to see.] One of five elected magistrates with supervisory power over the kings of Sparta.

**epi-** *or* **ep-** *pref.* [Gk. < *epi*, upon.] **1. a.** On : upon <*epi*phyte> **b.** Over : above <*epi*center> **c.** Around <*epi*carp> **d.** Close to : NEAR <*epi*calyx> **e.** Besides <*epi*phenomenon> **f.** After <*epi*crisis> **2.** A chemical substance related to a specified chemical substance <*epi*cholesterol>

**ep·i·blast** (ĕp'ə-blăst') *n.* Ectoderm. **—ep'i·blast'ic** *adj.*

**e·pib·o·ly** (ĭ-pĭb'ə-lē) *n.* [Gk. *epibolē*, addition < *epiballein*, to throw on : *epi-*, on + *ballein*, to throw.] Gastrulation by the differential growth of the cells of one embryonic part over and around another. **—ep'i·bol'ic** (ĕp'ə-bŏl'ĭk) *adj.*

**ep·ic** (ĕp'ĭk) *n.* [Lat. *epicus* < Gk. *epikos* < *epos*, song.] **1.** A long narrative poem celebrating the feats of a traditional or legendary hero. **2.** A literary or dramatic composition resembling an epic. **3.** A series of events considered suitable for an epic <the *epic* of Columbus' explorations>

**ep·i·ca·lyx** (ĕp'ĭ-kā'lĭks, -kăl'ĭks) *n.*, *pl.* **-ca·lyx·es** *or* **-ca·ly·ces** (-kā'lĭ-sēz', -kăl'ĭ-). A set of bracts close to and resembling a calyx.

**ep·i·can·thic fold** (ĕp'ĭ-kăn'thĭk) *n.* A fold of skin of the upper eyelid tending to cover the inner corner of the eye.

**ep·i·can·thus** (ĕp'ĭ-kăn'thəs) *n.*, *pl.* **-thi** (-thī, -thē) [NLat. : EPI- + Gk. *kanthos*, corner of the eye.] The epicanthic fold.

**ep·i·car·di·um** (ĕp'ĭ-kär'dē-əm) *n.*, *pl.* **-di·a** (-dē-ə) [NLat. : EPI- + Gk. *kardia*, heart.] The inner layer of the pericardium in actual contact with the heart. **—ep'i·car'di·al** *adj.*

**ep·i·carp** (ĕp'ĭ-kärp') *n.* Exocarp.

**epic drama** *n.* Modern narrative drama that asks the audience to examine social problems analytically rather than emotionally.

**ep·i·cene** (ĕp'ə-sēn') *adj.* [ME, having only one form of the noun for either gender < Lat. *epicoenus* < Gk. *epikoinos* : *epi-*, on + *koinos*, common.] **1. a.** Belonging to or with the characteristics of both sexes. **b.** Womanish : effeminate. **c.** Neuter : sexless. **2.** Having only one form of the noun for both male and female. **—ep'i·cene'** *n.* **—ep'i·cen'ism** *n.*

**ep·i·cen·ter** (ĕp'ĭ-sĕn'tər) *n.* **1.** The part of the earth's surface directly above the origin of an earthquake. **2.** A focal point <the *epicenter* of the controversy> **—ep'i·cen'tral** *adj.*

**ep·i·chlo·ro·hy·drin** (ĕp'ĭ-klôr'ə-hī'drĭn, -klōr'-) *n.* A colorless liquid, $C_3H_5OCl$, used as a solvent in making resins.

**ep·i·cot·yl** (ĕp'ĭ-kŏt'l) *n.* [EPI- + COTYL(EDON).] The part of the stem of a seedling or embryonic plant above the cotyledons and below the first true leaves.

**ep·i·crit·ic** (ĕp'ĭ-krĭt'ĭk) *adj.* [Gk. *epikritikos*, decisive < *epikritos*, decided on < *epikrinein*, to decide : *epi-*, over + *krinein*, to judge.] Relating to sensory nerve fibers that enable acute tactile and thermal sensitivity.

**epic theater** *n.* Epic drama.

**ep·i·cure** (ĕp'ĭ-kyŏŏr') *n.* [After *Epicurus* (341–270 B.C.).] **1.** One with refined taste esp. in food and wine. **2.** *Archaic.* One devoted to luxurious living and sensuous pleasure.

**ep·i·cu·re·an** (ĕp′ĭ-kyo͝o-rē′ən) *adj.* **1. Epicurean.** Of or pertaining to Epicurus or Epicureanism. **2.** Suited to the tastes of an epicure <an *epicurean* banquet> —*n.* **1. Epicurean.** A follower of Epicurus. **2.** An epicure.

**Ep·i·cu·re·an·ism** (ĕp′ĭ-kyo͝o-rē′ə-nĭz′əm) *n.* The philosophy of Epicurus, who sought freedom from pain and emotional disturbance and rejected a belief in the afterlife or the influence of gods on human affairs.

**ep·i·cur·ism** (ĕp′ĭ-kyo͝o-rĭz′əm) *n.* The beliefs, tastes, or lifestyle of an epicure.

**ep·i·cu·ti·cle** (ĕp′ĭ-kyo͞o′tĭ-kəl) *n.* The outermost layer of cuticle of an insect exoskeleton, consisting chiefly of wax.

**ep·i·cy·cle** (ĕp′ĭ-sī′kəl) *n.* [ME *epicicle* < LLat. *epicyclus* < Gk. *epikuklos* : *epi-*, on + *kuklos*, circle.] A small circle, the center of which moves on the circumference of a larger circle at whose center is the earth and the circumference of which describes the orbit of one of the planets around the earth in Ptolemaic cosmology. **—ep′i·cy′clic** (-sī′klĭk, -sĭk′lĭk) *adj.*

**epicyclic train** *n.* A system of gears in which at least one wheel axis revolves around another.

**ep·i·cy·cloid** (ĕp′ĭ-sī′kloid′) *n.* The curve described by a point on the circumference of a circle as it rolls on the outside of the circumference of a fixed circle. **—ep′i·cy·cloid′al** (-kloid′l) *adj.*

**ep·i·dem·ic** (ĕp′ĭ-dĕm′ĭk) *also* **ep·i·dem·i·cal** (-ĭ-kəl) *adj.* [Fr. *épidémique* < *épidémie*, an epidemic < OFr. *espydymie* < LLat. *epidemia* < Gk. *epidēmia* < *epidēmos*, prevalent : *epi-*, on + *dēmos*, people.] **1.** Affecting many individuals throughout an area at the same time <Cholera was *epidemic.*> **2.** Widely prevalent <*epidemic* anxiety> —*n.* **epidemic. 1.** A rapidly spreading outbreak of contagious disease. **2.** A rapid spread, growth, or development <an *epidemic* of robberies> **—ep′i·dem′i·cal·ly** *adv.*

**ep·i·de·mi·ol·o·gy** (ĕp′ĭ-dē′mē-ŏl′ə-jē, -dĕm′ē-) *n.* [LLat. *epidemia*, an epidemic + -LOGY.] The branch of medicine that studies epidemics and epidemic diseases. **—ep′i·de′mi·o·log′ic** (-ə-lŏj′ĭk), **ep′i·de′mi·o·log′i·cal** (-ĭ-kəl) *adj.* **—ep′i·de′mi·ol′o·gist** *n.*

**ep·i·der·mis** (ĕp′ĭ-dûr′mĭs) *n.* [NLat. < Gk. : *epi-*, on + *derma*, skin.] **1. a.** *Anat.* The protective, outer, nonvascular layer of the skin. **b.** An integument or outer layer of various organisms. **2.** *Bot.* The outermost layer of cells or protective covering of a plant or plant part. **—ep′i·der′mal** (-məl) *adj.*

**ep·i·der·moid** (ĕp′ĭ-dûr′moid′) *also* **ep·i·der·moid·al** (-dər-moid′l) *adj.* [EPIDERM(IS) + -OID.] Of, pertaining to, or having the characteristics of the epidermis.

**ep·i·di·a·scope** (ĕp′ĭ-dī′ə-skōp′) *n.* A machine for projecting the images of opaque objects or transparencies on a screen.

**ep·i·did·y·mis** (ĕp′ĭ-dĭd′ə-mĭs) *n., pl.* **-mi·des** (-mĭ-dēz′) [Gk. : *epi-*, at + *didumos*, testicle.] A long, narrow, flattened convoluted body that is part of the spermatic duct system, lying on the lateral edge of the posterior border of the testis. **—ep′i·did′y·mal** *adj.*

**ep·i·dote** (ĕp′ĭ-dōt′) *n.* [Fr. *épidote* < Gk. *epididonai*, to increase : *epi-*, upon + *didonai*, to give.] A natural, yellow, green, or black mineral composed of a silicate of calcium, aluminum, and iron, commonly found in metamorphic rock. **—ep′i·dot′ic** (-dŏt′ĭk) *adj.*

**ep·i·du·ral** (ĕp′ĭ-do͝or′əl, -dyo͝or′-) *adj.* Located on or over the dura mater.

**ep·i·gas·tri·um** (ĕp′ĭ-găs′trē-əm) *n., pl.* **-tri·a** (-trē-ə) [NLat. < Gk. *epigastrion* : *epi-*, above + *gastēr*, stomach.] The upper middle abdominal region. **—ep′i·gas′tric** (-trĭk) *adj.*

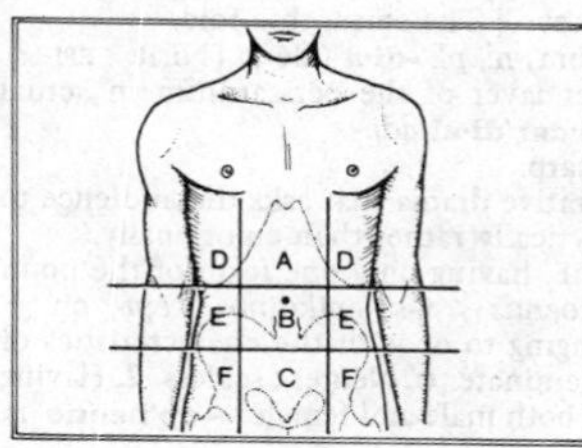


**epigastrium**
*The regions of the abdomen: A. epigastrium, B. umbilical region, C. hypogastrium, D. left and right hypochondrium, E. left and right lumbar, F. left and right iliac fossa*

**ep·i·ge·al** (ĕp′ə-jē′əl) *also* **ep·i·ge·an** (-ən) *or* **ep·i·ge·ous** (-əs) *adj.* [Gk. *epigaios*, on the earth : *epi-*, on + *gē*, earth.] **1.** *Biol.* Living or occurring on or near the surface of the ground. **2.** *Bot.* Designating or marked by cotyledons that appear aboveground.

**ep·i·gene** (ĕp′ə-jēn′) *adj.* [Fr. *épigène* < Gk. *epigenēs*, arising after : *epi-*, upon + *genos*, birth.] **1.** Occurring or originating on or just below the surface of the earth. **2.** Foreign to the material in which it is found. —Used of a crystal.

**ep·i·gen·e·sis** (ĕp′ə-jĕn′ĭ-sĭs) *n.* **1.** *Biol.* The theory that the individual is developed by structural elaboration of the unstructured egg rather than by a simple enlargement of a preformed entity. **2.** *Geol.* Change in the mineral characteristics of a rock due to external influence. **—ep′i·ge·net′ic** (-jə-nĕt′ĭk) *adj.*

**e·pig·e·nous** (ĭ-pĭj′ə-nəs) *adj.* *Bot.* Growing or developing on an upper surface, as fungi on leaves.

**ep·i·glot·tis** (ĕp′ĭ-glŏt′ĭs) *n., pl.* **-glot·tis·es** *or* **-glot·ti·des** (-glŏt′ĭ-dēz′) [Gk. *epiglōttis* : *epi-*, over + *glōttis*, glottis.] An elastic cartilage at the root of the tongue that folds over the glottis to prevent food from entering the windpipe during the act of swallowing. **—ep′i·glot′tal** (-glŏt′l), **ep′i·glot′tic** (-glŏt′ĭk) *adj.*

**ep·i·gone** (ĕp′ĭ-gōn′) *n.* [< Gk. *Epigonoi*, sons of the seven heroes against Thebes < pl. of *epigonos*, born after : *epi-*, after + *gignesthai*, to be born.] An imitator, esp. a second-rate follower of an artist or philosopher. **—ep′i·gon′ic** (-gŏn′ĭk) *adj.* **—e·pig′on·ism** (ĭ-pĭg′ə-nĭz′əm) *n.*

**ep·i·gram** (ĕp′ĭ-grăm′) *n.* [OFr. *epigramme* < Lat. *epigramma* < Gk. < *epigraphein*, to write on : *epi-*, on + *graphein*, to write.] **1.** A short poem tersely and wittily expressing a single thought or observation. **2.** A concise, cleverly worded and often paradoxical saying. **3.** Epigrammatic discourse or expression.

**ep·i·gram·mat·ic** (ĕp′ĭ-grə-măt′ĭk) *also* **ep·i·gram·mat·i·cal** (-ĭ-kəl) *adj.* **1.** Of or having the nature of an epigram. **2.** Given to or marked by the use of epigrams. **—ep′i·gram·mat′i·cal·ly** *adv.*

**ep·i·gram·ma·tism** (ĕp′ĭ-grăm′ə-tĭz′əm) *n.* Literary style marked by the use of epigrams. **—ep′i·gram′ma·tist** *n.*

**ep·i·gram·ma·tize** (ĕp′ĭ-grăm′ə-tīz′) *v.* **-tized, -tiz·ing, -tiz·es.** —*vt.* To express in an epigram. —*vi.* To create an epigram.

**ep·i·graph** (ĕp′ĭ-grăf′) *n.* [Gk. *epigraphē* < *epigraphein*, to write on. —see EPIGRAM.] **1.** An inscription, as on a statue or building. **2.** A quotation usu. at the beginning of a literary composition that suggests the theme. **—ep′i·graph′ic, ep′i·graph′i·cal** *adj.* **—ep′i·graph′i·cal·ly** *adv.*

**e·pig·ra·phy** (ĭ-pĭg′rə-fē) *n.* **1.** Inscriptions as a whole. **2. a.** Study of inscriptions. **b.** Decipherment esp. of ancient inscriptions. **—e·pig′ra·pher, e·pig′ra·phist** *n.*

**e·pig·y·nous** (ĭ-pĭj′ə-nəs) *adj.* Having floral parts or organs attached to or near the summit of the ovary. **—e·pig′y·ny** (-nē) *n.*

**ep·i·lep·sy** (ĕp′ə-lĕp′sē) *n.* [OFr. *epilepsie* < Lat. *epilepsia* < Gk. < *epilambanein*, to lay hold of : *epi-*, upon + *lambanein*, to take.] A disorder marked by recurring motor, sensory, or psychic malfunctions with or without unconsciousness or convulsive movements.

**ep·i·lep·tic** (ĕp′ə-lĕp′tĭk) *adj.* **1.** Afflicted with epilepsy. **2.** Of, typical of, or associated with epilepsy. **—ep′i·lep′tic** *n.*

**ep·i·lep·toid** (ĕp′ə-lĕp′toid′) *adj.* [EPILEPT(IC) + -OID.] Resembling epilepsy or any of its symptoms.

**ep·i·logue** *also* **ep·i·log** (ĕp′ə-lôg′, -lŏg′) *n.* [ME *epiloge* < OFr. *epilogue* < Lat. *epilogus* < Gk. *epilogos* < *epilegein*, to say more : *epi-*, in addition to + *legein*, to say.] **1. a.** A short speech spoken directly to the audience after the end of a play. **b.** The performer speaking such an epilogue. **2.** A short concluding section at the end of a literary work, often discussing the future of its characters.

**ep·i·mor·pho·sis** (ĕp′ə-môr′fə-sĭs) *n.* Regeneration of a part of an organ characterized by proliferation of new tissue.

**ep·i·mys·i·um** (ĕp′ə-mĭz′ē-əm, -mĭzh′ē-) *n., pl.* **-mys·i·a** (-mĭz′ē-ə, -mĭzh′ē-ə) [NLat. : EPI- + Gk. *mus*, muscle.] The fibrous sheath enclosing a muscle.

**ep·i·nas·ty** (ĕp′ə-năs′tē) *n., pl.* **-ties.** A downward bending of plant parts, as leaves, caused by excessive growth of the upper side. **—ep′i·nas′tic** (-tĭk) *adj.*

**ep·i·neph·rine** *also* **ep·i·neph·rin** (ĕp′ə-nĕf′rĭn) *n.* **1.** An adrenal hormone that stimulates autonomic nerve action. **2.** A white to brownish crystalline compound, $C_9H_{13}NO_3$, used in medicine esp. as a heart stimulant, muscle relaxant, and vasoconstrictor.

**ep·i·neu·ri·um** (ĕp′ə-no͝or′ē-əm, -nyo͝or′-) *n., pl.* **-neu·ri·a** (-no͝or′ē-ə, -nyo͝or′-) [NLat. : EPI- + Gk. *neuron*, nerve.] The connective tissue sheath of a nerve trunk, containing blood and lymph vessels. **—ep′i·neu′ri·al** (-ē-əl) *adj.*

**ep·i·pe·lag·ic** (ĕp′ə-pə-lăj′ĭk) *adj.* Of or relating to the part of the oceanic zone into which enough sunlight enters for photosynthesis.

**e·piph·a·ny** (ĭ-pĭf′ə-nē) *n., pl.* **-nies.** [ME *epiphanie* < OFr. < LLat. *epiphania* < Gk. *epiphaneia*, appearance < *epiphainein*, to manifest : *epi-*, to + *phainein*, to show.] **1. Epiphany.** A Christian festival held on Jan. 6 to celebrate the manifestation of the divine nature of Christ to the Gentiles as represented by the Magi. **2.** A revelatory manifestation of a divine being. **3. a.** A sudden manifestation of the meaning or essence of something. **b.** A sudden intuitive realization or perception of reality.

**ep·i·phe·nom·e·nal·ism** (ĕp′ə-fĭ-nŏm′ə-nə-lĭz′əm) *n.* *Philos.* The doctrine that mental activities are simply epiphenomena of the neural processes of the brain.

**ep·i·phe·nom·e·non** (ĕp′ə-fĭ-nŏm′ə-nŏn′) *n., pl.* **-na** (-nə). **1.** A secondary phenomenon resulting from and accompanying another. **2.** *Pathol.* A secondary condition in the course of a disease, not necessarily associated with the disease. **—ep′i·phe·nom′e·nal** (-nəl) *adj.* **—ep′i·phe·nom′e·nal·ly** *adv.*

**e·piph·y·sis** (ĭ-pĭf′ĭ-sĭs) *n., pl.* **-ses** (-sēz′) [Gk. *epiphusis* : *epi-*, upon + *phusis*, growth < *phuein*, to grow.] *Anat.* A part, often an end of a long bone, that initially develops separated from the main part by cartilage. **—ep′i·phys′i·al** (ĕp′ə-fĭz′ē-əl), **ep′i·phys′e·al** *adj.*

**ep·i·phyte** (ĕp′ə-fīt′) *n.* A plant, as a certain kind of orchid or fern, that grows on another plant on which it depends for mechanical support but not for nutrients. **—ep′i·phyt′ic** (-fĭt′ĭk), **ep′i·phyt′i·cal** *adj.* **—ep′i·phyt′i·cal·ly** *adv.*

**ep·i·phy·tot·ic** (ĕp′ə-fī-tŏt′ĭk) *adj.* Of, relating to, or characterizing a sudden or abnormally destructive outbreak of a plant disease, usu. over a wide geographic area. **—ep′i·phy·tot′ic** *n.*

**e·pis·co·pa·cy** (ĭ-pĭs′kə-pə-sē) *n., pl.* **-cies.** [< EPISCOPATE.] **1.** An episcopate. **2.** Church government in which bishops are the chief ministers.

**e·pis·co·pal** (ĭ-pĭs′kə-pəl) *adj.* [ME < LLat. *episcopalis* < *episcopus*, bishop < Gk. *episkopos*, overseer : *epi-*, over + *skopos*, watcher.] **1.** Of or relating to a bishop. **2.** Of, relating to, or involving church government by bishops. **3. Episcopal.** Designating or relating to the Protestant Episcopal Church. **—e·pis′co·pal·ly** *adv.*

**Episcopal Church** *n.* The Protestant Episcopal Church.

**E·pis·co·pa·lian** (ĭ-pĭs′kə-pā′lē-ən, -pāl′yən) *adj.* **1.** Of or belonging to the Protestant Episcopal Church. **2. episcopalian.** Of or advocating church government by bishops. **—E·pis′co·pa′lian** *n.*

**e·pis·co·pate** (ĭ-pĭs′kə-pĭt, -pāt′) *n.* [LLat. *episcopatus* < *episcopus*, bishop. —see EPISCOPAL.] **1.** The term, position, or office of a bishop. **2.** The jurisdiction of a bishop : DIOCESE. **3.** Bishops as a group.

**ep·i·si·ot·o·my** (ĭ-pē′zē-ŏt′ə-mē) *n., pl.* **-mies.** [Gk. *epision*, pubic region + -TOMY.] Surgical incision of the perineum during childbirth to facilitate delivery.

**ep·i·sode** (ĕp′ĭ-sōd′) *n.* [Gk. *epeisodion*, parenthetic story < *epeisodios*, coming in besides : *epi-*, in addition + *eisodios*, entering (*eis*, into + *hodos*, way).] **1. a.** An event or incident that is part of a narrative but forms a separate unit within the whole. **b.** One of a series of related events in the course of a continuous account. **2.** A part of a narrative that relates an event or series of connected events and forms a coherent story in itself : INCIDENT **3.** A separate part of a serialized work, as a novel or television program. **4.** A section of a classic Greek tragedy that occurs between two choric songs. **5.** *Mus.* A passage between statements of a main subject or theme, as in a rondo or fugue.

**ep·i·sod·ic** (ĕp′ĭ-sŏd′ĭk) *also* **ep·i·sod·i·cal** (-ĭ-kəl) *adj.* **1.** Relating to or like an episode. **2.** Made up of episodes. **3.** Limited to the duration of an episode : TEMPORARY. **—ep′i·sod′i·cal·ly** *adv.*

**ep·i·spas·tic** (ĕp′ĭ-spăs′tĭk) *adj.* [NLat. *epispasticus* < Gk. *epispastikos*, drawing toward oneself < *epispan*, to draw toward : *epi-*, toward + *span*, to draw.] Causing blisters. —*n.* A blistering agent.

**e·pis·ta·sis** (ĭ-pĭs′tə-sĭs) *n., pl.* **-ses** (-sēz′) [Gk., stoppage < *ephistanai*, to stop : *epi-*, upon + *histanai*, to place.] **1.** *Genetics.* A nonreciprocal interaction between nonalternative forms of gene in which one gene suppresses the expression of another affecting the same part of an organism. **2.** *Med.* Matter that rises to the surface of a body discharge. **—ep′i·stat′ic** (ĕp′ĭ-stăt′ĭk) *adj.*

**ep·i·stax·is** (ĕp′ĭ-stăk′sĭs) *n.* [Gk. < *epistazein*, to bleed from the nose : *epi-*, upon + *stazein*, to drip.] Nosebleed.

**ep·i·ste·mic** (ĕp′ĭ-stē′mĭk) *adj.* [Gk. *epistēmikos*, of knowledge < *epistēmē*, knowledge. —see EPISTEMOLOGY.] Of, pertaining to, or involving knowledge or the act of knowing. **—ep′i·ste′mi·cal·ly** *adv.*

**e·pis·te·mol·o·gy** (ĭ-pĭs′tə-mŏl′ə-jē) *n., pl.* **-gies.** [Gk. *epistēmē*, knowledge < *epistanai*, to understand (*epi-*, upon + *histanai*, to stand) + -LOGY.] **1.** The division of philosophy that investigates the nature and origin of knowledge. **2.** A theory of the nature of knowledge. **—e·pis′te·mo·log′i·cal** (-mə-lŏj′ĭ-kəl) *adj.* **—e·pis′te·mo·log′i·cal·ly** *adv.* **—e·pis′te·mol′o·gist** *n.*

**e·pis·tle** (ĭ-pĭs′əl) *n.* [ME < OFr. < Lat. *epistola* < Gk. *epistolē* < *epistellein*, to send to : *epi-*, to + *stellein*, to send.] **1.** A letter, esp. a formal one. **2. Epistle. a.** One of the letters included as a book in the New Testament. **b.** An excerpt from one of these letters, read as part of a religious service. **3.** A composition in the form of a letter.

**e·pis·tler** (ĭ-pĭs′lər) *n.* A writer of an epistle.

**e·pis·to·lar·y** (ĭ-pĭs′tə-lĕr′ē) *adj.* [Lat. *epistolaris* < *epistola*, epistle.] **1.** Of or associated with letters or letter writing. **2.** Being in the form of a letter. **3.** Carried on by or made up of letters <an *epistolary* acquaintanceship>

**ep·i·style** (ĕp′ĭ-stīl′) *n.* [Lat. *epistylium* < Gk. *epistulion* : *epi-*, upon + *stulos*, pillar.] ARCHITRAVE 1.

**ep·i·taph** (ĕp′ĭ-tăf′) *n.* [ME < OFr. *epitaphe* < Lat. *epitaphium* < Gk. *epitaphion* : *epi-*, upon + *taphos*, tomb.] **1.** An inscription on a tombstone in memory of the one buried there. **2.** A brief literary piece honoring a deceased person. **—ep′i·taph′ic** (-tăf′ĭk) *adj.*

**e·pit·a·sis** (ĭ-pĭt′ə-sĭs) *n., pl.* **-ses** (-sēz′) [Gk. < *epiteinein*, to intensify : *epi-*, upon + *teinein*, to stretch.] The middle part of a play that develops the main action leading to the catastrophe.

**ep·i·tha·la·mi·um** (ĕp′ə-thə-lā′mē-əm) *or* **ep·i·tha·la·mi·on** (-ən) *n., pl.* **-mi·ums** *or* **-mi·a** (-mē-ə) [Lat. < Gk. *epithalamion* : *epi-*, upon + *thalamos*, bridal chamber.] A poem or song in honor of a bride and bridegroom.

**ep·i·the·li·a** (ĕp′ə-thē′lē-ə) *n. pl. of* EPITHELIUM.

**ep·i·the·li·oid** (ĕp′ə-thē′lē-oid′) *adj.* Resembling epithelium.

**ep·i·the·li·o·ma** (ĕp′ə-thē′lē-ō′mə) *n., pl.* **-ma·ta** (-mə-tə) *or* **-mas.** A carcinoma derived from the epithelium. **—ep′i·the′li·om′a·tous** (-ŏm′ə-təs) *adj.*

**ep·i·the·li·um** (ĕp′ə-thē′lē-əm) *n., pl.* **-li·ums** *or* **-li·a** (-lē-ə) [NLat. : EPI- + Gk. *thēlē*, nipple.] Membranous tissue, usu. in a single layer, made up of closely arranged cells separated by little intercellular substance and forming the covering of most internal surfaces and organs and the outer surface of an animal body. **—ep′i·the′li·al** (-əl), **ep′i·the′li·oid′** (-oid′) *adj.*

**ep·i·thet** (ĕp′ə-thĕt′) *n.* [Lat. *epitheton* < Gk. < *epithetos*, added < *epitithenai*, to place upon : *epi-*, upon + *tithenai*, to place.] **1. a.** A term used to characterize a person or thing. **b.** A term used as a descriptive substitute for the name or title of a person. **2.** A derogatory or abusive word or phrase. **—ep′i·thet′ic, ep′i·thet′i·cal** *adj.*

**e·pit·o·me** (ĭ-pĭt′ə-mē) *n.* [Lat. < Gk. *epitomē* < *epitemnein*, to cut short : *epi-*, upon + *temnein*, to cut.] **1.** A brief summary, as of a book or article : ABSTRACT. **2.** An ideal example : EMBODIMENT.

**e·pit·o·mize** (ĭ-pĭt′ə-mīz′) *vt.* **-mized, -miz·ing, -miz·es. 1.** To make an epitome of : SUMMARIZE. **2.** To be an ideal example of.

**ep·i·zo·ic** (ĕp′ĭ-zō′ĭk) *adj.* Growing on the exterior of a living animal. **—ep′i·zo′ism** *n.* **—ep′i·zo′ite′** *n.*

**ep·i·zo·ot·ic** (ĕp′ĭ-zō-ŏt′ĭk) *adj.* Attacking a large number of animals simultaneously. —Used of a disease. —*n.* An epizootic disease. **—ep′i·zo·ot′i·cal·ly** *adv.*

**ep·och** (ĕp′ək, ē′pŏk′) *n.* [LLat. *epocha*, a point in time < Gk. *epokhē* < *epekhein*, to stop : *epi-*, upon + *ekhein*, hold.] **1. a.** A specific historical period, esp. one considered extraordinary. **b.** A notable event marking the beginning of such a period. **2.** *Geol.* A unit of geologic time that is a division of a period. **3.** *Astron.* An instant in time arbitrarily selected as a point of reference.

**ep·och·al** (ĕp′ə-kəl, -ŏk′əl) *adj.* **1.** Of, relating to, or typical of an epoch. **2.** Highly important or significant : MOMENTOUS.

**ep·ode** (ĕp′ōd′) *n.* [Lat. *epodos*, a type of lyric poem < Gk. *epōidos* < *epōidos*, sung after < *epaidein*, to sing after : *epi-*, upon + *aidein*, to sing.] **1.** The strophe that follows the strophe and the apostrophe to complete the triad that forms the basic compositional unit of a lyric ode. **2.** A lyric poem in which a long line of verse is followed by a short one.

**ep·o·nym** (ĕp′ə-nĭm′) *n.* [Gk. *epōnumos* < *epōnumos*, named after : *epi-*, to + *ōnuma*, name.] A person whose name is or is thought to be the source of the name of something. **—ep′o·nym′ic** *adj.*

**e·pon·y·mous** (ĭ-pŏn′ə-məs) *adj.* Of, pertaining to, or constituting an eponym.

**e·pon·y·my** (ĭ-pŏn′ə-mē) *n.* Derivation of a proper name, as of a city or institution, from that of a real or fictitious person.

**ep·o·pee** (ĕp′ə-pē′) *n.* [Fr. *épopée* < Gk. *epopoiia* < *epopoios*, epic poet : *epos*, epic + *poiein*, to make.] **1.** Epic poetry, esp. as a literary genre. **2.** An epic poem.

**ep·os** (ĕp′ŏs′) *n.* [Lat. < Gk.] **1.** Epic poetry handed down by word of mouth. **2.** An epic poem.

**ep·ox·y** (ĕp′ŏk′sē, ĭ-pŏk′-) *n., pl.* **-ies.** [EP(I)- + OXY(GEN).] One of various usu. thermosetting resins capable of forming tight cross-linked polymer structures marked by toughness, strong adhesion, and high corrosion and chemical resistance, used esp. in adhesives and surface coatings. —*vt.* **-ied, -y·ing, -ies.** To fasten together with epoxy.

**ep·si·lon** (ĕp′sə-lŏn′, -lən) *n.* [Gk. *e psilon*, simple e.] The fifth letter in the Greek alphabet. —See table at ALPHABET.

**Ep·som salts** *also* **Ep·som salt** (ĕp′səm) *n.* [After *Epsom*, England.] Hydrated magnesium sulfate used as a cathartic.

**eq·ua·ble** (ĕk′wə-bəl, ē′kwə-) *adj.* [Lat. *aequabilis* < *aequare*, to make even < *aequus*, even.] **1. a.** Steady : unvarying. **b.** Free from extremes. **2.** Not easily disturbed : SERENE <an *equable* temperament> **—eq′ua·bil′i·ty, eq′ua·ble·ness** *n.* **—eq′ua·bly** *adv.*

**e·qual** (ē′kwəl) *adj.* [Lat. *aequalis* < *aequus*, even.] **1.** Having the same measure, quantity, or value as another. **2.** *Math.* Being the same or identical to in value. **3. a.** Having the same rights, privileges, or status <*equal* in a court of law> **b.** Being the same for all members of a group <gave every employee an *equal* chance> **4. a.** Having the qualities, as strength, intelligence, or ability, needed for a situation or task. **b.** Sufficient in extent, amount, or degree. —*n.* One equal to another. —*vt.* **e·qualed, e·qual·ing, e·quals** *or* **e·qualled, e·qual·ling, e·quals. 1.** To be equal to, esp. in value. **2.** To do or produce something equal to. **—e′qual·ly** *adv.*

**e·qual·i·tar·i·an** (ĭ-kwŏl′ĭ-târ′ē-ən) *adj.* Egalitarian. **—e·qual′i·tar′i·an·ism** *n.*

**e·qual·i·ty** (ĭ-kwŏl′ĭ-tē) *n., pl.* **-ties.** [ME *equalite* < OFr. *equalite* < Lat. *aequalitas* < *aequalis*, equal.] **1.** The quality or state of being equal. **2.** A mathematical statement that one thing equals another.

**e·qual·ize** (ē′kwə-līz′) *v.* **-ized, -iz·ing, -iz·es.** —*vt.* **1.** To make equal. **2.** To make uniform. —*vi.* To constitute or induce equality, equilibrium, or balance. **—e′qual·i·za′tion** (ē′kwəl-ĭ-zā′shən) *n.*

**e·qual·iz·er** (ē′kwə-lī′zər) *n.* **1.** One that equalizes. **2.** A device for equalizing pressure or strain. **3.** A tone control system designed to compensate for frequency distortion in audio systems. **4.** *Slang.* A weapon, esp. a revolver.

**equal sign** *n.* The symbol (=) used to indicate logical or mathematical equivalence.

**equal temperament** *n. Mus.* The modification of the intervals of just intonation in the tuning of instruments of fixed intonation to permit the modulation of harmony.

**e·qua·nim·i·ty** (ē′kwə-nĭm′ĭ-tē, ĕk′wə-) *n.* [Lat. *aequanimitas* < *aequanimis*, calm : *aequus*, even + *animus*, mind.] The quality of being calm and even-tempered : COMPOSURE.

**e·quate** (ĭ-kwāt′) *v.* **e·quat·ed, e·quat·ing, e·quates.** [ME *equaten* < Lat. *aequare* < *aequus*, even.] —*vt.* **1.** To make equal : EQUALIZE. **2.** To adjust so as to reduce to a standard or average. **3.** To consider, treat, or depict as equal <*equates* power with money> —*vi.* To be or seem to be equal : CORRESPOND.

**e·qua·tion** (ĭ-kwā′zhən, -shən) *n.* **1.** The act or process of equating or of being equated. **2.** The state of being equal. **3.** *Math.* A linear array of mathematical symbols separated into left and right sides that are designated at least conditionally equal by an equal sign. **4.** *Chem.* A symbolic representation of a chemical reaction as a linear array of chemical symbols and signs. **5.** A complex of variable elements or factors. **—e·qua′tion·al** *adj.* **—e·qua′tion·al·ly** *adv.*

**e·qua·tor** (ĭ-kwā′tər) *n.* [ME < Med. Lat., equalizer < Lat. *aequare*, to equate.] **1. a.** The great circle circumscribing the earth's surface, the reckoning datum of latitudes and dividing boundary of Northern and Southern hemispheres, formed by the intersection of a plane passing through the earth's center perpendicular to its axis of rotation. **b.** A similar great circle drawn on the surface of a celestial body at right angles to the axis of rotation. **2.** The celestial equator.

**e·qua·to·ri·al** (ē′kwə-tôr′ē-əl, -tōr′-, ĕk′wə-) *adj.* **1.** Of, pertaining to, or like the equator. **2.** Pertaining to conditions that exist at the earth's equator <*equatorial* rainfall> **3.** Having or constituting a support with two perpendicular axles, one of which is parallel to the earth's rotational axis. —*n.* **—e′qua·to′ri·al·ly** *adv.*

**equatorial current** *n.* One of the surface currents drifting westward through the oceans at the equator.

**eq·uer·ry** (ĕk′wə-rē) *n., pl.* **-ries.** [Fr. *écurie*, stable < OFr. *escurie* < *escuier*, squire. —see ESQUIRE.] **1.** An officer charged with supervision of the horses belonging to a royal or noble household. **2.** A personal attendant to the British royal household.

**e·ques·tri·an** (ĭ-kwĕs′trē-ən) *adj.* [< Lat. *equester* < *equus*, horse.] **1.** Of or relating to horsemanship or horseback riding. **2.** Depicted or represented on horseback <an *equestrian* statue of Peter the Great> **3.** Of, relating to, or composed of horsemen or knights. —*n.* A person who rides a horse.

**e·ques·tri·enne** (ĭ-kwĕs′trē-ĕn′) *n.* [EQUESTRI(AN) + *-enne*, as in *comedienne*.] A woman who rides a horse.

**equi-** *pref.* [ME < Lat. *aequi-* < *aequus*, equal.] Equal : equally <*equiangular*>

**e·qui·an·gu·lar** (ē′kwē-ăng′gyə-lər, ĕk′wē-) *adj.* Having all angles equal. **—e′qui·an′gu·lar′i·ty** (-lâr′ĭ-tē) *n.*

**e·qui·dis·tant** (ē′kwē-dĭs′tənt, ĕk′wē-) *adj.* [OFr. < LLat. *aequidistans* : *aequus-*, equal + *distans*, distant.] Equally distant. **—e′qui·dis′tance** (-təns) *n.* **—e′qui·dis′tant·ly** *adv.*

**e·qui·lat·er·al** (ē′kwə-lăt′ər-əl, ĕk′wə-) *adj.* [LLat. *aequilateralis* : *aequus*, equal + *latus*, side.] Having all sides or faces equal. —*n.* **1.** A side exactly equal to others. **2.** A geometric figure having equal sides. **—e′qui·lat′er·al·ly** *adv.*

**e·quil·i·brant** (ĭ-kwĭl′ə-brənt) *n.* A force capable of balancing a system of forces to produce equilibrium.

**e·quil·i·brate** (ĭ-kwĭl′ə-brāt′) *v.* **-brat·ed, -brat·ing, -brates.** [EQUILIBR(IUM) + -ATE[1].] —*vi.* To be in or bring about equilibrium. —*vt.* To keep in or bring into equilibrium. **—e·quil′i·bra′tion** *n.*

**e·quil·i·bra·tor** (ĭ-kwĭl′ə-brā′tər) *n.* A device that brings about and maintains equilibrium. **—e·quil′i·bra·to′ry** (-brə-tôr′ē, -tōr′ē) *adj.*

**e·quil·i·brist** (ĭ-kwĭl′ə-brĭst) *n.* [Fr. *équilibriste* < Lat. *aequilibrium*, equilibrium.] A person who performs feats of balance, as tightrope walking. **—e·quil′i·bris′tic** *adj.*

**e·qui·lib·ri·um** (ē′kwə-lĭb′rē-əm, ĕk′wə-) *n.* [Lat. *aequilibrium* : *aequus*, level + *libra*, balance.] **1.** A condition in which all acting influences are canceled by others, resulting in a stable, balanced, or unchanging system. **2.** *Physics.* The condition of a system in which the resultant of all acting forces is zero and the sum of all torques about any axis is zero. **3.** *Chem.* The state of a reaction in which its forward and reverse reactions occur at equal rates so that the concentration of the reactants does not change with time. **4.** Emotional or mental balance : POISE.

**e·qui·mo·lar** (ē′kwə-mō′lər) *adj. Chem.* Having an equal number of moles.

**e·quine** (ē′kwīn′, ĕk′wīn′) *adj.* [Lat. *equinus* < *equus*, horse.] **1.** Of, relating to, or typical of a horse. **2.** Of or belonging to the family Equidae, including the horses, asses, and zebras. **—e′quine′** *n.*

**e·qui·noc·tial** (ē′kwə-nŏk′shəl, ĕk′wə-) *adj.* [ME *equinoxial* < OFr. < Lat. *aequinoctialis* < *aequinoctium*, equinox.] **1.** Relating to an equinox. **2.** Relating to the celestial equator. **3.** *Bot.* Having or designating flowers that open and close at specific times. —*n.* **1.** A violent windstorm and rainstorm held to occur at or near the time of the equinox. **2.** The celestial equator.

**equinoctial circle** *n.* The celestial equator.

**e·qui·nox** (ē′kwə-nŏks′, ĕk′wə-) *n.* [ME < OFr. *equinoxe* < Med Lat. *aequinoxium* < Lat. *aequinoctium* : *aequus*, equal + *nox* night.] **1.** Either of two points on the celestial sphere where the ecliptic intersects the celestial equator. **2.** Either of the two times during a year when the sun crosses the celestial equator and the length of day and night are approx. equal.

**e·quip** (ĭ-kwĭp′) *vt.* **e·quipped, e·quip·ping, e·quips.** [Fr. *équiper* of Germanic orig.] **1. a.** To furnish with necessities, as tools or provisions. **b.** To supply with the qualities necessary for performance <an experience that *equipped* them to deal with future problems> **2.** To dress up.

**eq·ui·page** (ĕk′wə-pĭj) *n.* [Fr. < *équiper*, to equip.] **1.** Equipment or furnishings : OUTFIT. **2. a.** A horse-drawn carriage. **b.** A carriage equipped with horses and attendants. **3.** *Archaic.* A retinue, as of a monarch. **4.** *Archaic.* A set of small household articles, as a dessert service. **5.** *Archaic.* A collection of small personal articles.

**e·quip·ment** (ĭ-kwĭp′mənt) *n.* **1.** The act of equipping or state of being equipped. **2.** Something with which one is equipped. **3. a.** The rolling stock of a railroad. **b.** A single piece of such equipment **4.** The qualities or traits that make up the mental and emotional resources of an individual.

**e·qui·poise** (ē′kwə-poiz′, ĕk′wə-) *n.* **1.** A state of equilibrium. **2.** A counterpoise : counterbalance.

**e·qui·pol·lence** (ē′kwə-pŏl′əns, ĕk′wə-) *n.* Equality, as in effectiveness or validity : EQUIVALENCE.

**e·qui·pol·lent** (ē′kwə-pŏl′ənt, ĕk′wə-) *adj.* [ME < OFr. < Lat. *aequipollens* : *aequus*, even + *pollens*, pr.part. of *pollēre*, to be strong.] **1.** Equal in power, effectiveness, or significance. **2.** *Logic.* Validly derived from each other : DEDUCIBLE. **3.** Equivalent. —*n.* An equivalent. **—e′qui·pol′lent·ly** *adv.*

**e·qui·pon·der·ance** (ē′kwə-pŏn′dər-əns, ĕk′wə-) *n.* [Med. Lat. *aequiponderans*, pr.part. of *aequiponderare*, to equiponderate.] Equality of weight : EQUIPOISE. **—e′qui·pon′der·ant** *adj.*

**e·qui·pon·der·ate** (ē′kwə-pŏn′də-rāt′, ĕk′wə-) *vt.* **-at·ed, -at·ing, -ates.** [Med. Lat. *aequiponderare*, *aequiponderat-* : Lat. *aequus*, equal + Lat. *ponderare*, to weigh.] **1.** To counterbalance. **2.** To give equal weight to.

**e·qui·po·ten·tial** (ē′kwə-pə-tĕn′shəl, ĕk′wə-) *adj.* **1.** Having equal potential. **2.** *Physics.* Having the same potential at every point.

**eq·ui·se·tum** (ĕk′wə-sē′təm) *n.* [NLat. *Equisetum*, genus name < Lat. *equisaetum*, horsetail : *equus*, horse + *saeta*, bristle.] A flowerless, seedless plant of the genus *Equisetum*, as the horsetail.

**eq·ui·ta·ble** (ĕk′wĭ-tə-bəl) *adj.* [Fr. *équitable* < OFr. < *equite*, equity.] **1.** Just and fair : IMPARTIAL. **2.** *Law.* Possessing existence or validity in equity, as distinguished from statute and common law. **—eq′ui·ta·ble·ness** *n.* **—eq′ui·ta·bly** *adv.*

**eq·ui·tant** (ĕk′wĭ-tənt) *adj.* [Lat. *equitans*, *equitant-*, pr.part. of *equitare*, to ride < *eques*, rider < *equus*, horse.] Overlapping at the base to form a flat fanlike arrangement, as the leaves of some irises.

**eq·ui·ta·tion** (ĕk′wĭ-tā′shən) *n.* [Lat. *equitatio* < *equitare*, to ride. —see EQUITANT.] The art and practice of riding a horse.

**eq·ui·ty** (ĕk′wĭ-tē) *n., pl.* **-ties.** [ME *equite* < OFr. < Lat. *aequitas* < *aequus*, even, fair.] **1.** The quality, state, or ideal of being just, fair, and impartial. **2.** Something just, fair, and impartial. **3.** The money value of a property beyond any mortgage or liabilities existing on it. **4.** *Law.* **a.** Justice applied in circumstances not covered by law. **b.** A system of jurisprudence supplementing common law. **c.** An equitable right or claim. **5.** *Law.* Equity of redemption.

**equity of redemption** *n. Law.* The right of one who has mortgaged his or her property to redeem that property on payment of the sum due within a reasonable amount of time after the due date.

**equity stock** *n.* Common stock.

**e·quiv·a·lence** (ĭ-kwĭv′ə-ləns) *n.* **1.** The state or condition of being equivalent : EQUALITY. **2.** *Math.* A reflexive, symmetric, and transitive relation between elements of a set that establishes any two elements in the set as equivalent or nonequivalent.

**equivalence relationship** *n.* EQUIVALENCE 2.

**e·quiv·a·len·cy** (ĭ-kwĭv′ə-lən-sē) *n., pl.* **-cies.** Equivalence.

**e·quiv·a·lent** (ĭ-kwĭv′ə-lənt) *adj.* [ME < LLat. *aequivalens*, pr.part. of *aequivalēre*, to be equal in value : *aequi-*, equi- + *valēre*, to be strong.] **1. a.** Equal, as in force, value, or meaning. **b.** Having identical or similar effects. **2.** Corresponding or practically equal in effect <a request that was *equivalent* to an order> **3.** *Math.* **a.** Capable of being put into a one-to-one relationship. —Used of two sets. **b.** Having virtually identical or corresponding parts. **4.** *Chem.* Having the same ability to combine. —*n.* **1.** Something equivalent. **2.** *Chem.* Equivalent weight. **—e·quiv′a·lent·ly** *adv.*

**equivalent weight** *n.* The number of parts by weight of any element combining with or replacing the equivalent of half the atomic weight of oxygen or with one atomic weight of hydrogen.

**e·quiv·o·cal** (ĭ-kwĭv′ə-kəl) *adj.* [LLat. *aequivocus* : *aequus*, same + *vocare*, to call.] **1.** Marked by often intentional ambiguity so as to mislead. **2.** Of uncertain significance or worth. **3.** Of an uncertain nature. **—e·quiv′o·cal·ly** *adv.* **—e·quiv′o·cal·ness** *n.*

**e·quiv·o·cate** (ĭ-kwĭv'ə-kāt') *vi.* **-cat·ed, -cat·ing, -cates.** [ME *equivocaten* < Med. Lat. *aequivocare* < LLat. *aequivocus,* equivocal.] **1.** To use vague or intentionally evasive language. **2.** To avoid making an explicit statement. **—e·quiv'o·ca'tor** *n.*
☆ ***syns:*** EQUIVOCATE, HEDGE, PUSSYFOOT, TERGIVERSATE, WAFFLE *v. core meaning* : to use evasive or deliberately vague language <a candidate who *equivocated* on the major issues>

**e·quiv·o·ca·tion** (ĭ-kwĭv'ə-kā'shən) *n.* **1.** Use of equivocal language. **2.** An equivocal expression or statement.

**eq·ui·voque** *also* **eq·ui·voke** (ĕk'wə-vōk', ē'kwə-) *n.* [LLat. *aequivocus,* equivocal.] **1.** EQUIVOCATION 2. **2.** A play on words : PUN. **3.** A double meaning.

**-er**[1] *suff.* [ME < OE *-ere* < Lat. *-arius,* -ary.] **1. a.** One that performs a specified action <swimm*er*> **b.** One that undergoes or is capable of undergoing a specified action <broil*er*> **c.** One that provides <pork*er*> **d.** One that has <ten-pound*er*> **2. a.** One associated or involved with <bank*er*> **b.** Native or resident of <New York*er*> **c.** One that is <foreign*er*>

**-er**[2] *suff.* [ME *-ere* < OE *-ra.*] —Used to form the comparative degree of adjectives and adverbs <dark*er*><fast*er*>

**Er** *symbol for* ERBIUM.

**e·ra** (ĭr'ə, ĕr'ə) *n.* [LLat. *aera* < Lat., counters, pl. of *aes,* money.] **1.** A time period reckoned from a specific date in history. **2. a.** A time period marked by particular circumstances, events, or personages. **b.** A point that marks the beginning of such a period. **3.** *Geol.* A division of geologic time comprising one or more periods.

**e·rad·i·cate** (ĭ-răd'ĭ-kāt') *vt.* **-cat·ed, -cat·ing, -cates.** [Lat. *eradicare, eradicat-* : *ex-,* out + *radix,* root.] **1.** To pull or tear up by or as if by the roots. **2.** To get rid of completely <Their goal was to *eradicate* political corruption.> **—e·rad'i·ca·ble** (-kə-bəl) *adj.* **—e·rad'i·ca'tion** *n.* **—e·rad'i·ca'tive** *adj.* **—e·rad'i·ca'tor** *n.*

**e·rase** (ĭ-rās') *vt.* **e·rased, e·ras·ing, e·ras·es.** [Lat. *eradere, eras-* : *ex-,* out + *radere,* to scratch.] **1. a.** To remove (e.g., something written) by rubbing or scraping. **b.** To remove recorded material from <*erase* a videotape> **2.** To remove all traces of. **3.** To remove or destroy as if by wiping out. **4.** *Computer Sci.* To replace all the binary digits in a storage device by binary zeros. **—e·ras'a·bil'i·ty** *n.* **—e·ras'a·ble** *adj.*

**e·ras·er** (ĭ-rā'sər) *n.* Something, as a piece of rubber, used to erase.

**e·ra·sure** (ĭ-rā'shər) *n.* An act or instance of erasing.

**Er·a·to** (ĕr'ə-tō') *n.* [Lat. < Gk. *Eratō* < *eratos,* loved < *eran,* to love.] *Gk. Myth.* The Muse of lyric poetry and mime.

**er·bi·um** (ûr'bē-əm) *n.* [After *Ytterby,* Sweden.] *Symbol* **Er** A soft, malleable, silvery element, used in metallurgy and nuclear research and to color glass and porcelain; atomic number 68; atomic weight 167.26.

**ere** (âr) [ME *er* < OE *ǣr* and ON *ār.*] *Archaic.* —*prep.* Previous to : BEFORE. —*conj.* **1.** Before. **2.** Sooner than : rather than.

**Er·e·bus** (ĕr'ə-bəs) *n.* [Lat. < Gk. *Erebos.*] *Gk. Myth.* The dark underworld region through which the dead must pass before they reach Hades.

**e·rect** (ĭ-rĕkt') *adj.* [ME < Lat. *erectus,* p.part. of *erigere,* to set up : *ex-,* up from + *regere,* to guide.] **1. a.** Not lying down : UPRIGHT. **b.** Vertical. **2.** Being in a stiff, rigid condition. **3.** *Archaic.* Wide-awake : alert. —*vt.* **e·rect·ed, e·rect·ing, e·rects. 1.** To construct by assembling materials and parts <*erect* a building> **2.** To raise to an upright or rigid condition. **3.** To fix in an upright position. **4.** To assemble or set up : ESTABLISH. **5.** *Math.* To construct (e.g., a perpendicular) from or on a given base. **—e·rect'a·ble** *adj.* **—e·rect'ly** *adv.* **—e·rect'ness** *n.*

**e·rec·tile** (ĭ-rĕk'təl, -tīl') *adj.* **1.** Capable of being erected. **2.** *Physiol.* Of or relating to vascular tissue capable of filling with blood and becoming rigid. **—e·rec·til'i·ty** (-tĭl'ĭ-tē) *n.*

**e·rec·tion** (ĭ-rĕk'shən) *n.* **1.** The act of erecting. **2.** Something erected : CONSTRUCTION. **3.** *Physiol.* **a.** The condition of erectile tissue when filled with blood. **b.** An erect penis.

**e·rec·tor** *also* **e·rect·er** (ĭ-rĕk'tər) *n.* **1.** One that erects. **2.** *Anat.* A muscle that causes or maintains the erection of a body part.

**E region** *n.* The E layer.

**ere·long** (âr-lông', -lŏng') *adv. Archaic.* Before long : SOON.

**er·e·mite** (âr'ə-mīt') *n.* [ME < LLat. *eremita* < LGk. *erēmitēs* < *erēmia,* desert < *erēmos,* uninhabited.] A hermit, esp. a religious recluse. **—er'e·mit'ic** (-mĭt'ĭk), **er'e·mit'i·cal** *adj.*

**er·e·mur·us** (ĕr'ə-myŏŏr'əs) *n.* [NLat. *Eremurus,* genus name : Gk. *erēmos,* solitary + Gk. *oura,* tail.] The foxtail lily.

**ere·now** (âr-nou') *adv. Archaic.* Before now : HERETOFORE.

**e·rep·sin** (ĭ-rĕp'sən) *n.* [Lat. *eripere,* to snatch away (*ex-,* away + *rapere,* to snatch) + (P)EPSIN.] A mixture of peptidases in the small intestines that produces amino acids.

**er·e·thism** (ĕr'ə-thĭz'əm) *n.* [Fr. *éréthisme* < Gk. *erethismos,* irritation < *erethizein,* to irritate.] Abnormal irritability and sensibility to stimulation in any part of the body. **—er'e·this'mic** (-mĭk) *adj.*

**ere·while** (âr-hwīl', -wīl') *also* **ere·whiles** (-hwīlz', -wīlz') *adv. Archaic.* Some time ago : FORMERLY.

**erg** (ûrg) *n.* [Gk. *ergon,* work.] A centimeter-gram-second unit of energy or work equal to the work done by a force of one dyne acting over a distance of one centimeter.

**er·go** (ûr'gō, âr'-) *conj.* [Lat.] Therefore : consequently. —*adv.* Consequently : hence.

**er·go·cal·cif·er·ol** (ûr'gō-kăl-sĭf'ə-rôl', -rōl') *n.* [ERGO(T) + CALCIFEROL.] Vitamin $D_2$.

**er·go·graph** (ûr'gə-grăf') *n.* [Gk. *ergon,* work + -GRAPH.] A device for determining the work capacity of a muscle or group of muscles by measuring the extent of movement.

**er·gom·e·ter** (ûr-gŏm'ĭ-tər) *n.* [Gk. *ergon,* work + -METER.] A device for measuring the amount of work done by a group of muscles under control conditions. **—er'go·met'ric** *adj.*

**er·go·nom·ics** (ûr'gə-nŏm'ĭks) *n.* [Gk. *ergon,* work + (ECO)NOMICS.] *(sing. in number).* Study of equipment design in order to reduce operator fatigue and discomfort. **—er'go·nom'ic** *adj.* **—er'go·nom'i·cal·ly** *adv.* **—er·gon'o·mist** (ûr-gŏn'ə-mĭst) *n.*

**er·gos·ter·ol** (ûr-gŏs'tə-rôl', -rōl') *n.* [ERGO(T) + STEROL.] A crystalline sterol, $C_{28}H_{44}O$, synthesized by yeast from sugars or derived from ergot and converted under ultraviolet irradiation to vitamin $D_2$.

**er·got** (ûr'gət, -gŏt') *n.* [Fr. < OFr. *argor,* cock's spur.] **1.** Any of various fungi of the genus *Claviceps,* infecting cereal plants, as rye, and forming black sclerotia that replace many of the seeds of the host plant. **2.** The disease caused by the ergot. **3.** The dried sclerotia of ergot, usu. obtained from rye seed and used as a source of several medicinally important alkaloids and as the basic source of lysergic acid.

**er·got·a·mine** (ûr-gŏt'ə-mēn', -mĭn) *n.* A crystalline alkaloid, $C_{33}H_{35}N_5O_5$, derived from ergot, that induces vasoconstriction and is used in the treatment of migraine.

**er·got·ism** (ûr'gə-tĭz'əm) *n.* Poisoning by ergot-infected grain or grain products, marked by lameness and necrosis of the extremities.

**E·rie** (ĭr'ē) *n., pl.* **Erie** *or* **E·ries. 1. a.** A tribe of Indians inhabiting the region around Lake Erie. **b.** A member of this tribe. **2.** The Iroquoian language of the Erie.

**E·ris** (ĭr'ĭs, ĕr'-) *n.* [Gk. < *eris,* strife.] *Gk. Myth.* The goddess of discord.

**e·ris·tic** (ĭ-rĭs'tĭk) *also* **e·ris·ti·cal** (-tĭ-kəl) *adj.* [Gk. *eristikos* < *erizein,* to wrangle < *eris,* strife.] **1.** Of or pertaining to controversy or argument. **2.** Given to argument or polemics. —*n.* **1.** A person given to argument or dispute. **2.** The art or practice of debate.

**Er·len·mey·er flask** (ûr'lən-mī'ər, âr'-) *n.* [After Emil *Erlenmeyer* (1825–1909).] A conical laboratory flask with a narrow neck and a broad flat bottom.

**er·mine** (ûr'mĭn) *n.* [ME *ermin* < OFr., perh. of Germanic orig.] **1.** A weasel, *Mustela erminea* of northern regions, with brownish fur that turns white in winter. **2.** The white fur of the ermine.

**erne** *also* **ern** (ûrn) *n.* [ME *ern* < OE *earn.*] A sea eagle, esp. *Haliaeetus albicella* of the Old World.

**e·rode** (ĭ-rōd') *v.* **e·rod·ed, e·rod·ing, e·rodes.** [Lat. *erodere,* to gnaw off : *ex-,* off + *rodere,* to gnaw.] —*vt.* **1.** To wear (something) away by or as if by abrasion <wind *eroding* the sand dunes> **2.** To eat into or corrode. **3.** To form by wearing away <The stream *eroded* a deep gully.> —*vi.* To become worn or eroded.

**e·rog·e·nous** (ĭ-rŏj'ə-nəs) *also* **er·o·gen·ic** (ĕr'ə-jĕn'ĭk) *adj.* [Gk. *erōs,* sexual love + -GENOUS.] Responsive to sexual stimulation.

**Er·os** (ĕr'ŏs', ĭr'-) *n.* [Lat. < Gk. *Erōs* < *erōs,* sexual love.] **1.** *Gk. Myth.* The god of love. **2.** *Psychoanal.* The sum of all self-preservative instincts.

**e·rose** (ĭ-rōs') *adj.* [Lat. *erosus,* p.part. of *erodere,* to gnaw off. —see ERODE.] Irregularly notched, toothed, or indented. **—e·rose'ly** *adv.*

**e·ro·sion** (ĭ-rō'zhən) *n.* **1.** The process of eroding or state of being eroded. **2.** Natural processes, as weathering or gravity, by which material is moved on the earth's surface. **—e·ro'sion·al** *adj.* **—e·ro'sion·al·ly** *adv.*

**e·ro·sive** (ĭ-rō'sĭv) *adj.* Causing erosion <the *erosive* effect of the wind> **—e·ro'sive·ness, e'ro·siv'i·ty** *n.*

**e·rot·ic** (ĭ-rŏt'ĭk) *adj.* [Gk. *erōtikos* < *erōs,* sexual love.] **1.** Of, relating to, or promoting sexual love and desire : AMATORY. **2.** Dominated by sexual love or desire. **—e·rot'ic** *n.* **—e·rot'i·cal·ly** *adv.*

**e·rot·i·ca** (ĭ-rŏt'ĭ-kə) *pl.n.* [Gk. *erōtika,* neuter pl. of *erōtikos,* erotic.] *(sing. or pl. in number).* Literature or art intended to arouse sexual desire.

**e·rot·i·cism** (ĭ-rŏt'ĭ-sĭz'əm) *also* **er·o·tism** (ĕr'ə-tĭz'əm) *n.* **1.** An erotic theme or quality. **2.** Sexual excitement. **3.** Abnormally persistent sexual excitement. **—e·rot'i·cist** *n.*

**err** (ûr, ĕr) *vi.* **erred, err·ing, errs.** [ME *erren* < OFr. *errer* < Lat. *errare.*] **1.** To make a mistake. **2.** To violate traditional moral standards : SIN. **3.** *Archaic.* To go astray.

**er·ran·cy** (ĕr'ən-sē) *n., pl.* **-cies.** The state of erring.

**er·rand** (ĕr'ənd) *n.* [ME *erand* < OE *ǣrend.*] **1. a.** A short trip taken to perform a specified task, usu. for another <went on an *errand* for the boss> **b.** The purpose of such a trip. **2.** *Archaic.* A mission.

**er·rant** (ĕr'ənt) *adj.* [ME *erraunt* < AN, partly < OFr. *errer,* to travel (< VLat. **iterare* < LLat. *itinerare*), and partly < OFr. *errer,* to err (< Lat. *errare*).] **1.** Traveling esp. in search of adventure <an *errant* knight> **2.** Straying from the standard or ordinary : ECCENTRIC.

3. Wandering outside the established limits. **4.** Moving aimlessly <an *errant* breeze> **—er′rant** *n.* **—er′rant·ly** *adv.*

**er·rant·ry** (ĕr′ən-trē) *n.* The condition of traveling or roving about, esp. in search of adventure.

**er·ra·ta** (ĭ-rä′tə, ĭ-rā′-) *n. pl. of* ERRATUM. ***usage:*** As the plural of *erratum, errata* usu. takes a plural verb: *The errata were noted in the appendix.*

**er·rat·ic** (ĭ-răt′ĭk) *adj.* [ME *erratik* < OFr. *erratique* < Lat. *erraticus* < *errare,* to wander.] **1.** Having no regular or fixed course : WANDERING. **2.** Lacking regularity, consistency, or uniformity. **3.** Unconventional or odd : ECCENTRIC. **—er·rat′i·cal·ly** *adv.*

**er·ra·tum** (ĭ-rä′təm, ĭ-rā′-) *n., pl.* **-ta** (-tə) [Lat. < neuter p.part. of *errare,* to stray.] An error in printing or writing, esp. such an error noted in a list of corrections and bound into a book.

**er·rhine** (ĕr′īn′) *adj.* [NLat. *errhinum,* an errhine medicine < Gk. *errhinon* : *en-,* in + *rhis,* nostril.] Promoting nasal discharge. *—n.* An errhine medication.

**er·ro·ne·ous** (ĭ-rō′nē-əs) *adj.* [ME < Lat. *erroneus* < *errare,* to err.] Containing or based on error : MISTAKEN. **—er·ro′ne·ous·ly** *adv.* **—er·ro′ne·ous·ness** *n.*

**er·ror** (ĕr′ər) *n.* [ME < OFr. < Lat. < *errare,* to err.] **1.** An act, assertion, or belief that unintentionally deviates from what is correct, right, or true. **2.** The state of having false knowledge. **3.** A deviation from an accepted code of behavior. **4.** A mistake. **5.** The difference between a computed or measured value and a correct value. **6.** *Baseball.* A defensive misplay when a normal play should have resulted in an out or prevented a base runner's advance.

☆ ***syns:*** ERROR, ERRATUM, LAPSE, MISCUE, MISSTEP, MISTAKE, SLIP, SLIP-UP, TRIP *n. core meaning* : an unintentional deviation from what is correct, right, or true <an *error* in judgment>

**er·satz** (ĕr′zäts′, ĕr-zäts′) *adj.* [< G. *Ersatz,* replacement < *ersetzen,* to replace < OHG *irsezzen.*] Being an artificial, usu. inferior, substitute. **—er′satz′** *n.*

**Erse** (ûrs) *n.* [ME *Erisch,* Irish.] **1.** Irish Gaelic. **2.** Scottish Gaelic. *—adj.* Of or relating to the Scottish or Irish Celts or their language.

**erst** (ûrst) [ME *erest* < OE *ǣrest.*] *Archaic. —adv.* **1.** At first. **2.** Formerly. *—adj.* First.

**erst·while** (ûrst′hwīl′, -wīl′) *adj.* Former. **—erst′while** *adv.*

**er·u·bes·cence** (ĕr′ə-bĕs′əns, ĕr′yə-) *n.* [< Lat. *erubescens,* pr.part. of *erubescere,* to blush : *ex-* (intensive) + *rubescere,* to redden < *rubere,* to be red.] A blush. **—er′u·bes′cent** *adj.*

**e·ruct** (ĭ-rŭkt′) *vt. & vi.* **e·ruct·ed, e·ruct·ing, e·ructs.** [Lat. *eructare* : *ex-,* out + *ructare,* to belch.] To belch.

**e·ruc·ta·tion** (ĭ-rŭk-tā′shən, ē′rŭk-) *n.* An act or instance of belching. **—e·ruc′ta·tive** (ĭ-rŭk′tə-tĭv) *adj.*

**er·u·dite** (ĕr′yə-dīt′, ĕr′ə-) *adj.* [ME *erudit* < Lat. *eruditus,* p.part. of *erudire,* to instruct : *ex-,* out + *rudis,* rude.] Learned <*erudite* scholars> **—er′u·dite′ly** *adv.* **—er′u·dite′ness** *n.*

**er·u·di·tion** (ĕr′yə-dĭsh′ən, ĕr′ə-) *n.* Deep, extensive learning.

**e·rum·pent** (ĭ-rŭm′pənt) *adj.* [Lat. *erumpens,* pr. part of *erumpere, erumpent-,* to burst. —see ERUPT.] Bursting through or as if through a surface or covering.

**e·rupt** (ĭ-rŭpt′) *v.* **e·rupt·ed, e·rupt·ing, e·rupts.** [Lat. *erumpere, erupt-* : *ex-,* out + *rumpere,* to break.] *—vi.* **1.** To emerge violently from restraint or limits : EXPLODE. **2.** To become violently active. **3.** To force out or release something, as steam, with sudden violence. **4. a.** To pierce the gum in developing. —Used of a tooth. **b.** To appear as a blemish on the skin. *—vt.* To eject or force out violently. **—e·rup′tive** *adj.* **—e·rup′tive·ly** *adv.*

**e·rup·tion** (ĭ-rŭp′shən) *n.* **1.** An act, process, or instance of erupting. **2.** A sudden, often violent outburst. **3.** A skin blemish or rash.

**-ery** *suff.* [ME *-erie* < OFr. : *-er,* -ary (< Lat. *-arius*) + *-ie,* -ia (< Lat. *-ia*).] **1.** A place for <bak*ery*> **2.** A collection or class <fin*ery*> **3.** A state or condition <slav*ery*> **4.** Act : practice <brib*ery*> **5.** Characteristics or qualities of <snobb*ery*>

**e·ryn·go** (ĭ-rĭng′gō) *n., pl.* **-goes.** [NLat. *Eryngium,* genus name < Lat. *eryngion,* sea holly < Gk. *ērungion.*] **1.** A plant of the genus *Eryngium,* with spiny leaves and dense clusters of small bluish flowers. **2.** *Obs.* The candied root of *E. maritimum,* the sea holly, used as an aphrodisiac.

**er·y·sip·e·las** (ĕr′ĭ-sĭp′ə-ləs, ĭr′-) *n.* [ME *erisipila* < Lat. *erysipelas* < Gk. *erusipelas* : *erisi-,* red + *-pelas,* skin.] An acute disease of the skin and subcutaneous tissue caused by a streptococcus and marked by spreading inflammation. **—er′y·si·pel′a·tous** (-sĭ-pĕl′ə-təs) *adj.*

**er·y·sip·e·loid** (ĕr′ĭ-sĭp′ə-loid′, ĭr′ə-) *n.* [ERYSIPEL(AS) + -OID.] An infectious disease of the hands marked by red lesions and caused by the bacterium *Erysipelothrix rhusiopathiae,* found in infected meat or fish.

**er·y·the·ma** (ĕr′ə-thē′mə) *n.* [Gk. *eruthēma* < *eruthainein,* to be red < *eruthros,* red.] A redness of the skin, as caused by chemical poisoning or sunburn. **—er′y·them′a·tous** (-thĕm′ə-təs, -thē′mə-təs), **er′y·the·mat′ic** (-thĭ-măt′ĭk), **er′y·the′mic** *adj.*

**er·y·thor·bic acid** (ĕr′ə-thôr′bĭk) *n.* [ERYTH(RO)- + (ASC)ORBIC ACID.] An optical isomer of ascorbic acid used as an antioxidant.

**erythr-** *pref. var. of* ERYTHRO-.

**er·y·thrism** (ĕr′ə-thrĭz′əm) *n.* Unusual redness of pigmentation, as of hair or plumage. **—er′y·thris′mal** (ĕr′ə-thrĭz′məl) *adj.*

**er·y·thrite** (ĕr′ə-thrīt′) *n.* A reddish hydrated arsenate of cobalt found in veins bearing cobalt and arsenic and used in coloring glass.

**erythro-** *or* **erythr-** *pref.* [Gk. *eruthros,* red.] **1.** Red <*erythro*cyte> **2.** Erythrocyte <*erythro*poiesis>

**e·ryth·ro·blast** (ĭ-rĭth′rə-blăst′) *n.* Any of the nucleated cells in bone marrow that develop into erythrocytes. **—e·ryth′ro·blas′tic** (-blăs′tĭk) *adj.*

**e·ryth·ro·blas·to·sis** (ĭ-rĭth′rō-blă-stō′sĭs) *n., pl.* **-ses** (-sēz′). Abnormal presence of erythroblasts in the blood.

**erythroblastosis fe·tal·is** (fə-tăl′əs) *n.* [NLat., fetal erythroblastosis.] A hemolytic disease of the fetus and newborn usu. caused by the production of antibodies from an Rh-negative mother against an Rh-positive fetus, characterized by anemia, jaundice, and the progressive destruction of circulating erythrocytes.

**e·ryth·ro·cyte** (ĭ-rĭth′rə-sīt′) *n.* The yellowish, nonnucleated, disk-shaped blood cell that contains hemoglobin and is responsible for the color of blood. **—e·ryth′ro·cyt′ic** (-sĭt′ĭk) *adj.*

**e·ryth·ro·cy·tom·e·ter** (ĭ-rĭth′rə-sī-tŏm′ĭ-tər) *n.* A hemacytometer.

**e·ryth·ro·my·cin** (ĭ-rĭth′rə-mī′sĭn) *n.* An antibiotic agent from cultures of the bacterium *Streptomyces erythreus,* effective esp. against Gram-positive bacteria.

**e·ryth·ro·poi·e·sis** (ĭ-rĭth′rō-poi-ē′sĭs) *n.* Formation and production of erythrocytes. **—e·ryth′ro·poi·et′ic** (-ĕt′ĭk) *adj.*

**e·ryth·ro·poi·e·tin** (ĭ-rĭth′rō-poi-ē′tĭn) *n.* [ERYTHROPOIET(IC) + -IN.] A hormone that regulates erythropoiesis.

**Es** *symbol for* EINSTEINIUM.

**-es**[1] *suff. var. of* -S[1]. —Used after *s, z, ch, sh,* and postconsonantal *y.*

**-es**[2] *suff. var. of* -S[2]. —Used after *s, z, ch, sh,* and postconsonantal *y.*

**es·ca·drille** (ĕs′kə-drĭl′, -drē′) *n.* [Fr. < Sp. *escuadrilla,* dim. of *escuadrón,* squadron < *escuadrar,* to square, of Lat. orig.] A unit of a European air command usu. made up of six airplanes.

**es·ca·lade** (ĕs′kə-lād′, -läd′) *n.* [Fr. < Ital. *scalata* < *scalare,* to climb < *scala,* ladder < Lat. < *scalae,* steps.] The act of scaling the walls of a fortification. *—vt.* **-lad·ed, -lad·ing, -lades.** To climb up and over (a fortified wall). **—es′ca·lad′er** *n.*

**es·ca·late** (ĕs′kə-lāt′) *v.* **-lat·ed, -lat·ing, -lates.** [Back-formation < ESCALATOR.] *—vt.* To enlarge, increase, or intensify. *—vi.* To increase in intensity or scope. **—es′ca·la′tion** *n.*

**es·ca·la·tor** (ĕs′kə-lā′tər) *n.* [Orig. a trademark.] **1.** A moving stairway consisting of steps attached to a continuously circulating belt. **2.** An escalator clause.

**escalator clause** *n.* A provision in a contract stipulating an increase or decrease, as in wages, benefits, or prices, under certain conditions, as changes in the cost of living.

**es·cal·lop** (ĭ-skŏl′əp, ĭ-skăl′-) *n. & v. var. of* SCALLOP.

**es·ca·pade** (ĕs′kə-pād′) *n.* [Fr. < OFr. < OSp. or OPort. *escapada* < *escapar,* to escape.] An adventurous action that usu. violates conventional standards of behavior.

**es·cape** (ĭ-skāp′) *v.* **-caped, -cap·ing, -capes.** [ME *escapen* < ONFr. *escaper* : Lat. *ex-,* out + Med. Lat. *cappa,* cloak.] *—vi.* **1.** To break free from confinement. **2.** To issue from confinement : leak out. **3.** To avoid capture, danger, or harm. **4.** To grow beyond a cultivated area. *—vt.* **1.** To get free of. **2.** To succeed in avoiding. **3.** To elude the memory or comprehension of <The exact date *escapes* me.> **4.** To issue involuntarily from. *—n.* **1.** An act or instance of escaping. **2.** A means of escaping. **3.** A means of obtaining temporary freedom from worry, care, or unpleasantness <The movies provided an *escape.*> **4.** A gradual effusion from an enclosure : LEAKAGE. **5.** A cultivated plant established away from cultivation. **—es·cap′a·ble** *adj.* **—es′cap·ee′** (ĕs′kā-pē′, ĭ-skā′pē′) *n.* **—es·cap′er** *n.*

☆ ***syns:*** ESCAPE, ABSCOND, DECAMP, FLEE, FLY, RUN AWAY *v. core meaning* : to break loose from confinement and leave suddenly <*escaped* from the prison cell>

**es·cape·ment** (ĭ-skāp′mənt) *n.* **1.** A mechanism having an escape wheel and anchor, used esp. in timepieces to control the wheel movement and provide energy impulses to a pendulum or balance. **2.** A mechanism, as in a typewriter, that controls the lateral movement of the carriage. **3.** An escape. **4.** A means of escape.

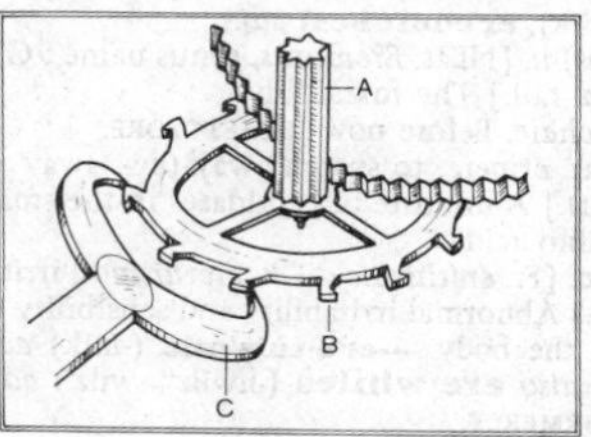


**escapement**
*A. escape wheel spindle, B. escape wheel, C. lever*

**escape velocity** *n.* The minimum velocity that a body must attain to overcome the gravitational attraction of another body.
**escape wheel** *n.* The rotating notched wheel periodically engaged and disengaged by the anchor in an escapement.
**es·cap·ism** (ĭ-skā'pĭz'əm) *n.* The tendency to escape from unpleasant realities by fantasy or entertainment.
**es·cap·ist** (ĭ-skā'pĭst) *n.* A person whose behavior is marked by escapism.
**es·cap·ol·o·gy** (ĕs'kā-pŏl'ə-jē) *n.* The art of escaping. **—es'cap·ol'o·gist** (-jĭst) *n.*
**es·car·got** (ĕs'kär-gō') *n., pl.* **-gots** (-gō') [Fr. < OFr. < OProv. *escaragol.*] An edible snail, esp. when prepared for eating.
**es·ca·role** (ĕs'kə-rōl') *n.* [Fr. < OFr. *scariole* < LLat. *escariola* < Lat. *escarius,* of food < *esca,* food < *edere,* to eat.] ENDIVE 1.
**es·carp** (ĭ-skärp') *n.* [Fr. *escarpe* < OFr. < OItal. *scarpa.*] **1.** A steep slope or cliff : ESCARPMENT. **2.** The inner wall of a trench or ditch dug around a fortification. —*vt.* **-carped, -carp·ing, -carps. 1.** To cause to form a steep slope. **2.** To furnish with an escarp or escarps.
**es·carp·ment** (ĭ-skärp'mənt) *n.* **1.** A steep slope or long cliff resulting from erosion or faulting and separating two relatively level areas of differing elevations. **2.** A steep slope in front of a fortification.
**-escence** *suff.* [OFr. < Lat. *-escentia* < *-escens,* -escent.] The process or quality of emitting or reflecting light in a specified way <fluor*escence*>
**-escent** *suff.* [OFr. < Lat. *-escens, -escent-,* pr.part. suffix of inchoative verbs in *-escere.*] **1.** Beginning to be : BECOMING <juven*escent*> **2.** Characterized by : RESEMBLING <opal*escent*>
**esch·a·lot** (ĕsh'ə-lŏt') *n.* [Obs. Fr. *eschallotte.* —see SHALLOT.] A shallot.
**es·char** (ĕs'kär') *n.* [ME *escare,* ult. < Lat. *eschara* < Gk. *eskhara.*] A dry scab formed esp. as a result of a burn.
**es·cha·rot·ic** (ĕs'kə-rŏt'ĭk) *adj.* Producing or capable of producing an eschar. —*n.* A caustic or corrosive agent.
**es·cha·tol·o·gy** (ĕs'kə-tŏl'ə-jē) *n.* [Gk. *eskhatos,* last + -LOGY.] The branch of theology concerned with final events, as death. **—es·chat'o·log'i·cal** (ĭ-skăt'l-ŏj'ĭ-kəl, ĕs'kə-tə-lŏj'-) *adj.* **—es·chat'o·log'i·cal·ly** *adv.* **—es'cha·tol'o·gist** *n.*
**es·cheat** (ĭs-chēt') *n.* [ME *eschete* < OFr. < *escheoir,* to fall out < VLat. **excadere* : Lat. *ex-,* out + Lat. *cadere,* to fall.] **1.** Reversion of land held under feudal tenure to the manor in the absence of legal heirs or claimants. **2.** *Law.* Reversion of property to the state in the absence of legal heirs or claimants. **3.** *Law.* Property that has reverted to the state in the absence of legal heirs or claimants. —*vi.* & *vt.* **-cheat·ed, -cheat·ing, -cheats.** To revert or cause to revert by escheat. **—es·cheat'a·ble** *adj.*
**es·cheat·age** (ĭs-chē'tĭj) *n. Law.* The state's right to acquire property by escheat.
**es·chew** (ĭs-choo') *vt.* **-chewed, -chew·ing, -chews.** [ME *eschewen* < OFr. *eschivir,* of Germanic orig.] To keep away from : AVOID. **—es·chew'al** (-əl) *n.*
**es·co·lar** (ĕs'kə-lär') *n., pl.* **escolar** or **-lars.** [Sp., student (from the spectaclelike rings around its eyes) < LLat. *scholaris,* of a school. —see SCHOLAR.] A slender fish of the family Gempylidae, esp. *Lepidocybium flavobrunneum* of warm marine waters.
**es·cort** (ĕs'kôrt') *n.* [Fr. *escorte* < OFr. < OItal. *scorta* < *scorgere,* to conduct < VLat. **excorrigere* : Lat. *ex-,* out + Lat. *corrigere,* to set right.] **1.** One or more persons accompanying another to guide, protect, or pay honor. **2.** A man who accompanies a woman, esp. on a social occasion. **3. a.** One or more vehicles accompanying another vehicle to guide, protect, or honor its passengers. **b.** A warship or aircraft or a group of warships or aircraft used to defend or protect other craft from enemy attack. **4.** The state of being accompanied by a person or protective guard. —*vt.* (ĭ-skôrt', ĕ-skôrt', ĕs'kôrt') **-cort·ed, -cort·ing, -corts.** To accompany as an escort.
**es·cri·toire** (ĕs'krĭ-twär') *n.* [Obs. Fr. < OFr. *escriptoire,* study < Med. Lat. *scriptorium* < Lat. *scribere,* to write.] **1.** A writing table or desk. **2.** A desk with a top section for books.
**es·crow** (ĕs'krō', ĕ-skrō') *n.* [AN *escrowe* < OFr. *escroe,* scroll, of Germanic orig.] Money, property, a deed, or a bond put into the custody of a third party for delivery to a grantee only after the fulfillment of specified conditions.
**es·cu·do** (ĭ-skoo'dō) *n., pl.* **-dos.** [Port. and Sp., shield, escudo < Lat. *scutum,* shield.] —See table at CURRENCY.
**es·cu·lent** (ĕs'kyə-lənt) *adj.* [Lat. *esculentus* < *esca,* food < *edere,* to eat.] Fit for eating : EDIBLE. **—es'cu·lent** *n.*
**es·cutch·eon** (ĭ-skŭch'ən) *n.* [ME *escochon* < OFr. *escuchon* < VLat. **scutio* < Lat. *scutum,* shield.] **1.** A shield or shield-shaped emblem bearing a coat of arms. **2.** A protective or ornamental plate, as for a keyhole. **3.** The plate on the stern of a ship inscribed with the ship's name. **—es·cutch'eoned** *adj.*
**Es·dras** (ĕz'drəs) *n.* —See table at BIBLE.
**-ese** *suff.* [OFr. *-eis* and Ital. *-ese* < Lat. *-ensis,* originating in.] **1.** Of, pertaining to, typical of, or originating in a given place <Vietnam*ese*> **2.** Native or resident of <Taiwan*ese*> **3. a.** Language or dialect of <Chin*ese*> **b.** Literary style or diction of <journal*ese*>
**es·er·ine** (ĕs'ə-rēn') *n.* [Fr. *ésère,* Calabar bean (< Kongo *anzadi*) + -INE[2].] *Biochem.* Physostigmine.
**es·ker** (ĕs'kər) *n.* [Ir. Gael. *eiscir* < OIr. *escir.*] A long, narrow ridge of coarse gravel deposited by a stream flowing in an ice-walled valley or tunnel in a decaying glacial ice sheet.
**Es·ki·mo** (ĕs'kə-mō') *n., pl.* **Eskimo** or **-mos.** [Dan. < Fr. *Esquimaux* (pl.), of Algonquian orig.] **1.** One of a people native to the Arctic coastal regions of North America and to parts of Greenland and northeastern Siberia. **2.** The language of the Eskimo people. **—Es'ki·mo'an** (ĕs'kə-mō'ən) *adj.*
**Eskimo dog** *n.* A large dog orig. bred in Greenland and Labrador, with a thick coat and bushy tail.
**e·soph·a·gus** (ĭ-sŏf'ə-gəs) *n., pl.* **-gi** (-jī') [ME *ysophagus* < Gk. *oisophagos.*] A muscular, membranous tube for the passage of food from the pharynx to the stomach. **—e·soph'a·ge'al** (-ə-jē'əl) *adj.*
**es·o·ter·ic** (ĕs'ə-tĕr'ĭk) *adj.* [Gk. *esōterikos* < *esōterō,* comp. of *esō,* within.] **1.** Intended for or understood only by a particular group. **2. a.** Known by a restricted number. **b.** Confined to a small group. **3.** Not publicly disclosed : CONFIDENTIAL. **—es'o·ter'i·cal·ly** *adv.*
**ESP** (ē'ĕs-pē') *n.* [E(XTRA)S(ENSORY) P(ERCEPTION).] Extrasensory perception.
**es·pa·drille** (ĕs'pə-drĭl') *n.* [Fr. < Prov. *espardilho,* dim. of *espart,* esparto < Lat. *spartum.*] A shoe having a flexible sole, as of twisted rope, and a canvas upper.
**es·pal·ier** (ĭ-spăl'yər, -yā') *n.* [Fr. < Ital. *spalliera,* stakes at shoulder's height < *spalla,* shoulder < Med. Lat. *spatula* < Lat., dim. of *spatho,* broad sword < Gk. *spathē,* broad blade.] **1.** A shrub or tree trained to grow in a flat plane against a wall, often in a symmetric pattern. **2.** A framework, as a trellis, on which an espalier is grown. —*vt.* **-iered, -ier·ing, -iers. 1.** To train on an espalier. **2.** To provide with an espalier.
**es·par·to** (ĭ-spär'tō) *n., pl.* **-tos.** [Sp. < Lat. *spartum* < Gk. *sparton,* rope.] A tough, wiry grass, *Stipa tenacissima* of northern Africa, yielding a fiber used to make paper and as cordage.
**es·pe·cial** (ĭ-spĕsh'əl) *adj.* [ME < OFr. < Lat. *specialis* < *species,* species.] **1.** Standing above or apart from others : EXCEPTIONAL. **2.** Relating to a particular person or thing. **—es·pe'cial·ly** *adv.*
**es·per·ance** (ĕs'pər-əns) *n. Obs.* [ME *esperaunce* < OFr. < Lat. *sperans,* pr. part. of *sperare,* to hope.] Hope.
**Es·pe·ran·to** (ĕs'pə-răn'tō, -rän'-) *n.* [After Dr. *Esperanto,* pseudonym of L.L. Zamenhof (1859–1917).] An artificial language with a vocabulary based on word roots common to many European languages and a regularized system of inflection.
**es·pi·al** (ĭ-spī'əl) *n.* [ME *espiaille* < OFr. < *espier,* to watch, of Germanic orig.] **1.** The act of noticing or observing. **2.** The fact of being seen or noticed.
**es·pi·o·nage** (ĕs'pē-ə-näzh', -nĭj) *n.* [Fr. *espionnage* < OFr. < *espionner,* to spy < *espion,* spy < OItal. *spione,* of Germanic orig.] The act or practice of spying or of using spies to obtain secret intelligence.
**es·pla·nade** (ĕs'plə-näd', -nād') *n.* [Fr. < Ital. *spianala* < *spianare,* to level < Lat. *explanare* : *ex-,* out + *planus,* level.] A flat, open stretch of pavement or grass, esp. one used as a shoreline promenade.
**es·pous·al** (ĭ-spou'zəl) *n.* **1. a.** A betrothal. **b.** A wedding ceremony. **2.** Adoption of a cause or idea.
**es·pouse** (ĭ-spouz') *vt.* **-poused, -pous·ing, -pous·es.** [ME *espousen* < OFr. *espouser* < Lat. *sponsare* < *spondere,* to betroth.] **1.** To take in marriage : MARRY. **2.** To give in marriage. **3.** To give one's support to : ADOPT. **—es·pous'er** *n.*
**es·pres·so** (ĭ-sprĕs'ō) *n., pl.* **-sos.** [Ital., p.part. of *esprimere,* to press out < Lat. *exprimere* : *ex-,* out + *premere,* to press.] A strong coffee brewed by forcing steam through darkly roasted, powdered coffee beans.
**es·prit** (ĕ-sprē') *n.* [Fr. < Lat. *spiritus,* spirit.] **1.** Esprit de corps. **2.** Liveliness of mind and expression : WIT.
**es·prit de corps** (ĕ-sprē' də kôr') *n.* [Fr. : *esprit,* spirit + *de,* of + *corps,* body.] A common spirit of enthusiasm and devotion to a cause among the members of a group.
**es·py** (ĭ-spī') *vt.* **-pied, -py·ing, -pies.** [ME *espien* < OFr. *espier,* to watch, of Germanic orig.] To catch sight of : GLIMPSE.
**-esque** *suff.* [Fr. < Ital. *-esco,* of Germanic orig.] In the manner of : RESEMBLING <Lincoln*esque*>
**es·quire** (ĕs'kwīr', ĭ-skwīr') *n.* [ME < OFr. *esquier* < LLat. *scutarius* < Lat. *scutum,* shield.] **1.** A candidate for medieval knighthood, serving a knight as attendant and shield bearer. **2.** A member of the English gentry ranking below a knight. **3.** *Archaic.* An English country gentleman : SQUIRE. **4.** —Used as a title of courtesy usu. in its abbreviated form after a person's full name, esp. an attorney or a consular officer <John Doe, *Esq.*><Jane Roe, *Esq.*> ***usage:*** *Esquire* and its abbreviation *Esq.,* once reserved for men, has come into use on correspondence addressed to women who are attorneys and consular officers.
**ess** (ĕs) *n.* The letter *s.*
**-ess** *suff.* [ME *-esse* < OFr. < LLat. *-issa* < Gk.] Female <lion*ess*>
**es·say** (ĕ-sā', ĕs'ā') *vt.* **-sayed, -say·ing, -says.** [OFr. *essaier* < *essai, assai,* trial < LLat. *exagium,* a weighing : Lat. *ex-,* out + Lat. *agere,* to drive.] **1.** To make an attempt at : TRY. **2.** To subject to a test. —*n.*

(ĕs′ā′, ĕ-sā′). **1.** An attempt : endeavor. **2.** A trial or test of the value or nature of a thing <an *essay* of our fortitude> **3.** (ĕs′ā′). **a.** A short literary composition on a single subject, usu. presenting the author's viewpoint. **b.** Something resembling a written essay <a photojournalistic *essay*> **—es·say′er** *n.*

**es·say·ist** (ĕs′ā′ĭst) *n.* A writer of essays.

**es·sence** (ĕs′əns) *n.* [ME *essencia* < Lat. *essentia* < *esse*, to be.] **1.** The indispensable or intrinsic properties that characterize or identify something. **2.** The most important element. **3.** The inherent, unchanging nature of a thing or class of things. **4. a.** An extract that has the fundamental properties of a substance in concentrated form. **b.** Such an extract in an alcohol solution. **c.** A perfume : scent. **5.** An existing thing, esp. a spiritual or incorporeal entity.

☆ ***syns:*** ESSENCE, BEING, ESSENTIALITY, NATURE, PITH, QUINTESSENCE, TEXTURE *n. core meaning* : a basic trait or set of traits that defines the character of something <Free enterprise is the *essence* of capitalism.>

**Es·sene** (ĕs′ēn′, ĭ-sēn′) *n.* A member of an ascetic Jewish sect existing in ancient Palestine from the 2nd cent. B.C. to the 3rd cent. A.D. **—Es·se′ni·an** (ĕ-sē′nē-ən), **Es·sen′ic** (ĕ-sĕn′ĭk) *adj.*

**es·sen·tial** (ĭ-sĕn′shəl) *adj.* **1.** Constituting or part of the nature of something : INHERENT. **2.** Indispensable or basic : NECESSARY <*essential* supplies> **—es·sen′tial** *n.* **—es·sen′ti·al′i·ty** (-shē-ăl′ĭ-tē), **es·sen′tial·ness** *n.* **—es·sen′tial·ly** *adv.*

**essential amino acid** *n.* An amino acid needed by the body for optimum growth and supplied by dietary protein.

**essential oil** *n.* A volatile oil, usu. possessing the characteristic odor or flavor of the plant from which it is obtained, used in perfumery and to make flavorings.

**es·so·nite** (ĕs′ə-nīt′) *n.* [Fr. < Gk. *hēssōn*, inferior, from its being softer than true hyacinth.] A brown or yellowish-brown garnet.

**-est¹** *suff.* [ME < OE *-est, -ast, -ost.*] —Used to form the superlative degree of adjectives and adverbs <great*est*><earli*est*>

**-est²** *suff.* [ME < OE *-est, -ast.*] —Used to form the archaic second person singular of English verbs <com*est*>

**es·tab·lish** (ĭ-stăb′lĭsh) *vt.* **-lished, -lish·ing, -lish·es.** [ME *establissen* < OFr. *establir, establiss-* < Lat. *stabilire* < *stabilis*, firm.] **1.** To make secure or firm. **2.** To set in a secure condition or position <*established* them in the wholesale market> **3.** To cause to be recognized and accepted <an invention that *established* their reputation> **4.** To found. **5.** To make a state institution of (a church). **6.** To introduce and enforce (e.g., a law). **7.** To prove the truth of. **—es·tab′lish·er** *n.*

**established church** *n.* A church officially recognized and supported by a government as a national institution.

**es·tab·lish·ment** (ĭ-stăb′lĭsh-mənt) *n.* **1. a.** The act of establishing. **b.** The fact or state of being established. **2. a.** A business firm, club, institution, or residence, including its members or occupants. **b.** A place of business, including its possessions and employees. **c.** An organized group, as a government, political party, or military force. **3.** An established church. **4.** *often* **Establishment. a.** An exclusive group of powerful people who rule a government or society. **b.** A group that controls a specified field of activity <the publishing *establishment*>

**es·ta·mi·net** (ĕ-stä′mē-nā′) *n.* [Fr.] A small café.

**es·tan·cia** (ĕ-stän′syä) *n.* [Am. Sp. < Sp., enclosure < VLat. *stantia < Lat. *stare*, to stand.] A large estate or ranch in Spanish America.

**es·tate** (ĭ-stāt′) *n.* [ME *estat*, condition < OFr. < Lat. *status* < *stare*, to stand.] **1.** A landed, usu. extensive property. **2.** All of one's possessions, esp. all of the property and debts left by a deceased person. **3.** *Law.* The nature and extent of an owner's rights with respect to his or her property. **4.** The situation or circumstances of one's life <pauper's *estate*> **5. a.** Social position or rank. **b.** *Obs.* High rank or status. **6.** *Archaic.* Display of power or wealth : POMP. **7.** A class, as the nobility, commons, or clergy, that once possessed specific political rights.

**Es·tates-Gen·er·al** (ĭ-stāts′jĕn′ər-əl) *n.* [Transl. of Fr. *états généraux.*] The States-General.

**es·teem** (ĭ-stēm′) *vt.* **-teemed, -teem·ing, -teems.** [ME *estemen*, to appraise < OFr. *estimer* < Lat. *aestimare.*] **1.** To regard with respect : PRIZE. **2.** To regard as : CONSIDER. *—n.* **1.** High regard : RESPECT <held in high *esteem* by all> **2.** *Archaic.* Judgment : opinion.

**es·ter** (ĕs′tər) *n.* [G., prob. short for *Essigäther* : *Essig*, vinegar (< MHG *ezzich* < OHG *ezzīh* < Lat. *acetum*) + *Äther*, ether (< Lat. *aether*).] Any of a class of organic compounds corresponding to the inorganic salts formed from an acid by the replacement of hydrogen by an alkyl radical.

**es·ter·ase** (ĕs′tə-rās′) *n.* An enzyme that catalyzes the hydrolysis of an ester.

**es·ter·i·fi·ca·tion** (ĕ-stĕr′ə-fĭ-kā′shən) *n.* A reaction resulting in the formation of at least one ester product.

**es·ter·i·fy** (ĕ-stĕr′ə-fī′) *vi.* & *vt.* **-fied, -fy·ing, -fies.** To change or cause to change to an ester.

**Es·ther** (ĕs′tər) *n.* [Heb. *Estēr* < Pers. *sitareh.*] —See table at BIBLE.

**es·the·sia** (ĕs-thē′zhə) *n. var. of* AESTHESIA.

**es·the·si·om·e·ter** (ĕs-thē′zē-ŏm′ə-tər) *n.* [ESTHESI(A) + -METER.] An instrument for determining tactile discrimination.

**es·thete** (ĕs′thēt) *n. var. of* AESTHETE.

**es·thet·ic** (ĕs-thĕt′ĭk) *adj. var. of* AESTHETIC.

**es·the·ti·cian** (ĕs′thĭ-tĭsh′ən) *n. var. of* AESTHETICIAN.

**es·thet·i·cism** (ĕs-thĕt′ĭ-sĭz′əm) *n. var. of* AESTHETICISM.

**es·thet·ics** (ĕs-thĕt′ĭks) *n. var. of* AESTHETICS.

**es·ti·ma·ble** (ĕs′tə-mə-bəl) *adj.* **1.** Capable of being estimated. **2.** Worthy of esteem. **—es′ti·ma·ble·ness** *n.* **—es′ti·ma·bly** *adv.*

**es·ti·mate** (ĕs′tə-māt′) *vt.* **-mat·ed, -mat·ing, -mates.** [Lat. *aestimare.*] **1.** To calculate approx. the amount or extent of. **2.** To form an opinion about : EVALUATE. *—n.* (ĕs′tə-mĭt). **1.** A tentative evaluation or rough calculation. **2. a.** A preliminary calculation of the cost of a project. **b.** The statement of such a calculation. **3.** A judgment : opinion <our *estimate* of the situation> **—es′ti·ma′tive** *adj.* **—es′ti·ma′tor** *n.*

**es·ti·ma·tion** (ĕs′tə-mā′shən) *n.* **1.** An act or instance of estimating. **2.** ESTIMATE 3. **3.** High regard : ESTEEM.

**es·ti·val** (ĕs′tə-vəl) *adj. var. of* AESTIVAL.

**es·ti·vate** (ĕs′tə-vāt′) *v. var. of* AESTIVATE.

**es·ti·va·tion** (ĕs′tə-vā′shən) *n. var. of* AESTIVATION.

**Es·to·ni·an** (ĕ-stō′nē-ən) *n.* **1.** The Finno-Ugric language of Estonia. **2.** A native or resident of Estonia. **—Es·to′ni·an** *adj.*

**es·top** (ĕ-stŏp′) *vt.* **-topped, -top·ping, -tops.** [ME *estoppen* < AN *estopper*, perh. < STOP.] **1.** *Law.* To impede or prohibit by estoppel. **2.** *Archaic.* To stop up. **—es·top′page** (ĕ-stŏp′ĭj) *n.*

**es·top·pel** (ĕ-stŏp′əl) *n.* [Perh. < OFr. *estouppail*, stopper < *estouper*, to stop up, ult. < Lat. *stuppa*, tow < Gk. *stuppē.*] *Law.* A bar against an assertion or denial contrary to one's previous assertion or denial of a fact.

**es·tra·di·ol** (ĕs′trə-dī′ôl′, -ōl′) *n.* [ESTR(US) + DI- + -OL.] An estrogenic hormone, $C_{18}H_{24}O_2$, usu. isolated commercially from sow ovaries or the urine of pregnant mares, used in treating estrogen deficiency.

**es·tral** (ĕs′trəl) *adj.* Estrous.

**estral cycle** *n.* Estrous cycle.

**es·trange** (ĭ-strānj′) *vt.* **-tranged, -trang·ing, -trang·es.** [OFr. *estranger* < Lat. *extraneare* < *extraneus*, strange.] **1. a.** To alienate the affections of. **b.** To make hostile or unsympathetic. **2.** To remove from a customary place or relation. **—es·trange′ment** *n.* **—es·trang′er** *n.*

☆ ***syns:*** ESTRANGE, ALIENATE, DISAFFECT *v. core meaning* : to make hostile, distant, or unsympathetic <political differences that had *estranged* two old friends>

**es·tray** (ĭ-strā′) *n.* [AFr. < *estraier*, to stray < OFr.] **1.** *Archaic.* A stray. **2.** *Law.* A stray domestic animal. *—vi.* **-trayed, -tray·ing, -trays.** *Archaic.* To stray.

**es·tri·ol** (ĕs′trī-ôl′, -ōl′, ĕ-strī′-) *n.* [ES(TRUS) + TRI- + -OL².] An estrogenic hormone, $C_{18}H_{24}O_3$, found in mammalian ovaries, obtained commercially from the urine of pregnant animals and used in treating estrogen deficiency.

**es·tro·gen** (ĕs′trə-jən) *n.* [ESTR(US) + -GEN.] One of several steroid hormones produced chiefly by the ovary and responsible for the regulation of certain female reproductive functions and the development and maintenance of female secondary sex characteristics. **—es′tro·gen′ic** (-jĕn′ĭk) *adj.* **—es′tro·gen′i·cal·ly** *adv.*

**es·trone** (ĕs′trōn′) *n.* [ESTR(US) + -ONE.] An estrogenic hormone, $C_{18}H_{22}O_2$, found in mammalian ovaries, isolated commercially from the urine of pregnant female animals and used in treating estrogen deficiency.

**es·trous** (ĕs′trəs) *adj.* **1.** Of or relating to estrus. **2.** In heat.

**estrous cycle** *n.* The series of chemical and physiological changes in female mammals from one estrus to the next.

**es·trus** (ĕs′trəs) *n.* [NLat. < Lat. *oestrus*, frenzy < Gk. *oistros.*] A regularly recurrent period of ovulation and sexual excitement in female mammals other than humans.

**es·tu·a·rine** (ĕs′cho͞o-ə-rīn′, -rēn′) *adj.* Of, relating to, or occurring in an estuary.

**es·tu·ar·y** (ĕs′cho͞o-ĕr′ē) *n., pl.* **-ies.** [Lat. *aestuarium* < *aestus*, tide.] **1.** The wide lower course of a river where its current is met by the tides. **2.** An arm of the sea that extends inland to meet the mouth of a river. **—es·tu·ar′i·al** (-âr′ē-əl) *adj.*

**e·su·ri·ent** (ĭ-so͝or′ē-ənt, ĭ-zo͝or′-) *adj.* [Lat. *esuriens, esurient-*, pr.part. of *esurire*, desiderative of *edere*, to eat.] Greedy. **—e·su′ri·ence** (-əns), **e·su′ri·en·cy** (-ən-sē) *n.* **—e·su′ri·ent·ly** *adv.*

**-et** *suff.* [ME < OFr.] **1.** Small <falcon*et*> **2.** Group <quart*et*>

**e·ta** (ā′tə, ē′tə) *n.* [Gk. *ēta*, of Phoenician orig.; akin to Heb. *hēth*, heth.] The seventh letter of the Greek alphabet. —See table at ALPHABET.

**é·ta·gère** *also* **e·ta·gere** (ā′tä-zhâr′) *n.* [Fr. < OFr. *estagiere*, estage, floor. —see STAGE.] A piece of furniture with open shelves for ornaments : WHATNOT.

**eta particle** *n.* A neutral, spinless elementary particle with a mass 1,074 times that of an electron.

**et cet·er·a** *also* **et·cet·er·a** (ĕt-sĕt′ər-ə, -sĕt′rə) [Lat., and the rest.] And other unspecified things of the same class : and so forth. ***usage:*** The use of *et cetera* and its abbreviation *etc.* is inappropriate

ă **pat** ā **pay** âr **care** ä **father** ĕ **pet** ē **be** hw **which** ĭ **pit**
ī **tie** îr **pier** ŏ **pot** ō **toe** ô **paw, for** oi **noise** o͝o **took**

to formal writing, but this use is acceptable in informal or technical contexts. —*n.* **et·cet·er·a. 1.** An unspecified number. **2. etceteras.** A miscellany of extras.

**etch** (ĕch) *v.* **etched, etch·ing, etch·es.** [Du. *etsen* < G. *ätzen* < MHG *etzen* < OHG *ezzen*, to be eaten.] —*vt.* **1. a.** To cut into the surface of (e.g., glass) by the action of acid. **b.** To make or create by this method <*etch* initials on glasses> **2.** To imprint or impress clearly. —*vi.* To practice etching. **—etch'er** *n.*

**etch·ing** (ĕch'ĭng) *n.* **1.** The art or technique of preparing etched metal plates and printing designs or pictures with them. **2.** A design or picture etched on such a plate. **3.** A print made from an etched plate.

**e·ter·nal** (ĭ-tûr'nəl) *adj.* [ME < OFr. < LLat. *aeternalis* < Lat. *aeternus* < *aevum*, age.] **1.** Without beginning or end : EVERLASTING. **2.** Having a beginning but without interruption or end <an *eternal* memorial> **3.** Not affected by time : TIMELESS <*eternal* struggle for justice> **4.** Seemingly endless : INTERMINABLE. **5.** Of or pertaining to spiritual communion with God, esp. after death <*eternal* life> —*n.* **1.** Something eternal. **2. the Eternal.** GOD 1a, b. **—e·ter'nal'i·ty** (ē'tər-năl'ĭ-tē), **e·ter'nal·ness** *n.* **—e·ter'nal·ly** *adv.*

**e·ter·nal·ize** (ĭ-tûr'nə-līz') *vt.* **-ized, -iz·ing, -iz·es.** To eternize.

**e·terne** (ĭ-tûrn') *adj.* [ME < OFr. < Lat. *aeternus* < *aevum*, age.] *Archaic.* Eternal.

**e·ter·ni·ty** (ĭ-tûr'nĭ-tē) *n., pl.* **-ties.** [ME *eternite* < OFr. *eternité* < Lat. *aeternitas* < *aeternus*, eternal.] **1.** The totality of time without beginning or end. **2.** The quality or state of being eternal. **3. a.** The endless period of time after death. **b.** The afterlife : immortality. **4.** A seemingly endless or very long time <waited an *eternity* for the traffic light to change>

**e·ter·nize** (ĭ-tûr'nīz') *vt.* **-nized, -niz·ing, -niz·es.** [OFr. *eterniser* < Med. Lat. *aeternizare* < *aeternus*, eternal < *aevum*, age.] **1.** To make eternal. **2.** To make perpetually famous : IMMORTALIZE.

**e·te·sian** (ĭ-tē'zhən) *adj.* [Lat. *etesius* < Gk. *etēsios* < *etos*, year.] Annually recurring. —Used of prevailing northerly summer winds of the Mediterranean. **—e·te'sian** *n.*

**eth** (ĕth) *n. var. of* EDH.

**-eth¹** *suff.* [ME < OE *-eð.*] —Used to form the archaic third person present singular indicative of verbs <lead*eth*>

**-eth²** *suff. var. of* -TH³.

**eth·a·cryn·ic acid** (ĕth'ə-krĭn'ĭk) *n.* [ETH(YL) + AC(ETIC) + (BUTY)RY(L) + (PHE)N(OL) + -IC.] A compound, $C_{13}H_{12}Cl_2O_4$, used as a diuretic in the treatment of edema.

**eth·ane** (ĕth'ān') *n.* [ETH(YL) + -ANE.] A colorless, odorless gas, $C_2H_6$, occurring as a constituent of natural gas and used as a fuel and refrigerant.

**eth·a·nol** (ĕth'ə-nôl', -nōl') *n.* [ETHAN(E) + -OL.] ALCOHOL 1.

**eth·a·nol·a·mine** (ĕth'ə-nŏl'ə-mēn', -nōl'-) *n.* A colorless liquid, $C_2H_7NO$, used in paints and as a dry-cleaning solvent.

**eth·ene** (ĕth'ēn') *n.* [ETH(YL) + -ENE.] *Chem.* Ethylene.

**e·ther** (ē'thər) *n.* [ME, upper air < Lat. *aethēr* < Gk. *aithēr.*] **1.** Any of a class of organic compounds in which two hydrocarbon groups are linked by an oxygen atom. **2.** A volatile, highly flammable liquid, $C_4H_{10}O$, derived from the distillation of ethyl alcohol with sulfuric acid and used chiefly in industry and as an anesthetic. **3.** The heavens. **4.** *Physics.* An all-pervading, infinitely elastic, massless medium formerly postulated as the medium of propagation of electromagnetic waves. **—e·ther'ic** (ĭ-thĕr'ĭk, -thîr'-) *adj.*

**e·the·re·al** (ĭ-thîr'ē-əl) *adj.* [Lat. *aetherius* < Gk. *aitherios* < *aithēr*, upper air.] **1.** Marked by lightness and insubstantiality : INTANGIBLE. **2.** Highly refined : DELICATE. **3. a.** Of the celestial spheres : HEAVENLY. **b.** Spiritual : unearthly. **4.** *Chem.* Of or relating to ether. **—e·the're·al'i·ty** (-ăl'ĭ-tē), **e·the're·al·ness** *n.* **—e·the're·al·ly** *adv.*

**e·the·re·al·ize** (ĭ-thîr'ē-ə-līz') *vt.* & *vi.* **-ized, -iz·ing, -iz·es.** To make or become ethereal. **—e·the're·al·i·za'tion** *n.*

**e·ther·i·fy** (ĭ-thĕr'ə-fī') *vt.* **-fied, -fy·ing, -fies.** To convert (an alcohol) into ether. **—e·ther'i·fi·ca'tion** *n.*

**e·ther·ize** (ē'thə-rīz') *vt.* **-ized, -iz·ing, -iz·es. 1.** To subject to the fumes of ether : ANESTHETIZE. **2.** *Chem.* To etherify. **—e'ther·i·za'tion** *n.* **—e'ther·iz'er** *n.*

**eth·ic** (ĕth'ĭk) *n.* [ME *ethik* < OFr. *ethique* < LLat. *ethica* < Lat. *ethice* < Gk. *ēthikē* < *ēthikos*, ethical < *ēthos*, character.] **1.** A principle of right or good behavior. **2.** A system of moral principles or values. **3. ethics** *(sing. in number).* The study of the general nature of morals and the specific moral choices an individual makes in relating to others. **4. ethics.** The rules or standards of conduct governing the members of a profession <medical *ethics*>

**eth·i·cal** (ĕth'ĭ-kəl) *adj.* **1.** Of, relating to, or dealing with ethics. **2.** Conforming to accepted principles of right and wrong that govern the conduct of a profession. **3.** Designating a drug dispensed only on a physician's prescription. **—eth'i·cal·ly** *adv.* **—eth'i·cal·ness, eth'i·cal'i·ty** (-kăl'ĭ-tē) *n.*

**eth·i·on** (ĕth'ē-ŏn') *n.* [Blend of ETHYL and THION-.] A highly toxic, liquid, organophosphate pesticide, $C_9H_{22}O_4P_2S_4$.

**E·thi·op** (ē'thē-ŏp') *also* **E·thi·ope** (-ōp') *n.* [Lat. *Aethiops* < Gk. *Aithiops.*] *Archaic.* Ethiopian.

**E·thi·o·pi·an** (ē'thē-ō'pē-ən) *adj.* **1.** Of or designating the zoogeographic region that includes Africa and most of Arabia. **2.** Of, pertaining to, or typical of Ethiopia or the inhabitants of Ethiopia.

**E·thi·op·ic** (ē'thē-ŏp'ĭk, -ō'pĭk) *n.* The Afro-Asiatic language of ancient Ethiopia still used as a liturgical language in the Christian Church in Ethiopia. —*adj.* **1.** Of or relating to Ethiopic. **2.** Ethiopian.

**eth·moid** (ĕth'moid') *also* **eth·moid·al** (ĕth-moid'l) *adj.* [Fr. *ethmoide* < Gk. *ēthmoeidēs*, sievelike < *ēthmos*, strainer < *ēthein*, to strain.] Of or relating to a light spongy bone situated between the orbits that forms part of the walls of the superior nasal cavity. —*n.* **ethmoid.** The ethmoid bone.

**eth·nic** (ĕth'nĭk) *also* **eth·ni·cal** (-nĭ-kəl) *adj.* [ME, heathen < LLat. *ethnicus* < Gk. *ethnikos* < *ethnos*, nation.] **1.** Of or relating to a religious, racial, national, or cultural group. **2.** Relating to a people not Christian or Jewish : HEATHEN. —*n.* **ethnic.** *Informal.* A member of an ethnic group. **—eth'ni·cal·ly** *adv.*

**eth·nic·i·ty** (ĕth-nĭs'ĭ-tē) *n.* **1.** The state of belonging to an ethnic group. **2.** Ethnic pride.

**ethno-** *pref.* [Fr. < Gk. *ethnos*, people.] Race : people <*ethno*logy>

**eth·no·cen·trism** (ĕth'nō-sĕn'trĭz'əm) *n.* **1.** Belief in the superiority of one's own ethnic group. **2.** Overriding concern with race. **—eth'no·cen'tric** (-trĭk) *adj.* **—eth'no·cen'tri·cal·ly** *adv.*

**eth·nog·ra·phy** (ĕth-nŏg'rə-fē) *n., pl.* **-phies. 1.** The descriptive anthropology of technologically primitive societies. **2.** Ethnology. **—eth·nog'ra·pher** *n.* **—eth'no·graph'ic** (ĕth'nə-grăf'ĭk), **eth'no·graph'i·cal** *adj.* **—eth'no·graph'i·cal·ly** *adv.*

**eth·nol·o·gy** (ĕth-nŏl'ə-jē) *n.* The anthropological study of cultural heritage and socioeconomic systems in technologically primitive societies, esp. the study of cultural origins and factors influencing cultural development. **—eth'no·log'ic** (ĕth'nə-lŏj'ĭk), **eth'no·log'i·cal** *adj.* **—eth'no·log'i·cal·ly** *adv.* **—eth·nol'o·gist** *n.*

**eth·no·mu·si·col·o·gy** (ĕth'nō-myōō'zĭ-kŏl'ə-jē) *n.* The study of the music of various cultures. **—eth'no·mu'si·col'o·gist** (-jĭst) *n.*

**e·thol·o·gy** (ĭ-thŏl'ə-jē, ē-) *n.* [Lat. *ethologia*, art of depicting character < Gk. : *ēthos*, ethos + *-logia*, -logy.] Scientific study of animal behavior. **—eth'o·log'i·cal** (ĕth'ə-lŏj'ĭ-kəl) *adj.* **—e·thol'o·gist** *n.*

**e·thos** (ē'thŏs') *n.* [Gk. *ēthos*, custom.] The character, disposition, or basic values peculiar to a specific people, culture, or movement <"The revolutionary *ethos* had become corrupted" —Irving Howe>

**eth·ox·yl** (ĭ-thŏk'səl) *also* **eth·ox·y** (ĭ-thŏk'sē) *n.* [ETH(YL) + OX- + -YL.] The univalent radical $C_2H_5O$.

**eth·yl** (ĕth'əl) *n.* [ETHER + -YL.] A univalent organic radical, $C_2H_5$. **—eth·yl'ic** (ĕ-thĭl'ĭk) *adj.*

**ethyl acetate** *n.* A colorless, volatile, flammable liquid, $CH_3COOC_2H_5$, used chiefly as a solvent.

**ethyl alcohol** *n. Chem.* ALCOHOL 1.

**eth·yl·a·mine** (ĕth'ə-lə-mēn') *n.* A colorless, volatile liquid, $C_2H_5NH_2$, used in petroleum refining, detergents, and organic synthesis.

**eth·yl·ate** (ĕth'ə-lāt') *vt.* **-at·ed, -at·ing, -ates.** *Chem.* To introduce the ethyl group into (a compound). **—eth'yl·a'tion** *n.*

**ethyl chloride** *n.* A chemical compound, $C_2H_5Cl$, a gas at ordinary temperatures and a colorless, volatile, flammable liquid when compressed, used in manufacturing tetraethyl lead and as a solvent and refrigerant.

**eth·yl·ene** (ĕth'ə-lēn') *n.* **1.** A colorless, flammable gas, $C_2H_4$, derived from natural gas and petroleum and used as a source of organic compounds, in welding and cutting metals, to color citrus fruits, and as an anesthetic. **2.** The bivalent organic radical $C_2H_4$. **—eth'yl·e'nic** (-ə-lē'nĭk, -lĕn'ĭk) *adj.*

**ethylene glycol** *n.* A colorless, syrupy alcohol, $C_2H_6O_2$, used as an antifreeze in cooling and heating systems.

**ethyl ether** *n.* ETHER 2.

**ethyl mercaptan** *n.* Mercaptan.

**-etic** *suff.* [Lat. *-eticus* < Gk. *-etikos* < *-etos*, verbal ending.] —Used to form adjectives usu. from nouns ending in *-esis*; e.g., *aphaeretic* from *aphaeresis.*

**e·ti·o·late** (ē'tē-ə-lāt') *v.* **-lat·ed, -lat·ing, -lates.** [Fr. *étioler* < Norman Fr. *étieuler*, to grow into haulm < *éteule*, stalk < OFr. *esteule* < Lat. *stipula.*] —*vt.* **1.** To cause (a plant) to develop without normal green coloring by preventing exposure to sunlight. **2.** To make weak by stunting the growth or development of. —*vi.* To become blanched or whitened, as when grown without sunlight. **—e'ti·o·la'tion** *n.*

**e·ti·ol·o·gy** (ē'tē-ŏl'ə-jē) *n., pl.* **-gies.** [LLat. *aetiologia* < Gk. *aitiologia* : *aitia*, cause + *-logia*, -logy.] **1.** The study of causes or origins. **2.** The branch of medicine that deals with the causes of disease. **3. a.** The assignment of an origin, cause, or reason for something. **b.** The cause of a disorder or disease as determined by medical diagnosis. **—e'ti·o·log'ic** (-ə-lŏj'ĭk), **e'ti·o·log'i·cal** *adj.* **—e'ti·o·log'i·cal·ly** *adv.* **—e'ti·ol'o·gist** *n.*

**et·i·quette** (ĕt'ĭ-kĕt', -kĭt) *n.* [Fr., etiquette, label < OFr. *estiquet*, label. —see TICKET.] The forms and practices prescribed by social convention or by authority.

**E·ton collar** (ēt'n) *n.* [After *Eton* College, England.] A broad white collar worn over the lapels of a jacket.
**Eton jacket** *n.* [After *Eton* College, England.] A waist-length jacket having wide lapels and cut square at the hips.
**E·tru·ri·an** (ĭ-tro͝or'ē-ən) *n.* Etruscan.
**E·trus·can** (ĭ-trŭs'kən) *n.* **1.** An inhabitant of ancient Etruria. **2.** The extinct language of the Etruscans, of unknown linguistic affiliation. **—E·trus'can** *adj.*
**-ette** *suff.* [ME < OFr., fem. of *-et*, -et.] **1.** Small : diminutive <di*nette*> **2.** Female <usher*ette*>
**é·tude** *also* **e·tude** (ā'to͞od', -tyo͞od') *n.* [Fr. *étude* < OFr. *estudie*, study.] **1.** A piece of music for the development of a given point of technique. **2.** An artistic composition embodying some point of technique but performed because of its artistic merit.
**é·tui** (ā-twē') *n., pl.* **é·tuis** (ā-twēz') [Fr. < OFr. *estui*, prison < *estuier*, to guard.] A small, usu. ornamental case.
**et·y·mol·o·gize** (ĕt'ə-mŏl'ə-jīz') *v.* **-gized, -giz·ing, -giz·es.** *—vt.* To trace and state the etymology of. *—vi.* To suggest or give the etymology of a word.
**et·y·mol·o·gy** (ĕt'ə-mŏl'ə-jē) *n., pl.* **-gies.** [ME *ethimologie* < OFr. < Med. Lat. *ethimologia* < Lat. *etymologia* < Gk. *etumologia* : *etumon*, true sense of a word (< *etumos*, true) + *-logia*, -logy.] **1.** The origin and historical development of a linguistic form as shown by determining its basic elements, discovering its earliest known use, recording its changes in form and meaning, tracing its transmission from one language to another, and identifying its cognates in other languages. **2.** The branch of linguistics that deals with etymologies. **—et'y·mo·log'i·cal, et'y·mo·log'ic** *adj.* **—et'y·mo·log'i·cal·ly** *adv.* **—et'y·mol'o·gist** *n.*
**et·y·mon** (ĕt'ə-mŏn') *n., pl.* **-mons** *or* **-ma** (-mə) [Lat. < Gk. *etumon*, true sense of a word < *etumos*, true.] **1.** An earlier form of a word in the same language or in an ancestor language. **2.** A morpheme or word from which derivatives and compounds are formed. **3.** A foreign word from which a particular loan-word is derived.
**eu-** *pref.* [ME < Lat. < Gk. < *eus*, good.] **1.** Good : well : true <*eu*plastic> **2.** A derivative of a specified substance <*eu*caine>
**Eu** *symbol for* EUROPIUM.
**eu·caine** (yo͞o'kān') *n.* A crystalline substance, $C_{15}H_{21}NO_2$, once used as a local anesthetic.
**eu·ca·lyp·tol** (yo͞o'kə-lĭp'tôl', -tōl') *also* **eu·ca·lyp·tole** (-tōl') *n.* A colorless oily liquid, $C_{10}H_{18}O$, derived from eucalyptus oil and used in pharmaceuticals, flavoring, and perfumery.
**eu·ca·lyp·tus** (yo͞o'kə-lĭp'təs) *n., pl.* **-tus·es** *or* **-ti** (-tī') [NLat. *Eucalyptus*, genus name : Gk. *eu-*, well + Gk. *kaluptos*, covered < *kaluptein*, to cover.] A tall native Australian tree of the genus *Eucalyptus*, with wood valued as timber and aromatic leaves that yield an oil used medicinally.
**eu·car·y·ote** *also* **eu·kar·y·ote** (yo͞o-kăr'ē-ōt', -ē-ət) *n.* [EU- + Gk. *karyōtos*, having nuts < *karyon*, nut.] An organism having one or more cells with well-defined nuclei. **—eu·car'y·ot'ic** (-ŏt'ĭk) *adj.*
**Eu·cha·rist** (yo͞o'kər-ĭst) *n.* [ME *eukarist* < OFr. *eucariste* < LLat. *eucharistia* < Gk. *eukharistia*, gratitude < *eukharistos*, grateful, thankful : *eu-*, well + *kharizesthai*, to show favor < *kharis*, grace.] **1.** COMMUNION 4a. **2.** *Christian Science.* Spiritual communion with God. **—Eu'cha·ris'tic, Eu'cha·ris'ti·cal** *adj.*
**eu·chre** (yo͞o'kər) *n.* [Orig. unknown.] **1.** A card game in which each player is dealt five cards and the player making the trump is required to take at least three tricks to win. **2.** The act of euchring an opponent. *—vt.* **-chred, -chring, -chres. 1.** To prevent (an opponent) from taking three tricks in euchre. **2.** *Informal.* To cheat : trick <*euchred* us out of our pension plan>
**eu·chro·ma·tin** (yo͞o-krō'mə-tĭn') *n.* Genetically active chromatin. **—eu'chro·mat'ic** (yo͞o'krō-măt'ĭk) *adj.*
**Eu·clid·e·an** *also* **Eu·clid·i·an** (yo͞o-klĭd'ē-ən) *adj.* Of or relating to Euclid's geometric principles.
**eu·de·mon** *also* **eu·dae·mon** (yo͞o-dē'mən) *n.* A good or benevolent spirit.
**eu·de·mon·ism** *also* **eu·dae·mon·ism** (yo͞o-dē'mə-nĭz'əm) *n.* A system of ethics that evaluates the morality of actions in terms of their capacity to produce happiness. **—eu·de'mo·nist** *n.* **—eu·de'mon·is'tic, eu·de'mon·is'ti·cal** *adj.*
**eu·gen·ic** (yo͞o-jĕn'ĭk) *adj.* **1.** Of or relating to eugenics. **2.** Relating or adapted to the production of good offspring.
**eu·gen·i·cist** (yo͞o-jĕn'ĭ-sĭst) *also* **eu·gen·ist** (yo͞o'jə-nĭst) *n.* An advocate of or specialist in eugenics.
**eu·gen·ics** (yo͞o-jĕn'ĭks) *n.* *(sing. in number).* The study of hereditary improvement by genetic control.
**eu·gen·ist** (yo͞o'jə-nĭst) *n. var. of* EUGENICIST.
**eu·ge·nol** (yo͞o'jə-nôl', -nōl') *n.* [< NLat. *Eugenia*, genus of the clove plant, after *Eugene*, Prince of Savoy (1663–1736).] A colorless aromatic oil, $C_{10}H_{12}O_2$, found in cloves and used mainly in perfumes and germicides.
**eu·gle·na** (yo͞o-glē'nə) *n.* [NLat. : Gk. *eu-*, good + Gk. *glēnē*, eyeball.] A minute unicellular freshwater organism of the genus *Euglena*, marked by the presence of chlorophyll, a reddish eyespot, and a single anterior flagellum.
**eu·glob·u·lin** (yo͞o-glŏb'yə-lĭn) *n.* A globulin soluble in dilute salt solutions and insoluble in distilled water.
**eu·he·mer·ism** (yo͞o-hē'mə-rĭz'əm, -hĕm'ə-) *n.* [After *Euhemerus*, a Greek philosopher of the 4th cent. B.C.] A theory attributing the origin of the gods to the deification of historical heroes. **—eu·he'mer·ist** *n.* **—eu·he'mer·is'tic** *adj.* **—eu·he'mer·is'ti·cal·ly** *adv.*
**eu·he·mer·ize** (yo͞o-hē'mə-rīz', -hĕm'ə-) *vt.* **-ized, -iz·ing, -iz·es.** To explain or interpret euhemeristically.
**eu·kar·y·ote** (yo͞o-kăr'ē-ōt', -ē-ət) *n. var. of* EUCARYOTE.
**eu·la·chon** (yo͞o'lə-kŏn') *n., pl.* **eulachon** *or* **-chons.** [Chinook *vlâkân.*] The candlefish.
**eu·lo·gize** (yo͞o'lə-jīz') *vt.* **-gized, -giz·ing, -giz·es.** To write or deliver a eulogy for. **—eu'lo·gist** (-jĭst), **eu'lo·giz'er** *n.*
**eu·lo·gy** (yo͞o'lə-jē) *n., pl.* **-gies.** [ME *euloge* < Med. Lat. *eulogium* < Gk. *eulogia*, praise : *eu-*, well + *-logia*, discourse.] **1.** A laudatory tribute, either oral or written. **2.** High praise. **—eu'lo·gis'tic** (-jĭs'tĭk) *adj.*
**eu·nuch** (yo͞o'nək) *n.* [ME *eunuk* < Lat. *eunuchus* < Gk. *eunoukhos* : *eunē*, bed + *ekhein*, to keep.] **1.** A castrated man employed as a harem attendant or functionary in certain Oriental courts. **2.** A man with undeveloped testes.
**eu·on·y·mus** (yo͞o-ŏn'ə-məs) *n.* [NLat. *Euonymus*, genus name < Lat. *euonymus* < Gk. *euōnumos*, of good name : *eu-*, good + *onoma*, name.] A shrub, tree, or vine of the genus *Euonymus*, cultivated for its decorative foliage or fruits.
**eu·pat·rid** (yo͞o-păt'rĭd, yo͞o'pə-trĭd) *n., pl.* **eu·pat·ri·dae** (yo͞o-păt'rĭ-dē') *or* **eu·pat·rids.** [Gk. *eupatridēs* : *eu-*, good + *patēr*, father + *-idēs*, patronymic suffix.] A member of the ancient Athenian hereditary aristocracy. **—eu·pat'rid** *adj.*
**eu·pep·si·a** (yo͞o-pĕp'sē-ə, -shə) *n.* [Gk. < *eupeptos*, eupeptic.] Good digestion.
**eu·pep·tic** (yo͞o-pĕp'tĭk) *adj.* [Gk. *eupeptos* : *eu-*, well + *peptein*, to digest.] **1.** Relating to or having good digestion. **2.** Conducive to digestion. **—eu·pep'ti·cal·ly** *adv.*
**eu·phe·mism** (yo͞o'fə-mĭz'əm) *n.* [Gk. *euphēmismos* < *euphēmos*, using auspicious words : *eu-*, good + *phēmē*, speech.] An act or example of the substitution of an inoffensive term for one considered offensive <"*Euphemisms* such as 'slumber room' . . . abound in the funeral business" —Jessica Mitford> **—eu'phe·mist** *n.* **—eu'phe·mis'tic** (-mĭs'tĭk) *adj.* **—eu'phe·mis'ti·cal·ly** *adv.*
**eu·phe·mize** (yo͞o'fə-mīz') *v.* **-mized, -miz·ing, -miz·es.** *—vt.* To speak of or refer to by means of a euphemism. *—vi.* To speak with euphemisms. **—eu'phe·miz'er** *n.*
**eu·phen·ics** (yo͞o-fĕn'ĭks) *n.* *(sing. in number).* The study of phenotypic improvement of humans after birth.
**eu·phon·ic** (yo͞o-fŏn'ĭk) *adj.* **1.** Pertaining to euphony. **2.** Euphonious. **—eu·phon'i·cal·ly** *adv.*
**eu·pho·ni·ous** (yo͞o-fō'nē-əs) *adj.* Pleasing to the ear. **—eu·pho'ni·ous·ly** *adv.*
**eu·pho·ni·um** (yo͞o-fō'nē-əm) *n.* [< Gk. *euphōnos*, sweet-voiced. —see EUPHONY.] A brass wind instrument similar to the tuba but with a somewhat higher pitch and a mellower sound.
**eu·pho·nize** (yo͞o'fə-nīz') *vt.* **-nized, -niz·ing, -niz·es.** To make euphonious.
**eu·pho·ny** (yo͞o'fə-nē) *n., pl.* **-nies.** [Fr. *euphonie* < LLat. *euphonia* < Gk. < *euphōnos*, sweet-voiced : *eu-*, good + *phōnē*, sound.] Agreeable sound, esp. in the phonetic quality of words.
**eu·phor·bi·a** (yo͞o-fôr'bē-ə) *n.* [NLat. *Euphorbia*, genus name < Lat. *euphorbea*, after *Euphorbus*, Greek physician of the 1st cent. A.D.] A plant of the genus *Euphorbia*, including the spurges.
**eu·pho·ri·a** (yo͞o-fôr'ē-ə, -fōr'-) *n.* [NLat. < Gk. < *euphoros*, healthy : *eu-*, well + *pherein*, to bear.] A feeling of elation or well-being. **—eu·phor'ic** (-fôr'ĭk, -fŏr'-) *adj.* **—eu·phor'i·cal·ly** *adv.*
**eu·phot·ic** (yo͞o-fŏt'ĭk) *adj.* Relating to, designating, or characterizing the uppermost layer of a body of water that receives sufficient light for the growth of green plants.
**Eu·phros·y·ne** (yo͞o-frŏs'ə-nē) [Lat. *Euphrosyne* < Gk. *Euphrosunē* < *euphrōn*, cheerful : *eu-*, good + *phrēn*, mind.] *Gk. Myth.* One of the Graces.
**eu·phu·ism** (yo͞o'fyo͞o-ĭz'əm) *n.* [After *Euphues*, a character in *Euphues: the Anatomy of Wit* and *Euphues and his England* by John Lyly (1554?–1606).] **1.** A pretentiously elegant literary style of the late 16th and early 17th cent., marked by alliteration, antitheses, and similes. **2.** Affected elegance in language. **—eu'phu·ist** *n.* **—eu'phu·is'tic, eu'phu·is'ti·cal** *adj.* **—eu'phu·is'ti·cal·ly** *adv.*
**eu·plas·tic** (yo͞o-plăs'tĭk) *adj.* Readily healing.
**eu·ploid** (yo͞o'ploid') *adj.* Having a chromosome complement that is an exact multiple of the haploid complement. **—eu'ploid** *n.* **—eu'ploi'dy** *n.*
**eup·ne·a** (yo͞op-nē'ə) *n.* [NLat. < Gk. *eupnoia* < *eupnoos*, breathing well : *eu-*, well + *pnein*, to breathe.] Normal, unlabored breathing. **—eup·ne'i·cal·ly** *adv.*
**eu·re·ka** (yo͞o-rē'kə) *interj.* [Gk. *heurēka*, I have found (it), supposedly exclaimed by Archimedes upon discovering how to measure the volume of an irregular solid and thereby determine the purity of a gold object.] —Used to express triumph on a discovery.

**eu·rhyth·mics** (yo͞o-rĭth′mĭks) *n. var. of* EURYTHMICS.
**eu·rhyth·my** (yo͞o-rĭth′mē) *n. var. of* EURYTHMY.
**eu·ri·pus** (yo͞o-rī′pəs) *n., pl.* **-pi** (-pī′) [Lat. < Gk. *euripos* : *eu-*, good + *ripē*, rush < *riptein*, to throw.] A sea channel whose currents are unpredictable and turbulent in either direction.
**eu·ro** (yo͝or′ō) *n., pl.* **-ros.** [Native word in Australia.] The wallaroo.
**Eu·ro·bond** (yo͝or′ō-bŏnd′) *n.* [EURO(PE) + BOND.] A bond of a U.S. corporation issued in Europe.
**eu·ro·cur·ren·cy** (yo͝or′ō-kûr′ən-sē, -kŭr′-) *n., pl.* **-cies.** [EURO(PE) + CURRENCY.] Currency, as of the United States or Japan, used in the European money market.
**Eu·ro·dol·lar** (yo͝or′ō-dŏl′ər) *n.* [EURO(PE) + DOLLAR.] A U.S. dollar on deposit with a bank abroad, esp. in Europe.
**Eu·ro·pa** (yo͞o-rō′pə) *n.* [Lat. < Gk. *Eurōpē.*] *Gk. Myth.* A Phoenician princess abducted to Crete by Zeus, who had taken on the form of a white bull.
**Eu·ro·pe·an** (yo͝or′ə-pē′ən) *adj.* Of, pertaining to, or derived from Europe. **—Eu′ro·pe′an** *n.*
**Eu·ro·pe·an·ize** (yo͝or′ə-pē′ə-nīz′) *vt.* **-ized, -iz·ing, -iz·es.** To make European. **—Eu′ro·pe′an·i·za′tion** *n.*
**European plan** *n.* A hotel plan in which the rates include only the charges for a room and services and not for meals.
**eu·ro·pi·um** (yo͞o-rō′pē-əm) *n.* [After *Europe.*] *Symbol* **Eu** A silvery-white, soft element used to absorb neutrons in research; atomic number 63; atomic weight 151.96.
**Eu·rus** (yo͝or′əs) *n.* [Lat. < Gk. *Euros.*] *Gk. Myth.* The god of the east or southeast wind.
**eury-** *pref.* [NLat. < Gk. *eurus*, wide.] Wide : broad <*eury*thermal>
**eu·ry·bath·ic** (yo͝or′ə-băth′ĭk) *adj.* Capable of dwelling at the bottom of a body of water in a wide range of depths.
**Eu·ryd·i·ce** (yo͞o-rĭd′ĭ-sē) *n.* [Lat. < Gk. *Eurudike* : *eurus*, wide + *dikē*, justice.] *Gk. Myth.* The wife of Orpheus, whom he failed to rescue from Hades when he looked back at her and thus violated Pluto's command on their journey back to the world of the living.
**eu·ry·ha·line** (yo͝or′ə-hā′lĭn′, -hăl′-īn′) *adj.* Capable of tolerating a wide range of saltwater concentrations.
**eu·ryp·ter·id** (yo͞o-rĭp′tər-ĭd) *n.* [NLat. *Eurypterida*, order name < *Eurypterus*, genus name : EURY- + Gk. *pteron*, wing.] Any of various large, extinct, segmented aquatic arthropods of the order Eurypterida, existing from the Ordovician to the Permian periods.
**eu·ry·ther·mal** (yo͝or′ə-thûr′məl) *also* **eu·ry·ther·mic** (-mĭk) *or* **eu·ry·ther·mous** (-məs) *adj.* Adaptable to a wide range of temperatures. —Used of an organism. **—eu′ry·therm′** *n.*
**eu·ryth·mics** *also* **eu·rhyth·mics** (yo͞o-rĭth′mĭks) *n.* *(sing. in number).* The choreographic art of interpreting musical composition by a rhythmical, graceful free-style body movement in response to the rhythm of the music. **—eu·ryth′mic** *adj.*
**eu·ryth·my** *also* **eu·rhyth·my** (yo͞o-rĭth′mē) *n.* [Lat. *eurythmia* < Gk. *euruthmia* < *euruthmos*, graceful : *eu-*, good + *ruthmos*, proportion.] **1.** Harmony of proportion in architecture. **2.** A system of rhythmical body movements in harmony with the rhythm of spoken words.
**Eu·sta·chian tube** (yo͞o-stā′shən, -shē-ən, -stā′kē-ən) *n.* [After Bartolommeo *Eustachio* (1524?–1574).] A bony and cartilaginous tube joining the tympanic cavity to the nasal part of the pharynx.
**eu·tec·tic** (yo͞o-tĕk′tĭk) *adj.* [< Gk. *eutēktos*, easily melted : *eu-*, well + *tēkein*, to melt.] **1.** Of, relating to, or formed at the lowest possible temperature of solidification for any mixture of specified constituents. —Used esp. of alloys. **2.** Exhibiting the constitution or properties of a eutectic solid. —*n.* **1.** A eutectic solution, mixture, or alloy. **2.** The eutectic temperature.
**Eu·ter·pe** (yo͞o-tûr′pē) *n.* [Gk. *Euterpē* : *eu-*, well + *terpein*, to please.] *Gk. Myth.* The Muse of music and lyric poetry.
**eu·tha·na·sia** (yo͞o′thə-nā′zhə, -zhē-ə) *n.* [Gk. : *eu-*, good + *thanatos*, death.] The intentional causing of a painless and easy death to a patient suffering from an incurable or painful disease.
**eu·then·ics** (yo͞o-thĕn′ĭks) *n.* [< Gk. *euthenein*, to flourish.] *(sing. in number).* The study of the improvement of human functioning and well-being by improvement of environment.
**eu·the·ri·an** (yo͞o-thîr′ē-ən) *adj.* [< NLat. *Eutheria*, a subdivision of the class Mammalia.] Of or relating to the Eutheria, a division of mammals to which all the placental mammals belong.
**eu·troph·ic** (yo͞o-trŏf′ĭk, -trō′fĭk) *adj.* [Prob. < G. *Eutroph* < Gk. *eutrophos*, well-nourished < *eutrophein*, to thrive : *eu-*, well + *trephein*, to nourish.] Designating a body of water in which the increase of mineral and organic nutrients has reduced the dissolved oxygen, producing an environment that favors plant over animal life. **—eu·troph′i·ca′tion** *n.* **—eu′tro·phy** (yo͞o′trə-fē) *n.*
**eux·e·nite** (yo͞ok′sə-nīt′) *n.* [G. *Euxenit* < Gk. *euxenos*, kind to strangers : *eu-*, good + *xenos*, stranger (so called because the mineral contains unusual elements).] A lustrous blackish-brown mineral consisting of cerium, erbium, titanium, uranium, and yttrium.

**e·vac·u·ant** (ĭ-văk′yo͞o-ənt) *adj.* Causing evacuation of an organ, esp. the bowels. **—e·vac′u·ant** *n.*
**e·vac·u·ate** (ĭ-văk′yo͞o-āt′) *v.* **-at·ed, -at·ing, -ates.** [Lat. *evacuare, evacuat-*, to empty out : *ex-*, out + *vacuus*, empty < *vacare*, to be empty.] —*vt.* **1. a.** To empty or remove the contents of. **b.** To create a vacuum in. **2.** To discharge or excrete (waste matter), esp. from the bowels. **3. a.** To relinquish military possession or occupation of (e.g., a town). **b.** To withdraw (troops or inhabitants) from a dangerous area. **4.** To depart from **:** VACATE <hurriedly *evacuated* the building> —*vi.* **1.** To vacate a place or area, esp. a threatened area. **2.** To excrete waste matter from the body. **—e·vac′u·a′tor** *n.*
**e·vac·u·a·tion** (ĭ-văk′yo͞o-ā′shən) *n.* **1.** The act of evacuating or state of being evacuated. **2. a.** Excretion of waste materials from the excretory passages, esp. from the bowels. **b.** The material thus discharged.
**e·vac·u·ee** (ĭ-văk′yo͞o-ē′) *n.* One evacuated from a hazardous area.
**e·vade** (ĭ-vād′) *v.* **e·vad·ed, e·vad·ing, e·vades.** [OFr. *evader* < Lat. *evadere* : *ex-*, out + *vadere*, to go.] —*vt.* **1.** To escape or avoid by cunning. **2. a.** To avoid the performance or fulfillment of <*evaded* their responsibilities> **b.** To fail to make payment of <*evade* import duty> **3.** To avoid giving a direct answer to. **4.** To elude or baffle <The error *evades* explanation.> —*vi.* To use cunning in avoiding or escaping. **—e·vad′a·ble, e·vad′i·ble** *adj.* **—e·vad′er** *n.*
**e·vag·i·nate** (ĭ-văj′ə-nāt′) *v.* **-nat·ed, -nat·ing, -nates.** [Lat. *evaginare, evaginat-*, to unsheath : *ex-*, from + *vagina*, sheath.] —*vi.* To turn inside out by eversion of an inner surface of a bodily part or organ. —*vt.* To cause (a bodily part) to turn inside out. **—e·vag′i·na′tion** *n.*
**e·val·u·ate** (ĭ-văl′yo͞o-āt′) *vt.* **-at·ed, -at·ing, -ates.** [Back-formation < E. *evaluation* < Fr. *évaluation* < OFr. *evaluation* < *evaluer*, to evaluate : *e-*, out (< Lat. *ex-*) + *value*, value. —see VALUE.] **1.** To determine or fix the value of. **2.** To examine carefully **:** APPRAISE. **3.** *Math.* To calculate or set down the numerical value of. **—e·val′u·a′tion** *n.*
**ev·a·nesce** (ĕv′ə-nĕs′) *vi.* **-nesced, -nesc·ing, -nesc·es.** [Lat. *evanescere*, to vanish : *ex-* (intensive) + *vanescere*, to disappear < *vanus*, empty.] To dissipate like vapor. **—ev′a·nes′cence** *n.*
**ev·a·nes·cent** (ĕv′ə-nĕs′ənt) *adj.* Vanishing or apt to vanish **:** FLEETING. **—ev′a·nes′cent·ly** *adv.*
**e·van·gel** (ĭ-văn′jəl) *n.* [ME *evangelie* < LLat. *evangelium* < Gk. *euangelion*, good news < *euangelos*, bringing good news : *eu-*, good + *angelos*, messenger.] **1.** The Christian gospel. **2.** An evangelist.
**e·van·gel·i·cal** (ē′văn-jĕl′ĭ-kəl, ĕv′ən-) *also* **e·van·gel·ic** (-jĕl′ĭk) *adj.* **1.** Of, relating to, or in accordance with the Christian gospel, esp. the four Gospels of the New Testament. **2.** Protestant. **3.** Of, relating to, or being a Protestant group emphasizing the authority of the gospel and holding that salvation is from faith and grace rather than from good works and sacraments alone. **4. Evangelical.** Of or relating to the Evangelical Church in Germany. **5.** Relating or belonging to the Low Church party of the Church of England. —*n.* **Evangelical.** A member of an evangelical church or party. **—e′van·gel′i·cal·ly** *adv.*
**e·van·gel·i·cal·ism** (ē′văn-jĕl′ĭ-kə-lĭz′əm, ĕv′ən-) *n.* **1.** Evangelical beliefs or doctrines. **2.** Adherence to a church or party professing evangelical beliefs or doctrines.
**e·van·gel·ism** (ĭ-văn′jə-lĭz′əm) *n.* **1.** Zealous preaching and dissemination of the gospel, as through missionary work. **2.** Militant zeal for a cause.
**e·van·gel·ist** (ĭ-văn′jə-lĭst) *n.* **1.** *often* **Evangelist.** An author of any of the four New Testament Gospels. **2.** A person who practices evangelism, esp. a Protestant preacher or missionary. **—e·van′gel·is′tic** *adj.* **—e·van′gel·is′ti·cal·ly** *adv.*
**e·van·gel·ize** (ĭ-văn′jə-līz′) *v.* **-ized, -iz·ing, -iz·es.** —*vt.* **1.** To preach the gospel to. **2.** To convert to Christianity. —*vi.* To preach the gospel. **—e·van′gel·i·za′tion** *n.* **—e·van′gel·iz′er** *n.*
**e·vap·o·ra·ble** (ĭ-văp′ər-ə-bəl) *adj.* Capable of being evaporated. **—e·vap′o·ra·bil′i·ty** *n.*
**e·vap·o·rate** (ĭ-văp′ə-rāt′) *v.* **-rat·ed, -rat·ing, -rates.** [ME *evaporaten* < Lat. *evaporare* : *ex-*, out + *vapor*, steam.] —*vt.* **1. a.** To convert into a vapor. **b.** To draw off in the form of vapor. **2.** To draw moisture from, leaving only the dry solid portion. **3.** To deposit (a metal) on a substrate by vacuum sublimation. —*vi.* **1. a.** To change into vapor. **b.** To pass off in or as if in vapor. **2.** To produce vapor. **3.** To disappear **:** vanish <Our financial worries *evaporated.*> **—e·vap′o·ra′tion** *n.* **—e·vap′o·ra′tive** *adj.* **—e·vap′o·ra′tive·ly** *adv.* **—e·vap′o·ra·tiv′i·ty** (-rə-tĭv′ĭ-tē) *n.* **—e·vap′o·ra′tor** *n.*
**evaporated milk** *n.* Concentrated, unsweetened milk made by evaporating some of the water from whole milk.
**e·vap·o·rite** (ĭ-văp′ə-rīt′) *n.* [EVAPOR(ATION) + -ITE.] A sedimentary deposit resulting from the evaporation of sea water. **—e·vap′o·rit′ic** (-rĭt′ĭk) *adj.*
**e·va·sion** (ĭ-vā′zhən) *n.* [ME *evasioun* < OFr. *evasion* < LLat. *evasio* < Lat. *evadere*, to evade.] An act or means of evading.
**e·va·sive** (ĭ-vā′sĭv) *adj.* **1.** Marked by or displaying evasion. **2.** Intentionally ambiguous or vague **:** EQUIVOCAL <an *evasive* proposition> **—e·va′sive·ly** *adv.* **—e·va′sive·ness** *n.*
**eve** (ēv) *n.* [ME < OE *ǣfen.*] **1.** The evening or day preceding a

special day, as a holiday. **2.** The period just preceding a certain event. **3.** Evening.

**e·vec·tion** (ĭ-vĕk′shən) *n.* [Lat. *evectio,* a going up < *evehere,* to raise up : *ex-,* up from + *vehere,* to carry.] Solar perturbation of the lunar orbit. **—e·vec′tion·al** *adj.*

**e·ven**[1] (ē′vən) *adj.* [ME < OE *efen.*] **1. a.** Having a horizontal surface : FLAT <an *even* road> **b.** Having no irregularities, roughness, or indentations : SMOOTH. **2.** Having the same plane or line : LEVEL. **3.** Having no variations or fluctuations : UNIFORM <an *even* rhythm> **4.** Of uniform thickness or distribution <an *even* coat of paint> **5.** Equally matched or balanced <an *even* contest> **6.** Equal or identical in degree, extent, or amount. **7.** Having equal probability <an *even* chance of winning> **8. a.** Having an equal score <The Red Sox and the Yankees are *even.*> **b.** Being equal for each opponent. —Used of a score. **9.** Neither owing nor being owed <paid me back and we were *even*> **10.** Having exacted full revenge. **11. a.** *Math.* Exactly divisible by 2. **b.** Marked or indicated by a number exactly divisible by 2. **12. a.** Having an even number in a series. **b.** Having an even number of members. **13.** Having an exact extent, amount, or number <an *even* dozen> —*adv.* **1.** To a greater extent or degree. —Used as an intensive <an *even* worse insult> **2.** At the same time as : JUST. **3.** In spite of : NOTWITHSTANDING <*Even* with all their experience, I won the match.> **4.** Indeed : moreover. —Used as an intensive <looked sad, *even* depressed> **5.** To a degree that extends to <courageous *even* unto death> **6.** *Nonstandard.* Smoothly : evenly. —*vt. & vi.* **e·vened, e·ven·ing, e·vens.** To make or become even. **—break even.** *Informal.* To have neither gains nor losses. **—get even.** To exact a full measure of revenge. **—e′ven·ly** *adv.* **—e′ven·ness** *n.*

**e·ven**[2] (ē′vən) *n.* [ME < OE *æfen.*] *Archaic.* Evening : twilight.

**e·ven·fall** (ē′vən-fôl′) *n.* The beginning of evening : TWILIGHT.

**e·ven·hand·ed** (ē′vən-hăn′dĭd) *adj.* Dealing equitably with all : IMPARTIAL. **—e′ven·hand′ed·ly** *adv.* **—e′ven·hand′ed·ness** *n.*

**eve·ning** (ēv′nĭng) *n.* [ME < OE *æfnung* < *æfnian,* to become evening < *æfen,* evening.] **1.** The period of decreasing daylight between afternoon and night. **2. a.** The period between sunset and bedtime. **b.** This period occupied in a given manner <an *evening* at the movies> **3.** A time or period of decline <in the *evening* of my life> —*adv.* **evenings.** Regularly or habitually in the evening.

**evening dress** *n.* Clothing, esp. formal clothing, worn for evening social events.

**evening gown** *n.* A woman's usu. long formal dress worn esp. in the evening.

**Evening Prayer** *n.* A daily evening service in the Anglican Church.

**evening primrose** *n.* A North American plant of the genus *Oenothera,* with four-petaled flowers that open in the evening.

**evening star** *n.* A planet, esp. Venus or Mercury, that crosses the local meridian before midnight and is prominent in the west shortly after sunset.

**evening stock** *n.* A Eurasian plant, *Mathiola bicornis,* with sweet-smelling purple flowers that bloom at night.

**e·ven·song** (ē′vən-sông′, -sŏng′) *n.* **1.** A song sung in the evening. **2.** A vesper service. **3.** *Archaic.* Evening. **4.** Evening Prayer.

**e·vent** (ĭ-vĕnt′) *n.* [Lat. *eventus* < *evenire,* to happen : *ex-,* out + *venire,* to come.] **1. a.** Something that takes place : OCCURRENCE. **b.** A significant occurrence <a literary *event*> **2.** The actual outcome or final result. **3.** One of the items in a program of sports. **4.** *Physics.* A coincidence of two or more point objects at a particular position in space at a particular instant of time, regarded as the fundamental observational entity in relativity theory. **—at all events.** In any case. **—in any event.** In any case. **—in the event.** If it should happen.

**e·vent·ful** (ĭ-vĕnt′fəl) *adj.* **1.** Full of events <an *eventful* year> **2.** Momentous : important <an *eventful* occurrence> **—e·vent′ful·ly** *adv.* **—e·vent′ful·ness** *n.*

**e·ven·tide** (ē′vən-tīd′) *n.* [ME < OE *æfentid* : *æfen,* evening + *tīd,* time.] Evening.

**e·ven·tu·al** (ĭ-vĕn′cho͞o-əl) *adj.* [< EVENT.] **1.** Taking place at an unspecified future time : ULTIMATE <their *eventual* success> **2.** Dependent on circumstance : CONTINGENT. **—e·ven′tu·al·ly** *adv.*

**e·ven·tu·al·i·ty** (ĭ-vĕn′cho͞o-ăl′ĭ-tē) *n., pl.* **-ties.** Something that may take place : POSSIBILITY.

**e·ven·tu·ate** (ĭ-vĕn′cho͞o-āt′) *vi.* **-at·ed, -at·ing, -ates.** To result ultimately.

**ev·er** (ĕv′ər) *adv.* [ME < OE *æfre.*] **1.** At all times : ALWAYS <was *ever* polite> **2.** At any time <Have you *ever* been arrested?> **3.** In any possible way or case. —Used for emphasis <was *ever* so happy> **—ever and again** (or **anon**). Now and then : OCCASIONALLY.

**ev·er·glade** (ĕv′ər-glād′) *n.* [After the *Everglades,* Florida.] Marshland, esp. of southern Florida, usu. under water and covered in places with tall grass.

**ev·er·green** (ĕv′ər-grēn′) *adj.* **1.** Having foliage that persists and remains green throughout the year. **2.** Remaining fresh. —*n.* **1.** An evergreen shrub, tree, or plant. **2. evergreens.** Twigs or branches of evergreen plants used for decorating.

**ev·er·last·ing** (ĕv′ər-lăs′tĭng) *adj.* **1.** Lasting forever : ETERNAL. **2. a.** Continuing for a long time or indefinitely. **b.** Persisting too long : TEDIOUS <their *everlasting* complaining> **3.** Retaining color and form for a long time when cut or dried, as certain plants. —*n.* **1. the Everlasting.** GOD 1a, b. **2.** Eternal duration : ETERNITY. **3.** A plant, as the strawflower or one of the genus *Anaphalis,* that retains form and color long after it is dry. **—ev′er·last′ing·ly** *adv.* **—ev′er·last′ing·ness** *n.*

**ev·er·more** (ĕv′ər-môr′, -mōr′) *adv.* **1.** *Archaic.* Forever : always. **2.** In the future.

**e·vert** (ĭ-vûrt′) *vt.* **e·vert·ed, e·vert·ing, e·verts.** [Lat. *evertere,* to overturn : *ex-,* out + *vertere,* to turn.] To turn outward or inside out. **—e·ver′si·ble** (ĭ-vûr′sə-bəl) *adj.* **—e·ver′sion** (-zhən, -shən) *n.*

**ev·e·ry** (ĕv′rē) *adj.* [ME < OE *æfre ælc* : *æfre,* ever + *ælc,* each.] **1. a.** Constituting each and all members of a class with no exception. **b.** Being each and all within a limited range or class. **2.** Being each of a designated succession of objects or intervals <*every* fifth car> <*every* six hours> **3.** Being the highest degree or expression of <gave them *every* opportunity to succeed> **—every bit.** *Informal.* In all ways : EQUALLY <We are *every bit* as competent as they are.> **—every now and then** (or **again**). Occasionally. **—every so often.** Occasionally. **—every which way.** *Informal.* In disorder.

**ev·e·ry·bod·y** (ĕv′rē-bŏd′ē, -bŭd′ē) *pron.* Every person : EVERYONE.

**ev·e·ry·day** (ĕv′rē-dā′) *adj.* **1.** Suitable for ordinary days or routine occasions <an *everyday* shirt> **2.** Common : ordinary <*everyday* concerns>

**ev·e·ry·one** (ĕv′rē-wŭn′) *pron.* Every person : EVERYBODY.

**ev·e·ry·place** (ĕv′rē-plās′) *adv.* Everywhere.

**ev·e·ry·thing** (ĕv′rē-thĭng′) *pron.* **1.** All factors or things that exist or relate to a given instance. **2.** The essential fact or consideration <Health meant *everything* to them.>

**ev·e·ry·where** (ĕv′rē-hwâr′, -wâr′) *adv.* In all places.

**e·vict** (ĭ-vĭkt′) *vt.* **e·vict·ed, e·vict·ing, e·victs.** [ME *evicten* < Lat. *evincere,* to vanquish : *ex-* (intensive) + *vincere,* to defeat.] **1.** To expel or put out (a tenant) by legal process. **2.** To force out : EJECT. **3.** To recover (e.g., property) by a superior claim or legal process. **—e·vic′tion** *n.* **—e·vic′tor** *n.*

**ev·i·dence** (ĕv′ĭ-dəns) *n.* [ME < OFr. < LLat. *evidentia* < Lat. *evidens,* evident.] **1.** The data on which a conclusion or judgment may be based <glacial *evidence* of climatic change> **2.** Something that indicates <Your reaction was *evidence* of innocence.> **3.** *Law.* The documentary or verbal statements and material objects admissible as testimony in a court of law. —*vt.* **-denced, -denc·ing, -denc·es.** **1.** To indicate clearly : EXEMPLIFY. **2.** To support by testimony : ATTEST. **—in evidence.** Present and plainly visible : CONSPICUOUS <Foreign cars are *in evidence* everywhere.>

**ev·i·dent** (ĕv′ĭ-dənt) *adj.* [ME < OFr. < Lat. *evidens* : *ex-,* out + *videns,* pr.part. of *vidēre,* to see.] Easily seen or understood.

**ev·i·den·tial** (ĕv′ĭ-dĕn′shəl) *adj.* Relating to, providing, or of the nature of evidence. **—ev′i·den′tial·ly** *adv.*

**ev·i·den·tia·ry** (ĕv′ĭ-dĕn′shə-rē, -shē-ĕr′ē) *adj.* Evidential.

**ev·i·dent·ly** (ĕv′ĭ-dənt-lē, ĕv′ĭ-dĕnt′lē) *adv.* **1.** Clearly : obviously. **2.** According to the available evidence <We're *evidently* too early.>

**e·vil** (ē′vəl) *adj.* [ME < OE *yfel.*] **1.** Morally wrong or bad : WICKED. **2.** Causing injury, ruin, or pain : HARMFUL. **3.** Marked by or indicating future misfortune : OMINOUS <*evil* portents> **4.** Reputedly bad or blameworthy : INFAMOUS <a person of *evil* reputation> **5.** Marked by anger or spite : MALICIOUS <an *evil* disposition> —*n.* **1.** A cause of harm, misfortune, or destruction. **2.** Something morally reprehensible : WICKEDNESS. **3.** An evil power or force. **4.** A source or cause of suffering, harm, or destruction <the *evil* of injustice> —*adv. Archaic.* In an evil manner. **—e′vil·ly** *adv.* **—e′vil·ness** *n.*

▲ word history: The word *evil* is ultimately related to the words *up* and *over* and to the prefix *hypo-,* "under, beneath." The basic sense of *evil,* which is now lost, was therefore probably "exceeding proper bounds" or "overreaching," and the word did not signify merely the absence of good.

**e·vil·do·er** (ē′vəl-do͞o′ər) *n.* A perpetrator of evil. **—e′vil·do′ing** *n.*

**evil eye** *n.* **1.** A look or a stare believed to be injurious to others. **2.** One believed to have the power of the evil eye.

**e·vil-mind·ed** (ē′vəl-mīn′dĭd) *adj.* Having evil thoughts or intentions. **—e′vil-mind′ed·ly** *adv.* **—e′vil-mind′ed·ness** *n.*

**e·vince** (ĭ-vĭns′) *vt.* **e·vinced, e·vinc·ing, e·vinc·es.** [Lat. *evincere,* to prove. —see EVICT.] To show or demonstrate convincingly : MANIFEST. **—e·vinc′i·ble** *adj.*

**e·vis·cer·ate** (ĭ-vĭs′ə-rāt′) *v.* **-at·ed, -at·ing, -ates.** [Lat. *eviscerare, eviscerat-,* to disembowel : *ex-,* out + *viscera,* internal organs.] —*vt.* **1.** To remove the entrails of : DISEMBOWEL. **2.** To take away a critical or essential part of. **3.** *Med.* **a.** To remove the contents of (an eyeball). **b.** To remove an organ, as an eye, from (a patient). —*vi. Med.* To protrude through an incision after an operation. **—e·vis′cer·a′tion** *n.*

**ev·i·ta·ble** (ĕv′ĭ-tə-bəl) *adj.* [Lat. *evitabilis* < *evitare,* to shun : *ex-* (intensive) + *vitare,* to avoid.] Avoidable.

**ev·o·ca·tion** (ĕv′ə-kā′shən, ē′və-) *n.* **1.** An act of evoking. **2.** Creation anew by way of the imagination or memory. **—ev′o·ca′tor** *n.*

**e·voc·a·tive** (ĭ-vŏk′ə-tĭv) *adj.* Tending or having the power to evoke. **—e·voc′a·tive·ly** *adv.* **—e·voc′a·tive·ness** *n.*
**e·voke** (ĭ-vōk′) *vt.* **e·voked, e·vok·ing, e·vokes.** [Lat. *evocare* : *ex-*, out + *vocare*, to call.] **1.** To call forth or summon <*evoke* dark spirits> **2. a.** To call to mind or memory <The incident *evoked* childhood dreams.> **b.** To create anew, esp. via the imagination. **—ev′o·ca·ble** (ĕv′ə-kə-bəl, ĭ-vō′kə-) *adj.*
**ev·o·lute** (ĕv′ə-lo͞ot′, ē′və-) *n.* [< Lat. *evolutus*, p.part. of *evolvere*, to unroll. —see EVOLVE.] The locus of the centers of curvature of a given curve.
**ev·o·lu·tion** (ĕv′ə-lo͞o′shən, ē′və-) *n.* [Lat. *evolutio* < *evolvere*, to unroll.—see EVOLVE.] **1.** A gradual process in which something changes into a different and usu. better or more complex form. **2.** *Biol.* **a.** The theory that groups of organisms, as species, may change over time so that descendants differ morphologically and physiologically from their ancestors. **b.** Historical development of a related group of organisms. **3.** Gradual growth or development of something, as a social institution. **4.** A movement that is part of a set of ordered movements. **5.** *Math.* Extraction of a root of a quantity. **—ev′o·lu′tion·al, ev′o·lu′tion·ar′y** (-shə-nĕr′ē) *adj.* **—ev′o·lu′tion·ar′i·ly** *adv.* **—ev′o·lu′tion·ism** *n.* **—ev′o·lu′tion·ist** *n.*
**e·volve** (ĭ-vŏlv′) *v.* **e·volved, e·volv·ing, e·volves.** [Lat. *evolvere*, to unroll : *ex-*, out of + *volvere*, to roll.] *—vt.* **1. a.** To achieve or develop gradually. **b.** To work out : DEVELOP. **2.** *Biol.* To develop by evolutionary processes from a primitive to a more highly organized form. **3.** To give off : EMIT. *—vi.* **1.** To undergo evolutionary change. **2.** To undergo any change. **—e·volv′a·ble** *adj.* **—e·volve′ment** *n.*
**e·vul·sion** (ĭ-vŭl′shən) *n.* [Lat. *evulsio* < *evellere*, to pull out : *ex-*, out + *vellere*, to pull.] A forcible extraction.
**ev·zone** (ev′zōn′) *n.* [Mod. Gk. *euzōnos* < Gk., dressed for exercise : *eu-*, good + *zōnē*, girdle.] A member of a special Greek infantry unit.
**ewe** (yo͞o) *n.* [ME < OE *ēowu*.] A female sheep, esp. when fully grown.
▲ word history: The Old English form of the word *ewe* was *ēowu*, derived from prehistoric Germanic *awiz*. The Indo-European form *owis* is the ancestor of *awiz* as well as of Latin *ovis*, "sheep." *Ewe* and *ovis* are therefore cognate words, which means that they have a common ancestor.
**E·we** (ā′wā′, ā′vā′) *n.* **1. a.** A Negro people of Togo, Ghana, and parts of Dahomey. **b.** A member of this people. **2.** The Niger-Congo language of the Ewe people.
**ewe-neck** (yo͞o′nĕk′) *n.* A defect in a horse or dog in which the neck is thin and has a concave arch. **—ewe′-necked′** *adj.*
**ew·er** (yo͞o′ər) *n.* [ME *ever* < AN < OFr. *eviere* < VLat. **aquaria* < Lat. *aquarius*, relating to water < *aqua*, water.] A large, wide-mouthed jug or pitcher.

**ewer**
*19th-century American*

**ex**[1] (ĕks) *prep.* [Lat.] **1.** Out of : FROM. **2.** Free of charge to the purchaser until removed from (a particular place or thing).
**ex**[2] (ĕks) *n.* The letter x.
**ex**[3] (ĕks) *n. Slang.* A former spouse.
**ex-** *pref.* [ME < OFr. < Lat. *ex*, out of.] **1.** Outside : away from <exodontia> **2.** Not : without <excaudate> **3.** Former <ex-president>
**ex·ac·er·bate** (ĭg-zăs′ər-bāt′) *vt.* **-bat·ed, -bat·ing, -bates.** [Lat. *exacerbare, exacerbat-* : *ex-* (intensive) + *acerbare*, to make harsh < *acerbus*, harsh.] **1.** To increase the severity of : AGGRAVATE <*exacerbate* a quarrel> **2.** To irritate : embitter. **—ex·ac′er·ba′tion** *n.*
**ex·act** (ĭg-zăkt′) *adj.* [Lat. *exactus*, p.part. of *exigere*, to demand : *ex-*, out + *agere*, to impel.] **1.** Precise and accurate. **2.** Strictly and completely in accord with fact. **3.** Meticulously observing of or adhering to a standard. *—vt.* **-act·ed, -act·ing, -acts. 1.** To force the payment or yielding of : EXTORT. **2.** To demand and obtain by or as if by force or authority. **—ex·act′a·ble** *adj.* **—ex·ac′ti·tude** *n.* **—ex·act′ness** *n.* **—ex·ac′tor, ex·act′er** *n.*
**ex·act·a** (ĭg-zăk′tə) *n.* [< Am. Sp. *quiniela exacta*, exact quiniela (a game of chance).] A method of betting, as on a horse race, in which the bettor must correctly pick those finishing in the first and second places in precisely that sequence.
**ex·act·ing** (ĭg-zăk′tĭng) *adj.* **1.** Making rigorous or severe demands <an *exacting* teacher> **2.** Requiring great accuracy or effort <an *exacting* problem> **—ex·act′ing·ly** *adv.* **—ex·act′ing·ness** *n.*
**ex·ac·tion** (ĭg-zăk′shən) *n.* **1. a.** The act of exacting. **b.** Extortion. **2.** Something exacted.
**ex·act·ly** (ĭg-zăkt′lē) *adv.* **1.** In an exact way : ACCURATELY. **2.** In all respects : JUST <*exactly* what we wanted> **3.** As you say. —Used to indicate agreement.
**ex·ag·ger·ate** (ĭg-zăj′ə-rāt′) *v.* **-at·ed, -at·ing, -ates.** [Lat. *exaggerare* : *ex-* (intensive) + *aggerare*, to pile up < *agger*, pile.] *—vt.* **1.** To increase or enlarge to an abnormal degree. **2.** To make greater than is actually the case : OVERSTATE <*exaggerated* their own importance> *—vi.* To distort through overstatement. **—ex·ag′ger·at′ed·ly** *adv.* **—ex·ag′ger·a′tion** *n.* **—ex·ag′ger·a′tive, ex·ag′ger·a·to′ry** (ĭg-zăj′ər-ə-tôr′ē, -tōr′ē) *adj.* **—ex·ag′ger·a′tor** *n.*
**ex·alt** (ĭg-zôlt′) *vt.* **-alt·ed, -alt·ing, -alts.** [ME *exalten* < Lat. *exaltare* : *ex-*, up + *altus*, high.] **1.** To raise in rank, character, or status : ELEVATE. **2.** To glorify : PRAISE. **3.** To increase the intensity or effect of : HEIGHTEN. **4.** *Obs.* To fill with an intensified feeling, as joy or pride : ELATE. **—ex·alt′er** *n.*
**ex·al·ta·tion** (ĕg′zôl-tā′shən) *n.* **1.** The act of exalting or state of being exalted. **2.** Intense exhilaration and well-being : ELATION.
**ex·alt·ed** (ĭg-zôl′tĭd) *adj.* **1.** Elevated in rank, character, or status. **2.** Lofty : sublime. **—ex·alt′ed·ly** *adv.* **—ex·alt′ed·ness** *n.*
**ex·am** (ĭg-zăm′) *n.* An examination : test.
**ex·a·men** (ĭg-zā′mən) *n.* [Lat. < *exigere*, to examine.] Examination.
**ex·am·i·nant** (ĭg-zăm′ə-nənt) *n.* **1.** A person who examines. **2.** An examinee.
**ex·am·i·na·tion** (ĭg-zăm′ə-nā′shən) *n.* **1.** The act of examining or state of being examined. **2.** An exercise testing skill or knowledge. **3.** Formal interrogation. **—ex·am′i·na′tion·al** *adj.*
**ex·am·ine** (ĭg-zăm′ĭn) *vt.* **-ined, -in·ing, -ines.** [ME *examinen* < OFr. *examiner* < Lat. *examinare* < *examen*, a weighing < *exigere*, to weigh. —see EXACT.] **1. a.** To inspect in detail. **b.** To analyze or observe carefully. **2.** To test the state of. **3.** To determine the aptitude, qualifications, or skills of by questions or exercises. **4.** To question formally. **—ex·am′in·a·ble** *adj.* **—ex·am′in·er** *n.*
**ex·am·in·ee** (ĭg-zăm′ə-nē′) *n.* A person who is examined.
**ex·am·ple** (ĭg-zăm′pəl) *n.* [ME < OFr. *example, essample* < Lat. *exemplum* < *eximere*, to take out : *ex-*, out + *emere*, to take.] **1.** One representative of a group. **2.** One serving as a specific kind of pattern <a good *example*> **3.** A case or situation serving as a precedent or model for another one that is similar. **4. a.** A punishment given as a warning to others. **b.** The recipient of such a punishment. **5.** A problem or exercise that illustrates a method or principle. **—for example.** Serving as an illustration, model, or instance.
**ex·an·the·ma** (ĕg′zăn-thē′mə) *also* **ex·an·them** (ĭg-zăn′thəm) *n., pl.* **-them·a·ta** (-thĕm′ə-tə) *or* **-the·mas** *also* **-thems.** [LLat. *exanthema* < Gk. *exanthēma*, eruption < *exanthein*, to burst forth : *ex-*, out + *anthein*, to blossom < *anthos*, flower.] **1.** A skin eruption. **2.** A disease, as measles, accompanied by a skin eruption. **—ex·an′the·mat′ic** (ĭg-zăn′-thə-măt′ĭk), **ex′an·them′a·tous** (ĕg′zăn-thĕm′ə-təs) *adj.*
**ex·arch** (ĕk′särk′) *n.* [LLat. < Gk. *exarkhos* < *exarkhein*, to lead : *ex-*, out + *arkhein*, to rule.] **1.** The ruler of a Byzantine Empire province. **2.** An Eastern Orthodox bishop ranking just below a patriarch. **—ex·arch′al** *adj.* **—ex′ar′chate** (ĕk′sär′kāt), **ex′ar′chy** (ĕk′sär′kē) *n.*
**ex·as·per·ate** (ĭg-zăs′pə-rāt′) *vt.* **-at·ed, -at·ing, -ates.** [Lat. *exasperare, exasperat-* : *ex-* (intensive) + *asperare*, to make rough < *asper*, rough.] **1.** To make angry : IRRITATE. **2.** To increase the gravity or intensity of. **—ex·as′per·at′er** *n.* **—ex·as′per·at′ing·ly** *adv.* **—ex·as′per·a′tion** (-rā′shən) *n.*
**Ex·cal·i·bur** (ĭk-skăl′ə-bər) *n.* [ME < OFr. *Escalibor* < Med. Lat. *Caliburnus* < Welsh *Caledvwlch*.] The legendary sword belonging to King Arthur.
**ex ca·the·dra** (ĕks′ kə-thē′drə) *adj.* & *adv.* [Lat. : *ex*, from + *cathedra*, chair. —see CATHEDRA.] With the authority derived from one's office or position.
**ex·cau·date** (ĕk-skô′dāt′) *adj.* Lacking a tail : TAILLESS.
**ex·ca·vate** (ĕk′skə-vāt′) *v.* **-vat·ed, -vat·ing, -vates.** [Lat. *excavare, excavat-*, to hollow out : *ex-*, out + *cavare*, to hollow < *cavus*, hollow.] *—vt.* **1.** To make a hole or cavity in. **2.** To form by hollowing out. **3.** To remove by scooping or digging out. **4.** To expose by or as if by digging. *—vi.* To engage in digging, hollowing out, or removing. **—ex′ca·va′tion** *n.*
**ex·ca·va·tor** (ĕk′skə-vā′tər) *n.* One that excavates, esp. a power shovel.
**ex·ceed** (ĭk-sēd′) *vt.* **-ceed·ed, -ceed·ing, -ceeds.** [ME *exceden* < Lat. *excedere* : *ex-*, out + *cedere*, to go.] **1.** To be greater than : SURPASS. **2.** To go beyond the limits of.
**ex·ceed·ing** (ĭk-sē′dĭng) *adj.* Extraordinary : exceptional. *—adv. Archaic.* Exceedingly.
**ex·ceed·ing·ly** (ĭk-sē′dĭng-lē) *adv.* To an unusual or advanced degree : EXTREMELY.
**ex·cel** (ĭk-sĕl′) *v.* **-celled, -cel·ling, -cels.** [ME *excellen* < Lat. *excellere*.] *—vt.* To be greater than : SURPASS. *—vi.* To surpass or do better than others.

**ex·cel·lence** (ĕk'sə-ləns) *n.* [ME < OFr. < Lat. *excellentia* < *excellens,* excellent.] **1.** The quality or state of excelling : SUPERIORITY. **2.** Something in which one excels. **3. Excellence.** Excellency.

**Ex·cel·len·cy** (ĕk'sə-lən-sē) *n., pl.* **-cies.** —Used as a form of address or title for high officials, as ambassadors or bishops.

**ex·cel·lent** (ĕk'sə-lənt) *adj.* [ME < OFr. < Lat. *excellens,* pr.part. of *excellere,* to surpass.] **1.** Exceptionally good of its kind. **2.** *Archaic.* Superior. **—ex'cel·lent·ly** *adv.*

☆ ***syns:*** EXCELLENT, CAPITAL, CHAMPION, DANDY, FINE, FIRST-CLASS, FIRST-RATE, GREAT, PRIME, SPLENDID, SUPER, SUPERB, SUPERIOR, SWELL, TERRIFIC, TOP, TOPFLIGHT, TOPNOTCH *adj. core meaning* : exceptionally good of its kind <an *excellent* film>

**ex·cel·si·or** (ĭk-sĕl'sē-ər) *n.* [Lat., comp. of *excelsis,* high < *excellere,* to rise.] Fine, curved wood shavings used esp. for packing.

**ex·cept** (ĭk-sĕpt') *prep.* [ME < Lat. *excipere* : *ex-,* out + *capere,* to take.] With the exclusion of : BUT <everyone *except* you> *—conj.* **1.** If it were not for the fact that : ONLY <They would buy the boat, *except* that it is too expensive.> **2.** Otherwise than <didn't open your mouth *except* to complain> **3.** *Archaic.* Unless. *—v.* **-cept·ed, -cept·ing, -cepts.** *—vt.* To leave out : EXCLUDE. *—vi.* To object.

**ex·cept·ing** (ĭk-sĕp'tĭng) *prep.* Excluding : except. *—conj. Archaic.* Except : unless.

**ex·cep·tion** (ĭk-sĕp'shən) *n.* **1.** The act of excepting or state of being excepted. **2.** One that is excepted, esp. a case not conforming to normal rules. **3.** A criticism or objection. **4.** *Law.* A formal objection taken in the course of an action or proceeding. **—take exception. 1.** To object to <I *take exception* to your remarks.> **2.** *Archaic.* To take offense : RESENT.

**ex·cep·tion·a·ble** (ĭk-sĕp'shə-nə-bəl) *adj.* Open to objection. **—ex·cep'tion·a·bil'i·ty** *n.* **—ex·cep'tion·a·bly** *adv.*

**ex·cep·tion·al** (ĭk-sĕp'shə-nəl) *adj.* **1.** Being an exception : UNUSUAL. **2.** Being above average. **3. a.** Relating to or describing one whose ability deviates from the norm. **b.** Being below average. **—ex·cep'tion·al·ly** *adv.*

**ex·cep·tive** (ĭk-sĕp'tĭv) *adj.* **1.** Of, being, or containing an exception. **2.** *Archaic.* Captious.

**ex·cerpt** (ĕk'sûrpt') *n.* [Lat. *excerptum* < *excerpere,* to pick out : *ex-,* out + *carpere,* to pluck.] A scene or passage, as from a book. *—vt.* (ĭk-sûrpt') **-cerpt·ed, -cerpt·ing, -cerpts.** To select, quote, or take out, as from a book : EXTRACT.

**ex·cess** (ĭk-sĕs', ĕk'sĕs') *n.* [ME < OFr. < Lat. *excessus,* p.part. of *excedere,* to exceed.] **1.** The state of exceeding what is normal or sufficient. **2.** An amount beyond the normal, sufficient, required, or appropriate. **3.** The amount or degree by which one quantity exceeds another : REMAINDER. **4.** Overindulgence : intemperance. *—adj.* Being more than is usual, required, or allowed. **—in excess of.** Greater or more than. **—to excess.** To an extreme degree or extent.

☆ ***syns:*** EXCESS, GLUT, OVERAGE, OVERFLOW, OVERSTOCK, OVERSUPPLY, SUPERFLUITY, SURPLUS *n. core meaning* : an amount beyond the needed or appropriate <an *excess* of grain> ***ant:*** dearth, deficiency, shortfall

**ex·ces·sive** (ĭk-sĕs'ĭv) *adj.* Exceeding what is proper, normal, or reasonable. **—ex·ces'sive·ly** *adv.* **—ex·cess'ive·ness** *n.*

**ex·change** (ĭks-chānj') *v.* **-changed, -chang·ing, -chang·es.** [ME *eschaungen* < AN *eschaungier* < OFr. *eschangier* < VLat. **excambiare* : Lat. *ex-,* out + LLat. *cambiare,* to barter, prob. of Celt. orig.] *—vt.* **1.** To take or give in return for something else <*exchange* recipes> **2.** To give up (one thing) for another <*exchange* a career in teaching for an editorship> **3.** To turn in for replacement by other merchandise. **4.** To provide in return for something of equal value : TRADE. *—vi.* **1.** To trade for something of equal value. **2.** To take part in a mutual trade, as of goods or services. *—n.* **1.** An act or instance of exchanging. **2.** One that is exchanged. **3.** A place where things are exchanged, esp. a center where securities and commodities are bought and sold. **4.** A telephone exchange. **5. a.** A system of payments using instruments, as negotiable drafts, instead of money. **b.** The fee or percentage charged for participating in such a system of payment. **6.** A bill of exchange. **7.** Rate of exchange. **8.** The amount of difference in the actual value of two or more currencies, or between values of the same currency at two or more places. **—ex·change'a·bil'i·ty** *n.* **—ex·change'a·ble** *adj.* **—ex·chang'er** *n.*

**exchange force** *n.* A force arising between two elementary particles due to the continuous interchange of space or spin coordinates.

**exchange rate** *n.* Rate of exchange.

**ex·cheq·uer** (ĭks'chĕk'ər, ĭks-chĕk'ər) *n.* [ME *escheker* < OFr. *eschequier,* counting table, chessboard < *eschec,* check in chess. —see CHECK.] **1. Exchequer.** The British governmental department charged with the collection and care of the national revenue. **2. Exchequer.** The Court of Exchequer. **3.** A treasury, as of a nation or organization. **4.** Financial resources : FUNDS.

▲ word history: The Exchequer got its name from the checkered cloth resembling a chessboard that covered the counting tables of the Norman and Angevin kings of England. The word *exchequer* is derived from the Old French word *eschequier,* "chessboard." The *x* in *exchequer* is an etymological phantom. Many Old French words beginning with *es-* come from Latin words beginning with *ex-,* but *eschequier* is derived from Latin *scaccus,* "check in chess," which is ultimately from Persian *shāh,* "king," the most important piece in chess. When other English words like *exchange* (from Old French *eschangier* and Vulgar Latin *excambiare*) were respelled to reflect their Latin origins, *exchequer* was erroneously altered in the same way.

**ex·ci·mer** (ĕk'sə-mər) *n.* [EXC(ITED) + (D)IMER.] A dimer existing in an energy level above the ground state.

**ex·cip·i·ent** (ĭk-sĭp'ē-ənt) *n.* [Lat. *excipiens, excipient-,* pr.part. of *excipere,* to take out. —see EXCEPT.] An inert substance used as a diluent or vehicle for a drug.

**ex·cis·a·ble** (ĭk-sī'zə-bəl) *adj.* Subject to an excise.

**ex·cise**[1] (ĕk'sīz') *n.* [Obs. Du. *excijs* < MDu., prob. < OFr. *acceis* < LLat. *accensare,* to tax : Lat. *ad-,* to + Lat. *census,* tax.] **1.** An internal tax levied on the sale, production, or consumption of certain commodities, as liquor, within a country. **2.** A tax often levied in the form of a licensing charge or a fee for certain privileges. *—vt.* **-cised, -cis·ing, -cis·es.** To levy an excise on.

**ex·cise**[2] (ĭk-sīz') *vt.* **-cised, -cis·ing, -cis·es.** [Lat. *excidere, excis-* : *ex-,* out + *caedere,* to cut.] To remove by or as if by cutting. **—ex·ci'sion** (-sĭzh'ən) *n.*

**ex·cise·man** (ĭk'sīz'mən, ĭk-sīz'-) *n. Chiefly Brit.* An excise tax officer or collector.

**ex·cit·a·ble** (ĭk-sī'tə-bəl) *adj.* **1.** Capable of being easily excited. **2.** Capable of responding to stimuli. **—ex·cit'a·bil'i·ty, ex·cit'a·ble·ness** *n.* **—ex·cit'a·bly** *adv.*

**ex·ci·tant** (ĭk-sīt'nt) *also* **ex·ci·ta·tive** (-sī'tə-tĭv) *or* **ex·ci·ta·to·ry** (-sī'tə-tôr'ē, -tōr'ē) *adj.* Capable of exciting. **—ex·ci'tant** *n.*

**ex·ci·ta·tion** (ĕk'sī-tā'shən) *n.* Excitement.

**ex·ci·ta·tive** (ĭk-sī'tə-tĭv) *or* **ex·ci·ta·to·ry** (-sī'tə-tôr'ē, -tōr'ē) *adj. vars. of* EXCITANT.

**ex·cite** (ĭk-sīt') *vt.* **-cit·ed, -cit·ing, -cites.** [ME *exciten* < Lat. *excitare,* freq. of *exciēre* : *ex-,* out + *ciēre,* to call.] **1.** To stir to activity or motion. **2.** To elicit, as a reaction or emotion : INDUCE <*excited* their interest> **3.** To arouse strong feeling in : PROVOKE. **4.** *Biol.* To produce increased activity in (an organism or part) : STIMULATE. **5.** *Physics.* **a.** To increase the energy of. **b.** To raise (e.g., an atom) to a higher energy level. **—ex·cite'ment** *n.*

**ex·cit·ed** (ĭk-sī'tĭd) *adj.* **1.** Being in a state of excitement : STIRRED. **2.** *Physics.* Being at an energy level higher than the ground state. **—ex·cit'ed·ly** *adv.*

**ex·cit·er** (ĭk-sī'tər) *n.* **1.** One that excites. **2.** *Elect.* **a.** An auxiliary generator used to provide field current for a larger generator or alternator. **b.** An oscillator used to generate the carrier frequency of a transmitter.

**ex·cit·ing** (ĭk-sī'tĭng) *adj.* Creating excitement : ROUSING. **—ex·cit'ing·ly** *adv.*

**ex·ci·ton** (ĕk'sĭ-tŏn', -sī-) *n.* [EXCIT(ATION) + -ON[1].] An electrically neutral excited state of a crystal, often regarded as a bound state of an electron and a hole.

**ex·ci·ton·ics** (ĕk'sə-tŏn'ĭks, -sī-) *n. (sing. in number).* Study of excitons and their behavior in semiconductors and dielectrics.

**ex·ci·tor** (ĭk-sī'tər) *n.* A stimulant.

**ex·claim** (ĭk-sklām') *vt. & vi.* **-claimed, -claim·ing, -claims.** [OFr. *exclamer* < Lat. *exclamare* : *ex-,* out + *clamare,* to call.] To cry out or speak suddenly, as from emotion. **—ex·claim'er** *n.*

**ex·cla·ma·tion** (ĕk'sklə-mā'shən) *n.* **1.** A sudden forceful utterance. **2.** An outcry, as of protest.

**exclamation point** *n.* A punctuation mark (!) used after an exclamation.

**ex·clam·a·to·ry** (ĭk-sklăm'ə-tôr'ē, -tōr'ē) *adj.* Of, being, containing, or using an exclamation.

**ex·clave** (ĕk'sklāv') *n.* [EX- + (EN)CLAVE.] An area of a country isolated from the main part and constituting an enclave in alien territory.

**ex·clude** (ĭk-sklōōd') *vt.* **-clud·ed, -clud·ing, -cludes.** [ME *excluden* < Lat. *excludere* : *ex-,* out + *claudere,* to shut.] **1.** To keep out : BAR. **2.** To omit from notice or consideration : DISREGARD. **3.** To put out : EXPEL. **—ex·clud'a·bil'i·ty** *n.* **—ex·clud'a·ble, ex·clud'i·ble** *adj.* **—ex·clud'er** *n.*

**ex·clu·sion** (ĭk-sklōō'zhən) *n.* [Lat. *exclusio* < *excludere,* to shut out. —see EXCLUDE.] The act of excluding or state of being excluded. **—ex·clu'sion·ar'y** (-zhə-nĕr'ē) *adj.*

**ex·clu·sion·ist** (ĭk-sklōō'zhə-nĭst) *n.* An advocate of excluding others from privileges or rights. **—ex·clu'sion·ism** *n.* **—ex·clu'sion·is'tic** *adj.*

**exclusion principle** *n.* The principle that no two particles of a given type, as protons, electrons, or neutrons, can occupy a particular quantum state.

**ex·clu·sive** (ĭk-sklōō'sĭv) *adj.* **1.** Relating to or marked by exclusion. **2.** Not shared with others <*exclusive* rights to publish a book> **3.** Independent or single : SOLE. **4.** Complete : whole. **5.** Excluding certain people, as from membership or participation. **6.** Catering to a wealthy clientele <*exclusive* salons> *—n.* **1.** A news item released to only one person or publication. **2.** An exclusive right

or privilege, as to sell a product. **—ex·clu'sive·ly** *adv.* **—ex·clu'sive·ness** *n.* **—ex'clu·siv'i·ty** (ĕk'sklōō-sĭv'ĭ-tē) *n.*
**xclusive of** *prep.* Not including : BESIDES.
**x·cog·i·tate** (ĭk-skŏj'ĭ-tāt') *vt.* **-tat·ed, -tat·ing, -tates.** [Lat. *excogitare, excogitat-*, to find out by thinking : *ex-*, out + *cogitare*, to think. —see COGITATE.] To think out in great detail : DEVISE. **—ex·cog'i·ta'tion** *n.* **—ex·cog'i·ta'tive** *adj.*
**x·com·mu·ni·ca·ble** (ĕks'kə-myōō'nĭ-kə-bəl) *adj.* Liable to or punishable by excommunication.
**x·com·mu·ni·cate** (ĕks'kə-myōō'nĭ-kāt') *vt.* **-cat·ed, -cat·ing, -cates.** [ME *excommunicaten* < LLat. *excommunicare* : Lat. *ex-*, out + Lat. *communis*, common.] **1.** To deprive of the right of church membership by ecclesiastical authority. **2.** To exclude from membership in a group. —*n.* (ĕks'kə-myōō'nĭ-kĭt). One who has been excommunicated. —*adj.* (ĕks'kə-myōō'nĭ-kĭt, -kāt'). Excommunicated. **—ex·com·mu'ni·ca'tion** *n.* **—ex'com·mu'ni·ca'tive** (ĭks'kə-myōō'nĭ-kā'tĭv, -kə-tĭv), **ex·com·mu'ni·ca·to'ry** (-kə-tôr'ē, -tōr'ē) *adj.* **—ex'com·mu'ni·ca'tor** *n.*
**x·co·ri·ate** (ĭk-skôr'ē-āt', -skōr'-) *vt.* **-at·ed, -at·ing, -ates.** [ME *excoriaten* < Lat. *excoriare*, to strip of its skin : *ex-*, off + *corium*, skin.] **1.** To wear off or tear the skin of : ABRADE. **2.** To censure strongly : UPBRAID. **—ex·co'ri·a'tion** *n.*
**x·cre·ment** (ĕk'skrə-mənt) *n.* [Lat. *excrementum* < *excernere*, to excrete.] Waste material, esp. fecal matter, expelled from the body after digestion. **—ex'cre·ment'al** (ĕk'skrə-mĕn'tl) *adj.* **—ex'cre·men·ti'tious** (-mĕn-tĭsh'əs) *adj.*
**x·cres·cence** (ĭk-skrĕs'əns) *n.* [ME < Lat. *excrescentia* < *excrescere*, to grow out : *ex-*, out + *crescere*, to grow.] An abnormal outgrowth or enlargement.
**x·cres·cen·cy** (ĭk-skrĕs'ən-sē) *n., pl.* **-cies.** An excrescence.
**x·cres·cent** (ĭk-skrĕs'ənt) *adj.* **1.** Growing out abnormally or excessively. **2.** Epenthetic. **—ex·cres'cent·ly** *adv.*
**x·cre·ta** (ĭk-skrē'tə) *pl.n.* [Lat. < *excernere*, to discharge. —see EXCREMENT.] Waste matter, as sweat, urine, or feces, excreted from the body. **—ex·cre'tal** *adj.*
**x·crete** (ĭk-skrēt') *vt.* **-cret·ed, -cret·ing, -cretes.** [Lat. *excernere, excret-* : *ex-*, out + *cernere*, to separate.] To eliminate (waste matter) from the blood, tissues, or organs.
**x·cre·tion** (ĭk-skrē'shən) *n.* **1.** The act or process of excreting. **2.** Matter excreted.
**x·cre·to·ry** (ĕk'skrĭ-tôr'ē, -tōr'ē) *adj.* **1.** Of or relating to excretion. **2.** Having the function of excreting <*excretory* organs>
**x·cru·ci·ate** (ĭk-skrōō'shē-āt') *vt.* **-at·ed, -at·ing, -ates.** [Lat. *excruciare* : *ex-* (intensive) + *cruciare*, to crucify < *crux*, cross.] **1.** To inflict with severe pain : TORTURE. **2.** To subject to great mental distress. **—ex·cru'ci·a'tion** *n.*
**x·cru·ci·at·ing** (ĭk-skrōō'shē-ā'tĭng) *adj.* **1.** Extremely painful : AGONIZING. **2.** Intensely distressing. **—ex·cru'ci·at'ing·ly** *adv.*
**x·cul·pate** (ĕk'skəl-pāt', ĭk-skŭl'-) *vt.* **-pat·ed, -pat·ing, -pates.** [Med. Lat. *exculpare* : Lat. *ex-*, away + *culpa*, guilt.] To clear of blame. **—ex·cul'pa·ble** (ĭk-skŭl'pə-bəl) *adj.* **—ex'cul·pa'tion** *n.*
**x·cul·pa·to·ry** (ĭk-skŭl'pə-tôr'ē, -tōr'ē) *adj.* Acting or tending to exculpate.
**x·cur·rent** (ĭk-skûr'ənt, -skŭr'-) *adj.* [Lat. *excurrens, excurrent-*, pr.part. of *excurrere*. —see EXCURSION.] **1. a.** Flowing or running out. **b.** Characterized by an outward flow of current. **2.** *Bot.* **a.** Having a single, undivided trunk with lateral branches, as many coniferous trees. **b.** Extending beyond the apex of a leaf, as a midrib or vein.
**ex·cur·sion** (ĭk-skûr'zhən) *n.* [Lat. *excursio* < *excurrere*, to run out : *ex-*, out + *currere*, to run.] **1.** A short journey : OUTING. **2. a.** A pleasure tour, esp. one of limited duration and at a special reduced fare. **b.** The party on such a tour. **3.** Deviation from the main topic : DIGRESSION. **4. a.** A movement from a mean position or axis in an oscillating or alternating motion. **b.** The distance traversed in such a movement. **—ex·cur'sion·ist** *n.*
**ex·cur·sive** (ĭk-skûr'sĭv) *adj.* Given to, marked by, or of the nature of digression. **—ex·cur'sive·ly** *adv.* **—ex·cur'sive·ness** *n.*
**ex·cur·sus** (ĭk-skûr'səs) *n., pl.* **-sus·es.** [Lat. < *excurrere*, to run out. —see EXCURSION.] **1.** A lengthy, appended exposition of a point. **2.** A digression.
**ex·cus·a·to·ry** (ĭk-skyōō'zə-tôr'ē, -tōr'ē) *adj.* Tending or serving to excuse.
**ex·cuse** (ĭk-skyōōz') *vt.* **-cused, -cus·ing, -cus·es.** [ME *excusen* < OFr. *excuser* < Lat. *excusare* : *ex-*, away from + *causa*, accusation.] **1. a.** To apologize for. **b.** To seek to remove the blame from <*excused* myself for my late arrival> **2. a.** To grant pardon to : FORGIVE. **b.** To make allowance for : OVERLOOK. **3.** To serve as justification for <Nothing can *excuse* your rudeness.> **4.** To free, as from an obligation or duty : EXEMPT. **5.** To give permission to leave : RELEASE <*excused* the class early> —*n.* (ĭk-skyōōs'). **1.** An explanation offered to elicit or justify forgiveness. **2.** The reason for excusing. **3.** An act of excusing. **4.** A note explaining an absence. **5.** *Informal.* An inferior example <a poor *excuse* for a governor> **—ex·cus'a·ble** *adj.* **—ex·cus'a·ble·ness** *n.* **—ex·cus'a·bly** *adv.* **—ex·cus'er** *n.*

**ex·e·cra·ble** (ĕk'sĭ-krə-bəl) *adj.* [ME < Lat. *execrabilis* < *execrari*, to execrate.] **1.** Deserving execration. **2.** Inferior : bad <*execrable* accommodations> **—ex'e·cra·ble·ness** *n.* **—ex'e·cra·bly** *adv.*
**ex·e·crate** (ĕk'sĭ-krāt') *vt.* **-crat·ed, -crat·ing, -crates.** [Lat. *execrari, execrat-* : *ex-*, away from + *sacrare*, to consecrate < *sacer*, sacred.] **1.** To declare to be abhorrent : DENOUNCE. **2.** To feel loathing for : DETEST. **3.** *Archaic.* To invoke a curse on. **—ex'e·cra'tive** *adj.* **—ex'e·cra'tor** *n.* **—ex'e·cra·to'ry** (-krə-tôr'ē, -tōr'ē) *adj.*
**ex·e·cra·tion** (ĕk'sĭ-krā'shən) *n.* **1.** The act of cursing. **2.** A curse. **3.** Something cursed or hated.
**ex·ec·u·tant** (ĭg-zĕk'yə-tənt) *n.* A performer, esp. a skilled one.
**ex·e·cute** (ĕk'sĭ-kyōōt') *vt.* **-cut·ed, -cut·ing, -cutes.** [ME *executen* < OFr. *executer* < Med. Lat. *executare* < Lat. *exequi* : *ex-*, out + *sequi*, to follow.] **1.** To put into effect : CARRY OUT <*execute* orders> **2.** To perform. **3.** To create (e.g., a work of art) in agreement with a prescribed design. **4.** To make valid or legal, as by signing <*execute* a license> **5.** To carry out what is required by <*execute* a contract> **6.** To put to death, esp. by carrying out a capital sentence. **—ex'e·cut'a·ble** *adj.* **—ex'e·cut'er** *n.*
**ex·e·cu·tion** (ĕk'sĭ-kyōō'shən) *n.* **1.** The act of executing or state of being executed. **2.** Result or manner of performance. **3.** An act or instance of putting or being put to death as a legal penalty. **4.** *Law.* **a.** The carrying into effect of a court judgment. **b.** A writ empowering an officer to enforce a judgment. **5.** *Law.* Validation of a legal document by the performance of the necessary formalities. **6.** Effective, punitive, or destructive action.
**ex·e·cu·tion·er** (ĕk'sĭ-kyōō'shən-ər) *n.* **1.** A person who executes. **2.** A person who puts another to death.
**execution time** *n. Computer Sci.* The time needed for a computer to decode and perform an instruction.
**ex·ec·u·tive** (ĭg-zĕk'yə-tĭv) *n.* **1.** An administrator or manager in an organization. **2.** The chief officer of a government, state, or political division. **3.** The branch of government empowered with the responsibility of executing a country's laws and administering its functions. **4.** *Computer Sci.* A set of coded instructions designed to control and process other coded instructions. —*adj.* **1.** Of, relating to, or capable of executing or carrying out. **2.** Of or relating to the branch of government charged with the execution and administration of the nation's laws. **3.** Of or relating to an executive.
**executive agreement** *n.* An agreement between heads of state without senatorial ratification.
**executive council** *n.* **1.** A council that assists or advises a political executive. **2.** A council having the highest executive power.
**executive officer** *n.* **1.** The officer second in command of a military or naval establishment. **2.** A person holding executive power in an organization.
**executive order** *n.* REGULATION 3.
**executive routine** *n. Computer Sci.* A set of coded instructions designed to utilize a computer to control or develop other routines.
**executive secretary** *n.* A secretary with administrative duties.
**executive session** *n.* A usu. closed legislative session.
**ex·ec·u·tor** (ĭg-zĕk'yōō-tər, ĕk'sĭ-kyōō'tər) *n.* **1.** One who performs something. **2.** *Law.* One appointed to execute a will by a testator. **—ex·ec'u·to'ri·al** (-tôr'ē-əl, -tōr'-) *adj.* **—ex·ec'u·tor·ship'** *n.*
**ex·ec·u·to·ry** (ĭg-zĕk'yə-tôr'ē) *adj.* **1.** Administrative. **2.** In effect : OPERATIVE. **3.** *Law.* Meant to go into effect or having the potential of becoming effective at a future time : CONTINGENT.
**ex·ec·u·trix** (ĭg-zĕk'yə-trĭks') *n., pl.* **-trix·es** *or* **-tri·ces** (-trī'sēz'). *Law.* A woman appointed to execute a will by a testator.
**ex·e·dra** (ĕk'sĭ-drə, ĭk-sē'-) *n.* [Lat. < Gk. : *ex-*, out + *hedra*, seat.] **1.** An often semicircular portico with seats, used in ancient Greece and Rome for discussions. **2.** A usu. curved outdoor bench with a high back.
**ex·e·ge·sis** (ĕk'sə-jē'sĭs) *n., pl.* **-ses** (-sēz') [Gk. *exēgēsis* < *exēgeisthai*, to interpret : *ex-*, out + *hēgeisthai*, to lead.] Critical explanation or interpretation, esp. of a text.
**ex·e·gete** (ĕk'sə-jēt') *also* **ex·e·get·ist** (ĕk'sə-jĕt'ĭst) *n.* [Gk. *exēgētēs* < *exēgeisthai*, to interpret. —see EXEGESIS.] One who is skilled in exegesis.
**ex·e·get·ic** (ĕk'sə-jĕt'ĭk) *also* **ex·e·get·i·cal** (-ĭ-kəl) *adj.* Of or relating to exegesis : ANALYTIC. **—ex'e·get'i·cal·ly** *adv.*
**ex·em·pla** (ĭg-zĕm'plə) *n. pl. of* EXEMPLUM.
**ex·em·plar** (ĭg-zĕm'plär', -plər) *n.* [ME < LLat. *exemplarium* < Lat. *exemplum*, example.] **1.** One worthy of imitation : MODEL. **2.** A typical example. **3.** An ideal serving as a pattern : ARCHETYPE. **4.** A copy, as of a book.
**ex·em·pla·ry** (ĭg-zĕm'plə-rē) *adj.* **1.** Worthy of imitation : COMMENDABLE <*exemplary* scholarship> **2.** Serving as a model. **3.** Serving as an illustration : TYPICAL. **4.** Serving as a warning : CAUTIONARY. **—ex'em·plar'i·ly** (ĕg'zəm-plâr'ə-lē) *adv.* **—ex·em'pla·ri·ness, ex'em·plar'i·ty** (ĕg'zəm-plăr'ĭ-tē) *n.*
**ex·em·pli·fi·ca·tion** (ĭg-zĕm'plə-fĭ-kā'shən) *n.* **1.** An act of exemplifying. **2.** One that exemplifies : EXAMPLE. **3.** *Law.* A certified copy of a document.
**ex·em·pli·fy** (ĭg-zĕm'plə-fī') *vt.* **-fied, -fy·ing, -fies.** [ME *exemplifien* < Med. Lat. *exemplificare* : Lat. *exemplum*, example + *facere*, to make.] **1. a.** To illustrate by example. **b.** To serve as an

example of. **2.** *Law.* To make a certified copy of (a document.) **—ex·em′pli·fi′a·ble** *adj.* **—ex·em′pli·fi′er** *n.*

**ex·em·pli gra·ti·a** (ĭg-zĕm′plē′ grä′tē-ä′) *adv.* [Lat.] For example.

**ex·em·plum** (ĭg′zĕm′pləm) *n., pl.* **-pla** (-plə) [Lat. —see EXAMPLE.] **1.** An example. **2.** A short story that illustrates a moral or makes a point in an argument.

**ex·empt** (ĭg-zĕmpt′) *vt.* **-empt·ed, -empt·ing, -empts.** [ME *exempten* < Lat. *eximere, exempt-.* —see EXAMPLE.] **1.** To free from a duty or obligation required of others. **2.** To isolate. *—adj.* **1.** Freed from a duty or obligation required of others. **2.** *Obs.* Set apart : ISOLATED. *—n.* One exempted from a duty. **—ex·empt′i·ble** *adj.*

**ex·emp·tion** (ĭg-zĕmp′shən) *n.* **1.** The act of exempting or state of being exempt : IMMUNITY. **2. a.** Something exempted, esp. from taxation. **b.** One so exempted or a source of such an exemption.

**ex·en·do·sperm·ous** (ĕk-sĕn′də-spûr′məs) *adj. Bot.* Lacking an endosperm.

**ex·en·ter·ate** (ĭk-zĕn′tə-rāt′) *vt.* **-at·ed, -at·ing, -ates.** [Lat. *exenterare, exenterat-,* to disembowel : *ex,* from + Gk. *enteron,* intestines.] **1.** To disembowel : eviscerate. **2.** *Med.* To remove the contents of (an organ). **—ex·en′ter·a′tion** *n.*

**ex·er·cise** (ĕk′sər-sīz′) *n.* [ME < OFr. *exercice* < Lat. *exercitium* < *exercēre,* to exercise : *ex-,* out of + *arcēre,* to restrain.] **1.** An act of putting into use. **2.** Discharge of a function, duty, or office. **3.** Activity requiring physical or mental exertion, esp. when performed to maintain or develop fitness. **4.** Something practiced so as to increase one's skill <a piano *exercise*> **5. exercises.** A secular or religious ceremony that includes traditional rites, as speeches and awards <commencement *exercises*> *—v.* **-cised, -cis·ing, -cis·es.** *—vt.* **1.** To put into operation : USE. **2.** To bring to bear : EXERT <*exercised* police power> **3. a.** To subject to practice or exertion so as to train or develop <*exercise* balancing skill> **b.** To put through exercises <*exercise* the dogs> **4.** To carry out the functions of <*exercise* the role of parent> **5. a.** To absorb the attentions of, esp. by anxiety or worry. **b.** To arouse the anger of <was *exercised* by their slovenliness> *—vi.* To take exercise. **—ex′er·cis′a·ble** *adj.*

**ex·er·cis·er** (ĕk′sər-sī′zər) *n.* **1.** One that exercises. **2.** A device for exercising the body.

**ex·er·ci·ta·tion** (ĭg-zûr′sĭ-tā′shən) *n.* [ME *exercitacioun* < Lat. *exercitatio* < *exercitare,* to exercise often, freq. of *exercēre,* to exercise.] An act or instance of exercising.

**ex·er·gon·ic** (ĕk′sər-gŏn′ĭk) *adj.* [EX(O)- + Gk. *ergon,* work + -IC.] Releasing energy.

**ex·ergue** (ĕk′sûrg′, ĕg′zûrg′) *n.* [Fr. < NLat. *exergum* : Gk. *ex-,* out of + Gk. *ergon,* work.] The space on the reverse of a coin or medal, usu. below the central design and often giving the date and place of engraving.

**ex·ert** (ĭg-zûrt′) *vt.* **-ert·ed, -ert·ing, -erts.** [Lat. *exserere, exsert-,* to put forth : *ex-,* out + *serere,* to bring forth.] **1.** To put forth (e.g., strength). **2.** To bring to bear : EXERCISE <*exert* clout> **3.** To put (oneself) to a strenuous effort. **4.** To make use of : EMPLOY. **—ex·er′tion** (-zûr′shən) *n.*

**ex·e·unt** (ĕk′sē-ənt, -o͝ont′) [Lat., 3rd pers. pl. of *exire,* to go out : *ex-,* out + *ire,* to go.] —Used as a stage direction to indicate that two or more actors leave the stage.

**ex·fo·li·ate** (ĕks-fō′lē-āt′) *v.* **-at·ed, -at·ing, -ates.** [Lat. *exfoliare, exfoliat-,* to strip of leaves : *ex-,* off + *folium,* leaf.] *—vt.* **1.** To remove (e.g., bark) in scales or flakes : PEEL. **2.** To cast off in scales, flakes, or splinters. *—vi.* To come off or separate as scales, flakes, sheets, or layers. **—ex·fo′li·a′tion** *n.* **—ex·fo′li·a′tive** *adj.*

**ex·ha·lant** *also* **ex·ha·lent** (ĕks-hā′lənt, ĕk-sā′-) *adj.* Capable of or functioning in exhalation.

**ex·hale** (ĕks-hāl′, ĕk-sāl′) *v.* **-haled, -hal·ing, -hales.** [ME *exhalen* < Lat. *exhalare* : *ex-,* out + *halare,* to breathe.] *—vi.* **1. a.** To breathe out. **b.** To emit air or vapor. **2.** To be given off or emitted. *—vt.* **1.** To blow forth or breathe out. **2.** To give off : EMIT. **—ex′ha·la′tion** (ĕks′hə-lā′shən, ĕk′sə-) *n.*

**ex·ha·lent** (ĕks-hā′lənt, ĕk-sā′-) *adj. var. of* EXHALANT.

**ex·haust** (ĭg-zôst′) *v.* **-haust·ed, -haust·ing, -hausts.** [Lat. *exhaurire, exhaust-* : *ex-,* out + *haurire,* to draw.] *—vt.* **1.** To let out or draw off. **2.** To draw out the contents of : DRAIN. **3.** To use up : CONSUME <*exhaust* one's food supply> **4.** To wear out completely : TIRE. **5.** To drain of properties or resources : DEPLETE <plant matter that *exhausted* the lake's oxygen supply> **6.** To deal with comprehensively <*exhausted* every possible lead> *—vi.* To escape, as steam. *—n.* **1. a.** The escape of vaporous waste, as from an engine. **b.** The gases so released. **2.** A conduit through which vaporous gases are emitted. **3.** An apparatus for drawing out noxious air or waste gases by means of a partial vacuum. **—ex·haust′er** *n.* **—ex·haust′i·bil′i·ty** *n.* **—ex·haust′i·ble** *adj.* **—ex·haus′tion** (-zôs′chən) *n.*

**ex·haus·tive** (ĭg-zô′stĭv) *adj.* **1.** Tending to exhaust. **2.** Comprehensive : thorough <an *exhaustive* analysis> **—ex·haus′tive·ly** *adv.* **—ex·haus′tive·ness** *n.* **—ex·haus·tiv′i·ty** *n.*

**ex·haust·less** (ĭg-zôst′lĭs) *adj.* Impossible to exhaust : INEXHAUSTIBLE. **—ex·haust′less·ly** *adv.* **—ex·haust′less·ness** *n.*

**ex·hib·it** (ĭg-zĭb′ĭt) *v.* **-it·ed, -it·ing, -its.** [ME *exhibiten* < Lat. *exhibere* : *ex-,* out + *habēre,* to hold.] *—vt.* **1.** To show externally : DISPLAY. **2. a.** To present for public view. **b.** To show or enter in an exhibition or contest. **3.** To give an instance or evidence of : DEMONSTRATE. **4.** *Law.* **a.** To submit (documents or evidence) in a cou[rt]. **b.** To present or introduce officially. *—vi.* To put something on pu[b]lic display. *—n.* **1.** An act of exhibiting. **2.** Something exhibite[d]. **3.** *Law.* Something, as a document, formally introduced as evidenc[e] in court. **—ex·hib′it·er, ex·hib′i·tor** *n.* **—ex·hib′i·to′ry** (-ĭ-tôr′ē, -tōr′ē) *adj.*

**ex·hi·bi·tion** (ĕk′sə-bĭsh′ən) *n.* **1.** An act of exhibiting. **2.** Som[e]thing exhibited. **3.** A public display, as of art objects or agricultura[l] products. **4.** *Chiefly Brit.* A grant awarded by a school or universit[y] to scholars.

**ex·hi·bi·tion·er** (ĕk′sə-bĭsh′ə-nər) *n. Chiefly Brit.* A student wh[o] is awarded an exhibition.

**ex·hi·bi·tion·ism** (ĕk′sə-bĭsh′ə-nĭz′əm) *n.* **1.** The act or practic[e] of attracting attention to oneself by egotistical behavior. **2.** *Psycho[l].* Compulsive exposure of the sexual organs in public. **—ex′hi·bi′tion·ist** *n.* **—ex′hi·bi′tion·is′tic** *adj.*

**ex·hib·i·tive** (ĭg-zĭb′ĭ-tĭv) *adj.* Tending to exhibit or serving as a[n] exhibition. **—ex·hib′i·tive·ly** *adv.*

**ex·hil·a·rant** (ĭg-zĭl′ər-ənt) *adj.* Exhilarating.

**ex·hil·a·rate** (ĭg-zĭl′ə-rāt′) *vt.* **-rat·ed, -rat·ing, -rates.** [Lat. *exh[i]larare, exhilarat-* : *ex-* (intensive) + *hilarare,* to make cheerful < *hilaris,* cheerful < Gk. *hilaros.*] **1.** To make happy : ELATE. **2.** T[o] vitalize or refresh : STIMULATE. **—ex·hil′a·ra′tion** *n.* **—ex·hil′a·rat′ive** *adj.* **—ex·hil′a·ra′tor** *n.*

**ex·hort** (ĭg-zôrt′) *v.* **-hort·ed, -hort·ing, -horts.** [ME *exhorten* < Lat. *exhortari* : *ex-* (intensive) + *hortari,* to encourage.] *—vt.* T[o] urge by strong argument, advice, or appeal. *—vi.* To make urge[nt] appeal. **—ex·hort′er** *n.*

**ex·hor·ta·tion** (ĕg′zôr-tā′shən, ĕk′sôr-) *n.* **1.** An act of exhorting. **2.** The practice of exhorting. **3.** An exhortative speech.

**ex·hor·ta·tive** (ĭg-zôr′tə-tĭv) *also* **ex·hor·ta·to·ry** (-tôr′ē, -tōr′ē) *adj.* **1.** Relating to exhortation. **2.** Serving to exhort.

**ex·hume** (ĭg-zo͞om′, -zyo͞om′, ĭk-syo͞om′, -hyo͞om′) *vt.* **-humed, -hum·ing, -humes.** [Fr. *exhumer* < Med. Lat. *exhumare* : Lat. *ex* out of + Lat. *humus,* ground.] **1.** To remove from a grave : DISINTER. **2.** To bring to light, esp. after a time of obscurity. **—ex′hu·ma′tion** *n.* **—ex·hum′er** *n.*

**ex·i·gence** (ĕk′sə-jəns) *n.* Exigency.

**ex·i·gen·cy** (ĕk′sə-jən-sē, ĭg-zĭj′ən-) *n., pl.* **-cies. 1.** The quality o[r] state of being exigent. **2.** An urgent situation. **3.** *often* **exigencies** Pressing needs.

**ex·i·gent** (ĕk′sə-jənt) *adj.* [Lat. *exigens,* pr.part. of *exigere,* to de[mand]. —see EXACT.] **1.** Needing immediate attention or remedy : URGENT. **2.** Excessively demanding. **—ex′i·gent·ly** *adv.*

**ex·ig·u·ous** (ĭg-zĭg′yo͞o-əs, ĭk-sĭg′-) *adj.* [Lat. *exiguus,* measured < *exigere,* to weigh. —see EXACT.] Extremely meager : SCANTY. **—ex′i·gu′i·ty** (ĕk′sĭ-gyo͞o′ĭ-tē) *n.* **—ex·ig′u·ous·ly** *adv.* **—ex·ig′u·ous·ness** *n.*

**ex·ile** (ĕg′zĭl′, ĕk′sĭl′) *n.* [ME *exil* < OFr. < Lat. *exilium* < *exul,* exiled person.] **1. a.** Enforced removal from one's native country. **b.** Self-imposed absence from one's country. **2. a.** The state or cir[]cumstance of being in exile. **b.** A time period spent in exile. **3. a.** One sent into exile by official decree. **b.** One who has voluntar[]ily left one's country. *—vt.* **-iled, -il·ing, -iles.** To send into exile : BANISH. **—ex·il′ic** (ĭg-zĭl′ĭk, ĭk-sĭl′-), **ex·il′ian** (-zĭl′yən, -zĭl′ē-ən, ĭk-sĭl′yən, -sĭl′ē-ən) *adj.*

**ex·ine** (ĕk′sēn′, -sīn′) *n.* [EX(O)- + Gk. *is, in-,* tendon.] *Bot.* The outer layer of the wall of a spore or pollen grain.

**ex·ist** (ĭg-zĭst′) *vi.* **-ist·ed, -ist·ing, -ists.** [Lat. *existere* : *ex-,* out + *sistere,* to stand.] **1.** To have material or spiritual being or actuality. **2. a.** To have life : LIVE. **b.** To continue to be. **3.** To live at a substan[]dard socioeconomic level <the inner-city slums whose inhabitants only *exist*> **4.** To occur under certain circumstances.

**ex·is·tence** (ĭg-zĭs′təns) *n.* **1.** The fact or state of existing : BEING. **2.** The fact or state of continued being : LIFE. **3. a.** All that exists. **b.** Something that exists : ENTITY. **4.** Manner of existing <day-by-day *existence*> **5.** Specific presence <the *existence* of extraterrestrial life>

**ex·is·tent** (ĭg-zĭs′tənt) *adj.* **1.** Having life or being : EXISTING. **2.** Oc[]curring or present at the moment : CURRENT. *—n.* One that exists.

**ex·is·ten·tial** (ĕg′zĭ-stĕn′shəl, ĕk′sĭ-) *adj.* **1.** Of or relating to exis[]tence. **2.** Based on experience : EMPIRICAL. **3.** Relating to existential[]ism. **—ex′is·ten′tial·ly** *adv.*

**ex·is·ten·tial·ism** (ĕg′zĭ-stĕn′shə-lĭz′əm, ĕk′sĭ-) *n.* A philosophy that emphasizes the uniqueness and isolation of the individual expe[]rience in a hostile or indifferent universe, regards human existence as unexplainable, and stresses freedom of choice and responsibility for the consequences of one's acts. **—ex′is·ten′tial·ist** *n.*

**ex·it[1]** (ĕg′zĭt, ĕk′sĭt) [Lat., 3rd. person sing. of *exire,* to go out : *ex-,* out + *ire,* to go.] —Used as a stage direction for a specified actor to leave the stage.

**ex·it[2]** (ĕg′zĭt, ĕk′sĭt) *n.* [Lat. *exire, exit-* : *ex-,* out + *ire,* to go.] **1. a.** The act of going away or out. **b.** Death. **2.** A passage or way out

3. Departure of a performer from the stage. —*vi.* **-it·ed, -it·ing, -its.** To make one's exit.

**ex li·bris** (ĕks lī′brĭs, lē′-) *n., pl.* **ex libris.** [Lat., from the books.] A bookplate.

**ex ni·hi·lo** (ĕks nē′ə-lō′, nī′-, nĭ′-) *adj. & adv.* [Lat.] Out of nothing.

**exo-** *pref.* [Gk. *exō*, outside < *ex*, out of.] Outside : external <*exo*skeleton>

**ex·o·bi·ol·o·gy** (ĕk′sō-bī-ŏl′ə-jē) *n.* **1.** A branch of biology that studies or searches for extraterrestrial living organisms. **2.** A branch of biology dealing with the effects of extraterrestrial space on living organisms. **—ex′o·bi′o·log′i·cal** (-ə-lŏj′ĭ-kəl) *adj.* **—ex′o·bi·ol′o·gist** *n.*

**ex·o·carp** (ĕk′sō-kärp′) *n. Bot.* The outermost layer of the pericarp of fruit.

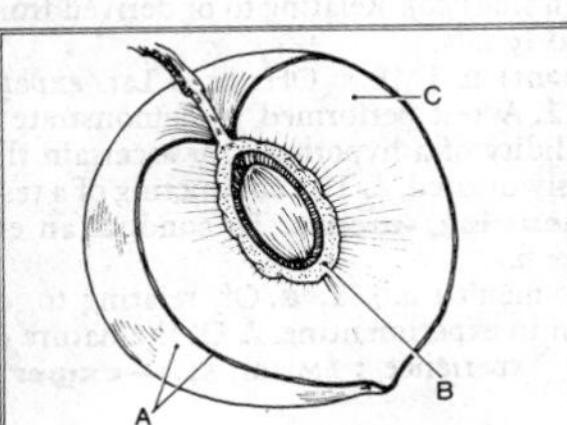


**exocarp**
*A longitudinal section of a peach showing: A. exocarp, B. endocarp, C. mesocarp*

**ex·o·crine** (ĕk′sə-krĭn, -krēn, -krīn′) *adj.* [EXO- + Gk. *krinein*, to separate.] **1.** Having or secreting through a duct. —Used of a gland. **2.** Of or relating to the secretion of a gland having a duct.

**ex·o·cy·clic** (ĕk′sō-sī′klĭk, -sĭk′lĭk) *adj.* Occurring outside a chemical ring structure.

**ex·o·don·tia** (ĕk′sə-dŏn′shə, -shē-ə) *n.* Dentistry specializing in extraction of teeth. **—ex′o·don′tist** *n.*

**ex·o·dus** (ĕk′sə-dəs) *n.* [LLat. < Gk. *exodos* : *ex-*, out + *hodos*, way.] **1.** A large-scale departure of people. **2. Exodus.** The Israelites' departure from Egypt. **3. Exodus.** —See table at BIBLE.

**ex·o·en·zyme** (ĕk′sō-ĕn′zīm′) *n.* An enzyme, as a digestive enzyme, that functions outside a cell.

**ex·o·er·gic** (ĕk′sō-ûr′jĭk) *adj.* [EXO- + Gk. *ergon*, work + -IC.] Exothermic.

**ex of·fi·ci·o** (ĕks′ ə-fĭsh′ē-ō′) *adj. & adv.* [Lat.] By virtue of office or position.

**ex·og·a·my** (ĕk-sŏg′ə-mē) *n.* **1.** The custom of marrying outside one's social unit, as a tribe or family. **2.** *Biol.* Reproduction by fusion of gametes of different ancestries. **—ex′o·gam′ic** (ĕk′sə-găm′ĭk), **ex·og′a·mous** (ĕk-sŏg′ə-məs) *adj.*

**ex·og·e·nous** (ĕk-sŏj′ə-nəs) *adj.* [Fr. *exogène* : exo-, exo- + *gène*, born < Gk. *-genēs*.] **1.** *Biol.* Derived or developed from external causes. **2.** *Bot.* Characterized by the addition of layers of woody tissue. **3.** Having a cause external to the body. —Used of diseases. **—ex·og′e·nous·ly** *adv.*

**ex·on·er·ate** (ĭg-zŏn′ə-rāt′) *vt.* **-at·ed, -at·ing, -ates.** [ME *exoneraten* < Lat. *exonerare, exonerat-*, to free from a burden : *ex-*, off + *onus*, burden.] **1.** To free from blame. **2.** To release from a responsibility or obligation. **—ex·on′er·a′tion** *n.* **—ex·on′er·a′tive** *adj.*

**ex·o·nu·cle·ase** (ĕk′sō-nōō′klē-ās′, -āz′, -nyōō′-) *n.* One of a group of enzymes that remove nucleotides sequentially from the end of a DNA chain.

**ex·oph·thal·mic goiter** (ĕk′səf-thăl′mĭc) *n.* A disease caused by excessive production of thyroid hormone and characterized by an enlarged thyroid gland, protrusion of the eyeballs, tachycardia, and nervous excitability.

**ex·oph·thal·mos** *also* **ex·oph·thal·mus** (ĕk′səf-thăl′məs) *n.* [Gk., with prominent eyes : *ex-*, out + *ophthalmos*, eye.] Abnormal protrusion of the eyeball. **—ex′oph·thal′mic** *adj.*

**ex·or·bi·tance** (ĭg-zôr′bĭ-təns) *n.* **1.** Excessiveness, as of price or demand : EXTRAVAGANCE. **2.** An exorbitant act, esp. one deviating from what is proper or right.

**ex·or·bi·tant** (ĭg-zôr′bĭ-tənt) *adj.* [ME, aberrant < OFr. < Med. Lat. *exorbitans*, pr.part. *exorbitare*, to deviate : *ex*, out of + *orbita*, path < *orbis*, ring.] **1.** Exceeding appropriate limits or bounds : IMMODERATE. **2.** *Law.* Exceeding established limits of right or propriety. **—ex·or′bi·tant·ly** *adv.*

**ex·or·cise** (ĕk′sôr-sīz′, -sər-) *vt.* **-cised, -cis·ing, -cis·es.** [ME *exorcisen* < Med. Lat. *exorcizare* < Gk. *exorkizien* : *ex-*, out + *horkos*, oath.] **1.** To expel (an evil spirit) by or as if by adjuration. **2.** To free from evil spirits. **—ex′or·cis′er** *n.*

**ex·or·cism** (ĕk′sôr-sĭz′əm, -sər-) *n.* **1.** An act of exorcising. **2.** A formula used in exorcising. **—ex′or·cist** *n.*

**ex·or·di·um** (ĭg-zôr′dē-əm, ĭk-sôr′-) *n., pl.* **-di·ums** *or* **-di·a** (-dē-ə) [Lat. < *exordiri*, to begin : *ex-* (intensive) + *ordiri*, to begin.] A beginning or introduction, as of a speech. **—ex·or′di·al** *adj.*

**ex·o·skel·e·ton** (ĕk′sō-skĕl′ĭ-tən) *n.* An external protective or supporting structure of many invertebrates, as insects and crustaceans. **—ex′o·skel′e·tal** *adj.*

**ex·os·mo·sis** (ĕk′sŏz-mō′sĭs, -sŏs-) *n.* [EX(O)- + OSMOSIS.] The flow of a fluid through a permeable membrane into a less dense fluid. **—ex′os·mot′ic** (-mŏt′ĭk) *adj.*

**ex·o·sphere** (ĕk′sō-sfîr′) *n.* The outermost portion of the atmosphere, estimated to begin 300 to 600 miles or approx. 480 to 960 kilometers above the earth. **—ex′o·spher′ic** *adj.*

**ex·o·spore** (ĕk′sō-spôr′, -spōr′) *n. Bot.* The outermost layer of a spore in some algae and fungi.

**ex·o·spor·i·um** (ĕk′sō-spôr′ē-əm, -spōr′-) *n., pl.* **-i·a** (ē-ə) [NLat. : EXO- + *spora*, spore.] *Bot.* Exine.

**ex·os·to·sis** (ĕk′sŏ-stō′sĭs) *n., pl.* **-ses** (-sēz′) [Gk. *exostōsis* : *ex-*, out of + *osteon*, bone.] A bony tumor on the surface of a bone.

**ex·o·ter·ic** (ĕk′sə-tĕr′ĭk) *adj.* [Lat. *exotericus*, external < Gk. *exōterikos* < *exōterō*, comp. of *exō*, outside < *ex*, out.] **1.** Not confined to a select few. **2.** Comprehensible to the public : POPULAR. **3.** Relating to the outside : EXTERNAL. **—ex′o·ter′i·cal·ly** *adv.*

**ex·o·ther·mic** (ĕk′sō-thûr′mĭk) *also* **ex·o·ther·mal** (-məl) *adj.* Releasing heat. **—ex′o·ther′mi·cal·ly** *adv.*

**ex·ot·ic** (ĭg-zŏt′ĭk) *adj.* [Lat. *exoticus* < Gk. *exōtikos* < *exō*, outside < *ex*, out.] **1.** From another part of the world : FOREIGN. **2.** Strikingly and charmingly different. **3.** Of or relating to striptease. —*n.* **1.** One that is exotic. **2.** A striptease performer. **—ex·ot′i·cal·ly** *adv.* **—ex·ot′ic·ness** *n.*

**ex·ot·i·cism** (ĭg-zŏt′ĭ-sĭz′əm) *n.* The quality or condition of being exotic.

**ex·o·tox·in** (ĕk′sō-tŏk′sĭn) *n.* A toxin excreted by a microorganism into a surrounding medium and recoverable from a culture without destruction of the producing agent.

**ex·pand** (ĭk-spănd′) *v.* **-pand·ed, -pand·ing, -pands.** [ME *expanden*, to spread out < Lat. *expandere* : *ex-*, out + *pandere*, to spread.] —*vt.* **1.** To increase the volume, size, or scope of. **2.** To express at length or in detail. **3.** *Math.* To write (a quantity) as a sum of terms, a continued product, or another extended form. —*vi.* **1.** To open up or out. **2.** To become greater in volume, size, or scope. **3.** To speak or write at length or in detail. **4.** To feel outgoing or generous. **—ex·pand′a·ble** *adj.* **—ex·pand′er** *n.*

**expanding universe theory** *n.* **1.** The interpretation of the shifts of the lines in the spectra of galaxies as resulting from a Doppler effect, with the experimental results that all galaxies are retreating from each other at speeds proportional to the distance separating them and that the universe is expanding. **2.** The cosmological theory in which violent eruption from a point source leads to formation of elementary particles, subsequent formation of hydrogen and helium, and dispersion of the galaxies that develop from this matter.

**ex·pan·dor** (ĭk-spăn′dər) *n.* A transducer for a given range of input voltages that produces a larger range of output voltages.

**ex·panse** (ĭk-spăns′) *n.* [Lat. *expansum* < *expandere*, to spread out. —see EXPAND.] A wide, open extent <the *expanse* of sky>

**ex·pan·si·ble** (ĭk-spăn′sə-bəl) *adj.* Capable of expanding or of being expanded. **—ex·pan′si·bil′i·ty** *n.*

**ex·pan·sile** (ĭk-spăn′səl, -sīl′) *adj.* Of, relating to, or capable of expanding.

**ex·pan·sion** (ĭk-spăn′shən) *n.* **1. a.** The act or process of expanding. **b.** The state of being expanded. **2. a.** An expanded part. **b.** A product of expanding. **3.** Degree or extent of expansion. **4.** *Math.* **a.** A quantity written in an extended form, as a sum of terms or a continued product. **b.** The process of obtaining this form.

**ex·pan·sion·ar·y** (ĭk-spăn′shə-nĕr′ē) *adj.* Tending or directed toward expansion.

**expansion bolt** *n.* A bolt with an attachment that expands when the bolt is driven into a surface.

**ex·pan·sion·ism** (ĭk-spăn′shə-nĭz′əm) *n.* A national practice or policy of territorial or economic expansion. **—ex·pan′sion·ist** *n.*

**ex·pan·sive** (ĭk-spăn′sĭv) *adj.* **1.** Capable of expanding or tending to expand. **2.** Broad : comprehensive. **3.** Inclined to be open and generous : OUTGOING. **4.** Characterized by euphoria and delusions of self-importance. **5.** Grand in scale. **—ex·pan′sive·ly** *adv.* **—ex·pan′sive·ness** *n.* **—ex′pan·siv′i·ty** (ĕk′spăn-sĭv′ĭ-tē) *n.*

**ex par·te** (ĕks pär′tē) *adj. & adv.* [Lat., in part.] **1.** *Law.* From or on one side only. **2.** One-sided : partisan.

**ex·pa·ti·ate** (ĭk-spā′shē-āt′) *vi.* **-at·ed, -at·ing, -ates.** [Lat. *expatiari, expatiat-* : *ex-*, out + *spatiari*, to spread < *spatium*, space.] **1.** To speak or write at length <*expatiated* on the probability of an immediate economic recovery> **2.** *Archaic.* To wander freely. **—ex·pa′ti·a′tion** *n.*

**ex·pa·tri·ate** (ĕk-spā′trē-āt′) *v.* **-at·ed, -at·ing, -ates.** [Med. Lat. *expatriare, expatriat-* : Lat. *ex-*, out of + Lat. *patria*, native land < *pater*, father.] —*vt.* **1.** To send into exile. **2.** To remove (oneself) from residence in one's native land. —*vi.* **1.** To give up residence in one's native land. **2.** To renounce one's allegiance to one's native land. —*n.* (-ĭt, -āt′). **1.** One who lives in a foreign country. **2.** One who has renounced his or her native land. —*adj.* Residing in a foreign country. **—ex·pa′tri·a′tion** *n.*

**ex·pect** (ĭk-spĕkt′) *vt.* **-pect·ed, -pect·ing, -pects.** [Lat. *expectare* : *ex-*, out + *spectare*, to look at, freq. of *specere*, to see.] **1.** To look forward to the probable occurrence or appearance of <*expected* a raise> **2.** To consider likely or certain <*expected* the rain to stop> **3.** To consider reasonable or due <*expected* civility from sales clerks> **4.** To consider obligatory : REQUIRE. **5.** *Informal.* To presume : suppose. **—ex·pect′a·ble** *adj.* **—ex·pect′a·bly** *adv.* **—ex·pect′ed·ly** *adv.* **—ex·pect′ed·ness** *n.*

**ex·pec·tan·cy** (ĭk-spĕk′tən-sē) *n., pl.* **-cies. 1.** EXPECTATION 1. **2. a.** EXPECTATION 3b. **b.** An expected amount based on statistical probability. **—ex·pec′tance** *n.*

**ex·pec·tant** (ĭk-spĕk′tənt) *adj.* **1.** Having or marked by expectation <*expectant* of victory> **2.** With child : PREGNANT. **—ex·pec′tant** *n.* **—ex·pec′tant·ly** *adv.*

**ex·pec·ta·tion** (ĕk′spĕk-tā′shən) *n.* **1.** The act of expecting or state of being expected. **2.** Eager anticipation. **3. a. expectations.** Future prospects <great *expectation*> **b.** Something expected. **4.** The expected value of a random variable, esp. the mean. **5.** EXPECTANCY 2b.

**ex·pec·ta·tive** (ĭk-spĕk′tə-tĭv) *adj.* Of, relating to, or marked by expectation.

**ex·pect·ing** (ĭk-spĕk′tĭng) *adj.* Pregnant.

**ex·pec·to·rant** (ĭk-spĕk′tər-ənt) *adj.* Promoting or facilitating expulsion of mucous from the respiratory tract. —*n.* An expectorant medication.

**ex·pec·to·rate** (ĭk-spĕk′tə-rāt′) *v.* **-rat·ed, -rat·ing, -rates.** [Lat. *expectorare, expectorat-*, to drive from the breast : *ex-*, out of + *pectus*, breast.] —*vt.* **1.** To eject from the mouth : SPIT. **2.** To cough up and eject by spitting. —*vi.* **1.** To spit. **2.** To clear the chest and lungs by coughing up and ejecting matter. **—ex·pec′to·ra′tion** *n.*

**ex·pe·di·ence** (ĭk-spē′dē-əns) *n.* Expediency.

**ex·pe·di·en·cy** (ĭk-spē′dē-ən-sē) *n., pl.* **-cies. 1.** Appropriateness to the purpose at hand. **2.** Adherence to self-serving means. **3.** An expedient. **4.** *Obs.* Speed.

**ex·pe·di·ent** (ĭk-spē′dē-ənt) *adj.* [ME < Lat. *expediens*, pr.part. of *expedire*, to make ready. —see EXPEDITE.] **1.** Appropriate to a given purpose. **2. a.** Serving to promote one's interest. **b.** Based on or marked by a concern for policy rather than principle. **3.** *Obs.* Speedy : expeditious. —*n.* **1.** A means to an end. **2.** A contrivance for meeting an urgent need. **—ex·pe′di·ent·ly** *adv.*

**ex·pe·di·en·tial** (ĭk-spē′dē-ĕn′shəl) *adj.* Of or relating to what is expedient. **—ex·pe′di·en′tial·ly** *adv.*

**ex·pe·dite** (ĕk′spĭ-dīt′) *vt.* **-dit·ed, -dit·ing, -dites.** [Lat. *expedire, expedit-*, to free from entanglement : *ex-*, out + *pes*, foot.] **1.** To speed up the progress of : FACILITATE. **2.** To perform efficiently and quickly. **3.** To issue officially : DISPATCH. **—ex′pe·dit′er, ex′pe·di′tor** *n.*

☆ ***syns:*** EXPEDITE, ACCELERATE, HASTEN, HURRY, QUICKEN, STEP UP *v. core meaning* : to speed up the progress of <*expedite* a delivery><*expedite* tax reform measures through the legislature>

**ex·pe·di·tion** (ĕk′spĭ-dĭsh′ən) *n.* [ME *expedicioun*, military campaign < OFr. *expedition* < Lat. *expeditio* < *expedire*, to extricate. —see EXPEDITE.] **1. a.** A journey undertaken for a definite purpose. **b.** The group undertaking such a journey. **2.** Speed in performance : PROMPTNESS.

**ex·pe·di·tion·ar·y** (ĕk′spĭ-dĭsh′ə-nĕr′ē) *adj.* Relating to or constituting an esp. military expedition.

**ex·pe·di·tious** (ĕk′spə-dĭsh′əs) *adj.* Acting or carried out with speed and efficiency. **—ex′pe·di′tious·ly** *adv.* **—ex′pe·di′tious·ness** *n.*

**ex·pel** (ĭk-spĕl′) *vt.* **-pelled, -pel·ling, -pels.** [ME *expellen* < Lat. *expellere* : *ex-*, out + *pellere*, to drive.] **1.** To force or drive out. **2.** To discharge from or as if from a receptacle. **3.** To dismiss, as from a school or society. **—ex·pel′la·ble** *adj.* **—ex·pel′ler** *n.*

**ex·pel·lant** *also* **ex·pel·lent** (ĭk-spĕl′ənt) *adj.* Expelling or tending to expel. **—ex·pel′lant** *n.*

**ex·pel·lee** (ĕk′spĕl-lē′) *n.* One who is expelled.

**ex·pel·lent** (ĭk-spĕl′ənt) *adj. var. of* EXPELLANT.

**ex·pend** (ĭk-spĕnd′) *vt.* **-pend·ed, -pend·ing, -pends.** [ME *expenden* < Lat. *expendere* : *ex*, out + *pendere*, to pay.] **1.** To lay out : SPEND. **2.** To use up : CONSUME.

**ex·pend·a·ble** (ĭk-spĕn′də-bəl) *adj.* **1.** Subject to use or consumption. **2.** Considered unworthy of retention or maintenance. **—ex·pend′a·ble** *n.*

**ex·pen·di·ture** (ĭk-spĕn′də-chər) *n.* **1.** The act or process of expending. **2. a.** An amount expended. **b.** An expense.

**ex·pense** (ĭk-spĕns′) *n.* [ME < AN *expense* < Lat. *expensa* < fem. p.part. of Lat. *expendere*, to expend.] **1. a.** Something paid out to attain a goal or accomplish a purpose. **b.** Something given up for the sake of something else : SACRIFICE. **2. expenses. a.** Charges incurred by an employee in the performance of assigned duties. **b.** *Informal.* Money allotted for payment of such charges. **3.** Something requiring the expenditure of money. **4.** *Archaic.* An expenditure. —*vt.* **-pensed, -pens·ing, -pens·es. 1.** To charge with expenses. **2.** To write off as an expense.

**expense account** *n.* An account of expenses for repayment to an employee.

**ex·pen·sive** (ĭk-spĕn′sĭv) *adj.* Bringing a large price : COSTLY. **—ex·pen′sive·ly** *adv.* **—ex·pen′sive·ness** *n.*

☆ ***syns:*** EXPENSIVE, COSTLY, DEAR, HIGH, PRICEY *adj. core meaning* : bringing a large price <*expensive* jewels> ***ant:*** cheap

**ex·pe·ri·ence** (ĭk-spîr′ē-əns) *n.* [ME < OFr. or < Lat. *experientia* < *experiens*, pr.part. of *experiri*, to try.] **1.** Apprehension or perception of an object, thought, emotion, or event through the senses or mind. **2. a.** Active participation in events or activities, leading to accumulation of knowledge or skill. **b.** The knowledge or skill so derived. **3. a.** An event or series of events participated in. **b.** The totality of such events in the past of an individual or group. —*vt.* **-enced, -enc·ing, -enc·es.** To participate in personally : UNDERGO <*experienced* a sense of elation>

**experience table** *n.* A table compiled from life-insurance statistics to indicate longevity.

**ex·pe·ri·en·tial** (ĭk-spîr′ē-ĕn′shəl) *adj.* Relating to or derived from experience. **—ex·pe′ri·en′tial·ly** *adv.*

**ex·per·i·ment** (ĭk-spĕr′ə-mənt) *n.* [ME < OFr. or < Lat. *experimentum* < *experiri*, to try.] **1.** A test performed to demonstrate a known truth, examine the validity of a hypothesis, or ascertain the efficacy of something previously untried. **2.** The conducting of a test. —*vi.* (-mĕnt′) **-ment·ed, -ment·ing, -ments.** To conduct an experiment. **—ex·per′i·ment′er** *n.*

**ex·per·i·men·tal** (ĭk-spĕr′ə-mĕn′tl) *adj.* **1. a.** Of, relating to, or based on experiment. **b.** Given to experimenting. **2.** Of the nature of an experiment. **3.** Proven by experience : EMPIRICAL. **—ex·per′i·men′tal·ly** *adv.*

**ex·per·i·men·tal·ism** (ĭk-spĕr′ə-mĕn′tl-ĭz′əm) *n.* Use of experimental methods in determining the validity of an idea. **—ex·per′i·men′tal·ist** *n.*

**ex·per·i·men·ta·tion** (ĭk-spĕr′ə-mĕn-tā′shən) *n.* The act, process, or practice of experimenting.

**experiment station** *n.* An establishment in which scientific experiments are conducted in a specific field, as agriculture, and practical applications are developed.

**ex·pert** (ĕk′spûrt′) *n.* [ME < OFr., experienced < Lat. *expertus*, p.part. of *experiri*, to try.] **1.** A person with a high degree of skill in or knowledge of a specific subject. **2. a.** The highest grade that can be achieved in marksmanship. **b.** One who has achieved this grade. —*adj.* (ĕk′spûrt, ĭk-spûrt′). Having or displaying great skill, dexterity, or knowledge as the result of experience. **—ex′pert′ly** *adv.* **—ex′pert′ness** *n.*

**ex·per·tise** (ĕk′spûr-tēz′) *n.* [Fr. < OFr. < *expert*, experienced. —see EXPERT.] **1.** Expert opinion or advice. **2.** Specialized knowledge or skill : MASTERY.

**ex·pi·a·ble** (ĕk′spē-ə-bəl) *adj.* Capable of being expiated.

**ex·pi·ate** (ĕk′spē-āt′) *v.* **-at·ed, -at·ing, -ates.** [Lat. *expiare, expiat-* : *ex-* (intensive) + *piare*, to atone < *pius*, devout.] —*vt.* To make atonement for. —*vi.* To make expiation. **—ex′pi·a′tor** *n.*

**ex·pi·a·tion** (ĕk′spē-ā′shən) *n.* **1.** The act of expiating : ATONEMENT. **2.** Means of atonement. **—ex′pi·a·to′ry** (-ə-tôr′ē, -tōr′ē) *adj.*

**ex·pi·ra·tion** (ĕk′spə-rā′shən) *n.* **1.** The act of coming to a close : TERMINATION. **2.** The act of breathing out. **3.** *Obs.* Death.

**ex·pi·ra·to·ry** (ĭk-spī′rə-tôr′ē, -tōr′ē) *adj.* Of, relating to, or involving the expiration of air from the lungs.

**ex·pire** (ĭk-spīr′) *v.* **-pired, -pir·ing, -pires.** [ME *expiren* < Lat. *exspirare* : *ex-*, out + *spirare*, to breathe.] —*vi.* **1.** To come to an end : TERMINATE <My subscription has *expired*.> **2.** To die. **3.** To breathe out : EXHALE. —*vt.* **1.** To breathe out from or as if from the lungs. **2.** *Archaic.* To give off.

**ex·pi·ry** (ĭk-spīr′ē) *n., pl.* **-ries. 1.** An expiration, esp. of a contract or agreement. **2.** Death.

**ex·plain** (ĭk-splān′) *v.* **-plained, -plain·ing, -plains.** [ME *explanen* < Lat. *explanare* : *ex-* (intensive) + *planus*, clear.] —*vt.* **1.** To make understandable. **2.** To define : expound <We *explained* our plan.> **3.** To offer reasons for or a cause of : JUSTIFY <*explain* an absence> —*vi.* To provide an explanation. **—explain away. 1.** To dismiss by or as if by explaining. **2.** To minimize by explanation. **—ex·plain′a·ble** *adj.*

**ex·pla·na·tion** (ĕk′splə-nā′shən) *n.* **1.** The act or process of explaining. **2.** Something that explains. **3.** Mutual clarification of misunderstandings : RECONCILIATION.

**ex·plan·a·tive** (ĭk-splăn′ə-tĭv) *adj.* Explanatory. **—ex·plan′a·tive·ly** *adv.*

**ex·plan·a·to·ry** (ĭk-splăn′ə-tôr′ē, -tōr′ē) *adj.* Serving or intended to explain. **—ex·plan′a·to′ri·ly** *adv.*

**ex·plant** (ĕk-splănt′) *vt.* **-plant·ed, -plant·ing, -plants.** To take (living tissue) from the natural site of growth and place in a medium or culture. —*n.* Material explanted. **—ex′plan·ta′tion** *n.*

**ex·ple·tive** (ĕk′splĭ-tĭv) *n.* [< LLat. *expletivus*, serving to fill out < Lat. *expletus* < *explēre*, to fill out : *ex-*, out + *plēre*, to fill.] **1.** An often profane or obscene exclamation. **2. a.** An added word or phrase that does not contribute meaning but serves to fill out a sentence or metrical line. **b.** A word standing in place of and anticipating a fol-

ă **pat** ā **pay** âr **care** ä **father** ĕ **pet** ē **be** hw **which** ĭ **pit** ī **tie** îr **pier** ŏ **pot** ō **toe** ô **paw, for** oi **noise** ŏŏ **took**

lowing word or phrase; e.g., in the sentence *"There are many books on the table,"* the word *there* is an expletive. —*adj.* Added to fill out something, as a metrical line or sentence.

**ex·ple·to·ry** (ĕk'splĭ-tôr'ē) *adj.* Expletive.

**ex·pli·ca·ble** (ĕk'splĭ-kə-bəl) *adj.* Capable of being explained. **—ex'pli·ca·bly** *adv.*

**ex·pli·cate** (ĕks'plĭ-kāt') *vt.* **-cat·ed, -cat·ing, -cates.** [Lat. *explicare,* to unfold : *ex-,* out + *plicare,* to fold.] To make clear the meaning of : EXPLAIN. **—ex·pli·ca'tion** *n.* **—ex'pli·ca'tor** *n.*

**ex·pli·ca·tion de texte** (ĕk-splē-kä-syôɴ' də tĕkst') *n., pl.* **ex·pli·ca·tions de texte** (ĕk-splē-kä-syôɴ' də tĕkst') [Fr. : *explication,* explanation + *de,* of + *texte,* text.] A method of literary criticism involving intense analysis and exhaustive interpretation of each part of the work.

**ex·pli·ca·tive** (ĕk'splĭ-kə-tĭv) *adj.* Explanatory. **—ex'pli·ca·tive** *n.* **—ex'pli·ca·tive·ly** *adv.*

**ex·plic·it** (ĭk-splĭs'ĭt) *adj.* [Fr. *explicite* < Lat. *explicitus,* p.part. of *explicare,* to unfold. —see EXPLICATE.] **1. a.** Expressed without vagueness or ambiguity : SPECIFIC. **b.** Clearly formulated or defined. **2.** Forthright and unreserved in expression. **—ex·plic'it·ly** *adv.* **—ex·plic'it·ness** *n.*

**ex·plode** (ĭk-splōd') *v.* **-plod·ed, -plod·ing, -plodes.** [Lat. *explodere,* to drive out by clapping : *ex-,* out + *plaudere,* to clap.] —*vi.* **1.** To release mechanical, chemical, or nuclear energy in an explosion. **2.** To burst violently from internal pressure. **3.** To burst forth suddenly and often violently. **4.** To increase suddenly, sharply, and without control. —*vt.* **1.** To cause to explode or burst violently and noisily. **2.** To show to be unreliable or false <*explode* a theory> **3.** *Obs.* To drive off the stage by the unrestrained expression of dissatisfaction. **—ex·plod'er** *n.*

☆ ***syns:*** EXPLODE, BLAST, BLOW UP, BURST, DETONATE, GO OFF *v. core meaning* : to release energy violently and suddenly, esp. with a loud report <a bomb that *exploded* in midair>

**exploded view** *n.* An illustration or diagram of a construction that shows its parts separately but in positions that indicate their proper relationships to the whole.

**ex·ploit** (ĕk'sploit', ĭk-sploit') *n.* [ME < OFr. < Lat. *explicitum,* neuter p.part. of *explicare,* to explicate.] An act or deed, esp. a brilliant or heroic feat. —*vt.* (ĭk-sploit', ĕk'sploit') **-ploit·ed, -ploit·ing, -ploits. 1.** To utilize to the greatest possible advantage. **2.** To make use of unethically or selfishly <*exploiting* the employees> **—ex·ploit'a·ble** *adj.* **—ex·ploit'a·tive** *adj.* **—ex·ploit'er** *n.*

**ex·ploi·ta·tion** (ĕk'sploi-tā'shən) *n.* **1.** An act of exploiting. **2.** Utilization of another person for selfish purposes. **3.** A publicity or advertising program.

**ex·plore** (ĭk-splôr', -splōr') *v.* **-plored, -plor·ing, -plores.** [Lat. *explorare.*] —*vt.* **1.** To investigate systematically : EXAMINE <*explore* every suggestion given> **2.** To search into or range over for the purpose of discovery. **3.** *Med.* To examine for diagnostic purposes. —*vi.* To make a careful search or examination. **—ex'plo·ra'tion** (ĕk'splə-rā'shən) *n.* **—ex·plor'a·to'ry** (ĭk-splôr'ə-tôr'ē, -splōr'ə-tōr'ē) *adj.*

**ex·plor·er** (ĭk-splôr'ər, -splōr'-) *n.* **1.** One who explores, esp. one who explores a geographic area. **2.** An implement used for exploring : PROBE. **3. Explorer.** Any of a series of early U.S. satellites, two of which were instrumental in the discovery of the Van Allen belts.

**ex·plo·sion** (ĭk-splō'zhən) *n.* [Lat. *explosio* < *explodere,* to drive out by clapping. —see EXPLODE.] **1. a.** An act or instance of exploding. **b.** The loud, sharp sound made by an explosion. **2.** A sudden, often vehement outburst, as of emotion. **3.** A sudden and great increase <the population *explosion*> **4.** Plosion.

**ex·plo·sive** (ĭk-splō'sĭv) *adj.* [< Lat. *explodere, explos-,* to drive out by clapping. —see EXPLODE.] **1.** Relating to or of the nature of an explosion. **2.** Tending to explode. —*n.* **1.** A substance, esp. a prepared chemical, that explodes or causes explosion. **2.** STOP 12. **—ex·plo'sive·ly** *adv.* **—ex·plo'sive·ness** *n.*

**ex·po·nent** (ĭk-spō'nənt, ĕk'spō'nənt) *n.* [Lat. *exponens, exponent-,* pr.part. of *exponere,* to put forward : *ex-,* out + *ponere,* to put.] **1.** One that expounds or interprets. **2.** One that speaks for, represents, or advocates. **3.** *Math.* A number or symbol, as 3 in $(x+y)^3$, placed to the right of and above another number, symbol, or expression and denoting the power to which the latter is to be raised. —*adj.* Expository : explanatory.

**ex·po·nen·tial** (ĕk'spə-nĕn'shəl) *adj.* **1.** *Math.* **a.** Containing, involving, or expressed as an exponent. **b.** Expressed in terms of a designated power of *e,* the base of natural logarithms. **2.** Of or relating to an exponent. **—ex'po·nen'tial·ly** *adv.*

**ex·po·nen·ti·a·tion** (ĕk'spə-nĕn'shē-ā'shən) *n. Math.* The act of raising a quantity to a power.

**ex·port** (ĭk-spôrt', -spōrt', ĕk'spôrt', -spōrt') *v.* **-port·ed, -port·ing, -ports.** [Lat. *exportare* : *ex-,* out + *portare,* to carry.] —*vt.* To send or transport (e.g., a commodity) abroad, esp. for sale or trade. —*vi.* To send abroad merchandise, esp. for trade or sale. —*n.* (ĕk'spôrt', -spōrt'). **1.** An act of exporting. **2.** A commodity exported. **—ex·port'a·bil'i·ty** *n.* **—ex·port'a·ble** *adj.* **—ex'por·ta'tion** (ĕk'spôr-tā'shən, -spōr-) *n.* **—ex·port'er** *n.*

**ex·pose** (ĭk-spōz') *vt.* **-posed, -pos·ing, -pos·es.** [ME *exposen* < OFr. *exposer* < Lat. *exponere.* —see EXPOUND.] **1. a.** To remove shelter or protection from. **b.** To lay open, as to something undesirable or injurious. **2.** To subject (e.g., a photographic film) to the action of light. **3.** To make visible <removed the paneling to *expose* the brickwork> **4. a.** To make known (e.g., a crime). **b.** To reveal the guilt or wrongdoing of <*expose* an embezzler> **5.** To abandon or put out without shelter or food. **—ex·pos'er** *n.*

**ex·po·sé** (ĕk'spō-zā') *n.* [Fr. < p.part. of *exposer,* to expose < OFr.] **1.** An exposure of something disreputable. **2.** An exposition of facts.

**ex·po·si·tion** (ĕk'spə-zĭsh'ən) *n.* [ME *exposicioun* < OFr. *exposition* < Lat. *expositio* < *exponere,* to expound.] **1.** A statement of meaning or intent. **2.** A definitive statement intended to give an explanation of difficult material. **3.** *Mus.* **a.** The first part of a composition in sonata form that introduces the themes. **b.** The opening section of a fugue. **4.** The part of a play that introduces the theme and chief characters. **5.** The act or an example of exposing. **6.** A public exhibition or show, as of artistic or agricultural developments. **—ex·pos'i·tive** (ĭk-spŏz'ĭ-tĭv), **ex·pos'i·to'ry** (-tôr'ē, -tōr'ē) *adj.* **—ex·pos'i·tor** *n.*

**ex post fac·to** (ĕks' pōst făk'tō) *adj.* [Med. Lat. *ex postfacto,* from what is done afterward.] Enacted, formulated, or operating retroactively, as a law.

**ex·pos·tu·late** (ĭk-spŏs'chə-lāt') *vi.* **-lat·ed, -lat·ing, -lates.** [Lat. *expostulare, expostulat-,* to demand strongly : *ex-* (intensive) + *postulare,* to demand.] To reason thoughtfully with someone in an effort to dissuade or correct : INVEIGH. **—ex·pos'tu·la'tion** *n.* **—ex·pos'tu·la'tor** *n.* **—ex·pos'tu·la·to'ry** (-lə-tôr'ē, -tōr'ē), **ex·pos'tu·la'tive** *adj.*

**ex·po·sure** (ĭk-spō'zhər) *n.* **1.** The act or an instance of exposing. **2.** The state of being exposed, esp. to the forces of nature. **3.** A position in relation to climatic conditions or points of the compass <a northern *exposure*> **4. a.** The act of exposing sensitized photographic film or plate. **b.** A film or plate so exposed. **c.** The amount of radiant energy needed to expose a photographic film. **5.** An act of abandoning without shelter or food.

**exposure meter** *n.* A photoelectric instrument that measures light intensity in a given area and, in photographic use, indicates proper exposure settings.

**ex·pound** (ĭk-spound') *v.* **-pound·ed, -pound·ing, -pounds.** [ME *expounden* < AN *espoundre* < Lat. *exponere* : *ex-,* out + *ponere,* to place.] —*vt.* **1.** To give a detailed statement of : SET FORTH. **2.** To explain by presenting in detail. —*vi.* To make a detailed statement. **—ex·pound'er** *n.*

**ex·press** (ĭk-sprĕs') *vt.* **-pressed, -press·ing, -press·es.** [ME *expressen* < OFr. *expresser* < Med. Lat. *expressare* : Lat. *ex-,* out + Lat. *pressare,* to press, freq. of *premere,* to press.] **1.** To make known in words. **2.** To communicate or manifest, as by a gesture : SHOW. **3. a.** To make known (e.g., one's opinion) <*expressed* their concerns intelligently> **b.** To communicate (e.g., one's feelings) esp. through artistic activity. **4.** To make a representation of : DEPICT. **5.** To represent by a symbol or sign : SYMBOLIZE. **6.** To press out, as juice from a fruit. **7.** To send by express. —*adj.* **1.** Firmly and explicitly stated <their *express* disapproval> **2.** Particular : specific <an *express* purpose> **3. a.** Sent out with or moving at high speed. **b.** Direct, rapid, and usu. nonstop <*express* parcel delivery> **c.** Of, relating to, or appropriate for rapid travel <an *express* lane> —*adv.* By express transport or delivery. —*n.* **1. a.** A special messenger. **b.** A message delivered by special courier. **2. a.** A rapid, efficient system for the delivery of goods and mail. **b.** Goods and mail conveyed by such a system. **3.** A rapid means of transportation. **—ex·press'er** *n.* **—ex·press'i·ble** *adj.*

**ex·press·age** (ĭk-sprĕs'ĭj) *n.* **1.** Conveyance of goods by express. **2.** The amount charged for such conveyance.

**ex·pres·sion** (ĭk-sprĕsh'ən) *n.* **1.** An act of expressing, conveying, or depicting in words, art, music, or movement : MANIFESTATION. **2.** Something that communicates or expresses. **3.** *Math.* A designation of any symbolic form, as an equation. **4.** The way in which one expresses oneself, esp. in speaking, depicting, or performing. **5.** A specific phrase or word <a regional *expression*> **6.** Outward manifestation of an emotional state. **7.** A facial aspect or look conveying a special feeling <an *expression* of disapproval> **8.** The act of pressing or squeezing out.

**ex·pres·sion·ism** (ĭk-sprĕsh'ə-nĭz'əm) *n.* An artistic movement during the latter part of the 19th and early part of the 20th cent. that emphasized subjective expression of the artist's inner experiences. **—ex·pres'sion·ist** *n.* **—ex·pres'sion·is'tic** *adj.* **—ex·pres'sion·is'ti·cal·ly** *adv.*

**ex·pres·sion·less** (ĭk-sprĕsh'ən-lĭs) *adj.* Devoid of expression.

**ex·pres·sive** (ĭk-sprĕs'ĭv) *adj.* **1.** Of, relating to, or marked by expression. **2.** Serving to indicate or express <a tone of voice *expressive* of anger> **3.** Full of expression : SIGNIFICANT <an *expressive* smile> **—ex·pres'sive·ly** *adv.* **—ex·pres'sive·ness** *n.*

**ex·pres·siv·i·ty** (ĕk'sprĕs-ĭv'ĭ-tē) *n.* **1.** The quality of being expressive. **2.** *Genetics.* The degree to which a particular gene can affect the phenotype of an organism.

**ex·press·ly** (ĭk-sprĕs′lē) *adv.* **1.** In an express manner : EXPLICITLY <*expressly* invited them> **2.** Especially : particularly <designated *expressly* for handicapped people>

**ex·press·man** (ĭk-sprĕs′mən) *n.* An express agency employee.

**ex·press·way** (ĭk-sprĕs′wā′) *n.* A multilane highway designed for high-speed travel.

**ex·pro·pri·ate** (ĕk-sprō′prē-āt′) *vt.* **-at·ed, -at·ing, -ates.** [Med. Lat. *expropriare, expropriat-* : Lat. *ex-*, away + Lat. *proprius,* one's own.] **1.** To deprive of possession. **2.** To transfer (another's property) to oneself. **—ex·pro′pri·a′tion** *n.* **—ex·pro′pri·a′tor** *n.* **—ex·pro′pri·a·to′ry** (-ə-tôr′ē, -tōr′ē) *adj.*

**ex·pul·sion** (ĭk-spŭl′shən) *n.* The act of expelling or state of being expelled.

**ex·punc·tion** (ĭk-spŭngk′shən, -spŭng′shən) *n.* [< Lat. *expungere, expunct-,* to strike out. —see EXPUNGE.] The act of expunging or state of being expunged : DELETION.

**ex·punge** (ĭk-spŭnj′) *vt.* **-punged, -pung·ing, -pung·es.** [Lat. *expungere,* to strike out : *ex-,* out + *pungere,* to prick.] **1.** To erase : delete. **2.** To obliterate totally : ANNIHILATE. **—ex·pung′er** *n.*

**ex·pur·gate** (ĕk′spər-gāt′) *vt.* **-gat·ed, -gat·ing, -gates.** [Lat. *expurgare, expurgat-,* to purify : *ex-* (intensive) + *purgare,* to cleanse.] To remove objectionable, obscene, or erroneous material from. **—ex′pur·ga′tion** *n.* **—ex′pur·ga′tor** *n.* **—ex·pur′ga·to′ry** (-spûr′gə-tôr′ē, -tōr′ē), **ex·pur′ga·to′ri·al** (-tôr′ē-əl, -tōr′ē-əl) *adj.*

**ex·qui·site** (ĕk′skwĭ-zĭt, ĭk-skwĭz′ĭt) *adj.* [ME *exquisit* < Lat. *exquisitus,* choice < *exquirere,* to search out : *ex-,* out + *quaerere,* to seek.] **1.** Marked by intricately beautiful design or execution <an *exquisite* carving> **2.** Of such delicacy or beauty as to arouse delight <an *exquisite* display of fireworks> **3.** Acutely perceptive : DISCRIMINATING <an *exquisite* intuitive sense> **4.** Intense : keen <*exquisite* agony> **5.** *Obs.* Ingeniously devised. *—n.* One who is overly fastidious in dress, manners, or taste. **—ex′qui·site·ly** *adv.* **—ex′qui·site·ness** *n.*

**ex·san·gui·nate** (ĕks-săng′gwə-nāt′) *vt.* **-nat·ed, -nat·ing, -nates.** [< Lat. *exsanguinatus,* bloodless : *ex-,* without + *sanguis,* blood.] To drain of blood. **—ex·san′gui·na′tion** *n.*

**ex·scind** (ĭk-sĭnd′) *vt.* **-scind·ed, -scind·ing, -scinds.** [Lat. *exscindere* : *ex-,* out + *scindere,* to cut.] To cut out : EXCISE.

**ex·sert** (ĭk-sûrt′) *vt.* **-sert·ed, -sert·ing, -serts.** [Lat. *exserere, exsert-.*] To thrust forth. *—adj. also* **ex·sert·ed** (-sûr′tĭd). *Biol.* Thrust outward : PROTRUDING. **—ex·ser′tion** *n.*

**ex·sic·cate** (ĕk′sĭ-kāt′) *vi.* & *vt.* **-cat·ed, -cat·ing, -cates.** [Lat. *exsiccare, exsiccat-,* to dry out : *ex-,* out + *siccare,* to dry < *siccus,* dry.] To dry up or cause to dry up. **—ex′sic·ca′tion** *n.* **—ex′sic·ca′tive** *adj.* **—ex′sic·ca′tor** *n.*

**ex·stip·u·late** (ĕks-stĭp′yə-lĭt) *adj. Bot.* Having no stipules.

**ex·tant** (ĕk′stənt, ĕk-stănt′) *adj.* [Lat. *extans, extant-,* pr.part. of *extare,* to stand out : *ex-,* out + *stare,* to stand.] **1. a.** Still in existence. **b.** Not destroyed or lost <*extant* writings> **2.** *Archaic.* Standing out : PROJECTING.

**ex·tem·po·ral** (ĭk-stĕm′pər-əl) *adj. Archaic.* Extemporaneous.

**ex·tem·po·ra·ne·ous** (ĭk-stĕm′pə-rā′nē-əs) *adj.* [LLat. *extemporaneus* < Lat. *ex tempore,* of the time.] **1.** Done with little or no prior preparation or practice : IMPROMPTU <an *extemporaneous* performance> **2.** Prepared in advance but given without notes or text <an *extemporaneous* lecture> **3.** Skilled at or given to unrehearsed speech or performance. **4.** Provided, made, or adapted as an expedient : MAKESHIFT. **—ex·tem′po·ra′ne·ous·ly** *adv.* **—ex·tem′po·ra′ne·ous·ness** *n.*

**ex·tem·po·rar·y** (ĭk-stĕm′pə-rĕr′ē) *adj.* [< Lat. *ex tempore,* of the time.] Extemporaneous. **—ex·tem′po·rar′i·ly** (-râr′ə-lē) *adv.*

**ex·tem·po·re** (ĭk-stĕm′pə-rē) *adj.* [Lat. *ex tempore* : *ex,* of + *tempus,* time.] Extemporaneous. **—ex·tem′po·re** *adv.*

**ex·tem·po·rize** (ĭk-stĕm′pə-rīz′) *v.* **-rized, -riz·ing, -riz·es.** *—vt.* To perform (something) extemporaneously. *—vi.* To perform or utter extemporaneously. **—ex·tem′po·ri·za′tion** *n.* **—ex·tem′po·riz′er** *n.*

**ex·tend** (ĭk-stĕnd′) *v.* **-tend·ed, -tend·ing, -tends.** [ME *extenden* < Lat. *extendere* : *ex-,* out + *tendere,* to stretch.] *—vt.* **1.** To open or straighten out : UNBEND. **2.** To stretch or spread out to full length. **3. a.** To exert (oneself) completely. **b.** To cause to move at full gallop. **4. a.** To increase in bulk or quantity by adding a cheaper substance. **b.** To adulterate. **5. a.** To expand the area or scope of. **b.** To increase the influence of. **c.** To make more comprehensive or inclusive. **6. a.** To offer <*extend* one's congratulations> **b.** To make available : PROVIDE. **7.** To prolong the time of repayment of. **8.** *Law.* **a.** *Chiefly Brit.* To assess or evaluate : APPRAISE. **b.** To seize or make a levy on for the purpose of settling a debt. *—vi.* To stretch or reach <political clout that *extended* throughout the government> **—ex·tend′i·bil′i·ty** *n.* **—ex·tend′i·ble** *adj.*

**ex·tend·ed** (ĭk-stĕn′dĭd) *adj.* **1.** Stretched or spread out. **2.** Continued for a long period of time : PROTRACTED. **3.** Extensive in meaning, scope, or influence. **—ex·tend′ed·ly** *adv.*

**extended family** *n.* A family unit making up one household that consists of parents, children, and other close relatives.

**ex·tend·er** (ĕk-stĕn′dər) *n.* A substance added to another substance to modify, dilute, or adulterate.

**ex·ten·si·ble** (ĭk-stĕn′sə-bəl) *adj.* [< Lat. *extendere, extens-,* to hold out. —see EXTEND.] Capable of being protruded or extended **—ex·ten′si·bil′i·ty** *n.*

**ex·ten·sile** (ĭk-stĕn′sĭl) *adj.* Extensible.

**ex·ten·sion** (ĭk-stĕn′shən) *n.* [ME *extensioun* < Lat. *extensio* < *extensus,* p.part. of *extendere,* to extend.] **1. a.** The act of extending or state of being extended. **b.** Something extended. **2.** The amount, degree, or range to which something extends or is extensible. **3. a.** The extending of a limb. **b.** The position assumed by an extended limb. **4.** *Med.* Application of traction to a fractured or dislocated limb to restore its normal position. **5.** An addition to a main structure. **6.** An additional telephone connected to the main line. **7. a.** An allowance of extra time, esp. for repayment of a debt. **b.** The period of this extra time. **8.** The property of an object by which it occupies space. **9.** *Logic.* The class of objects designated by a specific concept or term : DENOTATION. **10.** *Math.* A set that includes a given and similar set as a subset. **—ex·ten′sion·al** *adj.*

**ex·ten·si·ty** (ĭk-stĕn′sĭ-tē) *n.* **1.** The quality of having extension or being extensive. **2.** The attribute of sensation enabling one to perceive space or size.

**ex·ten·sive** (ĭk-stĕn′sĭv) *adj.* **1.** Great in extent, range, or amount. **2.** Of or relating to cultivation of vast land areas with minimal labor or expense. **3.** *Physics.* Having a value that is equal to the sum of the values for the subdivisions of a thermodynamic system. —Used of volume. **—ex·ten′sive·ly** *adv.* **—ex·ten′sive·ness** *n.*

**ex·ten·som·e·ter** (ĕk′stĕn-sŏm′ĭ-tər) *n.* [EXTENS(ION) + -METER.] An instrument for measuring minute deformations in a test specimen of a material.

**ex·ten·sor** (ĭk-stĕn′sər) *n.* [NLat. < Lat. *extendere,* to stretch out. —see EXTEND.] A muscle that extends or stretches a limb.

**ex·tent** (ĭk-stĕnt′) *n.* [ME *extente,* assessment < AN < Lat. *extenta,* fem. p.part. of *extendere,* to extend.] **1. a.** The magnitude, range, or distance over which a thing extends. **b.** The degree to which such a thing extends. **2.** An extensive area or space <an *extent* of grassland> **3.** *Archaic.* An assessment or valuation, as of land, esp. for taxation. **4.** *Law.* A writ allowing a creditor to seize a debtor's property temporarily.

**ex·ten·u·ate** (ĭk-stĕn′yōō-āt′) *vt.* **-at·ed, -at·ing, -ates.** [Lat. *extenuare, extenuat-,* to diminish : *ex-,* out + *tenuare,* to make thin < *tenuis,* thin.] **1.** To lessen or try to lessen the seriousness or magnitude of by offering excuses. **2.** *Obs.* To belittle. **3.** To make emaciated. **4.** To reduce the strength of. **—ex·ten′u·a′tive** *adj.* & *n.* **—ex·ten′u·a′tor** *n.* **—ex·ten′u·a·to′ry** (-ə-tôr′ē, -tōr′ē) *adj.*

**ex·ten·u·a·tion** (ĭk-stĕn′yōō-ā′shən) *n.* **1.** The act of extenuating or state of being extenuated. **2.** An excuse.

**ex·te·ri·or** (ĭk-stîr′ē-ər) *adj.* [Lat., comp. of *exterus,* outward.] **1.** External : outer. **2.** Originating or acting from the outside. **3.** Appropriate for outdoor use <an *exterior* varnish> *—n.* **1.** A surface or part that is outside. **2.** An outward or external appearance <a hostile *exterior*> **—ex·te′ri·or·ly** *adv.*

**exterior angle** *n.* **1.** The angle between any side of a polygon and an extended adjacent side. **2.** Any of the four angles that do not include a region of the space between two lines intersected by a transversal.

**ex·te·ri·or·i·ty** (ĭk-stîr′ē-ôr′ĭ-tē, -ŏr′-) *n.* Externality.

**ex·te·ri·or·ize** (ĭk-stîr′ē-ə-rīz′) *vt.* **-ized, -iz·ing, -iz·es.** To externalize.

**ex·ter·mi·nate** (ĭk-stûr′mə-nāt′) *vt.* **-nat·ed, -nat·ing, -nates.** [Lat. *exterminare, exterminat-,* to drive out : *ex-,* out of + *terminus,* boundary.] To destroy completely. **—ex·ter′mi·na′tion** *n.* **—ex·ter′mi·na′tive, ex·ter′mi·na·to′ry** (-nə-tôr′ē, -tōr′ē) *adj.*

**ex·ter·mi·na·tor** (ĭk-stûr′mə-nā′tər) *n.* One that exterminates, esp. one whose occupation is the extermination of vermin.

**ex·tern** *or* **ex·terne** (ĕk′stûrn′) *n.* [OFr. *externe* < Lat. *externus,* external.] A person associated with but not officially residing in an institution, as a nonresident physician on a hospital staff.

**ex·ter·nal** (ĭk-stûr′nəl) *adj.* [ME < Lat. *externus,* outward.] **1.** Relating to, existing on, or connected with the outside or an outer part : EXTERIOR. **2.** Appropriate for exterior application. **3.** Existing independently of the mind <*external* reality> **4.** Acting or coming from the outside <*external* influences> **5.** Of or relating to outward appearance : SUPERFICIAL. **6.** Of or relating to foreign affairs or countries. *—n.* **1.** An exterior surface or part. **2. externals. a.** External circumstances. **b.** Outward appearances. **—ex·ter′nal·ly** *adv.*

**ex·ter·nal-com·bus·tion engine** (ĭk-stûr′nəl-kəm-bŭs′chən) *n.* An engine, as a steam engine, in which the fuel is burned outside the engine cylinder.

**external ear** *n.* The part of the ear including the auricle and the external acoustic meatus.

**ex·ter·nal·ism** (ĭk-stûr′nə-lĭz′əm) *n.* Inordinate concern with externals. **—ex·ter′nal·ist** *n.*

**ex·ter·nal·i·ty** (ĕk′stər-năl′ĭ-tē) *n., pl.* **-ties. 1.** The quality or state of being external or externalized. **2.** Something external.

**ex·ter·nal·ize** (ĭk-stûr′nə-līz′) *vt.* **-ized, -iz·ing, -iz·es. 1. a.** To make external. **b.** To manifest externally. **2.** To attribute to outside causes : RATIONALIZE. **—ex·ter′nal·i·za′tion** *n.*

**ex·ter·o·cep·tor** (ĕk′stə-rō-sĕp′tər) *n.* [Lat. *exter,* outside + RECEPTOR.] A sense organ receiving and responding to external stimuli. **—ex′ter·o·cep′tive** *adj.*

**ex·ter·ri·to·ri·al** (ĕks′tĕr-ĭ-tôr′ē-əl, -tōr′-) *adj.* Extraterritorial. **—ex′ter·ri·to′ri·al′i·ty** (-ăl′ĭ-tē) *n.* **—ex′ter·ri·to′ri·al·ly** *adv.*

**ex·tinct** (ĭk-stĭngkt′) *adj.* [ME < Lat. *extinguere,* to extinguish.] **1.** Inactive <an *extinct* volcano> **2.** No longer existing <an *extinct* species> **3.** Lacking a claimant <an *extinct* title> **4.** No longer in use <an *extinct* practice> **—ex·tinc′tion** (-stĭngk′shən) *n.*

**ex·tinc·tive** (ĭk-stĭngk′tĭv) *adj.* Tending to extinguish or make extinct.

**ex·tin·guish** (ĭk-stĭng′gwĭsh) *vt.* **-guished, -guish·ing, -guish·es.** [Lat. *extinguere* : *ex-,* out + *stinguere,* to quench.] **1.** To put out (e.g., a fire) : QUENCH. **2.** To put an end to : DESTROY. **3.** To eclipse : obscure. **4.** *Law.* **a.** To settle or discharge (a debt). **b.** To nullify. **—ex·tin′guish·a·ble** *adj.* **—ex·tin′guish·ment** *n.*

☆ ***syns:*** EXTINGUISH, DOUSE, PUT OUT, QUENCH, SNUFF (out) *v.* *core meaning* : to cause to stop burning <*extinguish* a fire> ***ant:*** ignite, light

**ex·tin·guish·er** (ĭk-stĭng′gwĭ-shər) *n.* One that extinguishes, esp. a fire extinguisher.

**ex·tir·pate** (ĕk′stər-pāt′) *vt.* **-pat·ed, -pat·ing, -pates.** [Lat. *extirpare, extirpat-,* to root out : *ex-,* out + *stirps,* root.] **1.** To pull up by the roots. **2.** To destroy : exterminate. **3.** To remove surgically. **—ex′tir·pa′tion** *n.* **—ex′tir·pa′tive** *adj.* **—ex′tir·pa′tor** *n.*

**ex·tol** *also* **ex·toll** (ĭk-stōl′) *vt.* **-tolled, -tol·ling, -tols** *also* **-tolled, -toll·ing, -tolls.** [ME *extollen* < Lat. *extollere,* to lift up : *ex-,* up from + *tollere,* to lift.] To praise lavishly. **—ex·tol′ler** *n.* **—ex·tol′ment** *n.*

**ex·tort** (ĭk-stôrt′) *vt.* **-tort·ed, -tort·ing, -torts.** [Lat. *extorquēre, extort-,* to wrench out : *ex-,* out + *torquēre,* to twist.] To obtain by coercive means, as threats or intimidation. **—ex·tort′er** *n.* **—ex·tor′tive** *adj.*

**ex·tor·tion** (ĭk-stôr′shən) *n.* **1.** The act or an instance of extorting. **2.** Illegal use of one's official position or powers to obtain property, funds, or patronage. **3.** An exorbitant charge. **4.** Something extorted. **—ex·tor′tion·ar′y** (-shə-nĕr′ē) *adj.* **—ex·tor′tion·ist, ex·tor′tion·er** *n.*

**ex·tor·tion·ate** (ĭk-stôr′shə-nĭt) *adj.* **1.** Marked by extortion. **2.** Exorbitant <*extortionate* interest> **—ex·tor′tion·ate·ly** *adv.*

**ex·tra** (ĕk′strə) *adj.* [Prob. short for EXTRAORDINARY.] **1.** Being beyond the expected, usual, or necessary. **2.** Being better than ordinary : SUPERIOR <*extra* quality> **3.** Liable to an additional charge. —*n.* **1.** Something more than the expected, usual, or necessary. **2.** Something for which an additional charge is made, as an accessory on a motor vehicle. **3.** A special edition of a newspaper. **4.** An additional or alternate worker. **5.** A performer hired to play a minor film part, as in a crowd scene. **6.** Something of exceptional quality. —*adv.* Exceptionally : unusually <*extra* firm>

**extra-** *or* **extro-** *pref.* [Lat. *extra,* outside < *exterus,* outward.] Outside : beyond <*extra*ordinary>

**ex·tra-base hit** (ĕk′strə-bās′) *n. Baseball.* A double, a triple, or a home run.

**ex·tra·cel·lu·lar** (ĕk′strə-sĕl′yə-lər) *adj.* Occurring or found outside a cell. **—ex′tra·cel′lu·lar·ly** *adv.*

**ex·tra·code** (ĕk′strə-kōd′) *n. Computer Sci.* A sequence of machine code instructions used to simulate hardware functions.

**ex·tra·cra·ni·al** (ĕk′strə-krā′nē-əl) *adj.* Situated or occurring outside the cranium.

**ex·tract** (ĭk-străkt′) *vt.* **-tract·ed, -tract·ing, -tracts.** [ME *extracten* < Lat. *extrahere,* to draw out : *ex-,* out + *trahere,* to draw.] **1.** To draw out forcibly <*extract* a tooth> **2. a.** To obtain despite resistance <*extract* a confession> **b.** To draw forth by great effort. **3.** To obtain from a substance by chemical or mechanical action, as by pressure or distillation. **4.** To remove for separate consideration or publication : EXCERPT. **5.** *Math.* To calculate or determine (the root of a number). —*n.* (ĕk′străkt′). **1.** A passage from a literary work : EXCERPT. **2.** A concentrated preparation of the basic elements of a substance : CONCENTRATE <beef *extract*> **—ex·tract′a·ble, ex·tract′i·ble** *adj.* **—ex·tract′or** *n.*

**ex·trac·tion** (ĭk-străk′shən) *n.* **1.** The act of extracting or state of being extracted. **2.** Something obtained by extracting. **3.** Origin : ancestry <of German *extraction*>

**ex·trac·tive** (ĭk-străk′tĭv) *adj.* **1.** Of, involving, or used in extraction. **2.** Capable of being extracted. —*n.* Something that can be extracted. **—ex·trac′tive·ly** *adv.*

**ex·tra·cur·ric·u·lar** (ĕk′strə-kə-rĭk′yə-lər) *adj.* **1.** Being outside the regular curriculum of a school or college. **2.** Being outside the usual duties of a job or profession.

**ex·tra·dit·a·ble** (ĕk′strə-dī′tə-bəl) *adj.* **1.** Subject to extradition. **2.** Making liable to extradition <an *extraditable* offense>

**ex·tra·dite** (ĕk′strə-dīt′) *vt.* **-dit·ed, -dit·ing, -dites.** [Back-formation < EXTRADITION.] **1.** To surrender to extradition. **2.** To obtain as a result of the extradition of.

**ex·tra·di·tion** (ĕk′strə-dĭsh′ən) *n.* [Fr. : Lat. *ex-,* out + *traditio,* surrender.] Legal surrender of an alleged criminal to the jurisdiction of another state, country, or government for trial.

**ex·tra·dos** (ĕk′strə-dŏs′, -dō′, -dōs′) *n., pl.* **ex·tra·dos** (-dōz′) or **ex·tra·dos·es.** [Fr. : Lat. *extra,* outside + Fr. *dos,* back.] The upper or exterior curve of an arch.

**ex·tra·ga·lac·tic** (ĕk′strə-gə-lăk′tĭk) *adj.* Situated or originating beyond the galaxy.

**ex·tra·he·pat·ic** (ĕk′strə-hĭ-păt′ĭk) *adj.* Situated or occurring outside the liver.

**ex·tra·ju·di·cial** (ĕk′strə-jōō-dĭsh′əl) *adj.* **1.** Outside the authority of a court. **2.** Outside the usual judicial proceedings. **—ex′tra·ju·di′cial·ly** *adv.*

**ex·tra·le·gal** (ĕk′strə-lē′gəl *adj.* Not governed or sanctioned by law. **—ex′tra·le′gal·ly** *adv.*

**ex·tra·mar·i·tal** (ĕk′strə-măr′ĭ-tl) *adj.* Adulterous.

**ex·tra·mun·dane** (ĕk′strə-mŭn-dān′, -mŭn′dān′) *adj.* Occurring or existing outside the physical world or universe.

**ex·tra·mu·ral** (ĕk′strə-myŏŏr′əl) *adj.* Occurring or located outside the walls or boundaries, as of a community.

**ex·tra·ne·ous** (ĭk-strā′nē-əs) *adj.* [Lat. *extraneus* < *extra,* outside.] **1.** Coming from the outside. **2.** Not vital or essential. **3.** Irrelevant. **—ex·tra′ne·ous·ly** *adv.* **—ex·tra′ne·ous·ness** *n.*

**ex·tra·nu·cle·ar** (ĕk′strə-nōō′klē-ər, -nyōō′-) *adj.* Located or occurring outside a nucleus.

**ex·traor·di·nar·y** (ĭk-strôr′dn-ĕr′ē, ĕk′strə-ôr′-) *adj.* **1.** Beyond what is common or usual <*extraordinary* integrity> **2.** Very exceptional : REMARKABLE <*extraordinary* talent> **3.** Used for a special service or occasion. **—ex·traor′di·nar′i·ly** (-âr′ə-lē) *adv.*

**ex·trap·o·late** (ĭk-străp′ə-lāt′) *v.* **-lat·ed, -lat·ing, -lates.** [EXTRA- + (INTER)POLATE.] —*vt.* **1.** *Math.* To estimate (a value or values of a function) for values of the argument not used in the process of estimation : infer (a value or values) from known values. **2.** To infer or estimate by projecting or extending known information. —*vi.* To engage in the process of extrapolating. **—ex·trap′o·la′tion** *n.* **—ex·trap′o·la′tive** *adj.* **—ex·trap′o·la′tor** *n.*

**ex·tra·sen·so·ry** (ĕk′strə-sĕn′sə-rē) *adj.* Being outside the normal range of the senses.

**extrasensory perception** *n.* Perception by means other than normal sense perceptions, esp. by supernatural means.

**ex·tra·sys·to·le** (ĕk′strə-sĭs′tə-lē) *n. Med.* A premature cardiac contraction.

**ex·tra·ter·res·tri·al** (ĕk′strə-tə-rĕs′trē-əl) *adj.* Originating, situated, or occurring outside the earth or its atmosphere. **—ex′tra·ter·res′tri·al** *n.*

**ex·tra·ter·ri·to·ri·al** (ĕk′strə-tĕr′ĭ-tôr′ē-əl, -tōr′-) *adj.* **1.** Situated outside territorial limits. **2.** Of or relating to persons exempt from the legal jurisdiction of the country in which they reside. **—ex′tra·ter′ri·to′ri·al·ly** *adv.*

**ex·tra·ter·ri·to·ri·al·i·ty** (ĕk′strə-tĕr′ĭ-tôr′ē-ăl′ĭ-tē, -tōr′-) *n.* Exemption from local jurisdiction, as that granted to foreign diplomats.

**ex·tra·u·ter·ine** (ĕk′strə-yōō′tər-ĭn, -tə-rīn′) *adj.* Situated or occurring outside the uterus <*extrauterine* pregnancy>

**ex·trav·a·gance** (ĭk-străv′ə-gəns) *n.* **1.** The quality of being extravagant. **2.** Immoderate display or expense. **3.** Something extravagant <A mink coat is an *extravagance* I can ill afford.>

**ex·trav·a·gan·cy** (ĭk-străv′ə-gən-sē) *n., pl.* **-cies.** Extravagance.

**ex·trav·a·gant** (ĭk-străv′ə-gənt) *adj.* [ME *extravagaunt,* unusual < Med. Lat. *extravagans,* pr.part. of *extravagari,* to wander : Lat. *extra,* outside + Lat. *vagari,* to wander.] **1.** Given to imprudent or lavish spending. **2.** Exceeding reasonable limits : UNRESTRAINED <*extravagant* requests> **3.** Extremely abundant : PROFUSE <*extravagant* vegetation> **4.** Unreasonably high : EXORBITANT <*extravagant* prices> **5.** *Archaic.* Straying beyond limits or bounds : WANDERING. **—ex·trav′a·gant·ly** *adv.* **—ex·trav′a·gant·ness** *n.*

**ex·trav·a·gan·za** (ĭk-străv′ə-găn′zə) *n.* [Ital. *estravaganza* < *estravagante,* extravagant < Med. Lat. *extravagans.*] **1.** A literary or musical work marked by diversity and freedom of form, often with burlesque elements. **2.** A spectacular event or production.

**ex·trav·a·gate** (ĭk-străv′ə-gāt′) *vi.* **-gat·ed, -gat·ing, -gates.** [Med. Lat. *extravagari, extravagat-,* to wander. —see EXTRAVAGANT.] **1.** To roam or wonder : STRAY. **2.** To exceed reasonable limits.

**ex·trav·a·sate** (ĭk-străv′ə-sāt′) *v.* **-sat·ed, -sat·ing, -sates.** —*vt.* **1.** *Pathol.* To force the flow of (blood or lymph) out into surrounding tissue. **2.** *Geol.* To cause (molten lava) to pour forth from a volcanic vent. —*vi.* **1.** *Pathol.* To exude into the surrounding tissues. **2.** *Geol.* To erupt. **—ex·trav′a·sa′tion** (-sā′shən) *n.*

**ex·tra·vas·cu·lar** (ĕk′strə-văs′kyə-lər) *adj.* **1.** Occurring or situated outside a blood vessel or the vascular system. **2.** Lacking vessels : NONVASCULAR.

**ex·tra·ve·hic·u·lar activity** (ĕk′strə-vē-hĭk′yə-lər) *n.* Activity performed by an astronaut outside a spacecraft.

**ex·tra·ver·sion** (ĕk′strə-vûr′zhən) *n. var. of* EXTROVERSION.

**ex·tra·vert** (ĕk′strə-vûrt′) *n. var. of* EXTROVERT.

**ex·treme** (ĭk-strēm′) *adj.* [ME < OFr. < Lat. *extremus.*] **1.** Outermost or farthest from a center <the *extreme* edge of the forest>

**2.** Final : last. **3.** Being in the highest degree <*extreme* happiness><*extreme* poverty> **4.** Extending far beyond the norm <an *extreme* liberal> **5.** Of the greatest severity : DRASTIC <*extreme* measures to curb inflation> —*n.* **1.** The greatest or utmost degree or point. **2.** Either of the two things situated at opposite ends of a range <*extremes* of rain and drought> **3.** An extreme condition. **4.** A drastic or immoderate expedient <*Extremes* are necessary only occasionally.> **5.** *Math.* The last or first term of a series or ratio. **6.** *Logic.* The minor or major term of a syllogism. **—ex·treme'ly** *adv.* **—ex·treme'ness** *n.*

**extremely high frequency** *n.* A radio-frequency band with a range of 30,000 to 300,000 megahertz.

**extreme unction** *n. Rom. Cath. Ch.* The sacrament in which a priest anoints and prays for one in danger of death.

**ex·trem·ist** (ĭk-strē'mĭst) *n.* One who advocates or resorts to extreme measures, esp. in politics. **—ex·trem'ism** *n.*

**ex·trem·i·ty** (ĭk-strĕm'ĭ-tē) *n., pl.* **-ties. 1.** The outermost or farthest point or portion. **2.** The highest or utmost degree <the *extremity* of exertions> **3.** Extreme danger, necessity, or distress. **4.** A moment at which death or destruction is imminent. **5.** EXTREME 4. **6.** A bodily limb or appendage. **7.** A hand or a foot.

**ex·tri·cate** (ĕk'strĭ-kāt') *vt.* **-cat·ed, -cat·ing, -cates.** [Lat. *extricare, extricat-* : *ex-*, out + *tricae*, hindrances, perplexities.] **1.** To free from an entanglement or difficulty : DISENGAGE. **2.** *Archaic.* To distinguish from something related. **—ex'tri·ca·ble** (-kə-bəl) *adj.* **—ex'tri·ca'tion** *n.*

**ex·trin·sic** (ĭk-strĭn'sĭk, -zĭk) *adj.* [LLat. *extrinsecus*, from outside.] **1.** Not forming an essential part : EXTRANEOUS. **2.** Not inherent : ACCESSORY. **3.** Originating from the outside : EXTERNAL. **—ex·trin'si·cal·ly** *adv.*

**extrinsic factor** *n.* Vitamin $B_{12}$.

**extro-** *pref. var. of* EXTRA-.

**ex·trorse** (ĕk'strôrs') *adj.* [LLat. *extrorsus*, turned outward : *extra*, outside + *versus*, p.part of *vertere*, to turn.] *Bot.* Facing outward : turned away from the axis. —Used esp. of anthers.

**ex·tro·ver·sion** *also* **ex·tra·ver·sion** (ĕk'strə-vûr'zhən) *n.* **1.** Selfless interest in one's environment or in others. **2.** A turning inside out, as of an organ or part. **—ex'tro·ver'sive** *adj.* **—ex'tro·ver'sive·ly** *adv.*

**ex·tro·vert** *also* **ex·tra·vert** (ĕk'strə-vûrt') *n.* [EXTRO- + Lat. *vertere*, to turn.] An individual whose interests are selfless and lie in others or the environment. **—ex'tro·vert'ed** *adj.*

**ex·trude** (ĭk-strōōd') *v.* **-trud·ed, -trud·ing, -trudes.** [Lat. *extrudere*, to thrust out : *ex-*, out + *trudere*, to thrust.] —*vt.* **1.** To push or thrust out. **2.** To shape (e.g., metal or plastic) by forcing through a die. —*vi.* To project or protrude.

**ex·tru·sion** (ĭk-strōō'zhən) *n.* [Med. Lat. *extrusio* < Lat. *extrudere*, to thrust out. —see EXTRUDE.] **1.** The act or process of extruding. **2.** Something produced by extruding.

**ex·tru·sive** (ĭk-strōō'sĭv, -zĭv) *adj.* **1.** Tending to extrude. **2.** *Geol.* Derived from magma. —Used of rock.

**ex·u·ber·ance** (ĭg-zōō'bər-əns) *n.* **1.** The quality or state of being exuberant. **2.** An exuberant act or expression.

**ex·u·ber·ant** (ĭg-zōō'bər-ənt) *adj.* [ME, overabundant < Lat. *exuberans*, pr.part. of *exuberare*, to exuberate.] **1.** Full of unrestrained enthusiasm or joy. **2.** Excessive : overflowing. **3.** Growing or producing abundantly : LUXURIANT. **—ex·u'ber·ant·ly** *adv.*

**ex·u·ber·ate** (ĭg-zōō'bə-rāt') *vi.* **-at·ed, -at·ing, -ates.** [Lat. *exuberare* : *ex-* (intensive) + *uberare*, to be fruitful (< *uber*, fertile).] **1.** To be exuberant. **2.** *Archaic.* To overflow or abound.

**ex·u·date** (ĕks'yōō-dāt') *n.* [Lat. *exudatum*, neuter p.part. of *exudare*, to exude.] An exuded substance.

**ex·u·da·tion** (ĕks'yōō-dā'shən) *n.* **1.** The act or an instance of exuding. **2.** An exudate. **—ex'u'da·tive** *adj.*

**ex·ude** (ĭg-zōōd', ĭk-sōōd') *v.* **-ud·ed, -ud·ing, -udes.** [Lat. *exudare* : *ex-*, out + *sudare*, to sweat.] —*vi.* To ooze forth. —*vt.* **1.** To discharge or emit gradually. **2.** To exhibit conspicuously <a dancer *exuding* eroticism>

**ex·ult** (ĭg-zŭlt') *vi.* **-ult·ed, -ult·ing, -ults.** [Lat. *exultare*, freq. of *exsilire*, to spring out : *ex-*, out + *salire*, to leap.] **1.** To rejoice greatly and jubilantly. **2.** *Obs.* To leap upward, esp. for joy. **—ex·ul'tance, ex·ul'tan·cy** *n.* **—ex·ult'ing·ly** *adv.*

**ex·ul·tant** (ĭg-zŭl'tənt) *adj.* Marked by great joy and jubilation. **—ex·ul'tant·ly** *adv.*

**ex·ul·ta·tion** (ĕk'səl-tā'shən, ĕg'zəl-) *n.* The act of exulting or state of being exultant.

**ex·urb** (ĕk'sûrb') *n.* [EX- + (SUB)URB).] A region beyond the suburbs of a city, inhabited chiefly by the well-to-do. **—ex·ur'ban** *adj.*

**ex·ur·ban·ite** (ĕk-sûr'bə-nīt, ĕg-zûr'-) *n.* A resident of an exurb.

**ex·ur·bi·a** (ĕk-sûr'bē-ə, ĕg-zûr'-) *n.* An exurban area.

**ex·u·vi·ae** (ĭg-zōō'vē-ē') *pl.n.* [Lat. < *exuere*, to take off.] The cast-off skins or coverings of various animals, esp. the larvae and nymphs of insects. **—ex·u'vi·al** (-vē-əl) *adj.*

**ex·u·vi·ate** (ĭg-zōō'vē-āt') *v.* **-at·ed, -at·ing, -ates.** [< EXUVIAE.] —*vt.* To shed or cast off (a covering). —*vi.* To shed or cast off exuviae. **—ex·u'vi·a'tion** *n.*

**-ey** *suff. var. of* -Y¹.

**ey·as** (ī'əs) *n.* [Alteration of ME *a nias*, an eyas < OFr. *niais* < Lat. *nidus*, nest.] A nestling hawk or falcon, esp. one to be trained for falconry.

**eye** (ī) *n.* [ME < OE *ēage*.] **1.** An organ of vision or of light sensitivity. **2. a.** The vertebrate organ of vision, one of a pair of hollow structures located in fixed bony sockets of the skull, each with a lens capable of focusing incident light on an internal photosensitive retina. **b.** The external, visible portion of this organ together with its associated structures, as the eyelids, eyelashes, and eyebrows. **c.** The pigmented iris of this organ. **3.** The faculty of seeing : VISION. **4.** The ability to make aesthetic or intellectual judgments <a good *eye* for character> **5. a.** A look : gaze. **b.** A point of view. **6.** Something suggestive of an eye, esp.: **a.** An opening in a needle. **b.** A circular marking on a peacock feather. **c.** A loop, as in a hook. **7.** *Bot.* **a.** A bud on a twig or tuber. **b.** The often differently colored center of the corolla of some flowers. **8.** *Meteorol.* The circular area of relative calm at the center of a cyclone. **9.** Something construed as a center or focal point. **10.** *Informal.* A detective. —*vt.* **eyed, eye·ing** *or* **ey·ing, eyes. 1.** To concentrate the eyes on. **2.** To watch closely. **3.** To supply with an eye. **—an eye for an eye.** Punishment requiring that the offender suffer what the victim has suffered. **—catch (someone's) eye.** *Informal.* To attract another's attention. **—eye to eye.** In agreement. **—give (someone) the eye.** *Informal.* To look at admiringly or invitingly. **—in a pig's eye.** *Slang.* Under no condition : NEVER. **—my eye.** *Slang.* In no way : not at all. **—with an eye to.** With a view to.

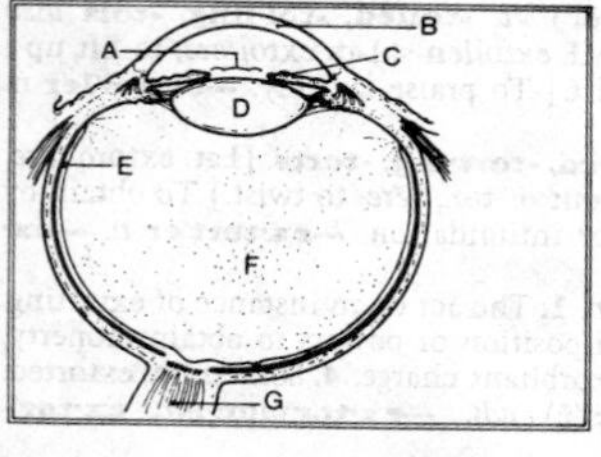


**eye**
*A. pupil, B. cornea, C. iris, D. lens, E. retina, F. vitreous humor, G. optic nerve*

**eye·ball** (ī'bôl') *n.* **1.** The ball-shaped portion of the eye enclosed by the socket and eyelids. **2.** The eye itself. —*vt.* **-balled, -ball·ing, -balls.** *Informal.* To look over intently : SCRUTINIZE.

**eye·ball-to-eye·ball** (ī'bôl'tə-ī'bôl') *adj. & adv. Informal.* Face to face, as in a confrontational posture.

**eye bank** *n.* A place at which corneas taken from human cadavers immediately after death are stored and preserved for subsequent transplantation to individuals with corneal defects.

**eye bath** *n.* An eyecup.

**eye·bolt** (ī'bōlt') *n.* A bolt having a looped head designed to receive a hook or rope.

**eye·bright** (ī'brīt') *n.* A plant of the genus *Euphrasia*, esp. *E. officinalis* of the Old World, with small white and purple flowers.

**eye·brow** (ī'brou') *n.* **1.** The bony ridge extending over the eye. **2.** The arch of short hairs covering this ridge.

**eyebrow pencil** *n.* A cosmetic pencil for the eyebrows.

**eye contact** *n.* Direct visual contact with another's eyes.

**eye·cup** (ī'kŭp') *n.* A small cup with a rim contoured to fit the orbit of the eye, used to apply a liquid medicine or wash to the eye.

**eyed** (īd) *adj.* Having eyes of a specified number or kind <green-*eyed*>

**eye dialect** *n.* The use of misspellings, as *sez* for *says*, to represent dialectal or nonstandard speech.

**eye·drop·per** (ī'drŏp'ər) *n.* A dropper for administering liquid eye medicines.

**eye·ful** (ī'fōōl') *n.* **1.** A complete or satisfying view. **2.** One pleasing to the sight, esp. a beautiful person.

**eye·glass** (ī'glăs') *n.* **1. a. eyeglasses.** GLASS 4b. **b.** A monocle. **2.** An eyepiece. **3.** An eyecup.

**eye·hole** (ī'hōl') *n.* **1.** The orbit of the eye. **2.** A peephole.

**eye·hook** (ī'hōōk') *n.* A hook attached to a ring at the end of a rope or chain.

**eye·lash** (ī'lăsh') *n.* **1.** One of a row of short hairs fringing the edge of the eyelid. **2.** A row of the hairs fringing the eyelid.

**eye·let** (ī'lĭt) *n.* [ME *oilet* < OFr. *oillet*, dim. of *oil*, eye < Lat. *oculus*.] **1. a.** A small hole or perforation, usu. rimmed with metal, cord, fabric, or leather, used for fastening with a cord or hook. **b.** A metal ring designed to reinforce such a hole : GROMMET. **2.** A small hole edged with embroidered stitches as part of a design. **3.** A peephole. **4.** A small eye.

**eye·lid** (ī'lĭd') *n.* Either of two folds of skin and muscle that can be closed over an eye.

**eye·lin·er** (ī'lī'nər) *n.* Make-up used to outline the eyes.

**eye opener** *n.* **1.** A shocking or surprising revelation. **2.** A drink of alcohol, taken to stimulate, esp. on awakening.
**eye·piece** (ī′pēs′) *n.* The lens or lens group closest to the eye in an optical instrument : OCULAR.
**eye rhyme** *n.* A false rhyme having words, as *lint* and *pint,* with similar spellings but different sounds.
**eye shadow** *n.* A colored cosmetic applied esp. to the eyelids.
**eye·shot** (ī′shŏt′) *n.* Range of vision : SIGHT.
**eye·sight** (ī′sīt′) *n.* **1.** The faculty of sight : VISION. **2.** Range of vision.
**eyes-on·ly** (īz′ōn′lē) *adj.* Of or relating to private or top-secret information <"a secret *eyes-only* memo" —Jeff Kamen>
**eye·sore** (ī′sôr′, ī′sōr′) *n.* Something ugly.
**eye·spot** (ī′spŏt′) *n.* **1.** A light-sensitive pigmented area in certain organisms, as algae or protozoans. **2.** A rounded eyelike marking, as on the tail of a peacock.
**eye·stalk** (ī′stôk′) *n.* A movable stalklike structure bearing at its tip one of the eyes of a crab or similar crustacean.
**eye·strain** (ī′strān′) *n.* Fatigue of the ciliary muscle or of the extrinsic muscles of the eyeball caused by refractive errors or imbalance of the ocular muscles and marked by pain in the eyes, lacrimation, headache, nausea, dizziness, or other reflex symptoms.
**eye·tooth** (ī′to͞oth′) *n.* A canine of the upper jaw.
**eye·wash** (ī′wŏsh′, ī′wôsh′) *n.* **1.** A medicated solution applied as a wash for the eyes. **2.** *Slang.* Meaningless or misleading language.
**eye·wink** (ī′wĭngk′) *n.* **1.** A wink of the eye. **2.** An instant. **3.** *Obs.* A glance.
**eye·wit·ness** (ī′wĭt′nəs) *n.* One who has personally seen someone or something and can bear witness to the fact.
**ey·ing** (ī′ĭng) *v. var. p.p. of* EYE.
**ey·ra** (âr′ə) *n.* [Am. Sp. (South America) *eirá,* a kind of fox < Guarani *eirara.*] A reddish-brown color phase of the jaguarondi.
**eyre** (âr) *n.* [ME < OFr. *eire* < Lat. *iter,* journey.] *Obs.* **1.** A circuit : itineration. **2.** A circuit court held by itinerant royal justices in medieval England.
**ey·rie** (âr′ē, îr′ē) *n. var. of* AERIE.
**ey·rir** (ā′rîr′) *n., pl.* **au·rar** (ou′rär′, œ′rär′) [Icel. < ON, money, prob. < Lat. *aurum,* gold.] —See table at CURRENCY.
**E·ze·ki·el** (ĭ-zē′kē-əl) *n.* [Heb. *Yĕḥezqēl.*] —See table at BIBLE.
**Ez·ra** (ĕz′rə) *n.* [Heb. *'Ezrā.*] —See table at BIBLE.

# Ff

**f** or **F** (ĕf) *n., pl.* **f's** or **F's.** **1.** The sixth letter of the English alphabet. **2.** A speech sound represented by the letter *f.* **3.** The sixth in a series. **4. F** *Mus.* **a.** The fourth tone in the scale of C major or the sixth tone in the relative minor scale. **b.** The key or a scale in which F is the tonic. **c.** A written or printed note representing this tone. **d.** A string, key, or pipe tuned to the pitch of this tone. **5. F** A failing grade.
**F** *symbol for* FLUORINE.
**fa** (fä) *n.* [ME < Med. Lat. —see GAMUT.] The fourth tone of the diatonic scale in solmization.
**Fa·bi·an** (fā′bē-ən) *adj.* **1. a.** Of or relating to the caution and avoidance of direct confrontation typical of the Roman general Quintus Fabius Maximus, who defeated Hannibal by the use of this strategy. **b.** Dilatory : cautious. **2.** Of, relating to, or being a member of the Fabian Society, a group committed to a gradual rather than revolutionary spread of socialism. **—Fa′bi·an** *n.* **—Fa′bi·an·ism** *n.* **—Fa′bi·an·ist** *n.*
**fa·ble** (fā′bəl) *n.* [ME < OFr. < Lat. *fabula* < *fari,* to speak.] **1.** A fictitious story making a moral or cautionary point and often using animal characters that speak and act like humans. **2.** A legend. **3.** A falsehood : lie. —*v.* **-bled, -bling, -bles.** —*vt.* To recount as if true. —*vi. Archaic.* To compose fables. **—fa′bler** *n.*
**fa·bled** (fā′bəld) *adj.* **1.** Renowned through fables : LEGENDARY. **2.** Fictitious.
**fab·li·au** (făb′lē-ō′) *n., pl.* **-li·aux** (-lē-ō′, -ōz′) [Fr. < OFr. *fabliaux,* pl. of *fablel,* dim. of *fable,* fable.] A medieval verse tale marked by comic, ribald treatment of worldly themes.
**fab·ric** (făb′rĭk) *n.* [ME *fabryke* < OFr. *fabrique* < Lat. *fabrica* < *faber,* workman, artificer.] **1.** A material structure of connected parts. **2.** A complex underlying structure <the *fabric* of our society> **3.** A style or method of construction. **4. a.** A cloth produced esp. by knitting, weaving, or felting fibers. **b.** The texture or quality of this kind of cloth.
**fab·ric·a·ble** (făb′rĭ-kə-bəl) *adj.* Capable of being molded <*fabricable* alloys> **—fab′ric·a·bil′i·ty** *n.*
**fab·ri·cant** (făb′rĭ-kənt) *n.* A manufacturer : builder.
**fab·ri·cate** (făb′rĭ-kāt′) *vt.* **-cat·ed, -cat·ing, -cates.** [ME *fabricaten* < Lat. *fabricari,* to make < *fabrica,* fabric.] **1.** To construct by combining or assembling : MAKE. **2.** To make up in order to deceive <*fabricate* excuses> **—fab′ri·ca′tion** *n.* **—fab′ri·ca′tor** *n.*
**Fa·bry's disease** (făb′rēz) *n.* [After Johannes *Fabry* (1860–1930).] A hereditary disease of fat metabolism marked by impaired renal functioning.
**fab·u·list** (făb′yə-lĭst) *n.* [OFr. *fabuliste* < Lat. *fabula,* fable.] **1.** A composer of fables. **2.** A liar.
**fab·u·lous** (făb′yə-ləs) *adj.* [ME < Lat. *fabulosus* < *fabula,* fable.] **1.** Of the nature of a fable or myth : LEGENDARY. **2.** Told of or celebrated in fables. **3.** Barely credible : ASTONISHING. **4.** *Informal.* Very pleasing or successful <had a *fabulous* time> **—fab′u·lous·ly** *adv.* **—fab′u·lous·ness** *n.*
☆ ***syns:*** FABULOUS, AMAZING, ASTONISHING, FANTASTIC, INCREDIBLE, MARVELOUS, MIRACULOUS, PHENOMENAL, PRODIGIOUS, STUPENDOUS, UNBELIEVABLE, WONDERFUL, WONDROUS *adj. core meaning* : so remarkable as to cause disbelief <the *fabulous* endurance of a long-distance runner>
**fa·çade** also **fa·cade** (fə-säd′) *n.* [Fr. < Ital. *facciata* < *faccia,* face < Lat. *facies.*] **1.** The face of a building. **2.** An artificial or deceptive outward appearance.
☆ ***syns:*** **1.** FAÇADE, FACE, FRONT, FRONTAL, FRONTISPIECE *n. core meaning* : the forward outer surface of a building <the famous *façade* of the Supreme Court building> **2.** FAÇADE, CLOAK, FACE, FRONT, GUISE, MASK, PRETENSE, PUT-ON, SHOW, VENEER, WINDOW-DRESSING *n. core meaning* : a deceptive outward appearance <a *façade* of respectability>
**face** (fās) *n.* [ME < OFr. < Lat. *facies.*] **1.** The surface of the front of the head from ear to ear and from the top of the forehead to the base of the chin. **2.** The expression of the features of the countenance. **3.** A contorted facial expression : GRIMACE. **4.** Outward appearance <the *face* of the neighborhood> **5.** Standing in the eyes of others : PRESTIGE <didn't want to lose *face*> **6.** Effrontery : impudence <had the *face* to dispute me> **7.** The most significant or prominent surface of an object, esp.: **a.** The surface presented to view : FRONT. **b.** A façade. **c.** The outer surface <on the *face* of the earth> **d.** A marked side <the *face* of a watch> **e.** The right side, as of fabric. **8.** A planar surface bounding a solid. **9.** Any of the surfaces of a rock or crystal. **10.** The end, as of a mine or tunnel, at which work is progressing. **11.** Appearance and geologic surface features of an area of land : TOPOGRAPHY. **12.** Typeface. —*v.* **faced, fac·ing, fac·es.** —*vt.* **1.** To occupy a position with the face toward. **2.** To front on. **3. a.** To confront with complete awareness <*faced* the issue> **b.** To overcome by confronting boldly or bravely. **4.** To be certain to encounter <An uneducated youth *faces* a tough life.> **5.** To cause (troops) to change direction by giving a command. **6.** To turn (a playing card) so that the face is up. **7.** To furnish with a surface of a different material. **8.** To line or trim the edge of, esp. with contrasting material. **9.** To treat the surface of so as to make even. —*vi.* **1.** To be turned or placed with the front toward a specified direction. **2.** To turn the face in a given direction. **—face down.** To overcome by or as if by a hard stare. **—face off.** To start play, as in hockey or lacrosse, by releasing the puck or ball between two opponents. **—face up to.** To recognize the existence of and confront <*faced up* to the problems> **—face′less** *adj.*
**face angle** *n.* The angle formed between two edges of a polyhedral angle.
**face card** *n.* A playing card, as a king, queen, or jack, that bears a stylized image of a person.
**face cloth** *n.* A washcloth.
**face-hard·en** (fās′här′dn) *vt.* **-ened, -en·ing, -ens.** To harden the surface of (a metal) : CASEHARDEN.
**face·lift·ing** (fās′lĭf′tĭng) also **face-lift** (-lĭft′) *n.* **1.** Plastic sur-

gery for tightening facial tissues and improving the appearance of facial skin. **2.** Modernization, as of a building.

**face-off** (fās′ôf′, -ŏf′) *n.* **1.** A way of starting play in ice hockey in which an official drops the puck between two opponents who contend for its control. **2.** A confrontation <a *face-off* between a mob and the police>

**face·plate** (fās′plāt′) *n.* **1.** A disk attached to the mandrel of a lathe to hold flat or irregularly shaped work. **2.** The glass front of a cathode-ray tube on which the image is projected.

**fac·er** (fā′sər) *n.* **1.** One that faces. **2.** A device used in smoothing or dressing a surface. **3.** *Chiefly Brit.* An unexpected blow or defeat.

**face-sav·er** (fās′sā′vər) *n.* Something that preserves one's dignity or self-esteem. **—face′-sav′ing** *n.*

**fac·et** (făs′ĭt) *n.* [Fr. *facette,* dim. of *face,* face < OFr.] **1.** One of the flat polished surfaces cut on a gemstone. **2.** A small planar or rounded smooth surface on a bone or tooth. **3.** One of the lenslike divisions of a compound eye, as of an insect. **4.** An aspect : phase <several *facets* to the problem> **—fac′et·ed, fac′et·ted** *adj.*

**fa·ce·ti·ae** (fə-sē′shē-ē′) *pl.n.* [Lat., pl. of *facetia,* jest < *facetus,* witty.] Witty writings and sayings.

**fa·ce·tious** (fə-sē′shəs) *adj.* [OFr. *facetieux* < *facetie,* jest < Lat. *facetia* < *facetus,* witty.] Playfully jocular : HUMOROUS. **—fa·ce′tious·ly** *adv.* **—fa·ce′tious·ness** *n.*

**face value** *n.* **1.** The value printed or written on the face, as of a bill or bond. **2.** Apparent significance <had to take the comments at *face value*>

**fa·cial** (fā′shəl) *adj.* Of or having to do with the face. *—n.* A treatment for the face, usu. consisting of a massage and application of cosmetic creams. **—fa′cial·ly** *adv.*

**facial index** *n.* The ratio of facial length to facial width multiplied by 100.

**facial nerve** *n.* Either of the seventh pair of cranial nerves that supply motor fibers to the facial muscles and sensory fibers to the taste buds of the anterior portion of the tongue.

**-facient** *suff.* [< Lat. *faciēns, facient-,* p.part. of *facere,* to do.] **1.** Bringing about : CAUSING <somni*facient*> **2.** Something that causes or brings about <aborti*facient*>

**fa·ci·es** (fā′shē-ēz′, -shēz′) *n., pl.* **facies.** [Lat. *facies,* form, appearance.] **1.** General aspect or outward appearance, as of a particular growth of flora. **2.** *Geol.* **a.** A part differentiated from other parts in a rock by appearance or composition. **b.** A rock distinguished from related or similar rocks. **c.** A stratigraphic body distinguished from others by appearance or composition.

**fac·ile** (făs′əl) *adj.* [OFr. < Lat. *facilis* < *facere,* to do.] **1.** Carried out with little difficulty or effort : EASY. **2.** Working, performing, or speaking effortlessly : FLUENT <a *facile* orator> **3.** Arrived at without due care, effort, or examination : SUPERFICIAL. **4.** Relaxed or easygoing in manner : POISED. **5.** Yielding : compliant. **—fac′ile·ly** *adv.* **—fac′ile·ness** *n.*

**fa·cil·i·tate** (fə-sĭl′ĭ-tāt′) *vt.* **-tat·ed, -tat·ing, -tates.** [< Fr. *faciliter* < Ital. *facilitare* < *facile,* facile < Lat. *facilis.*] To make easier <*facilitate* the transport of students> **—fa·cil′i·ta′tion** *n.*

**fa·cil·i·ty** (fə-sĭl′ĭ-tē) *n., pl.* **-ties. 1.** Ease in moving, performing, or doing : APTITUDE <"an extreme *facility* in acquiring new dialects" —W.H. Hudson> **2.** Readiness to be persuaded : PLIABILITY. **3.** *often* **facilities.** Something that facilitates an action or process. **4.** Something created to serve a particular function <a new mental health *facility*>

**fac·ing** (fā′sĭng) *n.* **1. a.** A piece of material sewn to the edge of a garment as lining or decoration. **b.** Material used for this. **2.** An outer protective or decorative layer applied to a surface.

**fac·sim·i·le** (făk-sĭm′ə-lē) *n.* [Lat. *fac simile,* make (it) similar.] **1.** An exact copy, as of a document. **2. a.** An electronic method of transmitting images or printed matter. **b.** An image transmitted electronically. *—adj.* **1.** Of or used to produce facsimiles. **2.** Exactly reproduced : DUPLICATE.

**facsimile modulation** *n.* A method for varying in time the physical properties of a wave in facsimile transmission.

**fact** (făkt) *n.* [Lat. *factum,* deed < *factus,* p.part. of *facere,* to do.] **1.** Something put forth as objectively real. **2.** Something objectively verified. **3. a.** Something with real, demonstrable existence <Travel to the moon is now a *fact.*> **b.** The quality of being real or actual. **4.** Something carried out or performed. **5.** *Law.* **a.** The aspect of a case at law comprising events determined by evidence as distinguished from interpretation of law <The jury made a finding of *fact.*> **b.** A crime <an accessory before the *fact*>

**fact-find·ing** (făkt′fīn′dĭng) *n.* Discovery or determination of facts. **—fact′-find′er** *n.*

**fac·tion** (făk′shən) *n.* [OFr. < Lat. *factio* < *factus,* p.part. of *facere,* to do.] **1.** A group of persons forming a cohesive, usu. contentious minority within a larger group. **2.** Conflict within an organization or nation <a country afflicted with *faction* and civil war> **—fac′tion·al** *adj.* **—fac′tion·al·ism** *n.* **—fac′tion·al·ly** *adv.*

**-faction** *suff.* [ME *-faccioun* < OFr. *-faction* < Lat. *-factio.*] Production : making <petri*faction*>

**fac·tious** (făk′shəs) *adj.* **1.** Produced or marked by faction. **2.** Creating faction : DIVISIVE. **—fac′tious·ly** *adv.* **—fac′tious·ness** *n.*

**fac·ti·tious** (făk-tĭsh′əs) *adj.* [Lat. *facticius* < *facere,* to make. **1.** Produced artificially. **2.** Lacking authenticity or genuineness <th *factitious* value of some highly speculative stocks> **—fac·ti′tious ly** *adv.* **—fac·ti′tious·ness** *n.*

**fac·ti·tive** (făk′tĭ-tĭv) *adj.* [NLat. *factitivus* < Lat. *facere,* to do.] C or being a transitive verb, as *elect,* that sometimes takes an objectiv complement to modify its direct object. **—fac′ti·tive·ly** *adv.*

**fact of life** *n.* **1. facts of life.** The basic physiological function involved in sex and reproduction. **2.** An unavoidable situation tha must be recognized and dealt with.

**fac·tor** (făk′tər) *n.* [ME *factour* < OFr. *facteur* < Lat. *factor,* make < *facere,* to make.] **1. a.** One who acts for someone else : AGENT **b.** One that accepts accounts receivable as security for short-term loans. **2.** One that actively contributes to an accomplishment, result or process. **3.** *Math.* One of two or more quantities that when multi plied together yield a given product <2 and 3 are *factors* of 6.> **4.** A gene. *—vt.* **-tored, -tor·ing, -tors.** *Math.* To determine or indicat explicitly the factors of. **—fac′tor·a·ble** *adj.* **—fac′tor·ship′** *n.*

**fac·tor·age** (făk′tər-ĭj) *n.* **1.** The business of a factor. **2.** A factor′ commission or fee.

**fac·to·ri·al** (făk-tôr′ē-əl, -tōr′-) *n.* The product of all the positive integers from 1 to a given number; e.g., 4 *factorial,* usu. written 4!, is equal to 24 (1·2·3·4 = 24). **—fac·to′ri·al** *adj.*

**fac·tor·ize** (făk′tə-rīz′) *vt.* **-ized, -iz·ing, -iz·es.** To factor. **—fac′ tor·i·za′tion** *n.*

**fac·to·ry** (făk′tə-rē) *n., pl.* **-ries.** [Med. Lat. *factoria,* establishmen for factors < Lat. *factor,* factor.] **1.** A place where goods are manufac tured : PLANT. **2.** A business establishment for commercial agents o factors in a foreign country.

**factory ship** *n.* An oceangoing vessel equipped with devices fo processing and storing the catch of a fishing fleet.

**fac·to·tum** (făk-tō′təm) *n.* [Med. Lat. *factotum* : Lat. *fac,* imper. of *facere,* to do + Lat. *totum,* everything < *totus,* all.] An employee o assistant who performs a wide range of functions.

**fac·tu·al** (făk′cho͞o-əl) *adj.* **1.** Of the nature of fact. **2.** Containin facts. **—fac′tu·al′i·ty** (-ăl′ĭ-tē) *n.* **—fac′tu·al·ly** *adv.* **—fac′tu·al ness** *n.*

**fac·tu·al·ism** (făk′cho͞o-ə-lĭz′əm) *n.* Devotion or adherence to fact **—fac′tu·al·ist** *n.*

**fac·u·la** (făk′yə-lə) *n., pl.* **-lae** (-lē′) [Lat., small torch, dim. of *fax* torch.] Any of various large bright spots or streaks on the sun's pho tosphere, most conspicuous at the solar edge or near sunspots.

**fac·ul·ta·tive** (făk′əl-tā′tĭv) *adj.* **1.** Of or relating to a mental fac ulty. **2. a.** Capable of happening or not happening : CONTINGENT **b.** Not required or compulsory : OPTIONAL. **3.** Granting permission o authority. **4.** *Biol.* Capable of adaptive response to varying environ ments. **—fac′ul·ta·tive·ly** *adv.*

**fac·ul·ty** (făk′əl-tē) *n., pl.* **-ties.** [ME *faculte* < OFr. *faculte* < Lat *facultas* < *facilis,* facile.] **1.** An inborn ability or power. **2.** Any of the powers or capacities possessed by the human mind <an extraor dinary *faculty* for observation> **3.** The ability to perform or act. **4.** *Obs.* Occupation : trade. **5. a.** A division or comprehensive branch of learning at a college or university <the *faculty* of medicine> **b.** The instructors within such a division. **c.** A body of instructors as distinguished from their students. **6.** The members of a learned profession <the medical *faculty*> **7.** Authorization granted by authority : conferred power.

**fad** (făd) *n.* [Orig. unknown.] A transitory fashion adopted with wide enthusiasm. **—fad′dism** *n.* **—fad′dist** *n.* **—fad′dy** *adj.*

**fad·dish** (făd′ĭsh) *adj.* **1.** Of the nature of a fad. **2.** Given to fads. **—fad′dish·ly** *adv.* **—fad′dish·ness** *n.*

**fade** (fād) *v.* **fad·ed, fad·ing, fades.** [ME *faden* < OFr. *fader* < *fade,* faded < VLat. **fatidus,* prob. < Lat. *fatuus,* insipid.] *—vi.* **1.** To lose brightness, loudness, or brilliance gradually : DIM. **2.** To lose freshness : WITHER. **3.** To lose vitality or strength : WANE. **4.** To disappear gradually : VANISH <hopes that *faded* away> *—vt.* **1.** To cause to fade. **2.** *Football.* To move back from the scrimmage line. —Used of a quarterback. **3.** *Slang.* To meet the bet of (an opposing player) in dice. **—fade in. 1.** To appear gradually. **2.** To cause to appear gradually. —Used of a motion-picture or television image or of a sound. **—fade out. 1.** To disappear gradually. **2.** To cause to disappear gradually. —Used of a motion-picture or television image or of a sound. *—n.* **1.** A gradual diminution in the brightness or visibility of an image in motion pictures or television. **2.** A periodic reduction in the received strength of a radio transmission.

☆ ***syns:*** FADE, DECLINE, DETERIORATE, FAIL, FLAG, LANGUISH, WANE, WEAKEN *v. core meaning* : to lose strength, vitality, or power <a patient who was slowly *fading*>

**fade-in** (fād′ĭn′) *n.* **1.** The gradual coming or bringing into full visibility of an image in motion pictures or television. **2.** The gradual coming or bringing into audibility of a sound, as in broadcasting.

**fade·less** (fād′lĭs) *adj.* Not fading or not subject to fading. **—fade′less·ly** *adv.*

**fade-out** (fād′out′) *n.* **1.** The gradual disappearance of a motion-picture or television image. **2.** A gradual lessening of broadcast sound.

**fad·ing** (fā′dĭng) *n.* **1.** A waning or decline <"The final factor in the *fading* of the Renaissance was the Counter Reformation" —Will Durant> **2.** Fluctuation in the strength of received radio signals because of variations in the transmission medium.

**fa·do** (fä′thŏŏ, făth′ō) *n., pl.* **-dos.** [Port. < Lat. *fatum,* fate.] A mournful Portuguese folk song.

**fae·ces** (fē′sēz) *pl.n. var. of* FECES.

**fa·e·na** (fä-ā′nä) *n.* [Sp., manual labor < Catalan *feyna* < Lat. *facienda,* things to be done, neuter pl. gerund. of *facere,* to do.] The series of final passes performed by a matador before killing the bull.

**fa·er·ie** *also* **fa·er·y** (fā′ə-rē, fâr′ē) [ME *fairie.* —see FAIRY.] *Archaic.* —*n., pl.* **-ies. 1.** FAIRY 1. **2.** The land of the fairies. —*adj.* **1.** Of or like a fairy or fairies. **2.** Enchanted : visionary.

**Faer·o·ese** (fâr′ō-ēz′, -ēs′) *n. var. of* FAROESE.

**Faf·nir** (fäv′nər, -nîr′) *n.* [ON *Fāfnir.*] *Norse Myth.* The dragon that guarded the treasure of the Nibelungs and was slain by Sigurd.

**fag¹** (făg) *n.* [Orig. unknown.] **1. a.** Tedious or fatiguing work : DRUDGERY. **b.** A drudge. **2.** *Chiefly Brit.* A student at a public school who is required to perform menial tasks for an upperclassman. —*v.* **fagged, fag·ging, fags.** —*vi.* **1.** To work to exhaustion. **2.** *Chiefly Brit.* To serve as the fag of an upperclassman. —*vt.* To exhaust : weary <was *fagged* out after the long trip>

**fag²** (făg) *n.* [Short for FAG END.] *Chiefly Brit.* A cigarette.

**fag end** *n.* [ME *fag.*] **1.** The frayed end of a length of cloth or rope. **2. a.** An inferior or worn-out remnant. **b.** The last part.

**fag·got** (făg′ət) *n.* & *v. var. of* FAGOT.

**fag·got·ing** (făg′ə-tĭng) *n. var. of* FAGOTING.

**fag·ot** *also* **fag·got** (făg′ət) [ME < OFr.] —*n.* **1.** A bundle of twigs, sticks, or branches bound together. **2.** A bundle of pieces of iron or steel to be welded or hammered into bars. —*vt.* **-ot·ed, -ot·ing, -ots** *also* **-got·ed, -got·ing, -gots. 1.** To collect or bind into a fagot : BUNDLE. **2.** To decorate with fagoting.

**fag·ot·ing** *also* **fag·got·ing** (făg′ə-tĭng) *n.* **1.** A method of decorating cloth by pulling out horizontal threads and tying the remaining vertical threads into hourglass-shaped bunches. **2.** A method of joining hemmed edges by crisscrossing thread over an open seam.

**Fahr·en·heit** (făr′ən-hīt′) *adj.* [After Gabriel D. *Fahrenheit* (1686–1736).] Of or relating to a temperature scale that registers the freezing point of water as 32°F and the boiling point as 212°F under standard atmospheric pressure.

**fa·ience** *also* **fa·ïence** (fī-äns′, fā-, -äɴs′) *n.* [Fr. < *Fayence,* Faenza, Italy.] **1.** Earthenware decorated with colorful, opaque glazes. **2.** A moderate to strong greenish blue.

**fail** (fāl) *v.* **failed, fail·ing, fails.** [ME *failen* < OFr. *faillir* < VLat. **fallire* < Lat. *fallere,* to deceive.] —*vi.* **1.** To prove so deficient as to be totally ineffective. **2.** To be unsuccessful. **3.** To receive an academic grade below the acceptable minimum. **4.** To prove insufficient in quantity or duration : GIVE OUT. **5.** To decline in strength or effectiveness. **6.** To cease functioning properly. **7.** To become bankrupt <a business that *failed*> —*vt.* **1.** To disappoint or prove undependable to <Our attorneys *failed* us.> **2.** To abandon : forsake <My strength *failed* me.> **3.** To omit or neglect <*failed* to answer the summons> **4. a.** To receive an academic grade below the acceptable minimum in (e.g., a course). **b.** To give such a grade of failure to (a student). —*n.* **1.** Failure. —Used to intensify the force esp. of a command or promise <Do it without *fail.*> **2. a.** A failure to deliver securities to a purchaser within a specified time. **b.** A failure to receive the proceeds of a transaction, as a sale of stock or securities, by a specified date.

☆ ***syns:*** FAIL, FLUNK, WASH OUT *v. core meaning* : to receive less than a passing grade <*flunked* the course> ***ant:*** pass

**fail·ing** (fā′lĭng) *n.* **1.** The act of one that fails : FAILURE. **2.** A minor fault or defect. —*prep.* In the absence of : WITHOUT <*Failing* a blizzard, we will arrive on Christmas eve.>

**faille** (fīl) *n.* [Fr. < OFr.] A slightly ribbed fabric woven of silk, cotton, or rayon.

**fail-safe** (fāl′sāf′) *adj.* **1.** Capable of compensating automatically for a mechanical failure. **2.** Acting according to a variety of predetermined conditions to stop a military attack, esp. a nuclear attack. **3.** Guaranteed not to fail. —*vi.* **-safed, -saf·ing, -safes.** To compensate automatically for failure. **—fail′-safe′** *n.*

**fail-soft** (fāl′sôft′, -sŏft) *adj.* Capable of compensating automatically for a partial failure. —Used of an electronic device.

**fail·ure** (fāl′yər) *n.* [ME *failer,* one who fails < *failen,* to fail.] **1.** The condition or fact of not achieving the desired end or ends <the *failure* of a rocket test> **2.** One that fails. **3.** Insufficiency or inadequacy. **4.** A cessation of proper functioning <a power *failure*> **5.** Nonperformance of what is requested or expected : OMISSION <*failure* to report to duty on time> **6.** The act or fact of failing to pass a course, test, or assignment. **7.** A decline in strength or effectiveness. **8.** The act or fact of becoming bankrupt.

**fain** (fān) [ME < *fain,* glad < OE *fægen.*] *Archaic.* —*adv.* **1.** Preferably : rather. **2.** Happily : gladly. —*adj.* **1.** Ready : willing. **2.** Pleased : happy. **3.** Obliged or required by circumstances.

**fai·né·ant** (fā′nā-äɴ′) *adj.* [Fr., alteration of OFr. *faignant,* idler, pr.part. of *faindre,* to feign.] Idle : lazy. —*n.* An idler.

**faint** (fānt) *adj.* **-er, -est.** [ME < OFr., p.part. of *faindre,* to feign. —see FEIGN.] **1.** Having little strength or vigor : FEEBLE. **2.** Lacking conviction, boldness, or courage : TIMID. **3.** Lacking clarity and brightness : DIM <a *faint* light> **4.** Likely to swoon <felt *faint*> —*n.* An abrupt, usu. brief loss of consciousness, gen. associated with failure of normal blood circulation. —*vi.* **faint·ed, faint·ing, faints. 1.** To fall into a swoon. **2.** *Archaic.* To weaken in purpose or spirit : LANGUISH. **—faint′er** *n.* **—faint′ly** *adv.* **—faint′ness** *n.*

**faint·heart** (fānt′härt′) *n.* A timid person.

**faint-heart·ed** (fānt′här′tĭd) *adj.* Lacking conviction or courage : TIMID. **—faint′-heart′ed·ly** *adv.* **—faint′-heart′ed·ness** *n.*

†**fair¹** (fâr) *adj.* **-er, -est.** [ME < OE *fæger.*] **1.** Visually pleasing : LOVELY. **2.** Of light color, as: **a.** Blond <*fair* hair> **b.** Not dark or ruddy <*fair* skin> **3.** Clear and sunny <*fair* skies> **4.** Free of blemishes or stains : PURE <one's *fair* name> **5.** Regular and even. **6.** Free of obstacles : OPEN <*fair* sailing> **7.** Promising <in a *fair* way to win> **8.** Impartial <a *fair* judge> **9.** Just to all parties <a *fair* compromise> **10.** Being in accordance with rules, logic, or ethics <*fair* tactics> **11.** Moderately good <a *fair* performance> **12.** Superficially true or good : SPECIOUS. **13.** Lawful to hunt or attack <*fair* game> —*adv.* **1.** In a proper or legal way <played *fair*> **2.** Directly : straight <a blow caught *fair* in the chest> —*n. Archaic.* **1.** Loveliness : beauty. **2.** A beautiful or beloved woman. —*v.* **faired, fair·ing, fairs.** —*vt.* To join so as to be smooth, even, or regular. —*vi. Regional.* To become clear. **—fair and square.** Just and honest. **—fair′ness** *n.*

☆ ***syns:*** FAIR, DISPASSIONATE, EQUITABLE, FAIR-MINDED, IMPARTIAL, JUST, NONPARTISAN, OBJECTIVE, UNBIASED, UNPREJUDICED *adj. core meaning* : free from bias in judgment <a *fair* appraisal of the issue> ***ant:*** biased, partisan, prejudiced, unfair

**fair²** (fâr) *n.* [ME *faire* < OFr. *feire* < Med. Lat. *feria* < Lat. *feriae,* holidays.] **1.** A gathering for the buying and selling of goods : MARKET. **2.** An exhibition, as of farm products or manufactured goods, usu. accompanied by competitions and entertainments <a state *fair*> **3.** An exhibition intended to inform the public about a product <a book *fair*> **4.** An event, usu. for the benefit of a charity or public institution, including entertainment and the sale of goods : BAZAAR <a church *fair* at Christmas>

**fair ball** *n. Baseball.* A batted ball that first strikes the ground or leaves the field beyond first or third base within the foul lines or that is within the foul lines as it bounces past first or third base or that comes to rest or is touched by a fielder in front of first or third base within the foul lines.

**fair catch** *n. Football.* A catch of a punt on the fly by a defensive player who has signaled that he will not run with the ball and who thus may not be tackled.

**fair copy** *n.* A copy of a document made after all corrections and revisions have been made.

**fair·ground** (fâr′ground′) *also* **fair·grounds** (-groundz′) *n.* Open land where fairs or exhibitions are held.

**fair-haired** (fâr′hârd′) *adj.* **1.** Having blond hair. **2.** Favorite <the *fair-haired* child of the family>

**fair·ing¹** (fâr′ĭng) *n.* An auxiliary structure or the external surface of an aircraft functioning to reduce drag.

**fair·ing²** (fâr′ĭng) *n. Chiefly Brit.* A gift, esp. one bought or given at a fair.

**fair·ish** (fâr′ĭsh) *adj.* Of moderately good size or quality.

**fair-lead** (fâr′lēd′) *also* **fair-lead·er** (-lē′dər) *n. Naut.* A device, as a ring or block of wood with a hole in it, through which rigging is passed to hold it in place or prevent it from snagging.

**fair·ly** (fâr′lē) *adv.* **1. a.** In a fair or just way. **b.** Legitimately : suitably. **2.** Clearly : distinctly. **3.** Actually : fully <The walls *fairly* shook with screams.> **4.** Moderately : rather <a *fairly* good time> **5.** *Obs.* **a.** Gently. **b.** Courteously.

**fair-mind·ed** (fâr′mīn′dĭd) *adj.* Just and impartial. **—fair′-mind′ed·ly** *adv.* **—fair′-mind′ed·ness** *n.*

**fairness doctrine** *n.* The principle and practice in the broadcast media of affording opposing candidates equal time to air their views on controversial issues.

**fair play** *n.* Conformity to established rules and ethics.

**fair sex** *n.* Women as a group.

**fair shake** *n. Slang.* A fair chance.

**fair-spo·ken** (fâr′spō′kən) *adj.* Civil, courteous, and gentle in speech.

**fair trade** *n.* Trade conforming to a fair-trade agreement.

**fair-trade** (fâr′trād′) *vt.* **-trad·ed, -trad·ing, -trades.** To sell (a product) at a price conforming to a fair-trade agreement.

**fair-trade agreement** *n.* A commercial agreement under which distributors sell products of a given class at no less than a minimum price set by the manufacturer.

**fair·way** (fâr′wā′) *n.* **1.** An obstacle-free stretch of ground. **2.** The part of a golf course covered with short grass and extending from the tee to the putting green. **3.** *Naut.* **a.** A navigable deep-water channel in a river or harbor or along a coastline. **b.** The usual course taken by vessels through a harbor or coastal waters.

**fair-weath·er** (fâr′wĕth′ər) *adj.* **1.** Suitable or used only during fair weather. **2.** Dependable only when times are good <*fair-weather* friends>

**fair·y** (fâr′ē) *n., pl.* **-ies.** [ME *fairie* < OFr. *faerie* < *fae* < Lat. *fata,* the Fates < *fatum,* fate.] A tiny imaginary being in human form, depicted as mischievous, clever, and having magical powers.

**fair·y·land** (fâr′ē-lănd′) *n.* **1.** The land of the fairies. **2.** A charming, enchanting place.

**fairy lily** *n.* The atamasco lily.

**fairy ring** *n.* [From the belief that it is a dancing place for fairies.] A circle of mushrooms in a grassy area, marking the periphery of underground mycelial growth.

**fairy shrimp** *n.* A freshwater crustacean of the order Anostraca.

**fairy tale** *n.* **1.** A fictitious tale of legendary deeds and fanciful creatures, usu. intended for children. **2.** A fictitious, highly fanciful explanation or story.

**fait ac·com·pli** (fā′tä-kôN-plē′, fĕt′ä-) *n., pl.* **faits ac·com·plis** (fā′tä-kôN-plē′, -plēz′, fĕt′ä-) [Fr.] An accomplished and presumably irreversible deed or fact.

**faith** (fāth) *n.* [ME < AN *fed* < OFr. *feid* < Lat. *fides* < *fidere,* to trust.] **1.** Confident belief in the truth, value, or trustworthiness of a person, idea, or thing. **2.** Belief not based on logical proof or material evidence. **3.** Loyalty to a person or thing : ALLEGIANCE <keeping *faith* with the President> **4. a.** Belief and trust in God. **b.** Religious conviction. **5.** A system of religious beliefs. **6.** A set of principles or beliefs.

**faith·ful** (fāth′fəl) *adj.* **1.** Firmly and devotedly supportive : LOYAL. **2.** Worthy of trust or belief : RELIABLE. **3.** Consistent with truth or fact <a *faithful* reproduction of the document> **4.** Having faith. —*n., pl.* **faithful** or **-fuls. 1.** The practicing members of a religious faith. **2.** A steadfast adherent of a faith or cause. **—faith′ful·ly** *adv.* **—faith′ful·ness** *n.*

☆ ***syns:*** FAITHFUL, ALLEGIANT, CONSTANT, FAST, FIRM, LOYAL, LIEGE, RESOLUTE, STAUNCH, STEADFAST, STEADY, TRUE *adj. core meaning* : firmly and devotedly supportive <a *faithful* member of the party><a *faithful* spouse> ***ant:*** faithless, unfaithful

**faith healer** *n.* One who treats disease with prayer. **—faith healing** *n.*

**faith·less** (fāth′lĭs) *adj.* **1.** Not true to duty or obligation : DISLOYAL. **2.** Lacking religious faith. **3.** Unworthy of faith or trust : UNRELIABLE. **—faith′less·ly** *adv.* **—faith′less·ness** *n.*

☆ ***syns:*** FAITHLESS, DISLOYAL, FALSE, FALSE-HEARTED, PERFIDIOUS, RECREANT, TRAITOROUS, TREACHEROUS, UNFAITHFUL, UNTRUE *adj. core meaning* : not true to duty or obligation <a *faithless* spouse><a *faithless* defector> ***ant:*** faithful

**fai·tour** (fā′tər) *n.* [ME < OFr. *faiteor* < Lat. *factor,* maker < *facere,* to make.] *Obs.* An impostor.

**fake**[1] (fāk) *adj.* [Orig. unknown.] Having a false or misleading appearance : FRAUDULENT. —*n.* **1.** One that is not genuine or authentic : SHAM. **2.** A brief feint or aborted change of direction intended to mislead one's opponent or the opposing team in certain sports. —*v.* **faked, fak·ing, fakes.** —*vt.* **1.** To contrive and present as genuine : COUNTERFEIT. **2.** To feign : simulate <*faked* their grief> **3.** To improvise (a musical passage). —*vi.* **1.** To engage in faking. **2.** To perform a fake in certain sports.

**fake**[2] (fāk) *n.* [ME *faken,* to coil a rope.] One loop or winding of a coiled rope or cable. —*vt.* **faked, fak·ing, fakes.** To coil (a rope or cable).

**fak·er** (fā′kər) *n.* One who fakes or who produces fakes : SWINDLER. **—fak′ery** (-kə-rē) *n.*

**fa·kir** (fə-kîr′, fä-, fă-) *n.* [Ar. *faqīr* < *faqura,* he was poor.] **1.** A Moslem religious mendicant. **2.** A Hindu ascetic or religious mendicant, esp. one who performs feats of magic or endurance.

**fa·la·fel** or **fe·la·fel** (fə-lä′fəl) *n.* [Ar. *falafil.*] **1.** Ground spiced chickpeas and fava beans shaped into balls and fried. **2.** A sandwich filled with falafel.

**Fa·lange** (fā′lănj′, fə-lănj′) *n.* [Sp. < *falange,* phalanx < Lat. *phalanx.* —see PHALANX.] A fascist organization constituting the official ruling party of Spain after the civil war of 1936–39. **—Fa·lan′gist** (fə-lăn′gĭst, fā′lăn′-) *n.*

**fal·cate** (făl′kāt′) *also* **fal·cat·ed** (-kā′tĭd) *adj.* [Lat. *falcatus* < *falx,* sickle.] Curved and tapering to a point : SICKLE-SHAPED.

**fal·ces** (făl′sēz′, fôl′-) *n. pl. of* FALX.

**fal·chion** (fôl′chən) *n.* [ME *fauchoun* < OFr. < Lat. *falx,* sickle.] **1.** A short, broad medieval sword with a convex cutting edge and a sharp point. **2.** *Archaic.* A sword of any kind.

**fal·ci·form** (făl′sə-fôrm′) *adj.* [Lat. *falx, falc-,* sickle + -FORM.] Falcate.

**fal·con** (făl′kən, fôl′-, fô′kən) *n.* [ME *faucoun* < OFr. *faucon* < LLat. *falco.*] **1. a.** A bird of prey of the family Falconidae, and esp. of the genus *Falco,* having long, pointed, powerful wings adapted for swift flight. **b.** Any of these or related birds such as hawks, trained to hunt small game. **2.** A small 15th–17th cent. cannon.

**fal·con·er** (făl′kə-nər, fôl′-, fô′kə-) *n.* **1.** A breeder and trainer of falcons **2.** A hunter who uses falcons.

**fal·con·et** (făl′kə-nĕt′, fôl′-, fô′kə-) *n.* **1.** A small or young falcon. **2.** A small falcon of the genus *Microhierax* of tropical Asia.

**fal·con-gen·tle** (făl′kən-jĕn′tl, fôl′-, fô′kən-) *n.* [ME *faucon gentil* < OFr. *faucon gentil,* noble falcon.] The female peregrine falcon.

**fal·con·ry** (făl′kən-rē, fôl′-, fô′kən-) *n.* **1.** The sport of hunting with falcons. **2.** The training of falcons for hunting.

**fal·cu·late** (făl′kyə-lāt′) *adj.* [< Lat. *falcula,* small sickle, dim. of *falx,* sickle.] Falcate.

**fal·de·ral** (făl′də-răl′) *n. var. of* FOLDEROL.

**fald·stool** (fôld′stōōl′) *n.* [Partial transl. of Med. Lat. *faldistolium,* folding stool, of Germanic orig.] **1.** A folding or small desk stool at which worshipers kneel to pray, esp. one on which the British monarchs kneel at their coronations. **2.** A folding chair or stool, esp. one used by a bishop when not occupying his throne or when presiding away from his own cathedral. **3.** A desk at which the litany is recited in Anglican churches.

**fall** (fôl) *v.* **fell** (fĕl), **fall·en** (fô′lən), **fall·ing, falls.** [ME *fallen* < OE *feallan.*] —*vi.* **1.** To come down without restraint, due to weight or gravity. **2.** To drop oneself to a lower or less erect position <*fell* back onto the sofa> **3. a.** To lose an upright or erect position suddenly. **b.** To drop wounded or dead, esp. in battle. **4.** To descend from or as if from the sky <Night *falls* quickly in the desert.> **5.** To come to rest : SETTLE <The light *fell* on the page.> **6.** To hang down <hair *falling* in curls> **7.** To be cast down. **8.** To assume an upset or disappointed expression <The child's face *fell* upon hearing the news.> **9.** To undergo conquest or capture, esp. as the result of military attack. **10.** To experience defeat or ruin. **11.** To slope downward <The plain *falls* gently toward the coast.> **12.** To lessen in degree, amount, or value <air pressure *falling*> **13.** To diminish in volume or pitch. **14.** To decline in rank, status, or importance. **15. a.** To give in to temptation : SIN. **b.** To lose one's chastity. **16.** To pass into a specific state or situation <at last *fell* silent><*fell* in love> **17.** To happen at a given time <Christmas *falls* on a Monday this year.> **18.** To happen at a given place <The stress *falls* on the first syllable.> **19.** To come, as by chance. **20. a.** To be given, as by assignment <The hardest job *fell* to us.> **b.** To be given by right or inheritance. **21.** To be included within the scope or range of something <The specimens *fall* into four categories.> **22.** To come into contact : STRIKE <My gaze *fell* on a priceless piece of silver.> **23.** To come out : ISSUE. **24.** To begin with vigor. **25.** To be born. —Used chiefly of lambs. —*vt.* To cut down (a tree) : FELL. **—fall away. 1.** To decline. **2.** To withdraw friendship or support. **—fall back.** To give ground : RETREAT. **—fall back on** (or **upon**). To resort to. **—fall behind. 1.** To lag behind. **2.** To be in arrears <*fell behind* in their payments> **—fall down.** *Informal.* To fail or lag in performance. **—fall for.** *Informal.* **1.** To fall in love with. **2.** To be tricked or deceived by <Don't *fall for* that ploy.> **—fall in.** To take one's place in a military formation. **—fall in with. 1.** To meet accidentally. **2.** To come to an agreement. **3.** To be in harmony with. **—fall off. 1.** To become less : DECREASE. **2.** *Naut.* To change course to leeward. **—fall on** (or **upon**). To attack without warning. **—fall out. 1.** To leave a military formation. **2.** To quarrel. **—fall through.** To fail <The plan *fell through.*> —*n.* **1.** An act or instance of falling. **2.** A sudden drop from a relatively erect to a less erect position. **3.** Something that has fallen <a *fall* of snow> **4. a.** An amount that has fallen <a *fall* of two inches of snow> **b.** The distance that something falls <a *fall* of 40 floors> **5.** Autumn. **6. falls** (*sing. or pl. in number*). A waterfall. **7.** A downward movement or slope. **8.** A pendent article of dress, esp.: **a.** A veil hung from a woman's hat and down her back. **b.** An ornamental cascade of lace or trimming attached to a dress, usu. at the collar. **9.** A woman's hair piece with long, free-hanging hair. **10. a.** An overthrow : collapse <the *fall* of a dynasty> **b.** A military capture of an objective under siege. **11.** A reduction in value, amount, or degree. **12.** A decline in status, rank, or importance. **13. a.** A moral lapse. **b.** A loss of chastity. **14.** *often* **Fall.** Loss of innocence and grace resulting from Adam's eating the forbidden fruit in the Garden of Eden. **15. a.** The act of throwing or forcing a wrestling opponent down on his back. **b.** Any of various wrestling maneuvers used for this. **16.** *Naut.* A break or rise in the level of a deck. **17. falls.** *Naut.* The apparatus used to hoist and transfer cargo or lifeboats. **18.** The end of a cable, rope, or chain pulled by the power source in hoisting. **19. a.** The birth of an animal, esp. a lamb. **b.** All of the animals born at one birth : LITTER. **—fall flat.** To fail to achieve an intended result. **—fall foul** (or **afoul**). **1.** *Naut.* To collide. —Used of vessels. **2.** To quarrel. **—fall short.** To fail to attain the requisite amount, level, or degree.

☆ ***syns:*** **1.** FALL, DESCEND, DROP *v. core meaning* : to move downward in response to gravity <apples *falling* from trees> **2.** FALL, DROP, PITCH, PLUNGE, SPILL, SPRAWL, TOPPLE, TUMBLE *v. core meaning* : to come to the ground suddenly and involuntarily <stumbled and *fell*>

**fal·la·cious** (fə-lā′shəs) *adj.* **1.** Containing fundamental errors in reasoning. **2.** Misleading : deceptive. **—fal·la′cious·ly** *adv.*

☆ ***syns:*** FALLACIOUS, FALSE, ILLOGICAL, INVALID, SOPHISTIC, SPE-

CIOUS, SPURIOUS *adj. core meaning* : containing fundamental errors in reasoning <a *fallacious* argument><*fallacious* logic>

**fal·la·cy** (făl′ə-sē) *n., pl.* **-cies.** [Lat. *fallacia*, deceit < *fallax*, deceitful < *fallere*, to deceive.] **1.** A false notion. **2.** A statement or argument based on a false or invalid inference. **3.** Incorrectness of reasoning or belief. **4.** The quality of being deceptive.

**fall·back** (fôl′băk′) *n.* **1.** A mechanism for carrying forth programmed instructions despite malfunction or failure of the primary device. **2.** Something to which one can retreat or resort. **3.** A retreat. **4.** Something that falls back.

**fall·en** (fô′lən) *v. p.p. of* FALL.

**fall·fish** (fôl′fĭsh′) *n., pl.* **fallfish** *or* **-fish·es.** A small, silvery freshwater fish, *Semotilus corporalis*, found in eastern North American streams and rivers.

**fall guy** *n. Slang.* **1.** SCAPEGOAT 1. **2.** A gullible victim : DUPE.

**fal·li·ble** (făl′ə-bəl) *adj.* [ME < Med. Lat. *fallibilis* < Lat. *fallere*, to deceive.] **1.** Capable of making an error. **2.** Apt to be erroneous. **—fal′li·bil′i·ty, fal′li·ble·ness** *n.* **—fal′li·bly** *adv.*

**fall·ing-out** (fô′lĭng-out′) *n., pl.* **fall·ings-out** or **fall·ing-outs.** A disagreement : quarrel.

**falling rhythm** *n.* Rhythm in which the stress regularly falls on the first syllable of each foot.

**falling star** *n.* An object, as a meteoroid, often visible as a result of being ignited by atmospheric friction.

**fall line** *n.* **1.** A line connecting the waterfalls of nearly parallel rivers that marks a drop in land level. **2.** The natural line of descent, as for skiing, between two points on a slope.

**fall·off** (fôl′ôf′, -ŏf′) *n.* A decrease <a *falloff* in first-quarter book sales>

**Fal·lo·pi·an tube** (fə-lō′pē-ən) *n.* [After Gabriello *Fallopio* (1523–1562).] Either of a pair of slender ducts connecting the uterus to the region of each of the ovaries in the female reproductive system of human beings and higher vertebrates.

**Fal·lot's tetralogy** (fă-lōz′) *n.* [After Etienne *Fallot* (1850–1911).] A congenital heart condition marked by narrowing of the pulmonary artery, enlargement of the right ventricle, malpositioning of the aorta, and a defective ventricular septum.

**fall·out** (fôl′out′) *n.* **1. a.** Slow descent of minute particles of radioactive debris in the atmosphere following a nuclear explosion. **b.** Particles of radioactive debris that descend in this way. **2.** An incidental side effect <the technological *fallout* of the space program>

**fal·low** (făl′ō) *adj.* [ME *falow* < OE *fealg*, plowed land.] **1.** Plowed but left unseeded during a growing season. **2.** Marked by inactivity <talents wasted or lying *fallow*> —*n.* **1.** Land left fallow. **2.** The act of plowing land and leaving it fallow. **3.** The state or period of being fallow. —*vt.* **-lowed, -low·ing, -lows. 1.** To make (land) fallow by plowing. **2.** To plow and till (land), esp. to get rid of weeds. **—fal′low·ness** *n.*

**fallow deer** *n.* [Obs. *fallow*, reddish-yellow (< ME *falwe* < OE *fealo*) + DEER.] A Eurasian deer, *Dama dama* or *D. mesopotamica*, having a yellowish coat spotted with white in summer and broad, flattened antlers in the male.

**false** (fôls) *adj.* **fals·er, fals·est.** [ME *fals* < OFr. < Lat. *falsus*, p.part. of *fallere*, to deceive.] **1.** Being contrary to truth or fact. **2.** Arising from mistaken ideas <*false* dreams> **3.** Deliberately untrue. **4.** Intentionally deceptive <*false* promises> **5.** Not keeping faith : TREACHEROUS <a *false* friend> **6.** Not real or natural : ARTIFICIAL <*false* teeth> **7.** Erected temporarily, as for support during construction. **8.** Resembling but not accurately or properly designated as such. **9.** *Mus.* Of incorrect pitch. —*adv.* Treacherously or faithlessly <played me *false*> **—false′ly** *adv.* **—false′ness** *n.*

☆ ***syns:*** FALSE, ERRONEOUS, INACCURATE, INCORRECT, SPECIOUS, TRUTHLESS, UNTRUTHFUL, WRONG *adj. core meaning* : devoid of truth and accuracy <a *false* story> ***ant:*** true

**false alarm** *n.* **1.** An emergency alarm, as a fire alarm, that is activated unnecessarily. **2.** A groundless warning.

**false arrest** *n. Law.* Unlawful or unjustifiable arrest.

**false bottom** *n.* A partition made to conceal a compartment between it and the bottom of a container, as a trunk.

**false-heart·ed** (fôls′här′tĭd) *adj.* Deceitful : treacherous. **—false′heart′ed·ness** *n.*

**false·hood** (fôls′ho͝od′) *n.* **1.** Lack of conformity to truth or fact : INACCURACY. **2.** The act of lying. **3.** A lie.

**false imprisonment** *n. Law.* Unlawful arrest or detention.

**false indigo** *n.* **1.** A shrub, *Amorpha fruticosa* of eastern North America, having compound leaves with numerous leaflets and long purplish flower clusters. **2.** A plant, *Baptisia australis* of the southeastern United States, having compound leaves with three leaflets and deep-blue or purplish flowers.

**false keel** *n.* A protective strip fixed below a ship's main keel.

**false miterwort** *n.* The foamflower.

**false pretense** *n. Law.* Calculated misrepresentation of fact for purposes of fraud, as through forged documents.

**false rib** *n.* Any of the five lower pairs of ribs that do not unite directly with the sternum.

**false Solomon's seal** *n.* A plant of the genus *Smilacina*, esp. *S. racemosa*, with a plumelike cluster of small greenish-white flowers.

**fal·set·to** (fôl-sĕt′ō) *n., pl.* **-tos.** [Ital., dim. of *falso*, false < Lat. *falsus*.] **1.** A male singing voice marked by artificially produced tones in an upper register beyond the normal range esp. of a tenor. **2.** One that sings falsetto. **—fal·set′to** *adv.*

**fals·ie** (fôl′sē) *n. often* **falsies.** *Informal.* Padding in a brassiere.

**fal·si·fy** (fôl′sə-fī′) *v.* **-fied, -fy·ing, -fies.** [ME *falsifien*, to show to be untrue < OFr. *falsifier*, to falsify < Med. Lat. *falsificare*, to pervert : Lat. *falsus*, false + Lat. *facere*, to make.] —*vt.* **1.** To give an untruthful account of <*falsify* testimony> **2. a.** To make false by altering or adding to. **b.** To counterfeit : forge <*falsify* transit visas> **3.** To prove to be false. —*vi.* To make untrue statements : LIE. **—fal′si·fi·ca′tion** *n.* **—fal′si·fi′er** *n.*

**fal·si·ty** (fôl′sĭ-tē) *n., pl.* **-ties. 1.** The quality or state of being false. **2.** A lie.

**Fal·staff·i·an** (fôl-stăf′ē-ən) *adj.* [After John *Falstaff*, a character in *Henry IV* and *Merry Wives of Windsor*, by William Shakespeare (1564–1616).] Jovial and convivial.

**falt·boat** (fält′bōt′, fôlt′-) *n.* [Partial transl. of G. *Faltboot*, folding boat : *falten*, to fold (< OHG *falden*) + *Boot*, boat.] A foldboat.

**fal·ter** (fôl′tər) *vi.* **-tered, -ter·ing, -ters.** [ME *falteren*, to stagger.] **1.** To waver : hesitate. **2.** To speak hesitatingly : STAMMER. **3. a.** To move ineptly or haltingly : STUMBLE. **b.** To operate or perform unsteadily. —*n.* **1.** Unsteadiness in speech or action. **2.** A faltering sound. **—fal′ter·er** *n.* **—fal′ter·ing·ly** *adv.*

**falx** (fălks, fôlks) *n., pl.* **fal·ces** (făl′sēz′, fôl′-) [Lat., sickle.] *Anat.* A sickle-shaped structure.

**fame** (fām) *n.* [ME < OFr. < Lat. *fama.*] **1. a.** Great renown. **b.** Public esteem. **2.** *Archaic.* Rumor. —*vt.* **famed, fam·ing, fames. 1.** To make renowned. **2.** To report to be.

**fa·mil·ial** (fə-mĭl′yəl) *adj.* **1.** Of or relating to a family <*familial* relationships> **2.** Passed on in a family : HEREDITARY.

**fa·mil·iar** (fə-mĭl′yər) *adj.* [ME < OFr. *familier*, familial < Lat. *familiāris* < *familia*, family.] **1.** Frequently encountered : COMMON <a *familiar* sight> **2.** Having good knowledge of something <*familiar* with that city> **3.** Of established friendship : INTIMATE <on *familiar* terms with them> **4.** Natural and unstudied : INFORMAL <lectured to the class in a *familiar* style> **5.** Arrogantly self-confident : PRESUMPTUOUS. **6.** *Archaic.* Familial. **7.** Domesticated : tame. —Used of animals. —*n.* **1.** A close friend or associate. **2.** A spirit, often assuming animal form, believed to serve esp. a witch. **3.** A domestic servant esp. in the household of a high church official. **4.** One who frequents a place. **—fa·mil′iar·ly** *adv.*

☆ ***syns:*** FAMILIAR, ACQUAINTED, CONVERSANT, UP (on) *adj. core meaning* : having good knowledge of <*familiar* with the roads here><*familiar* with that case> ***ant:*** unfamiliar

**fa·mil·iar·i·ty** (fə-mĭl′yăr′ĭ-tē, -mĭl′ē-ăr′-) *n., pl.* **-ties. 1.** Acquaintance with or knowledge of something. **2.** Established friendship. **3. a.** An excessively intimate or informal act : IMPROPRIETY. **b.** A sexual advance. **4.** The quality or state of being familiar.

**fa·mil·iar·ize** (fə-mĭl′yə-rīz′) *vt.* **-ized, -iz·ing, -iz·es. 1.** To make familiar, known, or recognized. **2.** To make acquainted with. **—fa·mil′iar·i·za′tion** *n.* **—fa·mil′iar·iz′er** *n.*

**fam·i·ly** (făm′ə-lē, făm′lē) *n., pl.* **-lies.** [ME *familie* < Lat. *familia* < *famulus*, servant.] **1.** A fundamental social group in society consisting esp. of a man and woman and their offspring. **2.** A group of people sharing common ancestry. **3.** Distinguished lineage. **4.** All the members of a household living under one roof. **5. a.** A group of like things : CLASS. **b.** A group of individuals derived from a common stock. **6.** *Biol.* A taxonomic category above a genus and below an order. **7.** A language group, the members of which are derived from the same parent language. **8.** *Math.* A set of functions that can be generated by varying the parameters of a general form.

**family Bible** *n.* A Bible with pages for recording births, deaths, and marriages.

**family circle** *n.* A section of moderately priced theater seats.

**family man** *n.* **1.** A man having a wife and children. **2.** A man devoted to his family.

**family name** *n.* A surname.

**family planning** *n.* Planning of the number of one's children by use of birth-control techniques.

**family room** *n.* A recreation room for family members.

**family tree** *n.* **1.** A genealogical diagram of a family. **2.** The ancestors and descendants of a family.

**fam·ine** (făm′ĭn) *n.* [ME < OFr. < Lat. *fames*, hunger.] **1.** A drastic, wide-ranging food shortage. **2.** A drastic shortage : DEARTH. **3.** Severe hunger : STARVATION. **4.** *Archaic.* Extreme appetite.

**fam·ish** (făm′ĭsh) *v.* **-ished, -ish·ing, -ish·es.** [ME *famishen*, prob. < AN < VLat. **affamare* : Lat. *ad-*, to + Lat. *fames*, hunger.] —*vt.* **1.** To cause to endure severe hunger : STARVE. **2.** *Archaic.* To cause to starve to death. —*vi.* **1.** To endure severe deprivation, esp. of food. **2.** To starve to death. **—fam′ish·ment** *n.*

**fa·mous** (fā′məs) *adj.* [ME < AN < Lat. *famosus* < *fama*, fame.] **1.** Widely known. **2.** *Informal.* Excellent. **3.** *Archaic.* Notorious. **—fa′mous·ly** *adv.* **—fa′mous·ness** *n.*

☆ ***syns:*** FAMOUS, CELEBRATED, DISTINGUISHED, EMINENT, FAMED, ILLUSTRIOUS, NOTABLE, NOTED, PRE-EMINENT, PROMINENT, RENOWNED *adj. core meaning* : widely known <a *famous* trial lawyer><a *famous* concert pianist>

**fam·u·lus** (făm'yə-ləs) *n., pl.* **-li** (-lī') [G. < Lat.] **1.** A medieval scholar's assistant. **2.** A private secretary.

**fan¹** (făn) *n.* [ME < OE *fann* < Lat. *vannus.*] **1.** A device for creating a current of air or a breeze, esp.: **a.** A collapsible, usu. wedge-shaped device of a light material. **b.** A machine using an electric motor to rotate thin, rigid vanes so as to move air for cooling. **2.** A machine for winnowing. **3.** Something like an open fan. —*v.* **fanned, fanning, fans.** —*vt.* **1.** To move or cause a movement of (air) with or as if with a fan. **2.** To direct a current of air or a breeze on, esp. so as to cool <*fan* one's face> **3.** To stir up by or as if by fanning <*fan* the flames of resentment> **4.** To open out to a fan shape. **5. a.** To fire (an automatic firearm) in a continuous sweep by depressing the trigger. **b.** To fire (a nonautomatic gun) rapidly by chopping the hammer with the palm. **6.** To winnow. **7.** *Baseball.* To strike out (a batter). —*vi.* **1.** To spread like a fan <The group *fanned* out in a southerly direction.> **2.** *Baseball.* To strike out.

**fan²** (făn) *n.* [Short for FANATIC.] *Informal.* An ardent devotee.

**fa·nat·ic** (fə-năt'ĭk) *n.* [Lat. *fanaticus,* inspired by a god < *fanum,* temple.] One having excessive zeal for and irrational attachment to a cause or position. **—fa·nat'ic, fa·nat'i·cal** (-ĭ-kəl) *adj.* **—fa·nat'i·cal·ly** *adv.* **—fa·nat'i·cal·ness** *n.*

**fa·nat·i·cism** (fə-năt'ĭ-sĭz'əm) *n.* Excessive, irrational zeal.

**fa·nat·i·cize** (fə-năt'ĭ-sīz') *vt.* & *vi.* **-cized, -ciz·ing, -ciz·es.** To make fanatical or act as a fanatic.

**fan belt** *n.* A taut rubber belt that transfers torque from the crankshaft to the shaft of the cooling fan on an engine.

**fan·ci·er** (făn'sē-ər) *n.* **1.** One who has a special enthusiasm for or interest in something. **2.** A breeder of a plant or animal for those features held to be desirable <a cat *fancier*>

**fan·ci·ful** (făn'sĭ-fəl) *adj.* **1.** Created in the fancy : UNREAL. **2.** Tending to indulge in fancy <a *fanciful* mind> **3.** Exhibiting invention or whimsy in design : IMAGINATIVE. **—fan'ci·ful·ly** *adv.* **—fan'ci·ful·ness** *n.*

☆ ***syns:*** FANCIFUL, FANTASTIC, WHIMSICAL *adj. core meaning* : appealing to fancy <*fanciful* baroque fountains in the park>

**fan·cy** (făn'sē) *n., pl.* **-cies.** [ME *fansy.* —see FANTASY.] **1.** Imagination, esp. of a fantastic or whimsical nature. **2.** An image or fantastic invention created by the mind. **3.** A capricious notion : WHIM. **4.** A capricious tendency or liking. **5.** Critical sensibility : TASTE. **6.** *Obs.* Amorous or romantic attachment. **7. a.** The fans or enthusiasts of a sport or pursuit. **b.** The sport or pursuit holding the interest of such a group. —*adj.* **-ci·er, -ci·est. 1.** Highly decorated <a *fancy* collar> **2.** Arising in the fancy : CAPRICIOUS. **3.** Skillfully executed. **4.** Of superior grade : FINE <*fancy* honey> **5.** Exorbitant <paid a *fancy* price> **6.** Bred for unusual qualities or special points. —*vt.* **-cied, -cy·ing, -cies. 1.** To visualize : imagine. **2.** To take a fancy to : LIKE. **3.** To suppose : guess. —*interj.* —Used to express surprise <*Fancy* those clothes!> **—fan'ci·ly** *adv.* **—fan'ci·ness** *n.*

☆ ***syns:*** FANCY, CAPRICE, CONCEIT, HUMOR, IMPULSE, MEGRIM, NOTION, VAGARY, WHIM, WHIMSY *n. core meaning* : an impulsive, sometimes illogical turn of mind <had a sudden *fancy* to take up hang gliding>

**fancy dress** *n.* A masquerade costume.

**fan·cy-free** (făn'sē-frē') *adj.* **1.** Carefree. **2.** Not in love : UNATTACHED.

**fan·cy·work** (făn'sē-wûrk') *n.* Decorative needlework.

**fan·dan·go** (făn-dăng'gō) *n., pl.* **-gos.** [Sp.] **1.** An animated Spanish or Spanish-American dance in triple time. **2.** Music for a fandango.

**fan·fare** (făn'fâr') *n.* [Fr.] **1.** A loud trumpet flourish. **2.** *Informal.* A spectacular public display.

**fan·far·o·nade** (făn'făr-ə-nād', -näd') *n.* [Fr. *fanfaronnade* < Sp. *fanfarronada,* bluster < *fanfarrón,* a blusterer.] Bragging or blustering behavior.

**fang** (făng) *n.* [ME, capture < OE.] **1.** A long, pointed tooth, esp.: **a.** One of the hollow, grooved teeth with which a venomous snake injects its poison. **b.** One of the teeth of a carnivorous animal, with which it seizes and tears its prey. **2.** A fanglike structure, as a chelicera of a venomous spider. **—fanged** *adj.*

**fang**
(Left) *of a cat and* (right) *of a snake*

**fan-in** (făn'ĭn') *n. Computer Sci.* The number of inputs available to a given function or logic stage.

**fan-jet** (făn'jĕt') *n.* **1.** A jet engine providing additional thrust by means of a ducted fan in its forward end that draws in extra air. **2.** An aircraft equipped with a fan-jet engine or engines.

**fan letter** *n.* A piece of fan mail.

**fan·light** (făn'līt') *n.* **1.** A half-circle window, often with sash bars arranged like the ribs of a fan. **2.** *Chiefly Brit.* A transom.

**fan mail** *n.* Mail sent to a public figure by admirers.

**fan·ny** (făn'ē) *n., pl.* **-nies.** [< *Fanny,* a nickname for *Frances.*] *Slang.* The buttocks.

**fan-out** (făn'out') *n. Computer Sci.* The number of circuits fed input signals from an output terminal.

**fan palm** *n.* A palm tree with palmate leaves in a fanlike pattern.

**fan·tail** (făn'tāl') *n.* **1.** One of a breed of domestic pigeons having a rounded, fan-shaped tail. **2.** A goldfish of a breed having a wide, fanlike double tail fin. **3.** A bird of the genus *Rhipidura* of east Asia and Australia, having a long, fan-shaped tail. **4.** A fanlike tail or end. **5.** The stern overhang of a ship. **—fan'tailed'** *adj.*

**fan-tan** (făn'tăn') *n.* [Chin. *fan¹ tan¹* : *fan¹,* division + *tan¹,* to spread out.] **1.** A Chinese betting game in which the players lay wagers on the number of counters that will remain when a hidden pile of them has been divided by four. **2.** A card game in which sevens and their equivalent are played in sequence and the first player out of cards is the winner.

**fan·ta·sia** (făn-tā'zhə, -zhē-ə, făn'tə-zē'ə) *n.* [Ital. < Lat. *phantasia,* fantasy.] *Mus.* **1.** A free composition structured according to the composer's fancy. **2.** A medley of familiar themes, with variations and interludes.

**fan·ta·sist** (făn'tə-sĭst) *n.* A creator of fantasy or a fantasia.

**fan·ta·size** (făn'tə-sīz') *v.* **-sized, -siz·ing, -siz·es.** —*vt.* To imagine. —*vi.* To indulge in fantasies.

**fan·tast** (făn'tăst') *n.* [G. < Med. Lat. *phantasta* < *phantastes,* boaster < Gk. *phantazein,* to make visible. —see FANTASY.] A visionary.

**fan·tas·tic** (făn-tăs'tĭk) *adj.* [ME *fantastik,* imagined < OFr. *fantastique* < Lat. *fantasticus* < Gk. *phantastikos,* creating mental images < *phantazein,* to make visible. —see FANTASY.] **1.** Totally unrelated to reality or common sense and existing only in the fancy. **2. a.** Unrestrainedly fanciful : EXTRAVAGANT <*fantastic* desires> **b.** Bizarre : strange. **3.** Capriciously or fancifully eccentric. **4.** *Informal.* Wonderful or superb. —*n. Archaic.* One fancifully eccentric in behavior or appearance. **—fan·tas'ti·cal'i·ty** (-tĭ-kăl'ĭ-tē) *n.* **—fan·tas'ti·cal·ly** *adv.* **—fan·tas'ti·cal·ness** *n.*

☆ ***syns:*** FANTASTIC, ANTIC, BIZARRE, FAR-FETCHED, GROTESQUE *adj. core meaning* : having no reference to reality or common sense <a *fantastic* tale>

**fan·tas·ti·cate** (făn-tăs'tĭ-kāt') *vt.* **-cat·ed, -cat·ing, -cates.** To make fantastic. **—fan·tas'ti·ca'tion** *n.*

**fan·ta·sy** (făn'tə-sē, -zē) *n., pl.* **-sies.** [ME *fantasie, fansy* < OFr. *phantasie* < Lat. *phantasia* < Gk., appearance < *phantazein,* to make visible < *phainein,* to show.] **1.** Creative imagination. **2.** A creation of the fancy. **3.** A capricious or fantastic idea : CONCEIT. **4. a.** Literary or dramatic fiction marked by highly fanciful or supernatural elements. **b.** An example of such fiction. **5.** *Psychol.* An imagined event or condition fulfilling a wish. **6.** *Mus.* A fantasia. **7.** A coin issued esp. by a questionable authority and not intended for use as currency. **8.** *Obs.* An illusion : hallucination. —*vt.* **-sied, -sy·ing, -sies.** To imagine.

**fan·toc·ci·ni** (făn'tə-chē'nē) *pl.n.* [Ital., pl. of *fantoccino,* dim. of *fantoccio,* puppet, aug. of *fante,* child, short for *infante* < Lat. *infans,* infant. —see INFANT.] **1.** Puppets animated by moving wires or mechanical means. **2.** A play or puppet show employing fantoccini.

**fan·tod** (făn'tŏd') *n.* [Orig. unknown.] **1. fantods. a.** Nervous irritability. **b.** Nervous movements. **2.** An emotional outburst : FIT.

**fan·tom** (făn'təm) *n.* & *adj. var. of* PHANTOM.

**fan vaulting** *n.* An intricate style of traceried late English Gothic vaulting in which ribs arch out like a fan.

**fan·wort** (făn'wûrt', -wôrt') *n.* An aquatic plant of the genus *Cabomba,* with fanlike submerged leaves.

**far** (fär) *adv.* **far·ther** (fär'thər) *or* **fur·ther** (fûr'thər), **far·thest** (fär'thĭst) *or* **fur·thest** (fûr'thĭst) [ME (adj.) < OE *feor.*] **1.** To, from, or at considerable distance. **2.** To or at a specific distance, degree, or position <Just how *far* are you taking this complaint?> **3.** To a considerable degree : MUCH <felt *far* better on Friday> **4.** Not at all : anything but <is *far* from happy> **5.** To an advanced point or stage <a brilliant medical student who will go *far*> —*adj.* **far·ther** *or* **fur·ther, far·thest** *or* **fur·thest. 1.** At considerable distance <a *far* land> **2.** More distant <the *far* corner of the room> **3.** Extensive or lengthy <a *far* climb> **4.** Having extreme political views <the *far* right> **—by far.** To a considerable degree. **—far and away.** By a great margin <*far and away* the better swimmer> **—far and wide.** Everywhere. **—so far. 1.** Up to the present moment <*So far* we have heard nothing.> **2.** To a limited degree <You can only go *so far* on $15.>

☆ **syns:** FAR, DISTANT, FARAWAY, FAR-OFF, REMOTE, REMOVED *adj. core meaning* : widely separated from others in space, time, or relationship <the *far* north><a *far* country><the *far* past>

**far·ad** (făr′əd, -ăd′) *n.* [After Michael *Faraday* (1791–1867).] A unit of capacitance equal to the capacitance of a capacitor having a charge of one coulomb on each plate and a potential difference of one volt between the plates.

**far·a·da·ic** (făr′ə-dā′ĭk) *adj. var. of* FARADIC.

**far·a·day** (făr′ə-dā′, -dē) *n.* [After Michael *Faraday* (1791–1867).] The quantity of electricity capable of depositing or dissolving 1 gram equivalent weight of a substance in electrolysis, approx. $9.6494 \times 10^4$ coulombs.

**Far·a·day effect** (făr′ə-dā′, -dē) *n.* [After Michael *Faraday* (1791–1867), its discoverer.] The rotation of the plane of polarization of either a plane-polarized light beam passed through a transparent isotropic medium or a plane-polarized microwave passing through a magnetic field along the lines of that field.

**Faraday rotation** *n.* Faraday effect.

**fa·rad·ic** (fə-răd′ĭk) *also* **far·a·da·ic** (făr′ə-dā′ĭk) *adj.* [After Michael *Faraday* (1791–1867).] Of, relating to, or employing an intermittent asymmetric alternating electric current produced by an induction coil.

**far·a·dism** (făr′ə-dĭz′əm) *n.* Faradization.

**far·a·di·za·tion** (făr′ə-dĭ-zā′shən) *n.* Medical therapy by the application of faradic currents.

**far·a·dize** (făr′ə-dīz′) *vt.* **-dized, -diz·ing, -diz·es.** To treat (e.g., an organ) medically with faradic currents.

**far·an·dole** (făr′ən-dōl′) *n.* [Fr. < Prov. *farandoulo.*] **1.** A spirited circle dance of Provençal origin. **2.** The music for a farandole.

**far·a·way** (fär′ə-wā′) *adj.* **1.** Very distant : REMOTE. **2.** Dreamy <a *faraway* look>

**farce** (färs) *n.* [Fr. < OFr., stuffing, interpolation, interlude < *farcir*, to stuff < Lat. *farcire.*] **1.** A theatrical composition in which highly improbable plots and humorous characterizations are used for effect. **2.** A ludicrous, empty show : MOCKERY. **3.** A seasoned stuffing. —*vt.* **farced, farc·ing, farc·es. 1.** To pad or fill out (e.g., a speech) with jokes or witticisms. **2.** To stuff, as for roasting.

**far·ceur** (fär-sœr′) *n.* [Fr.] **1.** One who writes or acts in a farce. **2.** A comic : wag.

**far·ci** *also* **far·cie** (fär-sē′) *adj.* [Fr., p.part. of *farcir*, to stuff < OFr. < Lat. *farcire.*] Stuffed, esp. with finely ground meat.

**far·ci·cal** (fär′sĭ-kəl) *adj.* **1.** Of or relating to farce. **2.** Resembling farce : LUDICROUS. **3.** Ridiculously clumsy : ABSURD. **—far′ci·cal′i·ty** (-kăl′ĭ-tē) *n.* **—far′ci·cal·ly** *adv.* **—far′ci·cal·ness** *n.*

**far·cie** (fär-sē′) *adj. var. of* FARCI.

**far·cy** (fär′sē) *n.* [ME *farsi* < OFr. *farcin* < LLat. *farciminum* < Lat. *farcire*, to stuff.] Chronic cutaneous glanders.

**farcy bud** *n.* A craterlike ulcer typical of farcy.

**far·del** (fär′dl) *n.* [ME < OFr., dim. of *farde*, package < Ar. *fardah.*] *Archaic.* **1.** A pack : bundle. **2.** A burden.

**fare** (fâr) *vi.* **fared, far·ing, fares.** [ME *faren* < OE *faran.*] **1.** To get along <How is the project *faring*?> **2.** To turn out : GO. **3.** To feed on : EAT. **4.** *Archaic.* To travel : wander. —*n.* **1.** A transportation charge, as for a bus. **2.** A passenger transported for a fee. **3.** Food and drink : DIET <peasant *fare*> **—far′er** *n.*

**fare-thee-well** (fâr′thē-wĕl′) *n.* **1.** Perfection. **2.** The most extreme degree.

**fare·well** (fâr-wĕl′) *interj.* [ME *fare wel.*] —Used to say good-by. —*n.* **1.** A good-by. **2.** A departure.

**far·fel** or **far·fal** (fär′fəl) *n.* [Yiddish *farfl* < MHG *varveln.*] Noodles shaped like small grains or pellets.

**far-fetched** (fär′fĕcht′) *adj.* Highly improbable <a *far-fetched* alibi>

**far-flung** (fär′flŭng′) *adj.* **1.** Widely distributed. **2.** Distant : remote.

**fa·ri·na** (fə-rē′nə) *n.* [Lat. < *far*, a kind of grain.] Fine meal prepared from cereal grain and other plant products, often used as a cooked cereal or in puddings.

**far·i·na·ceous** (făr′ə-nā′shəs) *adj.* [LLat. *farinaceus*, mealy < Lat. *farina*, farina.] **1.** Made from, rich in, or composed of starch. **2.** Mealy or powdery in texture.

**far·i·nose** (făr′ə-nōs′) *adj.* [LLat. *farinosus*, mealy < Lat. *farina*, farina.] **1.** Similar to or yielding farina. **2.** *Biol.* Covered with mealy dust or powder.

**far·kle·berry** (fär′kəl-bĕr′ē) *n.* [*Farkle*, poss. alteration of SPARKLE + BERRY.] A shrub or small tree, *Vaccinium arboreum* of the southeastern United States, having leathery leaves and hard black berries.

**farm** (färm) *n.* [ME, lease < OFr. *ferme* < Med. Lat. *firma*, fixed payment < Lat. *firmare*, to establish < *firmus*, firm.] **1.** Land cultivated for agricultural production. **2. a.** Land devoted to the raising and breeding of domestic animals. **b.** An area of water devoted to the raising and breeding of a particular kind of aquatic animal <a trout *farm*> **3.** *Baseball.* A minor-league club affiliated with a major-league club for training recruits and maintaining temporarily unneeded players. **4.** *Obs.* **a.** The system of leasing out the rights of collecting and retaining taxes in a certain district. **b.** A district so leased. —*v.* **farmed, farm·ing, farms.** —*vt.* **1.** To cultivate or produce a crop on. **2.** To pay a fixed sum in order to have the right to collect and retain profits from (e.g., a business). **3.** To turn over (e.g., a business) to another in return for the payment of a fixed sum. —*vi.* To engage in farming. **—farm out. 1.** To send (work) from a central point to be done elsewhere. **2.** *Baseball.* To assign (a player) to a minor-league team.

▲ word history: The word *farm* is derived from Latin *firmus*, "fixed, firm," whose Medieval Latin form, *firma*, meant "fixed payment." The earliest sense of English *farm* is that of a fixed amount payable as a tax or rent. It was a common practice to lease agricultural lands for a fixed annual rent instead of for a percentage of the crop. The word *farm* was also applied to land occupied on such terms. From the 16th century *farm* was used to indicate any cultivated agricultural land, regardless of the circumstances of its tenancy or ownership.

**farm·er** (fär′mər) *n.* **1.** A farm operator, owner, or worker. **2.** One who has paid for the right to collect and retain certain revenues or profits.

**farmer cheese** *n.* An unripened cheese similar to cottage cheese but drier and firmer in texture.

**farmer's lung** *n.* A pulmonary condition caused by continued exposure to moldy hay.

**farm hand** *n.* A hired farm laborer.

**farm·house** (färm′hous′) *n.* A farm dwelling.

**farm·land** (färm′lănd′, -lənd) *n.* Land suitable for farming.

**farm·stead** (färm′stĕd′) *n.* A farm and its buildings.

**farm team** *n. Baseball.* A minor-league team.

**farm·yard** (färm′yärd′) *n.* An area surrounded by or adjacent to farm buildings.

**far·ne·sol** (fär′nĭ-sôl′, -sōl′) *n.* [G. < NLat. *farnesiana*, specific epithet of *Acacia farnesiana* (whose flowers are used in making perfume), after Odoardo *Farnese* (1573–1626).] A compound, $C_{15}H_{26}O$, extracted from the flowers and essential oils of various plants and used in perfumery.

**far·o** (fâr′ō) *n.* [Var. of PHARAOH.] A card game in which the players bet on the top card of the dealer's pack.

**Far·o·ese** *also* **Faer·o·ese** (fâr′ō-ēz′, -ēs′) *n., pl.* **Faroese** *also* **Faeroese. 1.** One of the Germanic people inhabiting the Faeroe Islands. **2.** The North Germanic language spoken by the residents of the Faeroe Islands. **—Fa′ro·ese′** *adj.*

**far-off** (fär′ôf′, -ŏf′) *adj.* Remote : distant.

**far-out** (fär′out′) *adj. Slang.* Very unconventional.

**far point** *n.* The farthest point at which an object can be seen distinctly by the eye at rest.

**far·rag·i·nous** (fə-răj′ə-nəs) *adj.* [< Lat. *farrago, farragin-*, mixture < *far*, a kind of grain.] Made up of a variety of substances.

**far·ra·go** (fə-rä′gō, -rā′-) *n., pl.* **-goes.** [Lat. *farrāgo* < *far*, a kind of grain.] A medley : conglomeration.

**far-reach·ing** (fär′rē′chĭng) *adj.* Having a wide range, influence, or effect <*far-reaching* diplomatic negotiations>

**far-red** (fär′rĕd′) *adj.* **1.** Of, relating to, or being electromagnetic radiation with wavelengths between 30 and 1000 microns. **2.** Of, relating to, or being infrared light of wavelengths closest to those of visible red light, about .8 micron.

**far·ri·er** (făr′ē-ər) *n.* [OFr. *ferrier* < Lat. *ferrārius* < *ferrum*, iron.] One that shoes horses or treats them medically. **—far′ri·er·y** *n.*

**far·row¹** (făr′ō) *n.* [ME **farwes* < OE *fearh*, pig.] A litter of pigs. —*v.* **-rowed, -row·ing, -rows.** —*vt.* To give birth to (a farrow). —*vi.* To produce a farrow.

▲ word history: The Old English ancestor of *farrow¹*, *fearh*, meant "little pig." It is cognate with the Latin word *porcus*, "pig," from which English *pork* is derived. Cognate words have a common ancestor, which in the case of *fearh* and *porcus* is the Indo-European form *porkos*. The development of *fearh* from *porkos* illustrates a linguistic process peculiar to the Germanic languages called Grimm's Law. This law observes that the consonants that originally existed in Indo-European changed in a systematic way in the earliest Germanic speech. In the development of *porkos* to *fearh*, *p* changed to *f* and *k* changed to *h*, which in Old English represented not just the simple aspirate but also the sound of *ch* in the Scottish word *loch*. The Latin word *porcus*, in which *c* is a spelling of the sound of *k*, preserves most of the features of the ancestor form.

**far·row²** (făr′ō) *adj.* [ME *ferow.*] Not pregnant. —Used of a cow.

**far·see·ing** (fär′sē′ĭng) *adj.* **1.** Foresighted. **2.** Keen-sighted.

**far·sight·ed** (fär′sī′tĭd) *adj.* **1. a.** Able to see objects better from a distance than from short range. **b.** Hyperopic. **2.** Foresighted. **—far′sight′ed·ly** *adv.* **—far′sight′ed·ness** *n.*

**far·ther** (fär′thər) *adv.* [ME *ferther* < OE *furðor.*] **1.** To or at a more distant or remote point in space or time. **2.** In addition. **3.** To a greater extent or degree. —*adj.* **1.** More distant or remote. **2.** Additional. ***usage:*** According to traditional grammarians, *farther* should be used only in connection with physical distance, as in *They sailed farther down the coast. Further* should be used esp. in referring to degree, quantity, or time, as in *They fell further into debt.* But it should be noted that the distinction between the two terms has often been ignored by writers since the time of Shakespeare.

**far·ther·most** (fär′*th*ər-mōst′) *adj.* Farthest.

**far·thest** (fär′thĭst) *adj.* [ME *ferthest.*] Most remote or distant. —*adv.* **1.** To or at the most distant or remote point in space or time. **2.** To the most advanced stage or point. **3.** By the greatest extent or degree.

**far·thing** (fär′thĭng) *n.* [ME *ferthing* < OE *fēorðung.*] **1.** A former British coin worth one fourth of a penny. **2.** Something of very little value.

**far·thin·gale** (fär′thĭn-gāl′, -thĭng-) *n.* [Alteration of OFr. *verdugale* < OSp. *verdugadu* < *verdugo*, stick < *verde*, green < Lat. *viridis.* —see VERDANT.] A support, as a hoop, making a skirt extend horizontally from the waist, worn by 16th- and 17th-cent. women.

**farthingale**

**fas·ces** (făs′ēz′) *pl.n.* [Lat., pl. of *fascis*, bundle.] A bundle of rods bound together about an ax with the blade projecting, carried before ancient Roman magistrates as an emblem of authority.

**fas·ci·a** (făsh′ē-ə) *n., pl.* **-ci·ae** (-ē-ē′) [NLat. < Lat., band.] **1.** *Anat.* A sheet of fibrous tissue beneath the surface of the skin, enveloping the body, enclosing muscles or muscular groups, and separating muscular layers. **2.** A broad and distinct band of color. **3.** (*also* fā′shē-ə). A flat horizontal band or member between architectural moldings, esp. in a classical entablature. **4.** (fā′shə). *Chiefly Brit.* The dashboard, as of a car. **—fas′ci·al** *adj.*

**fas·ci·ate** (făsh′ē-āt′) *also* **fas·ci·at·ed** (-ā′tĭd) *adj.* [Lat. *fasciatus*, p.part. of *fasciare*, to swathe < *fascia*, band.] **1.** *Bot.* Abnormally flattened or coalesced, as certain stems. **2.** *Zool.* Marked by broad bands of color, as certain insects.

**fas·ci·a·tion** (făs′ē-ā′shən, făsh′ē-) *n.* **1.** The act of binding up or fastening, as with bandages. **2.** The way in which something is bound up or fastened. **3.** *Bot.* An abnormal flattening or coalescence of stems or leaf stalks.

**fas·ci·cle** (făs′ĭ-kəl) *n.* [Lat. *fasciculus*, dim. of *fascis*, bundle.] **1.** A small bundle. **2.** *also* **fas·ci·cule** (-kyōōl′). One of the parts of a book published in separate sections. **3.** *Bot.* A bundlelike cluster, as of stems, flowers, or leaves. **4.** *Anat.* A fasciculus. **—fas′ci·cled** *adj.*

**fas·cic·u·lar** (fə-sĭk′yə-lər) *adj.* Of, relating to, or made up of fascicles. **—fas·cic′u·lar·ly** *adv.*

**fas·cic·u·late** (fə-sĭk′yə-lĭt) *also* **fas·cic·u·lat·ed** (-lā′tĭd) *adj.* Relating to or resembling a fascicle. **—fas·cic′u·late·ly** *adv.* **—fas·cic′u·la′tion** *n.*

**fas·ci·cule** (făs′ĭ-kyōōl′) *n. var. of* FASCICLE 2.

**fas·cic·u·lus** (fə-sĭk′yə-ləs) *n., pl.* **-li** (-lī′) [Lat., fascicle.] A bundle of anatomical fibers, esp. a bundle of nerve fibers having common functions and connections.

**fas·ci·nate** (făs′ə-nāt′) *vt.* **-nat·ed, -nat·ing, -nates.** [Lat. *fascinare, fascinat-*, to enchant < *fascinum*, witchcraft.] —*vt.* **1.** To have a strong interest or attraction for. **2.** To hold motionless : SPELLBIND. **3.** *Obs.* To bewitch. —*vi.* To be irresistibly attractive.

**fas·ci·nat·ing** (făs′ə-nā′tĭng) *adj.* Having the power to charm or allure : CAPTIVATING. **—fas′ci·nat′ing·ly** *adv.*

**fas·ci·na·tion** (făs′ə-nā′shən) *n.* **1.** The power of fascinating. **2.** The state of being fascinated. **3.** A fascinating quality or trait.

**fas·ci·na·tor** (făs′ə-nā′tər) *n.* **1.** One that fascinates. **2.** A woman's head scarf.

**fas·cine** (fă-sēn′, fə-) *n.* [Fr. < Lat. *fascina* < *fascis*, bundle.] A bundle of sticks bound together for use in construction, as of fortresses and earthworks.

**fas·ci·o·li·a·sis** (fə-sē′ə-lī′ə-sĭs, -sī′-) *n.* [NLat. *Fasciola*, fluke genus + -IASIS.] Infestation with a trematode worm of the genus *Fasciola.*

**fas·cism** (făsh′ĭz′əm) *n.* [Ital. *fascismo* < *fascio*, group < Lat. *fascis*, bundle.] **1.** A philosophy or governmental system marked by stringent socioeconomic control, a strong central government usu. headed by a dictator, and often a belligerently nationalistic policy. **2.** Oppressive, dictatorial control. **—fas·cis′tic** (fə-shĭs′tĭk) *adj.*

**fas·cist** *or* **Fas·cist** (făsh′ĭst) *n.* [Ital. *fascista* < *fascio*, group. —see FASCISM.] An advocate or supporter of fascism.

**Fa·scis·ti** (fă-shē′stē) *pl.n.* [Ital., pl. of *fascista*, fascist.] The members of the Italian political organization led by Benito Mussolini.

**fash·ion** (făsh′ən) *n.* [ME *facioun* < OFr. *faceon* < Lat. *factio* < *factus*, p.part. of *facere*, to do.] **1.** The manner in which something is formed : CONFIGURATION. **2.** Kind or variety : SORT. **3.** A manner of performing : WAY. **4.** Current style or custom. **5.** A piece of clothing in the current style. **6.** The manners, customs, and mode of life typical of the upper classes. —*vt.* **-ioned, -ion·ing, -ions. 1. a.** To shape or form into. **b.** To train or influence into a specific state or character. **2.** To adapt, as to a purpose or occasion. **3.** *Obs.* To contrive. **—after (or in) a fashion.** In some way or other, esp. to a limited extent <They paint, *after a fashion.*> **—fash′ion·er** *n.*

☆ ***syns:*** FASHION, CRAZE, CRY, MODE, RAGE, STYLE, THING, TREND, VOGUE *n. core meaning* : the current custom <jeans—the *fashion* of the day>

**fash·ion·a·ble** (făsh′ə-nə-bəl) *adj.* **1.** Conforming to the current style. **2.** Of or associated with people who conform to the current fashion. —*n.* A stylish person. **—fash′ion·a·bil′i·ty** *n.* **—fash′ion·a·ble·ness** *n.* **—fash′ion·a·bly** *adv.*

☆ ***syns:*** FASHIONABLE, CHIC, IN, MODISH, SHARP, SMART, STYLISH, TONY, TRENDY, WITH-IT *adj. core meaning* : in accordance with the current fashion <a *fashionable* fur><a *fashionable* little restaurant> ***ant:*** unfashionable

**fash·ion·mon·ger** (făsh′ən-mŭng′gər, -mŏng′-) *n.* A person concerned with following or setting fashions.

**fashion plate** *n.* **1.** An illustration of current styles in dress. **2.** One who consistently wears the latest fashions.

**fast**[1] (făst) *adj.* **-er, -est.** [ME < OE *fæst*, firmly fixed.] **1.** Performing or moving or capable of performing or moving quickly : SWIFT. **2.** Accomplished in relatively little time <a *fast* visit> **3.** Indicating a time somewhat ahead of the actual time <a *fast* watch> **4.** Adapted to or suitable for rapid movement. **5. a.** Disposed to dissipation : WILD <led a *fast* life> **b.** Flouting conventional sexual standards <a *fast* crowd> **6.** Resistant <acid-*fast*> **7.** Firmly fixed or fastened <a *fast* grip> **8.** Fixed firmly in position : SECURE <blinds *fast* against the rain> **9.** Loyal : firm <a *fast* friendship> **10.** Resisting fading <*fast* dyes> **11.** Deep and undisturbed <a *fast* sleep> **12.** Designed for or compatible with a short exposure time <*fast* film> —*adv.* **1.** Securely : tightly. **2.** Deeply : soundly <*fast* asleep> **3.** Quickly : rapidly. **4.** Ahead of the correct or expected time. **5.** In a dissipated, immoderate way <living *fast*> **6.** *Archaic.* Close by : NEAR. **—fast and loose 1.** Deceitfully <played *fast and loose* with the facts in order to win> **2.** Irresponsibly <played *fast and loose* with their partner's money>

☆ ***syns:*** FAST, BREAKNECK, FLEET, HELL-FOR-LEATHER, QUICK, RAPID, SPEEDY, SWIFT *adj. core meaning* : marked by great celerity <a *fast* freight train><a *fast* pace> ***ant:*** slow

**fast**[2] (făst) *vi.* **fast·ed, fast·ing, fasts.** [ME *fasten* < OE *fæstan.*] **1.** To abstain from food. **2.** To eat very little or abstain from certain foods, esp. as a religious discipline. **—fast** *n.*

**fast·back** (făst′băk′) *n.* A car having a curving downward slope from roof to rear.

**fast-breed·er reactor** (făst′brē′dər) *n.* A breeder reactor that requires high-energy neutrons to produce fissionable material.

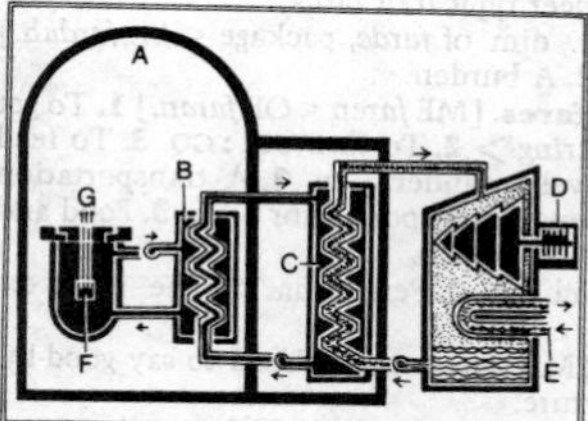


**fast-breeder reactor**
*A liquid metal fast-breeder reactor: A. containment structure, B. heat exchanger, C. steam generator, D. turbine generator, E. cooling water, F. core, G. control rods*

**fas·ten** (făs′ən) *v.* **-tened, -ten·ing, -tens.** [ME *fastnen* < OE *fæstnian.*] —*vt.* **1.** To attach firmly to : JOIN. **2. a.** To make fast or secure. **b.** To close, as by fixing firmly in place. **3.** To fix or direct steadily. **4. a.** To place : attribute. **b.** To impose (oneself) without welcome. —*vi.* **1.** To become attached, fixed, or joined. **2.** To cling fast. **3.** To fix or focus steadily. **—fas′ten·er** *n.*

**fas·ten·ing** (făs′ə-nĭng) *n.* Something, as a hook, used to fasten.

**fast food** *n.* Restaurant food prepared and served quickly. **—fast′-food′** (făst′fōōd′) *adj.*

**fas·tid·i·ous** (fă-stĭd′ē-əs, fə-) *adj.* [ME < OFr. *fastidieux* < *Lat. fastidiosus* < *fastidium*, loathing.] **1.** Meticulously attentive to detail. **2.** Difficult to please : EXACTING. **3.** Excessively scrupulous, esp. in matters of taste or propriety. **4.** Having complicated nutritional requirements. **—fas·tid′i·ous·ly** *adv.* **—fas·tid′i·ous·ness** *n.*

**fas·tig·i·ate** (fă-stĭj′ē-ĭt) *also* **fas·tig·i·at·ed** (-ē-ā′tĭd) *adj.* [Med. Lat. *fastigiatus*, high < Lat. *fastigium*, top.] **1.** Tapering to a point. **2.** *Bot.* Erect and almost parallel, as certain branches. **—fas·tig′i·ate·ly** *adv.*

**fas·tig·i·um** (fă-stĭj′ē-əm) *n.* [NLat. < Lat., extremity.] The period of maximum development of a disease.

**fast·ness** (făst′nĭs) *n.* **1. a.** A fortified place. **b.** A remote and secret place. **2.** The quality or state of being fast, esp.: **a.** Firmness : security. **b.** Rapidity : swiftness. **c.** Colorfastness.

**fast-talk** (făst′tôk′) *vt.* **-talked, -talk·ing, -talks.** To affect or persuade by esp. deceptive talk or means. **—fast′-talk′er** *n.*

**fast-track** (făst′trăk′) *adj.* Very high-powered and aggressive <a *fast-track* career><a *fast-track* manager>

**fat** (făt) *n.* [ME < *fat*, plump < OE *fǣtt*, fattened.] **1. a.** The glyceride ester of a fatty acid. **b.** Any of various soft solid or semisolid organic compounds comprising the glyceride esters of fatty acids and associated phosphatides, sterols, alcohols, hydrocarbons, ketones, and related compounds. **c.** A mixture of such compounds widely occurring in organic tissue, esp. in the subcutaneous connective tissue of animals and in the seeds, nuts, and fruits of plants. **d.** Organic tissue containing such substances. **e.** A solidified animal or vegetable oil. **2.** Obesity. **3.** The most desirable part. *—adj.* **fat·ter, fat·test. 1.** Having much or too much fat or flesh. **2.** Full of fat or oil : GREASY. **3.** Abounding in desirable elements <*Fat* pine yields much resin.> **4.** Fertile or productive : RICH. **5.** Having an abundant supply <a *fat* wallet> **6.** Yielding profit or plenty : LUCRATIVE <a *fat* commission> **7. a.** Thick : large. **b.** Puffed up : SWOLLEN. *—vt. & vi.* **fat·ted, fat·ting, fats.** To make or become fat. **—fat chance.** *Slang.* Little or no possibility of taking place. **—fat′ly** *adv.* **—fat′ness** *n.*

☆ ***syns:*** FAT, CORPULENT, FLESHY, GROSS, OBESE, OVERWEIGHT, PORCINE, PORTLY, STOUT, WEIGHTY *adj. core meaning* : having too much flesh <a *fat* person> ***ant:*** skinny, thin

**fa·tal** (făt′l) *adj.* [ME < OFr. < Lat. *fatalis* < *fatum*, fate.] **1.** Causing or capable of causing death. **2.** Causing ruin or destruction : DISASTROUS <a *fatal* error> **3.** Most decisive : FATEFUL. **4.** Controlling destiny. **5.** *Obs.* Destined. **—fa′tal·ly** (-ē) *adv.*

▲ word history: The Latin adjective *fatalis*, the source of English *fatal*, meant primarily "destined by fate," but in later Latin it also took on the malign senses of the noun *fatum*, "fate," from which it is derived, and meant "deadly, destructive." English adopted *fatal* with its original, neutral Latin meaning, though the pejorative sense, "causing ruin or death," has now completely supplanted the more neutral sense "destined."

**fa·tal·ism** (făt′l-ĭz′əm) *n.* **1.** The doctrine that all events are predetermined by fate and thus cannot be changed by human beings. **2.** Acceptance of the doctrine of fatalism. **—fa′tal·ist** *n.* **—fa′tal·is′tic** *adj.* **—fa′tal·is′ti·cal·ly** *adv.*

**fa·tal·i·ty** (fā-tăl′ĭ-tē, fə-) *n., pl.* **-ties. 1. a.** A death resulting from an unexpected occurrence. **b.** One that is killed as a result of such an occurrence. **2.** The ability to cause death or disaster. **3.** The quality or state of being decided or governed by fate. **4.** A decree made by fate : DESTINY. **5.** The quality or state of being doomed to disaster.

**fatality rate** *n.* Death rate.

**fa·ta mor·ga·na** (fä′tə môr-gä′nə) *n.* [Ital., mirage, Morgan le Fay (from the belief that the mirage was caused by her witchcraft).] MIRAGE 1.

**fat·back** (făt′băk′) *n.* The strip of fat taken from the upper part of a side of pork, usu. dried and salt-cured.

**fat body** *n.* A food reserve of fatty tissue in the larval stages of an insect.

**fat cat** *n. Slang.* **1.** A wealthy and privileged person. **2.** A rich person who is a heavy contributor to a political party.

**fat city** *n. Slang.* A condition of great wealth.

**fate** (fāt) *n.* [ME < OFr. < Lat. *fatum* < neuter p.part. of *fari*, to speak.] **1. a.** The force, principle, or power that is thought to predetermine events. **b.** Inevitable events predestined by this force. **2.** A final consequence : OUTCOME. **3.** Unfavorable destiny : DOOM <met an unhappy *fate*> **4. Fates.** *Gk. & Rom. Myth.* The three goddesses who govern human destiny.

☆ ***syns:*** FATE, DESTINY, KISMET, LOT, PORTION, PREDESTINATION *n. core meaning* : that which is inevitably destined to occur <a person whose *fate* it was to become a monarch>

▲ word history: Human beings seem to find it difficult to keep an open mind about the future: they expect either the best or the worst to happen. The word *fate* exemplifies this tendency. *Fatum*, the Latin ancestor of *fate*, started out neutrally as the past participle of *fari*, "to say." It originally meant "an utterance," especially an oracle or a prophecy, but even in classical times *fatum* had come to mean "calamity" and especially "death."

**fat·ed** (fā′tĭd) *adj.* **1.** Governed by fate. **2.** Condemned to death or destruction : DOOMED.

**fat farm** *n. Slang.* A spa for dieters.

**fate·ful** (fāt′fəl) *adj.* **1.** Affecting one's destiny or future <a *fateful* decision> **2.** Governed by or as if by fate : PREDETERMINED. **3.** Causing death or destruction. **4.** Portentous : ominous. **—fate′ful·ly** *adv.* **—fate′ful·ness** *n.*

**fat·head** (făt′hĕd′) *n. Slang.* A stupid person. **—fat′head′ed** *adj.* **—fat′head′ed·ly** *adv.* **—fat′head′ed·ness** *n.*

**fa·ther** (fä′thər) *n.* [ME *fader* < OE *fæder.*] **1.** A male parent of a child. **2.** One who functions in a paternal capacity. **3.** A male ancestor : FOREFATHER. **4.** A man who creates or originates something <a *father* of the space program> **5. Father. a.** GOD 1a, b. **b.** The first person of the Trinity. **6.** An old or venerable man. —Used as a title of respect. **7.** A member of the ancient Roman senate. **8.** *often* **Father.** Any of various authoritative early writers in the Christian Church who formulated doctrines and codified religious observances. **9.** A priest or clergyman in the Roman Catholic, Anglican, or Eastern Orthodox churches. —Used as a title. *—v.* **-thered, -ther·ing, -thers.** *—vt.* **1.** To beget. **2.** To act or serve as a father to. **3.** To create, found, or originate. **4.** To acknowledge responsibility for. **5. a.** To attribute the paternity, creation, or origin of. **b.** To assign falsely or unjustly : FOIST. *—vi.* To act or serve as a father. **—fa′ther·hood′** *n.* **—fa′ther·less** *adj.*

**Father Christmas** *n. Chiefly Brit.* Santa Claus.

**father confessor** *n.* **1.** A priest who hears confessions. **2.** A confidant.

**father figure** *n.* An influential or powerful person who is the object of emotions usu. reserved for a father.

**fa·ther-in-law** (fä′thər-ĭn-lô′) *n., pl.* **fa·thers-in-law. 1.** The father of one's spouse. **2.** A stepfather.

**fa·ther·land** (fä′thər-lănd′) *n.* **1.** One's native land. **2.** The native land of one's ancestors.

**fa·ther·ly** (fä′thər-lē) *adj.* **1.** Of, relating to, or suitable for a father. **2.** Showing the tenderness or affection of a father. **—fa′ther·li·ness** *n.* **—fa′ther·ly** *adv.*

**Father's Day** *n.* A day honoring fathers observed annually on the third Sunday in June.

**fath·om** (făth′əm) *n., pl.* **fathom** or **-oms.** [ME *fathme* < OE *fæðm*, outstretched arms.] A unit of length equal to 6 feet or approx. 1.83 meters, used mainly to measure and specify marine depths. *—vt.* **-omed, -om·ing, -oms. 1.** To determine the depth of : SOUND. **2.** To get to the bottom of and understand. **—fath′om·a·ble** *adj.*

**Fa·thom·e·ter** (fă-thŏm′ĭ-tər) A trademark for a sonic depth finder.

**fath·om·less** (făth′əm-lĭs) *adj.* **1.** Too deep to be measured. **2.** Too difficult to understand.

**fa·tid·ic** (fə-tĭd′ĭk) *also* **fa·tid·i·cal** (-ĭ-kəl) *adj.* [Lat. *fatidicus* : *fatum*, fate + *dicere*, to say.] Relating to or marked by prophecy.

**fat·i·ga·ble** (făt′ĭ-gə-bəl) *adj.* [Fr. < LLat. *fatigabilis* < Lat. *fatigare*, to fatigue.] Easily tired. **—fat′i·ga·bil′i·ty** *n.*

**fa·tigue** (fə-tēg′) *n.* [Fr. < OFr. < *fatiguer*, to fatigue < Lat. *fatigare.*] **1.** Physical or mental weariness due to exertion. **2.** Exhausting effort or activity : LABOR. **3.** *Physiol.* Decreased capacity or complete inability of an organism, organ, or part to function normally due to excessive stimulation or prolonged exertion. **4.** Weakness in a material, as metal or wood, resulting from prolonged stress. **5. a.** Manual or menial labor, as barracks cleaning, assigned to enlisted military personnel. **b. fatigues.** Clothing designed for fatigue or field duty. *—v.* **-tigued, -tigu·ing, -tigues.** *—vt.* **1.** To tire out : EXHAUST. **2.** To create fatigue in. *—vi.* To feel fatigue.

☆ ***syns:*** FATIGUE, DRAIN, EXHAUST, TIRE, WEAR OUT, WEARY *v. core meaning* : to diminish the strength and energy of <These endless meetings *fatigue* me.>

**fat·ling** (făt′lĭng) *n.* A young animal, as a lamb or calf, fattened for slaughter.

**fat·so** (făt′sō) *n., pl.* **-soes.** *Slang.* A fat person.

**fat-sol·u·ble** (făt′sŏl′yə-bəl) *adj.* Soluble in fats or fat solvents.

**fat·ten** (făt′n) *v.* **-tened, -ten·ing, -tens.** *—vt.* **1.** To make fat. **2.** To fertilize (land). **3.** To increase the amount or size of. *—vi.* To grow fat or fatter. **—fat′ten·er** *n.*

**fat·tish** (făt′ĭsh) *adj.* Somewhat fat. **—fat′tish·ness** *n.*

**fat·ty** (făt′ē) *adj.* **-ti·er, -ti·est. 1. a.** Containing fat. **b.** Containing too much fat. **2.** Full of fat : GREASY. **3.** Derived from or chemically related to fat. *—n., pl.* **-ties.** A fat person. **—fat′ti·ly** *adv.* **—fat′ti·ness** *n.*

**fatty acid** *n.* Any of a large group of monobasic acids having the general formula $C_nH_{2n+1}COOH$, esp. any of a commercially important subgroup obtained from animals and plants, the most abundant of which contain 16 or 18 carbon atoms and include palmitic, stearic, and oleic acids.

**fatty alcohol** *n.* Any of various alcohols derived from plant and animal oils and fats that are used in plastics and pharmaceuticals.

**fa·tu·i·ty** (fə-to͞o′ĭ-tē, -tyo͞o′-, fă-) *n., pl.* **-ties.** [OFr. *fatuite* < Lat. *fatuitas* < *fatuus*, fatuous.] **1.** Stupidity : folly. **2.** A fatuous act, remark, or sentiment.

**fat·u·ous** (făch′o͞o-əs) *adj.* [Lat. *fatuus.*] Smugly or foolishly stupid <a *fatuous* grin> **—fat′u·ous·ly** *adv.* **—fat′u·ous·ness** *n.*

**fau·bourg** (fō′bo͝org′, fō-bo͞or′) *n.* [ME < OFr., alteration of *forsborc* : *fors*, outside (< Lat. *foris*) + *borc*, town (< LLat. *burgus*, of Germanic orig.).] **1.** A suburb. **2.** An inner city district or quarter.

**fau·ces** (fô′sēz′) *pl.n.* [Lat.] (*sing. or pl. in number*). The space between the mouth and pharynx, bounded by the soft palate, the base of the tongue, and the palatine arches. **—fau′cal** (-kəl), **fau′cial** (-shəl) *adj.*

**fau·cet** (fô′sĭt) *n.* [ME < OFr. *fausset* < *fausser*, to break in < LLat. *falsare*, to falsify < Lat. *falsus*, false.] A device for drawing a flow of a liquid, as from a pipe or drum.

**fault** (fôlt) *n.* [ME *faulte* < OFr. < VLat. **fallita* < Lat. *fallere*, to fail.] **1. a.** A weakness : defect. **b.** A mistake : error. **c.** A minor offense. **2.** Responsibility for a mistake or offense : CULPABILITY. **3.** *Geol.* A break in the continuity of a rock formation caused by a

o͞o **boot** ou **out** th **thin** *th* **this** ŭ **cut** ûr **urge** y **young** yo͞o **abuse** zh **vision** ə **about,** item, edible, gallop, circus

shifting or dislodging of the earth's crust, in which adjacent surfaces are differentially displaced parallel to the plane of fracture. **4.** *Elect.* A defect in a circuit or wiring caused by imperfect connections, poor insulation, grounding, or shorting. **5.** A bad service, as in tennis. **6.** *Obs.* A lack or deficiency. —*v.* **fault·ed, fault·ing, faults.** —*vt.* **1.** To criticize. ***usage:*** This transitive sense of *fault* has recently come into widespread use; however, its occurrence actually dates from the 16th cent. **2.** *Geol.* To produce a fault in : FRACTURE. —*vi.* **1.** To commit a fault or error. **2.** *Geol.* To shift so as to produce a fault. **—at fault. 1.** Deserving blame : GUILTY. **2.** Confused and perplexed. **—find fault.** To criticize. **—to a fault.** Excessively <fastidious *to a fault*>

**fault·find·er** (fôlt′fīn′dər) *n.* A petty critic : chronic complainer. **—fault′find′ing** *adj.* & *n.*

**fault·less** (fôlt′lĭs) *adj.* Without fault : PERFECT. **—fault′less·ly** *adv.* **—fault′less·ness** *n.*

**fault plane** *n. Geol.* The plane along which the break or shear of a fault occurs.

**fault·y** (fôl′tē) *adj.* **-i·er, -i·est. 1.** Containing a fault or faults : IMPERFECT. **2.** *Obs.* Deserving blame. **—fault′i·ly** *adv.* **—fault′i·ness** *n.*

**faun** (fôn) *n.* [ME *faun* < Lat. *Faunus,* Faunus.] *Rom. Myth.* One of a group of rural deities depicted as having the body of a man and the horns, ears, tail, and occas. the legs of a goat.

**fau·na** (fô′nə) *n., pl.* **-nas** *or* **-nae** (-nē′) [NLat. < Lat. *Fauna,* sister of Faunus.] Animals as a whole, esp. those of a specific region or period. **—fau′nal** *adj.* **—fau′nal·ly** *adv.*

**fau·nis·tic** (fô-nĭs′tĭk) *adj.* Of or relating to the geographic distribution of fauna. **—fau·nis′ti·cal·ly** *adv.*

**Fau·nus** (fô′nəs) *n.* [Lat.] *Rom. Myth.* A god of nature and fertility.

**Faust** (foust) *also* **Faus·tus** (fou′stəs, fô′-) *n.* [G.] A legendary German magician who sold his soul to the devil in exchange for power and worldly experience. **—Faust′i·an** (fou′stē-ən) *adj.*

**fau·vism** (fō′vĭz′əm) *n.* [Fr. *fauvisme* < *fauve,* wild beast < *fauve,* wild, of Germanic orig.] An early 20th-cent. artistic movement characterized by the use of bold, often distorted forms and vivid colors on canvas.

**faux pas** (fō pä′) *n., pl.* **faux pas** (fō päz′) [Fr. : *faux,* false + *pas,* step.] A social blunder.

**fa·va bean** (fä′və) *n.* [Ital. *fava* (< Lat. *faba,* bean) + BEAN.] The broad bean.

**fa·ve·o·late** (fə-vē′ə-lāt′) *adj.* [< NLat. *faveolus,* dim. of Lat. *favus,* honeycomb.] Pitted with cavities or cells.

**fa·vo·ni·an** (fə-vō′nē-ən) *adj.* [Lat. *favonianus* < *Favonius,* the west wind.] **1.** Of or relating to the west wind. **2.** Mild : gentle.

**fa·vor** (fā′vər) *n.* [ME < OFr. < Lat. < *favēre,* to be favorable.] **1. a.** A kind, gracious, or friendly attitude. **b.** An act exhibiting such an attitude. **2. a.** Friendly regard shown esp. by a superior : PARTIALITY. **b.** A state of being held in such regard. **3.** Approval or support : SANCTION. **4. a.** Something given as a token of love, affection, or remembrance. **b.** A small, decorative gift, as a paper hat, given to each guest at a party. **5.** Advantage : benefit <a decision in our *favor*> **6.** *Obs.* A communication. **7.** *Obs.* **a.** Aspect : appearance. **b.** Countenance : face. **c.** A facial feature. —*vt.* **-vored, -vor·ing, -vors. 1.** To perform a kindness for : OBLIGE. **2.** To treat or look upon with favor. **3.** To be partial to : INDULGE. **4.** To be or tend to be in support of. **5.** To make easier or more possible : AID. **6.** To resemble <children who *favor* their parents> **7.** To treat with care <*favored* my sprained ankle> **—in favor of. 1.** In support of. **2.** To the advantage of. **—fa′vor·er** *n.* **—fa′vor·ing·ly** *adv.*

☆ ***syns:*** FAVOR, ADMIRATION, ESTEEM, ESTIMATION, HONOR, REGARD, RESPECT *n. core meaning* : a feeling of deference, approval, and liking <a judge held in great *favor* by the bar> ***ant:*** disfavor

**fa·vor·a·ble** (fā′vər-ə-bəl, fāv′rə-) *adj.* **1.** Advantageous. **2.** Indicating future success. **3.** Expressing approval. **4.** Embodying or conceding what was desired or requested <a *favorable* response> **5.** Indulgent : partial. **—fa′vor·a·ble·ness** *n.* **—fa′vor·a·bly** *adv.*

☆ ***syns:*** FAVORABLE, AUSPICIOUS, BRIGHT, PROPITIOUS *adj. core meaning* : indicative of future success <an economic climate *favorable* for investment><a *favorable* omen> ***ant:*** inauspicious

**fa·vored** (fā′vərd) *adj.* **1.** Treated or looked upon with special kindness or liking <a *favored* grandchild> **2.** Having special talents, gifts, or beauty. **3.** Having a specified type of physical appearance <well-*favored*><ill-*favored*>

**fa·vor·ite** (fā′vər-ĭt, fāv′rĭt) *n.* [OFr. *favorit* < OItal. *favorita,* p.part. of *favorire,* to favor < *favore,* favor < Lat. *favor.*] **1. a.** One enjoying special favor or regard. **b.** One trusted, indulged, or preferred above all others, esp. by a superior <a *favorite* of the President> **2.** A contestant or competitor considered most likely to win. —*adj.* Liked or preferred above all others.

**favorite son** *n.* **1.** A candidate favored by his own state delegates for nomination at a national political convention. **2.** A person, as a celebrity, viewed with much favor by his hometown.

**fa·vor·it·ism** (fā′vər-ĭ-tĭz′əm, fāv′rĭ-) *n.* **1.** A display of partiality toward a favored person or group. **2.** The state of being held in special favor.

**fa·vour** (fā′vər) *n.* & *v. Chiefly Brit. var. of* FAVOR.

**fa·vus** (fā′vəs) *n.* [Lat., honeycomb.] A chronic fungous infection of the scalp and nails.

**fawn**[1] (fôn) *vi.* **fawned, fawn·ing, fawns.** [ME *faunen* < OE *fagnian,* to rejoice.] **1.** To display obsequious affection. **2.** To seek favor supporting slavishly every opinion and suggestion of a superior. **—fawn′er** *n.* **—fawn′ing·ly** *adv.*

☆ ***syns:*** FAWN, APPLE-POLISH, BOOTLICK, GROVEL, KOWTOW, SLAVER, TRUCKLE *v. core meaning* : to seek favor by supporting slavishly every opinion and suggestion of a superior <an assistant who *fawned* on the general manager>

**fawn**[2] (fôn) *n.* [ME < OFr. *foun,* young animal < Lat. *fetus,* offspring.] **1.** A young deer, esp. one under a year old. **2.** A grayish yellowish brown.

**fawn lily** *n.* A North American plant of the genus *Erythronium,* esp. *E. grandiflorum* of western North America, with nodding yellow flowers.

**fax** (făks) *n.* [Shortening and alteration of FACSIMILE.] FACSIMILE 2.

**fay**[1] (fā) *vt.* **fayed, fay·ing, fays.** [ME *feien* < OE *fēgan.*] To join or fit closely or tightly.

**fay**[2] (fā) *n.* [ME *faie* < OFr. *fae,* enchanted < Lat. *fata,* the Fates, pl. of *fatum,* fate.] FAIRY 1.

**fay**[3] (fā) *n.* [ME *fai.*] *Obs.* Faith.

**fay·a·lite** (fā′ə-līt′) *n.* [G. *Fayalit* < *Faial,* Faial, an island in the Azores.] A yellowish to black mineral, $FeSiO_4$, of the olivine group.

**faze** (fāz) *vt.* **fazed, faz·ing, faz·es.** [Var. of FEEZE.] To disturb the composure of : DISCONCERT.

**Fe** [Lat. *ferrum,* iron.] *symbol for* IRON.

**fe·al·ty** (fē′əl-tē) *n., pl.* **-ties.** [ME *fealtye* < OFr. *fealte* < Lat. *fidelitas,* faithfulness < *fidelis,* faithful < *fides,* faith.] **1. a.** The loyalty of a vassal to a feudal lord. **b.** The obligation of such loyalty. **2.** Faithfulness : allegiance.

**fear** (fîr) *n.* [ME *fer* < OE *fǣr.*] **1. a.** Alarm and agitation caused by the expectation or realization of danger. **b.** An instance of this. **2.** Extreme reverence or awe. **3.** A ground for dread or apprehension : DANGER. —*v.* **feared, fear·ing, fears.** —*vt.* **1.** To be frightened of. **2.** To be apprehensive about. **3.** To be in awe of : REVERE <*fear* God> **4.** To suspect <I *fear* you are wrong.> **5.** *Archaic.* To feel fear within (oneself). —*vi.* **1.** To be frightened. **2.** To feel apprehensive. **—fear′er** *n.*

**fear·ful** (fîr′fəl) *adj.* **1.** Causing or capable of causing fear. **2.** Experiencing fear : FRIGHTENED. **3.** Feeling apprehensive. **4.** Feeling reverence, dread, or awe. **5.** Indicating fear. **6.** *Informal.* Extreme, as in degree or extent <a *fearful* error> **—fear′ful·ly** *adv.* **—fear′ful·ness** *n.*

**fear·less** (fîr′lĭs) *adj.* Having no fear : BRAVE. **—fear′less·ly** *adv.* **—fear′less·ness** *n.*

**fear·some** (fîr′səm) *adj.* **1.** FEARFUL 1. **2.** Timid : afraid. **—fear′some·ly** *adv.* **—fear′some·ness** *n.*

**fea·si·ble** (fē′zə-bəl) *adj.* [ME *fesable* < OFr. *faisible* < *faire,* to do < Lat. *facere.*] **1.** Capable of being accomplished or brought about : POSSIBLE <a *feasible* plan> **2.** Capable of being utilized or dealt with successfully : SUITABLE. **3.** Logical and likely. **—fea′si·bil′i·ty, fea′si·ble·ness** *n.* **—fea′si·bly** *adv.*

**feast** (fēst) *n.* [ME *feste* < OFr. < Lat. *festum* < *festus,* joyous.] **1.** A periodic religious festival commemorating an event or honoring a deity, event, or person. **2.** A large elaborate meal, usu. for many people and often having entertainment : BANQUET. **3.** Something providing great pleasure or satisfaction <a *feast* for the eyes> —*v.* **feast·ed, feast·ing, feasts.** —*vt.* **1.** To entertain or feed sumptuously. **2.** To provide with pleasure <*feasted* my eyes on the gorgeous vista> —*vi.* **1.** To partake of a feast. **2.** To experience something with delight. **—feast′er** *n.*

**Feast of Lights** *n.* Chanukah.

**feat**[1] (fēt) *n.* [ME *fet* < AN < OFr. *fait* < Lat. *factum* < neuter p.part. of *facere,* to do.] **1. a.** An act : deed. **b.** A courageous, daring act : EXPLOIT. **2.** An act of skill, endurance, imagination, or strength : ACHIEVEMENT. **3.** *Obs.* A specialized skill : KNACK.

☆ ***syns:*** FEAT, ACHIEVEMENT, EXPLOIT, GEST, MASTERSTROKE, STUNT, TOUR DE FORCE *n. core meaning* : a great or very clever deed <acrobatic *feats*><a *feat* of bureaucratic manipulation>

**feat**[2] (fēt) *adj.* **-er, -est.** [ME *fete,* suitable < OFr. *fait* < Lat. *factum,* deed. —see FEAT[1].] *Archaic.* **1.** Adroit : dexterous. **2.** Neat : trim. **—feat′ly** *adv.*

**feath·er** (fĕth′ər) *n.* [ME *fether* < OE *feðer.*] **1.** One of the light, horny structures forming birds' plumage, made up of many slender, closely arranged parallel barbs forming a vane on either side of a tapering hollow shaft. **2. feathers.** Plumage. **3. feathers.** Clothing. **4.** A tuft or fringe of hair, as on the legs or tail of some dogs. **5.** Character : nature. **6.** Something trivial or inconsequential. **7. a.** A strip, wedge, or flange used for strengthening something. **b.** A wedge or key that fits into a groove to make a joint. **8.** The vane of an arrow. **9.** A feather-shaped flaw, as in a gem. **10.** The wake made by a submarine's periscope. **11. a.** The act of feathering the blade of an oar in rowing. **b.** The act of feathering a propeller. —*v.* **-ered, -er·ing,**

**-ers.** —*vt.* **1.** To cover, dress, or decorate with or as if with feathers. **2.** To fit (an arrow) with a feather. **3. a.** To thin, reduce, or fringe the edge of by cutting, shaving, or wearing away. **b.** To shorten and taper (hair) by cutting and thinning. **4.** To connect with a tongue-and-groove joint. **5.** To turn (an oar blade) horizontal to the surface of the water between strokes. **6.** To change the pitch of (a propeller) so that the blade chords are parallel with the line of flight. —*vi.* **1.** To grow feathers or become feathered. **2.** To move, spread, or grow in a way suggestive of feathers. **3.** To feather an oar. **4.** To feather a propeller. **—a feather in (one's) cap.** An act or deed to one's credit. **—feather (one's) nest.** To grow wealthy by making use of property or funds left in one's trust. **—in fine feather.** In excellent form, health, or humor. **—feath'er·less** *adj.*

**feather bed** *n.* **1.** A mattress stuffed with feathers. **2.** A bed with a feather mattress.

**feath·er·bed·ding** (fĕth'ər-bĕd'ĭng) *n.* The practice of requiring an employer to hire more workers than needed for a given purpose or amount of production because of safety regulations or union rules. **—feath'er·bed'** *v.* **(-bed·ded, -bed·ding, -beds).**

**feath·er·bone** (fĕth'ər-bōn') *n.* A lightweight corset bone orig. made from the quills of domestic fowl.

**feath·er·brain** (fĕth'ər-brān') *n.* A silly, flighty person. **—feath'-er·brained'** *adj.*

**feath·er·edge** (fĕth'ər-ĕj') *n.* A thin fragile edge, esp. a tapering edge of a board.

**feather grass** *n.* A grass of the genus *Stipa*, with clusters of featherlike spikelets.

**feather rot** *n.* A disease of tree trunks caused by the fungus *Poria subacida* that makes the trunk become stringy or spongy.

**feather star** *n.* A crinoid of the genus *Antedon* or related genera, having a free-moving, stalkless adult stage with branched, feathery arms.

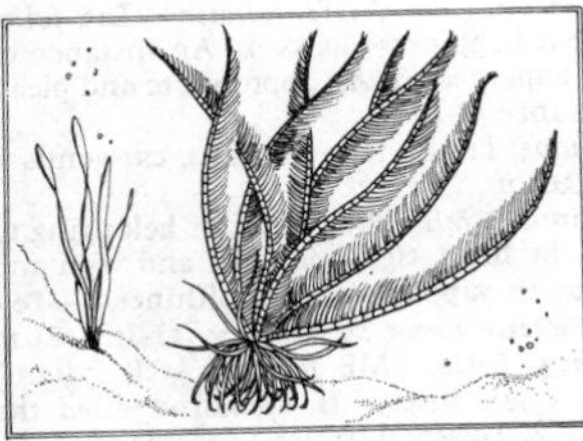
**feather star**

**feath·er·stitch** (fĕth'ər-stĭch') *n.* An embroidery stitch that produces a decorative zigzag line. **—feath'er·stitch'** *v.* **(-stitched, -stitch·ing, -stitch·es).**

**feath·er·weight** (fĕth'ər-wāt') *n.* **1.** A boxer or wrestler weighing between 118 and 127 pounds or approx. 54 and 57 kilograms. **2.** One of little weight or size. **3.** An insignificant person.

**feath·er·y** (fĕth'ə-rē) *adj.* **1.** Covered with or made of feathers. **2.** Resembling a feather. **—feath'er·i·ness** *n.*

**fea·ture** (fē'chər) *n.* [ME *feture* < OFr. < Lat. *factura* < *factus*, p.part. of *facere*, to make.] **1.** The make-up, shape, proportions, form, or outward appearance, esp. of a person. **2. a.** The make-up or appearance of the face or its parts. **b.** Any of the distinct parts of the face. **3.** A prominent aspect, quality, or characteristic. **4.** The main presentation at a movie theater. **5.** A prominent article in a newspaper or periodical. **6.** Something, as a sale item in a store, that is advertised as particularly attractive or as an inducement. **7.** *Archaic.* Form : shape. —*vt.* **-tured, -tur·ing, -tures. 1.** To give special attention to. **2.** To have as a characteristic. **3.** To draw or otherwise portray the features of. **4.** *Informal.* To resemble in features : FAVOR. **5.** *Informal.* To imagine.

**fea·tured** (fē'chərd) *adj.* **1. a.** Given particular attention or publicity. **b.** Starring <a *featured* performer> **2.** Having a specified kind of facial feature. **3.** Formed or given a particular appearance by facial features.

**fea·ture-length** (fē'chər-lĕngkth', -lĕngth') *adj.* Being of normal duration or full length <a *feature-length* movie>

**fe·bric·i·ty** (fĭ-brĭs'ĭ-tē) *n.* [Med. Lat. *febricitas* < Lat. *febricitare*, to have a fever < *febris*, fever.] The state of having a fever.

**feb·ri·fa·cient** (fĕb'rə-fā'shənt) *n.* [Lat. *febris*, fever + -FACIENT.] A fever-producing agent. —*adj.* Causing or producing fever.

**fe·brif·ic** (fĭ-brĭf'ĭk) *adj.* [Lat. *febris*, fever + -FIC.] **1.** Febrifacient. **2.** Having a fever : FEVERISH.

**feb·ri·fuge** (fĕb'rə-fyōōj') *n.* [Fr. *fébrifuge* : Lat. *febris*, fever + Lat. *fugare*, to drive away.] A fever-reducing agent. —*adj.* Reducing fever.

**feb·rile** (fĕb'rəl, fē'brəl) *adj.* [Fr. *fébrile* < Lat. *febris*, fever.] Of, relating to, or marked by fever : FEVERISH.

**Feb·ru·ar·y** (fĕb'rōō-ĕr'ē, fĕb'yōō-) *n., pl.* **-ies** *or* **-ys.** [ME < Lat. *Februarius* < *februa*, festival of purification, of Sabine orig.] The second month of the year according to the Gregorian calendar. —See table at CALENDAR.

**fe·cal** (fē'kəl) *adj.* Of, relating to, containing, or constituting feces.

**fe·ces** *also* **fae·ces** (fē'sēz) *pl.n.* [ME < Lat. *faeces*, pl. of *faex*, dregs.] Waste excreted from the bowels : EXCREMENT.

**feck·less** (fĕk'lĭs) *adj.* [Sc., short for EFFECT + -LESS.] **1.** Lacking purpose or vitality : INEFFECTIVE. **2.** Careless : irresponsible <*feckless* youth> **—feck'less·ly** *adv.* **—feck'less·ness** *n.*

**fec·u·lent** (fĕk'yə-lənt) *adj.* [ME < Lat. *faeculentus* < *faex*, dregs.] Full of foul matter : FECAL. **—fec'u·lence** *n.*

**fe·cund** (fē'kənd, fĕk'ənd) *adj.* [ME < OFr. *fecond* < Lat. *fecundus*.] **1.** Capable of producing offspring or vegetation : FRUITFUL. **2.** Intellectually productive. **—fe·cun'di·ty** (fĭ-kŭn'dĭ-tē) *n.*

**fe·cun·date** (fē'kən-dāt', fĕk'ən-) *vt.* **-dat·ed, -dat·ing, -dates.** [Lat. *fecundare*, *fecundat-* < *fecundus*, fecund.] **1.** To make fecund. **2.** To impregnate : FERTILIZE. **—fe'cun·da'tion** *n.*

**fed** (fĕd) *v. p.t. & p.p. of* FEED.

**Fe·da·yee** (fĭ-dä'yē', -dä'ē', -dä'-) *n., pl.* **-yeen** (-yēn', -ēn') [Ar. *fe-dā'yūn* < *fidā'ī*, one who sacrifices himself for his country < *fidā'*, redemption.] An Arab commando operating esp. against Israel.

**fed·er·al** (fĕd'ər-əl, fĕd'rəl) *adj.* [< Lat. *foedus, foeder-*, league.] **1.** Of, relating to, or designating a form of government in which a union of states recognizes the sovereignty of a central authority while retaining certain residual powers of government. **2.** Of, constituting, or marked by a form of government in which sovereign power is divided between a central authority and a number of constituent political units. **3.** Of or relating to the central government of a federation as distinct from the governments of its member units. **4.** Favorable to or advocating federation. **5.** Of or relating to a league, treaty, or compact. **6. Federal. a.** Of, relating to, or characterizing Federalism. **b.** Of, relating to, or supporting the Union during the American Civil War. —*n.* **Federal.** A supporter of the Union during the American Civil War, esp. a Union soldier. **—fed'er·al·ly** *adv.*

**federal district** *n.* An area reserved as the site of the national capital of a federation, as the District of Columbia.

**fed·er·al·ism** (fĕd'ər-ə-lĭz'əm, fĕd'rə-) *n.* **1. a.** The doctrine or system of federal government. **b.** Advocacy of this system of government. **2. Federalism.** The doctrine of the Federalists.

**fed·er·al·ist** (fĕd'ər-ə-lĭst, fĕd'rə-) *n.* **1.** An advocate of federalism. **2. Federalist.** A member or supporter of a political party founded in the United States in 1787 and favoring a strong federal government. —*adj.* **1.** Of or relating to federalism or its advocates. **2. Federalist.** Of or relating to Federalism or Federalists.

**fed·er·al·ize** (fĕd'ər-ə-līz', fĕd'rə-) *vt.* **-ized, -iz·ing, -iz·es. 1.** To unite in a federal union. **2.** To subject to the authority of a federal government. **—fed'er·al·i·za'tion** *n.*

**Federal Reserve System** *n.* A U.S. banking system consisting of 12 banks, with each one serving member banks in its own district.

**fed·er·ate** (fĕd'ə-rāt') *v.* **-at·ed, -at·ing, -ates.** [Lat. *foederare, foederat-*, to enter into a league < *foedus*, league.] —*vt.* To join or bring together in a league or federal union. —*vi.* To become united in a federal union. —*adj.* (fĕd'ər-ĭt, fĕd'rĭt). United under a central government.

**fed·er·a·tion** (fĕd'ə-rā'shən) *n.* **1.** The act of federating, esp. so as to create a federal union. **2.** A league formed by federating.

**fed·er·a·tive** (fĕd'ə-rā'tĭv, fĕd'ər-ə-, fĕd'rə-) *adj.* Forming, belonging to, or of the nature of a federation : FEDERAL. **—fed'er·a'tive·ly** *adv.*

**fe·do·ra** (fĭ-dôr'ə, -dōr'ə) *n.* [After *Fédora*, a play by Victorien Sardou (1831–1908).] A soft felt hat with a brim that can be turned up or down and a low crown creased lengthwise.

**fed up** *adj.* Extremely tired or disgusted <*fed up* with incompetent management>

**fee** (fē) *n.* [ME *fe* < AN *fee*, inherited estate < OFr. *fie*, fief, of Germanic orig.] **1.** A fixed charge <tuition *fees*> **2.** A charge for a professional service <a lawyer's *fee*> **3.** A tip : gratuity. **4.** *Law.* An inherited or heritable estate in land. **5. a.** A feudal estate in land held from a lord on condition of homage and service. **b.** The land so held. —*vt.* **feed, fee·ing, fees.** To give a tip to.

**fee·ble** (fē'bəl) *adj.* **-bler, -blest.** [ME *feble* < OFr., var. of *fleible* < Lat. *flebilis*, lamentable < *flere*, to weep.] **1. a.** Lacking strength : WEAK. **b.** Indicating weakness. **2. a.** Lacking vigor or force. **b.** Inadequate <a *feeble* answer> **—fee'ble·ness** *n.* **—fee'bly** *adv.*

**fee·ble-mind·ed** (fē'bəl-mīn'dĭd) *adj.* **1.** Intellectually subnormal. **2.** Dull-witted. **3.** *Archaic.* Weak-willed. **—fee'ble-mind'ed·ly** *adv.* **—fee'ble-mind'ed·ness** *n.*

**feed** (fēd) *v.* **fed** (fĕd), **feed·ing, feeds.** [ME *feden* < OE *fēdan*.] —*vt.* **1. a.** To supply with nourishment <*feed* livestock> **b.** To provide as food or nourishment <*feed* grain to livestock> **2. a.** To produce serve as food for <enough hay to *feed* my horse> **b.** To produce food for <The valley *feeds* an entire town.> **3.** To supply for consumption or utilization <*feed* ammunition to a gun crew> **4. a.** To minister to : GRATIFY <*fed* their appetite for violence> **b.** To support or promote <*feed* doubts> **5.** To supply as a cue <*feed* lines to a performer> **6.** To pass a ball or puck to (a teammate), esp. in order to score. —*vi.* **1.** To consume as food. **2.** To draw support or satisfaction from <Your ego *feeds* on flattery.> —*n.* **1. a.** Food for animals or birds : FODDER. **b.** A single allowance of fodder. **2.** *Informal.* A meal. **3. a.** Material or an amount of material supplied to a machine. **b.** The act of supplying material to a machine. **4. a.** The apparatus that supplies material to a machine <a paper *feed*> **b.** The aperture through which such material enters a machine.

**feed·back** (fēd′băk′) *n.* **1. a.** Return of a portion of the output of a process or system to the input, esp. to maintain performance or to control a system or process. **b.** The part of the output returned. **2.** Return of data about the result of a process. —*adj.* Of, relating to, or being a device or process that relies on feedback for its operation.

**feedback inhibition** *n.* A control mechanism in a cell in which the excessive accumulation of the end product of a series of biosynthetic reactions inhibits the action of an enzyme that occurs early in the reaction series.

**feed·bag** (fēd′băg′) *n.* A bag that fits over a horse's muzzle and holds its feed.

**feed·er** (fē′dər) *n.* **1.** One that supplies food. **2.** One that feeds materials into a machine for processing. **3.** Something contributing to the operation, maintenance, or supply of something else, esp.: **a.** A tributary. **b.** A branch line of a transport system. **c.** An animal being fattened. **4.** Any of the medium-voltage lines used to distribute electric power from a substation to consumers or smaller substations.

**feed·hole** (fēd′hōl′) *n. Computer Sci.* One of a noninformational series of holes in a paper tape that engages a driving sprocket to carry the tape through a reading or punching device.

**feed·lot** (fēd′lŏt′) *n.* A tract of land on which livestock are fattened for market.

**feed·stock** (fēd′stŏk′) *n.* Raw materials required for an industrial process.

**feed·stuff** (fēd′stŭf′) *n.* FEED 1a.

**feed·through** (fēd′thro͞o′) *n.* A conductor joining two circuits on opposite sides of a nonconducting surface.

**feel** (fēl) *v.* **felt** (fĕlt), **feel·ing, feels.** [ME *felen* < OE *fēlan.*] —*vt.* **1. a.** To perceive through the sense of touch. **b.** To perceive as a physical sensation <*feel* a dull pain><*feel* the heat> **2. a.** To touch. **b.** To examine by touching. **c.** To test or explore with caution <*feel* one's way in a new job> **3. a.** To undergo the experience of <*felt* my interest waning> **b.** To be aware of : SENSE. **c.** To be emotionally affected by <*feel* the loss of friendship> **4.** To believe : consider <Your answer was *felt* to be impolite.> —*vi.* **1.** To experience tactile sensations. **2.** To appear to be, esp. to the tactile sense <The satin *felt* smooth.> **3.** To be conscious of an impression or a physical or mental state. **4.** To search or be guided by or as if by the tactile sense. **5.** To have compassion <*feel* for them in their time of sorrow> **6.** To have a sentiment or subjective view <*feel* strongly about the candidate> —**feel out.** To sound out in order to ascertain the viewpoint or opinion of. —*n.* **1. a.** Perception by touching <the *feel* of velvet> **b.** An act or instance of touching <a *feel* of this grass> **2.** The sense of touch. **3.** The nature, state, or quality of something perceived through or as if through the tactile sense <the *feel* of a sports car> **4.** Emotional quality : ATMOSPHERE. **5.** Intuitive awareness, knowledge, or skill <a *feel* for the job>
☆ ***syns:*** **1.** FEEL, EXPERIENCE, HAVE *v. core meaning* : to be physically aware of through the senses <*felt* a sharp pain> **2.** FEEL, EXPERIENCE, HAVE, KNOW, SAVOR, TASTE *v. core meaning* : to undergo an emotional reaction <*felt* the first flush of victory>

**feel·er** (fē′lər) *n.* **1.** One that feels. **2.** Something, as a hint or question, designed to elicit the attitude or intention of another. **3.** A sensory or tactile organ, as an antenna, tentacle, or barbel.

**feel·ing** (fē′lĭng) *n.* **1. a.** The sensation involving tactile perception. **b.** A sensation perceived by touch. **c.** A physical sensation. **2.** An affective state of consciousness, as that resulting from emotions, sentiments, or desires <a *feeling* of horror> **3.** An awareness <a *feeling* of being watched> **4. a.** An emotional state or disposition <deep *feeling* expressed in the music> **b.** A tender emotion : FONDNESS. **5. a.** The ability to experience and react to the emotions : SENSIBILITY <a person of *feeling*> **b. feelings.** Sensitivities. **6.** Opinion as distinguished from reason. **7.** An impression produced <The guests gave the *feeling* of forced gaiety.> **8. a.** Appreciative regard <a *feeling* for propriety> **b.** Intuitive awareness or aptitude <a *feeling* for language> —*adj.* **1. a.** Capable of reacting or feeling emotionally : SENSITIVE. **b.** Easily moved emotionally. **2.** Having sensibility : SYMPATHETIC. **3.** Expressive of sensibility or emotion <a *feeling* look> —**feel′ing·ly** *adv.*

**fee simple** *n., pl.* **fees simple.** *Law.* An estate in land of which the inheritor has unqualified ownership and power of disposition.

**fee splitting** *n.* The practice of paying commissions to professional colleagues, as physicians, on the fees received from those who have been referred by these colleagues.

**feet** (fēt) *n. pl. of* FOOT.

**fee tail** *n., pl.* **fees tail.** *Law.* An estate in land limited in inheritance to a given individual, group, or class of heirs.

**feet of clay** *n.* A defect in an apparently sound character.

**†feeze** (fēz, fāz) [ME *fese* < *fesen,* to drive away.] *Regional.* —*n.* **1.** A heavy impact. **2.** Vexation. —*vt.* **feezed, feez·ing, feez·es. 1.** To put to flight. **2.** To faze.

**Feh·ling's solution** (fā′lĭngz) *n.* [After Hermann *Fehling* (1812–1885).] A solution of copper sulfate, sodium hydroxide, and Rochelle salt used in testing for the presence of sugars and aldehydes.

**feign** (fān) *v.* **feigned, feign·ing, feigns.** [ME *feinen* < OFr. *feindre* < Lat. *fingere,* to form.] —*vt.* **1. a.** To give a false appearance of <*feign* sickness> **b.** To represent falsely <*feign* authorship of story> **2.** *Archaic.* **a.** To invent. **b.** To fabricate. —*vi.* To dissemble.

**feigned** (fānd) *adj.* **1.** Not real : PRETENDED. **2.** Fictitious. —**feign′ed·ly** (fā′nĭd-lē) *adv.*

**feint** (fānt) *n.* [Fr. *feinte* < OFr. < p.part. of *feindre,* to feign.] **1.** A misleading movement or attack directed toward one part to draw defensive action away from the actual target or objective. **2.** A pretense. —*vi.* **feint·ed, feint·ing, feints.** To make a feint.

**†feist** (fīst) *also* **fice** (fīs) *n.* [Var. of obs. *fist,* short for *fisting dog* < ME *fist,* a foul smell < *fisten,* to break wind.] *Regional.* A small dog of mixed ancestry : MONGREL.

**†feist·y** (fī′stē) *adj.* **-i·er, -i·est.** [< FEIST.] **1.** *Regional.* Touchy : quarrelsome. **2.** Spirited : gutsy. —**feist′i·ness** *n.*

**fe·la·fel** (fə-lä′fəl) *n. var. of* FALAFEL.

**feld·spar** (fĕld′spär′, fĕl′-) *also* **fel·spar** (fĕl′-) *n.* [Partial transl. of obs. G. *Feldspath* : *Feld,* field (< OHG *feld*) + *Spath,* spar.] Any of a group of abundant rock-forming minerals occurring principally in igneous, plutonic, and some metamorphic rocks and composed of silicates of aluminum with potassium, sodium, calcium, and occas. barium.

**feld·spath·ic** (fĕld-spăth′ĭk, fĕl-) *adj.* [< obs. G. *Feldspath,* feldspar.] Of, relating to, or containing feldspar.

**fe·li·cif·ic** (fē′lĭ-sĭf′ĭk) *adj.* [Lat. *felicificus* : *felix, felici-,* happy + *-ficus,* -fic.] Producing or intended to produce happiness.

**fe·lic·i·tate** (fĭ-lĭs′ĭ-tāt′) *vt.* **-tat·ed, -tat·ing, -tates.** [Lat. *felicitare, felicitat-,* to make happy < *felix, felici-,* happy.] **1.** To wish happiness to : CONGRATULATE. **2.** *Archaic.* To make happy. —*adj. Obs.* Made happy. —**fe·lic′i·ta′tion** *n.* —**fe·lic′i·ta′tor** *n.*

**fe·lic·i·tous** (fĭ-lĭs′ĭ-təs) *adj.* **1.** Well-chosen : apt <a *felicitous* comparison> **2.** Yielding great pleasure or delight. —**fe·lic′i·tous·ly** *adv.* —**fe·lic′i·tous·ness** *n.*

**fe·lic·i·ty** (fĭ-lĭs′ĭ-tē) *n., pl.* **-ties.** [ME < OFr. *felicite* < Lat. *felicitas* < *felix,* happy.] **1. a.** Great happiness : BLISS. **b.** An instance of bliss. **2.** Something producing happiness. **3.** An appropriate and pleasing manner or style or an instance of it.

**fe·lid** (fē′lĭd) *adj.* [NLat. *Felidae,* family name < *Felis,* cat genus < Lat. *felis,* cat.] FELINE 1. —**fe′lid** *n.*

**fe·line** (fē′līn′) *adj.* [Lat. *felinus* < *felis,* cat.] **1.** Of or belonging to the family Felidae, including the lions, tigers, jaguars, and wild and domestic cats. **2.** Like a cat, as in suppleness or stealthiness. —**fe′line′** *n.* —**fe′line′ly** *adv.* —**fe′line′ness, fe·lin′i·ty** (fĭ-lĭn′ĭ-tē) *n.*

**fell**[1] (fĕl) *vt.* **felled, fell·ing, fells.** [ME *fellen* < OE *fyllan.*] **1. a.** To cut or knock down <*fell* trees> **b.** To kill <*felled* the opponent with a single bullet> **2.** To sew or finish (a seam) with the raw edges flattened, turned under, and stitched down. —**fell′a·ble** *adj.* —**fell′er** *n.*

**fell**[2] (fĕl) *adj.* [ME *fel* < OFr.] **1.** Inhumanly cruel : FIERCE. **2.** Capable of destroying : LETHAL. **3.** *Scot.* Sharp and biting. —**at one fell swoop.** All at once. —**fell′ness** *n.*

**fell**[3] (fĕl) *n.* [ME *fel* < OE.] The hide of an animal : PELT.

**fell**[4] (fĕl) *v. p.t. of* FALL.

**fel·lah** (fĕl′ə, fə-lä′) *n., pl.* **fel·lahs** *also* **fel·la·hin** *or* **fel·la·heen** (fĕl′ə-hēn′, fə-lä′hēn′) [Ar. *fellāḥ,* dial. var. of *fallāḥ* < *falaḥa,* to cultivate.] An Arab peasant or agricultural laborer.

**fell·mon·ger** (fĕl′mŭng′gər, -mŏng′) *n. Chiefly Brit.* A person who prepares hides for making leather. —**fell′mon′ger·ing, fell′mon′ger·y** (-gə-rē) *n.*

**fel·loe** (fĕl′ō) *n. var. of* FELLY.

**fel·low** (fĕl′ō) *n.* [ME < OE *fēolaga* < ON *fēlagi,* business partner : *fē,* money + *lag,* laying down.] **1. a.** A man or boy. **b.** *Informal.* A boyfriend. **2.** A comrade : associate. **3. a.** A person of equal rank, position, or background : PEER. **b.** One of a pair : MATE. **4.** A member of a learned society. **5.** A graduate student appointed to a position granting financial aid and providing for further study. **6.** *Chiefly Brit.* A member of an incorporated college or university. **7.** *Obs.* A person of a lower social class. —*adj.* Of the same kind, group, occupation, society, or locality <*fellow* students>

**fellow feeling** *n.* **1.** Sympathetic awareness of others. **2.** Community of interest.

**fellow man** *also* **fel·low·man** (fĕl′ō-măn′) *n.* A kindred person.

**fellow servant** *n. Law.* One of a group of employees working together under such circumstances that the employer cannot be expected to protect against or be liable for harm caused by the negligence of one employee to another.

**fel·low·ship** (fĕl′ō-shĭp′) *n.* **1. a.** The state of being together or of sharing similar interests or experiences : COMPANIONSHIP. **b.** Companionship of individuals in a congenial atmosphere and on equal terms. **2.** A union of friends or equals sharing similar interests : FRATERNITY. **3.** Friendship : comradeship. **4. a.** A scholarship or grant awarded a graduate student in a college or university. **b.** The state of having been awarded such a scholarship or grant. **c.** A foundation established for the awarding of such a scholarship or grant.

**fellow traveler** *n.* One who sympathizes with the tenets and programs of an organized group, as the Communist Party, without actually being a member of it.

**fel·ly** (fĕl′ē) *also* **fel·loe** (fĕl′ō) *n., pl.* **-lies** *also* **-loes.** [ME *felies* < OE *felg.*] The rim or a section of the rim of a wheel supported by spokes.

**fel·o-de-se** (fĕl′ō-dĭ-sā′, -sē′) *n., pl.* **fe·lo·nes-de-se** (fə-lō′nēz-) *or* **fel·os-de-se** (fĕl′ōz-) [Med. Lat., felon of himself.] *Law.* **1.** Suicide. **2.** One who commits suicide.

**fel·on**[1] (fĕl′ən) *n.* [ME *feloun* < *feloun,* wicked < OFr. *felon* < Med. Lat. *fello,* villain.] **1.** *Law.* One who has perpetrated a felony. **2.** *Archaic.* An evil person. —*adj. Archaic.* **1.** Evil. **2.** Cruel.

**fel·on**[2] (fĕl′ən) *n.* [ME *feloun,* prob. < Lat. *fel,* gall, bile.] A purulent infection at the distal end of a finger near or around the nail or the bone.

**fe·lo·nes-de-se** (fə-lō′nēz-dĭ-sā′, -sē′) *n. var. pl. of* FELO-DE-SE.

**fe·lo·ni·ous** (fə-lō′nē-əs) *adj.* **1.** *Law.* **a.** Of, relating to, or concerning a felony. **b.** Marked by or of the nature of a felony. **2.** *Archaic.* Evil : wicked. **—fe·lo′ni·ous·ly** *adv.* **—fe·lo′ni·ous·ness** *n.*

**fel·on·ry** (fĕl′ən-rē) *n.* Felons as a group.

**fel·o·ny** (fĕl′ə-nē) *n., pl.* **-nies.** *Law.* **1.** A crime, as murder, rape, or burglary, considered more serious than a misdemeanor and punishable by a stronger sentence. **2.** Any of several crimes in early English law that were punishable by forfeiture of land or goods and by possible loss of life or a bodily part.

**fel·os-de-se** (fĕl′ōz-dĭ-sā′, -sē′) *n. var. pl. of* FELO-DE-SE.

**fel·site** (fĕl′sīt′) *n.* [FELS(PAR) + -ITE.] A fine-grained igneous rock, essentially feldspar and quartz. **—fel·sit′ic** (-sĭt′ĭk) *adj.*

**fel·spar** (fĕl′spär′) *n. var. of* FELDSPAR.

**felt**[1] (fĕlt) *n.* [ME < OE.] **1.** A fabric of matted, compressed animal fibers, as wool or fur, occas. mixed with vegetable or synthetic fibers. **2.** A fabric or material resembling felt. **3.** Something made of felt. —*adj.* **1.** Made of felt. **2.** Relating to or like felt. —*v.* **felt·ed, felt·ing, felts.** —*vt.* **1.** To make into felt. **2.** To cover with felt. **3.** To press or mat together. —*vi.* To become like felt.

**felt**[2] (fĕlt) *v. p.t. & p.p. of* FEEL.

**felt·ing** (fĕl′tĭng) *n.* **1.** The process of making felt. **2.** The materials from which felt is made. **3.** Felted fabric.

**fe·luc·ca** (fə-lo͞o′kə, -lŭk′ə) *n.* [Ital. *feluca* < Ar. *fulk,* ship.] A narrow, swift sailing vessel chiefly of the Mediterranean, propelled by lateen sails.

**fel·wort** (fĕl′wûrt′, -wôrt′) *n.* [ME **feldwort* < OE *feldwyrt* : *feld,* field + *wyrt,* wort.] A plant of the genus *Gentiana* or related genera, esp. *G. amarella,* with small, purplish flowers.

**fe·male** (fē′māl′) *adj.* [ME, var. of *femelle* < OFr. < Lat. *femella,* dim. of *femina,* woman.] **1.** Of, relating to, or designating the sex that produces ova or bears young. **2.** Typical of or appropriate to the female sex : FEMININE. **3.** Consisting of members of the female sex. **4.** *Bot.* **a.** Relating to or designating an organ, as a pistil or ovary, that functions in producing seeds or spores after fertilization. **b.** Bearing pistils but not stamens <*female* flowers> **5.** Indicating or having a part, as a receptacle, designed to receive a complementary male part, as a plug. —*n.* **1.** A member of the sex that produces ova or bears young. **2.** One that is female. **3.** A woman or girl. **4.** *Bot.* A plant having only pistillate flowers. **—fe′male′ness** *n.*

▲ word history: The word *female* is unrelated to the word *male.* *Female* is a respelling of *femelle,* which is ultimately from Latin *femella,* a diminutive of *femina,* "woman." After its adoption into English from French in the 15th century, *femelle* was used primarily as an adjective. This circumstance led to the respelling of the word as *femal,* where *-al* was regarded as the adjectival suffix. The spelling *femal,* as well as the word's obvious correlation with the word *male,* suggested an etymological association with the word *male.* This error has been preserved in the now standard modern spelling *female.*

**female suffrage** *n.* Woman suffrage.

**fem·i·nine** (fĕm′ə-nĭn) *adj.* [ME < OFr. < Lat. *feminina* < *femina,* woman.] **1.** Of or belonging to the female sex. **2.** Marked by or having qualities gen. attributed to a woman. **3.** Effeminate : womanish. **4.** Being, designating, or belonging to the gender of words or grammatical forms that refer chiefly to females. —*n.* **1.** The feminine gender. **2.** A word or form belonging to the feminine gender. **—fem′i·nine·ly** *adv.* **—fem′i·nine·ness** *n.*

**feminine ending** *n.* **1.** Termination of a line of verse in an unaccented syllable. **2.** A final syllable or termination marking or forming words in the feminine gender.

**feminine rhyme** *n.* A rhyme with a final unaccented syllable.

**fem·i·nin·i·ty** (fĕm′ə-nĭn′ĭ-tē) *n., pl.* **-ties. 1.** The quality or state of being feminine. **2.** A female trait or characteristic. **3.** Women as a group. **4.** Womanishness : effeminacy.

**fem·i·nism** (fĕm′ə-nĭz′əm) *n.* **1.** A doctrine advocating for women the same rights granted men, as in political and economic status. **2.** The movement supporting feminism. **—fem′i·nist** *n.* **—fem′i·nis′tic** *adj.*

**fe·min·i·ty** (fĕ-mĭn′ĭ-tē) *n.* Femininity.

**fem·i·nize** (fĕm′ə-nīz′) *vt. & vi.* **-nized, -niz·ing, -niz·es.** To make or become feminine. **—fem′i·ni·za′tion** *n.*

**femme fa·tale** (fĕm′ fə-tăl′, -täl′, făm′) *n., pl.* **femmes fa·tales** (fĕm′ fə-tăl′, -tălz′, -täl′, -tälz′, făm′). [Fr. : *femme,* woman + *fatale,* fatal.] **1.** An attractive woman who may lead a man into a compromising or dangerous situation. **2.** A charming, mysterious woman.

**fem·o·ra** (fĕm′ər-ə) *n. var. pl. of* FEMUR.

**fem·o·ral** (fĕm′ər-əl) *adj.* [< Lat. *femur, femor-,* femur.] Of or relating to the thigh or the femur.

**femoral artery** *n.* The main artery of the thigh.

**femto-** *pref.* [Dan. or Norw. *femten,* fifteen < ON *fimmtān.*] One quadrillionth ($10^{-15}$) <*femto*meter>

**fem·to·joule** (fĕm′tə-jo͞ol′, -joul′) *n.* One quadrillionth ($10^{-15}$) of a joule.

**fem·tom·e·ter** (fĕm-tŏm′ĭ-tər) *n.* One quadrillionth ($10^{-15}$) of a meter.

**fem·to·sec·ond** (fĕm′tə-sĕk′ənd) *n.* One quadrillionth ($10^{-15}$) of a second.

**fe·mur** (fē′mər) *n., pl.* **fe·murs** *or* **fem·o·ra** (fĕm′ər-ə) [Lat.] **1. a.** The proximal bone of the lower or hind limb in vertebrates, located between the pelvis and knee in humans. **b.** THIGH 1. **2.** The usu. stout third segment of an insect's leg.

**fen** (fĕn) *n.* [ME < OE *fenn.*] Low, flat, marshy land : BOG.

**fence** (fĕns) *n.* [ME *fens,* short for *defense,* defense. —see DEFENSE.] **1.** A structure functioning as a boundary or barrier, usu. made of posts, boards, wire, or rails. **2.** *Archaic.* A means of defense : PROTECTION. **3.** FENCING 1. **4.** *Slang.* **a.** One who receives and sells stolen goods. **b.** A place where such goods are received and sold. —*v.* **fenced, fenc·ing, fenc·es.** —*vt.* **1.** To surround or close in by or as if by a fence. **2.** To separate or close off by or as if by a fence. **3.** *Archaic.* To defend or ward off. **4.** *Slang.* To receive and sell (stolen goods). —*vi.* **1.** To practice the art of fencing. **2.** To avoid giving direct answers : HEDGE. **3.** *Slang.* To act as a fence for stolen goods. **—on the fence.** *Informal.* **1.** Undecided regarding which side to support, as in a controversy. **2.** Neutral. **—fenc′er** *n.*

**fence sitter** *n.* One taking a neutral or undecided position, as in a controversial matter. **—fence′-sit′ting** (fĕns′sĭt′ĭng) *n.*

**fenc·ing** (fĕn′sĭng) *n.* **1.** The art, practice, or sport of using a foil, épée, or saber as a means of attack or defense. **2.** FENCE 1. **3.** Material, as wire or stakes, used in building fences. **4.** Fences in general.

**fend** (fĕnd) *v.* **fend·ed, fend·ing, fends.** [ME *fenden,* short for *defenden,* to defend. —see DEFEND.] —*vt.* **1.** *Archaic.* To defend. **2.** To keep or ward off <*fend* off an assault> —*vi.* **1.** *Chiefly Brit. Regional.* To make an effort to do something. **2.** To provide or earn a living <children *fending* for themselves>

**fend·er** (fĕn′dər) *n.* **1.** One that fends or keeps off. **2.** A usu. metal guard over the wheel of an automotive vehicle. **3.** A device at the front end of a locomotive or streetcar for pushing aside obstructions. **4.** A screen or metal framework placed in front of a fireplace to keep embers and debris from falling out. **5.** A device, as a bundle of rope or piece of timber, on the side of a vessel or dock for absorbing impact or friction.

**fe·nes·tra** (fə-nĕs′trə) *n., pl.* **-trae** (-trē′) [Lat., window.] **1.** *Anat.* A small opening, esp. either of two apertures in the medial wall of the middle ear. **2.** A windowlike opening. **3.** *Biol.* A transparent spot or marking, as on an insect's wing. **—fe·nes′tral** *adj.*

**fen·es·trate** (fĕn′ĭ-strāt′) *adj.* Fenestrated.

**fen·es·trat·ed** (fĕn′ĭ-strā′tĭd) *adj.* [< Lat. *fenestratus,* p.part. of *fenestrare,* to furnish with windows < *fenestra,* window.] **1.** Having windows or windowlike openings. **2.** *Biol.* Having fenestrae.

**fen·es·tra·tion** (fĕn′ĭ-strā′shən) *n.* **1.** Design and position of windows in a building. **2.** A structural opening. **3.** Surgical cutting of an artificial opening from the external auditory canal to the labyrinth of the internal ear to restore normal hearing.

**Fe·ni·an** (fē′nē-ən) *n.* [Blend of OIr. *fēinne,* pl. of *fīann,* legendary band of warriors, and *Fēne,* name of ancient inhabitants of Ireland.] **1.** One of a legendary group of heroic Irish warriors of the 2nd and 3rd cent. A.D. **2.** A member of a secret organization, founded in New York City in the mid-19th cent., with the goal of overthrowing British rule in Ireland. **—Fe′ni·an** *adj.* **—Fe′ni·an·ism** *n.*

**fen·nec** (fĕn′ĭk) *n.* [Ar. *fenek,* furry animal.] A small fox, *Fennecus zerda* of desert regions of northern Africa, with fawn-colored fur and large pointed ears.

**fen·nel** (fĕn′əl) *n.* [ME *fenel* < OE *finul* < Med. Lat. *fenuculum* < Lat. *faeniculum,* dim. of *faenum,* hay.] **1.** A native Eurasian plant, *Foeniculum vulgare,* with finely dissected leaves, small yellow flower clusters, and aromatic seeds used as flavoring. **2.** The seeds or edible stalks of the fennel.

**fen·thi·on** (fĕn-thī′ŏn′, -ən) *n.* [E. *fen-,* alteration of PHEN- + THI(O)- + -ON.] An organophosphorous insecticide, $C_{10}H_{15}O_3PS_2$, used for ornamental plants.

**fen·u·greek** (fĕn′yə-grēk′) *n.* [ME *fenigrek* < OFr. *fenugrek* < Lat. *fenugraecum* < *fenum graecum,* Greek hay.] **1.** A cloverlike Eurasian plant, *Trigonella foenum-graecum,* with white flowers and pungent, aromatic seeds used as flavoring. **2.** The seeds of the fenugreek.

**fen·u·ron** (fĕn′yə-rŏn′) *n.* [*fen-,* alteration of PHEN- + U(REA) + -ON[3].] A white compound, $C_9H_{12}N_2O$, used as a herbicide.

**feoff·ee** (fĕ-fē′, fē-) *n.* One to whom a feoffment is granted.

**feoff·er** *also* **feof·for** (fĕf′ər, fē′fər) *n.* One who grants a feoffment.

**feoff·ment** (fĕf′mənt, fēf′-) *n.* The act of granting a feudal estate or fee.

**feof·for** (fĕf′ər, fē′fər) *n. var. of* FEOFFER.

**-fer** *suff.* [Lat. < *ferre,* to carry.] One that bears <aqui*fer*>

**fe·ral** (fîr′əl, fĕr′-) *adj.* [< Lat. *fera,* wild animal < *ferus,* wild.] **1. a.** Existing in an untamed state. **b.** Having returned to an untamed state from domestication. **2.** Of or like a wild animal : SAVAGE <a *feral* grin>

**fer·bam** (fûr′băm′) *n.* [FER(RIC) (DIMETHYL-DITHIOCAR)BAM(ATE).] A black compound, $C_9H_{18}FeN_3S_6$, used as an agricultural fungicide.

**fer-de-lance** (fĕr′də-läns′, -läns′) *n., pl.* **fer-de-lance.** [Fr.] A venomous tropical American snake, *Bothrops atrox,* whose markings are brown and gray.

**fere** (fîr) *n.* [ME < OE *fēra.*] *Archaic.* **1.** A companion. **2.** A spouse.

**fe·ri·a** (fîr′ē-ə, fĕr′-) *n., pl.* **-ri·as** *or* **-ri·ae** (-ē-ē′) [Med. Lat. < LLat., festal day < Lat. *feriae.*] A weekday on a church calendar on which no feast is observed. **—fe′ri·al** *adj.*

**fe·rine** (fîr′īn′) *adj.* [Lat. *ferinus* < *fera,* wild animal. —see FERAL.] FERAL 1.

**fer·i·ty** (fĕr′ĭ-tē) *n.* [Lat. *feritas* < *ferus,* wild.] The quality or state of being feral.

**fer·ma·ta** (fĕr-mä′tə) *n.* [Ital. < fem. p.part. of *fermare,* to stop < Lat. *firmare,* to make firm < *firmus,* firm.] *Mus.* **1.** Prolongation of a tone, chord, or rest beyond its nominal value. **2.** The sign indicating a fermata.

**fer·ment** (fûr′mĕnt′) *n.* [ME < OFr. < Lat. *fermentum.*] **1.** Something, as a yeast, bacterium, mold, or enzyme, that causes fermentation. **2.** FERMENTATION 1. **3.** FERMENTATION 2. —*v.* (fər-mĕnt′) **-ment·ed, -ment·ing, -ments.** —*vt.* **1.** To produce by or as if by fermentation. **2.** To cause to undergo fermentation. **3.** To make turbulent : EXCITE. —*vi.* **1.** To undergo fermentation. **2.** To be turbulent : SEETHE. **—fer·ment′a·bil′i·ty** *n.* **—fer·ment′a·ble** *adj.* **—fer·ment′er** *n.*

**fer·men·ta·tion** (fûr′mən-tā′shən, -mĕn-) *n.* **1.** Any of a group of chemical reactions induced by living or nonliving ferments that split complex organic compounds into relatively simple substances, esp. the anaerobic conversion of sugar to carbon dioxide and alcohol by yeast. **2.** Great unrest : AGITATION.

**fer·men·ta·tive** (fər-mĕn′tə-tĭv) *adj.* **1. a.** Causing fermentation. **b.** Capable of causing or undergoing fermentation. **2.** Relating to or characteristic of fermentation.

**fer·mi** (fûr′mē, fĕr′-) *n.* [After Enrico *Fermi* (1901–1954).] A unit of length equal to $10^{-15}$ meter.

**fer·mi·on** (fûr′mē-ŏn′, fĕr′-) *n.* [After Enrico *Fermi* (1901–1954).] A particle, as an electron, proton, or neutron, having half-integral spin and obeying statistical rules requiring that not more than one in a set of identical particles may occupy a particular quantum state.

**fer·mi·um** (fûr′mē-əm, fĕr′-) *n.* [After Enrico *Fermi* (1901–1954).] *Symbol* **Fm** A synthetic transuranic metallic element; atomic number 100; longest-lived isotope Fm 257.

**fern** (fûrn) *n.* [ME < OE *fearn.*] Any of numerous flowerless, seedless, vascular plants of the class Filicinae, having fronds with divided leaflets and reproducing by spores. **—fern′y** *adj.*

**fern seed** *n.* The minute spores of ferns, once believed to be seeds and supposed to have the power of making the possessor invisible.

**fe·ro·cious** (fə-rō′shəs) *adj.* [Lat. *ferox, feroci-,* fierce.] **1.** Extremely savage : FIERCE. **2.** Unrelenting in intensity : EXTREME <a *ferocious* storm at sea> **—fe·ro′cious·ly** *adv.* **—fe·ro′cious·ness, fe·roc′i·ty** (-rŏs′ĭ-tē) *n.*

**-ferous** *suff.* [-FER + -OUS.] Bearing or producing : CONTAINING <carboni*ferous*>

**ferr-** *pref. var. of* FERRO-.

**fer·rate** (fĕr′āt′) *n.* FERRITE 1.

**fer·re·dox·in** (fĕr′ĭ-dŏk′sĭn) *n.* [FER(RO)- + REDOX + -IN.] An electron-transferring iron-containing plant protein.

**fer·ret¹** (fĕr′ĭt) *n.* [ME < OFr. *fuiret* < Lat. *fur,* thief.] **1.** A domesticated, usu. albino form of the Old World polecat, often trained to hunt rats or rabbits. **2.** A weasellike mammal, *Mustela nigripes* of central North America, related to the ferret and having yellowish fur and dark feet. —*v.* **-ret·ed, -ret·ing, -rets.** —*vt.* **1.** To hunt with ferrets. **2.** To drive out : EXPEL. **3.** To uncover and bring to light by searching <*ferreted* out my secrets> —*vi.* **1.** To hunt with ferrets. **2.** To search intensively. **—fer′ret·er** *n.* **—fer′ret·y** *adj.*

**fer·ret²** (fĕr′ĭt) *also* **fer·ret·ing** (-ĭ-tĭng) *n.* [Prob. alteration of Ital. *fioretti,* floss silk, pl. of *fioretto,* dim. of *fiore,* flower < Lat. *flos,* flower.] A narrow piece of tape for binding or edging fabric.

**ferri-** *pref.* [Lat. *ferrum,* iron.] Iron, esp. ferric iron <*ferri*cyanide>

**fer·ri·age** (fĕr′ē-ĭj) *n.* **1.** The act or business of ferrying. **2.** The toll charged for ferrying.

**fer·ric** (fĕr′ĭk) *adj.* Of, relating to, or containing iron, esp. iron with a valence of 3 or with a valence higher than in a corresponding ferrous compound.

**ferric ammonium citrate** *n.* An iron-containing salt used in treating some forms of anemia.

**ferric chloride** *n.* A salt, $FeCl_3$, used as an astringent and as a hematinic.

**ferric oxide** *n.* A dark compound, $Fe_2O_3$, occurring naturally as hematite ore and rust and used in metallurgy and in making pigments, polishing compounds, and magnetic tapes.

**fer·ri·cy·an·ic acid** (fĕr′ĭ-sī-ăn′ĭk, fĕr′ī-) *n.* A reddish-brown solid compound, $H_3Fe(CN)_6$.

**fer·ri·cy·a·nide** (fĕr′ĭ-sī′ə-nīd′, fĕr′ī-) *n.* Any of various salts derived from ferricyanic acid and used in making blue pigments.

**fer·rif·er·ous** (fə-rĭf′ər-əs, fĕ-) *adj.* Containing or yielding iron.

**Fer·ris wheel** *also* **fer·ris wheel** (fĕr′ĭs) *n.* [After George W. G. *Ferris* (1859–1896).] A large upright, rotating wheel with suspended cars in which passengers ride for amusement.

**fer·rite** (fĕr′īt′) *n.* **1.** Any of a group of nonmetallic, ceramiclike, usu. ferromagnetic compounds of ferric oxide with other oxides, esp. such a compound with spinel crystalline structure, marked by high electrical resistivity and used in computer memory elements, permanent magnets, and solid-state devices. **2.** Iron with a body-centered cubic crystalline form, found usu. in steel, cast iron, and pig iron below 910°C.

**ferrite core** *n.* A magnetic core used in computer core storage.

**fer·ri·tin** (fĕr′ĭ-tĭn) *n.* An iron-containing protein complex that functions as a form of iron storage in the tissues.

**ferro-** *or* **ferr-** *pref.* [Lat. *ferrum,* iron.] **1.** Iron <*ferro*magnetic> **2.** Ferrous iron <*ferro*cyanide>

**fer·ro·al·loy** (fĕr′ō-ăl′oi′, -ə-loi′) *n.* An alloy of iron and one or more other elements, as manganese or silicon, used as a raw material in steel production.

**fer·ro·con·crete** (fĕr′ō-kŏn′krēt′, -kŏn-krēt′) *n.* Concrete reinforced by steel bars or metal netting.

**fer·ro·cy·an·ic acid** (fĕr′ō-sī-ăn′ĭk) *n.* A solid white compound, $H_4Fe(CN)_6$.

**fer·ro·cy·a·nide** (fĕr′ō-sī′ə-nīd′) *n.* A salt derived from ferrocyanic acid, the sodium and potassium salts of which are used in making blue pigments, blueprint paper, and ferricyanide.

**fer·ro·e·lec·tric** (fĕr′ō-ĭ-lĕk′trĭk) *adj.* Of or relating to a crystalline dielectric that can be given a permanent electric polarization by application of an electric field. **—fer′ro·e·lec′tric** *n.* **—fer′ro·e·lec·tric′i·ty** *n.*

**fer·ro·mag·ne·sian** (fĕr′ō-măg-nē′zhən, -shən) *adj.* Containing iron and magnesium.

**fer·ro·mag·net** (fĕr′ō-măg′nĭt) *n.* **1. a.** A ferromagnetic substance. **b.** A substance with magnetic properties similar to those of iron. **2.** A ferromagnetic magnet.

**fer·ro·mag·net·ic** (fĕr′ō-măg-nĕt′ĭk) *adj.* Relating to or typical of substances, as iron, nickel, cobalt, and various alloys, that exhibit high magnetic permeability, the ability to acquire high magnetization in relatively weak magnetic fields, a characteristic saturation point, and magnetic hysteresis. **—fer′ro·mag′ne·tism** *n.*

**fer·ro·man·ga·nese** (fĕr′ō-măng′gə-nēz′, -nēs′) *n.* A ferroalloy of iron and manganese, used in making steel.

**fer·ro·sil·i·con** (fĕr′ō-sĭl′ĭ-kən, -kŏn′) *n.* A ferroalloy of iron and silicon used in making carbon steel.

**fer·ro·type** (fĕr′ə-tīp′) *n.* **1.** A positive photograph made directly on an iron plate varnished with a thin sensitized film. **2.** The process by which ferrotypes are made.

**fer·rous** (fĕr′əs) *adj.* Of, relating to, or containing iron, esp. with valence 2.

**ferrous oxide** *n.* A black powdery compound, FeO, used in making steel, green heat-absorbing glass, and enamels.

**ferrous sulfate** *n.* A greenish crystalline compound, $FeSO_4 \cdot 7H_2O$, used in sewage and water treatment and as a pigment, fertilizer, and feed additive.

**ferrous sulfide** *n.* A black to brown sulfide of iron, FeS, used in the manufacture of hydrogen sulfide.

**fer·ru·gi·nous** (fə-ro͞o′jə-nəs, fĕ-) *adj.* [Lat. *ferruginus* < *ferrugo,* iron rust < *ferrum,* iron.] **1.** Of, containing, or similar to iron. **2.** Having the color of iron rust.

**fer·rule** (fĕr′əl) *n.* [Var. of obs. *verrel* < ME *verrele* < OFr. *virelle* < Med. Lat. *virolla* < Lat. *viriola,* little bracelets, dim. of *viriae,* brace-

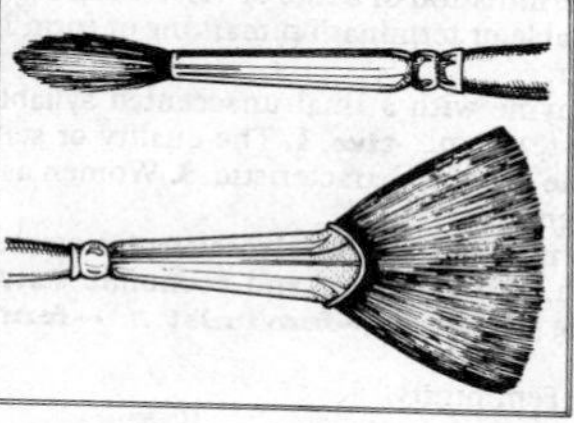

ferrule

lets.] **1.** A metal ring or cap placed around a pole or shaft to reinforce it or prevent splitting. **2.** A bushing for securing a pipe joint. —*vt.* **-ruled, -rul·ing, -rules.** To furnish with a ferrule.

**fer·ry** (fĕr′ē) *v.* **-ried, -ry·ing, -ries.** [ME *ferien* < OE *ferian.*] —*vt.* **1.** To transport by boat across a body of water. **2.** To cross. **3.** To deliver (a vehicle, esp. an aircraft) under its own power to its eventual user. **4.** To transport (people or goods) esp. by aircraft. —*vi.* To cross a body of water on or as if on a ferry. —*n., pl.* **-ries. 1. a.** A ferryboat. **b.** The embarkation point for a ferryboat. **2.** A franchise to operate a ferrying service for a fee. **3.** A service for transporting esp. an aircraft under its own power to its eventual user.

**fer·ry·boat** (fĕr′ē-bōt′) *n.* A boat used to ferry passengers or goods.

**fer·ry·man** (fĕr′ē-mən) *n.* One who owns, runs, or operates a ferry.

**fer·tile** (fûr′tl) *adj.* [ME *fertil* < OFr. *fertile* < Lat. *fertilis* < *ferre,* to bear.] **1.** *Biol.* **a.** Capable of reproducing. **b.** Capable of initiating, sustaining, or supporting reproduction. **c.** Capable of growing and developing. **2.** *Bot.* Bearing reproductive structures or material, as spores or pollen. **3.** Rich in material required to sustain plant growth. **4.** Characterized by great productivity. **—fer′tile·ly** *adv.* **—fer′tile·ness, fer·til′i·ty** (fər-tĭl′ĭ-tē) *n.*

☆ ***syns:*** FERTILE, FECUND, FRUITFUL, PRODUCTIVE, PROLIFIC, RICH *adj. core meaning:* characterized by great productivity <*fertile* farm land> <a brain *fertile* with new ideas> ***ant:*** barren, infertile

**fer·til·i·za·tion** (fûr′tl-ĭ-zā′shən) *n.* **1.** The act or process of initiating biological reproduction. **2.** The process in which two gametes unite to form a zygote. **3.** The act or process of applying a fertilizer. **—fer′til·i·za′tion·al** *adj.*

**fer·til·ize** (fûr′tl-īz′) *v.* **-ized, -iz·ing, -iz·es.** —*vt.* **1.** To cause fertilization of (e.g., an ovum), esp. to provide with sperm or pollen, thereby causing fertilization. **2.** To spread fertilizer on. —*vi.* To spread fertilizer. **—fer′til·iz′a·ble** *adj.*

**fer·til·iz·er** (fûr′tl-ī′zər) *n.* One that fertilizes, esp. any of a large number of natural and synthetic materials, including manure and nitrogen, phosphorus, and potassium compounds, spread on or worked into soil to increase fertility.

**fer·ule** (fĕr′əl) *n.* [Lat. *ferula.*] A baton, cane, or stick used in punishing children.

**fe·ru·lic acid** (fə-ro͞o′lĭk) *n.* [< NLat. *Ferula,* plant genus.] A compound, $C_{10}H_{10}O_4$, related to vanillin and obtained from certain plants.

**fer·vent** (fûr′vənt) *adj.* [ME < OFr. < Lat. *fervens,* pr.part. of *fervēre,* to boil.] **1.** Having or exhibiting great emotion or warmth : ARDENT. **2.** Very hot. **—fer′ven·cy** (-sē) *n.* **—fer′vent·ly** *adv.* **—fer′vent·ness** *n.*

**fer·vid** (fûr′vĭd) *adj.* [Lat. *fervidus* < *fervor,* passion < *fervēre,* to boil.] **1.** Intensely fervent or zealous : IMPASSIONED. **2.** Extremely hot. **—fer′vid·ly** *adv.* **—fer′vid·ness** *n.*

**fer·vor** (fûr′vər) *n.* [ME *fervour* < OFr. < Lat. *fervor.* —see FERVID.] **1.** Intensity of emotion : ARDOR. **2.** Intense heat.

**fer·vour** (fûr′vər) *n. Chiefly Brit. var. of* FERVOR.

**fes·cen·nine** (fĕs′ə-nīn′, -nēn′) *adj.* [Lat. *Fescinninus,* of Fescennia, a town in ancient Etruria noted for its licentious poetry.] Licentious : obscene.

**fes·cue** (fĕs′kyo͞o) *n.* [ME *festu,* stalk < OFr. < Lat. *festuca.*] A grass of the genus *Festuca,* often cultivated as pasturage.

**fess** *also* **fesse** (fĕs) *n.* [ME *fesse* < OFr. < Lat. *fascia,* band.] *Heraldry.* A wide horizontal band forming the middle section of an escutcheon.

**fess point** *n. Heraldry.* The center point of an escutcheon.

**-fest** *suff.* [< G. *Fest,* festival < Lat. *festum.*] A gathering or occasion marked by a specified activity <*slugfest*>

**fes·tal** (fĕs′təl) *adj.* [OFr. < Lat. *festum,* feast.] FESTIVE 1. **—fes′tal·ly** *adv.*

**fes·ter** (fĕs′tər) *v.* **-tered, -ter·ing, -ters.** [ME *festren* < *festre,* fistula < OFr. < Lat. *fistula.*] —*vi.* **1.** To generate pus : SUPPURATE. **2.** To form an ulcer. **3.** To decay or rot. **4.** To be or become a source of irritation : RANKLE <old resentments *festering*> —*vt.* To infect, inflame, or corrupt. —*n.* A small festering sore or ulcer.

**fes·ti·nate** (fĕs′tə-nĭt) *adj.* [Lat. *festinatus,* p.part. of *festinare,* to hasten.] Hasty. —*v.* (-nāt′) **-nat·ed, -nat·ing, -nates.** To hasten. **—fes′ti·nate·ly** *adv.*

**fes·ti·val** (fĕs′tə-vəl) *n.* [ME, festive < OFr. < Med. Lat. *festivalis* < Lat. *festivus* < *festus.*] **1.** An occasion for feasting or celebration, esp. a day or time of religious significance that recurs at regular intervals. **2.** An often regularly recurring program of performances, exhibitions, or competitions. **3.** Conviviality : revelry. —*adj.* Festive.

**fes·tive** (fĕs′tĭv) *adj.* [Lat. *festivus* < *festus.*] **1.** Of, relating to, or appropriate to a feast or festival. **2.** Merry : joyous. **—fes′tive·ly** *adv.* **—fes′tive·ness** *n.*

**fes·tiv·i·ty** (fĕ-stĭv′ĭ-tē) *n., pl.* **-ties. 1.** A festival. **2.** The pleasure, joy, and gaiety of a festival or celebration. **3. festivities.** The proceedings or events of a festival.

**fes·toon** (fĕ-sto͞on′) *n.* [Fr. *feston* < Ital. *festone* < *festa,* feast < Lat. *festus,* festive.] **1.** A string or garland hung in a loop between two points. **2.** A representation of a festoon, as in sculpture. —*vt.* **-tooned, -toon·ing, -toons. 1.** To decorate with or as if with a festoon. **2.** To form or make into a festoon. **—fes·toon′er·y** *n.*

**fest·schrift** (fĕst′shrĭft′) *n., pl.* **-schrif·ten** (-shrĭf′tən) *or* **-schrifts.** [G. : *Fest,* festival + *Schrift,* writing.] A written tribute or memorial to a scholar, consisting of a volume of learned articles or essays by his or her colleagues and admirers.

**fet-** *pref. var. of* FETO-.

**fet·a** (fĕt′ə, fē′tə) *n.* [Mod. Gk. *(turi) pheta,* (cheese) slice < Ital. *fetta,* slice < Lat. *offa,* morsel of food.] A white Greek cheese made usu. of goat's or ewe's milk and preserved in brine.

**fe·tal** *also* **foe·tal** (fēt′l) *adj.* Of, relating to, or having the nature of a fetus.

**fetal alcohol syndrome** *n.* A complex of birth defects, including retarded growth and cardiac abnormalities, occurring in infants born to alcoholic mothers.

**fetal position** *n.* [From its resemblance to the position of a fetus in the womb.] A bodily position at rest in which the spine is curved, the head is bowed forward, and the arms and legs are drawn in toward the chest.

**fe·ta·tion** (fē-tā′shən) *n.* Fetal development : PREGNANCY.

**fetch¹** (fĕch) *v.* **fetched, fetch·ing, fetch·es.** [ME *fecchen* < OE *feccean.*] —*vt.* **1.** To go or come after and return with. **2.** To cause to come. **3. a.** To draw in (breath) : INHALE. **b.** To bring forth (e.g., a sigh). **4.** To bring in as a price <*fetched* $200 at auction> **5.** *Informal.* To strike or deal (a blow). **6.** *Naut.* To arrive at : REACH. —*vi.* **1.** To go after and return with something. **2.** To retrieve killed game. **3.** *Naut.* **a.** To hold a course. **b.** To turn about : VEER. **—fetch up. 1.** To reach a place and halt. **2.** To make up (e.g., lost time). —*n.* **1.** An act or instance of fetching. **2.** *Computer Sci.* A program routine that brings a phase of the program from storage for immediate use. **3.** A trick or stratagem. **—fetch′er** *n.*

**fetch²** (fĕch) *n.* [Orig. unknown.] *Chiefly Brit.* **1.** A ghost. **2.** A doppelgänger.

**fetch·ing** (fĕch′ĭng) *adj.* Very attractive. **—fetch′ing·ly** *adv.*

**fete** *also* **fête** (fāt, fĕt) [Fr. *fête* < OFr. *feste.* —see FEAST.] —*n.* **1.** A feast or festival. **2.** An elaborate outdoor entertainment. **3.** An elaborate party. —*vt.* **fet·ed, fet·ing, fetes** *also* **fêt·ed, fêt·ing, fêtes. 1.** To celebrate with a fete. **2.** To pay honor to.

**fet·er·i·ta** (fĕt′ə-rē′tə) *n.* [Ar.] A sorghum, *Sorghum vulgare caudatum,* grown in warm regions for its grain and as forage.

**feti-** *pref. var. of* FETO-.

**fet·ich** (fĕt′ĭsh, fē′tĭsh) *n. var. of* FETISH.

**fet·ich·ism** (fĕt′ĭ-shĭz′əm, fē′tĭ-) *n. var. of* FETISHISM.

**fe·ti·cide** (fē′tĭ-sīd′) *n.* Intentional destruction of a human fetus.

**fet·id** (fĕt′ĭd, fē′tĭd) *also* **foe·tid** (fē′tĭd) *adj.* [ME < Lat. *fetidus* < *fetēre,* to stink.] Having an offensive odor. **—fet′id·ly** *adv.* **—fet′id·ness** *n.*

**fet·ish** *also* **fet·ich** (fĕt′ĭsh, fē′tĭsh) *n.* [Fr. *fétiche* < Port. *feitiço,* charm < Lat. *facticus,* factitious. —see FACTITIOUS.] **1.** An object believed to have magical powers, esp. of protection. **2.** An object of unreasonably excessive attention or reverence <made a *fetish* of cleanliness>

**fet·ish·ism** *also* **fet·ich·ism** (fĕt′ĭ-shĭz′əm, fē′tĭ-) *n.* **1.** Worship of or belief in magical fetishes. **2.** Excessive attachment or regard. **—fet′ish·ist** *n.* **—fet′ish·is′tic** *adj.*

**fet·lock** (fĕt′lŏk′) *n.* [ME *fitlok.*] **1. a.** A projection on the lower part of the leg of a horse or related animal, above and behind the hoof. **b.** A tuft of hair on such a projection. **2.** The joint marked by the fetlock.

**feto-** *or* **feti-** *or* **fet-** *pref.* [< FETUS.] Fetus : fetal <*fetology*>

**fe·tol·o·gy** (fē-tŏl′ə-jē) *n.* Medical study and treatment of a fetus. **—fe·tol′o·gist** *n.*

**fe·tor** (fē′tər, -tôr′) *also* **foe·tor** (fē′tər) *n.* [ME *fetour* < Lat. *fetor* < *fętēre,* to stink.] An offensive odor : STENCH.

**fe·tos·co·py** (fē-tŏs′kə-pē) *n.* Examination of a fetus in the uterus by insertion of a fiber-optic device into the amniotic cavity. **—fe′to·scope′** (fē′tə-skōp′) *n.*

**fet·ter** (fĕt′ər) *n.* [ME *feter* < OE.] **1.** A chain or shackle attached to the ankles to restrain movement. **2.** A restriction : restraint. —*vt.* **-tered, -ter·ing, -ters. 1.** To put fetters on : SHACKLE. **2.** To restrict the freedom of.

**fet·ter·bush** (fĕt′ər-bo͝osh′) *n.* **1.** A shrub, *Lyonia lucida* of the southeastern United States, with evergreen leaves and white flower clusters. **2.** A shrub related to or resembling the fetterbush, esp. one of the genus *Leucothoe.*

**fet·tle** (fĕt′l) *vt.* **-tled, -tling, -tles.** [ME *fetlen,* to shape, prob. < OE *fetel,* girdle.] *Metallurgy.* To line (the hearth of a reverberatory furnace) with loose sand or ore before pouring molten metal. —*n.* **1.** Material used to line a furnace in fettling. **2.** Proper or sound condition or mental state <in fine *fettle*>

**fet·tling** (fĕt′lĭng) *n.* Material, as loose ore and sand, used to line a reverberatory furnace.

**fet·tuc·ci·ne** (fĕt′ə-chē′nē) *n.* [Ital., pl. of *fettuccina,* dim. of *fettuccia,* ribbon, dim. of *fetta,* slice. —see FETA.] **1.** Narrow strips of pasta. **2.** A dish made with fettuccine.

**fettuccine Al·fre·do** (ăl-frā′dō) *n.* [After the owner of a restaurant in Rome noted for this dish.] A dish made of fettuccine, butter, Parmesan cheese, cream, and seasonings.

**fe·tus** *also* **foe·tus** (fē′təs) *n., pl.* **-tus·es.** [Lat., offspring.] The unborn young of a viviparous vertebrate; in humans, the unborn young from the end of the eighth week to the moment of birth.

**feud¹** (fyōōd) *n.* [ME *fede* < OFr. *faide,* of Germanic orig.] A bitter prolonged quarrel or state of enmity, as between two clans or individuals. —*vi.* **feud·ed, feud·ing, feuds.** To engage in a feud.

**feud²** (fyōōd) *n.* [Med. Lat. *feudum,* of Germanic orig.] FEE 5a.

**feu·dal** (fyōōd′l) *adj.* [Med. Lat. *feudalis* < *feudum,* feud.] **1.** Of, relating to, or typical of feudalism. **2.** Of or relating to lands held in fee or to the holding of such lands. **—feu′dal·ly** *adv.*

**feu·dal·ism** (fyōōd′l-ĭz′əm) *n.* A European political and economic system from the 9th to about the 15th cent., based on the relation of lord to vassal held on condition of homage and service. **—feu′dal·ist** *n.* **—feu′dal·is′tic** *adj.*

**feu·dal·i·ty** (fyōō-dăl′ĭ-tē) *n., pl.* **-ties. 1.** The quality or state of being feudal. **2.** A feudal holding, regime, or system.

**feu·dal·ize** (fyōōd′l-īz′) *vt.* **-ized, -iz·ing, -iz·es.** To make feudal. **—feu′dal·i·za′tion** *n.*

**feu·da·to·ry** (fyōō′də-tôr′ē, -tōr′ē) *n., pl.* **-ries.** [Med. Lat. *feudatarius,* of a feudatory < *feudare,* to enfeoff < *feudum,* feud.] **1.** One who holds a feudal fee : VASSAL. **2.** A feudal fee. —*adj.* **1.** Of or typical of the feudal relationship between vassal and lord. **2.** Owing feudal homage or allegiance.

**feud·ist¹** (fyōō′dĭst) *n.* A participant in a feud.

**feud·ist²** (fyōō′dĭst) *n.* A specialist in feudal law.

**Feul·gen reaction** (foil′gən) *n.* [After Robert *Feulgen* (1884–1955).] A DNA-specific staining reaction based on the formation of a reddish-purple color upon contact with a reagent containing fuchsin and sulfuric acid.

**fe·ver** (fē′vər) *n.* [ME < OE *fefor* and OFr. *fievre,* both < Lat. *febris.*] **1.** Abnormally high body temperature. **2.** A disease marked by abnormally high body temperatures. **3.** A condition of heightened excitement or activity <a *fever* of selling on the stock market> **4.** A contagious, usu. short-lived enthusiasm.

**fever blister** *n.* A cold sore.

**fe·ver·few** (fē′vər-fyōō′) *n.* [ME *feverfu* < OE *feferfuge* < Lat. *febrifugia* : *febris,* fever + *fugare,* to drive away.] An aromatic plant indigenous to Eurasia, *Chrysanthemum parthenium,* with buttonlike, white-rayed flower clusters.

**fe·ver·ish** (fē′vər-ĭsh) *also* **fe·ver·ous** (-əs) *adj.* **1. a.** Having a fever. **b.** Of or like a fever. **c.** Causing or tending to cause fever. **2.** Marked by strong agitation, emotion, or activity <working at a *feverish* pace> **—fe′ver·ish·ly** *adv.* **—fe′ver·ish·ness** *n.*

**fever pitch** *n.* Extreme excitement or disturbance.

**fever therapy** *n.* Treatment of disease involving artificially induced fever.

**fever tree** *n.* A tree, as one of certain species of eucalyptus or *Pinckneya pubens* of the southeastern United States, having leaves or bark used to allay fever.

**fe·ver·weed** (fē′vər-wēd′) *n.* Any of various plants believed to have medicinal properties.

**fe·ver·wort** (fē′vər-wûrt′, -wôrt′) *n.* A plant, as the horse gentian or boneset, believed to have medicinal properties.

**few** (fyōō) *adj.* **-er, -est.** [ME *fewe* < OE *fēawe.*] Amounting to or made up of a small number <spoke for only a *few* seconds> —*n.* *(pl. in number).* **1.** An indefinitely small number <borrowed a *few* of their toys><spoke to a *few* of my colleagues> **2.** A limited number of people <the discerning *few*> ***usage:*** Traditionally *few* and *fewer* are used only before a plural noun (*few books; fewer people*) and *less* is used before a mass noun (*less sugar*). —*pron.* *(pl. in number).* A small number of persons or things <"Many are called, but *few* are chosen" —Matthew 22:14> **—few′ness** *n.*

**fey** (fā) *adj.* [ME *feie* < OE *fǣge.*] **1.** *Scot.* **a.** Fated to die soon. **b.** Sensing imminent death. **2.** Having visionary power : CLAIRVOYANT. **3.** Appearing as if under a spell : ENCHANTED.

**fez** (fĕz) *n., pl.* **fez·zes.** [Fr. < Turk. < *Fez,* a town in Morocco.] A man's felt cap shaped like a flat-topped cone, usu. red with a black tassel hanging from the crown, worn primarily in the eastern Mediterranean region.

**fi·a·cre** (fē-ä′krə) *n.* [Fr., after the Hotel de St. *Fiacre,* Paris.] A small hackney coach.

**fi·an·cé** (fē′än-sā′, fē-än′sā′) *n.* [Fr., p.part. of *fiancer,* to betroth < OFr. *fiancier* < *fiance,* trust < *fier* to trust < Lat. *fidare.*] A man engaged to be married.

**fi·an·cée** (fē′än-sā′, fē-än′sā′) *n.* [Fr., fem. of *fiancé,* fiancé.] A woman engaged to be married.

**fi·as·co** (fē-ăs′kō, -ä′skō) *n., pl.* **-coes** *or* **-cos.** [Fr. < Ital.] A total failure.

**fi·at** (fē′ət, -ăt′, -ät′, fī′ăt′, -ət) *n.* [Lat., let it be done.] **1.** An arbitrary decree. **2.** Authorization or sanction.

**fiat money** *n.* Paper money decreed to be legal tender, not backed by gold or silver and not necessarily redeemable in coin.

**fib** (fĭb) *n.* [Orig. unknown.] An inconsequential lie. —*vi.* **fibbed, fib·bing, fibs.** To tell a fib. **—fib′ber** *n.*

**fi·ber** (fī′bər) *n.* [Fr. *fibre* < Lat. *fibra.*] **1.** A slender, elongated structure. **2.** One of the elongated, thick-walled cells giving strength and support to plant tissue. **3.** Any of the filaments constituting the intracellular matrix of connective tissue. **4.** Any of the elongated contractile cells of muscle tissue. **5.** The long threadlike process of a neuron. **6. a.** A natural or synthetic filament, as of cotton or nylon, capable of being spun into yarn. **b.** Material made of such filaments. **7.** Essential substance or character <disgusted down to the *fiber* of my being> **8.** Internal strength <lacked moral *fiber*>

**fi·ber·board** (fī′bər-bôrd′, -bōrd′) *n.* A building material made of plant fibers, as wood, bonded together and compressed into rigid sheets.

**Fi·ber·fil** (fī′bər-fĭl′). A trademark for a synthetic resin used as quilt filling.

**Fi·ber·glas** (fī′bər-glăs′). A trademark for a type of fiber glass.

**fiber glass** *n.* A composite material of glass fibers in resin.

**fi·ber·ize** (fī′bə-rīz′) *vt.* **-ized, -iz·ing, -iz·es.** To break into fibers. **—fi′ber·i·za′tion** *n.*

**fiber optics** *n.* *(sing. in number).* The optics of light transmission through very fine, flexible glass rods by internal reflection. **—fi′ber-op′tic** (fī′bər-ŏp′tĭk) *adj.*

**fi·ber·scope** (fī′bər-skōp′) *n.* A flexible fiber-optic instrument used to view otherwise inaccessible objects.

**Fi·bo·nac·ci number** (fē′bə-nä′chē) *n.* A number in the Fibonacci sequence.

**Fibonacci sequence** *n.* [After Leonardo *Fibonacci* (d. c. 1250).] A series of numbers, 1, 1, 2, 3, 5, 8, 13, . . . , in which each successive number is equal to the sum of the two preceding numbers.

**fibr-** *pref. var. of* FIBRO-.

**fi·bre** (fī′bər) *n. Chiefly Brit. var. of* FIBER.

**fi·bri·form** (fī′brə-fôrm′) *adj.* [FIBRE + -FORM.] Similar to a fiber in form.

**fi·bril** (fī′brəl, fĭb′rəl) *n.* [NLat. *fibrilla,* dim. of Lat. *fibra,* fiber.] A small, slender fiber. **—fi′bril·lar** (-lər), **fi′bril·lar′y** (-lĕr′ē) *adj.*

**fi·bril·late** (fĭb′rə-lāt′, fī′brə-) *vi.* & *vt.* **-lat·ed, -lat·ing, -lates.** [Back-formation < FIBRILLATION.] To undergo or cause to undergo fibrillation.

**fib·ril·la·tion** (fĭb′rə-lā′shən, fī′brə-) *n.* [NLat. *fibrilla,* fibril + -TION.] **1.** Formation of fibers. **2.** Uncoordinated twitching of individual muscular fibers with little or no movement of the muscle as a whole. **3.** Fine, rapid fibrillar cardiac movements replacing the normal contraction of the ventricular muscle.

**fi·bril·li·form** (fī-brĭl′ə-fôrm′, fə-) *adj.* Having the form of a fibril.

**fi·bril·lose** (fī′brə-lōs′, fĭb′rə-) *adj.* Having or made up of fibrils.

**fi·brin** (fī′brĭn) *n.* An elastic, insoluble protein derived from the interaction of fibrinogen with thrombin and forming a fibrous network in the coagulation of blood.

**fi·brin·o·gen** (fī-brĭn′ə-jən) *n.* A blood plasma protein that is converted to fibrin by the action of thrombin in the presence of ionized calcium.

**fi·bri·nol·y·sin** (fī′brə-nŏl′ĭ-sĭn) *n.* An enzyme capable of dissolving fibrin.

**fi·bri·nol·y·sis** (fī′brə-nŏl′ĭ-sĭs) *n., pl.* **-ses** (-sēz′). Breakdown of fibrin by the action of fibrinolysin. **—fi′bri·no·lyt′ic** (-nə-lĭt′ĭk) *adj.*

**fi·bri·nous** (fī′brə-nəs) *adj.* Of, relating to, or having the nature of fibrin.

**fibro-** *or* **fibr-** *pref.* [Lat. *fibra,* fiber.] **1.** Fiber <*fibrous*> **2.** Fibrous tissue <*fibroma*>

**fi·bro·blast** (fī′brə-blăst′) *n.* A cell that generates to connective tissue. **—fi′bro·blas′tic** *adj.*

**fi·bro·car·ti·lage** (fī′brō-kär′tl-ĭj) *n.* Cartilage having numerous thick bundles of collagen fibers.

**fi·broid** (fī′broid′) *adj.* Resembling or made up of fibrous tissue. —*n.* A benign neoplasm of smooth muscle, esp. in the uterine wall.

**fi·bro·in** (fī′brō-ĭn) *n.* [Fr. *fibroïne* < *fibre,* fiber.] A white protein, the essential component of raw silk and spider-web filaments.

**fi·bro·ma** (fī-brō′mə) *n., pl.* **-mas** *or* **-ma·ta** (-mə-tə). A benign neoplasm derived from fibrous tissue. **—fi·brom′a·tous** (-brŏm′ə-təs, -brō′mə-) *adj.*

**fi·bro·sis** (fī-brō′sĭs) *n.* Formation of fibrous tissue, as in a reparative or reactive process, in excess of amounts ordinarily present. **—fi·brot′ic** (-brŏt′ĭk) *adj.*

**fi·bro·si·tis** (fī′brə-sī′tĭs) *n.* [NLat. *fibrosus,* fibrous + -ITIS.] Inflammatory hyperplasia of white fibrous connective tissue.

**fi·brous** (fī′brəs) *adj.* Having, composed of, or like fibers. **—fi′brous·ly** *adv.* **—fi′brous·ness** *n.*

**fi·bro·vas·cu·lar** (fī′brō-văs′kyə-lər) *adj.* Having fibrous tissue and vascular tissue, as in the woody tissue of plants.

**fib·u·la** (fĭb′yə-lə) *n., pl.* **-lae** (-lē′) *or* **-las.** [NLat. < Lat., clasp, perh. < *figere,* to fix.] The outer and smaller of two bones of the human leg or the hind leg of an animal, between the knee and ankle.

**-fic** *suff.* [Lat. *-ficus* < *facere,* to do.] Causing : making <*soporific*>

**-fication** *suff.* [ME *-ficacioun* < OFr. *-fication* < Lat. *-ficatio* < *-ficare,* to make < *-ficus,* -fic.] Production : making <*jollification*>

**fice** (fīs) *n. var. of* FEIST.

**fiche** (fēsh) *n.* A microfiche.

**fich·u** (fĭsh′ōō, fē-shōō′) *n.* [Fr. < p.part. of *ficher,* to fix < Lat.

*figere.*] A woman's triangular scarf of lightweight fabric, worn over the shoulders and crossed or tied in a loose knot at the breast.

**fick·le** (fĭk′əl) *adj.* [ME *fikel* < OE *ficol,* deceitful.] Erratically changeable or unstable : CAPRICIOUS. **—fick′le·ness** *n.*

**fic·tile** (fĭk′təl, -tīl′) *adj.* [Lat. *fictilis,* made of clay < *fictus,* p.part. of *fingere,* to mold.] **1.** Capable of being molded : PLASTIC. **2.** Formed of a moldable substance, as clay or earth. **3.** Of or relating to earthenware or pottery.

**fic·tion** (fĭk′shən) *n.* [ME *ficcioun* < OFr. *fiction* < Lat. *fictio* < *fictus,* p.part. of *fingere,* to form.] **1.** An imaginative creation or a pretense. **2.** The act of inventing an imaginative creation or pretense. **3.** A lie. **4. a.** A literary work whose content is produced by the imagination and is not necessarily based on fact. **b.** The category of literature comprising works of this kind, including novels, short stories, and plays. **5.** *Law.* Something accepted as fact without real justification, merely for the sake of convenience. **—fic′tion·al** *adj.* **—fic′tion·al·ly** *adv.*

**fic·tion·al·ize** (fĭk′shə-nə-līz′) *vt.* **-ized, -iz·ing, -iz·es.** To treat as or make into fiction. **—fic′tion·al·i·za′tion** *n.*

**fic·tion·eer** (fĭk′shə-nîr′) *n.* A writer of fiction, esp. a prolific writer of pulp fiction.

**fic·tion·ize** (fĭk′shə-nīz′) *vt.* **-ized, -iz·ing, -iz·es.** To fictionalize. **—fic′tion·i·za′tion** *n.*

**fic·ti·tious** (fĭk-tĭsh′əs) *adj.* [Lat. *ficticius* < *fictus,* p.part. of *fingere,* to form.] **1.** Of, relating to, or marked by fiction : IMAGINARY. **2.** Assumed in order to deceive <a *fictitious* name> **3.** Not genuine : SHAM <*fictitious* enthusiasm> **—fic·ti′tious·ly** *adv.* **—fic·ti′tious·ness** *n.*

**fic·tive** (fĭk′tĭv) *adj.* **1.** Of or relating to the creation of fiction. **2.** Fictitious. **—fic′tive·ly** *adv.*

**fid** (fĭd) *n.* [Orig. unknown.] *Naut.* **1.** A square bar used as a support for a topmast. **2.** A large tapering pin for opening the strands of a rope prior to splicing.

**-fid** *suff.* [Lat. *-fidus* < *findere,* to split.] Divided into parts or lobes <pinnati*fid*>

**fid·dle** (fĭd′l) *n.* [ME *fidle* < OE *fiðele* < LLat. *vitula* < Lat. *vitulari,* to celebrate a victory < *Vitula,* goddess of victory.] **1.** *Informal.* **a.** A violin. **b.** A member of the violin family. **2.** *Naut.* A guardrail used on a table during rough weather to prevent objects from slipping off. **3.** *Informal.* Trifling nonsense. *—v.* **-dled, -dling, -dles.** *—vi.* **1.** *Informal.* To play a violin. **2.** To move one's fingers or hands nervously. **3.** To putter or tamper with something <always *fiddling* with the TV> *—vt. Informal.* To play (a tune) on a violin. **—fiddle away.** To fiddle-faddle **—fid′dler** *n.*

**fid·dle-de-dee** (fĭd′l-dē-dē′) *interj.* —Used to express mild annoyance or impatience.

**fid·dle-fad·dle** (fĭd′l-făd′l) *interj.* [Redup. of FIDDLE.] —Used to express mild annoyance or impatience. *—vi.* **-dled, -dling, -dles.** To fritter away one's time : DALLY. **—fid′dle-fad′dler** *n.*

**fid·dle·head** (fĭd′l-hĕd′) *n.* **1.** A curved scroll-like ornamentation at the top of a ship's bow that resembles the neck of a violin. **2.** The coiled young frond of any of various ferns, considered a delicacy when cooked.

**fiddler crab** *n.* A burrowing crab of the genus *Uca* of coastal areas, the males of which have a greatly enlarged anterior claw.

**fid·dle·sticks** (fĭd′l-stĭks′) *interj.* —Used to express mild annoyance or impatience.

**fi·del·i·ty** (fĭ-dĕl′ĭ-tē, fī-) *n., pl.* **-ties.** [ME *fidelite* < OFr. < Lat. *fidelitas* < *fidelis,* faithful < *fides,* faith.] **1.** Faithfulness to obligations, duties, or observances. **2.** Exact correspondence with fact or a given quality, condition, or event : ACCURACY. **3.** The degree to which an electronic system accurately reproduces the sound or image of its input signal.

**fidg·et** (fĭj′ĭt) *v.* **-et·ed, -et·ing, -ets.** [Obs. *fidge,* to move restlessly, perh. < ME *fiken.*] *—vi.* **1.** To behave or move nervously or restlessly <*fidgeted* about the room> **2.** FIDDLE 3. *—vt.* To cause to behave or move nervously or restlessly. *—n.* **1.** *often* **fidgets.** Restlessness. **2.** One who fidgets.

**fidg·et·y** (fĭj′ĭ-tē) *adj.* **1.** Habitually fidgeting. **2.** Unnecessarily fussy. **—fidg′et·i·ness** *n.*

**fi·do** (fī′dō) *n., pl.* **-dos.** [F(REAKS) + I(RREGULARS) + D(EFECTS) + O(DDITIES).] A coin containing a minting error.

**fi·du·cial** (fĭ-dōō′shəl, -dyōō′-, fī-) *adj.* [LLat. *fiducialis* < Lat. *fiducia,* trust < *fidĕre,* to trust.] **1.** Based on or relating to faith or trust. **2.** Relating to a legal trust : FIDUCIARY. **3.** Regarded or used as a standard of reference, as in surveying. **—fi·du′cial·ly** *adv.*

**fi·du·ci·ar·y** (fĭ-dōō′shē-ĕr′ē, -shə-rē, -dyōō′-, fī-) *adj.* [Lat. *fiduciarius* < *fiducia,* trust. —see FIDUCIAL.] **1.** Of, relating to, or involving one that holds something in trust for another <a *fiduciary* heir><a *fiduciary* contract> **2. a.** Of, relating to, or designating a trustee or trusteeship. **b.** Held in trust. **3.** Of, relating to, or consisting of fiat money. **4.** Of, relating to, or being a system of marking in the reticule of an optical instrument used as a reference point or a measuring scale. *—n., pl.* **-ies.** One that acts in a fiduciary capacity or holds a fiduciary relation to others.

**fie** (fī) *interj.* —Used to express distaste or shock.

**fief** (fēf) *n.* [Fr. < OFr.] FEE 5a.

**field** (fēld) *n.* [ME < OE *feld.*] **1.** A broad, level, open expanse of land. **2.** A meadow <a *field* of tulips> **3.** A cultivated expanse of land, esp. one set aside for a particular crop. **4.** A part of land or a geologic formation containing a specified natural resource <a natural gas *field*> **5.** A background area, as on a flag, painting, or coin <a red insignia on a *field* of blue> **6.** *Heraldry.* The background area of a shield or one of the divisions of the background. **7. a.** An area in which a sports event takes place : GROUND. **b.** The part of a playing field having specific dimensions on which the action of a game takes place. **c.** All the contestants or participants in an event. **d.** All the contestants except those specified. **e.** The members of a team participating in active play. **f.** The body of riders following a pack of hounds. **8. a.** An area of human activity or interest <a *field* of endeavor> **b.** A topic, subject, or area of academic interest or specialization. **c.** One's profession, employment, or business. **9.** An area or setting of practical activity or application <spent their sabbatical working in the *field*> **10. a.** The land where a battle is or has been fought. **b.** A battle. **11.** *Math.* A set with two binary operations, designated addition and multiplication, satisfying the conditions that the set is a commutative group with respect to addition, that the set with the identity of the additive group omitted is a commutative group with respect to multiplication, and that multiplication distributes over addition for all elements in the set. **12.** *Physics.* A region of space marked by a physical property, as gravitational or electromagnetic force or fluid pressure, having a determinable value at every point in the region. **13.** The usu. circular area in which the image is rendered by the lens system of an optical instrument. **14.** *Computer Sci.* A region, as a set of adjacent columns on a punched card, used to consistently record related information. *—v.* **field·ed, field·ing, fields.** *—vt.* **1.** To retrieve (a ball) and perform the required maneuver, esp. in baseball. **2.** To respond to adequately <*fielded* the questions expertly> **3. a.** To place or be equipped to place in a sports contest <*field* a baseball team> **b.** To put into action. *—vi.* To play as a fielder in a sports contest. **—play the field.** To indulge in a wide range of interests or possibilities. **—take the field.** To begin or resume activity, as in military operations or in a sport.

**field artillery** *n.* Artillery, excluding antiaircraft artillery, light enough to be mounted for use in the field.

**field capacity** *n.* The maximum amount of water a particular kind of soil can hold.

**field coil** *n.* An electric coil used to generate a magnetic field, as in a motor or direct-current generator.

**field corn** *n.* Any of several varieties of corn used chiefly as feed for livestock.

**field day** *n.* **1.** A day designated for a specific activity, as an athletic contest, nature study, or public demonstration. **2.** A festive day. **3.** *Informal.* An opportunity for expressing or asserting oneself with the greatest pleasure or triumph <Watergate proved to be a *field day* for the media.>

**field effect transistor** *n.* A transistor in which the output current is controlled by a variable electric field.

**field emission** *n.* Emission of electrons from the surface of a conductor, caused by a strong electric field.

**field·er** (fēl′dər) *n.* A sports player positioned in the field, esp. an outfielder in baseball.

**fielder's choice** *n. Baseball.* A play made by an infielder on a ground ball in which he or she chooses to put out an advancing base runner, thus allowing the batter to reach first base safely.

**field event** *n.* A throwing and jumping event of a track meet.

**field·fare** (fēld′fâr′) *n.* [ME *feldfare* < OE *feldeware* : *felde,* field + *ware,* dweller.] An Old World thrush, *Turdus pilaris,* with gray and brown plumage.

**field glass** *n. often* **field glasses.** A portable binocular instrument used for viewing distant objects.

**field goal** *n.* **1.** *Football.* A score worth three points made on an ordinary down by place-kicking or drop-kicking the ball over the crossbar and between the goal posts. **2.** *Basketball.* A score of two points made by throwing the ball through the basket.

**field hand** *n.* A hired farm laborer.

**field hockey** *n.* HOCKEY 2.

**field hospital** *n.* A hospital set up on a temporary basis to serve troops in a combat zone.

**field house** *n.* **1.** A building at an athletic field having locker rooms and storage and training facilities. **2.** A building having one or more areas for different athletic events and usu. grandstands.

**field ion microscope** *n.* A microscope that produces an image of the atoms on a metal surface by means of ions formed in a high-voltage electric field.

**field lens** *n.* The lens positioned farthest from the eye in a compound eyepiece.

**field magnet** *n.* A magnet used to provide a magnetic field in an electrical device, as a generator or motor.

**field marshal** *n.* An officer in some European armies, usu. ranking just below the commander-in-chief.

**field mint** *n.* The corn mint.
**field mouse** *n.* A small mouse, as of the genera *Apodemus* or *Microtus,* inhabiting meadows and often causing crop damage.
**field officer** *n.* A military officer, as a major, lieutenant colonel, or colonel, ranking above a captain and below a brigadier general.
**field of force** *n.* A region of space throughout which the force produced by a single agent, as an electric current, is operative.
**field of honor** *n.* **1.** The scene of a duel. **2.** A battlefield.
**field of view** *n.* FIELD 13.
**field of vision** *n.* A visual field.
**fields·man** (fēldz'mən) *n.* A fielder in cricket.
**field·stone** (fēld'stōn') *n.* A stone naturally occurring in fields, often used as a building material.
**field·strip** (fēld'strĭp') *vt.* **-stripped, -strip·ping, -strips. 1.** To disassemble (a weapon) for cleaning, repair, and inspection. **2.** To tear (a cigarette) into tiny pieces.
**field·test** (fēld'tĕst') *vt.* **-test·ed, -test·ing, -tests.** To test in natural operating conditions. **—field test** *n.*
**field trial** *n.* A test for young, untried hunting dogs to ascertain their competence in pointing and retrieving.
**field trip** *n.* A group excursion for firsthand observation, as to a museum, woods, or historical place.
**field winding** *n.* The electrically conducting winding of a field magnet producing electrical excitation, as of a motor or generator.
**field·work** (fēld'wûrk') *n.* **1.** A temporary military fortification erected in the field. **2.** Work carried out or observations made in the field.
**fiend** (fēnd) *n.* [ME < OE *fēond.*] **1.** An evil spirit : DEMON. **2.** *often* **Fiend.** DEVIL 1. **3.** A diabolically evil person. **4.** *Informal.* **a.** An addict <a dope *fiend*> **b.** One totally absorbed in or obsessed with something <a sports car *fiend*> **5.** *Informal.* One particularly adept at something <a *fiend* with calculators>
☆ ***syns:*** FIEND, BEAST, GHOUL, MONSTER, OGRE, VAMPIRE *n. core meaning* : a diabolically cruel or wicked person <a *fiend* who gassed thousands of helpless prisoners>
**fiend·ish** (fēn'dĭsh) *adj.* **1.** Of, relating to, or like a fiend : DIABOLICAL. **2.** Extremely wicked or cruel. **3.** Difficult <a *fiendish* problem> **—fiend'ish·ly** *adv.* **—fiend'ish·ness** *n.*
**fierce** (fîrs) *adj.* **fierc·er, fierc·est.** [ME *fiers* < OFr. < Lat. *ferus.*] **1.** Savage and violent in nature : FEROCIOUS. **2.** Extremely severe : TERRIBLE <a *fierce* storm> **3.** Extremely intense or ardent <*fierce* pride> **4.** *Informal.* Very difficult or unpleasant <a *fierce* English exam> **—fierce'ly** *adv.* **—fierce'ness** *n.*
☆ ***syns:*** FIERCE, BESTIAL, CRUEL, FELL, FERAL, FEROCIOUS, INHUMAN, SAVAGE, TRUCULENT, VICIOUS, WOLFISH *adj. core meaning* : violently destructive without scruples or restraint <a *fierce* interrogator><*fierce* cannibals>
**fi·e·ri fa·ci·as** (fī'ə-rē fā'shē-əs, fā'shəs) *n.* [ME < Lat., cause (it) to be done.] *Law.* A writ of execution commanding a sheriff to lay a claim to and seize the goods and chattels of a debtor to fulfill a judgment.
**fier·y** (fīr'ē, fī'ə-rē) *adj.* **-i·er, -i·est.** [ME < *fier,* fire < OE *fȳr.*] **1.** Consisting of or containing fire. **2.** Of, relating to, or resembling a fire <a *fiery* sunset> **3.** Torridly hot <*fiery* sunlight> **4.** Flammable. **5.** Emitting or appearing to emit sparks : GLOWING. **6.** Emotionally volatile : TEMPESTUOUS <had a *fiery* temper> **7.** Inflamed and usu. painful <a *fiery* sunburn> **—fier'i·ly** *adv.* **—fier'i·ness** *n.*
**fi·es·ta** (fē-ĕs'tə) *n.* [Sp. < Lat. *festa,* neuter pl. of *festus,* joyous.] A festival or religious holiday, esp. one honoring a saint's day in Spanish-speaking countries.
**fife** (fīf) *n.* [Fr. *fifre* or G. *Pfeife* < OHG *pfīffa.*] A musical instrument similar to a flute but higher in range, used chiefly to accompany drums in military music.
**fife rail** *n.* A rail around the lower part of a ship's mast to which the belaying pins for the rigging are secured.
**fif·teen** (fĭf-tēn') *n.* [ME *fiftene* < OE *fīftēne.*] **1.** The cardinal number equal to 14 + 1. **2.** The 15th in a set or sequence. **—fif·teen'** *adj. & pron.*
**fif·teenth** (fĭf-tēnth') *n.* **1.** The ordinal number matching the number 15 in a series. **2.** One of 15 equal parts. **—fif·teenth'** *adj. & adv.*
**fifth** (fĭfth) *n.* [ME < OE *fīfta.*] **1.** The ordinal number matching the number five in a series. **2.** One of five equal parts. **3.** One fifth of a gallon or four fifths of a quart of liquor. **4. a.** A musical interval encompassing five diatonic tones, as C, D, E, F, and G. **b.** Either of the two tones constituting the extremities of such an interval. **c.** The dominant of a tonality. **5. Fifth.** The Fifth Amendment to the U.S. Constitution. **—fifth** *adj. & adv.*
**Fifth Amendment** *n.* An amendment to the U.S. Constitution, ratified in 1791, that deals with the rights of accused criminals by providing for due process of law, forbidding double jeopardy, and stating that no one may be forced to testify as a witness against oneself.
**fifth column** *n.* [First applied in 1936 to rebel sympathizers inside Madrid when four columns of rebel troops were attacking the city.] A clandestine subversive organization working within a country to further an invading enemy's military and political aims. **—fifth columnist** *n.*
**fifth wheel** *n.* **1.** A wheel or portion of a wheel placed horizontally over the forward axle of a carriage to provide support and stability during turns. **2.** An additional wheel carried on a four-wheeled vehicle as a spare. **3.** A superfluous person or thing.
**fif·ti·eth** (fĭf'tē-ĭth) *n.* **1.** The ordinal number matching the number 50 in a series. **2.** One of 50 equal parts. **—fif'ti·eth** *adj. & adv.*
**fif·ty** (fĭf'tē) *n.* [ME *fifti* < OE *fīftig.*] The cardinal number equal to 5 × 10. **—fif'ty** *adj. & pron.*
**fif·ty-fif·ty** (fĭf'tē-fĭf'tē) *adj. Informal.* Divided or shared in two equal portions <a *fifty-fifty* split><a *fifty-fifty* chance of recovery> **—fif'ty-fif'ty** *adv.*
**fig**[1] (fĭg) *n.* [ME < OFr. *fige* < OProv. *figa* < Lat. *ficus.*] **1.** A tree or shrub of the genus *Ficus,* esp. *F. carica,* native to the Mediterranean region, cultivated for its edible fruit. **2.** The sweet, pear-shaped, many-seeded fruit of the fig tree. **3. a.** A plant bearing fruit similar to the fig. **b.** The fruit of such a plant. **4.** A trivial amount <didn't care a *fig* about the outcome>
**fig**[2] (fĭg) *n.* [Orig. unknown.] *Informal.* **1.** Dress : array <decked out in full *fig*> **2.** Physical condition : SHAPE <in poor *fig*>
**fight** (fīt) *v.* **fought** (fôt), **fight·ing, fights.** [ME *fighten* < OE *feohtan.*] *—vi.* **1.** To take part in combat. **2.** To participate in boxing or wrestling. **3.** To argue. **4.** To resist something or assert oneself. *—vt.* **1.** To contend with in battle. **2.** To box or wrestle against in a ring. **3.** To contend with or struggle against <*fight* the new zoning ordinance> **4.** To try to prevent or undo the development of <*fight* a serious infection> **5.** To wage (a battle). **6.** To contend for by or as if by combat <"I now resolved that Calais should be *fought* to the death" —Winston Churchill> **7.** To make (one's way) by or as if by combat. **8.** To set in combat with another. **—fight off.** To defend against or drive back (a hostile force). *—n.* **1.** A battle between opposing groups. **2.** A quarrel. **3. a.** A physical conflict between two or more individuals. **b.** A boxing or wrestling match. **4.** A struggle to gain an objective. **5.** The power or inclination to fight.
**fight·er** (fī'tər) *n.* **1.** One engaged in fighting. **2.** A boxer : pugilist. **3.** A pugnacious, unyielding, or determined person. **4.** A fast, maneuverable combat aircraft.
**fighting chance** *n.* A slight chance to win.
**fig leaf** *n.* A stylized representation of the leaf of a fig, used esp. to conceal the genitalia of male statues.
**fig marigold** *n.* A native South African plant of the genus *Mesembryanthemum,* with fleshy leaves and variously colored flowers.
**fig·ment** (fĭg'mənt) *n.* [ME < Lat. *figmentum* < *fingere,* to form.] An invention or fabrication <a *figment* of one's imagination>
**fig·ur·al** (fĭg'yər-əl) *adj.* Consisting of or forming a pictorial composition or design of human or animal figures.
**fig·u·rant** (fĭg'yə-ränt', -rănt', -räN') *n.* [Fr. < pr.part. of *figurer,* to represent < OFr. < Lat. *figurare,* to form < *figura,* figure.] **1.** A member of a corps de ballet who does not perform solos. **2.** A nonspeaking stage performer.
**fig·u·ra·tion** (fĭg'yə-rā'shən) *n.* **1.** The act of forming something into a given shape. **2.** A shape, form, or outline. **3.** The act of representing with figures. **4.** A figurative representation. **5.** *Mus.* Embellishment : ornamentation.
**fig·u·ra·tive** (fĭg'yər-ə-tĭv) *adj.* **1. a.** Based on or using figures of speech. **b.** Containing many figures of speech : ORNATE. **2.** Represented by a figure or figures. **3.** Of or relating to representation by means of animal or human figures. **—fig'u·ra·tive·ly** *adv.* **—fig'u·ra·tive·ness** *n.*
**fig·ure** (fĭg'yər) *n.* [ME < OFr. < Lat. *figura* < *fingere,* to form.] **1.** A written symbol representing something other than a letter, esp. a number. **2. figures.** Mathematical calculation involving the use of figures. **3.** An amount represented in numbers <a painting sold for a large *figure*> **4.** Outline, form, or silhouette. **5.** Human shape or form. **6.** An individual, esp. a well-known person. **7.** The impression an individual makes through his or her behavior or appearance <a dashing *figure*> **8.** One that symbolizes something. **9.** A pictorial or sculptural representation, esp. of the human body. **10. a.** A diagram. **b.** A design or pattern. **11.** An illustration printed from an engraved plate or block. **12.** A configuration or distinct group of steps in a dance. **13.** *Mus.* A brief melodic or harmonic unit often forming the base for a larger musical phrase or structure. **14.** *Logic.* One of the forms that a syllogism can take, depending on the position of the middle term. **15. a.** An indistinct object. **b.** *Obs.* An illusion : phantasm. *—v.* **-ured, -ur·ing, -ures.** *—vt.* **1.** To calculate with numbers. **2.** To make a likeness of : DEPICT. **3.** To adorn with a design or figures. **4.** *Mus.* To indicate the chordal structure of (a bass line of single notes) with a sequence of conventionalized numbers. **5.** *Informal.* **a.** To believe <didn't *figure* that it would happen> **b.** To interpret or regard <*figured* them for cheats> *—vi.* **1.** To calculate : compute. **2.** To be pertinent or involved <*figured* prominently in Paris society> **—figure on** (or **upon**). *Informal.* **1.** To depend on. **2.** To take into consideration : EXPECT. **—figure out.** *Informal.* To solve or comprehend.
**fig·ured** (fĭg'yərd) *adj.* **1.** Shaped or fashioned. **2.** Decorated with a

design <a richly *figured* cloth> **3.** Represented, as in graphic art or sculpture.

**figured bass** *n.* A continuo.

**figure eight** *n.* A form or representation having the shape of the number 8, as a knot or an ice-skating maneuver.

**fig·ure·head** (fĭg′yər-hĕd′) *n.* **1.** One in a position of nominal leadership but having no real authority. **2.** A carved figure on the prow of a ship.

**figure of speech** *n.* An expression, as a metaphor or hyperbole, in which a nonliteral and intensive sense of a word or words is used to create a forceful, dramatic, or illuminating image.

**figure skating** *n.* Ice skating in which the skater traces prescribed, usu. elaborate figures.

**figures shift** *n.* **1.** A shift in a typewriter to upper case. **2.** A data control character after which characters are interpreted as having been typed in the upper-case mode.

**fig·u·rine** (fĭg′yə-rēn′) *n.* [Fr. < Ital. *figurina,* dim. of *figura,* figure < Lat. *figura.*] A small molded or sculptured figure **:** STATUETTE.

**fig wasp** *n.* A small wasp of the genus *Blastophaga* that is the vehicle for caprification.

**fig·wort** (fĭg′wûrt′, -wôrt′) *n.* [FIG[1], piles (obs.) + WORT (from its use as a folk medicine).] A plant of the genus *Scrophularia,* with loose, branching clusters of small greenish or purple flowers.

**Fi·ji·an** (fē′jē-ən) The Austronesian language of Fiji. **—Fi′ji·an** *adj.*

**fil** (fĭl) *n.* [Alteration of Ar. *fils.*] —See table at CURRENCY.

**fi·la** (fī′lə) *n. pl. of* FILUM.

**fil·a·ment** (fĭl′ə-mənt) *n.* [NLat. *filamentum* < LLat. *filare,* to spin < Lat. *filum,* thread.] **1.** A fine or thinly spun thread, fiber, or wire. **2.** A slender, threadlike appendage, part, or structure, as the slender stalk of a stamen on which the anther or a chainlike series of cells is borne. **3. a.** A fine wire heated electrically to incandescence in an electric lamp. **b.** *Electron.* A high-resistance wire or ribbon forming the cathode in some thermionic tubes. **—fil′a·men′tous** (-mĕn′təs), **fil′a·men′ta·ry** (-mĕn′tə-rē, -mĕn′trē) *adj.*

**fi·lar** (fī′lər) *adj.* [< Lat. *filum,* thread.] **1.** Of or relating to a thread. **2.** Having fine threads across the field of view for measuring small distances, as in a telescope eyepiece.

**fil·a·ree** (fĭl′ə-rē′) *n.* [Mex. Sp. *alfilerillo.* —see ALFILARIA.] The alfilaria.

**fi·lar·i·a** (fə-lâr′ē-ə) *n., pl.* **-i·ae** (-ē-ē′) [NLat. *Filaria,* former genus name < Lat. *filum,* thread.] Any of various slender filamentous parasitic nematode worms of the superfamily Filarioidea that infest the blood or tissues of vertebrates and are often transmitted by biting insects. **—fi·lar′i·al** (-ē-əl), **fi·lar′i·an** (-ē-ən) *adj.*

**fil·a·ri·a·sis** (fĭl′ə-rī′ə-sĭs) *n.* [FILAR(IA) + -IASIS.] Infestation of tissue, esp. with filariae.

**fil·a·ture** (fĭl′ə-cho͝or′, -chər) *n.* [Fr. < LLat. *filare,* to spin. —see FILAMENT.] **1.** The act or process of spinning, drawing, or twisting into threads. **2.** The act or process of reeling raw silk from cocoons. **3.** A reel used in drawing silk from cocoons. **4.** An establishment where silk is reeled.

**fil·bert** (fĭl′bərt) *n.* [ME < AN, after St. *Philbert* (d. A.D. 684), whose feast day in late Aug. coincides with the ripening of the nut.] **1.** A Eurasian shrub or tree, *Corylus maxima,* a species of hazel, cultivated for its edible nuts. **2.** The rounded smooth-shelled nut of the filbert.

**filbert**

**filch** (fĭlch) *vt.* **filched, filch·ing, filch·es.** [ME *filchen.*] To steal (something). **—filch′er** *n.*

**file[1]** (fīl) *n.* [< ME *filen,* to put documents on a thread < OFr. *filer* < *fil,* thread < Lat. *filum.*] **1.** A receptacle used for keeping loose objects, esp. papers, in useful order. **2.** A collection of objects kept or arranged in or as if in a file. **3. a.** A line of persons, animals, or things positioned one behind another. **b.** A line of soldiers or military vehicles so positioned. **4.** Any of the rows of squares that run vertically or between players on a playing board in chess or checkers. **5.** *Obs.* A list **:** roll. —*v.* **filed, fil·ing, files.** —*vt.* **1.** To put or keep (e.g., papers) in useful order. **2.** To enter (e.g., a legal document) on public official record <*file* a complaint> **3.** To send or submit (copy) to a publication. —*vi.* **1.** To march or walk in a line. **2.** To apply <*file* for a job> **3.** To enter one's name in a political contest. **—on file.** Recorded and ready for reference. **—fil′er** *n.*

▲ word history: The various senses of the noun *file[1]* are derived from two different French words that ultimately have the same etymology. French *file,* "line of soldiers," is the source of English *file* meaning "line" and "row of squares on a chessboard." French *fil,* "thread," is the source of English *file* meaning "a receptacle for papers or documents." This sense developed in English from the practice of keeping papers in order by threading them on string or wire. The word has been extended to various filing systems for information and objects of all kinds. Both French words, *fil* and *file,* are derived from Latin *filum,* "thread."

**file[2]** (fīl) *n.* [ME < OE *fīl.*] Any of several steel tools with hardened ridged surfaces, used in smoothing, polishing, grinding down, or boring. —*vt.* **filed, fil·ing, files.** To smooth, polish, grind, bore, or remove with or as if with a file. **—fil′er** *n.*

**file[3]** (fīl) *vt.* **filed, fil·ing, files.** [ME *filien* < OE *fȳlen.*] *Obs.* To sully.

**file clerk** *n.* An employee who keeps the files and records of an office.

**file·fish** (fīl′fĭsh′) *n., pl.* **filefish** *or* **-fish·es.** Any of various chiefly tropical marine fishes of the family Balistidae, resembling the triggerfishes.

**file lockout** *n. Computer Sci.* A condition occurring when two simultaneously running programs request access to the same two files, with each gaining access to one file but not the other, with the result that neither program can proceed and the files block the function of the computer.

**fi·let[1]** (fĭ-lā′, fĭl′ā′) *n.* [Fr.] Net or lace with a simple pattern of squares.

**fi·let[2]** (fĭ-lā′, fĭl′ā′) *n. & v. var. of* FILLET 2.

**fi·let mi·gnon** (fĭl′ā mēn-yôɴ′, fĭ-lā′) *n.* [Fr. : *filet,* fillet + *mignon,* dainty.] A small, round, tender cut of beef from the loin.

**fil·i·al** (fĭl′ē-əl) *adj.* [ME < LLat. *filialis* < Lat. *filius,* son.] Of, relating to, or befitting a son or daughter. **—fil′i·al·ly** *adv.*

**filial generation** *n. Genetics.* A set of offspring from a specific mating that follow the parental generation.

**fil·i·ate** (fĭl′ē-āt′) *vt.* **-at·ed, -at·ing, -ates.** [Med. Lat. *filiare, filiat-,* to acknowledge as a son < Lat. *filius,* son.] **1.** To affiliate. **2.** *Law.* To assign paternity to (e.g., a bastard child).

**fil·i·a·tion** (fĭl′ē-ā′shən) *n.* **1.** The fact or state of being the child of a certain parent. **2.** A line of descent **:** DERIVATION. **3. a.** The act or fact of forming a new branch, as of a society or language group. **b.** The branch thus formed. **4.** *Law.* Assignment of paternity to someone, as a bastard child.

**fil·i·bus·ter** (fĭl′ə-bŭs′tər) *n.* [< Sp. *filibustero,* freebooter < Fr. *flibustier* < Du. *vrijbuiter,* pirate. —see FREEBOOTER.] **1. a.** Use of obstructionist tactics, esp. prolonged speechmaking, to delay legislative action. **b.** An instance of the use of such delaying tactics. **2.** An adventurer who engages in a private military action in a foreign country. —*v.* **-tered, -ter·ing, -ters.** —*vi.* **1.** To use obstructionist tactics in a legislative body. **2.** To engage in a private military action in a foreign country. —*vt.* To use obstructionist tactics against (legislative action). **—fil′i·bus′ter·er** *n.*

▲ word history: A *freebooter* and a *filibuster* may not share many attributes but they do share a common linguistic ancestor: both are derived from Dutch *vrijbuiter. Freebooter,* which appeared in the 16th century, was a direct borrowing of the Dutch word. *Filibuster,* however, had a more checkered career. The first instance of *filibuster* in English occurred in the 18th century with the spelling *flibustier,* which was probably a borrowing of the French form of *vrijbuiter.* The French word was also adopted by the Spanish, and it is the Spanish form, *filibustero,* that is the immediate source of *filibuster.* The development of the senses of *filibuster* also reflects its Spanish origins. At first, *flibustier* or *filibustero* meant "pirate," but in the 19th century in the United States it was used to denote adventurers who tried to foment revolution in the Spanish colonies of Central America and the Caribbean. It was also used as a verb to indicate the activities of a *filibuster.* The opprobrium conveyed by *filibuster* was probably uppermost in the minds of those who first applied the term to the practice of obstructing legislative action.

**fil·i·form** (fĭl′ə-fôrm′, fī′lə-) *adj.* [Lat. *filum,* thread + -FORM.] Resembling or having the form of a thread.

**fil·i·gree** (fĭl′ĭ-grē′) *n.* [Fr. *filigrane* < Ital. *filigrana* : Lat. *filum,* thread + Lat. *granum,* grain.] **1.** Delicate and intricate ornamental work made from gold, silver, or other fine twisted wire. **2.** Intricate, delicate, or whimsical ornamentation. —*vt.* **-greed, -gree·ing, -grees.** To decorate with or as if with filigree.

**fil·ing** (fī′lĭng) *n.* **1.** The act of using a file. **2.** A shaving removed by a file.

**Fil·i·pi·no** (fĭl′ə-pē′nō) *n., pl.* **-nos.** [Sp. < *(Islas) Filipinas,* Philippine (Islands).] A native, citizen, or resident of the Philippines. —*adj.* **1.** Of or relating to the Philippines. **2.** Of or relating to the Filipinos.

**fill** (fĭl) *v.* **filled, fill·ing, fills.** [ME *fillen* < OE *fyllan.*] —*vt.* **1.** To put into as much as can be held <*fill* a theater><*fill* a sack with grain> **2. a.** To plug up (e.g., an opening). **b.** To repair a cavity of (a tooth). **3.** To supply fully or completely **:** SATISFY <*fill* all requirements> **4.** To supply the materials for <*fill* a prescription> **5.** To complete by insertion or addition <*fill* out a form><*fill* in a questionnaire> **6.** To supply (an empty space) with material, as writing, an inscription, or an illustration. **7.** To put someone into <*fill* a vacant position> **8.** To discharge the duties of : HOLD <*fill* a post> **9.** To occupy the whole of <Laughter *filled* the room.> **10.** To occupy completely <a mind *filled* with original ideas> **11.** To add a foreign substance to. **12.** *Naut.* **a.** To cause (a sail) to swell. **b.** To adjust (a yard) so that wind will cause a sail to swell. *—vi.* To become full. **—fill in. 1.** To provide with information <*filled* us *in* on the latest news> **2.** To act as a substitute <an understudy *filling in* for the star> **—fill out.** To fill in what is lacking and make perfect : COMPLEMENT. *—n.* **1.** An amount required to make full, complete, or satisfied <eat one's *fill*> **2.** A built-up piece of land or the material, as earth or gravel, used for it. **—fill the bill.** *Informal.* To serve a purpose.

**filled gold** *n.* An inexpensive metal, as brass, with a relatively thick surface layer of bonded gold.

**filled milk** *n.* Skim milk with vegetable oils added as a substitute for butter fat.

**fill·er** (fĭl′ər) *n.* **1.** One that fills. **2.** Something added to augment weight or size or to fill space. **3.** A composition, esp. a semisolid that hardens on drying, used for filling pores, cracks, or holes in a wood, plaster, or other construction surface before it is finished. **4.** Tobacco used to form the body of a cigar. **5. a.** A short item used to fill space in a publication. **b.** Something, as a news item, public-service message, or music, used to fill time in a broadcast presentation. **6.** A sheaf of loose papers for filling a notebook or binder. **7.** An architectural element, as a plate, for filling the space between two supporting members.

**fil·lér** (fĭl′âr′) *n., pl.* **fillér** *or* **-lérs.** [Hung.] —See table at CURRENCY.

**fil·let** (fĭl′ĭt) *n.* [ME *filet* < OFr., dim. of *fil,* thread < Lat. *filum.*] **1.** A narrow strip of material, as ribbon, often worn as a headband. **2.** *also* **fi·let** (fĭ-lā′, fĭl′ā′). **a.** A strip or compact piece of boneless meat or fish. **b.** A boneless strip of meat rolled and tied, as for roasting. **3. a.** A flat, thin molding used as separation between or ornamentation for larger moldings. **b.** A ridge between the indentations of a fluted column. **4.** A narrow decorative line impressed on a book cover. **5.** *Heraldry.* A narrow horizontal band placed in the lower fourth area of the chief. **6.** *Anat.* A loop-shaped band of fibers, as the lemniscus. *—vt.* **-let·ed, -let·ing, -lets. 1.** To bind or decorate with or as if with a fillet. **2.** *also* **fi·let** (fĭ-lā′, fĭl′ā′). To slice, bone, or make into fillets.

**fill-in** (fĭl′ĭn′) *n.* One that fills in <a *fill-in* for the regular quarterback>

**fill·ing** (fĭl′ĭng) *n.* **1.** Something used to fill a space, cavity, or container. **2.** An edible mixture used to fill sandwiches, cakes, or pastries. **3.** The horizontal threads crossing the warp in weaving : WEFT.

**filling station** *n.* A retail establishment at which vehicles are serviced, esp. with gasoline, oil, air, and water.

**fil·lip** (fĭl′əp) *n.* [Imit.] **1.** A snap or light blow made by pressing a fingertip against the thumb and suddenly releasing it. **2.** Something that arouses or excites : STIMULUS. *—vt.* **-liped, -lip·ing, -lips. 1.** To strike or propel with a fillip. **2.** To arouse or excite.

**fil·ly** (fĭl′ē) *n., pl.* **-lies.** [ME *filli* < ON *fylja.*] **1.** A young female horse. **2.** *Informal.* A lively, high-spirited girl.

**film** (fĭlm) *n.* [ME < OE *filmen.*] **1.** A thin skin or membranous coating. **2.** An abnormal thin, opaque coating on the cornea of the eye. **3. a.** A thin coating <a *film* of dust> **b.** A thin, gen. flexible transparent sheet, as of plastic used in wrapping or packaging. **4.** A thin sheet or strip of flexible cellulose material coated with a photosensitive emulsion, used to make photographic negatives or transparencies. **5. a.** A movie. **b.** Movies as a whole. **6.** *Computer Sci.* A coating of magnetic alloys on glass used in making storage devices. *—v.* **filmed, film·ing, films.** *—vt.* **1.** To cover with or as if with a film. **2.** To make a movie of or based on. *—vi.* **1.** To become coated or obscured with or as if with a film. **2.** To make a movie.

**film·card** (fĭlm′kärd′) *n.* A microfiche.

**film·dom** (fĭlm′dəm) *n.* **1.** The movie industry. **2.** Employees of the movie industry.

**film·go·er** (fĭlm′gō′ər) *n.* One who frequently goes to see movies.

**film·ic** (fĭl′mĭk) *adj.* Of, relating to, or resembling movies. **—film′i·cal·ly** *adv.*

**film·mak·er** (fĭlm′mā′kər) *n.* A person who produces or directs movies.

**fil·mog·ra·phy** (fĭl-mŏg′rə-fē) *n., pl.* **-phies.** Writings, as lists or books, about films or film figures.

**film pack** *n.* A pack of photographic sheet films that can be exposed in succession and withdrawn from the exposure position for storage at the rear of the pack.

**film·set·ting** (fĭlm′sĕt′ĭng) *n.* Photocomposition.

**film·strip** (fĭlm′strĭp′) *n.* A length of film containing graphic material, as photographs or diagrams, prepared for still projection.

**film·y** (fĭl′mē) *adj.* **-i·er, -i·est. 1.** Of, like, or composed of film : GAUZY <a *filmy* gown> **2.** Covered by or as if by a film : HAZY. **—film′i·ly** *adv.* **—film′i·ness** *n.*

**fil·o·plume** (fĭl′ə-plo͞om′, fī′lə-) *n.* [Lat. *filum,* thread + PLUME.] A hairlike feather having few or no barbs.

**fi·lose** (fī′lōs′) *adj.* [< Lat. *filum,* thread.] **1.** Threadlike. **2.** Having or ending in a threadlike part.

**fil·ter** (fĭl′tər) *n.* [ME *filtre* < OFr. < Med. Lat. *filtrum,* of Germanic orig.] **1. a.** A porous substance through which a liquid or gas is passed in order to remove constituents such as suspended matter. **b.** A device containing or composed of such a substance so used. **2.** An electric, electronic, acoustic, or optical device used to reject signals, vibrations, or radiations of certain frequencies while allowing others to pass. *—v.* **-tered, -ter·ing, -ters.** *—vt.* **1.** To pass (a liquid or gas) through a filter. **2.** To remove by passing through a filter. *—vi.* **1.** To pass through or as if through a filter. **2.** To come or go in gradually and in small groups. **—fil′ter·er** *n.*

**fil·ter·a·ble** *also* **fil·tra·ble** (fĭl′tər-ə-bəl, fĭl′trə-) *adj.* **1. a.** Capable of being filtered. **b.** Capable of being removed by filtering. **2.** Being sufficiently minute to pass through a fine filter, thus maintaining the infectivity of the filtrate. —Used of certain viruses and bacteria. **—fil′ter·a·bil′i·ty** *n.*

**filterable virus** *n.* A virus small enough to pass through a very fine filter, thus maintaining the infectivity of the filtrate.

**filter bed** *n.* A layer of sand or gravel on the bottom of a reservoir or tank used to filter water or sewage.

**filter feeder** *n.* An organism that utilizes filtering mechanisms to gain nourishment from particulate organic material suspended in water.

**filter paper** *n.* Porous paper suitable for use as a filter.

**filth** (fĭlth) *n.* [ME < OE *fȳlð.*] **1. a.** Foul matter. **b.** Repellent refuse. **2.** A corrupt condition. **3.** Something, as language, regarded as obscene, prurient, or immoral.

**filth·y** (fĭl′thē) *adj.* **-i·er, -i·est. 1.** Covered with or full of filth. **2.** Obscene. **3.** So objectionable as to elicit despisal. **—filth′i·ly** *adv.* **—filth′i·ness** *n.*

☆ ***syns:*** FILTHY, ABHORRENT, CONTEMPTIBLE, DESPICABLE, DISGUSTING, FOUL, LOATHSOME, MEAN, NASTY, ODIOUS, REPUGNANT, ROTTEN, SHABBY, SORRY, VILE, WRETCHED *adj. core meaning* : so objectionable as to elicit despisal <a *filthy* trick>

**fil·tra·ble** (fĭl′tər-ə-bəl, fĭl′trə-) *adj. var. of* FILTERABLE.

**fil·trate** (fĭl′trāt′) *vt. & vi.* **-trat·ed, -trat·ing, -trates.** [NLat. *filtrare, filtrat-,* to filter < Med. Lat. *filtrum,* filter.] To put or go through a filter. *—n.* The portion of material subjected to filtration that passes through the filter. **—fil·tra′tion** (-trā′shən) *n.*

**fi·lum** (fī′ləm) *n., pl.* **-la** (-lə) [Lat., thread.] A threadlike anatomical structure : FILAMENT.

**fim·bri·a** (fĭm′brē-ə) *n., pl.* **-bri·ae** (-brē-ē′) [LLat., fringe < Lat. *fimbriae.*] A fringelike part or structure, as at the opening of a mammalian oviduct. **—fim′bri·al** *adj.*

**fim·bri·ate** (fĭm′brē-ĭt, -āt′) *also* **fim·bri·at·ed** (-ā′tĭd) *adj.* [LLat. *fimbriatus* < *fimbria,* fringe.] Fringed, as the edge of a petal or opening of a duct. **—fim′bri·a′tion** *n.*

**fin**[1] (fĭn) *n.* [ME < OE *finn.*] **1.** A membranous appendage extending from the body of an aquatic animal, as a fish, used for locomotion, steering, or maintaining balance. **2.** Something resembling a fin. **3.** A fixed or movable vane or airfoil used to stabilize an aircraft or missile in flight. **4.** An appendage on a seagoing vessel, as a submarine. **5.** A projecting vane for cooling, as on a radiator or engine cylinder. **6.** An ornamental projection, as on the rear fender of a car. *—v.* **finned, fin·ning, fins.** *—vt.* To equip with fins. *—vi.* To emerge with the fins above water.

**fin**[2] (fĭn) *n.* [Yiddish *finf,* five.] *Slang.* A five-dollar bill.

**fi·na·gle** (fə-nā′gəl) *v.* **-gled, -gling, -gles.** [Prob. < dial. *fainaigue,* to cheat.] *Informal —vt.* **1.** To make, gain, or achieve by indirect, often crafty methods : WANGLE. **2.** To get by trickery or deceit. *—vi.* To use crafty, deceitful methods. **—fi·na′gler** *n.*

**fi·nal** (fī′nəl) *adj.* [ME < OFr. < Lat. *finalis* < *finis,* end.] **1.** Forming or occurring at the end : LAST. **2.** Of, relating to, or constituting the last element in a series, process, or procedure. **3.** Ultimate and definitive : UNALTERABLE <The judges' decision is *final.*> *—n.* Something coming at or forming the end, esp.: **a.** The last or one of the last of a series of athletic contests. **b.** The last examination of an academic course. **—fi′nal·ly** *adv.*

**fi·nal·e** (fə-năl′ē, -nä′lē) *n.* [Ital. < Lat. *finalis,* final.] The concluding part, esp. of a musical composition.

**fi·nal·ist** (fī′nə-lĭst) *n.* A contestant in the final part of a contest.

**fi·nal·i·ty** (fī-năl′ĭ-tē, fə-) *n., pl.* **-ties. 1.** The fact or state of being final. **2.** A final, conclusive, or decisive act or utterance.

**fi·nal·ize** (fī′nə-līz′) *vt.* **-ized, -iz·ing, -iz·es.** To put into final form. ***usage:*** Because they associate it with the language of bureaucracy, many object to the use of *finalize.* There is no one word that is an exact synonym for *finalize,* however, so that those who wish to avoid this term must substitute an expression such as *make final* or *put in final form.* **—fi′nal·i·za′tion** *n.*

**fi·nance** (fə-năns′, fī-, fī′năns′) *n.* [ME *finaunce,* money supply < OFr. *finance,* gift < *finer,* to pay ransom < *fin,* end < Lat. *finis.*]

ă **pat** ā **pay** âr **care** ä **father** ĕ **pet** ē **be** hw **which** ĭ **pit**
ī **tie** îr **pier** ŏ **pot** ō **toe** ô **paw, for** oi **noise** o͝o **took**

**1.** The science of money management. **2.** The management of money, banking, investments, and credit. **3. finances.** Monetary resources : FUNDS. —*vt.* **-nanced, -nanc·ing, -nanc·es. 1.** To supply the funds or capital for <*financed* a new publishing venture> **2.** To supply funds to <*financing* us in a new business> **3.** To provide credit to. **—fi·nan'cial** (fə-năn'shəl, fī-) *adj.* **—fi·nan'cial·ly** *adv.*

**finance bill** *n.* A legislative act designed to raise public revenues.

**finance company** *n.* A loan company.

**fin·an·cier** (fĭn'ən-sîr', fə-năn'-, fī'nən-) *n.* [Fr. < OFr. < *finance,* gift. —see FINANCE.] An expert in large-scale financial affairs.

**fin·back** (fĭn'băk') *n.* The rorqual.

**finch** (fĭnch) *n.* [ME < OE *finc.*] A relatively small bird of the family Fringillidae, as a goldfinch, bullfinch, cardinal, grosbeak, or canary, having a short stout bill adapted for cracking seeds.

**find** (fīnd) *v.* **found** (found), **find·ing, finds.** [ME *finden* < OE *findan.*] —*vt.* **1.** To come upon, often by accident. **2.** To come upon after a search <*find* the cause of the trouble> **3.** To come upon through observation, experience, or study <*found* the answer at last> **4.** To succeed in reaching <The knife *found* its mark.> **5.** To gain or acquire by great effort <*found* the money by scrimping and saving> **6.** To regard : consider <*found* the children irresistible> **7.** To recover (something lost) <*found* my wallet> **8.** To recover the use of : REGAIN <*found* my voice> **9.** To decide on and make a declaration about <The jury deliberated and *found* a verdict.> **10.** To furnish : supply. **11. a.** To bring (oneself) to an awareness of what one truly wishes to be and do in life. **b.** To perceive (oneself) to be in a specific state or place. —*vi.* To come to a legal decision or verdict <The jury *found* for the plaintiff.> **—find out. 1.** To discover in a misdeed, as a crime <The thief was *found out* at once.> **2.** To detect the true nature of : EXPOSE. —*n.* **1.** An act of finding. **2.** Something found, esp. an unexpectedly valuable discovery <an archaeological *find*> **—find'a·ble** *adj.*

☆ ***syns:*** FIND, LOCATE, PINPOINT, SPOT *v. core meaning* : to discover by looking for <Help me *find* the needle.>

**find·er** (fīn'dər) *n.* **1.** One that finds. **2.** A device on a camera that indicates to the photographer what will appear in the field of view of the lens. **3.** *Astron.* A small telescope attached to the body of a larger one for locating an object to be observed with the larger telescope.

**finder's fee** *n.* A fee paid to the finder of financial backing for a venture or to a party that brings the principals together in a venture.

**fin de siè·cle** (făN' də sē-ĕk'lə) *adj.* [Fr. : *fin,* end + *de,* of + *siècle,* century.] Of or typical of the latter part of the 19th cent., esp. with reference to its artistic climate.

**find·ing** (fīn'dĭng) *n.* **1.** Something found. **2.** *often* **findings.** A conclusion reached after investigation or examination. **3. findings.** Small tools and materials used by an artisan.

**fine¹** (fīn) *adj.* **fin·er, fin·est.** [ME *fin* < OFr. < Med. Lat. *finus,* prob. < *finire,* to finish < Lat.] **1.** Of superior quality, skill, or appearance <a *fine* porcelain><a *fine* musician> **2.** Very small in size, weight, or bulk <*fine* type> **3. a.** Free from impurities. **b.** Containing pure metal in a given proportion or amount. **4.** Very sharp : KEEN <a knife with a *fine* edge> **5.** Thin : slender <*fine* hair> **6.** Showing skill or great delicacy <*fine* ornamentation> **7.** Consisting of very small particles <*fine* dust> **8. a.** Subtle or precise <a *fine* distinction> **b.** Able to make or detect subtle or precise effects : SENSITIVE <a *fine* eye for color> **9.** Trained to the highest degree of physical efficiency <a *fine* craftsman> **10.** Marked by refinement or elegance. **11.** Being in good health <I'm *fine.* And you?> —*adv.* **1.** Finely. **2.** *Informal.* Very well <The patient is doing *fine.*> —*vt.* & *vi.* **fined, fin·ing, fines.** To make or become finer, purer, or cleaner. **—fine'ness** *n.*

**fine²** (fīn) *n.* [ME *fin* < OFr. < Lat. *finis,* end.] **1.** A sum of money imposed as a penalty for an offense. **2.** *Law.* A forfeiture or penalty to be paid to the offended party in a civil action. **3.** *Law.* An amicable settlement of a suit over land ownership. **4.** *Obs.* An end : termination. —*vt.* **fined, fin·ing, fines.** To impose a fine on. **—fin'a·ble, fine'a·ble** *adj.*

**fi·ne³** (fē'nā) *n.* [Ital. < Lat. *finis,* end.] *Mus.* The end.

**fine art** *n.* **1. a.** Art produced or intended chiefly for beauty rather than utility. **b.** *often* **fine arts.** Any of such arts, including sculpture, painting, and music. **2.** Something requiring highly developed techniques and skills <the *fine art* of diplomacy>

**fine-drawn** (fīn'drôn') *adj.* **1.** Extended to a slender threadlike state, as wire. **2.** Subtly or precisely fashioned <a *fine-drawn* theory> **3.** Delicately formed <*fine-drawn* facial features>

**fine-grained** (fīn'grānd') *adj.* Having a fine, smooth, even grain <*fine-grained* wood>

**fine·ly** (fīn'lē) *adv.* **1.** In a fine way : SPLENDIDLY. **2.** To a fine point : DISCRIMINATINGLY <*finely* drawn distinctions> **3.** In small pieces : MINUTELY <*finely* chopped scallions>

**fine print** *n.* Something, as limitations in a contract, presented in an intentionally ambiguous or cryptic way.

**fin·er·y** (fī'nə-rē) *n., pl.* **-ies.** Elaborate adornment.

**fines herbes** (fēn zĕrb', fēn ĕrb') *pl.n.* [Fr.] Finely chopped herbs, as parsley, chives, tarragon, and thyme, used as a seasoning.

**fine-spun** (fīn'spŭn') *adj.* **1.** Developed to extreme fineness or subtlety : ELABORATE. **2.** Developed to excessive fineness or subtlety.

**fi·nesse** (fə-nĕs') *n.* [OFr., fineness < *fin,* fine.] **1.** Delicacy and refinement of performance, execution, or workmanship. **2.** Tact, subtlety, or skill in handling a situation. **3.** The playing of a card in a suit in which one holds a nonsequential higher card either to induce an opponent to play an intermediate card that one's partner can then top or to win the trick economically. —*v.* **-nessed, -ness·ing, -ness·es.** —*vt.* **1.** To accomplish by the use of finesse. **2.** To handle with deceptive or evasive strategy. **3.** To play (a card) as a finesse. —*vi.* To make a finesse in or as if in a card game.

**fine-toothed comb** (fīn'tōōtht', -tōōthd') *also* **fine-tooth comb** (-tōōth') *n.* **1.** A comb with closely set teeth. **2.** A method of searching or investigating in minute detail <examined the figures with a *fine-toothed comb*>

**fine-tune** (fīn'tyōōn', -tōōn') *vt.* **-tuned, -tun·ing, -tunes.** To make precise, minute adjustments in <"advertising agencies kept *fine-tuning* the coolly calculated machinery of merchandising and hype" —*New Yorker*> **—fine'-tun'er** *n.*

**fin·fish** (fĭn'fĭsh') *n.* An aquatic vertebrate of the superclass Pisces.

**fin·ger** (fĭng'gər) *n.* [ME < OE.] **1.** One of the five digits of the hand, esp. one other than the thumb. **2.** The part of a glove covering a finger. **3.** Something, as a peninsula, that resembles a finger. **4.** The length or width of a finger. —*v.* **-gered, -ger·ing, -gers.** —*vt.* **1.** To touch with the fingers : HANDLE. **2.** *Mus.* **a.** To mark (a score) with indications of which fingers are to play the notes. **b.** To play (an instrument) by using the fingers in a particular way or order. **3.** *Slang.* **a.** To inform on. **b.** To designate as an intended victim. —*vi.* **1.** To handle something with the fingers. **2.** *Mus.* To use the fingers in playing an instrument. **—twist** (*or* **wrap**) **around (one's) little finger.** *Informal.* To dominate easily and completely. **—fin'ger·er** *n.*

▲ word history: The ancestor of English *five* underlies the word *finger* and probably the word *fist* as well. *Five* is descended from the Indo-European form *penkwe.* The suffixed form *penkweros* meant "one of five" and from this form *finger* is derived. A variant of *penkwe* with a different suffix, *pnksti,* is possibly the source of *fist.* The surprising thing about the words *hand, finger,* and *fist* is that, despite their reference to such a quintessentially human appendage, the first two words have no cognates outside the Germanic languages and *fist* has only one.

**fin·ger·board** (fĭng'gər-bôrd', -bōrd') *n.* A strip of wood on the neck of a stringed instrument against which the strings are pressed in playing.

**finger bowl** *n.* A small bowl holding water for rinsing the fingers at the table.

**fin·ger·breadth** (fĭng'gər-brĕdth') *n.* The breadth of a finger.

**fin·gered** (fĭng'gərd) *adj.* Having a finger or fingers, esp. of a specific number, kind, or appearance <five-*fingered*><fat-*fingered*>

**fin·ger·ing** (fĭng'gər-ĭng) *n.* *Mus.* **1.** The technique used in playing an instrument with the fingers. **2.** The indication on a score of which fingers are to be used in playing.

**fin·ger·ling** (fĭng'gər-lĭng) *n.* A young or small fish.

**fin·ger·nail** (fĭng'gər-nāl') *n.* A thin, horny, transparent plate covering the dorsal surface of the tip of each finger.

**fin·ger-paint** (fĭng'gər-pānt') *vt.* & *vi.* **-paint·ed, -paint·ing, -paints.** To make by or engage in finger painting.

**finger painting** *n.* **1.** The technique of painting by applying color to moistened paper with the fingers. **2.** A picture made by finger painting.

**finger post** *n.* A guidepost shaped like a pointing hand.

**fin·ger·print** (fĭng'gər-prĭnt') *n.* An ink impression of the curves formed by the system of ridges on the skin surface of the distal phalanx of a finger, esp. such an impression used as a way of identifying persons. —*vt.* **-print·ed, -print·ing, -prints.** To take the fingerprints of.

**finger tip** *also* **fin·ger·tip** (fĭng'gər-tĭp') *n.* The extreme end of a finger.

**finger wave** *n.* A wave set into dampened hair using only the fingers and a comb.

**fin·i·al** (fĭn'ē-əl) *n.* [ME < *finial,* last, var. of *final.*] **1.** An ornament attached to the peak of an arch or arched structure. **2.** An ornamental terminating part, as the screw on top of a lampshade.

**fin·i·cal** (fĭn'ĭ-kəl) *adj.* [Prob. < FINE¹.] Finicky. **—fin'i·cal·ly** *adv.* **—fin'i·cal·ness** *n.*

**fin·ick·y** (fĭn'ĭ-kē) *adj.* **-i·er, -i·est.** [< *finick,* a finical person < FINICAL.] Difficult to please : FASTIDIOUS.

**fin·is** (fĭn'ĭs, fī'nĭs) *n.* [ME < Lat.] The end.

**fin·ish** (fĭn'ĭsh) *v.* **-ished, -ish·ing, -ish·es.** [ME *finishen* < OFr. *finir, finiss-,* to complete < Lat. *finire* < *finis,* end.] —*vt.* **1.** To arrive at or reach the end of <*finish* a road race> **2.** To bring to an end : TERMINATE <*finished* painting the garage> **3.** To consume all of : USE UP <*finish* a cake> **4.** To bring to a desired or required state <*finish* a drawing> **5.** To give (a surface) a desired or particular texture. **6.** To destroy : kill <*finish* an enemy> **7.** To bring about the ruin of <Bankruptcy *finished* them.> —*vi.* **1.** To come to an end : STOP. **2.** To reach the end of a task, course, or relationship. —*n.* **1. a.** The final part : END <a close *finish* in the road race> **b.** The reason for one's ruin : DOWNFALL. **2.** A completing, concluding, or perfecting part, material, or element, esp.: **a.** The last treatment or coating of a surface. **b.** The surface texture produced. **c.** The material used in surfacing or finishing. **3.** Completeness, thoroughness, or smoothness of execution : PERFECTION. **—fin'ish·er** *n.*

**fin·ished** (fĭn'ĭsht) *adj.* Highly accomplished or skilled : POLISHED.

**finishing school** *n.* A private girls' school that stresses training in cultural subjects and preparation for life in society.

**finish line** *n.* A line marking the end of a racecourse.

**fi·nite** (fī'nīt') *adj.* [ME *finit* < Lat. *finitum*, p.part. of *finire*, to limit < *finis*, end.] **1. a.** Having bounds <a *finite* list of options> **b.** Existing or enduring for a limited time only. **2.** Being neither infinite nor infinitesimal. **3.** *Math.* **a.** Bounded in an interval. —Used of a quantity defined in an interval. **b.** Incapable of being put into one-to-one correspondence with a part of itself. —Used of a set. **c.** Real or complex. —Used of a number. **4.** Limited by person, number, tense, and mood and capable of serving as a predicate. —Used of verbs. —*n.* Finite entities as a whole. **—fi'nite'ly** *adv.* **—fi'nite'ness, fin'i·tude'** (fĭn'ĭ-to͞od', -tyo͞od', fī'nĭ-) *n.*

**fink** (fĭngk) [Orig. unknown.] *Slang.* —*n.* **1.** A hired strikebreaker. **2.** An informer. **3.** An undesirable person. —*vi.* **finked, fink·ing, finks. 1.** To inform against another person. **2.** To withhold support or participation.

**Fin·land·ize** (fĭn'lən-dīz') *vt.* **-ized, -iz·ing, -izes.** [After *Finland*, a country that adopted such a policy.] To cause (a country or political unit) to adopt a neutral or conciliatory posture and policy in its relations with the U.S.S.R. **—Fin'land·i·za'tion** *n.*

**Finn** (fĭn) *n.* [Swed. *Finne.*] **1.** A native or resident of Finland. **2.** One who speaks Finnish or a Finnic language.

**fin·nan had·die** (fĭn'ən hăd'ē) *also* **finnan had·dock** (hăd'ək) *n.* [Obs. *findhorn haddock*, prob. after the *Findhorn* River, Scotland.] Smoked haddock.

**Fin·nic** (fĭn'ĭk) *adj.* Of or relating to Finland or the Finns. —*n.* A branch of Finno-Ugric including Finnish, Estonian, and Lapp.

**Finn·ish** (fĭn'ĭsh) *adj.* Of or relating to Finland, its language, or its people. —*n.* The Finno-Ugric language of the Finns.

**Fin·no-U·gric** (fĭn'ō-o͞o'grĭk, -yo͞o'-) *also* **Fin·no-U·gri·an** (-o͞o'grē-ən, -yo͞o'-) *n.* A subfamily of the Uralic language family that includes Finnish, Hungarian, and other languages of eastern Europe and northwestern U.S.S.R. —*adj.* **1.** Relating to the Finns and the Ugrians. **2.** Relating to Finno-Ugric.

**fin·ny** (fĭn'ē) *adj.* **-ni·er, -ni·est. 1.** Having fins. **2.** Resembling a fin. **3.** Of, relating to, or typical of fish.

**fi·no·chi·o** *also* **fi·noc·chi·o** (fə-nō'kē-ō') *n.* [Ital. *finocchio* < Lat. *feniculum*, fennel, dim. of *fenum*, hay.] A variety of fennel, *Foeniculum vulgare dulce*, whose stalks are eaten as a vegetable.

**fin rot** *n.* A bacterial disease of fish characterized by progressive deterioration of the fin tissue.

**fiord** (fyôrd, fyōrd) *n. var. of* FJORD.

**fip·ple** (fĭp'əl) *n.* [Orig. unknown.] **1.** A wooden block forming a flue at the mouth end of some wind instruments. **2.** An object similar to a fipple in an organ pipe.

**fipple flute** *n.* A flute, as a recorder, with a fipple.

**fir** (fûr) *n.* [ME *firre* < OE *fyrh.*] **1. a.** An evergreen tree of the genus *Abies*, having flat needles and erect cones. **b.** A similar or related tree, as the Douglas fir. **2.** The wood of a fir. **—fir'ry** *adj.*

**fire** (fīr) *n.* [ME *fir* < OE *fȳr.*] **1. a.** A rapid, persistent chemical reaction that releases heat and light, esp. the exothermic combination of a combustible substance with oxygen. **b.** A fire distinguished by magnitude, destructive power, or utility <a field *fire*><a stove *fire*> **2. a.** Great intensity : ardor. **b.** Enthusiasm. **3.** Luminosity or brilliance, as of a cut and polished gemstone. **4.** Liveliness and vivacity of imagination : INSPIRATION <the *fire* of the poet's verse> **5.** Torment, trial, or tribulation. **6.** Discharge of firearms. —*v.* **fired, fir·ing, fires.** —*vt.* **1.** To cause to burn : IGNITE. **2. a.** To add fuel to (something burning). **b.** To maintain or intensify a fire in. **3.** To bake in a kiln <*fire* a vase> **4.** To arouse the emotions of. **5.** To detonate or discharge (a firearm, explosives, or a projectile) <*fire* a pistol><*fire* a rocket> **6.** *Informal.* To project or hurl suddenly and forcefully <*fired* a ball at the batter> **7.** *Informal.* To discharge from a position : DISMISS. —*vi.* **1.** To become ignited. **2. a.** To become excited or ardent. **b.** To become angry or annoyed. **3.** To tend a fire. **4.** To detonate a weapon <*fired* at the oncoming troops> **5.** *Informal.* To hurl or project a missile. **—on fire. 1.** Ignited : ablaze. **2.** Filled with excitement or enthusiasm. **—under fire. 1.** Exposed or subjected to enemy attack. **2.** Exposed or subjected to critical attack. **—fir'er** *n.*

**fire alarm** *n.* A device, as a siren, used to signal the outbreak of a fire.

**fire ant** *n.* An ant of the genus *Solenopsis*, esp. *S. geminata* or *S. saevissima* of the southern United States and tropical America, that builds conspicuous mounds and inflicts a painful sting.

**fire·arm** (fīr'ärm') *n.* A weapon capable of firing a missile, esp. a pistol or rifle using an explosive charge as a propellant.

**fire·ball** (fīr'bôl') *n.* **1.** A brilliantly burning sphere. **2.** A particularly bright meteor. **3.** A very luminous, intensely hot spherical cloud of dust, gas, and vapor caused by a nuclear explosion. **4.** *Informal.* An energetic person.

**fire·base** (fīr'bās') *n.* A military site from which heavy fire is directed against an enemy.

**fire beetle** *n.* A tropical American click beetle of the genus *Pyrophorus*, esp. *P. noctilucus*, with brightly luminous spots.

**fire·bird** (fīr'bûrd') *n.* A bird, as the Baltimore oriole, with bright scarlet or orange plumage.

**fire blight** *n.* A destructive disease of apples, pears, and related trees and plants, caused by a bacterium, *Erwinia amylovora.*

**fire·boat** (fīr'bōt') *n.* A boat equipped to fight fires along waterfronts and on ships.

**fire bomb** *n.* An incendiary bomb.

**fire·box** (fīr'bŏks') *n.* **1.** A box having a device for sounding a fire alarm. **2.** A chamber, as the furnace of a steam locomotive, in which fuel is burned.

**fire·brand** (fīr'brănd') *n.* **1.** A piece of burning wood. **2.** One who stirs up trouble.

**fire·brat** (fīr'brăt') *n.* A small, wingless insect, *Thermobia domestica*, frequenting warm areas of dwellings and often destructive to paper.

**fire·break** (fīr'brāk') *n.* A strip of cleared land used to stop the spread of a fire.

**fire·brick** (fīr'brĭk') *n.* A refractory brick, esp. of fire clay, used for lining furnaces, fireboxes, chimneys, or fireplaces.

**fire brigade** *n.* An organized body of firefighters.

**fire·bug** (fīr'bŭg') *n.* *Informal.* One who deliberately sets fires : PYROMANIAC.

**fire clay** *n.* A heat-resistant clay used in making firebricks, crucibles, and other objects that are exposed to high temperatures.

**fire control** *n.* The control of the delivery of artillery fire on targets.

**fire·crack·er** (fīr'krăk'ər) *n.* A small explosive charge in a cylinder of heavy paper, used to make noise, as at celebrations.

**fire-cure** (fīr'kyo͝or') *vt.* **-cured, -cur·ing, -cures.** To cure (tobacco) by exposing it to the heat and smoke of a wood fire.

**fire·damp** (fīr'dămp') *n.* **1.** A combustible gas, chiefly methane, occurring naturally in coal mines and forming explosive mixtures with air. **2.** The explosive mixture of firedamp and air.

**fire department** *n.* A department, esp. of a municipal government, whose purpose is preventing and putting out fires.

**fire·dog** (fīr'dôg', -dŏg') *n.* An andiron.

**fire·drake** (fīr'drāk') *n.* [ME *firdrake* < OE *fȳrdraca* : *fȳr*, fir + *draca*, dragon.] A fiery dragon of Germanic mythology.

**fire drill** *n.* A practice exercise in the use of fire-fighting equipment or the exit procedure to be followed in case of a fire.

**fire-eat·er** (fīr'ē'tər) *n.* **1.** A performer who pretends to swallow fire. **2.** A belligerent person. **—fire'-eat'ing** *adj.*

**fire engine** *n.* A large motor vehicle that carries firefighters and equipment to a fire and supports extinguishing operations, as by pumping water.

**fire escape** *n.* A structure or device, as an outside stairway attached to a building, for emergency exit in case of fire.

**fire extinguisher** *n.* A portable apparatus containing chemicals that can be discharged in a jet to extinguish a small fire.

**fire·fight·er** (fīr'fī'tər) *n.* One who is employed by a fire department to fight fires.

**fire·flood** (fīr'flŭd') *or* **fire·flood·ing** (-ĭng) *n.* A procedure for extracting additional oil from producing wells by injecting compressed air into the petroleum reservoir and burning some of the oil to increase flow.

**fire·fly** (fīr'flī') *n.* Any of various nocturnal beetles of the family Lampyridae, with luminous abdominal organs that produce a flashing light.

**fire·guard** (fīr'gärd') *n.* **1.** A metal screen placed in front of an open fireplace to catch sparks. **2.** A firebreak.

**fire·house** (fīr'hous') *n.* A fire station.

**firehouse dog** *n.* A Dalmatian.

**fire hydrant** *n.* A hydrant.

**fire irons** *pl.n.* Equipment, including tongs, a shovel, and a poker, used to tend a fireplace.

**fire·light** (fīr'līt') *n.* The light from a fire, as in a fireplace.

**fire·lock** (fīr'lŏk') *n.* A flintlock.

**fire·man** (fīr'mən) *n.* **1.** A firefighter. **2.** One who tends fires : STOKER. **3.** An enlisted person in the U.S. Navy engaged in the operation of the engineering machinery. **4.** *Baseball.* A relief pitcher.

**fire opal** *n.* An opal with brilliant flamelike yellow, orange, and red colors.

**fire pink** *n.* A plant, *Silene virginica* of eastern North America, having red flowers with narrow notched petals.

**fire·place** (fīr'plās') *n.* **1.** An open recess for holding a fire at the base of a chimney : HEARTH. **2.** A structure, usu. of stone or brick, for holding an outdoor fire.

**fire·plug** (fīr′plŭg′) *n.* A large pipe at which water may be drawn from a water main for use in extinguishing a fire : HYDRANT.
**fire·pow·er** (fīr′pou′ər) *n.* The capacity, as of a weapon, military unit, or ship, for discharging fire.
**fire·proof** (fīr′pro͞of′) *adj.* Impervious or resistant to damage by fire. —*vt.* **-proofed, -proof·ing, -proofs.** To make fireproof.
**fire sale** *n.* A sale of fire-damaged commodities.
**fire screen** *n.* FIREGUARD 1.
**fire ship** *n.* A military vessel loaded with explosives and combustible material and set adrift among enemy ships or fortifications to destroy them.
**fire·side** (fīr′sīd′) *n.* **1.** The area immediately surrounding a fireplace or hearth. **2.** HOME 1.
**fire station** *n.* A building for fire equipment and firefighters.
**fire·stone** (fīr′stōn′) *n.* **1.** A flint or pyrites used to strike a fire. **2.** A fire-resistant stone, as certain sandstones.
**fire thorn** *n.* A thorny Asian shrub of the genus *Pyracantha*, often cultivated for its evergreen foliage and reddish or orange berries.
**fire tower** *n.* A tower in which a lookout for forest fires is posted.
**fire·trap** (fīr′trăp′) *n.* A building susceptible to fire or difficult to escape from in case of fire.
**fire wall** *n.* A fireproof wall used as a barrier to prevent the spread of a fire.
**fire·wa·ter** (fīr′wô′tər, -wŏt′ər) *n. Slang.* Strong liquor.
**fire·weed** (fīr′wēd′) *n.* **1.** A species of willow herb, *Epilobium angustifolium*, having terminal pinkish-purple flower clusters. **2.** A weedy North American plant, *Erechtites hieracifolia*, with small white or greenish flowers.
**fire·wood** (fīr′wo͝od′) *n.* Wood used for fuel.
**fire·work** (fīr′wûrk′) *n.* **1. a.** A device consisting of various combinations of explosives and combustibles used to generate colored lights, smoke, and noise for amusement. **b. fireworks.** A display of such devices. **2. fireworks. a.** An exciting or dramatic display, as of musical virtuosity. **b.** A display of rage.
**fir·ing** (fīr′ĭng) *n.* **1.** Application of fire or heat, as in the hardening or glazing of ceramics. **2.** Fuel for fires.
**firing line** *n.* **1.** The line of positions from which fire is directed against a target. **2.** The vanguard of an activity or pursuit.
**firing pin** *n.* The part of the bolt or breech of a firearm that strikes the primer and explodes the charge of a projectile.
**firing squad** *n.* **1.** A detachment assigned to shoot condemned persons. **2.** A detachment of troops chosen to fire a salute at a military funeral.
**fir·kin** (fûr′kĭn) *n.* [ME < MDu. *verdelkijn < veerdel, one-fourth.] **1.** A small wooden barrel or keg. **2.** Any of several British units of capacity, usu. equal to about 34 liters, 1/4 of a barrel, or 9 gallons.
**firm¹** (fûrm) *adj.* **-er, -est.** [ME *ferm* < OFr. < Lat. *firmus.*] **1.** Unyielding to pressure : SOLID. **2.** Marked by the tone and resiliency of healthy tissue <*firm* muscles> **3.** Securely fixed in place. **4.** Indicating resolution or determination <spoken with a *firm* voice> **5.** Constant and steadfast <a *firm* friend> **6. a.** Fixed : definite <a *firm* deal> **b.** Unfluctuating : steady <*firm* prices> **7.** Strong and sure <a *firm* grip> —*vt. & vi.* **firmed, firm·ing, firms.** To make or become firm. —*adv.* Without wavering : RESOLUTELY <stood *firm*> **—firm′ly** *adv.* **—firm′ness** *n.*
**firm²** (fûrm) *n.* [Ital. *firma* < *firmare*, to ratify by signature < LLat. *firmare* < Lat., to confirm < *firmus*, firm.] **1.** A commercial partnership of two or more persons. **2.** The name or designation under which a firm carries on business.
**fir·ma·ment** (fûr′mə-mənt) *n.* [ME < OFr. < LLat. *firmamentum* < Lat., support < *firmare*, to strengthen < *firmus*, firm.] The expanse of the heavens : SKY. **—fir′ma·men′tal** (-mĕn′tl) *adj.*

▲ word history: *Firmament* is a word that English owes to the long tradition of Biblical translation. The Latin Vulgate is at least two removes from the original text of the Old Testament. *Firmament* is from Latin *firmamentum*, "a support," which was used in the Vulgate to translate Hebrew *rāqī'a*. The Hebrew word literally means "expanse," but the verb from which it is derived means "to make firm or solid" in Syriac, a language closely related to Hebrew. The Greek word used to translate Hebrew *rāqī'a* was *stereōma*, "solid body, framework," chosen probably because of the translator's knowledge of the Syriac sense of the verb. The Latin translator in turn was influenced by the Greek interpretation of the Hebrew word.

**fir·mer chisel** (fûr′mər) *n.* [Fr. *fermoir* (< OFr. *formoir* < *former*, to form < Lat. *formare* < *forma*, form) + CHISEL.] A thin-bladed chisel or gouge used in shaping and finishing wood.
**firm·ware** (fûrm′wâr′) *n. Computer Sci.* Programming functions implemented through a small special-purpose memory unit.
**firn** (fîrn) *n.* [G. < dial. G., of last year < OHG *firni*, old.] Snow partially consolidated by thawing and freezing but not yet converted to glacial ice.
**first** (fûrst) *adj.* [ME < OE *fyrst.*] **1.** Corresponding in order to the number one. **2.** Coming before all others. **3.** Taking place or acting prior to all others : EARLIEST. **4.** Ranking above all others in importance or quality : FOREMOST. **5.** Highest in pitch or carrying the principal musical part <*first* horn> **6.** Of, relating to, or being the transmission gear or corresponding gear ratio used to produce the range of lowest drive speeds in an automotive vehicle. —*adv.* **1.** Before or above all others in time, order, rank, or importance. **2.** For the first time. **3.** Preferably : rather. —*n.* **1.** The ordinal number matching the number one in a series. **2.** The first in a set or sequence. **3.** The one coming, taking place, or ranking before or above all others. **4.** The beginning <from the *first*><at *first*> **5.** The voice or instrument highest in pitch or carrying the principal musical part. **6.** The transmission gear or corresponding gear ratio used to produce the range of lowest drive speeds in an automotive vehicle. **7.** The winning position in a contest. **—first′ly** *adv.*

☆ ***syns:*** FIRST, INITIAL, MAIDEN, ORIGINAL, PIONEER, PRIME *adj. core meaning* : preceding all others in time <America's *first* space flight> ***ant:*** last

**first aid** *n.* Emergency treatment administered to injured victims or sick people before professional medical care is available.
**first base** *n.* **1.** *Baseball.* **a.** The first of the bases in the infield counterclockwise from home plate. **b.** The fielding position occupied by the first baseman. **2.** *Informal.* The first stage or step <a reform bill that never got to *first base*> **—first baseman** *n.*
**first-born** (fûrst′bôrn′) *adj.* First in order of birth : ELDEST. **—first′-born′** *n.*
**first class** *n.* **1.** The first, highest, or best group of a specified category. **2.** The most luxurious and expensive class of accommodations on a train, ship, or aircraft. **3.** A class of mail including letters, post cards, and packages sealed against inspection.
**first-class** (fûrst′klăs′) *adj.* **1.** Indicating the first, highest, or best group of a specified category. **2.** Of the foremost excellence : FIRST-RATE <a *first-class* mind> **—first′-class′** *adv.*
**first cousin** *n.* COUSIN 1.
**first-de·gree burn** (fûrst′dĭ-grē′) *n.* A mild burn causing redness of the skin.
**first edition** *n.* **1. a.** The first published copies of a literary work printed from the same type and distributed at the same time. **b.** A single copy from a first edition. **2.** The day's first pressrun of a newspaper.
**first floor** *n.* **1.** The ground floor of a building. **2.** *Chiefly Brit.* The floor just above the ground floor.
**first·hand** (fûrst′hănd′) *adj.* Received from the original source <*firsthand* evidence> **—first′hand′** *adv.*
**First International** *n.* An international organization formed in 1864 by Karl Marx and Friedrich Engels to associate the trade unions of all nations.
**first lady** *n.* **1.** *often* **First Lady.** The wife or hostess of the chief executive of a country, state, or city. **2.** The foremost woman of a given profession or art.
**first lieutenant** *n.* A commissioned officer in the U.S. Army, Air Force, and Marine Corps ranking above a second lieutenant and below a captain.
**first·ling** (fûrst′lĭng) *n.* **1.** The first of a kind or category. **2.** A first-born offspring.
**first mate** *n.* A ship's officer ranking below the captain.
**first name** *n.* The name occurring first in a person's full name.
**first night** *n.* **1.** The opening performance of a theatrical production. **2.** The performance presented on a first night.
**first night·er** (nī′tər) *n.* A member of the audience on a first night.
**first offender** *n.* One convicted legally for the first time.
**first papers** *pl.n.* The documents first filed by one applying for U.S. citizenship.
**first person** *n.* **1. a.** A category of linguistic forms, as verbs or pronouns, designating the speaker or writer of the sentence in which they appear. **b.** One of the forms of this category. **2.** A discourse or literary style in which forms in the first person are used.
**first-rate** (fûrst′rāt′) *adj.* Foremost in quality, rank, or importance. **—first′-rate′** *adv.*
**first sergeant** *n.* The highest-ranking noncommissioned officer of a company or other military unit in the U.S. Army.
**first-string** (fûrst′strĭng′) *adj.* **1.** Being a regular member, as of a football team, rather than a substitute. **2.** First-rate.
**first water** *n.* [Prob. transl. of Ar. *mā'*, water luster.] **1.** The highest degree of quality or purity in diamonds or pearls. **2.** The foremost rank or quality.
**First World War** *n.* World War I.
**firth** (fûrth) *n.* [ME *furth* < ON *fjörðr.*] *Chiefly Scot.* A long, narrow inlet of the sea : FJORD.
**fisc** (fĭsk) *n.* [OFr. < Lat. *fiscus.*] A kingdom's or state's treasury.
**fis·cal** (fĭs′kəl) *adj.* [OFr. < Lat. *fiscalis* < *fiscus*, treasury.] **1.** Of or relating to the treasury or finances of a nation or branch of government. **2.** Of or relating to finances. **—fis′cal·ly** *adv.*
**fiscal year** *n.* A 12-month period for which an organization plans the use of its funds.
**fish** (fĭsh) *n., pl.* **fish** *or* **fish·es.** [ME < OE *fisc.*] **1.** Any of numerous cold-blooded aquatic vertebrates of the superclass Pisces, having fins, gills, and a streamlined body and including: **a.** Any of the class

Osteichthyes, having a bony skeleton. **b.** Any of the class Chondrichthyes, having a cartilaginous skeleton and including the sharks, rays, and skates. **c.** Any of the class Agnatha, lacking jaws and including the lampreys and hagfishes. **2.** Any of various unrelated aquatic animals, as a jellyfish, cuttlefish, or crayfish. **3.** *Informal.* A person <a strange *fish*> **4. Fish.** PISCES 2. —*v.* **fished, fish·ing, fish·es.** —*vi.* **1.** To catch or try to catch fish. **2.** To look for something by feeling one's way : GROPE. **3.** To seek something in a sly or indirect way <*fishing* for compliments> —*vt.* **1.** To catch or try to catch fish in. **2.** To catch or pull in the manner of one who fishes <*fished* the keys out of my pocket>

**fish and chips** *pl.n.* Fried fillets of fish and French-fried potatoes.

**fish·bowl** *also* **fish bowl** (fĭsh′bōl′) *n.* **1.** A transparent bowl in which live fish are kept. **2.** Lack of privacy.

**fish cake** *n.* A fried cake or patty of chopped fish.

**fish crow** *n.* A crow, *Corvus ossifragus* of the coast and rivers of the eastern United States.

**fish·er** (fĭsh′ər) *n.* **1.** One that fishes. **2. a.** A carnivorous mammal, *Martes pennanti* of northern North America, with thick dark-brown fur. **b.** The fur of the fisher.

**fish·er·man** (fĭsh′ər-mən) *n.* **1.** One who fishes as an occupation or sport. **2.** A commercial fishing vessel.

**fisherman's bend** *n.* A knot used for securing the end of a line to a ring or spar, made by two turns with the end passing back under both.

**fisherman's knot** *n.* A knot used for joining two lines, made by securing either end to the opposite standing part by an overhand knot.

**fish·er·y** (fĭsh′ə-rē) *n.*, *pl.* **-ies. 1.** The industry or occupation of catching, processing, or selling fish, shellfish, or similar aquatic products. **2.** A place where fish can be caught. **3.** A fish hatchery. **4.** The legal right to fish in specified waters or areas.

**fish·eye** (fĭsh′ī′) *adj.* Of, relating to, or being a wide-angle photographic lens that covers an angle of about 180°, producing a circular image with barrel distortion.

**fish farm** *n.* A facility consisting of tanks or ponds in which food fish are raised commercially.

**fish flour** *n.* A flour made of dried and powdered fish.

**fish fry** *n.* **1.** A cookout or other meal at which fried fish is the main course. **2.** Fried fish.

**fish·gig** (fĭsh′gĭg′) *n.* [Alteration of obs. *fisgig* < Sp. *fisga*, ult. < Lat. *fixus*, fixed.] A pronged instrument for spearing fish.

**fish hawk** *n.* OSPREY 1.

**fish·hook** (fĭsh′ho͝ok′) *n.* A barbed metal hook for catching fish.

**fish·ing** (fĭsh′ĭng) *n.* **1.** The act or practice of catching fish. **2.** FISHERY 2.

**fishing rod** *n.* A rod of wood, steel, or fiber glass used with a line for catching fish.

**fish joint** *n.* A joint formed by bolting fishplates to either side of two rails, timbers, or beams.

**fish·meal** (fĭsh′mēl′) *n.* A nutritive mealy substance produced from fish or fish parts and used as animal feed and fertilizer.

**fish·mon·ger** (fĭsh′mŭng′gər, -mŏng′-) *n. Chiefly Brit.* A seller or purveyor of fish.

**fish·net** (fĭsh′nĕt′) *n.* **1.** Netting used to catch fish. **2.** A mesh fabric resembling fishnet.

**fish·plate** (fĭsh′plāt′) *n.* [Prob. < OFr. *fiche*, peg (< *ficher*, to fix < Lat. *figere*) + PLATE.] One of the connecting metal plates bolted along the side of two rails or beams placed end to end, used esp. in laying railroad track.

**fish·pond** (fĭsh′pŏnd′) *n.* A small body of water where fish are found.

**fish protein concentrate** *n.* A flour or paste rich in protein, prepared from ground fish and used as a nutritional additive to foods.

**fish·skin disease** (fĭsh′skĭn′) *n.* Ichthyosis.

**fish stick** *n.* An oblong piece of breaded fish fillet.

**fish story** *n.* [From the fact that fishermen traditionally exaggerate the size of their catch.] *Informal.* An implausible and boastful story.

**fish·tail** (fĭsh′tāl′) *adj.* Resembling the tail of a fish. —*vi.* **-tailed, -tail·ing, -tails.** To swing the tail of an aircraft or the rear end of a motor vehicle from side to side while moving forward.

**fish·wife** (fĭsh′wīf′) *n.* **1.** A woman who sells fish. **2.** A coarse, abusive woman : SHREW.

**fish·y** (fĭsh′ē) *adj.* **-i·er, -i·est. 1.** Resembling fish, as in taste or odor. **2.** Cold or expressionless <a *fishy* gaze> **3.** *Informal.* Inspiring suspicion <a *fishy* excuse> **—fish′i·ly** *adv.* **—fish′i·ness** *n.*

**fissi-** *pref.* [< Lat. *fissus*, p.part. of *findere*, to split.] **1.** Fission <*fissiparous*> **2.** Split : cleft <*fissipalmate*>

**fis·sile** (fĭs′əl, -īl′) *adj.* [Lat. *fissilis* < *fissus*, split. —see FISSI-.] **1.** Capable of being split. **2.** *Physics.* Fissionable, esp. by neutrons of all energies. **—fis·sil′i·ty** (fĭ-sĭl′ĭ-tē) *n.*

**fis·sion** (fĭsh′ən) *n.* [Lat. *fissio*, a cleaving < *fissus*, split. —see FISSI-.] **1.** The act or process of splitting into parts. **2.** *Physics.* A nuclear reaction in which an atomic nucleus splits into fragments, usu. two fragments of comparable mass, with the evolution of approx. 100 million to several hundred million electron volts of energy. **3.** *Biol.* An asexual reproductive process in which a unicellular organism splits into two or more independently maturing daughter cells.

**fis·sion·a·ble** (fĭsh′ə-nə-bəl) *adj.* Capable of undergoing fission, esp. capable of being induced to undergo nuclear fission by slow neutrons. **—fis′sion·a·bil′i·ty** *n.*

**fission bomb** *n.* An atomic bomb.

**fis·si·pal·mate** (fĭs′ə-păl′māt′) *adj.* Having lobed or partially webbed separated toes, as the feet of certain birds.

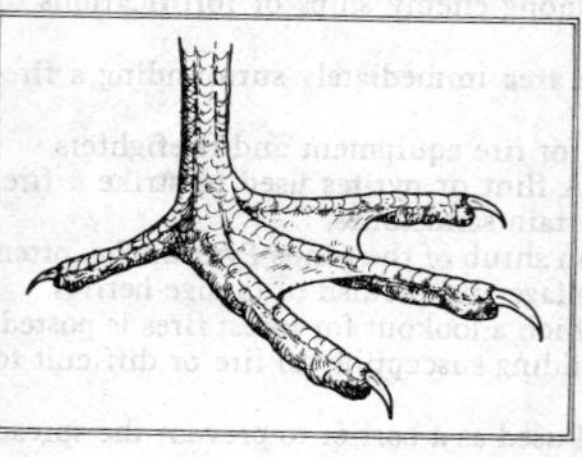

**fissipalmate**
*Fissipalmate foot of an ibis*

**fis·sip·a·rous** (fĭ-sĭp′ər-əs) *adj.* Reproducing by biological fission. **—fis·sip′a·rous·ly** *adv.* **—fis·sip′a·rous·ness** *n.*

**fis·si·ped** (fĭs′ə-pĕd′) *adj.* [LLat. *fissipes*, *fissiped-*, cloven-footed : Lat. *fissus*, split + Lat. *pes*, foot.] Having the toes separated from one another, as certain carnivorous mammals. —*n.* A fissiped carnivorous mammal.

**fis·sure** (fĭsh′ər) *n.* [ME, cut < OFr. < Lat. *fissura* < *fissus*, split. —see FISSI-.] **1.** A narrow crack or cleft, as in a rock face. **2.** Separation or division. **3.** A schism : split. **4.** *Anat.* A groove or furrow, as in the liver or brain, that divides an organ into lobes or separates it into areas. —*vi.* & *vt.* **-sured, -sur·ing, -sures.** To form a fissure or cause a fissure in : CRACK.

**fist** (fĭst) *n.* [ME < OE *fȳst.*] **1.** The hand closed tightly with the fingers bent against the palm. **2.** *Informal.* A grasping hand : CLUTCH. **3.** INDEX 3. —*vt.* **fist·ed, fist·ing, fists. 1.** To clench into a fist. **2.** To grasp with the fist.

**fist·fight** (fĭst′fīt′) *n.* A fight involving use of the fists.

**fist·ful** (fĭst′fo͝ol′) *n.*, *pl.* **-fuls.** A handful.

**fist·ic** (fĭs′tĭk) *adj.* Of or relating to boxing : PUGILISTIC.

**fist·i·cuffs** (fĭs′tĭ-kŭfs′) *pl.n.* **1.** A fistfight. **2.** Boxing. **—fist′i·cuff′er** *n.*

**fis·tu·la** (fĭs′chə-lə) *n.*, *pl.* **-las** or **-lae** (-lē′) [ME < Lat.] An abnormal duct or passage from an abscess, cavity, or hollow organ to the body surface or another hollow organ.

**fis·tu·lous** (fĭs′chə-ləs) *adj.* **1.** Of or like a fistula. **2.** Tubular and hollow : REEDLIKE.

**fit¹** (fĭt) *v.* **fit·ted** or **fit, fit·ted, fit·ting, fits.** [ME *fitten*, to be suitable.] —*vt.* **1. a.** To be the proper size and shape for. **b.** To cause to fit. **2.** To be suitable or appropriate to. **3.** To conform or agree with <observations that *fit* my theory> **4.** To make suitable. **5.** To make ready : PREPARE <education *fitting* one for a good job> **6.** To equip : outfit <*fit* out a boat> **7.** To provide a place or time for <The dentist can *fit* you in today.> **8.** To insert or adjust so as to be properly in place <*fit* a knob on a door> —*vi.* **1.** To be the proper size and shape. **2.** To be suited : BELONG <doesn't *fit* in with that crowd> **3.** To be in harmony : AGREE <a mood *fitting* in with the happy event> ***usage:*** Either *fitted* or *fit* is acceptable as the past tense of *fit* except when the sense is "to cause to fit by adjusting and altering"; then only *fitted* is acceptable as the past tense, as in *The tailor fitted the suit in a few minutes.* —*adj.* **fit·ter, fit·test. 1.** Suited, adapted, or acceptable for a given circumstance or purpose <This is not a *fit* time for further discussion.> **2.** Appropriate : proper <Do as you see *fit*.> **3.** Physically sound : HEALTHY. —*n.* **1.** The quality, state, or way of being fitted. **2.** The way in which garments fit. **3.** The degree of precision with which surfaces are adjusted or adapted to each other in a machine or collection of parts. **—fit′ly** *adv.* **—fit′ness** *n.*

**fit²** (fĭt) *n.* [ME, hardship.] **1.** *Med.* **a.** A sudden acute attack of disease. **b.** The sudden appearance of a symptom <a *fit* of coughing> **c.** A convulsion. **2.** A sudden emotional outburst <a *fit* of crying> **3.** A sudden period of vigorous activity. **—by (**or **in) fits and starts.** With irregular intervals of action and inaction : INTERMITTENTLY.

**fit³** (fĭt) *n.* [ME < OE.] *Archaic.* A section of a poem or ballad.

**fitch** (fĭch) *n.* [ME *fiche.*] The fur of the Old World polecat.

**fitch·ew** (fĭch′o͞o) *also* **fitch·et** (fĭch′ĭt) *n.* [ME *ficheux* < OFr. *ficheau* < MDu. *vitsau.*] *Archaic.* The Old World polecat or its fur.

**fit·ful** (fĭt′fəl) *adj.* Occurring in or marked by intermittent bursts of activity : IRREGULAR. **—fit′ful·ly** *adv.* **—fit′ful·ness** *n.*

**fit·ting** (fĭt′ĭng) *adj.* Suitable to the circumstances. —*n.* **1.** The act of trying on clothes whose fit is being adjusted. **2.** A small detachable

ă pat ā pay âr care ä father ĕ pet ē be hw which ĭ pit
ī tie îr pier ŏ pot ō toe ô paw, for oi noise o͝o took

part for a machine or an apparatus. **3. fittings.** *Chiefly Brit.* Furnishings : fixtures. **—fit'ting·ly** *adv.* **—fit'ting·ness** *n.*

**five** (fīv) *n.* [ME < OE *fīf*; akin to G. *fünf*, Lat. *quinque*, Gk. *pente*, and Skt. *pañca*.] **1.** The cardinal number equal to 4 + 1. **2.** The fifth in a set or sequence. **3.** Something having five parts, units, or members, esp. a basketball team. **4.** *Informal.* A five-dollar bill. **—five** *adj.* & *pron.*

**five-and-dime** (fīv'ən-dīm') *n.* A five-and-ten.

**five-and-ten** (fīv'ən-tĕn') *n.* A variety store selling inexpensive merchandise.

**five-fin·ger** (fīv'fĭng'gər) *n.* A plant having compound leaves with five leaflets, as the cinquefoil.

**five·fold** (fīv'fōld') *adj.* **1.** Consisting of five parts. **2.** Five times as many or as much. **—five'fold'** *adv.*

**fiv·er** (fī'vər) *n.* **1.** *Informal.* A five-dollar bill. **2.** *Chiefly Brit.* A five-pound note.

**†fix** (fĭks) *v.* **fixed, fix·ing, fix·es.** [ME *fixen* < *fix*, fixed in position < Lat. *fixus*, p.part. of *figere*, to fasten.] *—vt.* **1. a.** To fasten or place securely. **b.** To make fast to : ATTACH. **2.** To put into a stable or unalterable form, as: **a.** *Chem.* To make (a substance) nonvolatile or solid. **b.** *Biol.* To convert (nitrogen) into stable, biologically assimilable compounds. **c.** To kill and keep (a specimen) intact for microscopic study. **d.** To prevent discoloration of (a photographic image) by washing or coating with a chemical preservative. **3.** To direct steadily <*fixed* my eyes on the horizon> **4.** To establish definitely <*fix* a time to meet> **5.** To assign <*fixing* the blame on us> **6.** To set right : ADJUST. **7.** *Computer Sci.* To convert data from floating-point notation to fixed-point notation. **8.** To restore to proper state or working order : REPAIR <*fix* a broken watch> **9.** To make ready : PREPARE. **10.** To spay or castrate (an animal). **11.** *Informal.* To take revenge upon. **12.** To influence or arrange the outcome of by unlawful means. *—vi.* **1.** To become concentrated, directed, or attached. **2.** To become stable or firm : HARDEN. **3.** *Regional.* To intend <was *fixing* to go home> *—n.* **1.** A difficult or embarrassing position. **2.** The position of a ship, aircraft, etc., as determined by observations or radio. **3.** An instance of arranging for special consideration or exemption from a requirement, esp. by means of bribery. **4.** *Slang.* An intravenous injection of a narcotic. **—fix'a·ble** *adj.* **—fix'er** *n.*

**fix·ate** (fĭk'sāt') *v.* **-at·ed, -at·ing, -ates.** *—vt.* **1.** To make fixed, stable, or stationary. **2.** To focus one's eyes or concentrate one's attention on. **3.** *Psychol.* To attach (oneself) to a person or thing in an immature or neurotic way. *—vi.* **1.** To focus or concentrate one's attention. **2.** *Psychol.* **a.** To form a fixation. **b.** To be arrested at an immature stage of psychosexual development.

**fix·a·tion** (fĭk-sā'shən) *n.* **1.** The act or process of fixing or fixating. **2.** *Psychol.* A strong attachment to a person or thing, esp. such an attachment formed in childhood or infancy and persisting in immature or neurotic behavior.

**fix·a·tive** (fĭk'sə-tĭv) *n.* Something that fixes, protects, or preserves, esp.: **a.** A liquid preservative applied to artwork, as water-color paintings or charcoal drawings. **b.** A solution used to preserve fresh tissue for microscopic examination. **c.** A liquid mixed with perfume to prevent rapid evaporation. **—fix'a·tive** *adj.*

**fixed** (fĭkst) *adj.* **1.** Being set firmly in position : STATIONARY. **2.** *Chem.* **a.** Nonvolatile. **b.** Being in a stable combined form. **3.** Not subject to change or variation : CONSTANT <a *fixed* time for the meeting> **4.** Firmly, often dogmatically held <*fixed* notions> **5.** Illegally prearranged as to outcome <a *fixed* football game> **—fix'ed·ly** (fĭk'sĭd-lē) *adv.* **—fix'ed·ness** (-sĭd-nĭs) *n.*

**fixed head** *n.* A stationary device, as a tape-recording head, that reads and imprints information on a single track of magnetic tape.

**fixed oil** *n.* A nonvolatile oil, esp. a fatty oil.

**fixed-point** (fĭkst'point') *adj.* Of, relating to, or being a method of writing numerical quantities with a predetermined number of digits and with the decimal placed at a single, unchanging position.

**fixed star** *n.* A star so distant from the earth that its movements can be measured only by precise observations over long time periods.

**fix·ings** (fĭk'sĭngz) *pl.n. Informal.* Accessories : trimmings.

**fix·i·ty** (fĭk'sĭ-tē) *n., pl.* **-ties.** **1.** The quality or state of being fixed : STABILITY. **2.** A fixed or immovable object.

**fix·ture** (fĭks'chər) *n.* [Var. of obs. *fixure* < LLat. *fixura* < Lat. *fixus.* —see FIX.] **1.** Something securely fixed in place. **2.** A permanently attached appendage, appliance, or device <plumbing *fixtures*> **3.** *Law.* A chattel bound to realty. **4.** One long associated with, established in, or restricted to a given place, position, or function <a *fixture* of Paris society> **5. a.** The act or process of fixing. **b.** The state of being fixed.

**fizz** (fĭz) *vi.* **fizzed, fizz·ing, fizz·es.** [Imit.] To make a hissing or bubbling sound. *—n.* **1.** A hissing or bubbling sound. **2.** Effervescence. **3.** An effervescent beverage.

**fiz·zle** (fĭz'əl) *vi.* **-zled, -zling, -zles.** [Prob. < obs. *fist*, to break wind < ME *fisten*.] **1.** To make a hissing or sputtering sound. **2.** *Informal.* To fail or die out, esp. after a positive beginning. *—n. Informal.* A fiasco : failure.

**fjeld** (fyĕld) *n.* [Dan. < ON *fjall*, mountain.] A high, barren plateau in Scandinavia.

**fjord** or **fiord** (fyôrd, fyōrd) *n.* [Norw. < ON *fjörðr*.] A long, narrow, often deep inlet from the sea between steep cliffs and slopes.

**flab** (flăb) *n.* [Back-formation < FLABBY.] Loose, flaccid body tissue.

**flab·ber·gast** (flăb'ər-găst') *vt.* **-gast·ed, -gast·ing, -gasts.** [Orig. unknown.] To overwhelm with astonishment.

**flab·by** (flăb'ē) *adj.* **-bi·er, -bi·est.** [Alteration of *flappy*, tending to flap < FLAP.] **1.** FLACCID 1. **2.** Lacking force or vitality : INEFFECTUAL. **—flab'bi·ly** *adv.* **—flab'bi·ness** *n.*

**fla·bel·la** (flə-bĕl'ə) *n. pl. of* FLABELLUM.

**fla·bel·late** (flə-bĕl'ĭt, flăb'ə-lāt') *adj.* [< Lat. *flabellum*, small fan.] Fan-shaped.

**fla·bel·li·form** (flə-bĕl'ə-fôrm') *adj.* Flabellate.

**fla·bel·lum** (flə-bĕl'əm) *n., pl.* **-bel·la** (-bĕl'ə) [Lat. *flabellum*, small fan.] A fan-shaped anatomical structure.

**flac·cid** (flăk'sĭd, flăs'ĭd) *adj.* [Fr. *flaccide* < Lat. *flaccidus* < *flaccus*, flabby.] **1.** Lacking firmness or resilience <*flaccid* cheeks> **2.** Devoid of vigor or energy. **—flac·cid'i·ty** (-sĭd'ĭ-tē), **flac'cid·ness** *n.* **—flac'cid·ly** *adv.*

**flack** (flăk) *n.* [Orig. unknown.] A press agent. *—vi.* **flacked, flack·ing, flacks.** To function as a flack. **—flack'er·y** *n.*

**flac·on** (flăk'ən, -ŏn') *n.* [Fr. < OFr., flagon.] A small, often decorative bottle with a tight-fitting cap or stopper.

**flag¹** (flăg) *n.* [Orig. unknown.] **1.** A piece of cloth having a distinctive size, color, and design, used as a symbol, standard, signal, or emblem. **2.** *Mus.* A cross stroke added to a note that is less than a quarter note in value. **3.** FLAGSHIP 1. **4.** The masthead of a newspaper. **5.** A distinctively shaped or marked tail, as of a deer. *—vt.* **flagged, flag·ging, flags.** **1.** To mark with a flag for identification or ornamentation. **2. a.** To signal with or as if with a flag. **b.** To signal to stop <*flagged* the train between stations> **3.** *Computer Sci.* A bit or series of bits with two stable states, used in software to indicate a single bit of information. **—flag'ger** *n.*

**flag²** (flăg) *n.* [ME *flagge*, reed.] A plant having long bladelike leaves, as an iris or cattail.

**flag³** (flăg) *vi.* **flagged, flag·ging, flags.** [Orig. unknown.] **1.** To hang limply : DROOP. **2.** To decline in strength or vigor : WEAKEN <My appetite began to *flag*.> **3.** To decline in interest <The conversation *flagged*.>

**flag⁴** (flăg) *n.* [ME *flagge*, piece of turf < ON *flaga*, slab of stone.] **1.** A slab of flagstone. **2.** Flagstone. *—vt.* **flagged, flag·ging, flags.** To pave with flags.

**Flag Day** *n.* Jun. 14, commemorating adoption of the official U.S. flag in 1777.

**fla·gel·la** (flə-jĕl'ə) *n. pl. of* FLAGELLUM.

**flag·el·lant** (flăj'ə-lənt, flə-jĕl'ənt) *n.* [Lat. *flagellans*, *flagellant-*, pr.part. of *flagellare*, to flagellate.] One who whips, esp. one who scourges oneself by way of religious discipline or public penance. **—flag'el·lant·ism** *n.*

**fla·gel·lar** (flə-jĕl'ər) *adj.* Of or relating to a flagellum.

**flag·el·late** (flăj'ə-lāt') *vt.* **-lat·ed, -lat·ing, -lates.** [Lat. *flagellare*, *flagellat-*, to whip < *flagellum*, little whip, dim. of *flagrum*, whip.] **1.** To whip or flog. **2.** To punish or force as if by whipping. *—adj.* (-lĭt, -lāt', flə-jĕl'ĭt). **1.** Having a flagellum or flagella, as unicellular organisms of the class Flagellata or Magistophora. **2.** Whiplike. *—n.* (-lĭt, -lāt', flə-jĕl'ĭt). A flagellate organism.

**flag·el·la·tion** (flăj'ə-lā'shən) *n.* **1.** The act or practice of flagellating. **2.** The flagellar arrangement on an organism.

**fla·gel·li·form** (flə-jĕl'ə-fôrm') *adj.* [Lat. *flagellum*, little whip + FORM.] Long, thin, and tapering <*flagelliform* appendages>

**fla·gel·lin** (flə-jĕl'ĭn) *n.* A protein component of flagella.

**fla·gel·lum** (flə-jĕl'əm) *n., pl.* **-gel·la** (-jĕl'ə) [Lat., little whip.] **1.** *Biol.* A long, filamentous process, esp. one of the whiplike extensions of certain cells or unicellular organisms, usu. functioning in locomotion. **2.** A whip.

**flag·eo·let** (flăj'ə-lĕt', -lā') *n.* [Fr., dim. of OFr. *flajol*, flute.] A small flutelike instrument with a cylindrical mouthpiece, four finger holes, and two thumbholes.

**flag·ging¹** (flăg'ĭng) *adj.* **1.** Drooping : languid. **2.** Declining : weakening <a *flagging* economy> **—flag'ging·ly** *adv.*

**flag·ging²** (flăg'ĭng) *n.* A pavement laid with flagstones.

**fla·gi·tious** (flə-jĭsh'əs) *adj.* [ME *flagicious*, wicked < Lat. *flagitiosus* < *flagitium*, shameful act < *flagitare*, to incite to lewdness.] Marked by brutal or shocking crimes : VICIOUS. **—fla·gi'tious·ly** *adv.* **—fla·gi'tious·ness** *n.*

**flag·man** (flăg'mən) *n.* One who signals with or carries a flag.

**flag officer** *n.* A U.S. Navy or Coast Guard officer holding the rank of rear admiral, vice admiral, or admiral.

**flag of truce** *n.* A white flag brought or displayed to an enemy as an invitation to a conference or a signal of surrender.

**flag·on** (flăg'ən) *n.* [ME < OFr. *flacon* < LLat. *flasco*, flask. —see FLASK.] **1.** A large vessel for holding wine or other liquors, usu. made of metal or pottery and having a handle and spout and often a lid. **2.** The quantity of liquid a flagon holds.

**flag·pole** (flăg'pōl') *n.* A pole on which a flag is raised.

**fla·grant** (flā'grənt) *adj.* [Lat. *flagrans*, *flagrant-*, pr.part. of *flagare*, to burn.] **1.** Conspicuously bad or offensive <a *flagrant* miscarriage of justice> **2.** *Obs.* Flaming : blazing. **—fla'grance, fla'gran·cy** *n.* **—fla'grant·ly** *adv.*

☆ ***syns:*** FLAGRANT, ARRANT, CAPITAL, EGREGIOUS, GLARING, GROSS, RANK *adj. core meaning* : conspicuously bad or offensive <a *flagrant* violation of their civil rights>

**fla·gran·te de·lic·to** (flə-grăn'tē dĭ-lĭk'tō) *adv.* [Med. Lat., while the crime is blazing.] In the very act : RED-HANDED.

**flag·ship** (flăg'shĭp') *n.* **1.** A ship that carries a fleet or squadron commander and bears his or her flag. **2.** The chief one of a related group <the *flagship* of a grocery chain>

**flag·staff** (flăg'stăf') *n.* A flagpole.

**flag·stone** (flăg'stōn') *n.* A flat, fine-grained, hard, evenly layered stone split into slabs for use in paving.

**flag-wav·ing** (flăg'wā'vĭng) *n.* Fanatical or overzealous patriotism.

**flail** (flāl) *n.* [ME, partly < OE **flegil*, and partly < OFr. *flaiel*, both < LLat. *flagellum* < *flagrum*, whip.] A manual threshing device consisting of a long wooden handle or staff and a shorter free-swinging stick attached to its end. —*v.* **flailed, flail·ing, flails.** —*vt.* **1.** To thresh with a flail. **2.** To beat or strike with or as if with a flail. —*vi.* **1.** To thresh grain. **2.** To move with a flailing motion.

**flair** (flâr) *n.* [Fr. < OFr., odor < *flairer*, to smell < LLat. *flagrare* < Lat. *fragrare*, to emit an odor.] **1.** A natural talent or aptitude : KNACK <a *flair* for lexicography> **2.** Instinctive discernment : KEENNESS <a *flair* for using the right words> **3.** Distinctive elegance or style <served us with *flair*>

**flak** (flăk) *n.* [G. *Flak*, contraction of *Fliegerabwehrkanone.*] **1. a.** Antiaircraft artillery. **b.** The bursting shells fired from such artillery. **2.** *Informal.* **a.** Excessive or abusive criticism. **b.** Dissension : opposition.

**flake[1]** (flāk) *n.* [ME, of Scand. orig.] **1.** A flat, thin piece or layer : CHIP. **2.** A small piece : BIT. **3.** A small crystal of snow. **4.** *Slang.* One who is somewhat eccentric : ODDBALL. —*v.* **flaked, flak·ing, flakes.** —*vt.* **1.** To break flakes from : CHIP. **2.** To cover, mark, or overlay with or as if with flakes. —*vi.* To come off in flakes. **—flake out.** *Slang.* To collapse or fall asleep from fatigue or exhaustion. **—flak'er** *n.*

**flake[2]** (flāk) *n.* [ME *fleke* < ON *fleki.*] **1.** A frame or platform for drying fish or produce. **2.** A scaffold lowered over the side of a ship to support workers or caulkers.

**flake white** *n.* A pigment made of flakes of white lead.

**flak·y** (flā'kē) *adj.* **-i·er, -i·est. 1.** Made of or like flakes. **2.** Forming or tending to form flakes <a *flaky* pie crust> **3.** *Slang.* Somewhat eccentric. **—flak'i·ly** *adv.* **—flak'i·ness** *n.*

**flam[1]** (flăm) *n.* [Short for FLIMFLAM.] *Informal.* **1.** A hoax or deception. **2.** Nonsense.

**flam[2]** (flăm) *n.* [Prob. imit.] A drumbeat consisting of two almost simultaneous strokes of which the first is a very rapid grace note.

**flam·bé** (fläm-bā', fläN-) *vt.* **-béed, -bé·ing, -bés.** [< Fr., p.part. of *flamber*, to flame < OFr. < *flambe*, flame.] To drench with a liquor, as brandy, and ignite <*flambéed* the steak> —*adj.* Served flaming in ignited liquor <steak *flambé*>

**flam·beau** (flăm'bō') *n., pl.* **-beaux** (-bōz') or **-beaus.** [Fr. < OFr. < *flambe*, flame < Lat. *flammula*, dim. of *flamma*, flame.] **1.** A lighted torch. **2.** A large decorative candlestick.

**flam·boy·ant** (flăm-boi'ənt) *adj.* [Fr. < OFr., pr.part. of *flamboyer*, to blaze < *flambe*, flame.] **1.** Extremely elaborate : ORNATE. **2.** Richly colored : RESPLENDENT. **3.** Relating to or having waved lines and flamelike forms typical of 15th- and 16th-cent. French Gothic architecture. —*n.* The royal poinciana. **—flam·boy'ance, flam·boy'an·cy** *n.* **—flam·boy'ant·ly** *adv.*

**flame** (flām) *n.* [ME < AN *flaumbe* < OFr. *flambe* < Lat. *flamma.*] **1.** The zone of burning gases and fine suspended matter associated with combustion of a substance. **2.** *often* **flames.** Active, blazing combustion <burst into *flames*> **3.** Something like a flame in motion, brilliance, intensity, or shape. **4.** Violent passion. **5.** *Informal.* A sweetheart. —*v.* **flamed, flam·ing, flames.** —*vi.* **1.** To burn brightly : BLAZE. **2.** To color suddenly <cheeks *flaming* with embarrassment> **3.** To break out passionately <*flaming* with jealousy> —*vt.* **1.** To burn, ignite, or scorch with a flame. **2.** *Obs.* To foment : incite. **—flam'er** *n.*

**flame cell** *n.* A hollow ciliated cell in the excretory system of some platyhelminths and rotifers.

**fla·men** (flā'mən) *n., pl.* **-mens** or **flam·i·nes** (flăm'ə-nēz') [ME *flamin* < Lat. *flamen.*] A priest, esp. of a Roman deity.

**fla·men·co** (flə-mĕng'kō) *n.* [Sp., Flemish < MDu. *Vlāming*, Fleming.] **1.** A dance style of the Andalusian Gypsies marked by forceful, often improvised rhythms. **2.** Music for a flamenco dance.

**flame·out** (flām'out') *n.* In-flight failure of a jet aircraft engine.

**flame·proof** (flām'prōōf') *adj.* Flame-retardant. —*vt.* **-proofed, -proof·ing, -proofs.** To make flameproof. **—flame'proof'er** *n.*

**flame-re·tard·ant** (flām'rĭ-tär'dənt) *adj.* Resistant to catching fire. **—flame'-re·tard'ant** *n.*

**flame thrower** *n.* A weapon that projects ignited incendiary fuel, as napalm, in a steady stream.

**flam·i·nes** (flăm'ə-nēz') *n. var. pl. of* FLAMEN.

**flam·ing** (flā'mĭng) *adj.* **1.** On fire : ABLAZE. **2.** Like a flame in brilliance, color, or form. **3.** Intense : ardent. **—flam'ing·ly** *adv.*

**fla·min·go** (flə-mĭng'gō) *n., pl.* **-gos** or **-goes.** [Port. *flamengo* or Sp. *flamenco*, both prob. < OProv. *flamenc* < *flama*, flame < Lat. *flamma.*] **1.** Any of several large, gregarious tropical wading birds of the family Phoenicopteridae, with reddish or pinkish plumage, long legs, a long, flexible neck, and a bill turned downward at the tip. **2.** A moderate reddish orange.

**flam·ma·ble** (flăm'ə-bəl) *adj.* [< Lat. *flammare*, to flame < *flamma*, flame.] Easily ignited and capable of burning with great rapidity : INFLAMMABLE. ***usage:*** *Flammable* and *inflammable* are identical in meaning and are equally acceptable in standard English at all levels. **—flam'ma·bil'i·ty** *n.* **—flam'ma·ble** *n.*

**flam·y** (flā'mē) *adj.* **-i·er, -i·est.** FLAMING 2.

**flan** (flăn, flän, fläN) *n.* [Fr. < OFr. *flaon* < LLat. *flado*, flat cake, of Germanic orig.] **1.** A tart with a filling of custard, fruit, or cheese. **2.** A metal disk to be stamped as a coin : BLANK.

**flange** (flănj) *n.* [Prob. < *flanch*, to widen out.] A protruding rim, edge, rib, or collar, as on a wheel or a pipe shaft, used to strengthen an object, hold it in place, or attach it to another object. —*vt.* **flanged, flang·ing, flang·es.** To furnish with a flange.

**flank** (flăngk) *n.* [ME < OFr. *flanc*, of Germanic orig.] **1.** The section of flesh between the last rib and the hip. **2.** A cut of meat from the flank of an animal. **3.** A side or lateral part. **4. a.** The right or left side of a military formation <attacked on the right *flank*> **b.** The right or left side of a bastion. —*vt.* **flanked, flank·ing, flanks. 1.** To protect or guard the flank of. **2.** To attack or menace the flank of. **3.** To be situated at the flank or side of. **4.** To put something on each side of.

**flank·er** (flăng'kər) *n.* **1.** One that flanks. **2.** A flankerback.

**flank·er·back** (flăng'kər-băk') *n.* *Football.* The halfback of the offensive team positioned just behind the line of scrimmage and to the right of his team's right end.

**flan·nel** (flăn'əl) *n.* [ME, a kind of woolen cloth or garment.] **1.** A soft woven cloth of wool or a blend of wool and cotton or synthetics. **2. a. flannels.** Outer clothing, esp. trousers, made of flannel. **b.** Underclothing made of flannel. **—flan'nel·ly** *adj.*

**flannel bush** *n.* A shrub or small tree, *Fremontia californica* of California and northern Mexico, having downy, lobed leaves and yellow flowers.

**flannel cake** *n.* A pancake.

**flan·nel·ette** (flăn'ə-lĕt') *n.* A soft cotton cloth with a nap, used chiefly for infants' garments and underclothes.

**flan·nel-leaf** (flăn'əl-lēf') *n.* MULLEIN 1.

**flap** (flăp) *n.* [ME *flappe*, slap < *flappen*, to beat.] **1.** A flat, usu. thin piece attached at only one side. **2.** A projecting or hanging piece usu. intended to double over and protect or cover <the *flap* of an envelope> **3.** An act of waving or fluttering. **4.** The sound produced by the motion of a flap. **5.** A blow given with something flat : SLAP. **6.** A variable control surface on the trailing edge of an aircraft wing, used primarily to increase lift or drag. **7.** Partially detached tissue used in plastic surgery to fill an adjacent defect or cover the cut end of a bone after amputation. **8.** *Slang.* Agitation : confusion. —*v.* **flapped, flap·ping, flaps.** —*vt.* **1.** To wave (e.g., the arms) up and down. **2.** To cause to move or sway with a flap. **3.** To hit with something broad and flat : SLAP. **4.** *Informal.* To fling down : TOSS. —*vi.* **1.** To move or sway while fixed at one edge : FLUTTER. **2.** To wave arms or wings up and down. **3.** To fly by beating the air with the wings.

**flap·doo·dle** (flăp'dōōd'l) *n.* [Orig. unknown.] *Slang.* Nonsense.

**flap·jack** (flăp'jăk') *n.* A pancake.

**flap·pa·ble** (flăp'ə-bəl) *adj.* *Slang.* Easily excited or upset.

**flap·per** (flăp'ər) *n.* **1.** One that flaps. **2.** A broad, flexible part, as a flipper. **3.** *Informal.* A young woman, esp. one of the 1920's who showed disdain for conventional dress and behavior.

**flare** (flâr) *v.* **flared, flar·ing, flares.** [Orig. unknown.] —*vi.* **1.** To flame up with a bright light. **2.** To burst into intense, sudden flame. **3. a.** To erupt or intensify suddenly. **b.** To become suddenly angry. **4.** To expand or open outward in shape, as a skirt. —*vt.* **1.** To cause to flare. **2.** To signal with a blaze of light. —*n.* **1.** A blaze of light. **2.** A device producing a bright light for signaling, illumination, or identification. **3.** An outbreak, as of emotion or activity. **4.** An expanding or opening outward. **5.** A lens reflection or the resultant film fogging. **6.** A brief intense eruption from the sun's chromosphere, associated with sunspots. **7.** *Football.* A quick pass made to a back running toward the sideline.

**flare·back** (flâr'băk') *n.* A flame produced in a gun breech by ignition of residual gases.

**flare-up** (flâr'ŭp') *n.* **1.** A sudden outbreak of flame or light. **2.** An outburst or eruption <a *flare-up* of anger> **3.** An intensification <a *flare-up* of old hostilities>

**flash** (flăsh) *v.* **flashed, flash·ing, flash·es.** [ME *flashen*, to splash.] —*vi.* **1.** To burst forth into or as if into flame. **2.** To appear or occur suddenly. **3.** To emit light suddenly or intermittently. **4.** To proceed rapidly. —*vt.* **1. a.** To cause (light) to appear suddenly or intermittently. **b.** To cause to burst into flame. **c.** To reflect (light). **d.** To cause to reflect light from (a surface). **2.** To make known by flashing lights. **3.** To communicate or display at great speed. **4.** To exhibit briefly. **5.** To display ostentatiously : FLAUNT. **6.** To fill suddenly with water. **7.** To cover with a thin protective layer. —*n.* **1.** A sudden, brief, intense display of light. **2.** A sudden perception <a *flash* of insight> **3.** An instant <came in a *flash*> **4.** A brief news transmission or dispatch. **5.** *Obs.* The language or cant of thieves, tramps, or underworld figures. **6.** A flashlight. **7. a.** Instantaneous illumination for photography. **b.** A device, as a flash bulb or flash gun, used to produce such illumination. **8.** A flurry of activity.

**flash·back** (flăsh′băk′) *n.* **1.** A literary or dramatic device in which an earlier event is inserted into the normal chronological order of a narrative. **2.** The episode or scene depicted by a flashback.

**flash·board** (flăsh′bôrd′, -bōrd′) *n.* Boarding that extends above a dam to increase the depth of water held.

**flash bulb** *n.* A glass bulb filled with finely shredded aluminum or magnesium foil that is ignited by electricity to produce a short-duration high-intensity light flash for taking photographs.

**flash burn** *n.* A burn resulting from brief exposure to intense radiation.

**flash card** *n.* A card printed with words or numbers and briefly displayed as part of a learning drill.

**flash·cube** (flăsh′kyo͞ob′) *n.* A small cube containing four flash bulbs that rotates automatically when a picture is taken with a camera to which it is attached.

**flash·er** (flăsh′ər) *n.* **1.** One that flashes. **2.** A device that automatically switches an electric lamp off and on.

**flash flood** *n.* A sudden, violent flood after a heavy rain.

**flash-for·ward** (flăsh′fôr′wərd) *n.* A literary or dramatic device in which a future event is inserted into the normal chronological order of a narrative.

**flash gun** *n.* A dry-cell powered photographic device that holds and electrically triggers a flash bulb.

**flash·ing** (flăsh′ĭng) *n.* Sheet metal for reinforcing and weather-proofing the joints and angles of a roof.

**flash lamp** *n.* An electric lamp used to produce a high-intensity light of very short duration in photography.

**flash·light** (flăsh′līt′) *n.* **1.** A small, portable lamp usu. powered by batteries. **2.** A brief, brilliant flood of light from a photographic lamp. **3.** A bright flashing beam of light, as of a beacon or signal lamp.

**flash·o·ver** (flăsh′ō′vər) *n.* An unintended electric arc, as between two pieces of apparatus.

**flash photolysis** *n.* A method of investigating photochemical reactions that involves the breakdown of a chemical through exposure to a very brief, intense flash of light.

**flash point** *n.* **1.** The lowest temperature at which the vapor of a combustible liquid can be made to ignite momentarily in air. **2.** The brink of crisis or open warfare <international tensions reaching a *flash point*>

**flash tube** *n.* A gas discharge tube used in an electronic flash to produce a brief, intense flash of light.

**flash unit** *n.* **1.** An electronic flash system containing both power supply and flash tube in a single compact unit. **2. a.** A flash gun. **b.** A flash gun and reflector.

**flash·y** (flăsh′ē) *adj.* **-i·er, -i·est. 1.** Conveying a superficial or momentary impression of brilliance. **2.** Cheap and showy : GAUDY <a *flashy* satin shirt> **—flash′i·ly** *adv.* **—flash′i·ness** *n.*

**flask** (flăsk) *n.* [OFr. *flasque* < LLat. *flasco*, prob. of Germanic orig.] **1.** A small container, as a bottle, with a narrow neck and usu. a cap, esp.: **a.** A flat container for liquor. **b.** A container for carrying ammunition. **c.** A vial or round long-necked bottle for laboratory use. **2.** A frame for holding a sand mold in a foundry.

**flat[1]** (flăt) *adj.* **flat·ter, flat·test.** [ME < ON *flatr.*] **1.** Having a horizontal surface without a slope, tilt, or curvature. **2.** Having a smooth, even surface. **3.** Having a relatively broad surface in relation to thickness or depth <a *flat* board> **4.** Stretched out at full length along the ground : PRONE. **5.** Devoid of qualification : ABSOLUTE <a *flat* refusal> **6.** Fixed and unvarying <a *flat* rate> **7.** Neither more nor less : EXACT <15 minutes *flat*> **8.** Lacking interest or excitement : DULL. **9. a.** Lacking flavor. **b.** Having lost effervescence or sparkle <*flat* beer> **10.** Deflated. —Used of a tire. **11.** Commercially inactive : SLUGGISH <a *flat* market> **12.** Monotonous : unmodulated <*flat* tones> **13.** Lacking variety in tint or shading : UNIFORM. **14.** Having a dull finish : MAT. **15.** *Mus.* **a.** Being below the correct pitch. **b.** Being a semitone lower than the corresponding natural key. **16.** Designating the vowel *a* as pronounced in *bad* or *cat*. **17.** Taut. —Used of a sail. —*adv.* **1. a.** Level with the ground : HORIZONTALLY. **b.** Prostrate. **2.** So as to be flat. **3.** Directly : completely <went *flat* against the rules> **4.** *Mus.* Below the intended pitch. **5.** Without interest charge. —*n.* **1.** A flat surface or part. **2.** *often* **flats.** A stretch of level ground <raced on the salt *flats*> **3.** A shallow frame or box for seeds, seedlings, or semimature plants. **4.** Stage scenery on a movable frame. **5.** A flatcar. **6.** A deflated tire. **7.** A shoe with a flat heel. **8.** *Mus.* **a.** A sign (♭) affixed to a note to indicate it is to be lowered by a semitone. **b.** A note lowered a semitone. —*v.* **flat·ted, flat·ting, flats.** —*vt.* **1.** To make flat. **2.** *Mus.* To lower (a note) a semitone. —*vi. Mus.* To sing or play below the proper pitch. **—flat′ly** *adv.* **—flat′ness** *n.*

**flat[2]** (flăt) *n.* [Alteration of Sc. *flet*, inner part of a house < ME < OE.] **1.** An apartment on one floor. **2.** *Archaic.* STORY[2] 1.

**flat·bed** (flăt′bĕd′) *n.* A truck or trailor having a rear platform without sides.

**flat-bed press** (flăt′bĕd′) *n.* A printing press in which the type, locked into a chase, is supported by a flat surface or bed and the paper is applied to the type either by a flat platen or by a cylinder against which the bed moves.

**flat·boat** (flăt′bōt′) *n.* A boat with a flat bottom and square ends used for transporting freight on inland waterways.

**flat·car** (flăt′kär′) *n.* A railroad freight car with no sides or roof.

**flat·fish** (flăt′fĭsh′) *n., pl.* **flatfish** or **-fish·es.** Any of numerous chiefly marine fishes of the order Pleuronectiformes or Heterosomata, including the flounders, soles, and other fishes having a compressed body in which, at an early stage of development, one eye moves to the same side of the body as the other and the fish swims with its eyeless side downward.

**flat·foot** (flăt′fo͝ot′) *n.* **1.** *pl.* **-feet** (-fēt′). A condition in which the arch of the foot is flattened down so that the entire sole makes contact with the ground. **2.** *pl.* **-foots. a.** One afflicted with flatfoot. **b.** *Slang.* A police officer. —*vi.* **-foot·ed, -foot·ing, -foots.** To walk in a flat-footed way.

**flat-footed** (flăt′fo͝ot′ĭd) *adj.* **1.** Of or afflicted with flatfoot. **2. a.** Steady on the feet. **b.** *Informal.* Without reservation : FORTHRIGHT. **3.** Unable to react quickly : UNPREPARED <caught *flat-footed*> **—flat′-foot′ed·ly** *adv.* **—flat′-foot′ed·ness** *n.*

**Flat·head** (flăt′hĕd′) *n., pl.* **Flathead** or **-heads. 1. a.** One of several tribes of Indians living in the northwestern coast area of America who practiced head-flattening. **b.** A member of one of these tribes. **2. a.** A Salishan tribe of western Montana. **b.** A member of this tribe. **3. flathead.** *Slang.* A fool.

**flat·i·ron** (flăt′ī′ərn) *n.* A heated device, usu. with a flat metal base, used for pressing clothes.

**flat·land** (flăt′lănd′, -lənd) *n.* **1.** Land that varies little in elevation. **2. flatlands.** A geographic area composed chiefly of flatland. **—flat′land′er** *n.*

**flat·let** (flăt′lĭt) *n. Chiefly Brit.* An efficiency apartment.

**flat·ling** (flăt′lĭng) *also* **flat·lings** (-lĭngz) *adv. Archaic.* **1.** At full length : FLAT. **2.** With the flat of a sword.

**flat out** *adv.* **1.** In a direct, blunt way. **2.** At top speed.

**flat-out** (flăt′out′) *adj.* Thoroughgoing : out-and-out.

**flat silver** *n.* Eating utensils, as knives, forks, or spoons, made of silver or silver plate.

**flat·ten** (flăt′n) *v.* **-tened, -ten·ing, -tens.** —*vt.* **1.** To make flat or flatter. **2.** To knock down. —*vi.* To become flat or flatter. **—flat′-ten·er** *n.*

**flat·ter[1]** (flăt′ər) *v.* **-tered, -ter·ing, -ters.** [ME *flateren* < OFr. *flater*, of Germanic orig.] —*vt.* **1.** To compliment excessively and often insincerely, esp. so as to win favor. **2.** To gratify the vanity of : PLEASE. **3. a.** To portray favorably. **b.** To show off becomingly or advantageously. —*vi.* To practice flattery. **—flat′ter·er** *n.* **—flat·ter·ing·ly** *adv.*

☆ ***syns:*** FLATTER, ADULATE, BLANDISH, BUTTER UP, SOFT-SOAP, SWEET-TALK *v. core meaning* : to compliment excessively and ingratiatingly, often as a means to an end <sycophants who constantly *flattered* the President>

**flat·ter[2]** (flăt′ər) *n.* One that flattens, esp.: **a.** A blacksmith's flat-faced swage or hammer. **b.** A die plate for flattening metal into strips, as in making watch springs.

**flat·ter·y** (flăt′ə-rē) *n., pl.* **-ies. 1.** The act or practice of flattering. **2.** Excessive or insincere praise.

**flat·tish** (flăt′ĭsh) *adj.* Rather flat.

**flat·top** (flăt′tŏp′) *n. Informal.* **1.** An aircraft carrier. **2.** A man's short haircut with a flattish, brushlike crown.

**flat·u·lence** (flăch′ə-ləns) *also* **flat·u·len·cy** (-lən-sē) *n.* **1.** Excessive gas in the digestive tract. **2.** Pomposity : self-importance.

**flat·u·lent** (flăch′ə-lənt) *adj.* [OFr. < Lat. *flatus*, fart.] **1.** Of, relating to, or afflicted with flatulence. **2.** Producing or tending to produce flatulence. **3.** Pompous : pretentious. **—flat′u·lent·ly** *adv.*

**fla·tus** (flā′təs) *n.* [Lat., fart < p.part. of *flare*, to blow.] Gas generated in the stomach or intestines.

**flat·ware** (flăt′wâr′) *n.* **1.** Tableware that is fairly flat and fashioned usu. of a single piece, as plates. **2.** Table utensils, as knives, forks, and spoons.

**flat·wise** (flăt′wīz′) *also* **flat·ways** (-wāz′) *adv.* With the flat side down or in contact with a surface.

**flat·work** (flăt′wûrk′) *n.* Laundry that can be ironed on a mangle.

**flat·worm** (flăt′wûrm′) *n.* A platyhelminth.

**flaunt** (flônt) *v.* **flaunt·ed, flaunt·ing, flaunts.** [Orig. unknown.] —*vt.* **1.** To display ostentatiously : SHOW OFF <*flaunt* one's riches> **2.** *Nonstandard.* To flout. ***usage:*** *Flaunt* and *flout* are often confused. As a transitive verb, *flaunt* means "to display ostentatiously" (*flaunted their diamonds*); *flout* means "to show open contempt or scorn for" (*They flouted the conventions of their society*).

—*vi.* **1.** To show oneself off. **2.** To wave grandly <streamers *flaunting* in the wind> **—flaunt** *n.* **—flaunt′er** *n.* **—flaunt′ing·ly** *adv.*

**flaunt·y** (flôn′tē) *adj.* **-i·er, -i·est.** Inclined to flaunt : OSTENTATIOUS. **—flaunt′i·ly** *adv.* **—flaunt′i·ness** *n.*

**flau·tist** (flô′tĭst, flou′-) *n.* [Ital. *flautista* < *flauto*, flute < OProv. *flaut.*] A flutist.

**flav-** *pref. var. of* FLAVO-.

**fla·va·none** (flā′və-nōn′) *n.* A colorless crystalline compound, $C_{15}H_{12}O_2$, derived from flavone.

**fla·ves·cent** (flə-vĕs′ənt) *adj.* [Lat. *flavescens, flavescent-*, p.part. of *flavescere*, to turn yellow, inchoative of *flavere*, to be yellow < *flavus*, yellow.] Turning yellow.

**fla·vin** (flā′vĭn) *n.* **1.** Any of various water-soluble yellow pigments, including riboflavin, found in plant and animal tissue as coenzymes of flavoprotein. **2.** A compound, $C_{10}H_6N_4O_2$, the nucleus of various natural yellow pigments.

**flavin adenine di·nu·cle·o·tide** (dī-nōō′klē-ə-tīd′, -nyōō′-) *n.* A coenzyme, $C_{27}H_{33}N_9O_{15}P_2$, that contains riboflavin and acts as a hydrogen carrier in certain oxidative systems of the body.

**fla·vine** (flā′vēn′) *n.* **1.** A brownish-red crystalline powder, $C_{14}H_{15}N_3Cl_2$, used as an antiseptic. **2.** Flavin.

**flavo-** *or* **flav-** *pref.* [Lat. *flavus*, yellow.] **1.** Yellow <*flavin*> **2.** Flavin <*flavo*protein>

**fla·vone** (flā′vōn′) *n.* A crystalline compound, $C_{15}H_{10}O_2$, the parent substance of a number of important yellow pigments.

**fla·vo·noid** (flā′və-noid′) *n.* Any of a large group of plant substances that includes the anthocyanins, a class of flower pigments.

**fla·vo·pro·tein** (flā′vō-prō′tēn′, -tē-ĭn) *n.* Any of a class of enzymes containing protein-bound flavin and acting as dehydrogenation catalysts in biological reactions.

**fla·vor** (flā′vər) *n.* [ME *flavour*, aroma < OFr. *flaor* < VLat. **flator* < Lat. *flare*, to blow.] **1.** Distinctive taste <a *flavor* of garlic in the casserole> **2.** A distinctive yet intangible quality felt to be characteristic of a given thing <the *flavor* of the Middle East> **3.** A flavoring <artificial *flavors*> **4.** *Archaic.* Aroma : fragrance. —*vt.* **-vored, -vor·ing, -vors.** To give flavor to. **—fla′vor·er** *n.* **—fla′vor·less** *adj.* **—fla′vor·ous** (-əs), **fla′vor·some** (-səm) *adj.*

**fla·vor·ful** (flā′vər-fəl) *adj.* Full of flavor. **—fla′vor·ful·ly** *adv.*

**fla·vor·ing** (flā′vər-ĭng) *n.* A substance, as an extract or spice, that imparts flavor.

**fla·vour** (flā′vər) *n.* & *v. Chiefly Brit. var. of* FLAVOR.

**flaw**[1] (flô) *n.* [ME *flaue*, splinter, perh. < ON *flaga*, slab of stone.] **1.** A physical, often concealed imperfection <a structural *flaw*> **2.** A defect in something intangible <a *flaw* in one's character> **3.** A defect in a legal document that can render it invalid. —*vt.* & *vi.* **flawed, flaw·ing, flaws.** To make or become defective. **—flaw′less** *adj.* **—flaw′less·ly** *adv.* **—flaw′less·ness** *n.*

**flaw**[2] (flô) *n.* [Prob. of Scand. orig.] **1. a.** A brief gust of wind. **b.** A passing storm : SQUALL. **2.** *Obs.* A burst of passion. **—flaw′y** *adj.*

**flax** (flăks) *n.* [ME < OE *fleax.*] **1. a.** A plant of the genus *Linum*, esp. a widely cultivated species, *L. usitatissimum*, having blue flowers, seeds that yield linseed oil, and slender stems from which a fine, light-colored textile fiber is obtained. **b.** The textile fiber obtained from flax. **2.** A plant resembling flax. **3.** A grayish yellow.

**flax·en** (flăk′sən) *adj.* **1.** Made of or like flax. **2.** Having the pale-yellow color of flax fiber.

**flax·seed** (flăks′sēd′) *n.* The seed of flax.

**flax·y** (flăk′sē) *adj.* **-i·er, -i·est.** Similar to flax, as in texture.

**flay** (flā) *vt.* **flayed, flay·ing, flays.** [ME *flen* < OE *flēan.*] **1.** To strip off the skin of : DECORTICATE. **2.** To strip of money or goods : PLUNDER. **3.** To criticize harshly : EXCORIATE. **—flay′er** *n.*

**F layer** *n.* **1.** The highest zone of the ionosphere, extending continuously at night from 120 to 250 miles or approx. 190 to 400 kilometers. **2.** Either of two layers into which the F layer is divided during the day, esp. in summer, usu. designated $F_1$ and $F_2$ and extending respectively from 90 to 150 miles or approx. 145 to 240 kilometers and from 150 miles or 240 kilometers upward.

**flea** (flē) *n.* [ME *fle* < OE *flēah.*] **1.** Any of various small, wingless, bloodsucking insects of the order Siphonaptera, whose legs are adapted for jumping and that are parasitic on warm-blooded animals. **2.** Any of various small crustaceans resembling or moving like fleas.

**flea·bag** (flē′băg′) *n.* An inferior lodging place.

**flea·bane** (flē′bān′) *n.* A plant of the genus *Erigeron*, with variously colored, many-rayed daisylike flowers.

**flea beetle** *n.* Any of a group of small beetles of the family Chrysomelidae whose legs are adapted for jumping.

**flea·bite** (flē′bīt′) *n.* **1. a.** The bite of a flea. **b.** The small red lesion caused by a flea's bite. **2.** A trifling annoyance or loss.

**flea-bit·ten** (flē′bĭt′n) *adj.* **1.** Covered with fleas or fleabites. **2.** *Informal.* Shabby <a *flea-bitten* hotel> **3.** Having a pale coat with reddish-brown flecks. —Used of horses.

**flea collar** *n.* A collar, as for a cat or dog, containing an agent that kills fleas.

**flea market** *n.* A market where antiques, used household goods, and curios are sold.

**flèche** (flĕsh, flāsh) *n.* [Fr. < OFr., arrow.] A slender spire, esp. one on a church above the intersection of the nave and transepts.

flèche

**flé·chette** (flā-shĕt′, flĕ-) *n.* [Fr., dim of *flèche*, arrow.] A steel missile or dart dropped from an aircraft.

**fleck** (flĕk) *n.* [Prob. < ME *flekked*, spotted < ON *flekkr.*] **1.** A tiny spot or mark. **2.** A small flake or bit. —*vt.* **flecked, fleck·ing, flecks.** To spot or streak : SPECKLE.

**flec·tion** (flĕk′shən) *n. var. of* FLEXION 1.

**fled** (flĕd) *v. p.t.* & *p.p. of* FLEE.

**fledge** (flĕj) *v.* **fledged, fledg·ing, fledg·es.** [Prob. < obs. *fledge*, feathered < ME *flegge* < OE *flycge.*] —*vt.* **1.** To take care of (a young bird) until it is ready to fly. **2.** To cover with or as if with feathers. **3.** To equip (an arrow) with feathers. —*vi.* To grow the plumage necessary for flight.

**fledg·ling** *also* **fledge·ling** (flĕj′lĭng) *n.* **1.** A young bird that has recently acquired its flight feathers. **2.** A person who is young or inexperienced.

**flee** (flē) *v.* **fled** (flĕd), **flee·ing, flees.** [ME *flen* < OE *flēon.*] —*vt.* **1.** To run away. **2.** To pass swiftly away : VANISH. —*vt.* To run away from <*flee* an accident scene> **—fle′er** *n.*

**fleece** (flēs) *n.* [ME *fles* < OE *flēos.*] **1. a.** The coat of wool, as of a sheep. **b.** The yield of wool shorn from a sheep at one time. **2.** A soft, woolly covering or mass. **3.** Fabric with a soft, deep pile. —*vt.* **fleeced, fleec·ing, fleec·es. 1.** To shear the fleece from. **2.** To defraud of money or property : SWINDLE. **3.** To cover with or as if with fleece. **—fleec′er** *n.*

**fleec·y** (flē′sē) *adj.* **-i·er, -i·est.** Of, like, or covered with fleece <*fleecy* white clouds> **—fleec′i·ly** *adv.* **—fleec′i·ness** *n.*

**fleer** (flîr) *vi.* **fleered, fleer·ing, fleers.** [ME *flerien*, of Scand. orig.] To smirk or laugh contemptuously or derisively. —*n.* A contemptuous or derisive look or act. **—fleer′ing·ly** *adv.*

**fleet**[1] (flēt) *n.* [ME *flete* < OE *flēot* < *flēotan*, to float.] **1.** A number of warships operating together under one command. **2.** A group of vehicles, as taxicabs or fishing boats, owned or operated as a unit.

**fleet**[2] (flēt) *adj.* **-er, -est.** [Poss. < ME *fleten*, to drift < OE *flēotan.*] **1.** Moving swiftly. **2.** *Archaic.* Fleeting. —*v.* **fleet·ed, fleet·ing, fleets.** —*vi.* **1.** To move or pass swiftly. **2.** To fade out : VANISH. **3.** *Obs.* To drift. **4.** *Obs.* To flow. —*vt.* **1.** To cause (time) to pass quickly. **2.** *Naut.* To alter the position of : SHIFT. **—fleet′ly** *adv.* **—fleet′ness** *n.*

**Fleet Admiral** *n.* An Admiral of the Fleet.

**fleet·ing** (flē′tĭng) *adj.* Passing quickly : EPHEMERAL. **—fleet′ing·ly** *adv.* **—fleet′ing·ness** *n.*

**Fleet Street** *n.* [After *Fleet Street*, London, England, the center of British newspaper journalism.] London journalism.

**flei·shig** (flā′shĭk) *adj.* [Yiddish < MHG *vleischic*, meaty < *vleisch*, meat < OHG *fleisk*, flesh.] Composed of, prepared with, or relating to meat or meat products.

**Flem·ing** (flĕm′ĭng) *n.* [ME < MDu. *Vlāming.*] **1.** A native or resident of Flanders. **2.** A Flemish-speaking Belgian.

**Flem·ish** (flĕm′ĭsh) *adj.* Of or relating to Flanders, the Flemings, or their language. —*n.* **1.** The Low German language of the Flemings. **2.** The Flemings.

**flense** (flĕns) *vt.* **flensed, flens·ing, flens·es.** [Dan.] To strip the blubber or skin from (e.g., a whale). **—flens′er** *n.*

**flesh** (flĕsh) *n.* [ME < OE *flǣsc.*] **1. a.** Soft bodily tissue, esp. skeletal muscle distinguished from bone and viscera. **b.** Excess fat : AVOIRDUPOIS. **c.** The skin of the human body. **2.** The meat of animals as distinguished from the edible tissue of fish or fowl. **3.** The pulpy part of a fruit or vegetable. **4. a.** The human body itself. **b.** One's physical or carnal nature. **5.** Humanity in general. **6.** One's family : KIN. —*v.* **fleshed, flesh·ing, flesh·es.** —*vt.* **1.** To encourage (a hunting dog falcon) to take part in a chase by feeding it flesh from a kill. **2.** To fill out (e.g., a structure) <*fleshed* out the short story with a subplot> **3.** To clean (a hide) of adhering flesh. —*vi.* To gain weight. **—in the flesh.** Alive and in person.

**flesh and blood** *n.* **1.** Human nature or physical existence, together with its weaknesses. **2.** FLESH 6.

**flesh fly** *n.* A fly of the genus *Sarcophaga*, whose larva is parasitic in animal tissue.

**flesh·ly** (flĕsh′lē) *adj.* **-li·er, -li·est. 1.** Of or relating to the body : CORPOREAL. **2.** Of, relating to, or inclined to carnality : SENSUAL.

3. Not spiritual : WORLDLY. **4.** Tending to plumpness : FLESHY. **—flesh'li·ness** *n.*

**flesh·pot** (flĕsh'pŏt') *n.* **1.** A place in which physical comfort or gratification is obtained. **2.** Physical well-being and gratification.

**flesh wound** *n.* A wound penetrating the flesh but not damaging bones or vital organs.

**flesh·y** (flĕsh'ē) *adj.* **-i·er, -i·est. 1. a.** Relating to, composed of, or like flesh. **b.** Having much flesh : CORPULENT. **2.** Juicy or pulpy in texture. **—flesh'i·ness** *n.*

**fleshy fruit** *n.* A fruit, as a drupe, whose pericarp is soft and pulpy.

**fletch** (flĕch) *vt.* **fletched, fletch·ing, fletch·es.** [Prob. back-formation < FLETCHER.] To feather (an arrow).

**fletch·er** (flĕch'ər) *n.* [ME *fleccher* < OFr. *flechier* < *flèche,* arrow.] A maker of arrows.

**fleur-de-lis** *or* **fleur-de-lys** (flûr'də-lē', flŏŏr'-) *n., pl.* **fleurs-de-lis** *or* **fleurs-de-lys** (flûr'də-lēz', flŏŏr'-) [ME *flour de lice* < OFr. *flor de lis,* flower of the lily.] **1.** An iris, esp. a white-flowered form of *Iris germanica.* **2.** A heraldic device composed of a stylized three-petaled iris flower, once used as the armorial emblem of French sovereigns.

**flew** (flo͞o) *v. p.t. of* FLY[1].

**flews** (flo͞oz) *pl.n.* [Orig. unknown.] The pendulous corners of the upper lip of some dogs, as bloodhounds.

**flex** (flĕks) *v.* **flexed, flex·ing, flex·es.** [Lat. *flectere, flex-,* to bend.] *—vt.* **1. a.** To bend repeatedly. **b.** To bend (a joint). **c.** To bend (a joint) repeatedly. **2.** To contract (a muscle). *—vi.* To bend <hands *flexing* nervously> *—n. Chiefly Brit.* An electric cord.

**flex·a·gon** (flĕk'sə-gŏn') *n.* [FLEX + *-agon,* as in *pentagon.*] A folded paper construction capable of being flexed along its folds to alternately reveal and conceal its faces.

**flex·i·ble** (flĕk'sə-bəl) *adj.* **1.** Capable of being bent or flexed : PLIABLE. **2.** Susceptible to influence or persuasion : TRACTABLE. **3.** Responsive to change : ADAPTABLE <a *flexible* work schedule> **—flex'i·bil'i·ty, flex'i·ble·ness** *n.* **—flex'i·bly** *adv.*

☆ ***syns:*** **1.** FLEXIBLE, ELASTIC, PLASTIC, RESILIENT, SPRINGY, SUPPLE *adj. core meaning* : capable of withstanding stress without structural injury <a *flexible* girder> ***ant:*** inflexible, rigid. **2.** FLEXIBLE, ELASTIC, MALLEABLE, PLASTIC, PLIABLE, PLIANT *adj. core meaning* : easily influenced <a *flexible* mind><took a *flexible* approach to the issue> ***ant:*** inflexible, rigid

**flex·ile** (flĕk'səl, -sīl') *adj.* Flexible.

**flex·ion** (flĕk'shən) *n.* [Lat. *flexio,* a bending < *flectere,* to bend.] **1.** *also* **flec·tion.** *Anat.* The act of bending or state of being bent. **2.** A part that is bent. **—flex'ion·al** *adj.*

**flex·or** (flĕk'sər) *n.* [NLat. < Lat. *flexus,* p.part. of *flectere,* to bend.] A muscle that flexes a joint.

**flex·time** (flĕks'tīm') *n.* [FLEX(IBLE) + TIME.] An arrangement by which employees may set their own work schedules.

**flex·u·ous** (flĕk'sho͞o-əs) *adj.* [Lat. *flexuosus* < *flexus,* p.part. of *flectere,* to bend.] Bending or winding alternately from side to side. **—flex'u·ous·ly** *adv.*

**flex·ure** (flĕk'shər) *n.* **1.** A bend, curve, or turn. **2.** An act or instance of bending. **—flex'ur·al** *adj.*

**fley** (flā) *vt.* **fleyed, fley·ing, fleys.** [ME *fleien* < OE *flēgan.*] *Scot.* To frighten.

**flib·ber·ti·gib·bet** (flĭb'ər-tē-jĭb'ĭt) *n.* [ME *flibergebet.*] A silly, scatterbrained person.

**flick[1]** (flĭk) *n.* [Imit.] **1. a.** A light, quick blow, jerk, or touch. **b.** The sound accompanying a flick. **2.** A light splash, dash, or daub. *—v.* **flicked, flick·ing, flicks.** *—vt.* **1.** To touch or hit with a light, quick blow. **2.** To cause to move with a light blow <*flicked* the light switch off> **3.** To remove with a light, quick blow <*flicked* lint off the coat> *—vi.* To flutter or twitch.

**flick[2]** (flĭk) *n.* [Back-formation < FLICKER[1].] *Informal.* A movie.

**flick·er[1]** (flĭk'ər) *v.* **-ered, -er·ing, -ers.** [ME *flikeren* < OE *flicerian.*] *—vi.* **1.** To move waveringly : FLUTTER <shadows *flickering* on the wall> **2.** To burn unsteadily : GUTTER. *—vt.* To cause to flicker. *—n.* **1.** A brief movement : TREMOR. **2.** A wavering light. **3.** A brief or slight sensation. **4.** *Slang.* A movie.

**flick·er[2]** (flĭk'ər) *n.* [Perh. < FLICK[1].] A large North American woodpecker of the genus *Colaptes,* with a brownish back and a spotted breast.

**flick·er·tail** (flĭk'ər-tāl') *n.* A ground squirrel, *Citellus richardsoni* of western North America.

**flied** (flīd) *v. p.t. & p.p. of* FLY[1] 7.

**fli·er** *also* **fly·er** (flī'ər) *n.* **1.** One that flies, esp. an aircraft pilot. **2.** A step in a straight stairway. **3.** *Informal.* A daring venture. **4.** A circular or pamphlet for mass distribution.

**flight[1]** (flīt) *n.* [ME < OE *flyht.*] **1. a.** The motion of an object in or through a medium, esp. through the earth's atmosphere or through space. **b.** An instance of such motion. **c.** The distance covered in such motion. **2. a.** The act or process of flying through the air by means of wings. **b.** The ability to fly. **3.** A swift passage or movement. **4.** A scheduled airline run or trip. **5.** A group flying together <a *flight* of eagles><a *flight* of bombers> **6.** A number of aircraft in the U.S. Air Force forming a subdivision of a squadron. **7.** A brilliant, extraordinary effort or display. **8.** A series of stairs rising from one landing to another. *—vi.* **flight·ed, flight·ing, flights.** To fly or migrate in flocks. —Used of birds.

**flight[2]** (flīt) *n.* [ME < OE *flyht.*] An act or an instance of fleeing <unlawful *flight* to escape arrest>

**flight attendant** *n.* One employed to assist passengers in an aircraft.

**flight bag** *n.* A lightweight, flexible piece of luggage with zippered outside pockets.

**flight deck** *n.* **1.** The upper deck of an aircraft carrier, used as a runway. **2.** An elevated compartment in certain aircraft, used by the pilot, copilot, and flight engineer.

**flight engineer** *n.* The crew member responsible for the mechanical performance of an aircraft in flight.

**flight feather** *n.* One of the rather large, stiff feathers of a bird's wing or tail required for flight.

**flight·less** (flīt'lĭs) *adj.* Incapable of flying <*flightless* birds>

**flight surgeon** *n.* A specialist in aviation medicine.

**flight-test** (flīt'tĕst') *vt.* **-test·ed, -test·ing, -tests.** To test (e.g., an aircraft) during flight.

**flight·y** (flī'tē) *adj.* **-i·er, -i·est. 1. a.** Given to capricious or impulsive behavior. **b.** Marked by irresponsible behavior. **2.** Easily excited : SKITTISH. **—flight'i·ly** *adv.* **—flight'i·ness** *n.*

**flim·flam** (flĭm'flăm') [Prob. of Scand. orig.] *Informal. —n.* **1.** Nonsense. **2.** A swindle. *—vt.* **-flammed, -flam·ming, -flams.** To swindle.

**flim·sy** (flĭm'zē) *adj.* **-si·er, -si·est.** [Orig. unknown.] **1.** Light, thin, and insubstantial <*flimsy* lace> **2.** Lacking solidity or strength <a *flimsy* chair> **3.** Lacking plausibility <a *flimsy* alibi> *—n., pl.* **-sies. 1.** Thin paper for making multiple copies. **2.** Something written on flimsy. **—flim'si·ly** *adv.* **—flim'si·ness** *n.*

**flinch** (flĭnch) *vi.* **flinched, flinch·ing, flinch·es.** [OFr. *flenchir.*] **1.** To wince involuntarily, as from surprise. **2.** To draw back or away : RETREAT. **—flinch** *n.* **—flinch'er** *n.* **—flinch'ing·ly** *adv.*

**flin·ders** (flĭn'dərz) *pl.n.* [ME *flendris,* perh. of Scand. orig.] Fragments or splinters.

**fling** (flĭng) *v.* **flung** (flŭng), **fling·ing, flings.** [ME *flingen,* of Scand. orig.] *—vt.* **1.** To throw violently <*flung* the dish against the wall> **2.** To put or send suddenly <troops *flung* into battle> **3.** To throw (oneself) into an activity with abandon and energy. **4.** To toss aside : DISCARD <*fling* caution to the winds> *—vi.* To move quickly, violently, or impulsively. *—n.* **1.** An act of flinging. **2.** A brief period of indulging one's impulses : SPREE. **3.** *Informal.* A usu. brief attempt or effort <take a *fling* at flying a plane>

**flink·ite** (flĭng'kīt') *n.* [After Gustaf *Flink* (d. 1931).] A brownish-green mineral of manganese arsenate.

**flint** (flĭnt) *n.* [ME < OE.] **1.** A hard, fine-grained quartz that sparks when struck with steel. **2.** A piece of flint used as a tool by primitive human beings. **3.** A small solid cylinder of a spark-producing alloy, used in lighters to ignite the fuel. **4.** Something resembling flint.

**flint corn** *n.* A variety of corn, *Zea mays indurata,* with small, hard seeds.

**flint disease** *n.* Pneumonoconiosis caused by prolonged inhalation of stone dust.

**flint glass** *n.* A soft, fusible, lustrous, brilliant lead-oxide optical glass with high refraction and low dispersion.

**flint·head** (flĭnt'hĕd') *n.* The wood ibis.

**flint·lock** (flĭnt'lŏk') *n.* **1.** An obsolete gunlock in which a flint embedded in the hammer produces a spark that ignites the charge. **2.** A firearm equipped with a flintlock.

**flint·y** (flĭn'tē) *adj.* **-i·er, -i·est. 1.** Containing, composed of, or like flint. **2.** Stern <a *flinty* glare> **—flint'i·ly** *adv.* **—flint'i·ness** *n.*

**flip** (flĭp) *v.* **flipped, flip·ping, flips.** [Perh. imit.] *—vt.* **1.** To throw briskly <*flipped* me the ball> **2.** To toss in the air, imparting a spin <*flip* a coin> **3.** To strike or move with a light, quick blow : FLICK. **4.** To reverse or overturn quickly and effortlessly. *—vi.* **1.** To strike quickly or lightly, as with a snap of the fingers. **2.** To move suddenly or jerkily. **3.** To turn a somersault in the air. **4.** To turn or skim quickly <*flip* through the magazine> **5.** *Slang.* **a.** To go crazy. **b.** To react strongly and esp. enthusiastically <I *flipped* over the new house.> *—n.* **1.** An act of flipping, esp.: **a.** A fillip : tap. **b.** A quick, jerky movement. **c.** A somersault. **2.** A mixed drink made with an alcoholic beverage and often including beaten eggs. *—adj. Informal.* **1.** Disrespectful. **2.** Unconcerned : indifferent.

**flip chart** *n.* A chart with sheets hinged at the top that can be flipped over to present information sequentially.

**flip-flop** (flĭp'flŏp') *n.* **1.** The movement or sound of repeated flapping <the *flip-flop* of sandals on a stone floor> **2.** A backward somersault. **3.** *Informal.* A reversal, as of opinion. **4.** *Electron.* An electronic circuit or mechanical device capable of assuming either of two stable states and used in computers to store a single bit of information. **—flip'-flop'** *v.* **(-flopped, -flop·ping, -flops).**

**flip·pant** (flĭp'ənt) *adj.* [Prob. < FLIP.] **1.** Marked by disrespectful levity or indifference. **2.** *Archaic.* Talkative : voluble. **—flip'pan·cy** *n.* **—flip'pant·ly** *adv.*

**flip·per** (flĭp′ər) *n.* **1.** One that flips. **2.** A wide, flat limb, as of a seal, adapted esp. for swimming. **3.** A rubber foot covering with a flat, flexible portion that widens as it extends forward from the toes, used in swimming.
**flip side** *n.* The reverse side, as of a phonograph record.
**flirt** (flûrt) *v.* **flirt·ed, flirt·ing, flirts.** [Orig. unknown.] —*vi.* **1.** To make playfully romantic or sexual overtures. **2.** To act so as to attract or provoke <*flirting* with death in the bullring> **3.** To move abruptly or jerkily. —*vt.* **1.** To toss or flip suddenly. **2.** To move quickly. —*n.* **1.** One given to flirting. **2.** An abrupt, jerky movement.
**flir·ta·tion** (flûr-tā′shən) *n.* **1.** The practice of flirting. **2.** A superficial, usu. temporary romance. **3.** A brief involvement.
**flir·ta·tious** (flûr-tā′shəs) *adj.* **1.** Tending to flirt. **2.** Full of playful allure. **—flir·ta′tious·ly** *adv.* **—flir·ta′tious·ness** *n.*
**flit** (flĭt) *vi.* **flit·ted, flit·ting, flits.** [ME *flitten* < ON *flytja.*] **1.** To move about rapidly and nimbly. **2.** To move quickly from one state or location to another <*flitted* from one project to another> —*n.* A fluttering or darting movement. **—flit′ter** *n.*
**flitch** (flĭch) *n.* [ME *flicche* < OE *flicce.*] **1.** A salted and cured side of bacon. **2.** A longitudinal cut from a tree trunk. **3.** One of several planks secured together to form a single beam.
**flit·ter** (flĭt′ər) *vi.* **-tered, -ter·ing, -ters.** [Freq. of FLIT.] To flutter. **—flit′ter·y** *adj.*
**fliv·ver** (flĭv′ər) *n.* [Orig. unknown.] An old or cheap car.
**float** (flōt) *v.* **float·ed, float·ing, floats.** [ME *floten* < OE *flotian.*] —*vi.* **1. a.** To remain suspended within or on the surface of a fluid without sinking. **b.** To be suspended in or move through space as if supported by a liquid. **2.** To move from position to position, esp. at random. **3.** To find a level in relationship to other currencies solely in response to the law of supply and demand <allowed the dollar to *float*> —*vt.* **1.** To cause to remain suspended without sinking or falling. **2.** To flood (land), as for irrigation. **3.** To launch or establish (e.g., a business venture). **4.** To release (a security) for sale. **5.** To arrange for (a loan). **6.** To make the surface of (e.g., plaster) level or smooth. **7.** To allow (the exchange value of a currency) to find freely its real level in relationship to other currencies. **8.** *Computer Sci.* To convert data from fixed-point notation to floating-point notation. —*n.* **1.** Something that floats, as: **a.** A raft. **b.** A buoy. **c.** A life preserver. **d.** A floating object, as a cork, on a fishing line. **e.** A landing platform attached to a wharf and floating on the water. **f.** A hollow ball attached to a lever to regulate the water level in a tank. **g.** An air- or gas-filled organ or sac enabling an organism to remain suspended in water. **2.** A large, flat vehicle bearing an exhibit in a parade. **3.** A sum of money representing outstanding checks. **4.** A tool for smoothing the surface of plaster or cement. **5.** A soft drink with ice cream floating in it. **—float′a·ble** *adj.*
**float·age** (flō′tĭj) *n. var. of* FLOTAGE.
**floa·ta·tion** (flō-tā′shən) *n. var. of* FLOTATION.
**float·er** (flō′tər) *n.* **1.** One that floats. **2.** One who wanders : DRIFTER. **3.** An employee reassigned from job to job or shift to shift within an operation. **4.** One who votes illegally in different polling places. **5.** An insurance policy protecting movable property that is in transit or is regularly subject to use in a variety of places.
**float·ing** (flō′tĭng) *adj.* **1.** Buoyed on or suspended in or as if in a fluid. **2.** Not secured in place. **3.** Tending to move about. **4. a.** Being in circulation. —Used of capital. **b.** Short-term and usu. unfunded. —Used of a debt. **5.** Designed or constructed to operate smoothly and without vibration. **6.** Designating a bodily organ that is out of normal position <a *floating* kidney>
**floating dock** *n.* A structure that can be submerged to permit the entry and docking of a ship and then raised to lift the ship from the water for repairs.
**floating island** *n.* A dessert of soft custard with mounds of beaten egg whites or whipped cream floating on its surface.
**float·ing-point** (flō′tĭng-point′) *adj.* Of or relating to a method of writing numeric quantities with a mantissa representing the value of the digits and a characteristic indicating the power of the number base.
**floating rib** *n.* One of the four lower ribs that, unlike the other ribs, are not attached at the front.

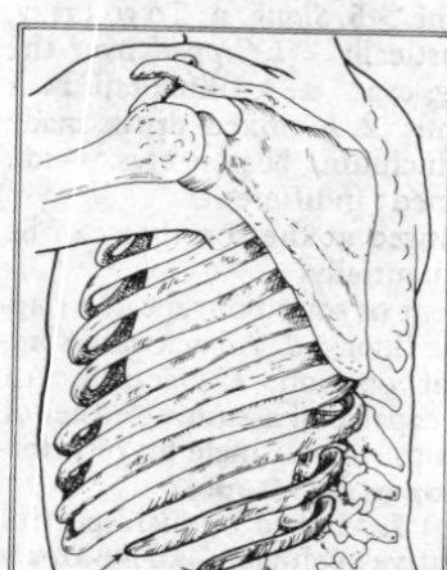
floating rib

**floc** (flŏk) *n.* [Short for FLOCCULUS.] A flocculent mass formed in certain serologic precipitin tests.
**floc·cu·late** (flŏk′yə-lāt′) *v.* **-lat·ed, -lat·ing, -lates.** —*vt.* **1.** To cause (soil) to form lumps or masses. **2.** To cause (clouds) to form fluffy masses. —*vi.* To form lumpy or fluffy masses. **—floc′cu·late′** *n.* **—floc′cu·la′tion** *n.*
**floc·cule** (flŏk′yo͞ol) *n.* [NLat. *flocculus,* flocculus.] A small, loosely held mass or aggregate of fine particles suspended in or precipitated from a solution.
**floc·cu·lent** (flŏk′yə-lənt) *adj.* **1.** Having a fluffy or woolly appearance. **2.** Composed of or containing woolly masses. **3.** Flaky, waxy, and woollike, as the secretion covering some insects. **—floc′cu·lence** *n.* **—floc′cu·lent·ly** *adv.*
**floc·cu·lus** (flŏk′yə-ləs) *n., pl.* **-li** (-lī′) [NLat., dim. of Lat. *floccus,* tuft of wool.] **1.** A small, fluffy mass or tuft. **2.** *Anat.* Either of two small lobes on the lower posterior border of each lobe of the cerebellum. **3.** *Astron.* Any of various masses of gases appearing as bright or dark patches on the sun's surface.
**flock**[1] (flŏk) *n.* [ME *flok* < OE *floc.*] **1.** A group of animals living, traveling, or feeding together. **2.** A group of people under the leadership of a single person, esp. the members of a church. **3.** A large number <a *flock* of queries> —*vi.* **flocked, flock·ing, flocks.** To congregate or travel in or as if in a flock.
**flock**[2] (flŏk) *n.* [ME *flok* < OFr. *floc* < Lat. *floccus.*] **1.** A tuft, as of fiber or hair. **2.** Waste wool or cotton for stuffing furniture and mattresses. **3.** An inferior grade of wool added to cloth for extra weight. **4.** Pulverized wool or felt applied to paper, cloth, or metal to produce a texture or pattern. **5.** A floccule. —*vt.* **flocked, flock·ing, flocks. 1.** To stuff with flock. **2.** To texture or pattern with flock.
**floe** (flō) *n.* [Prob. < Norw. *flo,* layer < ON *flō.*] **1.** A large, flat mass of ice formed on the surface of a body of water. **2.** A segment separated from an ice floe.
**flog** (flŏg, flôg) *vt.* **flogged, flog·ging, flogs.** [Perh. < Lat. *flagellare.* —see FLAGELLATE.] To beat severely with a whip or rod. **—flog′ger** *n.*
**flood** (flŭd) *n.* [ME *flod* < OE *flōd.*] **1.** An overflowing of water onto normally dry land. **2.** Flood tide. **3.** An abundant flow or outpouring <a *flood* of speculation> **4.** A floodlight. **5. Flood.** The universal deluge recorded in the Old Testament as having occurred during the life of Noah. —*v.* **flood·ed, flood·ing, floods.** —*vt.* **1.** To cover or submerge with or as if with a flood : INUNDATE. **2.** To fill with an abundance or an excess <*flood* the market with computers> —*vi.* **1.** To become inundated or submerged. **2.** To pour forth.
**flood·gate** (flŭd′gāt′) *n.* **1.** A gate for controlling the flow of a body of water. **2.** Something that restrains a flood or outpouring.
**flood·light** (flŭd′līt′) *n.* **1.** Artificial light in an intensely bright and broad beam. **2.** A unit producing a beam of intense light. —*vt.* **-light·ed** *or* **-lit** (-lĭt′), **-light·ing, -lights.** To illuminate with a floodlight.
**flood plain** *n.* A plain bordering a river subject to flooding.
**flood tide** *n.* **1.** The incoming tide. **2.** A high point : CLIMAX.
**floor** (flôr, flōr) *n.* [ME *flor* < OE *flōr.*] **1.** The surface of a room on which one stands. **2.** The lower or supporting surface of a structure. **3.** The surface of a structure on which vehicles travel. **4.** The ground or lowermost surface, as of a forest or ocean. **5.** The lower part of a room or hall, as a legislative chamber, where business is conducted. **6. a.** The right to address an assembly, as granted under parliamentary procedure. **b.** The body of assembly members <a motion from the *floor*> **7. a.** A story or level of a building. **b.** The occupants of such a story. **8.** A lower limit or base. —*vt.* **floored, floor·ing, floors. 1.** To provide with a floor. **2.** To knock down. **3.** To floorboard. **4.** To stun. **—floor′er** *n.*
**floor·age** (flôr′ĭj, flōr′-) *n.* Floor space.
**floor·board** (flôr′bôrd′, flōr′bōrd′) *n.* **1.** A board in a floor. **2.** The floor of a motor vehicle. —*vt.* **-board·ed, -board·ing, -boards.** To depress (the accelerator of a motor vehicle) to the floor.
**floor exercise** *n.* A competitive gymnastics event consisting of tumbling maneuvers performed on a mat.
**floor·ing** (flôr′ĭng, flōr′-) *n.* **1. a.** A floor. **b.** Floors in general. **2.** Material, as lumber or tile, used in making floors.
**floor lamp** *n.* A tall lamp with a base standing on the floor.
**floor leader** *n.* The member of a legislature chosen by fellow party members to be in charge of the party's activities on the floor.
**floor manager** *n.* **1.** A floorwalker. **2.** One in charge of directing something, as activities at a political convention, from the floor.
**floor plan** *n.* A scale diagram of a room or building drawn as if seen from above.
**floor sample** *n.* Merchandise sold at a reduced price because it has been a display or demonstration model.
**floor show** *n.* A series of nightclub entertainments.
**floor·walk·er** (flôr′wô′kər, flōr′-) *n.* A department store employee who supervises salespeople and assists customers.
**floo·zy** *also* **floo·zie** (flo͞o′zē) *n., pl.* **-zies.** [Orig. unknown.] *Slang.* A sleazy or vulgar woman, esp. a prostitute.

**flop** (flŏp) *v.* **flopped, flop·ping, flops.** [Alteration of FLAP.] —*vi.* **1.** To fall down heavily and noisily. **2.** To move about in a clumsy or relaxed way. **3.** *Informal.* To fail utterly. **4.** *Slang.* To go to bed. —*vt.* To cause to fall down heavily and noisily. —*n.* **1.** An act of flopping. **2.** The sound of flopping. **3.** *Informal.* An utter failure. **—flop'per** *n.*

**flop·house** (flŏp'hous') *n.* A cheap hotel.

**flop·py** (flŏp'ē) *adj.* **-pi·er, -pi·est.** Loose and flexible <*floppy* hats><*floppy* sandals> **—flop'pi·ly** *adv.* **—flop'pi·ness** *n.*

**floppy disk** *n. Computer Sci.* A flexible plastic disk coated with magnetic material used to store computer data.

**flo·ra** (flôr'ə, flōr'ə) *n., pl.* **-ras** *or* **flo·rae** (flôr'ē', flōr'ē') [< FLORA.] **1.** Plants as a whole, esp. those of a specific region or season. **2.** A treatise describing the plants of a region or season.

**Flo·ra** (flôr'ə, flōr'ə) *n.* [Lat. *Flora* < *flos*, flower.] *Rom. Myth.* The goddess of flowers.

**flo·rae** (flôr'ē', flōr'ē') *n. var. pl. of* FLORA.

**flo·ral** (flôr'əl, flōr'-) *adj.* Of, relating to, or like a flower or flora. **—flo'ral·ly** *adv.*

**floral envelope** *n.* The perianth of a flower.

**floral tube** *n.* A tube in some plants formed by the basal fusion of the sepals, petals, and stamens.

**flo·re·at·ed** (flôr'ē-ā'tĭd, flōr'-) *adj. var. of* FLORIATED.

**Flor·en·tine** (flôr'ən-tēn', -tīn', flŏr'-) *adj.* [Lat. *Florentinus* < *Florentia*, Florence, Italy.] **1.** Of or relating to the style of art and architecture that flourished in Florence, Italy, during the Renaissance. **2.** *often* **florentine.** Having a dull chased or rubbed finish. —Used of gold. **3.** Prepared or served with spinach.

**flo·res·cence** (flô-rĕs'əns, flə-) *n.* [NLat. *florescentia* < Lat. *florescens*, pr.part. of *florescere*, inchoative of *florere*, to bloom < *flos*, flower.] A state or time of blossoming. **—flo·res'cent** *adj.*

**flo·ret** (flôr'ĭt, flōr'-) *n.* [ME *flouret* < OFr. *florete*, dim. of *flor*, flower. —see FLOWER.] A small flower, esp. one of the disk or ray flowers of a composite plant, as a daisy.

**flo·ri·at·ed** *also* **flo·re·at·ed** (flôr'ē-ā'tĭd, flōr'-) *adj.* [< Lat. *flos*, *flor-*, flower.] Ornamented with floral designs.

**flo·ri·bun·da** (flôr'ə-bŭn'də, flōr'-) *n.* [NLat., fem. of *floribundus*, blossoming freely < Lat. *flos*, flower.] Any of various hybrid roses bearing many single or double flowers.

**flo·ri·cul·ture** (flôr'ĭ-kŭl'chər, flōr'-) *n.* [Lat. *flos*, *flor-*, flower + CULTURE.] Cultivation of flowering plants. **—flo'ri·cul'tur·al** *adj.* **—flo'ri·cul'tur·al·ly** *adv.* **—flo'ri·cul'tur·ist** *n.*

**flor·id** (flôr'ĭd, flŏr'-) *adj.* [Fr. *floride* < Lat. *floridus* < *florere*, to bloom < *flos*, flower.] **1.** Flushed with rosy color : RUDDY. **2.** Heavily embellished : ORNATE <*florid* prose> **3.** *Archaic.* Healthy : blooming. **4.** *Obs.* Abounding in or covered with flowers. **—flo·rid'i·ty** (flə-rĭd'ĭ-tē, flô-), **flor'id·ness** *n.* **—flor'id·ly** *adv.*

**flo·rif·er·ous** (flô-rĭf'ər-əs) *adj.* [Lat. *florifer*, bearing flowers : *flos*, flower + *ferre*, to bear + -OUS.] Bearing flowers, esp. in abundance. **—flo·rif'er·ous·ly** *adv.* **—flo·rif'er·ous·ness** *n.*

**flor·i·gen** (flôr'ə-jən, flōr'-) *n.* [Lat. *flos*, *flor-*, flower + -GEN.] A hormone stimulating the flowering of plant buds. **—flor'i·gen'ic** (-jĕn'ĭk) *adj.*

**flor·in** (flôr'ĭn, flŏr'-) *n.* [ME < OFr. < OItal. *fiorino* < *fiore*, flower < Lat. *flos*, flower.] **1.** A former British coin worth two shillings. **2.** A guilder. **3. a.** A gold coin first issued at Florence, Italy, in 1252. **b.** Any of several former European gold coins similar to the Florentine florin.

**flo·rist** (flôr'ĭst, flōr'-, flŏr'-) *n.* [Lat. *flos*, *flor-*, flower + -IST.] One whose business is the cultivation or sale of flowers and ornamental plants.

**flo·rist·ics** (flô-rĭs'tĭks, flō-) *n.* (*sing. in number*). Study of the numerical distribution of plants and plant groups. **—flo·rist'ic** *adj.* **—flo·rist'ic·al·ly** *adv.*

**-florous** *suff.* [LLat. *-florus* < Lat. *flos*, flower.] Having a given kind or number of flowers <tubuli*florous*>

**flos·cu·lus** (flŏs'kyə-ləs) *n.* [Lat., dim. of *flos*, flower.] A floret.

**floss** (flôs, flŏs) *n.* [Perh. < Fr. *floche*, down < OFr. *flosche*.] **1.** Short or waste silk fibers. **2.** A soft, loosely twisted thread used in embroidery. **3.** A soft, silky fibrous substance, as corn silk. **4.** Dental floss. —*v.* **flossed, floss·ing, floss·es.** —*vt.* To clean between (teeth) with dental floss. —*vi.* To use dental floss.

**floss·y** (flô'sē, flŏs'ē) *adj.* **-i·er, -i·est.** **1.** Made of or resembling floss. **2.** *Slang.* Ostentatiously stylish : FLASHY.

**flo·tage** *also* **float·age** (flō'tĭj) *n.* **1.** Flotation. **2.** Floating material or objects.

**flo·ta·tion** *also* **floa·ta·tion** (flō-tā'shən) *n.* **1.** The act or state of floating. **2.** An act or instance of launching or financing a business venture by selling an issue of stocks or bonds. **3.** A process in which different materials, chiefly minerals, are separated by agitation of a pulverized mixture of the materials with water, oil, and chemicals that cause differential wetting of the suspended particles, the unwetted particles being carried by air bubbles to the surface for collection.

**flo·til·la** (flō-tĭl'ə) *n.* [Sp., dim. of *flota*, fleet < OFr. *flote* < ON *floti*.] **1. a.** A fleet of small ships. **b.** A small fleet of ships. **2.** A group resembling a small fleet <*a flotilla* of bicycles>

**flot·sam** (flŏt'səm) *n.* [AN *floteson* < OFr. *floter*, to float, of Germanic orig.] **1.** Wreckage remaining afloat after a ship has sunk. **2. a.** Discarded odds and ends. **b.** Vagrant people : DRIFTERS.

**flounce[1]** (flouns) *n.* [Alteration of *frounce* < ME, pleat < OFr. *fronce*.] A strip of gathered or pleated material attached along its upper edge to another surface, as on a garment or curtain. —*vt.* **flounced, flounc·ing, flounc·es.** To trim with a flounce.

**flounce[2]** (flouns) *vi.* **flounced, flounc·ing, flounc·es.** [Poss. of Scand. orig.] **1.** To move with exaggerated movements <*flounced* out of the house in a huff> **2.** To struggle or flounder. **—flounce** *n.*

**floun·der[1]** (floun'dər) *vi.* **-dered, -der·ing, -ders.** [Prob. alteration of FOUNDER.] **1.** To make clumsy attempts to move or to regain one's balance. **2.** To proceed awkwardly and in great confusion. **—floun'der** *n.*

**floun·der[2]** (floun'dər) *n.* [ME, of Scand. orig.] Any of various marine flatfishes of the families Bothidae and Pleuronectidae, including important food fishes.

**flour** (flour) *n.* [ME. —see FLOWER.] **1.** A soft, fine, powdery substance obtained by grinding and sifting the meal of a grain, esp. wheat. **2.** A fine, soft powder. —*vt.* **floured, flour·ing, flours.** **1.** To cover or coat with flour. **2.** To make into flour. **—flour'y** *adj.*

▲ word history: The word *flour* is simply a specialized use of the word *flower* in its medieval spelling. The *flower* of wheat or any other grain is the finest part of the meal that is left after the bran has been sifted out. The distinction in spelling between *flour* and *flower* did not become standard until the 19th century.

**flour·ish** (flûr'ĭsh, flŭr'-) *v.* **-ished, -ish·ing, -ish·es.** [ME *florishen* < OFr. *florir*, *floriss-*, to bloom < VLat. **florire* < Lat. *florere* < *flos*, flower.] —*vi.* **1.** To grow luxuriantly : THRIVE. **2.** To fare well : PROSPER. **3.** To be in one's prime. **4.** To make bold, sweeping movements. —*vt.* To wield, wave, or exhibit dramatically <*flourish* a swagger stick> —*n.* **1.** An act or instance of waving or brandishing. **2.** An embellishment or ornamentation, esp. in writing. **3.** A dramatic act or gesture. **4.** A musical fanfare. **—flour'ish·er** *n.*

**flout** (flout) *v.* **flout·ed, flout·ing, flouts.** [Prob. < ME *flouten*, to play the flute < OFr. *flauter* < *flaute*, flute.] —*vt.* To show contempt for : SCORN <*flout* convention> —*vi.* To be scornful : JEER. —*n.* A contemptuous action or remark : INSULT. **—flout'er** *n.* **—flout'ing·ly** *adv.*

**flow** (flō) *v.* **flowed, flow·ing, flows.** [ME *flouen* < OE *flōwan*.] —*vi.* **1.** To move freely like a fluid. **2.** To circulate, as blood. **3.** To discharge a stream. **4.** To move with a continual shifting of the component particles <grain *flowing* into a silo> **5.** To proceed steadily and easily. **6.** To appear smooth, harmonious, or graceful. **7.** To rise. —Used of the tide. **8.** To derive <Several conclusions *flow* from this theory.> **9. a.** To be plentiful : ABOUND. **b.** To overflow : flood. **10.** To hang loosely and gracefully. **11.** To undergo plastic deformation without cracking or breaking. —*vt.* **1.** To release as a flow. **2.** To cause to flow. —*n.* **1. a.** The smooth motion of fluids. **b.** An act of flowing. **2.** A stream. **3. a.** A continuous outpouring <a *flow* of words> **b.** A continuous movement or circulation <the steady *flow* of traffic> **4.** The amount that flows in a given time period. **5.** A rising and overflowing, esp. of dry land, of a body of water. **6.** Continuity and smoothness of appearance. **7.** Menstrual discharge. **8.** The rising of the tide. **—flow'ing·ly** *adv.*

**flow·age** (flō'ĭj) *n.* **1.** An act of flowing or overflowing. **2.** The state of being flooded. **3.** A liquid that flows or overflows. **4.** The gradual plastic deformation of a solid body, as by heat.

**flow chart** *n.* A schematic representation of a sequence of operations, as for a computer program.

**flow·er** (flou'ər) *n.* [ME *flour* < OFr. *flor* < Lat. *flos*.] **1. a.** The reproductive structure of a seed-bearing plant, having specialized male and female organs as stamens and a pistil enclosed in an outer envelope of petals and sepals. **b.** Such a structure having large or colorful parts : BLOSSOM. **c.** A reproductive organ of other plants, as mosses. **2.** A plant cultivated or conspicuous for its blossoms. **3.** The period of highest development : PEAK. **4.** The highest example or best representative of something <the *flower* of our generation> **5.** An embellishment. **6. flowers.** *Chem.* A fine powder produced by condensation or sublimation. —*v.* **-ered, -er·ing, -ers.** —*vi.* **1.** To produce a flower. **2.** To develop fully : PEAK. —*vt.* To decorate with flowers or with a floral pattern. **—flow'er·er** *n.* **—flow'er·less** *adj.*

**flow·er·age** (flou'ər-ĭj) *n.* **1.** Flowers as a whole. **2.** The act of flowering or state of being in flower.

**flower bug** *n.* Any of a group of bugs from the family Anthocoridae, feeding on insects that infest flowers.

**flow·er·et** (flou'ər-ĭt) *n.* A small flower.

**flower girl** *n.* A young girl who carries flowers in a procession, esp. at a wedding.

**flowering dogwood** *n.* The dogwood.

**flowering maple** *n.* A tropical shrub of the genus *Abutilon*, esp. *A. hybridum*, having lobed leaves resembling those of the maple and variously colored flowers.

**flowering plant** *n.* An angiosperm.

**flowering quince** *n.* An Asian shrub of the genus *Chaenomeles*, having spiny branches and red or pink flowers.

**flow·er·pot** (flou′ər-pŏt′) *n.* A pot in which plants are grown.

**flow·er·y** (flou′ə-rē) *adj.* **-i·er, -i·est. 1.** Abounding in or bedecked with flowers. **2.** Like flowers. **3.** Full of figurative, ornate expressions <*flowery* prose> **—flow′er·i·ness** *n.*

**flow meter** *n.* A device for monitoring, measuring, or recording fluid flow, as of a gaseous fuel.

**flown[1]** (flōn) *adj.* [Obs. p.part. of FLOW.] *Archaic.* Filled to excess.

**flown[2]** (flōn) *v. p.p. of* FLY[1].

**flu** (flo͞o) *n. Informal.* Influenza.

**flub** (flŭb) *vt.* **flubbed, flub·bing, flubs.** To botch. *—n.* An act or instance of botching. **—flub′ber** *n.*

**flub·dub** (flŭb′dŭb′) *n.* [Orig. unknown.] *Slang.* Pretentious nonsense : BUNKUM.

**fluc·tu·ant** (flŭk′cho͞o-ənt) *adj.* [Lat. *fluctuans, fluctuant-,* pr.part. of *fluctuare,* to fluctuate.] **1.** Variable or unstable <*fluctuant* temperatures> **2.** Rising and falling in waves.

**fluc·tu·ate** (flŭk′cho͞o-āt′) *v.* **-at·ed, -at·ing, -ates.** [Lat. *fluctuare, fluctuat-* < *fluctus,* a flowing < p.part. of *fluere,* to flow.] *—vi.* **1.** To vary irregularly <stock prices *fluctuating* wildly> **2.** To rise and fall like waves. *—vt.* To cause to fluctuate. **—fluc′tu·a′tion** *n.*

**flue** (flo͞o) *n.* [Orig. unknown.] **1.** A conduit, as a pipe, tube, or channel through which hot air, gas, steam, or smoke may pass. **2. a.** A flue pipe. **b.** The air passage in a flue pipe.

**flu·ent** (flo͞o′ənt) *adj.* [Lat. *fluens, fluent-,* pr.part. of *fluere,* to flow.] **1.** Having facility in language use <*fluent* in Russian and German> **2. a.** Flowing effortlessly : POLISHED <*fluent* prose> **b.** Flowing smoothly : GRACEFUL <*fluent* curves> **3.** Flowing or capable of flowing : FLUID. **—flu′en·cy** *n.* **—flu′ent·ly** *adv.*

**flue pipe** *n.* An organ pipe sounded by a current of air striking a lip in the side of the pipe and causing the air within to vibrate.

**flu·er·ic** (flo͞o-ĕr′ĭk) *adj.* [Lat. *fluere,* to flow + -IC.] Fluidic.

**flu·er·ics** (flo͞o-ĕr′ĭks) *n. (sing. in number).* Fluidics.

**fluff** (flŭf) *n.* [Orig. unknown.] **1.** Light down or nap. **2.** Something having a light, soft, or frothy consistency or appearance. **3.** Something of little importance. **4.** *Informal.* An error, esp. in the delivery of lines, as by a performer or announcer. *—v.* **fluffed, fluff·ing, fluffs.** *—vt.* **1.** To make light and puffy by shaking or patting into a soft, loose mass. **2.** *Informal.* **a.** To mar or ruin by an error. **b.** To misread or forget (one's lines). *—vi.* **1.** To become soft and puffy or feathery. **2.** *Informal.* To make an error, esp. to forget one's lines.

**fluff·y** (flŭf′ē) *adj.* **-i·er, -i·est. 1.** Of, like, or covered with fluff. **2.** Light and airy : SOFT. **—fluff′i·ly** *adv.* **—fluff′i·ness** *n.*

**flü·gel·horn** (flo͞o′gəl-hôrn′, flü′-) *n.* [G. : *Flügel,* flank (from its use to summon flanks during a battle) + *Horn.*] A bugle with valves.

**flu·id** (flo͞o′ĭd) *n.* [Fr. *fluide* < Lat. *fluidus,* flowing < *fluere,* to flow.] A substance that exists or is held to exist as a continuum marked by low resistance to flow and the tendency to assume the shape of its container. *—adj.* **1.** Flowing easily like a fluid. **2.** Readily deformed : PLIABLE. **3.** Smooth and effortless. **4.** Easily changed or tending to change <a *fluid* diplomatic juncture fraught with uncertainty> **5.** Convertible into cash <*fluid* assets> **—flu·id′i·ty** (-ĭd′ĭ-tē), **flu′id·ness** *n.* **—flu′id·ly** *adv.*

**fluid dram** *n.* One eighth of a fluid ounce.

**flu·id·ex·tract** (flo͞o′ĭd-ĕk′străkt′) *n.* A concentrated alcohol solution of a vegetable drug containing the equivalent of one gram in powdered form of the active principle in each milliliter.

**flu·id·ic** (flo͞o-ĭd′ĭk) *adj.* Of, relating to, or being an apparatus operated by fluids. **—flu·id′ic** *n.*

**flu·id·ics** (flo͞o-ĭd′ĭks) *n. (sing. in number).* The technology of fluids used as nonmoving, nonelectrical components of control and sensing systems.

**fluid ounce** *n.* **1.** A unit of volume or capacity in the U.S. Customary System, used in liquid measure, equal to 1.804 cubic inches, 1/16 of a pint, or 29.574 milliliters. **2.** A unit of volume or capacity in the British Imperial System, used in liquid and dry measure, equal to 1.734 cubic inches or 3.552 cubic centimeters.

**fluke[1]** (flo͞ok) *n.* [ME < OE *flōc.*] **1.** *pl.* **fluke.** A flatfish, esp. a flounder of the genus *Paralichthys.* **2.** A trematode.

**fluke[2]** (flo͞ok) *n.* [Poss. < FLUKE[1].] **1.** The triangular blade at the end of either arm of an anchor, designed to catch in the ground. **2.** A barbed head, as on an arrow. **3.** One of the two horizontally flattened divisions of the tail of a whale or related animal.

**fluke[3]** (flo͞ok) *n.* [Orig. unknown.] **1.** An accidentally good or successful billiards or pool stroke. **2.** A stroke of good fortune.

**fluk·y** *also* **fluk·ey** (flo͞o′kē) *adj.* **-i·er, -i·est.** [< FLUKE[3].] **1.** Resulting merely from chance. **2.** Constantly shifting : UNCERTAIN <a *fluky* wind>

**flume** (flo͞om) *n.* [ME *flum,* river < OFr. < Lat. *flumen* < *fluere,* to flow.] **1.** A narrow gorge usu. with a stream flowing through it. **2.** An artificial channel or chute for a stream of water.

**flum·mer·y** (flŭm′ə-rē) *n., pl.* **-ies.** [Welsh *llymru.*] **1. a.** Any of several soft, light, bland foods, as a custard. **b.** A soft jelly made by straining boiled, slightly fermented oatmeal or flour. **2.** Nonsense.

**flum·mox** (flŭm′əks) *vt.* **-moxed, -mox·ing, -mox·es.** [Orig. unknown.] *Slang.* To confuse : perplex.

**flung** (flŭng) *v. p.t. & p.p. of* FLING.

**flunk** (flŭngk) *v.* **flunked, flunk·ing, flunks.** [Orig. unknown.] *Informal. —vi.* To fail esp. in a course or examination. *—vt.* **1.** To fail (a course or examination). **2.** To give a failing grade to. **—flunk out.** To expel or be expelled from a school or course because of work below the required standards. *—n.* A failing grade.

**flun·ky** *also* **flun·key** (flŭng′kē) *n., pl.* **-kies** *also* **-keys.** [Sc.] **1.** A liveried servant. **2.** An obsequious person : TOADY. **3.** One who does menial or trivial work. **—flun′ky·ism** *n.*

**flu·or** (flo͞o′ôr′, -ər) *n.* [NLat., mineral belonging to a group used as fluxes < Lat., a flowing < *fluere,* to flow.] Fluorite.

**fluor-** *pref. var. of* FLUORO-.

**flu·o·resce** (flo͞o′ə-rĕs′, flo͞o-rĕs′) *vi.* **-resced, -resc·ing, -resc·es.** [Back-formation < FLUORESCENCE.] To undergo, produce, or exhibit fluorescence.

**flu·o·res·ce·in** (flo͞o′ə-rĕs′ē-ĭn, flo͞o-rĕs′-) *n.* An orange-red compound, $C_{20}H_{12}O_5$, that exhibits intense fluorescence in alkaline solution and is used to dye sea water for spotting or tracing operations.

**flu·o·res·cence** (flo͞o′ə-rĕs′əns, flo͞o-rĕs′-) *n.* [FLUOR + -ESCENCE.] **1.** Emission of electromagnetic radiation, esp. of visible light, resulting from the absorption of incident radiation and persisting only as long as the stimulating radiation is continued. **2.** Radiation emitted by fluorescence.

**fluorescence microscopy** *n.* Microscopy in which the specimens are stained with a fluorescent dye and viewed by illumination with ultraviolet light.

**flu·o·res·cent** (flo͞o′ə-rĕs′ənt, flo͞o-rĕs′-) *adj.* Exhibiting or capable of exhibiting fluorescence.

**fluorescent lamp** *n.* A lamp generating visible light by fluorescence, esp. a glass tube whose inner wall is coated with a material that fluoresces when bombarded with secondary radiation generated by a gaseous discharge within the tube.

**fluor·i·date** (flo͝or′ĭ-dāt′, flôr′-, flōr′-) *vt.* **-dat·ed, -dat·ing, -dates.** To add a fluorine compound to (e.g., a water supply) so as to prevent tooth decay. **—fluor′i·da′tion** *n.*

**flu·o·ride** (flo͞o′ə-rīd′, flo͝or′īd′, flôr′-, flōr′-) *n.* A binary compound of fluorine.

**fluor·i·na·tion** (flo͝or′ĭ-nā′shən, flôr′-, flōr′-) *n.* Chemical introduction of fluorine into a compound.

**flu·o·rine** (flo͞o′ə-rēn′, -rĭn, flo͝or′ēn′, -ĭn, flôr′-, flōr′-) *n.* [FLUOR + -INE[2].] *Symbol* F A pale-yellow, highly corrosive, poisonous gaseous element, the most electronegative and reactive of all the elements; atomic number 9; atomic weight 18.9984.

**fluor·ite** (flo͝or′īt′, flôr′-, flōr′-) *n.* [Ital.] A gen. light-colored green, blue, violet, yellow, brown, or colorless mineral, essentially $CaF_2$, often fluorescent in ultraviolet light.

**fluoro-** *or* **fluor-** *pref.* **1.** Fluorine <*fluorosis*> **2.** Fluorescence <*fluoroscope*>

**flu·o·ro·car·bon** (flo͞o′ə-rō-kär′bən, flo͝or′ō-) *n.* Any of various inert organic compounds in which fluorine replaces hydrogen, used as aerosol propellants, refrigerants, solvents, and lubricants and in manufacturing plastics and resins.

**fluor·o·chrome** (flo͝or′ə-krōm′, flôr′-) *n.* Any of a group of fluorescent dyes used in staining microorganisms for examination by fluorescence microscopy.

**flu·o·rom·e·ter** (flo͞o′ə-rŏm′ĭ-tər, flo͞o-rŏm′-) *n.* An instrument for detecting and measuring fluorescence. **—flu′o·rom′e·try** *n.*

**fluor·o·scope** (flo͝or′ə-skōp′, flôr′-, flōr′-, flo͞o′ər-ə-) *n.* A fluorescent screen on which the internal structure of an optically opaque object, as of the human body, may be continuously viewed by transmission of x-rays through the object. *—vt.* **-scoped, -scop·ing, -scopes.** To examine the interior of with a fluoroscope. **—fluor′o·scop′ic** (-skŏp′ĭk) *adj.* **—fluor′o·scop′i·cal·ly** *adv.*

**flu·o·ros·co·py** (flo͞o′ə-rŏs′kə-pē, flo͞o-rŏs′-) *n.* Examination with the use of a fluoroscope.

**flu·o·ro·sis** (flo͞o′ə-rō′sĭs, flo͞o-rō′-) *n.* An abnormal condition caused by excessive fluorine intake, characterized by mottling of the teeth. **—flu′o·rot′ic** (-rŏt′ĭk) *adj.*

**flu·o·ro·u·ra·cil** (flo͞o′ə-rō-yo͝or′ə-sĭl, flo͝or′ō-) *n.* A pyrimidine, $C_4H_3FN_2O_2$, used as an anticancer drug.

**flu·or·spar** (flo͞o′ər-spär′, flo͝or′spär′) *n.* Fluorite.

**flur·ry** (flûr′ē, flŭr′ē) *n., pl.* **-ries.** [< obs. *flurr,* to scatter.] **1.** A sudden gust of wind. **2.** A light snowfall. **3.** A sudden burst of confusion, excitement, or bustling activity. **4.** A short period of active trading, as on the stock exchange. *—v.* **-ried, -ry·ing, -ries.** *—vt.* To confuse or make nervous : FLUSTER. *—vi.* To move or come down in a flurry.

**flush[1]** (flŭsh) *v.* **flushed, flush·ing, flush·es.** [Prob. < *flush,* to dart out.] *—vi.* **1.** To flow and spread out suddenly and abundantly. **2.** To turn red in the face : BLUSH. **3.** To glow, esp. with a reddish color. **4. a.** To be cleaned by a rapid, brief gush of water. **b.** To function by means of a flushing mechanism, as a toilet. *—vt.* **1.** To cause to redden. **2.** To excite, as with a feeling of pride or accomplishment <*flushed* with the first win> **3.** To wash, empty, or purify with a sudden, rapid flow of water. *—n.* **1.** A brief, copious gushing, as of water. **2.** A blush : glow. **3.** Redness of the skin, as with fever. **4.** Animation or exhilaration. **5.** Freshness or vigor <the *flush*

of youth> *—adj.* **-er, -est. 1.** Having a healthy reddish color. **2.** Abundant : plentiful. **3.** Having much money : AFFLUENT. **4.** Lively and vigorous : LUSTY. **5. a.** Having surfaces in the same plane : EVEN. **b.** Arranged with adjacent sides, surfaces, or edges close together <a bench *flush* against the wall> **c.** Having margins aligned with no indentations. **6.** Direct, straightforward, or solid, as a blow. *—adv.* **1.** So as to be even, in one plane, or aligned with a margin. **2.** Squarely or solidly <The punch hit me *flush* on the jaw.>

**flush[2]** (flŭsh) *n.* [OFr. *flus* < Lat. *fluxus,* flux.] A hand in which all the cards are of the same suit, rated above a straight and below a full house in a game such as poker.

**flush[3]** (flŭsh) *v.* **flushed, flush·ing, flush·es.** [ME *flusshen.*] *—vt.* To frighten (e.g., a game bird) from cover. *—vi.* To dart out or fly from cover. *—n.* A bird or a flock of birds suddenly taking flight.

**flus·ter** (flŭs′tər) *vt. & vi.* **-tered, -ter·ing, -ters.** [Prob. of Scand. orig.] To make or become nervous or upset. **—flus′ter** *n.*

**flute** (flo͞ot) *n.* [ME *floute* < OFr.] **1.** A high-pitched woodwind instrument, tubular in shape and with finger holes and keys on the side and a reedless mouthpiece either at the end or on the side. **2.** An organ stop whose flue pipe produces a flutelike tone. **3.** One of the rounded, parallel grooves incised on the shaft of a column as a decorative motif. **4. a.** A groove in cloth, as in a pleated ruffle. **b.** A similar short groove, as in a pie crust. *—v.* **flut·ed, flut·ing, flutes.** *—vt.* **1.** To play (a tune) on a flute. **2.** To produce in a flutelike tone. **3.** To make flutes in. *—vi.* **1.** To play a flute. **2.** To sing or whistle with a flutelike tone. **—flut′er** *n.*

**flut·ing** (flo͞o′tĭng) *n.* **1.** A decorative motif consisting of a series of long, rounded, parallel grooves, as those incised in the surface of a column. **2. a.** The grooves formed by narrow pleats in cloth, as in a ruffle. **b.** Similar short grooves, as in a pie crust. **3.** The act of incising or making grooves.

**flut·ist** (flo͞o′tĭst) *n.* A flute player.

**flut·ter** (flŭt′ər) *v.* **-tered, -ter·ing, -ters.** [ME *floteren* < OE *flotorian.*] *—vi.* **1.** To wave or flap rapidly and irregularly <draperies *fluttering* in the breeze> **2. a.** To fly by a quick, light flapping of the wings. **b.** To flap the wings without flying. **3.** To beat rapidly or erratically <My heart *fluttered* wildly.> **4.** To move quickly in a nervous, restless, or excited way : FLIT. *—vt.* To cause to flutter. *—n.* **1.** An act of fluttering. **2.** A state of nervous agitation : TIZZY. **3.** FLURRY 3. **4.** *Pathol.* Abnormal pulsation, as of the heart. **5.** A distortion in reproduced sound due to frequency deviations created by faulty recording or reproduction techniques. **—flut′ter·er** *n.* **—flut′ter·y** *adj.*

**flutter kick** *n.* A swimming kick in which the legs are held horizontally and alternately moved up and down in rapid strokes without bending the knees.

**flu·vi·al** (flo͞o′vē-əl) *adj.* [ME < Lat. *fluvialis* < *fluvius,* river < *fluere,* to flow.] **1.** Of, relating to, or living in a stream or river. **2.** Caused by the action of flowing water.

**flu·vi·o·ma·rine** (flo͞o′vē-ō-mə-rēn′) *adj.* [Lat. *fluvius,* river < *fluere,* to flow + MARINE.] Relating to deposits, esp. near the mouth of a river, formed by the joint action of the sea and a river.

**flux** (flŭks) *n.* [ME < OFr. < Lat. *fluxus* < p.part. of *fluere,* to flow.] **1. a.** A flow or flowing. **b.** A continued flow : FLOOD. **2.** *Physics.* **a.** A flow of matter or energy as a fluid or considered to be a fluid. **b.** Flux density. **c.** The lines of force of a magnetic field. **3.** *Med.* Discharge of large amounts of fluid from a bodily surface or cavity. **4.** Change : fluctuation <Our plans are in a state of *flux.*> **5.** *Chem. & Metallurgy.* A substance that aids, induces, or otherwise actively participates in a flowing, as: **a.** A mineral added to a furnace charge to promote fusing of metals or prevent the formation of oxides. **b.** A substance applied in soldering and brazing to parts of a surface to be joined, acting on application of heat to prevent oxide formation and facilitate the flowing of solder. **c.** A readily fusible glass or enamel used as a base in ceramic work. *—v.* **fluxed, flux·ing, flux·es.** *—vt.* **1.** To melt : fuse. **2.** To apply a flux to. *—vi.* **1.** To become fluid. **2.** To flow : stream.

**flux density** *n. Physics.* The quantity of flux per unit area.

**flux gate** *n.* A detector for indicating the direction of the terrestrial magnetic field.

**flux·ion** (flŭk′shən) *n.* [OFr. < Lat. *fluxio* < *fluxus,* flux.] **1.** Continual, constant change. **2.** *Math. Archaic.* **a.** A derivative. **b. fluxions.** Differential calculus. **—flux′ion·al, flux′ion·ar′y** (-shə-nĕr′ē) *adj.* **—flux′ion·al·ly** *adv.*

**fly[1]** (flī) *v.* **flew** (flo͞o), **flown** (flōn), **fly·ing, flies.** [ME *flien* < OE *flēogan.*] *—vi.* **1.** To move through the air by means of wings or winglike parts. **2. a.** To travel by air. **b.** To pilot an aircraft. **3. a.** To rise in or be carried through the air by the wind. **b.** To float in the air <pennants *flying*> **4.** To be sent or driven through the air with great speed or force <bullets *flying*> **5. a.** To rush : hasten <*flew* down the street> **b.** To escape. **6.** To pass by swiftly <a holiday *flying* by> **7.** *p.t. & p.p.* **flied.** *Baseball.* To hit a fly ball. **8.** To react explosively <*flew* into a towering rage> *—vt.* **1.** To cause to float in the air <*fly* a pennant> **2. a.** To pilot (an aircraft). **b.** To transport in an aircraft. **c.** To pass over in an aircraft <*fly* the ocean> **d.** To perform in an aircraft <*flew* 50 combat missions> *—n., pl.* **flies. 1.** An act of flying. **2.** A fold of cloth covering a garment fastening, esp. one on the front of trousers. **3.** A cloth flap covering an entrance or forming a roof extension for a tent or wagon. **4.** A flyleaf. **5.** *Baseball.* A fly ball. **6. a.** The span of a flag from the staff to the outer edge. **b.** The outer edge of a flag. **7. a.** A flywheel. **b.** An analogous device, esp. one used to regulate the speed of clockwork. **8. flies.** The area directly over the stage of a theater, containing overhead lights and other equipment. **9.** *Chiefly Brit.* A one-horse carriage : HACKNEY. **—fly blind.** To fly an aircraft in poor visibility with the aid of instruments. **—fly high.** To be elated. **—fly in the face** (or **teeth**) **of.** To defy or resist openly. **—fly off the handle.** To become suddenly angry. **—fly the coop.** *Slang.* To get away : ESCAPE. **—on the fly. 1.** In a great hurry. **2.** In flight.

**fly[2]** (flī) *n., pl.* **flies.** [ME *flie* < OE *flēoge.*] **1. a.** Any of numerous winged insects of the order Diptera, esp. one of the family Muscidae, including the housefly and the tsetse. **b.** Any of various other flying insects, as the caddis fly. **2.** A fishing lure simulating a fly. **—fly in the ointment.** A drawback.

**fly[3]** (flī) *adj.* [Prob. < FLY[1].] *Chiefly Brit.* Mentally alert : SHARP.

**fly agaric** *n.* A poisonous mushroom, *Amanita muscaria,* usu. having a red or orange cap with white patches.

**fly·a·way** (flī′ə-wā′) *adj.* **1.** Blown by the wind <*flyaway* hair> **2.** Given to frivolity : FLIGHTY. **3.** Ready for flight.

**fly ball** *n. Baseball.* A ball batted in a high arc, usu. to the outfield.

**fly·blow** (flī′blō′) *n.* The egg or larva of a blowfly, usu. deposited on food. *—vt.* **-blew** (-blo͞o′), **-blown** (-blōn′), **-blow·ing, -blows. 1.** To deposit (flyblows) in. **2.** To contaminate.

**fly·blown** (flī′blōn′) *adj.* Contaminated with or as if with flyblows.

**fly book** *n.* A case in which artificial flies for fishing are carried.

**fly·boy** (flī′boi′) *n. Slang.* A pilot, esp. in the U.S. Air Force.

**fly·by** (flī′bī′) *n., pl.* **-bys.** A flight passing close to a specified target or position, esp. a maneuver in which a spacecraft passes sufficiently close to a planet to make relatively detailed observations without landing.

**fly-by-night** (flī′bī-nīt′) *adj.* **1.** Of unreliable business character. **2.** Passing : temporary. *—n.* **1.** One who cheats creditors, as by absconding. **2.** Something dubiously transitory.

**fly·catch·er** (flī′kăch′ər, -kĕch′-) *n.* Any of various birds characterized by the habit of flying suddenly from a perch to catch flying insects, esp. a member of the New World family Tyrannidae or the Old World family Muscicapidae.

**fly·er** (flī′ər) *n. var. of* FLIER.

**fly-fish** (flī′fĭsh′) *vi.* **-fished, -fish·ing, -fish·es.** To angle using artificial flies for bait.

**fly front** *n.* A garment front with a fly concealing the fastenings.

**fly gallery** *n.* A platform at the side of a theater stage from which a stagehand works the ropes controlling equipment in the flies.

**fly·ing** (flī′ĭng) *adj.* **1.** Swiftly moving <*flying* fingers on the keyboard> **2.** Brief : hurried <a *flying* visit> **3.** Of, relating to, or concerned with aviation <a *flying* corps> *—n.* **1.** Flight in an aircraft. **2.** The piloting or navigation of an aircraft.

**flying boat** *n.* A large seaplane kept afloat by its hull rather than by pontoons.

**flying bomb** *n.* A robot bomb.

**flying buttress** *n.* A masonry prop that springs from a solid pier and abuts against another part of the structure to receive thrust.

**flying colors** *n.* Complete success or victory : TRIUMPH <passed the exam with *flying colors*>

**flying dragon** *n.* A flying lizard.

**Flying Dutchman** *n.* **1.** A legendary Dutch mariner condemned to sail the seas against the wind until Judgment Day. **2.** A spectral ship said to appear in storms near the Cape of Good Hope.

**flying fatigue** *n.* Aeroneurosis.

**flying field** *n.* A field graded for aircraft landings and takeoffs.

**flying fish** *n.* Any of various marine fishes of the family Exocoetidae, having enlarged pectoral or pelvic fins capable of sustaining them in brief, gliding flight over the water.

**flying fox** *n.* **1.** A fruit-eating bat of the genus *Pteropus,* chiefly of tropical Africa, Asia, and Australia, having a foxlike muzzle and ears. **2.** A mammal related to or resembling the flying fox.

**flying frog** *n.* An arboreal frog, *Rhacophorus reinwardtii* of southeastern Asia, having toes connected by broad webbing and capable of gliding considerable distances.

**flying gurnard** *n.* A chiefly tropical marine fish of the family Dactylopteridae, having winglike, much enlarged pectoral fins.

**flying jib** *n. Naut.* A light sail that extends beyond the jib and is attached to an extension of the jib boom.

**flying lemur** *n.* Either of two tropical Asian mammals, *Cynocephalus volans* or *C. variegatus,* sustained in gliding leaps by a wide fur-covered membrane extending from each side of the body.

**flying lizard** *n.* A small tropical Asian lizard of the genus *Draco,* capable of gliding by spreading the winglike membranes on either side of the body.

**flying machine** *n.* A machine designed for flight, esp. an early experimental type of aircraft.

**flying mare** *n.* A throw in which a wrestler grabs an opponent's wrist, turns around quickly, and flips the opponent over one shoulder onto the ground.
**flying phalanger** *n.* Any of several small marsupials of the family Phalangeridae, esp. one of the genus *Petaurus* of Australia, New Guinea, and Tasmania, capable of gliding through the air sustained by large folds of skin between the forelegs and hind legs.
**flying saucer** *n.* Any of various unidentified flying objects typically reported and described as luminous disks.
**flying squirrel** *n.* A squirrel of the genera *Pteromys, Glaucomys*, or related genera, with membranes between the forelegs and hind legs enabling it to glide through the air.
**flying start** *n.* **1.** The crossing of the starting line of a race at full speed. **2.** A quick start.
**fly·leaf** (flī'lēf') *n.* A blank page at the beginning or end of a book.
**fly-o·ver** (flī'ō'vər) *n.* **1.** A flight of aircraft at low altitude over a specific location, usu. as a military display. **2.** *Chiefly Brit.* An overpass on a highway.
**fly·pa·per** (flī'pā'pər) *n.* Paper coated with a sticky, occas. poisonous substance to catch flies.
**fly-poi·son** (flī'poi'zən) *n.* A poisonous plant, *Amianthium muscaetoxicum* of the southeastern United States, having narrow basal leaves and a terminal cluster of small white or greenish flowers.
**fly·speck** (flī'spĕk') *n.* **1.** A small, dark speck or stain made by the excrement of a fly. **2.** A minute spot.
**fly swatter** *n.* A device typically consisting of a flat square of plastic or wire mesh attached to a long handle and used to kill flies or other insects.
**fly·trap** (flī'trăp') *n.* **1.** A trap for catching flies. **2.** A plant, as the Venus's-flytrap, that traps insects.
**fly·weight** (flī'wāt') *n.* A boxer of the lightest weight class, weighing 112 pounds or less.
**fly·wheel** (flī'hwēl', -wēl') *n.* A heavy-rimmed rotating wheel used to minimize speed variation in a machine subject to fluctuation in drive and load.
**Fm** *symbol for* FERMIUM.
**f-num·ber** (ĕf'nŭm'bər) *n.* [F(OCAL LENGTH) + NUMBER.] The ratio of focal length to the effective aperture diameter in a lens or lens system.
**foal** (fōl) *n.* [ME *fole* < OE *fola.*] The young offspring of an equine animal, as a horse, esp. when under a year old. —*vi.* **foaled, foal·ing, foals.** To give birth to a foal.
**foam** (fōm) *n.* [ME *fom* < OE *fām.*] **1. a.** A mass of gas bubbles in a liquid-film matrix, esp. a light, bubbly gas and liquid mass formed by agitating a liquid containing certain soaps or detergents. **b.** A thick chemically produced froth, as shaving cream or certain fire-fighting substances. **2. a.** Frothy saliva from the mouth. **b.** The frothy sweat of an equine animal. **3.** The sea. **4.** Any of various light, bulky, more or less rigid materials used as thermal or mechanical insulators esp. in packaging and containers. —*vi.* & *vt.* **foamed, foam·ing, foams.** To form or cause to form foam. **—foam'ing·ly** *adv.*
**foam·flow·er** (fōm'flou'ər) *n.* A woodland plant, *Tiarella cordifolia* of eastern North America, with a narrow cluster of small white flowers.
**foam rubber** *n.* A light, firm, spongy rubber made by beating air into latex with subsequent curing and used as an upholstery material and insulating medium.
**foam·y** (fō'mē) *adj.* **-i·er, -i·est. 1.** Relating to or like foam. **2.** Consisting of or covered with foam. **—foam'i·ly** *adv.* **—foam'i·ness** *n.*
**fob¹** (fŏb) *n.* [Prob. of Germanic orig.] **1.** A small pocket at the front waistline of a pair of trousers or in the front of a vest, used esp. to hold a watch. **2.** A short chain or ribbon attached to a pocket watch and worn hanging in front of the vest or waist. **3.** An ornament or seal attached to a watch chain.
**fob²** (fŏb) *vt.* **fobbed, fob·bing, fobs.** [ME *fobben* < *fob*, trickster, prob. < *fob*, froth.] **1.** To dispose of (goods) by fraud or deception. **2.** To put off by deceitful or evasive means.
**fo·cal** (fō'kəl) *adj.* **1.** Of or relating to a focus. **2.** Positioned at or measured from a focus. **—fo'cal·ly** *adv.*
**focal infection** *n.* A localized infection.
**fo·cal·ize** (fō'kə-līz') *vt.* & *vi.* **-ized, -iz·ing, -iz·es. 1.** To adjust or come to a focus. **2.** To bring or be brought to a focus. **3.** To localize. **—fo'cal·i·za'tion** *n.*
**focal length** *n.* The distance of a focal point from the surface of a lens or mirror.
**focal point** *n.* A point on the axis of symmetry of an optical system, as of a mirror or lens, to which parallel incident rays converge or from which they appear to diverge after passing through the system.
**fo·ci** (fō'sī') *n. var. pl. of* FOCUS.
**fo'c's'le** (fōk'səl) *n. var. of* FORECASTLE.
**fo·cus** (fō'kəs) *n., pl.* **-cus·es** *or* **-ci** (-sī') [NLat. < Lat., hearth.] **1. a.** A focal point. **b.** Focal length. **c.** The distinctness or clarity with which an optical system renders an image. **d.** Adjustment for distinctness or clarity. **2.** A center of interest or activity. **3.** *Pathol.* The region of a localized bodily infection. **4.** *Geol.* The point of origin of an earthquake. **5.** *Math.* A point that together with a directrix determines a conic section. —*v.* **-cused, -cus·ing, -cus·es** *or* **-cussed, -cus·sing, -cus·ses.** —*vt.* **1. a.** To produce a clear image of (e.g., photographed material) by adjustment of optical equipment, as a projection lens. **b.** To adjust (e.g., a lens) to produce a clear image. **2.** To concentrate <*focused* all my attention on finishing the job> —*vi.* To converge at a point of focus. **—in focus.** Sharply or clearly defined : DISTINCT. **—out of focus.** Not sharply or clearly defined : INDISTINCT.
**fod·der** (fŏd'ər) *n.* [ME < OE *fōdor.*] **1.** Feed for livestock, often consisting of coarsely chopped stalks and leaves of corn mixed with hay. **2. a.** Raw material, as for artistic creation. **b.** Masses of people considered as raw material for achieving a particular goal <cannon *fodder*> —*vt.* **-dered, -der·ing, -ders.** To feed with fodder.
**foe** (fō) *n.* [ME *fo* < OE *gefā*, foe, and *fāh*, hostile.] **1.** A personal enemy. **2.** A wartime enemy. **3.** An opponent <a *foe* of tax increases> **4.** One that opposes, injures, or impedes.
**foehn** *also* **föhn** (fœn, fān) *n.* [G. *Föhn* < OHG *phōno* < Lat. *favonius*, the west wind.] A warm, dry wind coming off the lee slopes of a mountain range.
**foe·tal** (fēt'l) *adj. var. of* FETAL.
**foe·tid** (fĕ'tĭd) *adj. var. of* FETID.
**foe·tor** (fē'tər) *n. var. of* FETOR.
**foe·tus** (fē'təs) *n. var. of* FETUS.
**fog¹** (fôg, fŏg) *n.* [Perh. of Scand. orig.] **1.** Condensed water vapor in cloudlike masses that lie close to the ground and limit visibility. **2.** A mass of floating material, as dust or smoke, that forms an obscuring haze. **3.** Mental confusion or bewilderment. **4.** A dark blur on a developed photographic negative. —*v.* **fogged, fog·ging, fogs.** —*vt.* **1.** To cover or envelop with or as if with fog. **2.** To cause to be obscured : BLUR. **3.** To make uncertain or unclear : BEWILDER. **4.** To obscure or dim (a photographic negative) with a dark blur. —*vi.* **1.** To be covered with or as if with fog. **2.** To be blurred or obscured. **3.** To be dimmed or obscured with a dark blur. —Used of a photographic print or negative.
**fog²** (fôg, fŏg) *n.* [ME *fogge*, tall grass.] **1.** A second growth of grass appearing on a mown or grazed field. **2.** Tall, decaying grass left standing after the cutting or grazing season.
**fog bank** *n.* An opaque mass of fog sharply defined in contrast to surrounding clearer air, esp. such a fog occurring at sea.
**fog·bound** (fôg'bound', fŏg'-) *adj.* **1.** Immobilized by heavy fog. **2.** Clouded or obscured by fog.
**fog·bow** (fôg'bō', fŏg'-) *n.* A faint white or yellowish arc-shaped light, similar to a rainbow, often seen opposite the sun in a fog bank.
**fog·dog** (fôg'dôg', fŏg'dŏg') *n.* A bright or clear spot in a fog bank.
**fo·gey** (fō'gē) *n. var. of* FOGY.
**fog·gy** (fô'gē, fŏg'ē) *adj.* **-gi·er, -gi·est. 1. a.** Full of or covered by fog. **b.** Like fog. **2.** Clouded or blurred : INDISTINCT. **3.** Bewildered and perplexed. **—fog'gi·ly** *adv.* **—fog'gi·ness** *n.*
**fog·horn** (fôg'hôrn', fŏg'-) *n.* **1.** A horn used, as by ships, to sound warning signals in fog or darkness. **2.** A resounding, insistent voice.
**fo·gy** *also* **fo·gey** (fō'gē) *n., pl.* **-gies** *also* **-geys.** [Orig. unknown.] An old-fashioned person. **—fo'gy·ish** *adj.* **—fo'gy·ism** *n.*
**föhn** (fœn, fān) *n. var. of* FOEHN.
**foi·ble** (foi'bəl) *n.* [Obs. Fr. < obs. *foible*, weak < OFr. *feble.* —see FEEBLE.] **1.** A minor weakness or failing. **2.** The weaker section of a sword blade, from the middle to the tip.
**foil¹** (foil) *vt.* **foiled, foil·ing, foils.** [ME *foilen*, alteration of *fullen*, to trample, and *filen*, to pollute, defile.] **1.** To prevent from being successful : THWART. **2.** To obscure or confuse (a trail or scent) so as to evade pursuers. —*n.* **1.** *Archaic.* A repulse : setback. **2.** An animal's trail or scent.
**foil²** (foil) *n.* [ME < OFr. < Lat. *folium*, leaf.] **1.** A thin, flexible leaf or sheet of metal. **2.** A thin layer of bright metal placed under a displayed gem to lend it brilliance. **3.** One that by strong contrast underscores the distinctive characteristics of another. **4.** The metal coating applied to the back of a plate of glass to form a mirror. **5.** A leaflike design or space worked in stone or glass, found esp. in Gothic window tracery. **6.** *Naut.* A hydrofoil. —*vt.* **foiled, foil·ing, foils. 1.** To back or cover with foil. **2.** To set off by contrast.
**foil³** (foil) *n.* [Orig. unknown.] **1.** A fencing sword with a flat guard for the hand and a thin four-sided blade tipped with a blunt point to prevent injury. **2.** *often* **foils.** The art of fencing with foils.
**foils·man** (foilz'mən) *n.* One who fences with a foil : FENCER.
**foin** (foin) *vi.* **foined, foin·ing, foins.** [ME *foinen* < *foin*, a thrust < OFr. *foine*, three-pronged fish spear < Lat. *fuscina.*] *Archaic.* To thrust with a pointed weapon. **—foin** *n.*
**foi·son** (foi'zən) *n.* [ME *foisoun* < OFr. *foison* < Lat. *fusio*, a pouring < *fusus*, p.part. of *fundere*, to pour.] **1.** *Archaic.* A plentiful harvest. **2.** *Scot.* Physical strength. **3. foisons.** *Obs.* Reserves of power.
**foist** (foist) *vt.* **foist·ed, foist·ing, foists.** [Dial. Du. *vuisten*, to introduce a palmed die surreptitiously < *vuist*, fist.] **1.** To pass off as real, valuable, or worthy. **2.** To impose upon another by coercion or trickery. **3.** To insert fraudulently or deceitfully <*foisted* unfair provisions into the contract>
**fo·late** (fō'lāt') *n.* [FOL(IC ACID) + -ATE.] Folic acid.

**fold¹** (fōld) *v.* **fold·ed, fold·ing, folds.** [ME *folden* < OE *fealdan.*] —*vt.* **1.** To bend over or double up so that one part lies on top of another part <*fold* a sheet of paper> **2.** To make compact by successively bending over parts <*folded* up the cot> **3.** To bring from an extended to a closed position <The eagle *folded* its wings.> **4.** To place together and intertwine <*fold* one's arms> **5.** To clasp or entwine : EMBRACE. **6.** To mix in (a cooking ingredient) by slowly and gently turning one part over another <*folded* the egg whites into the batter> —*vi.* **1.** To become folded. **2.** *Informal.* To fail financially <Their business *folded.*> **3.** *Informal.* **a.** To give in : YIELD. **b.** To weaken or collapse from exertion. —*n.* **1.** An act or instance of folding. **2.** One part folded over another. **3.** The space at the junction of two folded parts. **4.** A hollow or dale in hilly or mountainous country. **5.** *Geol.* A bend in a stratum of rock. **6.** A coil, as of a snake. **7.** *Anat.* A crease apparently formed by folding, as of a membrane. **—fold'a·ble** *adj.*

**fold²** (fōld) *n.* [ME < OE *fald.*] **1. a.** A fenced enclosure for domestic animals, esp. sheep. **b.** The animals enclosed in such a pen. **2.** A flock of sheep. **3.** A church and its members. **4.** A group of people united by common aims and beliefs. —*vt.* **fold·ed, fold·ing, folds.** To place or keep (e.g., sheep) in a fold.

**-fold** *suff.* [ME < OE *-feald.*] **1.** Divided into a given number of parts <five*fold*> **2.** Multiplied by a given number <fifty*fold*>

**fold·boat** (fōld'bōt') *n.* [Transl. of G. *Faltboot.*] A small boat of rubberized canvas stretched over a collapsible frame.

**fold·er** (fōl'dər) *n.* **1.** One that folds. **2.** A booklet or pamphlet made of one or more folded sheets of paper. **3.** A sheet of cardboard or thick paper folded in the center and used as a holder for loose paper.

**fol·de·rol** (fŏl'də-rŏl') *also* **fal·de·ral** (făl'də-răl') *n.* [From a refrain in some old songs.] **1.** Nonsense. **2.** A gewgaw.

**fold-out** (fōld'out') *n.* A gatefold.

**fo·li·a** (fō'lē-ə) *n. pl. of* FOLIUM.

**fo·li·a·ceous** (fō'lē-ā'shəs) *adj.* [Lat. *foliāceus* < *folium,* leaf.] **1.** Of, pertaining to, or resembling the leaf of a plant. **2.** Having leaves or leaflike structures. **3.** Composed of thin laminated layers, as certain rocks.

**fo·li·age** (fō'lē-ĭj, fō'lĭj) *n.* [ME *foilage* < OFr. *foillage* < *foille,* leaf < Lat. *folium.*] **1. a.** Plant leaves as a whole. **b.** A cluster of leaves. **2.** An ornamental representation of leaves, branches, and flowers. **—fo'li·aged** *adj.*

**foliage plant** *n.* A plant cultivated mainly for its ornamental leaves.

**fo·li·ar** (fō'lē-ər) *adj.* [NLat. *foliaris* < Lat. *folium,* leaf.] FOLIATE 1.

**fo·li·ate** (fō'lē-ĭt, -āt') *adj.* [Lat. *foliatus,* leafy < *folium,* leaf.] **1.** Of or relating to a leaf or leaves. **2.** Shaped like a leaf. —*v.* (-āt') **-at·ed, -at·ing, -ates.** —*vt.* **1.** To hammer or cut (metal) into thin leaf or foil. **2. a.** To coat (e.g., glass) with metal foil. **b.** To furnish or adorn with metal foil. **3.** To separate into thin layers. **4.** To decorate with foliage. **5.** To number the leaves of (e.g., a book). —*vi.* **1.** To grow foliage. **2.** To split into thin layers.

**-foliate** *suff.* [< FOLIATE.] Having a specified kind or number of leaves <tri*foliate*>

**fo·li·a·tion** (fō'lē-ā'shən) *n.* [FOLIA(TE) + -TION.] **1.** The state of being in leaf. **2.** Decoration with foliage. **3.** Decoration of an archway, window, etc., with cusps and foils, as in Gothic tracery. **4.** The act or process of foliating metal or glass. **5. a.** The process of numbering the leaves of a book consecutively. **b.** The leaves of a book that are so numbered.

**fo·lic acid** (fō'lĭk) *n.* [Lat. *folium,* leaf + -IC + ACID.] A yellowish-orange compound, $C_{19}H_{19}N_7O_6$, a member of the vitamin B complex, occurring in green plants, fresh fruit, liver, and yeast and used to treat pernicious anemia.

**fo·lie à deux** (fô-lē' ä dœ', fŏl'ē) *n.* [Fr.] A condition in which the same delusional ideas or beliefs are shared by two people having a close relationship.

**fo·li·ic·o·lous** (fō'lē-ĭk'ə-ləs) *adj.* [Lat. *folium,* leaf + -COLOUS.] Thriving on or parasitic to leaves.

**fo·li·o** (fō'lē-ō') *n., pl.* **-os.** [ME < Lat., ablative of *folium,* leaf.] **1. a.** A large sheet of paper folded once in the middle, making two leaves or four pages of a book or manuscript. **b.** A book or manuscript of the largest common size, consisting of such folded sheets. **2.** A leaf of a book numbered only on the front side. **3.** A page number in a book or magazine. **4.** A page in an accounting ledger or two facing pages assigned a single number. **5.** *Law.* A specific number of words used as a unit for measuring the length of the text of a document. —*vt.* **-oed, -o·ing, -os.** To number consecutively the pages of (e.g., a book).

**-foliolate** *suff.* [< *foliole,* leaflet < Fr. < NLat. *foliolum,* dim. of Lat. *folium,* leaf.] Having a specified kind or number of leaflets <bi*foliolate*>

**fo·li·ose** (fō'lē-ōs') *adj.* [Lat. *foliosus* < *folium,* leafy.] **1.** Bearing numerous leaves or leaflets : LEAFY. **2.** Of, like, or pertaining to a leaf. **3.** Of or pertaining to a lichen with a flat and leafy thallus.

**fo·li·um** (fō'lē-əm) *n., pl.* **-li·a** (-lē-ə) [NLat. < Lat., leaf.] **1.** *Geol.* A thin layer or stratum occurring esp. in metamorphic rock. **2.** *Math.* A plane cubic curve having a single loop, a node, and two ends asymptotic to the same line.

**folk** (fōk) *n., pl.* **folk** *or* **folks.** [ME < OE *folc.*] **1.** An ethnic group : PEOPLE. **2.** *often* **folks.** People of a specified kind or group <town *folks*> **3. folks.** *Informal.* The members of one's family or childhood household : RELATIVES. **4. folks.** *Informal.* People in general <*Folks* gossip.> —*adj.* Of, occurring in, or originating among the common people <*folk* art>

**folk dance** *n.* **1. a.** A traditional dance originating among the common people of a nation or region. **b.** The music for a folk dance. **2.** A social gathering at which folk dances are performed. **—folk dancing** *n.*

**folk etymology** *n.* An alteration in form of a word or phrase so that it resembles a more familiar term mistakenly taken to be analogous, as *sparrowgrass* for *asparagus.*

**folk·lore** (fōk'lôr', -lōr') *n.* **1.** The traditional beliefs, practices, legends, and tales of a people, transmitted orally. **2.** The comparative study of folk knowledge and culture. **3.** A body of widely accepted but specious notions about a place, group, or institution <the *folklore* of Hollywood> **—folk'lor'ic** *adj.* **—folk'lor'ist** *n.*

**folk mass** *n.* A Mass in which folk music accompanies the service.

**folk medicine** *n.* Traditional medicine practiced by people who have no access to professional medical services and usu. involving natural remedies.

**folk·mote** (fōk'mōt') *also* **folk·moot** (-mōōt') *n.* [OE *folcmōt* : *folc,* folk + *mōt,* meeting.] A general assembly of the people of a medieval English town, district, or shire.

**folk music** *n.* Music originating among the common people of a nation or region and marked by a tradition of oral transmission.

**folk rock** *n.* Popular music mixing elements of rock 'n' roll and folk music and often conveying themes of social protest. **—folk'-rock'** *adj.*

**folk singer** *n.* A singer of folk songs. **—folk singing** *n.*

**folk song** *n.* **1.** A song belonging to the folk music of a people or area, marked chiefly by the directness and simplicity of the feelings expressed. **2.** A song composed in imitation of folk songs.

**folk·sy** (fōk'sē) *adj.* **-si·er, -si·est.** *Informal.* **1.** Simple and unpretentious in behavior. **2.** Marked by congeniality and affability. **—folk'si·ly** *adv.* **—folk'si·ness** *n.*

**folk·tale** (fōk'tāl') *n.* A traditional, usu. anonymous story transmitted orally from generation to generation.

**folk·way** (fōk'wā') *n.* A way of thinking or acting adopted by the members of a group as part of their shared culture.

**fol·li·cle** (fŏl'ĭ-kəl) *n.* [Lat. *folliculus,* little bag, dim. of *follis,* bellows.] **1.** *Anat.* **a.** An approx. spherical group of cells containing a cavity. **b.** A vascular body in the ovary containing ova. **2.** *Bot.* A dry single-chambered fruit that splits along only one seam to release its seeds.

**fol·li·cle-stim·u·lat·ing hormone** (fŏl'ĭ-kəl-stĭm'yə-lā'tĭng) *n.* A gonadotropic hormone of the anterior pituitary gland that stimulates the growth of follicles in the ovary and induces spermatogenesis in the testis.

**fol·lic·u·lar** (fə-lĭk'yə-lər) *adj.* **1.** Pertaining to, having, or like a follicle. **2.** Affecting or growing out of follicles.

**fol·lic·u·late** (fə-lĭk'yə-lĭt) *also* **fol·lic·u·lat·ed** (-lā'tĭd) *adj.* Having or consisting of a follicle or follicles.

**fol·lic·u·li·tis** (fə-lĭk'yə-lī'tĭs) *n.* [Lat. *folliculus,* follicle + -ITIS.] Inflammation of a follicle.

**fol·low** (fŏl'ō) *v.* **-lowed, -low·ing, -lows.** [ME *folowen* < OE *folgian.*] —*vt.* **1.** To come or go after <*followed* the leader> **2.** To move behind with the intention of overtaking : PURSUE <Detectives *followed* the suspect.> **3.** To move or go along the course of <We *followed* a road to the town.> **4.** To accept the guidance or leadership of : EMULATE. **5.** To be the result of <A fight *followed* the argument.> **6. a.** To act in agreement with : OBEY <*follow* the law> **b.** To keep to or stick to <*followed* the guidelines> **7.** To come after in order, time, or position <Night *follows* day.> **8.** To engage in (a trade or occupation). **9.** To be evident as a consequence of <Your conclusion does not *follow* your premise.> **10.** To listen to or watch closely <too sleepy to *follow* the program> **11.** To grasp the meaning or logic of : UNDERSTAND <Do you *follow* my argument?> **12.** To inform oneself of the course or progress of <*follow* the company's performance> —*vi.* **1.** To come, move, or take place after another person or thing in order or time. **2.** To occur or be evident as a consequence : RESULT <If you ignore your diet, trouble will *follow.*> **3.** To grasp the meaning or reasoning of what is said : UNDERSTAND. **—follow through. 1.** To carry a stroke to natural completion after hitting the ball, as in golf or bowling. **2.** To carry an act or project to completion. **—follow up. 1.** FOLLOW THROUGH 2. **2.** To increase the effectiveness of by repetition or further action. —*n.* **1.** An act or instance of following. **2.** A billiards shot in which the cue ball is struck in such a way that it follows the path of the object ball after impact.

**fol·low·er** (fŏl'ō-ər) *n.* **1.** An adherent to the teachings or methods of another. **2.** A pursuer. **3.** An attendant, servant, or subordinate. **4.** A machine element moved by another machine element.

**fol·low·ing** (fŏl′ō-ĭng) *adj.* **1.** Coming next in time or order <in the *following* section> **2.** Now to be enumerated <The *following* troops will report for duty.> —*n.* A group or gathering of admirers, adherents, or disciples <a professor with a large *following*>

**fol·low-through** (fŏl′ō-thrōō′) *n.* **1.** The act or process of following through. **2.** The concluding part of a stroke, as in golf or tennis, after a ball has been hit.

**fol·low-up** (fŏl′ō-ŭp′) *n.* **1.** The act or process of following up, esp. so as to increase effectiveness. **2.** A means, as a letter or visit, used to follow up. —*adj.* Designed to follow up, esp. to reinforce previous action <a *follow-up* memorandum>

**fol·ly** (fŏl′ē) *n., pl.* **-lies.** [ME *folie* < OFr. < *fol,* foolish < Lat. *follis,* bellows.] **1.** A lack of good sense, understanding, or foresight. **2. a.** An act or instance of foolishness. **b.** A costly undertaking having an absurd or ruinous outcome. **3. follies** *(sing. in number).* An elaborate theatrical revue consisting of a series of musical or dance skits. **4.** *Archaic.* Action or behavior regarded as immoral or criminal. **5.** *Obs.* Evil : wickedness.

**Fol·som** (fōl′səm) *adj.* [After *Folsom,* New Mexico.] Of or pertaining to an early North American culture of the Pleistocene period flourishing east of the Rocky Mountains and notable primarily for the use of leaf-shaped flint implements.

**Fo·mal·haut** (fō′məl-hôt′) *n.* [Ar. *fum′l-hūt,* mouth of the fish.] The brightest star in the constellation Piscis Austrinus.

**fo·ment** (fō-mĕnt′) *vt.* **-ment·ed, -ment·ing, -ments.** [ME *fomenten,* to apply warm liquids to the skin < Lat. *fomentare.*] **1.** To instigate <*foment* rebellion> **2.** To treat (e.g., the skin) by fomentation. **—fo·ment′er** *n.*

**fo·men·ta·tion** (fō′mən-tā′shən, -mĕn-) *n.* **1.** The act of promoting discontent or strife : INSTIGATION. **2. a.** A warm, moist medicinal compress. **b.** Therapeutic application of warmth and moisture.

**fo·mite** (fō′mīt′) *n.* [Back-formation < NLat. *fomites,* pl. of Lat. *fomes,* tinder.] An inanimate object or substance that functions to transfer infectious organisms from one individual to another.

**fond¹** (fŏnd) *adj.* **-er, -est.** [ME *fonned,* foolish < *fonnen,* to be foolish < *fonne,* fool.] **1.** Having or expressing feelings of affection : TENDER <a *fond* embrace> **2.** Having a strong liking or affection <*fond* of music> **3.** Immoderately or irrationally affectionate : DOTING. **4.** Cherished : dear <my *fondest* dreams> **5.** *Archaic.* Naively credulous or dependent : FOOLISH. **—fond′ly** *adv.*

▲ word history: *Fond* is an example of a word that has undergone *melioration,* or a shift in connotation from bad to neutral or good. *Fond* was originally *fonned,* meaning "insipid or tasteless" and "foolish." It later meant "foolishly affectionate, doting," but even the mild reproach conveyed by this sense was lost, and *fond* now means only "having a strong liking for." The verb *fon,* of which *fond* was once a past participle, is probably the same word as the word *fun.*

**fond²** (fŏnd) *n.* [Fr. < Lat. *fons* < Lat. *fundus,* bottom.] **1.** A foundation : basis. **2.** The background of a design in lace.

**fon·dant** (fŏn′dənt) *n.* [Fr. < pr.part. of *fondre,* to melt < Lat. *fundere.*] **1.** A sweet, creamy sugar paste used in candies and icings. **2.** A candy containing fondant.

**fon·dle** (fŏn′dl) *v.* **-dled, -dling, -dles.** [Freq. of obs. *fond,* to show fondness for.] —*vt.* **1.** To handle, stroke, or caress lovingly. **2.** *Obs.* To treat indulgently and solicitously : PAMPER. —*vi.* To show fondness or affection by caressing. **—fon′dler** *n.*

**fond·ness** (fŏnd′nĭs) *n.* **1.** Warm affection. **2.** Strong inclination : RELISH. **3.** *Archaic.* Naive trustfulness : CREDULITY.

**fon·due** *also* **fon·du** (fŏn-dōō′, -dyōō′) *n.* [Fr. < fem. p.part. of *fondre,* to melt. —see FONDANT.] A hot dish of melted cheese and wine that is eaten with bread.

**font¹** (fŏnt) *n.* [ME < OE < LLat. *fons,* water receptacle for baptism < Lat., fountain.] **1.** A basin holding baptismal water in a church. **2.** A receptacle for holy water : STOUP. **3.** The oil reservoir in an oil-burning lamp. **4.** An abundant source : FOUNT <a *font* of knowledge> **—font′al** (fŏn′tl) *adj.*

**font²** (fŏnt) *n.* [OFr., casting < *fondre,* to melt. —see FONDANT.] A complete set of printing type of one size and face.

**fon·ta·nel** *also* **fon·ta·nelle** (fŏn′tn-ĕl′) *n.* [ME *fontinel* < OFr. *fontanele,* dim. of *fontaine,* fountain.] Any of the soft membranous intervals between the incompletely ossified cranial bones of fetuses and infants.

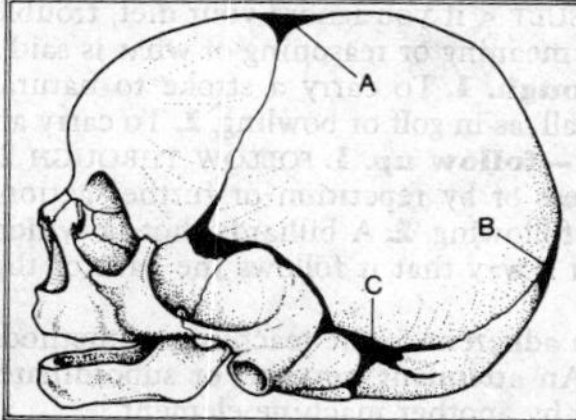


**fontanel**
*A. major, B. minor, C. mastoid*

**fon·ti·na** (fŏn-tē′nə) *n.* [Ital.] A ripened cheese that originated in Italy.

**food** (fōōd) *n.* [ME *fode* < OE *fōda.*] **1.** Material, usu. of plant or animal origin, containing or consisting of essential body nutrients, as carbohydrates, fats, proteins, vitamins, or minerals, taken in and assimilated by an organism to maintain growth and life. **2.** A specified kind of nourishment <breakfast *food*><plant *food*><cat *food*> **3.** Nourishment eaten in solid form. **4.** Something that nourishes in a way likened to physical nourishment.

**food chain** *n.* A succession of organisms in a community that constitutes a feeding chain in which food energy is transferred from one organism to another as each consumes a lower member and in turn is preyed upon by a higher member.

**food poisoning** *n.* Poisoning caused by ingesting food contaminated by natural toxins or bacteria, esp. bacteria of the genus *Staphylococcus,* and marked by vomiting, diarrhea, and prostration.

**food processor** *n.* An appliance having a container with interchangeable metal or plastic blades for processing food, as by slicing or shredding, at high speed.

**food stamp** *n.* A stamp issued by the government and sold or given to low-income persons to be redeemed for food.

**food·stuff** (fōōd′stŭf′) *n.* A substance that can be used or prepared for use as food.

**food vacuole** *n.* A vacuole in which phagocytized food is digested.

**food web** *n.* A complex of interrelated food chains in an ecological community.

**foo·fa·raw** (fōō′fə-rô′) *n.* [Orig. unknown.] **1.** Excessive or flashy ornamentation. **2.** A to-do over nothing.

**fool** (fōōl) *n.* [ME *fol* < OFr. < Lat. *follis,* bellows.] **1.** One deficient in judgment, sense, or understanding. **2.** One who acts unwisely <I was a *fool* to have taken this job.> **3.** A member of a royal or noble household who once entertained the court with jests and mimicry : JESTER. **4.** One who has been or can be easily deceived or imposed upon : DUPE <made a *fool* of me> **5.** *Chiefly Brit.* A dessert made of crushed, stewed fruit mixed with cream or custard and served cold. **6.** *Informal.* A person with a talent or enthusiasm for a certain activity <a *fool* for antique cars> **7.** *Obs.* A feeble-minded person : IDIOT. —*v.* **fooled, fool·ing, fools.** —*vt.* **1.** To deceive or trick : DUPE. **2.** To take unawares <We were sure they would fail, but they *fooled* us.> —*vi.* **1.** *Informal.* To engage in or amuse oneself with useless or trifling activity <just *fooling* around> **2.** To toy or tamper with aimlessly <shouldn't *fool* with matches> **3.** To act or speak in jest : JOKE. **4.** To act or speak without but as if with purposeful or harmful intent <I sounded angry, but I was only *fooling.*> **—fool away.** To waste (time or money) foolishly : SQUANDER.

**fool·er·y** (fōō′lə-rē) *n., pl.* **-ies. 1.** Foolish behavior or speech. **2.** An instance of foolery : JEST.

**fool·har·dy** (fōōl′här′dē) *adj.* **-di·er, -di·est.** [ME *folhardi* < OFr. *fol hardi,* bold fool.] Unwisely bold or venturesome : RASH. **—fool′har′di·ly** *adv.* **—fool′har′di·ness** *n.*

**fool·ish** (fōō′lĭsh) *adj.* **1.** Lacking good sense or judgment : SILLY <*foolish* remarks> **2. a.** Resulting from stupidity or misinformation : UNWISE <a *foolish* decision> **b.** So senseless as to be laughable. **3.** Devoid of meaning or coherence : INANE <a *foolish* smile> **4.** Abashed : embarrassed. **5.** *Archaic.* Insignificant : worthless. **—fool′ish·ly** *adv.* **—fool′ish·ness** *n.*

☆ ***syns:*** FOOLISH, ABSURD, BALMY, CRAZY, DIPPY, IDIOTIC, INSANE, LOONY, LUNATIC, MAD, NONSENSICAL, PREPOSTEROUS, SILLY, TOMFOOL, WACKY, ZANY *adj. core meaning* : so senseless as to be laughable <a *foolish* attempt to cross the country on a tricycle>

**fool·proof** (fōōl′prōōf′) *adj.* **1.** Designed so as to be impervious to human incompetence, error, or misuse <a *foolproof* detonator> **2.** Always effective : INFALLIBLE <a *foolproof* ploy>

**fools·cap** (fōōlz′kăp′) *n.* [From the watermark of a fool's cap with bells orig. used for this paper.] **1.** A sheet of writing or printing paper approx. 13 x 16 inches. **2.** A fool's cap.

**fool's cap** *n.* **1.** A gaily decorated cap, usu. with a number of loose peaks tipped with bells, formerly worn by court jesters and clowns. **2.** A dunce cap.

**fool's errand** *n.* A fruitless errand or undertaking.

**fool's gold** *n.* A mineral, as pyrite, found in gold-colored veins or nuggets and occas. mistaken for gold.

**fool's paradise** *n.* A state of delusive contentment or false hope.

**fool's-pars·ley** (fōōlz′pär′slē) *n.* A poisonous plant, *Aethusa cynapium,* indigenous to Eurasia, having finely divided leaves, small white flower clusters, and a foul odor.

**foot** (fōōt) *n., pl.* **feet** (fēt) [ME *fot* < OE *fōt.*] **1.** The lower extremity of the vertebrate leg that is in direct contact with the ground in standing or walking. **2.** A structure, as the muscular organ extending from the ventral side of a mollusk, used in an invertebrate animal for locomotion or attachment. **3.** Something resembling a foot <the *foot* of a mountain> **4.** The lower end of an object or the end opposite the head, as of a bed. **5.** The lowest part or rank. **6.** The part of a stocking or high-topped boot enclosing the foot. **7.** A manner of moving : STEP <runs with a light *foot*> **8.** Foot soldiers. **9.** The attachment on a sewing machine that clamps down and guides the cloth. **10.** A metrical unit consisting of a stressed or unstressed syllable or syllables. **11.** A unit of length in the U.S. Customary and British Imperial systems, equal to 1/3 yard or 12 inches. **12. foots.** Sediment that forms during the refining of oil and other liquids : DREGS. —*v.* **foot·ed, foot·ing, foots.** —*vi.* **1.** To go on foot : WALK. **2.** To dance. —*vt.* **1.** To go by foot over, on, or through : TREAD. **2.** To provide (e.g., a stocking) with a foot. **3.** To add (a column of numbers) and write the total at the bottom : TOTAL <*foot* up the bill> **4.** *Informal.* To pay <Can you *foot* the bill?> **—on foot.** Walking or standing. **—put (one's) best foot forward.** *Informal.* To make a good beginning or favorable first impression. **—put (one's) foot down.** *Informal.* To assert one's will emphatically. **—put (one's) foot in (one's) mouth.** *Informal.* To make a tactless or embarrassing blunder in speech.

**foot·age** (fo͝ot'ĭj) *n.* **1.** Length or extent, expressed in feet. **2.** A segment of motion-picture film, esp. a segment depicting a particular event or type of action <news *footage*><convention *footage*>

**foot-and-mouth disease** (fo͝ot'ən-mouth') *n.* An acute, highly contagious degenerative but usu. nonfatal viral disease of cattle and other cloven-hoofed animals, marked by fever and the eruption of vesicles around the mouth and hoofs.

**foot·ball** (fo͝ot'bôl') *n.* **1. a.** A game played by two teams of 11 players each on a rectangular 100-yard-long field with goal lines and posts at either end, the object being to gain possession of the ball and advance it in running or passing plays across the opponent's goal line. **b.** A similar game, popular in Canada, played by 12-man teams on a field 110 yards long. **c.** The inflated oval ball used in either of these games. **2.** *Chiefly Brit.* **a.** Rugby football. **b.** The ball used in Rugby football. **3.** *Chiefly Brit.* **a.** Soccer. **b.** The ball used in soccer. **4.** A problem or issue passed about among groups or persons without being settled <The court appointment became a political *football*.>

**foot·board** (fo͝ot'bôrd', -bōrd') *n.* **1.** A board or small raised platform on which to support or rest the feet, as in a carriage. **2.** An upright board across the foot of a bedstead.

**foot brake** *n.* A brake operated by foot pressure on a pedal.

**foot·bridge** (fo͝ot'brĭj') *n.* A pedestrian bridge.

**foot-can·dle** (fo͝ot'kăn'dl) *n. Physics.* The illumination of a surface one foot distant from a source of one candela, equal to one lumen per square foot.

**foot·cloth** (fo͝ot'klôth', -klŏth') *n.* **1.** A richly ornamented cloth draped over the back of a horse and touching the ground. **2.** *Obs.* A carpet or rug.

**foot-drag·ging** (fo͝ot'drăg'ĭng) *n.* Failure to take prompt or required action. **—foot'-drag'ger** *n.*

**foot·ed** (fo͝ot'ĭd) *adj.* **1.** Having a foot or feet. **2.** Having a given kind or number of feet <web-*footed*>

**foot·er** (fo͝ot'ər) *n.* **1.** A pedestrian. **2.** One measured by an indicated number of feet in height or length <a five-*footer*>

**foot·fall** (fo͝ot'fôl') *n.* **1.** A footstep. **2.** The sound of a footstep.

**foot fault** *n.* A fault against the server, as in tennis, called for failure to keep both feet behind the base line.

**foot·gear** (fo͝ot'gîr') *n.* Footwear.

**foot·hill** (fo͝ot'hĭl') *n.* A low hill close to the base of a mountain or mountain range.

**foot·hold** (fo͝ot'hōld') *n.* **1.** A place providing support for the foot in climbing or standing. **2.** A firm or secure position providing a base for further advancement.

**foot·ing** (fo͝ot'ĭng) *n.* **1.** Secure placement of the feet in standing or moving. **2.** A surface or the condition of a surface with respect to the ease with which one may walk or run on it <poor *footing* on the ice> **3.** The supporting base or groundwork of a structure, as for a monument or wall. **4. a.** A basis : foundation <a business begun on a good *footing*> **b.** A basis for social or business transactions with others : STANDING <on equal *footing* with other executives> **5.** The sum of a column of figures.

**foot-lam·bert** (fo͝ot'lăm'bərt) *n. Physics.* A unit of luminance equal to $1/\pi$ candela per square foot.

**foo·tle** (fo͞ot'l) [Orig. unknown.] *Informal.* —*vi.* **-tled, -tling, -tles.** **1.** To waste time : TRIFLE. **2.** To talk nonsense. —*n.* Nonsense. **—foo'tler** *n.*

**foot·less** (fo͝ot'lĭs) *adj.* **1.** Without feet. **2.** Lacking a firm support or basis. **3.** *Informal.* Without thought or skill : INEPT. **—foot'less·ly** *adv.* **—foot'less·ness** *n.*

**foot·lights** (fo͝ot'līts') *pl.n.* **1.** Lights placed in a row along the front of a stage floor. **2.** The theater as a profession.

**foot·lock·er** (fo͝ot'lŏk'ər) *n.* A small trunk for storing personal belongings and small items, esp. one kept at the foot of a bed.

**foot·loose** (fo͝ot'lo͞os') *adj.* Free to do as one pleases.

**foot·man** (fo͝ot'mən) *n.* **1.** A manservant employed in a house to wait at table, attend the door, and run various errands. **2.** *Archaic.* A foot soldier. **3.** *Archaic.* A pedestrian.

**foot·mark** (fo͝ot'märk') *n.* A footprint.

**foot·note** (fo͝ot'nōt') *n.* **1.** A note placed at the bottom of a page or in a special separate section of a book or manuscript that comments on or cites a reference for a designated part of the text. **2.** Something related to but of lesser importance than a larger work or occurrence. —*vt.* **-not·ed, -not·ing, -notes.** To furnish with or comment on in footnotes.

**foot·pace** (fo͝ot'pās') *n.* **1.** A walking pace. **2.** A raised platform in a room, as for a lecturer : DAIS.

**foot·pad** (fo͝ot'păd') *n.* [FOOT + obs. *pad*, highwayman.] *Archaic.* A highwayman or street robber who operates on foot.

**foot·path** (fo͝ot'păth', -päth') *n.* A narrow path for persons on foot.

**foot-pound** (fo͝ot'pound') *n.* A unit of work equal to the work done by a force of one pound acting through a distance of one foot in the direction of the force.

**foot-pound·al** (fo͝ot'poun'dl) *n.* A unit of work equal to the work done by a force of one poundal acting through a distance of one foot in the direction of the force.

**foot-pound-sec·ond** (fo͝ot'pound'sĕk'ənd) *adj.* Of, relating to, or typical of a system of units based on the foot, the pound, and the second as the fundamental units of length, weight, and time.

**foot·print** (fo͝ot'prĭnt') *n.* An outline or indentation left by a foot on a surface.

**foot·race** (fo͝ot'rās') *n.* A race run on foot. **—foot'rac'ing** *n.*

**foot·rest** (fo͝ot'rĕst') *n.* A support on which to rest the feet.

**foot·rope** (fo͝ot'rōp') *n.* **1.** A rope attached to the lower border of a sail. **2.** A rope, rigged beneath a yard, for crew members to stand on during the reefing or furling of sail.

**foot-rot** (fo͝ot'rŏt') *n.* **1.** A degenerative infection of the feet in hoofed animals, esp. cattle or sheep, often causing loss of the hoof. **2.** A disease of plants in which the stem or trunk rots at its base.

**foot soldier** *n.* An infantry soldier.

**foot·sore** (fo͝ot'sôr', -sōr') *adj.* Having sore or tired feet from much walking. **—foot'sore'ness** *n.*

**foot·stalk** (fo͝ot'stôk') *n. Bot.* A supporting stalk, as a peduncle or pedicel.

**foot·stall** (fo͝ot'stôl') *n.* The pedestal or plinth of a pillar.

**foot·step** (fo͝ot'stĕp') *n.* **1. a.** A step with the foot. **b.** The distance covered by one step <a *footstep* away> **c.** The sound of a foot stepping. **2.** A footprint. **3.** A step up or down. **—follow in (someone's) footsteps.** To carry on the behavior, work, or tradition of (a predecessor).

**foot·stone** (fo͝ot'stōn') *n.* A marking stone at the foot of a grave.

**foot·stool** (fo͝ot'sto͞ol') *n.* A low stool.

**foot·wall** (fo͝ot'wôl') *n.* The mass of rock underlying a mineral deposit in a mine.

**foot·way** (fo͝ot'wā') *n.* A walk for pedestrians.

**foot·wear** (fo͝ot'wâr') *n.* Covering for the feet, as shoes or boots.

**foot·work** (fo͝ot'wûrk') *n.* **1.** The way in which the feet are used or maneuvered, as in boxing. **2.** Work involving moving around on foot : LEGWORK.

**fop** (fŏp) *n.* [ME, fool.] A man preoccupied with his clothes and manners : DANDY.

**fop·per·y** (fŏp'ə-rē) *n., pl.* **-ies.** The dress or manner of a fop.

**fop·pish** (fŏp'ĭsh) *adj.* Of, relating to, or typical of a fop : DANDIFIED. **—fop'pish·ly** *adv.* **—fop'pish·ness** *n.*

**for** (fôr; fər *when unstressed*) *prep.* [ME < OE.] **1. a.** —Used to indicate the object, aim, or purpose of an act or activity <trained *for* the priesthood><put the car up *for* sale><plans to run *for* President> **b.** —Used to indicate a destination <were headed *for* town> **2.** —Used to indicate the object of a desire, intention, or perception <an eye *for* good-looking things><eager *for* success> **3. a.** —Used to indicate the recipient or beneficiary of an act <prepared brunch *for* us> **b.** On behalf of <spoke *for* all the crew members> **c.** In favor of <Were they *for* or against the legislation?> **4. a.** —Used to indicate equivalence or equality <paid $30 *for* a ticket><read the letter word *for* word> **b.** —Used to indicate correlation or correspondence <took two steps back *for* every step forward> **5.** —Used to indicate amount, extent, or duration <a bill *for* $30><walked *for* miles><was put on "hold" *for* several minutes> **6.** As being <take *for* granted><mistook me *for* the manager> **7.** As a result of <crying *for* joy> **8.** —Used to indicate appropriateness or suitability <It will be *for* the court to decide.> **9.** Notwithstanding : despite <were inefficient *for* all their experience> **10. a.** As regards : CONCERNING <a gift *for* organization> **b.** Considering the nature or usual character of <unusually warm *for* February> —*conj.* Because : since.

**for-** *pref.* [ME < OE.] Completely : excessively, esp. with destructive or detrimental effect <*for*worn>

**fo·ra** (fôr'ə, fōr'ə) *n. var. pl. of* FORUM.

**for·age** (fôr'ĭj, fŏr'-) *n.* [ME < OFr. *fourrage* < *feurre*, fodder, of Germanic orig.] **1.** FODDER 1. **2.** The act of looking or searching for forage or provisions. —*v.* **-aged, -ag·ing, -ag·es.** —*vi.* **1.** To search for forage or provisions. **2.** To make a raid, as for food. —*vt.* **1.** To wander or rummage through, esp. in search of provisions. **2.** To obtain by foraging. **—for'ag·er** *n.*

**for·am** (fôr′əm, fŏr′-) *n.* A foraminiferan.

**fo·ra·men** (fə-rā′mən) *n., pl.* **-ram·i·na** (-răm′ə-nə) *or* **-ra·mens.** [Lat. *foramen, foramin-*, an opening < *forare*, to bore.] An opening in a bone or through a membranous anatomical structure. **—fo·ram′i·nal** (-răm′ə-nəl), **fo·ram′i·nous** (-nəs) *adj.*

**foramen mag·num** (măg′nəm) *n.* [NLat.] The large orifice in the base of the skull through which the spinal cord passes and becomes continuous with the medulla oblongata.

**foramen o·val·e** (ō-văl′ē, -vă′lē, -vä′-) *n.* [NLat., oval opening.] An opening in the septum between the right and left atria in the heart of a fetus.

**fo·ram·i·na** (fə-răm′ə-nə) *n. var. pl. of* FORAMEN.

**for·a·min·i·fer·an** (fôr′ə-mĭn′ə-fər-ən, fŏr′-) *also* **for·a·min·i·fer** (-mĭn′ə-fər) *n.* [< NLat. *Foraminifera*, order name : Lat. *foramen*, an opening + Lat. *-fer*, -fer.] A unicellular microorganism of the order Foraminifera, having a calcareous shell with perforations through which numerous pseudopodia protrude. **—fo·ram′i·nif′er·ous** (fə-răm′ə-nĭf′ər-əs), **fo·ram′i·nif′er·al** (-əl) *adj.*

**for·as·much as** (fôr′əz-mŭch′ əz) *conj.* Inasmuch as : SINCE.

**for·ay** (fôr′ā′, fŏr′ā′, fôr′ā′) *n.* [ME *forrai* < *forraien*, to plunder.] **1.** A sudden raid or military advance. **2.** An initial attempt or first venture <made an unsuccessful *foray* into politics> *—vi. & vt.* **-ayed, -ay·ing, -ays.** To make a raid or make a raid against.

**forb** (fôrb) *n.* [Gk. *phorbē*, fodder < *pherbein*, to graze.] A herbaceous plant other than a grass, esp. one growing in a field or meadow.

**for·bad** (fər-băd′, fôr-) *v. var. p.t. of* FORBID.

**for·bade** (fər-băd′, -bād′, fôr-) *v. var. p.t. of* FORBID.

**for·bear¹** (fôr-bâr′) *v.* **-bore** (-bôr′, -bōr′), **-borne** (-bôrn′, -bōrn′), **-bear·ing, -bears.** [ME *forberen* < OE *forberan.*] *—vt.* **1.** To refrain from : RESIST <*forbear* replying> **2.** To desist from : CEASE. **3.** *Obs.* To avoid : shun. *—vi.* **1.** To hold back : REFRAIN. **2.** To be tolerant or patient in spite of provocation. **—for·bear′er** *n.*

**for·bear²** (fôr′bâr′, fōr′-) *n. var. of* FOREBEAR.

**for·bear·ance** (fôr-bâr′əns) *n.* **1.** An act of forbearing. **2.** Tolerance and restraint in the face of provocation : PATIENCE. **3.** *Law.* The act of a creditor who refrains from enforcing a debt when it falls due.

**for·bid** (fər-bĭd′, fôr-) *vt.* **-bade** (-băd′, -bād′) *or* **-bad** (-băd′), **-bid·den** (-bĭd′n) *or* **-bid, -bid·ding, -bids.** [ME *forbidden* < OE *forbēodan.*] **1.** To command (someone) not to do something <I *forbid* you to go.> **2.** To prohibit : interdict <Smoking is *forbidden.*> **3.** To have the effect of preventing : PRECLUDE. **—for·bid′dance** *n.*

☆ ***syns:*** FORBID, BAN, DISALLOW, ENJOIN, INTERDICT, OUTLAW, PROHIBIT, PROSCRIBE *v. core meaning* : to refuse to allow <The law *forbids* tax evasion.> *ant:* permit

**for·bid·den** (fər-bĭd′n, fôr-) *adj.* [< p.part of FORBID.] Having a low probability of occurrence. —Used of quantum phenomena <a *forbidden* transition>

**for·bid·ding** (fər-bĭd′ĭng, fôr-) *adj.* **1.** Tending or threatening to impede progress. **2.** Unfriendly : disagreeable <a *forbidding* glare> **—for·bid′ding·ly** *adv.*

**for·bore** (fôr-bôr′, -bōr′) *v. p.t. of* FORBEAR.

**for·borne** (fôr-bôrn′, -bōrn′) *v. p.p. of* FORBEAR.

**force** (fôrs, fōrs) *n.* [ME < OFr. < Lat. *fortis*, strong.] **1.** Capacity to work or cause change <the *force* of an explosion><a personality of great *force*> **2. a.** Power made operative against resistance : EXERTION <use *force* in driving a nail> **b.** The use of such power or exertion <a confession extracted by *force*> **3.** Intellectual power or vigor, as of a statement. **4. a.** A capacity for affecting, influencing, or persuading the mind or behavior : EFFICACY <the *force* of logical argumentation> **b.** One that possesses such capacity <the *forces* of evil> **5.** A group organized or available for a certain purpose <a large labor *force*><an armed *force*> **6.** *Law.* Legal validity. **7.** *Physics.* A vector quantity that tends to produce an acceleration of a body in the direction of its application. *—vt.* **forced, forc·ing, forc·es.** **1.** To compel through pressure or necessity. **2.** To obtain by the use of force or coercion <*force* a confession> **3.** To produce with effort and against one's will <*force* a smile> **4.** To move against resistance : PUSH. **5.** To move, open, or clear by force <*forced* my way through the crowd> **6.** To break down or open by force <*force* a lock> <*force* a door> **7.** RAPE¹ 1. **8.** To inflict : impose <*force* one's will on someone> **9.** To place undue strain on <*force* one's voice> **10.** To cause to grow by artificially accelerating the normal processes <*force* flowers in a greenhouse> **11.** *Baseball.* **a.** To put (a runner) out by tagging the base to which he or she must advance. **b.** To allow (a run) to be scored by walking a batter when the bases are loaded. **—in force. 1.** In full strength. **2.** In effect : OPERATIVE <a law no longer in *force*> **—force′a·ble** *adj.* **—forc′er** *n.*

**forced** (fôrst, fōrst) *adj.* **1.** Imposed or enforced : INVOLUNTARY <a *forced* landing><*forced* labor> **2.** Not spontaneous : UNNATURAL <*forced* laughter>

**force-feed** (fôrs′fēd′, fōrs′-) *vt.* **-fed** (-fĕd′), **-feed·ing, -feeds.** **1.** To feed forcibly. **2.** To force to assimilate <prisoners of war being *force-fed* the party line>

**force field** *n.* A field of force.

**force·ful** (fôrs′fəl, fōrs′-) *adj.* Marked by or full of force. **—force′ful·ly** *adv.* **—force′ful·ness** *n.*

**force ma·jeure** (fôrs′ mä-zhûr′, fōrs′) *n.* [Fr. : *force*, force + *majeure*, greater.] An unexpected or uncontrollable event.

**force·meat** (fôrs′mēt′, fōrs′-) *n.* [*Force*, var. of FARCE + MEAT.] Finely ground meat, fish, or poultry used in stuffing or as a garnish.

**force of habit** *n.* Automatic behavior, as from long practice or frequent repetition.

**for·ceps** (fôr′səps) *n., pl.* **forceps.** [Lat.] **1.** An instrument similar to a pair of tongs, used for grasping, manipulating, or extracting. **2.** A pincerlike clasping organ at the posterior end of the abdomen in certain insects.

**force pump** *n.* A pump with a solid piston and valves used to raise a liquid or expel it under pressure.

**forc·i·ble** (fôr′sə-bəl, fōr′-) *adj.* **1.** Achieved by use of force <a *forcible* entry into the house> **2.** Marked by force : FORCEFUL. **—forc′i·ble·ness** *n.* **—forc′i·bly** *adv.*

**for·ci·pate** (fôr′sə-pāt′) *adj.* [Lat. *forceps, forcip-*, pincers + -ATE¹.] Shaped like a forceps.

**ford** (fôrd, fōrd) *n.* [ME < OE.] A shallow place in a body of water, as a stream, where a crossing can be made without a boat. *—vt.* **ford·ed, ford·ing, fords.** To cross (a body of water) at a ford. **—ford′a·ble** *adj.*

**for·do** *also* **fore·do** (fôr-dōō′, fōr-) *vt.* **-did** (-dĭd′), **-done** (-dŭn′), **-do·ing, -does** (-dŭz′) [ME *fordon* < OE *fordōn* : *for-*, destruction + *dōn*, to do.] *Archaic.* **1.** To bring to ruin. **2.** To exhaust utterly.

**fore** (fôr, fōr) *adj.* [ME < *fore*, beforehand < OE.] Located at or toward the front : FORWARD. *—n.* **1.** Something situated at or near the front. **2.** The front part. *—adv.* At, toward, or near the front : FORWARD. *—prep. also* **'fore.** *Archaic.* Before. *—interj.* —Used by a golfer to warn those ahead that a ball is about to be driven in their direction. **—to the fore.** In, into, or toward a prominent position <A new piano virtuoso has come *to the fore.*>

**fore-** *pref.* [ME < OE < *fore*, in front.] **1.** Before : earlier <*foredoom*> **2.** In front of : FRONT <*foredeck*>

**fore and aft** *adv.* **1.** From the bow of a ship to the stern : LENGTHWISE. **2.** In, at, or toward both ends of a ship.

**fore-and-aft** (fôr′ən-ăft′, fōr′-) *adj.* Parallel with a ship's keel.

**fore-and-aft·er** (fôr′ən-ăf′tər, fōr′-) *n.* A sailing ship, as a schooner, with a fore-and-aft rig.

**fore-and-aft rig** *n.* A sailing ship rig with quadrilateral and triangular sails set to the fore-and-aft line and able to be trimmed to the leeward.

**fore-and-aft sail** *n.* A sail set parallel with the keel of a vessel, having the foremost edge or luff attached to the mast with travelers and the upper edge set on a gaff or stay.

**fore·arm¹** (fôr-ärm′, fōr-) *vt.* **-armed, -arm·ing, -arms.** To prepare or arm prior to a conflict.

**fore·arm²** (fôr′ärm′, fōr′-) *n.* The part of the arm extending from the wrist to the elbow.

**fore·bear** *also* **for·bear** (fôr′bâr′, fōr′-) *n.* [ME : *fore-*, fore- + *been*, to be.] An ancestor.

**fore·bode** (fôr-bōd′, fōr-) *vt.* **-bod·ed, -bod·ing, -bodes.** **1.** To indicate the likelihood of : PORTEND <food riots that *foreboded* revolution> **2.** To have a premonition of (a future misfortune).

**fore·bod·ing** (fôr-bō′dĭng, fōr-) *n.* **1.** A sense of impending misfortune or evil : PREMONITION. **2.** An evil omen. *—adj.* Ominous. **—fore·bod′ing·ly** *adv.*

**fore·brain** (fôr′brān′, fōr′-) *n.* **1.** The anterior region of the embryonic brain from which the telencephalon and diencephalon develop. **2.** The segment of the adult brain that develops from the embryonic forebrain and includes the cerebrum, thalamus, and hypothalamus.

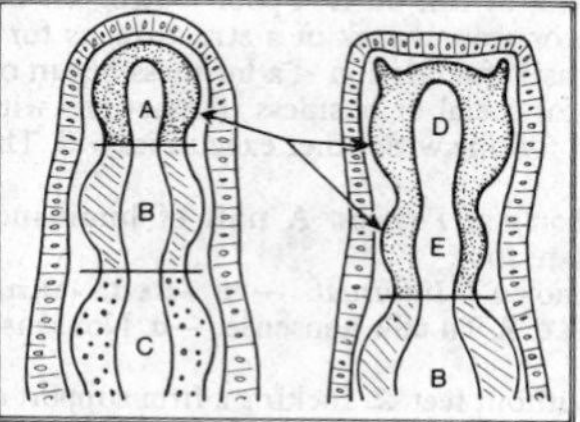


**forebrain**
(Left) *A. forebrain, B. midbrain, C. hindbrain,* (right, in enlarged view) *D. telencephalon, E. diencephalon, B. midbrain*

**fore·cast** (fôr′kăst′, fōr′-) *v.* **-cast** *or* **-cast·ed, -cast·ing, -casts.** [ME *forecasten*, to plan beforehand : *fore-*, fore- + *casten*, to contrive.] *—vt.* **1. a.** To estimate or calculate in advance <*forecast* next year's sales> **b.** To predict (weather conditions) by analysis of meteorological data. **2.** To serve as an advance indication of : FORESHADOW <price rises that *forecast* inflation> *—vi.* To make an estimation or calculation in advance. *—n.* A prediction, as of coming events. **—fore′cast′er** *n.*

**fore·cas·tle** (fōk′səl, fôr′kăs′əl, fōr′-) *also* **fo'c's'le** (fōk′səl) *n.* [ME < AN.] **1.** The section of a ship's upper deck situated at the bow forward of the foremast. **2.** A superstructure at the bow of a merchant ship where the crew is housed.

**fore·close** (fôr-klōz′, fōr-) *v.* **-closed, -clos·ing, -clos·es.** [ME *forclosen,* to exclude from an inheritance < OFr. *forclore* (p.part. *forclos*) : *fors-,* outside (< Lat. *foris*) + *clore,* to close (< Lat. *claudere*).] —*vt.* **1.** *Law.* **a.** To deprive (a mortgagor) of the right to redeem mortgaged property, as when payments have not been made. **b.** To bar an equity or right to redeem (a mortgage). **2.** To rule out : BAR. **3.** To settle or resolve ahead of time. —*vi.* To foreclose a mortgage. **—fore·clos′a·ble** *adj.*

**fore·clo·sure** (fôr-klō′zhər, fōr-) *n.* The act of foreclosing, esp. a legal proceeding by which a mortgage is foreclosed.

**fore·court** (fôr′kôrt′, fōr′kōrt′) *n.* **1.** A courtyard in front of a building. **2.** The area of a playing court nearest the net or wall, as in tennis or handball.

**fore·deck** (fôr′dĕk′, fōr′-) *n.* The forward part of a ship's deck, usu. the main deck.

**fore·do** (fôr-do͞o′, fōr-) *v. var. of* FORDO.

**fore·doom** (fôr-do͞om′, fōr-) *vt.* **-doomed, -doom·ing, -dooms.** To doom beforehand <*foredoomed* them to failure>

**fore·fa·ther** (fôr′fä′thər, fōr′-) *n.* [ME *forefader* < *fader,* father.] **1.** An ancestor. **2.** One from an earlier time and common tradition.

**fore·fend** (fôr-fĕnd′, fōr-) *v. var. of* FORFEND.

**fore·fin·ger** (fôr′fĭng′gər, fōr′-) *n.* The index finger.

**fore·foot** (fôr′fo͝ot′, fōr′-) *n.* **1.** One of the front feet of an animal. **2.** The part of a ship at which the prow joins the keel.

**fore·front** (fôr′frŭnt′, fōr′-) *n.* **1.** The foremost part. **2.** The position of greatest importance, prominence, or responsibility : VANGUARD <in the *forefront* of national politics>

**fore·gath·er** (fôr-găth′ər, fōr-) *v. var. of* FORGATHER.

**fore·go**[1] (fôr-gō′, fōr-) *vt.* **-went** (-wĕnt′), **-gone** (-gôn′, -gŏn′), **-go·ing, -goes** (-gōz′) [ME *forgon* < OE *forgān.*] To go before, as in time or place : PRECEDE. **—fore·go′er** *n.*

**fore·go**[2] (fôr-gō′, fōr-) *v. var. of* FORGO.

**fore·go·ing** (fôr-gō′ĭng, fōr-, fôr′gō′ĭng, fōr′-) *adj.* Said, written, or encountered just before : PREVIOUS <the *foregoing* comment>

**fore·gone** (fôr′gôn′, -gŏn′, fōr′-) *adj.* [< p.part. of FOREGO[1].] Past : previous <*foregone* remarks>

**foregone conclusion** *n.* An inevitable end or result.

**fore·ground** (fôr′ground′, fōr′-) *n.* **1.** The part of a view or picture that is or is depicted as nearest to the viewer. **2.** FOREFRONT 2.

**fore·gut** (fôr′gŭt′, fōr′-) *n.* The anterior part of the embryonic digestive tract from which the pharynx lining, lungs, esophagus, stomach, and small intestine develop.

**fore·hand** (fôr′hănd′, fōr′-) *adj.* **1.** Made or performed with the hand moving palm forward <a *forehand* stroke in tennis> **2.** *Archaic.* Taking place or performed beforehand : PRIOR. —*n.* **1.** A forehand stroke, as in tennis. **2.** The part of a horse in front of the rider. —*adv.* With a forehand motion or stroke.

**fore·hand·ed** (fôr′hăn′dĭd, fōr′-) *adj.* **1.** Forehand, as in tennis. **2.** Looking or planning ahead : CIRCUMSPECT. **3.** Having ample financial resources. **—fore′hand′ed·ly** *adv.* **—fore′hand′ed·ness** *n.*

**fore·head** (fôr′ĭd, fŏr′-, fôr′hĕd′, fōr′-) *n.* [ME *forhed* < OE *forhēafod.*] The part of the head or face between the eyebrows and the normal hairline.

**for·eign** (fôr′ĭn, fŏr′-) *adj.* [ME *forein* < OFr. *forain* < Lat. *foras,* outside.] **1.** Located away from one's native country <*foreign* cities> **2.** Of, from, or typical of a country other than one's own <*foreign* customs> **3.** Carried on or involved with other nations or governments <*foreign* trade agreements> **4.** Located in an abnormal or improper place in the body <a *foreign* object in one's eye> **5.** Not natural : ALIEN <Envy is *foreign* to my nature.> **6.** Not appropriate or essential : IRRELEVANT. **7.** *Law.* Subject to the jurisdiction of another political unit. **—for′eign·ness** *n.*

☆ ***syns:*** FOREIGN, ALIEN, EXOTIC *adj. core meaning* : of, from, or typical of another place or part of the world <a *foreign* species of fern><*foreign* languages>

**foreign bill** *n.* A draft for a sum of money to be paid in another country.

**foreign correspondent** *n.* A print or broadcast journalist who sends news reports or commentary from a foreign country for publication or airing.

**for·eign·er** (fôr′ə-nər, fŏr′-) *n.* A person from a foreign country.

**foreign exchange** *n.* **1.** Transaction of international monetary business, as between governments or business people of different countries. **2.** Negotiable bills drawn in one country to be paid in another country.

**foreign mission** *n.* **1.** A group sent to a foreign country to serve as missionaries. **2.** A group sent abroad for diplomatic service.

**foreign office** *n.* The governmental department in charge of foreign affairs in several countries.

**foreign policy** *n.* The diplomatic policy of a nation in its interactions with other nations.

**foreign service** *n.* A nation's diplomatic and consular staff.

**fore·judge** *also* **for·judge** (fôr-jŭj′, fōr-) *vt.* & *vi.* **-judged, -judg·ing, -judg·es.** To judge beforehand : PREJUDGE. **—fore·judg′ment** *n.*

**fore·know** (fôr-nō′, fōr-) *vt.* **-knew** (-no͞o′, -nyo͞o′), **-known** (-nōn′), **-know·ing, -knows.** To know ahead of time.

**fore·knowl·edge** (fôr-nŏl′ĭj, fōr-) *n.* Prior knowledge or awareness of something : PRESCIENCE.

**fore·known** (fôr-nōn′, fōr-) *v. p.p. of* FOREKNOW.

**fore·la·dy** (fôr′lā′dē, fōr′-) *n.* A forewoman.

**fore·land** (fôr′lənd, fōr′-) *n.* A projecting land mass : PROMONTORY.

**fore·leg** (fôr′lĕg′, fōr′-) *n.* One of an animal's front legs.

**fore·limb** (fôr′lĭm′, fōr′-) *n.* An anterior appendage, as a leg, wing, or flipper.

**fore·lock**[1] (fôr′lŏk′, fōr′-) *n.* A lock of hair growing or falling on the forehead.

**fore·lock**[2] (fôr′lŏk′, fōr′-) *n.* A cotter pin : LINCHPIN.

**fore·man** (fôr′mən, fōr′-) *n.* **1.** One in charge of a group of workers, as at a factory. **2.** The chairman and spokesman for a jury. **—fore′man·ship′** *n.*

**fore·mast** (fôr′məst, -măst′, fōr′-) *n.* The forward mast on a sailing vessel.

**fore·milk** (fôr′mĭlk′, fōr′-) *n.* Colostrum.

**fore·most** (fôr′mōst′, fōr′-) *adj.* [Alteration of ME *formest* < *forme,* first < OE *forma.*] **1.** Being first in time or place. **2.** Being ahead of all others, esp. in position or rank : PARAMOUNT. **—fore′most** *adv.*

**fore·moth·er** (fôr′mŭth′ər, fōr′-) *n.* A woman ancestor.

**fore·name** (fôr′nām′, fōr′-) *n.* A first name.

**fore·noon** (fôr′no͞on′, fōr′-, fôr-no͞on′, fōr-) *n.* The period between sunrise and noon.

**fo·ren·sic** (fə-rĕn′sĭk, -zĭk) *adj.* [Lat. *forensis* < *forum,* forum.] **1.** Of, relating to, or used in legal proceedings or argumentation <*forensic* medicine> **2.** Of, relating to, or used in debate or argument : RHETORICAL. **—fo·ren′si·cal·ly** *adv.*

**fo·ren·sics** (fə-rĕn′sĭks, -zĭks) *n. (sing. in number).* The study or practice of formal debate : ARGUMENTATION.

**fore·or·dain** (fôr′ôr-dān′, fōr′-) *vt.* **-dained, -dain·ing, -dains.** To appoint or ordain beforehand : PREDESTINE. **—fore′or·dain′ment, fore·or′di·na′tion** (-ôr′dn-ā′shən) *n.*

**fore·part** (fôr′pärt′, fōr′-) *n.* **1.** The first or earliest time period. **2.** An anterior part.

**fore·paw** (fôr′pô′, fōr′-) *n.* The paw of an animal's foreleg.

**fore·peak** (fôr′pēk′, fōr′-) *n.* The section of a ship's hold within the angle formed by the bow.

**fore·play** (fôr′plā′, fōr′-) *n.* Sexual stimulation before intercourse.

**fore·quar·ter** (fôr′kwôr′tər, fōr′-) *n.* **1.** The front section of a side of meat. **2.** The foreleg, shoulder, and adjacent lateral parts of an animal, esp. a horse.

**fore·ran** (fôr-răn′, fōr-) *v. p.t. of* FORERUN.

**fore·reach** (fôr-rēch′, fōr-) *v.* **-reached, -reach·ing, -reach·es.** —*vt.* To get ahead of, esp. in a sailing vessel. —*vi.* To gain ground, esp. on a sailing vessel.

**fore·run** (fôr-rŭn′, fōr-) *vt.* **-ran** (-răn′), **-run, -run·ning, -runs.** **1.** To run in front of. **2.** To be the precursor of : FORESHADOW. **3.** To forestall : prevent.

**fore·run·ner** (fôr′rŭn′ər, fōr′-) *n.* **1.** One that precedes, as in time : PREDECESSOR. **2.** An ancestor. **3.** One that provides advance notice of the coming of others : HARBINGER.

**fore·said** (fôr′sĕd′, fōr′-) *adj.* Aforesaid.

**fore·sail** (fôr′səl, -sāl′, fōr′-) *n. Naut.* **1.** The principal square sail hung to the foremast of a square-rigged vessel. **2.** The principal triangular sail hung to the mast of a fore-and-aft-rigged vessel. **3.** The triangular sail hung to the forestay of a cutter or sloop.

**fore·see** (fôr-sē′, fōr-) *vt.* **-saw** (-sô′), **-seen** (-sēn′), **-see·ing, -sees.** To see or know beforehand <*foresaw* the rapid inflationary spiral> **—fore·see′a·ble** *adj.* **—fore·se′er** *n.*

**fore·shad·ow** (fôr-shăd′ō, fōr-) *vt.* **-owed, -ow·ing, -ows.** To present an indication of beforehand : PRESAGE.

**fore·sheet** (fôr′shēt′, fōr′-) *n. Naut.* **1.** A rope used in trimming a foresail. **2. foresheets.** The space near the bow of an open boat.

**fore·shock** (fôr′shŏk′, fōr′-) *n.* A minor tremor occurring before an earthquake.

**fore·shore** (fôr′shôr′, fōr′shōr′) *n.* **1.** The part of a shore covered at high tide. **2.** The part of a shore between the water and inhabited or cultivated land.

**fore·short·en** (fôr-shôr′tn, fōr-) *vt.* **-ened, -en·ing, -ens.** **1.** To represent the long axis of (an object) by contracting its lines to produce an illusion of projection or extension in space. **2.** To shorten beforehand : CURTAIL.

**fore·show** (fôr-shō′, fōr-) *vt.* **-showed, -shown** (-shōn′) *or* **-showed, -show·ing, -shows.** To show in advance : PREFIGURE.

**fore·side** (fôr′sīd′, fōr′-) *n.* The front or upper side.

**fore·sight** (fôr′sīt′, fōr′-) *n.* **1. a.** An act of foreseeing. **b.** Ability to foresee. **2.** The act of looking forward. **3.** Concern or prudence regarding the future. **—fore'sight'ed** *adj.* **—fore'sight'ed·ly** *adv.* **—fore'sight'ed·ness** *n.*

**fore·skin** (fôr'skĭn', fōr'-) *n.* PREPUCE 1.

**fore·speak** (fôr-spēk', fōr-) *vt.* **-spoke** (-spōk'), **-spo·ken** (-spō'-kən), **-speak·ing, speaks. 1.** To predict. **2.** To arrange for ahead of time.

**for·est** (fôr'ĭst, fŏr'-) *n.* [ME < OFr. < Med. Lat. *foresta* < Lat. *foris,* outside.] **1.** A dense growth of trees, together with other plants, covering a large area. **2.** Something resembling a forest in density, quantity, or profusion <a *forest* of tall buildings> **3.** A defined area of land once set aside in England as a royal hunting ground. —*vt.* **-est·ed, -est·ing, -ests.** To plant trees on. **—for'est·al, fo·res'tial** (fə-rĕs'chəl) *adj.* **—for'es·ta'tion** *n.*

**fore·stall** (fôr-stôl', fōr-) *vt.* **-stalled, -stall·ing, -stalls.** [ME *forestallen,* to waylay and rob < *forestal,* highway robbery < OE *foresteall* : *fore,* in front of + *steall,* position.] **1.** To prevent, delay, or take precautionary measures against beforehand. **2.** To deal with or think of beforehand : ANTICIPATE. **3.** To hinder or prevent normal sales of by buying up merchandise, discouraging persons from bringing their goods to market, or encouraging an increase in prices of goods already on the market. **—fore·stall'er** *n.*

**fore·stay** (fôr'stā', fōr'-) *n. Naut.* A stay extending from the head of the foremast to the bowsprit.

**fore·stay·sail** (fôr'stā'səl, -sāl', fōr'-) *n. Naut.* A triangular sail set on the forestay.

**for·est·er** (fôr'ĭ-stər, fŏr'-) *n.* **1.** A specialist in forestry. **2.** An inhabitant of a forest. **3.** Any of various chiefly tropical moths of the family Agaristidae.

**for·est·land** (fôr'ĭst-lănd', fŏr'-) *n.* Land covered with forest.

**forest ranger** *n.* An officer in charge of managing or protecting a public forest or section of a public forest.

**for·est·ry** (fôr'ĭ-strē, fŏr'-) *n.* **1.** The art and science of cultivating, maintaining, and developing forests. **2.** Management of forestland. **3.** A forestland.

**fore·swear** (fôr-swâr', fōr-) *v. var. of* FORSWEAR.

**fore·taste** (fôr'tāst', fōr'-) *n.* An advance taste or realization <a *foretaste* of disaster> —*vt.* (fôr-tāst', fōr-, fôr'tāst', fōr'-) **-tast·ed, -tast·ing, -tastes.** To anticipate.

**fore·tell** (fôr-tĕl', fōr-) *vt.* **-told** (-tōld'), **-tell·ing, -tells.** To tell of or indicate beforehand : PREDICT. **—fore·tell'er** *n.*

**fore·thought** (fôr'thôt', fōr'-) *n.* **1.** Prior deliberation, consideration, or planning. **2.** Preparation or thought for the future : ANTICIPATION. **—fore'thought'ful** *adj.* **—fore'thought'ful·ly** *adv.* **—fore'thought'ful·ness** *n.*

**fore·to·ken** (fôr-tō'kən, fōr-) *vt.* **-kened, -ken·ing, -kens.** To foreshow : presage. —*n.* (fôr'tō'kən, fōr'-). An advance warning.

**fore·told** (fôr-tōld', fōr-) *v. p.t. & p.p. of* FORETELL.

**fore·top** (fôr'tŏp', fōr'-) *n.* **1.** (*also* -təp). A platform at the top of a ship's foremast. **2.** A forelock, esp. of a horse.

**fore·top·gal·lant** (fôr'tŏp-găl'ənt, fōr'-, fôr'tə-, fōr'tə-) *adj. Naut.* Of or pertaining to the mast directly above the foremast.

**fore·top·mast** (fôr'tŏp'məst, fōr'-, fôr'təp-măst', fōr'təp-) *n. Naut.* The mast above the foretop.

**fore·top·sail** (fôr'tŏp'səl, fōr'-, fôr'təp-, fōr'təp-) *n. Naut.* The sail hung from the foretopmast.

**for·ev·er** (fôr-ĕv'ər, fər-) *adv.* **1.** For everlasting time : ETERNALLY. **2.** At all times : INCESSANTLY <*forever* whining>

**for·ev·er·more** (fôr-ĕv'ər-môr', -mōr', fər-) *adv.* Forever.

**fore·warn** (fôr-wôrn', fōr-) *vt.* **-warned, -warn·ing, -warns.** To warn ahead of time.

**fore·went** (fôr-wĕnt') *v. p.t. of* FOREGO.

**fore·wing** (fôr'wĭng', fōr'-) *n.* One of a pair of anterior wings, as in certain insects.

**fore·wom·an** (fôr'wŏŏm'ən, fōr'-) *n.* **1.** A woman in charge of a group of workers, as at a factory. **2.** The chairwoman and spokeswoman for a jury.

**fore·word** (fôr'wərd, fōr'-) *n.* A preface or introductory note, esp. at the beginning of a book.

**fore·worn** (fôr-wôrn', fōr-wōrn') *adj. var. of* FORWORN.

**fore·yard** (fôr'yärd', fōr'-) *n. Naut.* The lowest yard on a foremast.

**for·feit** (fôr'fĭt) *n.* [ME < *forfet,* forfeited < OFr., p.part. of *forsfaire,* to commit a crime : *fors,* beyond (< Lat. *foris,* outside) + *faire,* to do (< Lat. *facere*).] **1.** Something surrendered as punishment for a crime, offense, error, or breach of contract. **2.** Something placed in escrow and then redeemed after payment of a fine. **3.** A forfeiture. **4. forfeits.** A game in which forfeits are required. —*adj.* Surrendered or alienated for a crime, offense, error, or breach of contract. —*vt.* **-feit·ed, -feit·ing, -feits. 1.** To surrender or be forced to surrender as a forfeit. **2.** To subject to forfeiture. **—for'feit·a·ble** *adj.* **—for'feit·er** *n.*

**for·fei·ture** (fôr'fĭ-chŏŏr', -chər) *n.* **1.** The act of surrendering something as a forfeit. **2.** Something forfeited.

**for·fend** *also* **fore·fend** (fôr-fĕnd', fōr-) *vt.* **-fend·ed, -fend·ing, -fends.** [ME *forfenden.*] **1.** To ward off : AVERT. **2.** *Archaic.* To forbid. **3.** To defend : protect.

**for·fi·cate** (fôr'fĭ-kĭt, -kāt') *adj.* [Lat. *forfex, forfic-,* scissors + -ATE[1].] Deeply forked or notched, as the tail of some birds.

**forficate**
*The forficate tail of a flycatcher*

**for·gath·er** *also* **fore·gath·er** (fôr-găth'ər, fōr-) *vi.* **-ered, -er·ing, -ers.** [Sc.] **1.** To gather together : ASSEMBLE. **2.** To meet accidentally. **3.** To keep company : CONSORT.

**for·gave** (fər-gāv', fôr-) *v. p.t. of* FORGIVE.

**forge[1]** (fôrj, fōrj) *n.* [ME < OFr. < Lat. *fabrica* < *faber,* worker.] **1.** A furnace or hearth where metals are heated or wrought : SMITHY. **2.** A shop in which pig iron is converted into wrought iron. —*v.* **forged, forg·ing, forg·es.** —*vt.* **1.** To form (metal) by heating in a forge and beating or hammering it into shape. **2.** To give form or shape to <*forge* an agreement> **3.** To fashion or reproduce for fraudulent purposes : COUNTERFEIT <*forge* a signature> —*vi.* **1.** To work at a forge. **2.** To make a forgery. **—forge'a·bil'i·ty** *n.* **—forge'a·ble** *adj.* **—forg'er** *n.*

**forge[2]** (fôrj, fōrj) *vi.* **forged, forg·ing, forg·es.** [Orig. unknown.] **1.** To advance gradually, yet firmly. **2.** To advance with an abrupt increase of speed <*forged* into the lead in the final lap of the race>

**for·ger·y** (fôr'jə-rē, fōr'-) *n., pl.* **-ies. 1.** The act of forging, esp. the illegal production of counterfeit material. **2.** Something counterfeit.

**for·get** (fər-gĕt', fôr-) *v.* **-got** (-gŏt'), **-got·ten** (-gŏt'n) *or* **-got, -get·ting, -gets.** [ME *forgeten* < OE *forgetan.*] —*vt.* **1.** To be unable to remember. **2.** To lack concern for : NEGLECT <*forget* one's duty> **3.** To leave behind unintentionally. **4.** To fail to mention. **5.** To banish from one's thoughts <*forget* an insult> —*vi.* **1.** To stop remembering. **2.** To fail or neglect to become aware at the proper or specified moment <*forget* about paying one's taxes> **—forget (oneself).** To lose one's reserve or self-restraint. **—for·get'ta·ble** *adj.* **—for·get'ter** *n.*

**for·get·ful** (fər-gĕt'fəl, fôr-) *adj.* **1.** Tending to forget. **2.** Neglectfully or thoughtlessly inattentive. **—for·get'ful·ly** *adv.* **—for·get'ful·ness** *n.*

**for·ge·tive** (fôr'jĭ-tĭv, fōr'-) *adj.* [Poss. FORGE[1] + (INVEN)TIVE.] *Archaic.* Capable of imagining or inventing.

**for·get-me-not** (fər-gĕt'mē-nŏt', fôr-) *n.* [Transl. of OFr. *ne m'oubliez mie.*] **1.** A low-growing plant of the genus *Myosotis,* having small blue flower clusters. **2.** A plant related to or resembling the forget-me-not.

**for·give** (fər-gĭv', fôr-) *v.* **-gave** (-gāv'), **-giv·en** (-gĭv'ən), **-giv·ing, -gives.** [ME *forgiven* < OE *forgifan.*] —*vt.* **1.** To excuse for a fault or offense : PARDON. **2.** To renounce anger or resentment against. **3.** To absolve from payment of. —*vi.* To accord forgiveness. **—for·giv'a·ble** *adj.* **—for·give'ness** (-gĭv'nĭs, fôr-) *n.* **—for·giv'er** *n.*

**for·go** *also* **fore·go** (fôr-gō', fōr-) *vt.* **-went** (-wĕnt'), **-gone** (-gôn', -gŏn'), **-go·ing, -goes.** [ME *forgon* < OE *forgān.*] To abstain from : RELINQUISH <*forgo* drinking> **—for·go'er** *n.*

**for·got** (fər-gŏt', fôr-) *v. p.t. & var. p.p. of* FORGET.

**for·got·ten** (fər-gŏt'n, fôr-) *v. var. p.p. of* FORGET.

**fo·rint** (fôr'ĭnt') *n.* [Hung. < Ital. *fiorino,* florin. —see FLORIN.] —See table at CURRENCY.

**for·judge** (fôr-jŭj', fōr-) *v. var. of* FOREJUDGE.

**fork** (fôrk) *n.* [ME *forke* < OE *forca* < Lat. *furca.*] **1.** An implement or piece of equipment with two or more prongs used for lifting, carrying, pitching, piercing, or digging. **2.** A forked utensil for serving or eating food. **3. a.** A separation into two or more branches or parts. **b.** The point at which such a bifurcation or separation occurs <a *fork* in a road> **c.** One of the branches of such a separation <the right *fork*> —*v.* **forked, fork·ing, forks.** —*vt.* **1.** To lift, carry, pitch, pierce, or dig with a fork. **2.** To give the shape of a fork to. **3.** To launch an attack on (two chesspieces). —*vi.* **1.** To divide into two or more branches. **2.** *Informal.* To hand over : PAY <*forked* over $1,000 as a down payment>

**forked** (fôrkt, fôr'kĭd) *adj.* **1.** Having or marked by a fork <a *forked* creek> **2.** Shaped like a fork <a *forked* tail>

**fork lift** *n.* An industrial vehicle with a power-operated pronged platform that can be raised and lowered for insertion under a load that is to be lifted and carried.

**for·lorn** (fər-lôrn', fôr-) *adj.* [ME *forloren,* p.part. of *forlesen,* to abandon < OE *forlēosan.*] **1.** Seeming to be sad or lonely due to desertion or abandonment. **2.** Suffering extreme want : DESTITUTE <*forlorn* of all hope> **3.** Wretched or pitiful <a *forlorn* war refugee> **4.** Practically hopeless. **—for·lorn'ly** *adv.* **—for·lorn'ness** *n.*

**forlorn hope** *n.* [By folk etymology < Du. *verloren hoop* : *verlo-*

*ren,* lost + *hoop,* troop.] **1.** A hopeless or arduous undertaking. **2.** An advance guard of troops sent on a hazardous mission.

**form** (fôrm) *n.* [ME *forme* < Lat. *forma.*] **1.** The shape and structure of something. **2.** The body or outward appearance of a person or animal taken separately from the face or head : FIGURE. **3.** The essence of something. **4.** The mode in which a thing exists, acts, or manifests itself : KIND <a *form* of plant life> **5.** Procedure as determined or governed by custom or regulation. **6.** Manners or conduct as governed by decorum, etiquette, or custom. **7.** Performance considered with regard to acknowledged criteria <a good skater with stylish *form*> **8.** Fitness, as of an animal, with regard to health or training. **9.** A fixed order of words or procedures, as in a ceremony. **10.** A document with blanks for the insertion of details or information. **11.** Style or manner of presenting ideas or concepts in literary or musical composition or in organized discourse <a treatise in the *form* of a dialogue> **12.** The design, structure, or pattern of a work of art <symphonic *form*> **13.** A model for making a mold. **14.** A copy of the human figure used for modeling clothes. **15.** Linotype assembled and locked up in a chase for printing. **16.** A grade in a British school or in some U.S. private schools <the sixth *form*> **17. a.** A linguistic form. **b.** The external aspect of words with regard to their inflections, pronunciation, or spelling <verb *forms*> **18.** *Chiefly Brit.* A bench. **19.** A hare's resting place. —*v.* **formed, form·ing, forms.** —*vt.* **1.** To give form to : SHAPE. **2.** To shape or mold into a given form. **3.** To fashion, train, or develop by instruction <*form* a child's mind> **4.** To come to have : DEVELOP <*form* an attachment to the child> **5.** To constitute or be an element, part, or characteristic of. **6.** To develop mentally : CONCEIVE <*form* an opinion> **7. a.** To produce (e.g., a tense) by assuming an inflection <*form* the past perfect> **b.** To make (a word) by derivation or composition. **8.** To put in order : ARRANGE. —*vi.* **1.** To become formed or shaped. **2.** To come into being : ARISE. **3.** To assume a given shape or pattern. **—form'a·bil'i·ty** *n.* **—form'a·ble** *adj.*

**-form** *suff.* Having the form of <plexi*form*>

**for·mal** (fôr'məl) *adj.* [ME < Lat. *formalis* < *forma,* shape.] **1.** Relating to the outward aspect of something. **2.** Being or relating to essential form or constitution <a *formal* principle> **3.** Adhering to accepted forms, conventions, or regulations <a *formal* requirement> **4.** Performed in regular or proper form <a *formal* rebuke> **5.** Marked by strict or meticulous observation of forms : METHODICAL. **6.** Stiff or cold <a *formal* manner> **7.** Having the outward appearance but lacking in substance <a purely *formal* farewell> —*n.* **1.** An occasion requiring formal attire, esp. evening clothes. **2.** Formal attire. **—for'mal·ly** *adv.*

**for·mal·de·hyde** (fôr-măl'də-hīd') *n.* [FORM(IC ACID) + ALDEHYDE.] A colorless, gaseous compound, HCHO, used in making melamine and phenolic resins, fertilizers, dyes, and embalming fluids and in aqueous solution as a preservative and disinfectant.

**for·ma·lin** (fôr'mə-lĭn) *n.* [Orig. a trademark.] A 37% by weight aqueous solution of formaldehyde with some methanol.

**for·mal·ism** (fôr'mə-lĭz'əm) *n.* **1.** Rigorous or excessive adherence to accepted or recognized forms, as in religion or art. **2.** An instance of formalism. **—for'mal·ist** *n.* **—for'mal·is'tic** *adj.*

**for·mal·i·ty** (fôr-măl'ĭ-tē) *n., pl.* **-ties. 1.** The quality or state of being formal. **2.** Rigorous or ceremonious adherence to established forms, rules, or customs. **3.** An established form, rule, or custom.

**for·mal·ize** (fôr'mə-līz') *vt.* **-ized, -iz·ing, -iz·es. 1.** To give definite form or shape to. **2.** To make formal. **3.** To give formal endorsement to. **—for'mal·i·za'tion** *n.*

**formal logic** *n.* The study of the properties of propositions and deductive reasoning by abstraction and analysis of the form rather than the content of propositions under consideration.

**form·am·id·ase** (fôr-măm'ĭ-dās', -dāz') *n.* [FORM(IC ACID) + AMID(E) + -ASE.] An enzyme participating in the catabolism of the amino acid tryptophan.

**for·mant** (fôr'mənt) *n.* [G. < Lat. *formans,* pr.part. of *formare,* to form < *forma,* form.] Any of several frequency regions of relatively great intensity in a sound spectrum, which together determine the characteristic quality of a vowel sound.

**for·mat** (fôr'măt') *n.* [Fr. < G. < Lat. *formatus,* p.part. of *formare,* to form < *forma,* form.] **1.** A plan for the organization and arrangement of a specified production. **2.** The form or layout of a publication. —*vt.* **-mat·ted, -mat·ting, -mats.** *Computer Sci.* To produce (e.g., data) in a specified form.

**for·mate** (fôr'māt') *n.* [FORM(IC ACID) + -ATE[2].] A salt or ester of formic acid.

**for·ma·tion** (fôr-mā'shən) *n.* **1.** The process of forming or producing. **2.** Something formed. **3.** The manner in which something is formed. **4.** An arrangement or deployment, as of troops. **5.** *Geol.* The primary unit of lithostratigraphy, having a succession of strata useful for mapping or description. **—for·ma'tion·al** *adj.*

**for·ma·tive** (fôr'mə-tĭv) *adj.* **1.** Forming or capable of forming. **2.** Susceptible of transformation via growth and development. **3.** Of or relating to formation, growth, or development <a *formative* stage of plant life> **4.** Relating to the formation or inflection of words. —*n.* The element of a word that is not contained in the base and gives the word a suitable form. **—for'ma·tive·ly** *adv.*

**form class** *n.* A set of linguistic forms whose exact substitutability in a given construction is determined by their common embodiment of one or more morphological or syntactic features.

**form·er**[1] (fôr'mər) *n.* One that forms.

**for·mer**[2] (fôr'mər) *adj.* [ME, comp. of *forme,* first < OE *forma.*] **1.** Happening earlier in time. **2.** Coming before in place or order. **3.** Being the first mentioned of two.

**for·mer·ly** (fôr'mər-lē) *adv.* At a former time : ONCE.

**form·fit·ting** (fôrm'fĭt'ĭng) *adj.* Closely fitted to the body.

**for·mic** (fôr'mĭk) *adj.* [< Lat. *formica,* ant.] **1.** Of or relating to ants. **2.** Of or derived from formic acid.

**For·mi·ca** (fôr-mī'kə). A trademark for any of various high-pressure laminated plastic sheets of melamine and phenolic materials used esp. for chemical and heat-resistant surfaces.

**formic acid** *n.* A colorless caustic fuming liquid, HCOOH, used in dyeing and finishing textiles and paper and in manufacturing fumigants, insecticides, and refrigerants.

**for·mi·car·y** (fôr'mĭ-kĕr'ē) *n., pl.* **-ies.** [Med. Lat. *formicarium* < Lat. *formica,* ant.] A nest of ants.

**for·mi·civ·o·rous** (fôr'mĭ-sĭv'ər-əs) *adj.* [Lat. *formica,* ant + -VOROUS.] Feeding on ants.

**for·mi·da·ble** (fôr'mĭ-də-bəl) *adj.* [OFr. < Lat. *formidabilis* < *formidare,* to fear < *formido,* fear.] **1.** Arousing fear or dread. **2.** Inspiring awe : ADMIRABLE <a *formidable* mind> **3.** Difficult to overcome, defeat, or undertake : AWESOME <*formidable* problems> **—for'mi·da·bil'i·ty, for'mi·da·ble·ness** *n.* **—for'mi·da·bly** *adv.*

**form·less** (fôrm'lĭs) *adj.* **1.** Having no specified form : SHAPELESS. **2.** Lacking order. **—form'less·ly** *adv.* **—form'less·ness** *n.*

**form letter** *n.* A usu. impersonal letter in a standardized format, sent to different people or large numbers of people.

**for·mu·la** (fôr'myə-lə) *n., pl.* **-las** *or* **-lae** (-lē') [Lat., dim. of *forma,* form.] **1.** An established form of words or symbols for use in a ceremony or procedure. **2.** A method of doing or treating something that relies on an established, uncontroversial model or approach <a new comedy that uses an old *formula* for laughs> **3.** An utterance of conventional notions or beliefs. **4.** *Chem.* **a.** A symbolic representation of the composition or the composition and structure of a chemical compound. **b.** The chemical compound so represented. **5.** A prescription of ingredients in fixed proportion : RECIPE. **6.** A liquid food prescribed for an infant and containing most required nutrients. **7.** A mathematical statement, esp. an equation, of a rule, principle, answer, or other logical relation. **—for'mu·la'ic** (-lā'ĭk) *adj.* **—for'mu·la'i·cal·ly** *adv.*

**for·mu·la·rize** (fôr'myə-lə-rīz') *vt.* **-rized, -riz·ing, -riz·es.** To formulate. **—for'mu·la·ri·za'tion** *n.*

**for·mu·lar·y** (fôr'myə-lĕr'ē) *n., pl.* **-ies. 1.** A book of formulas, as prayers. **2.** A statement expressed in formulas. **3.** A formula. **4.** A book listing the names of pharmaceuticals and their applications.

**for·mu·late** (fôr'myə-lāt') *vt.* **-lat·ed, -lat·ing, -lates. 1.** To state as a formula. **2.** To express in systematic terms or concepts. **3.** To devise : invent <*formulate* new strategy> **4.** To prepare as per a specified formula. **—for'mu·la'tion** *n.* **—for'mu·la'tor** *n.*

**formula weight** *n.* Molecular weight.

**for·mu·lize** (fôr'myə-līz') *vt.* **-lized, -liz·ing, -liz·es.** To formulate. **—for'mu·li·za'tion** *n.* **—for'mu·liz'er** *n.*

**for·myl** (fôr'mĭl') *n.* [FORM(IC ACID) + -YL.] The univalent radical CHO.

**For·nax** (fôr'năks') *n.* [Lat. *fornax,* furnace.] A constellation in the Southern Hemisphere.

**for·ni·cate** (fôr'nĭ-kāt') *vi.* **-cat·ed, -cat·ing, -cates.** [LLat. *fornicari, fornicat-* < *fornix,* brothel.] To commit fornication. **—for'ni·ca'tor** *n.*

**for·ni·ca·tion** (fôr'nĭ-kā'shən) *n.* Sexual intercourse between a man and woman not married to each other.

**for·nix** (fôr'nĭks) *n., pl.* **-ni·ces** (-nĭ-sēz') [NLat. < Lat., vault.] *Anat.* Either of a pair of bands of white fibers beneath the corpus callosum of the brain.

**for·sake** (fôr-sāk', fər-) *vt.* **-sook** (-so͝ok'), **-sak·en** (-sā'kən), **-sak·ing, -sakes.** [ME *forsaken* < OE *forsacan.*] **1.** To give up : RENOUNCE <*forsook* cigars> **2.** To leave altogether : ABANDON <*forsook* New York theater and returned to Hollywood>

**for·sooth** (fôr-so͞oth', fər-) *adv.* [ME *forsoth* < OE *forsōð.*] *Archaic.* In truth : INDEED.

**for·spent** (fôr-spĕnt', fər-) *adj.* *Archaic.* Worn out : EXHAUSTED.

**for·swear** *also* **fore·swear** (fôr-swâr', fōr-) *v.* **-swore** (fôr-swôr', fōr-swōr'), **-sworn** (fôr-swôrn', fōr-swōrn'), **-swear·ing, -swears.** [ME *forsweren* < OE *forswerian.*] —*vt.* **1.** To forsake unalterably : RENOUNCE. **2.** To disavow unalterably. **3.** To perjure (oneself). —*vi.* To commit perjury.

**for·syth·i·a** (fôr-sĭth'ē-ə, fər-) *n.* [NLat. (genus name), after William *Forsyth* (1737–1804).] An Asian shrub of the genus *Forsythia,* widely cultivated for its early-blooming yellow flowers.

**fort** (fôrt, fōrt) *n.* [ME < OFr. < *fort,* strong < Lat. *fortis.*] **1.** A fortified place occupied by troops. **2.** A permanent army post.

**for·ta·lice** (fôr′tə-lĭs) *n.* [ME < Med. Lat. *fortalitia* < Lat. *fortis,* strong.] A minor defensive position or structure, as a small fort.

**forte**[1] (fôrt, fōrt, fôr′tā′) *n.* [OFr. *fort* < *fort,* strong < Lat. *fortis.*] **1.** An activity in which one excels. **2.** The strong part of a sword blade, between the middle and the hilt.

**for·te**[2] (fôr′tā′) *adv.* [Ital. < *forte,* strong < Lat. *fortis.*] *Mus.* Loudly : forcefully. —Used as a direction. **—for′te** *n. & adj.*

**for·te·pi·an·o** (fôr′tā-pē-ăn′ō, -ä′nō) *adv.* [Ital. : *forte,* loud + *piano,* soft.] *Mus.* Loudly and then softly. —Used as a direction. **—for′te·pi·an′o** *adj.*

**forth** (fôrth, fōrth) *adv.* [ME < OE *forð.*] **1.** Forward in time, place, or order : ONWARD <from this moment *forth*> **2.** Into view <a stranger who came *forth* from the crowd> **3.** Away from a given place : ABROAD. —*prep. Archaic.* Out of : forth from.

**forth·com·ing** (fôrth-kŭm′ĭng, fōrth-) *adj.* **1.** About to appear or happen : APPROACHING <the *forthcoming* election> **2. a.** Available when required or as promised <Funds were no longer *forthcoming.*> **b.** Affable : outgoing <a considerate, *forthcoming* person> —*n.* (fôrth′kŭm′ĭng, fōrth′-). An act or instance of coming forth.

**forth·right** (fôrth′rīt′, fōrth′-) *adj.* **1.** Direct : straightforward <a *forthright* refusal> **2.** *Archaic.* Proceeding straight ahead. —*adv.* **1.** Directly : frankly. **2.** *Archaic.* Directly ahead. **3.** *Archaic.* At once. **—forth′right′ly** *adv.* **—forth′right′ness** *n.*

**forth·with** (fôrth-wĭth′, -wĭth′, fōrth-) *adv.* At once.

**for·ti·eth** (fôr′tē-ĭth) *n.* **1.** The ordinal number matching the number 40 in a series. **2.** One of 40 equal parts. **—for′ti·eth** *adj. & adv.*

**for·ti·fi·ca·tion** (fôr′tə-fĭ-kā′shən) *n.* **1.** The act, science, or art of fortifying. **2.** Something, esp. a military defensive work, that defends, strengthens, or fortifies.

**fortified wine** *n.* A wine, as sherry, to which alcohol, usu. in the form of grape brandy, has been added.

**for·ti·fy** (fôr′tə-fī′) *v.* **-fied, -fy·ing, -fies.** [ME *fortifien* < OFr. *fortifier* < LLat. *fortificare* < Lat. *fortis,* strong.] —*vt.* **1.** To strengthen and secure (a position) with fortifications. **2.** To add strength to (a structure) by reinforcement. **3.** To impart physical strength to : INVIGORATE <The hot tea *fortified* me for the long meeting.> **4.** To give moral or mental strength to : ENCOURAGE. **5.** To corroborate <*fortified* the allegations with new evidence> **6.** To strengthen or enrich (a substance), as by adding vitamins to food or alcohol to wine. —*vi.* To build fortifications. **—for′ti·fi′a·ble** *adj.* **—for′ti·fi′er** *n.*

**for·tis** (fôr′tĭs) *adj.* [NLat. < Lat., strong.] Pronounced with tension and strong articulation. —Used of certain consonants, as *f* and *p.* —*n.* A fortis consonant.

**for·tis·si·mo** (fôr-tĭs′ə-mō′) *adv.* [Ital., superl. of *forte,* strong.] *Mus.* Very loudly. —Used as a direction. **—for·tis′si·mo′** *n. & adj.*

**for·ti·tude** (fôr′tĭ-tōōd′, -tyōōd′) *n.* [ME < Lat. *fortitudo* < *fortis,* strong.] Strength of mind allowing one to endure pain or adversity courageously. **—for′ti·tu′di·nous** (-tōōd′n-əs, -tyōōd′-) *adj.*

**fort·night** (fôrt′nīt′) *n.* [ME *fourtenight,* alteration of *fourtene night,* 14 nights.] Two weeks.

**fort·night·ly** (fôrt′nīt′lē) *adj.* Taking place or appearing once in or every two weeks. —*n., pl.* **-lies.** A publication issued fortnightly. **—fort′night′ly** *adv.*

**FOR·TRAN** (fôr′trăn′) *n.* [FOR(MULA) + TRAN(SLATION).] *Computer Sci.* A programming language for problems expressible in algebraic terms.

**for·tress** (fôr′trĭs) *n.* [ME *forteress* < OFr. < Med. Lat. *fortalitia* < Lat. *fortis,* strong.] A fortified place, esp. a large and permanent military stronghold, often including a town.

**for·tu·i·tous** (fôr-tōō′ĭ-təs, -tyōō′-) *adj.* [Lat. *fortuitus* < *forte,* by chance, ablative of *fors,* chance.] **1.** Happening by accident or chance. **2.** Lucky : fortunate. ***usage:*** *Fortuitous* in the sense of "happening by chance" occurs commonly in contexts carrying an implication of lucky rather than unlucky chance, as in *a fortuitous meeting that led to a renewal of our friendship.* As a result *fortuitous* has come into use as an equivalent to *fortunate.* It should be noted, however, that many careful users of the language deplore this practice and insist upon maintaining the distinction between the two terms. **—for·tu′i·tous·ly** *adv.* **—for·tu′i·tous·ness** *n.*

**for·tu·i·ty** (fôr-tōō′ĭ-tē, -tyōō′-) *n., pl.* **-ties. 1.** An accidental occurrence : CHANCE. **2.** The quality or state of being fortuitous.

**For·tu·na** (fôr-tōō′nə, -tyōō′-) *n.* [Lat. < *fortūna,* fortune.] *Rom. Myth.* The goddess of fortune.

**for·tu·nate** (fôr′chə-nĭt) *adj.* [Lat. *fortunatus* < p.part. of *fortunare,* to prosper < *fortuna,* fortune.] **1.** Bringing something good and unforeseen : AUSPICIOUS. **2.** Having or characterized by unexpected good fortune. **—for′tu·nate·ly** *adv.* **—for′tu·nate·ness** *n.*

☆ ***syns:*** FORTUNATE, HAPPY, LUCKY, PROVIDENTIAL *adj. core meaning* : characterized by luck or good fortune <a *fortunate* turn of events> ***ant:*** unfortunate

**for·tune** (fôr′chən) *n.* [ME < OFr. < Lat. *fortuna* < *fors,* chance.] **1.** *often* **Fortune.** A hypothetical, often personified force or power that favorably or unfavorably controls the events of one's life <*Fortune* is not on our side.> **2.** The good or bad luck that is to befall someone : FATE. **3.** Success, esp. when at least partially resulting from luck. **4. a.** A person's condition or standing in life determined by material possessions or financial wealth. **b.** Extensive amounts of material possessions or money : WEALTH. **5.** A large amount of money <made a *fortune* on the commodities market> —*v.* **-tuned, -tun·ing, -tunes.** —*vt. Obs.* To ascribe good or bad fortune to. —*vi. Archaic.* To happen by chance.

▲ word history: A fortunate person possesses a good fortune, not an evil one. The same lack of equanimity in facing the future that led to the restriction of *fatal* to an evil fate is evident in the development of *fortune* and *fortunate.* Latin *fortuna,* the source of *fortune,* meant "chance" or "luck," either good or bad. But *fortunae,* the plural, meant "possessions, goods," and the verb *fortunare* meant "to prosper, make happy." *Fortunatus,* the past participle of *fortunare,* is the ancestor of *fortunate* and meant only "happy, prosperous."

**fortune cookie** *n.* A cookie made from a thin layer of dough folded and baked around a slip of paper bearing a prediction of fortune or a maxim.

**fortune hunter** *n.* One who seeks wealth, esp. through marriage.

**for·tune·tell·er** (fôr′chən-tĕl′ər) *n.* One who claims to be able to predict future events. **—for′tune·tell′ing** *n.*

**for·ty** (fôr′tē) *n., pl.* **-ties.** [ME < OE *fēowertig* : *fēower,* four + *-tig,* -ty.] The cardinal number equal to 4 × 10. **—for′ty** *adj. & pron.*

**for·ty-five** (fôr′tē-fīv′) *n.* **1.** A .45-caliber pistol. **2.** A phonograph record designed to be played at 45 revolutions per minute.

**for·ty-nin·er** (fôr′tē-nī′nər) *n.* One who took part in the 1849 California gold rush.

**forty winks** *n. (sing. or pl. in number). Informal.* A brief nap.

**fo·rum** (fôr′əm, fōr′-) *n., pl.* **fo·rums** *or* **fo·ra** (fôr′ə, fōr′ə) [ME < Lat.] **1.** The public square or marketplace of an ancient Roman city that was the place of assembly for judicial and other public activity. **2. a.** A public meeting place for open discussion. **b.** A medium for open discussion, as a radio or television program. **3.** A court of law : TRIBUNAL. **4.** A public meeting or presentation involving a discussion usu. among experts and often including audience participation.

**for·ward** (fôr′wərd) *adj.* [ME < OE *foreweard.*] **1. a.** At, near, or belonging to the front. **b.** Located in advance. **2.** Going, tending, or moving toward a position in front <a *forward* fall down the stairs> **3. a.** Ardently inclined : EAGER. **b.** Lacking restraint or modesty : BOLD. **4.** Progressive, esp. technologically, politically, or economically <a *forward* nation> **5.** Mentally, physically, socially, or biologically advanced : PRECOCIOUS. **6.** Completed or made in advance <bidding on *forward* contracts for grain> —*adv.* **1.** Toward or tending to the front : FRONTWARD <step *forward*> **2.** In or toward the future <looking *forward* to seeing you> **3.** Into view or prominence : FORTH <Come *forward* out of the shadows so that I can see you.> ***usage:*** As adverbs *forward* and *forwards* are interchangeable only in the sense of "toward the front," as in *They moved forward* (or *forwards*). —*n.* **1.** A player in certain games, as basketball, who is part of the front line of offense or defense. **2.** The position played by a forward. —*vt.* **-ward·ed, -ward·ing, -wards. 1.** To send on to a subsequent destination or address. **2.** To help advance : PROMOTE. **—for′ward·ly** *adv.* **—for′ward·ness** *n.*

**for·ward·er** (fôr′wər-dər) *n.* One that forwards, esp. an agent who facilitates the passage of received goods to their destination.

**forward pass** *n. Football.* A pass thrown in the direction of the opponent's goal.

**for·wards** (fôr′wərdz, fōr′-) *adv.* FORWARD 1.

**for·went** (fôr-wĕnt′) *v. p.t. of* FORGO.

**for·worn** *also* **fore·worn** (fôr-wôrn′, fōr-wōrn′) *adj. Archaic.* Worn out.

**for·zan·do** (fôrt-sän′dō) *adj. & adv. & n. var. of* SFORZANDO.

**foss** (fŏs) *n. var. of* FOSSE.

**fos·sa** (fŏs′ə) *n., pl.* **fos·sae** (fŏs′ē′) [Lat., ditch < fem. p.part. of *fodere,* to dig.] A hollow or depression, as in a bone. **—fos′sate′** (fŏs′āt′) *adj.*

**fosse** *also* **foss** (fŏs) *n.* [ME < OFr. and Lat. *fossa.*] A ditch or moat.

**fos·sick** (fŏs′ĭk) *v.* **-sicked, -sick·ing, -sicks.** [Orig. unknown.] *Austral.* —*vi.* **1.** To search for gold, esp. by reworking washings or waste piles. **2.** To rummage or search, esp. for a possible profit. —*vt.* To search for by or as if by rummaging. **—fos′sick·er** *n.*

**fos·sil** (fŏs′əl) *n.* [< Lat. *fossilis,* dug up < *fossus,* p.part. of *fodere,* to dig.] **1.** A remnant or trace of an organism of a past geologic age, as a skeleton or leaf imprint, embedded in the earth's crust. **2.** One that is outdated or antiquated, as a rigid theory or a person with outmoded ideas. **3.** An obsolete word or word element used only in an idiom, as *fro* in *to and fro.*

**fossil fuel** *n.* A hydrocarbon fuel, as petroleum, derived from living matter of a previous geologic time.

**fos·sil·if·er·ous** (fŏs′ə-lĭf′ər-əs) *adj.* Containing fossils.

**fos·sil·ize** (fŏs′ə-līz′) *v.* **-ized, -iz·ing, -iz·es.** —*vt.* **1.** To convert into a fossil. **2.** To make outmoded or rigid. —*vi.* To become a fossil. **—fos′sil·i·za′tion** *n.*

**fos·so·ri·al** (fŏ-sôr′ē-əl, -sōr′-) *adj.* [Med. Lat. *fossorius* < Lat. *fossus,* p.part. of *fodere,* to dig.] *Zool.* Adapted for or used in burrowing or digging.

**fos·ter** (fô′stər, fŏs′tər) *vt.* **-tered, -ter·ing, -ters.** [ME *fostren* < OE *fōstrian,* to nourish < *fōstor,* food.] **1.** To bring up : NURTURE. **2.** To promote the development or growth of : ENCOURAGE. **3.** To nurse : cherish. —*adj.* Receiving, sharing, or affording parental care and nurture although not related through legal or blood ties <*a foster* child>

**fos·ter·ling** (fô′stər-lĭng, fŏs′tər-) *n.* A foster child.

**Fou·cault pendulum** (fōō-kō′) *n.* [After Jean B.L. *Foucault* (1819–1868).] A simple pendulum suspended so the plane of motion is not fixed, is set into motion along a meridian, and appears to turn clockwise in the Northern Hemisphere or counterclockwise in the Southern Hemisphere, showing the axial rotation of the earth.

**fou·droy·ant** (fōō-droi′ənt, fōō′drwä-yäN′) *adj.* [Fr., pr.part of *foudroyer,* to strike with lightning < *foudre,* lightning < OFr. *fouldre* < Lat. *fulgur* < *fulgēre,* to flash.] Dazzling : stunning.

**fought** (fôt) *v. p.t. & p.p. of* FIGHT.

**foul** (foul) *adj.* **-er, -est.** [ME < OE *fūl.*] **1.** So offensive to the senses as to be revolting. **2.** Having an offensive odor : FETID. **3.** Rotten : putrid <*foul* meat> **4.** Full of dirt. **5.** Morally detestable. **6.** Vulgar or obscene <*foul* language> **7.** *Archaic.* Ugly : unattractive. **8.** *Informal.* Disagreeable or displeasing. **9.** Bad or unfavorable <*foul* weather> **10.** Not in accordance with accepted standards or rules or honor : DISHONORABLE <used *foul* means to achieve their aim> **11.** Contrary to the rules of a game or sport. **12.** *Baseball.* Being outside the foul line <a *foul* ball> **13.** Entangled or twisted <a *foul* anchor> **14.** Clogged or obstructed : BLOCKED <a *foul* ventilator shaft> —*n.* **1.** An infraction or violation of the rules of play in a sport. **2.** *Baseball.* A foul ball. **3.** An entanglement or collision. **4.** An instance of clogging or obstructing. —*adv.* In a foul way. —*v.* **fouled, foul·ing, fouls.** —*vt.* **1.** To make dirty. **2.** To bring into dishonor : BESMIRCH. **3.** To clog : obstruct. **4.** To entangle or catch (e.g., a rope). **5.** To encrust (a ship's hull) with foreign matter, as barnacles. **6.** To commit a sports foul against. **7.** *Baseball.* To hit (a ball) outside the foul lines. —*vi.* **1.** To become foul. **2.** To commit a sports foul. **3.** *Baseball.* **a.** To hit a ball outside the foul lines. **b.** To make an out by hitting a foul ball caught before it touches the ground. **4.** To become entangled or twisted <an anchor *fouling* on a rock> **5.** To become clogged. **—foul out.** To be put out of play by exceeding the number of permissible fouls. **—foul up.** *Informal.* To blunder or cause to blunder. **—foul′ly** *adv.* **—foul′ness** *n.*

**fou·lard** (fōō-lärd′) *n.* [Fr.] **1.** A lightweight twill or plain-woven fabric of silk or silk and cotton, usu. having a small printed design. **2.** An article of dress, esp. a necktie or scarf, made of foulard.

**foul ball** *n. Baseball.* A batted ball that touches the ground outside of fair territory.

**foul·brood** (foul′brōōd′) *n.* A disease of honeybee larvae caused by one of several types of bacteria, including *Bacillus alvei.*

**foul line** *n.* **1.** *Baseball.* Either of two straight lines running from the rear of home plate to the boundary of the playing field to indicate the area in which a fair ball can be hit. **2.** *Basketball.* A line from which a player makes a foul shot. **3.** A boundary limiting the playing area, esp. in bowling and tennis.

**foul-mouthed** (foul′mouthd′, -moutht′) *adj.* Using obscene or vulgar language.

**foul play** *n.* Unfair or treacherous behavior, esp. when involving violence.

**foul shot** *n. Basketball.* An unguarded throw to the basket from the foul line awarded to a fouled player and scored as one point if successful.

**foul tip** *n. Baseball.* A pitched ball slightly deflected off the bat into the foul zone.

**foul-up** (foul′ŭp′) *n. Informal.* **1.** Confusion due to bungling. **2.** Mechanical trouble.

**found**[1] (found) *vt.* **found·ed, found·ing, founds.** [ME *founden* < OFr. *fonder* < Lat. *fundare* < *fundus,* bottom.] **1.** To originate or establish (e.g., a college). **2.** To establish the basis of. **—found′er** *n.*

**found**[2] (found) *vt.* **found·ed, found·ing, founds.** [ME *founden* < OFr. *fondre* < Lat. *fundere.*] **1.** To melt (metal) and pour into a mold. **2.** To make (objects) by pouring molten material into a mold. **—found′er** *n.*

**found**[3] (found) *v. p.t. & p.p. of* FIND.

**foun·da·tion** (foun-dā′shən) *n.* **1.** The act of establishing, esp. the founding of an institution with provisions for future maintenance. **2.** The basis on which a thing stands, is founded, or is supported. **3.** Funds for the perpetual support of an institution : ENDOWMENT. **4.** An institution founded and supported by an endowment. **5.** A foundation garment. **6.** A cosmetic used as a base for facial make-up. **—foun·da′tion·al** *adj.*

**foundation garment** *n.* A woman's supporting undergarment, as a corset or girdle.

**foun·der** (foun′dər) *v.* **-dered, -der·ing, -ders.** [ME *foundren,* to sink to the ground < OFr. *fondrer* < VLat. *fundorare* < Lat. *fundus,* bottom.] —*vi.* **1.** To become disabled, esp. to become lame. —Used of horses. **2.** To break down : FAIL. **3.** *Naut.* To sink beneath the water. **4.** To cave in : SINK. —Used of ground or buildings. —*vt.* To cause to founder. —*n.* Laminitis.

**Found·ing Father** (foun′dĭng) *n.* **1.** A member of the American Constitutional Convention of 1787. **2. founding father.** One who founds or establishes something : ORIGINATOR.

**found·ling** (found′lĭng) *n.* [ME.] An abandoned child whose parents are unknown.

**found object** *n.* [Transl. of French *objet trouvé.*] Any of various objects picked up by chance and included in a work of art.

**foun·dry** (foun′drē) *n., pl.* **-dries. 1.** An establishment in which the founding of metals is done. **2. a.** The art or operation of founding. **b.** Castings made by founding.

**foundry proof** *n.* A proof taken from composed printing type for a final check before plates are made.

**fount**[1] (fount) *n.* [OFr. *font* < Lat. *fons,* fountain.] **1.** A fountain. **2.** A source <a *fount* of knowledge>

**fount**[2] (fount) *n. Chiefly Brit. var. of* FONT[2].

**foun·tain** (foun′tən) *n.* [ME < OFr. *fontaine* < LLat. *fontana* < Lat. *fontanus,* of a spring < *fons,* spring.] **1.** A spring, esp. the source of a stream. **2.** A point of origin : SOURCE. **3. a.** An artificially created stream of water. **b.** A device that produces and contains such a stream. **4.** A reservoir or chamber containing a supply of liquid that can be siphoned off as required. **5.** A soda fountain.

**foun·tain·head** (foun′tən-hĕd′) *n.* **1.** A spring that is the source of a stream. **2.** A principal origin or source.

**fountain pen** *n.* A pen filled from an external source and containing an ink reservoir that automatically feeds the writing point.

**four** (fôr, fōr) *n.* [ME < OE *fēower;* akin to G. *vier,* Lat. *quattuor,* Gk. *tettares,* and Skt. *catur.*] **1.** The cardinal number equal to 3 + 1. **2.** The fourth in a set or sequence. **3.** Something having four parts, units, or members. **—four** *adj. & pron.*

**four-bag·ger** (fôr′băg′ər, fōr′-) *n. Baseball. Informal.* A home run.

**four·chette** (fōōr-shĕt′) *n.* [Fr. < OFr. *forchete,* fork, dim. of *forche,* pitchfork < Lat. *furca.*] A narrow, forked strip of material that joins the front and back parts of the fingers of gloves.

**four-col·or** (fôr′kŭl′ər) *adj.* Of or relating a color printing or photographic process in which three primary colors and black are transferred by four different plates or filters to a surface, thereby reproducing the colors of the subject matter.

**four-cy·cle** (fôr′sī′kəl, fōr′-) *adj.* Designating an internal-combustion engine requiring four strokes of the piston for a cycle.

**four-di·men·sion·al** (fôr′dĭ-mĕn′shə-nəl, fōr′-) *adj.* Exhibiting or being specified by four dimensions, esp. the three spatial dimensions and single temporal dimension of relativity theory.

**Four·drin·i·er** (fōōr-drĭn′ē-ər) *adj.* [After Henry *Fourdrinier* (1766–1854) and Sealy *Fourdrinier* (d. 1847).] Designating a paper-making machine used to produce paper in a continuous web.

**four-eyed fish** (fôr′īd′, fōr′-) *n.* A freshwater fish, *Anableps anableps* or *A. microlepis* of tropical America, having bulging eyes divided longitudinally, with the upper part adapted for aerial vision and the lower part for underwater vision.

**four flush** *n.* A five-card poker hand containing four cards in the same suit.

**four-flush** (fôr′flŭsh′, fōr′-) *vi.* **-flushed, -flush·ing, -flush·es. 1.** To bluff in poker with a four flush. **2.** *Slang.* To make a pretense : BLUFF. **—four′-flush′er** *n.*

**four·fold** (fôr′fōld′, fōr′-) *adj.* **1.** Having four units or aspects : QUADRUPLE. **2.** Being four times as much or as many. —*adv.* (fôr′-fōld′, fōr′-). In quadruple measure.

**four-foot·ed** (fôr′fōōt′ĭd, fōr′-) *adj.* Having four feet.

**four·hand·ed** (fôr′hăn′dĭd, fōr′-) *adj.* **1.** Involving or requiring four players, as in cards. **2.** Intended for four hands, as a piano duet.

**Four-H Club** (fôr′āch′, fōr′-) *n.* [From its four goals to improve head, heart, hands, and health.] A youth organization sponsored by the Department of Agriculture and offering instruction in agriculture and home economics.

**four hundred** *also* **Four Hundred** *n.* The richest and most exclusive social set of a community.

**Fou·ri·er analysis** (fōōr′ē-ā′) *n.* Approximation of a function through application of a Fourier series to periodic data.

**Fou·ri·er·ism** (fōōr′ē-ə-rĭz′əm) *n.* A social reform system advocated by Charles Fourier in the early 19th cent., proposing that society be organized into small self-sustaining communal groups. **—Fou′ri·er·ist, Fou′ri·er·ite′** (-ə-rīt′) *n.*

**Fourier series** *n.* [After Jean B.J. *Fourier* (1768–1830).] An infinite series of sine and cosine functions capable if uniformly convergent of approximating numerous mathematical functions.

**four-in-hand** (fôr′ĭn-hănd′, fōr′-) *n.* **1.** A horse-drawn vehicle driven by one person. **2.** A team of four horses. **3.** A necktie tied in a slipknot with the ends left hanging and overlapping.

**four-leaf clover** (fôr′lēf′, fōr′-) *n.* A clover leaf having four leaflets instead of the normal three, regarded as a sign of good luck.

**four-let·ter word** (fôr′lĕt′ər, fōr′-) *n.* Any of several short English words gen. considered vulgar or obscene.

**four-o'clock** (fôr′ə-klŏk′, fōr′-) *n.* A plant of the genus *Mirabilis,* esp. *M. jalapa,* indigenous to tropical America and widely cultivated for its variously colored flowers that open in the late afternoon.

**four·pence** (fôr′pəns, fōr′-) *n. Chiefly Brit.* **1.** A sum of money equal to four pence. **2.** A small silver coin formerly worth four pence.

**four-post·er** (fôr′pō′stər, fōr′-) *n.* A bed with tall corner posts orig. intended to support curtains or a canopy.

**four·ra·gère** (fōōr′ə-zhâr′) *n.* [Fr. < *fourrage,* forage < OFr. —see FORAGE.] An ornamental braided cord usu. looped around the left shoulder, esp. such a cord awarded to a military unit.

**four·score** (fôr′skôr′, fōr′skōr′) *adj.* That is four times 20.

**four·some** (fôr′səm, fōr′-) *n.* [ME *four-som* < OE *fēowra sum,* one of four.] **1.** A group of four, esp. two couples. **2. a.** A game, esp. a golf match, played by four persons, two on each side. **b.** The players in such a game.

**four·square** (fôr′skwâr′, fōr′-) *adj.* **1.** Having four equal sides and four right angles : SQUARE. **2.** Marked by firm, unwavering conviction. **3.** Forthright : candid. —*adv.* Forthrightly : squarely.

**four-star** (fôr′stär′, fōr′-) *adj.* **1.** Pertaining to or being a general or an admiral whose insignia and rank carries four stars. **2.** Of superlative quality <a *four-star* Parisian restaurant>

**four·teen** (fôr-tēn′, fōr-) *n.* [ME *fourtene* < OE *fēowertēne.*] **1.** The cardinal number equal to 13 + 1. **2.** The 14th in a set or sequence. **3.** Something having 14 parts, units, or members. **—four·teen′** *adj. & pron.*

**four·teenth** (fôr-tēnth′, fōr-) *n.* **1.** The ordinal number matching the number 14 in a series. **2.** One of 14 equal parts. **—four·teenth′** *adj. & adv.*

**fourth** (fôrth, fōrth) *n.* [ME *fourthe* < OE *fēorða.*] **1.** The ordinal number matching the number four in a series. **2.** One of four equal parts. **3.** *Mus.* **a.** A tone four degrees above or below a given tone in a diatonic scale. **b.** The interval between two such tones. **c.** The harmonic combination of these tones. **d.** The subdominant in a scale. **4.** The fourth forward gear of a motor vehicle. **5. Fourth.** The Fourth of July. **—fourth** *adj. & adv.*

**fourth-class** (fôrth′klăs′, fōrth′-) *adj.* Designating a class of mail consisting of merchandise or printed matter weighing over eight ounces and not sealed against inspection. **—fourth′-class′** *adv.*

**fourth dimension** *n.* Time considered as a coordinate dimension and required by relativistic geometry, along with three spatial dimensions, to specify the location of any event.

**fourth estate** *n.* The public press.

**Fourth of July** *n.* Independence Day.

**Fourth World** *n.* The least-developed emerging countries of the Third World, esp. in Africa and Asia.

**four-wheel** (fôr′hwēl′, -wēl′, fōr′-) *adj.* **1.** Having four wheels. **2.** Of, relating to, or designating an automotive drive mechanism in which all four wheels are linked to the source of driving power.

**fo·ve·a** (fō′vē-ə) *n., pl.* **-ve·ae** (-vē-ē′) [NLat. < Lat., small pit.] A shallow cuplike depression or pit, as in a bone. **—fo′ve·al** (-əl), **fo′ve·ate′** (-āt′) *adj.* **—fo′ve·i·form′** (-ə-fôrm′) *adj.*

**fovea cen·tra·lis** (sĕn-trā′lĭs) *n.* [NLat., central fovea.] A small depression in the macula lutea of the retina, constituting the area of most distinct vision.

**fo·ve·ae** (fō′vē-ē′) *n. pl. of* FOVEA.

**fowl** (foul) *n., pl.* **fowl** *or* **fowls.** [ME *foul* < OE *fugol.*] **1.** A bird of the order Galliformes, esp. the common widely domesticated chicken, *Gallus gallus.* **2. a.** A bird used for food or hunted as game. **b.** The edible meat of such a bird. **3.** *Archaic.* Any bird. —*vi.* **fowled, fowl·ing, fowls.** To hunt, trap, or shoot wild fowl. **—fowl′er** *n.*

**fowling piece** *n.* A light shotgun esp. for shooting birds.

**fox** (fŏks) *n.* [ME < OE.] **1.** A carnivorous mammal of the genus *Vulpes* or related genera, related to the dogs and wolves and having upright ears, a pointed snout, and a long, bushy tail. **2.** The fur of a fox. **3.** A crafty or sly person. **4.** *Archaic.* A sword. **5.** *Naut.* Small cordage made by twisting together two or more strands of tarred yarn. —*v.* **foxed, fox·ing, fox·es.** —*vt.* **1. a.** To trick by ingenuity or cunning : OUTWIT. **b.** To baffle : confuse. **2.** *Archaic.* To intoxicate. **3.** To make (beer) sour by fermenting. **4.** To repair (a shoe) by adding a new upper. —*vi.* **1.** To act slyly or craftily. **2.** To turn sour in fermenting, as beer.

**Fox** (fŏks) *n., pl.* **Fox·es** *or* **Fox. 1. a.** A tribe of Indians once living in southwestern Wisconsin. **b.** A member of this tribe. **2.** The Algonquian language of the Fox.

**foxed** (fŏkst) *adj.* Discolored with yellowish-brown stains, as an old book or print.

**fox·fire** (fŏks′fīr′) *n.* [ME.] A phosphorescent glow, esp. one caused by certain fungi found on rotting wood.

**fox·glove** (fŏks′glŭv′) *n.* [ME.] **1.** A plant of the genus *Digitalis,* esp. *D. purpurea,* indigenous to Europe, having a long cluster of large, tubular, pinkish-purple flowers and leaves that are the source of the drug digitalis. **2.** A plant related to the foxglove.

**fox grape** *n.* A climbing woody vine, *Vitis labrusca* of the eastern United States, that bears purplish-black fruit and is the source of many cultivated grapes.

**fox·hole** (fŏks′hōl′) *n.* A shallow pit dug by a combat soldier for refuge against enemy fire.

**fox·hound** (fŏks′hound′) *n.* A dog bred and developed for fox hunting, esp. the English foxhound and the American foxhound.

**fox squirrel** *n.* A squirrel, *Sciurus niger* of the United States, with grayish or rusty fur.

**fox·tail** (fŏks′tāl′) *n.* A grass of the genus *Alopecurus,* with dense silky or bristly flowering spikes.

**foxtail lily** *n.* A native Asian plant of the genus *Eremurus* bearing a tall spirelike cluster of small bell-shaped white, yellowish, or pink flowers.

**fox terrier** *n.* A small dog orig. bred in England, having a white coat with dark markings and developed in both wire-haired and smooth-coated varieties.

**fox·trot** (fŏks′trŏt′) *vi.* **-trot·ted, -trot·ting, -trots.** To dance the fox trot.

**fox trot** *n.* **1. a.** A ballroom dance in 2/4 or 4/4 time, composed of various slow and fast steps. **b.** The music for this dance. **2.** The slow broken gait of a horse between a trot and a walk.

**fox·y** (fŏk′sē) *adj.* **-i·er, -i·est. 1.** Of or like a fox. **2.** Shrewdly clever : CUNNING. **3.** Having a reddish-brown color. **4.** Foxed. **5.** Having the distinctive sharp flavor of some American grapes <a *foxy* wine> **6.** *Informal.* Sensually attractive. **—fox′i·ly** *adv.* **—fox′i·ness** *n.*

**foy** (foi) *n. Scot.* [Dial. Du. *fooi,* prob. < OFr. *voie,* journey < Lat. *via,* road.] A farewell feast, drink, or gift, as at a wedding.

**foy·er** (foi′ər, foi′ā′, fwä′yā′) *n.* [Fr., hearth < Lat. *focus.*] **1.** A lobby or anteroom, as of a theater or hotel. **2.** A vestibule or entrance hall.

**Fr** *symbol for* FRANCIUM.

**Fra** (frä) *n.* [Ital., short for *frate,* brother < Lat. *frater.*] Brother. —Used as a title with the name of a friar.

**fra·cas** (frā′kəs, frăk′əs) *n.* [Fr. < Ital. *fracasso* < *fracassare,* to make an uproar.] A noisy row : BRAWL.

**fract·ed** (frăk′tĭd) *adj.* [Lat. *fractus.* —see FRACTION.] *Obs.* Broken.

**frac·tion** (frăk′shən) *n.* [ME *fraccioun,* a breaking < AN < LLat. *fractio,* a breaking of bread < Lat. *fractus,* p.part. of *frangere,* to break.] **1.** A small part : BIT <moved a *fraction* of a step> **2.** A disconnected piece : FRAGMENT. **3.** *Math.* An indicated quotient of two quantities. **4.** *Chem.* A component separated by fractionation.

**frac·tion·al** (frăk′shə-nəl) *adj.* **1.** Of, relating to, or being a fraction. **2.** Very small : INSIGNIFICANT <*fractional* assistance only> **3.** Being in fractions or pieces. **—frac′tion·al·ly** *adv.*

**fractional currency** *n.* **1.** A currency in a denomination less than the standard monetary unit. **2.** A coin worth less than a dollar in the United States.

**fractional distillation** *n.* **1.** Distillation with rectification to get the purest possible product. **2.** Distillation in which the product is collected in a series of separate fractions.

**frac·tion·ate** (frăk′shə-nāt′) *vt.* **-at·ed, -at·ing, -ates.** To separate (a chemical compound) into components, as by distillation or crystallization. **—frac′tion·a′tion** *n.* **—frac′tion·a′tor** *n.*

**frac·tion·ize** (frăk′shə-nīz′) *vt. & vi.* **-ized, -iz·ing, -iz·es.** To divide into fractions. **—frac′tion·i·za′tion** *n.*

**frac·tious** (frăk′shəs) *adj.* [< FRACTION.] **1.** Apt to cause trouble : UNRULY. **2.** Peevish : cranky. **—frac′tious·ly** *adv.* **—frac′tious·ness** *n.*

**frac·ture** (frăk′chər) *n.* [ME < OFr. < Lat. *fractura* < *fractus,* p.part of *frangere,* to break.] **1. a.** The act or process of breaking. **b.** The state of being broken. **2.** A break, rupture, or crack, as in bone. **3.** *Mineral.* **a.** The manner in which a mineral breaks. **b.** The appearance of a broken mineral. —*v.* **-tured, -tur·ing, -tures.** —*vt.* **1.** To break <*fracture* an arm> **2.** To disrupt as if by breaking up or bursting. **3.** To break or manipulate the restrictions or rules of. —*vi.* To undergo fracture.

**frae** (frā) *prep.* [ME *fra* < ON *frā.*] *Scot.* From.

**frag·ile** (frăj′əl, -īl′) *adj.* [OFr. < Lat. *fragilis* < *frangere,* to break.] **1.** Easily damaged or broken : BRITTLE. **2.** FRAIL[1] 1. **3.** Tenuous or flimsy <a *fragile* claim to the estate> **—frag′ile·ly** *adv.* **—fra·gil′i·ty** (frə-jĭl′ĭ-tē), **frag′ile·ness** *n.*

**frag·ment** (frăg′mənt) *n.* [ME < Lat. *fragmentum* < *frangere,* to break.] **1.** A part detached or broken off. **2.** Something incomplete <heard *fragments* of the conversation> —*v.* (-mĕnt′) **-ment·ed, -ment·ing, -ments.** —*vt.* To break or separate (something) into fragments. —*vi.* To break into fragments.

**frag·men·tal** (frăg-mĕn′tl) *adj.* **1.** Fragmentary. **2.** *Geol.* Consisting of broken material moved from its place of origin. **—frag·men′tal·ly** *adv.*

**frag·men·tar·y** (frăg′mən-tĕr′ē) *adj.* Made up of fragments. **—frag′men·tar′i·ly** (-târ′ə-lē) *adv.* **—frag′men·tar′i·ness** *n.*

**frag·men·ta·tion** (frăg′mən-tā′shən, -mĕn-) *n.* **1.** The act or process of breaking into fragments. **2.** The scattering of the fragments of an exploding grenade, bomb, or shell.

**fragmentation bomb** *n.* An aerial antipersonnel bomb that scatters shrapnel over a wide area upon explosion.

**frag·men·tize** (frăg′mən-tīz′) *vt. & vi.* **-tized, -tiz·ing, -tiz·es.** To break or become broken into fragments. **—frag′men·tiz′er** *n.*

**fra·grance** (frā′grəns) *n.* **1.** The quality or state of being fragrant. **2.** A pleasant odor : SCENT.

ă **pat** ā **pay** âr **care** ä **father** ĕ **pet** ē **be** hw **which** ĭ **pit** ī **tie** îr **pier** ŏ **pot** ō **toe** ô **paw, for** oi **noise** ōō **took**

**fra·grant** (frā'grənt) *adj.* [ME < Lat. *fragrans,* p.part. of *fragrare,* to emit an odor.] Having a pleasant odor. **—fra'grant·ly** *adv.*

**frail[1]** (frāl) *adj.* **-er, -est.** [ME < OFr. *fraile* < Lat. *fragilis* < *frangere,* to break.] **1.** Having a delicate constitution <a *frail* child> **2.** Slight or unsubstantial. **3.** FRAGILE 1. **4.** Easily led into evil. **—frail'ly** *adv.* **—frail'ness** *n.*

**frail[2]** (frāl) *n.* [ME *fraiel* < OFr.] **1.** A rush basket for holding fruit, esp. dried fruit. **2.** The quantity of fruit contained in a frail.

**frail·ty** (frāl'tē) *n., pl.* **-ties. 1.** The quality or state of being frail. **2.** A human fault, esp. weakness of resolution.

**fraise** (frāz) *n.* [Fr.] **1.** A barrier or defense of pointed, inclined stakes or of barbed wire. **2.** A ruff for the neck worn in the 16th cent.

**frak·tur** (fräk-tŏŏr') *n.* [G. < Lat. *fractura,* fracture.] A style of letter formerly used in German manuscripts and printing.

fraktur

**fram·be·sia** (frăm-bē'zhə, -zhē-ə) *n.* [NLat. < Fr. *framboise,* raspberry (from the appearance of the excrescence).] Yaws.

**frame** (frām) *v.* **framed, fram·ing, frames.** [ME *framen* < *frame,* structure.] *—vt.* **1.** To construct by putting together the various parts of. **2.** To conceive or design <*framed* a marketing proposal> **3.** To arrange or adjust for a given purpose <a question *framed* to draw only one answer> **4. a.** To put into words <*frame* a response> **b.** To form (words) silently with the lips. **5.** To enclose or encircle with or as if with a frame. **6.** *Slang.* **a.** To rig events or evidence so as to incriminate (a person) falsely. **b.** To fix (a contest) so as to ensure a desired fraudulent outcome <*frame* a football game> *—vi.* **1.** *Archaic.* To resort : proceed. **2.** *Obs.* To manage or contrive to do something. *—n.* **1.** Something made up of parts fitted and joined together. **2.** A skeletal structure designed to shape or support <the *frame* of a house> **3.** An open structure or rim for encasing, holding, or bordering something <window *frames*> **4.** The human body : PHYSIQUE <a robust *frame*> **5.** A cold frame. **6.** A machine built on or utilizing a frame. **7.** General structure : SYSTEM <the *frame* of democratic government> **8. a.** A round or period of play in some games, as bowling or billiards. **b.** *Baseball.* An inning. **9.** A single exposure on a roll of movie film. **10.** The total area of a television picture formed by a single traverse of the scanning spot. **11.** *Slang.* A frame-up. **12.** A minimal step in a sequence of programmed instruction. **13.** *Obs.* Shape. **—fram'er** *n.*

**frame of reference** *n.* **1.** *Physics.* A set of coordinate axes in terms of which position or movement may be specified or with reference to which physical laws may be mathematically stated. **2.** A set or system of ideas, as of philosophical or religious doctrine, in terms of which other ideas are interpreted or assigned meaning.

**frame·shift** (frām'shĭft') *n. Genetics.* Insertion or deletion of a pair of nucleotides in a gene that causes an alteration in the codon, resulting in the incorrect reading of the codon sequence as messenger RNA is being formed.

**frame-up** (frām'ŭp') *n. Informal.* **1.** A prearranged or fraudulent scheme. **2.** A scheme for incriminating an innocent victim.

**frame·work** (frām'wûrk') *n.* **1.** A supporting or enclosing structure, esp. a skeletal support used as the basis in an object being constructed. **2.** An external work platform : RIG. **3.** A basic arrangement, form, or system <"social structure is a stronger *framework* for behavior than national feeling" —Stanley Kauffman>

**franc** (frăngk) *n.* [Fr. < OFr. *franc* < Lat. *Francorum rex,* king of the Franks (from the legend on the first of these coins).] —See table at CURRENCY.

**fran·chise** (frăn'chīz') *n.* [ME *fraunchise,* freedom < OFr. *franchise* < *franche,* free.] **1.** A right or privilege officially granted a person or a group by a government, esp.: **a.** Constitutional or statutory right to vote. **b.** Establishment of a corporation's existence. **c.** The granting of certain rights and powers to a corporation. **d.** Legal immunity from certain burdens, servitude, or other restrictions, formerly granted to a person or group. **2.** Authorization granted by a manufacturer to a distributor or dealer to sell its products. **3.** The territory or limits within which a privilege, right, or immunity may be exercised. *—vt.* **-chised, -chis·ing, -chis·es.** To endow with a franchise.

**fran·chis·ee** (frăn'chī-zē') *n.* One enfranchised to run a unit of a business chain, as a restaurant.

**fran·chis·er** (frăn'chī'zər) *n.* One granting a franchise.

**Fran·cis·can** (frăn-sĭs'kən) *n.* A member of a religious mendicant order founded by Saint Francis of Assisi in 1209. **—Fran·cis'can** *adj.*

**fran·ci·um** (frăn'sē-əm) *n.* [NLat., after *France.*] *Symbol* **Fr** A highly unstable radioactive metallic element; atomic number 87; longest-lived isotope Fr 223.

**Franco-** *pref.* [< Med. Lat. *Francus,* Frenchman < LLat., Frank.] French <*Francophone*>

**Fran·co-A·mer·i·can** (frăng'kō-ə-mĕr'ĭ-kən) *n.* An American of French descent, esp. a French-Canadian. **—Fran'co-A·mer'i·can** *adj.*

**fran·co·lin** (frăng'kə-lĭn) *n.* [Fr. < Ital. *francolino.*] An Old World bird of the genus *Francolinus,* resembling a quail or partridge.

**Fran·co·phile** (frăng'kə-fīl') *also* **Fran·co·phil** (-fĭl') *n.* One who admires France, its people, and its culture.

**Fran·co·phobe** (frăng'kə-fōb') *n.* One who dislikes or fears France, its people, and its culture.

**Fran·co·phone** (frăng'kə-fōn') *adj.* French-speaking. *—n.* A French-speaking person.

**fran·gi·ble** (frăn'jə-bəl) *adj.* [ME < OFr. < Med. Lat. *frangibilis* < Lat. *frangere,* to break.] Easily broken. **—fran'gi·bil'i·ty, fran'gi·ble·ness** *n.*

**fran·gi·pan·i** (frăn'jə-păn'ē, -pä'nē) *n.* [Fr. *frangipane,* after Muzio *Frangipane,* a 16th-cent. Italian marquis who prepared a perfume similar to that of the shrub.] **1.** A tropical American shrub of the genus *Plumeria,* with milky juice and fragrant, variously colored flowers. **2.** A perfume derived from or similar in scent to the flowers of the frangipani. **3.** *also* **fran·gi·pane** (frăn'jə-pān'). A creamy pastry filling flavored with almonds.

**Fran·glais** (fräN-glā') *n.* [Blend of Fr. *Français,* French, and *Anglais,* English.] French marked by numerous borrowings from English.

**frank[1]** (frăngk) *adj.* **-er, -est.** [ME, free < OFr. *franc* < Med. Lat. *francus* < LLat. *Francus,* Frank.] **1.** Open and sincere : STRAIGHTFORWARD <a *frank* discussion> **2.** Clear : evident <*frank* pleasure> *—vt.* **franked, frank·ing, franks. 1. a.** To put an official mark on (a piece of mail) so it can be sent free of postage. **b.** To send (mail) free of charge. **2.** To place a stamp or mark on (a piece of mail) to show the payment of postage. **3.** To enable (a person) to come and go easily. *—n.* **1. a.** A mark or signature placed on a piece of mail to indicate the right to send it free of postage. **b.** The right to send mail free. **2.** A franked piece of mail. **—frank'ly** *adv.* **—frank'ness** *n.*

▲ word history: The word *frank,* "free," was originally the same word as *Frank,* a member of the Germanic people that conquered Gaul around A.D. 500. As the dominant group in the newly conquered territory, only the Franks possessed full freedom. The idea of political freedom was later extended to include freedom of behavior as well.

**frank[2]** (frăngk) *n. Informal.* A frankfurter.

**Frank** (frăngk) *n.* [ME < OE *Franca* and OFr. *Franc* < LLat. *Francus,* of Germanic orig.] A member of one of the Germanic tribes of the Rhine region in the early Christian era, esp. one of the Salian Franks who conquered Gaul about A.D. 500 and established an extensive empire that reached its greatest power in the 9th cent.

**Frank·en·stein** (frăng'kən-stīn') *n.* [After the protagonist of the novel *Frankenstein* by Mary W. Shelley (1797–1851).] **1.** A creation or agency that slips from the control of and destroys its creator. **2.** A monster having the appearance of a man.

**frank·furt·er** *also* **frank·fort·er** (frăngk'fər-tər) *or* **frank·furt** *or* **frank·fort** (-fərt) *n.* [After *Frankfurt am Main,* West Germany.] A smoked sausage of beef or beef and pork made in long reddish links.

**frank·in·cense** (frăng'kĭn-sĕns') *n.* [ME *frank encens* < OFr. *franc encens* : *franc,* superior + *encens,* incense.] An aromatic gum resin obtained from African and Asian trees of the genus *Boswellia* and used primarily as incense.

**Frank·ish** (frăng'kĭsh) *adj.* Of or relating to the Franks or their language. *—n.* The West Germanic language of the Franks.

**frank·lin** (frăng'klĭn) *n.* [ME *frankelein,* prob. < Med. Lat. *francus,* free. —see FRANK[1].] A medieval English freeholder of common birth but holding extensive property.

**frank·lin·ite** (frăng'klĭ-nīt') *n.* [After *Franklin,* New Jersey.] A black, slightly magnetic mineral of zinc, iron, and manganese that is a valuable source of zinc.

**Frank·lin stove** (frăng'klĭn) *n.* [After Benjamin *Franklin* (1706–1790), its inventor.] A cast-iron stove shaped like a fireplace but utilizing metal baffles to increase its energy efficiency.

**frank·pledge** (frăngk'plĕj') *n.* [ME *frankplegge* < AN *fraunc-pledge* : OFr. *franc,* free, frank + OFr. *plege,* pledge. —see PLEDGE.] **1.** A system in old English law in which units of ten households were formed, in each of which members were held responsible for one another's conduct. **2.** A member of a unit in frankpledge.

**fran·se·ri·a** (frăn-sîr'ē-ə) *n.* [NLat. *Franseria,* genus name, after Antonio *Franseri,* 18th-cent. Spanish botanist.] An herb or shrub of the genus *Franseria.*

**fran·tic** (frăn'tĭk) *adj.* [ME *frantik* < OFr. *frenetique* < Lat. *phreneticus.* —see FRENETIC.] **1.** Emotionally desperate <*frantic* with fear> **2.** Marked by rapid, disordered, or nervous action <*frantic* last-minute packing> **3.** *Archaic.* Crazy : insane. **—fran'ti·cal·ly, fran'tic·ly** *adv.* **—fran'tic·ness** *n.*

**frap** (frăp) *vt.* **frapped, frap·ping, fraps.** [ME *frapen,* to strike < OFr. *fraper.*] *Naut.* **1.** To make secure by lashing <*frap* a sail> **2.** To take up the slack of : TIGHTEN.

**frap·pé** (fră-pā', frăp) *n.* [Fr., chilled < p.part. of *frapper,* to chill < OFr. *fraper,* to strike.] **1.** A frozen fruit-flavored mixture served as an appetizer or dessert. **2.** A beverage, as a liqueur, poured over shaved ice. **3.** A milk shake containing ice cream.

**frat** (frăt) *n. Informal.* FRATERNITY 3.

**fra·ter·nal** (frə-tûr'nəl) *adj.* [ME < Med. Lat. *fraternalis* < Lat. *fraternus* < *frater,* brother.] **1. a.** Of or relating to brothers. **b.** Comradely : brotherly. **2.** Relating to or constituting a fraternity. **3.** *Biol.* Of or relating to a twin or twins developed from separately fertilized ova. **—fra·ter'nal·ism** *n.* **—fra·ter'nal·ly** *adv.*

**fra·ter·ni·ty** (frə-tûr'nĭ-tē) *n., pl.* **-ties.** [ME *fraternite* < OFr. *fraternité* < Lat. *fraternitas* < *fraternus,* fraternal.] **1.** A body of people associated for a common interest or purpose, as a guild. **2.** A group of people united by similar backgrounds, interests, or occupations <a *fraternity* of skydivers> **3.** A chiefly social organization of male college students, usu. designated by Greek letters. **4.** The quality or state of being brothers.

**frat·er·nize** (frăt'ər-nīz') *vi.* **-nized, -niz·ing, -niz·es.** [Fr. *fraterniser* < Med. Lat. *fraternizare* < Lat. *fraternus,* fraternal.] **1.** To associate with others in a congenial or brotherly way. **2.** To mix intimately with the people of an enemy or alien group, often in violation of military law. **—frat'er·ni·za'tion** *n.* **—frat'er·niz'er** *n.*

**frat·ri·cide** (frăt'rĭ-sīd') *n.* [Lat. *frater, fratr-,* brother + -CIDE.] **1.** The murder of one's brother or sister. **2.** One who has murdered his or her brother or sister. **—frat'ri·cid'al** (-sīd'l) *adj.*

**Frau** (frou) *n., pl.* **Frau·en** (frou'ən) [G. < MHG *vrowe* < OHG *frouwa.*] A married woman in a German-speaking area. —Used as a title corresponding to *Mrs.*

**fraud** (frôd) *n.* [ME *fraude* < OFr. < Lat. *fraus,* deceit.] **1.** A deliberate deception practiced so as to secure unfair or unlawful gain. **2.** Trickery. **3. a.** One who defrauds : CHEAT. **b.** One who pretends to be what he or she is not : IMPOSTOR.

**fraud·u·lent** (frô'jə-lənt) *adj.* [ME < OFr. < Lat. *fraudulentus* < *fraus,* deceit.] **1.** Engaging in fraud : DECEITFUL. **2.** Marked by, constituting, or gained by fraud <a *fraudulent* business contract> **—fraud'u·lence** *n.* **—fraud'u·lent·ly** *adv.*

**fraught** (frôt) *adj.* [ME, p.part. of *fraughten,* to load < MDu. *vrachten* < *vracht,* freight.] **1.** Filled or charged with : accompanied by <an event *fraught* with danger> **2.** Fully laden or provided. —*n. Obs.* Freight : cargo.

**Fräu·lein** (froi'līn') *n., pl.* **Fräulein.** [G., dim. of *Frau,* wife.] **1.** An unmarried girl or woman in a German-speaking area. —Used as a title corresponding to *Miss.* **2.** *Chiefly Brit.* A German governess.

**Fraun·ho·fer lines** (froun'hō'fər) *pl.n.* [After Joseph von *Fraunhofer* (1787–1826).] A set of several hundred dark lines appearing against the bright background of the continuous solar spectrum and produced by absorption of light by the cooler gases in the sun's outer atmosphere at frequencies corresponding to the atomic transition frequencies of these gases.

**frax·i·nel·la** (frăk'sə-nĕl'ə) *n.* [NLat., dim. of Lat. *fraxinus,* ash tree.] The gas plant.

**fray**[1] (frā) *n.* [ME *frai* < *affrai* < OFr. *effrei.*] **1.** A scuffle or brawl. **2.** A heated contest or dispute. —*v.* **frayed, fray·ing, frays** *Obs.* —*vt.* **1.** To alarm : frighten. **2.** To drive away. —*vi.* To fight.

**fray**[2] (frā) *v.* **frayed, fray·ing, frays.** [ME *fraien* < OFr. *fraier,* to rub < Lat. *fricare.*] —*vt.* **1.** To wear away (e.g., the edges of fabric) by rubbing. **2.** To strain <nerves *frayed* by the heavy traffic> —*vi.* To become frayed along the edges. —*n.* A frayed spot, as on fabric.

**fraz·zle** (frăz'əl) *Informal.* —*v.* **-zled, -zling, -zles.** [Perh. a blend of FRAY[2] and dial. *fazzle,* to tangle.] —*vt.* **1.** To fray, ravel, or tatter, as by rubbing. **2.** To exhaust physically or emotionally. —*vi.* To become frazzled. —*n.* **1.** A frayed condition. **2.** Fatigue or nervous exhaustion <worn to a *frazzle* by Friday>

**freak**[1] (frēk) *n.* [Orig. unknown.] **1.** A markedly unusual or irregular thing or occurrence <A *freak* of nature produced the summer hailstorm.> **2.** An abnormally formed organism, esp. a person or animal considered to be a curiosity or monstrosity. **3.** A sudden capricious turn of mind : WHIM. **4.** *Slang.* **a.** A drug user or addict. **b.** A hippie. **c.** A fan : enthusiast. —*vi.* & *vt.* **freaked, freak·ing, freaks.** *Slang.* **1.** To undergo or cause to undergo a negative reaction, as frightening hallucinations or feelings of paranoia, due to taking a drug, esp. a hallucinogen <*freaking* out on acid> **2.** To behave or cause to behave irrationally and uncontrollably <*freaked* completely when I was fired> **3. a.** To become or cause to become surprised or shocked <*freaked* out by the good news> **b.** To become or cause to become highly excited or elated.

**freak**[2] (frēk) *n.* [Orig. unknown.] A fleck or streak of color. —*vt.* **freaked, freak·ing, freaks.** To speckle or streak with color.

**freak·ish** (frē'kĭsh) *adj.* **1.** Unusual or abnormal. **2.** Relating to or typical of a freak. **3.** Capricious or whimsical. **—freak'ish·ly** *adv.* **—freak'ish·ness** *n.*

**freak-out** (frēk'out') *n. Slang.* **1.** An act or instance of freaking. **2.** A person who freaks.

**freak·y** (frē'kē) *adj.* **-i·er, -i·est. 1.** Freakish. **2.** *Slang.* Frightening : terrifying. **—freak'i·ly** *adv.*

**freck·le** (frĕk'əl) *n.* [ME *frakles* (pl.), alteration of *fraknes* < ON *freknur.*] A small precipitation of pigment in the skin, often brought out by the sun. —*vt.* & *vi.* **-led, -ling, -les.** To dot or become dotted with freckles. **—freck'ly** *adj.*

**free** (frē) *adj.* **fre·er, fre·est.** [ME *fre* < OE *frēo.*] **1.** Not imprisoned or enslaved : at liberty. **2.** Not controlled by obligation or the will of another <felt *free* to leave> **3. a.** Politically independent <a *free* nation> **b.** Governed by consent and possessing civil liberties <*free* citizenry> **c.** Immune to arbitrary governmental interference <a *free* press> **4. a.** Not affected or limited by a specified condition or circumstance <*free* from want><*free* of jealousy> **b.** Not subject to a specified condition : EXEMPT <a tax-*free* income> **5.** Not subject to external restraint <*free* speech> **6.** Not literal or exact <a *free* translation from the Latin> **7. a.** Costing nothing <a *free* dinner> **b.** Publicly supported <*free* education> **8. a.** Not occupied <a *free* table> **b.** Unobstructed : clear <a *free* lane> **9.** Guileless : frank. **10.** Taking undue liberties : OVERFAMILIAR. **11. a.** Liberal : lavish <*free* with one's money> **b.** Making use of something unstintingly <*free* with one's expense account> **12.** Uninhibited and outspoken. **13. a.** Given, made, or done of one's own accord : SPONTANEOUS <*free* advice> **b.** Determined according to one's own wishes <*free* choices> **14.** *Chem. & Physics.* **a.** Unconstrained : unconfined <*free* expansion> **b.** Capable of relatively unrestricted motion. **c.** Not chemically bound in a molecule <*free* oxygen> **d.** Involving no collisions or interactions <a *free* path> **e.** Empty <a *free* space> **15.** *Naut.* Favorable, as a wind. **16.** Not bound, fastened, or attached <the *free* end of a rope> **17.** Designating a vowel in an open syllable unchecked by a consonant, as *o* in *go.* —*adv.* **1.** In a free way : FREELY. **2.** Without charge <gave me the tire *free*> —*vt.* **freed, free·ing, frees. 1.** To set at liberty. **2.** To rid or release <a people *freed* from despotism> **3.** To clear or untangle <*free* an anchor line> **—free'ly** *adv.* **—free'ness** *n.*

☆ ***syns:*** FREE, EMANCIPATE, LIBERATE, MANUMIT, RELEASE, SPRING *v. core meaning* : to set at liberty <*free* the prisoners> FREE, LIBERATE, and RELEASE are the most general and are often interchangeable <*free* an innocent person><*liberate* hostages><*release* POW's> EMANCIPATE and MANUMIT usu. apply more narrowly to setting free from bondage or restraint <*emancipate* serfs> <*manumit* slaves> SPRING is a slang term usu. restricted to release from prison <a convict hoping his lawyer could *spring* him>

**free agent** *n.* A professional sports player who is free to sign a contract with any team.

**free alongside ship** *adj.* & *adv.* Delivered to the pier at no extra charge.

**free-as·so·ci·ate** (frē'ə-sō'shē-āt', -sē-) *vi.* **-at·ed, -at·ing, -ates.** To engage in free association.

**free association** *n.* **1.** A spontaneous, logically unconstrained association of ideas and feelings. **2.** A psychoanalytic technique in which a patient's articulation of free associations is encouraged in order to elicit repressed thoughts and emotions.

**free·base** (frē'bās') *vi.* **-based, -bas·ing, -bas·es.** To smoke the boiled residue of cocaine powder previously mixed with a solvent such as ether.

**free·bie** *also* **free·bee** (frē'bē) *n.* [< FREE.] *Slang.* Something given or received gratis <"offering cash rebates and *freebies* such as carpeting and landscaping" —*Newsweek*> **—free'bie** *adj.*

**free·board** (frē'bôrd', -bōrd') *n.* **1.** *Naut.* The distance between the water line and the uppermost full deck. **2.** The distance between the ground and the undercarriage of a car.

**freeboard deck** *n. Naut.* The uppermost deck officially considered completely watertight.

**free·boot·er** (frē'bōō'tər) *n.* [Du. *vrijbuiter* < *vrijbuit,* plunder : *vrij,* free + *buit,* booty.] One who pillages and plunders, esp. a pirate. **—free'boot'** *v.* **(-boot·ed, -boot·ing, -boots).**

**free·born** (frē'bôrn') *adj.* **1.** Born as a free person. **2.** Relating or appropriate to a person born free.

**free city** *n.* A city governed as an autonomous political unit under international auspices.

**freed·man** (frēd'mən) *n.* A man freed from bondage.

**free·dom** (frē'dəm) *n.* [ME *fredom* < OE *frēodom.*] **1.** The state of being free of restraints. **2.** Liberty of the person from slavery, oppression, or incarceration. **3. a.** Political independence. **b.** Possession of civil rights. **4.** Exemption from unpleasant or onerous conditions. **5.** Free will <the *freedom* to do as one wishes> **6.** Ease or facility in movement. **7.** Boldness or frankness <the new *freedom* in films and novels> **8. a.** Unrestricted access or use <was given the *freedom* of the entire military base> **b.** The right of enjoying all of the privileges of membership or citizenship.

☆ ***syns:*** FREEDOM, AUTONOMY, INDEPENDENCE, LIBERTY, SOVER-

EIGNTY *n. core meaning* : the condition of being politically free <granted the colony its *freedom*>

**free·dom of the seas** *n.* **1.** The doctrine that ships of any nation may travel through international waters unhampered. **2.** The right of neutral shipping in wartime to trade at will except where blockades have been established.

**freed·wom·an** (frēd′wo͝om′ən) *n.* A woman freed from bondage.

**free electron** *n.* An electron not bound to an atom, esp. an electron in a conductor that is available to move in a current.

**free energy** *n.* **1.** A thermodynamic quantity that is the difference between the internal energy and the product of the absolute temperature and entropy of a system. **2.** A thermodynamic quantity that is the difference between the enthalpy and the product of the absolute temperature and entropy of a system.

**free enterprise** *n.* The freedom of private businesses to operate competitively for profit with little government regulation.

**free fall** *n.* **1.** The fall of a body within the atmosphere without a drag-producing device, as a parachute. **2.** Unrestricted movement of a body in a gravitational field.

**free flight** *n.* Flight, as of an aircraft or spacecraft, after termination of powered flight.

**free-float·ing** (frē′flō′tĭng) *adj.* **1.** Not committed or decided. **2.** Experienced without obvious cause <*free-floating* anxiety>

**free-for-all** (frē′fər-ôl′) *n.* A brawl, argument, or competition in which everyone participates.

**free·form** (frē′fôrm′) *adj.* Designating a usu. flowing asymmetric shape or outline free of formal conventions.

**free form** *n.* A morpheme capable of standing alone and retaining meaning.

**free·hand** (frē′hănd′) *adj.* Drawn by hand without the aid of tracing or drafting devices <*freehand* sketches> **—free′hand′** *adv.*

**free hand** *n.* Freedom to do as one sees fit.

**free·hand·ed** (frē′hăn′dĭd) *adj.* Generous : openhanded. **—free′hand′ed·ly** *adv.* **—free′hand′ed·ness** *n.*

**free·heart·ed** (frē′här′tĭd) *adj.* **1.** Unreserved : open. **2.** Generous : liberal. **—free′heart′ed·ly** *adv.* **—free′heart′ed·ness** *n.*

**free·hold** (frē′hōld′) *n.* [ME *frehold,* transl. of Norman Fr. *fraunc tenement.*] **1.** *Law.* **a.** An estate held in fee or for life. **b.** The tenure by which such an estate is held. **2.** Life tenure of an office or a dignity. **—free′hold′er** *n.*

**free lance** *also* **free·lance** (frē′lăns′) *n.* **1.** One, esp. a writer or an artist, who sells his or her services to employers without a long-term commitment. **2.** One who remains uncommitted to a party and proceeds as an independent. **3.** A mercenary.

**free-lance** *also* **free·lance** (frē′lăns′) *vi.* & *vt.* **-lanced, -lanc·ing, -lanc·es.** To work as or produce and sell as a free lance. *—adj.* Of, relating to, or produced by a free lance. **—free′lanc′er** *n.*

**free-liv·ing** (frē′lĭv′ĭng) *adj.* **1.** Given to self-indulgence. **2.** *Biol.* Living or moving independently.

**free·load** (frē′lōd′) *vi.* **-load·ed, -load·ing, -loads.** *Slang.* To take advantage of the charity, generosity, or hospitality of others. **—free′load′er** *n.*

**free·man** (frē′mən) *n.* **1.** A person not in bondage or serfdom. **2.** One having all rights and privileges of a citizen.

**free·mar·tin** (frē′mär′tn) *n.* [Orig. unknown.] A sterile or otherwise sexually deficient female calf born as the twin of a bull calf.

**free·ma·son** (frē′mā′sən) *n.* **1.** A member of a guild of skilled, itinerant masons of the Middle Ages. **2. Freemason.** A member of the Free and Accepted Masons, an international secret fraternity.

**free·ma·son·ry** (frē′mā′sən-rē) *n.* **1.** Spontaneous fellowship and sympathy among numerous people. **2. Freemasonry.** The institutions, precepts, and rites of the Freemasons.

**free on board** *adj.* & *adv.* Without fee for delivery on board or into a carrier at a given point or location.

**free port** *n.* A port or an area of a port in which imported goods can be held or processed free of customs duties before re-export.

**fre·er** (frē′ər) *adj. compar. of* FREE.

**free radical** *n.* An atom or group of atoms with at least one unpaired electron.

**free rein** *n.* Unlimited freedom to act.

**free ride** *n.* Something acquired without the usual cost or effort. **—free rider** *n.*

**free·sia** (frē′zhə, -zhē-ə, -zē-ə) *n.* [NLat., after Friedrich H. T. *Freese* (d. 1876).] A plant of the genus *Freesia,* indigenous to southern Africa, having fragrant, variously colored flower clusters.

**free silver** *n.* The free coinage of silver, esp. at a fixed ratio to gold.

**free soil** *n.* U.S. territory in which slavery was prohibited before the Civil War.

**free-soil** (frē′soil′) *adj.* **1.** Prohibiting slavery <a *free-soil* state> **2. a.** Opposing the extension of slavery prior to the U.S. Civil War. **b. Free-Soil.** Relating to or designating a U.S. political party founded in 1848 to oppose the extension of slavery into U.S. territories and the admission of slave states into the Union.

**free-spo·ken** (frē′spō′kən) *adj.* Candid : outspoken <a *free-spoken* person> **—free′-spo′ken·ness** *n.*

**fre·est** (frē′ĭst) *adj. superl. of* FREE.

**free·stand·ing** (frē′stăn′dĭng) *adj.* Standing independently of attachment or support <a *freestanding* sculpture><a *freestanding* medical clinic>

**free·stone** (frē′stōn′) *n.* **1.** A stone, as limestone, soft enough to be cut easily without shattering or splitting. **2.** A fruit, esp. a peach, having a stone that does not cling to the pulp.

**free·style** (frē′stīl′) *n.* A competitive event, as swimming, in which the contestant chooses his or her own style.

**free-swim·ming** (frē′swĭm′ĭng) *adj.* Capable of swimming freely <*free-swimming* larva> **—free′-swim′mer** *n.*

**free·think·er** (frē′thĭng′kər) *n.* One who has rejected dogma and authority, esp. in religious thinking, in favor of rational inquiry and speculation. **—free′think′ing** *adj.* & *n.*

**free thought** *n.* Unorthodox thought : FREETHINKING.

**free throw** *n.* *Basketball.* A foul shot.

**free-throw line** (frē′thrō′) *n.* *Basketball.* FOUL LINE 2.

**free trade** *n.* Tariff-free trade between nations or states. **—free trader** *n.*

**free university** *n.* An independent unaccredited organization set up by students within a university studying nontraditional subjects.

**free verse** *n.* [Trans. of Fr. *vers libre.*] Verse not following a conventional metrical or stanzaic pattern and having either an irregular rhyme or no rhyme at all.

**free·way** (frē′wā′) *n.* **1.** A multilane highway and no intersections or stoplights : EXPRESSWAY. **2.** A highway without tolls.

**free·wheel** (frē′hwēl′, -wēl′) *vi.* **-wheeled, -wheel·ing, -wheels.** To live or move freely, aimlessly, or irresponsibly.

**free wheel** *n.* **1.** An automotive transmission device allowing the drive shaft to continue turning when its speed is greater than that of the engine shaft. **2.** A device in the rear-wheel hub of a bicycle that permits the wheel to turn without pedal action, as in coasting.

**free·wheel·ing** (frē′hwē′lĭng, -wē′-) *adj.* **1.** Relating to or equipped with free wheel. **2.** *Informal.* **a.** Free of rules or restraints in organization, methodology, or procedures. **b.** Not taking heed of consequences : CAREFREE.

**free·will** (frē′wĭl′) *adj.* Done of one's own accord : VOLUNTARY.

**free will** *n.* [Transl. of LLat. *liberum arbitrium.*] **1.** The ability or discretion to choose freely. **2.** The power, attributed esp. to human beings, of making free choices unconstrained by external circumstances or necessity.

**free world** *n.* The part of the world having democratic and capitalistic or socialistic political and economic systems rather than Communist or totalitarian ones.

**freeze** (frēz) *v.* **froze** (frōz), **fro·zen** (frō′zən), **freez·ing, freez·es.** [ME *fresen* < OE *frēosan.*] *—vi.* **1. a.** To change from a liquid to a solid state by losing heat. **b.** To acquire a surface of ice from cold. **2.** To become clogged or jammed due to formation of ice <plumbing that *froze* during the cold night> **3.** To become rigid and inflexible : SOLIDIFY <opinions that had *frozen* into dogma> **4.** To be at that degree of temperature at which ice forms <It may *freeze* tomorrow.> **5.** To be hurt, ruined, or killed by cold or frost <The citrus crops *froze.*> **6.** To feel the cold acutely <I'm *freezing* in this wind.> **7.** To become fixed, stuck, or attached by or as if by frost <The car door had *frozen.*> **8.** To become incapable of acting or reacting, as from fear <*froze* during the presentation> **9.** To become icily silent <*froze* at the insult> *—vt.* **1. a.** To convert into ice. **b.** To cause ice to form on. **c.** To cause to congeal or stiffen from extreme cold <winter cold that *froze* the marshes> **2.** To preserve (e.g., food) by subjecting to freezing temperatures. **3.** To hurt, ruin, or kill by cold or formation of ice. **4.** To make very cold : CHILL. **5.** To chill with an icy or formal manner <*froze* me with one look> **6.** To cause to become rigid and inflexible. **7. a.** To fix (prices or wages) at a given or current level. **b.** To prohibit further use or manufacture of. **c.** To prevent or restrict the exchange, liquidation, or granting of by law <banks that agreed to *freeze* investment loans> **8.** To anesthetize by freezing. *—n.* **1.** An act of freezing or the state of being frozen. **2.** A cold snap : FROST.

**freeze-dry** (frēz′drī′) *vt.* **-dried, -dry·ing, -dries.** To preserve (e.g., food) by rapid freezing and drying in a high vacuum.

**freeze-etch·ing** (frēz′ĕch′ĭng) *n.* A method of specimen preparation for electron microscopy in which a replica is made from a sample that has been rapidly frozen and then cut. **—freeze′-etched′, freeze′-etch′** *adj.*

**freeze frame** *n.* A single frame of a motion picture that is repeated to create the effect of a still photograph.

**freez·er** (frē′zər) *n.* One that freezes, esp. a thermally insulated cabinet or room that maintains a subfreezing temperature for the rapid freezing and storing of perishable food.

**freezing point** *n.* **1.** The temperature at which a liquid of specified composition solidifies under a specified pressure. **2.** The temperature at which the liquid and solid phases of a substance of specified composition are in equilibrium at atmospheric pressure.

**free zone** *n.* An area at a port or city where goods may be received and held without the payment of duty.

**F region** *n.* The F layer.

**freight** (frāt) *n.* [ME *fraught* < MDu. *vracht.*] **1.** Goods transported by a vessel or vehicle, esp. goods transported as cargo by a commercial carrier. **2.** A burden : load. **3.** Commercial transportation of goods. **4.** The charge for transporting goods by commercial carrier. **5.** A railroad train transporting goods only. —*vt.* **freight·ed, freight·ing, freights. 1.** To transport commercially as cargo. **2.** To load with goods that are to be transported. **3.** To load : charge.

**freight·age** (frā'tĭj) *n.* **1. a.** Commercial transportation of goods. **b.** The charge for such transportation. **2.** Cargo.

**freight car** *n.* A railroad car for transporting freight.

**freight·er** (frā'tər) *n.* **1.** A vehicle, esp. a ship, for transporting freight. **2.** A shipper of cargo.

**freight train** *n.* A railroad train composed of freight cars.

**frem·i·tus** (frĕm'ĭ-təs) *n., pl.* **fremitus.** [Lat., a murmuring < *fremere,* to murmur.] A palpable vibration, as felt by the hand placed on the chest during coughing or speaking.

**fre·na** (frē'nə) *n. var. pl. of* FRENUM.

**french** (frĕnch) *vt.* **frenched, french·ing, french·es. 1.** To cut into thin strips before cooking. **2.** To trim fat or bone from (e.g., a pork chop).

**French** (frĕnch) *adj.* [ME < OE *frencisc* < *Franca,* Frank.] Of, relating to, or typical of France or its people, language, or culture. —*n.* **1.** The Romance language of France, Switzerland, parts of Belgium, and other countries formerly under French influence or control. **2.** *(pl. in number).* The people of France.

**French bulldog** *n.* A small, compact dog orig. bred in France from toy English bulldogs and native breeds.

**French-Ca·na·di·an** *also* **French Canadian** (frĕnch'kə-nā'dē-ən) *n.* A Canadian of French descent. **—French'-Ca·na'di·an** *adj.*

**French chalk** *n.* Chalk of a soft, white variety of talc, used for marking fabrics and removing grease spots.

**French chop** *n.* A rib chop with the meat and fat trimmed from the end of the rib.

**French cuff** *n.* A wide cuff that is folded back and fastened with a cuff link.

**French curve** *n.* A flat drafting instrument with curved edges and scroll-shaped cutouts, used as a guide in connecting a set of individual points with a smooth curve.

**French door** *n.* A usu. paired door with glass panes extending the full length.

**French dressing** *n.* **1.** A seasoned oil and vinegar salad dressing. **2.** A commercially prepared creamy dressing, usu. pinkish and often sweet.

**French fries** *pl.n.* Strips of potatoes fried in deep fat.

**French-fry** (frĕnch'frī') *vt.* **-fried, -fry·ing, -fries.** To fry (e.g., potato strips) in deep fat.

**French heel** *n.* A curved, usu. high heel on women's shoes.

**French horn** *n.* A valved brass wind instrument with a circular shape, tapering from a narrow mouthpiece to a flaring bell at the other end and producing a mellow tone.

**French·i·fy** (frĕn'chə-fī') *vt.* **-fied, -fy·ing, -fies.** To make French in quality or character. **—French'i·fi·ca'tion** *n.*

**French knot** *n.* An embroidery stitch fashioned by looping thread two or more times around the needle, which is then inserted into the fabric.

**French leave** *n.* [From the 18th-cent. French custom of leaving without saying good-by to the host or hostess.] An abrupt or unannounced departure.

**French·man** (frĕnch'mən) *n.* A man who is a native or resident of France.

**French marigold** *n.* A native Mexican plant, *Tagetes patula,* having divided leaves and yellow flowers with reddish markings.

**French pastry** *n.* Rich and elaborate pastry prepared in individual portions.

**French provincial** *n.* A style of furniture or architecture characteristic of the 17th- and 18th-cent. French provinces.

**French seam** *n.* A seam stitched first on the right side and then turned in and stitched on the wrong side so that the raw edges are enclosed in and hidden by the seam.

**French telephone** *n.* A telephone whose receiver and transmitter are contained in a single hand piece.

**French toast** *n.* Sliced bread soaked in a milk and egg batter and then lightly fried.

**French window** *n.* **1.** A usu. paired window similar to a French door. **2.** A casement window.

**French·wom·an** (frĕnch'wŏŏm'ən) *n.* A woman who is a native or resident of France.

**fre·net·ic** (frə-nĕt'ĭk) *also* **fre·net·i·cal** (-ĭ-kəl) *adj.* [ME *frenetik* < OFr. *frenetique* < Lat. *phreneticus* < Gk. *phrenitikos* < *phrenitis,* brain disease < *phrēn,* mind.] Agitated or confused : FRANTIC. **—fre·net'i·cal·ly** *adv.*

**fren·u·lum** (frĕn'yə-ləm) *n., pl.* **-la** (-lə) [NLat., dim. of Lat. *frenum,* bridle.] **1.** A frenum. **2.** A spiny projection on moths' posterior wings that joins with a process on their anterior wings to fasten the wings together.

**fre·num** (frē'nəm) *n., pl.* **-nums** *or* **-na** (-nə) [Lat., bridle.] A membranous structure, as the fold under the tongue, that restrains movement or supports a part.

**fren·zied** (frĕn'zēd) *adj.* Affected with or marked by frenzy : FRANTIC <a *frenzied* search for the keys> **—fren'zied·ly** *adv.*

**fren·zy** (frĕn'zē) *n., pl.* **-zies.** [ME *frenesie* < OFr. < Lat. *phrenesis* < Gk. < *phrēn,* mind.] **1.** A violently agitated or wildly excited state, often accompanied by manic activity. **2.** Temporary madness or delirium. **3.** A craze. —*vt.* **-zied, -zy·ing, -zies.** To drive into a frenzy.

**Fre·on** (frē'ŏn'). A trademark for any of various nonflammable gaseous or liquid fluorocarbons that are used mainly as working fluids in refrigeration and air conditioning and as aerosol propellants.

**fre·quence** (frē'kwəns) *n.* [ME, multitude < Lat. *frequentia* < *frequens,* crowded.] Frequency.

**fre·quen·cy** (frē'kwən-sē) *n., pl.* **-cies.** [Lat. *frequentia,* multitude < *frequens,* crowded.] **1.** The property or state of occurring frequently. **2.** *Math. & Physics.* The number of times a specified phenomenon occurs within a given interval, as: **a.** The number of repetitions of a complete sequence of values of a periodic function per unit variation of an independent variable. **b.** The number of complete cycles of a periodic process occurring per unit time. **c.** The number of repetitions per unit time of a complete waveform, as of an electric current. **3.** *Statistics.* **a.** The number of measurements in an interval of a frequency distribution. **b.** The ratio of the number of times an event occurs in a series of trials of a chance experiment to the number of trials of the experiment performed.

**frequency curve** *n. Statistics.* A graphic approximation of the frequency polygon of a statistical distribution, usu. a smooth curve with the property that its abscissa is never negative and such that the area under the curve cut off by a class is the approximate fraction of the total frequency contained in that class.

**frequency distribution** *n. Statistics.* A set of intervals, usu. adjacent and of equal width, into which the range of a statistical distribution is divided, each associated with a frequency indicating the number of measurements in that interval.

**frequency modulation** *n.* The encoding of a carrier wave by variation of its frequency in accordance with an input signal.

**frequency polygon** *n. Statistics.* A graphic representation of a frequency distribution consisting of a set of points each obtained by plotting class frequency as ordinate and class mark as abscissa, together with line segments joining points of adjacent classes.

**fre·quent** (frē'kwənt) *adj.* [ME, ample < Lat. *frequens,* numerous.] Happening or appearing often or at close intervals <*frequent* errors of judgment> —*vt.* (frē-kwĕnt', frē'kwənt) **-quent·ed, -quent·ing, -quents.** To be in or at often. **—fre'quen·ta'tion** *n.* **—fre·quent'er** *n.* **—fre'quent·ly** *adv.* **—fre'quent·ness** *n.*

**fre·quen·ta·tive** (frē-kwĕn'tə-tĭv) *adj.* Expressing or denoting repeated action. —Used of a verb or verb form. **—fre·quen'ta·tive** *n.*

**fres·co** (frĕs'kō) *n., pl.* **-coes** *or* **-cos.** [Ital., fresh, of Germanic orig.] **1.** The art of painting on fresh, moist plaster with earth colors dissolved in water. **2.** A painting executed in fresco. —*vt.* **-coed, -co·ing, -coes.** To paint in fresco. **—fres'co·er, fres'co·ist** *n.*

**fresh** (frĕsh) *adj.* **-er, -est.** [ME < OE *fersc* and < OFr. *fres,* of Germanic orig.] **1.** New to one's experience. **2.** Novel : different <a *fresh* slant to the story> **3.** Recently made, produced, or harvested <*fresh* bread> **4.** Not preserved, as by canning, smoking, or freezing <*fresh* fruit> **5.** Not saline <*fresh* water> **6.** Not yet used or soiled : CLEAN <a *fresh* sheet of notepaper> **7.** Additional : new <*Fresh* evidence turned up.> **8.** Not dull or faded. **9.** Having the appearance of youth <a *fresh,* rosy complexion> **10.** Untried : inexperienced <*fresh* Marine recruits> **11.** Having just arrived <students *fresh* from Paris> **12.** Revived : refreshed <felt *fresh* as a daisy after the nap> **13.** Free from impurity or pollution : PURE <*fresh,* clean air> **14.** Fairly strong : BRISK <a *fresh* wind> **15.** Having recently calved and therefore with milk <a *fresh* cow> **16.** *Informal.* Bold and saucy : IMPUDENT <a *fresh* kid> —*adv.* Recently : newly <*fresh*-baked bread> —*n.* **1.** The early part <the *fresh* of the morning> **2.** A freshet. **—fresh'ly** *adv.* **—fresh'ness** *n.*

☆ ***syns:*** FRESH, INNOVATIVE, INVENTIVE, NEW, NEWFANGLED, NOVEL, ORIGINAL, UNPRECEDENTED *adj. core meaning* : showing a marked departure from previous practice <a *fresh* style of writing><a *fresh* approach to an old problem>

**fresh breeze** *n.* A wind with speeds of approx. 19 to 24 miles or 31 to 39 kilometers per hour.

**fresh·en** (frĕsh'ən) *v.* **-ened, -en·ing, -ens.** —*vi.* **1.** To become fresh, as in appearance <*freshened* up after the day's work> **2.** To increase in strength. —Used of the wind. **3.** To lose saltiness. **4.** To calve and produce milk. —*vt.* To make fresh. **—fresh'en·er** *n.*

**fresh·et** (frĕsh'ĭt) *n.* **1.** A sudden overflow of a stream resulting from a heavy rain or a thaw. **2.** A stream of fresh water emptying into a body of salt water.

**fresh gale** *n.* A wind with speeds of approx. 39 to 46 miles or 63 to 74 kilometers per hour.

**fresh·man** (frĕsh'mən) *n.* **1.** A student in the first-year class of a high school, college, or university. **2.** A beginner.

**fresh·wa·ter** (frĕsh′wô′tər, -wŏt′ər) *adj.* **1.** Of, relating to, living in, or composed of fresh water. **2.** Situated away from the sea : INLAND. **3.** Accustomed to sailing only on inland waters <a *freshwater* sailor>

**fret[1]** (frĕt) *v.* **fret·ted, fret·ting, frets.** [ME *freten* < OE *fretan,* to devour.] —*vt.* **1.** To cause to be anxious. **2. a.** To wear away : ERODE. **b.** To produce a worn spot or hole in : CORRODE. **3.** To form (a channel or passage) by erosion. **4.** To disturb the surface of (water) : AGITATE. —*vi.* **1.** To be vexed or troubled : WORRY. **2.** To be eaten away. **3.** To move agitatedly. **4.** To gnaw with the teeth as a rodent does. —*n.* **1.** An act or instance of fretting. **2.** A hole or worn spot made by abrasion or erosion. **3.** Irritation of mind : AGITATION.

**fret[2]** (frĕt) *n.* [Orig. unknown.] One of several ridges set across the fingerboard of a stringed instrument, as a guitar. —*vt.* **fret·ted, fret·ting, frets.** To provide with frets.

**fret[3]** (frĕt) *n.* [ME < OFr. *frete.*] **1.** An ornamental design composed of repeated and symmetric figures, often in relief, contained within a band or border. **2.** A headdress, worn by women during the Middle Ages, composed of interlaced wire. —*vt.* **fret·ted, fret·ting, frets.** To provide with a fret.

**fret·ful** (frĕt′fəl) *adj.* Inclined to fret. **—fret′ful·ly** *adv.* **—fret′ful·ness** *n.*

**fret saw** *n.* A long, narrow-bladed saw with fine teeth, used in making ornamental work in thin wood or metal.

**fret·work** (frĕt′wûrk′) *n.* **1.** Ornamental work composed of three-dimensional frets. **2.** Fretwork represented two-dimensionally by chiaroscuro.

**Freu·di·an** (froi′dē-ən) *adj.* Relating to or being in accord with the psychoanalytic theories of Sigmund Freud. **—Freu′di·an** *n.* **—Freu′di·an·ism** *n.*

**Freudian slip** *n.* A verbal slip caused by and indicating an unconscious thought, belief, or emotion.

**Freund's adjuvant** (froindz) *n.* [After Jules T. *Freund* (1890–1960).] A substance of killed microorganisms, as mycobacteria, in an oil and water emulsion that intensifies antigenicity.

**Frey** (frā) *also* **Freyr** (frâr) *n.* [ON *Freyr.*] *Norse Myth.* The god of peace, good weather, prosperity, and bountiful crops.

**Frey·a** *also* **Frey·ja** (frā′ə) *n.* [ON *Freyja.*] *Norse Myth.* The goddess of love and beauty.

**Freyr** (frâr) *n. var. of* FREY.

**fri·a·ble** (frī′ə-bəl) *adj.* [OFr. < Lat. *friabilis* < *friare,* to crumble.] Readily crumbled : BRITTLE. **—fri′a·bil′i·ty, fri′a·ble·ness** *n.*

**fri·ar** (frī′ər) *n.* [ME *frere* < OFr. < Lat. *frater,* brother.] A member of a usu. mendicant Roman Catholic order.

**fri·ar·bird** (frī′ər-bûrd′) *n.* A bird of the genus *Philemon* of Australia and adjacent regions, whose head is partially naked.

**friar's lantern** *n.* IGNIS FATUUS 1.

**fri·ar·y** (frī′ə-rē) *n., pl.* **-ies.** A monastery of friars.

**frib·ble** (frĭb′əl) *v.* **-bled, -bling, -bles.** [Orig. unknown.] —*vt.* To waste (e.g., time). —*vi.* To waste time : TRIFLE. —*n.* **1.** A frivolity : trifle. **2.** A frivolous person. **—frib′bler** *n.*

**fric·an·deau** (frĭk′ən-dō′) *n.* [Fr. < *fricasser,* to fricassee.] A cut of larded and braised veal.

**fric·as·see** (frĭk′ə-sē′, frĭk′ə-sē′) *n.* [Fr. *fricassée* < fem. p.part. of *fricasser,* to fricassee.] Meat cut into pieces and stewed in a gravy. —*vt.* **-seed, -see·ing, -sees.** To prepare as a fricassee.

**fric·a·tive** (frĭk′ə-tĭv) *n.* [NLat. *fricativus* < *fricare,* to rub.] A consonant, as *f* or *s* in English, produced by forcing breath through a constricted passage. **—fric′a·tive** *adj.*

**fric·tion** (frĭk′shən) *n.* [OFr. < Lat. *frictio* < *frictus,* p.part. of *fricare,* to rub.] **1.** The rubbing of one object or surface against another. **2.** Conflict, as between persons having dissimilar ideas or interests. **3.** *Physics.* A force tangential to the common boundary of two bodies in contact that resists the motion or tendency to motion of one relative to the other. **—fric′tion·al** *adj.* **—fric′tion·al·ly** *adv.*

**friction clutch** *n.* A clutch in which axial pressure with resultant friction between the clutch faces transmits torque.

**friction drive** *n.* An automotive transmission system in which motion is transmitted from one part to another by the surface friction of rolling contact.

**friction match** *n.* A match that ignites when struck on an abrasive surface.

**friction tape** *n.* A sturdy moisture-resistant adhesive tape used chiefly to insulate electrical conductors.

**Fri·day** (frī′dē, -dā′) *n.* [ME *Fridai* < OE *Frīgedæg,* Freya's day.] The sixth day of the week.

**fridge** (frĭj) *n. Informal.* A refrigerator.

**fried** (frīd) *v. p.t. & p.p. of* FRY[1].

**fried cake** *n.* A small pastry fried in deep fat, as a doughnut.

**Fried·land·er's bacillus** (frēd′lĕn′dərz) *n.* [After Karl *Friedländer* (1847–1887).] A species of pathogenic bacteria, *Klebsiella pneumoniae,* that causes pneumonia.

**Fried·man·ite** (frēd′mə-nīt′) *n.* [After Milton *Friedman* (b. 1912).] One supporting the theory that the government should directly regulate the money supply.

**friend** (frĕnd) *n.* [ME < OE *frēond.*] **1.** A person whom one knows, likes, and trusts. **2.** An acquaintance. **3.** A person with whom one is allied in a struggle or cause : COMRADE. **4.** A supporter, sympathizer, or patron of a group, cause, or movement. **5. Friend.** A member of the Society of Friends : QUAKER. —*vt.* **friend·ed, friend·ing, friends.** *Archaic.* To befriend. **—friend′less** *adj.* **—friend′less·ness** *n.* **—friend′ship′** *n.*

**friend·ly** (frĕnd′lē) *adj.* **-li·er, -li·est. 1.** Of, relating to, or worthy of a friend. **2.** Favorably disposed. **3.** Warm : comforting. —*n., pl.* **-lies.** One fighting on or favorable to one's own side. **—friend′li·ness** *n.* **—friend′ly** *adv.*

☆ ***syns:*** FRIENDLY, AMIABLE, AMICABLE, CHUMMY, CONGENIAL, CONVIVIAL, WARMHEARTED *adj. core meaning* : of or befitting a friend or friends <a *friendly* letter><*friendly* conversation> ***ant:*** unfriendly

**fri·er** (frī′ər) *n. var. of* FRYER.

**Frie·sian** (frē′zhən) *n. var. of* FRISIAN.

**frieze[1]** (frēz) *n.* [OFr. *frise* < Lat. *Phrygium (opus),* Phrygian (work), after *Phrygia,* an ancient country in Asia Minor.] **1.** A plain or decorated horizontal part of an entablature between the architrave and cornice. **2.** A decorative horizontal band, as along the upper part of a wall in a room.

**frieze[2]** (frēz) *n.* [ME *frise* < OFr. < MDu.] A coarse, shaggy woolen cloth with an uncut nap.

**frig·ate** (frĭg′ĭt) *n.* [Fr. *frégate* < Ital. *fregata.*] **1.** A high-speed, medium-sized sailing war vessel of the 17th, 18th, and 19th cent. **2.** A U.S. warship of approx. 5,000 to 7,000 tons, intermediate between a cruiser and a destroyer, used chiefly for escort duty. **3.** *Archaic.* A fast, light vessel, as a sailboat.

**frigate bird** *n.* A tropical sea bird of the genus *Fregata,* having long, powerful wings and dark plumage, that snatches food from other birds in flight.

**Frigg** (frĭg) *also* **Frig·ga** (frĭg′ə) *n.* [ON.] *Norse Myth.* The goddess of the heavens.

**fright** (frīt) *n.* [ME < OE *fyrhto.*] **1.** Sudden, intense fear : ALARM. **2.** *Informal.* Something very unsightly, alarming, or strange. —*vt.* **fright·ed, fright·ing, frights.** *Archaic.* To frighten.

**fright·en** (frīt′n) *v.* **-ened, -en·ing, -ens.** [ME < OE *fyrhtan.*] —*vt.* **1.** To make afraid. **2.** To drive or force by arousing fear <was *frightened* into confessing guilt> —*vi.* To become afraid. **—fright′en·er** *n.* **—fright′en·ing·ly** *adv.*

☆ ***syns:*** FRIGHTEN, ALARM, PANIC, SCARE, STARTLE, TERRIFY, TERRORIZE *v. core meaning* : to fill with fear <an explosion that *frightened* us>

**fright·ful** (frīt′fəl) *adj.* **1.** Eliciting shock or disgust : HORRIFYING. **2.** Producing fright : TERRIFYING. **3.** *Informal.* **a.** Excessive : extreme <a *frightful* boor> **b.** Disagreeable <a *frightful* climate> **—fright′ful·ly** *adv.* **—fright′ful·ness** *n.*

**frig·id** (frĭj′ĭd) *adj.* [Lat. *frigidus* < *frigēre,* to be cold < *frigus,* the cold.] **1.** Extremely cold. **2.** Lacking warmth of feeling. **3.** Stiff and formal <a *frigid* glance> **—fri·gid′i·ty** (frĭ-jĭd′ĭ-tē) *n.* **—frig′id·ness** *n.* **—frig′id·ly** *adv.*

**Frig·i·daire** (frĭj′ĭ-dâr′). A trademark for a refrigerator.

**frig·i·do·re·cep·tor** (frĭj′ĭ-dō-rĭ-sĕp′tər) *n.* A sensory receptor that responds to cold stimuli.

**frigid zone** *n.* **1.** The area within the Arctic Circle. **2.** The area within the Antarctic Circle.

**frig·o·rif·ic** (frĭg′ə-rĭf′ĭk) *also* **frig·o·rif·i·cal** (-ĭ-kəl) *adj.* [Lat. *frigorificus* : *frigus,* the cold + *-ficus,* -fic.] Causing coldness.

**†fri·jol** (frē-hōl′, frē′hōl′) *also* **fri·jo·le** (frē-hō′lē) *n., pl.* **fri·jo·les** (frē-hō′lēz, frē′hō′-). [Sp., var. of *frejol* < Catalan *fesol* < Lat. *phaseolus* < Gk. *phaselos.*] *Southwestern U.S.* A cultivated bean used for food.

**frill** (frĭl) *n.* [Orig. unknown.] **1.** A ruffled, gathered, or pleated projection or border, as a fabric edge used to trim clothing or a curled paper strip for decorating the end of the bone of a piece of meat. **2.** *Zool.* A ruff of hair or feathers about the neck of an animal or bird. **3.** A wrinkling of the edge of a photographic film. **4.** *Informal.* A superfluous item. —*v.* **frilled, frill·ing, frills.** —*vt.* **1.** To make into a ruffle or frill. **2.** To add a ruffle or frill to. —*vi.* To become wrinkled along the edge, as photographic film does. **—frill′y** *adj.*

**frilled lizard** *n.* An Australian lizard, *Chlamydosaurus kingi,* with a broad, contractile membrane extending from the neck and throat.

**fringe** (frĭnj) *n.* [ME *frenge* < OFr. < LLat. *fimbria.* —see FIMBRIA.] **1.** A decorative border or edging of hanging threads, cords, or strips, often attached to a separate band. **2.** Something like a fringe. **3.** A marginal, peripheral, or secondary part <on the *fringes* of the crowd> **4.** Extremist members of a group or political party <a lunatic *fringe*> **5.** Any of the light or dark bands produced by the diffraction or interference of light. —*vt.* **fringed, fring·ing, fring·es.** **1.** To decorate with or as if with a fringe. **2.** To serve as a fringe to <a pool *fringed* with lush ferns> **—fring′y** *adj.*

**fringe area** *n.* A zone just outside of the range of a broadcasting station in which signals are weakened and distorted.

**fringe benefit** *n.* An employment benefit given in addition to one's wages or salary.
**fringed gentian** *n.* A plant, *Gentiana crinita* of eastern North America, having blue, tubular flowers with fringed petals.
**fringed orchis** *n.* An orchid of the genus *Habenaria,* having variously colored flowers with fringed lips.
**fringed polygala** *n.* A plant, *Polygala paucifolia* of eastern North America, with fringed reddish-purple flowers.
**fringe tree** *n.* A shrub or small tree, *Chionanthus virginicus* of the southeastern United States, having drooping white flower clusters and dark blue fruit.
**frip·per·y** (frĭp′ə-rē) *n., pl.* **-ies.** [OFr. *freperie,* old clothes < *frepe,* rag.] **1.** Pretentious or showy finery. **2.** Pretentious elegance : OSTENTATION. **3.** A trivial or nonessential item.
**Fris·bee** (frĭz′bē). A trademark for a disk-shaped plastic toy that players throw and catch.
**fri·sé** (frē-zā′) *n.* [Fr. < p.part. of *friser,* to curl.] FRIEZE².
**fri·sette** (frĭ-zĕt′) *n. var. of* FRIZETTE.
**fri·seur** (frē-zûr′) *n.* [Fr. < *friser,* to curl.] A hairdresser.
**Fri·sian** (frĭzh′ən, frē′zhən) *also* **Frie·sian** (frē′zhən) *n.* [< Lat. *Frisii,* the Frisians.] **1.** A native or resident of the Frisian Islands or Friesland. **2.** The West Germanic language of the Frisians. **—Fri′sian** *adj.*
**frisk** (frĭsk) *v.* **frisked, frisk·ing, frisks.** [< obs. *frisk,* lively < OFr. *frisque.*] *—vi.* To move about briskly and playfully : FROLIC. *—vt.* To search (a person) for something concealed, esp. a weapon, by passing the hands quickly over the clothes. *—n.* **1.** An energetic, playful movement : GAMBOL. **2.** An act of frisking. **—frisk′er** *n.*
**frisk·y** (frĭs′kē) *adj.* **-i·er, -i·est.** [Obs. *frisk,* lively < OFr. *frisque.*] Energetic, lively, and playful. **—frisk′i·ly** *adv.* **—frisk′i·ness** *n.*
**frit** (frĭt) *n.* [Ital. *fritta* < fem. p.part. of *friggere,* to fry < Lat. *frigere.*] **1.** The fused or partially fused materials used in glassmaking. **2.** A vitreous substance used in making porcelain or glazes. *—vt.* **frit·ted, frit·ting, frits.** To make into frit.
**frit fly** *n.* [Orig. unknown.] A small fly of the family Chloropidae or Oscinidae, esp. *Oscinella frit,* having larvae destructive to cereal plants.
**frith** (frĭth) *n.* [Var. of FIRTH.] *Scot.* An estuary.
**frit·il·lar·y** (frĭt′l-ĕr′ē) *n., pl.* **-ies.** [NLat. *Fritillaria* (genus) < Lat. *fritillus,* dice box.] **1.** A bulbous plant of the genus *Fritillaria,* with nodding, variously colored, often spotted or checkered flowers. **2.** Any of various butterflies of the family Nymphalidae, and esp. of the genera *Speyeria* and *Boloria,* having brownish wings marked with black or silvery spots.
**frit·ter¹** (frĭt′ər) *vt.* **-tered, -ter·ing, -ters.** [< obs. *fitter,* fragment.] **1.** To reduce or squander a little at a time <*fritter* away one's money on trifles> **2.** To break, tear, or cut into bits : SHRED.
**frit·ter²** (frĭt′ər) *n.* [ME *friture* < OFr. < Lat. *frigere,* to fry.] A small cake made of batter, often containing fruit, vegetables, or fish, sautéed or fried in deep fat.
**friv·ol** (frĭv′əl) *vi.* **-oled, -ol·ing, -ols** *or* **-olled, -ol·ling, -ols.** [Back-formation < FRIVOLOUS.] *Informal.* To behave frivolously. **—friv′ol·er** *n.*
**friv·o·lous** (frĭv′ə-ləs) *adj.* [ME < *frivol,* trifle < OFr. *frivole* < Lat. *frivolum.*] **1.** Unworthy of serious attention : TRIVIAL. **2.** Inappropriately silly. **—fri·vol′i·ty** (frĭ-vŏl′ĭ-tē) *n.* **—friv′o·lous·ly** *adv.* **—friv′o·lous·ness** *n.*
**fri·zette** *also* **fri·sette** (frĭ-zĕt′) *n.* [Fr. *frisette* < *friser,* to curl.] A curled fringe of hair, worn on the forehead by a woman.
**frizz¹** (frĭz) *vt.* & *vi.* **frizzed, frizz·ing, frizz·es.** [Fr. *friser.*] To form or be formed into small, tight curls or tufts. *—n.* **1.** The state of being frizzed. **2.** A small, tight curl or tuft. **—frizz′er** *n.*
**frizz²** (frĭz) *v.* **frizzed, frizz·ing, frizz·es.** [Prob. < FRY.] *—vt.* To fry or burn with a sizzling noise. *—vi.* To make a sizzling noise while frying or searing.
**friz·zle¹** (frĭz′əl) *v.* **-zled, -zling, -zles.** [Perh. blend of FRY and SIZZLE.] *—vt.* **1.** To fry until crisp and curled. **2.** To scorch or sear with heat. *—vi.* To fry or sear with a sizzling noise.
**friz·zle²** (frĭz′əl) *vi.* & *vt.* **-zled, -zling, -zles.** [Orig. unknown.] To frizz or cause to frizz. *—n.* A small, tight curl.
**friz·zly** (frĭz′lē) *adj.* **-zli·er, -zli·est.** Frizzy.
**friz·zy** (frĭz′ē) *adj.* **-zi·er, -zi·est.** Tightly curled. **—friz′zi·ly** *adv.* **—friz′zi·ness** *n.*
**fro** (frō) *adv.* [ME < OE *fram* and < ON *frām.*] Away : back <running to and *fro*> *—prep. Scot.* From.
**frock** (frŏk) *n.* [ME *frok* < OFr. *froc* < Med. Lat. *froccus,* of Germanic orig.] **1.** A long, loose outer garment, as that worn by artists and craftspeople : SMOCK. **2.** A woolen garment once worn by sailors. **3.** A robe worn by monks, friars, and clerics : HABIT. **4.** A woman's dress. *—vt.* **frocked, frock·ing, frocks. 1.** To clothe in a frock. **2.** To invest with clerical office.
**frock coat** *n.* A man's dress coat with knee-length skirts.
**froe** *also* **frow** (frō) *n.* [Orig. unknown.] A cleaving tool with a heavy blade set at right angles to the handle.
**frog** (frôg, frŏg) *n.* [ME *frogge* < OE *frogga.*] **1.** Any of numerous tailless, chiefly aquatic amphibians of the order Salientia, and esp. of the family Ranidae, having a smooth, moist skin, webbed feet, and long hind legs adapted for jumping. **2.** A horny wedge-shaped prominence in the sole of a horse's hoof. **3.** A loop fastened to a belt to hold a tool or weapon. **4.** An ornamental looped braid or cord with a button or knot for fastening the front of a garment. **5.** A device on intersecting railroad tracks that allows wheels to cross the junction. **6.** A spiked or perforated device used to support stems in a flower arrangement. **7.** *Informal.* Hoarseness in the throat.
**frog·eye** (frôg′ī′, frŏg′ī′) *n.* A plant disease caused by fungi in which rounded spots appear on the leaves.
**frog·fish** (frôg′fĭsh′, frŏg′-) *n., pl.* **frogfish** *or* **-fish·es.** Any of various anglerfishes of the family Antennariidae of tropical and temperate seas, covered with fleshy or filamentous processes.
**frog·hop·per** (frôg′hŏp′ər, frŏg′-) *n.* The spittlebug.
**frog kick** *n.* A swimming kick in which the legs are drawn up close beneath one, then thrust outward and together vigorously.
**frog·man** (frôg′măn′, -mən, frŏg′-) *n.* A swimmer equipped with breathing equipment and other devices to execute underwater maneuvers, esp. military maneuvers.
**frog·mouth** (frôg′mouth′, frŏg′-) *n.* A brown or gray nocturnal bird of the genus *Podargus* or *Batrachostomus* of southeastern Asia and Australia, having a hooked bill and a wide mouth.
**frog spit** *also* **frog spittle** *n.* **1.** Cuckoo spit. **2.** A foamlike aggregation of small aquatic plants, as algae, on the surface of a pond.
**frol·ic** (frŏl′ĭk) *n.* [Du. *vrolijk,* merry < MDu. *vrolijc* < *vro,* happy.] **1.** Merriment. **2.** A gay, carefree time. **3.** A playful antic. *—vi.* **-icked, -ick·ing, -ics. 1.** To behave playfully and uninhibitedly : ROMP. **2.** To engage in merrymaking, joking, or teasing. *—adj. Archaic.* Merry. **—frol′ick·er** *n.*
**frol·ic·some** (frŏl′ĭk-səm) *adj.* Full of high-spirited fun : PLAYFUL.
**from** (frŭm, frŏm; frəm *when unstressed*) *prep.* [ME < OE.] **1. a.** —Used to indicate a particular time or place as a starting point <*from* seven o'clock on><ran home *from* the store> **b.** —Used to indicate a specific point as the first of two limits <*from* grades six to eight> **2.** —Used to indicate a source, cause, agent, or instrument <a note *from* the school principal><take a book *from* the library> **3.** —Used to indicate separation, removal, or exclusion <keep me *from* making an error><liberation *from* slavery> **4.** —Used to indicate differentiation <know good *from* evil> **5.** Because of <dizzy *from* the height>
**frond** (frŏnd) *n.* [Lat. *frons, frond-,* foliage.] **1.** A fern's usu. compound leaf. **2.** A large compound leaf of another plant, as a palm. **3.** A leaflike thallus, as of a seaweed or lichen. **—frond′ed** *adj.*
**fron·des·cent** (frŏn-dĕs′ənt) *adj.* [Lat. *frondescens, frondescent-,* pr.part. of *frondescere,* to become leafy < *frondēre,* to put forth leaves < *frons,* foliage.] Having, resembling, or having many leaves or fronds : LEAFY. **—fron·des′cence** *n.*
**fron·dose** (frŏn′dōs′) *adj.* [Lat. *frondōsus* : *frons,* foliage + *-osus,* -ose.] **1.** Bearing fronds. **2.** Like a frond. **—fron′dose′ly** *adv.*
**front** (frŭnt) *n.* [ME < OFr. < Lat. *frons,* façade.] **1.** The forward part or surface, as of a building. **2.** The area or position directly before or ahead. **3.** A position of leadership or superiority. **4.** *Archaic.* The first part : BEGINNING. **5.** The forehead, esp. of an animal. **6.** *Archaic.* The human face. **7.** Demeanor or bearing, esp. in the face of danger or difficulty <put up a brave *front*> **8.** An outward, often feigned appearance or manner. **9.** Land bordering a body of water or a street. **10.** A promenade along the water at a resort. **11.** The detachable part of a man's dress shirt covering the chest. **12. a.** The most forward line of a combat force. **b.** The area of contact between opposing combat forces. **13.** *Meteorol.* The interface between air masses at different temperatures. **14.** A group or movement uniting individuals or groups for the achievement of a common goal : COALITION <a revolutionary *front*> **15.** A nominal leader lacking true authority : FIGUREHEAD. **16.** A seemingly respectable person, group, or business used as a cover for secret or illegal activities <a *front* for a smuggling operation> **17.** A field of activity <the economic *front*> *—adj.* **1.** Of, relating to, aimed at, or located in the front. **2.** Produced at or toward the front of the oral cavity <a *front* vowel> *—v.* **front·ed, front·ing, fronts.** *—vt.* **1.** To look out on : FACE. **2.** To meet in opposition : CONFRONT. **3.** To provide a front for. **4.** To serve as a front for. *—vi.* To have a front : FACE <The property *fronts* on the beach.>
**front·age** (frŭn′tĭj) *n.* **1.** The front part of a piece of property. **2.** The land between a building and a street. **3.** The direction in which something faces. **4.** Land that lies adjacent to a building, street, body of water, etc.
**fron·tal¹** (frŭn′tl) *adj.* **1.** Of, relating to, or located at the front. **2.** Of or relating to the forehead. **3.** Of or relating to the frontal plane. **4.** Of or relating to a weather front. **—fron′tal·ly** *adv.*
**fron·tal²** (frŭn′tl) *n.* [ME *frontel* < OFr. < Med. Lat. *frontellum* < Lat. *frons,* façade.] **1.** A drapery covering the front of an altar. **2.** A building façade.
**frontal bone** *n.* A cranial bone having a vertical portion corresponding to the forehead and an orbital or horizontal portion that is part of the roofs of the orbital and nasal cavities.

**frontal lobe** *n.* The largest part of the anterior portion of the cerebral cortex.
**frontal lobotomy** *n.* A prefrontal lobotomy.
**frontal plane** *n.* A plane parallel to the long axis of the body and perpendicular to the sagittal plane.
**front·court** (frŭnt′kôrt′, -kōrt′) *n. Basketball.* The offensive half of the court.
**front-end** (frŭnt′ĕnd′) *adj.* Of or relating to the initial phase of a project <a *front-end* investment>
**front-end load** *n.* The amount deducted from early payments made to a mutual fund purchase plan that covers expenses such as sales commissions.
**fron·ten·is** (frŭn-tĕn′ĭs, frŏn′tĕn′ĭs) *n.* [Am. Sp., blend of Sp. *fronton,* gable, jai alai court (< *frenta,* forehead < Lat. *frons*), and *tenis,* tennis (< E. TENNIS).] A Latin-American tennis game played on a three-walled court.
**fron·tier** (frŭn-tîr′, frŏn-, frŭn′tîr′, frŏn′-) *n.* [ME *frountier* < OFr. *frontier* < *front,* front.] **1. a.** An international border. **b.** The area bordering on a frontier. **2.** A region just beyond or at the edge of a settled area. **3.** An undeveloped field for research or discovery.
▲ word history: *Frontier* and *front* are both derived from Latin *frons,* "forehead, front, façade." *Frontier* was borrowed into English from French in the 15th century with the meaning "borderland," the region of a country that fronts on another country. The use of *frontier* to mean "a region at the edge of a settled area" is a special American development. During most of American history the edge of the settled country was the place where unlimited cheap land was available to anyone willing to live the hard but independent life of the pioneer farmer. It has long been recognized that the experience of frontier life had an important part in shaping American society and character. This sense of *frontier* has also been extended to other areas of achievement and conquest.
**fron·tiers·man** (frŭn-tîrz′mən, frŏn-) *n.* A man living in a frontier area.
**fron·tis·piece** (frŭn′tĭ-spēs′) *n.* [OFr. *frontispice,* a building's principal façade < Med. Lat. *frontispicium* : Lat. *frons,* façade + Lat. *specere,* to look at.] **1.** An illustration that faces or immediately precedes the title page of a book, book section, or magazine. **2.** *Archaic.* A title page. **3.** An architectural façade, esp. an ornamental one. **4.** A small ornamental pediment, as on top of a door or window.
**front·let** (frŭnt′lĭt) *n.* [ME.] **1.** An ornament or band worn on the forehead, as a phylactery. **2.** The forehead of an animal, esp. when distinctively marked. **3.** The ornamental border of a frontal.
**front-line** (frŭnt′līn′) *adj.* **1.** Located or used at a military front. **2.** Of or pertaining to the most advanced or important position or activity in a field or undertaking.
**front man** *n.* FRONT 15.
**front matter** *n.* The material, as the frontispiece, preface, and title page, preceding the text in a book.
**front money** *n.* **1.** Money paid in advance for goods or services. **2.** Money spent during the initial part of a project.
**front office** *n.* An organization's policy-making officers.
**fron·to·gen·e·sis** (frŭn′tō-jĕn′ĭ-sĭs) *n.* Development or intensification of a weather front.
**fron·tol·y·sis** (frŭn-tŏl′ĭ-sĭs) *n.* Disintegration of a weather front.
**front-page** (frŭnt′pāj′) *adj.* So newsworthy as to require coverage on the front page of a newspaper. —*vt.* **-paged, -pag·ing, -pag·es.** To report or place on the front page of a newspaper.
**front room** *n.* A living room.
**front-run·ner** (frŭnt′rŭn′ər) *n.* **1.** A leading contender in a competition, as a race <the presidential *front-runner*> **2.** A competitor who performs best when in the lead. **—front′-run′ning** *adj.*
**front·ward** (frŭnt′wərd) *adj. & adv.* At or toward the front. **—front′wards** *adv.*
**frore** (frôr, frōr) *adj.* [ME, p.part. of *fresen,* to freeze < OE *frēosan.*] *Archaic.* Very cold : FROSTY.
**frosh** (frŏsh) *n., pl.* **frosh.** [Shortening and alteration of FRESHMAN.] *Informal.* A freshman.
**frost** (frôst, frŏst) *n.* [ME < OE.] **1.** A deposit of tiny ice crystals formed from frozen water vapor. **2.** The atmospheric conditions when the temperature is below the freezing point of water. **3.** The process of freezing. **4.** A cold or icy manner. **5.** *Informal.* A failure. —*v.* **frost·ed, frost·ing, frosts.** —*vt.* **1.** To cover with frost. **2.** To damage or kill by frost. **3.** To cover (e.g., glass) with a roughened or speckled decorative surface. **4.** To cover or decorate with icing. —*vi.* To become covered with or as if with frost.
**frost·bite** (frôst′bīt′, frŏst′-) *n.* Local tissue destruction resulting from freezing. —*vt.* **-bit** (-bĭt′), **-bit·ten** (-bĭt′n), **-bit·ing, -bites.** To injure or damage with frost.
**frost heave** *also* **frost heaving** *n.* An uplifting of a surface, as pavement or soil, caused by subsurface freezing.
**frost·ing** (frô′stĭng, frŏs′tĭng) *n.* **1.** ICING 1. **2.** A roughened or speckled surface imparted to glass or metal.
**frost line** *n.* The limit to which frost penetrates the earth.
**frost·work** (frôst′wûrk′, frŏst′-) *n.* **1.** Intricate patterns produced by frost, as on a windowpane. **2.** Ornamental patterns similar to frostwork, produced artificially, as on metal or glass.
**frost·y** (frô′stē, frŏs′tē) *adj.* **-i·er, -i·est. 1.** Producing or marked by frost : FREEZING. **2.** Covered with or as if with frost. **3.** Silvery white : HOARY. **4.** Cold in manner. **—frost′i·ly** *adv.* **—frost′i·ness** *n.*
**froth** (frôth, frŏth) *n.* [ME < ON *froða.*] **1.** A mass of bubbles in or on a liquid : FOAM. **2.** A salivary foam released as a result of disease or exhaustion. **3.** Something trivial or unsubstantial. —*v.* **frothed, froth·ing, froths.** —*vt.* **1.** To cover with foam. **2.** To cause to foam. —*vi.* To expel froth : EXUDE.
**froth·y** (frô′thē, frŏth′ē) *adj.* **-i·er, -i·est. 1.** Made of, covered with, or like froth : FOAMY. **2.** Playfully frivolous <a *frothy* French comedy> **—froth′i·ly** *adv.* **—froth′i·ness** *n.*
**frot·tage** (frô-täzh′) *n.* [Fr. < *frotter,* to rub.] **1.** A method of making a design by placing a piece of paper on top of an object and then rubbing over it, as with a pencil or charcoal. **2.** A design made by frottage.
**frou-frou** (frōō′frōō) *n.* [Fr.] **1.** A rustling sound, as of silk. **2.** Fussy dress or ornamentation.
**frow** (frō) *n. var. of* FROE.
**fro·ward** (frō′wərd, frō′ərd) *adj.* [ME : *fro,* fro + *-ward,* -ward.] Stubbornly contrary and disobedient : OBSTINATE. **—fro′ward·ly** *adv.* **—fro′ward·ness** *n.*
**frown** (froun) *v.* **frowned, frown·ing, frowns.** [ME *frounen* < OFr. *froigner,* of Celt. orig.] —*vi.* **1.** To wrinkle the brow, as in thought or displeasure. **2.** To regard with disapproval or distaste <*frowned* on smoking> —*vt.* To express (e.g., disapproval) by wrinkling the brow. **—frown** *n.* **—frown′er** *n.* **—frown′ing·ly** *adv.*
**frows·ty** (frou′stē) *adj.* **-ti·er, -ti·est.** [Orig. unknown.] *Chiefly Brit.* Having a stale smell : MUSTY.
**frow·zy** *also* **frow·sy** (frou′zē) *adj.* **-zi·er, -zi·est** *also* **-si·er, -si·est.** [Orig. unknown.] **1.** Unkempt : slovenly. **2.** Having an unpleasant smell. **—frow′zi·ness** *n.*
**froze** (frōz) *v. p.t. of* FREEZE.
**fro·zen** (frō′zən) *adj.* [< p.part. of FREEZE.] **1.** Made into, covered with, or surrounded by ice. **2.** Very cold <a *frozen* Arctic night> **3.** Preserved by freezing. **4.** Rendered immobile. **5.** Coldly unfriendly. **6. a.** Kept at a fixed level <Wages were *frozen.*> **b.** Incapable of being withdrawn, sold, or liquidated <*frozen* assets>
**frozen food** *n.* Food preserved by freezing, esp. commercially.
**fruc·tif·er·ous** (frŭk-tĭf′ər-əs, frŏŏk-) *adj.* [Lat. *fructifer* (*fructus,* fruit + *-fer,* -fer) + -OUS.] Bearing fruit.
**fruc·ti·fi·ca·tion** (frŭk′tə-fĭ-kā′shən, frŏŏk′-) *n.* **1.** Production of fruit. **2.** A seed- or spore-bearing structure.
**fruc·ti·fy** (frŭk′tə-fī′, frŏŏk′-) *v.* **-fied, -fy·ing, -fies.** [ME *fructifien* < OFr. *fructifier* < Lat. *fructificare* : *fructus,* fruit + *facere,* to make.] —*vt.* To make fruitful. —*vi.* To bear fruit.
**fruc·tose** (frŭk′tōs′, frŏŏk′-) *n.* [Lat. *fructus,* fruit + -OSE.] A very sweet sugar, $C_6H_{12}O_6$, occurring in many fruits and honey and used as a food preservative and an intravenous nutrient.
**fruc·tu·ous** (frŭk′chōō-əs, frŏŏk′-) *adj.* [ME < OFr. < Lat. *fructuosus* < *fructus,* fruit.] Fruitful : productive.
**fru·gal** (frōō′gəl) *adj.* [Lat. *frugalis* < *frux,* success.] **1.** Avoiding unnecessary monetary expenditure : THRIFTY <a *frugal* New England farmer> **2.** Costing little : INEXPENSIVE <a *frugal* dinner> **—fru·gal′i·ty** (frōō-găl′ĭ-tē), **fru′gal·ness** *n.* **—fru′gal·ly** *adv.*
**fru·giv·o·rous** (frōō-jĭv′ər-əs) *adj.* [Lat. *frux, frug-,* fruit + -VOROUS.] Feeding on fruit.
**fruit** (frōōt) *n., pl.* **fruit** *or* **fruits.** [ME < OFr. < Lat. *fructus* < p.part. of *frui,* to enjoy.] **1. a.** The ripened ovary or ovaries of a seed-bearing plant, along with its accessory parts, containing the seeds and occurring in numerous forms. **b.** An edible, usu. sweet and fleshy form of such a structure. **c.** A part or amount of such a plant product, served as food. **2.** The fertile, often spore-bearing structure of a plant that does not bear seeds. **3.** A plant crop or product. **4.** Result : outcome <the *fruit* of their labor> **5.** Offspring : progeny. —*vi. & vt.* **fruit·ed, fruit·ing, fruits.** To produce or cause to produce fruit.
**fruit·age** (frōō′tĭj) *n.* **1.** The time, process, or state of bearing fruit. **2.** Fruit in general. **3.** Produce from the land. **4.** FRUIT 4.
**fruit bat** *n.* Any of various tropical and subtropical Old World fruit-eating bats of the family Pteropodidae.

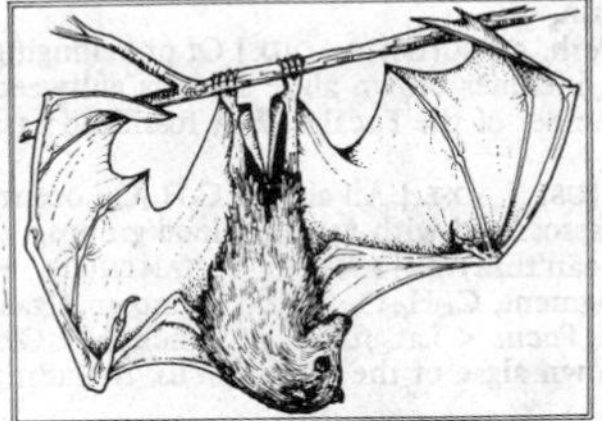

**fruit bat**
*May reach a wingspan of 5 feet*

**fruit·cake** (fro͞ot′kăk′) *n.* **1.** A heavy, spiced cake containing nuts and candied or dried fruits. **2.** *Slang.* An insane person.

**fruit cup** *n.* A mixture of fresh or preserved fruits cut into pieces and served as an appetizer or dessert.

**fruit·er·er** (fro͞o′tər-ər) *n.* A grower or seller of fruit.

**fruit fly** *n.* **1.** A small fly of the family Drosophilidae, having larva that feeds on ripening or fermenting fruit, esp. a common species, *Drosophila melanogaster.* **2.** Any of various flies of the family Tripetidae or Tephritidae, whose larvae hatch in and harm plant tissue.

**fruit·ful** (fro͞ot′fəl) *adj.* **1.** Producing fruit. **2.** Producing abundantly : PROLIFIC. **3.** Conducive to productivity. **4.** Producing results : PROFITABLE. **—fruit′ful·ly** *adv.* **—fruit′ful·ness** *n.*

**fruiting body** *n.* A specialized spore-producing structure, esp. of a fungus.

**fru·i·tion** (fro͞o-ĭsh′ən) *n.* [ME *fruicioun* < OFr. *fruition* < Med. Lat. *fruitio*, enjoyment < Lat. *frui*, to enjoy.] **1.** Enjoyment derived from use or possession. **2. a.** The condition of bearing fruit. **b.** Achievement of something desired or worked for.

**fruit·less** (fro͞ot′lĭs) *adj.* **1.** Not producing fruit. **2.** Unproductive <a *fruitless* search for the wallet> **—fruit′less·ly** *adv.* **—fruit′less·ness** *n.*

**fruit sugar** *n.* Fructose.

**fruit·y** (fro͞o′tē) *adj.* **-i·er, -i·est. 1.** Of, containing, or relating to fruit. **2.** Tasting and smelling of fruit. **3.** Overly sentimental or sweet. **4.** *Slang.* Crazy : nutty. **—fruit′i·ness** *n.*

**fru·men·ta·ceous** (fro͞o′mən-tā′shəs, -mĕn-) *adj.* [LLat. *frumentaceus* < Lat. *frumentum*, grain, perh. < *frui*, to enjoy.] Resembling or consisting of grain, esp. wheat.

**fru·men·ty** (fro͞o′mən-tē) *n.* [ME *frumente* < OFr. *frumentée* < *frument*, grain < Lat. *frumentum*, perh. < *frui*, to enjoy.] Hulled wheat boiled in milk and flavored with sugar and spices.

**frump** (frŭmp) *n.* [Orig. unknown.] **1.** A plain, unfashionable woman or girl. **2.** A colorless, prim person. **—frump′i·ly** *adv.* **—frump′i·ness** *n.* **—frump′y** *adj.*

**frump·ish** (frŭm′pĭsh) *adj.* **1.** Dull or plain. **2.** Prim and sedate. **—frump′ish·ly** *adv.* **—frump′ish·ness** *n.*

**frus·ta** (frŭs′tə) *n. var. pl. of* FRUSTUM.

**frus·trate** (frŭs′trāt′) *vt.* **-trat·ed, -trat·ing, -trates.** [ME *frustraten* < Lat. *frustrare*, to disappoint < *frustra*, in error.] **1. a.** To prevent from attaining a goal or fulfilling a desire : THWART. **b.** To cause feelings of discouragement or bafflement in. **2.** To make ineffectual or invalid : NULLIFY. **—frus′trat′er** *n.* **—frus·tra′tion** *n.*

**frus·tule** (frŭs′cho͞ol, -tyo͞ol) *n.* [Fr. < Lat. *frustulum*, dim. of *frustum*, piece broken off.] A diatom's hard, siliceous shell.

**frus·tum** (frŭs′təm) *n., pl.* **-tums** *or* **-ta** (-tə) [Lat., piece broken off.] A part of a solid, as a cone or pyramid, between two parallel planes cutting the solid, esp. the section between the base and a plane parallel to the base.

**fru·tes·cent** (fro͞o-tĕs′ənt) *adj.* [Lat. *frutex*, bush + -ESCENT.] Relating to, like, or assuming the form of a shrub. **—fru·tes′cence** *n.*

**fru·ti·cose** (fro͞o′tĭ-kōs′) *adj.* [Lat. *fruticosus* < *frutex*, bush.] Shrublike, esp. in form.

**fry**[1] (frī) *v.* **fried, fry·ing, fries.** [ME *frien* < OFr. *frire* < Lat. *frigere.*] *—vt.* To cook over direct heat in hot fat or oil. *—vi.* **1.** To be cooked in a pan over direct heat in hot fat or oil. **2.** *Slang.* To undergo execution in an electric chair. *—n., pl.* **fries. 1.** A dish of a fried food. **2.** A gathering at which food is fried and eaten <an oyster *fry*>

**fry**[2] (frī) *n., pl.* **fry.** [ME *fri*, prob. < AN.] **1. a.** A small fish, esp. a young, recently hatched one. **b.** The young of certain other animals. **2.** Individuals : persons <the young *fry*>

**fry·er** *also* **fri·er** (frī′ər) *n.* **1.** One that fries. **2.** A small, young chicken suitable for frying.

**frying pan** *n.* A shallow, long-handled pan for frying food.

**f-stop** (ĕf′stŏp′) *n.* [F(OCAL LENGTH) + STOP.] **1.** A camera lens aperture setting that is calibrated to a corresponding f-number. **2.** An f-number.

**f-sys·tem** (ĕf′sĭs′təm) *n.* A way of indicating the relative aperture of a camera lens based on the f-number.

**fuch·sia** (fyo͞o′shə) *n.* [NLat. *Fuchsia*, genus name, after Leonard *Fuchs* (1501–1566).] **1.** A chiefly tropical shrub of the genus *Fuchsia*, cultivated widely for its drooping purplish, white, or reddish flowers. **2.** A strong, vivid purplish red.

**fuch·sin** *also* **fuch·sine** (fyo͝ok′sĭn, -sēn′) *n.* [FUCHS(IA) + -IN.] A dark-green synthetic aniline dyestuff, the hydrochloride of rosaniline, used to make a purple-red dye employed in coloring textiles and leather and as a bacterial stain.

**fu·coid** (fyo͞o′koid′) *adj.* [Perh. < FUC(US) + -OID.] Of or belonging to the order Fucales, which includes brown algae such as gulfweed and rockweed. *—n.* **1.** A member of the Fucales. **2.** A fossilized cast or impression of a fucoid.

**fu·cose** (fyo͞o′kōs′) *n.* [FUC(US) + -OSE.] An aldose, $C_6H_{12}O_5$, occurring in the polysaccharides associated with various blood groups.

**fu·co·xan·thin** (fyo͞o′kō-zăn′thĭn) *n.* [FUC(US) + XANTH(O)- + -IN.] A brown carotenoid pigment, $C_{40}H_{60}O_6$, found in brown algae.

**fu·cus** (fyo͞o′kəs) *n.* [NLat. *Fucus* < Lat. *fucus*, rock lichen < Gk. *phukos.*] Any of various brown algae of the genus *Fucus*, including many of the rockweeds.

**fud·dle** (fŭd′l) *v.* **-dled, -dling, -dles.** [Orig. unknown.] *—vt.* **1.** To make drunk : INTOXICATE. **2.** To make confused : BEFUDDLE. *—vi.* To drink : tipple. *—n.* **1.** A state of drunkenness. **2.** Confusion.

**fud·dy-dud·dy** (fŭd′ē-dŭd′ē) *n., pl.* **-dies.** [Orig. unknown.] A fussy, old-fashioned person.

**fudge** (fŭj) *n.* [Orig. unknown.] **1.** A rich soft candy of sugar, milk, butter, flavoring, and usu. chocolate. **2.** Nonsense. *—v.* **fudged, fudg·ing, fudg·es.** *—vt.* **1.** To falsify <*fudged* the expense figures> **2.** To evade (e.g., an issue) : DODGE. *—vi.* **1.** To behave indecisively. **2. a.** To exceed the proper limits. **b.** To act dishonestly : CHEAT.

**fueh·rer** (fyo͝or′ər) *n. var. of* FÜHRER.

**fu·el** (fyo͞o′əl) *n.* [ME *feuel* < OFr. *fouaille* < Lat. *focus*, fire.] **1.** Matter consumed to generate energy, esp.: **a.** A material, as wood, coal, gas, or oil, burned to generate heat. **b.** Fissionable material used in a nuclear reactor. **c.** Nutritive material metabolized by a living organism. **2.** Something that stimulates or maintains an activity or emotion. *—v.* **-eled, -el·ing, -els** *or* **-elled, -el·ling, -els.** *—vt.* **1.** To provide with fuel. **2. a.** To support the activity or existence of. **b.** To stimulate <comments that *fueled* the argument> *—vi.* To take in fuel. **—fu′el·er** *n.*

**fuel cell** *n.* An electrochemical cell in which the energy of a reaction between a fuel, as liquid hydrogen, and an oxidant, as liquid oxygen, is converted directly and continuously into electrical energy.

**fuel injection** *n.* A method or mechanical system by which fuel is vaporized and sprayed directly into the cylinders of an internal-combustion engine.

**fuel oil** *n.* A liquid or liquefiable petroleum product for generating power or heat.

**fug** (fəg) *n.* [Orig. unknown.] A malodorous emanation <"In spite of the open windows the stench had become a reeking *fug*" —Colleen McCullough>

**fu·ga·cious** (fyo͞o-gā′shəs) *adj.* [Lat. *fugax*, *fugac-*, swift < *fugere*, to flee.] **1.** Passing away quickly : EVANESCENT. **2.** *Bot.* Withering or dropping off early. **—fu·ga′cious·ly** *adv.* **—fu·ga′cious·ness, fu·gac′i·ty** (-găs′ĭ-tē) *n.*

**-fuge** *suff.* [< Lat. *fugare*, to expel < *fuga*, flight.] One that expels or drives away <vermi*fuge*>

**fu·gi·tive** (fyo͞o′jĭ-tĭv) *adj.* [ME *fugitif* < OFr. < Lat. *fugitivus* < *fugere*, to flee.] **1.** Fleeing, as from the law <a *fugitive* bank robber> **2. a.** Passing quickly : FLEETING <*fugitive* hours> **b.** Difficult to understand or retain : ELUSIVE <*fugitive* facts and figures> **c.** Given to change or disappearance : PERISHABLE. **3.** Tending to wander : VAGABOND. **4.** Relating to topics of temporary interest : EPHEMERAL. *—n.* **1.** One who flees <a *fugitive* from justice> **2.** Something ephemeral. **—fu′gi·tive·ly** *adv.* **—fu′gi·tive·ness** *n.*

**fu·gle** (fyo͞o′gəl) *vi.* **-gled, -gling, -gles.** [Back-formation < FUGLEMAN.] *Archaic.* **1.** To act as a fugleman. **2.** To make signals.

**fu·gle·man** (fyo͞o′gəl-mən) *n.* [Alteration of G. *Flügelmann* : *Flügel*, wing + *Mann*, man.] **1.** *Archaic.* A soldier serving as a guide and model for his company. **2.** A leader, esp. a political leader.

**fugue** (fyo͞og) *n.* [Fr. < Ital. *fuga* < Lat., flight.] **1.** *Mus.* A polyphonic composition in which a theme or themes stated successively by a number of voices in imitation are developed contrapuntally. **2.** A pathological condition during which one is apparently conscious of one's actions but has no recollection of them after returning to a normal state. **—fu′gal** (fyo͞o′gəl) *adj.* **—fu′gal·ly** *adv.*

**füh·rer** *also* **fueh·rer** (fyo͝or′ər) *n.* [G. < MHG *vüerer*, bearer < *vüeren*, to bear < OHG *fuoren*, to lead.] **1.** An all-powerful ruler, esp. a tyrant. **2. Führer.** Adolf Hitler's title as the leader of the German Nazis.

**Fu·jian** (fo͞o′jĕn′) *n.* A dialect of Chinese spoken in northern Fujian and parts of Taiwan.

**Fu·kien** (fo͞o′kyĕn′) *n.* Fujian.

**-ful** *suff.* [ME < OE < *full*, full.] **1. a.** Full of <play*ful*> **b.** Characterized by : RESEMBLING <master*ful*> **c.** Tending, given, or able to <use*ful*> **2.** A quantity that fills <arm*ful*>

**Fu·la** *also* **Fu·lah** (fo͞o′lə) *n., pl.* **Fula** *or* **-las** *also* **Fulah** *or* **-lahs. 1.** A chiefly Moslem people of northwestern and central Africa, of mixed Hamitic and Negroid stock. **2.** A member of the Fula people.

**Fu·la·ni** (fo͞o′lä′nē, fo͞o-lä′-) *n., pl.* **Fulani** *or* **-nis. 1.** One of the Fula. **2.** The language of the Fula.

**ful·crum** (fo͝ol′krəm, fŭl′-) *n., pl.* **-crums** *or* **-cra** (-krə) [Lat., bedpost < *fulcire*, to support.] **1.** The point or support on which a lever turns. **2.** A centralized means of exerting influence or pressure.

**ful·fill** *also* **ful·fil** (fo͝ol-fĭl′) *vt.* **-filled, -fill·ing, -fills** *also* **-filled, -fil·ling, -fils.** [ME *fulfillen* < OE *fulfyllan* : *ful*, full + *fyllan*, to fill.] **1.** To bring into actuality : EFFECT. **2.** To carry out (e.g., an order). **3.** To measure up to : SATISFY <*fulfill* all requirements for the job> **4.** To bring to an end : COMPLETE. **—ful·fill′er** *n.* **—ful·fill′ment, ful·fil′ment** *n.*

**ful·gent** (fo͝ol′jənt, fŭl′-) *adj.* [ME < Lat. *fulgens*, pr.part. of *fulgēre*, to shine.] Brilliant or shining : RADIANT. **—ful′gent·ly** *adv.*

**ful·gu·rant** (fo͝ol′gyər-ənt, -gər-, fŭl′-) *adj.* [Lat. *fulgurans*, *fulgu-*

*rant-*, pr.part. of *fulgurare*, to lighten. —see FULGURATE.] Flashing like lightning.

**ful·gu·rate** (fo͝ol'gyə-rāt', -gə-, fŭl'-) *vi.* **-rat·ed, -rat·ing, -rates.** [Lat. *fulgurare, fulgurat-*, to lighten < *fulgur*, lightning < *fulgēre*, to flash.] To emit in flashes. **—ful'gu·ra'tion** *n.*

**ful·gu·rite** (fo͝ol'gyə-rīt', -gə-, fŭl'-) *n.* [Lat. *fulgur*, lightning + -ITE.] A tubular body of glassy rock produced by lightning striking exposed surfaces.

**ful·gu·rous** (fo͝ol'gyər-əs, -gər-, fŭl'-) *adj.* [Lat. *fulgur*, lightning + -OUS.] **1.** Emitting lightning flashes. **2.** Emitting in flashes resembling lightning.

**fu·lig·i·nous** (fyo͞o-lĭj'ə-nəs) *adj.* [LLat. *fuliginosus* < Lat. *fuligo*, soot.] **1.** Sooty. **2.** Colored by or as if by soot. **—fu·lig'i·nous·ly** *adv.*

**full**[1] (fo͝ol) *adj.* **-er, -est.** [ME *ful* < OE.] **1.** Containing all that is possible or normal <a *full* bucket> **2.** Complete in every detail <a *full* account of the incident> **3. a.** Of maximum degree <drove at *full* speed> **b.** At the peak of growth and development <roses in *full* bloom> **4.** Having many <a garden *full* of flowers> **5.** FULL-FLEDGED 3. **6. a.** Rounded in shape : plump <a *full* figure> **b.** Having or made with a generous amount of fabric <a *full* skirt> **7. a.** Satiated, esp. with food or drink. **b.** Providing an abundance, esp. of food. **8. a.** Having depth and body : RICH. **b.** Resonant <in *full* voice> **9.** Totally preoccupied. **10.** Possessing both parents in common. —*adv.* **1.** To a complete extent <*full*-grown><knowing *full* well> **2.** Exactly : directly <*full* in the path of the car> —*v.* **fulled, full·ing, fulls.** —*vt.* To make (a garment) full, as by pleating or gathering. —*vi.* To become full. —Used of the moon. —*n.* **1.** The maximum or complete size or amount. **2.** The highest degree or state. **—full'ness, ful'ness** *n.* **—ful'ly** *adv.*

**full**[2] (fo͝ol) *vt.* **fulled, full·ing, fulls.** [ME *fullen* < OFr. *fuler, fouler* < Med. Lat. *fullare* < Lat. *fullo*, fuller.] To increase the weight and bulk of (cloth) by shrinking and beating or pressing.

**full·back** (fo͝ol'băk') *n.* **1. a.** *Football.* A backfield player whose position is behind the quarterback and halfbacks and who performs offensive blocking and line plunges and defensive linebacking. **b.** A similar player in field hockey, soccer, and rugby. **2.** The position played by a fullback.

**ull blood** *n.* **1.** Relationship established through sharing the same set of parents. **2.** An individual of unmixed race or breed.

**ull-blood·ed** (fo͝ol'blŭd'ĭd) *adj.* **1. a.** Of unmixed ancestry : PUREBRED. **b.** Related through having the same parents. **2. a.** Not pale or anemic. **b.** Vigorous and vital. **3.** FULL-DRESS 1. **—full'-blood'ed·ness** *n.*

**ull-blown** (fo͝ol'blōn') *adj.* **1.** Having blossomed completely <*full-blown* roses> **2.** Fully developed or matured. **3.** Having or displaying all the characteristics necessary for completeness.

**ull-bod·ied** (fo͝ol'bŏd'ēd) *adj.* Having richness and intensity of flavor <*full-bodied* wines>

**ull dress** *n.* Attire appropriate for ceremonial or formal occasions.

**ull-dress** (fo͝ol'drĕs') *adj.* **1.** Complete in every respect <a *full-dress* rehearsal> **2.** Thorough <a *full-dress* inquest>

**ull·er**[1] (fo͝ol'ər) *n.* [ME *fullere* < OE and < OFr. *fouleor*, both < Lat. *fullo*.] One who fulls cloth.

**ul·ler**[2] (fo͝ol'ər) *n.* [Orig. unknown.] **1.** A hammer used by a blacksmith for grooving or spreading iron. **2.** A groove made by a fuller.

**uller's earth** *n.* A highly absorbent claylike substance used chiefly in fulling woolen cloth, in talcum powders, and as a filter and a catalyst.

**uller's teasel** *n.* A European plant, *Dipsacus fullonum*, having bristly flower heads used by fullers to raise the nap on cloth.

**ull-fash·ioned** (fo͝ol'făsh'ənd) *adj.* Knitted so as to conform closely to body lines.

**ull-fledged** (fo͝ol'flĕjd') *adj.* **1.** Having fully developed adult plumage. **2.** Having reached full development : MATURE. **3.** Having full status or rank <a *full-fledged* member>

**ull gainer** *n.* A forward dive in which the diver executes a full back somersault before entering the water.

**ull house** *n.* A poker hand made up of three of a kind and a pair.

**ull-length** (fo͝ol'lĕngkth', -lĕngth') *adj.* **1.** Showing or fitted to the entire length, esp. of the human body <a *full-length* mirror> **2.** Being of standard or normal length <a *full-length* historical novel>

**ull moon** *n.* **1.** The phase of the moon when visible as a fully lighted disk. **2.** The time of the month when a full moon occurs.

**ull-mouthed** (fo͝ol'mouthd', -moutht') *adj.* **1.** Having a complete set of teeth. —Used of livestock. **2.** Uttered loudly or noisily.

**ull nelson** *n.* A wrestling hold in which both hands are thrust under the opponent's arms from behind and then pressed against the back of the opponent's neck.

**ull rhyme** *n.* Perfect rhyme.

**ull-scale** (fo͝ol'skāl') *adj.* **1.** Of actual or full size <a *full-scale* model of the boat> **2.** Utilizing all resources <a *full-scale* promotional campaign>

**ull stop** *n.* A period indicating the end of a sentence.

**ull tilt** *adv.* At high or top speed <ran *full tilt* into the wall>

**full-time** (fo͝ol'tīm') *adj.* Used for or involving a standard number of working hours <a *full-time* manager> **—full'-time'** *adv.*

**ful·mar** (fo͝ol'mər, -mär') *n.* [Of Scand. orig.] **1.** A gull-like bird, *Fulmarus glacialis* of Arctic regions, having smoky gray plumage. **2.** A similar or related bird.

**ful·mi·nant** (fo͝ol'mə-nənt, fŭl'-) *adj.* [Lat. *fulminans, fulminant-*, pr.part. of *fulminare*, to strike with lightning. —see FULMINATE.] **1.** Fulminating. **2.** *Pathol.* Taking place suddenly, rapidly, and with great intensity.

**ful·mi·nate** (fo͝ol'mə-nāt', fŭl'-) *v.* **-nat·ed, -nat·ing, -nates.** [ME *fulminaten* < Lat. *fulminare*, to strike with lightning < *fulmen*, lightning < *fulgēre*, to flash.] —*vi.* **1.** To issue a thunderous verbal attack <*fulminate* against political corruption> **2.** To explode with sudden violence. —*vt.* **1.** To thunder out or issue (e.g., a denunciation). **2.** To cause to explode. —*n.* An explosive salt of fulminic acid, esp. fulminate of mercury. **—ful'mi·na'tion** *n.* **—ful'mi·na'tor** *n.* **—ful'mi·na·to'ry** (-nə-tôr'ē, -tōr'ē) *adj.*

**fulminate of mercury** *n.* A gray crystalline powder, $Hg(CNO)_2$, that explodes when dry under the slightest friction or shock and is used as a high explosive.

**ful·mine** (fo͝ol'mĭn, fŭl'-) *vt. & vi.* **-mined, -min·ing, -mines.** [OFr. *fulminer* < Lat. *fulminare*, to strike with lightning. —see FULMINATE.] *Archaic.* To fulminate.

**ful·min·ic acid** (fo͝ol-mĭn'ĭk, fŭl-) *n.* [Lat. *fulmen, fulmin-*, lightning (see FULMINATE) + -IC.] An unstable acid, HONC, that forms highly explosive salts.

**ful·some** (fo͝ol'səm) *adj.* [ME *fulsom*, loathsome < *ful*, full.] **1.** Offensively flattering or insincere. **2.** Offensive to the taste or sensibilities. **3.** *Archaic.* Copious or abundant in supply. ***usage:*** *Fulsome* is sometimes mistakenly used in the sense of "full or abundant." This sense is no longer current, however; in modern usage *fulsome* combines the ideas of fullness or abundance with that of an offensive or insincere excess. **—ful'some·ly** *adv.* **—ful'some·ness** *n.*

**fu·ma·rate** (fyo͞o'mə-rāt') *n.* [FUMAR(IC ACID) + -ATE.] A salt or ester of fumaric acid.

**fu·mar·ic acid** (fyo͞o-măr'ĭk) *n.* [< NLat. *Fumaria*, genus of herbs (< LLat. *fumaria*, fumitory < Lat. *fumus*, smoke) + -IC.] An acid, $C_4H_4O_4$, found in various plants and produced synthetically, used mainly in resins, paints, and varnishes.

**fu·ma·role** (fyo͞o'mə-rōl') *n.* [Ital. *fumarola* < LLat. *fumariolum* < Lat. *fumarium*, smoke chamber < *fumus*, smoke.] A hole in a volcanic area from which smoke and gases arise. **—fu'ma·rol'ic** *adj.*

**fu·ma·to·ri·um** (fyo͞o'mə-tôr'ē-əm, -tōr'-) *n., pl.* **-to·ri·ums** or **-to·ri·a** (-tôr'ē-ə, -tōr'-) [NLat. < Lat. *fūmatus*, p.part. of *fumare*, to smoke < *fumus*, smoke.] An airtight fumigation chamber in which chemical vapors are used to destroy insects and fungi on plants.

**fu·ma·to·ry** (fyo͞o'mə-tôr'ē, -tōr'ē) *adj.* [< Lat. *fumare, fumat-*, to smoke < *fumus*, smoke.] Of or relating to smoke or fumigating. —*n., pl.* **-ries.** A fumatorium.

**fum·ble** (fŭm'bəl) *v.* **-bled, -bling, -bles.** [Perh. of Scand. orig.] —*vi.* **1.** To handle nervously or idly. **2.** To grope awkwardly <*fumble* for a key> **3.** To proceed awkwardly and uncertainly : BLUNDER <*fumble* through a speech> **4. a.** *Baseball.* To mishandle a ground ball. **b.** *Football.* To drop a ball that is in play. —*vt.* **1.** To handle nervously or idly. **2.** To make a botch of : BUNGLE. **3.** To feel or make (one's way) awkwardly. **4. a.** *Baseball.* To mishandle (a ground ball). **b.** *Football.* To drop (a ball) while in play. —*n.* **1.** An act or instance of fumbling. **2.** A fumbled ball. **—fum'bler** *n.*

**fume** (fyo͞om) *n.* [ME < OFr. *fum* < Lat. *fumus*.] **1.** A usu. irritating or disagreeable exhalation, as of smoke, vapor, or gas. **2.** A strong or acrid odor. **3.** Irritation or anger. —*v.* **fumed, fum·ing, fumes.** —*vt.* **1.** To subject to or treat with fumes. **2.** To give off in or as if in fumes. —*vi.* **1.** To emit fumes. **2.** To rise in fumes. **3.** To experience or exhibit irritation or anger.

**fu·mi·gate** (fyo͞o'mĭ-gāt') *v.* **-gat·ed, -gat·ing, -gates.** [Lat. *fumigare, fumigat-*, to smoke : *fumus*, smoke + *agere*, to make.] —*vt.* To subject to smoke or fumes, usu. in order to exterminate vermin or insects or to disinfect. —*vi.* To use smoke or fumes as an exterminating or disinfecting agent. **—fu'mi·ga'tion** *n.* **—fu'mi·ga'tor** *n.*

**fu·mi·to·ry** (fyo͞o'mĭ-tôr'ē, -tōr'ē) *n., pl.* **-ries.** [ME *fumetere* < OFr. *fumeterre* < Med. Lat. *fumus terre* : Lat. *fumus*, smoke + Lat. *terra*, earth.] A native European climbing plant, *Fumaria officinalis*, with spurred purplish flowers and finely divided leaves.

**fun** (fŭn) *n.* [Perh. < ME *fonnen*, to fool < *fon*, fool.] **1.** A source of amusement, enjoyment, or pleasure <Clowns are *fun*.> **2.** Enjoyment or amusement : PLEASURE <have *fun* at a party> **3.** Playful, often noisy activity. —*vi.* **funned, fun·ning, funs.** To behave playfully : JOKE. **—for (or in) fun.** As a joke : PLAYFULLY. **—like fun.** *Slang.* Absolutely not : of course not.

▲ word history: The word *fun* meaning "amusement" was probably quite new in the 18th century, for Dr. Johnson records it with disapproval in his dictionary, which was published in 1755. *Fun* is very likely a borrowing of a dialectal form of *fon*, "to make a fool of, to be foolish," which has become obsolete in the standard language. The past participle of *fon*, originally spelled *fonned*, has become the Modern English word *fond*.

**fu·nam·bu·list** (fyo͞o-năm'byə-lĭst) *n.* [Lat. *funambulus* : *funis*, rope + *ambulare,* to walk.] A performer on a tightrope or a slack rope. **—fu·nam'bu·lism** *n.*

**func·tion** (fŭngk'shən) *n.* [Lat. *functio,* performance < *fungi,* to perform.] **1.** The activity for which one is specifically fitted or employed. **2. a.** Assigned duty or activity. **b.** Specific occupation or role <in your *function* as the family's physician> **3.** An official ceremony or formal social occasion. **4.** Something closely related to another thing and dependent on it for its existence, value, or significance <growth as a *function* of nutrition> **5.** *Math.* **a.** A variable so related to another that for each value assumed by one there is a value determined for the other. **b.** A rule of correspondence between two sets such that there is a unique element in one set assigned to each element in the other. *—vi.* **-tioned, -tion·ing, -tions.** To have or perform a function. **—func'tion·less** *adj.*

**func·tion·al** (fŭngk'shə-nəl) *adj.* **1.** Of or relating to a function. **2.** Designed for or adapted to a specific function or use <*functional* sculpture> **3.** Capable of performing : OPERATIVE. **4.** *Pathol.* Involving functions rather than a physiological or structural cause. **5.** *Math.* Of, pertaining to, or indicating a function or functions. **—func'tion·al·ly** *adv.*

**functional group** *n.* A distinctive reactive constituent, as a carboxyl group, of a chemical compound.

**functional illiterate** *n.* One with some education but below a minimum literacy standard.

**func·tion·al·ism** (fŭngk'shə-nə-lĭz'əm) *n.* **1.** The doctrine that the function of an object should determine its design and materials. **2.** A doctrine stressing purpose, practicality, and utility.

**functional shift** *n.* A shift in a word's syntactic function, as when a noun functions as a verb.

**func·tion·ar·y** (fŭngk'shə-nĕr'ē) *n., pl.* **-ies.** One who holds an office or performs a particular function : OFFICIAL.

**function word** *n.* A word, as a preposition or a conjunction, chiefly indicating a grammatical relationship in a sentence or phrase.

**fund** (fŭnd) *n.* [Lat. *fundus,* piece of land.] **1.** A source of supply : STOCK. **2. a.** A sum of money set aside for a specific purpose. **b. funds.** Available money. **3.** An organization established to administer a fund. *—vt.* **fund·ed, fund·ing, funds. 1.** To provide money for paying off the interest or principal of. **2.** To convert into a long-term or floating debt with fixed interest payments. **3.** To place in a fund. **4.** To furnish a fund for.

**fun·da·ment** (fŭn'də-mənt) *n.* [ME *foundement* < OFr. *fondement* < Lat. *fundamentum* < *fundare,* to lay the foundation < *fundus,* bottom.] **1. a.** The buttocks. **b.** The anus. **2.** The natural features of a land surface unaltered by human beings. **3.** A foundation. **4.** An underlying theoretical basis or principle.

**fun·da·men·tal** (fŭn'də-mĕn'tl) *adj.* **1. a.** Constituting or functioning as an essential component of a system or structure : BASIC. **b.** Of major significance : CENTRAL. **2.** *Physics.* **a.** Of or relating to the component of lowest frequency of a periodic wave or quantity. **b.** Of or relating to the lowest possible frequency of a vibrating element or system. **3.** *Mus.* Having the root in the bass. *—n.* **1.** An essential part. **2.** *Physics.* The lowest frequency of a periodically varying quantity or of a vibrating system. **—fun'da·men'tal·ly** *adv.*

**fun·da·men·tal·ism** (fŭn'də-mĕn'tl-ĭz'əm) *n.* **1. a.** *often* **Fundamentalism.** A Protestant movement characterized by a belief in the literal truth of the Bible. **b.** Adherence to this belief. **2.** A movement or point of view marked by rigid adherence to fundamental or basic principles. **—fun'da·men'tal·ist** *n.* **—fun'da·men'tal·is'tic** *adj.*

**fundamental particle** *n. Physics.* An elementary particle.

**fun·di** (fŭn'dī') *n. pl. of* FUNDUS.

**fund raiser** *n.* **1.** One who raises funds, as for an organization or a campaign. **2.** A social function, as a dinner, for raising funds.

**fund-rais·ing** (fŭnd'rā'zĭng) *adj.* Intended to raise funds <a *fund-raising* drive>

**fun·dus** (fŭn'dəs) *n., pl.* **-di** (-dī') [Lat., bottom.] *Anat.* The inner basal surface of an organ farthest away from the opening, as in the eye or uterus. **—fun'dic** *adj.*

**fu·ner·al** (fyōō'nər-əl) *n.* [ME *funerelles* < OFr. *funerailles* < Med. Lat. *funeralia* < Lat. *funus,* death rites.] **1.** The ceremonies held in connection with the burial or cremation of the dead. **2.** The persons accompanying a body to the grave. **3.** An end of existence. **4.** *Informal.* A source of great concern or care <If you don't meet the deadline, that's your *funeral.*>

**funeral home** *n.* An establishment where the dead are prepared for burial or cremation and where wakes and funerals may be held.

**fu·ner·ar·y** (fyōō'nə-rĕr'ē) *adj.* [LLat. *funerarius* < Lat. *funus,* funeral.] Of or appropriate for a funeral or burial.

**fu·ne·re·al** (fyōō-nîr'ē-əl) *adj.* [Lat. *funereus* < *funus,* funeral.] **1.** Of or relating to a funeral. **2.** Appropriate for or like a funeral <*funereal* gloom after their departure> **—fu·ne're·al·ly** *adv.*

**fun·gal** (fŭng'gəl) *adj.* Fungous.

**fun·gi** (fŭn'jī') *n. pl. of* FUNGUS.

**fun·gi·ble** (fŭn'jə-bəl) *adj.* [< Lat. *fungi,* to perform.] *Law.* Being of such a nature or kind that one unit or part may be exchanged or substituted for another unit or equal part to discharge an obligation. *—n. often* **fungibles.** Something fungible, as money or grain.

**fun·gi·cide** (fŭn'jĭ-sīd', fŭng'gĭ-) *n.* A substance that destroys or inhibits the growth of fungi. **—fun'gi·cid'al** (-sīd'l) *adj.* **—fun'gi·cid'al·ly** *adv.*

**fun·gi·form** (fŭn'jə-fôrm', fŭng'gə-) *adj.* [FUNG(US) + -FORM.] Shaped like a mushroom.

**fun·gi·stat** (fŭn'jĭ-stăt', fŭng'gĭ-) *n.* A substance inhibiting the growth of fungi.

**fun·giv·or·ous** (fŭn-jĭv'ər-əs, fŭng-gĭv'-) *adj.* Feeding on fungi.

**fun·go** (fŭng'gō) *n., pl.* **-goes.** [Orig. unknown.] *Baseball.* A practice fly ball hit to a fielder with a specially designed bat.

**fun·gous** (fŭng'gəs) *adj.* [ME, tender < Lat. *fungosus* < *fungus,* fungus.] **1.** Of, relating to, like, or characteristic of a fungus. **2.** Caused by a fungus.

**fun·gus** (fŭng'gəs) *n., pl.* **fun·gi** (fŭn'jī') *or* **fun·gus·es.** [Lat., perh. < Gk. *sphongos,* sponge.] Any of numerous plants of the division or subkingdom Thallophyta, lacking chlorophyll, ranging in form from a single cell to a body mass of branched filamentous hyphae that often produce specialized fruiting bodies and including the yeasts, molds, smuts, and mushrooms.

**fun house** *n.* A building or an attraction in an amusement park featuring devices intended to surprise, frighten, or amuse.

**fu·ni·cle** (fyōō'nĭ-kəl) *n.* A funiculus.

**fu·nic·u·lar** (fyōō-nĭk'yə-lər, fə-) *adj.* **1.** Of, relating to, or resembling a rope or cord. **2.** Operated by a cable. **3.** Of, relating to, or constituting a funiculus. *—n.* A cable railway on a steep incline, esp. such a railway with simultaneously ascending and descending cars counterbalancing one another.

**fu·nic·u·lus** (fyōō-nĭk'yə-ləs, fə-) *n., pl.* **-li** (-lī') [Lat. *funiculus,* slender rope, dim. of *funis,* rope.] **1.** *Anat.* A slender cordlike strand or band, esp.: **a.** A bundle of nerve fibers in the nerve trunk. **b.** The umbilical cord. **2.** *Bot.* A stalk connecting an ovule or seed with the placenta.

**funk** (fŭngk) [Prob. < obs. Flem. *fonck.*] *Informal. —n.* **1. a.** Cowardly fright : PANIC. **b.** A state of depression. **2.** A cowardly, fearful person. *—v.* **funked, funk·ing, funks.** *—vt.* **1.** To take fright and shrink from. **2.** To be afraid of. *—vi.* To shrink in fright.

**funk·y¹** (fŭng'kē) *adj.* **-i·er, -i·est.** Frightened : panicky.

**funk·y²** (fŭng'kē) *adj.* **-i·er, -i·est.** [< *funk,* strong smell.] **1.** Having a moldy or musty smell. **2.** *Slang.* **a.** Having an earthy quality characteristic of the blues <*funky* music> **b.** Earthy and uncomplicated. **3.** Characterized by faddish self-expression or originality <wore *funky* clothes from the Roaring Twenties> **—funk'i·ness** *n.*

**fun·nel** (fŭn'əl) *n.* [ME *fonel* < Prov. *fonilh* < LLat. *fundibulum* < Lat. *infundibulum* < *infundere,* to pour in. —see INFUSE.] **1.** A conical utensil with a small hole or narrow tube at the apex used to channel the flow of a substance into a container. **2.** Something shaped like a funnel. **3.** A shaft, flue, or stack for ventilation or the passage of smoke. *—v.* **-neled, -nel·ing, -nels** *or* **-nelled, -nel·ling, -nels.** *—vi.* **1.** To take on the shape of a funnel. **2.** To move through or as if through a funnel <passengers *funneling* slowly through the departure gate> *—vt.* **1.** To cause to assume the shape of a funnel. **2.** To cause to move through or as if through a funnel.

**fun·ny** (fŭn'ē) *adj.* **-ni·er, -ni·est.** [< FUN.] **1. a.** Eliciting laughter or amusement. **b.** Intended to amuse. **2.** Strangely or suspiciously odd : CURIOUS <something *funny* going on> **3.** Tricky or deceitful. *—n., pl.* **-nies.** *Informal.* **1.** A joke. **2. funnies.** Comic strips. **—fun'ni·ly** *adv.* **—fun'ni·ness** *n.*

**funny bone** *n. Informal.* **1.** The point near the elbow where a nerve may be pressed against bone to produce a tingling sensation. **2.** A sense of humor.

**funny book** *n.* A comic book.

**fun·ny·man** (fŭn'ē-măn') *n.* A humorous individual, esp. a professional comedian.

**funny paper** *n.* A newspaper section or supplement containing comic strips.

**fur** (fûr) *n.* [ME *furre,* prob. short for *furrer* < OFr. *forreure,* of Germanic orig.] **1. a.** The thick coat of soft hair covering the body of an animal, as a fox or beaver. **b.** The hair-covered, dressed pelt of such an animal, used in making garments and as trimming or decoration. **2.** A garment made of or lined with fur. **3.** A coating similar to fur. *—vt.* **furred, fur·ring, furs. 1.** To cover or line with or as if with fur. **2.** To provide fur garments for. **3.** To line (a wall or floor) with furring.

**fu·ran** (fyŏŏr'ăn', fyŏŏ-răn') *n.* [FUR(FURAL) + -AN.] A colorless, volatile, liquid heterocyclic compound, $C_4H_4O$, derived from dehydration of certain carbohydrates and used in synthesizing organic compounds, esp. nylon.

**fu·ra·nose** (fyŏŏr'ə-nōs') *n.* A sugar, the cyclic structure of which is composed of four carbon atoms and one oxygen atom.

**fur·be·low** (fûr'bə-lō') *n.* [Alteration of *falbala.*] **1.** A flounce or ruffle on a garment. **2.** A piece of showy ornamentation. *—vt.* **-lowed, -low·ing, -lows.** To decorate with a furbelow.

**fur·bish** (fûr'bĭsh) *vt.* **-bished, -bish·ing, -bish·es.** [ME *furbishen* < OFr. *fourbir, fourbiss-,* of Germanic orig.] **1.** To brighten by cleaning or rubbing: POLISH. **2.** To restore to serviceable or attractive state: RENOVATE. **—fur'bish·er** *n.*

**fur·cate** (fûr'kāt') *vi.* **-cat·ed, -cat·ing, -cates.** [LLat. *furcatus* < Lat. *furca,* fork.] To divide into branches: FORK. *—adj.* Forked. **—fur'cate'ly** *adv.* **—fur·ca'tion** *n.*

**fur·cu·la** (fûr'kyə-lə) *n., pl.* **-lae** (-lē') [NLat. < Lat., forked, prob. dim. of *furca,* fork.] A forked bone process, esp. the wishbone of a bird. **—fur'cu·lar** *adv.*

**fur·fur** (fûr'fər) *n., pl.* **-fu·res** (-fyə-rēz') [Lat.] An epidermal scale, as in dandruff.

**fur·fu·ra·ceous** (fûr'fə-rā'shəs, -fyə-) *adj.* [LLat. *furfuraceus,* like bran < Lat. *furfur,* bran.] **1.** Made of or covered with scaly particles, as dandruff. **2.** Relating to or like bran.

**fur·fu·ral** (fûr'fə-răl', -fyə-) *n.* [FURFUR + AL(DEHYDE).] A colorless mobile liquid, $C_4H_3OCHO$, used as a solvent for nitrocellulose and in manufacturing plastics and dyes.

**fur·fu·ran** (fûr'fə-răn', -fyə-) *n.* Furan.

**fur·fu·res** (fûr'fyə-rēz') *n. pl. of* FURFUR.

**fu·ri·o·so** (fyŏŏr'ē-ō'sō, -zō) *adj. & adv.* [Ital. < Lat. *furiosus,* furious.] *Mus.* In a vigorous manner. —Used as a direction.

**fu·ri·ous** (fyŏŏr'ē-əs) *adj.* [ME < OFr. < Lat. *furiosus* < *furia,* fury.] **1. a.** Full of, exhibiting, or marked by rage. **b.** Ragingly violent. **2.** Marked by violent or turbulent activity. **—fu'ri·ous·ly** *adv.*

**furl** (fûrl) *v.* **furled, furl·ing, furls.** [OFr. *ferlier* : *fer,* firm (< Lat. *firmus*) + *lier,* to bind (< Lat. *ligare*).] *—vt.* To roll up and secure (e.g., a flag) to an object. *—vi.* To be rolled up. *—n.* **1.** An act of furling. **2.** A single roll or rolled section of furled material.

**fur·long** (fûr'lông', -lŏng') *n.* [ME < OE *furlang* : *furh,* furrow + *lang,* long.] A unit for measuring distance, equal to 1/8 mile, 220 yards, or approx. 201 meters.

▲ word history: Since Old English times the word *furlong* has been used to indicate a unit of length, literally "a furrow's length." A *furlong* was originally defined as the length of a furrow in a square field of ten acres, but as the size of an acre varied during medieval times so did the length of a furrow. *Furlong* was also used to denote one eighth of both a Roman and an English mile. A *furlong way,* first recorded in Chaucer's poetry, was a measure of time—the length of time it took to walk one eighth of a mile.

**fur·lough** (fûr'lō) *n.* [Du. *verlof.*] **1.** A leave of absence from duty granted esp. to personnel of the armed services. **2.** The papers granting a furlough. *—vt.* **-loughed, -lough·ing, -loughs.** To grant a furlough to.

**fur·nace** (fûr'nĭs) *n.* [ME < OFr. *furnais* < Lat. *fornax,* oven.] **1.** An enclosure in which energy in a nonthermal form is converted to heat. **2.** An intensely hot enclosed place. **3.** A severe test or trial.

**fur·nish** (fûr'nĭsh) *vt.* **-nished, -nish·ing, -nish·es.** [ME *furnisshen* < OFr. *furnir, furniss-,* of Germanic orig.] **1.** To equip with what is needed, esp. to provide furniture for. **2.** To supply: give <*furnish* an example> **—fur'nish·er** *n.*

☆ **syns:** FURNISH, ACCOUTER, APPOINT, EQUIP, FIT (out), GEAR, OUTFIT, RIG, TURN OUT *v. core meaning:* to supply what is needed for an activity or purpose <troops *furnished* with winter uniforms><*furnish* a new apartment>

**fur·nish·ing** (fûr'nĭ-shĭng) *n.* **1.** A piece of equipment necessary or useful for convenience or comfort. **2. furnishings.** The furniture, appliances, and other movable articles in a home or office. **3. furnishings.** Wearing apparel and accessories <men's *furnishings*>

**fur·ni·ture** (fûr'nĭ-chər) *n.* [OFr. *fourniture* < *furnir,* to furnish.] **1.** The movable articles in a room or establishment that make it fit for use. **2.** *Archaic.* The necessary equipment for a horse. **3.** Blank strips of wood or metal placed between and around type on a page to hold it in place.

**fu·ror** (fyŏŏ'ôr', -ŏr) *n.* [Lat. < *furere,* to rage.] **1.** Violent anger. **2.** Intense excitement or ecstasy. **3. a.** General commotion. **b.** Public uproar: RUMPUS <a *furor* over the tax bill> **4.** A fashion enthusiastically adopted by the public: FAD.

**fu·rore** (fyŏŏr'ôr', -ōr') *n. Chiefly Brit.* FUROR 3b, 4.

**fu·ro·se·mide** (fyŏŏ-rō'sə-mīd') *n.* [FUR(FURAL) + S(ULF)- + *-emide,* alteration of AMIDE.] A compound, $C_{12}H_{11}ClN_2O_5S$, used as a diuretic.

**furred** (fûrd) *adj.* **1.** Bearing fur. **2.** Made, covered, or trimmed with or as if with fur. **3.** Wearing fur. **4.** Provided with furring, as a wall.

**fur·ri·er** (fûr'ē-ər) *n.* [ME *furrer* < AN.] One whose business is the dressing, selling, or repairing of furs.

**fur·ri·er·y** (fûr'ē-ə-rē) *n., pl.* **-ies. 1.** Fur garments and trimmings as a whole. **2.** The business of a furrier.

**fur·ring** (fûr'ĭng) *n.* **1. a.** A fur trimming or lining. **b.** Fur trimmings and linings as a whole. **2.** A furlike coating, as on the tongue. **3. a.** The act of preparing a wall or floor with strips of wood or metal to provide a level surface or an air space. **b.** Strips of material used for furring.

**fur·row** (fûr'ō, fŭr'ō) *n.* [ME *forwe* < OE *furh.*] **1.** A long, narrow, shallow trench made in the ground by a plow. **2.** A rut, groove, or narrow depression. **3.** A deep wrinkle in the skin, as on the forehead. *—v.* **-rowed, -row·ing, -rows.** *—vt.* **1.** To make furrows in: PLOW. **2.** To form deep wrinkles in. *—vi.* To become furrowed or wrinkled.

**fur·ry** (fûr'ē) *adj.* **-ri·er, -ri·est. 1.** Composed of or like fur. **2.** Covered with fur or a furlike coating. **3.** Resembling fur in thickness or softness. **—fur'ri·ness** *n.*

**fur seal** *n.* An eared seal of the genera *Callorhinus* or *Arctocephalus,* having thick, soft underfur valued commercially.

**fur·ther** (fûr'thər) *adj.* [ME, earlier < OE *furðor.*] **1.** More distant in degree, time, or space <was *further* from the truth><the *further* street light> **2.** Additional. *—adv.* **1.** To a greater extent: MORE. **2.** In addition: FURTHERMORE. **3.** At or to a more distant point in space or time. *—vt.* **-thered, -ther·ing, -thers.** To help the progress of: ADVANCE. **—fur'ther·er** *n.*

**fur·ther·ance** (fûr'thər-əns) *n.* The act of furthering, advancing, or helping forward.

**fur·ther·more** (fûr'thər-môr', -mōr') *adv.* In addition: MOREOVER.

**fur·ther·most** (fûr'thər-mōst') *adj.* Most distant or remote.

**fur·thest** (fûr'thĭst) *adj.* [ME < *further,* earlier. —see FURTHER.] Most distant in degree, time, or space. *—adv.* **1.** To the greatest extent or degree. **2.** At or to the most distant point in space or time.

**fur·tive** (fûr'tĭv) *adj.* [Fr. < OFr. *furtif* < Lat. *furtivus* < *furtum,* theft < *fur,* thief.] **1.** Stealthy: surreptitious. **2.** Expressive of stealth: SHIFTY. **—fur'tive·ly** *adv.* **—fur'tive·ness** *n.*

**fu·run·cle** (fyŏŏr'ŭng'kəl) *n.* [Lat. *furunculus,* knob on a vine, dim. of *fur,* thief.] BOIL². **—fu·run'cu·lar** (fyŏŏ-rŭng'kyə-lər), **fu·run'cu·lous** (-ləs) *adj.*

**fu·run·cu·lo·sis** (fyŏŏ-rŭng'kyə-lō'sĭs) *n.* [Lat. *furunculus,* furuncle + -OSIS.] A condition in which recurring furuncles develop.

**fu·ry** (fyŏŏr'ē) *n., pl.* **-ries.** [ME *furie* < OFr. < Lat. *furia* < *furere,* to rage.] **1.** Violent, unrestrained anger: RAGE. **2.** Violent, uncontrolled action: TURBULENCE. **3. Fury.** *Gk. Myth.* One of three winged deities held to have pursued and punished evildoers. **4.** An angry or spiteful woman.

☆ **syns:** FURY, IRE, RAGE, WRATH *n. core meaning:* violent or unrestrained anger <smashed the glass in a black *fury*>

**furze** (fûrz) *n.* [ME *furse* < OE *fyrs.*] Gorse.

**fuse¹** (fyŏŏz) *n.* [Ital. *fuso* < Lat. *fusus,* spindle.] **1.** A length of readily combustible material that is lighted at one end to carry a flame to and detonate an explosive at the other. **2.** *var. of* FUZE.

**fuse²** (fyŏŏz) *v.* **fused, fus·ing, fus·es.** [Lat. *fundere, fus-,* to melt.] *—vt.* **1.** To reduce to a liquid or plastic state by heating: MELT. **2.** To mix together by or as if by melting: BLEND. *—vi.* **1.** To become liquefied from heat. **2.** To become mixed or united by or as if by melting together. *—n.* A device containing an element that protects an electric circuit by melting when overloaded, thus opening the circuit.

**fu·see** *also* **fu·zee** (fyŏŏ-zē') *n.* [Fr. *fusée,* spindleful of thread < OFr. < *fus,* spindle < Lat. *fusus.*] **1.** A large-headed friction match capable of burning in a wind. **2.** A colored flare used as a railway warning signal. **3.** A grooved cone-shaped pulley in old-style clocks, used to equalize the force of the mainspring by maintaining a differential winding and unwinding of the cord or chain from the spring container. **4.** FUSE¹ 1.

**fu·se·lage** (fyŏŏ'sə-läzh', -zə-) *n.* [Fr. < *fuselé,* spindle-shaped < OFr. *fusel,* dim. of *fus,* spindle < Lat. *fusus.*] The central body of an aircraft that holds passengers, cargo, and crew and to which the wings and tail assembly are attached.

**fu·sel oil** (fyŏŏ'zəl) *n.* [G. *Fusel,* bad liquor.] A clear, colorless, poisonous liquid mixture of amyl alcohols, obtained as a by-product of the fermentation of starch-containing and sugar-containing plant materials and used as a solvent for fats, oils, resins, and waxes and in making explosives and pure amyl alcohols.

**fu·si·ble** (fyŏŏ'zə-bəl) *adj.* Capable of being fused or melted by heating. **—fu'si·bil'i·ty** *n.* **—fu'si·ble·ness** *n.*

**fusible metal** *n.* A metal alloy having a low melting point and used as solder and for safety plugs and fuses.

**fu·si·form** (fyŏŏ'zə-fôrm') *adj.* [Lat. *fusus,* spindle + -FORM] Tapering at each end.

**fu·sil** (fyŏŏ'zəl) *n.* [Fr. < OFr., steel for a tinderbox < Lat. *focus,* hearth.] A light flintlock musket.

**fu·sile** (fyŏŏ'zəl, -zīl') *also* **fu·sil** (-zəl) *adj.* [Lat. *fusilis,* molten < *fusus,* p.part. of *fundere,* to melt.] **1.** Formed by melting or casting. **2.** Capable of being fused.

**fu·sil·ier** *also* **fu·sil·eer** (fyŏŏ'zə-lîr') *n.* [Fr. < *fusil,* fusil.] **1.** A soldier armed with a fusil. **2.** A soldier in a British army regiment.

**fu·sil·lade** (fyŏŏ'sə-läd', -lād', -zə-, fyŏŏ'sə-läd', -lād', -zə-) *n.* [Fr. < *fusiller,* to shoot < *fusil,* fusil.] **1.** A discharge from a number of firearms, fired simultaneously or in rapid succession. **2.** A rapid outburst or barrage. *—vt.* **-lad·ed, -lad·ing, -lades.** To attack or shoot down with a fusillade.

**fu·sion** (fyŏŏ'zhən) *n.* [Lat. *fusio,* a melting < *fusus,* p.part. of *fundere,* to melt.] **1.** The act or procedure of liquefying or melting together by heat. **2.** The liquid or melted state induced by heat. **3.** A union resulting from fusing. **4.** The merging of different elements into a union. **5.** *Physics.* A nuclear reaction in which nuclei combine to form more massive nuclei with the release of energy.

**fusion bomb** *n.* An atomic bomb, esp. a hydrogen bomb, that derives its energy output mainly from fusion reactions among light nuclei.

**fu·sion·ism** (fyōō′zhə-nĭz′əm) *n.* The theory or practice of forming coalitions, esp. of political factions.

**fuss** (fŭs) *n.* [Orig. unknown.] **1.** Needlessly nervous or useless activity : COMMOTION. **2. a.** Excessive, unwarranted concern over an unimportant issue. **b.** An objection : protest. **3.** A quarrel. —*v.* **fussed, fuss·ing, fuss·es.** —*vi.* **1.** To worry over trifles. **2.** To be excessively careful or solicitous <*fussed* over their children> **3.** To get into or be in a state of nervous or useless activity. **4.** To object : complain. —*vt. Informal.* To disturb with unimportant issues. **—fuss′er** *n.*

**fuss-budg·et** (fŭs′bŭj′ĭt) *n.* One who fusses over trifles.

**fuss·y** (fŭs′ē) *adj.* **-i·er, -i·est. 1.** Tending to fuss. **2.** Paying great attention to details : FASTIDIOUS. **3.** Calling for or requiring great attention to trivial details : METICULOUS. **4.** Full of superfluous details. **—fuss′i·ly** *adv.* **—fuss′i·ness** *n.*

**fus·tian** (fŭs′chən) *n.* [ME < OFr. *fustaigne* < Med. Lat. *fustaneum*, perh. after *Fostat*, Egypt.] **1. a.** A coarse sturdy cloth of cotton and flax. **b.** A thick twilled cotton fabric with a short nap. **2.** Pretentious language.

**fus·tic** (fŭs′tĭk) *n.* [ME *fustik* < OFr. *fustoc* < Ar. *fustuq* < Gk. *pistakē*, pistachio.] **1.** A small tropical American tree, *Chlorophora tinctoria*, having wood yielding a yellow dyestuff. **2.** The wood of the fustic. **3.** A dyestuff obtained from fustic wood.

**fus·ti·gate** (fŭs′tĭ-gāt′) *vt.* **-gat·ed, -gat·ing, -gates.** [Lat. *fustigare, fustigat-* < *fustis*, club.] **1.** To beat with a club : CUDGEL. **2.** To criticize harshly. **—fus′ti·ga′tion** *n.*

**fus·ty** (fŭs′tē) *adj.* **-ti·er, -ti·est.** [ME < OFr. *fuste*, wine cask < Lat. *fustis*, club.] **1.** Smelling of mildew or decay : MOLDY. **2.** Old-fashioned <*fusty* notions> **—fus′ti·ly** *adv.* **—fus′ti·ness** *n.*

**fu·thark** (fōō′thärk′) *also* **fu·thorc** *or* **fu·thork** (-thôrk′) *n.* [From the first six letters of the alphabet : *f, u, (th), a, r, c.*] The runic alphabet.

**fu·tile** (fyōōt′l, fyōō′tīl′) *adj.* [Lat. *futilis.*] **1.** Having no useful result : INEFFECTUAL. **2.** Trifling and frivolous. **—fu′tile·ly** *adv.* **—fu′tile·ness** *n.* **—fu·til′i·ty** (fyōō-tĭl′ĭ-tē) *n.*

☆ ***syns:*** FUTILE, FRUITLESS, UNAVAILING, USELESS, VAIN *adj. core meaning* : having no useful or effectual result <*futile* attempts to reconciliate>

**fu·til·i·tar·i·an** (fyōō-tĭl′ĭ-târ′ē-ən) *n.* [Blend of FUTILE and UTILITARIAN.] A believer in the futility of human endeavor. **—fu·til′i·tar′i·an·ism** *n.*

**fut·tock** (fŭt′ək) *n.* [ME *fottek.*] *Naut.* One of the curved timbers forming a rib in a ship's frame.

**futtock plate** *n. Naut.* One of the iron plates attached to the top of a mast to hold the ends of the futtock shrouds.

**futtock shroud** *n. Naut.* One of the iron rods extending from the futtock plate, used to brace the base of a ship's mast.

**fu·ture** (fyōō′chər) *n.* [ME < OFr. *futur* and < Lat. *futurus.*] **1.** The time yet to come. **2.** Something that will occur in time yet to come. **3.** A prospective or expected condition, esp. with regard to growth, advancement, or development <a relationship with no *future*> **4. futures.** Commodities or stocks bought or sold upon agreement of delivery in time to come. **5. a.** The future tense. **b.** A verb form in the future tense. —*adj.* That is to be <*future* plans> **—fu′ture·less** *adj.*

**future perfect** *n.* A verb tense expressing action completed by a specified time in the future that is formed in English by combining *will have* or *shall have* with a past participle.

**future shock** *n.* [From the book *Future Shock* by Alvin Toffler (b. 1928).] Distress and disorientation engendered by an inability to cope with rapid technological and societal changes.

**future tense** *n.* A verb tense expressing future time.

**fu·tur·ism** (fyōō′chə-rĭz′əm) *n.* An artistic movement originating in Italy about 1910 and marked by an attempt to depict vividly the energetic and dynamic quality of contemporary life esp. by the motion and force of modern machinery. **—fu′tur·ist** *n.* **—fu′tur·is′tic** *adj.*

**fu·tur·ist·ics** (fyōō′chə-rĭs′tĭks) *n.* (*sing. in number*). Futurology. **—fu′tur·ist** *n.*

**fu·tu·ri·ty** (fyōō-tŏŏr′ĭ-tē, -tyŏŏr′-, -chŏŏr′-) *n., pl.* **-ties. 1.** FUTURE 1. **2.** The quality or state of being in or of the future. **3.** A future event or possibility. **4.** A futurity race.

**futurity race** *n.* **1.** A race for which entries are made well in advance of the event. **2.** A race for horses entered as competitors at or before their birth.

**futurity stakes** *pl.n.* **1.** The stakes awarded to the winner or winners in a futurity race. **2.** A futurity race.

**fu·tu·rol·o·gy** (fyōō′chə-rŏl′ə-jē) *n.* The study or forecast of potential developments, as in science and technology, using current conditions or trends as a point of departure. **—fu′tu·rol′o·gist** *n.*

**fuze** *also* **fuse** (fyōōz) *n.* [Var. of FUSE[1].] A mechanical, electrical, or electronic mechanism used to detonate an explosive, as a grenade or bomb.

**fu·zee** (fyōō-zē′) *n. var. of* FUSEE.

**fuzz**[1] (fŭz) *n.* [Perh. back-formation < FUZZY.] A mass of fine, light particles, fibers, or hairs : DOWN <peach *fuzz*> —*vt.* **fuzzed, fuzz·ing, fuzz·es. 1.** To cover with fuzz. **2.** To make indistinct.

**fuzz**[2] (fŭz) *n.* [Orig. unknown.] *Slang.* The police.

**fuzz·y** (fŭz′ē) *adj.* **-i·er, -i·est.** [Perh. < LG *fussig*, spongy.] **1.** Covered with fuzz. **2.** Of or like fuzz. **3.** Not clear : INDISTINCT. **4.** Not clearly worked out : CONFUSED. **—fuzz′i·ly** *adv.* **—fuzz′i·ness** *n.*

**-fy** *suff.* [ME *-fien* < OFr. *-fier* < Lat. *-ficare* < *-ficus*, -fic.] Make : cause to become <basi*fy*>

**fyke** (fīk) *n.* [Du. *fuik* < MDu. *fūke.*] A long bag-shaped net held open by hoops, used for catching fish.

**fyke**

**fyl·fot** (fĭl′fŏt′) *n.* [ME.] A swastika.

# Gg

**g** *or* **G** (jē) *n., pl.* **g's** *or* **G's. 1.** The seventh letter of the English alphabet. **2.** A speech sound represented by the letter g. **3.** The seventh in a series. **4.** *Mus.* **a.** The fifth tone in the scale of C major or the seventh tone in the relative minor scale. **b.** The key or a scale in which G is the tonic. **c.** A written or printed note representing this tone. **5. G.** *Slang.* One thousand dollars. **6. G.** A unit of force equal to the gravity exerted on a body at rest.

**G** (jē) *adj.* [Short for GENERAL.] Of or designating a motion-picture rating of such nature that all ages may be granted admission.

**Ga** *symbol for* GALLIUM.

**gab** (găb) *vi.* **gabbed, gab·bing, gabs.** [Perh. var. of dial. *gob* < GOB[2].] *Informal.* To chat idly. **—gab** *n.* **—gab′ber** *n.*

**gab·ar·dine** (găb′ər-dēn′, găb′ər-dēn′) *n.* [Alteration of GABERDINE.] **1.** A worsted cotton, wool, or rayon twill. **2.** GABERDINE 1, 2.

**gab·ble** (găb′əl) *v.* **-bled, -bling, -bles.** [MDu. *gabbelen.*] —*vi.* **1.** To speak incoherently or rapidly : JABBER. **2.** To make rapid, repeated cackling noises, as a duck. —*vt.* To utter incoherently or rapidly. —*n.* **1.** Rapid incoherent speech. **2.** A jumble of meaningless noises. **—gab′bler** *n.*

**gab·bro** (găb′rō) *n., pl.* **-bros.** [Ital. < Lat. *glaber*, smooth.] A usu. coarse-grained igneous rock made up of calcic plagioclase and pyroxene. **—gab·bro′ic** (gă-brō′ĭk) *adj.*

**gab·broid** (găb′roid′) *adj.* Like gabbro.

**gab·by** (găb′ē) *adj.* **-bi·er, -bi·est.** *Informal.* Tending to talk too much <a *gabby* neighbor> **—gab′bi·ness** *n.*

**ga·belle** (gə-bĕl′) *n.* [ME < OFr. < OItal. *gabella* < Ar. *qabāla*, tribute < *qabala*, he received.] A tax, esp. the salt tax imposed in France before 1790.

**gab·er·dine** (găb'ər-dēn', găb'ər-dēn') *n.* [OFr. *gauvardine* < MHG *wallevart*, pilgrimage : *wallen*, to roam (< OHG *wallōn*) + *vart*, journey < OHG < *faran*, to go.] **1.** A long coarse cloak or frock worn esp. by Jews during the Middle Ages. **2.** *Chiefly Brit.* A loose smock worn by laborers. **3.** GABARDINE 1.

**gab·fest** (găb'fĕst') *n.* *Slang.* An informal gathering during which news and gossip are exchanged.

**ga·bi·on** (gā'bē-ən) *n.* [OFr. < OItal. *gabbione*, aug. of *gabbia*, cage < Lat. *cavea* < *cavus*, hollow.] **1.** A cylindrical wicker basket filled with earth and stones, once used in building fortifications. **2.** A hollow metal cylinder used esp. in building dams and foundations.

**ga·ble** (gā'bəl) *n.* [ME < ON *gafl* and OFr. *gable*, of Germanic orig.] **1. a.** The triangular wall section at the ends of a pitched roof, bounded by the two roof slopes and the ridge pole. **b.** An end of a building with a gable in the roof section. **2.** A usu. ornamental triangular architectural section. **—ga'bled** *adj.*

**gable roof** *n.* A pitched roof ending in a gable.

**Ga·bri·el** (gā'brē-əl) *n.* [Heb. *Gabhrī'ēl.*] An archangel described in the Bible who acts as God's messenger.

**gad**[1] (găd) *vi.* **gad·ded, gad·ding, gads.** [ME *gadden.*] To roam about restlessly and with little purpose : ROVE. **—gad'der** *n.*

**gad**[2] (găd) *n.* [ME < ON *gaddr.*] **1.** A pointed tool, as a spike, for working or breaking rock or ore. **2.** A goad, as for prodding cattle. —*vt.* **gad·ded, gad·ding, gads.** To break up (e.g., ore) with a gad.

**gad·a·bout** (găd'ə-bout') *n.* One who roams about, esp. to seek gossip or excitement.

**gad·fly** (găd'flī') *n.* **1.** Any of various flies, esp. of the family Tabanidae, that bite or annoy livestock. **2.** One acting as a provocative stimulus. **3.** One who habitually criticizes existing institutions.

**gadg·et** (găj'ĭt) *n.* [Orig. unknown.] *Informal.* A small specialized mechanical device : CONTRIVANCE. **—gadg'et·y** *adv.*

☆ ***syns:*** GADGET, CONTRAPTION, CONTRIVANCE, DOODAD, DOOHICKEY, GIMMICK, GISMO, JIGGER, THING, THINGAMABOB, THINGAMAJIG *n. core meaning* : a small specialized mechanical device <a *gadget* for peeling potatoes>

**gadg·e·teer** (găj'ĭ-tîr') *n.* *Informal.* One who designs, builds, or delights in the use of gadgets.

**gadg·et·ry** (găj'ĭ-trē) *n.* **1.** Gadgets as a whole. **2.** Design or construction of gadgets.

**ga·doid** (gā'doid', găd'oid') *adj.* [NLat. *Gadus*, fish genus (< Gk. *gados*, a kind of fish) + -OID.] Of or belonging to the family Gadidae, including fish such as the cod and the hake. —*n.* A member of the Gadidae.

**gad·o·lin·ite** (găd'l-ə-nīt') *n.* [G. *Gadolinit*, after Johann *Gadolin* (1760–1852).] A dark-colored silicate mineral containing several of the rare earths combined with iron.

**gad·o·lin·i·um** (găd'l-ĭn'ē-əm) *n.* [After Johann *Gadolin* (1760–1852).] *Symbol* **Gd** A silvery-white, malleable, ductile metallic element, used to improve the high-temperature characteristics of iron, chromium, and related metallic alloys; atomic number 64; atomic weight 157.25.

**ga·droon** (gə-dro͞on') *n.* [Fr. *godron* < OFr. *goderon.*] **1.** A band of convex molding ornamentally carved with beading or reeding. **2.** An ornamental band, used esp. in silverwork, embellished with fluting or reeding. **—ga·droon'ing** *n.*

**gad·wall** (găd'wôl') *n.* [Orig. unknown.] A widely distributed duck, *Anas strepera*, with grayish or brown plumage.

**Gae·a** (jē'ə) *n.* [Gk. *Gaia* < *gaia*, earth.] *Gk. Myth.* The goddess of the earth and mother of the Titans.

**Gael** (gāl) *n.* [Sc. Gael. *Gaidheal.*] **1.** A Gaelic-speaking Celt of Scotland, Ireland, or the Isle of Man. **2.** A Scottish Highlander.

**Gael·ic** (gā'lĭk) *adj.* Of or relating to the Gaels or their languages. —*n.* **1.** Goidelic. **2.** Any of the Goidelic languages.

**gaff** (găf) *n.* [ME *gaffe* < OFr., of Celt. orig.] **1.** A metal hook fastened to a pole, used to land large fish. **2.** A spar for extending the top edge of a fore-and-aft sail. **3. a.** A metal spur for the leg of a gamecock. **b.** A climbing hook used by linemen. **4.** A hoax : trick. **5.** *Slang.* Harsh treatment : ABUSE. —*vt.* **gaffed, gaf·fing, gaffs. 1.** To hook with a gaff. **2.** To equip (a gamecock) with a gaff. **3.** *Slang.* **a.** To take in : CHEAT. **b.** To alter (e.g., dice) so as to cheat.

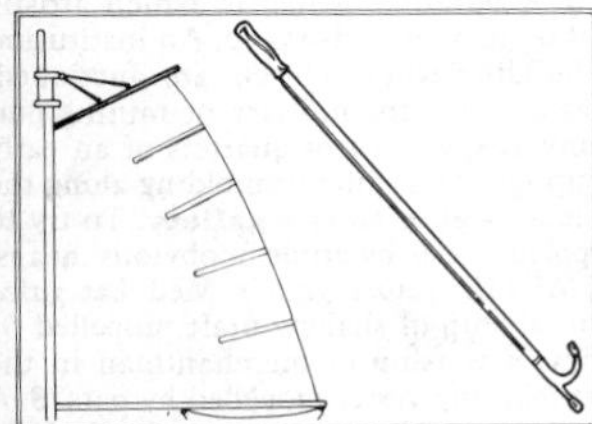

**gaff**
*Two types of gaffs:* (left) *a fore-and-aft sail gaff and* (right) *a fishing gaff*

**gaffe** (găf) *n.* [Fr.] A faux pas.

**†gaf·fer** (găf'ər) *n.* [Perh. alteration of GODFATHER.] **1.** *Regional.* **a.** An old man. **b.** A rustic. **2.** An electrician who handles lighting on a film or television stage set.

**gaff rig** *n.* *Naut.* A rig with a fore-and-aft sail that has its upper edge supported by a gaff.

**gaff-top·sail** (găf'tŏp'səl, -sāl') *n.* *Naut.* A light triangular or quadrilateral sail set over a gaff.

**gag** (găg) *n.* [< ME *gaggen*, to suffocate.] **1.** Something forced into or put over the mouth to prevent speaking or crying out. **2.** An obstacle to or censoring of free speech. **3.** A device placed in the mouth to keep it open, as in dentistry. **4.** *Informal.* **a.** A practical joke. **b.** A comic remark or effect. —*v.* **gagged, gag·ging, gags.** —*vt.* **1.** To prevent from speech or outcry by using a gag. **2.** To repress or censor (free speech). **3.** To keep (the mouth) open by using a gag. **4.** To block off or obstruct. **5.** To cause to choke or retch. —*vi.* **1.** To retch from nausea. **2.** *Informal.* To make jokes or quips.

**ga·ga** (gä'gä') *adj.* [Fr. < *gaga*, old fool.] *Slang.* **1.** Silly : crazy. **2.** Totally absorbed, enthusiastic, or infatuated.

**gage**[1] (gāj) *n.* [ME < OFr., of Germanic orig.] **1.** Something deposited or given as security against an obligation : PLEDGE. **2.** Something, as a glove, offered or thrown down as a pledge or challenge to fight. **3.** A challenge. —*vt.* **gaged, gag·ing, gag·es.** *Archaic.* **1.** To pledge as security. **2.** To offer as a stake in a bet : WAGER.

**gage**[2] (gāj) *n.* [After Sir William *Gage* (1777–1864).] A variety of plum, as the greengage.

**gage**[3] (gāj) *n.* & *v.* [Orig. unknown.] *var. of* GAUGE.

**gag·er** (gā'jər) *n.* *var. of* GAUGER.

**gag·ger** (găg'ər) *n.* **1.** One that gags. **2.** A piece of iron for keeping the core in position in a foundry mold.

**gag·gle** (găg'əl) *n.* [ME *gagel* < *gagelen*, to cackle.] **1.** A flock of geese. **2.** A cluster : group <a *gaggle* of news reporters>

**gag·man** (găg'măn') *n.* A writer or user of jokes or comedy routines.

**gag order** *n.* A court order prohibiting public reporting or commentary, as by the news media, on a case currently before a court.

**gag rule** *n.* A rule, as in a legislative body, restricting debate on a particular issue.

**gahn·ite** (gä'nīt') *n.* [G. *Gahnit*, after Johan G. *Gahn* (1745–1818).] A dark-green to brown or black mineral, $ZnAl_2O_4$.

**gai·e·ty** *also* **gay·e·ty** (gā'ĭ-tē) *n.*, *pl.* **-ties.** [Fr. *gaieté* < OFr. *gai*, cheerful. —see GAY.] **1.** A state of joyful exuberance. **2.** Festive or joyful activity. **3.** Gay color or showiness, as of dress.

☆ ***syns:*** **1.** GAIETY, GLEE, GLEEFULNESS, HILARITY, JOCULARITY, JOCUNDITY, JOVIALITY, MERRIMENT, MIRTH *n. core meaning* : a state of joyful exuberance <the *gaiety* of the holiday season> **2.** GAIETY, FESTIVITY, FUN, MERRYMAKING, REVEL, REVELRY *n. core meaning* : joyful, exuberant activity <invited the guests to join in the *gaiety*>

**gail·lar·di·a** (gə-lär'dē-ə) *n.* [NLat. *Gaillardia*, genus name, after *Gaillard* de Marentonneau, 18th-cent. French botanist.] Any of several plants of the genus *Gaillardia* of western North America, with yellow or reddish rayed flowers.

**gai·ly** *also* **gay·ly** (gā'lē) *adv.* In a gay manner : MERRILY.

**gain**[1] (gān) *v.* **gained, gain·ing, gains.** [OFr. *gaaignier*, of Germanic orig.] —*vt.* **1.** To become the owner of : ACQUIRE <*gained* wealth in the computer industry> **2.** To acquire in competition : WIN <*gained* a decisive win> **3.** To earn as profit or payment <*gain* a living> **4.** To develop an increase of : BUILD UP <a political movement that *gained* momentum><*gained* 20 pounds> **5.** To come to : REACH <*gained* the top of the hill> —*vi.* **1.** To become better or greater : PROGRESS <*gaining* in strength> **2.** To come nearer <The other team is *gaining* on us.> **3.** To increase in weight. —*n.* **1. a.** Something gained <territorial *gains*> **b.** Progress : advancement <financial *gains*> **2.** The act of acquiring something. **3.** An increase in degree or amount. **4.** *Electron.* **a.** An increase in signal power. **b.** The ratio of output to input, as of output power to input power in an antenna or of output voltage to input voltage in an amplifier.

**gain**[2] (gān) *n.* [Orig. unknown.] A notch cut into a board to receive another part. —*vt.* **gained, gain·ing, gains. 1.** To cut out a gain in. **2.** To join by or fit into a gain.

**gain·er** (gā'nər) *n.* **1.** One that gains. **2.** A dive in which the diver leaves the board facing forward, does a back somersault, and enters the water feet first.

**gain·ful** (gān'fəl) *adj.* Providing a gain : PROFITABLE <*gainful* employment> **—gain'ful·ly** *adv.* **—gain'ful·ness** *n.*

**gain·less** (gān'lĭs) *adj.* Profitless. **—gain'less·ness** *n.*

**gain·ly** (gān'lē) *adj.* **-li·er, -li·est.** [ME *geinli*, gracious < *gein*, helpful < ON *gegn.*] Graceful. **—gain'li·ness** *n.*

**gain·say** (gān-sā') *vt.* **-said** (-sād', -sĕd'), **-say·ing, -says** (-sāz', -sĕz') [ME *gainsayen* : *gain-*, against (< OE *gegn-*) + *sayen*, to say < OE *secgan.*] **1.** To declare false : DENY. **2.** To oppose, esp. by contradiction. **—gain·say'er** *n.*

**'gainst** *also* **gainst** (gĕnst, gānst) *prep.* Against.

**gait** (gāt) *n.* [ME *gate*, path < ON *gata.*] **1.** A particular way of moving on foot. **2.** Any of the ways a horse can move by lifting the feet in different order or rhythm, as a canter, trot, or walk. —*vt.* **gait·ed, gait·ing, gaits.** To teach a certain gait or gaits to (a horse).

**gait·ed** (gā′tĭd) *adj.* Having a specified gait <fast-*gaited*>

**gai·ter** (gā′tər) *n.* [Fr. *guêtre.*] **1.** A leather or cloth covering for the legs extending from the instep to the ankle or knee. **2.** An ankle-high shoe with elastic sides. **3.** An overshoe with a cloth top.

**gal** (găl) *n.* [Alteration of GIRL.] *Informal.* A girl.

**ga·la** (gā′lə, găl′ə, gä′lə) *n.* [Ital. < OSp. < OFr. *gale,* rejoicing.] A festive occasion. —*adj.* Characterized by or worthy of celebration.

**galact-** *pref. var. of* GALACTO-.

**ga·lac·tic** (gə-lăk′tĭk) *adj.* **1.** Of or relating to a galaxy, esp. the Milky Way. **2.** Very large : IMMENSE.

**galactic equator** *n.* The great circle of the celestial sphere that lies in the plane bisecting the band of the Milky Way, inclined at an angle of approx. 62° to the celestial equator.

**galactic nebula** *n.* A nebula lying within the Milky Way.

**galactic noise** *n.* Radio-frequency radiation originating within the Milky Way.

**galacto-** or **galact-** *pref.* [< Gk. *gala, galakt-,* milk.] Milk <*ga-lactose*>

**ga·lac·to·poi·e·sis** (gə-lăk′tə-poi-ē′sĭs) *n.* Secretion and continued production of milk. **—ga·lac′to·poi·et′ic** (-ĕt′ĭk) *adj.*

**gal·ac·tos·am·ine** (găl′ək-tō′sə-mēn′, gə-lăk-) *n.* An amino-acid derivative of galactose.

**ga·lac·tose** (gə-lăk′tōs′) *n.* A simple sugar, $C_6H_{12}O_6$, typically occurring in lactose.

**ga·lac·to·se·mi·a** (gə-lăk′tə-sē′mē-ə) *n.* A congenital metabolic disorder caused by the inherited absence of an enzyme that catalyzes galactose and marked by mental retardation and cataracts. **—ga·lac′to·se′mic** *adj.*

**ga·lac·to·side** (gə-lăk′tə-sīd′) *n.* Any of a group of glycosides that yield galactose on hydrolysis.

**ga·la·go** (gə-lā′gō, -lä′-) *n., pl.* **-gos.** [NLat. *Galago,* genus name, perh. < Wolof *golokh,* monkey.] A small African primate of the genera *Galago* or *Euoticus,* having dense woolly fur, large round eyes, prominent ears, and a long tail.

**Gal·a·had** (găl′ə-hăd′) *n.* **1.** The purest knight of King Arthur's Round Table who alone succeeded in the quest for the Holy Grail. **2.** A man regarded as pure or chivalrous.

**ga·lan·gal** (gə-lăng′gəl) *n.* [Alteration of GALINGALE.] **1.** A plant, *Alpinia officinarum* of eastern Asia, having pungent aromatic roots used medicinally and as seasoning. **2.** The dried roots of the galangal.

**gal·an·tine** (găl′ən-tēn′) *n.* [ME *galauntine,* a kind of sauce < OFr. *galantine,* alteration of *galatine* < Med. Lat. *galatina.*] A dish of boned, stuffed meat or poultry cooked and served cold with aspic.

**ga·lan·ty show** (gə-lăn′tē) *n.* [Perh. < Ital. *galante,* a gallant < OFr. *galant.* —see GALLANT.] A play performed by casting the shadows of miniature figures on a screen or wall.

**gal·a·te·a** (găl′ə-tē′ə) *n.* [After the *Galatea,* a 19th-cent. English warship (from its having been used for children's sailor suits).] A durable, often striped cotton fabric used in making garments.

**Gal·a·te·a** (găl′ə-tē′ə) *n.* [Lat. < Gk. *Galateia.*] *Gk. Myth.* An ivory statue of a maiden, brought to life by Aphrodite in response to the pleas of its sculptor, Pygmalion.

**Ga·la·tians** (gə-lā′shənz) *n. (sing. in number).* —See table at BIBLE.

**gal·a·vant** (găl′ə-vănt′) *v. var. of* GALLIVANT.

**ga·lax** (gā′lăks′) *n.* [NLat. *Galax,* genus name.] A plant, *Galax aphylla* of the southeastern United States, having glossy evergreen leaves and a cluster of small white flowers.

**gal·ax·y** (găl′ək-sē) *n., pl.* **-ies.** [ME *galaxie,* the Milky Way < LLat. *galaxias* < Gk. *galaxias,* milky < *gala,* milk.] **1.** *Astron.* **a.** Any of numerous large-scale aggregates of stars, gas, and dust, having one of a group of more or less definite overall structures, containing an average of 100 billion ($10^{11}$) solar masses, and ranging in diameter from 1,500 to 300,000 light-years. **b.** *often* **Galaxy.** The Milky Way. **2.** A brilliant or distinguished assembly.

**gal·ba·num** (găl′bə-nəm, gôl-) *n.* [ME < Lat. < Gk. *khalbanē,* of Semitic orig.] A bitter aromatic gum resin extracted from an Asiatic plant, *Ferula galbaniflua,* used in incense and medicinally as a counterirritant.

**gale¹** (gāl) *n.* [Orig. unknown.] **1. a.** A very powerful wind. **b.** A wind having a speed between 32 and 63 miles or approx. 51.5 and 101.4 kilometers per hour. **2.** *Archaic.* A breeze. **3.** A forceful outburst <*gales* of laughter>

**gale²** (gāl) *n.* [ME *gail* < OE *gagel.*] The sweet gale.

**ga·le·a** (gā′lē-ə) *n., pl.* **-le·ae** (-lē-ē′) [NLat. < Lat., helmet.] A helmet-shaped part, as the upper petal of certain plants or part of the maxilla of an insect.

**ga·le·ate** (gā′lē-āt′) *also* **ga·le·at·ed** (-ā′tĭd) *adj.* [Lat. *galeatus,* p.part. of *galeare,* to cover with a helmet < *galea,* helmet.] **1.** *Biol.* Having a galea. **2.** Helmet-shaped.

**ga·le·i·form** (gā′lē-ə-fôrm′, gə-lē′-) *adj.* [Lat. *galea,* helmet + -FORM.] Helmet-shaped.

**ga·le·na** (gə-lē′nə) *n.* [Lat., lead ore.] A gray mineral, essentially PbS, the principal ore of lead.

**Ga·len·ism** (gā′lə-nĭz′əm) *n.* The medical theories or practices advanced by the Greek physician Galen. **—Ga·len′ic** (gā-lĕn′ĭk), **Ga·len′i·cal** *adj.*

**gal Friday** *n.* A girl Friday.

**Ga·li·bi** (gə-lē′bē) *n., pl.* **Galibi** *or* **-bis.** [Carib.] **1.** A member of the Carib people of French Guiana. **2.** The language of the Galibi.

**Ga·li·cian** (gə-lĭsh′ən) *adj.* Of or relating to Spanish Galicia, its people, or their language. —*n.* **1.** A native or resident of Spanish Galicia. **2.** The Portuguese dialect spoken in Spanish Galicia.

**Gal·i·le·an¹** *also* **Gal·i·lae·an** (găl′ə-lē′ən) *n.* **1.** A Christian. **2.** Jesus.

**Gal·i·le·an²** (găl′ə-lē′ən, -lā′-) *adj.* Of, relating to, or being in accord with the work of the Italian scientist Galileo.

**gal·i·lee** (găl′ə-lē′) *n.* [ME *galile* < Med. Lat. *galilaea* < Lat. *Galilea,* Galilee.] A small chapel or porch at the western end of some medieval English churches.

**gal·i·ma·ti·as** (găl′ə-mā′shē-əs, -măt′ēəs) *n.* [Fr.] Nonsensical talk.

**gal·in·gale** (găl′ĭn-găl′) *n.* [ME, a kind of root < OFr. *galingal* < Ar. *khalanjān.*] A sedge of the genus *Cyperus,* esp. *C. longus* of Europe, having rough-edged leaves, reddish spikelets, and aromatic roots.

**gal·i·ot** *also* **gal·li·ot** (găl′ē-ət) *n.* [ME < OFr. < Med. Lat. *galiota,* dim. of *galea,* galley. —see GALLEY.] **1.** A light, swift galley used at one time on the Mediterranean. **2.** A light, single-masted, flat-bottomed Dutch merchant ship.

**gal·i·pot** (găl′ə-pŏt′, -pō′) *n.* Crude turpentine obtained from a pine tree, *Pinus pinaster* of southern Europe.

**gall¹** (gôl) *n.* [ME *galle* < OE *gealla.*] **1. a.** Liver bile. **b.** The gallbladder. **2. a.** Bitter feeling : RANCOR. **b.** Something bitter to undergo. **3.** Impudence : effrontery <had the *gall* to show up uninvited>

**gall²** (gôl) *n.* [ME *galle* < OE *gealla.*] **1.** A skin sore caused by friction and abrasion <a saddle *gall*> **2. a.** Exasperation : vexation. **b.** The cause of such vexation. —*v.* **galled, gall·ing, galls.** —*vt.* **1.** To make (the skin) sore by abrasion : CHAFE. **2.** To damage or break the surface of by or as if by friction. **3.** To exasperate : vex. —*vi.* To become irritated or sore.

**gall³** (gôl) *n.* [ME *galle* < Lat. *galla,* gallnut.] An abnormal swelling of plant tissue caused by insects, microorganisms, or external injury.

**Gal·la** (găl′ə) *n., pl.* **Galla** *or* **-las.** [Perh. < Ar. *ghalīẓ,* rough.] **1.** A member of a pastoral Hamitic people of southern Ethiopia and the Somali Republic. **2.** The Cushitic language of the Galla. **—Gal′la** *adj.*

**gal·lant** (găl′ənt) *adj.* [ME *galaunt* < OFr. *galant,* pr.part. of *galer,* to rejoice < *gale,* rejoicing.] **1.** Showy and gay : DASHING <a *gallant* feathered hat> **2.** Stately : majestic. **3.** High-spirited and courageous <*gallant* troops> **4.** (gə-lănt′, -länt′). **a.** Attentive to women : CHIVALROUS. **b.** Flirtatious. —*n.* (gə-lănt′, -länt′, găl′ənt). **1.** A stylish young man. **2. a.** A man courteously attentive to women. **b.** A woman's lover : PARAMOUR. —*v.* (gə-lănt′, -länt′) **-lant·ed, -lant·ing, -lants.** —*vt.* To pay court to (a lady). —*vi.* To play the gallant. **—gal′lant·ly** *adv.*

**gal·lant·ry** (găl′ən-trē) *n., pl.* **-ries. 1.** Nobility of spirit or action. **2.** Chivalrous attention to women. **3.** An act or instance of gallantry. **4.** *Archaic.* A bold or colorful display or appearance.

**gall·blad·der** *also* **gall bladder** (gôl′blăd′ər) *n.* A small pear-shaped muscular sac, located under the right lobe of the liver, in which bile secreted by the liver is stored.

**gal·le·ass** (găl′ē-ăs′, -əs) *n.* [OFr. *galeasse* < OItal. *galeaza,* aug. of *galea,* galley < Med. Lat. —see GALLEY.] A large heavily armed 16th–17th-cent. three-masted Mediterranean galley.

**gal·le·on** (găl′ē-ən) *n.* [Sp. *galeon* < OFr. *galion* < *galie,* galley. —see GALLEY.] A large three-masted sailing ship gen. having two or more decks, used during the 15th and 16th cent. esp. by Spain as a merchantman or warship.

**†gal·ler·y** (găl′ə-rē) *n., pl.* **-ies.** [ME *galerie* < OFr., portico < OItal. *galleria* < Med. Lat. *galeria.*] **1.** A roofed promenade, esp. one extending along the wall of a building and supported by arches on the outer side. **2. a.** A narrow enclosed passageway, as a hall or corridor. **b.** An establishment gen. like such a corridor in length and used for a specified purpose <a shooting *gallery*> **3.** *Regional.* A porch : verandah. **4. a.** An upper floor projecting over the rear part of the main floor of a theater and usu. providing cheaper seats than those in the orchestra. **b.** The seats in such a section. **c.** The audience occupying these seats. **d.** A similar floor in a large building such as a church. **5. a.** A large group of spectators, as at a tennis match. **b.** The general public when regarded as typifying a lack of artistic discrimination or sophistication. **6. a.** A building or hall in which artistic work is displayed. **b.** A collection of artistic works. **c.** An institution selling artistic works. **d.** A building where objects are auctioned. **7.** An underground tunnel, as one dug for military or mining purposes. **8.** A platform or balcony at the stern or quarters of an early sailing ship. **9.** A decorative upright trimming or molding along the edge of a table top, tray, or shelf. **—play to the gallery.** To try to gain general public favor or applause, esp. by crude or obvious means.

**gal·ley** (găl′ē) *n., pl.* **-leys.** [ME *galei* < OFr. *galie* < Med. Lat. *galea* < Med. Gk.] **1.** A large medieval ship of shallow draft propelled by sails and oars that was used as a warship or merchantman in the Mediterranean. **2.** An ancient seagoing vessel propelled by oars. **3.** A large rowboat once used in England. **4.** The kitchen of a ship or large passenger aircraft. **5. a.** A long, usu. metal tray for holding composed printing type. **b.** A galley proof.

**galley**

**galley proof** *n.* A printer's proof taken from composed type prior to page composition to permit detection and correction of errors.
**galley slave** *n.* **1.** A slave or convict forced to man an oar of a galley. **2.** A drudge.
**gall·fly** (gôl'flī') *n.* Any of various small insects of the family Cecidomyiidae that deposit their eggs on plant stems or in the bark of trees, causing the formation of galls in which their larvae grow.
**gal·liard** (găl'yərd) *n.* [ME *gaillard* < OFr. *gaillart.*] **1.** A spirited dance popular in France in the 16th and 17th cent. **2.** The triple-time music for a galliard. —*adj. Archaic.* Spirited, lively, and gay.
**Gal·lic** (găl'ĭk) *adj.* [Lat. *Gallicus* < *Galli,* the Gauls.] Of or relating to Gaul or France : FRENCH.
**gal·lic acid** (găl'ĭk, gôl'ĭk) *n.* A colorless crystalline compound, $C_7H_6O_5 \cdot H_2O$, derived from tannin and used in photography, as a tanning agent, and in ink and paper manufacture.
**Gal·li·can** (găl'ĭ-kən) *adj.* **1.** Relating to or characteristic of Gallicanism. **2.** Gallic. —*n.* A supporter of Gallicanism.
**Gal·li·can·ism** (găl'ĭ-kə-nĭz'əm) *n.* A movement originating among the French Roman Catholic clergy that favored restriction of papal control and achievement by each nation of individual administrative autonomy.
**Gal·li·cism** (găl'ĭ-sĭz'əm) *n.* **1.** A French idiom or phrase appearing in another language. **2.** A French trait.
**Gal·li·cize** (găl'ĭ-sīz') *vt.* & *vi.* **-cized, -ciz·ing, -ciz·es.** To make or become like the French. **—Gal'li·ci·za'tion** *n.*
**†gal·li·gas·kins** (găl'ĭ-găs'kĭnz) *pl.n.* [Perh. alteration of OFr. *garguesque,* var. of *greguesque* < OItal. *grechesca,* fem. of *grechesco,* Greek < *greco* < Lat. *Graecus.* —see GREEK.] **1.** Full-length, loosely fitting hose or breeches worn in the 16th and 17th cent. **2.** Loose breeches. **3.** *Regional.* Leggings.
**gal·li·mau·fry** (găl'ə-mô'frē) *n., pl.* **-fries.** [Fr. *galimafrée* < OFr. *calimafree.*] A hodgepodge.
**gal·li·na·ceous** (găl'ə-nā'shəs) *adj.* [Lat. *gallinaceus,* of poultry < *gallina,* hen, fem. of *gallus,* cock.] Of, belonging to, or characteristic of the order Galliformes, including the common domestic fowl, pheasants, turkeys, and grouse.
**gall·ing** (gô'lĭng) *adj.* Causing acute irritation or exasperation : VEXING <a *galling* series of delays> **—gall'ing·ly** *adv.*
**gal·li·nip·per** (găl'ə-nĭp'ər) *n.* [Orig. unknown.] A large insect, as a mosquito, capable of inflicting a painful bite.
**gal·li·nule** (găl'ə-no͞ol', -nyo͞ol') *n.* [NLat. *Gallinula,* genus name < Lat. *gallinula,* pullet, dim. of *gallina,* hen. —see GALLINACEOUS.] A wading bird of the genera *Gallinula, Porphyrio,* or *Porphyrula,* frequenting swampy regions and having dark iridescent plumage.
**gal·li·ot** (găl'ē-ət) *n. var. of* GALIOT.
**gal·li·pot** (găl'ə-pŏt') *n.* [ME *galy pott.*] A small glazed earthenware jar once used by druggists for medicaments.
**gal·li·um** (găl'ē-əm) *n.* [< Lat. *gallus,* cock, transl. of *Lecoq* de Boisbaudran (1838–1912), its discoverer.] *Symbol* **Ga** A rare metallic element used in semiconductor technology and as a component of various low-melting alloys; atomic number 31; atomic weight 69.72.
**gallium arsenide** *n.* A dark-gray crystalline compound, GaAs, used in transistors, solar cells, and semiconducting lasers.
**gal·li·vant** *also* **gal·a·vant** (găl'ə-vănt') *vi.* **-vant·ed, -vant·ing, -vants.** [Perh. alteration of GALLANT.] **1.** To roam about in search of amusement : GAD. **2.** To flirt.
**gal·li·wasp** (găl'ə-wŏsp', -wôsp') *n.* [Orig. unknown.] Any of several long-bodied lizards of the genera *Diploglossus* or *Celestus* of Central America and the West Indies.
**gall midge** *n.* A gallfly.
**gall·nut** (gôl'nŭt') *n.* The nutgall.
**Gal·lo·ma·ni·a** (găl'ō-mā'nē-ə, -mān'yə) *n.* [Fr. *gallomanie* : *gallo-,* France (< Lat. *Gallus,* Gaul) + *-manie,* -mania.] A strong predilection for anything French.
**gal·lon** (găl'ən) *n.* [ME, a liquid measure < ONFr. *galon* and < Med. Lat. *galona.*] **1. a.** A unit of volume or capacity in the U.S. Customary System, used in liquid measure, equal to 4 quarts or 3.785 liters. **b.** A unit of volume in the British Imperial System, used in liquid and dry measure, equal to 4.546 liters. **2.** A container with a capacity of one gallon.
**gal·lon·age** (găl'ə-nĭj) *n.* An amount calculated in gallons.
**gal·loon** (gə-lo͞on') *n.* [Fr. *galon* < OFr. *galonner,* to decorate the hair with ribbons.] A narrow lace, metallic thread, or embroidered band or braid used as trimming. **—gal·looned'** *adj.*
**gal·loot** (gə-lo͞ot') *n. var. of* GALOOT.
**gal·lop** (găl'əp) *n.* [< ME *galopen,* to go at a gallop < OFr. *galoper,* of Germanic orig.] **1. a.** A natural three-beat gait of a horse, faster than a canter and slower than a run. **b.** A rapid running motion of other quadrupeds. **2.** A ride taken at the gallop. **3.** A rapid pace. —*v.* **-loped, -lop·ing, -lops.** —*vt.* **1.** To cause to gallop. **2.** To transport at or as if at a gallop. —*vi.* **1.** To ride a horse at a gallop. **2.** To move or progress rapidly. **—gal'lop·er** *n.*
**gal·lo·pade** (găl'ə-pād', -päd') *n. var. of* GALOP.
**gal·lop·ing** (găl'ə-pĭng) *adj.* **1.** Of or like a gallop, esp. in rhythm or speed. **2.** Developing at a faster rate and leading to death. —Used of certain diseases.
**gal·low·glass** (găl'ō-glăs') *n.* [Ir. Gael. *galloglach* : *gall,* foreigner + *oglach,* soldier.] An armed retainer or mercenary in the service of an Irish chieftain.
**gal·lows** (găl'ōz) *n., pl.* **gallows** or **-lows·es.** [ME *galwes,* pl. of *galwe,* gallows < OE *galga.*] **1. a.** A device usu. composed of two upright beams supporting a crossbeam from which a noose is suspended and used for execution by hanging. **b.** A similar structure used for supporting or suspending. **2.** Execution on a gallows or by hanging.
**gallows humor** *n.* Humorous treatment of a frightening or very serious situation.
**gal·lows·tree** (găl'ōz-trē') *n.* GALLOWS 1a.
**gall·stone** (gôl'stōn') *n.* A small hard pathological concretion of cholesterol crystals, formed in the gallbladder or a bile duct.
**gal·lus·es** (găl'ə-sĭz) *pl.n.* [Pl. of *gallus,* suspenders, alter. of GALLOWS.] *Informal.* Suspenders for trousers.
**gall wasp** *n.* Any of various wasps of the family Cynipidae that produce galls on oaks and other plants.
**Ga·lois theory** (găl-wä') *n.* [After Évariste *Galois* (1811–1832).] The portion of mathematical group theory concerned with the conditions under which a polynomial equation of power *n* with coefficients in a given mathematical field can be solved by repeating given operations and extracting the *n*th roots.
**ga·loot** *also* **gal·loot** (gə-lo͞ot') *n.* [Orig. unknown.] *Slang.* A clumsy, uncouth, or sloppy person.
**gal·op** (găl'əp) *also* **gal·lo·pade** (găl'ə-pād', -päd') *n.* [Fr.] **1.** A lively dance in duple rhythm, popular in the 19th cent. **2.** The music for the galop.
**ga·lore** (gə-lôr', -lōr') *adj.* [Ir. Gael. *go leór,* to sufficiency.] *Informal.* Abundant in number <shoes *galore*>
**ga·losh** (gə-lŏsh') *n.* [ME *galoche,* wooden-sole shoe < OFr., prob. < LLat. *gallicula,* dim. of Lat. *gallica (solea),* Gaulish (sandals) < *Galli,* Gauls.] **1.** A waterproof overshoe. **2.** *Obs.* A sturdy heavy-soled boot or shoe.
**gal·van·ic** (găl-văn'ĭk) *adj.* [GALVAN(ISM) + -IC.] **1.** Of or relating to direct-current electricity, esp. when produced chemically. **2. a.** Having the effect of an electric shock. **b.** Produced as if by an electric shock. **—gal·van'i·cal·ly** *adv.*
**galvanic cell** *n.* A primary cell.
**galvanic couple** *n.* A voltaic couple.
**gal·va·nism** (găl'və-nĭz'əm) *n.* [Fr. *galvanisme* or < Ital. *galvanismo,* after Luigi *Galvani* (1737–1798).] Direct-current electricity, esp. when produced chemically.
**gal·va·nize** (găl'və-nīz') *vt.* **-nized, -niz·ing, -niz·es. 1.** To stimulate or shock with an electric current. **2.** To arouse to awareness or action. **3.** To coat (iron or steel) with rust-resistant zinc. **—gal'va·ni·za'tion** *n.* **—gal'va·niz'er** *n.*
**galvanized iron** *n.* Iron coated with zinc to prevent rust.
**gal·va·nom·e·ter** (găl'və-nŏm'ĭ-tər) *n.* [GALVAN(ISM) + -METER.] A device for detecting or measuring electric currents by way of mechanical effects produced by those currents. **—gal'va·no·met'ric** (-nō-mĕt'rĭk), **gal'va·no·met'ri·cal** *adj.* **—gal'va·nom'e·try** *n.*
**gal·van·o·scope** (găl-văn'ə-skōp', găl'və-nə-) *n.* [GALVAN(ISM) + -SCOPE.] A galvanometer used to detect the presence and direction of electric currents by the deflection of a magnetic needle. **—gal·van'o·scop'ic** (-skŏp'ĭk) *adj.* **—gal'va·nos'co·py** (găl'və-nŏs'kə-pē) *n.*
**gal·yak** (găl'yăk') *n.* [Dial. R. *galyak.*] A flat, glossy fur made from the pelt of a stillborn lamb or kid.
**gam[1]** (găm) *n.* [Perh. short for GAMMON[2].] **1.** A school of whales. **2.** A social visit or friendly conversation, esp. between whalers. —*v.* **gammed, gam·ming, gams.** —*vi.* To visit, esp. while at sea. —*vt.* **1.** To visit with. **2.** To spend (time) in visiting.
**gam[2]** (găm) *n.* [Prob. < *gamb,* an animal leg on a coat of arms < ONFr. *gambe,* leg < LLat. *gamba,* hoof. —see GAMBOL.] *Slang.* A person's leg.
**gam-** *pref. var. of* GAMO-.
**gam·ba·do[1]** (găm-bā'dō) *n., pl.* **-does** or **-dos.** [Sp. *gambeta* < Ital. *gambata* < OItal. —see GAMBOL.] **1.** A low leap of a horse in which all four feet leave the ground. **2.** A leaping movement.

**gam·ba·do²** (găm-bā′dō) *n., pl.* **-does** or **-dos.** [< Ital. *gamba,* leg < OItal. —see GAMBOL.] **1.** Either of a pair of protective leather gaiters attached to a saddle. **2.** A rider's legging.

**gam·bier** *also* **gam·bir** (găm′bîr) *n.* [Malay.] The resinous, astringent extract of a woody vine, *Uncaria gambier* of south-central Asia, used medicinally and in tanning and dyeing.

**gam·bit** (găm′bĭt) *n.* [Ital. *gambetto* < *gamba,* leg < OItal. —see GAMBOL.] **1.** A chess opening in which one or more pawns are offered in exchange for a favorable position. **2.** A remark intended to open a conversation. **3.** A carefully evaluated strategy.

**gam·ble** (găm′bəl) *v.* **-bled, -bling, -bles.** [Prob. < obs. *gamel,* to play games < ME *gamen,* to play < OE *gamian.*] —*vi.* **1. a.** To bet money on the outcome of a game or contest. **b.** To play a game of chance for stakes. **2.** To take a risk in the hope of gaining an advantage : SPECULATE. —*vt.* **1.** To put up in gambling : WAGER. **2.** To expose to hazard : RISK. —*n.* **1.** A bet, wager, or other gambling venture. **2.** An undertaking of questionable outcome. **—gam′bler** (-blər) *n.*

**gam·boge** (găm-bōj′, -bo͞ozh′) *n.* [NLat. *gambogium,* alteration of *cambugium,* after *Cambodia.*] **1.** A brownish or orange resin obtained from a tree of the genus *Garcinia* of south-central Asia, yielding a yellow pigment. **2.** A strong yellow. **—gam·boge′** *adj.*

**gam·bol** (găm′bəl) *vi.* **-boled, -bol·ing, -bols** or **-bolled, -bol·ling, -bols.** [OFr. *gambade,* horse's jump < OItal. *gambata* < *gamba,* leg < LLat., hoof, perh. < Gk. *kampē,* bend.] To frolic. **—gam′bol** *n.*

**gam·brel** (găm′brəl) *n.* [ONFr. *gamberel* < *gambe,* leg. —see GAM².] **1.** The hock of a horse or other animal. **2.** A frame used by butchers for hanging carcasses by the legs.

**gambrel roof** *n.* A ridged roof with two slopes on each side, the lower slope having the steeper pitch.

**gam·bu·sia** (găm-byo͞o′zhə) *n.* [NLat. *Gambusia,* genus name < Am. Sp. *gambusino,* gambusia.] Any of a genus, *Gambusia,* of topminnows that feed on mosquito larvae.

**game¹** (gām) *n.* [ME < OE *gamen.*] **1.** A way of amusing oneself : DIVERSION. **2.** A set of rules completely specifying a competition, including the permissible actions of and information available to each participant, the mathematical probabilities with which chance events may occur, the criteria for termination of the competition, and the distribution of payoffs. **3. a.** A competitive activity, as a sport, governed by specific rules <the *game* of badminton> **b.** An instance of such an activity <We lost the third *game.*> **4. a.** The total number of points needed to win a game. **b.** The score accumulated at any time in a game <At half time the *game* was 22 to 14.> **5.** The equipment needed for playing certain games. **6.** A particular style or way of playing a game <played a passable bridge *game*> **7.** A calculated action or approach : SCHEME <You'll never see through their *game.*> **8. a.** Wild animals, birds, or fish hunted for food or sport. **b.** The flesh of game, eaten as food. **9. a.** One hunted or fit to be hunted : QUARRY. **b.** An object of teasing, ridicule, or scorn <make *game* of someone> —*v.* **gamed, gam·ing, games.** —*vt. Archaic.* To waste or lose by gambling. —*vi.* To play for stakes, esp. for money. —*adj.* **gam·er, gam·est. 1.** Plucky and unyielding : RESOLUTE. **2.** *Informal.* Ready and willing <Are you *game* for a jog?> **—game′ly** *adv.* **—game′ness** *n.*

**game²** (gām) *adj.* **gam·er, gam·est.** [Orig. unknown.] Crippled : lame.

**game·cock** (gām′kŏk′) *n.* A rooster trained for cockfighting.

**game fowl** *n.* **1.** A bird sought as game. **2.** Any of several breeds of domestic fowl raised esp. for cockfighting.

**game·keep·er** (gām′kē′pər) *n.* A person employed to protect and maintain wildlife, esp. on an estate or game preserve.

**gam·e·lan** (găm′ə-lăn′) *n.* [Javanese.] An orchestra common to Southeast Asia, consisting chiefly of tuned metal or wooden chimes and other percussion instruments.

**game plan** *n.* **1.** The strategy devised before or used during a sports event. **2.** A strategy for reaching an objective.

**game show** *n.* A television show in which contestants vie for prizes usu. by playing a competitive game, as a quiz.

**games·man·ship** (gāmz′mən-shĭp′) *n.* The skill or practice of winning a game, contest, or struggle by dubious or unsportsmanlike methods and strategies not actually breaking the rules.

**game·some** (gām′səm) *adj.* Playful : frolicsome. **—game′some·ly** *adv.* **—game′some·ness** *n.*

**game·ster** (gām′stər) *n.* A habitual gambler.

**gam·e·tan·gi·um** (găm′ĭ-tăn′jē-əm) *n., pl.* **-gi·a** (-jē-ə) [GAMET(E) + Gk. *angeion,* dim. of *angos,* vessel.] An organ or cell in which gametes are produced, esp. in primitive plant forms. **—gam′e·tan′gi·al** (-əl) *adj.*

**gam·ete** (găm′ēt′, gə-mēt′) *n.* [NLat. *gameta* < Gk. *gametēs,* husband < *gamein,* to marry < *gamos,* marriage.] A germ cell possessing the haploid number of chromosomes, esp. a mature sperm or egg capable of participating in fertilization. **—ga·met′ic** (-mĕt′ĭk) *adj.* **—ga·met′i·cal·ly** *adv.*

**game theory** *n.* Mathematical analysis of abstract models of strategic competition with the determination of best strategy as a goal, having applications in linear programming, statistical decision making, operations research, and military and economic planning.

**gameto-** *pref.* [< NLat. *gameta,* gamete.] Gamete <*gameto*cyte>

**ga·me·to·cyte** (gə-mē′tə-sīt′) *n.* A cell from which gametes are developed by division : spermatocyte or oocyte.

**ga·me·to·gen·e·sis** (gə-mē′tə-jĕn′ĭ-sĭs) *also* **gam·e·tog·e·ny** (găm′ĭ-tŏj′ə-nē) *n.* Production of gametes. **—ga·me′to·gen′ic, gam′e·tog′e·nous** (găm′ĭ-tŏj′ə-nəs) *adj.*

**ga·me·to·phore** (gə-mē′tə-fôr′, -fōr′) *n.* A structure on which gametangia are borne. **—ga·me′to·phor′ic** (-fôr′ĭk, -fōr′-) *adj.*

**ga·me·to·phyte** (gə-mē′tə-fīt′) *n.* The generation or form that reproduces sexually in a plant characterized by alternation of generations. **—ga·me′to·phyt′ic** (-fĭt′ĭk) *adj.*

**gam·ic** (găm′ĭk) *adj.* [< Gk. *gamos,* marriage.] Requiring fertilization in reproduction : SEXUAL.

**gam·in** (găm′ĭn′) *n.* [Fr.] A boy who roams about the streets : URCHIN.

**ga·mine** (gă-mēn′) *n.* [Fr., fem. of *gamin,* gamin.] **1.** A girl who roams about the streets : URCHIN. **2.** A girl with impish appeal.

**gam·ing** (gā′mĭng) *n.* The playing of games of chance : GAMBLING.

**gam·ma** (găm′ə) *n.* [ME < Gk., of Phoenician orig.; akin to Heb. *gīmel,* gimel.] **1.** The third letter of the Greek alphabet. —See table at ALPHABET. **2.** A gamma ray.

**gamma decay** *n.* **1.** A radioactive process in which an atomic nucleus loses energy by emitting a gamma ray without a change in its atomic or mass numbers. **2.** Decay of an unstable elementary particle by photon emission.

**gamma globulin** *n.* Any of several globulin fractions of blood serum closely associated with immune bodies and used to treat infectious diseases, as measles, poliomyelitis, and infectious hepatitis.

**gamma ray** *n.* **1.** Electromagnetic radiation emitted by radioactive decay and having energies in a range overlapping that of the highest energy x-rays, extending up to several hundred thousand electron volts. **2.** Electromagnetic radiation with energy greater than several hundred thousand electron volts. **3.** A high-energy photon.

**gam·ma-ray astronomy** (găm′ə-rā′) *n.* Astronomy dealing with the origin and nature of periodic gamma-ray emissions from extraterrestrial sources.

**†gam·mer** (găm′ər) *n.* [Alteration of GRANDMOTHER.] *Regional.* An elderly woman.

**gam·mon¹** (găm′ən) *n.* [Prob. < ME *gamen,* game < OE.] A victory in backgammon occurring before the loser has removed a single man. —*vt.* **-moned, -mon·ing, -mons.** To defeat by scoring a gammon.

**gam·mon²** (găm′ən) [Orig. unknown.] *Chiefly Brit.* —*n.* Misleading or nonsensical talk. —*v.* **-moned, -mon·ing, -mons.** —*vt.* To mislead by deceptive talk. —*vi.* To talk gammon. **—gam′mon·er** *n.*

**gam·mon³** (găm′ən) *n.* [ONFr. *gambon* < *gambe,* leg. —see GAM².] **1.** A cured or smoked ham. **2.** The lower part of a side of bacon.

**gam·mon⁴** (găm′ən) *vt.* **-moned, -mon·ing, -mons.** [Orig. unknown.] To fasten (a bowsprit) to the stem of a ship.

**gamo-** or **gam-** *pref.* [< Gk. *gamos,* marriage.] **1.** United : joined <*gamo*petalous> **2.** Sexual <*gamo*genesis>

**gam·o·gen·e·sis** (găm′ə-jĕn′ə-sĭs) *n.* Sexual reproduction. **—gam′o·ge·net′ic** (-jə-nĕt′ĭk) *adj.* **—gam′o·ge·net′i·cal·ly** *adv.*

**gam·o·pet·al·ous** (găm′ə-pĕt′l-əs) *adj.* Having a corolla with the petals fused or partially fused.

**gam·o·phyl·lous** (găm′ə-fĭl′əs) *adj.* Having or designating united leaves or leaflike parts.

**gam·o·sep·al·ous** (găm′ə-sĕp′ə-ləs) *adj.* Having the sepals united or partly united.

**-gamous** *suff.* [< Gk. *gamos,* marriage.] **1. a.** Having a specified number of marriages <mono*gamous*> **b.** Practicing a specified kind of marriage <exo*gamous*> **2.** Having a specified kind of reproductive organs <hetero*gamous*>

**gamp** (gămp) *n.* [After Mrs. Sarah *Gamp,* a character in the novel *Martin Chuzzlewit* by Charles Dickens (1812–1870).] *Chiefly Brit.* A large baggy umbrella.

**gam·ut** (găm′ət) *n.* [ME, the musical scale < Med. Lat. *gamma ut* : *gamma,* first note of the medieval scale (< Gk. *gamma,* gamma) + *ut,* first note of the diatonic scale.] **1.** The complete range or extent <a face expressing the *gamut* of emotion, from anger to contentment> **2.** *Mus.* The entire series of recognized notes.

**gam·y** (gā′mē) *adj.* **-i·er, -i·est. 1.** Having the flavor or odor of game, esp. that which is slightly spoiled. **2.** Showing a plucky spirit. **3. a.** Disreputable. **b.** Scandalous. **—gam′i·ly** *adv.* **—gam′i·ness** *n.*

**-gamy** *suff.* [Gk. *-gamia* < *gamos,* marriage.] **1.** Marriage <exo*gamy*> **2.** Procreative or propagative union <allo*gamy*> **3.** The possession of a specified mode of fertilization or specified reproductive organs <apo*gamy*>

**gan·der** (găn′dər) *n.* [ME < OE *gandra.*] **1.** A male goose. **2.** *Informal.* A halfwit. **3.** *Slang.* A quick look : GLANCE <took a *gander* at the display and walked off>

**gan·dy dancer** (găn′dē) *n.* [Poss. after the now defunct *Gandy* Manufacturing Company of Chicago, which made tools.] *Slang.* **1.** A railroad worker. **2.** An itinerant laborer.

**ga·nef** or **ga·nof** (gä′nəf) *also* **gon·if** (gŏn′ĭf) *n.* [Yiddish < Heb. *gannābh.*] A thief or scoundrel.

**gang[1]** (găng) *n.* [ME, band of men < OE, journey.] **1.** A group of people who socialize regularly <saw the whole *gang* at the picnic> **2.** A group of criminals or hoodlums who band together for mutual protection and profit. **3.** A group of adolescents, esp. juvenile delinquents, who band together. **4.** A group of laborers organized together on one job or under one foreman <a road *gang*> **5.** A set, as of matched tools <a *gang* of chisels> **6. a.** A pack of wolves or wild dogs. **b.** A herd, esp. of buffalo or elk. —*v.* **ganged, gang·ing, gangs.** —*vi.* To form into or band together as a gang. —*vt.* **1.** To group together or arrange (e.g., pages of type) into a gang. **2.** To attack as a gang. **—gang up.** *Informal.* To harass or attack as a group.

**gang[2]** (găng) *n. var. of* GANGUE.

**gang·bus·ter** (găng′bŭs′tər) *n. Slang.* A law officer who fights to break up organized criminal groups. **—like gangbusters.** *Slang.* With great force or zeal <came on *like gangbusters* at the beginning of the campaign>

**gang·er** (găng′ər) *n. Chiefly Brit.* A gang foreman.

**gang hook** *n.* A multiple fishhook consisting of two or more hooks joined shank to shank.

**gan·gli·a** (găng′glē-ə) *n. pl. of* GANGLION.

**gan·gli·at·ed** (găng′glē-ā′tĭd) *also* **gan·gli·ate** (-ĭt, -āt′) *adj.* Having ganglia.

**gan·gling** (găng′glĭng) *adj.* [Perh. < dial. *gang,* to go < ME *gangen* < OE *gangan.*] Tall and ungracefully thin.

**gan·gli·on** (găng′glē-ən) *n., pl.* **-gli·a** (-glē-ə) *or* **-gli·ons.** [Gk., cystlike tumor.] **1.** *Anat.* A group of nerve cells, as one found outside the brain or spinal cord, in vertebrates. **2.** A center of activity, power, or energy. **3.** *Pathol.* A cystic lesion similar to a tumor, occurring in a tendon sheath or joint capsule. **—gan′gli·on′ic** (-ŏn′ĭk) *adj.*

**gan·gli·on·at·ed** (găng′-glē-ə-nā′tĭd) *adj.* Gangliated.

**gan·gli·o·side** (găng′glē-ə-sīd′) *n.* [GANGLI(ON) + -OS(E) + -IDE.] Any of a group of glycosphingolipids found in ganglionic cells.

**gan·gly** (găng′glē) *adj.* **-gli·er, -gli·est.** Gangling.

**gang·plank** (găng′plăngk′) *n.* A board or ramp used as a removable footway between a ship and a pier.

▲ word history: The element *gang-* in *gangplank* and *gangway* is the same as the word *gang[1]*, but it preserves an older meaning of that word. *Gang* in Old and Middle English denoted the action of walking, with specific applications such as "way, passage," and "journey." A *gangplank* thus provides passage between a ship and a landing place. *Gangway* denotes a gangplank as well as various other kinds of passageways, as aisles. *Gang[1]* is related to the Old English verb *gangan,* "to walk, go," which has now been replaced by the verb go.

**gang·plow** (găng′plou′) *n.* A plow equipped with several blades that make parallel furrows.

**gang·punch** (găng′pŭnch′) *vt.* **-punched, -punch·ing, -punch·es.** To duplicate data from a punched card onto succeeding cards.

**gan·grel** (găng′rəl) *n.* [ME.] *Scot.* A drifter : vagabond.

**gan·grene** (găng′grēn′, găng-grēn′) *n.* [Lat. *gangraena* < Gk. *gangraina.*] Death and decay of tissue in a bodily part, usu. a limb, due to failure of blood supply, injury, or disease. —*vt.* & *vi.* **-grened, -gren·ing, -grenes.** To affect or become affected with gangrene. **—gan′gre·nous** (găng′grə-nəs) *adj.*

**gang·ster** (găng′stər) *n.* A member of a criminal organization : RACKETEER. **—gang′ster·ism** (-stə-rĭz′əm) *n.*

**gangue** *also* **gang** (găng) *n.* [Fr. < G. *Gang,* lode < OHG, a going.] Worthless material, as rock, in which valuable minerals are found.

**gang·way** (găng′wā′) *n.* **1.** A passageway, as through a crowd or an obstructed area. **2.** *Naut.* **a.** A passage along either side of a ship's upper deck. **b.** A gangplank. **c.** An opening in a ship's bulwark through which passengers may board. **3.** *Chiefly Brit.* **a.** The aisle dividing the front and rear seating sections of the House of Commons. **b.** An aisle between seating sections, as in a theater. **4.** The main level of a mine.

**gan·is·ter** *also* **gan·nis·ter** (găn′ĭ-stər) *n.* [Orig. unknown.] **1.** A silicon-rich sedimentary rock used for refractory furnace linings. **2.** A mixture of fire clay and ground quartz, used for furnace linings.

**gan·ja** (gän′jə) *n.* [Hindi *gājā* < Skt. *ganjā.*] A highly resinous form of marijuana, prepared by collecting only the flowering tops and leaves of carefully selected and cultivated plants.

**gan·net** (găn′ĭt) *n.* [ME *ganet* < OE *ganot.*] A large sea bird of the family Sulidae, esp. *Morus bassanus* of northern coastal regions, having white plumage with black wing tips.

**gan·nis·ter** (găn′ĭ-stər) *n. var. of* GANISTER.

**ga·nof** (gä′nəf) *n. var. of* GANEF.

**gan·oid** (găn′oid′) *adj.* [< Gk. *ganos,* brightness.] Of or characteristic of certain bony fishes, as the sturgeon and gar, having armorlike scales consisting of bony plates covered with layers of dentine and enamel. **—gan′oid′** *n.*

**gant·let[1]** (gônt′lĭt, gänt′-) *n.* [< GAUNTLET[2].] A section of overlapping but independent railroad track where two sets of tracks are overlapped to afford passage at a narrow place without switching. —*vt.* **-let·ed, -let·ing, -lets.** To overlap (railroad tracks) to form a gantlet.

**gant·let[2]** (gônt′lĭt, gänt′-) *n. var. of* GAUNTLET[1].

**gant·let[3]** (gônt′lăt, gänt′-) *n. var. of* GAUNTLET[2].

**gant·line** (gănt′līn′, -lĭn) *n.* [Alteration of *girtline,* gantline.] A rope passed through a single block at the top of a mast, used for hoisting.

**gan·try** (găn′trē) *n., pl.* **-tries.** [Prob. dial. *gawn,* gallon + TREE.] **1.** *Aerospace.* A massive vertical frame used in assembling or servicing rockets. **2.** A support for a barrel lying on its side. **3. a.** A bridgelike frame over which a traveling crane moves. **b.** A similar frame supporting a group of railway signals over several tracks.

**Gantt chart** (gănt) *n.* [After Henry Laurence *Gantt* (1861–1919).] A chart designed for comparing rates, as of planned production versus actual production.

**Gan·y·mede** (găn′ə-mēd′) *n.* [Gk. *Ganumēdēs.*] **1.** *Gk. Myth.* A very handsome Trojan boy whom Zeus carried away to be cupbearer to the gods. **2.** A young man who serves liquors. **3.** The fourth moon of Jupiter.

**gaol** (jāl) *n.* & *v. Chiefly Brit. var. of* JAIL.

**gap** (găp) *n.* [ME < ON, chasm.] **1.** An opening, as in a wall : CLEFT. **2.** A pass through mountains. **3.** A suspension of continuity : HIATUS <a *gap* in the news report> **4.** A conspicuous disparity <a *gap* between receipts and expenses> **5.** *Elect.* A space traversed by an electric spark. **6.** *Computer Sci.* An absence of data on a recording medium, often used to signal the end of a segment of data. **7.** *Electron.* The distance between the head of a recording device and the surface of the recording medium. —*v.* **gapped, gap·ping, gaps.** —*vt.* To make an opening or gap in. —*vi.* To be or become open.

**gape** (gāp, găp) *vi.* **gaped, gap·ing, gapes.** [ME *gapen* < ON *gapa.*] **1.** To open the mouth wide : YAWN. **2.** To stare wonderingly, as with the mouth open. **3.** To become widely open or separated <curtains *gaping* in the wind> —*n.* **1.** An act or instance of gaping. **2.** A large opening. **3.** *Zool.* The width of the space between the open jaws or mandibles of a vertebrate. **4. gapes** *(sing. in number).* A disease of birds, esp. young domesticated chickens and turkeys, caused by gapeworms and resulting in obstructed breathing. **5. gapes.** A fit of yawning. **—gap′er** *n.*

**gape·worm** (gāp′wûrm′, găp′-) *n.* A nematode worm of the genus *Syngamus,* esp. *S. trachea,* infecting the trachea of certain birds and causing gapes.

**gap·ing** (gā′pĭng) *adj.* Cavernous <a *gaping* chest wound> **—gap′ing·ly** *adv.*

**gar[1]** (gär) *n.* [Short for GARFISH.] **1.** A ganoid fish of the genus *Lepisosteus* of fresh and brackish waters of North and Central America, with an elongated body and long snout. **2.** A fish resembling or related to the gar, as the needlefish.

**gar[2]** (gär) *vt.* **garred, gar·ring, gars.** [ME *geren* < ON *gera,* to make.] *Scot.* To compel.

**ga·rage** (gə-räzh′, -räj′) *n.* [Fr. < *garer,* to shelter < OFr., to protect, of Germanic orig.] **1.** A building or wing of a building in which to park a car. **2.** A commercial establishment where cars are repaired, serviced, or parked. —*vt.* **-raged, -rag·ing, -rag·es.** To put in or take to a garage.

**garage sale** *n.* A sale of used household goods held at the seller's home.

**garb** (gärb) *n.* [OFr. *garbe,* grace, or OItal. *garbo,* grace, both of Germanic orig.] **1.** Clothing, esp. when distinctive <sailors' *garb*> **2.** An outward appearance : GUISE. —*vt.* **garbed, garb·ing, garbs.** To cover or furnish with or as if with clothing : DRESS.

**gar·bage** (gär′bĭj) *n.* [ME, refuse of fowls.] **1.** Food wastes, as from a kitchen. **2.** Worthless matter : TRASH. **3.** *Computer Sci.* Unwanted or incorrect data in a device's input, output, or storage.

**gar·ban·zo** (gär-bän′zō) *n., pl.* **-zos.** [Sp.] A seed of the chickpea.

**gar·ble** (gär′bəl) *vt.* **-bled, -bling, -bles.** [ME *garbelen,* to inspect and remove refuse from spices < OItal. *garbellare* < Ar. *gharbala,* he selected, poss. ult. < LLat. *cribellum,* dim. of Lat. *cribrum,* sieve.] **1.** To distort so as to be unintelligible : SCRAMBLE <a *garbled* radio message> **2.** To sort out : CULL. **—gar′ble** *n.* **—gar′bler** (-blər) *n.*

**gar·board** (gär′bôrd′, -bōrd′) *n.* [Obs. Du. *gaarboord.*] The first range or strake of planks laid next to a ship's keel.

**gar·boil** (gär′boil′) *n.* [OFr. *garbouil* < OItal. *garbuglio* < Lat. *bullire,* to boil.] *Archaic.* Confusion : uproar.

**gar·çon** (gär-sôN′) *n., pl.* **-çons** (-sôN′) [Fr. < OFr. *garçun,* servant.] A restaurant waiter.

**gar·dant** (gär′dnt) *adj. var. of* GUARDANT.

**gar·den** (gär′dn) *n.* [ME *gardin* < ONFr., of Germanic orig.] **1.** Land for the cultivation of flowers, vegetables, or fruit. **2. gardens.** Grounds adorned with flowers, shrubs, and trees for public enjoyment. **3.** A yard : lawn. **4.** A fertile well-cultivated region. —*v.* **-dened, -den·ing, -dens.** —*vt.* **1.** To cultivate (ground) as a garden. **2.** To furnish with a garden. —*vi.* To work as a gardener. —*adj.* **1.** Of, intended for, or found in a garden. **2.** Provided with open areas and greenery <*garden* condominiums> **3.** Ordinary : usual.

**garden cress** *n.* An annual herb of the mustard family, *Lepidium sativum.*

**gar·den·er** (gärd′nər, gär′dn-ər) *n.* One who works in or takes care of a garden for pleasure or profit.

**garden heliotrope** *n.* A species of valerian, *Valeriana officinalis,* with small purplish, pink, or white flower clusters.

**gar·de·nia** (gär-dēn′yə) *n.* [NLat., genus name, after Alexander *Garden* (1731–1790).] **1.** A shrub or tree of the genus *Gardenia,* esp. *G. jasminoides,* native to China, having glossy evergreen leaves and large, fragrant, usu. white flowers. **2.** The flower of the gardenia.
**gar·den-va·ri·e·ty** (gär′dn-və-rī′ĭ-tē) *adj.* Common : unremarkable <*a garden-variety* crook>
**garde·robe** (gärd′rōb′) *n.* [ME < OFr. : *garder,* to keep + *robe,* robe.] *Archaic.* **1. a.** A chamber for storing clothes : WARDROBE. **b.** The contents of a wardrobe. **2.** A private chamber.
**Gar·eth** (găr′ĭth) *n.* A nephew of King Arthur and a knight of the Round Table.
**gar·fish** (gär′fĭsh′) *n., pl.* **garfish** *or* **-fish·es.** [ME.] GAR[1].
**gar·ga·ney** (gär′gə-nē) *n., pl.* **-neys.** [Dial. Ital. *gargenei.*] An Old World duck, *Anas querquedula,* having the head prominently striped with white in the male.
**Gar·gan·tu·a** (gär-găn′cho͞o-ə) *n.* A giant king noted for his enormous physical and intellectual appetites, the hero of Rabelais' satire *Gargantua and Pantagruel.*
**gar·gan·tu·an** *also* **Gar·gan·tu·an** (gär-găn′cho͞o-ən) *adj.* Immense in size or volume : COLOSSAL.
**gar·get** (gär′gĭt) *n.* [Perh. < ME, throat < OFr. *garguette.*] Mastitis of domestic animals, esp. cattle.
**gar·gle** (gär′gəl) *v.* **-gled, -gling, -gles.** [OFr. *gargouiller.*] *—vi.* **1.** To force exhaled air through a liquid held in the back of the mouth in order to cleanse or medicate the mouth or throat. **2.** To produce the sound of gargling when speaking or singing. *—vt.* **1.** To rinse or medicate. **2.** To circulate or apply (e.g., a medicine) by gargling. **3.** To utter with a gargling sound. *—n.* **1.** A medicated solution for gargling. **2.** A gargling sound.
**gar·goyle** (gär′goil′) *n.* [ME *gargoile* < OFr. *gargole.*] **1.** A roof spout carved to represent a grotesque human or animal figure and projected from a gutter to carry rainwater clear of the wall. **2.** A grotesque ornamental figure.
**gar·i·bal·di** (găr′ə-bôl′dē) *n.* A loose high-necked blouse styled after the red shirts worn by the Italian nationalist leader Garibaldi and his soldiers.
**gar·ish** (gâr′ĭsh) *adj.* [Orig. unknown.] **1. a.** Marred by excessive color or ornamentation : GAUDY. **b.** Loud and flashy <*garish* make-up> **2.** Glaring and dazzling. **—gar′ish·ly** *adv.* **—gar′ish·ness** *n.*
**gar·land** (gär′lənd) *n.* [ME < OFr. *garlande.*] **1. a.** A wreath, circlet, or festoon, esp. one of flowers or leaves. **b.** A wreath worked in metal for ornamentation or as a heraldic device. **2.** *Naut.* A ring or collar of rope used to hoist spars or prevent fraying. **3.** An anthology <*a garland* of verse> *—vt.* **-land·ed, -land·ing, -lands. 1.** To embellish or deck with a garland. **2.** To form into a garland.
**gar·lic** (gär′lĭk) *n.* [ME < OE *gārlēac* : *gār,* spear + *lēac,* leek.] **1.** A plant, *Allium sativum,* related to the onion, having a bulb with a strong characteristic flavor and odor. **2.** The bulb of the garlic, divisible into separate cloves and used as a seasoning.
**gar·lick·y** (gär′lĭ-kē) *adj.* Smelling or tasting of or containing garlic.
**garlic mustard** *n.* A European weedy plant, *Alliaria officinalis,* having small white flowers and a garlicky odor.
**gar·ment** (gär′mənt) *n.* [ME < OFr. *garnement* < *garnir,* to equip, of Germanic orig.] An article of clothing, esp. of outer clothing. *—vt.* **-ment·ed, -ment·ing, -ments.** To clothe : dress.
**garment bag** *n.* A long bag in which hanging clothes are held when one is traveling, that often folds in half and has a center handle to facilitate carrying.
**gar·ner** (gär′nər) *vt.* **-nered, -ner·ing, -ners.** [< ME, granary < OFr. *gernier* < Lat. *granarium* < *granum,* grain.] **1.** To gather and store in or as if in a granary. **2.** To amass : acquire. *—n.* A granary.
**gar·net**[1] (gär′nĭt) *n.* [ME < OFr. *grenate* < *grenat,* pomegranate-red < *pome grenate,* pomegranate. —see POMEGRANATE.] **1.** Any of several common widespread silicate minerals, occurring in two internally isomorphic series, gen. crystallized, often imbedded in igneous and metamorphic rocks, colored red, brown, black, green, yellow, or white, and used as gemstones and abrasives. **2.** A dark red.
**gar·net**[2] (gär′nĭt) *n.* [ME *garnett,* prob. < MDu. *garnaat.*] *Naut.* A tackle for hoisting light cargo.
**gar·net·if·er·ous** (gär′nə-tĭf′ər-əs) *adj.* Containing garnets.
**gar·ni·er·ite** (gär′nē-ə-rīt′) *n.* [After Jules *Garnier* (1839?–1904).] An earthy apple-green mineral, $(Ni,Mg)_6(OH)_6Si_4O_{11}\cdot H_2O$, an important nickel ore.
**gar·nish** (gär′nĭsh) *vt.* **-nished, -nish·ing, -nish·es.** [ME *garnishen* < OFr. *garnir, garniss-,* of Germanic orig.] **1. a.** To furnish with beautifying details : EMBELLISH. **b.** To provide (food) with a garnish. **2.** *Law.* To garnishee. *—n.* **1. a.** Ornamentation : embellishment. **b.** An embellishment added to a food or drink for extra flavor or color. **2.** *Slang.* An unwarranted fee, as one extorted from a new prisoner by a jailer.
**gar·nish·ee** (gär′nĭ-shē′) *Law. —n.* **1.** A debtor against whom a plaintiff has instituted a process of garnishment. **2.** A third party who has been warned that money or property in his or her control but due or belonging to a defendant has been attached. *—vt.* **-eed, -ee·ing, -ees. 1.** To attach (e.g., a debtor's pay) by garnishment. **2.** To serve with a garnishment.
**gar·nish·ment** (gär′nĭsh-mənt) *n.* **1.** *Law.* **a.** A legal proceeding whereby money or property due or belonging to a debtor but in the possession of another is applied to the payment of the debt to the plaintiff. **b.** A court order directing a third party who owes a defendant money or holds property belonging to him or her to withhold such money or property and to appear in court to answer inquiries. **2.** Ornamentation : embellishment.
**gar·ni·ture** (gär′nĭ-chər) *n.* [OFr. < *garnir,* to garnish, of Germanic orig.] Something that garnishes : EMBELLISHMENT.
**gar·pike** (gär′pīk′) *n.* **1.** GAR[1]. **2.** A European marine fish, *Belone belone,* having green bones.
**gar·ret** (găr′ĭt) *n.* [ME < OFr. *garite,* watchtower < *garir,* to protect, of Germanic orig.] A room on the top floor of a house, typically right under a sloping roof : ATTIC.
**gar·ri·son** (găr′ĭ-sən) *n.* [ME *garison,* fortified place < OFr. < *garir,* to protect, of Germanic orig.] **1.** A usu. permanently established military post. **2.** Troops assigned at a garrison. *—vt.* **-soned, -son·ing, -sons. 1.** To assign (troops) to a military post. **2.** To supply (a post) with troops. **3.** To occupy as or convert into a garrison.
**garrison cap** *n.* A soft visorless cloth cap worn as a dress headgear chiefly by Army and Air Force personnel.
**Garrison finish** *n.* [After Edward H. *Garrison* (1868–1930).] A finish in a contest or race in which the winner comes from behind at the last moment.
**gar·rote** *or* **gar·rotte** (gə-rŏt′, -rōt′) *n.* [Sp. *garrote,* cudgel, instrument of torture < OFr. *garrot.*] **1. a.** Execution by strangulation or by breaking the neck with an iron collar screwed tight. **b.** A collar used for this. **2.** Strangulation, esp. in order to rob. *—vt.* **-rot·ed, -rot·ing, -rots** *or* **-rot·ted, -rot·ting, -rottes. 1.** To execute by garrote. **2.** To strangle in order to rob. **—gar·rot′er** *n.*
**gar·ru·lous** (găr′ə-ləs, găr′yə-) *adj.* [Lat. *garrulus* < *garrire,* to chatter.] Habitually, often excessively talkative. **—gar·ru′li·ty** (gə-ro͞o′lĭ-tē) *n.* **—gar′ru·lous·ly** *adv.* **—gar′ru·lous·ness** *n.*
**gar·ter** (gär′tər) *n.* [ME, band to support socks < OFr. < *garet,* bend of knee, prob. of Celt. orig.] **1. a.** An elasticized band worn around the leg to support hose. **b.** A suspender strap with a fastener attached to a girdle or belt for supporting hose. **c.** An elasticized band worn around the arm to keep the sleeve pushed up. **2. Garter. a.** The badge of the Order of the Garter. **b.** The order itself. **c.** Membership in this order. *—vt.* **-tered, -ter·ing, -ters. 1.** To fasten and hold with a garter. **2.** To put a garter on.
**garter snake** *n.* Any of various nonvenomous North American snakes of the genus *Thamnophis,* with longitudinal stripes.
**garth** (gärth) *n.* [ME, enclosed yard < ON *garðr.*] **1.** A grassy quadrangle surrounded by cloisters. **2.** *Archaic.* A yard or paddock.
**gas** (găs) *n., pl.* **gas·es** *or* **gas·ses.** [Du., an occult physical principle < Gk. *khaos,* chaos, coined by J.B. van Helmont, (1577–1644).] **1. a.** The state of matter differentiated from the solid and liquid states by very low density and viscosity, rather great expansion and contraction with changes in pressure and temperature, the ability to diffuse readily, and the spontaneous tendency to become distributed uniformly throughout a container. **b.** A substance in the gaseous state. **2.** A gaseous fuel, as natural gas. **3. a.** Gasoline. **b.** The speed control of a gasoline engine <let up on the *gas*> **4.** A gaseous asphyxiant, irritant, or poison. **5.** A gaseous anesthetic. **6.** *Slang.* Idle boastful talk. **7.** *Slang.* One providing great fun and excitement <The party was a *gas.*> *—v.* **gassed, gas·sing, gas·es** *or* **gass·es.** *—vt.* **1.** To supply with gas or gasoline. **2.** To treat chemically with gas. **3.** To poison with gas. *—vi.* **1.** To give off gas. **2.** *Slang.* To talk excessively. **3.** *Informal.* To fill the tank of a vehicle with gas <*gassed* up before leaving>
**gas·bag** (găs′băg′) *n.* **1.** An expansible bag for holding gas. **2.** *Slang.* An idle boastful talker.
**gas burner** *n.* A nozzle or jet on a fitting through which combustible gas is released to burn.
**gas chamber** *n.* A sealed enclosure in which prisoners are executed by a poisonous gas.
**gas chromatograph** *n.* A device used to separate a sample into its components for analysis in gas chromatography.
**gas chromatography** *n.* Chromatography in which the substance to be analyzed is vaporized and diffused along with a carrier gas through a liquid or solid adsorbent for differential adsorption.
**Gas·con** (găs′kən) *n.* **1.** A native or resident of Gascony. **2.** The French dialect of the Gascons. **3. gascon.** A boastful person. *—adj.* Of or relating to Gascony or the Gascons.
**gas·con·ade** (găs′kə-nād′) *n.* Boastfulness : bravado. **—gas′con·ade′** *v.* **(-ad·ed, -ad·ing, -ades). —gas′con·ad′er** *n.*
**gas·dy·nam·ics** (găs′dī-năm′ĭks) *n. (sing. in number).* The branch of dynamics that deals with thermal gaseous fluids. **—gas′dy·nam′ic** *adj.* **—gas′dy·nam′i·cist** *n.*
**gas·e·ous** (găs′ē-əs, găsh′əs) *adj.* **1.** Of, relating to, or existing as a gas. **2.** Lacking concreteness : TENUOUS. **—gas′e·ous·ness** *n.*
**gas fitter** *n.* A worker who installs or repairs gas pipes, fixtures, or appliances.
**gas gangrene** *n.* Gangrene occurring in a wound infected with bacteria of the genus *Clostridium,* esp. with *C. welchi* or *C. oedema-*

*tiens,* and marked by the presence of gas in the affected tissue and constitutional septic symptoms.

**gas-guz·zler** (găs′gŭz′lər) *n. Informal.* A car that consumes excessive amounts of gasoline.

**gas-guz·zling** (găs′gŭz′lĭng) *adj. Informal.* Using excessive amounts of gasoline <*gas-guzzling* old heaps>

**gash** (găsh) *vt.* **gashed, gash·ing, gash·es.** [ME *garsen,* to cut < ONFr. *garser,* prob. LLat. *charaxare* < Gk. *kharassein.*] To make a deep long slash in. —*n.* A deep long cut or wound.

**gas·hold·er** (găs′hōl′dər) *n.* A storage container for fuel gas, esp. a large telescoping cylindrical tank.

**gas·house** (găs′hous′) *n.* A gasworks.

**gas·i·form** (găs′ə-fôrm′) *adj.* Being in gaseous form.

**gas·i·fy** (găs′ə-fī′) *vt. & vi.* **-fied, -fy·ing, -fies.** To convert into or become gas. **—gas′i·fi′a·ble** *adj.* **—gas′i·fi·ca′tion** *n.*

**gas jet** *n.* **1.** A gas burner. **2.** The flame of burning gas from a gas burner.

**gas·ket** (găs′kĭt) *n.* [Fr. *garcette,* dim. of *garce,* girl.] **1.** A seal or packing used between matched machine parts or around pipe joints to prevent the escape of a gas or fluid. **2.** *Naut.* A cord or canvas strap used to secure a furled sail to a yard boom or gaff. **—blow a gasket.** *Slang.* To erupt in anger.

**gas·kin** (găs′kĭn) *n.* [Prob. short for GALLIGASKINS.] **1.** The part of the hind leg of a horse or related animal between the stifle and the hock. **2. gaskins.** *Obs.* Galligaskins.

gaskin

**gas·light** (găs′līt′) *n.* **1.** Light generated by burning illuminating gas. **2.** A gas burner or lamp.

**gas log** *n.* A fireplace gas heater designed to look like a log.

**gas main** *n.* A major pipeline conveying gas to smaller pipes for distribution to consumers.

**gas mask** *n.* A respirator covering the face and having a chemical air filter to protect against poisonous gases.

**gas·o·hol** (găs′ə-hôl) *n.* [GAS(OLINE) + (ALC)OHOL.] A fuel composed of a blend of ethanol and unleaded gasoline, esp. a blend of 10% ethanol and 90% gasoline.

**gas·o·line** *also* **gas·o·lene** (găs′ə-lēn′, găs′ə-lēn′) *n.* A volatile mixture of flammable liquid hydrocarbons derived chiefly from crude petroleum and used principally as a fuel for internal-combustion engines and as a solvent, illuminant, and thinner.

**gas·om·e·ter** (găs-ŏm′ĭ-tər) *n.* [Fr. *gazomètre* : *gaz,* gas + *-mètre,* -meter.] **1.** A device for measuring gases. **2.** A gasholder.

**gasp** (găsp) *v.* **gasped, gasp·ing, gasps.** [ME *gaspen,* to gape < ON *geispa,* to yawn.] —*vi.* **1.** To draw in or catch the breath sharply, as from shock. **2.** To breathe laboriously or convulsively. —*vt.* To say in a breathless way. —*n.* A short convulsive intake of the breath.

**gasp·er** (găs′pər) *n. Chiefly Brit.* A cigarette.

**gas plant** *n.* A Eurasian plant, *Dictamnus albus,* having aromatic foliage and white flowers and emitting a flammable vapor.

**gas·ser** (găs′ər) *n.* **1.** A well or drilling that yields natural gas. **2.** *Slang.* GAS 7.

**gas station** *n.* A filling station.

**gas·sy** (găs′ē) *adj.* **-si·er, -si·est. 1.** Containing, full of, or resembling gas. **2.** *Slang.* Bombastic : boastful. **—gas′si·ness** *n.*

**gast** (găst) *vt.* **gast·ed, gast·ing, gasts.** [ME *gasten* < OE *gǣstan.*] *Obs.* To frighten : scare.

**gas·tight** (găs′tīt′) *adj.* Preventing the escape or entry of gas. **—gas′tight′ness** *n.*

**gastr-** *pref. var. of* GASTRO-.

**gas·trec·to·my** (gă-strĕk′tə-mē) *n., pl.* **-mies.** Surgical excision of part or all of the stomach.

**gas·tric** (găs′trĭk) *adj.* Of or relating to the stomach.

**gastric juice** *n.* The colorless, watery, acidic digestive fluid secreted by the stomach glands and containing hydrochloric acid, pepsin, rennin, and mucin.

**gas·trin** (găs′trĭn) *n.* A secretion of the gastric mucosa that stimulates production of gastric juice.

**gas·tri·tis** (gă-strī′tĭs) *n.* Inflammation of the stomach.

**gastro-** *or* **gastr-** *pref.* [< Gk. *gastēr, gastr-,* belly.] **1. a.** Belly <*gastropod*> **b.** Stomach <*gastritis*> **2.** Gastric <*gastrin*>

**gas·tro·en·ter·i·tis** (găs′trō-ĕn′tə-rī′tĭs) *n.* Inflammation of the mucous membrane of the stomach and intestine.

**gas·tro·en·ter·ol·o·gy** (găs′trō-ĕn′tə-rŏl′ə-jē) *n.* Medical study of the stomach and the intestines. **—gas′tro·en·ter′ic** (-ĕn-tĕr′ĭk) *adj.* **—gas′tro·en′ter·o·log′i·cal** *adj.* **—gas′tro·en′ter·ol′o·gist** *n.*

**gas·tro·in·tes·ti·nal** (găs′trō-ĭn-tĕs′tə-nəl) *adj.* Of or relating to the stomach and intestines.

**gas·tro·lith** (găs′trə-lĭth′) *n.* A small stony pathological mass formed in the stomach.

**gas·trol·o·gy** (gă-strŏl′ə-jē) *n.* Medical study of the stomach and its diseases. **—gas·trol′o·gist** *n.*

**gas·tro·nome** (găs′trə-nōm′) *also* **gas·tron·o·mer** (gă-strŏn′ə-mər) *n.* [Fr., back-formation < *gastronomie,* gastronomy.] A connoisseur of good food and drink : GOURMET.

**gas·tro·nom·ic** (găs′trə-nŏm′ĭk) *also* **gas·tro·nom·i·cal** (-ĭ-kəl) *adj.* Relating to gastronomy. **—gas′tro·nom′i·cal·ly** *adv.*

**gas·tron·o·mist** (gă-strŏn′ə-mĭst) *n.* A gastronome.

**gas·tron·o·my** (gă-strŏn′ə-mē) *n.* [Fr. *gastronomie.*] **1.** The art of good eating. **2.** Cooking, as of a specific region of the world.

**gas·tro·pod** (găs′trə-pŏd′) *n.* [NLat. *Gastropoda,* class name : GASTRO- + Gk. *pous,* foot.] A mollusk of the class Gastropoda, as a snail, slug, cowry, or limpet, having a single, usu. coiled shell and a ventral muscular mass serving as an organ of locomotion. —*adj.* Of or belonging to the Gastropoda. **—gas·trop′o·dan** (gă-strŏp′ə-dən), **gas·trop′o·dous** (-dəs) *adj.*

**gas·tro·scope** (găs′trə-skōp′) *n.* An instrument for examining the interior of the stomach. **—gas′tro·scop′ic** (-skŏp′ĭk) *adj.* **—gas·tros′co·pist** (gă-strŏs′kə-pĭst) *n.* **—gas·tros′co·py** (-strŏs′kə-pē) *n.*

**gas·tros·to·my** (gă-strŏs′tə-mē) *n., pl.* **-mies.** Surgical construction of a permanent opening from the external surface of the body into the stomach, usu. for inserting a feeding tube.

**gas·trot·o·my** (gă-strŏt′ə-mē) *n., pl.* **-mies.** Surgical incision into the stomach.

**gas·tro·vas·cu·lar** (găs′trō-văs′kyə-lər) *adj.* Having both digestive and circulatory functions.

**gas·tru·la** (găs′trə-lə) *n., pl.* **-las** *or* **-lae** (-lē′) [NLat. < Gk. *gastēr,* belly.] An embryo at the stage following the blastula and consisting of ectoderm, endoderm, and archenteron. **—gas′tru·lar** (-lər) *adj.*

**gas·tru·late** (găs′trə-lāt′) *vi.* **-lat·ed, -lat·ing, -lates.** To form or become a gastrula. **—gas′tru·la′tion** *n.*

**gas turbine** *n.* An air-breathing internal-combustion engine composed of an air compressor, a combustion chamber, and a turbine wheel, used esp. for propulsion.

**gas·works** (găs′wûrks′) *n.* (*sing. in number*). A factory where gas for heating and lighting is produced.

**gat¹** (găt) *n.* [Prob. < Du.] A narrow passage extending inland from a shore : CHANNEL.

**gat²** (găt) *n.* [GAT(LING GUN).] *Slang.* A pistol.

**gat³** (găt) *v. Archaic. var. p.t. of* GET.

**gate¹** (gāt) *n.* [ME < OE *geat.*] **1.** A structure that can be swung, drawn, or lowered to block an entrance or passageway. **2. a.** An opening in a wall or fence for entrance or exit. **b.** The structure surrounding such an opening, as the monumental or fortified entrance to a palace. **3.** Something providing access <the *gate* to immense fortune> **4.** A device for controlling the passage of water or gas through a dam or conduit. **5.** The total admission receipts or attendance at a public spectacle. **6.** The channel through which molten metal flows into the shaped cavity of a mold. **7.** *Electron.* **a.** A circuit widely used in computers that has an output dependent on some function of its input. **b.** Such a circuit having an output when any or all of a designated set of inputs are received within a specified time interval. —*vt.* **gat·ed, gat·ing, gates.** *Chiefly Brit.* **1.** To punish (a student) by confinement within the college gates after a certain hour. **2.** *Electron.* To select (part of a wave) for transmission, reception, or processing by magnitude or time interval. **—get the gate.** *Slang.* To be ejected. **—give (someone) the gate.** *Slang.* To eject.

**†gate²** (gāt) *n.* [ME < ON *gata.*] **1.** *Archaic.* A path or road : WAY. **2.** *Regional.* A particular way of acting or doing : MANNER.

**gate·crash·er** (gāt′krăsh′ər) *n. Slang.* One who gains admittance without being invited or enters without paying admission.

**gate·fold** (gāt′fōld′) *n.* A folded insert in a publication whose full size exceeds that of the regular page.

**gate·keep·er** (gāt′kē′pər) *n.* One in charge of a gate.

**gate-leg table** (gāt′lĕg′) *n.* A drop-leaf table with movable legs arranged in pairs.

**gate·post** (gāt′pōst′) *n.* An upright post on which a gate is hung or against which a gate is closed.

**gate·way** (gāt′wā′) *n.* **1.** An opening or a structure framing an opening that may be closed by a gate. **2.** GATE¹ 3.

**gath·er** (găth′ər) *v.* **-ered, -er·ing, -ers.** [ME *getheren* < OE *gadrian.*] —*vt.* **1.** To cause to come together : CONVENE. **2. a.** To amass gradually. **b.** To harvest or pick <*gather* raspberries> **c.** To gain or increase by degrees <*gather* speed> **3. a.** To collect into one place : ASSEMBLE. **b.** To arrange (signatures) in sequence for bookbinding. **4.** To run a thread through (cloth) so as to draw it up into small folds or puckers. **5.** To draw (e.g., a garment) about or closer to something. **6.** To conclude : infer <I *gather* the project has been canceled.> **7. a.** To summon up : MUSTER <*gather* strength> **b.** To collect (one's wits or powers). **8.** To attract or be a center of attraction for. —*vi.* **1.** To come together in a group : ASSEMBLE <An angry mob *gathered.*> **2.** To accumulate. **3.** To grow or increase by degrees. **4.** To come to a head, as a boil does : FESTER. —*n.* **1. a.** An act or instance of gathering. **b.** A quantity gathered. **2.** A small fold or pucker made in cloth by gathering it. **—gath'er·er** *n.*

☆ ***syns:*** **1.** GATHER, CALL, CONVENE, CONVOKE, MUSTER, SUMMON *v. core meaning* : to bring together <*gathered* the faculty for a meeting> **2.** GATHER, ACCUMULATE, AMASS, COLLECT *v. core meaning* : to come together <a crowd that *gathered* to watch the parade> GATHER is the most general term and COLLECT is often interchangeable with it <crowds *collecting* along the parade route>, but it is often used to imply the careful selection of related things <paintings *collected* by an avid art lover> ACCUMULATE and AMASS refer to the gradual increase of something over a period of time <snow *accumulating* on the sidewalks> <a fortune *amassed* over the years> ASSEMBLE suggests a convening for a common purpose <legislators *assembling* for a joint session of Congress>

**gath·er·ing** (găth'ər-ĭng) *n.* **1.** Something amassed : COLLECTION. **2.** An assembly : meeting. **3.** A gather in cloth. **4.** A suppurated swelling : BOIL.

**Gat·ling gun** (găt'lĭng) *n.* [After Richard J. *Gatling* (1818–1903), its inventor.] A machine gun with a cluster of barrels that turn simultaneously and fire once per each revolution.

**Gat·or·ade** (gā'tər-ād'). A trademark for a thirst-quenching beverage used esp. by athletes.

**gauche** (gōsh) *adj.* [Fr. < OFr. *gauchir,* to turn aside, of Germanic orig.] Socially awkward. **—gauche'ly** *adv.* **—gauche'ness** *n.*

**gau·che·rie** (gō'shə-rē') *n.* [Fr. < *gauche,* gauche.] **1.** An awkward or tactless action, manner, or expression. **2.** Social awkwardness.

**gau·cho** (gou'chō) *n., pl.* **-chos.** [Am. Sp. (South America), prob. < Quechua *wáhcha,* vagabond.] A South American cowboy.

**gaud** (gôd) *n.* [ME *gaude,* trinket.] A gaudy ornament.

**gaud·er·y** (gô'də-rē) *n., pl.* **-ies.** Showy things : FINERY.

**gaud·y¹** (gô'dē) *adj.* **-i·er, -i·est.** Marked by tasteless or showy ornaments : GARISH. **—gaud'i·ly** *adv.* **—gaud'i·ness** *n.*

☆ ***syns:*** GAUDY, CHINTZY, FLASHY, GARISH, GLARING, LOUD, MERETRICIOUS, TACKY, TAWDRY, TINSEL *adj. core meaning* : tastelessly showy <*gaudy* clothes>

**gaud·y²** (gô'dē) *n., pl.* **-ies.** [ME, ornamental rosary bead < Lat. *gaudium,* joy < *gaudēre,* to rejoice.] *Chiefly Brit.* A feast, esp. an annual university dinner.

**gauf·fer** (gŏf'ər, gô'fər) *v.* & *n. var. of* GOFFER.

**gauge** *also* **gage** (gāj) [ME < ONFr.] —*n.* **1. a.** A standard or scale of measurement. **b.** A standard dimension, quantity, or capacity. **2.** A device for measuring or testing. **3.** A means of estimating or evaluating : TEST <a *gauge* of one's true character> **4.** The position of a vessel in relation to another vessel and the wind. **5. a.** The distance between the two railroad rails. **b.** The distance between two wheels on an axle. **6.** The diameter of a shotgun barrel as measured by the number of lead balls of a size exactly fitting the barrel that can be made from a pound of lead. **7.** The amount of plaster of Paris mixed with common plaster to speed its setting. **8.** Thickness or diameter, as of sheet metal or wire. **9.** The fineness of knitted cloth as determined by the number of loops per 1½ inches. —*vt.* **gauged, gaug·ing, gaug·es** *also* **gaged, gag·ing, gag·es. 1.** To measure precisely. **2.** To determine the capacity, volume, or contents of. **3.** To evaluate : judge <*gauge* a student's ability> **4.** To adapt to a specified measurement. **5.** To mix (plaster) in specific proportions. **6.** To chip or rub (e.g., bricks) to size. **—gauge'a·ble** *adj.*

**gaug·er** *also* **gag·er** (gā'jər) *n.* **1.** One that gauges. **2.** *Chiefly Brit.* A revenue officer who inspects bulk goods subject to duty.

**Gaul** (gôl) *n.* **1.** A Celt of ancient Gaul. **2.** A Frenchman.

**Gaul·ish** (gô'lĭsh) *n.* The Celtic language of ancient Gaul.

**Gaull·ism** (gō'lĭz'əm, gô'-) *n.* **1.** The political movement supporting General Charles de Gaulle as leader of the French government in exile during World War II. **2. a.** A political movement headed by de Gaulle after World War II. **b.** The political principles and goals of de Gaulle. **—Gaull'ist** *n.*

**†gaum** (gôm) *vt.* **gaumed, gaum·ing, gaums.** [Alteration of GUM¹.] *Regional.* To smudge : smear.

**gaunt** (gônt) *adj.* **-er, -est.** [ME.] **1.** Thin and bony : ANGULAR. **2.** Emaciated and haggard : DRAWN. **3.** Bleak and desolate : BARREN <a *gaunt* landscape> **—gaunt'ly** *adv.* **—gaunt'ness** *n.*

**gaunt·let¹** *also* **gant·let** (gônt'lĭt, gänt'-) *n.* [ME < OFr. *gantelet,* dim. of *gant,* glove, of Germanic orig.] **1.** A protective glove worn with medieval armor. **2.** A protective glove with a flaring cuff used in manual labor. **3.** A challenge. **4.** A dress glove cuffed above the wrist.

**gaunt·let²** *also* **gant·let** (gônt'lĭt, gänt'-) *n.* [Alteration of *gantelope* < Swed. *gatlopp* : *gata,* lane + *lopp,* course.] **1.** Two lines of men facing each other and armed with sticks or other weapons with which they beat a person forced to run between them. **2.** A severe trial : ORDEAL <The embattled President ran the *gauntlet* of hostile questions from the press.>

▲ word history: The two English words spelled *gauntlet* are unrelated in origin and meaning; they share only the fate of now being restricted, in the living language, to idiomatic phrases. *Gauntlet¹*, denoting a type of glove, is derived from Old French *gantelet,* a diminutive of *gant,* "glove." *Gauntlet²*, denoting a kind of punishment or ordeal, is an alteration of the earlier English form *gantlope,* from Swedish *gatlopp,* a compound of *gata,* "lane," and *lopp,* "course." It is not clear exactly why the first syllable of *gantlope* was altered so radically from the original Swedish form, but the change occurred soon after the word was borrowed in the 16th century. The form *gantlet,* used at various times for both *gauntlet¹* and *gauntlet²*, is merely a spelling variant with no apparent significance.

**gaur** (gour) *n.* [Hindi < Skt. *gauraḥ.*] A large dark-coated bovine mammal, *Bos gaurus,* of hilly areas of southeastern Asia.

**gauss** (gous) *n., pl.* **gauss** *or* **gauss·es.** [After Karl F. *Gauss* (1777–1855).] The centimeter-gram-second electromagnetic unit of magnetic flux density, equal to one maxwell per square centimeter.

**Gauss·i·an distribution** (gou'sē-ən) *n.* [After Karl F. *Gauss* (1777–1855).] Normal distribution.

**gauze** (gôz) *n.* [OFr. *gaze.*] **1. a.** A thin transparent fabric with a loose open weave, used for curtains or clothing. **b.** A cotton surgical dressing. **c.** A thin plastic or metal woven mesh. **2.** A mist or haze. **—gauz'i·ly** *adv.* **—gauz'i·ness** *n.* **—gauz'y** *adj.*

**ga·vage** (gə-väzh') *n.* [Fr.] Introduction of material into the stomach via a tube.

**gave** (gāv) *v. p.t. of* GIVE.

**gav·el¹** (găv'əl) *n.* [Orig. unknown.] **1.** The mallet or hammer used by a presiding officer or auctioneer to signal for attention or order. **2.** A maul used by masons in fitting stones. —*vt.* **-eled, -el·ing, -els** *also* **-elled, -el·ling, -els.** To cause or compel by using a gavel.

**gav·el²** (găv'əl) *n.* [ME < OE *gafol.*] Ancient and medieval English tribute or rent.

**gav·el·kind** (găv'əl-kīnd') *n.* [ME *gavelkinde* : OE *gafol,* gavel + OE *gecynd,* kind.] An English system of land tenure from Anglo-Saxon times to 1926 that provided for equal division of an intestate's estate among all the sons or other heirs.

**gav·el-to-gav·el** (găv'əl-tə-găv'əl) *adj.* Extending from beginning to end <*gavel-to-gavel* coverage of the Congressional hearings>

**ga·vi·al** (gā'vē-əl) *n.* [Fr. < Hindi *ghaṛiyāl.*] A large reptile, *Gavialis gangeticus* of southern Asia, related to and resembling the crocodiles and having a long slender snout.

**ga·votte** (gə-vŏt') *n.* [Fr. < Prov. *gavoto* < *Gavot,* native of the Alps.] **1.** A French peasant dance similar to the minuet. **2.** Music for the gavotte in moderately quick 4/4 time.

**Ga·wain** (gə-wān', gä'wān', gou'ən) *n.* A nephew of King Arthur and a knight of the Round Table.

**gawk** (gôk) *n.* [Perh. alteration of obs. *gaw,* to gape < ME *gawen* < ON *gā,* to heed.] An oaf. —*vi.* **gawked, gawk·ing, gawks.** *Informal.* To stare or gape stupidly. **—gawk'er** *n.*

**gawk·y** (gô'kē) *adj.* **-i·er, -i·est.** Clumsy : awkward.

**gay** (gā) *adj.* **-er, -est.** [ME *gai* < OFr., of Germanic orig.] **1.** Showing or marked by exuberance or happy excitement : MERRY. **2.** Bright, esp. in color. **3.** Full of or given to social pleasures. **4.** Dissolute. **5.** Homosexual. —*n.* A homosexual. **—gay** *adv.* **—gay'ness** *n.*

**ga·yal** (gə-yäl') *n.* [Bengali *gayāl,* prob. < Skt. *gauḥ,* ox.] A domesticated bovine mammal, *Bos frontalis* of India and Burma, having thick pointed horns, a dark coat, and a tufted tail.

**gay·e·ty** (gā'ĭ-tē) *n. var. of* GAIETY.

**gay·ly** (gā'lē) *adv. var. of* GAILY.

**gaze** (gāz) *vi.* **gazed, gaz·ing, gaz·es.** [ME *gasen,* prob. of Scand. orig.] To look fixedly <*gazed* at me across the room> **—gaze** *n.* **—gaz'er** *n.*

☆ ***syns:*** GAZE, EYE, GAPE, GAWK, GOGGLE, OGLE, PEER, STARE *v. core meaning* : to look fixedly <*gazed* in awe at the crown jewels>

**ga·ze·bo** (gə-zā'bō, -zē'bō) *n., pl.* **-bos** *or* **-boes.** [Orig. unknown.] **1.** A freestanding, roofed, usu. open-sided structure providing a shady resting place. **2.** A belvedere.

**ga·zelle** (gə-zĕl') *n.* [Fr. < OFr., prob. < Sp. *gacela* < Ar. *ghazāl.*] An African or Asian hoofed mammal of the genus *Gazella* or related genera, having a slender neck and ringed lyrate horns.

**ga·zette** (gə-zĕt') *n.* [Fr. < Ital. *gazetta,* prob. < *gazeta,* a coin for which such a newspaper sold.] **1.** A newspaper. **2.** An official journal. **3.** *Chiefly Brit.* An announcement in an official journal. —*vt.* **-zet·ted, -zet·ting, -zettes.** *Chiefly Brit.* To announce or publish in a gazette.

**gaz·et·teer** (găz'ĭ-tîr') *n.* **1.** A geographic dictionary or index. **2.** *Archaic.* A journalist.

**gaz·pa·cho** (gə-spä'chō, gəz-pä'-) *n.* [Sp.] A chilled soup of tomatoes, onions, green peppers, and herbs.

**G clef** *n. Mus.* The treble clef.

**Gd** *symbol for* GADOLINIUM.

**Ge** *symbol for* GERMANIUM.

**ge-** *pref. var. of* GEO-.

**ge·an·ti·cline** (jē-ăn'tĭ-klīn') *n.* A large upward fold of the earth's crust. **—ge·an'ti·cli'nal** *adj.*

**gear** (gîr) *n.* [ME *gere,* equipment < ON *gervi.*] **1. a.** A machine part, as a toothed wheel or cylinder that meshes with another toothed part to transmit motion or to change direction or speed. **b.** A complete assembly that performs a specific function in a larger machine. **c.** A transmission configuration for a specific ratio of engine to axle torque in a motor vehicle. **2.** Equipment, as tools or clothing, required for an activity : PARAPHERNALIA. **3.** The harness for a horse. **4.** The rigging of a ship. **5.** A sailor's personal effects. **6.** *Chiefly Brit. Regional.* **a.** Nonsense. **b.** Doings. —*v.* **geared, gear·ing, gears.** —*vt.* **1. a.** To provide with gears. **b.** To connect by gears. **c.** To put into gear. **2. a.** To adjust or adapt <*geared* the program to learning-disabled students> **b.** To prepare for action <*gear* oneself up for a big game> **3.** To provide with gear. —*vi.* **1.** To be or become in gear. **2.** To become adjusted so as to fit or blend.

**gear·box** (gîr'bŏks') *n.* An automotive transmission.

**gear·ing** (gîr'ĭng) *n.* **1.** A system of gears and associated elements by which motion is transferred within a machine. **2.** The act or technique of providing with gears.

**gear·shift** (gîr'shĭft') *n.* A mechanism for changing from one transmission gear to another.

**gear train** *n.* A system of interconnected gears.

**gear·wheel** *also* **gear wheel** (gîr'hwēl') *n.* A wheel configured with a toothed rim.

**geck·o** (gĕk'ō) *n., pl.* **-os** *or* **-oes.** [Malay *ge'kok.*] Any of various usu. small lizards of the family Gekkonidae of warm regions, having toes with adhesive pads enabling them to climb on vertical surfaces.

**gecko**
*4–5 inches long*

**gee¹** (jē) *n.* The letter g.

**gee²** (jē) *interj.* —Used to express a command, as to a horse, to turn to the right or to go forward. —*vi.* **geed, gee·ing, gees.** To turn to the right.

**gee³** (jē) [Alteration of JESUS.] *interj.* —Used as a mild expletive or exclamation of surprise.

**gee⁴** (jē) *n.* [< GEE¹, from the first letter of GRAND.] *Slang.* A thousand dollars.

**geek** (gēk) *n.* [Perh. < dial. *geek,* fool < MLG *geck.*] *Slang.* A carnival performer whose act usu. consists of biting the head off a live chicken or snake.

**gee·pound** (jē'pound') *n.* [GEE¹ (from the first letter of GRAVITY) + POUND¹.] SLUG¹ 6.

**geese** (gēs) *n. pl. of* GOOSE.

**gee whiz** *interj.* GEE³.

**gee-whiz** (jē'hwĭz', -wĭz') *adj. Slang.* **1.** Having a sensational aspect or quality <The microchip industry was once in the *gee-whiz* stage of development.> **2.** Marked by wide-eyed amazement or enthusiasm <a *gee-whiz* attitude about European travel>

**gee·zer** (gē'zər) *n.* [Prob. alteration of dial. *guiser,* masquerader < GUISE.] *Slang.* An old man.

**ge·fil·te fish** (gĕ-fĭl'tə) *n.* [Yiddish, filled fish.] Chopped fish mixed with crumbs, eggs, and seasonings, cooked in a broth and usu. served chilled in the form of balls or oval-shaped cakes.

**ge·gen·schein** (gā'gən-shīn') *n.* [G. : *gegen,* against (< OHG *gegin*) + *Schein,* light < *scheinen,* to shine < OHG *scīnan.*] A faint glowing spot in the sky, exactly opposite the position of the sun.

**Ge·hen·na** (gĭ-hĕn'ə) *n.* [LLat. < Gk. *Geenna* < Heb. *Ge' Hinnōm,* Valley of Hinnom, a valley south of Jerusalem.] **1.** A place or state of torment or suffering. **2.** Hell.

**Gei·ger counter** (gī'gər) *n.* An instrument consisting of a Geiger tube and associated electronic equipment, used to detect, measure, and record nuclear emanations, cosmic rays, and artificially produced subatomic particles.

**Geiger tube** *n.* [After Hans *Geiger* (1882–1945).] A gas-filled tube containing coaxial cylindrical electrodes between which a potential difference slightly below the breakdown voltage is maintained, so that production of a pair of ions in the gas by passage of a charged particle or by ionizing radiation causes a breakdown throughout the volume of the tube.

**gei·sha** (gā'shə) *n., pl.* **geisha** *or* **-shas.** [J. : *gei,* art + *sha,* person.] A Japanese girl trained to entertain men, as by singing, dancing, or talking amusingly.

**gel** (jĕl) *n.* [Short for GELATIN.] A colloid in which the disperse phase has combined with the continuous phase to produce a semisolid material, as a jelly. —*vi.* **gelled, gel·ling, gels.** To become a gel. **—gel'a·ble** *adj.*

**ge·län·de·sprung** (gə-lĕn'də-shprŏŏng') *n.* [G.: *Gelände,* ground (< *Land,* land < OHG *lant*) + *Sprung,* jump < *springan,* to jump < OHG.] A skiing jump performed from a crouch with the use of both poles.

**gel·ate** (jĕl'āt') *vi.* **-at·ed, -at·ing, -ates.** To gel.

**gel·a·tin** *also* **gel·a·tine** (jĕl'ə-tən) *n.* [Fr. *gélatine* < Ital. *gelatina,* dim. of *gelata,* jelly < Lat., p.part. of *gelare,* to freeze < *gelu,* frost.] **1.** A colorless or slightly yellow, transparent, brittle protein formed by boiling the specially prepared skin, bones, and connective tissue of animals, used in foods, drugs, and photographic film. **2.** A substance similar to gelatin. **3.** A jelly made with gelatin, used esp. as a salad base or dessert. **4.** A thin membrane used over a theatrical light to color it.

**ge·lat·i·nize** (jə-lăt'n-īz', jĕl'ə-tə-nīz') *v.* **-nized, -niz·ing, -niz·es.** —*vt.* **1.** To convert to gelatin or jelly. **2.** To coat with gelatin. —*vi.* To become gelatinous. **—ge·lat'i·ni·za'tion** *n.*

**ge·lat·i·nous** (jə-lăt'n-əs) *adj.* **1.** Like gelatin in consistency : VISCOUS. **2.** Of, relating to, containing, or resembling gelatin. **—ge·lat'i·nous·ly** *adv.* **—ge·lat'i·nous·ness** *n.*

**ge·la·tion** (jĕ-lā'shən) *n.* [Lat. *gelatio* < *gelare,* to freeze < *gelu,* frost.] **1.** Solidification by cooling or freezing. **2.** Formation of a gel.

**geld¹** (gĕld) *vt.* **geld·ed** *or* **gelt** (gĕlt), **geld·ing, gelds.** [ME *gelden* < ON *gelda.*] To castrate (e.g., a horse).

**geld²** (gĕld) *n.* [Med. Lat. *geldum* < OE *gield.*] A tax paid to the crown by English landholders under Anglo-Saxon and Norman kings.

**geld·ing** (gĕl'dĭng) *n.* [ME < ON *geldingr* < *gelda,* to geld.] A castrated animal, esp. a male horse.

**gel·id** (jĕl'ĭd) *adj.* [Lat. *gelidus* < *gelu,* frost.] Very cold <*gelid* Arctic waters> **—ge·lid'i·ty** (jə-lĭd'ĭ-tē), **gel'id·ness** *n.* **—gel'id·ly** *adv.*

**gel·ig·nite** (jĕl'ĭg-nīt') *n.* [GEL(ATIN) + Lat. *ignis,* fire + -ITE.] An explosive mixture of nitroglycerin, guncotton, wood pulp, and potassium nitrate.

**gelt¹** (gĕlt) *n.* [Yiddish < OHG, recompense.] *Slang.* Money.

**gelt²** (gĕlt) *v. var. p.t. & p.p. of* GELD¹.

**gem** (jĕm) *n.* [ME *gemme* < Lat. *gemma.*] **1.** A precious or semiprecious stone, esp. one cut and polished. **2. a.** Something valued for its beauty or perfection. **b.** A beloved or highly prized person. —*vt.* **gemmed, gem·ming, gems.** To adorn with or as if with gems.

**Ge·ma·ra** (gə-mär'ə, -môr'ə) *n.* [Aram. *gəmārā,* completion < *gəmār,* he finished.] The second part of the Talmud, consisting chiefly of commentary on the Mishnah. **—Ge·ma'ric** *adj.* **—Ge·ma'rist** *n.*

**gem·i·nate** (jĕm'ə-nāt') *v.* **-nat·ed, -nat·ing, -nates.** [Lat. *geminare, geminat-* < *geminus,* twin.] —*vt.* To arrange in pairs or to double. —*vi.* To occur in pairs. —*adj.* (jĕm'ə-nĭt, -nāt'). Forming a pair : DOUBLED. **—gem'i·na'tion** *n.*

**Gem·i·ni** (jĕm'ə-nī', -nē') *n.* [Lat. < pl. of *geminus,* twin.] **1.** *Astron.* A constellation in the Northern Hemisphere. **2. a.** The third sign of the zodiac. **b.** One born under this sign.

**gem·ma** (jĕm'ə) *n., pl.* **gem·mae** (jĕm'ē') [NLat. < Lat., bud.] An asexual reproductive structure, as in liverworts or the hydra, consisting of a cell or group of cells capable of developing into a new individual : BUD.

**gem·mate** (jĕm'āt') *adj.* [< Lat. *gemmare, gemmat-,* to bud < *gemma,* bud.] Having or reproducing by gemmae. —*vi.* **-mat·ed, -mat·ing, -mates.** To produce gemmae or reproduce by gemmae. **—gem·ma'ceous** (jĕ-mā'shəs) *adj.* **—gem·ma'tion** (jĕ-mā'shən) *n.*

**gem·mip·a·rous** (jĕ-mĭp'ər-əs) *adj.* [NLat. *gemmiparus* : Lat. *gemma,* bud + Lat. *parere,* to bring forth.] Reproducing by gemmae. **—gem·mip'a·rous·ly** *adv.*

**gem·mol·o·gy** (jĕ-mŏl'ə-jē) *n. var. of* GEMOLOGY.

**gem·mu·la·tion** (jĕm'yə-lā'shən) *n.* Production of or reproduction by gemmules.

**gem·mule** (jĕm'yōōl) *n.* [Fr. < Lat. *gemmula,* dim. of *gemma,* bud.] **1.** A small gemma, esp. a reproductive structure in some sponges that remains dormant through the winter and later develops into a new individual. **2.** A hypothetical particle of heredity postulated in the theory of pangenesis. **—gem'mu·lif'er·ous** (-yōō-lĭf'ər-əs) *adj.*

**gem·my** (jĕm'ē) *adj.* **-mi·er, -mi·est. 1.** Full of or set with gems. **2.** Glittering, as a gem.

**gem·ol·o·gy** *or* **gem·mol·o·gy** (jĕ-mŏl'ə-jē) *n.* Study of gems. **—gem'o·log'i·cal** (jĕm'ə-lŏj'ĭ-kəl) *adj.* **—gem·ol'o·gist** *n.*

**ge·mot** *also* **ge·mote** (gə-mōt') *n.* [OE *gemōt* : *ge-,* together + *mōt,* assembly.] An English public meeting or local judicial assembly before the Norman Conquest.

**gems·bok** (gĕmz'bŏk') *n.* [Afr. < Du. < G. *Gemsbock* : *Gemse,* chamois + *Bock,* buck < OHG *boc.*] An antelope, *Oryx gazella* of arid regions of southern Africa, having long, sharp, straight horns.

**gem·stone** (jĕm'stōn') *n.* A precious or semiprecious stone that may be used as a jewel when cut and polished.

**ge·müt·lich** (gə-müt'lĭKH) *adj.* [G. < *gemüt,* spirit.] Having a feeling of warmth or congeniality : FRIENDLY.

**ge·müt·lich·keit** (gə-mōōt'lĭKH-kīt') *n.* [G. < *gemütlich,* gemütlich.] Cordial regard : AMICABILITY.

**-gen** or **-gene** *suff.* [Fr. *-gène* < Gk. *-genēs,* born.] **1.** Producer <an*drogen*> **2.** One that is produced <phos*gene*>
**gen·darme** (zhän′därm′) *n.* [Fr. < *gens d'armes,* men of arms.] **1.** A member of a French national police organization forming a branch of the armed forces with responsibility for general law enforcement. **2.** A police officer.
**gen·dar·me·rie** *also* **gen·dar·mer·y** (zhän-där′mə-rē) *n.* [Fr. < *gendarme,* gendarme.] A group of gendarmes.
**gen·der** (jĕn′dər) *n.* [ME *gendre* < OFr., kind < Lat. *genus.*] **1. a.** A set of two or more categories, as masculine, feminine, and neuter, into which words are divided according to sex, animation, psychological associations, or other characteristic and that govern agreement with or the selection of modifiers, referents, or grammatical forms. **b.** One category of such a set. **c.** The classification of a word or grammatical form in such a category. **d.** The distinguishing form or forms used. **2.** Classification of sex. —*vt.* **-dered, -der·ing, -ders.** *Archaic.* To engender.
**gene** (jēn) *n.* [G. *Gen,* short for *Pangen* : Gk. *pan,* all + Gk. *-genēs,* born.] A functional hereditary unit that occupies a fixed location on a chromosome, has a specific influence on phenotype, and is capable of undergoing mutation to various allelic forms.
**-gene** *suff. var. of* -GEN.
**ge·ne·al·o·gy** (jē′nē-ŏl′ə-jē, -ăl′ə-jē, jĕn′ē-) *n., pl.* **-gies.** [ME *genealogie* < OFr. < LLat. *genealogia* < Gk. : *genea,* family + *-logia,* -logy.] **1.** A record or table of familial descent. **2.** Direct descent from an ancestor : LINEAGE. **3.** The study of family histories. **—ge′ne·a·log′i·cal** (-ə-lŏj′ĭ-kəl) *adj.* **—ge′ne·a·log′i·cal·ly** *adv.* **—ge′ne·al′o·gist** *n.*
**gen·er·a** (jĕn′ər-ə) *n. pl. of* GENUS.
**gen·er·a·ble** (jĕn′ər-ə-bəl) *adj.* [ME *generabill* < Lat. *generabilis* < *generare,* to produce < *genus,* birth.] Capable of being generated.
**gen·er·al** (jĕn′ər-əl) *adj.* [ME < Lat. *generalis* < *genus,* kind.] **1.** Pertaining to, concerned with, or applicable to the whole or every member of a class or category. **2.** Affecting or characteristic of the majority : PREVALENT <*general* unrest> **3.** Being usu. the case. **4. a.** Not limited in scope, area, or application <as a *general* rule> **b.** Not limited to one class or set <*general* studies> **5.** Involving only the main features <a *general* description> **6.** Highest or superior in rank <the *general* manager> —*n.* **1. a.** An officer in the U.S. Army, Air Force, or Marine Corps holding a rank above colonel. **b.** A military officer holding a rank just below field marshal in England, Canada, and certain other countries. **2.** Something, as a condition, principle, or fact, that embraces or is applicable to the whole. **3.** *Archaic.* The public <"'twas caviare to the *general*" —Shakespeare> **—in general.** Generally. **—gen′er·al·ness** *n.*

☆ ***syns:*** GENERAL, ALL-INCLUSIVE, ALL-ROUND, BROAD, BROAD-SPECTRUM, COMPREHENSIVE, EXPANSIVE, EXTENSIVE, GLOBAL, INCLUSIVE, OVERALL, SWEEPING, WIDESPREAD *adj. core meaning* : covering a wide scope <*general* discontent>

▲ word history: The Latin adjective *generalis,* the ultimate ancestor of the English word *general,* meant literally "pertaining to the whole *genus* (kind or class)." The English noun *general* meaning "a military officer" preserves something of the Latin sense. It originally designated a general officer, the commander of the general, or entire, army. Since such an officer was of a very high rank, the term *general* was also prefixed to the titles of other ranks of officers to indicate superiority of rank and command.

**general anesthetic** *n.* An anesthetic that anesthetizes the entire body and induces unconsciousness.
**general assembly** *n.* **1.** A legislative body, esp. a U.S. state legislature. **2. General Assembly.** The chief deliberative body of the United Nations. **3.** The supreme governing body of some religious denominations.
**General Court** *n.* **1.** A legislative body with judicial powers in Colonial times. **2.** The state legislature in Massachusetts and New Hampshire.
**general court-martial** *n.* A court-martial for trying major offenses and composed of at least five officers.
**gen·er·al·cy** (jĕn′ər-əl-sē) *n., pl.* **-cies.** GENERALSHIP 1.
**general delivery** *n.* **1.** A post office department that holds mail for addressees until it is called for. **2.** Mail sent to general delivery.
**general election** *n.* An election involving most or all constituencies of a state or nation in the choice of candidates.
**gen·er·al·is·si·mo** (jĕn′ər-ə-lĭs′ə-mō′) *n., pl.* **-mos.** [Ital., superl. of *generale,* a general < Lat. *generalis,* general < *genus,* kind.] The commander in chief of all the armed forces in certain countries.
**gen·er·al·ist** (jĕn′ər-ə-lĭst) *n.* One with broad general knowledge and skills in several fields.
**gen·er·al·i·ty** (jĕn′ə-răl′ĭ-tē) *n., pl.* **-ties. 1.** The quality or state of being general. **2.** GENERALIZATION 2. **3.** A vague, imprecise statement or idea. **4.** The greater portion or number : MAJORITY.
**gen·er·al·i·za·tion** (jĕn′ər-ə-lə-zā′shən) *n.* **1.** An act or instance of generalizing. **2.** A general principle, statement, or idea.
**gen·er·al·ize** (jĕn′ər-ə-līz′) *v.* **-ized, -iz·ing, -iz·es.** —*vt.* **1. a.** To reduce to a general form, class, or law. **b.** To make indefinite or unspecific. **2. a.** To infer from many particulars. **b.** To draw inferences or a general conclusion from. **3. a.** To make generally applicable. **b.** To popularize. —*vi.* **1. a.** To form a concept inductively. **b.** To form general conclusions. **2.** To speak or think in generalities. **3.** *Med.* To spread through the body, as a usu. localized disease.
**gen·er·al·ized** (jĕn′ər-ə-līzd′) *adj.* Not well-adapted to a specific environment or function : UNDIFFERENTIATED.
**gen·er·al·ly** (jĕn′ər-ə-lē) *adv.* **1.** For the most part : WIDELY <*generally* admired> **2.** As a rule : USUALLY. **3.** In disregard of particular instances and details <*generally* speaking>
**general officer** *n.* GENERAL 1a.
**General of the Air Force** *n.* A general having the highest rank in the U.S. Air Force and an insignia of five stars.
**General of the Army** *n.* A general having the highest rank in the U.S. Army and an insignia of five stars.
**general paresis** *n.* A syphilitic brain disorder characterized by mental deterioration, speech disturbances, and progressive muscular weakness.
**general practitioner** *n.* A physician who treats a wide variety of medical problems.
**gen·er·al-pur·pose** (jĕn′ər-əl-pûr′pəs) *adj.* Having many uses.
**general relativity** *n.* The geometric theory of gravitation developed by Albert Einstein, incorporating and extending the special theory of relativity to accelerated frames of reference and introducing the principle that gravitational and inertial forces are equivalent.
**general semantics** *n. (sing. or pl. in number).* A doctrine proposed by Alfred Korzybski (1879–1950) that presents a method of improving human behavior through a more critical use of words and symbols.
**gen·er·al·ship** (jĕn′ər-əl-shĭp′) *n.* **1.** The rank, office, or tenure of a general. **2.** Leadership or skill in the conduct of a war. **3.** Skillful management.
**general staff** *n.* A group of officers charged with assisting the commander of a division or larger unit in planning and supervising military operations.
**general store** *n.* A retail store that sells a wide variety of merchandise but is not subdivided into departments.
**general term** *n. Math.* An expression that represents the successive terms of a sequence or series by some combination of variables and constants which, when successive integers are substituted for one of the variables, yields back the sequence or series.
**gen·er·ate** (jĕn′ə-rāt′) *vt.* **-at·ed, -at·ing, -ates.** [Lat. *generare, generat-,* to produce < *genus,* birth.] **1.** To bring into existence : PRODUCE. **2.** To engender (offspring) : BEGET. **3.** To form (a geometric figure) by describing a curve or surface. **4.** *Computer Sci.* To produce a program by instructing a computer to follow given parameters with a skeleton program. **—gen′er·a′tive** (jĕn′ə-rā′tĭv, -ər-ə-tĭv) *adj.*
**gen·er·a·tion** (jĕn′ə-rā′shən) *n.* **1.** The act or process of generating, esp. procreation, origination, or production. **2.** Offspring sharing a common parent or parents and forming a single stage of descent. **3.** A class of objects derived from a preceding class <the new *generation* of microcomputers> **4. a.** A group of contemporaneous individuals. **b.** A group of individuals considered as sharing a common contemporaneous cultural or social attribute <the beat *generation*> **5.** The average time interval between the birth of parents and the birth of their offspring. **6.** *Computer Sci.* The technique of generating programs. **—gen·er·a′tion·al** *adj.*
**generation gap** *n.* The differences in values and attitudes between one generation and the next, esp. a generation of adolescents and that of their parents.
**generative grammar** *n.* A system of rules intended to produce all the well-formed sentences of a language when applied to its lexicon, esp. such a system whose syntactic component is generated successively by rules for construction of phrases containing a semantic component, deep structure, and by rules for production of a phonological component, surface structure, by transforming one grammatical structure into another semantically equivalent structure.
**gen·er·a·tor** (jĕn′ə-rā′tər) *n.* **1.** One that generates. **2.** A machine that converts mechanical energy into electrical energy. **3.** An apparatus that generates vapor or gas. **4.** A generatrix. **5.** *Computer Sci.* A routine that performs a generating function.
**gen·er·a·trix** (jĕn′ə-rā′trĭks) *n., pl.* **-er·a·tri·ces** (-ə-rā′trĭ-sēz′, -ər-ə-trī′sēz). A geometric element that generates a geometric figure, esp. a line that generates a surface by moving in a specified way.
**ge·ner·ic** (jə-nĕr′ĭk) *adj.* [Fr. *générique* < Lat. *genus,* kind.] **1.** Pertaining to or describing an entire group or class : GENERAL. **2.** *Biol.* Of or pertaining to a genus. **3.** Not bearing a trademark or trade name <*generic* raisins> **—ge·ner′i·cal·ly** *adv.*
**gen·er·os·i·ty** (jĕn′ə-rŏs′ĭ-tē) *n., pl.* **-ties. 1.** The quality of being generous, esp. liberality in giving : MUNIFICENCE. **2.** Nobility of thought or behavior. **3.** Amplitude : abundance. **4.** A generous act.
**gen·er·ous** (jĕn′ər-əs) *adj.* [OFr. *genereux,* of noble birth < Lat. *generosus* < *genus,* birth.] **1.** Noble and forbearing in thought or behavior : MAGNANIMOUS. **2. a.** Liberal in giving or sharing. **b.** Marked by bounteous giving. **3.** Marked by abundance : AMPLE.

4. Rich-flavored, as wine. 5. *Obs.* Of noble lineage. **—gen'er·ous·ly** *adv.* **—gen'er·ous·ness** *n.*

☆ ***syns:*** **1.** GENEROUS, BIG, BIG-HEARTED, GREATHEARTED, MAGNANIMOUS, SELFLESS, UNSELFISH *adj. core meaning* **:** willing to give of oneself and one's possessions <a *generous* contributor to charity> **2.** GENEROUS, HANDSOME, LIBERAL, MUNIFICENT, UNSPARING *adj. core meaning* **:** marked by bounteous giving <a *generous* allowance for the children> **3.** GENEROUS, ABUNDANT, AMPLE, BOUNTEOUS, BOUNTIFUL, COPIOUS, LIBERAL, PLENTEOUS, PLENTIFUL *adj. core meaning* **:** marked by abundance <a *generous* serving of peas>

**gen·e·sis** (jĕn'ĭ-sĭs) *n., pl.* **-ses** (-sēz') [Lat. < Gk.] **1.** A coming into being **:** ORIGIN. **2. Genesis.** —See table at BIBLE.

**-genesis** *suff.* [NLat. < Lat. < Gk., birth, origin.] Origin **:** production <*abiogenesis*>

**gene-splic·ing** (jēn'splī'sĭng) *n.* The process in which DNA fragments from one or more different organisms are combined and made to function within the cells of a host organism.

**gen·et[1]** (jĕn'ĭt, jə-nĕt') *n.* [ME *genete* < OFr. < Ar. *jarnayṭ.*] Any of several Old World carnivorous mammals of the genus *Genetta,* with dark-spotted yellowish or grayish fur and a long ringed tail.

**gen·et[2]** (jĕn'ĭt) *n. var. of* JENNET.

**ge·net·ic** (jə-nĕt'ĭk) *also* **ge·net·i·cal** (-ĭ-kəl) *adj.* [< GENESIS.] **1.** Of or relating to origin or development. **2. a.** Of or relating to genetics. **b.** Affecting or affected by genes. **—ge·net'i·cal·ly** *adv.*

**genetic code** *n.* The information coded within the nucleotide sequences of RNA and DNA that specifies the amino acid sequence in the synthesis of proteins and on which heredity is based. **—genetic coding** *n.*

**genetic counseling** *n.* Counseling of prospective parents on the probabilities of inherited diseases occurring in offspring and on the diagnosis and treatment of such diseases. **—genetic counselor** *n.*

**genetic engineering** *n.* Intentional alteration of genetic material or gene-controlled processes to prevent or ameliorate hereditary defects.

**ge·net·i·cist** (jə-nĕt'ĭ-sĭst) *n.* A specialist in genetics.

**ge·net·ics** (jə-nĕt'ĭks) *n.* **1.** *(sing. in number).* The biology of heredity, esp. the study of mechanisms of hereditary transmission and variation of organismal characteristics. **2.** *(pl. in number).* Genetic constitution of an individual, group, or class.

**Ge·ne·va bands** (jə-nē'və) *pl.n.* [After *Geneva,* Switzerland.] Two strips of white cloth hanging from the collar of some clerical or academic vestments.

**Geneva Convention** *n.* An agreement first formulated at an international convention held in Geneva, Switzerland, in 1864, establishing rules for the treatment of prisoners of war and the sick or wounded.

**Geneva cross** *n.* A red Greek or St. George's cross on a white ground used as a symbol by the Red Cross and a sign of neutrality.

**Geneva gown** *n.* [After *Geneva,* Switzerland.] A loose black academic or clerical gown with wide sleeves.

**Ge·ne·van** (jə-nē'vən) *also* **Gen·e·vese** (jĕn'ə-vēz', -vēs') *—adj.* **1.** Of or pertaining to Geneva, Switzerland. **2.** Of or pertaining to Geneva during the time of Calvin **:** CALVINISTIC. —*n.* **1.** A native or inhabitant of Geneva, Switzerland. **2.** A Calvinist.

**gen·ial[1]** (jēn'yəl) *adj.* [Lat. *genialis,* festive < *genius,* spirit of festivity.] **1.** Pleasant or friendly **:** KINDLY. **2.** Conducive to life, growth, or comfort **:** MILD <"the *genial* sunshine . . . saturating his miserable body with its warmth" —Jack London> **3.** *Obs.* Characteristic of or pertaining to genius. **4.** *Obs.* Of or pertaining to marriage **:** NUPTIAL. **—ge'ni·al'i·ty** (jē'nē-ăl'ĭ-tē), **gen'ial·ness** *n.* **—gen'ial·ly** *adv.*

**ge·ni·al[2]** (jĭ-nī'əl) *n.* [< Gk. *geneion,* chin < *genus,* jaw.] Of or relating to the chin.

**gen·ic** (jē'nĭk, jĕn'ĭk) *adj.* Of, pertaining to, produced by, or being a gene. **—gen'ic·al·ly** *adv.*

**-genic** *suff.* [-GEN + -IC.] **1.** Producing **:** generating <*dysgenic*> **2.** Produced or generated by <*cryptogenic*> **3.** Suitable for production or reproduction by a specified medium <*telegenic*>

**ge·nic·u·late** (jə-nĭk'yə-lĭt) *also* **ge·nic·u·lat·ed** (-lā'tĭd) *adj.* [Lat. *geniculatus,* with bended knee < *geniculum,* dim. of *genu,* knee.] **1.** Bent at an abrupt angle. **2.** Jointed so as to be capable of bending at an abrupt angle. **—ge·nic'u·late·ly** *adv.* **—ge·nic'u·la'tion** *n.*

**ge·nie** (jē'nē) *n.* [Fr. *génie,* spirit < Lat. *genius,* guardian spirit.] **1.** A supernatural creature who does one's bidding. **2.** *var. of* JINNI.

**gen·ip** (jĕn'əp) *n.* [Sp. *genipa,* a kind of palm, prob. of Carib orig.] **1. a.** A tropical American tree, *Melicocca bijuga,* with small greenish-white flowers and small yellow fruit. **b.** The sweet, edible fruit of the genip. **2.** The genipap.

**gen·i·pap** (jĕn'ə-păp') *n.* [Portuguese *genipapo,* from Tupi.] **1.** An evergreen tree, *Genipa americana* of the West Indies, with yellowish-white flowers and bearing edible fruit. **2.** The reddish-brown fruit of the genipap.

**gen·i·tal** (jĕn'ĭ-təl) *adj.* [ME < Lat. *genitalis* < *gignere,* to beget.] **1.** Of or relating to biological reproduction. **2.** Of or relating to the genitalia. *—pl.n.* **genitals.** The genitalia. **—gen'i·tal·ly** *adv.*

**gen·i·ta·li·a** (jĕn'ĭ-tā'lē-ə, -tāl'yə) *pl.n.* [Lat., neuter pl. of *genitalis,* generative < *gignere,* to beget.] The reproductive organs, esp. the external sex organs.

**gen·i·ti·val** (jĕn'ĭ-tī'vəl) *adj.* Of, pertaining to, or in the genitive case. **—gen'i·ti'val·ly** *adv.*

**gen·i·tive** (jĕn'ĭ-tĭv) *adj.* [ME *genetif* < Lat. *genetivus* < *gignere,* to beget.] **1.** Of, relating to, or designating a grammatical case that expresses possession, measurement, or source. **2.** Of or relating to an affix or a construction, as a prepositional phrase, characteristic of the genitive case. —*n.* **1.** The genitive case. **2.** A genitive construction or word form.

**gen·i·tor** (jĕn'ĭ-tər) *n.* [ME *genitour* < OFr. *genitor* < Lat. < *gignere,* to beget.] **1.** One who begets or creates. **2.** A natural father as opposed to the socially responsible foster father in certain cultures.

**gen·i·to·u·ri·nar·y** (jĕn'ĭ-tō-yŏŏr'ə-nĕr'ē) *adj.* [GENIT(AL) + URINARY.] Relating to the genital and urinary organs or their functions.

**gen·ius** (jēn'yəs) *n., pl.* **-ius·es.** [Lat., guardian spirit.] **1. a.** Exceptional intellectual and creative power. **b.** One gifted with such power. **2. a.** A natural inclination or talent <had a *genius* for singing> **b.** The possessor of an inclination or talent <a *genius* at diplomacy> **3.** The prevailing spirit or character, as of a place, person, time, or group <the *genius* of the Elizabethan poets> **4.** *pl.* **ge·ni·i** (jē'nē-ī'). *Rom. Myth.* A tutelary deity or guardian spirit of a person or place. **5.** One with great influence over another. **6.** A jinni or a demon.

**ge·ni·us lo·ci** (jē'nē-əs lō'sī') *n.* [Lat.] **1.** A guardian deity of a specific locality. **2.** The distinctive atmosphere or particular character of a place.

**gen·o·cide** (jĕn'ə-sīd') *n.* [Gk. *genos,* race + -CIDE.] Systematic, planned annihilation of a racial, political, or cultural group. **—gen'o·cid'al** (-sīd'l) *adj.* **—gen'o·cid'al·ly** *adv.*

**ge·nome** (jē'nōm') *also* **ge·nom** (-nŏm) *n.* [G. *Genom* **:** *Gen,* gene + *(Chromos)om,* chromosome.] A complete haploid set of chromosomes. **—ge·nom'ic** (-nŏm'ĭk) *adj.*

**gen·o·type** (jĕn'ə-tīp', jē'nə-) *n.* [Gk. *genos,* race + -TYPE.] **1.** The genetic constitution of an organism. **2.** A group or class of organisms having the same genetic constitution. **3.** The type species of a genus. **—gen'o·typ'ic** (-tĭp'ĭk), **gen'o·typ'i·cal** *adj.* **—gen'o·typ'i·cal·ly** *adv.* **—gen'o·ty·pic'i·ty** (-tī-pĭs'ĭ-tē) *n.*

**-genous** *suff.* [-GEN + -OUS.] **1.** Producing **:** generating <hematogenous> **2.** Produced by or in a specified manner <hypogenous>

**gen·re** (zhän'rə) *n.* [Fr. < OFr., kind < Lat. *genus.*] **1.** Type **:** class. **2. a.** An artistic category marked by a distinctive style, form, or content, esp. a painting style depicting common everyday scenes and subjects. **b.** A distinctive category of literary composition.

**gen·ro** (gĕn'rō') *n., pl.* **-ros.** [J. *genrō* **:** Chin. *yuan*[2], first + Chin. *lao*[3], elder.] A group of elder statesmen, once functioning as advisers to the Japanese emperor.

**gens** (jĕnz) *n., pl.* **gen·tes** (jĕn'tēz') [Lat.] **1.** The patrilinear clan forming the basic unit of the Roman tribe and having orig. a common name, land, cult, and burial ground. **2.** An exogamous patrilineal clan.

**gent[1]** (jĕnt) *adj.* [ME < OFr. < Lat. *genitus,* p.part. of *gignere,* to beget.] *Obs.* Graceful **:** elegant.

**gent[2]** (jĕnt) *n.* [Short for GENTLEMAN.] *Informal.* A fellow.

**gen·ta·mi·cin** (jĕn'tə-mī'sĭn) *n.* [Alteration of *gentamycin* **:** GENT(I)A(N VIOLET) + MYCIN.] A broad-spectrum antibiotic derived from an actinomycete of the genus *Micromonospora.*

**gen·teel** (jĕn-tēl') *adj.* [OFr. *gentil.* —see GENTLE.] **1.** Refined **:** polite. **2.** Free from rudeness or vulgarity. **3.** Elegantly fashionable. **4. a.** Striving to convey a refined, respectable appearance. **b.** Characterized by affected, rather prudish refinement. **—gen·teel'ly** *adv.* **—gen·teel'ness** *n.*

**gen·teel·ism** (jĕn-tēl'ĭz'əm) *n.* A word or expression considered by its user to be genteel.

**gen·tes** (jĕn'tēz') *n. pl. of* GENS.

**gen·tian** (jĕn'shən) *n.* [ME *gencian* < Lat. *gentiana.*] **1.** Any of numerous plants of the genus *Gentiana,* with showy blue flowers. **2.** The dried rhizome and roots of a yellow-flowered European gentian, *G. lutea,* occas. used as a tonic.

**gentian violet** *n.* A purple dye used chiefly as a biological stain and bactericide.

**gen·tile** (jĕn'tīl') *n.* [ME *gentil* < LLat. *gentilis,* pagan < Lat., of the same clan < *gens,* clan.] **1. Gentile. a.** One not of the Jewish faith. **b.** One of a non-Jewish nation. **2. Gentile.** A Christian. **3.** A pagan **:** heathen. **4. Gentile.** One who is not a Mormon. **5.** A member of a gens. *—adj.* **1.** Of or pertaining to a Gentile. **2.** Of or pertaining to tribal society, esp. of the gens.

**gen·ti·lesse** (jĕn'tə-lĕs') *n.* [ME < OFr. < *gentil,* noble. —see GENTLE.] *Archaic.* Refinement and courtesy developed by good breeding.

**gen·til·i·ty** (jĕn-tĭl'ĭ-tē) *n.* [ME *gentilete,* nobility of birth < OFr. < *gentil,* noble. —see GENTLE.] **1.** The state of being genteel. **2.** The state of being born to the gentry. **3.** Persons of the upper class as a whole. **4.** A usu. obsessive attempt to convey or maintain a refined, respectable appearance.

**gen·tle** (jĕn'tl) *adj.* **-tler, -tlest.** [ME *gentil,* noble < OFr. < Lat.

*gentilis*, of the same clan < *gens*, clan.] **1.** Considerate or kindly : AMIABLE. **2.** Not harsh, severe, or violent <a *gentle* breeze> **3.** Easily managed or handled : DOCILE. **4.** Not steep or sudden : GRADUAL <a *gentle* incline to the shore> **5.** Well-born. **6.** *Archaic.* Noble : chivalrous. —*n.* **1.** *Archaic.* One of gentle birth or station. **2.** The larva of a bluebottle fly. —*vt.* **-tled, -tling, -tles. 1.** To make gentle : MOLLIFY. **2.** To tame or break (a horse). **3.** *Obs.* To raise to the status of a noble. **—gen'tle·ness** *n.* **—gen'tly** *adv.*

☆ ***syns:* 1.** GENTLE, MILD, SOFT, SOFTHEARTED, TENDER, TENDERHEARTED *adj. core meaning* : kind and considerate <a *gentle* parent> **2.** GENTLE, DOCILE, MEEK, TAME *adj. core meaning* : easily controlled, led, or handled <a *gentle* mare>

▲ word history: The now archaic sense of *gentle*, "noble," is the first to be recorded, having been borrowed from French *gentil* in the 12th or 13th century. In the 16th century the French word was borrowed again as *genteel*, meaning "well-born," although not necessarily "noble." The French word *gentil* is derived from Latin *gentilis*, "belonging to the same nation or clan." The Latin word was also used to indicate those belonging to a nation or clan different from one's own, and hence it came to mean a non-Roman and, in Christian writings, a non-Jew or non-Christian. The Latin word *gentilis* in this extended sense is the source of English *gentile*.

**gentle breeze** *n.* A wind having a speed between 8 and 12 miles or approx. 12.9 and 19.3 kilometers per hour.

**gen·tle·folk** (jĕn'tl-fōk') *also* **gen·tle·folks** (-fōks') *pl.n.* Well-bred persons of good family background.

**gen·tle·man** (jĕn'tl-mən) *n.* **1.** A man of gentle or noble birth or superior social position. **2.** A polite, gracious, or considerate man having high standards of propriety or correct behavior. **3.** A man of independent means who does not or need not work. **4.** A man who considers manual labor to be beneath him. **5. a.** A man. **b. gentlemen.** —Used as a written or oral form of address esp. for a group of men. **6.** A manservant. **—gen'tle·man·ly** *adj.*

**gen·tle·man-at-arms** (jĕn'tl-mən-ət-ärmz') *n., pl.* **gen·tle·men-at-arms** (-mĭn-ət-ärmz'). Any of a military corps of 40 gentlemen who attend the British sovereign as a ceremonial guard on state occasions.

**gentleman farmer** *n., pl.* **gentlemen farmers.** A man who farms chiefly for pleasure rather than income.

**gentleman's agreement** *n.* An agreement guaranteed only by the honor of the participants.

**gentleman's gentleman** *n.* GENTLEMAN 6.

**gen·tle·wom·an** (jĕn'tl-wo͝om'ən) *n.* **1.** A woman of gentle or noble birth or superior social position. **2.** A polite, gracious, or considerate woman. **3.** A woman acting as a personal attendant to a lady of rank.

**Gen·too** (jĕn-to͞o') *n., pl.* **-toos.** [Port. *gentio* < *gentio*, pagan < LLat. *gentilis.* —see GENTILE.] *Archaic.* A Hindu.

**gen·tri·fi·ca·tion** (jĕn'trə-fĭ-kā'shən) *n.* [GENTRY + -FICATION.] Restoration of deteriorated urban property esp. in working-class neighborhoods by the middle and upper classes. **—gen'tri·fy'** *v.* **(-fied, -fy·ing, -fies).**

**gen·try** (jĕn'trē) *n.* [ME *gentri*, nobility of birth < OFr. *genterie* < *gentil*, noble. —see GENTLE.] **1.** People of gentle birth, good breeding, or high social position. **2.** The English upper middle classes. **3.** People of a specified class or group <the suburban *gentry*>

**gen·u·flect** (jĕn'yə-flĕkt') *vi.* **-flect·ed, -flect·ing, -flects.** [LLat. *genuflectere* : Lat. *genu*, knee + Lat. *flectere*, to bend.] **1.** To bend the knee in a kneeling or half-kneeling position, as in worship. **2.** To exhibit a deferential or obsequious manner or attitude. **—gen'u·flec'tion** (-flĕk'shən) *n.*

**gen·u·ine** (jĕn'yo͞o-ĭn) *adj.* [Lat. *genuinus*, natural.] **1. a.** Actually possessing the attributes reputed <a *genuine* French wine> **b.** Actually produced by or arising from the reputed source <a *genuine* signature> **2.** Not counterfeit <*genuine* money> **3.** Free from hypocrisy or dishonesty <a *genuine* liking> **4.** Of pure or original stock <a *genuine* Hawaiian> **—gen'u·ine·ly** *adv.* **—gen'u·ine·ness** *n.*

**ge·nus** (jē'nəs) *n., pl.* **gen·er·a** (jĕn'ər-ə) [Lat., kind.] **1.** *Biol.* A taxonomic category ranking below a family and above a species, used in taxonomic nomenclature, either alone or followed by a Latin adjective or epithet, to form the name of a species. **2.** *Logic.* A class of objects divided into subordinate species having certain common attributes.

**-geny** *suff.* [Gk. *-geneia* < *-genēs*, born.] Production : generation : origin <onto*geny*>

**geo-** *or* **ge-** *pref.* [Gk. *geō-* < *gē*, earth.] **1.** Earth <*geo*centric> **2.** Geography <*geo*political>

**ge·o·bot·a·ny** (jē'ō-bŏt'n-ē) *n.* Phytogeography. **—ge'o·bo·tan'ic** (-bə-tăn'ĭk), **ge'o·bo·tan'i·cal** *adj.* **—ge'o·bo·tan'i·cal·ly** *adv.* **—ge'o·bot'a·nist** (-bŏt'n-ĭst) *n.*

**ge·o·cen·tric** (jē'ō-sĕn'trĭk) *adj.* **1.** Relating to, measured from, or observed from the center of the earth. **2.** Having the earth as a center. **—ge'o·cen'tri·cal·ly** *adv.*

**ge·o·chem·is·try** (jē'ō-kĕm'ĭ-strē) *n.* The chemistry of the composition and alterations of the earth's crust. **—ge'o·chem'i·cal** *adj.* **—ge'o·chem'i·cal·ly** *adv.* **—ge'o·chem'ist** *n.*

**ge·o·chro·nol·o·gy** (jē'ō-krə-nŏl'ə-jē) *n.* The chronology of the earth's history as governed by geological events. **—ge'o·chron'o·log'ic** (-krŏn'ə-lŏj'ĭk), **ge'o·chron'o·log'i·cal** (-ĭ-kəl) *adj.* **—ge'o·chron'o·log'i·cal·ly** *adv.* **—ge'o·chro·nol'o·gist** *n.*

**ge·o·chro·nom·e·try** (jē'ō-krə-nŏm'ə-trē) *n.* Measurement of geologic time, as through isotopic radioactive decay. **—ge'o·chron'o·met'ric** (-krŏn'ə-mĕt'rĭk) *adj.*

**ge·o·co·ro·na** (jē'ō-kə-rō'nə) *n.* The outermost part of the earth's atmosphere, consisting chiefly of ionized hydrogen.

**ge·ode** (jē'ōd') *n.* [Lat. *geodes*, a precious stone < Gk. *geōdēs*, earthlike : *gē*, earth + *eidos*, shape.] A small, hollow, usu. spheroidal rock with crystals lining the inside wall.

**ge·o·des·ic** (jē'ə-dĕs'ĭk) *adj.* **1.** *Math.* Of or relating to the geometry of geodesics. **2.** Geodetic. —*n. Math.* A curve whose principal normal at any point is the normal to the surface on which the curve occurs in three-dimensional Euclidean space : the shortest line between two points on a mathematically derived surface.

**geodesic dome** *n.* A domed or vaulted structure of lightweight straight elements that form interlocking polygons.

**ge·od·e·sy** (jē-ŏd'ĭ-sē) *n.* [Ult. < Gk. *geōdaisia* : *gē*, earth + *daiesthai*, to divide.] The geologic science of the size and shape of the earth. **—ge·od'e·sist** *n.*

**ge·o·det·ic** (jē'ə-dĕt'ĭk) *also* **ge·o·det·i·cal** (-ĭ-kəl) *adj.* **1.** Of or relating to geodesy. **2.** Geodesic. **—ge'o·det'i·cal·ly** *adv.*

**geo·duck** (go͞o'ē-dŭk') *n.* [Chinook Jargon *go-duck.*] A very large edible clam, *Panope generosa*, of the northwestern American Pacific coast.

**ge·og·no·sy** (jē-ŏg'nə-sē) *n.* [GEO- + Gk. *-gnōsia*, knowledge < *gnōsis* < *gignōskein*, to know.] Scientific study of the organization and structure of the earth and its materials.

**ge·o·graph·ic** (jē'ə-grăf'ĭk) *also* **ge·o·graph·i·cal** (-ĭ-kəl) *adj.* **1.** Of or relating to geography. **2.** Concerning the topography of a specific region. **—ge'o·graph'i·cal·ly** *adv.*

**geographic mile** *n.* A nautical mile.

**ge·og·ra·phy** (jē-ŏg'rə-fē) *n., pl.* **-phies.** [Lat. *geographia* < Gk. *geōgraphia* : *gē*, earth + *graphein*, to write.] **1.** Study of the earth and its features and of the distribution of life on the earth, including human life and the effects of human activity. **2.** Geographic characteristics of an area. **3.** A book on geography. **4.** An ordered arrangement of constituent parts <charting a *geography* of the mind> **—ge·og'ra·pher** *n.*

**ge·oid** (jē'oid') *n.* [G. < Gk. *geoeidēs*, earthlike : *gē*, earth + *-eidēs*, -oid.] The hypothetical surface of the earth that coincides everywhere with mean sea level. **—ge·oid'al** (-oid'l) *adj.*

**geologic time** *n.* The period covering the earth's geologic history.

**ge·ol·o·gize** (jē-ŏl'ə-jīz') *vi.* **-gized, -giz·ing, -giz·es.** To study geology or make geologic investigations.

**ge·ol·o·gy** (jē-ŏl'ə-jē) *n., pl.* **-gies. 1.** Scientific study of the origin, history, and structure of the earth. **2.** Structure of a specific region of the earth's surface. **3.** A book on geology. **4.** Scientific study of the origin, history, and structure of the solid matter of a celestial body. **—ge'o·log'ic** (jē'ə-lŏj'ĭk), **ge'o·log'i·cal** *adj.* **—ge'o·log'i·cal·ly** *adv.* **—ge·ol'o·gist, ge·ol'o·ger** *n.*

**ge·o·mag·net·ic equator** (jē'ō-măg-nĕt'ĭk) *n.* The great circle on the earth's surface formed by the intersection of a plane passing through the earth's center perpendicular to the axis connecting the north and south magnetic poles.

**geomagnetic storm** *n.* Magnetic storm.

**ge·o·mag·ne·tism** (jē'ō-măg'nĭ-tĭz'əm) *n.* **1.** The earth's magnetism. **2.** Study of the earth's magnetic field. **—ge'o·mag·net'ic** (-nĕt'ĭk) *adj.* **—ge'o·mag·net'i·cal·ly** *adv.*

**ge·o·man·cy** (jē'ə-măn'sē) *n.* [ME *geomancie* < OFr. < Med. Lat. *geomantia* < LGk. *geōmanteia*, divination by signs from the earth : Gk. *gē*, earth + Gk. *manteia*, divination. —see -MANCY.] Divination by lines and figures. **—ge'o·man'cer** *n.* **—ge'o·man'tic** *adj.*

**ge·o·met·ric** (jē'ə-mĕt'rĭk) *also* **ge·o·met·ri·cal** (-rĭ-kəl) *adj.* **1.** Of or relating to geometry and its methods and principles. **2.** Utilizing simple geometric forms in design and decoration. **3.** Increasing or decreasing in a geometric progression. **—ge'o·met'ri·cal·ly** *adv.*

**geometric isomer** *n.* ISOMER 1e.

**geometric mean** *n.* The *n*th root, usu. the positive *n*th root, of a product of *n* factors.

**geometric pace** *n.* PACE[1] 2d.

**geometric progression** *n.* A sequence of terms, as 1, 3, 9, 27, 81, each of which is a constant multiple of the preceding term.

**ge·om·e·trid** (jē-ŏm'ĭ-trĭd) *n.* [NLat. *Geometridae*, family name < Lat. *geometres*, geometer < Gk. *geōmetrēs* < *geōmetrein*, to measure land. —see GEOMETRY.] A moth of the family Geometridae, having caterpillars that move by looping the body in alternate contractions and expansions. —*adj.* Of or belonging to the Geometridae.

**ge·om·e·trize** (jē-ŏm'ĭ-trīz') *vi.* **-trized, -triz·ing, -triz·es. 1.** To study geometry. **2.** To apply geometric methods.

**ge·om·e·try** (jē-ŏm'ĭ-trē) *n., pl.* **-tries.** [ME *geometrie* < OFr. < Lat. *geometria* < Gk. *geōmetria* < *geōmetrein*, to measure land : *gē*, earth + *metron*, measure.] **1. a.** The mathematics of the properties,

## GEOLOGIC TIME SCALE

| ERAS | PERIODS | | EPOCHS | APPROX. NO. OF YEARS AGO |
|---|---|---|---|---|
| Cenozoic | Quaternary | | Holocene (Recent) | 11.000 |
| | | | Pleistocene (Glacial) | 500.000 to 2.000.000 |
| | Tertiary | | Pliocene | 13.000.000 |
| | | | Miocene | 25.000.000 |
| | | | Oligocene | 36.000.000 |
| | | | Eocene | 58.000.000 |
| | | | Paleocene | 63.000.000 |
| Mesozoic | Cretaceous | | | 135.000.000 |
| | Jurassic | | | 180.000.000 |
| | Triassic | | | 230.000.000 |
| Paleozoic | Permian | | | 280.000.000 |
| | Carboniferous | Pennsylvanian (Upper Carboniferous) | | 310.000.000 |
| | | Mississippian (Lower Carboniferous) | | 345.000.000 |
| | Devonian | | | 405.000.000 |
| | Silurian | | | 425.000.000 |
| | Ordovician | | | 500.000.000 |
| | Cambrian | | | 600.000.000 |
| Precambrian | | | | |

measurement, and relationships of points, lines, angles, surfaces, and solids. **b.** A particular system of geometry <Euclidean *geometry*> **c.** A geometry restricted to a class of problems or objects <plane *geometry*> **2.** Configuration. **3.** A surface shape. **4.** Physical arrangement resembling geometric lines or forms. **—ge·om'e·tri'cian** (jē-ŏm'ĭ-trĭsh'ən, jē'ə-mĭ-), **ge·om'e·ter** *n.*

**ge·o·mor·phic** (jē'ə-môr'fĭk) *adj.* Of or like the earth, its shape, or surface configuration.

**ge·o·mor·phol·o·gy** (jē'ō-môr-fŏl'ə-jē) *n.* Geologic study of the configuration and evolution of land forms. **—ge'o·mor'pho·log'ic** (-môr'fə-lŏj'ĭk), **ge'o·mor'pho·log'i·cal** *adj.* **—ge'o·mor'pho·log'i·cal·ly** *adv.* **—ge'o·mor·phol'o·gist** *n.*

**ge·oph·a·gy** (jē-ŏf'ə-jē) *n.* Ingestion of earthy substances, as clay. **—ge·oph'a·gism** *n.* **—ge·oph'a·gist** *n.*

**ge·o·phone** (jē'ə-fōn') *n.* An electronic receiver for picking up seismic vibrations.

**ge·o·phys·ics** (jē'ō-fĭz'ĭks) *n. (sing. in number).* The physics of geologic phenomena, including fields such as meteorology, oceanography, geodesy, and seismology. **—ge'o·phys'i·cal** *adj.* **—ge'o·phys'i·cal·ly** *adv.* **—ge'o·phys'i·cist** *n.*

**ge·o·phyte** (jē'ə-fīt') *n.* A perennial plant propagated by underground buds.

**ge·o·pol·i·tics** (jē'ō-pŏl'ĭ-tĭks) *n. (sing. in number).* **1.** Study of the relationship between politics and geography. **2.** A Nazi expansionist doctrine that focused on reallocation of geographic, economic, and political boundaries. **—ge'o·po·lit'i·cal** (-pə-lĭt'ĭ-kəl) *adj.*

**ge·o·pon·ic** (jē'ə-pŏn'ĭk) *adj.* [Gk. *geōponikos* < *geōponein,* to till : *gē,* earth + *ponein,* to toil.] Of or relating to agriculture.

**ge·o·pon·ics** (jē'ə-pŏn'ĭks) *n. (sing. in number).* The study or science of agriculture.

**ge·o·pres·sured** (je'ō-prĕsh'ərd) *adj.* Being under high pressure within the earth.

**ge·o·pres·sur·ized** (jē'ō-prĕsh'ə-rīzd') *adj.* Geopressured.

**ge·o·probe** (jē'ō-prōb') *n.* A spacecraft designed to explore and study space near the earth.

**George** (jôrj) *n.* **1.** A jeweled figure of Saint George killing the dragon, used as an insignia of the Knights of the Garter. **2.** An English coin during the reign of Henry VIII, imprinted with a representation of Saint George.

**Geor·gette crepe** (jôr-jĕt') *n.* [Orig. a trademark.] A sheer strong silk or silklike fabric having a dull creped surface.

**Geor·gian** (jôr'jən) *adj.* **1.** Of, relating to, or characteristic of the reigns of the four Georges who ruled Great Britain from 1714 to 1830. **2.** Of or relating to the U.S. state of Georgia or its residents. **3.** Of or relating to the Georgian S.S.R., its people, or their language. *—n.* **1.** A native or resident of the state of Georgia. **2. a.** A native or resident of the Georgian S.S.R. **b.** The language of the Soviet Georgians. **3.** One belonging to or whose style is imitative of the period of the first four Georges of England.

**geor·gic** (jôr'jĭk) *also* **geor·gi·cal** (-jĭ-kəl) *adj.* [Lat. *georgicus* < Gk. *geōrgikos* < *geōrgos,* farmer : *gē,* earth + *ergon,* work.] Of or relating to rural life or agriculture. *—n.* A poem about rural life.

**ge·o·sci·ence** (jē'ō-sī'əns) *n.* A science, as geology or geochemistry, that deals with the earth. **—ge'o·sci'en·tist** *n.*

**ge·o·sta·tion·ar·y** (jē'ō-stā'shə-nĕr'ē) *adj.* Of or relating to a satellite that travels about the earth's equator at an altitude of at least 35,000 kilometers at a speed matching that of the earth's rotation, thereby maintaining a constant relation to points on the earth.

**ge·o·stroph·ic** (jē'ə-strŏf'ĭk) *adj.* [GEO- + Gk. *strophē,* a turning < *strephein,* to turn.] Of or relating to force caused by the earth's rotation. **—ge'o·stroph'i·cal·ly** *adv.*

**ge·o·syn·chro·nous** (jē'ō-sĭng'krə-nəs, -sĭn'-) *adj.* Geostationary. **—ge'o·syn'chro·nous·ly** *adv.*

**ge·o·syn·cline** (jē'ō-sĭn'klīn') *n.* An extensive, usu. linear depression in the earth's crust in which a succession of sedimentary strata has accumulated. **—ge'o·syn·cli'nal** (-sĭn-klī'nəl) *adj.*

**ge·o·tax·is** (jē'ō-tăk'sĭs) *n. Biol.* Movement of an organism in response to gravity. **—ge'o·tac'tic** (-tĭk) *adj.* **—ge'o·tac'ti·cal·ly** *adv.*

**ge·o·tec·ton·ic** (jē'ō-tĕk-tŏn'ĭk) *adj.* Of or pertaining to the shape, structure, and arrangement of the rock masses forming the earth's crust.

**ge·o·ther·mal** (jē'ō-thûr'məl) *also* **ge·o·ther·mic** (-mĭk) *adj.* Of or relating to the earth's internal heat. **—ge'o·ther'mal·ly** *adv.*

**ge·ot·ro·pism** (jē-ŏt'rə-pĭz'əm) *n. Biol.* Response of a living organism to gravity. **—ge'o·tro'pic** (jē'ə-trō'pĭk, -trŏp'ĭk) *adj.* **—ge'o·tro'pi·cal·ly** *adv.*

**ge·rah** (gîr'ə) *n.* [Heb. *gērāh.*] An ancient Hebrew coin and unit of weight.

**Ge·raint** (jə-rānt') *n.* A knight of the Round Table and husband of Enid in Arthurian legend.

**ge·ra·ni·ol** (jə-rā'nē-ôl', -ōl') *n.* [GERANI(UM) + -OL[1].] A fragrant pale-yellow liquid, $C_{10}H_{18}O$, derived chiefly from the oils of geranium and citronella and used in flavorings and cosmetics.

**ge·ra·ni·um** (jə-rā'nē-əm) *n.* [NLat. *Geranium,* genus name < Lat. *geranium,* cranesbill < Gk. *geranion* < *geranos,* crane.] **1.** Any of various plants of the genus *Geranium,* with divided leaves and purplish or pink flowers. **2.** A plant of the genus *Pelargonium,* native chiefly to southern Africa, esp. *P. domesticum,* widely cultivated for its rounded, often variegated leaves and red, pink, or white flower clusters. **3.** A strong to vivid red.

▲ word history: The word *geranium* refers to plants of two different but related genera. One genus is called *Geranium,* which is ultimately from Greek *geranos,* "crane." The genus is so named because the fruits of the plants that belong to it are similar in shape to the bill of the crane. The popular name for these plants is *cranesbill.* The plant commonly called the *geranium* in English belongs to the genus *Pelargonium,* from *pelargos,* the Greek word for "stork." The name for this genus also alludes to the shape of its fruit, which resembles the bill of a stork.

**ger·bil** (jûr'bəl) *n.* [NLat. *Gerbillus,* genus name < Fr. *gerbo,* jerboa < Ar. *yerbō'.*] A small mouselike rodent of the genus *Gerbillus* and related genera of arid regions of Africa and Asia Minor, with long hind legs and a long tail.

**ge·rent** (jîr′ənt) *n.* [< Lat. *gerens, gerent-*, pr.part. of *gerere*, to manage.] *Archaic.* A manager : overseer.

**ger·e·nuk** (gĕr′ə-no͝ok′) *n.* [Somali *garanūg*.] An African gazelle, *Litocranius walleri*, with long legs, a long slender neck, and backward-curving horns in the male.

**gerenuk**
*Approximately 5 feet long*

**ger·fal·con** (jûr′făl′kən, -fôl′-, -fô′-) *n. var. of* GYRFALCON.

**ger·i·at·ric** (jĕr′ē-ăt′rĭk) *adj.* [Gk. *gēras*, old age + -IATRIC.] **1.** Of or relating to geriatrics. **2.** Of or relating to the aged or their characteristic afflictions. —*n.* An aged person considered as a recipient of geriatric care.

**ger·i·at·rics** (jĕr′ē-ăt′rĭks) *n. (sing. in number).* Medical study of the physiology and pathology of old age. **—ger′i·a·tri′cian** (-ə-trĭsh′ən), **ger′i·at′rist** (-rĭst, jə-rī′ə-trĭst) *n.*

**germ** (jûrm) *n.* [Fr. *germe* < Lat. *germen*, bud.] **1.** *Biol.* A small organic structure or cell from which a new organism may develop. **2.** Something that can serve as the basis of further growth or development <the *germ* of a new research project> **3.** *Med.* A microorganism, esp. one that is pathogenic.

**ger·man¹** (jûr′mən) *n.* [Short for *German cotillion*.] **1.** An intricate dance for many couples. **2.** A party for dancing at which the german is featured.

**ger·man²** (jûr′mən) *adj.* [ME *germain* < OFr. < Lat. *germanus* < *germen*, offshoot.] Having the same parents or having the same grandparents on one side <cousin-*german*>

**Ger·man** (jûr′mən) *adj.* [Med. Lat. *Germanus* < Lat., perh. of Celt. orig.] Of, relating to, or characteristic of Germany, its people, or their language. —*n.* **1. a.** A native or resident of Germany. **b.** A person of German descent. **2.** The West Germanic language of Germany, Austria, and part of Switzerland.

**German cockroach** *n.* The Croton bug.

**ger·man·der** (jər-măn′dər) *n.* [ME *germandre* < OFr. *germandree* < Med. Lat. *germandra*, alteration of *gamandrea* < LGk. *khamandrua* < Gk. *khamaidrus* : *khamai*, on the ground + *drus*, oak.] A usu. aromatic plant of the genus *Teucrium*, having purplish or reddish flowers.

**ger·mane** (jər-mān′) *adj.* [ME *germain*, having the same parents. —see GERMAN².] Relevant to a point at hand : PERTINENT <a question *germane* to the issue at hand>

**Ger·man·ic** (jər-măn′ĭk) *adj.* **1. a.** Of, relating to, or characteristic of Germany or of the German people or their culture. **b.** Of or relating to Teutons. **c.** Of or relating to speakers of a Germanic language. **2.** Of, relating to, or constituting the Germanic languages. —*n.* A branch of the Indo-European language family that comprises North Germanic, West Germanic, and East Germanic.

**Ger·man·ism** (jûr′mə-nĭz′əm) *n.* **1.** An attitude, custom, or feature that seems to be characteristically German. **2.** A German idiom or expression appearing in another language. **3.** High regard for Germany and emulation of German ways.

**ger·ma·ni·um** (jər-mā′nē-əm) *n.* [< Lat. *Germanus*, German.] *Symbol* **Ge** A brittle, crystalline, gray-white semiconducting element, widely used as a semiconductor and as an alloying agent and catalyst; atomic number 32; atomic weight 72.59.

**Ger·man·ize** (jûr′mə-nīz′) *v.* **-ized, -iz·ing, -iz·es.** —*vt.* **1.** To give a German character to. **2.** *Archaic.* To translate into German. —*vi.* To adopt German attitudes or customs. **—Ger′man·i·za′tion** *n.* **—Ger′man·iz′er** *n.*

**German measles** *n.* A mild, contagious, eruptive disease caused by a virus and capable of causing congenital defects in infants born to mothers infected during the first three months of pregnancy.

**Ger·man·o·phile** (jər-măn′ə-fīl′) *n.* An admirer of Germany and German things.

**Ger·man·o·phobe** (jər-măn′ə-fōb′) *n.* One who hates or fears Germany and German things.

**German shepherd** *n.* A large dog orig. bred in Germany, having a dense brownish or black coat and often trained to assist the police and the blind.

**German silver** *n.* Nickel silver.

**germ cell** *n.* A cell having reproduction as its principal function, esp. an egg or sperm cell.

**ger·mi·cide** (jûr′mĭ-sīd′) *n.* A germ-killing agent. **—ger′mi·ci′dal** (-sīd′l) *adj.*

**ger·mi·nal** (jûr′mə-nəl) *adj.* [Fr. < Lat. *germen*, seed.] **1.** Of, relating to, or having the nature of a germ cell. **2.** Of, in, or relating to the earliest stage of development <the *germinal* stages of the space program> **—ger′mi·nal·ly** *adv.*

**germinal disc** *n.* A disklike region from which the embryo begins to develop in certain ova.

**germinal vesicle** *n.* The nucleus of an oocyte.

**ger·mi·nate** (jûr′mə-nāt′) *vi.* & *vt.* **-nat·ed, -nat·ing, -nates.** [Lat. *germinare, germinat-*, to sprout < *germen*, seed.] To begin to or cause to grow : SPROUT. **—ger′mi·na·ble** (-nə-bəl), **ger′mi·na′tive** *adj.* **—ger′mi·na′tion** *n.* **—ger′mi·na′tor** *n.*

**germ layer** *n.* One of three cellular layers, the ectoderm, endoderm, or mesoderm, into which most animal embryos differentiate.

**germ plasm** *n.* **1.** The cytoplasm of a germ cell. **2.** Germ cells as a whole. **3.** Hereditary material : GENES.

**germ theory** *n.* The doctrine that infectious diseases are caused by microorganismic activity within the body.

**germ warfare** *n.* Use of disease germs as weapons in war.

**geronto-** *or* **geront-** *pref.* [Fr. *géronto-* < Gk. *gerōn*, old man.] Old age : aged one <*gerontology*>

**ger·on·toc·ra·cy** (jĕr′ən-tŏk′rə-sē) *n., pl.* **-cies. 1.** Government in which the elders rule. **2.** A ruling group of elders. **—ge·ron′to·crat′ic** (jə-rŏn′tə-krăt′ĭk) *adj.*

**ger·on·tol·o·gy** (jĕr′ən-tŏl′ə-jē) *n.* Study of the physiological and pathological phenomena associated with aging. **—ge·ron′to·log′i·cal** (jə-rŏn′tə-lŏj′ĭ-kəl), **ge·ron′to·log′ic** (-lŏj′ĭk) *adj.* **—ger′on·tol′o·gist** *n.*

**ger·ry·man·der** (jĕr′ē-măn′dər, gĕr′-) *vt.* **-dered, -der·ing, -ders.** [After Elbridge *Gerry* (1744–1814) + (SALA)MANDER (from the shape of an election district created while Gerry was governor of Massachusetts).] To divide (a geographic area) into voting districts in order to give unfair advantage to one party in elections. **—ger′ry·man′der** *n.*

**ger·und** (jĕr′ənd) *n.* [LLat. *gerundium* < Lat. *gerundum*, var. of *gerendum*, gerund of *gerere*, to carry on.] **1.** A Latin verbal form that functions as a noun. **2.** A form, as English *cooking*, analogous to the Latin gerund. **—ge·run′di·al** (jə-rŭn′dē-əl) *adj.*

**ge·run·dive** (jə-rŭn′dĭv) *n.* [ME *gerundif* < LLat. *gerundivus* < *gerundium*. —see GERUND.] **1.** The Latin future passive participle that functions adjectivally and expresses the idea of fitness or necessity. **2.** A verbal adjective analogous to the Latin gerundive.

**Ge·ry·on** (jîr′ē-ən, gĕr′-) *n.* [Lat. < Gk. *Gēruōn*.] *Gk. Myth.* A winged monster with three bodies that was slain by Hercules.

**ges·so** (jĕs′ō) *n.* [Ital. < *gesso*, gypsum < Lat. *gypsum*. —see GYPSUM.] **1.** A preparation of plaster of Paris and glue used as a base for low relief or as a surface for painting. **2.** A surface of gesso.

**gest** *or* **geste** (jĕst) *n.* [ME *geste*. —see JEST.] *Archaic.* **1.** A notable feat. **2. a.** A verse romance or tale. **b.** A prose romance.

**ge·stalt** or **Ge·stalt** (gə-shtält′, -shtôlt′) *n., pl.* **-stalts** *or* **-stalt·en** (-shtält′n, -shtôlt′n) [G., shape < MHG, shaped, p.part. of *stellen*, to place < OHG.] A physical, psychological, or symbolic configuration or pattern so unified as a whole that its properties cannot be derived from its parts.

**Gestalt psychology** *n.* A psychological school or doctrine holding that psychological phenomena are irreducible gestalts.

**Ge·sta·po** (gə-stä′pō, -shtä′-) *n.* [G. *Ge(heime) Sta(ats)po(lizei)*, secret state police.] The Nazi internal security police, infamous for its terrorist methods directed against those suspected of treason or questionable loyalty. —*adj.* Behaving like the Gestapo <*Gestapo* interrogation tactics>

**ges·tate** (jĕs′tāt′) *vt.* **-tat·ed, -tat·ing, -tates.** [Back-formation < GESTATION.] **1.** To carry (unborn young) within the uterus for a period following conception. **2.** To conceive and develop mentally.

**ges·ta·tion** (jĕ-stā′shən) *n.* [Lat. *gestatio*, a carrying < *gestare*, freq. of *gerere*, to carry.] **1.** The period of carrying developing offspring in the uterus after conception : PREGNANCY. **2.** Mental conception and development of an idea. **—ges′ta·to′ry** (jĕs′tə-tôr′ē, -tōr′ē), **ges·ta′tion·al** *adj.*

**geste** (jĕst) *n. var. of* GEST.

**ges·tic** (jĕs′tĭk) *adj.* [< obs. *gest*, bearing < Fr. *geste* < Lat. *gestus* < *gerere*, to bear.] Relating to bodily movement, esp. in dancing.

**ges·tic·u·late** (jĕ-stĭk′yə-lāt′) *v.* **-lat·ed, -lat·ing, -lates.** [Lat. *gesticulari, gesticulat-* < *gesticulus*, gesticulation, dim. of *gestus*, gesture < p.part. of *gerere*, to behave.] —*vi.* To make gestures, esp. to do so to add force or emphasis during speech. —*vt.* To say or express by gestures. **—ges·tic′u·la′tion** (-lā′shən) *n.* **—ges·tic′u·la′tive** *adj.* **—ges·tic′u·la′tor** *n.* **—ges·tic′u·la·to′ry** (-lə-tôr′ē, -tōr′ē) *adj.*

**ges·ture** (jĕs′chər) *n.* [ME < Med. Lat. *gestura*, bearing < Lat. *gestus* < p.part. of *gerere*, to behave.] **1.** Movement of the limbs or body to express or help express thought or to emphasize speech. **2.** The act of moving the limbs or body as an expression of thought or emphasis. **3.** An act or expression made as an often formal sign of attitude or intent. —*v.* **-tured, -tur·ing, -tures.** —*vi.* To make gestures. —*vt.* To express or direct by gestures. **—ges′tur·er** *n.*

**Ge·sund·heit** (gə-zo͝ont′hīt′) [G.] *interj.* —Used to wish good health to a person who has just sneezed.

**†get** (gĕt) *v.* **got** (gŏt), **got** *or* **got·ten** (gŏt'n), **get·ting, gets.** [ME *geten* < ON *geta.*] —*vt.* **1.** To come into possession of : OBTAIN. **2.** To go after and obtain <*get* a book from the library> **3.** To take esp. by force : SEIZE. **4.** To acquire as a result of action or effort <*got* a prize for a high academic record> **5.** To acquire involuntarily : CATCH <*get* the measles> **6.** To have current possession of. —Used in the present perfect with the meaning of the present <has *got* a large collection of English silver> **7.** To beget. **8. a.** To cause to assume or be in a specified state <*got* the kids all tired and cross> **b.** To make ready : PREPARE <*get* dinner for a group> **9.** To cause to come or go. **10.** To cause to undertake or perform <*got* the guide to give us a good tour of the mansion> **11.** To have as an obligation. —Used in the present perfect with the meaning of the present <We have *got* to leave now.> **12. a.** To evoke an emotional response in <Violent movies really *get* me.> **b.** To annoy : irritate. **13. a.** To take revenge on. **b.** To kill in revenge for a wrong. **14. a.** To hit : strike <*got* me square on the mouth> **b.** To receive as retribution or punishment. **c.** To inflict ruin or destruction on. **15. a.** To gain or have understanding of. **b.** To learn by heart : MEMORIZE. **c.** To find or reach by calculating <*get* a final total> **16.** To present a difficult problem to : PUZZLE <What *gets* me is how easy it seems.> **17.** To perceive by hearing <didn't *get* their names> **18.** To make contact with, as by telephone. **19.** *Baseball.* To put out. —*vi. Informal.* **1.** To reach a given state or condition <*got* better> **2.** To arrive <When will we *get* to Charleston?> **3.** To start or come to be doing something <*Get* moving!> **4.** (gĭt). *Regional & Informal.* To depart at once. **5.** To make money. ***usage:*** Some uses of *get,* although not incorrect, are felt to be inappropriate to a formal written style. To be avoided in such contexts are the use of *get* in place of *be* or *become* (*We got arrested*) and the use of *get* or *get to* in place of *start* or *begin* (*They got to talking about the old days and I thought they'd never stop*). **—get across. 1.** To make understandable or clear. **2.** To be convincing or understandable. **—get along. 1.** To be in a state of harmony. **2.** To manage or fare with reasonable success. **3.** To advance, esp. in years. **—get around. 1.** To evade : circumvent. **2.** *Informal.* To convince or win over by flattering or cajoling. **—get away with.** *Informal.* To be successful in avoiding retribution or criticism for. **—get back at.** *Informal.* To take revenge on. **—get by. 1.** To succeed with minimal effort. **2.** To manage : survive <We'll *get by* if we're careful.> **—get off. 1.** To write and send, as a letter. **2.** To escape, as from punishment or danger <*got off* with a slap on the wrist> **3.** To obtain a release or lesser penalty for <*got* my client *off* with a suspended sentence> **4.** *Slang.* To get high, as from a drug. **5.** *Slang.* To feel great pleasure. **—get on. 1.** To get along. **2.** To advance <is *getting on* in life> **—get up.** To act as the originator or organizer of <*got up* a petition against video game parlors>—*n.* **1.** The act of begetting. **2.** Progeny : offspring. **3.** A return in tennis on a shot that seems impossible to reach. **—get it.** *Informal.* To be scolded or punished. **—get nowhere.** To make no progress. **—get on the stick.** *Slang.* To begin working at once and energetically. **—get there.** To achieve success. **—get'a·ble, get'ta·ble** *adj.*

**get·a·way** (gĕt'ə-wā') *n.* **1.** An act or instance of escaping. **2.** The start, as of a race : TAKEOFF.

**Geth·sem·a·ne** (gĕth-sĕm'ə-nē) *n.* [Gk. *Gethsēmanē.*] **1.** The garden outside Jerusalem that was the scene of Jesus' agony and arrest. **2. gethsemane.** A place or instance of great suffering.

**get·ter** (gĕt'ər) *n.* **1.** One that gets. **2.** A material that is added in small amounts during a chemical or metallurgical process to absorb impurities.

**get-to·geth·er** (gĕt'tə-gĕth'ər) *n. Informal.* A small party.

**get-up** (gĕt'ŭp') *n.* **1.** An outfit or costume. **2.** The arrangement and production style, as of a magazine or book. **3.** *also* **get-up-and-go.** Energy and ambition : SPUNK.

**gew·gaw** (gyōō'gô') *n.* [Orig. unknown.] A decorative trinket.

**gey·ser** (gī'zər) *n.* [Icel. *Geysir,* name of a hot spring in Iceland < *geysa,* to gush < ON.] **1.** A natural hot spring that intermittently ejects a column of water and steam into the air. **2.** (gē'zər). *Chiefly Brit.* A gas-operated hot-water heater.

**gey·ser·ite** (gī'zə-rīt') *n.* An opaline siliceous deposit formed around natural hot springs.

**ghast·ly** (găst'lē) *adj.* **-li·er, -li·est.** [ME *gastli* < *gasten,* to terrify < OE *gæstan.*] **1.** Shockingly repellent <a *ghastly* car accident> **2.** Gruesomely suggestive of ghosts or death. **3.** Highly unpleasant or bad <a *ghastly* little book> **4.** Very serious or great <a *ghastly* error> **—ghast'li·ness** *n.*

☆ ***syns:*** **1.** GHASTLY, GRIM, GRISLY, GRUESOME, HIDEOUS, HORRIBLE, HORRID, LURID, MACABRE *adj. core meaning* : shockingly repellent <the *ghastly* sight of starving children> **2.** GHASTLY, CADAVEROUS, DEATHLY, GHOSTLY, SPECTRAL *adj. core meaning* : gruesomely suggestive of ghosts or death <a *ghastly* Halloween costume><a *ghastly,* livid face>

▲ word history: The words *ghastly* and *ghost* are related, but the former is not derived from the latter. *Ghastly* is formed from the Middle English verb *gasten,* which meant "to frighten, terrify." *Ghost* is the modern descendent of Old English *gāst,* which meant "spirit, soul, spiritual being." Both *ghastly* and *ghost* are derived from a prehistoric Germanic root *gaist-.* The presence of the *h* in the modern spelling of both words is a relatively recent development. It was introduced into *ghost* in the 15th century, perhaps from the influence of Flemish *gheest,* "ghost," and appeared in *ghastly* in the 16th century, probably as a result of the semantic association of the two English words.

**ghat** *also* **ghaut** (gôt, gät) *n.* [Hindi *ghāt* < Skt. *ghaṭṭaḥ.*] A broad flight of steps down to the bank of a river.

**gher·kin** (gûr'kĭn) *n.* [Du. *agurkje,* pickled gherkin, ult. < LGk. *angourion.*] **1. a.** A tropical American vine, *Cucumis anguria,* bearing edible prickly fruit. **b.** The fruit of the gherkin. **2.** A small cucumber, esp. one used for pickling.

**ghet·to** (gĕt'ō) *n., pl.* **-tos** *or* **-toes.** [Ital.] **1.** A city slum inhabited by a minority group who live there due to social or economic pressure. **2.** A section or quarter in a European city to which Jews were restricted. **3.** Something resembling the isolation or restriction of a ghetto.

**ghet·to·ize** (gĕt'ō-īz') *vt.* **-ized, -iz·ing, -iz·es.** To set apart in or as if in a ghetto : ISOLATE <"The cities will become . . . *ghettoized,* and the flight of business and the middle class will accelerate . . . ." —*Newsweek*> **—ghet'to·i·za'tion** *n.*

**Ghib·el·line** (gĭb'ə-lēn', -līn', -lĭn) *n.* [Ital. *Ghibellino.*] A member of the aristocratic political faction who fought during the Middle Ages for German imperial control of Italy.

**ghil·lie** *also* **gil·lie** (gĭl'ē) *n., pl.* **-lies.** [Sc. Gael. *gille,* servant.] A low-cut sports shoe with fringed laces.

**ghost** (gōst) *n.* [ME *gost* < OE *gāst.*] **1.** The spirit of a deceased person, believed to haunt living people or former habitats. **2.** *Archaic.* The animus : soul. **3.** A demon : spirit. **4.** A recurring or haunting memory or image. **5.** A slight trace <not a *ghost* of a chance for success> **6.** A faint, false, occas. secondary photographic or television image. **7.** A ghostwriter. **8.** A nonexistent publication listed in bibliographies. —*vi. & vt.* **ghost·ed, ghost·ing, ghosts.** *Informal.* To ghostwrite.

**ghost crab** *n.* Any of several light-colored burrowing crabs of the genus *Ocypoda,* frequenting the tide line along sandy shores.

**ghost dance** *n.* Either of two religious dances practiced by certain North American Indians during the latter half of the 19th cent. to invoke a return of their former condition.

**ghost·ly** (gōst'lē) *adj.* **-li·er, -li·est. 1.** Relating to or like a ghost : SPECTRAL. **2.** Relating to the spirit or to religion. **—ghost'li·ness** *n.*

**ghost town** *n.* A town, esp. a Western boom town, that has been totally abandoned.

**ghost word** *n.* A word that has come into a language through the perpetuation of a misreading of a manuscript, a typographical error, or a misunderstanding.

**ghost·write** (gōst'rīt') *vi. & vt.* **-wrote** (-rōt'), **-writ·ten** (-rĭt'n), **-writ·ing, -writes.** To work as a ghostwriter or to write (something) as a ghostwriter.

**ghost·writ·er** (gōst'rī'tər) *n.* One who writes for and gives credit of authorship to another person.

**ghoul** (gōōl) *n.* [Ar. *ghūl* < *ghāla,* he took suddenly.] **1.** An evil spirit or demon in Moslem folklore held to plunder graves and feed on corpses. **2.** A grave robber. **3.** One who delights in the loathsome. **—ghoul'ish** *adj.* **—ghoul'ish·ly** *adv.* **—ghoul'ish·ness** *n.*

**GI** (jē'ī') *n., pl.* **GIs** *or* **GI's.** [Abbrev. for *government issue.*] An enlisted person in or veteran of the U.S. armed forces. —*adj.* **1.** Relating to or characteristic of a GI. **2.** Conforming to or in accordance with U.S. military rules, regulations, or procedures. **3.** Issued by an official U.S. military supply department <GI blankets>

**gi·ant** (jī'ənt) *n.* [ME *geaunt* < OFr. *geant* < VLat. **gagante* < Lat. *gigas* < Gk.] **1. a.** One of extraordinary size. **b.** One of extraordinary power or importance. **2.** *Gk. Myth.* One of a race of enormous humanlike creatures who warred with the Olympians and by whom they were finally destroyed. **3.** A being in folklore or myth similar to a giant.

☆ ***syns:*** GIANT, COLOSSAL, ELEPHANTINE, ENORMOUS, GARGANTUAN, GIGANTIC, HERCULEAN, HUGE, IMMENSE, JUMBO, MAMMOTH, MASSIVE, MIGHTY, MONSTROUS, MONUMENTAL, MOUNTAINOUS, PRODIGIOUS, STUPENDOUS, TITAN, TITANIC, TREMENDOUS, VAST *adj. core meaning* : of extraordinary size and power <a *giant* corporation>

**gi·ant·ism** (jī'ən-tĭz'əm) *n.* **1.** The state of being a giant. **2.** GIGANTISM 1.

**giant panda** *n.* PANDA 1.

**giant sequoia** *n.* A very tall evergreen tree, *Sequoia gigantea* of mountainous regions of southern California, having a massive trunk and light-colored reddish wood.

**giant star** *n.* Any of a class of highly luminous, massive stars.

**giaour** (jour) *n.* [Ult. < Pers. *gaur,* infidel, var. of *gebr,* fire worshiper.] A nonbeliever from the viewpoint of Islam, esp. a Christian : INFIDEL.

**gib** (gĭb) *n.* [Orig. unknown.] A plain or notched, often wedge-shaped piece of wood or metal intended to hold parts of a machine or structure in place or to provide a bearing surface, usu. adjusted by a screw or key.

**gib·ber** (jĭb′ər) *vi.* **-bered, -ber·ing, -bers.** [Imit.] To chatter unintelligibly. —*n.* GIBBERISH 1.

**gib·ber·el·lic acid** (jĭb′ə-rĕl′ĭk) *n.* [< GIBBERELLIN.] A substance, $C_{19}H_{22}O_6$, produced from a fungus, *Gibberella fujikuroi*, and used to promote plant growth, esp. the growth of seedlings.

**gib·ber·el·lin** (jĭb′ə-rĕl′ĭn) *n.* [< NLat. *Gibberella* (*fujikoroi*), the fungus from which gibberellin was first isolated.] A substance of plant origin, as gibberellic acid, used to promote plant stem growth.

**gib·ber·ish** (jĭb′ər-ĭsh) *n.* **1.** Rapid nonsensical chatter. **2.** Unintelligible writing.

**gib·bet** (jĭb′ĭt) *n.* [ME *gibet* < OFr., dim. of *gibe*, staff, perh. of Germanic orig.] **1.** A gallows. **2.** An upright post with a crosspiece, forming a T-shaped structure from which executed criminals were hung for public viewing. —*vt.* **-bet·ed, -bet·ing, -bets** *or* **-bet·ted, -bet·ting, -bets. 1.** To execute by hanging. **2.** To hang on a gibbet for public viewing. **3.** To expose to infamy or public ridicule.

**gib·bon** (gĭb′ən) *n.* [Fr.] An arboreal ape of the genera *Hylobates* or *Symphalangus* of tropical Asia, having a slender body and long arms.

**gib·bos·i·ty** (gĭ-bŏs′ĭ-tē) *n., pl.* **-ties. 1.** The state of being gibbous. **2.** A rounded protuberance.

**gib·bous** (gĭb′əs) *adj.* [ME, bulging < Lat. *gibbus*, hump.] **1.** Marked by convexity : PROTUBERANT. **2.** More than half but less than fully illuminated. —Used of a celestial body, as the moon. <"the *gibbous* moon, its light reflecting whitely" —John Barth> **3.** Having a hump : HUMPBACKED. **—gib′bous·ly** *adv.* **—gib′bous·ness** *n.*

**gibe** (jīb) *v.* **gibed, gib·ing, gibes.** [Poss. < OFr. *giber*, to handle roughly.] —*vi.* To make mocking remarks : HECKLE. —*vt.* To reproach by mocking and heckling. **—gibe** *n.* **—gib′er** *n.*

**gib·let** (jĭb′lĭt) *n.* [ME *gibelet* < OFr.] The heart, liver, or gizzard of a fowl, esp. when treated as meat.

**Gib·son** (gĭb′sən) *n.* [< the name *Gibson*.] A dry martini with a small pickled onion.

**Gibson girl** *n.* The young American woman of the 1890's as idealized in sketches by Charles Dana Gibson.

**gid** (gĭd) *n.* [Back-formation < GIDDY.] A disease of sheep caused by the presence of the larva of a tapeworm, *Taenia caenurus*, in the brain and resulting in a staggering gait.

**gid·dy** (gĭd′ē) *adj.* **-di·er, -di·est.** [ME *gidi*, crazy < OE *gidig*.] **1. a.** Having a reeling, lightheaded sensation : DIZZY. **b.** Causing or capable of causing dizziness. **2.** Frivolous and lighthearted : FLIGHTY. —*vt. & vi.* **-died, -dy·ing, -dies.** To make or become giddy. **—gid′di·ly** *adv.* **—gid′di·ness** *n.*

☆ ***syns:*** **1.** GIDDY, BIRDBRAINED, DIZZY, EMPTY-HEADED, FEATHER-BRAINED, FLIGHTY, FRIVOLOUS, GAGA, HAREBRAINED, LIGHTHEADED, SCATTERBRAINED, SILLY, SKITTISH *adj. core meaning* : given to lighthearted silliness <*giddy* youngsters> **2.** GIDDY, DIZZY, DIZZYING, VERTIGINOUS *adj. core meaning* : producing dizziness or vertigo <a *giddy* height>

**Gid·e·on** (gĭd′ē-ən) *n.* [Heb. *Gidh'ōn* < *gādha*, he hewed.] A judge of Israel and conqueror of the Midianites in the Old Testament.

**gie** (gē) *v. Scot. var. of* GIVE.

**gift** (gĭft) *n.* [ME < ON.] **1.** Something bestowed voluntarily and without compensation. **2.** The act, right, or power of giving <Your request is not in my *gift*.> **3.** A talent, endowment, aptitude, or bent. —*vt.* **gift·ed, gift·ing, gifts. 1.** To present with a gift. **2.** To endow with : INVEST. ***usage:*** Although unacceptable to some, the use of *gift* as a verb meaning "to present with a gift" has a long history in English.

**gift·ed** (gĭf′tĭd) *adj.* **1.** Endowed with natural ability or talent <*gifted* children> **2.** Displaying talent <a *gifted* opera singer> **—gift′ed·ly** *adv.* **—gift′ed·ness** *n.*

**gift of tongues** *n.* An ecstatic utterance partially or entirely unintelligible to hearers, esp. such an utterance as practiced liturgically in certain Christian congregations.

**gig¹** (gĭg) *n.* [< obs. *gig*, spinning top.] **1.** A light two-wheeled one-horse carriage. **2. a.** A long, light ship's boat with oars, sails, or a motor, usu. reserved for use by a ship's captain. **b.** A light, fast rowboat. —*vi.* **gigged, gig·ging, gigs.** To ride in a gig.

**†gig²** (gĭg) *n.* [Short for FISHGIG.] **1.** An arrangement of barbless hooks dragged through a school of fish to hook them. **2.** A pronged fishing spear. —*v.* **gigged, gig·ging, gigs.** —*vt.* **1.** To catch with a gig. **2.** *Regional.* To goad : prod. —*vi.* To fish with a gig.

**gig³** (gĭg) [Orig. unknown.] *Slang.* —*n.* A demerit, esp. one assigned as punishment to military personnel. —*vt.* **gigged, gig·ging, gigs.** To give a demerit to.

**gig⁴** (gĭg) *n.* [Orig. unknown.] *Slang.* A job, esp. a booking for musicians.

**giga-** *pref.* [< Gk. *gigas*, giant.] One billion ($10^9$) <*gigahertz*>

**gig·a·bit** (jĭg′ə-bĭt′, gĭg′-) *n. Computer Sci.* One billion ($10^9$) bits.

**gig·a·cy·cle** (jĭg′ə-sī′kəl, gĭg′-) *n.* One billion ($10^9$) cycles.

**gig·a·hertz** (jĭg′ə-hûrtz, gĭg′-) *n.* One billion ($10^9$) hertz.

**gi·gan·tesque** (jī′găn-tĕsk′) *adj.* [Fr. < Ital. *gigantesco* < Gk. *gigas*, giant.] GIGANTIC 1.

**gi·gan·tic** (jī-găn′tĭk) *adj.* [< Lat. *gigas, gigant-*, giant < Gk.] **1.** Relating to or suggesting a giant. **2.** Exceedingly large or extensive <a *gigantic* radio network> **—gi·gan′ti·cal·ly** *adv.*

**gi·gan·tism** (jī-găn′tĭz′əm) *n.* **1.** Excessive bodily growth due to oversecretion of the pituitary growth hormone. **2.** Abnormal size.

**gig·gle** (gĭg′əl) *vi.* **-gled, -gling, -gles.** [Imit.] To laugh with repeated short high-pitched sounds. —*n.* A high-pitched spasmodic laugh. **—gig′gler** *n.* **—gig′gling·ly** *adv.* **—gig′gly** *adj.*

**gig·o·lo** (jĭg′ə-lō′, zhĭg′-) *n., pl.* **-los.** [Fr.] **1.** A man kept by a woman as a lover and supported financially. **2.** A man hired as a woman's escort or dancing partner.

**gig·ot** (jĭg′ət, zhē-gō′) *n.* [OFr., dim. of *gigue*, leg, of Germanic orig.] **1.** A leg of mutton, lamb, or veal for cooking. **2.** A leg-of-mutton sleeve.

**gigue** (zhēg) *n.* [Fr. < E. JIG.] *Mus.* **1.** JIG 1a. **2. a.** A lively dance form in 6/8, 9/8, or 12/8 time, often forming the final movement of a suite. **b.** The music for this dance.

**Gi·la monster** (hē′lə) *n.* [After *Gila* River, Arizona.] A venomous lizard, *Heloderma suspectum* of the southwestern United States, covered with black and pinkish or yellowish scales.

**gil·bert** (gĭl′bərt) *n.* [After William *Gilbert* (1836–1911).] The centimeter-gram-second electromagnetic unit of magnetomotive force, equal to $^{10}/_{4\pi}$ ampere-turn.

**gild¹** (gĭld) *vt.* **gild·ed** *or* **gilt** (gĭlt), **gild·ing, gilds.** [ME *gilden* < OE *gyldan*.] **1.** To cover with or as if with a thin coating of gold. **2.** To give an often deceptively attractive or improved appearance to. **3.** *Archaic.* To smear with blood. **—gild the lily.** To adorn needlessly something that is already beautiful. **—gild′er** *n.*

**gild²** (gĭld) *n. var. of* GUILD.

**gild·ing** (gĭl′dĭng) *n.* **1.** The art or process of applying gilt to a surface. **2.** Gold leaf or a paint containing or simulating gold : GILT. **3.** Something used to provide something else with a superficially attractive appearance.

**Gil·ga·mesh** (gĭl′gə-mĕsh′) *n.* [Of Sumerian orig.] *Myth.* The semidivine king of Erech and hero of an epic collection of mythic Babylonian tales, one of which tells of a flood that covered the earth.

**gill¹** (gĭl) *n.* [ME *gile*, of Scand. orig.] **1.** *Zool.* The respiratory organ of fishes, larval amphibians, and many aquatic invertebrates. **2.** *often* **gills.** A bird's wattle. **3.** *often* **gills.** *Informal.* The area around the human chin and neck. **4.** *Bot.* One of the thin platelike structures on the underside of the cap of a mushroom or a similar fungus. —*vt.* **gilled, gill·ing, gills. 1.** To catch (fish) in a gill net. **2.** To gut or clean (fish).

**gill²** (jĭl) *n.* [ME *gille* < Med. Lat. *gillo*, a pot.] **1.** A unit of volume or capacity in the U.S. Customary System, used in liquid measure, equal to 4 fluid ounces, 1/4 pint, or 23.656 milliliters. **2.** A unit of volume or capacity in the British Imperial System, used in dry and liquid measure, equal to 5 fluid ounces, 1/4 pint, or 28.423 milliliters.

**gill³** (gĭl) *n.* [ME *gille* < ON *gil*.] *Chiefly Brit.* **1.** A ravine. **2.** A narrow stream.

**gill⁴** *also* **Gill** (jĭl) *n.* [ME *gille* < *Gille*, a woman's name.] A girlfriend.

**gill fungus** (gĭl) *n.* A fleshy fungus having a cap with gills on the underside.

**gil·lie** *also* **gil·ly** (gĭl′ē) *n., pl.* **-lies.** [Sc. Gael. *gille*.] **1.** *Scot.* A professional guide for deerstalking and fishing. **2.** *var. of* GHILLIE.

**gill net** (gĭl) *n.* A fishing net set vertically in the water so that fish swimming into it are entangled in its mesh by the gills.

**gill slit** (gĭl) *n.* One of several narrow external openings connecting with the pharynx, present in all vertebrates during embryonic development and characteristic of sharks and related fishes.

**gil·ly·flow·er** (gĭl′ē-flou′ər) *n.* [By folk etymology < ME *gilofre* < OFr. *girofle* < Med. Lat. *corophylum*, clove < Gk. *karuophullon* : *karuon*, nut + *phullon*, leaf.] **1.** A plant, as the carnation, of the genus *Dianthus*. **2.** A plant having fragrant flowers, as the stock or wallflower.

**Gil·son·ite** (gĭl′sə-nīt′). A trademark for a natural black bitumen found in Utah and Colorado, used in manufacturing acid, alkali, and waterproof coatings.

**gilt¹** (gĭlt) *adj.* [P.part. of GILD¹] **1.** Covered with gold or a simulation of gold. **2.** Resembling gold. —*n.* **1.** A thin layer of gold or a simulation of it, applied in gilding. **2. a.** Shining brilliance : GLITTER. **b.** Superficial brilliance.

**gilt²** (gĭlt) *n.* [ME, young sow < ON *gylvt*.] A young sow not yet farrowed.

**gilt-edged** (gĭlt′ĕjd′) *also* **gilt-edge** (-ĕj′) *adj.* **1.** Having gilded edges. **2.** Of the highest quality or value <*gilt-edged* securities>

**gim·bal** (gĭm′bəl, jĭm′-) *n.* [OFr. *gemel*. —see GIMMAL.] *often* **gimbals.** A device consisting of two rings mounted on axes at right angles to each other so that an object, as a ship's compass, will remain suspended in a horizontal plane between them regardless of any motion of its support.

**gim·crack** (jĭm′krăk′) *n.* [Orig. unknown.] A cheap, showy object of little or no use : GEWGAW. —*adj.* Cheap and shoddy : FLIMSY. **—gim′crack′ery** *n.*

**gim·el** (gĭm′əl) *n.* [Heb. *gīmel*.] The third letter of the Hebrew alphabet. —See table at ALPHABET.

**gim·let** (gĭm′lĭt) *n.* [ME < AN *guimblet*.] **1.** A small hand tool for boring holes, having a spiraled shank, a screw tip, and a cross handle.

2. A cocktail made with vodka or gin and lime juice, garnished with a slice of lime. —*vt.* **-let·ed, -let·ing, -lets.** To penetrate with or as if with a gimlet. —*adj.* Penetrating : piercing <*gimlet* eyes>

**gim·mal** (gĭm′əl, jĭm′-) *n.* [OFr. *gemel* < Lat. *gemelus*, dim. of *geminus*, twin.] A ring made of two narrower rings interlocked.

**gim·mick** (gĭm′ĭk) *n.* [Orig. unknown.] **1.** A device often used illegally to cheat or trick, esp. a mechanism for the secret control of a gambling wheel or other apparatus. **2.** A stratagem employed to promote a project <a sales *gimmick*> **3.** A significant feature that is obscured or misrepresented : CATCH. **4.** A trivial or unnecessary innovation, as a gadget, added to enhance appeal. **5.** A gismo. —*vt.* **-micked, -mick·ing, -micks.** To add gimmicks to <*gimmick* up a car> **—gim′mick·ry** *n.* **—gim′mick·y** *adj.*

**gimp**[1] (gĭmp) *n.* [Prob. < Du.] A narrow braid or cord of fabric, occas. stiffened, used for trimmings.

**gimp**[2] (gĭmp) [Orig. unknown.] *Slang.* —*n.* **1.** A limping gait. **2.** A person who limps. —*vi.* **gimped, gimp·ing, gimps.** To limp. **—gimp′y** *adj.*

**gimp**[3] (gĭmp) *n.* [Orig. unknown.] Spirit : pep.

**gin**[1] (jĭn) *n.* [Short for *geneva* < MDu. *genever* < OFr. *geneivre* < Lat. *juniperus*, juniper.] **1.** A strong alcoholic beverage made by distilling rye or other grains with juniper berries. **2.** A liquor similar to gin but flavored with some other aromatic substance, as aniseed.

**gin**[2] (jĭn) *n.* [ME, short for OFr. *engin* < Lat. *ingenium*, skill.] **1.** A machine or device, esp.: **a.** A machine for hoisting or moving heavy objects. **b.** A pile driver. **c.** A snare or trap for game. **d.** A pump operated by a windmill. **2.** A cotton gin. —*vt.* **ginned, gin·ning, gins. 1.** To remove the seeds from (cotton) with a gin. **2.** To trap in a gin.

**gin**[3] (jĭn) *n.* Gin rummy.

**gin·ger** (jĭn′jər) *n.* [ME *gingivere* < OE *gingifer* and OFr. *gingivre*, both < Med. Lat. *gingiber* < Lat. *zinziberi* < Gk. *zingiberis*.] **1. a.** A tropical Asian plant, *Zingiber officinale*, having yellowish-green flowers and a pungent aromatic rootstock. **b.** The rootstock of the ginger, often dried and powdered and used as a spice. **2. a.** Any of various plants of the family Zingiberaceae, having variously colored, often fragrant flowers. **b.** The wild ginger. **3.** A deep brown. **4.** *Informal.* Liveliness : vigor. —*vt.* **-gered, -ger·ing, -gers. 1.** To spice with ginger. **2.** *Informal.* To make more lively <*gingered* up the evening> **—gin′ger·y** *adj.*

**ginger ale** *n.* A ginger-flavored effervescent soft drink.

**ginger beer** *n.* A nonalcoholic drink similar to ginger ale but flavored with fermented ginger.

**gin·ger·bread** (jĭn′jər-brĕd′) *n.* **1. a.** A dark molasses cake flavored with ginger. **b.** A soft molasses and ginger cookie cut in various shapes, occas. decorated elaborately. **2. a.** Elaborate ornamentation. **b.** Superfluous or tasteless embellishment, esp. in architecture.

**gingerbread palm** *n.* The doom palm.

**gin·ger·ly** (jĭn′jər-lē) *adv.* [Poss. < OFr. *gensor*, comp. of *gent*, gentle. —see GENT[1].] With great delicacy or care : CAUTIOUSLY. **—gin′ger·li·ness** *n.* **—gin′ger·ly** *adj.*

**gin·ger·snap** (jĭn′jər-snăp′) *n.* A flat, brittle cookie spiced with ginger and sweetened with molasses.

**ging·ham** (gĭng′əm) *n.* [Malay *ginggang* < *ginggang*, striped.] Yarn-dyed cotton woven in stripes, checks, plaids, or solid colors.

**gin·gi·va** (jĭn′jə-və, jĭn-jī′-) *n., pl.* **-vae** (-vē′). GUM[2].

**gin·gi·val** (jĭn′jə-vəl, jĭn-jī′-) *adj.* [< Lat. *gingiva*, gum.] **1.** Of or relating to the gums. **2.** Alveolar.

**gin·gi·vi·tis** (jĭn′jə-vī′tĭs) *n.* [Lat. *gingiva*, gum + -ITIS.] Inflammation of the gums.

**gink·go** *also* **ging·ko** (gĭng′kō) *n., pl.* **-goes** *also* **-koes.** [J. *ginkyō*, of Chin. orig.] A tree indigenous to China, *Ginkgo biloba*, having fan-shaped leaves and fleshy yellowish fruit, often used as an ornamental street tree.

**gin mill** *n. Slang.* A saloon.

**gin rummy** *n.* Rummy for two or more players in which a player may win by matching all his or her cards or may end the game by melding when his or her unmatched cards add up to ten points or less.

**gin·seng** (jĭn′sĕng′) *n.* [Chin. (Mandarin) *ren*[2] *shen*[1] : *ren*[2], man + *shen*[1], ginseng.] **1.** A plant of the genus *Panax*, esp. *P. schinseng* of eastern Asia or *P. quinquefolium* of North America, with small greenish flowers and a forked root thought to have medicinal properties. **2.** The root of a ginseng plant.

**gip** (jĭp) *v.* & *n. var. of* GYP.

**Gip·sy** (jĭp′sē) *n. var. of* GYPSY.

**gi·raffe** (jə-răf′) *n., pl.* **-raffes** or **giraffe.** [NLat. *Giraffa*, genus name < Ital. *giraffa* < Ar. *zirāfah*.] An African ruminant mammal, *Giraffa camelopardis*, having a very long neck and legs, a tan coat with brown blotches, and short horns.

**gir·an·dole** (jĭr′ən-dōl′) *n.* [Fr. < Ital. *girandola* < *girare*, to turn < Lat. *gyrare*. —see GYRATE.] **1.** A composition or structure in radiating form or arrangement, as a rotating display of fireworks. **2.** A branched candleholder, sometimes backed by a mirror.

**girandole**

**gir·a·sol** *also* **gir·o·sol** (jĭr′ə-sôl′, -sōl′, -sŏl′) *n.* [Ital. *girasole* : *girare*, to turn (< Lat. *gyrare* < *gyrus*, circle < Gk. *guros*) + *sole*, sun < Lat. *sol*.] A fire opal.

**gird**[1] (gûrd) *vt.* **gird·ed** or **girt** (gûrt), **gird·ing, girds.** [ME *girden* < OE *gyrdan*.] **1.** To encircle or fasten with or as if with a belt or band. **2.** To equip : endow. **3.** To prepare (oneself) for action.

**gird**[2] (gûrd) *vi.* & *vt.* **gird·ed, gird·ing, girds.** [ME *girden*, to strike.] To jeer or jeer at. —*n.* A jeering remark. **—gird′er** *n.*

**gird·er** (gûr′dər) *n.* A horizontal beam, as of steel or wood, used as a main support for a structure, as a building.

**gir·dle** (gûr′dl) *n.* [ME *girdel* < OE *gyrdel*.] **1. a.** A belt or sash worn around the waist. **b.** Something that encircles like a belt. **2.** An elasticized flexible undergarment worn over the hips and waist. **3.** A band made around the trunk of a tree by the removal of a strip of bark. **4.** The edge of a cut gem held by the setting. **5.** *Anat.* The pelvic or pectoral arch. —*vt.* **-dled, -dling, -dles. 1.** To encircle with or as if with a belt. **2.** To put a girdle on. **3.** To remove a band of bark completely from the circumference of (a tree), usu. to kill it.

**gird·ler** (gûrd′lər) *n.* **1.** One that makes girdles. **2.** Any of various insects that chew circular bands around twigs or stems in preparing nesting sites.

**girl** (gûrl) *n.* [ME *girle*.] **1.** A female who has not yet attained womanhood. **2.** A female child. **3.** An unmarried young woman. **4.** A daughter. **5.** A sweetheart. **6.** A woman servant, employee, or clerk. **—girl′hood′** *n.*

▲ word history: In Modern English *girl* is the ordinary word denoting a female child. In Middle English times, however, from the 13th to the 15th century, *girl* indicated a child or youth of either sex. Nothing is known of the history of *girl* before its appearance in English, and it appears that no related forms exist in any other language.

**girl Friday** *n.* [GIRL + (MAN) FRIDAY.] *Informal.* A woman employee, esp. one having multiple responsibilities.

**girl·friend** *also* **girl friend** (gûrl′frĕnd′) *n.* **1.** A woman friend. **2.** A close woman friend of a boy or man, esp. one seen romantically.

**Girl Guide** *n.* A member of the Girl Guides, a British youth organization founded in 1910.

**girl·ie** *also* **girl·y** (gûr′lē) *adj. Informal.* Featuring minimally clothed young women <*girlie* calendars>

**girl·ish** (gûr′lĭsh) *adj.* Relating to, characteristic of, or suitable for a girl. **—girl′ish·ly** *adv.* **—girl′ish·ness** *n.*

**Girl Scout** *n.* A member of the Girl Scouts, a youth organization founded in the United States in 1912 on the plan of the Girl Guides.

**girl·y** (gûr′lē) *adj. var. of* GIRLIE.

**Gi·ronde** (jə-rŏnd′, zhĭ-) *n.* [After *Gironde*, a department in France.] A moderate republican political party of revolutionary France from 1791 to 1793. **—Gi·rond′ist** *n.*

**gir·o·sol** (jĭr′ə-sôl′, -sōl′, -sŏl′) *n. var. of* GIRASOL.

**girt**[1] (gûrt) *v.* **girt·ed, girt·ing, girts.** [Var. of GIRD.] —*vt.* **1.** To gird. **2.** To measure the girth of. —*vi.* To measure in girth.

**girt**[2] (gûrt) *v. var. p.t.* & *p.p. of* GIRD[1].

**girth** (gûrth) *n.* [ME *gerth*, strap passing under a horse's belly to secure a load < ON *gyǫrð*, girdle.] **1.** The distance around an object : CIRCUMFERENCE. **2.** Size : bulk. **3.** A strap encircling an animal's body to secure a load or saddle on its back : CINCH. —*vt.* **girthed, girth·ing, girths. 1.** To measure the circumference of. **2.** To encircle. **3.** To secure with a girth.

**gi·sarme** (gĭ-zärm′) *n.* [ME < OFr.] A halberd with a long shaft and a two-sided blade, carried by medieval foot soldiers.

**gis·mo** *also* **giz·mo** (gĭz′mō) *n., pl.* **-mos.** [Orig. unknown.] *Slang.* A mechanical device or part whose name is forgotten or unknown.

**gist** (jĭst) *n.* [< OFr., it lies < *gesir*, to lie < Lat. *jacēre*.] **1.** The most central and material part : HEART. **2.** *Law.* The grounds for action in a suit.

**give** (gĭv) *v.* **gave** (gāv), **giv·en** (gĭv′ən), **giv·ing, gives.** [ME *given* < OE *giefan*.] —*vt.* **1. a.** To make a present of <*gave* them money for their birthdays> **b.** To deliver in exchange or in recompense : PAY <will *give* $15 for the book> **c.** To put temporarily at the disposal of <*gave* us the beach house for a week> **d.** To place in the hands of <*Give* me the spoon.> **2. a.** To convey or offer for conveyance <*Give* them my warmest wishes.> **b.** To bestow, esp. officially : CONFER <*give* authority> **c.** To accord to another <*Give* me your trust.> **3.** To donate <*give* one's time to a worthy cause> **4. a.** To be a source of <a remark that *gave* offense> **b.** To cause to have or be subject to <*gave* them the mumps> **5.** To produce <a cow that *gives* three gallons of milk> **6.** To provide (something expected or required) <*Give* your name and age.> **7. a.** To inflict as punishment <*gave* me a tongue-lashing> **b.** To administer <*gave* me antibiotics> **8.** To relinquish : yield <*give* ground> **9.** To emit : utter <*give* a sigh of relief> **10. a.** To allot : assign <*gave* us 20 minutes to finish> **b.** To designate : specify <*give* a departure time> **11.** To award as due <*gave* them third prize> **12.** To ascribe : attribute <*gave* us the blame> **13. a.** To cause to take place, esp. for entertainment <*give* a luncheon> **b.** To proffer or offer <*give* a toast> **c.** To manifest : show <*gives* promise of success> **d.** To perform for an audience <*give* a concert> **14.** To submit for consideration <*give* an opinion> **15.** To allow or lead <*gave* me to think I was talented> **16.** To devote : apply <They *give* themselves to teaching.> **17.** To sacrifice <*gave* sons to the war> **18.** To care to the extent of <don't *give* a damn> —*vi.* **1.** To make gifts or donations <*gives* generously to charity> **2.** To yield, as to pressure : COLLAPSE <The bridge *gave* under the trucks' weight.> **3.** To afford a view of or access to : OPEN <The French doors *give* onto a deck.> **4.** *Slang.* To be in progress : HAPPEN <What *gives*?> **—give away. 1.** To make a gift of. **2.** To present (a bride) to the bridegroom at a wedding ceremony. **3.** To reveal or make known, often accidentally. **—give back.** To return. **—give in. 1.** To hand in : SUBMIT. **2.** To cease opposition : YIELD. **—give off.** To send forth : EMIT <chemical changes that *give off* energy> **—give out. 1.** To let (something) be known <*gave out* the good news> **2.** To distribute : ISSUE <*gave out* surplus food to the destitute> **3.** To stop functioning : FAIL. **4.** To become used up : RUN OUT. **—give over. 1.** To entrust. **2.** To make available for a given purpose : DEVOTE. **3.** To surrender totally and unrestrainedly, as to an emotion. **—give up. 1.** To surrender <*gave* themselves *up* to the FBI> **2.** To leave off <*give up* drinking> **3.** To part with : RELINQUISH <*gave up* the idea> **4.** To abandon hope for <*gave* us *up* as lost> **5.** To admit defeat. **6.** To abandon what one is engaging in or planning to engage in. —*n. Informal.* Resilient springiness <a foam mattress with a lot of *give*> **—give way. 1. a.** To withdraw. **b.** To make room for <*give way* to an oncoming truck> **2.** To collapse. **3.** To abandon oneself <*give way* to a flood of tears>

**give-and-take** (gĭv′ən-tāk′) *n.* **1.** The practice of compromise. **2.** Lively exchange of conversation or ideas.

**give·a·way** (gĭv′ə-wā′) *n. Informal.* **1.** Something given away free. **2.** Something that betrays or exposes, often accidentally.

**giv·en** (gĭv′ən) *adj.* **1. a.** Specified <a *given* hour> **b.** Issued on a specified date. —Used of legal documents. **2.** Granted as a supposition : ACKNOWLEDGED <*Given* their superiority, they can't expect to lose.> **3.** Having a tendency : INCLINED <*given* to crying> **4.** Bestowed : presented. **—giv′en** *n.*

**given name** *n.* A name given to a person at birth or baptism.

**giz·mo** (gĭz′mō) *n. var. of* GISMO.

**giz·zard** (gĭz′ərd) *n.* [ME *giser* < OFr. *gisier* < VLat. **gicerium* < Lat. *gigeria*, cooked entrails of poultry.] **1.** An enlargement of the alimentary canal in birds, often having dense muscular walls and containing fine grit eaten to aid in the digestion of seeds. **2.** A digestive organ similar to the gizzard in certain invertebrates, as the earthworm.

**gla·bel·la** (glə-bĕl′ə) *n., pl.* **-bel·lae** (-bĕl′ē′) [NLat. < Lat. *glabellus*, hairless < *glaber*.] The smooth area between the eyebrows just above the nose. **—gla·bel′lar** *adj.*

**gla·brous** (glā′brəs) *adj.* [Lat. *glaber*, bald.] Having no hairs or pubescence : SMOOTH. **—gla′brous·ness** *n.*

**gla·cé** (glă-sā′) *adj.* [Fr. < p.part. of *glacer*, to glaze < *glace*, ice < Lat. *glacies*.] **1.** Having a glazed surface. **2.** Coated with a sugar glaze. —*vt.* **-céed, -cé·ing, -cés. 1.** To glaze. **2.** To candy.

**gla·cial** (glā′shəl) *adj.* [Lat. *glacialis*, icy < *glacies*, ice.] **1. a.** Of, relating to, or caused by a glacier. **b.** *often* **Glacial.** Characterized or dominated by the existence of glaciers. —Used esp. of the Pleistocene. **2.** Extremely cold : ICY <*glacial* waters> **3.** Looking like ice. **4.** Cold and forbidding <a *glacial* stare> **—gla′cial·ly** *adv.*

**glacial acetic acid** *n.* Acetic acid that is at least 99.8% pure.

**glacial epoch** *n.* **1.** Any of several periods during the Pleistocene epoch up to 1,000,000 years ago when much of the earth's surface was covered by glaciers. **2.** The Pleistocene epoch.

**gla·ci·ate** (glā′shē-āt′, -sē-) *vt.* **-at·ed, -at·ing, -ates.** [Lat. *glaciare*, *glaciat-*, to freeze < *glacies*, ice.] **1.** To subject to glacial action. **2.** To freeze. **—gla′ci·a′tion** *n.*

**gla·cier** (glā′shər) *n.* [Fr. < *glace*, ice < Lat. *glacies*.] A huge mass of laterally limited, moving ice originating from compacted snow.

**glacier lily** *n.* The fawn lily.

**gla·ci·ol·o·gy** (glā′shē-ŏl′ə-jē, -sē-) *n.* [GLACI(ER) + -LOGY.] Scientific study of glaciers. **—gla′ci·o·log′ic** (-ə-lŏj′ĭk), **gla′ci·o·log′i·cal** *adj.* **—gla′ci·ol′o·gist** *n.*

**gla·cis** (glă-sē′, glăs′ē, glā′sĭs) *n.* [Fr. < OFr. *glacier*, to slide < *glace*, ice < Lat. *glacies*.] **1.** A gentle slope : INCLINE. **2.** A slope extending downward from a fortification.

**glad** (glăd) *adj.* **glad·der, glad·dest.** [ME < OE *glæd*.] **1.** Feeling or displaying joy and pleasure. **2.** Providing joy and pleasure. **3.** Pleased : willing <*glad* to come> **4.** *Archaic.* Having a cheerful disposition. —*vt.* & *vi.* **glad·ded, glad·ding, glads.** *Obs.* To gladden. **—glad′ly** *adv.* **—glad′ness** *n.*

☆ ***syns:* 1.** GLAD, CHEERFUL, CHEERY, FESTIVE, GAY, JOYFUL, JOYOUS *adj. core meaning* : providing joy and pleasure <*glad* tidings> **2.** GLAD, HAPPY, READY *adj. core meaning* : eagerly compliant <*glad* to help>

**glad·den** (glăd′n) *vt.* & *vi.* **-dened, -den·ing, -dens.** To make or become glad.

**glade** (glād) *n.* [Perh. < GLAD, shining (obs.).] An open space in a forest.

**glad hand** *n. Informal.* A hearty, often insincere and offensively familiar welcome or greeting.

**glad-hand** (glăd′hănd′) *v.* **-hand·ed, -hand·ing, -hands.** *Informal.* —*vt.* To extend a glad hand to. —*vi.* To extend a glad hand. **—glad′-hand′er** *n.*

**glad·i·ate** (glăd′ē-āt′, -ĭt, glā′dē-) *adj.* [NLat. *gladiatus* < Lat. *gladius*, sword.] Sword-shaped <a *gladiate* leaf>

**glad·i·a·tor** (glăd′ē-ā′tər) *n.* [ ME < Lat. < *gladius*, sword.] **1.** One trained to entertain the public by engaging in mortal combat in the ancient Roman arena. **2.** One engaged in a controversy or dispute, esp. in public : COMBATANT. **3.** A prizefighter. **—glad′i·a·to′ri·al** (-ə-tôr′ē-əl, -tōr′-) *adj.*

**glad·i·o·lus** (glăd′ē-ō′ləs) *n., pl.* **-li** (-lī′, -lē′) *or* **-lus·es.** [Lat., dim. of *gladius*, sword.] **1.** *also* **glad·i·o·la** (glăd′ē-ō′lə). Any of various plants of the genus *Gladiolus*, native to tropical regions but widely cultivated elsewhere, having sword-shaped leaves and a showy, variously colored flower spike. **2.** *Anat.* The large middle section of the sternum.

**glad rags** *pl.n. Informal.* One's best clothes.

**glad·some** (glăd′səm) *adj.* **1.** Glad : joyful. **2.** Producing joy. **—glad′some·ly** *adv.* **—glad′some·ness** *n.*

**Glad·stone** (glăd′stōn′, -stən) *n.* [After William E. *Gladstone* (1809–1898).] **1.** A light four-wheeled convertible carriage with two interior seats and places outside for a driver and footman. **2.** A Gladstone bag.

**Gladstone bag** *n.* [After William E. *Gladstone* (1809–1898).] A piece of light hand luggage composed of two hinged compartments.

**glair** *also* **glaire** (glâr) *n.* [ME *glaire* < OFr. < Lat. *clarus*, clear.] **1.** Raw egg white used in glazing or sizing. **2.** A glaze or sizing made of egg white.

**glair·y** (glâr′ē) *adj.* **-i·er, -i·est. 1.** Resembling glair. **2.** Coated with glair. **—glair′i·ness** *n.*

**glaive** (glāv) *n.* [ME < OFr. < Lat. *gladius*.] *Archaic.* A sword, esp. a broadsword.

**glam·or** (glăm′ər) *n. var. of* GLAMOUR.

**glam·or·ize** *also* **glam·our·ize** (glăm′ə-rīz′) *vt.* **-ized, -iz·ing, -iz·es. 1.** To make glamorous. **2.** To treat or portray in a romantic or idealistic way. **—glam′or·i·za′tion** *n.* **—glam′or·iz′er** *n.*

**glam·or·ous** *also* **glam·our·ous** (glăm′ər-əs) *adj.* Marked by glamour. **—glam′or·ous·ly** *adv.* **—glam′or·ous·ness** *n.*

**glam·our** *also* **glam·or** (glăm′ər) *n.* [Sc., alteration of GRAMMAR.] **1.** An air of compelling charm, romance, and excitement, esp. when delusively alluring. **2.** *Archaic.* Magic : enchantment. ***usage:*** The preferred spelling of *glamour* makes it an exception to the usual American spelling practice illustrated by such words as *honor* and *labor*. In both British and American English the adjective is more often spelled *glamorous*.

▲ word history: *Glamour* as a word in standard Modern English is a borrowing from Scots. It is a doublet of *grammar*, since both words are descended from Greek *grammatikē* through Latin *grammatica*. The Greek word originally meant "pertaining to letters or literature," from *gramma*, "letter, written character." The Latin word in medieval times denoted not just literacy but learning in general, including knowledge of such occult sciences as astrology and magic. The extended sense of Latin *grammatica* was preserved by its Old French descendant *gramaire*, which was borrowed into English and became the modern word *grammar*. A variant of the Old French form in which *l* was substituted for *r* must have existed because this form, probably spelled *glomerie* or *glamorie*, was also borrowed into English where it survived in the dialect of Scotland as *glamour*. The Scottish word preserved only the sense "magic, magic spell." *Glamour* was introduced into standard English with the meaning "magic spell" by Sir Walter Scott. The current sense of *glamour*, "alluring charm," developed in English later in the 19th century.

**glam·our·ize** (glăm′ə-rīz′) *v. var. of* GLAMORIZE.

**glam·our·ous** (glăm′ər-əs) *adj. var. of* GLAMOROUS.

**glance**[1] (glăns) *v.* **glanced, glanc·ing, glanc·es.** [ME *glenchen*, *glansen*, to strike obliquely.] —*vi.* **1.** To strike a surface at such an angle as to be deflected <A small rock *glanced* off the windshield.> **2.** To direct the gaze briefly <*glance* at a menu> **3.** To shine briefly : GLINT. **4.** To make a passing reference. —*vt.* **1.** To strike (a surface) at an angle : GRAZE. **2.** To cause to strike a surface at an angle : SKIP <*glance* a pebble over the lake> —*n.* **1.** An oblique movement following impact : DEFLECTION. **2.** A brief look. **3.** A quick flash of light : GLEAM.

ă pat ā pay âr care ä father ĕ pet ē be hw which ĭ pit ī tie îr pier ŏ pot ō toe ô paw, for oi noise ŏŏ took

**glance²** (glăns) *n.* [G. *Glanz* < OHG *glanz,* bright.] Any of various minerals with a brilliant luster.

**gland** (glănd) *n.* [Fr. *glande* < OFr., gland, acorn < Lat. *glans,* acorn.] **1. a.** An organ that extracts specific substances from the blood and concentrates or alters them for subsequent secretion. **b.** Any of various nonsecretory or excretory organs that resemble such organs. **2.** *Bot.* An organ or structure that secretes a substance. **3.** A sliding machine part designed to hold something in place.

**glan·ders** (glăn′dərz) *n.* [< OFr. *glandre,* glandular swelling < Lat. *glandula,* dim. of *glans,* acorn.] *(sing. or pl. in number).* A contagious, often chronic, occas. fatal disease of horses and other animals, caused by a bacillus, *Actinobacillus mallei,* and marked by a nasal discharge and ulcers in the lungs, respiratory tract, and skin. **—glan′der·ous** *adj.*

**glan·des** (glăn′dēz′) *n. pl. of* GLANS.

**glan·du·lar** (glăn′jə-lər) *adj.* [Fr. *glandulaire* < *glandule,* small gland < Lat. *glandula,* glandular swelling. —see GLANDERS.] **1.** Of, relating to, affecting, or resembling a gland or its secretion. **2.** Functioning as a gland. **3.** Having glands. **4.** Resulting from abnormal gland function. **5.** Innate : instinctive <a *glandular* distrust of foreigners> **—glan′du·lar·ly** *adv.*

**glandular fever** *n.* Infectious mononucleosis.

**glans** (glănz) *n., pl.* **glan·des** (glăn′dēz′) [Lat. < *glans,* acorn (from its shape).] **1.** The glans penis. **2.** The glans clitoridis.

**glans cli·tor·i·dis** (klĭ-tôr′ĭ-dĭs, klī-) *n.* The small mass of erectile tissue at the tip of the clitoris.

**glans penis** *n.* The head or tip of the penis.

**glare¹** (glâr) *v.* **glared, glar·ing, glares.** [ME *glaren,* to shine brightly.] —*vi.* **1.** To stare fixedly and angrily. **2.** To shine intensely and blindingly <sun *glaring* down on us> **3.** To be obtrusively conspicuous. —*vt.* To express by staring fixedly and angrily <*glared* their dislike> —*n.* **1.** A fixed, angry stare. **2.** An intense, blinding light. **3.** Showy brilliance : GAUDINESS.

**glare²** (glâr) *n.* [Prob. < GLARE¹.] A sheet of ice.

**glar·ing** (glâr′ĭng) *adj.* **1.** Staring fixedly and angrily. **2.** Shining intensely and blindingly. **3.** Gaudy : garish. **4.** Very conspicuous <a *glaring* mistake> **—glar′ing·ly** *adv.*

**glar·y** (glâr′ē) *adj.* **-i·er, -i·est.** GLARING 2.

**glass** (glăs) *n.* [ME *glas* < OE *glæs.*] **1.** Any of a large class of materials with highly variable mechanical and optical properties that solidify from the molten state without crystallization, are typically based on silicon dioxide, boric oxide, aluminum oxide, or phosphorus pentoxide, are gen. transparent or translucent, and are regarded physically as supercooled liquids rather than true solids. **2.** Objects made of glass. **3.** Something made of glass, esp.: **a.** A drinking vessel. **b.** A mirror. **c.** A barometer. **d.** A windowpane. **4. a.** A device containing a lens or lenses, used as an aid to vision. **b. glasses.** A pair of lenses mounted in a light frame, used to correct faulty vision or protect the eyes. **5.** The quantity contained by a drinking vessel. —*v.* **glassed, glass·ing, glass·es.** —*vt.* **1.** To put within glass or a glass container. **2.** To provide with glass or glass parts. **3.** To make glassy : GLAZE. **4. a.** To see reflected, as in a mirror. **b.** To mirror : reflect. —*vi.* To become glassy.

**glass blowing** *n.* The process or art of shaping objects from molten glass by blowing air into them through a tube. **—glass blower** *n.*

**glass eel** *n.* An eel in its transparent postlarval stage.

**glass·ful** (glăs′fo͝ol′) *n.* GLASS 5.

**glass harmonica** *n.* A musical instrument composed of a set of graduated glass bowls that produce tones when a moistened finger is moved over their rims.

**glass·house** (glăs′hous′) *n.* **1.** A glassworks. **2.** *Chiefly Brit.* A greenhouse.

**glass·ine** (glă-sēn′) *n.* An almost transparent, resilient glazed paper resistant to the passage of air and grease.

**glass·mak·er** (glăs′mā′kər) *n.* A maker of glass.

**glass·man** (glăs′mən, -măn′) *n.* **1.** A seller of glass. **2.** A glassmaker.

**glass snake** *n.* [From the brittleness of its tail.] Any of several slender, limbless, snakelike lizards of the genus *Ophisaurus,* having a tail that breaks or snaps off readily.

**glass·ware** (glăs′wâr′) *n.* GLASS 2.

**glass wool** *n.* Fine-spun fibers of glass used esp. in air filters and for insulation.

**glass·work** (glăs′wûrk′) *n.* **1. a.** The making of glassware or glass. **b.** The cutting and fitting of glass panes : GLAZIERY. **2.** GLASS 2. **3. glassworks** *(sing. in number).* A place where glass is made.

**glass·wort** (glăs′wûrt′, -wôrt′) *n.* [From its former use in making glass.] Any of various plants of the genus *Salicornia,* growing in salt marshes and having fleshy stems and rudimentary scalelike leaves.

**glass·y** (glăs′ē) *adj.* **-i·er, -i·est. 1.** Like glass. **2.** Lifeless : expressionless <a fixed *glassy* grin> **—glass′i·ly** *adv.* **—glass′i·ness** *n.*

**Glau·ber's salts** *also* **Glau·ber's salt** (glou′bərz) *n.* [After Johann R. *Glauber* (1604–1668).] A hydrated sodium sulfate, $Na_2SO_4 \cdot 10H_2O$, used in making paper and glass and as a cathartic.

**glau·co·ma** (glou-kō′mə, glô-) *n.* [Lat., cataract < Gk. *glaukōma* < *glaukos,* gray.] A disease of the eye marked by high intraocular pressure, damaged optic disk, hardening of the eyeball, and partial or total vision loss. **—glau·co′ma·tous** (-kō′mə-təs) *adj.*

**glau·co·nite** (glô′kə-nīt′) *n.* [Gk. *glaukon,* neuter of gray + -ITE.] A hydrous silicate of potassium, iron, aluminum, or magnesium, $K_2(Mg,Fe)_2Al_6(Si_4O_{10})_3(OH)_{12}$, occurring in greensand and used as a water softener and fertilizer. **—glau′co·nit′ic** (-nĭt′ĭk) *adj.*

**glau·cous** (glô′kəs) *adj.* [Lat. *glaucus* < Gk. *glaukos.*] *Bot.* Grayish green or bluish green due to a fine, whitish, powdery coating <*glaucous* grapes> **—glau′cous·ness** *n.*

**glaze** (glāz) *n.* [ME *glasen* < *glas,* glass < OE *glæs.*] **1.** A smooth, thin, shiny coating. **2.** A thin, glassy ice coating. **3. a.** A coating of colored, opaque, or transparent material applied to ceramics before firing. **b.** A coating, as of syrup, applied to food. **c.** A transparent coating applied to the surface of a painting to modify color tones. **4.** A glassy film, as over the eyes. —*v.* **glazed, glaz·ing, glaz·es.** —*vt.* **1.** To furnish or fit with glass <*glaze* the broken windows> **2.** To apply a glaze to <*glazing* a dozen doughnuts><*glaze* a set of pottery dishes> **3.** To give a smooth, lustrous surface to. —*vi.* **1.** To be or become glazed or glassy <eyes *glazing* over with fatigue> **2.** To form a glaze. **—glaz′er** *n.*

**gla·zier** (glā′zhər) *n.* [ME *glasier* < *glas,* glass < OE *glæs.*] One who cuts and fits window glass. **—gla′zier·y** (-zhə-rē) *n.*

**glaz·ing** (glā′zĭng) *n.* **1. a.** Glasswork. **b.** Glass set or made to be set in frames. **2. a.** A glaze. **b.** The act or process of applying a glaze.

**gleam** (glēm) *n.* [ME *glem* < OE *glæm.*] **1.** A brief beam or flash of light. **2.** A subdued but steady shining : GLOW <the *gleam* of lights across the bay> **3.** A brief or dim indication : TRACE <a *gleam* of comprehension> —*v.* **gleamed, gleam·ing, gleams.** —*vi.* **1.** To emit a gleam : FLASH. **2.** To be manifested or indicated faintly or briefly. —*vt.* To cause to emit a gleam. **—gleam′er** *n.*

**glean** (glēn) *v.* **gleaned, glean·ing, gleans.** [ME *glenen* < OFr. *glener* < LLat. *glennare.*] —*vi.* To gather grain left by reapers. —*vt.* **1.** To gather (grain) left by reapers. **2.** To collect little by little <"records from which historians *glean* their knowledge" —Kemp Malone> **—glean′er** *n.*

**glean·ings** (glē′nĭngz) *pl.n.* Things gathered bit by bit <the *gleanings* of assiduous researchers>

**gle·ba** (glē′bə) *n., pl.* **-bae** (-bē′) [NLat. < Lat., clod.] The inner spore-bearing mass of a puffball.

**glebe** (glēb) *n.* [Lat. *gleba,* clod.] **1.** *Chiefly Brit.* Land granted to a cleric as part of his benefice during his tenure of office. **2.** *Archaic.* The soil or earth : LAND.

**glede** (glēd) *n.* [ME < OE *glida.*] A predatory bird, as the kite.

**glee** (glē) *n.* [ME *gle,* entertainment < OE *glēo.*] **1.** Jubilant gaiety : JOY. **2.** An unaccompanied part song for three or more usu. male voices.

**glee club** *n.* A group of singers who perform usu. short pieces of choral music.

**gleed** (glēd) *n.* [ME *glede* < OE *glēd.*] *Chiefly Brit. Regional.* A glowing coal : EMBER.

**glee·ful** (glē′fəl) *adj.* Full of glee. **—glee′ful·ly** *adv.* **—glee′ful·ness** *n.*

**glee·man** (glē′mən) *n.* [ME *gleman* < OE *glēoman* : *glēo,* minstrelsy + *mann,* man.] *Archaic.* A medieval itinerant singer.

**glee·some** (glē′səm) *adj. Archaic.* Gleeful.

**gleet** (glēt) *n.* [ME *glet,* slime < OFr. *glete* < Lat. *glittus,* sticky.] **1.** Chronic inflammation of the urethra, marked by mucopurulent discharge. **2.** The discharge symptomatic of gleet. **—gleet′y** *adj.*

**gleg** (glĕg) *adj.* [ME, clear-sighted < ON *glöggr.*] *Scot.* Alert.

**glen** (glĕn) *n.* [ME < Sc. Gael. *gleann* < OIr. *glend.*] A valley.

**Glen·gar·ry** (glĕn-găr′ē) *n., pl.* **-ries.** [After *Glengarry,* a valley in Scotland.] A Scottish woolen cap creased lengthwise with short ribbons at the back.

**gley** (glā) *n.* [R. *gleĭ,* clay.] A sticky, bluish-gray clay layer formed under the influence of high soil moisture.

**gli·a·din** (glī′ə-dĭn) *n.* [Ital. *gliadina* < Med. Gk. *glia,* glue.] Any of several simple proteins obtained from rye or wheat gluten.

**glib** (glĭb) *adj.* **glib·ber, glib·best.** [Poss. of Low German orig.] **1. a.** Performed with a natural, offhand ease <a *glib* talker> **b.** Showing little thought, preparation, or concern <*glib* answers> **2.** Marked by a fluency or quickness that often suggests or stems from deception or insincerity. **—glib′ly** *adv.* **—glib′ness** *n.*

**glide** (glīd) *v.* **glid·ed, glid·ing, glides.** [ME *gliden* < OE *glīdan.*] —*vi.* **1.** To move in a smooth, effortless way <dancers *gliding* around the ballroom> **2.** To move silently and furtively. **3.** To occur or pass imperceptibly. **4.** To fly without propulsion. —Used of an aircraft. **5.** *Mus.* To blend one tone into the next : SLUR. **6.** To articulate a glide in speech. —*vt.* To cause to glide. —*n.* **1.** The act of gliding. **2.** *Mus.* A slur. **3. a.** The transitional sound produced in moving from the articulatory position of one speech sound to that of another. **b.** A semivowel.

**glid·er** (glī′dər) *n.* **1.** One that glides. **2.** A light engineless aircraft designed to glide after being towed aloft or launched from a catapult. **3.** A swinging couch hung from a vertical frame. **4.** A device that aids gliding.

**glim** (glĭm) *n.* [Perh. short. for GLIMMER.] A light source, as a candle or lamp.

**glim·mer** (glĭm′ər) *n.* [< ME *glimeren,* to shine.] **1.** A dim or intermittent light : FLICKER. **2.** A faint manifestation or vague idea : TRACE <a *glimmer* of hope> —*vi.* **-mered, -mer·ing, -mers. 1.** To give off a dim or intermittent light. **2.** To appear or be indicated faintly.

**glimpse** (glĭmps) *n.* [< ME *glimsen,* to glance.] **1.** A quick, incomplete view or look. **2.** *Archaic.* A brief flash of light. —*v.* **glimpsed, glimps·ing, glimps·es.** —*vt.* To obtain a quick, incomplete view of <*glimpsed* my friend through the window> —*vi.* To look quickly : GLANCE <*glimpsed* at the morning newspaper> **—glimps′er** *n.*

**glint** (glĭnt) *n.* [ME *glent,* of Scand. orig.] **1.** A momentary flash of light : SPARKLE. **2.** A faint or fleeting indication : TRACE <a *glint* of compassion> —*v.* **glint·ed, glint·ing, glints.** —*vi.* **1.** To gleam or flash. **2.** *Archaic.* To move abruptly : DART. —*vt.* To cause to gleam or flash.

**glis·sade** (glĭ-säd′, -säd′) *n.* [Fr. < *glisser,* to slide < OFr. *glier,* to glide, of Germanic orig.] **1.** A gliding ballet step. **2.** A controlled slide, in either a standing or a sitting position, used in descending a steep icy or snowy incline. —*vi.* **-sad·ed, -sad·ing, -sades.** To perform a glissade. **—glis·sad′er** *n.*

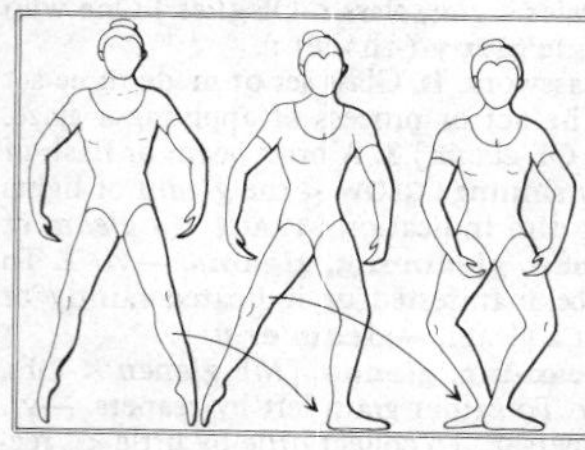
**glissade**

**glis·san·do** (glĭ-sän′dō) *n., pl.* **-di** (-dē) *or* **-dos.** [Prob. alteration of Fr. *glissade,* sliding motion. —see GLISSADE.] *Mus.* A rapid slide through a series of consecutive tones in a scalelike passage.

**glis·ten** (glĭs′ən) *vi.* **-tened, -ten·ing, -tens.** [ME *glisnen* < OE *glisnian.*] To shine or reflect with glittering radiance : GLEAM. **—glis′ten** *n.*

**glis·ter** (glĭs′tər) *vi.* **-tered, -ter·ing, -ters.** [ME *glisteren* < MDu. *glinsteren* and MLG *glisteren.*] To glisten. —*n.* Glitter : brilliance.

**glitch** (glĭch) *n.* [Prob. < Yiddish *glitsh,* slippery area < *glitshn,* to slide < G. *glitschen* < MHG *glīten,* to glide < OHG *glitan.*] **1.** A minor mishap, malfunction, or technical problem. **2.** *Electron.* A false or spurious electronic signal caused by a brief, unwanted surge of electric power. **3.** *Astron.* A sudden change in the period of rotation of a neutron star. **—glitch′y** *adj.*

**glit·ter** (glĭt′ər) *n.* [ME *gliteren* < ON *glitra.*] **1.** A sparkling light or brightness. **2.** Brilliant or showy attractiveness. **3.** Small bits of light-reflecting decorative material. —*vi.* **-tered, -ter·ing, -ters. 1. a.** To sparkle brilliantly : GLISTEN. **b.** To sparkle malevolently or coldly <eyes *glittering* with hate> **2.** To be brilliantly, often misleadingly attractive. **—glit′ter·ing·ly** *adv.* **—glit′ter·y** *adj.*

**glit·te·ra·ti** (glĭt′ə-rä′tē) *pl.n.* [Blend of GLITTER and LITERATI.] *Informal.* Highly fashionable celebrities.

**glitz** (glĭts) *n.* [Yiddish, glitter.] *Slang.* Excessive showiness : FLASHINESS. **—glitz′y** *adj.*

**gloam** (glōm) *n. Archaic.* Twilight : gloaming.

**gloam·ing** (glō′mĭng) *n.* [ME *gloming* < OE *glōmung* < *glōm,* dusk.] Dusk : twilight.

**gloat** (glōt) *vi.* **gloat·ed, gloat·ing, gloats.** [Perh. of Scand. orig.] To feel or express great, often malicious joy or self-satisfaction. —*n.* **1.** An act of gloating. **2.** A feeling of great, often malicious joy or self-satisfaction. **—gloat′er** *n.*

**glob** (glŏb) *n.* [ME *globbe,* large mass < Lat. *globus,* globular mass.] **1.** A small drop : GLOBULE. **2.** A rounded, usu. large mass or lump <a *glob* of sticky dough>

**glob·al** (glō′bəl) *adj.* **1.** Having the shape of a globe : SPHERICAL. **2.** Of, pertaining to, or involving the entire earth : WORLDWIDE <a *global* agreement on outer space> **3.** Total : comprehensive <a *global* approach to problem solving> **—glob′al·ly** *adv.*

**glob·al·ism** (glō′bə-lĭz′əm) *n.* **1.** Globalization. **2.** A policy fostering globalization. **—glob′al·ist** *n.*

**glob·al·i·za·tion** (glō′bə-lĭ-zā′shən) *n.* The act, process, or policy of making something worldwide in scope or application.

**glob·al·ize** (glō′bə-līz′) *vt.* **-ized, -iz·ing, -iz·es.** To make worldwide in application or scope. **—glob′al·iz′er** *n.*

**glo·bate** (glō′bāt′) *also* **glo·bat·ed** (-bā′tĭd) *adj.* [Lat. *globatus,* p.part. of *globare,* to form into a ball < *globus,* ball.] Shaped like a globe : GLOBULAR.

**globe** (glōb) *n.* [ME < Lat. *globus.*] **1.** A body shaped like a sphere, esp. a representation of the earth or heavens as a hollow ball. **2. a.** The earth. **b.** A planet. **3.** An object resembling a globe, esp. a rounded container, as a glass sphere covering a light bulb. **4.** A sphere emblematic of sovereignty : ORB. —*vt.* & *vi.* **globed, glob·ing, globes.** To form into or assume the shape of a globe.

**globe·fish** (glōb′fĭsh′) *n., pl.* **globefish** *or* **-fish·es.** A fish, as the ocean sunfish, having or capable of assuming a globular shape.

**globefish**
*Approximately 9 inches long*

**globe·flow·er** (glōb′flou′ər) *n.* Any of several plants of the genus *Trollius,* with globe-shaped, usu. yellow flowers.

**globe·trot·ter** (glōb′trŏt′ər) *n.* One who travels widely and often, esp. for sightseeing. **—globe′trot′ting** *n.*

**glo·bin** (glō′bĭn) *n.* [Back-formation < HEMOGLOBIN.] A simple protein derived from hemoglobin.

**glo·boid** (glō′boid′) *adj.* Having a globelike shape : SPHEROID. **—glo′boid′** *n.*

**glo·bose** (glō′bōs′) *also* **glo·bous** (-bəs) *adj.* [Lat. *globosus* < *globus,* sphere.] Globular : spherical. **—glo′bose′ly** *adv.* **—glo′bose′ness, glo·bos′i·ty** (-bŏs′ĭ-tē) *n.*

**glob·u·lar** (glŏb′yə-lər) *adj.* **1.** Shaped like a globe or globule : SPHERICAL. **2.** Consisting of globules. **3.** Global : worldwide. **—glob′u·lar·ly** *adv.* **—glob′u·lar·ness** *n.*

**globular cluster** *n.* A system of stars, gen. smaller than a galaxy, more or less globular in conformation.

**glob·ule** (glŏb′yo͞ol) *n.* [Lat. *globulus,* dim. of *globus,* sphere.] A small, often minute spherical mass, esp. a small drop of liquid.

**glob·u·lif·er·ous** (glŏb′yə-lĭf′ər-əs) *adj.* Composed of or producing globules.

**glob·u·lin** (glŏb′yə-lĭn) *n.* [GLOBULE + -IN.] Any of a class of simple proteins found extensively in blood, milk, muscle, and plant seeds, insoluble in pure water, soluble in dilute salt solution, and coagulable by heat.

**glo·chid·i·um** (glō-kĭd′ē-əm) *n., pl.* **-i·a** (-ē-ə) [NLat. < Gk. *glōkhis,* barb of an arrow.] **1.** *Zool.* A parasitic larva of certain freshwater mussels of the family Unionidae, with hooks for attaching to a host fish. **2.** *also* **glo·chid** (glō′kĭd). *Bot.* A barbed hair or bristle on a plant, as the prickly pear. **—glo·chid′i·ate** (-ĭt, -āt′) *adj.*

**glock·en·spiel** (glŏk′ən-spēl′, -shpēl′) *n.* [G. : *Glocke,* bell (< OHG *glocka*) + *Spiel,* play.] A percussion instrument with a series of metal bars tuned to the chromatic scale and played with two light hammers.

**glogg** (glŏg) *also* **glögg** (glœg) *n.* [Sw. *glögg* < *glödga,* to mull < OSw.] A hot punch of red wine, sherry, and brandy flavored with almonds, raisins, and orange peel.

**glom·er·ate** (glŏm′ər-ĭt) *adj.* [Lat. *glomeratus,* p.part. of *glomerare,* to wind into a ball < *glomus,* ball.] Formed into a rounded, compact mass : CLUSTERED.

**glom·er·ule** (glŏm′ə-ro͞ol′, glŏm′yə-) *n.* [NLat. *glomerulus* < Lat. *glomus,* ball.] **1.** *Bot.* A compact flower cluster. **2.** *Anat.* A glomerulus. **—glo·mer′u·late** (glō-mĕr′yə-lĭt) *adj.*

**glo·mer·u·lus** (glō-mĕr′yə-ləs) *n., pl.* **-li** (-lī′) [NLat. < Lat. *glomus,* ball.] **1.** A tuft of capillaries located at the origin of a vertebrate kidney. **2.** The twisted secretory portion of a sweat gland.

**gloom** (glo͞om) *n.* [Prob. < ME *gloumen,* to become dark.] **1. a.** Partial or complete darkness : DIMNESS. **b.** A partially or completely dark place. **2. a.** An atmosphere of depression or melancholy. **b.** A state of depression or melancholy : DESPONDENCY. —*v.* **gloomed, gloom·ing, glooms.** —*vi.* **1.** To be or become dark, shaded, or obscure. **2.** To feel, appear, or act despondent. —*vt.* **1.** To make dark, shaded, or obscure. **2.** *Archaic.* To make despondent : SADDEN.

**gloom·y** (glo͞o′mē) *adj.* **-i·er, -i·est. 1.** Partially or wholly dark, esp. dreary and dismal. **2.** Showing or filled with gloom <*gloomy* expressions> **3. a.** Causing or producing gloom : DEPRESSING <*gloomy* reports> **b.** Marked by hopelessness : PESSIMISTIC <*gloomy* expectations> **—gloom′i·ly** *adv.* **—gloom′i·ness** *n.*

☆ ***syns:*** **1.** GLOOMY, BLEAK, CHEERLESS, DISMAL, DREARY, GLUM, JOYLESS, SOMBER *adj. core meaning* : dark and depressing <a *gloomy,* rainy day> **2.** GLOOMY, BLUE, DEPRESSED, DOWNHEARTED, LOW, MELANCHOLY, SAD, SADDENED, UNHAPPY *adj. core meaning* : in poor spirits <felt *gloomy* after the holidays> **3.** GLOOMY, DARK, DISMAL, PESSIMISTIC *adj. core meaning* : marked by little hopefulness <a *gloomy* economic future>

**glop** (glŏp) [Imit. of the sound of food being mixed.] *Slang.* —*n.* **1.** A messy mixture, as of food. **2.** Something considered worthless <wrote terrible *glop*> —*vt.* **glopped, glop·ping, glops.** To cover with glop or put glop on. **—glop'py** *adj.*

**Glo·ri·a** (glôr'ē-ə, glōr'-) *n.* [ME < Lat. *gloria*, glory.] **1. a.** A Christian hymn of praise to God that begins with the word *Gloria*. **b.** The music for such a hymn. **2. gloria.** A halo or nimbus.

**Gloria in ex·cel·sis De·o** (ĭn ĕk-sĕl'sĭs dā'ō, dē'ō) *n.* [LLat., Glory to God in the highest.] A Latin doxology forming part of the Ordinary of the Mass, beginning with the words *Gloria in excelsis Deo.*

**Gloria Pa·tri** (păt'rē, pä'trē) *n.* [LLat., Glory to the Father.] A Latin doxology beginning with the words *Gloria Patri.*

**glo·ri·fy** (glôr'ə-fī', glōr'-) *vt.* **-fied, -fy·ing, -fies.** [ME *glorifien* < OFr. *glorefiier* < LLat. *glorificare* : Lat. *gloria*, glory + Lat. *facere*, to make.] **1.** To give glory, honor, or high praise to : EXALT. **2.** To cause to be or seem better than is actually the case <*glorified* the banal little story into an epic> **3.** To give glory to, esp. through worship. **—glo'ri·fi·ca'tion** *n.* **—glo'ri·fi'er** *n.*

**glo·ri·ole** (glôr'ē-ōl', glōr'-) *n.* [Fr. < Lat. *gloriola*, dim. of *gloria*, glory.] GLORY 8.

**glo·ri·ous** (glôr'ē-əs, glōr'-) *adj.* **1.** Having or deserving glory : FAMOUS. **2.** Conferring or advancing glory <*glorious* accomplishments> **3.** Marked by great beauty and splendor : MAGNIFICENT <*glorious* mountain scenery> **4.** Wonderful : delightful <had a *glorious* time> **—glo'ri·ous·ly** *adv.* **—glo'ri·ous·ness** *n.*

**glo·ry** (glôr'ē, glōr'ē) *n., pl.* **-ries.** [ME *glorie* < OFr. < Lat. *gloria.*] **1.** Great honor, praise, or distinction given by common consent : RENOWN. **2.** Something that brings honor or renown. **3.** A highly praiseworthy asset, esp. a physical asset, as beautiful hair. **4.** Adoration, praise, and thanksgiving offered in worship. **5.** Majestic beauty and splendor : RESPLENDENCE <the *glory* of a stately pine forest> **6.** The splendor and perfect happiness of heaven : BLISS. **7.** A peak of achievement, enjoyment, or prosperity. **8.** A halo, nimbus, or aureole. —*vi.* **-ried, -ry·ing, -ries.** To rejoice triumphantly : EXULT <*gloried* in success>

**gloss**[1] (glôs, glŏs) *n.* [Perh. of Scand. orig.] **1.** A surface shininess or luster. **2.** A superficially or deceptively attractive appearance. —*vt.* **glossed, gloss·ing, gloss·es. 1.** To give a bright luster or sheen to. **2.** To make attractive or acceptable by deception or superficial treatment. —*vi.* To become lustrous or shiny.

**gloss**[2] (glôs, glŏs) *n.* [ME *glose* < Med. Lat. *glosa* < Lat. *glossa*, word requiring explanation < Gk. *glōssa.*] **1. a.** A brief explanatory note or translation of a difficult or technical expression usu. inserted in the margin or between lines of a text or manuscript. **b.** A collection of such notes : GLOSSARY. **2.** A deliberately misleading explanation or interpretation. **3.** An in-depth commentary, often accompanying a text or publication. —*vt.* **glossed, gloss·ing, gloss·es. 1.** To provide (a text) with glosses. **2.** To give a false interpretation to. **—gloss'er** *n.*

**glos·sal** (glŏ'səl, glŏs'əl) *adj.* [< Gk. *glōssa*, tongue.] Of or relating to the tongue.

**glos·sa·ry** (glô'sə-rē, glŏs'ə-) *n., pl.* **-ries.** [Lat. *glossarium* < Lat. *glossa*, word requiring explanation. < Gk. *glōssa.*] A list of words with their definitions, often placed at the back of a book. **—glos·sar'i·al** (glô-sâr'ē-əl, glŏ-) *adj.* **—glos'sa·rist** *n.*

**glos·sog·ra·phy** (glô-sŏg'rə-fē, glŏ-) *n.* The writing and compilation of glosses or glossaries. **—glos·sog'ra·pher** *n.*

**glos·so·la·li·a** (glô'sə-lā'lē-ə, glŏs'ə-) *n.* [Gk. *glōssa*, tongue + Gk. *lalein*, to babble.] **1.** Fabricated and nonsensical speech, esp. the speech associated with certain schizophrenic syndromes. **2.** The gift of tongues.

**gloss·y** (glô'sē, glŏs'ē) *adj.* **-i·er, -i·est. 1.** Having a smooth, shiny, lustrous surface <*glossy* laurel leaves> **2.** Superficially and often artificially attractive : SHOWY. —*n., pl.* **-ies.** A photographic print on smooth, shiny paper. **—gloss'i·ly** *adv.* **—gloss'i·ness** *n.*

**glot·tal** (glŏt'l) *adj.* Pertaining to or articulating in the glottis.

**glottal stop** *n.* A speech sound produced by a momentary complete closure of the glottis, followed by an explosive release.

**glot·tis** (glŏt'ĭs) *n., pl.* **-tis·es** or **-ti·des** (-ĭ-dēz') [Gk. *glōttis* < *glōtta, glōssa*, tongue.] **1.** The space between the vocal cords at the upper part of the larynx. **2.** The vocal structures of the larynx.

**glove** (glŭv) *n.* [ME < OE *glōf.*] **1. a.** A fitted covering for the hand with a separate sheath for the thumb and each finger. **b.** GAUNTLET[1]. **2. a.** *Baseball.* An oversized padded leather covering for the hand, used in catching balls, esp. one with more finger sheath than the catcher's or first baseman's mitt. **b.** A boxing glove. —*vt.* **gloved, glov·ing, gloves. 1.** To furnish with gloves. **2.** To cover with or as if with a glove.

**glove compartment** *n.* A small storage container in the dashboard of a motor vehicle.

**glov·er** (glŭv'ər) *n.* A maker or seller of gloves.

**glow** (glō) *vi.* **glowed, glow·ing, glows.** [ME *glouen* < OE *glōwan.*] **1.** To shine brightly and steadily, esp. without a flame <city lights *glowing* on the horizon> **2. a.** To have a bright, warm, usu. reddish color <faces *glowing* with anticipation> **b.** To flush : blush. **3.** To be exuberant or radiant <*glowing* with joy> —*n.* **1.** A light produced by a body heated to luminosity : INCANDESCENCE. **2.** Brilliance or warmth of color, esp. redness. **3.** A sensation of physical warmth. **4.** A warm feeling of emotion or passion : ARDOR.

**glow·er** (glou'ər) *vi.* **-ered, -er·ing, -ers.** [ME *gloren.*] To stare or look angrily or sullenly. **—glow'er** *n.* **—glow'er·ing·ly** *adv.*

**glow plug** *n.* A small heating element in a diesel engine cylinder for facilitating start-up.

**glow·worm** (glō'wûrm') *n.* A firefly, esp. the luminous larva or wingless grublike female of a firefly.

**glox·in·i·a** (glŏk-sĭn'ē-ə) *n.* [NLat., after Benjamin P. *Gloxin*, 18th-cent. German botanist.] A tropical South American plant of the genus *Sinningia*, esp. *S. speciosa*, cultivated as a house plant for its variously colored flowers.

**gloze** (glōz) *v.* **glozed, gloz·ing, gloz·es.** [ME *glosen*, to interpret < OFr. *gloser* < *glose*, gloss < Med. Lat. *glosa.* —see GLOSS[2].] —*vt.* To underplay or minimize : GLOSS <*glozed* over the confidential material> —*vi. Archaic.* To use flattery or cajolery.

**gluc-** *pref. var. of* GLUCO-.

**glu·ca·gon** (glo͞o'kə-gŏn') *n.* [GLUC(O)- + Gk. *agōn*, pr. part. of *agein*, to lead, drive.] A proteinaceous pancreatic hormone that increases blood sugar.

**gluco-** or **gluc-** *pref.* [< GLUCOSE.] Glucose <*glu*cagon>

**glu·cose** (glo͞o'kōs') *n.* [Fr. < Gk. *gleukos*, sweet wine.] **1.** Dextrose. **2.** A colorless to yellowish syrupy mixture of dextrose, maltose, and dextrins with about 20% water, used in treating tobacco and in confectionery, alcoholic fermentation, and tanning.

**glu·co·side** (glo͞o'kə-sīd') *n.* A glycoside whose sugar component is glucose. **—glu'co·sid'ic** (-sĭd'ĭk) *adj.* **—glu'co·sid'i·cal·ly** *adv.*

**glue** (glo͞o) *n.* [ME *gleu* < OFr. *glu* < LLat. *glus* < Lat. *gluten.*] **1.** A thick, sticky liquid used to hold things together. **2.** An adhesive obtained by boiling certain animal proteins. —*v.* **glued, glu·ing, glues.** —*vt.* To stick or fasten with glue. —*vi.* To fasten on something attentively <All eyes were *glued* on the candidate.> **—glu'ey** *adj.* **—glu'i·ness** *n.*

**glum** (glŭm) *adj.* **glum·mer, glum·mest.** [Orig. unknown.] **1.** Being in low spirits : DEJECTED. **2.** Dismal : gloomy. **—glum'ly** *adv.* **—glum'ness** *n.*

**glu·ma·ceous** (glo͞o-mā'shəs) *adj.* Having or like a glume or glumes.

**glume** (glo͞om) *n.* [NLat. *gluma* < Lat., husk.] A chaffy basal bract on the spikelet of a grass.

**glu·on** (glo͞o'ŏn) *n.* [GLU(E) + -ON[1].] A massless, neutral elementary particle held to mediate the strong interaction that binds quarks together.

**glut** (glŭt) *v.* **glut·ted, glut·ting, gluts.** [ME *glotten* < OFr. *glotoiier*, to eat greedily < Lat. *gluttire.*] —*vt.* **1.** To fill beyond capacity, esp. with food : SATIATE. **2.** To flood (a market) with so many goods that supply exceeds demand. —*vi.* To eat or indulge in something excessively. —*n.* An oversupply <a *glut* of heating oil>

**glu·tam·ic acid** (glo͞o-tăm'ĭk) *n.* [GLUT(EN) + AM(IDE) + -IC.] An amino acid in all complete proteins, found widely in plant and animal tissue and having a salt, sodium glutamate, that is used as a flavor-intensifying seasoning.

**glu·ta·mine** (glo͞o'tə-mēn') *n.* [GLUT(EN) + AMINE.] A white crystalline amino acid, $C_5H_{10}N_2O_3$, found in plant and animal tissue and used in medicine and biochemical research.

**glu·tar·al·de·hyde** (glo͞o'tə-răl'də-hīd') *n.* [Blend of *glutaric acid* (< GLUTEN) and ALDEHYDE.] A water-soluble oily liquid, $C_5H_8O_2$, containing two aldehyde groups, used in tanning leather and as a fixative for biological tissues.

**glu·te·i** (glo͞o'tē-ī', glo͞o-tē'ī') *n. pl. of* GLUTEUS.

**glu·ten** (glo͞ot'n) *n.* [Lat., glue.] A mixture of plant proteins found in cereal grains, mainly corn and wheat, and used as an adhesive and flour substitute. **—glu'ten·ous** *adj.*

**gluten bread** *n.* Bread made from flour with a high gluten but low starch content.

**glu·te·us** (glo͞o'tē-əs, glo͞o-tē'-) *n., pl.* **glu·te·i** (glo͞o'tē-ī', glo͞o-tē'ī') [NLat. < Gk. *gloutos*, buttock.] Any of three large muscles of the buttocks. **—glu'te·al** *adj.*

**glu·ti·nous** (glo͞ot'n-əs) *adj.* [Lat. *glutinosus* < *gluten*, glue.] Like glue : STICKY. **—glu'ti·nous·ly** *adv.* **—glu'ti·nous·ness, glu'ti·nos'i·ty** (-ŏs'ĭ-tē) *n.*

**glut·ton** (glŭt'n) *n.* [ME *glotoun* < OFr. *gloton* < Lat. *glutto.*] **1.** One who consumes immoderate amounts of food and drink. **2.** One with an inordinate capacity to undergo or withstand something <a *glutton* for punishment> **3.** The wolverine. **—glut'ton·ous** *adj.* **—glut'ton·ous·ly** *adv.* **—glut'ton·y** (glŭt'n-ē) *n.*

**glyc-** *pref. var. of* GLYCO-.

**glyc·er·al·de·hyde** (glĭs'ə-răl'də-hīd') *n.* [GLYCER(IN) + ALDEHYDE.] A sweet colorless solid, $C_3H_6O_3$, an intermediate compound in carbohydrate metabolism.

**gly·cer·ic acid** (glĭ-sĕr'ĭk) *n.* [< GLYCERIN.] A syrupy colorless compound, $C_3H_6O_4$.

**glyc·er·ide** (glĭs'ə-rīd') *n.* [GLYCER(IN) + -IDE.] An ester of glycerol and fatty acids.

**glyc·er·in** (glĭs′ər-ĭn) *n.* [Fr. < Gk. *glukeros,* sweet.] Glycerol.

**glyc·er·ol** (glĭs′ə-rôl′, -rōl′, -rŏl′) *n.* [GLYCER(IN) + -OL.] A syrupy, sweet, colorless or yellowish liquid, $C_3H_8O_3$, derived from fats and oils as a by-product of the manufacture of soaps and fatty acids and used as a solvent, antifreeze and antifrost fluid, plasticizer, and sweetener and in production of dynamite, cosmetics, liquid soaps, inks, and lubricants.

**glyc·er·yl** (glĭs′ər-əl) *n.* [GLYCER(IN) + -YL.] The trivalent glycerol radical $CH_2CHCH_2$.

**gly·cin** (glī′sĭn) *also* **gly·cine** (-sēn′, -sĭn) *n.* [< GLYCINE.] A poisonous compound, $C_8H_9NO_3$, used as a photographic developer.

**gly·cine** (glī′sēn′, -sĭn) *n.* **1.** A white, extremly sweet crystalline amino acid, $C_2H_5NO_2$, the principal amino acid found in sugar cane, derived by alkaline hydrolysis of gelatin and used in biochemical research and medicine. **2.** *var. of* GLYCIN.

**glyco–** *or* **glyc–** *pref.* [< Gk. *glukus,* sweet.] **1.** Sugar <*glyco*protein> **2.** Glycogen <*glyco*genesis>

**gly·co·gen** (glī′kə-jən) *n.* A white, sweet-tasting powder, $(C_6H_{10}O_5)_n$, occurring as the chief animal storage carbohydrate, chiefly in the liver. **—gly′co·gen′ic** *adj.*

**gly·co·gen·e·sis** (glī′kə-jĕn′ĭ-sĭs) *n.* **1.** Formation of glycogen. **2.** Formation of sugar from glycogen. **—gly′co·ge·net′ic** (-jə-nĕt′ĭk) *adj.*

**gly·col** (glī′kôl′, -kōl′, -kŏl′) *n.* **1.** Ethylene glycol. **2.** Any of several dihydric alcohols.

**gly·col·ic acid** (glī-kŏl′ĭk) *n.* A colorless crystalline compound, $C_2H_4O_3$, found in sugar beets, cane sugar, and unripe grapes and used in leather dyeing and tanning and in pharmaceuticals, pesticides, adhesives, and plasticizers.

**gly·co·pro·tein** (glī′kō-prō′tēn′, -tē-ĭn) *n.* Any of several conjugated proteins that contain carbohydrates as prosthetic groups.

**gly·co·side** (glī′kə-sīd′) *n.* [*Glycose* (var. of GLUCOSE) + -IDE.] Any of a group of organic compounds, occurring abundantly in plants, that produce sugars and related substances on hydrolysis. **—gly′co·sid′ic** (-sĭd′ĭk) *adj.*

**gly·co·su·ri·a** (glī′kə-sŏŏr′ē-ə, -shŏŏr′-) *n.* [*Glycose* (var. of GLUCOSE) + -URIA.] Excretion of abnormal quantities of sugar in the urine. **—gly′co·su′ric** *adj.*

**gly·ox·a·line** (glī-ŏk′sə-lēn′, -lĭn) *n.* [GLY(COL) + OXAL(IC ACID) + -INE.] Imidazole.

**glyph** (glĭf) *n.* [Gk. *gluphē,* carving < *gluphein,* to carve.] **1.** A vertical groove, esp. in a Doric column or frieze. **2.** A symbolic figure usu. incised or engraved. **3.** A symbol, as figures of people on a road sign, that communicates nonverbally. **—glyph′ic** *adj.*

**glyp·tic** (glĭp′tĭk) *adj.* [Gk. *gluptikos* < *gluphein,* to carve.] Of or pertaining to engraving or carving, esp. on precious stones.

**glyp·tics** (glĭp′tĭks) *n.* *(sing. in number).* The art of engraving or carving, esp. on precious stones : GLYPTOGRAPHY.

**glyp·to·graph** (glĭp′tə-grăf′) *n.* [Gk. *gluptos,* carved (< *gluphein,* to carve) + -GRAPH.] An inscription engraved on a precious stone.

**glyp·tog·ra·phy** (glĭp-tŏg′rə-fē) *n.* The art or process of engraving or carving on precious stones. **—glyp·tog′ra·pher** *n.* **—glyp′to·graph′ic** (-tə-grăf′ĭk), **glyp′to·graph′i·cal** *adj.*

**G-man** (jē′măn′) *n.* [G(OVERNMENT) + MAN.] An FBI agent.

**gnar** *also* **gnarr** (när) *vi.* **gnarred, gnar·ring, gnars.** [Imit.] GNARL[1].

**gnarl[1]** (närl) *vi.* **gnarled, gnarl·ing, gnarls.** [Freq. of GNAR.] To growl.

**gnarl[2]** (närl) *n.* [Back-formation < GNARLED.] A protruding knot on a tree. *—vt.* **gnarled, gnarl·ing, gnarls.** To make knotted : TWIST.

**gnarled** (närld) *adj.* [Prob. var. of KNURLED.] **1.** Having gnarls : KNOTTY <*gnarled* old apple trees> **2.** Crabbed in dispostion. **3.** Rugged and roughened, as from old age or work <the *gnarled* hands of a farmer>

**gnarr** (när) *v. var. of* GNAR.

**gnash** (năsh) *vt.* **gnashed, gnash·ing, gnash·es.** [Alteration of ME *gnasten, gnaisten,* of Scand. orig.] *—vt.* **1.** To strike or grind (e.g., the teeth) together. **2.** To bite by grinding the teeth. **—gnash** *n.*

**gnat** (năt) *n.* [ME < OE *gnæt.*] A small winged insect, esp. one that bites.

**gnat·catch·er** (năt′kăch′ər, -kĕch′-) *n.* Any of several small New World birds of the genus *Polioptila,* with grayish and white plumage and a long tail.

**gna·thal** (nā′thəl, năth′əl) *adj.* [< Gk. *gnathos,* jaw.] Gnathic.

**gnath·ic** (năth′ĭk) *adj.* [< Gk. *gnathos,* jaw.] Of or relating to the jaw.

**gna·thite** (nā′thīt′, năth′īt′) *n.* [Gk. *gnathos,* jaw + -ITE.] A jaw or jawlike appendage of an insect or other arthropod.

**-gnathous** *suff.* [NLat. *-gnathus,* < Gk. *gnathos,* jaw.] Having a specified kind of jaw <*metagnathous*>

**gnaw** (nô) *v.* **gnawed, gnaw·ing, gnaws.** [ME *gnauen* < OE *gnagan.*] *—vt.* **1. a.** To chew on, bite, or erode with the teeth. **b.** To produce by gnawing <*gnaw* an opening in the wall> **c.** To erode or diminish gradually as if by gnawing <rough tree branches *gnawing* the side of the barn> **2.** To afflict or trouble persistently <hunger *gnawed* their stomachs> *—vi.* **1.** To bite or chew persistently <Rats *gnawed* at the ship's lines.> **2.** To cause erosion or gradual diminishment. **—gnaw′er** *n.*

**gneiss** (nīs) *n.* [G. *Gneis.*] A banded or foliated metamorphic rock, usu. of the same composition as granite, with the minerals arranged in layers. **—gneiss′ic** (nī′sĭk), **gneiss′oid′** (nī′soid′), **gneiss′ose′** (nī′sōs′) *adj.*

**gnoc·chi** (nyô′kē) *pl.n.* [Ital., pl. of *gnocco,* alteration of *nocchio,* knot in wood.] Dumplings made of flour, semolina, or potatoes, baked or boiled and served with grated cheese or sauces.

**gnome[1]** (nōm) *n.* [Fr. < NLat. *gnomus.*] **1.** One of a fabled race of dwarflike underground creatures who guard treasure hoards. **2.** A shriveled old man. **—gnom′ish** *adj.*

**gnome[2]** (nōm) *n.* [Gk. *gnōmē* < *gignōskein,* to know.] A pithy saying expressing a fundamental principle or general truth.

**gno·mic** (nō′mĭk) *adj.* [< GNOME[2].] Marked by aphorisms : APHORISTIC <essays written in *gnomic* style>

**gno·mon** (nō′mŏn′, -mən) *n.* [Lat. < Gk. *gnōmōn* < *gignōskein,* to know.] **1.** An object, as the style of a sundial, that makes a shadow used as an indicator. **2.** The figure remaining after a parallelogram has been removed from a similar but larger parallelogram with which it has a common corner. **—gno·mon′ic, gno·mon′i·cal** *adj.*

**gno·sis** (nō′sĭs) *n.* [Gk. *gnōsis,* knowledge < *gignōskein,* to know.] Intuitive apprehension of spiritual truths.

**gnos·tic** (nŏs′tĭk) *adj.* [Lat. *Gnosticus,* a Gnostic < Gk. *Gnōstikos* < *gignōskein,* to know.] **1.** Of, relating to, or possessing spiritual or intellectual knowledge. **2. Gnostic.** Of or relating to Gnosticism. *—n.* **Gnostic.** A believer in Gnosticism.

**Gnos·ti·cism** (nŏs′tĭ-sĭz′əm) *n.* The doctrines of certain early Christian sects that valued inquiry into spiritual truth above faith and considered salvation attainable only by the few whose faith enabled them to transcend matter.

**gnu** (nōō, nyōō) *n.* [Xhosa *i-gnu.*] A large African antelope, *Connochaetes gnou* or *C. taurinus,* with a drooping beard, a long tufted tail, and curved horns in both sexes.

**go[1]** (gō) *v.* **went** (wĕnt), **gone** (gôn, gŏn), **go·ing, goes.** [ME *gon* < OE *gān.*] *—vi.* **1.** To proceed along. **2.** To move away from a place : DEPART <*Go* before I cry.> **3. a.** To pursue a particular course. **b.** To resort to another, as for aid <*went* to the fire department> **4. a.** To extend between two points or in a given direction : RUN <curtains *going* from floor to ceiling> **b.** To give entry : LEAD <a stairway *going* to the first floor> **5.** To function properly <The bike won't *go.*> **6. a.** To have currency. **b.** To pass from one person to another, as a story : CIRCULATE. **7.** To pass as the result of a sale <a figurine that *went* to the highest bidder> **8.** *Informal.* —Used as an intensifier when joined by *and* to a coordinate verb <*went* and complained to personnel> **9.** —Used in the progressive tense with an infinitive to indicate future intent or expectation <is *going* to learn how to sing> **10. a.** To be in a given state <The vote *went* against them.> **b.** To come to be in a given state <*go* nuts><*go* to sleep> **c.** To continue in a given state or continue an activity <The party was still *going* strong at midnight.> **d.** To carry out an action to a certain point or extent <*went* to great lengths> **11. a.** To be customarily located : BELONG <Where do the sheets *go?*> **b.** To be capable of entering or fitting <Will the case *go* into the trunk of your car?> **12. a.** To pass into one's possession <The estate *went* to the children.> **b.** To be allotted <One quarter of my salary *goes* for rent.> **13.** To be a contributing factor that leads to <It all *goes* to show us that it can be done.> **14.** To have a specific form <as the saying *goes*> **15. a.** To pass by : ELAPSE. **b.** To be used up. **c.** To be discarded or abolished <Luxuries will have to *go.*> **16. a.** To become weak : FAIL <My hearing started to *go.*> **b.** To come apart : break up. **17.** To cease living : DIE. **18. a.** To get along : FARE <How are things *going?*> **b.** To have a successful outcome <made the ad campaign really *go*> **19. a.** To be suitable or appropriate as an accessory or accompaniment <That color *goes* well with dark hair.> **b.** To be as a general rule <was well-behaved as dogs *go*> **20. a.** To have authority <Whatever I say *goes.*> **b.** To be acceptable, valid, or adequate. **21.** *Informal.* To excrete waste from the bladder or bowels. **22.** *Obs.* To walk. *—vt. Informal.* To say. —Used chiefly in verbal narration <First I go "thank you," then my cousin *goes* "for what?"> **—go about.** To undertake. **—go after.** To pursue : seek. **—go along.** To cooperate. **—go around.** To satisfy a demand or requirement <just enough meat to *go around*> **—go at.** To attack, esp. energetically. **—go back on. 1.** To fail to keep <*go back on* a promise> **2.** To betray. **—go down. 1.** To sink below the horizon : SET. **2.** To undergo ruin or defeat. **3.** *Chiefly Brit.* To leave a university. **4.** *Slang.* To occur <A robbery is *going down.*> **—go for. 1.** To have a special liking for <really *goes for* chocolate> **2.** To attack <*go for* the jugular> **—go in.** To take part in a cooperative venture. **—go in for. 1.** To have an interest in <*goes in for* offbeat clothes> **2.** To take part in <*go in for* water-skiing> **—go off. 1.** To explode. **2.** To make a noise : SOUND. **—go on.** To happen <What's *going on?*> **—go out.** To become extinguished. **—go out for.** To seek to become a participant in <*go out for* track> **—go over. 1.** To gain approval or acceptance. **2.** To examine <*go over* test papers> **—go under. 1.** To suffer destruction or defeat. **2.** To lose consciousness.

**—go with.** To date regularly. *—n., pl.* **goes. 1.** The act or an instance of going. **2.** An attempt : effort <had a go at acting> **3.** The time or period of an activity. **4.** *Informal.* Energy : vitality <had lots of go> **5.** A go-ahead. *—adj. Informal.* Functioning correctly and ready for action <All systems are *go.*> **—go one better.** To surpass by one degree. **—go places.** To be on the way to success. **—go steady.** To date exclusively. **—go straight.** To reform <former addicts now *going straight*> **—go the distance.** To carry a course of action through to completion. **—go to (one's) head. 1.** To make one excited or dizzy. **2.** To make one vain or overconfident. **—go to pieces. 1.** To lose one's self-control. **2.** To suffer the loss of one's health. **—no go.** *Informal.* Ineffective : useless. **—on the go.** Constantly active. **—to go.** To be taken off the premises, as restaurant food or drink <coffee and sandwiches *to go*>

**go²** (gō) *n.* [J.] A Japanese game for two, played with counters on a board ruled with 19 vertical and 19 horizontal lines.

**go·a** (gō'ə) *n.* [Tibetan *dgoba.*] A gazelle, *Procapra picticaudata* of eastern Asia, the male of which has backward-curving horns.

**goad** (gōd) *n.* [ME *gode* < OE *gād.*] **1.** A long stick with a pointed end for prodding animals. **2.** Something that urges or prods : STIMULUS. *—vt.* **goad·ed, goad·ing, goads. 1.** To prod with a goad. **2.** To give impetus to as if with a goad : INCITE.

**go-a·head** (gō'ə-hĕd') *n. Informal.* Permission to proceed.

**goal** (gōl) *n.* [ME *gol,* boundary.] **1.** The objective toward which an endeavor is directed. **2.** The finish line of a race. **3. a.** A specified structure or area into or over which players try to advance a ball or puck. **b.** The score awarded for such an act.

**goal·ie** (gō'lē) *n.* A goalkeeper.

**goal·keep·er** (gōl'kē'pər) *n.* A player who protects the goal in various sports.

**goal post** *n.* One of a pair of posts with or without a crossbar forming the goal in various games.

**goal·tend·er** (gōl'tĕn'dər) *n.* A goalkeeper.

**goal·tend·ing** (gōl'tĕn'dĭng) *n.* **1.** An act of protecting the goal in a sport, as hockey. **2.** *Basketball.* An illegal play in which a player deflects a ball that is on the downward path to the basket or has already broken the plane of the cylinder.

**go-a·round** (gō'ə-round') *n.* **1.** An argument. **2.** RUN-AROUND 1.

**goat** (gōt) *n.* [ME *got* < OE *gāt.*] **1.** A bearded horned ruminant mammal of the genus *Capra,* orig. of Old World mountainous regions, esp. a domesticated form of *C. hircus.* **2. Goat.** CAPRICORN 1, 2a. **3.** A lecher. **4.** A scapegoat. **—get (someone's) goat.** *Informal.* To make annoyed or angry. **—goat'ish** *adj.*

**goat antelope** *n.* A ruminant mammal, as the mountain goat or the chamois, with characteristics both of goats and antelopes.

**goat·ee** (gō-tē') *n.* A small pointed chin beard.

**goat·fish** (gōt'fĭsh') *n., pl.* **goatfish** or **-fish·es.** Any of various brightly colored fishes of the family Mullidae of warm seas, with two sensory barbels on the chin.

**goats·beard** *also* **goat's-beard** (gōts'bîrd') *n.* **1.** A native European plant, *Tragopogon pratensis,* with grasslike leaves and yellow dandelionlike flowers. **2.** A tall plant, *Aruncus dioicus,* bearing compound leaves and small white flowers in branching clusters.

**goat·skin** (gōt'skĭn') *n.* **1. a.** The skin of a goat. **b.** Leather made from this. **2.** A container, as for wine, made from a goatskin.

**goat's-rue** (gōts'ro͞o') *n.* **1.** A North American plant, *Tephrosia virginiana,* with yellow and pink flowers. **2.** A Eurasian plant, *Galega officinalis,* cultivated for its variously colored flowers.

**goat·suck·er** (gōt'sŭk'ər) *n.* [From the belief that the bird sucked milk from goats.] Any of various chiefly nocturnal birds of the family Caprimulgidae, including the whippoorwill and nighthawk.

**gob¹** (gŏb) *n.* [ME *gobbe* < *gobet.* —see GOBBET.] **1.** A small piece or lump. **2.** *often* **gobs.** *Informal.* A large quantity <*gobs* of money>

**gob²** (gŏb) *n.* [Sc. and Ir. Gael.] *Slang.* The mouth.

**gob³** (gŏb) *n.* [Orig. unknown.] *Slang.* A sailor.

**gob·bet** (gŏb'ĭt) *n.* [ME *gobet* < OFr.] **1.** A chunk or piece, esp. of raw meat. **2.** A lump : mass. **3.** A drop of liquid.

**gob·ble¹** (gŏb'əl) *v.* **-bled, -bling, -bles.** [Perh. < GOB¹.] *—vt.* **1.** To devour in large, greedy gulps. **2.** To take greedily : GRAB <*gobbled* up the only remaining jobs> *—vi.* To eat rapidly or greedily.

**gob·ble²** (gŏb'əl) *n.* [Imit.] The chortling, guttural sound of a male turkey. *—vi.* **-bled, -bling, -bles.** To make a gobble.

**gob·ble·dy·gook** *also* **gob·ble·de·gook** (gŏb'əl-dē-go͝ok') *n.* [Coined by Maury Maverick (1895–1954).] Unclear, wordy jargon.

**gob·bler** (gŏb'lər) *n.* A male turkey.

**Go·be·lin** (gō'bə-lĭn, gŏb'ə-) *n.* A tapestry of a kind woven at the Gobelin works in Paris, France, noted for rich pictorial design.

**go-be·tween** (gō'bĭ-twēn') *n.* One who acts as a messenger between two sides : INTERMEDIARY.

**gob·let** (gŏb'lĭt) *n.* [ME *gobelet* < OFr., dim. of *gobel,* cup.] **1.** A drinking vessel, as a glass, with a stem and base. **2.** *Archaic.* A drinking bowl without handles.

**gob·lin** (gŏb'lĭn) *n.* [ME *gobelin* < OFr. < Med. Lat. *gobelinus.*] A grotesque elfin creature of folklore, thought to work mischief or evil.

**go·by** (gō'bē) *n., pl.* **goby** or **-bies.** [Lat. *gobius,* gudgeon < Gk. *kōbios.*] Any of numerous usu. small freshwater and marine fishes of the family Gobiidae, with the pelvic fins joined to form a sucking disk.

**go-by** (gō'bī') *n. Informal.* An intentional slight : SNUB.

**go-cart** (gō'kärt') *n.* **1.** A small wagon for children. **2.** A small frame on casters that supports a child learning to walk. **3.** A handcart. **4.** A stroller.

**god** (gŏd) *n.* [ME < OE.] **1. God. a.** A being conceived as the perfect, omnipotent, omniscient originator and ruler of the universe, the principal object of faith and worship in monotheistic religions. **b.** The force, effect, or a manifestation or aspect of this being. **c.** *Christian Science.* —Used to refer to <"Infinite Mind; Spirit; Soul; Principle; Life; Truth; Love"—Mary Baker Eddy> **2.** A being of supernatural powers or attributes, believed in and worshiped by a people, esp. a male deity held to control some part of nature or reality. **3.** An image of a supernatural being : IDOL. **4.** Something worshiped or idealized <In ancient Egypt the sun was a *god.*> **5.** A very handsome man.

**God-aw·ful** (gŏd'ô'fəl) *adj. Slang.* Particularly disagreeable.

**god·child** (gŏd'chīld') *n.* One for whom another serves as sponsor at baptism.

**god·daugh·ter** (gŏd'dô'tər) *n.* A girl who is a godchild.

**god·dess** (gŏd'ĭs) *n.* **1.** A female being of supernatural powers or attributes, believed in and worshiped by a people. **2.** An image of a supernatural being : IDOL. **3.** Something worshiped or idealized. **4.** An extremely beautiful woman.

**go-dev·il** (gō'dĕv'əl) *n.* **1.** A logging sled. **2.** A railway handcar. **3.** A tool for cleaning an oil pipeline and disengaging obstructions. **4.** An iron dart dropped down an oil well to explode a charge of dynamite.

**god·fa·ther** (gŏd'fä'*th*ər) *n.* **1.** A man who sponsors a person at baptism. **2.** *Informal.* A mentor.

**god·for·sak·en** *also* **God·for·sak·en** (gŏd'fər-sā'kən) *adj.* **1.** Situated in a remote or dismal area. **2.** Forlorn : desolate.

**god·head** (gŏd'hĕd') *n.* [ME *godhede* : *god,* god (< OE) + *-hede,* -hood.] **1.** Divinity : godhood. **2. Godhead. a.** GOD 1a, b. **b.** The essential and divine nature of God, regarded abstractly.

**god·hood** (gŏd'ho͝od') *n.* [ME *godhode* : *god,* god (< OE) + *-hode,* -hood.] The quality or state of being a god : DIVINITY.

**god·less** (gŏd'lĭs) *adj.* Worshiping or recognizing no god. **—god'less·ly** *adv.* **—god'less·ness** *n.*

**god·like** (gŏd'līk') *adj.* Like or of the nature of a god or God : DIVINE. **—god'like'ness** *n.*

**god·ling** (gŏd'lĭng) *n.* A minor god.

**god·ly** (gŏd'lē) *adj.* **-li·er, -li·est. 1.** Having great reverence for God : PIOUS. **2.** Divine. **—god'li·ness** *n.*

**god·moth·er** (gŏd'mŭ*th*'ər) *n.* A woman who sponsors a person at baptism.

**god·par·ent** (gŏd'pâr'ənt, -păr'-) *n.* A godfather or godmother.

**God's acre** *n.* [Transl. of G. *Gottesacker.*] A churchyard or burial ground.

**god·send** (gŏd'sĕnd') *n.* Something wanted or needed that comes or happens unexpectedly.

**god·son** (gŏd'sŭn') *n.* A boy who is a godchild.

**God·speed** (gŏd'spēd') *n.* [From the phrase *God speed you.*] Success or good fortune <They boarded the ship, and we bade them *Godspeed.*>

**god·wit** (gŏd'wĭt') *n.* [Orig. unknown.] Any of various wading birds of the genus *Limosa,* with a long, slender, slightly upturned bill.

**goe·thite** (gō'thīt', gœ'tīt') *n.* [After Johann W. von *Goethe* (1749–1832).] A brown mineral, essentially $HFeO_2$, used as an iron ore.

**go·fer** *also* **go-fer** (gō'fər) *n.* [Alteration of *go for,* from that person's having to go for or after things.] *Slang.* An employee who runs errands in addition to performing regular duties.

**gof·fer** *also* **gauf·fer** (gŏf'ər, gô'fər) [Fr. *gaufrer,* to emboss < OFr. *gaufre,* honeycomb < MLG *wāfel.*] *—vt.* **-fered, -fer·ing, -fers.** To press narrow pleats or ridges into (e.g., a frill). *—n.* **1.** An iron used for goffering. **2.** Pleated or ridged ornamentation made by goffering.

**go-get·ter** (gō'gĕt'ər) *n. Informal.* An enterprising person : HUSTLER.

**gog·gle** (gŏg'əl) *v.* **-gled, -gling, -gles.** [ME *gogelen,* to squint.] *—vi.* **1.** To stare with bulging eyes. **2.** To roll or bulge. —Used of the eyes. *—vt.* To roll or bulge (the eyes). *—n.* **1.** A stare or leer. **2. goggles.** A pair of large, usu. tinted spectacles with shielding side pieces worn as protection against wind, dust, or glare. **—gog'gly** *adj.*

**gog·gle-eyed** (gŏg'əl-īd') *adj.* Having prominent or rolling eyes.

**go-go** *also* **go·go** (gō'gō') *adj.* [< À GOGO.] *Informal.* **1. a.** Of or relating to discotheques or to the energetic music and dancing performed at discotheques. **b.** Engaged to perform at a discotheque <a go-go dancer> **2.** Lively : energetic. **3.** Of, relating to, or engaging in a type of speculative and risky stock-market operation <go-go investments>

**Goi·del·ic** (goi-dĕl'ĭk) *n.* [< OIr. *Goidel,* Gael.] A branch of the Celtic languages that includes Irish Gaelic, Scottish Gaelic, and Manx. *—adj.* **1.** Of or relating to the Gaels. **2.** Of, relating to, or typifying of Goidelic.

**go·ing** (gō'ĭng) *n.* **1.** An act or instance of going : DEPARTURE. **2.** The condition underfoot as it affects one's progress in walking or riding <The *going* was rough.> **3.** *Informal.* Progress toward a goal : HEADWAY. —*adj.* **1.** Working : running <in *going* order> **2.** In full operation : FLOURISHING <a *going* enterprise> **3.** Prevailing : current <The *going* rate on oil deliveries has declined.> **4.** Available : existing <the best ball team *going*>

**go·ing-o·ver** (gō'ĭng-ō'vər) *n., pl.* **go·ings-o·ver.** *Informal.* **1.** An inspection : examination. **2. a.** A severe beating : THRASHING. **b.** A harsh reprimand.

**go·ings-on** (gō'ĭngz-ŏn') *pl.n. Informal.* Behavior or proceedings, esp. when looked upon with disapproval.

**goi·ter** *also* **goi·tre** (goi'tər) *n.* [Fr. *goitre* < Prov. *goitron* < Lat. *guttur,* throat.] A chronic, noncancerous enlargement of the thyroid gland, visible as a swelling at the front of the neck, occurring without hyperthyroidism and associated with iodine deficiency. **—goi'trous** (-trəs) *adj.*

**Gol·con·da** (gŏl-kŏn'də) *n.* [After *Golconda,* India.] A source of great riches, as a mine.

**gold** (gōld) *n.* [ME < OE.] **1.** *Symbol* **Au** A soft, yellow, corrosion-resistant, highly malleable and ductile metallic element that is used as an international monetary standard, in jewelry, for decoration, and as a plated coating on a wide variety of electrical and mechanical components; atomic number 79; atomic weight 196.967. **2. a.** Gold coinage. **b.** A gold standard. **3.** Riches : money. **4.** A moderate to vivid yellow. **5.** Something highly valued, as for purity or goodness <a heart of *gold*> —*adj.* Of the color of gold.

**gold·beat·er's skin** (gōld'bē'tərz) *n.* Treated animal membrane for separating sheets of gold being hammered into gold leaf.

**gold·beat·ing** (gōld'bē'tĭng) *n.* The act, art, or process of beating sheets of gold into gold leaf. **—gold'beat'er** *n.*

**gold brick** *n.* **1.** A bar of gilded metal that appears to be genuine gold. **2.** A fraudulent and worthless substitute.

**gold·brick** (gōld'brĭk') *Slang.* —*n.* One, esp. a soldier, who avoids assigned tasks : SHIRKER. —*v.* **-bricked, -brick·ing, -bricks.** —*vi.* To evade one's assigned tasks. —*vt.* To cheat : swindle. **—gold'brick'er** *n.*

**gold bug** *n.* **1.** A North American beetle, *Metriona bicolor,* having a metallic luster. **2.** A supporter of the gold standard.

**gold certificate** *n.* A monetary note issued at one time by the U.S. Treasury to the public and redeemable in gold but now issued to Federal Reserve Banks to certify compliance with their legal reserve requirements.

**gold digger** *n. Slang.* A woman who seeks money and gifts from men.

**gold dust** *n.* Gold in powder form.

**gold·en** (gōl'dən) *adj.* **1.** Of, relating to, composed of, or containing gold. **2. a.** Having a gold color or a yellow color resembling gold. **b.** Radiant : lustrous <the *golden* marigold> **c.** Suggestive of gold, as in richness or splendor <a *golden* age of opera> **3.** Of the greatest value or importance : PRECIOUS. **4.** Marked by peace, prosperity, and often creativeness <*golden* years> **5.** Very advantageous or favorable : EXCELLENT <a *golden* opportunity> **6.** Having a promising future <*golden* youth> **7.** Of or relating to a 50th anniversary. **—gold'en·ly** *adv.* **—gold'en·ness** *n.*

**golden age** *n.* **1.** *Gk. & Rom. Myth.* The first age of the world, a prosperous and untroubled era when people lived in ideal happiness. **2.** An era of peace, prosperity, and happiness.

**golden ager** *n.* An elderly person.

**golden Al·ex·an·ders** (ăl'ĭg-zăn'dərz) *n. (sing. or pl. in number).* A plant, *Zizia aurea* of eastern North America, bearing small yellow flower clusters.

**golden aster** *n.* Any of various North American plants of the genus *Chrysopsis,* with yellow rayed flowers.

**golden bantam** *n.* A variety of corn with large bright-yellow kernels on a rather small ear.

**golden calf** *n.* **1.** A golden image of a sacrificial calf wrought by Aaron and worshiped by the Israelites. **2.** Money as an object of worship : MAMMON.

**golden club** *n.* An aquatic plant, *Orontium aquaticum* of the eastern United States, bearing small golden-yellow flowers that cover a clublike spadix.

**golden eagle** *n.* An eagle, *Aquila chrysaetos* of mountainous areas of the Northern Hemisphere, bearing dark plumage with yellowish feathers on the neck and head.

**gol·den·eye** (gōl'dən-ī') *n.* [From their golden-yellow eyes.] A duck, *Bucephala clangula* or *B. islandica* of northern regions, with a short black bill and black and white plumage.

**Golden Fleece** *n. Gk. Myth.* The fleece of the golden ram, stolen from the king of Colchis by Jason and the Argonauts.

**golden glow** *n.* A tall plant, *Rudbeckia laciniata hortensis,* with many-rayed double yellow flowers.

**Golden Horde** *n.* [From the golden tent of their commander.] The Mongol army that swept across eastern Europe in the 13th cent. and controlled Russia until 1486.

**golden mean** *n.* A course between extremes.

**golden oldie** *n. Informal.* A recording, movie, or other form of entertainment that had great popularity in the past.

**golden parachute** *n. Slang.* A lucrative termination agreement with an executive who is fired or demoted following a corporate takeover.

**golden pheasant** *n.* A pheasant, *Chrysolophus pictus* of China and Tibet, with a long tail and brilliantly colored plumage.

**gold·en·rod** (gōl'dən-rŏd') *n.* Any of numerous chiefly North American plants of the genus *Solidago,* bearing small yellow flower clusters in late summer or fall.

**golden rule** *n.* The Biblical teaching that one ought to behave toward others as one would have others behave toward oneself.

**gold·en·seal** (gōl'dən-sēl') *n.* A woodland plant, *Hydrastis canadensis* of eastern North America, with small greenish-white flowers and a yellow root occas. used medicinally.

**golden section** *n.* A ratio, observed esp. in the fine arts, between the two dimensions of a plane figure or the two divisions of a line such that the smaller is to the larger as the larger is to the sum of the two, approx. a ratio of three to five.

**gold-filled** (gōld'fĭld') *adj.* Made of a hard base metal with an outer layer of gold.

**gold·finch** (gōld'fĭnch') *n.* **1.** A small New World bird of the genus *Spinus,* esp. *S. tristis,* of which the male has yellow plumage with a black forehead, wings, and tail. **2.** A small Old World bird, *Carduelis carduelis,* having brownish plumage with red, yellow, and black markings.

**gold·fish** (gōld'fĭsh') *n., pl.* **goldfish** *or* **-fish·es.** A freshwater fish native to eastern Asia, *Carassius auratus,* with brassy or reddish coloring and bred as an aquarium fish.

**gold foil** *n.* Gold that is beaten or rolled into thin sheets thicker than gold leaf.

**gold·i·locks** (gōl'dē-lŏks') *n.* [Obs. *goldy,* golden (< GOLD) + LOCKS.] *(sing. in number).* A European plant, *Linosyrus vulgaris,* with narrow leaves and small yellow flower clusters.

**gold leaf** *n.* Gold beaten into very thin sheets used esp. for gilding.

**gold mine** *n. Informal.* A lucrative source, as of information.

**gold-of-pleas·ure** (gōld'əv-plĕzh'ər) *n.* An Old World plant, *Camelina sativa,* bearing small yellow flowers and seeds rich in oil.

**gold point** *n.* **1.** The point in foreign-exchange rates at which it is no more expensive to export or import gold bullion in settling international accounts than to buy or sell bills of exchange. **2.** A fixed point on the international temperature scale that is equal to the melting point of gold, 1064.43°C, at a pressure of one atmosphere.

**gold reserve** *n.* The gold bullion reserve held by a government or central bank for redeeming its notes.

**gold rush** *n.* A rush of migrants to a region where gold has been found.

**gold·smith** (gōld'smĭth') *n.* **1.** An artisan who creates objects of gold. **2.** A trader in gold articles.

**goldsmith beetle** *n.* A scarabaeid beetle, *Cotalpa lanigera* or *Cetonia aurata,* with metallic greenish-yellow coloring.

**gold standard** *n.* A monetary standard under which the basic unit of currency equals the value of and is exchangeable for a specified amount of gold.

**gold·stone** (gōld'stōn') *n.* An aventurine with particles of gold-colored material.

**gold·thread** (gōld'thrĕd') *n.* A low-growing woodland plant, *Coptis trifolia,* bearing white flowers and slender yellow roots.

**go·lem** (gō'ləm) *n.* [Yiddish *goylem* < Heb. *gōlem* < *gālam,* he wrapped up.] An artificially created human being supernaturally endowed with life.

**golf** (gŏlf, gôlf) *n.* [ME.] An outdoor game played on a large course with a series of 9 or 18 holes spaced far apart, the object being to drive a small hard ball, using special clubs, into each hole with as few strokes as possible. —*vi.* **golfed, golf·ing, golfs.** To play golf. **—golf'er** *n.*

**golf club** *n.* **1.** One of a set of clubs with a slender shaft and a head of wood or iron, used in golf. **2.** An organization of golfers.

**golf course** *n.* A large tract of land laid out for playing golf.

**golf·links** (gŏlf'lĭngks') *pl.n.* A golf course.

**Gol·gi apparatus** *also* **Gol·gi complex** (gôl'jē) *n.* [After Camillo *Golgi* (1844–1926).] A network of fibrils, granules, and membranous structures present in living cells and held to function in the formation of secretions within the cell.

**gol·go·tha** (gŏl'gə-thə) *n.* [After *Golgotha,* the hill near Jerusalem where Jesus was crucified.] A place or occasion of great suffering.

**gol·iard** (gōl'yərd, -yär') *n.* [OFr., glutton < *gole,* throat < Lat. *gula.*] A wandering student in medieval Europe inclined to conviviality, license, and the making of satirical and ribald Latin songs. **—gol·iar'dic** (gōl-yär'dĭk) *adj.*

**Go·li·ath** (gə-lī'əth) *n.* [Heb. *Golyath.*] **1.** A giant Philistine warrior in the Old Testament slain by David with a sling and stone. **2.** One of colossal power or achievement <a *Goliath* of the oil business>

**gol·li·wog** *or* **gol·li·wogg** (gŏl'ē-wŏg') *n.* [After *Golliwog,* a character in books by Bertha Upton (d. 1912).] **1.** A grotesque black male doll. **2.** A grotesque person.

**gol·ly** (gŏl'ē) *interj.* [Alteration of GOD.] —Used to express mild surprise or wonder.

**gom·pho·sis** (gŏm-fō'sĭs) *n., pl.* **-ses** (-sēz') [Gk. *gomphōsis* < *gomphoun,* to fasten with bolts < *gomphos,* bolt.] An immovable peg and rigid socket articulation, as of a tooth and its bony socket.

**gon-** *pref. var. of* GONOR-.

**-gon** *suff.* [Gk. *-gōnon* < *gōnia,* angle.] A figure having a specified kind or number of angles <isogon>

**go·nad** (gō'năd') *n.* [NLat. < Gk. *gonos,* procreation.] A bodily organ that produces gametes. **—go·nad'al** (gō-năd'l), **go·nad'ic** *adj.*

**go·nad·o·trop·ic** (gō-năd'ə-trŏp'ĭk, -trō'pĭk) *also* **go·nad·o·troph·ic** (-trŏf'ĭk, -trō'fĭk) *adj.* Acting on or stimulating the gonads, as a hormone.

**go·nad·o·tro·pin** (gō-năd'ə-trō'pĭn, -trŏp'ĭn) *also* **go·nad·o·tro·phin** (-trō'fĭn, -trō'pĭn) *n.* A gonadotropic substance.

**Gond** (gŏnd) *n.* One of a people of Dravidian stock of central India.

**Gon·di** (gŏn'dē) *n.* The Dravidian language of the Gonds.

**gon·do·la** (gŏn'dl-ə, gŏn-dō'lə) *n.* [Ital.] **1.** A narrow lightweight barge with ends that curve up into a point and often a small cabin in the middle, propelled from the stern with a single oar, used on the canals of Venice. **2.** A flat-bottomed river boat. **3.** A gondola car. **4.** A basket, enclosure, or instrument sling suspended from and carried aloft by a balloon. **5.** An enclosed car suspended from a cable, used for transporting passengers, as to and from a ski slope.

**gondola**
(Left) *a Venetian gondola and* (right) *a cable gondola*

**gondola car** *n.* An shallow, open, freight car.

**gon·do·lier** (gŏn'dl-îr') *n.* [Fr. < Ital. *gondoliere* < *gondola,* gondola.] The boatman of a gondola.

**gone** (gôn, gŏn) *adj.* [P.part. of GO¹] **1.** Bygone : past. **2.** Advanced beyond recall or hope. **3.** Dying or dead. **4.** Lost : ruined. **5.** Carried away : absorbed. **6.** Used up : EXHAUSTED. **7.** *Slang.* Infatuated.

**gon·er** (gô'nər, gŏn'ər) *n.* [< GONE.] *Slang.* One doomed or ruined.

**gon·fa·lon** (gŏn'fə-lŏn', -lən) *n.* [Ital. *gonfalone,* of Germanic orig.] A banner hung from a crosspiece, esp. as a standard in an ecclesiastical procession or as the ensign of a medieval Italian republic.

**gon·fa·lon·ier** (gŏn'fə-lə-nîr') *n.* [Fr. < Ital. *gonfaloniere* < *gonfalon,* gonfalon.] A bearer of a gonfalon.

**gong** (gông, gŏng) *n.* [Malay *gōng.*] **1.** A rimmed metal disk that resounds loudly and sonorously when struck with a padded mallet. **2.** A usu. saucer-shaped bell struck with a mechanically operated hammer. *—vi.* **gonged, gong·ing, gongs.** To make the sound of a gong.

**Gon·gor·ism** (gŏng'gə-rĭz'əm) *n.* [After Luis de *Góngora* y Argote (1561–1627).] Florid and deliberately obscure literary style. **—Gon'·gor·is'tic** (gŏng'gə-rĭs'tĭk) *adj.*

**go·nid·i·um** (gō-nĭd'ē-əm) *n., pl.* **-i·a** (-ē-ə) [NLat. < Gk. *gonos,* birth.] **1.** An asexually produced reproductive cell that separates from the parent body, as in certain colonial microorganisms. **2.** An algal cell in the thallus of a lichen.

**gon·if** (gŏn'ĭf) *n. var. of* GANEF.

**go·ni·om·e·ter** (gō'nē-ŏm'ĭ-tər) *n.* [Gk. *gōnia,* angle + -METER.] **1.** An optical instrument for measuring crystal angles. **2.** A radio receiver and directional antenna used as a system for determining the angular direction of incoming radio signals. **—go'ni·o·met'ric** (-nē-ə-mĕt'rĭk), **go'ni·o·met'ri·cal** *adj.* **—go'ni·om'e·try** *n.*

**go·ni·on** (gō'nē-ŏn') *n.* [< Gk. *gonia,* angle.] The point of the angle on either side of the lower jaw.

**gono-** or **gon-** *pref.* [NLat. < Gk. < *gonos,* seed, procreation.] Sexual : reproductive <gonophore>

**gon·o·coc·cus** (gŏn'ə-kŏk'əs) *n., pl.* **-coc·ci** (-kŏk'sī', -kŏk'ī'). A bacterium, *Neisseria gonorrhoeae,* causing gonorrhea. **—gon'o·coc'·cal** (-kŏk'əl), **gon'o·coc'cic** (-kŏk'ĭk, -kŏk'sĭk) *adj.*

**go-no-go** (gō-nō'gō) *adj.* Of, relating to, or requiring the outcome of a parameter in order to stop or continue a course of action <a go-no-go space launch>

**gon·o·phore** (gŏn'ə-fôr', -fōr') *n.* A structure bearing or consisting of a reproductive organ or part, as a reproductive cell or bud in a hydroid colony. **—gon'o·pho'ric** (-fôr'ĭk, -fōr'-), **go·noph'o·rous** (gə-nŏf'ər-əs) *adj.*

**gon·o·pore** (gŏn'ə-pôr', -pōr') *n.* A reproductive aperture or pore.

**gon·or·rhe·a** (gŏn'ə-rē'ə) *n.* [LLat., spermatorrhea < Gk. *gonorrhoia* : *gonos,* semen + *rhoia,* flow.] An infectious disease of the genitourinary tract, rectum, and cervix, caused by gonococci, transmitted chiefly by sexual intercourse, and marked by acute purulent urethritis with dysuria. **—gon'or·rhe'al, gon'or·rhe'ic** *adj.*

**-gony** *suff.* [Lat. *-gonia* < Gk. < *gonos,* offspring.] Generation : reproduction : manners of origin <heterogony>

**goo** (gōō) *n.* [Orig. unknown.] *Informal.* **1.** A wet, sticky substance. **2.** Sentimental drivel. **—goo'ey** *adj.*

**†goo·ber** (gōō'bər) *n.* [Kongo *nguba.*] *Regional.* A peanut.

**good** (go͝od) *adj.* **bet·ter** (bĕt'ər), **best** (bĕst) [ME *god* < OE *gōd.*] **1.** Having desirable or positive qualities. **2.** Serving the desired end : SUITABLE <a *good* auto wax> **3. a.** Not ruined or spoiled <The fruit is still *good.*> **b.** Being in excellent condition : SOUND <a *good* tooth> **4. a.** Better than average <a *good* report> **b.** Designating the U.S. Government grade of meat higher than standard and lower than choice. **5. a.** Of high quality <*good* music> **b.** Discriminating <*good* judgment> **6.** Handsome : attractive <*good* appearance> **7.** Beneficial : salutary <a *good* square meal> **8.** Skilled : competent <a *good* sailor> **9.** Thorough : complete <a *good* rest> **10. a.** Safe : sure <*good* stocks and bonds> **b.** Valid or true <a *good* explanation> **c.** Real : genuine <a *good* early American antique> **11. a.** Ample : substantial <a *good* annual dividend> **b.** Bountiful <a *good* wheat crop> **12.** Full <a *good* distance from Cheyenne> **13. a.** Enjoyable : pleasant <Have a *good* day!> **b.** Propitious : favorable <*good* flying weather><a *good* omen> **14. a.** Of moral excellence : UPRIGHT. **b.** Benevolent : kind. **c.** Staunch : loyal <a *good* citizen> **15. a.** Obedient : well-behaved <a *good* and gentle dog> **b.** Socially correct : PROPER <*good* manners> ***usage:*** *Good* is properly used as an adjective after linking verbs such as *be, seem,* or *appear*: *I look and feel good.* It should not be used as an adverb with other verbs: *My car runs well* (not *good*). *—n.* **1. a.** Something that is good. **b.** A valuable or useful aspect or part. **2.** Benefit : welfare <laws for the *good* of all> **3.** Virtue : goodness <the *good* in everyone> **4. goods. a.** Commodities : wares <a shop specializing in dry *goods*> **b.** Portable personal property. **c.** (*sing. or pl. in number*). Fabric : material <stock of drapery *goods*> **5. goods.** *Slang.* Incriminating information or evidence <The police had the *goods* on the gang.> *—adv. Informal.* Well. **—as good as.** Practically : nearly <*as good as* finished> **—for good.** Forever : permanently <settled down *for good*> **—good and.** *Informal.* Thoroughly : very <soup that is *good and* hot.> **—no good.** *Informal.* **1.** Worthless. **2.** Futile <It's *no good* discussing it any more.> **—to the good. 1.** For the best : ADVANTAGEOUS. **2.** In an advantageous financial position.

**good book** *also* **Good Book** *n.* The Bible.

**good-by** or **good-bye** (go͝od-bī') [Alteration of *God be with you.*] *—interj.* —Used to express farewell. *—n., pl.* **-bys** or **-byes.** An expression of farewell.

**good-fel·low·ship** (go͝od'fĕl'ō-shĭp') *n.* Pleasant comradeship and sociability.

**good-for-noth·ing** (go͝od'fər-nŭth'ĭng) *n.* One of little worth or usefulness. **—good'-for-noth'ing** *adj.*

**Good Friday** *n.* The Friday before Easter, observed by Christians to commemorate the Crucifixion.

**good·heart·ed** (go͝od'här'tĭd) *adj.* Generous and kind. **—good'·heart'ed·ly** *adv.* **—good'heart'ed·ness** *n.*

**good-hu·mored** (go͝od'hyo͞o'mərd) *adj.* Amiable : cheerful. **—good'hu'mored·ly** *adv.* **—good'-hu'mored·ness** *n.*

**good·ish** (go͝od'ĭsh) *adj.* **1.** Rather good. **2.** GOODLY 2.

**good-look·ing** (go͝od'lo͝ok'ĭng) *adj.* Handsome : attractive.

**good·ly** (go͝od'lē) *adj.* **-li·er, -li·est. 1.** Of pleasing appearance : COMELY. **2.** Rather large : CONSIDERABLE <a *goodly* fee> **—good'li·ness** *n.*

**good·man** (go͝od'mən) *n. Archaic.* **1. a.** The male head of a household : MASTER. **b.** A husband. **2.** A courteous title of address for a man not of noble birth.

**good nature** *n.* An obliging, cheerful disposition.

**good-na·tured** (go͝od'nā'chərd) *adj.* Easygoing and cheerful in disposition. **—good'-na'tured·ly** *adv.* **—good'-na'tured·ness** *n.*

**good·ness** (go͝od'nĭs) *n.* **1.** The quality or state of being good. **2.** The good part : ESSENCE. *—interj.* —Used to express surprise.

**Good Samaritan** *n.* **1.** The only passer-by in a New Testament parable to help a man who had been robbed and beaten. **2.** An unselfish helper of others.

**good-sized** (go͝od'sīzd') *adj.* GOODLY 2.

**good-tem·pered** (go͝od'tĕm'pərd) *adj.* Having an even temper : AMIABLE. **—good'-tem'pered·ly** *adv.* **—good'-tem'pered·ness** *n.*

**good·wife** (go͝od'wīf') *n. Archaic.* **1.** The female head of a household : MISTRESS. **2.** A courteous title of address for a woman not of noble birth.

**good will** *also* **good·will** (go͝od'wĭl') *n.* **1.** An attitude of friendliness or kindness : BENEVOLENCE. **2.** Cheerful acquiescence or willingness. **3.** A good relationship, as of a business with its customers or a country with other counties.

**good·y[1]** (gŏŏd'ē) *Informal.* —*n., pl.* **-ies.** Something attractive or delectable, esp. something sweet to eat. —*interj.* —Used to express delight.

**good·y[2]** (gŏŏd'ē) *n., pl.* **-ies.** [Short for GOODWIFE.] *Archaic.* A polite title usu. used with the surname of a married woman of humble rank.

**good·y-good·y** (gŏŏd'ē-gŏŏd'ē) *adj.* Affectedly sweet, good, or virtuous. **—good'y-good'y** *n.*

**goof** (gōōf) [Poss. alteration of obs. *goff,* fool.] *Slang.* —*n.* **1.** An incompetent, foolish, or stupid person : KLUTZ. **2.** A careless mistake : SLIP. —*v.* **goofed, goof·ing, goofs.** —*vi.* **1.** To make a silly mistake : BLUNDER. **2.** To kill or waste time <*goofing* off all morning> —*vt.* To spoil : bungle <*goof* up the project>

**goof-off** (gōōf'ôf', -ŏf) *n.* One who shuns work or responsibility : SHIRKER.

**goof·y** (gōō'fē) *adj.* **-i·er, -i·est.** *Slang.* Silly : ridiculous <a *goofy* new style> **—goof'i·ly** *adv.* **—goof'i·ness** *n.*

**goo·gol** (gōō'gôl') *n.* [Coined by Milton Sirotta, nephew of Edward Kasner (1878–1955).] The number equivalent to $10^{100}$.

**goo·gol·plex** (gōō'gôl-plĕks') *n.* [GOOGOL + (DU)PLEX.] The number 10 raised to the power googol.

**goo-goo** (gōō'gōō') *adj. Slang.* Amorous, often humorously so <They are *goo-goo* about each other.>

**gook** (gŏŏk, gōōk) *n.* [Poss. alteration of GOO.] *Slang.* A dirty, sludgy, or slimy substance.

**goon** (gōōn) *n.* [Perh. alteration of dial. *gooney,* fool.] **1.** *Informal.* A hired thug. **2.** *Slang.* A stupid or oafish person.

**goo·ney bird** *also* **goo·ny bird** (gōō'nē) *n.* [< dial. *gooney,* fool.] An albatross, esp. *Diomedea nigripes,* that is common on Pacific islands.

**goos·an·der** (gōō-săn'dər) *n.* [Orig. unknown.] *Chiefly Brit.* The common merganser, *Mergus merganser.*

**goose[1]** (gōōs) *n., pl.* **geese** (gēs) [ME *goos* < OE *gōs.*] **1. a.** Any of various wild or domesticated water birds of the family Anatidae, and esp. of the genera *Anser* and *Branta,* with a shorter neck than that of a swan and a shorter, more pointed bill than that of a duck. **b.** A female goose. **c.** The flesh of the goose used as food. **2.** *Informal.* A silly person. **3.** *pl.* **goos·es.** A tailor's pressing iron with a long curved handle. **—cook one's goose.** *Informal.* To ruin one's chances.

**goose[2]** (gōōs) *vt.* **goosed, goos·ing, goos·es.** [Orig. unknown.] *Slang.* To poke (a person) between the buttocks.

**goose·ber·ry** (gōōs'bĕr'ē, -bə-rē, gōōz'-) *n.* **1. a.** A spiny shrub native to Eurasia, *Ribes grossularia,* bearing lobed leaves, greenish flowers, and edible greenish berries. **b.** The fruit of the gooseberry plant. **2.** A plant bearing fruit similar to the gooseberry.

**goose bumps** *pl.n.* Goose flesh.

**goose egg** *n. Slang.* Zero, esp. when written as a numeral to indicate that no points have been scored, as in a game.

**goose·fish** (gōōs'fĭsh') *n., pl.* **goosefish** *or* **-fish·es.** An anglerfish of the genus *Lophius,* esp. *L. americanus* of North American Atlantic waters.

**goose flesh** *n.* Brief roughness of skin caused by erection of the papillae in response to fear or cold.

**goose·foot** (gōōs'fŏŏt') *n., pl.* **-foots.** [From the shape of its leaves.] Any of various usu. weedy plants of the genus *Chenopodium,* bearing small greenish flowers.

**goose grass** *n.* Cleavers.

**goose·neck** (gōōs'nĕk') *n.* A slender curved part or object, as the flexible shaft of a kind of desk lamp. **—goose'necked'** *adj.*

**goose pimples** *pl.n.* Goose flesh.

**goose step** *n.* A military parade step executed by swinging the legs sharply from the hips while keeping the knees locked.

**goose-step** (gōōs'stĕp') *vi.* **-stepped, -step·ping, -steps.** To march in or execute a goose step.

**goos·y** *also* **goos·ey** (gōō'sē) *adj.* **-i·er, -i·est. 1.** Relating to or resembling a goose. **2.** Scatterbrained or foolish. **3.** Nervous and keyed up.

**go·pher** (gō'fər) *n.* [Orig. unknown.] **1.** Any of various short-tailed burrowing mammals of the family Geomyidae of North America, with fur-lined external cheek pouches. **2.** A ground squirrel. **3.** A burrowing tortoise of the genus *Gopherus,* esp. *G. polyphemus* of the southeastern United States.

**gopher ball** *n. Baseball.* A pitched ball hit for a home run.

**gopher snake** *n.* A bull snake.

**go·pher·wood** (gō'fər-wŏŏd') *n.* YELLOWWOOD 1.

**Gor·di·an knot** (gôr'dē-ən) *n.* **1.** An intricate knot tied by King Gordius of Phrygia and cut by Alexander the Great with his sword after hearing an oracle promise that whoever could undo it would be the next ruler of Asia. **2.** A highly complex problem or a deadlock.

**Gordon setter** *n.* [After Alexander, 4th Duke of *Gordon* (d. 1827).] A hunting dog orig. bred in Scotland, with a silky black and tan coat.

**gore[1]** (gôr) *vt.* **gored, gor·ing, gores.** [ME *goren* < *gore,* spear < OE *gār.*] To stab or pierce with a tusk or horn.

**gore[2]** (gôr) *n.* [ME < OE *gāra,* triangular piece of land.] **1.** A triangular or tapering piece of cloth, as in a skirt or sail. **2.** A small triangular piece of land. —*vt.* **gored, gor·ing, gores. 1.** To provide with a gore. **2.** To cut into a gore.

**gore[3]** (gôr) *n.* [ME, filth < OE *gor.*] Blood, esp. coagulated blood from a wound.

**gorge** (gôrj) *n.* [ME, throat < OFr. < Lat. *gurges,* gulf.] **1.** A deep narrow passage with steep rocky sides, enclosed on each side by mountains. **2.** A narrow entrance into the outwork of a fortification. **3.** The throat : gullet. **4.** An instance of gluttonous eating. **5.** A mass obstructing a passage <a harbor entrance closed by an ice *gorge*> —*v.* **gorged, gorg·ing, gorg·es.** —*vt.* **1.** To stuff : glut. **2.** To devour greedily. —*vi.* To eat gluttonously <*gorged* on chocolate cake> **—gorg'er** *n.*

**gor·geous** (gôr'jəs) *adj.* [ME *gorgeouse,* prob. < OFr. *gorrias,* elegant.] **1. a.** Dazzlingly beautiful or magnificent <dressed in *gorgeous* French gowns> **b.** Characterized by brilliance or magnificence <*gorgeous* coloratura singing> **2.** *Informal.* Wonderful or delightful. **—gor'geous·ly** *adv.* **—gor'geous·ness** *n.*

**gor·ger·in** (gôr'jər-ĭn) *n.* [Fr. < *gorge,* throat < OFr. —see GORGE.] The necking of an architectural column.

**gor·get** (gôr'jĭt) *n.* [ME < OFr. *gorgete,* dim. of *gorge,* throat < Lat. *gurges,* gulf.] **1.** Armor protecting the throat. **2.** An ornamental collar. **3.** The scarflike part of a wimple hanging over the neck and shoulders. **4.** A distinctively colored band or patch on the throat esp. of a bird.

**Gor·gon** (gôr'gən) *n.* [ME < Lat. *Gorgo* < Gk. *Gorgō* < *gorgos,* terrible.] **1.** *Gk. Myth.* Any of three sisters who had snakes for hair and eyes that turned the beholder into stone. **2. gorgon.** A repulsively ugly or terrifying woman. **—Gor·go'ni·an** (-gō'nē-ən) *adj.*

**gor·go·ni·an** (gôr-gō'nē-ən) *n.* [Lat. *gorgonia,* coral < *Gorgo,* Gorgon. —see GORGON.] Any of various corals of the order Gorgonacea, with flexible, often branching skeletons of horny material. —*adj.* Of or belonging to the Gorgonacea.

**gor·gon·ize** (gôr'gə-nīz') *vt.* **-ized, -iz·ing, -iz·es.** [< GORGON.] To have a paralyzing or stupefying effect on.

**Gor·gon·zo·la** (gôr'gən-zō'lə) *n.* [After *Gorgonzola,* Italy.] A pungent, blue-veined, cream-colored Italian cheese made of pressed cow's milk.

**go·ril·la** (gə-rĭl'ə) *n.* [< Gk. *Gorillai,* a tribe of hairy women.] **1.** A large anthropoid ape, *Gorilla gorilla* of equatorial Africa, with a stocky body and dark, coarse hair. **2.** A brutish or thuglike man.

**gor·mand·ize** (gôr'mən-dīz') *v.* **-ized, -iz·ing, -iz·es.** [< obs. *gormandise,* gluttony < Fr. *gourmandise* < *gourmand,* glutton.] —*vi.* To eat gluttonously : GORGE. —*vt.* To devour (food) gluttonously. **—gor'mand·iz'er** *n.*

**go-round** (gō'round') *n.* A go-around.

**gorp** (gôrp) *n.* [Perh. < slang *gorp,* to eat greedily.] A mixture of foods, as dried fruit, nuts, seeds, and chocolate, eaten esp. as a high-energy snack.

**gorse** (gôrs) *n.* [ME *gorst* < OE.] Any of several spiny, thickset shrubs of the genus *Ulex,* esp. the native European variety *U. europeaus,* bearing fragrant yellow flowers.

**go·ry** (gôr'ē, gōr'ē) *adj.* **-ri·er, -ri·est. 1.** Covered or stained with gore : BLOODY. **2.** Full of or marked by violence and bloodshed <a *gory* battle> **—gor'i·ly** *adv.* **—gor'i·ness** *n.*

**gosh** (gŏsh) *interj.* [Alteration of GOD.] —Used to express mild surprise or delight.

**gos·hawk** (gŏs'hôk') *n.* [ME *goshauk* < OE *goshafoc* : *gos,* goose + *hafoc,* hawk.] **1.** A large hawk, *Accipiter gentilis,* with broad rounded wings and gray or brownish plumage. **2.** A hawk similar or related to the goshawk.

**gos·ling** (gŏz'lĭng) *n.* [ME < ON *gæslingr.*] **1.** A young goose. **2.** An inexperienced or naive young person.

**gos·pel** (gŏs'pəl) *n.* [ME < OE *gōdspel* : *gōd,* good + *spel,* tidings.] **1.** *often* **Gospel.** Jesus' and the Apostles' teachings. **2. a. Gospel.** One of the first four books of the New Testament, telling of the life, death, and resurrection of Jesus Christ. **b.** A similar narrative. **3.** *often* **Gospel.** A lection from any of the Gospels read as part of a religious service. **4.** A religious teacher's doctrine or teaching. **5.** Gospel music. **6.** Something, as an idea or principle, accepted as undoubtedly true. **—gos'pel** *adj.*

▲ word history: The ancestor of *gospel* is the Old English compound *gōdspel,* literally "good tidings." It originated as a loan translation of Latin *evangelium,* a borrowing of Greek *euangelion,* which in Christian contexts meant "good tidings," specifically the good news of the kingdom of God brought by Jesus.

**gos·pel·er** *also* **gos·pel·ler** (gŏs'pə-lər) *n.* **1.** One who teaches or professes faith in a gospel. **2.** One who reads or sings the Gospel as part of a church service.

**gospel music** *n.* American religious music associated with evangelism and based on simple folk melodies blended with rhythmic and melodic elements of spirituals and jazz.

**gospel side** *also* **Gospel side** *n.* [So called from the practice in some churches of reading the Gospel and the Epistle from different sides.] The left side of a chancel or altar.

**gos·port** (gŏs′pôrt′, -pōrt′) *n.* [After *Gosport,* England.] A flexible speaking tube used for communicating between individual compartments or cockpits of an aircraft.

**gos·sa·mer** (gŏs′ə-mər) *n.* [ME *gossomer* : *gos,* goose (< OE *gōs*) + *somer,* summer < OE *sumor.*] **1.** A fine film of cobwebs often seen floating in the air or caught on grass or bushes. **2.** A soft, sheer, gauzy fabric. **3.** Something delicate, light, or flimsy. **—gos′sa·mery** (-mə-rē) *adj.*

**gos·sip** (gŏs′əp) *n.* [ME *godsib,* godparent < OE *godsibb* : *god,* god + *sibb,* kinsman.] **1.** Rumor or talk of a personal, intimate, and often sensational nature. **2.** One who habitually repeats intimate or private rumors or facts. **3.** Trivial, chatty writing or talk. **4.** *Archaic.* A close friend or companion. **5.** *Archaic.* A godparent. *—vi.* **-siped, -sip·ing, -sips.** To engage in or spread gossip. **—gos′sip·er** *n.* **—gos′sip·ry** *n.* **—gos′sip·y** *adj.*

▲ word history: The ordinary senses of *gossip* current in Modern English are the latest to have developed. *Gossip* was originally a compound of *god* and *sib,* meaning "godparent." At first denoting only the relationship of godparent to godchild, *gossip* was later used to indicate the relationship of godparent to parent and the relationship between the godparents of the same child. *Gossip* thus designated a relationship among peers as much as one between generations and from the extended senses the meaning "friend" evolved. The derogatory use of *gossip* to mean a person who engages in idle chatter and rumor-mongering first appears in the 16th century, but the use of *gossip* to mean the conversation of a gossip is not recorded until the early 18th century.

**gos·sip·mong·er** (gŏs′əp-mŭng′gər, -mŏng′-) *n.* A spreader of gossip : GOSSIPER.

**got** (gŏt) *v. p.t. & var. p.p. of* GET.

**Goth** (gŏth) *n.* [< ME *Gothes,* Goths < LLat. *Gothi,* of Goth. orig.] A member of a Germanic people who invaded the Roman Empire during the early Christian era.

**Goth·ic** (gŏth′ĭk) *adj.* **1. a.** Of or relating to the Goths or their language. **b.** Germanic : Teutonic. **2.** Of or relating to the Middle Ages : MEDIEVAL. **3. a.** Of or relating to an architectural style prevalent in 12th–15th-cent. western Europe and typified by pointed arches, rib vaulting, and flying buttresses. **b.** Of or relating to painting, sculpture, or other art forms prevalent in northern Europe from the 12th through the 15th cent. **c.** Of or relating to an architectural style derived from medieval Gothic. **4.** *often* **gothic.** Of or relating to a style of fiction emphasizing the grotesque, mysterious, and desolate. **5. gothic.** Barbarous and crude : UNCOUTH. *—n.* **1.** The extinct East Germanic language of the Goths. **2.** Gothic art or architecture. **3.** *often* **gothic. a.** Black letter. **b.** Sans serif. **4.** A novel written in the Gothic style. **—Goth′i·cal·ly** *adv.*

**Gothic arch** *n.* A pointed arch, esp. one with a jointed apex.

**Goth·i·cism** (gŏth′ĭ-sĭz′əm) *n.* **1.** Use of or imitation of Gothic style, as in architecture. **2.** A barbarous or crude style or manner.

**Goth·i·cize** (gŏth′ĭ-sīz′) *vt.* **-cized, -ciz·ing, -ciz·es.** To make Gothic.

**GO TO** *n. Computer Sci.* A programming instruction directing a computer to leave the current sequence of instructions for another sequence at another point in the program.

**got·ten** (gŏt′n) *v. var. p.p. of* GET.

**gouache** (gwăsh) *n.* [Fr. < Ital. *guazzo* < Lat. *aquatio,* watering < *aquari,* to fetch water < *aqua,* water.] **1.** A method of painting with opaque water colors mixed with a gum preparation. **2.** An opaque pigment used in gouache. **3.** A picture painted by gouache.

**Gou·da** (gou′də, gōō′-) *n.* [After *Gouda,* The Netherlands.] A mild, close-textured, pale-yellow cheese made from whole or partially skimmed milk.

**gouge** (gouj) *n.* [ME < OFr. *goi* < LLat. *gubia,* of Celt. orig.] **1.** A chisel having a rounded troughlike blade. **2.** A digging or scooping action, as with a gouge. **3.** A hole or groove scooped with or as if with a gouge. **4.** *Informal.* An excessive amount exacted or extorted. *—vt.* **gouged, goug·ing, goug·es. 1.** To scoop or cut out with or as if with a gouge. **2.** To force out (an eye). **3.** To exact exorbitantly or extort from. **4.** *Slang.* To swindle. **—goug′er** *n.*

**gou·lash** (gōō′läsh′, -lăsh) *n.* [Hung. *gulyás* (*hus*), herdsman's (meat) < *gulya,* herd.] A beef or veal and vegetable stew seasoned mostly with paprika.

**gou·ra·mi** (gōō-rä′mē, gōōr′ə-) *n., pl.* **-mis.** [Malay *gurami.*] Any of various freshwater fishes of the family Anabantidae of southeastern Asia, with many species being brightly colored and popular in home aquariums.

**gourd** (gôrd, gōrd, gōōrd) *n.* [ME *gourde* < AN, ult. < Lat. *cucurbita.*] **1.** Any of several vines of the family Cucurbitaceae, related to the pumpkin, squash, and cucumber and bearing fruits with a hard rind. **2. a.** The fruit of a gourd, as a calabash, often irregular and unusual in shape. **b.** The dried hollowed-out shell of a gourd, used as a drinking vessel.

**gourde** (gōōrd) *n.* [Haitian < Fr. *gourd,* dull < Lat. *gurdus.*] —See table at CURRENCY.

**gour·mand** (gōōr-mänd′, gōōr′mənd) *n.* [ME *gourmant,* glutton < OFr. *gormant.*] One who delights in eating heartily and well. ***usage:*** A *gourmand* is one who loves good food and drink and sometimes partakes of them to excess; a *gourmet* or an *epicure,* on the other hand, has discriminating and cultivated taste in food and wine.

**gour·man·dise** (gōōr′mən-dēz′) *n.* [ME *gromandise,* gluttony < OFr. *gormandise* < *gormant,* glutton.] A taste and relish for good food.

**gour·met** (gōōr-mā′, gōōr′mā′) *n.* [Fr. < OFr., wine merchant's servant.] A connoisseur of fine food and drink.

**gout** (gout) *n.* [ME *goute* < OFr. *goute,* gout, drop < Lat. *gutta,* drop (from the belief that gout was caused by drops of morbid humors).] **1.** A disturbance of the uric-acid metabolism occurring chiefly in men, characterized by painful inflammation of the joints, esp. of the hands and feet, and arthritic attacks and having the potential of becoming chronic and causing deformity. **2.** A large blob or clot <*gouts* of blackish blood> **—gout′i·ness** *n.* **—gout′y** *adj.*

**gov·ern** (gŭv′ərn) *v.* **-erned, -ern·ing, -erns.** [ME *governen* < OFr. *governer* < Lat. *gubernare* < Gk. *kubernan.*] *—vt.* **1.** To set forth and administer the public policy and affairs of. **2.** To control the speed or magnitude of : REGULATE <a device that *governs* temperature in the boiler> **3.** To control the actions or behavior of. **4.** To keep under control : RESTRAIN. **5.** To exercise a determining influence on <Interest rates *govern* the real estate market.> **6. a.** To require (a verb or noun) to be in a particular mood or case. **b.** To require the use of (a particular mood or case). *—vi.* **1.** To exercise political authority. **2.** To have or exercise a determining influence. **—gov′ern·a·ble** *adj.*

**gov·er·nance** (gŭv′ər-nəns) *n.* **1.** The act, process, or power of governing : GOVERNMENT. **2.** The state of being governed.

**gov·ern·ess** (gŭv′ər-nĭs) *n.* A woman employed in a private home to train and educate the children of the owners.

**gov·ern·ment** (gŭv′ərn-mənt) *n.* **1.** The act or process of governing, esp. the administration and control of public policy in a political unit. **2.** The office, function, or authority of one who governs or of a governing body. **3.** Exercise of authority in a political unit : RULE. **4.** The agency or apparatus through which a governing individual or body exercises authority and performs the required duties. **5. a.** A governing body or organization. **b.** The individuals comprising a governing body. **6.** A system or policy for governing a political unit. **7.** Management or administration of an organization, business, or institution. **8.** Political science. ***usage:*** *Government* always takes a singular verb in American usage. In British usage *government* is construed as a plural collective and takes a plural verb. **—gov′ern·ment′al** (-mĕn′tl) *adj.* **—gov′ern·ment′al·ly** *adv.*

**Government Issue** *n.* Items, as U.S. military equipment, issued by the government.

**gov·er·nor** (gŭv′ər-nər) *n.* [ME *governour* < OFr. *governeor* < Lat. *gubernator* < *gubernare,* to govern. —see GOVERN.] **1.** One who governs, esp.: **a.** The chief executive of a U.S. state. **b.** An official appointed to govern a territory or colony. **c.** A member of a governing body. **2.** The administrative manager of an organization, business, or institution. **3.** A military commandant. **4.** *Chiefly Brit.* Mister : sir. —Used in direct address. **5.** A feedback device on a machine or engine for providing automatic control, as of speed, pressure, or temperature. **—gov′er·nor·ship′** *n.*

**gov·er·nor-gen·er·al** (gŭv′ər-nər-jĕn′ər-əl) *n., pl.* **gov·er·nors-gen·er·al** *or* **gov·er·nor-gen·er·als.** A governor of a large territory having jurisdiction over subordinate governors. **—gov′er·nor-gen′er·al·ship′** *n.*

**gow·an** (gou′ən) *n.* [Prob. alteration of ME *gollan* < ON *gullinn,* golden < *gull,* gold.] *Scot.* A yellow or white wildflower, esp. the Old World daisy.

**gown** (goun) *n.* [ME *goune* < OFr. < Med. Lat. *gunna,* fur robe.] **1.** A long, loose, flowing garment, as a nightgown or robe. **2.** A long, usu. formal woman's dress. **3.** A distinctive outer robe worn on ceremonial occasions, as by members of the clergy or scholars. **4.** The faculty and student body of a university <general harmony between town and *gown*> *—vt.* **gowned, gown·ing, gowns.** To dress in or invest with a gown.

**Graaf·i·an follicle** (grä′fē-ən, grăf′ē-) *n.* [After Regnier de *Graaf* (1641–1673).] Any of the follicles in the mammalian ovary containing a maturing ovum.

**grab**[1] (grăb) *v.* **grabbed, grab·bing, grabs.** [MLG *grabben.*] *—vt.* **1.** To grasp or take suddenly. **2.** To capture or restrain : ARREST. **3.** To obtain or appropriate forcibly or unscrupulously. **4.** To take hurriedly <*grab* a sandwich> **5.** *Slang.* To capture the attention of <new television programs that do not *grab* the viewers> *—vi.* To make a snatch <*grabbed* for the knife> *—n.* **1.** A sudden seizing. **2.** Something grabbed. **3.** A mechanical device for gripping an object. **—up for grabs.** *Informal.* Available for anyone to win or take. **—grab′ber** *n.*

**grab**[2] (grăb) *n.* [Ar. *ghurāb.*] A two- or three-masted Oriental coastal vessel.

**grab bag** *n.* **1.** A container full of articles, as party favors, to be drawn sight unseen. **2.** A miscellaneous assortment.

**grab·ble** (grăb′əl) *vi.* **-bled, -bling, -bles.** [Du. *grabbelen* < MDu., freq. of *grabben,* to grab.] **1.** To feel around with the hands : GROPE. **2.** To fall down : SPRAWL.

**grab·by** (grăb′ē) *adj.* **-bi·er, -bi·est.** Inclined to greediness : ACQUISITIVE. **—grab′bi·ness** *n.*

**gra·ben** (grä′bən) *n.* [G., trench < OHG *grabo* < *graban,* to dig.] A usu. elongated depression of the earth's crust between two parallel faults.

**grace** (grās) *n.* [ME, divine love < OFr. < Lat. *gratia,* good will < *gratus,* pleasing.] **1.** Apparently effortless charm or beauty of movement, form, or proportion. **2.** A characteristic or quality pleasing for its charm or refinement. **3.** A sense of fitness or propriety. **4. a.** A disposition to be generous or helpful : GOOD WILL. **b.** Clemency : mercy. **5.** A favor rendered voluntarily : INDULGENCE. **6.** Temporary immunity or exemption : REPRIEVE. **7. a.** Divine love and protection bestowed freely on human beings. **b.** Protection or sanctification by the favor of God. **c.** An excellence or power granted by God. **8.** A short prayer of blessing or thanksgiving said before or after a meal. **9.** —Used with *his, her,* or *your* as a title of courtesy for a duke, duchess, or archbishop. **10.** *Mus.* An embellishment, as a trill or appoggiatura. —*vt.* **graced, grac·ing, grac·es. 1.** To favor or honor. **2.** To impart beauty, elegance, or charm to. **3.** *Mus.* To embellish with grace notes. **—in good graces.** In favor with. **—with good grace.** Willingly.

**grace cup** *n.* **1.** A cup used at the end of a meal, usu. after grace, for the last toast. **2.** The last toast of a meal.

**grace·ful** (grās′fəl) *adj.* Displaying grace of movement, form, or proportion. **—grace′ful·ly** *adv.* **—grace′ful·ness** *n.*

**grace·less** (grās′lĭs) *adj.* **1.** Lacking grace : CLUMSY. **2.** Having no sense of propriety or decency : UNCOUTH. **3.** Inferior or clumsy in treatment or performance : AWKWARD. **—grace′less·ly** *adv.* **—grace′less·ness** *n.*

**grace note** *n. Mus.* A note, esp. an appoggiatura, added as an embellishment.

**Grac·es** (grā′sĭz) *pl.n. Gk. Myth.* Three sister goddesses who give charm and beauty.

**grac·ile** (grăs′əl, -īl′) *adj.* [Lat. *gracilis.*] **1.** Gracefully slender. **2.** Graceful. **—gra·cil′i·ty** (grə-sĭl′ĭ-tē) *n.*

**gra·ci·o·so** (grä′sē-ō′sō, -zō) *n., pl.* **-sos.** [Sp. < Lat. *gratiosus,* agreeable. —see GRACIOUS.] A clown or buffoon in Spanish comedies.

**gra·cious** (grā′shəs) *adj.* [ME < OFr. *gracieus* < Lat. *gratiosus,* agreeable < *gratia,* good will < *gratus,* pleasing.] **1. a.** Marked by kindness and warm courtesy. **b.** Marked by elaborate, usu. formal courtesy. **2.** Marked by tact and propriety <reacted to the insult with *gracious* humor> **3.** Of a compassionate or merciful nature. **4.** Condescendingly courteous : INDULGENT. **5.** Characterized by charm or beauty : GRACEFUL. **6.** Marked by good taste and elegance. **7.** *Obs.* Fortunate or prosperous. —*interj.* —Used to express surprise or mild emotion. **—gra′cious·ly** *adv.* **—gra′cious·ness** *n.*

☆ ***syns:*** **1.** GRACIOUS, AFFABLE, CONGENIAL, CORDIAL, GENIAL, SOCIABLE *adj. core meaning* : characterized by kindness and warm, unaffected courtesy <a *gracious* attitude toward everyone> ***ant:*** ungracious **2.** GRACIOUS, CHIVALROUS, COURTLY, GALLANT, KNIGHTLY, STATELY *adj. core meaning* : characterized by elaborate but usu. formal courtesy <*gracious* old-world manners>

**grack·le** (grăk′əl) *n.* [Lat. *graculus,* jackdaw.] **1.** Any of several New World blackbirds of the family Icteridae, esp. of the genera *Quiscalus* or *Cassidix,* with iridescent blackish plumage. **2.** An Asian myna of the genus *Gracula.*

**grad** (grăd) *n. Informal.* A school or college graduate.

**gra·date** (grā′dāt′) *v.* **-dat·ed, -dat·ing, -dates.** [Back-formation < GRADATION.] —*vi.* To pass imperceptibly from one degree, shade, or tone to another. —*vt.* **1.** To cause to pass imperceptibly from one degree, shade, or tone to another. **2.** To arrange by or in grades.

**gra·da·tion** (grā-dā′shən) *n.* [Lat. *gradatio* < *gradus,* step.] **1. a.** A series of gradual, successive stages. **b.** A stage or degree in such a series. **2.** Advancement by successive stages, tones, or shades, as from one color to another. **3.** The act of gradating or arranging in grades. **4.** An ablaut. **—gra·da′tion·al** *adj.* **—gra·da′tion·al·ly** *adv.*

**grade** (grād) *n.* [Fr. < Lat. *gradus,* step.] **1.** A stage or degree in a process. **2.** A position in a scale of size or quality, as of eggs or beef. **3.** A group of persons or things all falling in the same specified limits : CLASS. **4. a.** A class at an elementary school or the pupils in it. **b. grades.** Elementary school. **5.** A mark showing a student's level of achievement. **6.** A military, naval, or civil service rank. **7.** Degree of inclination, as of a slope or road. **8.** A gradual inclination, esp. of a road or railroad track. **9.** A domestic animal produced by crossbreeding one of purebred stock with one of ordinary stock. —*v.* **grad·ed, grad·ing, grades.** —*vt.* **1.** To arrange in degrees or steps. **2.** To arrange in a series or according to a scale. **3. a.** To determine the quality of (e.g., academic work) : EVALUATE. **b.** To give a grade to (e.g., a student). **4.** To smooth or level to a desired or horizontal gradient. **5.** To gradate. **6.** To improve the quality of (livestock) by crossbreeding with purebred stock. —*vi.* **1.** To hold a particular rank or position. **2.** To change or progress gradually. **—make the grade. 1.** To reach a goal or standard : SUCCEED. **2.** To reach the highest point of an inclined slope.

**grade crossing** *n.* An intersection of roads, railroad tracks, or a road and a railroad track at the same level.

**grade point** *also* **grade index** *n.* A point assigned to a course credit, as in a college, that corresponds to the letter grade made in a course.

**grade point average** *also* **grade point index** *n.* The average grade earned by a student, computed by dividing the grade points earned by the number of credits attempted.

**grade school** *n.* An elementary school.

**gra·di·ent** (grā′dē-ənt) *n.* [Perh. < GRADE.] **1.** A rate of inclination : SLOPE. **2.** An ascending or descending part : INCLINE. **3.** *Physics.* The maximum rate at which a variable physical quantity changes in value per unit change in position. **4.** *Math.* A vector having coordinate components that are the partial derivatives of a function with respect to its variables.

**gra·din** (grād′n) *also* **gra·dine** (grā′dēn′, grə-dēn′) *n.* [Fr. < Ital. *gradino,* dim. of *grado,* step < Lat. *gradus.*] One of a series of steps or tiered seats, as in an amphitheater.

**grad·u·al** (grăj′o͞o-əl) *adj.* [Med. Lat. *gradualis* < Lat. *gradus,* step.] Progressing by regular or continuous degrees <*gradual* erosion><a *gradual* incline> —*n. Rom. Cath. Ch.* **1.** A book containing the choral portions of the Mass. **2.** The antiphon sung between the Epistle and the Gospel. **—grad′u·al·ly** *adv.* **—grad′u·al·ness** *n.*

**grad·u·al·ism** (grăj′o͞o-ə-lĭz′əm) *n.* The policy of or belief in advancing toward a goal by gradual, often slow stages. **—grad′u·al·ist** *n.* **—grad′u·al·is′tic** *adj.*

**grad·u·ate** (grăj′o͞o-āt′) *v.* **-at·ed, -at·ing, -ates.** [ME *graduaten,* to confer a degree < Med. Lat. *graduare* < Lat. *gradus,* degree.] —*vi.* **1.** To be granted an academic degree or diploma. **2.** To change gradually. —*vt.* **1. a.** To grant an academic degree or diploma to. **b.** *Nonstandard.* To receive an academic degree from. **2.** To arrange or divide into categories, steps, or grades. **3.** To divide into marked intervals, esp. for use in measurement. ***usage:*** A traditionalist would insist that *I was graduated from the university* is the only correct usage. But the usage *I graduated from college* is equally acceptable; the variant without the preposition, as in *I graduated college* is rapidly gaining ground but has not yet achieved full acceptability. —*n.* (-ĭt). **1.** One who has received an academic degree or diploma. **2.** A graduated container, as a flask or beaker. —*adj.* (-ĭt). **1.** Holding an academic degree or diploma. **2.** Of, for, or pertaining to studies beyond a bachelor's degree <*graduate* school> **—grad′u·a·tor** *n.*

**grad·u·a·tion** (grăj′o͞o-ā′shən) *n.* **1. a.** The conferring or receipt of an academic degree or diploma marking successful completion of studies. **b.** A ceremony at which degrees or diplomas are conferred : COMMENCEMENT. **2. a.** A division or interval on a graduated scale. **b.** A mark indicating the boundary of such an interval. **3.** An arrangement in or division into stages or degrees.

**Graeco-** *pref. var. of* GRECO-.

**graf·fi·to** (grə-fē′tō, grä-, grä-) *n., pl.* **-ti** (-tē) [Ital., dim. of *graffio,* a scratching < *graffiare,* to scratch.] A crude inscription or drawing scratched, painted, or sprayed on a surface, as a wall, usu. so as to be seen by the public.

**graft[1]** (grăft) *v.* **graft·ed, graft·ing, grafts.** [ME *graffen* < *graffe,* graft < OFr. *graife,* stylus (from its shape) < Lat. *graphium* < Gk. *graphion* < *graphein,* to write.] —*vt.* **1. a.** To unite (a shoot or bud) with a growing plant by insertion or placing in close contact. **b.** To join (a plant or plants) in this way. **2.** To transplant or implant (e.g., tissue) surgically into a bodily part to compensate for a defect. **3.** To join or unite closely. —*vi.* **1.** To make a graft. **2.** To be or become grafted. —*n.* **1. a.** A detached bud or shoot united or to be united with a growing plant. **b.** The union or point of union of a detached bud or shoot with a growing plant by attachment or insertion. **c.** A plant produced by such union. **2. a.** Material, esp. tissue or an organ, surgically attached to or inserted into a bodily part to compensate for a defect. **b.** Surgical implantation or transplantation of such material. **c.** The configuration or condition resulting from such a procedure. **—graft′er** *n.*

**graft[2]** (grăft) *n.* [Perh. < GRAFT[1].] **1.** Unscrupulous use of one's position to derive profit or advantages. **2.** Money or an advantage gained or yielded by unscrupulous means. **—graft** *v.* **(graft·ed, graft·ing, grafts). —graft′er** *n.*

**graft·age** (grăf′tĭj) *n.* The process of making a horticultural graft.

**gra·ham** (grā′əm) *n.* [After Sylvester *Graham* (1794–1851).] Whole-wheat flour.

**graham cracker** *n.* A somewhat sweet cracker made of whole-wheat flour.

**grail** *also* **Grail** (grāl) *n.* [ME *gral* < OFr. *graal* < Med. Lat. *gradalis,* dish.] **1.** The cup or chalice in medieval legend used by Christ at the Last Supper and subsequently the object of many chivalrous quests. **2.** An objective of a prolonged endeavor.

**grain** (grān) *n.* [ME < OFr. *graine* < Lat. *granum.*] **1. a.** A small hard fruit or seed, esp. that produced by a cereal grass. **b.** The seeds of such plants as a whole, esp. after harvest. **2.** Cereal grasses as a whole <many acres of *grain*> **3. a.** A relatively small discrete particulate or crystalline mass <*grains* of sugar> **b.** A small amount or the smallest amount possible <without a *grain* of decency> **4.** *Aerospace.* A mass of solid propellant formed from a number of smaller pieces. **5.** A unit of weight in the U.S. Customary System, an avoirdupois unit equal to 0.002285 ounce or .065 gram. **6.** Arrangement, direction, or pattern of the fibrous tissue in wood. **7. a.** The side of a hide or piece of leather from which the fur or hair is removed. **b.** The pattern or markings on this side of leather. **8.** The pattern produced, as in stone, by arrangement of particulate constituents. **9.** The relative size of the particles composing a substance or pattern <a fine *grain*> **10.** A painted, stamped, or printed design imitating a pattern found in wood, leather, or stone. **11.** The direction or texture of fibers in a woven fabric. **12.** Fine crystallization. **13. a.** Basic temperament : DISPOSITION. **b.** An essential characteristic or quality. **14.** *Archaic.* Color : tint. **15.** *Obs.* **a.** Cochineal or kermes. **b.** Red dye made from cochineal or kermes. **c.** A fast dye. —*v.* **grained, graining, grains.** —*vt.* **1.** To cause to form into grains : GRANULATE. **2.** To paint, stamp, or print with a design imitating the grain of wood, leather, or stone. **3.** To give a granular texture to. **4.** To remove the hair or fur from (hides) in preparation for tanning. —*vi.* To form grains. **—against the (**or **one's) grain.** In contradiction to one's natural disposition or character. **—with a grain of salt.** With skepticism. **—grain'er** *n.*

**grain alcohol** *n.* Alcohol.

**grain elevator** *n.* A tall building equipped with mechanical lifting devices and used for grain storage.

**grains of paradise** *pl.n.* **1.** The pungent aromatic seeds of a tropical African plant, *Aframomum melegueta,* used medicinally. **2.** Cardamom seeds.

**grain·y** (grā'nē) *adj.* **-i·er, -i·est. 1.** Made or suggestive of grain : GRANULAR. **2.** Resembling the grain of wood. **—grain'i·ness** *n.*

**gram**[1] (grăm) *n.* [Fr. *gramme* < LLat. *gramma,* a small weight < Gk.] A metric unit of mass and weight, equal to one thousandth ($10^{-3}$) of a kilogram.

**gram**[2] (grăm) *n.* [Obs. Port. < Lat. *granum,* seed.] **1.** A plant, as the chickpea, bearing seeds used widely as food esp. in tropical Asia. **2.** The seeds of a gram.

**-gram** *suff.* [Lat. *-gramma* < Gk. < *gramma,* letter.] Something written or drawn : RECORD <cardio*gram*>

**gra·ma** (grä'mə, grăm'ə) *n.* [Sp. *grama* < Lat. *gramen,* grass.] A grass of the genus *Bouteloua* of western North America and South America, forming dense tufts and often used as pasturage.

**gram·a·rye** (grăm'ə-rē) *n.* [ME *gramarie* < OFr. *gramaire,* grammar. —see GRAMMAR.] Occult learning : MAGIC.

**gram atom** *n.* The mass in grams of an element numerically equal to the atomic weight.

**gram calorie** *n.* CALORIE 1.

**gra·mer·cy** (grə-mûr'sē, grăm'ər-) *interj.* [ME *gramerci* < OFr. *grand merci,* great thanks.] *Archaic.* —Used to express surprise or gratitude.

**gram·i·ci·din** (grăm'ĭ-sīd'n) *n.* [GRAM(-POSITIVE) + -CID(E) + -IN.] An antibiotic produced by a bacterium, *Bacillus brevis,* and used against most Gram-positive pathogenic bacteria.

**gra·min·e·ous** (grə-mĭn'ē-əs) *adj.* [Lat. *gramineus,* grassy < *gramen,* grass.] **1.** Of, relating to, or characteristic of grasses. **2.** Of or belonging to the family Gramineae, including the grasses. **—gra·min'e·ous·ness** *n.*

**gram·i·niv·o·rous** (grăm'ə-nĭv'ər-əs) *adj.* [Lat. *gramen,* grass + -VOROUS.] Eating grasses, grain, or seeds.

**gram·mar** (grăm'ər) *n.* [ME *gramere* < OFr. *gramaire* < Lat. *grammatica* < Gk. *grammatikē* < *grammatikos,* of letters < *gramma,* letter.] **1.** The study of language as a systematically composed body of words with discernible regularity of structure and arrangement into sentences and occas. including such aspects of language as pronunciation, meanings, and etymology. **2. a.** The phenomena with which grammar deals as exhibited by a specific language at a specific time. **b.** The system of rules implicit in a language, regarded as a mechanism for generating all sentences possible in that language. **3.** A normative or prescriptive system of rules delineating the current standard of usage for pedagogical or reference purposes. **4.** Writing or speech evaluated with regard to the rules or practice of grammar <Their *grammar* is excellent.> **5.** A book containing the morphologic, syntactic, and semantic rules for a given language. **6. a.** The basic principles of an area of knowledge <the *grammar* of dance> **b.** A book dealing with such principles.

**gram·mar·ian** (grə-mâr'ē-ən) *n.* A specialist in grammar.

**grammar school** *n.* **1.** ELEMENTARY SCHOOL 2. **2.** *Chiefly Brit.* A secondary or preparatory school. **3.** A school emphasizing study of classical languages.

**gram·mat·i·cal** (grə-măt'ĭ-kəl) *adj.* [LLat. *grammaticalis* < Lat. *grammaticus* < Gk. *grammatikos,* of letters < *gramma,* letter.] **1.** Of or pertaining to grammar. **2.** In conformity with the rules of grammar. **—gram·mat'i·cal'i·ty** (-kăl'ĭ-tē) *n.* **—gram·mat'i·cal·ly** *adv.*

**grammatical gender** *n.* The gender assigned to a word in the grammar of a language.

**gram·ma·tol·o·gy** (grăm'ə-tŏl'ə-jē) *n.* [Gk. *gramma, grammat-,* letter + -LOGY.] The study and science of systems of graphic script. **—gram'ma·to·log'ic** (-tə-lŏj'ĭk), **gram'ma·to·log'i·cal** *adj.* **—gram'ma·tol'o·gist** *n.*

**gramme** (grăm) *n. Chiefly Brit. var. of* GRAM[1].

**gram-mo·lec·u·lar weight** (grăm'mə-lĕk'yə-lər) *n.* MOLE[5].

**gram molecule** *n.* MOLE[5].

**Gram-neg·a·tive** (grăm'nĕg'ə-tĭv) *adj.* Of, relating to, or being a microorganism that does not retain the purple dye used in Gram's method.

**gram·o·phone** (grăm'ə-fōn') *n.* [Orig. a trademark.] A phonograph.

**Gram-pos·i·tive** (grăm'pŏz'ĭ-tĭv) *adj.* Of, relating to, or being a microorganism that retains the purple dye used in Gram's method.

**gram·pus** (grăm'pəs) *n.* [Alteration of ME *graspeis* < OFr. *craspois* : *cras,* fat (< Lat. *crassus*) + *pois,* fish < Lat. *piscis.*] **1.** A marine mammal, *Grampus griseus,* related to and resembling the dolphin but without a beaklike snout. **2.** A cetacean, as the killer whale, similar to the grampus.

**Gram's method** (grămz) *n.* [After Hans C.J. *Gram* (1855–1938).] A differential staining technique using the retention or lack of retention of a purple dye to classify bacteria.

**gran·a·dil·la** (grăn'ə-dĭl'ə, -dē'yə) *n.* [Sp., dim of *granada,* pomegranate < Lat. *granatus,* seedy < *granum,* grain.] **1.** A tropical American passionflower, esp. *Passiflora quadrangularis,* bearing edible fruit. **2.** The fleshy egg-shaped fruit of the granadilla.

**gran·a·ry** (grăn'ə-rē, grā'nə-) *n., pl.* **-ries.** [Lat. *granarium* < *granum,* seed.] **1.** A storage building for threshed grain. **2.** A region yielding copious amounts of grain.

**grand** (grănd) *adj.* **-er, -est.** [Fr. < OFr. < Lat. *grandis.*] **1.** Having higher rank than others of the same group or class <a *grand* admiral> **2.** More important than others : PRINCIPAL. **3.** Including or covering all units or aspects <a *grand* total of our expenditures> **4.** Large and impressive in scope, size, or extent : MAGNIFICENT. **5. a.** Rich and sumptuous <a *grand* feast> **b.** Solemn, stately, or splendid in nature, as certain kinds of public ceremonies. **6. a.** Dignified or noble in appearance or effect <*grand* old houses> **b.** Noble or admirable in conception or intent <a *grand* purpose> **c.** Lofty or sublime <opera in the *grand* style> **7.** *Informal.* Wonderful or terrific <a *grand* party> —*n.* **1.** A grand piano. **2.** *Slang.* A thousand dollars. **—grand'ly** *adv.* **—grand'ness** *n.*

☆ ***syns:*** GRAND, BARONIAL, GRANDIOSE, IMPOSING, KINGLY, LORDLY, MAGNIFICENT, MAJESTIC, NOBLE, PRINCELY, REGAL, ROYAL, STATELY, SUBLIME *adj. core meaning* : large and impressive in size, scope, or extent <the *grand* architecture of Buckingham Palace>

**gran·dad·dy** (grăn'dăd'ē) *n. var. of* GRANDDADDY.

**gran·dam** (grăn'dăm', -dəm) *also* **gran·dame** (-dām', -dəm) *n.* [ME *graundame* < OFr. *grand dame,* great lady.] **1.** A grandmother. **2.** An old woman.

**grand·aunt** (grănd'ănt', -änt') *n.* A sister of one's grandparent.

**grand·child** (grănd'chīld', grăn'-) *n.* A child of one's son or daughter <my eldest *grandchild*>

**grand·dad** (grăn'dăd') *n. Informal.* A grandfather.

**grand·dad·dy** *also* **gran·dad·dy** (grăn'dăd'ē) *n., pl.* **-dies. 1.** GRANDFATHER 1. **2.** One that is the oldest, first, or most respected of its kind <the *granddaddy* of modern rockets>

**grand·daugh·ter** (grăn'dô'tər) *n.* The daughter of one's son or daughter.

**grand duchess** *n.* **1.** The wife or widow of a grand duke. **2.** A woman who rules a grand duchy. **3.** A daughter of a czar or of one of his male descendants.

**grand duchy** *n.* A territory ruled by a grand duke or a grand duchess.

**grand duke** *n.* **1.** A nobleman ranking below a king and ruling a grand duchy. **2.** A son of a czar or of one of his male descendants.

**gran·dee** (grăn-dē') *n.* [Sp. *grande* < Lat. *grandis,* great.] **1.** A nobleman of the highest rank in Portugal or Spain. **2.** An eminent or high-ranking personage.

**gran·deur** (grăn'jər, -jŏŏr') *n.* [OFr. < *grand,* great < Lat. *grandis*] **1.** The condition or quality of being grand : MAGNIFICENCE <the *grandeur* of Versailles> **2.** Nobility or greatness of character.

**grand·fa·ther** (grănd'fä'*th*ər, grăn'-) *n.* **1.** The father of one's mother or father. **2.** A male ancestor : FOREFATHER.

**grandfather clause** *n.* **1.** A clause in the constitutions of several Southern states prior to 1915 meant to disfranchise blacks by exempting from strict voting requirements all descendants of persons who voted before 1867. **2.** A clause in some laws creating exemption because of conditions existing before enactment of the legislation.

**grandfather clock** *n.* [From the song *My Grandfather's Clock* by Henry C. Work (1832–1884).] A pendulum clock in a tall narrow cabinet.

**grandfather file** *n. Computer Sci.* A magnetic tape or disk containing the original data for a programming system.

**grand·fa·ther·ly** (grănd'fä'*th*ər-lē, grăn'-) *adj.* **1.** Characteristic of or befitting a grandfather. **2.** Having the qualities traditionally associated with a grandfather : BENEVOLENT.

**grandfather tape** *n. Computer Sci.* A grandfather file.

**gran·dil·o·quence** (grăn-dĭl'ə-kwəns) *n.* [< Lat. *grandiloquus,* speaking loftily : *grandis,* great + *loqui,* to speak.] Bombastic or pompous expression. **—gran·dil'o·quent** *adj.* **—gran·dil'o·quent·ly** *adv.*

**gran·di·ose** (grăn'dē-ōs', grăn'dē-ōs') *adj.* [Fr. < Ital. *grandioso* < *grande,* great < Lat. *grandis.*] **1.** Marked by breadth of scope or intent : GRAND. **2.** Marked by affected grandeur : POMPOUS. **—gran'di·ose'ly** *adv.* **—gran'di·os'i·ty** (-ŏs'ĭ-tē), **gran'di·ose'ness** *n.*

**gran·di·o·so** (grän'dē-ō'sō, -zō, grăn'-) *adv.* [Ital. —see GRANDIOSE.] *Mus.* In a grand and noble style. —Used as a direction. **—gran·di·o'so** *adj.*

**grand jury** *n.* A jury of 12 to 23 persons convened in private session to evaluate accusations against persons charged with crime and to decide if the evidence warrants a bill of indictment.

**Grand La·ma** (grănd' lä'mə) *n.* The Dalai Lama.

**grand larceny** *n.* Theft of property of a value exceeding the amount constituting petit larceny.

**grand·ma** (grănd'mä', grăn'mä', grăm'mä', grăm'ə) *n. Informal.* GRANDMOTHER 1.

**grand mal** (grăN mäl', grăn, grănd' măl') *n.* [Fr. : *grand,* great + *mal,* illness.] Epilepsy marked by severe seizures involving spasms and loss of consciousness.

**grand·moth·er** (grănd'mŭ*th*'ər, grăn'-) *n.* **1.** The mother of one's father or mother. **2.** A woman ancestor.

**grand·moth·er·ly** (grănd'mŭ*th*'ər-lē, grăn'-) *adj.* **1.** Characteristic of or befitting a grandmother. **2.** Having the qualities traditionally associated with a grandmother : SOLICITOUS.

**grand·neph·ew** (grănd'nĕf'yōō, grăn'-) *n.* A son of one's nephew or niece.

**grand·niece** (grănd'nēs', grăn'-) *n.* A daughter of one's nephew or niece.

**grand opera** *n.* A drama with the complete text set to music.

**grand·pa** (grănd'pä', grăn'pä', grăm'pä', grăm'pə) *n. Informal.* GRANDFATHER 1.

**grand·par·ent** (grănd'pâr'ənt, -păr'-, grăn'-) *n.* A parent of one's mother or father : grandmother or grandfather.

**grand piano** *n.* A piano with strings in a horizontal harp-shaped frame supported usu. on three legs.

**Grand Prix** (grăN' prē') *n., pl.* **Grand Prix** (prēz', prē') [Fr., short for *Grand Prix de Paris,* great prize of Paris.] Any of several competitive international road races for sports cars of specific engine size.

**grand·sire** (grănd'sīr', grăn'-) *also* **grand·sir** (-sər) *n. Archaic.* **1.** GRANDFATHER 1. **2.** An old man.

**grand slam** *n.* **1.** SLAM[2]. **2.** Victory in all the major or specified events in a contest, esp. sporting events on a professional circuit. **3.** *Baseball.* A home run hit when three runners are on base.

**grand·son** (grănd'sŭn', grăn'-) *n.* The son of one's son or daughter.

**grand·stand** (grănd'stănd', grăn'-) *n.* **1.** A roofed stand for spectators at a stadium or racetrack. **2.** The audience at an event. *—vi.* **-stand·ed, -stand·ing, -stands.** To perform ostentatiously so as to impress an audience. **—grand'stand'er** *n.*

**grand tour** *n.* **1.** An extended tour of continental Europe once considered a finishing course in the education of upper-class young Englishmen. **2.** A comprehensive tour or survey.

**grand·un·cle** (grănd'ŭng'kəl) *n.* The uncle of one's father or mother.

**grange** (grānj) *n.* [ME *graunge,* granary < OFr. *grange* < Med. Lat. *granica* < Lat. *granum,* seed.] **1. Grange. a.** An association of farmers founded in the United States in 1867. **b.** One of the branch lodges of this association. **2.** *Chiefly Brit.* A farm, esp. the residence and outbuildings of a gentleman farmer. **3.** *Archaic.* A granary. **—grang'er** *n.*

**grang·er·ize** (grān'jə-rīz') *vt.* **-ized, -iz·ing, -iz·es.** [After James *Granger* (1723–1776).] **1.** To illustrate (a book) with pictures taken from other books. **2.** To mutilate (a book) by removing illustrative material for similar use in another book. **—grang'er·ism** *n.* **—grang'er·i·za'tion** *n.* **—grang'er·iz'er** *n.*

**grani-** *pref.* [Lat. < *granum,* seed.] Grain : seed <*granivorous*>

**gran·ite** (grăn'ĭt) *n.* [Ital. *granito* < *granito,* grained, p.part. of *granire,* to make grainy < *grano,* grain < Lat. *granum.*] **1.** A common coarse-grained, light-colored, hard igneous rock consisting mostly of quartz, orthoclase or microcline, and mica, used in monuments and for building. **2.** Unyielding endurance : STEADFASTNESS <a will of *granite*> **—gra·nit'ic** (grə-nĭt'ĭk), **gran'it·oid'** (grăn'ĭ-toid') *adj.*

**granite paper** *n.* Paper that has a low proportion of colored mottling fibers.

**gran·ite·ware** (grăn'ĭt-wâr') *n.* **1.** Enameled iron utensils. **2.** Earthenware with a speckled glaze suggestive of granite.

**gra·niv·o·rous** (grə-nĭv'ər-əs) *adj.* Eating grain and seeds.

**†gran·ny** *or* **gran·nie** (grăn'ē) *n., pl.* **-nies.** [Short for GRANDMOTHER.] **1.** A grandmother. **2.** A fussy person. **3.** *Southern U.S.* A midwife.

**granny knot** *n.* An insecure knot resembling a square knot but with the second tie crossed incorrectly.

granny knot

**grano-** *pref.* [G. < *Granit,* granite < Ital. *granito.* —see GRANITE.] Granite <*grano*lith>

**gra·no·la** (grə-nō'lə) *n.* [Orig. a trademark.] Rolled oats mixed with additional ingredients, as dried fruit, brown sugar, and nuts, used esp. as a cereal. **—gra·no'la** *adj.*

**gran·o·lith** (grăn'ə-lĭth') *n.* A paving stone of cement and crushed granite. **—gran'o·lith'ic** *adj.*

**gran·o·phyre** (grăn'ə-fīr') *n.* [G. *Granophyr* : *grano-,* grano- + *Porphyr,* porphyry < Med. Lat. *porphyrium.* —see PORPHYRY.] A fine-grained granite porphyry having a groundmass with irregular intergrowths of quartz and feldspar. **—gran'o·phyr'ic** (-fĭr'ĭk) *adj.*

**grant** (grănt) *vt.* **grant·ed, grant·ing, grants.** [ME *graunter* < OFr. *granter,* var. of *creanter,* to assure < VLat. **credentare* < Lat. *credens,* pr.part. of *credere,* to believe.] **1.** To agree to the fulfillment of <*grant* a wish> **2.** To permit or accord, as a favor or privilege <civil rights *granted* to everyone> **3. a.** To bestow : confer <*grant* support> **b.** To transfer (property) by a deed. **4.** To concede : acknowledge <I *grant* the suitability of the house.> *—n.* **1.** An act of granting. **2.** Something granted. **3.** *Law.* **a.** Transfer of property by deed. **b.** The property so transferred. **c.** The deed that transfers the property. **4.** One of several tracts of land in New Hampshire, Maine, and Vermont orig. granted to an individual or group. **—grant'a·ble** *adj.* **—grant'er** *n.*

**grant·ee** (grăn-tē') *n. Law.* One to whom a grant is made.

**grant-in-aid** (grănt'ĭn-ād') *n., pl.* **grants-in-aid. 1.** A federal grant of funds to a state or local government to subsidize a public project. **2.** A grant of funds to an institution or an individual to subsidize a project or program.

**gran·tor** (grăn'tər, -tôr') *n. Law.* One that makes a grant.

**grants·man·ship** (grănts'mən-shĭp') *n.* [GRANT + (GAME)SMANSHIP).] The technique of obtaining grants-in-aid.

**gran·u·lar** (grăn'yə-lər) *adj.* **1.** Composed or appearing to be composed of granules or grains. **2.** Grainy in texture. **—gran'u·lar'i·ty** (-lăr'ĭ-tē) *n.* **—gran'u·lar·ly** *adv.*

**gran·u·late** (grăn'yə-lāt') *v.* **-lat·ed, -lat·ing, -lates.** *—vt.* **1.** To form into granules or grains. **2.** To make rough and grainy. *—vi.* To become granular or grainy. **—gran'u·la'tive** *adj.* **—gran'u·la'tor** *n.*

**gran·u·la·tion** (grăn'yə-lā'shən) *n.* **1.** An act or process of granulating or the state of being granulated. **2.** *Physiol.* **a.** Formation of small, fleshy, beadlike protuberances on the surface of a healing wound. **b.** One of these protuberances.

**gran·ule** (grăn'yōōl) *n.* [LLat. *granulum,* dim. of *granum,* grain.] **1.** A small grain or pellet : PARTICLE. **2.** *Astron.* Any of the smallest brilliant transient markings visible in the sun's photosphere.

**gran·u·lite** (grăn'yə-līt') *n.* A granular, often banded metamorphic rock composed mostly of feldspar, quartz, and garnet. **—gran'u·lit'ic** (-lĭt'ĭk) *adj.*

**gran·u·lo·cyte** (grăn'yə-lō-sīt') *n.* A granular leukocyte. **—gran'u·lo·cyt'ic** (-sĭt'ĭk) *adj.*

**gran·u·lo·ma** (grăn'yə-lō'mə) *n., pl.* **-mas** *or* **-ma·ta** (-mə-tə). One of numerous granulated nodules of inflamed tissue, usu. occurring with ulcerated infections. **—gran'u·lo'ma·tous** (-mə-təs) *adj.*

**gran·u·lose** (grăn'yə-lōs') *adj.* Covered with granules.

**grape** (grāp) *n.* [ME < OFr., bunch of grapes.] **1.** A woody vine of the genus *Vitis,* bearing edible fruit clusters and widely cultivated in many species and varieties. **2.** The fleshy, smooth-skinned purple, red, or green fruit of a grape, eaten raw or dried and widely used in winemaking. **3.** A dark violet to dark grayish purple. **4.** Grapeshot.

**grape fern** *n.* Any of various ferns of the genus *Botrychium,* with a fertile frond bearing small grapelike clusters of spore cases.

**grape·fruit** (grāp'frōōt') *n.* [So called because the fruit grows in clusters.] **1.** An evergreen tropical or semitropical tree, *Citrus paradisi,* cultivated for its edible fruit. **2.** The large round fruit of the grapefruit, with a yellow rind and juicy, acidic pulp.

**grape hyacinth** *n.* A native Eurasian plant of the genus *Muscari,* with narrow leaves and rounded, usu. blue flowers in dense terminal clusters.

**grape·shot** (grāp'shŏt') *n.* [From its resemblance to a cluster of grapes.] A cluster of small iron balls once used as a cannon charge.

**grape sugar** *n.* Dextrose.

**grape·vine** (grāp′vīn′) *n.* **1.** A vine on which grapes grow. **2. a.** Informal transmission of information, gossip, or rumor from person to person. **b.** A secret information source.

**graph** (grăf) *n.* [Short for *graphic formula.*] **1.** A drawing that shows an often functional relationship between two sets of numbers as a set of points having coordinates determined by the relationship. **2.** A pictorial device, as a pie chart or bar graph, for displaying numerical relationships. **3.** A representation of a quantity, as of a complex number, by a geometric object such as a point in a plane. —*vt.* **graphed, graph·ing, graphs. 1.** To represent by a graph. **2.** To plot (a function) on a graph.

**-graph** *suff.* [Fr. *-graphe* < Lat. *-graphum* < Gk. *-graphon* < *graphein,* to write.] **1.** Something written or drawn <mono*graph*> **2.** An instrument for writing, drawing, or recording <seismo*graph*>

**graph·eme** (grăf′ēm′) *n.* [Gk. *graphein,* to write + -EME.] **1.** A letter of an alphabet. **2.** The sum of letters and letter combinations that represent a single phoneme. **—gra·phe′mic** (gră-fē′mĭk) *adj.* **—gra·phe′mi·cal·ly** *adv.*

**-grapher** *suff.* [< LLat. *-graphus* < Gk. *-graphos* < *graphein,* to write.] One who writes about a specified subject or in a specified manner <steno*grapher*>

**graph·ic** (grăf′ĭk) *also* **graph·i·cal** (-ĭ-kəl) *adj.* [Lat. *graphicus* < Gk. *graphikos* < *graphē,* writing < *graphein,* to write.] **1. a.** Of or relating to written representation. **b.** Of or relating to pictorial representation. **2.** Of, relating to, or represented by or as if by a graph. **3. a.** Described in vivid detail. **b.** Sharply outlined or set forth. **4.** Of or relating to the graphic arts. **5.** Of or relating to graphics. **6.** *Geol.* Having crystals resembling printed characters. —*n.* **1.** A work of graphic art. **2.** A pictorial device, as an illustration or chart, used for exemplifying something, as in a lecture. **3.** A graphic display generated by a computer or imaging device. **—graph′i·cal·ly** *adv.* **—graph′ic·ness** *n.*

**graphic arts** *pl.n.* **1.** Any of the fine or applied visual arts that involves the application of lines and strokes to a two-dimensional surface. **2.** Reproductions made from blocks, plates, or type, as engravings and lithographs.

**graph·ics** (grăf′ĭks) *n.* (*sing. or pl. in number*). **1. a.** The making of drawings in accordance with the rules of mathematics, as in architecture or engineering. **b.** Calculations, as of structural stress, from such drawings. **2.** The process by which a computer displays graphics for operator manipulation.

**graph·ite** (grăf′īt′) *n.* [Gk. *graphein,* to write + -ITE[1].] The soft, steel-gray to black, hexagonally crystallized allotrope of carbon, used in lead pencils, lubricants, paints and coatings, and various fabricated forms including molds, bricks, electrodes, crucibles, and rocket nozzles. **—gra·phit′ic** (gră-fĭt′ĭk) *adj.*

**graph·i·tize** (grăf′ĭ-tīz′) *vt.* **-tized, -tiz·ing, -tiz·es. 1.** To convert into graphite by a heating process. **2.** To impregnate or coat with graphite. **—graph′i·ti·za′tion** *n.*

**gra·phol·o·gy** (gră-fŏl′ə-jē) *n.* [Gk. *graphē,* writing (< *graphein,* to write) + -LOGY.] The study of handwriting, esp. when used as a way of analyzing character. **—graph′o·log′i·cal** (grăf′ə-lŏj′ĭ-kəl) *adj.* **—gra·phol′o·gist** *n.*

**graph paper** *n.* Paper ruled into small squares of equal size for use in drawing charts, graphs, or diagrams.

**-graphy** *suff.* [Lat. *-graphia* < Gk. < *graphein,* to write.] **1.** A writing or representation produced in a specified manner or by a specified process <photo*graphy*> **2. a.** A writing about a specified subject <oceano*graphy*> **b.** A representation of a specified object <phono*graphy*>

**grap·nel** (grăp′nəl) *n.* [ME *grapenel,* prob. ult. < OFr. *grapin,* hook, of Germanic orig.] A small anchor with three or more flukes.

**grap·pa** (grä′pə) *n.* [Ital.] An Italian brandy distilled from the residue of pressed wine.

**grap·ple** (grăp′əl) *n.* [ME *grapel* < OFr. *grape,* hook.] **1. a.** An iron shaft with claws at one end for grasping and holding, esp. one for drawing and holding an enemy ship alongside. **b.** A grapnel. **2.** The act of grappling. **3. a.** A contest in which the participants try to grip or clutch each other. **b.** A grip or grasp in such a contest. —*v.* **-pled, -pling, -ples.** —*vt.* **1.** To seize and hold with a grapple. **2.** To seize firmly with the hands. —*vi.* **1.** To hold on to something with or as if with a grapple. **2.** To use a grapple, as for dragging. **3.** To struggle or come to grips in hand-to-hand combat : WRESTLE. **4.** To attempt to cope <*grappling* with unemployment> **—grap′pler** *n.*

**grap·pling** (grăp′lĭng) *n.* **1.** A grappling iron. **2.** A grapnel.

**grappling iron** *also* **grappling hook** *n.* An iron bar with claws at one end, used for raising sunken objects or securing a ship alongside.

**grap·to·lite** (grăp′tə-līt′) *n.* [Gk. *graptos,* written (< *graphein,* to write) + -LITE (from the resemblance of the fossils' impressions on shale to markings on a slate).] Any of numerous extinct colonial marine animals chiefly of the orders Dendroidea and Graptoloidea, of the Cambrian to the Mississippian periods.

**grap·y** (grā′pē) *adj.* **-i·er, -i·est.** Of or suggestive of grapes.

**grasp** (grăsp) *v.* **grasped, grasp·ing, grasps.** [ME *graspen.*] —*vt.* **1.** To seize or take hold of firmly with or as if with the hand. **2.** To clasp firmly with or as if with the hand. **3.** To take hold of intellectually : COMPREHEND. —*vi.* **1.** To make a seizing, snatching, or clutching motion. **2.** To show prompt and eager willingness or acceptance <*grasps* at any chance> —*n.* **1.** The act of grasping. **2. a.** A firm hold or grip. **b.** An embrace. **3.** The power or ability to seize or attain : REACH. **4.** Comprehension : understanding.

**grasp·ing** (grăs′pĭng) *adj.* Very desirous of material gain : AVARICIOUS. **—grasp′ing·ly** *adv.* **—grasp′ing·ness** *n.*

**grass** (grăs) *n.* [ME *gras* < OE *græs.*] **1. a.** Any of numerous plants of the family Gramineae, with narrow leaves, hollow, jointed stems, and spikes or clusters of membranous flowers borne in smaller spikelets. **b.** Such plants as a group. **2.** Any of various plants with slender leaves like those of the true grasses. **3.** An expanse of ground, as a lawn, covered with grass or similar plants. **4.** Grazing land : PASTURE. **5.** *Slang.* Marijuana. **6.** *Electron.* The small variations in amplitude of an oscilloscope display due to electrical noise. —*v.* **grassed, grass·ing, grass·es.** —*vt.* **1. a.** To cover with grass. **b.** To grow grass on. **2.** To feed (livestock) with grass. —*vi.* **1.** To become covered with grass. **2.** To graze.

**grass green** *n.* A moderate yellow-green to strong or dark yellowish-green. **—grass′-green′** *adj.*

**grass·hop·per** (grăs′hŏp′ər) *n.* **1.** Any of numerous insects of the families Locustidae or Acrididae and Tettigoniidae, often destructive to plants and having long hind legs adapted for jumping. **2.** A light, usu. unarmed aircraft used for reconnaissance and liaison. **3.** A cocktail of crème de menthe, crème de cacao, and cream.

**grass·land** (grăs′lănd′) *n.* An area, as a meadow or prairie, of grass or grasslike vegetation.

**grass·roots** (grăs′ro͞ots′, -ro͝ots′) *n.* (*sing. or pl. in number*). **1.** People or society at a local level. **2.** The groundwork or fundamental source of something.

**grass skiing** *n.* The sport of skiing down straw- or grass-covered hills on skis equipped with rollers.

**grass snake** *n.* Any of various greenish snakes, esp. *Opheodrys vernalis* of eastern North America.

**grass snipe** *n.* The pectoral sandpiper.

**grass tree** *n.* A woody-stemmed Australian plant of the genus *Xanthorrhoea,* bearing stiff grasslike leaves and a small white flower spike.

**grass widow** *n.* **1.** A woman who is separated or divorced from her husband. **2.** A woman whose husband is temporarily absent. **3.** An abandoned mistress. **4.** The mother of an illegitimate child.

**grass·y** (grăs′ē) *adj.* **-i·er, -i·est. 1.** Covered with or abounding in grass. **2.** Resembling grass.

**grate**[1] (grāt) *v.* **grat·ed, grat·ing, grates.** [ME *graten* < OFr. *grater,* to scrape, of Germanic orig.] —*vt.* **1.** To reduce to fragments, shreds, or powder by rubbing against an abrasive surface. **2.** To cause (e.g., teeth) to make a harsh grinding or rasping sound through friction. **3.** To annoy or irritate persistently. **4.** *Archaic.* To wear away. —*vi.* **1.** To make a harsh rasping sound by or as if by scraping or grinding. **2.** To cause annoyance or irritation <a shrill whistle *grating* on our nerves> —*n.* A harsh, rasping sound made by rubbing or scraping.

**grate**[2] (grāt) *n.* [ME < Med. Lat. *grata* < Lat. *cratis,* wickerwork.] **1.** A framework of latticed or parallel bars for blocking an opening. **2.** A framework of metal bars for holding fuel in a stove, furnace, or fireplace. **3.** A fireplace. **4.** A perforated iron plate or screen for sieving and grading crushed ore. —*vt.* **grat·ed, grat·ing, grates.** To equip with a grate.

**grate·ful** (grāt′fəl) *adj.* [< obs. *grate,* pleasing < Lat. *gratus.*] **1.** Thankful : appreciative. **2.** Expressing gratitude. **3.** Affording pleasure or comfort <*grateful* warmth> **—grate′ful·ly** *adv.* **—grate′ful·ness** *n.*

**grat·er** (grā′tər) *n.* **1.** One that grates. **2.** An implement with sharp-edged slits and perforations for grating foods.

**grat·i·fy** (grăt′ə-fī′) *vt.* **-fied, -fy·ing, -fies.** [ME *gratifien,* to favor < Lat. *gratificare,* to oblige : *gratus,* pleasing + *facere,* to make.] **1.** To please or satisfy. **2.** To give what is desired to : INDULGE <*gratified* my curiosity> **3.** *Archaic.* To reward. **—grat′i·fi·ca′tion** (-fĭ-kā′shən) *n.* **—grat′i·fi′er** *n.*

**gra·tin** (grät′n, grăt′n) *n.* [Fr. < OFr. < *grater,* to scrape, of Germanic orig.] A crust of browned crumbs and butter, often with grated cheese.

**grat·ing** (grā′tĭng) *n.* **1.** A grill or network of bars fixed in a window or door or used as a partition : GRATE. **2.** *Physics.* Diffraction grating.

**grat·is** (grăt′ĭs, grā′tĭs) *adv.* & *adj.* [ME < Lat., alteration of *gratiis,* out of kindness, ablative pl. of *gratia,* kindness < *gratus,* pleasing.] Without charge : FREE.

**grat·i·tude** (grăt′ĭ-to͞od′, -tyo͞od′) *n.* [ME < Med. Lat. *gratitudo* < Lat. *gratus,* pleasing.] The state of being grateful.

**gra·tu·i·tous** (grə-to͞o′ĭ-təs, -tyo͞o′-) *adj.* [Lat. *gratuitus,* voluntary < *gratia,* favor < *gratus,* pleasing.] **1.** Given without recompense or return : UNEARNED. **2.** Given or received without cost or obligation : FREE. **3.** Unnecessary or unwarranted : UNJUSTIFIED <*gratuitous* criticism> **—gra·tu′i·tous·ly** *adv.* **—gra·tu′i·tous·ness** *n.*

**gra·tu·i·ty** (grə-to͞o′ĭ-tē, -tyo͞o′-) *n., pl.* **-ties.** [OFr. *gratuite* < Med.

Lat. *gratuitas,* gift < Lat. *gratia, favor* < *gratus,* pleasing.] A favor or gift, usu. in the form of money, given in return for service.

**grat·u·late** (grăch′ə-lāt′) *vt.* **-lat·ed, -lat·ing, -lates.** [Lat. *gratulari, gratulat-* < *gratus,* pleasing.] *Archaic.* To congratulate. **—grat′u·la′tion** *n.* **—grat′u·la·to′ry** (-lə-tôr′ē, -tōr′ē) *adj.*

**grau·pel** (grou′pəl) *n.* [G., dim. of *Graupe,* barley.] Precipitation consisting of snow pellets.

**gra·va·men** (grə-vā′mən) *n., pl.* **-va·mens** *or* **-vam·i·na** (-văm′ə-nə) [LLat. *gravamen* < Lat. *gravare,* to burden < *gravis,* heavy.] *Law.* The part of a charge or accusation that weighs most heavily against the accused.

**grave**[1] (grāv) *n.* [ME < OE *græf.*] **1.** An excavation for interring a corpse. **2.** A burial place. **3.** Death or extinction.

**grave**[2] (grāv) *adj.* **grav·er, grav·est.** [OFr. < Lat. *gravis.*] **1.** Extremely serious : IMPORTANT <*grave* consequences of such an act> **2.** Fraught with danger or harm <a *grave* illness> **3.** Dignified in conduct or character <a *grave* occasion> **4.** Somber : dark. —Used of colors. **5.** (*also* gräv). **a.** Written with or modified by the mark (`) as the è in *Sèvres.* **b.** Articulated toward the back of the oral cavity. —*n.* (*also* gräv). The grave accent (`) indicating a pronounced *e* for the sake of meter in the usu. nonsyllabic ending *-ed* in English poetry. **—grave′ly** *adv.* **—grave′ness** *n.*

**grave**[3] (grāv) *vt.* **graved, grav·en** (grā′vən) *or* **graved, grav·ing, graves.** [ME *graven* < OE *grafan.*] **1.** To carve or sculpt : ENGRAVE. **2.** To impress or stamp deeply, as ideas or words.

**grave**[4] (grāv) *vt.* **graved, grav·ing, graves.** [ME *graven.*] To clean and coat with pitch (the bottom of a wooden ship).

**gra·ve**[5] (grä′vā) *adv.* [Ital. < Lat. *gravis,* ponderous.] *Mus.* Slowly and solemnly. —Used as a direction. **—gra′ve** *adj.*

**grave·dig·ger** (grāv′dĭg′ər) *n.* A digger of graves.

**grav·el** (grăv′əl) *n.* [ME < OFr. *gravele,* dim. of *grave,* coarse sand, of Celt. orig.] **1.** An unconsolidated mixture of pebbles or rock fragments. **2.** *Pathol.* The sandlike granular material of urinary calculi. —*vt.* **-eled, -el·ing, -els** *or* **-elled, -el·ling, -els. 1.** To apply a surface of gravel to. **2.** To perplex : confuse. **3.** *Informal.* To irritate.

**grav·el·ly** (grăv′ə-lē) *adj.* **1.** Of, full of, or covered with gravel. **2.** Having a harsh rasping sound <replied in a *gravelly* voice>

**grav·en** (grā′vən) *v. var. p.p. of* GRAVE[3].

**graven image** *n.* An idol carved in wood or stone.

**grav·er** (grā′vər) *n.* **1.** An engraver or carver. **2.** An engraver's cutting tool.

**grave robber** *n.* One who steals valuables from graves or tombs or who steals corpses, as for illicit dissection.

**Graves' disease** (grāvz) *n.* [After Robert J. Graves (1796–1853).] Exophthalmic goiter.

**grave·stone** (grāv′stōn′) *n.* A stone marking a grave.

**grave·yard** (grāv′yärd′) *n.* A burial ground : CEMETERY.

**graveyard shift** *n.* **1.** A work shift that begins late at night, as 11:00 P.M. or midnight. **2.** The workers on a graveyard shift.

**grav·id** (grăv′ĭd) *adj.* [Lat. *gravidus* < *gravis,* heavy.] Pregnant. **—gra·vid′i·ty** (grə-vĭd′ĭ-tē), **grav′id·ness** *n.* **—grav′id·ly** *adv.*

**gra·vim·e·ter** (grə-vĭm′ĭ-tər, grăv′ə-mē-) *n.* [Fr. *gravimètre* : Lat. *gravis,* heavy + *-mètre,* -meter.] An instrument for determining specific gravity. **—gra·vim′e·try** (grə-vĭm′ĭ-trē) *n.*

**grav·i·met·ric** (grăv′ə-mĕt′rĭk) *also* **grav·i·met·ri·cal** (-rĭ-kəl) *adj.* [Lat. *gravis,* heavy + METRIC.] Of or relating to measurement by weight. **—grav′i·met′ri·cal·ly** *adv.*

**graving dock** *n.* A dry dock for repairing ships and cleaning their hulls.

**grav·i·sphere** (grăv′ĭ-sfîr′) *n.* [GRAVI(TY) + SPHERE.] The spherical region of space dominated by the gravitational influence of a celestial body.

**grav·i·tate** (grăv′ĭ-tāt′) *vi.* **-tat·ed, -tat·ing, -tates. 1.** To move in response to the force of gravity. **2.** To move downward. **3.** To be attracted by or as if by an irresistible force <teen-agers *gravitating* toward rock concerts> **—grav′i·tat′er** *n.*

**grav·i·ta·tion** (grăv′ĭ-tā′shən) *n.* **1.** *Physics.* **a.** Natural attraction between massive bodies. **b.** The degree of such attraction. **c.** The action or process of moving under the influence of this attraction. **2.** A movement toward a source of attraction <*gravitation* of the very rich to the southern coast of France> **—grav′i·ta′tion·al** *adj.* **—grav′i·ta′tion·al·ly** *adv.* **—grav′i·ta′tive** *adj.*

**gravitational collapse** *n.* The process by which stars, star clusters, and galaxies form from dilute interstellar gas under the influence of gravity.

**gravitational interaction** *n.* A hypothesized weak fundamental interaction between elementary particles.

**gravitational wave** *n.* A wave hypothesized to propagate the force of gravity and travel at the speed of light.

**grav·i·ton** (grăv′ĭ-tŏn′) *n.* [GRAVIT(ATION) + -ON[1].] A particle postulated to be the quantum of gravitational interaction and presumed to have zero electric charge, zero rest mass, and spin 2.

**grav·i·ty** (grăv′ĭ-tē) *n.* [OFr. *gravite,* heaviness < Lat. *gravitas* < *gravis,* heavy.] **1.** *Physics.* **a.** The force of gravitation, which for any two sufficiently massive bodies is directly proportional to the product of their masses and inversely proportional to the square of the distance between them, esp. the attractive central gravitational force exerted by a celestial body such as the earth. **b.** GRAVITATION 1. **c.** Weight. **2.** Grave or serious consequence : IMPORTANCE <the *gravity* of the urban problem> **3.** Solemnity or dignity.

**gravity wave** *n.* A gravitational wave.

**gra·vure** (grə-vyŏŏr′) *n.* [Fr. < *graver,* to engrave < OFr., of Germanic orig.] **1. a.** A method of printing with etched plates or cylinders : INTAGLIO. **b.** Photogravure. **2.** A plate or reproduction produced by or used in the gravure process.

**gra·vy** (grā′vē) *n., pl.* **-vies.** [ME *grave* < OFr.] **1. a.** The juices that drip from cooking meat. **b.** A sauce made by thickening and seasoning these juices. **2.** *Slang.* Money, profit, or benefit easily or unexpectedly gained.

**gravy boat** *n.* An elongated dish or pitcher for serving gravy.

**gravy train** *n. Slang.* A highly profitable occupation or job requiring little effort.

**gray** (grā) *also* **grey** [ME *grei* < OE *græg.*] —*adj.* **-er, -est. 1.** Of or relating to an achromatic color of any lightness between the extremes of black and white. **2. a.** Dull or dark, as from lack of light <a cold, *gray* January day> **b.** Lacking in cheer : GLOOMY. **3. a.** Having gray hair : HOARY. **b.** Old, venerable, or ancient. **4.** Intermediate in character or position, esp. in morality or propriety <political deals that lay in a *gray* area of questionable honesty> —*n.* **1.** An achromatic color of any lightness between the extremes of black and white. **2.** An object or animal of the color gray. **3.** *often* **Gray. a.** A member of the Confederate Army in the American Civil War. **b.** The Confederate Army. —*vt.* & *vi.* **grayed, gray·ing, grays** *also* **greyed, grey·ing, greys.** To make or become gray.

**gray·beard** (grā′bîrd′) *n.* An old man.

**gray·fish** (grā′fĭsh′) *n., pl.* **grayfish** *or* **-fish·es.** DOGFISH 1.

**gray·ish** (grā′ĭsh) *adj.* Rather gray.

**gray·lag goose** (grā′lăg′) *n.* [Poss. GRAY + dial. *lag,* last.] A gray goose, *Anser anser* of Old World marshy areas.

**gray·ling** (grā′lĭng) *n., pl.* **grayling** *or* **-lings.** Any of several freshwater food fishes of the genus *Thymallus* of the Northern Hemisphere, with a small mouth and a large dorsal fin.

**gray matter** *n.* **1.** The brownish-gray nerve tissue of the brain and spinal cord, composed of nerve cells and fibers and some supportive tissue. **2.** *Informal.* Intellect : brains.

**gray mullet** *n.* The mullet.

**gray panther** *n.* [GRAY + (BLACK) PANTHER.] A member of a militant organization of elderly people.

**gray squirrel** *n.* A common squirrel, *Sciurus carolinensis* of eastern North America, with gray or blackish fur.

**gray·wacke** (grā′wăk′, -wăk′ə) *n.* [Partial transl. of G. *Grauwacke* : *grau,* gray + *Wacke,* boulder.] Any of various shale-containing dark-gray sandstones.

**gray whale** *n.* A whalebone whale, *Eschrichtius glaucus* of Pacific waters, having grayish coloring with white blotches.

**gray wolf** *n.* The timber wolf.

**graze**[1] (grāz) *v.* **grazed, graz·ing, graz·es.** [ME *grasen* < OE *grasian* < *græs,* grass.] —*vi.* To feed on growing grasses and herbage. —*vt.* **1.** To feed on (herbage) in a field or on pasture land. **2.** To feed on the herbage of (land). **3.** To put (livestock) out to feed. **4.** To afford adequate herbage for <This pasture will *graze* a large flock of sheep.> **5.** To tend (feeding livestock) in a pasture. **—graz′er** *n.*

**graze**[2] (grāz) *v.* **grazed, graz·ing, graz·es.** [Perh. < GRAZE[1].] —*vt.* **1.** To touch lightly in passing : BRUSH. **2.** To scrape or scratch slightly : ABRADE. —*vi.* To scrape or touch something lightly in passing. —*n.* **1.** An act of scraping or brushing along a surface. **2.** A slight scratch or abrasion. **—graz′ing·ly** *adv.*

**gra·zier** (grā′zhər) *n.* One who grazes cattle.

**gra·zi·o·so** (grät′sē-ō′sō, -zō) *adv.* [Ital. < Lat. *gratiosus,* agreeable. —see GRACIOUS.] *Mus.* Gracefully : smoothly. —Used as a direction. **—gra′zi·o′so** *adj.*

**grease** (grēs) *n.* [ME *grese* < AN *grece* < OFr. *graisse* < Lat. *crassus,* fat.] **1.** Animal fat when melted or soft. **2.** A thick oil or viscous substance, esp. when used as a lubricant. **3. a.** The oily substance present in raw wool : SUINT. **b.** Raw wool that has not been cleansed. —*vt.* (grēs, grēz) **greased, greas·ing, greas·es. 1.** To coat, smear, or soil with grease or lard. **2.** To lubricate with grease. **—grease (someone's) palm.** *Slang.* To bribe. **—grease the wheels.** To facilitate or expedite the progress of an operation.

**grease monkey** *n. Slang.* A mechanic.

**grease paint** *n.* Theatrical make-up.

**grease·wood** (grēs′wŏŏd′) *n.* **1.** A spiny shrub, *Sarcobatus vermiculatus* of western North America, bearing small alternate leaves, white stems, and small greenish flowers. **2.** A plant resembling or related to the greasewood, as the creosote bush.

**greas·y** (grē′sē, -zē) *adj.* **-i·er, -i·est. 1.** Soiled or coated with grease. **2.** Containing grease, esp. too much grease <*greasy* fried chicken> **3.** Suggestive of grease, as in slipperiness or slickness. **—greas′i·ly** *adv.* **—greas′i·ness** *n.*

**greasy spoon** *n. Slang.* A cheap diner or restaurant.

**great** (grāt) *adj.* **-er, -est.** [ME *grete* < OE *grēat,* thick, coarse.] **1.** Extremely large : BIG. **2.** Larger than others of the same kind.

**3.** Large in quantity or number <A *great* throng was at the gate.> **4.** Extensive in distance or time. **5.** Remarkable or outstanding in magnitude, degree, or extent <a *great* disaster> **6.** Of outstanding importance or significance <a *great* exhibition of paintings> **7.** Chief or principal <the *great* hall of the manor house> **8.** Superior in quality or character : NOBLE. **9.** Powerful : influential <a *great* nation> **10.** Eminent : distinguished <a *great* musician> **11.** Grand : aristocratic. **12.** *Archaic.* Pregnant <*great* with child> **13.** *Informal.* Enthusiastic <a *great* yachtsman> **14.** *Informal.* Very skillful <*great* at math> **15.** *Informal.* Very good : FIRST-RATE. **16.** Being one generation removed from the relative specified <a *great*-uncle> —*n.* One that is great. —*adv. Informal.* Very well. **—great'ly** *adv.* **—great'ness** *n.*

**great auk** *n.* A large flightless sea bird, *Pinguinus impennis,* once flourishing along northern Atlantic coasts but now extinct.

**great-aunt** (grāt'ănt', -änt') *n.* A grandaunt.

**Great Bear** *n.* Ursa Major.

**great circle** *n.* A circle that is the intersection of the surface of a sphere with a plane passing through the center of the sphere.

**great·coat** (grāt'kōt') *n.* A heavy overcoat.

**Great Dane** *n.* A large and powerful dog orig. bred in Germany, with a short smooth coat and a narrow head.

**great divide** *n.* **1.** A major watershed. **2.** A major turning point.

**great·en** (grāt'n) *vt.* & *vi.* **-ened, -en·ing, -ens.** *Archaic.* To make or become great or greater.

**great·er** *also* **Great·er** (grā'tər) *adj.* Designating a city and its populous suburbs <*greater* Springfield>

**great·heart·ed** (grāt'här'tĭd) *adj.* **1.** Noble or courageous. **2.** Magnanimous : generous. **—great'heart'ed·ly** *adv.* **—great'heart'ed·ness** *n.*

**great horned owl** *n.* A large North American owl, *Bubo virginianus,* with brownish plumage.

**great laurel** *n.* ROSEBAY 1.

**Great Powers** *pl.n.* Those nations with the greatest social, political, and economic influence in international affairs.

**Great Pyr·e·nees** (pĭr'ə-nēz') *n.* A large heavy-boned dog of an ancient breed, with a thick white coat.

**Great Russian** *n.* A member of the Russian-speaking people living in the central and northeastern U.S.S.R.

**great seal** *n.* The principal seal of a government or state, for stamping official documents.

**Great Spirit** *n.* The principal deity in the religion of many North American Indian tribes.

**great-un·cle** (grāt'ŭng'kəl) *n.* A granduncle.

**Great War** *n.* World War I.

**greave** (grēv) *n.* [ME *greve* < OFr.] *often* **greaves.** Leg armor worn below the knee.

**greaves** (grēvz) *pl.n.* [LG *greven.*] The unmelted residue left after animal fat or tallow has been rendered.

**grebe** (grēb) *n.* [Fr. *grèbe.*] Any of various diving birds of the family Podicipedidae, with fleshy lobed membranes along each toe and a pointed bill.

**Gre·cian** (grē'shən) *adj.* [< Lat. *Graecia,* Greece < *Graecus,* Greek. —see GREEK.] Greek. —*n.* A native or resident of Greece.

**Gre·cism** (grē'sĭz'əm) *n.* **1.** The style or spirit of Greek culture, art, or thought. **2.** Something imitative of Greek style or spirit. **3.** A Greek idiom.

**Gre·cize** (grē'sīz') *vt.* **-cized, -ciz·ing, -ciz·es.** [Lat. *graecizare* < *Graecus,* Greek. —see GREEK.] To make Greek or Hellenic in form or style.

**Greco-** *or* **Graeco-** *pref.* [Lat. *Graeco-* < *Graecus,* Greek.] Greece : Greek <*Greco*-Roman>

**Grec·o-Ro·man** (grĕk'ō-rō'mən, grē'kō-) *adj.* Of or relating to both Greece and Rome <*Greco-Roman* statuary>

**gree** (grē) *n.* [ME *gre* < OFr., step < Lat. *gradus.*] *Scot.* Superiority.

**greed** (grēd) *n.* [Back-formation < GREEDY.] An overwhelming desire to acquire or have, as wealth or power, in excess of what one requires or deserves.

**greed·y** (grē'dē) *adj.* **-i·er, -i·est.** [ME *gredi* < OE *grædig.*] **1.** Excessively eager to acquire or possess something, esp. in quantity. **2.** Wanting to eat or drink more than one can reasonably consume. **3.** Extremely desirous. **—greed'i·ly** *adv.* **—greed'i·ness** *n.*

☆ ***syns:*** **1.** GREEDY, ACQUISITIVE, AVARICIOUS, AVID, COVETOUS, DESIROUS, GRABBY, GRASPING, HUNGRY, RAPACIOUS *adj. core meaning* : having a strong urge to obtain or possess something, esp. material wealth, in quantity <*greedy* politicians> **2.** GREEDY, EDACIOUS, GLUTTONOUS, HOGGISH, PIGGISH, RAVENOUS, VORACIOUS *adj. core meaning* : wanting to eat or drink more than is reasonable <*greedy* people at a buffet>

**Greek** (grēk) *n.* [ME *Grek* < OE *Grecas,* the Greeks < Lat. *Graecus,* Greek < Gk. *Graikos,* tribal name.] **1.** The Indo-European language of the Greeks. **2. a.** An indigenous inhabitant of Greece. **b.** A descendant of an indigenous inhabitant of Greece. **3.** *Slang.* A member of a fraternity or sorority having a name composed of Greek letters. **4.** Something unintelligible <This document is *Greek* to me.> —*adj.* **1.** Of, relating to, or designating Greece, the Hellenes, their language, or their culture. **2.** Of, relating to, or designating the Greek Orthodox Church.

**Greek Catholic** *n.* **1.** A member of the Eastern Orthodox Church. **2.** A member of a Uniat Church.

**Greek cross** *n.* A cross formed by two bars of equal length crossing in the middle at right angles to each other.

**Greek fire** *n.* An incendiary material used by the Byzantine Greeks for burning enemy ships.

**green** (grēn) *n.* [ME *grene* < OE *grēne.*] **1.** Any of a group of colors that may vary in lightness and saturation and whose hue is that of the emerald or somewhat less yellow than that of growing grass; the hue of that portion of the spectrum lying between yellow and blue; one of the additive or light primaries; one of the psychological primary hues, evoked in the normal observer by radiant energy having a wavelength of approx. 530 nanometers. **2.** Something green in color. **3. greens.** Green foliage or growth, esp.: **a.** Branches and leaves of plants used for decoration. **b.** Leafy plants or plant parts eaten as vegetables. **4. a.** A grassy lawn or plot. **b.** A putting green. **5. greens.** Garments that are green in color <hospital *greens*> —*adj.* **-er, -est. 1.** Of the color green. **2.** Abounding in or covered with green growth or foliage. **3.** Made with green or leafy vegetables <a *green* salad> **4.** Mild or temperate in climate. **5.** Youthful : vigorous. **6.** Fresh : brand-new. **7.** Not mature or ripe : YOUNG <*green* tomatoes> **8.** Pale and sickly : WAN. **9.** Not yet fully processed, esp.: **a.** Not aged <*green* wood> **b.** Not cured or tanned <*green* pelts> **10.** Lacking experience or training. **11. a.** Naive : unsophisticated. **b.** Easily duped : GULLIBLE. —*vt.* & *vi.* **greened, green·ing, greens.** To make or become green. **—green'ly** *adv.* **—green'ness** *n.*

**green alga** *n.* An alga of the division Chlorophyta, with pronounced green coloring, as spirogyra or sea lettuce.

**green·back** (grēn'băk') *n.* A legal-tender note of U.S. currency.

**Greenback Party** *n.* A former U.S. political party, organized in 1874, that advocated the use of inconvertible paper money.

**green bean** *n.* STRING BEAN 2.

**green·belt** (grēn'bĕlt') *n.* An area of recreational parks, farmland, or uncultivated land surrounding a community.

**Green Beret** *n.* [From the green beret that is part of the uniform.] A member of the U.S. Army Special Forces.

**green·bri·er** (grēn'brī'ər) *n.* The catbrier.

**green corn** *n.* Tender young ears of sweet corn.

**green dragon** *n.* A plant, *Arisaema dracontium* of eastern North America, with divided leaves and minute flowers at the base of a long slender spadix projecting from a narrow green spathe.

**green·er·y** (grē'nə-rē) *n., pl.* **-ies. 1.** Green foliage. **2.** A place where plants are grown.

**green-eyed** (grēn'īd') *adj.* Jealous.

**green·finch** (grēn'fĭnch') *n.* A Eurasian bird, *Carduelis chloris* or *Chloris chloris,* with green and yellow plumage.

**green·fly** (grēn'flī') *n.* A greenish insect, esp. an aphid.

**green·gage** (grēn'gāj') *n.* A variety of plum with yellowish-green skin and sweet flesh.

**green·gro·cer** (grēn'grō'sər) *n. Chiefly Brit.* A retailer of fresh fruit and vegetables. **—green'gro'cer·y** *n.*

**green·head** (grēn'hĕd') *n.* A male mallard duck.

**green·heart** (grēn'härt') *n.* **1.** A tropical American tree, *Ocotea rodioei* or *Nectandra rodioei,* with durable greenish wood. **2.** A tree similar to the greenheart. **3.** The wood of the greenheart.

**green·horn** (grēn'hôrn') *n.* [Obs. *greynhorne,* animal with immature horns.] **1.** An immature or inexperienced person. **2.** A newcomer, esp. an immigrant unfamiliar with local customs. **3.** A gullible person.

**green·house** (grēn'hous') *n.* **1.** A usu. glass-enclosed structure for cultivating plants that must have controlled temperature and humidity. **2.** *Slang.* A part of an aircraft enclosed in a clear plastic shell or bubble.

**greenhouse effect** *n.* The sequence of phenomena comprising the absorption of solar radiation by the earth, its conversion and re-emission in the infrared, and the absorption of this radiation, esp. in the wavelength region from 5 to 17 microns, by atmospheric ozone, water vapor, and carbon dioxide, preventing its dissipation into space and resulting in a steady, gradual rise in the temperature of the atmosphere.

**green·ish** (grē'nĭsh) *adj.* Rather green.

**Green·land spar** (grēn'lənd) *n.* Cryolite.

**green·let** (grēn'lĭt) *n.* Any of various greenish birds of the genus *Hylophilus* of Central and South America, related to the vireos.

**green light** *n.* **1.** The green-colored light that signals traffic to proceed. **2.** Permission to proceed.

**green·ling** (grēn'lĭng) *n.* Any of various food fishes of the family Hexagrammidae, of the northern Pacific.

**green manure** *n.* A growing crop, as a clover or grass, plowed under the soil to enhance fertility.

**green monkey** *n.* An African monkey of the genus *Cercopithecus,* esp. *C. aethiops sabaeus,* having yellowish-gray fur with a greenish tinge.

**gree·nock·ite** (grē'nə-kīt') *n.* [After Charles Cathcart, Lord

*Greenock* (d. 1859).] A yellow to brown or red mineral, essentially CdS, used as a cadmium ore.

**green pepper** *n.* The unripened green fruit of various pepper plants.

**green plover** *n.* The lapwing.

**green revolution** *n.* Significant increase in agricultural productivity following introduction of high-yield varieties of grains and improved management techniques.

**green·room** (grēn′ro͞om′, -ro͝om′) *n.* [So called because such rooms were orig. painted green.] A waiting room or lounge in a theater or concert hall for performers to use when off stage.

**green·sand** (grēn′săndʹ) *n.* A sand or sediment given a dark greenish color by grains of glauconite.

**green·shank** (grēn′shăngk′) *n.* An Old World wading bird, *Tringa nebularia*, with greenish legs and a long bill.

**green·sick·ness** (grēn′sĭk′nĭs) *n.* CHLOROSIS 2. **—green′sick′** *adj.*

**green snake** *n.* A nonvenomous North American snake of the genus *Opheodrys*, with a slender yellow-green body.

**green soap** *n.* A translucent, yellowish-green soft or liquid soap made from vegetable oils and used to treat chronic skin disorders.

**green·stone** (grēn′stōn′) *n.* Any of various altered basic igneous rocks colored green by chlorite, hornblende, and epidote.

**green·sward** (grēn′swôrd′) *n.* Turf on which green grass grows.

**green tea** *n.* Tea made from leaves that are not fermented before being dried.

**green thumb** *n.* A knack for growing plants.

**green turtle** *n.* A large marine turtle, *Chelonia mydas*, with greenish flesh valued as food.

**Green·wich time** *also* **Greenwich mean time** (grĭn′ĭj, -ĭch, grĕn′-) *n.* Mean solar time for the meridian at Greenwich, England, used as a basis for calculating time throughout most of the world.

**greet** (grēt) *vt.* **greet·ed, greet·ing, greets.** [ME *greten* < OE *grētan.*] **1.** To address in a friendly and respectful way : WELCOME. **2.** To receive with a specified reaction <*greet* the guests with respect> **3.** To be perceived by <A horrible sight *greeted* our eyes.> **—greet′er** *n.*

**greet·ing** (grē′tĭng) *n.* A gesture or word of salutation or welcome.

**greeting card** *n.* A card imprinted with a greeting and a suitable illustration for use on a special occasion or holiday.

**greg·a·rine** (grĕg′ə-rīn′) *n.* [NLat. *Gregarina*, genus name < Lat. *gregarius*, belonging to a flock < *grex*, flock.] Any of various sporozoan protozoans of the order Gregarinida that are parasitic within invertebrates such as arthropods and annelids. *—adj.* Of or belonging to the Gregarinida. **—greg′a·rin′i·an** (-rĭn′ē-ən) *adj.*

**gre·gar·i·ous** (grĭ-gâr′ē-əs) *adj.* [Lat. *gregarius*, belonging to a flock < *grex*, flock.] **1.** Tending to move in or form a group, as a herd, pack, or flock, with others of the same kind. **2.** Seeking and enjoying the company of others : SOCIABLE. **3.** *Bot.* Growing in groups that are close together but not densely clustered or matted. **—gre·gar′i·ous·ly** *adv.* **—gre·gar′i·ous·ness** *n.*

**Gre·go·ri·an calendar** (grĭ-gôr′ē-ən, -gōr′-) *n.* The calendar used throughout most of the world, sponsored by Pope Gregory XIII in 1582.

**Gregorian chant** *n.* [After Pope Gregory I (540?–604).] Unharmonized, unaccompanied liturgical plainsong of the Roman Catholic Church, introduced by Pope Gregory I.

**grei·sen** (grī′zən) *n.* [G. < *greissen*, to split.] A granitic rock composed primarily of quartz and mica.

**grem·lin** (grĕm′lĭn) *n.* [Orig. unknown.] **1.** An imaginary gnomelike creature often blamed for mechanical problems, esp. in aircraft. **2.** A maker of mischief.

**gre·nade** (grə-nād′) *n.* [Fr. < OFr. *grenate*, pomegranate (from its shape), ult. < Lat. *granatum* < *granatus*, having many seeds < *granum*, seed.] **1.** A weapon containing priming and bursting charges, designed to be thrown by hand or fired from a rifle equipped with a launcher. **2.** A glass container filled with a volatile chemical or a liquid that is dispersed when the glass is hurled and smashed <a tear gas *grenade*>

**gren·a·dier** (grĕn′ə-dîr′) *n.* [Fr. < *grenade*, grenade.] **1.** A member of a regiment formerly bearing grenades. **2.** Any of various fishes of the family Macrouridae, chiefly of the deep ocean, with a long tapering tail and no tail fin.

**gren·a·dine** (grĕn′ə-dēn′, grĕn′ə-dēn′) *n.* [Fr. < *grenade*, pomegranate < OFr. *grenate.* —see GRENADE.] **1.** A thick, sweet syrup made from pomegranates or red currants and used as flavoring, esp. in beverages. **2.** A thin openwork fabric of silk, cotton, or a synthetic.

**Gresh·am's law** (grĕsh′əmz) *n.* [After Sir Thomas Gresham (1519?–1579).] The theory that if two kinds of money in circulation have the same denominational value but different intrinsic values, the money with higher intrinsic value will be hoarded and eventually forced out of circulation by the money with lesser intrinsic value.

**gres·so·ri·al** (grĕ-sôr′ē-əl, -sōr′-) *adj.* [NLat. *gressorius* < Lat. *gradi*, to walk.] **1.** Adapted for walking. **2.** Having legs adapted for walking.

**grew** (gro͞o) *v. p.t. of* GROW.

**grey** (grā) *adj.* & *n.* & *v. var. of* GRAY.

**grey·hen** (grā′hĕn′) *n.* The female of the black grouse.

**grey·hound** (grā′hound′) *n.* [ME *grehound* < OE *grīghund.*] A large slender dog of an ancient breed, with a smooth coat, a narrow head, long legs, and the ability to run swiftly.

**grey·lag** (grā′lăg′) *n.* The graylag goose.

**grib·ble** (grĭb′əl) *n.* [Poss. dim. of GRUB.] A small wood-boring marine crustacean of the genus *Limnoria*, esp. *L. lignorum*, which often damages underwater wooden structures.

**grid** (grĭd) *n.* [Short for GRIDIRON.] **1.** A framework of parallel or crisscrossed bars : GRIDIRON. **2.** A pattern of horizontal and vertical lines forming squares of uniform size on a map, chart, aerial photograph, or optical device, used as a reference for locating points. **3.** A football field. **4.** *Elect.* **a.** An interconnected system of electric cables and power stations that distributes electricity over a large area. **b.** A corrugated or perforated conducting plate in a storage battery. **c.** A network or coil of fine wires located between the plate and the filament in an electron tube. **5.** The starting positions of cars on a racecourse. **6.** A device in a phototypesetting machine on which are etched the characters used in composition.

**grid·dle** (grĭd′l) *n.* [ME *gridel*, gridiron < OFr. *gridil* < Lat. *craticula*, dim. of *cratis*, wickerwork.] A flat pan or surface for cooking by dry heat. *—vt.* **-dled, -dling, -dles.** To cook on a griddle.

**grid·dle·cake** (grĭd′l-kāk′) *n.* A pancake.

**grid·i·ron** (grĭd′ī′ərn) *n.* [ME *gridere*, alteration of *gridel.* —see GRIDDLE.] **1.** A flat framework of parallel metal bars for broiling meat or fish. **2.** A framework or network suggestive of a gridiron. **3.** A football field. **4.** A metal structure high above the stage of a theater from which ropes or cables are strung to scenery and lights.

**grid·lock** (grĭd′lŏk′) *n.* A massive, city-wide traffic jam in which streets are so clogged with vehicles that all intersections become blocked and traffic in all directions is stopped.

**grief** (grēf) *n.* [ME < OFr. < *grever*, to grieve. —see GRIEVE.] **1.** Deep mental anguish, as over a loss : SORROW. **2.** A source of grief. **3.** *Obs.* A grievance.

**griev·ance** (grē′vəns) *n.* [ME *grevaunce* < OFr. *grevance* < *grever*, to harm. —see GRIEVE.] **1. a.** A circumstance regarded as just cause for protest. **b.** A complaint based on such a circumstance. **2.** Indignation or resentment aroused by a feeling of having been wronged. **3.** *Obs.* **a.** The act of inflicting hardship or harm. **b.** A cause of harm or hardship.

**grieve** (grēv) *v.* **grieved, griev·ing, grieves.** [ME *grevan* < OFr. *grever*, to harm < Lat. *gravare*, to burden < *gravis*, heavy.] *—vt.* **1.** To cause to be sorrowful : DISTRESS. **2.** *Archaic.* To hurt : harm. *—vi.* To feel grief : MOURN. **—griev′ing·ly** *adv.*

**griev·ous** (grē′vəs) *adj.* **1.** Causing grief, pain, or anguish <a *grievous* injury> **2.** Serious or dire : GRAVE <a *grievous* wrong> **—griev′ous·ly** *adv.* **—griev′ous·ness** *n.*

**grif·fin** *also* **grif·fon** *or* **gryph·on** (grĭf′ən) *n.* [ME *griffoun* < OFr. *griffon* < Lat. *gryphus* < Gk. *grups.*] A fabulous beast with the head and wings of an eagle and the body of a lion.

griffin

**grif·fon** (grĭf′ən) *n.* [Alteration of GRIFFIN.] **1.** A breed of dog with a wiry coat, esp. a small dog orig. bred in Belgium, with a short bearded muzzle. **2.** *var. of* GRIFFIN.

**grift** (grĭft) [Var. of GRAFT².] *Slang.* *—n.* **1.** Money made dishonestly. **2.** A swindle or confidence game. *—vi.* **grift·ed, grift·ing, grifts.** To practice swindling or cheating. **—grift′er** *n.*

**grig** (grĭg) *n.* [ME, dwarf.] *Archaic.* A lively, bright person.

**gri·gri** *also* **gris-gris** (grē′grē) *n.* [Of African orig.] An African charm, fetish, or amulet.

**grill** (grĭl) *vt.* **grilled, grill·ing, grills.** [Fr. *griller* < *gril*, gridiron < OFr. *greille* < Lat. *craticula*, dim. of *cratis*, wickerwork.] **1.** To broil on a gridiron. **2.** To torture as if by broiling. **3.** *Informal.* To question relentlessly : CROSS-EXAMINE. **4.** To mark or emboss with a gridiron. *—n.* **1.** A cooking utensil with parallel metal bars : GRIDIRON. **2.** Food cooked by broiling or grilling. **3.** A grillroom. **4.** A series of marks grilled or embossed on a surface. **5.** *var. of* GRILLE. **—grill′er** *n.*

**gril·lage** (grĭl′ĭj) *n.* [Fr. < *grille*, gridiron < OFr. *greille.* —see GRILL.] A network or frame of crossed timbers serving as a foundation, usu. on treacherous soil.

**grille** *also* **grill** (grĭl) *n.* [Fr. —see GRILLAGE.] **1.** A metal grating serving as a screen, divider, barrier, or decorative element, as in a window or gateway. **2.** A square opening at the back of the hazard side of a court-tennis court.

**grill·room** (grĭl′ro͞om′, -ro͝om′) *n.* A place where grilled foods are served.

**grilse** (grĭls) *n., pl.* **grilse.** [ME *grills.*] A young salmon on its first return from the sea to fresh or brackish waters.

**grim** (grĭm) *adj.* **grim·mer, grim·mest.** [ME < OE, fierce.] **1.** Unrelenting : rigid <*grim* determination> **2.** Uninviting or unnerving in aspect : FORBIDDING. **3.** Sinister : ghastly. **4.** Dismal : gloomy. **5.** Savage : ferocious. **—grim′ly** *adv.* **—grim′ness** *n.*

**grim·ace** (grĭm′ĭs, grĭ-mās′) *n.* [Fr. < OFr., of Germanic orig.] A sharp facial contortion expressing pain, contempt, or disgust. *—vi.* **-aced, -ac·ing, -ac·es.** To make a grimace. **—grim′ac·er** *n.*

**gri·mal·kin** (grĭ-môl′kĭn, -măl′-) *n.* [Var. of *graymalkin* : GRAY + obs. *Malkin*, a woman's name.] **1.** An old female cat. **2.** A shrewish old woman.

**grime** (grīm) *n.* [ME < MLG.] Black dirt or soot, esp. such dirt clinging to or ingrained in a surface. *—vt.* **grimed, grim·ing, grimes.** To cover with grime : BEGRIME.

**Grimm's Law** (grĭmz) *n.* [After Jakob *Grimm* (1785–1863).] A formula describing the regular changes undergone by Indo-European stop consonants represented in Germanic, essentially stating that Indo-European *p, t,* and *k* became Germanic *f, th,* and *h;* Indo-European *b, d,* and *g,* Germanic *p, t,* and *k;* and Indo-European *bh, dh,* and *gh,* Germanic *b, d,* and *g.*

**grim·y** (grī′mē) *adj.* **-i·er, -i·est.** Covered with grime. **—grim′i·ly** *adv.* **—grim′i·ness** *n.*

**grin** (grĭn) *v.* **grinned, grin·ning, grins.** [ME *grennan*, to grimace < OE *grennian.*] *—vi.* To smile broadly, showing the teeth. *—vt.* To express with a grin. *—n.* **1.** An act of grinning. **2.** The facial expression produced by grinning. **—grin′ner** *n.* **—grin′ning·ly** *adv.*

**grind** (grīnd) *v.* **ground** (ground), **grind·ing, grinds.** [ME *grinden* < OE *grindan.*] *—vt.* **1. a.** To crush, pulverize, or powder by friction, esp. by rubbing between two hard surfaces. **b.** To shape, sharpen, or refine with friction <*grind* a crystal> **2.** To rub (two surfaces) together : GNASH. **3.** To bear down on harshly : CRUSH <*grind* the cigarette out> **4. a.** To operate by turning a crank <*grind* a pepper mill> **b.** To produce by turning a crank <*grind* a special blend of coffee> **5.** To produce mechanically <"The production line *grinds* out a uniform product" —Dwight Macdonald> **6.** To instill or teach by persistent repetition <*grind* the lesson into their heads> *—vi.* **1.** To perform the operation of grinding. **2.** To become crushed, pulverized, or powdered by friction. **3.** To move with noisy friction : GRATE <a pully *grinding* on a squeaky wheel> **4.** *Informal.* To devote oneself to study or work <*grinding* over the law books> **5.** *Slang.* To rotate the pelvis in the manner of a stripteaser. *—n.* **1.** The act of grinding. **2.** A crunching or grinding noise. **3.** A specific grade or degree of pulverization, as of coffee beans <filter *grind*> **4.** *Informal.* A laborious task, routine, or study <the *grind* of daily chores> **5.** *Informal.* A student who studies or works excessively. **6.** *Slang.* Erotic rotation of the pelvis. **—grind′ing·ly** *adv.*

**grind·er** (grīn′dər) *n.* **1.** One that grinds, esp.: **a.** One who sharpens cutting edges. **b.** A machine that grinds <a food *grinder*> **2.** A molar. **3. grinders.** *Informal.* The teeth. **4.** *Slang.* HERO 5.

**grind·stone** (grīnd′stōn′) *n.* **1.** A stone disk turned on an axle for grinding, polishing, or sharpening tools. **2.** A millstone.

**grip¹** (grĭp) *n.* [ME < OE *gripe*, grasp.] **1. a.** A tight hold : GRASP. **b.** The strength or pressure of such a grasp. **c.** A way of grasping and holding. **2.** Mastery : command <a firm *grip* on the language> **3. a.** A mechanical device that grasps and holds. **b.** A part designed to be grasped and held : HANDLE. **4.** A suitcase or valise. **5. a.** A stagehand who helps in shifting scenery. **b.** A member of a film production crew who adjusts props and sets and sometimes assists the camera operator. *—v.* **gripped, grip·ping, grips.** *—vt.* **1.** To secure and keep a tight hold on : GRASP. **2.** To hold the interest or attention of <scenes that *gripped* the audience> *—vi.* To hold securely. **—grip′ping·ly** *adv.*

**grip²** (grĭp) *n. var. of* GRIPPE.

**gripe** (grīp) *v.* **griped, grip·ing, gripes.** [ME *gripen*, to seize < OE *gripan.*] *—vt.* **1.** To cause sharp pain in the bowels of. **2.** *Informal.* To irritate : annoy. **3.** To grasp : seize. **4.** *Archaic.* To oppress or afflict. *—vi.* **1.** *Informal.* To complain naggingly or petulantly : GRUMBLE. **2.** To have sharp pains in the bowels. *—n.* **1.** *Informal.* A complaint. **2. gripes.** Sharp, persistent pains in the bowels. **3.** A grip : grasp. **4.** A handle. **—grip′er** *n.*

**grippe** *also* **grip** (grĭp) *n.* [Fr. < *gripper*, to seize < OFr., of Germanic orig.] Influenza. **—grip′py** *adj.*

**gri·saille** (grĭ-zī′, -zāl′) *n.* [Fr. < *gris*, gray < OFr., of Germanic orig.] **1.** A style of monochromatic painting in shades of gray. **2.** A painting or design in grisaille.

**gris·e·ous** (grĭz′ē-əs, grĭs′-) *adj.* [Med. Lat. *griseus*, of Germanic orig.] Grizzled or mottled with gray.

**gri·sette** (grĭ-zĕt′) *n.* [Fr. < *grisette*, a cheap gray dress fabric < *gris*, gray. —see GRISAILLE.] A young French working-class woman.

**gris-gris** (grē′grē) *n. var. of* GRIGRI.

**gris·ly** (grĭz′lē) *adj.* **-li·er, -li·est.** [ME *grisli* < OE *grislīc.*] Gruesome. **—gris′li·ness** *n.*

**gri·son** (grī′sən, grĭz′ən) *n.* [Fr. < OFr., gray animal < *gris*, gray. —see GRISAILLE.] A carnivorous mammal, *Grison vittatus* or *G. cuja* of Central and South America, with grizzled fur, a slender body, and short legs.

**grist** (grĭst) *n.* [ME < OE.] **1.** Grain or a quantity of grain for grinding. **2.** Ground grain. **—grist for (*or* to) the mill.** Something that can be turned to one's profit.

**gris·tle** (grĭs′əl) *n.* [ME *gristel* < OE *gristle.*] Cartilage, esp. in meat.

**gris·tly** (grĭs′lē) *adj.* **-tli·er, -tli·est. 1.** Composed of or containing gristle. **2.** Like gristle. **—gris′tli·ness** *n.*

**grist·mill** (grĭst′mĭl′) *n.* A mill for grinding grain.

**grit** (grĭt) *n.* [ME *gret*, sand < OE *grēot.*] **1.** Minute rough granules, as of sand. **2.** The texture or structure of stone to be used in grinding. **3.** A coarse hard sandstone for making grindstones and millstones. **4.** *Informal.* Indomitable spirit : PLUCK. *—v.* **grit·ted, grit·ting, grits.** *—vt.* **1.** To clamp (the teeth) together. **2.** To cover or treat with grit. *—vi.* To make a grinding noise.

**grith** (grĭth) *n.* [ME < OE *griđ.*] Protection or sanctuary provided individuals by Old English law in certain circumstances, as when in a church or traveling on the monarch's highway.

**grits** (grĭts) *pl.n.* [< ME *grutla*, coarse meal < OE *grytla.*] Coarsely ground grain, esp. corn.

**grit·ty** (grĭt′ē) *adj.* **-ti·er, -ti·est. 1.** Containing or resembling grit. **2.** Demonstrating resolution and fortitude : PLUCKY <*gritty* determination> **—grit′ti·ly** *adv.* **—grit′ti·ness** *n.*

**griv·et** (grĭv′ĭt) *n.* [Fr.] A long-tailed African monkey, *Cercopithecus aethiops*, with a greenish-gray coat.

**griz·zle** (grĭz′əl) *vt. & vi.* **-zled, -zling, -zles.** [ME *grisel*, gray < OFr. < *gris*, gray, of Germanic orig.] To make or become gray. *—n.* **1.** *Archaic.* Gray hair. **2. a.** The color of a roan animal. **b.** A roan animal. *—adj.* **1.** Gray. **2.** Roan.

**griz·zly** (grĭz′lē) *adj.* **-zli·er, -zli·est.** Grayish or flecked with gray. *—n., pl.* **-zlies.** A grizzly bear.

**grizzly bear** *n.* The grayish form of the brown bear, *Ursus arctos* of northwestern North America, often considered a separate species, *U. horribilis.*

**groan** (grōn) *v.* **groaned, groan·ing, groans.** [ME *gronen* < OE *grānian.*] *—vi.* **1.** To give voice to a prolonged deep, wordless sound expressive of pain, grief, annoyance, or disapproval. **2.** To make a sound indicating stress or strain <The old floorboards *groaned* under our weight.> *—vt.* To utter or convey with groaning. **—groan** *n.* **—groan′er** *n.* **—groan′ing·ly** *adv.*

**groat** (grōt) *n.* [ME *grot* < MDu. *groot*, a small coin.] A British silver fourpence piece used from the 14th to the 17th cent.

**groats** (grōts) *pl.n.* [ME *grotes* < OE *grotan.*] Hulled, usu. crushed grain, esp. oats.

**gro·cer** (grō′sər) *n.* [ME, wholesaler < AN *grasser* < OFr. *grossier* < Med. Lat. *grossarius* < *grossus*, gross < LLat., thick.] A storekeeper who sells foodstuffs and household supplies.

**gro·cer·y** (grō′sə-rē) *n., pl.* **-ies. 1.** A store selling foodstuffs and household supplies. **2. groceries.** Commodities sold by a grocer.

**grog** (grŏg) *n.* [After Old *Grog*, nickname of Admiral Edward Vernon (1684–1757), who ordered that diluted rum be served to his sailors.] Alcoholic liquor, esp. rum diluted with water.

**grog·gy** (grŏg′ē) *adj.* **-gi·er, -gi·est.** [< GROG.] Dazed and unsteady : SHAKY. **—grog′gi·ly** *adv.* **—grog′gi·ness** *n.*

**grog·ram** (grŏg′rəm, grō′grəm) *n.* [Alteration of GROSGRAIN.] A coarse, often stiffened fabric of silk, mohair, or wool or a blend of these materials.

**groin** (groin) *n.* [ME *grinde*, perh. < OE *grynde*, abyss.] **1.** *Anat.* The crease at the junction of the thigh and the trunk, together with the adjacent region. **2.** The curved edge at the junction of two intersecting architectural vaults. **3.** A structure projecting out from a shoreline into the water as protection against beach erosion. *—vt.* **groined, groin·ing, groins.** To provide or build with groins.

**grom·met** (grŏm′ĭt) *also* **grum·met** (grŭm′-) *n.* [Prob. < obs. Fr. *gormette*, chain joining the ends of a bit < *gourmer*, bridle.] **1. a.** A reinforced eyelet, as in leather or cloth, through which a fastener may be passed. **b.** A small metal or plastic ring used to reinforce such an eyelet. **2.** *Naut.* A rope or metal ring for securing the edge of a sail.

**grom·well** (grŏm′wəl, -wĕl′) *n.* [ME *gromil* < OFr.] **1.** A plant of the genus *Lithospermum*, bearing small yellow or white flowers. **2.** Any of several plants resembling or related to the gromwell.

**groom** (gro͞om, gro͝om) *n.* [ME *grom.*] **1.** A person employed to take care of horses. **2.** A bridegroom. **3.** An officer in an English royal household. **4.** *Archaic.* **a.** A man. **b.** A manservant. *—vt.* **groomed, groom·ing, grooms. 1.** To make clean, neat, and pleasing in appearance. **2.** To clean and brush (an animal). **3.** To train, as for a specific position : PREPARE <*groom* someone else for the new position>

▲ word history: The word *groom* has come to mean "bridegroom" by the process of folk etymology. In this process an unfamiliar word element is replaced by, or refashioned to resemble, a more familiar word. The Old English form of *bridegroom* was *brȳdguma*, literally "bride's man." *Guma,* cognate with Latin *homo,* "human being, man," was an Old English word for "man" that did not survive the 16th century. About the time that *guma* disappeared from the language the word *groom* came into general use as a word for "man" or "youth," and *guma* in the compound *brȳdguma* was replaced by the more familiar word *groom.* The modern use of *groom* to mean "bridegroom" results from shortening *bridegroom* to *groom.*

**grooms·man** (gro͞omz′mən, gro͞omz′-) *n.* A bridegroom's attendant at his wedding.

**groove** (gro͞ov) *n.* [ME *groof,* mining shaft, prob. < MDu. *groove,* ditch.] **1.** A long, narrow furrow or channel. **2.** *Slang.* An activity or situation to which one is esp. well suited. **3.** *Slang.* A settled, humdrum routine : RUT. **4.** *Slang.* A very pleasurable experience. —*v.* **grooved, groov·ing, grooves.** —*vt.* To cut a groove in. —*vi. Slang.* **1.** To take great pleasure or satisfaction : enjoy oneself. **2.** To react or come together harmoniously.

**groov·y** (gro͞o′vē) *adj.* **-i·er, -i·est.** *Slang.* Deeply satisfying : PLEASING. **—groov′i·ness** *n.*

**grope** (grōp) *v.* **groped, grop·ing, gropes.** [ME *gropen* < OE *grāpian.*] —*vi.* **1.** To reach about or feel one's way uncertainly <*groped* for the light switch> **2.** To search blindly or uncertainly <*grope* for a solution> —*vt.* To make (one's way) by groping. —*n.* The act of groping. **—grop′er** *n.* **—grop′ing·ly** *adv.*

**gros·beak** (grōs′bēk′) *n.* [Partial transl. of Fr. *grosbec* : *gros,* thick + *bec,* beak.] A finch of the genera *Hesperiphona, Pinicola,* or related genera, with a thick rounded bill.

**gro·schen** (grō′shən) *n., pl.* **groschen.** [G. < MHG *grosse* < Med. Lat. (*denarius*) *grossus,* thick (denarius) < LLat. *grossus,* thick.] —See table at CURRENCY.

**gros·grain** (grō′grān′) *n.* [Fr. *gros grain,* coarse grain.] **1.** A heavy, horizontally ribbed silk or rayon fabric. **2.** A grosgrain ribbon.

**gros point** (grō) *n.* [Fr. : *gros,* large + *point,* point.] **1.** A large needlepoint stitch covering two vertical and two horizontal threads. **2.** Work done in gros point.

**gross** (grōs) *adj.* **-er, -est.** [ME *gros,* large < OFr. < LLat. *grossus,* thick.] **1.** Exclusive of deductions : TOTAL <*gross* income> **2. a.** Unmitigated in any way : UTTER <*gross* negligence> **b.** Glaringly obvious : FLAGRANT <*gross* unfairness> **3.** *Slang.* **a.** Vulgar : coarse. **b.** Offensive : disgusting. **c.** Lacking sensitivity or discernment : UNREFINED. **d.** Carnal : sensual. **4. a.** Overweight : corpulent. **b.** Dense : profuse. **5.** Broad : general <the *gross* outlines of a project> **6.** *Pathol.* Visible to the naked eye <a *gross* lesion> —*n.* **1.** *pl.* **gross·es.** The entire body or amount : TOTAL. **2.** *pl.* **gross.** A group of 144 or 12 dozen items. —*vt.* **grossed, gross·ing, gross·es.** To earn as a total profit or income before deductions. **—gross out.** *Slang.* To disgust : nauseate. **—gross′ly** *adv.* **—gross′ness** *n.*

**gross index** *n. Computer Sci.* The general index first consulted in locating particular records.

**gross national product** *n.* The total market value of all the goods and services produced by a nation during a specified period.

**gros·su·la·rite** (grŏs′yə-lə-rīt′) *n.* [G. *Grossularit* < NLat. *Grossularia,* a former genus of gooseberry (from the color of some garnets) < Fr. *groseille,* gooseberry < OFr. *groselle.*] A light-green, pink, gray, or brown garnet with composition $Ca_3Al_2(SiO_4)_3$, found alone or as a constituent of the common garnet.

**grosz** (grôsh) *n., pl.* **gro·szy** (grô′shē) [Pol. < Czech *gros* < Med. Lat. (*denarius*) *grossus,* thick (denarius) < LLat. *grossus,* thick.] —See table at CURRENCY.

**grot** (grŏt) *n.* A grotto.

**gro·tesque** (grō-tĕsk′) *adj.* [< Fr., a fanciful style of decorative art < Ital. *grottesca* < *grottesco,* of a grotto < *grotta,* grotto.] **1.** Characterized by incongruous or ludicrous distortion. **2.** Bizarre : outlandish. **3.** Of or designating the grotesque in an artistic work executed in this style. —*n.* **1.** One that is grotesque. **2. a.** An art style developed in 16th-cent. Italy, marked by incongruous combinations of monstrous or natural forms. **b.** A work of art executed in this style. **—gro·tesque′ly** *adv.* **—gro·tesque′ness** *n.*

**gro·tes·que·ry** *also* **gro·tes·que·rie** (grō-tĕs′kə-rē) *n., pl.* **-ries.** **1.** The state of being grotesque. **2.** Something grotesque.

**grot·to** (grŏt′ō) *n., pl.* **-toes** *or* **-tos.** [Ital. *grotta* < OItal. < VLat. **grupta* < Lat. *crypta,* vault. —see CRYPT.] **1.** A small cave or cavern. **2.** An artificial structure or excavation made to look like a cave or cavern.

**grot·ty** (grŏt′ē) *adj.* **-ti·er, -ti·est.** [Alteration of GROTESQUE.] *Chiefly Brit.* Wretched : miserable.

**grouch** (grouch) *vi.* **grouched, grouch·ing, grouch·es.** [Prob. alteration of obs. *grutch,* to complain < ME *grucchen* < OFr. *grouchier.*] To sulk or grumble. —*n.* **1.** A sulky or gumbling mood. **2.** A complaint. **3.** A habitually irritable or complaining person.

**grouch·y** (grou′chē) *adj.* **-i·er, -i·est.** Tending to complain and grumble : PEEVISH. **—grouch′i·ly** *adv.* **—grouch′i·ness** *n.*

**ground¹** (ground) *n.* [ME < OE *grund.*] **1. a.** The solid surface of the earth. **b.** The floor of a body of water, esp. the sea. **2.** Earth : soil. **3.** *often* **grounds.** An area of land designated for a given purpose <picnic *grounds*> **4. grounds.** The land around or forming part of a house or other building <the capitol *grounds*> **5.** *often* **grounds.** The foundation for an argument, belief, or action : BASIS. **6.** *often* **grounds.** The underlying condition prompting an action : CAUSE <*grounds* for a lawsuit> **7.** An area of reference : SUBJECT. **8.** A surrounding area : BACKGROUND. **9.** The preparatory coat of paint on which a picture is to be painted. **10. grounds.** Sediment at the bottom of a liquid, esp. coffee. **11.** *Elect.* **a.** The position or portion of an electric circuit at zero potential with respect to the earth. **b.** A conducting connection to such a position or to the earth. **c.** A large conducting body, as the earth, used as a return for electric currents and as an arbitrary zero of potential. —*v.* **ground·ed, ground·ing, grounds.** —*vt.* **1.** To set or place on the ground. **2.** To provide a basis for (e.g., a theory) : JUSTIFY. **3.** To supply with basic information. **4. a.** To prevent (an aircraft or pilot) from flying. **b.** *Informal.* To restrict, esp. to a certain place, as a punishment. **5.** *Elect.* To connect (an electric circuit) to a ground. **6.** *Naut.* To run (a vessel) aground. **7.** *Baseball.* To hit (a ball) on the ground. **8.** *Football.* To throw (a ball) to the ground to halt play and avoid being tackled behind the line of scrimmage. —*vi.* **1.** To hit or reach the ground. **2.** *Baseball.* To hit a ground ball. **3.** *Naut.* To run aground. **—break ground. 1.** To dig or cut into the soil, as in plowing or excavating. **2.** To start an undertaking. **—cover ground. 1.** To move about or travel, esp. for a great distance and at a good speed. **2.** To accomplish a lot. **—from the ground up.** Omitting nothing : THOROUGHLY. **—gain ground. 1.** To make progress. **2.** To gain popularity or favor. **—give ground.** To give way : YIELD. **—hold (***or* **stand) (one's) ground.** To maintain one's position despite opposition.

**ground²** (ground) *v. p.t. & p.p. of* GRIND.

**ground ball** *n. Baseball.* A batted ball that bounces or rolls along the ground.

**ground bass** *n.* A short musical passage constantly repeated in the bass under the changing harmonies and melodies of the upper range.

**ground beetle** *n.* Any of numerous chiefly brown or black beetles of the family Carabidae that often crawl under stones, logs, or debris.

**ground cherry** *n.* Any of various chiefly New World plants of the genus *Physalis,* bearing round fleshy fruit enclosed in a papery bladderlike husk.

**ground cloth** *n.* A ground sheet.

**ground cover** *n.* Low-growing plants that form a dense, extensive growth and tend to prevent weeds and soil erosion.

**ground crew** *n.* A team of technicians and mechanics who maintain and service aircraft or spacecraft on the ground.

**ground-ef·fect machine** (ground′ĭ-fĕkt′) *n.* A vehicle designed for traveling over land or water by means of an air cushion.

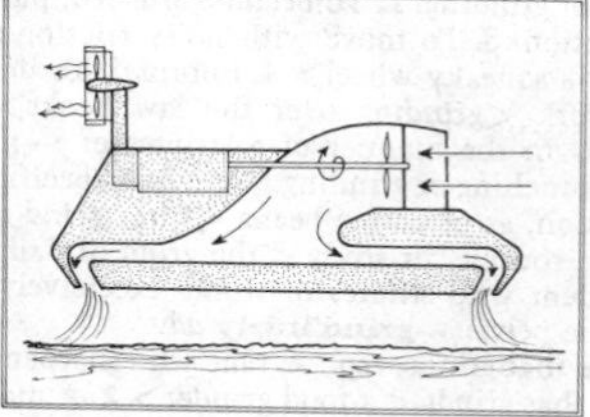

**ground-effect machine**

**ground·er** (groun′dər) *n. Baseball.* A ground ball.

**ground floor** *n.* The floor of a building at or nearly at ground level.

**ground glass** *n.* Glass that has been ground or etched to create a roughened, nontransparent surface.

**ground hemlock** *n.* A low-growing yew, *Taxus canadensis* of northeastern North America.

**ground hog** *n.* The woodchuck.

**ground-hog day** (ground′hôg′, -hŏg′) *n.* [From the legend that the ground hog emerges from hibernation on this day and returns to its burrow if it sees its shadow, presaging prolonged winter weather.] Feb. 2, the date that traditionally indicates an early or late spring.

**ground ivy** *n.* A creeping or trailing aromatic plant, *Glechoma hederacea,* native to Eurasia, with rounded scalloped leaves and small purplish flowers.

**ground·less** (ground′lĭs) *adj.* Having no basis or foundation : UNSUBSTANTIATED <*groundless* expectations> **—ground′less·ly** *adv.* **—ground′less·ness** *n.*

**ground·ling** (ground′lĭng) *n.* **1. a.** A plant or animal living on or close to the ground. **b.** A fish living at the bottom of the water. **2.** One with uncultivated tastes. **3.** A spectator in the cheaper part of an Elizabethan theater.

**ground loop** *n.* A sharp, uncontrollable turn of an aircraft in taxiing, landing, or taking off.

**ground·mass** (ground′măs′) *n.* The fine-grained crystalline base of porphyritic rock in which phenocrysts are embedded.

**ground·nut** (ground′nŭt′) *n.* **1. a.** A climbing vine, *Apios tuberosa* of eastern North America, with compound leaves, fragrant brownish flower clusters, and small edible tubers. **b.** Any of several plants with underground tubers or nutlike parts. **c.** The tuber or nutlike part of such a plant. **2.** *Chiefly Brit.* PEANUT 1, 2.

**ground pine** *n.* **1.** A club moss, esp. *Lycopodium obscurum* or a similar species. **2.** A low-growing Old World plant, *Ajuga chamaepitys,* with narrow leaves, yellow flowers, and a resinous odor.

**ground plan** *n.* **1.** A plan of a floor of a building as if seen from above. **2.** A preliminary plan.

**ground plum** *n.* **1.** A plant, *Astragalus crassicarpus* of the central and western United States, bearing compound leaves, purple or white flowers, and edible green plumlike fruit. **2.** The fruit of the ground plum.

**ground rent** *n. Chiefly Brit.* Rent paid for land to be used primarily for building.

**ground robin** *n.* The towhee.

**ground rule** *n.* **1.** A rule affecting the playing of a game on a particular field, course, or court. **2.** A basic rule, as of procedures.

**ground·sel**[1] (ground′səl, groun′-) *n.* [ME *groundeswille* < OE *grundeswylige,* perh. : *gund,* pus + *swelgan,* to swallow (from its use in reducing abscesses).] Any of various plants of the genus *Senecio,* bearing rayed, usu. yellow flowers.

**ground·sel**[2] (ground′səl, groun′-) *n. var. of* GROUNDSILL.

**ground sheet** *n.* **1.** A waterproof cover for protecting an area of ground, as a baseball field. **2.** A waterproof sheet laid under camp bedding to protect against dampness.

**ground·sill** (ground′sĭl′) *also* **ground·sel** (ground′səl, groun′-) *n.* The horizontal timber closest to the ground in the frame of a building.

**ground speed** *also* **ground·speed** (ground′spēd′) *n.* The speed of an airborne aircraft computed in terms of the ground distance traversed in a specified time interval.

**ground squirrel** *n.* A rodent of the genus *Citellus* or *Spermophilus* and related genera, related to and resembling the chipmunk.

**ground state** *n. Physics.* The energy level, as of a system of interacting elementary particles, having the least energy of all its possible states.

**ground swell** *n.* **1.** An undulation of the ocean with deep rolling waves, often the result of a distant storm or earthquake. **2.** A sudden gathering of force, as of public opinion.

**ground water** *also* **ground·wa·ter** (ground′wô′tər, -wŏt′ər) *n.* Water beneath the earth's surface between saturated soil and rock that supplies wells and springs.

**ground wave** *n.* A radio wave that travels along the earth's surface.

**ground·work** (ground′wûrk′) *n.* Preliminary work : BASIS.

**ground zero** *n.* **1.** The target of a missile, bomb, or other projectile. **2.** The site of a nuclear explosion.

**group** (grōōp) *n.* [Fr. *groupe* < Ital. *gruppo,* of Germanic orig.] **1.** An assemblage of persons or objects located or gathered together : AGGREGATION. **2.** Two or more figures comprising a unit or a design, as in sculpture. **3.** A number of things or individuals considered together because of similarities. **4.** A subdivision of a linguistic family, less inclusive than a branch. **5. a.** A military unit of two or more battalions and a headquarters. **b.** A unit of two or more U.S. Air Force squadrons, smaller than a wing. **6.** A class or collection of related objects or entities, as: **a.** Two or more atoms behaving or regarded as behaving as a single chemical unit. **b.** A vertical column in the periodic table of elements. **c.** A geologic stratigraphic unit, esp. a unit of two or more formations. **7.** *Math.* A set together with a binary operation under which the set is closed and associative and for which the set contains an identity element and an inverse for every element in the set. ***usage:*** As a collective noun, *group* may take either a singular or plural verb. The verb is singular when the members of the group are considered to be a unit, as in *The group is ready to leave.* When its members are considered individually *group* takes a plural verb, as in *The group were divided in their sympathies.* —*v.* **grouped, group·ing, groups.** —*vt.* To place or arrange in a group. —*vi.* To form or belong to a group.

**grou·per**[1] (grōō′pər) *n., pl.* **grouper** *or* **-pers.** [Port. *garoupa.*] A large fish of the genera *Epinephelus, Mycteroperca,* or related genera, indigenous to warm seas.

**group·er**[2] (grōō′pər) *n.* GROUPIE 2.

**group·ie** (grōō′pē) *n. Slang.* **1.** A young woman who is a fan esp. of a rock group and who follows the group around on tours. **2.** A member of a group sharing the same weekend house, usu. at a beach or ski resort.

**group insurance** *n.* Insurance covering members of a group under one contract or under individual contracts.

**group theory** *n.* The branch of mathematics dealing with the properties of groups.

**group therapy** *n.* Psychotherapy involving sessions guided by a therapist and attended by several patients who discuss their emotional problems with each other.

**group·think** (grōōp′thĭngk′) *n.* **1.** The act or practice of decision- and policy-making by a group, as a board of directors or a research team. **2.** Conformity to group values or ethical standards.

**grouse**[1] (grous) *n., pl.* **grouse** *or* **grous·es.** [Orig. unknown.] Any of various plump birds of the family Tetraonidae, chiefly of the Northern Hemisphere, with mottled brown or grayish plumage.

**grouse**[2] (grous) [Orig. unknown.] *Informal.* —*vi.* **groused, grous·ing, grous·es.** To grumble : complain. —*n.* A cause for complaint : GRIEVANCE. **—grous′er** *n.*

**grout** (grout) *n.* [ME, grain used for making malt, mud < OE *grūt,* coarse meal.] **1. a.** A thin mortar for filling cracks and crevices in masonry. **b.** A thin plaster for finishing walls and ceilings. **2.** *often* **grouts.** *Chiefly Brit.* Sediment : lees. —*vt.* **grout·ed, grout·ing, grouts.** To fill or finish with grout. **—grout′er** *n.*

**grove** (grōv) *n.* [ME < OE *grāf.*] A small wood or stand of trees without dense undergrowth.

**grov·el** (grŏv′əl, grŭv′-) *vi.* **-eled, -el·ing, -els** *also* **-elled, -el·ling, -els.** [Back-formation < obs. *groveling,* prone, face downward < ME < *grufe* < ON *á grūfu* < *grūfa,* to lie face down.] **1.** To behave in a demeaning or servile way : CRINGE. **2.** To lie or creep in a prostrate position, often as a token of humility or subservience. **3.** To surrender oneself to base pleasures. **—grov′el·er** *n.* **—grov′el·ing·ly** *adv.*

**grow** (grō) *v.* **grew** (grōō), **grown** (grōn), **grow·ing, grows.** [ME *growen* < OE *grōwan.*] —*vi.* **1.** To increase in size by a natural process. **2. a.** To expand : gain <The population *grew* after World War II.> **b.** To increase in amount or degree : INTENSIFY <My fears *grew.*> **3.** To develop and attain maturity. **4.** To be capable of growth : THRIVE <corn that will *grow* in intense heat> **5.** To become attached by or as if by the process of growth <lichen that had *grown* on stones> **6.** To come into existence from a source : DEVELOP <friendship that *grew* out of acquaintance> **7.** To come to be by a gradual process or by degrees : BECOME <*grow* sad><*grow* hot><*grow* wealthy> —*vt.* **1.** To cause to grow : RAISE <*grow* mushrooms> **2.** To let grow <*grow* long hair> **—grow on** (*or* **upon).** To become gradually more pleasurable or more acceptable to <The unusual taste of truffles *grew on* them.> **—grow up.** To become an adult. **—grow′er** *n.*

**growing pains** *pl.n.* **1.** Pains in the limbs and joints of children, often erroneously attributed to rapid growth. **2.** Problems arising in the early stages of an enterprise.

**growl** (groul) *n.* [Prob. imit.] **1.** The low, guttural, threatening sound made by a dog or other animal. **2.** A gruff, surly utterance. —*v.* **growled, growl·ing, growls.** —*vi.* **1.** To utter a growl. **2.** To make a sound like a growl <artillery *growling* in the distance> **3.** To speak in an angry or surly way. —*vt.* To utter by growling <*growled* complaints at everyone>

**growl·er** (grou′lər) *n.* **1.** One that growls. **2.** A small iceberg. **3.** *Elect.* An electromagnetic device with two poles, used for magnetizing, demagnetizing, and finding short-circuited coils.

**grown** (grōn) *adj.* [P.part. of GROW.] **1.** Having full growth : MATURE. **2.** Produced or cultivated in a certain way or place <shade-*grown* tobacco>

**grown-up** (grōn′ŭp′) *adj.* Mature and adult. **—grown′-up′** *n.*

**growth** (grōth) *n.* **1. a.** The process of growing. **b.** A stage in the process of growing : SIZE. **c.** Full development : MATURITY. **2.** Development from a lower or simpler to a higher or more complex form : EVOLUTION. **3.** An increase, as in size, number, value, or strength : EXPANSION <urban *growth*> **4.** Something that grows or has grown <luxuriant *growth* of hair> **5.** An abnormal mass of tissue growing in or on a living organism. **6.** The result of growth : PRODUCTION.

**growth company** *n.* A company whose rate of growth markedly exceeds that of the average in its field or the overall rate of economic growth.

**growth fund** *n.* A mutual fund whose goal is capital appreciation.

**grub** (grŭb) *v.* **grubbed, grub·bing, grubs.** [ME *grubben.*] —*vt.* **1.** To clear of roots and stumps by digging. **2.** To dig up by the roots. **3.** *Slang.* To obtain by cadging <*grub* a free meal> —*vi.* **1.** To dig in the earth <*grub* for turnips> **2. a.** To search laboriously : RUMMAGE. **b.** To toil arduously : DRUDGE <*grub* for a meager existence> —*n.* **1.** The thick, wormlike larva of certain beetles and other insects. **2.** A drudge. **3.** *Slang.* Food. **—grub′ber** *n.*

**grub·by** (grŭb′ē) *adj.* **-bi·er, -bi·est.** **1.** Unkempt : dirty. **2.** Infested with grubs. **3.** Contemptible. **—grub′bi·ly** *adv.* **—grub′bi·ness** *n.*

**grub·stake** (grŭb′stāk′) *n.* Supplies or funds advanced to a mining prospector or a person beginning a business in return for a promised share of the profits. —*vt.* **-staked, -stak·ing, -stakes.** To supply with a grubstake. **—grub′stak′er** *n.*

**Grub Street** (grŭb) *n.* [After *Grub Street,* London.] The world of impoverished writers and literary hacks.

**grudge** (grŭj) *vt.* **grudged, grudg·ing, grudg·es.** [ME *grucchen,* to complain < OFr. *grouchier.*] To be reluctant to admit or give : BEGRUDGE. —*n.* Deep-seated resentment or rancor. **—grudg′er** *n.* **—grudg′ing·ly** *adv.*

**gru·el** (grōō′əl) *n.* [ME < OFr. < *gru,* groats, of Germanic orig.] **1.** Thin watery porridge. **2.** *Chiefly Brit.* Severe punishment.

**gru·el·ing** *also* **gru·el·ling** (grōō′ə-lĭng) *adj.* Demanding and exhausting <a *grueling* march> **—gru′el·ing·ly** *adv.*

**grue·some** (grōō′səm) *adj.* [Obs. *grue,* to shudder (< ME *gruen*) +

-SOME[1].] Causing horror and repugnance : SHOCKING <a *gruesome* series of crimes> **—grue′some·ly** *adv.* **—grue′some·ness** *n.*

**gruff** (grŭf) *adj.* **-er, -est.** [Du. *grof* < MDu.] **1.** Brusque and unfriendly : STERN. **2.** Harsh : hoarse <a *gruff* command> **—gruff′ly** *adv.* **—gruff′ness** *n.*

☆ ***syns:*** GRUFF, BLUFF, BLUNT, BRUSQUE, CURT *adj. core meaning* : abrupt and sometimes markedly impolite in manner or speech <a *gruff* retort> GRUFF implies rough and often harsh speech, but does not necessarily suggest intentional rudeness. BRUSQUE emphasizes rude abruptness of manner <a *brusque* refusal to help> BLUNT stresses utter frankness and usu. a disconcerting directness <a *blunt* criticism> BLUFF refers to unpolished, unceremonious manner but usu. implies good nature <a *bluff* old sea dog> CURT refers to briefness and abruptness of speech and manner and usu. implies rudeness <a *curt* dismissal>

**grum** (grŭm) *adj.* **grum·mer, grum·mest.** [Perh. blend of GRIM and GLUM.] Glum : morose.

**grum·ble** (grŭm′bəl) *v.* **-bled, -bling, -bles.** [Freq. of ME *grummen,* to grumble.] *—vi.* **1.** To mutter in a surly way <*grumbling* about the weather> **2.** To rumble or growl. *—vt.* To express in a discontented way. *—n.* **1.** A muttered complaint. **2.** RUMBLE 1. **—grum′bler** *n.* **—grum′bling·ly** *adv.* **—grum′bly** *adj.*

**grum·met** (grŭm′ĭt) *n. var. of* GROMMET.

**grump** (grŭmp) *n.* [Perh. imit. of discontented muttering.] **1. grumps.** A fit of ill temper. **2.** A cranky, complaining person. *—vi.* **grumped, grump·ing, grumps. 1.** To complain and mutter. **2.** To behave in a grumpy manner.

**grump·y** (grŭm′pē) *adj.* **-i·er, -i·est.** Fretful and peevish : CRANKY. **—grump′i·ly** *adv.* **—grump′i·ness** *n.*

**grun·gy** (grŭn′jē) *adj.* **-gi·er, -gi·est.** [Orig. unknown.] *Slang.* Being in a dirty, run-down, or inferior condition <*grungy* old running shoes>

**grun·ion** (grŭn′yən) *n.* [Perh. < Sp. *gruñón,* grumbler < *gruñir,* to grumble < Lat. *grunnire,* to grunt.] A small fish, *Leuresthes tenuis* of coastal waters of California and Mexico, that spawns along beaches during high spring tides at full moon.

**grunt** (grŭnt) *v.* **grunt·ed, grunt·ing, grunts.** [ME *grunten* < OE *grunnettan.*] *—vi.* **1.** To utter a deep guttural sound, as a hog does. **2.** To utter a sound like a grunt, as in disgust. *—vt.* To utter or express with a grunt <*grunted* agreement> *—n.* **1.** A deep guttural sound. **2.** A chiefly tropical marine fish of the genus *Haemulon* or related genera that produces grunting sounds. **3.** *Slang.* A U.S. infantry soldier, esp. in the Vietnam War. **4.** *Slang.* One who performs routine or mundane tasks. **—grunt′er** *n.* **—grunt′ing·ly** *adv.*

**Grus** (grŭs) *n.* [Lat., crane.] A constellation in the Southern Hemisphere.

**Gru·yère** (grōō-yâr′, grē-) *n.* [After *Gruyère,* a district in Switzerland.] A pale-yellow firm-textured cheese, with or without holes, made from whole milk.

**gryph·on** (grĭf′ən) *n. var. of* GRIFFIN.

**G-string** (jē′strĭng′) *n.* [Orig. unknown.] **1.** A narrow loincloth supported by a waistband : BREECHCLOTH. **2.** A garment similar to a G-string worn esp. by stripteasers.

**G-suit** (jē′sōōt′) *n.* [G(RAVITY) + SUIT.] A flight garment designed to counteract the effects of high acceleration by exerting pressure on parts of the body below the chest.

**gua·ca·mo·le** (gwä′kə-mō′lē) *n.* [Mex. Sp. < Nahuatl *ahuacamolli* : *ahuacatl,* avocado + *molli,* sauce.] A spread of mashed avocado, tomato pulp, mayonnaise, and seasonings.

**gua·cha·ro** (gwä′chə-rō′) *n., pl.* **-ros.** [Sp. (South America) < Quechua *guacho,* orphan.] A nocturnal bird, *Steatornis caripensis* of tropical America, whose young have a layer of fat that yields an oil used in cooking and for lighting.

**gua·co** (gwä′kō) *n., pl.* **-cos.** [Am. Sp.] A tropical American plant, esp. *Mikania guaco* or *Aristolochia serpentina,* used as an antidote against snakebite.

**guai·ac** (gwī′ăk′) *n.* GUAIACUM 3.

**guai·a·col** (gwī′ə-kôl′, -kōl′) *n.* [GUAIAC(UM) + -OL[2].] A yellowish, oily, aromatic liquid, $C_7H_8O_2$, used primarily as an expectorant and a local anesthetic.

**guai·a·cum** (gwī′ə-kəm) *n.* [NLat. < Sp. *guayacan* < Taino.] **1.** A tree of the genus *Guaiacum* : LIGNUM VITAE. **2.** The wood of a guaiacum. **3.** A greenish-brown resin obtained from the lignum vitae and used medicinally and in varnishes.

**guan** (gwän) *n.* [South American Sp.] A bird of the genus *Penelope* or related genera of the jungles of tropical America, related to and resembling the curassow.

**gua·na·co** (gwə-nä′kō) *n., pl.* **-cos** or **guanaco.** [Sp. < Quechua *huanaco.*] A brownish South American mammal, *Lama guanicoe,* related and similar to the domesticated llama.

**gua·neth·i·dine** (gwä-nĕth′ĭ-dēn′) *n.* [Blend of GUANIDINE and ETHYL.] A drug, $C_{10}H_{22}N_4$, used in treating of hypertension.

**gua·ni·dine** (gwä′nĭ-dēn′) *n.* [GUAN(INE) + -ID(E) + -INE.] A strong crystalline base, $CH_5N_3$, found in plant and animal tissues and used for organic syntheses.

**gua·nine** (gwä′nēn′) *n.* [< GUANO, in which it is found.] A purine, $C_5H_5N_5O$, that is a constituent of both ribonucleic and deoxyribonucleic acids.

**gua·no** (gwä′nō) *n., pl.* **-nos.** [Sp. < Quechua *huanu,* dung.] A substance composed chiefly of sea bird or bat dung, accumulated in certain coastal regions or in caves and used as fertilizer.

**gua·no·sine monophosphate** (gwä′nə-sēn′) *n.* [GUAN(INE) + (RIB)OSE + -INE + MONOPHOSPHATE.] Cyclic GMP.

**guar** (gwär) *n.* [Hindi *guār.*] A legume, *Cyamopsis tetragonoloba,* adapted to semiarid regions and grown for its seeds and as forage.

**gua·ra·ni** (gwä′rə-nē′) *n., pl.* **guarani** or **-nis.** [Sp. *guaraní,* Guarani.] —See table at CURRENCY.

**Gua·ra·ni** (gwä′rə-nē′) *n., pl.* **-nis** or **Guarani.** [Sp. *guaraní.*] **1. a.** A Tupi-Guaranian group of South American Indians of Paraguay, Bolivia, and southern Brazil. **b.** A member of one of these tribes. **2.** The language of the Guaranis.

**guar·an·tee** (găr′ən-tē′) *n.* [Perh. alteration of GUARANTY.] **1.** Something that ensures a particular outcome or condition <Hard work is not a *guarantee* of success.> **2.** A promise or assurance, esp. as to the durability or quality of a product or service. **3.** A guaranty. **4.** Something given or held as security : PLEDGE. **5.** A guarantor. *—vt.* **-teed, -tee·ing, -tees. 1.** To assume responsibility for the debt, default, or miscarriage of. **2.** To assume responsibility for the quality or execution of. **3.** To undertake to accomplish or secure something <*guaranteed* to win the contest> **4.** To furnish security for. **5.** To give a guarantee for <The company *guarantees* all of its products.> **6.** To make certain : ENSURE.

**guar·an·tor** (găr′ən-tôr′, găr′ən-tər) *n.* **1.** One that makes or gives a guarantee. **2.** One that makes or gives a guaranty.

**guar·an·ty** (găr′ən-tē) *n., pl.* **-ties.** [OFr. *garantie* < *garant,* warrant, of Germanic orig.] **1.** An agreement by which one party assumes the responsibility of assuring payment or fulfillment of another party's debts or obligations. **2.** Something that guarantees. **3. a.** Something held or provided as security for the execution, completion, or existence of something else. **b.** The act of providing such security. **4.** A guarantor. *—vt.* **-tied, -ty·ing, -ties.** To guarantee.

▲ word history: *Guaranty* is derived from Old French *garant* or *guarant,* "warrant," which is ultimately of Germanic origin. The Germanic source of *guarant* had an initial *w.* A number of Germanic words with initial *w* were borrowed into French in medieval times. In most Old French dialects the *w* was changed to *g* or *gu,* as in words such as modern French *guerre,* "war," *garde,* "guard," and the name *Guillaume,* "William." In the northeastern French dialects, as that of Normandy, initial *w* did not change. Sometimes the same word was borrowed into English from the two different groups of dialects and the two forms have come to be differentiated into two English words. The doublet of *guaranty* derived from northeastern French is *warranty.*

**guard** (gärd) *v.* **guard·ed, guard·ing, guards.** [Ult. < OFr. *garden,* of Germanic orig.] *—vt.* **1.** To shield from danger or harm, esp. by careful watching : PROTECT <*guard* the payroll truck> **2.** To watch over to prevent escape, violence, or indiscretion <*guarded* the captives> **3.** To keep watch at (e.g., a door) to supervise entries and exits. **4.** To furnish (a device or object) with a protective piece. **5.** *Archaic.* To escort. *—vi.* To take precautions <*guard* against spread of the disease> *—n.* **1. a.** One that acts as a sentinel or stands watch. **b.** One who supervises prisoners. **c.** A body of persons forming an escort or performing drill exhibitions on ceremonial occasions <an honor *guard*> **2.** *Chiefly Brit.* A railway employee in charge of a train. **3.** *Football.* One of the two players on either side of the center. **4.** *Basketball.* Either of the two players who initiate plays from the center of the court. **5.** A defensive position or stance in certain sports, as boxing or fencing. **6. a.** The act or duty of guarding. **b.** Protection : watch <detainees under close *guard*> **7.** A safeguard <a *guard* against sunburn> **8.** A device that prevents injury, damage, or loss, esp.: **a.** An attachment or covering put on a machine to protect the operator. **b.** A chain or band for safeguarding something, as a bracelet, from loss. **c.** A ring for preventing a more valuable ring from sliding off the finger. **9.** *Electron.* A signal that prevents accidental activation of a device or ambiguous interpretation of data. **—off (one's) guard.** Not alert : UNPREPARED. **—on (one's) guard.** Alert and watchful : CAUTIOUS. **—guard′er** *n.*

**guar·dant** *also* **gar·dant** (gär′dnt) *adj.* [OFr. *gardant,* pr.part. of *garder,* to guard, of Germanic orig.] *Heraldry.* Indicating an animal shown in full face, turned toward the viewer.

**guard cell** *n.* One of the paired epidermal cells controlling the opening and closing of a stoma in plant tissue.

**guard·ed** (gär′dĭd) *adj.* Restrained : cautious <*guarded* enthusiasm> **—guard′ed·ly** *adv.* **—guard′ed·ness** *n.*

**guard hair** *n.* Any of the coarse hairs covering the underfur of certain mammals.

**guard·house** (gärd′hous′) *n.* **1.** A building that accommodates a military guard. **2.** A military jail for detaining those guilty of minor offenses.

**guard·i·an** (gär′dē-ən) *n.* [ME *gardein* < OFr. < *garder,* to guard, of Germanic orig.] **1.** One that guards. **2.** One legally responsible for the care and management of the person or property of one, as a minor child, whom the law regards as incompetent to manage his or her own affairs. **3.** A superior in a Franciscan monastery. **—guard'i·an·ship'** *n.*

**guard·rail** (gärd'rāl') *n.* A protective rail, as on a highway.

**guard·room** (gärd'ro͞om', -ro͝om') *n.* **1.** A room used by guards on duty. **2.** A room for confining military prisoners.

**guards·man** (gärdz'mən) *n.* **1.** A member of the U.S. National Guard. **2.** *Chiefly Brit.* A soldier in a household guard regiment.

**Guar·ne·ri·us** (gwär-nâr'ē-əs, -nîr'-) *n.* A violin made by a member of the Guarneri family of Italy in the 17th and 18th cent.

**gua·va** (gwä'və) *n.* [Sp. *guayaba* < a native word in the Caribbean islands.] **1.** A tropical American shrub or tree of the genus *Psidium,* esp. *P. guajava,* bearing white flowers and edible fruit. **2.** The fruit of a guava.

**gua·yu·le** (gwī-o͞o'lē) *n.* [Am. Sp. < Nahuatl *cuauhuli* : *cuahuitl,* tree + *uli,* latex gum.] A woody plant or shrub, *Parthenium argentatum* of the southwestern United States and Mexico, producing sap occas. used as a source of rubber.

**gu·ber·na·to·ri·al** (go͞o'bər-nə-tôr'ē-əl, -tōr'-, gyo͞o'-) *adj.* [< Lat. *gubernator,* governor.] Of or relating to a governor.

**guck** (gŭk, go͝ok) *n.* [Poss. G(OO) + (M)UCK.] *Slang.* A messy substance, as sludge.

**gudg·eon¹** (gŭj'ən) *n.* [ME *gojoun* < OFr. *goujon* < Lat. *gobius* < Gk. *kōbios.*] **1. a.** A small Eurasian freshwater fish, *Gobio gobio.* **b.** A fish similar to the gudgeon. **2.** *Slang.* One who is easily duped.

**gudg·eon²** (gŭj'ən) *n.* [ME *gojoun* < OFr. *gojon,* peg < *goi,* gouge. —see GOUGE.] **1.** A metal pivot or journal at the end of a shaft or axle, around which a wheel or other device turns. **2.** The socket of a hinge into which the pin fits. **3.** *Naut.* The socket for the pintle of a rudder. **4.** A metal pin that joins two pieces of stone.

**gudgeon pin** *n.* A wrist pin.

**Gud·run** (go͝od'ro͞on') *also* **Guth·run** (go͝oth'-) *n.* [ON *Guðrun.*] *Norse Myth.* The daughter of the king of the Nibelungs and wife of Sigurd in the *Volsunga Saga.*

**guel·der rose** (gĕl'dər) *n.* [After *Guelderland,* a province in the Netherlands.] A native Eurasian shrub, *Viburnum opulus,* bearing white flower clusters and small red fruit.

**Guelph** *also* **Guelf** (gwĕlf) *n.* [Ital. *Guelf.*] A member of a faction in medieval Italy that supported the pope and the city-states in a struggle against the German emperors and the Ghibellines.

**Guen·e·vere** (gwĕn'ə-vîr') *n. var. of* GUINEVERE.

**gue·non** (gə-nōn') *n.* [Fr.] Any of various African monkeys of the genus *Cercopithecus,* with long hind legs and a long tail.

**guenon**
*3–5 feet long including tail*

**guer·don** (gûr'dn) *n.* [ME < OFr. < Med. Lat. *widerdonum,* alteration of OHG *widarlōn* : *widar,* back + *lōn,* reward.] A reward : requital. *—vt.* **-doned, -don·ing, -dons.** To reward.

**gue·ril·la** (gə-rĭl'ə) *n. var. of* GUERRILLA.

**Guern·sey** (gûrn'zē) *n., pl.* **-seys.** One of a breed of brown and white dairy cattle orig. bred on the Isle of Guernsey.

**guer·ril·la** *or* **gue·ril·la** (gə-rĭl'ə) *n.* [Sp. *guerilla,* dim. of *guerra,* war, of Germanic orig.] **1.** A member of an irregular military force operating usu. in small independent groups capable of great speed and mobility. **2.** *Archaic.* Warfare carried out by guerrillas.

**guerrilla theater** *n.* Street theater.

**guess** (gĕs) *v.* **guessed, guess·ing, guess·es.** [ME *gessen,* to infer.] *—vt.* **1.** To make a judgment about without adequate information. **2.** To find the correct answer to by surmise or inference. **3.** To suppose. *—vi.* To make a conjecture. **—guess** *n.* **—guess'er** *n.*

**guess·ti·mate** (gĕs'tə-mĭt) *n.* [Blend of GUESS and ESTIMATE.] *Informal.* An approximate estimate. **—guess'ti·mate'** (-māt') *v.* **(-mat·ed, -mat·ing, -mates).**

**guess·work** (gĕs'wûrk') *n.* **1.** The process of making guesses. **2.** A result or answer obtained by guessing.

**guest** (gĕst) *n.* [ME *gest* < ON *gesvt.*] **1.** The recipient of hospitality at the home or table of another. **2.** One to whom entertainment or hospitality has been extended. **3.** A patron of an establishment such as a restaurant or hotel. **4.** A visiting participant in a program. **5.** *Zool.* A commensal organism, esp. an insect living in the nest or burrow of another species. *—vt. & vi.* **guest·ed, guest·ing, guests.** To entertain as or to be a guest.

**guff** (gŭf) *n.* [< obs. *guff,* puff.] *Slang.* Nonsense.

**guf·faw** (gə-fô') *n.* [Imit.] A coarse or hearty burst of laughter. **—guf·faw'** *v.* **(-fawed, -faw·ing, -faws).**

**guid·ance** (gīd'ns) *n.* **1.** An act or instance of guiding. **2.** Counseling, as on vocational, educational, or personal problems. **3.** A process for guiding the path of a missile using built-in equipment.

**guide** (gīd) *n.* [ME < OFr. < OProv. *guida,* of Germanic orig.] **1.** One who leads the way, directs, or advises. **2.** One employed to point out and give information about objects of interest. **3.** A guidebook. **4. a.** Something serving to indicate or direct. **b.** A device acting to regulate or direct a motion or operation. **5.** A soldier at the right or left of a column who controls the alignment of the marchers, shows the direction, or marks the point of pivot. *—v.* **guid·ed, guid·ing, guides.** *—vt.* **1.** To serve as a guide for : CONDUCT. **2.** To direct the course of. **3.** To exert influence or control over. *—vi.* To serve as a guide. **—guid'a·ble** *adj.* **—guid'er** *n.*

☆ ***syns:*** GUIDE, CONDUCT, DIRECT, ESCORT, LEAD, PILOT, ROUTE, SHEPHERD, SHOW, STEER, USHER *v. core meaning* : to show the way to <*guided* them to safety>

**guide·book** (gīd'bo͝ok') *n.* A handbook of information.

**guided missile** *n.* A missile capable of being guided in flight.

**guided wave** *n.* An electromagnetic or acoustic wave transmitted by a process that limits its physical dispersion along the length of its transmission.

**guide·line** (gīd'līn') *n.* A statement of policy or procedure.

**guide·post** (gīd'pōst') *n.* **1.** A post with a directional sign. **2.** One that gives direction.

**guide rope** *n.* A rope fastened to another rope that is lifting a load, to guide the rope and steady the load.

**guide·word** (gīd'wûrd') *n.* A word or term at the top of a page of a reference book, as a dictionary, indicating the first or last entry on the page.

**gui·don** (gī'dŏn', gīd'n) *n.* [Fr. < Ital. *guidone* < *guida,* guide < OProv., of Germanic orig.] **1.** A small flag or pennant carried as a standard by a military unit. **2.** A soldier bearing a guidon.

**guild** *also* **gild** (gĭld) *n.* [ME < ON *gildi.*] **1. a.** An association of persons of the same trade or pursuits. **b.** A medieval association or society of merchants, craftsmen, or artisans. **2.** *Ecol.* One of four groups of plants, the lianas, epiphytes, saprophytes, and parasites, with a characteristic mode of existence involving some dependence upon other plant life.

**guil·der** (gĭl'dər) *n.* [Alteration of Du. *gulden.* —see GULDEN.] —See table at CURRENCY.

**guild·hall** (gĭld'hôl') *n.* **1.** The meeting hall of a guild. **2.** A town hall.

**guilds·man** (gĭldz'mən) *n.* **1.** A member of a guild. **2.** An adherent or advocate of guild socialism.

**guild socialism** *n.* An early 20th-cent. English socialist doctrine by which industry was to be owned by the state but managed by a council of workers.

**guile** (gīl) *n.* [ME < OFr., of Germanic orig.] **1.** Insidious, treacherous cunning : DECEIT. **2.** *Obs.* A trick : stratagem. *—vt.* **guiled, guil·ing, guiles.** *Archaic.* To beguile : deceive. **—guile'ful** *adj.* **—guile'ful·ly** *adv.* **—guile'ful·ness** *n.*

**guile·less** (gīl'lĭs) *adj.* Free from guile : ARTLESS. **—guile'less·ly** *adv.* **—guile'less·ness** *n.*

**guil·le·mot** (gĭl'ə-mŏt') *n.* [Fr. < *Guillaume,* William.] Any of several small sea birds of the genus *Cepphus* of northern regions, bearing black plumage with white markings.

**guil·loche** (gĭ-lōsh', gē-yōsh') *n.* [Fr. *guillochis.*] An ornamental border of two or more bands interlaced in such a way as to repeat a rounded design.

**guil·lo·tine** (gĭl'ə-tēn', gē'ə-) *n.* [After Joseph I. *Guillotin* (1738–1814).] **1.** A machine with a heavy blade that falls freely between upright guides to behead a condemned prisoner. **2.** A cutting instrument, as a paper cutter, similar to a guillotine. **3.** A method of limiting debate in a legislative body by fixing beforehand a time for voting. *—vt.* **-tined, -tin·ing, -tines.** To behead with a guillotine.

**guilt** (gĭlt) *n.* [ME *gilt* < OE *gylt.*] **1.** The fact of being responsible for an offense or wrongdoing. **2.** *Law.* **a.** Culpability for a crime or lesser breach of regulations. **b.** The disposition to break the law. **3.** Guilty behavior. **4.** Remorseful awareness of having done something wrong or of having failed to do something required or expected.

**guilt·less** (gĭlt'lĭs) *adj.* Free from guilt : INNOCENT. **—guilt'less·ly** *adv.* **—guilt'less·ness** *n.*

**guilt·y** (gĭl'tē) *adj.* **-i·er, -i·est.** **1.** Responsible for or chargeable with a reprehensible act <*guilty* of cheating> **2.** *Law.* Having committed a crime <a *guilty* defendant> **3.** Being at fault : CULPABLE <the *guilty* party> **4.** Burdened with, prompted by, or showing a sense of guilt <a *guilty* look> **—guilt'i·ly** *adv.* **—guilt'i·ness** *n.*

**guimpe** (gămp, gĭmp) *n.* [Fr. < OFr. *guimple,* of Germanic orig.] **1.** A blouse worn with a jumper. **2.** A yoke insert for a low-necked dress. **3.** A starched cloth covering the neck and shoulders as part of a nun's habit. **4.** GIMP¹.

**guin·ea** (gĭn′ē) *n.* [After the *Guinea* coast of Africa, the source of the gold from which it was first made.] **1.** A British gold coin worth one pound and five pence, no longer in circulation. **2.** The sum of one pound and five pence.

**guinea fowl** *n.* [After the *Guinea* coast of Africa.] A pheasantlike bird of the family Numididae, native to Africa, esp. a widely domesticated species, *Numida meleagris*, bearing blackish plumage marked with many small white spots.

**guinea hen** *n.* **1.** A female guinea fowl. **2.** The guinea fowl.

**guinea pig** *n.* [Prob. alteration of *Guiana.*] **1.** Any of various South American burrowing rodents of the genus *Cavia*, with variously colored hair and no visible tail and widely domesticated as pets and as experimental animals. **2.** One used as a subject for experimentation.

**guinea worm** *n.* [Prob. after the *Guinea* coast of Africa.] A long threadlike nematode worm, *Dracunculus medinensis* of tropical Asia and Africa, that is a subcutaneous parasite of humans and other animals.

**Guin·e·vere** (gwĭn′ə-vîr′) *also* **Guen·e·vere** (gwĕn′-) *n.* The wife of King Arthur and the mistress of Lancelot according to Arthurian legend.

**gui·pure** (gĭ-po͝or′, -pyo͝or′) *n.* [Fr. < OFr. < *guiper*, to cover with silk, of Germanic orig.] A coarse large-patterned lace without a net ground.

**guise** (gīz) *n.* [ME, fashion < OFr., of Germanic orig.] **1.** Outward appearance : ASPECT. **2.** Mode of dress : GARB. **3.** *Obs.* Habit : custom.

**gui·tar** (gĭ-tär′) *n.* [Fr. *guitare* < Sp. *guitarra* < Ar. *gītār* < Gk. *kithara*, cithara.] A musical instrument with a large flat-backed sound box similar in shape to a violin, a long fretted neck, and usu. six strings, played by strumming or plucking. **—gui·tar′ist** *n.*

**gui·tar·fish** (gĭ-tär′fĭsh′) *n., pl.* **guitarfish** or **-fish·es.** Any of several marine fishes of the family Rhinobatidae, related to the skates and rays and having a guitar-shaped body.

**Gu·ja·ra·ti** (go͞o′jə-rä′tē, go͞oj′ə-) *n., pl.* **Gujarati. 1.** The Indic language of Gujarat. **2.** A native or inhabitant of Gujarat.

**gul** (go͞ol) *n.* [Pers., rose.] A stylized rose motif in an Oriental rug.

**gu·lag** (go͞o′läg′) *n.* [R., acronym for *glavnoe upravlenie lagery*, Chief Administration of Collective Labor Camps.] The Soviet penal system and its administration.

**gu·lar** (go͞o′lər, gyo͞o′-) *adj.* [< Lat. *gula*, throat.] Of, relating to, or located on the throat.

**gulch** (gŭlch) *n.* [Orig. unknown.] A small shallow canyon having smoothly inclined slopes and steep sides : RAVINE.

**gul·den** (go͞ol′dən, go͞ol′-) *n., pl.* **-dens** or **gulden.** [ME < MDu. *gulden (florijn)*, golden (florin).] A guilder.

**gules** (gyo͞olz) *n.* [ME *goules* < OFr., pl. of *gole*, throat < Lat. *gula.*] *Heraldry.* The color red, indicated on a blazon by engraved vertical lines.

**gulf** (gŭlf) *n.* [ME *goulf* < OFr. *golf* < OItal. *golfo*, ult. < Gk. *kolphos.*] **1.** A large area of a sea or ocean partially enclosed by land, esp. a long landlocked portion of sea opening through a strait. **2.** A deep, wide chasm : ABYSS. **3.** A distance that separates : GAP <the *gulf* between the generations> **4.** A whirlpool : eddy. *—vt.* **gulfed, gulf·ing, gulfs.** To engulf.

**gulf·weed** (gŭlf′wēd′) *n.* [After the *Gulf* of Mexico, where it is found.] Any of several brownish seaweeds of the genus *Sargassum* of tropical Atlantic waters, that have rounded air bladders and often form dense floating masses.

**gull[1]** (gŭl) *n.* [ME *gull*, of Celt. orig.] Any of various chiefly coastal aquatic birds of the subfamily Larinae, with long wings, webbed feet, and usu. gray and white plumage.

**gull[2]** (gŭl) *n.* [Orig. unknown.] A gullible person : DUPE. *—vt.* **gulled, gull·ing, gulls.** To take advantage of : CHEAT.

**Gul·lah** (gŭl′ə) *n.* **1.** One of a group of blacks inhabiting the Sea Islands and coastal area of South Carolina, Georgia, and northern Florida. **2.** The language of the Gullahs, based on English but including vocabulary elements and grammatical features from several African languages.

**gul·let** (gŭl′ĭt) *n.* [ME *golet* < OFr., dim. of *gole*, throat < Lat. *gula.*] **1.** *Anat.* The esophagus. **2.** The throat.

**gul·li·ble** (gŭl′ə-bəl) *adj.* [< GULL[2].] Easily duped or deceived. **—gul′li·bil′i·ty** *n.* **—gul′li·bly** *adv.*

**gul·ly[1]** (gŭl′ē) *n., pl.* **-lies.** [Prob. alteration of GULLET.] A deep ditch or channel cut in the earth by running water after a downpour. *—vt. & vi.* **-lied, -ly·ing, -lies.** To wear a gully in or to form a gully.

**gul·ly[2]** (gŭl′ē) *n., pl.* **-lies.** [Short for dial. *gully knife*, a large knife.] *Chiefly Brit. Regional.* A large knife.

**gulp** (gŭlp) *v.* **gulped, gulp·ing, gulps.** [ME *gulpen.*] *—vt.* **1.** To swallow rapidly or greedily in large amounts. **2.** To stifle by or as if by swallowing. *—vi.* To gasp or choke, as in swallowing. *—n.* **1.** An act of gulping. **2.** A large mouthful <a *gulp* of hot tea> **3.** *Computer Sci.* A small group of bytes that may be either data or instruction. **—gulp′er** *n.* **—gulp′ing·ly** *adv.*

**gum[1]** (gŭm) *n.* [ME *gomme* < OFr. *gome* < Lat. *gummi* < Gk. *kommi*, of Egypt. orig.] **1. a.** Any of various viscous substances exuded by certain plants and trees and drying into water-soluble, noncrystalline, brittle solids. **b.** A similar plant exudate, as a resin. **2.** A substance resembling a plant gum. **3. a.** A tree, as one of the genera *Eucalyptus, Liquidambar*, or *Nyssa*, that is a source of gum. **b.** The wood of such a tree. **4.** Chewing gum. *—v.* **gummed, gum·ming, gums.** *—vt.* To cover, smear, seal, fill, or fix in place with or as if with gum. *—vi.* **1.** To exude or form gum. **2.** To become clogged or sticky with or as if with gum. **—gum up.** *Slang.* To bungle : ruin <*gummed up* the whole project>

**gum[2]** (gŭm) *n.* [ME *goma* < OE *gōma*, palate.] The firm connective tissue covered by mucous membrane that envelops the alveolar arches of the jaw and surrounds the bases of the teeth.

**gum ammoniac** *n.* Ammoniac.

**gum arabic** *n.* A gum exuded by an African tree of the genus *Acacia*, esp. *A. senegal*, used in preparing pills and emulsions, in manufacturing mucilage and candies, and as a thickener and colloidal stabilizer.

**gum benjamin** *n.* Benzoin.

**gum benzoin** *n.* Benzoin.

**gum·bo** (gŭm′bō) *n., pl.* **-bos.** [Louisiana Fr. *gombo*, of African orig.] **1.** OKRA 1, 2. **2.** A soup or stew thickened with okra. **3.** A fine silty soil, common in the southern and western United States, that forms an unusually sticky mud when wet. **4. Gumbo.** A patois spoken by some blacks and Creoles in Louisiana and the French West Indies.

**gum·boil** (gŭm′boil′) *n.* A small boil or abscess on the gum.

**gum·bo-lim·bo** (gŭm′bō-lĭm′bō) *n., pl.* **-bos.** [GUMBO + *limbo*, of unknown orig.] An aromatic tree, *Bursera simaruba* of Florida and the West Indies, with compound leaves and small white flowers.

**gum·drop** (gŭm′drŏp′) *n.* A firm jellylike piece of candy made of sweetened, colored, and flavored gum arabic or gelatin and coated with coarse granulated sugar.

**gum·ma** (gŭm′ə) *n., pl.* **-mas** or **-ma·ta** (-ə-tə) [NLat. < Lat. *gummi*, gum. —see GUM.] A small rubbery tumor formed at an advanced stage of syphilis. **—gum′ma·tous** *adj.*

**gum·mose** (gŭm′ōs′) *adj. var. of* GUMMOUS.

**gum·mo·sis** (gŭ-mō′sĭs) *n.* [Lat. *gummi*, gum + -OSIS.] Pathological formation of patches of gum on certain plants, as sugar cane and certain fruit trees, caused by insects, microorganisms, or adverse weather conditions.

**gum·mous** (gŭm′əs) *also* **gum·mose** (gŭm′ōs′) *adj.* Like gum : GUMMY.

**gum·my** (gŭm′ē) *adj.* **-mi·er, -mi·est. 1.** Consisting of or containing gum. **2.** Yielding gum. **3.** Sticky. **—gum′mi·ness** *n.*

**gum plant** *n.* A North American plant of the genus *Grindelia*, esp. *G. squarosa*, with sticky leaves and bracts and yellow rayed flowers.

**gump·tion** (gŭmp′shən) *n.* [Orig. unknown.] *Informal.* **1.** Common sense. **2.** Initiative : boldness.

**gum resin** *n.* A mixture of gum and resin that exudes from some plants or trees.

**gum·shoe** (gŭm′sho͞o′) *Slang.* *—n.* A detective. *—vi.* **-shoed, -shoe·ing, -shoes.** To perform or engage in the work of a detective.

**gum tree** *n.* GUM[1] 3a.

**gum·wood** (gŭm′wo͝od′) *n.* GUM[1] 3b.

**gun** (gŭn) *n.* [ME *gonne*, cannon, perh. < ON *gunnr*, battle.] **1.** A weapon consisting of a metal tube that fires a projectile at high velocity into a flat trajectory. **2.** A cannon. **3.** A portable firearm. **4.** A device that shoots a projectile <a dart *gun*> **5.** A discharge of a gun as a signal or salute. **6.** One who uses or carries a gun, esp.: **a.** A hunter. **b.** One skilled in the use of a gun. **7.** A mechanism controlling the flow of fuel to an engine : THROTTLE. *—v.* **gunned, gun·ning, guns.** *—vt.* **1.** To fire upon : SHOOT. **2.** To open the throttle of so as to accelerate <*gun* the motor> *—vi.* To shoot or hunt with a gun. **—gun for. 1.** To seek to catch, overcome, or destroy <a posse *gunning for* cattle rustlers> **2.** To seek with tenacity <*gun for* a quick promotion>

**gun·boat** (gŭn′bōt′) *n.* A small armed vessel used esp. on deep rivers and in coastal waters.

**gun carriage** *n.* A frame or structure upon which a gun is mounted for maneuvering or firing.

**gun·cot·ton** (gŭn′kŏt′n) *n.* Nitrocellulose.

**gun dog** *n.* A dog trained to assist hunters, as in flushing or retrieving game.

**gun·fight** (gŭn′fīt′) *n.* A battle with guns. **—gun′fight′er** *n.*

**gun·fire** (gŭn′fīr′) *n.* Discharge of guns.

**gun·flint** (gŭn′flĭnt′) *n.* A piece of flint for striking the igniting spark in a flintlock.

**gung ho** (gŭng′ hō′) *adj.* [Pidgin E., prob. < Chin. (Mandarin) *gong[1] he[2]*, to work together : *gong[1]*, work + *he[2]*, together.] *Slang.* **1.** Unswervingly loyal. **2.** Extremely enthusiastic.

**gunk** (gŭngk) *n.* [< *Gunk*, a trademark for liquid soap.] *Informal.* A filthy, slimy, or greasy substance.

**gun·lock** (gŭn′lŏk′) *n.* A device for igniting the charge of a firearm.

**gun·man** (gŭn′mən) *n.* **1. a.** A man armed with a gun. **b.** A professional killer. **2.** A man of great skill in using a gun.

**gun·met·al** (gŭn′mĕt′l) *n.* **1.** An alloy of copper with 10% tin. **2.** Metal used for guns. **3.** A dark gray. **—gun′met′al** *adj.*

**gun moll** *n.* *Slang.* A gangster's girlfriend and partner in crime.

**Gun·nar** (gōōn′är′, -ər) *n. Norse Myth.* The husband of Brynhild, brother-in-law of Sigurd, and brother of Gudrun.
**Gunn effect** (gŭn) *n.* [After J. B. *Gunn* (b. 1928).] *Electron.* Production of high-speed current fluctuations when voltage in excess of a critical level is applied to a semiconductor device, resulting in microwave generation.
**gun·nel**[1] (gŭn′əl) *n.* [Orig. unknown.] Any of various long eellike fishes of the family Pholidae, inhabiting northern seas.
**gun·nel**[2] (gŭn′əl) *n. var. of* GUNWALE.
**gun·ner** (gŭn′ər) *n.* **1.** A soldier, sailor, or airman who aims or fires a gun. **2.** One who hunts with a gun. **3.** A warrant officer in charge of ordnance.
**gun·ner·y** (gŭn′ə-rē) *n.* **1.** The art and science of constructing and operating guns. **2.** The use of guns.
**gunnery sergeant** *n.* A noncommissioned U.S. Marine officer ranking above a staff sergeant and below a master sergeant.
**gun·ny** (gŭn′ē) *n.* [Hindi *gonī* < Skt. *goṇī*, sack.] A coarse fabric of jute or hemp.
**gun·ny·sack** (gŭn′ē-săk′) *n.* A sack made of gunny.
**gun·pow·der** (gŭn′pou′dər) *n.* An explosive powder for propelling projectiles from guns, esp. a black explosive mixture of potassium nitrate, charcoal, and sulfur.
**gunpowder tea** *n.* A green tea whose leaves are rolled into pellets.
**gun·room** (gŭn′rōōm′, -rŏŏm′) *n.* The quarters occupied by midshipmen and junior officers on a British warship.
**gun·run·ner** (gŭn′rŭn′ər) *n.* One that smuggles firearms and ammunition. **—gun′run′ning** *n.*
**gun·shot** (gŭn′shŏt′) *n.* **1.** Shot fired from a gun. **2.** A particular gun's range.
**gun-shy** (gŭn′shī′) *adj.* **1.** Afraid of loud noise, as that made by gunfire. **2.** Extremely distrustful : WARY.
**gun·sling·er** (gŭn′slĭng′ər) *n.* A gunman. **—gun′sling′ing** *n.*
**gun·smith** (gŭn′smĭth′) *n.* A maker or repairer of firearms.
**gun·stock** (gŭn′stŏk′) *n.* A handle on a gun : STOCK.
**Gun·ter's chain** (gŭn′tərz) *n.* [After Edmund *Gunter* (1581–1626).] CHAIN 9a.
**Gun·ther** (gōōn′tər) *n.* [G.] A king of Burgundy and husband of Brunhild in the *Nibelungenlied.*
**gun·wale** *also* **gun·nel** (gŭn′əl) *n.* The upper edge of a ship's side.
**Guo·yu** (kōō′yōō′) *n.* [Chin. : *guo*[2], nation + *yu*[3], language.] MANDARIN 3.
**gup·py** (gŭp′ē) *n., pl.* **-pies.** [After R.J.L. *Guppy* (1836–1916).] A small, brightly colored freshwater fish, *Poecilia reticulata* or *Lebistes reticulatus* of northern South America and adjacent islands of the West Indies, popular in home aquariums.
**gur·gi·ta·tion** (gûr′jĭ-tā′shən) *n.* [< LLat. *gurgitare*, to engulf < *gurges*, whirlpool.] A whirling motion : EBULLITION.
**gur·gle** (gûr′gəl) *v.* **-gled, -gling, -gles.** [Prob. imit.] *—vi.* **1.** To flow in an uneven, broken current making intermittent low sounds. **2.** To make a gurgling sound. *—vt.* To pronounce or express with a gurgling sound. **—gur′gle** *n.* **—gur′gling·ly** *adv.*
**Gur·kha** (gŏŏr′kə) *n.* **1.** A member of a Rajput ethnic group predominant in Nepal. **2.** A soldier from Nepal serving in the British or Indian armies.
**gur·nard** (gûr′nərd) *n., pl.* **-nards** or **gurnard.** [ME < OFr. *gornart.*] **1.** A marine fish of the family Triglidae, and esp. of the Old World genus *Trigla*, with large fanlike pectoral fins. **2.** The flying gurnard.
**gur·ry** (gûr′ē) *n.* [Orig. unknown.] Fish offal.
**gu·ru** (gōō′rōō, gōō-rōō′) *n., pl.* **-rus.** [Hindi *gurū* < Skt. *guru-*, venerable.] **1.** A personal spiritual teacher. **2. a.** A recognized guide or leader. **b.** An acknowledged advocate, as of a movement or idea.
**gush** (gŭsh) *v.* **gushed, gush·ing, gush·es.** [ME *gushen*, prob. of Scand. orig.] *—vi.* **1.** To flow forth suddenly and violently. **2.** To issue or emit a copious flow. **3.** To make an exaggerated display of enthusiasm or sentiment. *—vt.* To emit abundantly. *—n.* **1.** A sudden, violent, or copious outflow <a *gush* of hysterical laughter> **2.** Something emitted by gushing. **3.** An exaggerated display of enthusiasm or sentiment. **—gush′er** *n.*
**gush·y** (gŭsh′ē) *adj.* **-i·er, -i·est.** Marked by exaggerated displays of enthusiasm or sentiment. **—gush′i·ly** *adv.* **—gush′i·ness** *n.*
**gus·set** (gŭs′ĭt) *n.* [ME < OFr. *gosset.*] A triangular insert, as in a garment, for enlarging or strengthening.
**gus·sy** (gŭs′ē) *vt.* **-sied, -sy·ing, -sies.** [Orig. unknown.] To dress up : DECORATE <all *gussied* up in silks and satins>
**gust**[1] (gŭst) *n.* [Prob. < ON *gusvt.*] **1.** A violent, abrupt rush of wind. **2.** A sudden outburst. *—vi.* **gust·ed, gust·ing, gusts.** To blow in gusts. **—gust′i·ly** *adv.* **—gust′i·ness** *n.* **—gust′y** *adj.*
**gust**[2] (gŭst) *n.* [ME *guste*, taste < Lat. *gustus.*] **1.** *Archaic.* Relish : gusto. **2.** *Obs.* Personal taste or inclination : LIKING.
**gus·ta·tion** (gŭ-stā′shən) *n.* [Lat. *gustatio*, a tasting < *gustare*, to taste < *gustus*, taste.] The act or faculty of tasting.
**gus·ta·to·ry** (gŭs′tə-tôr′ē, -tōr′ē) *adj.* Of or relating to the sense of taste. **—gus′ta·to′ri·ly** *adv.*
**gus·to** (gŭs′tō) *n., pl.* **-toes.** [Ital. < Lat. *gustus*, taste.] **1.** A specialized or individual taste. **2.** Vigorous enjoyment : ZEST. **3.** *Archaic.* Artistic style.
**gut** (gŭt) *n.* [< ME *gutles*, entrails < OE *gutlas.*] **1.** The alimentary canal or a section thereof, esp. the stomach or intestine. **2. guts.** The bowels : entrails. **3. guts.** The essential components <the *guts* of old washing machines> **4.** The intestines of some animals used as strings for musical instruments or as surgical sutures. **5. guts.** *Slang.* Fortitude : courage. **6.** A narrow channel or passage. **7.** Fibrous material removed from the silk gland of a silkworm before it spins a cocoon, used for fishing tackle. *—vt.* **gut·ted, gut·ting, guts. 1.** To remove the intestines or entrails of : EVISCERATE. **2.** To destroy the interior of <wreckers *gutting* the hotel> *—adj. Slang.* **1.** Arousing or involving basic emotions : VISCERAL <a *gut* issue><a *gut* response> **2.** Easy <*gut* courses> **—gut′ty** *adj.*
**Guth·run** (gōōth′rōōn′) *n. var. of* GUDRUN.
**gut·less** (gŭt′lĭs) *adj.* Lacking courage. **—gut′less·ness** *n.*
**guts·y** (gŭt′sē) *adj.* **-i·er, -i·est.** *Slang.* Full of courage. **—guts′i·ly** *adv.* **—guts′i·ness** *n.*
**gut·ta** (gŭt′ə) *n., pl.* **gut·tae** (gŭt′ē′) [Lat., drop.] **1.** One of a group of small droplike ornaments on a Doric entablature. **2.** *Med.* A drop.
**gut·ta-per·cha** (gŭt′ə-pûr′chə) *n.* [Malay *gĕtah percha* : *gĕtah*, sap + *percha*, strip of cloth.] A rubbery substance derived from the latex of tropical trees of the genera *Palaquium* and *Payena* and used for electrical insulation and waterproofing.
**gut·tate** (gŭt′āt′) *also* **gut·tat·ed** (-ā′tĭd) *adj.* [Lat. *guttatus*, speckled < *gutta*, drop.] **1. a.** Being in the form of drops. **b.** Having drops. **2.** Spotted as if by drops.
**gut·ter** (gŭt′ər) *n.* [ME *goter* < OFr. *gotier* < VLat. **guttarie* < Lat. *gutta*, drop.] **1.** A channel for draining off water at the edge of a street or road. **2.** A pipe or trough for draining off water under the eaves of a roof. **3.** A furrow or groove made by running water. **4.** The trough on either side of a bowling alley. **5.** The white space between the facing pages of a book. **6.** The lowest state of human existence. *—v.* **-tered, -ter·ing, -ters.** *—vt.* To form gutters or furrows in. *—vi.* **1.** To flow in channels or rivulets. **2.** To melt away through the channel in the side of the hollow formed by a burning wick. —Used of a candle. **3.** To burn with a low flame : FLICKER. *—adj.* Filthy : foul <*gutter* language>
**gut·ter·snipe** (gŭt′ər-snīp′) *n.* **1.** A street urchin. **2.** One who is of the lowest class.
**gut·tur·al** (gŭt′ər-əl) *adj.* [OFr. < Lat. *guttur*, throat.] **1.** Of or relating to the throat. **2.** Produced in the throat <a *gutteral* moan> **3.** Velar. **—gut′tur·al** *n.* **—gut′tur·al·ism** *n.* **—gut′tur·al′i·ty** (-ə-răl′ĭ-tē) *n.* **—gut′tur·al·ly** *adv.* **—gut′tur·al·ness** *n.*
**guy**[1] (gī) *n.* [Prob. of LG orig.] A rope, cord, or cable used for steadying, guiding, or holding. *—vt.* **guyed, guy·ing, guys.** To steady, guide, or hold with a guy.
**guy**[2] (gī) *n.* [After *Guy* Fawkes (1570–1606).] **1.** *Informal.* A fellow. **2.** *Informal.* **guys.** Persons <Where have you *guys* been?> **3.** *Chiefly Brit.* A person of odd or grotesque appearance or dress. **4.** *often* **Guy.** An effigy of Guy Fawkes paraded through the streets of English towns and burned on Guy Fawkes Day. *—vt.* **guyed, guy·ing, guys.** To make fun of : MOCK.
**Guy Fawkes Day** (gī′ fôks′) *n.* Nov. 5, celebrated in England in commemoration of the 1605 attempt led by Guy Fawkes to burn down the houses of Parliament.
**guz·zle** (gŭz′əl) *v.* **-zled, -zling, -zles.** [Orig. unknown.] *—vt.* To drink greedily or habitually <*guzzling* martinis> *—vi.* To drink esp. alcoholic beverages greedily or habitually. **—guz′zler** *n.*
**gybe** (jīb) *v. & n. var. of* JIBE[1].
**gym** (jĭm) *n. Informal.* **1.** A gymnasium. **2.** Physical education. **3.** A frame supporting structures used in outdoor play.
**gym·kha·na** (jĭm-kä′nə) *n.* [Prob. alteration of Hindi *gend-khānā*, racket court.] A contest involving display of skill, as automobile racing or horseback riding.
**gym·na·si·um** (jĭm-nā′zē-əm) *n., pl.* **-si·ums** *or* **-si·a** (-zē-ə) [Lat., school < Gk. *gumnasion* < *gumnazein*, to exercise naked < *gumnos*, naked.] **1.** A room or building equipped for gymnastics and sports. **2.** (gĭm-nä′zē-ōōm′). An academic high school in various European countries, esp. Germany, that prepares students for university.
**gym·nast** (jĭm′năst′) *n.* [Gk. *gumnastēs* < *gumnazein*, to exercise naked < *gumnos*, naked.] One skilled in gymnastic exercises.
**gym·nas·tics** (jĭm-năs′tĭks) *n. (sing. or pl. in number).* Body-building exercises, esp. those performed with special equipment in a gymnasium. **—gym·nas′tic** *adj.*
**gym·nos·o·phist** (jĭm-nŏs′ə-fĭst) *n.* [Lat. *gymnosophista* < Gk. *gumnosophistēs* : *gumnos*, naked + *sophistēs*, expert. —see SOPHIST.] One of an ancient sect of naked Hindu ascetics, as described in classical antiquity.
**gym·no·sperm** (jĭm′nə-spûrm′) *n.* [NLat. *Gymnospermae*, class name : Gk. *gumnos*, naked + Gk. *sperma*, seed.] A plant of the class Gymnospermae, including the coniferous trees and other plants bearing seeds not enclosed within an ovary. **—gym′no·sper′mous** *adj.*
**gyn-** *pref. var. of* GYNO-.

**gynaeco-** *or* **gynaec-** *pref. vars. of* GYNECO-.
**gy·nan·dro·morph** (jĭ-năn′drə-môrf′, gī-) *n.* An individual with male and female characteristics. **—gy·nan′dro·mor′phic, gy·nan′dro·mor′phous** *adj.* **—gy·nan′dro·mor′phism, gy·nan′dro·mor′phy** *n.*
**gy·nan·drous** (jĭ-năn′drəs, gī-) *n.* **1.** Having the stamens and pistil united to form a column. **2.** Hermaphroditic.
**gyn·ar·chy** (jĭn′är′kē, jĭ′när′-, gī′-) *n., pl.* **-chies.** Government by women. **—gyn·ar′chic** *adj.*
**-gyne** *suff.* [< Gk. *gunē,* woman.] Female reproductive organ <tri*chogyne*>
**gyneco-** *or* **gynec-** *or* **gynaeco-** *or* **gynaec-** *pref.* [Gk. *gunaiko-* < *gunē,* woman.] Woman <*gynecology*>
**gyn·e·coc·ra·cy** (jĭn′ĭ-kŏk′rə-sē, gī′-nĭ-) *n., pl.* **-cies.** [Gk. *gunaikokratia* : *gunē,* woman + *-kratia,* -cracy.] Political dominance by women.
**gy·ne·col·o·gy** (gī′nĭ-kŏl′ə-jē, jĭn′ĭ-) *n.* The branch of medicine concerned with disease, reproductive physiology, and endocrinology in females. **—gy′ne·co·log′i·cal** (-kə-lŏj′ĭ-kəl), **gy′ne·co·log′ic** *adj.* **—gy′ne·col′o·gist** *n.*
**gyn·e·cop·a·thy** (jĭn′ĭ-kŏp′ə-thē, gī′nĭ-) *n.* Any of various diseases peculiar to women.
**gyn·i·at·rics** (jĭn′ē-ăt′rĭks, gī′nē-) *n. (sing. in number).* Treatment of diseases specific to women.
**gyno-** *or* **gyn-** *pref.* [Gk. *guno-* < *gunē,* woman.] **1.** Woman <*gyn*iatrics> **2.** Female reproductive organ : PISTIL <*gyno*phore>
**gy·noe·ci·um** (jĭ-nē′sē-əm, gī-) *n., pl.* **-ci·a** (-sē-ə) [NLat., alteration of Lat. *gynaeceum,* women's apartments < Gk. *gunaikeion* < *gunē,* woman.] The female reproductive organs of a flower.
**gyn·o·phore** (jĭn′ə-fôr′, -fōr′, gī′nə-) *n.* The stalk of a pistil.
**-gynous** *suff.* [NLat. *-gynus* < Gk. *-gunos* < *gunē, woman.*] **1.** Of, relating to, or having a specified number of females <heterogy*nous*> **2. a.** Of, pertaining to, or situated in a specified place with respect to female organs of a plant <epi*gynous*> **b.** Having a specified number or kind of female organs of a plant <tetra*gynous*>
**-gyny** *suff.* [< Gk. *gunē,* woman.] **1.** The state or condition of having a specified number of women or females <mono*gyny*> **2.** The condition of being situated in a specified place with respect to female plant organs <epi*gyny*>
**gyp** *also* **gip** (jĭp) [Poss. short for GYPSY.] *Informal.* —*vt.* **gypped, gyp·ping, gyps** *also* **gipped, gip·ping, gips.** To swindle, cheat, or defraud. —*n.* **1.** An act or instance of gypping : SWINDLE. **2.** One who gyps : SWINDLER. **—gyp′per** *n.*
**gyp joint** *n. Slang.* An establishment that deliberately and regularly overcharges or defrauds its clientele.
**gyp·lure** (jĭp′lŏŏr′) *n.* [GYP(SY MOTH) + LURE.] A synthetic form of the sex attractant of the female gypsy moth for trapping male gypsy moths.
**gyp·soph·i·la** (jĭp-sŏf′ə-lə) *n.* [NLat. *Gypsophila,* genus name : Gk. *gupsos,* chalk + Gk. *philos,* loving.] A plant of the genus *Gypsophila,* as the baby's-breath, bearing small white or pink flowers.
**gyp·sum** (jĭp′səm) *n.* [Lat. < Gk. *gupsos,* of Semitic orig.] A white mineral, $CaSO_4 \cdot 2H_2O$, used to make plaster of Paris, gypsum plaster and plasterboard, Portland cement, wallboards, and fertilizers. **—gyp′se·ous** (-sē-əs), **gyp·sif′er·ous** (-sĭf′ər-əs) *adj.*
**gypsum board** *n.* Plasterboard.
**Gyp·sy** *also* **Gip·sy** (jĭp′sē) *n., pl.* **-sies.** [Shortening and alteration of EGYPTIAN.] **1.** One of a nomadic Caucasoid people orig. migrating from the border region between Iran and India to Europe in the 14th or 15th cent. and now living mostly in Europe and the United States. **2.** ROMANY 2. **3. gypsy.** One held to resemble a Gypsy in appearance or behavior.
**gypsy cab** *n.* A taxicab that is licensed only to respond to calls but that cruises the streets for passengers.
**gypsy moth** *n.* **1.** A moth, *Porthetria dispar,* native to the Old World, having hairy caterpillars that eat foliage and are very destructive to trees. **2.** *Informal.* A politically moderate Republican member of the U.S. House of Representatives from a northeastern or midwestern urban area.
**gy·ral** (jī′rəl) *adj.* **1.** Moving in a circular or spiral path : GYRATORY. **2.** Relating to a gyrus. **—gy′ral·ly** *adv.*
**gy·rate** (jī′rāt′) *vi.* **-rat·ed, -rat·ing, -rates.** [Lat. *gyrare, gyrat-* < *gyrus,* circle < Gk. *guros.*] **1.** To revolve around or on a center or axis. **2.** To circle or spiral. —*adj. Biol.* In rings : COILED. **—gy·ra′tion** *n.* **—gy′ra·tor** *n.* **—gy′ra·to′ry** (-rə-tôr′ē, -tōr′ē) *adj.*
**gyre** (jīr) *n.* [Lat. *gyrus* < Gk. *guros.*] **1. a.** A ring or circle. **b.** A spiral. **2.** A circular or spiral motion.
**gy·rene** (jī-rēn′) *n.* [Prob. alteration of MARINE.] *Slang.* A U.S. Marine.
**gyr·fal·con** *also* **ger·fal·con** (jûr′făl′kən, -fôl′-, -fô′-) *n.* [ME *gerfaucoun* < OFr. *girfaut,* of Germanic orig.] A large falcon, *Falco rusticolus* of northern regions, with various color phases ranging from black to white.
**gy·ri** (jī′rī′) *n. pl. of* GYRUS.
**gy·ro**[1] (jī′rō) *n., pl.* **-ros. 1.** A gyroscope. **2.** A gyrocompass.
**gy·ro**[2] (jī′rō) *n.* [Mod. Gk. *gurō* < Gk. *guros,* turn.] A sandwich of pita bread filled with roasted meat, esp. lamb, and often vegetables such as onions and tomatoes.
**gyro-** *pref.* [Lat. *guro-,* circle < Gk. *guro-* < *guros.*] **1.** Spinning <*gyro*magnetic> **2.** Circle : spiral <*gyro*plane> **3.** Gyroscope <*gyro*stabilizer>
**gy·ro·com·pass** (jī′rō-kŭm′pəs, -kŏm′-) *n.* A navigational device in which the interaction of a gyroscope's angular momentum with the force produced by the earth's rotation maintains a north-south orientation of the gyroscopic spin axis, thereby providing a stable directional reference.
**gy·ro·cop·ter** (jī′rō-kŏp′tər) *n.* [GYRO- + (HELI)COPTER.] A rotary-wing aircraft driven forward by a conventional propeller.
**gy·ro·mag·net·ic** (jī′rō-măg-nĕt′ĭk) *adj.* Of, relating to, or resulting from the magnetic properties of a spinning, electrically charged particle.
**gyromagnetic ratio** *n.* The ratio of the magnetic moment to the intrinsic angular momentum of a spinning particle.
**gyro pilot** *n.* An automatic pilot incorporating a gyroscope to initiate corrections to aircraft control surfaces and thus maintain a preset course and altitude.
**gy·ro·plane** (jī′rə-plān′) *n.* An aircraft, as a helicopter or autogiro, with wings that rotate about an approx. vertical axis.
**gy·ro·scope** (jī′rə-skōp′) *n.* **1.** A device consisting of a spinning mass, usu. a disk or wheel, the spin axis of which turns between two low-friction supports and maintains its angular orientation with respect to inertial coordinates when not subjected to external torques. **2.** A spinning mass. **—gy′ro·scop′ic** (-skŏp′ĭk) *adj.* **—gy′ro·scop′i·cal·ly** *adv.*
**gy·ro·sta·bi·liz·er** (jī′rō-stā′bə-lī′zər) *n.* A device having a heavy gyroscope whose axis spins in a vertical plane to reduce the side-to-side rolling of a ship or aircraft.
**gy·ro·stat** (jī′rə-stăt′) *n.* A gyrostabilizer. **—gy′ro·stat′ic** *adj.* **—gy′ro·stat′i·cal·ly** *adv.*
**gy·rus** (jī′rəs) *n., pl.* **-ri** (-rī′) [Lat., circle < Gk. *guros.*] Any of the prominent, rounded, elevated convolutions at the surfaces of the cerebral hemispheres.
**gyve** (jīv) *n.* [ME *give.*] *Archaic.* A fetter or shackle, esp. for the leg. —*vt.* **gyved, gyv·ing, gyves.** To shackle or fetter.

# Hh

**h** *or* **H** (āch) *n., pl.* **h's** *or* **H's. 1.** The eighth letter of the English alphabet. **2.** A speech sound represented by the letter *h.* **3.** The eighth in a series. **4.** Something shaped like the letter H.
**H** *symbol for* HYDROGEN.
**ha** *also* **hah** (hä) *interj.* —Used to express surprise, laughter, wonder, triumph, puzzlement, or pique.
**Ha·bak·kuk** (hăb′ə-kŭk′, hə-băk′ək) *n.* [Heb. *Ḥăbhaqqūq,* prob. < *hābhaq,* he embraced.] —See table at BIBLE.
**ha·ba·ne·ra** (hä′bə-nâr′ə, ä′bə-) *n.* [Sp. *(danza) habanera,* Havanan (dance) < *la Habana,* Havana, Cuba.] **1.** A slow dance originating in Cuba. **2.** The music for a habanera, in duple time with a repetitive rhythmic pattern.
**hab·da·lah** *also* **Hab·da·lah** (häv′lä-lä′) *n.* [Heb. *habdālāh,* separation.] A Jewish religious ceremony observed at the end of the Sabbath or a holy day.
**ha·be·as corpus** (hā′bē-əs) *n.* [ME < Med. Lat., you should have the body (from the first words of the writ).] *Law.* A writ issued to bring a person before a court or judge in order to release that person from unlawful restraint or detention.
**hab·er·dash·er** (hăb′ər-dăsh′ər) *n.* [ME < AN *haberdassher.*] **1.** A

dealer in men's furnishings, as shirts, ties, and socks. **2.** *Chiefly Brit.* A dealer in sewing notions and small wares.

**hab·er·dash·er·y** (hăb'ər-dăsh'ə-rē) *n., pl.* **-ies. 1.** The articles sold by a haberdasher. **2.** A haberdasher's shop.

**hab·er·geon** (hăb'ər-jən) *n.* [ME < OFr. *hauberjon* < *hauberc,* hauberk. —see HAUBERK.] **1.** A short, sleeveless coat of mail worn in medieval times. **2.** A hauberk.

**hab·ile** (hăb'ĭl) *adj.* [Fr. < Lat. *habilis* < *habēre,* to handle.] Adroit.

**ha·bil·i·ment** (hə-bĭl'ə-mənt) *n.* [ME *habylement* < OFr. *habillement* < *habiller,* to clothe < *habile,* habile.] **1.** *often* **habiliments. a.** The attire associated with an office, rank, or occasion. **b.** Clothes. **2. habiliments.** Characteristic equipment or furnishings : TRAPPINGS <surrounded by the *habiliments* of the theater>

**ha·bil·i·tate** (hə-bĭl'ĭ-tāt') *v.* **-tat·ed, -tat·ing, -tates.** [LLat. *habilitare, habilitat-,* to enable < Lat. *habilitas,* ability < *habilis,* able < *habēre,* to handle.] —*vt.* **1.** To clothe. **2.** To fit out or equip (a mine) for operation. **3.** *Obs.* To give an ability or capacity to : QUALIFY. —*vi.* To qualify oneself for a position or office. **—ha·bil'i·ta'tion** *n.*

**hab·it** (hăb'ĭt) *n.* [ME, clothing < OFr., custom < Lat. *habitus,* condition < p.part. of *habēre,* to have.] **1. a.** A continual, often involuntary or unconscious inclination to perform an activity, acquired through frequent repetition. **b.** An established disposition of the mind or character. **2.** Customary practice or manner <a person of unselfish *habits*> **3.** An addiction. **4.** Physical constitution. **5.** Typical appearance, form, or manner of growth, esp. of a plant. **6. a.** Distinctive clothing, esp. that worn by members of a religious order. **b.** A riding habit. —*vt.* **-it·ed, -it·ing, -its.** To clothe.

☆ ***syns:*** HABIT, CHARACTERISTIC, PATTERN, TRAIT *n. core meaning* : an activity done without thinking <a bad *habit* of interrupting others>

**hab·it·a·ble** (hăb'ĭ-tə-bəl) *adj.* [ME < OFr. < Lat. *habitabilis* < *habitare,* to dwell, freq. of *habēre,* to have.] Fit to live in. **—hab'it·a·bil'i·ty, hab'it·a·ble·ness** *n.* **—hab'it·a·bly** *adv.*

**hab·i·tant** (hăb'ĭ-tənt) *n.* [OFr. < pr.part. of *habiter,* to dwell < Lat. *habitare.* —see HABITABLE.] **1.** An inhabitant. **2.** *also* **ha·bi·tan** (ä'-bē-täN'). A person of French descent in Canada or Louisiana who belongs to the small farmer class.

**hab·i·tat** (hăb'ĭ-tăt') *n.* [Lat., it dwells < *habitare,* to dwell. —see HABITABLE.] **1.** The environment in which an organism or biological population usu. lives or grows. **2.** The place where something is most apt to be found.

**hab·i·ta·tion** (hăb'ĭ-tā'shən) *n.* [ME *habitacioun* < Lat. *habitatio* < *habitatus,* p.part. of *habitare,* to dwell. —see HABITABLE.] **1.** The act of inhabiting or state of being inhabited. **2. a.** Natural environment or locality. **b.** Dwelling place.

**hab·it-form·ing** (hăb'ĭt-fôr'mĭng) *adj.* **1.** Leading to or causing physiological addiction <a *habit-forming* substance> **2.** Tending to become habitual.

**ha·bit·u·al** (hə-bĭch'o͞o-əl) *adj.* **1.** Of the nature of a habit. **2.** Acting in a certain way by habit : INVETERATE. **3.** Established by long use : USUAL <one's *habitual* chair by the fireplace> **—ha·bit'u·al·ly** *adv.* **—ha·bit'u·al·ness** *n.*

**ha·bit·u·ate** (hə-bĭch'o͞o-āt') *v.* **-at·ed, -at·ing, -ates.** [LLat. *habituare, habituat-,* to bring into a condition < Lat. *habitus,* condition. —see HABIT.] —*vt.* To make familiar by frequent repetition or prolonged exposure : ACCUSTOM. —*vi.* To develop a tolerance or psychological dependence through frequent use. **—ha·bit'u·a'tion** *n.*

**hab·i·tude** (hăb'ĭ-to͞od', -tyo͞od') *n.* [ME < Lat. *habitudo,* condition < *habitus.* —see HABIT.] A usual behavior or manner.

**ha·bit·u·é** (hə-bĭch'o͞o-ā') *n.* [Fr. < p.part. of *habituer,* to frequent < LLat. *habituare,* to be in a condition. —see HABITUATE.] One who frequents a particular place or type of place.

**hab·i·tus** (hăb'ĭ-təs) *n., pl.* **habitus.** [Lat., condition. —see HABIT.] Physical and constitutional characteristics, esp. in regard to susceptibility to a disease.

**ha·ček** (hä'chĕk') *n.* [Czech *háček.*] A diacritical mark (ˇ) that looks like an inverted circumflex and is used over certain letters to indicate quality of pronunciation.

**ha·chure** (hă-sho͝or', hăsh'o͝or) *n.* [Fr. < *hacher,* to crosshatch. —see HASH[1].] One of the short lines used to shade or indicate slopes on maps and also their degree and direction. —*vt.* (hă-sho͝or') **-chured, -chur·ing, -chures.** To make hatching on (a map).

**ha·ci·en·da** (hä'sē-ĕn'də) *n.* [Sp. < Lat. *facienda,* things to be done, neuter pl. gerund. of *facere,* to do.] **1.** A large estate or plantation in Spanish-speaking countries. **2.** The residence of a hacienda owner.

**hack[1]** (hăk) *v.* **hacked, hack·ing, hacks.** [ME *hakken* < OE *haccian.*] —*vt.* **1.** To cut or chop with repeated irregular blows. **2.** To break up the surface of (land) into clods or ridges. **3.** To cut or mutilate as if by hacking <*hacked* the manuscript in half> **4.** *Informal.* To cope with successfully : MANAGE <couldn't *hack* the extra responsibility> —*vi.* **1.** To chop or cut by hacking. **2.** To cough roughly or harshly. —*n.* **1.** A rough, irregular cut made by hacking. **2.** A tool, as a hoe, used for chopping or breaking up something. **3.** A blow made by hacking. **4.** A rough, dry cough. **—hack'er** *n.*

**hack[2]** (hăk) *n.* [Short for HACKNEY.] **1.** A horse used for riding or driving : HACKNEY. **2.** An old or worn-out horse. **3. a.** One who does unpleasant or distasteful tasks for money or reward : HIRELING. **b.** A writer hired to produce routine, often formulaic writing. **4.** A coach or carriage for hire. **5.** *Informal.* **a.** A taxicab. **b.** The driver of a taxicab. —*v.* **hacked, hack·ing, hacks.** —*vt.* **1.** To let out (a horse) for hire. **2.** To make banal or trite with indiscriminate use. —*vi.* **1.** *Informal.* To work as the driver of a taxicab. **2.** *Informal.* To work for hire as a writer. **3.** To ride on horseback at an ordinary pace. —*adj.* **1.** By, typical of, or designating a hack <*hack* writing> **2.** Banal : trite.

**hack·a·more** (hăk'ə-môr', -mōr') *n.* [Alteration of Sp. *jaquima,* halter < OSp. *xaquima* < Ar. *shakīmah,* bit of a bridle.] A rope or rawhide halter with a wide band that can be lowered over a horse's eyes, used in breaking horses to a bridle.

**hack·ber·ry** (hăk'bĕr'ē) *n.* [Alteration of obs. *hagberry,* of Scandinavian orig.] **1.** A tree or shrub of the genus *Celtis,* with berrylike, often edible fruit. **2.** The fruit of a hackberry. **3.** The soft, yellowish wood of a hackberry.

**hack·but** (hăk'bŭt') *also* **hag·but** (hăg'-) *n.* [OFr. *haquebute* < MLG *hakebusse.* —see HARQUEBUS.] A harquebus. **—hack'but·eer'** (-bə-tîr'), **hack'but'ter** *n.*

**hack·er** (hăk'ər) *n. Slang.* One who gains unauthorized, usu. nonfraudulent access to another's computer system.

**hack·ie** (hăk'ē) *n. Slang.* A taxicab driver.

**hack·le[1]** (hăk'əl) *n.* [ME *hakell,* of OE orig.] **1.** One of the long, slender, often glossy feathers on the neck of a bird, esp. a male domestic fowl. **2. hackles.** The erectile hairs at the back of the neck, esp. of a dog. **3.** A tuft of feathers on an artificial fishing fly. —*vt.* **-led, -ling, -les.** To trim (a fly) with a hackle. **—get (one's) hackles up.** To be ready to fight.

**hack·le[2]** (hăk'əl) *v.* **-led, -ling, -les.** [Freq. of HACK[1].] —*vt.* To chop roughly. —*vi.* To hack.

**hack·ly** (hăk'lē) *adj.* [< HACKLE[2].] Nicked or notched : JAGGED.

**hack·man** (hăk'mən) *n.* The driver of a hack or hired carriage.

**hack·ma·tack** (hăk'mə-tăk') *n.* [Of Algonquian orig.] The tamarack.

**hack·ney** (hăk'nē) *n., pl.* **-neys.** [ME *hakenei,* prob. after *Hakenei,* Hackney, a borough in England where such horses were raised.] **1.** *often* **Hackney.** A horse of a breed developed in England, with a gait marked by pronounced flexion of the knee. **2.** A horse fit for routine riding or driving. **3.** A coach or carriage for hire. —*vt.* **-neyed, -ney·ing, -neys. 1.** To cause to become banal or trite through overuse. **2.** To hire out. —*adj.* **1.** Banal : trite. **2.** Hired.

**hack·neyed** (hăk'nēd) *adj.* Overused : trite.

**hack·saw** (hăk'sô') *n.* A saw having a strong, fine-toothed blade stretched taut in a frame, used for cutting metal.

**hack·work** (hăk'wûrk') *n.* Commissioned work, as writing, usu. done quickly and by formula.

**had** (hăd) *v. p.t. & p.p. of* HAVE.

**ha·dal** (hād'l) *adj.* [Fr. < *Hadès,* Hades.] Of or pertaining to the ocean depths below 6,000 meters or approx. 20,000 feet.

**had·dock** (hăd'ək) *n., pl.* **haddock** *or* **-docks.** [ME *haddok.*] A food fish, *Melanogrammus aeglefinus* of northern Atlantic waters, resembling the cod.

**hade** (hād) *n.* [Orig. unknown.] The angle of inclination from the vertical of a fault, vein, or lode.

**Ha·des** (hā'dēz) *n.* [Gk. *Haidēs.*] **1.** *Gk. Myth.* The god of the netherworld and distributor of earthly riches. **2.** *Gk. Myth.* The netherworld kingdom of Hades, the abode of the dead. **3.** *also* **hades.** Hell.

**hadj** (hăj) *n. var. of* HAJ.

**hadj·i** (hăj'ē) *n. var. of* HAJI.

**had·n't** (hăd'nt). Had not.

**had·ron** (hăd'rŏn') *n.* [Gk. *hadros,* thick + -ON[1].] Any of a class of elementary particles that participate in strong interactions. **—had·ron'ic** *adj.*

**hadst** (hădst) *v. Archaic. 2nd person sing. p.t. of* HAVE.

**hae** (hā, hă) *vt.* **haed, haen** (hān, hăn), **hae·ing, haes.** *Scot.* To have.

**-haemia** *suff. var. of* -EMIA.

**haet** (hāt) *n.* [Sc., short for *hae it,* take it.] *Scot.* A tiny amount.

**ha·fiz** (hä'fĭz) *n.* [Pers. < Ar. *hāfiz* < *hafiza,* he memorized.] **1.** A Moslem who has memorized the Koran. **2.** A title of respect used with the name of a Moslem who has memorized the Koran.

**haf·ni·um** (hăf'nē-əm) *n.* [NLat. < *Hafnia,* Lat. name for Copenhagen, Denmark.] *Symbol* **Hf** A brilliant, silvery, metallic element used in nuclear reactor control rods and in the manufacture of tungsten filaments; atomic number 72; atomic weight 178.49.

**haft** (hăft) *n.* [ME < OE *hæft.*] A handle or hilt, esp. a handle of a tool or weapon. —*vt.* **haft·ed, haft·ing, hafts.** To fit into or provide with a haft.

**haf·ta·rah** *or* **haf·to·rah** (häf'tə-rä', häf-tôr'ə) *n. vars. of* HAPHTARAH.

**hag[1]** (hăg) *n.* [ME *hagge,* perh. short for OE *hægtesse,* witch.] **1.** An ugly old woman. **2.** A witch : sorceress. **3.** *Obs.* A woman demon.

**hag[2]** (hăg) *n.* [Of Scandinavian orig.] *Chiefly Brit.* **1.** A bog : quagmire. **2.** A spot in boggy land softer or more solid than the surrounding area. **3.** A cutting in a peat bog.

**hag·born** (hăg′bôrn′) *adj.* Born of a witch or hag.
**hag·but** (hăg′bŭt′) *n. var. of* HACKBUT.
**hag·fish** (hăg′fĭsh′) *n., pl.* **hagfish** *or* **-fish·es.** [HAG¹ + FISH.] Any of various primitive, eel-shaped marine fishes of the family Myxinidae, having a jawless sucking mouth with rasping teeth.
**Hag·ga·dah** *also* **Hag·ga·da** (hə-gä′də, -gô′də) *n., pl.* **-doth** (-dōt′, -dōth′) [Heb. *haggādāh* < *higgidh,* he narrated.] **1.** Traditional Jewish literature, esp. the nonlegal part of the Talmud. **2.** The book having the story of the Exodus and the ritual of the Seder, read at the Passover Seder. **—hag·gad′ic** (-găd′ĭk, -gä′dĭk, -gô′dĭk) *adj.*
**hag·ga·dist** (hə-gä′dĭst, -gô′-) *n.* **1.** A haggadic writer. **2.** A student of haggadic literature. **—hag′ga·dis′tic** (hăg′ə-dĭs′tĭk) *adj.*
**Hag·ga·i** (hăg′ē-ī′, hăg′ī′) *n.* [Heb.] —See table at BIBLE.
**hag·gard** (hăg′ərd) *adj.* [OFr. *hagard,* wild.] **1. a.** Worn, pale, and exhausted **:** GAUNT. **b.** Wild and unruly **:** UNCONTROLLED. **2.** Wild and intractable. —Used of a hawk in falconry. —*n.* An adult hawk captured for training. **—hag′gard·ly** *adv.* **—hag′gard·ness** *n.*
☆ ***syns:*** HAGGARD, CAREWORN, GAUNT, WAN, WORN *adj. core meaning* **:** pale and exhausted because of worry, disease, hunger, or fatigue <the *haggard* appearance of the hobo>
**hag·gis** (hăg′ĭs) *n.* [ME *hagese.*] A Scottish dish made of a mixture of the minced heart, lungs, and liver of a sheep or calf blended with suet, onions, oatmeal, and seasonings and boiled in the stomach of the animal.
**hag·gish** (hăg′ĭsh) *adj.* Of or typical of a hag. **—hag′gish·ly** *adv.* **—hag′gish·ness** *n.*
**hag·gle** (hăg′əl) *v.* **-gled, -gling, -gles.** [Freq. of dial. *hag,* to cut < ME *haggen* < ON *hǫggva.*] —*vi.* **1.** To bargain, as over the price of something **:** DICKER. **2.** To argue in an attempt to come to terms. —*vt.* **1.** To cut in a rough, unskillful manner **:** HACK. **2.** *Archaic.* To pester or worry by wrangling. —*n.* An instance of haggling. **—hag′gler** *n.*
**hagi-** *pref. var. of* HAGIO-.
**hag·i·ar·chy** (hăg′ē-är′kē, hā′jē-) *also* **hag·i·oc·ra·cy** (hăg′ē-ŏk′rə-sē, hā′jē-) *n., pl.* **-chies** *also* **-cies.** Government by holy men, as priests or saints.
**hagio-** *or* **hagi-** *pref.* [LLat. < Gk. < *hagios,* holy.] **1.** Saint <*hagiography*> **2.** Holy <*hagioscope*>
**hag·i·oc·ra·cy** (hăg′ē-ŏk′rə-sē, hā′jē-) *n. var. of* HAGIARCHY.
**Hag·i·og·ra·pha** (hăg′ē-ŏg′rə-fə, hā′jē-) *pl.n.* [LLat. < Gk. : *hagio-,* sacred + *-graphos,* written.] *(sing. or pl. in number).* The third of the three ancient Jewish divisions of the Old Testament, containing those books not in the Law (Torah) or the Prophets and usu. including the Psalms, Proverbs, Job, the Song of Solomon, Ruth, Lamentations, Ecclesiastes, Esther, Daniel, Ezra, Nehemiah, and Chronicles.
**hag·i·og·ra·phy** (hăg′ē-ŏg′rə-fē, hā′jē-) *n., pl.* **-phies. 1.** Biography of saints. **2.** A worshipful or idealizing biography. **—hag′i·og′rapher** *n.* **—hag′i·o·graph′ic** (-ə-grăf′ĭk), **hag′i·o·graph′i·cal** *adj.*
**hag·i·ol·o·gy** (hăg′ē-ŏl′ə-jē, hā′jē-) *n., pl.* **-gies. 1.** Literature treating the lives of saints. **2.** A collection of sacred writings. **3.** An authoritative list of saints. **—hag′i·o·log′ic** (-ə-lŏj′ĭk), **hag′i·o·log′i·cal** *adj.* **—hag′i·ol′o·gist** *n.*
**hag·i·o·scope** (hăg′ē-ə-skōp′, hā′jē-) *n.* A small opening in an interior wall of a church giving those in the transept a view of the main altar. **—hag′i·o·scop′ic** (-skŏp′ĭk) *adj.*
**hag·rid·den** (hăg′rĭd′n) *adj.* **1.** Harassed by or as if by a witch. **2.** Tormented or harassed, as by unreasonable fears.
**hah** (hä) *interj. var. of* HA.
**ha-ha¹** (hä′hä′) *n.* A sound made in imitation of laughter.
**ha-ha²** (hä′hä′) *also* **haw-haw** (hô′hô′) *n.* [Fr.] A moat, walled ditch, or hedge that serves as a fence without impairing the view.
**Hai·da** (hī′də) *n., pl.* **Haida** *or* **-das. 1. a.** Any of the Indian tribes inhabiting the Queen Charlotte Islands, British Columbia, and Prince of Wales Island, Alaska. **b.** A member of any of these tribes. **2. a.** A language family of the Na-dene phylum. **b.** The language of the Haida and the only surviving language of the Haida language family. **—Hai′dan** *adj.*
**haik** (hīk, hāk) *n.* [Ar. *ḥā'ik* < *ḥāka,* he wove.] A large piece of cotton, silk, or wool cloth worn by Arabs as an outer garment.

**haik**

**hai·ku** (hī′ko͞o) *n., pl.* **haiku.** [J. : *hai,* amusement (< Chin. *pa²,* parallel) + *ku,* sentence (< Chin. *ju⁴*).] An unrhymed Japanese lyric poem having a fixed three-line form containing five, seven, and five syllables respectively.
**hail¹** (hāl) *n.* [ME < OE *hægel.*] **1.** Precipitation of small pellets of ice and hard snow. **2.** Something that falls or pours out with the force of a shower of hail <a *hail* of bullets> —*v.* **hailed, hail·ing, hails.** —*vi.* **1.** To precipitate. **2.** To fall like hail. —*vt.* To pour down or forth <*hail* oaths at one's tormentor>
**hail²** (hāl) *v.* **hailed, hail·ing, hails.** [ME *heilen* < *(wæs) hæil,* (be) healthy. —see WASSAIL.] —*vt.* **1. a.** To salute or greet **:** WELCOME. **b.** To greet or acclaim enthusiastically. **2.** To call out to or signal in order to catch the attention of <*hail* a taxi> —*vi.* To signal or call to a passing ship as a greeting or identification of oneself. **—hail from.** To come or originate from. —*n.* **1.** The act of hailing. **2.** A shout made to greet or catch the attention of someone. **3.** Hearing distance <commanded us to stay within *hail*> —*interj.* —Used to express a greeting or tribute.
**hail·er** (hā′lər) *n.* **1.** One that hails. **2.** A bullhorn.
**hail-fel·low** (hāl′fĕl′ō) *also* **hail-fel·low-well-met** (hāl′fĕl′ō-wĕl′mĕt′) *adj.* [From the obs. greeting *hail, fellow!*] Heartily friendly and sociable. **—hail′-fel′low** *n.*
**Hail Mary** *n.* A Roman Catholic prayer based on the greetings of Gabriel and Saint Elizabeth to the Virgin Mary.
**hail·stone** (hāl′stōn′) *n.* A hard pellet of snow and ice.
**hail·storm** (hāl′stôrm′) *n.* A storm with hail.
**hair** (hâr) *n.* [ME *her* < OE *hǣr.*] **1. a.** One of the fine, cylindrical, often pigmented filaments growing from the epidermis of a mammal. **b.** A growth of such filaments, as that forming an animal's coat or covering a human being's scalp. **2.** A filamentous projection or bristle resembling a hair, as a seta of an arthropod or an epidermal process of a plant. **3.** Fabric woven from the hair of certain animals <a coat of camel's *hair*> **4.** A tiny distance or narrow margin <win the election by a *hair*> **5.** An exact degree <calculated the distance to a *hair*> **—let (one's) hair down.** To drop or lose one's reserve or inhibitions. **—split hairs.** To make fine, usu. petty distinctions.
**hair·ball** (hâr′bôl′) *n.* A small mass of hair ingested by an animal, often causing indigestion or convulsions.
**hair·breadth** (hâr′brĕdth′) *adj.* Extremely close <a *hairbreadth* escape> —*n. var. of* HAIRSBREADTH.
**hair·brush** (hâr′brŭsh′) *n.* A brush for the hair.
**hair·cloth** (hâr′klôth′, -klŏth′) *n.* A wiry fabric woven esp. from horsehair or camel's hair, used in upholstery and for stiffening garments.
**hair·cut** (hâr′kŭt′) *n.* **1.** An act or instance of cutting or trimming the hair. **2.** The style in which hair is cut.
**hair·do** (hâr′do͞o′) *n., pl.* **-dos.** The style in which hair is arranged.
**hair·dress·er** (hâr′drĕs′ər) *n.* One who cuts or arranges hair.
**hair·dress·ing** (hâr′drĕs′ĭng) *n.* **1.** The occupation of a hairdresser. **2.** The act or an instance of dressing or arranging the hair. **3.** A cosmetic or medicinal preparation used on the hair.
**haired** (hârd) *adj.* Having a specified type of hair <a short-*haired* breed of dogs>
**hair·less** (hâr′lĭs) *adj.* Having little or no hair.
**hair·line** (hâr′līn′) *n.* **1.** The outline of the growth of hair on the scalp, esp. across the front. **2.** A very thin line. **3. a.** A fine line on a typeface. **b.** A style of type using such lines. **4. a.** A textile design with thin, hairlike stripes. **b.** A usu. worsted fabric with such stripes.
**hair piece** *n.* A covering or bunch of human or artificial hair used to cover baldness or give shape to a hairdo.
**hair·pin** (hâr′pĭn′) *n.* **1.** A thin, cylindrical strip of metal or other material bent in a long U shape, used to secure a hairdo or a headdress. **2.** Something shaped like a hairpin, esp. a sharp turn in a road.
**hair-rais·er** (hâr′rā′zər) *n.* Something causing great excitement, terror, or thrills.
**hair-rais·ing** (hâr′rā′zĭng) *adj.* Causing excitement, horror, or thrills.
**hairs·breadth** *or* **hair's-breadth** (hârz′brĕdth′) *also* **hair·breadth** (hâr′brĕdth′) *n.* A small space, extent, or margin.
**hair seal** *n.* A seal of the family Phocidae, having a stiff hairlike coat and ears visible only as small indentations.
**hair shirt** *n.* A coarse haircloth garment worn next to the skin as penance.
**hair space** *n.* The narrowest space used in printing to separate words or letters.
**hair·split·ting** (hâr′splĭt′ĭng) *n.* The act or process of making unreasonably fine distinctions **:** QUIBBLING. **—hair′split′ter** *n.*
**hair spray** *n.* A commercial product sprayed on the hair to keep it in place.
**hair·spring** (hâr′sprĭng′) *n.* A fine coiled spring regulating the movement of the balance wheel in a watch or clock.
**hair·streak** (hâr′strēk′) *n.* A butterfly of the subfamily Theclinae, with transverse streaks on the underwings and fine, hairlike projections on the hind wings.
**hair stroke** *n.* A fine line in writing or printing.
**hair style** *n.* Hairdo. **—hair stylist** *n.*
**hair trigger** *n.* A gun trigger adjusted to respond to a very slight pressure.

**hair-trig·ger** (hâr′trĭg′ər) *adj.* Responding to the least provocation <*a hair-trigger* temper>
**hair-weav·ing** (hâr′wē′vĭng) *n.* The act or process of interweaving a hair piece of human hair with a balding person's own hair.
**hair·worm** (hâr′wûrm′) *n.* **1.** A slender, parasitic nematode worm of the genus *Trichostrongylus* that infests the stomach and small intestine of animals, as cattle. **2.** A horsehair worm.
**hair·y** (hâr′ē) *adj.* **-i·er, -i·est. 1.** Covered with hair or hairlike projections. **2.** Of or resembling hair <a *hairy* coat> **3.** *Slang.* Fraught with difficulties : HAZARDOUS <a *hairy* escape> **—hair′i·ness** *n.*
**Hai·tian** (hā′shən, -tē-ən) *adj.* Of or pertaining to Haiti, its people, or its dialect. —*n.* **1.** A native or resident of Haiti. **2.** Haitian Creole.
**Haitian Creole** *n.* A language spoken by the majority of Haitians that is based on French and various African languages.
**haj** *or* **hajj** *also* **hadj** (hăj) *n.* [Ar. *ḥajj* < *ḥajja*, he went on a pilgrimage.] A pilgrimage to Mecca made during Ramadan.
**haj·i** *or* **haj·ji** *also* **hadj·i** (hăj′ē) *n.* [Ar. *ḥājjī* < *ḥajj*, pilgrimage. —see HAJ.] One who has made a pilgrimage to Mecca during Ramadan. —Often used as a title of address.
**hake** (hāk) *n., pl.* **hake** *or* **hakes.** [ME, poss. < OE *haca*, hook (from the shape of its lower jaw).] A marine food fish of the genera *Merluccius* or *Urophycis*, resembling the cod.
**Ha·ken·kreuz** (hä′kən-kroits′) *n.* [G. : *Haken*, hook + *Kreuz*, cross.] A swastika used as a symbol of Nazi Germany or anti-Semitism.
**ha·kim¹** (hä′kēm) *n.* [Ar. *ḥakīm* < *ḥakama*, he was wise.] A Moslem doctor.
**ha·kim²** (hä′kĭm) *n., pl.* **hakim** *or* **-kims.** [Ar. *ḥākim* < *ḥakama*, to exercise authority.] A Moslem ruler, provincial governor, or judge.
**hal-** *pref. var. of* HALO-.
**Ha·la·kah** *also* **Hal·la·cha** (hä′lä-KHä′, hä-lä′KHə) *n.* [Heb. *halākhāh*, tradition < *hālakh*, he went.] The legal part of Talmudic literature that is an interpretation of the laws of the Scriptures. **—Ha·lak′ic** (hə-lăk′ĭk) *adj.*
**ha·la·tion** (hā-lā′shən) *n.* [HAL(O) + -ATION.] **1.** A blurring or spreading of light around bright objects or areas on a photographic image. **2.** A ring of light around a bright object on a television screen.
**ha·la·vah** (hä′lə-vä′) *n. var. of* HALVAH.
**hal·berd** (hăl′bərd, hôl′-) *also* **hal·bert** (-bərt) *n.* [OFr. *hallebarde* < MHG *helmbarde* : *helm*, handle + *barte*, ax < OHG *barta*.] A 15th- and 16th-cent. weapon with an axlike blade and a steel spike mounted on the end of a long shaft. **—hal′ber·dier′** (-bər-dîr′) *n.*
**hal·cy·on** (hăl′sē-ən) *n.* [ME *alcioun* < Lat. *alcyon* < Gk. *alkuōn*, a mythical bird.] **1.** A fabled bird, identified with the kingfisher, regarded as having the power to calm the wind and the waves during the winter solstice while it nested on the sea. **2.** A kingfisher. —*adj.* **1.** Calm and peaceful : TRANQUIL. **2.** Prosperous : golden <our *halcyon* days>
**Hal·cy·o·ne** (hăl-sī′ə-nē′) *n. var. of* ALCYONE.
**hale¹** (hāl) *adj.* **hal·er, hal·est.** [ME < OE *hāl*.] Not infirm : HEALTHY. **—hale′ness** *n.*
**hale²** (hāl) *vt.* **haled, hal·ing, hales.** [ME *halen* < OFr. *haler*, of Germanic orig.] **1.** To force to go <*haled* us into court> **2.** *Archaic.* To pull, drag, draw, or hoist.
**ha·ler** (hä′lər, -lĕr′) *n., pl.* **-lers** *or* **-le·ru** (-lə-rōō′) [Czech < MHG *haller*, a coin < *Hall*, a town in Germany where it was once minted.] —See table at CURRENCY.
**half** (hăf, häf) *n., pl.* **halves** (hăvz, hävz) [ME < OE *healf*.] **1. a.** One of two equal parts that together make up a whole. **b.** A part of something approx. equal to the remainder. **2.** *Informal.* A half dollar. **3.** One of the two playing periods into which certain games are divided. **4.** *Chiefly Brit.* A school term : SEMESTER. **5.** Half an hour <at *half* past one> ***usage:*** The phrases *a half, half of a*, and *half a* are all correct; although the phrase *a half a* occurs frequently, esp. in speech, it is unacceptable to some. —*adj.* **1.** Being one of two equal parts. **2.** Being approx. a half. **3.** Partial or incomplete <a *half* wave to the onlookers> **4.** Having only one parent in common with another person. —*adv.* **1.** To the extent of exactly or nearly 50% <a *half*-empty glass> **2.** Not entirely or sufficiently : PARTLY <*half* asleep>
**half-and-half** (hăf′ənd-hăf′, häf′ənd-häf′) *adj.* Being half one thing and half another. —*adv.* In equal portions. —*n.* **1.** A blend of two things in equal portions, esp. such a mixture of milk and cream. **2.** *Chiefly Brit.* A blend of malt liquors, esp. porter and ale.
**half·back** (hăf′băk′, häf′-) *n.* **1.** *Football.* One of the two players positioned near the flanks behind the line of scrimmage. **2.** One of several players stationed behind the forward line in various sports. **3.** The position played by a halfback.
**half-baked** (hăf′bākt′, häf′-) *adj.* **1.** Only partially baked. **2.** *Informal.* Not sufficiently thought out : FOOLISH <a *half-baked* idea> **3.** *Informal.* Lacking common sense <a *half-baked* visionary>
**half binding** *n.* A book binding in which the back and often the corners of the volume are bound in a material differing from the rest of the cover.
**half blood** *also* **half-blood** (hăf′blŭd′, häf′-) *n.* **1. a.** The relationship between persons having only one parent in common. **b.** A person having such a relationship. **2.** A half-blooded animal.
**half-blood·ed** (hăf′blŭd′ĭd, häf′-) *adj.* **1.** Having only one parent in common. **2.** Having parents of different ethnic types. **3.** Having one parent of pedigreed stock and the other of unknown or mixed ancestry. —Used of animals.
**half boot** *n.* A low boot extending just above the ankle.
**half-bound** (hăf′bound′, häf′-) *adj.* Having a half binding.
**half-bred** (hăf′brĕd′, häf′-) *adj.* Having only one purebred parent. —Used of animals.
**half-breed** (hăf′brēd′, häf′-) *adj.* Half-blooded : hybrid.
**half brother** *n.* A brother related through one parent only.
**half-caste** (hăf′kăst′, häf′-) *n.* A person of mixed racial descent. —*adj.* Of mixed racial descent.
**half cock** *n.* The position of a firearm's hammer when it is raised halfway and locked by a catch so the trigger cannot be pulled.
**half-cocked** (hăf′kŏkt′, häf′-) *adj.* **1.** At the position of half cock. **2.** *Informal.* Inadequately prepared or conceived.
**half-crown** (hăf′kroun′, häf′-) *n.* A former British coin worth two shillings and sixpence.
**half dime** *n.* An obsolete U.S. silver coin worth five cents.
**half dollar** *n.* A U.S. silver coin worth 50 cents.
**half eagle** *n.* An obsolete U.S. gold coin worth five dollars.
**half gainer** *n.* A dive in which the diver springs from the board facing forward, rotates backward in the air in a half backward somersault, and enters the water headfirst, facing the diving board.
**half-heart·ed** (hăf′här′tĭd, häf′-) *adj.* Done with or having little interest, enthusiasm, or warmth : UNINSPIRED. **—half′heart′ed·ly** *adv.* **—half′heart′ed·ness** *n.*
**half hitch** *n.* A hitch made by looping a rope or strap around an object and then back around itself, bringing the end of the rope through the loop.
**half-hour** (hăf′our′, häf′-) *n.* **1.** A period of 30 minutes. **2.** The middle point of an hour. **—half′-hour′ly** *adj. & adv.*
**half-in·te·gral** (hăf′ĭn′tĭ-grəl, häf′-) *adj.* Having an integer as a numerator and 2 as a denominator, as a fraction.
**half-length** (hăf′lĕngkth′, -lĕngth′, häf′-) *n.* A portrait depicting only the upper half and hands of a person. —*adj.* **1.** Of or denoting a half-length portrait. **2.** Of half the full length.
**half-life** (hăf′līf′, häf′-) *n.* **1.** *Physics.* The time necessary for half the nuclei in a sample of a specific isotopic species to undergo radioactive decay. **2.** *Biol.* **a.** The time required by living tissue, an organ, or an organism to eliminate by biological processes half the quantity of a substance taken in. **b.** The time required for the radioactivity of material taken in by a living organism to be reduced to half its initial value by a combination of biological elimination processes and radioactive decay.
**half-light** (hăf′līt′, häf′-) *n.* Soft, subdued light, as that found at dusk or dawn or in dimly lit interiors.
**half-line** (hăf′līn′, häf′-) *n.* A straight line extending in only one direction from a given point.
**half-mast** (hăf′măst′, häf′-) *n.* The position approx. halfway up a pole at which a flag is flown as a symbol of mourning or a signal of distress. —*vt.* **-mast·ed, -mast·ing, -masts.** To place (a flag) at half-mast.
**half-moon** (hăf′mōōn′, häf′-) *n.* **1.** The moon when only half its disk is illuminated. **2.** Something shaped like a crescent.
**half nelson** *n.* A wrestling hold in which one arm is passed under the opponent's arm from behind to the back of his neck.
**half note** *n.* *Mus.* A note having one half the value of a whole note.
**half·pen·ny** (hā′pə-nē, hāp′nē) *n.* **1.** *pl.* **half·pence** (hā′pəns) *or* **half·pen·nies.** A British coin worth one half of a penny. **2.** The sum of one half of a penny.
**half pint** *n.* *Slang.* A small person or animal.
**half relief** *n.* Sculptural relief having modeled forms that project approx. halfway from the background.
**half sister** *n.* A sister related through one parent only.
**half-slip** (hăf′slĭp′, häf′-) *n.* A woman's underskirt extending from the waist to the hem of an outer garment.
**half sole** *n.* A shoe sole covering the shank to the toe.
**half-sole** (hăf′sōl′, häf′-) *vt.* **-soled, -sol·ing, -soles.** To fit or repair with a half sole.
**half sovereign** *n.* A former British gold coin worth ten shillings.
**half-staff** (hăf′stăf′, häf′-) *n.* Half-mast.
**half step** *n.* **1.** A semitone. **2.** A marching step of 15 inches at quick time and 18 at double time.
**half-tim·bered** (hăf′tĭm′bərd, häf′-) *also* **half-tim·ber** (-bər) *adj.* Having a wooden framework with plaster or masonry filling the spaces.
**half time** *n.* The intermission between halves in certain sports.
**half title** *n.* The title of a book printed at the top of the first page of the text or on a full page before the main title page.
**half·tone** (hăf′tōn′, häf′-) *n.* **1.** A tone or value halfway between a highlight and a dark shadow. **2. a.** A picture in which the gradations

of light are shown by the relative darkness and density of tiny dots produced by photographing the subject through a fine screen. **b.** A picture made by such a process.

**half tone** *n.* A semitone.

**half-track** (hăf′trăk′, häf′-) *n.* A military motor vehicle, usu. lightly armored, with caterpillar treads instead of wheels. **—half′-track′, half′-tracked′** *adj.*

**half-truth** (hăf′trōōth′, häf′-) *n.* A statement, esp. one meant to deceive, that omits some of the facts necessary for a truthful description or account.

**half volley** *n.* A stroke in certain games, as tennis, in which the ball is hit immediately after it bounces off the ground.

**half·way** (hăf′wā′, häf′-) *adj.* **1.** Midway between two points or conditions. **2.** Reaching or including only half or a portion : PARTIAL <*halfway* measures> **—half′way′** *adv.*

**halfway house** *n.* **1.** An inn or other stopping place at the midpoint of a journey. **2.** A rehabilitation center where people who have left an institution, as a prison or hospital, are helped to readjust to the outside world.

**half-wit** (hăf′wĭt′, häf′-) *n.* **1.** A mentally deficient person. **2.** A foolish or stupid person. **—half′-wit′ted** *adj.* **—half′-wit′ted·ly** *adv.* **—half′-wit′ted·ness** *n.*

**hal·i·but** (hăl′ə-bət, hŏl′-) *n., pl.* **halibut** *or* **-buts.** [ME : *hali*, holy (from its being eaten on holy days) + *butte*, flatfish < MDu.] A large edible flatfish of the genus *Hippoglossus* or related genera of northern Atlantic or Pacific waters.

**hal·ide** (hăl′īd′, hā′līd′) *n.* A binary chemical compound of a halogen with a more electropositive element or group.

**hal·i·dom** (hăl′ĭ-dəm) *n.* [ME < OE *haligdom* : *halig*, holy + *-dom*, -dom.] *Obs.* **1.** Something regarded as holy. **2.** A sanctuary.

**hal·ite** (hăl′īt′, hā′līt′) *n.* Rock salt.

**hal·i·to·sis** (hăl′ĭ-tō′sĭs) *n.* [Lat. *halitus*, breath (< *halare*, to breathe) + -OSIS.] A condition marked by stale or bad-smelling breath.

**hall** (hôl) *n.* [ME *halle* < OE *heall.*] **1.** A corridor or passageway in a building. **2.** A large entrance room or vestibule in a building : LOBBY. **3. a.** A building for public meetings or entertainments. **b.** The large room in which such events are held. **4.** A building used for the meetings, entertainments, or living quarters of a fraternity or other social or religious organization. **5. a.** A school, college, or university building that provides classroom, dormitory, or dining facilities. **b.** A large room in such a building. **c.** The group of students occupying such a building. **d.** *Chiefly Brit.* A meal served in such a building. **6.** The main house on an estate. **7. a.** The house or castle of a medieval monarch or noble. **b.** The large principal room in such a house or castle.

**hal·lah** (KHä′lə, hä′-) *n. var. of* CHALLAH.

**Hall effect** *n.* [After Edwin H. *Hall* (1855–1938).] The generation of an electric potential perpendicular to both an electric current flowing along a thin conducting material and an external magnetic field applied at right angles to the current upon application of the magnetic field.

**hal·lel** (hä-lāl′) *n.* [Heb. *hallēl* < *hĭllēl*, he praised.] A chant of praise used during Passover and on certain other Jewish holidays.

**hal·le·lu·jah** (hăl′ə-lōō′yə) *interj.* [Heb. *hallelūyāh*, praise God : *hallelū*, pl. imper. of *hĭllēl*, he praised + *Yāh*, God.] —Used to express praise or joy. *—n.* **1.** The exclamation of "hallelujah." **2.** A joyful musical composition based on the word "hallelujah."

**Hal·ley's comet** (hăl′ēz) *n.* [After Edmund *Halley* (1656–1742), from his prediction of its return after observing it in 1682.] A comet, last seen in 1910, with a period of approx. 76 years.

**hal·liard** (hăl′yərd) *n. var. of* HALYARD.

**hall·mark** (hôl′märk′) *n.* [After Goldsmith's *Hall* in London, England, where gold and silver articles were appraised and stamped.] **1.** A mark used to stamp esp. gold and silver articles that meet established standards of purity or quality. **2.** A mark indicating quality or excellence. **3.** A distinguishing feature or characteristic <"The sense of guilt is the *hallmark* of civilized humanity" —Theodor Reik> *—vt.* **-marked, -mark·ing, -marks.** To mark with a hallmark.

**hall of fame** *also* **Hall of Fame** *n.* **1.** A building housing memorial items honoring outstanding persons. **2.** A group of persons judged outstanding, as in a sport or profession.

**hal·loo** (hə-lōō′) *also* **hal·loa** (hə-lō′) [Alteration of obs. *holla*, stop! —see HELLO.] *—interj.* **1.** —Used to gain someone's attention. **2.** —Used to urge on hounds in a hunt. *—n.* A shout or call of "halloo." *—v.* **-looed, -loo·ing, -loos** *also* **-loaed, -loa·ing, -loas.** *—vi.* To shout "halloo." *—vt.* **1.** To urge on or pursue by shouting or calling "halloo." **2.** To call out to. **3.** To shout or yell.

**hal·low** (hăl′ō) *vt.* **-lowed, -low·ing, -lows.** [ME *halwen* < OE *hālgian.*] **1.** To make or set apart as holy : CONSECRATE. **2.** To honor or respect : REVERE.

**hal·lowed** (hăl′ōd) *adj.* **1.** Sanctified : consecrated. **2.** Greatly venerated : SACROSANCT.

**Hal·low·een** *also* **Hal·low·e'en** (hăl′ə-wēn′, hŏl′-) *n.* [Short for *All Hallow Even.*] Oct. 31, the eve of All Saints' Day, celebrated by children who beg treats or play pranks.

**Hal·low·mas** *also* **Hal·low·mass** (hăl′ō-məs, -măs′) *n.* [Short for ALLHALLOWMAS.] *Archaic.* The feast of All Saints' Day or Allhallowmas on Nov. 1.

**Hall·statt** (hôl′stăt′, häl′shtät′) *adj.* [After the type-site at *Hallstatt*, Austria.] Of or relating to a dominant Iron Age culture of central and western Europe, prob. chiefly Celtic, that flourished from the 9th to the 5th cent. B.C.

**hal·lu·ces** (hăl′yə-sēz′, hăl′ə-) *n. pl. of* HALLUX.

**hal·lu·ci·nate** (hə-lōō′sə-nāt′) *v.* **-nat·ed, -nat·ing, -nates.** [Lat. *hallucinari, hallucinat-*, to dream.] *—vi.* To experience hallucinations. *—vt.* To cause to have hallucinations.

**hal·lu·ci·na·tion** (hə-lōō′sə-nā′shən) *n.* **1. a.** False or distorted perception of objects or events with an overwhelming sense of their reality, usu. as a product of mental disorder or as a response to a drug. **b.** The material so perceived. **2.** A mistaken or false idea : DELUSION. **—hal·lu′ci·na′tion·al** *adj.* **—hal·lu′ci·na′tive** *adj.*

**hal·lu·ci·na·to·ry** (hə-lōō′sə-nə-tôr′ē, -tōr′ē) *adj.* **1.** Of, characterizing, or marked by hallucination. **2.** Inducing hallucination.

**hal·lu·cin·o·gen** (hə-lōō′sə-nə-jən) *n.* [HALLUCIN(ATION) + -GEN.] A drug inducing hallucination. **—hal·lu′cin·o·gen′ic** (-jĕn′ĭk) *adj.*

**hal·lu·ci·no·sis** (hə-lōō′sə-nō′sĭs) *n.* [HALLUCIN(ATION) + -OSIS.] An abnormal condition or mental state marked by hallucination.

**hal·lux** (hăl′əks) *n., pl.* **hal·lu·ces** (hăl′yə-sēz′, hăl′ə-) [Lat.] **1.** The inner or first digit on the hind foot of a mammal; in humans, the big toe. **2.** The often backward-directed toe of a bird.

**hall·way** (hôl′wā′) *n.* HALL 1, 2.

**halm** (hôm) *n. var. of* HAULM.

**ha·lo** (hā′lō) *n., pl.* **-los** *or* **-loes.** [Med. Lat. < Lat. *halos* < Gk. *halōs.*] **1.** A circular band of colored light around a light source, as around the sun or moon, produced by the refraction and reflection of light by ice particles suspended in the intervening atmosphere. **2. a.** Something resembling a halo. **b.** A nimbus. **3.** The aura of glory surrounding one regarded with reverence, awe, or sentiment. *—vt.* **-loed, -lo·ing, -los** *or* **-loes.** To encircle with or as if with a halo.

**halo-** *or* **hal-** *pref.* [Fr. < Gk. < *hals*, sea, salt.] **1.** Salt <*halophyte*> **2.** Halogen <*halocarbon*>

**hal·o·bi·ont** (hăl′ō-bī′ŏnt′) *n.* An organism existing or growing in a saline environment.

**hal·o·car·bon** (hăl′ə-kär′bən) *n.* A compound made up of carbon and a halogen.

**hal·o·cline** (hăl′ə-klīn′) *n.* A vertical gradient in ocean salinity.

**hal·o·gen** (hăl′ə-jən) *n.* Any of a group of five chemically related nonmetallic elements including fluorine, chlorine, bromine, iodine, and astatine. **—ha·log′e·nous** (hă-lŏj′ə-nəs) *adj.*

**hal·o·ge·nate** (hăl′ə-jə-nāt′) *vt.* **-nat·ed, -nat·ing, -nates.** To treat or mix with a halogen. **—hal′o·ge·na′tion** *n.*

**hal·o·phile** (hăl′ə-fīl′) *n.* An organism requiring a saline environment. **—hal′o·phil′ic** (-fĭl′ĭk), **ha·loph′i·lous** (hă-lŏf′ə-ləs) *adj.*

**hal·o·phyte** (hăl′ə-fīt′) *n.* A plant growing in saline soil. **—hal′o·phyt′ic** (-fĭt′ĭk) *adj.*

**hal·o·thane** (hăl′ə-thān′) *n.* [HALO- + (E)THANE.] A colorless, nonflammable liquid, $C_2HBrClF_3$, used as an inhalational anesthetic.

**halt¹** (hôlt) *n.* [G. < MHG < imper. of *halten*, to stop < OHG *haltan.*] A stoppage of movement or progress, esp. a temporary one. *—v.* **halt·ed, halt·ing, halts.** *—vt.* To cause to stop. *—vi.* To stop : pause.

**halt²** (hôlt) *vi.* **halt·ed, halt·ing, halts.** [ME *halten*, to limp < OE *healtian.*] **1.** To be defective or proceed poorly, as in the development of a logical argument. **2.** To proceed or act uncertainly or indecisively : HESITATE. **3.** To limp or hobble. *—adj. Archaic.* Lame : crippled.

**hal·ter¹** (hôl′tər) *n.* [ME < OE *hælftre.*] **1.** A device of rope or leather straps that is placed around the head or neck of an animal in order to lead or secure it. **2.** A rope with a noose for execution by hanging. **3.** Death or execution by hanging. **4.** A bodice for women that ties behind the neck and across the back, leaving the arms, shoulders, and back bare. *—vt.* **-tered, -ter·ing, -ters. 1. a.** To put a halter on. **b.** To restrain or control with or as if with a halter. **2.** To hang (someone).

**hal·ter²** (hôl′tər, häl′-) *also* **hal·tere** (-tîr′) *n., pl.* **hal·ter·es** (hôl-tîr′ēz, hăl-) [Lat., lead weights used in leaping exercises < Gk. < *hallesthai*, to jump.] Either of the small, clublike balancing organs that are the rudimentary hind wings of dipterous insects, as the flies or mosquitoes.

**halt·ing** (hôl′tĭng) *adj.* **1.** Limping : lame. **2.** Imperfect : defective <a *halting* argument> **3.** Hesitant or indecisive <a *halting* voice>

**hal·vah** *or* **hal·va** (häl-vä′, häl′vä′) *also* **ha·la·vah** (hä′lə-vä′) *n.* [Yiddish *halva* < Rum. < Turk. *helva* < Ar. *ḥalwā.*] A confection of Turkish origin made of crushed sesame seeds, honey, and often other ingredients.

**halve** (hăv, häv) *vt.* **halved, halv·ing, halves.** [ME *halven* < *half*, half. —see HALF.] **1.** To divide or separate into two equal parts. **2.** To reduce by half. **3.** *Informal.* To share equally. **4.** To play (a golf game or hole) in the same number of strokes as an opponent.

**halves** (hăvz, hävz) *n. pl. of* HALF.

**hal·yard** *also* **hal·liard** (hăl'yərd) *n.* [Alteration of ME *halier* < *halen,* to pull. —see HALE².] A rope used to raise or lower a sail, flag, or yard.

**ham** (hăm) *n.* [ME *hamme* < OE *hamm.*] **1.** The thigh of the hind leg of certain animals, esp. a hog. **2.** The meat of the thigh of a hog. **3.** The back of the knee. **4.** The back of the thigh. **5. hams.** The buttocks. **6.** *Slang.* A performer who overacts or exaggerates. **7.** *Informal.* A licensed amateur radio operator. —*v.* **hammed, ham·ming, hams.** —*vi.* To overact. —*vt.* To exaggerate (e.g., a role or line).

**ham·a·dry·ad** (hăm'ə-drī'əd) *n., pl.* **-ads** *or* **-a·des** (-ə-dēz') [Lat. *hamadryas, hamadryad-,* < Gk. *Hamadruas* : *hama,* together with + *druas,* dryad < *drus,* tree.] **1.** *Gk. & Rom. Myth.* A wood nymph who lives only as long as the tree of which she is the spirit lives. **2.** The king cobra.

**ham·a·dry·as** (hăm'ə-drī'əs) *n.* [Lat., hamadryad. —see HAMADRYAD.] A baboon, *Comopithecus hamadryas* of northern Africa and Arabia, the adult male of which has a heavy mane.

**ha·mal** *also* **ham·mal** (hə-mäl') *n.* [Ar. *ḥammāl* < *ḥamala,* he carried.] A porter or bearer in certain Islamic countries.

**ha·mate** (hā'māt') *adj.* [Lat. *hamatus* < *hamus,* hook.] Hooked at the tip.

**ham·burg·er** (hăm'bûr'gər) *also* **ham·burg** (-bûrg') *n.* [Short for *Hamburger steak,* after *Hamburg,* Germany.] **1. a.** Ground meat, usu. beef. **b.** A cooked patty of such meat. **2.** A sandwich made with a hamburger patty, usu. in a roll or bun.

**hame** (hām) *n.* [ME < MDu.] One of the two curved wooden or metal pieces of a harness that fit around the neck of a draft animal and to which the traces are attached.

**Ham·il·to·ni·an** (hăm'əl-tō'nē-ən) *n.* [After William R. *Hamilton* (1805–1865).] A mathematical function that can be used systematically to generate the equations of motion of a dynamic system, equal for many such systems to the sum of the kinetic and potential energies of the system expressed in terms of the system's coordinates and momenta treated as independent variables.

**Ham·ite** (hăm'īt') *n.* [After HAM.] A member of a group of related peoples inhabiting northern and northeastern Africa, including the Berbers and the descendants of the ancient Egyptians.

**Ha·mit·ic** (hă-mĭt'ĭk) *adj.* Of or pertaining to the Hamites or their language. —*n.* A subfamily of the Afro-Asiatic language family that includes Berber, Egyptian, Coptic, and the various Cushitic languages of Ethiopia.

**Ham·i·to-Se·mit·ic** (hăm'ĭ-tō-sə-mĭt'ĭk) *n.* [HAMIT(IC) + SEMITIC.] Afro-Asiatic.

**ham·let** (hăm'lĭt) *n.* [ME *hamelet* < OFr., dim. of *ham,* village, of Germanic orig.] A small town or village.

**ham·mal** (hə-mäl') *n. var. of* HAMAL.

**ham·mer** (hăm'ər) *n.* [ME *hamer* < OE *hamor.*] **1.** A hand tool consisting of a handle with a perpendicularly attached head of a relatively heavy, rigid material, used mainly to drive in nails or pound and shape metals. **2.** A device similar to a hammer in function or action, as: **a.** The part of a gunlock that hits the primer or firing pin or explodes the percussion cap causing the gun to go off. **b.** One of the padded wooden pieces of a piano that strike the strings. **c.** A part of an apparatus that strikes a bell or gong as in a clock. **3.** *Anat.* The malleus. **4.** A metal ball weighing 16 pounds with a long wire or wooden handle by which it is thrown in track-and-field competition. **5.** A small mallet used by auctioneers. —*v.* **-mered, -mer·ing, -mers.** —*vt.* **1.** To hit with or as if with a hammer : POUND. **2.** To beat into a shape or flatten with or as if with a hammer <*hammered* out a new union contract> **3.** To put together, fasten, or seal, esp. with nails, by hammering. **4.** To force upon by constant repetition. —*vi.* **1.** To deal repeated blows with or as if with a hammer : PUMMEL <"wind *hammered* at us violently in gusts" —Thor Heyerdahl> **2.** To beat in the manner of a hammer. **3.** *Informal.* To keep at something constantly <*hammered* away at the problem> **—hammer and tongs.** With great energy or effort. **—ham'mer·er** *n.*

**hammer and sickle** *n.* A Communist symbol consisting of a crossed hammer and sickle representing the alliance of workers and peasants.

**ham·mered** (hăm'ərd) *adj.* Shaped or worked with a metalworker's hammer and often showing the marks of such a tool <*hammered* steel>

**ham·mer·head** (hăm'ər-hĕd') *n.* **1.** The head of a hammer. **2.** A large predatory shark of the genus *Sphyrna,* having the sides of the head elongated into fleshy extensions with the eyes at the ends. **3.** A bird, *Scopus umbretta* of Africa and southwestern Asia, with brown plumage, a bladelike bill, and a backward-pointing crest.

**hammer lock** *n.* A wrestling hold in which the opponent's arm is pulled behind his back and twisted upward.

**ham·mer·toe** (hăm'ər-tō') *n.* A toe, usu. the second, congenitally bent downward.

**ham·mock¹** (hăm'ək) *n.* [Sp. *hamaca* < Taino.] An easily swung cot or lounge of canvas or netting suspended between two trees or other supports.

**ham·mock²** (hăm'ək) *n. var. of* HUMMOCK.

**ham·my** (hăm'ē) *adj.* **-mi·er, -mi·est.** Marked by overacting or exaggeration. **—ham'mi·ly** *adv.* **—ham'mi·ness** *n.*

**ham·per¹** (hăm'pər) *vt.* **-pered, -per·ing, -pers.** [ME *hamperen.*] To restrict or impede the free movement, action, or progress of. —*n.* *Naut.* Necessary but encumbering equipment on a ship.

**ham·per²** (hăm'pər) *n.* [ME < OFr. *hanapier,* a case for holding goblets < *hanap,* goblet, of Germanic orig.] A large basket or receptacle, usu. with a cover.

**Hamp·shire** (hămp'shîr, -shər) *n.* [After *Hampshire,* England.] **1.** A large sheep orig. bred in England. **2.** A pig of a breed developed in the United States, having a white-banded black body.

**ham·ster** (hăm'stər) *n.* [G. < OHG *haumstro,* of Slavic orig.] Any of several Eurasian rodents of the family Cricetidae, esp. *Mesocricetus auratus,* with large cheek pouches and a short tail, used in laboratory research and kept as a pet.

**ham·string** (hăm'strĭng') *n.* **1.** Either of two tendons at the rear hollow of the human knee. **2.** The large sinew in the rear of the hock of a quadruped. —*vt.* **-strung** (-strŭng'), **-string·ing, -strings. 1.** To cripple by cutting the hamstring. **2.** To destroy or restrict the free movement or efficiency of : FRUSTRATE.

**ham·u·lus** (hăm'yə-ləs) *n., pl.* **-li** (-lī') [Lat., dim. of *hamus,* hook.] A small hooklike projection or process, as at the end of a bone.

**ham·za** *also* **ham·zah** (hăm'zə) *n.* [Ar. < *hamaza,* he compressed.] A sign in Arabic orthography used to represent the sound of a glottal stop, transliterated in English as an apostrophe.

**hand** (hănd) *n.* [ME < OE.] **1. a.** The terminal part of the human arm below the wrist, including the palm, four fingers, and an opposable thumb and used for grasping and holding. **b.** A homologous part in other animals. **2.** A unit of length equal to 4 inches or 10.16 centimeters, used esp. to measure the height of a horse. **3.** Something suggesting the shape or function of the human hand. **4. a.** Any of the rotating pointers on the face of a mechanical clock. **b.** A pointer, as on a gauge or dial. **5.** INDEX 3. **6.** Lateral direction indicated according to the way one is facing <at my left *hand*> **7.** A style or sample of writing : PENMANSHIP. **8.** A round of applause. **9.** Physical assistance : HELP <gave me a *hand* with the dishes> **10. a.** The cards held by a given player at any time. **b.** The number of cards dealt each player. **c.** A player or participant. **d.** A portion of a game during which all the cards dealt out are played <a *hand* of poker> **11. a.** One who does manual labor <a farm *hand*> **b.** One who is part of a group or crew. **12. a.** A participant in an activity. **b.** One who specializes in a specific activity or pursuit <an old *hand* at diplomatic maneuvering> **13. a.** The immediacy of a source of information <got the story at first *hand*> **b.** The strength or force of one's position <negotiated from a strong *hand*> **14. a.** *often* **hands.** Possession, ownership, or keeping <The check should be in your *hands* before Thursday.> **b.** *often* **hands.** Control : jurisdiction <in good *hands*> **15. a.** Involvement or participation <"In all this was evident the *hand* of the counterrevolutionaries" —John Reed> **b.** An effect or influence <had a *hand* in all the big decisions> **16.** An ability or aptitude <tried my *hand* at oil painting> **17.** Evidence of craftsmanship or skill <can see the *hand* of a master even in the early novels> **18.** A way of doing something <applied make-up with a heavy *hand*> **19. a.** Permission or a promise, esp. a pledge to wed. **b.** A commitment or agreement, esp. when sealed by a handshake : WORD. —*vt.* **hand·ed, hand·ing, hands. 1.** To give or pass with or as if with the hands : TRANSMIT <*Hand* me the butter, please.> **2.** To aid, direct, or conduct with the hands <The usher *handed* me to my seat in the third row.> **3.** *Naut.* To roll up and secure (a sail) : FURL. **—at hand. 1.** Close by : NEAR. **2.** Soon in time : IMMINENT. **—at the hand (***or* **hands) of.** Done by someone or through the agency of someone. **—bite the hand that feeds (one).** To repay kindness with malice. **—by hand.** Done manually. **—come to hand.** To become apparent. **—force (one's) hand.** To compel or pressure one to act prematurely or unwillingly in a given situation. **—hand and foot.** With total effort or fidelity <waited on me *hand and foot*> **—hand down. 1.** To bequeath as an inheritance to one's heirs. **2.** To deliver an official decision, esp. a court verdict. **—hand in (***or* **and) glove.** In close association. **—hand in hand.** In cooperation : JOINTLY. **—hand it to.** *Slang.* To give due credit to. **—hand on.** To turn over to another. **—hand out. 1.** To distribute freely : DISSEMINATE. **2.** To administer or deal out. **—hand over.** To give up or relinquish to another. **—hand over fist.** *Slang.* At a tremendous rate <making money *hand over fist*> **—hands down.** With no trouble : EASILY. **—in hand. 1.** Under control. **2.** Currently available or accessible. **3.** In preparation. **—off (one's) hands.** Out of one's responsibility or care. **—on hand.** Available. **—on (***or* **upon) (one's) hands.** In one's care or possession, often as a responsibility or burden. **—on the one hand.** In one respect. **—on the other hand.** As another or opposite point of view. **—out of hand. 1.** Out of control. **2.** Immediately. **3.** Finished. **4.** Uncalled for or improper : INDISCREET. **—show (one's) hand.** To reveal something previously hidden, as a plan or intention. **—throw up (one's) hands.** To concede. **—tip (one's) hand.** To reveal something unwittingly. **—to hand. 1.** Nearby. **2.** In one's possession. **—with a heavy hand. 1.** In a clumsy or awkward manner. **2.** With great emphasis or harshness.

**hand·bag** (hănd′băg′) *n.* **1.** A bag for carrying small personal articles : POCKETBOOK. **2.** A small suitcase.

**hand·ball** (hănd′bôl′) *n.* **1.** A game played by two or more players batting a ball against a wall with their hands, usu. with a special glove. **2.** The small rubber ball used in handball.

**hand·bar·row** (hănd′băr′ō) *n.* A flat framework or litter with carrying poles at each end.

**hand·bill** (hănd′bĭl′) *n.* A printed sheet or pamphlet distributed by hand.

**hand·book** (hănd′bo͝ok′) *n.* **1.** A concise manual or reference book giving specific information or instruction about a particular subject. **2. a.** A book in which off-track bets are recorded. **b.** A place where off-track bets are taken.

**hand·breadth** (hănd′brĕdth′) *also* **hand's-breadth** or **hand's breadth** (hăndz′-) *n.* A linear measurement, approx. the width of the palm of the hand, ranging from 2½ to 4 inches or 6.25 to 10 centimeters.

**hand·car** (hănd′kär′) *n.* A small open railroad car propelled by hand or a small motor.

**hand·cart** (hănd′kärt′) *n.* A small, usu. two-wheeled cart pushed or pulled by hand.

**hand·clasp** (hănd′klăsp′) *n.* The act or an instance of clasping the hand of another, esp. in friendship.

**hand·craft** (hănd′krăft′) *n. var. of* HANDICRAFT. *—vt.* **-craft·ed, -craft·ing, -crafts.** To make by hand. **—hand′crafts′man, hand′craft′man** *n.*

**hand·cuff** (hănd′kŭf′) *n. often* **handcuffs.** A device consisting of a pair of strong, connected hoops that can be locked about one or both wrists of a prisoner in custody : MANACLE. *—vt.* **-cuffed, -cuff·ing, -cuffs. 1.** To put handcuffs on. **2.** To make ineffective.

**hand·ed** (hăn′dĭd) *adj.* **1.** Using or designed for use by one hand in preference to the other <a right-*handed* baseball player> <a left-*handed* golf club> **2.** Pertaining to a specified number of people <a four-*handed* card game>

**hand·ed·ness** (hăn′dĭd-nĭs) *n.* A tendency to use one hand in preference to the other <left-*handedness*>

**hand·fast** (hănd′făst′) *n. Archaic.* A handclasp used to signify a pledge, as a marriage.

**hand·ful** (hănd′fo͝ol′) *n., pl.* **-fuls. 1.** The amount or number that can be held in the hand. **2.** A small, undefined amount or number <a *handful* of change> **3.** *Informal.* One too difficult to control or handle easily.

**hand glass** *n.* **1.** A small magnifying glass held in the hand. **2.** A mirror with a handle.

**hand·grip** (hănd′grĭp′) *n.* **1.** A grip by the hand. **2.** Something suited to a grip by the hand, as a handle. **3. handgrips.** Hand-to-hand combat.

**hand·gun** (hănd′gŭn′) *n.* A firearm that can be used with one hand : PISTOL.

**hand·hold** (hănd′hōld′) *n.* **1.** A grip by the hand. **2.** Something that can be held on to for support.

**hand·i·cap** (hăn′dē-kăp′) *n.* [< obs. *hand in cap,* a game in which forfeits were held in a cap.] **1. a.** A race or competition in which advantages or penalties are given to individual contestants in order to equalize the chances of winning. **b.** Such an advantage or penalty. **2.** A disadvantage or deficiency, esp. a physical or mental disability that prevents or restricts normal achievement. **3.** A hindrance. *—vt.* **-capped, -cap·ping, -caps. 1.** To assign a handicap or handicaps to (a contestant). **2.** To be a disadvantage to : IMPEDE.

**hand·i·cap·per** (hăn′dē-kăp′ər) *n.* **1.** One who assigns handicaps. **2.** One who predicts the winners in a horse race, esp. one who publishes such predictions as a guide for bettors.

**hand·i·craft** (hăn′dē-krăft′) *also* **hand·craft** (hănd′krăft′) *n.* [ME *handiecraft,* var. of *handcraft.*] **1.** Skill and facility with the hands : WORKMANSHIP. **2.** A trade, craft, or occupation requiring skilled use of the hands. **3.** Work produced by skilled hands.

**hand·i·crafts·man** (hăn′dē-krăfts′mən) *n.* One who is skilled in handicraft.

**hand·i·ly** (hăn′dĭ-lē) *adv.* **1.** In a handy or easy way. **2.** Conveniently.

**hand·i·work** (hăn′dē-wûrk′) *n.* [ME *handiwerk* < OE *handgeweorc* : *hand,* hand + *geweorc,* work.] **1.** Work done by hand. **2.** Something accomplished by a single person's efforts. **3.** The product of a person's work or actions.

**hand·ker·chief** (hăng′kər-chĭf) *n.* **1.** A small square of cloth used esp. for wiping the nose or mouth. **2.** A usu. square piece of cloth worn as a decorative article : SCARF.

**han·dle** (hăn′dl) *v.* **-dled, -dling, -dles.** [ME *handelen* < OE *handlian.*] *—vt.* **1.** To touch, lift, or hold with the hands. **2.** To operate with the hands : MANIPULATE. **3.** To deal with or have responsibility for : CONDUCT <*handle* criminal law> **4. a.** To direct, execute, or dispose of <*handle* a stock portfolio> **b.** To manage, administer to, or represent <*handle* a prizefighter> **5.** To confront or cope with <*handle* a difficult situation> **6.** To deal or trade in the purchase or sale of <a store that *handles* computer software> *—vi.* To act or function under operation <a car that *handles* well at high speeds> *—n.* **1.** A part designed to be held or operated with the hand. **2.** An opportunity or means for achieving a purpose. **3.** *Slang.* A person's name. **4.** The total amount of money bet on an event or over a set period of time.

☆ ***syns:*** HANDLE, MANIPULATE, PLY, WIELD *v. core meaning* : to use with or as if with the hands <*handles* the golf club like a pro>

**han·dle·bar** (hăn′dl-bär′) *n. often* **handlebars.** A curved metal steering bar, as on a bicycle.

**handlebar mustache** *n.* A long, curved mustache.

**han·dler** (hănd′lər) *n.* **1.** One that handles something. **2. a.** One who trains or shows an animal, as a dog. **b.** One who acts as the trainer or second of a boxer.

**han·dling** (hănd′lĭng) *n.* **1.** An act or instance of one that handles something. **2.** The way in which a matter, esp. a delicate one, is taken care of. **3.** The way in which a presentation, esp. an artistic or theatrical work, is treated.

**hand·made** (hănd′mād′) *adj.* Made or prepared by hand.

**hand·maid** (hănd′mād′) *also* **hand·maid·en** (-mād′n) *n.* **1.** A woman who is a servant or attendant. **2.** Something serving as an aid.

**hand-me-down** (hănd′mē-doun′) *adj.* **1.** Handed down to one person after being used by another. **2.** Inferior in quality : SHABBY. *—n.* Something passed on from one person to another.

**hand·off** (hănd′ôf′, -ŏf′) *n. Football.* A play in which one player hands the ball to another.

**hand organ** *n.* A barrel organ operated by turning a crank.

**hand·out** (hănd′out′) *n.* **1.** Food, clothing, or money given to the needy. **2.** A folder or leaflet circulated free. **3.** A prepared news or publicity release.

**hand-pick** (hănd′pĭk′) *vt.* **-picked, -pick·ing, -picks. 1.** To assemble, gather, or pick by hand. **2.** To choose personally. **—hand′-picked′** *adj.*

**hand·print** (hănd′prĭnt′) *n.* An outline or an indentation left by a hand on a surface.

**hand·rail** (hănd′rāl′) *n.* A narrow rail to be grasped with the hand for support.

**hand·saw** (hănd′sô′) *n.* A small saw operated by hand.

**hand's-breadth** or **hand's breadth** (hăndz′brĕdth′) *n. vars. of* HANDBREADTH.

**hand·sel** (hănd′səl) *also* **han·sel** (hăn′-) [ME *hanselle* < OE *handselen* and ON *handsal,* transfer.] *Chiefly Brit.* *—n.* **1.** A gift to express good wishes at the beginning of a new year or enterprise. **2.** The first money or barter taken in, as by a new business or on the opening day of business, esp. when regarded as a token of good luck. **3. a.** A first payment : earnest money. **b.** A foretaste of what is to come. *—vt.* **-seled, -sel·ing, -sels** or **-selled, -sel·ling, -sels. 1.** To give a handsel to. **2.** To launch with a ceremonial gesture or gift. **3.** To perform or use for the first time.

**hand·set** (hănd′sĕt′) *n.* A portable telephone transmitter and receiver module.

**hand·shake** (hănd′shāk′) *n.* The grasping of hands by two people, as in greeting or leave-taking.

**hands-off** (hăndz′ôf′, -ŏf′) *adj.* Marked by nonintervention or noninterference.

**hand·some** (hăn′səm) *adj.* **-som·er, -som·est.** [ME *handsom,* handy.] **1.** Pleasing and dignified. **2.** Bounteous or copious : LIBERAL <a *handsome* reward> **3.** Marked by or requiring dexterity or skill <*handsome* gymnastic routines > **4.** Appropriate or fitting. **5.** Fairly large. **—hand′some·ly** *adv.* **—hand′some·ness** *n.*

**hands-on** (hăndz′ŏn′, -ôn′) *adj.* Applied as opposed to theoretical <"We're involved in *hands-on* operations, pulling levers, pushing buttons" —Arthur R. Taylor>

**hand·spike** (hănd′spīk′) *n.* A bar serving as a lever.

**hand·spring** (hănd′sprĭng′) *n.* A gymnastic feat in which the body is flipped totally forward or backward from an upright position, landing first on the hands, then on the feet.

**hand·stand** (hănd′stănd′) *n.* The act of balancing on the hands with one's feet in the air.

**hand-to-hand** (hănd′tə-hănd′) *adj.* At close quarters.

**hand-to-mouth** (hănd′tə-mouth′) *adj.* Having or supplying only the bare essentials.

**hand·work** (hănd′wûrk′) *n.* Work done by hand.

**hand·wo·ven** (hănd′wō′vən) *adj.* **1.** Produced on a hand-operated loom. **2.** Woven by hand <*handwoven* sweaters>

**hand·writ·ing** (hănd′rī′tĭng) *n.* **1.** Writing done with the hand. **2.** The writing typical of a particular person. **—hand′writ′ten** (-rĭt′n) *adj.*

**hand·y** (hăn′dē) *adj.* **-i·er, -i·est.** [< HAND.] **1.** Adept in using one's hands, esp. in a number of ways. **2.** Within easy reach. **3.** Useful : convenient <a *handy* gadget> **4.** Easy to use or handle <a *handy* recipe file> **—hand′i·ness** *n.*

**hand·y·man** (hăn′dē-măn′) *n.* One who does various jobs or small tasks.

**hang** (hăng) *v.* **hung** (hŭng), **hang·ing, hangs.** [ME *hongen,* partly < OE *hangian,* to hang, and partly < OE *hōn,* to hang.] *—vt.*

**1.** To fasten from above with no support from below : SUSPEND. **2.** To suspend or fasten so as to permit free movement at or about the point of suspension <*hang* a louvered door> **3.** *p.t. & p.p.* **hanged** or **hung.** To execute by suspending by the neck. ***usage:*** When the sense intended is "to execute by suspending by the neck," *hanged* is the preferred form for the past tense and the past participle of *hang.* **4.** To fix or join at an appropriate angle <*hang* a scythe to its handle> **5.** To alter the hem of (a garment) so as to fall evenly at a particular height. **6.** To decorate, furnish, or appoint by suspending objects around or about <*hang* a room with draperies> **7.** To hold or incline downward <*hang* one's head in shame> **8.** To attach to a wall <*hang* wallpaper> **9.** To deadlock (a jury) by preventing a unanimous verdict. **10.** *Baseball.* To throw (a pitch) in such a way so as to fail to break. —*vi.* **1.** To be attached from above with no support from below. **2.** To die as a result of hanging. **3.** To remain stationary over a place or object : HOVER. **4.** To attach oneself as an impediment or dependent : cling. **5.** To incline downward : DROOP. **6.** To depend <The final decision *hangs* on your vote.> **7.** To pay close attention <*hang* on every word> **8.** To remain unresolved or uncertain <a future *hanging* in the balance> **9.** To fit the body in loose lines <a coat that *hangs* well> **10.** To be on exhibition, as in a gallery. **11.** *Baseball.* To fail to break or move in the intended way, as a curve ball. **—hang around.** *Informal.* **1.** To spend time idly : LOITER. **2.** To keep company. **—hang back (*or* off).** To hold back : HESITATE. **—hang on. 1.** To cling firmly to something. **2.** To continue persistently : PERSEVERE. **3.** To keep a telephone connection open. **—hang onto.** To hold or cling firmly to. **—hang out.** *Slang.* To spend one's leisure time in a certain place. **—hang together. 1.** To stand united. **2.** To make up a coherent totality. **—hang up. 1.** To suspend on a hook or hanger. **2. a.** To replace (a telephone receiver) on its cradle. **b.** To end a telephone conversation. **3.** To impede or hinder <*hang up* a million-dollar deal> **4.** To halt the movement or action of. **5.** To become halted or snagged. —*n.* **1.** The way in which something hangs. **2.** A downward inclination or slope. **3.** Particular meaning or significance. **4.** *Informal.* The proper method of doing, using, or handling something <get the *hang* of it> **5.** A suspension of motion. **—give (*or* care) a hang.** To be concerned or worried. **—hang fire. 1.** To be slow in firing, as a gun. **2.** To delay. **—hang in there.** *Informal.* To persevere despite difficulties : PERSIST. **—hang loose.** *Slang.* To stay calm or relaxed. **—hang tough.** *Informal.* To remain firmly resolved <"We are going to *hang tough* on this" —Donald T. Regan> **—let it all hang out.** *Slang.* **1.** To be totally relaxed. **2.** To be totally candid.

**han·gar** (hăng′ər, hăng′gər) *n.* [Fr. < OFr., prob. < Med. Lat. *angarium,* shed for shoeing horses.] A structure esp. for housing or repairing aircraft.

**hang·dog** (hăng′dôg′, -dŏg′) *adj.* **1.** Ashamed or guilty. **2.** Downcast : intimidated. —*n.* A sneaky or despicable person.

**hang·er** (hăng′ər) *n.* **1.** One that hangs. **2.** A device or structure to which something hangs or by which something is hung. **3.** A device around which a garment is draped for hanging from a hook or rod. **4.** A loop or strap by which something is hung. **5.** A bracket on an automobile's spring shackle that is designed to hold it to the chassis. **6.** A decorative strip of cloth hung on a wall or garment.

**hang·er-on** (hăng′ər-ŏn′, -ôn′) *n., pl.* **hang·ers-on** (hăng′ərz-). One who depends on another for support : PARASITE.

**hang glider** *n.* A device resembling a kite from which a harnessed rider hangs while gliding from a height. **—hang gliding** *n.*

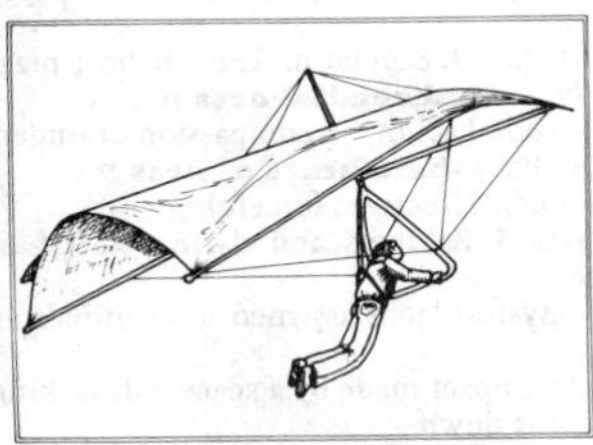

**hang glider**

**hang·ing** (hăng′ĭng) *n.* **1.** An execution on a gallows. **2.** Something hung. **3.** A descending slope or inclination. —*adj.* **1.** Situated on a sharp declivity. **2.** Projecting downward : OVERHANGING. **3.** Suited for holding something that hangs. **4. a.** Deserving death by hanging <a *hanging* offense> **b.** Tending to impose the sentence of death by hanging <a *hanging* judge>

**hanging indention** *n.* The indention of every line in a paragraph except the first.

**hang·man** (hăng′mən) *n.* One employed to execute condemned prisoners by hanging.

**hang·nail** (hăng′nāl′) *n.* [Alteration of AGNAIL.] A small piece of partially detached dead skin at the side or the base of a fingernail. ▲ word history: The word *hangnail* is an example of a word that has acquired its modern form by the process of folk etymology. By this process an unfamiliar element in a word is made to resemble a more familiar word that is often semantically associated with the word being refashioned. The Old English form of *hangnail* was *angnægel,* a compound of *ang–,* "tight, painful," and *nægel,* "nail, peg." The Old English word originally meant "corn" on the foot and later came to denote various kinds of painful conditions of the fingers and toes, including what is now denoted by *hangnail.* Because hangnails consist of partially detached skin, the unfamiliar element *ang–* was refashioned in modern times as *hang.* The normal modern development of Old English *angnægel,* and a form still current, is *agnail.*

**hang·out** (hăng′out′) *n.* A frequently visited place.

**hang·o·ver** (hăng′ō′vər) *n.* **1.** Unpleasant physical effects following the heavy use of alcohol. **2.** A letdown, as after a period of excitement. **3.** A vestige : holdover <*hangovers* from a past generation>

**hang·tag** (hăng′tăg′) *n.* A tag attached to a piece of merchandise giving information about its composition and proper care and use.

**hang-up** (hăng′ŭp′) *n. Informal.* **1.** A psychological or emotional difficulty or inhibition. **2.** An impediment to progress or growth.

**hank** (hăngk) *n.* [ME < ON *hǫnk.*] **1.** A coil or loop, esp. of hair. **2.** *Naut.* A ring on a stay attached to the head of a jib or staysail. **3.** A looped bundle, as of yarn : SKEIN.

**han·ker** (hăng′kər) *vi.* **-kered, -ker·ing, -kers.** [Perh. < dial. Du. *hankeren.*] To have a longing : CRAVE. **—hank′er·er** *n.*

**han·kie** *also* **han·ky** (hăng′kē) *n., pl.* **-kies.** A handkerchief.

**han·ky-pan·ky** (hăng′kē-păng′kē) *n.* [Perh. alteration of HOCUS-POCUS.] *Slang.* **1.** Devious or mischievous activity. **2.** Foolish talk or action.

**Han·o·ve·ri·an** (hăn′ō-vîr′ē-ən) *adj.* Of or pertaining to the kingdom or province of Hanover, the electoral house of Hanover, or the royal family of Hanover.

**Han·sard** (hăn′sərd) *n.* [After Luke *Hansard* (1752–1828).] The official report of the proceedings and debates of the British or Canadian Parliament.

**hanse** (hăns) *n.* [ME < OFr. < MLG < OHG *hansa,* military troop.] A medieval merchant guild or trade association. **—han′se·at′ic** (hăn′sē-ăt′ĭk) *adj.*

**Hanseatic League** *n.* A protective commercial association of free towns in northern Germany and adjoining areas, formally organized in 1358 and dissolved in the 17th cent.

**han·sel** (hăn′səl) *n. & v. var. of* HANDSEL.

**Han·sen's disease** (hăn′sənz) *n.* [After A.G.H. *Hansen* (1841–1912).] Leprosy.

**han·som** (hăn′səm) *n.* [After Joseph A. *Hansom* (1803–1882).] A two-wheeled covered carriage with the driver's seat at the back.

**Ha·nuk·kah** *or* **Ha·nu·kah** (KHä′nə-kə, hä′-) *n. vars. of* CHANUKAH.

**han·u·man** (hŭn′o͞o-män′) *n., pl.* **-mans.** [Skt. < *hanu,* jaw.] A monkey, *Presbytis entellus* of southern Asia, with bristly hairs on the crown and sides of the face.

**hao** (hou) *n.* [Vietnamese < Chin. *hao²,* a unit of weight.] —See table at CURRENCY.

**hao·le** (hou′lē, -lā) *n.* [Hawaiian.] A person, esp. a Caucasian, who is not a native Hawaiian.

**hap** (hăp) *n.* [ME < ON *happ.*] **1.** Fortune : chance. **2.** An occurrence. —*vi.* **happed, hap·ping, haps.** To take place.

**ha·pax le·go·me·non** (hā′păks′ lĭ-gŏm′ə-nŏn′) *n., pl.* **ha·pax le·go·me·na** (-gŏm′ə-nə) [Gk., once said.] A word or form occurring only once in the recorded corpus of a given language.

**hap·haz·ard** (hăp-hăz′ərd) *adj.* Dependent on or marked by chance. —*n.* Chance : fortuity. —*adv.* By chance : CASUALLY. **—hap·haz′ard·ly** *adv.* **—hap·haz′ard·ness** *n.*

**haph·ta·rah** *also* **haf·ta·rah** *or* **haf·to·rah** (häf′tə-rä′, häf-tôr′ə) *n., pl.* **-ta·roth** (-tə-rōt′, -rōs′, -tôr′ōt′, -ōs′) [Heb. *haphṭārāh* < *haphṭēr,* he concluded < *pāṭar,* he separated.] A reading from the Prophets that follows each lesson from the Torah in the synagogue service on the Sabbath.

**hap·less** (hăp′lĭs) *adj.* Without luck : UNFORTUNATE.

**hap·lite** (hăp′līt′) *n. var. of* APLITE.

**hap·loid** (hăp′loid′) *adj.* [Gk. *haploeidēs,* single : *haplous,* single + *-eidēs,* -oid.] *Genetics.* Having the number of chromosomes of a germ cell equal to half the number in the normal somatic cell. —*n.* A haploid individual.

**hap·loi·dy** (hăp′loi′dē) *n. Genetics.* The state or condition of being haploid.

**hap·lol·o·gy** (hăp-lŏl′ə-jē) *n.* [Gk. *haplous,* single, simple + *logos,* speech < *legein,* to speak.] The shortening of a word by contraction of similar sounds or syllables in its pronunciation.

**hap·lo·sis** (hăp-lō′sĭs) *n.* [Gk. *haplous,* single + -OSIS.] *Genetics.* Reduction of the diploid number of chromosomes by one half to the haploid number by meiosis.

**hap·pen** (hăp′ən) *vi.* **-pened, -pen·ing, -pens.** [ME *happenen* < *hap,* hap.] **1. a.** To come to pass. **b.** To come into being. **2.** To take place by chance. **3.** To come upon something by chance. **4.** To appear by chance.

**hap·pen·chance** (hăp′ən-chăns′) *n. var. of* HAPPENSTANCE.
**hap·pen·ing** (hăp′ə-nĭng) *n.* **1.** An event, esp. a significant one. **2.** An improvised, often spontaneous spectacle or performance, esp. one involving audience participation.
**hap·pen·stance** (hăp′ən-stăns′) *also* **hap·pen·chance** (-chăns′) *n.* [HAPPEN + (CIRCUM)STANCE.] An unexpected circumstance : ACCIDENT.
**hap·py** (hăp′ē) *adj.* **-pi·er, -pi·est.** [ME < *hap,* hap.] **1.** Marked by good luck : FORTUNATE. **2.** Enjoying, displaying, or characterized by pleasure or joy. **3.** Appropriate : felicitous <a *happy* choice of words> **4.** Cheerful : willing <*happy* to be of help> **5. a.** Marked by an unpredictable or obsessive inclination to use something <trigger-*happy*> **b.** Enthusiastic about, esp. to a disproportionate degree <power-*happy*> **—hap′pi·ly** *adv.* **—hap′pi·ness** *n.*
**hap·py-go-luck·y** (hăp′ē-gō-lŭk′ē) *adj.* Carefree : lighthearted.
**happy hour** *n.* A period of time, usu. in late afternoon and early evening, during which a bar offers drinks at reduced prices.
**hap·ten** (hăp′tĕn′) *also* **hap·tene** (-tēn′) *n.* [G. : Gk. *haptein,* to fasten + G. *-en,* -ene.] An incomplete antigen that cannot by itself cause antibody formation but can neutralize specific antibodies in an artificial environment outside the body.
**hap·tic** (hăp′tĭk) *or* **hap·ti·cal** (-tĭ-kəl) *adj.* [< Gk. *haptesthai,* to touch.] Of or pertaining to the sense of touch.
**ha·ra-ki·ri** (hăr′ĭ-kîr′ē) *n.* [J.] Ritual suicide by disembowelment, formerly practiced by the Japanese samurai and upper classes.
**ha·rangue** (hə-răng′) *n.* [ME *arang,* a speech to an assembly < OFr. *arenge* < Med. Lat. *harenga.*] **1.** A long, pompous speech. **2.** A speech or literary work marked by strong feeling or expression : TIRADE. —*v.* **-rangued, -rangu·ing, -rangues.** —*vt.* To deliver a harangue to. —*vi.* To deliver a harangue. **—ha·rangu′er** *n.*
**ha·rass** (hə-răs′, hăr′əs) *vt.* **-rassed, -rass·ing, -rass·es.** [Fr. *harasser,* prob. < OFr. *harer,* to set a dog on < *hare,* interjection used to set a dog on, of Germanic orig.] **1.** To annoy or torment repeatedly and persistently. **2.** To wear out : EXHAUST. **3.** To impede (an enemy) by repeated attacks or raids. **—ha·rass′er** *n.* **—ha·rass′ment** *n.*
**har·bin·ger** (här′bĭn-jər) *n.* [ME *herbengar,* person sent ahead to prepare lodging < OFr. *herbergeor* < *herbergier,* to provide lodgings for < *herberge,* lodgings, of Germanic orig.] One that indicates what is to come : FORERUNNER. —*vt.* **-gered, -ger·ing, -gers.** To signal the approach of : PRESAGE.
**har·bor** (här′bər) *n.* [ME *herberwe.*] **1.** A sheltered part of a body of water deep enough to provide anchorage for ships. **2.** A place of shelter : REFUGE. —*vt.* **-bored, -bor·ing, -bors. 1.** To give shelter to. **2.** To provide with often temporary lodging. **3.** To have or nourish (a given thought or feeling) <*harbor* a grudge> **—har′bor·er** *n.*
**har·bor·age** (här′bər-ĭj) *n.* **1.** Anchorage and protection for ships. **2.** Shelter : refuge.
**har·bor·mas·ter** (här′bər-măs′tər) *n.* One who oversees and enforces the regulations of a harbor.
**harbor seal** *n.* A hair seal, *Phoca vitulina* of coastal waters of the Northern Hemisphere, having a spotted coat.
**har·bour** (här′bər) *n.* & *v. Chiefly Brit. var. of* HARBOR.
**hard** (härd) *adj.* **-er, -est.** [ME < OE *heard.*] **1.** Resistant to pressure : RIGID. **2. a.** Physically toughened : RUGGED. **b.** Mentally toughened : STRONG-MINDED. **3. a.** Demanding great effort or endurance <a project that took years of *hard* work> **b.** Performed with or marked by great diligence or energy. **4. a.** Intense in force or degree <a *hard* blow> **b.** Inclement : bleak <a long, *hard* winter> **5. a.** Stern or strict <a *hard* taskmaster> **b.** Resistant to persuasion : STUBBORN. **c.** Making few or no concessions <drives a *hard* bargain> **6. a.** Difficult to endure <a *hard* life> **b.** Oppressive or unfair in nature or effect <made demands that were *hard* on the new students> **c.** Unsympathetic : callous. **7. a.** Harsh or severe in intention or effect <said some *hard* things that I won't forget> **b.** Bitter : rancorous <Please don't have *hard* feelings about this.> **8. a.** Causing damage or premature wear <Snow and ice are *hard* on a car's finish.> **b.** Bad : adverse <*hard* luck> **9.** Difficult to resolve or complete : TROUBLESOME <ran into some *hard* problems> **10.** Difficult to understand or explain <Physics was the *hardest* course for me.> **11. a.** Based on facts <*hard* evidence> **b.** Definite : firm <a *hard* commitment> **c.** Searching : penetrating <need to take a *hard* look at the situation> **d.** Free from illusions : PRACTICAL <brought some *hard* common sense to the discussion> **12. a.** Marked by sharp outline, rigidity, or stiffness. **b.** Lacking in delicacy or nuance. **13. a.** Metallic as opposed to paper <*hard* money> **b.** Backed by gold rather than by credit. —Used of currency. **c.** High and stable <*hard* prices> **14.** Durable : lasting <*hard* merchandise> **15.** Having high alcoholic content <*hard* liquor> **16.** Containing dissolved substances, as salts of calcium or magnesium, that prevent soap from forming a lather. —Used of water. **17.** Pronounced as the *c* in *cold* and the *g* in *get.* **18.** *Physics.* Of relatively high energy <*hard* x-rays> **19.** Having a high gluten content <*hard* wheat> **20.** Not liable to biodegradation <a *hard* detergent> **21.** Causing physical addiction and damage to the health <*hard* drugs> —*adv.* **1.** With strenuous effort : INTENTLY <worked *hard* all day> **2.** With great force, vigor, or energy <pressed *hard* on the brake> **3.** In such a manner as to cause great damage or hardship <industrial areas hit *hard* by unemployment> **4.** With great distress, grief, or bitterness <took the separation *hard*> **5.** Firmly : securely <held *hard* to the rope> **6.** Toward or into a solid condition <cement that sets *hard* within a day> **7.** Near to in time or space : CLOSE <a diner *hard* by the railroad tracks> **8.** *Naut.* Completely : fully <*hard* alee> **—hard and fast.** Fixed : invariable <*hard and fast* rules> **—hard up.** *Informal.* In need : POOR.
☆ ***syns:*** HARD, FIRM, SOLID *adj. core meaning* : unyielding to physical pressure <*hard* ground><a *hard* surface> ***ant:*** soft
**hard·back** (härd′băk′) *adj.* Bound in hard covers. —Used of a book. **—hard′back′** *n.*
**hard·ball** (härd′bôl′) *n.* **1.** Baseball. **2.** Strong, uncompromising measures taken to achieve a desired end <played *hardball* politics in the race for mayor>
**hard-bit·ten** (härd′bĭt′n) *adj.* Made hard or tough by experience.
**hard·board** (härd′bôrd′, -bōrd′) *n.* Construction board made by compressing fibers of wood chips usu. with a binder at a high temperature.
**hard-boiled** (härd′boild′) *adj.* **1.** Cooked by boiling in the shell to a solid state <*hard-boiled* eggs> **2.** *Informal.* Callous : unfeeling. **3.** *Informal.* Unsentimental and tough.
**hard·bound** (härd′bound′) *adj.* & *n.* Hardback.
**hard cider** *n.* Fermented cider.
**hard coal** *n.* Anthracite.
**hard copy** *n.* Readable printed copy of the output of a machine, as a computer.
**hard core** *n.* The most intractable or resistant nucleus of a group or organization <the *hard core* of the revolutionary movement>
**hard-core** *also* **hard·core** (härd′kôr′, -kōr′) *adj.* **1.** Stubbornly resistant to change or improvement <*hard-core* poverty> **2.** Extremely graphic or explicit <*hard-core* pornography>
**hard·cov·er** (härd′kŭv′ər) *adj.* & *n.* Hardback.
**hard·edge** (härd′ĕj′) *n.* A form of abstract painting characterized by clearly defined geometric shapes and often bright colors.
**hard·en** (här′dn) *v.* **-ened, -en·ing, -ens.** —*vt.* **1.** To make hard or harder. **2.** To toughen mentally or physically : INURE. **3.** To make unsympathetic or callous <Selfishness had *hardened* her heart.> **4.** To define or outline sharply. —*vi.* **1.** To become hard or harder. **2. a.** To rise. —Used of prices. **b.** To become stable. **3.** To become toughened physically or mentally.
☆ ***syns:*** HARDEN, CAKE, CONCRETE, CONGEAL, DRY, INDURATE, PETRIFY, SET, SOLIDIFY *v. core meaning* : to make or become physically hard <glue that *hardens* in ten seconds>***ant:*** soften
**hard·en·er** (här′dn-ər) *n.* **1.** One that hardens. **2.** A substance added (e.g., to paint) to give a harder finish.
**hard-fist·ed** (härd′fĭs′tĭd) *adj.* Not generous : STINGY.
**hard·hack** (härd′hăk′) *n.* A woody plant, *Spiraea tomentosa* of eastern North America, having rusty down on the undersides of the leaves and pyramidal clusters of small pink flowers.
**hard-hand·ed** (härd′hăn′dĭd) *adj.* **1.** Having hands made calloused or hard by work. **2.** Tyrannical. **—hard′-hand′ed·ness** *n.*
**hard·hat** (härd′hat′) *adj.* **1.** Relating to construction work or demolition. **2.** *Informal.* Of, relating to, or marked by extreme conservatism. **3.** *Slang.* Extremely patriotic.
**hard hat** *n.* **1.** A helmet, usu. of metal or reinforced plastic, worn for protection by construction workers. **2.** *Informal.* A construction worker. **3.** *Informal.* An ultraconservative. **4.** *Slang.* An extremely patriotic person.
**hard·head** (härd′hĕd′) *n.* **1.** A practical and realistic person. **2.** A stubborn, unyielding person. **3.** *pl.* **hardhead** *or* **-heads.** Any of several fishes having a bony head, esp. a common croaker, *Micropogon undulatus* of Atlantic waters.
**hard·head·ed** (härd′hĕd′ĭd) *adj.* **1.** Stubborn. **2.** Realistic : pragmatic. **—hard′head′ed·ly** *adv.* **—hard′head′ed·ness** *n.*
**hard·heart·ed** (härd′här′tĭd) *adj.* Lacking in compassion or understanding. **—hard′heart′ed·ly** *adv.* **—hard′heart′ed·ness** *n.*
**hard-hit·ting** (härd′hĭt′ĭng) *adj.* Effective : forceful.
**har·di·hood** (här′dē-hŏŏd′) *n.* **1.** Boldness and daring. **2.** Self-assured impudence or insolence.
**hard labor** *n.* Compulsory physical labor assigned to criminals as part of a prison term.
**hard landing** *n.* A landing by impact made by a spacecraft lacking devices, as retrorockets, to slow it down.
**hard line** *n.* A firm and uncompromising position.
**hard-line** *also* **hard·line** (härd′līn′) *adj.* Marked by a firm, uncompromising position. **—hard′-lin′er** *n.*
**hard·ly** (härd′lē) *adv.* [ME *hardli,* hardily < OE *heardlīce.*] **1.** Barely : just. **2.** Almost surely not. **3. a.** With severity : HARSHLY. **b.** With great force or vigor. **4.** With great difficulty : PAINFULLY.
***usage:*** In formal style a clause following *hardly* is introduced by *when* or sometimes by *before,* but not by *than,* as in *were hardly seated before* (not *than*) *the fire broke out; had hardly finished when you arrived.*
**hard maple** *n.* The sugar maple.
**hard·ness** (härd′nĭs) *n.* **1.** The quality or state of being hard. **2.** The

relative resistance of a mineral to scratching, as measured on the Mohs scale.

**hard news** *n.* News dealing with significant or serious topics and events. **—hard'-news'** (härd'no͞oz', -nyo͞oz') *adj.*

**hard-nosed** (härd'nōzd') *adj.* Hardheaded.

**hard palate** *n.* The relatively hard, bony anterior portion of the palate that forms the roof of the mouth.

**hard·pan** (härd'păn') *n.* **1.** A layer of hard subsoil or clay. **2.** Hard, unbroken ground. **3.** A foundation : bedrock.

**hard rock** *n.* A style of rock music characterized by a harsh, amplified sound and often making use of electronic modulations, distortion, and feedback.

**hard rubber** *n.* A relatively firm rubber containing 30–50% sulfur and usu. lime or magnesia as a filler.

**hards** (härdz) *pl.n.* [ME < OE *heordan.*] *(used with a sing. verb).* The coarse refuse of flax or similar fiber.

**hard sauce** *n.* A dessert sauce made of butter creamed together with sugar and rum, brandy, or vanilla, served esp. on puddings.

**hard·scrab·ble** (härd'skrăb'əl) *adj.* Earning a bare living <the sharecropper's *hardscrabble* life> —*n.* Barren or marginal farmland.

**hard sell** *n. Informal.* Aggressive high-pressure sales or promotional methods.

**hard·set** (härd'sĕt') *adj.* **1.** In a difficult or ticklish position. **2.** Rigid and unyielding.

**hard-shell** (härd'shĕl') *also* **hard-shelled** (-shĕld') *adj.* **1.** Having a thick, heavy, or hardened shell. **2.** Uncompromising : confirmed. —*n.* A hard-shell clam or crab.

**hard-shell clam** *n.* The quahog.

**hard-shell crab** *n.* A marine crab with a fully hardened shell, esp. the edible species *Callinectes sapidus* of eastern North America in this stage.

**hard·ship** (härd'shĭp') *n.* **1.** Extreme privation : SUFFERING. **2.** Something that is a source or cause of privation or suffering.

**hard-spun** (härd'spŭn') *adj.* Twisted tightly in spinning, often to the point of curling and looping. —Used of yarn.

**hard·stand** (härd'stănd') *n.* An area with a hard surface, used for parking planes or ground vehicles.

**hard·tack** (härd'tăk') *n.* A hard biscuit or bread made only with flour and water.

**hard·top** (härd'tŏp') *n.* A car having a rigidly fixed, hard top, designed to resemble a convertible.

**hard·ware** (härd'wâr') *n.* **1.** Metal goods and utensils, as locks, tools, and machine parts. **2. a.** A computer and its associated physical apparatus directly involved in the performance of communications or data-processing functions. **b.** Machines and the devices or equipment used in the performance of an industrial, technological, or military function. **3.** *Informal.* Weapons, esp. military weapons.

**hard·wired** (härd'wīrd') *adj.* Of, pertaining to, or effected by means of logic circuitry that is permanently connected within a computer or calculator. **—hard'wire'** *v.* **(-wired, -wir·ing, -wires).**

**hard·wood** (härd'wo͝od') *n.* **1.** The wood of a broad-leaved flowering tree as distinguished from the wood of a conifer. **2.** A broad-leaved flowering tree.

**har·dy** (här'dē) *adj.* **-di·er, -di·est.** [ME < OFr. *hardi* < *hardir,* to harden, of Germanic orig.] **1.** Stalwart and rugged : ROBUST. **2.** Courageous : bold. **3.** Brazenly daring : AUDACIOUS. **4.** Capable of surviving unfavorable conditions, as cold weather or lack of moisture <*hardy* perennials> **—har'di·ly** *adv.* **—har'di·ness** *n.*

**hare** (hâr) *n.* [ME < OE *hara.*] Any of various mammals of the family Leporidae, esp. of the genus *Lepus,* related to the rabbits but with longer ears, large hind feet, and long, powerful legs for jumping.

**hare and hounds** *n.* A game in which one group of players leaves a trail of paper scraps that a pursuing group must try to follow.

**hare·bell** (hâr'bĕl') *n.* [ME *harebelle* : *hare,* hare + *belle,* bell.] A plant, *Campanula rotundifolia,* with slender stems and leaves and bell-shaped blue flowers.

**hare·brained** (hâr'brānd') *adj.* Foolish : flighty <a *harebrained* notion>

**Ha·re Krish·na** (hä'rē krĭsh'nə) *n., pl.* **Hare Krish·nas.** [Hindi *hare,* invocation of God + *Krishna,* Krishna.] A member of a religious group devoted to the Hindu god Krishna.

**hare·lip** (hâr'lĭp') *n.* A congenital fissure in the upper lip. **—hare'-lipped'** *adj.*

**har·em** (hâr'əm, hăr'-) *n.* [Ar. *ḥarīm.*] **1.** A house or a part of a house set aside for the women of a Moslem household. **2.** The wives, concubines, women relatives, and servants who live in a harem.

**har·i·cot** (hăr'ĭ-kō') *n.* [Fr.] The edible pod or seed of any of several beans, esp. the string bean.

**hark** (härk) *vi.* **harked, hark·ing, harks.** [ME *herken* < *herkenen,* to hark < OE *hercnian.*] To listen attentively or closely. **—hark back.** To return to a previous subject or circumstance.

**har·ken** (här'kən) *v. var. of* HEARKEN.

**harl** (härl) *n.* [ME, fiber, filament, perh. of MLG orig.] Filaments or fibers, as of hemp or flax.

**har·le·quin** (här'lĭ-kwĭn, -kĭn) *n.* [Obs. Fr. < OItal. *arlecchino* < OFr. *Helquin,* a demon, poss. of Germanic orig.] **1. Harlequin.** A conventional buffoon of the commedia dell'arte, usu. wearing a mask and parti-colored tights. **2.** A clown : buffoon. —*adj.* Having a pattern of brightly colored diamond shapes.

**har·le·quin·ade** (här'lĭ-kwə-nād') *n.* A comedy or pantomime in which Harlequin plays a leading part.

**harlequin bug** *n.* A flat-bodied, brightly colored insect, *Murgantia histrionica,* destructive to cabbage and other plants.

**har·lot** (här'lət) *n.* [ME < OFr. *arlot,* vagabond.] A prostitute. **—har'lot·ry** (här'lə-trē) *n.*

**harm** (härm) *n.* [ME < OE *hearm.*] **1.** Physical or psychological injury or damage. **2.** Mischief : evil. —*vt.* **harmed, harm·ing, harms.** To cause damage to.

☆ ***syns:*** HARM, DAMAGE, DETRIMENT, HURT, INJURY, OUTRAGE *n. core meaning* : the action or result of inflicting loss or pain <did *harm* to the hostages>

**har·mat·tan** (här'mə-tăn', här-măt'n) *n.* [Twi *haramata.*] A dry, dusty wind that occurs in certain seasons on the northwestern coast of Africa.

**harm·ful** (härm'fəl) *adj.* Causing or capable of causing harm. **—harm'ful·ly** *adv.* **—harm'ful·ness** *n.*

**harm·less** (härm'lĭs) *adj.* Incapable of harming. **—harm'less·ly** *adv.* **—harm'less·ness** *n.*

**har·mon·ic** (här-mŏn'ĭk) *adj.* [Lat. *harmonicus* < Gk. *harmonikos* < *harmonia,* harmony.] **1. a.** Of or relating to musical harmony as distinguished from melody or rhythm. **b.** Of or relating to a harmonic or harmonics. **2.** Marked by agreeable harmony. —*n.* **1.** A tone in the harmonic series of overtones produced by a fundamental tone. **2.** A tone produced on a stringed instrument by lightly touching a vibrating string at a given fraction of its length so that both segments vibrate. **3.** A wave whose frequency is a whole-number multiple of that of another. **4. harmonics.** *(used with a sing. verb).* The theory or study of the physical properties and characteristics of musical sound. **—har·mon'i·cal·ly** *adv.*

**har·mon·i·ca** (här-mŏn'ĭ-kə) *n.* [Obs. *armonica,* glass bowl instrument < Ital. *armonica,* harmonious < Lat. *harmonicus,* harmonic. —see HARMONIC.] **1.** A small rectangular musical instrument having a row of free reeds set back in air holes, played by exhaling or inhaling. **2.** A musical instrument having a series of glass bowls of varying sizes played by rubbing the finger along the wet rims.

**harmonic analysis** *n.* Expression of mathematical functions by means of linear operations, as summation or integration on characteristic sets of functions, esp. such expression by means of a Fourier series.

**harmonic mean** *n.* The reciprocal of the arithmetic mean of the reciprocals of a specified set of numbers.

**harmonic progression** *n.* A sequence of quantities whose reciprocals form an arithmetic progression, as 1, 1/3, 1/5, 1/7, . . . .

**harmonic series** *n.* **1.** *Math.* A series whose terms are in harmonic progression, as 1 + 1/3 + 1/5 + 1/7 + . . . . **2.** A series of tones consisting of a fundamental tone and the overtones produced by it, whose frequencies are consecutive integral multiples of the frequency of the fundamental.

**har·mo·ni·ous** (här-mō'nē-əs) *adj.* **1.** Displaying accord and good will in feeling or action <a *harmonious* relationship> **2.** Having component elements pleasingly or appropriately combined <a *harmonious* structure> **3.** Marked by harmony of sound : MELODIOUS. **—har·mo'ni·ous·ly** *adv.* **—har·mo'ni·ous·ness** *n.*

**har·mo·nist** (här'mə-nĭst) *n.* One adept in musical harmony. **—har'mo·nis'tic** *adj.* **—har'mo·nis'ti·cal·ly** *adv.*

**har·mo·ni·um** (här-mō'nē-əm) *n.* [Fr. < *harmonie,* harmony < OFr. *armonie* < Lat. *harmonia.* —see HARMONY.] An organlike keyboard instrument producing tones with free metal reeds actuated by air forced from a bellows.

**har·mo·nize** (här'mə-nīz') *v.* **-nized, -niz·ing, -niz·es.** —*vt.* **1.** To bring into harmony or agreement. **2.** To provide harmony for (a melody). —*vi.* **1.** To be in agreement. **2.** To sing or play in harmony. **—har'mo·ni·za'tion** *n.* **—har'mo·niz'er** *n.*

**har·mo·ny** (här'mə-nē) *n., pl.* **-nies.** [ME *armonie* < Lat. *harmonia* < Gk. < *harmos,* joint.] **1.** Agreement in feeling or opinion : ACCORD. **2.** A pleasing combination of the elements forming a whole <*harmony* of color and design> **3.** *Mus.* **a.** The study of the structure, progression, and relation of chords. **b.** The simultaneous combination of notes in a chord. **c.** The structure of a musical work or passage as regarded from the point of view of its chordal characteristics and relationships. **4.** A combination of musical sounds thought to be pleasing. **5.** A collation of parallel passages, esp. from the Gospels, with a commentary demonstrating their consonance.

☆ ***syns:*** HARMONY, CONCORD, RAPPORT, UNITY *n. core meaning* : the state of individuals who are in total agreement <family *harmony*> ***ant:*** discord

**har·ness** (här'nĭs) *n.* [ME *harnes* < OFr., poss. of Germanic orig.] **1.** The gear or tackle, other than a yoke, of a draft animal. **2.** Something resembling a harness, as the arrangement of straps used to hold a parachute to the body. **3.** A device that raises and lowers the warp threads on a loom. **4.** *Archaic.* Armor for a man or a horse. —*vt.* **-nessed, -ness·ing, -ness·es. 1. a.** To put a harness on (a draft

o͞o **boot** ou **out** th **thin** *th* **this** ŭ **cut** ûr **urge** y **young**
yo͞o ab**u**se zh vi**si**on ə **a**bout, it**e**m, edible, gall**o**p, circ**u**s

animal). **b.** To fasten by means of a harness. **2.** To control and direct : UTILIZE <*harness* water power> **—in harness.** At one's usual job. **—in harness with.** In close association. **—har'ness·er** *n.*

**harness race** *n.* A horse race between trotters or pacers harnessed to sulkies.

**harp** (härp) *n.* [ME < OE *hearpe* and < OFr. *harpe* (of Germanic orig).] **1.** A musical instrument having an upright, open triangular frame with 46 strings of graded lengths played by plucking with the fingers. **2.** Something similar to a harp. *—vi.* **harped, harp·ing, harps.** To play a harp. **—harp on (***or* **upon).** To dwell on excessively and tediously. **—harp'er** *n.* **—harp'ist** *n.*

**har·pins** (här'pĭnz) *also* **har·pings** (-pĭngz) *pl.n.* [Perh. < HARP.] Extensions of the ribbands of a ship being built.

**har·poon** (här-po͞on') *n.* [Prob. < Du. *harpoen* < OFr. *harpon* < *harper,* to seize.] A spearlike weapon with a barbed head used in hunting whales and large fish. *—vt.* **-pooned, -poon·ing, -poons.** To strike or kill with or as if with a harpoon. **—har·poon'er** *n.*

**harpoon gun** *n.* A small cannonlike apparatus for firing harpoons.

**harp·si·chord** (härp'sĭ-kôrd', -kōrd') *n.* [Ital. *arpicordo* : *arpi,* harp + *corda,* string < Lat. *chorda* < Gk. *khordē.*] A keyboard instrument resembling a small piano whose strings are plucked by quills or leather plectrums. **—harp'si·chord'ist** *n.*

**Har·py** (här'pē) *n., pl.* **-pies.** [Lat. *Harpyia* < Gk. *Harpuiai.*] **1.** *Gk. Myth.* One of several loathsome voracious monsters with the head and trunk of a woman and the tail, wings, and talons of a bird. **2. harpy.** A predatory person. **3. harpy.** A shrewish woman.

**har·que·bus** (här'kwə-bəs, -kə-) *n.* [OFr. *harquebuse* < MLG *hakebusse* : *hake,* hook + *busse,* gun.] A heavy, portable matchlock gun invented during the 15th cent.

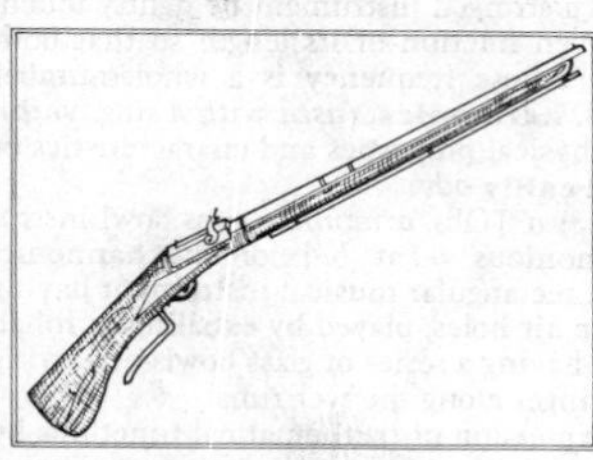
**harquebus**

**har·ri·dan** (hăr'ĭ-dn) *n.* [Poss. < Fr. *haridelle,* gaunt woman.] A shrewish woman : SCOLD.

**har·ri·er**[1] (hăr'ē-ər) *n.* **1.** One that harries. **2.** A slender narrow-winged hawk of the genus *Circus,* which preys on small animals.

**har·ri·er**[2] (hăr'ē-ər) *n.* [Prob. < HARE.] **1.** One of a breed of small hounds orig. used in hunting hares. **2.** A cross-country runner.

**har·row**[1] (hăr'ō) *n.* [ME *harwe.*] A farm implement having a heavy frame with sharp teeth or upright disks, used to break up and level off plowed ground. *—vt.* **-rowed, -row·ing, -rows. 1.** To work (soil or land) with a harrow. **2.** To inflict great distress or torment on. **—har'row·er** *n.*

**har·row**[2] (hăr'ō) *vt.* **-rowed, -row·ing, -rows.** [ME *herwen,* var. of *harien,* to harry.] *Archaic.* To plunder : sack.

**har·row·ing** (hăr'ō-ĭng) *adj.* Extremely distressing : TORMENTING <a *harrowing* illness>

**har·ry** (hăr'ē) *vt.* **-ried, -ry·ing, -ries.** [ME *harien* < OE *hergian.*] **1.** To pillage, as in war. **2.** To annoy or disturb by or as if by constant attacks : HARASS.

**harsh** (härsh) *adj.* **-er, -est.** [ME *harsk,* of Scand. orig.] **1.** Unpleasant to the senses, esp. to the ear. **2.** Extremely stern and severe <*harsh* punishment> **—harsh'ly** *adv.* **—harsh'ness** *n.*

☆ ***syns:*** HARSH, DISCORDANT, GRATING, HOARSE, RASPING, RASPY, ROUGH, SQUAWKY, SQUEAKY, STRIDENT *adj. core meaning* : disagreeable to the ear <a *harsh* metallic clanking>

**harsh·en** (här'shən) *vt. & vi.* **-ened, -en·ing, -ens.** To make or become harsh.

**hars·let** (här'slĭt) *n. var. of* HASLET.

**hart** (härt) *n., pl.* **harts** *or* **hart.** [ME < OE *heorot.*] A male deer, esp. a male red deer more than five years old.

**har·te·beest** (här'tə-bēst', härt'bēst') *n., pl.* **-beests** *or* **hartebeest.** [Obs. Afr. < Du. : *hart,* deer + *beest,* beast.] Either of two African antelopes, *Alcelaphus buselaphus* or *A. lichtensteini,* with ridged outward-curving horns and a brownish coat.

**hart's-tongue** (härts'tŭng') *n.* [So called from the shape of its fronds.] An evergreen fern, *Phyllitis scolopendrium,* with narrow undivided fronds.

**har·um-scar·um** (hâr'əm-skâr'əm, hăr'əm-skăr'əm) *adj.* [Perh. < HARE + SCARE.] Lacking a sense of responsibility : RECKLESS. **—har'um-scar'um** *adv.*

**ha·rus·pex** (hə-rŭs'pĕks', hăr'ə-spĕks') *n., pl.* **ha·rus·pi·ces** (hə-rŭs'pĭ-sēz') [Lat.] A Roman priest who practiced divination by the inspection of the entrails of animals.

**har·vest** (här'vĭst) *n.* [ME < OE *hærfest.*] **1.** The act or process of gathering a crop. **2.** The crop that ripens or is gathered in a season <the spring *harvest*> **3. a.** The amount or measure of the crop gathered in one season. **b.** The time or season of such gathering. **4.** The consequence or result of an action. *—v.* **-vest·ed, -vest·ing, -vests.** *—vt.* **1.** To gather (a crop). **2.** To gather a crop from. **3.** To receive the consequences of (an action). *—vi.* To gather a crop.

**harvest bug** *n.* CHIGGER 1.

**har·vest·er** (här'vĭ-stər) *n.* **1.** One who gathers a crop. **2.** REAPER 2.

**harvest fly** *n.* A cicada of the genus *Tibicen,* producing a shrill sound heard in late summer.

**harvest home** *n.* **1.** The completion of a harvest. **2. a.** The time of completing a harvest. **b.** A festival held at the completion of harvest. **c.** A song sung at this time.

**har·vest·man** (här'vĭst-mən) *n.* **1.** One who harvests. **2.** The daddy longlegs.

**harvest mite** *n.* CHIGGER 1.

**harvest moon** *n.* The full moon occurring nearest to the time of the autumnal equinox.

**has** (hăz) *v. 3rd person sing. present tense of* HAVE.

**has-been** (hăz'bĭn') *n. Informal.* One that is no longer successful, useful, or in favor.

**ha·sen·pfef·fer** (hä'zən-fĕf'ər) *n.* [G. : *Hase,* rabbit (< MHG < OHG *haso*) + *Pfeffer,* pepper (< OHG *pfeffar* < Lat. *piper*).] A highly seasoned stew of marinated rabbit meat.

**hash**[1] (hăsh) *n.* [Fr. *hachis* < *hacher,* to chop up < OFr. *hachier* < *hache,* ax, of Germanic orig.] **1.** A usu. baked or browned dish of chopped meat, potatoes, and sometimes vegetables. **2.** A hodgepodge : jumble. **3.** A reworking of already familiar material. *—vt.* **hashed, hash·ing, hash·es. 1.** To chop into pieces : MINCE. **2.** *Informal.* To make a mess of. **3.** *Informal.* To discuss carefully <*hash* over the itinerary for the trip> **—make a hash of. 1.** To make a mess of : BUNGLE. **2.** To defeat decisively. **—settle (someone's) hash.** To silence or subdue.

**hash**[2] (hăsh) *n. Slang.* Hashish.

**hash·eesh** (hăsh'ēsh') *n. var. of* HASHISH.

**hash house** *n. Slang.* A cheap restaurant.

**hash·ish** (hăsh'ēsh', -ĭsh) *also* **hash·eesh** (-ēsh') *n.* [Ar. *ḥashīsh.*] A purified extract derived from the dried flowers of the hemp plant, smoked or chewed as a mild narcotic.

**hash mark** *n. Slang.* A service stripe on the sleeve of an enlisted person's uniform.

**Ha·sid** (KHä'sĭd) *n. var. of* CHASSID.

**has·let** (hăs'lĭt, hăz'-) *also* **hars·let** (här'slĭt) *n.* [ME *hastelet* < OFr., dim. of *haste,* roast meat, spit, perh. < Lat. *hasta,* spear.] The heart, liver, and other edible viscera of an animal, esp. a hog.

**has·n't** (hăz'ənt). Has not.

**hasp** (hăsp) *n.* [ME < OE *hæpse.*] A metal fastener with a hinged, slotted part that fits over a staple and is secured by a pin, bolt, or padlock. *—vt.* **hasped, hasp·ing, hasps.** To fasten or lock with a hasp.

**Has·sid** (KHä'sĭd) *n. var. of* CHASSID.

**has·sle** (hăs'əl) [Poss. HA(GGLE) + (TU)SSLE.] *Informal. —n.* **1.** An argument or fight. **2.** Bother : trouble. *—v.* **-sled, -sling, -sles.** *—vi.* To argue or fight. *—vt.* To bother or harass <Don't *hassle* me with trivial questions.>

**has·sock** (hăs'ək) *n.* [ME *hassok,* clump of grass < OE *hassuc.*] **1.** A thick cushion used as a footstool or for kneeling. **2.** A dense clump of grass.

**hast** (hăst) *v. Archaic. 2nd person sing. present tense of* HAVE.

**has·tate** (hăs'tāt') *adj.* [NLat. *hastatus* < Lat. *hasta,* spear.] Shaped like a spearhead, as certain leaves. **—has'tate'ly** *adv.*

**haste** (hāst) *n.* [ME < OFr., of Germanic orig.] **1.** Rapidity : swiftness. **2.** Overeagerness to act. **3.** Rash or impetuous action : PRECIPITATENESS. *—vi. & vt.* **hast·ed, hast·ing, hastes.** To hasten or cause to hasten. **—make haste.** To hasten : hurry.

☆ ***syns:*** HASTE, BUSTLE, CELERITY, DISPATCH, EXPEDITION, HURRY, HUSTLE, RAPIDITY, SPEED, SPEEDINESS, SWIFTNESS *n. core meaning* : rapidness of movement or activity <left the room in great *haste*> *ant* : deliberateness

**has·ten** (hā'sən) *v.* **-tened, -ten·ing, -tens.** *—vi.* To move or act swiftly. *—vt.* **1.** To hurry. **2.** To expedite the progress of.

**hast·y** (hā'stē) *adj.* **-i·er, -i·est. 1.** Rapid : fast. **2.** Made or done too quickly to be accurate or wise : IMPETUOUS <a *hasty* judgment> **3.** Easily angered : IRRITABLE. **—hast'i·ly** *adv.* **—hast'i·ness** *n.*

**hasty pudding** *n.* **1.** Cornmeal mush served with a sweetening, as maple syrup or brown sugar. **2.** *Chiefly Brit.* A mush of flour or oatmeal.

**hat** (hăt) *n.* [ME < OE *hæt.*] **1.** A covering for the head, esp. one with a shaped crown and brim. **2. a.** A distinctive hat worn as a symbol of office. **b.** The office symbolized by the wearing of such a hat. **3.** A role or office symbolized by or as if by the wearing of special hats. *—vt.* **hat·ted, hat·ting, hats.** To cover or provide with a hat. **—at the drop of a hat.** At the slightest provocation or pretext. **—pass the hat.** To take up a collection of money. **—take (one's) hat off to.** To respect, admire, or congratulate. **—talk through (one's) hat. 1.** To talk nonsense. **2.** To bluff. **—throw (**or

**toss) (one's) hat into the ring.** To enter a political race. **—under (one's) hat.** As a secret or in confidence.

**hat·band** (hăt′bănd′) *n.* A band worn on a hat just above the brim.

**hat·box** (hăt′bŏks′) *n.* A usu. round box or case for a hat.

**hatch[1]** (hăch) *n.* [ME, small door < OE *hæc.*] **1. a.** An opening, as in the deck of a ship or the roof or floor of a building. **b.** The cover for such an opening. **c.** A hatchway. **d.** A ship's compartment. **2.** A floodgate.

**hatch[2]** (hăch) *v.* **hatched, hatch·ing, hatch·es.** [ME *hacchen.*] —*vi.* To emerge from or break out of an egg. —*vt.* **1.** To produce (young) from an egg. **2.** To cause (an egg) to produce young. **3.** To create or devise, esp. in secret <*hatch* a plot> —*n.* **1.** The act or an instance of hatching. **2.** The young hatched at one time : BROOD. **—hatch′er** *n.*

**hatch[3]** (hăch) *vt.* **hatched, hatch·ing, hatch·es.** [ME *hachen,* to inlay < OFr. *hachier,* to draw lines, to cut < *hache,* ax, of Germanic orig.] To shade by drawing or etching fine crossed or parallel lines on. —*n.* A fine line used in hatching.

**hatch·back** (hăch′băk′) *n.* An automobile, as a coupe, having a hatch that opens upward in a sloping back.

**hatch·el** (hăch′əl) *n.* [ME *hechele.*] A comb for separating flax fibers. —*vt.* **-eled, -el·ing, -els** *also* **-elled, -el·ling, els.** To separate (flax fibers) with a hatchel.

**hatch·er·y** (hăch′ə-rē) *n., pl.* **-ies.** A place where eggs, esp. those of fish or poultry, are hatched.

**hatch·et** (hăch′ĭt) *n.* [ME *hachet* < OFr. *hachete,* dim. of *hache,* ax, of Germanic orig.] **1.** A small short-handled ax for use in one hand. **2.** A tomahawk. **—bury the hatchet.** To make peace.

**hatchet face** *n.* A long, gaunt, sharp-featured face. **—hatch′et-faced′** *adj.*

**hatchet job** *n.* A usu. malicious verbal attack.

**hatchet man** *n. Slang.* **1.** A hired killer. **2.** One hired or assigned to carry out a disagreeable task or unscrupulous order.

**hatch·ing** (hăch′ĭng) *n.* **1.** The fine lines used in graphic arts for shading. **2.** The act or process of decorating with hatching.

**hatch·ment** (hăch′mənt) *n.* [Prob. alteration of ACHIEVEMENT.] A panel bearing the coat of arms of a deceased person.

**hatch·way** (hăch′wā′) *n.* **1.** A hatch leading to a hold, compartment, or cellar. **2.** A stairway or ladder within a hatchway.

**hate** (hāt) *v.* **hat·ed, hat·ing, hates.** [ME *haten* < OE *hatian.*] —*vt.* **1.** To feel great hostility or animosity toward. **2.** To feel dislike or distaste for <*hates* driving in traffic> —*vi.* To feel hatred. —*n.* **1.** Intense dislike or animosity : HATRED. **2.** An object of detestation or hatred <My pet *hate* is pollution.> **—hat′er** *n.*

☆ ***syns:*** HATE, ABOMINATE, DESPISE, DETEST, EXECRATE, LOATHE *v. core meaning* : to feel great hostility and dislike for <*hated* the occupying troops> ***ant*** : love

**hate·ful** (hāt′fəl) *adj.* **1.** Exciting hatred : ABOMINABLE. **2.** Feeling or exhibiting hatred : MALEVOLENT. **—hate′ful·ly** *adv.* **—hate′ful·ness** *n.*

**hath** (hăth) *v. Archaic. 3rd person sing. present tense of* HAVE.

**ha·tred** (hā′trĭd) *n.* [ME : *hate,* hate + OE *rǣden,* condition.] Intense animosity or hostility.

**hat·ter** (hăt′ər) *n.* One whose occupation is the manufacture, selling, or repair of hats.

**hat trick** *n.* [From the hat that was a reward for the feat.] **1.** Three wickets taken in cricket by a bowler in three consecutive balls. **2.** Three consecutive wins, hits, or goals made by one player in one game, as in ice hockey.

**hau·berk** (hô′bərk) *n.* [ME < OFr. *hauberc,* of Germanic orig.] A long tunic of chain mail.

**haugh** (hôKH) *n.* [ME *hawch* < OE *healh,* corner of land.] *Scot.* A low-lying meadow in a river valley.

**haugh·ty** (hô′tē) *adj.* **-ti·er, -ti·est.** [< ME *haut* < OFr. < Lat. *altus,* high.] Arrogantly and inordinately proud. **—haugh′ti·ly** *adv.* **—haugh′ti·ness** *n.*

**haul** (hôl) *v.* **hauled, haul·ing, hauls.** [ME *haulen* < OFr. *haler,* of Germanic orig.] —*vt.* **1.** To pull or drag forcibly. **2.** To transport, as with a truck or cart. **3.** To change the course of (a ship), esp. so as to sail closer into the wind. —*vi.* **1.** To pull or drag something. **2.** To provide transportation : CART. **3. a.** To shift direction, as the wind. **b.** To change one's mind. **4.** To change the course of a ship. **—haul off.** *Informal.* To draw back slightly, as in preparing for action <*hauled off* and belted the guy> **—haul up.** To come to a halt. —*n.* **1.** The act of pulling or dragging. **2.** The act of transporting or carting. **3.** A distance, esp. the distance over which something is hauled. **4.** Something hauled : LOAD. **5.** The total collected or acquired by a single effort <a *haul* of lobsters> **—haul′er** *n.*

**haul·age** (hô′lĭj) *n.* **1.** The act or process of hauling. **2.** A charge made for hauling.

**haulm** *also* **halm** (hôm) *n.* [ME *halm,* straw < OE.] *Chiefly Brit.* The stems of peas, beans, potatoes, or grasses.

**haunch** (hônch, hŏnch) *n.* [ME *haunche* < OFr. *hanche* < Med. Lat. *hancha,* of Germanic orig.] **1.** The hip, buttock, and upper thigh in humans and animals. **2.** The loin and leg of a four-footed animal, esp. used for food. **3.** Either of the sides of an arch, curving down from the apex to an impost.

**†haunt** (hônt, hŏnt) *v.* **haunt·ed, haunt·ing, haunts.** [ME *haunten,* to frequent < OFr. *hanter.*] —*vt.* **1.** To inhabit, visit, or appear to in the form of an apparition. **2.** To visit often : FREQUENT <*haunted* the poolroom> **3.** To recur continually to the mind of : OBSESS <The memory *haunted* me.> **4.** To be continually present in : PERVADE <despair that *haunts* the atmosphere of the conquered city> —*vi.* To recur or visit often, esp. as a ghost or other supernatural being. —*n.* **1.** A place much frequented. **2.** (hănt). *Regional.* A supernatural being, as a ghost.

**haunt·ing** (hôn′tĭng, hŏn′-) *adj.* Recurring continually to the mind : UNFORGETTABLE <a *haunting* theme song> **—haunt′ing·ly** *adv.*

**Hau·sa** (hou′sə, -zə) *n., pl.* **Hausa** *or* **-sas. 1.** One of a Negroid people of Niger and northern Nigeria. **2.** The language of the Hausa, widely employed as a trade language in western Africa.

**haus·frau** (hous′frou′) *n.* [G.] A housewife.

**haus·tel·lum** (hô-stĕl′əm) *n., pl.* **-tel·la** (-stĕl′ə) [NLat. < Lat. *haustus,* p.part. of *haurire,* to draw up.] A proboscis adapted as a sucking organ, as in many insects. **—haus·tel′late** (hô-stĕl′ĭt, hô′stə-lāt′) *adj.*

**haus·to·ri·um** (hô-stôr′ē-əm, -stōr′-) *n., pl.* **-to·ri·a** (-stôr′ē-ə, -stōr′-) [NLat. < Lat. *haustus,* p.part. of *haurire,* to draw up.] *Bot.* A specialized branch of hyphae or a similar structure by which parasitic plants such as fungi obtain food from a host plant. **—haus·to′ri·al** *adj.*

**haut·boy** *also* **haut·bois** (hō′boi′, ō′boi′) *n., pl.* **-boys** *also* **-bois** (-boiz′) [Fr. *hautbois* : *haut,* high (< Lat. *altus*) + *bois,* wood, of Germanic orig.] An oboe.

**haute cou·ture** (ōt′ ko͞o-to͝or′) *n.* [Fr. : *haute,* high + *couture,* sewing.] **1.** The leading designers who create exclusive fashions for women. **2.** The fashions created by these designers.

**haute cui·sine** (ōt′ kwĭ-zēn′) *n.* [Fr. : *haute,* high + *cuisine,* cooking.] **1.** Skillfully prepared and often elaborate cuisine. **2.** The food prepared in the style of haute cuisine.

**haute é·cole** (ōt′ ā-kôl′) *n.* [Fr. : *haute,* high + *école,* school.] The art, techniques, or practice of expert horsemanship.

**hau·teur** (hō-tûr′, ō-tûr′) *n.* [Fr. < *haut,* high < Lat. *altus.*] Disdainful arrogance or pride.

**have** (hăv) *v.* **had** (hăd), **hav·ing** (hăv′ĭng), **has** (hăz) [ME *haven* < OE *habban.*] —*vt.* **1. a.** To be in possession of <*has* a house> **b.** To possess as a characteristic, quality, or function <*has* ambition><*had* blue eyes> **c.** To possess or contain as a constituent part <a book that *has* a happy ending> **2.** To occupy a particular relation to <*had* many new students> **3.** To possess knowledge of or facility in <*has* little math> **4.** To hold in the mind : ENTERTAIN <*had* my suspicions> **5.** To exhibit in action <*have* pity> **6. a.** To come into possession of : ACQUIRE <the finest bicycle to be *had*> **b.** To receive : get <*had* good news> **c.** To accept : TAKE <I'll *have* the chocolate cake.> **7. a.** To suffer from <*has* a bad back> **b.** To be subject to the experience of <*had* a lot of trouble> **8. a.** To cause to be done or performed <*have* the car fixed> **b.** To cause to, as by persuasion or compulsion <*had* me give a talk> **c.** To cause to be <*had* us baffled> **9.** To permit : allow <won't *have* such language here> **10.** To carry on, perform, or execute <*have* a fight> **11. a.** To place at a disadvantage <I *have* my competitor now.> **b.** *Informal.* To trick or deceive <was *had* by a con man> **12. a.** To beget : bear. **b.** To give birth to <*had* twins> **13.** To partake of <*have* a snack> **14.** To be obliged to : MUST <I *have* to leave now.> **15.** To engage in sexual intercourse with. **16.** To influence by dishonest means : BRIBE <could not be *had* for any payoff> —*aux.v.* —Used with a past participle to form the following tenses indicating completed action: **a.** Present perfect <*has* left for Hawaii> **b.** Past perfect <thought I *had* lost the keys> **c.** Future perfect <will *have* arrived by the time you finish> **—have at.** To attack. **—have on. 1.** To wear. **2.** To be scheduled <*have* a meeting *on* for 10 A.M.> —*n.* One that possesses esp. material wealth. **—had better (*or* best).** Ought to <I *had better* call the doctor.> **—have done with.** To stop : cease <*Have done with* this silly arguing!> **—have had it. 1.** To have done everything that will be permitted. **2.** To have endured all one can. **3.** To be in a state beyond remedy, repair, or salvage. **—have it in for (someone).** To intend to harm, esp. because of a grudge. **—have it out.** To settle decisively, esp. through discussion or argument. **—have (one's) eye on. 1.** To look at, esp. attentively and continuously. **2.** To have as one's goal. **—have (something) coming.** To deserve whatever one receives. **—have to do with.** To be concerned or associated with.

**have·lock** (hăv′lŏk′, -lək) *n.* [After Henry *Havelock* (1795–1857).] A cloth covering for a cap, with a flap to cover and protect the back of the neck.

**ha·ven** (hā′vən) *n.* [ME < OE *hæfen.*] **1.** A harbor or anchorage : PORT. **2.** A sanctuary : refuge. —*vt.* **-vened, -ven·ing, -vens.** To put into or provide with a haven.

**have-not** (hăv′nŏt′) *n.* One having little or no material wealth.

**have·n't** (hăv′ənt). Have not.

**hav·er·sack** (hăv′ər-săk′) *n.* [Fr. *havresac* < G. *Habersack* : *Haber,* oats (< OHG *habaro*) + *Sack*, sack (< OHG *sac* < Lat. *saccus*).] A bag for supplies worn over one shoulder.

**hav·oc** (hăv′ək) *n.* [ME *havok* < AN.] **1.** Widespread destruction : DEVASTATION. **2.** Confusion, disorder, or chaos. —*vt.* **-ocked, -ock·ing, -ocs.** To pillage or destroy. **—cry havoc.** To sound an alarm.

**haw[1]** (hô) *n.* [Imit.] An utterance used by a speaker who is groping for words. —*vi.* **hawed, haw·ing, haws.** To hesitate in speaking.

**haw[2]** (hô) *n.* [ME < OE *haga.*] **1.** The fruit of a hawthorn. **2.** A hawthorn or similar tree or shrub.

**haw[3]** (hô) *n.* [Orig. unknown.] A nictitating membrane, esp. of a domesticated animal.

**haw[4]** (hô) *interj.* —Used to command an animal to turn left. —*vi.* **hawed, haw·ing, haws.** To turn left.

**Ha·wai·ian** (hə-wä′yən) *n.* **1.** A native or resident of Hawaii. **2.** The Polynesian language of Hawaii. **—Ha·wai′ian** *adj.*

**Hawaiian guitar** *n.* An electric guitar having a long sounding board and six to eight steel strings that are plucked while being pressed with a steel bar.

**haw·finch** (hô′fĭnch′) *n.* [HAW[2] + FINCH.] **1.** An Old World bird, *Coccothraustes coccothraustes*, with a thick bill, brown, white, and black plumage, and a short tail. **2.** Any of various birds similar to the hawfinch.

**haw-haw** (hô′hô′) *n. var. of* HA-HA[2].

**hawk[1]** (hôk) *n.* [ME *hauk* < OE *hafoc.*] **1. a.** Any of various birds of prey of the order Falconiformes, and esp. of the genera *Accipiter* or *Buteo,* with a short, hooked bill and strong claws adapted for seizing. **b.** Any of various similar birds. **2.** A predatory person. **3. a.** *Informal.* One who advocates military force or action in order to carry out foreign policy. **b.** One whose attitude is aggressive or combative, as in an argument. —*vi.* **hawked, hawk·ing, hawks. 1.** To hunt with trained hawks. **2.** To swoop and strike in the manner of a hawk. **—hawk′ish** *adj.* **—hawk′ish·ly** *adv.* **—hawk′ish·ness** *n.*

**hawk[2]** (hôk) *v.* **hawked, hawk·ing, hawks.** [Back-formation < HAWKER.] —*vi.* To peddle goods by calling out. —*vt.* To peddle (goods) by hawking.

**hawk[3]** (hôk) *v.* **hawked, hawk·ing, hawks.** [Imit.] —*vi.* To clear or attempt to clear the throat by or as if by coughing up phlegm. —*vt.* To clear the throat of (phlegm). —*n.* An audible effort to clear the throat by coughing up phlegm.

**hawk·er** (hô′kər) *n.* [Prob. < LG *höker* < MLG *hōker* < *hōken,* to peddle.] One who hawks goods : PEDDLER.

**hawk-eyed** (hôk′īd′) *adj.* Having keen vision.

**hawk moth** *n.* Any of various moths of the family Sphingidae, with a large body and long, narrow forewings and feeding while in flight on nectar from flowers.

**hawk's-beard** (hôks′bîrd′) *n.* A plant of the genus *Crepis,* related to the dandelion.

**hawks·bill** (hôks′bĭl′) *n.* A tropical sea turtle, *Eretmochelys imbricata,* that is a source of tortoiseshell.

**hawk·weed** (hôk′wēd′) *n.* An often hairy plant of the genus *Hieracium,* bearing yellow or orange dandelionlike flowers.

**hawse** (hôz) *n.* [ME *hals,* bow strake < ON, bow.] **1.** The part of a ship where the hawseholes are located. **2.** A hawsehole. **3.** The space between the bows and anchors of an anchored ship. **4.** The arrangement of the anchor cables of a ship when both starboard and port anchors are secured.

**hawse·hole** (hôz′hōl′) *n.* An opening in the bow of a ship through which a hawser is passed.

**haw·ser** (hô′zər) *n.* [ME < AN *hauceur* < OFr. *haucier,* to hoist < Lat. *altus,* high.] A cable or rope used in mooring or towing a ship.

**haw·thorn** (hô′thôrn′) *n.* [ME < OE *hagaðorn* : *haga,* haw + *ðorn,* thorn.] A thorny tree or shrub of the genus *Crataegus,* bearing white or pinkish flowers and reddish fruit.

**hay** (hā) *n.* [ME < OE *hīeg.*] **1.** Plants, as grass, clover, or alfalfa, that are cut and dried for fodder. **2.** *Slang.* A trifling amount of money <An outlay of $2 million is not *hay.*> —*v.* **hayed, hay·ing, hays.** —*vi.* To mow and cure herbage for hay. —*vt.* **1.** To make into hay. **2.** To feed with hay. **—hay′er** *n.*

**hay·cock** (hā′kŏk′) *n. Chiefly Brit.* A conical mound of hay.

**hay fever** *n.* An acute, allergic catarrhal condition marked by a running nose, sneezing, conjunctivitis, and headaches and often caused by an abnormal sensitivity to certain airborne pollens, esp. of the ragweed : POLLINOSIS.

**hay·fork** (hā′fôrk′) *n.* **1.** A fork-shaped hand tool for pitching hay. **2.** A machine-operated fork for moving hay.

**hay·loft** (hā′lôft′, -lŏft′) *n.* A loft for storing hay.

**hay·mak·er** (hā′mā′kər) *n.* **1.** One that hays. **2.** *Slang.* A forceful blow with the fist.

**hay·mow** (hā′mou′) *n.* **1.** A hayloft. **2.** The hay stored in a hayloft. **3.** *Archaic.* A haystack.

**hay·rack** (hā′răk′) *n.* **1.** A rack from which livestock feed. **2. a.** A rack fitted to a wagon for carrying hay. **b.** A wagon fitted with such a rack.

**hay·seed** (hā′sēd′) *n.* **1.** Grass seed shaken out of hay. **2.** Pieces of chaff or straw that fall from hay. **3.** *Slang.* A bumpkin : yokel.

**hay·stack** (hā′stăk′) *n.* A large stack of hay esp. for winter storage.

**hay·wire** (hā′wīr′) *n.* Wire used in baling hay. —*adj. Informal.* **1.** Failing to function properly <The TV went *haywire.*> **2.** Mentally or emotionally confused or upset.

**haz·ard** (hăz′ərd) *n.* [ME, dice game < OFr. *hasard* < OSp. *azar* < Ar. *azzahr,* gaming die.] **1.** A chance happening : ACCIDENT. **2.** A chance of being harmed or injured : DANGER. **3.** A possible source of danger <a highway *hazard*> **4.** A dice game similar to craps. **5.** An obstacle, as a sandtrap, on a golf course. —*vt.* **-ard·ed, -ard·ing, -ards. 1.** To expose to harm or danger. **2.** To attempt or risk : VENTURE <*hazard* a rough estimate>

**haz·ard·ous** (hăz′ər-dəs) *adj.* **1.** Dangerous. **2.** Depending on chance : RISKY. **—haz′ard·ous·ly** *adv.* **—haz′ard·ous·ness** *n.*

**haze[1]** (hāz) *n.* [Prob. back-formation < HAZY.] **1. a.** Atmospheric moisture, dust, smoke, and vapor suspended to form a partly opaque condition. **b.** The atmospheric condition so formed. **2.** A vague or muddled state of mind. —*vi.* **hazed, haz·ing, haz·es.** To become hazy or misty : BLUR.

☆ ***syns:*** HAZE, BRUME, FILM, FOG, MIST, SMAZE, SMOG *n. core meaning* : a thick, heavy atmospheric condition offering reduced visibility due to the presence of suspended particles <a *haze* of cigarette smoke><a cityscape obscured by *haze*>

**†haze[2]** (hāz) *vt.* **hazed, haz·ing, haz·es.** [Orig. unknown.] **1.** *Naut.* To harass or persecute with meaningless, difficult, or demeaning tasks. **2.** To initiate, as into a college fraternity, by playing rough practical jokes on or exacting demeaning performances from. **3.** *Regional.* To drive (e.g., cattle) with saddle horses. **—haz′er** *n.*

**ha·zel** (hā′zəl) *n.* [ME *hasel* < OE *hæsel.*] **1. a.** A shrub or small tree of the genus *Corylus,* esp. *C. avellana* of Europe or *C. americana* of North America, bearing edible nuts enclosed in a leafy husk. **b.** The nut of a hazel, with a smooth brown shell. **2.** A light to strong brown or yellowish brown. **—ha′zel** *adj.*

**ha·zel·nut** (hā′zəl-nŭt′) *n.* The edible hard-shelled nut of a hazel.

**haz·y** (hā′zē) *adj.* **-i·er, -i·est.** [Orig. unknown.] **1.** Marked by haze : MISTY. **2.** Not clearly defined. **—haz′i·ly** *adv.* **—haz′i·ness** *n.*

**haz·zan** (KHä′zən) *n. var. of* CHAZAN.

**H-bomb** (āch′bŏm′) *n.* A hydrogen bomb.

**he[1]** (hē) *pron.* [ME < OE *hē.*] **1.** The male that is neither the speaker nor the hearer. **2.** —Used to refer to a person whose sex is not specified <*He* goes quickest who goes alone.> —*n.* A male animal or person <Is the cat a *he?*>

**he[2]** (hā) *n.* [Heb. *hē.*] The fifth letter of the Hebrew alphabet. —See table at ALPHABET.

**He** *symbol for* HELIUM.

**head** (hĕd) *n.* [ME < OE *hēafod.*] **1. a.** The uppermost or forwardmost part of the body of a vertebrate, containing the brain or principal ganglia and the eyes, ears, nose, mouth, and jaws. **b.** The analogous part of an invertebrate organism. **2.** The seat of the faculty of reason : MIND <Use your *head!*> **3.** Mental ability or aptitude <a good *head* for arithmetic> **4.** Freedom of choice or of action <give him his *head*> **5.** *Slang.* **a.** A habitual drug user <an acid *head*> **b.** Fan : devotee <an ice-cream *head*> **6.** A portrait or representation of a person's head. **7.** *often* **heads** *(sing. in number).* The side of a coin having the principal design and date. **8.** *Informal.* A headache. **9. a.** An individual <tickets at $10 a *head*>> **b.** *pl.* **head.** One of a number, as of cattle or sheep. **10.** One who leads, rules, or is in charge of something <at the *head* of the line> **11.** The foremost or leading position. **12.** A headwater. **13. a.** The difference in depth of a liquid at two given points. **b.** The measure of pressure at the lower point expressed in terms of this difference. **c.** The pressure exerted by a liquid or gas <built up a *head* of steam> **14.** The foam or froth that rises to the top in pouring an effervescent liquid, as beer. **15.** The pus-filled tip of an abscess, boil, or pimple. **16.** A turning point : CRISIS <The situation came to a *head.*> **17. a.** A projection, weight, or fixture at the end of an elongated object <the *head* of a tack> **b.** The working end of a tool or implement, as of a hammer. **c.** The part of an explosive device carrying the explosive : WARHEAD. **18.** An attachment to or part of a machine that holds or contains the operative device. **19.** A rounded, compact mass of leaves, buds, or flowers <a *head* of lettuce> **20.** *Bot.* A dense, compact cluster of flowers, as of composite plants or clover. **21.** The uppermost part : TOP <My name was at the *head* of the list.> **22.** The end regarded as the most important <the *head* of the table> **23.** Either end of an object whose two ends are interchangeable, as a drum. **24.** *Naut.* **a.** The forward part of a vessel. **b.** The toilet on a ship. **c.** The top part or upper edge of a sail. **25.** A passage or gallery in a coal mine. **26. a.** The top of a book or of a page. **b.** *Informal.* A heading or headline. **c.** A distinct topic or category. **27.** Progress : headway. **28.** A word in a construction having the same grammatical function as the construction as a whole, as *child* in *a lovable, affectionate child.* —*v.* **head·ed, head·ing, heads.** —*vt.* **1.** To be in charge of : LEAD <I *headed* the task force.> **2.** To be in the first or foremost position of <*heads* the list> **3.** To aim, point, or turn in a certain direction <*headed* the car toward town> **4.** To remove the head or top of. **5.** To hit (a soccer ball) in the air with one's head. **6.** To place a heading on. —*vi.* **1.** To go or proceed in a certain direc-

ă **pat** ā **pay** âr **care** ä **father** ĕ **pet** ē **be** hw **which** ĭ **pit** ī **tie** îr **pier** ŏ **pot** ō **toe** ô **paw, for** oi **noise** ŏŏ **took**

tion <*head* for home> **2.** To form a head, as lettuce or cabbage. **3.** To originate, as a stream or river : RISE. **—go to (one's) head. 1.** To make one lightheaded or drunk. **2.** To make conceited. **—head and shoulders above.** Far superior to <*head and shoulders above* one's competitors in figures> **—head off.** To block the progress of : INTERCEPT <*head* the outlaws *off* at the pass> **—head over heels. 1.** Rolling, as in a somersault <fell *head over heels*> **2.** Deeply : completely <*head over heels* in love> **—keep (one's) head.** To remain calm. **—lose (one's) head.** To lose one's poise or self-control. **—off (or out of) (one's) head.** Delirious or crazy. **—over (one's) head. 1.** Beyond one's ability to understand or deal with. **2.** To a higher-ranking person <went *over* my *head* and complained to the boss> **—put heads together.** To consult and plan together.

**head·ache** (hĕd′āk′) *n.* **1.** A pain in the head. **2.** *Informal.* Something causing trouble or annoyance. **—head′ach′y** (-ā′kē) *adj.*

**head·band** (hĕd′bănd′) *n.* **1.** A band worn around the head. **2.** An ornamental strip at the top of a page or beginning of a chapter or paragraph. **3.** A cloth band attached to the top of the spine of a book.

**head·board** (hĕd′bôrd′, -bōrd′) *n.* A board or panel forming the head, as of a bed.

**head·cheese** (hĕd′chēz′) *n.* A jellied loaf or sausage made from edible parts of the feet, head, and sometimes the tongue and heart of an animal, esp. a hog.

**head cold** *n.* Coryza.

**head·dress** (hĕd′drĕs′) *n.* **1.** A covering or ornament for the head. **2.** A hairdo.

**head·ed** (hĕd′ĭd) *adj.* **1.** Growing or grown into a head. **2.** Having a head or heading. **3.** Having a specified kind or number of heads <a three-*headed* monster><a level-*headed* person>

**head·er** (hĕd′ər) *n.* **1.** One that fits a head on an object. **2.** One that removes a head from an object, esp. a machine that reaps and passes the heads of grain into a wagon or receptacle. **3.** A pipe that serves as a central connection for two or more smaller pipes. **4.** A beam placed between two long beams that supports the ends of the tailpieces. **5.** A brick laid across rather than parallel with a wall. **6.** *Informal.* A headlong fall or dive.

**head·first** (hĕd′fûrst′) *also* **head·fore·most** (-fôr′mōst′, -məst, -fōr′-) *adv.* **1.** With the head leading <*headfirst* down the steps> **2.** HEADLONG 2.

**head gate** *n.* **1.** A control gate upstream of a lock or canal. **2.** A floodgate controlling the flow of water, as in a sluice or channel.

**head·gear** (hĕd′gîr′) *n.* **1.** Something, as a hat or helmet, that covers the head. **2.** The part of a harness that fits about a horse's head. **3.** The hauling or lifting rigging at the head of a mine shaft.

**head-hunt·ing** (hĕd′hŭn′tĭng) *n.* **1.** The custom of cutting off and preserving the heads of enemies as trophies. **2.** *Slang.* The process of attempting to deprive usu. political enemies of position or power. **3.** *Slang.* The attempt to recruit esp. executive personnel, as for a corporation. **—head′-hunt′er** *n.*

**head·ing** (hĕd′ĭng) *n.* **1.a.** The title, subtitle, or topic at the top or beginning, as of a text. **b.** The address and date at the beginning of a business letter. **2.** The course or direction in which a ship or aircraft is moving. **3.** A gallery or drift in a mine.

**head·lamp** (hĕd′lămp′) *n.* A headlight.

**head·land** (hĕd′lənd, -lănd′) *n.* **1.** A point of land, usu. high and with a steep drop, extending out into a body of water : PROMONTORY. **2.** The unplowed land at the end of a plowed furrow.

**head·less** (hĕd′lĭs) *adj.* **1. a.** Without a head. **b.** Decapitated. **2.** Lacking a leader. **3.** Foolish. **—head′less·ness** *n.*

**head·light** (hĕd′līt′) *n.* A lamp mounted on the front of a vehicle.

**head·line** (hĕd′līn′) *n.* **1.** The title or caption of a published article, usu. set in large type. **2.** A line at the head of a page or passage giving information such as the title, author, and page number. *—vt.* **-lined, -lin·ing, -lines. 1.** To supply (e.g., a page) with a headline. **2. a.** To present as a headliner <The theater management *headlined* a classical guitarist.> **b.** To serve as the headliner of <the impressionist who *headlines* the bill>

**head·lin·er** (hĕd′lī′nər) *n.* A performer who receives prominent billing : STAR.

**head·lock** (hĕd′lŏk′) *n.* A wrestling hold in which the head of one wrestler is locked under the arm of the other.

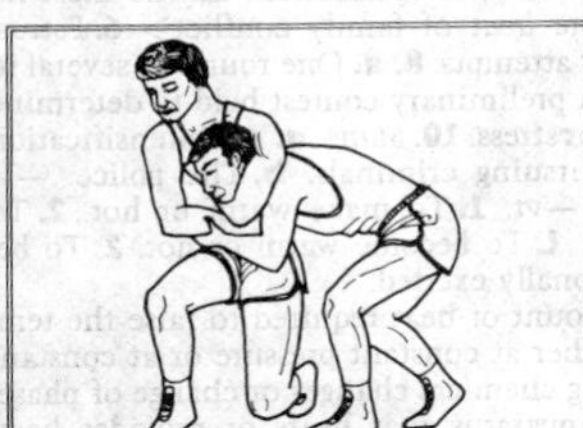
headlock

**head·long** (hĕd′lông′, -lŏng′) *adv.* [ME *hedling* < *hed*, head.] **1.** HEADFIRST 1. **2.** Impetuously : rashly. *—adj.* (hĕd′lông′, -lŏng′). **1.** Done with the head leading <a *headlong* plunge> **2.** Impetuous : rash. **3.** *Archaic.* Steep : sheer.

**head·mas·ter** *also* **head master** (hĕd′măs′tər) *n.* A man who is principal esp. of a private school.

**head·mis·tress** *also* **head mistress** (hĕd′mĭs′trĭs) *n.* A woman who is principal esp. of a private school.

**head·most** (hĕd′mōst′, -məst) *adj.* Leading : foremost.

**head-on** (hĕd′ŏn′, -ôn′) *adj.* **1.** Facing forward : FRONTAL. **2.** With the front end foremost <a *head-on* crash> **—head′on′** *adv.*

**head·phone** (hĕd′fōn′) *n.* A receiver, as for a radio, held to the ear by a headband.

**head·piece** (hĕd′pēs′) *n.* **1.** A protective covering for the head. **2.** A pair of headphones : HEADSET. **3.** A headstall. **4.** An ornamental design, esp. at the top of a page.

**head pin** *n.* KINGPIN 1.

**head·quar·ter** (hĕd′kwôr′tər) *vt.* & *vi.* **-tered, -ter·ing, -ters.** *Informal.* To provide with or establish headquarters.

**head·quar·ters** (hĕd′kwôr′tərz) *pl.n.* (*sing. or pl. in number*). **1.** The offices of a commander, as of a military unit, from which official orders are issued. **2.** A center of administration or operations.

***usage:*** The noun *headquarters*, which may occur with a singular or plural verb, usu. takes a plural verb when referring to a place, as in *The headquarters are in Boston*, and a singular verb when referring to authority, as in *Headquarters has refused to O.K. the project.*

**head·race** (hĕd′rās′) *n.* A watercourse that channels water into a mill, water wheel, or turbine.

**head·rest** (hĕd′rĕst′) *n.* **1.** A support for the head, as at the back of a chair. **2.** A padded cushion attached to the top of the back of an automobile seat, esp. to prevent whiplash injury.

**head restraint** *n.* HEADREST 2.

**head·sail** (hĕd′səl, -sāl′) *n.* A sail, as a jib, set forward of a foremast.

**head·set** (hĕd′sĕt′) *n.* A set of headphones.

**head·ship** (hĕd′shĭp′) *n.* The position or office of a leader.

**head shop** *n. Slang.* A specialty shop selling drug paraphernalia.

**head shrinker** *n. Slang.* A psychiatrist.

**heads·man** (hĕdz′mən) *n.* An executioner who beheads condemned prisoners.

**head·spring** (hĕd′sprĭng′) *n.* A principal source : FOUNTAINHEAD.

**head·stall** (hĕd′stôl′) *n.* The part of a bridle that fits over the head.

**head start** *n.* **1.** A start before other contestants in a race. **2.** An early start that confers an advantage <got a *head start* in preschool>

**head·stock** (hĕd′stŏk′) *n.* A stationary part of a machine or powered tool supporting a revolving part, as the spindle of a lathe.

**head·stone** (hĕd′stōn′) *n.* **1.** A memorial stone at the head of a grave. **2.** *also* **head stone.** A keystone.

**head·strong** (hĕd′strông′, -strŏng′) *adj.* **1.** Willful : obstinate. **2.** Resulting from willfulness and obstinacy.

**heads up** *interj.* —Used as a warning to be on the aware for a potential source of danger, as at a construction site.

**head·trip** (hĕd′trĭp′) *n. Slang.* **1.** An experience that stimulates the mind. **2.** An exploration of one's own perceptions or emotions.

**head·wait·er** (hĕd′wā′tər) *n.* A waiter in charge of the other waiters in a restaurant and often responsible for taking reservations and seating guests.

**head·wa·ter** (hĕd′wô′tər, -wŏt′ər) *n.* often **headwaters.** The water from which a river rises.

**head·way** (hĕd′wā′) *n.* **1.** Forward movement. **2.** Progress toward an objective. **3.** The clear vertical space beneath a ceiling or archway : CLEARANCE. **4.** The amount of distance or time separating two vehicles traveling the same route.

**head wind** *n.* A wind blowing directly opposite to the course of an aircraft or ship.

**head·work** (hĕd′wûrk′) *n.* Mental work. **—head′work′er** *n.*

**head·y** (hĕd′ē) *adj.* **-i·er, -i·est. 1.** Tending to make dizzy : INTOXICATING <a *heady* victory><a *heady* floral fragrance> **2.** Headstrong : obstinate. **—head′i·ly** *adv.* **—head′i·ness** *n.*

**heal** (hēl) *v.* **healed, heal·ing, heals.** [ME *helen* < OE *hǣlan.*] *—vt.* **1.** To restore to health or soundness : CURE. **2.** To set right : MEND <*healed* the break between us> **3.** To restore to spiritual wholeness. *—vi.* To return to health. **—heal′a·ble** *adj.* **—heal′er** *n.*

**heal-all** (hēl′ôl′) *n.* The self-heal.

**health** (hĕlth) *n.* [ME *helthe* < OE *hǣlð.*] **1.** The overall condition of an organism at a given time <in excellent *health*> **2.** Soundness, esp. of body or mind. **3.** A condition of optimal well-being <economic *health*> **4.** A wish for someone's good health, often expressed as a toast.

**health food** *n.* A food held to be highly beneficial to health.

**health·ful** (hĕlth′fəl) *adj.* **1.** Conducive to good health : SALUBRIOUS. **2.** Healthy. **—health′ful·ly** *adv.* **—health′ful·ness** *n.*

**health insurance** *n.* Insurance against expenses incurred through illness of the insured.

**health spa** *n.* A business establishment with equipment and facilities to help customers lose weight.

**health·y** (hĕl′thē) *adj.* **-i·er, -i·est. 1.** Having good health. **2.** Conducive to good health : HEALTHFUL <*healthy* food> **3.** Indicative of good health : SOUND <a *healthy* outlook> **4.** Of considerable size : LARGE <a *healthy* serving> ***usage:*** Although the term *healthful* is largely restricted to the sense "conducive to health," *healthy* may be used both in that sense and also in the sense "full of health." Therefore one may properly refer to both *a healthy person* and a *healthy climate.* **—health'i·ly** *adv.* **—health'i·ness** *n.*

☆ ***syns:*** HEALTHY, FIT, HALE, HARDY, HEARTY, RIGHT, ROBUST, SOUND, VIGOROUS, WELL, WHOLE, WHOLESOME *adj. core meaning* : possessing good health <a *healthy* baby> ***ant:*** unhealthy

**heap** (hēp) *n.* [ME < OE *hēap.*] **1.** A group of things piled haphazardly or in disorder. **2.** *Informal.* A great deal. **3.** *Slang.* An old or dilapidated car. *—vt.* **heaped, heap·ing, heaps. 1.** To put or pile up in a heap. **2.** To fill to overflowing. **3.** To bestow lavishly or in abundance.

**hear** (hîr) *v.* **heard** (hûrd), **hear·ing, hears.** [ME *hearen* < OE *hīeran.*] *—vt.* **1.** To perceive by means of the ear. **2.** To listen to attentively. **3. a.** To learn by hearing. **b.** To acquire, as information or news. **4.** To listen to in an official, professional, or formal capacity <The judge *heard* the evidence.> **5.** To attend (e.g., a lecture or religious service). *—vi.* **1.** To be capable of perceiving sound. **2.** To receive information or news : LEARN <We *heard* about your illness.> **3.** To consider something. —Used in the negative <They won't *hear* of a party.> **—hear'er** *n.*

**hear·ing** (hîr'ĭng) *n.* **1.** The sense by which sound is perceived. **2.** Range of audibility : EARSHOT. **3.** A chance to be heard. **4.** *Law.* **a.** A preliminary examination of an accused person. **b.** The trial of an equity case. **5.** A session, as of an investigatory committee, at which witnesses submit testimony.

**hearing aid** *n.* An electronic apparatus that amplifies sound and is worn to compensate for poor hearing.

**hear·ken** *also* **har·ken** (här'kən) *v.* **-kened, -ken·ing, -kens.** [ME *herknen* < OE *hercnian.*] *—vi.* To listen attentively. *—vt. Archaic.* To listen to : HEAR.

**hear·say** (hîr'sā') *n.* **1.** Information heard from another. **2.** *Law.* Evidence based on the reports of others rather than on the personal knowledge of a witness and therefore gen. not admissible.

**hearse** (hûrs) *n.* [ME *herse,* a harrow-shaped structure for holding candles over a coffin < OFr. *herce* < Med. Lat. *hercia* < Lat. *hirpex,* harrow.] **1.** A vehicle for conveying a dead person to a church or cemetery. **2.** *Rom. Cath. Ch.* A triangular candelabrum used at Tenebrae during Holy Week. **3.** A framelike structure over a coffin or tomb on which to hang epitaphs.

▲ word history: The common meaning of *hearse* denoting a vehicle was the latest to develop. The Latin word *hirpex,* the ancestor of *hearse,* meant "harrow," an agricultural implement that in earlier times consisted of a framework with teeth that was used to break up plowed land. The emergence of the sense "vehicle" occurred by a series of logical steps. The first sense of English *hearse,* recorded in the 13th century, was that of a framework used for holding candles, which no doubt resembled an agricultural harrow. This framework was used at the Tenebrae service during Holy Week, at which all the candles in the church were extinguished in commemoration of Christ's death and descent into hell. Such hearses were also used to hold candles over a coffin at a funeral. Later extensions of the meaning of *hearse* preserved the idea of a framework and the funereal circumstances of its use. In the 16th century the word denoted a support for the funeral pall, and in the 17th century it indicated the bier or coffin itself. Finally, late in the 17th century, *hearse* was used for the vehicle that carries the coffin at a funeral.

**heart** (härt) *n.* [ME *hert* < OE *heorte.*] **1. a.** *Anat.* The hollow, muscular organ in vertebrates that pumps blood received from the veins into the arteries, thereby supplying the entire circulatory system. **b.** A similarly functioning structure in invertebrates. **2.** The approximate bodily area in which the heart is located : BREAST. **3.** The vital center of one's being, emotions, and sensibilities. **4. a.** Emotional constitution, disposition, or mood <a hard *heart*> **b.** Capacity for concern or sympathy : COMPASSION <a teacher with *heart*> **c.** Love : affection <The kitten won my *heart.*> **5.** Courage : fortitude <Don't lose *heart!*> **6.** The most central and material part. **7.** A conventionalized two-lobed representation of the heart. **8. a.** A playing card bearing a red, heart-shaped symbol. **b. hearts.** The suit of cards marked with this symbol. **9. hearts** *(sing. in number).* A card game in which the object is either to avoid hearts when taking tricks or to take all the hearts. *—vt.* **heart·ed, heart·ing, hearts.** *Archaic.* To encourage : hearten. **—by heart.** By memory or rote. **—heart and soul.** Completely : entirely. **—to (one's) heart's content.** To one's unlimited satisfaction. **—with all (one's) heart. 1.** With great willingness or pleasure. **2.** With the deepest feeling or devotion.

☆ ***syns:*** HEART, CORE, ESSENCE, GIST, KERNEL, MARROW, MEAT, NITTY-GRITTY, NUB, PITH, QUINTESSENCE, ROOT, SOUL, STUFF *n. core meaning* : the most central and material part <Let's get to the *heart* of the matter.>

**heart·ache** (härt'āk') *n.* Emotional sorrow or anguish.

**heart attack** *n.* **1.** The condition or an instance of heart failure. **2.** An acute episode of abnormal heart functioning, as a coronary thrombosis.

**heart·beat** (härt'bēt') *n.* **1.** A single complete pulsation of the heart. **2.** The vital force or driving impulse <the *heartbeat* of the enterprise>

**heart block** *n.* Reduction or complete lack of coordination in the beating of the atria and ventricles of the heart.

**heart·break** (härt'brāk') *n.* Intense grief or sorrow.

**heart·break·ing** (härt'brā'kĭng) *adj.* **1.** Causing heartbreak. **2.** Causing great difficulty or distress. **—heart'break'ing·ly** *adv.*

**heart·bro·ken** (härt'brō'kən) *adj.* Suffering from or displaying great grief or despair. **—heart'bro'ken·ly** *adv.* **—heart'bro'ken·ness** *n.*

**heart·burn** (härt'bûrn') *n.* A burning sensation in the stomach and esophagus, often accompanied by the eructation of small quantities of acidic fluid, caused by stomach hyperacidity.

**heart disease** *n.* Organic or functional abnormality of the heart.

**heart·en** (här'tn) *vt.* **-ened, -en·ing, -ens.** To encourage.

**heart failure** *n.* Inability of the heart to pump blood at an adequate rate, resulting in congestion in the tissues, shortness and wheezing of breath, pitting edema, and enlarged and tender liver.

**heart·felt** (härt'fĕlt') *adj.* Deeply or earnestly felt : SINCERE.

**hearth** (härth) *n.* [ME *herth* < OE *heorth.*] **1.** The floor of a fireplace, usu. extending into a room and paved with brick, flagstone, or cement. **2.** The fireside as a symbol of home or family life. **3.** *Metallurgy.* **a.** The lowest part of a blast furnace or cupola, from which the molten metal flows. **b.** The bottom of a reverberatory furnace, where ore is exposed to the flame. **4.** The fireplace of a blacksmith's forge.

**hearth money** *n.* Peter's pence.

**hearth·stone** (härth'stōn') *n.* **1.** Stone used in the construction of a hearth. **2.** The home. **3.** A soft stone or powder used for scouring and whitening hearths or doorsteps.

**heart·i·ly** (här'tl-ē) *adv.* **1.** In a hearty manner. **2.** With sincere good will : WARMLY. **3.** Thoroughly : entirely <wished *heartily* that I were elsewhere> **4.** With great appetite or gusto <dined *heartily*>

**heart·land** (härt'lănd') *n.* A central region, esp. one held in geopolitical theory to be economically or militarily vital to a nation.

**heart·less** (härt'lĭs) *adj.* **1.** Lacking compassion. **2.** *Archaic.* Lacking enthusiasm. **—heart'less·ly** *adv.* **—heart'less·ness** *n.*

**heart-rend·ing** (härt'rĕn'dĭng) *adj.* Evoking anguish or distress.

**hearts·ease** *also* **heart's-ease** (härts'ēz') *n.* [ME *herts ease.*] **1.** Peace of mind. **2.** A Eurasian plant, *Viola tricolor,* bearing small, spurred, variously colored flowers.

**heart·sick** (härt'sĭk') *adj.* Profoundly despondent. **—heart'sick'·ness** *n.*

**heart-strick·en** (härt'strĭk'ən) *also* **heart-struck** (-strŭk') *adj.* Overwhelmed with strong emotion, as grief or remorse.

**heart·string** (härt'strĭng') *n.* **1. heartstrings.** The deepest feelings or affections. **2.** One of the nerves or tendons once thought to brace and sustain the heart.

**heart-struck** (härt'strŭk') *adj. var. of* HEART-STRICKEN.

**heart·throb** (härt'thrŏb') *n.* **1.** A heartbeat. **2. a.** Tender or sentimental emotion. **b.** A sweetheart.

**heart-to-heart** (härt'tə-härt') *adj.* Candid : frank.

**heart·warm·ing** (härt'wôrm'ĭng) *adj.* Prompting feelings of inspiration or encouragement.

**heart·wood** (härt'wo͝od') *n.* The older, inactive central wood of a tree or woody plant, usu. darker and harder than the sapwood.

**heart·worm** (härt'wûrm') *n.* A nematode worm, *Dirofilaria immitis,* parasitic in the heart and bloodstream of mammals, as dogs.

**heart·y** (här'tē) *adj.* **-i·er, -i·est. 1.** Marked by unrestrained warmth of feeling <a *hearty* greeting> **2.** Thoroughgoing : unequivocal <a *hearty* dislike> **3.** Robust : vigorous. **4. a.** Enjoying or requiring much food. **b.** Providing abundant nourishment. **c.** Substantial : satisfying <a *hearty* breakfast> *—n., pl.* **-ies. 1.** A good fellow : COMPANION. **2.** A sailor.

**heat** (hēt) *n.* [ME *hete* < OE *hǣtu.*] **1.** A form of energy associated with the motion of atoms or molecules in solids and capable of being transmitted through solid and fluid media by conduction, through fluid media by convection, and through empty space by radiation. **2.** The physiological sensation of being hot. **3. a.** A pathologically high body temperature. **b.** Intense or excessive warmth. **4.** The condition of being hot. **5. a.** Intensity, as of emotion. **b.** The most intense or active stage <in the *heat* of family conflict> **6.** Estrus. **7.** One of a series of efforts or attempts. **8. a.** One round of several in a competition, as a race. **b.** A preliminary contest held to determine finalists. **9.** *Informal.* Pressure : stress. **10.** *Slang.* **a.** An intensification esp. of police activity in pursuing criminals. **b.** The police. *—v.* **heat·ed, heat·ing, heats.** *—vt.* **1.** To make warm or hot. **2.** To inflame the feelings of. *—vi.* **1.** To become warm or hot. **2.** To become intellectually or emotionally excited.

**heat capacity** *n.* The amount of heat required to raise the temperature of a body by 1°, either at constant pressure or at constant volume and without inducing chemical changes or change of phase.

**heat·er** (hē'tər) *n.* **1.** An apparatus that heats or provides heat.

2. One who heats something or tends a heating apparatus. 3. *Slang.* A pistol.

**heat exchanger** *n.* A device, as a radiator, used to transfer heat from one fluid to another without mixture.

**heat exhaustion** *n.* A reaction to excessive heat, marked by profuse sweating, weakness, dizziness, and collapse.

**heath** (hēth) *n.* [ME, uncultivated land < OE *hǣð.*] 1. An Old World, usu. low-growing shrub of the genus *Erica* and related genera, bearing small evergreen leaves and small, urn-shaped pink or purplish flowers. 2. An extensive tract of open, uncultivated land covered with herbage and low shrubs.

**hea·then** (hē'thən) *n., pl.* **-thens** *or* **heathen.** [ME *hethen* < OE *hǣðen.*] 1. a. One who belongs to a tribe or nation that does not acknowledge the God of Judaism, Christianity, or Islam. b. Such persons as a group. 2. a. One considered irreligious or uncivilized. b. Such persons as a group. **—hea'then** *adj.* **—hea'then·dom** (-dəm), **hea'then·ism, hea'then·ry** *n.*

**hea·then·ish** (hē'thə-nĭsh) *adj.* 1. Of or pertaining to the heathen. 2. Uncivilized : barbarous. **—hea'then·ish·ly** *adv.*

**heath·er** (hĕth'ər) *n.* [ME *hathir.*] 1. A low-growing Eurasian shrub, *Calluna vulgaris*, growing in dense masses and bearing small evergreen leaves and pinkish-purple flower clusters. 2. HEATH 1. 3. A grayish purple to purplish red. **—heath'er** *adj.*

**heath·er·y** (hĕth'ə-rē) *adj.* 1. Of, pertaining to, or resembling heather. 2. Flecked with various colors.

**heath hen** *n.* A prairie chicken, *Tympanuchus cupido*, now extinct in eastern North America.

**heat island** *n.* An industrial or urban area in which a greater amount of heat is retained, as by buildings and streets, than in nearby areas.

**heat lightning** *n.* Intermittent flashes of light across the horizon on a hot summer evening, unaccompanied by thunder and held to be cloud reflections of distant lightning.

**heat of fusion** *n.* The amount of heat needed to melt a unit mass of a solid at a specified temperature.

**heat of vaporization** *n.* The amount of heat needed to convert a unit liquid mass at a specified temperature into vapor.

**heat prostration** *n.* Heat exhaustion.

**heat pump** *n.* An engine that transfers heat from a relatively low-temperature reservoir to one at a higher temperature.

**heat rash** *n.* Miliaria.

**heat shield** *n.* A barrier that prevents the heating of a space by absorbing, reflecting, or dissipating external heat, esp. a protective structure on a spacecraft or missile that dissipates heat on atmospheric re-entry by melting and vaporizing.

**heat sink** *n.* 1. An environment having a greater heat capacity and at a lower temperature than an object with which it is in thermal contact. 2. A device by means of which heat is absorbed or stored in or removed from a thermal system.

**heat stroke** *n.* Severe illness caused by exposure to excessively high temperatures and marked by cessation of sweating, high fever, tachycardia, and in serious cases collapse and coma.

**heat-treat** (hēt'trēt') *vt.* **-treat·ed, -treat·ing, -treats.** To treat (e.g., metal) by alternate heating and cooling in order to produce desired characteristics. **—heat treater** *n.* **—heat treatment** *n.*

**heat wave** *n.* A period of unusually hot weather.

**heave** (hēv) *v.* **heaved, heav·ing, heaves.** [ME *heven* < OE *hebban.*] —*vt.* 1. To hoist with great force or effort. 2. To throw with great effort : HURL. 3. To utter effortfully <*heaved* a sigh of relief> 4. *p.t. & p.p.* **hove** (hōv). *Naut.* To raise or haul by a rope. 5. a. To cause to rise. b. To displace or move (e.g., a stratum). —*vi.* 1. To rise up or swell : BULGE. 2. *Informal.* To retch. 3. *p.t. & p.p.* **hove** (hōv). *Naut.* a. To move to a specified position. b. To move a ship in a particular manner or direction. **—heave to.** To bring or come to a stop. —Used of a ship. —*n.* 1. The effort of heaving. 2. An act of hurling : THROW. 3. A horizontal dislocation, as of a rock stratum, at a fault. 4. a. An upward movement. b. **heaves.** *Informal.* The act or an instance of retching. 5. **heaves** *(sing. or pl. in number).* A pulmonary disease of horses, manifested esp. in cold weather or after exercise and characterized by respiratory irregularities, as coughing. **—heav'er** *n.*

**heave-ho** (hēv'hō') *n.* [< *heave ho!*, an interjection used by sailors when pulling on a rope.] *Informal.* Dismissal, esp. from employment.

**heav·en** (hĕv'ən) *n.* [ME *heven* < OE *heofan.*] 1. *often* **heavens.** The sky or universe as seen from the earth : FIRMAMENT. 2. The abode of God, the angels, and the souls of those granted salvation. 3. **Heaven.** GOD 1a, b. 4. A condition or place of supreme happiness.

**heav·en·ly** (hĕv'ən-lē) *adj.* 1. Delightful : sublime. 2. Of or pertaining to heaven or the heavens : CELESTIAL. 3. Of or pertaining to the abode of God. **—heav'en·li·ness** *n.*

**heav·en·ward** (hĕv'ən-wərd) *adj. & adv.* Toward heaven. **—heav'en·wards** (-wərdz) *adv.*

**heav·i·er-than-air** (hĕv'ē-ər-thən-âr') *adj.* Being an aircraft heavier than the air it displaces.

**heav·i·ly** (hĕv-ə-lē) *adv.* 1. In a heavy manner. 2. Slowly and laboriously. 3. Greatly or severely <*heavily* armed>

**Heav·i·side layer** (hĕv'ĭ-sīd') *n.* [After Oliver *Heaviside* (1850–1925).] E layer.

**heav·y** (hĕv'ē) *adj.* **-i·er, -i·est.** [ME *hevy* < OE *hefig.*] 1. Having great weight. 2. Having relatively high density or a high specific gravity. 3. a. Large, as in number or quantity <*heavy* snowfall><*heavy* traffic> b. Large in capacity, yield, or output. 4. a. Dense or thick <*heavy* smog> b. Of great intensity or depth <*heavy* thoughts> 5. a. Having great power or force, as stormy seas. b. Difficult or awkward to move, esp. due to great weight. c. Stoutly built. 6. a. Indulging to a great degree <a *heavy* eater> b. Involved or participating on a large scale <a *heavy* bettor> 7. a. Of great import or seriousness : GRAVE. b. Sad or painful <*heavy* tidings> 8. a. Hard to do or accomplish : ARDUOUS. b. Not easily borne : OPPRESSIVE <*heavy* debts> 9. Too rich to digest easily or quickly. 10. Having large or marked physical features : COARSE. 11. Weighed down with concern or sadness <a *heavy* spirit> 12. Lacking vitality. 13. Deficient in vivacity or grace. 14. Strong and pervasive : PUNGENT <a *heavy* perfume> 15. a. Bent down : BURDENED <trees *heavy* with apples> b. Showing or marked by drowsiness or weariness <eyes *heavy* with sleep> 16. Of, pertaining to, or involving the production of basic products, as steel. 17. Pregnant. 18. Of or relating to a serious dramatic role. 19. *Physics.* a. Designating an isotope with a mass greater than that of others found in the same element. b. Designating an atomic particle having a mass between that of pi mesons and protons. 20. Bearing weighty arms or armor. 21. *Slang.* Of great import or profundity. 22. *Slang.* Very popular or important. —*adv.* Heavily. —*n., pl.* **-ies.** 1. A villain in a story or play. 2. A villain. 3. a. A serious or tragic dramatic role. b. An actor performing a heavy role. 4. *Slang.* One that is very important or influential. **—heav'i·ness** *n.*

☆ ***syns:*** 1. HEAVY, BURDENSOME, HEFTY, MASSIVE, PONDEROUS, WEIGHTY *adj. core meaning* : having great physical weight <a *heavy* boulder> ***ant*** : light 2. HEAVY, CUMBERSOME, PONDEROUS *adj. core meaning* : unwieldy or awkward, esp. because of excess weight <*heavy* sandbags><a *heavy* grand piano>

**heav·y-dut·y** (hĕv'ē-dōō'tē, -dyōō'-) *adj.* Made to withstand hard use or wear <*heavy-duty* machinery>

**heav·y-foot·ed** (hĕv'ē-fŏŏt'ĭd) *adj.* Having a heavy, lumbering gait.

**heav·y-hand·ed** (hĕv'ē-hăn'dĭd) *adj.* 1. Awkward : clumsy. 2. Oppressive : harsh. **—heav'y-hand'ed·ness** *n.*

**heav·y-heart·ed** (hĕv'ē-här'tĭd) *adj.* Melancholy and sad : DEPRESSED. **—heav'y-heart'ed·ly** *adv.* **—heav'y-heart'ed·ness** *n.*

**heavy hydrogen** *n.* An isotope of hydrogen with mass number greater than 1 : DEUTERIUM.

**heavy metal** *n. Slang.* Hard rock music.

**heav·y·set** (hĕv'ē-sĕt') *adj.* Having a heavy, compact build.

**heavy spar** *n.* Barite.

**heavy water** *n.* Any of several isotopic varieties of water, esp. deuterium oxide, consisting primarily or exclusively of molecules containing hydrogen with mass number greater than 1 and used as a moderator in certain nuclear reactors.

**heav·y·weight** (hĕv'ē-wāt') *n.* 1. One of above average weight. 2. One competing in the heaviest class, esp. a boxer weighing more than 175 pounds or 81 kilograms. 3. *Informal.* One of great importance or influence : BIGWIG.

**heb·do·mad** (hĕb'də-măd') *n.* [Lat. *hebdomas, hebdomad-*, the number seven < Gk. < *hepta*, seven.] 1. A group of seven. 2. A period of seven days : WEEK.

**heb·dom·a·dal** (hĕb-dŏm'ə-dəl) *adj.* Occurring weekly. **—heb·dom'a·dal·ly** *adv.*

**He·be** (hē'bē) *n.* [Gk. *Hēbē* < *hēbē*, youthful.] *Gk. Myth.* The goddess of youth and spring.

**he·be·phre·ni·a** (hē'bə-frē'nē-ə, -frĕn'ē-) *n.* [Gk. *hēbē*, youth + -PHRENIA.] A schizophrenia marked by foolish mannerisms, delusions, hallucinations, and regressive behavior. **—he'be·phren'ic** (-frĕn'ĭk, -frē'nĭk) *adj.*

**heb·e·tate** (hĕb'ĭ-tāt') *vt.* **-tat·ed, -tat·ing, -tates.** [Lat. *hebetare, hebetat-* < *hebes*, blunt.] To make dull or obtuse. —*adj. Bot.* Having a blunt point <*hebetate* leaves> **—heb'e·ta'tion** *n.* **—heb'e·ta'tive** *adj.*

**heb·e·tude** (hĕb'ĭ-tōōd', -tyōōd') *n.* [LLat. *hebetudo* < Lat. *hebes*, dull.] Mental lethargy. **—heb'e·tu'di·nous** (-tōōd'n-əs, -tyōōd'-) *adj.*

**He·bra·ic** (hĭ-brā'ĭk) *also* **He·bra·i·cal** (-ĭ-kəl) *adj.* [ME *Ebraik* < LLat. *Hebraicus* < Gk. *Hebraikos* < *Hebraios.* —see HEBREW.] Of, relating to, or typical of the Hebrews or their language or culture. **—He·bra'i·cal·ly** *adv.*

**He·bra·ism** (hē'brā-ĭz'əm) *n.* [< HEBRAIC.] 1. A custom or manner typical of the Hebrews. 2. A linguistic feature typical of Hebrew occurring esp. in another language. 3. The character, culture, or spirit of the Hebrew people. 4. Judaism.

**He·bra·ist** (hē'brā'ĭst) *n.* A scholar who studies Hebrew. **—He'bra·is'tic, He'bra·is'ti·cal** *adj.* **—He'bra·is'ti·cal·ly** *adv.*

**He·bra·ize** (hē'brā-īz') *v.* **-ized, -iz·ing, -iz·es.** —*vt.* To make Hebraic, as in idiom. —*vi.* To use or adopt Hebraisms. **—He'bra·i·za'tion** *n.*

**He·brew** (hē′bro͞o) *n.* [ME *Ebreu* < OFr. < Lat. *Hebraeus,* Hebraic < Gk. *Hebraios* < Aram. *'ibhray* < Heb. *'ibhrī* < *'ēbher,* region across < *ābhar,* he crossed over.] **1.** A member or descendant of a northern Semitic people : ISRAELITE. **2. a.** The Semitic language of the ancient Hebrews. **b.** Any of the various later forms of this language, esp. the language of the Israelis. **3. Hebrews** *(sing. in number).* —See table at BIBLE. **—He′brew** *adj.*

**Hebrew Scriptures** *pl.n.* The Pentateuch, the Prophets, and the Hagiographa, forming of the covenant between God and the Jewish people that is the foundation of Judaism. —See table at BIBLE.

**Hec·a·te** (hĕk′ə-tē) *n.* [Lat. < Gk. *Hekatē.*] *Gk. Myth.* A goddess identified with the underworld and with witches and sorcery.

**hec·a·tomb** (hĕk′ə-tōm′) *n.* [Lat. *hecatombe* < Gk. *hekatombē* : *hekaton,* hundred + *bous,* ox.] **1.** A sacrifice to the gods in ancient Greece and Rome consisting orig. of 100 oxen. **2.** A large-scale sacrifice or slaughter.

**heck** (hĕk) *interj.* & *n.* Hell.

**heck·le** (hĕk′əl) *vt.* **-led, -ling, -les.** [ME *hekelen,* to comb with a hatchel < *hekel,* hatchel < MDu.] **1.** To attempt to annoy or embarrass by questions, satirical comments, or objections : BADGER. **2.** To comb (flax or hemp) with a hatchel. **—heck′ler** *n.*

**hect-** *pref. var. of* HECTO-.

**hec·tare** (hĕk′târ′) *n.* A metric unit of area equal to 100 ares or 2.471 acres.

**hec·tic** (hĕk′tĭk) *adj.* [ME *etik,* recurring, consumptive < OFr. *etique* < LLat. *hecticus* < Gk. *hektikos* < *hexis,* habit.] **1.** Marked by feverish activity, haste, or confusion <a *hectic* schedule> **2.** Of, pertaining to, or having an undulating fever, as in diseases such as septicemia or tuberculosis. **3.** Feverish : consumptive. **4.** Suffused with blood : FLUSHED.

**hecto-** *or* **hect-** *pref.* [Fr. < Gk. *hekaton,* hundred.] One hundred ($10^2$) <*hectare*>

**hec·to·cot·y·lus** (hĕk′tō-kŏt′l-əs) *n., pl.* **-cot·y·li** (-kŏt′l-ī′) [NLat. : HECTO- + Gk. *kotulē,* small cup.] A modified arm of the male of certain cephalopods, as the octopus, containing sperm and functioning as a reproductive organ.

**hec·to·gram** *also* **hec·to·gramme** (hĕk′tə-grăm′) *n.* A metric unit of mass equal to 100 grams or 3.527 avoirdupois ounces.

**hec·to·graph** (hĕk′tə-grăf′) *n.* [G. *Hektograph* : *hekto-,* hecto- + *-graph,* -graph.] A machine using a glycerin-coated layer of gelatin to make copies of typed, written, or drawn material. —*vt.* **-graphed, -graph·ing, -graphs.** To copy by means of a hectograph. **—hec′to·graph′ic** *adj.* **—hec′to·graph′i·cal·ly** *adv.*

**hec·to·li·ter** *also* **hec·to·li·tre** (hĕk′tə-lē′tər) *n.* A metric unit of capacity or volume, used in dry measure and equal to 100 liters or 2.8378 bushels.

**hec·to·me·ter** *also* **hec·to·me·tre** (hĕk′tə-mē′tər, hĕk-tŏm′ĭ-tər) *n.* A metric unit of length equal to 100 meters or 328 feet.

**hec·tor** (hĕk′tər) *n.* [Gk. *Hektōr.*] **1. Hector.** *Gk. Myth.* A Trojan prince killed by Achilles in Homer's *Iliad.* **2.** A bully. —*v.* **-tored, -tor·ing, -tors.** —*vt.* To intimidate or dominate in a blustering manner. —*vi.* To act like a bully : SWAGGER.

▲ word history: The word *hector,* which is both a noun meaning "a bully" and a verb meaning "to intimidate," is derived from *Hector,* the name of the Trojan hero who figures so largely in the *Iliad.* In the 17th century the name of this hero was used as a generic term for a swaggering braggart or bully. The behavior of such persons is denoted by the verb *hector,* which appeared at the same time as the noun.

**Hec·u·ba** (hĕk′yə-bə) *n.* [Lat. < Gk. *Hekabē.*] *Gk. Myth.* The wife of Priam in Homer's *Iliad.*

**he'd** (hēd). **1.** He had. **2.** He would.

**hed·dle** (hĕd′l) *n.* [Prob. alteration of ME *helde* < OE *hefeld.*] One of a set of parallel wires or cords in a loom used to separate and guide the warp threads and make a path for the shuttle.

**hedge** (hĕj) *n.* [ME < OE *hecg.*] **1.** A row of closely planted shrubs or low-growing trees forming a boundary or fence. **2.** A means of defense or protection esp. against financial loss. **3.** A deliberately noncommittal or unexplicit statement. —*v.* **hedged, hedg·ing, hedg·es.** —*vt.* **1.** To bound or enclose with or as if with hedges. **2.** To enclose or encumber with or as if with a hedge or barrier <*hedged* about by bureaucratic red tape> **3.** To minimize or protect against the loss of by counterbalancing one transaction, as a bet, against another. —*vi.* **1.** To plant or cultivate hedges. **2.** To take compensatory measures so as to counterbalance possible loss. **3.** To avoid making an unequivocal statement or response. **—hedg′er** *n.* **—hedg′y** *adj.*

▲ word history: Hedges have been used in England for hundreds of years as fences to enclose fields and mark the boundaries between them. They also provide shelter for birds and small animals and will, in a pinch, provide the same for human beings. This property of hedges probably gave rise to the sense "a means of defense or protection" for *hedge,* which has been current since the 14th century. In the 17th century that meaning of *hedge* was applied to the case of financial loss, especially in the areas of investment and gambling.

**hedge fund** *n.* A limited-partnership fund that invests private capital speculatively to maximize capital appreciation.

**hedge·hog** (hĕj′hôg′, -hŏg′) *n.* [ME *hedge hogge.*] **1.** Any of several small Old World mammals of the family Erinaceidae, and esp. of the genus *Erinaceus,* bearing dense, erectile spines on the back and rolling into a ball for protection. **2.** Any of several spiny animals similar to the hedgehog.

**hedge·hop** (hĕj′hŏp′) *vi.* **-hopped, -hop·ping, -hops.** To fly an airplane close to the ground, rising above objects as they appear, as in spraying crops. **—hedge′hop′per** *n.*

**hedge hyssop** *n.* A plant of the genus *Gratiola,* growing in damp places and bearing small whitish or yellow flowers.

**hedge·row** (hĕj′rō′) *n.* A row of bushes, shrubs, or trees forming a hedge.

**he·don·ic** (hĭ-dŏn′ĭk) *adj.* [Gk. *hēdonikos* < *hēdonē,* pleasure.] **1.** Of, pertaining to, or characterized by pleasure. **2.** Of or pertaining to hedonism or hedonists. **—he·don′i·cal·ly** *adv.*

**he·don·ics** (hĭ-dŏn′ĭks) *n. (sing. in number).* **1.** *Psychol.* The study of pleasant and unpleasant sensations. **2.** *Philos.* A branch of ethics dealing with the relation of pleasure to duty.

**he·don·ism** (hēd′n-ĭz′əm) *n.* [Gk. *hēdonē,* pleasure + -ISM.] **1.** Pursuit of or devotion to pleasure. **2.** The ethical doctrine that only that which is pleasant or has pleasant consequences is intrinsically good. **3.** *Psychol.* The doctrine that behavior is motivated by the desire for pleasure and the avoidance of pain.

**he·don·ist** (hēd′n-ĭst) *n.* One who holds that pleasure is the chief good. **—he′don·is′tic** *adj.* **—he′don·is′ti·cal·ly** *adv.*

**-hedral** *suff.* [< -HEDRON.] Having a given kind or number of surfaces <di*hedral*>

**-hedron** *suff.* [Gk. *-edron* < *hedra,* base.] A crystal or geometric figure with a given kind or number of surfaces <hepta*hedron*>

**hee·bie-jee·bies** (hē′bē-jē′bēz) *pl.n.* [Coined by Billy De Beck (1890–1942), in his comic strip *Barney Google.*] *Slang.* A feeling of uneasiness or nervousness : JITTERS.

**heed** (hēd) *v.* **heed·ed, heed·ing, heeds.** [ME *heden* < OE *hēdan.*] —*vt.* To pay attention to : MIND <*Heed* what I say!> —*vi.* To pay attention. —*n.* Close attention.

**heed·ful** (hēd′fəl) *adj.* Paying close attention. **—heed′ful·ly** *adv.* **—heed′ful·ness** *n.*

**heed·less** (hēd′lĭs) *adj.* Paying little attention : THOUGHTLESS. **—heed′less·ly** *adv.* **—heed′less·ness** *n.*

**hee·haw** (hē′hô′) *n.* [Imit.] **1.** The braying sound made by a donkey. **2.** A noisy laugh : GUFFAW. **—hee′haw′** *v.* **(-hawed, -haw·ing, -haws).**

**heel**[1] (hēl) *n.* [ME < OE *hēla.*] **1. a.** The rounded posterior portion of the human foot under and behind the ankle. **b.** The corresponding part of the hind foot of other vertebrates. **c.** A similar anatomical part, as the fleshy rounded base of the human palm or the hind toe of a bird. **2. a.** The part that covers the heel <the *heel* of a stocking> **b.** The built-up portion of a shoe or boot, supporting the heel. **3. a.** A crusty end of a loaf of bread. **b.** The first or last piece of a sliced loaf of bread. **4.** *Naut.* **a.** The lower end of a mast. **b.** The after end of a ship's keel. **5.** The basal end of a plant cutting or tuber used in propagation. **6.** *Slang.* A dishonorable man : CAD. —*v.* **heeled, heel·ing, heels.** —*vt.* **1.** To furnish with a heel. **2.** *Slang.* To provide esp. with money. —*vi.* To follow at one's heels, as a dog. **—on (or upon) the heels of. 1.** Directly behind. **2.** Immediately following <*On the heels of* the earthquake came a flood.> **—take to (one's) heels.** To run away. **—to heel. 1.** Close behind. **2.** Under control or discipline. **—heel′less** *adj.*

**heel**[2] (hēl) *v.* **heeled, heel·ing, heels.** [Alteration of ME *helden* < OE *hieldan.*] —*vi.* To tilt to one side : LIST. —*vt.* To cause (a ship) to list. —*n.* An inclination or tilt, as of a boat, to one side.

**heel-and-toe** (hēl′ən-tō′) *adj.* Marked by a stride in which the heel of one foot touches ground before the toe of the other foot is lifted.

**heel bone** *n.* The calcaneus.

**heel·er** (hē′lər) *n.* **1.** One who heels shoes. **2.** *Informal.* A ward heeler.

**heel·post** (hēl′pōst′) *n.* The post to which a door or gate is hinged.

**heel·tap** (hēl′tăp′) *n.* A small amount of liquor remaining in a glass or drinking container.

**heft** (hĕft) *n.* [Alteration of HEAVE.] Heaviness : weight. —*v.* **heft·ed, heft·ing, hefts.** —*vt.* **1.** To lift in order to determine the weight of. **2.** To hoist up : HEAVE. —*vi.* To weigh.

**heft·y** (hĕf′tē) *adj.* **-i·er, -i·est. 1.** Of considerable weight. **2.** Rugged : powerful. **3.** Sizable. **—heft′i·ly** *adv.* **—heft′i·ness** *n.*

**He·ge·li·an·ism** (hā-gā′lē-ə-nĭz′əm) *n.* The idealist philosophy of Georg Wilhelm Friedrich Hegel in which the dialectic is used as an analytic tool for approaching a higher unity. **—He·ge′li·an** *adj.* & *n.*

**he·gem·o·ny** (hĭ-jĕm′ə-nē, hĕj′ə-mō′nē) *n., pl.* **-nies.** [Gk. *hēgemonia* < *hēgemōn,* leader < *hēgeisthai,* to lead.] The influence of one state over others. **—heg′e·mon′ic** (hĕj′ə-mŏn′ĭk) *adj.*

**he·gi·ra** (hĭ-jī′rə, hĕj′ər-ə) *n.* [Med. Lat. < Ar. *alhijrat* < *hajara,* he left.] **1. Hegira.** The flight of Mohammed from Mecca to Medina in A.D. 622. **2.** A journey esp. to escape danger : FLIGHT.

**Hei·del·berg man** (hīd'l-bûrg') *n.* An extinct early member of the human species, a form of *Homo erectus*, known primarily from a fossil jawbone found near Heidelberg, West Germany, in 1907.
**heif·er** (hĕf'ər) *n.* [ME < OE *hēahfore*.] A young cow, esp. one that has not yet given birth to a calf.
**heigh-ho** (hī'hō', hā'-) *interj.* —Used to express fatigue, boredom, disappointment, or mild surprise.
**height** (hīt) *n.* [ME < OE *hēahðu*.] **1.** The uppermost or highest point : SUMMIT. **2. a.** The highest or most advanced degree : ZENITH <at the *height* of one's popularity> **b.** The point of highest intensity : CLIMAX <the *height* of confusion> **3. a.** The distance from the base to the top of something. **b.** Elevation above a given level : ALTITUDE. **4. a.** The condition or attribute of being high or tall. **b.** Stature, esp. of the human body. **5. a.** An eminence, as a mountain or hill. **b.** A high point, position, or degree. **6.** *Obs.* High rank, estate, or degree.
**height·en** (hīt'n) *vt. & vi.* **-ened, -en·ing, -ens. 1.** To increase or cause to increase in quantity or degree. **2.** To make or become high or higher. **—height'en·er** *n.*
**height-to-pa·per** (hīt'tə-pā'pər) *n.* The height of type from foot to face, standardized at 0.9186 inch or 2.296 centimeters.
**Heim·lich maneuver** (hīm'lĭkh', -lĭk') *n.* [After Henry J. *Heimlich*, 20th-cent. American surgeon.] A maneuver designed to dislodge an object, as food, from a choking person's windpipe in which the victim is clasped from behind, a closed fist is placed below the rib cage, and air is forced out of the lungs with a hard upward thrust.
**hei·nous** (hā'nəs) *adj.* [ME < OFr. *hainos* < *haine*, hatred < *hair*, to hate, of Germanic orig.] Grossly wicked or deserving strong condemnation : ABOMINABLE. **—hei'nous·ly** *adv.* **—hei'nous·ness** *n.*
**heir** (âr) *n.* [ME < AN < Lat. *heres*.] **1.** *Law.* One who inherits or is entitled by law or by the terms of a will to inherit the estate of another. **2.** One who succeeds or is in line to succeed to a hereditary rank, title, or office. **3.** One who receives or is expected to receive an endowment or heritage, as of talent, from a parent or predecessor.
**heir apparent** *n., pl.* **heirs apparent.** *Law.* An heir whose right to inheritance is indefeasible by law if he survives his ancestor.
**heir·dom** (âr'dəm) *n.* **1.** Succession by right of blood. **2.** An inheritance.
**heir·ess** (âr'ĭs) *n.* A woman who is an heir, esp. to great wealth.
**heir·loom** (âr'lo͞om') *n.* [ME *heirlome* : *heir*, heir + *lome*, implement.] **1.** A valued family possession handed down from generation to generation. **2.** *Law.* An article of personal property included in an inherited estate.
**heir presumptive** *n., pl.* **heirs presumptive.** *Law.* An heir whose right to inheritance can be defeated by the birth of a closer relative.
**heir·ship** (âr'shĭp') *n.* **1.** The condition of being an heir. **2.** The right to inheritance : HEIRDOM.
**heist** (hīst) [Alteration of HOIST.] *Slang.* —*vt.* **heist·ed, heist·ing, heists. 1.** To perpetrate a robbery on. **2.** To steal. —*n.* A robbery.
**Hel** (hĕl) *n.* [ON.] *Norse Myth.* **1.** The daughter of Loki and the goddess of death. **2.** The underworld of the dead not killed in battle.
**He·La cell** (hĕl'ə) *n.* [After *He(len) La(ne)*, who donated such cells in 1951.] Any of the human cancer cells of a continuously cultured strain used in biomedical research.
**held** (hĕld) *v. p.t. & p.p. of* HOLD.
**hel·den·te·nor** *also* **Hel·den·te·nor** (hĕl'dən-tə-nôr', -nōr') *n.* [G. : *Held*, hero + *Tenor*, tenor.] A tenor with a strikingly dramatic or brilliant voice that is suitable for heroic roles, as those in Wagnerian opera.
**Hel·en of Troy** (hĕl'ən) *n.* [Gk. *Helenē*.] *Gk. Myth.* The daughter of Zeus and Leda and wife of Menelaus whose abduction by Paris caused the Trojan War.
**heli-**[1] *pref.* [< HELICOPTER.] Helicopter <*heli*port><*heli*pad>
**heli-**[2] *pref. var. of* HELIO-.
**he·li·a·cal** (hĭ-lī'ə-kəl) *adj.* [LLat. *heliacus* < Gk. *hēliakos* < *hēlios*, sun.] Of or relating to the sun, esp. rising and setting with the sun, as a star. **—he·li'a·cal·ly** *adv.*
**helic-** *pref. var. of* HELICO-.
**hel·i·cal** (hĕl'ĭ-kəl, hē'lĭ-) *adj.* **1.** Of or shaped like a helix. **2.** Having a shape approx. like that of a helix. **—hel'i·cal·ly** *adv.*
**hel·i·ces** (hĕl'ĭ-sēz', hē'lĭ-) *n. var. pl. of* HELIX.
**he·lic·i·ty** (hē-lĭs'ĭ-tē, hĕ-) *n.* The component of the spin of a particle along its direction of motion.
**helico-** or **helic-** *pref.* [Gk. *heliko-* < *helix*, spiral.] Helix : spiral <*helic*ity>
**hel·i·coid** (hĕl'ĭ-koid', hē'lĭ-) *adj.* [Gk. *helikoeidēs* : *helix*, spiral + *-oeidēs*, -oid.] Arranged in or having a shape approx. like that of a flattened spiral. —*n. Math.* A surface generated by a plane curve or a twisted curve that is rotated about a linear axis and at the same time is translated in the direction of the axis so that the two rates have a constant ratio.
**hel·i·con** (hĕl'ĭ-kŏn', -kən) *n.* [Prob. < Gk. *helix*, *helik-*, spiral.] A large circular brass tuba that encircles the player's shoulder.

helicon

**hel·i·cop·ter** (hĕl'ĭ-kŏp'tər) *n.* [Fr. *hélicoptère* : Gk. *helix*, spiral + Gk. *pteron*, wing.] An aircraft deriving its lift from blades that rotate about an approx. vertical central axis. —*vi. & vt.* **-tered, -ter·ing, -ters.** To travel or convey by helicopter.
**helio-** or **heli-** *pref.* [< Gk. *hēlios*, sun.] Sun <*helio*gram>
**he·li·o·cen·tric** (hē'lē-ō-sĕn'trĭk) *also* **he·li·o·cen·tri·cal** (-trĭ-kəl) *adj.* **1.** Relative or referred to the sun. **2.** Having the sun as a center. **—he'li·o·cen·tric'i·ty** (-sĕn-trĭs'ĭ-tē) *n.*
**he·li·o·gram** (hē'lē-ə-grăm') *n.* A message sent by heliograph.
**he·li·o·graph** (hē'lē-ə-grăf') *n.* **1.** An apparatus once used to photograph the sun. **2.** A signaling apparatus that reflects sunlight with a movable mirror to flash coded messages. —*vt. & vi.* **-graphed, -graph·ing, -graphs.** To signal or communicate by heliograph. **—he'li·og'raph·er** (-ŏg'rə-fər) *n.* **—he'li·o·graph'ic** *adj.* **—he'li·og'raph·y** *n.*
**he·li·o·gra·vure** (hē'lē-ō-grə-vyo͝or') *n.* Photogravure.
**he·li·om·e·ter** (hē'lē-ŏm'ĭ-tər) *n.* A telescope for measuring small angular distances between celestial bodies. **—he'li·o·met'ric** (-ə-mĕt'rĭk), **he'li·o·met'ri·cal** *adj.* **—he'li·o·met'ri·cal·ly** *adv.* **—he'li·om'e·try** *n.*
**He·li·os** (hē'lē-ŏs') *n.* [Gk. *Hēlios* < *hēlios*, sun.] *Gk. Myth.* The sun god, depicted as driving a chariot across the sky from east to west.
**he·li·o·stat** (hē'lē-ə-stăt') *n.* An instrument in which a mirror is automatically moved so as to reflect sunlight in a constant direction.
**he·li·o·tax·is** (hē'lē-ō-tăk'sĭs) *n. Biol.* The movement of an organism in response to sunlight.
**he·li·o·ther·a·py** (hē'lē-ō-thĕr'ə-pē) *n. Med.* Therapy involving exposure to sunlight.
**he·lio·trope** (hēl'yə-trōp') *n.* [Lat. *heliotropium* < Gk. *hēliotropion* : *hēlios*, sun + *tropos*, turn.] **1.** A South American plant of the genus *Heliotropium*, esp. *H. arborescens*, with small, fragrant purplish flowers. **2.** The garden heliotrope. **3.** A plant that turns toward the sun. **4.** Bloodstone. **5.** A moderate, light, or brilliant violet to moderate or deep reddish purple. **—he'lio·trope'** *adj.*
**he·lio·tro·pin** (hēl'yə-trō'pĭn, hē'lē-ŏt'rə-pĭn) *n.* Piperonal.
**he·li·ot·ro·pism** (hē'lē-ŏt'rə-pĭz'əm) *n.* Growth or movement of an organism toward or away from sunlight. **—he'li·o·trop'ic** (-ə-trŏp'ĭk) *adj.* **—he'li·o·trop'i·cal·ly** *adv.*
**he·li·o·type** (hē'lē-ə-tīp') *n.* **1.** A photomechanically produced plate for pictures or type made by exposing a gelatin film under a negative, hardening it with chrome alum, and printing from it directly. **2.** *also* **he·li·o·typ·y** (-tī'pē). The process of producing a heliotype. —*vt.* **-typed, -typ·ing, -types.** To produce a heliotype of. **—he'li·o·typ'ic** (-tĭp'ĭk) *adj.*
**he·li·o·zo·an** (hē'lē-ə-zō'ən) *n.* Any of various aquatic protozoans of the order Heliozoa, with numerous stiff, radiating pseudopodia. **—he'li·o·zo'ic** (-zō'ĭk) *adj.*
**hel·i·pad** (hĕl'ə-păd') *n.* A heliport.
**hel·i·port** (hĕl'ə-pôrt', -pōrt') *n.* A place for helicopters to land and take off.
**hel·i·stop** (hĕl'ĭ-stŏp') *n.* A heliport.
**he·li·um** (hē'lē-əm) *n.* [NLat. < Gk. *hēlios*, sun (so called because its existence was deduced from the solar spectrum).] *Symbol* **He** A colorless, odorless, tasteless inert gaseous element used to provide lift for balloons and as an inert component of various artificial atmospheres; atomic number 2; atomic weight 4.0026.
**helium I** *n.* Liquid helium existing as a normal fluid between the superfluid transition point of approx. 2.178°K at 1 atmosphere pressure and its boiling point of 4.2°K.
**helium II** *n.* Liquid helium existing as a superfluid below the transition point of approx. 2.178°K at 1 atmosphere and having extremely low viscosity and high thermal conductivity.
**he·lix** (hē'lĭks) *n., pl.* **-lix·es** *or* **hel·i·ces** (hĕl'ĭ-sēz', hē'lĭ-) [Lat. < Gk.] **1.** *Math.* A three-dimensional curve that lies on a cylinder or cone and cuts the elements at a constant angle. **2.** A spiral form or structure. **3.** *Anat.* The folded rim of skin and cartilage around the outer ear. **4.** A volute on a Corinthian or Ionic capital.
**hell** (hĕl) *n.* [ME *helle* < OE.] **1.** The abode of the dead in ancient traditions : UNDERWORLD. **2.** *often* **Hell.** The abode of devils and condemned souls in many religions. **3.** A place or situation of evil, misery, discord, or destruction. **4.** *Informal.* **a.** Anguish : torment <They put us through *hell*.> **b.** One that causes anguish, trouble, or annoyance <The boss is *hell* when we're behind schedule.> **5. Hell.** *Christian Science.* Mortal belief : sin or error. **6.** *Informal.* A

sharp scolding <gave me *hell* for being late> **7. a.** A tailor's receptacle for discarded material. **b.** A hellbox. **8.** —Used as an intensive <How the *hell* did it happen?> *—vi.* **helled, hell·ing, hells.** *Informal.* To behave riotously : CAROUSE. *—interj. Slang.* —Used to express impatience, anger, or disgust. **—hell or (or and) high water.** *Informal.* Troubles or difficulties of whatever magnitude <I'm going, come *hell or high water.*> **—hell to pay.** *Informal.* Trouble to be faced <There'll be *hell to pay* if they're embezzling.>

▲ word history: The word *hell* has existed since Old English times as a name for an abode of the dead, both Christian and pagan. *Hell* is derived from an Indo-European form that probably had the basic sense of "hide" or "conceal," and has many relatives in the Germanic and other Indo-European languages. English words related to *hell* are *helm²*, *hull*, *hole*, *hollow*, *hill*, and *hall*. The Old Norse word *hel* indicated the abode of the dead, like English *hell*, and also denoted the goddess who ruled it. A Latin relative of *hell* is the verb *occulere*, "to cover over," from which English *occult* is derived. A related Greek word is the verb *kaluptein*, "to conceal." *Calypso*, the name of the goddess who hid Odysseus on an island for seven years, is derived from *kaluptein*. Another derivative of *kaluptein* is Greek *apokalupsis*, literally "uncovering," which is the source of English *apocalypse*.

**he'll** (hēl). **1.** He will. **2.** He shall.

**hell·bend·er** (hĕl'bĕn'dər) *n.* A large aquatic salamander, *Cryptobranchus alleganiensis* of eastern and central North America.

**hell-bent** (hĕl'bĕnt') *adj.* Obstinately or impetuously determined <is *hell-bent* on buying a new car>

**hell·box** (hĕl'bŏks') *n.* A printer's box for discarded or broken type.

**hell·cat** (hĕl'kăt') *n.* **1.** An ill-humored and wicked woman : VIXEN. **2.** One who torments others.

**hell·div·er** (hĕl'dī'vər) *n. Informal.* A New World grebe, *Podilymbus podiceps.*

**Hel·le** (hĕl'ē) *n.* [Gk. *Hellē.*] *Gk. Myth.* The daughter of a Greek king who, while fleeing with her brother from their stepmother, drowned in the Hellespont, thereafter named for her.

**hel·le·bore** (hĕl'ə-bôr', -bōr') *n.* [ME *ellebre* < OFr. < Lat. *elleborus* < Gk. *helleboros*, perh. : *hellos*, fawn + *-boros*, eaten < *bibrōskein*, to eat.] **1.** Any of various Eurasian plants of the genus *Helleborus*, most species of which are poisonous. **2.** A North American plant of the genus *Veratrum*, esp. *V. viride*, bearing large leaves and greenish flowers and yielding a toxic alkaloid used medicinally.

**hel·le·bo·rin** (hĕl'ə-bôr'ĭn, -bōr'-) *n.* A poisonous compound, $C_{28}H_{36}O_6$, extracted from *Helleborus viridis*, a species of hellebore.

**Hel·lene** (hĕl'ēn') *also* **Hel·le·ni·an** (hĕ-lē'nē-ən) *n.* [Gk. *Hellēn.*] A Greek.

**Hel·len·ic** (hĕ-lĕn'ĭk) *adj.* Of or pertaining to the ancient Greeks or their language. *—n.* The branch of the Indo-European language family consisting only of Greek.

**Hel·le·nism** (hĕl'ə-nĭz'əm) *n.* **1.** An idiom peculiar to the Greeks. **2.** The civilization and culture of ancient Greece. **3.** Adoption of Greek ideas, style, or culture.

**Hel·le·nist** (hĕl'ə-nĭst) *n.* **1.** One in classical times who adopted the Greek language and culture, esp. a Jew of the Diaspora. **2.** A student or devotee of Greek civilization, language, or literature.

**Hel·le·nis·tic** (hĕl'ə-nĭs'tĭk) *also* **Hel·le·nis·ti·cal** (-tĭ-kəl) *adj.* **1.** Pertaining to the Hellenists. **2. a.** Of or pertaining to Greek history and culture from the time of Alexander the Great into the 1st cent. B.C. **b.** Pertaining to or in the style of the Greek art or architecture of this period.

**Hel·le·nize** (hĕl'ə-nīz') *v.* **-nized, -niz·ing, -niz·es.** *—vi.* To adopt Greek or Hellenistic speech and ways. *—vt.* To make Greek or Hellenistic. **—Hel'le·ni·za'tion** *n.* **—Hel'le·niz'er** *n.*

**†hell·er** (hĕl'ər) *n.* [< HELL.] *Regional.* One who behaves wildly or rashly.

**hell-for-leath·er** (hĕl'fər-lĕth'ər) *adv. & adj. Informal.* At breakneck speed.

**hell·gram·mite** (hĕl'grə-mīt') *n.* [Orig. unknown.] The large, brownish aquatic larva of the dobson fly, often used as fishing bait.

**hell·hole** (hĕl'hōl') *n.* **1.** A hellish place. **2.** *Obs.* The pit of hell.

**hell·hound** (hĕl'hound') *n.* **1.** Cerberus. **2.** A fiendish person : DEVIL.

**hel·lion** (hĕl'yən) *n.* [Prob. alteration of dial. *hallion*, worthless person.] *Informal.* A mischievous, troublesome person.

**hell·ish** (hĕl'ĭsh) *adj.* Of, resembling, or worthy of hell : FIENDISH. **—hell'ish·ly** *adv.* **—hell'ish·ness** *n.*

**hel·lo** (hĕ-lō', hə-) *also* **hul·lo** (hə-) [Alteration of obs. *holla*, stop!, prob. < OFr. *hola* : *ho*, ho! + *la*, there.] *—interj.* —Used to greet someone, answer the telephone, or express surprise. *—n., pl.* **-los.** An expression or greeting of "hello." *—vi.* **-loed, -lo·ing, -loes.** To say or call "hello."

**helm¹** (hĕlm) *n.* [ME < OE *helma.*] **1.** The steering gear of a ship, esp. the tiller or wheel. **2.** A position of control or leadership. *—vt.* **helmed, helm·ing, helms.** To take the helm of : STEER.

**helm²** (hĕlm) *n.* [ME < OE and < ON *hjālmr.*] *Archaic.* A helmet. *—vt.* **helmed, helm·ing, helms.** To cover or supply with a helmet.

**hel·met** (hĕl'mĭt) *n.* [ME.] **1.** A usu. metal piece of armor designed to protect the head. **2. a.** A head covering of sturdy material, as metal or plastic, worn to protect the head. **b.** The headgear with a glass mask worn by deep-sea divers. **c.** A pith helmet : TOPI. **d.** A helmet-shaped hat. **3.** *Bot.* The hood-shaped sepal or corolla of some flowers. *—vt. & vi.* **-met·ed, -met·ing, -mets.** To supply with or don a helmet. **—hel'met·ed** *adj.*

**hel·minth** (hĕl'mĭnth') *n.* [Gk. *helmins, helminth-.*] A worm, esp. a parasitic intestinal nematode or trematode worm.

**hel·min·thi·a·sis** (hĕl'mĭn-thī'ə-sĭs) *n.* A disease caused by an infestation of parasitic worms.

**hel·min·thic** (hĕl-mĭn'thĭk) *adj.* **1.** Of or pertaining to worms, esp. parasitic intestinal worms. **2.** Tending to expel worms : ANTHELMINTIC. *—n.* A vermifuge or anthelmintic.

**hel·min·thol·o·gy** (hĕl'mĭn-thŏl'ə-jē) *n.* The scientific study of worms, esp. parasitic worms. **—hel'min·thol'o·gist** *n.*

**helms·man** (hĕlmz'mən) *n.* A person who steers a ship. **—helms'man·ship'** *n.*

**hel·ot** (hĕl'ət) *n.* [Lat. *Helotes*, Helots < Gk. *Heilōtes.*] **1. Helot.** One of a class of serfs in ancient Sparta. **2.** A bondsman : serf. **—hel'ot·ry** (-ə-trē) *n.*

**hel·ot·ism** (hĕl'ə-tĭz'əm) *n.* **1.** The state or condition of a helot. **2.** *Biol.* A symbiotic relationship in which one organism dominates and uses another to its own advantage.

**help** (hĕlp) *v.* **helped, help·ing, helps.** [ME *helpen* < OE *helpan* and < ON *hjalpa.*] *—vt.* **1.** To give aid to : ASSIST. **2.** To further the progress or advancement of <*help* a worthy cause> **3.** To give relief to <*help* the disadvantaged> **4.** To alleviate <aspirin to *help* your headache> **5.** To be able to prevent, change, or rectify <can't *help* my shyness> **6.** To refrain from : AVOID <couldn't *help* crying> **7.** To wait on, as in a store or restaurant. *—vi.* To give assistance. *—n.* **1. a.** An act or instance of helping. **b.** Assistance : aid. **2.** Remedy : relief. **3.** One that helps. **4. a.** One employed to help, esp. a farm worker or domestic servant. **b.** Such employees as a group. **—cannot help but.** To be unable to resist or avoid <*cannot help but* eat too much> **—help (oneself) to. 1.** To serve oneself. **2.** To take without asking permission <*helped* myself *to* the car keys> **—help'er** *n.*

☆ ***syns:*** HELP, AID, ASSIST, SUCCOR *v. core meaning* : to provide assistance <*helped* the homeless>

**help·ful** (hĕlp'fəl) *adj.* Providing help. **—help'ful·ly** *adv.* **—help'ful·ness** *n.*

**help·ing** (hĕl'pĭng) *n.* A portion of food for one person.

**help·less** (hĕlp'lĭs) *adj.* **1.** Unable to manage alone. **2.** Lacking strength or power. **—help'less·ly** *adv.* **—help'less·ness** *n.*

**help·mate** (hĕlp'māt') *n.* A helper and companion, esp. a spouse.

**help·meet** (hĕlp'mēt') *n.* [HELP + MEET².] A helpmate.

▲ word history: The word *helpmeet* owes its existence to a misreading of a passage in the King James Bible. In Genesis 2:18 God promises Adam "to make him an help meet for him," that is, a helper suitable for him. The words *help meet* were even in the 17th century misread as one word, and since the "help" turned out to be Eve, the new compound was interpreted as meaning "spouse." *Helpmate* is an alteration of *helpmeet* that substitutes *mate*, "spouse," for the now obscure *meet* of *helpmeet*.

**hel·ter-skel·ter** (hĕl'tər-skĕl'tər) *adv.* [Orig. unknown.] **1.** In disorderly haste : PELL-MELL. **2.** In a haphazard way. *—adj.* **1.** Carelessly hurried and confused. **2.** Haphazard. *—n.* Confusion : disorder.

**helve** (hĕlv) *n.* [ME < OE *hielfe.*] A handle of a tool, as an ax, chisel, or hammer.

**Hel·ve·tian** (hĕl-vē'shən) *adj.* [Lat. *Helvetius.*] **1.** Of or pertaining to the Helvetii. **2.** Swiss. *—n.* **1.** One of the Helvetii. **2.** A Swiss.

**Hel·ve·ti·i** (hĕl-vē'shē-ī') *pl.n.* [Lat.] A Celtic people inhabiting Helvetia during the time of Julius Caesar.

**hem¹** (hĕm) *n.* [ME < OE.] An edge or border of a piece of cloth, esp. a finished edge, as for a garment, made by folding the selvage or raw edge under and stitching it down. *—vt.* **hemmed, hem·ming, hems. 1.** To fold back and stitch down the edge of. **2.** To enclose and shut in : CONFINE <My car was *hemmed* in by a big truck.> **—hem'mer** *n.*

**hem²** (hĕm) *n.* [Imit.] A short cough or clearing of the throat made esp. to gain attention, hide embarrassment, or fill a pause in speech. *—vi.* **hemmed, hem·ming, hems. 1.** To utter a hem. **2.** To hesitate in speech. **—hem and haw.** To be indecisive and hesitant : EQUIVOCATE <*hemmed and hawed* over making a decision>

**hem-** or **hema-** *pref. var. of* HEMO-.

**he·ma·cy·tom·e·ter** (hē'mə-sī-tŏm'ĭ-tər) *n.* An instrument for estimating the number of blood cells in a measured volume of blood.

**he·mag·glu·ti·nate** (hē'mə-glōōt'n-āt') *vt.* **-nat·ed, -nat·ing, -nates.** To cause agglutination of (red blood cells). **—he'mag·glu'ti·na'tion** *n.*

**he·mag·glu·ti·nin** (hē'mə-glōōt'n-ĭn) *n.* An antibody that causes agglutination of red blood cells containing or coated with the corresponding antigen.

**he·mal** (hē'məl) *adj.* **1.** Of or relating to the blood or blood vessels. **2.** Relating to or located on or in the side of the body containing the heart.

**he-man** (hē′măn′) *n. Informal.* A strong, robust, virile man.
**hemat-** *pref. var. of* HEMATO-.
**he·ma·te·in** (hē′mə-tē′ĭn, hē′mə-tēn′) *n.* A dark-purple crystalline compound, $C_{16}H_{12}O_6$, used as a biological stain.
**he·mat·ic** (hĭ-măt′ĭk) *adj.* [Gk. *haimatikos* < *haima*, blood.] Of, relating to, like, containing, or acting on blood. —*n. Med.* A remedy for a blood disease, as anemia.
**he·ma·tin** (hē′mə-tĭn) *n.* A blue to blackish-brown powder, $C_{34}H_{32}N_4O_4 \cdot FeOH$, the hydroxide of heme, containing ferric iron.
**he·ma·tin·ic** (hē′mə-tĭn′ĭk) *adj.* Stimulating red blood cell or hemoglobin production. —*n.* A hematinic drug.
**he·ma·tite** (hē′mə-tīt′) *n.* [Lat. *haemites* < Gk. (*lithos*) *haimatitēs*, bloodlike (stone) < *haima*, blood.] A blackish-red to brick-red mineral, primarily $Fe_2O_3$, the chief ore of iron. **—he′ma·tit′ic** (-tĭt′ĭk) *adj.*
**hemato-** *or* **hemat-** *pref.* [Gk. *haimato-* < *haima*, blood.] Blood <*hematology*>
**he·ma·to·blast** (hē′mə-tə-blăst′, hĭ-măt′ə-) *n.* **1.** A platelet of the blood. **2.** An immature blood cell. **—he′ma·to·blas′tic** *adj.*
**he·mat·o·crit** (hĭ-măt′ə-krĭt′) *n.* [HEMATO- + Gk. *kritēs*, judge < *krinein*, to judge.] A centrifuge for separating the cellular and other particulate matter of blood from the plasma.
**he·ma·to·gen·e·sis** (hē′mə-tə-jĕn′ĭ-sĭs, hĭ-măt′ə-) *n.* Hematopoiesis. **—he′ma·to·gen′ic, he′ma·to·ge·net′ic** (-jə-nĕt′ĭk) *adj.*
**he·ma·tog·e·nous** (hē′mə-tŏj′ə-nəs) *adj.* **1.** Producing blood. **2.** Originating in or involving the blood.
**he·ma·toid** (hē′mə-toid′) *adj.* **1.** Bloody. **2.** Like blood.
**he·ma·tol·o·gy** (hē′mə-tŏl′ə-jē) *n.* The science encompassing the generation, anatomy, physiology, pathology, and therapeutics of blood. **—he′ma·to·log′ic** (-tə-lŏj′ĭk), **he′ma·to·log′i·cal** *adj.* **—he′ma·to·log′i·cal·ly** *adv.* **—he′ma·tol′o·gist** *n.*
**he·ma·tol·y·sis** (hē′mə-tŏl′ĭ-sĭs) *n.* Hemolysis.
**he·ma·to·ma** (hē′mə-tō′mə) *n., pl.* **-mas** *or* **-ma·ta** (-mə-tə). A localized blood-filled swelling or tumor.
**he·ma·to·poi·e·sis** (hē′mə-tō-poi-ē′sĭs, hĭ-măt′ə-) *n.* The formation of blood in the body. **—he′ma·to·poi·et′ic** (-ĕt′ĭk) *adj.*
**he·ma·to·sis** (hē′mə-tō′sĭs) *n.* Oxygenation of venous blood in the lungs.
**he·ma·tox·y·lin** (hē′mə-tŏk′sə-lĭn) *n.* [< NLat. *Haematoxylon*, a genus of plants.] A red or yellow crystalline compound, $C_{16}H_{14}O_6 \cdot 3H_2O$, the coloring agent of logwood, used in dyes, inks, and biological stains.
**he·ma·to·zo·on** (hē′mə-tō-zō′ŏn′, hĭ-măt′ə-) *n., pl.* **-zo·a** (-zō′ə). A parasitic protozoan or similar organism living in the blood. **—he′ma·to·zo′al, he′ma·to·zo′ic** *adj.*
**he·ma·tu·ri·a** (hē′mə-tŏŏr′ē-ə, -tyŏŏr′-) *n.* The presence of blood or red blood cells in the urine.
**heme** (hēm) *n.* [< HEMATIN.] The nonprotein, ferrous-iron-containing component of hemoglobin, $C_{34}H_{32}FeN_4O_4$.
**he·mel·y·tron** (hĕ-mĕl′ĭ-trŏn′) *n., pl.* **-tra** (-trə). An insect forewing thickened at the base and membranous at the apex, typical of the true bugs.
**hem·er·a·lo·pi·a** (hĕm′ər-ə-lō′pē-ə) *n.* [NLat. < Gk. *hēmeralōps* : *hēmera*, day + *alaos*, blind + *ōps*, eye.] Inability to see as clearly in bright light as in dim light.
**hemi-** *pref.* [Lat. *hemi-* < Gk. *hēmi-*.] **1.** Half <*hemi*hedral> **2.** Partial : partially <*hemi*parasite>
**-hemia** *suff. var. of* -EMIA.
**hem·i·al·gi·a** (hĕm′ē-ăl′jē-ə) *n.* Pain affecting half the human body.
**he·mic** (hē′mĭk) *adj.* Of, relating to, or produced by blood.
**hem·i·cel·lu·lose** (hĕm′ĭ-sĕl′yə-lōs′, -lōz′) *n.* Any of several polysaccharides that are more complex than a sugar and less complex than cellulose, derived from plants and produced commercially from corn grain hulls.
**hem·i·chor·date** (hĕm′ĭ-kôr′dāt′, -dĭt) *n.* Any of various wormlike marine animals of the phylum or subphylum Hemichordata, with a primitive notochord and gill slits. —*adj.* Of or belonging to the Hemichordata.
**hem·i·cy·cle** (hĕm′ĭ-sī′kəl) *n.* [Fr. *hémicycle* < Lat. *hemicyclium* < Gk. *hēmikuklion* : *hēmi-*, half + *kuklos*, circle.] A semicircular arrangement or structure.
**hem·i·dem·i·sem·i·qua·ver** (hĕm′ē-dĕm′ē-sĕm′ē-kwā′vər) *n. Chiefly Brit.* A sixty-fourth note.
**hem·i·he·dral** (hĕm′ĭ-hē′drəl) *adj.* [HEMI- + -HEDR(ON) + -AL.] Displaying only half the faces needed for total symmetry <a *hemihedral* crystal> **—hem′i·he′dral·ly** *adv.*
**hem·i·hy·drate** (hĕm′ĭ-hī′drāt′) *n.* A hydrate in which the molecular ratio of water molecules to anhydrous compound is 1:2. **—hem′i·hy′drat′ed** *adj.*
**hem·i·mor·phic** (hĕm′ĭ-môr′fĭk) *adj.* Asymmetric at the axial ends.
**hem·i·mor·phite** (hĕm′ĭ-môr′fīt′) *n.* [HEMIMORPH(IC) + -ITE.] CALAMINE 1.

ŏŏ **boo**t ou **ou**t th **th**in *th* **th**is ŭ **cu**t ûr **ur**ge y **y**oung
yōō ab**u**se zh vi**si**on ə **a**bout, it**e**m, ed**i**ble, gall**o**p, circ**u**s

**he·min** (hē′mĭn) *n.* A brown or blue crystalline compound, $C_{34}H_{32}N_4O_4FeCl$, the chloride of heme, used in identifying blood stains.
**hem·i·par·a·site** (hĕm′ĭ-păr′ə-sīt′) *n.* A partially parasitic plant or organism. **—hem′i·par′a·sit′ic** (-sĭt′ĭk) *adj.*
**hem·i·ple·gia** (hĕm′ĭ-plē′jə, -jē-ə) *n.* [MGk. *hēmiplēgia* : Gk. *hēmi-*, half + Gk. *plēgē*, stroke.] Paralysis of one side of the body. **—hem′i·ple′gic** (-plē′jĭk) *adj.*
**he·mip·ter·an** (hĭ-mĭp′tər-ən) *adj.* Hemipterous. —*n.* A hemipterous insect. **—he·mip′ter·on′** (-tə-rŏn′) *n.*
**he·mip·ter·ous** (hĭ-mĭp′tər-əs) *adj.* Of or belonging to the Hemiptera, a large group of insects including the true bugs of the order Heteroptera and their allies of the order Homoptera.
**hem·i·sphere** (hĕm′ĭ-sfîr′) *n.* **1. a.** A half of a sphere bounded by a great circle. **b.** A half of a symmetric, approx. spherical object as divided by a plane of symmetry. **2.** Either half of the celestial sphere as divided by the ecliptic, the celestial equator, or the horizon. **3.** Either the northern or southern half of the earth as divided by the equator or the eastern or western half as divided by a meridian. **—hem′i·spher′ic** (-sfîr′ĭk, -sfĕr′-), **hem′i·spher′i·cal** *adj.* **—hem′i·spher′i·cal·ly** *adv.*
**hem·i·stich** (hĕm′ĭ-stĭk′) *n.* [Lat. *hemistichium* < Gk. *hēmistikhion* : *hēmi-*, half + *stikhos*, line.] Half a line of verse, esp. when divided rhythmically from the rest of the line by a caesura.
**hem·line** (hĕm′līn′) *n.* The line formed by the hem of a garment, as a dress.
**hem·lock** (hĕm′lŏk′) *n.* [ME *hemlok*, poisonous hemlock < OE *hymlīc*.] **1. a.** An evergreen tree of the genus *Tsuga* of North America and eastern Asia, with short, flat needles and small cones. **b.** The wood of the hemlock, used as a source of lumber, wood pulp, and tannic acid. **2. a.** Any of several poisonous plants of the genera *Conium* and *Cicuta*, as the poison hemlock. **b.** A poison obtained from the poison hemlock.
**hemo-** *or* **hema-** *or* **hem-** *pref.* [< Gk. *haima*, blood.] Blood <*hema*cytometer>
**he·mo·chro·ma·to·sis** (hē′mə-krō′mə-tō′sĭs) *n.* A disorder of iron metabolism marked by accumulation of iron-containing pigments in the skin, liver enlargement, and diabetes.
**he·mo·cy·a·nin** (hē′mō-sī′ə-nĭn) *n.* A bluish, oxygen-bearing, copper-containing substance similar to hemoglobin, found in the blood of certain insects and crustaceans.
**he·mo·cyte** (hē′mə-sīt′) *n.* A cell or similar formation in the blood.
**he·mo·di·al·y·sis** (hē′mō-dī-ăl′ĭ-sĭs) *n.* Dialysis of the blood.
**he·mo·flag·el·late** (hē′mō-flăj′ə-lāt′, -lĭt, -flə-jĕl′ĭt) *n.* A flagellate protozoan, as a trypanosome, parasitic in the blood.
**he·mo·glo·bin** (hē′mə-glō′bĭn) *n.* [Short for obs. *hematoglobulin*.] The oxygen-bearing, iron-containing conjugated protein in vertebrate red blood cells, consisting of approx. 6% heme and 94% globin and having as a typical formula $(C_{738}H_{1166}FeN_{203}O_{208}S_2)_4$.
**he·mo·leu·ko·cyte** (hē′mə-lōō′kə-sīt′) *n.* A white blood cell.
**he·mo·ly·sin** (hē′mə-lī′sĭn) *n.* An agent or substance that causes lysis of red blood cells, thereby liberating hemoglobin.
**he·mol·y·sis** (hĭ-mŏl′ĭ-sĭs, hē′mə-lī′sĭs) *n.* Lysis of red blood cells. **—he′mo·lyt′ic** (hē′mə-lĭt′ĭk) *adj.*
**he·mo·phil·i·a** (hē′mə-fĭl′ē-ə) *n.* A hereditary plasma-coagulation disorder principally affecting males but transmitted by females and marked by excessive, sometimes spontaneous bleeding.
**he·mo·phil·i·ac** (hē′mə-fĭl′ē-ăk′) *n.* One who has hemophilia.
**he·mo·phil·ic** (hē′mə-fĭl′ĭk) *adj.* **1.** Of or relating to hemophilia. **2.** Growing well in blood or in a culture containing blood <*hemophilic* bacteria>
**he·mo·pho·bi·a** (hē′mə-fō′bē-ə) *n.* A morbid fear of blood. **—he′mo·pho′bic** *adj.*
**he·mo·poi·e·sis** (hē′mə-poi-ē′sĭs) *n.* Hematopoiesis. **—he′mo·poi·et′ic** (-ĕt′ĭk) *adj.*
**he·mop·ty·sis** (hĭ-mŏp′tĭ-sĭs) *n.* [HEMO- + Gk. *ptusis*, a spitting < *ptuein*, to spit.] Expectoration of blood from the lungs or bronchial tubes.
**hem·or·rhage** (hĕm′ər-ĭj) *n.* [Obs. *hemoragie* < OFr. *hemorragie* < Lat. *haemorrhagia* < Gk. *haimorrhagia* : *haimo*, hemo- + *rhēgnunai*, to break.] Bleeding, esp. abnormally great discharge of blood from the blood vessels. —*vi.* **-rhaged, -rhag·ing, -rhag·es.** To bleed in or as if in a hemorrhage. **—hem′or·rhag′ic** (hĕm′ə-răj′ĭk) *adj.*
**hem·or·rhoid** (hĕm′ə-roid′) *n.* [ME *emoroides*, hemorrhoids < Lat. *haemorrhoidae* < Gk. *haimorrhoides* < *haimorrhos*, flowing with blood : *haima*, blood + *rhein*, to flow.] **1.** An itching or painful mass of dilated veins in swollen anal tissue. **2. hemorrhoids.** The pathological condition in which hemorrhoids occur.
**hem·or·rhoid·al** (hĕm′ə-roid′l) *adj.* **1.** Of, relating to, or involving hemorrhoids. **2.** *Anat.* Supplying the region of the rectum and anus. —Used of certain arteries.
**hem·or·rhoid·ec·to·my** (hĕm′ə-roi-dĕk′tə-mē) *n., pl.* **-mies.** Surgical excision of hemorrhoids.
**he·mo·sta·sis** (hē′mə-stā′sĭs) *also* **he·mo·sta·sia** (-zhə, -zhē-ə, -zē-ə) *n.* Stoppage of a flow or circulation of blood.
**he·mo·stat** (hē′mə-stăt′) *n.* **1.** An agent, as a chemical, that stops

bleeding. **2.** A clamplike surgical instrument for preventing or reducing bleeding.

**he·mo·stat·ic** (hē′mə-stăt′ĭk) *adj.* Acting to stop the flow of blood. —*n.* A hemostatic agent.

**hemp** (hĕmp) *n.* [ME < OE *hænep.*] **1. a.** A tall Asian plant, *Cannabis sativa,* with stems that yield a fiber used in cordage. **b.** The fiber of the hemp. **c.** A narcotic, as hashish, derived from hemp. **2. a.** A plant related to or resembling hemp, esp. one yielding a fiber similar to that of *Cannabis sativa.* **b.** The fiber of such a plant.

**hemp agrimony** *n.* A Eurasian plant, *Eupatorium cannabinum,* bearing small reddish-purple flower clusters.

**hemp·en** (hĕm′pən) *adj.* Of, relating to, or like hemp.

**hemp nettle** *n.* A Eurasian plant of the genus *Galeopsis,* esp. *G. tetrahit,* with bristly stems and white or reddish flowers.

**hemp·weed** (hĕmp′wēd′) *n.* Climbing hempweed.

**hem·stitch** (hĕm′stĭch′) *n.* **1.** A decorative stitch usu. bordering a hem, as on a handkerchief, made by drawing out several parallel threads and catching together the cross threads in uniform groups. **2.** Needlework incorporating the hemstitch. —*vt.* **-stitched, -stitch·ing, -stitch·es.** To embroider or ornament with hemstitch. **—hem′stitch′er** *n.*

**hen** (hĕn) *n.* [ME < OE.] **1.** A female bird, esp. the adult female of the domestic fowl. **2.** The female of certain aquatic animals, as a lobster. **3.** *Slang.* A woman, esp. a fussy old woman.

**hen-and-chick·ens** (hĕn′ən-chĭk′ənz) *n., pl.* **hens-and-chick·ens** (hĕnz′-). Any of several plants with many runners or offshoots, esp. the houseleek.

**hen·bane** (hĕn′bān′) *n.* A poisonous, malodorous plant, *Hyoscyamus niger,* native to the Mediterranean region, bearing clammy leaves and funnel-shaped greenish-yellow flowers and yielding a juice used medicinally.

**hen·bit** (hĕn′bĭt′) *n.* [HEN + BIT[1].] A European plant, *Lamium amplexicaule,* bearing toothed leaves and small white or purplish-red flowers.

**hence** (hĕns) *adv.* [ME, from this place < *henne* < OE *heonon.*] **1. a.** For this reason : THEREFORE <one of a kind and *hence* valuable> **b.** From this source <I worked in the Peace Corps; *hence* my interest in Ghanaian culture.> **2.** From this time : from now <A month *hence* we will have graduated.> **3. a.** From this place. **b.** From this life.

**hence·forth** (hĕns′fôrth′) *also* **hence·for·ward** (hĕns-fôr′wərd) *adv.* From now on.

**hench·man** (hĕnch′mən) *n.* [ME *hencheman,* servant to a person of rank < *hengest,* horse < OE + *man,* man.] **1. a.** A loyal and trusted follower or subordinate. **b.** One who supports a political figure chiefly out of self-interest. **2.** A member of a criminal gang. **3.** *Obs.* A page to a person of high rank, esp. a prince.

**hen·dec·a·syl·lab·ic** (hĕn-dĕk′ə-sĭ-lăb′ĭk) *adj.* [< Lat. *hendecasyllabus,* a line of 11 syllables : Gk. *hendeka,* eleven (*hen,* one + *deka,* ten) + Gk. *sullabē,* syllable. —see SYLLABLE.] Having 11 syllables. —*n.* A verse composed of 11 syllables. **—hen·dec′a·syl′la·ble** (-sĭl′ə-bəl) *n.*

**hen·di·a·dys** (hĕn-dī′ə-dĭs) *n.* [Med. Lat. < Gk. *hen dia duoin,* one by means of two.] A figure of speech in which two words connected by a conjunction are used to express a single notion that would usu. be expressed by an adjective and a substantive, as *grace and favor* instead of *gracious favor.*

**hen·e·quen** *also* **hen·e·quin** (hĕn′ĭ-kwĭn) *n.* [Sp. *henequén,* perh. of Taino orig.] **1.** A tropical American plant, *Agave fourcroydes,* with thick leaves that yield a coarse reddish fiber used in making rope and twine. **2.** The fiber obtained from the henequen.

**hen harrier** *n.* The marsh hawk.

**hen·na** (hĕn′ə) *n.* [Ar. *ḥinnā′′.*] **1. a.** A tree or shrub, *Lawsonia inermis* of Asia and northern Africa, bearing fragrant white or reddish flowers. **b.** A reddish dyestuff obtained from the leaves of the henna, used as a cosmetic dye and for coloring leather. **2.** A moderate or strong reddish brown to strong brown. —*vt.* **-naed, -na·ing, -nas.** To dye (e.g, hair) with henna. **—hen′na** *adj.*

**hen·ner·y** (hĕn′ə-rē) *n., pl.* **-ies. 1.** A poultry farm. **2.** A pen for domestic fowl.

**hen·o·the·ism** (hĕn′ə-thē-ĭz′əm) *n.* [G. *Henotheismus* : Gk. *heis,* one + Gk. *theos,* god.] Belief in one god without denying the existence of others. **—hen′o·the′ist** *n.* **—hen′o·the·is′tic** *adj.*

**hen party** *n.* A party or get-together for women only.

**hen·peck** (hĕn′pĕk′) *vt.* **-pecked, -peck·ing, -pecks.** *Informal.* To dominate or harass (one's husband) by scolding.

**hen·ry** (hĕn′rē) *n., pl.* **-ries** *or* **-rys.** [After Joseph *Henry* (1797–1878).] The unit of inductance in which an induced electromotive force of one volt is produced when the current is varied at the rate of one ampere per second.

**hent** (hĕnt) *vt.* **hent·ed, hent·ing, hents.** [ME *henten* < OE *hentan.*] *Obs.* To take hold of : SEIZE.

**hen track** *n.* Illegible handwriting.

**hep** (hĕp) *adj.* & *n. var. of* HIP[2].

**hep·a·rin** (hĕp′ər-ĭn) *n.* [Gk. *hēpar,* liver + -IN.] A complex organic acid found esp. in lung and liver tissue that is used medically to prevent the clotting of blood.

**hepat-** *pref. var. of* HEPATO-.

**he·pat·ic** (hĭ-păt′ĭk) *adj.* [ME *epatic* < Lat. *hepaticus* < Gk. *hēpatikos* < *hēpar,* liver.] **1.** Of, relating to, or like the liver. **2.** Occurring in or acting on the liver. **3.** Of or belonging to the Hepaticae, a class of mosslike plants including the liverworts. —*n.* **1.** A drug used in treating liver diseases. **2.** A plant of the class Hepaticae.

**he·pat·i·ca** (hĭ-păt′ĭ-kə) *n.* [NLat. *Hepatica,* genus name < Med. Lat., liverwort < Lat. *hepaticus,* hepatic.] A woodland plant of the genus *Hepatica,* esp. *H. americana* of eastern North America, with three-lobed leaves and white or lavender flowers.

**hep·a·ti·tis** (hĕp′ə-tī′tĭs) *n.* Inflammation of the liver, marked by jaundice and usu. fever and caused by infectious or toxic agents.

**hepato-** *or* **hepat-** *pref.* [< Gk. *hēpar, hēpat-,* liver.] Liver <*hepatitis*><*hepatotoxin*>

**hep·a·to·gen·ic** (hĕp′ə-tō-jĕn′ĭk) *also* **hep·a·tog·e·nous** (-tŏj′ə-nəs) *adj.* Originating in or produced by the liver.

**hep·a·to·tox·ic·i·ty** (hĕp′ə-tō-tŏk-sĭs′ĭ-tē) *n.* **1.** The quality or condition of being toxic to the liver. **2.** The degree to which something causes hepatotoxicity. **—hep′a·to·tox′ic** (-tŏk′sĭk) *adj.*

**hep·a·to·tox·in** (hĕp′ə-tō-tŏk′sĭn) *n.* A substance that causes hepatotoxicity.

**hep·cat** (hĕp′kăt′) *n. Slang.* A devotee or performer of swing and jazz during the 1940's.

**He·phaes·tus** (hĭ-fĕs′təs) *n.* [Lat. < Gk. *Hēphaistos.*] *Gk. Myth.* The god of fire and metalworking.

**Hep·ple·white** (hĕp′əl-hwīt′, -wīt′) *adj.* [After George *Hepplewhite* (d. 1786).] Of or designating an English style of furniture of the late 18th cent., noted for its light, graceful lines.

**hepta-** *or* **hept-** *pref.* [< Gk. *hepta,* seven.] Seven <*heptarchy*>

**hep·tad** (hĕp′tăd′) *n.* [Gk. *heptas, heptad-,* the number seven < *hepta,* seven.] A series or group of seven.

**hep·ta·gon** (hĕp′tə-gŏn′) *n.* [Gk. *heptagōnos,* having seven angles : *hepta,* seven + *gōnos,* -gon.] A polygon having seven sides. **—hep·tag′o·nal** (-tăg′ə-nəl) *adj.*

**hep·ta·he·dron** (hĕp′tə-hē′drən) *n.* A polyhedron with seven faces. **—hep′ta·he′dral** (-drəl) *adj.*

**hep·tam·e·ter** (hĕp-tăm′ĭ-tər) *n.* **1.** A unit of verse having seven feet. **2.** A line of verse written in heptameter.

**hep·tane** (hĕp′tān′) *n.* A volatile, colorless, highly flammable liquid hydrocarbon, $CH_3(CH_2)_5CH_3$, obtained in the fractional distillation of petroleum and used as an anesthetic or solvent and as a standard in determining octane ratings.

**hep·tar·chy** (hĕp′tär′kē) *n., pl.* **-chies. 1. a.** Government by seven persons. **b.** A state so governed. **2.** *often* **Heptarchy.** The informal confederation of the Anglo-Saxon kingdoms from the 5th to the 9th cent., consisting of Kent, Sussex, Wessex, Essex, Northumbria, East Anglia, and Mercia.

**hep·ta·stich** (hĕp′tə-stĭk′) *n.* A stanza or strophe of seven lines.

**Hep·ta·teuch** (hĕp′tə-to͞ok′, -tyo͞ok′) *n.* [Gk. *heptateukhos,* volume containing seven books : *hepta,* seven + *teukhos,* book.] The first seven books of the Old Testament.

**her** (hər, ər; hûr *when stressed*) *pron.* [ME < OE *hire.*] *objective case of* SHE. —Used: **1.** As the direct object of a verb <I met *her* at the beach.> **2.** As the indirect object of a verb <I sold *her* a car.> **3.** As the object of a preposition <This dress belongs to *her.*> —*adj.* —Used as a modifier before a noun <*her* hat><*her* achievements><*her* first romance>

**He·ra** (hîr′ə) *n.* [Gk. *Hēra.*] *Gk. Myth.* The sister and wife of Zeus.

**Her·a·cles** *or* **Her·a·kles** (hĕr′ə-klēz′) *n. vars. of* HERCULES 1.

**her·ald** (hĕr′əld) *n.* [ME < AN, of Germanic orig.] **1.** One who carries or announces important news : MESSENGER. **2.** One that gives a sign or indication of something to come : HARBINGER. **3.** *Chiefly Brit.* An official whose specialty is heraldry. **4. a.** An official formerly charged with making royal proclamations and bearing messages of state between sovereigns. **b.** An official who formerly made proclamations and conveyed challenges at a tournament. —*vt.* **-ald·ed, -ald·ing, -alds.** To announce : proclaim <was *heralded* by thunderous applause>

**he·ral·dic** (hə-răl′dĭk) *adj.* Of or relating to heralds or heraldry. **—he·ral′di·cal·ly** *adv.*

**her·ald·ry** (hĕr′əl-drē) *n., pl.* **-ries. 1.** The study or art of tracing genealogies, determining, designing, and granting coats of arms, and settling questions of rank or protocol. **2.** Armorial ensigns or devices. **3.** Pageantry.

**herb** (ûrb, hûrb) *n.* [ME *herbe* < OFr. *erbe* < Lat. *herba.*] **1.** A plant with a fleshy stem as distinguished from the woody tissue of shrubs and trees and that gen. dies back at the end of each growing season. **2.** Any of various often aromatic plants used esp. in medicine or cookery.

**her·ba·ceous** (hûr-bā′shəs, ûr-) *adj.* [Lat. *herbaceus* < *herba,* herb.] **1.** Relating to or typical of an herb as distinguished from a woody plant. **2.** Green and leaflike in appearance or texture.

**herb·age** (ûr′bĭj, hûr′-) *n.* [ME < OFr. *erbage* < Med. Lat. *herba-*

*gium* < Lat. *herba.*] **1.** Grass or similar herbaceous vegetation used for pasturage. **2.** The fleshy, often edible parts of plants.

**herb·al** (hûr′bəl, ûr′-) *adj.* Of, pertaining to, or containing herbs. —*n.* A book about plants and herbs, esp. those that are useful to humans.

**herb·al·ist** (hûr′bə-lĭst, ûr′-) *n.* One who grows, collects, or specializes in the use of herbs, esp. medicinal herbs.

**her·bar·i·um** (hûr-bâr′ē-əm, ûr-) *n., pl.* **-i·ums** or **-i·a** (-ē-ə) [LLat. < Lat. *herba,* vegetation.] **1.** A collection of dried plants mounted and labeled for use in scientific examination. **2.** A place or institution where a herbarium is maintained.

**herb bennet** *n.* [ME *herbe benet* < AN < Med. Lat. *herba benedicta.*] A hairy Eurasian plant, *Geum urbanum,* with small yellow flowers and an astringent root once used medicinally.

**her·bi·cide** (hûr′bĭ-sīd′, ûr′-) *n.* A substance used to destroy plants, esp. weeds. **—her′bi·cid′al** (-sīd′l) *adj.*

**her·bi·vore** (hûr′bə-vôr′, -vōr′, ûr′-) *n.* [NLat. *Herbivora,* former mammalian group < *herbivorus,* herbivorous.] An animal that is herbivorous.

**her·biv·o·rous** (hûr-bĭv′ər-əs, ûr-) *adj.* [NLat. *herbivorus* : Lat. *herba,* vegetation + *vorare,* to swallow up.] Feeding on plants. **—her·biv′o·rous·ly** *adv.*

**herb Par·is** (păr′ĭs) *n.* [Prob. Med. Lat. *herba paris.*] A European plant, *Paris quadrifolia,* bearing a whorl of four leaves and a solitary greenish or yellow flower.

**herb Rob·ert** (rŏb′ərt) *n.* [ME *herbe Robert* < OFr. *erbe Robert* < Med. Lat. *herba Roberti.*] A low-growing plant, *Geranium robertianum,* with divided leaves and small reddish-purple flowers.

**her·cu·le·an** (hûr′kyə-lē′ən, hûr-kyo͞o′lē-) *adj.* **1.** Of unusual size, force, or difficulty. **2.** *often* **Herculean.** Of or like Hercules <*Herculean* power> **3. Herculean.** Of or pertaining to Hercules.

**Her·cu·les** (hûr′kyə-lēz′) *n.* [Lat. < Gk. *Hēraklēs.*] **1.** *also* **Hera·cles** or **Her·a·kles** (hĕr′ə-klēz′). *Gk. & Rom. Myth.* The son of Zeus and Alcmene, a hero of extraordinary strength who won immortality by performing 12 extremely difficult labors demanded by Hera. **2.** *Astron.* A constellation in the Northern Hemisphere.

**Her·cu·les′-club** (hûr′kyə-lēz-klŭb′) *n.* **1.** A tree or shrub, *Aralia spinosa* of the southeastern United States, bearing prickly compound leaves and large clusters of small white flowers. **2.** A spiny tree, *Zanthoxylum clava-herculis* of the southeastern United States, bearing small greenish-yellow flower clusters.

**herd** (hûrd) *n.* [ME < OE *heord.*] **1. a.** A group of domestic animals of a single kind, as cattle or sheep, kept together for a particular use. **b.** A number of wild animals of one species that remain together as a group <a *herd* of antelope> **2. a.** A large number of people : CROWD. **b.** The common people : the masses. —*v.* **herd·ed, herd·ing, herds.** —*vi.* **1.** To come together in a herd. **2.** To keep company : ASSOCIATE. —*vt.* **1.** To gather, keep, or drive (animals) in a herd. **2.** To gather and lead in a group <*herded* the tourists into the museum> **3.** To tend (sheep or cattle).

**herd·er** (hûr′dər) *n.* **1.** A person who drives or tends a herd. **2.** A herdsman.

**her·dic** (hûr′dĭk) *n.* [After Peter *Herdic* (1824–1888).] A small horse-drawn cab with two wheels, side seats, and a back entrance.

**herds·man** (hûrdz′mən) *n.* One who owns or breeds livestock.

**here** (hîr) *adv.* [ME < OE *hēr.*] **1.** At or in this place <Sit *here.*> **2.** At this time : NOW <We'll end the game *here.*> **3.** At or on this point, detail, or item <*Here* we must object.> **4.** In the present life or condition. **5.** To this place : HITHER <Come *here.*> ***usage:*** In constructions introduced by *here,* the number of the verb is governed by the subject, which appears after the verb (*Here* are the tickets). —*adj.* **1.** —Used for emphasis after a demonstrative pronoun <Which street? This one *here.*> **2.** —Used for emphasis after a noun modified by a demonstrative pronoun <this street *here*> **3.** *Nonstandard.* —Used for emphasis between a demonstrative pronoun and a noun <this *here* street> —*interj.* —Used to respond to a roll call, attract attention, command an animal, or rebuke or admonish. **—neither here nor there.** Insignificant and immaterial.

**here·a·bout** (hîr′ə-bout′) *also* **here·a·bouts** (-bouts′) *adv.* In this general vicinity.

**here·af·ter** (hîr-ăf′tər) *adv.* **1.** Immediately following this in time, order, or place. **2.** In a future time or state. —*n.* The afterlife.

**here and now** *n.* The present <rewards in the *here and now*>

**here·by** (hîr-bī′) *adv.* By this means.

**her·e·dit·a·ment** (hĕr′ĭ-dĭt′ə-mənt) *n.* [Med. Lat. *hereditamentum* < LLat. *hereditare,* to inherit. —see HERITAGE.] *Law.* Property that can be inherited.

**he·red·i·tar·y** (hə-rĕd′ĭ-tĕr′ē) *adj.* [Lat. *hereditarius* < *hereditas,* inheritance < *heres,* heir.] **1.** *Law.* **a.** Descending from an ancestor to a legal heir. **b.** Having title or possession through inheritance. **2.** Genetically transmitted or transmissible. **3. a.** Appearing in or typical of successive generations. **b.** Derived from or fostered by one's ancestors. **4.** Ancestral : traditional <their *hereditary* country estate> **5.** Of or relating to heredity or inheritance. **—he·red′i·tar′i·ly** (-târ′ə-lē) *adv.* **—he·red′i·tar′i·ness** *n.*

**he·red·i·tist** (hə-rĕd′ĭ-tĭst) *n.* One who advocates the theory that heredity rather than environment determines personality.

**he·red·i·ty** (hə-rĕd′ĭ-tē) *n., pl.* **-ties.** [OFr. *heredite,* inheritance < Lat. *hereditas* < *heres,* heir.] **1.** The genetic transmission of characteristics from parents to offspring. **2.** The characteristics and related potentialities transmitted to an individual organism by heredity.

**Here·ford** (hûr′fərd, hĕr′ə-fərd) *n.* Any of a breed of beef cattle developed in Herefordshire, England, with a reddish coat and white markings.

**here·in** (hîr-ĭn′) *adv.* In or into this.

**here·in·af·ter** (hîr′ĭn-ăf′tər) *adv.* In a following part of this document or written material.

**here·in·be·fore** (hîr′ĭn-bĭ-fôr′, -fōr′) *adv.* In a preceding part of this document or written material.

**here·in·to** (hîr-ĭn′to͞o) *adv.* Into this matter, circumstance, situation, or place.

**here·of** (hîr-ŭv′, -ŏv′) *adv.* Relating to or concerning this.

**here·on** (hîr-ŏn′, -ôn′) *adv.* Hereupon.

**he·re·si·arch** (hə-rē′zē-ärk′, hĕr′ĭ-sē-) *n.* [LLat. *haeresiarcha* < LGk. *hairesiarkhēs* : Gk. *hairesis,* sect + Gk. *-arkhēs,* -arch.] The founder or chief proponent of a heresy.

**her·e·sy** (hĕr′ĭ-sē) *n., pl.* **-sies.** [ME *heresie* < OFr. < LLat. *haeresis* < LGk. *hairesis* < Gk., faction < *haireisthai,* to choose.] **1. a.** An opinion or doctrine in conflict with established religious beliefs, esp. dissension from or denial of Roman Catholic dogma by a professed believer or baptized church member. **b.** Adherence to such dissenting opinion or doctrine. **2. a.** An unorthodox or controversial opinion or doctrine, as in philosophy, science, or politics. **b.** Adherence to such unorthodox or controversial opinion.

**her·e·tic** (hĕr′ĭ-tĭk) *n.* [ME *heretik* < OFr. *eretique* < LLat. *haereticus* < Gk. *hairetikos,* factious < *hairesthai,* to choose.] One who holds or advocates controversial opinions, esp. one who publicly opposes the officially accepted dogma of the Roman Catholic Church.

**he·ret·i·cal** (hə-rĕt′ĭ-kəl) *adj.* **1.** Of or relating to heresy or heretics. **2.** Marked by, revealing, or approaching deviation from established standards or beliefs. **—he·ret′i·cal·ly** *adv.* **—he·ret′i·cal·ness** *n.*

**here·to** (hîr-to͞o′) *adv.* To this document, matter, or proposition.

**here·to·fore** (hîr′tə-fôr′, -fōr′) *adv.* [ME : *here,* here (< OE *hēr*) + *to fore,* previously (< OE *toforan*).] Up to the present time.

**here·un·to** (hîr-ŭn′to͞o) *adv.* Hereto.

**here·up·on** (hîr′ə-pŏn′, -pôn′) *adv.* Immediately after this.

**here·with** (hîr-wĭth′, -wĭth′) *adv.* **1.** Along with this. **2.** By this means.

**her·i·ot** (hĕr′ē-ət) *n.* [ME < OE *heregeatu* : *here,* army + *geatwe,* equipment.] A tribute or service rendered to a feudal lord on the death of a tenant.

**her·i·ta·ble** (hĕr′ĭ-tə-bəl) *adj.* [ME < OFr. < *heriter,* to inherit. —see HERITAGE.] **1.** Capable of being inherited : HEREDITARY. **2.** Capable of inheriting or of taking by inheritance. **—her′i·ta·bil′i·ty** *n.*

**her·i·tage** (hĕr′ĭ-tĭj) *n.* [ME < OFr. < *heriter,* to inherit < LLat. *hereditare* < Lat. *heres,* heir.] **1.** Property that is or can be inherited : INHERITANCE. **2.** Something passed down from preceding generations : TRADITION. **3.** The status gained by a person through birth : BIRTHRIGHT <a *heritage* of wealth and power>

**her·i·tor** (hĕr′ĭ-tər) *n.* [ME *heriter* < AN < Lat. *hereditarius,* hereditary. —see HEREDITARY.] An inheritor.

**her·i·tress** (hĕr′ĭ-trĭs) *n.* A woman who is an inheritor.

**her·ma** (hûr′mə) *also* **herm** (hûrm) *n.* [Lat. < Gk. *hermēs* (after *Hermēs,* Hermes).] A statue consisting of the head of the Greek god Hermes mounted on a square stone post.

**her·maph·ro·dite** (hər-măf′rə-dīt′) *n.* [ME *hermofrodite* < Med. Lat. *hermofroditus* < Lat. *hermaphroditus* < Gk. *hermaphroditos* < *Hermaphroditos,* Hermaphroditus.] **1.** One who has the sex organs and many of the secondary sex characteristics of both male and female. **2.** *Biol.* An organism, as an earthworm, with male and female reproductive organs in the same individual. **3.** Something composed of different or opposing elements. **—her·maph′ro·dit′ic** (-dĭt′ĭk) *adj.* **—her·maph′ro·dit′i·cal·ly** *adv.*

**hermaphrodite brig** *n.* A two-masted vessel with a square-rigged foremast and a schooner-rigged mainmast.

**her·maph·ro·dit·ism** (hər-măf′rə-dī-tĭz′əm) *also* **her·maph·ro·dism** (-rə-dĭz′əm) *n.* The condition of being a hermaphrodite.

**Her·maph·ro·di·tus** (hər-măf′rə-dī′təs) *n.* [Gk. *Hermaphroditos* : *Hermēs,* Hermes + *Aphroditē,* Aphrodite.] *Gk. Myth.* The son of Hermes and Aphrodite, who became united in one body with the nymph Salmacis.

**her·me·neu·tic** (hûr′mə-no͞o′tĭk, -nyo͞o′-) *also* **her·me·neu·ti·cal** (-tĭ-kəl) *adj.* [Gk. *hermeneutikos* < *hermēneutēs,* interpreter < *hermēnuenein,* to interpret < *hermēneus,* interpreter.] Interpretive : explanatory. **—her′me·neu′ti·cal·ly** *adv.*

**her·me·neu·tics** (hûr′mə-no͞o′tĭks, -nyo͞o′-) *n.* [< HERMENEUTIC.] *(sing. in number).* The science and methodology of interpretation, esp. of the Bible.

**Her·mes** (hûr′mēz) *n.* [Gk. *Hermēs.*] *Gk. Myth.* The god of commerce, invention, and theft and messenger for the other gods.

**Hermes Tris·me·gis·tus** (trĭs′mə-jĭs′təs) *n.* [Lat. < Gk. *Hermēs trismegistos.*] *Myth.* The Egyptian god Thoth, the legendary author of works on alchemy, astrology, and magic.

**her·met·ic** (hər-mĕt′ĭk) *also* **her·met·i·cal** (-ĭ-kəl) *adj.* [NLat. *hermeticus,* after *Hermes* Trismegistus.] **1. a.** Totally sealed, esp. against the escape or entry of air. **b.** Impervious to outside interference or influence. **c.** Reclusive <a *hermetic* existence on a remote island> **2.** *often* **Hermetic. a.** Of or pertaining to Hermes Trismegistus or the works ascribed to him. **b.** Having to do with the occult sciences, esp. alchemy : MAGICAL. **—her·met′i·cal·ly** *adv.*

▲ word history: An airtight seal is called a *hermetic* seal by a roundabout chain of circumstances. *Hermetic* is an adjective derived from the name *Hermes,* specifically Hermes Trismegistus, the Greek name for the Egyptian god Thoth, who was regarded as the originator of the science of alchemy. In the 17th century the adjective *hermetic* meant "pertaining to alchemy" and the occult sciences in general. Alchemy, and later chemistry, was itself variously known as the *hermetic* art, philosophy, or science. A *hermetic* seal was a kind of seal used by alchemists that involved melting closed an opening in a glass vessel. Since the resulting seal was airtight, any airtight seal has come to be called *hermetic.*

**her·mit** (hûr′mĭt) *n.* [ME *heremite* < Med. Lat. *heremita* < LLat. *eremita* < Gk. *erēmitēs* < *erēmia,* desert < *erēmos,* solitary.] **1.** One who has withdrawn from society and lives a solitary existence : RECLUSE. **2.** A spiced cookie with molasses, raisins, and nuts. **—her·mit′ic, her·mit′i·cal** *adj.* **—her·mit′i·cal·ly** *adv.*

**her·mit·age** (hûr′mĭ-tĭj) *n.* [ME < OFr. *hermitaige* < Med. Lat. *heremita,* hermit. —see HERMIT.] **1. a.** The habitation of a hermit. **b.** A monastery or abbey. **2.** A retreat where one can live in seclusion. **3.** The condition or way of life of a hermit.

**Her·mi·tage** (ĕr′mĭ-täzh′) *n.* [After Tain *l'Ermitage,* a commune in France.] A rich, full-bodied, red or white wine produced in southeastern France.

**hermit crab** *n.* Any of various crustaceans of the section Anomura within the order Decapoda, with a soft, unprotected abdomen and occupying and carrying about the empty shell of a univalve mollusk, as a snail.

**hermit thrush** *n.* A North American songbird, *Hylocichla guttata,* with brownish plumage and a spotted breast.

**†hern** (hûrn) *n.* [Var. of HERON.] *Archaic & Regional.* A heron.

**her·ni·a** (hûr′nē-ə) *n., pl.* **-ni·as** *or* **-ni·ae** (-nē-ē′) [ME < Lat.] The protrusion of an organ, organic part, or other bodily structure through the wall that usu. contains it : RUPTURE. **—her′ni·al** *adj.*

**her·ni·ate** (hûr′nē-āt′) *vi.* **-at·ed, -at·ing, -ates.** [HERNI(A) + -ATE.] To protrude so as to form a hernia. **—her′ni·a′tion** *n.*

**he·ro** (hîr′ō) *n., pl.* **-roes.** [Lat. *heros* < Gk. *hērōs.*] **1.** A mythological or legendary figure, often of divine ancestry, who is favored by the gods, endowed with great courage and strength, and celebrated for his bold exploits. **2.** A man noted for courageous acts or nobility of purpose, esp. one who has risked or sacrificed his life. **3.** A man noted for special achievements in a particular field <the *heroes* of science> **4.** The principal male character in a literary work or dramatic presentation. **5.** *Slang.* A large sandwich consisting of a long, split roll having a variety of fillings, as meats, cheeses, lettuce, tomatoes, and onions.

**He·ro** (hîr′ō) *n.* [Lat. < Gk. *Hērō.*] *Gk. Myth.* A priestess of Aphrodite, beloved by Leander.

**he·ro·ic** (hĭ-rō′ĭk) *also* **he·ro·i·cal** (-ĭ-kəl) *adj.* **1.** Of, pertaining to, or like the heroes of myth, legend, or literature. **2.** Having, exhibiting, or marked by the qualities appropriate to a hero : COURAGEOUS. **3. a.** Impressive in size or scope : GRAND <a *heroic* achievement> **b.** Of a size or scale that is larger than life <creates *heroic* paintings> *—n.* **1.** A heroic verse or poem. **2. heroics.** Melodramatic language or behavior <"We trust the House . . . will come up with answers without all the political *heroics*" *—Atlanta Constitution*> **—he·ro′i·cal·ly** *adv.* **—he·ro′i·cal·ness** *n.*

**heroic couplet** *n.* A verse unit having two rhymed lines in iambic pentameter.

**heroic meter** *n.* Heroic verse.

**heroic play** *n.* A Restoration tragedy written in rhymed couplets and gen. marked by extravagant declamatory rhetoric.

**heroic verse** *n.* One of several verse forms suitable for and traditionally used in epic and dramatic poetry, esp.: **a.** The dactylic hexameter in Greek and Latin. **b.** The iambic pentameter in English.

**her·o·in** (hĕr′ō-ĭn) *n.* [G. < Gk. *hērōs,* hero (from the feeling of omnipotence experienced while taking the drug).] A white, odorless, bitter crystalline compound, $C_{17}H_{17}NO(C_2H_3O_2)_2$, a highly addictive narcotic derived from morphine. **—her′o·in·ism** *n.*

**her·o·ine** (hĕr′ō-ĭn) *n.* [Lat. *heroina* < Gk. *hērōinē,* fem. of *hērōs,* hero.] **1.** A woman noted for courageous acts. **2.** A woman noted for special achievements in a particular field. **3.** The principal female character in a literary work or dramatic presentation.

**her·o·ism** (hĕr′ō-ĭz′əm) *n.* **1.** The quality or state of being heroic. **2.** Heroic conduct. **3.** Heroic characteristics : COURAGE.

☆ ***syns:*** HEROISM, GALLANTRY, PROWESS, VALIANCE, VALOR, VALOROUSNESS *n. core meaning* : the quality or state of being heroic <the rescue was an act of sheer *heroism*> ***ant*** : pusillanimity

**her·on** (hĕr′ən) *n.* [ME < OFr., of Germanic orig.] Any of various long-necked, long-legged wading birds of the family Ardeidae, with a long pointed bill.

**hero worship** *n.* Strong or excessive admiration for a hero or a person regarded as a hero.

**he·ro-wor·ship** (hîr′ō-wûr′shĭp) *vt.* **-shiped, -ship·ing, -ships** *or* **-shipped, -ship·ping, -ships.** To feel hero worship for. **—he′ro-wor′ship·er** *n.*

**her·pes** (hûr′pēz) *n.* [Lat. < Gk. *herpēs* < *herpein,* to creep.] A viral disease causing eruptions of the skin or mucous membrane, esp. herpes simplex or herpes zoster. **—her·pet′ic** (hər-pĕt′ĭk) *adj.*

**herpes sim·plex** (sĭm′plĕks′) *n.* [NLat., simple herpes.] A viral infection resulting in blistering of the lips, external nares, glans, prepuce, or vulva.

**her·pes·vi·rus** (hûr′pēz-vī′rəs) *n.* Any of various DNA-containing animal viruses that produce herpes.

**herpes zos·ter** (zŏs′tər, zō′stər) *n.* [NLat., girdle herpes.] A viral infection causing eruption of vesicles along a nerve path on one side of the body, often accompanied or followed by severe neuralgia.

**her·pe·tol·o·gy** (hûr′pĭ-tŏl′ə-jē) *n.* [Gk. *herpeton,* reptile < *herpein,* to creep.] The zoological study of reptiles and amphibians. **—her′pe·to·log′ic** (-tə-lŏj′ĭk), **her′pe·to·log′i·cal** *adj.* **—her′pe·to·log′i·cal·ly** *adv.* **—her′pe·tol′o·gist** *n.*

**Herr** (hĕr) *n., pl.* **Her·ren** (hĕr′ən) [G.] A courtesy title prefixed to the name or professional title of a German, equivalent to *Mister.*

**Her·ren·volk** (hĕr′ən-fōk′, -fôlk′) *n.* [G.] A people purportedly endowed with the right to dominate and exploit other peoples.

**her·ring** (hĕr′ĭng) *n., pl.* **herring** *or* **-rings.** [ME < OE *hǣring.*] A fish of the family Clupeidae, esp. *Clupea harengus,* a commercially important food fish of Atlantic and Pacific waters.

**her·ring·bone** (hĕr′ĭng-bōn′) *n.* **1.** A pattern made up of rows of short, slanted parallel lines, with the direction of the slant alternating row by row. **2.** A woven, twilled fabric in a herringbone pattern. **3.** A method of climbing a ski slope with the tips of the skis pointed outward. *—v.* **-boned, -bon·ing, -bones.** *—vt.* To arrange or ornament with a herringbone pattern. *—vi.* **1.** To produce a herringbone pattern. **2.** To climb a ski slope by executing a herringbone.

**herring gull** *n.* A common, widely distributed gull, *Larus argentatus,* with gray and white plumage and black wing tips.

**hers** (hûrz) *pron.* [ME < *hire,* her.] —Used to indicate the one or ones belonging to her <The idea for the fund raiser was *hers.*>

**her·self** (hûr-sĕlf′) *pron.* [ME *hire selfe* < OE *hire selfre.*] **1.** That one identical with her. —Used: **a.** Reflexively as the direct or indirect object of a verb or as the object of a preposition <She cut *herself.*> **b.** For emphasis <She *herself* isn't sure.> **c.** In an absolute construction <In debt *herself,* she was unable to contribute.> **2.** Her normal or healthy condition or state <She hasn't been *herself* for the past two weeks.>

**hertz** (hûrts) *n.* [After Heinrich R. *Hertz* (1857-1894).] A unit of frequency equal to one cycle per second.

**Hertz·i·an wave** (hûrt′sē-ən, hĕrt′-) *n.* [After Heinrich R. *Hertz* (1857–1894).] An electromagnetic wave, usu. of radio frequency, created by the oscillation of electricity in a conductor.

**Hertz·sprung-Rus·sell diagram** (hĕrts′sprŭng-rŭs′əl) *n.* [After Ejnar *Hertzsprung* (1873–1967) and Henry N. *Russell* (1877–1957).] A graph of the logarithms of the luminosities of stars plotted against the logarithms of their surface temperatures.

**he's** (hēz). **1.** He has. **2.** He is.

**Hesh·van** *also* **Hesh·wan** (KHĕsh′vən) *n.* [Heb. *ḥeshwān.*] The second month of the Hebrew calendar. —See table at CALENDAR.

**hes·i·tant** (hĕz′ĭ-tənt) *adj.* Tending to hesitate. **—hes′i·tan·cy** (-tən-sē) *n.* **—hes′i·tant·ly** *adv.*

**hes·i·tate** (hĕz′ĭ-tāt′) *vi.* **-tat·ed, -tat·ing, -tates.** [Lat. *haesitare, haesitat-,* to hesitate, freq. of *haerēre,* to hold fast.] **1. a.** To be slow to speak, act, or decide. **b.** To pause in uncertainty : WAVER. **2.** To be reluctant <*hesitated* to bother you> **3.** To speak haltingly : FALTER. **—hes′i·tat′er** *n.* **—hes′i·tat′ing·ly** *adv.*

☆ ***syns:*** HESITATE, FALTER, HALT, PAUSE, SHILLY-SHALLY, STAGGER, VACILLATE, WAVER *v. core meaning* : to be irresolute in acting or doing <*hesitated* before answering the question>

**hes·i·ta·tion** (hĕz′ĭ-tā′shən) *n.* **1.** An act or instance of hesitating. **2.** The state of being hesitant. **3.** A pause or fumbling in speech.

**Hes·pe·ri·an** (hĕ-spîr′ē-ən) *adj.* [< Lat. *Hesperius* < Gk. *hesperios* < *hesperos,* evening.] Of or relating to the west.

**Hes·per·i·des** (hĕ-spĕr′ĭ-dēz′) *pl.n.* [Lat. *Hesperides* < Gk., pl. of *hesperis,* western < *hesperios* < *hesperos,* evening.] *Gk. Myth.* **1.** The nymphs who together with a dragon watch over a garden that produces golden apples. **2.** *(sing. in number).* A garden at the western end of the earth, in which golden apples grow. **—Hes′per·id′i·an, Hes′per·id′e·an** (hĕs′pə-rĭd′ē-ən) *adj.*

**hes·per·id·i·a** (hĕs′pə-rĭd′ē-ə) *n. pl. of* HESPERIDIUM.

**hes·per·i·din** (hĕ-spĕr′ĭ-dĭn) *n.* [HESPERID(IUM) + -IN.] A white or colorless crystalline compound, $C_{28}H_{34}O_{15}$, found in citrus fruit.

**hes·per·id·i·um** (hĕs′pə-rĭd′ē-əm) *n., pl.* **-i·a** (-ē-ə) [NLat., after the *Hesperides,* where golden apples grow.] A form of berry with a thickened leathery rind and juicy pulp divided into segments, as an orange or grapefruit.

**Hes·per·us** (hĕs′pər-əs) *n.* [Lat. < Gk. *Hesperos.*] The planet Venus in its appearance as the evening star.

**Hes·sian** (hĕsh′ən) *n.* **1.** An inhabitant of Hesse. **2.** A German mercenary in the British army in America during the Revolutionary War. **3.** A mercenary.

**Hessian boot** *n.* A high, tasseled man's boot.

**Hessian fly** *n.* A small fly, *Mayetiola destructor,* with larvae that infest and destroy grain plants, as wheat.

**hes·so·nite** (hĕs′ə-nīt′) *n. var. of* ESSONITE.

**hest** (hĕst) *n. Archaic.* [ME < OE *hǣs.*] Command : behest.

**Hes·ti·a** (hĕs′tē-ə) *n.* [Gk.] *Gk. Myth.* The goddess of the hearth.

**he·tae·ra** (hĭ-tîr′ə) *also* **he·tai·ra** (-tīr′ə) *n., pl.* **-tae·rae** (-tîr′ē) *or* **-tae·ras** *also* **-tai·rai** (-tīr′ī′) *or* **-tai·ras.** [Gk. *hetaira,* fem. of *hetairos,* companion.] **1.** A concubine or courtesan in ancient Greece, esp. one of a class of highly cultivated female companions. **2.** An adventuress.

**heter-** *pref. var. of* HETERO-.

**het·er·e·cious** (hĕt′ə-rē′shəs) *adj. var. of* HETEROECIOUS.

**het·ero** (hĕt′ə-rō′) *n., pl.* **-os.** A heterosexual. **—het′ero** *adj.*

**hetero-** *or* **heter-** *pref.* [< Gk. *heteros,* other.] **1.** Other : different <*hetero*chromatic> **2.** Containing different kinds of atoms <*het-erocyclic*>

**het·er·o·cer·cal** (hĕt′ə-rō-sûr′kəl) *adj.* [HETERO- + Gk. *kerkos,* tail + -AL.] Relating to, designating, or marked by a tail fin having two unequal lobes, with the vertebral column extending into the upper, usu. larger lobe, as in sharks.

**het·er·o·chro·mat·ic** (hĕt′ə-rō-krō-măt′ĭk) *adj.* **1.** Of, relating to, or having different colors : VARICOLORED. **2.** Having different wavelengths or frequencies. **3.** Of or relating to heterochromatin. **—het′er·o·chro′ma·tism** (-krō′mə-tĭz′əm) *n.*

**het·er·o·chro·ma·tin** (hĕt′ə-rō-krō′mə-tĭn) *n.* Chromosomal material exhibiting maximal staining in the nuclear meiotic interphase and lacking specific genetic activity.

**het·er·o·chro·mo·some** (hĕt′ə-rō-krō′mə-sōm′) *n.* A chromosome made up essentially of heterochromatin.

**het·er·o·cy·clic** (hĕt′ə-rō-sī′klĭk, -sĭk′lĭk) *adj.* Containing more than one kind of atom joined in a ring. **—het′er·o·cy′cle** (-sī′kəl) *n.* **—het′er·o·cy′clic** *n.*

**het·er·o·dox** (hĕt′ər-ə-dŏks′) *adj.* [LLat. *heterodoxus* < Gk. *heterodoxos* : *heteros,* other + *doxa,* opinion < *dokein,* to think.] **1.** Not agreeing with accepted beliefs, esp. deviating from church doctrine or dogma. **2.** Holding unorthodox opinions : UNCONVENTIONAL. **—het′er·o·dox′y** (-dŏk′sē) *n.*

**het·er·o·dyne** (hĕt′ər-ə-dīn′) *adj.* Having alternating currents of two different frequencies combined to generate a current that has sum and difference frequencies, either of which may be used in radio or television receivers by proper tuning or filtering. —*vt.* **-dyned, -dyn·ing, -dynes.** To combine (a radio-frequency wave) with a locally generated wave of different frequency in order to produce a new frequency equal to the sum or difference of the two.

**het·er·oe·cious** *also* **het·er·e·cious** (hĕt′ə-rē′shəs) *adj.* [HETERO- + Gk. *oikos,* house + -OUS.] Spending alternate stages of a life cycle on different unrelated hosts. —Used of parasites, as rusts and tapeworms. **—het′er·oe′cism** (-sĭz′əm) *n.*

**het·er·o·gam·ete** (hĕt′ə-rō-găm′ēt′, -gə-mēt′) *n.* Either of two conjugating gametes, as the small, motile male spermatozoon and the larger, nonmotile female ovum, that differ in size, form, or behavior.

**het·er·o·ga·met·ic** (hĕt′ə-rō-gə-mĕt′ĭk) *adj.* Having a dissimilar pair of sex chromosomes, as in human males, or one unpaired sex chromosome, as in some male insects.

**het·er·og·a·mous** (hĕt′ə-rŏg′ə-məs) *adj.* **1.** *Biol.* Marked by the fusion of unlike gametes in the reproductive process. **2.** *Bot.* Having flowers of different kinds, esp. both male and female flowers.

**het·er·og·a·my** (hĕt′ə-rŏg′ə-mē) *n.* **1.** Alternation of generations, one sexual, the other parthenogenetic, as in some aphids. **2.** A state in which uniting gametes are dissimilar in structure, size, and function. **—het′er·o·gam′ic** (-rō-găm′ĭk) *adj.*

**het·er·o·ge·ne·ous** (hĕt′ər-ə-jē′nē-əs, -jēn′yəs) *adj.* [Gk. *heterogenēs* (*heteros,* other + *genos,* kind) + -OUS.] **1.** Consisting of or involving dissimilar elements or parts : MIXED. **2.** Completely different : INCONGRUOUS. **—het′er·o′ge·ne′i·ty** (hĕt′ə-rō′jə-nē′ĭ-tē) *n.* **—het′er·o·ge′ne·ous·ly** *adv.* **—het′er·o·ge′ne·ous·ness** *n.*

**het·er·o·gen·e·sis** (hĕt′ə-rō-jĕn′ĭ-sĭs) *n.* Metagenesis. **—het′ero·ge·net′ic** (-jə-nĕt′ĭk) *adj.*

**het·er·o·gen·ic** (hĕt′ə-rō-jĕn′ĭk) *adj.* HETEROGENOUS[1].

**het·er·og·e·nous[1]** (hĕt′ə-rŏj′ə-nəs) *adj.* Originating from a source outside the body, esp. from a different species. **—het′er·og′e·ny** *n.*

**het·er·og·e·nous[2]** (hĕt′ə-rŏj′ə-nəs) *adj.* HETEROGENEOUS 1.

**het·er·og·o·nous** (hĕt′ə-rŏg′ə-nəs) *adj.* [HETERO- + -GON(Y) + -OUS.] Marked by the alternation of sexual and asexual generations. **—het′er·og′o·ny** *n.*

**het·er·o·graft** (hĕt′ə-rō-grăft′) *n.* Tissue taken from one species and grafted onto another.

**het·er·og·y·nous** (hĕt′ə-rŏj′ə-nəs) *adj.* Having two types of females, one able to reproduce sexually, the other infertile, as in ants.

**het·er·o·lec·i·thal** (hĕt′ə-rō-lĕs′ə-thəl) *adj.* [HETERO- + Gk. *lekithos,* yolk + -AL.] Having nonhomogeneous nutrient distribution in an ovum.

**het·er·ol·o·gous** (hĕt′ə-rŏl′ə-gəs) *adj.* [HETERO- + Gk. *logos,* word + -OUS.] **1.** Derived from a different species, as a tissue graft. **2.** Of or relating to histological or cytologic elements not normally found in a designated part of the body. **—het′er·ol′o·gous·ly** *adv.*

**het·er·ol·o·gy** (hĕt′ə-rŏl′ə-jē) *n.* [HETERO- + (ANA)LOGY.] Lack of structural, organizational, or developmental correspondence between bodily parts, arising from differences in origin.

**het·er·ol·y·sis** (hĕt′ə-rŏl′ĭ-sĭs, -ə-rō-lī′sĭs) *n., pl.* **-ses** (-sēz′). **1.** *Biol.* Dissolution of cells or protein components in one species by lytic agents of another. **2.** *Chem.* An organic reaction in which the breaking of bonds leads to the formation of ion pairs. **—het′er·o·lyt′ic** (-rō-lĭt′ĭk) *adj.*

**het·er·om·er·ous** (hĕt′ə-rŏm′ər-əs) *adj.* Having differing or unequal parts within the same structure or similar structures.

**het·er·o·mor·phic** (hĕt′ə-rō-môr′fĭk) *adj.* **1.** Having different forms at different periods of the life cycle. **2.** Heteromorphous. **—het′er·o·mor′phism** *n.*

**het·er·o·mor·phous** (hĕt′ə-rō-môr′fəs) *adj.* Having an atypical or irregular form or forms, as in stages of insect metamorphosis.

**het·er·on·o·mous** (hĕt′ə-rŏn′ə-məs) *adj.* [HETERO- + Gk. *nomos,* law + -OUS.] **1.** Subject to external or foreign laws or control. **2.** Differing in development or manner of specialization, as the dissimilar segments of certain arthropods. **—het′er·on′o·mous·ly** *adv.*

**het·er·o·nym** (hĕt′ər-ə-nĭm′) *n.* [Back-formation < HETERONYMOUS.] One of two or more words having identical spelling but different meanings and pronunciations, as *sow* (to plant) and *sow* (a pig).

**het·er·on·y·mous** (hĕt′ə-rŏn′ə-məs) *adj.* [LGk. *heterōnumos* : Gk. *hetero-,* different (< *heteros,* other) + Gk. *onoma,* name.] **1.** Relating to, being, or of the nature of a heteronym. **2.** Being different names or terms but having correspondence or interrelationship, as *uncle* and *aunt.*

**het·er·o·phyl·lous** (hĕt′ə-rō-fĭl′əs) *adj. Bot.* Having leaves of different form on one plant. **—het′er·o·phyl′ly** *n.*

**het·er·o·phyte** (hĕt′ər-ə-fīt′) *n.* A plant, as a saprophyte or parasite, that derives its nourishment from living or dead organic sources. **—het′er·o·phyt′ic** (-fĭt′ĭk) *adj.*

**het·er·o·plas·ty** (hĕt′ər-ə-plăs′tē) *n., pl.* **-ties.** The surgical grafting of tissue obtained from another person or from a lower animal. **—het′er·o·plas′tic** *adj.*

**het·er·o·ploid** (hĕt′ə-rə-ploid′) *adj.* Having a chromosome number that is not a whole-number multiple of the haploid chromosome number. **—het′er·o·ploid′** *n.* **—het′er·o·ploi′dy** (-ploi′dē) *n.*

**het·er·op·ter·ous** (hĕt′ə-rŏp′tər-əs) *adj.* Of or belonging to the insect order Heteroptera, marked by forewings and hind wings that differ from one another and including the true bugs.

**het·er·o·sex** (hĕt′ə-rō-sĕks′) *n.* Heterosexuality.

**het·er·o·sex·u·al** (hĕt′ə-rō-sĕk′sho͞o-əl) *adj.* **1.** Marked by attraction to the opposite sex. **2.** Of or relating to different sexes. —*n.* One who is heterosexual. **—het′er·o·sex′u·al′i·ty** (-ăl′ĭ-tē) *n.* **—het′er·o·sex′u·al·ly** *adv.*

**het·er·o·sis** (hĕt′ə-rō′sĭs) *n.* [Gk. *heteroiōsis,* alternation < *heteroioun,* to alter < *heteroios,* different in kind < *heteros,* other.] Increased vigor or capacity for growth arising from the crossbreeding of genetically different plants or animals. **—het′er·ot′ic** (-rŏt′ĭk) *adj.*

**het·er·o·spo·rous** (hĕt′ər-ə-spôr′əs, -spōr′-, hĕt′ə-rŏs′pər-əs) *adj.* Producing megaspores and microspores. **—het′er·o·spo′ry** *n.*

**het·er·o·tax·is** (hĕt′ə-rō-tăk′sĭs) *also* **het·er·o·tax·y** (hĕt′ər-ə-tăk′sē) *or* **het·er·o·tax·i·a** (hĕt′ə-rō-tăk′sē-ə) *n.* Abnormal structural arrangement. **—het′er·o·tac′tic** (-tăk′tĭk), **het′er·o·tac′tous** (-tăk′təs) *adj.*

**het·er·o·thal·lic** (hĕt′ə-rō-thăl′ĭk) *adj.* [HETERO- + THALL(US) + -IC.] Producing male gametangia in one structure or plant and female gametangia in a different structure or plant, as in some algae and fungi. **—het′er·o·thal′lism** *n.*

**het·er·o·tro·phic** (hĕt′ər-ə-trō′fĭk, -trŏf′ĭk) *adj.* Deriving nourishment from organic substances, as do some plants and all animals. **—het′er·o·tro′phi·cal·ly** *adv.* **—het′er·ot′ro·phy** (-ə-rŏt′rə-fē) *n.*

**het·er·o·typ·ic** (hĕt′ə-rō-tĭp′ĭk) *also* **het·er·o·typ·i·cal** (-ĭ-kəl) *adj.* [HETERO- + TYPIC(AL).] **1.** *Biol.* Designating or pertaining to the first reduction division of meiosis. **2.** Of a different type or form.

**het·er·o·zy·go·sis** (hĕt′ə-rō-zī-gō′sĭs) *n.* **1.** Derivation from or union between genetically different gametes. **2.** The condition of being a heterozygote.

**het·er·o·zy·gote** (hĕt′ə-rō-zī′gōt′) *n.* A zygote with inherited different alleles at one or more loci. **—het′er·o·zy′gous** *adj.*

**heth** (KHāt, KHāth, KHĕt, KHĕth) *n.* [Heb. *hēth.*] The eighth letter of the Hebrew alphabet. —See table at ALPHABET.

**heu·land·ite** (hyo͞oʹlən-dītʹ) *n.* [After Henry *Heuland* (1777–1856).] A white, red, or yellow zeolite mineral with composition $(Ca,Na,K)_6Al_{10}(Al,Si)_4\ Si_{29}O_{80}{\cdot}25H_2O$.

**heu·ris·tic** (hyo͞o-rĭsʹtĭk) *adj.* [< Gk. *heuriskein*, to find.] **1.** Of or pertaining to a usu. speculative formulation serving as a guide in the study or solution of a problem. **2.** Of, pertaining to, or constituting an educational method in which discoveries made by the investigating student serve as the vehicle of learning. **3.** *Computer Sci.* Pertaining to or using a problem-solving technique in which the best solution of several found by alternative methods is selected at successive stages of a program for use in the next step of the program. —*n.* A heuristic method or process. **—heu·risʹti·cal·ly** *adv.*

**hew** (hyo͞o) *v.* **hewed, hewn** (hyo͞on) *or* **hewed, hew·ing, hews.** [ME *hewen* < OE *hēawan.*] —*vt.* **1.** To make or shape with or as if with an ax <*hew* a way through the woods> **2.** To cut down with an ax : FELL. **3.** To strike or cleave with a heavy cutting instrument. —*vi.* **1.** To cut by repeated blows, as of an ax. **2.** To adhere or conform strictly : HOLD. —Used esp. in the phrase *hew to the line.* **—hewʹer** *n.*

**hex¹** (hĕks) *n.* [Pennsylvania Du. < G. *Hexe*, witch < MHG *hecse.*] **1.** An evil spell : CURSE. **2.** One believed to bring bad luck. —*vt.* **hexed, hex·ing, hex·es. 1.** To put a hex on. **2.** To bring or wish bad luck to. **—hexʹer** *n.*

**hex²** (hĕks) *adj.* Hexagonal. —Used of hardware.

**hexa-** or **hex-** *pref.* [Gk. < *hex*, six.] **1.** Six <*hexa*chord> **2.** Containing six atoms, molecules, or groups <*hex*ose>

**hex·a·chlo·ro·eth·ane** (hĕkʹsə-klôrʹō-ĕthʹānʹ, -klōrʹ-) *also* **hex·a·chlor·eth·ane** (-klôr-ĕthʹānʹ, -klōr-) *n.* A colorless crystalline compound, $Cl_3CCCl_3$, used in pyrotechnics, explosives, and veterinary medicine and as a substitute for camphor.

**hex·a·chlo·ro·phene** (hĕkʹsə-klôrʹə-fēnʹ, -klōrʹ-) *n.* [HEXA- + CHLORO- + PHEN(OL).] An almost odorless white powder, $(C_6HCl_3OH)_2CH_2$, used as a bactericidal agent in soaps, cosmetics, and skin medications.

**hex·a·chord** (hĕkʹsə-kôrdʹ) *n.* [HEXA- + Gk. *khordē*, string.] A sequence in medieval music of six whole tones with a semitone in the middle.

**hex·ad** (hĕkʹsădʹ) *n.* [LLat. *hexas, hexad-*, the number six < Gk. < *hex*, six.] A series or group of six. **—hex·adʹic** (-sădʹĭk) *adj.*

**hex·a·gon** (hĕkʹsə-gŏnʹ) *n.* [Gk. *hexagōnon* < *hexagōnos*, having six angles : *hexa-*, six + *gōnos*, -gon.] A polygon with six sides.

**hex·ag·o·nal** (hĕk-săgʹə-nəl) *adj.* **1.** Having six sides. **2.** Shaped like or having a hexagon. **3.** *Mineral.* Having three equal axes intersecting at 60° in one plane and one axis of variable length at right angles to the others. **—hex·agʹo·nal·ly** *adv.*

**hex·a·gram** (hĕkʹsə-grămʹ) *n.* **1.** A six-pointed star formed by extending each of the sides of a regular hexagon into equilateral triangles. **2.** A figure of six lines or sides.

**hex·a·he·dron** (hĕkʹsə-hēʹdrən) *n., pl.* **-drons** *or* **-dra** (-drə) [Gk. *hexaedron* < *hexaedros*, having six sides : *hexa-*, six + *hedra*, side.] A polyhedron with six faces. **—hexʹa·heʹdral** (-drəl) *adj.*

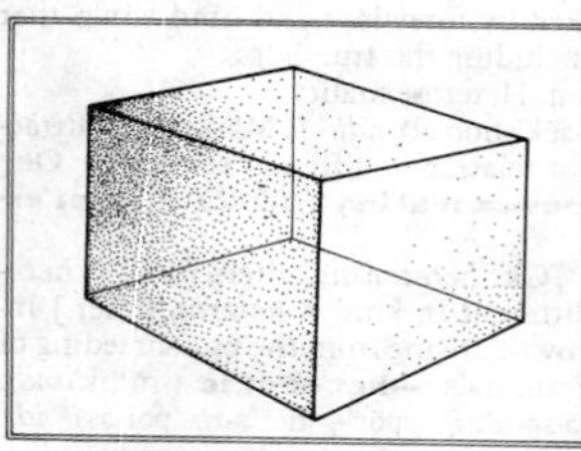

**hexahedron**

**hex·am·er·ous** (hĕk-sămʹər-əs) *adj.* **1.** Having six similar divisions or parts. **2.** *Bot.* Bearing flower parts, as petals, sepals, and stamens, in sets of six. **—hex·amʹer·ism** *n.*

**hex·am·e·ter** (hĕk-sămʹĭ-tər) *n.* [Lat. < Gk. *hexametron* : *hexa-*, six (< *hex*) + *metron*, meter.] A dactylic line including five dactyls and a trochee or spondee. **—hexʹa·metʹric** (-sə-mĕtʹrĭk), **hexʹa·metʹri·cal** *adj.*

**hex·a·meth·yl·ene·tet·ra·mine** (hĕkʹsə-mĕthʹə-lēn-tĕtʹrə-mēnʹ) *n.* Methenamine.

**hex·ane** (hĕkʹsānʹ) *n.* A flammable liquid, $CH_3(CH_2)_4CH_3$, derived from the fractional distillation of petroleum and used primarily as a solvent.

**hex·a·pod** (hĕkʹsə-pŏdʹ) *n.* [NLat. *Hexapoda*, class name : Gk. *hexa-*, six + Gk. *pous*, foot.] A member of the class Insecta or Hexapoda : INSECT. —*adj.* **1.** Of or belonging to the Hexapoda. **2.** Having six legs or feet. **—hex·apʹo·dous** (hĕk-săpʹə-dəs) *adj.*

**Hex·a·teuch** (hĕkʹsə-to͞okʹ, -tyo͞okʹ) *n.* [HEXA- + Gk. *teukos*, book.] The first six books of the Old Testament.

**hex·o·san** (hĕkʹsə-sănʹ) *n.* Any of several polysaccharides that form a hexose on hydrolysis.

**hex·ose** (hĕkʹsōsʹ) *n.* A simple sugar having six carbon atoms per molecule.

**hex·yl** (hĕkʹsəl) *n.* The univalent hydrocarbon radical $C_6H_{13}$.

**hex·yl·re·sor·ci·nol** (hĕkʹsəl-rĭ-zôrʹsə-nôlʹ, -nōlʹ) *n.* A yellowish-white phenol, $C_{12}H_{18}O_2$, used as an antiseptic and vermifuge.

**hey** (hā) *interj.* [ME.] —Used to attract attention or to express surprise, appreciation, wonder, or pleasure.

**hey·day** (hāʹdāʹ) *n.* [Orig. unknown.] The period of greatest power, success, or popularity : PRIME.

**Hf** *symbol for* HAFNIUM.

**Hg** [NLat. *hydrargyrum*, mercury < Lat. *hydrargyrus*, artificial mercury < Gk. *hudrarguros* : *hudr-*, hydro- + *arguros*, silver.] *symbol for* MERCURY.

**hi** (hī) *interj. Informal.* —Used to express greeting.

**hi·a·tus** (hī-āʹtəs) *n., pl.* **-tus·es** *or* **hiatus.** [Lat. < *hiare*, to gape.] **1.** A gap or interruption in space, time, or continuity : BREAK <a *hiatus* in radio programming> **2.** A slight pause occurring when two adjacent vowels in consecutive syllables are pronounced, as in *duality.* **3.** *Anat.* A separation, aperture, or fissure.

**hi·ba·chi** (hĭ-băʹchē) *n., pl.* **-chis.** [J. : *hi*, fire + *bachi*, bowl.] A portable charcoal-burning brazier with a grill, used esp. for cooking.

**hi·ber·nac·u·lum** (hīʹbər-năkʹyə-ləm) *n., pl.* **-la** (-lə) [Lat., winter residence < *hibernare*, to winter < *hibernus*, wintry.] *Biol.* **1.** A case, covering, or structure in which an organism remains dormant for the winter. **2.** The shelter of a hibernating animal.

**hi·ber·nal** (hī-bûrʹnəl) *adj.* [Lat. *hibernalis* < *hibernus*, wintry.] Of or relating to winter.

**hi·ber·nate** (hīʹbər-nātʹ) *vi.* **-nat·ed, -nat·ing, -nates.** [Lat. *hibernare, hibernat-*, to winter < *hibernus*, wintry.] **1.** To pass the winter in a dormant or torpid state. **2.** To be in an inactive or dormant state or period. **—hiʹber·naʹtion** *n.* **—hiʹber·naʹtor** *n.*

▲ word history: The English word *hibernate* is ultimately derived from Latin *hibernus*, "wintry." *Hibernus* descends from the Indo-European root *ghiem-*, from which the Sanskrit word *himā-*, "snow," is also derived. The name *Himalaya* is a Sanskrit compound of *himā-*, "snow," and *ālaya*, "abode, place," and means "the place of snow."

**Hi·ber·ni·an** (hī-bûrʹnē-ən) *n.* [< Lat. *Hibernia*, Ireland.] A native or resident of Ireland. **—Hi·berʹni·an** *adj.*

**hi·bis·cus** (hī-bĭsʹkəs) *n.* [NLat., genus name < Lat., marsh mallow < Gk. *hibiskos.*] A chiefly tropical plant, shrub, or tree of the genus *Hibiscus*, bearing large, variously colored flowers.

**hic·cup** *also* **hic·cough** (hĭkʹəp) [Imit.] —*n.* **1.** A spasm of the diaphragm resulting in a sudden, abortive inhalation stopped by a spasmodic glottal closure. **2. hiccups.** An attack of hiccups. —*vi.* **-cupped, -cup·ping, -cups** *also* **-coughed, -cough·ing, -coughs. 1.** To make a hiccup. **2.** To have an attack of hiccups.

**hick** (hĭk) [After *Hick*, a nickname for Richard.] *Informal.* —*n.* A gullible, provincial person : YOKEL. —*adj.* Marked by a lack of sophistication : PROVINCIAL <comes from a *hick* town>

**hick·ey** (hĭkʹē) *n., pl.* **-eys.** [Orig. unknown.] *Informal.* **1.** A device or contrivance : GADGET. **2. a.** A pimple. **b.** A reddish mark on the skin caused by kissing, biting, or sucking, as in lovemaking. **3.** A pipe-bending apparatus. **4.** A threaded electrical fitting for connecting a fixture to an outlet box.

**hick·o·ry** (hĭkʹə-rē) *n., pl.* **-ries.** [Earlier *pohickery* < *pawcohiccora*, food made from crushed hickory nuts, of Algonquian orig.] **1.** A chiefly North American deciduous tree of the genus *Carya*, having smooth or shaggy bark and compound leaves and bearing hard, smooth nuts with an edible kernel. **2. a.** The hard, tough, heavy wood of a hickory tree. **b.** A walking stick or switch made of hickory.

**hid** (hĭd) *v. p.t. & var. p.p. of* HIDE¹.

**hi·dal·go** (hĭ-dălʹgō, ē-thälʹ-) *n., pl.* **-gos.** [Sp. < OSp. *hijo dalgo* : *hijo*, son (< Lat. *filius*) + *de*, of (< Lat.) + *algo*, something < Lat. *aliquid* (*alius*, some + *quid*, something).] A member of the lesser nobility in Spain.

**hid·den** (hĭdʹn) *v. var. p.p. of* HIDE¹.

**hid·den·ite** (hĭdʹn-ītʹ) *n.* [After William E. *Hidden* (1853–1918).] A transparent emerald-green variety of spodumene, used as a gemstone.

**hide¹** (hīd) *v.* **hid** (hĭd), **hid·den** (hĭdʹn) *or* **hid, hid·ing, hides.** [ME *hiden* < OE *hȳdan.*] —*vt.* **1.** To put or keep out of sight : CACHE. **2.** To prevent the recognition or disclosure of : CONCEAL <attempted to *hide* the truth> **3.** To cut off from sight : cover up <Trees *hid* the view.> **4.** To avert (one's gaze), esp. in shame or grief. —*vi.* **1.** To keep oneself out of sight. **2.** To seek refuge. **—hidʹer** *n.*

☆ **syns:** HIDE, BURY, CACHE, CONCEAL, ENSCONCE, PLANT, SECRETE, STASH *v. core meaning* : to put (something) out of sight and keep it there <*hid* the stolen money in an abandoned mine>

**hide²** (hīd) *n.* [ME < OE *hȳd.*] The skin of an animal, esp. the thick, tough skin of a large animal. —*vt.* **hid·ed, hid·ing, hides.** To beat severely : THRASH. **—hide or (or nor) hair.** The smallest indication of someone or something: TRACE <haven't seen *hide or hair* of the tenants>

**hide[3]** (hīd) *n.* [ME < OE *hīd.*] An old English measure of land, usu. the amount held adequate for one free family and its dependents.

**hide-and-seek** (hīd'n-sēk') *n.* **1.** A children's game in which one player tries to find and catch others who are hiding. **2.** A game or action involving evasion.

**hide·a·way** (hīd'ə-wā') *n.* **1.** A hide-out. **2.** A secluded place.

**hide·bound** (hīd'bound') *adj.* **1.** Having abnormally dry, stiff skin that adheres closely to the underlying flesh. —Used of domestic animals, as cattle. **2.** Having the bark so contracted and unyielding as to hinder growth. —Used of trees. **3.** Stubbornly narrow-minded, inflexible, or biased.

**hid·e·ous** (hĭd'ē-əs) *adj.* [ME < AN *hidous* < OFr. *hide,* fear.] **1.** Repulsive, esp. to the sight : UGLY. **2.** Repugnant to the moral sense : DESPICABLE. **—hid'e·ous·ly** *adv.* **—hid'e·ous·ness** *n.*

**hide-out** (hīd'out') *n.* A place of shelter or concealment.

**hid·ing** (hī'dĭng) *n. Informal.* A severe beating : THRASHING.

**hi·dro·sis** (hĭ-drō'sĭs) *n.* [Gk. *hidrōsis,* sweating < *hidros,* sweat.] Perspiration, esp. in excessive or abnormal amounts. **—hi·drot'ic** (-drŏt'ĭk) *adj.*

**hie** (hī) *vi.* **hied, hie·ing** *or* **hy·ing, hies.** [ME *hien* < OE *hīgian.*] To go quickly : HURRY.

**hi·e·mal** (hī'ə-məl) *adj.* [Lat. *hiemalis* < *hiems,* winter.] Of or relating to winter : HIBERNAL.

**hier-** *pref. var. of* HIERO-.

**hi·er·arch** (hī'ə-rärk', hī'rärk') *n.* [OFr. *hierarche* < Med. Lat. *hierarcha* < Gk. *hierarkhēs,* high priest : *hieros,* sacred + *arkhos,* leader.] **1.** One having a position of high authority in an ecclesiastical hierarchy. **2.** One having a high position in a hierarchy. **—hi'er·ar'chal** *adj.*

**hi·er·ar·chy** (hī'ə-rär'kē, hī'rär'-) *n., pl.* **-chies.** [ME *iherarchie* < OFr. *ierarchie* < Med. Lat. *hierarchia* < Gk. *hierarkhia,* rule of a high priest < *hierarkhēs,* high priest. —see HIERARCH.] **1. a.** A group of persons organized or classified according to authority or rank. **b.** A ranked or graded series <a *hierarchy* of professional goals> **2. a.** A body of clergy organized into successively higher ranks or grades. **b.** Ecclesiastical rule by a hierarchy. **3.** A division of angels. **—hi'er·ar'chi·cal, hi'er·ar'chic** *adj.* **—hi'er·ar'chi·cal·ly** *adv.*

**hi·er·at·ic** (hī'ə-răt'ĭk, hī-răt'-) *adj.* [Lat. *hieraticus* < Gk. *hieratikos* < *hierasthai,* to be a priest < *hieros,* sacred.] **1.** Of or associated with priests or the priesthood. **2.** Of or relating to a simplified cursive style of Egyptian hieroglyphics. **—hi'er·at'i·cal·ly** *adv.*

**hiero-** *or* **hier-** *pref.* [Gk. < *hieros.*] Sacred : holy <*hierology*>

**hi·er·oc·ra·cy** (hī'ə-rŏk'rə-sē, hī-rŏk'-) *n., pl.* **-cies.** Government by the clergy. **—hi'er·o·crat'ic** (hī'ər-ə-krăt'ĭk, hī'rə-krăt'-), **hi'ero·crat'i·cal** *adj.*

**hi·er·o·dule** (hī'ər-ə-do͞ol', -dyo͞ol') *n.* [LLat. *hierodulus* < Gk. *hierodoulos* : *hieron,* temple + *doulos,* slave.] A temple slave in the service of a given deity. **—hi'er·o·du'lic** (-do͞o'lĭk, -dyo͞o'-) *adj.*

**hi·er·o·glyph** (hī'ər-ə-glĭf', hī'rə-) *n.* A hieroglyphic.

**hi·er·o·glyph·ic** (hī'ər-ə-glĭf'ĭk, hī'rə-) *also* **hi·er·o·glyph·i·cal** (-ĭ-kəl) *adj.* [Fr. *hiéroglyphique* < LLat. *hieroglyphicus* < Gk. *hierogluphikos* : *hieros,* sacred + *gluphē,* carving < *gluphein,* to carve.] **1.** Of or relating to a system of writing, as that of ancient Egypt, in which pictorial symbols are used to represent words or sounds. **2.** Written with hieroglyphic symbols. **3.** Difficult to read or decipher <*hieroglyphic* handwriting> —*n.* **hieroglyphic. 1.** A picture or symbol used in hieroglyphic writing. **2. hieroglyphics. a.** Hieroglyphic writing. **b.** Illegible or undecipherable writing. **3.** Something resembling a hieroglyphic esp. in being indecipherable. **—hi'er·o·glyph'i·cal·ly** *adv.*

**hi·er·ol·o·gy** (hī'ə-rŏl'ə-jē, hī-rŏl'-) *n., pl.* **-gies.** The sacred literature of a given people.

**hi·er·o·phant** (hī'ər-ə-fănt', hī'rə-, hī-ĕr'ə-fənt) *n.* [LLat. *hierophanta* < Gk. *hierophantēs* : *hieros,* sacred + *phainein,* to reveal.] **1.** A chief priest of the Eleusinian mysteries. **2.** An interpreter of sacred mysteries or arcane knowledge. **—hi'er·o·phan'tic** *adj.*

**hi·fa·lu·tin** (hī'fə-lo͞ot'n) *adj. var. of* HIGHFALUTIN.

**hi-fi** (hī'fī') *n.* [HI(GH) FI(DELITY).] **1.** High fidelity. **2.** An electronic system for reproducing high-fidelity sound.

**hig·gle** (hĭg'əl) *vi.* **-gled, -gling, -gles.** [Prob. alteration of HAGGLE.] To haggle. **—hig'gler** *n.*

**hig·gle·dy-pig·gle·dy** (hĭg'əl-dē-pĭg'əl-dē) *adv.* [Orig. unknown.] In complete confusion or disorder. —*adj.* Jumbled : topsy-turvy.

**high** (hī) *adj.* **-er, -est.** [ME < OE *hēah.*] **1. a.** Having a relatively great elevation <a *high* hill> **b.** Extending a specified distance upward <a bookshelf six feet *high*> **2.** Being at or near its peak <*high* tide> **3.** Beginning to decompose, as meat : GAMY. **4.** Far removed in time : REMOTE. **5. a.** Having a musical pitch that corresponds to a relatively great number of cycles per second <can't hit the *high* notes> **b.** Not soft or hushed : PIERCING <a child's *high* voice> **6.** Situated far from the equator, as a latitude. **7.** Extremely important <a *high* objective> **8.** Eminent in rank or status <a *high* authority> **9.** Serious : grave <*high* transgressions> **10.** Lofty or exalted in quality or character <*high* standards> **11. a.** Greater than usual or expected <a cheese that is *high* in calcium><*high* taxes> **b.** Favorable <*high* esteem> **12.** Of great force or violence, as strong winds. **13.** Elated : excited. **14.** *Informal.* Intoxicated by or as if by alcohol or a narcotic. **15.** Advanced in development or complexity <*higher* math> **16.** Luxurious : extravagant <*high* style> **17.** Pronounced with part of the tongue close to the palate <a *high* vowel> —*adv.* **-er, -est.** At, in, or to a high position, price, or level <soaring *high* above the trees> —*n.* **1.** A high place or region. **2.** A high level or degree <Sales reached a new *high.*> **3.** The transmission gear of an automotive vehicle producing maximum speed. **4.** A center of high atmospheric pressure : ANTICYCLONE. **5.** *Informal.* Intoxication or euphoria induced by or as if by a stimulant or narcotic. **—high and dry. 1.** Abandoned : stranded <left us *high and dry*> **2.** Out of water, as a ship. **—high and low.** Here and there : EVERYWHERE <searched *high and low* for the book> **—high and mighty.** Domineering : arrogant. **—high'ly** *adv.*

**high·ball** (hī'bôl') *n.* **1.** A beverage of alcoholic liquor and water or a carbonated liquid served in a tall glass. **2.** A railroad signal indicating full speed ahead. —*vi.* **-balled, -ball·ing, -balls.** To move ahead at full speed.

**high beam** *n.* A high-intensity headlight on a vehicle.

**high·bind·er** (hī'bīn'dər) *n.* [After the *Highbinders,* a group of ruffians in New York City ca. 1806.] **1.** A member of a Chinese-American secret society of paid assassins and blackmailers. **2.** A corrupt politician.

**high blood pressure** *n.* Hypertension.

**high·born** (hī'bôrn') *adj.* Of noble birth.

**high·boy** (hī'boi') *n.* A tall chest of drawers mounted on four legs.

**high·bred** (hī'brĕd') *adj.* Of superior breed or stock.

**high·brow** (hī'brou') *n. Informal.* One who has or affects superior learning or culture. **—high'brow', high'browed'** *adj.* **—high'-brow'ism** *n.*

**high-bush cranberry** (hī'bo͝osh') *n.* A North American shrub, *Viburnum trilobum,* bearing broad white flower clusters and yielding scarlet fruit.

**high·chair** (hī'châr') *n.* A baby's feeding chair mounted on tall legs and often having a detachable tray.

**High-Church** (hī'chûrch') *adj.* Favoring or characterized by the incorporation of elements, as the liturgy, usu. associated with Roman Catholicism into the forms of worship of the Anglican Church. **—High'-Church'man** *n.*

**high-class** (hī'klăs') *adj.* Of high quality or character : FIRST-CLASS.

**high comedy** *n.* Comedy marked by sophisticated characterizations and clever dialogue.

**high court** *n.* A supreme court.

**high-en·er·gy** (hī'ĕn'ər-jē) *adj.* **1.** Of or pertaining to elementary particles with energies exceeding hundreds of thousands of electron volts. **2.** Yielding a large amount of energy upon undergoing chemical reaction. **3.** Dynamic : vigorous.

**higher criticism** *n.* Critical study of Biblical texts with regard to questions of their character, composition, editing, and collection.

**higher education** *n.* Education beyond the secondary level, esp. at the college level.

**higher learning** *n.* Education or scholastic attainment at the college or university level.

**high·er-up** (hī'ər-ŭp') *n. Informal.* One who has a superior rank, status, or position.

**high explosive** *n.* A powerful fast-acting explosive.

**high·fa·lu·tin** *or* **hi·fa·lu·tin** (hī'fə-lo͞ot'n) *also* **high·fa·lu·ting** (-lo͞ot'n, -lo͞o'tĭng) *adj.* [Orig. unknown.] *Informal.* **1.** Affected : pretentious <*highfalutin* airs> **2.** Pompous : grandiose <a *highfalutin* speech>

**high fashion** *n.* The latest in trend-setting fashion or design.

**high fidelity** *n.* The electronic reproduction of sound, esp. from broadcast, recorded, or taped sources, with minimal distortion. **—high'-fi·del'i·ty** *adj.*

**high-flown** (hī'flōn') *adj.* **1.** Lofty : exalted. **2.** Pretentious.

**high·fly·ing** (hī'flī'ĭng) *adj.* **1.** Rising to a great height. **2.** Lofty in form or ambition.

**high frequency** *n.* A radio frequency between 3 and 30 megahertz.

**high gear** *n.* **1.** HIGH 3. **2.** A state of maximum activity or energy.

**High German** *n.* **1.** German as indigenously spoken and written in central and southern Germany. **2.** GERMAN 2.

**high-grade** (hī'grād') *adj.* Of superior quality <*high-grade* eggs>

**high·hand·ed** (hī'hăn'dĭd) *adj.* Arrogant or domineering. **—high'hand'ed·ly** *adv.* **—high'hand'ed·ness** *n.*

**high-hat** (hī'hăt') *vt.* **-hat·ted, -hat·ting, -hats.** *Slang.* To treat in a supercilious or condescending way. **—high'-hat'** *adj.* & *n.*

**High Holiday** *n.* **1.** Rosh Hashanah. **2.** Yom Kippur.

**high·jack** (hī'jăk') *v.* & *n. var. of* HIJACK.

**high jinks** *pl.n.* Mischievous pranks.

**high jump** *n.* **1.** A jump for height in a track and field contest. **2.** A contest in which high jumps are made.

high jump

**high·land** (hī′lənd) *n.* **1.** Elevated land. **2. highlands.** A mountainous region. —*adj.* **1.** Of, pertaining to, or typical of a highland. **2. Highland.** Of or pertaining to the Highlands of Scotland.

**high·land·er** (hī′lən-dər) *n.* **1.** An inhabitant of a highland area. **2. Highlander.** An inhabitant of the Highlands of Scotland.

**Highland fling** *n.* A folk dance of the Highlands of Scotland.

**high-lev·el** (hī′lĕv′əl) *adj.* **1.** Situated, occurring, or carried out at a high level. **2.** Being at a high level of importance <a *high-level* administrator>

**high·light** (hī′līt′) *n.* **1.** A light or brilliantly lighted area, as in a photograph or painting. **2.** An outstanding or major event. —*vt.* **-light·ed, -light·ing, -lights. 1.** To give a highlight to. **2.** To make prominent : EMPHASIZE. **3.** To be the highlight of.

**high·light·er** (hī′lī′tər) *n.* **1.** One that highlights. **2.** A cosmetic for emphasizing facial features.

**High Mass** *n.* A sung mass celebrated by a priest or prelate, often assisted by a deacon and a subdeacon.

**high-mind·ed** (hī′mīn′dĭd) *adj.* **1.** Marked by elevated conduct or ideals : NOBLE. **2.** *Archaic.* Disdainfully proud : HAUGHTY. **—high′-mind′ed·ly** *adv.* **—high′-mind′ed·ness** *n.*

**high muckamuck** *n.* [Chinook Jargon *hiu muckamuck.*] *Slang.* A muckamuck.

**high·ness** (hī′nĭs) *n.* **1.** The quality of being high or tall : HEIGHT. **2. Highness.** A title of honor for royalty.

**high noon** *n.* **1.** Exactly noon. **2.** The highest or most advanced stage or period <at the *high noon* of artistic genius>

**high-oc·tane** (hī′ŏk′tān′) *adj.* Having a high octane number.

**high-per·form·ance** (hī′-pər-fôr′məns) *adj.* Operating or capable of operating at maximum power and efficiency, as a car engine.

**high-pitched** (hī′pĭcht′) *adj.* **1.** High in pitch, as a voice or musical tone. **2.** Steeply sloped, as a roof.

**high place** *n.* A place of worship on top of a hill.

**high-pow·ered** (hī′pou′ərd) *also* **high-pow·er** (-pou′ər) *adj.* Having great power or energy : DYNAMIC.

**high-pres·sure** (hī′prĕsh′ər) *adj.* **1.** Of or relating to pressures higher than normal, esp. higher than atmospheric pressure. **2.** *Informal.* Using aggressive and often annoyingly persistent methods of persuasion. —*vt.* **-sured, -sur·ing, -sures.** *Informal.* To influence or persuade by using high-pressure methods.

**high priest** *n.* **1.** A chief priest, esp. of the ancient Levitical priesthood. **2.** *Mormon Ch.* A priest of the Melchizedek order. **3.** The head or chief proponent, as of a movement or doctrine <the *high priest* of contemporary filmmaking> **—high priesthood** *n.*

**high priestess** *n.* A chief priestess.

**high relief** *n.* Sculptural relief in which the modeled forms project from the background by at least half their depth.

**high-rise** (hī′rīz′) *adj.* **1.** Designating or being a multistoried building equipped with elevators. **2.** Of, pertaining to, or being a bicycle with handlebars that are higher and longer than average. —*n.* **1.** A high-rise building. **2.** A high-rise bicycle.

**high-ris·er** (hī′rī′zər) *n.* **1.** A high-rise building : SKYSCRAPER. **2.** A high-rise bicycle.

**high-risk** (hī′rĭsk′) *adj.* **1.** Of, pertaining to, or marked by risk <a *high-risk* enterprise> **2.** Being particularly subject to potential danger or hazard <a *high-risk* teen-age pregnancy>

**high·road** (hī′rōd′) *n.* **1.** *Chiefly Brit.* A main road : HIGHWAY. **2.** A simple, direct, or sure path <the *highroad* to success>

**high roller** *n.* *Informal.* **1.** A big spender. **2.** One who gambles, esp. for high stakes. **—high′-roll′ing** (hī′rō′lĭng) *adj.*

**high school** *n.* A secondary school of grades 9 or 10 through 12 or grades 7 through 12. **—high′-school′er** (hī′sko͞o′lər) *n.*

**high seas** *n.* The open waters of an ocean or sea beyond the limits of the territorial jurisdiction of a country.

**high-sound·ing** (hī′soun′dĭng) *adj.* HIGH-FLOWN 2.

**high-spir·it·ed** (hī′spĭr′ĭ-tĭd) *adj.* **1.** Having a proud or unbroken spirit : BRAVE. **2.** Exuberant. **—high′-spir′it·ed·ly** *adv.* **—high′-spir′it·ed·ness** *n.*

**high street** *n.* *Chiefly Brit.* A main street.

**high-strung** (hī′strŭng′) *adj.* Inclined to be extremely nervous and sensitive <a *high-strung* thoroughbred animal>

**high style** *n.* High fashion.

**high·tail** (hī′tāl′) *vi.* **-tailed, -tail·ing, -tails.** [From those animals who raise their tails before fleeing.] *Slang.* To leave in great haste.

**high tea** *n.* *Chiefly Brit.* A substantial meal served in the late afternoon or early evening.

**high tech** *n.* **1.** A style of interior design incorporating industrial materials or motifs. **2.** High technology.

**high technology** *n.* Technology involving highly advanced or specialized systems or devices.

**high-ten·sion** (hī′tĕn′shən) *adj.* Having a high voltage.

**high-test** (hī′tĕst′) *adj.* **1.** Meeting the most exacting requirements. **2.** Of or relating to highly volatile high-octane gasoline.

**high tide** *n.* **1. a.** The tide at its full, when the water reaches its highest level. **b.** The time at which high tide occurs. **2.** A point of culmination : CLIMAX.

**high-toned** (hī′tōnd′) *adj.* **1.** Intellectually, morally, or socially superior. **2.** *Informal.* Pretentiously elegant or fashionable.

**high water** *n.* **1.** High tide. **2.** The state of a body of water that has reached its highest level.

**high-wa·ter mark** (hī′wô′tər, -wŏt′ər) *n.* **1.** A mark showing the highest level reached by a body of water. **2.** The highest point, as of achievement : CULMINATION.

**high·way** (hī′wā′) *n.* A main public road, esp. one connecting towns and cities.

**high·way·man** (hī′wā′mən) *n.* A robber who waylays travelers on a public road.

**highway robbery** *n.* **1.** Robbery usu. of travelers on a highway. **2.** Undue profit or benefit from something, as a business deal.

**hi·jack** *also* **high·jack** (hī′jăk′) [Orig. unknown.] *Informal.* —*vt.* **-jacked, -jack·ing, -jacks. 1.** To stop and rob (a vehicle in transit). **2.** To steal (goods) from a vehicle in transit. **3.** To seize control of (e.g., an aircraft), esp. in order to go somewhere other than the scheduled destination. **4.** To steal from. **5.** To force or coerce. —*n.* An act or instance of hijacking. **—hi′jack′er** *n.*

**hike** (hīk) *v.* **hiked, hik·ing, hikes.** [Orig. unknown.] —*vi.* **1.** To go on an extended walk for pleasure or exercise. **2.** To be raised or caught up <a jacket that *hiked* up when I sat down> —*vt.* **1.** To raise or increase in amount <*hiked* prices for Christmas> **2.** To pull up or raise with a sudden motion : HITCH. —*n.* **1.** A long walk or march. **2.** An upward movement : RISE. **—hik′er** *n.*

**hi·la** (hī′lə) *n.* *pl. of* HILUM.

**hi·lar·i·ous** (hĭ-lâr′ē-əs, -lăr′-) *adj.* [Lat. *hilarus,* cheerful + -IOUS. —see HILARITY.] Boisterously funny. **—hi·lar′i·ous·ly** *adv.* **—hi·lar′i·ous·ness** *n.*

**hi·lar·i·ty** (hĭ-lăr′ĭ-tē, -lâr′-) *n.* [OFr. *hilarite,* cheerfulness < Lat. *hilaritas* < *hilarus,* cheerful < Gk. *hilaros.*] Boisterous merriment.

**hill** (hĭl) *n.* [ME *hille* < OE *hyl.*] **1.** A well-defined natural elevation smaller than a mountain. **2.** A small heap, pile, or mound. **3. a.** A mound of earth piled around and over a plant. **b.** A plant thus covered. **4.** An incline, esp. in a road : SLOPE. —*vt.* **hilled, hill·ing, hills. 1.** To form into a hill, pile, or heap. **2.** To cover (a plant) with a mound of soil. **—hill′er** *n.*

**hill·bil·ly** (hĭl′bĭl′ē) *n., pl.* **-lies.** [HILL + *Billy,* a nickname for William.] *Informal.* One from the mountains or backwoods.

**hill·ock** (hĭl′ək) *n.* [ME *hillok* < *hille,* hill.] A small hill.

**hill·side** (hĭl′sīd′) *n.* The side or slope of a hill.

**hill·top** (hĭl′tŏp′) *n.* The top or crest of a hill.

**hill·y** (hĭl′ē) *adj.* **-i·er, -i·est. 1.** Having many hills. **2.** Like a hill, as in steepness. **—hill′i·ness** *n.*

**hilt** (hĭlt) *n.* [ME < OE.] The handle of a weapon or tool, esp. of a sword or dagger. **—to the hilt.** To the limit : COMPLETELY.

**hi·lum** (hī′ləm) *n., pl.* **-la** (-lə) [Lat., trifle.] **1.** *Bot.* **a.** The scarlike mark on a seed, as a bean, formed at the point where it was joined to the stalk connecting it to the placenta. **b.** The nucleus of a starch grain. **2.** *Anat.* A small aperture or notch through which ducts, nerves, or vessels enter or leave an organ.

**him** (hĭm) *pron.* [ME < OE.] *objective case of* HE. —Used: **1.** As the direct object of a verb <I met *him* at the beach.> **2.** As the indirect object of a verb <I sold *him* a car.> **3.** As the object of a preposition <This shirt belongs to *him.*>

**hi·mat·i·on** (hĭ-măt′ē-ŏn′) *n., pl.* **-i·a** (-ē-ə) [Gk., dim. of *hima,* garment < *hennunai,* to clothe.] A long loose outer garment worn by men and women in ancient Greece.

**him·self** (hĭm-sĕlf′) *pron.* [ME < OE *himselfum.*] **1.** That one identical with him. —Used: **a.** Reflexively as the direct or indirect object of a verb or the object of a preposition <He cut *himself.*> **b.** For emphasis <He *himself* isn't sure.> **c.** In an absolute construction <In debt *himself,* he was unable to contribute.> **2.** His normal or healthy condition or state <He hasn't been *himself.*>

**Him·yar·ite** (hĭm′yə-rīt′) *n.* [After *Himyar,* a legendary king in Yemen.] **1.** A member of an ancient tribe of southwestern Arabia. **2.** The Semitic language of the ancient Himyarites. —*adj.* Of, pertaining to, or typical of the Himyarite people, their language, or their culture. **—Him′yar·it′ic** (-rĭt′ĭk) *adj.*

**hin** (hĭn) *n.* [Heb. *hīn,* of Egyptian orig.] A unit of liquid measure used by the ancient Hebrews, equal to approx. a half gallon.

ă **pat** ā **pay** âr **care** ä **father** ĕ **pet** ē **be** hw **which** ĭ **pit**
ī **tie** îr **pier** ŏ **pot** ō **toe** ô **paw, for** oi **noise** o͝o **took**

**Hi·na·ya·na** (hē′nə-yä′nə) *n.* [Skt. *hīnayāna*, lesser vehicle.] A small, conservative branch of Buddhism following the Pali scriptures and the nontheistic ideal of self-purification to nirvana. **—Hi′na·ya′nist** *n.* **—Hi′na·ya·nis′tic** (-yä-nĭs′tĭk) *adj.*

**hind**[1] (hīnd) *adj.* [ME *hinde*, short for *bihinde* < OE *bihindan*.] Forming or located at the back or rear : POSTERIOR.

**hind**[2] (hīnd) *n.* [ME < OE.] **1.** A female red deer. **2.** An Atlantic fish of the genus *Epinephelus*, related to the grouper.

**hind**[3] (hīnd) *n.* [ME < OE *hīwan*.] *Chiefly Brit.* **1.** A farm laborer, esp. a skilled worker. **2.** *Archaic.* A bumpkin : yokel.

**hind·brain** (hīnd′brān′) *n.* The rhombencephalon.

**hin·der** (hĭn′dər) *vt.* **-dered, -der·ing, -ders.** [ME *hindren* < OE *hindrian*.] **1.** To get in the way of : HAMPER. **2.** To impede or delay the progress of. **—hin′der·er** *n.*

☆ ***syns:*** HINDER, BOG (down), ENCUMBER, HAMPER, HOLD BACK, IMPEDE, OBSTRUCT, RETARD *v. core meaning* : to interfere with the progress of <Unreasonable demands *hindered* the truce negotiations.> ***ant:*** expedite, facilitate

**hind·gut** (hīnd′gŭt′) *n.* The posterior portion of the embryonic alimentary canal.

**Hin·di** (hĭn′dē) *n.* [Hindi *Hindī* < *Hind*, India < Pers. < OPers. *Hindu*, the Indus River.] **1. a.** A group of vernacular Indic dialects spoken in northern India. **b.** The literary and official language of northern India. **2.** A member of a cultural group of northern India speaking a Hindi dialect. **—Hin′di** *adj.*

**hind·most** (hīnd′mōst′) *adj.* Farthest to the rear : LAST.

**Hin·doo** (hĭn′do͞o) *n.* & *adj. Archaic. var. of* HINDU.

**Hin·doo·ism** (hĭn′do͞o-ĭz′əm) *n. Archaic. var. of* HINDUISM.

**hind·quar·ter** (hīnd′kwôr′tər) *n.* **1.** The posterior portion of a side of beef, lamb, veal, or mutton, including a hind leg and one or two ribs. **2. hindquarters.** The hind part of a quadruped, adjacent to the hind legs.

**hin·drance** (hĭn′drəns) *n.* [ME *hindraunce*, harm < *hindren*, to hinder.] **1.** The act of hindering or state of being hindered. **2.** One that hinders.

**hind·sight** (hīnd′sīt′) *n.* **1.** The rear sight of a firearm. **2.** Perception of the nature and import of events after they have occurred.

**Hin·du** (hĭn′do͞o) *n.* [Pers. *Hindū* < *Hind*, India. —see HINDI.] **1.** A native of India, esp. northern India. **2.** A believer in Hinduism. *—adj.* **1.** Of or relating to the Hindus or their culture. **2.** Of or relating to Hinduism.

**Hindu calendar** *n.* The lunisolar calendar of the Hindus.

**Hin·du·ism** (hĭn′do͞o-ĭz′əm) *n.* A diverse body of religion, philosophy, and cultural practice native to and predominant in India, marked by a belief in reincarnation and a supreme being of many forms and natures, by a desire for liberation from earthly evils, and by the view that opposing theories are aspects of one eternal truth.

**Hin·du·sta·ni** (hĭn′do͞o-stä′nē, -stăn′ē) *n.* **1.** A native of Hindustan. **2.** A group of Indic dialects including Urdu and Hindi. *—adj.* Of or relating to Hindustani or Hindustan.

**hinge** (hĭnj) *n.* [ME.] **1.** A jointed or flexible device that allows the turning or pivoting of a part, as a door or lid, on a stationary frame. **2.** A part or structure like a hinge, as one enabling the valves of a bivalve mollusk to open and close. **3.** A small folded piece of paper gummed on one side, used esp. to fasten stamps in an album. **4.** A point or circumstance upon which later events depend. *—v.* **hinged, hing·ing, hing·es.** *—vt.* To equip with or attach by a hinge. *—vi.* To be contingent <Your raise *hinges* on your job performance.>

**hin·ny** (hĭn′ē) *n., pl.* **-nies.** [Lat. *hinnus* < Gk. *innos*.] The hybrid offspring of a male horse and a female ass.

**hint** (hĭnt) *n.* [Poss. < obs. *hent*, the act of seizing < obs. *hent*, to seize < ME *henten* < OE *hentan*.] **1.** A slight intimation or indication. **2. a.** A brief or indirect suggestion : TIP. **b.** A statement conveying information indirectly. **3.** A barely perceivable amount <just a *hint* of lemon> **4.** *Obs.* An occasion : opportunity. *—v.* **hint·ed, hint·ing, hints.** *—vt.* To make known by a hint : INTIMATE <*hint* one's doubts> *—vi.* To give a hint <*hinted* about my birthday> **—hint′er** *n.*

**hin·ter·land** (hĭn′tər-lănd′) *n.* [G. : *hinter-*, behind + *Land*, land.] **1.** The land directly adjacent to and inland from a coast. **2.** A region remote from urban areas.

**hip**[1] (hĭp) *n.* [ME < OE *hype*.] **1. a.** The laterally projecting prominence of the pelvis or pelvic region extending from the waist to the thigh. **b.** A homologous posterior part in quadrupeds. **2.** The hip joint. **3.** The external angle formed by the meeting of two adjacent slopes of a roof.

**hip**[2] (hĭp) *also* **hep** (hĕp) [Orig. unknown.] *Slang.* *—adj.* **hip·per, hip·pest** *also* **hep·per, hep·pest. 1.** Aware of and interested in the latest trends. **2. a.** Cognizant : wise <*hip* to what's happening> **b.** Comprehending. *—n.* The quality or state of being hip. **—hip′ness** *n.*

**hip**[3] (hĭp) *n.* [ME *hipe* < OE *hēopa*.] The fleshy, berrylike, often brightly colored seed receptacle of a rose.

**hip**[4] (hĭp) *interj.* —Used as a cheer or a signal for a cheer.

**hip·bone** (hĭp′bōn′) *n.* The innominate bone.

**hip-hug·gers** (hĭp′hŭg′ərz) *pl.n.* Tight-fitting pants whose waistline rests at hip level.

**hip joint** *n.* The joint between the innominate bone and femur.

**hipped** (hĭpt) *adj.* [Shortening and alteration of HYPOCHONDRIAC.] **1.** Sad : depressed. **2.** *Slang.* Absorbed to an extreme degree <*hipped* on golf>

**hip·pie** *also* **hip·py** (hĭp′ē) *n., pl.* **-pies.** [< HIP[2].] One who opposes and rejects many of the traditional standards and customs of society, esp. one who advocates extreme liberalism in sociopolitical attitudes and lifestyles. **—hip′pie·dom** (-dəm) *n.*

**hip·po** (hĭp′ō) *n., pl.* **-pos.** *Informal.* A hippopotamus.

**hip·po·cam·pus** (hĭp′ə-kăm′pəs) *n., pl.* **-pi** (-pī′) [NLat. *Hippocampus*, sea horse genus < LLat., a mythical sea creature < Gk. *hippokampos* : *hippos*, horse + *kampos*, sea monster.] *Anat.* A ridge along each lateral ventricle of the brain. **—hip′po·cam′pal** *adj.*

▲ word history: The anatomical structure called the *hippocampus* derives its name from its resemblance to the sea horse, which belongs to the genus *Hippocampus*. The word *hippocampus*, a compound of the Greek words for "horse" and "sea monster," was the name of the creature ridden by Poseidon and other sea gods. According to legend, this beast had a horse's body and a fish's tail.

**hip·po·cras** (hĭp′ə-krăs′) *n.* [ME *ipocras* < OFr. < *Ypocras*, Hippocrates (460?–377? B.C.).] A spiced cordial made from wine, once used as a medicine.

**Hip·po·crat·ic oath** (hĭp′ə-krăt′ĭk) *n.* An oath of ethical professional behavior sworn by new physicians, attributed to Hippocrates.

**Hip·po·crene** (hĭp′ə-krēn′, hĭp′ə-krē′nē) *n.* [Lat. < Gk. *Hippokrēnē* : *hippos*, horse (from the myth that Pegasus' hoof created it) + *krēnē*, fountain.] *Gk. Myth.* A fountain on Mount Helicon, Greece, that was held sacred to the Muses and was regarded as a source of poetic inspiration.

**hip·po·drome** (hĭp′ə-drōm′) *n.* [OFr. < Lat. *hippodromos* < Gk. : *hippos*, horse + *dromos*, race.] **1.** An open-air stadium with an oval course for horse and chariot races in ancient Greece and Rome. **2.** An arena for horse shows.

**hip·po·griff** *also* **hip·po·gryph** (hĭp′ə-grĭf′) *n.* [Fr. *hippogriffe* < Ital. *ippogrifo* : *ippo-*, horse (< Lat. *hippos* < Gk.) + *grifo*, griffin (< LLat. *gryphus*).] A mythological monster with the wings, claws, and head of a griffin and the body and hindquarters of a horse.

**Hip·pol·y·ta** (hĭ-pŏl′ĭ-tə) *n.* [Lat. < Gk. *Hippolutē*.] *Gk. Myth.* A queen of the Amazons slain by Hercules as the last of his labors.

**Hip·pol·y·tus** (hĭ-pŏl′ĭ-təs) *n.* [Lat. < Gk. *Hippolutos*.] *Gk. Myth.* A son of Theseus who spurned the advances of his stepmother, Phaedra, and was killed by Poseidon.

**hip·po·pot·a·mus** (hĭp′ə-pŏt′ə-məs) *n., pl.* **-mus·es** *or* **-mi** (-mī′) [Lat. < LGk. *hippopotamos* : Gk. *hippos*, horse + Gk. *potamos*, river.] **1.** A large, short-legged, chiefly aquatic African mammal, *Hippopotamus amphibius*, with dark, thick, almost hairless skin and a broad wide-mouthed muzzle. **2.** An animal, *Choeropsis liberiensis*, similar to but smaller than the hippopotamus.

**hip·py** (hĭp′ē) *n. var. of* HIPPIE.

**hip roof** *n.* A roof having sloping edges and sides.

**hip·ster** (hĭp′stər) *n. Slang.* One who is highly aware of or interested in the latest trends, esp. in jazz or drugs. **—hip′ster·ism** *n.*

**hi·ra·ga·na** (hîr′ə-gä′nə) *n.* [J. : *hira*, flat + *kana*, kana.] One of two sets of Japanese syllabaries, having a cursive form.

**hir·cine** (hûr′sīn′, -sĭn) *adj.* [Lat. *hircinus* < *hircus*, goat.] Of or typical of a goat, esp. in strong odor.

**hire** (hīr) *vt.* **hired, hir·ing, hires.** [ME *hiren* < OE *hȳrian*.] **1.** To engage the services of for a fee <*hire* a caterer> **2.** To engage the temporary use of for a fee : RENT <*hire* a car> **—hire out. 1.** To grant one's services for a fee. **2.** To allow the use of for a fee <*hired out* the summer cottage> *—n.* **1.** Payment for services or for the use of something. **2.** The act of hiring. **3.** The fact or state of being hired. **—hir′a·ble, hire′a·ble** *adj.* **—hir′er** *n.*

**hire·ling** (hīr′lĭng) *n.* One who offers services solely for compensation, esp. one willing to perform menial or unpleasant tasks for a fee.

**hiring hall** *n.* A union-operated placement center where jobs from various employers are allotted to registered applicants according to an order based usu. on rotation or seniority.

**hir·sute** (hûr′so͞ot′, hîr′-, hər-so͞ot′) *adj.* [Lat. *hirsutus*.] Covered with hair. **—hir′sute′ness** *n.*

**hir·sut·ism** (hûr′so͞o-tĭz′əm, hîr′-, hər-so͞o′-) *n.* Heavy and often abnormally distributed growth of hair.

**hir·u·din** (hîr-o͞od′n, hîr′ə-dən, -yə-) *n.* [Lat. *hirudo*, *hirudin-*, leech.] A substance obtained from the salivary glands of leeches and used as an anticoagulant.

**his** (hĭz) *adj.* [ME < OE.] —Used to indicate that the male previously referred to is the possessor or the agent or recipient of an action <*his* hat><*his* achievements><*his* first romance> *—pron. (sing. or pl. in number).* That or those belonging to him <The cats are *his*.><The idea for the fund raiser was *his*.>

**His·pan·ic** (hĭ-spăn′ĭk) *adj.* [Lat. *Hispanicus*, after *Hispania* (Spain).] Of or relating to the language, people, or culture of Spain, Portugal, or Latin America. *—n.* An American of Spanish or Latin-American origin or descent. **—His·pan′i·cism** (-ĭ-sĭz′əm) *n.* **—His·pan′i·cist** (-ĭ-sĭst) *n.*

**his·pid** (hĭs′pĭd) *adj.* [Lat. *hispidus.*] Covered with bristly hairs, as a leaf or stem. **—his·pid′i·ty** (hĭ-spĭd′ĭ-tē) *n.*

**hiss** (hĭs) *n.* [ME *hissen.*] **1.** A sharp, sibilant sound similar to a sustained *s.* **2.** An expression of disapproval, dissatisfaction, or contempt conveyed by a hiss. —*v.* **hissed, hiss·ing, hiss·es.** —*vi.* To utter a hiss. —*vt.* **1.** To utter with a hiss. **2.** To express by hissing <*hissed* our disapproval> **—hiss′er** *n.*

**hist-** *pref. var. of* HISTO-.

**his·tam·i·nase** (hĭ-stăm′ə-nās′, -nāz′, hĭs′tə-mə-) *n.* A digestive enzyme that converts histidine to histamine.

**his·ta·mine** (hĭs′tə-mēn′, -mĭn) *n.* A white crystalline compound, $C_5H_9N_3$, occurring in plant and animal tissue, that is formed from histidine by the action of putrefactive bacteria, is a stimulant of gastric secretion, and is used in medicine as a vasodilator. **—his′ta·min′ic** (-mĭn′ĭk) *adj.*

**his·ti·dine** (hĭs′tĭ-dēn′, -dĭn) *n.* [HIST(O)- + -ID(E) + -INE.] A colorless, crystalline amino acid, $C_6H_9N_3O_2$, used as a feed additive and dietary supplement.

**his·ti·o·cyte** (hĭs′tē-ə-sīt′) *n.* [Gk. *histion,* web (dim. of *histos* < *histanai,* to cause to stand) + -CYTE.] A fixed phagocyte that is part of the reticuloendothelial system and ingests foreign particles in the blood. **—his′ti·o·cyt′ic** (-sĭt′ĭk) *adj.*

**histo-** *or* **hist-** *pref.* [< Gk. *histos,* web < *histanai,* to cause to stand.] Body tissue <*histogenesis*>

**his·to·chem·is·try** (hĭs′tō-kĕm′ĭ-strē) *n.* The chemistry of tissues and cells. **—his′to·chem′i·cal** (-kəl) *adj.* **—his′to·chem′i·cal·ly** *adv.*

**his·to·com·pat·i·bil·i·ty** (hĭs′tō-kəm-păt′ə-bĭl′ĭ-tē) *n.* A state in which the absence of immunological interference permits the transfusion of blood or grafting of tissue. **—his′to·com·pat′i·ble** *adj.*

**his·to·gen·e·sis** (hĭs′tō-jĕn′ĭ-sĭs) *n.* Formation and development of bodily tissues. **—his′to·ge·net′ic** (-jə-nĕt′ĭk), **his′to·gen′ic** *adj.* **—his′to·ge·net′i·cal·ly, his′to·gen′i·cal·ly** *adv.*

**his·to·gram** (hĭs′tə-grăm′) *n.* [Gk. *histos,* beam, mast + -GRAM.] *Statistics.* A graphic representation of a frequency distribution in which the widths of contiguous vertical bars are proportional to the class widths of the variable and the heights of the bars are proportional to the class frequencies.

**his·tol·o·gy** (hĭ-stŏl′ə-jē) *n.* [Fr. *histologie* : *histo-,* histo- + *-logie,* -logy.] **1.** Anatomical study of the microscopic structure of animal and plant tissues. **2.** The microscopic structure of tissue. **3.** A treatise on histology. **—his′to·log′i·cal** (hĭs′tə-lŏj′ĭ-kəl) *adj.* **—his′to·log′i·cal·ly** *adv.* **—his·tol′o·gist** *n.*

**his·tol·y·sis** (hĭ-stŏl′ĭ-sĭs) *n.* Disintegration of organic tissue. **—his′to·lyt′ic** (hĭs′tə-lĭt′ĭk) *adj.* **—his′to·lyt′i·cal·ly** *adv.*

**his·tone** (hĭs′tōn′) *n.* A simple water-soluble protein occurring esp. in glandular tissue that can release on hydrolysis a high proportion of basic amino acids.

**his·to·pa·thol·o·gy** (hĭs′tō-pə-thŏl′ə-jē, -pă-) *n.* The pathology of changes in diseased tissue. **—his′to·path′o·log′ic** (-păth′ə-lŏj′ĭk), **his′to·path′o·log′i·cal** *adj.* **—his′to·path′o·log′i·cal·ly** *adv.* **—his′to·pa·thol′o·gist** *n.*

**his·to·phys·i·ol·o·gy** (hĭs′tō-fĭz′ē-ŏl′ə-jē) *n.* The physiology of the microscopic functioning of bodily tissues. **—his′to·phys′i·o·log′ic** (-ē-ə-lŏj′ĭk), **his′to·phys′i·o·log′i·cal** *adj.*

**his·to·ri·an** (hĭ-stôr′ē-ən, -stōr′-, -stŏr′-) *n.* **1.** A writer, student, or scholar of history. **2.** One who records proceedings.

**his·tor·ic** (hĭ-stôr′ĭk, -stŏr′-) *adj.* **1.** Having importance in or influence on history. **2.** Historical. ***usage:*** Although *historic* and *historical* overlap in meaning and may be used interchangeably in certain contexts (*historic* or *historical times*), in other contexts they are differentiated. *Historic* refers to whatever is important in history (*the historic voyage of Columbus*); *historical* refers to anything in the past, whether important or not (*a historical character*).

**his·tor·i·cal** (hĭ-stôr′ĭ-kəl, -stŏr′-) *adj.* **1.** Of, pertaining to, or of the character of history. **2.** Based on or concerned with events in history. **3.** Having importance or influence in history : HISTORIC. **4.** Diachronic. **—his·tor′i·cal·ly** *adv.* **—his·tor′i·cal·ness** *n.*

**historical linguistics** *n.* (*sing. in number*). The scientific study of chronological language development with emphasis on evolutionary development.

**historical materialism** *n.* A major tenet in the Marxist theory of history that regards material economic forces as the base on which sociopolitical institutions and ideas are built.

**historical present** *n.* The present tense used to narrate or recount events set in the past.

**historical school** *n.* A school of theorists, as in economics or law, emphasizing the influence of historical conditions.

**his·tor·i·cism** (hĭ-stôr′ĭ-sĭz′əm, -stŏr′-) *n.* A theory of history holding that events are determined by conditions and inherent processes beyond the control of humans. **—his·tor′i·cist** *adj. & n.*

**his·to·ric·i·ty** (hĭs′tə-rĭs′ĭ-tē) *n.* Historical authenticity.

**his·tor·i·cize** (hĭ-stôr′ĭ-sīz′, -stŏr′-) *v.* **-cized, -ciz·ing, -ciz·es.** —*vt.* To make or make appear historical. —*vi.* To use historical materials or details.

**his·to·ri·og·ra·pher** (hĭ-stôr′ē-ŏg′rə-fər, -stōr′-) *n.* **1.** One specializing in historiography. **2.** A historian, esp. one designated officially by a group or public institution.

**his·to·ri·og·ra·phy** (hĭ-stôr′ē-ŏg′rə-fē, -stōr′-) *n.* [OFr. *historiographie* < Gk. *historiographia* : *historia,* history + *-graphia,* writing.] **1.** The principles or methodology of historical study. **2.** The writing of history. **3.** Historical literature. **—his·to′ri·o·graph′ic** (-ē-ə-grăf′ĭk), **his·to′ri·o·graph′i·cal** *adj.* **—his·to′ri·o·graph′i·cal·ly** *adv.*

**his·to·ry** (hĭs′tə-rē) *n., pl.* **-ries.** [Lat. *historia* < Gk. < *histōr,* learned man.] **1.** A narrative of past events. **2.** A chronological, often explanatory or commentary record of events, as of the life or development of a people or institution. **3.** The branch of knowledge that records and analyzes past events. **4.** The events and details forming the subject matter of history. **5.** An interesting past <a wedding gown with a *history*> **6.** Something no longer of current concern <My childhood is now *history.*> **7.** A drama based on historical events. **8.** A record of a patient's medical background.

**his·tri·on·ic** (hĭs′trē-ŏn′ĭk) *also* **his·tri·on·i·cal** (-ĭ-kəl) *adj.* [LLat. *histrionicus* < Lat. *histrio,* actor, of Etruscan orig.] **1.** Of or relating to actors or acting. **2.** Excessively dramatic or emotional : MELODRAMATIC. **—his′tri·on′i·cal·ly** *adv.*

**his·tri·on·ics** (hĭs′trē-ŏn′ĭks) *n.* **1.** (*sing. in number*). Theatrical arts. **2.** (*pl. in number*). Exaggerated emotional behavior calculated for effect.

**hit** (hĭt) *v.* **hit, hit·ting, hits.** [ME *hitten* < OE *hyttan* < ON *hitta.*] —*vt.* **1.** To come in contact with forcefully : STRIKE. **2.** To cause to come into contact <*hit* the ball against the fence> **3.** To deal a blow to. **4.** To strike with a projectile. **5. a.** To reach (a player) with a propelled object in a sport. **b.** To score in this way. **c.** To perform (a shot or maneuver) successfully. **6.** *Baseball.* **a.** To make (a base hit) effectively. **b.** To bat against (a pitcher or kind of pitch) successfully. **7.** To affect adversely <*hit* hard by the death of my spouse> **8.** To come upon or find, often by chance. **9.** To attain or reach <Sales *hit* an all-time low.> **10.** To accord with : SUIT <The restaurant *hit* our fancies.> **11.** To propel with a blow. **12.** *Informal.* To request or obtain from. —Often used with up. **13.** *Informal.* To resort to frequently or excessively. —*vi.* **1.** To strike or deal a blow. **2.** To come in contact : COLLIDE. **3. a.** To attack. **b.** To happen or occur <All at once the storm *hit.*> **4.** To achieve or find something desired or sought <finally *hit* on a compromise> **5.** *Baseball.* To bat. —*n.* **1.** A collision or impact. **2.** A successfully executed shot, blow, thrust, or throw. **3.** A successful or popular venture. **4.** An apt or effective remark. **5.** *Baseball.* A base hit. **6.** *Slang.* A dose of a narcotic drug. **7.** *Slang.* A murder planned and carried out usu. by a member of an underworld syndicate. **—hit it off.** To get along well together. **—hit (one's) stride.** To attain optimum speed or effectiveness. **—hit the books.** To study, esp. hard. **—hit the hay** (*or* **sack**). To go to bed. **—hit the jackpot.** To suddenly become very successful. **—hit the nail on the head.** To be exactly right. **—hit the road.** To set out, as on a trip : LEAVE. **—hit the roof** (*or* **ceiling**). To express strong anger. **—hit the spot.** To give satisfaction, as food or drink.

☆ ***syns:*** HIT, BASH, BELT, BOP, CLIP, CLOBBER, CLOUT, KNOCK, PASTE, SLAM, SLOG, SLUG, SMACK, SMASH, SMITE, SOCK, STRIKE, SWAT, WALLOP, WHACK, WHAM *v. core meaning* : to deliver (a powerful blow) suddenly and sharply <*hit* the other boxer in the jaw>

**hit-and-run** (hĭt′n-rŭn′) *adj.* **1.** Designating or involving the driver of a motor vehicle who drives on after striking a pedestrian or another vehicle. **2.** *Baseball.* Relating to or designating a play in which a player on base runs on the pitch and the batter tries to hit the ball.

**hitch** (hĭch) *v.* **hitched, hitch·ing, hitch·es.** [Orig. unknown.] —*vt.* **1.** To fasten temporarily with or as if with a loop, hook, or noose. **2.** To connect or attach <*hitched* the trailer to the car> **3.** *Informal.* To join in marriage. **4.** To raise by pulling or jerking. **5.** *Informal.* To hitchhike. —*vi.* **1.** To move haltingly : HOBBLE. **2.** To become entangled, snarled, or fastened. **3.** *Slang.* To be united in marriage. **4.** *Informal.* To hitchhike. —*n.* **1.** Any of various knots used as a temporary fastening. **2.** A device used to connect one thing to another. **3.** A short jerking motion : TUG. **4.** A hobble or limp. **5.** An impediment or delay <The meeting went without a *hitch.*> **6.** A term of service, esp. of military service.

**hitch·hike** (hĭch′hīk′) *v.* **-hiked, -hik·ing, -hikes.** —*vi.* To travel by soliciting free rides along a road. —*vt.* To solicit or obtain (a free ride) along a road. **—hitch′hik′er** *n.*

**hitching post** *n.* A post for temporarily tying up an animal, esp. a horse.

**hith·er** (hĭth′ər) *adv.* [ME < OE *hider.*] To or toward this place. —*adj.* Located on the near side.

**hith·er·most** (hĭth′ər-mōst′) *adj.* Nearest to this place or side.

**hith·er·to** (hĭth′ər-tōō′, hĭth′ər-tōō′) *adv.* Until this time.

**hith·er·ward** (hĭth′ər-wərd) *also* **hith·er·wards** (-wərdz) *adv.* Hither.

**Hit·ler·i·an** (hĭt-lîr′ē-ən) *adj.* Of, relating to, or typical of Adolf Hitler, his beliefs, or his regime in Germany.

**Hit·ler·ism** (hĭt′lə-rĭz′əm) *n.* The fascistic and nationalistic theo-

ries and methods of Adolf Hitler and the Nazis. **—Hit'ler·ite'** (-lə-rīt') *n.*
**hit list** *n. Slang.* **1.** A list of the potential murder victims of a crime syndicate. **2.** A list designating a target, esp. for attack or elimination <"had a *hit list* of executives he wanted fired" —*New York*>
**hit man** *n. Slang.* HATCHET MAN 1.
**hit-or-miss** (hĭt'ər-mĭs') *adj.* Lacking care or precision.
**hit·ter** (hĭt'ər) *n.* **1.** One who hits something. **2.** *Baseball.* A batter.
**Hit·tite** (hĭt'īt') *n.* [Heb. *Ḥittī* < Hittite *Hatti.*] **1.** A member of an ancient people living in Asia Minor and northern Syria about 2000–1200 B.C. **2.** The Indo-European language of the Hittites. —*adj.* Of or relating to the Hittites, their culture, or their language.
**hive** (hīv) *n.* [ME < OE *hȳf.*] **1.** A structure for housing bees, esp. honeybees. **2.** A colony of bees living in a hive. **3.** A place swarming with people. —*v.* **hived, hiv·ing, hives.** —*vt.* **1.** To collect (bees) into a hive. **2.** To store (honey) in a hive. **3.** To store up : ACCUMULATE. —*vi.* **1.** To enter a hive. **2.** To dwell in close quarters.
**hives** (hīvz) *pl.n.* [Orig. unknown.] *(sing. or pl. in number).* Urticaria.
**ho** (hō) *interj.* [ME < OFr. and < ON *hō.*] —Used esp. to attract attention to something sighted or to urge onward.
**Ho** *symbol for* HOLMIUM.
**hoa·gie** (hō'gē) *n.* [Orig. unknown.] HERO 5.
**hoar** (hôr, hōr) *adj.* [ME *hor* < OE *hār.*] Hoary. —*n.* Hoarfrost.
**hoard** (hôrd, hōrd) *n.* [ME *hord* < OE.] A secret fund or supply stored for future use : CACHE. —*v.* **hoard·ed, hoard·ing, hoards.** —*vi.* To gather or accumulate a hoard. —*vt.* To gather or accumulate by saving or hiding. **—hoard'er** *n.*
☆ **syns:** HOARD, LAY UP, SQUIRREL, STASH, STOCKPILE, TREASURE *v. core meaning* : to store up (e.g., supplies or money), usu. well beyond one's current needs <criticized for *hoarding* food>
**hoard·ing** (hôr'dĭng, hōr'-) *n.* [Obs. *hoard* < AN *hurdis* < OFr. *hourd,* scaffold, of Germanic orig.] *Chiefly Brit.* **1.** A temporary wooden fence around a building or structure under construction or repair. **2.** A billboard.
**hoar·frost** (hôr'frôst', -frŏst', hōr'-) *n.* Frozen dew forming a white covering on a surface.
**hoarse** (hôrs, hōrs) *adj.* **hoars·er, hoars·est.** [ME *hors* < OE *hās.*] **1.** Having a harsh, raspy quality <a *hoarse* voice> **2.** Having a hoarse voice <yelled myself *hoarse*> **—hoarse'ly** *adv.* **—hoarse'ness** *n.*
**hoars·en** (hôr'sən, hōr'-) *vt.* & *vi.* **-ened, -en·ing, -ens.** To make or become hoarse.
**hoar·y** (hôr'ē, hōr'ē) *adj.* **-i·er, -i·est. 1.** Gray or white with or as if with age. **2.** Covered with grayish hair or pubescence, as some leaves. **3.** Very old : ANCIENT. **—hoar'i·ness** *n.*
**hoat·zin** (wät-sēn') *n.* [Am. Sp. < Nahuatl *uatzin,* pheasant.] A tropical South American brownish, crested bird, *Opisthocomus hoazin,* whose young have claws on the first and second digits of the wings.
**hoax** (hōks) *n.* [Perh. alteration of HOCUS.] **1.** An act meant to trick or deceive. **2.** Something established or accepted by fraudulent means. —*vt.* **hoaxed, hoax·ing, hoax·es.** To cheat or deceive by using a hoax. **—hoax'er** *n.*
**hob¹** (hŏb) *n.* [Orig. unknown.] **1.** A projection or shelf at the side or back of the inside of a fireplace, used for keeping things warm. **2.** A tool for cutting the teeth of machine parts.
**hob²** (hŏb) *n.* [After ME *Hobbe,* a nickname for Robert.] A hobgoblin, sprite, or elf.
**Hobb·ism** (hŏb'ĭz'əm) *n.* A political theory promulgated by Thomas Hobbes, advocating powerful governmental control as the only means of effectively dealing with the inevitable problems created by the inherently selfish, aggrandizing nature of human beings.
**hob·ble** (hŏb'əl) *v.* **-bled, -bling, -bles.** [ME *hobblen* < MDu. *hobbelen.*] —*vi.* To walk or move along haltingly or with difficulty : LIMP. —*vt.* **1.** To put a device around the legs of (an animal) so as to hamper but not prevent movement. **2.** To cause to limp. **3.** To impede the action or progress of. —*n.* **1.** An awkward, clumsy, or irregular walk or gait. **2.** A device, as a rope or strap, for hobbling an animal. **3.** *Archaic.* An awkward situation. **—hob'bler** *n.*
**hob·ble·bush** (hŏb'əl-bo͝osh') *n.* A shrub, *Viburnum alnifolium* of northeastern North America, bearing flat clusters of white flowers with the marginal flowers larger than the others.
**hob·ble·de·hoy** (hŏb'əl-dē-hoi') *n., pl.* **-hoys.** [Orig. unknown.] An awkward adolescent boy.
**hobble skirt** *n.* A long skirt, very narrow below the knees, that was popular around 1911.
**hob·by¹** (hŏb'ē) *n., pl.* **-bies.** [ME *hobi,* small horse < *Hobyn,* nickname for *Robert.*] An activity or interest pursued outside of one's regular work primarily for pleasure. **—hob'by·ist** *n.*
▲ word history: The word *hobby¹* meaning "an activity pursued for pleasure" is a shortened form of *hobbyhorse,* which originally denoted a small horse. *Hobbyhorse* also developed other senses denoting representations of horses, such as the horse in the morris dance and a toy horse for children to ride. It is this last sense that led to the development of the modern meaning of the word *hobby.* A person with a hobby was thought to pursue a favorite pastime or obsession with the single-minded zeal of a child riding a hobbyhorse.
**hob·by²** (hŏb'ē) *n., pl.* **-bies.** [ME *hobi* < OFr. *hobe.*] Any of several small falcons of the genus *Falco,* once trained for hawking.
**hob·by·horse** (hŏb'ē-hôrs') *n.* **1.** A child's toy having a long stick with an imitation horse's head on one end. **2.** A rocking horse. **3. a.** A figure of a horse worn around the waist of a mummer pretending to ride a horse. **b.** One wearing a hobbyhorse. **4. a.** A favorite topic or hobby. **b.** An obsession : fixation.
**hob·gob·lin** (hŏb'gŏb'lĭn) *n.* **1.** An ugly, mischievous goblin or elf. **2.** An object of extreme dislike : BUGBEAR.
**hob·nail** (hŏb'nāl') *n.* A short nail with a thick head used for protecting the soles of boots or shoes.
**hob·nob** (hŏb'nŏb') *vi.* **-nobbed, -nob·bing, -nobs.** [< the phrase *(drink) hob or nob,* (toast) one another alternately.] To associate familiarly <*hobnobs* with ambassadors and statesmen>
**ho·bo** (hō'bō) *n., pl.* **-boes** *or* **-bos.** [Orig. unknown.] **1.** A tramp or vagrant. **2.** A migratory, usu. unskilled worker. **—ho'bo·ism** *n.*
**Hob·son's choice** (hŏb'sənz) *n.* [After Thomas *Hobson* (1544–1631), English liveryman, from his requirement that customers take either the horse nearest the stable door or none.] An apparently free choice that offers no actual alternative.
**hock¹** (hŏk) *n.* [ME *hoche* < OE *hōh,* heel.] **1.** The tarsal joint of the hind leg of a digitigrade quadruped, as a horse, corresponding to the human ankle. **2.** A joint in the leg of a domestic fowl similar to the hock of a quadruped. —*vt.* **hocked, hock·ing, hocks.** To disable by cutting the tendons of the hock : HAMSTRING.
**hock²** (hŏk) *n.* [Alteration of G. *Hochheimer,* after *Hochheim,* West Germany.] *Chiefly Brit.* White Rhine wine.
**hock³** (hŏk) [< Du. *hok,* prison.] *Informal.* —*vt.* **hocked, hock·ing, hocks.** To pawn. —*n.* The state of being pawned. **—in hock.** In debt.
**hock·ey** (hŏk'ē) *n.* [Orig. unknown.] **1.** A game played on ice in which two opposing teams of skaters try to drive a puck into the opponents' goal using curved sticks. **2.** A form of hockey played on foot on a turf field and using a ball rather than a puck.
**hockey stick** *n.* A long-handled stick with a curved end, used in hockey.
**hock·shop** (hŏk'shŏp') *n. Informal.* A pawnshop.
**ho·cus** (hō'kəs) *vt.* **-cused, -cus·ing, -cus·es** *or* **-cussed, -cus·sing, -cus·ses.** [Short for HOCUS-POCUS.] **1.** To deceive : hoax. **2.** To put a drug into (food or drink).
**ho·cus-po·cus** (hō'kəs-pō'kəs) *n.* [Poss. < an alteration of Lat. *hoc est corpus,* this is the body (from its use in the Eucharist at the time of transubstantiation).] **1.** Nonsense words or phrases used as a formula by conjurers. **2.** A trick executed by a magician or juggler. **3.** A chicanery or deception. —*v.* **-cused, -cus·ing, -cus·es** *or* **-cussed, -cus·sing, -cus·ses.** —*vt.* To trick : deceive. —*vi.* To be deceptive.
**hod** (hŏd) *n.* [Perh. alteration of dial. *hot* < ME *hotte,* pannier < OFr., of Germanic orig.] **1.** A trough carried over the shoulder for transporting loads, as bricks. **2.** A coal scuttle.
**ho·dad** (hō'dăd') *also* **ho·dad·dy** (-dăd'ē) *n.* [Orig. unknown.] One who does not surf but who spends time at surfing beaches pretending to be a surfer.
**hodge·podge** (hŏj'pŏj') *n.* [ME *hochepot* < OFr. —see HOTCHPOT.] A mixture of diverse ingredients : MISHMASH.
**Hodg·kin's disease** (hŏj'kĭnz) *n.* [After Thomas *Hodgkin* (1798–1866).] A usu. chronic, progressive disease of unknown etiology, characterized by inflammatory enlargement of the lymph nodes, spleen, and often liver and kidneys.
**hoe** (hō) *n.* [ME *houe* < OFr., of Germanic orig.] A tool with a flat blade attached approx. at right angles to a long handle, used for weeding, cultivating, and digging. —*v.* **hoed, hoe·ing, hoes.** —*vt.* To weed, cultivate, or dig up with a hoe. —*vi.* To work with a hoe. **—ho'er** *n.*
**hoe·cake** (hō'kāk') *n.* A thin cake made of cornmeal.
**hoe-down** (hō'doun') *n.* **1.** A square dance. **2.** The music for a hoe-down. **3.** A party featuring hoe-downs.
**hog** (hôg, hŏg) *n.* [ME < OE *hogg,* poss. of Celtic orig.] **1.** Any of various mammals of the family Suidae, which includes the domesticated pig as well as wild species, as the boar and wart hog. **2.** A domesticated pig, esp. one weighing more than 120 pounds. **3.** A greedy, self-indulgent, or filthy person. **4.** *also* **hogg.** *Chiefly Brit.* A young sheep before its second shearing. —*v.* **hogged, hog·ging, hogs.** —*vt.* **1.** To take more than one's share of. **2.** To cause (the back) to arch like that of a hog. **3.** To shorten the mane of (a horse). —*vi. Naut.* To arch upward in the middle. —Used of a ship's keel. **—go hog wild.** To react excitedly, intemperately, or irrationally. **—high off** (*or* **on**) **the hog.** In a lavish or extravagant manner.
**ho·gan** (hō'gän', -gən) *n.* [Navaho.] A Navaho dwelling usu. built of logs and mud.
**hog·back** (hôg'băk', hŏg'-) *n.* A sharp ridge with steeply sloping sides, produced by the erosion of the broken edges of highly tilted strata.
**hog cholera** *n.* A highly infectious, often fatal viral disease of swine, marked by fever, loss of appetite, diarrhea, and exhaustion.

**hog·fish** (hôg′fĭsh′, hŏg′-) *n., pl.* **hogfish** *or* **-fish·es. 1.** A colorful fish, *Lachnolaimus maximus* of warm Atlantic waters, with a long snout in the adult male. **2.** The pigfish.

**hogg** (hôg, hŏg) *n. var. of* HOG 4.

**hog·gish** (hô′gĭsh, hŏg′ĭsh) *adj.* Coarsely self-indulgent, greedy, or filthy. **—hog′gish·ly** *adv.* **—hog′gish·ness** *n.*

**Hog·ma·nay** (hŏg′mə-nā′) *n.* [Orig. unknown.] *Scot.* New Year's Eve, when children go from house to house asking for presents.

**hog·nose snake** (hôg′nōz′, hŏg′-) *n.* A thick-bodied nonvenomous North American snake of the genus *Heterodon,* with an upturned snout.

**hognose snake**
*2–3 feet long*

**hog peanut** *n.* A twining North American vine, *Amphicarpa bracteata,* with pinkish or white flower clusters and bearing curving pods as well as basal or underground fleshy one-seeded fruit.

**hogs·head** (hôgz′hĕd′, hŏgz′-) *n.* **1.** Any of various units of volume or capacity ranging from 63 to 140 gallons or approx. 238 to 530 liters, esp. a unit of capacity used in liquid measure in the United States, equal to 63 gallons or approx. 238 liters. **2.** A large barrel or cask with the capacity for holding a hogshead.

**hog-tie** *also* **hog·tie** (hôg′tī′, hŏg′-) *vt.* **-tied, -ty·ing** *or* **-tie·ing, -ties. 1.** To tie together the legs of. **2.** To disrupt or impede in action or movement.

**hog·wash** (hôg′wôsh′, hŏg′wŏsh′) *n.* **1.** Garbage fed to hogs : SWILL. **2.** False, worthless, or ridiculous speech or writing.

**hog·weed** (hôg′wēd′, hŏg′-) *n.* Any of various coarse, weedy plants.

**ho hum** *interj.* —Used to express tiredness or boredom.

**ho-hum** (hō′hŭm′) *adj.* Boring and conventional <"a *ho-hum* speaker who couldn't capture the attention of the conventioneers" —*Chicago Tribune*>

**hoicks** (hoiks) *interj. var. of* YOICKS.

**hoi pol·loi** (hoi′ pə-loi′) *n.* [Gk., the many.] The common people.

**hoist** (hoist) *vt.* **hoist·ed, hoist·ing, hoists.** [Alteration of dial. *hoise.*] To raise or haul up with or as if with a mechanical apparatus. —*n.* **1.** An apparatus for raising heavy or unwieldy objects. **2.** The act of hoisting : LIFT. **3.** *Naut.* **a.** The height or vertical dimension of a flag or of any square sail other than a course. **b.** A group of flags raised together as a signal. **—hoist′er** *n.*

**hoi·ty-toi·ty** (hoi′tē-toi′tē) *adj.* [Redup. of dial. *hoit,* to romp.] **1.** Pretentiously self-important. **2.** Flighty : scatterbrained. **3.** Easily offended. —*n., pl.* **-ties. 1.** Arrogance : pomposity. **2.** Giddiness.

**hok·ey** (hō′kē) *adj.* **-i·er, -i·est.** *Slang.* **1.** Corny : trite. **2.** Artificial : phony. **—hok′i·ly** *adv.* **—hok′i·ness** *n.*

**ho·key-po·key** (hō′kē-pō′kē) *n.* **1.** Hocus-pocus : chicanery. **2.** Ice cream formerly sold by street venders.

**hok·ku** (hō′ko͞o) *n., pl.* **hokku.** Haiku.

**ho·kum** (hō′kəm) *n.* [Perh. HO(CUS-POCUS) + (BUN)KUM.] **1.** Something seemingly impressive but actually insincere or false. **2.** A stock technique for eliciting a response from an audience.

**hol-** *pref. var. of* HOLO-.

**Hol·arc·tic** (hō-lärk′tĭk, -lär′tĭk, hŏ-) *adj.* Of or designating the zoogeographic region that includes the northern areas of the earth and is divided into Nearctic and Palearctic regions.

**hold**[1] (hōld) *v.* **held** (hĕld), **hold·ing, holds.** [ME *holden* < OE *healdon.*] —*vt.* **1. a.** To have and keep in one's grasp. **b.** To keep up : SUPPORT <The table can't *hold* much weight.> **c.** To maintain in a certain position or relationship <*held* the camera steady> **d.** To keep in reserve or custody <*holding* the suspect without bail> **2. a.** To receive or be able to receive as contents : CONTAIN <The couch *holds* four people.> **b.** To have in store <Who knows what tomorrow *holds*?> **3. a.** To have and maintain in one's possession. **b.** To maintain control over <The cordons *held* the mob.> **c.** To retain or defend by force <The soldiers *held* the ridge.> **4. a.** To impose control or restraint upon : CHECK <*hold* one's anger> <tried to *hold* prices down> **b.** To stall or delay <Please *hold* all my calls.> **c.** To stop the movement or progress of <*Hold* the bus!> **5. a.** To maintain in a given condition or action <*held* the ship on course> **b.** To keep the attention or interest of <The film *held* me spellbound.> **6. a.** To be the legal possessor of. **b.** To bind by a contract. **c.** To adjudge or decree. **d.** To cause to keep : OBLIGATE <will *hold* you to your word> **7.** To comport : carry <*held* themselves straight and tall> **8. a.** To keep in the mind or heart : harbor, as a grudge. **b.** To consider or believe : JUDGE <*holds* it wrong to cheat on taxes> **c.** To assert : affirm <*holds* that this is the only workable solution> **9. a.** To have or occupy (a position) <*held* the mayoralty for two years> **b.** To regard in a given way <*held* them in esteem> **10. a.** To cause to take place <*hold* a dance> **b.** To convene or assemble <*held* a meeting of the executive council> —*vi.* **1.** To maintain a grasp or grip. **2.** To maintain a desired or accustomed position or condition. **3.** To adhere closely : KEEP <*held* to a northeasterly course> **4.** To stand up under stress, pressure, or opposition : ENDURE. **5.** To be valid, applicable, or true <Our assumption *holds.*> **6.** *Slang.* To have illicit or illegally obtained material or goods, esp. narcotics, in one's possession. **—hold back. 1.** To keep in one's possession or control. **2.** To impede the progress of. **—hold down.** To work at and keep (a job). **—hold off.** To stop or delay doing something. **—hold on. 1.** To maintain one's grip : CLING. **2.** To keep at : CONTINUE. **3.** To stop or wait for someone or something. **—hold out. 1.** To endure : last. **2.** To continue to resist. **3.** To refuse to reach or satisfy an agreement. **—hold over. 1.** To delay or postpone. **2.** To continue or prolong the term or engagement of. **—hold to.** To remain loyal or faithful to. **—hold up. 1.** To offer or present as an example <*held* the student's essay *up* as a model> **2.** To obstruct or delay. **3.** To rob. **4.** To hold out <How are you *holding up* in this heat?> **—hold with.** To agree with : SUPPORT. —*n.* **1.** The act or a means of grasping : GRIP. **2.** Something held onto, as for support. **3. a.** A bond or force that restrains, dominates, or affects someone or something. **b.** Complete control <need to take *hold* of my life> **4.** A prison cell. **5.** *Archaic.* A fortified place : STRONGHOLD. **6.** *Mus.* **a.** The prolongation of a note beyond its indicated time value. **b.** The symbol designating this pause : FERMATA. **7.** A temporary halt, as in a countdown. **—hold forth.** To talk at great length. **—hold the fort. 1.** To maintain a secure position. **2.** To assume responsibility esp. in another's absence. **—hold the line.** To keep something at an acceptable level by regulation or aggressive action. **—hold water.** To stand up to critical examination. **—on hold. 1.** In or into a condition or period of waiting <The receptionist put me *on hold.*> **2.** In or into a state of delay or suspended activity <put the wedding *on hold*>

**hold**[2] (hōld) *n.* [Alteration of ME *hole* < OE *hulu,* hull.] The lower interior part of a ship or airplane in which cargo is stored.

**hold·all** (hōld′ôl′) *n.* A case or bag for carrying miscellaneous items, as when traveling.

**hold·back** (hōld′băk′) *n.* A strap or iron bar between the shaft and the harness on a drawn wagon, permitting the horse to stop or back up.

**hold·en** (hōl′dən) *v. Archaic. var. p.p. of* HOLD[1].

**hold·er** (hōl′dər) *n.* **1.** One who holds something. **2.** A device for holding something. **3. a.** One who possesses something : OWNER. **b.** One who occupies or controls something. **4.** *Law.* One who legally possesses and is entitled to the payment of a check, bill, or promissory note.

**hold·fast** (hōld′făst′) *n.* **1.** A device for fastening something securely. **2.** *Biol.* An organ or structure of attachment, esp. the basal, rootlike formation by which certain seaweeds or other algae are attached to a surface.

**hold·ing** (hōl′dĭng) *n.* **1.** Land rented or leased from another. **2.** *often* **holdings.** Legally owned property, as land, capital, or stocks. **3.** A court ruling, esp. a ruling on a point of law raised in an official proceeding.

**holding company** *n.* A company controlling partial or complete interest in other companies.

**holding pattern** *n.* **1.** A usu. circular pattern flown by aircraft awaiting clearance to land at an airport. **2.** A state or period of waiting or delay.

**hold·out** (hōld′out′) *n. Informal.* One who withholds or delays cooperation or agreement.

**hold·o·ver** (hōld′ō′vər) *n.* One that remains from an earlier time, esp. an officeholder kept in office after his or her term is over.

**hold·up** (hōld′ŭp′) *n.* **1.** A delay. **2.** A usu. armed robbery.

**hole** (hōl) *n.* [ME < OE *hol.*] **1.** A cavity in a solid. **2.** An opening or perforation through something <a *hole* in one's sock> **3.** A deep place in a body of water. **4.** An animal's hollowed-out habitation, as a burrow. **5.** An ugly or squalid dwelling. **6.** A deep or isolated place of confinement : DUNGEON. **7.** A fault or flaw <*holes* in one's reasoning> **8.** A difficult situation : PREDICAMENT. **9. a.** The small pit lined with a cup into which the ball must be hit in golf. **b.** One of the divisions of a golf course, from tee to cup. **10.** *Electron.* A vacant electron energy state manifested as a charge defect in a crystalline solid, the defect behaving as a positive charge carrier with charge magnitude equal to that of the electron. —*v.* **holed, hol·ing, holes.** —*vt.* **1.** To put a hole in. **2.** To put or propel into a hole. —*vi.* To make a hole. **—hole in one.** The driving of a golf ball from the tee into the hole in only one stroke. **—hole out.** To hit a golf ball into the hole. **—hole up. 1.** To hibernate in or as if in a hole. **2.** To hide out or shut oneself up. **—hole′y** *adj.*

**hol·i·day** (hŏl′ĭ-dā′) *n.* [ME *holidai,* holy day < OE *hālig dæg.*] **1.** A day on which custom or the law dictates a cessation of general business activity to celebrate or commemorate a particular event. **2.** A religious feast day: HOLY DAY. **3.** A day free from work that one may spend at leisure. **4.** *often* **holidays.** *Chiefly Brit.* A vacation. —*vi.* **-dayed, -day·ing, -days.** *Chiefly Brit.* To pass a holiday or vacation.

**ho·li·er-than-thou** (hō′lē-ər-thən-thou′) *adj.* Self-righteously pious or virtuous.

**ho·li·ness** (hō′lē-nĭs) *n.* **1.** The quality or state of being holy: SANCTITY. **2. Holiness.** A title of address used for high ecclesiastical dignitaries and esp. for the pope.

**ho·lism** (hō′lĭz′əm) *n.* The theory that reality is made up of organic or unified wholes greater than the simple sum of their parts.

**ho·lis·tic** (hō-lĭs′tĭk) *adj.* **1.** Of or relating to holism. **2. a.** Emphasizing the importance of the whole and the interdependence of its parts. **b.** Concerned with wholes rather than with analysis or dissection into parts <*holistic* health care> **—ho·lis′ti·cal·ly** *adv.*

**hol·land** (hŏl′ənd) *n.* [ME *holand*, after *Holand*, a county in the Netherlands < MDu.] An often glazed cotton or linen fabric used esp. for window shades and upholstery.

**hol·lan·daise sauce** (hŏl′ən-dāz′) *n.* [Fr. *sauce Hollandaise.*] A creamy sauce of butter, egg yolks, and lemon or vinegar.

**Hol·lands** (hŏl′əndz) *n.* [Du. *Hollandsch* < *hollandsch genever*, Dutch gin.] Gin made in the Netherlands.

**hol·ler**[1] (hŏl′ər) *v.* **-lered, -ler·ing, -lers.** [< OFr. *hold*, stop! —see HELLO.] —*vi.* **1.** To yell or shout. **2.** *Informal.* To complain. —*vt.* To yell or shout out <*hollered* "Help!"> —*n.* A yell or shout: CALL.

†**hol·ler**[2] (hŏl′ər) *adj. & n. & v. Regional. var. of* HOLLOW.

**Hol·ler·ith** (hŏl′ə-rĭth′) *also* **Hollerith code** *n.* [After Herman *Hollerith* (1860–1929).] A code used for recording alphanumeric information on punch cards.

**Hollerith card** *n.* A punch card.

**hol·low** (hŏl′ō) *adj.* **-er, -est.** [ME *holwe* < *holgh*, hole < OE *holh.*] **1.** Having a cavity, hole, or space within <a *hollow* log> **2.** Being deeply indented or concave: SUNKEN <a *hollow* rock> **3.** Without substance, worth, or character <a *hollow* individual> **4.** Lacking validity or truth: SPECIOUS <a *hollow* victory> **5.** Having a reverberating, muffled sound. —*n.* **1.** A cavity, hole, or space within something. **2.** An indented or concave area. **3.** A void <The children went to college and left a *hollow* in our life.> —*v.* **-lowed, -low·ing, -lows.** —*vt.* **1.** To make hollow <*hollow* out a tomato> **2.** To scoop or form by making concave <canoes *hollowed* out from logs> —*vi.* To become hollow. **—hol′low·ly** *adv.* **—hol′low·ness** *n.*

**hol·low·ware** (hŏl′ō-wâr′) *n.* Articles for the table, as bowls, pitchers, or knife handles, that are tubular or bowl-shaped.

**hol·ly** (hŏl′ē) *n., pl.* **-lies.** [ME *holi* < OE *holen.*] **1. a.** A tree or shrub of the genus *Ilex*, often bearing bright-red berries and glossy evergreen leaves with spiny margins. **b.** Branches or leaves of the holly, traditionally used for Christmas decoration. **2.** Any of various trees or plants related or similar to the holly.

**hol·ly·hock** (hŏl′ē-hŏk′) *n.* [ME *holihocke*, marsh mallow : *holi*, holy (< OE *halig*) + *hoc*, mallow (< OE).] A tall plant native to China, *Althaea rosea*, widely cultivated for its tall spike of large, variously colored flowers.

**Hol·ly·wood** (hŏl′ē-wo͝od′) *n.* [After *Hollywood*, California.] The U.S. motion-picture industry or the atmosphere attributed to it.

**hollywood bed** *n.* A mattress on a box spring supported by a metal frame or attached low legs, often with an upholstered headboard.

**holm** (hōm, hōlm) *n.* [ME < ON *holmr.*] *Chiefly Brit.* An island in a river.

**hol·mic** (hōl′mĭk) *adj.* Relating to holmium in its trivalent state.

**hol·mi·um** (hōl′mē-əm) *n.* [NLat. < *Holmia* (Stockholm), Sweden.] *Symbol* **Ho** A relatively soft, malleable, stable metallic element; atomic number 67; atomic weight 164.930.

**holm oak** (hōm, hōlm) *n.* [Poss. ME *holm* < *holm*, holly < *holi.*] A tree native to the Mediterranean region, *Quercus ilex*, with prickly evergreen leaves.

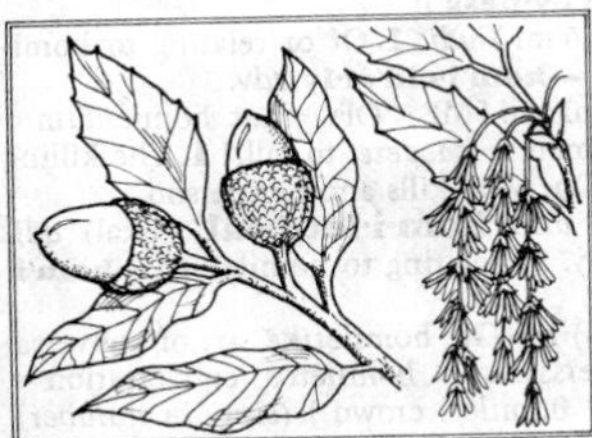

**holm oak**

**holo-** *or* **hol-** *pref.* [< Gk. *holos*, whole.] Whole: entirely <*holoblastic*>

**ho·lo·blas·tic** (hŏ′lə-blăs′tĭk, hŏl′ə-) *adj.* Displaying cleavage in which the entire egg divides into individual blastomeres. **—ho′lo·blas′ti·cal·ly** *adv.*

**ho·lo·caust** (hō′lə-kôst′, hŏl′ə-) *n.* [ME < Lat. *holocaustum* < Gk. *holokauston* < *holokaustos*, burnt whole : *holo-*, whole + *kaustos*, burnt < *kaein*, to burn.] **1.** Great or total destruction, esp. by fire. **2. a.** Widespread destruction **b.** A disaster. **3.** *often* **Holocaust.** A massive slaughter, esp. the genocide of European Jews by the Nazis during World War II. **4.** A sacrificial offering consumed entirely by flames. **—ho′lo·caus′tal, ho′lo·caus′tic** *adj.*

**Ho·lo·cene** (hō′lə-sēn′, hŏl′ə-) *adj.* Of, belonging to, or designating the geologic time, rock series, or sedimentary deposits of the more recent of the two epochs of the Quaternary period, extending from the end of the Pleistocene to the present. —*n.* The Holocene epoch or system of deposits.

**ho·lo·crine** (hō′lə-krĭn, -krīn′, -krēn′, hŏl′ə-) *adj.* [HOLO- + Gk. *krinein*, to separate.] Pertaining to a gland, as a sebaceous gland, whose secretion is formed by the degeneration of the gland's cells.

**ho·lo·en·zyme** (hō′lō-ĕn′zīm′, hŏl′ō-) *n.* A complete enzyme having an apoenzyme and a coenzyme.

**ho·lo·gram** (hō′lə-grăm′, hŏl′ə-) *n.* **1.** The pattern produced on a photosensitive medium that has been exposed by holography and then photographically developed. **2.** The photosensitive medium so exposed and developed.

**ho·lo·graph** (hō′lə-grăf′, hŏl′ə-) *n.* [< LLat. *holographus*, entirely written by the signer < Gk. *holographos* : *holos*, whole + *-graphos*, written.] **1. a.** A document written entirely in the handwriting of the individual whose signature it bears. **b.** The handwriting itself. **2.** A hologram. **—ho′lo·graph′ic, ho′lo·graph′i·cal** *adj.* **—ho′lo·graph′i·cal·ly** *adv.*

**ho·log·raph·y** (hō-lŏg′rə-fē) *n.* The technique of producing images by wave-front reconstruction, esp. by using lasers to record on a photographic plate the diffraction pattern from which a three-dimensional image can be projected. **—ho′lo·graph′** *v.* **(-graphed, -graph·ing, -graphs). —ho·log′ra·pher** *n.*

**ho·lo·he·dral** (hō′lə-hē′drəl, hŏl′ə-) *adj.* Having as many planes as needed for total symmetry in a given crystal system.

**ho·lo·me·tab·o·lism** (hō′lō-mə-tăb′ə-lĭz′əm, hŏl′ō-) *n.* Complete metamorphosis of a developing insect. **—ho′lo·me·tab′o·lous** *adj.*

**ho·lo·phras·tic** (hō′lə-frăs′tĭk, hŏl′ə-) *adj.* [HOLO- + Gk. *phrastikos*, expressive < *phrazein*, to show.] Polysynthetic.

**ho·lo·thu·ri·an** (hō′lə-tho͝or′ē-ən, -thyo͝or′-, hŏl′ə-) *n.* [NLat. *Holothuria*, genus name < Gk. *holothourion*, water polyp.] Any of various echinoderms of the class Holothuroidea, which includes the sea cucumbers. **—hol′o·thu′ri·an** *adj.*

**ho·lo·type** (hō′lə-tīp′, hŏl′ə-) *n.* The single specimen used as the basis of the first published description of a taxonomic species and later designated as the type specimen. **—ho′lo·typ′ic** (-tĭp′ĭk) *adj.*

**ho·lo·zo·ic** (hō′lə-zō′ĭk, hŏl′ə-) *adj.* Deriving nourishment by the ingestion of organic material, in the manner of most animals.

**holp** (hōlp) *v. Archaic. var. p.t. of* HELP.

**hol·pen** (hōl′pən) *v. Archaic. var. p.p. of* HELP.

**Hol·stein** (hōl′stīn′) *n.* [After *Holstein*, a region in West Germany.] Any of a breed of large black and white dairy cattle orig. developed in Friesland.

**hol·ster** (hōl′stər) *n.* [Du.] **1.** A leather case shaped to hold a pistol. **2.** A belt with loops or slots for carrying equipment, as small tools. **—hol′stered** *adj.*

**holt** (hōlt) *n.* [ME < OE.] *Archaic.* A wood or grove: COPSE.

**ho·lus-bo·lus** (hō′ləs-bō′ləs) *adv.* All at once.

**ho·ly** (hō′lē) *adj.* **-li·er, -li·est.** [ME < OE *hālig.*] **1.** Belonging to, derived from, or associated with a divine power: SACRED. **2.** Regarded with or deserving of worship or veneration: REVERED <*holy* Scripture> **3.** Living according to a strict or highly moral religious or spiritual system: SAINTLY. **4.** Specified or set apart for a religious purpose <a *holy* season> **5.** Solemnly undertaken: SACROSANCT <a *holy* vow> **6.** Deserving or regarded with special reverence or respect. **7.** —Used as an intensive or in combination as a mild expletive <The child was a *holy* terror.> <*Holy* cow!> **—ho′li·ly** *adv.* **—ho′li·ness** *n.*

☆ **syns: 1.** HOLY, DEVOUT, GODLY, PIETISTIC, PIOUS, RELIGIOUS *adj. core meaning*: deeply concerned with God and the beliefs and practice of religion <*holy* saints of the church> **2.** HOLY, BLESSED, HALLOWED, SACRED, SACROSANCT, SANCTIFIED *adj. core meaning*: regarded with particular reverence or respect <the *Holy* Grail>

**Holy Ark** *n.* The chest in a synagogue in which the scrolls of the Torah are kept.

**Holy Communion** *n.* The Eucharist.

**holy day** *also* **ho·ly·day** (hō′lē-dā′) *n.* A day specified or set apart for religious observance.

**Holy Father** *n.* One of the titles of the pope.

**Holy Ghost** *n.* The third person of the Christian Trinity: HOLY SPIRIT.

**Holy Grail** *n.* The Grail.

**Holy Office** *n.* A congregation of the Roman Catholic Church dealing with the protection of the faith and morals.

**holy of holies** *n.* **1.** The innermost shrine of a Jewish tabernacle and temple. **2.** An especially sacrosanct place.

**holy orders** *n.* **1.** The sacrament or rite of ordination. **2.** The rank of an ordained Christian minister or priest. **3.** The principal orders of the clergy in the Roman Catholic, Eastern Orthodox, and Anglican churches.

**Holy Saturday** *n.* The Saturday before Easter.

**Holy See** *n. Rom. Cath. Ch.* The court, office, or jurisdiction of the pope.

**Holy Spirit** *n.* The Holy Ghost.

**ho·ly·stone** (hō'lē-stōn') *n.* [Perh. from its being used while kneeling.] A piece of soft sandstone for scouring the wooden decks of a ship. —*vt.* **-stoned, -ston·ing, -stones.** To scour with a holystone.

**Holy Synod** *n.* The governing body of any of the Eastern Orthodox churches.

**Holy Thursday** *n.* **1.** Maundy Thursday. **2.** Ascension Day.

**holy water** *n.* Water blessed by a priest.

**Holy Week** *n.* The week before Easter, commemorating the last days of Christ.

**hom-** *pref. var. of* HOMO-.

**hom·age** (hŏm'ĭj, ŏm'-) *n.* [ME < OFr. < Med. Lat. *hominaticum* < Lat. *homo,* man.] **1.** Special honor or respect shown or expressed publicly <paid *homage* to the famous composer> **2.** Ceremonial acknowledgment by a vassal of allegiance to his lord.

**hom·bre**[1] (ŏm'brā', -brē) *n.* [Sp. < Lat. *homo.*] *Slang.* A fellow.

**hom·bre**[2] (hŏm'bər) *n. var. of* OMBRE.

**Hom·burg** *also* **hom·burg** (hŏm'bûrg') *n.* [After *Homburg,* West Germany.] A man's felt hat with a soft, dented crown and a shallow, slightly rolled brim.

**home** (hōm) *n.* [ME < OE *hām.*] **1.** A place where one lives : RESIDENCE. **2.** The physical structure within which one lives. **3.** A dwelling place together with the family or social unit that occupies it : HOUSEHOLD. **4. a.** An environment affording security and happiness. **b.** A valued place considered to be a refuge or place of origin. **5.** The place, as a country or town, where one was born or resided for a long period. **6.** Native habitat, as of a plant or animal. **7. a.** The place where something is discovered, originated, developed, or promoted : SOURCE <Several states claim to be the *home* of the potato.> **b.** Headquarters <the *home* of the opera company> **8.** *Baseball.* Home plate. **9.** HOME BASE 3. **10.** An institution where people are cared for <a convalescent *home*> —*adj.* **1.** Of or relating to a home, esp. to one's house or household <*home* decorating><*home* repairs> **2.** Of, relating to, or being a place of origin or headquarters <the *home* team> **3.** Taking place in the city where a team is franchised <a *home* ball game> —*adv.* **1.** At, to, or toward the direction of home. **2. a.** To the point at which something is directed <The bullet hit *home.*> **b.** To the center or heart of something : DEEPLY <Their insults struck *home.*> —*v.* **homed, hom·ing, homes.** —*vi.* **1.** To go or return home. **2.** To be guided to a target automatically, as by means of electronic signals. **3.** To move or lead toward a goal <The police were *homing* in on the gang.> —*vt.* To guide (a missile or aircraft) to a target automatically. **—at home. 1.** Available to receive visitors. **2.** Comfortable and relaxed. **3.** Feeling or exhibiting an easy competence and familiarity <*at home* in Russian> **—home free.** Free of tension or stress, usu. after expending considerable effort <signed the contract and was *home free*>

☆ **syns:** HOME, ABODE, DIGS, DWELLING, HABITATION, HOUSE, LODGINGS, PAD, PLACE *n. core meaning* : a building or shelter where one lives <a modest *home* on the edge of town>

**home base** *n.* **1.** *Baseball.* Home plate. **2.** A base of operations : HEADQUARTERS. **3.** An objective toward which players of a game, as baseball or backgammon, progress.

**home·bod·y** (hōm'bŏd'ē) *n. pl.* **-ies.** One whose interests and concerns center on the home.

**home·bound**[1] (hōm'bound') *adj.* [HOME + BOUND[4].] Heading toward home.

**home·bound**[2] (hōm'bound') *adj.* [HOME + BOUND[3].] Restricted or confined to home <*homebound* with the flu>

**home·bred** (hōm'brĕd') *adj.* Bred or reared at home.

**home-brew** (hōm'bro͞o') *n.* An alcoholic beverage, esp. beer, made at home. **—home'-brewed'** *adj.*

**home·com·ing** (hōm'kŭm'ĭng) *n.* **1.** An arrival or return home. **2.** An annual event for visiting alumni at colleges and universities.

**home economics** *n. (sing. or pl. in number).* The science and art of home management. **—home economist** *n.*

**home front** *n.* The civilian population of a nation or state at war.

**home·grown** (hōm'grōn') *adj.* Made or grown at home.

**home·land** (hōm'lănd') *n.* One's native land.

**home·less** (hōm'lĭs) *adj.* Having no home or refuge.

**home·like** (hōm'līk') *adj.* Like a home, esp. in coziness or wholesome simplicity.

**home·ly** (hōm'lē) *adj.* **-li·er, -li·est. 1.** Typical of the home or of domestic life. **2.** Of a simple or unpretentious nature : PLAIN <*homely* verities> **3.** Lacking elegance or refinement. **4.** Not attractive or good-looking. **—home'li·ness** *n.*

**home·made** (hōm'mād') *adj.* **1.** Made or prepared in the home. **2.** Made by oneself. **3.** Simply or crudely fashioned.

**home·mak·er** (hōm'mā'kər) *n.* One who manages a household. **—home'mak'ing** *n.* & *adj.*

**homeo-** *or* **homoio-** *pref.* [Lat. *homoeo-* < Gk. *homoio-* < *homoios,* similar < *homos,* same.] Like : similar <*homeostasis*>

**ho·me·o·mor·phism** (hō'mē-ə-môr'fĭz'əm) *n.* [HOMEO- + Gk. *morphē,* form + -ISM.] **1.** *Chem.* A close similarity in the crystal forms of dissimilar chemical compounds. **2.** *Math.* A one-to-one correspondence between the points of two geometric figures that is continuous in both directions. **—ho'me·o·mor'phous** *adj.*

**ho·me·op·a·thy** (hō'mē-ŏp'ə-thē) *n.* [G. *Homöopathie* : *homöo-,* homeo- + *-pathie,* -pathy.] A system of medical treatment based on the use of small quantities of remedies that in massive doses produce effects similar to those of the disease being treated. **—ho'me·o·path'** (-ə-păth'), **ho'me·op'a·thist** *n.* **—ho'me·o·path'ic** *adj.* **—ho'me·o·path'i·cal·ly** *adv.*

**ho·me·o·sta·sis** (hō'mē-ō-stā'sĭs) *n.* A state of physiological equilibrium produced by a balance of functions and chemical composition within an organism. **—ho'me·o·stat'ic** (-stăt'ĭk) *adj.*

**ho·me·o·therm** (hō'mē-ə-thûrm') *n. var. of* HOMOIOTHERM.

**ho·me·o·ther·mous** (hō'mē-ə-thûr'məs) *adj.* Homoiothermic.

**home plate** *n. Baseball.* A hard rubber slab at one of the corners of a diamond at which a batter stands when hitting and which a base runner must touch last in order to score.

**hom·er**[1] (hō'mər) *n.* **1.** *Baseball.* A home run. **2.** A homing pigeon.

**ho·mer**[2] (hō'mər) *n.* [Heb. *ḥomer.*] An ancient Hebrew measure of capacity equal to 10 ephahs, or about 10 or 11 bushels, in dry measure or 10 baths, or about 100 gallons, in liquid measure.

**Ho·mer·ic** (hō-mĕr'ĭk) *adj.* **1.** Of, relating to, or typical of the poet Homer, his works, or the legends and period of which he wrote. **2.** Of heroic proportions. **—Ho·mer'i·cal·ly** *adv.*

**home·room** (hōm'ro͞om', -ro͝om') *n.* A classroom where a group of pupils must report before morning and sometimes afternoon classes.

**home rule** *n.* The principle or practice of self-government in the internal affairs of a dependent political unit.

**home run** *n. Baseball.* A hit that allows the batter to make a complete circuit of the diamond and score a run.

**home screen** *n.* Television.

**home·sick** (hōm'sĭk') *adj.* Acutely desiring one's family and home. **—home'sick'ness** *n.*

**home·spun** (hōm'spŭn') *adj.* **1.** Spun, woven, or made in the home. **2.** Made of a homespun fabric. **3.** Simple and homely : UNPRETENTIOUS <*homespun* wisdom> —*n.* **1.** A plain coarse woolen cloth of homespun yarn. **2.** A sturdy fabric like homespun made on a power loom.

**home·stead** (hōm'stĕd') *n.* **1.** A house, esp. a farmhouse, with adjoining buildings and land. **2.** *Law.* Property designated by a householder as his or her home and protected by law from forced sale to meet debts. **3.** Land claimed by a settler or a squatter, esp. under the Homestead Act. **4.** The place where one's home is located. —*v.* **-stead·ed, -stead·ing, -steads.** —*vi.* To settle and work land, esp. under the Homestead Act. —*vt.* To claim and settle (land) as a homestead. **—home'stead'er** *n.*

**Homestead Act** *n.* An act passed by Congress in 1862 promising ownership of a 160-acre tract of public land to the head of a family after the land had been cleared and improved and lived on for five years.

**homestead law** *n.* Any of several laws passed in most states exempting a householder's homestead from attachment or forced sale to meet general debts.

**home·stretch** (hōm'strĕch') *n.* **1.** The part of a racetrack from the last turn to the finish line. **2.** The final stages of a venture.

**home·town** (hōm'toun') *n.* The town or city of one's birth or principal residence.

**home·ward** (hōm'wərd) *adj.* & *adv.* At or toward home. **—home'wards** (-wərdz) *adv.*

**home·work** (hōm'wûrk') *n.* **1.** Work, as schoolwork or piecework, done at home. **2.** Preparatory or preliminary work.

**hom·ey** *also* **hom·y** (hō'mē) *adj.* **-i·er, -i·est.** *Informal.* Having a homelike atmosphere. **—hom'ey·ness** *n.*

**hom·i·cid·al** (hŏm'ĭ-sīd'l, hō'mĭ-) *adj.* **1.** Of or relating to homicide. **2.** Inclined to homicide. **—hom'i·cid'al·ly** *adv.*

**hom·i·cide** (hŏm'ĭ-sīd', hō'mĭ-) *n.* [ME < OFr. < Lat. *homicidum* < *homicida,* murderer : *homo,* man + *caedere,* to kill.] **1.** The killing of one person by another. **2.** One who kills another person.

**hom·i·let·ic** (hŏm'ə-lĕt'ĭk) *also* **hom·i·let·i·cal** (-ĭ-kəl) *adj.* **1.** Relating to or like a homily. **2.** Relating to homiletics. **—hom'i·let'i·cal·ly** *adv.*

**hom·i·let·ics** (hŏm'ə-lĕt'ĭks) *n.* [Gk. *homilētikē,* art of conversation < *homilētikos,* of conversation < *homilētos,* conversation < *homilein,* to converse with < *homilos,* crowd.] *(sing. in number).* The art of preaching.

**hom·i·ly** (hŏm'ə-lē) *n., pl.* **-lies.** [ME *omelie* < OFr. < LLat. *homilia* < Gk., discourse < *homilos,* crowd.] **1.** A sermon, esp. one meant to edify a congregation on a practical matter. **2. a.** A tedious moralizing lecture or admonition. **b.** A platitude. **—hom'i·list** *n.*

**homing pigeon** *n.* A domestic pigeon trained to return to its home roost.
**hom·i·nid** (hŏm′ə-nĭd) *n.* [NLat. *Hominidae,* family name : < *Homo,* genus name < Lat. *homo,* man.] A primate of the family Hominidae, of which the modern human, *Homo sapiens,* is the only extant species. —*adj.* Of or belonging to the Hominidae.
**hom·i·noid** (hŏm′ə-noid′) *adj.* [NLat. *Hominoidea,* superfamily name < *Homo,* genus name < Lat. *homo,* man.] **1.** Of or belonging to the superfamily Hominoidea, which includes the apes and humans. **2.** Like a human being. —*n.* A member of the Hominoidea.
**hom·i·ny** (hŏm′ə-nē) *n.* [Perh. of Algonquian orig.] Hulled and dried kernels of corn, prepared as food by boiling.
**hominy grits** *pl.n.* Hominy ground into a coarse white meal : GRITS.
**hom·mos** (hŭm′əs, hŏm′-) *n. var. of* HUMMUS.
**ho·mo** (hō′mō) *n.* [NLat. *Homo,* genus name < Lat. *homo,* man.] A member of the genus *Homo,* which includes the extinct and extant species of human beings.
**homo-** *or* **hom-** *pref.* [Lat. < Gk. < *homos,* same.] Same : like <*homo*phone>
**ho·mo·cen·tric** (hō′mə-sĕn′trĭk, hŏm′ə-) *adj.* [NLat. *homocentricus* : Gk. *homos,* same + Gk. *kentron,* center.] Having the same center.
**ho·mo·cer·cal** (hō′mə-sûr′kəl, hŏm′ə-) *adj.* [HOMO- + Gk. *kerkos,* tail + -AL.] Relating to, designating, or marked by a tail fin having two symmetric lobes extending from the end of the vertebral column, as in most bony fishes.
**ho·mo·chro·mat·ic** (hō′mə-krō-măt′ĭk, hŏm′ə-) *adj.* Of or marked by one color : MONOCHROMATIC. **—ho′mo·chro′ma·tism** (-krō′mə-tĭz′əm) *n.*
**ho·mog·a·mous** (hō-mŏg′ə-məs) *adj. Bot.* **1.** Having flowers that are sexually alike in the same plant or inflorescence. **2.** Having stamens and pistils that mature simultaneously.
**ho·mo·ge·ne·i·ty** (hō′mə-jə-nē′ĭ-tē, -nā′-, hŏm′ə-) *n.* The quality or state of being homogeneous.
**ho·mo·ge·ne·ous** (hō′mə-jē′nē-əs, -jēn′yəs) *adj.* [Med. Lat. *homogeneus* < Gk. *homogenēs* : *homos,* same + *genos,* kind.] **1.** Of the same or similar nature or kind. **2.** Uniform throughout in structure or make-up. **3.** *Math.* Having terms of the same degree or elements of the same dimension. **—ho′mo·ge′ne·ous·ly** *adv.* **—ho′mo·ge′ne·ous·ness** *n.*
**ho·mog·e·nize** (hō-mŏj′ə-nīz′, hə-) *vt.* **-nized, -niz·ing, -niz·es.** [< HOMOGENEOUS.] **1.** To make homogeneous. **2. a.** To reduce to particles and disperse throughout a fluid. **b.** To make uniform in consistency, esp. to render (milk) homogeneous by emulsifying the fat content. **—ho·mog′e·ni·za′tion** *n.* **—ho·mog′e·niz′er** *n.*
**ho·mog·e·nous** (hō-mŏj′ə-nəs, hə-) *adj.* **1.** *Biol.* Of or displaying homogeny. **2.** Homogeneous.
**ho·mog·e·ny** (hō-mŏj′ə-nē, hə-) *n.* [Gk. *homogenia,* homogeneity < *homogenēs,* homogeneous.] *Biol.* Correspondence between parts or organs, possibly of dissimilar function, related by common descent.
**ho·mo·graft** (hō′mə-grăft′, hŏm′ə-) *n.* A graft of tissue obtained from a member of the same species as the individual receiving it.
**hom·o·graph** (hŏm′ə-grăf′, hō′mə-) *n.* One of two or more words having the same spelling but differing in origin, meaning, and sometimes pronunciation. **—hom′o·graph′ic** *adj.*
**homoio-** *pref. var. of* HOMEO-.
**ho·moi·o·therm** (hō-moi′ə-thûrm′) *also* **ho·me·o·therm** (hō′mē-) *n.* A homoiothermous organism, as a bird or mammal.
**ho·moi·o·ther·mic** (hō-moi′ə-thûr′mĭk) *or* **ho·moi·o·ther·mal** (-məl) *also* **ho·moi·o·ther·mous** (-məs) *adj.* Maintaining a relatively constant and warm body temperature regardless of environmental temperature : WARM-BLOODED.
**Ho·moi·ou·si·an** (hō′moi-ōō′zē-ən, -sē-) *n.* [< Gk. *homoiousios,* of similar substance : *homoios,* similar + *ousia,* substance.] A member of an Arian party in the 4th cent. who believed that Jesus the Son and God the Father were of similar but not of the same substance.
**ho·mol·o·gate** (hō-mŏl′ə-găt′, hə-) *vt.* **-gat·ed, -gat·ing, -gates.** [Med. Lat. *homolagare, homologat-* < Gk. *homologein,* to agree < *homologos,* agreeing. —see HOMOLOGOUS.] *Scot.* To approve, esp. to confirm officially.
**ho·mo·log·i·cal** (hō′mə-lŏj′ĭ-kəl, hŏm′ə-) *also* **ho·mo·log·ic** (-lŏj′ĭk) *adj.* Homologous. **—ho′mo·log′i·cal·ly** *adv.*
**ho·mol·o·gize** (hō-mŏl′ə-jīz′, hə-) *vt.* **-gized, -giz·ing, -giz·es. 1.** To make homologous. **2.** To show to be homologous.
**ho·mol·o·gous** (hō-mŏl′ə-gəs, hə-) *adj.* [Gk. *homologos,* agreeing : *homos,* same + *logos,* proportion.] **1.** Similar or corresponding in position, value, structure, or function. **2.** *Biol.* Corresponding in structure and evolutionary origin, as the flippers of a seal and the arms of a human being. **3.** *Genetics.* Having the same linear sequence of genes as another chromosome. **4.** *Chem.* Belonging to or being a series of organic compounds each successive member of which differs from the preceding member by a constant increment, esp. by an added $CH_2$ group.
**hom·o·lo·graph·ic** (hŏm′ə-lə-grăf′ĭk) *adj.* [Gk. *homalos,* even + GRAPHIC.] Maintaining the ratio of parts.
**homolographic projection** *n.* An equal-area projection reproducing the ratios of areas as they exist on the earth's surface.
**hom·o·logue** *also* **hom·o·log** (hŏm′ə-lôg′, -lŏg′, hō′mə-) *n.* A homologous part or organ.
**ho·mol·o·gy** (hō-mŏl′ə-jē, hə-) *n., pl.* **-gies.** [Gk. *homologia,* agreement < *homologos,* agreeing. —see HOMOLOGOUS.] **1.** The quality or state of being homologous. **2.** A homologous correspondence or relationship. **3.** *Math.* A topologic classification of configurations into distinct types that imposes an algebraic structure or hierarchy on sets of geometric figures.
**ho·mol·o·sine projection** (hō-mŏl′ə-sīn′) *n.* [Gk. *homalos,* even + -INE.] An equal-area map of the earth's surface laid out on the basis of sinusoidal curves, with the interruptions over ocean areas so the continents appear with minimal distortion.
**ho·mo·mor·phism** (hō′mə-môr′fĭz′əm, hŏm′ə-) *n.* The quality or state of being similar in external appearance, size, or form. **—ho′mo·mor′phic, ho′mo·mor′phous** *adj.*
**hom·o·nym** (hŏm′ə-nĭm′, hō′mə-) *n.* [Lat. *homonymum* < Gk. *homōnumon* < *homōnumos,* homonymous.] **1.** One of two or more words having the same sound and often the same spelling but differing in meaning. **2. a.** A word used to designate several different things. **b.** A namesake. **3.** *Biol.* One of two or more identical but conflicting taxonomic designations independently proposed for members of different categories. **—hom′o·nym′ic** *adj.*
**ho·mon·y·mous** (hō-mŏn′ə-məs) *adj.* [Lat. *homonymus* < Gk. *homōnumos* : *homos,* same + *onoma,* name.] **1.** Having the same name. **2.** Like a homonym : HOMONYMIC. **—ho·mon′y·mous·ly** *adv.* **—ho·mon′y·my** (-mē) *n.*
**Ho·mo·ou·si·an** (hō′mō-ōō′sē-ən, -zē-, hŏm′ō-) *n.* [LLat. *homousianus* < *homousius,* of same substance < Gk. *homoousios* : *homos,* same + *ousia,* substance.] A Christian supporting the Council of Nicaea's Trinitarian designation of Jesus the Son of God as consubstantial with God the Father.
**ho·mo·phile** (hō′mə-fīl′) *adj.* **1.** Homosexual. **2.** Being actively concerned with the welfare and rights of homosexuals. **—ho′mo·phile′** *n.*
**ho·mo·pho·bi·a** (hō′mə-fō′bē-ə) *n.* Unreasonable fear of homosexuals or homosexuality. **—ho′mo·pho′bic** (-bĭk) *adj.*
**hom·o·phone** (hŏm′ə-fōn′, hō′mə-) *n.* One of two or more words having the same sound but differing in spelling, origin, and meaning. **—ho·moph′o·nous** (hō-mŏf′ə-nəs) *adj.*
**hom·o·phon·ic** (hŏm′ə-fŏn′ĭk, hō′mə-) *adj.* [Gk. *homophōnos* : *homos,* same + *phōnē,* sound.] **1.** Having the same sound as another word. **2.** *Mus.* Having or marked by a single melodic line with accompaniment. **—ho·moph′o·ny** (hō-mŏf′ə-nē) *n.*
**ho·mo·phy·ly** (hō′mə-fī′lē, hŏm′ə-, hō-mŏf′ə-lē) *n.* [HOMO- + PHYL(UM) + -Y.] Resemblance caused by common ancestry. **—ho′mo·phyl′ic** (-fĭl′ĭk) *adj.*
**ho·mo·plas·tic** (hō′mə-plăs′tĭk, hŏm′ə-) *adj.* **1.** Of, relating to, or displaying homoplasy. **2.** Of, pertaining to, or derived from a different individual of the same species <a *homoplastic* bone graft> **—ho′mo·plas′ti·cal·ly** *adv.*
**ho·mo·pla·sy** (hō′mə-plā′sē, -plăs′ē, hŏm′ə-) *n.* Superficial structural similarity arising from convergence or parallel evolution.
**ho·mop·ter·an** (hō-mŏp′tər-ən) *n.* [NLat. *Homoptera,* order name : Gk. *homos,* same + Gk. *pteron,* wing.] A homopterous insect. —*adj.* Homopterous.
**ho·mop·ter·ous** (hō-mŏp′tər-əs) *adj.* [< NLat. *Homoptera,* order name. —see HOMOPTERAN.] Of or belonging to the order Homoptera, which includes the aphids, cicadas, and scale insects.
**Ho·mo sa·pi·ens** (hō′mō sā′pē-ənz, -ĕnz′) *n.* [NLat. *Homo sapiens,* specific name : *Homo,* genus name (< Lat. *homo,* man) + Lat. *sapiens,* pr.part. of *sapere,* to be wise.] The modern human being, the only extant species of the genus *Homo.*
**ho·mo·sex·u·al** (hō′mō-sĕk′shōō-əl) *adj.* Relating to, typical of, or displaying homosexuality. —*n.* A homosexual person.
**ho·mo·sex·u·al·i·ty** (hō′mō-sĕk′shōō-ăl′ĭ-tē) *n.* **1.** Sexual desire for others of one's own sex. **2.** Sexual activity with another member of the same sex.
**ho·mo·spo·rous** (hō′mə-spôr′əs, -spōr′-, hŏm′ə-, hō-mŏs′pər-əs) *adj. Bot.* Producing spores of one kind only. **—ho′mo·spo′ry** *n.*
**ho·mo·tax·is** (hō′mō-tăk′sĭs, hŏm′ō-) *n.* Similarity of arrangement and fossils in noncontemporaneous or widely separated geologic deposits. **—ho′mo·tax′ic** (-tăk′sĭk), **ho′mo·tax′i·al** (-tăk′sē-əl) *adj.*
**ho·mo·thal·lic** (hō′mō-thăl′ĭk, hŏm′ō-) *adj. Bot.* Having male and female reproductive structures in the same thallus, as in some algae and fungi.
**ho·mo·zy·go·sis** (hō′mō-zī-gō′sĭs, hŏm′ō-) *n.* The union of genetically identical gametes, causing the formation of a homozygote. **—ho′mo·zy·got′ic** (-gŏt′ĭk) *adj.*
**ho·mo·zy·gote** (hō′mō-zī′gōt′, hŏm′ō-) *n.* A zygote derived from the union of genetically identical gametes.
**ho·mo·zy·gous** (hō′mō-zī′gəs, hŏm′ō-) *adj.* Having identical alleles at corresponding chromosomal loci. **—ho′mo·zy′gous·ly** *adv.*
**ho·mun·cu·lus** (hō-mŭng′kyə-ləs) *n., pl.* **-li** (-lī′) [Lat., dim. of *homo,* man.] A diminutive or dwarfish man.

**hom·y** (hō′mē) *adj. var. of* HOMEY.

**ho·nan** *also* **Ho·nan** (hō′nän′) *n.* An evenly colored pongee fabric made orig. from silk produced by the silkworms of Honan (now Henan) Province, China.

**hon·cho** (hŏn′chō) [J., squad leader : *han,* squad + *chō,* chief.] *Slang.* —*n., pl.* **-chos.** One who is in charge, as a leader or manager <"Some of the big-name *honchos* . . . featured in the glossy . . . magazines" —*New Yorker*> —*vt.* **-choed, -cho·ing, -chos.** To undertake the direction and management of <"*honchoing* preparations for the forthcoming . . . economic summit" —*Newsweek*>

**hone[1]** (hōn) *n.* [ME < OE *hān.*] **1.** A fine-grained whetstone for imparting a sharp edge to a cutting tool. **2.** A tool with a rotating abrasive tip for enlarging holes to precise dimensions. —*vt.* **honed, hon·ing, hones. 1.** To sharpen on a hone. **2.** To make perfect or more effective <*hone* one's writing style>

**hone[2]** (hōn) *vi.* **honed, hon·ing, hones.** [OFr. *hoigner* < *hon,* cry of discontent.] *Informal.* **1.** To whine or moan. **2.** To long for : YEARN.

**hon·est** (ŏn′ĭst) *adj.* [ME < OFr. *honeste* < Lat. *honestus* < *honos,* honor.] **1.** Characterized by or exhibiting truthfulness and integrity : INCORRUPTIBLE <earning an *honest* wage> **2.** Not deceptive or fraudulent : GENUINE <an *honest* mistake> **3.** Equitable : fair <earning an *honest* wage> **4. a.** Marked by integrity and truth <an *honest* newscast> **b.** Sincere : frank <an *honest* editorial> **5. a.** Of good repute : RESPECTABLE. **b.** Without affectation : PLAIN <a simple, *honest* soul> **6.** Virtuous : chaste. **—hon′est·ly** *adv.*

**honest broker** *n.* A neutral agent, as in mediation.

**hon·es·ty** (ŏn′ĭ-stē) *n., pl.* **-ties. 1.** The quality or condition of being honest : INTEGRITY. **2.** Sincerity : truthfulness. **3.** *Archaic.* Chastity. **4.** A Eurasian plant, *Lunaria annua,* grown for its fragrant purplish flowers and round, flat, papery, silver-white seed pods.

☆ ***syns:*** HONESTY, INCORRUPTIBILITY, INCORRUPTIBLENESS, INTEGRITY, RECTITUDE, UPRIGHTNESS *n. core meaning* : the quality of truthfulness and probity combined with overall moral excellence <*honesty* in all business dealings> ***ant:*** dishonesty

**hone·wort** (hōn′wûrt′, -wôrt′) *n.* [*Hone-* (of unknown orig.) + WORT.] A plant of the genus *Cryptotaenia,* esp. *C. canadensis* of eastern North America, with small whitish flower clusters.

**hon·ey** (hŭn′ē) *n., pl.* **-eys.** [ME *honi* < OE *hunig.*] **1. a.** A sweet yellowish or brownish viscid fluid produced by various bees from the nectar of flowers and used as food. **b.** A similar substance made by various other insects. **2.** Sweetness. **3.** *Informal.* Dear. —Used as a term of affection. **4.** *Informal.* Something notably fine <a *honey* of a boat> —*vt.* **-eyed** *or* **-ied, -ey·ing, -eys. 1.** To sweeten with or as if with honey. **2.** To persuade with sweet talk.

**honey bear** *n.* The kinkajou.

**hon·ey·bee** (hŭn′ē-bē′) *n.* Any of various social bees of the genus *Apis* that produce honey, esp. *A. mellifera,* widely domesticated as a source of honey and beeswax.

**hon·ey·comb** (hŭn′ē-kōm′) *n.* **1.** A structure of hexagonal, thin-walled cells fabricated from beeswax by honeybees to contain honey and eggs. **2.** Something similar to a honeycomb in structure or pattern. —*vt.* **-combed, -comb·ing, -combs. 1.** To fill with holes : RIDDLE <a target *honeycombed* with bullets> **2.** To form in or cover with a honeycomb pattern.

**hon·ey·creep·er** (hŭn′ē-krē′pər) *n.* **1.** Any of various small, often brightly colored tropical American birds of the subfamily Dacninae, with a curved bill adapted for sucking nectar from flowers. **2.** Any of several Hawaiian birds of the family Drepanididae, similar to the honeycreepers.

**hon·ey·dew** (hŭn′ē-do͞o′, -dyo͞o′) *n.* **1.** A sweet, sticky substance excreted by various insects, esp. aphids, on the leaves of plants. **2.** A sweet exudate similar to honeydew on the leaves of plants. **3.** A honeydew melon.

**honeydew melon** *n.* A variety of melon, *Cucumis melo,* with a smooth whitish rind and green flesh.

**hon·ey·eat·er** (hŭn′ē-ē′tər) *n.* Any of various birds of the family Meliphagidae of Australia and adjacent regions, with a long extensible tongue adapted for sucking nectar from flowers.

**hon·eyed** *also* **hon·ied** (hŭn′ēd) *adj.* **1.** Having, full of, or sweetened with honey. **2.** Ingratiating : flattering <*honeyed* words of praise> **3.** Sweet : dulcet <*honeyed* tones>

**honey guide** *n.* Any of various tropical Old World birds of the family Indicatoridae, some species of which lead animals or people to the nests of wild honeybees, where they eat the wax remaining after the honey has been removed.

**honey locust** *n.* The mesquite.

**hon·ey·moon** (hŭn′ē-mo͞on′) *n.* [From the idea that the first month of marriage is sweet.] **1.** A holiday or trip taken by a newly married couple. **2.** The early harmonious period of a relationship. **—hon′ey·moon′** *v.* **(-mooned, -moon·ing, -moons). —hon′ey·moon′er** *n.*

**hon·ey·suck·le** (hŭn′ē-sŭk′əl) *n.* [ME *honisouke* < OE *hunisūce* : *hunig,* honey + *sūcan,* to suck.] **1.** A shrub or vine of the genus *Lonicera,* bearing tubular, often highly fragrant yellowish, white, or pink flowers. **2.** A plant related or similar to the honeysuckle.

**hong** (hông, hŏng) *n.* [Cantonese, a business establishment.] A factory, warehouse, or foreign trading house in China.

**hon·ied** (hŭn′ēd) *adj. var. of* HONEYED.

**honk** (hôngk, hŏngk) *n.* [Imit.] **1.** The raucous, resonant sound typical of a wild goose. **2.** A sound, as that of a horn, similar to a goose's honk. —*v.* **honked, honk·ing, honks.** —*vi.* & *vt.* To produce or cause to produce a honk. **—honk′er** *n.*

**hon·ky-tonk** (hông′kē-tôngk′, hŏng′kē-tŏngk′) *n.* [Orig. unknown.] *Slang.* A cheap, noisy bar or dance hall. —*adj. Mus.* Of or denoting a type of ragtime usu. played on a tinny old piano. —*vi.* **-tonked, -tonk·ing, -tonks.** *Slang.* To make the rounds of honky-tonks.

**hon·or** (ŏn′ər) *n.* [ME < OFr. < Lat.] **1.** Special esteem or respect : REVERENCE <show *honor* to the clergy> **2. a.** Good name : REPUTATION. **b.** A source or cause of credit <You're an *honor* to your community.> **3. a.** Recognition or distinction. **b.** A mark, sign, or gesture of respect or distinction <given the seat of *honor*> **c.** A military decoration. **d.** A title, as a knighthood, conferred for achievement. **4.** Nobility of mind : PROBITY. **5.** High rank. **6.** The dignity accorded to rank or position <the *honor* of the President's office> **7.** Great privilege <I have the *honor* to present the senator.> **8. Honor.** A title of address often accorded to mayors and judges <may it please your *Honor*> **9. a.** A code of chiefly male dignity, integrity, and pride, maintained in some societies, as in feudal Europe, by force of arms. **b.** Personal integrity maintained without legal or other obligation. **c.** A woman's chastity : PURITY. **10. honors.** Social courtesies offered to guests <Will you do the *honors* at tea?> **11. honors. a.** Special recognition for high academic achievement. **b.** A program of individual advanced study for exceptional students. **12.** The right of being first at the tee in golf. **13.** *often* **honors.** The four or five highest cards in trump or in all suits, esp. in bridge. —*vt.* **-ored, -or·ing, -ors. 1. a.** To hold in respect : ESTEEM. **b.** To display respect for. **2.** To confer distinction on <*honored* us with a gala feast> **3.** To accept or pay (e.g., a check) as valid. **—hon′or·er** *n.*

☆ ***syns:*** **1.** HONOR, DEFERENCE, HOMAGE, OBEISANCE *n. core meaning* : great respect accorded as a right or due <received the *honor* appropriate to their rank> ***ant:*** dishonor **2.** HONOR, DIGNITY, PRESTIGE, REPUTATION, REPUTE, STATUS *n. core meaning* : a person's high standing among others <went to court to defend their *honor*>

**hon·or·a·ble** (ŏn′ər-ə-bəl) *adj.* **1.** Worthy of or winning honor and respect. **2.** Bringing recognition or distinction <*honorable* diplomatic service> **3.** Having and marked by honor <an *honorable* individual> **4.** Consistent with honor or good name <the only *honorable* thing to do> **5.** Distinguished : illustrious <an *honorable* assembly> **6.** Accompanied by marks of honor and recognition <an *honorable* military discharge> **7. Honorable. a.** —Used as a title of respect for certain high officials. **b.** *Chiefly Brit.* —Used as a courtesy title of the children of barons and viscounts and the younger sons of earls. **—hon′or·a·ble·ness** *n.* **—hon′or·a·bly** *adv.*

**honorable mention** *n.* A citation to one who has performed well in a competition or exhibition but has not been awarded a prize.

**hon·o·rar·i·um** (ŏn′ə-râr′ē-əm) *n., pl.* **-i·ums** *or* **-i·a** (-ē-ə) [Lat. < *honorarius,* honorary.] A payment given to someone, as a consultant, for services for which fees are not legally required.

**hon·or·ar·y** (ŏn′ə-rĕr′ē) *adj.* [Lat. *honorarius* < *honor,* honor.] **1.** Held or awarded as a mark of honor, esp. conferred as an honor without the usual adjuncts <an *honorary* academic degree> **2. a.** Holding an unpaid office or title given as an honor. **b.** Voluntary <an *honorary* chairperson> **3.** Not legally enforceable, as a duty or obligation.

**hon·or·ee** (ŏn′ə-rē′) *n.* One who receives an honor.

**hon·or·if·ic** (ŏn′ə-rĭf′ĭk) *adj.* [Lat. *honorificus* : *honor,* honor + *-ficus,* -fic.] Conferring or showing honor or respect. —*n.* A title, phrase, or grammatical form conveying respect, used esp. when addressing a social superior. **—hon′or·if′i·cal·ly** *adv.*

**honor roll** *n.* A register of names of persons deserving honor, as for academic achievement or military service.

**honors of war** *pl.n.* Courtesies granted a surrendering foe, as the privilege of marching out bearing arms and colors.

**hon·our** (ŏn′ər) *n.* & *v. Chiefly Brit. var. of* HONOR.

**hooch[1]** *also* **hootch** (ho͞och) *n.* [Short for *hoochino,* after *Hoochino,* an Alaskan tribe that made a kind of distilled liquor.] *Slang.* **1.** Alcoholic liquor, esp. bootleg liquor. **2.** Marijuana.

**hooch[2]** (ho͞och) *n.* [Alteration of J. *uchi,* house.] *Slang.* A small dwelling, esp. a thatched hut.

**hood[1]** (ho͝od) *n.* [ME *hod* < OE *hōd.*] **1.** A loose pliable covering for the head and neck, either separate or attached to a garment, as a jacket or robe. **2.** An ornamental draping of cloth hung from the shoulders of an ecclesiastical or academic robe. **3.** A sack for covering a falcon's head to keep it quiet. **4.** Something shaped or functioning like a hood, as: **a.** A metal cover or cowl for a hearth or stove. **b.** A carriage top. **c.** The hinged metal lid covering an automobile engine. **d.** An expanded part, crest, or marking on or near the head of an animal. —*vt.* **hood·ed, hood·ing, hoods.** To provide or cover with a hood.

**hood[2]** (ho͝od) *n.* [Short for HOODLUM.] *Slang.* A hoodlum.

**-hood** *suff.* [ME *-hode* < OE *-hād.*] **1. a.** Quality or state <man*hood*> **b.** An instance of a given quality or state <false*hood*> **2.** A group sharing a given quality or state <sister*hood*>

**hood·ed** (hŏŏd'ĭd) *adj.* **1.** Having or covered with a hood. **2.** Shaped like a hood or cowl. **3.** *Zool.* Having a crest, coloration, or skin formation suggesting a hood.

**hooded seal** *n.* A seal, *Cystophora cristata* of northern seas, with a spotted grayish coat and an inflatable hoodlike or bladderlike pouch near the nose.

**hood·lum** (hŏŏd'ləm, hōōd'-) *n.* [Orig. unknown.] **1.** A thug : gangster. **2.** A tough, destructive youth. **—hood'lum·ism** *n.*

**hoo·doo** (hōō'dōō) *n., pl.* **-doos.** [Of African orig.] **1.** Voodoo. **2. a.** Bad luck. **b.** One thought to bring bad luck. —*vt.* **-dooed, -doo·ing, -doos.** To bring bad luck to. **—hoo'doo·ism** *n.*

**hood·wink** (hŏŏd'wĭngk') *vt.* **-winked, -wink·ing, -winks.** **1.** To trick or deceive. **2.** *Archaic.* To blindfold. **3.** *Obs.* To conceal. **—hood'wink'er** *n.*

**hoo·ey** (hōō'ē) *n.* [Orig. unknown.] *Slang.* Nonsense.

**hoof** (hŏŏf, hōōf) *n., pl.* **hoofs** *or* **hooves** (hōōvz, hŏŏvz) [ME *hof* < OE *hōf.*] **1. a.** The horny sheath covering the toes or lower part of the foot of a mammal of the orders Perissodactyla and Artiodactyla, as a horse, ox, or deer. **b.** The foot of such an animal, esp. a horse. **2.** *Slang.* The human foot. —*v.* **hoofed, hoof·ing, hoofs.** —*vt.* **1.** To trample with the hoofs. **2.** *Informal.* To walk. —*vi. Slang.* **1.** To dance. **2.** To go on foot : WALK.

**hoof-and-mouth disease** (hŏŏf'ən-mouth', hōōf'-) *n.* Foot-and-mouth disease.

**hoof·bound** (hŏŏf'bound', hōōf'-) *adj.* Afflicted with lameness as a result of drying and contraction of the hoof. —Used of a horse.

**hoofed** (hŏŏft, hōōft) *adj.* Having hoofs : UNGULATE.

**hoof·er** (hŏŏf'ər, hōō'fər) *n. Slang.* A professional dancer.

**hook** (hŏŏk) *n.* [ME *hok* < OE *hōc.*] **1.** A curved or sharply bent, usu. metal device used to catch, drag, suspend, attach, or close something. **2.** A fishhook. **3.** A catch : snag. **4.** Something shaped like a hook, esp.: **a.** A barbed or curved plant or animal part. **b.** A short curved or angled line on a letter. **c.** The lip of a breaking wave. **d.** A sickle. **5.** *Baseball.* A curve ball. **6.** A short swinging blow in boxing delivered with a crooked arm. **7.** A golf stroke sending the ball to the left of a right-handed player or to the right of a left-handed player. —*v.* **hooked, hook·ing, hooks.** —*vt.* **1. a.** To connect or catch with or as if with a hook. **b.** *Informal.* To snare. **c.** *Informal.* To please and make a fan of <was *hooked* on tennis> **d.** *Slang.* To cause to become addicted. **e.** *Slang.* To steal. **2.** To fasten by means of a hook. **3.** To gore or pierce as if with a hook. **4.** To make (a rug) by looping yarn through canvas with a type of crochet hook. **5.** *Baseball.* To pitch (a ball) with a curve. **6.** To hit with a hook in boxing. **7.** To hit (a golf ball) in a hook. —*vi.* **1.** To bend like a hook. **2.** To fasten by means of a hook or a hook and eye. **—by hook or (by) crook.** By whatever means possible. **—hook, line, and sinker.** *Slang.* Completely : entirely. **—hook up. 1.** To assemble or wire (a mechanism). **2.** To connect a mechanism to a source of power. **3.** *Slang.* To form a tie or connection <*hooked up* with a rough crowd> **—off the hook. 1.** *Slang.* Freed, as from blame or an obligation. **2.** Left off the cradle, as a telephone receiver. **—on (one's) own hook.** *Informal.* By one's own efforts.

**hook·ah** (hŏŏk'ə) *n.* [Urdu < Ar. *ḥuqqah,* the hookah's water urn.] A smoking pipe having a long tube passing through an urn of water that cools the smoke as it is drawn through.

**hook and eye** *n.* A clothes fastener having a small blunt metal hook with a corresponding loop.

**hook-and-lad·der truck** (hŏŏk'ən-lăd'ər) *n.* A fire engine equipped with extension ladders and hooked poles.

**hook·er**[1] (hŏŏk'ər) *n.* [Du., alteration of MDu. *hoeckboot* : *hoec,* fishhook + *boat,* boat.] **1.** A single-masted fishing smack used off the coast of Ireland. **2.** An old dilapidated or clumsy boat.

**hook·er**[2] (hŏŏk'ər) *n.* **1.** One that hooks. **2.** *Slang.* A prostitute.

**hook·nose** (hŏŏk'nōz') *n.* An aquiline nose. **—hook'nosed'** *adj.*

**hook shot** *n. Basketball.* A shot made by arcing the far hand upward while moving or being positioned sideways to the basket.

**hook·up** (hŏŏk'ŭp') *n.* **1.** A system of electric circuits and electrically powered equipment devised to operate together. **2. a.** A configuration of mechanical parts or devices acting as an integrated unit. **b.** A plan or schema of such a system or a configuration. **3.** *Informal.* A connection, often between unlikely factors or associates.

**hook·worm** (hŏŏk'wûrm') *n.* Any of numerous small, parasitic nematode worms of the family Ancylostomatidae, bearing hooked mouth parts with which they fasten themselves to the intestinal walls of various hosts, including humans, causing ancylostomiasis.

**hookworm disease** *n.* Ancylostomiasis.

**hook·y** (hŏŏk'ē) *n.* [Orig. unknown.] *Informal.* Truancy. —Used chiefly in the phrase *play hooky.*

**hoo·li·gan** (hōō'lĭ-gən) *n.* [Orig. unknown.] *Informal.* A young ruffian : HOODLUM. **—hoo'li·gan·ism** *n.*

**hoop** (hōōp, hŏŏp) *n.* [ME *hop.*] **1.** A circular band of metal or wood put around a cask or barrel to bind the staves together. **2.** A large wooden, plastic, or metal ring used as a plaything. **3.** One of the lightweight circular supports for a hoop skirt. **4.** A circular ringlike earring. **5.** One of a pair of circular wooden or metal frames used to hold material taut for needlework, as embroidery. **6.** *Basketball. Informal.* **a.** The basket. **b.** The game of basketball. **7.** A croquet wicket. —*vt.* **hooped, hoop·ing, hoops. 1.** To fasten or support with or as if with a hoop. **2.** To encircle.

**hoop·er** (hōō'pər, hŏŏp'ər) *n.* A cooper.

**hoop·la** (hōōp'lä', hŏŏp'-) *n.* [Fr. *houp-là,* oops!] *Slang.* **1.** Boisterous jovial commotion or excitement. **2.** Misleading or confusing talk.

**hoo·poe** (hōō'pōō, -pō) *n.* [Alteration of obs. *hoop* < OFr. *huppe* < Lat. *upupa.*] An Old World bird, *Upupa epops,* with a downward-curving bill, fanlike crest, and distinctive plumage.

**hoop skirt** *n.* A long full skirt belled out with a series of connected hoops.

**hoop snake** *n.* Any of several snakes, as the mud snake, that allegedly grasp the tail in the mouth and move with a rolling motion.

**hoo·ray** (hōō-rā') *interj. & n. & v. var. of* HURRAH.

**hoose·gow** (hōōs'gou') *n.* [Sp. *juzgado,* courtroom < p.part. of *juzgar,* to judge < Lat. *judicare* < *judex,* judge.] *Slang.* A jail.

**Hoo·sier** (hōō'zhər) *n.* [Orig. unknown.] A nickname for a native or resident of Indiana.

**hoot**[1] (hōōt) *v.* **hoot·ed, hoot·ing, hoots.** [ME *houten.*] —*vi.* **1.** To utter the typical cry of an owl. **2.** To make a loud raucous cry, esp. of contempt or derision. —*vt.* **1.** To force to leave by means of jeering cries <*hoot* the actor off the stage> **2.** To express or convey by hooting <*hooted* their disapproval> —*n.* **1. a.** The typical cry of an owl. **b.** A sound like an owl's cry, esp. the sound of an automobile horn. **2.** A jeer or taunt. **3.** *Chiefly Brit.* Something hilariously funny. **—not give (or care) a hoot.** To be completely indifferent <didn't *give a hoot* about the expense> **—hoot'er** *n.*

**hoot**[2] (hōōt, ōōt) *also* **hoots** (hōōts, ōōts) *interj. Chiefly Scot.* —Used to express objection or annoyance.

**hootch** (hōōch) *n. var. of* HOOCH[1].

**hoot·en·an·ny** (hōōt'n-ăn'ē) *n., pl.* **-nies.** [Orig. unknown.] **1.** An informal performance by folk singers. **2.** *Informal.* An unidentified or unidentifiable gadget.

**hoot owl** *n.* Any of various owls emitting a hooting cry.

**hoots** (hōōts, ōōts) *interj. var. of* HOOT[2].

**hooves** (hōōvz, hŏŏvz) *n. var. pl. of* HOOF.

**hop**[1] (hŏp) *v.* **hopped, hop·ping, hops.** [ME *hoppen* < OE *hoppian.*] —*vi.* **1.** To move with light bounding skips or leaps. **2.** To jump on one foot. **3.** To make a quick trip, esp. by plane. —*vt.* **1.** To move over by hopping <*hop* a puddle> **2.** To jump aboard <*hop* a bus> —*n.* **1. a.** A light springy jump or leap, esp. on one foot. **b.** A rebound. **2.** *Informal.* A dance. **3.** A short distance. **4.** A short trip, esp. by air. **5.** A free ride : LIFT. **—hop, skip, and (a) jump.** A short distance.

**hop**[2] (hŏp) *n.* [ME *hoppe* < MDu.] **1.** A twining vine of the genus *Humulus,* esp. *H. lupulus,* bearing lobed leaves and green conelike flowers. **2. hops.** The dried ripe flowers of the hop plant, having a bitter aromatic oil and used in brewing beer. **3.** *Slang.* Opium. —*vt.* **hopped, hop·ping, hops.** To flavor with hops. **—hop up.** *Slang.* **1.** To increase the power or energy of <*hop up* a car engine> **2.** To stimulate with or as if with a narcotic.

**hop clover** *n.* A Eurasian clover, *Trifolium agrarium* or one of a similar closely related species, with small yellow flower heads that resemble hops when withered.

**hope** (hōp) *v.* **hoped, hop·ing, hopes.** [ME *hopen* < OE *hopian.*] —*vi.* **1.** To wish for something with expectation of its fulfillment. **2.** *Archaic.* To have confidence : TRUST. —*vt.* **1.** To look forward to with confidence or expectation <*hoped* the children would go to college> **2.** To expect and desire <We *hope* it's a girl.> —*n.* **1.** A wish or desire accompanied by confident expectation of its fulfillment. **2.** Something hoped for or desired. **3.** One that is a cause of or reason for hope. **4.** *Archaic.* Trust : confidence. **—hope against hope.** To hope with little reason or justification. **—hop'er** *n.*

**hope chest** *n.* A chest used by a young woman for clothing and household goods, as linens and silver, in preparing for marriage.

**hope·ful** (hōp'fəl) *adj.* **1.** Having or expressing hope. **2.** Inspiring hope : PROMISING. —*n.* One who aspires to success or displays promise of succeeding, esp. as a political candidate. **—hope'ful·ness** *n.*

**hope·ful·ly** (hōp'fə-lē) *adv.* **1.** In a hopeful manner. **2.** It is to be hoped.

**hope·less** (hōp'lĭs) *adj.* **1.** Having no hope : DESPAIRING. **2.** Offering no hope. **3.** Incurable. **4.** Having no possibility of solution : IMPOSSIBLE. **—hope'less·ly** *adv.* **—hope'less·ness** *n.*

☆ ***syns:*** HOPELESS, CURELESS, IMPOSSIBLE, INCURABLE, IRREMEDIABLE, IRREPARABLE *adj. core meaning* : offering no hope or expectation of improvement <a *hopeless* financial position> ***ant:*** hopeful

**hop·head** (hŏp'hĕd') *n. Slang.* A drug addict.

**hop hornbeam** *n.* Any of several eastern North American trees of the genus *Ostrya,* esp. *O. virginiana,* yielding fruit resembling hops.

**Ho·pi** (hō'pē) *n., pl.* **Hopi** *or* **-pis.** [Hopi *hópi,* peaceful.] **1. a.** A tribe of Indians now inhabiting a reservation in northeastern Arizona. **b.** A member of this tribe. **2.** The Uto-Aztecan language of the Hopi.

**hop·lite** (hŏp′līt′) *n.* [Gk. *hoplitēs* < *hoplon,* weapon.] A heavily armed foot soldier of ancient Greece. **—hop·lit′ic** (-lĭt′ĭk) *adj.*

**hop-o′-my-thumb** (hŏp′ə-mī-thŭm′) *n.* [Alteration of *hop on my thumb.*] A tiny person.

**hop·per** (hŏp′ər) *n.* **1.** One that hops. **2. a.** A funnel-shaped container in which materials such as grain or fuel are temporarily stored before delivery. **b.** Any of various other receptacles in which something is stored temporarily. **c.** A freight car with a door in the floor through which materials are unloaded.

**hop·sack·ing** (hŏp′săk′ĭng) *also* **hop·sack** (-săk′) *n.* [From its being used for bags by hop growers.] A loosely woven, coarse cotton or wool fabric used in clothing.

**hop·scotch** (hŏp′skŏch′) *n.* A children's game in which players toss a small object into the numbered spaces of a pattern of rectangles outlined on the ground and then hop or jump through the spaces to retrieve the object.

**ho·ra** *also* **ho·rah** (hôr′ə, hōr′ə) *n.* [Heb. *hōrāh* < Rum. *horă.*] A traditional Rumanian and Israeli round dance.

**ho·ra·ry** (hôr′ə-rē, hōr′-) *adj.* [Med. Lat. *horarius* < Lat. *hora,* hour.] **1.** Of an hour or hours. **2.** Occurring once an hour : HOURLY.

**Ho·ra·tian** (hə-rā′shən) *adj.* Of, pertaining to, or typical of the Latin poet Horace or his works.

**horde** (hôrd, hōrd) *n.* [OFr. < G. < Pol. *horda* < Turk. *ordū,* camp.] **1.** A large group or crowd : SWARM <*hordes* of tourists> **2. a.** A nomadic Mongol tribe. **b.** A nomadic tribe or group.

**hore·hound** (hôr′hound′, hōr′-) *n.* [ME < OE *hārehūne* : *hār,* hoary + *hūne,* a kind of plant.] **1. a.** An aromatic Eurasian plant, *Marrubium vulgare,* bearing leaves covered with whitish pubescence and yielding a bitter extract used as flavoring and a cough remedy. **b.** A preparation or candy flavored with this extract. **2.** Any of various plants similar to the horehound, as the black horehound.

**ho·ri·zon** (hə-rī′zən) *n.* [ME *orizon* < OFr. *orizonte* < LLat. *horizon* < Gk. *horizōn (kuklos),* limiting (circle) < *horizein,* to limit < *horos,* boundary.] **1.** The apparent intersection of the earth and sky as seen by an observer. **2.** *Astron.* **a.** The circular intersection of a plane tangent to the earth at the observer's station with the celestial sphere. **b.** The intersection with the celestial sphere of a plane through the center of the earth and perpendicular to the line connecting the zenith and the nadir. **c.** The great circle of the celestial sphere at the intersection of the sensible and rational horizons at infinity, its plane passing through the center of the earth. **3.** The range of a person's interest, knowledge, or experience. **4.** *Geol.* **a.** A specific position in a stratigraphic column, as the location of one or more fossils, that identifies the stratum with a particular period. **b.** A specific layer of soil in a cross section of land.

**hor·i·zon·tal** (hôr′ĭ-zŏn′tl, hŏr′-) *adj.* [< LLat. *horizon,* horizon.] **1. a.** Of, pertaining to, or close to the horizon. **b.** Parallel to or in the plane of the horizon. **2.** Occupying or restricted to the same level in a hierarchy <a *horizontal* study of reading ability> —*n.* Something horizontal, as a line, plane, or object. **—hor′i·zon′tal·ly** *adv.*

**horizontal union** *n.* A craft union.

**hor·mone** (hôr′mōn′) *n.* [Gk. *hormōn,* pr.part. of *horman,* to urge on < *hormē,* impulse.] A substance produced by one organ and conveyed, as by the blood stream, to another, which it stimulates to function by means of its chemical activity. **—hor·mon′al** (-mō′nəl), **hor·mon′ic** (-mŏn′ĭk) *adj.* **—hor·mon′al·ly** *adv.*

**horn** (hôrn) *n.* [ME < OE.] **1.** One of the hard, usu. permanent structures projecting from the head of certain mammals, as cattle, sheep, deer, or goats, consisting of a bony core covered with a sheath of keratinous material. **2.** A hard protuberance similar to a horn, as an antler or a projection on the head of a giraffe or rhinoceros. **3. a.** The hard, smooth keratinous material forming the outer covering of the horns of cattle or related animals. **b.** A natural or synthetic substance resembling this material. **4.** A container made from a horn <a drinking *horn*> **5.** Something shaped like a horn, esp.: **a.** A cornucopia. **b.** Either of the ends of a new moon. **c.** The point of an anvil. **d.** The pommel of a saddle. **e.** An ear trumpet. **f.** A device for projecting sound waves, as in a loudspeaker. **g.** A hollow, metallic electromagnetic transmission antenna with a rectangular cross section. **6.** *Mus.* **a.** A wind instrument made of an animal horn. **b.** A wind instrument made of brass. **c.** A French horn. **d.** *Informal.* A trumpet. **7.** A usu. electrical signaling device producing a sound similar to that of a horn. **8.** *Slang.* A telephone. —*v.* **horned, horn·ing, horns.** *Slang.* To join without being invited : INTRUDE <*horned* in on our tête-à-tête> —*adj.* Made of horn. **—blow (***or* **toot) (one's) own horn.** To brag or boast about oneself.

**horn·beam** (hôrn′bēm′) *n.* **1.** A tree of the genus *Carpinus,* with smooth bark and hard whitish wood. **2.** The wood of a hornbeam.

**horn·bill** (hôrn′bĭl′) *n.* Any of various tropical Old World birds of the family Bucerotidae, with a very large bill, often surmounted by an enlarged protuberance at the base.

**horn·blende** (hôrn′blĕnd′) *n.* [G. : *Horn,* horn + *Blende,* blende.] An amphibole mineral, $CaNa(Mg,Fe)_4(Al,Fe,Ti)_3Si_6O_{22}(O,OH)_2$, commonly green or bluish-green to black, formed in the late stages of cooling in igneous rock.

**horn·book** (hôrn′bo͝ok′) *n.* **1.** An early primer having a single page protected by a transparent sheet of horn, once used in teaching children to read. **2.** A rudimentary text.

**horned** (hôrnd) *adj.* Having a horn.

**horned pout** *n.* A hornpout.

**horned toad** *n.* A short-tailed lizard of the genus *Phrynosoma* of western North America and Central America, with hornlike projections on the head and a flattened spiny body.

**horned viper** *n.* A venomous African snake, *Cerastes cornutus,* with a hornlike projection above each eye.

**hor·net** (hôr′nĭt) *n.* [ME < OE *hyrnet.*] Any of various large stinging wasps, chiefly of the genera *Vespa* and *Vespula,* that build a large papery nest.

**hor·ni·to** (hôr-nē′tō) *n., pl.* **-tos.** [Sp., dim. of *horno,* oven < Lat. *furnus.*] A low mound of volcanic origin, sometimes emitting vapor or smoke.

**horn of plenty** *n.* A cornucopia.

**horn·pipe** (hôrn′pīp′) *n.* **1.** A musical instrument with a single reed, finger holes, and a bell and mouthpiece made of horn. **2. a.** A spirited British folk dance orig. accompanied by a hornpipe. **b.** The music accompanying a hornpipe.

**horn·pout** (hôrn′pout′) *n.* A freshwater catfish native to eastern North America, *Ictalurus nebulosus* or *Ameiurus nebulosus,* with a large head bearing barbels.

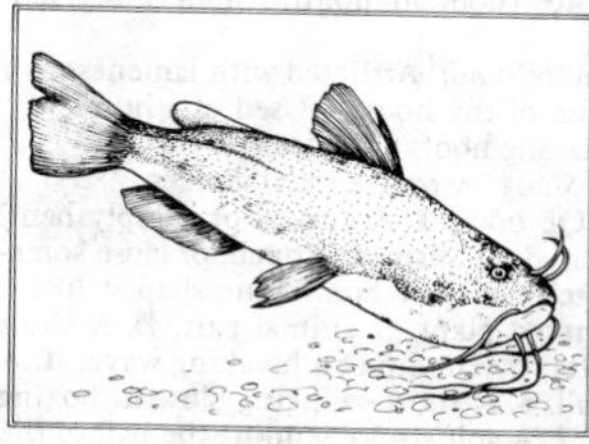

**hornpout**
*Approximately 10 inches long*

**†horn·swog·gle** (hôrn′swŏg′əl) *vt.* **-gled, -gling, -gles.** [Orig. unknown.] *Regional.* To bamboozle : deceive.

**horn·tail** (hôrn′tāl′) *n.* Any of various sawflies of the family Siricidae, the female of which has a long, stout ovipositor.

**horn·worm** (hôrn′wûrm′) *n.* The larva of the hawk moth, bearing a hornlike posterior segment.

**horn·wort** (hôrn′wûrt′, -wôrt′) *n.* An aquatic plant of the genus *Ceratophyllum,* forming submerged branching masses in slow-moving water.

**horn·y** (hôr′nē) *adj.* **-i·er, -i·est. 1.** Having horns or hornlike projections. **2.** Made of horn or a similar substance. **3.** Tough and calloused, as skin. **—horn′i·ness** *n.*

**hor·o·loge** (hôr′ə-lōj′, hŏr′-) *n.* [ME *orloge* < OFr. < Lat. *horologium* < Gk. *hōrologion* : *hōra,* hour + *legein,* to speak.] A timepiece, esp. an early or primitive one.

**Hor·o·lo·gi·um** (hôr′ə-lō′jē-əm, hŏr′-) *n.* [Lat. *horologium,* horologe.] A constellation in the Southern Hemisphere.

**ho·rol·o·gy** (hô-rŏl′ə-jē) *n.* [Gk. *hōra,* hour + -LOGY.] **1.** The science of measuring time. **2.** The art of making timekeeping instruments. **—hor′o·log′ic** (hôr′ə-lŏj′ĭk, hŏr′-), **hor′o·log′i·cal** (-ĭ-kəl) *adj.* **—ho·rol′o·gist, ho·rol′o·ger** *n.*

**hor·o·scope** (hôr′ə-skōp′, hŏr′-) *n.* [OFr. < Lat. *horoscopus* < Gk. *hōroskopos* : *hōra,* hour + *skopos,* observer.] **1. a.** The aspect of the planets and stars at a given moment, as the moment of a person's birth, used in astrology. **b.** A diagram of the signs of the zodiac based on such an aspect. **2.** A forecast based on a horoscope.

**hor·ren·dous** (hô-rĕn′dəs, hə-) *adj.* [Lat. *horrendus* < gerund. of *horrēre,* to tremble.] Dreadful : horrible. **—hor·ren′dous·ly** *adv.*

**hor·rent** (hôr′ənt, hŏr′-) *adj.* [Lat. *horrens, horrent-,* pr.part. of *horrēre,* to tremble.] *Archaic.* Covered with bristles : BRISTLING.

**hor·ri·ble** (hôr′ə-bəl, hŏr′-) *adj.* [ME < OFr. < Lat. *horribilis* < *horrēre,* to tremble.] **1.** Arousing or tending to arouse horror : DREADFUL <"War is beyond all words *horrible*" —Winston Churchill> **2.** Extremely unpleasant : DISAGREEABLE. **—hor′ri·ble·ness** *n.* **—hor′ri·bly** *adv.*

**hor·rid** (hôr′ĭd, hŏr′-) *adj.* [Lat. *horridus* < *horrēre,* to tremble.] **1.** Horrible. **2.** *Archaic.* Bristling : rough. **—hor′rid·ly** *adv.* **—hor′rid·ness** *n.*

**hor·rif·ic** (hô-rĭf′ĭk, hŏ-) *adj.* [OFr. *horrifique* < Lat. *horrificus* : *horrēre,* to tremble + *-ficus,* -fic.] HORRIBLE 1. **—hor·rif′i·cal·ly** *adv.*

**hor·ri·fy** (hôr′ə-fī′, hŏr′-) *vt.* **-fied, -fy·ing, -fies.** [Lat. *horrificare* < *horrificus,* horrific.] **1.** To cause to feel horror. **2.** To cause unpleasant surprise to. **—hor′ri·fi·ca′tion** *n.* **—hor′ri·fy′ing·ly** *adv.*

**hor·rip·i·la·tion** (hô-rĭp′ə-lā′shən, hŏ-) *n.* [LLat. *horripilatio* < Lat. *horripilare,* to bristle with hairs : *horrēre,* to tremble + *pilus,* hair.] The bristling of body hair, as from fear or cold : GOOSE FLESH. **—hor·rip′i·late′** *v.* **(-lat·ed, -lat·ing, -lates).**

**hor·ror** (hôr′ər, hŏr′-) *n.* [ME *horrour* < OFr. *horreur* < Lat. *horror* < *horrēre,* to tremble.] **1.** A strong and painful feeling of fear and repugnance. **2.** Intense dislike : ABHORRENCE <has a *horror* of snakes> **3.** One that causes horror. **4.** *Informal.* Something unpleasant, disagreeable, or ugly <That house is a *horror!*> **5.** *Slang.* Intense nervous depression or anxiety. —Usu. used with *the.*

**hors de com·bat** (ôr′ də kôN-bä′) *adj.* & *adv.* [Fr.] Out of action : INCAPACITATED.

**hors d'oeuvre** (ôr dûrv′) *n., pl.* **hors d'oeuvres** (ôr dûrvz′) *or* **hors d'oeuvre.** [Fr. : *hors,* outside + *de,* of + *oeuvre,* work.] An appetizer served usu. before a meal.

**horse** (hôrs) *n.* [ME < OE *hors.*] **1. a.** A large hoofed mammal, *Equus caballus,* with a short-haired coat, a long mane, and a long tail, domesticated since ancient times. **b.** An adult male horse : STALLION. **c.** Any of various equine mammals, as the wild Asian species, *E. przewalskii,* or certain extinct forms related ancestrally to the modern horse. **2.** Mounted soldiers : CAVALRY. **3.** A usu. four-legged frame or device used for holding or supporting. **4.** A piece of gymnastic equipment with an upholstered body used esp. for vaulting. **5.** *Slang.* Heroin. **6.** *often* **horses.** Horsepower. **7.** *Geol.* **a.** A block of rock interrupting a vein and containing no minerals. **b.** A large block of displaced rock caught along a fault. —*v.* **horsed, hors·ing, hors·es.** —*vt.* **1.** To supply with a horse. **2.** To haul or hoist energetically. —*vi.* To be in heat. —Used of mares. **—horse around.** To indulge in horseplay. —*adj.* **1.** Of or relating to a horse. **2.** Mounted on horses. **3.** Drawn or operated by a horse. **4.** Larger or cruder than other similar things. **—a horse of another (***or* **a different) color.** Another matter entirely. **—be (***or* **get) on (one's) high horse.** To be or become disdainful, supercilious, or conceited. **—hold (one's) horses.** To restrain oneself. **—the horse's mouth.** The original and unquestionably accurate source.

**horse·back** (hôrs′băk′) *n.* **1.** The back of a horse. **2.** A natural ridge : HOGBACK. —*adv.* On the back of a horse.

**horse balm** *n.* [From its former use for treating ailments of horses.] A plant, *Collinsonia canadensis* of eastern North America, with yellow lemon-scented flower clusters.

**horse bean** *n.* The broad bean.

**horse·car** (hôrs′kär′) *n.* **1.** A streetcar drawn by horses. **2.** A car for transporting horses.

**horse chestnut** *n.* **1.** A Eurasian tree of the genus *Aesculus,* esp. *A. hippocastanum,* with palmate leaves, erect clusters of white flowers tinged with red, and brown shiny nuts enclosed in a spiny bur. **2.** The nut of the horse chestnut.

**horse·flesh** (hôrs′flĕsh′) *n.* **1.** The flesh of a horse. **2.** Horses as a group, esp. for driving, riding, or racing.

**horse·fly** *also* **horse fly** (hôrs′flī′) *n.* Any of numerous large flies of the family Tabanidae, the females of which suck the blood of various mammals.

**horse gentian** *n.* A plant of the genus *Triosteum,* bearing small purplish-brown flowers and leathery orange-yellow fruit.

**horse·hair** (hôrs′hâr′) *n.* **1.** The hair of a horse, esp. from the mane or tail. **2.** Cloth made of horsehair.

**horsehair worm** *n.* Any of various slender aquatic worms of the phylum Nematomorpha, whose larvae are parasitic within insects.

**horse·hide** (hôrs′hīd′) *n.* **1. a.** The hide of a horse. **b.** Leather made from horsehide. **2.** *Informal.* A baseball.

**horse latitudes** *pl.n.* Either of two belts of latitudes located over the oceans at approx. 30° to 35° north and south, having high barometric pressure, calms, and light, variable winds.

**horse·laugh** (hôrs′lăf′, -läf′) *n.* A loud, raucous laugh.

**horse·leech** (hôrs′lēch′) *n.* A large leech of the genus *Haemopis.*

**horse·less carriage** (hôrs′lĭs) *n.* An automobile.

**horse mackerel** *n.* **1.** Any of various marine fishes of the genus *Trachurus.* **2.** Any of various tunas or related fishes.

**horse·man** (hôrs′mən) *n.* **1. a.** A man who rides a horse. **b.** A man skilled at horsemanship. **2.** A man who breeds and raises horses.

**horse·man·ship** (hôrs′mən-shĭp′) *n.* The art of equitation.

**horse marine** *n.* **1.** A marine assigned to the cavalry. **2.** A cavalryman assigned to a ship.

**horse·mint** (hôrs′mĭnt′) *n.* Any of several coarse aromatic plants, esp. *Monarda punctata* or *Mentha longifolia.*

**horse nettle** *n.* A prickly-stemmed plant, *Solanum carolinense* of eastern and central North America, bearing purplish or white star-shaped flowers and yellowish berries.

**horse opera** *n.* WESTERN 2.

**horse·play** (hôrs′plā′) *n.* Rowdy or unruly behavior.

**horse·pow·er** (hôrs′pou′ər) *n.* **1.** A unit of power in the U.S. Customary System, equal to 745.7 watts or 33,000 foot-pounds per minute. **2.** The power exerted by a horse in pulling.

**horse·rad·ish** (hôrs′răd′ĭsh) *n.* **1.** A coarse Eurasian plant, *Armoracia rusticana* or *A. lapathifolia,* with a thick, whitish, pungent root. **2.** The grated or shredded root of the horseradish, often used as a condiment.

**horse sense** *n. Informal.* Common sense.

**horse·shoe** (hôrs′shōō′, hôrsh′-) *n.* **1.** A narrow U-shaped iron plate fitted and nailed to a horse's hoof. **2.** Something shaped like a horseshoe. **3. horseshoes** *(sing. in number).* A game in which players try to toss horseshoes so that they encircle a stake. —*vt.* **-shoed, -shoe·ing, -shoes.** To fit with horseshoes.

**horseshoe crab** *n.* A marine arthropod of the class Merostomata, esp. *Limulus polyphemus* or *Xiphosura polyphemus* of eastern North America, with a large rounded body and a stiff pointed tail.

**horse·tail** (hôrs′tāl′) *n.* A nonflowering plant of the genus *Equisetum,* with a jointed hollow stem and narrow, sometimes much reduced leaves.

**horse trade** *n.* A transaction marked by vigorous and shrewd bargaining. **—horse trader** *n.*

**horse·weed** (hôrs′wēd′) *n.* A weedy North American plant, *Erigeron canadensis,* bearing narrow leaves and numerous small white or greenish flowers.

**horse·whip** (hôrs′hwĭp′, -wĭp′) *n.* A whip used to drive or control a horse. —*vt.* **-whipped, -whip·ping, -whips.** To beat with or as if with a horsewhip. **—horse′whip′per** *n.*

**horse·wom·an** (hôrs′wŏŏm′ən) *n.* **1. a.** A woman who rides a horse. **b.** A woman skilled at horsemanship. **2.** A woman who breeds and raises horses.

**hors·ey** (hôr′sē) *adj. var. of* HORSY.

**horst** (hôrst) *n.* [G. < MHG < OHG *hurst,* hill.] A massive, usu. elevated block of the earth's crust that lies between two faults.

**hors·y** *also* **hors·ey** (hôr′sē) *adj.* **-i·er, -i·est. 1.** Of, relating to, or typical of a horse. **2.** Devoted to horses and horsemanship. **3.** Large and clumsy. **—hors′i·ly** *adv.* **—hors′i·ness** *n.*

**hor·ta·tive** (hôr′tə-tĭv) *adj.* [LLat. *hortativus* < Lat. *hortari,* to exhort.] Hortatory. **—hor′ta·tive·ly** *adv.*

**hor·ta·to·ry** (hôr′tə-tôr′ē, -tōr′ē) *adj.* [LLat. *hortatorius* < *hortari,* to exhort.] Marked by or given to exhortation.

**hor·ti·cul·ture** (hôr′tĭ-kŭl′chər) *n.* [Lat. *hortus,* garden + (AGRI)CULTURE.] **1.** The science or art of cultivating fruits, vegetables, flowers, and plants. **2.** The cultivation of a garden. **—hor′ti·cul′tur·al** *adj.* **—hor′ti·cul′tur·al·ly** *adv.* **—hor′ti·cul′tur·ist** *n.*

**Ho·rus** (hôr′əs, hōr′-) *n.* [LLat. < Gk. *Hōros,* of Egypt. orig.] *Myth.* The ancient Egyptian sun god, represented as having the head of a hawk.

**ho·san·na** (hō-zăn′ə) *interj.* [ME *osanna* < Lat. < Gk. *hōsanna* < Heb. *hosha'nā,* short for *hosh' āhanna,* save (us).] —Used to express praise or veneration to God. —*n.* A cry of "hosanna."

**hose** (hōz) *n., pl.* **hose.** [ME, a stocking < OE.] **1. a.** Stockings. **b.** Socks. **2. a.** A man's garment that covers the legs and hips and fastens to a doublet by points. **b.** Short, knee-length breeches. **3.** *pl.* **hos·es.** A flexible tube for conveying liquids or gases under pressure. —*vt.* **hosed, hos·ing, hos·es.** To wash or water with a hose.

**Ho·se·a** (hō-zē′ə, -zā′ə) *n.* [Heb. *Hōshēa'.*] —See table at BIBLE.

**ho·sier** (hō′zhər) *n.* [ME < *hose,* a stocking < OE.] A manufacturer of or dealer in hosiery.

**ho·sier·y** (hō′zhə-rē) *n.* **1. a.** Stockings and socks : HOSE. **b.** *Chiefly Brit.* Stockings, socks, and underclothing. **2.** The business of a hosier.

**hos·pice** (hŏs′pĭs) *n.* [Fr. < OFr. < Lat. *hospitium,* hospitality < *hospes,* host.] **1.** A shelter or lodging, often maintained by monks, for travelers, children, or the needy. **2.** An establishment or program caring for the physical and emotional needs of terminally ill patients.

**hos·pi·ta·ble** (hŏs′pĭ-tə-bəl, hŏ-spĭt′ə-bəl) *adj.* [NLat. *hospitabilis* < Med. Lat. *hospitare,* to receive as a guest < Lat. *hospes,* host.] **1.** Cordial and generous to guests. **2.** Having an open mind : RECEPTIVE. **3.** Environmentally favorable to growth and development. **—hos′pi·ta·bly** *adv.*

**hos·pi·tal** (hŏs′pĭ-tl, -pĭt′l) *n.* [ME, hospice < OFr. *ospital* < Med. Lat. *hospitale* < Lat. *hospitalis,* of a guest < *hospes,* guest.] **1.** An institution providing medical or surgical care and treatment for the sick and injured. **2.** An often charitable home for old people, the infirm, or foundlings. **3.** A repair shop, as for dolls.

**Hos·pi·tal·er** *also* **Hos·pi·tal·ler** (hŏs′pĭt′l-ər) *n.* [ME *Hospiteler* < OFr. *hospitalier* < Med. Lat. *hospitale,* hospice. —see HOSPITAL.] A member of a military religious order founded among European Crusaders in 12th-cent. Palestine.

**hos·pi·tal·i·ty** (hŏs′pĭ-tăl′ĭ-tē) *n., pl.* **-ties.** [ME *hospitalite* < OFr. < Lat. *hospitalitas* < *hospitalis,* of a guest. —see HOSPITAL.] **1.** Cordial reception of guests. **2.** An instance of being hospitable.

**hos·pi·tal·ize** (hŏs′pĭt′l-īz′) *vt.* **-ized, -iz·ing, -iz·es.** To put into a hospital for care or treatment. **—hos′pi·tal·i·za′tion** *n.*

**Hos·pi·tal·ler** (hŏs′pĭt′l-ər) *n. var. of* HOSPITALER.

**host**[1] (hōst) *n.* [ME < OFr. < Lat. *hospes.*] **1. a.** One who entertains guests in a social or business capacity. **b.** One that furnishes facilities for a function or event. **c.** One who emcees a radio or television program. **2.** *Biol.* An organism that harbors and provides nourishment for a parasite. —*vt.* **host·ed, host·ing, hosts.** *Informal.* To serve as host for or at <"the garden party he had *hosted* last spring" —*Saturday Review*> **—host′ly** *adj.*

**host²** (hōst) *n.* [ME < OFr. < LLat. *hostis* < Lat., enemy.] **1.** An army. **2.** A great number : MULTITUDE <a *host* of angels>
**host³** *also* **Host** (hōst) *n.* [ME < Lat. *hostia*, sacrifice.] The consecrated bread or wafer of the Eucharist.
**hos·tage** (hŏs′tĭj) *n.* [ME < OFr. < *host*, guest, host. —see HOST¹.] **1.** One held as a pledge that certain terms will be fulfilled by the opposing party. **2.** The state of being held as a hostage.
**hos·tel** (hŏs′təl) *n.* [ME, lodging < OFr. < Med. Lat. *hospitale*, hospice. —see HOSPITAL.] **1.** A supervised, inexpensive lodging, esp. for youthful travelers. **2.** An inn. —*vi.* **-teled, -tel·ing, -tels.** To stay at hostels while traveling.
**hos·tel·er** (hŏs′tə-lər) *n.* **1.** *Archaic.* An innkeeper. **2.** A youthful traveler who stays at hostels.
**hos·tel·ry** (hŏs′təl-rē) *n., pl.* **-ries.** A lodging, as an inn or hotel.
**host·ess** (hō′stĭs) *n.* **1.** A woman who acts as host. **2.** A woman employed to greet and serve patrons, as in a restaurant or on an airplane.
**hos·tile** (hŏs′təl, -tīl′) *adj.* [OFr. < Lat. *hostilis* < *hostis*, enemy.] **1.** Of or relating to an enemy. **2.** Feeling or displaying enmity : ANTAGONISTIC. **3.** Not hospitable. —*n.* One that is hostile. **—hos′tile·ly** *adv.*
**hos·til·i·ty** (hŏ-stĭl′ĭ-tē) *n., pl.* **-ties. 1.** The state of being hostile : ANTAGONISM. **2. a.** A hostile act. **b. hostilities.** Open warfare.
**hos·tler** (hŏs′lər, ŏs′-) *n.* [ME < AN *hostiler* < OFr. *hostel*, lodging. —see HOSTEL.] **1.** One who takes charge of horses, as at an inn. **2.** One who services a large vehicle or engine, as a locomotive.
**hot** (hŏt) *adj.* **hot·ter, hot·test.** [ME < OE *hāt*.] **1. a.** Having or marked by great heat. **b.** Yielding much heat. **c.** Being at a high temperature. **2.** Higher in temperature than is normal or desirable : FEVERISH. **3.** Highly spiced <*hot* chili sauce> **4. a.** Charged or energized with electricity, as a wire. **b.** Radioactive, esp. to a dangerous degree. **5. a.** Characterized by intensity or warmth of emotion : FIERY. **b.** Having or displaying desire or enthusiasm. **c.** Arousing excited interest and attention <a *hot* debate> **6.** *Slang.* **a.** Obtained by stealing. **b.** Wanted by the police. **7.** Close to a successful solution or conclusion. **8.** *Informal.* **a.** Just made or released : FRESH <*hot* news> **b.** Extremely popular <a *hot* record album> **9.** *Slang.* Very good or impressive. **10.** *Slang.* **a.** Having or displaying unusual skill <a *hot* tennis player> **b.** Unusually lucky. **11.** *Mus.* Marked by strong rhythms and improvisation, as jazz. **—hot under the collar.** *Informal.* Angry. **—in hot water.** *Informal.* In trouble. **—make it hot for.** *Informal.* To make things uncomfortable or dangerous for. **—hot′ness** *n.*
☆ ***syns:*** HOT, BAKING, BLISTERING, BOILING, BROILING, BURNING, FIERY, HEATED, RED-HOT, SCALDING, SCORCHING, SIZZLING, SULTRY, SWELTERING, TORRID *adj. core meaning* : marked by much heat <the *hot* desert sun> ***ant:*** cold
**hot air** *n. Slang.* Empty, often boastful talk.
**hot·bed** (hŏt′bĕd′) *n.* **1.** A glass-covered bed of soil heated with fermenting manure or by electricity, used for germinating seeds or protecting tender plants. **2.** An environment conducive to rapid, vigorous growth and development, esp. of something bad <a *hotbed* of violence>
**hot-blood·ed** (hŏt′blŭd′ĭd) *adj.* Easily aroused or excited. **—hot′-blood′ed·ness** *n.*
**hot·box** (hŏt′bŏks′) *n.* An overheated axle or journal box, as on a railway car, caused by excessive friction.
**hot cake** *n.* A pancake. **—go (or sell) like hot cakes.** To be in great demand <Tickets are *going like hot cakes.*>
**hotch** (hŏch) *vi.* **hotched, hotch·ing, hotch·es.** [Perh. < OFr. *hocher*, to shake.] *Scot.* To fidget.
**hotch·pot** (hŏch′pŏt′) *n.* [An < OFr., mixture, stew: *hocher*, to shake together + *pot*, pot.] *Law.* The combination of properties to ensure an equal division of the total for distribution among the heirs of an intestate parent.
**hotch·potch** (hŏch′pŏch′) *n.* [Alteration of HOTCHPOT.] **1.** A hodgepodge. **2.** A hotchpot.
**hot cross bun** *n.* A sweet bun often made with raisins and marked on top with a cross of frosting, traditionally eaten during Lent.
**hot dog** *n.* A frankfurter served heated, usu. in a long soft roll. —*interj. Informal.* —Used to express enthusiasm or satisfaction.
**hot-dog** (hŏt′dôg′, -dŏg′) *vi.* **-dogged, -dog·ging, -dogs.** *Slang.* To do acrobatic stunts, esp. while surfing or skiing. **—hot′-dog′ger** *n.*
**ho·tel** (hō-tĕl′) *n.* [Fr. *hôtel* < OFr. *hostel*, hostel.] A public house that provides lodging and usu. meals and various services.
**hot flash** *n.* A transient vasomotor symptom of menopause involving dilation of the skin capillaries and the sensation of heat over the whole body.
**hot·foot** (hŏt′fo͝ot′) *vi.* **-foot·ed, -foot·ing, -foots.** To go hastily <*hotfoot* to the office> —*adv.* In haste. —*n., pl.* **-foots.** A practical joke in which a match is inserted into the side of someone's shoe and lit.
**hot·head** (hŏt′hĕd′) *n.* A hotheaded person.
**hot·head·ed** (hŏt′hĕd′ĭd) *adj.* **1.** Easily excited or angered. **2.** Impetuous : rash. **—hot′head′ed·ly** *adv.* **—hot′head′ed·ness** *n.*
**hot·house** (hŏt′hous′) *n.* A heated greenhouse for plants requiring an even, relatively warm temperature. —*adj.* Resembling or having qualities like those of a plant grown in a hothouse : DELICATE.
**hot line** *n.* **1.** A direct communications link, as a telephone line, esp. one between heads of state, for use in a crisis or emergency. **2.** A telephone facility enabling a caller to talk confidentially with a sympathetic listener about a personal problem or crisis.
**hot·ly** (hŏt′lē) *adv.* In an intense or fiery fashion <argued *hotly*>
**hot pepper** *n.* **1.** The pungent fruit of any of several varieties of *Capsicum frutescens.* **2.** PEPPER 4.
**hot plate** *n.* **1.** An electrically heated plate for cooking or warming food. **2.** A table-top cooking device with one or two burners.
**hot rod** *also* **hot·rod** *or* **hot-rod** (hŏt′rŏd′) *n. Slang.* An automobile modified or rebuilt for increased power and speed. **—hot′-rod′** *v.* **(-rod·ded, -rod·ding, -rods). —hot rodder** *n.*
**hot seat** *n.* **1.** *Slang.* The electric chair. **2.** *Informal.* A situation or position of stress, embarrassment, or uneasiness.
**hot·shot** (hŏt′shŏt′) *n.* **1.** *Slang.* An ostentatiously skillful person. **2.** A nonstop freight train.
**hot spring** *n.* A natural spring discharging water above 98°F.
**Hot·ten·tot** (hŏt′n-tŏt′) *n., pl.* **Hottentot** *or* **-tots.** [Afr.] **1. a.** A people of southern Africa thought to be related to the Bantu and Bushmen. **b.** A member of this people. **2.** The language of the Hottentot.
**hot toddy** *n.* TODDY 1.
**hot tub** *n.* A large usu. wooden tub in which one or more people may soak.
**hou·dah** (hou′də) *n. var. of* HOWDAH.
**Hou·dan** (ho͞o′dăn′) *n.* [Fr., after *Houdan*, France.] A domesticated fowl with a V-shaped comb and black and white plumage.
**hound** (hound) *n.* [ME < OE *hund*.] **1. a.** A dog of any of various breeds used for hunting, with drooping ears, a short coat, and a deep, resonant voice. **b.** A dog. **2.** A contemptible person : SCOUNDREL. **3.** An enthusiast or addict <a chocolate *hound*> —*vt.* **hound·ed, hound·ing, hounds. 1.** To pursue persistently and stubbornly. **2.** To urge insistently : NAG. **—hound′er** *n.*
**hound's-tongue** (houndz′tŭng′) *n.* Any of several plants of the genus *Cynoglossum*, esp. the Eurasian variety *C. officinale*, with hairy leaves, small reddish-purple flowers, and prickly, clinging fruit.
**hound's-tooth check** (houndz′to͞oth′) *n.* A small checkered textile design.
**hour** (our) *n.* [ME < OFr. *houre* < Lat. *hora* < Gk. *hōra*.] **1.** One of the 24 parts of a day. **2. a.** One of the points on a timepiece marking off 12 or 24 successive intervals of 60 minutes, from midnight to noon and noon to midnight or from midnight to midnight. **b.** The time of day shown by a 12-hour clock. **c. hours.** The time of day determined on a 24-hour basis <1600 *hours*> **3. a.** A customary time <the cocktail *hour*> **b. hours.** A specified period of time <business *hours*> **c.** A particular time <my *hours* of glory> **4. a.** The work that can be accomplished in an hour. **b.** The distance that can be traveled in an hour. **5.** A class session. **6. a.** A time set aside for daily liturgical devotion. **b. hours.** The canonical hours.
**hour angle** *n.* The angle measured westward along the celestial equator from the celestial meridian of the observer to the hour circle passing through a celestial body.
**hour circle** *n.* A great circle passing through the poles of the celestial sphere and intersecting the celestial equator at right angles.
**hour·glass** (our′glăs′) *n.* An instrument for measuring time consisting of two connected glass chambers and containing water, sand, or mercury requiring exactly one hour to trickle from one chamber to the other.
**hou·ri** (ho͝or′ē, ho͞o′rē) *n., pl.* **-ris.** [Fr. < Pers. *ḥūrī* < Ar. *ḥaurā′*.] **1.** A beautiful and voluptuous woman. **2.** One of the beautiful virgins of the Koranic paradise.
**hour·ly** (our′lē) *adj.* **1. a.** Every hour. **b.** Frequent : continual <in *hourly* fear of losing one's job> **2.** By the hour as a unit <*hourly* wages> —*adv.* **1.** At or during every hour. **2.** Frequently : often.
**house** (hous) *n., pl.* **hous·es** (hou′zĭz, -sĭz) [ME < OE *hūs*.] **1. a.** A building used as a dwelling by one or more families. **b.** A household. **2.** Something serving as a shelter or habitation for an animal. **3. a.** A building used for a particular purpose, as entertainment. **b.** The audience or patrons of such an establishment <an appreciative *house*> **4.** A dwelling for students or a religious community. **5. House.** A family line, esp. of a noble family <the *House* of Anjou> **6.** A commercial firm <a fashion *house*> **7. a.** A legislative or deliberative assembly. **b.** The hall in which such an assembly meets. **c.** A quorum of such an assembly. **8. a.** One of the 12 parts into which the heavens are divided in astrology. **b.** The sign of the zodiac indicating the seat or station of a planet in the heavens. —*v.* (houz) **housed, hous·ing, hous·es.** —*vt.* **1.** To provide living quarters for : LODGE <*housed* us in the guest cottage> **2.** To shelter, keep, or store in or as if in a house. **3.** To serve as cover or shelter for <The garage *housed* our boat.> —*vi.* To reside : dwell. **—clean house. 1.** To take care of and clean a house. **2.** To eliminate or discard some-

one or something undesirable. **—on the house.** At the expense of the management : FREE.

**house arrest** *n.* Confinement to one's domicile rather than prison by administrative or judicial order.

**house·boat** (hous′bōt′) *n.* A barge used as a home or cruiser.

**house·bound** (hous′bound′) *adj.* Confined to the house.

**house·break·ing** (hous′brā′kĭng) *n.* The act of unlawfully breaking into another's residence for the purpose of committing a felony. **—house′break′er** *n.*

**house·bro·ken** (hous′brō′kən) *adj.* **1.** Trained in habits of excretion appropriate for a house pet. **2.** Trained to be compliant : DOCILE.

**house·carl** (hous′kärl′) *n.* [OE *hūscarl* < ON *hūskarl* : *hūs*, house + *karl*, man.] A member of the household troops or bodyguard of a Danish or early English king or noble.

**house·coat** (hous′kōt′) *n.* A woman's often long-skirted robe, for wear at home.

**house·fly** (hous′flī′) *n.* A common, widely distributed fly, *Musca domestica*, that breeds in moist or decaying organic matter, frequents human dwellings, and transmits a wide variety of diseases.

**house·hold** (hous′hōld′) *n.* [ME : *house*, house + *hold*, possession.] A domestic establishment including the members of a family and others who live under the same roof.

**household arts** *n.* (*sing. or pl. in number*). Home economics.

**house·hold·er** (hous′hōl′dər) *n.* **1.** One who occupies or owns a house. **2.** The head of a household.

**household word** *n.* A commonly used word, phrase, or name.

**house·hus·band** (hous′hŭz′bənd) *n.* A husband who manages the household while his wife earns all or most of the family income.

**house·keep·er** (hous′kē′pər) *n.* **1.** One hired to perform the domestic tasks in a household. **2.** HOUSEWIFE 1.

**house·keep·ing** (hous′kē′pĭng) *n.* **1.** The management of a household. **2.** The management of internal affairs, as of a business. **3.** The necessary routine tasks that enable a system to operate.

**hou·sel** (hou′zəl) [ME < OE *hūsel.*] *Archaic.* —*n.* The Eucharist. —*vt.* **-seled, -sel·ing, -sels.** To administer the Eucharist to.

**house·leek** (hous′lēk′) *n.* An Old World plant of the genus *Sempervivum*, esp. *S. tectorum*, with a basal rosette of fleshy leaves and a branching cluster of pinkish or purplish flowers.

**house·lights** (hous′līts′) *pl.n.* The lights illuminating the audience section of a concert hall, theater, or auditorium.

**house·maid** (hous′mād′) *n.* A woman employed to do housework.

**housemaid's knee** *n.* A chronic inflammatory swelling of the bursa in front of the kneecap.

**house martin** *n.* An Old World bird, *Delichon urbica*, having a forked tail and blue-black plumage with white markings.

**house·mas·ter** (hous′măs′tər) *n.* A teacher who supervises a residence hall of a boys' school.

**house·mate** (hous′māt′) *n.* One that shares a house with another.

**house·moth·er** (hous′mŭth′ər) *n.* A woman employed as supervisor or housekeeper of a residence hall or dormitory for young people.

**House of Burgesses** *n.* The lower house of the legislature of colonial Virginia.

**House of Commons** *n.* The lower house of Parliament in the United Kingdom and Canada.

**house of correction** *n.* An institution for persons convicted of minor criminal offenses.

**House of Delegates** *n.* The lower house of the state legislature of Maryland, Virginia, and West Virginia.

**House of Lords** *n.* The upper house of Parliament in the United Kingdom, composed of members of the nobility and high-ranking clergy.

**House of Representatives** *n.* The lower branch of the U.S. Congress and of most state legislatures.

**house organ** *n.* A periodical published by a business organization for its employees or clients.

**house party** *n.* A party at which guests stay overnight or for several days in a private home or other residence, as a fraternity house.

**house physician** *n.* **1.** A physician employed by and residing in a hospital. **2.** A physician employed by an establishment, as a hotel.

**house-rais·ing** (hous′rā′zĭng) *n.* The construction of a house or its framework by a group of neighbors.

**house·room** (hous′ro͞om′, -ro͝om′) *n.* Room for accommodation in or as if in a house.

**house·sit** (hous′sĭt′) *vi.* **-sat** (-săt′), **-sit·ting, -sits.** To act as a house sitter.

**house sitter** *n.* One who lives in and cares for a house while the occupant is away.

**house snake** *n.* The milk snake.

**house sparrow** *n.* A small bird, *Passer domesticus*, native to the Old World but widely naturalized elsewhere, with brown and gray plumage and a black throat in the male.

**house·top** (hous′tŏp′) *n.* The roof of a house.

**house·warm·ing** (hous′wôr′mĭng) *n.* A party to celebrate the recent occupancy of a dwelling, esp. a house.

**house·wife** (hous′wīf′) *n.* **1.** *pl.* **-wives** (-wīvz′). A married woman who manages the affairs of a household. **2.** (hŭz′ĭf), *pl.* **house·wifes** (hŭz′ĭfs) *or* **house·wives** (hŭz′ĭvz). A small container for sewing equipment.

**house·wife·ly** (hous′wīf′lē) *adj.* Of, relating to, or suited to a housewife : DOMESTIC. **—house′wife′li·ness** *n.*

**house·wif·er·y** (hous′wī′fə-rē, -wīf′rē) *n.* The domestic function or duties of a housewife.

**house·work** (hous′wûrk′) *n.* The tasks of housekeeping, as cleaning or cooking. **—house′work′er** *n.*

**hous·ing¹** (hou′zĭng) *n.* **1. a.** Residences or dwelling places for people. **b.** A place to live : DWELLING. **2. a.** Something that covers, protects, or supports. **b.** A frame, bracket, or box for holding or protecting a mechanical part. **c.** A frame in which a shaft revolves. **3.** A hole, groove, or slot in a piece of wood for the insertion of another piece. **4.** A niche for a statue. **5.** *Naut.* The part of a mast that is below deck or of a bowsprit that is inside the hull.

**hous·ing²** (hou′zĭng) *n.* [ME *house* < OFr. *housse* < Med. Lat. *hucia.*] **1.** A decorative or protective covering for a saddle. **2.** *often* **housings.** Trappings.

**housing development** *n.* A group of similarly designed houses or apartment buildings, usu. under a single management.

**housing project** *n.* A publicly funded and administered housing development, usu. for low-income families.

**hove** (hōv) *v. p.t. & p.p. of* HEAVE.

**hov·el** (hŭv′əl, hŏv′-) *n.* [ME, hut.] **1.** A small, wretched dwelling. **2.** A low, open shed.

**hov·er** (hŭv′ər, hŏv′-) *vi.* **-ered, -er·ing, -ers.** [ME *hoveren*, freq. of *hoven*, to hover.] **1.** To remain floating or suspended in the air over a particular place <a hawk *hovering* high above> **2.** To stay or linger in or near a place <nervous mothers *hovering* around backstage> **3.** To be in a state of uncertainty or suspense : WAVER <*hovering* between life and death> —*n.* An act or instance of hovering. **—hov′er·er** *n.* **—hov′er·ing·ly** *adv.*

**Hov·er·craft** (hŭv′ər-krăft′, hŏv′-). A trademark for a vehicle used in low-level flight over land or water.

**how** (hou) *adv.* [ME *howe* < OE *hū.*] **1.** In what manner or way <*How* do you play this game?> **2.** In what state or condition <*How* do I look with short hair?> **3.** To what extent, amount, or degree <*How* do they like the new car?> **4.** For what reason or purpose : WHY <*How* did you happen to buy it?> **5.** With what meaning <*How* should I take that remark?> **6.** By what name. **7.** What <*How* is that again?> —*conj.* **1.** In what way or manner <can't remember *how* it was done> **2.** That <reminded me *how* they had places to go> **3.** In whatever way or manner <sang it *how* we wanted> —*n.* A manner or method of doing or performing <knew the *how* but not the why> **—how about.** What do you think or feel about <*How about* a sandwich?> **—how come.** For what reason : WHY <*How come* you lost?> ***usage:*** The expressions *seeing as how* and *being as how*, while occurring frequently in spoken English, are inappropriate to formal style.

**how·be·it** (hou-bē′ĭt) *Archaic.* —*adv.* Be that as it may : NEVERTHELESS. —*conj.* Although.

**how·dah** *also* **hou·dah** (hou′də) *n.* [Urdu and Pers. *haudah* < Ar. *haudaj.*] A seat, usu. fitted with a canopy and railing, on the back of an elephant or camel.

**†how·dy** (hou′dē) *interj.* [< *how do you do.*] *Regional.* —Used to express greeting.

**how·ev·er** (hou-ev′ər) *adv.* **1.** By whatever manner or means <*However* it happened, we aren't to blame.> **2.** To whatever degree or extent <never asked for money, *however* broke I was> —*conj.* **1.** Nevertheless : yet <It looks like rain; *however*, the picnic is still on.> **2.** *Archaic.* Although : notwithstanding that. ***usage:*** The use of *however* as the first word of a sentence is entirely acceptable at all levels of spoken or written English.

**how·it·zer** (hou′ĭt-sər) *n.* [Du. *houwitser* < G. *Haubitze* < Czech *houfnice*, catapult.] A short cannon that discharges shells with medium velocities, usu. at a high trajectory.

**howl** (houl) *v.* **howled, howl·ing, howls.** [ME *houlen.*] —*vi.* **1.** To emit or utter a long, mournful, plaintive sound. **2.** To cry loudly, as in pain, grief, or anger. **3.** *Slang.* To laugh heartily. —*vt.* **1.** To express or utter with a howl. **2.** To effect, drive, or force by or as if by howling. —*n.* **1.** A long, wailing cry. **2.** *Slang.* Something uproariously funny or preposterous.

**howl·er** (hou′lər) *n.* **1.** One that howls. **2.** *also* **howler monkey.** A tropical American monkey of the genus *Alouatta*, with a long prehensile tail and a loud howling call. **3.** *Slang.* An amusing or stupid blunder.

**howl·ing** (hou′lĭng) *adj.* **1.** Marked by howling. **2.** *Slang.* Extremely great : TREMENDOUS <The play was a *howling* success.>

**how·so·ev·er** (hou′sō-ev′ər) *adv.* **1.** To whatever extent or degree. **2.** By whatever means.

**hoy** (hoi) *n.* [ME *hoie* < MDu. *hoey.*] **1.** A small sloop-rigged coasting ship. **2.** A heavy barge used for freight.

**hoy·den** (hoid′n) *n.* [Prob. MDu. *heiden*, heathen.] A high-spirited, often boisterous girl or woman. —*adj.* High-spirited : boisterous. **—hoy′den·ish** *adj.*

**Hoyle** (hoil) *n.* [After Edmund *Hoyle* (1672–1769).] A reference book of rules for indoor games, esp. for card games.

**hua·ra·che** (wə-rä'chē, hə-) *n.* [Mex. Sp.] A flat-heeled sandal with woven leather strips on the upper part.

**hub** (hŭb) *n.* [Prob. alteration of HOB¹.] **1.** The center portion of a wheel, fan, or propeller. **2.** A center of interest or activity.

**hub·ble-bub·ble** (hŭb'əl-bŭb'əl) *n.* [Redup. of *bubble.*] **1.** An uproar. **2.** A hookah.

**Hub·ble's constant** (hŭb'əlz) *n.* [After Edwin P. *Hubble* (1889–1953).] The ratio of the speed at which a distant galaxy is receding from the earth to its distance from the earth, approx. 100 kilometers per second per million parsecs.

**hub·bub** (hŭb'ŭb') *n.* [Of Ir. orig.] **1.** A confused babble of voices and loud sounds : UPROAR. **2.** Tumult : confusion.

**hub·by** (hŭb'ē) *n., pl.* **-bies.** *Informal.* A husband.

**hub·cap** (hŭb'kăp') *n.* A round metal covering attached to the hub of an automobile wheel.

**hu·bris** (hyōō'brĭs) *n.* [Gk., violence.] Excessive pride : ARROGANCE. **—hu·bris'tic** (-brĭs'tĭk) *adj.*

**huck** (hŭk) *n.* Huckaback.

**huck·a·back** (hŭk'ə-băk') *n.* [Orig. unknown.] A coarse absorbent cotton or linen fabric used esp. for toweling.

**huck·le·ber·ry** (hŭk'əl-bĕr'ē) *n.* [Prob. alteration of obs. *hurtleberry,* whortleberry.] **1.** A New World shrub of the genus *Gaylussacia,* related to the blueberries and bearing edible fruit. **2.** The glossy, blackish, many-seeded berry of a huckleberry.

**huck·ster** (hŭk'stər) *n.* [ME *hukster.*] **1.** One who sells goods in the street : PEDDLER. **2.** *Slang.* A writer of advertising copy. *—v.* **-stered, -ster·ing, -sters.** *—vt.* **1.** To sell by peddling. **2.** To promote (e.g., a commercial product). **3.** To haggle over. *—vi.* To haggle. **—huck'ster·ism** *n.*

**hud·dle** (hŭd'l) *n.* [Orig. unknown.] **1.** A densely packed group, as of people or animals. **2.** *Football.* A brief gathering of a team's players behind the line of scrimmage to prepare for the next play. **3.** A small private conference. *—v.* **-dled, -dling, -dles.** *—vi.* **1.** To crowd together. **2.** To draw oneself together : CURL UP. **3.** *Football.* To gather in a huddle. **4.** To gather for conference or consultation. *—vt.* **1.** To cause to crowd together. **2.** To draw (oneself) together : HUNCH <*huddled* myself against the cold> **3.** *Chiefly Brit.* To arrange, do, or make hastily or carelessly. **—hud'dler** *n.*

**Hu·di·bras·tic** (hyōō'də-brăs'tĭk) *adj.* [After *Hudibras,* a satiric epic by Samuel Butler (1612–1680).] Having a mock-heroic style.

**Hud·son seal** (hŭd'sən) *n.* [After *Hudson* Bay, inland sea in Canada.] Imitation seal made from processed muskrat fur.

**hue** (hyōō) *n.* [ME, color < OE *hīw.*] **1.** The dimension of color referred to a scale of perceptions ranging from red through yellow, green, and blue, and circularly back to red. **2.** A particular color as distinguished from other colors : SHADE. **3.** Color <rainbow *hues*> **4.** Appearance : aspect <gave the party an exotic *hue*>

**hue and cry** *n.* [ME *heu,* a loud outcry (< OFr. *hu* < *huer,* to shout) + CRY.] **1.** A public clamor, as of protest or demand <raised a great *hue and cry* about double-digit unemployment> **2.** The loud shout once used to announce the pursuit of a felon.

**hued** (hyōōd) *adj.* Having a specified hue, aspect, or character <sickly-*hued*>

**huff** (hŭf) *n.* [Imit. of the sound of puffing.] A fit of anger or annoyance : PIQUE. *—v.* **huffed, huff·ing, huffs.** *—vi.* **1.** To blow or puff. **2.** To speak or act with loud, empty threats : BLUSTER. **3.** To act or react indignantly. *—vt.* **1.** To puff up : INFLATE. **2.** *Archaic.* To treat with insolence : BULLY. **3.** To anger or annoy.

**huff·ish** (hŭf'ĭsh) *adj.* **1.** Ill-humored : peevish. **2.** HUFFY 3. **—huff'ish·ly** *adv.* **—huff'ish·ness** *n.*

**huff·y** (hŭf'ē) *adj.* **-i·er, -i·est. 1.** Easily offended : TOUCHY. **2.** Vexed : annoyed. **3.** Haughty : arrogant. **—huff'i·ly** *adv.* **—huff'i·ness** *n.*

**hug** (hŭg) *v.* **hugged, hug·ging, hugs.** [Prob. of Scand. orig.] *—vt.* **1.** To clasp or hold closely, esp. in the arms : EMBRACE. **2.** To hold steadfastly to : CHERISH. **3.** To keep, remain, or be situated near to <a footpath that *hugged* the slope> *—vi.* To embrace or be in physical contact. *—n.* An affectionate clasp or embrace. **—hug'ga·ble** (hŭg'ə-bəl) *adj.* **—hug'ger** *n.*

**huge** (hyōōj) *adj.* **hug·er, hug·est.** [ME < OFr. *ahuge.*] Of great size, extent, or amount. **—huge'ly** *adv.* **—huge'ness** *n.*

**huge·ous** (hyōō'jəs) *adj. Informal.* Huge. **—huge'ous·ly** *adv.* **—huge'ous·ness** *n.*

**hug·ger·mug·ger** *or* **hug·ger-mug·ger** (hŭg'ər-mŭg'ər) *n.* [Orig. unknown.] **1.** Disorderly confusion : MUDDLE. **2.** Concealment : secrecy. *—adj.* **1.** Disorderly. **2.** Secret. *—v.* **-gered, -ger·ing, -gers.** *—vt.* To keep secret : CONCEAL. *—vi.* To act surreptitiously.

**hug-me-tight** (hŭg'mē-tīt') *n.* A woman's close-fitting, sleeveless, usu. knitted jacket.

**Hu·gue·not** (hyōō'gə-nŏt') *n.* [Fr., alteration (influenced by gate of Roi-*Hugon,* where the Protestants of Tours assembled at night) of dial. Fr. *eyguenot* < dial. G. *Eidgenossen,* confederates.] A French Protestant of the 16th and 17th cent. **—Hu'gue·not'ic** (-nŏt'ĭk) *adj.* **—Hu'gue·not'ism** *n.*

**huh** (hŭ) *interj.* —Used to express interrogation, surprise, contempt, or indifference.

**hu·la** (hōō'lə) *also* **hu·la-hu·la** (hōō'lə-hōō'lə) *n.* [Hawaiian.] **1.** A Polynesian dance marked by undulating movements of the hips, arms, and hands and often accompanied by rhythmic drumbeats and chants. **2.** The music for a hula.

**hulk** (hŭlk) *n.* [ME < OE *hulc.*] **1.** A heavy, unwieldy ship. **2. a.** The hull of an old, unseaworthy, or wrecked ship. **b.** An old or unseaworthy ship used as a prison or warehouse. **3.** One that is large, clumsy, or awkward. *—vi.* **hulked, hulk·ing, hulks. 1.** To be or seem impressively or exaggeratedly large. **2.** To move awkwardly.

**hulk·ing** (hŭl'kĭng) *also* **hulk·y** (hŭl'kē) *adj.* Bulky : massive.

**hull** (hŭl) *n.* [ME *hulle,* husk < OE *hulu.*] **1. a.** The enlarged, easily detached, usu. green calyx of a fruit, as a strawberry. **b.** The dry outer covering of a fruit, seed, or nut : HUSK. **2. a.** The framework or body of a ship, exclusive of masts, sails, yards, and rigging. **b.** The fuselage of a flying boat. **c.** *Aerospace.* The outer casing of a rocket, guided missile, or spaceship. *—vt.* **hulled, hull·ing, hulls.** To remove the hulls of (fruit or seeds). **—hull'er** *n.*

**hul·la·ba·loo** *also* **hul·la·bal·loo** (hŭl'ə-bə-lōō') *n., pl.* **-loos.** [Orig. unknown.] Great noise or excitement : UPROAR.

**hul·lo** (hə-lō') *interj. & n. & v. var. of* HELLO.

**hum** (hŭm) *v.* **hummed, hum·ming, hums.** [ME *hummen.*] *—vi.* **1. a.** To make or emit a low droning sound like that of the speech sound (m) when prolonged. **b.** To emit the continuous droning sound of an insect on the wing. **c.** To give forth a low drone blended of many sounds <The courtroom *hummed* with excitement.> **2.** To be in a state of busy activity <The classroom was *humming.*> **3.** To sing a tune without opening the lips or articulating words. *—vt.* To sing by humming. *—n.* **1.** The noise produced by humming. **2.** The act of humming. *—interj.* —Used to indicate hesitation, displeasure or surprise. **—hum'mer** *n.*

**hu·man** (hyōō'mən) *adj.* [ME *humain* < OFr. < Lat. *humanus.*] **1.** Of, pertaining to, or typical of humans or humankind. **2.** Having or exhibiting the form, nature, or qualities typical of humans. **3.** Prone to or characterized by the frailties and weaknesses associated with humans as imperfect beings. **4.** Made up of people. *—n.* A human being. **—hu'man·ness** *n.*

☆ ***syns:*** HUMAN, HUMANE, HUMANITARIAN *adj. core meaning* : having or exhibiting the nature or qualities typical of humans <the alleviation of poverty—a *human* concern> HUMAN is essentially a classifying term relating to individuals or people collectively <*human* kindness>, while HUMANE stresses the qualities of kindness and compassion <*humane* treatment> HUMANITARIAN applies to what actively promotes the needs and welfare of people <*humanitarian* considerations in the dispensation of charity money>

**human being** *n.* A member of the genus *Homo* and esp. of the species *Homo sapiens.*

**hu·mane** (hyōō-mān') *adj.* [ME *humain,* human.] **1.** Characterized by kindness, mercy, or compassion <a *humane* attorney> **2.** Characterized by an emphasis on humanistic values and concerns <a *humane* educational background> **—hu·mane'ly** *adv.* **—hu·mane'ness** *n.*

**human engineering** *n.* **1.** The industrial management of labor. **2.** The technology of the efficient use of machines by human beings.

**hu·man·ism** (hyōō'mə-nĭz'əm) *n.* **1.** A doctrine or attitude concerned chiefly with human beings and their values, capacities, and achievements. **2.** Absorption in or study of the humanities. **3.** *often* **Humanism.** A cultural and intellectual movement of the Renaissance that emphasized secular concerns as a result of the study of the literature, art, and civilization of ancient Greece and Rome. **—hu'man·ist** *n.* **—hu'man·is'tic** *adj.* **—hu'man·is'ti·cal·ly** *adv.*

**hu·man·i·tar·i·an** (hyōō-măn'ĭ-târ'ē-ən) *n.* One devoted to the promotion of human welfare and the advancement of social reforms : PHILANTHROPIST.

**hu·man·i·tar·i·an·ism** (hyōō-măn'ĭ-târ'ē-ə-nĭz'əm) *n.* Concern for human welfare, esp. as manifested through philanthropy.

**hu·man·i·ty** (hyōō-măn'ĭ-tē) *n., pl.* **-ties.** [ME *humanite* < OFr. < Lat. *humanitas* < *humanus,* human.] **1.** Human beings as a group. **2.** The quality, condition, or fact of being human : HUMANNESS. **3.** The quality of being humane : BENEVOLENCE. **4.** A humane quality or action. **5. humanities.** Those subjects, as philosophy, literature, and the fine arts, that are concerned with human beings and their culture, as distinguished from the sciences.

**hu·man·ize** (hyōō'mə-nīz') *vt.* **-ized, -iz·ing, -iz·es. 1.** To represent or endow with human characteristics or attributes. **2.** To make humane. **—hu'man·i·za'tion** *n.* **—hu'man·iz'er** *n.*

**hu·man·kind** (hyōō'mən-kīnd') *n.* The human race.

**hu·man·ly** (hyōō'mən-lē) *adv.* **1. a.** In a human way. **b.** In a humane way. **2.** Within the scope of human means, capabilities, or powers <not *humanly* possible> **3.** According to human experience or knowledge <*humanly* speaking>

**hu·man·oid** (hyōō'mə-noid') *adj.* Resembling a human being in appearance. *—n.* An android.

**hum·ble** (hŭm′bəl) *adj.* **-bler, -blest.** [ME < OFr. < Lat. *humilis* < *humus,* ground.] **1.** Characterized by modesty or meekness in behavior, attitude, or spirit. **2.** Exhibiting deferential or submissive respect. **3.** Unpretentious <a *humble* dwelling> —*vt.* **-bled, -bling, -bles. 1.** To humiliate. **2.** To make lower in condition or status. **—hum′ble·ness** *n.* **—hum′bler** *n.* **—hum′bly** *adv.*

**hum·ble·bee** (hŭm′bəl-bē′) *n.* [ME *humbulbe.*] A bumblebee.

**humble pie** *n.* [Obs. *humble,* edible animal organs (< ME *hombuls* < OFr. *nombles* < Lat. *lumbulus* < *lumbus,* loin) + PIE[1].] A pie formerly made from the edible organs of a deer or hog. **—eat humble pie.** To apologize abjectly in humiliating circumstances.

▲ word history: *Humble pie* has no etymological connection with humility, although the spelling of the phrase has probably been influenced by the adjective *humble. Humble pie* is a dish made of an animal's *numbles,* that is, its entrails and other internal organs. *Numbles* was borrowed from French in the 14th century. In the 15th century a variant form *umbles* appeared and existed alongside *numbles,* especially in the compound *umble pie.* The idiom "to eat humble pie" probably arose in a dialect that dropped initial *h.* In such a dialect *umble* and the adjective *humble* would be pronounced alike and the semantic confusion, or perhaps pun, involved in the idiom would be possible. The form *humble pie* did not appear in standard English until the 19th century.

**Hum·boldt Current** (hŭm′bōlt′) *n.* [After F.H. Alexander von *Humboldt* (1769–1859).] A cold ocean current of the South Pacific, flowing northward from the northern coast of Chile and Peru to southern Ecuador.

**hum·bug** (hŭm′bŭg′) *n.* [Orig. unknown.] **1.** A deception : hoax. **2.** One who attempts to trick or deceive. **3.** Nonsense. —*v.* **-bugged, -bug·ging, -bugs.** —*vt.* To trick or deceive. —*vi.* To practice deception or trickery. **—hum′bug′ger** *n.* **—hum′bug′gery** *n.*

**hum·ding·er** (hŭm′dĭng′ər) *n.* [Orig. unknown.] *Slang.* Something superior or extraordinary.

**hum·drum** (hŭm′drŭm′) *adj.* [Orig. unknown.] Lacking change, variety, or excitement : MONOTONOUS. —*n.* One that is boring.

**hu·mec·tant** (hyōō-mĕk′tənt) *n.* [< Lat. *humectans, humectant-,* pr.part. of *humectare,* to moisten < *humectus,* moist < *humēre,* to be moist.] A substance promoting moisture retention. —*adj.* Promoting moisture retention.

**hu·mer·al** (hyōō′mər-əl) *adj.* **1.** Of, relating to, or situated in the region of the humerus or the shoulder. **2.** Relating to or being a bodily part analogous to the humerus. **—hu′mer·al** *n.*

**humeral veil** *n. Rom. Cath. Ch.* A silk veil covering the shoulders of a subdeacon or priest at a High Mass.

**hu·mer·us** (hyōō′mər-əs) *n., pl.* **-meri** (-mə-rī′) [NLat. < Lat., upper arm.] The long bone of the upper part of the arm, extending from the shoulder to the elbow.

**hu·mic** (hyōō′mĭk) *adj.* Of, relating to, or derived from humus.

**hu·mid** (hyōō′mĭd) *adj.* [OFr. *humide* < Lat. *humidus* < *humēre,* to be moist.] Having or marked by a large amount of water or water vapor <a *humid* atmosphere> **—hu′mid·ly** *adv.*

**hu·mid·i·fi·er** (hyōō-mĭd′ə-fī′ər) *n.* An apparatus for increasing the humidity in an enclosed space, as a room or greenhouse.

**hu·mid·i·fy** (hyōō-mĭd′ə-fī′) *vt.* **-fied, -fy·ing, -fies.** To make humid. **—hu·mid′i·fi·ca′tion** *n.*

**hu·mid·i·stat** (hyōō-mĭd′ĭ-stăt′) *n.* An instrument that indicates or controls the relative humidity of the air.

**hu·mid·i·ty** (hyōō-mĭd′ĭ-tē) *n.* [ME *humidite* < OFr. < Med. Lat. *humiditas* < Lat. *humidus,* humid.] Dampness, esp. of the air.

**hu·mi·dor** (hyōō′mĭ-dôr′) *n.* [< HUMID.] A container for cigars having a device for keeping the humidity level constant.

**hu·mil·i·ate** (hyōō-mĭl′ē-āt′) *vt.* **-at·ed, -at·ing, -ates.** [LLat. *humiliare, humiliat-,* to humble < *humilis,* humble.] To lower the pride or dignity of : ABASE. **—hu·mil′i·a′tion** *n.* **—hu·mil′i·a·to′ry** (-ē-ə-tôr′ē, -tōr′ē) *adj.*

**hu·mil·i·ty** (hyōō-mĭl′ĭ-tē) *n., pl.* **-ties.** [ME *humilite* < OFr. < LLat. *humilitas* < *humilis,* humble. —see HUMBLE.] The quality or condition of being humble.

**hum·ming·bird** (hŭm′ĭng-bûrd′) *n.* Any of numerous very small, chiefly tropical New World birds of the family Trochilidae, with a long slender bill, wings capable of beating very rapidly, and often brilliantly colored plumage.

**hum·mock** (hŭm′ək) *also* **ham·mock** (hăm′-) *n.* [Orig. unknown.] **1.** A low mound or ridge of earth : KNOLL. **2.** A ridge or hill of ice in an ice field. **—hum′mock·y** *adj.*

**hum·mus** (hŭm′əs) *also* **hom·mos** (hŭm′əs, hŏm′-) *n.* [Ar. *ḥummuṣ,* chickpea.] A smooth, thick mixture of puréed chickpeas and tahini used esp. as a sandwich spread or a dip.

**hu·mon·gous** (hyōō-mŏng′gəs, -mŭng′-) *or* **hu·mun·gous** (-mŭng′-) *adj.* [Perh. alteration of HUGE + TREMENDOUS.] *Slang.* Extremely large : ENORMOUS.

**hu·mor** (hyōō′mər) *n.* [ME, fluid < OFr. *umor* < Lat. *humor.*] **1.** The quality of being amusing or comical <couldn't see the *humor* of the remark> **2.** The ability to perceive, enjoy, or express what is comical or funny <a keen sense of *humor*> **3.** One of the four fluids of the body in medieval physiology, blood, phlegm, choler, and black bile, the dominance of which was thought to determine one's character and general health. **4. a.** A state of mind : MOOD <in a bad *humor*> **b.** Characteristic disposition : TEMPERAMENT <a child of cheerful *humor*> **5.** A sudden, unexpected whim. **6.** *Physiol.* **a.** A clear or hyaline bodily fluid, as blood, lymph, or bile. **b.** Aqueous humor. —*vt.* **-mored, -mor·ing, -mors. 1.** To comply with the ideas or desires of : INDULGE. **2.** To accommodate oneself to.

▲ word history: The Latin word *umor* or *humor,* from which English *humor* is derived, meant "a liquid, fluid; moisture." The word first appeared in English during the Middle Ages with that meaning. *Humor* came to denote mental qualities and dispositions through its use in medieval physiology. A *humor* was one of the four principal fluids of the body (blood, phlegm, choler, and black bile) that were mixed in different proportions in each person. The preponderance of one or another humor gave everyone a characteristic temperament or disposition, which was also called a *humor. Humor* also came to indicate temporary and changing moods and states of mind, particularly whimsical and capricious fancies which, when revealed in actions as humorous traits or eccentricities, afford amusement to others. *Humor* at last came to denote the ability both to amuse and to perceive what is amusing.

**hu·mor·al** (hyōō′mər-əl) *adj.* Of, relating to, or arising from any of the bodily humors.

**hu·mor·esque** (hyōō′mə-rĕsk′) *n.* [G. *Humoreske* < *Humor,* humor < E. HUMOR.] A light-spirited musical composition.

**hu·mor·ist** (hyōō′mər-ĭst) *n.* **1.** One who has a good sense of humor. **2.** A writer or performer who specializes in humor.

**hu·mor·less** (hyōō′mər-lĭs) *adj.* **1.** Lacking a sense of humor. **2.** Devoid of humor <a *humorless* smile> **—hu′mor·less·ly** *adv.* **—hu′mor·less·ness** *n.*

**hu·mor·ous** (hyōō′mər-əs) *adj.* **1.** Having or marked by humor <a *humorous* anecdote> **2.** Utilizing or expressing humor <a *humorous* essayist> **3.** *Obs.* Damp and moist. **—hu′mor·ous·ly** *adv.* **—hu′mor·ous·ness** *n.*

☆ ***syns:*** HUMOROUS, FACETIOUS, FUNNY, JESTING, JOCOSE, JOCULAR, WITTY *adj. core meaning* : intended to excite laughter or amusement <*humorous* political comments>

**hu·mour** (hyōō′mər) *n.* & *v. Chiefly Brit. var. of* HUMOR.

**hump** (hŭmp) *n.* [Orig. unknown.] **1.** A rounded mass or protuberance, as the fleshy structure on the back of a camel or some cattle. **2.** A deformity of the back, caused in human beings by an abnormal curvature of the spine. **3.** A low mound of earth : HUMMOCK. **4.** *Chiefly Brit.* A depressed feeling. —*v.* **humped, hump·ing, humps.** —*vt.* **1.** To make or bend into a hump. **2.** *Slang.* To subject (oneself) to exertion. —*vi.* **1.** To arch so as to become a hump. **2.** *Slang.* To exert oneself. **—over the hump.** Past the hardest or worst part of something.

**hump·back** (hŭmp′băk′) *n.* **1.** HUNCHBACK 1, 3. **2.** HUNCHBACK 2. **3.** A whalebone whale, *Megaptera novaeangliae,* with a rounded back and long, knobby flippers. **—hump′backed′** *adj.*

**humph** (hŭmf) *interj.* —Used to express contempt, displeasure, or doubt. **—humph** *v.* **(humphed, humph·ing, humphs).**

**hump·y** (hŭm′pē) *adj.* **-i·er, -i·est. 1.** Covered with or having humps. **2.** Resembling a hump.

**hu·mun·gous** (hyōō-mŭng′gəs) *adj. var. of* HUMONGOUS.

**hu·mus** (hyōō′məs) *n.* [Lat., soil.] A brown or black organic substance consisting of decayed vegetable matter that provides nutrients for plants and increases the water-retention of soil.

**Hun** (hŭn) *n.* [LLat. *Hunni,* the Huns < Turki *Hunyü.*] **1.** One of a fierce barbaric race of Asiatic nomads who ravaged Europe in the 4th and 5th cent. A.D. **2.** *often* **hun.** A savage, barbaric, or destructive person. **3.** *Slang.* A German.

**hunch** (hŭnch) *n.* [Orig. unknown.] **1.** An intuitive feeling or guess <had a *hunch* I was right> **2.** A hump. **3.** A chunk or lump. —*v.* **hunched, hunch·ing, hunch·es.** —*vt.* **1.** To bend or draw up into a hump <*hunch* one's shoulders> **2.** To push or shove. —*vi.* **1.** To draw oneself up closely into a crouched or cramped posture <The terrified puppy *hunched* against the wall.> **2.** To thrust oneself forward.

**hunch·back** (hŭnch′băk′) *n.* **1.** One having an abnormally curved or hunched back. **2.** Kyphosis. **3.** An abnormally curved or hunched back. **—hunch′backed′** *adj.*

**hun·dred** (hŭn′drĭd) *n., pl.* **hundred** *or* **-dreds.** [ME < OE.] **1.** The cardinal number equal to 10 x 10 or $10^2$. **2.** A currency note worth 100 dollars. **3.** The number in the third position left of the decimal point in an Arabic numeral. **4. hundreds.** The numbers between 100 and 999. **5.** An administrative division of some English and American counties. **—hun′dred** *adj.* & *pron.*

**hun·dredth** (hŭn′drĭdth) *n.* **1.** The ordinal number matching the number 100 in a series. **2.** One of 100 equal parts. **—hun′dredth** *adj.* & *adv.*

**hun·dred·weight** (hŭn′drĭd-wāt′) *n., pl.* **hundredweight** *or* **-weights. 1.** A unit of weight in the U.S. Customary System equal to 100 pounds or approx. 45.6 kilograms. **2.** A unit of weight in the

British Imperial System equal to 112 pounds or approx. 50.8 kilograms.
**hung** (hŭng) *v. var. p.t. & p.p. of* HANG.
**Hun·gar·i·an** (hŭng-gâr'ē-ən) *adj.* Of or relating to Hungary or its people, language, or culture. —*n.* **1.** A native or resident of Hungary. **2.** MAGYAR 2.
**hun·ger** (hŭng'gər) *n.* [ME < OE *hungor.*] **1. a.** A strong need or desire for food. **b.** The discomfort, weakness, or pain caused by a lack of food. **2.** A strong craving or desire <a *hunger* for love> —*v.* **-gered, -ger·ing, -gers.** —*vi.* **1.** To have a need or desire for food. **2.** To have a strong craving or desire. —*vt.* To make hungry.
**hunger strike** *n.* A voluntary fast undertaken as a protest against something. **—hunger striker** *n.*
**hung jury** *n.* A jury unable to agree on a verdict.
**hung over** *adj.* Suffering from a hangover.
**hun·gry** (hŭng'grē) *adj.* **-gri·er, -gri·est.** [ME *hungri* < OE *hungrig.*] **1.** Experiencing hunger. **2.** Extremely desirous : AVID <*hungry* for affection> **3.** Marked by or expressing hunger or craving <a *hungry* stare> **4.** Lacking richness or fertility <*hungry* garden soil> **—hun'gri·ly** *adv.* **—hun'gri·ness** *n.*
**hung up** *adj. Informal.* **1.** Delayed <was *hung up* at the airport> **2.** Having a hang-up : DISTRESSED <*hung up* about their debts> **3.** Involved : preoccupied <*hung up* on physical fitness>
**hunk** (hŭngk) *n.* [Perh. < Flem. *hunke,* a piece of food.] **1.** *Informal.* A large piece : CHUNK. **2.** *Slang.* An attractive man.
**†hun·ker** (hŭng'kər) *vi.* **-kered, -ker·ing, -kers.** [Prob. of Scand. orig.] To crouch close to the ground : SQUAT. —*n.* **hunkers.** *Regional.* The haunches.
**hunks** (hŭngks) *n., pl.* **hunks.** [Orig. unknown.] **1.** A grouchy or disagreeable old person. **2.** A stingy person : MISER.
**hun·ky** (hŭng'kē) *n., pl.* **-kies.** *Slang.* A bohunk.
**hun·ky-do·ry** (hŭng'kē-dôr'ē, -dōr'ē) *adj.* [Obs. *hunk,* goal + *-dory,* of unknown orig.] *Slang.* Perfectly fine.
**Hun·nish** (hŭn'ĭsh) *adj.* **1.** Of, relating to, or like the Huns. **2.** Barbarous. **—Hun'nish·ness** *n.*
**hunt** (hŭnt) *v.* **hunt·ed, hunt·ing, hunts.** [ME *hunten* < OE *huntian.*] —*vt.* **1. a.** To pursue (game) for food or sport. **b.** To seek out : look for. **2.** To search through, as for game or prey. **3.** To utilize (e.g., hounds) in pursuing game. **4.** To drive out forcibly. —*vi.* **1.** To pursue game or other animals in order to capture or kill them. **2.** To make a search : SEEK. **3.** *Aerospace.* **a.** To yaw back and forth about a flight path, as if seeking a new direction or another angle of attack. —Used of aircraft, rockets, and space vehicles. **b.** To rotate up and down or back and forth without being deflected by the pilot. —Used of a control surface or a rocket motor in gimbals. **4. a.** To oscillate about a selected value. —Used of a control system. **b.** To swing back and forth or oscillate. —Used of an indicator on a display or instrument panel. —*n.* **1.** The act or sport of hunting. **2. a.** A hunting expedition or outing, usu. with horses and hounds. **b.** The participants in a hunt. **3.** A diligent and thorough pursuit or search <the *hunt* for the lost child>
**hunt·er** (hŭn'tər) *n.* **1.** One who hunts. **2.** A dog bred or trained for use in hunting. **3.** A horse, esp. a fast, strong jumper, bred or trained for hunting.
**hunt·ing** (hŭn'tĭng) *n.* **1.** The sport or activity of pursuing game. **2.** *Elect.* The periodic variation in speed of a synchronous motor with respect to the current.
**Hun·ting·ton's chorea** (hŭn'tĭng-tənz) *n.* [After George *Huntington,* 1850–1916.] A hereditary chorea in adults, marked by progressive mental deterioration and dementia.
**hunt·ress** (hŭn'trĭs) *n.* A woman who hunts : HUNTER.
**hunts·man** (hŭnts'mən) *n.* **1.** A man who hunts : HUNTER. **2.** A man who manages the hounds in the hunting field.
**hur·dle** (hûr'dl) *n.* [ME *hurdel,* portable panel for temporary fences < OE *hyrdel.*] **1. a.** A light, portable barrier used in obstacle races that usu. has two uprights between which a horizontal bar may be hung at varying heights. **b.** *often* **hurdles.** A race in which hurdles must be jumped. **2.** An obstacle or difficulty. **3.** *Chiefly Brit.* A portable section of fencing used chiefly for penning sheep. **4.** *Chiefly Brit.* A frame or sledge once used to carry condemned persons to their executions. —*v.* **-dled, -dling, -dles.** —*vt.* **1.** To jump or leap over (a barrier) in or as if in a race. **2.** To deal with or overcome successfully. —*vi.* To jump or leap over a barrier in or as if in a race. **—hur'dler** *n.*
**hur·dy-gur·dy** (hûr'dē-gûr'dē, hûr'dē-gûr'dē) *n., pl.* **-dies.** [Prob. imit.] A musical instrument, as a barrel organ, played by turning a crank.
**hurl** (hûrl) *v.* **hurled, hurl·ing, hurls.** [ME *hurlen.*] —*vt.* **1.** To throw with or as if with great force. **2.** To send with great vigor : THRUST <*hurled* the troops against the invaders> **3.** To throw down : OVERTHROW <a dictator *hurled* from power> —*vi.* **1.** To move with great speed, force, or violence : HURTLE. **2.** To pitch a baseball. —*n.* An act or instance of hurling. **—hurl'er** *n.*
**hurl·ing** (hûr'lĭng) *n.* An Irish game similar to lacrosse but played with a broad-bladed netless stick.
**hur·ly-bur·ly** (hûr'lē-bûr'lē) *n., pl.* **-lies.** [Alteration and redup. of *hurling,* gerund of HURL.] Tumultuous uproar : COMMOTION.
**Hu·ron** (hyŏŏr'ən, -ŏn') *n., pl.* **Huron** *or* **-rons.** [Fr. < OFr. *hure,* fur.] **1. a.** A confederation of four Indian tribes formerly inhabiting the region east of Lake Huron and the St. Lawrence Valley. **b.** A member of any of these tribes. **2.** The Iroquoian language of the Huron. **—Hu'ron** *adj.*
**hur·rah** (hŏŏ-rä', -rô') *also* **hoo·ray** *or* **hur·ray** (-rā') [Alteration of HUZZAH.] —*interj.* —Used to express pleasure, approval, elation, or victory. —*n.* A shout of "hurrah." —*v.* **-rahed, -rah·ing, -rahs** *also* **-rayed, -ray·ing, -rays.** —*vt.* To applaud, cheer, or approve by shouting "hurrah." —*vi.* To shout "hurrah."
**hur·ri·cane** (hûr'ĭ-kān') *n.* [Sp. *huracan* < Carib.] A severe tropical cyclone with winds exceeding approx. 74 miles or 119 kilometers per hour, originating in the tropical regions of the Atlantic Ocean or Caribbean Sea, traveling north, northwest, or northeast from its point of origin, and usu. producing heavy rain.
**hurricane deck** *n.* The upper deck on a passenger steamship.
**hurricane lamp** *n.* A lamp having a candle or electric bulb protected by a glass chimney.
**hur·ried** (hûr'ēd, hŭr'-) *adj.* **1.** Moving or acting rapidly : RUSHED. **2.** Done in great haste. **—hur'ried·ly** *adv.* **—hur'ried·ness** *n.*
**hur·ry** (hûr'ē, hŭr'-) *v.* **-ried, -ry·ing, -ries.** [Perh. ME *horien.*] —*vi.* To move or act with haste or speed <*hurried* to school> —*vt.* **1.** To cause to move or act more rapidly : HASTEN. **2.** To cause to move or act too quickly : RUSH <Don't *hurry* me!> **3.** To speed the progress or completion of : EXPEDITE. —*n., pl.* **-ries.** **1.** The act of hurrying. **2.** The need or wish to hurry <We're in a *hurry* to get home.> **—hur'ri·er** *n.*
**hur·ry-scur·ry** *also* **hur·ry-skur·ry** (hûr'ē-skûr'ē, hŭr'ē-skŭr'ē) [Redup. of HURRY.] —*vi.* **-ried, -ry·ing, -ries.** To move or act with excessive hurry and confusion. —*n., pl.* **-ries.** Confused haste : COMMOTION. **—hur'ry-scur'ry** *adj. & adv.*
**hurt** (hûrt) *v.* **hurt, hurt·ing, hurts.** [ME *hurten,* perh. < OFr. *hurter,* to bang into, of Germanic orig.] —*vt.* **1.** To cause physical pain or damage to : INJURE. **2.** To cause mental or emotional suffering to : AFFLICT. **3.** To impair <*hurt* one's chances for a promotion> —*vi.* **1.** To have a feeling of pain or discomfort <My head *hurts.*> **2.** To cause suffering, distress, or damage <The rent increase *hurts.*> —*n.* **1.** Something that hurts, as a pain, injury, or wound. **2.** Mental suffering : ANGUISH. **3.** A wrong : harm. **—hurt'er** *n.*
**hurt·ful** (hûrt'fəl) *adj.* Causing hurt or injury : DAMAGING. **—hurt'ful·ly** *adv.* **—hurt'ful·ness** *n.*
**hur·tle** (hûr'tl) *v.* **-tled, -tling, -tles.** [ME *hurtlen,* to collide, freq. of *hurten,* to hurt.] —*vi.* To move with or as if with great speed and a rushing noise. —*vt.* To throw or fling with great force : HURL.
**hurt·less** (hûrt'lĭs) *adj.* **1.** Causing no hurt : HARMLESS. **2.** Not hurt.
**hus·band** (hŭz'bənd) *n.* [ME *huseband* < OE *hūsbōnda* < ON *hūsbōndi* : *hūs,* house + *bōndi,* dwelling < *būa,* to dwell.] **1.** A married man. **2.** *Archaic.* A manager or steward, as of a household. **3.** A prudent and thrifty manager. —*vt.* **-band·ed, -band·ing, -bands.** **1.** To use or manage economically : CONSERVE <*husbanded* our resources> **2.** *Archaic.* To find a husband for. **—hus'band·ly** *adj.*
**hus·band·man** (hŭz'bənd-mən) *n.* [ME *husbondman.*] One whose occupation is husbandry : FARMER.
**hus·band·ry** (hŭz'bən-drē) *n.* [ME *husbondri* < *huseband,* husband.] **1. a.** The cultivation of crops and the breeding and raising of livestock : AGRICULTURE. **b.** The application of scientific principles, esp. to animal breeding. **2.** Careful management of resources.
**hush** (hŭsh) *v.* **hushed, hush·ing, hush·es.** [Prob. back-formation < ME *husht,* silent.] —*vt.* **1.** To make quiet or silent. **2.** To soothe or calm. **3.** To keep from public knowledge : SUPPRESS. —*vi.* To be or become silent or still. —*n.* A stillness or silence, esp. after noise. —*adj. Archaic.* Silent : quiet.
**hush-hush** (hŭsh'hŭsh') *adj. Informal.* Confidential : secret.
**hush money** *n. Informal.* A bribe or payment made to keep something secret.
**hush·pup·py** (hŭsh'pŭp'ē) *n.* [From its use as food for dogs.] A fried cornmeal fritter.
**husk** (hŭsk) *n.* [ME.] **1.** The membranous or green outer envelope of many fruits and seeds, as of an ear of corn or a nut. **2.** An often worthless outer layer. **3.** A framework used as a support. —*vt.* **husked, husk·ing, husks.** To remove the husk from.
**husking bee** *n.* CORNHUSKING 2.
**husk·y¹** (hŭs'kē) *adj.* **-i·er, -i·est.** [< HUSK.] **1.** Hoarse or deep, as from emotion <a *husky,* tearful voice> **2. a.** Like a husk. **b.** Having husks. **—husk'i·ly** *adv.* **—husk'i·ness** *n.*
**husk·y²** (hŭs'kē) *adj.* **-i·er, -i·est.** [Perh. < HUSK.] *Informal.* Strong and rugged : BURLY. —*n., pl.* **-ies.** A husky person. **—husk'i·ness** *n.*
**hus·ky³** (hŭs'kē) *n., pl.* **-kies.** [Prob. shortening and alteration of ESKIMO.] **1.** *often* **Husky.** A dog of a breed developed in Siberia for pulling sleds, with a dense, furry, variously colored coat. **2.** A dog orig. bred in the Arctic, similar to a husky.
**hus·sar** (hə-zär', -sär') *n.* [Hung. *huszár* < Serbian *husar* < OItal. *corsaro.* —see CORSAIR.] **1.** A horseman of the Hungarian light cav-

alry organized during the 15th cent. **2.** A member of any of various European units of light cavalry.

**Huss·ite** (hŭs′īt′, hōōs′-) *n.* A follower of the Bohemian religious reformer John Huss. —*adj.* Of or relating to John Huss or his religious theories. **—Huss′it′ism** *n.*

**hus·sy** (hŭz′ē, hŭs′ē) *n., pl.* **-sies.** [Alteration of HOUSEWIFE.] **1.** A saucy or mischievous girl. **2.** An immoral woman.

▲ word history: The word *hussy* represents the normal phonetic development of the Middle English compound *hūswīf,* which is also the ancestor of Modern English *housewife.* The development of *hūswīf* into *hussy* was completed by the 17th century and had the following states: *huswif* (shortening of the vowels), *hussif* (elision of *w*), *hussi* (loss of the final consonant), and finally *hussy* (respelling of final *i* as *y*). The form *housewife* preserves the long vowels of the Middle English form, which were changed to diphthongs in the 15th and 16th centuries. *Hussy* and *housewife* were originally synonymous, both meaning "mistress of a household." As *hussy* developed pejorative senses, *housewife* became restricted, when referring to persons, to the original sense of the word.

**hust·ings** (hŭs′tĭngz) *pl.n.* [ME *husting,* court of common pleas < OE *hūsting,* meeting < ON *hūsđing* : *hūs,* house + *đing,* assembly.] *(sing. or pl. in number).* **1.** A court formerly held in some English cities and still held at rare intervals in London. **2.** *Chiefly Brit.* **a.** A platform on which candidates for Parliament formerly stood to address the electors. **b.** The proceedings at a parliamentary election. **3. a.** A place where political speeches are made. **b.** An act or instance of political campaigning.

**hus·tle** (hŭs′əl) *v.* **-tled, -tling, -tles.** [Du. *husselen,* to shake < MDu. *hustelen,* freq. of *hutsen.*] —*vt.* **1.** To shove or jostle roughly. **2.** *Informal.* To move hurriedly or urgently <*hustled* the culprit out of the room> **3.** *Informal.* To urge forward : hurry along. **4.** *Slang.* **a.** To obtain by energetic effort. **b.** To sell or obtain by questionable means. —*vi.* **1.** To jostle : push. **2.** *Informal.* To perform or move quickly and energetically. **3.** *Slang.* To obtain something by deceitful and underhand methods. **4.** *Slang.* To solicit customers for or as a prostitute. —*n. Informal.* **1.** An act or instance of hustling. **2.** Busy activity. **—hus′tler** *n.*

**hut** (hŭt) *n.* [Fr. *hutte,* of Germanic orig.] **1.** A crudely built dwelling or shelter. **2.** A temporary structure for sheltering troops. —*vt.* & *vi.* **hut·ted, hut·ting, huts.** To shelter or take shelter in a hut.

**hutch** (hŭch) *n.* [ME *huche* < OFr., fishpond < Med. Lat.] **1.** A pen or coop for small animals, esp. rabbits. **2.** A cupboard with drawers for storage and usu. open shelves above. **3.** A chest or bin for storage. **4.** A hut.

**hutz·pah** (hōōt′spə, KHōōt′-) *n. var. of* CHUTZPAH.

**Huy·gens′ principle** (hī′gənz) *n.* [After Christiaan *Huygens* (1629–1695).] The principle that any point on a wave front may be regarded as the source of a secondary wave and that the position of the wave front at any time is determined by the envelope at that time of the secondary waves arising from a previous wave front.

**huz·za** *also* **huz·zah** (hə-zä′) [Orig. unknown.] *Archaic.* —*n.* A shout of encouragement or triumph : CHEER. —*interj.* —Used to express joy, encouragement, or triumph.

**hy·a·cinth** (hī′ə-sĭnth) *n.* [Lat. *hyacinthus* < Gk. *huakinthos,* wild hyacinth.] **1. a.** A bulbous plant of the genus *Hyacinthus,* native to the Mediterranean region, bearing narrow leaves and a terminal cluster of variously colored, usu. very fragrant flowers, esp. the widely cultivated species *H. orientalis.* **b.** Any of various related or similar plants. **2.** *Gk. Myth.* A plant, perhaps a lily, gladiolus, or iris, that sprang from the blood of the slain Hyacinthus. **3.** A deep purplish blue to vivid violet. **4. a.** A reddish or cinnamon-colored variety of transparent zircon, used as a gemstone. **b.** A blue precious stone, perhaps a sapphire, known in antiquity. **—hy′a·cin′thine** (-sĭn′thĭn, -thīn′) *adj.*

**hyacinth bean** *n.* A twining vine, *Dolichos lablab* of the Old World tropics, with purple or white flowers and edible pods and seeds.

**Hy·a·cin·thus** (hī′ə-sĭn′thəs) *n.* [Lat. < Gk. *Huakinthos.*] *Gk. Myth.* A beautiful youth, loved but accidentally killed by Apollo, from whose blood Apollo caused the hyacinth to grow.

**Hy·a·des** (hī′ə-dēz′) *pl.n.* [Lat. < Gk. *Huades.*] **1.** *Gk. Myth.* The five daughters of Atlas and sisters of the Pleiades, placed by Zeus in the heavens. **2.** *Astron.* An asterism of five stars in the constellation Taurus, believed by ancient astronomers to indicate rain when they rose with the sun.

**hy·ae·na** (hī-ē′nə) *n. var. of* HYENA.

**hy·a·lin** (hī′ə-lĭn) *also* **hy·a·line** (-lĭn, -līn′) *n.* [Gk. *hualos,* glass + -IN.] **1.** *Physiol.* The uniform matrix of hyaline cartilage. **2.** *Pathol.* A transparent substance occurring in certain degenerative skin conditions.

**hy·a·line** (hī′ə-lĭn, -līn′) *adj.* [LLat. *hyalinus* < Gk. *hualinos* < *hualos,* glass.] Translucent or transparent like glass. —*n.* **1.** A glassy or transparent appearance. **2.** Something translucent or transparent. **3.** *var. of* HYALIN.

**hyaline cartilage** *n.* Cartilage having a glassy translucent appearance and a bluish color, which in the adult is composed of cells in a seemingly homogeneous, translucent matrix, as in joints, and which in the fetus forms most of the skeleton.

**hy·a·lite** (hī′ə-līt′) *n.* [G. *Hyalit* < Gk. *hualos,* glass.] A clear colorless opal.

**hy·a·loid** (hī′ə-loid′) *adj.* [Gk. *hualoeidēs* : *hualos,* glass + *-eidēs,* -oid.] Transparent or glassy in appearance : HYALINE.

**hy·a·lo·plasm** (hī′ə-lō-plăz′əm) *n.* [G. *Hyaloplasma* : Gk. *hualos,* glass + G. *-plasma,* -plasm.] The clear fluid portion of cytoplasm as distinguished from its granular and netlike components.

**hy·brid** (hī′brĭd) *n.* [Lat. *hybrida.*] **1.** *Genetics.* The offspring of genetically dissimilar parents or stock, esp. the offspring produced by breeding plants or animals of different varieties, species, or races. **2.** Something of mixed origin or composition. **3.** A word whose elements are derived from different languages. **—hy′brid·ism** *n.* **—hy·brid′i·ty** (hī-brĭd′ĭ-tē) *n.*

**hy·brid·ize** (hī′brĭ-dīz′) *vi.* & *vt.* **-ized, -iz·ing, -iz·es.** To produce or cause to produce hybrids : CROSSBREED. **—hy′brid·i·za′tion** *n.* **—hy′brid·iz′er** *n.*

**hy·brid·o·ma** (hī′brĭ-dō′mə) *n.* A cell resulting from the hybridization of a lymphocyte and tumor cell, used to produce a specific antibody.

**hybrid vigor** *n.* Heterosis.

**hy·da·thode** (hī′də-thōd′) *n.* [Gk. *hudōr, hydat-,* water + *hodos,* road.] A microscopic epidermal structure in many plants through which water is excreted.

**hy·da·tid** (hī′də-tĭd) *n.* [Gk. *hudatis, hudatid-,* watery vesicle < *hudōr,* water.] **1.** A cyst formed as a result of infestation by a larval form of a tapeworm, *Echinococcus granulosus.* **2.** The encysted larva of *E. granulosus.*

**hydr-** *pref. var. of* HYDRO-.

**Hy·dra** (hī′drə) *n.* [ME *Idra* < Lat. *Hydra* < Gk. *Hudra.*] **1.** *Gk. Myth.* A many-headed monster slain by Hercules. **2.** A constellation in the equatorial region of the southern sky. **3. hydra** *pl.* **-dras** or **-drae** (-drē). A small freshwater polyp of the genus *Hydra* or related genera, with a naked cylindrical body and an oral opening surrounded by tentacles. **4. hydra.** A multifarious source of evil or destruction that cannot be eradicated by a single attempt.

**hydra**
*Up to one-half inch in length*

**hy·dral·a·zine** (hī-drăl′ə-zēn′) *n.* [HYDR(O)- + (PHTH)AL(IC ACID) + AZINE.] An antihypertensive drug, $C_8H_8N_4$.

**hy·dran·gea** (hī-drān′jə, -drăn′-) *n.* [NLat. *Hydrangea,* genus name : Gk. *hudōr,* water + Gk. *angos,* vessel.] A shrub or tree of the genus *Hydrangea,* bearing large flat-topped or rounded clusters of white, pink, or blue flowers.

**hy·drant** (hī′drənt) *n.* An upright pipe with a spout or nozzle for drawing water from a water main.

**hy·dranth** (hī′drănth′) *n.* [HYDR(O)- + Gk. *anthos,* flower.] The oral opening and tentacles of a feeding polyp in a hydroid colony.

**hy·drase** (hī′drās′, -drāz′) *n.* An enzyme that catalyzes the addition or removal of water from a substrate.

**hy·dras·tine** (hī-drăs′tēn′, -tĭn) *n.* [NLat. *Hydrastis,* plant genus + -INE².] A poisonous white alkaloid, $C_{21}H_{21}NO_6$, derived from the root of the goldenseal, *Hydrastis canadensis,* and used to treat catarrhal inflammation of mucous membranes.

**hy·drate** (hī′drāt′) *n.* A compound containing water combined in a definite ratio, the water being retained or thought of as being retained in its molecular state. —*v.* **-drat·ed, -drat·ing, -drates.** —*vt.* To combine with water, esp. to form a hydrate. —*vi.* To become a hydrate. **—hy·dra′tion** *n.* **—hy′dra′tor** *n.*

**hy·drat·ed** (hī′drā′tĭd) *adj.* Chemically combined with water, esp. in the form of a hydrate.

**hy·drau·lic** (hī-drô′lĭk) *adj.* [Lat. *hydraulicus* < Gk. *hudraulis,* water organ : *hudor,* water + *aulos,* pipe.] **1.** Of, involving, moved, or operated by a pressurized fluid, esp. water. **2.** Setting and hardening under water, as Portland cement. **3.** Of or relating to hydraulics. **—hy·drau′li·cal·ly** *adv.*

**hydraulic brake** *n.* A brake in which the braking force is transmitted to the braking surface by a compressed fluid.

**hydraulic cement** *n.* A cement capable of setting and hardening under water.

**hydraulic press** *n.* A machine in which a large force is exerted on the larger of two pistons in a pair of hydraulically coupled cylinders by means of a relatively small force applied to the smaller piston.

**hydraulic ram** *n.* **1.** A water pump in which the downward flow of naturally running water is intermittently stopped by a valve so that the flow is forced upward through an open pipe into a reservoir. **2.** The large output piston of a hydraulic press.

**hy·drau·lics** (hī-drô′lĭks) *n. (sing. in number).* The physical science and technology of the static and dynamic behavior of fluids.

**hy·dra·zine** (hī′drə-zēn′, -zĭn) *n.* A colorless, fuming, corrosive hygroscopic liquid, $H_2NNH_2$, used in jet and rocket fuels.

**hy·dric** (hī′drĭk) *adj.* **1.** Of, having, or relating to hydrogen. **2.** Relating to, marked by, or requiring considerable moisture.

**hy·dride** (hī′drīd′) *n.* A compound of hydrogen with another, more electropositive element or group.

**hy·dri·od·ic acid** (hī′drē-ŏd′ĭk) *n.* A clear colorless or pale-yellow aqueous solution of hydrogen iodide, HI, a strong acid and reducing agent.

**hydro-** *or* **hydr-** *pref.* [< Gk. *hudōr,* water.] **1. a.** Water <*hydro*electric> **b.** Liquid <*hydro*dynamics> **2.** Hydrogen <*hydro*chloride>

**hy·dro·bi·ol·o·gy** (hī′drō-bī-ŏl′ə-jē) *n.* Limnology. **—hy′dro·bi′o·log′i·cal** (-bī′ə-lŏj′ĭ-kəl) *adj.* **—hy′dro·bi·ol′o·gist** *n.*

**hy·dro·bro·mic acid** (hī′drə-brō′mĭk) *n.* A clear colorless or faintly yellow highly acidic and corrosive aqueous solution of hydrogen bromide, HBr, used to make bromides.

**hy·dro·car·bon** (hī′drə-kär′bən) *n.* An organic compound, as methane or benzene, that contains only carbon and hydrogen. **—hy′dro·car′bo·na′ceous** (-bə-nā′shəs), **hy′dro·car·bon′ic** (-bŏn′ĭk), **hy′dro·car′bon·ous** (-bə-nəs) *adj.*

**hy·dro·cele** (hī′drə-sēl′) *n.* [Lat. < Gk. *hudrokēlē* : *hudōr,* water + *kēlē,* tumor.] *Pathol.* An accumulation of serous fluid in a bodily cavity, esp. in the testes.

**hy·dro·ceph·a·lus** (hī′drō-sĕf′ə-ləs) *also* **hy·dro·ceph·a·ly** (-lē) *n.* [NLat. < Gk. *hudrokephalon* : *hudōr,* water + *kephalē,* head.] A usu. congenital condition in which an abnormal accumulation of fluid in the cerebral ventricles results in enlargement of the skull and compression of the brain. **—hy′dro·ce·phal′ic** (-sə-făl′ĭk), **hy′dro·ceph′a·loid′, hy′dro·ceph′a·lous** *adj.*

**hy·dro·chlo·ric acid** (hī′drə-klôr′ĭk, -klōr′-) *n.* A clear, colorless, fuming, poisonous, highly acidic aqueous solution of hydrogen chloride, HCl, used as a chemical intermediate, in petroleum production, and in food processing, pickling, metal cleaning, and ore reduction.

**hy·dro·chlo·ride** (hī′drə-klôr′īd′, -klōr′-) *n.* A compound resulting or thought of as resulting from the reaction of hydrochloric acid with an organic base.

**hy·dro·col·loid** (hī′drə-kŏl′oid′) *n.* A substance that forms a gel with water. **—hy′dro·col·loid′al** (-kə-loid′l) *adj.*

**hy·dro·cor·al** (hī′drə-kôr′əl, -kŏr′-) *n.* Any of various colonial marine hydrozoans of the order Hydrocorallinae, having a limestone skeleton and thus resembling the true corals.

**hy·dro·cor·ti·sone** (hī′drə-kôr′tĭ-sōn′, -zōn′) *n.* A bitter crystalline hormone, $C_{21}H_{30}O_5$, derived from the adrenal cortex and used in medicines like cortisone.

**hy·dro·cy·an·ic acid** (hī′drō-sī-ăn′ĭk) *n.* A colorless, volatile, extremely toxic, flammable aqueous solution of hydrogen cyanide, HCN, used in making dyes, fumigants, and plastics.

**hy·dro·dy·nam·ic** (hī′drō-dī-năm′ĭk) *also* **hy·dro·dy·nam·i·cal** (-ĭ-kəl) *adj.* **1.** Of or relating to hydrodynamics. **2.** Of, relating to, or operated by the force of moving liquid. **—hy′dro·dy·nam′i·cal·ly** *adv.*

**hy·dro·dy·nam·ics** (hī′drō-dī-năm′ĭks) *n. (sing. in number).* The dynamics of fluids, esp. incompressible fluids, in motion. **—hy′dro·dy·nam′i·cist** *n.*

**hy·dro·e·lec·tric** (hī′drō-ĭ-lĕk′trĭk) *adj.* **1.** Generating electricity by conversion of the energy of running water. **2.** Of, relating to, or using electricity so generated. **—hy′dro·e·lec′tri·cal·ly** *adv.* **—hy′dro·e·lec·tric′i·ty** (-ĭ-lĕk-trĭs′ĭ-tē) *n.*

**hy·dro·flu·or·ic acid** (hī′drō-flōō-ôr′ĭk, -ŏr′-, -flôr′ĭk) *n.* A colorless, fuming, corrosive, dangerously poisonous aqueous solution of hydrogen fluoride, HF, used to clean masonry, pickle certain metals, and etch or polish glass.

**hy·dro·foil** (hī′drə-foil′) *n.* **1.** One of a set of blades affixed to the hull of a boat and aligned in the water at a small angle to the horizontal so that when the boat is in motion the fluid striking each blade's underside creates a high-pressure region below the blade, low pressure above it, and a resultant lift that raises the craft out of the water for efficient high-speed operation. **2.** A boat with hydrofoils.

**hy·dro·form·ing** (hī′drə-fôr′mĭng) *n.* A process in which naphthas are converted to high-octane aromatics in the presence of hydrogen and a catalyst under pressure and heat. **—hy′dro·form′er** *n.*

**hy·dro·gen** (hī′drə-jən) *n.* [Fr. *hydrogène* : Gk. *hudōr,* water + *-gène,* -gen.] *Symbol* **H** A colorless, highly flammable gaseous element used in producing synthetic ammonia and methanol, in petroleum refining, as a reducing atmosphere, and in oxyhydrogen torches and rocket fuels; atomic number 1; atomic weight 1.00797. **—hy·drog′e·nous** (-drŏj′ə-nəs) *adj.*

**hy·drog·e·nase** (hī-drŏj′ə-nās′, -nāz′) *n.* An enzyme that catalyzes the formation of hydrogen in certain microorganisms.

**hy·dro·gen·ate** (hī′drə-jə-nāt′, hī-drŏj′ə-) *vt.* **-at·ed, -at·ing, -ates.** To combine with or subject to the action of hydrogen, esp. to combine (an unsaturated compound) with hydrogen. **—hy′dro·gen·a′tion** *n.* **—hy′dro·gen·a′tor** *n.*

**hydrogen bomb** *n.* An explosive weapon of great destructive power produced by the fusion of nuclei of various hydrogen isotopes in the formation of helium nuclei.

**hydrogen bond** *n.* A primarily ionic chemical bond between a strongly electronegative atom and a hydrogen atom already bonded to another strongly electronegative atom.

**hydrogen bromide** *n.* An irritating colorless gas, HBr, used to make synthetic hormones and barbiturates.

**hydrogen chloride** *n.* A colorless, fuming, corrosive, suffocating gas, HCl, used in making plastics.

**hydrogen cyanide** *n.* Hydrocyanic acid.

**hydrogen fluoride** *n.* A colorless, fuming, mobile, corrosive liquid or a highly soluble corrosive gas, HF, used in making hydrofluoric acid, as a reagent, catalyst, and fluorinating agent, and in the refining of uranium and the preparation of many compounds.

**hydrogen iodide** *n.* A corrosive, colorless, suffocating gas, HI, used to make hydriodic acid.

**hydrogen ion** *n.* The positively charged ion of hydrogen, $H^+$, formed by removal of the electron from atomic hydrogen.

**hy·dro·gen·ol·y·sis** (hī′drō-jə-nŏl′ĭ-sĭs) *n.* The breaking of a chemical bond in an organic molecule with the simultaneous addition of a hydrogen atom to each of the resulting molecular fragments.

**hydrogen peroxide** *n.* An unstable, colorless, heavy, strongly oxidizing liquid, $H_2O_2$, capable of reacting explosively with combustibles and used chiefly in aqueous solution as an antiseptic, bleaching or oxidizing agent, and laboratory reagent.

**hydrogen sulfide** *n.* A colorless, flammable, poisonous compound, $H_2S$, having a characteristic rotten-egg odor and used as a precipitator, purifier, and reagent.

**hy·drog·ra·phy** (hī-drŏg′rə-fē) *n., pl.* **-phies.** [OFr. *hydrographie* : *hydro-,* hydro- + *-graphie,* -graphy.] **1.** The scientific description and analysis of the physical conditions, boundaries, flow, and related characteristics of surface waters, as oceans, lakes, and rivers. **2.** The mapping of bodies of water. **3.** A book on hydrography. **—hy·drog′ra·pher** *n.* **—hy′dro·graph′ic** (hī′drə-grăf′ĭk) *adj.* **—hy′dro·graph′i·cal·ly** *adv.*

**hy·droid** (hī′droid′) *n.* [NLat. *Hydra,* hydra genus + -OID.] **1.** Any of numerous colonial hydrozoan coelenterates having a polyp rather than a medusoid form as the dominant stage of the life cycle. **2.** The asexual polyp in the life cycle of a hydrozoan. *—adj.* Of, relating to, or typical of a hydroid.

**hy·dro·ki·net·ic** (hī′drō-kĭ-nĕt′ĭk, -kī-) *also* **hy·dro·ki·net·i·cal** (-ĭ-kəl) *adj.* **1.** Of or relating to hydrokinetics. **2.** Of or relating to the kinetic energy and motion of fluids.

**hy·dro·ki·net·ics** (hī′drō-kĭ-nĕt′ĭks, -kī-) *n. (sing. in number).* The kinetics of fluids, esp. incompressible fluids, in motion.

**hy·drol·o·gy** (hī-drŏl′ə-jē) *n.* The scientific study of the properties, distribution, and effects of water in the atmosphere, on the earth's surface, and in soil and rocks. **—hy′dro·log′ic** (-drə-lŏj′ĭk), **hy′dro·log′i·cal** *adj.* **—hy′dro·log′i·cal·ly** *adv.* **—hy·drol′o·gist** *n.*

**hy·drol·y·sate** (hī-drŏl′ĭ-sāt′, hī′drə-lī′-) *also* **hy·drol·y·zate** (-zāt′) *n.* [HYDROLYS(IS) + -ATE.] A product of hydrolysis.

**hy·drol·y·sis** (hī-drŏl′ĭ-sĭs) *n.* Decomposition of a chemical compound by reaction with water, as the dissociation of a dissolved salt or the catalytic conversion of glucose to starch. **—hy′dro·lyte′** (-līt′) *n.* **—hy′dro·lyt′ic** (-drə-lĭt′ĭk) *adj.* **—hy′dro·lyt′i·cal·ly** *adv.*

**hy·dro·lyze** (hī′drə-līz′) *vt.* & *vi.* **-lyzed, -lyz·ing, -lyz·es.** To subject to or undergo hydrolysis. **—hy′dro·lyz′a·ble** *adj.* **—hy′dro·ly·za′tion** *n.*

**hy·dro·mag·net·ics** (hī′drō-măg-nĕt′ĭks) *n.* Magnetohydrodynamics. **—hy′dro·mag·net′ic** *adj.*

**hy·dro·man·cy** (hī′drə-măn′sē) *n.* [ME *ydromancy* < OFr. *hydromancie* < Lat. *hydromantia* < Gk. *hudromanteia* : *hudōr,* water + *manteia,* divination. —see -MANCY.] Divination by water.

**hy·dro·me·chan·ics** (hī′drō-mĭ-kăn′ĭks) *n. (sing. in number).* Mechanics dealing with the motion and equilibrium states of fluids. **—hy′dro·me·chan′i·cal** *adj.*

**hy·dro·me·du·sa** (hī′drō-mĭ-dōō′sə, -dyōō′-) *n., pl.* **-sas** *or* **-sae** (-sē) [HYDRO(ID) + MEDUSA.] A hydrozoan in its medusoid stage. **—hy′dro·me·du′soid′** (-dōō′soid′, -dyōō′-) *adj.*

**hy·dro·mel** (hī′drə-mĕl′) *n.* [ME *ydromel* < OFr. < Lat. *hydromel* < Gk. *hudromeli* : *hudōr,* water + *meli,* honey.] MEAD¹.

**hy·dro·met·al·lur·gy** (hī′drō-mĕt′l-ûr′jē) *n.* The separation of metal from ores and ore concentrates by chemical reactions in aqueous solution, as leaching, extraction, and precipitation. **—hy′dro·met′al·lur′gi·cal** *adj.*

**hy·dro·me·te·or** (hī′drō-mē′tē-ər, -ôr′) *n.* A precipitation body, as rain, snow, sleet, or hail, derived from the condensation of water in the atmosphere.

**hy·dro·me·te·or·ol·o·gy** (hī′drō-mē′tē-ə-rŏl′ə-jē) *n.* The meteorology of the occurrence, motion, and changes of state of atmospheric water. **—hy′dro·me′te·or′o·log′i·cal** (-ôr′ə-lŏj′ĭ-kəl, -ŏr′-) *adj.* **—hy′dro·me′te·or·ol′o·gist** *n.*

**hy·drom·e·ter** (hī-drŏm′ĭ-tər) *n.* An instrument for determining specific gravity, esp. a sealed, graduated tube, weighted at one end, that sinks in a fluid to a depth used as a measure of the fluid's specific gravity. **—hy′dro·met′ric** (hī′drə-mĕt′rĭk), **hy′dro·met′ri·cal** *adj.* **—hy′dro·met′ri·cal·ly** *adv.* **—hy·drom′e·try** *n.*

**hy·dro·ni·um** (hī-drō′nē-əm) *n.* [HYDR(O)- + (AMM)ONIUM.] A hydrated hydrogen ion, $H_3O^+$.

**hy·drop·a·thy** (hī-drŏp′ə-thē) *n.* Therapeutic use of water. **—hy′dro·path′ic** (hī′drə-păth′ĭk), **hy′dro·path′i·cal** *adj.* **—hy·drop′a·thist, hy′dro·path′** *n.*

**hy·dro·phane** (hī′drə-fān′) *n.* An opal almost opaque when dry but transparent when wet. **—hy·droph′a·nous** (hī-drŏf′ə-nəs) *adj.*

**hy·dro·phil·ic** (hī′drə-fĭl′ĭk) *adj.* [NLat. *hydrophilus,* hydrophilous + -IC.] Having an affinity for, absorbing, wetting smoothly with, tending to combine with, or capable of dissolving in water. **—hy′dro·phile′** (hī′drə-fīl′) *n.* **—hy′dro·phi·lic′i·ty** (-fə-lĭs′ĭ-tē) *n.*

**hy·droph·i·lous** (hī-drŏf′ə-ləs) *adj. Bot.* Growing or thriving in water. **—hy·droph′i·ly** *n.*

**hy·dro·pho·bi·a** (hī′drə-fō′bē-ə) *n.* **1.** Fear of water. **2.** Rabies.

**hy·dro·pho·bic** (hī′drə-fō′bĭk, -fŏb′ĭk) *adj.* **1.** Antagonistic to, shedding, tending not to combine with, or incapable of dissolving in water. **2.** Of or exhibiting hydrophobia.

**hy·dro·phone** (hī′drə-fōn′) *n.* An electrical instrument for detecting or monitoring sound under water.

**hy·dro·phyte** (hī′drə-fīt′) *n.* A plant growing in and adapted to an aquatic or very wet environment. **—hy′dro·phyt′ic** (-fĭt′ĭk) *adj.*

**hy·dro·plane** (hī′drə-plān′) *n.* **1.** A seaplane. **2.** A motorboat designed so the prow and much of the hull lift out of the water and skim the surface at high speeds. **3.** HYDROFOIL 2. **4.** A horizontal rudder on a submarine. *—vi.* **-planed, -plan·ing, -planes. 1.** To pilot or ride in a hydroplane. **2. a.** To skim along on the surface of the water. **b.** To go out of control by skimming along the surface of a wet road. —Used of a car. **—hy′dro·plan′er** *n.*

**hy·dro·pon·ics** (hī′drə-pŏn′ĭks) *n.* [HYDRO- + (GEO)PONICS.] *(sing. in number).* The cultivation of plants in water containing dissolved inorganic nutrients. **—hy′dro·pon′ic** *adj.* **—hy′dro·pon′i·cal·ly** *adv.* **—hy′dro·pon′i·cist, hy′dro·pon′ist** *n.*

**hy·dro·qui·none** (hī′drō-kwĭ-nōn′, -kwĭn′ōn′) *also* **hy·dro·quin·ol** (-kwĭn′ôl′, -ōl′) *n.* A white crystalline compound, $C_6H_4(OH)_2$, used as a photographic developer, antioxidant, stabilizer, and reagent.

**hy·dro·scope** (hī′drə-skōp′) *n.* An optical instrument for viewing objects deep in the water. **—hy′dro·scop′ic** (-skŏp′ĭk) *adj.*

**hy·dro·sol** (hī′drə-sôl′, -sōl′) *n.* [HYDRO- + SOL(UTION).] A sol with water as the dispersing medium. **—hy′dro·sol′ic** (-sŏl′ĭk) *adj.*

**hy·dro·sphere** (hī′drə-sfîr′) *n.* The waters of the earth as distinguished from the lithosphere and the atmosphere. **—hy′dro·spher′ic** (-sfîr′ĭk, -sfĕr′-) *adj.*

**hy·dro·stat·ic** (hī′drə-stăt′ĭk) *also* **hy·dro·stat·i·cal** (-ĭ-kəl) *adj.* Of or pertaining to hydrostatics. **—hy′dro·stat′i·cal·ly** *adv.*

**hy·dro·stat·ics** (hī′drə-stăt′ĭks) *n. (sing. in number).* The statics of fluids, esp. incompressible fluids.

**hy·dro·sul·fate** (hī′drə-sŭl′fāt′) *n.* A salt formed by the union of sulfuric acid with an alkaloid or other organic base.

**hy·dro·sul·fide** (hī′drə-sŭl′fīd′) *n.* A chemical compound derived from hydrogen sulfide by replacement of one of the hydrogen atoms with a basic radical or base.

**hy·dro·sul·fite** (hī′drə-sŭl′fīt′) *n.* **1.** A salt of hyposulfurous acid. **2.** Sodium hydrosulfite.

**hy·dro·sul·fu·rous acid** (hī′drō-sŭl-fyŏor′əs, -sŭl′fər-əs) *n.* Hyposulfurous acid.

**hy·dro·tax·is** (hī′drə-tăk′sĭs) *n. Biol.* Movement of an organism in response to moisture. **—hy′dro·tac′tic** (-tăk′tĭk) *adj.*

**hy·dro·ther·a·peu·tics** (hī′drə-thĕr′ə-pyōō′tĭks) *n. (sing. in number).* Hydrotherapy. **—hy′dro·ther′a·peu′tic** *adj.*

**hy·dro·ther·a·py** (hī′drə-thĕr′ə-pē) *n., pl.* **-pies.** The medical use of water in the treatment of disease.

**hy·dro·ther·mal** (hī′drə-thûr′məl) *adj.* **1.** Of or pertaining to hot water. **2.** *Geol.* **a.** Of or pertaining to hot magmatic emanations that are rich in water. **b.** Of or pertaining to the rocks, ore deposits, and springs produced by such emanations. **—hy′dro·ther′mal·ly** *adv.*

**hy·dro·tho·rax** (hī′drə-thôr′ăks′, -thōr′-) *n.* The presence of serous fluid from the blood in one or both pleural cavities, often associated with cardiac failure.

**hy·drot·ro·pism** (hī-drŏt′rə-pĭz′əm) *n.* Growth or movement of an organism in response to water. **—hy′dro·tro′pic** (hī′drə-trō′pĭk, -trŏp′ĭk) *adj.* **—hy′dro·tro′pi·cal·ly** *adv.*

**hy·drous** (hī′drəs) *adj.* Containing water, esp. that of crystallization or hydration.

**hy·drox·ide** (hī-drŏk′sīd′) *n.* A chemical compound containing the hydroxyl group.

**hydroxide ion** *n.* The ion OH−, typical of basic hydroxides.

**hy·drox·y** (hī-drŏk′sē) *adj.* [< HYDROXYL.] Containing the hydroxyl group, as an acid.

**hy·drox·yl** (hī-drŏk′sĭl) *n.* [HYDR(O)- + OX(YGEN) + -YL.] The univalent radical or group OH, typical of bases, certain acids, phenols, alcohols, carboxylic and sulfonic acids, and amphoteric compounds. **—hy′drox·yl′ic** (hī′drŏk-sĭl′ĭk) *adj.*

**hy·drox·yl·a·mine** (hī-drŏk′sə-lə-mēn′, hī′drŏk-sĭl′ə-mēn′) *n.* A colorless crystalline compound, $NH_2OH$, explosive when heated above 130°C, used as a reducing agent and in organic synthesis.

**hy·drox·yl·ate** (hī-drŏk′sə-lāt′) *vt.* **-at·ed, -at·ing, -ates.** To introduce the hydroxyl radical into. **—hy·drox′y·la′tion** *n.*

**hydroxyl ion** *n.* A hydroxide ion.

**hy·dro·zo·an** (hī′drə-zō′ən) *n.* [< NLat. *Hydrozoa,* class name : HYDRO- + Gk. *zōia,* pl. of *zōion,* animal.] Any of numerous coelenterates of the class Hydrozoa, which includes the hydroids, hydrocorals, and siphonophores. *—adj.* Of, relating to, or belonging to the class Hydrozoa.

**Hy·drus** (hī′drəs) *n.* [Lat. < Gk. *hudros,* water serpent.] A constellation in the Southern Hemisphere.

**hy·e·na** *also* **hy·ae·na** (hī-ē′nə) *n.* [ME *hiena* < Med. Lat. < Lat. *hyaena* < Gk. *huaina* < *hus,* hog.] Any of several carnivorous mammals of the genera *Hyaena* or *Crocuta* of Africa and Asia, with powerful jaws, coarse hair, and relatively short hind limbs.

**hy·e·tal** (hī′ĭ-tl) *adj.* [< Gk. *huetos,* rain.] Of or pertaining to rain or to rainy areas.

**Hy·ge·ia** (hī-jē′ə) *n.* [Gk. *Hugieia* < *hugiēs,* healthy.] *Gk. Myth.* The goddess of health.

**hy·giene** (hī′jēn′) *n.* [Fr. *hygiène* and NLat. *hygiena,* both < Gk. *hugieinos,* healthful < *hugiēs,* healthy.] **1.** The science of health and the prevention of disease. **2.** Conditions and practices promoting or preserving health. **—hy·gien′ist** (hī-jē′nĭst, hī′jē′-, hī-jĕn′ĭst) *n.*

**hy·gi·en·ic** (hī′jē-ĕn′ĭk, hī-jĕn′-) *adj.* **1.** Of or relating to hygiene. **2.** Tending to promote or preserve health. **3.** Sanitary. **—hy′gi·en′i·cal·ly** *adv.*

**hy·gi·en·ics** (hī′jē-ĕn′ĭks, hī-jĕn′-) *n. (sing. in number).* HYGIENE 1.

**hygro-** *pref.* [< Gk. *hugros,* moist.] Moisture : humidity <hygroscope>

**hy·gro·graph** (hī′grə-grăf′) *n.* An automatic hygrometer that records variations in atmospheric humidity.

**hy·grom·e·ter** (hī-grŏm′ĭ-tər) *n.* An instrument measuring atmospheric humidity. **—hy′gro·met′ric** (hī′grə-mĕt′rĭk) *adj.* **—hy·grom′e·try** *n.*

**hy·gro·scope** (hī′grə-skōp′) *n.* An instrument measuring changes in atmospheric moisture.

**hy·gro·scop·ic** (hī′grə-skŏp′ĭk) *adj.* Readily absorbing moisture, as from the atmosphere. **—hy′gro·scop′i·cal·ly** *adv.* **—hy′gro·sco·pic′i·ty** (-skō-pĭs′ĭ-tē) *n.*

**hy·ing** (hī′ĭng) *v. var. p.p. of* HIE.

**hy·lo·zo·ism** (hī′lə-zō′ĭz′əm) *n.* [Gk. *hulē,* matter + Gk. *zoē,* life + -ISM.] The philosophical doctrine that life is a property or derivative of matter, that life and matter are inseparable, or that matter possesses a spiritual component. **—hy′lo·zo′ic** *adj.* **—hy′lo·zo′ist** *n.* **—hy′lo·zo·is′tic** (-zō-ĭs′tĭk) *adj.*

**hy·men** (hī′mən) *n.* [LLat. < Gk. *humēn,* membrane.] A membranous fold of tissue partially or completely occluding the external vaginal orifice. **—hy′men·al** *adj.*

**Hy·men** (hī′mən) *n.* [Lat. < Gk. *Humēn.*] *Gk. Myth.* The god of marriage.

**hy·me·ne·al** (hī′mə-nē′əl) *adj.* [Lat. *hymenaeus* < Gk. *humēnaios* < *Humēn,* Hymen.] Of or relating to a wedding or marriage : NUPTIAL. *—n.* A wedding song or poem. **—hy′me·ne′al·ly** *adv.*

**hy·me·ni·um** (hī-mē′nē-əm) *n., pl.* **-ni·a** (-nē-ə) *or* **-ni·ums.** [NLat. < Gk. *humēn,* membrane.] The spore-bearing layer of the fruiting body of certain fungi, containing asci or basidia. **—hy·me′ni·al** (-əl) *adj.*

**hy·me·nop·ter·an** (hī′mə-nŏp′tər-ən) *also* **hy·me·nop·ter·on** (-tə-rŏn′) *n.* [NLat. *Hymenoptera,* order name : Gk. *humēn,* membrane + Gk. *pteron,* wing.] An insect of the order Hymenoptera, which includes the bees, wasps, and ants, having two pairs of membranous wings. *—adj.* **hymenopteran.** Of or belonging to the Hymenoptera. **—hy′me·nop′ter·ous** (-tər-əs) *adj.*

**hymn** (hĭm) *n.* [ME *imne* < OFr. *ymne* < Lat. *hymnus,* song of praise < Gk. *humnos.*] **1.** A song of praise or thanksgiving, esp. to God. **2.** A hymnlike song of praise or joy : PAEAN. *—v.* **hymned, hymn·ing, hymns.** *—vt.* To praise, glorify, or worship in or as if in a hymn. *—vi.* To sing hymns.

**hym·nal** (hĭm′nəl) *n.* [ME *hymnale* < Med. Lat. *hymnale* < Lat. *hymnus,* hymn.] A book or collection of hymns.

**hymn·book** (hĭm′bŏŏk′) *n.* A hymnal.

**hym·no·dy** (hĭm′nə-dē) *n., pl.* **-dies.** [Med. Lat. *hymnodia* < Gk. *humnōidia* : *humnos,* hymn + *ōidē,* song.] **1.** The singing of hymns. **2.** Composition of hymns. **3.** The hymns of a particular place, time, or church. **—hym′no·dist** (-dĭst) *n.*

**hym·nol·o·gy** (hĭm-nŏl′ə-jē) *n.* [Gk. *humnologia,* singing of hymns : *humnos,* hymn + *-logia,* -logy.] Hymnody. **—hym′no·log′ic** (hĭm′nə-lŏj′ĭk), **hym′no·log′i·cal** *adj.* **—hym·nol′o·gist** *n.*

**hy·oid** (hī′oid′) *adj.* [NLat. *hyoides,* the hyoid bone < Gk. *huoeidēs* : *hu,* name of the letter upsilon + *-eidēs,* -oid.] Of or pertaining to the hyoid bone. —*n.* The hyoid bone.

**hyoid bone** *n.* A U-shaped bone between the mandible and the larynx at the base of the tongue.

**hy·o·scine** (hī′ə-sēn′) *n.* [G. *Hyoscin* < NLat. *Hyoscyamus,* henbane genus < Gk. *huoskuamus,* henbane : *hus,* pig + *kuamos,* bean.] Scopolamine.

**hy·o·scy·a·mine** (hī′ə-sī′ə-mēn′) *n.* [< NLat. *Hyoscyamus,* henbane genus. —see HYOSCINE.] A poisonous, white crystalline alkaloid, $C_{17}H_{23}NO_3$, isometric with atropine and used as an antispasmodic, analgesic, and sedative.

**hyp-** *pref. var. of* HYPO-.

**hyp·a·bys·sal** (hĭp′ə-bĭs′əl, hī′pə-) *adj.* Solidifying primarily as a minor intrusion, esp. as a dike or sill, before reaching the earth's surface. —Used of rocks. **—hyp′a·bys′sal·ly** *adv.*

**hy·pae·thral** (hĭ-pē′thrəl) *adj.* [Lat. *hypaethrus* < Gk. *hupaithros* : *hupo,* under + *aithēr,* sky.] **1.** Open to the sky. **2.** Lacking a roof, as an ancient temple.

**hy·pan·thi·um** (hī-păn′thē-əm) *n., pl.* **-thi·a** (-thē-ə). The modified, often enlarged floral receptacle of various plants, having a tubular or cup-shaped form. **—hy·pan′thi·al** *adj.*

**hype**[1] (hīp) [Shortening and alteration of HYPODERMIC.] *Slang.* —*n.* **1.** A hypodermic injection, syringe, or needle. **2.** A drug addict. —*vt.* **hyped, hyp·ing, hypes.** To stimulate with or as if with a hypodermic injection.

**hype**[2] (hīp) [Orig. unknown.] *Slang.* —*n.* **1.** Something intentionally misleading. **2.** Exaggerated or extravagant claims made esp. in advertising or promotional material <"The Christmas *hype* starts at Thanksgiving" —*Vogue*> —*vt.* **hyped, hyp·ing, hypes.** To advertise or promote by misleading or exaggerated claims <"The most dazzling dancers don't get *hyped* into stardom until their late teens" —*New York Times*>

**hyped-up** (hīpt′ŭp′) *adj. Slang.* **1.** Stimulated with or as if with a hypodermic injection. **2.** Excited : enthusiastic.

**hy·per** (hī′pər) *adj.* [Short for HYPERACTIVE.] *Slang.* **1.** Extremely nervous or excitable : HIGH-STRUNG. **2.** Emotionally stimulated or overexcited.

**hyper-** *pref.* [< Gk. *huper,* over, beyond.] **1.** Over : above : beyond <*hyper*charge> **2.** Excessive : excessively <*hyper*critical>

**hy·per·ac·id** (hī′pər-ăs′ĭd) *adj.* Containing excessive acid. **—hy′per·a·cid′i·ty** (-ə-sĭd′ĭ-tē) *n.*

**hy·per·ac·tive** (hī′pər-ăk′tĭv) *adj.* Excessively or abnormally active <a *hyperactive* child> **—hy′per·ac·tiv′i·ty** (-ăk-tĭv′ĭ-tē) *n.*

**hy·per·ae·mi·a** (hī′pə-rē′mē-ə) *n. var. of* HYPEREMIA.

**hy·per·aes·the·sia** (hī′pər-ĭs-thē′zhə) *n. var. of* HYPERESTHESIA.

**hy·per·bar·ic** (hī′pər-băr′ĭk) *adj.* Of, relating to, producing, operating, or occurring at pressures higher than normal atmospheric pressure. **—hy′per·bar′i·cal·ly** *adv.*

**hy·per·bo·la** (hī-pûr′bə-lə) *n., pl.* **-las** or **-lae** (-lē) [NLat. < Gk. *huperbolē.* —see HYPERBOLE.] A plane curve having two branches, formed by: **a.** A conic section intersecting both halves of a right circular cone. **b.** The locus of points related to two given points such that the difference in the distances of each point from the two given points is a constant.

**hy·per·bo·le** (hī-pûr′bə-lē) *n.* [Lat. *hyperbole* < Gk. *huperbolē* < *huperballein,* to exceed : *huper,* beyond + *ballein,* to throw.] An exaggeration or extravagant statement used as a figure of speech, e.g., *I'm so hungry I could eat a horse.*

**hy·per·bol·ic** (hī′pər-bŏl′ĭk) *also* **hy·per·bol·i·cal** (-ĭ-kəl) *adj.* **1.** Of, relating to, or using hyperbole. **2.** *Math.* **a.** Of, relating to, or having the form of a hyperbola. **b.** Based on or having a metric that is a hyperbola <*hyperbolic* geometry> **c.** Of or pertaining to a hyperbolic function <*hyperbolic* tangent> **—hy′per·bol′i·cal·ly** *adv.*

**hyperbolic function** *n.* Any of a set of six functions related, for a real variable *z,* to the hyperbola in a manner analogous to the relationship of the trigonometric functions to a circle, including: **a.** The hyperbolic sine, defined by the equation $\sinh z = \frac{1}{2}(e^z - e^{-z})$. **b.** The hyperbolic cosine, defined by the equation $\cosh z = \frac{1}{2}(e^z + e^{-z})$. **c.** The hyperbolic tangent, defined by the equation $\tanh z = \sinh z/\cosh z$. **d.** The hyperbolic cotangent, defined by the equation $\coth z = \cosh z/\sinh z$. **e.** The hyperbolic secant, defined by the equation $\operatorname{sech} z = 1/\cosh z$. **f.** The hyperbolic cosecant, defined by the equation $\operatorname{csch} z = 1/\sinh z$.

**hyperbolic paraboloid** *n.* A surface of which all sections parallel to one coordinate plane are hyperbolas and all sections parallel to another coordinate plane are parabolas.

**hy·per·bo·lism** (hī-pûr′bə-lĭz′əm) *n.* **1.** The use of hyperbole. **2.** A hyperbole.

**hy·per·bo·lize** (hī-pûr′bə-līz′) *vi.* & *vt.* **-lized, -liz·ing, -liz·es.** To use or express with hyperbole : EXAGGERATE.

**hy·per·bo·loid** (hī-pûr′bə-loid′) *n.* Either of two quadric surfaces having a finite center with plane sections that are hyperbolas or either ellipses or circles. **—hy·per′bo·loid′al** (-loid′l) *adv.*

**Hy·per·bo·re·an** (hī′pər-bôr′ē-ən, -bōr′-, -bə-rē′ən) *n.* [Lat. *Hyperborei,* the Hyperboreans < Gk. *Huperboreoi* : *huper,* extreme + *boreios,* northern < *Boreas,* Boreas.] *Gk. Myth.* One of a people known to the ancient Greeks from the earliest times, living in an unidentified country in the far north. —*adj.* **1.** Of or relating to the Hyperboreans. **2. hyperborean. a.** Of or relating to the far north : ARCTIC. **b.** Very cold : FRIGID.

**hy·per·cat·a·lex·is** (hī′pər-kăt′l-ĕk′sĭs) *n.* [HYPER- + NLat. *catalexis,* omission in the last foot of a line < Gk. *katalēxis,* ending < *katalēgein,* to leave off.] The addition of one or more syllables more than the normal number in a verse or metric line. **—hy′per·cat′a·lec′tic** (-ĕk′tĭk) *adj.*

**hy·per·charge** (hī′pər-chärj′) *n.* A quantum number numerically equal to twice the average electric charge of a particle multiplet or to the sum of the strangeness and the baryon number.

**hy·per·crit·ic** (hī′pər-krĭt′ĭk) *n.* One who is overcritical.

**hy·per·crit·i·cal** (hī′pər-krĭt′ĭ-kəl) *adj.* Excessively critical : CAPTIOUS. **—hy′per·crit′i·cal·ly** *adv.* **—hy′per·crit′i·cism** *n.*

**hy·per·e·mi·a** *also* **hy·per·ae·mi·a** (hī′pə-rē′mē-ə) *n.* Abnormally large blood supply. **—hy′per·e′mic** (-mĭk) *adj.*

**hy·per·es·the·sia** *also* **hy·per·aes·the·sia** (hī′pər-ĭs-thē′zhə) *n.* Abnormal sensitivity of the senses. **—hy′per·es·thet′ic** (-thĕt′ĭk) *adj.*

**hy·per·eu·tec·tic** (hī′pər-yo͞o-tĕk′tĭk) *adj. Chem.* Having the minor component present in a larger amount than in the eutectic composition of the same components.

**hy·per·ex·ten·sion** (hī′pər-ĭk-stĕn′shən) *n.* Extension of a bodily limb beyond normal limits.

**hy·per·fine structure** (hī′pər-fīn′) *n.* The splitting of a spectral line into two or more components as a result of the spin or magnetic moment of the atomic nucleus.

**hy·per·gly·ce·mi·a** (hī′pər-glī-sē′mē-ə) *n.* An abnormally high level of glucose in the blood. **—hy′per·gly·ce′mic** *adj.*

**hy·per·gol·ic** (hī′pər-gŏl′ĭk) *adj.* [G. *Hypergol,* a hypergolic fluid propellant (*hyp-,* hyper- + Gk. *ergon,* work + *-ol,* -ole) + -IC.] Igniting spontaneously when mixed together, as a rocket fuel. **—hy′per·gol′** (hī′pər-gôl′, -gōl′) *n.* **—hy′per·gol′i·cal·ly** *adv.*

**hy·per·in·fla·tion** (hī′pər-ĭn-flā′shən) *n.* Very high inflation.

**hy·per·in·su·lin·ism** (hī′pər-ĭn′sə-lə-nĭz′əm) *n.* The presence of abnormally excessive quantities of insulin in the blood, resulting in hypoglycemia.

**Hy·pe·ri·on** (hī-pîr′ē-ən) *n.* [Lat. < Gk. *Huperiōn.*] *Gk. Myth.* A Titan, the son of Gaea and Uranus and father of Helios, the sun god.

**hy·per·ir·ri·ta·bil·i·ty** (hī′pər-ĭr′ĭ-tə-bĭl′ĭ-tē) *n.* Excessive sensitivity to stimuli. **—hy′per·ir′ri·ta·ble** *adj.*

**hy·per·ker·a·to·sis** (hī′pər-kĕr′ə-tō′sĭs) *n.* [HYPER- + Gk. *keras, kerat-,* horn + -OSIS.] Hypertrophy of the horny layer of the skin. **—hy′per·ker′a·tot′ic** (-tŏt′ĭk) *adj.*

**hy·per·ki·ne·sia** (hī′pər-kĭ-nē′zhə) *also* **hy·per·ki·ne·sis** (-sĭs) *n.* [HYPER- + Gk. *kinēsis,* movement + -IA.] Pathologically excessive motion. **—hy′per·ki·net′ic** (-nĕt′ĭk) *adj.*

**hy·per·mar·ket** (hī′pər-mär′kĭt) *n. Chiefly Brit.* A large commercial establishment having a department store and supermarket.

**hy·per·me·ter** (hī-pûr′mĭ-tər) *n.* Hypercatalexis. **—hy′per·met′ric** (hī′pər-mĕt′rĭk), **hy′per·met′ri·cal** *adj.*

**hy·per·me·tro·pi·a** (hī′pər-mĭ-trō′pē-ə) *n.* [Gk. *hupermetros,* beyond measure (*huper,* beyond + *metron,* measure) + -OPIA.] Hyperopia. **—hy′per·me·tro′pic** (-trō′pĭk, -trŏp′ĭk), **hy′per·me·tro′pi·cal** *adj.* **—hy′per·met′ro·py** (-mĕt′rə-pē) *n.*

**hy·perm·ne·sia** (hī′pərm-nē′zhə) *n.* [HYPER- + (A)MNESIA.] Unusually accurate or vivid memory. **—hy′perm·ne′sic** (-zĭk, -sĭk) *adj.*

**hy·per·on** (hī′pə-rŏn′) *n.* A subatomic particle with mass greater than the nucleon, decaying into a nucleon or another hyperon and lighter particles and having $2I + 1$ charge states, where $I$ is the isospin of the particle multiplet.

**hy·per·o·pi·a** (hī′pə-rō′pē-ə) *n.* A pathological condition of the eye in which parallel rays are focused behind the retina because of a refractive error or because of flattening of the globe of the eye, so that vision is better for distant than near objects. **—hy′per·ope′** (hī′pə-rōp′) *n.* **—hy′per·o′pic** (-ō′pĭk, -ŏp′ĭk) *adj.*

**hy·per·os·to·sis** (hī′pər-ŏ-stō′sĭs) *n.* [HYPER- + OST(EO) + -OSIS.] Excessive or abnormal thickening or growth of bone tissue. **—hy′per·os·tot′ic** (-stŏt′ĭk) *adj.*

**hy·per·pi·tu·i·ta·rism** (hī′pər-pĭ-to͞o′ĭ-tə-rĭz′əm, -tyo͞o′-) *n.* Pathologically excessive production of anterior pituitary hormones, esp. growth hormones, causing acromegaly or gigantism. **—hy′per·pi·tu′i·tar′y** (-tĕr′ē) *adj.*

**hy·per·pla·sia** (hī′pər-plā′zhə) *n.* A nontumorous increase in the number of cells in an organ or tissue with consequent enlargement of the affected part. **—hy′per·plas′tic** (-plăs′tĭk) *adj.*

**hy·per·ploid** (hī′pər-ploid′) *adj.* Having a chromosome number in excess of but not an exact multiple of the normal diploid number. **—hy′per·ploid′** *n.* **—hy′per·ploi′dy** *n.*

**hy·perp·ne·a** (hī′pərp-nē′ə, hī′pər-nē′ə) *n.* [HYPER- + Gk. *pnoē,*

breath < *pnein*, to breathe.] Abnormally deep or rapid breathing. **—hy'perp·ne'ic** (-ĭk) *adj.*
**hy·per·py·rex·i·a** (hī'pər-pī-rĕk'sē-ə) *n.* Abnormally high fever. **—hy'per·py·rex'i·al, hy'per·py·ret'ic** (-rĕt'ĭk) *adj.*
**hy·per·sen·si·tive** (hī'pər-sĕn'sĭ-tĭv) *adj.* Excessively or abnormally sensitive. **—hy'per·sen'si·tive·ness, hy'per·sen'si·tiv'i·ty** (-tĭv'ĭ-tē) *n.*
**hy·per·sex·u·al** (hī'pər-sĕk'shōō-əl) *adj.* Excessively interested or involved in sexual activity. **—hy'per·sex'u·al'i·ty** (-ăl'ĭ-tē) *n.*
**hy·per·son·ic** (hī'pər-sŏn'ĭk) *adj.* Of speed equal to or exceeding five times the speed of sound. **—hy'per·son'i·cal·ly** *adv.*
**hy·per·sthene** (hī'pərs-thēn') *n.* [Fr. *hypersthène* : *hyper-*, hyper- + Gk. *sthenos*, strength.] A green, brown, or black splintery, cleavable pyroxene mineral, chiefly $(Fe,Mg)_2Si_2O_6$. **—hy'per·sthen'ic** (-thĕn'ĭk) *adj.*
**hy·per·ten·sion** (hī'pər-tĕn'shən) *n.* Abnormally high arterial blood pressure. **—hy'per·ten'sive** *adj.* & *n.*
**hy·per·ther·mi·a** (hī'pər-thûr'mē-ə) *n.* Unusually high fever, esp. when induced for therapeutic purposes. **—hy'per·ther'mal** *adj.*
**hy·per·thy·roid** (hī'pər-thī'roid') *adj.* Of, relating to, or having hyperthyroidism.
**hy·per·thy·roid·ism** (hī'pər-thī'roi-dĭz'əm) *n.* Pathologically excessive production of thyroid hormones.
**hy·per·to·ni·a** (hī'pər-tō'nē-ə) *also* **hy·per·to·nic·i·ty** (-tō-nĭs'ĭ-tē) *n.* The state of being hypertonic.
**hy·per·ton·ic** (hī'pər-tŏn'ĭk) *adj.* **1.** *Pathol.* Having extreme arterial or muscular tension <a *hypertonic* bladder> **2.** *Chem.* Having the higher osmotic pressure of two solutions.
**hy·per·tro·phy** (hī-pûr'trə-fē) *n.* A nontumorous increase in the size of an organ or part due to the enlargement without increase in number of constituent cells. —*vi.* & *vt.* **-phied, -phy·ing, -phies.** To grow or cause to grow abnormally large. **—hy'per·tro'phic** (-trō'fĭk) *adj.*
**hy·per·ven·ti·la·tion** (hī'pər-vĕn'tl-ā'shən) *n.* Abnormally rapid or deep respiration in which excessive quantities of air are taken in, causing buzzing in the ears, tingling of extremities, and sometimes fainting. **—hy'per·ven'ti·late'** (-vĕnt'əl-āt') *v.* **(-lat·ed, -lat·ing, -lates).**
**hy·per·vi·ta·min·o·sis** (hī'pər-vī'tə-mə-nō'sĭs) *n.* Any of various abnormal conditions in which the physiological effect of a vitamin is produced to a pathological degree by excessive use of the vitamin.
**hy·pes·the·sia** (hī'pĭs-thē'zhə) *n. var. of* HYPOESTHESIA.
**hy·pha** (hī'fə) *n., pl.* **-phae** (-fē) [NLat. < Gk. *huphē*, web.] Any of the threadlike filaments forming the mycelium of a fungus. **—hy'phal** *adj.*
**hy·phen** (hī'fən) *n.* [LLat. < LGk. *huphen*, a sign indicating a compound < Gk., together : *hupo*, under + *hen*, one.] A punctuation mark ( - ) used to connect the parts of a compound word or between syllables of a word, esp. of a word divided at the end of a line. —*vt.* **-phened, -phen·ing, -phens.** To hyphenate.
**hy·phen·ate** (hī'fə-nāt') *vt.* **-at·ed, -at·ing, -ates.** To connect or divide (word elements or syllables) with a hyphen. **—hy'phen·a'tion** *n.*
**hypn-** *pref. var. of* HYPNO-.
**hyp·na·gog·ic** *also* **hyp·no·gog·ic** (hĭp'nə-gŏj'ĭk, -gō'jĭk) *adj.* [Fr. *hypnagogique* : Gk. *hupnos*, sleep + Gk. *agōgos*, leading < *agein*, to lead.] **1.** Inducing sleep. **2.** Of or relating to the period or state of drowsiness preceding sleep.
**hypno-** *or* **hypn-** *pref.* [< Gk. *hupnos*, sleep.] **1.** Sleep <*hypno*phobia> **2.** Hypnosis <*hypno*analysis>
**hyp·no·a·nal·y·sis** (hĭp'nō-ə-năl'ĭ-sĭs) *n.* Psychoanalytic treatment involving the use of hypnosis.
**hyp·no·gen·e·sis** (hĭp'nō-jĕn'ĭ-sĭs) *n.* The process of inducing or entering a hypnotic state. **—hyp'no·ge·net'ic** (-jə-nĕt'ĭk) *adj.* **—hyp'no·ge·net'i·cal·ly** *adv.*
**hyp·no·gog·ic** (hĭp'nə-gŏj'ĭk, -gō'jĭk) *adj. var. of* HYPNAGOGIC.
**hyp·noid** (hĭp'noid') *also* **hyp·noi·dal** (hĭp-noid'l) *adj.* Of or like hypnosis or sleep.
**hyp·no·pho·bi·a** (hĭp'nə-fō'bē-ə) *n.* Abnormal fear of sleep. **—hyp'no·pho'bic** *adj.*
**hyp·no·pom·pic** (hĭp'nə-pŏm'pĭk) *adj.* [HYPNO- + Gk. *pompē*, a sending away + -IC.] Of or relating to the partially conscious period or state preceding complete awakening.
**Hyp·nos** (hĭp'nŏs') *n.* [Gk.] *Gk. Myth.* The god of sleep.
**hyp·no·sis** (hĭp-nō'sĭs) *n., pl.* **-ses** (-sēz'). **1.** An artificially induced sleeplike condition in which an individual is receptive to suggestions made by the hypnotist. **2.** Hypnotism. **3.** A sleeplike condition.
**hyp·no·ther·a·py** (hĭp'nō-thĕr'ə-pē) *n.* Therapy based on or using hypnosis. **—hyp'no·ther'a·pist** *n.*
**hyp·not·ic** (hĭp-nŏt'ĭk) *adj.* [Fr. *hypnotique* < LLat. *hypnoticus* < Gk. *hupnōtikos* < *hupnoun*, to put to sleep < *hupnos*, sleep.] **1. a.** Of or relating to hypnosis. **b.** Of or relating to hypnotism. **2.** Inducing or tending to induce sleep : SOPORIFIC. —*n.* **1. a.** One who is hypnotized. **b.** One who can be hypnotized. **2.** An agent causing sleep : SOPORIFIC. **—hyp·not'i·cal·ly** *adv.*
**hyp·no·tism** (hĭp'nə-tĭz'əm) *n.* **1.** The theory or practice of inducing hypnosis. **2.** An act of inducing hypnosis. **—hyp'no·tist** *n.*
**hyp·no·tize** (hĭp'nə-tīz') *vt.* **-tized, -tiz·ing, -tiz·es. 1.** To put in a state of hypnosis. **2.** To fascinate by or as if by hypnosis <*hypnotized* by the movement of the windshield wipers> **—hyp'no·tiz'a·ble** *adj.* **—hyp'no·ti·za'tion** *n.* **—hyp'no·tiz'er** *n.*
**hy·po**[1] (hī'pō) *n.* [Short for HYPOSULFITE.] Sodium thiosulfate.
**hy·po**[2] (hī'pō) *n., pl.* **-pos.** *Informal.* **1.** A hypodermic syringe. **2.** A hypodermic injection.
**hypo-** *or* **hyp-** *pref.* [Gk. *hupo-* < *hupo*, under, beneath.] **1.** Below : beneath <*hypo*dermic> **2.** Less than normal : deficient <*hypo*esthesia> **3.** In the lowest state of oxidation <*hypo*xanthine>
**hy·po·a·cid·i·ty** (hī'pō-ə-sĭd'ĭ-tē) *n.* **1.** *Chem.* Slight acidity. **2.** *Med.* Acidity that is below normal.
**hy·po·bar·ic** (hī'pə-băr'ĭk) *adj.* Below normal pressure. **—hy'po·bar'ism** *n.*
**hy·po·blast** (hī'pə-blăst') *n.* The endoblast. **—hy'po·blas'tic** *adj.*
**hy·po·caust** (hī'pə-kôst') *n.* [Lat. *hypocaustum* < Gk. *hupokauston* < *hupokaiein*, to light a fire beneath : *hupo*, beneath + *kaiein*, to burn.] A space under the floor where heat from a furnace was accumulated to heat a room or bath in ancient Rome.
**hy·po·cen·ter** (hī'pə-sĕn'tər) *n.* The surface position directly beneath the center of a nuclear explosion. **—hy'po·cen'tral** *adj.*
**hy·po·chlo·rite** (hī'pə-klôr'īt', -klōr'-) *n.* A salt or ester of hypochlorous acid.
**hy·po·chlo·rous acid** (hī'pə-klôr'əs, -klōr'-) *n.* A weak, unstable acid, HOCl, occurring only in solution and used as a deodorant, disinfectant, bleach, and oxidizer.
**hy·po·chon·dri·a** (hī'pə-kŏn'drē-ə) *n.* [LLat., abdomen (the seat of melancholy) < Gk. *hupokhondria*, pl. of *hupokhondrion*, abdomen < *hupokhondrios*, under the cartilage of the breastbone : *hupo*, under + *khondros*, cartilage.] **1.** The persistent neurotic conviction that one is or is likely to become ill, often involving experiences of actual pain, when illness is neither present nor likely. **2.** *pl. of* HYPOCHONDRIUM.
**hy·po·chon·dri·ac** (hī'pə-kŏn'drē-ăk') *n.* One afflicted with hypochondria. —*adj.* **1.** Relating to or having hypochondria. **2.** *Anat.* Relating to or located in the hypochondrium. **—hy'po·chon·dri'a·cal** (-kŏn-drī'ə-kəl) *adj.* **—hy'po·chon·dri'a·cal·ly** *adv.*
**hy·po·chon·dri·a·sis** (hī'pə-kən-drī'ə-sĭs) *n.* [HYPOCHONDR(IA) + -IASIS.] Hypochondria.
**hy·po·chon·dri·um** (hī'pə-kŏn'drē-əm) *n., pl.* **-dri·a** (-drē-ə) [NLat. < Gk. *hupokhondrion*, abdomen. —see HYPOCHONDRIA.] The upper lateral region of the abdomen, below the lowest ribs.
**hy·poc·o·rism** (hī-pŏk'ə-rĭz'əm, hī'pə-kôr'ĭz'əm, -kōr'-) *n.* [LLat. *hypocorisma* < Gk. *hupokorisma* < *hupokorizesthai*, to call by endearing names : *hypo*, below + *korizesthai*, to caress < *koros*, boy, and *korē*, girl.] **1.** A name of endearment : pet name. **2.** The use of hypocorisms. **—hy'po·co·ris'tic** (hī'pə-kə-rĭs'tĭk), **hy'po·co·ris'ti·cal** *adj.* **—hy'po·co·ris'ti·cal·ly** *adv.*
**hy·po·cot·yl** (hī'pə-kŏt'l) *n.* [HYPO- + COTYL(EDON).] The part of the axis of a plant embryo or seedling plant below the cotyledons.
**hy·poc·ri·sy** (hĭ-pŏk'rĭ-sē) *n., pl.* **-sies.** [ME *ipocrisie* < OFr. < LLat. *hypocrisis* < Gk. *hupokrisis*, pretense < *hupokrinesthai*, to pretend : *hupo-*, from under + *krinesthai*, to explain.] **1.** The practice of expressing feelings, beliefs, or virtues one does not hold or possess : INSINCERITY. **2.** An act or instance of hypocrisy.
**hyp·o·crite** (hĭp'ə-krĭt') *n.* [ME *ipocrite* < OFr. < LLat. *hypocrita* < Gk. *hupocritēs*, actor < *hupokrinein*, to play a part. —see HYPOCRISY.] One given to hypocrisy or dissemblance.
**hy·po·crit·i·cal** (hĭp'ə-krĭt'ĭ-kəl) *adj.* **1.** Marked by hypocrisy. **2.** Being a hypocrite. **—hyp'o·crit'i·cal·ly** *adv.*
**hy·po·cy·cloid** (hī'pō-sī'kloid') *n.* The plane locus of a point fixed on a circle that rolls on the inside circumference of a fixed circle.
**hy·po·derm** (hī'pə-dûrm') *n. var. of* HYPODERMIS.
**hy·po·der·mal** (hī'pə-dûr'məl) *adj.* **1.** Of or relating to the hypodermis. **2.** Lying beneath the epidermis.
**hy·po·der·mic** (hī'pə-dûr'mĭk) *adj.* [HYPO- + DERM(ATO)- + -IC.] **1.** Of or relating to the layer just beneath the epidermis. **2.** Relating to the hypodermis. **3.** Injected beneath the skin. —*n.* **1.** A hypodermic injection. **2.** A hypodermic needle. **3.** A hypodermic syringe. **—hy'po·der'mi·cal·ly** *adv.*
**hypodermic injection** *n.* A subcutaneous, intramuscular, or intravenous injection by means of a hypodermic syringe and needle.
**hypodermic needle** *n.* **1.** A hollow needle used with a hypodermic syringe. **2.** A hypodermic syringe.
**hypodermic syringe** *n.* A syringe fitted with a hypodermic needle for hypodermic injections.
**hy·po·der·mis** (hī'pə-dûr'mĭs) *also* **hy·po·derm** (hī'pə-dûrm') *n.* **1.** *Zool.* An epidermal layer of cells that secretes an overlying chitinous cuticle, as in arthropods. **2.** *Bot.* A layer of cells lying immediately beneath the epidermis.

**hy·po·es·the·sia** (hī'pō-ĭs-thē'zhə) *also* **hy·pes·the·sia** (hī'-pĭs-) *n.* [HYPO- + (AN)ESTHESIA.] *Pathol.* Partial loss of sensation.

**hy·po·eu·tec·tic** (hī'pō-yōō-tĕk'tĭk) *adj. Chem.* Having the minor component present in a smaller amount than in the eutectic composition of the same components.

**hy·po·gas·tri·um** (hī'pə-găs'trē-əm) *n., pl.* **-tri·a** (-trē-ə) [NLat. < Gk. *hypogastrion* : *hupo*, below + *gastēr*, belly.] The lowest of the three median regions of the abdomen. **—hy'po·gas'tric** *adj.*

**hy·po·ge·a** (hī'pə-jē'ə) *n. pl. of* HYPOGEUM.

**hy·po·ge·al** (hī'pə-jē'əl) *also* **hy·po·ge·an** (-ən) *or* **hy·po·ge·ous** (-əs) *adj.* [< LLat. *hypogeus* < Gk. *hupogaios* : *hupo*, below + *gaia*, earth.] **1.** Located beneath the earth's surface : UNDERGROUND. **2.** *Bot.* Designating or marked by cotyledons remaining below the surface of the ground. **—hy'po·ge'al·ly** *adv.*

**hy·po·gene** (hī'pə-jēn') *adj.* [HYPO- + (EPI)GENE.] Formed or situated below the earth's surface <*hypogene* rocks>

**hy·pog·e·nous** (hī-pŏj'ə-nəs) *adj. Bot.* Developing or growing on a lower surface, as fungi on leaves.

**hy·po·ge·ous** (hī'pə-jē'əs) *adj. var. of* HYPOGEAL.

**hy·po·ge·um** (hī'pə-jē'əm) *n., pl.* **-ge·a** (-jē'ə) [Lat. < Gk. *hupogaion* < *hupogaios*, hypogeal.] **1.** A subterranean chamber of an ancient building. **2.** An ancient subterranean burial chamber, as a catacomb.

**hy·po·glos·sal** (hī'pə-glŏs'əl) *adj.* [< NLat. *hypoglossus*, hypoglossal nerve : HYPO- + Gk. *glōssa*, tongue.] *Anat.* Of or relating to the hypoglossal nerve. —*n.* The hypoglossal nerve.

**hypoglossal nerve** *n.* A motor nerve attached to the medulla oblongata and innervating the muscles of the tongue.

**hy·po·gly·ce·mi·a** (hī'pō-glī-sē'mē-ə) *n.* An abnormally low level of glucose in the blood. **—hy'po·gly·ce'mic** *adj.*

**hy·pog·y·nous** (hī-pŏj'ə-nəs) *adj. Bot.* Having or characterizing floral organs or parts that are below and not in contact with the ovary. **—hy·pog'y·ny** (-nē) *n.*

**hy·po·ma·ni·a** (hī'pə-mā'nē-ə, -măn'yə) *n.* A mild state of mania involving slightly abnormal overactivity and elation. **—hy'po·man'ic** (-măn'ĭk) *adj.*

**hy·po·nas·ty** (hī'pə-năs'tē) *n.* An upward bending of plant parts, as leaves, caused by growth of the lower side. **—hy'po·nas'tic** *adj.*

**hy·po·phos·phite** (hī'pō-fŏs'fīt') *n.* A salt of hypophosphorous acid.

**hy·po·phos·pho·rous acid** (hī'pō-fŏs'fər-əs, -fŏs-fôr'əs, -fōr'-) *n.* A clear, colorless or slightly yellow liquid, $H_3PO_2$, used in preparing hypophosphites.

**hy·poph·y·sis** (hī-pŏf'ĭ-sĭs) *n., pl.* **-ses** (-sēz') [NLat. < Gk. *hupophusis*, attachment underneath < *hupophein*, to grow beneath : *hupo*, beneath + *phuein*, to grow.] The pituitary gland. **—hy·poph'y·se'al** (-sē'əl), **hy'po·phys'i·al** (hī'pə-fĭz'ē-əl) *adj.*

**hy·po·pi·tu·i·ta·rism** (hī'pō-pĭ-tōō'ĭ-tə-rĭz'əm, -tyōō'-) *n.* Deficient or decreased production of pituitary hormones. **—hy'po·pi·tu'i·tar'y** (-tĕr'ē) *adj.*

**hy·po·pla·sia** (hī'pō-plā'zhə) *n. Pathol.* Incomplete or arrested development of an organ or part. **—hy'po·plas'tic** (-plăs'tĭk) *adj.*

**hy·po·ploid** (hī'pō-ploid') *adj. Genetics.* Having a chromosome number less by only a few chromosomes than the normal diploid number. **—hy'po·ploi'dy** *n.*

**hy·po·pne·a** (hī'pō-nē'ə) *n.* [HYPO- + Gk. *pnoē*, breath < *pnein*, to breathe.] Abnormally slow and shallow breathing.

**hy·po·sen·si·tiv·i·ty** (hī'pō-sĕn'sĭ-tĭv'ĭ-tē) *n.* Less than normal sensitivity. **—hy'po·sen'si·tive** *adj.*

**hy·po·sen·si·tize** (hī'pō-sĕn'sĭ-tīz') *vt.* **-tized, -tiz·ing, -tiz·es.** To make less sensitive. **—hy'po·sen'si·ti·za'tion** *n.*

**hy·pos·ta·sis** (hī-pŏs'tə-sĭs) *n., pl.* **-ses** (-sēz') [LLat. < Gk. *hupostasis* : *hupo*, beneath + *stasis*, a standing.] **1.** *Philos.* The substance or essence of something. **2. a.** Any of the persons of the Trinity. **b.** The essential person of Christ in which His human and divine natures are united. **3.** An entity that has been hypostatized. **4. a.** A settling of solid particles in a fluid. **b.** Something that settles to the bottom of a fluid : SEDIMENT. **5.** A condition in which the action of one gene conceals or suppresses the action of another gene that is not its allele but that affects the same organ, part, or state of the body. **—hy'po·stat'ic** (hī'pə-stăt'ĭk), **hy'po·stat'i·cal** *adj.* **—hy'po·stat'i·cal·ly** *adv.*

**hy·pos·ta·tize** (hī-pŏs'tə-tīz') *vt.* **-tized, -tiz·ing, -tiz·es.** [< Gk. *hupostatos*, standing under < *hyphistasthai*, to stand under : *hupo*, beneath + *histasthai*, to stand.] **1.** To symbolize (a concept) in a material form. **2.** To ascribe material existence to. **—hy·pos'ta·ti·za'tion** *n.*

**hy·po·sthe·ni·a** (hī'pəs-thē'nē-ə) *n.* [HYPO- + Greek *sthenos*, strength.] Abnormal weakness. **—hy'po·sthen'ic** (-thĕn'ĭk) *adj.*

**hy·po·style** (hī'pə-stīl') *n.* [< Gk. *hupostulos*, resting upon pillars : *hupo*, beneath + *stulos*, pillar.] A building having a roof or ceiling supported by rows of columns. **—hy'po·style'** *adj.*

**hy·po·sul·fite** (hī'pō-sŭl'fīt') *n.* Sodium thiosulfate.

**hy·po·sul·fu·rous acid** (hī'pō-sŭl-fyŏŏr'əs, -sŭl'fər-əs) *n.* An unstable acid, $H_2S_2O_4$, known only in aqueous solution and used as a bleaching and reducing agent.

**hy·po·tax·is** (hī'pə-tăk'sĭs) *n.* [Gk. *hupotaxis*, subjection < *hupotassein*, to arrange under : *hupo*, under + *tattein*, to arrange.] The dependent or subordinate relationship of clauses with connectives. **—hy'po·tac'tic** (-tăk'tĭk) *adj.*

**hy·po·ten·sion** (hī'pō-tĕn'shən) *n.* Abnormally low blood pressure.

**hy·pot·e·nuse** (hī-pŏt'n-ōōs', -yōōs') *also* **hy·poth·e·nuse** (-pŏth'ə-nōōs', -nyōōs') *n.* [Lat. *hypotenusa* < Gk. *hupoteinousa* < *hupoteinein*, to stretch under : *hupo*, under + *teinein*, to stretch.] The side of a right triangle opposite the right angle.

**hy·po·thal·a·mus** (hī'pō-thăl'ə-məs) *n.* The part of the brain that lies below the thalamus, forming the major portion of the ventral region of the diencephalon and functioning to regulate autonomic activities, as bodily temperature and certain metabolic processes. **—hy'po·tha·lam'ic** (-thə-lăm'ĭk) *adj.*

**hy·poth·e·cate** (hī-pŏth'ĭ-kāt') *vt.* **-cat·ed, -cat·ing, -cates.** [Med. Lat. *hypothecare, hypothecat-* < LLat. *hypotheca*, pledge < Gk. *hupothēkē* < *hupotithenai*, to give as a pledge : *hupo*, beneath + *tithenai*, to place.] To pledge (property) as security or collateral for a debt without transfer of title or possession. **—hy·poth'e·ca'tion** *n.* **—hy·poth'e·ca'tor** (-kā'tər) *n.*

**hy·poth·e·nuse** (hī-pŏth'ə-nōōs', -nyōōs') *n. var. of* HYPOTENUSE.

**hy·po·ther·mal** (hī'pō-thûr'məl) *adj. Geol.* Of, relating to, or being high-temperature deposits derived from magmatic emanations forced under pressure into place in pre-existing rock openings.

**hy·po·ther·mi·a** (hī'pō-thûr'mē-ə) *n.* [NLat. : HYPO- + Gk. *thermē*, heat.] Abnormally low body temperature. **—hy'po·ther'mic** *adj.*

**hy·poth·e·sis** (hī-pŏth'ĭ-sĭs) *n., pl.* **-ses** (-sēz') [Gk. *hupothesis* < *hupotithenai*, to suppose : *hupo*, beneath + *tithenai*, to place.] **1.** An explanation accounting for a set of facts that can be tested by further investigation : THEORY. **2.** Something considered to be true for the purpose of investigation or argument : ASSUMPTION.

**hy·poth·e·size** (hī-pŏth'ĭ-sīz') *vt.* & *vi.* **-sized, -siz·ing, -siz·es.** To assert as or form a hypothesis.

**hy·po·thet·i·cal** (hī'pə-thĕt'ĭ-kəl) *also* **hy·po·thet·ic** (-thĕt'ĭk) *adj.* [Gk. *hupothetikos* < *hupothesis*, hypothesis.] **1.** Of, relating to, or based on a hypothesis. **2. a.** Conjectural : suppositional. **b.** Contingent : conditional. **—hy'po·thet'i·cal·ly** *adv.*

**hy·po·thy·roid** (hī'pō-thī'roid') *adj.* Affected by or exhibiting hypothyroidism.

**hy·po·thy·roid·ism** (hī'pō-thī'roi-dĭz'əm) *n.* **1.** Insufficient production of thyroid hormones. **2.** A pathological condition, esp. cretinism or myxedema, resulting from severe thyroid insufficiency.

**hy·po·ton·ic** (hī'pō-tŏn'ĭk) *adj.* **1.** *Pathol.* Having less than normal tone or tension. **2.** *Chem.* Having the lower osmotic pressure of two fluids. **—hy'po·to·nic'i·ty** (-tə-nĭs'ĭ-tē) *n.*

**hy·pot·ro·phy** (hī-pŏt'rə-fē) *n.* Less than normal growth. **—hy'po·tro'phic** (hī'pə-trō'fĭk) *adj.*

**hy·po·xan·thine** (hī'pō-zăn'thēn') *n.* A white powder, $C_5H_4N_4O$, an intermediate in the metabolism of animal purines.

**hy·pox·i·a** (hī-pŏk'sē-ə, hĭ-) *n.* Deficiency in the amount of oxygen reaching bodily tissues.

**hypso-** *or* **hyps-** *pref.* [< Gk. *hupsos*, height.] Height <*hypsometer*><*hypsography*>

**hyp·sog·ra·phy** (hĭp-sŏg'rə-fē) *n., pl.* **-phies. 1. a.** The scientific study of the earth's topologic configuration above sea level, esp. the measurement and mapping of land elevations. **b.** A representation or description of such features, as on a map. **2.** Hypsometry. **—hyp'so·graph'ic** (hĭp'sə-grăf'ĭk), **hyp'so·graph'i·cal** *adj.*

**hyp·som·e·ter** (hĭp-sŏm'ĭ-tər) *n.* An instrument using the altitude-pressure dependence of boiling points to determine land elevations.

**hyp·som·e·try** (hĭp-sŏm'ĭ-trē) *n.* The measurement of elevation relative to sea level. **—hyp'so·met'ric** (hĭp'sə-mĕt'rĭk), **hyp'so·met'ri·cal** *adj.* **—hyp'so·met'ri·cal·ly** *adv.* **—hyp·som'e·trist** *n.*

**hy·rax** (hī'răks') *n., pl.* **-rax·es** *or* **-ra·ces** (-rə-sēz') [Gk. *hurax*, shrew mouse.] Any of several herbivorous mammals of the family Procaviidae within the order Hyraoidea, of Africa and adjacent Asia, resembling woodchucks or similar rodents but more closely related to the hoofed mammals.

**hyrax**
*Approximately 18 inches long*

**hy·son** (hī′sən) *n.* [Chin. (Mandarin) *xi*[1] *chun*[1] : *xi*[1], warm, sunny + *chun*[1], springlike.] A Chinese green tea with curled or twisted leaves.

**hys·sop** (hĭs′əp) *n.* [ME *ysop* < OE *hysope* and OFr. *ysope,* both < Lat. *hyssōpus* < Gk. *hussōpos,* of Semitic orig.] **1. a.** A woody Asian plant, *Hyssopus officinalis,* bearing spikes of small blue flowers and aromatic leaves used as a condiment and in perfumery. **b.** Any of various related plants. **2.** An unidentified plant mentioned in the Bible as the source of twigs used for sprinkling in certain Hebraic purificatory rites.

**hyster-** *pref. var. of* HYSTERO-.

**hys·ter·ec·to·my** (hĭs′tə-rĕk′tə-mē) *n., pl.* **-mies.** Partial or complete surgical removal of the uterus.

**hys·ter·e·sis** (hĭs′tə-rē′sĭs) *n., pl.* **-ses** (-sēz′) [Gk. *husterēsis,* a shortcoming < *husterein,* to come late < *husteros,* late.] The failure of a property that has been altered by an external agent to return to its original value when the cause of the alteration is removed. **—hys′ter·et′ic** (-rĕt′ĭk) *adj.*

**hys·ter·i·a** (hĭ-stĕr′ē-ə, -stîr′-) *n.* [NLat. < HYSTERIC.] **1.** A neurosis marked by conversion symptoms, a calm mental attitude, and episodes of hallucination, somnambulism, amnesia, and other mental aberrations. **2.** Excessive or uncontrollable emotion, as panic or fear.

**hys·ter·ic** (hĭ-stĕr′ĭk) *n.* [Lat. *hystericus,* hysterical < Gk. *husterikos* < *hustera,* womb (from the former idea that disturbances in the womb caused hysteria).] **1.** One suffering from hysteria. **2. hysterics** *(sing. or pl. in number).* **a.** A fit of uncontrollable laughing and crying. **b.** An attack of hysteria. —*adj.* Hysterical.

**hys·ter·i·cal** (hĭ-stĕr′ĭ-kəl) *adj.* **1.** Of, marked by, or caused by hysteria. **2.** Having or tending to have hysterics. **—hys·ter′i·cal·ly** *adv.*

**hystero-** *or* **hyster-** *pref.* [< Gk. *hustera,* womb.] **1.** Uterus <*hyster*ectomy> **2.** Hysteria <*hyster*oid>

**hys·ter·o·gen·ic** (hĭs′tə-rō-jĕn′ĭk) *adj.* Causing hysteria.

**hys·ter·oid** (hĭs′tə-roid′) *adj.* Resembling hysteria.

**hys·ter·on prot·er·on** (hĭs′tə-rŏn prŏt′ə-rŏn′) *n.* [LLat. < Gk. *husteron proteron,* latter first.] **1.** A figure of speech in which the natural or rational order of its terms is reversed, as *true and tried* instead of *tried and true.* **2.** The logical fallacy of assuming as a premise a proposition following something yet to be proved.

**hys·ter·ot·o·my** (hĭs′tə-rŏt′ə-mē) *n., pl.* **-mies.** Surgical incision of the uterus.

# Ii

**i** *or* **I** (ī) *n., pl.* **i's** *or* **I's. 1.** The ninth letter of the English alphabet. **2.** A speech sound represented by the letter *i.* **3.** The ninth in a series. **4.** A grade indicating that a student's work is incomplete. **5.** Something shaped like the letter *I.* **6. I** The Roman numeral for one.

**I**[1] (ī) *pron.* [ME < OE *ic.*] The one who is the speaker or writer. —*n., pl.* **I's.** The self : ego.

**I**[2] *symbol for* IODINE.

**-i-** [ME < OFr. < Lat., stem vowel of nouns and adjectives used in combination.] —Used as a connective to join word elements <brut*i*fy>

**-ia**[1] *suff.* [NLat. < Lat. and Gk., n. suffix.] **1.** Disease : pathological condition <anox*ia*> **2.** Territory : country <Manchur*ia*>

**-ia**[2] *suff.* [Partly < Lat., neuter pl. of *-ius,* and partly < Gk., neuter pl. of *-ios,* n. and adj. suffixes.] Things derived from, pertaining to, or belonging to <personal*ia*>

**-ial** *suff.* [ME < OFr. < Lat. *-ialis.*] Of, pertaining to, or characterized by <baron*ial*>

**i·amb** (ī′ămb′, -ăm′) *also* **i·am·bus** (ī-ăm′bəs) *n., pl.* **i·ambs** *also* **-bus·es** *or* **-bi** (-bī′) [Fr. *iambe* < Lat. *iambus* < Gk. *iambos.*] A metrical foot made up of a short syllable followed by a long syllable or an unstressed syllable followed by a stressed syllable.

**i·am·bic** (ī-ăm′bĭk) *adj.* **1.** Composed of iambs or distinguished by their predominance <*iambic* verse> **2.** Employing iambic rhythm, esp. in the various genres associated with its use <the *iambic* poetry of Shakespeare> —*n.* **1.** An iamb. **2.** *often* **iambics.** A verse, stanza, or poem written in iambs <the *iambics* of Greek satirists>

**i·am·bus** (ī-ăm′bəs) *n. var. of* IAMB.

**-ian** *suff.* [OFr. *-ien* < Lat. *-ianus,* adj. and n. suffix.] **1.** Of, pertaining to, or like <Boston*ian*> **2.** One pertaining to, belonging to, or like <academic*ian*>

**-iana** *suff. var. of* -ANA

**-iasis** *suff.* [NLat. < Gk., n. suffix.] A pathological condition characterized or produced by <ten*iasis*>

**-iatric** *suff.* [< Gk. *iatrikos,* medical < *iatros,* physician < *iasthai,* to heal.] Of or relating to a specified kind of medical treatment or healing <ger*iatric*>

**-iatrics** *suff.* [< -IATRIC.] Medical treatment <bar*iatrics*>

**i·at·ro·gen·ic** (ī-ăt′rə-jĕn′ĭk) *adj.* [Gk. *iatros,* physician (< *iasthai,* to heal) + -GENIC.] Induced in a patient by a physician's actions or words. —Used esp. of imagined illnesses. **—i·at′ro·gen′i·cal·ly** *adv.*

**-iatry** *suff.* [Fr. *-iatrie* < NLat. *-iatria* < Gk. *-iatreia,* art of healing < *iatros,* physician < *iasthai,* to heal.] Medical treatment <psych*iatry*>

**I-beam** (ī′bēm′) *n.* A steel girder or beam with a cross section shaped like the letter I.

**I·be·ri·an** (ī-bîr′ē-ən) *adj.* **1. a.** Of or relating to the ancient ethnologic group or groups that lived in the Iberian Peninsula. **b.** Of or relating to the language or culture of these groups. **2.** Of or relating to the Iberian Peninsula. **3.** Of or relating to ancient Iberia in the Caucasus, its inhabitants, their language, or their culture. —*n.* **1. a.** A member of the ancient Caucasoid people that lived in the Iberian Peninsula. **b.** Any of the languages spoken by the ancient peoples of the Iberian Peninsula. **2.** An inhabitant of the Iberian Peninsula. **3.** An inhabitant of ancient Iberia in the Caucasus.

**i·bex** (ī′bĕks′) *n.* [Lat.] A wild goat of the genus *Capra* of mountainous regions of the Old World, esp. *C. ibex,* with long, ridged, backward-curving horns.

**I·bib·i·o** (ĭ-bĭb′ē-ō) *n., pl.* **Ibibio** *or* **-os. 1. a.** A people of southeastern Nigeria. **b.** A member of this people. **2.** The Niger-Congo language of the Ibibio.

**i·bi·dem** (ĭb′ĭ-dĕm′, ĭ-bī′dəm) *adv.* [Lat.] In the same place. —Used in footnotes and bibliographies to refer to the book, chapter, article, or page cited just before.

**-ibility** *suff. var. of* -ABILITY.

**i·bis** (ī′bĭs) *n.* [Lat. < Gk., of Egypt. orig.] **1.** A long-billed wading bird of the family Threskiornithidae. **2.** The wood ibis.

**-ible** *suff. var. of* -ABLE.

**I·bo** (ē′bō) *n., pl.* **Ibo** *or* **I·bos. 1.** A member of one of various Negroid tribes of Nigeria. **2.** The Kwa language of the Ibo.

**-ic** *suff.* [ME < Lat. *-icus.*] **1.** Of, relating to, or marked by <seism*ic*> **2.** Having a valence higher than that of a specified element in compounds or ions identified by adjectives ending in *-ous* <sulfur*ic* acid> **3.** One relating to or marked by <academ*ic*>

**Ic·a·rus** (ĭk′ər-əs) *n.* [Lat. < Gk. *Ikaros.*] **1.** *Gk. Myth.* The son of Daedalus, who in escaping from Crete on artificial wings made by his father, flew too near the sun so that the wax that fastened his wings melted, and he fell into the Aegean Sea. **2.** An asteroid with an eccentric orbit that passes closest to the sun at a distance of within 30 million kilometers or approx. 19 million miles.

**ice** (īs) *n.* [ME *ise* < OE *īs.*] **1.** Water frozen solid. **2.** A layer, surface, or mass of frozen water. **3.** Something resembling frozen water. **4.** A dessert of flavored and sweetened crushed ice. **5.** Cake frosting : ICING. **6.** *Slang.* Diamonds. **7.** The playing field in ice hockey : RINK. **8.** *Informal.* Extreme unfriendliness or reserve. —*v.* **iced, ic·ing, ic·es.** —*vt.* **1.** To coat or slick with ice. **2.** To cause to become ice : FREEZE. **3.** To cool by setting in or as if in ice. **4.** To cover or decorate (e.g., a cake) with icing. **5.** *Slang.* To guarantee victory, as in a game : CLINCH. **6.** To shoot (the puck) far out of defensive territory in ice hockey. —*vi.* To turn into or become coated with ice : FREEZE <The road *iced* over during the night.> **—break the ice. 1.** To relax an unduly formal or tense social situation or atmosphere. **2.** To make a beginning : START. **—on ice. 1.** *Informal.* **a.** In reserve or readiness. **b.** Held incommunicado. **2.** Sure to be won or accomplished.

**ice age** *n.* **1.** A cold period characterized by extensive glaciation. **2. Ice Age.** The Pleistocene or glacial epoch.

**ice ax** *n.* A tool used in mountain climbing for cutting steps in ice.

**ice bag** *n.* A waterproof bag used as an ice pack.

**ice·berg** (īs′bûrg′) *n.* [Partial transl. of Dan. and Norw. *isberg* : *is,* ice + *berg,* mountain.] **1.** A mass of floating ice broken away from a glacier. **2.** *Informal.* A cold or aloof person.
**ice·blink** (īs′blĭngk′) *n.* **1.** A yellowish glare in the sky above an ice field. **2.** A coastal ice cliff.
**ice·boat** (īs′bōt′) *n.* **1.** A boatlike vehicle on runners that sails on ice. **2.** ICEBREAKER 1. **—ice′boat′ing** *n.*
**ice·bound** (īs′bound′) *adj.* Covered over or locked in by ice.
**ice·box** (īs′bŏks′) *n.* **1.** An insulated chest or box in which ice cools and preserves food. **2.** A refrigerator.
**ice·break·er** (īs′brā′kər) *n.* **1.** A sturdy ship for breaking a channel through icebound waters. **2.** A pier or dock apron used as a buffer against floating ice.
**ice cap** *n.* An extensive perennial cover of ice and snow, esp. an ice sheet with an outward flow from a relatively level central area.
**ice cream** *n.* A smooth, sweet, cold food made from a frozen mixture of milk products, flavoring, and occas. small amounts of colloidal materials and emulsifiers.
**ice-cream cone** (īs′krēm′) *n.* **1.** A conical wafer for holding a scoop of ice cream. **2.** An ice-cream cone filled with ice cream.
**ice-cream soda** *n.* A refreshment consisting of ice cream scoops in a mixture of soda water and flavoring syrup.
**iced** (īst) *adj.* **1.** Covered over with ice. **2.** Cooled with ice. **3.** Coated or decorated with icing.
**ice·fall** (īs′fôl′) *n.* **1.** The face or sheer side of a glacier, resembling a frozen waterfall. **2.** An avalanche of ice.
**ice field** *n.* A large level expanse of floating ice.
**ice floe** *n.* A flat expanse of floating ice, smaller than an ice field.
**ice fog** *n.* Pogonip.
**ice foot** *n.* A belt or ledge of ice that forms along shorelines in polar regions.
**ice hockey** *n.* HOCKEY 1.
**ice·house** (īs′hous′) *n.* A place where ice is made, stored, or sold.
**Ice·land·er** (īs′lən-dər) *n.* A native or resident of Iceland.
**Ice·land·ic** (īs-lăn′dĭk) *adj.* Of or relating to Iceland, its residents, their language, or their culture. —*n.* The North Germanic language of the Icelanders.
**Iceland moss** *n.* A brittle, grayish-brown edible lichen, *Cetraria islandica* of northern regions.
**Iceland spar** *n.* A doubly refracting transparent calcite used in optical instruments.
**ice milk** *n.* A smooth, sweet, cold food similar to ice cream but containing less butterfat.
**ice needle** *n.* Any of the thin ice crystals that float high in the atmosphere in cold clear weather.
**ice-out** (īs′out′) *n.* The thawing of ice on the surface of a body of water, as a pond or lake.
**ice pack** *n.* **1.** A floating mass of compacted ice fragments. **2.** A folded cloth or waterproof bag filled with ice and applied to sore or swollen parts of the body.
**ice pick** *n.* A pointed awl for chipping or breaking ice.
**ice plant** *n.* A plant native to southern Africa, *Mesembryanthemum crystallinum,* bearing white or pink flowers and fleshy leaves and stems coated with shiny encrustations.
**ice point** *n.* The temperature at which pure water and ice are in equilibrium in a mixture at one atmosphere of pressure.
**ice show** *n.* An entertainment with dances, stunts, and buffoonery performed by ice skaters.
**ice skate** *n.* **1.** A metal blade or runner fitted to the sole of a shoe for skating on ice. **2.** A shoe or light boot with a permanently fixed runner for skating on ice.
**ice-skate** (īs′skāt′) *vi.* **-skat·ed, -skat·ing, -skates.** To skate on ice. **—ice skater** *n.*
**ice storm** *n.* A storm in which precipitation freezes on contact.
**ice water** *n.* Very cold or chilled water.
**ich** (ĭk) *n.* [Short for NLat. *Ichthyophthirius,* genus name : ICHTHYO- + Gk. *phtheir,* louse.] A contagious disease of tropical aquarium fishes, characterized by small white pustules on the body.
**ich·neu·mon** (ĭk-nōō′mən, -nyōō′-) *n.* [Lat. < Gk. *ikhneumōn,* weasel < *ikhneuein,* to track < *ikhnos,* track.] A mongoose of the genus *Herpestes,* esp. *H. ichneumon* of Africa.
**ichneumon fly** *n.* Any of various wasplike insects of the family Ichneumonidae, with larvae that are parasitic on the larvae of other insects.

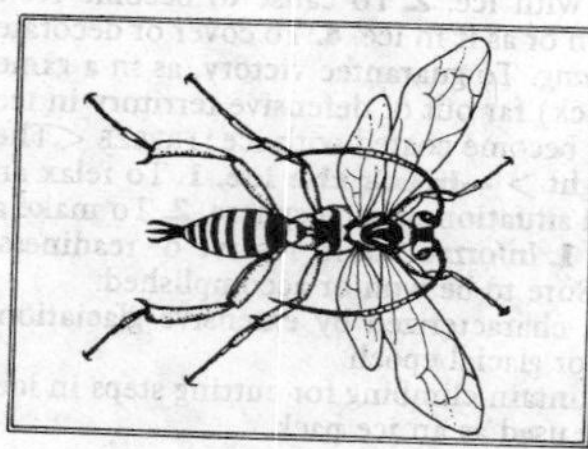
**ichneumon fly**
*Up to 1½ inches long*

**ichneumon wasp** *n.* An ichneumon fly.
**ich·nite** (ĭk′nīt′) *n.* [Gk. *iknos,* track + -ITE.] A fossilized footprint.
**i·chor** (ī′kôr′, ī′kər) *n.* [Gk. *ikhōr.*] **1.** *Gk. Myth.* The rarefied fluid said to run in the veins of the gods. **2.** *Pathol.* A watery acrid discharge from an ulcer or wound. **—i′chor·ous** (ī′kər-əs) *adj.*
**ichthy-** *pref. var. of* ICHTHYO-.
**ich·thy·ic** (ĭk′thē-ĭk) *adj.* Of, relating to, or typical of fishes.
**ichthyo-** or **ichthy-** *pref.* [Lat. < Gk. *ikhthuo-* < *ikhthus,* fish.] Fish <*ichthyo*phagous>
**ich·thy·o·fau·na** (ĭk′thē-ə-fô′nə) *n.* The fish of a given region.
**ich·thy·oid** (ĭk′thē-oid′) *also* **ich·thy·oi·dal** (ĭk′thē-oid′l) *adj.* Typical of or like a fish. —*n.* A fish or fishlike vertebrate.
**ich·thy·ol·o·gy** (ĭk′thē-ŏl′ə-jē) *n.* The zoological study of fishes. **—ich′thy·o·log′ic** (-ə-lŏj′ĭk), **ich′thy·o·log′i·cal** *adj.* **—ich′thy·ol′o·gist** *n.*
**ich·thy·oph·a·gous** (ĭk′thē-ŏf′ə-gəs) *adj.* Feeding on fish.
**ich·thy·or·nis** (ĭk′thē-ôr′nĭs) *n.* [NLat. *Ichthyornis,* genus name : ICHTHY(O) + Gk. *ornis,* bird.] An extinct, toothed, fish-eating bird of the genus *Ichthyornis* that existed during the Cretaceous period.
**ich·thy·o·saur** (ĭk′thē-ə-sôr′) *also* **ich·thy·o·sau·rus** (ĭk′-thē-ə-sôr′əs) *n., pl.* **-saurs** *also* **-sau·ri** (-sôr′ī′) [ICHTHYO- + Gk. *sauros,* lizard.] An extinct fishlike marine reptile of the order Ichthyosauria of the Triassic to the Cretaceous periods.
**ich·thy·o·sis** (ĭk′thē-ō′sĭs) *n.* A congenital skin disease, marked by dry, thickened, scaly skin.
**-ician** *suff.* [ME < OFr. *-icien.*] One who practices : SPECIALIST <techn*ician*>
**i·ci·cle** (ī′sĭ-kəl) *n.* [ME *isikel* : *is,* ice + *ikel,* icicle < OE *gicel.*] **1.** A tapering spike of ice made by the freezing of dripping water. **2.** *Informal.* A cold or aloof person.
**ic·ing** (ī′sĭng) *n.* **1.** A sweet glaze made of sugar, butter, and egg whites or milk, often flavored and cooked, for coating or decorating cakes, cookies, and other baked goods. **2.** The act of intentionally shooting the puck far out of defensive territory in ice hockey.
**ick·y** (ĭk′ē) *adj.* **-i·er, -i·est.** [Perh. alteration of STICKY.] *Informal.* **1.** Disagreeably sticky : GUMMY <*icky* candy> **2.** Offensively sentimental : SACCHARINE <*icky* greeting cards>
**i·con** *also* **i·kon** (ī′kŏn′) *n.* [Lat. < Gk. *eikōn,* likeness, image.] **1. a.** An image : representation. **b.** A simile or symbol. **2.** A representation or picture of a sacred Christian personage, traditional to the Eastern Churches.
**icon-** *pref. var. of* ICONO-.
**i·con·ic** (ī-kŏn′ĭk) *adj.* **1.** Relating to or like an icon. **2.** Having a conventional formulaic style <an *iconic* memorial statue>
**icono-** or **icon-** *pref.* [Gk. *eikono-* < *eikōn,* image.] Image : icon <*icono*latry>
**i·con·o·clasm** (ī-kŏn′ə-klăz′əm) *n.* [< ICONOCLAST.] **1.** The doctrine or practice of destroying religious images. **2.** The attacking of established institutions, practices, or attitudes.
**i·con·o·clast** (ī-kŏn′ə-klăst′) *n.* [Med. Lat. *iconoclastes* < Med. Gk. *eikonoklastēs* : Gk. *eikōn,* image + *-klastēs,* breaker < Gk. *klan,* to break.] **1. a.** One who destroys sacred images. **b.** One who opposes the use or worship of sacred images. **2.** One who attacks and seeks to overthrow popular or traditional ideas or institutions. **—i·con′o·clas′tic** (-klăs′tĭk) *adj.*
**i·co·nog·ra·phy** (ī′kə-nŏg′rə-fē) *n., pl.* **-phies.** [Gk. *eikonographia,* description, sketch : *eikōn,* image, likeness + *-graphia,* -graphy.] **1. a.** Pictorial illustration of a specific subject. **b.** The collected representations illustrating a subject. **2. a.** A set of traditional or specified symbolic forms associated with the subject or theme of a stylized work of art. **b.** The conventions defining such forms and governing their interrelationship. **3.** A treatise or book about iconography. **—i′co·nog′ra·pher** *n.* **—i·con′o·graph′ic** (ī-kŏn′ə-grăf′ĭk), **i·con′o·graph′i·cal** *adj.*
**i·co·nol·a·try** (ī′kə-nŏl′ə-trē) *n.* Worship of sacred images or icons. **—i′co·nol′a·ter** *n.*
**i·co·nol·o·gy** (ī′kə-nŏl′ə-jē) *n.* The branch of art history embracing description, analysis, and interpretations of icons. **—i·con′o·log′i·cal** (ī-kŏn′ə-lŏj′ĭ-kəl) *adj.* **—i′co·nol′o·gist** *n.*
**i·con·o·scope** (ī-kŏn′ə-skōp′) *n.* A television camera tube designed for rapid scanning of an information-storing photoactive mosaic.
**i·co·nos·ta·sis** (ī′kə-nŏs′tə-sĭs) *n., pl.* **-ses** (-sēz′) [< LGk. *eikonostasion,* shrine : *eikōn,* image + Gk. *stasis,* a standing.] The screen separating and concealing the sanctuary from the main area of an Eastern Orthodox church.
**i·co·sa·he·dron** (ī-kō′sə-hē′drən) *n., pl.* **-dra** (-drə) or **-drons.** [Gk. *eikosaedron* : *eikosi,* twenty + *-edron,* -hedron.] A polyhedron with 20 faces. **—i·co′sa·he′dral** *adj.*
**-ics** *suff.* [-IC + -S (n.pl. suffix), transl. of Gk. *-ika* < neuter pl. of *-ikos,* adj. suffix.] **1.** Science : art : study : knowledge : skill <graph*ics*> **2.** Actions, activities, or practices of <athlet*ics*> **3.** Qualities or operations of <mechan*ics*>

**ic·ter·ic** (ĭk-tĕr'ĭk) *adj.* [ICTER(US) + -IC.] **1.** Relating to or having jaundice. **2.** Used for treating jaundice. *—n.* A remedy for jaundice.

**ic·ter·o·gen·ic** (ĭk'tər-ə-jĕn'ĭk) *adj.* [ICTER(US) + -GENIC.] Causing jaundice.

**ic·ter·us** (ĭk'tər-əs) *n.* [NLat. < Gk. *ikteros.*] Jaundice.

**ic·tus** (ĭk'təs) *n., pl.* **-tus·es** *or* **ictus.** [Lat., stroke < p.part. of *icere,* to strike.] *Pathol.* A sudden attack : STROKE.

**i·cy** (ī'sē) *adj.* **i·ci·er, i·ci·est. 1.** Having or covered with ice. **2.** Resembling ice. **3.** Bitterly cold : FREEZING <an *icy* season> **4.** Lacking warmth or friendliness <an *icy* gaze> **—i'ci·ly** *adv.* **—i'ci·ness** *n.*

**id** (ĭd) *n.* [NLat. < Lat., it.] The division of the psyche associated with instinctual impulses and demands for immediate gratification of primitive needs.

**I'd** (īd). **1.** I had. **2.** I should. **3.** I would.

**-id** *suff.* [< Lat. *-is, -id-,* fem. patronymic suffix.] Body : particle <chromat*id*>

**ID card** (ī'dē') *n.* A card often having a photograph that gives identifying data, as name, age, or organizational association, about an individual.

**-ide** *suff.* [G. *-id* < Fr. *-ide* (as in *oxide*) < *acide,* acid.] Chemical compound <cyan*ide*>

**i·de·a** (ī-dē'ə) *n.* [Lat. < Gk., form, class, notion.] **1.** Something existing in the mind, actually or potentially, as a result of mental activity, such as a thought, image, or conception. **2.** An opinion, conviction, or principle <has startling economic *ideas*> **3.** A plan, scheme, or method. **4.** Significance : import. **5.** A fancy : notion. **6.** *Obs.* A mental image of something remembered. **7.** *Mus.* A theme or motif. **8.** *Philos.* **a.** The Platonic archetype of which a corresponding being in phenomenal reality is assertedly an imperfect replica. **b.** The Kantian concept of reason as transcendent but nonempirical. **c.** The Hegelian absolute truth as the complete and ultimate product of reason.

**i·de·al** (ī-dē'əl, ī-dēl') *n.* [Fr. < LLat. *idealis* < Lat. *idea,* idea.] **1.** A conception of something in its absolute perfection. **2.** One considered to be a standard of perfection or excellence and worthy of imitation. **3.** An ultimate object of endeavor : GOAL. **4.** An honorable or worthy aim or principle. *—adj.* **1. a.** Of, relating to, or embodying an ideal. **b.** Conforming to an ultimate standard or form of excellence or perfection. **2.** Regarded as the best of its kind. **3.** Completely or highly satisfactory <Their marriage is *ideal.*> **4. a.** Existing only in the mind : IMAGINARY. **b.** Lacking practicality or the possibility of realization. **5.** Of, relating to, or made up of ideas or mental images. **6.** *Philos.* **a.** Existing as an archetype or pattern, esp. as a Platonic idea or perception. **b.** Of or relating to idealism.

☆ ***syns:*** IDEAL, EXAMPLE, EXEMPLAR, MODEL, STANDARD *n. core meaning* : one worthy of imitation <a person who was the *ideal* of virtue>

**i·de·al·ism** (ī-dē'ə-lĭz'əm) *n.* **1.** The act or practice of envisioning things in ideal form. **2.** Pursuit of one's ideals. **3.** An idealizing treatment of a subject in art or literature. **4.** *Philos.* The theory that the object of external perception, in itself or as perceived, is made up of ideas.

**i·de·al·ist** (ī-dē'ə-lĭst) *n.* **1.** One who pursues ideals, esp. when they conflict with practical considerations. **2.** One who is impractical and unrealistic. **3.** An artist or writer whose work is permeated with idealism. **4.** *Philos.* An adherent of a system of idealism.

**i·de·al·is·tic** (ī-dē'ə-lĭs'tĭk) *adj.* Relating to or typical of an idealist or idealism. **—i'de·al·is'ti·cal·ly** *adv.*

**i·de·al·i·ty** (ī'dē-ăl'ĭ-tē) *n., pl.* **-ties. 1.** The quality or state of being ideal. **2.** Existence in idea only.

**i·de·al·ize** (ī-dē'ə-līz') *v.* **-ized, -iz·ing, -iz·es.** *—vt.* **1.** To regard as ideal. **2.** To make or envision as ideal. *—vi.* **1.** To render something as an ideal. **2.** To conceive an ideal or ideals. **—i·de'al·i·za'tion** *n.* **—i·de'al·iz'er** *n.*

**i·de·al·ly** (ī-dē'ə-lē) *adv.* **1.** In conformity with an ideal : PERFECTLY. **2.** In imagination or theory : THEORETICALLY.

**i·de·ate** (ī'dē-āt') *v.* **-at·ed, -at·ing, -ates.** *—vt.* To form an idea of : IMAGINE. *—vi.* To conceive mental images : THINK. **—i'de·a'tion** *n.* **—i'de·a'tion·al** *adj.*

**i·dée fixe** (ē-dā fēks') *n., pl.* **i·dées fixes** (ē-dā fēks') [Fr.] A fixed idea : OBSESSION.

**i·dem** (ī'dĕm') *pron.* [Lat., the same < *id,* it.] —Used to indicate a previously mentioned reference.

**i·den·tic** (ī-dĕn'tĭk) *adj.* [Med. Lat. *identicus,* identical.] **1.** Designating diplomatic language or action in which two or more governments agree to use the same forms in their relations with other governments. **2.** *Archaic.* Identical.

**i·den·ti·cal** (ī-dĕn'tĭ-kəl) *adj.* [Med. Lat. *identicus* < LLat. *identitas,* identity.] **1.** Being the same <wore *identical* clothes> **2.** Exactly equal and alike. **3.** Having such similarity or near resemblance as to be fundamentally equal or interchangeable. **4.** *Biol.* Of or relating to a twin or twins developed from the same ovum. ***usage:*** Either *with* or *to* is acceptable as a preposition after *identical.* **—i·den'ti·cal·ly** *adv.* **—i·den'ti·cal·ness** *n.*

**i·den·ti·fi·ca·tion** (ī-dĕn'tə-fĭ-kā'shən) *n.* **1.** The act of identifying or state of being identified. **2.** Evidence of identity. **3.** *Psychol.* **a.** One's recognition of a personal or group identity. **b.** Transferal of response to an object regarded as identical to another.

**identification card** *n.* An ID card.

**i·den·ti·fi·er** (ī-dĕn'tə-fī'ər) *n.* **1.** One that identifies. **2.** *Computer Sci.* A symbol that identifies, indicates, or names a body of data.

**i·den·ti·fy** (ī-dĕn'tə-fī') *v.* **-fied, -fy·ing, -fies.** [Med. Lat. *identificare* : LLat. *identitas,* identity + Lat. *facere,* to make.] *—vt.* **1. a.** To establish the identity of. **b.** To find out the origin, nature, or definitive elements of. **2.** To determine the taxonomic classification of. **3.** To consider as similar or identical : EQUATE. **4.** *Psychol.* To associate or affiliate (oneself) closely with a person or group. *—vi.* To establish an identification with another or others. ***usage:*** When used in the sense of "to see oneself as similar or identical to," the verb *identify* may be used with or without the reflexive pronoun, as in *I identified myself with the hero* or *I identified with the hero.* **—i·den'ti·fi'a·ble** *adj.* **—i·den'ti·fi'er** *n.*

☆ ***syns:*** IDENTIFY, EMPATHIZE, RELATE (to), SYMPATHIZE *v. core meaning* : to associate or affiliate oneself closely with a person or group <I couldn't *identify* with the book's central character.>

**i·den·ti·ty** (ī-dĕn'tĭ-tē) *n., pl.* **-ties.** [LLat. *identitas* < Lat. *idem,* the same < *id,* it.] **1.** The collective aspect of the characteristics by which a thing is distinctly recognizable or known. **2.** The set of behavioral or personal characteristics by which an individual is recognizable as a member of a group. **3.** The quality or condition of being the same as something else. **4.** The distinct personality of an individual regarded as a continuing entity : INDIVIDUALITY. **5.** *Math.* **a.** An equality satisfied by all values of the variables for which the expressions involved in the equality are defined. **b.** A unity.

**identity crisis** *n.* **1.** A psychosocial state of disorientation and role confusion occurring esp. in adolescents because of conflicting pressures and expectations. **2.** A state of disorientation and role confusion occurring in a social structure, as in an institution.

**identity element** *n.* The element of a set of numbers that when combined with another number in an operation leaves that number unchanged; e.g., 0 is the identity element under addition for the real numbers, since if $a$ is any real number, $a + 0 = 0 + a = a$; similarly, 1 is the identity element under multiplication for the real numbers, since $a \times 1 = 1 \times a = a$.

**identity matrix** *n.* A square matrix with numeral 1's along the diagonal from upper left to lower right and 0's in all other positions.

**identity sign** *n.* A mathematical symbol (≡) used to denote identity rather than equality.

**ideo-** *pref.* [Fr. *idéo-* < Gk. *idea,* form, idea.] Idea <*ideo*graphy>

**id·e·o·gram** (ĭd'ē-ə-grăm', ī'dē-) *also* **id·e·o·graph** (-grăf') *n.* **1.** A character or symbol representing an idea or thing without expressing a particular word or phrase for it, as Chinese characters. **2.** A graphic symbol, as &, $, or @.

**id·e·og·ra·phy** (ĭd'ē-ŏg'rə-fē, ī'dē-) *n.* **1.** Representation of ideas by graphic symbols. **2.** Use of ideograms for expressing ideas. **—id'e·o·graph'ic** (-ə-grăf'ĭk) *adj.*

**i·de·o·log·i·cal** (ī'dē-ə-lŏj'ĭ-kəl, ĭd'ē-) *also* **i·de·o·log·ic** (-lŏj'ĭk) *adj.* **1.** Of or pertaining to ideology. **2.** Of or concerned with ideas.

**i·de·ol·o·gist** (ī'dē-ŏl'ə-jĭst, ĭd'ē-) *n.* **1.** An advocate or adherent of a particular ideology. **2.** *Archaic.* A visionary : theorist.

**i·de·o·logue** (ī'dē-ə-lŏg, ĭd'ē-) *n.* [Fr. *idéologue,* back-formation < *idéologie,* ideology.] An advocate of a particular ideology, esp. one of its official exponents.

**i·de·ol·o·gy** (ī'dē-ŏl'ə-jē, ĭd'ē-) *n., pl.* **-gies.** [Fr. *idéologie* : *idéo-,* ideo- + *-logie,* -logy.] The body of ideas reflecting the social needs and aspirations of an individual, group, class, or culture.

**i·de·o·mo·tor** (ī'dē-ə-mō'tər, ĭd'ē-) *adj.* Of or being a motor response to an ideational rather than a sensory stimulus.

**ides** (īds) *n.* [ME *idus* < OFr. *ides* < Lat. *idus.*] *(sing. in number).* The 15th day of Mar., May, Jul., or Oct. or the 13th day of the other months in the ancient Roman calendar.

**idio-** *pref.* [Gk. < *idios,* personal, private.] **1.** Private : personal <*idio*lect> **2.** Distinct : separate <*idio*plasm>

**id·i·o·blast** (ĭd'ē-ə-blăst') *n.* A plant cell that differs markedly in form from neighboring cells. **—id'i·o·blas'tic** *adj.*

**id·i·o·cy** (ĭd'ē-ə-sē) *n., pl.* **-cies.** [< IDIOT.] **1.** *Psychol.* A condition of subnormal intellectual development or ability, marked by intelligence in the lowest measurable range. **2.** Extreme foolishness or stupidity. **3.** A foolish or stupid deed or utterance.

**id·i·o·lect** (ĭd'ē-ə-lĕkt') *n.* [IDIO- + (DIA)LECT.] The speech of an individual, considered as a linguistic pattern unique among speakers of his or her language or dialect. **—id'i·o·lec'tal, id'i·o·lec'tic** *adj.*

**id·i·om** (ĭd'ē-əm) *n.* [OFr. *idiome* < LLat. *idioma* < Gk. *idiōma* < *idiousthai,* to make one's own < *idios,* own, personal, private.] **1.** A speech form or expression of a language that is peculiar to itself grammatically or that cannot be understood from the individual meanings of its elements. **2.** The particular grammatical, syntactic, and structural character of a given language. **3.** A regional speech or dialect. **4.** A specialized vocabulary used by a group of people : JARGON <diplomatic *idiom*> **5.** A style of artistic expression typical of a particular individual, school, period, or medium <the *idiom* of the minimalist painters>

▲ word history: *Idiom*, like *idiot*, is derived ultimately from Greek *idios*, "private." Greek *idiōma* meant basically any peculiarity or unique feature, but especially a peculiarity of language or literary style. In English the word *idiom* was used to mean "language" in general, then "dialect," and finally to denote a peculiarity of expression, phrase, or grammatical construction in a particular language.

**id·i·o·mat·ic** (ĭd′ē-ə-măt′ĭk) *adj.* **1.** Peculiar to or typical of a particular language. **2.** Like or having the nature of an idiom. **3.** Using many idioms. **—id′i·o·mat′i·cal·ly** *adv.*

**id·i·o·mor·phic** (ĭd′ē-ə-môr′fĭk) *adj.* [< Gk. *idiomorphos*, having one's own form : *idios*, own + *morphē*, shape.] Having a characteristic configuration. —Used of well-crystallized minerals. **—id′i·o·mor′phi·cal·ly** *adv.*

**id·i·op·a·thy** (ĭd′ē-ŏp′ə-thē) *n.* [Gk. *idiopathia*, disease having its own origin : *idios*, own + *pathos*, suffering.] *Med.* **1.** A disease of unknown origin or cause. **2.** A disease for which no etiology is known. **—id′i·o·path′ic** (-ō-păth′ĭk) *adj.* **—id′i·o·path′i·cal·ly** *adv.*

**id·i·o·plasm** (ĭd′ē-ə-plăz′əm) *n.* A hypothetical structural unit of germ plasm. **—id′i·o·plas′mic, id′i·o·plas·mat′ic** (-ō-plăz-măt′ĭk) *adj.*

**id·i·o·syn·cra·sy** (ĭd′ē-ō-sĭng′krə-sē) *n., pl.* **-sies.** [Gk. *idiosunkrasia* : *idios*, own + *sunkrasis*, mixture, temperament (*sun*, together + *krasis*, mixture).] **1.** A structural or behavioral characteristic unique to an individual or group. **2.** A physiological or temperamental peculiarity. **3.** Hypersensitivity to a drug. **—id′i·o·syn·crat′ic** (-sĭn-krăt′ĭk) *adj.* **—id′i·o·syn·crat′i·cal·ly** *adv.*

**id·i·ot** (ĭd′ē-ət) *n.* [ME, ignorant person < OFr. *idiote* < Lat. *idiota* < Gk. *idiōtēs*, private person, layman < *idios*, own, private.] **1.** *Psychol.* A mentally deficient person, with intelligence in the lowest measurable range, being unable to guard against common dangers and incapable of learning connected speech. **2.** A stupid person : FOOL.

▲ word history: The development of the pejorative senses of *idiot* occured in ancient Greek, although the meaning "a mentally deficient person" is a more modern refinement. Greek *idiōtēs*, the source of *idiot*, is derived from *idios*, "private," and originally meant a private citizen in contrast to a public official. The use of *idiōtēs* was extended to other pairs of opposites such as layman/professional, layman/priest, common person/distinguished person, and unskilled worker/craftsman. In general, a person of no special status, knowledge, or skill was *idiōtēs*, and the term became one of abuse.

**idiot box** *n. Slang.* Television.

**id·i·ot·ic** (ĭd′ē-ŏt′ĭk) *adj.* **1.** Exhibiting idiocy. **2.** Foolish or stupid. **—id′i·ot′i·cal·ly** *adv.*

**idiot light** *n.* A light on the instrument panel of an automobile that gives advance warning, as of low oil pressure.

**i·dle** (īd′l) *adj.* **i·dler, i·dlest.** [ME *idel* < OE *īdel*.] **1. a.** Not in use. **b.** Without a job : UNEMPLOYED. **c.** Not scheduled to compete, as a sports team. **2.** Shiftless : lazy. **3.** Lacking basis in fact. *—v.* **i·dled, i·dling, i·dles.** *—vi.* **1.** To pass time without working or in avoiding work. **2.** To move lazily. **3.** To run at a slow speed or out of gear. *—vt.* **1.** To pass (time) without working or in avoiding work : WASTE <*idle* the whole week away> **2.** To make or cause to be unemployed or inactive <A slowdown in orders *idled* many workers.> ***usage:*** This transitive sense of *idle* is now completely acceptable at all levels of speech and writing. **3.** To cause (e.g., a motor) to idle. **—i′dle·ness** *n.* **—i′dler** (-īd′lər) *n.* **—i′dly** *adv.*

**idle character** *n.* An alphanumeric or digital character that is transmitted over a communications line but does not appear in the output of the receiving terminal.

**idle pulley** *also* **idler pulley** *n.* A pulley on a shaft that presses against or rests on a drive belt to guide it or take up slack.

**idle wheel** *n.* **1.** A gear, wheel, or roller interposed between two similar parts for conveying motion from one to the other without change in speed or direction of motion. **2.** An idle pulley.

**i·do·crase** (ī′də-krās′, -krāz′, ĭd′ə-) *n.* [Fr. : Gk. *eidos*, form + *krasis*, mixture.] A green, brown, yellow, or blue mineral, chiefly $Ca_{10}Al_4(Mg,Fe)_2Si_9O_{34}(OH)_4$.

**i·dol** (īd′l) *n.* [ME < OFr. *idole* < LLat. *idolum* < Gk. *eidōlon*, image < *eidos*, form.] **1. a.** An image regarded as an object of worship. **b.** A false god. **2.** A person or thing blindly or excessively adored. **3.** *Archaic.* Something visible but lacking substance.

**i·dol·a·ter** (ī-dŏl′ə-tər) *n.* [ME *idolatrer* < OFr. *idolatre* < LLat. *idololatres* < Gk. *eidōlolatrēs* : *eidōlon*, idol + *-latrēs*, worshiper.] **1.** One who worships idols. **2.** One who blindly or excessively adores a person or thing.

**i·dol·a·trous** (ī-dŏl′ə-trəs) *adj.* **1.** Of or relating to idolatry. **2.** Given to idolatry. **3.** Constituting idolatry. **—i·dol′a·trous·ly** *adv.* **—i·dol′a·trous·ness** *n.*

**i·dol·a·try** (ī-dŏl′ə-trē) *n., pl.* **-tries.** [ME *idolatrie* < OFr. < Med. Lat. *idolatria* < LLat. *idololatria* < Gk. *eidōlolatria* : *eidolōn*, idol + *latreia*, service.] **1.** Worship of idols. **2.** Blind or excessive adoration or devotion.

**i·dol·ize** (īd′l-īz′) *vt.* **-ized, -iz·ing, -iz·es.** **1.** To regard with unquestioning or excessive admiration or devotion. **2.** To worship as an idol. **—i′dol·i·za′tion** *n.* **—i′dol·iz′er** *n.*

**i·dyll** *also* **i·dyl** (īd′l) *n.* [Lat. *idyllium* < Gk. *eidullion*, dim. of *eidos*, form, figure.] **1.** A short literary work describing a picturesque episode or pleasant scene of country life. **2.** An event or scene of rural simplicity. **3.** A narrative poem on a tragic, epic, or romantic theme. **4. a.** A carefree experience. **b.** A romantic interlude.

**i·dyl·list** (īd′l-ĭst) *n.* A writer of idylls.

**i·dyl·lic** (ī-dĭl′ĭk) *adj.* **1.** Of, relating to, or like an idyll. **2.** Naturally charming and picturesque. **—i·dyl′li·cal·ly** *adv.*

**-ie** *suff. var. of* -Y3.

**if** (ĭf) *conj.* [ME < OE *gif*.] **1. a.** In the event that <*If* I were to go, I would rent a car.> **b.** Granting that <*If* that's correct, then the problem is solved.> **c.** On condition that <They will cater the dinner only *if* they are paid in advance.> **2.** Although possibly : even though <a handsome *if* impractical purchase> **3.** Whether <Ask *if* the plane will arrive on time.> **4.** —Used to introduce an exclamatory clause, indicating a wish <*If* you had done what I asked!> ***usage:*** There is a growing tendency to use *would have* in contrary-to-fact clauses begining with *if*, but this usage is still considered incorrect. Correct models are *if I had been successful* (not *if I would have been successful*) or *if I were responsible* (not *if I would have been responsible*). *—n.* A possibility, condition, or stipulation <no *ifs*, ands, or buts>

**if·fy** (ĭf′ē) *adj.* [< IF.] *Informal.* Marked by doubt, uncertainty, or chance.

**I formation** *n. Football.* An alignment of the offensive team in which all the backs line up in single file behind the center.

**ig·loo** (ĭg′lo͞o) *n., pl.* **-loos.** [Eskimo *iglu*, house.] **1.** A usu. dome-shaped Eskimo dwelling built of blocks of ice or hard snow or occas. wood, stone, or sod. **2.** A dome-shaped structure.

**ig·ne·ous** (ĭg′nē-əs) *adj.* [Lat. *igneus* < *ignis*, fire.] **1.** Of, pertaining to, or typical of fire. **2.** *Geol.* **a.** Formed by solidification from a molten or partially molten state. —Used of rocks. **b.** Of or relating to rock so formed.

**ig·nis fat·u·us** (ĭg′nĭs făch′o͞o-əs) *n., pl.* **ig·nes fat·u·i** (ĭg′nēz făch′o͞o-ī′) [Med. Lat., foolish fire.] **1.** A phosphorescent light hovering or flitting over swampy ground at night, possibly caused by spontaneous combustion of gases given off by rotting organic matter. **2.** Something that deludes or misleads : ILLUSION.

**ig·nite** (ĭg-nīt′) *v.* **-nit·ed, -nit·ing, -nites.** [LLat. *ignire*, *ignit-* < *ignis*, fire.] *—vt.* **1. a.** To cause to burn. **b.** To set fire to. **2.** To subject to great heat, esp. to make luminous by heat. **3.** To arouse or kindle the passions of : EXCITE <Their rudeness *ignited* my indignation.> *—vi.* **1.** To start to burn. **2.** To start to glow. **—ig·nit′a·ble, ig·nit′i·ble** *adj.* **—ig·nit′er, ig·ni′tor** *n.*

☆ ***syns:*** IGNITE, ENKINDLE, KINDLE, LIGHT, TORCH *v. core meaning* : to cause to burn or undergo combustion <*ignite* a fire> ***ant:*** extinguish

**ig·ni·tion** (ĭg-nĭsh′ən) *n.* **1.** The raising of a substance to its ignition point, as by friction, electric current, or mechanical shock. **2. a.** An electrical system, usu. operated by a magneto or battery, that provides the spark to ignite the fuel mixture in an internal-combustion engine. **b.** A switch activating this system.

**ignition point** *n.* The minimum temperature at which a substance will continue burning without additional application of external heat.

**ig·ni·tron** (ĭg-nī′trŏn′, ĭg′nĭ-) *n.* [Lat. *ignis*, fire + -TRON.] A single-anode mercury-vapor rectifier in which current passes as an arc between the anode and a mercury-pool cathode, used in power rectification.

**ig·no·ble** (ĭg-nō′bəl) *adj.* [Lat. *ignobilis* : *in-*, not + *nobilis*, noble.] **1.** Not having a noble character or purpose : BASE. **2.** Not being a member of the nobility : COMMON. **—ig′no·bil′i·ty** (-bĭl′ĭ-tē), **ig·no′ble·ness** *n.* **—ig·no′bly** *adv.*

**ig·no·min·i·ous** (ĭg′nə-mĭn′ē-əs) *adj.* **1.** Marked by shame or disgrace : DISHONORABLE. **2.** Deserving disgrace or shame : DESPICABLE. **3.** Degrading : debasing <an *ignominious* knockout in the first round> **—ig′no·min′i·ous·ly** *adv.* **—ig′no·min′i·ous·ness** *n.*

**ig·no·min·y** (ĭg′nə-mĭn′ē, -mə-nē) *n., pl.* **-ies.** [Lat. *ignominia* : *in-*, not + *nomen*, name, reputation.] **1.** Great personal humiliation or dishonor. **2.** Disgraceful or shameful action, conduct, or character.

**ig·no·ra·mus** (ĭg′nə-rā′məs) *n.* [NLat. < Lat., we do not know < *ignorare*, to be ignorant.] An ignorant person.

**ig·no·rant** (ĭg′nər-ənt) *adj.* [ME *ignoraunt* < OFr. *ignorant* < Lat. *ignorans*, pr.part. of *ignorare*, to be ignorant.] **1.** Without knowledge or education. **2.** Displaying lack of knowledge or education. **3.** Unaware or uninformed : OBLIVIOUS. **—ig′no·rance** (ĭg′nər-əns) *n.* **—ig′no·rant·ly** *adv.*

☆ ***syns:*** **1.** IGNORANT, ILLITERATE, NESCIENT, UNEDUCATED, UNLEARNED, UNSCHOOLED, UNTAUGHT *adj. core meaning* : without knowledge or education <*ignorant* children> **2.** IGNORANT, OBLIVIOUS, UNAWARE, UNCONSCIOUS, UNFAMILIAR, UNKNOWING, UNWITTING *adj. core meaning* : not aware or informed <*ignorant* of the dangers around them> ***ant:*** informed

**ig·nore** (ĭg-nôr′, -nōr′) *vt.* **-nored, -nor·ing, -nores.** [Fr. *ignorer* < Lat. *ignorare*, not to know, to be ignorant.] To pay no attention to. **—ig·nor′a·ble** *adj.* **—ig·nor′er** *n.*

☆ **syns:** IGNORE, DISREGARD, NEGLECT, SLIGHT, SNUB *v. core meaning* : to refuse to pay attention to <*ignored* those who didn't agree>

**I·go·rot** (ĭg'ə-rōt', ē'gə-) *n., pl.* **Igorot** or **-rots. 1.** A member of any of several related tribes of northern Luzon, Philippines. **2.** The Austronesian language of the Igorot.

**i·gua·na** (ĭ-gwä'nə) *n.* [Sp. < Arawak *iwana.*] Any of various large tropical American lizards of the family Iguanidae, often with spiny projections along the back.

**i·guan·o·don** (ĭ-gwä'nə-dŏn') *n.* [NLat. *Iguanodon,* genus name : IGUANA + Gk. *odōn,* tooth.] Any of various large dinosaurs of the genus *Iguanodon* of the Jurassic and Cretaceous periods.

**ih·ram** (ē-räm') *n.* [Ar. *iḥrām,* prohibition < *ḥarama,* he prohibited.] **1.** The sacred dress worn by Moslem pilgrims, made of two lengths of white cotton. **2.** The sacred state in which Moslem pilgrims exist while wearing the ihram.

**IHS** *n.* [Short for Gk. *IHSOUS,* Jesus.] A graphic symbol for Jesus.

**i·ke·ba·na** (ē'kā-bä'nä, ĭk'ə-) *n.* [J. : *ikeru,* to arrange + *hana,* flower.] The Japanese art of flower arrangement with special regard to form, balance, and harmony.

**i·kon** (ī'kŏn') *n. var. of* ICON.

**il-**[1] *pref. var. of* IN-[1]. —Used before *l.*

**il-**[2] *pref. var. of* IN-[2]. —Used before *l.*

**i·lang-i·lang** (ē'läng'ē'läng') *n. var. of* YLANG-YLANG.

**-ile**[1] *suff.* [ME < OFr. < Lat. *-ilis.*] Of, relating to, or capable of <aud*ile*>

**-ile**[2] *suff.* [Perh. < Lat. *-ilis,* adj. suffix.] A division of a specified size in the range of a statistic <percent*ile*>

**il·e·a** (ĭl'ē-ə) *n. pl. of* ILEUM.

**il·e·ac** (ĭl'ē-ăk') *adj.* **1.** Of or relating to ileus. **2.** Of or relating to the ileum.

**il·e·i·tis** (ĭl'ē-ī'tĭs) *n.* [ILE(UM) + -ITIS.] Inflammation of the ileum.

**il·e·o·ce·cal** (ĭl'ē-ō-sē'kəl) *adj.* [ILE(UM) + CECUM.] Of or relating to the ileum and the cecum.

**il·e·os·to·my** (ĭl'ē-ŏs'tə-mē) *n., pl.* **-mies.** [ILE(UM) + -STOMY.] Surgical construction of an artificial excretory opening through the abdominal wall into the ileum.

**il·e·um** (ĭl'ē-əm) *n., pl.* **-e·a** (-ē-ə) [NLat. < Lat., groin, flank.] The section of the small intestine between the jejunum and the cecum. **—il'e·al** *adj.*

**il·e·us** (ĭl'ē-əs) *n.* [Lat. *ileos* < Gk. *eileos,* intestinal obstruction < *eilein,* to squeeze, turn.] Intestinal obstruction inducing colic, vomiting, and toxemia.

**i·lex** (ī'lĕks') *n.* [Lat., holm oak.] **1.** A tree or shrub of the genus *Ilex* : HOLLY. **2.** The holm oak.

**il·i·um** (ĭl'ē-əm) *n., pl.* **-i·a** (-ē-ə) [NLat. < Lat. *ileum,* groin, flank.] The uppermost and widest of three bones comprising one of the lateral halves of the pelvis. **—il'i·ac'** (-ăk') *adj.*

**ilk** (ĭlk) *n.* [ME, same < OE *ilca.*] Type or sort. *—pron. Scot.* The same. —Used following a name to indicate that the one named resides in an area bearing the same name <Duncan of that *ilk*> *—adj. var. of* ILKA.

**il·ka** (ĭl'kə) *also* **ilk** (ĭlk) *adj.* [ME *ilk a,* each one : *ilk,* var. of *ech,* each + *a,* one, a.] *Scot.* Every : each.

**ill** (ĭl) *adj.* **worse** (wûrs), **worst** (wûrst) [ME < ON *illr,* bad.] **1.** Not healthy : SICK <hoped I wouldn't be *ill*> **2.** Not normal : UNSOUND <*ill* health> **3.** Resulting in suffering : DISTRESSING <*ill* effects caused by the drought> **4.** Having evil intentions : HOSTILE <did us an *ill* turn> **5.** Not favorable : UNPROPITIOUS <*ill* luck> **6.** Not up to recognized standards of excellence or conduct. **7.** Cruel : harmful. *—adv.* **worse, worst. 1.** In an ill manner : not well. **2.** Scarcely or with difficulty. *—n.* **1.** Sin : evil. **2.** Disaster, distress, or harm. **3.** Something causing suffering : TROUBLE <persistent social *ills* of the city> **—ill at ease.** Nervous and uncomfortable.

**I'll** (īl). **1.** I will. **2.** I shall.

**ill-ad·vised** (ĭl'əd-vīzd') *adj.* Done without careful thought or wise counsel. **—ill'-ad·vis'ed·ly** (-vī'zĭd-lē) *adv.*

**il·la·tion** (ĭ-lā'shən) *n.* [LLat. *illatio* < Lat. *illatus,* p.part. of *inferre,* to carry in, infer : *in-,* in + *ferre,* to carry.] **1.** The act or process of inferring or drawing conclusions. **2.** A conclusion : deduction.

**il·la·tive** (ĭl'ə-tĭv, ĭ-lā'-) *adj.* **1.** Of, relating to, or like an illation. **2.** Expressing or preceding an inference. —Used of a word. *—n.* **1.** An illative word or phrase, as *hence* or *for that reason.* **2.** ILLATION 2.

**ill-being** (ĭl'bē'ĭng) *n.* Lack of health, prosperity, or happiness.

**ill-bod·ing** (ĭl'bō'dĭng) *adj.* Portending evil : INAUSPICIOUS.

**ill-bred** (ĭl'brĕd') *adj.* **1.** Badly brought up. **2.** Not thoroughbred.

**il·le·gal** (ĭ-lē'gəl) *adj.* **1.** Forbidden by law. **2.** Forbidden by official rules. **3.** *Computer Sci.* That is unacceptable to or not performable by a computer <an *illegal* operation> *—n.* An illegal emigrant. **—il'le·gal'i·ty** (ĭl'ē-găl'ĭ-tē) *n.* **—il·le'gal·ly** *adv.*

☆ **syns:** ILLEGAL, ILLICIT, UNLAWFUL, WRONGFUL *adj. core meaning* : prohibited by law <*illegal* tax deductions> **ant:** legal

**il·leg·i·ble** (ĭ-lĕj'ə-bəl) *adj.* Not legible or decipherable. **—il·leg'i·bil'i·ty, il·leg'i·ble·ness** *n.* **—il·leg'i·bly** *adv.*

**il·le·git·i·ma·cy** (ĭl'ĭ-jĭt'ə-mə-sē) *n.* **1.** The quality or condition of being illegitimate. **2.** Bastardy.

**il·le·git·i·mate** (ĭl'ĭ-jĭt'ə-mĭt) *adj.* **1.** Against the law : ILLEGAL. **2.** Born out of wedlock. **3.** Not in proper grammatical usage. **4.** Incorrectly deduced : ILLOGICAL. **—il'le·git'i·mate·ly** *adv.*

**ill-fat·ed** (ĭl'fā'tĭd) *adj.* **1.** Destined for misfortune : DOOMED. **2.** Characterized by or causing misfortune : UNLUCKY.

**ill-fa·vored** (ĭl'fā'vərd) *adj.* **1.** Having an unattractive or ugly face. **2.** Arousing disapproval : OFFENSIVE. **3.** ILL-FATED 1.

**ill-got·ten** (ĭl'gŏt'n) *adj.* Obtained in an evil or dishonest way.

**ill humor** *n.* An irritable state of mind : SURLINESS.

**ill-hu·mored** (ĭl'hyōō'mərd) *adj.* Being irritable and surly. **—ill'-humored·ly** *adv.*

**il·lib·er·al** (ĭ-lĭb'ər-əl) *adj.* [Lat. *illiberalis* : *in-,* not + *liberalis,* liberal.] **1.** Bigoted : narrow-minded. **2.** *Archaic.* Ungenerous or stingy. **3.** *Archaic.* **a.** Lacking liberal culture. **b.** Ill-bred : vulgar. **—il·lib'er·al'i·ty, il·lib'er·al·ness** *n.* **—il·lib'er·al·ly** *adv.*

**il·lic·it** (ĭ-lĭs'ĭt) *adj.* [Lat. *illicitus* : *in-,* not + *licitus,* lawful. —see LICIT.] Not permitted by custom or law : UNLAWFUL. **—il·lic'it·ly** *adv.* **—il·lic'it·ness** *n.*

**il·lim·it·a·ble** (ĭ-lĭm'ĭ-tə-bəl) *adj.* Incapable of being limited or circumscribed : LIMITLESS. **—il·lim'it·a·bil'i·ty, il·lim'it·a·ble·ness** *n.* **—il·lim'it·a·bly** *adv.*

**Il·li·noi·an** (ĭl'ə-noi'ən) *adj.* [After the state of *Illinois.*] Of or relating to the third glacial stage in North America.

**Il·li·nois** (ĭl'ə-noi', -noiz') *n., pl.* **Illinois.** [Fr., of Algonquian orig.] **1. a.** A confederacy of Indian tribes that inhabited Illinois and parts of Iowa, Wisconsin, and Missouri. **b.** A member of this confederacy or one of the member tribes. **2.** The Algonquian language of the Illinois and Miami peoples.

**il·liq·uid** (ĭ-lĭk'wĭd) *adj.* **1.** Incapable of being readily or easily converted into cash <*illiquid* property> **2.** Lacking in cash or liquid assets <brokerage houses that were *illiquid*> **—il'li·quid'i·ty** (ĭl'ĭ-kwĭd'ĭ-tē) *n.*

**il·lit·er·a·cy** (ĭ-lĭt'ər-ə-sē) *n., pl.* **-cies. 1.** The quality or condition of being unable to read and write. **2.** An error caused by or thought to be typical of illiteracy.

**il·lit·er·ate** (ĭ-lĭt'ər-ĭt) *adj.* [Lat. *illiteratus* : *in-,* not + *literatus,* literate.] **1. a.** Unable to read and write. **b.** Having little or no formal education. **2. a.** Characterized by inferiority to an expected standard of familiarity with language and literature. **b.** Violating prescribed standards of speech or writing. **3.** Ignorant of the fundamentals of a given art or branch of knowledge <mathematically *illiterate*> *—n.* One who is illiterate. **—il·lit'er·ate·ly** *adv.* **—il·lit'er·ate·ness** *n.*

**ill-man·nered** (ĭl'măn'ərd) *adj.* Lacking or showing a lack of good manners : RUDE. **—ill'-man'nered·ly** *adv.*

**ill-na·tured** (ĭl'nā'chərd) *adj.* **1.** Having or exhibiting a disagreeable disposition : SURLY. **2.** Spiteful and nasty. **—ill'-na'tured·ly** *adv.* **—ill'-na'tured·ness** *n.*

**ill·ness** (ĭl'nĭs) *n.* **1. a.** Sickness of body or mind. **b.** A particular sickness : AILMENT. **2.** *Obs.* Evil : wickedness.

**il·log·ic** (ĭ-lŏj'ĭk) *n.* Lack of logic.

**il·log·i·cal** (ĭ-lŏj'ĭ-kəl) *adj.* **1.** Contradicting or ignoring the principles of logic. **2.** Without logic : SENSELESS. **—il·log'i·cal'i·ty** (-kăl'ĭ-tē), **il·log'i·cal·ness** *n.* **—il·log'i·cal·ly** *adv.*

**ill-sort·ed** (ĭl'sôr'tĭd) *adj.* Badly matched.

**ill-starred** (ĭl'stärd') *adj.* Unlucky or ill-fated : UNFORTUNATE.

**ill-tem·pered** (ĭl'tĕm'pərd) *adj.* Having a bad temper : IRRITABLE. **—ill'-tem'pered·ly** *adv.*

**ill-timed** (ĭl'tīmd') *adj.* Done or happening at an inappropriate time : UNTIMELY.

**ill-treat** (ĭl'trēt') *vt.* **-treat·ed, -treat·ing, -treats.** To maltreat. **—ill'-treat'ment** *n.*

**il·lume** (ĭ-lōōm') *vt.* **-lumed, -lum·ing, -lumes.** [Short for ILLUMINE.] To illuminate.

**il·lu·mi·nance** (ĭ-lōō'mə-nəns) *n.* ILLUMINATION 7.

**il·lu·mi·nant** (ĭ-lōō'mə-nənt) *n.* Something that emits light.

**il·lu·mi·nate** (ĭ-lōō'mə-nāt') *v.* **-nat·ed, -nat·ing, -nates.** [Lat. *illuminare, illuminat-* : *in-,* in + *luminare,* to light up < *lumen,* light.] *—vt.* **1.** To supply or brighten with light. **2.** To decorate or hang with lights. **3.** To make understandable : CLARIFY. **4.** To enlighten intellectually or spiritually : EDIFY. **5.** To endow with splendor or fame : CELEBRATE. **6.** To adorn (e.g., a book page) with decorative designs, miniatures, or lettering in bright colors or precious metals. **7.** To reveal by or expose to radiation. *—vi.* **1.** To become lighted : GLOW. **2.** To be revealed by or exposed to radiation. *—n.* (-nĭt). One who has or professes to have a high degree of enlightenment.

**il·lu·mi·na·ti** (ĭ-lōō'mə-nä'tē) *pl.n.* [Lat., pl. of *illuminatus,* p.part. of *illuminare,* to illuminate.] **1.** Persons claiming to be highly enlightened with respect to some subject. **2. Illuminati.** Any of various groups that claimed special religious enlightenment.

**il·lu·mi·na·tion** (ĭ-lōō'mə-nā'shən) *n.* **1.** An act of illuminating or the state of being illuminated. **2.** A source of light. **3.** Lighting used as decoration. **4.** Spiritual or intellectual enlightenment. **5.** Clarification : elucidation. **6. a.** The art or act of adorning a text, page, or initial letter with decorative designs, miniatures, or lettering. **b.** An example of this art. 7. *Physics.* The luminous flux per unit area at any point on a surface exposed to incident light.

**il·lu·mi·na·tive** (ĭ-lo͞o′mə-nā′tĭv) *adj.* Of, producing, or capable of producing illumination.

**il·lu·mi·na·tor** (ĭ-lo͞o′mə-nā′tər) *n.* **1.** One that illuminates. **2.** A device for producing, concentrating, or reflecting light. **3.** One who illuminates manuscripts or other objects.

**il·lu·mine** (ĭ-lo͞o′mĭn) *vt.* **-mined, -min·ing, -mines.** [ME *illuminen* < Lat. *illuminare,* to illuminate.] To give light to : ILLUMINATE. **—il·lu′mi·na·ble** *adj.*

**il·lu·mi·nism** (ĭ-lo͞o′mə-nĭz′əm) *n.* [ILLUMIN(ATI) + -ISM.] **1.** Belief in or proclamation of a special personal enlightenment. **2. Illuminism.** The principles and ideas of various groups of Illuminati. **—il·lu′mi·nist** *n.*

**ill-use** (ĭl′yo͞oz′) *vt.* **-used, -us·ing, -us·es.** To maltreat. —*n.* (ĭl′yo͞os′) *also* **ill-us·age** (-yo͞o′sĭj, -zĭj). Bad or unjust treatment.

**il·lu·sion** (ĭ-lo͞o′zhən) *n.* [ME < OFr. < LLat. *illusio* < Lat., a mocking, irony < *illusus,* p.part. of *illudere,* to mock : *in-,* against + *ludere,* to play.] **1. a.** A mistaken perception of reality. **b.** A mistaken belief or concept. **2.** The condition of being deceived by an erroneous belief or perception. **3.** Something, as a fantastic desire or plan, that causes an erroneous perception or belief. **4.** ILLUSIONISM 2. **5.** A fine transparent cloth used for dresses or trimmings. **—il·lu′sion·al, il·lu′sion·ar′y** *adj.*

**il·lu·sion·ism** (ĭ-lo͞o′zhə-nĭz′əm) *n.* **1.** The doctrine stating that the material world is an immaterial product of the senses. **2.** Use of illusionary devices and techniques in art or decoration. **—il·lu′sion·is′tic** *adj.*

**il·lu·sion·ist** (ĭ-lo͞o′zhə-nĭst) *n.* **1.** An adherent of the doctrine of illusionism. **2.** A magician or ventriloquist. **3.** An artist whose work is characterized by illusionism.

**il·lu·sive** (ĭ-lo͞o′sĭv) *adj.* Of, relating to, or like an illusion : ILLUSORY. **—il·lu′sive·ly** *adv.* **—il·lu′sive·ness** *n.*

**il·lu·so·ry** (ĭ-lo͞o′sə-rē, -zə-rē) *adj.* Produced by, based on, or having the nature of an illusion.

☆ ***syns:*** ILLUSORY, DELUSIVE, DELUSORY, ILLUSIVE *adj. core meaning* : tending to deceive <*illusory* hopes>

**il·lus·trate** (ĭl′ə-strāt′, ĭ-lŭs′trāt′) *v.* **-trat·ed, -trat·ing, -trates.** [Lat. *illustrare, illustrat-* : *in-,* in + *lustrare,* to make bright < *lustrum,* purification.] —*vt.* **1. a.** To clarify, as by using examples or making comparisons. **b.** To clarify by serving as an example or comparison. **2.** To provide (a publication) with explanatory or decorative features. **3.** *Obs.* To illuminate. —*vi.* To present a clarification, example, or explanation. **—il′lus·tra′tor** *n.*

**il·lus·tra·tion** (ĭl′ə-strā′shən) *n.* **1.** An act of clarifying or explaining or the state of being clarified or explained. **2.** Something used to clarify or explain. **3.** Visual matter for clarifying or decorating a text. **4.** *Obs.* Illumination. **—il′lus·tra′tion·al** *adj.*

**il·lus·tra·tive** (ĭ-lŭs′trə-tĭv, ĭl′ə-strā′tĭv) *adj.* Acting as an illustration. **—il·lus′tra·tive·ly** *adv.*

**il·lus·tri·ous** (ĭ-lŭs′trē-əs) *adj.* [< Lat. *illustris,* distinguished, bright < *illustrare,* to illustrate.] **1.** Renowned or celebrated. **2.** *Obs.* Shining brightly. **—il·lus′tri·ous·ly** *adv.* **—il·lus′tri·ous·ness** *n.*

**il·lu·vi·ate** (ĭ-lo͞o′vē-āt′) *vi.* **-at·ed, -at·ing, -ates.** [Back-formation < ILLUVIATION.] To undergo illuviation.

**il·lu·vi·a·tion** (ĭ-lo͞o′vē-ā′shən) *n.* [IL-² + (AL)LUVI(UM) + -ATION.] The deposition in an underlying soil layer of colloids, soluble salts, and mineral particles leached out of an overlying soil layer.

**ill will** *n.* Unfriendly feeling : ENMITY.

**il·ly** (ĭl′lē) *adv.* Badly : ill.

**Il·lyr·i·an** (ĭ-lĭr′ē-ən) *n.* **1.** One of a people inhabiting ancient Illyria. **2.** The Indo-European language of the Illyrians. —*adj.* Of, relating to, or typical of the Illyrians or their language.

**il·men·ite** (ĭl′mə-nīt′) *n.* [G. *Ilmenit,* after Ilmen, a range in the Ural Mountains, where it was first found.] A lustrous black-to-brownish titanium ore, chiefly $FeTiO_3$.

**I·lo·ca·no** *also* **I·lo·ka·no** (ē′lō-kä′nō) *n., pl.* **Ilocano** *also* **Ilokano** or **-nos.** [Sp. *Ilócano* < *Iloko,* a Malayan people in the Philippines.] **1.** One of a people inhabiting northwestern Luzon, Philippines. **2.** The Austronesian language of the Ilocano. —*adj.* Of, relating to, or typical of the Ilocano or their language.

**I'm** (īm). I am.

**im-¹** *pref. var. of* IN-¹. —Used before *b, m,* and *p.*

**im-²** *pref. var. of* IN-². —Used before *b, m,* and *p.*

**im·age** (ĭm′ĭj) *n.* [ME < OFr. < Lat. *imago.*] **1.** A reproduction of the form of someone or something, esp. a sculptured likeness. **2.** An optically formed duplicate, counterpart, or other representative reproduction of an object, esp. an optical reproduction of an object by a mirror or lens. **3.** A close or exact resemblance to another : DOUBLE. **4. a.** The idea of someone or something that is held by the public. **b.** The character projected by someone or something to the public, esp. by the mass media : REPUTATION. **5.** A personification of something specified <You are the *image* of happiness.> **6.** A mental picture of something unreal or not present. **7. a.** A vivid description or representation. **b.** A figure of speech. **8.** *Obs.* An apparition. **9.** *Math.* A set of values of a function corresponding to a particular subset of a domain. **10.** *Computer Sci.* An exact duplication of data in a file onto another medium. —*vt.* **-aged, -ag·ing, -ag·es. 1.** To make a likeness of. **2.** To reflect. **3.** To symbolize or typify. **4.** To picture mentally : IMAGINE. **5.** To describe, esp. to describe so vividly as to call up a mental picture of.

**image orthicon** *n.* An orthicon.

**im·age·ry** (ĭm′ĭj-rē) *n., pl.* **-ries.** [ME *imagerie* < OFr. < *image,* image.] **1.** Mental pictures or images. **2. a.** The use of vivid descriptions or figures of speech in speaking or writing to produce mental images. **b.** A metaphoric representation, as in music, art, or drama. **3. a.** Representative images, esp. statues or icons. **b.** The art of making such images.

**i·mag·i·na·ble** (ĭ-măj′ə-nə-bəl) *adj.* Capable of being conceived of by the imagination. **—i·mag′i·na·bly** *adj.*

**i·ma·gi·nal** (ĭ-mā′gə-nəl, ĭ-mä′-) *adj.* [< NLat. *imago, imagin-,* imago.] Of or pertaining to an insect imago.

**i·mag·i·nar·y** (ĭ-măj′ə-nĕr′ē) *adj.* **1.** Existing only in the imagination : UNREAL. **2.** *Math.* **a.** Of, relating to, or being the coefficient of the imaginary unit in a complex number. **b.** Of, relating to, involving, or being an imaginary number. **c.** Involving only a complex number of which the real part is zero. —*n., pl.* **-ies.** *Math.* An imaginary number. **—i·mag′i·nar′i·ly** *adv.* **—i·mag′i·nar′i·ness** *n.*

**imaginary number** *n.* A complex number in which the real part is zero and the coefficient of the imaginary unit is not zero.

**imaginary unit** *n.* The positive square root of −1.

**i·mag·i·na·tion** (ĭ-măj′ə-nā′shən) *n.* **1. a.** The power of the mind to form a mental image or concept of something that is unreal or not present. **b.** Such power of the mind used creatively. **2.** The ability to confront and cope with reality by using the creative power of the mind : RESOURCEFULNESS. **3.** *Archaic.* **a.** An unrealistic idea or notion : FANCY. **b.** A plan or scheme. **4.** A traditional or widely held belief or opinion. **—i·mag′i·na′tion·al** *adj.*

☆ ***syns:*** IMAGINATION, IMAGINATIVENESS, FANCY, FANTASY *n. core meaning* : the power of the mind to form images <the writer's lively *imagination*>

**i·mag·i·na·tive** (ĭ-măj′ə-nə-tĭv, -nā′tĭv) *adj.* **1.** Having a lively, creative imagination. **2.** Inclined to indulge in the fanciful or in make-believe. **3. a.** Created by, indicative of, or marked by imagination or creativity. **b.** False : untrue. **—i·mag′i·na·tive·ly** *adv.* **—i·mag′i·na·tive·ness** *n.*

**i·mag·ine** (ĭ-măj′ĭn) *v.* **-ined, -in·ing, -ines.** [ME *imaginen* < OFr. *imaginer* < Lat. *imaginari* < *imago,* image.] —*vt.* **1.** To form a mental image or picture of. **2.** To think : suppose <I *imagine* they'll all go.> **3.** To have a notion without adequate foundation : FANCY <*imagine* themselves to be artists> —*vi.* **1.** To use the imagination. **2.** To guess : conjecture. **—i·mag′in·er** *n.*

**im·a·gism** (ĭm′ə-jĭz′əm) *n.* An early 20th-cent. literary movement that promoted free verse and precise imagery. **—im′a·gist** *n.* **—im′a·gis′tic** *adj.*

**i·ma·go** (ĭ-mā′gō, ĭ-mä′-) *n., pl.* **-goes** or **-gi·nes** (-gə-nēz′) [NLat. *imago, imagin-* < Lat. image.] **1.** An insect in its sexually mature adult stage after metamorphosis. **2.** *Psychoanal.* An often idealized image of a person, usu. a parent, formed in childhood and persisting into adulthood.

**i·mam** (ĭ-mäm′) *n.* [Ar. *imām,* leader < *amma,* he led.] **1.** A prayer leader of Islam. **2.** A Moslem scholar, esp. an authority on Islamic law. **3. Imam. a.** A title accorded to Mohammed and his four immediate successors. **b.** One of the leaders considered by the Shiites to be successors of Mohammed. **c.** Any of various religious and temporal leaders claiming descent from Mohammed.

**i·mam·ate** (ĭ-mä′māt′) *n.* **1.** The office of an Imam. **2.** A country or region governed by an Imam.

**i·ma·ret** (ĭ-mä′rĕt) *n.* [Turk. < Ar. *imārah,* hospice < *amara,* he built.] An inn or hospice for pilgrims in Turkey.

**im·bal·ance** (ĭm-băl′əns) *n.* A lack of balance, as in distribution, proportion, or functioning.

**im·be·cile** (ĭm′bə-sĭl, -səl) *n.* [OFr. *imbecille,* feeble < Lat. *imbecillus* : *in-,* not + *bacillum,* staff, dim. of *baculum,* rod.] **1.** A feeble-minded person. **2.** A stupid person. —*adj. also* **im·be·cil·ic** (ĭm′bə-sĭl′ĭk). **1.** Deficient in mental ability. **2.** Stupid : idiotic. **—im′be·cile·ly** *adv.*

**im·be·cil·i·ty** (ĭm′bə-sĭl′ĭ-tē) *n., pl.* **-ties. 1.** The quality or condition of being an imbecile. **2. a.** Great foolishness or stupidity. **b.** Something imbecilic.

**im·bed** (ĭm-bĕd′) *v. var. of* EMBED.

**im·bibe** (ĭm-bīb′) *v.* **-bibed, -bib·ing, -bibes.** [ME *embiben,* to soak, saturate < OFr. *embiber* < Lat. *imbibere,* to drink in : *in-,* in + *bibere,* to drink.] —*vt.* **1.** To drink. **2.** To absorb or take in as if by drinking <"the whole body . . . *imbibes* delight through every pore" —Thoreau> **3.** To receive and absorb into the mind <"Gladstone had . . . *imbibed* a strong prejudice against Americans" —Philip Magnus> **4.** *Obs.* To permeate : saturate. —*vi.* To drink. **—im·bib′er** *n.*

**im·bi·bi·tion** (ĭm′bə-bĭsh′ən) *n.* **1.** The act of imbibing. **2.** Absorption of fluid by a solid or colloid that results in swelling.

**im·bri·cate** (ĭm′brĭ-kāt′) *adj.* [Lat. *imbricatus,* p.part. of *imbricare,* to cover with roof tiles < *imbrex,* roof tile < *imber,* rain.] With edges

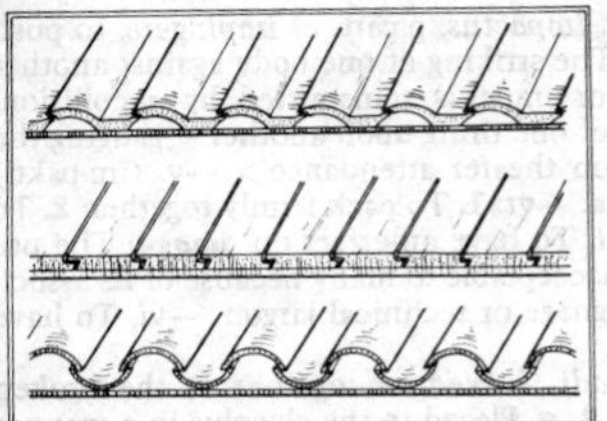

**imbricate**
*Three styles of imbricate roofing: (top) Spanish, (middle) English, (bottom) mission*

overlapping in a regular arrangement, as roof tiles or the scales of a fish. **—im'bri·cate'** *v.* **(-cat·ed, -cat·ing, -cates).**

**im·bri·ca·tion** (ĭm'brĭ-kā'shən) *n.* **1.** A regular overlapping of edges. **2.** A pattern or design with overlapping edges.

**im·bro·glio** (ĭm-brōl'yō) *n., pl.* **-glios.** [Ital., prob. < Fr. *embrouiller,* to confuse. —see EMBROIL.] **1. a.** A difficult or intricate situation : ENTANGLEMENT. **b.** A complicated or confused disagreement. **2.** A confused heap : TANGLE.

**im·brue** (ĭm-brōō') *vt.* **-brued, -bru·ing, -brues.** [ME *enbrewen* < OFr. *embreuver,* to moisten : *en-,* in + *breu,* broth, of Germanic orig.] To saturate or stain.

**im·brute** (ĭm-brōōt') *vt. & vi.* **-brut·ed, -brut·ing, -brutes.** To make or become brutal.

**im·bue** (ĭm-byōō') *vt.* **-bued, -bu·ing, -bues.** [Lat. *imbuere,* to moisten, stain.] **1.** To saturate, as with stain or dye. **2.** To inspire, permeate, or pervade <work *imbued* with creative energy>

**im·id·az·ole** (ĭm'ĭd-ăz'ōl') *n.* [IMID(E) + AZOLE.] Any of a group of heterocyclic compounds, esp. the white crystalline base $C_3H_4N_2$.

**im·ide** (ĭm'ĭd') *n.* [Alteration of AMIDE.] A compound derived from ammonia containing the divalent NH group combined with two acid radicals.

**im·ine** (ĭm'ēn', ĭ-mēn') *n.* [Alteration of AMINE.] A compound derived from ammonia containing the divalent NH group combined with alkyl or other nonacid radicals.

**i·mip·ra·mine** (ĭ-mĭp'rə-mēn', ĭm'ə-prā'mēn) *n.* [IMI(DE) + PR(OPYL) + AMINE.] A water-soluble compound, $C_{19}H_{24}N_2$, used medically as an antidepressant.

**im·i·ta·ble** (ĭm'ĭ-tə-bəl) *adj.* Capable or worthy of being imitated.

**im·i·tate** (ĭm'ĭ-tāt') *vt.* **-tat·ed, -tat·ing, -tates.** [Lat. *imitari, imitat-.*] **1.** To model oneself after the behavior or actions of (another). **2. a.** To copy the actions, appearance, or mannerisms of (another) <*imitated* their singsong speech> **b.** To use the literary, artistic, or musical style of (another). **3.** To duplicate exactly : REPRODUCE. **4.** To appear like : RESEMBLE. **—im'i·ta'tor** *n.*

☆ ***syns:*** IMITATE, APE, BURLESQUE, MIMIC, MOCK, PARODY *v. core meaning* : to copy the manner or expression of (another), often mockingly <a comedian who *imitated* the President>

**im·i·ta·tion** (ĭm'ĭ-tā'shən) *n.* **1.** An act of imitating. **2.** Something imitative. **3.** *Mus.* The repetition of a phrase or sequence often with variations in key, rhythm, and voice. **—im'i·ta'tion·al** *adj.*

**im·i·ta·tive** (ĭm'ĭ-tā'tĭv) *adj.* **1.** Of or involving imitation. **2.** Not original : DERIVATIVE. **3.** Tending to imitate. **4.** Onomatopoeic. **—im'i·ta'tive·ly** *adv.* **—im'i·ta'tive·ness** *n.*

**im·mac·u·late** (ĭ-măk'yə-lĭt) *adj.* [ME *immaculat* < Lat. *immaculatus* : *in-,* not + *maculatus,* p.part. of *maculare,* to blemish < *macula,* spot.] **1.** Free from blemish or stain : PURE. **2.** Free from error or fault <an *immaculate* report> **3.** Impeccably clean : SPOTLESS. **4.** Bearing no markings. **—im·mac'u·late·ly** *adv.* **—im·mac'u·late·ness** *n.*

**Immaculate Conception** *n. Rom. Cath. Ch.* The doctrine that the Virgin Mary was conceived in her mother's womb free from all stain of original sin.

**im·ma·nent** (ĭm'ə-nənt) *adj.* [LLat. *immanens, immanent-,* pr.part. of *immanēre,* to remain in : Lat. *in-,* in + Lat. *manēre,* to remain.] **1.** Existing or remaining within : INHERENT <believed in a God *immanent* in human beings> **2.** Restricted completely to the mind : SUBJECTIVE. **—im'ma·nence, im'ma·nen·cy** *n.* **—im'ma·nent·ly** *adv.*

**im·ma·nent·ism** (ĭm'ə-nən-tĭz'əm) *n.* A religious theory postulating that a deity or abstract spirit is immanent in the world.

**im·ma·te·ri·al** (ĭm'ə-tîr'ē-əl) *adj.* **1.** Without material body or form. **2.** Of no importance or relevance : INCONSEQUENTIAL. **—im'ma·te'ri·al·ly** *adv.* **—im'ma·te'ri·al·ness** *n.*

**im·ma·te·ri·al·ism** (ĭm'ə-tîr'ē-ə-lĭz'əm) *n. Philos.* A metaphysical doctrine asserting the nonexistence of corporeal reality. **—im'ma·te'ri·al·ist** *n.*

**im·ma·te·ri·al·i·ty** (ĭm'ə-tîr'ē-ăl'ĭ-tē) *n., pl.* **-ties. 1.** The quality or state of being immaterial. **2.** Something immaterial.

**im·ma·te·ri·al·ize** (ĭm'ə-tîr'ē-ə-līz') *vt.* **-ized, -iz·ing, -iz·es.** To render immaterial.

**im·ma·ture** (ĭm'ə-tyŏŏr', -tŏŏr', -chŏŏr') *adj.* [Lat. *immaturus* : *in-,* not + *maturus,* mature.] **1.** Not fully grown or developed : UNRIPE. **2.** Characterized by or suggesting a lack of normal maturity. **—im'ma·ture'ly** *adv.* **—im'ma·tur'i·ty, im'ma·ture'ness** *n.*

**im·meas·ur·a·ble** (ĭ-mĕzh'ər-ə-bəl) *adj.* **1.** Incapable of being measured. **2.** Without limit : VAST. **—im·meas'ur·a·bil'i·ty, im·meas'ur·a·ble·ness** *n.* **—im·meas'ur·a·bly** *adv.*

**im·me·di·a·cy** (ĭ-mē'dē-ə-sē) *n., pl.* **-cies. 1.** The quality or state of being immediate. **2.** Something immediate, as in importance.

**im·me·di·ate** (ĭ-mē'dē-ĭt) *adj.* [LLat. *immediatus* : Lat. *in-,* not + *mediatus,* p.part. of *mediare,* to be in the middle < Lat. *medius,* middle.] **1.** Acting or taking place without the interposition of another agency or object : DIRECT <an *immediate* change> **2.** Directly apprehended or perceived <*immediate* consciousness> **3.** Next in line or relation <the *immediate* heir> **4.** Occurring at once. **5. a.** Of or near the present time <the *immediate* past> **b.** Of or relating to the present <concentrating first on our *immediate* needs> **6.** Close at hand : NEAR. **7.** Directly affecting someone or something <problems of the *immediate* community> **—im·me'di·ate·ness** *n.*

**immediate constituent** *n.* A meaningful constituent, as a word, that enters directly into the formation of a linguistic construction, as a phrase.

**im·me·di·ate·ly** (ĭ-mē'dē-ĭt-lē) *adv.* **1.** Without intermediary : DIRECTLY <dealt with those *immediately* concerned> **2.** Without delay. —*conj.* As soon as <I phoned *immediately* I reached home.>

**im·med·i·ca·ble** (ĭ-mĕd'ĭ-kə-bəl) *adj.* Incurable.

**Im·mel·mann turn** (ĭm'əl-mən) *n.* [After Max *Immelmann* (1890–1916).] An airplane maneuver in which half a loop and then half a roll are completed in order to simultaneously gain altitude and change direction in flight.

**im·me·mo·ri·al** (ĭm'ə-môr'ē-əl, -mōr'-) *adj.* [Med. Lat. *immemorialis* : Lat. *in-,* not + Lat. *memorialis,* memorial < *memoria,* memory < *memor,* mindful.] Beyond the limits of memory, tradition, or recorded history. **—im'me·mo'ri·al·ly** *adv.*

**im·mense** (ĭ-mĕns') *adj.* [OFr. < Lat. *immensus* : *in-,* not + *mensus,* p.part. of *metiri,* to measure.] **1.** Exceptionally large : HUGE. **2.** Immeasurably vast. **3.** *Slang.* Very good. **—im·mense'ly** *adv.* **—im·mense'ness** *n.* **—im·men'si·ty** (-mĕn'sĭ-tē) *n.*

**im·men·sur·a·ble** (ĭ-mĕn'shər-ə-bəl) *adj.* [LLat. *immensurabilis* : Lat. *in,* not + *mensurabilis,* measurable < *mensura,* measure.] Immeasurable.

**im·merge** (ĭ-mûrj') *vi.* **-merged, -merg·ing, -merg·es.** [Lat. *immergere.* —see IMMERSE.] To submerge or disappear in or as if in a liquid. **—im·mer'gence** *n.*

**im·merse** (ĭ-mûrs') *vt.* **-mersed, -mers·ing, -mers·es.** [< Lat. *immersus,* p.part. of *immergere,* to immerse : *in-,* in + *mergere,* to dip.] **1.** To cover completely in a liquid : SUBMERGE. **2.** To baptize by submerging in water. **3.** To involve profoundly : ABSORB.

**im·mers·i·ble** (ĭ-mûr'sə-bəl) *adj.* Capable of immersion in water without suffering damage, as some electrical heating elements.

**im·mer·sion** (ĭ-mûr'zhən, -shən) *n.* **1.** An act of immersing. **2.** The condition of being immersed. **3.** Baptism by total submersion of a person in water. **4.** *Astron.* The obscuring of a celestial body by another or by the shadow of another.

**im·mesh** (ĭ-mĕsh') *v. var. of* ENMESH.

**im·me·thod·i·cal** (ĭm'ə-thŏd'ĭ-kəl) *adj.* Not methodical. **—im'me·thod'i·cal·ly** *adv.*

**im·mi·grant** (ĭm'ĭ-grənt) *n.* **1.** A person who leaves one country to settle permanently in another. **2.** An organism that appears where it was previously unknown.

**im·mi·grate** (ĭm'ĭ-grāt') *v.* **-grat·ed, -grat·ing, -grates.** [Lat. *immigrare, immigrat-,* to go into : *in-,* in + *migrare,* to depart.] —*vi.* To enter and settle in a country or region to which one is not native. —*vt.* To send or introduce as immigrants. **—im'mi·gra'tion** *n.*

**im·mi·nent** (ĭm'ə-nənt) *adj.* [Lat. *imminens,* pr.part. of *imminēre,* to overhang.] About to occur at any moment : IMPENDING. **—im'mi·nent·ly** *adv.* **—im'mi·nence** (-nəns), **im'mi·nen·cy** (-sē), **im'mi·nent·ness** *n.*

**im·min·gle** (ĭ-mĭng'gəl) *vi. & vt.* **-gled, -gling, -gles.** To intermingle.

**im·mis·ci·ble** (ĭ-mĭs'ə-bəl) *adj.* Incapable of blending or mixing. **—im·mis'ci·bil'i·ty** *n.* **—im·mis'ci·bly** *adv.*

**im·mit·i·ga·ble** (ĭ-mĭt'ĭ-gə-bəl) *adj.* Incapable of being mitigated. **—im·mit'i·ga·bly** *adv.*

**im·mit·tance** (ĭ-mĭt'ns) *n.* [IM(PEDANCE) + (AD)MITTANCE.] Electrical impedance or admittance.

**im·mix** (ĭ-mĭks') *vt.* **-mixed, -mix·ing, -mix·es.** [Back-formation < ME *immixte,* mixed in < Lat. *immixtus,* p.part. of *immiscere,* to blend : *in-,* in + *miscere,* to mix.] To commingle : blend. **—im·mix'ture** *n.*

**im·mo·bile** (ĭ-mō'bəl, -bēl', -bīl') *adj.* [ME *immobil* < OFr. *immobile* < Lat. *immobilis* : *in-,* not + *mobilis,* mobile.] **1.** Firmly in position : FIXED. **2.** Not moving : MOTIONLESS. **—im'mo·bil'i·ty** *n.*

**im·mo·bi·lize** (ĭ-mō'bə-līz') *vt.* **-lized, -liz·ing, -liz·es. 1.** To render immobile. **2.** To impede movement or use of <heavy snow *immobilizing* trucks> **3. a.** To withdraw (specie) from circulation and reserve as security for other money. **b.** To convert (floating capital) into fixed capital. **—im·mo'bi·li·za'tion** *n.* **—im·mo'bi·liz'er** *n.*

**im·mod·er·ate** (ĭ-mŏd'ər-ĭt) *adj.* [ME < Lat. *immoderatus* : *in-*, not + *moderatus*, moderate.] Exceeding normal or appropriate bounds : EXTREME <*immoderate* drinking> **—im·mod'er·a·cy** (-mŏd'ər-ə-sē) *n.* **—im·mod'er·ate·ly** *adv.* **—im·mod'er·ate·ness, im·mod'er·a'tion** *n.*

**im·mod·est** (ĭ-mŏd'ĭst) *adj.* [Lat. *immodestus* : *in-*, not + *modestus*, modest.] **1.** Without modesty. **2.** Morally offensive. **3.** Arrogant or boastful. **—im·mod'est·ly** *adv.* **—im·mod'es·ty** *n.*

**im·mo·late** (ĭm'ə-lāt') *vt.* **-lat·ed, -lat·ing, -lates.** [Lat. *immolare, immolat-*, to sacrifice, sprinkle with sacrificial meal : *in-*, on + *mola*, meal.] **1.** To kill as a sacrifice. **2.** To destroy completely. **—im'mo·la'tion** *n.* **—im'mo·la'tor** *n.*

**im·mor·al** (ĭ-môr'əl, ĭ-mŏr'-) *adj.* Contrary to accepted moral principles. **—im·mor'al·ly** *adv.*

**im·mor·al·ist** (ĭ-môr'ə-lĭst, -mŏr'-) *n.* An advocate of immorality.

**im·mor·al·i·ty** (ĭm'ô-răl'ĭ-tē, ĭm'ə-) *n.*, *pl.* **-ties. 1.** The quality or state of being immoral. **2.** An immoral act or practice.

**im·mor·tal** (ĭ-môr'tl) *adj.* [ME < OFr. *immortel* < Lat. *immortalis* : *in-*, not + *mortalis*, mortal.] **1.** Not subject to death. **2.** Lasting forever, as in fame : IMPERISHABLE. **3.** Of or relating to immortality. —*n.* **1.** One not subject to death. **2.** One whose fame endures <*immortals* such as Mozart and Beethoven> **3. immortals.** The gods of ancient Greece and Rome. **—im·mor'tal·ly** *adv.*

**im·mor·tal·i·ty** (ĭm'ôr-tăl'ĭ-tē) *n.* **1.** The quality or state of being immortal. **2.** Endless life. **3.** Enduring fame.

**im·mor·tal·ize** (ĭ-môr'tl-īz') *vt.* **-ized, -iz·ing, -iz·es.** To make immortal <a leader *immortalized* in history>

**im·mor·telle** (ĭm'ôr-tĕl') *n.* [Fr. < fem. of *immortel*, immortal < OFr.] A plant bearing flowers that retain their color when dried.

**im·mo·tile** (ĭ-mōt'l, ĭ-mō'tīl') *adj.* **1.** Not moving. **2.** IMMOVABLE 1. **—im'mo·til'i·ty** *n.*

**im·mov·a·ble** (ĭ-mo͞o'və-bəl) *adj.* **1.** Not capable of moving or of being moved. **2.** Not capable of alteration. **3.** Unyielding in principle, purpose, or adherence : STEADFAST. **4.** Incapable of being moved emotionally : IMPASSIVE. **5.** *Law.* Not liable to be removed <*immovable* dwellings> —*n.* **1.** One that is incapable of movement. **2.** *often* **immovables.** Property, as real estate, that cannot be moved. **—im·mov'a·bil'i·ty, im·mov'a·ble·ness** *n.* **—im·mov'a·bly** *adv.*

**im·mune** (ĭ-myo͞on') *adj.* [Lat. *immunis*, exempt from service : *in-*, not + *munia*, duties.] **1. a.** Exempt <*immune* from prosecution> **b.** Not affected or responsive : RESISTANT <*immune* to all appeals> **2.** *Med.* Having immunity. **—im·mune'** *n.*

**im·mu·ni·ty** (ĭ-myo͞o'nĭ-tē) *n.*, *pl.* **-ties. 1.** The quality or condition of being immune. **2.** An inherited, acquired, or induced resistance to a particular pathogen.

**im·mu·nize** (ĭm'yə-nīz') *vt.* **-nized, -niz·ing, -niz·es.** To render immune. **—im'mu·ni·za'tion** *n.*

**immuno-** *pref.* [< IMMUNE.] Immunity <*immunogenic*>

**im·mu·no·as·say** (ĭm'yə-nō-ăs'ā, ĭm-yo͞o'-) *n.* Analysis and identification of a substance based on its antigenic actions.

**im·mu·no·chem·is·try** (ĭm'yə-nō-kĕm'ĭ-strē) *n.* The chemistry of immunologic phenomena, as of antigen stimulation of tissue or of antigen-antibody reactions.

**im·mu·no·e·lec·tro·pho·re·sis** (ĭm'yə-nō-ĭ-lĕk'trə-fə-rē'sĭs, ĭm-yo͞o'-) *n.* Separation of antigens by electrophoresis with identification through specific immunological reactions.

**im·mu·no·ge·net·ics** (ĭm'yə-nō-jə-nĕt'ĭks) *n.* *(sing. in number).* The study of the interrelation between immunity to disease and genetic make-up.

**im·mu·no·gen·ic** (ĭm'yə-nō-jĕn'ĭk) *adj.* Causing immunity.

**im·mu·no·glob·u·lin** (ĭm'yə-nō-glŏb'yə-lĭn, ĭm-yo͞o'-) *n.* One of a group of blood serum proteins capable of acting as an antibody.

**im·mu·nol·o·gy** (ĭm'yə-nŏl'ə-jē) *n.* The medical study of immunity. **—im'mu·no·log'ic** (-nə-lŏj'ĭk), **im'mu·no·log'i·cal** *adj.* **—im'mu·no·log'i·cal·ly** *adv.*

**im·mu·no·sup·pres·sive** (ĭm'yə-nō-sə-prĕs'ĭv) *adj.* Tending to suppress a natural immune response of an organism to an antigen <an *immunosuppressive* drug>

**im·mu·no·ther·a·py** (ĭm'yə-nō-thĕr'ə-pē, ĭm-yo͞o'-) *n.* **1.** Treatment of disease by use of antigenic preparations. **2.** Treatment of disease by immunosuppressive techniques. **—im'mu·no·ther'a·pist** *n.*

**im·mure** (ĭ-myo͝or') *vt.* **-mured, -mur·ing, -mures.** [Med. Lat. *immurare* : Lat. *in-*, in + Lat. *murus*, wall.] **1.** To confine within or as if within walls : IMPRISON. **2.** To build into a wall, esp. to entomb in a wall. **—im·mure'ment** *n.*

**im·mu·ta·ble** (ĭ-myo͞o'tə-bəl) *adj.* [ME < Lat. *immutabilis* : *in-*, not + *mutabilis*, mutable.] Not subject or susceptible to change. **—im·mu'ta·bil'i·ty, im·mu'ta·ble·ness** *n.* **—im·mu'ta·bly** *adv.*

**imp** (ĭmp) *n.* [ME *impe*, scion, sprig, offspring < OE *impa*, young shoot < *impian*, to graft, ult. < Gk. *emphuein*, to implant : *en-*, in + *phuein*, to grow.] **1.** A mischievous child. **2.** A small demon. **3.** *Archaic.* A graft. —*vt.* **imped, imp·ing, imps. 1.** To graft (new feathers) onto the wing or tail of a falcon to repair damage or increase flying capacity. **2.** *Archaic.* To furnish with wings.

**im·pact** (ĭm'păkt') *n.* [Lat. *impactus*, p.part. of *impingere*, to push against. —see IMPINGE.] **1.** The striking of one body against another : COLLISION. **2.** The force or impetus transmitted by a collision. **3.** The effect or impression of one thing upon another <gauging the *impact* of cable television on theater attendance> —*v.* (ĭm-păkt') **-pact·ed, -pact·ing, -pacts.** —*vt.* **1.** To pack firmly together. **2.** To strike forcefully. **3.** *Informal.* To have an effect on. ***usage:*** The use of *impact* in this sense is unacceptable to many because of its association with bureaucratic language or technical jargon. —*vi.* To have an effect. **—im·pac'tion** *n.*

**im·pact·ed** (ĭm-păk'tĭd) *adj.* **1.** Wedged together at the broken ends <an *impacted* bone> **2. a.** Placed in the alveolus in a manner prohibiting eruption into a normal position. —Used of a tooth. **b.** Driven upward into the alveolar process or surrounding tissue. —Used of a tooth.

**im·pair** (ĭm-pâr') *vt.* **-paired, -pair·ing, -pairs.** [ME *empairen* < OFr. *empeirer* < VLat. **impejorare* : Lat. *in-* (intensive) + LLat. *pejorare*, to worsen < Lat. *pejor*, worse.] To decrease in strength, value, amount, or quality. **—im·pair'ment** *n.*

**im·pa·la** (ĭm-pä'lə) *n.* [Zulu *im-pala.*] A large brownish African antelope, *Aepyceros melampus*, the male of which has ridged, slender, curved horns.

**im·pale** (ĭm-pāl') *vt.* **-paled, -pal·ing, -pales.** [Med. Lat. *impalare* : Lat. *in-*, in + Lat. *palus*, stake.] **1. a.** To pierce with a sharp stake or point. **b.** To torture or kill by impaling. **2.** To render helpless as if by impaling <*impaled* by grief> **—im·pale'ment** *n.* **—im·pal'er** *n.*

**im·pal·pa·ble** (ĭm-păl'pə-bəl) *adj.* **1.** Not perceptible to the touch : INTANGIBLE. **2.** Not easily perceived or understood by the mind. **—im·pal'pa·bil'i·ty** *n.* **—im·pal'pa·bly** *adv.*

**im·pan·el** (ĭm-păn'əl) *vt.* **-eled, -el·ing, -els** *or* **-elled, -el·ling, -els.** To enroll (a jury) upon a panel or list. **—im·pan'el·ment** *n.*

**im·par·i·ty** (ĭm-păr'ĭ-tē) *n.*, *pl.* **-ties.** [LLat. *imparitas* < Lat. *impar*, not equal : *in-*, not + *par*, equal.] Disparity : inequality.

**im·part** (ĭm-pärt') *vt.* **-part·ed, -part·ing, -parts.** [Lat. *impartire* : *in-*, in + *partire*, to share < *pars*, part.] **1.** To grant a share of : BESTOW <*impart* wisdom> **2.** To make known : COMMUNICATE <*impart* the bad news>

**im·par·tial** (ĭm-pär'shəl) *adj.* Not partial or biased. **—im'par·ti·al'i·ty** (-shē-ăl'ə-tē), **im·par'tial·ness** *n.* **—im·par'tial·ly** *adv.*

**im·part·i·ble** (ĭm-pär'tə-bəl) *adj.* [LLat. *impartibilis* : Lat. *in-*, not + Lat. *partibilis*, partible.] Not capable of being separated : INDIVISIBLE. **—im·part'i·bil'i·ty** *n.* **—im·part'i·bly** *adv.*

**im·pass·a·ble** (ĭm-păs'ə-bəl) *adj.* Impossible to travel over or across. **—im·pass'a·bil'i·ty, im·pass'a·ble·ness** *n.* **—im·pass'a·bly** *adv.*

**im·passe** (ĭm'păs') *n.* [Fr. : *in-*, not (< Lat. *in-*) + *passer*, to pass < OFr.] **1.** A road having no exit : CUL-DE-SAC. **2. a.** A difficult situation offering no practical escape. **b.** A deadlock, as in negotiations.

**im·pas·si·ble** (ĭm-păs'ə-bəl) *adj.* [ME < OFr. < LLat. *impassibilis* : *in-*, not + *passibilis*, passible.] **1.** Not subject to pain. **2.** Impassive. **—im·pas'si·bil'i·ty, im·pas'si·ble·ness** *n.* **—im·pas'si·bly** *adv.*

**im·pas·sion** (ĭm-păsh'ən) *vt.* **-sioned, -sion·ing, -sions.** [Ital. *impassionare* : *in-*, in (< Lat.) + *passione*, passion < LLat. *passio*, suffering. —see PASSION.] To arouse the passions of.

**im·pas·sioned** (ĭm-păsh'ənd) *adj.* Filled with passion.

**im·pas·sive** (ĭm-păs'ĭv) *adj.* [IM-[1] + Lat. *passivus*, capable of feeling. —see PASSIVE.] **1.** Lacking or not subject to emotion. **2.** Displaying no emotion : EXPRESSIONLESS. **3.** Incapable of physical sensation. **4.** Without motion : STILL. **—im·pas'sive·ly** *adv.* **—im·pas'sive·ness, im'pas·siv'i·ty** *n.*

**im·paste** (ĭm-păst') *vt.* **-past·ed, -past·ing, -pastes.** [Ital. *impastare* : *in-*, in (< Lat.) + *pasta*, paste < LLat.] **1.** To make into a paste. **2.** To apply pigment thickly to.

**im·pas·to** (ĭm-păs'tō, -päs'-) *n.* [Ital. < *impastare*, to impaste.] **1.** The application of thick layers of pigment to a canvas or other surface in painting. **2.** The pigment so applied.

**im·pa·tiens** (ĭm-pā'shənz, -shəns) *n.* [NLat. *Impatiens*, genus name < Lat. *impatiens*, impatient.] A plant of the genus *Impatiens*, which includes the jewelweed.

**im·pa·tient** (ĭm-pā'shənt) *adj.* [ME *impacient* < OFr. *impatient* < Lat. *impatiens* : *in-*, not + *patiens*, pr.part. of *pati*, to suffer, endure.] **1.** Unable to wait patiently or tolerate delay : RESTLESS. **2.** Unable to endure opposition or irritation : INTOLERANT. **3.** Expressing or caused by impatience <uttered an *impatient* oath> **4.** Restively desirous or eager <*impatient* to depart> **—im·pa'tience** (-pā'shəns) *n.* **—im·pa'tient·ly** *adv.*

**im·peach** (ĭm-pēch') *vt.* **-peached, -peach·ing, -peach·es.** [ME *empechen*, to impede, molest, accuse < AN *empecher* < LLat. *impedicare*, to entangle : Lat. *in-*, in + Lat. *pedica*, fetter.] **1. a.** To make an accusation against. **b.** To charge with misconduct in office before a proper tribunal. **2.** To challenge or discredit : ATTACK. **—im·peach'a·ble** *adj.* **—im·peach'er** *n.* **—im·peach'ment** *n.*

**im·pearl** (ĭm-pûrl') *vt.* **-pearled, -pearl·ing, -pearls. 1.** To form into pearls. **2.** To adorn with or as if with pearls.

**im·pec·ca·ble** (ĭm-pĕk′ə-bəl) *adj.* [Lat. *impeccabilis* : *in-*, not + *peccare*, to sin.] **1.** Having no flaws : PERFECT. **2.** Incapable of sin or wrongdoing. **—im·pec′ca·bil′i·ty** *n.* **—im·pec′ca·bly** *adv.*
**im·pe·cu·ni·ous** (ĭm′pĭ-kyo͞o′nē-əs) *adj.* [IM-[1] + obs. *pecunious*, rich < ME < Lat. *pecuniosus* < *pecunia*, money, wealth.] Having no money. **—im′pe·cu′ni·ous·ly** *adv.* **—im′pe·cu′ni·os′i·ty** (-ŏs′ĭ-tē), **im′pe·cu′ni·ous·ness** *n.*
**im·pe·dance** (ĭm-pēd′ns) *n.* [< IMPEDE.] A measure of the total opposition to current flow in an alternating-current circuit, equal to the ratio of the rms electromotive force in the circuit to the rms current produced by it, and usu. represented in complex notation as $Z = R + iX$, where $R$ is the ohmic resistance and $X$ is the reactance.
**impedance matching** *n.* The use of electric circuits, transmission lines, and other devices to make the impedance of a load equal to the internal impedance of the source of power, thereby making possible an efficient transfer of power.
**im·pede** (ĭm-pēd′) *vt.* **-ped·ed, -ped·ing, -pedes.** [Lat. *impedire* : *in-*, in + *pes*, foot.] To obstruct or delay the progress of : HINDER. **—im·ped′er** *n.*
**im·ped·i·ment** (ĭm-pĕd′ə-mənt) *n.* [Lat. *impedimentum* < *impedire*, to impede.] **1.** A hindrance or obstruction. **2.** Something that impedes, esp.: **a.** An organic defect preventing clear articulation or perception <a speech *impediment*> <a hearing *impediment*> **b.** *Law.* Something obstructing the making of a legal contract. **—im·ped′i·men′tal** (-mĕn′tl), **im·ped′i·men′ta·ry** *adj.*
**im·ped·i·men·ta** (ĭm-pĕd′ə-mĕn′tə) *pl.n.* [Lat. *impedimenta*, pl. of *impedimentum*, impediment.] Objects, as baggage or provisions, that impede or encumber.
**im·pel** (ĭm-pĕl′) *vt.* **-pelled, -pel·ling, -pels.** [Lat. *impellere* : *in-*, against + *pellere*, to drive.] **1.** To urge to action : COMPEL. **2.** To drive forward : PROPEL.
**im·pel·ler** (ĭm-pĕl′ər) *n.* **1.** One that impels. **2. a.** A rotor or rotor blade. **b.** A rotating device for forcing a gas in a given direction under pressure.
**im·pend** (ĭm-pĕnd′) *vi.* **-pend·ed, -pend·ing, -pends.** [Lat. *impendēre* : *in-*, against + *pendēre*, to hang.] **1.** To hang or hover in a threatening manner. **2.** To be about to take place. **3.** *Archaic.* To overhang.
**im·pen·dent** (ĭm-pĕn′dənt) *adj.* Close at hand : IMPENDING.
**im·pen·e·tra·bil·i·ty** (ĭm-pĕn′ĭ-trə-bĭl′ĭ-tē) *n.* **1.** The quality or state of being impenetrable. **2.** The inability of two bodies to occupy the same space at the same time.
**im·pen·e·tra·ble** (ĭm-pĕn′ĭ-trə-bəl) *adj.* [ME < OFr. < Lat. *impenetrabilis* : *in-*, not + *penetrabilis*, penetrable.] **1.** Incapable of being penetrated or entered. **2.** Incapable of being understood : UNFATHOMABLE. **3.** Impervious to reason or persuasion. **—im·pen′e·tra·ble·ness** *n.* **—im·pen′e·tra·bly** *adv.*
**im·pen·i·tent** (ĭm-pĕn′ĭ-tənt) *adj.* [LLat. *impaenitens* : Lat. *in-*, not + Lat. *paenitens*, penitent.] Not penitent. **—im·pen′i·tence** *n.* **—im·pen′i·tent** *n.* **—im·pen′i·tent·ly** *adv.*
**im·per·a·tive** (ĭm-pĕr′ə-tĭv) *adj.* [LLat. *imperativus* < Lat. *imperare*, to command. —see EMPEROR.] **1.** Expressing a plea or command : PEREMPTORY. **2.** Having the authority or power to control or command. **3.** Of, relating to, or constituting the mood of a verb that expresses a command or request. **4.** Urgent : obligatory. *—n.* **1. a.** The imperative mood. **b.** A verb form of the imperative mood. **2. a.** A command or order. **b.** An obligation. **—im·per′a·tive·ly** *adv.* **—im·per′a·tive·ness** *n.*
**im·pe·ra·tor** (ĭm′pə-rä′tôr′, -tər) *n.* [Lat. —see EMPEROR.] A supreme commander in ancient Rome. **—im′pe·ra·to′ri·al** *adj.*
**im·per·cep·ti·ble** (ĭm′pər-sĕp′tə-bəl) *adj.* **1.** Not perceptible by the mind or senses. **2.** So subtle, small, or gradual as to be barely perceptible. **—im′per·cep′ti·bil′i·ty, im′per·cep′ti·ble·ness** *n.* **—im′per·cep′ti·bly** *adv.*

☆ ***syns:*** IMPERCEPTIBLE, IMPALPABLE, INDISTINGUISHABLE, INSENSIBLE, INTANGIBLE, UNAPPRECIABLE, UNNOTICEABLE *adj. core meaning* : incapable of being seen, measured, or detected by the mind or senses <the *imperceptible* movement of a distant planet> ***ant:*** perceptible

**im·per·cep·tive** (ĭm′pər-sĕp′tĭv) *adj.* Not perceptive. **—im′per·cep·tiv′i·ty, im′per·cep′tive·ness** *n.*
**im·per·fect** (ĭm-pûr′fĭkt) *adj.* [ME *imparfit* < OFr. *imparfait* < Lat. *imperfectus* : *in-*, not + *perfectus*, perfect.] **1.** Not perfect. **2.** Of or being the tense of a verb that shows, usu. in the past, an action or condition as incomplete, continuous, or coincident with another action. **3.** *Bot.* Having either stamens or a pistil only <*imperfect* flowers> **4.** Not enforceable by law. *—n.* **1.** The imperfect tense. **2.** A verb in the imperfect tense. **3.** An item of merchandise, esp. clothing, with a minor defect that does not impair its use, usu. sold at a discount. **—im·per′fect·ly** *adv.* **—im·per′fect·ness** *n.*

☆ ***syns:*** IMPERFECT, DEFECTIVE, FAULTY *adj. core meaning* : having a defect or defects <*imperfect* glassware> ***ant:*** perfect

**imperfect fungus** *n.* Any of various fungi of the order Fungi Imperfecti, which reproduce only by asexual means.

o͞o **boot** ou **out** th **thin** *th* **this** ŭ **cut** ûr **urge** y **young** yo͞o **abuse** zh **vision** ə **a**bout, it**e**m, edib**i**le, gall**o**p, circ**u**s

**im·per·fec·tion** (ĭm′pər-fĕk′shən) *n.* **1.** The quality or condition of being imperfect. **2.** A defect.
**im·per·fec·tive** (ĭm′pər-fĕk′tĭv) *adj.* Denoting a verb aspect or form that expresses action without regard to its beginning or end.
**im·per·fo·rate** (ĭm-pûr′fər-ĭt) *adj.* **1.** Having no opening : not perforated. **2.** Not perforated into perforated rows. —Used of stamps and sheets of stamps. *—n.* An imperforate stamp.
**im·pe·ri·al¹** (ĭm-pîr′ē-əl) *adj.* [ME < OFr. < LLat. *imperialis* < Lat. *imperium*, command. —see EMPIRE.] **1.** Of or relating to an empire or emperor. **2.** Designating a nation or government holding sovereign rights over colonies or dependencies. **3. a.** *Obs.* Having supreme authority : SOVEREIGN. **b.** Regal : majestic. **4.** Outstanding in quality or size. **5.** Of or pertaining to the British Imperial System of weights and measures. *—n.* **1. Imperial.** A supporter or soldier of the Holy Roman Empire. **2.** An emperor or empress. **3.** The top of a carriage. **4.** Something outstanding in size or quality. **5.** A variable size of paper, usu. 23 by 33 inches. **—im·pe′ri·al·ly** *adv.*
**im·pe·ri·al²** (ĭm-pîr′ē-əl) *n.* [Fr. *impériale*, imperial, after Emperor Napoleon III of France.] A pointed beard on the lower lip and chin.
**im·pe·ri·al·ism** (ĭm-pîr′ē-ə-lĭz′əm) *n.* **1.** The policy of extending a nation's authority by acquisition of territory or by the establishment of economic and political hegemony over other nations. **2.** The system, policies, or practices of an imperial government. **—im·pe′ri·al·ist** *n.* **—im·pe′ri·al·is′tic** *adj.* **—im·pe′ri·al·is′ti·cal·ly** *adv.*
**imperial moth** *n.* A large New World moth, *Eacles imperialis*, with yellow wings having purplish or brownish markings.

**imperial moth**
*Wingspan approximately 4½ inches*

**im·per·il** (ĭm-pĕr′əl) *vt.* **-iled, -il·ing, -ils** *or* **-illed, -il·ling, -ils.** To put in peril : ENDANGER. **—im·per′il·ment** *n.*
**im·pe·ri·ous** (ĭm-pîr′ē-əs) *adj.* [Lat. *imperiosus* < *imperium*, imperium.] **1.** Domineering or overbearing. **2.** *Obs.* Regal : imperial. **3.** Urgent. **—im·pe′ri·ous·ly** *adv.* **—im·pe′ri·ous·ness** *n.*
**im·per·ish·a·ble** (ĭm-pĕr′ĭ-shə-bəl) *adj.* Not perishable : INDESTRUCTIBLE. **—im·per′ish·a·bil′i·ty, im·per′ish·a·ble·ness** *n.* **—im·per′ish·a·bly** *adv.*
**im·pe·ri·um** (ĭm-pîr′ē-əm) *n., pl.* **-ri·a** (-ē-ə) [Lat. —see EMPIRE.] **1.** Absolute rule : supreme power. **2.** A sphere of dominion or power : EMPIRE. **3.** *Law.* The right or power of a state to enforce the law.
**im·per·ma·nent** (ĭm-pûr′mə-nənt) *adj.* Not permanent : TRANSITORY. **—im·per′ma·nence, im·per′ma·nen·cy** *n.*
**im·per·me·a·ble** (ĭm-pûr′mē-ə-bəl) *adj.* [LLat. *impermeabilis* : Lat. *in-*, not + *permeabilis*, permeable.] Not permeable. **—im·per′me·a·bil′i·ty, im·per′me·a·ble·ness** *n.* **—im·per′me·a·bly** *adv.*
**im·per·mis·si·ble** (ĭm′pər-mĭs′ə-bəl) *adj.* Not permissible. **—im′per·mis′si·bil′i·ty** *n.* **—im′per·mis′si·bly** *adv.*
**im·per·son·al** (ĭm-pûr′sə-nəl) *adj.* **1. a.** Having no personal reference or connection <an *impersonal* comment> **b.** Displaying no emotion or personality <*impersonal* actions of the committee> **c.** Not responsive to or expressive of human personalities <the *impersonal* atmosphere of the assembly line> **2. a.** Denoting a verb that expresses the action of an unspecified agent and is used in the third person singular with no subject expressed, as *methought*, or with a formal subject, as in *it snowed*. **b.** Indefinite. —Used of pronouns. **—im·per′son·al′i·ty** *n.* **—im·per′son·al·ly** *adv.*
**im·per·son·al·ize** (ĭm-pûr′sə-nə-līz′) *vt.* **-ized, -iz·ing, -iz·es.** To make impersonal.
**im·per·son·ate** (ĭm-pûr′sə-nāt′) *vt.* **-at·ed, -at·ing, -ates.** **1.** To play the character or part of. **2.** *Archaic.* To embody : personify. **—im·per′son·a′tion** *n.* **—im·per′son·a′tor** *n.*
**im·per·ti·nence** (ĭm-pûr′tn-əns) *n.* **1.** The quality or state of being impertinent, as: **a.** Insolence or discourtesy. **b.** Irrelevance. **2.** An impertinent act or statement.
**im·per·ti·nen·cy** (ĭm-pûr′tn-ən-sē) *n., pl.* **-cies.** Impertinence.
**im·per·ti·nent** (ĭm-pûr′tn-ənt) *adj.* [ME, irrelevant < OFr. < LLat. *impertinens* : Lat. *in-*, not + *pertinens*, pertinent.] **1.** Not constrained within established or proper limits, esp. of manners or good taste : INSOLENT <an *impertinent* answer> **2.** Not pertinent : IRRELEVANT. **—im·per′ti·nent·ly** *adv.*
**im·per·turb·a·ble** (ĭm′pər-tûr′bə-bəl) *adj.* Unshakably calm. **—im′per·turb′a·bil′i·ty, im′per·turb′a·ble·ness** *n.* **—im′per·turb′a·bly** *adv.*
**im·per·vi·ous** (ĭm-pûr′vē-əs) *adj.* [Lat. *impervius* : *in-*, not + *pervius*, pervious.] **1.** Incapable of being penetrated <a raincoat *imper-*

*vious* to water> **2.** Incapable of being affected <*impervious* to anxiety> **—im·per'vi·ous·ly** *adv.* **—im·per'vi·ous·ness** *n.*

**im·pe·ti·go** (ĭm'pĭ-tī'gō) *n.* [Lat. < *impetere,* to attack. —see IMPETUS.] A contagious skin disease characterized by superficial pustules that burst and typically form thick yellow crusts.

**im·pe·trate** (ĭm'pĭ-trāt') *vt.* **-trat·ed, -trat·ing, -trates.** [Lat. *impetrare, impetrat-,* to obtain : *in-* (intensive) + *patrare,* to bring about.] **1.** To gain by entreaty or petition. **2.** To beseech. **—im'pe·tra'tion** *n.* **—im'pe·tra'tor** *n.*

**im·pet·u·os·i·ty** (ĭm-pĕch'o͞o-ŏs'ĭ-tē) *n., pl.* **-ties. 1.** The quality or state of being impetuous. **2.** An impetuous act.

**im·pet·u·ous** (ĭm-pĕch'o͞o-əs) *adj.* [ME < OFr. *impetueux* < Lat. *impetuosus* < *impetus,* impetus.] **1.** Marked by sudden energy, action, or emotion : IMPULSIVE <an *impetuous* move that led to a series of problems> **2.** Having or characterized by violent force <*impetuous,* roaring north winds> **—im·pet'u·ous·ly** *adv.* **—im·pet'u·ous·ness** *n.*

**im·pe·tus** (ĭm'pĭ-təs) *n., pl.* **-tus·es.** [Lat., attack < *impetere,* to attack : *in-,* against + *petere,* to go toward, seek.] **1. a.** A driving force : IMPULSE. **b.** Something that incites : STIMULUS. **2.** The force or energy associated with a moving body.

**im·pi·e·ty** (ĭm-pī'ĭ-tē) *n., pl.* **-ties. 1.** The quality or state of being impious. **2.** An impious act. **3.** Undutifulness.

**im·pinge** (ĭm-pĭnj') *vi.* **-pinged, -ping·ing, -ping·es.** [Lat. *impingere,* to push against : *in-,* against + *pangere,* to fasten.] **1.** To collide or strike <brilliant colors *impinging* on the optic nerve> **2.** To encroach <*impinging* on my independence> **—im·pinge'ment** *n.* **—im·ping'er** *n.*

**im·pi·ous** (ĭm'pē-əs, ĭm-pī'-) *adj.* [Lat. *impius,* irreverent, undutiful : *in-,* not + *pius,* reverent, dutiful.] **1.** Not pious : IRREVERENT. **2.** Lacking proper respect. **—im'pi·ous·ly** *adv.* **—im'pi·ous·ness** *n.*

**imp·ish** (ĭm'pĭsh) *adj.* Of or befitting an imp : MISCHIEVOUS. **—imp'ish·ly** *adv.* **—imp'ish·ness** *n.*

**im·plac·a·ble** (ĭm-plăk'ə-bəl, -plā'kə-) *adj.* [Lat. *implacabilis* : *in-,* not + *placabilis,* placable.] Incapable of mitigation or appeasement : INEXORABLE <*implacable* opponents> **—im·plac'a·bil'i·ty, im·plac'a·ble·ness** *n.* **—im·plac'a·bly** *adv.*

**im·plant** (ĭm-plănt') *vt.* **-plant·ed, -plant·ing, -plants. 1.** To set in firmly, as in the ground. **2.** To establish decisively, as in the mind or consciousness : INSTILL. **3.** *Med.* To insert or embed surgically, as in grafting. *—n.* (ĭm'plănt'). Something implanted, esp. surgically. **—im'plan·ta'tion** *n.*

**im·plau·si·ble** (ĭm-plô'zə-bəl) *adj.* Difficult to believe. **—im·plau'si·bil'i·ty, im·plau'si·ble·ness** *n.* **—im·plau'si·bly** *adv.*

**im·plead** (ĭm-plēd') *vt.* **-plead·ed, -plead·ing, -pleads.** [ME *empleden* < AN *empleder* : *en-* (intensive < Lat. *in-*) + *pleder,* to plead. —see PLEAD.] To sue in a court of law.

**im·ple·ment** (ĭm'plə-mənt) *n.* [ME < LLat. *implementum,* a filling up < Lat. *implēre,* to fill up : *in-* (intensive) + *plēre,* to fill.] **1.** A tool, utensil, or instrument for doing a task. **2.** An article used to outfit or equip. **3.** A means for accomplishing an end : AGENT. *—vt.* (ĭm'plə-mĕnt') **-ment·ed, -ment·ing, -ments. 1.** To put into effect : CARRY OUT <*implement* the combat plan> **2.** To supply with implements. **—im'ple·men·ta'tion** (ĭm'plə-mən-tā'shən, -mĕn-) *n.*

**im·pli·cate** (ĭm'plĭ-kāt') *vt.* **-cat·ed, -cat·ing, -cates.** [Lat. *implicare, implicat-,* to entangle, unite : *in-,* in + *plicare,* to fold.] **1.** To connect or involve intimately or incriminatingly, esp. in illegal activity. **2.** To imply. **3.** *Archaic.* To entangle or interweave : ENTWINE.

**im·pli·ca·tion** (ĭm'plĭ-kā'shən) *n.* **1.** An act of implicating or the state of being implicated. **2.** An act of implying or the state of being implied. **3.** Something implied, esp.: **a.** An indirect indication. **b.** An inference. **—im'pli·ca'tive** *adj.* **—im'pli·ca'tive·ly** *adv.*

**im·plic·it** (ĭm-plĭs'ĭt) *adj.* [Lat. *implicitus,* var. of *implicatus,* p.part. of *implicare,* to entangle. —see IMPLICATE.] **1.** Implied or understood although not directly expressed <an *implicit* argument> **2.** Contained in the nature of someone or something although not readily apparent <could see a great park *implicit* in the carefully arranged plantings of young trees> **3.** Without doubts or reservations. **—im·plic'it·ly** *adv.* **—im·plic'it·ness** *n.*

☆ ***syns:*** **1.** IMPLICIT, IMPLIED, TACIT, UNDERSTOOD, UNEXPRESSED, UNSAID, UNSPOKEN, UNUTTERED *adj. core meaning* : conveyed indirectly without words or speech <an *implicit* approval> ***ant:*** explicit **2.** IMPLICIT, UNCONDITIONAL *adj. core meaning* : having no reservations <*implicit* trust>

**implicit differentiation** *n.* The process of isolating the derivative of a dependent variable of an implicit function by differentiating each term of the function separately, expressing the desired derivative as a symbol, and solving the resulting expression for the symbol.

**implicit function** *n.* A mathematical expression in which the function of concern is not directly expressed but must be arrived at by manipulation of the expression.

**im·plode** (ĭm-plōd') *v.* **-plod·ed, -plod·ing, -plodes.** [IN-[2] + (EX)PLODE.] *—vi.* To undergo implosion. *—vt.* **1.** To express by implosion. **2.** To cause to undergo implosion.

**im·plore** (ĭm-plôr', -plōr') *vt.* **-plored, -plor·ing, -plores.** [Lat. *implorare* : *in-,* in + *plorare,* to weep.] **1.** To appeal to in supplication : ENTREAT. **2.** To plead or beg for urgently. **—im'plo·ra'tion** *n.* **—im·plor'er** *n.* **—im·plor'ing·ly** *adv.*

**im·plo·sion** (ĭm-plō'zhən) *n.* [IN-[2] + (EX)PLOSION.] **1.** A violent collapse inward, as of a highly evacuated glass vessel. **2.** Violent compression. **3.** The stopping of the breath in the formation of a stop consonant.

**im·plo·sive** (ĭm-plō'sĭv) *adj.* Expression by implosion. *—n.* A consonant expression by implosion.

**im·ply** (ĭm-plī') *vt.* **-plied, -ply·ing, -plies.** [ME *implien,* to enfold < OFr. *emplier* < Lat. *implicare.* —see IMPLICATE.] **1.** To involve or suggest by logical necessity : ENTAIL. **2.** To say or express indirectly. **3.** *Obs.* To entangle.

**im·po·lite** (ĭm'pə-līt') *adj.* [Lat. *impolitus,* unpolished, inelegant : *in-,* not + *politus,* polished, p.part. of *polire,* to polish.] Not polite : DISCOURTEOUS. **—im'po·lite'ly** *adv.* **—im'po·lite'ness** *n.*

**im·pol·i·tic** (ĭm-pŏl'ĭ-tĭk) *adj.* **1.** Not expedient or wise. **2.** Lacking sensitivity : TACTLESS. **—im·pol'i·tic·ly** *adv.*

**im·pon·der·a·ble** (ĭm-pŏn'dər-ə-bəl) *adj.* Incapable of being weighed or evaluated with precision. **—im·pon'der·a·bil'i·ty, im·pon'der·a·ble·ness** *n.* **—im·pon'der·a·bly** *adv.*

**im·pone** (ĭm-pōn') *vt.* **-poned, -pon·ing, -pones.** [Lat. *imponere,* to place upon : *in-,* on + *ponere,* to place.] *Obs.* To wager or stake.

**im·port** (ĭm-pôrt', -pōrt', ĭm'pôrt', -pōrt') *v.* **-port·ed, -port·ing, -ports.** [ME *importen* < Lat. *importare* : *in-,* in + *portare,* to carry.] *—vt.* **1.** To bring or carry in from an outside source, esp. to bring in (goods) from a foreign country for trade or sale. **2.** To carry or hold the meaning of : SIGNIFY <fluctuating energy costs *importing* changes in manufacturers' overhead> **3.** To imply. **4.** *Archaic.* To have meaning for. *—vi.* To be significant. *—n.* (ĭm'pôrt', -pōrt'). **1.** Something imported. **2.** The occupation of importing. **3.** Meaning : signification. **4.** Importance : significance <a corporate decision with long-term *import*> **—im·port'a·bil'i·ty** *n.* **—im·port'a·ble** *adj.* **—im·port'er** *n.*

**im·por·tance** (ĭm-pôr'tns) *n.* **1.** The quality or condition of being important : SIGNIFICANCE. **2.** Personal status : STANDING. **3.** *Obs.* An important matter. **4.** *Obs.* Meaning : import. **5.** *Obs.* Importunity.

**im·por·tant** (ĭm-pôr'tnt) *adj.* [OFr. < OItal. *importante* < Med. Lat. *importans,* pr.part. of *importare,* to mean < Lat., to import.] **1.** Having or characterized by great value, significance, or consequence. **2.** Having or suggesting an air of great weight or moment : AUTHORITATIVE. **3.** *Obs.* Importunate. **—im·por'tant·ly** *adv.*

☆ ***syns:*** IMPORTANT, BIG, CONSEQUENTIAL, HISTORIC, MEANINGFUL, MOMENTOUS, SIGNIFICANT, WEIGHTY *adj. core meaning* : having great significance <an *important* development in medicine> ***ant:*** insignificant, unimportant

**im·por·ta·tion** (ĭm'pôr-tā'shən, -pōr-) *n.* **1.** The act or business of importing. **2.** Something imported.

**im·por·tu·nate** (ĭm-pôr'chə-nĭt) *adj.* Stubbornly or unreasonably persistent in pressing demands or requests. **—im·por'tu·nate·ly** *adv.* **—im·por'tu·nate·ness** *n.*

**im·por·tune** (ĭm'pôr-to͞on', -tyo͞on', ĭm-pôr'chən) *vt.* **-tuned, -tun·ing, -tunes.** [Med. Lat. *importunari,* to be troublesome < Lat. *importunus,* inconvenient: *in-,* not + *portus,* harbor.] **1.** To press with insistent demands or requests. **2.** *Obs.* To ask for insistently. **3.** *Obs.* To annoy. *—adj.* Importunate. **—im'por·tune'ly** *adv.* **—im'por·tun'er** *n.* **—im'por·tu'ni·ty** (-to͞o'nĭ-tē, -tyo͞o'-) *n.*

**im·pose** (ĭm-pōz') *v.* **-posed, -pos·ing, -pos·es.** [OFr. < Lat. *impositus,* p.part. of *imponere,* to place upon: *in-,* on + *ponere,* to place.] *—vt.* **1.** To enact or apply as compulsory : LEVY <*impose* license fees> **2.** To apply or make prevail by or as if by authority <*impose* strike arbitration> **3.** To obtrude or force (e.g., oneself) upon another or others. **4.** To arrange (printing type or plates) on an imposing stone. **5.** To pass off on others <*imposed* a deception on the people> *—vi.* To take unfair advantage of <always *imposing* on their kindness> **—im·pos'er** *n.*

**im·pos·ing** (ĭm-pō'zĭng) *adj.* Awesome : impressive.

**imposing stone** *n.* A stone or metal slab on which material for printing is arranged.

**im·po·si·tion** (ĭm'pə-zĭsh'ən) *n.* **1.** The act of imposing. **2.** Something imposed, as a tax, unfair burden, or fraud. **3.** An unfair or burdensome demand, as upon someone's time or patience. **4.** The arrangement of printed matter to form a sequence of pages.

☆ ***syns:*** IMPOSITION, INFLICTION, INTRUSION *n. core meaning* : an excessive, unwelcome burden <The unexpected dinner guests were an *imposition.*>

**im·pos·si·ble** (ĭm-pŏs'ə-bəl) *adj.* [ME < OFr. < Lat. *impossibilis* : *in-,* not + *possibilis,* possible.] **1.** Incapable of existing or taking place. **2.** Unlikely to happen or be accomplished. **3.** Unacceptable. **4.** Extremely difficult to tolerate or deal with. **—im·pos'si·bil'i·ty** (-bĭl'ĭ-tē) *n.* **—im·pos'si·bly** *adv.*

**im·post[1]** (ĭm'pōst') *n.* [OFr. < Med. Lat. *impostum* < Lat. *impostus,* p.part. of *imponere,* to place upon. —see IMPOSE.] **1.** Something imposed or levied, as a tax or duty. **2.** The weight a horse must carry in a handicap race.

**im·post²** (ĭm′pōst′) *n.* [Fr. *imposte* < Ital. *imposta* < Lat., fem. p.part. of *imponere,* to place upon. —see IMPOSE.] The topmost section of a column or pillar supporting an arch.

**im·pos·tor** (ĭm-pŏs′tər) *n.* [OFr. *imposteur* < LLat. *impositor* < Lat. *impositus,* p.part. of *imponere,* to place upon. —see IMPOSE.] One who is not what one claims to be.

☆ ***syns:*** IMPOSTOR, CHARLATAN, FAKE, FRAUD, PHONY, PRETENDER, QUACK *n. core meaning* : one who is not what one claims to be <not a doctor, just an *impostor*>

**im·pos·ture** (ĭm-pŏs′chər) *n.* [LLat. *impostura* < Lat. *impositus,* p.part. of *imponere,* to place upon. —see IMPOSE.] The act or practice of deceiving by the assumption of a false identity.

**im·po·tent** (ĭm′pə-tənt) *adj.* [ME < OFr. < Lat. *impotens* : *in-,* not + *potens,* potent.] **1.** Lacking strength or vigor : WEAK. **2.** Powerless : ineffectual. **3.** *Obs.* Lacking self-restraint. **—im′po·tence** (-təns), **im′po·ten·cy** (-tən-sē) *n.* **—im′po·tent·ly** *adv.*

**im·pound** (ĭm-pound′) *vt.* **-pound·ed, -pound·ing, -pounds. 1.** To confine in or as if in a pound. **2.** To seize and keep in legal custody. **3.** To accumulate (water) in a reservoir. **—im·pound′age, im·pound′ment** *n.* **—im·pound′er** *n.*

**im·pov·er·ish** (ĭm-pŏv′ər-ĭsh) *vt.* **-ished, -ish·ing, -ish·es.** [ME *empoverishen* < OFr. *empovrir, empovriss-* : *en-* (causative) + *povre,* poor < Lat. *pauper.*] **1.** To make poor. **2.** To deprive of natural richness or strength. **—im·pov′er·ish·ment** *n.*

**im·prac·ti·ca·ble** (ĭm-prăk′tĭ-kə-bəl) *adj.* **1.** Not capable of being done or carried out. ***usage:*** *Impracticable* applies to that which is not capable of being carried out or put into practice. *Impractical* refers to that which is not sensible or prudent. A plan may be *impractical* because it involves undue cost or effort and yet it may not be *impracticable.* **2.** Unfit for passage, as a road or bridge. **3.** *Archaic.* Unmanageable : intractable. **—im·prac′ti·ca·bil′i·ty, im·prac′ti·ca·ble·ness** *n.* **—im·prac′ti·ca·bly** *adv.*

**im·prac·ti·cal** (ĭm-prăk′tĭ-kəl) *adj.* **1.** Unwise to put into effect or maintain in practice <an *impractical* plan> **2.** Incapable of dealing efficiently with practical matters, esp. money <an *impractical* idealist> **3.** Not a part of experience, fact, or practice : THEORETICAL. **—im·prac′ti·cal′i·ty, im·prac′ti·cal·ness** *n.*

**im·pre·cate** (ĭm′prĭ-kāt′) *vt.* **-cat·ed, -cat·ing, -cates.** [Lat. *imprecari, imprecat-* : *in-,* on + *precari,* to pray, ask.] To call down evil upon : CURSE. **—im′pre·ca′tion** (-kā′shən) *n.* **—im′pre·ca′tor** *n.* **—im′pre·ca·to′ry** (-kə-tôr′ē, -tōr′ē) *adj.*

**im·pre·cise** (ĭm′prĭ-sīs′) *adj.* Not precise. **—im′pre·cise′ly** *adv.* **—im′pre·ci′sion** (-sĭzh′ən) *n.*

**im·preg·na·ble¹** (ĭm-prĕg′nə-bəl) *adj.* [ME < OFr. : *in-,* not (< Lat. *in-*) + *pregnable,* pregnable.] **1.** Incapable of being taken or entered by force <an *impregnable* fortress> **2.** Unable to be changed or criticized, as a conviction.

**im·preg·na·ble²** (ĭm-prĕg′nə-bəl) *adj.* Capable of being impregnated, as an egg.

**im·preg·nate** (ĭm-prĕg′nāt′) *vt.* **-nat·ed, -nat·ing, -nates.** [LLat. *impregnare, impregnat-* : Lat. *in-,* in + *praegnans,* pregnant.] **1.** To make pregnant : INSEMINATE. **2.** To fertilize (e.g., an ovum). **3.** To fill throughout or saturate. **4.** To permeate or imbue. —*adj.* Saturated or filled. **—im′preg·na′tion** *n.* **—im′preg·na′tor** *n.*

**im·pre·sa** (ĭm-prā′zə) *n.* [Ital., undertaking. —see IMPRESARIO.] An emblem or device with a motto.

**im·pre·sa·ri·o** (ĭm′prĭ-sär′ē-ō′, -sâr′-) *n., pl.* **-os.** [Ital. < *impresa,* undertaking < *impreso,* p.part.of *imprendere,* to undertake < VLat. —see EMPRISE.] **1.** One who produces or sponsors entertainment, esp. the director of an opera company. **2.** A manager or producer.

**im·press¹** (ĭm-prĕs′) *vt.* **-pressed, -press·ing, -press·es.** [ME *impressen* < Lat. *impressus,* p.part. of *imprimere* : *in-,* in + *premere,* to press.] **1.** To produce or apply with pressure. **2.** To mark or stamp with or as if with pressure. **3.** To produce a vivid perception or image of. **4.** To affect or influence deeply or strongly. **5.** To transmit a force or motion to. —*n.* (ĭm′prĕs′). **1.** The act of impressing. **2.** A mark or pattern made by impressing. **3.** A stamp or seal for impressing.

**im·press²** (ĭm-prĕs′) *vt.* **-pressed, -press·ing, -press·es. 1.** To compel (someone) to serve in a military force. **2.** To confiscate (property). —*n.* Impressment.

**im·press·i·ble** (ĭm-prĕs′ə-bəl) *adj.* Capable of being impressed. **—im·press′i·bil′i·ty** *n.* **—im·press′i·bly** *adv.*

**im·pres·sion** (ĭm-prĕsh′ən) *n.* **1.** The act or process of impressing. **2.** The effect, mark, or imprint made on a surface by pressure. **3.** An effect, feeling, or image retained as a result of experience. **4.** An indistinct notion, recollection, or belief. **5. a.** All the copies of a publication printed at one time from the same set of type. **b.** A single copy of this printing. **6.** A humorous or satiric imitation of a famous personality done esp. by a professional entertainer. **7.** An initial or single coat of color or paint. **8.** A wax or plaster imprint of the teeth and surrounding tissue, used in dentistry as a mold in making dentures or inlays.

**im·pres·sion·a·ble** (ĭm-prĕsh′ə-nə-bəl) *adj.* **1.** Capable of receiving an impression : PLASTIC. **2.** Easily influenced : SUGGESTIBLE. **—im·pres′sion·a·bil′i·ty, im·pres′sion·a·ble·ness** *n.*

**im·pres·sion·ism** (ĭm-prĕsh′ə-nĭz′əm) *n.* **1.** *often* **Impressionism.** A theory or style of painting developed in France during the 1870's, characterized chiefly by concentration on the general impression created by a scene or object and by applying unmixed primary colors in small strokes to simulate reflected light. **2.** A literary style marked gen. by the use of details and mental associations to inspire subjective and sensory impressions rather than the re-creation of objective reality. **3.** A musical style of the late 19th and early 20th cent., using rich and diversified tones and harmonies to call forth suggestions of mood, place, and natural phenomena. **4.** The practice or habit of expressing or developing one's subjective response to a work of art or actual experience.

**im·pres·sion·ist** (ĭm-prĕsh′ə-nĭst) *n.* **1.** One, esp. an artist or a composer, who practices or advocates the theories of impressionism. **2.** An entertainer who does impressions.

**im·pres·sion·is·tic** (ĭm-prĕsh′ə-nĭs′tĭk) *adj.* **1.** Of, relating to, or made up of impressionism. **2.** Of, relating to, involving, or based on impression rather than reason or fact. **3.** Impressionable. **—im·pres′sion·is′ti·cal·ly** *adv.*

**im·pres·sive** (ĭm-prĕs′ĭv) *adj.* Making a strong impression : commanding attention. **—im·pres′sive·ly** *adv.* **—im·pres′sive·ness** *n.*

**im·press·ment** (ĭm-prĕs′mənt) *n.* The act or policy of seizing people or property for public service or use.

**im·prest¹** (ĭm-prĕst′) *n.* [< Ital. *impresto,* lent, advanced < *imprestare,* to lend : *in-,* toward (< Lat.) + *prestare,* to lend < Lat. *praestare,* to give < *praesto,* at hand.] An advance or loan of government funds for service to the government.

**im·prest²** (ĭm-prĕst′) *v. Archaic. var. p.t. & p.p. of* IMPRESS.

**im·pri·ma·tur** (ĭm′prə-mä′tər, -mā′tər) *n.* [NLat., let it be printed.] **1.** Official approval to print or publish, esp. under a situation of censorship. **2.** Official sanction.

**im·pri·mis** (ĭm-prī′mĭs) *adv.* [ME < Lat. *in primis,* among the first (things).] In the first place.

**im·print** (ĭm-prĭnt′) *vt.* **-print·ed, -print·ing, -prints.** [ME *emprenten* < OFr. *empreinter* < *empreinte,* impression < *empreindre,* to print < Lat. *imprimere,* to impress.] **1.** To make or impress (a mark or design) on a surface. **2.** To make or stamp a mark on. **3. a.** To produce a strong, often favorable effect on the mind or emotions. **b.** To establish firmly in the mind. —*n.* (ĭm′prĭnt′). **1.** A mark or pattern made by imprinting. **2.** A distinguishing influence or effect <a harsh climate leaving its *imprint* on their culture> **3.** A publisher's name, often with the date, address, and edition of a publication, printed at the bottom of a title page.

**im·print·ing** (ĭm′prĭn′tĭng) *n.* A learning process occurring early in the life of a social animal in which a behavior pattern is established through association with a parent or other role model.

**im·pris·on** (ĭm-prĭz′ən) *vt.* **-oned, -on·ing, -ons.** [ME *emprisonen* < OFr. *emprisoner* : *en-,* in (< Lat. *in-*) + *prison,* prison.] **1.** To put in prison. **2.** To restrain, limit, or confine as if in a prison. **—im·pris′on·ment** *n.*

☆ ***syns:*** IMPRISON, CLOSET, CONFINE, ISOLATE, SHUT IN *v. core meaning* : to restrain, limit, or confine as if in a prison <was *imprisoned* in the cabin during the blizzard>

**im·prob·a·ble** (ĭm-prŏb′ə-bəl) *adj.* [Lat. *improbabilis* : *in-,* not + *probabilis,* probable.] Not likely to occur or to be true <an *improbable* outcome><an *improbable* story> **—im·prob′a·bil′i·ty** (-bĭl′ĭ-tē) *n.* **—im·prob′a·ble·ness** *n.* **—im·prob′a·bly** *adv.*

**im·pro·bi·ty** (ĭm-prō′bĭ-tē) *n.* [ME *improbite* < Lat. *improbitas* < *improbus,* dishonest : *in-,* not + *probus,* honest, good.] Lack of probity or integrity : DISHONESTY.

**im·promp·tu** (ĭm-prŏmp′to͞o, -tyo͞o) *adj.* [Fr. < Lat. *in promptu,* at hand : *in,* in + *promptus,* ready. —see PROMPT.] Performed or devised without rehearsal or preparation <an *impromptu* song and dance> —*adv.* Spontaneously. —*n.* **1.** Something made or done impromptu, as a speech. **2.** *Mus.* A short lyrical composition esp. for the piano.

**im·prop·er** (ĭm-prŏp′ər) *adj.* [OFr. *impropre* < Lat. *improprius* : *in-,* not + *proprius,* proper.] **1.** Not suited to circumstances or needs : UNSUITABLE <*improper* punishment for the crime> **2.** Not conforming to conventional mores : INDECOROUS. **3.** Not consistent with truth, fact, or rule : INCORRECT. **4.** Irregular or abnormal, as in form. **—im·prop′er·ly** *adv.* **—im·prop′er·ness** *n.*

**improper fraction** *n.* A fraction with the numerator larger than or equal to the denominator.

**improper integral** *n.* An integral with at least one nonfinite limit or an integrand that becomes infinite between the limits of integration.

**im·pro·pri·e·ty** (ĭm′prə-prī′ĭ-tē) *n., pl.* **-ties. 1.** The quality or state of being improper. **2.** An improper act or statement. **3.** An improper or unacceptable usage in speech or writing.

☆ ***syns:*** IMPROPRIETY, INDECENCY, INDECORUM, INDELICACY *n. core meaning* : an improper act or statement <barred from the club after numerous *improprieties*>

**im·prove** (ĭm-pro͞ov′) *v.* **-proved, -prov·ing, -proves.** [ME *improven,* to enclose land for cultivation < AN *emprouwer* : OFr. *en-* (causative) + *prou,* profit < LLat. *prode,* advantageous. —see

PROUD.] —*vt.* **1.** To advance to a better quality or state. **2.** To increase the productivity or value of (land). —*vi.* **1.** To become or get better. **2.** To make advantageous additions or changes.

**im·prove·ment** (ĭm-pro͞ov′mənt) *n.* **1.** The act or process of improving. **2.** The state of being improved. **3.** Something that improves.

**im·prov·i·dent** (ĭm-prŏv′ĭ-dənt) *adj.* **1.** Not providing for the future : UNTHRIFTY. **2.** Incautious : reckless. **—im·prov′i·dence** *n.* **—im·prov′i·dent·ly** *adv.*

☆ ***syns:*** IMPROVIDENT, THRIFTLESS, UNTHRIFTY *adj. core meaning* : reckless, esp. in the use of material resources <the *improvident* use of our forest lands> ***ant:*** provident

**im·pro·vi·sa·tion** (ĭm-prŏv′ĭ-zā′shən, ĭm′prə-vĭ-) *n.* **1.** The act or process of improvising. **2.** Something improvised, as a dramatic skit.

**im·prov·i·sa·tor** (ĭm-prŏv′ĭ-zā′tər) *n.* One who improvises.

**im·prov·i·sa·to·ry** (ĭm-prŏv′ĭ-zə-tôr′ē, ĭm′prə-vĭ′zə-tôr′ē, -tōr′ē) *also* **im·prov·i·sa·to·ri·al** (ĭm-prŏv′ĭ-zə-tôr′ē-əl, -tōr′ē-əl) *adj.* **1.** Of or relating to improvisation. **2.** Of or relating to an improviser.

**im·pro·vise** (ĭm′prə-vīz′) *v.* **-vised, -vis·ing, -vis·es.** [Fr. *improviser* < Ital. *improvvisare* < *improvviso,* unforeseen < Lat. *improvisus* : *in-,* not + *provisus,* p.part. of *providere,* to foresee. —see PROVIDE.] —*vt.* **1.** To make up, compose, recite, or do without preparation. **2.** To make or provide from available materials. —*vi.* To invent, compose, recite, or perform something offhand. **—im′pro·vis′er** *n.*

☆ ***syns:*** IMPROVISE, AD-LIB, EXTEMPORIZE, FAKE *v. core meaning* : to compose or recite without preparation <*improvise* a speech>

**im·pru·dent** (ĭm-pro͞od′ənt) *adj.* [ME < Lat. *imprudens* : *in-,* not + *prudens,* prudent.] Not prudent. **—im·pru′dence** *n.*

**im·pu·dent** (ĭm′pyə-dənt) *adj.* [ME < Lat. *impudens* : *in-,* not + *pudens,* pr.part. of *pudēre,* to be ashamed.] **1.** Marked by rude behavior or impertinent disrespect. **2.** *Obs.* Immodest. **—im′pu·dence, im′pu·den·cy** *n.* **—im′pu·dent·ly** *adv.*

☆ ***syns:*** IMPUDENT, AUDACIOUS, BOLD, BRAZEN, CHEEKY, CONTUMELIOUS, FLIP, FORWARD, FRESH, IMPERTINENT, INSOLENT, PRESUMPTUOUS, SASSY, SAUCY, SMART, SMART-ALECKY, WISE *adj. core meaning* : rude and disrespectful <was expelled from school for being *impudent*>

**im·pu·dic·i·ty** (ĭm′pyo͞o-dĭs′ĭ-tē) *n.* [OFr. *impudicite* < Lat. *impudicus,* immodest : *in-,* not + *pudicus,* modest < *pudēre,* to be ashamed.] Shamelessness : immodesty.

**im·pugn** (ĭm-pyo͞on′) *vt.* **-pugned, -pugn·ing, -pugns.** [ME *impugnen* < OFr. *impugner* < Lat. *impugnare,* to fight against : *in-,* against + *pugnare,* to fight.] To oppose or attack as untrue, esp. to censure or refute by argumentation. **—im·pugn′a·ble** *adj.* **—im·pugn′er** *n.*

**im·pu·is·sance** (ĭm-pyo͞o′ĭ-səns, ĭm-pwĭs′əns) *n.* Lack of power or effectiveness : WEAKNESS. **—im·pu′is·sant** *adj.*

**im·pulse** (ĭm′pŭls′) *n.* [Lat. *impulsus* < p.part. of *impellere,* to impel. —see IMPEL.] **1. a.** An impelling force. **b.** The motion such a force produces. **2. a.** A sudden spontaneous urge or inclination. **b.** A driving force : INCENTIVE. **3. a.** An inherent, usu. nonrational propensity <"Respect for the liberty of others is not a natural *impulse* in most men" —Bertrand Russell> **b.** A general tendency or spirit : CURRENT <"The Romantic *impulse* led in Germany to a technical recasting of the novel form" —Leslie Fiedler> **4.** *Electron.* A short-term change in the intensity of a medium. **5.** *Physics.* The product of the average value of a force with the time during which it acts, equal in general to the change in momentum produced by the force in this time interval. **6.** *Physiol.* An instance of the transmission of energy from one neuron to another.

**im·pul·sion** (ĭm-pŭl′shən) *n.* **1.** An act of impelling or the state of being impelled. **2.** An impelling force : THRUST. **3.** The motion an impelling force produces : MOMENTUM. **4.** An urging : compulsion <"I do not move . . . unless it be under the *impulsion* of a third party" —Samuel Beckett>

**im·pul·sive** (ĭm-pŭl′sĭv) *adj.* **1.** Tending to act on impulse rather than thought. **2.** Produced as a result of impulse : PRECIPITATE <an *impulsive* comment> **3.** Capable of impelling or inciting : FORCEFUL. **4.** *Physics.* Acting within brief time intervals. —Used esp. of a force. **—im·pul′sive·ly** *adv.* **—im·pul′sive·ness** *n.*

**im·pu·ni·ty** (ĭm-pyo͞o′nĭ-tē) *n., pl.* **-ties.** [Lat. *impunitas* < *impunis,* not punished : *in-,* not + *poena,* penalty < Gk. *poinē.*] Exemption from punishment or penalty.

**im·pure** (ĭm-pyo͝or′) *adj.* **1.** Not pure or clean : CONTAMINATED. **2.** Not purified by a religious ritual : DEFILED. **3.** Unchaste or obscene. **4.** Mixed with another and usu. inferior substance : ADULTERATED. **5.** Being a composite of more than one color or mixed with black or white. —Used of color. **6.** Deriving from more than one source, style, or convention. —Used esp. of the arts. **7.** Not correct or consistent in grammar, vocabulary, idiom, or other usage. **—im·pure′ly** *adv.* **—im·pure′ness** *n.*

☆ ***syns:*** IMPURE, ADULTERATED, ALLOYED, DEGRADED *adj. core meaning* : mixed with other, usu. inferior substances <*impure* gold> ***ant:*** pure

**im·pu·ri·ty** (ĭm-pyo͝or′ĭ-tē) *n., pl.* **-ties. 1.** The quality or state of being impure, esp.: **a.** Contamination or pollution. **b.** Lack of consistency or homogeneity : ADULTERATION. **c.** A state of immorality : SIN. **2.** Something that adulterates or contaminates.

**im·put·a·ble** (ĭm-pyo͞o′tə-bəl) *adj.* Capable of being ascribed : ATTRIBUTABLE **—im·put′a·bly** *adv.*

**im·pu·ta·tion** (ĭm′pyo͞o-tā′shən) *n.* **1.** The act of imputing. **2.** Something that is imputed. **—im·pu′ta·tive** (ĭm-pyo͞o′tə-tĭv) *adj.* **—im·pu′ta·tive·ly** *adv.*

**im·pute** (ĭm-pyo͞ot′) *vt.* **-put·ed, -put·ing, -putes.** [ME *imputen* < OFr. *emputer* < Lat. *imputare,* to charge : *in-,* in + *putare,* to reckon, compute.] **1.** To ascribe (e.g., a crime or defect) to another. **2.** To attribute to a source or cause. **3.** To attribute (e.g., credit for a good turn) to a person other than the one who initiated the act.

**in** (ĭn) *prep.* [ME < OE.] **1. a.** Within the bounds or area of <*in* the winter><went walking *in* the woods> **b.** From the outside to a point within : INTO <threw the log *in* the fireplace> **2.** To or at a situation or condition of <was divided *in* two><*in* poverty><*in* love> **3. a.** Having the activity, occupation, or function of <*in* business><*in* charge> **b.** During the act or process of <an old letter discovered *in* looking through a book> **4. a.** With the arrangement or order of <hanging *in* soft pleats><*in* equal payments> **b.** After the style or form of <*in* Spenserian meter> **5. a.** With the characteristic, attribute, or property of <a tall person *in* a short overcoat> **b.** —Used to indicate a material or element from which something is made <a painting *in* oil> **6.** With the aim or purpose of <went *in* pursuit of the thief> **7.** By the instrumentality or means of <paneled the office *in* walnut> **8.** With reference to <four feet *in* height> **9.** —Used to indicate the second and larger term of a ratio or proportion <picked only one *in* ten> —*adv.* **1.** To or toward the inside <They walked *in.*> **2.** To or toward a destination or goal <The advance force closed *in.*> **3.** In a customary place, as of business or residence <They're not *in.*> **4. a.** In a position of success or favor. **b.** In a relationship <*in* favor with the president> **5. a.** In fashion. **b.** In season. —*adj.* **1.** Fashionable <the *in* restaurant this year> **2.** Extremely concerned with or aware of the latest fashion <one of the *in* crowd> **3.** Incoming <the *in* bus> **4.** Having power : INCUMBENT <the *in* clique> —*n.* **1.** One having position, influence, or power <people who are the *ins* this year> **2.** *Informal.* Power : influence <have an *in* with the boss> **—in for.** Guaranteed to get or have <*in for* a surprise> **—ins and outs. 1.** The twists and turns, as of a roadway. **2.** The characteristic difficulties and features. **—in that.** For the reason that.

**In** *symbol for* INDIUM.

**in-**[1] *or* **il-** *or* **im-** *or* **ir-** *pref.* [ME < OFr. < Lat.] Not <*in*articulate>

**in-**[2] *or* **il-** *or* **im-** *or* **ir-** *pref.* [ME < OFr. < Lat. < *in,* in, within.] **1.** In : into : within <*in*tubation> **2.** EN-[1].

**-in** *suff.* [Fr. *-ine.* —see -INE[2].] **1.** *or* **-ine.** A neutral chemical compound <globul*in*> **2.** Enzyme <pancreat*in*> **3. a.** A pharmaceutical <niac*in*> **b.** An antibiotic <penicill*in*> **4.** *var. of* -INE[2] 3.

**in·a·bil·i·ty** (ĭn′ə-bĭl′ĭ-tē) *n.* Lack of ability or means.

**in ab·sen·tia** (ĭn ăb-sĕn′shə, -shē-ə) *adv.* [Lat.] In absence.

**in·ac·ces·si·ble** (ĭn′ăk-sĕs′ə-bəl) *adj.* Not accessible : UNAPPROACHABLE. **—in′ac·ces′si·bil′i·ty** *n.* **—in′ac·ces′si·bly** *adv.*

☆ ***syns:*** INACCESSIBLE, UNAPPROACHABLE, UNATTAINABLE, UNREACHABLE *adj. core meaning* : unable to be reached <an *inaccessible* mountain aerie> ***ant:*** accessible

**in·ac·cu·ra·cy** (ĭn-ăk′yər-ə-sē) *n., pl.* **-cies. 1.** The quality or state of being inaccurate. **2.** An error.

**in·ac·cu·rate** (ĭn-ăk′yər-ĭt) *adj.* **1.** Not accurate. **2.** Incorrect or mistaken. **—in·ac′cu·rate·ly** *adv.* **—in·ac′cu·rate·ness** *n.*

**in·ac·tion** (ĭn-ăk′shən) *n.* Lack of action.

**in·ac·ti·vate** (ĭn-ăk′tə-vāt′) *vt.* **-vat·ed, -vat·ing, -vates.** To render inactive. **—in·ac′ti·va′tion** *n.*

**in·ac·tive** (ĭn-ăk′tĭv) *adj.* **1.** Not active or inclined to be active. **2. a.** Out of use. **b.** Retired from duty or service. **3. a.** *Chem.* Not readily participating in chemical reactions. **b.** *Biol.* Having no effect on or interaction with living organisms. **c.** *Med.* Quiescent. —Used esp. of a disease. **d.** *Physics.* Exhibiting little or no radioactivity. **4.** Not functioning or operating. **—in·ac′tive·ly** *adv.* **—in·ac′tive·ness, in′ac·tiv′i·ty** *n.*

☆ ***syns:*** INACTIVE, DORMANT, IDLE, INERT, INOPERATIVE *adj. core meaning* : not involved in or tending toward action or movement <led an *inactive* life> INACTIVE is very neutral and does not imply favorable or unfavorable judgment. DORMANT refers mainly to suspended activity that may be renewed <*dormant* cancer> IDLE refers to human inactivity <unemployed, *idle* people> INERT implies mental or spiritual lethargy <an *inert* person unwilling to try> INOPERATIVE often refers to mechanical inactivity <a temporarily *inoperative* elevator> ***ant:*** active

**in·ad·e·qua·cy** (ĭn-ăd′ĭ-kwə-sē) *n., pl.* **-cies. 1.** The quality or state of being inadequate. **2.** A shortcoming or failing.

**in·ad·e·quate** (ĭn-ăd′ĭ-kwĭt) *adj.* **1.** Not adequate : INSUFFICIENT. **2.** Not able : INCAPABLE. **—in·ad′e·quate·ly** *adv.*

**in·ad·mis·si·ble** (ĭn′əd-mĭs′ə-bəl) *adj.* Not admissible : OBJECTIONABLE. **—in′ad·mis′si·bil′i·ty** *n.* **—in′ad·mis′si·bly** *adv.*

**in·ad·ver·tence** (ĭn′əd-vûr′tns) *n.* [Med. Lat. *inadvertentia* : Lat.

*in-*, not + Lat. *advertens*, pr.part. of *advertere*, to turn toward. —see ADVERSE.] **1.** The quality or state of being inadvertent. **2.** An instance of being inadvertent.

**in·ad·ver·ten·cy** (ĭn'əd-vûr'tn-sē) *n.*, *pl.* **-cies.** Inadvertence.

**in·ad·ver·tent** (ĭn'əd-vûr'tnt) *adj.* **1.** Not duly attentive. **2.** Unintentional : accidental. **—in'ad·ver'tent·ly** *adv.*

**in·ad·vis·a·ble** (ĭn'əd-vī'zə-bəl) *adj.* Not recommended : UNWISE. **—in'ad·vis'a·bil'i·ty** *n.*

**in ae·ter·num** (ĭn ē-tûr'nəm) *adv.* [Lat.] To eternity : FOREVER.

**in·al·ien·a·ble** (ĭn-āl'yə-nə-bəl, -ā'lē-ə-nə-) *adj.* Not capable of being transferred to another <*inalienable* liberties> **—in·al'ien·a·bil'i·ty** *n.* **—in·al'ien·a·bly** *adv.*

**in·al·ter·a·ble** (ĭn-ôl'tər-ə-bəl) *adj.* Not changeable. **—in·al'ter·a·bil'i·ty** *n.* **—in·al'ter·a·bly** *adv.*

**in·am·o·ra·ta** (ĭn-ăm'ə-rä'tə) *n.*, *pl.* **-tas.** [Ital., fem. of *inamorato*, inamorato.] A woman with whom one is in love or has an intimate relationship.

**in·am·o·ra·to** (ĭn-ăm'ə-rä'tō) *n.*, *pl.* **-tos.** [Ital. < p.part. of *inamore*, to enamor : *in-*, in (< Lat.) + *amore*, love < Lat. *amor* < *amare*, to love.] A man with whom one is in love or has an intimate relationship.

**in-and-in** (ĭn'ənd-ĭn') *adj.* & *adv.* Repeatedly within the same or closely related stocks <horses showing effects of *in-and-in* breeding> **—in'-and-in'** *adj.*

**in-and-out** (ĭn'ənd-out') *adj.* Involving the buying and selling of a single security within a short period.

**in·ane** (ĭn-ān') *adj.* [Lat. *inanis*, empty, vain.] Without sense or substance : EMPTY <an *inane* comment> **—in·ane'ly** *adv.*

**in·an·i·mate** (ĭn-ăn'ə-mĭt) *adj.* **1.** Not having the qualities associated with active, living organisms. **2.** Appearing lifeless or dead. **3.** Lacking animation or energy : DULL. **—in·an'i·mate·ly** *adv.* **—in·an'i·mate·ness** *n.*

**in·a·ni·tion** (ĭn'ə-nĭsh'ən) *n.* [ME *inanisioun* < LLat. *inanitio* < *inanire*, to make empty < *inanis*, empty.] **1.** Exhaustion, as from lack of nourishment. **2.** The quality or state of being empty.

**in·an·i·ty** (ĭn-ăn'ĭ-tē) *n.*, *pl.* **-ties. 1.** The quality or state of being inane. **2.** Something fatuous or absurd.

**in·ap·peas·a·ble** (ĭn'ə-pē'zə-bəl) *adj.* Incapable of being appeased.

**in·ap·pe·tence** (ĭn-ăp'ĭ-təns) *also* **in·ap·pe·ten·cy** (-tən-sē) *n.* Lack of appetite. **—in·ap'pe·tent** *adj.*

**in·ap·pli·ca·ble** (ĭn-ăp'lĭ-kə-bəl) *adj.* Not applicable. **—in·ap'pli·ca·bil'i·ty** *n.* **—in·ap'pli·ca·bly** *adv.*

**in·ap·po·site** (ĭn-ăp'ə-zĭt) *adj.* Not pertinent : UNSUITABLE. **—in·ap'po·site·ly** *adv.*

**in·ap·pre·cia·ble** (ĭn'ə-prē'shə-bəl) *adj.* Too small to be noticed or make a significant difference : NEGLIGIBLE. **—in'ap·pre'cia·bly** *adv.*

**in·ap·pre·cia·tive** (ĭn'ə-prē'shə-tĭv, -shē-ā'tĭv) *adj.* Showing or feeling no appreciation : UNAPPRECIATIVE. **—in'ap·pre'cia·tive·ly** *adv.* **—in'ap·pre'cia·tive·ness** *n.*

**in·ap·proach·a·ble** (ĭn'ə-prō'chə-bəl) *adj.* Not approachable. **—in'ap·proach'a·bil'i·ty** *n.* **—in'ap·proach'a·bly** *adv.*

**in·ap·pro·pri·ate** (ĭn'ə-prō'prē-ĭt) *adj.* Not suited or appropriate : IMPROPER. **—in'ap·pro'pri·ate·ly** *adv.* **—in'ap·pro'pri·ate·ness** *n.*

**in·apt** (ĭn-ăpt') *adj.* **1.** Inappropriate. **2.** Inept. **—in·apt'ly** *adv.* **—in·apt'ness** *n.*

**in·ap·ti·tude** (ĭn-ăp'tĭ-tōōd', -tyōōd') *n.* **1.** Inappropriateness. **2.** Lack of skill : INEPTITUDE.

**in·ar·tic·u·late** (ĭn'är-tĭk'yə-lĭt) *adj.* **1.** Expressed without the use of normal words or syllables : INCOMPREHENSIBLE <"a cry . . . that . . . sank down into an *inarticulate* whine" —Jack London> **2.** Unable to speak : SPEECHLESS <*inarticulate* with embarrassment> **3.** Unable to speak with eloquence or clarity. **4.** Unable to be expressed <*inarticulate* grieving> **5.** *Biol.* Lacking joints or segments. **—in'ar·tic'u·late·ly** *adv.* **—in'ar·tic'u·late·ness** *n.*

**in·ar·tis·tic** (ĭn'är-tĭs'tĭk) *adj.* **1.** Not conforming to artistic principles. **2.** Lacking interest or taste in art. **—in·ar·tis'tic·al·ly** *adv.*

**in·as·much as** (ĭn'əz-mŭch') *conj.* **1.** Because of the fact that : SINCE. **2.** To the extent that : INSOFAR AS.

**in·at·ten·tion** (ĭn'ə-tĕn'shən) *n.* Lack of attention or notice.

**in·at·ten·tive** (ĭn'ə-tĕn'tĭv) *adj.* Displaying a lack of attention : NEGLIGENT. **—in'at·ten'tive·ly** *adv.* **—in'at·ten'tive·ness** *n.*

**in·au·di·ble** (ĭn-ô'də-bəl) *adj.* Not capable of being heard. **—in·au'di·bly** *adv.*

**in·au·gu·ral** (ĭn-ô'gyər-əl) *adj.* **1.** Of, pertaining to, or typical of an inauguration. **2.** Initial : first <the *inaugural* event of the centennial> —*n.* **1.** An inaugural speech. **2.** An inauguration.

**in·au·gu·rate** (ĭn-ô'gyə-rāt') *vt.* **-rat·ed, -rat·ing, -rates.** [Lat. *inaugurare, inaugurat-* : *in-*, in + *augurare*, to auger < *augur*, soothsayer.] **1.** To induct into office by a formal ceremony. **2.** To begin officially. **3.** To open or begin use of formally with a ceremony : DEDICATE. **—in·au'gu·ra'tor** *n.*

**in·au·gu·ra·tion** (ĭn-ô'gyə-rā'shən) *n.* **1.** A formal beginning or introduction. **2.** Formal induction into an office.

**in·aus·pi·cious** (ĭn'ô-spĭsh'əs) *adj.* Not auspicious : UNFAVORABLE. **—in'aus·pi'cious·ly** *adv.* **—in'aus·pi'cious·ness** *n.*

**in between** *adv.* & *prep.* Between.

**in-be·tween** (ĭn'bĭ-twēn') *adj.* Intermediate <The president issued a vague *in-between* statement.> —*n.* An intermediate or intermediary <Some enjoy sports, some dislike them, and some are *in-between.*>

**in·board** (ĭn'bôrd', -bōrd') *adj.* **1.** Within the hull or toward the center of a ship. **2.** Near to the fuselage of an aircraft <the *inboard* engines> **3.** Toward the center of a machine. —*n.* A motor attached to the inside of the hull of a boat. **—in'board'** *adv.*

**in·born** (ĭn'bôrn') *adj.* **1.** Possessed by an organism at birth. **2.** Inherited or hereditary.

**in·bound** (ĭn'bound') *adj.* **1.** Incoming. **2.** Homeward bound.

**in·bounds** (ĭn'boundz') *adj.* *Basketball.* Of or relating to a means of putting the ball in play by having one player standing out of bounds pass it to another player on the court.

**in·breathe** (ĭn'brēth') *vt.* **-breathed, -breath·ing, -breathes.** To breathe (something) in : INHALE.

**in·bred** (ĭn'brĕd') *adj.* **1.** Produced by inbreeding. **2.** Deep-seated : innate <an *inbred* love of music>

**in·breed** (ĭn'brēd') *vt.* **-bred** (-brĕd'), **-breed·ing, -breeds. 1.** To produce by the continued breeding of closely related individuals. **2.** To breed or develop within : ENGENDER. **—in·breed'er** *n.*

**in·breed·ing** (ĭn'brē'dĭng) *n.* Breeding of closely related individuals or animals.

**In·ca** (ĭng'kə) *n.*, *pl.* **Inca** or **-cas.** [Sp. < Quechua *inka*, ruler.] **1.** An Indian of the group of Quechuan peoples who ruled Peru before the Spanish conquest. **2.** A king or other member of the royal family of the Inca.

**In·ca·ic** (ĭn-kā'ĭk) *adj.* & *n.* Incan.

**in·cal·cu·la·ble** (ĭn-kăl'kyə-lə-bəl) *adj.* **1.** Not capable of being calculated : INDETERMINATE. **2.** Not certain : UNPREDICTABLE <an *incalculable* temper> **—in·cal'cu·la·bil'i·ty, in·cal'cu·la·ble·ness** *n.* **—in·cal'cu·la·bly** *adv.*

**in·ca·les·cent** (ĭn'kə-lĕs'ənt) *adj.* [Lat. *incalescens, incalescent-*, pr.part. of *incalescere*, to grow warm : *in-* (intensive) + *calescere*, to grow warm, inceptive of *calēre*, to be warm.] Growing warmer or more ardent. **—in·ca·les'cence** *n.*

**in cam·er·a** (ĭn kăm'ər-ə) *adv.* [Lat., in the chamber.] **1.** In secret : PRIVATELY. **2.** *Law.* In private with a judge instead of in open court.

**In·can** (ĭng'kən) *adj.* Of or pertaining to the Incas or their civilization or language. —*n.* **1.** An Inca. **2.** Quechua.

**in·can·desce** (ĭn'kən-dĕs') *v.* **-desced, -desc·ing, -desc·es.** [Lat. *incandescere*, to glow : *in-* (intensive) + *candescere*, to glow, inceptive of *candēre*, to shine.] —*vi.* To become incandescent. —*vt.* To cause to become incandescent.

**in·can·des·cence** (ĭn'kən-dĕs'əns) *n.* **1.** Emission of visible light by a hot object. **2.** Light emitted by an incandescent object. **3.** A high degree of emotion, intensity, or brilliance.

**in·can·des·cent** (ĭn'kən-dĕs'ənt) *adj.* **1.** Emitting visible light as a result of being heated. **2.** Shining brilliantly. **3.** Marked by ardent emotion, intensity, or brilliance <an *incandescent* performance in the leading role> **—in'can·des'cent·ly** *adv.*

**incandescent lamp** *n.* An electric lamp in which a filament is heated to incandescence by an electric current.

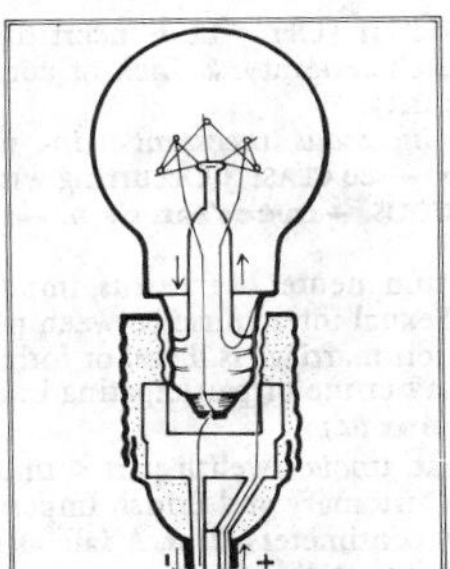

**incandescent lamp**

**in·can·ta·tion** (ĭn'kăn-tā'shən) *n.* [ME *incantacioun* < OFr. *incantation* < LLat. *incantatio*, spell < Lat. *incantore*, to enchant. —see ENCHANT.] **1.** Ritual recitation of charms or spells to produce a magical effect. **2. a.** The formulaic words, phrases, or sounds of an incantation. **b.** Conventional words and slogans used and repeated in a manner similar to the utterance of spells <the tiresome *incantations* of desperate advertisers> **3.** The casting of spells. **—in'can·ta'tion·al** *adj.* **—in·can'ta·to'ry** (-tə-tôr'ē, -tōr'ē) *adj.*

**in·ca·pa·ble** (ĭn-kā'pə-bəl) *adj.* **1. a.** Lacking the needed ability or power <*incapable* of playing three sets of tennis> **b.** Not being in a state or of a kind to admit <*incapable* of fear> **2.** *Law.* Lacking legal qualifications or requirements : INELIGIBLE. **—in·ca'pa·bil'i·ty, in·ca'pa·ble·ness** *n.* **—in·ca'pa·bly** *adv.*

**in·ca·pac·i·tant** (ĭn′kə-păs′ĭ-tənt) *n.* A drug or chemical agent designed to incapacitate a person or animal temporarily by causing dizziness, drowziness, and disorientation.

**in·ca·pac·i·tate** (ĭn′kə-păs′ĭ-tāt′) *vt.* **-tat·ed, -tat·ing, -tates. 1.** To deprive of strength or ability : DISABLE. **2.** *Law.* To make ineligible : DISQUALIFY. **—in′ca·pac′i·ta′tion** *n.*

**in·ca·pac·i·ty** (ĭn′kə-păs′ĭ-tē) *n., pl.* **-ties. 1.** Inadequate strength or ability. **2.** A defect or handicap : DISABILITY. **3.** *Law.* Something that renders one legally ineligible : DISQUALIFICATION.

**in·cap·su·late** (ĭn-kăp′sə-lāt′) *v. var. of* ENCAPSULATE.

**in·car·cer·ate** (ĭn-kär′sə-rāt′) *vt.* **-at·ed, -at·ing, -ates.** [Lat. *incarcerare, incarcerat-* : *in-*, in + *carcer*, prison.] **1.** To jail. **2.** To shut in : CONFINE. **—in·car′cer·a′tion** *n.* **—in·car′cer·a′tor** *n.*

**in·car·na·dine** (ĭn-kär′nə-dīn′, -dēn′, -dĭn) *adj.* [OFr. *incarnadin* < OItal. *incarnatino* < *incarnato*, flesh-colored < Lat. *incarnatus*, incarnate.] **1.** Flesh-colored. **2.** Blood-red. —*vt.* **-dined, -din·ing, -dines.** To make incarnadine.

**in·car·nate** (ĭn-kär′nĭt) *adj.* [LLat. *incarnatus*, p.part. of *incarnare*, to make flesh : *in-*, in + *caro*, flesh.] **1. a.** Having a bodily nature and form. **b.** Embodied in human form : PERSONIFIED. **2.** Incarnadine. —*vt.* (ĭn-kär′nāt′) **-nat·ed, -nat·ing, -nates. 1. a.** To give bodily, esp. human, form to. **b.** To personify. **2.** To realize in action or fact : ACTUALIZE.

**in·car·na·tion** (ĭn′kär-nā′shən) *n.* **1.** The act of incarnating or state of being incarnated. **2. a. Incarnation.** The embodiment of God in the human form of Jesus. **b.** A bodily manifestation of a supernatural being. **3.** One regarded as personifying a given abstract quality or idea. **4.** Time passed in a given condition <hopes for a better life in another *incarnation*>

**in·case** (ĭn-kās′) *v. var. of* ENCASE.

**in·cau·tious** (ĭn-kô′shəs) *adj.* Not cautious : IRRESPONSIBLE. **—in·cau′tious·ly** *adv.* **—in·cau′tious·ness** *n.*

**in·cen·di·ar·y** (ĭn-sĕn′dē-ĕr′ē) *adj.* [Lat. *incendiarius* < *incendium*, fire < *incendere*, to set on fire.] **1. a.** Causing or capable of causing fire. **b.** Of or designating a device, esp. a missile, containing chemicals that produce intensely hot fire when exploded. **c.** Of or involving arson. **2.** Tending to inflame : INFLAMMATORY. —*n., pl.* **-ies. 1.** An arsonist. **2.** An incendiary device. **3.** One who causes factionalism or sedition : AGITATOR. **—in·cen′di·a·rism** (-ə-rĭz′əm) *n.*

**incendiary bomb** *n.* A bomb for starting a fire.

**in·cense¹** (ĭn-sĕns′) *vt.* **-censed, -cens·ing, -cens·es.** [ME *encensen* < OFr. *incenser* < Lat. *incensus*, p.part. of *incendere*, to set on fire.] To cause to be angry : INFURIATE.

**in·cense²** (ĭn′sĕns′) *n.* [ME *encens* < OFr. < LLat. *incensum* < Lat., neuter p.part. of *incendere*, to set on fire.] **1.** An aromatic substance, as a gum or wood, that burns with a fragrant odor. **2.** The smoke or odor produced by the burning of incense. **3.** A fragrant smell. **4.** Flattery or adulation. —*vt.* **-censed, -cens·ing, -cens·es. 1.** To perfume with incense. **2.** To burn incense to as a ritual offering.

**in·cen·tive** (ĭn-sĕn′tĭv) *n.* [ME < Lat. *incentivum* < *incentivus*, inciting < *incentus*, p.part. of *incinere*, to sound : *in-* (intensive) + *canere*, to sing.] Something, as the fear of punishment or the expectation of reward, that incites one to action or effort : STIMULUS. —*adj.* Inciting : motivating.

**in·cep·tion** (ĭn-sĕp′shən) *n.* [Lat. *inceptio* < *incipere*, to begin : *in-*, in + *capere*, to take.] A beginning or commencement.

**in·cep·tive** (ĭn-sĕp′tĭv) *adj.* **1.** Incipient. **2.** INCHOATIVE 2. —*n.* An inchoative verb.

**in·cer·ti·tude** (ĭn-sûr′tĭ-to͞od′, -tyo͞od′) *n.* [OFr. < LLat. *incertitudo* : *in-*, not + *certitudo*, certitude.] **1.** Uncertainty. **2.** Lack of confidence : DOUBT. **3.** Instability or insecurity.

**in·ces·sant** (ĭn-sĕs′ənt) *adj.* [LLat. *incessans, incessant-* : *in-*, not + *cessans*, pr.part. of *cessare*, to stop. —see CEASE.] Occurring without respite or interruption : CONTINUOUS. **—in·ces′san·cy** *n.* **—in·ces′sant·ly** *adv.*

**in·cest** (ĭn′sĕst′) *n.* [ME < Lat. *incestum*, neuter of *incestus*, impure : *in-*, not + *castus*, pure, chaste.] **1.** Sexual intercourse between persons who are so closely related that their marriage is illegal or forbidden by law or custom. **2.** The statutory crime of participating in an incestuous relationship. **—in·ces′tu·ous** *adj.*

**inch¹** (ĭnch) *n.* [ME < OE *ynce* < Lat. *uncia*, twelfth part < *unus*, one.] **1.** A unit of length in the U.S. Customary and British Imperial systems, equal to $^1/_{12}$ of a foot or 2.54 centimeters. **2. a.** A fall, as of rain or snow, sufficient to cover a surface to the depth of one inch. **b.** A unit or degree of atmospheric or other pressure as measured by a barometer or manometer that is equal to the pressure balanced by a one-inch column of liquid, usu. mercury, in the measuring device. **3.** A very small degree or amount <They wouldn't give an *inch*.> —*vi.* & *vt.* **inched, inch·ing, inch·es.** To move or cause to move slowly or by small degrees. **—every inch.** In all respects : ENTIRELY <"Ay, *every inch* a king!" —Shakespeare> **—inch by inch.** Very gradually or slowly. **—within an inch of.** Almost to the point of.

▲ word history: The Latin word *uncia* meaning "a one-twelfth part" was borrowed into English twice. *Uncia* was borrowed into Old English as *ynce*, denoting a linear measure of one twelfth of a foot. The word remains in Modern English as *inch*. *Uncia* also developed into Old French *unce* as a unit of weight equal to one twelfth of a pound. It was borrowed into Middle English and survives in Modern English as *ounce*. The system of troy weight still used for precious metals is based on a 12-ounce pound. The more common avoirdupois system contains a 16-ounce pound.

**inch²** (ĭnch) *n.* [ME < Sc. Gael. *innis*.] *Scot.* A small island.

**inch·meal** (ĭnch′mēl′) *adv.* [INCH + (PIECE)MEAL.] Little by little.

**in·cho·ate** (ĭn-kō′ĭt) *adj.* [Lat. *inchoatus*, p.part. of *inchoare, incohare*, to begin : *in-*, in + *colum*, part of a yoke harness.] **1.** In an early stage. **2.** Imperfectly formed or developed : SHAPELESS <an *inchoate* notion> **—in·cho′ate·ly** *adv.* **—in·cho′ate·ness** *n.*

**in·cho·a·tive** (ĭn-kō′ə-tĭv) *adj.* **1.** Beginning : initial. **2.** Of or pertaining to a verb or verbal form that denotes the start of an action, state, or event. **—in·cho′a·tive** *n.* **—in·cho′a·tive·ly** *adv.*

**inch·worm** (ĭnch′wûrm′) *n.* A measuring worm.

**in·ci·dence** (ĭn′sĭ-dəns) *n.* **1.** An act, instance, or manner of affecting : OCCURRENCE. **2.** The extent or frequency of occurrence <low *incidence* of street crime> **3.** *Physics.* The arrival of incident radiation or of an incident projectile at a surface.

**in·ci·dent** (ĭn′sĭ-dənt) *n.* [ME < OFr. < Lat. *incidens*, pr.part. of *incidere*, to happen : *in-*, on + *cadere*, to fall.] **1.** A separate and definite occurrence : EVENT. **2.** A usu. unimportant event or condition that is subordinate to another. **3.** Something contingent upon or related to something else. **4.** An occurrence or event that interrupts normal procedure or precipitates a crisis. —*adj.* **1.** Tending to arise or occur as a concomitant <crop losses *incident* to uncertain weather> **2.** *Law.* Contingent upon or related to something else. **3.** *Physics.* Falling upon : STRIKING <*incident* radiation>

**in·ci·den·tal** (ĭn′sĭ-dĕn′tl) *adj.* **1.** Occurring or apt to occur as an unpredictable or minor concomitant <tie-ups *incidental* to heavy evening traffic> **2.** Of a minor, casual, or subordinate nature <*incidental* costs> —*n. often* **incidentals.** A minor concomitant circumstance, event, item, or expense.

**in·ci·den·tal·ly** (ĭn′sĭ-dĕn′tl-ē) *adv.* **1.** By chance : CASUALLY. **2.** (*also* ĭn′sĭ-dĕnt′lē). Apart from the main subject : PARENTHETICALLY.

**incidental music** *n.* Music played as an accompaniment to a dramatic work, as a play or film.

**in·cin·er·ate** (ĭn-sĭn′ə-rāt′) *v.* **-at·ed, -at·ing, -ates.** [Med. Lat. *incinerare, incinerat-* : Lat. *in-*, in + *cinis*, ashes.] —*vt.* To consume by burning. —*vi.* To burn or burn up. **—in·cin′er·a′tion** *n.*

**in·cin·er·a·tor** (ĭn-sĭn′ə-rā′tər) *n.* One that consumes by burning, esp. an apparatus for incinerating waste.

**in·cip·i·ent** (ĭn-sĭp′ē-ənt) *adj.* [Lat. *incipiens, incipient-*, pr.part. of *incipere*, to begin. —see INCEPTION.] Beginning to exist or appear <signs of *incipient* illness> **—in·cip′i·ence, in·cip′i·en·cy** *n.* **—in·cip′i·ent·ly** *adv.*

**in·cise** (ĭn-sīz′) *vt.* **-cised, -cis·ing, -cis·es.** [OFr. *inciser* < Lat. *incisus*, p.part. of Lat. *incidere* : *in-*, in + *caedere*, to cut.] **1.** To mark or cut into with a sharp instrument. **2. a.** To engrave (e.g., designs or writing) into a surface : CARVE. **b.** To engrave designs, writing, or other patterns into.

**in·cised** (ĭn-sīzd′) *adj.* **1.** Cut into : ENGRAVED. **2.** Made with or as if with a sharp instrument. **3.** Deeply notched : TOOTHED <the *incised* edge of a handsaw>

**in·ci·sion** (ĭn-sĭzh′ən) *n.* **1.** The act of incising. **2. a.** A surgical cut into soft tissue. **b.** The scar resulting from such a cut. **3.** A notch, as in the edge of a leaf. **4.** The quality of being incisive : TRENCHANCY.

**in·ci·sive** (ĭn-sī′sĭv) *adj.* **1.** Having or suggesting sharp intellect. **2.** Direct and effective : TELLING <an *incisive* summary of the report> **—in·ci′sive·ly** *adv.* **—in·ci′sive·ness** *n.*

☆ ***syns:*** INCISIVE, ACUTE, BITING, CLEAR-CUT, PENETRATING, PERCEPTIVE, PROBING, SHARP, SHREWD, TRENCHANT *adj. core meaning* : having or suggesting keen, discerning intellect <an *incisive* analysis of the problem>

**in·ci·sor** (ĭn-sī′zər) *n.* A cutting tooth, at the apex of the dental arch.

**in·ci·ta·tion** (ĭn′sī-tā′shən) *n.* **1.** An act or instance of inciting : STIMULATION. **2.** Something that incites : INCENTIVE.

**in·cite** (ĭn-sīt′) *vt.* **-cit·ed, -cit·ing, -cites.** [OFr. *inciter* < Lat. *incitare*, to urge forward : *in-* (intensive) + *citare*, to stimulate, freq. of *ciēre*, to put in motion.] To provoke to action : GOAD. **—in·cite′ment** *n.* **—in·cit′er** *n.*

**in·ci·vil·i·ty** (ĭn′sĭ-vĭl′ĭ-tē) *n., pl.* **-ties. 1.** The quality or state of being uncivil. **2.** A rude or discourteous act.

**in·clasp** (ĭn-klăsp′) *v. var. of* ENCLASP.

**in·clem·ent** (ĭn-klĕm′ənt) *adj.* [Lat. *inclemens, inclement-*, harsh, severe : *in-*, not + *clemens*, mild.] **1.** Stormy. **2.** Severe : unmerciful. **—in·clem′en·cy** *n.* **—in·clem′ent·ly** *adv.*

**in·clin·a·ble** (ĭn-klī′nə-bəl) *adj.* **1.** Favorably disposed : AMENABLE. **2.** Having a tendency to do something.

**in·cli·na·tion** (ĭn′klə-nā′shən) *n.* **1.** An attitude or disposition toward something. **2.** A trend or general tendency toward a particular aspect, condition, or character. **3.** Something for which one has a preference or leaning <an *inclination* to eat too much chocolate> **4.** The act of inclining, as a bow. **5.** The state of being inclined : TILT.

**6. a.** A deviation from a horizontal or vertical direction. **b.** The degree of deviation from a horizontal or vertical.

**in·cline** (ĭn-klīn′) *v.* **-clined, -clin·ing, -clines.** [ME *enclinen* < OFr. *encliner* < Lat. *inclinare* : *in-*, toward + *clinare*, to lean.] —*vi.* **1.** To deviate from a horizontal or vertical : SLANT. **2.** To be disposed to a given preference, opinion, or disposition. **3.** To lower or bend the head or body, as in a nod or bow. —*vt.* **1.** To cause to lean, slant, or slope. **2.** To influence to have a certain tendency : DISPOSE <Previous experience *inclined* them to be wary.> **3.** To bend or lower in a nod or bow. —*n.* (ĭn′klīn′). An inclined surface. **—in·clin′er** *n.*

**in·clined** (ĭn-klīnd′) *adj.* **1.** Having a preference, disposition, or tendency. **2.** Sloping, slanting, or leaning.

**inclined plane** *n.* A plane slanted to the horizontal, a simple machine used to raise or lower a load by rolling or sliding.

**in·cli·nom·e·ter** (ĭn′klə-nŏm′ĭ-tər) *n.* **1.** An instrument for determining magnetic dip. **2.** An instrument for showing the inclination of an aircraft or ship relative to the horizontal. **3.** A clinometer.

**in·close** (ĭn-klōz′) *v. var. of* ENCLOSE.

**in·clude** (ĭn-klo͞od′) *vt.* **-clud·ed, -clud·ing, -cludes.** [ME *includen* < Lat. *includere*, to enclose : *in-*, in + *claudere*, to close.] **1.** To have or take in as a part or member : CONTAIN. **2.** To put into a group, class, or total. **—in·clud′a·ble, in·clud′i·ble** *adj.*

**in·clud·ed** (ĭn-klo͞o′dĭd) *adj.* **1.** *Bot.* Not protruding beyond a surrounding part, as stamens that do not project from a corolla. **2.** Formed by and between two intersecting straight lines <an *included* angle>

**in·clu·sion** (ĭn-klo͞o′zhən) *n.* [Lat. *inclusio* < *inclusus*, p.part. of *includere*, to enclose. —see INCLUDE.] **1.** The act of including or state of being included. **2.** Something included. **3.** A solid, liquid, or gaseous foreign body enclosed in a mineral or rock. **4.** *Biol.* A nonliving mass in cytoplasm. **5.** *Computer Sci.* A logical operation assuming that the second statement of a pair is true if the first one is true.

**inclusion body** *n.* An abnormal structure in a cell nucleus or cytoplasm having typical staining properties and associated esp. with the presence of filterable viruses.

**in·clu·sive** (ĭn-klo͞o′sĭv) *adj.* **1.** Taking much or everything within its scope : COMPREHENSIVE. **2.** Including the specified extremes or limits as well as the area between them <the numbers one to ten, *inclusive*> **—in·clu′sive·ly** *adv.* **—in·clu′sive·ness** *n.*

**inclusive of** *prep.* Taking into consideration or account.

**in·co·er·ci·ble** (ĭn′kō-ûr′sə-bəl) *adj.* Incapable of coercion.

**in·cog·i·tant** (ĭn-kŏj′ĭ-tənt) *adj.* [Lat. *incogitans*, *incogitant-* : *in-*, not + *cogitans*, pr.part. of *cogitare*, to think. —see COGITATE.] Inconsiderate : thoughtless.

**in·cog·ni·ta** (ĭn-kŏg′nĭ-tə, ĭn′kŏg-nē′tə) *adj. & adv.* [Ital., fem. of *incognito*, incognito.] With one's identity disguised or hidden. —Used of a woman. **—in·cog′ni·ta** *n.*

**in·cog·ni·to** (ĭn-kŏg′nĭ-tō′, ĭn′kŏg-nē′tō) *adj. & adv.* [Ital. < Lat. *incognitus*, unknown : *in-*, not + *cognitus*, p.part. of *cognoscere*, to learn. —see COGNITION.] With one's identity disguised or hidden. —*n.* **1.** One who is incognito. **2.** The state of being incognito.

**in·cog·ni·zant** (ĭn-kŏg′nĭ-zənt) *adj.* Lacking awareness or knowledge of something : UNAWARE.

**in·co·her·ent** (ĭn′kō-hîr′ənt) *adj.* **1.** Lacking order, connection, or harmony. **2.** Unable to think or express one's thoughts in a clear or orderly manner. **—in′co·her′ence, in′co·her′en·cy** *n.* **—in′co·her′ent·ly** *adv.* **—in′co·her′ent·ness** *n.*

**in·com·bus·ti·ble** (ĭn′kəm-bŭs′tə-bəl) *adj.* [ME < Med. Lat. *incombustibilis* : Lat. *in-*, not + Lat. *combustus*, p.part. of *comburere*, to burn up. —see COMBUSTION.] Incapable of burning. **—in′com·bus′ti·bil′i·ty** *n.* **—in′com·bus′ti·ble** *n.* **—in′com·bus′ti·bly** *adv.*

**in·come** (ĭn′kŭm′) *n.* [ME, enhance, arrival : *in*, in + *comen*, to come.] **1.** Money or its equivalent received during a time period in exchange for labor or services, from the sale of goods or property, or as profit from financial investments. **2.** *Archaic.* An influx.

**income tax** *n.* A tax levied on net income of a business or an individual.

**in·com·ing** (ĭn′kŭm′ĭng) *adj.* **1.** Coming in : ENTERING. **2.** About to come in <the *incoming* administration> —*n.* **1.** The act of coming in : ARRIVAL. **2.** *often* **incomings.** Revenue : income. **3.** *Slang.* An airborne missile or artillery shell about to strike one's position.

**in·com·men·su·ra·ble** (ĭn′kə-mĕn′sər-ə-bəl, -shər-) *adj.* **1. a.** Incapable of being measured. **b.** Without a common quality upon which to make a comparison. **2.** *Math.* Having no common measure. —*n.* Something incommensurable. **—in′com·men′su·ra·bil′i·ty** *n.* **—in′com·men′su·ra·bly** *adv.*

**in·com·men·su·rate** (ĭn′kə-mĕn′sər-ĭt, -shər-) *adj.* **1. a.** Not commensurate : DISPROPORTIONATE <punishment *incommensurate* with the crime> **b.** Inadequate. **2.** Incommensurable. **—in′com·men′su·rate·ly** *adv.* **—in′com·men′su·rate·ness** *n.*

**in·com·mode** (ĭn′kə-mōd′) *vt.* **-mod·ed, -mod·ing, -modes.** [Fr. *incommoder* < OFr. < Lat. *incommodare* < *incommodus*, inconvenient : *in-*, not + *commodus*, convenient. —see COMMODIOUS.] To cause to be inconvenienced : DISTURB.

**in·com·mo·di·ous** (ĭn′kə-mō′dē-əs) *adj.* Inconvenient or uncomfortable, as by not offering sufficient space. **—in′com·mo′di·ous·ly** *adv.* **—in′com·mo′di·ous·ness** *n.*

**in·com·mod·i·ty** (ĭn′kə-mŏd′ĭ-tē) *n., pl.* **-ties. 1.** Inconvenience. **2.** Something inconvenient.

**in·com·mu·ni·ca·ble** (ĭn′kə-myo͞o′nĭ-kə-bəl) *adj.* Not communicable. **—in′com·mu′ni·ca·bil′i·ty** *n.* **—in′com·mu′ni·ca·bly** *adv.*

**in·com·mu·ni·ca·do** (ĭn′kə-myo͞o′nĭ-kä′dō) *adj.* [Sp., p.part. of *incomunicar*, to deny communication : *in-*, not (< Lat.) + *comunicar*, to communicate < Lat. *communicare* < *communis*, common.] Without the right or means of communicating with others, as one held in solitary confinement. **—in′com·mu′ni·ca′do** *adv.*

**in·com·mu·ni·ca·tive** (ĭn′kə-myo͞o′nĭ-kā′tĭv, -kə-tĭv) *adj.* Not communicative. **—in′com·mu′ni·ca′tive·ly** *adv.* **—in′com·mu′ni·ca′tive·ness** *n.*

**in·com·mut·a·ble** (ĭn′kə-myo͞o′tə-bəl) *adj.* **1.** Incapable of being traded or exchanged. **2.** Unalterable. **—in′com·mut′a·bil′i·ty, in′com·mut′a·ble·ness** *n.* **—in′com·mut′a·bly** *adv.*

**in·com·pa·ra·ble** (ĭn-kŏm′pər-ə-bəl) *adj.* **1.** Incapable of being compared. **2.** Without equal or rival : UNIQUE. **—in·com′pa·ra·bil′i·ty, in·com′pa·ra·ble·ness** *n.* **—in·com′pa·ra·bly** *adv.*

**in·com·pat·i·bil·i·ty** (ĭn′kəm-păt′ə-bĭl′ĭ-tē) *n., pl.* **-ties. 1.** The quality or state of being incompatible. **2. incompatibilities.** Mutually exclusive or antagonistic qualities or things.

**in·com·pat·i·ble** (ĭn′kəm-păt′ə-bəl) *adj.* [Med. Lat. *incompatibilis* : *in-*, not (< Lat.) + *compatibilis*, compatible.] **1.** Not in harmony or agreement : INCONGRUOUS. **2.** Incapable of being held simultaneously by one person, as offices. **3.** *Logic.* Incapable of being simultaneously true : mutually exclusive. —*n. often* **incompatibles.** Incompatible persons or things. **—in′com·pat′i·ble·ness** *n.* **—in′com·pat′i·bly** *adv.*

**in·com·pe·tent** (ĭn-kŏm′pĭ-tənt) *adj.* Not competent : UNQUALIFIED. —*n.* An incompetent person. **—in·com′pe·tence, in·com′pe·ten·cy** *n.* **—in·com′pe·tent·ly** *adv.*

**in·com·plete** (ĭn′kəm-plēt′) *adj.* **1.** Not complete. **2.** *Football.* Not caught or not caught in bounds. —Used of a forward pass. **—in′com·plete′ly** *adv.* **—in′com·plete′ness, in′com·ple′tion** *n.*

**in·com·pli·ant** (ĭn′kəm-plī′ənt) *adj.* Not compliant : OBSTINATE. **—in′com·pli′ance, in′com·pli′an·cy** *n.* **—in′com·pli′ant·ly** *adv.*

**in·com·pre·hen·si·ble** (ĭn′kŏm-prĭ-hĕn′sə-bəl, ĭn-kŏm′-) *adj.* **1.** Incapable of being understood, as: **a.** Unintelligible <*incomprehensible* jargon> **b.** Unknowable : unfathomable. **2.** *Archaic.* Boundless. **—in′com·pre·hen′si·bil′i·ty, in′com·pre·hen′si·ble·ness** *n.* **—in′com·pre·hen′si·bly** *adv.*

☆ ***syns:*** INCOMPREHENSIBLE, IMPENETRABLE, UNFATHOMABLE, UNCOMPREHENSIBLE, UNINTELLIGIBLE *adj. core meaning* : incapable of being grasped by the intellect <the *incomprehensible* vastness of deep space> ***ant:*** comprehensible, understandable

**in·com·pre·hen·sion** (ĭn′kŏm-prĭ-hĕn′shən, ĭn-kŏm′-) *n.* Lack of understanding or comprehension.

**in·com·pre·hen·sive** (ĭn′kŏm-prĭ-hĕn′sĭv, ĭn-kŏm′-) *adj.* Limited in range or scope. **—in′com·pre·hen′sive·ly** *adv.* **—in′com·pre·hen′sive·ness** *n.*

**in·com·press·i·ble** (ĭn′kəm-prĕs′ə-bəl) *adj.* Incapable of being compressed : UNYIELDING. **—in′com·press′i·bil′i·ty** *n.*

**in·com·put·a·ble** (ĭn′kəm-pyo͞o′tə-bəl) *adj.* Incapable of being computed or calculated. **—in′com·put′a·bil′i·ty** *n.*

**in·con·ceiv·a·ble** (ĭn′kən-sē′və-bəl) *adj.* **1.** Incapable of being understood or grasped fully. **2.** So surprising or unlikely as to have been considered impossible. **—in′con·ceiv′a·bil′i·ty, in′con·ceiv′a·ble·ness** *n.* **—in′con·ceiv′a·bly** *adv.*

**in·con·cin·ni·ty** (ĭn′kən-sĭn′ĭ-tē) *n.* [Lat. *inconcinnitas*, awkwardness < *inconcinnus*, awkward : *in-*, not + *concinnus*, skillfully put together.] Lack of harmony or agreement.

**in·con·clu·sive** (ĭn′kən-klo͞o′sĭv) *adj.* Not conclusive. **—in′con·clu′sive·ly** *adv.* **—in′con·clu′sive·ness** *n.*

**in·con·den·sa·ble** *also* **in·con·den·si·ble** (ĭn′kən-dĕn′sə-bəl) *adj.* Incapable of being condensed. **—in′con·den′sa·bil′i·ty** *n.*

**in·con·dite** (ĭn-kŏn′dĭt, -dīt′) *adj.* [Lat. *inconditus* : *in-*, not + *conditus*, p.part. of *condere*, to put together.] Poorly constructed : CRUDE. **—in·con′dite·ly** *adv.*

**in·con·form·i·ty** (ĭn′kən-fôr′mĭ-tē) *n.* Nonconformity.

**in·con·gru·ent** (ĭn-kŏng′gro͞o-ənt, ĭn′kŏn-gro͞o′ənt) *adj.* **1.** Not congruent. **2.** Incongruous. **—in·con′gru·ence** *n.* **—in·con′gru·ent·ly** *adv.*

**in·con·gru·i·ty** (ĭn′kŏn-gro͞o′ĭ-tē) *n., pl.* **-ties. 1.** The quality or state of being incongruous. **2.** Something that is incongruous.

**in·con·gru·ous** (ĭn-kŏng′gro͞o-əs) *adj.* [Lat. *incongruus* : *in-*, not + *congruus*, congruous.] **1.** Not corresponding : DISAGREEING <ambitions *incongruous* with their talents> **2.** Made up of disparate, inconsistent, or discordant parts or qualities. **3.** Not consistent with what is logical, usual, or correct : INAPPROPRIATE <*incongruous* clothing for such activities> **—in·con′gru·ous·ly** *adv.* **—in·con′gru·ous·ness** *n.*

☆ ***syns:*** INCONGRUOUS, DISCONSONANT, DISCORDANT, DISCREPANT, DISSONANT, INCOMPATIBLE, INCONGRUENT, INCONSISTENT *adj.*

*core meaning* : made up of parts or qualities that are disparate or otherwise markedly lacking in consistency <a hodgepodge of *incongruous* literary styles> ***ant:*** congruent, congruous

**in·con·nec·tor** (ĭn'kə-nĕk'tər) *n.* A flow-chart symbol indicating continuation of a broken line of flow.

**in·con·se·quent** (ĭn-kŏn'sĭ-kwənt) *adj.* [LLat. *inconsequens, inconsequent-* : Lat. *in-*, not + *consequens*, consequent.] **1.** Not obtained as a result. **2.** Not derived from the premises or obtained by logic or reason : IRRELEVANT. **3.** Proceeding without logical sequence : HAPHAZARD. **4.** Out of character with the style or nature of something. **5.** Unimportant <*inconsequent* matters> **—in·con'se·quence** *n.* **—in·con'se·quent·ly** *adv.*

**in·con·se·quen·tial** (ĭn-kŏn'sĭ-kwĕn'shəl) *adj.* **1.** Without consequence or importance : PETTY. **2.** Inconsequent. —*n.* A trivial thing. **—in·con'se·quen'ti·al'i·ty** (-kwĕn'shē-ăl'ĭ-tē), **in·con'se·quen'tial·ness** *n.* **—in·con'se·quen'tial·ly** *adv.*

**in·con·sid·er·a·ble** (ĭn'kən-sĭd'ər-ə-bəl) *adj.* Too unimportant or small to merit consideration or attention : TRIVIAL. **—in'con·sid'er·a·ble·ness** *n.* **—in'con·sid'er·a·bly** *adv.*

**in·con·sid·er·ate** (ĭn'kən-sĭd'ər-ĭt) *adj.* [Lat. *inconsideratus* : *in-*, not + *consideratus*, considerate.] Not considerate : THOUGHTLESS. **—in'con·sid'er·ate·ly** *adv.* **—in'con·sid'er·ate·ness, in'con·sid'er·a'tion** *n.*

**in·con·sis·tent** (ĭn'kən-sĭs'tənt) *adj.* **1.** Not consistent, esp.: **a.** Not predictable or regular : ERRATIC. **b.** Lacking in correct logical relation : CONTRADICTORY <*inconsistent* reasons> **c.** Not in agreement or harmony : INCOMPATIBLE <words *inconsistent* with the theme of the speech> **2.** *Math.* Not solvable for the unknowns by the same set of values. **—in'con·sis'tence, in'con·sis'ten·cy** *n.* **—in'con·sis'tent·ly** *adv.*

**in·con·sol·a·ble** (ĭn'kən-sō'lə-bəl) *adj.* Incapable of being consoled. **—in'con·sol'a·bil'i·ty, in'con·sol'a·ble·ness** *n.* **—in'con·sol'a·bly** *adv.*

**in·con·so·nant** (ĭn-kŏn'sə-nənt) *adj.* Lacking harmony, agreement, or consistency : DISCORDANT. **—in·con'so·nance** *n.* **—in·con'so·nant·ly** *adv.*

**in·con·spic·u·ous** (ĭn'kən-spĭk'yo͞o-əs) *adj.* Not readily noticed or seen. **—in'con·spic'u·ous·ly** *adv.* **—in'con·spic'u·ous·ness** *n.*

☆ ***syns:*** INCONSPICUOUS, OBSCURE, UNCONSPICUOUS, UNNOTICEABLE, UNOBTRUSIVE *adj. core meaning* : not readily noticed or seen <*inconspicuous* attire><an *inconspicuous* position> ***ant:*** conspicuous, prominent

**in·con·stant** (ĭn-kŏn'stənt) *adj.* **1.** Likely to change, esp. often and without discernible reason or pattern : UNPREDICTABLE. **2.** Faithless : fickle. **—in·con'stan·cy** (-kŏn'stən-sē) *n.* **—in·con'stant·ly** *adv.*

**in·con·sum·a·ble** (ĭn'kən-so͞o'mə-bəl) *adj.* Incapable of being consumed. **—in'con·sum'a·bly** *adv.*

**in·con·test·a·ble** (ĭn'kən-tĕs'tə-bəl) *adj.* Incapable of being contested : UNQUESTIONABLE <*incontestable* evidence> **—in'con·test'a·bil'i·ty, in'con·test'a·ble·ness** *n.* **—in'con·test'a·bly** *adv.*

**in·con·ti·nent** (ĭn-kŏn'tə-nənt) *adj.* [ME < OFr. < Lat. *incontinens*, unrestrained : *in-*, not + *continens*, restrained. —see CONTINENT[2].] Not continent, esp.: **a.** Not restrained : UNCONTROLLED. **b.** Incapable of controlling the excretory functions. **—in·con'ti·nence** *n.* **—in·con'ti·nent·ly** *adv.*

**in·con·trol·la·ble** (ĭn'kən-trō'lə-bəl) *adj.* Not controllable.

**in·con·tro·vert·i·ble** (ĭn-kŏn'trə-vûr'tə-bəl) *adj.* Incapable of being disputed : UNQUESTIONABLE. **—in·con'tro·vert'i·bil'i·ty, in·con'tro·vert'i·ble·ness** *n.* **—in·con'tro·vert'i·bly** *adv.*

**in·con·ven·ience** (ĭn'kən-vēn'yəns) *n.* **1.** The quality or state of being inconvenient. **2.** Something inconvenient. —*vt.* **-ienced, -ienc·ing, -ienc·es.** To cause inconvenience to or for : BOTHER.

**in·con·ven·ient** (ĭn'kən-vēn'yənt) *adj.* [ME < OFr. < Lat. *inconveniens* : *in-*, not + *conveniens*, convenient.] Not convenient, esp.: **a.** Not handy or accessible. **b.** Awkward or difficult to perform. **c.** Inopportune : ill-timed. **—in'con·ven'ient·ly** *adv.*

**in·con·vert·i·ble** (ĭn'kən-vûr'tə-bəl) *adj.* **1.** Incapable of being converted or traded. **2.** Not redeemable for money in coin <paper currency *inconvertible* into gold> **—in'con·vert'i·bil'i·ty, in'con·vert'i·ble·ness** *n.* **—in'con·vert'i·bly** *adv.*

**in·con·vinc·i·ble** (ĭn'kən-vĭn'sə-bəl) *adj.* That is incapable of being convinced.

**in·co·or·di·nate** (ĭn'kō-ôr'dn-ĭt, -āt') *adj.* Lacking coordination. **—in'co·or'di·nate·ly** *adv.*

**in·co·or·di·na·tion** (ĭn'kō-ôr'dn-ā'shən) *n.* Lack of coordination, esp. lack of the ability to exercise normal voluntary control of relatively complex muscular movement.

**in·cor·po·rate** (ĭn-kôr'pə-rāt') *v.* **-rat·ed, -rat·ing, -rates.** [ME *incorporaten* < LLat. *incorporare*, to form into a body : Lat. *in-*, in + Lat. *corpus*, body.] —*vt.* **1.** To unite with or blend indistinguishably into an existing thing. **2.** To admit as a member to an organization, as a corporation. **3.** To cause to merge or combine together into a united whole. **4.** To cause to form into a legal corporation. **5.** To give material form or substance to : EMBODY. —*vi.* **1.** To become united or combined into an organized body. **2.** To form a legal corporation. —*adj.* (-pər-ĭt). **1.** Combined into one united body : MERGED. **2.** Formed into a legal corporation. **—in·cor'po·ra·ble** (-pə-rə-bəl) *adj.* **—in·cor'po·ra'tion** *n.* **—in·cor'po·ra'tive** *adj.* **—in·cor'po·ra'tor** *n.*

**in·cor·po·rat·ed** (ĭn-kôr'pə-rā'tĭd) *adj.* **1.** United as one body : COMBINED. **2.** That is organized and maintained as a legal business corporation.

**in·cor·po·re·al** (ĭn'kôr-pôr'ē-əl, -pōr'-) *adj.* [< Lat. *incorporeus* : *in-*, not + *corpus*, body.] **1.** Lacking material form or substance. **2.** *Law.* Intangible, as a patent or right. **—in'cor·po're·al·ly** *adv.*

**in·cor·po·re·i·ty** (ĭn-kôr'pə-rē'ĭ-tē) *n.* [< Lat. *incorporeus*, incorporeal.] The quality or state of being incorporeal : IMMATERIALITY.

**in·cor·rect** (ĭn'kə-rĕkt') *adj.* Not correct, esp.: **a.** Mistaken : wrong. **b.** Defective : faulty. **c.** Improper : inappropriate. **—in'cor·rect'ly** *adv.* **—in'cor·rect'ness** *n.*

**in·cor·ri·gi·ble** (ĭn-kôr'ĭ-jə-bəl, -kŏr'-) *adj.* [ME < LLat. *incorrigibilis* : Lat. *in-*, not + Lat. *corrigere*, to correct. —see CORRECT.] **1.** Incapable of being reformed or corrected <an *incorrigible* offender> **2.** Deeply rooted : INERADICABLE. **3.** Not manageable. —*n.* One that will not be reformed or corrected. **—in·cor'ri·gi·bil'i·ty, in·cor'ri·gi·ble·ness** *n.* **—in·cor'ri·gi·bly** *adv.*

**in·cor·rupt** (ĭn'kə-rŭpt') *adj.* [ME < Lat. *incorruptus* : *in-*, not + *corruptus*, corrupt.] **1.** Not corrupt. **2.** Not decayed. **3.** Not marred by faults or errors. **—in'cor·rupt'ly** *adv.* **—in'cor·rupt'ness** *n.*

**in·cor·rupt·i·ble** (ĭn'kə-rŭp'tə-bəl) *adj.* **1.** Incapable of being corrupted morally : HONEST. **2.** Not subject to corruption or decay. **—in'cor·rupt'i·bil'i·ty** *n.* **—in'cor·rupt'i·bly** *adv.*

**in·crease** (ĭn-krēs') *v.* **-creased, -creas·ing, -creas·es.** [ME *encresen* < OFr. *encreistre, encreiss-* < Lat. *increscere* : *in-*, in + *crescere*, to grow.] —*vi.* **1.** To become greater or larger. **2.** To multiply : reproduce. —*vt.* To make greater or larger. —*n.* (ĭn'krēs'). **1.** The act or an instance of increasing. **2.** The amount or rate by which something is increased <a large *increase* in wholesale prices> **3.** *Archaic.* Reproduction and spread : PROPAGATION. **—in·creas'a·ble** *adj.* **—in·creas'er** *n.* **—in·creas'ing·ly** *adv.*

☆ ***syns:*** INCREASE, AMPLIFY, AUGMENT, BEEF UP, BUILD UP, ENLARGE, EXPAND, EXTEND, GROW, MAGNIFY, MOUNT, MULTIPLY, SNOWBALL, SWELL *v. core meaning* : to become larger or greater <Difficulties continued to *increase*.> ***ant:*** decrease

**in·cre·ate** (ĭn'krē-āt', ĭn-krē'ĭt) *adj.* Existing without having been created, as divine presences or qualities. **—in'cre·ate'ly** *adv.*

**in·cred·i·ble** (ĭn-krĕd'ə-bəl) *adj.* [ME < Lat. *incredibilis* : *in-*, not + *credibilis*, credible.] **1.** Too implausible to be believed : UNBELIEVABLE. **2.** Astonishing. **—in·cred'i·ble·ness** *n.* **—in·cred'i·bly** *adv.*

**in·cre·du·li·ty** (ĭn'krĭ-do͞o'lĭ-tē, -dyo͞o'-) *n.* The quality or state of being incredulous : DISBELIEF.

**in·cred·u·lous** (ĭn-krĕj'ə-ləs) *adj.* [Lat. *incredulus* : *in-*, not + *credulus*, credulous.] **1.** Skeptical : disbelieving <*incredulous* of ghost stories> **2.** Expressing disbelief <an *incredulous* look on my face> **—in·cred'u·lous·ly** *adv.* **—in·cred'u·lous·ness** *n.*

**in·cre·ment** (ĭn'krə-mənt, ĭng'-) *n.* [ME < Lat. *incrementum* < neuter p.part. of *increscere*, to increase.] **1.** An increase in number, size, or extent. **2.** Something gained or added. **3.** A small increase in quantity. **4.** One of a series of regular additions or contributions. **5.** *Math.* A small positive or negative change in a variable. **—in'cre·men'tal** (-mĕn'tl) *adj.* **—in'cre·men'tal·ly** *adv.*

**in·cre·men·tal·ism** (ĭn'krə-mĕn'tl-ĭz'əm) *n.* Social or political gradualism. **—in'cre·men'tal·ist** *n.*

**in·cres·cent** (ĭn-krĕs'ənt) *adj.* [Lat. *increscens, increscent-*, pr.part. of *increscere*, to increase.] Exhibiting a progressively larger lighted surface : WAXING. —Used of the moon.

**in·cre·tion** (ĭn-krē'shən) *n.* [IN-[2] + (SE)CRETION.] **1.** The process of internal secretion typical of endocrine glands. **2.** The product of incretion : HORMONE.

**in·crim·i·nate** (ĭn-krĭm'ə-nāt') *vt.* **-nat·ed, -nat·ing, -nates.** [LLat. *incriminare, incriminat-* : Lat. *in-*, in + Lat. *crimen*, crime.] To involve in or charge with a wrongful act, as a crime. **—in·crim'i·na'tion** *n.* **—in·crim'i·na·to'ry** (-nə-tôr'ē, -tōr'ē) *adj.*

**in·crust** (ĭn-krŭst') *v. var. of* ENCRUST.

**in·cu·bate** (ĭn'kyə-bāt', ĭng'-) *v.* **-bat·ed, -bat·ing, -bates.** [Lat. *incubare, incubat-*, to lie down on : *in-*, on + *cubare*, to lie down.] —*vt.* **1.** To warm (eggs), as by bodily heat, so as to bring about embryonic development and the hatching of young : BROOD. **2.** To maintain (e.g., a bacterial culture) at optimum environmental conditions for development. **3.** To cause to develop : FOMENT. —*vi.* **1.** To brood eggs. **2.** To develop and hatch. **3.** To undergo the process of incubation. **—in'cu·ba'tive** *adj.*

**in·cu·ba·tion** (ĭn'kyə-bā'shən, ĭng'-) *n.* **1.** The act of incubating or state of being incubated. **2.** *Med.* The development of an infection from the time of its entry into or initiation within an organism up to the time of the first appearance of signs or symptoms. **—in'cu·ba'tion·al** *adj.* **—in'cu·ba'tive** *adj.*

**in·cu·ba·tor** (ĭn'kyə-bā'tər, ĭng'-) *n.* One that incubates, esp.: **a.** A cabinet in which a constant temperature can be maintained, used in growing bacterial cultures. **b.** An apparatus for maintaining an in-

fant, esp. a premature infant, in a controlled environment. **c.** A temperature-controlled enclosure for hatching eggs artificially.
**in·cu·bus** (ĭn′kyə-bəs, ĭng′-) *n., pl.* **-bus·es** *or* **-bi** (-bī′) [ME < LLat. < Lat. *incubare,* to lie down on. —see INCUBATE.] **1.** An evil spirit thought to descend upon and have sexual intercourse with sleeping women. **2.** A nightmare. **3.** Something nightmarishly burdensome.
**in·cu·des** (ĭng-kyōō′dēz) *n. pl. of* INCUS.
**in·cul·cate** (ĭn-kŭl′kāt′, ĭn′kŭl-) *vt.* **-cat·ed, -cat·ing, -cates.** [Lat. *inculcare, inculcat-,* to force upon : *in-,* in + *calcare,* to trample < *calx,* heel.] To teach or impress by emphasis or frequent repetition : INSTILL. **—in′cul·ca′tion** *n.* **—in·cul′ca′tor** *n.*
**in·cul·pa·ble** (ĭn-kŭl′pə-bəl) *adj.* Free from guilt : BLAMELESS.
**in·cul·pate** (ĭn-kŭl′pāt′, ĭn′kŭl-) *vt.* **-pat·ed, -pat·ing, -pates.** [LLat. *inculpare, inculpat-* : Lat. *in-,* on + Lat. *culpare,* to blame < *culpa,* fault.] To incriminate. **—in′cul·pa′tion** *n.* **—in·cul′pa·to′ry** (-pə-tôr′ē, -tōr′ē) *adj.*
**in·cult** (ĭn-kŭlt′) *adj.* [Lat. *incultus* : *in-,* not + *cultus,* p.part. of *colere,* to till.] Not cultured : COARSE.
**in·cum·ben·cy** (ĭn-kŭm′bən-sē) *n., pl.* **-cies. 1.** The quality or state of being incumbent. **2.** Something incumbent. **3. a.** The holding and administering of an office or ecclesiastical benefice. **b.** The term of such an office or benefice.
**in·cum·bent** (ĭn-kŭm′bənt) *adj.* [ME < Lat. *incumbens,* pr.part. of *incumbere,* to lean upon : *in-,* on + *cumbere,* to recline.] **1.** Lying, leaning, or resting on something else. **2.** Imposed as a duty or obligation : OBLIGATORY. **3.** Currently holding a given office <the *incumbent* governor> —*n.* One currently holding an office. **—in·cum′bent·ly** *adv.*
**in·cu·nab·u·lum** (ĭn′kyə-năb′yə-ləm, ĭng′-) *n., pl.* **-la** (-lə) [< Lat. *incunabula* (pl.), swaddling clothes, cradle : *in-,* in + *cunabula,* cradle, infancy < *cunae,* cradle.] **1.** A book printed prior to 1501. **2.** An artifact of an early period. **—in′cu·nab′u·lar** (-lər) *adj.*
**in·cur** (ĭn-kûr′) *vt.* **-curred, -cur·ring, -curs.** [Lat. *incurrere,* to run upon : *in-,* in + *currere,* to run.] To become liable or subject to, esp. because of one's own actions <People who buy pets *incur* great responsibilities.> **—in·cur′rence** *n.*
**in·cur·a·ble** (ĭn-kyŏŏr′ə-bəl) *adj.* Not curable : HOPELESS. **—in·cur′a·bil′i·ty, in·cur′a·ble·ness** *n.* **—in·cur′a·bly** *adv.*
**in·cu·ri·ous** (ĭn-kyŏŏr′ē-əs) *adj.* [Lat. *incuriosus,* careless : *in-,* not + *curiosus,* careful < *cura,* care.] Lacking interest : DETACHED. **—in·cu′ri·os′i·ty** (-ŏs′ĭ-tē), **in·cu′ri·ous·ness** *n.* **—in·cu′ri·ous·ly** *adv.*
**in·cur·rent** (ĭn-kûr′ənt) *adj.* [Lat. *incurrens, incurrent-,* pr.part. of *incurrere,* to run upon. —see INCUR.] Permitting passage to an inflowing current.
**in·cur·sion** (ĭn-kûr′zhən, -shən) *n.* [ME < OFr. < Lat. *incursio* < *incursus,* p.part. of *incurrere,* to run upon. —see INCUR.] **1.** A sudden attack on hostile territory : RAID. **2.** An act of entering another's territory.
**in·cur·vate** (ĭn-kûr′vāt′, ĭn′kûr-vāt′) *vt.* **-vat·ed, -vat·ing, -vates.** To cause to curve inward. —*adj.* (-kûr′vāt′, -vĭt). Curved inward. **—in′cur·va′tion, in·cur′va·ture′** (-chŏŏr′, -chər) *n.*
**in·curve** (ĭn-kûrv′, ĭn′kûrv′) *vt.* & *vi.* **-curved, -curv·ing, -curves.** [Lat. *incurvare* : *in-,* in + *curvus,* curve.] To bend into an inward curve. —*n.* (ĭn′kûrv′). An inward curve.
**in·cus** (ĭng′kəs) *n., pl.* **in·cu·des** (ĭng-kyōō′dēz) [Lat. *incus, incud-,* anvil < *incusus,* p.part. of *incudere,* to forge with a hammer : *in-,* in + *cudere,* to beat, forge.] An anvil-shaped bone in the mammalian middle ear.
**in·cuse** (ĭn-kyōōz′, -kyōōs′) *adj.* [Lat. *incusus,* p.part. of *incudere,* to forge with a hammer. —see INCUS.] Formed by hammering, stamping, or pressing. —Used of designs on coins and medals.
**in·da·ba** (ĭn-dä′bə) *n.* [Zulu *in-daba,* matter for discussion.] A conference of tribes in southern Africa.
**in·da·mine** (ĭn′də-mēn′) *n.* [IND(IGO) + AMINE.] Any of a group of organic bases that form unstable bluish or greenish salts used as dyes.
**in·debt·ed** (ĭn-dĕt′ĭd) *adj.* [ME *endetted* < OFr. *endette* < p.part. of *endetter,* to oblige : *en-,* in (< Lat. *in-*) + *dette,* debt. —see DEBT.] Obligated to another : BEHOLDEN. **—in·debt′ed·ness** *n.*
**in·de·cent** (ĭn-dē′sənt) *adj.* **1.** Offensive to good taste : UNSEEMLY. **2.** Offensive to accepted moral values : IMMODEST. **—in·de′cen·cy** *n.* **—in·de′cent·ly** *adv.*
**in·de·ci·pher·a·ble** (ĭn′dĭ-sī′fər-ə-bəl) *adj.* Incapable of being deciphered. **—in′de·ci′pher·a·bil′i·ty, in′de·ci′pher·a·ble·ness** *n.*
**in·de·ci·sion** (ĭn′dĭ-sĭzh′ən) *n.* Reluctance or inability to make up one's mind : IRRESOLUTION.
**in·de·ci·sive** (ĭn′dĭ-sī′sĭv) *adj.* **1.** Not decisive : INCONCLUSIVE. **2.** Prone to or marked by indecision : IRRESOLUTE. **3.** Not clearly defined : INDEFINITE. **—in′de·ci′sive·ly** *adv.* **—in′de·ci′sive·ness** *n.*
**in·de·clin·a·ble** (ĭn′dĭ-klī′nə-bəl) *adj.* Lacking grammatical inflections.
**in·de·com·pos·a·ble** (ĭn-dē′kəm-pō′zə-bəl) *adj.* Not capable of being separated into component parts.

**in·dec·o·rous** (ĭn-dĕk′ər-əs) *adj.* Lacking good taste or propriety. **—in·dec′o·rous·ly** *adv.* **—in·dec′o·rous·ness** *n.*
**in·de·co·rum** (ĭn′dĭ-kôr′əm, -kōr′-) *n.* **1.** Lack of decorum : IMPROPRIETY. **2.** An instance of indecorous behavior.
**in·deed** (ĭn-dēd′) *adv.* [ME *indede,* in fact : *in,* in + *dede,* deed.] **1.** Without a doubt : CERTAINLY. **2.** In reality : in fact. **3.** Unquestionably : admittedly. —*interj.* —Used to express surprise, skepticism, or irony.
**in·de·fat·i·ga·ble** (ĭn′dĭ-făt′ĭ-gə-bəl) *adj.* [Lat. *indefatigabilis* : *in-,* not + *defatigare,* to tire out (*de-,* thoroughly + *fatigare,* to weary).] Incapable of being fatigued : TIRELESS. **—in′de·fat′i·ga·bil′i·ty, in′de·fat′i·ga·ble·ness** *n.* **—in′de·fat′i·ga·bly** *adv.*
**in·de·fea·si·ble** (ĭn′dĭ-fē′zə-bəl) *adj.* Not capable of being nullified or made void <*indefeasible* claims to the land> **—in′de·fea′si·bil′i·ty** *n.* **—in′de·fea′si·bly** *adv.*
**in·de·fec·ti·ble** (ĭn′dĭ-fĕk′tə-bəl) *adj.* **1.** Capable of resisting defect or failure : LASTING. **2.** Without flaw : PERFECT. **—in′de·fec′ti·bil′i·ty** *n.* **—in′de·fec′ti·bly** *adv.*
**in·de·fen·si·ble** (ĭn′dĭ-fĕn′sə-bəl) *adj.* Not capable of being defended, esp.: **a.** Inexcusable. **b.** Invalid : untenable. **c.** Open to physical attack. **—in′de·fen′si·bil′i·ty, in′de·fen′si·ble·ness** *n.* **—in′de·fen′si·bly** *adv.*
**in·de·fin·a·ble** (ĭn′dĭ-fī′nə-bəl) *adj.* Not capable of being described or analyzed. —*n.* One that is indefinable. **—in′de·fin′a·ble·ness** *n.* **—in′de·fin′a·bly** *adv.*
**in·def·i·nite** (ĭn-dĕf′ə-nĭt) *adj.* [Lat. *indefinitus* : *in-,* not + *definitus,* definite.] Not definite, esp.: **a.** Unclear : vague. **b.** Lacking precise limits. **c.** Not decided : UNCERTAIN. **—in·def′i·nite·ly** *adv.* **—in·def′i·nite·ness** *n.*
☆ ***syns:*** **1.** INDEFINITE, INDETERMINABLE, INEXACT, UNDETERMINED *adj. core meaning* : lacking precise limits <closed for an *indefinite* time> ***ant:*** definite **2.** INDEFINITE, OPEN, UNCERTAIN, UNDECIDED, UNDETERMINED, UNRESOLVED, UNSETTLED, UNSURE, VAGUE *adj. core meaning* : marked by lack of firm decision or commitment and of questionable outcome <*indefinite* vacation plans> ***ant:*** definite
**indefinite article** *n.* An article, as English *a* or *an,* that does not fix the identity of the noun modified.
**indefinite integral** *n. Math.* The set of all functions of which a given function is the derivative, usu. represented by $\int f(x)dx + C$, where $\int f(x)dx$ is any member of the set and $C$ is an arbitrary constant.
**indefinite pronoun** *n.* A pronoun, as English *any* or *some,* that does not specify the identity of its object.
**in·de·his·cent** (ĭn′dĭ-hĭs′ənt) *adj.* Not splitting open at maturity <*indehiscent* seedpods> **—in′de·his′cence** *n.*
**in·del·i·ble** (ĭn-dĕl′ə-bəl) *adj.* [Lat. *indelebilis* : *in-,* not + *delebilis,* capable of being destroyed < *delēre,* to destroy.] **1.** Incapable of being removed : PERMANENT. **2.** Making a mark not easily erased or washed away <written with *indelible* ink> **—in·del′i·bil′i·ty, in·del′i·ble·ness** *n.* **—in·del′i·bly** *adv.*
**in·del·i·cate** (ĭn-dĕl′ĭ-kĭt) *adj.* **1. a.** Lacking in or offensive to propriety. **b.** Bordering on vulgarity : COARSE. **2.** Lacking sensitivity to the feelings of others : TACTLESS. **—in·del′i·ca·cy** (-ka-sē) *n.* **—in·del′i·cate·ly** *adv.* **—in·del′i·cate·ness** *n.*
**in·dem·ni·fi·ca·tion** (ĭn-dĕm′nə-fĭ-kā′shən) *n.* **1.** An act of indemnifying or the state of being indemnified. **2.** Something that indemnifies : INDEMNITY.
**in·dem·ni·fy** (ĭn-dĕm′nə-fī′) *vt.* **-fied, -fy·ing, -fies.** [< Lat. *indemnis,* uninjured : *in-,* not + *damnum,* harm.] **1.** To protect against damage, loss, or injury : INSURE. **2.** To make compensation to for damage, loss, or injury. **—in·dem′ni·fi′er** *n.*
**in·dem·ni·ty** (ĭn-dĕm′nĭ-tē) *n., pl.* **-ties.** [ME *indempnite* < AN < LLat. *indemnitas* < Lat. *indemnis,* uninjured. —see INDEMNIFY.] **1.** Security against damage, loss, or injury. **2.** A legal exemption from liability for damages. **3.** Compensation for damage, loss, or injury.
**in·de·mon·stra·ble** (ĭn′dĭ-mŏn′strə-bəl) *adj.* Incapable of being proved or demonstrated. **—in′de·mon′stra·bil′i·ty, in′de·mon′stra·ble·ness** *n.* **—in′de·mon′stra·bly** *adv.*
**in·dene** (ĭn′dēn′) *n.* [IND(OLE) + -ENE.] A colorless organic liquid, $C_9H_8$, derived from coal tar and used in making synthetic resins.
**in·dent**[1] (ĭn-dĕnt′) *v.* **-dent·ed, -dent·ing, -dents.** [ME *endenten,* to notch < OFr. *endenter* < Med. Lat. *indentare* : Lat. *in-,* in + Lat. *dens,* tooth.] —*vt.* **1. a.** To cut or tear (a document with two or more copies) along an irregular line so that the parts can later be matched to show authenticity. **b.** To draw up (a document) in duplicate or triplicate. **2. a.** To notch or make jagged the edge of : SERRATE. **b.** To make notches, grooves, or holes in (e.g., wood) for the purpose of mortising. **c.** To fit or join together by or as if by mortising. **3.** To set (e.g., the first line of a paragraph) in from the margin. —*vi.* To form an indentation. —*n.* (ĭn-dĕnt′, ĭn′dĕnt′). **1.** An indenture. **2.** A U.S. certificate issued at the end of the Revolutionary War for interest due on the public debt. **3.** An indention. **—in·dent′er** *n.*
**in·dent**[2] (ĭn-dĕnt′) *vt.* **-dent·ed, -dent·ing, -dents. 1.** To make a dent or depression in. **2.** To impress (e.g., a design) : STAMP. —*n.* (ĭn-dĕnt′, ĭn′dĕnt′). An indentation.
**in·den·ta·tion** (ĭn′dĕn-tā′shən) *n.* **1.** The act of indenting or state of being indented. **2.** A notch or jagged cut in an edge. **3.** A recess in

a border, coastline, or other boundary. **4.** The blank space between a margin and the beginning of an indented line.

**in·den·tion** (ĭn-dĕn′shən) *n.* **1.** INDENTATION 1. **2.** INDENTATION 4. **3.** *Archaic.* INDENTATION 2, 3.

**in·den·ture** (ĭn-dĕn′chər) *n.* [ME *endenture,* a written agreement < AN < OFr. *endenter,* to indent.] **1. a.** A document in duplicate with indented edges. **b.** A legal contract or deed executed between two or more parties. **c.** *often* **indentures.** A contract binding one party into the service of another for a specified term. **d.** An official or authenticated voucher or list. **2.** INDENTATION 2. —*vt.* **-tured, -tur·ing, -tures. 1.** To bind into the service of another by indenture. **2.** *Archaic.* To form an indentation in.

**in·de·pend·ence** (ĭn′dĭ-pĕn′dəns) *n.* **1.** The quality or state of being independent. **2.** *Archaic.* Sufficient income for self-support.

**Independence Day** *n.* Jul. 4, a U.S. legal holiday celebrating the anniversary of the adoption of the Declaration of Independence in Philadelphia in 1776.

**in·de·pend·en·cy** (ĭn′dĭ-pĕn′dən-sē) *n., pl.* **-cies. 1.** Independence. **2.** An independent territory or state. **3. Independency.** The Independent movement in 17th-cent. England.

**in·de·pend·ent** (ĭn′dĭ-pĕn′dənt) *adj.* **1.** Politically autonomous : SELF-GOVERNING. **2.** Free from the influence, guidance, or control of another or others : SELF-RELIANT <an *independent* thinker> **3.** Not influenced or determined by someone or something else <an *independent* poll of opinion> **4.** Associated with or loyal to no one political party or organization <an *independent* candidate> **5.** Not dependent on or affiliated with a larger or controlling group or system <an *independent* gasoline station> **6. a.** Financially self-sufficient : SELF-SUPPORTING. **b.** Providing or being enough income to enable one to live without working <a person of *independent* resources> **7.** *Math.* **a.** Not dependent on other variables. **b.** Of or relating to a system of equations no one of which is necessarily satisfied by a set of values of the independent variables that satisfy all the others. **c.** Of, relating to, describing, or being the result of a trial of a chance experiment the probability of which does not depend on the outcome of any other trial of the chance experiment. —*n.* **1.** One that is independent, esp. a voter who does not pledge loyalty to any one political party. **2. Independent.** A member of a 17th-cent. movement in England advocating political and religious independence of individual congregations. **3. Independent.** *Chiefly Brit.* A Congregationalist.

**independent clause** *n.* A clause containing a subject, a verb, and sometimes an object and modifiers and capable of standing alone as a complete sentence.

**in-depth** (ĭn′dĕpth′) *adj.* Thorough : detailed <an *in-depth* analysis of market trends>

**in·de·scrib·a·ble** (ĭn′dĭ-skrī′bə-bəl) *adj.* **1.** Incapable of being described. **2.** Exceeding description <*indescribable* wrath><*indescribable* beauty> **—in′de·scrib′a·bil′i·ty, in′de·scrib′a·ble·ness** *n.* **—in′de·scrib′a·bly** *adv.*

**in·de·struc·ti·ble** (ĭn′dĭ-strŭk′tə-bəl) *adj.* Incapable of being destroyed : UNBREAKABLE. **—in′de·struc′ti·bil′i·ty, in′de·struc′ti·ble·ness** *n.* **—in′de·struc′ti·bly** *adv.*

**in·de·ter·min·a·ble** (ĭn′dĭ-tûr′mə-nə-bəl) *adj.* **1.** Incapable of being ascertained, measured, or fixed. **2.** Incapable of being finally settled or decided. **—in′de·ter′min·a·bly** *adv.*

**in·de·ter·mi·nate** (ĭn′dĭ-tûr′mə-nĭt) *adj.* [ME *determinat* < LLat. *indeterminatus* : Lat. *in-,* not + *determinatus,* determinate.] **1. a.** Not precisely determined <old houses of *indeterminate* condition> **b.** Incapable of being determined. **c.** Lacking precision or clarity : VAGUE. **d.** Not known in advance. **2.** *Bot.* Not terminating in a flower and continuing to grow at the apex <an *indeterminate* inflorescence> **—in′de·ter′mi·na·cy** *n.* **—in′de·ter′mi·nate·ly** *adv.* **—in′de·ter′mi·nate·ness, in′de·ter′mi·na′tion** *n.*

**in·de·ter·min·ism** (ĭn′dĭ-tûr′mə-nĭz′əm) *n.* **1.** Unpredictability. **2.** *Philos.* The doctrine that in some circumstances volition occurs independent of physiological and psychological antecedents. **—in′de·ter′min·ist** *n.* **—in′de·ter′min·is′tic** *adj.*

**in·dex** (ĭn′dĕks′) *n., pl.* **-dex·es** *or* **-di·ces** (-dĭ-sēz′) [Lat. *index, indic-,* forefinger, pointer < *indicare,* to indicate.] **1.** Something serving to guide, point out, or otherwise aid reference, as: **a.** An alphabetized listing of names, places, and subjects in a printed work that gives the page on which each item is mentioned. **b.** A series of notches cut into the edge of a book for easy access to chapters or other divisions. **c.** A table, file, or catalogue. **2.** Something that reveals or indicates : SIGN. **3.** A printing character (☞) calling attention to a particular paragraph or section. **4.** An indicator or pointer, as in a scientific instrument. **5.** *Math.* **a.** A number or symbol, often written as a subscript or superscript to a mathematical expression, indicating an operation to be performed on, an ordering relation involving, or a use of the associated expression. **b.** A number derived from a formula for characterizing a set of data <the wholesale price *index*> **6. Index.** A list once published by the Roman Catholic Church authorities restricting or forbidding the reading of certain books. —*vt.* **-dexed, -dex·ing, -dex·es. 1.** To furnish with an index <*index* an encyclopedia> **2.** To enter in an index. **3.** To indicate or signal. **4.** To adjust through indexation. **—in′dex′er** *n.*

**in·dex·a·tion** (ĭn′dĕk-sā′shən) *n.* The linkage of economic factors, as wages, interest, or prices, to a cost-of-living index so they rise and fall within the rate of inflation.

**index finger** *n.* The finger next to the thumb.

**index number** *n.* A number indicating change in magnitude, as of price, wage, employment, or production shifts, relative to the magnitude at some specified point usu. taken as 100.

**index of refraction** *n.* The ratio of the speed of light in a vacuum to the speed of light in a medium under consideration.

**India ink** *n.* **1.** A black pigment made from lampblack mixed with a binding agent and molded into cakes or sticks. **2.** A liquid ink made from India ink.

**In·di·an** (ĭn′dē-ən) *n.* **1.** A native or resident of India or the East Indies. **2.** A member of any of the aboriginal peoples of North America, South America, or the West Indies. **3.** Any of the native languages of the American Indians. **4.** Indus. —*adj.* **1.** Of or relating to India or the East Indies, their culture, or their people. **2.** Of or relating to the aboriginal people of North America, South America, or the West Indies.

**Indian agent** *n.* An official representing the U.S. government in dealings with American Indians, esp. on reservations.

**Indian almond** *n.* A tree, *Terminalia catappa* of tropical Asia, bearing fruit with edible seeds.

**Indian bread** *n.* A plant, as the breadroot, with edible sections used by American Indians for food.

**Indian club** *n.* A bottle-shaped wooden club swung in the hand for gymnastic exercise.

**Indian corn** *n.* CORN[1] 1c.

**Indian file** *n.* Single file.

**Indian giver** *n.* *Informal.* One who gives something as a gift to another and then demands or takes it back.

**Indian hemp** *n.* Hemp.

**Indian licorice** *n.* The rosary pea.

**Indian meal** *n.* CORNMEAL 1.

**Indian paintbrush** *n.* A plant of the genus *Castilleja,* bearing flower spikes surrounded by brightly colored bracts.

**Indian pipe** *n.* A waxy white, occas. pinkish woodland plant, *Monotropa uniflora,* with scalelike leaves and a nodding flower.

**Indian pipe**

**Indian pudding** *n.* A pudding of cornmeal and milk sweetened with molasses.

**Indian red** *n.* An iron oxide used as a paint and cosmetic pigment.

**Indian summer** *n.* **1.** A period of mild weather in late autumn or early winter. **2.** A pleasant, peaceful, or flourishing period occurring at the end of something.

**Indian tobacco** *n.* A poisonous North American plant, *Lobelia inflata,* with light-blue flowers and rounded seedpods.

**Indian turnip** *n.* The jack-in-the pulpit.

**Indian wrestling** *n.* **1.** Arm wrestling. **2.** A form of wrestling in which two opponents lie supine in reversed position with their near arms and near raised legs locked and attempt to force each other's leg down. **3.** A form of wrestling in which two opponents stand facing each other with usu. right hands interlocked and the outsides of their near feet set together and attempt to unbalance each other.

**India paper** *n.* A thin, uncoated, delicate paper of vegetable fiber, used esp. for taking impressions of engravings.

**India rubber** *n.* RUBBER[1] 1.

**In·dic** (ĭn′dĭk) *adj.* **1.** Of or relating to India, its people, or their culture. **2.** Of, relating to, or constituting the Indic languages. —*n.* A branch of the Indo-European language family comprising the languages of the Indian subcontinent and Sri Lanka.

**in·di·can** (ĭn′dĭ-kăn′) *n.* [< Lat. *indicum,* indigo.] **1.** A potassium salt, $C_8H_6NOSOK$, occurring in sweat and urine and resulting from the conversion of tryptophan to indole by intestinal bacteria. **2.** A glucoside, $C_{14}H_{17}NO_6$, occurring in the indigo plant.

**in·di·cant** (ĭn′dĭ-kənt) *n.* Something that indicates.

**in·di·cate** (ĭn′dĭ-kāt′) *vt.* **-cat·ed, -cat·ing, -cates.** [Lat. *indicare, indicat-,* to show : *in-,* in + *dicare,* to proclaim.] **1.** To show or point out <*indicate* the quickest route> **2.** To serve as a sign, symptom, or token of : SIGNIFY <"The cracking and booming of the ice *indicate* a change of temperature" —Thoreau> **3.** To suggest or demonstrate the need, expedience, or advisability of <A sagging roof *indicates* immediate repairs.> **4.** To express briefly <*indicated* our hopes> **—in'di·ca·to'ry** (-kə-tôr'ē, -tōr'ē) *adj.*

**in·di·ca·tion** (ĭn'dĭ-kā'shən) *n.* **1.** An act of indicating. **2. a.** Something that indicates. **b.** Something indicated as required or expedient. **3.** The degree indicated by a measuring instrument.

**in·dic·a·tive** (ĭn-dĭk'ə-tĭv) *adj.* **1.** Serving to indicate <attitude *indicative* of disappointment> **2.** Relating to, designating, or being a verb mood for indicating that the denoted act or condition is an objective fact. *—n.* **1.** The indicative mood. **2.** A verb in the indicative mood. **—in·dic'a·tive·ly** *adv.*

**in·di·ca·tor** (ĭn'dĭ-kā'tər) *n.* **1.** One that indicates, as: **a.** A pointer or index. **b.** An instrument, as a meter or a gauge for monitoring the operation or condition of a physical system, as an engine, furnace, electrical network, or reservoir. **c.** The needle, dial, or other registering device on such an instrument. **2.** *Chem.* A substance, as litmus or phenolphthalein, that indicates the presence, absence, or concentration of a substance or the degree of reaction between two or more substances by means of a characteristic change, esp. in color. **3.** Any of various statistical values that collectively indicate the stability of an economic system.

**in·di·ces** (ĭn'dĭ-sēz') *n. var. pl. of* INDEX.

**in·di·cia** (ĭn-dĭsh'ə, -dĭsh'ē-ə) *pl.n.* [Lat., pl. of *indicium*, sign < *index*, index.] **1.** Identifying marks. **2.** Markings on bulk mailings used as a substitute for stamps or cancellations.

**in·dict** (ĭn-dīt') *vt.* **-dict·ed, -dict·ing, -dicts.** [Alteration of ME *enditen*, to accuse, write a document < AN *enditer*, to indite.] **1.** To accuse of a crime or other offense : CHARGE. **2.** *Law.* To make a formal accusation or indictment against by the findings of a jury, esp. a grand jury. **—in·dict'a·ble** *adj.* **—in'dict·ee'** (ĭn'dī-tē') *n.* **—in·dict'er, in·dict'or** *n.*

**in·dic·tion** (ĭn-dĭk'shən) *n.* [ME *indiccioun* < LLat. *indictio*, proclamation, period of 15 years < *indicere*, to proclaim. —see INDITE.] A 15-year cycle used as a chronological unit in ancient Rome and in some medieval systems.

**in·dict·ment** (ĭn-dīt'mənt) *n.* **1.** The act of indicting or state of being indicted. **2.** *Law.* A written statement charging a party with the commission of a crime or other offense, drawn up by a prosecuting attorney and found and presented by a grand jury.

**in·dif·fer·ent** (ĭn-dĭf'ər-ənt, -dĭf'rənt) *adj.* [ME < OFr. < Lat. *indifferens* : *in-*, not + *differens*, different.] **1.** Marked by a lack of bias. **2.** Not mattering one way or the other. **3.** Having no marked feeling one way or the other : without a preference. **4.** Having no particular interest or concern : APATHETIC. **5.** Neither too much nor too little : MODERATE. **6.** Neither good nor bad : MEDIOCRE. **7.** Neither right nor wrong. **8.** Not active or involved : NEUTRAL. **9.** *Biol.* Undifferentiated, as cells or tissue. **—in·dif·fer·ence** (-dĭf'ər-əns, -dĭf'rəns) *n.* **—in·dif'fer·ent·ly** *adv.*

**in·dif·fer·ent·ism** (ĭn-dĭf'ər-ən-tĭz'əm, -dĭf'rən-) *n.* The belief that religions are all of like validity. **—in·dif'fer·ent·ist** *n.*

**in·di·gen** (ĭn'dĭ-jən, -jĕn') *also* **in·di·gene** (-jēn') *n.* [Lat. *indigena.*] One native or indigenous to an area.

**in·di·gence** (ĭn'dĭ-jəns) *n.* Poverty.

**in·dig·e·nous** (ĭn-dĭj'ə-nəs) *adj.* [LLat. *indigenus* < *indigena*, a native.] **1.** Living or occurring naturally in a specific area or environment : NATIVE. **2.** Intrinsic : innate. **—in·dig'e·nous·ly** *adv.* **—in·dig'e·nous·ness** *n.*

**in·di·gent** (ĭn'dĭ-jənt) *adj.* [ME < OFr. < Lat. *indigens*, pr.part. of *indigēre*, to need : *indu*, in + *egēre*, to lack.] **1.** Without the means of subsistence : IMPOVERISHED. **2.** *Archaic.* Lacking or deficient. *—n.* A destitute or needy person. **—in'di·gent·ly** *adv.*

**in·di·gest·ed** (ĭn'dĭ-jĕs'tĭd, -dī-) *adj. Archaic.* **1. a.** Not carefully thought over or considered. **b.** Shapeless or chaotic. **2.** Not digested.

**in·di·gest·i·ble** (ĭn'dĭ-jĕs'tə-bəl, -dī-) *adj.* Difficult or impossible to digest. **—in'di·gest'i·bil'i·ty** *n.* **—in'di·gest'i·bly** *adv.*

**in·di·ges·tion** (ĭn'dĭ-jĕs'chən, -dī-) *n.* **1.** Inability to digest something, esp. food. **2.** Discomfort or illness caused by indigestion.

**in·dign** (ĭn-dīn') *adj.* [ME *indigne* < OFr. < Lat. *indignus* : *in-*, not + *dignus*, worthy.] *Obs.* **1.** Unworthy. **2.** Shameful : disgraceful.

**in·dig·nant** (ĭn-dĭg'nənt) *adj.* Marked by or filled with indignation. **—in·dig'nant·ly** *adv.*

**in·dig·na·tion** (ĭn'dĭg-nā'shən) *n.* [ME *indignacioun* < Lat. *indignatio* < *indignari*, to regard as unworthy : *in-*, not + *dignus*, worthy.] Anger aroused by one that is unjust, mean, or unworthy.

**in·dig·ni·ty** (ĭn-dĭg'nĭ-tē) *n., pl.* **-ties.** [Lat. *indignitas* < *indignus*, unworthy. —see INDIGN.] **1. a.** Humiliating, degrading, or abusive treatment. **b.** Something that offends one's pride or sense of dignity : AFFRONT. **2.** *Obs.* The lack of dignity or honor.

**in·di·go** (ĭn'dĭ-gō') *n., pl.* **-gos** or **-goes.** [Sp. *indigo* < Lat. *indicum* < Gk. *indikon (pharmakon)*, Indian (dye) < *India*, India.] **1. a.** A plant of the genus *Indigofera*, often yielding a blue dyestuff. **b.** Any of various plants similar or related to the indigo. **2.** A blue dye obtained from indigo or other plants or produced synthetically. **3.** Dark blue to grayish purplish blue.

**indigo bunting** *n.* A small bird, *Passerina cyanea* of North and Central America, the male of which has deep-blue plumage.

**indigo snake** *n.* A nonvenomous bluish-black snake, *Drymarchon corais* of the southern United States and northern Mexico.

**in·dig·o·tin** (ĭn-dĭg'ə-tĭn, ĭn'dĭ-gō'-) *n.* [INDIGO + -IN.] A dark-blue crystalline compound, $C_{16}H_{10}N_2O_2$, the primary coloring matter of indigo.

**in·di·rect** (ĭn'dĭ-rĕkt', -dī-) *adj.* **1.** Not proceeding straight to a destination. **2. a.** Not straight to the point, as in speaking : CIRCUMLOCUTORY. **b.** Not forthright and candid : DEVIOUS. **3.** Not directly planned for : SECONDARY <*indirect* results> **—in'di·rect'ly** *adv.* **—in'di·rect'ness** *n.*

☆ ***syns:*** INDIRECT, CIRCUITOUS, CIRCULAR, ROUNDABOUT *adj. core meaning* : not going straight to a destination or mark <an *indirect* route> ***ant:*** direct

**indirect discourse** *n.* Discourse reporting the words of another with consequent grammatical changes to conform the reported statement to the sentence in which it is included.

**in·di·rec·tion** (ĭn'dĭ-rĕk'shən, -dī-) *n.* **1.** The quality or state of being indirect. **2.** Lack of direction : AIMLESSNESS. **3.** Lack of straightforwardness : DEVIOUSNESS.

**indirect lighting** *n.* Illumination by reflected or diffused light.

**indirect object** *n.* A grammatical object indirectly affected by the action of a verb, as *me* in *sent me a telegram.*

**indirect tax** *n.* A tax levied on persons who ultimately pass on the burden of the tax to others, esp. a tax on goods passed on to the consumer in the form of higher prices.

**in·dis·creet** (ĭn'dĭ-skrēt') *adj.* Lacking discretion : INJUDICIOUS. **—in'dis·creet'ly** *adv.* **—in'dis·creet'ness** *n.*

**in·dis·crete** (ĭn'dĭ-skrēt') *adj.* Not divided or divisible into separate parts : UNIFIED.

**in·dis·cre·tion** (ĭn'dĭ-skrĕsh'ən) *n.* **1.** Lack of discretion : INJUDICIOUSNESS. **2.** An indiscreet act or remark.

**in·dis·crim·i·nate** (ĭn'dĭ-skrĭm'ə-nĭt) *adj.* **1.** Lacking in discrimination <*indiscriminate* application of force> **2.** Haphazard : random <*indiscriminate* errors> **3.** Not sorted out : CONFUSED. **4.** Not properly restrained : PROMISCUOUS. **—in'dis·crim'i·nate·ly** *adv.* **—in'dis·crim'i·nate·ness** *n.*

**in·dis·crim·i·na·tion** (ĭn'dĭ-skrĭm'ə-nā'shən) *n.* The quality or state of being indiscriminate. **—in'dis·crim'i·na'tive** *adj.*

**in·dis·pen·sa·ble** (ĭn'dĭ-spĕn'sə-bəl) *adj.* **1.** Incapable of being dispensed with : ESSENTIAL. **2.** Incapable of being escaped : INEVITABLE. *—n.* One that is indispensable. **—in'dis·pen'sa·bil'i·ty, in'dis·pen'sa·ble·ness** *n.* **—in'dis·pen'sa·bly** *adv.*

**in·dis·pose** (ĭn'dĭ-spōz') *vt.* **-posed, -pos·ing, -pos·es.** **1.** To make averse : DISINCLINE. **2.** To make unfit : DISQUALIFY. **3.** To cause to be or feel mildly ill : SICKEN.

**in·dis·posed** (ĭn'dĭ-spōzd') *adj.* **1.** Mildly ill. **2.** Disinclined : averse.

**in·dis·po·si·tion** (ĭn-dĭs'pə-zĭsh'ən) *n.* **1.** Disinclination : aversion. **2.** A minor ailment.

**in·dis·put·a·ble** (ĭn'dĭ-spyo͞o'tə-bəl) *adj.* Incapable of being disputed. **—in'dis·put'a·ble·ness** *n.* **—in'dis·put'a·bly** *adv.*

**in·dis·sol·u·ble** (ĭn'dĭ-sŏl'yə-bəl) *adj.* **1.** Impossible to break or undo : BINDING <an *indissoluble* deed> **2.** Incapable of being dissolved, disintegrated, or decomposed. **—in'dis·sol'u·bil'i·ty, in'dis·sol'u·ble·ness** *n.* **—in'dis·sol'u·bly** *adv.*

**in·dis·tinct** (ĭn'dĭ-stĭngkt') *adj.* **1.** Not clearly delineated. **2. a.** Dim : faint. **b.** Vague : unclear. **—in'dis·tinct'ly** *adv.* **—in'dis·tinct'ness** *n.*

**in·dis·tinc·tive** (ĭn'dĭ-stĭngk'tĭv) *adj.* Having no distinctive qualities. **—in'dis·tinc'tive·ly** *adv.* **—in'dis·tinc'tive·ness** *n.*

**in·dis·tin·guish·a·ble** (ĭn'dĭ-stĭng'gwĭ-shə-bəl) *adj.* Not distinguishable, esp.: **a.** Not easily perceptible. **b.** Indistinctive. **—in'dis·tin'guish·a·bil'i·ty, in'dis·tin'guish·a·ble·ness** *n.*

**in·dite** (ĭn-dīt') *vt.* **-dit·ed, -dit·ing, -dites.** [ME *enditen*, to write a document < AN *enditer* < VLat. **indictare* < Lat. *indicere*, to proclaim : *in-*, toward + *dicere*, to say.] **1.** To write : compose. **2.** To put in writing. **3.** *Obs.* To dictate. **—in·dite'ment** *n.* **—in·dit'er** *n.*

**in·di·um** (ĭn'dē-əm) *n.* [IND(IGO) + -IUM.] *Symbol* **In** A soft, malleable, silvery-white metallic element used as a silver-plating for mirrors and in making transistors; atomic number 49; atomic weight 114.82.

**in·di·vid·u·al** (ĭn'də-vĭj'o͞o-əl) *adj.* [ME, single, indivisible < Med. Lat. *individualis* < Lat. *individuus* : *in-*, not + *dividuus*, divisible < *dividere*, to divide.] **1. a.** Of or pertaining to a single human being. **b.** By or for one person <an *individual* serving> **2.** Existing as a distinct entity : SEPARATE <*individual* parts> **3.** Distinguished by specific attributes or identifying traits : DISTINCTIVE <an *individual* style of dancing> **4.** Indivisible as an entity : INSEPARABLE. *—n.* **1. a.** A human being regarded separately from a group or from society. **b.** An organism as distinguished from a group or colony. **2.** A particular person <a lucky *individual*> ***usage:*** Careful writers and stylists avoid the use of the term *individual* as a substitute for *person.* **—in'di·vid'u·al·ly** *adv.*

**in·di·vid·u·al·ism** (ĭn'də-vĭj'o͞o-ə-lĭz'əm) *n.* **1.** Individuality. **2.** The assertion of one's own will and personality. **3. a.** The theory that one should have freedom in one's economic pursuits and should succeed by one's own initiative. **b.** The practice of this theory. **4.** The doctrine that the interests of the individual should have preference over the interests of the state or social group.

**in·di·vid·u·al·ist** (ĭn'də-vĭj'o͞o-ə-lĭst) *n.* **1.** One who asserts individuality by independence of action or thought. **2.** An advocate of individualism. **—in'di·vid'u·al·is'tic** *adj.* **—in'di·vid'u·al·is'ti·cal·ly** *adv.*

**in·di·vid·u·al·i·ty** (ĭn'də-vĭj'o͞o-ăl'ĭ-tē) *n., pl.* **-ties. 1.** The quality of being individual : DISTINCTNESS. **2.** The aggregate of qualities and characteristics that distinguish one from others. **3.** A distinct entity. **4.** *Archaic.* Indivisibility.

**in·di·vid·u·al·ize** (ĭn'də-vĭj'o͞o-ə-līz') *vt.* **-ized, -iz·ing, -iz·es. 1.** To give individuality to. **2.** To consider or deal with individually : PARTICULARIZE. **3.** To modify to suit the needs or wishes of a specific individual. **—in'di·vid'u·al·i·za'tion** *n.*

**in·di·vid·u·ate** (ĭn'də-vĭj'o͞o-āt') *vt.* **-at·ed, -at·ing, -ates. 1.** To individualize. **2.** To form into a distinct and separate entity.

**in·di·vid·u·a·tion** (ĭn'də-vĭj'o͞o-ā'shən) *n.* **1.** The act or process of individuating, esp. the process by which social individuals become differentiated one from the other. **2.** Individuality.

**in·di·vis·i·ble** (ĭn'də-vĭz'ə-bəl) *adj.* **1.** Incapable of being divided. **2.** *Math.* Incapable of being divided exactly. **—in·di·vis'i·bil'i·ty, in'di·vis'i·ble·ness** *n.* **—in'di·vis'i·bly** *adv.*

**Indo-** *pref.* [Gk. < *Indos,* India.] **1.** India : East Indies <*Indochina*> **2.** Indo-European <*Indo-Hittite*>

**In·do-Ar·y·an** (ĭn'dō-âr'ē-ən, -ăr'-) *adj.* **1.** Belonging to or typical of any of the Indo-European-speaking peoples of the Indian subcontinent. **2.** Indo-Iranian. *—n.* **1.** One of the Indo-Aryan peoples. **2.** Indo-Iranian.

**in·doc·ile** (ĭn-dŏs'əl) *adj.* Difficult to control or instruct : UNRULY. **—in'do·cil'i·ty** (ĭn'dŏ-sĭl'ĭ-tē, -dō-) *n.*

**in·doc·tri·nate** (ĭn-dŏk'trə-nāt') *vt.* **-nat·ed, -nat·ing, -nates. 1.** To instruct in a body of doctrine. **2.** To teach to accept a system of thought uncritically <*indoctrinating* the people with false ideas> **—in·doc'tri·na'tion** *n.*

☆ ***syns:*** **1.** INDOCTRINATE, DRILL, INCULCATE, INSTILL *v. core meaning* : to instruct in a body of doctrine or belief <soldiers *indoctrinated* with lectures on patriotism> **2.** INDOCTRINATE, BRAINWASH, PROGRAM, PROPAGANDIZE *v. core meaning* : to teach to accept a system of thought uncritically <cultists *indoctrinated* by promises of immortality>

**In·do-Eu·ro·pe·an** (ĭn'dō-yo͝or'ə-pē'ən) *—n.* **1. a.** A family of languages comprising most of the languages of Europe as well as those of Iran, the Indian subcontinent, and other parts of Asia. **b.** Proto-Indo-European. **2.** A member of a people speaking an Indo-European language. **—In'do-Eu'ro·pe'an** *adj.*

**In·do-Ger·man·ic** (ĭn'dō-jər-măn'ĭk) *n.* Indo-European. **—In'do-Ger·man'ic** *adj.*

**In·do-Hit·tite** (ĭn'dō-hĭt'īt') *n.* **1.** A language family including Indo-European and Anatolian. **2.** The hypothetical parent language of Indo-European and Anatolian.

**In·do-I·ra·ni·an** (ĭn'dō-ĭ-rā'nē-ən) *n.* A subfamily of the Indo-European language family comprising the Indic and Iranian branches. **—In'do-I·ra'ni·an** *adj.*

**in·dole** (ĭn'dōl') *n.* [IND(IGO) + -OLE.] A white crystalline compound, $C_8H_7N$, derived from coal tar and used in perfumery and medicine and as a flavoring.

**in·dole·a·ce·tic acid** (ĭn'dō-lə-sē'tĭk) *n.* A plant hormone, $C_{10}H_9NO_2$, stimulating growth.

**in·dole·am·ine** (ĭn'dō-lăm'ēn, ĭn'dō-lə-mēn') *n.* Any of various derivatives of indole containing an amine group.

**in·do·lent** (ĭn'də-lənt) *adj.* [LLat. *indolens, indolent-,* painless : Lat. *in-,* not + Lat. *dolens,* pr.part. of *dolēre,* to feel pain.] **1. a.** Disinclined to exert oneself : SLOTHFUL. **b.** Conducive to inactivity or laziness : LANGUOROUS. **c.** Marked by indolence <an *indolent* shrug of the shoulders> **2.** *Pathol.* Resulting in little or no pain <an *indolent* tumor> **—in'do·lence** *n.* **—in'do·lent·ly** *adv.*

**in·do·meth·a·cin** (ĭn'dō-mĕth'ə-sĭn) *n.* [INDO(LE) + METH- + AC(ETIC ACID) + -IN.] An anti-inflammatory and analgesic drug, $C_{19}H_{16}ClNO_4$, used for treating rheumatoid arthritis.

**in·dom·i·ta·ble** (ĭn-dŏm'ĭ-tə-bəl) *adj.* [LLat. *indomitabilis,* untameable : Lat. *in-,* not + *domitare,* to tame, freq. of *domare,* to subdue.] Incapable of being subdued or vanquished : UNCONQUERABLE. **—in·dom'i·ta·ble·ness** *n.* **—in·dom'i·ta·bly** *adv.*

**In·do·ne·sian** (ĭn'də-nē'zhən, -shən) *n.* **1.** A native or resident of the Republic of Indonesia. **2.** A member of a hypothetical non-Malay race of Indonesia, Malaysia, and the Philippines, with both Mongoloid and Polynesian characteristics. **3. a.** A subfamily of Austronesian including Malay, Tagalog, and the languages of Indonesia. **b.** Bahasa Indonesia. *—adj.* Of or relating to Indonesia, its people, or their language.

**in·door** (ĭn'dôr', -dōr') *adj.* **1.** Of, relating to, or located inside a house or other building <an *indoor* tennis court> **2.** Carried on within doors <an *indoor* game>

**in·doors** (ĭn-dôrz', -dōrz') *adv.* In or into a house or other building.

**in·dorse** (ĭn-dôrs') *v. var. of* ENDORSE.

**In·dra** (ĭn'drə) *n.* [Skt. *Indraḥ.*] A chief Vedic deity associated with rain and thunder.

**in·draft** (ĭn'drăft', -dräft') *n.* **1.** A drawing or pulling inward. **2.** An inward current or flow, as of air.

**in·drawn** (ĭn'drôn') *adj.* **1.** Drawn in. **2.** Marked by reserve.

**in·du·bi·ta·ble** (ĭn-do͞o'bĭ-tə-bəl, -dyo͞o'-) *adj.* Too apparent to be doubted : UNQUESTIONABLE. **—in·du'bi·ta·bly** *adv.*

**in·duce** (ĭn-do͞os', -dyo͞os') *vt.* **-duced, -duc·ing, -duc·es.** [ME *inducen* < Lat. *inducere,* to bring in : *in-,* in + *ducere,* to lead.] **1.** To lead or move by persuasion or influence <*induced* us to make the trip> **2. a.** To bring about : CAUSE. **b.** To arouse by stimulating. **3.** To infer by inductive reasoning. **4.** *Physics.* To produce (an electric current or magnetic effect) by induction. **—in·duc'er** *n.* **—in·duc'i·ble** *adj.*

**in·duce·ment** (ĭn-do͞os'mənt, -dyo͞os'-) *n.* **1.** The act or process of inducing. **2.** Something that induces or leads to action : MOTIVE. **3.** *Law.* An introductory or background statement explaining the allegations in a legal proceeding.

**in·duct** (ĭn-dŭkt') *vt.* **-duct·ed, -duct·ing, -ducts.** [ME *inducten* < Lat. *inductus,* p.part. of *inducere,* to bring in. —see INDUCE.] **1. a.** To install ceremoniously or formally in an office or position. **b.** To admit as a member : INITIATE. **c.** To admit to military service. **2.** *Physics.* To induce. **—in·duc·tee'** *n.*

**in·duc·tance** (ĭn-dŭk'təns) *n.* A circuit element, usu. a conducting coil, in which electromagnetic induction generates electromotive force.

**in·duc·tion** (ĭn-dŭk'shən) *n.* **1.** The act of inducting or of being inducted. **2.** *Elect.* **a.** The generation of electromotive force in a closed circuit by a varying magnetic flux through the circuit. **b.** The charging of an isolated conducting object by momentarily grounding it while a charged body is nearby. **3. a.** The act or process of deriving general principles from particular instances or facts. **b.** *Math.* A deductive method of proof in which verification of a proposition consists of proving the first case and the case immediately following an arbitrary case for which the proposition is assumed to be correct. **4.** The act of adducing. **5.** *Archaic.* A preface or prologue esp. to a literary composition.

**induction coil** *n.* A transformer, often used in automotive ignition systems, in which an interrupted, low-voltage direct current in the primary is changed into an intermittent, high-voltage current in the secondary.

**in·duc·tive** (ĭn-dŭk'tĭv) *adj.* **1.** Of, pertaining to, or utilizing induction <the *inductive* process> **2.** *Elect.* Of or resulting from inductance <*inductive* reactance> **3.** Causing or influencing : INDUCING. **4.** Introductory. **—in·duc'tive·ly** *adv.* **—in·duc'tive·ness** *n.*

**inductive statistics** *n. (sing. in number).* The branch of statistics dealing with generalizations, predictions, estimations, and decisions from data initially presented.

**in·duc·tor** (ĭn-dŭk'tər) *n.* **1.** One that inducts. **2.** *Elect.* A device that functions by or introduces inductance into a circuit.

**in·due** (ĭn-do͞o', -dyo͞o') *v. var. of* ENDUE.

**in·dulge** (ĭn-dŭlj') *v.* **-dulged, -dulg·ing, -dulg·es.** [Lat. *indulgēre,* to be kind.] *—vt.* **1.** To accede to the desires and whims of, esp. to an excessive degree : HUMOR. **2. a.** To yield to : GRATIFY <*indulge* their desire for a new car> **b.** To allow (oneself) unrestrained gratification. **3.** To grant an ecclesiastical indulgence or dispensation to. *—vi.* To indulge oneself. **—in·dulg'er** *n.*

**in·dul·gence** (ĭn-dŭl'jəns) *n.* **1.** The act of indulging or state of being indulgent. **2.** Something indulged in <A season ticket at the opera is an expensive *indulgence.*> **3. a.** Something granted as a favor or privilege. **b.** Permission to extend the time of payment or performance. **4.** Liberal or lenient treatment : TOLERANCE. **5.** *Rom. Cath. Ch.* The remission of punishment still due for a sin that has been sacramentally absolved. *—vt.* **-genced, -genc·ing, -genc·es.** *Rom. Cath. Ch.* To attach an indulgence to.

**in·dul·gent** (ĭn-dŭl'jənt) *adj.* Displaying, marked by, or given to indulgence : LENIENT. **—in·dul'gent·ly** *adv.*

**in·dult** (ĭn-dŭlt') *n.* [ME < Med. Lat. *indultum* < Lat. *indultus,* p.part. of *indulgēre,* to be kind.] *Rom. Cath. Ch.* A usu. temporary dispensation.

**in·du·pli·cate** (ĭn-do͞o'plĭ-kĭt, -dyo͞o'-) *adj.* [IN- + Lat. *duplicatus,* doubled. —see DUPLICATE.] *Bot.* With the edges turned or folded inward.

**in·du·rate** (ĭn'də-rāt', -dyə-) *v.* **-rat·ed, -rat·ing, -rates.** [Lat. *indurare, indurat-* : *in-* (intensive) + *durus,* hard.] *—vt.* **1.** To make hard. **2.** To make hardy. **3.** To make callous. *—vi.* **1.** To harden. **2.** To become firmly established or fixed. *—adj.* (ĭn'do͞o-rĭt, -dyə-). Obstinate : unfeeling. **—in'du·ra'tion** *n.* **—in'du·ra'tive** *adj.*

**In·dus** (ĭn'dəs) *n.* A constellation in the Southern Hemisphere.

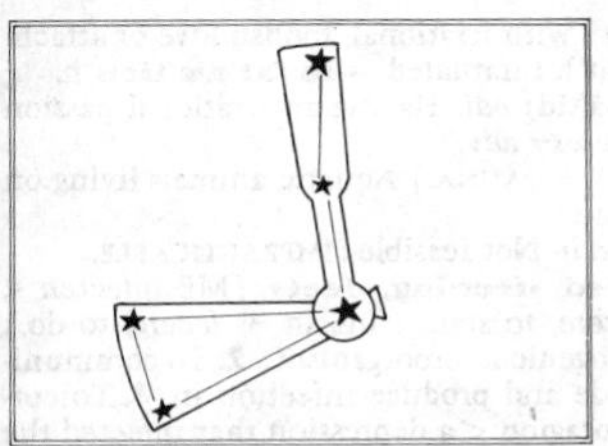
Indus

**in·du·si·um** (ĭn-do͞o′zē-əm, -zhē-, -dyo͞o′-) *n., pl.* **-si·a** (-zē-ə, -zhē-ə) [Lat., tunic < *induere,* to put on.] An enclosing membrane, as that covering the sorus of a fern.
**in·dus·tri·al** (ĭn-dŭs′trē-əl) *adj.* **1.** Of, relating to, or derived from industry. **2.** Having highly developed industries <an *industrial* culture> **3.** Employed, necessary, or used in industry <*industrial* machinery> —*n.* **1.** One employed in industry. **2.** An industrial firm. **3.** A stock or bond issued by an industrial enterprise.
**industrial arts** *pl.n. (sing. in number).* A subject of study in schools aimed at developing manual and technical skills for working with machinery or tools.
**in·dus·tri·al·ism** (ĭn-dŭs′trē-ə-lĭz′əm) *n.* A system in which industries are dominant. **—in·dus′tri·al·ist** *n.*
**in·dus·tri·al·ize** (ĭn-dŭs′trē-ə-līz′) *v.* **-ized, -iz·ing, -iz·es.** —*vt.* To cause to be or become industrial. —*vi.* To become industrial. **—in·dus′tri·al·i·za′tion** *n.*
**industrial park** *n.* An area usu. situated on the outskirts of a city and zoned for a group of industries and businesses.
**industrial psychology** *n.* Psychology applied to problems of industry, as personnel selection, training, and efficiency. **—industrial psychologist** *n.*
**industrial revolution** *n. often* **Industrial Revolution.** Radical social and economic changes, as those in late 18th-cent. England, brought about when extensive mechanization of production systems results in a shift from home manufacturing to large-scale factory production.
**industrial union** *n.* A labor union to which all the workers of a particular industry can belong regardless of their trade.
**in·dus·tri·ous** (ĭn-dŭs′trē-əs) *adj.* **1.** Diligently active : ASSIDUOUS. **2.** *Obs.* Skillful and clever. **—in·dus′tri·ous·ly** *adv.* **—in·dus′tri·ous·ness** *n.*
**in·dus·try** (ĭn′də-strē) *n., pl.* **-tries.** [ME *industrie,* skill < OFr. < Lat. *industria,* diligence.] **1.** The commercial production and sale of goods and services. **2.** A particular branch of manufacture and trade <the electronics *industry*> **3.** Industrial management as distinguished from labor. **4.** Diligence : assiduity.
**in·dwell** (ĭn-dwĕl′) *v.* **-dwelt** (-dwĕlt′), **-dwell·ing, -dwells.** —*vi.* To exist as an energizing inner spirit, force, or principle <creative powers *indwelling* throughout the world> —*vt.* To reside within as an energizing spirit, force, or principle. **—in′dwell′er** *n.*
**in·dwell·ing** (ĭn′dwĕl′ĭng) *adj. Med.* Left within a bodily organ or passage, esp. to facilitate drainage <an *indwelling* catheter>
**-ine**[1] *suff.* [ME < OFr. *-in,* partly < Lat. *-īnus,* adj. suffix, and partly < Lat. *-inus* < Gk. *-inos,* adj. suffix.] **1.** Of or relating to <Benedic*tine*> **2.** Made of : RESEMBLING <opal*ine*>
**-ine**[2] *suff.* [Fr. < Lat. *-ina,* fem. of *-inus,* of or belonging to.] **1.** A chemical substance <az*ine*> **2.** Halogen <brom*ine*> **3.** *or* **-in.** Alkaloid <quin*ine*> **4.** A mixture of compounds <gasol*ine*> **5.** Commercial material <glass*ine*> **6.** *var. of* -IN 1.
**in·e·bri·ant** (ĭn-ē′brē-ənt) *adj.* Intoxicating. —*n.* An intoxicant.
**in·e·bri·ate** (ĭn-ē′brē-āt′) *vt.* **-at·ed, -at·ing, -ates.** [Lat. *inebriare, inebriat-* : *in-* (intensive) + *ebriare,* to intoxicate < *ebrius,* drunk.] **1.** To make drunk : INTOXICATE. **2.** To stupefy or exhilarate as if with alcohol. —*adj.* (ĭn-ē′brē-ĭt). Intoxicated. —*n.* (ĭn-ē′brē-ĭt). An intoxicated person, esp. a drunkard. **—in·e′bri·a′tion** *n.*
**in·e·bri·e·ty** (ĭn′ĭ-brī′ĭ-tē) *n.* Intoxication : drunkenness.
**in·ed·i·ble** (ĭn-ĕd′ə-bəl) *adj.* Not edible.
**in·ef·fa·ble** (ĭn-ĕf′ə-bəl) *adj.* [ME < OFr. < Lat. *ineffabilis* : *in-,* not + *effabilis,* utterable < *effari,* to utter (*ex-,* out + *fari,* to speak).] **1.** Beyond expression : INDESCRIBABLE <*ineffable* happiness> **2.** Not to be uttered : TABOO <the *ineffable* name of God> **—in·ef′fa·bil′i·ty, in·ef′fa·ble·ness** *n.* **—in·ef′fa·bly** *adv.*
**in·ef·face·a·ble** (ĭn′ĭ-fā′sə-bəl) *adj.* Not effaceable : INDELIBLE. **—in′ef·face′a·bil′i·ty** *n.* **—in′ef·face′a·bly** *adv.*
**in·ef·fec·tive** (ĭn′ĭ-fĕk′tĭv) *adj.* **1.** Not causing an intended effect <an *ineffective* petition> **2.** Incapable of performing efficiently <an *ineffective* executive officer> **—in′ef·fec′tive·ly** *adv.* **—in′ef·fec′tive·ness** *n.*
**in·ef·fec·tu·al** (ĭn′ĭ-fĕk′cho͞o-əl) *adj.* **1.** INEFFECTIVE 1. **2.** Lacking the ability to do or perform effectively : POWERLESS. **—in′ef·fec′tu·al′i·ty, in′ef·fec′tu·al·ness** *n.* **—in′ef·fec′tu·al·ly** *adv.*
**in·ef·fi·ca·cious** (ĭn-ĕf′ĭ-kā′shəs) *adj.* INEFFECTIVE 1. **—in·ef′fi·ca′cious·ly** *adv.* **—in·ef′fi·ca′cious·ness** *n.* **—in·ef′fi·ca·cy** (-kə-sē) *n.*
**in·ef·fi·cien·cy** (ĭn′ĭ-fĭsh′ən-sē) *n., pl.* **-cies. 1.** The quality, state, or fact of being inefficient. **2.** Something inefficient.
**in·ef·fi·cient** (ĭn′ĭ-fĭsh′ənt) *adj.* **1.** Not efficient. **2.** Lacking in ability : INCOMPETENT. **3.** Wasteful of time, energy, or materials. **4.** Not causing the intended result. **—in′ef·fi′cient·ly** *adv.*
**in·e·las·tic** (ĭn′ĭ-lăs′tĭk) *adj.* **1.** Not elastic : UNYIELDING. **2.** Not responding or adapting readily to change. **—in′e·las·tic′i·ty** (-ĭ-lă-stĭs′ĭ-tē) *n.*
**in·el·e·gant** (ĭn-ĕl′ĭ-gənt) *adj.* [OFr. < Lat. *inelegans* : *in-,* not + *elegans,* elegant.] **1.** Lacking elegance. **2.** Without refinement or polish. **—in·el′e·gant·ly** *adv.* **—in·el′e·gance** *n.*
**in·el·i·gi·ble** (ĭn-ĕl′ĭ-jə-bəl) *adj.* **1.** Not qualified for election or appointment to an office or position. **2.** Not worthy of being chosen. **—in·el′i·gi·bil′i·ty** *n.* **—in·el′i·gi·ble** *n.* **—in·el′i·gi·bly** *adv.*
**in·el·o·quent** (ĭn-ĕl′ə-kwənt) *adj.* Not eloquent. **—in·el′o·quence** *n.* **—in·el′o·quent·ly** *adv.*
**in·e·luc·ta·ble** (ĭn′ĭ-lŭk′tə-bəl) *adj.* [Lat. *ineluctabilis* : *in-,* not + *eluctari,* to struggle out of (*ex-,* out + *luctari,* to struggle).] Not to be avoided or overcome : INESCAPABLE <*ineluctable* consequences> **—in′e·luc′ta·bil′i·ty** *n.* **—in′e·luc′ta·bly** *adv.*
**in·ept** (ĭn-ĕpt′) *adj.* [Lat. *ineptus* : *in-,* not + *aptus,* suitable. —see APT.] **1.** Not suitable to the situation or occasion : INAPPROPRIATE. **2.** Lacking in judgment, sense, or reason : FOOLISH. **3.** Incompetent. **—in·ep′ti·tude, in·ept′ness** *n.* **—in·ept′ly** *adv.*
**in·e·qual·i·ty** (ĭn′ĭ-kwŏl′ĭ-tē) *n., pl.* **-ties. 1.** An instance or the state of being unequal. **2.** Social or economic disparity. **3.** Lack of regularity : UNEVENNESS. **4.** Variability : changeability. **5.** *Math.* An algebraic statement that a quantity is greater than or less than another quantity.
☆ ***syns:*** INEQUALITY, DISPARITY, DISPROPORTION *n. core meaning* : the condition or fact of being unequal, as in age, rank, or degree <the *inequality* between the rich and the poor> ***ant:*** equality
**in·eq·ui·ta·ble** (ĭn-ĕk′wĭ-tə-bəl) *adj.* Not equitable : UNFAIR. **—in·eq′ui·ta·bly** *adv.*
**in·eq·ui·ty** (ĭn-ĕk′wĭ-tē) *n., pl.* **-ties. 1.** Injustice : unfairness. **2.** An instance of injustice or unfairness.
**in·e·rad·i·ca·ble** (ĭn′ĭ-răd′ĭ-kə-bəl) *adj.* Incapable of being eradicated. **—in′e·rad′i·ca·bly** *adv.*
**in·er·rant** (ĭn-ĕr′ənt) *adj.* Free from errors. **—in·er′ran·cy** *n.*
**in·ert** (ĭn-ûrt′) *adj.* [Lat. *iners, inert-,* inactive : *in-,* not + *ars,* skill.] **1.** Unable to move or act. **2.** Moving or acting very slowly : SLUGGISH. **3.** *Chem.* **a.** Displaying no chemical activity. **b.** Displaying chemical activity only under special or extreme conditions. **—in·ert′ly** *adv.* **—in·ert′ness** *n.*
**in·er·tia** (ĭn-ûr′shə) *n.* [NLat. < Lat., idleness < *iners,* inert.] **1.** *Physics.* The tendency of a body to resist acceleration, as the tendency of a body at rest to remain at rest or of a body in motion to stay in motion in a straight line unless disturbed by an external force. **2.** Resistance to change or motion. **—in·er′tial** *adj.* **—in·er′tial·ly** *adv.*
**inertial frame** *n.* A frame of reference relative to which the Newtonian law of motion, that a mass *m* subjected to a force *F* moves in accordance with the equation $F = ma$ where *a* is the acceleration, is valid.
**inertial guidance** *n.* Guidance in which gyroscopic and accelerometer data are used by a computer to maintain a predetermined course.
**inertial platform** *n.* The devices used in inertial guidance and their mounting platform.
**inertia welding** *n.* A welding of metals caused by the heat of friction from pressing a spinning metallic piece against a stationary one.
**in·es·cap·a·ble** (ĭn′ĭ-skā′pə-bəl) *adj.* Inevitable : unavoidable. **—in′es·cap′a·bly** *adv.*
**in·es·sen·tial** (ĭn′ĭ-sĕn′shəl) *adj.* **1.** Not required : UNESSENTIAL. **2.** Without essence. **—in·es·sen′tial** *n.* **—in′es·sen′ti·al′i·ty** *n.*
**in·es·ti·ma·ble** (ĭn-ĕs′tə-mə-bəl) *adj.* **1.** Incapable of being estimated or computed : INDETERMINABLE <floods causing *inestimable* damage> **2.** Of incalculable value. **—in·es′ti·ma·bly** *adv.*
**in·ev·i·ta·ble** (ĭn-ĕv′ĭ-tə-bəl) *adj.* Incapable of being avoided or prevented. **—in·ev′i·ta·bil′i·ty** *n.* **—in·ev′i·ta·bly** *adv.*
**in·ex·act** (ĭn′ĭg-zăkt′) *adj.* **1.** Not true, accurate, or precise. **2.** Not rigorous. **—in′ex·act′ly** *adv.* **—in′ex·act′ness** *n.*
**in·ex·act·i·tude** (ĭn′ĭg-zăk′tĭ-to͞od′, -tyo͞od′) *n.* Lack of exactitude.
**in·ex·cus·a·ble** (ĭn′ĭk-skyo͞o′zə-bəl) *adj.* That cannot be excused. **—in′ex·cus′a·ble·ness** *n.* **—in′ex·cus′a·bly** *adv.*
**in·ex·haust·i·ble** (ĭn′ĭg-zô′stə-bəl) *adj.* **1.** Incapable of being used up or consumed. **2.** Incapable of being tired. **—in′ex·haust′i·bil′i·ty, in′ex·haust′i·ble·ness** *n.* **—in′ex·haust′i·bly** *adv.*
**in·ex·is·tent** (ĭn′ĭg-zĭs′tənt) *adj.* Lacking existence. **—in′ex·is′tence** *n.*
**in·ex·o·ra·ble** (ĭn-ĕk′sər-ə-bəl) *adj.* [Lat. *inexorabilis* : *in,* not + *exorabilis,* pliant < *exorare,* to prevail upon (*ex-,* intensive + *orare,* to argue).] Not capable of being persuaded : UNYIELDING. **—in·ex′o·ra·bil′i·ty, in·ex′o·ra·ble·ness** *n.* **—in·ex′o·ra·bly** *adv.*

**in·ex·pe·di·ent** (ĭn′ĭk-spē′dē-ənt) *adj.* Inadvisable. **—in′ex·pe′di·ence, in′ex·pe′di·en·cy** *n.* **—in′ex·pe′di·ent·ly** *adv.*
**in·ex·pen·sive** (ĭn′ĭk-spĕn′sĭv) *adj.* Not high in price : CHEAP. **—in′ex·pen′sive·ly** *adv.* **—in′ex·pen′sive·ness** *n.*
**in·ex·pe·ri·ence** (ĭn′ĭk-spîr′ē-əns) *n.* **1.** Lack of experience. **2.** Lack of the knowledge gained from experience.
**in·ex·pe·ri·enced** (ĭn′ĭk-spîr′ē-ənst) *adj.* Lacking experience.
☆ ***syns:*** INEXPERIENCED, GREEN, INEXPERT, RAW, UNPRACTICED, UNSEASONED, UNTRIED, UNVERSED *adj. core meaning* : lacking experience and the knowledge gained from it <Flying a plane is not for the *inexperienced* person.> ***ant:*** experienced
**in·ex·pert** (ĭn-ĕk′spûrt′) *adj.* Not experienced or expert : UNSKILLED. **—in·ex′pert′ly** *adv.* **—in·ex′pert′ness** *n.*
**in·ex·pi·a·ble** (ĭn-ĕk′spē-ə-bəl) *adj.* **1.** Not capable of being expiated or atoned for. **2.** *Obs.* Implacable. **—in·ex′pi·a·bly** *adv.*
**in·ex·plain·a·ble** (ĭn′ĭk-splā′nə-bəl) *adj.* Inexplicable. **—in′ex·plain′a·bly** *adv.*
**in·ex·pli·ca·ble** (ĭn-ĕk′splĭ-kə-bəl, ĭn′ĭk-splĭk′ə-bəl) *adj.* Incapable of being explained or interpreted. **—in·ex′pli·ca·bil′i·ty, in·ex′pli·ca·ble·ness** *n.* **—in·ex′pli·ca·bly** *adv.*
**in·ex·plic·it** (ĭn′ĭk-splĭs′ĭt) *adj.* Not explicit : VAGUE.
**in·ex·press·i·ble** (ĭn′ĭk-sprĕs′ə-bəl) *adj.* Incapable of being expressed : INDESCRIBABLE. **—in′ex·press′i·bil′i·ty, in′ex·press′i·ble·ness** *n.* **—in′ex·press′i·bly** *adv.*
**in·ex·pug·na·ble** (ĭn′ĭk-spŭg′nə-bəl, -spyōō′nə-) *adj.* Not capable of being attacked or overthrown : IMPREGNABLE. **—in′ex·pug′na·bil′i·ty** *n.* **—in′ex·pug′na·bly** *adv.*
**in·ex·ten·si·ble** (ĭn′ĭk-stĕn′sə-bəl) *adj.* Not extensible.
**in ex·ten·so** (ĭn ĕk-stĕn′sō) *adv.* [Lat.] At full length.
**in·ex·tin·guish·a·ble** (ĭn′ĭk-stĭng′gwĭ-shə-bəl) *adj.* Not capable of being extinguished. **—in′ex·tin′guish·a·bly** *adv.*
**in·ex·tir·pa·ble** (ĭn′ĭk-stûr′pə-bəl) *adj.* Incapable of being destroyed or eradicated.
**in ex·tre·mis** (ĭn ĕk-strē′mĭs) *adv.* [Lat., in extreme (circumstances).] At the point of death.
**in·ex·tri·ca·ble** (ĭn-ĕk′strĭ-kə-bəl) *adj.* **1. a.** Incapable of being untangled or untied. **b.** Too complicated or intricate to solve. **2.** That is impossible to escape or get free of <an *inextricable* web of circumstances> **—in·ex′tri·ca·bil′i·ty, in·ex′tri·ca·ble·ness** *n.* **—in·ex′tri·ca·bly** *adv.*
**in·fal·li·ble** (ĭn-făl′ə-bəl) *adj.* [Fr. < Med. Lat. *infallibilis* : Lat. *in-*, not + *fallibilis*, fallible.] **1.** Incapable of erring <an *infallible* leader> **2.** Incapable of failing : CERTAIN <an *infallible* method> **3.** *Rom. Cath. Ch.* Incapable of error in promulgating doctrine on faith or morals. **—in·fal′li·bil′i·ty, in·fal′li·ble·ness** *n.* **—in·fal′li·bly** *adv.*
**in·fa·mous** (ĭn′fə-məs) *adj.* [ME < Med. Lat. *infamosus* < Lat. *infamis* : *in-*, not + *fama*, renown, fame.] **1.** Having a bad reputation : NOTORIOUS. **2.** Causing or deserving infamy. **3.** *Law.* Convicted of a crime, as treason or felony, that results in infamy. **—in′fa·mous·ly** *adv.* **—in′fa·mous·ness** *n.*
**in·fa·my** (ĭn′fə-mē) *n., pl.* **-mies.** [ME *infamie*, dishonor < OFr. < Lat. *infamia* < *infamis*, infamous.] **1.** Evil fame or reputation. **2.** The state of being infamous. **3.** A publicly known criminal or evil act.
**in·fan·cy** (ĭn′fən-sē) *n., pl.* **-cies. 1.** The state or period of being an infant. **2.** An early stage of existence. **3.** *Law.* The state or period of being a minor.
**in·fant** (ĭn′fənt) *n.* [ME < OFr. *enfant* < Lat. *infans* : *in-*, not + *fans*, pr.part. of *fari*, to speak.] **1.** A child in the earliest period of its life. **2.** *Law.* A person under the legal age of majority : MINOR.
**in·fan·ta** (ĭn-făn′tə, -fän′-) *n.* [Sp. and Port., fem. of *infante*, infante.] A daughter of a Spanish or Portuguese monarch.
**in·fan·te** (ĭn-făn′tē, -fän′tā) *n.* [Sp. and Port. < Lat. *infans*, infant.] A son of a Spanish or Portuguese monarch other than the heir to the throne.
**in·fan·ti·cide** (ĭn-făn′tĭ-sīd′) *n.* [LLat. *infanticidium*, the killing of a child, and *infanticida*, killer of a child : Lat. *infans*, infant + Lat. *caedere*, to kill.] **1.** The killing of an infant. **2.** One who kills an infant.
**in·fan·tile** (ĭn′fən-tīl′, -tĭl) *adj.* [Fr. < Lat. *infantilis* < *infans*, infant.] **1.** Of or pertaining to infants or infancy. **2.** Lacking maturity : CHILDISH <*infantile* behavior by so-called mature adults>
**infantile autism** *n.* AUTISM 2.
**infantile paralysis** *n.* Poliomyelitis.
**in·fan·til·ism** (ĭn′fən-tə-lĭz′əm, ĭn-făn′tə-) *n.* A state of arrested development in an adult, marked by a retention of infantile mentality accompanied by stunted growth and sexual immaturity.
**in·fan·try** (ĭn′fən-trē) *n., pl.* **-tries.** [Fr. *infanterie* < Ital. *infanteria* < *infante*, youth, foot soldier < Lat. *infans*, infant.] The branch of an army composed of units trained to fight on foot.
**in·fan·try·man** (ĭn′fən-trē-mən) *n.* A soldier in the infantry.
**infant school** *n. Chiefly Brit.* A kindergarten.
**in·farct** (ĭn′färkt′, ĭn-färkt′) *n.* [Lat. *infarctus*, p.part. of *infarcire*, to cram : *in-*, in + *farcire*, to stuff.] An area of necrotic tissue caused by insufficient blood supply. **—in·farct′ed** *adj.* **—in·farc′tion** *n.*
**in·fat·u·ate** (ĭn-făch′ōō-āt′) *vt.* **-at·ed, -at·ing, -ates.** [Lat. *infatuare*, *infatuat-* : *in-* (causative) + *fatuus*, foolish.] **1.** To cause to behave foolishly. **2.** To inspire with irrational, foolish love or attachment. —*adj.* (ĭn-făch′ōō-ĭt, -āt′). Infatuated. **—in·fat′u·a′tion** *n.*
**in·fat·u·at·ed** (ĭn-făch′ōō-ā′tĭd) *adj.* Having an irrational passion or attraction. **—in·fat′u·at′ed·ly** *adv.*
**in·fau·na** (ĭn′fô′nə) *n.* [IN-2 + FAUNA.] Aquatic animals living on the substrate of a body of water.
**in·fea·si·ble** (ĭn-fē′zə-bəl) *adj.* Not feasible : IMPRACTICABLE.
**in·fect** (ĭn-fĕkt′) *vt.* **-fect·ed, -fect·ing, -fects.** [ME *infecten* < Lat. *infectus*, p.part. of *inficere*, to stain : *in-*, in + *facere*, to do.] **1.** To contaminate with pathogenic microorganisms. **2.** To communicate a disease to. **3.** To invade and produce infection in. **4.** To corrupt. **5.** To affect as if by contagion <a depression that *infected* the entire group>
**in·fec·tion** (ĭn-fĕk′shən) *n.* **1. a.** Invasion by pathogenic microorganisms of a bodily part in which conditions are favorable for growth, production of toxins, and resulting injury to tissue. **b.** An instance of such invasion. **c.** The pathological state caused by such invasion. **d.** An agent or substance responsible for such invasion. **2.** An infectious disease. **3.** Communication of an emotion or attitude by contact or example.
**in·fec·tious** (ĭn-fĕk′shəs) *adj.* **1.** Capable of causing infection. **2.** Capable of being transmitted by infection without actual contact. **3.** Caused by a microorganism. **4.** Easily communicated <*infectious* joy> **—in·fec′tious·ly** *adv.* **—in·fec′tious·ness** *n.*
**infectious mononucleosis** *n.* MONONUCLEOSIS 2.
**in·fec·tive** (ĭn-fĕk′tĭv) *adj.* Capable of producing infection : INFECTIOUS. **—in·fec′tive·ness, in′fec·tiv′i·ty** *n.*
**in·fe·lic·i·tous** (ĭn′fĭ-lĭs′ĭ-təs) *adj.* **1.** Not happy : UNFORTUNATE. **2.** Inopportune : inappropriate. **—in′fe·lic′i·tous·ly** *adv.*
**in·fe·lic·i·ty** (ĭn′fĭ-lĭs′ĭ-tē) *n., pl.* **-ties.** [ME *infelicite*, unhappiness < Lat. *infelicitas* < *infelix*, unhappy : *in-*, not + *felix*, happy.] **1.** The quality or state of being infelicitous. **2.** Something inopportune or inappropriate.
**in·fer** (ĭn-fûr′) *v.* **-ferred, -fer·ring, -fers.** [OFr. *inferer* < Lat. *inferre*, to bring in, deduce : *in-*, in- + *ferre*, to bear.] —*vt.* **1.** To conclude from certain premises or evidence. **2.** To have as a logical consequence. **3.** To lead to as a result or conclusion. —*vi.* To draw inferences. **—in·fer′a·ble** *adj.* **—in·fer′a·bly** *adv.* **—in·fer′rer** *n.*
**in·fer·ence** (ĭn′fər-əns) *n.* **1. a.** The act or process of inferring or deriving a conclusion from facts or premises. **b.** The premises and conclusions of an act of inferring. **2.** Something inferred, esp. a conclusion derived by inference.
**in·fer·en·tial** (ĭn′fə-rĕn′shəl) *adj.* **1.** Of, pertaining to, or involving inference. **2.** Derived or capable of being derived by inference. **—in′fer·en′tial·ly** *adv.*
**in·fe·ri·or** (ĭn-fîr′ē-ər) *adj.* [ME < Lat., comp. of *inferus*, low.] **1.** Situated under or beneath. **2.** Low or lower in order, degree, or rank. **3.** Low or lower in quality, value, or estimation. **4.** *Bot.* Located below the perianth and other floral parts. —Used of an ovary. **5.** Set below a usual line of printing type. **—in·fe′ri·or** *n.* **—in·fe′ri·or′i·ty** (-ôr′ĭ-tē, -ŏr′-) *n.*
**inferiority complex** *n.* A sense of inadequacy or a tendency to self-diminishment.
**in·fer·nal** (ĭn-fûr′nəl) *adj.* [ME < OFr. < LLat. *infernalis* < *infernus*, hell < Lat., lower, underground.] **1.** Of or pertaining to a lower world of the dead. **2.** Of or pertaining to hell. **3. a.** Devilish : fiendish. **b.** Damnable : abominable. **—in·fer′nal·ly** *adv.*
**infernal machine** *n.* An explosive device malevolently designed to harm or destroy.
**in·fer·no** (ĭn-fûr′nō) *n., pl.* **-nos.** [Ital., hell < LLat. *infernus*. —see INFERNAL.] **1.** A place or condition suggestive of hell, esp. causing extreme suffering or having intolerable heat. **2.** A hellish place.
**in·fer·tile** (ĭn-fûr′tl) *adj.* Not fertile : UNPRODUCTIVE. **—in′fer·til′i·ty** (ĭn′fər-tĭl′ĭ-tē) *n.*
**in·fest** (ĭn-fĕst′) *vt.* **-fest·ed, -fest·ing, -fests.** [ME *infesten*, to distress < OFr. *infester* < Lat. *infestare*, to attack < *infestus*, hostile.] To spread in or overrun in large numbers so as to be harmful or unpleasant. **—in′fes·ta′tion** *n.* **—in·fest′er** *n.*
**in·fi·del** (ĭn′fĭ-dəl, -dĕl′) *n.* [ME *infidele*, heathen < OFr. < Lat. *infidelis*, unbelieving : *in-*, not + *fides*, faith.] **1.** One without religious beliefs. **2.** An unbeliever in a religion, esp. Christianity or Islam. **3.** A skeptic.
**in·fi·del·i·ty** (ĭn′fĭ-dĕl′ĭ-tē) *n., pl.* **-ties. 1.** Lack of religious belief. **2.** Lack of fidelity or loyalty : UNFAITHFULNESS. **3. a.** Marital unfaithfulness. **b.** An act of marital unfaithfulness.
**in·field** (ĭn′fēld′) *n.* **1.** A field close to a farmhouse. **2.** *Baseball.* **a.** The area of a baseball field bounded by home plate and first, second, and third bases. **b.** The defensive positions of first base, second base, third base, and shortstop considered as a unit. **3.** The area inside a racetrack or running track.
**in·field·er** (ĭn′fēl′dər) *n. Baseball.* A player in the infield.
**in·fight·ing** (ĭn′fī′tĭng) *n.* **1.** Fighting or boxing at close range. **2.** Contention among members of a group or organization. **—in′fight′er** *n.*

**in·fil·trate** (ĭn-fĭl′trāt′, ĭn′fĭl-) *v.* **-trat·ed, -trat·ing, -trates.** —*vt.* **1.** To cause (e.g., a liquid) to permeate a substance by passing through its interstices or pores. **2.** To permeate by passing a liquid or gas through the interstices of. **3.** To pass (e.g., troops) stealthily into enemy-held territory. **4.** To enter or take up positions in gradually or stealthily. —*vi.* To gain entrance gradually or stealthily. —*n.* A substance that accumulates gradually in bodily tissues. **—in′fil·tra′tion** (-trā′shən) *n.* **—in·fil′tra·tive** (-trə-tĭv) *adj.*

**in·fi·nite** (ĭn′fə-nĭt) *adj.* [ME *infinit* < OFr. < Lat. *infinitus* : *in-*, not + *finitus*, finite.] **1.** Without boundaries or limits. **2.** Immeasurably great, as in extent or duration. **3.** *Math.* **a.** Existing beyond or being greater than any arbitrarily large value. **b.** Unlimited in spatial extent. **c.** Of or relating to a set capable of being put into one-to-one correspondence with a proper subset of itself. —*n.* Something infinite. **—in′fi·nite·ly** *adv.* **—in′fi·nite·ness** *n.*

**in·fin·i·tes·i·mal** (ĭn′fĭn-ĭ-tĕs′ə-məl) *adj.* [< NLat. *infinitesimus*, infinite in rank < Lat. *infinitus*, infinite.] **1.** Immeasurably or incalculably small. **2.** *Math.* Capable of having values arbitrarily close to zero. —*n.* **1.** An infinitesimal amount or quantity. **2.** *Math.* A function with values arbitrarily close to zero. **—in′fin·i·tes′i·mal·ly** *adv.*

**infinitesimal calculus** *n.* Differential and integral calculus.

**in·fin·i·ti·val** (ĭn′fĭn-ĭ-tī′vəl) *adj.* Pertaining to the infinitive.

**in·fin·i·tive** (ĭn-fĭn′ĭ-tĭv) *n.* [LLat. *infinitivus*, unlimited, indefinite < Lat. *infinitus*, infinite.] A verb form functioning as a substantive while retaining certain verbal characteristics, as modification by adverbs, and that in English may be preceded by *to*, as in *To proceed cautiously is not always to be wise* or *I asked them to stay longer*, or may also occur without *to*, as in *I had them read the manuscript.*

**in·fin·i·tude** (ĭn-fĭn′ĭ-to͞od′, -tyo͞od′) *n.* **1.** INFINITY 1. **2.** An infinite quantity, number, or extent.

**in·fin·i·ty** (ĭn-fĭn′ĭ-tē) *n., pl.* **-ties. 1.** The quality or state of being infinite. **2.** Unbounded space, time, or quantity. **3.** An indefinitely large number. **4.** *Math.* The limit that a function $f$ is said to approach at $x = a$ when for $x$ close to $a$, $f(x)$ is larger than any preassigned number.

**in·firm** (ĭn-fûrm′) *adj.* [ME *infirme* < Lat. *infirmus* : *in-*, not + *firmus*, strong, firm.] **1.** Physically weak, esp. from old age : FEEBLE. **2.** Lacking moral firmness : IRRESOLUTE. **3.** Not sound or valid : INSECURE. **—in·firm′ly** *adv.*

**in·fir·ma·ry** (ĭn-fûr′mə-rē) *n., pl.* **-ries.** [Med. Lat. *infirmaria* < Lat. *infirmus*, infirm.] An institution for the care of the sick or injured, esp. a small hospital or dispensary.

**in·fir·mi·ty** (ĭn-fûr′mĭ-tē) *n., pl.* **-ties. 1.** Lack of power : DISABILITY. **2.** Physical debilitation : FRAILTY. **3.** Moral weakness. **4.** A failing or defect in a person's character.

**in·fix** (ĭn-fĭks′) *vt.* **-fixed, -fix·ing, -fix·es.** [Lat. *infigere, infix-* : *in-*, in + *figere*, to fasten.] **1.** To fix into another. **2.** To fix in the mind : INSTILL. **3.** To insert (a morphological element) into the body of a word. —*n.* (ĭn′fĭks′). An inflectional or derivational element appearing in the body of a word, as Tagalog *sinulat*, "written," in which the infix *-in-* appears as the indicator of a passive form that contrasts with the active form *sulat*, "write."

**in·flame** (ĭn-flām′) *v.* **-flamed, -flam·ing, -flames.** [ME *inflaumen* < OFr. *enflammer* < Lat. *inflammare* : *in-* (intensive) + *flammare*, to set on fire < *flamma*, flame.] —*vt.* **1.** To set on fire : KINDLE. **2.** To arouse to strong emotion. **3.** To make more violent : INTENSIFY <"*inflamed* to madness an already savage nature" —Robert Graves> **4.** To produce inflammation in. —*vi.* **1.** To catch fire. **2.** To become excited or aroused. **3.** To be affected by inflammation.

**in·flam·ma·ble** (ĭn-flăm′ə-bəl) *adj.* [Fr. < Med. Lat. *inflammabilis* < Lat. *inflammare*, to inflame.] **1.** Tending to catch fire easily and burn rapidly : FLAMMABLE. **2.** Easily aroused to strong emotion. **—in·flam′ma·bil′i·ty** *n.* **—in·flam′ma·ble** *n.* **—in·flam′ma·bly** *adv.*

▲ word history: Hydrogen was once called the *inflammable* gas not because it does not burn but because it is so easily ignited. *Inflammable* is derived ultimately from the Latin prefix *in-*, "in," and the noun *flamma*, "flame." There is another prefix *in-*, however, which English also borrowed from Latin, that means "not," and the word *inflammable* can be misunderstood as meaning "not capable of burning." In order to eliminate possibly dangerous confusion about the combustibility of various materials, safety officials in the 20th century have adopted the term *flammable*, which had a brief life in the early 19th century, to mean "able to burn." Materials that do not burn are unambiguously labeled *nonflammable.*

**in·flam·ma·tion** (ĭn′flə-mā′shən) *n.* **1.** The act of inflaming or state of being inflamed. **2.** Redness, heat, swelling, and pain caused by irritation, injury, or infection.

**in·flam·ma·to·ry** (ĭn-flăm′ə-tôr′ē, -tōr′ē) *adj.* **1.** Tending to arouse strong emotion, as passion, anger, or violence. **2.** Marked or caused by inflammation. **—in·flam′ma·to′ri·ly** *adv.*

**in·flate** (ĭn-flāt′) *v.* **-flat·ed, -flat·ing, -flates.** [Lat. *inflare, inflat-* : *in-*, in + *flare*, to blow.] —*vt.* **1.** To swell and fill with a gas. **2.** To cause to expand abnormally. **3.** To raise (e.g., prices) to an abnormally high level. —*vi.* To become inflated. **—in·flat′a·ble** *adj.*

**in·flat·ed** (ĭn-flā′tĭd) *adj.* **1.** Distended or expanded by or as if by gas or air. **2.** Unduly increased or puffed up <an *inflated* opinion of yourself> **3.** Full of empty or pretentious language : BOMBASTIC. **4.** Increased or raised to abnormal levels <*inflated* fees> **5.** Hollow and enlarged <*inflated* sepals>

**in·fla·tion** (ĭn-flā′shən) *n.* **1.** The act of inflating or state of being inflated. **2.** An abnormal increase in available currency and credit beyond the proportion of available goods, causing a sharp and continuing rise in price levels. **—in·fla′tion·ar′y** (-shə-nĕr′ē) *adj.*

**in·fla·tion·ist** (ĭn-flā′shə-nĭst) *n.* One who advocates a policy of economic inflation. **—in·fla′tion·ism** *n.*

**in·flect** (ĭn-flĕkt′) *v.* **-flect·ed, -flect·ing, -flects.** [ME *inflecten*, to bend < Lat. *inflectere* : *in-*, intensive + *flectere*, to blend.] —*vt.* **1.** To turn from a course or alignment : BEND. **2.** To alter (the voice) in pitch or tone : MODULATE. **3.** To change (a word) by inflection. —*vi.* To be modified by inflection. —Used of a word. **—in·flec′tive** *adj.* **—in·flec′tor** *n.*

**in·flec·tion** (ĭn-flĕk′shən) *n.* **1.** The act of inflecting or state of being inflected. **2.** An alteration in pitch or tone of the voice. **3. a.** A change in the form of a word indicating grammatical features such as number, person, or tense. **b.** A word element involved in such a change. **c.** An inflected form of a word. **—in·flec′tion·al** *adj.* **—in·flec′tion·al·ly** *adv.*

**in·flexed** (ĭn-flĕkst′) *adj.* [< Lat. *inflexus*, p.part. of *inflectere*, to bend. —see INFLECT.] Bent or curved downward or inward, as petals or sepals.

**in·flex·i·ble** (ĭn-flĕk′sə-bəl) *adj.* **1.** Not flexible : RIGID. **2.** Incapable of being changed : UNALTERABLE. **3.** Adhering firmly to a particular intention or purpose. **—in·flex′i·bil′i·ty, in·flex′i·ble·ness** *n.* **—in·flex′i·bly** *adv.*

☆ ***syns:*** INFLEXIBLE, IMMUTABLE, INALTERABLE, INVARIABLE, IRONBOUND, IRONCLAD, STIFF, UNCHANGEABLE *adj. core meaning* : incapable of changing or of being changed <*inflexible* standards> ***ant:*** flexible

**in·flex·ion** (ĭn-flĕk′shən) *n. Chiefly Brit.* Inflection.

**in·flict** (ĭn-flĭkt′) *vt.* **-flict·ed, -flict·ing, -flicts.** [Lat. *infligere, inflitc-* : *in-*, on + *fligere*, to strike.] **1.** To cause or carry out by aggressive action, as physical assault. **2.** To impose something painful or unwelcome upon. **3.** To afflict. **—in·flict′er, in·flic′tor** *n.* **—in·flic′tion** *n.* **—in·flic′tive** *adj.*

**in-flight** (ĭn′flīt′) *adj.* **1.** Performed while in flight. **2.** Provided for use or enjoyment while in flight <*in-flight* movies>

**in·flo·res·cence** (ĭn′flə-rĕs′əns) *n.* [NLat. *inflorescentia* < LLat. *inflorescere*, to begin to flower : *in-* (intensive) + *florescere*, to begin to blossom, inceptive of *florēre*, to blossom < *flos*, flower.] **1.** *Bot.* A typical arrangement of flowers on a stalk or in a cluster. **2.** A flowering. **—in′flo·res′cent** *adj.*

**in·flow** (ĭn′flō′) *n.* **1.** The act or process of flowing in or into. **2.** Something that flows in or into.

**in·flu·ence** (ĭn′flo͞o-əns) *n.* [ME < OFr. < Med. Lat. *influentia* < Lat. *influens*, pr.part. of *influere*, to flow in : *in-*, in + *fluere*, to flow.] **1.** A power indirectly or intangibly affecting a person or event. **2. a.** Power to sway or affect based on prestige, wealth, ability, or status. **b.** One exercising such power. **c.** An effect produced by such power. **3. a.** An occult ethereal fluid thought to flow from the stars and affect the fate of people. **b.** An occult power thought to emanate from the stars. —*vt.* **-enced, -enc·ing, -enc·es. 1.** To have power over : AFFECT. **2.** To cause a change in the character, thought, or action of. **—in′flu·enc·er** *n.*

**in·flu·ent** (ĭn′flo͞o-ənt, ĭn-flo͞o′-) *adj.* [ME < Lat. *influens*, flowing in. —see INFLUENCE.] Flowing in or into. —*n.* Something that flows in or into, esp. a tributary.

**in·flu·en·tial** (ĭn′flo͞o-ĕn′shəl) *adj.* Having or exerting influence. **—in′flu·en′tial·ly** *adv.*

**in·flu·en·za** (ĭn′flo͞o-ĕn′zə) *n.* [Ital. < Med. Lat. *influentia*, influence.] An acute infectious viral disease marked by inflammation of the respiratory tract, fever, muscular pain, and irritation in the intestinal tract.

**in·flux** (ĭn′flŭks′) *n.* [LLat. *influxus* < Lat., p.part. of *influere*, to flow in. —see INFLUENCE.] INFLOW 1.

**in·fold** (ĭn-fōld′) *v.* **-fold·ed, -fold·ing, -folds.** —*vi.* To fold inward. —*vt.* To enfold. **—in·fold′er** *n.* **—in·fold′ment** *n.*

**in·form** (ĭn-fôrm′) *v.* **-formed, -form·ing, -forms.** [ME *enfourmen* < OFr. *enfourmer* < Lat. *informare* : *in-*, in + *forma*, form.] —*vt.* **1.** To give character or form to. **2.** To inspire or animate with a specific quality or character : IMBUE. **3.** To form or shape (the mind or character) by instruction or training. **4.** To impart information to. —*vi.* To give often incriminating information.

**in·for·mal** (ĭn-fôr′məl) *adj.* **1.** Not following or bound by prescribed regulations or forms <an *informal* debate> **2.** Of, for, or relating to ordinary everyday use : CASUAL <*informal* speech> **3.** More suitable for spoken than written language esp. in business, technical, or official communications. **—in·for′mal·ly** *adv.*

☆ ***syns:*** INFORMAL, CASUAL, EASY, EASYGOING, RELAXED *adj. core meaning* : not bound by rigid standards <an *informal* lifestyle> ***ant:*** formal

**in·for·mal·i·ty** (ĭn′fôr-măl′ĭ-tē) *n., pl.* **-ties. 1.** The quality or state of being informal. **2.** An informal act.

**in·form·ant** (ĭn-fôr′mənt) *n.* **1.** One who discloses information : INFORMER. **2.** One who furnishes cultural or linguistic information to a researcher.

**in·for·mat·ics** (ĭn′fər-măt′ĭks) *n.* [INFORMAT(ION) + -ICS.] (*sing. in number*). Information science.

**in·for·ma·tion** (ĭn′fər-mā′shən) *n.* **1.** The act of informing or state of being informed. **2.** Knowledge derived from study, experience, or instruction. **3.** Knowledge of a particular event or situation : NEWS. **4.** *Law.* A formal accusation of a crime made by a public officer rather than by indictment by a grand jury. **5.** A nonaccidental signal used as an input to a computer or communications system. **6.** A numerical measure of the uncertainty of an experimental outcome. **—in′for·ma′tion·al** *adj.*

**information science** *n.* The scientific study of the gathering, manipulation, classification, storage, and retrieval of recorded knowledge.

**information theory** *n.* The theory of the probability of transmission of messages with specified accuracy when the bits of information in the messages are subject, with certain probabilities, to transmission failure, distortion, and accidental additions.

**in·form·a·tive** (ĭn-fôr′mə-tĭv) *adj.* Providing or disclosing information : INSTRUCTIVE. **—in·form′a·tive·ly** *adv.* **—in·form′a·tive·ness** *n.*

**in·formed** (ĭn-fôrmd′) *adj.* **1.** Having, displaying, or utilizing information <the *informed* buyer> **2.** Based on factual knowledge <an *informed* opinion>

**in·form·er** (ĭn-fôr′mər) *n.* **1.** An informant. **2.** One who informs against others, often for compensation.

**infra-** *pref.* [Lat. *infra*, below.] Below : beneath <*infra*sonic>

**in·fract** (ĭn-frăkt′) *vt.* **-fract·ed, -fract·ing, -fracts.** [Lat. *infractus*, p.part. of *infringere*, to destroy. —see INFRINGE.] To infringe : violate. **—in·frac′tion** *n.* **—in·frac′tor** *n.*

**in·fra dig** (ĭn′frə dĭg′) *adj.* [Short for Lat. *infra dignitatem.*] Beneath one's dignity.

**in·fran·gi·ble** (ĭn-frăn′jə-bəl) *adj.* [OFr. < LLat. *infrangibilis* : Lat. *in-*, not + Lat. *frangere*, to break.] **1.** Not capable of being broken. **2.** Inviolable. **—in·fran′gi·bil′i·ty** *n.* **—in·fran′gi·bly** *adv.*

**in·fra·or·bit·al** (ĭn′frə-ôr′bĭ-təl) *adj. Anat.* Located or occurring beneath the orbit.

**in·fra·red** (ĭn′frə-rĕd′) *adj.* **1.** Of or relating to electromagnetic radiation with wavelengths greater than those of visible light and shorter than those of microwaves. **2.** Producing, using, or sensitive to infrared radiation.

**in·fra·son·ic** (ĭn′frə-sŏn′ĭk) *adj.* **1.** Producing or using waves with frequencies below that of audible sound. **2.** SUBSONIC 1.

**in·fra·sound** (ĭn′frə-sound′) *n.* A wave phenomenon sharing the physical nature of sound but with a range of frequencies below that of human hearing.

**in·fra·struc·ture** (ĭn′frə-strŭk′chər) *n.* **1.** An underlying base or foundation, esp. for an organization. **2.** The basic facilities, equipment, and installations needed for the functioning of a system.

**in·fre·quent** (ĭn-frē′kwənt) *adj.* **1.** Not frequent : RARE. **2.** Not occurring regularly : OCCASIONAL <an *infrequent* visitor> **—in·fre′quence, in·fre′quen·cy** *n.* **—in·fre′quent·ly** *adv.*

☆ ***syns:*** INFREQUENT, OCCASIONAL, RARE, SCARCE, SPORADIC, UNCOMMON, UNUSUAL *adj. core meaning* : rarely occurring or appearing <Their visits were *infrequent.*> ***ant:*** frequent

**in·fringe** (ĭn-frĭnj′) *v.* **-fringed, -fring·ing, -fring·es.** [Lat. *infringere* : *in-* (intensive) + *frangere*, to break.] *—vt.* **1.** To violate or go beyond the limits of (e.g., a law). **2.** *Obs.* To defeat : invalidate. *—vi.* To encroach upon something <extensive quoting that *infringed* upon the copyright> **—in·fring′er** *n.*

**in·fringe·ment** (ĭn-frĭnj′mənt) *n.* **1.** A violation, as of a law or agreement. **2.** An encroachment, as of a privilege or right.

**in·fun·dib·u·li·form** (ĭn′fən-dĭb′yə-lə-fôrm′) *adj.* Funnel-shaped.

**in·fun·dib·u·lum** (ĭn′fən-dĭb′yə-ləm) *n., pl.* **-la** (-lə) [Lat., funnel < *infundere*, to pour in. —see INFUSE.] *Physiol.* A funnel-shaped bodily passage or part. **—in′fun·dib′u·lar** (-lər), **in′fun·dib′u·late′** (-lāt′, -lĭt′) *adj.*

**in·fu·ri·ate** (ĭn-fyo͝or′ē-āt′) *vt.* **-at·ed, -at·ing, -ates.** [Med. Lat. *infuriare, infuriat-* : Lat. *in-* (intensive) + *furia*, fury < *furere*, to be mad.] To make angry : ENRAGE. *—adj.* (ĭn-fyo͝or′ē-ĭt). *Archaic.* Furious. **—in·fu′ri·at′ing·ly** *adv.* **—in·fu′ri·a′tion** *n.*

**in·fuse** (ĭn-fyo͞oz′) *vt.* **-fused, -fus·ing, -fus·es.** [ME *infusen* < OFr. *infuser* < Lat. *infusus*, p.part. of *infundere*, to pour in : *in-*, in + *fundere*, to pour.] **1.** To put into : INTRODUCE <*infused* new enthusiasm into the campaign> **2.** To cause to pervade : IMBUE <*infused* us with pride> **3.** To give an animating or motivating impulse to. **4.** To steep without boiling in order to extract soluble elements. **—in·fus′er** *n.*

**in·fus·i·ble** (ĭn-fyo͞o′zə-bəl) *adj.* Incapable of being fused or melted. **—in·fus′i·bil′i·ty, in·fus′i·ble·ness** *n.*

**in·fu·sion** (ĭn-fyo͞o′zhən) *n.* **1.** The act or process of infusing. **2.** An admixture. **3.** The liquid product obtained by infusing. **4.** Introduction of a solution into a vein.

**in·fu·so·ri·al** (ĭn′fyo͞o-sôr′ē-əl, -sōr′-, -zôr′-, -zōr′-) *adj.* **1.** Of or relating to infusorians. **2.** Containing or composed of infusorians.

**in·fu·so·ri·an** (ĭn′fyo͞o-sôr′ē-ən, -sōr′-, -zôr′-, -zōr′-) *n.* [NLat. *Infusoria*, class of protozoan < Lat. *infusus*, p.part. of *infundere*, to pour in. —see INFUSE.] Any of various microscopic organisms, esp. of the phylum Protozoa or order Rotifera, occurring in stagnant water or in infusions containing organic material. *—adj.* Of or relating to infusorians.

**-ing**[1] *suff.* [ME, alteration of *-end, -ind* < OE *-ende*, pr.part. suffix.] **1.** —Used to form the present participle of verbs <see*ing*> **2.** —Used to form adjectives resembling present participles but not derived from verbs <swashbuckl*ing*>

**-ing**[2] *suff.* [ME < OE *-ung.*] **1. a.** Action, process, or art <danc*ing*> **b.** An instance of an action, process, or art <a gather*ing*> **2.** An action or process connected with a given thing <berry*ing*> **3. a.** Something required to perform an action or process <moor*ing*> **b.** The result of an action or process <a draw*ing*> **c.** Something connected with a given thing or concept <sid*ing*> <off*ing*>

**-ing**[3] *suff.* [ME < OE, belonging to, descended from.] One having a given quality or nature <wild*ing*>

**in·gen·ious** (ĭn-jēn′yəs) *adj.* [Fr. *ingénieux* < Lat. *ingeniosus* < *ingenium*, inborn talent : *in-*, in + *gignere*, to beget.] **1.** Having or displaying great skill in creating or devising. **2.** Original and imaginative. **3.** *Obs.* Having genius : BRILLIANT. **—in·gen′ious·ly** *adv.* **—in·gen′ious·ness** *n.*

**in·gé·nue** (ăn′zhə-no͞o′) *n.* [Fr., fem. of *ingénu*, guileless < Lat. *ingenuus*, ingenuous.] **1.** An artless young woman. **2. a.** The dramatic role of an ingénue. **b.** An actress playing an ingénue.

**in·ge·nu·i·ty** (ĭn′jə-no͞o′ĭ-tē, -nyo͞o′-) *n., pl.* **-ties.** [Lat. *ingenuitas*, frankness < *ingenuus*, ingenuous.] **1.** Inventive imagination or skill : CLEVERNESS. **2.** Clever and imaginative design or construction. **3.** An ingenious or imaginative device. **4.** *Archaic.* Ingenuousness.

**in·gen·u·ous** (ĭn-jĕn′yo͞o-əs) *adj.* [Lat. *ingenuus*, honest, freeborn : *in-*, in + *gignere*, to beget.] **1.** Lacking worldliness or sophistication : ARTLESS. **2.** Straightforward and frank : CANDID. **3.** *Obs.* Ingenious. **—in·gen′u·ous·ly** *adv.* **—in·gen′u·ous·ness** *n.*

**in·gest** (ĭn-jĕst′) *vt.* **-gest·ed, -gest·ing, -gests.** [Lat. *ingerere, ingest-*, to carry in : *in-*, in + *gerere*, to carry.] To take in by or as if by swallowing. **—in·ges′tion** *n.* **—in·ges′tive** *adj.*

**in·ges·ta** (ĭn-jĕs′tə) *pl.n.* [NLat., neuter pl. of Lat. *ingestus*, p.part. of *ingerere*, to carry in. —see INGEST.] Ingested matter, esp. food.

**in·gle** (ĭng′gəl) *n.* **1.** A fire upon a hearth. **2.** A fireplace.

**in·gle·nook** (ĭng′gəl-no͝ok′) *n.* [Sc. Gael. *aingeal.*] **1.** A nook or corner next to an open fireplace. **2.** A bench in an inglenook.

**in·glo·ri·ous** (ĭn-glôr′ē-əs, -glōr′-) *adj.* [Lat. *inglorius* : *in-*, not + *gloria*, fame.] **1.** Not glorious. **2.** Dishonorable : ignominious. **—in·glo′ri·ous·ly** *adv.* **—in·glo′ri·ous·ness** *n.*

**in·got** (ĭng′gət) *n.* [ME, mold for casting metal, alteration of OFr. *lingot*, metal ingot.] **1.** A mass of metal shaped in a bar or block. **2.** A casting mold for metal.

**in·grain** (ĭn-grān′) *vt.* **-grained, -grain·ing, -grains.** **1.** To impress indelibly on the mind or nature : INFUSE. **2.** *Archaic.* To dye or stain into the fiber of. *—adj.* **1.** Deeply rooted : INSTILLED. **2.** Dyed in the yarn before weaving or knitting. **3.** Made of fiber or yarn dyed before weaving. —Used esp. of rugs. *—n.* **1.** Yarn or fiber dyed before manufacture. **2.** An article made of ingrain yarn, as a carpet.

**in·grained** (ĭn-grānd′) *adj.* **1.** Worked deeply into the texture or fiber. **2.** Firmly established : DEEP-SEATED.

**in·grate** (ĭn′grāt′) *n.* [ME *ingrat*, ungrateful < Lat. *ingratus* : *in-*, not + *gratus*, pleasing, thankful.] An ungrateful person.

**in·gra·ti·ate** (ĭn-grā′shē-āt′) *vt.* **-at·ed, -at·ing, -ates.** [IN-[2] + Lat. *gratia*, favor < *gratus*, pleasing.] To try to insinuate (oneself) into the good graces or favor of another. **—in·gra′ti·at′ing·ly** *adv.* **—in·gra′ti·a′tion** *n.* **—in·gra′ti·a·to′ry** (-shē-ə-tôr′ē, -tōr′ē) *adj.*

**in·grat·i·tude** (ĭn-grăt′ĭ-to͞od′, -tyo͞od′) *n.* [ME < OFr. < LLat. *ingratitudo* < *ingratus*, ungrateful. —see INGRATE.] Lack of gratitude.

**in·gre·di·ent** (ĭn-grē′dē-ənt) *n.* [ME < Lat. *ingrediens*, pr.part. of *ingredi*, to enter : *in-*, in + *gradi*, to step.] A constituent element of a mixture or compound.

**in·gress** (ĭn′grĕs′) *n.* [ME *ingresse* < Lat. *ingressus* < p.part. of *ingredi*, to enter. —see INGREDIENT.] **1.** *also* **in·gres·sion** (ĭn-grĕsh′ən). A going in or entering. **2.** Permission or right to enter.

**in·gres·sive** (ĭn-grĕs′ĭv) *adj.* **1.** Of or involving ingress. **2.** INCHOATIVE 2. **—in·gres′sive** *n.* **—in·gres′sive·ness** *n.*

**in-group** (ĭn′gro͞op′) *n. Informal.* A group united by common beliefs, attitudes, and interests and usu. excluding outsiders.

**in·grow·ing** (ĭn′grō′ĭng) *adj.* Growing inward.

**in·grown** (ĭn′grōn′) *adj.* **1.** Grown abnormally into the flesh <an *ingrown* toenail> **2.** Grown within : INNATE.

**in·growth** (ĭn′grōth′) *n.* **1.** The act of growing inward. **2.** Something growing inward.

**in·gui·nal** (ĭng′gwə-nəl) *adj.* [Lat. *inguinalis* < *inguen*, groin.] Of, pertaining to, or situated in the groin.

**in·gur·gi·tate** (ĭn-gûr′jĭ-tāt′) *vt.* **-tat·ed, -tat·ing, -tates.** [Lat. *ingurgitare, ingurgitat-* : *in-*, in + *gurges*, whirlpool.] To swallow greedily or in excessive amounts : GUZZLE. **—in·gur′gi·ta′tion** *n.*

**in·hab·it** (ĭn-hăb′ĭt) *v.* **-it·ed, -it·ing, -its.** [ME *enhabiten* < OFr. *enhabiter* < Lat. *inhabitare* : *in-*, in + *habitare*, to dwell, freq. of *habēre*, to have.] —*vt.* **1.** To reside in. **2.** To be present in. —*vi. Archaic.* To dwell. **—in·hab′it·a·bil′i·ty** *n.* **—in·hab′it·a·ble** *adj.* **—in·hab′i·ta′tion** *n.* **—in·hab′it·er** *n.*

**in·hab·i·tan·cy** (ĭn-hăb′ĭ-tən-sē) *n., pl.* **-cies.** Occupancy.

**in·hab·i·tant** (ĭn-hăb′ĭ-tənt) *n.* A resident.

**in·hab·it·ed** (ĭn-hăb′ĭ-tĭd) *adj.* Having inhabitants : POPULATED.

**in·ha·lant** (ĭn-hā′lənt) *adj.* Used in or for inhaling. —*n.* Something that is inhaled, as a medicine.

**in·ha·la·tor** (ĭn′hə-lā′tər) *n.* A device producing a vapor to ease breathing or to medicate by inspiration.

**in·hale** (ĭn-hāl′) *v.* **-haled, -hal·ing, -hales.** [Lat. *inhalare* : *in-*, in + *halare*, to breath.] —*vt.* To take in by breathing. —*vi.* To breathe in. **—in′ha·la′tion** *n.*

**in·hal·er** (ĭn-hā′lər) *n.* **1.** One that inhales. **2.** An inhalator. **3.** A respirator.

**in·har·mon·ic** (ĭn′här-mŏn′ĭk) *adj.* Not harmonic : DISCORDANT.

**in·har·mo·ni·ous** (ĭn′här-mō′nē-əs) *adj.* **1.** Harsh or unpleasant : DISCORDANT. **2.** Not in accord or agreement. **—in′har·mo′ni·ous·ly** *adv.* **—in′har·mo′ni·ous·ness** *n.*

**in·here** (ĭn-hîr′) *vi.* **-hered, -her·ing, -heres.** [Lat. *inhaerēre* : *in*, in + *haerēre*, to stick.] To be inherent or innate. **—in·her′ence** (-hîr′əns, -hĕr′-), **in·her′en·cy** *n.*

**in·her·ent** (ĭn-hîr′ənt, -hĕr′-) *adj.* [Lat. *inhaerens, inhaerent-*, pr.part. of *inhaerēre*, to inhere.] Existing as an essential constituent or characteristic : INTRINSIC. **—in·her′ent·ly** *adv.*

**in·her·it** (ĭn-hĕr′ĭt) *v.* **-it·ed, -it·ing, -its.** [ME *enheriten*, to make (someone) an heir < OFr. *enheriter* < LLat. *inhereditare* : Lat. *in-*, in + Lat. *heres*, heir.] —*vt.* **1.** To come into possession of : POSSESS. **2.** To receive (property) from a person by legal succession or will. **3.** *Biol.* To receive genetically from an ancestor. —*vi.* To hold or take possession of an inheritance. **—in·her′i·tor** *n.* **—in·her′i·trix** (-ĭ-trĭks) *n.*

**in·her·it·a·ble** (ĭn-hĕr′ĭ-tə-bəl) *adj.* **1.** Having the right to inherit. **2.** Capable of being inherited.

**in·her·i·tance** (ĭn-hĕr′ĭ-təns) *n.* **1.** The act of inheriting. **2.** Something inherited or to be inherited. **3.** Something regarded as a heritage <the cultural *inheritance* of Greece> **4.** *Biol.* **a.** Genetic transmission of characteristics. **b.** A characteristic so inherited.

**inheritance tax** *n.* A tax on inherited property.

**in·hib·it** (ĭn-hĭb′ĭt) *vt.* **-it·ed, -it·ing, -its.** [ME *inhibiten*, to forbid < Lat. *inhibitus*, p.part. of *inhibēre*, to restrain : *in-*, in + *habēre*, to have.] **1.** To restrict or hold back : RESTRAIN. **2.** To prohibit : FORBID. **—in·hib′it·a·ble** *adj.* **—in·hib′i·tive, in·hib′i·to′ry** (-tôr′ē, -tōr′ē) *adj.*

**in·hib·it·er** (ĭn-hĭb′ĭ-tər) *n. var. of* INHIBITOR.

**in·hi·bi·tion** (ĭn′hə-bĭsh′ən, ĭn′ə-) *n.* **1.** The act of inhibiting or state of being inhibited. **2. a.** Restraint of a behavioral process or the condition inducing such restraint. **b.** The process by which the superego prevents conscious expression of an instinct.

**in·hib·i·tor** *also* **in·hib·it·er** (ĭn-hĭb′ĭ-tər) *n.* **1.** A substance used to retard an undesirable reaction <a rust *inhibitor*> **2.** One that inhibits.

**in·hos·pi·ta·ble** (ĭn-hŏs′pĭ-tə-bəl, ĭn′hŏ-spĭt′ə-bəl) *adj.* **1.** Showing no hospitality : UNFRIENDLY. **2.** Not affording shelter or sustenance <the *inhospitable* regions of the Arctic> **—in·hos′pi·ta·ble·ness, in·hos′pi·tal′i·ty** *n.* **—in·hos′pi·ta·bly** *adv.*

**in-house** (ĭn′hous′) *adj.* Being or coming from within an organization <an *in-house* publication>

**in·hu·man** (ĭn-hyōō′mən) *adj.* [Lat. *inhumanus* : *in-*, not + *humanus*, human.] **1. a.** Lacking kindness or pity : BRUTAL. **b.** Lacking emotional warmth : COLD. **2.** Not in accord with human needs <an *inhuman* atmosphere> **3.** Not of ordinary human form : MONSTROUS. **—in·hu′man·ly** *adv.* **—in·hu′man·ness** *n.*

**in·hu·mane** (ĭn′hyōō-mān′) *adj.* Lacking pity or compassion : CRUEL. **—in′hu·mane′ly** *adv.*

**in·hu·man·i·ty** (ĭn′hyōō-măn′ĭ-tē) *n., pl.* **-ties. 1.** Lack of pity or compassion. **2.** An inhuman or cruel act.

**in·hume** (ĭn-hyōōm′) *vt.* **-humed, -hum·ing, -humes.** [Lat. *inhumare* : *in-*, in + *humus*, earth.] To bury in a grave : INTER. **—in′hu·ma′tion** *n.* **—in·hum′er** *n.*

**in·im·i·cal** (ĭn-ĭm′ĭ-kəl) *adj.* [LLat. *inimicalis* < Lat. *inimicus*, enemy. —see ENEMY.] **1.** Injurious or harmful in effect : ADVERSE <eating habits *inimical* to good nutrition> **2.** Hostile : unfriendly <a stern and *inimical* glare>

**in·im·i·ta·ble** (ĭn-ĭm′ĭ-tə-bəl) *adj.* Defying imitation : MATCHLESS. **—in·im′i·ta·bil′i·ty** *n.* **—in·im′i·ta·bly** *adv.*

**in·iq·ui·tous** (ĭ-nĭk′wĭ-təs) *adj.* Of or marked by wickedness : SINFUL. **—in·iq′ui·tous·ly** *adv.* **—in·iq′ui·tous·ness** *n.*

**in·iq·ui·ty** (ĭ-nĭk′wĭ-tē) *n., pl.* **-ties.** [ME *iniquite* < OFr. < Lat. *iniquitas* < *iniquus*, unjust, harmful : *in-*, not + *aequus*, equal.] **1.** Wickedness : sinfulness. **2.** A grossly immoral act : SIN.

**in·i·tial** (ĭ-nĭsh′əl) *adj.* [Lat. *initialis* < *initium*, beginning < *initus*, p.part. of *inire*, to enter : *in-*, in + *ire*, to go.] **1.** Happening or being at the very beginning : FIRST. **2.** Denoting the first letter or letters of a word. —*n.* **1.** *often* **initials.** The first letter or letters of a person's name or names, used as a shortened signature or for identification. **2.** The first letter of a word. **3.** A large, often highly decorated letter set at the opening of a chapter, verse, or paragraph. —*vt.* **-tialed, -tial·ing, -tials** *also* **-tialled, -tial·ling, -tails.** To sign or mark with initials. **—in·i′tial·ly** *adv.*

**in·i·tial·ize** (ĭ-nĭsh′ə-līz′) *vt.* **-ized, iz·ing, -iz·es.** *Computer Sci.* To set to a starting position or value. **—in·i′tial·i·za′tion** *n.* **—in·i′tial·iz′er** *n.*

**initial teaching alphabet** *n.* An alphabet with 44 symbols, each of which represents a single sound, used to teach beginning reading of English.

**in·i·ti·ate** (ĭ-nĭsh′ē-āt′) *vt.* **-at·ed, -at·ing, -ates.** [Lat. *initiare, initiat-* < *initium*, beginning. —see INITIAL.] **1.** To cause to begin <*initiated* the autumn music season> **2.** To introduce (a person) to a new field, interest, skill, or activity. **3.** To admit into membership, as with ceremonies or ritual. —*adj.* (-ĭt). Initiated. —*n.* (-ĭt). **1.** One who has been initiated. **2.** A novice : beginner. **—in·i′ti·a′tor** *n.*

**in·i·ti·a·tion** (ĭ-nĭsh′ē-ā′shən) *n.* **1. a.** An act or instance of initiating. **b.** The state of being initiated. **2.** A ceremony, ritual, test, or period of instruction with which an organization admits a new member to office or knowledge.

**in·i·tia·tive** (ĭ-nĭsh′ə-tĭv) *n.* **1.** The power, ability, or instinct to begin or to follow through energetically with a plan or task. **2.** The first step : opening move <opponents seized the *initiative*> **3. a.** The right or power to introduce a new legislative measure. **b.** The right and procedure by which citizens can propose a law by petition and ensure its submission to the electorate. —*adj.* **1.** Of or relating to initiation. **2.** Used to initiate. **—on (one's) own initiative.** Without prompting or direction from others. **—in·i′tia·tive·ly** *adv.*

**in·i·ti·a·to·ry** (ĭ-nĭsh′ē-ə-tôr′ē, -tōr′ē) *adj.* **1.** Introductory : initial. **2.** INITIATIVE 2.

**in·ject** (ĭn-jĕkt′) *vt.* **-ject·ed, -ject·ing, -jects.** [< Lat. *injectus*, p.part. of *inicere*, to put in : *in-*, in + *jacere*, to throw.] **1.** To force or drive (a fluid) into something <*inject* gasoline into the cylinder> **2. a.** *Med.* To introduce (a fluid) into the skin, subcutaneous tissue, muscle, blood vessels, or a bodily cavity. **b.** To introduce a fluid into <*injected* antibiotics into the muscle> **3.** To introduce into conversation or consideration <*inject* a touch of seriousness into the discussion> **4.** To place into an orbit, trajectory, or stream. **—in·jec′tor** *n.*

**in·jec·tion** (ĭn-jĕk′shən) *n.* **1.** The act of injecting. **2.** An injected fluid, esp. a dose of liquid medicine.

**in·ju·di·cious** (ĭn′jōō-dĭsh′əs) *adj.* Lacking or displaying a lack of judgment or discretion <*injudicious* schemes> **—in′ju·di′cious·ly** *adv.* **—in′ju·di′cious·ness** *n.*

**in·junc·tion** (ĭn-jŭngk′shən) *n.* [LLat. *injunctio*, command < Lat. *injunctus*, p.part. of *injungere*, to enjoin : *in-*, in + *jungere*, to join.] **1.** The act or an instance of enjoining. **2.** *Law.* A court order enjoining a party from a given course of action. **—in·junc′tive** *adj.*

**in·jure** (ĭn′jər) *vt.* **-jured, -jur·ing, -jures.** [Back-formation < INJURY.] **1.** To cause physical harm to : HURT. **2.** To cause damage to : IMPAIR. **3.** To cause distress to : WOUND <*injured* your pride> **4.** To commit an injustice or offense against : WRONG. **—in′jur·er** *n.*

**in·ju·ri·ous** (ĭn-jŏŏr′ē-əs) *adj.* **1.** Causing or tending to cause injury. **2.** Libelous : slanderous <*injurious* statements about our client's private life> **—in·ju′ri·ous·ly** *adv.* **—in·ju′ri·ous·ness** *n.*

**in·ju·ry** (ĭn′jə-rē) *n., pl.* **-ries.** [ME *injurie* < AN < Lat. *injuria*, a wrong < *injurius*, unjust : *in-*, not < *jus*, law.] **1.** Damage of or to a person, property, reputation, or thing. **2.** A wound or other specific damage. **3.** *Law.* A wrong or damage done to a person or to his or her property, reputation, or rights when caused by the wrongful act of another. **4.** *Obs.* An insult.

**in·jus·tice** (ĭn-jŭs′tĭs) *n.* [ME < OFr. < Lat. *injustitia* < *injustus*, unjust : *in-*, not + *justus*, just.] **1. a.** Lack of justice. **b.** Violation of another's rights or of what is right. **2.** A specific unjust act : WRONG.

**ink** (ĭngk) *n.* [ME *inke* < OFr. *enque* < LLat. *encaustum*, purple ink < Gk. *enkauston* < *enkaiein*, to paint in encaustic. —see ENCAUSTIC.] **1.** A pigmented liquid or paste used esp. for writing or printing. **2.** A dark liquid secreted by cuttlefish and other cephalopods. —*vt.* **inked, ink·ing, inks.** To mark or stain with ink. **—ink′i·ness** *n.* **—ink′y** *adj.*

**ink·ber·ry** (ĭngk′bĕr′ē) *n.* **1.** A shrub, *Ilex glabra* of eastern North America, bearing black berrylike fruit. **2.** Pokeweed. **3.** The fruit of an inkberry.

**ink·blot** (ĭngk′blŏt′) *n.* **1.** A blotted pattern of spilled ink. **2.** A pattern resembling an inkblot and used in the Rorschach test.

**ink·horn** (ĭngk′hôrn′) *n.* A small container made esp. of horn for holding writing ink. —*adj.* Bookish : recondite <*inkhorn* phrases>

**ink·ling** (ĭngk′lĭng) *n.* [Perh. < ME *inklen*, to mention.] **1.** A hint or slight suggestion. **2.** A vague notion or idea.

**ink sac** *n. Biol.* An organ containing ink, located near the rectum in some cephalopods.

**ink·stand** (ĭngk′stănd′) *n.* **1.** A tray or rack for holding writing implements, as pens and ink. **2.** An inkwell.

ōō **boot** ou **out** th **thin** *th* **this** ŭ **cut** ûr **urge** y **young**
yōō ab**u**se zh vi**si**on ə **a**bout, it**e**m, ed**i**ble, gall**o**p, circ**u**s

**ink·well** (ĭngk′wĕl′) *n.* A small reservoir for ink.
**inky cap** (ĭngk′ē) *n.* A mushroom of the genus *Coprinus*, with gills that dissolve into a dark liquid on maturing.

**inky cap**

**in·lace** (ĭn-lās′) *v. var. of* ENLACE.
**in·laid** (ĭn′lād′) *adj.* **1.** Set into a surface in a decorative design. **2.** Decorated with a pattern set into a surface.
**in·land** (ĭn′lənd) *adj.* **1.** Of, relating to, or situated in the interior part of a country or area. **2.** Operating or applying within the borders of a country : DOMESTIC. —*adv.* In, toward, or into the interior. —*n.* (-lănd′, -lənd). The interior of a country or area. **—in′land′er** *n.*
**in-law** (ĭn′lô′) *n.* [Back-formation < such compound words as *mother-in-law.*] A relative by marriage.
**in·lay** (ĭn-lā′, ĭn′lā′) *vt.* **-laid, -lay·ing, -lays. 1. a.** To set (e.g., pieces of wood) into a surface to form a design. **b.** To decorate by setting in such designs. **2.** To insert (e.g., a photograph) within a mat in a book. —*n.* (ĭn′lā′). **1.** Material set into a surface in pieces to form a design. **2.** A design, pattern, or decoration made by inlaying. **3.** A solid dental filling fitted to a cavity and cemented in place. **—in·lay′er** *n.*
**in·let** (ĭn′lĕt′, -lĭt) *n.* **1.** A recess, as a bay or cove, along a coastline. **2.** A stream or bay leading inland, as from the ocean : ESTUARY. **3.** A narrow passage of water, as between two islands. **4.** A drainage passage, as to a culvert. **5.** An opening providing a means of entrance.
**in·li·er** (ĭn′lī′ər) *n.* An older rock formation completely surrounded by newer strata.
**in-line** (ĭn′līn′) *adj.* Having the parts arranged in a straight line. —*adv.* Being arranged in an in-line manner.
**in lo·co pa·ren·tis** (ĭn lō′kō pə-rĕn′tĭs) *adv.* [Lat.] In the place or position of a parent.
**in·ly** (ĭn′lē) *adv.* Inwardly.
**in·mate** (ĭn′māt′) *n.* **1.** A resident in a building or dwelling. **2.** One confined to an institution, as a prison or hospital.
**in me·di·as res** (ĭn mē′dē-əs rās′) *adv.* [Lat., in the middle of things.] In or into the middle of a sequence of events, as of a plot.
**in me·mo·ri·am** (ĭn′ mə-môr′ē-əm, -mōr′-) *prep.* [Lat.] In memory of.—Used in epitaphs.
**in·most** (ĭn′mōst′) *adj.* Innermost.
**inn** (ĭn) *n.* [ME < OE.] **1.** A lodging house serving food and drink to travelers : HOTEL. **2.** A restaurant. **3.** *Chiefly Brit.* A residence hall formerly provided for students.
**in·nards** (ĭn′ərdz) *pl.n.* [Alteration of INWARDS.] *Informal.* **1.** Internal bodily organs : VISCERA. **2.** The inner parts, as of a structure.
**in·nate** (ĭ-nāt′, ĭn′āt′) *adj.* [ME *innat* < Lat. *innatus*, p.part. of *innasci*, to be born in : *in-*, in + *nasci*, to be born.] **1.** Possessed at birth : INBORN. **2.** Possessed as an essential characteristic : INHERENT. **3.** Of or produced by thought rather than experience. **—in·nate′ly** *adv.* **—in·nate′ness** *n.*
**in·ner** (ĭn′ər) *adj.* [ME < OE *innera.*] **1.** Occurring or located farther inside <an *inner* compartment> **2.** Less evident : DEEPER <the *inner* meaning of a novel> **3.** Of or relating to the spirit or mind <"Beethoven's manuscript looks like a bloody record of a tremendous *inner* battle" —Leonard Bernstein> **4.** More important or exclusive <the *inner* circles of politics>
**inner city** *n.* The older, central part of a city, esp. when marked by overpopulation and blight.
**in·ner-di·rect·ed** (ĭn′ər-dĭ-rĕk′tĭd, -dī-) *adj.* Guided, as in thought and action, by one's own set of values rather than by conventional norms.
**inner ear** *n.* The internal ear.
**in·ner·most** (ĭn′ər-mōst′) *adj.* **1.** Occurring or situated farthest within. **2.** Most intimate <*innermost* thoughts>
**inner product** *n. Math.* Scalar product.
**inner space** *n.* **1.** Space at or near the earth's surface, esp. space beneath the sea. **2.** The inner, spiritual, or subconscious part of the self.
**in·ner·vate** (ĭ-nûr′vāt′, ĭn′ər-) *vt.* **-vat·ed, -vat·ing, -vates. 1.** To supply (a bodily part) with nerves. **2.** To stimulate (a nerve or bodily part). **—in′ner·va′tion** *n.*
**in·nerve** (ĭ-nûrv′) *vt.* **-nerved, -nerv·ing, -nerves.** To give nervous energy to : STIMULATE.
**in·ning** (ĭn′ĭng) *n.* [< IN.] **1. a.** *Baseball.* One of nine divisions or periods of a regulation game, in which each team has a turn at bat as limited by three outs. **b. innings.** *(sing. in number).* The division or period of a cricket game during which one team is at bat. **2.** *often* **innings.** An opportunity to act or speak out : TURN. **3.** *Archaic.* Reclamation of flooded or marshy land.
**inn·keep·er** (ĭn′kē′pər) *n.* The manager or owner of an inn.
**in·no·cence** (ĭn′ə-səns) *n.* **1.** The quality, state, or virtue of being innocent. **2.** Bluets.
**in·no·cent** (ĭn′ə-sənt) *adj.* [ME < OFr. < Lat. *innocens* : *in-*, not + *nocens*, pr.part. of *nocēre*, to harm.] **1.** Uncorrupted by wickedness, malice, or wrongdoing : SINLESS <*innocent* children> **2.** Not guilty of a given crime : legally blameless <found *innocent* on all counts> **3.** Harmless : innocuous <an *innocent* remark> **4.** Inexperienced : naive <*innocent* newcomers> **5.** Not exposed to or familiar with something specified : IGNORANT. **6.** Betraying or suggesting no guile or deception : ARTLESS <an *innocent* look> **—in′no·cent** *n.* **—in′no·cent·ly** *adv.*
**in·noc·u·ous** (ĭ-nŏk′yo͞o-əs) *adj.* [Lat. *innocuus* : *in-*, not + *nocuus*, harmful < *nocēre*, to harm.] **1.** Having no adverse effect : HARMLESS. **2.** Lacking importance or distinction : INSIPID <an *innocuous* lecture> **—in·noc′u·ous·ly** *adv.* **—in·noc′u·ous·ness** *n.*
**in·nom·i·nate** (ĭ-nŏm′ə-nĭt) *adj.* [LLat. *innominatus* : Lat. *in-*, not + *nominatus*, p.part. of *nominare*, to name < *nomen*, name.] **1.** Having no name. **2.** Anonymous.
**innominate artery** *n.* An artery arising from the aortic arch and dividing into the right subclavian and right carotid arteries.
**innominate bone** *n.* A large flat bone forming the lateral half of the pelvis.
**innominate vein** *n.* One of a pair of veins each formed by the union of the internal jugular and subclavian veins that join to form the superior vena cava.
**in·no·vate** (ĭn′ə-vāt′) *v.* **-vat·ed, -vat·ing, -vates.** [Lat. *innovare*, *innovat*, to renew : *in-* (intensive) + *novus*, new.] —*vt.* To start or introduce (something new). —*vi.* To start or introduce something new : be creative. **—in′no·va′tive** *adj.* **—in′no·va′tor** *n.*
**in·no·va·tion** (ĭn′ə-vā′shən) *n.* **1.** The act of innovating. **2.** Something new or unusual. **—in′no·va′tion·al** *adj.*
**Inns of Court** *pl.n.* **1.** The four legal societies in England founded about the beginning of the 14th cent. and having the exclusive right to confer the degree of barrister on law students. **2.** The buildings housing the Inns of Court.
**in·nu·en·do** (ĭn′yo͞o-ĕn′dō) *n., pl.* **-does.** [Lat. *innuendo*, by hunting < *innuendum*, gerund of *innuere*, to nod to.] **1.** An indirect or subtle and usu. derogatory implication : INSINUATION. **2.** *Law.* **a.** A plaintiff's interpretation, in a libel suit, of allegedly libelous or slanderous material. **b.** An explanation of a word or charge.
**In·nu·it** (ĭn′yo͞o-ĭt) *n., pl.* **Innuit** *or* **-its.** [Eskimo, people, pl. of *innuk*, person.] **1.** An Eskimo of North America and Greenland as distinguished from one of Asia and the Aleutian Islands. **2.** The language of the Innuits.
**in·nu·mer·a·ble** (ĭ-no͞o′mər-ə-bəl, ĭ-nyo͞o′-) *adj.* Too many to be counted. **—in·nu′mer·a·ble·ness** *n.* **—in·nu′mer·a·bly** *adv.*
**in·nu·mer·ous** (ĭ-no͞o′mər-əs) *adj.* Innumerable.
**in·nu·tri·tion** (ĭn′no͞o-trĭsh′ən, -nyo͞o-) *n.* Poor nourishment. **—in′nu·tri′tious** *adj.*
**in·ob·ser·vance** (ĭn′əb-zûr′vəns) *n.* **1.** Lack of attention or heed. **2.** Nonobservance, as of a law or custom. **—in′ob·ser′vant** *adj.*
**in·ob·tru·sive** (ĭn′əb-tro͞o′sĭv) *adj.* Not noticeable : UNOBTRUSIVE.
**in·oc·u·la·ble** (ĭ-nŏk′yə-lə-bəl) *adj.* **1.** Transmissible by inoculation. **2.** Susceptible to a disease transmitted by inoculation. **—in·oc′u·la·bil′i·ty** *n.*
**in·oc·u·lant** (ĭ-nŏk′yə-lənt) *n.* Inoculum.
**in·oc·u·late** (ĭ-nŏk′yə-lāt′) *vt.* **-lat·ed, -lat·ing, -lates.** [ME *inoculaten*, to graft a scion < Lat. *inoculare* : *in-*, in + *oculus*, eye, bud.] **1.** To communicate a disease to by transferring its virus or other causative agent into the body. **2.** To introduce the virus of a disease or other antigenic material into in order to immunize, cure, or experiment. **3.** To implant microorganisms or infectious material into (a culture medium). **—in·oc′u·la′tive** *adj.* **—in·oc′u·la′tor** *n.*
**in·oc·u·la·tion** (ĭ-nŏk′yə-lā′shən) *n.* **1.** The act or process of inoculating. **2.** Inoculum.
**in·oc·u·lum** (ĭ-nŏk′yə-ləm) *n.* [NLat. < Lat. *inoculare*, to graft a scion. —see INOCULATE.] Material used in an inoculation.
**in·o·dor·ous** (ĭn-ō′dər-əs) *adj.* Having no odor.
**in·of·fen·sive** (ĭn′ə-fĕn′sĭv) *adj.* **1.** Giving no offense : UNOBJECTIONABLE. **2.** Causing no harm : HARMLESS. **—in′of·fen′sive·ly** *adv.* **—in′of·fen′sive·ness** *n.*
**in·of·fi·cious** (ĭn′ə-fĭsh′əs) *adj.* [Lat. *inofficiosus*, undutiful : *in-*, not + *officiosus*, dutiful < *officium*, duty.] *Law.* Contrary to natural affection or moral duty. —Used of a will in which the testator disinherits rightful heirs without sufficient reason.
**in·op·er·a·ble** (ĭn-ŏp′ər-ə-bəl, -ŏp′rə-) *adj.* **1.** Not operable. **2.** Not susceptible to surgery. **—in·op′er·a·bly** *adv.*
**in·op·er·a·tive** (ĭn-ŏp′ər-ə-tĭv, -ŏp′rə-) *adj.* Not working or functioning. **—in·op′er·a·tive·ness** *n.*

**in·o·per·cu·late** (ĭn′ō-pûr′kyə-lĭt) *adj. Biol.* Lacking an operculum. **—in′o·per′cu·late** *n.*

**in·op·por·tune** (ĭn-ŏp′ər-to͞on′, -tyo͞on′) *adj.* Inconvenient : ill-timed. **—in·op′por·tune′ly** *adv.* **—in·op′por·tune′ness** *n.*

**in·or·di·nate** (ĭn-ôr′dn-ĭt) *adj.* [ME *inordinat* < Lat. *inordinatus* : *in-*, not + *ordinatus*, p.part. of *ordinare*, to set in order < *ordo*, order.] **1.** Exceeding normal or reasonable limits : IMMODERATE. **2.** Not regulated : DISORDERLY. **—in·or′di·na·cy, in·or′di·nate·ness** *n.* **—in·or′di·nate·ly** *adv.*

**in·or·gan·ic** (ĭn′ôr-găn′ĭk) *adj.* **1. a.** Involving neither organic life nor the products of organic life. **b.** Not made up of organic matter. **2.** Of or pertaining to the chemistry of compounds not usu. classified as organic. **3.** Not produced by in normal growth : ARTIFICIAL. **4.** Lacking structure or system. **—in′or·gan′i·cal·ly** *adv.*

**in·os·cu·late** (ĭn-ŏs′kyə-lāt′) *v.* **-lat·ed, -lat·ing, -lates.** [IN-² + Lat. *osculare, osculat-*, to provide with an opening < *osculum*, dim. of *os*, mouth.] *—vt.* **1.** To unite (e.g., blood vessels) by small openings. **2.** To make continuous. *—vi.* **1.** To open into one another. **2.** To unite so as to be continuous : BLEND. **—in·os′cu·la′tion** *n.*

**in·o·si·tol** (ĭ-nō′sĭ-tôl′, -tōl′) *n.* [Gk. *is, in-*, sinew + -IT(E) + -OL.] Any of nine isomeric alcohols, $C_6H_6(OH)_6$, esp. one found in plant and animal tissue and classified as part of the vitamin B complex.

**in·o·tro·pic** (ē′nə-trō′pĭk, -trŏp′ĭk, ĭ′nə-) *adj.* [Gk. *is, in-*, tendon, sinew + -TROPIC.] Influencing muscular contractility.

**in·pa·tient** (ĭn′pā′shənt) *n.* A patient admitted to a hospital.

**in-person** (ĭn′pûr′sən) *adj.* Of or relating to the physical presence of a person or subject : LIVE <an *in-person* concert>

**in per·so·nam** (ĭn′ pər-sō′nəm) *adj.* & *adv.* [Lat.] *Law.* Against a person. —Used of a proceeding.

**in pet·to** (ĭn pĕt′ō) *adj.* & *adv.* [Ital., in the breast.] In secret or private. —Used of appointments of cardinals by the pope undisclosed in consistory.

**in·phase** (ĭn′fāz′) *adj. Elect.* Having the same phase.

**in pos·se** (ĭn pŏs′ē) *adj.* & *adv.* [Lat., in possibility.] In potential but not in actuality.

**in-print** (ĭn′prĭnt′) *adj.* Being in print.

**in pro·pri·a per·so·na** (ĭn prō′prē-ə pər-sō′nə) *adv.* [Lat.] In one's own person or self : PERSONALLY.

**in·put** (ĭn′po͝ot′) *n.* **1.** Something introduced into a system or expended in its operation to attain a result or output, esp.: **a.** Energy, work, or power for driving a machine. **b.** Current, electromotive force, or power supplied to an electric circuit, network, or device. **c.** Information put into a communications system for transmission or into a data-processing system for processing. **d.** The body of basic resources, including materials, equipment, and funds, needed to complete a project. **e.** A terminal or station at which input enters a system. **2. a.** The act of putting in : INFUSION. **b.** An amount put in. **3.** Contribution to or participation in a common effort <wanted *input* from all project supervisors> **4.** Information in general. **—in′put′** *v.* **(-put·ted** or **-put, -put·ting, -puts).**

**in·quest** (ĭn′kwĕst′) *n.* [ME *enqueste* < OFr., of Lat. orig.] **1. a.** A judicial inquiry held before a jury, esp. an inquiry into the cause of a death. **b.** A jury making such an inquiry. **2.** An investigation.

**in·qui·e·tude** (ĭn-kwī′ĭ-to͞od′, -tyo͞od′) *n.* [ME, disturbance < LLat. *inquietudo*, restlessness < Lat. *inquies*, restless : *in-*, not + *quietus*, quiet.] A state of uneasiness : DISQUIETUDE.

**in·qui·line** (ĭn′kwə-līn′, -lĭn, ĭng′-) *n.* [Lat. *inquilinus*, lodger, tenant : *in-*, in + *colere*, to dwell.] An animal that lives in the dwelling place or burrow of another kind of animal. *—adj.* Being or living as an inquiline. **—in′qui·lin·ism** (-lə-nĭz′əm), **in′qui·lin′i·ty** (-lĭn′ĭ-tē) *n.* **—in′qui·lin′ous** (-lī′nəs) *adj.*

**in·quire** (ĭn-kwīr′) *v.* **-quired, -quir·ing, -quires.** [ME *enquiren* < OFr. *enquerrer* < Med. Lat. *inquerere* < Lat. *inquirere* : *in-* (intensive) + *quaerere*, to seek.] *—vi.* **1. a.** To put a question. **b.** To request information <*inquire* about mortgage rates><*inquire* after a friend's condition> **2.** To make a study or search : INVESTIGATE. *—vt.* To ask about. **—in·quir′er** *n.* **—in·quir′ing·ly** *adv.*

**in·quir·y** (ĭn-kwīr′ē, ĭn′kwīr′ē, ĭn′kwə-rē, ĭng′-) *n., pl.* **-ies.** **1.** The act of inquiring. **2.** A question : query. **3.** A close examination of a matter for information or truth.

**in·qui·si·tion** (ĭn′kwĭ-zĭsh′ən, ĭng′-) *n.* [ME *inquisicioun* < OFr. *inquisicion* < Lat. *inquisitio* < *inquirere*, to inquire.] **1.** The act of inquiring into a matter : INVESTIGATION. **2. a.** An inquest. **b.** The verdict of a judicial inquiry. **3. a. Inquisition.** *Rom. Cath. Ch.* A tribunal formerly directed at suppression of heresy. **b.** An investigation violating individual rights. **c.** Rigorous, abusive interrogation. **—in′qui·si′tion·al** *adj.*

**in·quis·i·tive** (ĭn-kwĭz′ə-tĭv) *adj.* **1.** Unduly curious : SNOOPY <*inquisitive* neighbors> **2.** Eager to learn. **—in·quis′i·tive·ly** *adv.* **—in·quis′i·tive·ness** *n.*

**in·quis·i·tor** (ĭn-kwĭz′ĭ-tər) *n.* One who inquires, esp. an excessively rigorous or abrasive questioner.

**in·quis·i·to·ri·al** (ĭn-kwĭz′ĭ-tôr′ē-əl, -tōr′-) *adj.* **1.** Of, relating to, typical of, or having the function of an inquisitor. **2.** *Law.* **a.** Pertaining to a trial in which one party acts as both prosecutor and judge. **b.** Relating to a criminal proceeding conducted in secrecy. **—in·quis′i·to′ri·al·ly** *adv.*

**in re** (ĭn rā′, rē′) *prep.* [Lat.] In regard to : CONCERNING.

**in rem** (ĭn rĕm′) *adj.* & *adv.* [Lat.] *Law.* Against a thing, as property, status, or a right.

**in·road** (ĭn′rōd′) *n.* [IN + ROAD, raid (obs.).] **1.** A hostile intrusion : RAID. **2.** An advance, esp. at another's expense : ENCROACHMENT <foreign products making *inroads* into American markets>

**in·rush** (ĭn′rŭsh′) *n.* A sudden influx.

**in·sal·i·vate** (ĭn-săl′ə-vāt′) *vt.* **-vat·ed, -vat·ing, -vates.** To mix (food) with saliva in chewing. **—in·sal′i·va′tion** *n.*

**in·sa·lu·bri·ous** (ĭn′sə-lo͞o′brē-əs) *adj.* Not salubrious : UNHEALTHY.

**in·sane** (ĭn-sān′) *adj.* [Lat. *insanus* : *in-*, not + *sanus*, sane, healthy.] **1.** Of, displaying, or afflicted with insanity : mentally unsound. **2.** Typical of, used by, or for the insane. **3.** Extremely foolish : ABSURD. **—in·sane′ly** *adv.* **—in·sane′ness** *n.*

☆ ***syns:*** INSANE, BATTY, CRAZY, LOCO, LOONY, LUNATIC, MAD, MANIACAL, NUTS, SCREWY, TOUCHED, UNBALANCED *adj. core meaning* : suffering from or showing irrationality and mental unsoundness <diagnosed *insane* by two psychiatrists> ***ant:*** sane

**in·san·i·tar·y** (ĭn-săn′ĭ-tĕr′ē) *adj.* Not sanitary : CONTAMINATED.

**in·san·i·ty** (ĭn-săn′ĭ-tē) *n., pl.* **-ties.** **1.** Serious mental disorder or derangement impairing one's ability to function safely and normally. **2. a.** *Law.* Unsoundness of mind sufficient in the judgment of a civil court to render a person unfit to maintain a contractual or other legal relationship or to warrant commitment to a mental institution. **b.** *Law.* A degree of mental malfunctioning regarded in most criminal jurisdictions as being sufficient to prevent the accused from knowing right from wrong concerning the act he or she is charged with or to render him or her unaware of the nature of the act when committing it. **3. a.** Extreme folly. **b.** Something that is extremely foolish.

☆ ***syns:*** INSANITY, DEMENTIA, LUNACY, MADNESS, MANIA, UNBALANCE *n. core meaning* : serious mental illness or disorder impairing one's ability to function safely and normally <hallucinations and wild suspicions—often signs of *insanity*> ***ant:*** sanity

**in·sa·tia·ble** (ĭn-sā′shə-bəl, -shē-ə-) *adj.* [ME *insaciable* < OFr. < Lat. *insatiabilis* : *in-*, not + *satiare*, to fill < *satis*, enough.] Incapable of being satisfied <an *insatiable* desire to learn> **—in·sa′tia·bil′i·ty, in·sa′tia·ble·ness** *n.* **—in·sa′tia·bly** *adv.*

**in·sa·ti·ate** (ĭn-sā′shē-ĭt) *adj.* Not satisfied : INSATIABLE. **—in·sa′ti·ate·ly** *adv.* **—in·sa′ti·ate·ness** *n.*

**in·scribe** (ĭn-skrīb′) *vt.* **-scribed, -scrib·ing, -scribes.** [Lat. *inscribere* : *in-*, in + *scribere*, to write.] **1. a.** To write, print, or engrave (letters or words) on a surface. **b.** To engrave or mark (a surface) with letters or words. **2.** To enter (a name) on a list or in a register. **3.** To sign one's name or write a brief message in or on (a book or picture) when giving it as a gift. **4.** *Math.* To enclose (a polygon or polyhedron) within a closed configuration of lines, curves, or surfaces so that every vertex of the enclosed figure is incident on the enclosing configuration. **—in·scrib′er** *n.*

**in·scrip·tion** (ĭn-skrĭp′shən) *n.* [ME *inscripcioun* < Lat. *inscriptio* < *inscribere*, to inscribe.] **1.** An act or instance of inscribing. **2.** Something inscribed, as the wording on a coin, gravestone, or label. **3.** An enrollment or registration of names. **4.** A short signed message in a book or on a picture given as a gift. **—in·scrip′tion·al, in·scrip′tive** *adj.* **—in·scrip′tive·ly** *adv.*

**in·scru·ta·ble** (ĭn-skro͞o′tə-bəl) *adj.* Difficult to comprehend : ENIGMATIC <an *inscrutable* statement> **—in·scru′ta·bil′i·ty, in·scru′ta·ble·ness** *n.* **—in·scru′ta·bly** *adv.*

**in·sect** (ĭn′sĕkt′) *n.* [Lat. *insectum* < *insectus*, segmented, p.part. of *insecare*, to cut up : *in-*, in + *secare*, to cut.] **1. a.** Any of numerous usu. small invertebrate animals of the class Insecta or Hexapoda, with an adult stage marked by three pairs of legs, a segmented body with three major divisions, and usu. two pairs of wings. **b.** A similar invertebrate animal, as a spider, centipede, or tick. **2.** A small or contemptible person.

**in·sec·ta·ry** (ĭn′sĕk′tə-rē, ĭn-sĕk′-) *n., pl.* **-ries.** A place where living insects are kept or bred.

**in·sec·ti·cide** (ĭn-sĕk′tĭ-sīd′) *n.* An insect-killing agent. **—in·sec′ti·cid′al** (-sīd′l) *adj.* **—in·sec′ti·cid′al·ly** *adv.*

**in·sec·ti·vore** (ĭn-sĕk′tə-vôr′, -vōr′) *n.* [NLat. *Insectivora*, order name : Lat. *insectum*, insect + Lat. *vorare*, to devour.] **1.** A mammal of the order Insectivora, feeding on insects and including the shrew, mole, and hedgehog. **2.** An organism feeding on insects.

**in·sec·tiv·o·rous** (ĭn′sĕk-tĭv′ər-əs) *adj.* **1.** Feeding on insects. **2.** *Bot.* Capable of trapping and absorbing insects, as the Venus's-flytrap.

**in·se·cure** (ĭn′sĭ-kyo͝or′) *adj.* **1.** Not secure or safe <an *insecure* lock> **2.** Unsound : unstable <*insecure* footing> **3.** Lacking self-confidence : UNCERTAIN <*insecure* about the meeting> **—in′se·cure′ly** *adv.* **—in′se·cure′ness, in′se·cu′ri·ty** (-kyo͝or′ĭ-tē) *n.*

**in·sem·i·nate** (ĭn-sĕm′ə-nāt′) *vt.* **-nat·ed, -nat·ing, -nates.** [Lat. *inseminare, inseminat-* : *in-*, in + *seminare*, to plant < *semen*, seed.] **1.** To sow seed in. **2.** To introduce semen into the uterus of. **—in·sem′i·na′tion** *n.* **—in·sem′i·na′tor** *n.*

**in·sen·sate** (ĭn-sĕn'sāt', -sĭt) *adj.* **1. a.** Lacking sensation : INANIMATE. **b.** Unconscious. **2.** Lacking sensibility : CALLOUS. **3.** Lacking sense : FOOLISH. **—in·sen'sate'ly** *adv.* **—in·sen'sate'ness** *n.*

**in·sen·si·ble** (ĭn-sĕn'sə-bəl) *adj.* **1.** Inappreciable : imperceptible <an *insensible* variation> **2.** Deprived of the power of feeling : UNCONSCIOUS <lay *insensible* from the punch> **3. a.** Unsusceptible : unaffected <*insensible* to the heat> **b.** Unheeding : unmindful <not *insensible* of employee suggestions> **c.** Unfeeling : callous. **4.** Lacking intelligence : IRRATIONAL. **—in·sen'si·bil'i·ty, in·sen'si·ble·ness** *n.* **—in·sen'si·bly** *adv.*

**in·sen·si·tive** (ĭn-sĕn'sĭ-tĭv) *adj.* **1.** Not physically sensitive : NUMB. **2. a.** Lacking sensitivity to the feelings or circumstances of others : CALLOUS. **b.** Lacking responsiveness <*insensitive* to the client's needs> **—in·sen'si·tive·ly** *adv.* **—in·sen'si·tive·ness, in·sen'si·tiv'i·ty** *n.*

**in·sen·tient** (ĭn-sĕn'shənt) *adj.* Without consciousness or sensation : INANIMATE. **—in·sen'tience** *n.*

**in·sep·a·ra·ble** (ĭn-sĕp'ər-ə-bəl, -sĕp'rə-) *adj.* Incapable of being separated. **—in·sep'a·ra·bil'i·ty, in·sep'a·ra·ble·ness** *n.* **—in·sep'a·ra·bly** *adv.*

**in·sert** (ĭn-sûrt') *vt.* **-sert·ed, -sert·ing, -serts.** [Lat. *inserere, insert-* : *in-*, in + *serere*, to sow.] **1.** To put or set into, between, or among <*inserted* the pencil in the sharpener> **2.** To introduce into the body of something : INTERPOLATE <*insert* a gloss into a text> **3.** To place into an orbit, trajectory, or stream. *—n.* (ĭn'sûrt'). Something inserted or intended for insertion, as a flier in a newspaper. **—in·sert'er** *n.*

**in·ser·tion** (ĭn-sûr'shən) *n.* **1.** The act of inserting. **2.** Something inserted, as a strip of lace or embroidery inserted in a fabric. **3.** A point or mode of attachment. **4.** An extra section, as for advertising, inserted in a newspaper. **—in·ser'tion·al** *adj.*

**in-serv·ice** (ĭn'sûr'vĭs) *adj.* **1.** Of, relating to, or being a full-time employee <*in-service* firefighters> **2.** Occurring or continuing while one is a full-time employee <*in-service* career education>

**in·ses·so·ri·al** (ĭn'sĕ-sôr'ē-əl, -sōr'-) *adj.* [LLat. *insessor*, occupant < Lat. *insessus*, p.part. of *insidēre*, to sit upon —see INSIDIOUS.] Perching or adapted for perching.

**in·set** (ĭn'sĕt', ĭn-sĕt') *vt.* **-set, -set·ting, -sets.** To set in : INSERT. *—n.* (ĭn'sĕt'). **1.** Something set in, as: **a.** A small illustration or map set within a larger one. **b.** A page or group of pages inserted in a publication. **c.** Material set into a garment as trim. **2. a.** An inflow, as of water. **b.** A channel.

**in·shore** (ĭn'shôr', -shōr') *adj.* **1.** Close to a shore. **2.** Coming toward a shore. **—in'shore'** *adv.*

**in·shrine** (ĭn-shrīn') *v. var. of* ENSHRINE.

**in·side** (ĭn-sīd', ĭn'sīd') *n.* **1.** An interior part. **2.** An inner side or surface. **3.** The middle part or the part away from the edge. **4. insides.** *Informal.* **a.** The inner organs : ENTRAILS. **b.** The inner parts or workings. **5.** *Slang.* Confidential information : TIP. *—adj.* **1.** Inner : interior. **2.** For the interior. **3.** Relating to or emanating from those in authority. **4.** *Slang.* Known only to a select group <the *inside* story> **5.** *Baseball.* Passing too near the body of the batter. —Used of a pitch. *—adv.* Into or in the interior : WITHIN. *—prep.* **1.** Within <*inside* a box> **2.** Into <going *inside* the house> **—inside of.** *Informal.* Within <*inside* of the barn><*inside* of a week> **—inside out. 1.** With the inner surface turned out. **2.** *Informal.* Thoroughly. **—on the inside.** In a position of confidence or influence.

**in·sid·er** (ĭn-sī'dər) *n. Informal.* **1.** An accepted member of a group. **2.** One who has special knowledge or access to confidential information <White House *insiders*>

**inside track** *n.* **1.** The path next to the inner rail in a curved race track. **2.** An advantageous position in a competition.

**in·sid·i·ous** (ĭn-sĭd'ē-əs) *adj.* [Lat. *insidiosus* < *insidiae*, ambush < *insidēre*, to sit upon : *in-*, on + *sedēre*, to sit.] **1.** Working or spreading harmfully in a stealthy or subtle manner. **2.** Designed to entrap : TREACHEROUS. **3.** Beguiling but harmful : SEDUCTIVE. **—in·sid'i·ous·ly** *adv.* **—in·sid'i·ous·ness** *n.*

**in·sight** (ĭn'sīt') *n.* **1.** The capacity to discern the true nature of a situation : INSTINCT. **2.** An elucidating glimpse.

**in·sight·ful** (ĭn'sīt'fəl, ĭn-sīt'-) *adj.* Having or showing insight : PERCEPTIVE. **—in'sight'ful·ly** *adv.*

**in·sig·ni·a** (ĭn-sĭg'nē-ə) *also* **in·sig·ne** (-nē) *n., pl.* **insignia** *or* **-ni·as.** [Lat., pl. of *insigne*, badge of office < *insignis*, remarkable : *in-*, in + *signum*, sign.] **1.** A badge of office or rank : EMBLEM. **2.** A distinguishing sign. ***usage:*** Although it is technically a Latin plural form, *insignia* may be used with either a singular or plural verb in English, as in *The insignia was* (or *were*) *prominently displayed.* A second plural form, *insignias*, is entirely acceptable and is the standard form in the U.S. armed forces.

**in·sig·nif·i·cant** (ĭn'sĭg-nĭf'ĭ-kənt) *adj.* Not significant, esp.: **a.** Lacking in importance : TRIVIAL. **b.** Small in power, size, or value. **c.** Lacking in meaning : MEANINGLESS. **—in'sig·nif'i·cance, in'sig·nif'i·can·cy** *n.* **—in'sig·nif'i·cant·ly** *adv.*

**in·sin·cere** (ĭn'sĭn-sîr') *adj.* Not sincere : HYPOCRITICAL. **—in'sin·cere'ly** *adv.* **—in'sin·cer'i·ty** (-sĕr'ĭ-tē) *n.*

**in·sin·u·ate** (ĭn-sĭn'yo͞o-āt') *v.* **-at·ed, -at·ing, -ates.** [Lat. *insinuare, insinuat-* : *in-*, in + *sinuare*, to curve < *sinus*, curve.] *—vt.* **1. a.** To introduce (e.g., a thought) gradually or slyly. **b.** To introduce (oneself) by subtle or artful means. **2.** To convey by indirect means : HINT. *—vi.* To make insinuations. **—in·sin'u·a'tive** *adj.* **—in·sin'u·a'tor** *n.*

**in·sin·u·at·ing** (ĭn-sĭn'yo͞o-ā'tĭng) *adj.* **1.** Provoking gradual distrust or suspicion : SUGGESTIVE <*insinuating* statements> **2.** Ingratiating <a sweet, *insinuating* manner> **—in·sin'u·at'ing·ly** *adv.*

**in·sin·u·a·tion** (ĭn-sĭn'yo͞o-ā'shən) *n.* **1.** The act or practice of insinuating. **2.** Something insinuated, esp. an artful suggestion.

**in·sip·id** (ĭn-sĭp'ĭd) *adj.* [LLat. *insipidus*: Lat. *in-*, not + *sapidus*, savory < Lat. *sapere*, to taste.] **1.** Lacking zest or flavor : TASTELESS. **2.** Lacking stimulation, excitement, or interest : DULL. **—in'si·pid'i·ty** (ĭn'sĭ-pĭd'ĭ-tē), **in·sip'id·ness** *n.* **—in·sip'id·ly** *adv.*

**in·sip·i·ence** (ĭn-sĭp'ē-əns) *n.* [ME < OFr. < Lat. *insipientia* < *insipiens*, unwise : *in-*, not + *sapiens*, wise, pr.part. of *sapere*, to taste, have sense.] *Archaic.* Lack of wisdom.

**in·sist** (ĭn-sĭst') *v.* **-sist·ed, -sist·ing, -sists.** [Lat. *insistere*, to persist : *in-*, on + *sistere*, to stand.] *—vi.* To be firm in a course or demand <They *insisted* on coming along with us.> *—vt.* To demand or assert persistently <*insisted* they were innocent> **—in·sis'tence, in·sis'ten·cy** *n.*

**in·sis·tent** (ĭn-sĭs'tənt) *adj.* **1.** Adamant in asserting a demand or opinion : FIRM. **2.** Demanding attention <an *insistent* buzzing> **—in·sis'tent·ly** *adv.*

**in si·tu** (ĭn sī'to͞o) *adj. & adv.* [Lat.] In the original place.

**in·snare** (ĭn-snâr') *v. var. of* ENSNARE.

**in·so·bri·e·ty** (ĭn'sə-brī'ĭ-tē) *n.* Lack of sobriety : INTEMPERANCE.

**in·so·far** (ĭn'sō-fär') *adv.* To such an extent.

**insofar as** *conj.* To the extent that.

**in·so·late** (ĭn'sō-lāt', ĭn-sō'-) *vt.* **-lat·ed, -lat·ing, -lates.** [Lat. *insolare, insolat-* : *in-*, in + *sol*, sun.] To expose to sunlight, as for bleaching.

**in·so·la·tion** (ĭn'sō-lā'shən) *n.* **1. a.** Exposure to sunlight. **b.** Therapeutic exposure to sunlight. **2.** Sunstroke. **3. a.** The solar radiation incident on the earth or another planet. **b.** The rate of delivery of such radiation per unit area surface.

**in·sole** (ĭn'sōl') *n.* **1.** The inner sole of a shoe. **2.** An extra strip of material put inside a shoe for protection or comfort.

**in·so·lent** (ĭn'sə-lənt) *adj.* [ME < Lat. *insolens* : *in-*, not + *solens*, pr.part. of *solēre*, to be accustomed.] **1.** Presumptuous and abrasive in speech or manner : ARROGANT. **2.** Brazenly impudent : IMPERTINENT. **—in'so·lence** *n.* **—in'so·lent·ly** *adv.*

**in·sol·u·ble** (ĭn-sŏl'yə-bəl) *adj.* [ME *insolible*, unanswerable < Lat. *insolubilis*, irrefutable : *in-*, not + *solvere*, to loosen.] **1.** Incapable of being dissolved. **2.** Incapable of being explained or solved. **—in·sol'u·bil'i·ty, in·sol'u·ble·ness** *n.* **—in·sol'u·bly** *adv.*

**in·solv·a·ble** (ĭn-sŏl'və-bəl) *adj.* Incapable of being solved. **—in·solv'a·bly** *adv.*

**in·sol·vent** (ĭn-sŏl'vənt) *adj.* **1.** Unable to meet debts or discharge liabilities : BANKRUPT. **2.** Of or relating to insolvent or bankrupt persons. *—n.* One who is insolvent. **—in·sol'ven·cy** *n.*

**in·som·ni·a** (ĭn-sŏm'nē-ə) *n.* [Lat. < *insomnis*, sleepless : *in-*, not + *somnus*, sleep.] Chronic inability to sleep : SLEEPLESSNESS. **—in·som'ni·ac'** (-ăk') *adj. & n.*

**in·so·much as** (ĭn'sō-mŭch') *conj.* **1.** To such extent or degree as. **2.** Inasmuch as : SINCE.

**in·sou·ci·ance** (ĭn-so͞o'sē-əns, ăN'so͞o-syäNs') *n.* Blithe lack of concern : NONCHALANCE.

**in·sou·ci·ant** (ĭn-so͞o'sē-ənt, ăN'so͞o-syäN') *adj.* [Fr. : *in-*, not (< Lat.) + *souciant*, pr.part. of *soucier*, to trouble < Lat. *sollicitare*, to vex. —see SOLICIT.] Cheerfully nonchalant. **—in·sou'ci·ant·ly** *adv.*

**in·soul** (ĭn-sōl') *v. var. of* ENSOUL.

**in·spect** (ĭn-spĕkt') *vt.* **-spect·ed, -spect·ing, -spects.** [Lat. *inspectare*, freq. of *inspicere*, to look into : *in-*, in + *specere*, to look.] **1.** To examine in detail, esp. for flaws. **2.** To review or examine officially <*inspected* the safety procedures at the factory> **—in·spec'tive** *adj.*

**in·spec·tion** (ĭn-spĕk'shən) *n.* **1.** The act of inspecting. **2.** Official review or examination. **—in·spec'tion·al** *adj.*

**in·spec·tor** (ĭn-spĕk'tər) *n.* **1.** A person, esp. an official, who inspects. **2.** A police officer ranking next below superintendent. **—in·spec'to·ral, in'spec·to'ri·al** (-tôr'ē-əl, -tōr'-) *adj.* **—in·spec'tor·ship'** *n.*

**in·spec·tor·ate** (ĭn-spĕk'tər-ĭt) *n.* **1.** The duties or office of an inspector. **2.** A staff of inspectors. **3.** An inspector's district.

**inspector general** *n., pl.* **inspectors general.** An officer with general investigative powers within a civil or military organization.

**in·sphere** (ĭn-sfîr') *v. var. of* ENSPHERE.

**in·spi·ra·tion** (ĭn'spə-rā'shən) *n.* **1. a.** Stimulation of the mind or emotions to a high level of activity or feeling. **b.** The condition of being so stimulated. **2.** One that moves the intellect or emotions or prompts action or invention. **3.** A sudden creative idea or act. **4.** The quality of exalting or inspiring <a sculpture full of *inspiration*> **5.** Divine influence or guidance exerted directly on the mind and soul of man. **6.** The act of breathing in : INHALATION.

**in·spi·ra·tion·al** (ĭn′spə-rā′shə-nəl) *adj.* **1.** Of or relating to inspiration. **2.** Providing or meant to convey inspiration. **3.** Resulting from inspiration. **—in′spi·ra′tion·al·ly** *adv.*

**in·spi·ra·tor** (ĭn′spə-rā′tər) *n.* **1.** An inhaler. **2.** A respirator.

**in·spir·a·to·ry** (ĭn-spīr′ə-tôr′ē, -tōr′ē) *adj.* Of, relating to, or used for the drawing in of air.

**in·spire** (ĭn-spīr′) *v.* **-spired, -spir·ing, -spires.** [ME *enspiren* < OFr. *enspirer* < Lat. *inspirare* : *in-*, into + *spirare*, to breathe.] *—vt.* **1.** To guide, affect, or arouse by divine influence. **2.** To fill with high or reverent emotion : EXALT <works that *inspired* the gathering> **3.** To stimulate to creativity or action. **4.** To elicit : arouse. **5.** To be the cause or source of : OCCASION <a discovery that *inspired* new cancer research> **6.** To inhale (air). **7.** *Archaic.* **a.** To breathe on. **b.** To breathe life into. *—vi.* To inhale. **—in·spir′er** *n.* **—in·spir′ing·ly** *adv.*

**in·spir·it** (ĭn-spĭr′ĭt) *vt.* **-it·ed, -it·ing, -its.** To fill with spirit : ANIMATE.

**in·spis·sate** (ĭn-spĭs′āt′, ĭn′spĭ-sāt′) *vi. & vt.* **-sat·ed, -sat·ing, -sates.** [< LLat. *inspissatus*, thickened : Lat. *in-* (intensive) + *spissus*, thick.] To thicken or cause to thicken, as by evaporation or boiling : CONDENSE. **—in′spis·sa′tion** *n.* **—in′spis′sa′tor** *n.*

**in·sta·bil·i·ty** (ĭn′stə-bĭl′ĭ-tē) *n., pl.* **-ties.** Lack of stability.

**in·stall** *also* **in·stal** (ĭn-stôl′) *vt.* **-stalled, -stall·ing, -stalls** *also* **-stalled, -stal·ling, -stals.** [ME *installen* < OFr. *installer* < Med. Lat. *installare* : *in-*, in + *stallum*, place.] **1.** To set in position or adjust for use. **2.** To induct into an office or rank <a meeting to *install* the new general manager> **3.** To settle in a certain place or condition : ESTABLISH <*installed* myself in the guest room> **—in·stall′er** *n.*

**in·stal·la·tion** (ĭn′stə-lā′shən) *n.* **1.** The act of installing or state of being installed. **2.** An apparatus, as a system of machinery, set up for use. **3.** A military camp or base.

**in·stall·ment**[1] *also* **in·stal·ment** (ĭn-stôl′mənt) *n.* [Alteration of obs. *estallment* < OFr. *estaler*, to place, fix < *estal*, place, of Germanic orig.] **1.** One of several successive payments in settlement of a debt. **2.** A part of something issued at intervals. **3.** One chapter or section of a literary work presented serially.

**in·stall·ment**[2] *also* **in·stal·ment** (ĭn-stôl′mənt) *n.* INSTALLATION 1.

**in·stance** (ĭn′stəns) *n.* [ME *instaunce* < OFr. *instance* < Lat. *instantia*, presence < *instans*, present. —see INSTANT.] **1.** Something illustrative of a class or group : EXAMPLE. **2.** A legal proceeding or process : SUIT. **3.** A step in a process. **4. a.** Prompting : request <called at the *instance* of my supervisor> **b.** *Archaic.* Urgent solicitation. **5.** *Obs.* An impelling motive. *—vt.* **-stanced, -stanc·ing, -stanc·es. 1.** To offer as an example : CITE. **2.** To demonstrate or clarify with examples : EXEMPLIFY.

**in·stan·cy** (ĭn′stən-sē) *n.* **1.** Urgency. **2.** Immediateness.

**in·stant** (ĭn′stənt) *n.* [ME < OFr. < Lat. *instans*, present, pr.part. of *instare*, to approach : *in-*, on + *stare*, to stand.] **1.** A short time : MOMENT. **2.** A particular point in time <the *instant* I left> **3.** The current month. *—adj.* **1.** Immediate <an *instant* prize winner> **2.** Urgent : imperative <an *instant* request> **3.** Now under consideration : PRESENT. **4. a.** Designed for quick and easy preparation <*instant* coffee><*instant* soup> **b.** Appearing, done, or occurring with or as if with maximum quickness and ease. *—adv.* Instantly.

**in·stan·ta·ne·ous** (ĭn′stən-tā′nē-əs) *adj.* [Med. Lat. *instantaneus* < Lat. *instans*, present. —see INSTANT.] **1.** Occurring or completed without perceptible delay. **2.** Present or occurring at a given instant <*instantaneous* velocity> **—in′stan·ta′ne·ous·ly** *adv.* **—in′stan·ta′ne·ous·ness** *n.*

**in·stan·ter** (ĭn-stăn′tər) *adv.* [Med. Lat. < Lat., urgently < *instans*, present. —see INSTANT.] Without delay : INSTANTLY.

**in·stant·ly** (ĭn′stənt-lē) *adv.* **1.** At once : IMMEDIATELY. **2.** *Archaic.* Urgently. *—conj.* As soon as.

**instant replay** *n.* A recording of an event on videotape for playback immediately after completion of the event.

**in·star**[1] (ĭn-stär′) *vt.* **-starred, -star·ring, -stars.** To stud with or as if with stars.

**in·star**[2] (ĭn′stär′) *n.* [NLat. < Lat., image, form.] An insect, as an arthropod, between molts, as during metamorphosis.

**in·state** (ĭn-stāt′) *vt.* **-stat·ed, -stat·ing, -states.** To establish in office : INSTALL.

**in·stau·ra·tion** (ĭn′stô-rā′shən) *n.* [Lat. *instauratio* < *instaurare*, to renew.] *Archaic.* **1.** Renovation : restoration. **2.** Institution.

**in·stead** (ĭn-stĕd′) *adv.* [ME *in sted of*, in place of.] In the place of something previously mentioned : as an alternative or substitute <were going to fly, but drove *instead*>

**instead of** *prep.* In place of : rather than.

**in·step** (ĭn′stĕp′) *n.* **1.** The arched middle part of the human foot. **2.** The part of a shoe or stocking covering the instep.

**in·sti·gate** (ĭn′stĭ-gāt′) *vt.* **-gat·ed, -gat·ing, -gates.** [Lat. *instigare, instigat-*.] **1.** To urge on : GOAD. **2.** To stir up : FOMENT. **—in′sti·ga′tion** *n.* **—in′sti·ga′tive** *adj.* **—in′sti·ga′tor** *n.*

**in·still** *also* **in·stil** (ĭn-stĭl′) *vt.* **-stilled, -still·ing, -stills** *also* **-stils.** [Lat. *instillare*, to drip in : *in-*, in + *stillare*, to drip < *stilla*, drop.] **1.** To introduce by gradual, persistent means. **2.** To pour in drop by drop. **—in′stil·la′tion** (ĭn′stə-lā′shən) *n.* **—in·still′er** *n.*

**in·stinct** (ĭn′stĭngkt′) *n.* [ME < Lat. *instinctus*, impulse < p.part. of *instinguere*, to incite : *in-*, on + *stinguere*, to prick.] **1. a.** The innate, complex, and normally adaptive aspect of behavior. **b.** A strong impulse or motivation. **2.** A natural capability or aptitude. *—adj.* (ĭn-stĭngkt′). **1.** *Obs.* Impelled from within. **2.** Imbued : filled <a look *instinct* with concern>

**in·stinc·tive** (ĭn-stĭngk′tĭv) *adj.* **1.** Of, relating to, or resulting from instinct. **2.** Arising from a natural impulse : SPONTANEOUS <an *instinctive* distrust of strangers> **—in·stinc′tive·ly** *adv.*

**in·sti·tute** (ĭn′stĭ-to͞ot′, -tyo͞ot′) *vt.* **-tut·ed, -tut·ing, -tutes.** [ME *instituten* < Lat. *institutus*, p.part. of *instituere*, to establish : *in-*, in + *statuere*, to set up < *stare*, to stand.] **1. a.** To organize, establish, and set in operation. **b.** To begin : initiate. **2.** To establish in an office or position. *—n.* **1.** *Obs.* The act of instituting. **2.** Something instituted, esp. an authoritative precedent or rule. **3.** An organization established to promote a cause <an *institute* for the blind> **4. a.** An educational institution. **b.** The building or buildings housing such an institution. **5.** A usu. short, intensive seminar or workshop on a subject. **6. institutes.** A digest of the principles or rudiments of a subject, esp. a legal abstract. **—in′sti·tut′er, in′sti·tu′tor** *n.*

**in·sti·tu·tion** (ĭn′stĭ-to͞o′shən, -tyo͞o′shən) *n.* **1.** The act of instituting. **2. a.** An important custom, relationship, or behavioral pattern in a culture or society. **b.** *Informal.* A lasting feature : FIXTURE. **3. a.** An established organization or foundation, esp. one dedicated to public service. **b.** The building or buildings housing such an organization. **4.** A place of confinement, as a mental asylum. **—in′sti·tu′tion·al** *adj.* **—in′sti·tu′tion·al·ly** *adv.*

**in·sti·tu·tion·al·ism** (ĭn′stĭ-to͞o′shə-nə-lĭz′əm, -tyo͞o′-) *n.* **1.** Adherence to or belief in established forms, esp. those of organized religion. **2.** Use of institutions for those incapable of self-care. **—in′sti·tu′tion·al·ist** *n.*

**in·sti·tu·tion·al·ize** (ĭn′stĭ-to͞o′shə-nə-līz′, -tyo͞o′-) *vt.* **-ized, -iz·ing, -iz·es. 1.** To make into, treat as, or give the character of an institution to. **2.** To commit to an institution. **—in′sti·tu′tion·al·i·za′tion** *n.*

**in·struct** (ĭn-strŭkt′) *v.* **-struct·ed, -struct·ing, -structs.** [ME *instructen* < Lat. *instructus*, p.part. of *instruere*, to prepare : *in-*, in + *struere*, to build.] *—vt.* **1.** To furnish with knowledge : TEACH. **2.** To give orders to : DIRECT. *—vi.* To serve as an instructor.

**in·struc·tion** (ĭn-strŭk′shən) *n.* **1.** The act, practice, or profession of instructing. **2. a.** Imparted knowledge. **b.** An acquired or imparted item of knowledge : LESSON. **3.** *Computer Sci.* A machine code telling a computer to perform a particular operation. **4.** *often* **instructions.** An authoritative direction : ORDER. **—in·struc′tion·al** *adj.*

**in·struc·tive** (ĭn-strŭk′tĭv) *adj.* Conveying knowledge or data : ENLIGHTENING. **—in·struc′tive·ly** *adv.* **—in·struc′tive·ness** *n.*

**in·struc·tor** (ĭn-strŭk′tər) *n.* **1.** A person who instructs : TEACHER. **2. a.** An academic rank below assistant professor. **b.** A person holding such a rank. **—in·struc′tor·ship′** *n.*

**in·stru·ment** (ĭn′strə-mənt) *n.* [ME < Lat. *instrumentum*, tool < *instruere*, to prepare. —see INSTRUCT.] **1.** A means by which something is accomplished : AGENCY. **2. a.** One used to accomplish some purpose. **b.** A person used or controlled by another : DUPE. **3.** A utensil or implement. **4.** A device for recording or measuring, esp. such a device functioning as part of a control system. **5.** A device for producing music. **6.** A legal document. *—vt.* (-mĕnt′) **-ment·ed, -ment·ing, -ments. 1.** To equip with instruments. **2.** To address a legal document to. **3.** To arrange (music) for performance.

**in·stru·men·tal** (ĭn′strə-mĕn′tl) *adj.* **1.** Serving as a means or agency : IMPLEMENTAL. **2.** Of, relating to, or done with an instrument or tool. **3.** Performed on or written for a musical instrument. **4.** Of or designating a grammatical case typically used to express means, agency, or accompaniment. *—n.* **1.** The instrumental case. **2.** A word in the instrumental case. **—in′stru·men′tal·ly** *adv.*

**in·stru·men·tal·ism** (ĭn′strə-mĕn′tl-ĭz′əm) *n.* A pragmatic theory that ideas are instruments that function as guides of action, their validity being determined by the success of the action.

**in·stru·men·tal·ist** (ĭn-strə-mĕn′tl-ĭst) *n.* **1.** A person who plays a musical instrument. **2.** A student or advocate of instrumentalism.

**in·stru·men·tal·i·ty** (ĭn′strə-mĕn-tăl′ĭ-tē) *n., pl.* **-ties. 1.** The quality or state of being instrumental. **2.** Means : agency.

**in·stru·men·ta·tion** (ĭn′strə-mĕn-tā′shən) *n.* **1.** Application or use of instruments. **2. a.** Study and practice of arranging music for instruments. **b.** Arrangement or orchestration resulting from such practice. **3.** Study, development, and manufacture of instruments, as for scientific use. **4.** Instrumentality.

**instrument panel** *n.* A mounted array of instruments for operating a machine.

**in·sub·or·di·nate** (ĭn′sə-bôr′dn-ĭt) *adj.* Not submissive to authority : DISOBEDIENT <*insubordinate* behavior> **—in′sub·or′di·na′tion** *n.* **—in′sub·or′di·nate·ly** *adv.*

**in·sub·stan·tial** (ĭn′səb-stăn′shəl) *adj.* **1.** Lacking reality or substance : IMAGINARY. **2. a.** Not solid : FLIMSY. **b.** Delicate : fine. **—in′sub·stan′ti·al′i·ty** (-shē-ăl′ĭ-tē) *n.*

**in·suf·fer·a·ble** (ĭn-sŭf'ər-ə-bəl, -sŭf'rə-) *adj.* Incapable of being endured : INTOLERABLE <*insufferable* rudeness> **—in·suf'fer·a·ble·ness** *n.* **—in·suf'fer·a·bly** *adv.*

**in·suf·fi·cient** (ĭn'sə-fĭsh'ənt) *adj.* Not sufficient : INADEQUATE. **—in'suf·fi'cien·cy** *n.* **—in'suf·fi'cient·ly** *adv.*

**in·suf·flate** (ĭn'sə-flāt', ĭn-sŭf'lāt') *vt.* **-flat·ed, -flat·ing, -flates.** [LLat. *insufflare, insufflat-* : Lat. *in-*, on + Lat. *sufflare,* to inflate.] **1.** To blow or breathe into or on. **2.** To treat medically by blowing a powder, gas, or vapor into a bodily cavity. **—in'suf·fla'tor** *n.*

**in·suf·fla·tion** (ĭn'sə-flā'shən) *n.* **1.** The act or an instance of insufflating. **2.** A Christian rite of exorcism performed by breathing on a person.

**in·su·lar** (ĭn'sə-lər, ĭns'yə-) *adj.* [LLat. *insularis* < Lat. *insula,* island.] **1.** Of, relating to, or being an island. **2. a.** Typical or suggestive of the isolated life of an island. **b.** Circumscribed and detached in viewpoint and experience. **3.** *Anat.* Designating isolated tissue or an island of tissue. **—in'su·lar·ism, in'su·lar'i·ty** (-lăr'ĭ-tē) *n.* **—in'su·lar·ly** *adv.*

**insular sclerosis** *n.* Multiple sclerosis.

**in·su·late** (ĭn'sə-lāt', ĭns'yə-) *vt.* **-lat·ed, -lat·ing, -lates.** [< LLat. *insulatus,* made into an island < Lat. *insula,* island.] **1.** To cause to be in an isolated or detached position. **2.** *Physics.* To prevent the passage of heat, electricity, or sound into or out of, esp. by interposition of an insulator.

**in·su·la·tion** (ĭn'sə-lā'shən, ĭns'yə-) *n.* **1.** The act of insulating or state of being insulated. **2.** Material used in insulating.

**in·su·la·tor** (ĭn'sə-lā'tər, ĭns'yə-) *n.* **1.** An insulating material, esp. a nonconductor of heat or electricity. **2.** An insulating device.

**in·su·lin** (ĭn'sə-lĭn) *n.* [NLat. *insula,* islet (of Langerhans) < Lat., island + -IN.] **1.** A polypeptide hormone secreted by the islands of Langerhans and functioning to regulate carbohydrate metabolism by controlling blood glucose levels. **2.** A preparation derived from the pancreas of the pig or the ox for use in the medical treatment of diabetes.

**insulin shock** *n.* Hypoglycemia caused by excessive insulin in the blood.

**in·sult** (ĭn-sŭlt') *v.* **-sult·ed, -sult·ing, -sults.** [OFr. *insulter,* to triumph over < Lat. *insultare,* to revile : *in-*, on + *saltare,* to dance, freq. of *salire,* to jump.] —*vt.* **1. a.** To speak to or treat insolently or contemptuously. **b.** To cause damage or offense to <a condescending remark that *insulted* our intelligence> **2.** *Obs.* To make an attack upon. —*vi. Obs.* To behave arrogantly. —*n.* (ĭn'sŭlt'). **1.** An offensive action or remark. **2.** *Med.* An injury, irritation, or trauma. **—in·sult'er** *n.* **—in·sult'ing·ly** *adv.*

☆ ***syns:*** INSULT, AFFRONT, OFFEND, OUTRAGE *v. core meaning* : to cause resentment or hurt by rude, unfeeling behavior <foul language that *insulted* everyone>

**in·su·per·a·ble** (ĭn-so͞o'pər-ə-bəl) *adj.* Incapable of being overcome : INSURMOUNTABLE <*insuperable* complications> **—in·su'per·a·bil'i·ty, in·su'per·a·ble·ness** *n.* **—in·su'per·a·bly** *adv.*

**in·sup·port·a·ble** (ĭn'sə-pôr'tə-bəl, -pōr'-) *adj.* **1.** Unendurable : intolerable. **2.** Lacking validity : UNJUSTIFIABLE <*insupportable* allegations> **—in'sup·port'a·ble·ness** *n.* **—in'sup·port'a·bly** *adv.*

**in·sup·press·i·ble** (ĭn'sə-prĕs'ə-bəl) *adj.* Irrepressible. **—in'sup·press'i·bly** *adv.*

**in·sur·ance** (ĭn-sho͝or'əns) *n.* **1. a.** The act, business, or system of insuring. **b.** The state of being insured. **c.** A means of being insured. **2. a.** Contractural coverage binding a party to indemnify another against specified loss in return for premiums paid. **b.** The sum for which such a contract insures something. **c.** The premium paid for such coverage.

**in·sure** (ĭn-sho͝or') *v.* **-sured, -sur·ing, -sures.** [ME *ensuren,* to assure < OFr. *enseurer,* perh. var. of *assurer,* to assure. —see ASSURE.] —*vt.* **1.** To cover with insurance. **2.** To make certain or secure. —*vi.* To buy or sell insurance. **—in·sur'a·bil'i·ty** *n.* **—in·sur'a·ble** *adj.*

**in·sured** (ĭn-sho͝ord') *n.* One covered by insurance.

**in·sur·er** (ĭn-sho͝or'ər) *n.* One that insures, esp. an underwriter.

**in·sur·gence** (ĭn-sûr'jəns) *n.* Revolt : insurrection.

**in·sur·gen·cy** (ĭn-sûr'jən-sē) *n.* **1.** The quality or state of being insurgent. **2.** Insurgence.

**in·sur·gent** (ĭn-sûr'jənt) *adj.* [Lat. *insurgens, insurgent-*, pr.part. of *insurgere,* to rise up : *in-* (intensive) + *surgere,* to rise.] Rising in revolt against established authority, as a government. —*n.* **1.** A person who revolts against civil authority. **2.** A member of a political party who rebels against its leadership. **—in·sur'gent·ly** *adv.*

**in·sur·mount·a·ble** (ĭn'sər-moun'tə-bəl) *adj.* Insuperable <*insurmountable* difficulties> **—in'sur·mount'a·bly** *adv.*

**in·sur·rec·tion** (ĭn'sə-rĕk'shən) *n.* [ME *insurrecion* <OFr. < Lat. *insurrectio* < *insurgere,* to rise up. —see INSURGENT.] An act or instance of open revolt against established authority, as a government. **—in'sur·rec'tion·al** *adj.* **—in'sur·rec'tion·ar'y** *adj.* & *n.* **—in'sur·rec'tion·ism** *n.* **—in'sur·rec'tion·ist** *n.*

**in·sus·cep·ti·ble** (ĭn'sə-sĕp'tə-bəl) *adj.* Not susceptible : INSENSITIVE. **—in·sus·cep'ti·bil'i·ty** *n.* **—in'sus·cep'ti·bly** *adv.*

**in·tact** (ĭn-tăkt') *adj.* [ME *intacte* < Lat. *intactus* : *in-*, not + *tactus,* p.part. of *tangere,* to touch.] **1.** Not damaged in any way. **2.** Having all parts : WHOLE. **—in·tact'ness** *n.*

**in·tag·lio** (ĭn-tăl'yō, -tăl'-) *n., pl.* **-glios.** [Ital. < *intagliare,* to engrave : *in-*, in (< Lat.) + *tagliare,* to cut < VLat. ***talliare.* —see TAILOR.] **1. a.** A design or figure incised beneath the surface of hard metal or stone. **b.** The art or process of carving a design in this manner. **2.** A gemstone carved in intaglio. **3.** Printing done with a plate bearing an image in intaglio. **4.** A die incised so as to produce a design in relief.

**in·take** (ĭn'tāk') *n.* **1.** An opening by which a fluid is admitted into a container or conduit. **2. a.** The act of taking in. **b.** Something, esp. energy, taken in.

**in·tan·gi·ble** (ĭn-tăn'jə-bəl) *adj.* **1.** Incapable of being apprehended by the mind or senses. **2.** Incapable of being defined <an *intangible* premonition of disaster> —*n.* Something intangible, esp. an asset that cannot be apprehended by the mind or senses. **—in·tan'gi·bil'i·ty, in·tan'gi·ble·ness** *n.* **—in·tan'gi·bly** *adv.*

**in·tar·si·a** (ĭn-tär'sē-ə) *n.* [Ital. *intarsio* < *tarsia,* inlaid mosaic work < Ar. *tarṣi.*] **1.** A mosaic worked in wood. **2.** The art or practice of making intarsia.

**in·te·ger** (ĭn'tĭ-jər) *n.* [< Lat., whole, complete.] **1.** A member of the set of positive whole numbers (1, 2, 3, . . . ), negative whole numbers (−1, −2, −3, . . . ), and zero (0). **2.** A complete entity or unit.

**in·te·gra·ble** (ĭn'tĭ-grə-bəl) *adj.* Capable of being integrated. **—in'te·gra·bil'i·ty** *n.*

**in·te·gral** (ĭn'tĭ-grəl, ĭn-tĕg'rəl) *adj.* [LLat. *integralis,* making up a whole < Lat. *integer,* complete.] **1.** Essential for completeness. **2.** Having everything required : ENTIRE. **3.** (ĭn'tĭ-grəl). *Math.* **a.** Expressed or expressible as or in terms of integers. **b.** Expressed as or involving integrals. —*n.* **1.** A complete unit : WHOLE. **2.** (ĭn'tĭ-grəl). *Math.* **a.** A definite integral. **b.** An indefinite integral. **—in'te·gral'i·ty** *n.* **—in'te·gral·ly** *adv.*

**integral calculus** *n.* The mathematical study of integration, the properties of integrals, and their applications.

**integral domain** *n.* A commutative ring with unity having no proper divisors of zero, that is, having no nonzero elements *a, b* such that $a \cdot b = 0$, where 0 is the additive identity.

**in·te·grand** (ĭn'tĭ-grănd') *n.* [< Lat. *integrandus,* gerund. of *integrare,* to integrate.] A function or equation to be integrated.

**in·te·grant** (ĭn'tĭ-grənt) *adj.* Integral.

**in·te·grate** (ĭn'tĭ-grāt') *v.* **-grat·ed, -grat·ing, -grates.** [Lat. *integrare, integrat-*, to make whole < *integer,* complete.] —*vt.* **1.** To make into a whole by bringing all parts together : UNIFY. **2.** To join with something else : UNITE. **3.** To open without restriction to people of all races or ethnic groups : DESEGREGATE. **4.** *Math.* **a.** To calculate the integral of. **b.** To perform integration on. **5.** To bring about the integration of (personality traits). —*vi.* To become integrated or undergo integration. **—in'te·gra'tive** *adj.*

**integrated circuit** *n.* A tiny wafer of substrate material on which a complex of electronic components and their interconnections is etched or imprinted. **—integrated circuitry** *n.*

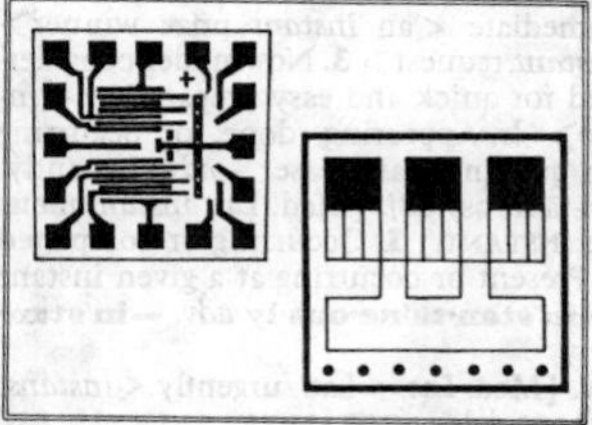

**integrated circuit**

**in·te·gra·tion** (ĭn'tĭ-grā'shən) *n.* **1. a.** An act or process of integrating. **b.** The state of becoming integrated. **c.** Desegregation. **2.** The organization of organic, psychological, or social traits and tendencies of a personality into a harmonious whole. **—in'te·gra'tion·ist** *n.*

**in·te·gra·tor** (ĭn'tĭ-grā'tər) *n.* **1.** One that integrates. **2.** An instrument for mechanically computing definite integrals.

**in·teg·ri·ty** (ĭn-tĕg'rĭ-tē) *n.* [ME *integrite* < OFr. < Lat. *integritas,* soundness < *integer,* whole, complete.] **1.** Firm adherence to a code or standard of values : PROBITY. **2.** The state of being unimpaired : SOUNDNESS. **3.** The quality or condition of being undivided : COMPLETENESS.

**in·te·gro·dif·fer·en·tial** (ĭn'tĭ-grō-dĭf'ə-rĕn'shəl, ĭn-tĕg'rō-) *adj.* [INTEGR(ATION) + DIFFERENTIAL.] Involving both mathematical differentiation and integration.

**in·teg·u·ment** (ĭn-tĕg'yo͞o-mənt) *n.* [Lat. *integumentum* < *integere,* to cover : *in-*, on + *tegere,* to cover.] An outer covering or coat, as the skin of an animal, the coat of a seed, or the membrane enclosing an organ. **—in·teg'u·men'ta·ry** (-mĕn'tə-rē, -mĕn'trē) *adj.*

**in·tel·lect** (ĭn'tl-ĕkt') *n.* [ME < OFr. < Lat. *intellectus,* perception < p.part. of *intellegere,* to perceive. —see INTELLIGENT.] **1. a.** The capacity for understanding and knowledge. **b.** The ability to think abstractly or profoundly. **2.** A person of great intellectual ability.

**in·tel·lec·tion** (ĭn′tl-ĕk′shən) *n.* [ME *intelleccioun,* understanding < OFr. < Lat. *intellectio* < *intellectus,* intellect.] **1.** The act or process of exercising the intellect. **2.** A thought or idea.

**in·tel·lec·tive** (ĭn′tl-ĕk′tĭv) *adj.* Of, relating to, or generated by the intellect. **—in′tel·lec′tive·ly** *adv.*

**in·tel·lec·tron·ics** (ĭn′tl-ĕk-trŏn′ĭks) *n.* [Blend of INTELLECT and ELECTRONICS.] *(sing. in number).* The use of electronic devices to extend human intellect.

**in·tel·lec·tu·al** (ĭn′tl-ĕk′cho͞o-əl) *adj.* **1. a.** Of or relating to the intellect. **b.** Rational rather than emotional. **2.** Appealing to or engaging the intellect. **3. a.** Intelligent. **b.** Given to exercise of the intellect. *—n.* An intellectual person. **—in′tel·lec′tu·al′i·ty** (-ăl′ĭ-tē) *n.* **—in′tel·lec′tu·al·ly** *adv.*

**in·tel·lec·tu·al·ism** (ĭn′tl-ĕk′cho͞o-ə-lĭz′əm) *n.* **1.** Exercise or application of the intellect. **2.** Devotion to development or exercise of the intellect. **—in′tel·lec′tu·al·ist** *n.* **—in′tel·lec′tu·al·is′tic** *adj.*

**in·tel·lec·tu·al·ize** (ĭn′tl-ĕk′cho͞o-ə-līz′) *vt.* **-ized, -iz·ing, -iz·es.** To give a rational structure or meaning for. **—in′tel·lec′tu·al·i·za′tion** *n.* **—in′tel·lec′tu·al·iz′er** *n.*

**in·tel·li·gence** (ĭn-tĕl′ə-jəns) *n.* **1. a.** The capacity to acquire and apply knowledge. **b.** The faculty of thought and reason. **c.** Superior mental powers. **2. a.** An intelligent, incorporeal being, esp. an angel. **b. Intelligence.** *Christian Science.* The primal, eternal quality of God. **3.** News : information. **4. a.** Secret information, esp. about an enemy. **b.** An agency employed in gathering such information.

**intelligence quotient** *n.* The ratio of tested mental age to chronological age, usu. expressed as a quotient multiplied by 100.

**in·tel·li·genc·er** (ĭn-tĕl′ə-jən-sər, -jĕn′-) *n.* **1.** One who conveys news : reporter. **2.** A secret agent : SPY.

**intelligence test** *n.* A standardized test used to determine the relative mental ability of an individual.

**in·tel·li·gent** (ĭn-tĕl′ə-jənt) *adj.* [Lat. *intellegens, intellegent-,* pr.part. of *intellegere,* to perceive : *inter-,* between + *legere,* to choose.] **1.** Having intelligence. **2.** Having a high degree of intelligence : mentally acute. **3.** Displaying sound judgment. **4.** Guided or motivated by the intellect : RATIONAL. **5.** Capable of performing certain computer functions <an *intelligent* terminal> **—in·tel′li·gen′tial** (ĭn-tĕl′ə-jĕn′shəl) *adj.* **—in·tel′li·gent·ly** *adv.*

**in·tel·li·gent·si·a** (ĭn-tĕl′ə-jĕnt′sē-ə, -gĕnt′-) *n.* [R. *intelligentsiya* < Lat. *intellegentia,* intelligence < *intelligens,* intelligent.] The intellectual elite of a society.

**in·tel·li·gi·ble** (ĭn-tĕl′ĭ-jə-bəl) *adj.* [ME < Lat. *intellegibilis* < *intellegere,* to perceive. —see INTELLIGENT.] **1.** Capable of being understood <*intelligible* radio transmissions> **2.** Capable of being apprehended by the mind alone. **—in·tel′li·gi·bil′i·ty, in·tel′li·gi·ble·ness** *n.* **—in·tel′li·gi·bly** *adv.*

**in·tem·per·ance** (ĭn-tĕm′pər-əns, -prəns) *n.* Lack of moderation, esp. in the excessive consumption of alcoholic beverages.

**in·tem·per·ate** (ĭn-tĕm′pər-ĭt, -tĕm′prĭt) *adj.* Not temperate : EXCESSIVE. **—in·tem′per·ate·ly** *adv.* **—in·tem′per·ate·ness** *n.*

**in·tend** (ĭn-tĕnd′) *v.* **-tend·ed, -tend·ing, -tends.** [ME *entenden* < OFr. *entendre* < Lat. *intendere* : *in-,* into + *tendere,* to stretch.] *—vt.* **1.** To have in mind : PLAN. **2. a.** To design for a particular purpose. **b.** To have in mind for a particular use. **3.** To signify : mean. *—vi.* To have a design or purpose in mind.

**in·ten·dance** (ĭn-tĕn′dəns) *n.* **1.** The function of an intendant : MANAGEMENT. **2.** An intendancy.

**in·ten·dan·cy** (ĭn-tĕn′dən-sē) *n., pl.* **-cies. 1.** The function or position of an intendant. **2.** Intendants as a group. **3.** The district supervised by an intendant, as in Latin America.

**in·ten·dant** (ĭn-tĕn′dənt) *n.* [Fr. < OFr., administrator < Lat. *intendens,* pr.part. of *intendere,* to intend.] **1.** An administrative official serving a French, Spanish, or Portuguese monarch. **2.** A district administrator in some Latin American countries.

**in·tend·ed** (ĭn-tĕn′dĭd) *adj.* **1.** Intentional : deliberate. **2.** Proposed for the future : PROSPECTIVE. *—n. Informal.* An engaged person.

**in·tend·ment** (ĭn-tĕnd′mənt) *n.* The true meaning or intention of something, esp. a law.

**in·ten·er·ate** (ĭn-tĕn′ə-rāt′) *vt.* **-at·ed, -at·ing, -ates.** [IN-2 + Lat. *tener,* tender + -ATE1.] To tender : soften. **—in·ten′er·a′tion** *n.*

**in·tense** (ĭn-tĕns′) *adj.* [ME < OFr. < Lat. *intensus,* stretched, p.part. of *intendere,* to intend.] **1.** Having or exhibiting a distinctive feature to an extreme degree <the *intense* heat of the desert> **2.** Extreme in degree. **3.** Involving or displaying strain or extreme effort. **4. a.** Deeply felt : PROFOUND. **b.** Tending to feel deeply. **—in·tense′ly** *adv.* **—in·tense′ness** *n.*

☆ ***syns:*** INTENSE, DESPERATE, FIERCE, FURIOUS, TERRIBLE, VEHEMENT, VIOLENT *adj. core meaning* : extreme in degree, strength, or effect <an *intense* storm>

**in·ten·si·fi·er** (ĭn-tĕn′sə-fī′ər) *n.* **1.** One that intensifies. **2.** An intensive.

**in·ten·si·fy** (ĭn-tĕn′sə-fī′) *v.* **-fied, -fy·ing, -fies.** *—vt.* **1.** To make intense or more intense. **2.** To increase the contrast of (a photographic image). *—vi.* To become intense or more intense. **—in·ten′si·fi·ca′tion** *n.*

**in·ten·sion** (ĭn-tĕn′shən) *n.* [Lat. *intensio,* an intensifying < *intensus,* stretched. —see INTENSE.] **1.** *Logic.* The properties connoted by a term. **2.** Intensity.

**in·ten·si·ty** (ĭn-tĕn′sĭ-tē) *n., pl.* **-ties. 1.** Exceptionally great concentration, power, or force. **2.** *Physics.* **a.** The measure of effectiveness of a force field given by the force per unit test element. **b.** The energy transferred by a wave per unit time across a unit area perpendicular to the direction of propagation.

**in·ten·sive** (ĭn-tĕn′sĭv) *adj.* **1.** Of, relating to, or marked by intensity. **2.** Tending to intensify or emphasize. **3.** Highly concentrated. **4.** Constituting or relating to a method esp. of land cultivation whose purpose is to increase the productivity of a given area by means of an increase in the capital and labor. **5.** *Physics.* Having the same value for any subdivision of a thermodynamic system. ***usage:*** *Intensive,* which is often used interchangeably with *intense,* also has the special meaning of "concentrated," so that one may speak of *intense heat* but *intensive study.* *—n.* A linguistic element, as the adverbs *extremely* or *awfully,* that add force or emphasis.

**intensive care** *n.* **1.** Special medical equipment and services provided for seriously ill patients. **2.** A hospital unit specializing in intensive care.

**in·tent** (ĭn-tĕnt′) *n.* [ME *entente* < OFr. < Med. Lat. *intentus* < Lat., attentive to, p.part. of *intendere,* to intend.] **1.** That which is intended : PURPOSE. **2.** The state of mind operative at the time of an action. **3. a.** Meaning : significance. **b.** Connotation. *—adj.* **1.** Firmly fixed : CONCENTRATED. **2.** Having the attention applied : ENGROSSED. **3.** Having the mind fixed on some purpose <*intent* on leaving>

**in·ten·tion** (ĭn-tĕn′shən) *n.* [ME *entencioun* < OFr. < Lat. *intentio,* attention < *intendere,* to intend.] **1.** A plan of action : DESIGN. **2. a.** An aim that guides action : OBJECT. **b. intentions.** Purpose in regard to marriage. **3.** The import, significance, or thrust of something. **4.** *Med.* The manner or course of healing of a surgical wound. **5.** A concept regarded as the product of attention directed to an object of knowledge.

☆ ***syns:*** INTENTION, AIM, DESIGN, END, GOAL, INTENT, MEANING, PLAN, POINT, PURPOSE, TARGET, VIEW *n. core meaning* : what one intends to do or achieve <was never my *intention* to get involved in their scheme>

**in·ten·tion·al** (ĭn-tĕn′shə-nəl) *adj.* **1.** Deliberately done <an *intentional* oversight> **2.** Relating to logical intention or connotation. **—in·ten′tion·al′i·ty** (-năl′ĭ-tē) *n.* **—in·ten′tion·al·ly** *adv.*

**in·ter** (ĭn-tûr′) *vt.* **-terred, -ter·ring, -ters.** [ME *enteren* < OFr. *enterrer* < Med. Lat. *interrare* : Lat. *in-,* in + Lat. *terra,* earth.] To place in a grave or tomb.

**inter-** *pref.* [ME < OFr. < Lat. < *inter,* between, among.] **1.** Between : among <*inter*national> **2.** In the midst of : WITHIN <*inter*tropical> **3.** Mutual : mutually <*inter*relate> **4.** Reciprocal : reciprocally <*inter*mingle>

**in·ter·a·bang** (ĭn-tĕr′ə-băng′) *n. var. of* INTERROBANG.

**in·ter·act** (ĭn′tər-ăkt′) *vi.* **-act·ed, -act·ing, -acts.** To act on each other. **—in′ter·ac′tion** *n.* **—in′ter·ac′tive** *adj.*

**in·ter·ac·tive terminal** (ĭn′tər-ăk′tĭv) *n.* A computer or data-processing terminal capable of providing a source of both input and output for the computer system to which it is connected.

**in·ter a·li·a** (ĭn′tər ā′lē-ə, ä′lē-ə) *adv.* [Lat.] Among other things.

**in·ter a·li·os** (ĭn′tər ā′lē-ōs′, ä′lē-ōs′) *adv.* [Lat.] Among other persons.

**in·ter·brain** (ĭn′tər-brān′) *n.* The diencephalon.

**in·ter·breed** (ĭn′tər-brēd′) *v.* **-bred** (-brĕd′), **-breed·ing, -breeds.** *—vi.* **1.** To breed with another kind or species. **2.** To breed within a narrow range or with closely related types or individuals : INBREED. *—vt.* To cause to interbreed.

**in·ter·ca·lar·y** (ĭn-tûr′kə-lĕr′ē, ĭn′tər-kăl′ə-rē) *adj.* [Lat. *intercalarius* < *intercalare,* to intercalate.] **1. a.** Inserted in the calendar to make the calendar year correspond to the solar year. —Used of a day or a month. **b.** Having an intercalary day or month inserted. —Used of a year. **2.** Interpolated.

**in·ter·ca·late** (ĭn-tûr′kə-lāt′) *vt.* **-lat·ed, -lat·ing, -lates.** [Lat. *intercalare, intercalat-* : *inter-,* among + *calare,* to proclaim.] **1.** To insert (a day or month) in a calendar. **2.** To insert, interpolate, or interpose. **—in·ter′ca·la′tion** *n.* **—in·ter′ca·la′tive** *adj.*

**in·ter·cede** (ĭn′tər-sēd′) *vi.* **-ced·ed, -ced·ing, -cedes.** [Lat. *intercedere,* to intervene : *inter-,* between + *cedere,* to go.] **1.** To argue on another's behalf. **2.** To act as mediator in a dispute. **—in′ter·ced′er** *n.*

**in·ter·cel·lu·lar** (ĭn′tər-sĕl′yə-lər) *adj.* Between or among cells.

**in·ter·cept** (ĭn′tər-sĕpt′) *vt.* **-cept·ed, -cept·ing, -cepts.** [Lat. *intercipere, intercept-* : *inter-,* between + *capere,* to seize.] **1. a.** To stop or interrupt the progress or course of. **b.** To take possession of by catching (an opponent's ball), esp. in football. **2.** To intersect. **3.** *Obs.* To cut off from access or communication. **4.** To prevent. *—n.* (ĭn′tər-sĕpt′). **1.** *Math.* The distance from the origin of coordinates along a coordinate axis to the point at which a line, curve, or surface intersects the axis. **2.** Interception <a radio *intercept*> **—in′ter·cep′tive** *adj.*

**in·ter·cep·tion** (ĭn′tər-sĕp′shən) *n.* **1.** The act of intercepting or state of being intercepted. **2. a.** Something intercepted. **b.** An intercepted pass, esp. a forward pass in football.

**in·ter·cep·tor** *also* **in·ter·cept·er** (ĭn′tər-sĕp′tər) *n.* **1.** One that intercepts. **2.** A fast-climbing, highly maneuverable fighter plane designed to intercept enemy aircraft.

**in·ter·ces·sion** (ĭn′tər-sĕsh′ən) *n.* [ME < OFr. < Lat. *intercessio,* intervention < *intercessus,* p.part. of *intercedere,* to intervene. —see INTERCEDE.] **1.** Entreaty in favor of another. **2.** Mediation in a dispute. **—in′ter·ces′sion·al** *adj.* **—in′ter·ces′sor** *n.* **—in′ter·ces′so·ry** *adj.*

**in·ter·change** (ĭn′tər-chānj′) *v.* **-changed, -chang·ing, -chang·es.** [ME *enterchaungen* < OFr. *entrechangier* : *inter-,* between (< Lat.) + *changier,* to change.] *—vt.* **1.** To switch each of (two things) into the place of the other. **2.** To give mutually **:** EXCHANGE. *—vi.* To change places with each other. *—n.* (ĭn′tər-chānj′). **1.** The act, process, or an instance of interchanging **:** EXCHANGE. **2.** A highway intersection of allowing traffic to move freely from one road to another without crossing another line of traffic. **—in′ter·chang′er** *n.*

**in·ter·change·a·ble** (ĭn′tər-chān′jə-bəl) *adj.* Capable of mutual exchange. **—in′ter·change′a·bil′i·ty, in′ter·change′a·ble·ness** *n.* **—in′ter·change′a·bly** *adv.*

**in·ter·clav·i·cle** (ĭn′tər-klăv′ĭ-kəl) *n.* A bone in front of the sternum and between the clavicles in most reptiles.

**in·ter·col·le·giate** (ĭn′tər-kə-lē′jĭt, -jē-ĭt) *adj.* Involving or representing two or more colleges.

**in·ter·co·lum·ni·a·tion** (ĭn′tər-kə-lŭm′nē-ā′shən) *n.* **1.** The open spaces between the columns in a colonnade. **2.** The system by which the columns of a colonnade are spaced.

**in·ter·com** (ĭn′tər-kŏm′) *n.* [Short for INTERCOMMUNICATION.] An intercommunication system, as between two rooms.

**in·ter·com·mu·ni·cate** (ĭn′tər-kə-myōō′nĭ-kāt′) *vi.* **-cat·ed, -cat·ing, -cates. 1.** To communicate with each other. **2.** To be connected or next to, as rooms. **—in′ter·com·mu′ni·ca′tion** *n.* **—in′ter·com·mu′ni·ca′tive** (-kā′tĭv, -kə-tĭv) *adj.*

**in·ter·con·nect** (ĭn′tər-kə-nĕkt′) *v.* **-nect·ed, -nect·ing, -nects.** *—vi.* To be connected one to the other. *—vt.* To connect with one another. **—in′ter·con·nec′tion** *n.*

**in·ter·con·ti·nen·tal** (ĭn′tər-kŏn′tə-nĕn′tl) *adj.* **1.** Extending or occurring between or among continents <*intercontinental* trade> **2.** Capable of traveling from one continent to another <*intercontinental* aircraft>

**in·ter·con·ver·sion** (ĭn′tər-kən-vûr′zhən, -shən) *n.* Mutual conversion <*interconversion* of two chemical compounds> **—in′ter·con·vert′i·bil′i·ty** *n.* **—in′ter·con·vert′i·ble** *adj.*

**in·ter·cos·tal** (ĭn′tər-kŏs′təl) *adj.* [NLat. *intercostalis* : INTER- + Lat. *costa,* rib.] Situated or occurring between the ribs.

**in·ter·course** (ĭn′tər-kôrs′, -kōrs′) *n.* [ME *entercours* < OFr. *entrecours* < Lat. *intercursus,* p.part. of *intercurrere,* to mingle with : *inter-,* between + *currere,* to run.] **1.** Exchange or communications between persons or groups. **2.** Sexual intercourse.

**in·ter·crop** (ĭn′tər-krŏp′) *v.* **-cropped, -crop·ping, -crops.** *—vi.* To grow a second crop between the rows of another. *—vt.* To plant a crop between the rows of (another crop). **—in′ter·crop′** *n.*

**in·ter·cur·rent** (ĭn′tər-kûr′ənt) *adj.* [Lat. *intercurrens, intercurrent-,* pr.part. of *intercurrere,* to mingle with. —see INTERCOURSE.] **1.** Occurring as an interruption or delay in a process. **2.** *Pathol.* Occurring during the course of an existing disease.

**in·ter·den·tal** (ĭn′tər-dĕn′tl) *adj.* **1.** Located between the teeth. **2.** Pronounced with the tip of the tongue between the teeth, as (*th*) in *these* or (th) in *thunder.* *—n.* An interdental consonant.

**in·ter·de·pend·ent** (ĭn′tər-dĭ-pĕn′dənt) *adj.* Mutually dependent. **—in′ter·de·pend′ence** *n.*

**in·ter·dict** (ĭn′tər-dĭkt′) *vt.* **-dict·ed, -dict·ing, -dicts.** [Alteration of ME *enterditen,* to place under a church ban < OFr. *entredit,* an interdict < p.part. of *entredire,* to forbid < Lat. *interdicere* : *inter-,* between + *dicere,* to say.] **1.** To forbid or place under an ecclesiastical or legal sanction. **2.** To destroy or interrupt (an enemy line of communication) by firepower so as to halt an enemy's advance. *—n.* (ĭn′tər-dĭkt′). **1.** A prohibition by court order. **2.** *Rom. Cath. Ch.* An ecclesiastical censure excluding a person or area from participation in most sacraments and from Christian burial. **—in′ter·dic′tion** *n.* **—in′ter·dic′tive, in′ter·dic′to·ry** *adj.* **—in′ter·dic′tive·ly** *adv.* **—in′ter·dic′tor** *n.*

**in·ter·est** (ĭn′trĭst, -tər-ĭst, -trĕst′) *n.* [ME, legal claim < OFr. < Lat., it is of importance, 3rd person sing. indicative of *interesse,* to be between : *inter-,* between + *esse,* to be.] **1. a.** A feeling of concern or curiosity about something. **b.** Something, as a quality, that causes such a feeling. **2.** Regard for one's advantage or benefit **:** SELF-INTEREST. **3. a.** A claim, right, or legal share in something. **b.** Something in which such a claim, right, or share is held. **c.** Involvement with or participation in something. **4. a.** A charge for a loan, usu. a percentage of the amount loaned. **b.** An excess beyond what is due. **5.** A group of persons sharing esp. a financial interest in an industry or enterprise. *—vt.* **-est·ed, -est·ing, -ests. 1.** To stimulate the curiosity or hold the attention of. **2.** To cause to become concerned or involved with. **3.** *Obs.* To concern or affect.

**in·ter·est·ed** (ĭn′trĭ-stĭd, -tər-ĭ-stĭd, -tə-rĕs′tĭd) *adj.* **1.** Having or displaying curiosity or concern. **2.** Possessing a claim, right, or share. **—in′ter·est·ed·ly** *adv.* **—in′ter·est·ed·ness** *n.*

**in·ter·est·ing** (ĭn′trĭ-stĭng, -tər-ĭ-stĭng, -tə-rĕs′tĭng) *adj.* Arousing or holding attention. **—in′ter·est·ing·ly** *adv.*

**in·ter·face** (ĭn′tər-fās′) *n.* **1.** A surface forming a common boundary between adjacent regions. **2. a.** A point at which independent systems or diverse groups interact. **b.** The device or system by which interaction at an interface is effected. *—v.* **-faced, -fac·ing, -fac·es.** *—vt.* **1.** To join by means of an interface. **2.** To serve as an interface for. *—vi.* **1.** To serve as an interface or to become interfaced. **2.** To coordinate or interact smoothly. **—in′ter·fa′cial** *adj.*

**in·ter·fere** (ĭn′tər-fîr′) *vi.* **-fered, -fer·ing, -feres.** [ME *enterferen,* to meddle < OFr. *entreferer.*] **1.** To come between so as to be an impediment **:** HINDER. **2.** To obstruct illegally the movement of the ball or of an opposing player, esp. to hinder illegally the catching of a pass in football. **3.** To intrude in the affairs of others **:** MEDDLE. **4.** To strike one hoof against the opposite hoof or leg while moving. —Used of a horse. **5.** *Physics.* To produce interference with (another wave). **6.** *Electron.* To prevent or inhibit clear reception of (broadcast signals). **—in′ter·fer′er** *n.* **—in′ter·fer′ing·ly** *adv.*

☆ ***syns:*** INTERFERE, MEDDLE, TAMPER *v. core meaning* **:** to involve oneself in the affairs of others <a neighbor who *interfered* in my life> Although INTERFERE and MEDDLE are interchangeable, MEDDLE is the stronger word, implying unwanted, unwarranted intrusion <*meddle* in the private lives of others> TAMPER refers to rash, harmful, or illegal intervention <a child who *tampered* with the power tools><attempted to *tamper* with the jury>

**in·ter·fer·ence** (ĭn′tər-fîr′əns) *n.* **1. a.** An act or process of interfering. **b.** Something that interferes. **2. a.** *Football.* The blocking of defensive tacklers to protect the ball carrier. **b.** Illegal hindrance or obstruction of the ball or of an opposing player. **3.** *Physics.* The phenomenon of two or more waves of the same frequency combining to form a wave in which the disturbance at any point is the algebraic or vector sum of the disturbances due to the interfering waves at that point. **4.** *Electron.* **a.** Inhibition or prevention of clear reception of broadcast signals. **b.** The distorted portion of a received signal. **—in′ter·fer·en′tial** (-fə-rĕn′shəl) *adj.*

**in·ter·fe·rom·e·ter** (ĭn′tər-fə-rŏm′ĭ-tər) *n.* Any of several optical, acoustical, or radio-frequency instruments that use interference phenomena between a reference wave and an experimental wave, or between two parts of an experimental wave, to determine wavelengths, wave velocities, distances, and directions. **—in′ter·fer′o·met′ric** *adj.* **—in′ter·fer′o·met′ri·cal·ly** *adv.* **—in′ter·fe·rom′e·try** *n.*

**in·ter·fer·on** (ĭn′tər-fîr′ŏn′) *n.* [INTERFER(E) + -ON[3].] A cellular protein produced in response to and acting to prevent replication of an infectious viral form within an infected cell.

**in·ter·fer·tile** (ĭn′tər-fûr′tl) *adj.* Capable of interbreeding.

**in·ter·ga·lac·tic** (ĭn′tər-gə-lăk′tĭk) *adj.* Between galaxies.

**in·ter·gen·er·a·tion·al** (ĭn′tər-jĕn′ə-rā′shə-nəl) *adj.* Being or occurring between generations <*intergenerational* differences>

**in·ter·gla·cial** (ĭn′tər-glā′shəl) *adj.* Between glacial epochs.

**in·ter·gov·ern·men·tal** (ĭn′tər-gŭv′ərn-mĕn′tl) *adj.* Existing or occurring between two or more governments or divisions of government.

**in·ter·grade** (ĭn′tər-grād′) *vi.* **-grad·ed, -grad·ing, -grades.** To merge into each other through a series of stages, forms, or types. *—n.* (ĭn′tər-grād′). A transitional grade, step, or form. **—in′ter·gra·da′tion** (-grā-dā′shən) *n.*

**in·ter·group** (ĭn′tər-grōōp′) *adj.* Existing or occurring between two or more social groups.

**in·ter·im** (ĭn′tər-ĭm) *n.* [< Lat., in the meantime < *inter,* at intervals, between.] An intervening period. *—adj.* Belonging to or taking place during an interim **:** TEMPORARY <an *interim* arrangement>

**in·ter·i·on·ic** (ĭn′tər-ī-ŏn′ĭk) *adj.* Located or occurring between ions.

**in·te·ri·or** (ĭn-tîr′ē-ər) *adj.* [Lat., comp. of *inter,* within.] **1.** Of, relating to, or located in the inside **:** INNER. **2.** Of or relating to the spiritual or mental life. **3.** Situated away from a shoreline or border **:** INLAND. *—n.* **1.** The internal part or area. **2.** One's spiritual or mental being. **3.** A representation of the inside of a room or structure, as in a painting. **4.** The inland part of a particular political or geographic entity. **—in·te′ri·or′i·ty** (-ôr′ĭ-tē, -ŏr′-) *n.* **—in·te′ri·or·ly** *adv.*

**interior angle** *n.* **1. a.** One of four angles formed between two straight lines cut by a transversal. **b.** A vertex angle measured wholly within a polygon. **2.** The angle formed inside a polygon by two adjacent sides.

**interior decoration** *n.* The planning and execution of the layout, decoration, and furnishing of an architectural interior.

**interior decorator** *n.* A person who practices or specializes in interior decoration.

**interior design** *n.* Interior decoration. **—interior designer** *n.*

**interior monologue** *n.* An often lengthy monologue representing a fictional character's thoughts and emotions.

**in·ter·ject** (ĭn'tər-jĕkt') *vt.* **-ject·ed, -ject·ing, -jects.** [Lat. *interjicere, interject-,* to put between : *inter-,* between + *jacere,* to throw.] To insert between other parts or elements : INTERPOSE. **—in·ter·jec'tor** *n.* **—in'ter·jec'to·ry** *adj.*

**in·ter·jec·tion** (ĭn'tər-jĕk'shən) *n.* **1.** An exclamation. **2. a.** A part of speech made up of exclamatory words or expressions capable of standing alone. **b.** A word, phrase, or utterance used exclamatorily to express emotion, as *Heavens!* or *Oh!* **—in'ter·jec'tion·al** *adj.* **—in'ter·jec'tion·al·ly** *adv.*

**in·ter·lace** (ĭn'tər-lās') *v.* **-laced, -lac·ing, -lac·es.** *—vt.* **1.** To connect by or as if by lacing together : INTERWEAVE. **2.** To intermix : intersperse. *—vi.* To intertwine. **—in'ter·lace'ment** *n.*

**in·ter·lam·i·nate** (ĭn'tər-lăm'ə-nāt') *vt.* **-nat·ed, -nat·ing, -nates.** **1.** To insert between layers. **2.** To arrange in alternate layers. **—in'ter·lam'i·nar** (-nər) *adj.* **—in'ter·lam'i·na'tion** *n.*

**in·ter·lard** (ĭn'tər-lärd') *vt.* **-lard·ed, -lard·ing, -lards.** [ME *interlarden,* to mix fat into < OFr. *entrelarder* : *entre-,* between (< Lat. *inter-*) + *lard,* lard < Lat. *laridum.*] To insert something foreign into <*interlarded* the book with drawings>

**in·ter·leaf** (ĭn'tər-lēf') *n.* A blank leaf inserted between two leaves of a book.

**in·ter·leave** (ĭn'tər-lēv') *vt.* **-leaved, -leav·ing, -leaves.** To provide with an interleaf.

**in·ter·line**[1] (ĭn'tər-līn') *vt.* **-lined, -lin·ing, -lines.** To insert between printed lines. **—in'ter·lin'e·a'tion** (-lĭn'ē-ā'shən) *n.*

**in·ter·line**[2] (ĭn'tər-līn') *vt.* **-lined, -lin·ing, -lines.** To provide (a garment) with an interlining.

**in·ter·lin·e·ar** (ĭn'tər-lĭn'ē-ər) *adj.* **1.** Inserted between the lines of a text. **2.** Written or printed with different languages or versions in alternating lines.

**in·ter·lin·ing** (ĭn'tər-lī'nĭng) *n.* An extra lining between the outer fabric and the regular lining of a garment.

**in·ter·lock** (ĭn'tər-lŏk') *vi.* **-locked, -lock·ing, -locks.** **1.** To lock together or join closely, as by hooking or dovetailing. **2.** *Computer Sci.* To prevent initiation of new operations until current operations are completed. *—vt.* To connect in such a way that no part can operate independently. **—in'ter·lock'** *n.*

**in·ter·lo·cu·tion** (ĭn'tər-lō-kyo͞o'shən) *n.* [Lat. *interlocutio* < *interlocutus,* p.part. of *interloqui,* to interrupt : *inter-,* between + *loqui,* to speak.] Conversation between two or more persons.

**in·ter·loc·u·tor** (ĭn'tər-lŏk'yə-tər) *n.* **1.** One who participates in a conversation. **2.** The performer in a minstrel show who is placed midway between the end men and engages in banter with them.

**in·ter·loc·u·to·ry** (ĭn'tər-lŏk'yə-tôr'ē, -tōr'ē) *adj.* Pronounced or decided during the course of a legal action and merely temporary or provisional in nature.

**in·ter·lope** (ĭn'tər-lōp', ĭn'tər-lōp') *vi.* **-loped, -lop·ing, -lopes.** [Back-formation < E. *interloper* : INTER- + Du. *loper,* running < MDu. < *loopen,* to run.] **1.** To violate the rights, esp. the trading rights, of others. **2.** To meddle in the affairs of others. **—in'ter·lop'er** *n.*

**in·ter·lude** (ĭn'tər-lo͞od') *n.* [ME *enterlude,* a dramatic entertainment < OFr. *entrelude* < Med. Lat. *interludium* : Lat. *inter-,* between + Lat. *ludus,* play.] **1.** An intervening feature, episode, or period of time. **2. a.** A short farcical entertainment performed between the acts of a medieval morality or mystery play. **b.** A 16th-cent. genre of comedy derived from such farces. **c.** An entertainment between the acts of a play. **3.** A short musical piece interposed between the parts of a longer composition.

**in·ter·lu·nar** (ĭn'tər-lo͞o'nər) *adj.* Of or pertaining to the period between the old and new moon when the moon is not visible.

**in·ter·mar·ry** (ĭn'tər-măr'ē) *vi.* **-ried, -ry·ing, -ries.** **1.** To marry a member of another group. **2.** To be bound together by the marriages of members. **3.** To marry within one's family, clan, or tribe. **—in'ter·mar'riage** *n.*

**in·ter·med·dle** (ĭn'tər-mĕd'l) *vi.* **-dled, -dling, -dles.** [ME *entermedlen* < OFr. *entremedler* : *entre-,* between (< Lat. *inter-*) + *medler,* to mix. —see MEDDLE.] To intrude in the affairs of others : MEDDLE. **—in'ter·med'dler** *n.*

**in·ter·me·di·ar·y** (ĭn'tər-mē'dē-ĕr'ē) *n., pl.* **-ies.** **1.** One who acts as a mediator. **2.** One that acts as an agent between persons or things : MEANS. **3.** An intermediate stage or state.

**in·ter·me·di·ate** (ĭn'tər-mē'dē-ĭt) *adj.* [ME < Med. Lat. *intermediatus* < Lat. *intermedius* : *inter-,* between + *medius,* middle.] **1.** Situated or occurring between two extremes or in a middle position or state. **2.** Of or relating to an intermediate school. *—n.* **1.** One that is intermediate. **2.** An intermediary. **3.** *Chem.* A substance formed as a necessary stage in the manufacture of a desired end-product. **4.** An automobile smaller than a full-sized model but larger than a compact. *—vi.* (ĭn'tər-mē'dē-āt') **-at·ed, -at·ing, -ates.** **1.** To act as an intermediary : MEDIATE. **2.** To intervene. **—in'ter·me'di·a·cy** *n.* **—in'ter·me'di·ate·ly** *adv.* **—in'ter·me'di·ate·ness** *n.* **—in'ter·me'di·a'tion** *n.* **—in'ter·me'di·a'tor** *n.*

**intermediate school** *n.* **1.** A junior high school. **2.** A school usu. including grades 4-6.

**in·ter·ment** (ĭn-tûr'mənt) *n.* The act or ritual of interring.

**in·ter·mez·zo** (ĭn'tər-mĕt'sō, -mĕd'zō) *n., pl.* **-zos** or **-zi** (-sē, -zē) [Ital. < Lat. *intermedius,* intermediate.] **1.** A brief entr'acte. **2. a.** A short movement separating the major sections of a lengthy musical composition or work. **b.** An independent instrumental composition having the character of such a movement.

**in·ter·mi·na·ble** (ĭn-tûr'mə-nə-bəl) *adj.* Boringly protracted <an *interminable* lecture> **—in·ter'mi·na·bly** *adv.*

**in·ter·min·gle** (ĭn'tər-mĭng'gəl) *vt.* & *vi.* **-gled, -gling, -gles.** To blend or become mixed together.

**in·ter·mis·sion** (ĭn'tər-mĭsh'ən) *n.* [Lat. *intermissio* < *intermissus,* p.part. of *intermittere,* to intermit.] **1.** The act of intermitting or state of being intermitted. **2.** A respite or recess. **3.** The period between the parts of a performance, as the acts of a play.

**in·ter·mit** (ĭn'tər-mĭt') *vt.* & *vi.* **-mit·ted, -mit·ting, -mits.** [Lat. *intermittere* : *inter-,* at intervals + *mittere,* to let go.] To suspend or cause to suspend some activity periodically or temporarily. **—in'ter·mit'ter** *n.*

**in·ter·mit·tent** (ĭn'tər-mĭt'nt) *adj.* Starting and stopping at intervals. **—in'ter·mit'tence** *n.* **—in'ter·mit'tent·ly** *adv.*

☆ ***syns:*** INTERMITTENT, FITFUL, OCCASIONAL, PERIODIC, SPORADIC *adj. core meaning* : starting and stopping at intervals <*intermittent* showers> ***ant:*** incessant

**intermittent current** *n.* A periodically interrupted one-directional electric current.

**in·ter·mix** (ĭn'tər-mĭks') *vt.* & *vi.* **-mixed, -mix·ing, -mix·es.** [Back-formation < *intermixt,* intermixed < Lat. *intermixtus,* p.part. of *intermiscēre,* to mix together : *inter-,* among + *miscēre,* to mix.] To mix or become mixed together.

**in·tern** *also* **in·terne** (ĭn'tûrn') [Fr. *interne,* house physician < OFr., internal < Lat. *internus.*] *—n.* **1.** An advanced student or recent graduate undergoing supervised practical training. **2.** One who is interned : INTERNEE. *—v.* **-terned, -tern·ing, -terns.** *—vi.* (ĭn'tûrn'). To train or work as an intern. *—vt.* (ĭn-tûrn'). To confine, esp. in wartime. *—adj.* (ĭn-tûrn'). *Archaic.* Internal. **—in'tern·ship'** *n.*

**in·ter·nal** (ĭn-tûr'nəl) *adj.* [NLat. *internalis* < Lat. *internus* < *inter,* within.] **1.** Of, relating to, or situated within the limits or surface of something : INNER. **2.** Residing in or dependent on the essential nature of something : INHERENT. **3.** Located, effective, or acting within the body. **4.** Of or pertaining to the domestic affairs of a country. **—in'ter·nal'i·ty** *n.* **—in·ter'nal·ly** *adv.*

**in·ter·nal-com·bus·tion engine** (ĭn-tûr'nəl-kəm-bŭs'chən) *n.* An engine, as an automotive gasoline piston engine or a diesel, in which fuel is burned within the engine proper rather than in an external furnace, as in a steam engine.

**internal ear** *n.* The portion of the ear including the semicircular canals, vestibule, and cochlea.

**in·ter·nal·ize** (ĭn-tûr'nə-līz') *vt.* **-ized, -iz·ing, -iz·es.** **1.** To make internal. **2.** To absorb (e.g., cultural values) and make an integral part of one's beliefs or attitudes. **—in·ter'nal·i·za'tion** *n.*

**internal medicine** *n.* Medical study and treatment of nonsurgical diseases in adults.

**internal rhyme** *n.* Rhyme between a word within a line and another word at the end of that line or between two words within two different lines.

**internal secretion** *n.* A secretion of an endocrine gland discharged directly into the blood.

**in·ter·na·tion·al** (ĭn'tər-năsh'ə-nəl) *adj.* **1.** Of, pertaining to, or involving two or more nations. **2.** Extending across two or more national borders. *—n.* **International.** Any of various socialist organizations of international scope formed during the late 19th and early 20th cent. **—in'ter·na'tion·al'i·ty** *n.* **—in'ter·na'tion·al·ly** *adv.*

**International Date Line** *n.* The date line.

**in·ter·na·tion·al·ism** (ĭn'tər-năsh'ə-nə-lĭz'əm) *n.* **1.** The quality or state of being international in principles, character, concern, or attitude. **2.** A policy or practice of cooperation among nations, esp. in politics and economy. **—in'ter·na'tion·al·ist** *n.*

**in·ter·na·tion·al·ize** (ĭn'tər-năsh'ə-nə-līz') *vt.* **-ized, -iz·ing, -iz·es.** **1.** To make international. **2.** To place under international control. **—in'ter·na'tion·al·i·za'tion** *n.*

**international law** *n.* A set of rules gen. regarded as binding in relations between states and nations.

**international Morse code** *n.* The continental code.

**International Phonetic Alphabet** *n.* A phonetic alphabet that includes a unique symbol for each speech sound.

**international pitch** *n.* A standard of tuning of 440 vibrations per second for A above middle C.

**International Red Cross** *n.* An organization formed according to the terms of the Geneva Convention of 1864 for the care of the wounded, sick, and homeless in wartime and now also during and following national disasters.

**International System** *n.* A complete, coherent system of units used for scientific work, based on the metric system with the addition of units of time, electric current, temperature, and luminous intensity.

**international unit** *n.* An internationally accepted quantity of a biological substance, as a vitamin or antibiotic, that produces a specific biological effect.

**in·terne** (ĭn'tûrn', ĭn-tûrn') *n.* & *v.* & *adj. var. of* INTERN.

**in·ter·nec·ine** (ĭn'tər-nĕs'ēn', -ĭn, -nē'sĭn') *adj.* [Lat. *internecinus* < *internecio*, massacre < *internecare*, to slaughter : *inter-* (intensive) + *necare*, to kill.] **1.** Mutually destructive. **2.** Marked by bloodshed or slaughter. **3.** Of or pertaining to struggle within a group.

**in·tern·ee** (ĭn'tûr-nē') *n.* A person who is interned.

**in·ter·neu·ron** (ĭn'tər-nŏŏr'ŏn', -nyŏŏr'-) *n.* An internuncial neuron. **—in'ter·neu'ro·nal** (-nŏŏr'ə-nəl, -nyŏŏr'-, -nŏŏ-rō'-, -nyŏŏ-) *adj.*

**in·ter·nist** (ĭn-tûr'nĭst) *n.* [INTERN(AL MEDICINE) + -IST.] A physician specializing in internal medicine.

**in·tern·ment** (ĭn-tûrn'mənt) *n.* The act of interning or state of being interned.

**in·ter·node** (ĭn'tər-nōd') *n.* A section or part between two nodes, as of a nerve or stem. **—in'ter·nod'al** (-nōd'l) *adj.*

**in·ter·nu·cle·ar** (ĭn'tər-nōō'klē-ər, -nyōō'-) *adj.* Situated or occurring between nuclei.

**in·ter·nun·cial** (ĭn'tər-nŭn'shəl, -sē-əl) *adj.* [INTERNUNCI(O) + -AL.] Linking two neurons in a neuronal pathway. **—in'ter·nun'cial·ly** *adv.*

**in·ter·nun·ci·o** (ĭn'tər-nŭn'sē-ō', -nōōn'-) *n., pl.* **-os.** [Ital. *internunzio* < Lat. *internuntius*, mediator : *inter-*, between + *nuntius*, messenger.] **1.** A Vatican diplomatic envoy or representative ranking just beneath a nuncio. **2.** An agent or messenger : GO-BETWEEN.

**in·ter·o·cep·tor** (ĭn'tər-ō-sĕp'tər) *n.* [INTER(IOR) + (RE)CEPTOR.] A specialized sensory nerve receptor responding to stimuli originating in internal organs. **—in'ter·o·cep'tive** *adj.*

**in·ter·of·fice** (ĭn'tər-ô'fĭs, -ŏf'ĭs) *adj.* Being or conducted between offices, esp. of an organization.

**in·ter·pel·late** (ĭn'tər-pĕl'āt', ĭn'tər-pə-lāt') *vt.* **-lat·ed, -lat·ing, -lates.** [Lat. *interpellare, interpellat-*, to interrupt, disturb.] To question (a government official) formally about government policy or action or about personal behavior. **—in'ter·pel·la'tion** *n.* **—in'ter·pel'la'tor** *n.*

**in·ter·pen·e·trate** (ĭn'tər-pĕn'ĭ-trāt') *v.* **-trat·ed, -trat·ing, -trates.** —*vt.* To penetrate between or throughout. —*vi.* To penetrate mutually. **—in'ter·pen'e·tra'tion** *n.*

**in·ter·per·son·al** (ĭn'tər-pûr'sə-nəl) *adj.* Of, pertaining to, involving, or being relations between persons. **—in'ter·per'son·al·ly** *adv.*

**in·ter·phase** (ĭn'tər-fāz') *n.* A stage or period between two successive mitotic divisions of a cell nucleus. **—in'ter·phase'** *v.* **(-phased, -phas·ing, -phas·es).**

**in·ter·plan·e·tar·y** (ĭn'tər-plăn'ĭ-tĕr'ē) *adj.* Between planets.

**in·ter·play** (ĭn'tər-plā') *n.* Reciprocal action and reaction : INTERACTION. —*vi.* **-played, -play·ing, -plays.** To act or react on each other.

**in·ter·plead** (ĭn'tər-plēd') *vi.* **-plead·ed, -plead·ing, -pleads.** [ME *enterpleden* < AN *enterpleder* : *enter-*, between (< Lat. *inter-*) + *pleder*, to plead < OFr. *plaidier.* —see PLEAD.] *Law.* To go to court together to establish a dispute involving a third party.

**in·ter·plead·er** (ĭn'tər-plē'dər) *n. Law.* A legal procedure to determine which of two persons bringing the same suit against a third person is the rightful claimant.

**in·ter·po·late** (ĭn-tûr'pə-lāt') *v.* **-lat·ed, -lat·ing, -lates.** [Lat. *interpolare, interpolat-* : *inter-*, between + *polire*, to embellish.] —*vt.* **1.** To insert or introduce between other parts or elements. **2. a.** To insert (material) into a text. **b.** To insert into a conversation. **3.** To change or falsify (a text) by introducing additional or untrue material. **4.** *Math.* To determine a value of (a function) between known values by a procedure or algorithm different from that specified by the function itself. —*vi.* To make insertions. **—in·ter'po·la'tion** *n.* **—in·ter'po·la'tive** *adj.* **—in·ter'po·la'tor** *n.*

**in·ter·pose** (ĭn'tər-pōz') *v.* **-posed, -pos·ing, -pos·es.** [OFr. *interposer* < Lat. *interpositus*, p.part. of *interponere*, to put between : *inter-*, between + *ponere*, to put.] —*vt.* **1. a.** To introduce or insert between parts. **b.** To place (oneself) between : INTRUDE. **2.** To introduce or interject (e.g., a comment) during a conversation or discourse. **3.** To exert (influence or authority) in order to intervene or interfere. —*vi.* **1.** To come between. **2.** To come between the parties in a dispute : INTERVENE. **3.** To insert a comment, question, or argument. **—in'ter·po'sal** *n.* **—in'ter·pos'er** *n.* **—in'ter·po·si'tion** (-pə-zĭsh'ən) *n.*

**in·ter·pret** (ĭn-tûr'prĭt) *v.* **-pret·ed, -pret·ing, -prets.** [ME *interpreten* < OFr. *interpreter* < Lat. *interpretari* < *interpres*, negotiator, explainer.] —*vt.* **1.** To explain to oneself the meaning of : ELUCIDATE. **2.** To expound the significance of. **3.** To represent or delineate the meaning of, esp. through artistic performance. —*vi.* **1.** To present an explanation. **2.** To serve as an interpreter between speakers of different languages. **—in·ter'pret·a·bil'i·ty, in·ter'pret·a·ble·ness** *n.* **—in·ter'pret·a·ble** *adj.*

**in·ter·pre·ta·tion** (ĭn-tûr'prĭ-tā'shən) *n.* **1.** The act, process, or result of interpreting : EXPLANATION. **2.** A representation of the meaning of an artistic work. **—in·ter'pre·ta'tion·al** *adj.*

**in·ter·pre·ta·tive** (ĭn-tûr'prĭ-tā'tĭv) *adj.* Explanatory : expository. **—in·ter'pre·ta'tive·ly** *adv.*

**in·ter·pret·er** (ĭn-tûr'prĭ-tər) *n.* **1.** One who translates orally from one language into another. **2.** One who gives an explanation. **3.** *Computer Sci.* A program that translates an instruction into a machine language and executes it before proceeding to the next one.

**in·ter·pre·tive** (ĭn-tûr'prĭ-tĭv) *adj.* Interpretative.

**in·ter·pu·pil·lar·y** (ĭn'tər-pyōō'pə-lĕr'ē) *adj.* Situated or occurring between the pupils of the eyes.

**in·ter·reg·num** (ĭn'tər-rĕg'nəm) *n., pl.* **-nums** *or* **-na** (-nə) [Lat. : *inter-*, between + *regnum*, reign < *rex*, king.] **1.** The period of time between the end of a sovereign's reign and the accession of a successor. **2.** A temporary suspension of the usual functions of control or government. **3.** A lapse in continuity. **—in'ter·reg'nal** *adj.*

**in·ter·re·late** (ĭn'tər-rĭ-lāt') *v.* **-lat·ed, -lat·ing, -lates.** —*vt.* To put in mutual relationship. —*vi.* To come into mutual relationship. **—in'ter·re·la'tion** *n.* **—in'ter·re·la'tion·ship'** *n.*

**in·ter·ro·bang** *also* **in·ter·a·bang** (ĭn-tĕr'ə-băng') *n.* [INTERRO(GATION POINT) + BANG, (printers' slang) exclamation point.] A punctuation mark used esp. to end a simultaneous question and exclamation.

**in·ter·ro·gate** (ĭn-tĕr'ə-gāt') *vt.* **-gat·ed, -gat·ing, -gates.** [Lat. *interrogare, interrogat-* : *inter-*, between + *rogare*, to ask.] **1.** To question formally. **2.** *Computer Sci.* To send out a signal to for producing an appropriate response. **—in·ter'ro·ga'tion** *n.* **—in·ter'ro·ga'tion·al** *adj.* **—in·ter'ro·ga'tor** *n.*

**interrogation point** *n.* A question mark.

**in·ter·rog·a·tive** (ĭn'tə-rŏg'ə-tĭv) *adj.* **1.** Of the nature of a question. **2.** Used to ask a question. —*n.* **1.** A word or form used to ask a question. **2.** An interrogative expression or sentence. **—in'ter·rog'a·tive·ly** *adv.*

**in·ter·rog·a·to·ry** (ĭn'tə-rŏg'ə-tôr'ē, -tōr'ē) *adj.* Interrogative. —*n., pl.* **-ries.** *Law.* A written question, as to a witness, usu. answered under oath. **—in'ter·rog'a·to'ri·ly** *adv.*

**in·ter·rupt** (ĭn'tə-rŭpt') *v.* **-rupt·ed, -rupt·ing, -rupts.** [ME *interrupten* < OFr. *interrupte*, interrupted < Lat. *interruptus*, p.part. of *interrumpere*, to break off : *inter-*, between + *rumpere*, to break.] —*vt.* **1.** To break the harmony or continuity of. **2.** To impede or stop by breaking in on. —*vi.* To break in on an action or discourse. —*n.* **1.** *Computer Sci.* A signal to a computer that stops the execution of a running program in order to run a program of higher priority. **2.** A circuit that transmits an interrupt signal. **—in'ter·rup'tion** *n.* **—in'ter·rup'tive** *adj.*

**in·ter·rupt·er** (ĭn'tə-rŭp'tər) *n.* **1.** One that interrupts. **2.** *Elect.* A device for automatically opening or closing an electric circuit.

**in·ter·scho·las·tic** (ĭn'tər-skə-lăs'tĭk) *adj.* Occurring or conducted between or among schools <*interscholastic* sports>

**in·ter se** (ĭn'tər sē', sā') *adv.* & *adj.* [Lat.] Between or among themselves.

**in·ter·sect** (ĭn'tər-sĕkt') *v.* **-sect·ed, -sect·ing, -sects.** [Lat. *intersecare, intersect-* : *inter-*, between + *secare*, to cut.] —*vt.* **1.** To cut through or across. **2.** To form an intersection with. —*vi.* **1.** To overlap or cut across each other. **2.** To form an intersection.

**in·ter·sec·tion** (ĭn'tər-sĕk'shən) *n.* **1. a.** The act or process of intersecting. **b.** (*also* ĭn'tər-sĕk'-). A place where things, as streets, intersect. **2.** *Math.* **a.** The point or locus of points common to two or more geometric figures. **b.** A set every member of which is an element of each of two or more given sets.

**in·ter·ses·sion** (ĭn'tər-sĕsh'ən) *n.* The period between two academic semesters or sessions. **—in'ter·ses'sion·al** *adj.*

**in·ter·sex** (ĭn'tər-sĕks') *n.* An intersexual individual.

**in·ter·sex·u·al** (ĭn'tər-sĕk'shōō-əl) *adj.* **1.** Occurring between the sexes. **2.** Having sexual characteristics intermediate between those of a typical male and a typical female. **—in'ter·sex'u·al'i·ty** (-ăl'ĭ-tē) *n.* **—in'ter·sex'u·al·ly** *adv.*

**in·ter·space** (ĭn'tər-spās') *vt.* **-spaced, -spac·ing, -spac·es.** To make or occupy a space between. —*n.* (ĭn'tər-spās'). An intervening space : INTERVAL. **—in'ter·spa'tial** (-spā'shəl) *adj.*

**in·ter·spe·cif·ic** (ĭn'tər-spĭ-sĭf'ĭk) *adj.* Arising between species.

**in·ter·sperse** (ĭn'tər-spûrs') *vt.* **-spersed, -spers·ing, -spers·es.** [< Lat. *interspersus*, interspersed : *inter-*, between + *sparsus*, p.part. of *spargere*, to scatter.] **1.** To scatter among other things. **2.** To diversify or supply with things distributed at intervals. **—in'ter·spers'ed·ly** (-spûr'sĭd-lē) *adv.* **—in'ter·sper'sion** (-spûr'zhən, -shən) *n.*

**in·ter·state** (ĭn'tər-stāt') *adj.* Relating to, existing between, or connecting two or more states <*interstate* commerce> —*n.* One of a system of highways connecting U.S. cities or states.

**in·ter·stel·lar** (ĭn'tər-stĕl'ər) *adj.* Among or between the stars.

**in·ter·stice** (ĭn-tûr'stĭs) *n., pl.* **-sti·ces** (-stĭ-sēz', -sĭz) [Fr. < LLat. *interstitium* < Lat. *interstitus*, p.part. of *intersistere*, to stand in the middle : *inter-*, between + *sistere*, to stand.] A space, esp. a small or narrow one, between things or parts.

**in·ter·sti·tial** (ĭn'tər-stĭsh'əl) *adj.* **1.** Of, relating to, or occurring in interstices. **2.** Affecting or based on interstices.

**in·ter·tex·ture** (ĭn'tər-tĕks'chər) *n.* **1.** The act of interweaving or state of being interwoven. **2.** Something interwoven.

**in·ter·tid·al** (ĭn'tər-tīd'l) *adj.* Of, relating to, or being the region between the extremes of high and low tide. **—in'ter·tid'al·ly** *adv.*

**in·ter·tri·bal** (ĭn′tər-trī′bəl) *adj.* Existing between tribes.

**in·ter·trop·i·cal** (ĭn′tər-trŏp′ĭ-kəl) *adj.* **1.** Located between or in the tropics. **2.** Of or relating to the tropics.

**in·ter·twine** (ĭn′tər-twīn′) *vt.* & *vi.* **-twined, -twin·ing, -twines.** To join by twining together. **—in′ter·twine′ment** *n.*

**in·ter·twist** (ĭn′tər-twĭst′) *vt.* & *vi.* **-twist·ed, -twist·ing, -twists.** To intertwine.

**in·ter·ur·ban** (ĭn′tər-ûr′bən) *adj.* Relating to or connecting urban areas <*interurban* transportation>

**in·ter·val** (ĭn′tər-vəl) *n.* [ME *intervalle* < OFr. < Lat. *intervallum* : *inter-*, between + *vallum*, rampart.] **1.** A space between objects, points, or units. **2.** The pause between two specified instants, events, or states. **3.** *Math.* **a.** A set containing all the numbers between a pair of given numbers. **b.** Such a set including the endpoints. **c.** Such a set not including the endpoints. **d.** A line segment representing such a set. **e.** A set of numbers greater than or less than a given number and excluding or including the given number. **4.** *Chiefly Brit.* An intermission. **5.** Difference in pitch between two musical tones.

**†in·ter·vale** (ĭn′tər-vāl′) *n.* [Obs. *intervale,* alteration of INTERVAL.] *Regional.* A tract of low-lying land, esp. along a river.

**in·ter·vene** (ĭn′tər-vēn′) *vi.* **-vened, -ven·ing, -venes.** [Lat. *intervenire* : *inter-*, between + *venire*, to come.] **1.** To enter or occur as an unnecessary condition or characteristic. **2.** To appear, come, or lie between two things. **3.** To occur or fall between two periods or points of time. **4.** To come in or between so as to modify or hinder <*intervened* in an argument> **5.** To interfere, usu. through force or threat of force, in the affairs of another nation. **6.** *Law.* To enter into a suit as a third party for the protection of an alleged interest. **—in′ter·ven′er** *n.* **—in′ter·ven′tion** (-vĕn′shən) *n.*

**in·ter·ven·tion·ism** (ĭn′tər-vĕn′shə-nĭz′əm) *n.* The policy or practice of intervening in the affairs of another sovereign state. **—in′ter·ven′tion·ist** *n.*

**in·ter·ver·te·bral** (ĭn′tər-vûr′tə-brəl, -vûr-tē′-) *adj.* Located between vertebrae. **—in′ter·ver′te·bral·ly** *adv.*

**intervertebral disk** *n.* A broad disk of fibrocartilage located between adjoining vertebrae of the spinal column.

**in·ter·view** (ĭn′tər-vyo͞o′) *n.* [Fr. *entrevue* < *entrevu*, p.part. of *entrevoir*, to see : *entre-*, between (< Lat. *inter-*) + *voir*, to see < Lat. *vidēre*.] **1.** A formal face-to-face meeting, esp. one arranged for evaluating the qualifications of an applicant, as for employment. **2. a.** A conversation, as one conducted by a reporter, in which information is elicited from another. **b.** An account or reproduction of such a conversation. *—v.* **-viewed, -view·ing, -views.** *—vt.* To gain an interview from. *—vi.* To have an interview. **—in′ter·view·ee′** *n.* **—in′ter·view′er** *n.*

**in·ter vi·vos** (ĭn′tər vē′vōs, vī′-) *adj.* [Lat.] Between living persons <*inter vivos* awards>

**in·ter·vo·cal·ic** (ĭn′tər-vō-kăl′ĭk) *adj.* Immediately preceded and followed by a vowel.

**in·ter·volve** (ĭn′tər-vŏlv′) *vt.* & *vi.* **-volved, -volv·ing, -volves.** To intertwine.

**in·ter·weave** (ĭn′tər-wēv′) *v.* **-wove** (-wōv′), **-wo·ven** (-wō′vən), **-weav·ing, -weaves.** *—vt.* **1.** To weave together. **2.** To blend together : INTERMIX. *—vi.* To intertwine.

**in·tes·tate** (ĭn-tĕs′tāt′, -tĭt) *adj.* [ME < OFr. *intestat* < Lat. *intestatus* : *in-*, not + *testatus*, testate.] **1.** Having made no legal will. **2.** Not disposed of by a legal will. *—n.* One who dies intestate. **—in·tes′ta·cy** (-tə-sē) *n.*

**in·tes·ti·nal** (ĭn-tĕs′tə-nəl) *adj.* Of, relating to, or constituting the intestine. **—in·tes′ti·nal·ly** *adv.*

**intestinal fortitude** *n.* Courage : perseverance.

**in·tes·tine** (ĭn-tĕs′tĭn) *n.* [Lat. *intestinum* < *intestinus*, internal < *intus*, within.] The portion of the alimentary canal from the stomach to the anus.

**in·thrall** (ĭn-thrôl′) *v. var. of* ENTHRALL.

**in·throne** (ĭn-thrōn′) *v. var. of* ENTHRONE.

**in·ti·ma** (ĭn′tə-mə) *n., pl.* **-mae** (-mē′) *or* **-mas.** [NLat. < Lat., fem. of *intimus*, innermost.] *Anat.* The innermost layer of a bodily organ or part, esp. the wall of a lymphatic vessel, an artery, or a vein. **—in′ti·mal** *adj.*

**in·ti·ma·cy** (ĭn′tə-mə-sē) *n., pl.* **-cies.** The state of being intimate.

**in·ti·mate¹** (ĭn′tə-mĭt) *adj.* [LLat. *intimatus*, p.part. of *intimare*, to intimate.] **1.** Marked by close association, acquaintance, or familiarity <an *intimate* understanding of foreign diplomacy> **2.** Relating to or characteristic of one's deepest nature. **3.** Essential : fundamental. **4.** Marked by privacy and informality <an *intimate* café> **5. a.** Very personal : PRIVATE. **b.** Of or having sexual relations. *—n.* A close friend. **—in′ti·mate·ly** *adv.* **—in′ti·mate·ness** *n.*

**in·ti·mate²** (ĭn′tə-māt′) *vt.* **-mat·ed, -mat·ing, -mates.** [LLat. *intimare, intimate-*, to make known < Lat. *intimus*, innermost.] **1.** To communicate indirectly or subtly : HINT. **2.** To announce : proclaim. **—in′ti·mat′er** *n.* **—in′ti·ma′tion** *n.*

**in·tim·i·date** (ĭn-tĭm′ĭ-dāt′) *vt.* **-dat·ed, -dat·ing, -dates.** [Med. Lat. *intimidare, intimidat-* : Lat. *in-* (intensive) + *timidus*, timid.] **1.** To make timid : FRIGHTEN. **2.** To inhibit or discourage by or as if by threats. **—in·tim′i·da′tion** *n.* **—in·tim′i·da′tor** *n.*

**in·tinc·tion** (ĭn-tĭngk′shən) *n.* [LLat. *intinctio*, a dipping in < Lat. *intingere*, to dip in : *in-*, in + *tingere*, to moisten.] The administration of the Eucharist by dipping the host into the wine and offering both simultaneously to the communicant.

**in·tine** (ĭn′tēn′) *n.* [G. < Lat. *intus*, within.] The inner wall layer of a spore or pollen grain.

**in·ti·tule** (ĭn-tĭch′o͞ol) *vt.* **-uled, -ul·ing, -ules.** [OFr. *intituler* < LLat. *intitulare* : Lat. *in-*, in + Lat. *titulus*, title.] *Chiefly Brit.* To give a designation or title to (e.g., a legislative act).

**in·to** (ĭn′to͞o) *prep.* [ME < OE : *in*, in + *to*, to.] **1.** To the inside or interior of. **2.** To the activity or occupation of <went *into* medicine> **3.** To the condition or form of <fall *into* ruin> **4.** So as to be in or be included in <enter *into* an association> **5.** To a point within the limits of a period of time or extent of space <carried over *into* the following month> **6.** Against <ran *into* a wall> **7.** In the direction of : TOWARD <look *into* the future> **8.** *Informal.* Interested in or involved with <*into* jogging this year>

**in·tol·er·a·ble** (ĭn-tŏl′ər-ə-bəl) *adj.* **1.** That cannot be tolerated : UNBEARABLE <*intolerable* pain> **2.** Extravagant : inordinate. **—in·tol′er·a·bil′i·ty, in·tol′er·a·ble·ness** *n.* **—in·tol′er·a·bly** *adv.*

**in·tol·er·ant** (ĭn-tŏl′ər-ənt) *adj.* Not tolerant, esp.: **a.** Unwilling to tolerate differences in opinions or beliefs, esp. religious beliefs. **b.** Unable or unwilling to endure <*intolerant* of changes> **—in·tol′er·ance** *n.* **—in·tol′er·ant·ly** *adv.*

**in·to·nate** (ĭn′tə-nāt′) *vt.* **-nat·ed, -nat·ing, -nates.** **1.** To intone. **2.** To utter with a particular tone of voice.

**in·to·na·tion** (ĭn′tə-nā′shən, -tō-) *n.* **1. a.** The act of intoning. **b.** An intoned utterance. **2.** A manner of producing or uttering tones, esp. with regard to accuracy of pitch. **3. a.** The use of pitch as an element of meaning in language. **b.** A use of pitch typical of a speaker or dialect. **—in′to·na′tion·al** *adj.*

**in·tone** (ĭn-tōn′) *v.* **-toned, -ton·ing, -tones.** [ME *entonen* < OFr. *entoner* < Med. Lat. *intonare* : Lat. *in-*, in + Lat. *tonus*, tone.] *—vt.* **1.** To recite in a singing or chanting voice. **2.** To utter in a monotone. *—vi.* To speak with a singing tone or with a given intonation. **—in·ton′er** *n.*

**in to·to** (ĭn tō′tō) *adv.* [Lat.] Totally : entirely.

**in·tox·i·cant** (ĭn-tŏk′sĭ-kənt) *n.* Something that intoxicates, esp. an alcoholic beverage. *—adj.* Intoxicating.

**in·tox·i·cate** (ĭn-tŏk′sĭ-kāt′) *vt.* **-cat·ed, -cat·ing, -cates.** [Med. Lat. *intoxicare, intoxicat-*, to poison : Lat. *in-*, in + Lat. *toxicum*, poison. —see TOXIC.] **1.** To bring about, esp. by the effect of ingested alcohol, any of a series of progressively deteriorating states ranging from exhilaration to stupefaction. **2.** To excite or stimulate <"a man whom life *intoxicates,* who has no need of wine" —Anaïs Nin> **3.** To poison. **—in·tox′i·ca′tion** *n.* **—in·tox′i·ca′tive** *adj.* **—in·tox′i·ca′tor** *n.*

**intra-** *pref.* [LLat. < Lat. *intra*, within.] Within <*intra*ocular>

**in·tra·ar·te·ri·al** (ĭn′trə-är-tîr′ē-əl) *adj.* Within an artery. **—in′tra·ar·te′ri·al·ly** *adv.*

**in·tra-a·tom·ic** (ĭn′trə-ə-tŏm′ĭk) *adj.* Within an atom.

**in·tra·car·di·ac** (ĭn′trə-kär′dē-ăk′) *adj.* Within a heart chamber.

**in·tra·car·ti·lag·i·nous** (ĭn′trə-kär′tl-ăj′ə-nəs) *adj.* Within cartilage.

**in·tra·cel·lu·lar** (ĭn′trə-sĕl′yə-lər) *adj.* Within a cell or cells. **—in′tra·cel′lu·lar·ly** *adv.*

**in·tra·cos·tal** (ĭn′trə-kŏs′tl) *adj.* [INTRA- + Lat. *costa*, rib + -AL.] On the inner surface of a rib or ribs.

**in·tra·cra·ni·al** (ĭn′trə-krā′nē-əl) *adj.* Within the skull. **—in′tra·cra′ni·al·ly** *adv.*

**in·trac·ta·ble** (ĭn-trăk′tə-bəl) *adj.* **1.** Difficult to govern or manage : OBSTINATE. **2.** Difficult to manipulate or mold. **3.** Difficult to ease, remedy, or cure <an *intractable* virus> **—in·trac′ta·bil′i·ty, in·trac′ta·ble·ness** *n.* **—in·trac′ta·bly** *adv.*

**in·tra·cu·ta·ne·ous** (ĭn′trə-kyo͞o-tā′nē-əs) *adj.* Within the skin. **—in′tra·cu·ta′ne·ous·ly** *adv.*

**in·tra·day** (ĭn′trə-dā′) *adj.* Occurring in the course of a single day.

**in·tra·der·mal** (ĭn′trə-dûr′məl) *adj.* Within the dermis of the skin.

**in·tra·dos** (ĭn′trə-dŏs′, -dō′, -ĭn-trā′dŏs′, -dōs′) *n., pl.* **-dos** (-dōz′) *or* **-dos·es** (-dŏs′ĭz) [Fr. : *intra-*, within (< Lat.) + *dos*, back < Lat. *dorsum*.] The inner curve of an architectural arch.

intrados

**in·tra·ga·lac·tic** (ĭn′trə-gə-lăk′tĭk) *adj.* Occurring or located within the space of a galaxy.

**in·tra·mo·lec·u·lar** (ĭn′trə-mə-lĕk′yə-lər) *adj.* Within a molecule. **—in′tra·mo·lec′u·lar·ly** *adv.*

**in·tra·mu·ral** (ĭn′trə-myŏŏr′əl) *adj.* **1.** Carried on or existing within the bounds of an institution, esp. a school <*intramural* sports> **2.** *Anat.* Within the wall of a cavity or organ. **—in′tra·mu′ral·ly** *adv.*

**in·tra·mus·cu·lar** (ĭn′trə-mŭs′kyə-lər) *adj.* Within a muscle. **—in′tra·mus′cu·lar·ly** *adv.*

**in·tran·si·gent** *also* **in·tran·si·geant** (ĭn-trăn′sə-jənt) *adj.* [Fr. *intransigeant* < Sp. *intransigente* : *in-*, not (< Lat.) + *transigente*, pr.part. of *transigir*, to compromise < Lat. *transigere*, to come to an agreement (*trans-*, through + *agere*, to drive).] Refusing to modify a position : UNCOMPROMISING. **—in·tran′si·gence, in·tran′si·gen·cy** *n.* **—in·tran′si·gent** *n.* **—in·tran′si·gent·ly** *adv.*

**in·tran·si·tive** (ĭn-trăn′sĭ-tĭv) *adj.* [LLat. *intransitivus* : *in-*, not + *transitivus*, transitive.] Designating a verb or verb construction that does not need a direct object to complete its meaning. —*n.* An intransitive verb. **—in·tran′si·tive·ly** *adv.* **—in·tran′si·tive·ness** *n.*

**in·tra·nu·cle·ar** (ĭn′trə-nōō′klē-ər, -nyōō′-) *adj.* Within a nucleus.

**in·tra·oc·u·lar** (ĭn′trə-ŏk′yə-lər) *adj.* Within the eyeball.

**in·tra·per·son·al** (ĭn′trə-pûr′sə-nəl) *adj.* Taking place within one's own mind or self. **—in′tra·per′son·al·ly** *adv.*

**in·tra·psy·chic** (ĭn′trə-sī′kĭk) *adj.* Taking place or existing within the psyche. **—in′tra·psy′chi·cal·ly** *adv.*

**in·tra·spe·cif·ic** (ĭn′trə-spĭ-sĭf′ĭk) *adj.* Occurring among members of the same species.

**in·tra·state** (ĭn′trə-stāt′) *adj.* Existing within a state's borders.

**in·tra·u·ter·ine** (ĭn′trə-yōō′tər-ĭn, -tə-rīn′) *adj.* Within the uterus.

**intrauterine device** *n.* A metal or plastic loop, ring, or spiral inserted into the uterus as a contraceptive.

**in·trav·a·sa·tion** (ĭn-trăv′ə-sā′shən) *n.* Entry of foreign material into a blood vessel.

**in·tra·vas·cu·lar** (ĭn′trə-văs′kyə-lər) *adj.* Within the blood vessels or lymphatics. **—in′tra·vas′cu·lar·ly** *adv.*

**in·tra·ve·na·tion** (ĭn′trə-vē-nā′shən) *n.* Entry of foreign material into a vein.

**in·tra·ve·nous** (ĭn′trə-vē′nəs) *adj.* Within a vein or veins. **—in′tra·ve′nous·ly** *adv.*

**in·treat** (ĭn-trēt′) *v. var. of* ENTREAT.

**in·trench** (ĭn-trĕnch′) *v. var. of* ENTRENCH.

**in·trench·ment** (ĭn-trĕnch′mənt) *n. var. of* ENTRENCHMENT.

**in·trep·id** (ĭn-trĕp′ĭd) *adj.* [Fr. *intrépide* < Lat. *intrepidus* : *in-*, not + *trepidus*, alarmed.] Outstandingly courageous : FEARLESS. **—in′tre·pid′i·ty** (-trə-pĭd′ĭ-tē), **in·trep′id·ness** *n.* **—in·trep′id·ly** *adv.*

**in·tri·cate** (ĭn′trĭ-kĭt) *adj.* [ME < Lat. *intricatus*, p.part. of *intricare*, to entangle, perplex : *in-*, in + *tricae*, perplexities.] **1.** Having many elaborately arranged elements : COMPLEX. **2.** Difficult to solve or comprehend. **—in′tri·ca·cy** *n.* **—in′tri·cate·ly** *adv.* **—in′tri·cate·ness** *n.*

**in·tri·gant** *also* **in·tri·guant** (ĭn′trĭ-gănt′, ăn′trē-gäN′) *n.* [Fr. < Ital. *intrigante*, pr.part. of *intrigare*, to perplex. —see INTRIGUE.] One who intrigues : INTRIGUER.

**in·trigue** (ĭn′trēg′, ĭn-trēg′) *n.* [Fr. < Ital. *intrigo* < *intrigare*, to perplex < Lat. *intricare*. —see INTRICATE.] **1. a.** A furtive scheme : PLOT. **b.** The practice of or involvement in such schemes. **2.** A clandestine love affair. —*v.* (ĭn-trēg′) **-trigued, -trigu·ing, -trigues.** —*vi.* To engage in intrigue : PLOT. —*vt.* **1.** To effect by intriguing. **2.** To stimulate the curiosity or interest of. ***usage:*** The use of the verb *intrigue* in the sense of "to stimulate the curiosity or interest of" has been objected to by some purists on the grounds that it is an unnecessary substitute for such available terms as *fascinate, interest, pique,* or *puzzle.* Nevertheless this use of *intrigue* is widespread and completely acceptable at all levels. **—in·trigu′er** *n.*

**in·trin·sic** (ĭn-trĭn′zĭk, -sĭk) *adj.* [OFr. *intrinseque*, inner < LLat. *intrinsecus*, inward < Lat., inwardly : *intra*, within + *secus*, alongside.] **1.** Of or relating to the fundamental nature of a thing : INHERENT. **2.** *Anat.* Located within or belonging solely to a body part, as certain nerves and muscles. **—in·trin′si·cal·ly** *adv.*

**in·trin·si·cal** (ĭn-trĭn′zĭ-kəl, -sĭ-) *adj. Archaic.* Intrinsic.

**intrinsic factor** *n.* A substance produced by gastric mucosa that combines and promotes the stomach's absorption of vitamin $B_{12}$.

**intro-** *pref.* [Lat. < *intro*, to the inside.] **1.** In : inward <*intro*jection> **2.** Inward <*intro*vert>

**in·tro·duce** (ĭn′trə-dōōs′, -dyōōs′) *vt.* **-duced, -duc·ing, -duc·es.** [Lat. *introducere*, to bring in : *intro*, within + *ducere*, to lead.] **1. a.** To present (a person) by name to another in order to establish an acquaintance. **b.** To present (e.g., a performer) to the public for the first time. **2.** To create, propose, or bring into use or fashion for the first time <*introduce* new reforms> **3.** To provide with an elementary knowledge or first experience of something. **4. a.** To bring or put in something different : ADD. **b.** To bring in and establish in a new environment. **5.** To inject or insert. **6.** To open or begin : PREFACE. **—in′tro·duc′er** *n.* **—in′tro·duc′i·ble** *adj.*

☆ ***syns:*** INTRODUCE, INAUGURATE, INITIATE, INSTITUTE, LAUNCH, ORIGINATE *v. core meaning* : to bring into use, fashion, or practice <*introduce* a new product>

**in·tro·duc·tion** (ĭn′trə-dŭk′shən) *n.* [ME *introduccioun* < OFr. *introduction* < Lat. *introductio* < *introducere*, to bring in. —see INTRODUCE.] **1. a.** The act or process of introducing. **b.** The fact of being introduced. **2.** A means of presenting one person to another, as a personal letter. **3.** Something recently introduced <"He loathed a fork; it is a modern *introduction* which has still scarcely reached common people" —D.H. Lawrence> **4.** Something that introduces, esp.: **a.** A preface, as to a book. **b.** A short preliminary movement in a musical work. **c.** A basic instructive text or course of study.

**in·tro·duc·to·ry** (ĭn′trə-dŭk′tə-rē) *adj.* **1.** Of, relating to, or constituting an introduction. **2.** Serving to introduce. **—in′tro·duc′to·ri·ly** *adv.*

**in·tro·it** *also* **In·tro·it** (ĭn′trō′ĭt, -troit′, ĭn-trō′ĭt) *n.* [ME, entrance < OFr. *introït* < Lat. *introitus* < p.part. of *introire*, to enter : *intro*, in + *ire*, to go.] **1.** A psalm or hymn sung at the beginning of a service, esp. in the Anglican Church. **2.** *Rom. Cath. Ch.* The beginning of the proper of the Mass, usu. including a psalm verse, antiphon, and the Gloria Patri.

**in·tro·jec·tion** (ĭn′trə-jĕk′shən) *n.* [INTRO- + (PRO)JECTION.] **1.** Ascription of living characteristics to inanimate objects. **2.** Unconscious incorporation into one's personality of the characteristics of another person or of an inanimate object.

**in·tro·mis·sion** (ĭn′trə-mĭsh′ən) *n.* [Med. Lat. *intromissio* < Lat. *intromissus*, p.part. of *intromittere*, to intromit.] The act or process of intromitting. **—in′tro·mis′sive** *adj.*

**in·tro·mit** (ĭn′trə-mĭt′) *vt.* **-mit·ted, -mit·ting, -mits.** [Lat. *intromittere*, to send in : *intro*, in + *mittere*, to send.] To cause or allow to enter. **—in′tro·mit′tent** *adj.* **—in′tro·mit′ter** *n.*

**in·trorse** (ĭn′trôrs′) *adj.* [Lat. *introrsus*, contraction of *introversus*, inwards : *intro*, to the inside + *versus*, p.part. of *vertere*, to turn.] *Bot.* Facing inward. —Used esp. of anthers.

**in·tro·spect** (ĭn′trə-spĕkt′, ĭn′trə-spĕkt′) *vi.* **-spect·ed, -spect·ing, -spects.** [Lat. *introspicere, introspect-*, to look into : *intro*, within + *specere*, to look.] To examine one's own feelings, thoughts, and sensations. **—in′tro·spec′tive** *adj.* **—in′tro·spec′tive·ly** *adv.* **—in′tro·spec′tive·ness** *n.*

**in·tro·spec·tion** (ĭn′trə-spĕk′shən) *n.* Contemplation of one's own thoughts, feelings, and sensations. **—in′tro·spec′tion·al** *adj.*

**in·tro·ver·sion** (ĭn′trə-vûr′zhən, -shən) *n.* **1.** The act of introverting or state of being introverted. **2.** The direction of or tendency to direct one's thoughts and interests mostly toward oneself. **3.** *Med.* The turning of one part within another. **—in′tro·ver′sive** *adj.*

**in·tro·vert** (ĭn′trə-vûrt′, ĭn′trə-vûrt′) *v.* **-vert·ed, -vert·ing, -verts.** [INTRO- + Lat. *vertere*, to turn.] —*vt.* **1.** To turn or direct inward. **2.** To concentrate (one's interests) on oneself. **3.** To turn (a tubular organ or part) inward upon itself. —*vi.* To exhibit introversion. —*n.* (ĭn′trə-vûrt′). **1.** One whose personality is marked by introversion. **2.** An anatomical structure, as the intestine, that is turned inward on itself.

**in·trude** (ĭn-trōōd′) *v.* **-trud·ed, -trud·ing, -trudes.** [Lat. *intrudere*, to thrust in : *in-* in + *trudere*, to thrust.] —*vt.* **1.** To push or force in, esp. without invitation or permission. **2.** *Geol.* To thrust (molten rock) into a stratum. —*vi.* To come in inappropriately or rudely. **—in·trud′er** *n.*

**in·tru·sion** (ĭn-trōō′zhən) *n.* **1. a.** The act of intruding. **b.** The fact of being intruded upon. **2.** An unwelcome or inappropriate addition. **3.** *Law.* Illegal entry upon or appropriation of another's property. **4.** *Geol.* **a.** The forcing of molten rock into an earlier formation. **b.** The intrusive mass so produced.

**in·tru·sive** (ĭn-trōō′sĭv, -zĭv) *adj.* **1.** Intruding or inclined to intrude. **2.** *Geol.* Designating igneous rock that is forced into another stratum while in molten state : IRRUPTIVE. **3.** Constituting an epenthesis. **—in·tru′sive·ly** *adv.* **—in·tru′sive·ness** *n.*

**in·trust** (ĭn-trŭst′) *v. var. of* ENTRUST.

**in·tu·bate** (ĭn′tōō-bāt′, -tyōō-) *vt.* **-bat·ed, -bat·ing, -bates.** *Med.* To insert a tube into (a bodily organ or passage). **—in′tu·ba′tion** *n.* **—in′tur·ba′tion·al** *adj.*

**in·tu·it** (ĭn-tōō′ĭt, -tyōō′-) *vt.* **-it·ed, -it·ing, -its.** [Back-formation < INTUITION.] To sense or know by intuition.

**in·tu·i·tion** (ĭn′tōō-ĭsh′ən, -tyōō-) *n.* [ME *intuicioun*, insight < Med. Lat. *intuitio* < LLat., view < Lat. *intueri*, to look at : *in-*, on + *tueri*, to look at.] **1. a.** The act or faculty of knowing without the use of rational processes : immediate cognition. **b.** Knowledge acquired by the use of this faculty. **2.** Acute insight. **—in′tu·i′tion·al** *adj.* **—in′tu·i′tion·al·ly** *adv.*

**in·tu·i·tion·al·ism** (ĭn′tōō-ĭsh′ə-nə-lĭz′əm, -tyōō-) *n.* Intuitionism. **—in′tu·i′tion·al·ist** *n.*

**in·tu·i·tion·ism** (ĭn′tōō-ĭsh′ə-nĭz′əm, -tyōō-) *n.* **1.** The theory that basic truths are known by intuition rather than reason. **2.** The theory that objects of perception are known to be real by intuition. **3.** The theory that ethical principles are known to be valid and universal through intuition. **—in′tu·i′tion·ist** *n.*

**in·tu·i·tive** (ĭn-tōō′ĭ-tĭv, -tyōō′-) *adj.* **1.** Of, relating to, or arising from intuition. 2. Perceived or known through intuition. 3. Having or showing intuition. **—in·tu'i·tive·ly** *adv.* **—in·tu'i·tive·ness** *n.*

**in·tu·mesce** (ĭn'tōō-mĕs', -tyōō-) *vi.* **-mesced, -mesc·ing, -mesc·es.** [Lat. *intumescere,* to swell up : *in-* (intensive) + *tumescere,* to begin to swell < *tumere,* to swell.] To swell or enlarge.

**in·tu·mes·cence** (ĭn'tōō-mĕs'əns, -tyōō-) *n.* **1. a.** The process of swelling. **b.** The condition of being swollen. **2.** A swollen organ or part. **—in'tu·mes'cent** *adj.*

**in·tus·sus·cept** (ĭn'tə-sə-sĕpt') *vi. & vt.* **-cept·ed, -cept·ing, -cepts.** [Prob. back-formation < INTUSSUSCEPTION.] To undergo or cause to undergo intussusception : INVAGINATE. **—in'tus·sus·cep'-tive** *adj.*

**in·tus·sus·cep·tion** (ĭn'tə-sə-sĕp'shən) *n.* [NLat. *intussusceptio* : Lat. *intus,* within + Lat. *susceptio,* a taking up < *suscipere,* to take up (*sub-,* under + *capere,* to take).] Invagination, esp. an infolding of one part of the intestine into another.

**in·twine** (ĭn-twīn') *v. var. of* ENTWINE.

**in·twist** (ĭn-twĭst') *v. var. of* ENTWIST.

**in·u·lase** (ĭn'yə-lās') *n.* [INUL(IN) + -ASE.] An enzyme that catalyzes the conversion of inulin to levulose.

**in·u·lin** (ĭn'yə-lĭn) *n.* [Prob. < G. *Inulin* < NLat. *Inula,* plant genus < Lat. *inula,* elecampane < Gk. *helenion.*] A carbohydrate, $(C_6H_{10}O_5)_3$ or $(C_6H_{10}O_5)_4$, found in the roots of many plants and used in making fructose.

**in·unc·tion** (ĭn-ŭngk'shən) *n.* [Lat. *inunctio,* an anointing < *inunguere,* to anoint : *in-,* on + *unguere,* to smear.] **1.** The process of applying an ointment. **2.** The act of anointing, as in a religious ceremony.

**in·un·date** (ĭn'ŭn-dāt') *vt.* **-dat·ed, -dat·ing, -dates.** [Lat. *inundare, inundat-* : *in-,* in + *undare,* to flow < *unda,* wave.] **1.** To cover with water, esp. flood water : OVERFLOW. **2.** To overwhelm as if with a flood : SWAMP <*inundated* with work> **—in'un·da'tion** *n.* **—in'un·da'tor** *n.* **—in·un'da·to'ry** (ĭ-nŭn'də-tôr'ē, -tōr'ē) *adj.*

**in·ure** (ĭn-yŏor') *vt.* **-ured, -ur·ing, -ures.** [Back-formation < ME *enured,* accustomed < the phrase *in ure,* customary : *in,* in + *ure,* use < AN **eure* < Lat. *opera,* pl. of *opus,* work.] To make accustomed to something undesirable : HARDEN. **—in·ure'ment** *n.*

**in·urn** (ĭn-ûrn') *vt.* **-urned, -urn·ing, -urns. 1.** To put in an urn, as the ashes of the cremated. **2.** To entomb : inter.

**in·u·tile** (ĭn-yōōt'l, -yōō'tīl') *adj.* [ME < OFr. < Lat. *inutilis* : *in-,* not + *utilis,* useful < *uti,* to use.] Having no utility : USELESS. **—in·u'tile·ly** *adv.* **—in'u·til'i·ty** (ĭn'yōō-tĭl'ĭ-tē) *n.*

**in·vade** (ĭn-vād') *v.* **-vad·ed, -vad·ing, -vades.** [Lat. *invadere* : *in-,* in + *vadere,* to go.] *—vt.* **1.** To enter by force to conquer or pillage. **2.** To encroach upon : VIOLATE. **3.** To overrun as if by invading : PERMEATE. **4.** To enter and spread harm through. *—vi.* To make an invasion. **—in·vad'er** *n.*

**in·vag·i·nate** (ĭn-văj'ə-nāt') *v.* **-nat·ed, -nat·ing, -nates.** [Med. Lat. *invaginare, invaginat-* : Lat. *in-,* in + Lat. *vagina,* sheath.] *—vt.* **1.** To enclose in or as if in a sheath. **2.** To infold so as to form a hollow space within a previously solid structure : INTROVERT. *—vi.* To become enclosed or turned within.

**in·vag·i·na·tion** (ĭn-văj'ə-nā'shən) *n.* **1. a.** The act or process of invaginating. **b.** The state of being invaginated. **2.** Something invaginated, as an organ or part. **3.** The infolding of a blastula to form a gastrula.

**in·va·lid**[1] (ĭn'və-lĭd) *n.* [Fr. *invalide,* sickly, infirm < Lat. *invalidus,* weak : *in-,* not + *validus,* strong < *valēre,* to be strong.] A chronically ill or disabled person. *—adj.* **1.** Disabled by disease or injury. **2.** Of, relating to, or for invalids. *—vt.* **-lid·ed, -lid·ing, -lids. 1.** To make an invalid of. **2.** *Chiefly Brit.* To exempt or release from duty because of poor health. **—in'va·lid'ism** (-lĭ-dĭz'əm) *n.*

**in·val·id**[2] (ĭn-văl'ĭd) *adj.* [Lat. *invalidus,* weak. —see INVALID[1].] **1.** Not factually or legally valid : NULL. **2.** Falsely based or reasoned : FAULTY. **—in'va·lid'i·ty** (-və-lĭd'ĭ-tē) *n.* **—in·val'id·ly** *adv.*

**in·val·i·date** (ĭn-văl'ĭ-dāt') *vt.* **-dat·ed, -dat·ing, -dates.** To make invalid : NULLIFY. **—in·val'i·da'tion** *n.* **—in·val'i·da'tor** *n.*

**in·val·u·a·ble** (ĭn-văl'yōō-ə-bəl) *adj.* **1.** Of great value : PRICELESS <*invaluable* antiques> **2.** Of inestimable help or use : INDISPENSABLE. **—in·val'u·a·bly** *adv.* **—in·val'u·a·ble·ness** *n.*

**in·var·i·a·ble** (ĭn-vâr'ē-ə-bəl) *adj.* Not changing or subject to change : CONSTANT. **—in·var'i·a·bil'i·ty, in·var'i·a·ble·ness** *n.* **—in·var'i·a·bly** *adv.*

**in·var·i·ant** (ĭn-vâr'ē-ənt) *adj.* **1.** Not varying : CONSTANT. **2.** Unaffected by a given mathematical operation, as a transformation of coordinates. *—n.* An invariant function, quantity, configuration, or system. **—in·var'i·ance** *n.*

**in·va·sion** (ĭn-vā'zhən) *n.* [ME *invasioun* < OFr. *invasion* < LLat. *invasio* < Lat. *invadere,* to invade.] **1.** The act of invading, esp. entrance of an army to pillage or conquer. **2.** The onset of something harmful, as a disease.

**in·va·sive** (ĭn-vā'sĭv) *adj.* **1.** Tending to spread, esp. tending to invade healthy tissue. **2.** Of, relating to, or given to armed aggression. **—in·va'sive·ness** *n.*

**in·vec·tive** (ĭn-vĕk'tĭv) *n.* [< ME *invectif,* denunciatory < OFr. < Lat. *invectivus,* reproachful < *invehere,* to inveigh.] **1.** An abusive or denunciatory expression. **2.** Vituperative or denunciatory language. *—adj.* Of, relating to, or marked by invective. **—in·vec'tive·ly** *adv.* **—in·vec'tive·ness** *n.*

**in·veigh** (ĭn-vā') *vi.* **-veighed, -veigh·ing, -veighs.** [Lat. *invehi,* to attack, inveigh, passive of *invehere,* to carry in : *in-,* in + *vehere,* to carry.] To protest angrily : RAIL. **—in·veigh'er** *n.*

**in·vei·gle** (ĭn-vā'gəl, ĭn-vē'-) *vt.* **-gled, -gling, -gles.** [Alteration of OFr. *aveugler,* to blind < *aveugle,* blind < Med. Lat. *ab oculis,* without eyes.] **1.** To win over by guile or persuasion. **2.** To obtain by cajolery. **—in·vei'gle·ment** *n.* **—in·vei'gler** *n.*

**in·vent** (ĭn-vĕnt') *vt.* **-vent·ed, -vent·ing, -vents.** [Lat. *invenire, invent-,* to find : *in-,* on + *venire,* to come.] **1.** To produce or contrive (something previously unknown) by the use of ingenuity or imagination. **2.** To make up : fabricate. **—in·vent'i·ble** *adj.* **—in·ven'tor** *n.*

☆ ***syns:*** INVENT, CONCOCT, CONTRIVE, COOK UP, DEVISE, DREAM UP, FABRICATE, FORMULATE, HATCH, MAKE UP, THINK (up) *v.* *core meaning* : to use ingenuity in making, developing, or achieving <always *inventing* excuses> <*invented* a new gadget>

**in·ven·tion** (ĭn-vĕn'shən) *n.* **1.** The act or process of inventing. **2.** A new method, device, or process developed from study and experimentation. **3.** A mental fabrication, esp. a lie. **4.** Expertise in inventing : INVENTIVENESS. **5.** *Mus.* A short composition developing a single theme contrapuntally. **6.** *Archaic.* A discovery : finding. **—in·ven'tion·al** *adj.*

**in·ven·tive** (ĭn-vĕn'tĭv) *adj.* **1.** Of, relating to, or marked by invention. **2.** Skillful or adept at inventing : CREATIVE. **—in·ven'tive·ly** *adv.* **—in·ven'tive·ness** *n.*

**in·ven·to·ry** (ĭn'vən-tôr'ē, -tōr'ē) *n., pl.* **-ries.** [Med. Lat. *inventorium,* list, alteration of LLat. *inventarium* < Lat. *invenire,* to find. —see INVENT.] **1. a.** A detailed list of items in one's view or possession, esp. a periodic survey of all goods and materials in stock. **b.** The process of making such a survey. **c.** The items listed in such a survey. **d.** The supply of goods and materials on hand : STOCK. **2.** A survey or evaluation, as of personal characteristics. *—vt.* **-ried, -ry·ing, -ries. 1.** To make an inventory of. **2.** To include in an inventory. **—in'ven·to'ri·al** *adj.* **—in'ven·to'ri·al·ly** *adv.*

**in·ve·rac·i·ty** (ĭn'və-răs'ĭ-tē) *n., pl.* **-ties. 1.** Lack of truthfulness. **2.** An untruth : falsehood.

**in·ver·ness** *also* **In·ver·ness** (ĭn'vər-nĕs') *n.* [After *Inverness,* a city and former county of Scotland.] **1.** A loose overcoat with a detachable cape. **2.** The cape of an inverness.

**in·verse** (ĭn-vûrs', ĭn'vûrs') *adj.* [Lat. *inversus,* p.part. of *invertere,* to invert.] **1.** Reversed in nature, order, or effect. **2.** Turned upside down : INVERTED. *—n.* (ĭn'vûrs', ĭn-vûrs'). **1.** Something that is opposite, as in sequence or character : REVERSE. **2.** *Math.* An element $x^*$ in a set *S* related to a designated element x in *S* such that $x^* \cdot x = x \cdot x^* = I$, where • is a binary operation defined in *S* and *I* is the identity element, esp.: **a.** The reciprocal of a designated quantity. **b.** The negative of a designated quantity. **—in·verse'ly** *adv.*

**in·ver·sion** (ĭn-vûr'zhən, -shən) *n.* [Lat. *inversio* < *invertere,* to invert.] **1.** The act of inverting or state of being inverted. **2.** An interchange of position, esp. of adjacent objects in a sequence, as: **a.** A change in normal word order, as the placement of a verb before its subject. **b.** *Mus.* A rearrangement or result of the rearrangement of tones in which upper and lower voices are transposed, as in counterpoint, or in which each interval in a single melody is applied in the opposite direction. **3.** Homosexuality. **4.** *Chem.* Conversion from the dextrorotatory to the levorotatory or from the levorotatory to the dextrorotatory form. **5.** A state in which air temperature increases with increasing altitude, holding surface air and pollutants down.

**in·vert** (ĭn-vûrt') *v.* **-vert·ed, -vert·ing, -verts.** [Lat. *invertere* : *in-,* in + *vertere,* to turn.] *—vt.* **1.** To turn upside down or inside out. **2.** To reverse the order, position, or condition of. **3.** To subject to inversion. *—vi.* To be subjected to inversion. *—n.* (ĭn'vûrt'). **1.** Something inverted. **2.** A homosexual. **—in·vert'i·ble** *adj.*

**in·ver·tase** (ĭn-vûr'tās'; ĭn'vər-tās', -tāz') *n.* An enzyme that catalyzes the conversion of sucrose to glucose and fructose.

**in·ver·te·brate** (ĭn-vûr'tə-brĭt, -brāt') *adj.* [NLat. *invertebratus* : Lat. *in-,* not + NLat. *vertebratus,* vertebrate.] Lacking a backbone or spinal column. *—n.* An invertebrate animal.

**inverted comma** *n. Chiefly Brit.* A quotation mark.

**in·vert·er** (ĭn-vûr'tər) *n.* **1.** One that inverts. **2.** A device for converting direct current into alternating current. **3.** A circuit that takes in a positive pulse and puts out a negative one.

**invert sugar** *n.* A hygroscopic mixture of equal parts of glucose and fructose resulting from the hydrolysis of sucrose and used chiefly in brewing and in medicine.

**in·vest** (ĭn-vĕst') *v.* **-vest·ed, -vest·ing, -vests.** [Ital. *investire* < Lat., to clothe, surround : *in-,* in + *vestis,* clothes.] *—vt.* **1.** To commit (money) in order to gain profit or interest. **2.** To utilize for future benefit or advantage <*invested* their energies wisely> **3.** To furnish with authority or power. **4.** To install in office : INAUGURATE. **5.** To provide with an enveloping or pervasive quality. **6.** To clothe or adorn. **7.** To cover completely : ENVELOP. **8.** To surround with troops or ships : BESIEGE. —*vi.* To make an investment. **—in·ves'tor** *n.*

**in·ves·ti·gate** (ĭn-vĕs'tĭ-gāt') *v.* **-gat·ed, -gat·ing, -gates.** [Lat. *investigare, investigat-* : *in-*, in + *vestigare*, to track < *vestigium*, footprint.] —*vt.* To observe or inquire into in detail. —*vi.* To make a systematic inquiry or examination. **—in·ves'ti·ga·ble** (-gə-bəl), **in·ves'ti·ga'tive, in·ves'ti·ga·to'ry** (-gə-tôr'ē, -tōr'ē) *adj.* **—in·ves'ti·ga'tion** *n.*

**in·ves·ti·ga·tor** (ĭn-vĕs'tĭ-gā'tər) *n.* A person who investigates, esp. a detective. **—in·ves'ti·ga·to'ri·al** (-gə-tôr'ē-əl, -tōr'-) *adj.*

**in·ves·ti·ture** (ĭn-vĕs'tə-cho͝or', -chər) *n.* [ME < Med. Lat. *investitura* < Lat. *investire*, to clothe. —see INVEST.] **1.** The act or formal ceremony of conferring the authority and symbols of a high office. **2.** Something that adorns or covers.

**in·vest·ment** (ĭn-vĕst'mənt) *n.* **1.** The act of investing. **2.** An amount invested. **3.** A possession, as property, acquired for future income or benefit. **4.** Investiture. **5.** *Archaic.* A garment : vestment. **6.** An outer layer or covering. **7.** A military siege.

**in·vet·er·ate** (ĭn-vĕt'ər-ĭt) *adj.* [ME, obstinate, long-established < Lat. *inveteratus*, p.part. of *inveterare*, to make old : *in-*, in, into + *vetus*, old.] **1.** Solidly established by long standing : DEEP-ROOTED. **2.** Persisting in an ingrained habit : HABITUAL <an *inveterate* cynic> **—in·vet'er·a·cy** (-ər-ə-sē), **in·vet'er·ate·ness** *n.* **—in·vet'er·ate·ly** *adv.*

**in·vid·i·ous** (ĭn-vĭd'ē-əs) *adj.* [Lat. *invidiosus*, envious, hostile < *invidia*, envy. —see ENVY.] **1.** Tending to cause animosity or resentment. **2.** Containing or implying a slight : DISCRIMINATORY. **3.** *Obs.* Envious. **—in·vid'i·ous·ly** *adv.* **—in·vid'i·ous·ness** *n.*

**in·vig·or·ate** (ĭn-vĭg'ə-rāt') *vt.* **-at·ed, -at·ing, -ates.** To give vigor, vitality, or strength to : ANIMATE. **—in·vig'or·at'ing·ly** *adv.* **—in·vig'or·a'tion** *n.* **—in·vig'or·a'tor** *n.*

**in·vin·ci·ble** (ĭn-vĭn'sə-bəl) *adj.* [ME < OFr. < Lat. *invincibilis* : *in-*, not + *vincibilis*, vincible.] Incapable of being conquered, overrun, or subjugated. **—in·vin'ci·bil'i·ty, in·vin'ci·ble·ness** *n.* **—in·vin'ci·bly** *adv.*

☆ ***syns:*** INVINCIBLE, IMPREGNABLE, INCONQUERABLE, INSUPERABLE, UNCONQUERABLE *adj. core meaning* : incapable of being conquered, overrun, or subjugated <the *invincible* will of a free people>

**in·vi·o·la·ble** (ĭn-vī'ə-lə-bəl) *adj.* **1.** Secure from violation or profanation. **2.** Impregnable to trespass or assault. **—in·vi'o·la·bil'i·ty, in·vi'o·la·ble·ness** *n.* **—in·vi'o·la·bly** *adv.*

**in·vi·o·late** (ĭn-vī'ə-lĭt) *adj.* [ME < Lat. *inviolatus* : *in-*, not + *violatus*, p.part. of *violare*, to violate.] Not violated : PURE. **—in·vi'o·la·cy** (-lə-sē), **in·vi'o·late·ness** *n.* **—in·vi'o·late·ly** *adv.*

**in·vis·cid** (ĭn-vĭs'ĭd) *adj.* **1.** Having no viscosity. **2.** Of or relating to an inviscid fluid.

**in·vis·i·ble** (ĭn-vĭz'ə-bəl) *adj.* **1.** Not visible. **2.** Not open to view : HIDDEN. **3.** Not easily detected or noticed : IMPERCEPTIBLE. **4.** Not published in financial statements. —*n.* One that is invisible. **—in·vis'i·bil'i·ty, in·vis'i·ble·ness** *n.* **—in·vis'i·bly** *adv.*

**invisible ink** *n.* Ink that is colorless and invisible until treated by a chemical, heat, or special light.

**in·vi·ta·tion** (ĭn'vĭ-tā'shən) *n.* **1.** The act of inviting. **2.** A spoken or written request for one's presence or participation. **3.** An enticement, allurement, or attraction.

**in·vi·ta·tion·al** (ĭn'vĭ-tā'shə-nəl) *adj.* Restricted to invited participants <an *invitational* tennis match>

**in·vi·ta·to·ry** (ĭn-vī'tə-tôr'ē, -tōr'ē) *n., pl.* **-ries.** [ME *invitatorie* < Med. Lat. *invitiatorium* < LLat. *invitiatorius*, inviting < Lat. *invitare*, to invite.] A psalm or antiphon sung as an invitation to prayer in church services. —*adj.* Constituting or having an invitation.

**in·vite** (ĭn-vīt') *vt.* **-vit·ed, -vit·ing, -vites.** [OFr. *inviter* < Lat. *invitare.*] **1.** To request the presence or participation of. **2.** To ask formally. **3.** To welcome : encourage <*invite* questions from the press corps> **4.** To tend to bring on : PROVOKE. **5.** To tempt or lure : ENTICE. —*n.* (ĭn'vīt'). *Informal.* An invitation.

**in·vit·ing** (ĭn-vī'tĭng) *adj.* Attractive : tempting <an *inviting* buffet> **—in·vit'ing·ly** *adv.*

**in vi·tro** (ĭn vē'trō) *adj.* & *adv.* [NLat., in glass.] In an artificial environment outside the living organism.

**in vi·vo** (ĭn vē'vō) *adj.* & *adv.* [NLat., in a living body.] Within a living organism.

**in·vo·ca·tion** (ĭn'və-kā'shən) *n.* [ME < OFr. < Lat. *invocatio* < *invocare*, to invoke.] **1.** The act or process of invoking, esp. an appeal to a higher power for assistance. **2.** An invocatory prayer, as at the opening of a religious service. **3. a.** An act of conjuring up a spirit by incantation. **b.** An incantation used in conjuring.

**in·voc·a·to·ry** (ĭn-vŏk'ə-tôr'ē, -tōr'ē) *adj.* Of, relating to, or like an invocation.

**in·voice** (ĭn'vois') *n.* [Alteration of obs. *invoyes*, pl. of *invoy*, invoice < Fr. *envoy* < *envoyer*, to send. —see ENVOY.] **1.** An itemized list of goods shipped or services rendered, with an account of all costs : BILL. **2.** The goods or services listed in an invoice. —*vt.* **-voiced, -voic·ing, -voic·es.** To make an invoice of or submit an invoice to : BILL.

**in·voke** (ĭn-vōk') *vt.* **-voked, -vok·ing, -vokes.** [OFr. *invoquer* < Lat. *invocare* : *in-*, in + *vocare*, to call.] **1.** To call on for aid, support, or inspiration. **2.** To appeal to or cite in support or justification. **3.** To call for earnestly : SOLICIT. **4.** To call forth with incantations : CONJURE. **5.** To resort to : IMPLEMENT <The dictator *invoked* martial law.> **—in·vok'er** *n.*

**in·vol·u·cel** (ĭn-vŏl'yə-sĕl') *n.* [NLat. *involucellum*, dim. of *involucrum*, involucre.] *Bot.* A secondary involucre, as at the base of an umbellule in a compound umbel.

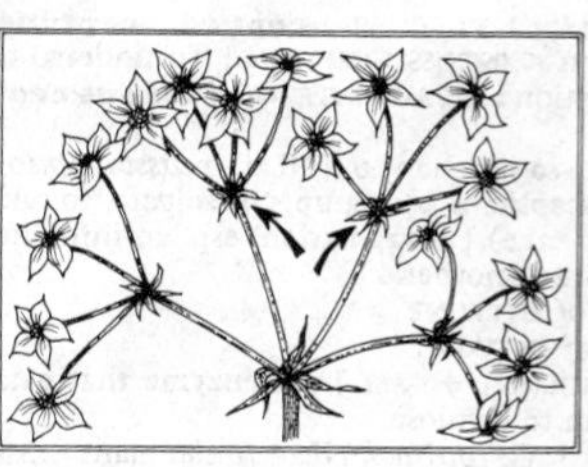

**involucel**

**in·vo·lu·cra** (ĭn'və-lo͞o'krə) *n. pl. of* INVOLUCRUM.

**in·vo·lu·cre** (ĭn'və-lo͞o'kər) *n.* [NLat. *involucrum*, involucrum.] A whorl of bracts beneath or around a flower or flower cluster. **—in'vo·lu'cral** (-krəl), **in'vo·lu'crate** (-krĭt, -krāt') *adj.*

**in·vo·lu·crum** (ĭn'və-lo͞o'krəm) *n., pl.* **-cra** (-krə) [NLat. < Lat., wrapper, envelope < *involvere*, to enwrap. —see INVOLVE.] An enveloping envelope or sheath.

**in·vol·un·tar·y** (ĭn-vŏl'ən-tĕr'ē) *adj.* **1. a.** Not done willingly or on purpose. **b.** Not involving or based on conscious choice. **2.** Not subject to control : AUTOMATIC. **—in·vol'un·tar'i·ly** (-târ'ə-lē) *adv.* **—in·vol'un·tar'i·ness** *n.*

☆ ***syns:*** INVOLUNTARY, AUTOMATIC, INSTINCTIVE *adj. core meaning* : not involving or based on conscious choice <an *involuntary* exclamation of pain> INVOLUNTARY refers to what is not subject to the control of the will <*involuntary* heart muscles> What is AUTOMATIC is done or produced by the body without conscious control or awareness <*automatic* reflexes> INSTINCTIVE actions are directed by unlearned inner drives <the *instinctive* migrations of birds><*instinctive* revulsion> ***ant:*** voluntary

**in·vo·lute** (ĭn'və-lo͞ot') *adj.* [Lat. *involutus*, p.part. of *involvere*, to enwrap. —see INVOLVE.] **1.** Intricate : complex. **2.** *Bot.* **a.** Having the margins rolled inward. **b.** Having whorls that obscure the axis or other volutions, as the shell of a cowry. —*n. Math.* **1.** The locus of a fixed point on a taut, inextensible string as it unwinds from a fixed plane curve. **2.** The locus of any point on a tangent line as it rolls but does not slide around a fixed curve. **—in'vo·lute'ly** *adv.*

**in·vo·lu·tion** (ĭn'və-lo͞o'shən) *n.* [Lat. *involutio* < *involvere*, to enwrap. —see INVOLVE.] **1.** The act of involving or state of being involved. **2. a.** Complexity : intricacy. **b.** Something intricate or complex, as a complicated grammatical construction. **3.** *Math.* Multiplication of a quantity by itself a specified number of times. **4.** *Biol.* Formation of a gastrula from a blastula by ingrowth of blastomeres at the dorsal lip. **—in'vo·lu'tion·al** *adj.*

**in·volve** (ĭn-vŏlv') *vt.* **-volved, -volv·ing, -volves.** [ME *involven* < OFr. *involver* < Lat. *involvere*, to enwrap : *in-*, in + *volvere*, to roll, turn.] **1.** To include or contain as a part. **2.** To have as an essential feature or consequence : ENTAIL. **3.** To draw in as a participant. **4.** To occupy or engross : ABSORB. **5.** To make intricate : COMPLICATE. **6.** To wrap : envelop. **7.** *Archaic.* To coil or wind about. **8.** *Math.* To raise (a number) to a specified power. **—in·volve'ment** *n.* **—in·volv'er** *n.*

**in·volved** (ĭn-vŏlvd') *adj.* **1.** Intricate : complicated. **2.** Involute : twisted. **3.** Confused : tangled. **—in·volv'ed·ly** (-vŏl'vĭd-lē) *adv.*

**in·vul·ner·a·ble** (ĭn-vŭl'nər-ə-bəl) *adj.* [Lat. *invulnerabilis* : *in-*, not + *vulnerare*, to wound < *vulnus*, wound.] **1.** Immune to attack : IMPREGNABLE. **2.** Incapable of being injured, damaged, or wounded. **—in·vul'ner·a·bil'i·ty, in·vul'ner·a·ble·ness** *n.* **—in·vul'ner·a·bly** *adv.*

**in·ward** (ĭn'wərd) *adj.* [ME < OE *inweard.*] **1.** Located inside : INNER. **2.** Moving or directed toward the interior. **3.** Of, relating to, or existing in the thoughts or mind. **4.** Closely acquainted : FAMILIAR. —*adv.* **1.** Toward the inside, center, or interior. **2.** Toward the mind or the self. —*n.* **1.** An inner or central part. **2.** An inner spirit or essence. **3. inwards.** Entrails : innards. **—in'wards** *adv.*

**in·ward·ly** (ĭn'wərd-lē) *adv.* **1.** On or in the inside : WITHIN. **2.** To oneself : PRIVATELY <*inwardly* laughing>

**in·ward·ness** (ĭn'wərd-nĭs) *n.* **1.** Intimacy : familiarity. **2.** Preoccupation with one's own thoughts or feelings : INTROSPECTION. **3.** Essential or fundamental nature. **4.** Internal quality or essence.

**in·weave** (ĭn-wēv') *vt.* **-wove** (-wōv'), **-wo·ven** (-wō'vən), **-weav·ing, -weaves.** To weave into a fabric or design.

**in·wind** (ĭn-wīnd') *v. var. of* ENWIND.

**in·wrap** (ĭn-răp′) *v. var. of* ENWRAP.
**in·wreathe** (ĭn-rēth′) *v. var. of* ENWREATHE.
**in·wrought** (ĭn-rôt′, ĭn′rôt′) *adj.* **1.** Worked or woven in. **2.** With a decorative pattern worked or woven in.
**I·o** (ī′ō) *n.* [Lat. < Gk. *Iō.*] *Gk. Myth.* A maiden loved by Zeus and transformed by Hera into a heifer.
**iod-** *pref. var. of* IODO-.
**i·o·date** (ī′ə-dāt′) *vt.* **-dat·ed, -dat·ing, -dates.** To iodize. —*n.* (ī′ə-dāt′, -dĭt). A salt of iodic acid. **—i′o·da′tion** *n.*
**i·od·ic acid** (ī-ŏd′ĭk) *n.* [Fr. *iodique* < *iode,* iodine.] A colorless or white crystalline powder, $HIO_3$, used as an antiseptic and deodorant.
**i·o·dide** (ī′ə-dīd′) *n.* A binary compound of iodine with a more electropositive atom or group.
**i·o·dine** (ī′ə-dīn′, -dĭn, -dēn′) *n.* [Fr. *iode,* iodine (< Gk. *iōdēs,* violet-colored < *ion,* violet) + -INE.] **1.** *Symbol* **I** A lustrous, grayish-black, corrosive, poisonous element having radioactive isotopes, esp. I 131, used as tracers, in thyroid disease diagnosis and therapy, and in compounds as germicides, antiseptics, and dyes; atomic number 53; atomic weight 126.9044. **2.** A tincture of iodine and sodium iodide, NaI, or potassium iodide, KI, used as an antiseptic.
**i·o·dize** (ī′ə-dīz′) *vt.* **-dized, -diz·ing, -diz·es.** To treat or combine with iodine or an iodide.
**iodo-** or **iod-** *pref.* [Fr. *iode,* iodine.] Iodine <*iodo*form>
**i·o·do·form** (ī-ō′də-fôrm′, ī-ŏd′ə-) *n.* [IODO- + FORM(YL).] A yellowish iodine compound, $CHI_3$, used as an antiseptic.
**i·o·do·phor** (ī-ō′də-fôr′) *n.* [IODO- + -PHOR(E).] A substance made up of iodine and a solubilizing agent that releases free iodine when in solution.
**i·o·dop·sin** (ī′ə-dŏp′sĭn) *n.* A light-sensitive pigment in the retinal cones of the eye.
**i·o moth** (ī′ō) *n.* [After *Io,* who was tormented by gadflies sent by Hera as a punishment.] A large yellowish moth, *Automeris io* of North America, with large eyelike spots on the hind wings.
**i·on** (ī′ən, ī′ŏn′) *n.* [< Gk. *ion,* something that goes, neuter pr.part. of *ienai,* to go.] An atom, group of atoms, or molecule that has acquired or is considered to have acquired a net electric charge by gaining electrons in or losing electrons from an initially electrically neutral configuration.
**-ion** *suff.* [ME < OFr. < Lat. *-io,* n. suffix.] **1. a.** Action or process <oxidizat*ion*> **b.** Result of an action or process <indent*ion*> **2.** State or condition <hydrat*ion*>
**ion engine** *n.* A rocket engine that develops thrust by expelling ions rather than gaseous combustion products.
**ion exchange** *n.* A reversible chemical reaction between a solid and a fluid mixture by means of which ions may be interchanged, used in softening water and separating radioactive isotopes.
**I·o·ni·an** (ī-ō′nē-ən) *adj.* IONIC 1. **—I·o′ni·an** *n.*
**i·on·ic** (ī-ŏn′ĭk) *adj.* Of, having, or involving ions.
**I·on·ic** (ī-ŏn′ĭk) *adj.* **1.** Of or relating to Ionia or the Ionians. **2.** Relating to or designating the Ionic order of architecture. —*n.* The ancient Greek dialect of Ionia.
**ionic bond** *n.* A chemical bond typical of salts, formed by the complete transfer of one or more electrons from one kind of atom to another.
**Ionic order** *n.* An order of classical Greek architecture marked by two opposed volutes in the capital.
**ionic propulsion** *n.* Propulsion by the reactive thrust of a high-speed beam of similarly charged ions ejected by an ion engine.
**i·on·i·za·tion** (ī′ə-nĭ-zā′shən) *n.* **1.** Formation of one or more ions by the addition of electrons to or removal of electrons from an electrically neutral atomic or molecular configuration by heat, electrical discharge, radiation, or chemical reaction. **2.** The state of being ionized.
**ionization chamber** *n.* A gas-filled enclosure fitted with electrodes between which electric current flows on ionization of the gas by incident radiation, the electrodes being maintained at a potential difference sufficient to collect ions thus produced without causing further ionization.
**ionization potential** *n.* The energy needed to remove entirely the weakest bound electron from its ground state in an atom or molecule so that the resulting ion is also in its ground state.
**i·on·ize** (ī′ə-nīz′) *vt.* & *vi.* **-ized, -iz·ing, -iz·es.** To convert completely or partially into ions.
**ionizing radiation** *n.* Radiation capable of producing ionization, including energetic charged particles such as alpha and beta rays, nonparticulate radiation such as x-rays, and neutrons.
**i·o·none** (ī′ə-nōn′) *n.* [Gk. *ion,* violet + -ONE.] Either of two yellowish to colorless liquid isomers, $C_{13}H_{20}O$, with a strong odor of violets, used in perfumes.
**i·on·o·sphere** (ī-ŏn′ə-sfîr′) *n.* [ION + -SPHERE.] An electrically conducting set of layers of the earth's atmosphere, extending from altitudes of approx. 30 to 250 miles or 50 to 400 kilometers, caused by ionization of rarefied atmospheric gases by incident solar radiation. **—i·on′o·spher′ic** (-sfîr′ĭk, -sfĕr′-) *adj.*
**ion propulsion** *n.* Ionic propulsion.
**ion rocket** *n.* **1.** A rocket using ionic propulsion. **2.** An ion engine.
**ion trap** *n.* A magnet mounted to the neck of a kinescope to prevent ions from striking the kinescope screen.
**i·o·ta** (ī-ō′tə) *n.* [Gk. *iōta,* of Phoenician orig.; akin to Hebrew *yōdh,* yod.] **1.** The ninth letter of the Greek alphabet. —See table at ALPHABET. **2.** A very small amount: BIT.
**i·o·ta·cism** (ī-ō′tə-sĭz′əm) *n.* [LLat. *iotacismus* < Gk. *iōtakismos* < *iōta,* iota.] The conversion of other vowel sounds in Greek to the sound of iota.
**IOU** (ī′ō-yo͞o′) *n., pl.* **IOU's** or **IOUs.** [< the pronunciation of *I owe you.*] A promise to pay a debt.
**-ious** *suff.* [ME, partly < Lat. *-ius,* and partly < OFr. *-ieus, -ieux* < Lat. *-iosus.*] Having: having the qualities of: full of <bil*ious*>
**I·o·wa** (ī′ə-wə) *n., pl.* **Iowa** or **-was.** [Dakota *Ayuhwa.*] **1. a.** A tribe of Indians once inhabiting the region of Minnesota, Iowa, and Missouri. **b.** A member of this tribe. **2.** The Siouan language of the Iowa. **—I′o·wa** *adj.*
**ip·e·cac** (ĭp′ĭ-kăk′) *n.* [< Port. *ipecacuanha* < Tupi *ipekaaguéne.*] **1.** A low-growing South American shrub, *Cephaelis ipecacuanha,* with roots used medicinally. **2.** The dried roots of the ipecac.
**Iph·i·ge·ni·a** (ĭf′ə-jə-nī′ə) *n.* [Lat. < Gk. *Iphigeneia.*] *Gk. Myth.* The daughter of Clytemnestra and Agamemnon, offered as a sacrifice to Artemis to enable the Greek fleet to sail for Troy.
**ip·se dix·it** (ĭp′sē dĭk′sĭt) *n.* [Lat., he himself said (it).] An arbitrary statement: DICTUM.
**ip·si·lat·er·al** (ĭp′sə-lăt′ər-əl) *adj.* [Alteration of Lat. *ipse,* self + LATERAL.] Affecting or located on the same side of the body. **—ip′si·lat′er·al·ly** *adv.*
**ip·sis·si·ma ver·ba** (ĭp-sĭs′ə-mə vûr′bə) *pl.n.* [Lat.] The very words.
**ip·so fac·to** (ĭp′sō făk′tō) *adv.* [Lat.] By the fact itself.
**ip·so ju·re** (ĭp′sō jo͝or′ē) *adv.* [Lat.] By the law itself.
**IQ** or **I.Q.** (ī′kyo͞o′) *n.* Intelligence quotient.
**Ir** *symbol for* IRIDIUM.
**ir-**[1] *pref. var. of* IN-[1]. —Used before *r.*
**ir-**[2] *pref. var. of* IN-[2]. —Used before *r.*
**I·ra·ni·an** (ĭ-rā′nē-ən) *adj.* Of or relating to Iran, its people, or their language. —*n.* **1.** A native or resident of Iran. **2.** A branch of the Indo-European language family including Persian, Kurdish, Pashto, and other languages of Iran, Afghanistan, and western Pakistan.
**I·ra·qi** (ĭ-rä′kē) *adj.* Of or relating to Iraq, its people, or their language. —*n., pl.* **Iraqi** or **-qis.** **1.** A native or resident of Iraq. **2.** The modern dialect of Arabic spoken in Iraq.
**i·ras·ci·ble** (ĭ-răs′ə-bəl, ī-răs′-) *adj.* [ME *irascible* < OFr. < LLat. *irascibilis* < Lat. *irasci,* to be angry < *ira,* anger.] **1.** Easily angered. **2.** Marked by or resulting from anger. **—i·ras′ci·bil′i·ty, i·ras′ci·ble·ness** *n.* **—i·ras′ci·bly** *adv.*
**i·rate** (ī-rāt′, ī′rāt′) *adj.* [Lat. *iratus,* p.part. of *irasci,* to be angry < *ira,* anger.] **1.** Extremely angry. **2.** Marked or occasioned by anger <an *irate* letter> **—i·rate′ly** *adv.*
**ire** (īr) *n.* [ME < OFr. < Lat. *ira.*] Anger: wrath.
**ire·ful** (īr′fəl) *adj.* Full of ire: WRATHFUL. **—ire′ful·ly** *adv.*
**i·ren·ic** (ī-rĕn′ĭk, ī-rē′nĭk) *adj.* also **i·ren·i·cal** (-ĭ-kəl, -nĭ-kəl) *adj.* [Gk. *eirēnikos* < *eirēnē,* peace.] Promoting or conducive to peace: CONCILIATORY. **—i·ren′i·cal·ly** *adv.*
**irid-** *pref. var. of* IRIDO-.
**ir·i·da·ceous** (ĭr′ĭ-dā′shəs) *adj.* [< NLat. *Iridacea,* iris family < *Iris,* type genus < Lat., iris.] Of or relating to the iris family.
**ir·i·dec·to·my** (ĭr′ĭ-dĕk′tə-mē, ī′rĭ-) *n., pl.* **-mies.** The surgical removal of part of the iris of the eye.
**ir·i·des·cent** (ĭr′ĭ-dĕs′ənt) *adj.* **1.** Producing a display of lustrous, rainbowlike colors. **2.** Brilliant, lustrous, or colorful in effect or appearance. **—ir′i·des′cence** (-dĕs′əns) *n.*
**i·rid·ic** (ĭ-rĭd′ĭk, ī-rĭd′-) *adj.* Of or relating to the iris of the eye.
**i·rid·i·um** (ĭ-rĭd′ē-əm) *n.* [Gk. *iris, irid-,* rainbow + -IUM (from the colors produced by dissolving it in hydrochloric acid).] *Symbol* **Ir** A very hard and brittle, exceptionally corrosion-resistant, whitish-yellow metallic element used to harden platinum and in high-temperature materials, electrical contacts, and wear-resistant bearings; atomic number 77; atomic weight 192.2.
**irido-** or **irid-** *pref.* [Lat. *iris, irid-,* rainbow.] **1.** Rainbow <*irid*escent> **2.** Iris of the eye <*irid*ectomy> **3.** Iridium <*irido*smine>
**ir·i·dos·mine** (ĭr′ĭ-dŏz′mēn′) *n.* [G. *Iridosmin.*] Osmiridium.
**i·ris** (ī′rĭs) *n., pl.* **i·ris·es.** [ME, rainbow < Lat. < Gk., rainbow, iris of the eye.] **1.** The pigmented, round, contractile membrane of the eye, located between the cornea and lens and perforated by the pupil. **2.** A plant of the genus *Iris,* with sword-shaped leaves and variously colored flowers. **3.** A rainbow or rainbowlike display of colors.
**I·ris** (ī′rĭs) *n.* [Lat. < Gk.] *Gk. Myth.* The goddess of the rainbow and messenger of the gods.
**iris diaphragm** *n.* A metallic diaphragm adjustable to vary the diameter of a central aperture, commonly used on cameras to control the amount of light admitted to a lens.
**I·rish** (ī′rĭsh) *adj.* [ME < OE *Iras,* the Irish.] Of or relating to Ireland, its people, or their language. —*n.* **1. a.** The inhabitants of Ireland. **b.** People of Irish descent. **2. a.** Irish Gaelic. **b.** Irish English.

**Irish coffee** *n.* A beverage of sweetened hot coffee and Irish whiskey, topped with whipped cream.
**Irish elk** *n.* A large extinct European deer of the genus *Megaceros* of the Pliocene and Pleistocene epochs, with palmate antlers.
**Irish English** *n.* English as spoken by the Irish.
**Irish Gaelic** *n.* The Goidelic language of Ireland.
**I·rish·ism** (ī′rĭsh-ĭz′əm) *n.* An Irish idiom or custom.
**I·rish·man** (ī′rĭsh-mən) *n.* **1.** A native or resident of Ireland. **2.** A person of Irish birth or descent.
**Irish moss** *n.* An edible North Atlantic seaweed, *Chondrus crispus,* that yields a mucilaginous substance used medicinally and in the making of jellies.
**Irish setter** *n.* A setter having a silky reddish-brown coat.
**Irish stew** *n.* A stew of meat and vegetables.
**Irish terrier** *n.* A terrier having a wiry brown coat.
**Irish whiskey** *n.* Whiskey made by the distillation of barley.
**Irish wolfhound** *n.* A large dog of an ancient breed, having a rough, shaggy coat.
**I·rish·wom·an** (ī′rĭsh-wo͝om′ən) *n.* A woman of Irish birth or descent.
**i·ri·tis** (ī-rī′tĭs) *n.* [IR(IS) + -ITIS.] Inflammation of the iris of the eye.
**irk** (ûrk) *vt.* **irked, irk·ing, irks.** [ME *irken,* to weary, prob. of Celt. orig.] To irritate : VEX.
**irk·some** (ûrk′səm) *adj.* Causing annoyance or bother. **—irk′some·ly** *adv.* **—irk′some·ness** *n.*
**i·ron** (ī′ərn) *n.* [ME *iren* < OE *īren.*] **1.** *Symbol* **Fe** A silvery-white, lustrous, malleable, ductile, magnetic or magnetizable metallic element used alloyed in many important structural materials; atomic number 26; atomic weight 55.847. **2.** Great strength or hardness : FIRMNESS. **3.** An implement made of iron alloy or similar metal, esp. a bar heated for use in branding, curling hair, or cauterizing. **4.** A golf club with a metal head, numbered from one to nine according to the degree of slant of the face of the club. **5.** An appliance with a handle and a weighted flat bottom, used when heated for pressing wrinkles from fabric. **6.** A harpoon. **7. irons.** Handcuffs : shackles. **8.** A medication, as a tonic, containing iron as a dietary supplement. *—v.* **i·roned, i·ron·ing, i·rons.** *—vt.* **1. a.** To smooth with a heated iron. **b.** To remove (creases) by pressing. **2.** To put in irons : FETTER. **3.** To fit or clad with iron. *—vi.* To iron clothes. **—iron out.** To settle through discussion or compromise : WORK OUT <*ironed out* their problems>
**Iron Age** *n.* The gen. prehistoric period succeeding the Bronze Age, marked by the introduction of iron metallurgy, in Europe beginning around the 8th cent. B.C.
**iron blue** *n.* A light- and heat-resistant, semitransparent blue pigment of powerful tinctorial strength, used mainly in permanent industrial finishes, printing inks, and artists' colors.
**i·ron·bound** (ī′ərn-bound′) *adj.* **1.** Bound with iron <*ironbound* kegs> **2.** IRONCLAD 2. **3.** Bound with rocks and cliffs, as a coast.
**i·ron·clad** (ī′ərn-klăd′) *adj.* **1.** Sheathed with iron plates for protection. **2.** Rigid : fixed <*an ironclad* law> *—n.* An armored 19th-cent. warship.
**Iron Curtain** *n.* A social, political, or military barrier preventing free exchange or communication, esp. the political and ideological barrier between the Soviet bloc and western Europe.
**iron gray** *n.* A dark gray with a slightly greenish tinge.
**iron hand** *n.* Despotic or rigorous control. **—i′ron·hand′ed** (ī′ərn-hăn′dĭd) *adj.* **—i′ron·hand′ed·ness** *n.*
**iron horse** *n. Informal.* A railroad locomotive.
**i·ron·ic** (ī-rŏn′ĭk) *also* **i·ron·i·cal** (ī-rŏn′ĭ-kəl) *adj.* **1.** Marked by or constituting irony. **2.** Given to the use of irony. **—i·ron′i·cal·ly** *adv.* **—i·ron′i·cal·ness** *n.*
**i·ron·ing** (ī′ər-nĭng) *n.* **1.** The act or process of pressing clothes with a heated iron. **2.** Clothing ironed or to be ironed.
**ironing board** *n.* A flat padded board on which clothes are ironed.
**i·ro·nist** (ī′rə-nĭst) *n.* A user of irony, esp. a writer.
**iron lung** *n.* A tank in which the entire body except the head is enclosed and by means of which pressure is regularly increased and decreased to provide artificial respiration.
**iron maiden** *n.* A medieval torture instrument having an iron frame in the form of a person in which a victim was enclosed and impaled on interior spikes.
**i·ron·mas·ter** (ī′ərn-măs′tər) *n. Chiefly Brit.* A maker of iron products.
**i·ron·mon·ger** (ī′ərn-mŭng′gər, -mŏng′-) *n. Chiefly Brit.* A hardware merchant. **—i′ron·mon′ger·y** (-mŭng′gə-rē, -mŏng′-) *n.*
**iron oxide** *n.* One of various oxides of iron, as ferrous oxide.
**iron pyrites** *n.* Pyrite.
**i·ron·smith** (ī′ərn-smĭth′) *n.* A worker in iron : BLACKSMITH.
**i·ron·stone** (ī′ərn-stōn′) *n.* **1.** One of various kinds of iron ore with admixtures of silica and clay. **2.** A hard white pottery.
**i·ron·ware** (ī′ərn-wâr′) *n.* Products made of iron.
**i·ron·weed** (ī′ərn-wēd′) *n.* A plant of the genus *Vernonia,* with purplish flower clusters.
**i·ron·wood** (ī′ərn-wo͝od′) *n.* **1.** One of various trees with very hard wood. **2.** The wood of an ironwood.
**i·ron·work** (ī′ərn-wûrk′) *n.* Work in iron, as gratings and rails.
**i·ron·works** (ī′ərn-wûrks′) *pl.n. (sing. or pl. in number).* A building or establishment where iron is smelted or where heavy iron products are made.
**i·ro·ny** (ī′rə-nē) *n., pl.* **-nies.** [Lat. *ironia* < Gk. *eirōneia,* feigned ignorance < *eirōn,* dissembler < *eirein,* to say.] **1. a.** Use of words to convey the opposite of their literal meaning. **b.** An expression or utterance marked by a deliberate contrast between apparent and intended meaning. **c.** A style of literature employing ironic contrasts for rhetorical or humorous effect. **2. a.** Incongruity between what might be expected and what actually happens <"Hyde noted the *irony* of Ireland's copying the nation she most hated" —Richard Kain> **b.** An occurrence, circumstance, or result notable for such incongruity. **3.** The dramatic effect attained by leading an audience to understand an incongruity between a situation and the accompanying speeches, while the characters in the play remain unaware of the incongruity. **4.** Feigned ignorance, as in the Socratic method of instruction.
**Ir·o·quoi·an** (ĭr′ə-kwoi′ən) *n.* **1.** A family of North American Indian languages of the eastern part of Canada and the United States that includes Iroquois, Cherokee, Conestoga, Erie, and Wyandot. **2.** A member of a tribe using a language of the Iroquoian family. *—adj.* Of or constituting the Iroquoian language family.
**Ir·o·quois** (ĭr′ə-kwoi′) *n., pl.* **Iroquois** (-kwoi′, -kwoiz′) [Fr. < Algonquin *Irinakhoiw.*] **1. a.** Any of several Indian tribes once inhabiting New York State and forming the confederacy known as the Five Nations, including the Cayuga, Mohawk, Oneida, Onondaga, and Seneca peoples. **b.** A member of any of these tribes. **2.** Any of the languages of the Iroquois. **—Ir′o·quois′** *adj.*
**ir·ra·di·ate** (ĭ-rā′dē-āt′) *v.* **-at·ed, -at·ing, -ates.** [Lat. *irradiare, irradiat-,* to illuminate : *in-,* on + *radiare,* to shine < *radius,* ray.] *—vt.* **1.** To expose to or treat with radiation. **2.** To send forth in a way analogous to the emission of light. *—vi. Archaic.* To send forth rays or become radiant. **—ir·ra′di·a′tive** *adj.* **—ir·ra′di·a′tor** *n.*
**ir·ra·di·a·tion** (ĭ-rā′dē-ā′shən) *n.* **1.** The act of irradiating or state of being irradiated. **2.** Treatment or therapy by exposure to radiation.
**ir·rad·i·ca·ble** (ĭ-răd′ĭ-kə-bəl) *adj.* [Med. Lat. *irradicabilis* : Lat. *in-,* not + Lat. *radix,* root.] Impossible to destroy : INVETERATE. **—ir·rad′i·ca·bly** *adv.*
**ir·ra·tion·al** (ĭ-răsh′ə-nəl) *adj.* **1. a.** Not capable of reasoning. **b.** Affected by loss of normal mental clarity : INCOHERENT. **c.** Contrary to reason : ILLOGICAL. **2. a.** Designating a syllable in Greek and Latin prosody whose length does not fit the metric pattern. **b.** Designating a metrical foot containing such a syllable. **3.** *Math.* Incapable of being expressed as an integer or a quotient of integers. **—ir·ra′tion·al·ly** *adv.* **—ir·ra′tion·al·ness** *n.*
**ir·ra·tion·al·ism** (ĭ-răsh′ə-nə-lĭz′əm) *n.* Irrational thought, expression, or behavior.
**ir·ra·tion·al·i·ty** (ĭ-răsh′ə-năl′ĭ-tē) *n., pl.* **-ties. 1.** The quality or state of being irrational. **2.** An irrational action or idea.
**irrational number** *n.* A member of the set of real numbers that is not a member of the set of rational numbers.
**ir·re·claim·a·ble** (ĭr′ĭ-klā′mə-bəl) *adj.* Incapable of reclamation <*irreclaimable* toxic waste dumps> **—ir′re·claim′a·bil′i·ty, ir′re·claim′a·ble·ness** *n.* **—ir′re·claim′a·bly** *adv.*
**ir·rec·on·cil·a·ble** (ĭ-rĕk′ən-sī′lə-bəl, ĭ-rĕk′ən-sī′-) *adj.* Not capable of being reconciled. *—n.* **1.** One who will not compromise, adjust, or submit. **2. irreconcilables.** Conflicting beliefs or ideas that cannot be brought into accord. **—ir·rec′on·cil′a·bil′i·ty** *n.* **—ir·rec′on·cil′a·bly** *adv.*
**ir·re·cov·er·a·ble** (ĭr′ĭ-kŭv′ər-ə-bəl) *adj.* Incapable of being recovered : IRREPARABLE <*irrecoverable* damages> **—ir′re·cov′er·a·ble·ness** *n.* **—ir′re·cov′er·a·bly** *adv.*
**ir·re·cu·sa·ble** (ĭr′ĭ-kyo͞o′zə-bəl) *adj.* [Fr. *irrécusable* < LLat. *irrecusabilis* : Lat. *in-,* not + *recusabilis,* deserving of rejection < Lat. *recusare,* to refuse. —see RECUSANT.] Not subject to challenge or objection. **—ir′re·cu′sa·bly** *adv.*
**ir·re·deem·a·ble** (ĭr′ĭ-dē′mə-bəl) *adj.* **1.** Incapable of being bought back or paid off. **2.** Not convertible into coin. **3.** Incapable of being remedied : IRREMEDIABLE. **4.** Incapable of being redeemed or reformed. **—ir′re·deem′a·bly** *adv.*
**ir·re·den·tist** (ĭr′ĭ-dĕn′tĭst) *n.* [Ital. *irredentista* < *(Italia) irredenta,* unredeemed (Italy), Italian-speaking areas subject to other countries.] An advocate of the recovery of lands of which his or her nation has been deprived or of territory historically or culturally related to his or her nation but now subject to a foreign government. **—ir′re·den′tism** *n.* **—ir′re·den′tist** *adj.*
**ir·re·duc·i·ble** (ĭr′ĭ-do͞o′sə-bəl, -dyo͞o′-) *adj.* Incapable of being reduced to a simpler or smaller form or amount. **—ir′re·duc′i·bil′i·ty, ir′re·duc′i·ble·ness** *n.* **—ir′re·duc′i·bly** *adv.*
**ir·ref·ra·ga·ble** (ĭ-rĕf′rə-gə-bəl) *adj.* [LLat. *irrefragabilis* : Lat. *in-,* not + Lat. *refragari,* to oppose.] Incapable of being refuted : INDISPUTABLE. **—ir·ref′ra·ga·bil′i·ty** *n.* **—ir·ref′ra·ga·bly** *adv.*

**ir·re·fran·gi·ble** (ĭr'ĭ-frăn'jə-bəl) *adj.* **1.** Incapable of being broken. **2.** *Physics.* Incapable of being refracted. **—ir're·fran'gi·bly** *adv.*

**ir·ref·u·ta·ble** (ĭ-rĕf'yə-tə-bəl, ĭr'ĭ-fyōō'-) *adj.* Incapable of being disproved. **—ir·ref'u·ta·bil'i·ty** *n.* **—ir·ref'u·ta·bly** *adv.*

**ir·re·gard·less** (ĭr'ĭ-gärd'lĭs) *adv. Nonstandard.* Regardless.

**ir·reg·u·lar** (ĭ-rĕg'yə-lər) *adj.* **1.** Not according to rule, traditional order, or general practice <*irregular* coffee breaks> **2.** Not conforming to moral law, legality, or social convention <*irregular* lifestyles> **3.** Not uniform, straight, or symmetric. **4.** Of uneven rate, occurrence, or duration <an *irregular* pulse> **5.** Asymmetrically arranged or atypical. **6.** *Bot.* Having differing floral parts, esp. petals. **7.** Of a quality below the manufacturer's standard or usual specifications : IMPERFECT. **8.** Departing from the usual set of inflectional forms <The verb "be" is *irregular.*> **9.** Not belonging to a permanent, organized military force <*irregular* volunteers> *—n.* **1.** One that is irregular. **2.** A soldier, as a guerrilla, who is not a member of a regular military force. **—ir·reg'u·lar·ly** *adv.*

**ir·reg·u·lar·i·ty** (ĭ-rĕg'yə-lăr'ĭ-tē) *n., pl.* **-ties. 1.** The quality or state of being irregular. **2.** Something irregular. **3.** Constipation.

**ir·rel·a·tive** (ĭ-rĕl'ə-tĭv) *adj.* **1.** Having no correlative relationship : UNCONNECTED. **2.** Irrelevant. **—ir·rel'a·tive·ly** *adv.*

**ir·rel·e·vant** (ĭ-rĕl'ə-vənt) *adj.* Having no applications or effects in a specified circumstance. **—ir·rel'e·vance** (-vəns), **ir·rel'e·van·cy** (-vən-sē) *n.* **—ir·rel'e·vant·ly** *adv.*

☆ ***syns:*** IRRELEVANT, EXTRANEOUS, IMMATERIAL, INAPPLICABLE *adj. core meaning* : not relating to the subject at hand <avoided *irrelevant* digressions> ***ant:*** relevant

**ir·re·lig·ion** (ĭr'ĭ-lĭj'ən) *n.* Hostility or indifference to religion.

**ir·re·lig·ious** (ĭr'ĭ-lĭj'əs) *adj.* Indifferent or hostile to religion. **—ir're·lig'ious·ly** *adv.* **—ir're·lig'ious·ness** *n.*

**ir·re·me·a·ble** (ĭ-rē'mē-ə-bəl) *adj.* [Lat. *irremeabilis* : *in-*, not + *remeare*, to return (*re-*, back + *meare*, to go).] *Archaic.* Affording no possibility of return.

**ir·re·me·di·a·ble** (ĭr'ĭ-mē'dē-ə-bəl) *adj.* Impossible to correct, remedy, or repair : HOPELESS. **—ir're·me'di·a·bly** *adv.*

**ir·re·mis·si·ble** (ĭr'ĭ-mĭs'ə-bəl) *adj.* Not remissible : UNPARDONABLE. **—ir're·mis'si·bil'i·ty** *n.* **—ir're·mis'si·bly** *adv.*

**ir·re·mov·a·ble** (ĭr'ĭ-mōō'və-bəl) *adj.* Not removable. **—ir're·mov'a·bil'i·ty** *n.* **—ir're·mov'a·bly** *adv.*

**ir·rep·a·ra·ble** (ĭ-rĕp'ər-ə-bəl) *adj.* Incapable of being rectified, repaired, or corrected. **—ir·rep'a·ra·bil'i·ty, ir·rep'a·ra·ble·ness** *n.* **—ir·rep'a·ra·bly** *adv.*

**ir·re·peal·a·ble** (ĭr'ĭ-pē'lə-bəl) *adj.* Not repealable.

**ir·re·place·a·ble** (ĭr'ĭ-plā'sə-bəl) *adj.* Incapable of being replaced. **—ir're·place'a·bly** *adv.*

**ir·re·pres·si·ble** (ĭr'ĭ-prĕs'ə-bəl) *adj.* Impossible to hold back or control <*irrepressible* joy> **—ir're·pres'si·bil'i·ty, ir're·pres'si·ble·ness** *n.* **—ir're·pres'si·bly** *adv.*

**ir·re·proach·a·ble** (ĭr'ĭ-prō'chə-bəl) *adj.* Beyond reproach. **—ir're·proach'a·ble·ness** *n.* **—ir're·proach'a·bly** *adv.*

**ir·re·sis·ti·ble** (ĭr'ĭ-zĭs'tə-bəl) *adj.* Impossible to resist <an *irresistible* desire> **—ir're·sis'ti·bil'i·ty, ir're·sis'ti·ble·ness** *n.* **—ir're·sis'ti·bly** *adv.*

**ir·re·sol·u·ble** (ĭr'ĭ-zŏl'yə-bəl) *adj.* Not capable of being solved.

**ir·res·o·lute** (ĭ-rĕz'ə-lōōt') *adj.* Lacking in resolution : INDECISIVE. **—ir·res'o·lute'ly** *adv.* **—ir·res'o·lute'ness, ir·res'o·lu'tion** *n.*

**ir·re·solv·a·ble** (ĭr'ĭ-zŏl'və-bəl) *adj.* **1.** Incapable of being resolved. **2.** Incapable of being separated into component parts : IRREDUCIBLE.

**ir·re·spec·tive** (ĭr'ĭ-spĕk'tĭv) *adj. Archaic.* Marked by disregard : HEEDLESS. **—ir're·spec'tive·ly** *adv.*

**irrespective of** *prep.* Without consideration of : REGARDLESS OF.

**ir·res·pi·ra·ble** (ĭ-rĕs'pər-ə-bəl, ĭr'ĭ-spĭr'-) *adj.* Not fit for breathing.

**ir·re·spon·si·ble** (ĭr'ĭ-spŏn'sə-bəl) *adj.* **1.** Not accountable to a higher authority. **2.** Not financially or mentally fit to assume responsibility. **3.** Lacking a sense of responsibility. **4.** Marked by a lack of responsibility <*irresponsible* allegations> *—n.* An irresponsible person. **—ir're·spon'si·bil'i·ty, ir're·spon'si·ble·ness** *n.* **—ir're·spon'si·bly** *adv.*

☆ ***syns:*** IRRESPONSIBLE, FECKLESS, INCAUTIOUS, RECKLESS *adj. core meaning* : lacking or showing a lack of a sense of responsibility <criticized for *irresponsible* behavior> ***ant:*** responsible

**ir·re·spon·sive** (ĭr'ĭ-spŏn'sĭv) *adj.* **1.** Not responsive, as to treatment or stimuli. **2.** Not answering or responding readily. **—ir're·spon'sive·ly** *adv.* **—ir're·spon'sive·ness** *n.*

**ir·re·triev·a·ble** (ĭr'ĭ-trē'və-bəl) *adj.* Impossible to retrieve or recover. **—ir're·triev'a·bil'i·ty, ir're·triev'a·ble·ness** *n.* **—ir're·triev'a·bly** *adv.*

**ir·rev·er·ence** (ĭ-rĕv'ər-əns) *n.* **1.** Lack of reverence. **2.** A disrespectful act or remark.

**ir·rev·er·ent** (ĭ-rĕv'ər-ənt) *adj.* **1. a.** Lacking in reverence : DISRESPECTFUL. **b.** Lightly or humorously sardonic. **2.** Proceeding from irreverence. **—ir·rev'er·ent·ly** *adv.*

**ir·re·vers·i·ble** (ĭr'ĭ-vûr'sə-bəl) *adj.* Impossible to reverse. **—ir're·vers'i·bil'i·ty, ir're·vers'i·ble·ness** *n.* **—ir're·vers'i·bly** *adv.*

**ir·rev·o·ca·ble** (ĭ-rĕv'ə-kə-bəl) *adj.* Incapable of being retracted or revoked : IRREVERSIBLE. **—ir·rev'o·ca·bil'i·ty, ir·rev'o·ca·ble·ness** *n.* **—ir·rev'o·ca·bly** *adv.*

**ir·ri·ga·ble** (ĭr'ĭ-gə-bəl) *adj.* Capable of being irrigated.

**ir·ri·gate** (ĭr'ĭ-gāt') *vt.* **-gat·ed, -gat·ing, -gates.** [Lat. *irrigare, irrigat-* : *in-*, in + *rigare*, to water.] **1.** To supply (dry land) with water by means of ditches, pipes, or streams. **2.** To wash out (a canal or wound) with water or a medicated fluid. **3.** To vitalize or make fertile. **—ir'ri·ga'tion** *n.* **—ir'ri·ga'tion·al** *adj.* **—ir'ri·ga'tor** *n.*

**ir·ri·ta·bil·i·ty** (ĭr'ĭ-tə-bĭl'ĭ-tē) *n.* **1.** The quality or state of being irritable : CRANKINESS. **2.** *Pathol.* Excessive sensitivity. **3.** *Biol.* The capacity to respond to stimuli.

**ir·ri·ta·ble** (ĭr'ĭ-tə-bəl) *adj.* [Lat. *irritabilis* < *irritare*, to irritate.] **1. a.** Easily annoyed or irritated. **b.** Having or showing a bad temper. **2.** *Pathol.* Abnormally sensitive. **3.** *Biol.* Responsive to stimuli. **—ir'ri·ta·ble·ness** *n.* **—ir'ri·ta·bly** *adv.*

☆ ***syns:*** IRRITABLE, CANTANKEROUS, CROSS, DISAGREEABLE, GROUCHY, GRUMPY, ILL-TEMPERED, IRASCIBLE, NASTY, PEEVISH, PETULANT, QUERULOUS, SURLY, TESTY *adj. core meaning* : having or showing a bad temper <an *irritable* coworker>

**ir·ri·tant** (ĭr'ĭ-tənt) *adj.* [Lat. *irritans, irritant-*, pr.part. of *irritare*, to irritate.] Causing irritation, esp. physical irritation. *—n.* Something that causes irritation.

**ir·ri·tate** (ĭr'ĭ-tāt') *vt.* **-tat·ed, -tat·ing, -tates.** [Lat. *irritare.*] **1. a.** To exasperate : VEX. **b.** To provoke. **2.** To inflame or chafe. **—ir'ri·tat'ing·ly** *adv.* **—ir'ri·ta'tor** *n.*

**ir·ri·ta·tion** (ĭr'ĭ-tā'shən) *n.* **1.** The act of irritating or the state of being irritated. **2.** Something that irritates. **3.** *Pathol.* Incipient inflammation, roughness, soreness, or irritability of a bodily part.

**ir·ri·ta·tive** (ĭr'ĭ-tā'tĭv) *adj.* Involving irritation.

**ir·ro·ta·tion·al** (ĭr'ō-tā'shə-nəl) *adj.* Not rotating.

**ir·rupt** (ĭ-rŭpt') *vi.* **-rupt·ed, -rupt·ing, -rupts.** [Lat. *irrumpere, irrupt-* : *in-*, in + *rumpere*, to burst.] **1.** To break or burst in. **2.** *Ecol.* To increase irregularly in number. **—ir·rup'tion** *n.*

**ir·rup·tive** (ĭ-rŭp'tĭv) *adj.* **1.** Irrupting or prone to irruption. **2.** *Geol.* Intrusive.

**is** (ĭz) *v.* [ME < OE.] *3rd person sing. present tense of* BE.

**is-** *pref. var. of* ISO-.

**I·sa·iah** (ī-zā'ə, ī-zī'ə) *n.* [Heb. *Yĕsha'yāhu.*] —See table at BIBLE.

**i·sal·lo·bar** (ī-săl'ə-bär') *n.* [IS(O)- + ALLO- + Gk. *baros*, weight.] A line on a weather map connecting places having equal changes in barometric pressure within a given period of time.

**-isation** *suff. var. of* -IZATION.

**is·che·mi·a** (ĭ-skē'mē-ə) *n.* [NLat. *ischaemia* < Gk. *iskhaimos*, a stopping of the blood : *iskhein*, to restrain + *haima*, blood.] A localized anemia resulting from mechanical obstruction of the blood supply. **—i·sche'mic** *adj.*

**is·chi·um** (ĭs'kē-əm) *n., pl.* **-chi·a** (-kē-ə) [Lat., hip joint < Gk. *iskhion.*] The lowest of three major bones comprising each half of the pelvis. **—is'chi·al** *adj.*

**-ise** *suff. var. of* -IZE.

**is·en·tro·pic** (ī'sən-trō'pĭk, -trŏp'ĭk) *adj.* [IS(O)- + ENTROP(Y) + -IC.] Being without change in entropy. **—is'en·tro'pi·cal·ly** *adv.*

**I·seult** (ĭ-sōōlt') *also* **I·sol·de** (ĭ-sōl'də, ĭ-zōl'-) *n.* A legendary Irish princess of Arthurian times who married the king of Cornwall and had a hopeless love affair with his knight Tristan.

**-ish** *suff.* [ME < OE *-isc.*] **1.** Of, pertaining to, or being <Swed*ish*> **2. a.** Typical of <girl*ish*> **b.** Having the esp. undesirable qualities of <child*ish*> **3.** Approximately : somewhat <green*ish*> **4.** Tending toward : preoccupied with <self*ish*>

**Ish·ma·el** (ĭsh'mē-əl) *n.* [LLat. *Ismaël* < Heb. *Yishmā'ēl* : *yishma'*, he will hear + *Ēl*, God.] An outcast.

**Ish·ma·el·ite** (ĭsh'mē-ə-līt') *n.* **1.** One of a group of desert-dwelling people thought by the ancient Hebrews to be descended from Ishmael. **2.** One at odds with society. **—Ish'ma·el·it'ism** *n.*

**Ish·tar** (ĭsh'tär') *n.* [Akkadian.] *Myth.* The ancient Assyrian and Babylonian goddess of love, fertility, and war.

**i·sin·glass** (ī'zən-glăs', ī'zĭng-) *n.* [By folk ety. < obs. Du. *huizenblas* < MDu. *huusblase* : *huus*, sturgeon + *blase*, bladder.] **1.** A transparent, almost pure gelatin prepared from the air bladder of certain fishes, as the sturgeon. **2.** The mineral muscovite.

**I·sis** (ī'sĭs) *n.* [Lat. < Gk.] *Myth.* An ancient Egyptian goddess of fertility and sister and wife of Osiris.

**Is·lam** (ĭs-läm', ĭz-, ĭs'läm', ĭz'-) *n.* [Ar. *islām*, submission (to God) < *aslama*, he surrendered < *salama*, he was safe.] **1.** A religion based on the teachings of the prophet Mohammed, believing in one God, Allah, and in Paradise and Hell, and having a body of law put forth in the Koran and the Sunna. **2. a.** Nations whose predominant populations and religion are Moslem. **b.** Islamic civilization. **3.** Moslems as a group. **—Is·lam'ic** *adj.*

**Is·lam·ism** (ĭs'lə-mĭz'əm, ĭz'-) *n.* The religious faith, tenets, or cause of Islam. **—Is·lam'ist** (-lä'mĭst) *n.*

**Is·lam·ize** (ĭs'lə-mīz', ĭz'-) *vt.* **-ized, -iz·ing, -iz·es.** To convert to Islam.

**is·land** (ī'lənd) *n.* [ME *iland* < OE *īgland.*] **1.** A land mass, esp. one smaller than a continent, completely surrounded by water. **2.** Something resembling an island, esp. in being completely isolated. **3.** *Anat.* A tissue or cluster of cells separated from surrounding tissue by a groove or differing from surrounding tissue in structure. —*vt.* **-land·ed, -land·ing, -lands.** To make into or as if into an island.

▲ word history: The words *island* and *isle* are synonymous but unrelated, in spite of their apparent phonetic and orthographic similarities. The word *isle* is derived from Latin *insula,* which developed into Old French *ile,* the form originally borrowed into English. The word *island* is a native English word whose earliest form was *īgland.* Until the 17th century the ordinary form of the word was *iland,* which was the spelling used in the King James Bible and Milton's *Paradise Lost.* In the 15th century the French respelled *ile* as *isle,* inserting an etymologically more "correct" *s* from Latin *insula.* The English adopted this spelling for *ile* and incorrectly respelled *iland* as *isle-land* and *island,* probably from a mistaken notion that the native word was a compound of French *ile* and English *land.* The modern spelling of *isle* and *island* became standard only in the 18th century.

**is·land·er** (ī′lən-dər) *n.* An inhabitant of an island.

**islands of Lang·er·hans** (läng′ər-häns′) *pl.n.* [After Paul *Langerhans* (1847–1888).] Irregular masses of small cells that lie in the interstitial tissue of the pancreas and secrete insulin.

**isle** (īl) *n.* [ME *ile* < OFr. *ile* < Lat. *insula.*] An island, esp. one that is small.

**is·let** (ī′lĭt) *n.* A little island.

**islets of Lang·er·hans** (läng′ər-häns′) *pl.n.* Islands of Langerhans.

**ism** (ĭz′əm) *n.* [< -ISM.] *Informal.* A distinctive system, doctrine, or theory.

**-ism** *suff.* [ME *-isme* < OFr. < Lat. *-ismus* < Gk. *-ismos,* n. suffix.] **1.** Action or process : PRACTICE <terror*ism*> **2.** Typical behavior or quality <hero*ism*> **3. a.** State : quality <pauper*ism*> **b.** State or condition resulting from an excess of something specified <strychnin*ism*> **4.** Distinctive or typical trait <Latin*ism*> **5.** System of principles : DOCTRINE <pacif*ism*>

**Is·ma·i·li** (ĭs′mä-ē′lē) *also* **Is·ma·i·li·an** (-ē′lē-ən) *n.* [Ar. *Isma′īlīy,* after *Isma′īl* (d. A.D. 760), son of the sixth Imam Jafar.] A Moslem of a Shiah sect.

**is·n't** (ĭz′ənt). Is not.

**iso-** *or* **is-** *pref.* [Gk. < *isos,* equal.] **1.** Equal : uniform <*iso*bar> **2.** Isomeric <*iso*propyl>

**i·so·ag·glu·ti·na·tion** (ī′sō-ə-glo͞ot′n-ā′shən) *n.* Agglutination of an agglutinogen by the serum of another individual of the same species.

**i·so·ag·glu·ti·nin** (ī′sō-ə-glo͞ot′n-ĭn) *n.* An isoantibody causing cell agglutination.

**i·so·ag·glu·tin·o·gen** (ī′sō-ăg′lo͞o-tĭn′ə-jən) *n.* [ISOAGGLUTIN(IN) + -GEN.] An isoantigen that on exposure to its isoantibody induces agglutination of cells to which it is attached.

**i·so·an·ti·bod·y** (ī′sō-ăn′tē-bŏd′ē) *n., pl.* **-ies.** An antibody occurring in only some individuals of a species and reacting specifically with the corresponding isoantigen.

**i·so·an·ti·gen** (ī′sō-ăn′tĭ-jən) *n.* An antigen occurring only in some individuals of a species and never in those having cells containing the corresponding isoantibody. **—i′so·an′ti·gen′ic** *adj.* **—i′so·an′ti·ge·nic′i·ty** *n.*

**i·so·bar** (ī′sə-bär′) *n.* [ISO- + Gk. *baros,* weight.] **1.** A line on a map connecting points of equal pressure. **2.** One of two or more nuclides with the same mass number but different atomic numbers. **—i′so·bar′ic** (-bär′ĭk, -băr′-) *adj.*

**i·so·chro·mat·ic** (ī′sə-krō-măt′ĭk) *adj.* ORTHOCHROMATIC 2.

**i·soch·ro·nal** (ī-sŏk′rə-nəl) *adj.* [< Gk. *isokhronos* : *isos,* equal + *khronos,* time.] **1.** Equal in duration. **2.** Marked by or occurring at equal time intervals. **—i·soch′ro·nal·ly** *adv.* **—i·soch′ro·nism** *n.*

**i·soch·ro·nize** (ī-sŏk′rə-nīz′) *vt.* **-nized, -niz·ing, -niz·es.** To make isochronal.

**i·soch·ro·nous** (ī-sŏk′rə-nəs) *adj.* Isochronal.

**i·soch·ro·ous** (ī-sŏk′rō-əs) *adj.* [ISO- + Gk. *khrōs,* flesh, color.] Having the same color throughout.

**i·so·cli·nal** (ī′sə-klī′nəl) *adj.* Having the same dip or inclination. —*n.* An isoclinic line. **—i′so·cli′nal·ly** *adv.*

**i·so·cline** (ī′sə-klīn′) *n.* An anticline or syncline with strata so tightly folded as to have the same dip.

**i·so·clin·ic** (ī′sə-klĭn′ĭk) *adj.* & *n.* Isoclinal.

**isoclinic line** *n.* A line on a map connecting points of equal magnetic dip.

**i·so·di·a·met·ric** (ī′sō-dī′ə-mĕt′rĭk) *adj.* Having equal diameters.

**i·so·di·mor·phism** (ī′sō-dī-môr′fĭz′əm) *n.* Isomorphism between crystalline forms of two dimorphic substances.

**i·so·dy·nam·ic** (ī′sō-dī-năm′ĭk) *adj.* Having equal strength or force.

**i·so·e·lec·tric** (ī′sō-ĭ-lĕk′trĭk) *adj.* Having equal electric potential.

**i·so·e·lec·tron·ic** (ī′sō-ĭ-lĕk-trŏn′ĭk) *adj.* Having equal numbers of electrons or the same electronic configuration.

**i·so·en·zyme** (ī′sō-ĕn′zīm′) *n.* One of two or more chemically distinct but functionally identical forms of an enzyme. **—i′so·en·zy′mic** *adj.*

**i·so·ga·mete** (ī′sō-gə-mēt′, -găm′ēt′) *n.* A gamete morphologically indistinguishable from one with which it unites.

**i·sog·a·my** (ī-sŏg′ə-mē) *n.* Conjugation of isogametes or of identical cells. **—i·sog′a·mous** *adj.*

**i·so·gloss** (ī′sə-glôs′, -glŏs′) *n.* [ISO- + Gk. *glossa,* language.] A geographic boundary delimiting the area in which a given linguistic form occurs. **—i′so·gloss′al** *adj.*

**i·so·gon** (ī′sə-gŏn′) *n.* An equiangular polygon.

**i·so·gon·ic** (ī′sə-gŏn′ĭk) *also* **i·sog·o·nal** (ī-sŏg′ə-nəl) *adj.* Having equal angles. —*n.* An isogonic line.

**isogonic line** *n.* A line on a map connecting points of equal magnetic declination.

**i·so·gram** (ī′sə-grăm′) *n.* A line on a map, chart, or graph connecting points of equal value.

**i·so·hel** (ī′sō-hĕl′) *n.* [ISO- + Gk. *helios,* sun.] A line on a map connecting points receiving equal sunlight.

**i·so·he·mo·ly·sin** (ī′sō-hē′mə-lī′sən, -hĕm′ə-, -hĭ-mŏl′ĭ-sĭn) *n.* Hemolysin derived from the serum of an individual injected with red blood cells from another individual of the same species.

**i·so·he·mol·y·sis** (ī′sō-hə-mŏl′ĭ-sĭs) *n.* Hemolysis due to the action of isohemolysin.

**i·so·hy·et** (ī′sō-hī′ĭt) *n.* [ISO- + Gk. *huetos,* rain.] A line on a map connecting points receiving equal rainfall.

**i·so·late** (ī′sə-lāt′) *vt.* **-lat·ed, -lat·ing, -lates.** [Back-formation < *isolated,* set apart < Fr. *isolé* < Ital. *isolato* < LLat. *insulatus,* made into an island < *insula,* island.] **1.** To set apart from a group or whole. **2.** To place in quarantine. **3.** *Chem.* To obtain (a substance) in an uncombined form. **4.** To render free of external influence : INSULATE. —*adj.* (-lĭt, -lāt′). Solitary : alone. **—i′so·la·ble** (-lə-bəl), **i′so·lat′a·ble** (-lāt′ə-bəl) *adj.* **—i′so·la′tion** *n.* **—i′so·la′tor** *n.*

**i·so·la·tion·ism** (ī′sə-lā′shə-nĭz′əm) *n.* A national policy of abstaining from economic or political entanglements with other countries. **—i′so·la′tion·ist** *n.*

**I·sol·de** (ĭ-sōl′də, ĭ-zōl′-) *n. var. of* ISEULT.

**i·so·lec·i·thal** (ī′sə-lĕs′ə-thəl) *adj.* [ISO- + LECITH(IN) + -AL.] Having the yolk evenly distributed throughout the egg.

**i·so·leu·cine** (ī′sə-lo͞o′sēn′) *n.* An essential amino acid, $C_6H_{13}NO_2$, isomeric with leucine.

**i·so·mag·net·ic** (ī′sō-măg-nĕt′ĭk) *adj.* Designating or relating to points of equal magnetic induction.

**i·so·mer** (ī′sə-mər) *n.* **1.** *Chem.* **a.** A compound with the same percentage composition and molecular weight as another compound but differing in chemical or physical properties. **b.** Such a compound so differing because of the manner of linkage of its constituent atoms. **c.** Such a compound so differing because of the manner of arrangement of its constituent atoms in space. **d.** A stereoisomer manifesting one of two structures that rotate the plane of polarization of polarized light either to the left or to the right. **e.** A stereoisomer having no effect on polarized light but exhibiting isomerism because of a structural asymmetry about a double bond in the molecule. **2.** *Physics.* An atom the nucleus of which can exist in any of several bound excited states for a measurable period of time. **—i′so·mer′ic** (-mĕr′ĭk) *adj.*

**i·som·er·ase** (ī-sŏm′ə-rās′) *n.* An enzyme that catalyzes isomerization reactions.

**i·som·er·ism** (ī-sŏm′ə-rĭz′əm) *n.* **1.** The phenomenon of the existence of isomers. **2.** The complex of chemical and physical phenomena typical of or attributable to isomers. **3.** The condition of being an isomer.

**i·som·er·ize** (ī-sŏm′ə-rīz′) *vi* & *vt.* **-ized, -iz·ing, -iz·es.** To change or cause to change into an isomeric form. **—i·som′er·i·za′tion** *n.*

**i·som·er·ous** (ī-sŏm′ər-əs) *adj.* **1.** Having an equal number of parts, as organs or markings. **2.** Having or designating floral whorls with equal numbers of parts.

**i·so·met·ric** (ī′sə-mĕt′rĭk) *also* **i·so·met·ri·cal** (-rĭ-kəl) *adj.* [< Gk. *isometros,* of equal measure : *isos,* equal + *metron,* measure.] **1.** Of or exhibiting equality in measurements or dimensions. **2.** Of or being a crystal system of three equal and mutually orthogonal axes. **3.** *Physiol.* Of or involving muscular contraction occurring when the ends of the muscle are fixed in place so that significant increases in tension occur without appreciable increases in length. —*n.* **isometric. 1.** A line connecting isometric points. **2. isometrics.** *(sing. in number).* Isometric exercise.

**isometric exercise** *n.* Exercise involving isometric contraction.

**i·so·me·tro·pi·a** (ī′sō-mĭ-trō′pē-ə) *n.* [Gk. *isometros,* isometric + -OPIA.] Equality of refraction in both eyes.

**i·som·e·try** (ī-sŏm′ĭ-trē) *n.* **1.** Equality of measure. **2.** Equality of elevation above sea level.

**i·so·morph** (ī′sə-môrf′) *n.* An object, organism, or group displaying isomorphism.

**i·so·mor·phic** (ī′sə-môr′fĭk) *adj.* Related by an isomorphism.

**i·so·mor·phism** (ī′sə-môr′fĭz′əm) *n.* **1.** *Biol.* Similarity in form, as in different kinds of organisms. **2.** *Math.* **a.** A one-to-one correspondence between the elements of two sets such that the result of an operation on elements of one set corresponds to the result of the analogous operation on their images in the other set. **b.** A mapping * of a group $G$ onto another group $H$ such that $(ab)^* = (a^*)(b^*)$ for all $a$, $b$ in $G$. **3.** The existence or an instance of the existence of two or more different substances with closely similar crystalline structure, crystalline dimensions, and chemical composition. **—i'so·mor'phous** *adj.*

**i·so·ni·a·zid** (ī'sə-nī'ə-zĭd) *n.* [ISONI(COTINIC ACID) + (HYDR)AZID(E).] A crystalline compound, $C_6H_7N_3O$, used for treating tuberculosis.

**i·so·oc·tane** (ī'sō-ŏk'tān') *n.* A highly flammable liquid, $C_8H_{18}$, used to determine the octane numbers of fuels.

**i·so·pi·es·tic** (ī'sō-pī-ĕs'tĭk, -pē-) *adj.* [ISO- + Gk. *piestos*, able to be compressed < *piezein*, to press tight.] Characterized by or registering equal pressure : ISOBARIC. —*n.* An isobar.

**i·so·pod** (ī'sə-pŏd') *n.* [NLat. *Isopoda*, order name : ISO- + Gk. *pous*, foot.] Any of various crustaceans of the order Isopoda, which includes the sow bugs and gribbles. **—i'so·pod'** *adj.*

**i·so·prene** (ī'sə-prēn') *n.* [ISO- + PR(OPYL) + -ENE.] A colorless volatile liquid, $C_5H_8$, used primarily in making synthetic rubber.

**i·so·pro·pyl alcohol** (ī'sə-prō'pəl) *n.* A clear, colorless, flammable, mobile liquid, $C_3H_8O$, used in antifreeze compounds, lotions, and cosmetics and as a solvent for shellac, gums, and essential oils.

**i·sos·ce·les** (ī-sŏs'ə-lēz') *adj.* [LLat. *isosceles* < Gk. *isoskelēs* : *isos*, equal + *skelos*, leg.] Having two equal sides.

**i·so·seis·mic** (ī'sə-sīz'mĭk) *also* **i·so·seis·mal** (-məl) *adj.* Of, relating to, or exhibiting equal seismic intensities.

**i·sos·mot·ic** (ī'sŏz-mŏt'ĭk, -sŏs-) *adj.* Of or exhibiting equal osmotic pressure.

**i·so·spin** (ī'sə-spĭn') *n.* [ISO(TOPIC) + SPIN.] A quantum number related to the number of charge states of a subatomic particle by the equation $2I + 1 = M$, where $M$ is the number of such states.

**i·sos·ta·sy** (ī-sŏs'tə-sē) *n.* [ISO- + Gk. *stasis*, a standstill.] Equilibrium caused by isotropic equalization of pressure.

**i·so·therm** (ī'sə-thûrm') *n.* [Fr. *isotherme*, having the same temperature : Gk. *isos*, equal + Gk. *thermē*, heat.] A line on a weather map or chart linking all points having identical mean temperature for a specified period or identical temperature at a specified time.

**i·so·ther·mal** (ī'sə-thûr'məl) *adj.* **1.** Of, relating to, or registering equal temperatures. **2.** Of or designating changes of pressure and volume at constant temperature. **3.** Of or relating to an isotherm. —*n.* An isotherm.

**i·so·tone** (ī'sə-tōn') *n.* [ISO- + Gk. *tonos*, tension, stretching.] One of two or more atoms whose nuclei have the same number of neutrons but different numbers of protons.

**i·so·ton·ic** (ī'sə-tŏn'ĭk) *adj.* [ISO- + Gk. *tonos*, tension.] **1.** Equal in tension. **2.** Isosmotic. **—i'so·ton'i·cal·ly** *adv.* **—i'so·to·nic'i·ty** (-tə-nĭs'ĭ-tē) *n.*

**i·so·tope** (ī'sə-tōp') *n.* [ISO- + Gk. *topos*, place.] One of two or more atoms whose nuclei have the same number of protons but different numbers of neutrons. **—i'so·top'ic** (-tŏp'ĭk) *adj.* **—i'so·top'i·cal·ly** *adv.*

**isotopic spin** *n.* Isospin.

**i·so·tro·pic** (ī'sə-trō'pĭk, -trŏp'ĭk) *adj.* Identical in all directions. **—i·sot'ro·py** (ī-sŏt'rə-pē), **i·sot'ro·pism** (-pĭz'əm) *n.*

**i·so·zyme** (ī'sə-zīm') *n.* [ISO- + (EN)ZYME.] Isoenzyme.

**Is·ra·el** (ĭz'rē-əl) *n.* [Lat. < Gk. *Israēl* < Heb. *Yisrā'ēl*.] **1.** The descendants of Jacob in the Old Testament. **2.** The whole Hebrew people, past, present, and future, thought to be the chosen people of Jehovah by virtue of the covenant of Jacob.

**Is·rae·li** (ĭz-rā'lē) *adj.* Of or relating to the state of Israel or its people. —*n., pl.* **Israeli** or **-lis.** A native or resident of the state of Israel.

**Is·ra·el·ite** (ĭz'rē-ə-līt') *n.* A Hebrew, esp. a descendant of Jacob. —*adj. also* **Is·ra·el·it·ic** (ĭz'rē-ə-lĭt'ĭk). Of or relating to Israel or the Israelites.

**Is·sa·char** (ĭs'ə-kär') *n.* [LLat. < Gk. < Heb. *Yissākhār*.] The Hebrew tribe descended from Issachar, son of Jacob and Leah.

**Is·sei** (ēs'sā') *n., pl.* **Issei** or **-seis.** [J., first generation < Chin. (Mandarin) *yi*[1] *shi*[4] : *yi*[1], first + *shi*[4], generation.] A Japanese immigrant to the United States or Canada.

**is·su·a·ble** (ĭsh'ōō-ə-bəl) *adj.* **1.** Capable of issuing or being issued. **2.** Open to debate or litigation. **3.** Authorized for issue.

**is·su·ance** (ĭsh'ōō-əns) *n.* An act of issuing : ISSUE.

**is·su·ant** (ĭsh'ōō-ənt) *adj.* **1.** *Archaic.* Emerging. **2.** *Heraldry.* Designating an animal with only the upper part depicted.

**is·sue** (ĭsh'ōō) *n.* [ME, exit, act of going out < OFr. < VLat. *exuta < Lat. *exitus*, p.part. of *exire*, to go out : *ex-*, out + *ire*, to go.] **1. a.** An act or instance of passing, flowing, or giving out. **b.** An act of circulating, distributing, or publishing by an office or official group. **2.** Something produced, published, or offered, as: **a.** An item or set of items, as stamps, made available at one time by an office or bureau. **b.** A single copy of a periodical. **c.** A set of copies of an edition of a book distinguished from others of that edition by variations in the printed matter. **d.** A final result or outcome. **e.** Proceeds from fines or estates. **f.** Something proceeding from a designated source <ideas that were the *issue* of a creative mind> **3.** Offspring : progeny. **4. a.** A point of debate, discussion, or dispute. **b.** A matter of public concern. **c.** The essential point : CRUX. **d.** A culminating point leading to a decision <brought the problem to an *issue*> **5.** A place of exit : OUTLET. **6.** *Pathol.* **a.** A discharge, as of blood. **b.** A suppurating sore. **7.** *Archaic.* Termination : close. —*v.* **-sued, -su·ing, -sues.** —*vi.* **1.** To go or come out. **2.** To accrue as profit or proceeds. **3.** To be born or be descended. **4.** To be published or circulated. **5.** To spring or result from. **6.** To result in or terminate. —*vt.* **1.** To cause to flow out : DISCHARGE. **2.** To distribute or circulate officially <*issued* weapons to the troops> **3.** To publish. **—at issue. 1.** In question or dispute. **2.** At variance. **—take issue.** To take an opposing point of view : DISAGREE. **—is'su·er** *n.*

**-ist** *suff.* [ME *-iste* < OFr. < Lat. *-ista, istes* < Gk. *-istēs*, agent n. suffix.] **1. a.** One that performs a given action <lobby*ist*> **b.** One that produces, makes, operates, plays, or is connected with a given thing <novel*ist*> **2.** A specialist in a given art, science, or skill <biolog*ist*> **3.** An adherent or advocate of a given doctrine, theory, or school of thought <anarch*ist*> **4.** One that is characterized by a given trait or quality <romantic*ist*>

**isth·mi** (ĭs'mī') *n. var. pl. of* ISTHMUS.

**isth·mi·an** (ĭs'mē-ən) *adj.* **1.** Of, relating to, or forming an isthmus. **2. Isthmian.** Of or relating to the Isthmus of Corinth, esp. with regard to the biennial pan-Hellenic games held there in antiquity.

**isth·mus** (ĭs'məs) *n., pl.* **-mus·es** or **-mi** (-mī') [Lat. < Gk. *isthmos*.] **1.** A narrow strip of land connecting two larger land masses. **2.** *Anat.* **a.** A narrow strip of tissue joining two larger organs or parts of an organ. **b.** A narrow passage connecting two larger cavities.

**is·tle** *also* **ix·tle** (ĭs'lē, ĭst'-) *n.* [Mex. Sp. *ixtle* < Nahuatl *ixtli*, fibrous stem.] A plant, pita, or its fiber.

**it** (ĭt) *pron.* [ME < OE *hit*.] **1.** That one previously mentioned. —Used as the subject, direct object, indirect object, or object of a preposition for a nonhuman entity, an animate being, as a person or animal whose sex is unknown, unspecified, or irrelevant, a group of objects or individuals, or an abstraction <scrubbed the floor until *it* was clean><couldn't figure out what *it* was><started the machine by turning *it* on><gave *it* a thorough examination> **2.** —Used as the subject of an impersonal verb <*It* is pouring outside.> **3. a.** —Used as an anticipatory subject <Is *it* known who will get the job?> **b.** —Used as an anticipatory subject to emphasize a term that is not itself a subject <*It* was on Tuesday that the fire occurred.> **4.** —Used to refer to a general condition or state of affairs <couldn't stand *it*> **5.** *Informal.* —Used to refer to something that is the best, the most desirable, or without equal <That performance was really *it!*> —*n.* A player in a game, as tag, who tries to find or catch the other players.

**it·a·col·u·mite** (ĭt'ə-kŏl'yə-mīt') *n.* [After *Itacolumi*, a mountain in Brazil where it is found.] A variety of sandstone that is flexible when cut into thin slabs.

**I·tal·ian** (ĭ-tăl'yən) *adj.* [ME < Lat. *Italianus* < *Italia*, Italy.] Relating to Italy, its people, or their language. —*n.* **1.** A native or resident of Italy or a person of Italian descent. **2.** The Romance language of Italy and one of the three official languages of Switzerland.

**I·tal·ian·ate** (ĭ-tăl'yə-nāt', -nĭt) *adj.* Italian in character.

**I·tal·ian·ism** (ĭ-tăl'yə-nĭz'əm) *n.* **1.** An Italian custom, trait, or expression. **2.** A quality typical of Italy or its people.

**I·tal·ian·ize** (ĭ-tăl'yə-nīz') *v.* **-ized, -iz·ing, -iz·es.** —*vt.* To give an Italian aspect to. —*vi.* To adopt Italian manners, speech, or customs. **—I·tal'ian·i·za'tion** *n.*

**Italian sandwich** *n.* HERO 5.

**Italian sonnet** *n.* A Petrarchan sonnet.

**I·tal·ic** (ĭ-tăl'ĭk, ī-tăl'-) *adj.* [Lat. *Italicus* < Gk. *Italikos* < *Italia*, Italy < Lat.] **1.** Of or relating to ancient Italy or its peoples. **2.** Of or relating to Italic. **3. italic.** Of or being a style of printing type patterned on a Renaissance script with the letters slanting to the right. —*n.* **1.** A branch of the Indo-European language family that includes Latin, Oscan, and Umbrian. **2.** *often* **italics.** Italic typeface or print.

**I·tal·i·cism** (ĭ-tăl'ĭ-sĭz'əm) *n.* An Italianism, esp. a word or idiom borrowed from or suggestive of the Italian language.

**i·tal·i·cize** (ĭ-tăl'ĭ-sīz', ī-tăl'-) *vt.* **-cized, -ciz·ing, -ciz·es. 1.** To print in italic type. **2.** To underscore (written matter) with a single line to indicate italics. **—i·tal'i·ci·za'tion** *n.*

**itch** (ĭch) *n.* [ME *icchen* < OE *giccan*.] **1.** A skin sensation resulting in a desire to scratch. **2.** A contagious skin disease characterized by intense irritation, eruptions, and itching. **3.** A restless desire or craving. —*v.* **itched, itch·ing, itch·es.** —*vi.* **1. a.** To feel, have, or produce an itch. **b.** To have a desire to scratch. **2.** To have a persistent restless craving. —*vt.* **1.** To cause to itch. **2.** To scratch an itch.

**itch mite** *n.* A parasitic mite, *Sarcoptes scabiei*, that causes scabies.

**itch·y** (ĭch'ē) *adj.* **-i·er, -i·est. 1.** Having or resulting in an itching sensation. **2.** Nervous or restless. **—itch'i·ness** *n.*

**-ite[1]** *suff.* [ME < OFr. < Lat. *-ita, -ites* < Gk. *-itēs*.] **1.** Native or resident of <New Jersey*ite*> **2. a.** Descendant of <Lev*ite*> **b.** Adherent or follower of <Ludd*ite*> **3.** A part of an organ, body, or bodily part <som*ite*> **4.** Rock : mineral <graph*ite*> **5.** A commercial product <ebon*ite*>

**-ite[2]** *suff.* [Fr., alteration of *-ate*, -ate < NLat. *-atum*. —see -ATE[2].] A salt or ester of an acid designated with an adjective ending in *-ous* <sulf*ite*>

ōō **boot** ou **out** th **thin** *th* **this** ŭ **cut** ûr **urge** y **young**
yōō ab**u**se zh vi**s**ion ə **a**bout, it**e**m, ed**i**ble, gall**o**p, circ**u**s

**i·tem** (īʹtəm) *n.* [< ME, also, moreover < Lat. < *ita,* so.] **1.** A single, separately specified article or unit included in an enumeration, collection, or series. **2.** A clause of a document, as a bill or charter. **3.** An entry in an account. **4. a.** A bit of information : DETAIL. **b.** A short piece in a magazine or newspaper. *—vt.* **i·temed, i·tem·ing, i·tems.** *Archaic.* To compute. *—adv.* Also : likewise. —Used to introduce each article in an enumeration or list.

**i·tem·ize** (īʹtə-mīzʹ) *vt.* **-ized, -iz·ing, -iz·es.** To set down item by item : LIST. **—iʹtem·i·zaʹtion** *n.* **—iʹtem·izʹer** *n.*

**it·er·ance** (ĭtʹər-əns) *n.* Iteration.

**it·er·ant** (ĭtʹər-ənt) *adj.* Characterized by iteration : REPEATING.

**it·er·ate** (ĭtʹə-rātʹ) *vt.* **-at·ed, -at·ing, -ates.** [Lat. *iterare, iterat-* < *iterum,* again.] To perform or say again : REPEAT. **—itʹer·aʹtion** *n.*

**it·er·a·tive** (ĭtʹə-rāʹtĭv, -ər-ə-tĭv) *adj.* **1.** Marked by or involving repetition. **2.** Frequentative. **3.** *Math.* Of, relating to, or being a computational procedure to produce a desired result by replication of a series of operations that successively better approximates the desired result.

**ith·y·phal·lic** (ĭthʹə-fălʹĭk) *adj.* [LLat. *ithyphallicus* < Gk. *ithuphallikos* < *ithuphallos,* erect phallus : *ithus,* straight + *phallos,* phallus.] **1.** Of or relating to the phallus carried in the ancient festival of Bacchus. **2.** Having the penis erect. —Used of graphic and sculptural representations. **3.** Lewd : salacious.

**i·tin·er·an·cy** (ī-tĭnʹər-ən-sē, ĭ-tĭnʹ-) *also* **i·tin·er·a·cy** (-ə-sē) *n.* A state or system of itinerating, esp. in the office or role of public speaker, minister, or judge.

**i·tin·er·ant** (ī-tĭnʹər-ənt, ĭ-tĭnʹ-) *adj.* [LLat. *itinerans, itinerant-,* pr.part. of *itinerari,* to travel < Lat. *iter,* journey.] Traveling from place to place, esp. to perform some work or duty. **—i·tinʹer·ant** *n.*

**i·tin·er·ar·y** (ī-tĭnʹə-rĕrʹē, ĭ-tĭnʹ-) *n., pl.* **-ies.** [ME *itinerarie* < LLat. *itinerarium,* course of travel < *itinerarius,* of traveling < Lat. *iter,* journey.] **1.** A route or proposed route of a journey. **2.** A record of a journey. **3.** A travelers' guidebook. *—adj.* **1.** Of or relating to a journey or to a route. **2.** Itinerant.

**i·tin·er·ate** (ī-tĭnʹə-rātʹ, ĭ-tĭnʹ-) *vi.* **-at·ed, -at·ing, -ates.** [LLat. *itinerari, itinerat-* < Lat. *iter,* journey.] To travel from place to place. **—i·tinʹer·aʹtion** *n.*

**-itis** *suff.* [NLat. < Gk., n. suffix.] **1.** Inflammation or disease of <laryng*itis*> **2.** Excessive preoccupation with, indulgence in, reliance on, or possession of the qualities of <television*itis*>

**it'll** (ĭtʹl). **1.** It will. **2.** It shall.

**-itol** *suff.* [-IT(E) + -OL.] An alcohol having more than one hydroxyl group <manni*tol*>

**its** (ĭts) *adj.* [Alteration of *it's* : IT + -'S[1].] —Used to indicate possession or the agent or recipient of an action <a dog wagging *its* tail><an agency that wouldn't reveal *its* secrets><a manuscript undergoing *its* fourth rewrite> ***usage:*** As the possessive form of *it, its* is never written with an apostrophe. The form *it's,* written with the apostrophe, is a contraction of *it is* or *it has.*

**it's** (ĭts). **1.** It is. **2.** It has.

**it·self** (ĭt-sĕlfʹ) *pron.* **1.** That one identical with it. —Used: **a.** Reflexively as the direct or indirect object of a verb or the object of a preposition <This microwave oven turns *itself* off.> **b.** For emphasis <The difficulty is not in the problem *itself.*> **c.** In an absolute construction <*Itself* no major work of art, it still reveals talent.> **2.** Its normal or healthy condition<The computer is acting *itself* again since the program was corrected.>

**it·ty-bit·ty** (ĭtʹē-bĭtʹē) *also* **it·sy-bit·sy** (ĭtʹsē-bĭtʹsē) *adj.* [Prob. alteration of *little bit.*] *Informal.* Extremely small.

**-ity** *suff.* [ME *-itie* < OFr. *-ite* < Lat. *-itas.*] State : quality <abnor*mality*>

**-ium** *suff.* [NLat. < Lat. < Gk. *-ion,* dim. suffix.] Chemical element or group <californ*ium*>

**I've** (īv). I have.

**-ive** *suff.* [ME < OFr. *-if, -ive* < Lat. *-ivus,* adj. suffix.] Performing or tending toward a specified action <demonstrat*ive*>

**i·vied** (īʹvēd) *adj.* Overgrown or covered with ivy.

**i·vo·ry** (īʹvə-rē, īvʹrē) *n., pl.* **-ries.** [ME *ivorie* < OFr. *ivoire* < Lat. *eboreus,* of ivory < *ebur,* ivory, of Egypt. orig.] **1. a.** The hard, smooth, yellowish-white dentine forming the main part of the tusks of the elephant. **b.** A similar substance forming the tusks or teeth of certain other animals, as the walrus. **2.** A tusk, esp. an elephant's tusk. **3.** A substance resembling ivory. **4.** A pale or grayish yellow to yellowish white. **5.** An article made of ivory. **6.** *often* **ivories. a.** Piano keys. **b.** Dice. **c.** The teeth. *—adj.* **1.** Made of ivory. **2.** Of the color ivory.

**i·vo·ry·bill** (īʹvə-rē-bĭlʹ, īvʹrē-) *n.* Ivory-billed woodpecker.

**i·vo·ry-billed woodpecker** (īʹvə-rē-bĭldʹ, īvʹrē-) *n.* A large, white-billed, prob. extinct North American woodpecker, *Campephilus principalis.*

**ivory black** *n.* A black pigment made from charred ivory.

**ivory nut** *n.* The hard seed of an American palm, *Phytelephas macrocarpa,* yielding an ivorylike substance.

**ivory tower** *n.* [Transl. of Fr. *tour d'ivoire.*] A place or attitude of retreat, esp. preoccupation with remote, lofty, or intellectual considerations rather than with mundane affairs.

**i·vy** (īʹvē) *n., pl.* **i·vies.** [ME *ivi* < OE *īfig.*] **1.** A woody climbing or trailing plant of the genus *Hedera,* native to the Old World, esp. *H. helix,* with lobed evergreen leaves and berrylike black fruit. **2.** Any of various creeping or climbing plants similar to the ivy.

**Ivy League** *n.* [So called because of the ivy that covers the older college buildings.] An association of eight colleges in the northeastern United States, usu. thought to be high in academic standards and social prestige. **—Ivy Leaguer** *n.*

**i·wis** (ĭ-wĭsʹ) *adv.* [ME < OE *gewis,* certain.] *Archaic.* Certainly : assuredly.

**Ix·i·on** (ĭk-sīʹən) *n.* [Lat. < Gk. *Ixiōn.*] *Gk. Myth.* A Thessalian king whom Zeus punished for his temerity in seeking Hera's love by having him bound to a perpetually revolving wheel in Hades.

**ix·tle** (ĭsʹlē, ĭstʹ-) *n. var. of* ISTLE.

**I·yar** *also* **Iy·yar** (ē-yärʹ, ēʹyärʹ) *n.* [Heb. *iyyār.*] The eighth month of the Hebrew year. —See table at CALENDAR.

**iz·ar** (ĭ-zärʹ) *n.* [Ar. *'izār,* veil.] A long, usu. white cotton outer garment worn by women in many Moslem countries.

**-ization** *or* **-isation** *suff.* [-IZ(E) + -ATION.] Action, process, or result of doing or making <colon*ization*>

**-ize** *or* **-ise** *suff.* [ME *-isen* < OFr. *-iser* < LLat. *-izare* < Gk. *-izein,* v. suffix.] **1. a.** To cause to be or to become <dramat*ize*> **b.** To cause to conform to or resemble <Hellen*ize*> **c.** To treat as <idol*ize*> **2. a.** To treat or affect with <anesthet*ize*> **b.** To subject to <tyrann*ize*> **3.** To treat according to or practice the method of <pasteur*ize*> **4.** To become or become like <material*ize*> **5.** To perform, engage in, or produce <botan*ize*> ***usage:*** The addition of *-ize* to nouns and adjectives to create a related verb, as *commercialize* or *specialize,* is an ancient and useful practice in English. It is a practice, however, that should be used with caution; new coinages such as *concretize* and *reprivatize* will become acceptable only when they have passed the tests of utility and permanence.

# Jj

**j** *or* **J** (jā) *n., pl.* **j's** *or* **J's 1.** The tenth letter of the English alphabet. **2.** A speech sound represented by the letter *j.* **3.** The tenth in a series. **4.** Something shaped like the letter J.

**jab** (jăb) *v.* **jabbed, jab·bing, jabs.** [Var. of JOB[2].] *—vt.* **1.** To poke abruptly, esp. with something sharp. **2.** To thrust into or against in a rough, abrupt manner. **3.** To punch with short blows. *—vi.* **1.** To make an abrupt jabbing motion. **2.** To deliver a quick punch. *—n.* **1.** A quick stab or blow. **2.** A short straight punch.

**jab·ber** (jăbʹər) *v.* **-bered, -ber·ing, -bers.** [Imit.] *—vi.* To speak rapidly, unintelligibly, or idly : GABBLE. *—vt.* To utter rapidly or unintelligibly. *—n.* Babbling talk : GIBBERISH. **—jabʹber·er** *n.*

**jab·ber·wock·y** (jăbʹər-wŏkʹē) *n.* [< *Jabberwocky,* a poem by Lewis Carroll (1832–1898).] Nonsensical speech or writing that appears to make sense.

**jab·i·ru** (jăbʹə-rōōʹ) *n.* [Port., perh. of Tupian orig.] A large tropical American wading bird, *Jabiru mycteria,* with white plumage and a naked neck.

**jab·o·ran·di** (jăbʹə-răn-dēʹ) *n., pl.* **-dis.** [Port. and Sp. (South America), perh. of Tupi-Guarani orig.] **1.** Either of two tropical American shrubs, *Pilocarpus jaborandi* or *P. microphyllus,* the dried leaves of which yield pilocarpine. **2.** The dried leaves of the jaborandi.

**ja·bot** (zhă-bōʹ, jăbʹō) *n.* [Fr.] A decorative cascade of ruffles down the front of a shirt or dress.

**ja·cal** (hä-kälʹ) *n., pl.* **ja·ca·les** (hä-käʹlās) *or* **ja·cals.** [Mex. Sp. <

Nahuatl *xacalli* : *xamitl*, adobe + *calli*, house.] A thatch-roofed hut in Mexico and the U.S. Southwest made of wattle and daub.

**jac·a·mar** (jăk'ə-mär') *n.* [Fr. < Tupi-Guarani *jacamaciri.*] Any of various insectivorous tropical American birds of the family Galbulidae, with iridescent plumage and a long bill.

**ja·ca·na** (zhä'sə-nä') *n.* [Port. *jaçanã*, perh. of Tupi-Guarani orig.] Any of several tropical marsh birds of the family Jacanidae, with long toes adapted for walking on floating vegetation.

**jac·a·ran·da** (jăk'ə-răn'də) *n.* [Port. and Sp. (South America) < Tupi *yacarandá.*] **1.** A tropical American tree of the genus *Jacaranda*, with compound leaves and pale-purple flower clusters. **2.** The hard, brown wood of the jacaranda tree.

jacaranda

**ja·cinth** (jā'sĭnth, jăs'ĭnth) *n.* [ME *jacinte* < OFr. *jacinte* or Med. Lat. *jacintus*, both < Lat. *hyacinthus*, a blue precious stone. —see HYACINTH.] HYACINTH 4a.

**jack** (jăk) *n.* [< the name JACK.] **1.** *often* **Jack. a.** A man : fellow. **b.** A jack-of-all-trades. **c.** A lumberjack. **d.** A sailor. **2.** A playing card showing the figure of a knave and ranking below a queen. **3. a.** A usu. portable device for lifting or moving heavy objects by means of force applied with a lever, screw, or hydraulic press. **b.** A wooden wedge for cleaving rock. **c.** A support or brace, esp. the iron crosstree on a topgallant masthead. **d.** A contrivance for turning a spit. **4.** The male of certain animals, esp. the ass. **5.** Any of various food and game fishes, chiefly of the genus *Caranx* of Atlantic and Pacific waters. **6.** A pin in some bowling games. **7. a. jacks.** *(sing. or pl. in number).* A game played with a set of small six-pointed metal pieces and a small ball, the object being to pick up the pieces in various combinations while tossing the ball. **b.** One of the metal pieces used in this game. **8.** A socket that accepts a plug at one end and attaches to electric circuitry at the other. **9.** A small flag displayed at the bow of a ship, usu. to indicate nationality. **10.** *Slang.* Money. **11.** Applejack. —*v.* **jacked, jack·ing, jacks.** —*vt.* **1.** To hunt or fish for with a jacklight. **2.** To hoist or move with or as if with a jack. **—jack up. 1.** To lift with a jack. **2.** To raise <*jack prices up*> **3.** To bolster confidence in : SUPPORT. —*vi.* To jacklight.

**jack·al** (jăk'əl, -ôl') *n.* [Turk. *chakāl* < Pers. *shagāl.*] **1.** A doglike carnivorous mammal of the genus *Canis* of Africa and Asia. **2. a.** One who aids another in committing base acts. **b.** One who does menial tasks for another.

**jack·a·napes** (jăk'ə-nāps') *n.* [ME *Jack Napis*, nickname of William de la Pole, first Duke of Suffolk (d. 1450).] **1.** An impudent person. **2.** A mischievous child. **3.** *Archaic.* A monkey or ape.

**jack·ass** (jăk'ăs') *n.* **1.** A male ass or donkey. **2.** A stupid person.

**jackass rig** *n. Naut.* A nonstandard combination of square rig and fore-and-aft rig on a sailing ship with two or more masts.

**jack bean** *n.* A tropical American plant, *Canavalia ensiformis*, with long pods and purple flower clusters.

**jack·boot** (jăk'bo͞ot') *n.* **1.** A stout military boot reaching above the knee. **2.** A ruthless bully. **3.** An aggressive or totalitarian, esp. military, policy or regime.

**jack·boot·ed** (jăk'bo͞o'tĭd) *adj.* **1.** Wearing jackboots. **2.** Cruelly and violently oppressive.

**jack·daw** (jăk'dô') *n.* A Eurasian black bird, *Corvus monedula*, related to and resembling the crow.

**jack·et** (jăk'ĭt) *n.* [ME *jaket*, dim. of OFr. *jaque*, short jacket.] **1.** A short coat, usu. reaching to the hip. **2.** An outer covering or casing, esp.: **a.** The skin of a potato. **b.** A fitted protective cover for the binding of a book. **c.** Insulation, as for a steam pipe, wire, or boiler. **d.** A paper or thin cardboard sleeve for a phonograph record. **e.** An open envelope or folder for filing papers. **f.** The shell casing of a bullet. —*vt.* **-et·ed, -et·ing, -ets.** To supply or cover with a jacket.

**Jack Frost** *n.* Frost or cold weather personified.

**jack·fruit** (jăk'fro͞ot') *n.* [Port. *jaca* (< Malayalam *chakka*) + FRUIT.] **1.** A tree, *Artocarpus heterophyllus* of tropical Asia, with large, edible fruit. **2.** The fruit of the jackfruit tree.

**jack·ham·mer** (jăk'hăm'ər) *n.* A hand-held pneumatic machine for drilling rock or breaking up concrete.

**jack-in-the-box** (jăk'ĭn-*th*ə-bŏks') *n.*, *pl.* **jack-in-the-box·es** or **jacks-in-the-box.** A toy in the shape of a small box from which a usu. grotesque puppet springs when the lid is released.

**jack-in-the-pul·pit** (jăk'ĭn-*th*ə-po͝ol'pĭt, -pŭl'-) *n.*, *pl.* **jack-in-the-pul·pits** or **jacks-in-the-pul·pit.** A plant, *Arisaema triphyllum* of eastern North America, with a leaflike spathe enclosing a clublike spadix.

**jack·knife** (jăk'nīf') *n.* **1.** A large clasp pocketknife. **2.** A dive executed by bending the body at the waist in midair and, with the knees unbent, touching the feet with the hands before straightening out to enter the water. —*v.* **-knifed, -knif·ing, -knifes.** —*vt.* **1.** To cut or stab with a jackknife. **2.** To fold or double like a jackknife. —*vi.* **1.** To bend or fold up like a jackknife. **2.** To form a 90° angle.

**jack·light** (jăk'līt') *n.* A light used as a lure in night hunting or fishing. —*vi.* **-light·ed, -light·ing, -lights.** To hunt or fish with a jacklight.

**jack mackerel** *n.* A food and game fish, *Trachurus symmetricus* of Pacific coastal waters.

**jack-of-all-trades** (jăk'əv-ôl'trādz') *n.*, *pl.* **jacks-of-all-trades.** A person who is handy at many kinds of work.

**jack-o'-lan·tern** (jăk'ə-lăn'tərn) *n.* **1.** A lantern made from a hollowed pumpkin into which a face has been cut. **2. a.** IGNIS FATUUS 1. **b.** Saint Elmo's fire.

**jack pine** *n.* An evergreen tree, *Pinus banksiana* of northern North America, with short twisted needles and soft wood.

**jack·plane** (jăk'plān') *n.* A bench plane for rough surfacing.

**jack·pot** (jăk'pŏt') *n.* **1. a.** The accumulated stakes in a poker game in which play can be opened only with a pair of jacks or better. **b.** A cumulative pool or kitty in various games and competitions. **2.** A top prize or reward. **—hit the jackpot.** *Informal.* To have great success or sudden good fortune.

**jack rabbit** *n.* [JACK(ASS) + RABBIT.] A large, long-eared, long-legged hare of the genus *Lepus* of western North America.

**jack·screw** (jăk'skro͞o') *n.* A jack for lifting, operated by a screw.

**jack·shaft** (jăk'shăft') *n.* A short shaft transmitting motion from a motor to a machine, esp. in an automobile.

**jack·snipe** (jăk'snīp') *n.*, *pl.* **jacksnipe** or **-snipes. 1.** A long-billed Old World wading bird, *Limnocryptes minima*, with brown plumage. **2.** Any of several American birds similar to the jacksnipe.

**Jack·so·ni·an** (jăk-sō'nē-ən) *adj.* Of or relating to Andrew Jackson, his concepts of government, or his Presidency. —*n.* A supporter of Andrew Jackson.

**jack·stay** (jăk'stā') *n. Naut.* **1.** A stay for racing or cruising vessels used to steady the mast against the strain of the gaff. **2.** A rope, rod, or batten along a ship's yard to which the sail is fastened. **3.** A rope or rod running up the forward side of the mast on which the yard moves.

**jack·straw** (jăk'strô') *n.* **1. jackstraws.** *(sing. in number).* A game in which each player in turn lets fall a pile of straws or thin sticks and then attempts to remove one without disturbing the others. **2.** A straw or stick used in jackstraws.

**jack-tar** *also* **Jack-tar** (jăk'tär') *n.* A sailor.

**Jac·o·be·an** (jăk'ə-bē'ən) *adj.* [NLat. *Jacobaeus* < *Jacobus*, James.] Of or relating to the reign of James I of England (1603–25) or his times. —*n.* A prominent figure of the Jacobean period.

**Jac·o·bin** (jăk'ə-bĭn) *n.* [Fr., after the *Jacobin* friars, in whose convent the Jacobins first met.] **1.** A member of a society of radical republicans during the French Revolution. **2.** A radical or extreme leftist. **3.** A Dominican friar. **—Jac'o·bin'ic, Jac'o·bin'i·cal** *adj.* **—Jac'o·bin·ism** *n.*

**Jac·o·bite** (jăk'ə-bīt') *n.* [< NLat. *Jacobus*, James.] A supporter of James II of England or the Stuart pretenders after 1688. **—Jac'o·bit'i·cal** (-bĭt'ĭ-kəl) *adj.* **—Jac'o·bit·ism** (-bī-tĭz'əm) *n.*

**Ja·cob's ladder** (jā'kəbz) *n.* [From the ladder seen by the patriarch Jacob in a dream.] **1.** *Naut.* A rope or chain ladder with rigid rungs. **2.** A plant of the genus *Polemonium*, with blue bell-shaped flowers and paired ladderlike leaflets.

**jac·o·net** (jăk'ə-nĕt') *n.* [Urdu *jagannāthī*, after *Jagannath*, India.] A lightweight cotton cloth similar to lawn that is used for clothing and bandages.

**jac·quard** *also* **Jac·quard** (jăk'ärd', jə-kärd') *adj.* [After Joseph M. *Jacquard* (1752–1834).] **1.** A loom or other apparatus used in the weaving of a figured fabric. **2.** A fabric with an intricately woven pattern made on a jacquard loom.

**Jac·que·rie** (zhä-krē') *n.* [Fr. < OFr. *jacqerie*, peasantry < *jacques*, peasant < the name *Jacques.*] **1.** The revolt of the French peasants against the nobility in 1358. **2. jacquerie.** A peasants' revolt, esp. an extremely bloody one.

**jac·ti·ta·tion** (jăk'tĭ-tā'shən) *n.* [Med. Lat. *jactitatio*, *jactitation-*, false declaration < *jactitare*, to utter, freq. of *jactare*, to boast, freq. of *jacere*, to throw.] **1.** A false boasting or claim, esp. one harmful to the interests of another. **2.** *Pathol.* Extreme restlessness or tossing in bed during illness.

**Ja·cuz·zi** (jə-ko͞o'zē, jă-). A trademark for a device that swirls the water in a bath or pool.

**jade¹** (jād) *n.* [Fr. < Sp. *ijada* < (*piedra de*) *ijada*, (stone of the) flank (from the belief that it cured renal colic) < VLat. **iliata* < Lat. *ilia*, pl. of *ilium*, flank.] Either of two distinct, usu. green minerals, nephrite and jadeite, used mainly as gemstones or in carving.

**jade[2]** (jād) *n.* [ME.] **1.** A broken-down or useless horse : NAG. **2. a.** A disreputable woman. **b.** A willful or coquettish girl. —*vt. & vi.* **jad·ed, jad·ing, jades.** To fatigue or become fatigued.

**jad·ed** (jā′dĭd) *adj.* **1.** Fatigued : worn-out. **2. a.** Made dull or insensitive as by excess or surfeit : SATED <"the sickeningly sweet life of the amoral, *jaded,* bored upper classes" —John Simon> **b.** Cynically or pretentiously callous. **—jad′ed·ly** *adv.* **—jad′ed·ness** *n.*

**jade·ite** (jā′dīt′) *n.* A rare, emerald to light-green, white, red-brown, yellow-brown, or violet jade, $NaAlSi_2O_6$, used as a gem and for ornamental carvings.

**jae·ger** (yā′gər) *n.* [G. *Jäger,* jaeger, hunter < OHG *jagāri* < *jagōn,* to hunt.] **1.** (*also* jā′gər). A large, strong sea bird of the genus *Stercorarius* that snatches food from other birds. **2.** A huntsman or hunting attendant.

**jag[1]** (jăg) *n.* [ME *jagge.*] **1.** A sharp projection : BARB. **2. a.** A hanging flap along the edge of a garment. **b.** A slash in a garment exposing material of a different color. —*vt.* **jagged, jag·ging, jags. 1.** To cut jags in : NOTCH. **2.** To cut raggedly. **3.** *Scot.* To prick or jab sharply.

**jag[2]** (jăg) *n.* [Orig. unknown.] **1.** *Slang.* A binge or spree <a drinking *jag*> **2.** A small portion.

**jag·ged** (jăg′ĭd) *adj.* **1. a.** Having a rough surface or edge. **b.** Having sharp or ragged projections on a surface or edge. **2.** Having a rough or harsh quality. **—jag′ged·ly** *adv.* **—jag′ged·ness** *n.*

**jag·ger·y** (jăg′ə-rē) *n.* [Port. *jagara* < Kanarese *sharkare* < Skt. *śarkarā,* sugar.] A coarse, dark sugar made from palm sap.

**jag·gy** (jăg′ē) *adj.* **-gi·er, -gi·est.** JAGGED 1b.

**jag·uar** (jăg′wär′, jăg′yoo-är′) *n.* [Sp. and Port. < Guarani *jaguá,* dog.] A large feline mammal, *Panthera onca* of tropical America, having a tawny coat with black rosettelike spots.

**jag·ua·ron·di** *also* **jag·ua·run·di** (jăg′wə-rŭn′dē) *n.* [Sp. and Port. < Guarani *jaguarundi,* var. of *jaguá,* dog.] A long-tailed grayish-brown wild cat, *Felis jaguarundi* of tropical America.

**Jah·veh** (yä′vā) *or* **Jah·weh** (yä′wā, -vā) *n. vars. of* YAHWEH.

**jai a·lai** (hī′ lī′, hī′ ə-lī′, hī′ ə-lī′) *n.* [Sp. < Basque : *jai,* festival + *alai,* joyous.] A court game similar to handball in which players use a long hand-shaped basket strapped to the wrist to catch and propel the ball.

jai alai

**jail** (jāl) *n.* [ME *jaiole* < OFr., ult. < Lat. *cavea,* cage.] **1.** A place for the lawful confinement of persons : PRISON. **2.** Detention in a jail. —*vt.* **jailed, jail·ing, jails.** To detain in custody.
☆ ***syns:*** JAIL, IMMURE, IMPRISON, INCARCERATE, INTERN, LOCK (up) *v. core meaning* : to put in jail <*jailed* the suspect overnight>

**jail·bait** (jāl′bāt′) *n. Slang.* A girl below the age of consent with whom sexual intercourse can constitute statutory rape.

**jail·bird** (jāl′bûrd′) *n. Informal.* A prisoner or ex-convict.

**jail·break** (jāl′brāk′) *n.* An escape from jail.

**jail delivery** *n.* **1.** *Chiefly Brit.* The clearing of a jail by bringing all prisoners to trial. **2.** A mass escape or forcible freeing of prisoners.

**jail·er** *also* **jail·or** (jā′lər) *n.* The keeper of a jail.

**jail·house** (jāl′hous′) *n.* JAIL 1.

**jailhouse lawyer** *n.* A prisoner who functions as his or her own attorney, as by filing appeals.

**Jain** (jīn) *also* **Jai·na** (jī′nə) *n.* [Hindi *jaina* < Skt. *jaina-,* relating to the saints < *jinaḥ,* saint, victor < *jayati,* he conquers.] A follower of Jainism.

**Jain·ism** (jī′nĭz′əm) *n.* An ascetic religion of India, founded in the 6th cent. B.C., that teaches the immortality and transmigration of the soul and denies the existence of a supreme being.

**†jakes** (jāks) *n.* [Orig. unknown.] (*sing. in number*). *Regional.* A privy.

**jal·ap** (jăl′əp, jä′ləp) *n.* [Fr. < Mex. Sp. *jalapa,* short for (*purga de*) *Jalapa,* (purgative of) Jalapa, after *Jalapa,* Mexico.] **1.** A Mexican plant, *Exogonium purga,* with a tuberous rootstock that is dried and powdered for use as a cathartic. **2.** A similar or related plant. **3.** The dried rootstock of the jalap or a related plant.

**ja·lop·y** (jə-lŏp′ē) *n., pl.* **-ies.** [Orig. unknown.] *Informal.* An old, run-down vehicle, esp. an automobile.

**jal·ou·sie** (jăl′ə-sē) *n.* [Fr. < *jalousie,* jealousy < OFr. *gelosie* < *gelos,* jealous. —see JEALOUS.] A blind or shutter with adjustable horizontal slats for regulating the passage of air and light.

**jam[1]** (jăm) *v.* **jammed, jam·ming, jams.** [Orig. unknown.] —*vt.* **1.** To drive or wedge forcibly into a tight position <*jammed* the lid on the box> **2.** To activate or apply suddenly, as automotive brakes. **3.** To cause to lock in an inoperable position <*jam* the typewriter keys> **4.** To fill to excess : pack tight <The fans *jammed* the arena.> **5.** To block or clog <The drain was *jammed* by debris.> **6.** To crush or bruise <*jam* a finger in the door> **7.** *Electron.* To interfere with or prevent the clear reception of (signals) by electronic means. —*vi.* **1.** To become wedged : STICK. **2.** To become inoperable because of jammed parts. **3.** To force into or through a tight space. **4.** *Mus.* To play jazz improvisations. —*n.* **1.** The act of jamming or state of being jammed. **2.** A crush or congestion in a limited space <a traffic *jam*> **3.** *Informal.* A difficult situation : PREDICAMENT.

**jam[2]** (jăm) *n.* [Poss. < JAM[1].] A preserve of whole fruit boiled to a pulp with sugar.

**jamb** (jăm) *n.* [ME *jambe* < OFr. < LLat. *gamba,* hoof. —see GAMBOL.] **1.** The vertical side piece or post of a door or window frame. **2.** A jambeau.

**jam·ba·lay·a** (jŭm′bə-lī′ə) *n.* [Louisiana Fr.] A Creole dish of rice cooked with shrimp, oysters, ham, or chicken and seasoned with spices and herbs.

**jam·beau** (jăm′bō) *n., pl.* **-beaux** (-bōz) [ME.] A piece of medieval leg armor worn below the knee.

**jam·bo·ree** (jăm′bə-rē′) *n.* [Orig. unknown.] **1.** A noisy, boisterous celebration. **2.** A large national or international assembly, as of Boy Scouts. **3.** A large, usu. festive assembly, as of a political party or association.

**James** (jāmz) *n.* —See table at BIBLE.

**James·i·an** (jām′zē-ən) *adj.* **1.** Relating to or typical of William James, his philosophy, or his teachings. **2.** Relating to or typical of Henry James or his writings.

**jam session** *n.* **1.** An impromptu gathering of jazz musicians to play improvisations. **2.** An impromptu or open discussion.

**jan·gle** (jăng′gəl) *v.* **-gled, -gling, -gles.** [ME *janglen,* to chatter < OFr. *jangler,* prob. of Germanic orig.] —*vi.* To make a harsh metallic sound, as a bell. —*vt.* **1.** To cause to make a discordant sound. **2.** To jar (the nerves). —*n.* A harsh metallic sound. **—jan′gler** *n.*

**jan·is·sar·y** (jăn′ĭ-sĕr′ē) *also* **jan·i·zar·y** (-zĕr′ē) *n., pl.* **-ies.** [Fr. *janissaire* < Turk. *yeniçeri* : *yeni,* new + *çeri,* soldier.] A soldier in an elite guard of Turkish troops organized about 1330 and abolished in 1826.

**jan·i·tor** (jăn′ĭ-tər) *n.* [Lat., doorkeeper < *janua,* door < *janus,* arch.] **1.** One who attends to maintenance, cleaning, and repairs in a building. **2.** A doorman. **—jan′i·to′ri·al** (-tôr′ē-əl, -tōr′-) *adj.*

**jan·i·zar·y** (jăn′ĭ-zĕr′ē) *n. var. of* JANISSARY.

**Jan·sen·ism** (jăn′sə-nĭz′əm) *n.* The theological principles of Cornelis Jansen, condemned by the Roman Catholic Church, which emphasize predestination, deny free will, and assert that human nature is incapable of good. **—Jan′sen·ist** *n.* **—Jan′sen·is′tic** *adj.*

**Jan·u·ar·y** (jăn′yoo-ĕr′ē) *n., pl.* **-ies** *or* **-ys.** [ME *januarie* < Lat. *Januarius (mensis),* (month of) Janus < *Janus,* Janus.] The first month of the year according to the Gregorian calendar. —See table at CALENDAR.

**Ja·nus** (jā′nəs) *n.* [Lat.] *Rom. Myth.* A Roman god of gates and doorways, depicted with two faces looking in opposite directions.

**Ja·nus-faced** (jā′nəs-fāst′) *adj.* Two-faced : hypocritical.

**ja·pan** (jə-păn′) *n.* [After *Japan.*] **1.** A black enamel or lacquer with a durable glossy finish. **2.** An object decorated and varnished in the Japanese manner. —*vt.* **-panned, -pan·ning, -pans. 1.** To enamel with japan. **2.** To coat with a glossy finish.

**Ja·pan clover** (jə-păn′) *n.* A leguminous plant native to Asia, *Lespedeza striata,* grown for forage and to improve the soil.

**Japan Current** *n.* A warm ocean current flowing northeast from the Philippine Sea past southeastern Japan to the North Pacific.

**Jap·a·nese** (jăp′ə-nēz′, -nēs′) *n., pl.* **Japanese. 1.** A native or resident of Japan. **2.** The language of the Japanese. **—Jap′a·nese′** *adj.*

**Japanese andromeda** *n.* A shrub native to Japan, *Pieris japonica,* with small, early-blooming white flowers.

**Japanese beetle** *n.* A metallic-green and brownish beetle indigenous to eastern Asia, *Popillia japonica,* the larvae and adults of which are serious plant pests in North America.

**Japanese cedar** *n.* A tall coniferous tree, the cryptomeria, or its wood.

**Japanese iris** *n.* A plant native to Asia, *Iris kaempferi,* cultivated in many horticultural varieties for its large flat flowers.

**Japanese ivy** *n.* Boston ivy.

**Japanese leaf** *n.* Chinese evergreen.

**Japanese maple** *n.* A shrub or small tree indigenous to eastern Asia, *Acer palmatum,* widely cultivated for its decorative, deeply lobed, often reddish foliage.

**Japanese quince** *n.* Japonica.

**Japanese river fever** *n.* Scrub typhus.

**Japanese spurge** *n.* Pachysandra.

**Japan wax** *n.* A pale-yellow solid wax obtained from the berries of certain plants of the genus *Rhus* and used in polishes, soaps, and food packaging and as a substitute for beeswax.

**jape** (jāp) *v.* **japed, jap·ing, japes.** [ME *japen,* prob. < OFr. *japer,* to chatter.] —*vi.* To joke or quip. —*vt.* To make sport of : MOCK. —*n.* A joke or quip. —**jap′er** *n.* —**jap′er·y** *n.*

**ja·pon·i·ca** (jə-pŏn′ĭ-kə) *n.* [NLat., specific epithet of the species *Chaenomeles japonica* < *Japonia*, Japan.] **1.** A Japanese shrub, *Chaenomeles japonica*, with red flowers. **2.** The camellia.

**jar[1]** (jär) *n.* [Fr. *jarre* < Prov. *jarra* < Ar. *jarrah*, earthen vessel.] **1.** A cylindrical glass or earthenware vessel with a wide mouth and usu. having no handles. **2.** The contents of a jar. —**jar′ful′** *n.*

**jar[2]** (jär) *v.* **jarred, jar·ring, jars.** [Perh. of imit. orig.] —*vi.* **1.** To make or utter a harsh sound. **2.** To have a disturbing effect : GRATE <The music *jarred* on my nerves.> **3.** To shake or shiver from impact. **4.** To clash or conflict <"We ourselves . . . often *jar* with the landscape" —Isak Dinesen> —*vt.* **1.** To bump or cause to move or shake from impact. **2.** To startle or unsettle. —*n.* **1.** A jolt or shock. **2.** A harsh grating sound. **3.** A clash of interests or opinions.

**jar·di·nière** (järd′n-îr′, zhär-dē-nyâr′) *n.* [Fr., fem. of *jardinier*, gardener < *jardin*, garden, of Germanic orig.] **1.** A large decorative stand or vessel for displaying plants or flowers. **2.** Diced cooked vegetables served as a garnish for meat.

**jar·gon[1]** (jär′gən) *n.* [ME *jargoun* < OFr. *jargon*.] **1.** Nonsensical, incoherent, or meaningless talk. **2.** A hybrid language or dialect : PIDGIN. **3.** The specialized language of a trade, profession, or similar group <computer *jargon*> —*vi.* **-goned, -gon·ing, -gons.** To speak in or use jargon.

**jar·gon[2]** (jär-gŏn′) *also* **jar·goon** (jär-go͞on′) *n.* [Fr.] A smoky yellow or colorless variety of zircon.

**jar·gon·ize** (jär′gə-nīz′) *v.* **-ized, -iz·ing, -iz·es.** —*vt.* To debase or translate into jargon. —*vi.* To talk in jargon.

**jar·goon** (jär-go͞on′) *n. var. of* JARGON[2].

**jarl** (yärl) *n.* [ON.] A great Scandinavian chieftain or nobleman.

**jas·mine** (jăz′mĭn) *also* **jes·sa·mine** (jĕs′ə-mĭn) *n.* [Fr. *jasmin* < Ar. *yāsamīn* < Pers. *yasmīn*.] **1.** A vine or shrub of the genus *Jasminum*, esp. *J. officinalis* of Asia, with fragrant white flowers used in making perfume. **2.** A woody vine of the genus *Gelsemium*, esp. *G. sempervirens* of the southeastern United States, with fragrant yellow flowers. **3.** Any of several plants or shrubs with fragrant flowers. **4.** A light to brilliant yellow.

**Ja·son** (jā′sən) *n.* [Gk. *Iasōn*.] *Gk. Myth.* The leader of the Argonauts in quest of the Golden Fleece and the husband of Medea.

**jas·per** (jăs′pər) *n.* [ME *jaspre* < AN < OFr. < Lat. *jaspis* < Gk. *iaspis*, of Semitic orig.] **1.** An opaque variety of quartz, usu. reddish, brown, or yellow in color. **2.** Chalcedony, esp. green chalcedony.

**jasper ware** *n.* A fine white porcelain or stoneware invented by Josiah Wedgwood, often colored by metallic oxides with raised designs remaining white.

**Jat** (jät) *n.* [Hindi *jāṭ*.] A member of an Indo-Aryan people of the Punjab and Uttar Pradesh.

**ja·to** (jā′tō) *n.* [< *JATO*, acronym for *jet-assisted takeoff*.] **1.** A takeoff aided by an auxiliary jet or rocket. **2.** An auxiliary unit giving thrust for a jato.

**jaun·dice** (jôn′dĭs, jän′-) *n.* [ME *jaunis* < OFr. *jaunice* < *jaune*, yellow < Lat. *galbinus*, yellowish < *galbus*, yellow.] **1.** Yellowish staining of the eyes, skin, and body fluids by bile pigment. **2.** A pathological condition in which the normal processing of bile is interrupted. **3.** Prejudice or jealousy.

**jaun·diced** (jôn′dĭst, jän′-) *adj.* **1.** Having jaundice. **2.** Yellow or yellowish. **3.** Feeling or showing envy, prejudice, or hostility.

**jaunt** (jônt, jänt) *n.* [Orig. unknown.] A short trip, made usu. for pleasure : OUTING. —*vi.* **jaunt·ed, jaunt·ing, jaunts.** To make a short journey.

**jaun·ty** (jôn′tē, jän′-) *adj.* **-ti·er, -ti·est.** [Fr. *gentil*, noble < OFr. —see GENTLE.] **1.** Crisp and dapper. **2.** Buoyantly carefree and self-confident : BRISK. **3.** *Archaic.* **a.** Genteel. **b.** Stylish. —**jaun′ti·ly** *adv.* —**jaun′ti·ness** *n.*

▲ word history: The English word *jaunty*, like *gentle* and *genteel*, is a borrowing of French *gentil*, "noble." *Jaunty* at one time did mean "genteel" or "well-bred," but it soon developed other senses such as "elegant" and "sprightly." These adjectives were used to characterize those who attempted to behave as if they were well-bred, whether they actually were or not.

**ja·va** (jăv′ə, jä′və) *n.* [After *Java*.] *Informal.* Brewed coffee.

**Ja·va man** (jä′və, jăv′ə) *n.* Pithecanthropus.

**Jav·a·nese** (jăv′ə-nēz′, -nēs′, jä′və-) *adj.* [*Java* + *-nese*, as in *Japanese*.] Of or relating to Java or to its people, language, or culture. —*n.*, *pl.* **Javanese. 1.** A native or resident of Java. **2.** The Indonesian language of Java.

**Java sparrow** *n.* A small, pink-billed grayish bird, *Padda oryzivora*, native to tropical Asia, often kept as a cage bird.

**jave·lin** (jăv′lĭn, jăv′ə-) *n.* [OFr. *javeline*, var. of *javelot*, of Celtic orig.] **1.** A light spear thrown as a weapon. **2. a.** A metal or metal-tipped spear, usu. at least 8½ feet long, used in contests of distance throwing. **b.** The athletic field event in which a javelin is thrown.

**Ja·velle water** *also* **Ja·vel water** (zhə-vĕl′) *n.* [After *Javel*, a former town in France.] An aqueous solution of potassium or sodium hypochlorite, used as a disinfectant or bleaching agent.

**jaw** (jô) *n.* [ME *jowe*.] **1.** Either of two bony or cartilaginous structures that form the framework of the mouth and hold the teeth in most vertebrates. **2. jaws.** The anatomical parts that form the walls of the mouth and serve to open and close it. **3.** Either of two opposed hinged parts in a mechanical device. **4. jaws.** The walls of a pass, canyon, or cavern. **5. jaws.** Something looming and ominous <the *jaws* of death> **6.** *Slang.* **a.** Impudent argument or back talk. **b.** Chatter. —*vi.* **jawed, jaw·ing, jaws.** *Slang.* To talk, esp. excessively or abusively.

**jaw·bone** (jô′bōn′) *n.* A bone of the jaw, esp. the lower jaw : MANDIBLE. —*v.* **-boned, -bon·ing, -bones.** *Informal.* —*vt.* To try to influence or pressure through strong persuasion, esp. to urge to comply voluntarily. —*vi.* To urge voluntary compliance with official wishes or guidelines. —**jaw′bon′er** *n.* —**jaw′bon′ing** *n.*

**jaw·break·er** (jô′brā′kər) *n.* **1.** A hard candy. **2.** *Slang.* A word difficult to pronounce.

**Jaws of Life.** A trademark for a pincerlike metal device used to provide access to persons trapped inside a crushed vehicle.

**jay[1]** (jā) *n.* The letter *j*.

**jay[2]** (jā) *n.* [ME *jai* < OFr. < LLat. *gaius*.] **1.** A usu. crested bird of the genera *Garrulus*, *Cyanocitta*, *Aphelocoma*, or related genera within the family Corvidae, often having a loud, harsh call. **2.** An overly talkative person : CHATTERBOX. **3.** *Slang.* A newcomer or inexperienced person.

**jay·bird** (jā′bûrd′) *n.* JAY[2] 1.

**Jay·cee** (jā′sē′) *n.* [From the initial letters in *junior chamber* (*of commerce*).] A member of a junior chamber of commerce.

**jay·hawk·er** (jā′hô′kər) *n.* [< *jayhawk*, a fictitious bird.] **1.** One of the free-soil guerrillas in Kansas and Missouri during the border disputes of 1857–59. **2.** A robber : bandit.

**jay·vee** (jā′vē′) *n.* [From the initial letters in JUNIOR VARSITY.] *Informal.* A junior varsity or a member of one.

**jay·walk** (jā′wôk′) *vi.* **-walked, -walk·ing, -walks.** [< JAY[2] (newcomer).] To cross a street illegally or recklessly. —**jay′walk′er** *n.*

**jazz** (jăz) *n.* [Orig. unknown.] **1.** A kind of American music first played extemporaneously by black bands in and around New Orleans at the turn of the century, marked by a strong but flexible rhythmic understructure with solo and ensemble improvisations on basic tunes and chord patterns and, in more recent styles, a highly sophisticated harmonic idiom. **2.** Big-band dance music, popular esp. in the 1930's and 1940's : SWING. **3.** *Slang.* Animation : enthusiasm. **4.** *Slang.* **a.** Pretentious talk <all that legal *jazz*> **b.** Nonsense. **5.** *Slang.* Miscellaneous and unspecified accompaniments <a salad made of spinach, sprouts, and the usual *jazz*> —*v.* **jazzed, jazz·ing, jazz·es.** —*vt.* **1.** To play in a jazz style. **2.** *Slang.* To lie or exaggerate to <Don't *jazz* me.> —*vi.* *Slang.* To lie or exaggerate. —**jazz up.** *Informal.* To make more interesting : ENLIVEN. —**jazz′er** *n.*

**jazz·man** (jăz′măn′, -mən) *n.* A jazz musician or composer.

**jazz·y** (jăz′ē) *adj.* **-i·er, -i·est. 1.** Resembling jazz : RHYTHMICAL. **2.** *Slang.* Showy : flashy. —**jazz′i·ly** *adv.* —**jazz′i·ness** *n.*

**jeal·ous** (jĕl′əs) *adj.* [ME *jelous* < OFr. *gelos*, jealous, zealous < Med. Lat. *zelosus* < LLat. *zelos*, zeal < Gk. *zēlos*.] **1.** Fearful or wary of being replaced by a rival, esp. in regard to another's affection. **2. a.** Resentful or bitter in rivalry : ENVIOUS <*jealous* of their friend's new car> **b.** Inclined to suspect rivalry. **3.** Vigilant in guarding something <*jealous* of one's civil rights> **4.** Arising from feelings of envy, apprehension, or bitterness <a *jealous* rage> **5.** Intolerant of disloyalty or infidelity <a *jealous* God> —**jeal′ous·ly** *adv.* —**jeal′ous·ness** *n.*

**jeal·ous·y** (jĕl′ə-sē) *n.*, *pl.* **-ies. 1.** A jealous attitude or disposition. **2.** Close vigilance.

**jean** (jēn) *n.* [Short for obs. *jene fustian*, Genoan fustian < ME *Jene*, Genoa.] **1.** A heavy, strong, twilled cotton, used for uniforms and work clothes. **2. jeans.** Pants made of jean or denim, usu. blue.

**Jeep** (jēp) A trademark for a civilian motor vehicle.

**jeer** (jîr) *v.* **jeered, jeer·ing, jeers.** [Origin unknown.] —*vi.* To speak or shout derisively : SCOFF. —*vt.* To deride openly : TAUNT. —*n.* A taunting remark or shout. —**jeer′er** *n.* —**jeer′ing·ly** *adv.*

**Jef·fer·so·ni·an** (jĕf′ər-sō′nē-ən) *adj.* Of, relating to, or typical of Thomas Jefferson or his politics. —*n.* A follower of Jefferson or a proponent of his politics. —**Jef′fer·so′ni·an·ism** *n.*

**Je·ho·vah** (jĭ-hō′və) *n.* [Alteration of Heb. *Yahweh*, Yahweh.] God, esp. in Christian translations of the Old Testament.

▲ word history: The form *Jehovah* did not exist as a Hebrew word. It is actually a conflation of two Hebrew forms that came about through a peculiarity of the Hebrew writing system. The Hebrew alphabet consists only of characters for consonants; vowels are indicated as dots or "points" written in characteristic positions above or below the consonants. The Hebrew name for God, the consonants of which are transliterated YHWH, was considered so sacred that it was never pronounced and its proper vowel points were never written. In some texts the vowel points for a completely different word, *Adonai*, "lord," were written with YHWH to indicated that the word *Adonai* was to be spoken whenever the reader came upon the word YHWH. YHWH was never intended to be pronounced with the vowels of *Ado-*

*nai,* but Christian scholars of the Renaissance made exactly that mistake, and the forms *Iehovah* (using the classical Latin equivalents of the Hebrew letters) and *Jehovah* (substituting, in English, *J* for consonantal *I*) came into common use.

**Jehovah's Witnesses** *n.* A religious sect founded in the United States during the late 19th cent., whose followers practice active evangelism, preach the imminent approach of the millennium, and are strongly opposed to war and the authority of organized government in matters of conscience.

**je·june** (jə-jo͞on′) *adj.* [< Lat. *jejunus,* hungry.] **1.** Lacking in nutrition : INSUBSTANTIAL. **2.** Containing nothing of interest : DULL <*jejune* philosophical writings> **3.** Immature : puerile <*jejune* behavior> **—je·june′ly** *adv.* **—je·june′ness** *n.*

**je·ju·num** (jə-jo͞o′nəm) *n., pl.* **-na** (-nə) [Med. Lat. < Lat. *jejunus,* fasting (so called because in dissection it was always found empty).] The section of the small intestine between the duodenum and the ileum.

**Je·kyll and Hyde** (jĕk′əl ən hīd′, jē′kəl) *n.* [After *The Strange Case of Dr. Jekyll and Mr. Hyde* by Robert L. Stevenson (1850–1894).] *Informal.* One with a dual personality that is alternately good and evil.

**jell** (jĕl) *v.* **jelled, jell·ing, jells.** [Back-formation < JELLY.] *—vi.* **1.** To become firm or gelatinous : CONGEAL. **2.** *Informal.* To take shape or fall into place : CRYSTALLIZE <Our vacation plans haven't *jelled* yet.> *—vt.* **1.** To cause to become jelly. **2.** To cause to take shape.

**Jell-O** (jĕl′ō). A trademark for a gelatin dessert.

**jel·ly** (jĕl′ē) *n., pl.* **-lies.** [ME *gele* < OFr. *gelee* < Lat. *gelata,* p.part. of *gelare,* to freeze < *gelu,* frost.] **1.** A soft, semisolid food substance with an elastic consistency, made by the setting of a liquid containing pectin or gelatin, or by the addition of gelatin to a liquid, esp. such a substance made of fruit juice containing pectin boiled with sugar. **2.** A substance with the consistency of jelly <petroleum *jelly*> **3.** Something that quivers like jelly <My knees turned to *jelly.*> *—v.* **-lied, -ly·ing, -lies.** *—vt.* To make or cause to become jelly. *—vi.* To become jelly.

**jel·ly·bean** (jĕl′ē-bēn′) *n.* A small ovoid candy with a hardened sugar coating over a chewy center.

**jel·ly·fish** (jĕl′ē-fĭsh′) *n., pl.* **jellyfish** *or* **-fish·es. 1.** Any of numerous usu. free-swimming marine coelenterates of the class Scyphozoa, having a gelatinous, tentacled, often bell-shaped medusoid stage as the dominant or only phase of its life cycle. **2.** Any of various similar or related coelenterates. **3.** *Informal.* A spineless weakling.

**jel·ly·roll** (jĕl′ē-rōl′) *n.* A thin sheet of sponge cake layered with jelly and then rolled up.

**je ne sais quoi** (zhə′ nə sā kwä′) *n.* [Fr., I know not what.] Something difficult to describe or express.

**jen·net** (jĕn′ĭt) *n.* [ME *genet* < OFr. < Sp. *jinete,* light horseman < Ar. *Zeneti,* a Berber tribe famed for horsemanship.] A small Spanish saddle horse.

**jen·ny** (jĕn′ē) *n., pl.* **-nies.** [< the name *Jenny.*] **1.** A female donkey. **2.** A female wren. **3.** A spinning jenny.

**jeop·ard·ize** (jĕp′ər-dīz′) *vt.* **-ized, -iz·ing, -izes.** To expose to loss or danger : IMPERIL.

**jeop·ard·y** (jĕp′ər-dē) *n., pl.* **-ies.** [ME *jupartie* < OFr. *jeu parti,* even game, uncertainty : *jeu,* game (< Lat. *jocus*) + *parti,* p.part. of *partir,* to divide < Lat. *partire* < *pars,* part.] **1.** Risk of loss or harm : PERIL <Your career is in *jeopardy.*> **2.** *Law.* The defendant's risk of conviction when put on trial.

▲ word history: The word *jeopardy* illustrates the human tendency to anticipate the worst in an uncertain situation. The French source of jeopardy, *jeu parti,* literally "divided game," originally denoted a chess problem and came to mean a position in any game for which the chances of either winning or losing were even. In English *jeopardy* retained the senses of the French word but extended them to mean "an uncertain or undecided situation." By Chaucer's time, the late 14th century, *jeopardy* had acquired its modern sense of "peril, danger."

**je·quir·i·ty bean** (jĭ-kwîr′ĭ-tē) *n.* [Ult. < Tupi-Guarani *jekirtí.*] Rosary pea.

**jer·bo·a** (jər-bō′ə) *n.* [Med. Lat. < Ar. *yerbō′.*] Any of various small leaping rodents of the family Dipodidae of Asia and northern Africa, with long hind legs and a long tufted tail.

**jerboa**
*Body approximately 3 inches, tail 7 inches, and hind foot 3 inches long*

**jer·e·mi·ad** (jĕr′ə-mī′əd) *n.* [Fr. *jérémiade,* after *Jérémie,* Jeremiah.] A long, elaborate lamentation or tale of woe.

**Jer·e·mi·ah** (jĕr′ə-mī′ə) *n.* [Heb. *Yirmayāhū.*] —See table at BIBLE.

**jerk[1]** (jûrk) *v.* **jerked, jerk·ing, jerks.** [Orig. unknown.] *—vt.* **1.** To give an abrupt thrust, push, pull, or twist to. **2.** To throw with a quick abrupt motion. **3.** To utter abruptly or sharply. **4.** To make and serve (e.g., ice cream sodas) at a soda fountain. *—vi.* **1.** To move quickly and abruptly : JOLT. **2.** To make spasmodic motions <legs *jerking* from fatigue> *—n.* **1.** A sudden abrupt motion. **2.** A jolting motion. **3.** *Physiol.* A sudden spasmodic muscular contraction. **4.** *Slang.* A dull, stupid, or fatuous person. **—jerk′er** *n.*

☆ ***syns:*** JERK, SNAP, TUG, WRENCH, YANK *v. core meaning* : to move (something) with a sudden abrupt motion <*jerked* the window open>

**jerk[2]** (jûrk) *vt.* **jerked, jerk·ing, jerks.** [Back-formation < JERKY[2].] To cut (meat) into long strips and dry in the sun or cure by exposing to smoke.

**jer·kin** (jûr′kĭn) *n.* [Orig. unknown.] A short, close-fitting, often sleeveless coat or jacket, usu. of leather.

**jerk·wa·ter** (jûrk′wô′tər, -wŏt′ər) *adj.* [< *jerkwater,* a branch-line train, so called because its small boiler had to be refilled often, requiring train crews to "jerk" or draw water from streams.] **1.** *Informal.* Remote, small, and insignificant <a *jerkwater* town> **2.** Contemptibly trivial.

**jerk·y[1]** (jûr′kē) *adj.* **-i·er, -i·est. 1.** Marked by jerks or jerking. **2.** *Slang.* Foolish. **—jerk′i·ly** *adv.* **—jerk′i·ness** *n.*

**jerk·y[2]** (jûr′kē) *n.* [Alteration of CHARQUI.] Meat cured by jerking.

**jer·o·bo·am** (jĕr′ə-bō′əm) *n.* [After *Jeroboam I* (d. 912? B.C.), king of northern Israel.] A large wine bottle holding about 4/5 of a gallon or 3.03 liters.

**Jer·ry** (jĕr′ē) *n., pl.* **-ries.** [Alteration of GERMAN.] *Chiefly Brit.* A German, esp. a German soldier.

**jer·ry-build** (jĕr′ē-bĭld′) *vt.* **-built** (-bĭlt′), **-build·ing, -builds.** [Orig. unknown.] To build poorly and cheaply. **—jer′ry-build′er** *n.*

**jer·sey** (jûr′zē) *n., pl.* **-seys.** [After *Jersey,* England.] **1.** A soft plain-knitted fabric used for clothing. **2.** A close-fitting knitted pullover shirt, sweater, or jacket. **3.** *often* **Jersey.** Any of a breed of fawn-colored, short-horned dairy cattle developed on the island of Jersey and yielding milk rich in butterfat.

**Je·ru·sa·lem artichoke** (jə-ro͞o′sə-ləm, -zə-ləm) *n.* [By folk ety. < Ital. *girasole,* sunflower. —see GIRASOL.] **1.** A North American sunflower, *Helianthus tuberosus,* with yellow rayed flowers and edible tuberous roots that resemble potatoes. **2.** The tuber of the Jerusalem artichoke, eaten as a vegetable.

**Jerusalem cherry** *n.* A small Old World shrub, *Solanum pseudocapsicum,* bearing inedible reddish fruit and used as an ornamental house plant.

**Jerusalem cross** *n.* A cross with four arms, each terminating in a crossbar.

**Jerusalem oak** *n.* A weedy North American plant, *Chenopodium botrys,* with lobed leaves and an odor suggestive of turpentine.

**Jerusalem thorn** *n.* A spiny tropical American tree, *Parkinsonia aculeata,* with yellow flower clusters.

**jess** (jĕs) *n.* [ME *ges* < OFr., pl. of *jet,* throw < VLat. **jectus* < Lat. *jacere,* to throw.] A short strap fastened around the leg of a hawk or falcon, to which a leash may be attached. *—vt.* **jessed, jess·ing, jess·es.** To fasten jesses on. **—jessed** (jĕst) *adj.*

**jes·sa·mine** (jĕs′ə-mĭn) *n. var. of* JASMINE.

**jest** (jĕst) *n.* [ME *geste,* tale < OFr. < Lat. *gesta,* deeds < *gerere,* to perform.] **1.** Something said or done to provoke amusement and laughter. **2.** A frolicsome mood or attitude <said only in *jest*> **3.** A jeering remark : TAUNT. **4.** An object of ridicule : LAUGHINGSTOCK. *—v.* **jest·ed, jest·ing, jests.** *—vi.* **1.** To act or speak playfully. **2.** To make witty or amusing remarks. **3.** To scoff : gibe.

**jest·er** (jĕs′tər) *n.* One given to jesting, esp. a fool or buffoon at medieval courts.

**Jes·u·it** (jĕzh′o͞o-ĭt, jĕz′yo͞o-) *n.* [Fr. *Jésuite* < *Jésus,* Jesus < LLat. *Jesus.* —see JESUS.] **1.** A member of the Society of Jesus, a Roman Catholic order founded by Saint Ignatius Loyola in 1534. **2.** *often* **jesuit.** One given to subtle casuistry. **—Jes′u·it′i·cal** *adj.* **—Jes′u·it′i·cal·ly** *adv.*

**Je·sus** (jē′zəs) *n.* [LLat. < Gk. *Iēsous* < Heb. *Yēshua′* < *Yĕhōshūa′,* Joshua.] **1.** The founder of Christianity, regarded by Christians as the son of God and the Messiah. **2.** *Christian Science.* —Used to refer to <"The highest human corporeal concept of the divine idea" —Mary Baker Eddy>

**Jesus freak** *n. Slang.* A member of a fundamentalist evangelical group, esp. of young people, devoted to the teachings of Jesus.

**jet[1]** (jĕt) *n.* [ME *get* < AN < OFr. *jayet* < Lat. *gagates* < Gk. *gagatēs,* after *Gagai,* a town in Asia Minor.] **1.** A dense black coal that takes a high polish and is used for jewelry. **2.** A deep black. *—adj.* **1.** Made of or resembling jet. **2.** Black as jet.

**jet[2]** (jĕt) *n.* [OFr. < *jeter,* to spout forth < VLat. **jectare* < Lat. *jactare,* freq. of *jacere,* to throw.] **1. a.** A high-velocity fluid stream emitted forcefully through a narrow nozzle or opening. **b.** A nozzle or other outlet for emitting such a stream. **2.** Something issued in or as if in a jetlike stream <"such myriad and such vivid *jets* of images" —Henry Roth> **3. a.** A jet airplane. **b.** A jet engine. —*v.* **jet·ted, jet·ting, jets.** —*vi.* **1.** To move quickly. **2.** To travel by jet airplane. —*vt.* To propel outward or squirt, as under pressure.

**jet airplane** *n.* An airplane driven by jet propulsion.

**jet boat** *n.* A boat propelled by a powerful jet of water created by a specially designed engine.

**je·té** (zhə-tā') *n.* [Fr. < p.part. of *jeter,* to throw.] A leap in ballet with one leg extending forward and the other backward.

**jet engine** *n.* **1.** An engine that develops thrust by ejecting a jet, esp. a jet of gaseous combustion products. **2.** An airplane engine equipped to consume atmospheric oxygen, as distinguished from rocket engines with self-contained fuel-oxidizer systems.

**jet lag** *n.* The psychological and physiological disruption of body rhythms caused by high-speed travel across several time zones in a jet airplane.

**jet·lin·er** (jĕt'lī'nər) *n.* A large passenger-carrying jet airplane.

**jet·port** (jĕt'pôrt', -pōrt') *n.* An airport equipped for jet aircraft.

**jet-pro·pelled** (jĕt'prə-pĕld') *adj.* Driven by jet propulsion.

**jet propulsion** *n.* Propulsion derived from the rearward expulsion of matter in a jet stream, esp. propulsion by jet engines.

**jet·sam** (jĕt'səm) *n.* [Alteration of JETTISON.] **1.** Cargo or equipment thrown overboard to lighten a ship in distress. **2.** Discarded cargo or equipment washed ashore. **3.** Discarded odds and ends.

**jet set** *n.* An international social set composed of wealthy people who travel from one fashionable place to another. **—jet setter** *n.*

**jet stream** *n.* **1.** A high-speed wind near the troposphere, gen. moving from a westerly direction at speeds often exceeding 250 miles or approx. 400 kilometers per hour. **2.** A high-speed stream : JET.

**jet·ti·son** (jĕt'ĭ-sən, -zən) *vt.* **-soned, -son·ing, -sons.** [< ME *jetteson,* a throwing overboard < AN *gettesone* < Lat. *jactatio* < *jactare,* freq. of *jacere,* to throw.] **1.** To cast off or overboard. **2.** To discard as useless or burdensome <*jettisoned* the whole diet plan> —*n.* **1.** The act of jettisoning. **2.** Jetsam.

**jet·ty**[1] (jĕt'ē) *n., pl.* **-ties.** [ME *gete* < OFr. *jete,* p.part. of *jeter,* to project. —see JET[2].] **1.** A pier or other structure built out into a body of water to influence the current or tide or protect a harbor or shoreline. **2.** A wharf.

**jet·ty**[2] (jĕt'ē) *adj.* **1.** Resembling jet. **2.** Black as jet. **—jet'ti·ness** *n.*

**jeu·nesse do·rée** (zhœ-nĕs' dô-rā') *n.* [Fr., gilded youth.] Fashionable and wealthy young people.

**Jew** (jo͞o) *n.* [ME *Jeu* < OFr. *giu* < Lat. *Judaeus* < Gk. *Ioudaios* < Aram. *Yəhūdāy* < Heb. *Yĕhūdī,* after *Yĕhūdāh,* Judah, son of Jacob and Leah.] **1.** An adherent of Judaism. **2.** A descendant of the ancient Hebrew people.

**jew·el** (jo͞o'əl) *n.* [ME *juel* < OFr. *jöel* < perh. < *jeu,* game < Lat. *jocus,* joke.] **1.** A costly ornament of precious metal or gems used as personal adornment. **2.** A precious stone : GEM. **3.** A small gem or gem substitute used as a bearing in a watch. **4.** One that is treasured or highly valued. —*vt.* **-eled, -el·ing, -els** or **-elled, -el·ling, -els. 1.** To adorn with jewels. **2.** To fit with jewels, as a watch.

**jew·el·er** *also* **jew·el·ler** (jo͞o'ə-lər) *n.* One who makes, repairs, or deals in jewelry.

**jew·el·fish** (jo͞o'əl-fĭsh') *n., pl.* **jewelfish** *or* **-fish·es.** A small, brilliantly colored freshwater fish, *Hemichromis bimaculatus* of tropical Africa, popular in home aquariums.

**jew·el·ler** (jo͞o'ə-lər) *n. var. of* JEWELER.

**jew·el·ry** (jo͞o'əl-rē) *n.* Jewels, esp. ornaments made of precious metals set with gems.

**jew·el·weed** (jo͞o'əl-wēd') *n.* A plant of the genus *Impatiens,* bearing yellowish spurred flowers and seed pods that burst open at a touch when ripe.

**Jew·ess** (jo͞o'ĭs) *n.* A Jewish woman.

**jew·fish** (jo͞o'fĭsh') *n., pl.* **jewfish** *or* **-fish·es.** Any of various large marine fishes of the family Serranidae, esp. *Epinephelus itajara* of tropical Atlantic waters.

**Jew·ish** (jo͞o'ĭsh) *adj.* Of, relating to, or typical of the Jews, their customs, or their religion. **—Jew'ish·ly** *adv.* **—Jew'ish·ness** *n.*

**Jewish calendar** *n.* A calendar used by the Jewish people that dates the creation of the world at 3761 B.C. —See table at CALENDAR.

**Jew·ry** (jo͞o'rē) *n.* **1.** The Jewish people as a group. **2.** A section of a medieval city inhabited by Jews : GHETTO.

**jew's-harp** *also* **jews'-harp** (jo͞oz'härp') *n.* A small lyre-shaped musical instrument that is held against the teeth and played by plucking a projecting flexible metal tongue.

**Jez·e·bel** (jĕz'ə-bĕl', -bəl) *n.* [Heb. *Izebhel.*] **1.** A Phoenician princess of the 9th cent. B.C., and a queen of Israel as the wife of Ahab. **2. jezebel.** A scheming, shameless woman.

**JHVH** or **JHWH** *n. vars. of* YHWH.

**ji·ao** (jē'ou') *n., pl.* **jiao.** [Chin. (Mandarin) *jiao*3.] —See table at CURRENCY.

**jib**[1] (jĭb) *n.* [Orig. unknown.] **1.** *Naut.* A triangular sail stretching from the foretopmast head to the jib boom and in small craft to the bowsprit or the bow. **2. a.** The arm of a mechanical crane. **b.** The boom of a derrick. —*v.* **jibbed, jib·bing, jibs.** *Naut.* —*vi.* To jibe. —*vt.* To make (a sail) jibe.

**jib**[2] (jĭb) *vi.* **jibbed, jib·bing, jibs.** [Orig. unknown.] To stop short, balk, or shy. —Used of an animal. —*n. also* **jib·ber** (jĭb'ər). An animal that jibs.

**jib boom** *n. Naut.* A spar beyond the bowsprit, on which the jib is spread.

**jibe**[1] (jīb) *v.* **jibed, jib·ing, jibes.** [Perh. < obs. Du. *gijben.*] —*vi.* To shift a fore-and-aft sail from one side of a vessel to the other while sailing before the wind. —*vt.* To cause to jibe.

**jibe**[2] (jīb) *vi.* **jibed, jib·ing, jibes.** [Orig. unknown.] *Informal.* To be in accord : AGREE <Your ideas *jibe* with mine.>

**jibe**[3] (jīb) *v.* & *n. var. of* GIBE.

**jif·fy** (jĭf'ē) *also* **jiff** (jĭf) *n.* [Orig. unknown.] *Informal.* A short period of time <I'll be out in a *jiffy.*>

**jig** (jĭg) *n.* [Orig. unknown.] **1. a.** Any of various lively dances in triple time. **b.** The music for such a dance. **2.** A trick or caper <The *jig* is up.>. **3.** A usu. metal fishing lure with one or more hooks, fished on or near the bottom with a jiggling retrieve. **4.** An apparatus for cleaning or separating ore by agitation in water. **5.** A device for guiding a tool or for holding machine work in place. —*v.* **jigged, jig·ging, jigs.** —*vi.* **1.** To dance a jig. **2.** To bob up and down jerkily and rapidly. **3.** To operate a jig. —*vt.* **1.** To jerk up and down or to and fro. **2.** To machine with the aid of a jig. **3.** To separate or clean (ore) by shaking a jig. **—in jig time.** Quickly.

**jig·ger**[1] (jĭg'ər) *n.* **1.** One who jigs or operates a jig. **2. a.** A small measure for liquor, usu. holding 1½ ounces. **b.** This amount of liquor. **3.** A mechanical device that operates with a jerking or jolting motion, as a drill. **4.** *Naut.* **a.** A light all-purpose tackle. **b.** A small sail set in the stern of a yawl or ketch. **5.** A tricky contrivance or trivial device whose name eludes one : GADGET.

**jig·ger**[2] (jĭg'ər) *n.* [Var. of CHIGOE.] **1.** CHIGGER 1. **2.** CHIGOE 1.

**jigger mast** *n.* **1.** The short after mast from which the jigger sail is set on a ketch or yawl. **2.** The fourth mast aft on a four-masted ship.

**jig·gle** (jĭg'əl) *v.* **-gled, -gling, -gles.** [Freq. of JIG.] —*vi.* To move or rock lightly up and down or to and fro in a jerky manner. —*vt.* To cause to jiggle. —*n.* A jiggling motion. **—jig'gly** *adj.*

**jig·saw** (jĭg'sô') *n.* A usu. power-driven saw with a narrow vertical reciprocating blade, used to cut sharp curves.

**jigsaw puzzle** *n.* **1.** A puzzle consisting of a mass of irregularly shaped pieces of cardboard, plastic, or wood that form a picture when fitted together. **2.** Anything resembling a jigsaw puzzle, as in complexity.

**ji·had** (jĭ-häd') *n.* [Ar. *jihād.*] **1.** A Moslem holy war against infidels. **2.** A crusade.

**jill** (jĭl) *n. var. of* GILL[4].

**jil·lion** (jĭl'yən) *n.* [Alteration of MILLION.] *Informal.* A large, indeterminate number.

**jilt** (jĭlt) *vt.* **jilt·ed, jilt·ing, jilts.** [Orig. unknown.] To discard (a lover) suddenly or callously. —*n.* A woman who discards a lover.

**Jim Crow** *or* **jim crow** *n.* [< *Jim Crow,* derogatory term for a black person, ult. < the title of a 19th-century song.] *Slang.* The systematic practice of discriminating against and suppressing black people. —*adj.* **1.** Upholding or practicing Jim Crow <*Jim Crow* laws><a *Jim Crow* town> **2.** Reserved or set aside for a racial or ethnic group that is being discriminated against <"I told them I wouldn't take a *Jim Crow* job" —Ralph Bunche> **—Jim'-Crow'ism** (jĭm'krō'ĭz'əm) *n.*

**jim-dan·dy** (jĭm'dăn'dē) [*Jim,* nickname for *James* + DANDY.] *Informal. n.* One that is very pleasing or excellent of its kind. —*adj.* Admirable : excellent.

**jim-jams** (jĭm'jămz') *pl.n.* [Orig. unknown.] *Slang.* **1.** The jitters. **2.** Delirium tremens.

**Jim·mies** (jĭm'ēz). A trademark for chocolate sprinkles for ice cream.

**jim·my** (jĭm'ē) *n., pl.* **-mies.** [< the name *Jimmy,* nickname for *James.*] A short crowbar, esp. one used by a burglar. —*vt.* **-mied, -my·ing, -mies.** To pry open with or as if with a jimmy.

**jim·son·weed** (jĭm'sən-wēd') *n.* [Alteration of E. *Jamestown weed,* after *Jamestown,* Virginia.] A coarse foul-smelling poisonous plant, *Datura stramonium,* with large, trumpet-shaped white or purplish flowers and prickly fruit.

**jin·gle** (jĭng'gəl) *v.* **-gled, -gling, -gles.** [ME *ginglen.*] —*vi.* **1.** To make a metallic clinking or ringing sound. **2.** To have the catchy sound of a poetic jingle. —*vt.* To cause to jingle. —*n.* **1. a.** The sound produced by bits of metal striking together. **b.** Something resembling this. **2.** A simple, repetitious, catchy rhyme or doggerel.

**jingle shell** *n.* The translucent, rounded, yellowish or grayish shell of any of several marine bivalve mollusks of the genus *Anomia.*

**jin·go** (jĭng'gō) *n., pl.* **-goes.** [From the phrase *by jingo,* used in the refrain of a bellicose English song.] One who vociferously supports his or her country, esp. one who supports a belligerent foreign policy : CHAUVINIST. —*adj.* **1.** Of or relating to a jingo. **2.** Marked by jingoism. **—jin'go·ish** *adj.*

**jin·go·ism** (jĭng'gō-ĭz'əm) *n.* Extreme nationalism or chauvinism marked esp. by a belligerent foreign policy. **—jin'go·ist** *n.* **—jin'go·is'tic** *adj.* **—jin'go·is'ti·cal·ly** *adv.*

**jink** (jĭngk) *vi.* **jinked, jink·ing, jinks.** [Orig. unknown.] To make a quick, evasive turn. —*n.* **1.** A sudden evasive turn. **2. jinks.** Rambunctious play : FROLIC. —Used esp. in the phrase *high jinks.*

**jin·ni** *also* **jin·nee** (jĭn'ē, jĭ-nē') *n., pl.* **jinn** (jĭn) [Ar. *jinnīy.*] A spirit with supernatural powers for good or evil in Moslem legend, capable of assuming human or animal form and enormous size.

**jin·rik·sha** *or* **jin·rick·sha** *also* **jin·rik·i·sha** (jĭn-rĭk'shô') *n.* [J. *jinrikisha* : *jin,* man + *riki,* strength + *sha,* vehicle (of Chin. orig.).] A small two-wheeled hooded carriage drawn by one or two men, formerly used in the Orient.

**jinx** (jĭngks) [Poss. < E. *jynx,* wryneck (from the use of the bird in witchcraft) < Lat. *iynx* < Gk. *iunx* < *iuzein,* to call.] *Informal.* —*n.* **1.** One thought to bring bad luck. **2.** A condition or period of bad luck thought to be caused by a jinx. —*vt.* **jinxed, jinx·ing, jinx·es.** To bring bad luck or misfortune to.

☆ ***syns:*** JINX, CURSE, HEX, HOODOO *n. core meaning* : one that is believed to bring bad luck <thought the new basketball coach was a *jinx*>

**ji·pi·ja·pa** (hē'pē-hä'pə) *n.* [Sp. (South America) after *Jipijapa,* Ecuador.] A palmlike plant, *Carludovica palmata* of Central and South America, with long-stalked, fanlike leaves used to make Panama hats.

**jit·ney** (jĭt'nē) *n., pl.* **-neys.** [Orig. unknown.] **1.** A small bus that transports passengers on a regular route for a low fare. **2.** *Slang.* A nickel.

**jit·ter** (jĭt'ər) *vi.* **-tered, -ter·ing, -ters.** [Orig. unknown.] To be nervous or uneasy : FIDGET. —*n.* **1.** A jittering movement. **2.** Nervous agitation <Exams give me the *jitters.*>

**jit·ter·bug** (jĭt'ər-bŭg') *n.* **1.** A strenuous dance performed to quick-tempo jazz or swing music and having various two-step patterns embellished with twirls and sometimes acrobatic maneuvers, popular esp. in the 1940's. **2.** One who dances the jitterbug. —*vi.* **-bugged, -bug·ging, -bugs.** To dance the jitterbug.

**jit·ter·y** (jĭt'ə-rē) *adj.* **-i·er, -i·est. 1.** Having the jitters. **2.** Characterized by jittering movements.

**jiu·jit·su** *or* **jiu·jut·su** (jo͞o-jĭt'so͞o) *n. vars. of* JUJITSU.

**jive** (jīv) [Orig. unknown.] *Slang.* —*n.* **1.** Jazz or swing music. **2.** The jargon of jazz musicians and enthusiasts. **3.** Deceptive, nonsensical, or glib talk. —*v.* **jived, jiv·ing, jives.** —*vi.* **1.** To play or dance to jive music. **2.** To talk nonsense : KID. —*vt.* To cajole or mislead. —*adj.* Misleading : phony. **—jiv'er** *n.*

**job¹** (jŏb) *n.* [Perh. < obs. *job,* piece.] **1.** An action that needs to be done : TASK. **2.** An activity performed regularly for payment, esp. a trade, occupation, or profession. **3. a.** A specific piece of work to be done for a set fee <a costly remodeling *job*> **b.** The object to be worked on <a plumbing *job*> **c.** Something resulting from or produced by work. **4.** A position in which one is employed. **5.** A specified duty <It was your *job* to rake the leaves.> **6.** *Computer Sci.* A program application to be performed as a single logical unit. **7.** *Informal.* A difficult or strenuous task <It was quite a *job* to convince them to leave.> **8.** *Informal.* A bad or unsatisfactory piece of work <The barber really did a *job* on your hair.> **9.** *Informal.* A state of affairs <Their arrangement turned out to be a bad *job.*> **10.** *Informal.* A criminal act, esp. a robbery <a bank *job*> **11.** *Chiefly Brit.* Something done ostensibly for the public welfare, but actually for improper private gain. —*v.* **jobbed, job·bing, jobs.** —*vi.* **1.** To do odd jobs or piecework. **2.** To act as a middleman or jobber. **3.** To exploit one's position for private profits. —*vt.* **1.** To buy (merchandise) from manufacturers and sell it to retailers. **2.** To arrange for (contracted work) to be done in portions by others. **3.** *Chiefly Brit.* To transact (official business) dishonestly for private profit.

**job²** (jŏb) [ME *jobben.*] *Archaic.* —*vt. & vi.* To jab or make a jab. —*n.* A jab.

**Job** (jōb) *n.* [LLat. < Gk. *Iōb* < Heb. *Iyyōbh.*] —See table at BIBLE.

**job action** *n.* A temporary action, as a strike or slowdown, by workers to exact demands or to protest a company decision.

**job·ber** (jŏb'ər) *n.* **1.** One that buys merchandise from manufacturers and sells it to retailers. **2.** One who works by the piece or at odd jobs. **3.** *Chiefly Brit.* A middleman in the exchange of stocks and securities among brokers.

**job·ber·y** (jŏb'ə-rē) *n.* [< JOB¹.] Corruption among public officials.

**job control language** *n. Computer Sci.* A language used for communication with a computer's operating system.

**job·hold·er** (jŏb'hōl'dər) *n.* One who has a regular job.

**job-hop·ping** (jŏb'hŏp'ĭng) *n.* The practice of changing jobs frequently, esp. as a means of quick financial gain. **—job'-hop'per** *n.*

**job·less** (jŏb'lĭs) *adj.* **1.** Being without a job. **2.** Of or pertaining to those who are without jobs. **—job'less·ness** *n.*

**job lot** *n.* **1.** Miscellaneous merchandise sold in one lot. **2.** An assortment of cheap items.

**job·name** (jŏb'nām') *n. Computer Sci.* A code assigned to a specific job instruction in a computer program for user reference.

**Job's comforter** *n.* [After *Job,* who was given false sympathy by his friends.] One who saddens or discourages while seeming to offer sympathy or comfort.

**Job's-tears** (jōbz'tîrz') *n. (sing. or pl. in number).* **1.** A grass, *Coix lacryma-jobi* of tropical Asia, with white beadlike seeds. **2.** The edible seeds of the Job's-tears, often used as beads.

**job stick** *n.* A composing stick in printing.

**Jo·cas·ta** (jō-kăs'tə) *n.* [Gk. *Iokastē.*] *Gk. Myth.* A Theban queen, widow of Laius, who unknowingly married her own son, Oedipus.

**jock¹** (jŏk) *n.* [Short for JOCKEY.] **1.** A jockey. **2.** A disc jockey.

**jock²** (jŏk) *n.* [Short for JOCKSTRAP.] **1.** A jockstrap. **2.** *Slang.* A male athlete, esp. in college.

**jock·ey** (jŏk'ē) *n., pl.* **-eys.** [Dim. of Sc. *Jock,* var. of *Jack,* nickname for *John.*] **1.** One who rides horses in races, esp. as a profession. **2.** *Slang.* One who operates a specified vehicle, machine, or device <a bus *jockey*> —*v.* **-eyed, -ey·ing, -eys.** —*vt.* **1.** To ride (a horse) as jockey. **2.** To direct or maneuver with skill or cunning <*jockeyed* them out of the company> **3.** To move gradually and adroitly <They *jockeyed* the piano down the stairs.> **4.** To cheat : trick. **5.** *Slang.* To drive : pilot. —*vi.* **1.** To ride a horse in a race. **2.** To maneuver for position or advantage. **3.** To utilize trickery.

**jock·strap** *also* **jock strap** (jŏk'străp') *n.* [Slang *jock,* penis + STRAP.] An athletic supporter.

**jo·cose** (jō-kōs') *adj.* [Lat. *jocosus* < *jocus,* joke.] **1.** Given to joking : PLAYFUL. **2.** Marked by joking : HUMOROUS. **—jo·cose'ly** *adv.* **—jo·cose'ness, jo·cos'i·ty** (jō-kŏs'ĭ-tē) *n.*

**joc·u·lar** (jŏk'yə-lər) *adj.* [Lat. *jocularis,* droll < *joculus,* dim. of *jocus,* joke.] **1.** Given to or marked by joking. **2.** Meant in jest : FACETIOUS. **—joc'u·lar'i·ty** (-lăr'ĭ-tē) *n.* **—joc'u·lar·ly** *adv.*

**joc·und** (jŏk'ənd, jō'kənd) *adj.* [ME *jocound* < OFr. *jocond* < LLat. *jocundus* < Lat. *jucundus* < *juvare,* to delight.] Having a cheerful disposition. **—jo·cun'di·ty** (jō-kŭn'dĭ-tē) *n.* **—joc'und·ly** *adv.*

**jodh·pur boots** (jŏd'pər) *n.* Ankle-high leather boots buckled at the side and worn with jodhpurs for riding.

**jodh·purs** (jŏd'pərz) *pl.n.* [After *Jodhpur,* a region in India.] Wide-hipped riding breeches, fitting tightly at the knees and ankles.

**Jo·el** (jō'əl) *n.* [Heb. *Yō'ēl.*] —See table at BIBLE.

**joe-pye weed** (jō'pī') *n.* [Orig. unknown.] A tall North American plant of the genus *Eupatorium,* with whorled leaves and terminal clusters of small pinkish or purplish flowers.

**jog¹** (jŏg) *v.* **jogged, jog·ging, jogs.** [Orig. unknown.] —*vt.* **1.** To jar or move by shoving, bumping, or jerking. **2.** To give a slight push or shake to : NUDGE. **3.** To stimulate or stir to activity <*jog* one's memory> —*vi.* **1. a.** To move or ride at a steady slow trot. **b.** To run in such a way for sport or exercise. **2.** To proceed in a leisurely, humdrum manner <"while his life was thus *jogging* easily along" —Duff Cooper> —*n.* **1.** A slight jolt or shake. **2.** A nudge. **3.** A slow steady pace. **—jog'ger** *n.*

**jog²** (jŏg) *n.* [Perh. var. of JAG¹.] **1.** A protruding or receding part in a surface or line. **2.** An abrupt shift in direction. —*vi.* **jogged, jog·ging, jogs.** To turn sharply : VEER.

**jog·ging** *n.* Running for sport or exercise at a slow, regular pace.

**jog·gle¹** (jŏg'əl) *v.* **-gled, -gling, -gles.** [Freq. of JOG¹.] —*vt.* To shake or jar slightly. —*vi.* To move with a shaking or jolting motion. —*n.* A slight shake or jolt.

**jog·gle²** (jŏg'əl) *n.* [< JOG².] **1.** A joint between two pieces of building material formed by a notch and a fitted projection. **2.** The notch or projecting piece used in a joggle. —*vt.* **-gled, -gling, -gles.** To join or attach by means of a joggle.

**jog trot** *n.* **1.** A slow, steady, jolting pace. **2.** A routine, humdrum way of living or of doing something.

**john** (jŏn) *n.* [< the name *John.*] *Slang.* A toilet.

**John** (jŏn) *n.* —See table at BIBLE.

**John Bar·ley·corn** (jŏn' bär'lē-kôrn') *n.* A personification of malt liquor or of alcoholic beverages in general.

**John Birch Society** (jŏn' bûrch') *n.* [After *John Birch* (d. 1945).] An ultraconservative anti-Communist organization established by Robert Welch in 1958. **—John Bircher** *n.*

**John Bull** (jŏn' bo͝ol') *n.* [After *John Bull,* a character in *Law is a Bottomless Pit* by John Arbuthnot (1667–1735).] **1.** A personification of England or the English. **2.** An Englishman.

**John Doe** (jŏn' dō') *n.* **1.** A *Law.* A fictitious or unidentified person. **2.** An average, undistinguished man.

**John Do·ry** (jŏn' dôr'ē, dōr'ē) *n.* Either of two fishes, *Zenopsis ocellata* of the western Atlantic or *Zeus faber* of the eastern Atlantic, with spiny fins and a laterally compressed body.

**John Han·cock** (jŏn' hăn'kŏk') *n.* [After *John Hancock* (1737–1793), from the prominence of his signature on the Declaration of Independence.] *Informal.* A person's signature.

**john·ny·cake** (jŏn'ē-kāk') *n.* A corncake.

**John·ny-come-late·ly** (jŏn'ē-kŭm-lāt'lē) *n., pl.* **-lies.** *Informal.* A newcomer or latecomer, esp. a recent adherent to a cause or fad <a *Johnny-come-lately* on the issue of environmental protection>

**John·ny-jump-up** (jŏn'ē-jŭmp'ŭp') *n.* [From its quick growth.] HEARTSEASE 2.

**John·ny-on-the-spot** (jŏn'ē-ŏn'thə-spŏt', -ôn'-) *n. Informal.* One who is available and ready to act when necessary.

**John·ny Reb** (jŏn'ē rĕb') *n. Informal.* A Confederate soldier.

**John·son grass** (jŏn′sən) *n.* [After William *Johnson* (d. 1859).] A coarse grass native to the Mediterranean area, *Sorghum halepense,* cultivated in the South for forage but often a troublesome weed.

**John·so·ni·an** (jŏn-sō′nē-ən) *adj.* Of, resembling, or pertaining to Samuel Johnson or his writings. —*n.* An admirer or student of Samuel Johnson or his work.

**joie de vi·vre** (zhwä′ də vē′vrə) *n.* [Fr., joy of living.] Carefree enjoyment of life.

**join** (join) *v.* **joined, join·ing, joins.** [ME *joinen* < OFr. *joindre* < Lat. *jungere.*] —*vt.* **1.** To bring or put together so as to make continuous or form a unit <*joined* the ends with tape> **2.** To bring or put into close association or relationship <*joined* in wedlock><*join* forces> **3.** *Math.* To connect (points), as with a straight line. **4.** To form a junction with <where the river *joins* the bay> **5.** To become a part or member of <*join* a club> **6.** To take a place among, in, or with <We'll *join* you at home.> **7.** *Informal.* To adjoin. —*vi.* **1.** To come together into a unit : CONNECT <Three lines *join* to make a triangle.> **2.** To become a member of a group. **3.** To take part : PARTICIPATE <We *joined* in the search.> —*n.* A joint : junction.

☆ ***syns:*** JOIN, CONNECT, LINK, UNITE *v. core meaning* : to bring or come together <*join* hands><*join* the links of a chain> JOIN has the widest application, in both literal and figurative use. CONNECT and LINK imply a looser relationship in which individual units retain their identity while coming together at some point <two rooms *connected* by a hallway><the Panama Canal that *links* the Atlantic and the Pacific> UNITE stresses the oneness that results from joining <a plan to *unite* the 13 Colonies under one government> ***ant:*** disjoin, part

**join·der** (join′dər) *n.* [< Fr. *joindre,* to join < Lat. *jungere.*] **1.** The act of joining. **2.** *Law.* **a.** A joining of causes of action or defense in a suit. **b.** A joining of parties in a suit. **c.** The formal acceptance of an issue offered.

**join·er** (joi′nər) *n.* **1.** One that joins. **2.** *Chiefly Brit.* A carpenter, esp. a cabinetmaker. **3.** *Informal.* One given to joining many clubs, organizations, or causes.

**joint** (joint) *n.* [ME < OFr. < p.part. of *joindre,* to join < Lat. *jungere.*] **1. a.** A point or position at which two or more things are joined. **b.** A configuration in or by which two or more things are joined. **c.** Manner of joining. **2.** *Anat.* A point of contact or articulation between more or less movable parts, as between bones or between segments in the leg of an arthropod. **3.** *Bot.* A point on a stem from which a branch or leaf may grow : NODE. **4.** *Geol.* A crack or fracture in a rock mass along which no perceptible movement has occurred. **5.** A large cut of meat for roasting. **6.** *Slang.* **a.** A cheap or disreputable gathering place, as a sleazy bar. **b.** Establishment : place <We ate at a fancy *joint.*> **7.** *Slang.* A marijuana cigarette. —*adj.* **1.** Shared by or common to two or more <a *joint* tax return> **2.** Sharing with another or others <*joint* owners> **3.** Formed or marked by cooperation <a *joint* effort> **4.** Involving both houses of a legislature <a *joint* session> **5.** *Law.* United in identity of interest or liability. **6.** *Math.* Involving two or more variables. —*vt.* **joint·ed, joint·ing, joints. 1.** To combine or attach at a joint or joints. **2.** To provide or construct with joints. **3.** To cut (meat) into joints. **—out of joint. 1.** Dislocated, as a bone. **2.** Not harmonious : INCONSISTENT <Their actions are *out of joint* with their words.> **3.** Out of order : UNSATISFACTORY <The times are *out of joint.*> **4.** In bad spirits or humor. **—joint′ed** *adj.* **—joint′ly** *adv.*

**Joint Chiefs of Staff** *n.* The principal military advisory group to the President of the United States, composed of the chiefs of the Army, Navy, and Air Force and the commandant of the Marine Corps.

**joint·er** (join′tər) *n.* One that joints, esp. a machine or tool used in making joints.

**joint probability** *n.* The probability that two or more specific outcomes will occur in an event.

**joint resolution** *n.* A resolution passed by both houses of a bicameral legislature and eligible to become a law if signed by the chief executive or passed over the chief executive's veto.

**joint stock** *n.* Stock or capital funds of a company held in common by its owners.

**joint-stock company** (joint′stŏk′) *n.* A business whose capital is held in transferable shares of stock by its joint owners.

**join·ture** (join′chər) *n.* [ME < OFr., a joining < Lat. *junctura* < *junctus,* p.part. of *jungere,* to join.] **1.** *Law.* **a.** An arrangement by which a husband sets aside property to be used for the support of his wife after his death. **b.** The property so designated. **2.** *Obs.* The act of joining or state of being joined.

**joint·worm** (joint′wûrm′) *n.* The larva of certain wasps of the family Eurytomidae, esp. of *Harmolita tritici,* that infests wheat and causes hard swellings in the stems.

**joist** (joist) *n.* [ME. *giste* < OFr., ult. < Lat. *jacēre,* to lie.] Any of the parallel beams set from wall to wall to support the boards of a floor or ceiling. —*vt.* **joist·ed, joist·ing, joists.** To construct with joists.

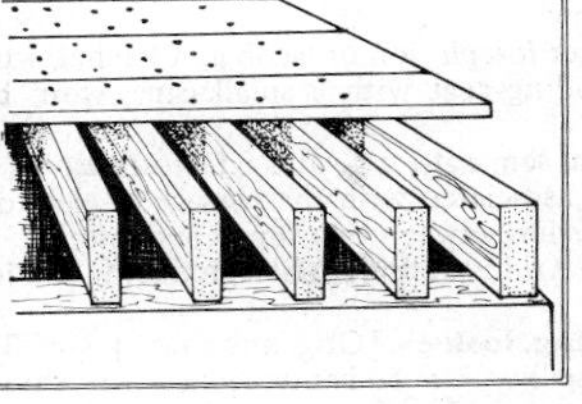
joist

**jo·jo·ba** (hə-hō′bə) *n.* [Mex. Sp.] A shrub, *Simmondsia Californica* of southwestern North America, with edible seeds that contain a valuable oil.

**joke** (jōk) *n.* [Lat. *jocus.*] **1.** A brief, amusing story, esp. one with a punch line. **2.** A jesting remark <made *jokes* at my expense> **3.** A mischievous trick : PRANK <can't take a *joke*> **4.** An amusing or ridiculous incident or situation. **5.** Something trivial or laughable <Driving on icy roads is no *joke.*> **6.** An object of amusement or laughter : LAUGHINGSTOCK. —*v.* **joked, jok·ing, jokes.** —*vi.* **1.** To tell or play jokes : JEST. **2.** To speak in fun. —*vt.* To make fun of : KID. **—jok′ing·ly** *adv.*

☆ ***syns:*** JOKE, GAG, JAPE, JEST, QUIP, WITTICISM *n. core meaning* : words or actions intended to cause laughter or amusement <opened the meeting with a *joke*>

**jok·er** (jō′kər) *n.* **1. a.** One who tells or plays jokes. **b.** An insolent person who seeks to make a show of cleverness : SMART ALECK. **c.** Fool : clown <Imagine that *joker* giving me advice.> **2.** An extra playing card in a deck, usu. printed with a picture of a jester, used in certain games as the highest ranking card or as a wild card. **3.** A seemingly innocent clause in a document or legislative bill that voids or changes its original or intended purpose. **4.** An unseen or unpredicted difficulty, fact, or circumstance. **5.** A deceptive way of getting the better of someone.

☆ ***syns:*** JOKER, CARD, CLOWN, COMEDIAN, COMIC, CUTUP, FARCEUR, HUMORIST, JESTER, JOKESTER, WAG, WIT, ZANY *n. core meaning* : one whose words or actions provoke or are intended to provoke amusement or laughter <Every party has a *joker.*>

**jol·li·fi·ca·tion** (jŏl′ə-fĭ-kā′shən) *n.* [< JOLLY.] Festivity : revelry.

**jol·li·ty** (jŏl′ĭ-tē) *n.* Merriment : gaiety.

**jol·ly** (jŏl′ē) *adj.* **-li·er, -li·est.** [ME *joli* < OFr.] **1.** Full of merriment and good spirits : FUN-LOVING. **2.** Expressing or eliciting happiness or mirth : CHEERFUL. **3.** Highly pleasing : ENJOYABLE. —*adv. Chiefly Brit.* Very : extremely <a *jolly* good story> —*v.* **-lied, -ly·ing, -lies.** —*vt.* To keep diverted or amused for one's own purposes : HUMOR. —*vi.* To engage in humorous or teasing banter. —*n., pl.* **-lies. 1.** *Chiefly Brit.* A good or festive time. **2. jollies.** *Slang.* Amusement : kicks <However you get your *jollies* is your business.> **—jol′li·ly** *adv.* **—jol′li·ness** *n.*

**jol·ly·boat** (jŏl′ē-bōt′) *n.* [Perh. alteration of obs. *jolywat.*] A medium-sized ship's boat used for rough work and minor tasks.

**Jolly Rog·er** (rŏj′ər) *n.* A black flag bearing the white skull and crossbones of a pirate ship.

**jolt** (jōlt) *v.* **jolt·ed, jolt·ing, jolts.** [Orig. unknown.] —*vt.* **1.** To bump into. **2.** To shake or knock about. **3.** To jar with or as if with a sudden, sharp blow <The bad news *jolted* us all.> **4.** To bring to a specified state by or as if by a blow <"now and then he *jolted* a nodding reader awake by inserting a witty paragraph" —Walter Blair> —*vi.* To proceed in an irregular, bumpy, or jerky manner. —*n.* **1.** A sudden jarring or jerking, as from a blow. **2.** A sudden or unexpected shock or reversal <The bank's failure came as a *jolt.*> **—jolt′er** *n.* **—jolt′i·ly** *adv.* **—jolt′i·ness** *n.* **—jolt′y** *adj.*

**Jo·ma·da** (jə-mä′dä) *n. var. of* JUMADA.

**Jo·nah** (jō′nə) *n.* [After *Jonah,* whose disobedience caused God to raise a great storm.] **1.** One thought to bring bad luck. **2.** —See table at BIBLE.

**Jon·a·than** (jŏn′ə-thən) *n.* [After *Jonathan* Hasbrouck (d. 1846).] A variety of red, late-ripening apple.

**jon·gleur** (zhôN-glœr′) *n.* [Fr. < OFr., var. of *joglere.* —see JUGGLER.] A wandering minstrel and storyteller in medieval France and England.

**jon·quil** (jŏng′kwəl, jŏn′-) *n.* [NLat. *jonquilla,* specific epithet of *Narcissus jonquilla* < Sp. *junquillo,* dim. of *junco,* reed < Lat. *juncus.*] A widely cultivated plant, *Narcissus jonquilla,* with long narrow leaves and short-tubed, fragrant yellow flowers.

**Jor·dan almond** (jôr′dn) *n.* [By folk ety. < ME *jardin almaund* : OFr. *jardin,* garden (of Germanic orig.) + ME *almaund,* almond. —see ALMOND.] A large variety of Spanish almond, covered with a hard, colored and flavored sugar coating and used as a confection.

**Jor·dan curve** (jôr′dn) *n.* [After Camille *Jordan* (1838–1922).] A curve, as a circle, that is closed and does not intersect itself.

**Jordan curve theorem** *n.* A basic theorem of topology that states that every simple closed curve divides a plane into two regions and acts as the common boundary between them.

**jo·rum** (jôr′əm, jōr′-) *n.* [Perh. after *Joram,* who brought vessels of silver, gold, and brass to King David.] **1.** A large drinking bowl. **2.** The contents of such a bowl.

**jo·seph** (jō′zəf, -səf) *n.* [After *Joseph,* son of Jacob and Rachel, who had a long coat.] A long riding coat with a small cape, worn by women in the 18th cent.

**Jo·seph·son effect** (jō′zəf-sən, -səf-) *n.* [After B.D. *Josephson,* (b. 1940).] The radiative effect associated with the passage of electron pairs across an insulating barrier between two superconductors.

**Josephson junction** *n.* An insulating barrier producing the Josephson effect.

**josh** (jŏsh) *v.* **joshed, josh·ing, josh·es.** [Orig. unknown.] —*vt.* To tease (someone) good-humoredly. —*vi.* To banter **:** joke. —*n.* A teasing or joking remark.

**Josh·u·a** (jŏsh′o͞o-ə) *n.* [Heb. *Yĕhōshūa.*] —See table at BIBLE.

**Joshua tree** *n.* [Prob. after JOSHUA, from the resemblance of the tree's greatly extended branches to Joshua's outstretched arm as he pointed with his spear to the city of Ai.] A treelike plant, *Yucca brevifolia* of the southwestern United States, with sword-shaped leaves and greenish-white flowers.

**joss** (jŏs) *n.* [Pidgin E. < Port. *deos,* god < Lat. *deus.*] A Chinese idol or image.

**joss house** *n.* A Chinese temple or shrine.

**joss stick** *n.* A stick of incense burned before a joss.

**jos·tle** (jŏs′əl) *v.* **-tled, -tling, -tles.** [Obs. *justle,* freq. of *just,* to joust < ME *justen* < OFr. *juster.* —see JOUST.] —*vi.* **1.** To come in contact or collide. **2.** To make one's way by pushing or elbowing. **3.** To contend for an advantage or position <*jostle* for a promotion> **4.** To be in close proximity. —*vt.* **1.** To come into close contact with. **2.** To force by pushing or elbowing. **3.** To vie with for an advantage or position. **4.** To be in close proximity with <"books written in all languages by men and women of all tempers, races, and ages *jostle* each other on the shelf" —Virginia Woolf> —*n.* **1.** A rough shove or push. **2.** The state of being crowded together.

**jot** (jŏt) *n.* [Lat. *iota* < Gk. *iōta,* iota.] The smallest bit **:** IOTA. —*vt.* **jot·ted, jot·ting, jots.** To write down briefly or hastily <*jot* down a phone number>

**Jo·tun·heim** (yô′to͞on-hām′) *also* **Jö·tunn·heim** (yœ′-) *n.* [ON *jötunheimar* : *jötunn,* giant + *heimr,* home.] *Norse Myth.* The home of a race of giants.

**joule** (jo͞ol, joul) *n.* [After James P. *Joule* (1818–1889).] **1.** The International System unit of energy, equal to the work done when a current of one ampere is passed through a resistance of one ohm for one second. **2.** A unit of energy, equal to the work done when the point of application of a force of one newton is displaced one meter in the direction of the force.

**jounce** (jouns) *vi.* & *vt.* **jounced, jounc·ing, jounc·es.** [ME *jouncen.*] To move or cause to move with bumps and jolts. **—jounce** *n.*

**jour·nal** (jûr′nəl) *n.* [ME < OFr. *jornel* < *jornel,* daily < LLat. *diurnalis.* —see DIURNAL.] **1. a.** A personal record of experiences and observations kept on a regular basis **:** DIARY. **b.** A record of daily events. **c.** An official record of daily transactions, as of a legislative body. **d.** A ship's log. **2. a.** DAYBOOK 1. **b.** A book of original entry in a double-entry system, listing all transactions and indicating the accounts to which they belong. **3.** A daily newspaper. **4.** A periodical containing articles of interest to a particular group <a dental *journal*> **5.** The part of a machine shaft or axle supported by a bearing.

**journal box** *n.* A housing in a machine enclosing a journal and its bearings.

**jour·nal·ese** (jûr′nə-lēz′, -lēs′) *n.* A slick, superficial style of writing considered typical of most newspapers.

**jour·nal·ism** (jûr′nə-lĭz′əm) *n.* **1.** Collection, writing, editing, and dissemination of news through the media. **2.** Material written for publication in the media. **3.** A style of writing used in newspapers and magazines, characterized by the direct presentation of facts or occurrences with little attempt at analysis or interpretation. **4.** Newspapers and magazines **:** the press. **5.** A course of study in the collection and preparation of the news. **6.** Written material of current interest or wide popular appeal.

**jour·nal·ist** (jûr′nə-lĭst) *n.* **1.** One whose occupation is journalism. **2.** One who keeps a journal.

**jour·nal·is·tic** (jûr′nə-lĭs′tĭk) *adj.* Of, relating to, or typical of journalism or journalists. **—jour′nal·is′ti·cal·ly** *adv.*

**jour·nal·ize** (jûr′nə-līz′) *v.* **-ized, -iz·ing, -iz·es.** —*vt.* To record in a journal. —*vi.* To keep a personal or financial journal.

**jour·ney** (jûr′nē) *n., pl.* **-neys.** [ME *journei,* day, day's travel, journey < OFr. *journee* < VLat. **diurnata* < Lat. *diurnus,* of a day < *dies,* day.] **1. a.** Travel from one place to another **:** TRIP. **b.** The distance to be traveled or the time required for such a trip. **2.** Any passage from one stage to another <the *journey* from cradle to grave> —*v.* **-neyed, -ney·ing, -neys.** —*vi.* To make a journey **:** TRAVEL. —*vt.* To travel over or through. **—jour′ney·er** *n.*

**jour·ney·man** (jûr′nē-mən) *n.* [ME *journeiman* < *journei,* a day's work. —see JOURNEY.] **1.** One who has served an apprenticeship in a trade or craft and is a qualified worker in another's employ. **2.** A competent but undistinguished worker.

▲ word history: A *journeyman,* a skilled craftsman, is not an itinerant worker. *Journeyman* preserves an older sense of the word *journey,* and one that reveals its origins. A *journeyman* was originally someone who worked for another for daily wages. He was distinguished from an apprentice, who was learning the trade, and a master artisan, who was in business for himself. *Journey* is derived from Old French *journee,* which meant "day," "day's work," and "day's travel." *Journee* is descended from Latin *diurnus,* "daily," from *dies,* "day." *Journee* in the sense "day's travel" developed in English into the word *journey,* "a trip."

**jour·ney·work** (jûr′nē-wûrk′) *n.* The work of a journeyman.

**joust** (joust, jŭst, jo͞ost) *also* **just** (jŭst) [ME < OFr. *juste* < *juster,* to joust < VLat. **juxtare* < Lat. *juxta,* close together.] —*n.* **1. a.** A combat with lances between two mounted knights **:** tilting match. **b. jousts.** A series of these matches **:** TOURNAMENT. **2.** Personal competition. —*vi.* **joust·ed, joust·ing, jousts** *also* **just·ed, just·ing, justs. 1.** To engage in combat on horseback, esp. with lances. **2.** To engage in personal competition.

**Jove** (jōv) *n.* [ME < Lat. *Jovis.*] JUPITER 1. —*interj.* **by Jove.** —Used to express surprise or emphasis. **—Jo′vi·an** (jō′vē-ən) *adj.*

**jo·vi·al** (jō′vē-əl) *adj.* [Fr. < Ital. *giovale,* born under the planet Jupiter < *Giove,* Jupiter.] Marked by hearty conviviality. **—jo′vi·al′i·ty** (-ăl′ĭ-tē) *n.* **—jo′vi·al·ly** *adv.*

▲ word history: The ultimate source of *jovial* is Latin *jovialis,* an adjective derived from *Jovis,* another name for the Roman god Jupiter. The meaning "jolly, convivial" for *jovial* was a development of astrological notions about the planets and other celestial objects. Astrologers believed that the planets had specific attributes and characteristics. If they were in certain positions in the heavens or in relation to each other they could affect persons or events on earth. The planet Jupiter in a person's horoscope had a very favorable influence and was regarded as a source of happiness. A *jovial* person was therefore literally one influenced astrologically by the planet Jupiter who as a result was characterized by mirth and conviviality.

**jowl**[1] (joul) *n.* [ME *chaule* < OE *ceafl.*] **1.** The jaw, esp. the lower jaw. **2.** The cheek.

**jowl**[2] (joul) *n.* [ME *cholle.*] **1.** The flesh under the lower jaw, esp. when flaccid. **2.** A fleshy part similar to a jowl, as a dewlap or a wattle.

**joy** (joi) *n.* [ME *joi* < OFr. < Lat. *gaudia,* pl. of *gaudium,* joy < *gaudere,* to rejoice.] **1. a.** Great pleasure or happiness **:** DELIGHT. **b.** The expression or display of this emotion. **2.** A source or object of pleasure or satisfaction. —*vt.* **joyed, joy·ing, joys.** *Archaic.* **1.** To fill with joy. **2.** To enjoy.

**joy·ful** (joi′fəl) *adj.* Feeling, causing, or displaying joy. **—joy′ful·ly** *adv.* **—joy′ful·ness** *n.*

**joy·less** (joi′lĭs) *adj.* Cheerless **:** unhappy. **—joy′less·ly** *adv.* **—joy′less·ness** *n.*

**joy·ous** (joi′əs) *adj.* Joyful. **—joy′ous·ly** *adv.* **—joy′ous·ness** *n.*

**joy·pop** (joi′pŏp′) *vi.* **-popped, -pop·ping, -pops.** *Slang.* To use narcotic drugs occas. without becoming addicted. **—joy′pop′per** *n.*

**joy ride** *n.* **1.** A ride taken for fun or for the thrill of reckless driving. **2.** A hazardous, reckless, often costly venture.

**joy·stick** (joi′stĭk′) *n. Slang.* **1.** The control stick of an airplane. **2.** A manual control device resembling an airplane's joystick.

**J particle** *n.* A neutral meson having an unusually long lifetime.

**ju·ba** (jo͞o′bə) *n.* [Orig. unknown.] A lively dance of blacks on Southern plantations, marked by complex rhythmic clapping and body movements.

**ju·bi·lant** (jo͞o′bə-lənt) *adj.* [Lat. *jubilans, jubilant-,* pr.part. of *jubilare,* to raise a shout of joy.] **1.** Exultingly joyful. **2.** Expressing joy. **—ju′bi·lance, ju′bi·lan·cy** *n.* **—ju′bi·lant·ly** *adv.*

**ju·bi·late** (jo͞o′bə-lāt′) *vi.* **-lat·ed, -lat·ing, -lates.** [Lat. *jubilare, jubilat-,* to raise a shout of joy.] To rejoice **:** exult. **—ju′bi·la′tion** *n.*

**Ju·bi·la·te** (yo͞o′bə-lä′tā, -tē, jo͞o′-) *n.* [Lat., imper. of *jubilare,* to raise a shout of joy (the first word of the psalm).] **1. a.** The 100th Psalm in the King James Bible and in most modern Catholic versions or the 99th in the Vulgate. **b.** A musical setting of the Jubilate. **2.** The third Sunday after Easter. **3. jubilate.** A song or outburst of joy and triumph.

**ju·bi·lee** (jo͞o′bə-lē′, jo͞o′bə-lē′) *n.* [ME *jubile* < OFr. < LLat. *jubilaeus,* the Jewish year of jubilee < LGk. *iōbēlaios* < Heb. *yōbhēl.*] **1. a.** A special anniversary, esp. a 50th anniversary. **b.** Celebration of such an anniversary. **2.** A season of joyful celebration. **3.** Jubilation **:** rejoicing. **4.** *often* **Jubilee.** A year of rest prescribed in the Old Testament to be observed by the Israelites every 50th year, during which slaves were to be set free, alienated property restored to the former owners, and the lands left untilled. **5.** *Rom. Cath. Ch.* A year during which plenary indulgence may be obtained by the performance of certain pious acts.

**Ju·dah** (jo͞o′də) *n.* [Heb. *Yĕhūdāh.*] One of the 12 tribes of Israel.

**Ju·da·ic** (jo͞o-dā′ĭk) *also* **Ju·da·i·cal** (-ĭ-kəl) *adj.* [Lat. *Judaicus* < Gk. *Ioudaikos* < Gk. *Ioudaios,* Jew.] Of, relating to, or typical of Jews or Judaism. **—Ju·da′i·cal·ly** *adv.*

**Ju·da·ism** (jo͞o′dē-ĭz′əm) *n.* [LLat. *Judaismus* < Gk. *Ioudaismos* < *Ioudaios,* Jew. —see JEW.] **1.** The monotheistic religion of the Jewish people, tracing its origins to Abraham and having its spiritual and ethical principles embodied chiefly in the Bible and the Talmud. **2.** Conformity to the traditional rites and ceremonies of the Jewish religion. **3.** The spiritual, cultural, and social way of life of the Jewish people. **4.** The Jewish people.

**Ju·da·ize** (jōō'dē-īz') *v.* **-ized, -iz·ing, -iz·es.** —*vt.* To bring into conformity with Judaism. —*vi.* To adopt Jewish customs and beliefs. **—Ju'da·i·za'tion** *n.* **—Ju'da·iz'er** *n.*

**Ju·das** (jōō'dəs) *n.* [After *Judas* Iscariot, the apostle who betrayed Jesus.] **1.** One who betrays under the appearance of friendship. **2. ju·das.** A one-way peephole in a door.

**Judas tree** *n.* [From the belief that Judas Iscariot hanged himself on such a tree.] The redbud.

**Jude** (jōōd) *n.* [< LLat. *Judas* < Gk. *Ioudas* < Heb. *Yĕhūdāh*, Judah.] —See table at BIBLE.

**Ju·de·o-Span·ish** (jōō-dā'ō-spăn'ĭsh) *n.* [Lat. *Judaeus*, Jewish + SPANISH.] Ladino.

**judge** (jŭj) *v.* **judged, judg·ing, judg·es.** [ME *jugen* < AN *juger* < OFr. *jugier* < Lat. *judicare* < *judex*, judge.] —*vt.* **1. a.** To pass judgment on in a court of law <*judged* them guilty> **b.** To sit in judgment on : TRY <*judge* a case> **2. a.** To decide authoritatively after deliberation. **b.** To appraise discriminatingly as an expert. **c.** To declare after determination <They *judged* me a good cook.> **3.** To form an opinion about <*judge* personality> **4.** To draw a general conclusion about : ESTIMATE <I *judge* the time to be noon.> **5.** To have as an opinion or assumption : SUPPOSE. **6.** *Obs.* To govern : rule. —*vi.* **1.** To act or decide as a judge. **2.** To form an opinion or evaluation. —*n.* **1. a.** A public official authorized to hear and decide cases brought before a court of law. **b.** An appointed arbiter in a contest or competition. **c.** One whose critical judgment or opinion is sought : CONNOISSEUR. **2. a.** A leader of the Israelites during a period of about 400 years between the death of Joshua and the accession of Saul. **b. Judges.** (*sing. in number*). —See table at BIBLE.
☆ ***syns:*** JUDGE, ADJUDGE, ADJUDICATE, ARBITRATE, DECIDE, DECREE, DETERMINE, REFEREE, RULE, UMPIRE *v. core meaning* : to make a decision about (a controversy or dispute) after deliberating about it <A jury *judged* the merits of the case.>

**judge advocate** *n., pl.* **judge advocates. 1.** A commissioned officer in the U.S. military assigned to the Judge Advocate General's Corps. **2.** A staff officer serving as legal adviser to a commander. **3.** An officer acting as prosecutor at a court-martial.

**Judge Advocate General** *n., pl.* **Judge Advocates General** *or* **Judge Advocate Generals.** The senior legal officer in the U.S. Army, Air Force, or Navy.

**judge·ment** (jŭj'mənt) *n. var. of* JUDGMENT.

**judge·ship** (jŭj'shĭp') *n.* The office or jurisdiction of a judge.

**judg·mat·ic** (jŭj-măt'ĭk) *also* **judg·mat·i·cal** (-ĭ-kəl) *adj.* [< JUDGE.] *Informal.* Judicious. **—judg·mat'i·cal·ly** *adv.*

**judg·ment** *also* **judge·ment** (jŭj'mənt) *n.* [ME *jugement* < OFr. < *jugier*, judge. —see JUDGE.] **1. a.** The ability to make a decision or form an opinion by discerning and evaluating. **b.** The capacity to make sound and reasonable decisions : good sense. **c.** Sound and reasonable decision-making. **d.** An opinion or estimate formed by sound and reasonable evaluation. **2.** A discriminating or authoritative appraisal or opinion. **3.** A rough guess or estimation <make a *judgment* of the cost> **4.** An assertion of something believed. **5.** A formal decision, as of an arbiter in a contest. **6.** *Law.* **a.** A determination of a court of law : judicial decision. **b.** A court act creating or affirming an obligation, as a debt. **c.** A writ in witness of such an act. **7.** A misfortune regarded as sent by God in punishment. **8. Judgment.** The final judgment of mankind by God. **—judg·men'tal** (-mĕn'tl) *adj.*

**Judgment Day** *n.* **1.** The day of God's final judgment in the teleology of Judaism, Christianity, and Islam. **2.** A day of reckoning or final judgment.

**ju·di·ca·ble** (jōō'dĭ-kə-bəl) *adj.* [LLat. *judicabilis* < Lat. *judicare*, to judge < *judex*, judge.] **1.** Capable of being judged. **2.** Liable to be judged.

**ju·di·ca·tor** (jōō'dĭ-kā'tər) *n.* [LLat. < Lat. *judicare*, to judge < *judex*, judge.] One that acts as judge.

**ju·di·ca·to·ry** (jōō'dĭ-kə-tôr'ē, -tōr'ē) *n., pl.* **-ries.** [Med. Lat. *judicatorium* < Lat. *judicare*, to judge < *judex*, judge.] **1.** A court of justice. **2.** JUDICIARY 2a. —*adj.* Of or relating to the administration of justice.

**ju·di·ca·ture** (jōō'dĭ-kə-chŏŏr') *n.* [Med. Lat. *judicatura* < Lat. *judicare*, to judge < *judex*, judge.] **1.** The administration of justice. **2.** The position, function, or authority of a judge. **3.** The jurisdiction of a judge or court. **4.** A court of law. **5.** Judges or courts as a whole.

**ju·di·cial** (jōō-dĭsh'əl) *adj.* [ME < Lat. *judicialis* < *judicium*, judgment < *judex*, judge.] **1.** Of, relating to, or befitting courts of law or the administration of justice. **2.** Decreed or enforced by a court of justice. **3.** Belonging or appropriate to the office of a judge. **4.** Relating to, marked by, or expressing judgment. **5.** Proceeding from a divine judgment. **—ju·di'cial·ly** *adv.*

**ju·di·ci·ar·y** (jōō-dĭsh'ē-ĕr'ē) *n., pl.* **-ies.** [Lat. *judiciarius* < *judicium*, judgment < *judex*, judge.] **1.** The judicial branch of government. **2. a.** A system of courts of law for the administration of justice. **b.** The judges of these courts.

**ju·di·cious** (jōō-dĭsh'əs) *adj.* [Fr. *judicieux* < OFr. < Lat. *judicium*, judgment < *judex*, judge.] Having or exhibiting sound judgment <a *judicious* investment> **—ju·di'cious·ly** *adv.* **—ju·di'cious·ness** *n.*

**Ju·dith** (jōō'dĭth) *n.* [Heb. *Yĕhūdīth.*] —See table at BIBLE.

**ju·do** (jōō'dō) *n.* [J. *jūdō* : *jū*, soft (< Chin. *rou*$^2$) + *dō*, way.] A Japanese martial art developed from jujitsu that applies principles of balance and leverage for self-defence, often used as a method of physical training. **—ju'do·ist** *n.*

**jug** (jŭg) *n.* [Poss. < *Jug*, nickname for *Joan.*] **1.** A small pitcher. **2. a.** A tall, often rounded vessel of earthenware, glass, or metal with a narrow neck, a handle, and usu. a stopper or cap. **b.** The contents of a jug. **3.** *Slang.* A jail. —*vt.* **jugged, jug·ging, jugs. 1.** To stew (e.g., a hare) in an earthenware vessel. **2.** *Slang.* To put in jail.

**ju·ga** (jōō'gə) *n. var. pl. of* JUGUM.

**ju·gate** (jōō'gāt', -gĭt) *adj.* [< Lat. *jugum*, yoke.] Joined in or forming a pair or pairs.

**jug band** *n.* A musical group that uses unconventional or improvised instruments, as jugs, kazoos, and washboards.

**Jug·ger·naut** (jŭg'ər-nôt') *n.* [Hindi *jagannath* < Skt. *jaganāthaḥ*, lord of the world : *jagat*, the world (< *gacchati*, it goes) + *nathaḥ*, lord.] **1.** A title of the Hindu deity Krishna, whose idol is drawn in an annual procession on a huge cart under the wheels of which worshipers are said to have thrown themselves to be crushed. **2. jug·gernaut.** Anything that elicits blind and destructive devotion or ruthless sacrifice. **3. juggernaut.** A relentless and overwhelming force or movement.

**jug·gle** (jŭg'əl) *v.* **-gled, -gling, -gles.** [ME *jogelen*, to entertain by performing tricks < OFr. *jogler* < Lat. *joculari*, to jest < *joculus*, dim. of *jocus*, joke.] —*vt.* **1.** To keep (several objects) in the air at one time by alternately tossing and catching them. **2.** To keep (e.g., several activities) in operation or progress at one time. **3.** To attempt to balance or otherwise cope with. **4.** To manipulate in order to deceive <*juggle* figures in a ledger> —*vi.* **1.** To perform tricks with sleight of hand. **2.** To make juggling motions. **3.** To use trickery to deceive. —*n.* **1.** An act of juggling. **2.** An act of manipulation.

**jug·gler** (jŭg'lər) *n.* [ME *jogelour*, jester < OFr. *joglere* < Lat. *joculator* < *joculari*, to jest. —see JUGGLE.] **1.** One who performs tricks of dexterity. **2.** One who uses trickery or deception.

**jug·gler·y** (jŭg'lə-rē) *n., pl.* **-ies. 1.** The art or performance of a juggler. **2.** Trickery : deception.

**jug·head** (jŭg'hĕd') *n. Slang.* A dull, slow-witted person.

**jug·u·lar** (jŭg'yə-lər) *adj.* [LLat. *jugularis* < Lat. *jugulum*, collarbone, dim. of *jugum*, yoke.] Of, relating to, or situated in the region of the neck or throat. —*n.* A jugular vein. **—go for the jugular.** To attempt to administer a finishing blow.

**jugular vein** *n.* Any of various large veins of the neck.

**ju·gum** (jōō'gəm) *n., pl.* **-ga** (-gə) *or* **-gums.** [Lat., yoke.] A paired or yokelike structure, as a pair of opposite leaflets or a lobe joining the bases of the forewings and hind wings of certain insects.

**juice** (jōōs) *n.* [ME *jus* < OFr. < Lat.] **1. a.** A fluid naturally contained in plant or animal tissue. **b.** A bodily secretion <gastric *juices*> **2.** *Slang.* **a.** Vigorous life and vitality. **b.** Inherent quality : ESSENCE. **c.** Power : clout. **3.** *Slang.* **a.** Electric current. **b.** Fuel for an engine. **4.** *Slang.* Liquor. **5.** *Slang.* An interesting fact or development, esp. of a private or scandalous nature. —*vt.* **juiced, juic·ing, juic·es.** To extract the juice from. **—juice up.** *Informal.* To add energy, power, or excitement to. **—juice'less** *adj.*

**juic·er** (jōō'sər) *n.* **1.** An appliance for extracting juice from fruits and vegetables. **2.** *Slang.* A heavy drinker.

**juic·y** (jōō'sē) *adj.* **-i·er, -i·est. 1.** Full of juice : SUCCULENT. **2. a.** Richly interesting <a *juicy* novel> **b.** Racy : titillating <*juicy* gossip> **3.** Profitable : lucrative <a *juicy* contract> **—juic'i·ly** *adv.* **—juic'i·ness** *n.*

**ju·jit·su** *also* **ju·jut·su** *or* **jiu·jit·su** *or* **jiu·jut·su** (jōō-jĭt'sōō) *n.* [J. *jūjitsu* : *jū*, soft (< Chin. *rou*$^2$) + *jitsu*, art (< Chin. *shu*$^4$).] A Japanese martial art using holds, throws, and stunning blows to disable or subdue an opponent.

**ju·ju** (jōō'jōō) *n.* [Perh. < Hausa *jùju*, evil spirit.] **1.** An object used as a fetish, charm, or amulet in West Africa. **2.** Magic power ascribed to a juju. **—ju'ju·ism'** *n.*

**ju·jube** (jōō'jōōb') *n.* [ME < Med. Lat. *jujuba* < Lat. *zizyphum* < Gk. *zizuphon.*] **1. a.** A spiny tree of the genus *Ziziphus*, esp. the Old World tree *Z. jujuba*, bearing small yellowish flowers and dark-red fruit. **b.** The fleshy, edible fruit of the jujube. **2.** (*also* jōō'jōō-bē') A small gelatinous fruit-flavored candy or lozenge.

**ju·jut·su** (jōō-jĭt'sōō) *n. var. of* JUJITSU.

**juke box** (jōōk) *n.* [< E. *jukehouse*, roadhouse with music and dancing, prob. < dial. E. *juke*, disorderly, poss. of African orig.] A coin-operated phonograph, usu. in a large lighted cabinet, equipped with push buttons for the selection of records.

**ju·lep** (jōō'lĭp) *n.* [ME, a sugar syrup < OFr. < Ar. *julāb* < Pers. *gulāb* : *gul*, rose + *āb*, water.] **1.** A mint julep. **2.** A sweet syrupy drink, esp. one to which medicine may be added.

**Jul·ian calendar** (jo͞ol′yən) *n.* A calendar introduced by Julius Caesar in Rome in 46 B.C. that established the 12-month year of 365 days with a leap year every 4 years, eventually replaced by the Gregorian calendar.

**ju·li·enne** (jo͞o′lē-ĕn′, zhü-lyĕn′) *adj.* [Fr.] Cut into long thin strips <*julienne* potatoes> —*n.* Consommé or broth garnished with julienne vegetables.

**Ju·ly** (jo͞o-lī′) *n., pl.* **-lies** *or* **-lys.** [ME *juil* < OFr. < Lat. *Julius,* after *Julius* Caesar.] The seventh month of the year according to the Gregorian calendar. —See table at CALENDAR.

**Ju·ma** (jo͞o′mä) *n.* The Islamic Sabbath, falling on Friday.

**Ju·ma·da** (jo͞o-mä′dä) *also* **Jo·ma·da** (jə-) *n.* [Ar. *Jumādā.*] Either the fifth or the sixth month of the Moslem year. —See table at CALENDAR.

**jum·ble** (jŭm′bəl) *v.* **-bled, -bling, -bles.** [Orig. unknown.] —*vi.* To move or mingle in a confused, disordered manner. —*vt.* **1.** To stir or mix in a disordered mass. **2.** To muddle : confuse. —*n.* **1.** A disordered mass. **2.** A confused state : MUDDLE.

**jum·bo** (jŭm′bō) *n., pl.* **-bos.** [After *Jumbo,* a large elephant exhibited by P.T. Barnum (1810–1891).] An unusually large person, animal, or thing. —*adj.* Larger than average <*jumbo* shrimp>

**jum·buck** (jŭm′bŭk′) *n.* [A native word in Australia.] *Austral.* A sheep.

**jump** (jŭmp) *v.* **jumped, jump·ing, jumps.** [Perh. of imit. orig.] —*vi.* **1. a.** To spring off the ground or other base by a muscular effort of the legs and feet : LEAP. **b.** To move suddenly and in one motion <*jumped* out of the car> **c.** To move involuntarily, as in surprise <*jumped* at the sound of the crash> **d.** To respond or act quickly <*Jump* when the officer gives you an order.> **2.** To spring at with the intent to assail or censure <*jumped* at me for not inviting them> **3. a.** To move or react eagerly so as to seize <*jump* at the chance><*jump* at a bargain> **b.** To arrive at hastily or haphazardly <*jump* to conclusions> **4.** To move randomly or aimlessly <*jumping* from job to job> **5. a.** To undergo a sudden and pronounced increase <Prices *jumped.*> **b.** To rise suddenly in position or rank <*jumped* to managing editor> **6. a.** To change abruptly <The speaker kept *jumping* from one subject to another.> **b.** To be displaced with a sudden jolt <The phonograph needle *jumped.*> **c.** To be displaced vertically or laterally because of improper alignment <The film *jumped* during projection.> **d.** *Computer Sci.* To move from one set of instructions in a program to another farther ahead or behind rather than moving sequentially. **7.** To move over an opponent's playing piece in checkers. **8.** To make a jump bid in bridge. **9.** To show enterprise and quickness. **10.** To have a lively, pulsating quality <a band that really *jumps*> —*vt.* **1.** To leap over or across <*jump* a puddle> **2.** To leap into <*jump* a taxi> **3.** To spring upon in sudden attack <Muggers *jumped* them in the alley.> **4.** To move or start before (e.g., a signal). **5.** To cause to leap <*jump* a horse over a fence> **6.** To cause to increase suddenly and markedly <The school *jumped* its fees.> **7.** To skip <The typewriter *jumped* a space.> **8.** To promote, esp. by more than one level <*jumped* me from instructor to associate professor> **9.** To take (an opponent's piece) in checkers by moving over it with one's own. **10.** To raise (a partner's bid) in bridge by more than is necessary. **11.** To leave (a course) through mishap <The train *jumped* the track.> **12.** *Slang.* **a.** To leave hastily : SKIP <*jumped* town> **b.** To leave (a position) in violation of a contract <*jumped* the team> —*n.* **1.** An act of jumping : LEAP. **2.** The space or distance covered by a leap. **3.** A hurdle, barrier, or span to be jumped. **4.** A track sport featuring skill in jumping <the high *jump*> **5. a.** A sudden, pronounced rise, as in price or salary. **b.** An impressive promotion. **6.** A step or level <managed to stay a *jump* ahead of inflation> **7.** A sudden or drastic transition, as from one career or subject to another. **8. a.** A short trip. **b.** One in a series of moves and stopovers, as with a circus or road show. **9.** A move in checkers made by jumping. **10.** *Computer Sci.* A movement from one set of instructions to another. **11. a.** An involuntary nervous movement, as when startled. **b. jumps.** The fidgets. **—get (or have) the jump on (someone).** *Informal.* To have a head start or early advantage over. **—jump a claim.** To take land or rights from another by violence or fraud. **—jump bail.** To forfeit one's bail by absconding. **—jump ship.** To desert. **—jump the gun.** To start something too soon.
☆ ***syns:*** JUMP, HURDLE, LEAP, VAULT *v. core meaning* : to spring into the air <*jumped* over the fence>

**jump ball** *n. Basketball.* A method of starting or resuming play in which an official tosses the ball up between two opposing players who must then jump and try to tap the ball to a teammate.

**jump bid** *n.* A bridge bid at a higher level than that required to exceed the preceding bid.

**jump·er¹** (jŭm′pər) *n.* **1.** One that jumps. **2.** A coasting sled. **3.** *Elect.* A short length of wire used temporarily to complete or bypass a circuit. **4.** *Basketball.* A jump shot.

**jum·per²** (jŭm′pər) *n.* [Prob. < E. dial. *jump,* loose jacket.] **1.** A sleeveless dress worn over a blouse or sweater. **2.** A loose, protective garment worn over other clothes. **3.** *often* **jumpers.** A child's garment consisting of straight-legged pants attached to a biblike bodice.

**jumper cable** *n.* A booster cable.

**jumping bean** *n.* A seed, as of certain Mexican shrubs or plants of the genera *Sebastiana* and *Sapium,* inhabited by the larva of a small moth, *Laspeyresia saltitans,* the movements of which cause the seed to jump about.

**jumping jack** *n.* **1.** A toy figure with jointed limbs that can be made to dance by pulling an attached string. **2.** A physical exercise performed by jumping to a position with legs spread wide and hands touching overhead and then back to a standing position with arms at the sides.

**jumping mouse** *n.* Any of various small rodents of the family Zapodidae, with a long tail and long hind legs.

**jumping mouse**
*8–9 inches long*

**jump·ing-off place** (jŭm′pĭng-ôf′, -ŏf′) *n.* **1.** A remote spot. **2.** A beginning point for an enterprise.

**jump-off** (jŭmp′ôf′, -ŏf′) *n.* The start of a race or a planned military attack.

**jump rope** *n.* A sturdy rope usu. with handles that is twirled and jumped over in children's games or in conditioning exercises.

**jump seat** *n.* **1.** A portable or collapsible seat, as in an automobile between the front and rear seats. **2.** A small rear seat in a sports car.

**jump shot** *n. Basketball.* A ball shot made by a player at the highest point of his or her jump.

**jump-start** (jŭmp′stärt′) *vt.* **-start·ed, -start·ing, -starts.** To start (an automobile engine) by pushing or rolling the automobile and suddenly releasing the clutch or by using a booster cable connected to the battery of another automobile. **—jump′-start′** *n.*

**jump suit** *n.* **1.** A one-piece uniform worn by parachutists. **2.** *also* **jump·suit** (jŭmp′so͞ot′). A one-piece garment combining shirt and pants.

**jump·y** (jŭm′pē) *adj.* **-i·er, -i·est. 1.** Marked by fitful, jerky movements. **2.** On edge : NERVOUS. **—jump′i·ness** *n.*

**jun** (jo͞on) *n., pl.* **jun.** [Korean.] —See table at CURRENCY.

**jun·co** (jŭng′kō) *n., pl.* **-cos** *or* **-coes.** [NLat. *Junco,* genus name < Sp. < Lat. *juncus,* reed.] A North American finch of the genus *Junco,* having primarily slate-colored plumage.

**junc·tion** (jŭngk′shən) *n.* [Lat. *junctio* < *junctus,* p.part. of *jungere,* to join.] **1. a.** The act or process of joining. **b.** The state of being joined. **2.** The place where two things meet, esp. the place where two roads or railway routes join or cross paths. **3.** A transition layer or boundary between two different materials or physically different regions in a single material, esp.: **a.** A connection between conductors or sections of a transmission line. **b.** The interface in a semiconductor between a region of predominantly positive-charge carriers and another of predominantly negative-charge carriers. **c.** A mechanical or alloyed contact between different metals or other materials, as in a thermocouple. **—junc′tion·al** *adj.*

**junction box** *n.* An enclosed panel used to connect or branch electric circuits without making permanent splices.

**junc·ture** (jŭngk′chər) *n.* [ME < Lat. *junctura* < *junctus,* p.part. of *jungere,* to join.] **1.** JUNCTION 1. **2.** The line or point where two things are joined : JOINT. **3.** A point in time, esp. a critical turning point. **4.** The transition or mode of transition from one sound to another in speech.

**June** (jo͞on) *n.* [ME *juin* < OFr. < Lat. *Junius,* after *Juno.*] The sixth month of the year according to the Gregorian calendar. —See table at CALENDAR.

**June beetle** *or* **June bug** *n.* Any of various North American beetles of the subfamily Melolonthinae, having larvae that live in the soil and are often destructive to crops.

**June·ber·ry** (jo͞on′bĕr′ē) *n.* The shadbush.

**June bug** *n.* A June beetle.

**Jung·i·an** (yo͝ong′ē-ən) *adj.* Of, relating to, or typical of Carl G. Jung or his theories of psychology. **—Jung′i·an** *n.*

**jun·gle** (jŭng′gəl) *n.* [Hindi *jaṅgal,* wasteland < Skt. *jāṅgala-,* wild, arid.] **1.** Land densely overgrown with tropical vegetation and trees. **2.** A dense thicket or growth. **3.** *Slang.* A hobo camp. **4.** An environment marked by intense, often ruthless competition or struggle for survival <the corporate *jungle*> **5.** A frustrating maze, entanglement, or confusion that leads nowhere. **—jun′gly** (-glē) *adj.*

**jungle fowl** *n.* A bird of the genus *Gallus* of southeastern Asia, esp. *G. gallus,* considered to be the ancestor of the common domestic fowl.
**jungle gym** *n.* [Orig. a trademark.] A structure of poles and bars on which children can play.
**jun·ior** (jo͞on′yər) *adj.* [Lat., compar. of *juvenis,* young.] **1.** Younger. —Used to distinguish the son from the father of the same name. **2.** Designed for or including young people. **3.** Lower in rank or shorter in length of tenure <the *junior* partner><a *junior* lien> **4.** Designating the third year of a U.S. high school or college. **5.** Lesser in scale than the usual. —*n.* **1.** One younger than another. **2.** One lesser in rank or time of participation or service : SUBORDINATE. **3.** A student in the third year of a U.S. high school or college. **4.** A clothing size for girls and women with slender figures.
**junior college** *n.* An educational institution offering a two-year course of study gen. equivalent to the first two years of a four-year undergraduate course.
**junior high school** *n.* A school in the U.S. system gen. including the seventh, eighth, and sometimes ninth grades.
**jun·ior·i·ty** (jo͞on-yôr′ĭ-tē, -yŏr′-) *n.* The rank or condition of being a junior.
**junior miss** *n.* **1.** A teenage girl. **2.** JUNIOR 4.
**junior varsity** *n.* A high-school or college team that competes in interschool sports on the level below varsity.
**ju·ni·per** (jo͞o′nə-pər) *n.* [ME < Lat. *juniperus.*] An evergreen tree or shrub of the genus *Juniperus,* having scalelike, often prickly foliage, fragrant wood, and bluish-gray berrylike fruit.
**juniper oil** *n.* An essential oil obtained from the fruit of the common juniper, used esp. for flavoring gin and liqueurs.
**junk**[1] (jŭngk) *n.* [ME *jonk,* an old cable or rope.] **1.** Scrapped materials, as glass, rags, paper, or metals, that can be converted into usable stock. **2.** *Informal.* **a.** Worn-out articles fit to be discarded. **b.** Something worthless or shoddy. **c.** Clutter : stuff. **d.** Nonsense. **3.** *Slang.* Heroin. **4.** *Naut.* **a.** Hard salted beef. **b.** Old cordage, reused for gaskets, oakum, and mats. —*vt.* **junked, junk·ing, junks.** To discard as useless : SCRAP.
**junk**[2] (jŭngk) *n.* [Port. *junco* < Javanese *djong.*] A Chinese flat-bottomed ship with a high poop and battened sails.

**junk**[2]
*A Chinese trading junk*

**junk art** *also* **junk sculpture** *n.* Three-dimensional art made from junked materials, as metal, glass, or wood.
**junk·er** (jŭng′kər) *n.* Something, esp. a motor vehicle, so old and dilapidated as to be beyond repair.
**Jun·ker** (yo͝ong′kər) *n.* [G. < OHG *junchērro* : *jung,* young + *hērro,* compar. of *hēr,* worthy.] A member of the Prussian landed aristocracy. **—Jun′ker·dom** *n.*
**jun·ket** (jŭng′kĭt) *n.* [ME *jonket,* a kind of food served on rushes < *jonket,* rush basket, ult. < Lat. *juncus,* rush.] **1.** A sweet custardlike food made from flavored milk set with rennet. **2.** A party, banquet, or outing. **3.** A trip or excursion, esp. one taken by an official and paid for with public funds. —*v.* **-ket·ed, -ket·ing, -kets.** —*vi.* **1.** To hold a party or banquet. **2.** To make an excursion. —*vt.* To fete at a party. **—jun′ket·er, jun′ket·eer′** *n.*
▲ word history: The transitional stage between the meanings "a food" and "a trip" for *junket* is found in the meaning "picnic." The most recent sense, "a trip taken by a public official at public expense," developed in the United States and has acquired decidedly negative connotations.
**junk food** *n.* Food having little nutritional value in relation to it's caloric content.
**junk·ie** *also* **junk·y** (jŭng′kē) *n., pl.* **-ies.** *Slang.* **1.** A narcotics addict, esp. one using heroin. **2.** One who has a consuming interest in or need for something <a soap opera *junkie*>
**junk mail** *n.* Third-class mail, as advertisements, mailed indiscriminately in large quantities.
**junk sculpture** *n. var. of* JUNK ART.
**junk·y** (jŭng′kē) *n. var. of* JUNKIE.
**junk·yard** (jŭngk′yärd′) *n.* A yard or lot used to store junk, as scrap metal, that can be resold.
**Ju·no** (jo͞o′nō) *n.* [Lat.] *Rom. Myth.* The principal goddess of the pantheon, wife and sister of Jupiter, and patroness of marriage and the well-being of women.
**Ju·no·esque** (jo͞o′nō-ĕsk′) *adj.* Having the stately bearing and regal beauty of the goddess Juno.
**jun·ta** (ho͝on′tə, jŭn′-) *n.* [Sp. and Port., conference < Lat. *juncta,* p.part. of *jungere,* to join.] **1.** A group of military officers holding state power in a country after a coup d'état. **2.** A council or small legislative body in a government, esp. in Central and South American countries. **3.** A junto.
**jun·to** (jŭn′tō) *n., pl.* **-tos.** [Var. of JUNTA.] A small, usu. secret group of persons united for a common purpose.
**Ju·pi·ter** (jo͞o′pĭ-tər) *n.* [Lat.] **1.** *Rom. Myth.* The supreme god of the pantheon, husband and brother of Juno, and patron of the Roman state. **2.** *Astron.* The fifth planet from the sun, largest in the solar system, having a diameter of approx. 88,000 miles or 142,000 kilometers, a mass approx. 318 times that of Earth, and a sidereal period of revolution about the sun of 11.86 years at a mean distance of 483 million miles or 777 million kilometers.
**ju·ral** (jo͝or′əl) *adj.* [< Lat. *jus, jur-,* law.] **1.** Of or relating to law. **2.** Of or relating to rights and obligations. **—ju′ral·ly** *adv.*
**Ju·ras·sic** (jo͞o-răs′ĭk) *adj.* [Fr. *jurassique,* after the *Jura* Mountains.] Of, belonging to, or designating the second period of the Mesozoic era, marked by the presence of dinosaurs and the appearance of primitive mammals and birds. —*n.* The Jurassic period.
**ju·rat** (jo͝or′ăt′) *n.* [< Lat. *juratum,* p.part. of *jurare,* to swear < *jus,* law.] A certification on an affidavit declaring when, where, and before whom it was sworn.
**ju·rid·i·cal** (jo͞o-rĭd′ĭ-kəl) *also* **ju·rid·ic** (-ĭk) *adj.* [Lat. *juridicus* : *jus,* law + *dicere,* to say.] Of or relating to the law and its administration. **—ju·rid′i·cal·ly** *adv.*
**ju·ris·con·sult** (jo͝or′ĭs-kŏn′sŭlt′) *n.* [Lat. *jurisconsultus* : *jus,* law + *consultus,* skilled, p.part. of *consulere,* to take counsel.] A person learned esp. in international and public law : JURIST.
**ju·ris·dic·tion** (jo͝or′ĭs-dĭk′shən) *n.* [ME *jurisdiccioun* < Med. Lat. *jurisdictio* < Lat. : *jus,* law + *dictio,* declaration. —see DICTION.] **1.** The right or power to interpret and apply the law. **2. a.** Authority or control. **b.** The sphere of authority or control. **3.** The territorial range over which any authority extends. **—ju′ris·dic′tion·al** *adj.* **—ju′ris·dic′tion·al·ly** *adv.*
**ju·ris doctor** (jo͝or′ĭs) *n.* [NLat. < Lat., doctor of law.] An academic degree equivalent to bachelor of laws.
**ju·ris·pru·dence** (jo͝or′ĭs-pro͞od′ns) *n.* [LLat. *jurisprudentia* : *jus,* law + *prudentia,* knowledge < *prudens,* knowing. —see PRUDENT.] **1.** The science or philosophy of law. **2.** A division or department of law. **—ju′ris·pru·den′tial** (-pro͞o-dĕn′shəl) *adj.* **—ju′ris·pru·den′tial·ly** *adv.*
**ju·ris·pru·dent** (jo͝or′ĭs-pro͞od′nt) *adj.* Versed in the law. —*n.* A jurist.
**ju·rist** (jo͝or′ĭst) *n.* [OFr. *juriste* < Med. Lat. *jurista* < Lat. *jus,* law.] One who is skilled in the law, esp. an eminent judge, lawyer, or legal scholar.
**ju·ris·tic** (jo͞o-rĭs′tĭk) *also* **ju·ris·ti·cal** (-tĭ-kəl) *adj.* **1.** Of or relating to a jurist or to jurisprudence. **2.** Of or relating to law or legality. **—ju·ris′ti·cal·ly** *adv.*
**ju·ror** (jo͝or′ər, -ôr′) *n.* [ME *jurour* < AN < Lat. *jurator,* swearer < *jurare,* to swear.] **1. a.** One who serves on a jury. **b.** One called or designated for jury duty. **2.** A member of any body of persons acting in the capacity of a jury.
**ju·ry**[1] (jo͝or′ē) *n., pl.* **-ries.** [ME *jure* < AN, ult. < Lat. *jurare,* to swear < *jus,* law.] **1.** A body of persons sworn to judge and give a verdict on a given matter, esp. a body of persons summoned by law and sworn to hear a case in court and render a verdict. **2.** A group of persons forming a committee to judge, as at a competition.
**ju·ry**[2] (jo͝or′ē) *adj.* [Orig. unknown.] *Naut.* Intended or designed for temporary use : MAKESHIFT <a *jury* mast>
**ju·ry-rigged** (jo͝or′ē-rĭgd′) *adj.* Constructed or designed in a makeshift manner.
**jus gen·ti·um** (yo͞os gĕn′tē-əm) *n.* [Lat.] The law of nations.
**jus·sive** (jŭs′ĭv) *n.* [< Lat. *jussus,* p.part. of *jubēre,* to command.] A word, mood, or form used to express command. **—jus′sive** *adj.*
**just**[1] (jŭst) *adj.* [ME *juste* < Lat. *justus.*] **1.** Honorable and fair in dealings and actions <a *just* king> **2.** Consistent with moral right : RIGHTEOUS <a *just* cause> **3.** Properly merited <*just* deserts> **4.** Valid within the law : LEGITIMATE <a *just* claim> **5.** Proper or suitable : FITTING <a *just* note of compassion> **6.** Based on fact or sound reason : WELL-FOUNDED <a *just* criticism> —*adv.* (jəst, jĭst; jŭst *when stressed*). **1.** Precisely : exactly <*just* right> **2.** At the exact moment of <It's *just* midnight.> **3.** Only a moment ago <*just* left> **4.** By a narrow margin : BARELY <You *just* missed the bus.> **5.** At a short distance <*just* down the street> **6.** Only : merely <I *just* said it in jest.> **7.** Simply : certainly <*just* delightful> **8.** Possibly : perhaps <We *just* might win.> **—just about.** Very nearly : ALMOST <I've *just about* finished.> **—just the same.** Nevertheless. **—just′ly** *adv.* **—just′ness** *n.*
☆ **syns:** JUST, DUE, RIGHT *adj. core meaning* : consistent with accepted standards and circumstances <*just* punishment>
**just**[2] (jŭst) *n. & v. var. of* JOUST.
**jus·tice** (jŭs′tĭs) *n.* [ME < OFr. < Lat. *justitia* < *justus,* just.]

1. a. The principle or ideal of moral rightness : EQUITY. **b.** Conformity to moral rightness in conduct or attitude : RIGHTEOUSNESS. **2.** The upholding of what is right and lawful, esp. fair treatment or punishment in accordance with honor, standards, or law : FAIRNESS. **3.** The quality of being fair or impartial. **4.** Sound reason <We're furious, and with *justice.*> **5.** The administration and procedure of law. **6. a.** A judge. **b.** A justice of the peace.

**justice of the peace** *n.* A magistrate of the lowest level of certain state court systems, empowered to act upon minor offenses, commit cases to a higher court for trial, perform marriages, and administer oaths.

**jus·ti·ci·a·ble** (jŭ-stĭsh'ə-bəl) *adj.* [OFr. < *justicier,* to try < Med. Lat. *justitiare* < Lat. *justitia,* justice < *justus,* just.] **1.** Appropriate for or subject to court trial. **2.** Liable for court decision.

**jus·ti·ci·ar·y** (jŭ-stĭsh'ē-ĕr'ē) *also* **jus·ti·ci·ar** (jŭ-stĭsh'ē-ər) *n., pl.* **-ies** *also* **-ars.** [Med. Lat. *justitiarius* < Lat. *justitia,* justice < *justus,* law.] A high judicial officer in medieval England.

**jus·ti·fi·a·ble** (jŭs'tə-fī'ə-bəl, jŭs'tə-fī'-) *adj.* That can be justified. **—jus'ti·fi'a·bil'i·ty, jus'ti·fi'a·ble·ness** *n.* **—jus'ti·fi'a·bly** *adv.*

**jus·ti·fi·ca·tion** (jŭs'tə-fĭ-kā'shən) *n.* **1.** The act of justifying or state of being justified. **2.** Something that justifies.

**jus·ti·fi·ca·tive** (jŭs'tə-fĭ-kā'tĭv) *also* **jus·tif·i·ca·to·ry** (jŭ-stĭf'ĭ-kə-tôr'ē, -tōr'ē) *adj.* Serving to justify.

**jus·ti·fy** (jŭs'tə-fī') *v.* **-fied, -fy·ing, -fies.** [ME *justifien* < OFr. *justifier,* to administer justice < LLat. *justificare,* to act justly toward : Lat. *justus,* just + Lat. *facere,* to do.] *—vt.* **1.** To show or prove to be just or valid. **2.** To declare free of blame : ABSOLVE. **3.** To free (a person) of the penalty attached to grievous sin. —Used only of God. **4.** *Law.* **a.** To show good reason for (an action taken). **b.** To prove to be qualified as a bondsman. **5.** To adjust or space (lines) to the proper length in printing. *—vi.* To be or become properly spaced and of the correct length. —Used of a line of type.

**jut** (jŭt) *vi.* **jut·ted, jut·ting, juts.** [Var. of JET[2].] To project beyond the limits of the main body : PROTRUDE <"He had a sharp crooked nose *jutting* out of a lean dancer's face" —Graham Greene> *—n.* Something that protrudes : PROJECTION.

**jute** (jo͞ot) *n.* [Bengali *jhuṭo* < Skt. *jūṭaḥ,* twisted hair.] **1.** Either of two Asian plants, *Corchorus capsularis* or *C. olitorius,* yielding a strong, coarse fiber used for sacking and cordage. **2.** The fiber obtained from the jute plant.

**Jute** (jo͞ot) *n.* [< ME *Iutes,* the Jutes < OE *Iotas.*] A member of a Germanic tribe that invaded Britain and settled in Kent in the 5th cent. A.D.

**ju·ve·nes·cent** (jo͞o'və-nĕs'ənt) *adj.* [Lat. *juvenescens, juvenescent-,* pr.part. of *juvenescere,* to reach the age of youth < *juvenis,* young.] Growing young or youthful. **—ju've·nes'cence** *n.*

**ju·ve·nile** (jo͞o'və-nīl', -nəl) *adj.* [Lat. *juvenilis* < *juvenis,* young.] **1.** Not fully developed : not yet adult. **2.** Typical of youth or children : IMMATURE <*juvenile* behavior> **3.** Intended or suitable for children or young persons <*juvenile* books> *—n.* **1. a.** A young person : CHILD. **b.** A young animal that has not reached sexual maturity. **2.** An actor who plays youthful roles. **3.** A children's book. **—ju've·nile'ly** *adv.* **—ju've·nile'ness** *n.*

**juvenile court** *n.* A court with jurisdiction over all cases involving children under a specified age, usu. 18 years.

**juvenile delinquency** *n.* **1.** Behavior by a juvenile that is so antisocial or criminal as to be beyond parental control and thus subject to legal action. **2.** A violation of the law committed by a juvenile and not punishable by life imprisonment or the death penalty. **—juvenile delinquent** *n.*

**juvenile hormone** *n.* A hormone that controls metamorphosis, secreted by the corpus allatum of arthropod larvae.

**ju·ve·nil·i·a** (jo͞o'və-nĭl'ē-ə) *pl.n.* [Lat., neuter pl. of *juvenilis,* juvenile < *juvenis,* young.] **1.** Written or artistic works produced in childhood or youth. **2.** Literary or artistic works for children.

**ju·ve·nil·i·ty** (jo͞o'və-nĭl'ĭ-tē) *n., pl.* **-ties. 1.** The quality or state of being foolishly immature. **2.** The quality or state of being young : YOUTHFULNESS. **3.** An instance of childishness. **4.** Young people.

**jux·ta·pose** (jŭk'stə-pōz') *vt.* **-posed, -pos·ing, -pos·es.** [Fr. *juxtaposer* : Lat. *juxta,* close together + Fr. *poser,* to place. —see POSE.] To place side by side, esp. for contrast or comparison. **—jux'ta·po·si'tion** (-pə-zĭsh'ən) *n.* **—jux'ta·po·si'tion·al** *adj.*

# Kk

**k** or **K** (kā) *n., pl.* **k's** or **K's. 1.** The 11th letter of the English alphabet. **2.** A speech sound represented by the letter *k.* **3.** The 11th in a series. **4. K** *Informal.* Thousand <a job that pays $40*K*> **5.** *Computer Sci.* A unit of storage capacity equal to $10^{24}$ bytes.

**K** [NLat. *kalium.*] *symbol for* POTASSIUM.

**Kaa·ba** (kä'bə) *n.* [Ar. *ka'bah* < *ka'b,* cube.] A Moslem shrine in Mecca, the goal of pilgrims, toward which Moslems turn to pray.

**kab** (kăb) *n. var. of* CAB[2].

**kab·a·la** or **kab·ba·la** (kăb'ə-lə, kə-bä'lə) *n. var. of* CABALA.

**ka·bob** (kə-bŏb') *n.* Shish kebab.

**ka·bu·ki** (kə-bo͞o'kē) *n.* [J., art of singing and dancing : *kabu,* singing and dancing (< Chin. *ge*[1] *wu*[3]) + *ki,* art, artist (< Chin. *ji*[4]).] A Japanese popular drama in which dances, gestures, and songs are performed in a formal and stylized way.

**Ka·byle** (kə-bīl') *n., pl.* **Kabyle** or **-byles.** [Fr. < Ar. *qabā'il,* pl. of *qabīlah,* tribe.] **1.** A Tunisian or Algerian Berber. **2.** The Berber language of the Kabyle.

**ka·chi·na** (kə-chē'nə) *n.* [Hopi *qacina,* supernatural.] A doll that symbolizes one of the rain-bringing ancestral spirits of the Hopi.

**Kad·dish** (kä'dĭsh) *n.* [Aram. *qaddīsh,* holy.] A prayer recited in daily synagogue services and by mourners after the death of a close relative.

**kaf·fee klatsch** (kô'fē klăch', kä'fē klăch') *n. var. of* COFFEE KLATCH.

**Kaf·fir** *also* **Kaf·ir** (kăf'ər) *n., pl.* **Kaffir** or **-firs** *also* **Kafir** or **-firs.** [Ar. *kāfir,* infidel.] **1. a.** A member of a Bantu-speaking tribe of South Africa : XHOSA. **b.** XHOSA 2. **2.** A non-Moslem. **3. kaffir** or **kafir.** A sorghum, *Sorghum vulgare caffrorum,* grown in dry regions as a source of grain and fodder.

**kaf·fi·yeh** (kä-fē'ə, kă-) *n.* [Ar. *kaffīyah.*] A cloth headdress fastened by a band around the crown and usu. worn by Arab men.

**Kaf·ir** (kăf'ər) *n., pl.* **Kafir** or **-firs.** [Ar. *kāfir,* infidel.] **1. a.** An Iranian people of northeastern Afghanistan. **b.** A member of these people. **2.** *var. of* KAFFIR.

**Kaf·i·ri** (kăf'ə-rē) *n.* The Indic language of the Iranian Kafir.

**Kaf·ka·esque** (käf'kə-ĕsk', kăf'-) *adj.* **1.** Of or typical of Franz Kafka or his writings. **2.** Marked by surreal distortion and usu. by a sense of impending danger <a *Kafkaesque* painting>

**kaf·tan** (kăf'tăn', kăf-tăn') *n. var. of* CAFTAN.

**kain** (kān) *n.* [ME *cain* < Sc. Gael. *cāin.*] Tax or rent payments made in kind.

**kai·nite** (kī'nīt', kā'-) *n.* [G. *Kainit* : Gk. *kainos,* new + *-it,* -ite.] A mineral, $KCL \cdot MgSo_4 \cdot 3H_2O$, used as fertilizer and as a source of potassium compounds.

**Kai·ser** (kī'zər) *n.* [ME < OE *cāsere,* ult. < Lat. *Caesar,* Caesar. —see CAESAR.] An emperor, esp. a German emperor from 1871 to 1918.

**Kai·ser·in** (kī'zər-ĭn) *n.* [G., fem. of *Kaiser,* Kaiser < OHG *Keisur* < Lat. *Caesar,* Caesar. —see CAESAR.] The wife of a Kaiser.

**ka·ka** (kä'kə) *n.* [Maori.] A brownish-green New Zealand parrot, *Nestor meridionalis.*

**ka·ka·po** (kä'kə-pō') *n., pl.* **-pos.** [Maori : *kaka,* parrot + *po,* night.] A ground-dwelling New Zealand parrot, *Strigops habroptilus,* having greenish plumage.

**ka·ke·mo·no** (kä'kə-mō'nō) *n., pl.* **-nos.** [J. : *kake,* hanging + *mono,* object.] A vertical Japanese scroll painting.

**kale** (kāl) *n.* [ME < OE *cāl,* cole < Lat. *caulis,* cabbage.] **1.** A cabbage, *Brassica oleracea acephala,* with crinkled leaves not forming a tight head. **2.** *Slang.* Money.

▲ word history: The word *kale* exhibits one of the most important features distinguishing the Scottish dialect from standard English, which is the use of long *a* where the standard language has long *o.* With regard to this feature Scots is more conservative. The Old English form of *kale* is *cāl.* In the 12th and 13th centuries most dialects of Middle English had changed all Old English long *a*'s to long *o*'s, but the northern dialects, of which Scots is the most important modern representative, did not. The word *kale* has a doublet *cole,* "cabbage," which represents the standard development of Old English *cāl.* Other pairs of words that exhibit the two different developments of Old English long *a* are *kame/comb, laird/lord,* and *hale/whole.*

**ka·lei·do·scope** (kə-lī'də-skōp') *n.* [Gk. *kalos,* beautiful + *eidos,* form + -SCOPE.] **1.** A tube-shaped optical instrument rotated to produce successive symmetric designs by means of mirrors reflecting the continuously changing patterns made by bits of colored glass at an end of the tube. **2.** A continuously changing set of colors. **3.** A series of changing phases or events. **—ka·lei′do·scop′ic** (-skŏp′ĭk), **ka·lei′do·scop′i·cal** *adj.* **—ka·lei′do·scop′i·cal·ly** *adv.*

**kal·ends** (kăl′əndz, kā′ləndz) *n. var. of* CALENDS.

**ka·lim·ba** (kə-lĭm′bə) *n.* [Of African orig.] An African musical instrument shaped like a wooden box and set with metal bars plucked with the fingers.

**Kal·muck** *also* **Kal·muk** (kăl′mŭk′, kăl-mŭk′) *or* **Kal·myk** (kăl′mĭk, kăl-mĭk′) *n.* [R. *Kalmyk.*] **1.** A member of a Buddhist Mongol people orig. of northwestern China. **2.** The Mongolian language of the Kalmucks.

**kal·so·mine** (kăl′sə-mīn′) *n. var. of* CALCIMINE.

**Ka·ma** (kä′mə) *n.* [Skt. *kāmaḥ.*] The Hindu god of love.

**ka·ma·la** (kä′mə-lə, kăm′ə-) *n.* [Skt. *kamalam,* lotus, prob. of Dravidian orig.] **1.** An Asian tree, *Mallotus philippinensis,* yielding a hairy, capsular fruit. **2.** A vermifugal powder obtained from the capsules of the kamala.

**Ka·ma·su·tra** (kä′mə-sōō′trə) *n.* [Skt. *kāmasūtram* : *kāmaḥ,* love + *sūtram,* manual.] A Sanskrit treatise explicating rules for love and marriage in accord with Hindu law.

**kame** (kām) *n.* [Sc. < ME *camb,* comb < OE.] A short ridge or mound of sand and gravel deposited during the melting of glacial ice.

**ka·mi·ka·ze** (kä′mĭ-kä′zē) *n.* [J. : *kami,* god + *kaze,* wind.] **1.** A Japanese pilot trained to make a suicidal crash attack in World War II. **2.** An aircraft loaded with explosives to be piloted in a suicide attack. *—adj.* Suicidal <*kamikaze* skiers>

**ka·na** (kä′nə) *n.* [J., pseudo-characters : *ka,* false (< Chin. *jia*³) + *na,* name (< Chin. *ming*²).] **1.** Hiragana. **2.** Katakana.

**kan·a·my·cin** (kăn′ə-mī′sĭn) *n.* [NLat. *kanamyceticus* (specific epithet of a species of actinomycete) + -IN.] A water-soluble broad-spectrum antibiotic, $C_{18}H_{36}O_{11}N_4$, from a soil actinomycete.

**Kan·a·rese** (kăn′ə-rēz′, -rēs′) *n., pl.* **Kanarese.** [After *Kanara,* an area of southwest India.] **1.** One of the Kannada-speaking peoples of Mysore, India. **2.** Kannada. **—Kan′ar·ese′** *adj.*

**kan·ga·roo** (kăng′gə-rōō′) *n.* [Prob. < a native word in Australia.] Any of various herbivorous marsupials of the family Macropodidae, of Australia and adjacent areas, having short forelegs, large hind legs adapted for leaping, and a long tapered tail.

**kangaroo court** *n.* **1.** A mock court set up in violation of established legal procedure. **2.** A dishonest or incompetent court.

**kangaroo rat** *n.* A long-tailed rodent of the genera *Dipodomys* or *Microdipodops* of arid areas of western North America, having long hind legs adapted for jumping.

**kangaroo vine** *n.* A native Australian climbing or trailing vine, *Cissus antarctica,* grown as a house plant for its glossy foliage.

**Kan·na·da** (kä′nə-də) *n.* The principal Dravidian language of Mysore, a state in southern India.

**ka·o·lin** *also* **ka·o·line** (kā′ə-lĭn) *n.* [Fr. < Chin. (Mandarin) *gao*¹ *ling*³, an area of Jiangxi Province where it was first obtained.] A fine clay used in ceramics and refractories and as a filler or coating for paper and textiles.

**ka·o·lin·ite** (kā′ə-lĭ-nīt′) *n.* A mineral, $Al_2O_3{\cdot}2SiO_2{\cdot}2H_2O$, the essential constituent of kaolin. **—ka′o·lin·i′tic** *adj.*

**ka·on** (kā′ŏn′) *n.* [*Ka,* pronunciation of the letter *K* + -ON¹.] An unstable meson produced in either an electrically charged form or a neutral form as the result of a high-energy particle collision.

**Ka·pell·meis·ter** (kə-pĕl′mī′stər, kä-) *n., pl.* **Kapellmeister.** [G. : *kapell,* choir + *Meister,* master.] A choir or orchestra conductor.

**kaph** (käf, kôf) *n.* [Heb. *kāph.*] The 11th letter of the Hebrew alphabet. —See table at ALPHABET.

**ka·pok** (kā′pŏk′) *n.* [Malay.] A silky fiber obtained from the fruit of the silk-cotton tree and used for insulation and as padding in mattresses, pillows, and life preservers.

**kap·pa** (kăp′ə) *n.* [Gk., of Phoenician orig., akin to Heb. *kāph.*] The tenth letter of the Greek alphabet. —See table at ALPHABET.

**ka·put** *also* **ka·putt** (kä-pŏŏt′, -pōōt′, kə-) *adj.* [G. *kaputt* < Fr. *capot,* not having won a trick at piquet.] *Informal.* **1.** Destroyed : wrecked. **2.** Incapacitated. **3.** Out of order.

**kar·a·bi·ner** (kăr′ə-bē′nər) *n.* [G.] An oblong steel ring snapped to the eye of a piton and through which a rope is run, used in mountaineering.

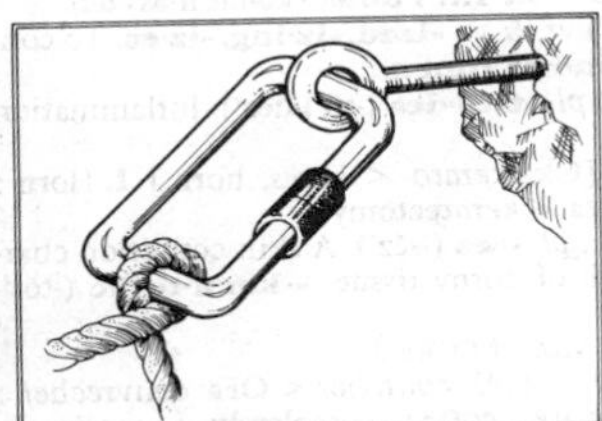

karabiner

**kar·a·kul** (kăr′ə-kəl) *n.* [After *Karakul,* a lake in Central Asian USSR.] **1.** One of a breed of Central Asian sheep with a pelt curled and glossy in the young but wiry and coarse in the adult. **2.** Fur made from the pelt of a karakul lamb.

**kar·at** (kăr′ət) *n.* [OFr. *carat,* unit of weight for gemstones. —see CARAT.] A unit of measure for the fineness of gold, equal to 1/24 of the total amount of pure gold in an alloy.

**ka·ra·te** (kə-rä′tē) *n.* [J. : *kara,* empty + *te,* hand.] A Japanese art of self-defense in which sharp blows and kicks are administered to pressure-sensitive points on an opponent's body. **—ka·ra′te·ist** *n.*

**kar·ma** (kär′mə, kûr′-) *n.* [Skt. *karman,* deed < *karoti,* he does.] **1.** The total effect of a person's actions and conduct during the successive phases of existence, held to determine destiny in Hinduism and Buddhism. **2.** Fate : destiny. **3.** *Informal.* A distinctive aura, atmosphere, or feeling <There's bad *karma* in the office today.> **—kar′mic** (-mĭk) *adj.*

**ka·roo** *also* **kar·roo** (kə-rōō′) *n.* [Afr. *karo,* of Hottentot orig.] An arid southern African plateau.

**karst** (kärst) *n.* [G.] An area of irregular limestone in which erosion has produced fissures, sinkholes, underground streams, and caverns. **—karst′ic** *adj.*

**kart** (kärt) *n.* [Prob. < *GoKart,* a trademark.] A small racing car.

**kart·ing** (kär′tĭng) *n.* The sport of racing miniature cars.

**karyo-** *also* **caryo-** *pref.* [NLat. < Gk. *karuon,* nut.] **1.** Cell nucleus <*karyo*gamy> **2.** Nut : kernel <*caryo*psis>

**kar·y·og·a·my** (kăr′ē-ŏg′ə-mē) *n.* The fusing of gamete nuclei.

**kar·y·o·ki·ne·sis** (kăr′ē-ō-kə-nē′sĭs) *n.* Mitosis.

**kar·y·o·lymph** (kăr′ē-ə-lĭmf′) *n.* The clear homogeneous liquid portion of nuclear protoplasm.

**kar·y·o·plasm** (kăr′ē-ə-plăz′əm) *n.* Nucleoplasm. **—kar′y·o·plas′mic** (-mĭk) *adj.*

**kar·y·o·some** (kăr′ē-ə-sōm′) *n.* A spherical aggregation of chromatin in a resting nucleus during mitosis.

**kar·y·o·type** (kăr′ē-ə-tīp′) *n.* **1.** The chromosomal complement of an individual or species. **2.** A photomicrograph of metaphase chromosomes in a standard array. **—kar′y·o·typ′ic** (-tĭp′ĭk), **kar′y·o·typ′i·cal** *adj.*

**ka·sha** (kä′shə) *n.* [R.] Buckwheat groats.

**Kash·mir goat** (kăzh′mîr′, kăsh′-) *n. var. of* CASHMERE GOAT.

**Kash·mir·i** (kăsh-mîr′ē, kăzh-) *n.* An Indic language of Jammu and Kashmir.

**ka·ta·ka·na** (kä′tä-kä′nä) *n.* [J. : *kata,* one + *kana,* kana.] A phonetic Japanese syllabary used for writing foreign words or documents.

**Ka·tha·rev·u·sa** (kä′thə-rĕv′ə-sä′) *n.* [Mod. Gk. *kathareuousa* < Gk. < *kathareuein,* to be pure < *katharos,* pure.] The official form of Modern Greek, containing morphological and lexical features borrowed from classical Greek.

**ka·ty·did** (kā′tē-dĭd′) *n.* [Imit. of its sound.] Any of various green insects related to the grasshoppers and the crickets, with specialized organs on the wings of the male that produce a shrill sound when rubbed together.

**katz·en·jam·mer** (kăt′sən-jăm′ər) *n.* [G. : *Katzen,* cats + *Jammer,* misery.] **1.** A loud discordant noise. **2.** A hangover. **3.** A state of bewilderment or depression.

**kau·ri** (kou′rē) *n.* [Maori *kawri.*] **1. a.** A coniferous tree of the genus *Agathis,* esp. *A. australis* of New Zealand. **b.** The white close-grained wood of this tree. **2.** A resin obtained from the kauri or deposits of its fossilized resin and used in enamels and varnishes.

**ka·va** (kä′və) *n.* [Tongan, kava, bitter.] **1.** An Australasian shrub, *Piper methysticum,* the roots of which are used to make an intoxicating beverage. **2.** The beverage made from kava.

**kay** (kā) *n.* The letter *k.*

**Kay** (kā) *n.* The foster brother and steward of King Arthur.

**kay·ak** (kī′ăk′) *n.* [Eskimo *qajaq.*] **1.** A watertight Eskimo canoe made of a light wooden frame covered with skins except for an opening in the center. **2.** A lightweight canoe similar to a kayak.

**kay·o** (kā-ō′, kā′ō′) [Pronunciation of *K.O.,* abbreviation of *knock out.*] *Slang.* *—n., pl.* **-os.** A knockout in boxing. *—vt.* **-oed, -o·ing, -os.** To knock out in boxing.

**ka·zoo** (kə-zōō′) *n., pl.* **-zoos.** [Imit. of its sound.] A toy musical instrument with a membrane that produces a sound when a player hums or sings into the mouthpiece.

**ke·a** (kē′ə) *n.* [Maori.] A large brownish-green New Zealand parrot, *Nestor notabilis,* that usu. eats insects but occas. kills sheep by slashing them and eating their flesh.

**ke·bab** *also* **ke·bob** (kə-bŏb′) *n.* Shish kebab.

**Kech·ua** (kĕch′wə, -wä′) *n. var. of* QUECHUA.

**kedge** (kĕj) [< *kedge,* to warp a vessel, poss. < ME *caggen,* to tie.] *Naut.* *—n.* A light anchor used for warping a vessel. *—v.* **kedged, kedg·ing, kedg·es.** *—vt.* To warp (a vessel) by means of a kedge. *—vi.* To move by means of a kedge.

**keel¹** (kēl) *n.* [ME *kele* < ON *kjölr.*] **1. a.** The main structural member of a ship, running lengthwise along the center line from bow to stern, to which the frames are attached. **b.** A corresponding structure on an aircraft. **2.** A ship. **3.** A structure, as the breastbone of a bird, that resembles a ship's keel. **4.** A pair of united petals in certain flowers, as those of the pea. *—vt. & vi.* **keeled, keel·ing, keels.** To capsize. **—keel over.** To collapse or fall in or as if in a faint.

**keel[2]** (kēl) *n.* [ME *kele* < MDu. *kiel.*] **1. a.** A freight barge, esp. one for carrying coal on the Tyne in England. **b.** The amount of coal this barge can hold. **2.** A British unit of weight once used for coal, equal to 21.2 long tons.

**†keel[3]** (kēl) *vt.* **keeled, keel·ing, keels.** [ME *kelen* < OE *cēlan,* to become cool.] *Regional.* To make cool.

**keel·boat** (kēl'bōt') *n.* A riverboat with a keel but without sails, used for carrying freight.

**keel·haul** (kēl'hôl') *vt.* **-hauled, -haul·ing, -hauls.** [Du. *kielhalen* : *kiel,* keel of a ship + *halen,* to haul.] **1.** To punish by dragging under the keel of a ship. **2.** To rebuke harshly.

**keel·son** (kēl'sən, kēl'-) *n.* [Poss. < LG *kielswin.*] *Naut.* A timber or girder fastened above and parallel to the keel for additional strength.

**keen[1]** (kēn) *adj.* **-er, -est.** [ME *kene* < OE *cēne,* brave.] **1.** Having a fine, sharp cutting edge or point. **2.** Intellectually acute. **3.** Highly sensitive. **4.** Sharp : vivid <"His entire body hungered for *keen* sensation, something exciting" —Richard Wright> **5.** Intense : piercing <a *keen* north wind> **6.** Pungent : acrid. **7. a.** Ardent : enthusiastic. **b.** Eagerly desirous <*keen* on going to the party> **8.** *Slang.* Great : splendid. **—keen'ly** *adv.* **—keen'ness** *n.*

**keen[2]** (kēn) *n.* [Ir. Gael. *caoine* < *caoninim,* I lament.] A loud, wailing lament for the dead. *—vi.* **keened, keen·ing, keens.** To wail or lament loudly. **—keen'er** *n.*

**keep** (kēp) *v.* **kept** (kĕpt), **keep·ing, keeps.** [ME *kepen* < OE *cēpan,* to observe.] *—vt.* **1.** To retain possession of. **2.** To have as a supply. **3.** To provide with maintenance and support <*kept* a large family on a small income> **4.** To put customarily <Where do you *keep* your forks?> **5. a.** To supply with room and board for a fee <*keep* student boarders> **b.** To raise <*keep* hens> **6.** To maintain for use or service <doesn't *keep* a car> **7.** To manage, tend, or have charge of <*keep* shop> **8.** To preserve (food). **9.** To cause to continue in a given state or course of action <tried to *keep* the injured person calm> **10. a.** To maintain records in <*keep* a yearly diary> **b.** To enter (data) in a book <*keep* sales records> **11. a.** To detain <was *kept* after class> **b.** To restrain <*kept* the dog away from the mail carrier> **c.** To refrain from divulging <*keep* secrets> **d.** To save in reserve <*keep* extra cash on hand> **12.** To maintain <*keep* late hours> **13.** To adhere to : FULFILL <*keep* a bargain> **14.** To celebrate : observe. *—vi.* **1.** To remain in a given state : STAY <*keep* in line><*keep* still> **2.** To continue to do <*keep* on arguing> **3.** To remain fresh or unspoiled <The fruit won't *keep.*> **—keep down.** To prevent from accomplishing or succeeding. **—keep off.** To stay away from. **—keep up. 1.** To maintain in good condition. **2.** To persevere in : CARRY ON. **3.** To continue at the same pace or level. **4.** To match one's competitors, colleagues, or neighbors in success or lifestyle <successfully *kept up* with my associates> *—n.* **1.** Care : charge <The kids are in my *keep* for the day.> **2.** The means by which one is supported <earn one's *keep*> **3. a.** A castle's stronghold. **b.** A jail. **—for keeps. 1.** For an indefinitely long time <gave it to me *for keeps*> **2.** Seriously and permanently <We're separating *for keeps.*> **—keep at it.** To persevere. **—keep (one's) eyes open (or peeled).** To be on the lookout. **—keep (one's) nose clean.** To stay out of trouble. **—keep pace.** To stay even. **—keep to.** To adhere to <*keep to* the original plan> **—keep to (oneself). 1.** To shun the company of others. **2.** To refrain from divulging <*kept* the secret *to* ourselves>

☆ ***syns:*** KEEP, RESERVE, RETAIN, WITHHOLD *v. core meaning* : to have and maintain in one's possession <*keeps* 40% of the profits> ***ant:*** relinquish

**keep·er** (kē'pər) *n.* **1.** One that keeps, esp.: **a.** An attendant, guard, or warden. **b.** One who has the charge or care of something. **2.** A device for keeping something in place.

**keep·ing** (kē'pĭng) *n.* **1.** The act of holding, guarding, or supporting. **2.** Custody : care. **3.** Harmony : conformity.

**keep·sake** (kēp'sāk') *n.* A memento.

**kees·hond** (kās'hônt') *n., pl.* **-hon·den** (-hôn'dən) *or* **-honds.** [Du., prob. : *kees,* nickname for *Cornelis,* Cornelius + *hond,* dog < MDu.] A dog orig. bred in the Netherlands, having a thick grayish-black coat.

**kef** (kĕf, kēf, käf) *n. var. of* KIF.

**ke·fir** (kĕ-fîr') *n.* [R., of Caucasian orig.] A creamy drink of fermented cow's milk.

**keg** (kĕg) *n.* [ME *kag* < ON *kaggi.*] A cask.

**keg·ler** (kĕg'lər) *n.* [G. < *kegeln,* to bowl < *Kegel,* bowling pin < OHG *kegil,* peg.] One who bowls : BOWLER.

**kel·ly green** (kĕl'ē) *n.* [< the name *Kelly.*] A strong yellowish green.

**ke·loid** (kē'loid') *n.* [Fr. *kéloïde* : Gk. *khēlē,* claw + *-oïde,* -oid.] Tissue scarring resulting from trauma or surgical incision. **—ke·loid'al** *adj.*

**kelp** (kĕlp) *n.* [ME *culp.*] **1.** Any of various brown, often very large seaweeds of the order Laminariales. **2.** The ash of kelp, used as a source of potash and iodine.

**kel·pie[1]** *also* **kel·py** (kĕl'pē) *n., pl.* **-pies.** [Prob. of Celt. orig.] A legendary malevolent Scottish water spirit.

**kel·pie[2]** (kĕl'pē) *n.* [< *Kelpie,* the name of an early specimen of the breed.] A sheep dog orig. bred in Australia.

**kel·py** (kĕl'pē) *n. var. of* KELPIE[1].

**Kelt** (kĕlt) *n. var. of* CELT.

**Kelt·ic** (kĕl'tĭk) *n. & adj. var. of* CELTIC.

**kel·vin** (kĕl'vĭn) *n.* [After William Thompson (1824–1907), first Baron *Kelvin.*] The unit of thermodynamic temperature equal to 1/273.16 of the thermodynamic temperature of the triple point of water.

**Kel·vin** (kĕl'vĭn) *adj.* Of or relating to an absolute scale of temperature whose zero point is approx. −273.16°C.

**kempt** (kĕmpt) *adj.* [Prob. back-formation < UNKEMPT.] Tidy : trim <a nicely *kempt* lawn>

**†ken** (kĕn) *v.* **kenned** *or* **kent** (kĕnt), **ken·ning, kens.** [ME *kennen* < OE *cennan,* to declare.] *—vt.* **1.** *Scot.* To know (a person or thing). **2.** *Regional.* To recognize. **3.** *Archaic.* To descry. *—vi. Scot.* To have an understanding of something. *—n.* **1.** Understanding. **2. a.** Range of vision. **b.** View : sight.

**Ken·dal green** (kĕn'dl) *n.* [After *Kendal,* England, where it was orig. made.] **1.** A coarse green woolen fabric similar to tweed. **2.** The color of Kendal green fabric.

**ken·do** (kĕn'dō) *n.* [J.] The Japanese art of fencing with bamboo sticks.

kendo

**Ken·il·worth ivy** (kĕn'əl-wûrth') *n.* [After *Kenilworth* Castle, Warwickshire, England.] A trailing or climbing vine, *Cymbalaria muralis,* with lobed leaves and pale-purple flowers.

**ken·nel[1]** (kĕn'əl) *n.* [ME *kenel* < OFr. *chenil* < VLat. ***canile** < Lat. *canis,* dog.] **1.** A shelter for a dog. **2.** A pack of dogs, esp. hounds. **3.** A place where dogs are bred, trained, or boarded. **4.** The lair of a wild animal, as a fox. *—v.* **-neled, -nel·ing, -nels** *or* **-nelled, -nel·ling, -nells.** *—vt.* To place or keep in or as if in a kennel. *—vi.* To take cover or lie in or as if in a kennel.

**ken·nel[2]** (kĕn'əl) *n.* [ME *cannel* < ONFr. *canel,* channel < Lat. *canalis.*] A street gutter.

**Ken·nel·ly-Heav·i·side layer** (kĕn'ə-lē-hĕv'ē-sīd') *n.* [After Arthur E. *Kennelly* (1861–1939) and Oliver *Heaviside* (1850–1925).] The E layer of the ionosphere.

**ken·ning** (kĕn'ĭng) *n.* [ON < *kenna,* to name with a kenning, to know.] A metaphorical, usu. compound expression used as a name, esp. in Old English and Old Norse poetry.

**ke·no** (kē'nō) *n.* [Orig. unknown.] A game of chance, similar to lotto, that uses balls rather than counters.

**kent** (kĕnt) *v. var. p.t. & p.p. of* KEN.

**kent·ledge** (kĕnt'lĭj) *n.* [Orig. unknown.] Pig iron used as permanent ballast.

**Ken·tuck·y bluegrass** (kən-tŭk'ē) *n.* BLUEGRASS 1.

**Kentucky coffee tree** *n.* A deciduous North American tree, *Gymnocladus dioica,* with flat pulpy pods containing seeds once used as a coffee substitute.

**Ke·ogh plan** (kē'ō) *n.* [After Eugene J. *Keogh* (b. 1907).] A retirement plan for the self-employed and their employees.

**ke·pi** (kā'pē, kĕp'ē) *n.* [Fr. *képi* < dial. G. *käppi,* dim. of G. *Kappe,* cap < OHG *kappa,* cloak, prob. < LLat. *cappa,* cloak.] A French military cap with a flat circular top and a visor.

**kept** (kĕpt) *v. p.t. & p.p. of* KEEP.

**kerat-** *pref. var. of* KERATO-.

**ker·a·tec·to·my** (kĕr'ə-tĕk'tə-mē) *n., pl.* **-mies.** Surgical removal of all or part of the cornea.

**ker·a·tin** (kĕr'ə-tĭn) *n.* [Gk. *keras, kerat-,* horn + -IN.] A tough, fibrous protein substance forming the outer layer of epidermal structures, as hair, nails, or horns. **—ke·rat'i·nous** (kə-răt'n-əs) *adj.*

**ker·a·tin·ize** (kĕr'ə-tə-nīz') *vt. & vi.* **-ized, -iz·ing, -iz·es.** To convert into keratin. **—ker'a·tin·i·za'tion** *n.*

**ker·a·ti·tis** (kĕr'ə-tī'tĭs) *n., pl.* **-tit·i·des** (-tĭt'ĭ-dēz'). Inflammation of the cornea.

**kerato-** *or* **kerat-** *pref.* [Gk. *kerato-* < *keras,* horn.] **1.** Horn : horny <*keratosis*> **2.** Cornea <*keratectomy*>

**ker·a·to·sis** (kĕr'ə-tō'sĭs) *n., pl.* **-ses** (-sēz'). A skin condition characterized by excessive growth of horny tissue. **—ker'a·tot'ic** (-tŏt'ĭk) *adj.*

**kerb** (kûrb) *n. Chiefly Brit. var. of* CURB 2.

**ker·chief** (kûr'chĭf, -chēf') *n.* [ME *courchef* < OFr. *couvrechef* : *covrir,* to cover (< Lat. *cooperire* : *co(m)-,* completely + *operire,* to cover) + *chef,* head < Lat. *caput.*] **1.** A woman's square scarf, often worn as a head covering. **2.** A handkerchief.

**kerf** (kûrf) *n.* [ME < OE *cyrf,* act of cutting.] **1.** A groove or notch made by a cutting tool, as an ax or saw. **2.** The width of a groove made by a cutting tool.

**ker·mes** (kûr'mēz) *n.* [Fr. *kermès* < Ar. *qirmiz.*] The dried bodies of the females of various scale insects of the genus *Kermes,* used as a red dyestuff.

**ker·mis** *also* **ker·mess** (kûr'mĭs) *n.* [Du. *kermis* < MDu. *kercmisse* : *kerc,* church + *misse,* mass < LLat. *missa* < Lat. *mittere,* to send.] **1.** An outdoor fair in the Low Countries. **2.** A fund-raising fair.

**kern**[1] *also* **kerne** (kûrn) *n.* [ME *kerne* < OIr. *ceithern,* band of foot soldiers, poss. < *cath,* battle.] **1.** A medieval Scottish or Irish foot soldier. **2.** A lout.

**kern**[2] (kûrn) *n.* [Fr. *carne,* corner < Lat. *cardo,* hinge.] The part of a typeface projecting beyond the shank of a character. —*vt.* **kerned, kern·ing, kerns.** To provide (type) with a kern.

**kerne** (kûrn) *n. var. of* KERN[1].

**ker·nel** (kûr'nəl) *n.* [ME *kirnel* < OE *cyrnel,* dim. of *corn,* seed.] **1.** A grain or seed, as of a cereal grass, enclosed in a hard husk. **2.** The inner, usu. edible part of a nut or fruit stone. **3.** The most material and central part : CORE. **4.** A tiny bit: MODICUM.

**kern·ite** (kûr'nīt') *n.* [After *Kern* County, California.] A colorless to white crystalline compound, $Na_2B_4O_7 \cdot 4H_2O$, that is a major source of borax and boron compounds.

**ker·o·gen** (kĕr'ə-jən) *n.* [Gk. *kēros,* wax + -GEN.] A bituminous material in shale that yields oil upon heating.

**ker·o·sene** *also* **ker·o·sine** (kĕr'ə-sēn', kĕr'ə-sēn', kăr'ə-sēn', kăr'ə-sēn') *n.* [Orig. a trademark.] A thin oil distilled from petroleum or shale oil, used as a fuel and alcohol denaturant.

**Ker·ry** (kĕr'ē) *n., pl.* **-ries.** [After County *Kerry,* Ireland.] One of a breed of small, black dairy cattle of Irish origin.

**Kerry blue terrier** *n.* [After County *Kerry,* Ireland.] One of a breed of terriers of Irish origin, with a dense, wavy bluish-gray coat.

**ker·sey** (kûr'zē) *n., pl.* **-seys.** [ME, prob. after *Kersey,* a village in Suffolk, England.] **1.** A woolen, often ribbed fabric once used for hose and trousers. **2.** A twilled woolen fabric, occas. with a cotton warp, used for coats. **3.** *often* **kerseys.** A garment made of kersey.

**ker·sey·mere** (kûr'zē-mîr') *n.* [Alteration of CASSIMERE.] A fine woolen cloth with a fancy twill weave.

**kes·trel** (kĕs'trəl) *n.* [ME *castrel,* prob. < OFr. *cresserelle.*] **1.** A small Old World falcon, *Falco tinnunculus,* having brown and gray plumage. **2.** Any of various Old World falcons.

**ket-** *pref. var. of* KETO-.

**ketch** (kĕch) *n.* [ME *cache.*] A two-masted fore-and-aft-rigged sailing vessel with a mizzen or jigger mast stepped aft of a taller mainmast but forward of the rudder.

**ketch·up** (kĕch'əp, kăch'-) *n.* [Chin. (Amoy) *ketsiap,* a kind of sauce.] A condiment consisting of a thick, smooth-textured, spicy sauce usu. made from tomatoes.

**ke·tene** (kē'tēn') *n.* A pungent, toxic, colorless gas, $H_2C_2O$, used primarily as an acetylation agent.

**keto-** *or* **ket-** *pref.* [< KETONE.] Ketone <*ketosis*>

**ke·to·gen·e·sis** (kē'tō-jĕn'ĭ-sĭs) *n.* Formation of ketone bodies, as in diabetes. **—ke'to·gen'ic** *adj.*

**ke·tone** (kē'tōn') *n.* [G. < *Aketon,* acetone.] An organic compound with a carbonyl group linked to a carbon atom in each of two hydrocarbon radicals and having the general formula $R_1(CO)R_2$, where $R_1$ may be the same as $R_2$. **—ke·ton'ic** (kē-tŏn'ĭk) *adj.*

**ketone body** *n.* Any of several substances, as acetoacetic acid, increasing in the blood in certain diabetic and other pathological conditions.

**ke·tose** (kē'tōs') *n.* Any of various carbohydrates containing a ketone group in each molecule.

**ke·to·sis** (kē-tō'sĭs) *n.* Pathological accumulation of ketone bodies in the body. **—ke·tot'ic** (-tŏt'ĭk) *adj.*

**ke·to·ste·roid** (kē'tō-stîr'oid', -stĕr'-) *n.* A steroid having in it a ketone group.

**ket·tle** (kĕt'l) *n.* [ME *ketel* < ON *ketill*] **1.** A metal pot, usu. with a lid, for boiling or stewing. **2.** A teakettle. **3.** A kettledrum. **4.** A depression left in a mass of glacial drift, possibly formed by the melting of an isolated block of glacial ice. **5.** A pothole. **—kettle of fish. 1.** A troublesomely embarrassing situation. **2.** A matter to be reckoned with.

**ket·tle·drum** (kĕt'l-drŭm') *n.* A large copper or brass hemispheric drum with a parchment head that can be tuned by adjusting the tension.

**kev·el** (kĕv'əl) *n.* [ME *kevile,* peg < ONFr. *keville* < Lat. *clavicula,* dim. of *clavis,* key.] A sturdy belaying pin for the heavier cables of a ship.

**kew·pie** (kyo͞o'pē) *n.* [Orig. a trademark.] A small, fat-cheeked, wide-eyed doll with a curl of hair on top of the head.

**key**[1] (kē) *n., pl.* **keys.** [ME < OE *cæg.*] **1. a.** A usu. metal notched and grooved implement turned to open or close a lock. **b.** A device functioning like a key <the *key* of a clock><opened the can with a *key*> **2.** A means of access, control, or possession. **3. a.** A vital element. **b.** A set of answers to a test. **c.** A table, gloss, or cipher for decoding or interpreting. **4.** A device, as a wedge or pin, inserted to lock together structural or mechanical parts. **5.** The keystone in the crown of an arch. **6. a.** A button or lever depressed with the finger to operate a machine. **b.** A button or lever depressed with the finger to produce or modulate the sound of a musical instrument, as a clarinet or a piano. **7.** *Mus.* **a.** A tonal system consisting of seven tones in fixed relationship to a tonic, having a characteristic key signature and being the structural foundation of the bulk of Western music : TONALITY. **b.** The principal tonality of a musical work <an etude in the *key* of E> **8.** The pitch, as of a voice <spoke in a low *key*> **9.** A characteristic tone or level of intensity, as of a speech, dramatic performance, or sales campaign. **10.** A samara. **11.** *Slang.* A kilogram of marijuana or heroin. —*adj.* Of crucial importance. —*vt.* **keyed, key·ing, keys. 1.** To lock with or as if with a key. **2.** To furnish (an arch) with a keystone. **3.** To regulate the musical pitch of. **4.** To bring into harmony : ADJUST. **5.** To supply an explanatory key for. **—key up.** To make intense or nervous.

**key**[2] (kē) *n., pl.* **keys.** [Sp. *cayo,* prob. < OFr. *quai,* quay, of Celtic orig.] A low offshore island or reef, esp. in the Gulf of Mexico.

**key·board** (kē'bôrd', -bōrd') *n.* A set of keys, as on a piano or word processor. —*vt.* **-board·ed, -board·ing, -boards. 1.** To set (copy) by means of a keyed typesetting machine. **2.** To generate (documents) by means of a word processor. **—key'board'er** *n.*

**key·card** (kē'kärd') *n.* A coded, usu. plastic card scanned in order to operate a mechanism, as a door or a cash-dispensing machine.

**key club** *n.* [From the key to the premises given to each member.] A private club featuring liquor and entertainment.

**key fruit** *n.* [From its shape.] A samara.

**key·hole** (kē'hōl') *n.* The hole in a lock into which a key fits.

**Keynes·i·an** (kān'zē-ən) *adj.* Of or relating to the economic theories of John M. Keynes, esp. those theories advocating government monetary and fiscal programs designed to increase employment. —*n.* A supporter of Keynes's economic views. **—Keynes'i·an·ism** *n.*

**key·note** (kē'nōt') *n.* **1.** The tonic of a musical key. **2.** A prime underlying theme <"the *keynote* of the revolution settlement was personal freedom under the law" —G.M. Trevelyan> —*vt.* **-not·ed, -not·ing, -notes. 1.** To give or set the keynote of. **2.** To give a keynote address at.

**keynote address** *n.* An opening speech, as at a convention, that outlines the issues under discussion.

**key·not·er** (kē'nō'tər) *n.* One who gives a keynote address.

**keynote speech** *n.* A keynote address.

**key·punch** (kē'pŭnch') *n.* A keyboard machine used to punch holes in cards or tapes for data-processing systems. —*vt.* & *vi.* **-punched, -punch·ing, -punch·es.** To process on a keypunch. **—key'punch'er** *n.*

**key signature** *n. Mus.* The group of sharps or flats placed to the right of the clef on a staff to identify the key.

**key·stone** (kē'stōn') *n.* **1.** The central wedge-shaped stone of an arch that locks its parts together. **2.** A central supporting element.

**key·stroke** (kē'strōk') *n.* A stroke of a key, as of a typewriter or word processor.

**key·way** (kē'wā') *n.* **1.** A slot in a wheel hub or shaft for a key. **2.** The keyhole of a cylinder lock.

**khak·i** (kăk'ē, kä'kē) *n.* [Urdu < *khāk,* dust < Pers.] **1.** A light olive brown to yellowish brown. **2.** A sturdy cloth of the color khaki. **3. khakis.** A uniform of khaki cloth. **—khak'i** *adj.*

**kham·sin** (kăm-sēn') *n.* [Ar. *rīḥ al-khamsīn,* wind of the 50 days.] A gen. southerly hot wind from the Sahara that blows across Egypt from late Mar. to early May.

**khan**[1] (kän, kăn) *n.* [ME *caan* < OFr. < Turk. *khān,* ruler.] **1.** A chieftain or an important person in India and some Central Asian countries. **2.** A medieval Mongolian, Tartar, or Turkish tribal ruler. **—khan'ate'** (kä'nāt', kăn'āt') *n.*

**khan**[2] (kän, kăn) *n.* [Ar. *khān.*] A caravansary or inn in certain Asian countries.

**khe·dive** (kə-dēv') *n.* [Fr. *khédive* < Turk. *hidiv* < Pers. *khidīw,* prince.] A Turkish viceroy ruling Egypt from 1867 to 1914.

**khi** (kī) *n. var. of* CHI.

**Khmer** (kə-mâr) *n., pl.* **Khmer** *or* **Khmers. 1. a.** A people of Cambodia. **b.** A member of this people. **2.** The Mon-Khmer language of the Khmer, the official language of Cambodia. **—Khmer'i·an** *adj.*

**Khoi·san** (koi'sän') *n.* A family of languages of southwestern Africa, including those of the Bushmen and the Hottentot.

**khoum** (ko͞om, ko͝om) *n.* [Native word in Mauritania.] —See table at CURRENCY.

**ki·ang** (kē-äng') *n.* [Tibetan *rkyan.*] A wild ass, *Equus hemionus kiang* of eastern Asian mountains.

**kiaugh** (kyäKH) *n.* [Prob. < Sc. Gael. *cabhag.*] *Scot.* Trouble or anxiety.

**kib·butz** (kĭ-bo͝ots', -bo͞ots') *n., pl.* **kib·but·zim** (kĭb'o͞ot-sēm', -o͞ot-) [Mod. Heb. *qibbūtz* < Heb., gathering < *qibbētz,* he gathered.] An Israeli collective farm or settlement.

**kibe** (kīb) *n.* [ME *kybe,* poss. < Welsh *cibi.*] An ulcerated chilblain, esp. one on the heel.

**kib·itz** (kĭb'ĭts, kĭ-bĭts') *vi.* **-itzed, -itz·ing, -itz·es.** [Yiddish *kibitsen* < G. *kiebitzen,* to look on < *Kiebitz,* kibitzer, lapwing.] *Informal.* **1.** To look on and offer unwanted, usu. meddlesome advice to others. **2.** To chat : converse. **—kib'itz·er** *n.*

**ki·bosh** (kī'bŏsh', kī-bŏsh') *n.* [Orig. unknown.] *Informal.* A check, restraint, or halt <put the *kibosh* on their idiotic plan>

**kick** (kĭk) *v.* **kicked, kick·ing, kicks.** [ME *kiken.*] *—vi.* **1.** To strike out with the foot or feet. **2. a.** To score or gain ground by kicking a ball. **b.** *Football.* PUNT[2]. **3.** To recoil, as a firearm when fired. **4.** *Informal.* **a.** To express negative feelings vigorously : COMPLAIN. **b.** To oppose by argument : PROTEST. *—vt.* **1.** To strike with the foot. **2.** To propel by striking with the foot. **3.** To recoil against, as a firearm does when fired. **4.** To score (a goal or point) in a sport by kicking a ball. **—kick around.** *Informal.* **1.** To treat badly : ABUSE. **2.** To move from place to place. **3.** To give thought to (an idea). **—kick back. 1.** To recoil violently and unexpectedly. **2.** *Slang.* To return (stolen items). **3.** *Slang.* To pay a kickback. **—kick in.** *Slang.* **1.** To contribute (one's share). **2.** To die. **—kick off. 1.** To begin or resume play in a sport with a kickoff. **2.** To begin : start. **3.** *Slang.* To die. **—kick out.** *Slang.* To throw out : DISMISS. **—kick up. 1.** To cause to be propelled upward forcefully <truck tires *kicking up* gravel> **2.** To stir up (trouble) <*kicked up* a controversy> **3.** To show signs of disorder <My ulcer has *kicked up* again.> *—n.* **1. a.** A vigorous blow with the foot. **b.** The thrusting motion of the legs in swimming. **2.** The recoil of a firearm. **3.** *Slang.* Complaint : protest. **4.** *Slang.* Power : force <still a lot of *kick* to that engine> **5.** *Slang.* **a.** Pleasurable stimulation <got a *kick* out of the film> **b. kicks.** Fun : thrills <did it just for *kicks*> **6.** *Slang.* Temporary, often obsessive interest <on a water-skiing *kick*> **7.** *Slang.* A sudden, striking surprise : TWIST. **8. a.** An act or instance of kicking a ball. **b.** A kicked ball. **c.** The distance spanned by a kicked ball. **—kick the bucket.** *Slang.* To die. **—kick the habit.** *Slang.* To free oneself of an addiction, as to drugs or cigarettes. **—kick up (one's) heels.** *Slang.* To shed one's inhibitions and have a good time. **—kick upstairs.** *Informal.* To promote to a higher yet less desirable position.

**Kick·a·poo** (kĭk'ə-po͞o') *n., pl.* **Kickapoo** or **-poos.** [Kickapoo *kiwĕgapawa.*] **1. a.** A tribe of Indians once living in northern Illinois and southern Wisconsin. **b.** A member of this tribe. **2.** The Algonquian language of the Kickapoo.

**kick·back** (kĭk'băk') *n.* **1.** *Slang.* A percentage payment to a person able to influence or control a source of income, as by confidential arrangement or coercion. **2.** A sharp reaction : REPERCUSSION.

**kick·er** (kĭk'ər) *n.* **1.** One that kicks. **2.** *Informal.* A sudden, surprising turn of events. **3.** A tricky or concealed condition : PITFALL. **4.** A condition that imposes an automatic increase, as in a pension plan.

**kick·off** (kĭk'ôf', -ŏf') *n.* **1.** A place kick in football or soccer with which play is begun. **2.** A beginning.

**kick plate** *n.* A protective metal sheet at the bottom of a door.

**kick·shaw** (kĭk'shô') *n.* [Alteration of Fr. *quelque chose,* something.] **1.** A fancy food : DELICACY. **2.** A trinket.

**kick·stand** (kĭk'stănd') *n.* A swiveling metal bar for holding a two-wheeled vehicle, as a motorcycle, upright when parked.

**kick·y** (kĭk'ē) *adj.* **-i·er, -i·est.** *Slang.* Providing thrills or pleasure by being unusual or unconventional.

**kid** (kĭd) *n.* [ME < ON *kið.*] **1. a.** A young goat. **b.** The young of an animal, as an antelope, that is similar to a young goat. **2.** The flesh of a young goat. **3.** Leather made from the skin of a young goat. **4.** An article of kidskin. **5.** *Informal.* **a.** A child. **b.** A young person. **6.** *Slang.* Pal. —Used as a term of familiar address, esp. for a young person. *—v.* **kid·ded, kid·ding, kids.** *—vt. Informal.* **1.** To mock playfully : TEASE. **2.** To deceive in fun : FOOL. *—vi.* **1.** *Informal.* To engage in teasing or good-humored fooling. **2.** To bear young, as a goat or an antelope. **—kid'der** *n.* **—kid'ding·ly** *adv.*

**Kid·der·min·ster** (kĭd'ər-mĭn'stər) *n.* [After *Kidderminster,* England, where it was originally made.] An ingrain carpet.

**kid·die** (kĭd'ē) *n. var. of* KIDDY.

**kid·do** (kĭd'ō) *n., pl.* **-os.** *Slang.* Pal. —Used as a term of familiar address.

**Kid·dush** (kĭd'əsh, kĭ-do͞osh') *n.* [Heb. *qiddūsh* < *qiddesh,* he sanctified.] The traditional Jewish blessing and prayer recited over bread or a cup of wine on the eve of the Sabbath or a festival.

**kid·dy** *also* **kid·die** (kĭd'ē) *n., pl.* **-dies.** *Slang.* A small child.

**kid glove** *n.* A glove of fine, soft leather, esp. kidskin. **—with kid gloves.** Cautiously and tactfully.

**kid·nap** (kĭd'năp') *vt.* **-naped, -nap·ing, -naps** or **-napped, -nap·ping, naps.** [Prob. back-formation < *kidnapper* : KID + obs. *napper,* thief.] To seize and detain (a person) unlawfully, usu. for ransom. **—kid'nap'er, kid'nap'per** *n.*

☆ ***syns:*** KIDNAP, ABDUCT, CARRY OFF, SNATCH, SPIRIT (away) *v. core meaning* : to seize and detain (a person) unlawfully, usu. for ransom <Terrorists *kidnaped* the ambassador.>

**kid·ney** (kĭd'nē) *n., pl.* **-neys.** [ME *kidenei.*] **1.** *Anat.* Either of a pair of structures in the dorsal region of the vertebrate abdominal cavity, functioning to maintain proper water balance, regulate acid-base concentration, and excrete metabolic wastes as urine. **2.** The kidney of certain animals, eaten as food. **3.** An excretory organ of certain invertebrates. **4.** Disposition : temperament.

**kidney bean** *n.* **1.** A bean, *Phaseolus vulgaris,* cultivated for its edible seeds. **2.** The reddish seed of the kidney bean.

**kidney stone** *n.* A small hard mass, usu. of mineral salts from urine, that forms in a kidney.

**kid·skin** (kĭd'skĭn') *n.* KID 3.

**kid stuff** *n. Informal.* **1.** Something suitable only for children. **2.** Something easy or uncomplicated.

**kid·vid** (kĭd'vĭd') *n.* [KID + VID(EO).] *Slang.* Television programs for children.

**kiel·ba·sa** (kēl-bä'sə, kĭl-, kēl-) *n.* [Pol.] A smoked Polish sausage.

**kie·sel·guhr** (kē'zəl-go͝or') *n.* [G. : *Kiesel,* pebble + *Guhr,* earthy deposit from water.] Diatomite.

**kie·ser·ite** (kē'zə-rīt') *n.* [G. *Kieserit,* after Dietrich G. *Kieser* (1779–1862).] A whitish to yellowish hydrous magnesium sulfate mineral.

**kif** (kĭf, kēf) *also* **kef** (kĕf, kēf, käf) *n.* [Ar. *kayf,* pleasure.] **1.** A smoking material, as Indian hemp, used esp. in the Maghreb. **2.** The euphoria caused by smoking kif.

**Ki·ku·yu** (kĭ-ko͞o'yo͞o) *n., pl.* **Kikuyu** or **-yus. 1.** A member of a Bantu people of Kenya. **2.** The Bantu language of the Kikuyu.

**kil·der·kin** (kĭl'dər-kĭn) *n.* [ME < MDu. *kindekijn,* dim. of *kintal,* hundredweight < Med. Lat. *quintale* < Ar. *qinṭār,* ult. < Lat. *centum,* hundred.] **1.** A cask. **2.** An English measure of capacity equal to approx. 68 liters or 18 gallons.

**kill** (kĭl) *v.* **killed, killing, kills.** [ME *killen.*] *—vt.* **1. a.** To put to death. **b.** To deprive of life <The Black Plague *killed* millions.> **2.** To put an end to : EXTINGUISH. **3. a.** To destroy a vitally essential quality in <Too much garlic *killed* the taste of the lamb.> **b.** To cause to cease operating <*killed* the engine> **4.** To pass (time) in aimless activity. **5.** To consume entirely <*kill* a bottle of champagne> **6.** To cause extreme pain or discomfort to <My new boots are *killing* me.> **7.** To mark for deletion. **8.** To thwart passage of : VETO <*kill* a legislative bill> **9.** To cause to stop : TURN OFF. **10. a.** To hit (a ball) with great force. **b.** To hit (a ball) with such force in a racket game as to make a return impossible. *—vi.* **1.** To be fatal. **2.** To commit murder. **—kill off.** To destroy in such large numbers as to render extinct. *—n.* **1.** An act of killing. **2.** An animal killed, esp. in hunting. **—in at the kill.** Present at the moment of triumph.

☆ ***syns:*** KILL, DESTROY, DISPATCH, DO IN, FINISH (off), SLAY, ZAP *v. core meaning* : to cause the death of <a disease that *killed* thousands>

**kill·deer** (kĭl'dîr') *n., pl.* **killdeer** or **-deers.** [Imit.] A New World bird, *Charadrius vociferus,* indigenous to inland waters and fields and having a distinctive cry.

**kill·er** (kĭl'ər) *n.* **1.** One that kills. **2.** The killer whale.

**killer whale** *n.* A black and white predatory whale, *Orcinus orca,* inhabiting cold seas.

**kil·lick** (kĭl'ĭk) *n.* [Orig. unknown.] A small anchor, esp. one made of a stone in a wooden frame.

**kil·li·fish** (kĭl'ĭ-fĭsh') *n., pl.* **killifish** or **-fish·es.** [Orig. unknown.] Any of numerous small fishes of the family Cyprinodontidae, inhabiting chiefly fresh and brackish waters in warm regions.

**kill·ing** (kĭl'ĭng) *n.* **1.** Murder. **2.** Quarry : kill. **3.** A sudden large profit. *—adj.* **1.** Designed or apt to kill : FATAL. **2.** Exhausting <a *killing* schedule> **3.** *Informal.* Hilarious. **—kill'ing·ly** *adv.*

**kill·joy** (kĭl'joi') *n.* One who spoils the enthusiasm or fun of others.

**kill ratio** *n.* The ratio of casualties to the total number of troops involved in a combat operation or to enemy casualties.

**kill shot** *n.* A shot in various games, esp. racket games, so forcefully hit or perfectly placed that it cannot be returned.

**kiln** (kĭln, kĭl) *n.* [ME *kilne* < OE *cyln* < Lat. *culina,* kitchen < *coquere,* to cook.] An oven for hardening, burning, or drying substances, as grain or clay, esp. a brick-lined oven for baking or firing ceramics. *—vt.* **kilned, kiln·ing, kilns.** To process in a kiln.

**ki·lo** (kē'lō) *n., pl.* **-los.** A kilogram.

**kilo-** *pref.* [Fr. < Gk. *khilioi,* thousand.] One thousand ($10^3$) <*kilo*watt><*kilo*cycle>

**kil·o·baud** (kĭl'ə-bôd') *n.* One thousand baud.

**kil·o·bit** (kĭl'ə-bĭt') *n.* One thousand binary digits.

**kil·o·cal·o·rie** (kĭl'ə-kăl'ə-rē) *n.* A kilogram calorie.

**kil·o·cy·cle** (kĭl'ə-sī'kəl) *n.* **1.** A unit equal to 1,000 cycles. **2.** One thousand cycles per second.

**kil·o·gram** (kĭl'ə-grăm') *n.* **1.** The fundamental unit of mass in the metric system, equal to approx. 2.2046 pounds. **2.** A force equal to a kilogram weight, or the product of a kilogram mass with the acceleration of gravity.

**kilogram calorie** *n.* CALORIE 3.

**kil·o·gram-me·ter** (kĭl'ə-grăm-mē'tər) *n.* A meter-kilogram-second unit of work equal to the work performed by a one-kilogram force acting through a distance of one meter.

**kil·o·hertz** (kĭl'ə-hûrts') *n.* One thousand hertz.
**kil·o·li·ter** (kĭl'ə-lē'tər) *n.* One thousand liters.
**kil·o·me·ter** (kĭl'ə-mē'tər, kĭ-lŏm'ĭ-tər) *n.* One thousand meters, approx. 0.62137 mile. **—kil'o·met'ric** (-mĕt'rĭk) *adj.*
**kil·o·par·sec** (kĭl'ə-pär'sĕk') *n.* One thousand parsecs.
**kil·o·rad** (kĭl'ə-răd') *n.* One thousand rads.
**kil·o·ton** (kĭl'ə-tŭn') *n.* **1.** One thousand tons. **2.** An explosive force equaling 1,000 tons of TNT.
**kil·o·volt** (kĭl'ə-vōlt') *n.* One thousand volts.
**kil·o·watt** (kĭl'ə-wŏt') *n.* One thousand watts.
**kil·o·watt-hour** (kĭl'ə-wŏt-our') *n.* A unit of electric power consumption indicating the total energy developed by a power of one kilowatt acting for one hour.
**kilt** (kĭlt) *n.* [< *kilt,* to tuck up < ME *kilten,* of Scand. orig.] A knee-length skirt with deep pleats, usu. of a tartan wool, worn esp. in the Scottish Highlands. *—vt.* **kilt·ed, kilt·ing, kilts.** To tuck up around the body.
**kil·ter** (kĭl'tər) *n.* [Orig. unknown.] Good condition : proper form <The TV was out of *kilter.*>
**Kim·bun·du** (kĭm-bo͞on'do͞o) *n.* A Bantu language of Angola.
**ki·mo·no** (kə-mō'nə, -nō) *n., pl.* **-nos.** [J. : *ki,* to wear + *mono,* object.] **1.** A long, loose, wide-sleeved Japanese robe worn with a broad sash. **2.** A loose robe worn chiefly by women.

**kimono**

**kin** (kĭn) *n.* [ME < OE *cyn.*] One's relatives : KINDRED. **—next of kin.** One's closest blood relatives.
☆ ***syns:*** KIN, KINDRED, KINSFOLK *n. core meaning* : one's relatives collectively <were finally united with their *kin*>
**-kin** *or* **-kins** *suff.* [ME < MDu.] Little one <devil*kin*>
**ki·na** (kē'nə) *n., pl.* **kina** *or* **-nas.** [Native word in Papua New Guinea.] —See table at CURRENCY.
**ki·nase** (kī'nās') *n.* [KIN(ETIC) + -ASE.] An enzyme catalyzing transfer of phosphate from ADP or ATP to an acceptor.
**kind¹** (kīnd) *adj.* **-er, -est.** [ME < OE *(ge)cynde,* natural.] **1.** Friendly, generous, or warm-hearted in nature. **2.** Exhibiting sympathy or understanding <a *kind* word> **3.** Humane : considerate <*kind* to animals> **4.** Forbearing : tolerant <was very *kind* about the bent fender> **5.** Generous : liberal <*kind* words of praise> **6.** Agreeable : beneficial <a climate *kind* to asthmatics> **—kind'ness** *n.*
☆ ***syns:*** KIND, BENIGN, GOODHEARTED, KINDHEARTED, KINDLY *adj. core meaning* : having or showing a tender, considerate, and helping nature <a *kind* person><a *kind* gesture> ***ant:*** unkind
▲ word history: *Kind¹* is a word that has undergone *melioration.* By this process a word with neutral or negative connotations develops favorable ones. The Old English ancestor of *kind, gecynde,* meant "natural, inborn, inherent" and was used of fearsome things such as death. In the 13th century, however, *kind* was used to mean "wellborn" and "of a good nature." These senses have passed from standard English, but the derived sense "possessing all the (good) qualities usually attributed to those of good birth" underlies the modern meanings of *kind.* Other English words that derive ultimately from words denoting social condition are *generous, gentle,* and *liberal.*
**kind²** (kīnd) *n.* [ME < OE *(ge)cynd,* nature.] **1.** A class of similar or related individuals. **2.** A specific type <What *kind* of car is that?> **3.** *Archaic.* Manner. **4.** A doubtful or peripheral member of a specified class <a *kind* of shelter><a *kind* of bluish color> ***usage:*** Traditionalists object to the use of the plurals *these* and *those* with *kind,* as in the phrase *these kind of books.* Strictly speaking, the modifier and the following noun and verb should agree in number with *kind: This kind of cookie is my favorite. Those kinds of books are for children.* **—all kinds of.** *Informal.* Plenty of : AMPLE <We have *all kinds of* time.> **—in kind.** In the same manner or with something equivalent <returned the slight *in kind*><was paid back *in kind*> **—kind of.** *Informal.* Rather : somewhat <I'm *kind of* hungry.>
☆ ***syns:*** KIND, BREED, FEATHER, ILK, LOT, SORT, SPECIES, STRIPE, TYPE, VARIETY *n. core meaning* : a class that is defined by the common attribute or attributes possessed by all of its members <flowers of all *kinds*><the *kind* of person who gets angry easily>

**kin·der·gar·ten** (kĭn'dər-gär'tn) *n.* [G. : *Kinder,* pl. of *Kind,* child (< OHG *kind*) + *Garten,* garden < MHG *garte* < OHG *garto.*] A program or class for four- to six-year-old children that is as an introduction to regular school.
**kin·der·gart·ner** (kĭn'dər-gärt'nər) *n.* [G. *Kindergärtner* < *Kindergarten,* kindergarten.] **1.** A child attending kindergarten. **2.** A kindergarten teacher.
**kind·heart·ed** (kīnd'här'tĭd) *adj.* Having or proceeding from a kind heart. **—kind'heart'ed·ly** *adv.* **—kind'heart'ed·ness** *n.*
**kin·dle¹** (kĭn'dl) *v.* **-dled, -dling, -dles.** [ME *kindelen* < ON *kynda.*] *—vt.* **1. a.** To build or fuel (a fire). **b.** To set fire to : IGNITE. **2.** To cause to glow <The sunset *kindled* the skies.> **3. a.** To make ardent : INFLAME <*kindled* their desire> **b.** To arouse : inspire. *—vi.* **1.** To catch fire : burst into flame. **2.** To become bright : GLOW. **3. a.** To become inflamed. **b.** To be stirred up : RISE. **—kin'dler** *n.*
**kin·dle²** (kĭn'dl) *vt. & vi.* **-dled, -dling, -dles.** [ME.] To give birth to young. —Used esp. of rabbits. **—kin'dle** *n.*
**kind·less** (kīnd'lĭs) *adj.* **1.** Heartless. **2.** *Obs.* Inhuman.
**kin·dling** (kĭnd'lĭng) *n.* Easily ignited material used as a fire starter.
**kind·ly** (kīnd'lē) *adj.* **-li·er, -li·est. 1.** Of a sympathetic, helpful, or benevolent nature <took a *kindly* interest in me><a *kindly* soul> **2.** Agreeable : pleasant <a *kindly* breeze> **3.** *Archaic.* Natural to its kind. *—adv.* **1.** Out of kindness <*kindly* overlooked their errors> **2.** In a kind way <spoke *kindly* to us> **3.** Pleasantly : agreeably <The sun shone *kindly.*> **4.** In an accommodating way <Would you *kindly* fill in your name?> **5.** *Obs.* In a way or course that is natural : FITTINGLY. **—take kindly to. 1.** To be receptive to <*take kindly to* new methods> **2.** To be naturally attracted or fitted to : thrive on. **—kind'li·ness** *n.*
**kin·dred** (kĭn'drĭd) *n.* [ME : *kin,* kin + OE *rǣden,* condition < *rǣdan,* to advise.] **1.** A group of related persons, as a clan or tribe. **2.** A person's relatives : KINFOLK. *—adj.* **1.** Of the same ancestry or family <*kindred* clans> **2.** Having a similar or related origin, nature, or character <*kindred* emotions> **—kin'dred·ness** *n.*
**kine** (kīn) *n.* [ME *kyn.*] *Archaic. var. pl. of* COW¹.
▲ word history: The word *kine,* an archaic plural of *cow¹,* is actually a double plural. The Old English form of *cow* was *cū,* respelled in Middle English as *ku.* The word *cū* formed its plural by a vowel change, like *foot* (plural *feet*) and was written *cy,* where *y* represented a sound like the vowel in French *tu.* In Middle English times the ordinary plural suffix *-en* was added to *cy* or *ky* to make it regular, and the form *kyn,* modern *kine,* persisted as the standard plural of *cow* into the 17th century. The plural *cows* did not appear until after 1600.
**kin·e·mat·ics** (kĭn'ə-măt'ĭks) *n.* [< Gk. *kinēma, kinēmat-,* motion < *kinein,* to move.] *(sing. in number).* The study of motion exclusive of the influences of mass and force. **—kin'e·mat'ic, kin'e·mat'i·cal** *adj.* **—kin'e·mat'i·cal·ly** *adv.*
**kin·e·scope** (kĭn'ĭ-skōp', kī'nĭ-) *n.* [Orig. a trademark.] **1.** A cathode-ray tube in a television receiver that translates received electrical signals into a visible picture on a luminescent screen. **2.** A film of a transmitted television program. *—vt.* **-scoped, -scop·ing, -scopes.** To make a kinescope of (a transmitted television program).
**ki·ne·si·ol·o·gy** (kə-nē'sē-ŏl'ə-jē, -zē-) *n.* [Gk. *kinēsis,* movement (< *kinein,* to move) + -LOGY.] The study of muscles and their movements, esp. for physical conditioning. **—ki·ne'si·ol'o·gist** *n.*
**-kinesis** *suff.* [NLat. < Gk. *kinēsis,* movement < *kinein,* to move.] **1.** Motion <photo*kinesis*> **2.** Division <cyto*kinesis*>
**kin·es·the·sia** (kĭn'ĭs-thē'zhə, kī'nĭs-) *n.* [Gk. *kinein,* to move + ESTHESIA.] The sensation of bodily position, presence, or movement resulting primarily from stimulation of sensory nerve endings in muscles, tendons, and joints. **—kin'es·thet'ic** (-thĕt'ĭk) *adj.* **—kin'es·thet'i·cal·ly** *adv.*
**ki·net·ic** (kĭ-nĕt'ĭk, kī-) *adj.* [Gk. *kinētikos* < *kinētos,* moving < *kinein,* to move.] Of, pertaining to, or produced by motion.
**kinetic art** *n.* An art form, as an assemblage or sculpture, composed of parts intended to be moved by an internal mechanism or external stimulus, as light or air. **—kinetic artist** *n.*
**kinetic energy** *n.* Energy associated with motion, equal for a body in pure translational motion at nonrelativistic speeds to one half of the product of its mass and the square of its speed.
**ki·net·i·cism** (kə-nĕt'ĭ-sĭz'əm) *n.* The theory or practice of kinetic art. **—ki·net'i·cist** *n.*
**ki·net·ics** (kĭ-nĕt'ĭks, kī-) *n.* *(sing. in number).* **1.** The study of all aspects of motion, comprising both kinematics and dynamics. **2.** The study of the relationship between motion and the forces affecting motion.
**kinetic theory** *n.* A theory of the thermodynamic behavior of matter, esp. of pressure-volume-temperature relationships in gases, based in its simplest form on the identification of heat with the kinetic energy of a substance's rapid, randomly moving molecules and on a classical dynamic analysis of molecular motion under simplifying assumptions, including the conservation of energy and momentum in all collisions and the applicability of statistical analysis for large numbers of molecules.
**ki·ne·tin** (kī'nə-tĭn) *n.* A plant growth substance that promotes cell division.

**kin·folk** (kĭn′fōk′) *also* **kins·folk** (kĭnz′fōk′) *or* **kin·folks** (kĭn′fōks′) *pl.n.* Kindred.

**king** (kĭng) *n.* [ME < OE *cyning.*] **1.** A man who is a monarch. **2.** The most powerful or eminent of a group, category, or place. **3. King. a.** GOD 1a, b. **b.** Christ. **4.** A playing card bearing a picture of a king. **5. a.** The principal chess piece, capable of being moved one square in any direction. **b.** A piece in checkers that has reached the opponent's side of the board and been crowned. **6. Kings** *(sing. in number).* —See table at BIBLE.

**king·bolt** (kĭng′bōlt′) *n.* A vertical bolt used to join the body of a wagon to the front axle and usu. acting as a pivot.

**King Charles spaniel** (chärlz) *n.* [After *King Charles* II of England (1630–1685).] A toy spaniel with long ears and a curly black and tan coat.

**king cobra** *n.* A large venomous snake, *Ophiophagus hannah* of tropical Asia.

**king crab** *n.* **1.** A large crab, *Paralithodes camtschatica,* inhabiting the coastal waters of Alaska, Japan, and Siberia and valued commercially for its edible flesh. **2.** The horseshoe crab.

**king·craft** (kĭng′krăft′) *n.* The manner used by a king in ruling.

**king-cup** (kĭng′kŭp′) *n.* A buttercup.

**king·dom** (kĭng′dəm) *n.* [ME < OE *cyningdom.*] **1.** A political or territorial unit ruled by a king or queen. **2. a.** God's eternal spiritual sovereignty. **b.** The realm over which God's sovereignty extends. **3.** An area, province, or realm in which one thing dominates <the *kingdom* of the imagination> **4.** One of the three main taxonomic divisions into which natural organisms and objects are classified <the animal *kingdom*>

**kingdom come** *n.* [From the phrase *thy kingdom come* in the Lord's Prayer.] The next world <tried to blow us to *kingdom come*>

**king·fish** (kĭng′fĭsh′) *n., pl.* **kingfish** *or* **-fish·es. 1. a.** Any of several food and game fishes of the genus *Menticirrhus,* indigenous to warm Atlantic waters. **b.** A similar or related fish. **2.** *Informal.* A powerful person, esp. a prominent political leader.

**king·fish·er** (kĭng′fĭsh′ər) *n.* [ME *kyngys fischare,* king's fisher.] Any of various birds of the family Alcedinidae, having crested heads.

**King James Bible** (jāmz) *n.* —See table at BIBLE.

**king·let** (kĭng′lĭt) *n.* A small grayish North American bird, *Regulus satrapa* or *R. calendula,* having a yellowish or reddish patch on the crown of the head.

**king·ly** (kĭng′lē) *adj.* **-li·er, -li·est. 1.** Having the rank or status of king. **2.** Pertaining to or suitable for a king : REGAL. —*adv.* As a king : ROYALLY. **—king′li·ness** *n.*

**king mackerel** *n.* A food and game fish, *Scomberomorus cavalla* of warm Atlantic waters.

**king·mak·er** (kĭng′mā′kər) *n.* A person with sufficient political clout to influence the selection of a candidate for office.

**king-of-arms** (kĭng′əv-ärmz′) *n., pl.* **kings-of-arms.** A high-ranking English heraldic officer.

**king·pin** (kĭng′pĭn′) *n.* **1.** The foremost pin in an arrangement of bowling pins. **2.** The most important or essential one. **3.** A kingbolt.

**king post** *n.* A supporting post extending vertically from a crossbeam to the apex of a triangular truss.

**king salmon** *n.* The Chinook salmon.

**King's bench** *n.* A division of the British superior courts system that hears criminal and civil cases. —Used when the monarch is a man.

**King's Counsel** *n.* A barrister appointed as counsel to the British crown. —Used when the monarch is a man.

**King's English** *n.* Spoken or written English considered as a standard of good usage.

**king·ship** (kĭng′shĭp′) *n.* **1.** The position, power, province, or prerogative of a king. **2.** The domain ruled by a king : KINGDOM. **3.** A king's tenure : REIGN. **4.** The style of a king : MAJESTY. **5.** A monarchy.

**king-size** (kĭng′sīz′) *also* **king-sized** (-sīzd′) *adj.* Larger than the standard or usual size <*king-size* cigarettes> <a *king-size* bed>

**king snake** *n.* A nonvenomous New World snake of the genus *Lampropeltis,* having yellow or reddish markings.

**king·wood** (kĭng′wo͝od′) *n.* **1.** A South American tree, *Dalbergia cearensis,* having hard, fine-textured, purplish-brown wood used in cabinetmaking. **2.** The wood of the kingwood tree.

**ki·nin** (kī′nĕn) *n.* [Gk. *kinein,* to move + -IN.] Any of various polypeptides that act in the contraction of smooth muscle.

**kink** (kĭngk) *n.* [LG *kinke* < MLG.] **1.** A tight curl or a sharp twist in a wirelike material, typically caused by the tensing of a looped section. **2.** A painful muscle spasm : CRICK. **3.** A slight difficulty or flaw. **4.** A physical or mental quirk. **5.** A clever idea. —*vt.* & *vi.* **kinked, kink·ing, kinks.** To form kinks.

**kink·a·jou** (kĭng′kə-jo͞o′) *n.* [Fr. *quincajou,* of Algonquian orig.] A tropical American arboreal mammal, *Potos flavus,* with brownish fur and a long prehensile tail.

**kink·y** (kĭng′kē) *adj.* **-i·er, -i·est. 1.** Tightly curled : FRIZZY. **2.** *Informal.* **a.** Marked by or engaging in a perverted eroticism. **b.** Sexually perverted. **—kink′i·ly** *adv.* **—kink′i·ness** *n.*

**kin·ni·kin·nick** *also* **kin·ni·kin·nic** (kĭn′ĭ-kĭ-nĭk′) *n.* [Of Algonquian orig.] **1.** A preparation made from dried leaves, bark, and occas. tobacco and smoked esp. by American Indians. **2.** A plant, as the bearberry, whose leaves or bark were used in kinnikinnick.

**ki·no** (kē′nō) *n., pl.* **-nos.** [Prob. < Mandingo *keno.*] A reddish resin obtained from several Old World tropical trees of the genera *Pterocarpus* and *Butea.*

**-kins** *suff. var. of* -KIN.

**kins·folk** (kĭnz′fōk′) *pl.n. var. of* KINFOLK.

**kin·ship** (kĭn′shĭp′) *n.* The state or quality of being kin.

**kins·man** (kĭnz′mən) *n.* **1.** A man relative. **2.** A man sharing the same racial, cultural, or national background as another.

**kins·wom·an** (kĭnz′wo͝om′ən) *n.* **1.** A woman relative. **2.** A woman sharing the same racial, cultural, or national background as another.

**ki·osk** (kē′ŏsk′, kē-ŏsk′) *n.* [Fr. *kiosque* < Turk. *kōshk* < Pers. *kūshk,* portico.] **1.** An open pavilion or gazebo. **2.** A small structure used as a newsstand or refreshment booth. **3.** A cylindrical structure on which advertisements are posted.

**Ki·o·wa** (kī′ə-wô′, -wä′, -wā′) *also* **Ki·o·way** (-wā′) *n., pl.* **Kiowa** *or* **-was** *also* **Kioway** *or* **-ways.** [Kiowa *Kâ-i-gwŭ.*] **1. a.** A tribe of Plains Indians once living in Colorado, Oklahoma, Kansas, New Mexico, and Texas. **b.** A member of this tribe. **2.** The Uto-Aztecan language of the Kiowa.

**kip**[1] (kĭp) *n., pl.* **kip.** [Thai.] —See table at CURRENCY.

**kip**[2] (kĭp) *n.* [Obs. Du.] The untanned hide of a small or young animal, as a calf.

**kip**[3] (kĭp) [Poss. < Dan. *kippe,* cheap inn.] *Chiefly Brit.* —*n.* **1.** A boarding house. **2.** A room or bed in a boarding house. **3.** A bed. **4.** Sleep. —*vi.* **kipped, kip·ping, kips.** To sleep.

**kip**[4] (kĭp) *n.* [KI(LO)- + P(OUND).] A 1,000-pound unit of weight.

**kip·per** (kĭp′ər) *n.* [ME *kypre* < OE *cypera.*] **1.** A male salmon or sea trout in the spawning season. **2.** A split, salted, and smoked herring. —*vt.* **-pered, -per·ing, -pers.** To cure (fish) by splitting, salting, and smoking.

**Kir·ghiz** (kîr-gēz′) *n., pl.* **Kirghiz** *or* **-ghiz·es. 1.** A member of a Turkic people residing chiefly in the Kirghiz S.S.R. of the Soviet Union. **2.** The Turkic language of the Kirghiz.

**kir·i·ga·mi** (kĭr′ĭ-gä′mē) *n.* [J. : *kiri,* to cut + *kami,* paper.] The Japanese art of making ornamental designs by cutting and folding paper.

**kirk** (kûrk) *n.* [ME < ON *kirkja* < OE *cirice,* church. —see CHURCH.] **1.** *Scot.* A church. **2. Kirk.** *Chiefly Brit.* The Presbyterian Church of Scotland.

**kirsch** (kîrsh) *n.* [G. *Kirsch(wasser),* cherry (water).] A colorless brandy made from fermented cherry juice.

**kir·tle** (kûr′tl) *n.* [ME < OE *cyrtel,* prob. ult. < Lat. *curtus,* shortened.] **1.** A knee-length tunic or coat once worn by men. **2.** A woman's long dress or skirt.

kirtle

**kish·ke** (kĭsh′kə) *n.* [Yiddish, of Slav. orig.] DERMA[2].

**Kis·lev** (kĭs′ləf) *n.* [Heb. *kislēw.*] The third month of the Hebrew year. —See table at CALENDAR.

**kis·met** (kĭz′mĕt′, -mĭt) *n.* [Turk. < Ar. *qismah,* portion < *qasama,* he allotted.] Fate : fortune.

**kiss** (kĭs) *v.* **kissed, kiss·ing, kiss·es.** [ME *kissen* < OE *cyssan.*] —*vt.* **1.** To caress or touch with the lips as a sign of sexual passion, affection, greeting, or respect. **2.** To touch lightly. —*vi.* To engage in mutual touching or caressing with the lips. **—kiss off.** *Informal.* To get rid of : DISMISS. —*n.* **1. a.** A caress or touch with the lips. **b.** A slight or gentle touch. **2. a.** A small piece of candy, esp. of chocolate. **b.** A baked confection of meringue. **—kiss good-by.** To resign oneself to the loss of.

**kiss·er** (kĭs′ər) *n.* **1.** A person who kisses. **2.** *Slang.* The human mouth. **3.** *Slang.* The human face.

**kissing bug** *n.* An assassin bug, *Melanolestes picipes,* that inflicts a painful bite, often on the lips of a sleeping person.

**kissing cousin** *n.* A distant relative known well enough to be kissed when greeted.

**kissing disease** *n.* *Informal.* Mononucleosis.

**kiss of death** *n.* [From the kiss by which Judas betrayed Jesus.] Something ultimately ruinous.

**kiss-off** (kĭs′ôf′, -ŏf′) *n. Informal.* A dismissal.
**kiss of life** *n. Chiefly Brit.* Mouth-to-mouth resuscitation.
**kiss of peace** *n.* A ceremonial gesture, as a kiss or a handclasp, used as a sign of brotherhood in Christian liturgies.
**kist** (kĭst) *n. var. of* CIST[2].
**kit[1]** (kĭt) *n.* [ME *kitt,* wooden tub.] **1.** A set of articles used for a particular purpose <a survival *kit*><a travel *kit*> **2.** A set of parts or materials to be assembled <a stereo *kit*> **3.** A packaged set of related materials <a sales *kit*> **4.** A container for a kit. **5.** *Chiefly Brit. Regional.* A tub. **—the (whole) kit and caboodle.** *Informal.* The entire lot or collection.
**kit[2]** (kĭt) *n.* [Short for KITTEN.] **1.** A kitten. **2.** A young, often undersized fur-bearing animal.
**kit[3]** (kĭt) *n.* [Orig. unknown.] A small three-stringed violin.
**kitch·en** (kĭch′ən) *n.* [ME *kichene* < OE *cycene,* ult. < LLat. *coquina* < fem. of Lat. *coquinus,* of cooking < *coquere,* to cook.] **1.** A place where food is cooked or prepared. **2.** The facilities and equipment used in preparing and serving food. **3.** A department that prepares, cooks, and serves food.
**kitchen cabinet** *n.* **1.** A cabinet or cupboard for kitchen use. **2.** An informal group of advisers to a head of state.
**kitch·en·er** (kĭch′ə-nər) *n.* **1.** A manager of a kitchen, esp. in a monastery. **2.** *Chiefly Brit.* A large cooking stove.
**kitch·en·ette** (kĭch′ə-nĕt′) *n.* A small kitchen.
**kitchen garden** *n.* A garden in which vegetables and fruits are grown for household use.
**kitchen midden** *n.* [Transl. of Dan. *køkkenmødding.*] A refuse heap or mound with artifacts, shells, and often bones indicating the site of a primitive human settlement.
**kitchen police** *n.* **1.** Enlisted military personnel assigned to kitchen work. **2.** The work of the kitchen police.
**kitch·en·ware** (kĭch′ən-wâr′) *n.* Utensils for kitchen use.
**kite** (kīt) *n.* [ME, bird of prey < OE *cyta.*] **1.** A light framework covered with cloth, plastic, or paper, designed to climb and fly in a steady breeze at the end of a long string. **2.** Any of the light sails of a ship, used in a light wind. **3.** Any of various predatory birds of the subfamilies Milvinae and Elaninae, having long, often forked tails. **4.** Negotiable paper, as a check, representing a fictitious financial transaction and used temporarily to sustain credit or raise money. *—v.* **kit·ed, kit·ing, kites.** *—vi.* **1.** To fly like a kite. **2.** To get money or credit with a kite. *—vt.* To use a kite to sustain credit or raise money.
**kith and kin** (kĭth′ ən kĭn′) *n.* [ME *kyth* < OE *cyð.*] Friends and neighbors.
**kitsch** (kĭch) *n.* [G. < *kitschen,* to put together sloppily.] **1.** Pretentious bad taste, esp. in the arts. **2.** Something that exemplifies kitsch. **—kitsch′y** *adj.*
**kit·ten** (kĭt′n) *n.* [ME *kitoun,* prob. < ONFr. **caton,* dim. of *cat,* cat < LLat. *cattus.*] A young cat. *—vi.* **-tened, -ten·ing, -tens.** To bear kittens.
**kit·ten·ish** (kĭt′n-ĭsh) *adj.* Playful and coy. **—kit′ten·ish·ly** *adv.* **—kit′ten·ish·ness** *n.*
**kit·ti·wake** (kĭt′ē-wāk′) *n.* [Imit. of its cry.] A gull, *Rissa tridactyla* or *R. brevirostris,* of northern regions.
**kit·tle** (kĭt′l) *adj.* [< Sc., to tickle < ME *kytyllen,* prob. of ON orig.] *Scot.* **1.** Unpredictable : capricious. **2.** Touchy.
**kit·ty[1]** (kĭt′ē) *n., pl.* **-ties.** [< KIT[1].] **1. a.** An extra hand or part of a hand to be used by the highest bidder in some card games. **b.** A fund made up of a portion of each player's winnings in a card game, used to pay the game expenses. **2.** A pool of money.
**kit·ty[2]** (kĭt′ē) *n., pl.* **-ties.** [Shortening and alteration of KITTEN.] *Informal.* A cat, esp. a kitten.
**kit·ty-cor·nered** (kĭt′ē-kôr′nərd) *adj.* Cater-cornered.
**Kitty Litter.** A trademark for absorbent material used in a box or pan for the waste of small domestic pets, esp. cats.
**ki·va** (kē′və) *n.* [Hopi.] An underground or partially underground room in a Pueblo Indian village, used by the men esp. for ceremonies or councils.
**ki·wi** (kē′wē) *n.* [Maori.] **1.** A flightless bird of the genus *Apteryx,* indigenous to New Zealand, having vestigial wings and a long slender bill. **2.** A native Asian vine, *Actinidia chinensis,* yielding fuzzy, edible fruit.
**Klan** (klăn) *n.* The Ku Klux Klan.
**Klans·man** (klănz′mən) *n.* A member of the Ku Klux Klan.
**klav·ern** (klăv′ərn) *n.* [KL(AN) + (C)AVERN.] A local unit of the Ku Klux Klan.
**Klax·on** (klăk′sən). A trademark for a loud horn formerly used on automobiles.
**Kleen·ex** (klē′nĕks′). A trademark for a soft cleansing tissue.
**Klein bottle** (klīn) *n.* [After Felix *Klein* (1849–1925).] A one-sided topologic surface having no inside or outside, formed by inserting the small open end of a tapered tube through the side of the tube and making it contiguous with the larger open end.
**klep·to·ma·ni·a** (klĕp′tə-mā′nē-ə, -mān′yə) *n.* [Gk. *kleptein,* to steal + -MANIA.] Obsessive desire to steal, esp. in the absence of economic necessity. **—klep′to·ma′ni·ac′** (-nē-ăk′) *n.*
**klieg light** (klēg) *n.* [After John H. *Kliegl* (1869–1959) and Anton T. *Kliegl* (1872–1927).] A powerful carbon-arc lamp generating an intense light and used esp. in making movies.
**klip·spring·er** (klĭp′sprĭng′ər) *n.* [Afr. : Du. *klip,* cliff + Du. *springer,* jumper < *springen,* to leap.] A small, hoofed, large-eared African mammal, *Oreotragus oreotragus.*
**kloof** (klo͞of) *n.* [Afr. < Du. < MDu. *clove.*] *So. Afr.* A deep ravine.
**klutz** (klŭts) *n.* [Yiddish < G. *Klotz,* wooden block < MHG *kloz.*] *Slang.* **1.** A clumsy, dull person. **2.** A bungler. **—klutz′i·ness** *n.* **—klutz′y** *adj.*
**kly·stron** (klī′strŏn′) *n.* [Orig. a trademark.] An electron tube for amplifying or generating radio waves of microwave range frequencies by means of velocity modulation.
**knack** (năk) *n.* [ME *knak.*] **1.** A clever, expedient, and specific way of doing something. **2.** A specific talent for doing something. **3.** *Archaic.* **a.** A cleverly designed device. **b.** A knickknack.
**knack·er** (năk′ər) *n.* [< Obs. *knacker,* saddler.] *Chiefly Brit.* **1.** One who buys worn-out or old livestock and sells the meat or hides. **2.** One who buys discarded structures and dismantles them to sell the materials. **—knack′er·y** (-ə-rē) *n.*
**knack·wurst** *also* **knock·wurst** (nŏk′wûrst′, -wo͝orst′) *n.* [G. : *knacken,* to crack (< MHG) + *Wurst,* sausage.] A short, thick, highly seasoned sausage similar to a frankfurter.
**knap[1]** (năp) *vt.* **knapped, knap·ping, knaps.** [ME *knappen.*] **1.** *Chiefly Brit.* To strike sharply : RAP. **2.** To break or chip flints with a sharp blow. **3.** *Chiefly Brit.* To snap at or bite. **—knap′per** *n.*
**†knap[2]** (năp) *n.* [ME < OE *cnæp.*] *Regional.* The crest of a hill : SUMMIT.
**knap·sack** (năp′săk′) *n.* [LG *knappsack* : *knappen,* to bite + *sack,* bag.] A supply or equipment bag, as of canvas or nylon, usu. strapped to the back.
**knap·weed** (năp′wēd′) *n.* [ME *knopwed* : *knop,* knob + *wed,* weed.] A plant of the genus *Centaurea,* having purplish thistlelike flowers.
**knar** *also* **knaur** (när) *n.* [ME *knarre.*] A knot on a tree.
**knave** (nāv) *n.* [ME < OE *cnafa,* boy.] **1.** An unprincipled, crafty person. **2.** *Archaic.* A manservant. **3.** JACK 2.
**knav·er·y** (nā′və-rē) *n.* **1.** Dishonest dealing. **2.** A piece of mischief or trickery.
**knav·ish** (nā′vĭsh) *adj.* Pertaining to or characteristic of a knave : ROGUISH. **—knav′ish·ly** *adv.* **—knav′ish·ness** *n.*
**knawel** (nôl) *n.* [G. *Knäuel,* ball of yarn < MHG *kliuwel,* dim. of *kliuwe,* ball of yarn < OHG *kliuwa.*] A low-growing, weedy Eurasian plant, *Scleranthus annuus,* having narrow leaves and inconspicuous green flowers.
**knead** (nēd) *vt.* **knead·ed, knead·ing, kneads.** [ME *kneden* < OE *cnedan.*] **1.** To mix and work (a substance) into a uniform mass, esp. to fold, press, and stretch (dough) with the hands. **2.** To make or shape by or as if by kneading. **—knead′er** *n.*
**knee** (nē) *n.* [ME < OE *cnēo.*] **1. a.** *Anat.* The joint or region of the human leg that is the articulation for the tibia, fibula, and patella. **b.** A joint of a vertebrate leg, as in the forelimb of a hoofed animal. **2.** One of the woody projections arising from the roots of some swamp-growing trees <cypress *knees*> *—vt.* **kneed, knee·ing, knees.** To push or strike with the knee.
**knee action** *n.* An automotive front-wheel suspension allowing independent vertical motion of each wheel.
**knee breeches** *pl.n.* Breeches extending to just below the knee.
**knee·cap** (nē′kăp′) *n.* PATELLA 1a.
**knee·cap** *vt.* **-capped, -cap·ping, -caps.** To shoot (a victim) in the knees so as to cripple or maim.
**knee-deep** (nē′dēp′) *adj.* **1.** Reaching to the knees. **2.** Submerged to the knees <*knee-deep* in snow> **3.** Deeply occupied or engaged<*knee-deep* in work>
**knee-high** (nē′hī′) *adj.* KNEE-DEEP 1. *—n.* (nē′hī′). A stocking that extends to just below the knee.
**knee·hole** (nē′hōl′) *n.* A space or opening for the knees, as under a desk or counter.
**knee jerk** *n.* A sudden involuntary kick forward produced by a smart tap to the tendon below the patella.
**knee-jerk** (nē′jûrk′) *adj.* **1.** Automatic <Civil unrest is often a *knee-jerk* reaction to totalitarianism.> **2.** Characterized by or reacting with unthinking predictability <"There is a *knee-jerk* quality to the simple assertion that reducing guns will reduce crime" *—New York Times*>
**kneel** (nēl) *vi.* **knelt** (nĕlt) or **kneeled, kneel·ing, kneels.** [ME *knelen* < OE *cnēowlian.*] To fall or rest on bent knees.
**kneel·er** (nē′lər) *n.* **1.** One who kneels. **2.** Something, as a cushion, to kneel on.
**knee·pad** (nē′păd′) *n.* A protective knee covering.
**knell** (nĕl) *v.* **knelled, knell·ing, knells.** [ME *knellen* < OE *cnyllan.*] *—vi.* **1.** To sound a bell, esp. for a funeral : TOLL. **2.** To sound mournfully or ominously. *—vt.* To signal, summon, or proclaim by tolling. *—n.* **1.** An act or instance of knelling : TOLL. **2.** A signal of disaster.

o͞o **boot** ou **out** th **thin** *th* **this** ŭ **cut** ûr **urge** y **young**
yo͞o **abuse** zh **vision** ə **about**, item, edible, gallop, circus

**knelt** (nělt) *v. var. p.t. & p.p. of* KNEEL.

**Knes·set** (kněs′ět′) *n.* [Heb. *Kěneseth*, assembly < *kanas*, he gathered.] The Israeli parliament.

**knew** (no͞o, nyo͞o) *v. p.t. of* KNOW.

**Knick·er·bock·er** (nĭk′ər-bŏk′ər) *n.* [After Diedrich *Knickerbocker*, fictitious author of *History of New York*, by Washington Irving.] **1. a.** A descendant of the Dutch settlers of New York. **b.** A New Yorker. **2. knickerbockers.** Full breeches gathered and banded just below the knee.

**knick·ers** (nĭk′ərz) *pl.n.* [Short for KNICKERBOCKERS.] **1.** Long bloomers once worn as underwear by women and girls. **2.** KNICKERBOCKERS 2.

**knick·knack** (nĭk′năk′) *n.* [Redup. of KNACK.] A trinket.

**knife** (nīf) *n., pl.* **knives** (nīvz) [ME *knif* < OE *cnīf*.] **1.** A cutting instrument having a sharp blade with a handle. **2.** A cutting edge : BLADE. —*v.* **knifed, knif·ing, knifes.** —*vt.* **1.** To use a knife on, esp. to cut, stab, or wound. **2.** *Informal.* To hurt, defeat, or betray by underhand means. —*vi.* To cut or slash a way with or as if with a knife. **—knif′er** *n.*

**knife-edge** (nīf′ěj′) *n.* **1.** The cutting edge of a blade. **2.** A sharp knifelike edge <felt the *knife-edge* of criticism> **3.** A metal wedge used as a low-friction fulcrum for a balancing beam or lever.

**knight** (nīt) *n.* [ME < OE *cniht*.] **1.** A medieval tenant giving military service as a mounted man-at-arms to a feudal landholder. **2.** A usu. high-born medieval gentleman-soldier raised by a sovereign to privileged military status after training as a page and squire. **3.** The holder of a nonhereditary dignity conferred by a sovereign in recognition of personal merit or services to the country. **4.** A member of an order or brotherhood designating its members knights. **5. a.** A zealous defender or champion of a principle or cause. **b.** A lady's devoted champion. **6.** A chess piece moved either two squares horizontally and one vertically or two vertically and one horizontally. —*vt.* **knight·ed, knight·ing, knights.** To raise (a person) to knighthood. **—knight′li·ness** *n.* **—knight′ly** *adj.*

**knight er·rant** (ěr′ənt) *n., pl.* **knights errant. 1.** A knight of medieval romance who wandered in search of adventure. **2.** One given to adventurous or quixotic conduct. **—knight′-er′rant·ry** (nīt′ěr′ən-trē) *n.*

**knight·hood** (nīt′ho͝od′) *n.* **1.** The rank, profession, or dignity of a knight. **2.** Behavior of or qualities worthy of a knight : CHIVALRY. **3.** Knights as a group.

**Knight of Co·lum·bus** (kə-lŭm′bəs) *n.* A member of a philanthropic fraternal society of Roman Catholic men.

**Knight of Pythias** *n.* A member of a secret philanthropic fraternal order.

**Knights of the Round Table** *pl.n.* The knights of the court of King Arthur in Arthurian legend.

**Knight Templar** *n., pl.* **Knights Templars.** A member of a 12th–14th cent. order of knights founded to protect pilgrims in the Holy Land during the Second Crusade.

**knish** (kə-nĭsh′) *n.* [Yiddish < R.] Dough stuffed with potato, meat, or cheese and baked or fried.

**knit** (nĭt) *v.* **knit** *or* **knit·ted, knit·ting, knits.** [ME *knitten* < OE *cnyttan*, to tie in a knot.] —*vt.* **1.** To make by intertwining yarn or thread in a series of connected loops. **2.** To unite securely and closely. **3.** To draw (the brows) together in wrinkles : FURROW. —*vi.* **1.** To make a fabric or garment by knitting. **2.** To come or grow together securely. **3.** To come together in wrinkles or furrows. **—knit′** *n.* **—knit′ter** *n.*

**knit·ting needle** (nĭt′ĭng) *n.* A long, thin, pointed rod used for knitting.

**knit·wear** (nĭt′wâr′) *n.* Knitted garments in general.

**knives** (nīvz) *n. pl. of* KNIFE.

**knob** (nŏb) *n.* [ME *knobbe*, prob. < MLG.] **1. a.** A rounded protuberance. **b.** A rounded dial. **2.** A prominent rounded hill or mountain. **—knobbed** *adj.* **—knob′by** *adj.*

**knob·ker·rie** (nŏb′kĕr′ē) *n.* [Afr. *knopkierie* : *knop*, knob (< MDu. *cnoppe*) + *kieri*, club < Hottentot *kirri*.] A short club with one knobbed end, used by South African tribesmen as a weapon.

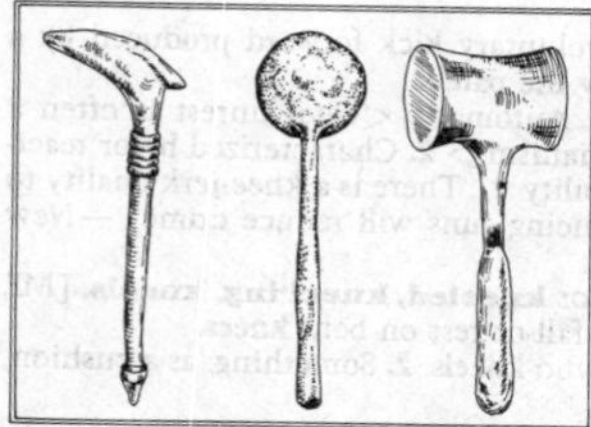

**knobkerrie**
*Three types of knobkerries*

**knock** (nŏk) *v.* **knocked, knock·ing, knocks.** [ME *knokken* < OE *cnocian*.] —*vt.* **1.** To strike with a hard blow. **2.** To cause to collide. **3.** To produce by hitting <*knocked* a hole in the fence> **4.** To instill as if with blows <*knocked* some sense into their heads> **5.** *Slang.* To criticize adversely : DISPARAGE. —*vi.* **1.** To strike a blow or series of blows. **2.** To collide. **3. a.** To make a clanking or pounding noise. **b.** To undergo engine knock. **—knock around** (*or* **about**). *Informal.* **1.** To be rough or brutal with : MALTREAT. **2.** To wander from place to place. **3.** To discuss or consider. **—knock back.** *Informal.* To gulp (an alcoholic drink). **—knock down. 1.** To disassemble into parts. **2.** To declare sold at an auction, as by striking a blow with a gavel. **3.** *Informal.* To reduce, as in price. **4.** *Slang.* To receive as wages : EARN. **—knock off. 1.** *Informal.* **a.** To take a break or rest from : STOP. **b.** To cease work. **2.** *Informal.* To make, accomplish, or consume hastily or easily. **3.** *Informal.* To eliminate : deduct <*knocked* 15% *off* the bill> **4.** *Slang.* To kill. **5.** *Slang.* To hold up or rob. **6.** *Informal.* To copy the design or production of. **—knock out. 1.** To render unconscious. **2.** To defeat by knocking down to the canvas for a count of ten in boxing. **3.** *Informal.* To render useless or inoperative <power *knocked out* by a storm> **4.** *Informal.* To exert or exhaust (oneself or another). **—knock together.** To make or assemble quickly or carelessly. **—knock up. 1.** *Chiefly Brit.* To gain the attention of or wake up by knocking at the door. **2.** To wear out : EXHAUST. —*n.* **1.** An instance of knocking : BLOW. **2.** The sound of a sharp tap on a hard surface : RAP. **3.** A clanking, pounding noise made by an engine, esp. one in poor operating condition. **4.** *Slang.* A cutting, often petty criticism. **—knock cold.** To knock out. **—knock dead.** *Slang.* To affect strongly, usu. positively <a virtuoso piano performance that *knocked us dead*> **—knock for a loop.** *Slang.* To surprise greatly : ASTONISH. **—knock out of the box.** *Baseball.* To force the removal of (an opposing pitcher) by heavy hitting.

**knock·a·bout** (nŏk′ə-bout′) *n.* A small sloop with a mainsail, a jib, and a keel but no bowsprit. —*adj.* **1.** Boisterous and rowdy. **2.** Appropriate for rough wear or use <*knockabout* clothes>

**knock·down** (nŏk′doun′) *adj.* **1.** Forceful enough to knock down or overwhelm : POWERFUL <a *knockdown* punch> **2.** Designed to be assembled and disassembled easily and quickly <*knockdown* office furniture> —*n.* **1.** An act of knocking down. **2.** An overwhelming blow. **3.** A device or mechanism designed to be assembled and disassembled quickly and easily.

**knock·down-drag·out** (nŏk′doun-drăg′out′) *adj.* Marked by roughness, violence, and acrimony.

**knock·er** (nŏk′ər) *n.* One that knocks, as a fixture for knocking on a door.

**knock-knee** (nŏk′nē′) *n.* An abnormal condition in which one knee is turned toward the other or in which each is turned toward the other. **—knock′-kneed′** *adj.*

**knock·off** (nŏk′ôf′, -ŏf′) *n. Informal.* A usu. inexpensive copy, as of a garment <a *knockoff* of a designer original>

**knock·out** (nŏk′out′) *n.* **1.** The act of knocking out or the state of being knocked out. **2.** The knocking out of an opponent in boxing. **3.** *Slang.* One that is very impressive or attractive.

**knockout drops** *pl.n. Slang.* A solution, as of chloral hydrate, put into a drink to render the drinker unconscious.

**knock·wurst** (nŏk′wûrst′, -wo͝orst′) *n. var. of* KNACKWURST.

**knoll[1]** (nōl) *n.* [ME *knolle* < OE *cnoll*.] A small rounded hill or mound : HILLOCK.

**knoll[2]** (nōl) [ME *knollen*, prob. alteration of *knellen*, to knell < OE *cnyllan*.] *Archaic.* —*vt. & vi.* **knolled, knoll·ing, knolls.** To ring or sound mournfully. **—knoll** *n.*

**knop** (nŏp) *n.* [ME *knoppe* < OE *cnop*.] A decorative knob.

**knot[1]** (nŏt) *n.* [ME < OE *cnotta*.] **1. a.** A compact intersection of interlaced material, as cord, ribbon, or rope. **b.** A fastening made by tying together lengths of material, as rope, in a prescribed way. **2.** A decorative bow of ribbon, fabric, or braid. **3.** A unifying bond, esp. a marriage bond. **4.** A tight group or cluster <*knots* of spectators> **5.** A difficult problem. **6. a.** A hard node, esp. on a tree, at a point from which a stem or branch grows. **b.** The circular, often darker cross section of such a node as it appears cross-grained on a piece of cut lumber. **7.** A protuberant growth in living tissue. **8.** *Naut.* **a.** A division on a log line used to measure the speed of a ship. **b.** A unit of speed, one nautical mile per hour, approx. 1.15 statute miles per hour. ***usage:*** *Knot* is a unit of nautical speed with the built-in meaning of "per hour." Therefore, a ship would properly be said to travel at ten *knots* (not at ten *knots per hour*). **c.** A distance of one nautical mile. —*v.* **knot·ted, knot·ting, knots.** —*vt.* **1.** To tie in or fasten with a knot. **2.** To entangle. **3.** To cause to form knots. —*vi.* **1.** To become entangled. **2.** To form a knot. **—knot′ted** *adj.*

**knot[2]** (nŏt) *n.* [Orig. unknown.] A shore bird, *Calidris canutus* or *C. tenvirostris*, related to the sandpiper.

**knot·grass** (nŏt′grăs′) *n.* **1.** A low-growing weedy plant, *Polygonum aviculare*, having tiny greenish flowers. **2.** A grass having jointed stems.

**knot·hole** (nŏt′hōl′) *n.* A hole in lumber where a knot used to be.

**knot·ty** (nŏt′ē) *adj.* **-ti·er, -ti·est. 1.** Tied or snarled in knots. **2.** Covered with knots or knobs : GNARLED. **3.** Difficult to compre-

ă **pat** ā **pay** âr **care** ä **father** ĕ **pet** ē **be** hw **which** ĭ **pit** ī **tie** îr **pier** ŏ **pot** ō **toe** ô **paw, for** oi **noise** o͝o **took**

hend or solve : PUZZLING <*knotty* financial problems> **—knot'ti·ness** *n.*

**knotty pine** *n.* Pine wood with a large number of knots, used esp. for furniture and paneling.

**knot·weed** (nŏt'wēd') *n.* A plant of the genus *Polygonum,* with jointed stems and inconspicuous flowers.

**knout** (nout) *n.* [Fr. < R. *knut,* prob. < *knūtr,* knot.] A leather scourge for flogging. *—vt.* **knout·ed, knout·ing, knouts.** To flog with a knout.

**know** (nō) *v.* **knew** (no͞o, nyo͞o), **known** (nōn), **know·ing, knows.** [ME *knowen* < OE *(ge)cnāwan.*] *—vt.* **1.** To perceive directly with the senses or mind. **2.** To believe to be true with absolute certainty. **3.** To have a practical understanding of or thorough experience with <*know* yacht racing> **4.** To be subjected to : EXPERIENCE <a person who had *known* no pain> **5. a.** To recognize (something) as being the same as something else previously known. **b.** To be familiar with. **6.** To be able to distinguish : RECOGNIZE. **7.** *Archaic.* To have sexual intercourse with <"And Adam *knew* Eve his wife; and she conceived" —Genesis 4:1> *—vi.* **1.** To possess knowledge. **2.** To be cognizant or aware. **—in the know.** Privy to restricted or secret data. **—know'a·ble** *adj.* **—know'er** *n.*

☆ ***syns:*** KNOW, APPREHEND, COMPREHEND, FATHOM, GRASP, UNDERSTAND *v. core meaning* : to perceive directly with the intellect <*knows* the job well>

**know-how** (nō'hou') *n. Informal.* The knowledge and skill required to do something right.

**know·ing** (nō'ĭng) *adj.* **1.** Possessing knowledge, intelligence, or comprehension. **2.** Possessing or exhibiting clever awareness and resourcefulness : SHREWD. **3.** Deliberate <*knowing* complicity> **—know'ing·ly** *adv.* **—know'ing·ness** *n.*

**know-it-all** (nō'ĭt-ôl') *n. Informal.* One who claims to know everything.

**knowl·edge** (nŏl'ĭj) *n.* [ME *knowlech* < *knowlechen,* to acknowledge < *knowen,* to know < OE *(ge)cnāwan.*] **1.** The state or fact of knowing. **2.** Familiarity, awareness, or comprehension acquired by experience or study. **3.** The sum or range of what has been perceived, discovered, or learned. **4.** Erudition <people of *knowledge*> **5.** Specific information. **6.** *Archaic.* Sexual intercourse.

☆ ***syns:*** **1.** KNOWLEDGE, INFORMATION, LORE, WISDOM *n. core meaning* : that which is known <new additions to our *knowledge* of the earth> **2.** KNOWLEDGE, EDUCATION, ERUDITION, INSTRUCTION, LEARNING, SCHOLARSHIP *n. core meaning* : facts, ideas, and skills that have been imparted <*knowledge* gained in the classroom>

**knowl·edge·a·ble** (nŏl'ĭ-jə-bəl) *adj.* Having or exhibiting knowledge. **—knowl'edge·a·bil'i·ty** (-bĭl'ĭ-tē), **knowl'edge·able·ness** *n.* **—knowl'edge·ably** *adv.*

**known** (nōn) *adj.* [P. part. of KNOW.] Proved, satisfactorily specified, or completely understood. **—known** *n.*

**know-noth·ing** (nō'nŭth'ĭng) *n.* **1. Know-Nothing.** A member of a mid-19th-cent. American political movement antagonistic toward immigrants and Roman Catholics. **2.** An ignorant person. **3.** An agnostic.

**knuck·le** (nŭk'əl) *n.* [ME *knokel.*] **1.** *Anat.* **a.** The prominence of the dorsal aspect of a joint of a finger, esp. one of the joints connecting the fingers to the hand. **b.** A rounded protuberance formed by the bones in a joint. **2.** A cut of meat centering on the carpal joint, as of a hog. **3.** The part of a hinge through which the pin passes. **4. knuckles.** Brass knuckles. *—vt.* **-led, -ling, -les. 1.** To press, rub, or hit with the knuckles. **2.** To shoot (a marble) with the thumb over the bent forefinger. **—knuckle down.** To devote oneself earnestly to a task. **—knuckle under.** To yield to pressure : cave in.

**knuckle ball** *n. Baseball.* A typically slow, randomly fluttering pitch thrown by gripping the ball with the knuckles of two or three fingers.

**knuck·le·bone** (nŭk'əl-bōn') *n.* A knobbed bone, as of a knuckle or joint.

**knuck·le-dust·ers** (nŭk'əl-dŭs'tərz) *pl.n. Slang.* Brass knuckles.

**knuck·le·head** (nŭk'əl-hĕd') *n.* A dumb person.

**knuckle joint** *n.* A hinged joint in which a pin fastens together two rods, one of which has an eye that fits between the two projections of the other.

**knuckle sandwich** *n. Slang.* A punch in the mouth.

**knur** (nûr) *n.* [ME *knor,* a swelling.] A gnarl.

**knurl** (nûrl) *n.* [Prob. < KNUR.] **1.** A protuberance, as a knob or knot. **2.** One of a series of small ridges, as along the edge of an object such as a thumbscrew. *—vt.* **knurled, knurl·ing, knurls.** To provide with knurls : MILL. **—knurl'y** *adj.*

**KO** (kā'ō') *Slang. —vt.* **KO'd, KO'ing, KO's.** KNOCK OUT 2. *—n.* (kā-ō', kā'ō') *pl.* **KO's.** KNOCKOUT 2.

**ko·a·la** (kō-ä'lə) *n.* [Native word in Australia.] An Australian arboreal marsupial, *Phascolarctos cinereus,* having dense grayish fur and feeding chiefly on eucalyptus leaves and bark.

**ko·an** (kō'än') *n.* [J. : *ko,* public + *an,* matter.] A riddle in the form of a paradox used in Zen Buddhism as an aid to meditation and a way of acquiring intuitive knowledge.

**ko·bo** (kô'bô') *n., pl.* **kobo.** [Alteration of COPPER.] —See table at CURRENCY.

**ko·bold** (kō'bôld') *n.* [G. < MHG *kobolt.*] **1.** A mischievous household elf in German folklore. **2.** A gnome that haunts underground places in German folklore.

**Ko·dak** (kō'dăk'). A trademark for a small hand camera and photographic film.

**Ko·di·ak bear** (kō'dē-ăk') *n.* [After *Kodiak* island, Alaska.] A brown bear, *Ursus arctos,* inhabiting Alaskan islands and coastal areas and occas. regarded as a separate species, *U. middendorffi.*

**kohl** (kōl) *n.* [Ar. *koḥl.*] A preparation used as eye make-up esp. by women.

**kohl·ra·bi** (kōl-rä'bē, -răb'ē) *n.* [G. < Ital. *cavolo rapa* : *cavolo,* cabbage (< Lat. *caulis*) + *rapa,* turnip < Lat.] A plant, *Brassica caulorapa,* with a thickened basal part eaten as a vegetable.

**Koi·ne** (koi-nā', koi'nā') *n.* [Gk. *koinē (dialektos),* common (language) < *koinos,* common.] **1.** A dialect of Greek that developed primarily from Attic and became the common language of the Hellenistic world from which later stages of Greek are descended. **2. koine.** A lingua franca.

**kok-sa·ghyz** (kōk'sə-gēz') *n.* [R. < Turk. *kok-sagïz* : *kok,* root + *sagïz,* rubber.] An Asian dandelion, *Taraxacum koksaghyz,* having fleshy roots that yield a form of rubber.

**ko·la** (kō'lə) *n.* [Of African orig.] An African tree, *Cola nitida* or *C. acuminata,* that bears nuts used in making beverages and medicines.

**kola nut** *n.* The nut of a kola tree, containing caffeine and theobromine and yielding an extract used in carbonated beverages and in pharmaceuticals.

**ko·lin·sky** (kə-lĭn'skē) *n., pl.* **-skies.** [R. *kolinskiĭ,* of Kola < *Kola,* peninsula in northwestern USSR.] **1.** A northern Eurasian mink, esp. *Mustela siberica.* **2.** The fur of the kolinsky.

**kol·khoz** (kŏl-kôz') *n.* [R. < *kollektivnoe khozyaistvo,* collective farm.] A Soviet collective farm.

**Kol Nid·re** (kōl nĭd'rā, -rə, kôl) *n.* [Aram. *kol nidhrē,* all the vows (the opening words of the prayer).] The opening prayer recited by Jewish worshipers on the eve of Yom Kippur.

**Ko·mo·do dragon** (kə-mō'dō) *n.* A large monitor lizard, *Varanus komodoensis,* indigenous to the Indonesian islands of Komodo and Flores.

**Kon·go** (kŏng'gō) *n.* A Bantu language spoken in the region of the lower Congo River.

**koo·doo** (ko͞o'do͞o) *n. var. of* KUDU.

**kook** (ko͝ok) *n.* [Poss. < CUCKOO.] *Slang.* An eccentric person. **—kook'i·ness** *n.* **—kook'y** *adj.*

**kook·a·bur·ra** (ko͝ok'ə-bûr'ə, -bŭr'ə) *n.* [Native word in Australia.] A large Australian kingfisher, *Dacelo novaeguineae* or *D. gigas,* having a call similar to raucous laughter.

**kookaburra**
*16 inches long*

**ko·peck** (kō'pĕk') *n.* [R. *kopeĭka.*] —See table at CURRENCY.

**kor** (kôr, kōr) *n.* [Heb. *kōr.*] HOMER².

**Ko·ran** (kə-răn', -rän', kô-, kō-) *n.* [Ar. *qur'ān* < *qara'a,* he read.] The sacred text of Islam, held to contain the revelations made by Allah to Mohammed. **—Ko·ran'ic** (-răn'ĭk) *adj.*

**Ko·re·an** (kə-rē'ən, kô-, kō-) *adj.* Of or pertaining to Korea or its people, language, or culture. *—n.* **1.** A native or resident of Korea. **2.** The language of the Koreans, of no known linguistic affiliation.

**Korean War.** *n.* A military action between North Korea and U.N. forces from 1950 to 1953.

**ko·sher** (kō'shər) *adj.* [Yiddish < Heb. *kāshēr,* proper.] **1.** Conforming to or prepared in accordance with Jewish dietary laws, as: **a.** Slaughtered or prepared for eating according to rabbinic law : ritually pure <*kosher* lamb> **b.** Specializing in the preparation or sale of such food <a *kosher* meat market> **2.** *Slang.* **a.** Proper : correct : permissible <*kosher* sales tactics> **b.** Genuine : legitimate. *—vt.* **-shered, -sher·ing, -shers.** To make kosher.

**ko·to** (kō'tō) *n., pl.* **-tos.** [J.] A Japanese musical instrument having 13 strings stretched over an oblong box.

**kou·miss** (ko͞o-mĭs', ko͞o'mĭs) *n. var. of* KUMISS.

**kow·tow** (kou-tou', kou'tou') *n.* [Chin. (Mandarin) *ke¹ tou²* : *ke¹,* to knock + *tou²,* head.] **1.** A Chinese salutation in which one touches the forehead to the ground as an expression of respect or submission. **2.** An obsequious act. —*vi.* **-towed, -tow·ing, -tows. 1.** To perform a kowtow. **2.** To show servile deference : FAWN.

**Kr** *symbol for* KRYPTON.

**kraal** (krôl, kräl) *n.* [Afr. < Port. *curral,* pen.] *So. Afr.* **1.** A native village typically consisting of huts surrounded by a stockade. **2.** An enclosure for livestock.

**kraft** (krăft) *n.* [G., strength < OHG.] A tough wrapping paper made from sulfate wood pulp.

**krait** (krīt) *n.* [Hindi *karait.*] Any of several venomous southeast Asian snakes of the genus *Bungarus.*

**kra·ken** (krä'kən) *n.* [Dial. Norw.] A legendary Scandanavian sea monster.

**K ration** *n.* A lightweight, emergency field ration used by the U.S. Army in World War II.

**kraut** (krout) *n.* [G. —see SAUERKRAUT.] Sauerkraut.

**Krebs cycle** (krĕbz) *n.* [After Sir Hans Adolf *Krebs* (1900–1981).] A series of enzymatic reactions in aerobic organisms involving oxidative metabolism of acetyl units, esp. during the process of respiration, to provide the main source of cellular energy in the form of phosphate-rich ATP.

**Krem·lin** (krĕm'lĭn) *n.* [Fr. < R. *kreml',* citadel.] **1. a.** The citadel of an ancient Russian city. **b.** The citadel of Moscow, housing the offices of the Soviet government. **2.** The Soviet government.

**Krem·lin·ol·o·gy** (krĕm'lə-nŏl'ə-jē) *n.* Study of the policies and strategy of the Soviet government. **—Krem'lin·o·log'i·cal** (-nə-lŏj'ĭ-kəl) *adj.* **—Krem'lin·ol'o·gist** *n.*

**kreu·zer** (kroit'sər) *n.* [G. < MHG *kriuzer* < *kriuze,* cross < OHG *kruzi* < Lat. *crux.*] A small coin of low value once used in Austria and Germany.

**Kriem·hild** (krēm'hĭld', -hĭlt') *also* **Kriem·hil·de** (krēm-hĭl'də) *n.* The wife of Siegfried and avenger of his murder in Germanic legend.

**krill** (krĭl) *pl.n.* [Norw., young of fish.] Small marine crustaceans of the order Euphausiacea, the principal food of whalebone whales.

**krim·mer** (krĭm'ər) *n.* [G. < *Krim,* Crimea.] Gray curly fur made from the pelts of lambs of the Crimean region.

**kris** (krēs) *n.* [Malay *kĕris.*] A Malayan sword with a wavy double-edged blade.

**Krish·na** (krĭsh'nə) *n.* [Hindi < Skt. *kṛṣṇaḥ* < *kṛṣṇa-,* black.] The eighth and principal avatar of Vishnu, often depicted as a handsome young man playing a flute. **—Krish'na·ism** *n.*

**Kriss Krin·gle** (krĭs krĭng'gəl) *n.* [G. *Christkindl,* Santa Claus, the Christ child : *Christ,* Christ + *Kindl,* dim. of *Kind,* child < OHG.] Santa Claus.

**kro·na**[1] (krō'nə) *n., pl.* **-nur** (-nər) [Icel. *krōna,* prob. < Lat. *corona,* wreath. —see CROWN.] —See table at CURRENCY.

**kro·na**[2] (krō'nə) *n., pl.* **-nor** (-nôr', -nər) [Swed., prob. ult. < Lat. *corona,* wreath. —see CROWN.]—See table at CURRENCY.

**kro·ne** (krō'nə) *n., pl.* **-ner** (-nər) [Norw., prob. ult. < Lat. *corona,* wreath. —see CROWN.] —See table at CURRENCY.

**kro·ner** (krō'nər) *n.* [Dan., prob. ult. < Lat. *corona,* wreath. —see CROWN.] **1.** —See table at CURRENCY. **2.** *pl. of* KRONE.

**kro·nor** (krō'nôr', -nər) *n. pl. of* KRONA[2].

**kro·nur** (krō'nər) *n. pl. of* KRONA[1].

**Kru·ger·rand** (kro͞o'gə-rănd', -ränd') *n.* [Afr., after S.J.P. *Kruger* (1825–1904) + *rand,* rand.] A gold coin issued by the Republic of South Africa.

**kryp·ton** (krĭp'tŏn') *n.* [Gk. *krupton,* neut. of *kruptos,* hidden < *kruptein,* to hide.] *Symbol* **Kr** A whitish, inert gaseous element used chiefly in gas-discharge lamps and fluorescent lamps; atomic number 36; atomic weight 83.80.

**Ksha·tri·ya** (kə-shăt'rē-ə, -chăt'-) *n.* [Skt. *kṣatriyaḥ* < *kṣatram,* rule, power < *kṣayati,* he rules.] **1.** A major Hindu caste, including the professional, governing, and military occupations. **2.** A member of the Kshatriya.

**ku·chen** (ko͞o'kən, -KHən) *n.* [G. *Kuchen* < MHG *kuoche,* cake < OHG *kuocho.*] A yeast-raised coffee cake.

**ku·dos** (kyo͞o'dŏs', -dōs', ko͞o'-) *n.* [Gk.] Acclaim or prestige resulting from notable achievement or high position <"the *kudos* of the Presidency of the United States" —Eric F. Goldman>

▲ word history: The word *kudos* is etymologically a singular form, a modern borrowing of Greek *kudos,* "glory, renown." In very recent times, however, *kudos* has been reanalyzed as a plural form and consequently a new singular *kudo* sometimes occurs. Certain features of *kudos* predispose it to this kind of treatment. In the first place, it is an unfamiliar word, drawn from the vocabulary of Homer by academic and learned persons. In their usage it did not often occur as the subject of a sentence, where the verb could provide a clue to whether *kudos* was singular or plural. Secondly, *kudos* has no recorded plural in English. A person unfamiliar with Homeric Greek who saw the form *kudos* in an English publication would be likely to interpret it as the regular plural of a noun ending in *o,* like *typos* for *typo* and *altos* for *alto.* Once *kudos* was treated as a plural, the linguistic pressure to supply a singular would have been very strong. Although the form *kudo* has not achieved general acceptance and the construction "kudos are" is often considered incorrect, the linguistic processes in the development of *kudo/kudos* as the singular and plural forms of an English noun are highly productive and have been going on in English and other languages since prehistoric times. The development of the singular *pea* from the earlier singular form *pease* is an example of the same kind of reanalysis.

**ku·du** *also* **koo·doo** (ko͞o'do͞o) *n., pl.* **kudu** *or* **-dus** *also* **koodoo** *or* **-doos.** [Afr. *koedoe,* of Xhosa orig.] An African hoofed mammal, *Tragelaphus strepsiceros* or *T. imberbis,* having a brownish coat with narrow white stripes and long spirally curved horns in the male.

**kud·zu** (ko͝od'zo͞o) *n.* [J. *kuzu.*] A Japanese vine, *Pueraria lobata,* with compound leaves and reddish-purple flower clusters, grown for fodder and forage and for containment of erosion.

kudzu

**Ku Klux·er** (ko͞o klŭk'sər, kyo͞o) *n.* A member of the Ku Klux Klan. **—Ku Klux'ism** *n.*

**Ku Klux Klan** (ko͞o' klŭks klăn', kyo͞o') *n.* [Orig. unknown.] **1.** A secret society organized in the South after the American Civil War to reassert white supremacy through terrorism. **2.** A secret fraternal organization founded in Georgia in 1915 and dedicated to maintaining legal and de facto segregation of blacks.

**ku·lak** (ko͞o-lăk', ko͞o'lăk', -läk') *n.* [R.] A prosperous Russian peasant during czarist times, the October Revolution, and the early 1920's and 1930's.

**Kul·tur** (ko͝ol-to͝or') *n.* [G. < Lat. *cultura,* culture. —see CULTURE.] **1.** Culture. **2.** An idealized view of German culture and civilization.

**Kul·tur·kampf** (ko͝ol-to͝or'kämpf') *n.* [G. : *Kultur,* Kultur + *Kampf,* struggle < OHG *kamph,* prob. ult. < Lat. *campus,* field.] A conflict between secular and religious authorities.

**ku·miss** *also* **kou·miss** (ko͞o-mĭs', ko͞o'mĭs) *n.* [R. *kumys,* of Tatar orig.] A drink made from the fermented milk of a mare or camel.

**küm·mel** (kĭm'əl, kü'məl) *n.* [G., cumin seed < OHG *kumīn,* ult. < Lat. *cuminum,* —see CUMIN.] A colorless liqueur flavored chiefly with caraway.

**kum·quat** (kŭm'kwŏt') *n.* [Cantonese *kam kwat,* golden orange.] **1.** A tree or shrub of the genus *Fortunella,* bearing a small, edible, orangelike fruit. **2.** The citrus fruit of the kumquat, with an acid pulp and a thin edible rind.

**kun·da·li·ni** (ko͞on'də-lē'nē) *n.* [Skt. *kuṇḍalinī* < *kuṇḍalin-,* coiled < *kundalam,* ring.] Spiritual energy in Yogic tradition that lies dormant at the base of the spine until it is activated and channeled upward to the brain to produce enlightenment.

**kung fu** (ko͝ong' fo͞o', go͝ong'-) *n.* [Chin. (Mandarin) *gong*[1] *fu*[5] : *gong*[1], skill + *fu*[5], distinguished person, artisan.] A Chinese art of self-defense similar to karate.

**kunz·ite** (ko͝ont'sīt') *n.* [After George F. *Kunz* (1856–1932).] A lilac-colored spodumene used as a gemstone.

**Kuo·yu** (kwô'yo͞o') *n. var. of* GUOYU.

**Kurd·ish** (kûr'dĭsh, ko͝or'-) *n.* The Iranian language of the Kurds.

**kur·ra·jong** (kûr'ə-jông', -jŏng', kŭr'-) *n.* [Native word in Australia.] An Australian tree, *Brachychiton populneum,* having evergreen leaves and yellowish or reddish flowers.

**kur·to·sis** (kər-tō'sĭs) *n.* [Gk. *kurtōsis,* curvature < *kurtos,* convex.] *Statistics.* The general form or a quantity indicative of the general form of a frequency curve near the distribution's mean.

**ku·ru**[1] (ko͞o'ro͞o) *n.* [Turk. *kuruş.*] —See table at CURRENCY.

**ku·ru**[2] (ko͞or'o͞o) *n.* [Native word in New Guinea.] A fatal neurological disease caused by a slow-acting virus, seen in New Guinea.

**kvass** (kə-väs') *n.* [R. *kvas.*] A fermented Russian beverage made from rye or barley.

**kvetch** (kə-vĕch') [Yiddish *kvetchen* < G. *quetschen,* to squeeze.] *Slang.* —*vi.* **kvetched, kvetch·ing, kvetch·es.** To complain persistently and querulously. —*n.* A chronic complainer.

**Kwa** (kwä) *n.* A branch of the Niger-Congo language family including Ibo, Yoruba, and other West African languages. **—Kwa** *adj.*

**kwa·cha** (kwä'chə) *n.* [Native word in Zambia.] —See table at CURRENCY.

**kwan·za** (kwän'zə) *n., pl.* **kwanza** *or* **-zas.** [Swahili.] —See table at CURRENCY.

**kwa·shi·or·kor** (kwä'shē-ôr'kôr') *n.* [Native word in Ghana.] Severe malnutrition esp. in children, marked by anemia, edema, potbelly, depigmentation of the skin, and loss of hair or change in hair color.
**ky·ack** (kī'ăk') *n.* [Orig. unknown.] A packsack that hangs on either side of a packsaddle.
**ky·a·nite** (kī'ə-nīt') *n.* [G. *Zyanit* : Gk. *kyanos*, dark blue enamel -*it*, -ite.] A usu. blue mineral, $Al_2SiO_5$, used as a refractory.
**kyat** (chät) *n.* [Burmese.] —See table at CURRENCY.
**ky·lix** (kī'lĭks, kĭl'ĭks) *n.*, *pl.* **ky·li·kes** (kī'lĭ-kēz', kĭl'ĭ-) [Gk. *kulix*.] A shallow, usu. tall-stemmed ancient Greek drinking cup.
**ky·mo·graph** (kī'mə-grăf') *n.* [Gk. *kuma*, something swollen (< *kuein*, to swell) + GRAPH.] An instrument for recording variations in pressure, as of the blood. **—ky'mo·graph'ic** *adj.*
**Kym·ric** (kĭm'rĭk) *adj.* & *n. var. of* CYMRIC.
**Kym·ry** (kĭm'rē) *n. var. of* CYMRY.
**ky·pho·sis** (kī-fō'sĭs) *n.* [Gk. *kuphōsis* < *kuphos*, bent.] Abnormal rearward curvature of the spine. **—ky·phot'ic** (-fŏt'ĭk) *adj.*
**Kyr·i·e** (kîr'ē-ā') *n.* [LLat. < Gk. *Kurie eleēson*, Lord, have mercy.] A liturgical prayer in the Christian church beginning with or composed of the words "Lord, have mercy."
**Kyrie e·le·i·son** (ĭ-lā'ĭ-sŏn', -sən) *n.* Kyrie.

# Ll

**l** or **L** (ĕl) *n.*, *pl.* **l's** or **L's.** **1.** The 12th letter of the English alphabet. **2.** A speech sound represented by the letter *l.* **3.** The 12th in a series. **4.** Something shaped like the letter L. **5.** The Roman numeral for 50.
**la**[1] (lä) *n.* [ME < Med. Lat. —see GAMUT.] *Mus.* **1.** The sixth tone of the diatonic scale. **2.** The tone A.
**†la**[2] (lä) *interj. Chiefly Regional.* —Used to indicate surprise.
**La** *symbol for* LANTHANUM.
**laa·ger** (lä'gər) *n.* [Obs. Afr. *lager.*] A defensive encampment surrounded by armored vehicles or by wagons. *—vi.* **-gered, -ger·ing, -gers.** To camp in a laager.
**lab** (lăb) *n.* A laboratory.
**lab·a·rum** (lăb'ər-əm) *n.*, *pl.* **-a·ra** (-ər-ə) [LLat.] **1.** A religious banner, esp. one carried in processions. **2.** The banner adopted by Constantine the Great after his conversion to Christianity.
**lab·da·num** (lăb'də-nəm) *also* **lad·a·num** (lăd'n-əm) *n.* [Med. Lat.] A resinous exudation of certain Old World plants of the genus *Cistus*, yielding a fragrant oil used in perfumes and flavorings.
**la·bel** (lā'bəl) *n.* [ME, ornamental strip of cloth < OFr., poss. of Germanic orig.] **1.** An object serving as a means of identification, esp. a small piece of cloth or paper attached to an article to indicate its origin, owner, contents, use, or destination. **2.** A descriptive term : EPITHET. **3.** *Computer Sci.* A symbol or set of symbols identifying the contents of a file, memory, tape, or record. **4.** A molding over a window or door. **5.** A figure in a heraldic field consisting of a narrow horizontal bar with several pendants. **6.** A distinctive name identifying a manufacturer or product. *—vt.* **-beled, -bel·ing, -bels** or **-belled, -bel·ling, -bels. 1.** To attach a label to. **2.** To identify or designate with a label. **3.** To describe or classify as. **—la'bel·er, la'bel·ler** *n.*
**la·bel·lum** (lə-bĕl'əm) *n.*, *pl.* **-bel·la** (-bĕl'ə) [NLat. < Lat., little lip, dim. of *labrum*, lip.] The often enlarged lip of an orchid. **—la·bel'late** (-ĭt) *adj.*
**la·bi·a** (lā'bē-ə) *n. pl. of* LABIUM.
**la·bi·al** (lā'bē-əl) *adj.* [Med. Lat. *labialis* < Lat. *labium*, lip.] **1.** Of or relating to the lips or labia. **2.** Articulated with one lip or both lips, as the sounds *b, m, v,* or *w.* **—la'bi·al** *n.* **—la'bi·al·ly** *adv.*
**la·bi·al·ize** (lā'bē-ə-līz') *vt.* **-ized, -iz·ing, -iz·es.** To make labial : ROUND. **—la'bi·al·i·za'tion** *n.*
**labia ma·jo·ra** (mə-jôr'ə, -jōr'ə) *pl.n.* [NLat. : Lat. *labia*, lips + Lat. *majora*, larger.] Two rounded folds of tissue forming the external lateral boundaries of the vulva.
**labia mi·no·ra** (mə-nôr'ə, -nōr'ə) *pl.n.* [NLat. : Lat. *labia*, lips + Lat. *minora*, smaller.] Two narrow folds of tissue enclosed within the cleft of the labia majora.
**la·bi·ate** (lā'bē-ĭt, -āt') *adj.* [NLat. *labiatus* < Lat. *labium*, lip.] **1.** Having lips or liplike parts. **2.** *Bot.* **a.** Bearing or characterizing flowers with the corolla divided into two liplike parts. **b.** Of or belonging to the family Labiatae. **—la'bi·ate** *n.*
**la·bile** (lā'bīl', -bəl) *adj.* [ME *labil*, forgetful, wandering < Lat. *labilis*, liable to slip < *labi*, to slip.] **1.** Open to change : ADAPTABLE. **2.** Constantly undergoing or likely to undergo chemical change : UNSTABLE. **—la·bil'i·ty** (-bĭl'ĭ-tē) *n.*
**labio-** *pref.* [< Lat. *labium*, lip.] Labial <*labio*velar>
**la·bi·o·den·tal** (lā'bē-ō-dĕn'tl) *adj.* Articulated with the lip and teeth, as the sounds *f* and *v.* *—n.* A labiodental sound.
**la·bi·o·ve·lar** (lā'bē-ō-vē'lər) *adj.* Being simultaneously labial and velar, as *kw* in *quick.* *—n.* A labiovelar sound.
**la·bi·um** (lā'bē-əm) *n.*, *pl.* **-bi·a** (-bē-ə) [NLat. < Lat., lip.] **1.** *Anat.* Any of four folds of tissue of the female external genitalia. **2.** *Zool.* A liplike structure, as that forming the floor of the mouth in insects. **3.** *Bot.* One of the liplike divisions of a labiate corolla.
**la·bor** (lā'bər) *n.* [ME < OFr. < Lat.] **1.** Physical or mental exertion : WORK. **2.** A specific task. **3.** A particular form of work or method of working <agricultural *labor*> **4.** Work for wages as opposed to work for profit. **5. a.** Workers as a whole. **b.** The trade-union movement, esp. its officials. **6. Labor.** A political party furthering the interests of workers, esp. in Great Britain. **7.** Something made by labor. **8.** The physical efforts of childbirth : PARTURITION. *—v.* **-bored, -bor·ing, -bors.** *—vi.* **1.** To work. **2.** To strive painstakingly. **3. a.** To proceed with great effort : PLOD. **b.** To pitch and roll, as a ship. **4.** To suffer from a burden or disadvantage <*labor* under a handicap> **5.** To undergo the efforts of childbirth. *—vt.* **1.** To deal with in exhaustive or too much detail <*labor* the question> **2.** To distress : burden. *—adj.* **1.** Of or relating to labor. **2. Labor.** Of or relating to a political party furthering the interests of the working class. **—la'bor·er** *n.*
☆ **syns:** LABOR, DRUDGERY, MOIL, SWEAT, TOIL, TRAVAIL, WORK *n. core meaning* : usu. difficult and exhausting physical exertion <found little joy in *labor*>
**lab·o·ra·to·ry** (lăb'rə-tôr'ē, -tōr'ē, lə-bŏr'ə-trē, -tə-rē) *n.*, *pl.* **-ries.** [Med. Lat. *laboratorium* < Lat. *laborare*, to labor < *labor*, labor.] **1.** A room or building equipped for scientific research. **2.** A place where drugs and chemicals are produced. **3.** A place for practice, observation, or testing. **4.** A period of the academic day devoted to work or study in a laboratory.
**Labor Day** *n.* The first Monday in Sept., a legal holiday in the United States and Canada honoring working people.
**la·bored** (lā'bərd) *adj.* **1.** Done or made with effort. **2.** Lacking natural ease : STRAINED.
**la·bor-in·ten·sive** (lā'bər-ĭn-tĕn'sĭv) *adj.* Requiring or having a large expenditure of labor in comparison to capital <"industries . . . like mining, trucking, and apparel . . . tend to be *labor-intensive* . . . some of which can barely afford to fund a pension plan" *—Forbes*>
**la·bo·ri·ous** (lə-bôr'ē-əs, -bōr'-) *adj.* [ME < OFr. *laborieux* < Lat. *laboriosus* < *labor*, labor.] **1.** Requiring or characterized by long, hard work : LABORED. **2.** Industrious : hard-working. **—la·bo'ri·ous·ly** *adv.* **—la·bo'ri·ous·ness** *n.*
**la·bor·ite** (lā'bə-rīt') *n.* **1.** A supporter or member of a labor movement or union. **2. Laborite.** A member of a political party furthering the interests of labor.
**la·bor·sav·ing** (lā'bər-sā'vĭng) *adj.* Designed to decrease or conserve the amount of human labor needed.
**labor union** *n.* An association of wage earners formed to help members secure satisfactory wages, benefits, and working conditions.
**la·bour** (lā'bər) *n.* & *v.* & *adj. Chiefly Brit. var. of* LABOR.
**la·bour·ite** (lā'bə-rīt') *n. Chiefly Brit. var. of* LABORITE.
**la·bra** (lā'brə) *n. pl. of* LABRUM.
**Lab·ra·dor Current** (lăb'rə-dôr') *n.* A cold ocean current flowing southward from Baffin Bay along the Labrador coast and turning east after joining the Gulf Stream.
**lab·ra·dor·ite** (lăb'rə-dôr'īt', -dô-rīt') *n.* [After *Labrador* peninsula, Canada.] A plagioclase feldspar occurring in igneous rocks and marked by brilliant colors in some specimens.
**Labrador retriever** *n.* [After *Labrador*, a region of Canada.] A dog orig. bred in Newfoundland and having a short, dense coat and a tapering tail.
**la·bret** (lā'brĭt) *n.* [Lat. *labrum*, lip + -ET.] A piece of jewelry inserted in a perforation in the lip.

**la·brum** (lā'brəm) *n., pl.* **-bra** (-brə) [NLat. < Lat., lip.] A lip or liplike structure, as that forming the roof of the mouth in insects.
**la·bur·num** (lə-bûr'nəm) *n.* [NLat. *Laburnum,* genus name < Lat. *laburnum,* broad-leaved bean-trefoil.] A tree or shrub of the genus *Laburnum,* esp. *L. anagyroides,* cultivated for its drooping yellow flower clusters.
**lab·y·rinth** (lăb'ə-rĭnth') *n.* [Lat. *labyrinthus* < Gk. *laburinthos.*] **1.** An intricate structure of interconnecting passages and dead-end alleys through which it is difficult to find one's way : MAZE. **2.** Something highly intricate or convoluted : PERPLEXITY <a *labyrinth* of legal precedures> **3.** *Anat.* **a.** A group of communicating anatomical cavities. **b.** The internal ear, composed of the semicircular canals, vestibule, and cochlea.
**lab·y·rin·thi·an** (lăb'ə-rĭn'thē-ən) *adj.* Labyrinthine.
**lab·y·rin·thine** (lăb'ə-rĭn'thĭn, -thēn') *adj.* Of, relating to, resembling, or constituting a labyrinth : INTRICATE.
**lac**[1] (lăk) *n.* [Du. *lac* or Fr. *laque* both < Hindi *lākh* < Skt. *lākṣā,* red dye, resin.] A resinous secretion of the lac insect used in shellac.
**lac**[2] *also* **lakh** (lăk) *n.* [Hindi *lākh* < Skt. *lakṣam,* mark.] **1.** The sum of 100,000 in India. **2.** A very large number.
**lac·co·lith** (lăk'ə-lĭth') *n.* [Gk. *lakkos,* cistern + -LITH.] A mass of igneous rock intruded between layers of sedimentary rock, causing uplift.

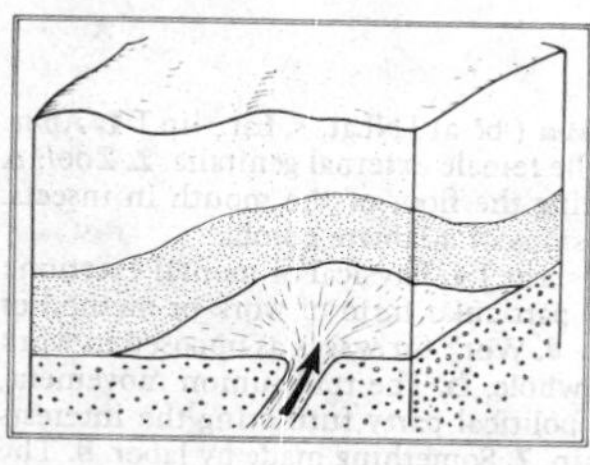

**laccolith**

**lace** (lās) *n.* [ME < OFr. < Lat. *laqueus,* noose.] **1.** A cord or ribbon for drawing and tying together two opposite edges, as of a shoe. **2.** A delicate fabric made of thread or yarn in an open weblike pattern. **3.** Gold or silver braid ornamenting a uniform. *—v.* **laced, lac·ing, lac·es.** *—vt.* **1.** To thread a cord through the eyelets or around the hooks of. **2. a.** To draw together and tie the laces of. **b.** To constrict or restrain by tightening laces, esp. of a corset. **3.** To pull or pass through : INTERTWINE. **4.** To decorate with or as if with lace. **5.** To add a touch of liquor to. **6.** To streak with color. **7.** To give a beating to : THRASH. **8.** *Computer Sci.* To punch holes in all the rows of (a punch-card column). *—vi.* To be fastened with a lace. **—lac'er** *n.*
**lace-cur·tain** (lās'kûr'tn) *adj.* Aspiring to the middle class.
**lac·er·ate** (lăs'ə-rāt') *vt.* **-at·ed, -at·ing, -ates.** [Lat. *lacerare, lacerat-* < *lacer,* lacerated.] **1.** To rip, cut, or tear. **2.** To cause deep emotional pain to : DISTRESS. *—adj.* (-rĭt, -rāt'). **1.** Mangled : torn. **2.** Wounded. **3.** Having jagged, deeply cut edges <a *lacerate* leaf> **—lac'er·a'tive** *adj.* **—lac'er·a'tion** *n.*
**La·cer·ta** (lə-sûr'tə) *n.* [NLat. < Lat. *lacerta,* lizard.] A constellation in the Northern Hemisphere.
**lace·wing** (lās'wĭng') *n.* Any of various greenish or brownish insects of the families Chrysopidae and Hemerobiidae, with four gauzy wings, threadlike antennae, and larvae that feed on insect pests, as aphids and scale insects.
**lach·es** (lăch'ĭz, lā'chĭz) *n., pl.* **laches.** [ME, negligence < AN < OFr. *lasche,* lax < Lat. *laxus.*] *Law.* Negligence, esp. delay in asserting a claim or a right.
**Lach·e·sis** (lăk'ĭ-sĭs) *n.* [Gk. *Lakhesis* < *lankhanein,* to obtain by lot.] *Gk. Myth.* One of the three Fates.
**lach·ry·mal** *also* **lac·ri·mal** (lăk'rə-məl) [Med. Lat. *lachrymalis* < Lat. *lacrima,* tear.] *adj.* **1.** Of or relating to tears. **2.** Of, relating to, or comprising the glands that produce tears. *—n.* **lachrymals.** The lachrymal glands.
**lach·ry·ma·tor** (lăk'rə-mā'tər) *n.* [< Lat. *lacrimatio,* a weeping < *lacrimare,* to cry < *lacrima,* tear.] Tear gas.
**lach·ry·mose** (lăk'rə-mōs') *adj.* [Lat. *lacrimosus* < *lacrima,* tear.] **1.** Weeping or given to weeping : TEARFUL. **2.** Causing or tending to cause tears : SORROWFUL. **—lach'ry·mose'ly** *adv.*
**lac·ing** (lā'sĭng) *n.* **1.** A cord or ribbon that laces : LACE. **2.** A beating or trouncing. **3.** A touch of liquor in a beverage or food. **4.** *Computer Sci.* A set of multiple holes in a punch card signifying to a device the end of a card run.
**la·cin·i·ate** (lə-sĭn'ē-ĭt, -āt') *adj.* [< Lat. *lacinia,* edge.] **1.** Fringed. **2.** With edges cut into narrow, fringelike segments or lobes <*laciniate* petals> **—la·cin'i·a'tion** *n.*
**lac insect** *n.* An insect of the subfamily Lacciferinae, esp. *Laccifer lacca* of southern Asia, the female of which secretes the resinous substance lac.
**lack** (lăk) *n.* [ME.] **1.** A deficiency or absence : WANT <a *lack* of companionship> **2.** Something needed. *—v.* **lacked, lack·ing, lacks.** *—vt.* **1.** To be entirely without or have very little of. **2.** To be in need of. *—vi.* **1.** To be wanting or deficient. **2.** To be in need of something.
☆ ***syns:*** LACK, NEED, REQUIRE, WANT *v. core meaning* : to be without what is needed, required, or essential <Their house *lacked* adequate heating.>
**lack·a·dai·si·cal** (lăk'ə-dā'zĭ-kəl) *adj.* [*Lackadaisy,* alteration of LACKADAY + -IC + -AL.] Lacking spirit, liveliness, or interest : LANGUID. **—lack'a·dai'si·cal·ly** *adv.* **—lack'a·dai'si·cal·ness** *n.*
**lack·a·day** (lăk'ə-dā') *interj.* [Alteration of *alack the day.*] *Archaic.* —Used to express regret or disapproval.
**lack·ey** (lăk'ē) *n., pl.* **-eys.** [OFr. *laquais* < Catalan *alacay.*] **1.** A liveried male servant : FOOTMAN. **2.** An obsequious follower : TOADY. *—v.* **-eyed, -ey·ing, -eys.** *—vt.* To wait on as a lackey : ATTEND. *—vi. Obs.* To act in a servile way : TOADY.
**lack·lus·ter** (lăk'lŭs'tər) *adj.* Without luster, brightness, or vitality : INSIPID.
**la·con·ic** (lə-kŏn'ĭk) *adj.* [Lat. *Laconicus,* Spartan < Gk. *Lakōnikos* (from the reputation of the Spartans for brevity of speech).] Using or marked by the use of few words : CONCISE. **—la·con'i·cal·ly** *adv.*
**lac·o·nism** (lăk'ə-nĭz'əm) *n.* Terse or succinct style or expression.
**lac·quer** (lăk'ər) *n.* [Obs. Fr. *lacre,* sealing wax < Port. < *laca,* resin of the lac insect < Hindi *lākh.* —see LAC[1].] **1.** Any of various clear or colored synthetic coatings made by dissolving nitrocellulose or other cellulose derivatives together with plasticizers and pigments in a mixture of volatile solvents and used to impart a high gloss to surfaces. **2.** A glossy, often resinous material, as the exudation of the lacquer tree, used for coating surfaces. **3.** A baked-on finish on the inside of food and beverage cans. *—vt.* **-quered, -quer·ing, -quers. 1.** To coat with lacquer. **2.** To give a glossy finish to. **—lac'quer·er** *n.*
**lacquer tree** *n.* A tree, *Rhus verniciflua* of eastern Asia, with a toxic exudation from which a black lacquer is obtained.
**lac·ri·mal** (lăk'rə-məl) *adj.* & *n. var. of* LACHRYMAL.
**lac·ri·ma·tion** (lăk'rə-mā'shən) *n.* Excessive secretion of tears.
**la·crosse** (lə-krôs', -krŏs') *n.* [Canadian Fr. *la crosse* < Fr. (*le jeu de*) *la crosse,* (the game of) the hooked stick < OFr., crosier, of Germanic orig.] A game played on a field by two teams of ten players each, in which participants maneuver a ball into the opposing team's goal by using a long-handled stick with a webbed pouch.
**lact-** *pref. var. of* LACTO-.
**lact·al·bu·min** (lăk'tăl-byōō'mĭn) *n.* The albumin in whey.
**lac·tase** (lăk'tās') *n.* An enzyme occurring in certain yeasts and in the intestinal juices of mammals that catalyzes the conversion of lactose into glucose and galactose.
**lac·tate** (lăk'tāt') *vi.* **-tat·ed, -tat·ing, -tates.** To secrete or produce milk. *—n.* A salt or ester of lactic acid. **—lac·ta'tion** *n.* **—lac·ta'tion·al** *adj.*
**lac·te·al** (lăk'tē-əl) *adj.* [< Lat. *lacteus* < *lac,* milk.] **1.** Of, relating to, or resembling milk. **2.** *Anat.* Of or relating to any of numerous minute lymph-carrying vessels conveying chyle from the intestine to the thoracic duct. *—n. Anat.* A lacteal vessel. **—lac'te·al·ly** *adv.*
**lac·tes·cent** (lăk-tĕs'ənt) *adj.* [Lat. *lactescens, lactescent-,* pr.part. of *lactescere,* inchoative of *lactēre,* to be milky < *lac,* milk.] **1.** Becoming milky. **2.** Milky. **3.** *Biol.* Secreting or yielding a milky juice, as certain plants and insects. **—lac·tes'cence** *n.*
**lac·tic** (lăk'tĭk) *adj.* Of, relating to, or derived from milk.
**lactic acid** *n.* A hygroscopic syrupy liquid, $C_3H_6O_3$, occurring in sour milk, molasses, various fruits, and wines and used in foods and beverages as an acidulant, flavoring, and preservative and in adhesives, plasticizers, and pharmaceuticals.
**lac·tif·er·ous** (lăk-tĭf'ər-əs) *adj.* [LLat. *lactifer,* bearing milk (*lac,* milk + *-fer,* -fer) + -OUS.] **1.** Secreting or conveying milk. **2.** *Bot.* Yielding latex or a similar milky juice. **—lac·tif'er·ous·ness** *n.*
**lacto-** *or* **lact-** *pref.* [Fr. < LLat. < Lat. *lac,* milk.] **1.** Milk <*lacto*protein> **2.** Lactose <*lact*ase> **3.** Lactic acid <*lact*ate>
**lac·to·ba·cil·lus** (lăk'tō-bə-sĭl'əs) *n.* Any of various bacilli of the genus *Lactobacillus* that ferment lactic acid from carbohydrates.
**lac·to·fla·vin** (lăk'tə-flā'vĭn, lăk'tə-flā'-) *n.* Riboflavin.
**lac·to·gen·ic** (lăk'tə-jĕn'ĭk) *adj.* Inducing lactation.
**lac·tone** (lăk'tōn') *n.* A cyclic ester of a hydroxyl acid, formed by removing the constituents of water from a molecule of the acid. **—lac·ton'ic** (-tŏn'ĭk) *adj.*
**lac·to·pro·tein** (lăk'tō-prō'tēn', -tē-ən) *n.* A protein normally found in milk.
**lac·tose** (lăk'tōs') *n.* A white crystalline disaccharide, $C_{12}H_{22}O_{11}$, made from whey and used in pharmaceuticals, infant foods, bakery products, and confections.
**la·cu·na** (lə-kyōō'nə) *n., pl.* **-nae** (-nē) *or* **-nas.** [Lat. —see LAGOON.] **1.** An empty space or missing part : GAP. **2.** *Anat.* A cavity or depression. **—la·cu'nal, la·cu'nar, la·cu'na·ry** *adj.*
▲ word history: The relationship between the meanings of *lacuna* and *lagoon* can be found in the notion of something hollow or empty that underlies both words. The Latin word *lacuna,* the immediate source of *lacuna* and the ultimate source of *lagoon,* meant basically "a hollow." In Latin the word also denoted a hollow where water

ă pat ā pay âr care ä father ĕ pet ē be hw which ĭ pit
ī tie îr pier ŏ pot ō toe ô paw, for oi noise ŏŏ took

collects, and meant "pond, sea." Latin *lacuna* developed into Italian *laguna*, which was used to denote the bodies of water characteristic of the area around Venice. This word was borrowed into English as *lagoon*, and was later extended to denote similar bodies of water formed by coral reefs. *Lacuna* is a direct borrowing of Latin *lacuna* in the sense "gap." It was first used in English to denote a gap in a manuscript or inscription, and this sense is still current. The anatomical sense "cavity, depression," is probably a scientific reborrowing of the Latin word.

**la·cu·nar** (lə-kyōō'nər) *n.* [Lat. < *lacuna*, hole. —see LAGOON.] **1.** A ceiling with recessed panels. **2.** *pl.* **lac·u·nar·i·a** (lăk'yə-nâr'ē-ə). A panel in a lacunar ceiling.

**la·cus·trine** (lə-kŭs'trĭn) *adj.* [< Fr. *lacustre* < Lat. *lacus*, lake.] **1.** Of or relating to lakes. **2.** Living or growing in lakes.

**lac·y** (lā'sē) *adj.* **-i·er, -i·est.** Of, relating to, or having the nature of lace. **—lac'i·ness** *n.*

**lad** (lăd) *n.* [ME.] **1.** A young man : YOUTH. **2.** *Informal.* A fellow.

**lad·a·num** (lăd'n-əm) *n. var. of* LABDANUM.

**lad·der** (lăd'ər) *n.* [ME < OE *hlǣder.*] **1.** A framework of two long structural members connected at regular intervals by parallel rungs for climbing or descending. **2.** Something resembling a ladder, esp. a run in a stocking. **3. a.** A means of ascent and descent <ascending the corporate *ladder*> **b.** A series of ranked stages or levels <low on the social *ladder*> —*vi.* **-dered, -der·ing, -ders.** To run, as a stocking does.

**lad·der-back** (lăd'ər-băk') *n.* **1.** A chair back consisting of two upright posts joined by horizontal slats. **2.** A chair having a ladder-back. **—lad'der-back'** *adj.*

**lad·die** (lăd'ē) *n.* A young lad.

**la-de-da** (lä'dē-dä') *adj. var. of* LA-DI-DA.

**lade** (lād) *v.* **lad·ed, lad·en** (lād'n) *or* **lad·ed, lad·ing, lades.** [ME *laden* < OE *hladan.*] —*vt.* **1. a.** To load with or as if with cargo. **b.** To place as a load for or as if for shipment. **2.** To burden or oppress : WEIGH DOWN. **3.** To take up or remove (water) with a ladle or dipper : BALE. —*vi.* **1.** To take on cargo. **2.** To ladle a liquid.

**lad·en** (lād'n) *adj.* [P.part. of LADE.] **1.** Weighed down with a load : HEAVY. **2.** Burdened : oppressed.

**la-di-da** *also* **la-de-da** (lä'dē-dä') *adj.* [Perh. imit. of affected speech.] *Informal.* Affectedly genteel : PRETENTIOUS.

**ladies' man** *n. var. of* LADY'S MAN.

**la·dies'-tress·es** *also* **la·dy's-tress·es** (lā'dēz-trĕs'ĭz) *n. (sing. or pl. in number).* An orchid of the genus *Spiranthes*, bearing a spike of small white flowers usu. in a spiral arrangement.

**La·din** (lə-dēn') *n.* [Rhaeto-Romanic < Lat. *Latinus*, Latin.] **1.** Romansch. **2.** A native speaker of Ladin.

**lad·ing** (lā'dĭng) *n.* **1.** An act of loading. **2.** Freight : cargo.

**La·di·no** (lə-dē'nō) *n.* [Sp., Latin < Lat. *Latinus.*] A Romance language with elements borrowed from Hebrew that is spoken by Sephardic Jews esp. in the Balkans.

**la·dle** (lād'l) *n.* [ME < OE *hlǣdel* < *hladan*, to lade.] A long-handled, deep-bowled spoon for serving liquids. —*vt.* **-dled, -dling, -dles.** To lift out and pour with a ladle.

**la·dy** (lā'dē) *n., pl.* **-dies.** [ME, female head of a household < OE *hlǣfdige.*] **1.** A woman with the refined habits and gentle manners often associated with good culture and breeding. **2. a.** A woman regarded as virtuous and proper. **b.** A well-behaved young girl. **3.** A woman head of a household. **4. a.** —Used in the sing. as a polite form of address <May I help you, *lady?*> **b.** —Used in the pl. as a polite form of group address <*Ladies* and gentlemen, please be seated.> **5. a.** A woman to whom a man is romantically attached. **b.** *Informal.* A wife. **6. Lady.** *Chiefly Brit.* The general feminine title of nobility and of other rank, used: **a.** For the wife of a knight or baronet. **b.** Semiformally for a marchioness, countess, or viscountess. **c.** As the usual style for the wife of a baron. **d.** Semiformally for a baroness in her own right. **e.** As a courtesy title for the daughter of a duke, marquis, or earl. **f.** As a courtesy title for the wife of a younger son of a duke or marquis. **g.** As a courtesy title with the name of a female member of an order of knighthood. **7.** *often* **Our Lady.** The Virgin Mary.

▲ word history: The word *lady* was not originally an honorific title but it did designate a woman of some social importance, at least within her own household. The Old English form of *lady, hlǣfdige*, denoted the mistress of a household, especially one who had authority over servants and other dependents. The word is ultimately a compound of *hlāf*, "bread," and *dīg-*, a root meaning basically "to knead," which is related to *dough* and *dairy.* As the "bread-kneader" of the household a lady was in a position of some authority and dominance; this circumstance is reflected in the later development of the word.

**lady beetle** *n.* A ladybug.

**la·dy·bird** (lā'dē-bûrd') *n.* A ladybug.

**la·dy·bug** (lā'dē-bŭg') *n.* [LADY, title of Mary, the mother of Jesus + BUG.] Any of numerous small beetles of the family Coccinellidae, often reddish with black spots, feeding on other insects, as aphids and scale insects.

**Lady Chapel** *also* **lady chapel** *n.* A cathedral chapel usu. situated behind the sanctuary and dedicated to the Virgin Mary.

**la·dy·fin·ger** (lā'dē-fĭng'gər) *also* **la·dys·fin·ger** (lā'dēz-) *n.* A small oval sponge cake resembling a finger.

**lady in waiting** *n., pl.* **ladies in waiting.** A lady of a court appointed to serve or attend a queen or princess.

**la·dy-kill·er** (lā'dē-kĭl'ər) *n. Slang.* A man reputed to be exceptionally successful and often ruthless with women.

**la·dy·like** (lā'dē-līk') *adj.* **1.** Characteristic of a lady : WELL-BRED. **2.** Appropriate for or becoming a lady. **3.** Oversensitive to matters of propriety or decorum. **4.** Lacking virility or strength.

**la·dy·love** (lā'dē-lŭv') *n.* A sweetheart.

**la·dys·fin·ger** (lā'dēz-fĭng'gər) *n. var. of* LADYFINGER.

**la·dy·ship** (lā'dē-shĭp') *n. often* **Ladyship.** —Used in speaking or referring to a woman holding the rank of lady.

**lady's man** *also* **ladies' man** *n.* A man who enjoys and attracts the company of women.

**la·dy's-slip·per** (lā'dēz-slĭp'ər) *n.* An orchid of the genus *Cypripedium*, bearing colored flowers with an inflated pouchlike lip.

**la·dy's-smock** (lā'dēz-smŏk') *n.* The cuckooflower.

**la·dy's-thumb** (lā'dēz-thŭm') *n.* A native European plant, *Polygonum persicaria*, with small pinkish flower clusters.

**la·dy's-tress·es** (lā'dēz-trĕs'ĭz) *n. var. of* LADIES'-TRESSES.

**La·er·tes** (lā-ûr'tēz) *n.* [Lat. < Gk. *Laertēs.*] *Gk. Myth.* The father of Odysseus.

**La·e·trile** (lā'ĭ-trĭl', -trəl). A trademark for an anticancer drug obtained by hydrolyzing amygdalin and oxidizing the resulting glycoside.

**lag**[1] (lăg) *v.* **lagged, lag·ging, lags.** [Orig. unknown.] —*vi.* **1.** To fail to keep up a pace : STRAGGLE. **2.** To develop or proceed slowly or with abnormal slowness. **3.** To fail, weaken, or slacken gradually : FLAG. **4.** To determine the order of play in billiards by successively hitting the cue ball against the end rail, with the ball rebounding closest to the head rail indicating the player to shoot first. —*vt.* **1.** To cause to fall or lag behind. **2.** To shoot, throw, or pitch (e.g., a coin) at a mark. —*n.* **1.** One that lags. **2.** The act, process, or state of lagging. **3.** Retardation or slowness. **4. a.** An extent or duration of lagging. **b.** An interval between phenomena or events. **—lag'ger** *n.*

**lag**[2] (lăg) *n.* [Prob. of Scand. orig.] **1.** A barrel stave. **2.** A strip, as of wood, that forms a part of the covering for a cylindrical object. —*vt.* **lagged, lag·ging, lags.** To cover or furnish with lags.

**lag**[3] (lăg) [Orig. unknown.] *Slang.* —*vt.* **lagged, lag·ging, lags.** **1.** To arrest. **2.** To send to prison. —*n.* **1. a.** A convict. **b.** An ex-convict. **2.** A period of imprisonment : SENTENCE.

**lag·an** (lăg'ən) *also* **li·gan** (lī'gən) *or* **lag·end** (lăg'ənd) *n.* [OFr., perh. of Scand. orig.] Cargo or equipment thrown into the sea but attached to a buoy so that it can be recovered.

**Lag b'O·mer** (läg' bō'mər, läg'bə ō'mər) *n.* [Heb.] A Jewish holiday celebrated on the 33rd day after the 2nd day of Passover, on the 18th day of Iyar.

**lag·end** (lăg'ənd) *n. var. of* LAGAN.

**la·ger** (lä'gər) *n.* [G. *Lagerbier* : *Lager*, stone + *bier*, beer.] A beer orig. brewed in Germany that contains a small amount of hops and is aged from six weeks to six months to allow sedimentation.

**lag·gard** (lăg'ərd) *n.* One that lags : STRAGGLER. —*adj.* Lagging or tending to lag. **—lag'gard·ly** *adv.* **—lag'gard·ness** *n.*

**lag·ging** (lăg'ĭng) *n.* [< LAG[2].] **1.** Insulation for preventing heat diffusion, as from a steam pipe. **2.** A wooden frame built esp. for supporting the sides of an arch until the keystone is positioned.

**la·gniappe** (lăn-yăp', lăn'yăp') *n.* [Louisiana Fr. < Am. Sp. *la ñapa*, the lagniappe < Quechua *yapay*, to give more.] **1.** A small gift presented by a store owner to a customer who has made a purchase. **2.** *Informal.* An extra or unexpected gift or benefit.

**lag·o·morph** (lăg'ə-môrf') *n.* [NLat. *Lagomorpha*, order name : Gk. *lagōs*, hare + Gk. *morphē*, shape.] Any of various gnawing mammals of the order Lagomorpha, including the rabbits, hares, and pikas. **—lag'o·mor'phic** (-môr'fĭk), **lag'o·mor'phous** (-fəs) *adj.*

**la·goon** (lə-gōōn') *n.* [Fr. *lagune* and Ital. *laguna*, both < Lat. *lacuna*, pool < *lacus*, lake.] A shallow body of water, esp. one separated from the sea by sandbars or coral reefs.

**lag screw** *n.* [From its orig. use for securing barrel staves.] A heavy wood screw with a square bolt head.

**la·ic** (lā'ĭk) *also* **la·i·cal** (-ĭ-kəl) *adj.* [LLat. *laicus.*] Of or relating to the laity : SECULAR. —*n.* A layman. **—la'i·cal·ly** *adv.*

**la·i·cize** (lā'ĭ-sīz') *vt.* **-cized, -ciz·ing, -ciz·es.** **1.** To free from church control. **2.** To change to lay status : SECULARIZE. **—la'i·ci·za'tion** *n.*

**laid** (lād) *v. p.t. & p.p. of* LAY[1].

**laid-back** (lād'băk') *adj. Informal.* Having a relaxed or casual atmosphere or character : EASYGOING <"group therapists appear to be *laid-back*, untroubled people" —*New Yorker*>

**laid paper** *n.* A paper made on wire molds that give it a characteristic watermark of close thin lines.

**lain** (lān) *v. p.p. of* LIE[1].

**lair** (lâr) *n.* [ME < OE *leger.*] **1.** The dwelling of a wild animal : DEN. **2.** A refuge or hideaway. **3.** *Obs.* A resting place : COUCH.

**laird** (lârd) *n.* [Sc. < ME *laverd,* lord < OE *hlāford.* —see LORD.] *Scot.* A landed estate owner.

**lais·sez faire** *also* **lais·ser faire** (lĕs'ā fâr') *n.* [Fr. < imper. of *laisser faire,* to let (people) do (as they choose).] **1.** An economic doctrine that opposes government regulation of or interference in commerce beyond the minimum necessary for a free-enterprise system to operate according to its own economic laws. **2.** *Informal.* Noninterference in the affairs of others. **—lais'sez-faire'** *adj.*

**lais·sez-pas·ser** (lĕs'ā-pä-sā') *n.* [Fr. : *laissez,* imper. of *laisser,* to let + *passer,* to pass.] A pass, esp. one used instead of a passport.

**la·i·ty** (lā'ĭ-tē) *n., pl.* **-ties.** [< LAY[2].] **1.** Lay people as a group. **2.** All those who are not members of a profession or specialized field.

**La·ius** (lā'əs) *n.* [Lat. < Gk. *Laios.*] *Gk. Myth.* The king of Thebes who was mistakenly killed by his own son, Oedipus.

**lake**[1] (lāk) *n.* [ME < OFr. *lac* < Lat. *lacus.*] **1.** A large inland body of fresh or salt water. **2.** A large pool of liquid.

**lake**[2] (lāk) *n.* [Var. of LAC[1].] **1.** A pigment made up of organic coloring matter with an inorganic base or carrier. **2.** A deep red. *—vt.* **laked, lak·ing, lakes.** To cause (blood plasma) to become red by releasing hemoglobin from erythrocytes, as by suspending the erythrocytes in water.

**lake dwelling** *n.* A dwelling, esp. a prehistoric one, standing on piles in a shallow lake.

**lake herring** *n.* A food fish, *Coregonus artedii* or *Leucichthys artedi* of the Great Lakes region, resembling the whitefish.

**lak·er** (lā'kər) *n.* **1.** A fish, as the lake trout, living in a lake. **2.** A ship used on lakes.

**lake trout** *n.* A freshwater food fish, *Salvelinus namaycush,* indigenous to the Great Lakes.

**lakh** (läk) *n. var. of* LAC[2].

**Lal·lan** (lăl'ən) *also* **Lal·lans** (lăl'ənz) *n.* [Sc., alteration of LOWLAND.] *Scot.* **1.** The Scottish Lowlands. **2.** Scots as spoken in southern and eastern Scotland. **—Lal'lan** *adj.*

**Lal·ly** (lă'lē). A trademark for a concrete-filled steel cylinder used as a supporting member in a building.

**lal·ly·gag** (lăl'ē-găg') *v. var. of* LOLLYGAG.

**lam**[1] (lăm) *v.* **lammed, lam·ming, lams.** [Prob. of Scand. orig.] *Slang. —vt.* To give a thorough beating to : THRASH. *—vi.* To wallop : strike.

**lam**[2] (lăm) [Orig. unknown.] *Slang. —vi.* **lammed, lam·ming, lams.** To escape, as from prison. *—n.* Flight, esp. from the law <murder suspects on the *lam*>

**la·ma** (lä'mə) *n.* [Tibetan *blama.*] A Buddhist monk of Tibet or Mongolia.

**La·marck·i·an** (lə-mär'kē-ən) *adj.* Of or relating to Lamarckism. *—n.* A supporter of Lamarckism.

**La·marck·ism** (lə-mär'kĭz'əm) *n.* [After Chevalier de *Lamarck* (1744–1829), its formulator.] The theory that adaptive responses to environment cause structural changes that can be inherited.

**La·maze** (lə-mäz') *adj.* [After Fernand *Lamaze* (1890–1957).] Pertaining to or being a method of childbirth in which the mother is prepared physically and psychologically to give birth without using drugs.

**lamb** (lăm) *n.* [ME < OE.] **1. a.** A young sheep, esp. one not yet weaned. **b.** The flesh of a young sheep used as meat. **c.** Lambskin. **2.** A sweet, mild-mannered person. **3.** One who can be fleeced, esp. in financial matters : DUPE. *—vi.* **lambed, lamb·ing, lambs.** To give birth to a lamb.

**lam·baste** (lăm-băst') *vt.* **-bast·ed, -bast·ing, -bastes.** [Perh. LAM[1] + BASTE[3].] *Slang.* **1.** To give a thrashing to : BEAT. **2.** To scold sharply.

**lamb·da** (lăm'də) *n.* [Gk., of Phoenician orig.; akin to Heb. *lāmedh,* lamed.] **1.** The 11th letter of the Greek alphabet. —See table at ALPHABET. **2.** An electrically neutral subatomic particle in the baryon family, having a mass 2,183 times that of the electron and a mean lifetime of approx. $2.5 \times 10^{-10}$ second.

**lambda point** *n.* **1.** The temperature at which the transition from helium I to superfluid helium II occurs, approx. 2.19°K. **2.** The temperature of a phase transition in which the specific heat regarded as a function of temperature has a logarithmic singularity.

**lam·bent** (lăm'bənt) *adj.* [Lat. *lambens, lambent-,* pr.part. of *lambere,* to lick.] **1.** Flickering lightly on or over a surface. **2.** Marked by effortless brilliance or lightness <*lambent* repartee> **3.** Having a gentle glow : LUMINOUS. **—lam'ben·cy** *n.* **—lam'bent·ly** *adv.*

**lam·bert** (lăm'bərt) *n.* [After Johann H. *Lambert* (1728–1777).] A unit of brightness equal to $1/\pi$ candle per square centimeter.

**lamb·kill** (lăm'kĭl') *n.* Sheep laurel.

**lam·bre·quin** (lăm'bər-kĭn, -brə-kĭn) *n.* [Fr. < Du. **lamperkin* < *lamper,* veil.] **1.** A short ornamental drapery for the top of a window or door or the edge of a shelf : VALANCE. **2.** A heavy protective cloth worn over a medieval helmet.

**lamb·skin** (lăm'skĭn') *n.* **1.** The hide of a lamb, esp. when dressed without removing the fleece. **2.** Leather made from the dressed hide of a lamb.

**lamb's-let·tuce** (lămz'lĕt'əs) *n.* Corn salad.

**lamb's-quar·ters** *also* **lamb's quarters** (lămz'kwôr'tərz) *n. (sing. or pl. in number).* PIGWEED 1.

**lamb's wool** *n.* **1.** Wool shorn from a lamb. **2.** Yarn or a fabric made of lamb's wool.

**lame**[1] (lām) *adj.* **lam·er, lam·est.** [ME < OE *lama.*] **1.** Disabled in one or more limbs, esp. in a leg or foot, so that walking or moving is easily hampered. **2.** Marked by pain or rigidness <a *lame* wrist> **3.** Weak and ineffectual : UNSATISFACTORY <a *lame* excuse for failing to finish the job> *—vt.* **lamed, lam·ing, lames.** To cause to become lame : CRIPPLE. **—lame'ly** *adv.* **—lame'ness** *n.*

**lame**[2] (lām) *n.* [OFr. < Lat. *lamina,* thin plate.] A thin metal plate, esp. one of the overlapping steel plates in medieval armor.

**la·mé** (lă-mā') *n.* [Fr. < *lamé,* spangled < OFr. *lame,* lame, thin metal plate.] A brocaded fabric with metallic threads in the warp or in the filling.

**lame·brain** (lām'brān') *n.* A thick-witted person : NINNY. **—lame'-brained'** (-brānd') *adj.*

**la·med** *also* **la·medh** (lä'mĕd') *n.* [Heb. *lāmedh.*] The 12th letter of the Hebrew alphabet. —See table at ALPHABET.

**lame duck** *n.* **1.** An elected officeholder or group remaining in office during the interval between failure to win an election and the inauguration of a successor. **2.** An ineffective person : WEAKLING.

**lamell-** *pref. var. of* LAMELLI-.

**la·mel·la** (lə-mĕl'ə) *n., pl.* **-mel·lae** (-mĕl'ē) *or* **-mel·las.** [NLat. < Lat., small thin plate < *lamina,* thin plate.] A thin scale, plate, or layer, as in the gills of a bivalve mollusk or forming one of the gills of a mushroom. **—la·mel'lar** *adj.* **—la·mel'lar·ly** *adv.*

**la·mel·late** (lə-mĕl'āt', lăm'ə-lāt') *adj.* **1.** Having, composed of, or arranged in thin layers or lamellae. **2.** Like a lamella. **—lam'el·la'ted** *adj.* **—lam'el·la'tion** *n.*

**lamelli-** or **lamell-** *pref.* [< LAMELLA.] Lamella <*lamelli*form>

**la·mel·li·branch** (lə-mĕl'ə-brănk') *n.* [NLat. *Lamellibranchia,* class name : LAMELLI- + Lat. *branchia,* gill < Gk. *brankhia,* gills.] Any of the mollusks of the class Pelecypoda or Lamellibranchia, with a hinged bivalve shell and including the clams, mussels, and oysters. *—adj.* Of or relating to lamellibranchs.

**la·mel·li·corn** (lə-mĕl'ĭ-kôrn') *adj.* [NLat. *Lamellicornia,* superfamily name : LAMELLI- + Lat. *cornu,* horn.] Of or belonging to the superfamily Lamellicornia or Scarabaeoidea, which includes the scarabs and other beetles whose antennae are tipped with movable leaflike plates. *—n.* A lamellicorn beetle.

**la·mel·li·form** (lə-mĕl'ə-fôrm') *adj.* Shaped like a lamella.

**la·ment** (lə-mĕnt') *v.* **-ment·ed, -ment·ing, -ments.** [OFr. *lamenter* < Lat. *lamentari* < *lamentum,* lament.] *—vt.* **1.** To express grief for or about : MOURN <*lament* the death of a leader> **2.** To regret deeply : DEPLORE. *—vi.* To grieve. *—n.* **1.** An expression or feeling of grief : LAMENTATION. **2.** A song or poem expressing grief : ELEGY. **—la·ment'er** *n.*

**la·men·ta·ble** (lə-mĕn'tə-bəl, lăm'ən-) *adj.* **1.** That is to be lamented : DEPLORABLE. **2.** Exhibiting sorrow : MOURNFUL. **—lam'en·ta·bly** *adv.*

**lam·en·ta·tion** (lăm'ən-tā'shən) *n.* **1.** The act or an instance of expressing grief. **2. Lam·en·ta·tions** *(sing. in number).* —See table at BIBLE.

**la·ment·ed** (lə-mĕn'tĭd) *adj.* Mourned for. **—la·ment'ed·ly** *adv.*

**la·mi·a** (lā'mē-ə) *n., pl.* **-as** *or* **-ae** (-ē') [ME < Lat. < Gk.] **1.** *Gk. Myth.* A monster depicted as a serpent with the head and breasts of a woman, reputed to prey on humans and suck the blood of children. **2.** A female vampire.

**lam·i·na** (lăm'ə-nə) *n., pl.* **-nae** (-nē') *or* **-nas.** [Lat.] **1.** A thin plate, sheet, or layer. **2.** *Bot.* The expanded area of a leaf. **3.** *Zool.* A scalelike or platelike structure, as one of the thin layers of sensitive tissue in a horse's hoof. **—lam'i·nal, lam'i·nar** *adj.*

**lam·i·nar flow** (lăm'ə-nər) *n.* Nonturbulent flow of a viscous fluid in layers near a boundary, as of lubricating oil in bearings.

**lam·i·nate** (lăm'ə-nāt') *v.* **-nat·ed, -nat·ing, -nates.** [LAMIN(A) + -ATE[1].] *—vt.* **1.** To compress or beat into a thin plate or sheet. **2.** To separate into thin layers. **3.** To make by uniting several layers. **4.** To cover with thin sheets. *—vi.* To split into thin layers or sheets. *—adj.* (-nĭt, -nāt'). Made up of, arranged in, or covered with laminae. *—n.* A laminated product, as plywood. **—lam'i·na'tor** *n.*

**lam·i·nat·ed** (lăm'ə-nā'tĭd) *adj.* **1.** Made up of bonded layers. **2.** Arranged in laminae : LAMINATE.

**lam·i·na·tion** (lăm'ə-nā'shən) *n.* **1.** The act of laminating or state of being laminated. **2.** Laminated material. **3.** A lamina.

**lam·i·ni·tis** (lăm'ə-nī'tĭs) *n.* Inflammation of the sensitive laminae in the hoof of a horse.

**Lam·mas** (lăm'əs) *n.* [ME < OE *hlāfmæss* : *hlāf,* loaf + *mæsse,* Mass.] A harvest festival once held in England on Aug. 1.

**lam·mer·gei·er** *also* **lam·mer·gey·er** (lăm'ər-gī'ər) *n.* [G. *Lämmergeier* : *Lamm,* lamb + *Geier,* vulture.] A large predatory bird, *Gypaetus barbatus* of mountainous regions of the Old World.

**lamp** (lămp) *n.* [ME < OFr. *lampe* < Lat. *lampas* < Gk. < *lampein,* to shine.] **1. a.** A device that generates heat, light, or therapeutic radiation. **b.** A vessel holding oil or alcohol burned through a wick for illumination. **2.** A celestial body, as a star, planet, or meteor. **3.** Mental or emotional enlightenment.

**lamp·black** (lămp′blăk′) *n.* A gray or black pigment made from soot obtained from the incomplete combustion of carbonaceous materials, used as a pigment and in matches, explosives, lubricants, and fertilizers.

**lam·pi·on** (lăm′pē-ən) *n.* [Fr. < Ital. *lampione*, aug. of *lampa*, lamp < OFr. *lampe*. —see LAMP.] An oil-burning lamp for outdoor use.

**lamp·light** (lămp′līt′) *n.* The light shed by a lamp.

**lamp·light·er** (lămp′lī′tər) *n.* One that lights lamps.

**lam·poon** (lăm-po͞on′) *n.* [Fr. *lampon* < *lampons*, let us drink (from the refrain of a drinking song) < *lamper*, to gulp down.] **1.** A broad satirical piece that ridicules a person, group, or institution. **2.** Light, good-humored satire. —*vt.* **-pooned, -poon·ing, -poons.** To satirize in or as if in a lampoon. **—lam·poon′er, lam·poon′ist** *n.* **—lam·poon′er·y** *n.*

**lamp·post** (lămp′pōst′) *n.* A post with a street lamp.

**lam·prey** (lăm′prē) *n., pl.* **-preys.** [ME *lamprei* < OFr. *lampreie* < Med. Lat. *lampreda.*] Any of various primitive elongated freshwater or anadromous fishes of the family Petromyzontidae, with a jawless sucking mouth and a rasping tongue.

**lamp shell** *n.* A brachiopod.

**la·nai** (lə-nī′, lä-) *n., pl.* **-nais.** [Hawaiian.] A roofed patio.

**la·nate** (lā′nāt′) *adj.* [Lat. *lanatus* < *lana*, wool.] Having or composed of woolly hairs.

**Lan·cas·tri·an** (lăng-kăs′trē-ən) *adj.* Of or relating to the royal house of Lancaster that ruled in England from 1399 to 1461.

**lance** (lăns) *n.* [ME < OFr. < Lat. *lancea.*] **1.** A thrusting weapon with a sharp metal head and a long wooden shaft. **2.** A sharp-pointed implement for spearing fish. **3.** LANCER 1. **4.** A lancet. —*vt.* **lanced, lanc·ing, lanc·es. 1.** To pierce with a lance. **2.** To make a surgical incision in <*lance* an abcess>

**lance corporal** *n.* **1.** An enlisted person in the U.S. Marine Corps, ranking above a private first class and below a corporal. **2.** A British Army private functioning as a corporal.

**lance·let** (lăns′lĭt) *n.* Any of various small, flattened marine organisms of the subphylum Cephalochordata, allied to the vertebrates but having a notochord rather than a true vertebral column.

**Lan·ce·lot** (lăn′sə-lət, -lŏt′, län′-) *n.* A knight of the Round Table in Arthurian legend whose love affair with Queen Guinevere resulted in a war with King Arthur.

**lan·ce·o·late** (lăn′sē-ə-lāt′) *adj.* [LLat. *lanceolatus* < Lat. *lanceola*, little lance, dim. of *lancea*, lance.] Narrow and tapering at each end, as leaves. **—lan′ce·o·late′ly** *adv.*

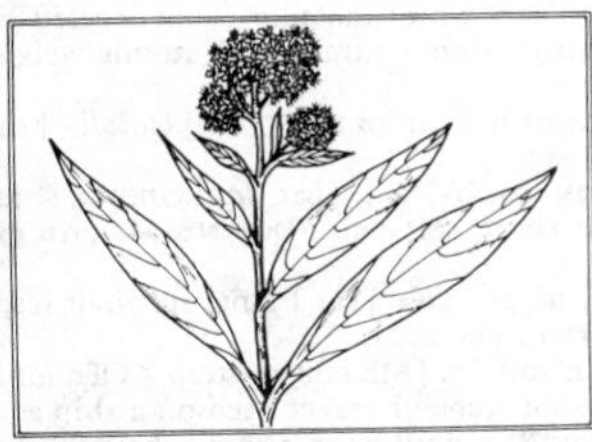

**lanceolate**
*Lanceolate leaves*

**lanc·er** (lăn′sər) *n.* [OFr. *lancier* < *lance*, lance.] **1.** A cavalry trooper armed with a lance. **2.** A member of a regiment orig. armed with lances. **3. lancers** *(sing. in number).* **a.** A set of five quadrilles, each having a different meter. **b.** The music for a quadrille.

**lan·cet** (lăn′sĭt) *n.* [ME < OFr., dim. of *lance*, lance.] **1.** A surgical knife with a short, wide, double-edged blade. **2. a.** A lancet arch. **b.** A lancet window.

**lancet arch** *n.* An arch narrow and pointed like a spearhead.

**lancet fish** *n.* A large marine fish, *Alepisaurus ferox* of the Atlantic or *A. richardsoni* of the Pacific, with long sharp teeth and a large dorsal fin.

**lancet window** *n.* A tall narrow window set in a lancet arch.

**lance·wood** (lăns′wo͝od′) *n.* **1.** A tropical American tree, as one of the genera *Calycophyllum* or *Mimusops*, with hard, durable, uniformly grained wood. **2.** The wood of a lancewood.

**land** (lănd) *n.* [ME < OE.] **1.** The solid ground of the earth. **2. a.** The ground or soil : EARTH <cultivate the *land*> **b.** A topographically or functionally distinct tract <rich *land*> **3. a.** A country : nation. **b.** The people of a nation, district, or region. **c. lands.** Territorial possessions or property. **4.** Public or private landed property : REAL ESTATE. **5.** An area or realm <the *land* beyond the sea> **6.** *Law.* **a.** A tract of land that may be owned, together with everything growing or built on it. **b.** A landed estate. **7.** The raised portion of a grooved surface, as any of the ridges in a rifle bore. —*v.* **land·ed, land·ing, lands.** —*vt.* **1. a.** To bring to and unload on land <*land* passengers> **b.** To set down on land or another surface <*land* a spacecraft> **2.** To cause to arrive in a place or condition <My misreading of the map *landed* me in the wrong city.> **3. a.** To catch and pull in (a fish). **b.** To win : secure <*land* a new position in the company> **4.** To deliver <*land* a series of rapid punches> —*vi.* **1. a.** To come to shore. **b.** To disembark. **2.** To descend toward and settle on the ground or another surface. **3.** To arrive in a place or condition <*landed* in Boise very early> **4.** To come to rest in a certain way or place <slipped and *landed* on my back>

**lan·dau** (lăn′dô′, -dou′) *n.* [After *Landau*, Bavaria, Germany.] **1.** A four-wheeled closed carriage with front and rear passenger seats facing each other and a roof in two sections that can be lowered or detached. **2.** A style of automobile with a similar roof.

**lan·dau·let** *or* **lan·dau·lette** (lăn′dl-ĕt′) *n.* **1.** A small landau. **2.** An automobile with a collapsible roof over the back seat and an open driver's seat.

**land bank** *n.* A bank that issues long-term loans on real estate in exchange for mortgages.

**land·ed** (lăn′dĭd) *adj.* **1.** Owning land <*landed* nobility> **2.** Consisting of land or real estate <a *landed* hunting preserve>

**land·er** (lăn′dər) *n.* **1.** One that lands. **2.** A space vehicle designed to land on a celestial body <a moon *lander*>

**land·fall** (lănd′fôl′) *n.* **1.** The act or an instance of sighting or reaching land after a voyage or flight. **2.** The land sighted or reached after a voyage or flight.

**land·fill** (lănd′fĭl′) *n.* **1.** A method of waste disposal in which garbage and trash are buried in low-lying ground. **2.** Land built up by landfill.

**land grant** *n.* A government grant of public land for a railroad, highway, or state college.

**land·hold·er** (lănd′hōl′dər) *n.* A person who owns land. **—land′hold′ing** *n.*

**land·ing** (lăn′dĭng) *n.* **1. a.** The act or process of coming to land or rest, esp. after a voyage or flight. **b.** A termination, esp. of a voyage or flight. **2.** A site for loading and unloading passengers and cargo. **3. a.** An intermediate platform on a flight of stairs. **b.** The area at the top or bottom of a staircase.

**landing craft** *n.* A naval craft designed to convey troops and equipment from ship to shore.

**landing field** *n.* A tract of land used by aircraft for landing and taking off.

**landing gear** *n.* An aircraft undercarriage, designed to support the weight of the craft and its load on the ground.

**landing strip** *n.* An aircraft runway without airport facilities.

**land·la·dy** (lănd′lā′dē) *n.* **1.** A woman who owns and rents land, buildings, or dwelling units. **2.** A woman who runs a rooming house or inn : INNKEEPER.

**land·less** (lănd′lĭs) *adj.* Owning or possessing no land.

**land·locked** (lănd′lŏkt′) *adj.* **1.** Surrounded by or almost entirely by land. **2.** Confined to inland waters <*landlocked* salmon>

**land·lord** (lănd′lôrd′) *n.* **1.** One who owns and rents land, buildings, or dwelling units. **2.** A man who runs a rooming house or inn : INNKEEPER.

**land·lord·ism** (lănd′lôr-dĭz′əm) *n.* Land management in which a private individual or group owns land and leases it to tenants, esp. tenants who cultivate it.

**land·lub·ber** (lănd′lŭb′ər) *n.* One unfamiliar with the sea or seamanship. **—land′lub′ber·ly** *adj.*

**land·mark** (lănd′märk′) *n.* **1.** A fixed marker, as a concrete block, indicating a boundary line. **2.** A conspicuous identifying feature of a landscape. **3.** An event marking an important phase of development or a decisive moment in history. **4.** A building or site that has historical and often aesthetic importance, esp. one marked for preservation by a municipal, state, or national government.

**land·mass** (lănd′măs′) *n.* A large land area.

**land mine** *n.* An explosive mine laid usu. just below the surface of the ground.

**land office** *n.* A government office that records the sales and transfers of public land.

**land-of·fice business** (lănd′ô′fĭs, -ŏf′ĭs) *n.* A thriving, extensive, or rapidly moving volume of trade.

**land·own·er** (lănd′ō′nər) *n.* A landholder. **—land′own′er·ship′** *n.* **—land′own′ing** *n.* & *adj.*

**land-poor** (lănd′po͝or′) *adj.* Owning much land but lacking capital to maintain or improve it.

**land reform** *n.* Measures, as the division of large properties into smaller ones, undertaken to effect a more equitable apportionment of agricultural land.

**land·scape** (lănd′skāp′) *n.* [Du. *landschap* : *land*, land + *-schap*, -ship.] **1.** A view or vista of scenery on land <a prairie *landscape*> **2.** A picture showing a landscape. **3.** The branch of art dealing with the depiction of natural scenery. —*v.* **-scaped, -scap·ing, -scapes.** —*vt.* To adorn or improve (a section of ground) by contouring the land and planting flowers, shrubs, or trees. —*vi.* To adorn or improve grounds artistically as a profession.

**landscape architect** *n.* One whose professional skill is the decorative and functional alteration and planting of grounds, esp. at or around a building site. **—landscape architecture** *n.*

**landscape gardener** *n.* One whose occupation is the decoration of land by designing gardens and planting trees and shrubs. **—landscape gardening** *n.*

**land·scap·ist** (lănd'skā'pĭst) *n.* A painter of landscapes.

**land·side** (lănd'sīd') *n.* The flat side of a plow opposite the furrow.

**lands·leit** (länts'līt') *n.* [Yiddish *landslayt*, pl. of *landsman*, landsman. —see LANDSMAN².] *pl. of* LANDSMAN².

**land·slide** (lănd'slīd') *n.* **1.** *also* **land·slip** (-slĭp'). **a.** The dislodging and fall of a mass of earth and rock. **b.** The dislodged mass. **2. a.** An overwhelming majority of votes for a political party or candidate. **b.** An election that sweeps a party or candidate into office. **c.** A very great victory.

**Lands·mål** (länts'môl') *n.* [Norw. : *land*, country + *mål*, speech.] An official and literary form of Norwegian based on the spoken dialects of Norway.

**lands·man¹** (lăndz'mən) *n.* One who lives and works on land.

**lands·man²** (länts'mən) *n., pl.* **lands·leit** (-līt') [Yiddish < MHG *lantsman* : OHG *lant*, land + MHG *man*, man.] A fellow Jew who comes from the same district or town, esp. in Eastern Europe.

**land·ward** (lănd'wərd) *adj. & adv.* To or toward land. **—land'wards** *adv.*

**lane** (lān) *n.* [ME < OE.] **1. a.** A narrow way or passage between walls, hedges, or fences. **b.** A narrow country road. **2.** A narrow passage, course, or track, as: **a.** A set course for ships or aircraft. **b.** A strip marked on a street or highway to accommodate one line of motor vehicles. **c.** One of a set of parallel courses marking the bounds for contestants in a race. **d.** A wood-surfaced passageway along which a bowling ball is rolled.

**lang** (lăng) *adj. Scot.* Long.

**lang·lauf** (läng'louf') *n.* [G. : *lang*, long + *lauf*, race.] A cross-country ski run. **—lang'lauf'er** *n.*

**lang·ley** (lăng'lē) *n.* [After Samuel *Langley* (1834–1906).] A unit of illumination used to measure temperature, as of a star, equal to one gram calorie per square centimeter of irradiated surface.

**Lan·go·bard** (lăng'gə-bärd') *n.* [Lat. *Langobardus.*] LOMBARD 1. **—Lan'go·bar'dic** *adj.*

**lan·gouste** (län-go͞ost') *n.* [Fr. < OFr. < OProv. *langosta.*] The spiny lobster.

**lang·syne** *also* **lang syne** (lăng-zīn') [ME *lang sine.*] *Scot.* —*adv.* Long ago : long since. —*n.* Time long past : times past.

**lan·guage** (lăng'gwĭj) *n.* [ME < OFr. < Lat. *lingua.*] **1. a.** Human use of voice sounds, and often of written symbols that represent these sounds, in organized combinations and patterns to express and communicate thoughts and feelings. **b.** A system of words formed from such combinations and patterns, used by the people of a particular country or by a group of people with a shared history or set of traditions. **2.** A nonverbal method of communication, as by a system of signs, symbols, or gestures <the *language* of the deaf> **3.** Body language. **4.** The special vocabulary and usages of a scientific, professional, or other group <the *language* of physics> **5.** A typical style of speech or writing <childish *language*> **6. a.** Abusive, violent, or profane utterance <"*language* that would make your hair curl" —W.S. Gilbert> **b.** A particular manner of utterance <harsh *language*> **7.** The manner or means of communication between living creatures other than humans <the *language* of monkeys> **8.** Language as a subject of study. **9.** The wording of a legal document or statute as distinct from the spirit. **10.** *Computer Sci.* Machine language.

☆ ***syns:*** **1.** LANGUAGE, DIALECT, SPEECH, TONGUE, VERNACULAR *n. core meaning* : a system of terms used by a people sharing a history and culture <Polish and Russian—two Slavic *languages*> **2.** LANGUAGE, CANT, IDIOM, JARGON, LEXICON, TERMINOLOGY, VOCABULARY *n. core meaning* : specialized expressions characteristic of a field, subject, trade, or subculture <the *language* of medicine><street *language*>

**langue d'oc** (läng dôk', läɴg) *n.* [Fr. < OFr., language of *oc*, from the use of the word *oc* for "yes" in Provençal.] Provençal.

**langue d'o·ïl** (läng doil', doi', läɴg' dô-ēl') *n.* [Fr. < OFr., language of *oïl*, from the use of the word *oïl* for "yes" in French.] FRENCH 1.

**lan·guet** (lăng'gwĭt, lăng-gwĕt') *n.* [ME < OFr. *languette*, dim. of *langue*, tongue < Lat. *lingua.*] Something resembling a tongue.

**lan·guid** (lăng'gwĭd) *adj.* [OFr. *languide* < Lat. *languidus* < *languēre*, to be languid.] **1.** Lacking energy or vitality : WEAK. **2.** Showing little or no animation. **3.** Lacking vigor or force : SLOW. **—lan'guid·ly** *adv.* **—lan'guid·ness** *n.*

**lan·guish** (lăng'gwĭsh) *vi.* **-guished, -guish·ing, -guish·es.** [ME *languishen* < OFr. *languir, languiss-* < Lat. *languēre*, to be languid.] **1. a.** To be or become weak or feeble : WASTE AWAY. **b.** To exist or continue in miserable or disheartening conditions. **2.** To remain unattended or be neglected <bills *languishing* in the governor's office> **3.** To become downcast : PINE. **4.** To affect a wistful or languid air, esp. in order to gain sympathy. **—lan'guish·er** *n.* **—lan'guish·ing·ly** *adv.* **—lan'guish·ment** *n.*

**lan·guor** (lăng'gər, lăng'ər) *n.* [ME < OFr. *langor* < Lat. *languor* < *languēre*, to be languid.] **1.** Lack of physical or mental energy : LISTLESSNESS. **2.** A dreamy, indolent mood or quality : LASSITUDE. **3.** Oppressive quiet or stillness. **—lan'guor·ous** *adj.* **—lan'guor·ous·ly** *adv.* **—lan'guor·ous·ness** *n.*

**lan·gur** (lŭng-go͝or') *n.* [Hindi *langūr*, perh. < Skt. *lāṅgūlam*, a hairy tail.] A slender long-tailed Asian monkey of the genus *Presbytis* or related genera.

**lan·iard** (lăn'yərd) *n. var. of* LANYARD.

**la·nif·er·ous** (lə-nĭf'ər-əs) *adj.* [Lat. *lanifer* (*lana*, wool + *-fer*, -fer) + -OUS.] Having wool or woollike hair.

**lank** (lăngk) *adj.* **-er, -est.** [ME *lank < OE *hlanc.*] **1.** Long and lean : GAUNT. **2.** Long, straight, and limp <*lank* hair> **—lank'ly** *adv.* **—lank'ness** *n.*

**lank·y** (lăng'kē) *adj.* **-i·er, -i·est.** Tall, thin, and ungainly. **—lank'i·ly** *adv.* **—lank'i·ness** *n.*

**lan·ner** (lăn'ər) *n.* [ME *laner* < AN < Med. Lat. *lanarius.*] **1.** A falcon, *Falco biarmicus* of Africa and the Mediterranean region. **2.** A female lanner, used in falconry.

**lan·ner·et** (lăn'ə-rĕt') *n.* [ME *laneret* < OFr., dim. of *lanier*, lanner < Med. Lat. *lanarius.*] A male lanner, smaller than the female, used in falconry.

**lan·o·lin** (lăn'ə-lĭn) *n.* [G. < Lat. *lana*, wool.] A yellowish-white fatty substance obtained from wool and used in soaps, cosmetics, and ointments.

**lan·ta·na** (lăn-tä'nə, -tăn'ə) *n.* [NLat. *Lantana*, genus name < dial. Ital. *lantana*, viburnum.] An aromatic, chiefly tropical shrub of the genus *Lantana*, bearing small variously colored flowers.

**lan·tern** (lăn'tərn) *n.* [ME < OFr. *lanterne* < Lat. *lanterna* < Gk. *lamptēr* < *lampein*, to shine.] **1. a.** An often portable case with transparent or translucent sides for holding and protecting a light. **b.** A decorative casing for a light, often of paper. **2. a.** *Obs.* A lighthouse. **b.** The room at the top of a lighthouse where the light is located. **3.** A structure built on a roof with open or windowed walls to let in air and light. **4.** A slide projector.

**lantern fish** *n.* Any of numerous small deep-sea fishes of the family Myctophidae, with phosphorescent light organs.

**lantern fly** *n.* Any of various chiefly tropical insects of the subfamily Fulgorinae, with an enlarged, elongated head.

**lantern jaw** *n.* A lower jaw protruding beyond the upper jaw. **—lan'tern-jawed'** *adj.*

**lantern pinion** *n.* A lantern wheel.

**lantern wheel** *n.* A small pinion having circular disks connected by cylindrical bars that serve as teeth, now used chiefly in inexpensive clocks.

**lan·tha·nide** (lăn'thə-nīd') *n.* [LANTHAN(UM) + -IDE.] A rare-earth element.

**lanthanide series** *n.* The set of chemically related elements with atomic numbers from 57 to 71 : rare-earth elements.

**lan·tha·num** (lăn'thə-nəm) *n.* [NLat. < Gk. *lanthanein*, to escape notice.] *Symbol* **La** A soft, silvery-white metallic element used in glass manufacture and in lighting; atomic number 57; atomic weight 138.91.

**lant·horn** (lănt'hôrn', lăn'tərn) *n.* [Var. of LANTERN.] *Chiefly Brit.* A lantern.

**la·nu·gi·nous** (lə-no͞o'jə-nəs, -nyo͞o'-) *adj.* [Lat. *lanuginosus* < *lanugo*, lanugo.] Covered with short, soft hair : DOWNY. **—la·nu'gi·nous·ness** *n.*

**la·nu·go** (lə-no͞o'gō, -nyo͞o'-) *n., pl.* **-gos.** [Lat.] Fine, soft hair, esp. that covering the fetus of certain mammals.

**lan·yard** *also* **lan·iard** (lăn'yərd) *n.* [ME *langer*, strap < OFr. *laniere*, of Germanic orig.] **1.** A short rope or gasket used on a ship as a fastener or for securing rigging. **2.** A cord worn around the neck for carrying something, as a knife. **3.** A cord with a hook at one end used to fire a cannon.

**Lao** (lou) *n., pl.* **Lao** *or* **Laos. 1. a.** One of a Buddhist people of Thai stock living in the area of the Mekong River in Laos and Thailand. **b.** The Lao people. **2.** The Tai language of the Lao. —*adj.* Of the Lao or their language.

**La·oc·o·on** (lā-ŏk'ō-ŏn') *n.* [Lat. < Gk. *Laokoōn.*] *Gk. Myth.* A Trojan priest of Apollo who was killed with his two sons by two serpents for having warned his people against the Trojan horse.

**La·od·i·ce·an** (lā-ŏd'ĭ-sē'ən) *adj.* [After *Laodicea*, an ancient city in Asia Minor.] Indifferent or lukewarm in religion or politics. **—La·od'i·ce'an** *n.*

**La·om·e·don** (lā-ŏm'ĭ-dŏn') *n.* [Lat. < Gk. *Laomedōn.*] *Gk. Myth.* The founder and king of Troy and father of Priam.

**La·o·tian** (lā-ō'shən, lou'shen) *n. & adj.* [Fr. *Laotien* < *Lao*, Lao.] Lao.

**lap¹** (lăp) *n.* [ME < OE *læppa*, lappet.] **1.** The front of a seated person from the lower trunk to the knees. **2.** The portion of a garment that covers the lower trunk to the knees. **3.** A place of nurture or control <*lap* of luxury> **—lap'ful'** (-fo͝ol') *n.*

**lap²** (lăp) *v.* **lapped, lap·ping, laps.** [ME *lappen* < *lap*, lap, flap of a garment.] —*vt.* **1.** To fold, wrap, or wind around or over something. **2.** To envelop : swathe. **3.** To place (a thing) so as to cover part of another : OVERLAP. **4.** To join by or as if by scarfing in cabinetwork. **5.** To get ahead of (an opponent) in a race by at least one complete circuit of the course. **6.** To polish until smooth. —*vi.* **1.** To fold or wind around something. **2.** To extend beyond or project onto an edge : OVERLAP. —*n.* **1. a.** An overlapping part. **b.** The amount by which a part overlaps another. **2. a.** One complete circuit, esp. of a racetrack. **b.** One complete length of a straight course, as of a swimming pool. **c.** A segment or stage, as of a trip. **d.** A length, as of rope, needed to make one complete turn, as around a wheel. **3.** A continuous band or layer of fiber, as cotton or flax. **4.** An either stationary or rotating wheel, disk, or slab of leather or metal for polishing and smoothing.

**lap³** (lăp) *v.* **lapped, lap·ping, laps.** [ME *lappen* < OE *lapian.*] —*vt.* **1.** To take in (a liquid or food) with the tongue. **2.** To wash against with a gentle intermittent slapping sound <waves *lapping* the hull of the boat> —*vi.* **1.** To take in a liquid or food with the tongue. **2.** To wash against a shore with a gentle intermittent slapping sound. **—lap up.** *Informal.* To take in eagerly or greedily <*lapping up* compliments> —*n.* **1. a.** The act or process of lapping. **b.** An amount taken into the mouth by lapping. **2.** A watery food or drink. **3.** The sound of lapping water. **—lap'per** *n.*

**lap·a·rot·o·my** (lăp'ə-rŏt'ə-mē) *n., pl.* **-mies.** [Gk. *lapara*, flank (< *laparos*, soft) + -TOMY.] Surgical incision into the abdominal wall.

**lap belt** *n.* A motor vehicle seat belt that fastens across the lap.

**lap·board** (lăp'bôrd', -bōrd') *n.* A flat board held on the lap as a substitute for a table or desk.

**lap dog** *n.* A small dog kept as a pet.

**la·pel** (lə-pĕl') *n.* [< LAP¹.] The part of a garment that is an extension of the collar and folds back against the breast.

**lap·i·dar·i·an** (lăp'ĭ-dâr'ē-ən) *adj.* [Lat. *lapidarius* < *lapis*, stone.] Cut in or inscribed on stone.

**lap·i·dar·y** (lăp'ĭ-dĕr'ē) *n., pl.* **-ies.** [Lat. *lapidarius* < *lapis*, stone.] **1.** One who cuts, polishes, or engraves gems. **2.** A dealer in precious or semiprecious stones. —*adj.* **1.** Of or relating to precious stones or the art of working with them. **2. a.** Engraved in stone. **b.** Suitable for inscription in stone <*lapidary* verses>

**la·pil·lus** (lə-pĭl'əs) *n., pl.* **-pil·li** (-pĭl'ī') [Lat., small stone, dim. of *lapis*, stone.] A small solidified fragment of lava.

**lap·in** (lăp'ĭn, lä-păN') *n.* [Fr.] **1.** A rabbit. **2.** Rabbit fur, esp. when sheared and dyed.

**lap·is laz·u·li** (lăp'ĭs lăz'yə-lē, lăzh'ə-) *n.* [ME < Med. Lat. : Lat. *lapis*, stone + *lazulum*, lapis lazuli < Ar. *lāzaward* < Pers. *lājward.*] **1.** An opaque azure-blue to deep-blue lazurite used as a gemstone. **2.** Lazurite.

**lap joint** *n.* A joint in which the ends or edges are overlapped and fastened together.

**Lapp** (lăp) *n.* [Swed.] **1.** One of a people of nomadic tradition who inhabit Lapland. **2.** The Finno-Ugric language of the Lapps.

**lap·pet** (lăp'ĭt) *n.* [< LAP¹.] **1.** A decorative flap or loose fold on a garment or headdress. **2.** A flaplike structure, as the wattle of a bird.

**lap robe** *n.* A small blanket to cover the lap, legs, and feet, as of a passenger in a car.

**lapse** (lăps) *v.* **lapsed, laps·ing, laps·es.** [Lat. *labi, laps-*, to lapse.] —*vi.* **1. a.** To fall from one level to a different, usu. less desirable one : BACKSLIDE. **b.** To slip gradually : DRIFT <*lapse* into a coma> **2.** To pass : elapse <Some time *lapsed* before the game began.> **3. a.** *Law.* To pass to another through omission or neglect. —Used of a right or privilege, a benefice, or an estate. **b.** To be no longer in force because of disuse, neglect, or the passage of time <The purchase option *lapsed.*> —*vt.* To allow to lapse. —*n.* **1. a.** A slip, error, or failure, esp. a slight or unimportant one <a *lapse* of judgment> **b.** A fall from rectitude. **2.** A slipping into a more unfavorable condition : DECLINE <a *lapse* into carelessness> **3. a.** Passage of time. **b.** A period of passing time. **4.** *Law.* The termination of a right or privilege through disuse, neglect, or death. **—laps'er** *n.*

**lap·strake** (lăp'strāk') *also* **lap·streak** (-strēk') *Naut.* —*adj.* Built with each strake overlapping the one below : CLINKER-BUILT. —*n.* A clinker-built boat.

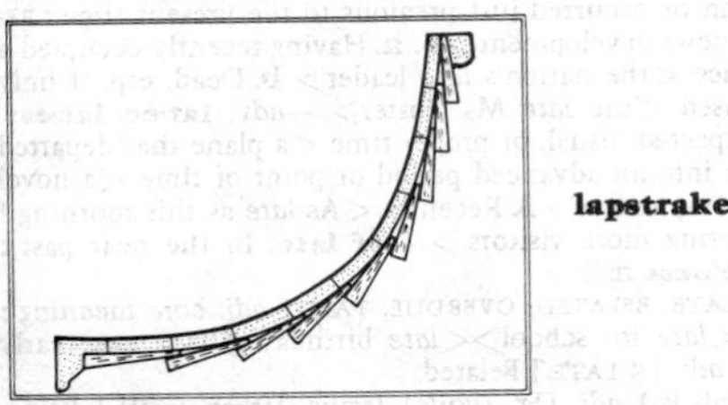

**lapstrake**

**La·pu·ta** (lə-pyōō'tə) *n.* A flying island in Swift's *Gulliver's Travels* inhabited by philosophers involved in absurdly impractical projects. **—La·pu'tan** (-pyōōt'n) *n. & adj.*

**lap·wing** (lăp'wĭng') *n.* [ME < OE *hlēapewince.*] An Old World bird of the genus *Vanellus*, related to the plover, esp. *V. vanellus*, with a narrow crest.

**Lar** (lär) *n., pl.* **Lar·es** (lâr'ēz) [Lat.] A tutelary deity or spirit of an ancient Roman household.

**lar·board** (lär'bərd) [ME *laddebord.*] *Naut.* —*n.* PORT². —*adj.* On the port side.

**lar·ce·nist** (lär'sə-nĭst) *also* **lar·ce·ner** (-nər) *n.* One who commits larceny.

**lar·ce·ny** (lär'sə-nē) *n., pl.* **-nies.** [ME < OFr. *larcin*, theft < Lat. *latrocinium*, robbery < *latro*, robber.] The unlawful taking and removing of another's personal property with the intent of permanently depriving the owner : THEFT. **—lar'ce·nous** *adj.*

**larch** (lärch) *n.* [MHG *larche* < Lat. *larix.*] **1.** A coniferous tree of the genus *Larix*, with deciduous needles and heavy durable wood. **2.** The wood of a larch.

**lard** (lärd) *n.* [ME < OFr. *larde* < Lat. *lardum.*] The white solid or semisolid rendered fat of a hog. —*vt.* **lard·ed, lard·ing, lards. 1.** To coat or cover with lard or a similar fat. **2.** To insert strips of fat in (meat) before cooking. **3. a.** To make richer with or as if with fat. **b.** To enrich with additions : EMBELLISH <*larded* the story with humorous anecdotes> **—lard'y** *adj.*

**lar·der** (lär'dər) *n.* [ME < AN < Med. Lat. *lardarium* < Lat. *lardum*, lard.] **1.** A place to store meat and other foods. **2.** A food supply.

**lar·don** (lär'dŏn') *also* **lar·doon** (lär-dōōn') *n.* [Fr. < OFr. < *larde*, lard.] A strip of fat for larding meat.

**lar·ee** (lâr'ē) *n.* [Ult. < Pers. *Lārī.*] —See table at CURRENCY.

**Lar·es** (lâr'ēz) *n. pl. of* LAR.

**lares and penates** *pl.n.* Household possessions.

**large** (lärj) *adj.* **larg·er, larg·est.** [ME < OFr. < Lat. *largus.*] **1.** Greater than average in size, extent, quantity, or amount : BIG. **2.** Greater than average in scope, breadth, or capacity : COMPREHENSIVE. **3.** Tolerant and understanding <a *large* and forgiving soul> **4. a.** Boastful : pretentious. —Used of speech or manners. **b.** *Obs.* Gross : coarse. —Used of speech or language. **5.** *Naut.* Designating a favorable wind. **—at large. 1.** At liberty <escaped bank robbers still *at large*> **2.** At length : COPIOUSLY. **3.** As a whole : in general <the television viewing audience *at large*> **4.** Representing a nation, state, or district as a whole. **5.** Not assigned to a specific country <an ambassador *at large*> **—large'ness** *n.*

**large calorie** *n.* CALORIE 3.

**large-heart·ed** (lärj'här'tĭd) *adj.* Generous in disposition. **—large'-heart'ed·ness** *n.*

**large intestine** *n.* The portion of the intestine that extends from the ileum to the anus, arching around the convolutions of the small intestine and including the cecum, colon, rectum, and anal canal.

**large·ly** (lärj'lē) *adv.* **1.** For the most part. **2.** On a large scale.

**large-mind·ed** (lärj'mīn'dĭd) *adj.* Marked by breadth or liberality of views : OPEN-MINDED. **—large'-mind'ed·ly** *adv.* **—large'-mind'ed·ness** *n.*

**large·mouth bass** (lärj'mouth') *n.* A North American freshwater food and game fish, *Micropterus salmoides.*

**large-scale** (lärj'skāl') *adj.* **1.** Large in scope : EXTENSIVE. **2.** Drawn or made large to show detail <a *large-scale* map><a *large-scale* model>

**lar·gess** *also* **lar·gesse** (lär-zhĕs', -jĕs', lär'jĕs') *n.* [ME *largesse* < OFr. *largece* < *large*, generous < Lat. *largus*, abundant.] **1. a.** Liberality in giving, esp. when attended by condescension. **b.** Money or gifts bestowed. **2.** Generosity of spirit or attitude.

**lar·ghet·to** (lär-gĕt'ō) [Ital. < *largo*, largo.] *Mus.* —*adv. & adj.* Moderately slow in tempo. —Used as a direction. —*n., pl.* **-tos.** A movement or passage played larghetto.

**larg·ish** (lär'jĭsh) *adj.* Rather large <a *largish* sofa>

**lar·go** (lär'gō) [Ital. < Lat. *largus*, large.] *Mus.* —*adv. & adj.* Slow and solemn in manner. —Used as a direction. —*n., pl.* **-gos.** A largo movement or passage.

**lar·i·at** (lăr'ē-ət) *n.* [Sp. *la reata* : *la*, the + *reatar*, to tie again (*re-*, again + *atar*, to tie).] **1.** A long rope with a running noose for catching livestock : LASSO. **2.** A rope for picketing grazing horses or mules.

**lark¹** (lärk) *n.* [ME < OE *lāwerce.*] **1.** Any of various chiefly Old World birds of the family Alaudidae, with a prolonged melodious song. **2.** A bird similar to the lark, as the meadowlark.

**lark²** (lärk) *n.* [Prob. < alteration of dial. *lake*, to play < ME *laken* < ON *leika.*] **1.** A carefree adventure. **2.** A harmless prank. —*vi.* **larked, lark·ing, larks.** To engage in harmless pranks.

**lark·spur** (lärk'spûr') *n.* A plant of the genus *Delphinium*, with spurred, variously colored flowers.

**lar·ri·gan** *also* **Lar·ri·gan** (lăr'ĭ-gən) *n.* [Orig. unknown.] A moccasin with knee-high leggings made of oiled leather.

**†lar·rup** (lăr'əp) [Orig. unknown.] *Regional.* —*vt.* **-ruped, -rup·ing, -rups.** To beat : flog. —*n.* A blow.

**lar·um** (lăr'əm) *n.* [Short for ALARUM.] *Archaic.* An alarm.

**lar·va** (lär'və) *n., pl.* **-vae** (-vē) [NLat. < Lat., ghost.] **1.** The wingless, often wormlike form of a newly hatched insect before metamorphosis. **2.** The newly hatched, earliest stage of any of various animals that undergo metamorphosis, differing noticeably in form and appearance from the adult. **—lar'val** *adj.*

**lar·vi·cide** (lär'vĭ-sīd') *n.* An insecticide for exterminating larval pests. **—lar'vi·cid'al** (-sīd'l) *adj.*

**laryng-** *pref. var. of* LARYNGO-.

**la·ryn·ge·al** (lə-rĭn′jē-əl, -jəl, lăr′ən-jē′əl) *also* **la·ryn·gal** (lə-rĭng′gəl) *adj.* [NLat. *laryngeus* < *larynx*, larynx.] **1.** Of, relating to, affecting, or situated near the larynx. **2.** Produced in or with the larynx : GLOTTAL. —*n.* **1.** A part of the larynx. **2.** A laryngeal sound. **3.** Any of a set of sounds of uncertain character reconstructed from indirect evidence for Proto-Indo-European.

**lar·yn·gec·to·my** (lăr′ən-jĕk′tə-mē) *n., pl.* **-mies.** Surgical removal of part or all of the larynx.

**la·ryn·ges** (lə-rĭn′jēz) *n. var. pl. of* LARYNX.

**lar·yn·gi·tis** (lăr′ən-jī′tĭs) *n.* Inflammation of the larynx. **—lar′yn·git′ic** (-jĭt′ĭk) *adj.*

**laryngo-** *or* **laryng-** *pref.* [NLat. < Gk. *larungo-* < *larunx*, larynx.] Larynx <*laryngitis*>

**lar·yn·gol·o·gy** (lăr′ən-gŏl′ə-jē) *n.* The branch of medicine concerned with the study and treatment of the larynx. **—lar′yn·gol′o·gist** *n.*

**la·ryn·go·scope** (lə-rĭng′gə-skōp′, -rĭn′jə-) *n.* A tubular instrument for observing the interior of the larynx. **—la·ryn′go·scop′ic** (-skŏp′ĭk), **la·ryn′go·scop′i·cal** *adj.* **—la·ryn′go·scop′i·cal·ly** *adv.* **—lar′yn·gos′co·py** (lăr′ən-gŏs′kə-pē) *n.*

**lar·ynx** (lăr′ĭngks) *n., pl.* **la·ryn·ges** (lə-rĭn′jēz) *or* **lar·ynx·es.** [NLat. < Gk. *larunx.*] The upper part of the respiratory tract between the pharynx and the trachea, having cartilaginous walls and containing the vocal cords enveloped in folds of mucous membrane attached to the sides.

**la·sa·gna** *also* **la·sa·gne** (lə-zän′yə) *n.* [Ital. < Lat. *lasanum*, cooking pot < Gk. *lasanon*, chamber pot.] **1.** Flat, wide noodles. **2.** A dish made by baking lasagna with layers of ground meat, tomato sauce, and cheese.

**las·car** (lăs′kər) *n.* [Hindi *lashkari* < *lashkar*, army < Pers. < Ar. *al-'askar*, the army.] An East Indian sailor.

**las·civ·i·ous** (lə-sĭv′ē-əs) *adj.* [ME < LLat. *lasciviosus* < Lat. *lascivia*, lasciviousness < *lascivus*, lascivious.] **1.** Of or marked by lust. **2.** Exciting sexual desires. **—las·civ′i·ous·ly** *adv.* **—las·civ′i·ous·ness** *n.*

**lase** (lāz) *vi.* **lased, las·ing, las·es.** [Back-formation < LASER.] **1.** To function as a laser. **2.** To emit coherent radiation by the action of a laser.

**la·ser** (lā′zər) *n.* [L(IGHT) A(MPLIFICATION BY) S(TIMULATED) E(MISSION OF) R(ADIATION).] **1.** Any of several devices that convert incident electromagnetic radiation of mixed frequencies to one or more discrete frequencies of highly amplified and coherent visible radiation. **2.** A device whose output is in an invisible region of the electromagnetic spectrum.

**lash**[1] (lăsh) *n.* [ME, prob. < *lashen*, to deal a blow.] **1. a.** A stroke or blow with or as if with a whip. **b.** A whip. **c.** The thongs of a whip. **2.** A cutting remark. **3.** An eyelash. —*v.* **lashed, lash·ing, lash·es.** —*vt.* **1.** To strike with or as if with a whip. **2.** To strike against with force or violence <heavy rain *lashing* the tent> **3.** To move or wave rapidly <A great marlin on the end of my line *lashed* its tail to and fro.> **4.** To make a scathing oral or written attack against. **5.** To incite or goad <speech that *lashed* the mob into a frenzy> —*vi.* **1.** To move rapidly or violently : DASH. **2.** To strike with or as if with a whip. **3.** To make a scathing verbal attack. **—lash′er** *n.*

**lash**[2] (lăsh) *vt.* **lashed, lash·ing, lash·es.** [ME *lashen*, to lace < OFr. *lachier* < Lat. *laqueare*, to ensnare < *laqueus*, snare.] To secure or bind, as with a rope, cord, or chain. **—lash′er** *n.*

**lash·ing** (lăsh′ĭng) *n.* Something used for securing or binding.

**lash·ings** (lăsh′ĭngz) *pl.n.* [< LASH[1], to lavish (obs.).] *Chiefly Brit.* Lavish quantities.

**lass** (lăs) *n.* [ME *las.*] **1.** A young woman or girl. **2.** A sweetheart.

**Las·sa fever** (lä′sə, lăs′ə) *n.* [After *Lassa*, a village in Nigeria.] An acute viral disease marked by high fever, headache, and ulcers of the mucous membranes and resulting in high mortality.

**las·sie** (lăs′ē) *n.* A lass.

**las·si·tude** (lăs′ĭ-to͞od′, -tyo͞od′) *n.* [Lat. *lassitudo* < *lassus*, weary.] Listless weakness or exhaustion.

**las·so** (lăs′ō, lă-so͞o′) *n., pl.* **-sos** *or* **-soes.** [Sp. < Lat. *laqueus*, snare.] A long rope or leather thong with a running noose at one end used esp. to catch horses and cattle : LARIAT. —*vt.* **-soed, -so·ing, -sos** *or* **-soes.** To catch with or as if with a lasso : ROPE. **—las′so·er** *n.*

**last**[1] (lăst) *adj.* [ME < OE *latost*, superl. of *læt*, late.] **1.** Being, coming, or placed after all others : FINAL <the *last* game in the series> **2.** Being the only one left <my *last* apple> **3.** Just past : most recent <*last* week> **4.** Most up-to-date : NEWEST <the *last* word in sportswear> **5.** Highest in extent or degree : UTMOST. **6.** Most valid, authoritative, or conclusive. **7.** Least likely or expected <the *last* thing anyone would have wanted> **8.** Being the latest possible <hesitated up to the *last* second> **9.** Lowest in size, rank, or importance. **10. a.** Of or relating to a terminal period or stage <their *last* years> **b.** Administered just before death <the *last* rites> —*adv.* **1.** After all others in chronology or sequence <happened *last*> **2.** At a time just previous to the present : most recently <telephoned *last* from Des Moines> **3.** At the end : FINALLY <And *last* came the clowns.> —*n.* **1.** One that is last <took every trick but the *last*> **2.** The end <They were there until the *last*.> **3.** The final mention or appearance <haven't seen the *last* of them> **—at last.** After a considerable length of time : FINALLY. **—at long last.** After a long time or wait. **—last′ly** *adv.*

☆ ***syns:*** **1.** LAST, CLOSING, CONCLUDING, FINAL, TERMINAL *adj. core meaning* : coming after all others <the *last* act> ***ant:*** first **2.** LAST, FINAL, ULTIMATE *adj. core meaning* : of or relating to a terminative condition, stage, or point <the *last* days of the war><the *last* rites> **3.** LAST, FOREGOING, PRECEDING, PREVIOUS *adj. core meaning* : next before the present one <*last* night>

**last**[2] (lăst) *v.* **last·ed, last·ing, lasts.** [ME *lasten* < OE *læsten.*] —*vi.* **1. a.** To continue in existence : go on <The tournament *lasted* six days.> **b.** To continue to live : SURVIVE <Sturdy old oaks really *last.*> **2.** To remain in good condition : ENDURE <A stone house *lasts* better than a wooden one.> **3.** To remain in adequate supply <Will the refreshments *last?*> —*vt.* To supply adequately <enough water to *last* the crew for two days> **—last′er** *n.*

**last**[3] (lăst) *n.* [ME < OE *læste* < *læst*, sole of the foot.] A block or form shaped like a human foot and used in making or repairing shoes. —*vt.* **last·ed, last·ing, lasts.** To mold or shape on a last.

**last**[4] (lăst) *n.* [ME, a kind of measure < OE *hlæst.*] *Chiefly Brit.* A unit of weight or volume varying for different commodities and in different districts, approx. 80 bushels, 640 gallons, or 2 tons.

**last-ditch** (lăst′dĭch′) *adj.* Done or made as a final measure, esp. to avoid a crisis or disaster <a *last-ditch* effort to prevent war>

**Las·tex** (lăs′tĕks′). A trademark for a yarn having a core of elastic rubber wound with rayon, nylon, silk, or cotton threads.

**last hurrah** *n.* A last appearance or effort, esp. at the end of a career.

**last·ing** (lăs′tĭng) *adj.* Continuing or enduring for a long time. —*n.* A durable twilled fabric. **—last′ing·ly** *adv.* **—last′ing·ness** *n.*

**Last Judgment** *n.* The final judgment by God of all humankind.

**last-min·ute** (lăst′mĭn′ĭt) *adj.* Pertaining to or being the moment just before a climactic, conclusive, or calamitous event <*last-minute* changes><a *last-minute* rescue>

**last straw** *n.* The last of a series of annoyances or disappointments that leads one to a final loss of patience, temper, trust, or hope.

**Last Supper** *n.* Christ's meal with His disciples on the night before the Crucifixion.

**last word** *n.* **1.** The final statement in a verbal argument. **2. a.** A conclusive or authoritative statement or treatment <the *last word* in automobile safety> **b.** The power or authority of ultimate decision. **3.** *Informal.* The newest or most up-to-date example of a category <the *last word* in video games>

**lat·a·ki·a** (lăt′ə-kē′ə) *n.* [After *Latakia*, Syria.] An aromatic Turkish smoking tobacco.

**latch** (lăch) *n.* [ME *latche* < *lacchen*, to seize < OE *læccan.*] A fastening or lock, usu. consisting of a bar that fits into a notch, slot, or cavity. —*vt.* **latched, latch·ing, latch·es.** To close or lock with or as if with a latch. **—latch on to** (*or* **onto**). *Informal.* **1.** To attach oneself to. **2.** To get possession of : OBTAIN.

**latch·et** (lăch′ĭt) *n.* [ME *lachet* < OFr., var. of *lacet*, shoestring < *lace*, lace. —see LACE.] A leather strap or thong for fastening a shoe or sandal.

**latch·key** (lăch′kē′) *n.* A key for opening a latch, esp. one on an outside door or gate.

**latchkey child** *n.* A young child who returns from school and remains at home unsupervised for an indefinite period until the parents arrive from work.

**latch·string** (lăch′strĭng′) *n.* A cord fastened to a latch and often passed through a hole in the door so that the latch can be lifted from the outside or pulled inside to thwart intruders.

**late** (lāt) *adj.* **lat·er, lat·est.** [ME < OE *læt.*] **1.** Coming, occurring, or remaining after the usual or proper time <*Late* arrivals will not be seated.> **2. a.** Beginning at or lasting until an advanced hour <a *late* show> **b.** Occurring, being, or continuing toward the end or more advanced part, as of a time period <during *late* December> **3.** Having begun or occurred just previous to the present time : RECENT <a *late* news development> **4. a.** Having recently occupied a position or place <the nation's *late* leader> **b.** Dead, esp. if only recently deceased <the *late* Ms. Foster> —*adv.* **lat·er, lat·est.** **1.** After the expected, usual, or proper time <a plane that departed *late*> **2.** At or into an advanced period or point of time <a novel written *late* in my career> **3.** Recently <As *late* as this morning I was still expecting more visitors.> **—of late.** In the near past : LATELY. **—late′ness** *n.*

☆ ***syns:*** LATE, BELATED, OVERDUE, TARDY *adj. core meaning* : not on time <*late* for school><*late* birthday gifts> ***ant:*** early

**lat·ed** (lā′tĭd) *adj.* [< LATE.] Belated.

**la·teen** (lə-tēn′, lă-) *adj.* [Fr. *(voile) latine*, lateen (sail).] Being, relating to, or rigged with a triangular sail on a long yard fastened at an angle to a short mast. —*n.* **1.** A lateen-rigged boat. **2.** A lateen sail.

**Late Greek** *n.* Greek as used from the 4th to the 9th cent.

**Late Latin** *n.* Latin as used from the 3rd to the 7th cent.

**late·ly** (lāt′lē) *adv.* Not long ago : RECENTLY.

**la·ten·cy** (lāt'n-sē) *n.* **1.** The quality or state of being latent. **2.** *Computer Sci.* The time required for a device to begin physical output of a desired piece of data once processing is complete.
**la·tent** (lāt'nt) *adj.* [Lat. *latens, latent-,* pr.part. of *latēre,* to lie hidden.] Present or potential but not evident or active <*latent* musical talent> —*n.* A fingerprint that is difficult to see but can be made visible for examination. **—la'tent·ly** *adv.*
**latent heat** *n.* The quantity of heat absorbed or released by a substance undergoing a change of state, as by ice changing to water or water to steam.
**latent period** *n.* **1.** The incubation period of an infectious disease. **2.** The interval between stimulus and response.
**lat·er·al** (lăt'ər-əl) *adj.* [Lat. *lateralis* < *latus,* side.] **1.** Of, pertaining to, or located at or on the side. **2.** Designating a sound produced by breath passing along one or both sides of the tongue. —*n.* **1.** A lateral part, projection, passage, or appendage. **2.** *Football.* A lateral pass. **3.** A lateral sound, as *l.* **—lat'er·al·ly** *adv.*
**lateral line** *n.* A linear series of sensory pores and tubes along the side of a fish or certain other aquatic animals.
**lateral pass** *n. Football.* A usu. underhand pass thrown sideways or backward.
**lat·er·ite** (lăt'ə-rīt') *n.* [Lat. *later,* brick + -ITE.] A red residual soil in humid tropical and subtropical regions, containing concentrations of iron and aluminum hydroxides and occas. used as an ore of iron, aluminum, manganese, or nickel. **—lat'er·it'ic** (-rĭt'ĭk) *adj.*
**la·tex** (lā'tĕks') *n., pl.* **la·ti·ces** (lā'tĭ-sēz', lăt'ĭ-) or **la·tex·es.** [NLat. *Latex, Latic-* < Lat., fluid.] **1.** The usu. milky, viscous sap of certain trees and plants, as the rubber tree, that coagulates when exposed to air. **2.** An emulsion of rubber or plastic globules in water, used in paints, adhesives, and various synthetic rubber products. **3.** A latex paint.
**latex paint** *n.* A paint with a latex binder.
**lath** (lăth) *n., pl.* **laths** (lăthz, lăths) [ME.] **1. a.** A narrow, thin strip of wood or metal, used esp. in making a supporting structure for plaster, shingles, slates, or tiles. **b.** A building material, as a sheet of metal mesh, used for similar purposes. **2. a.** Lathing. **b.** Work made with or from lathing. —*vt.* **lathed, lath·ing, laths.** To build, cover, or line with laths.
**lathe** (lāth) *n.* [Prob. < ME, supporting structure.] A machine on which a piece of wood, metal, or other material is spun and shaped by a fixed cutting or abrading tool. —*vt.* **lathed, lath·ing, lathes.** To shape or cut on a lathe.
**lath·er** (lăth'ər) *n.* [Ult. < OE *lēaðor.*] **1.** A light foam created by agitation of soap or detergent in water. **2.** Froth formed by profuse sweating, as on a horse. **3.** *Informal.* Impatient, troubled excitement : AGITATION. —*v.* **-ered, -er·ing, -ers.** —*vt.* **1.** To coat with lather. **2.** *Informal.* To give a beating to : WHIP. —*vi.* **1.** To produce lather : FOAM. **2.** To become coated with lather, as a horse. **—lath'er·er** *n.* **—lath'er·y** *adj.*
**lath·ing** (lăth'ĭng, lăth'-) *n.* **1.** The act or process of building with laths. **2.** Work made of laths. **3.** A quantity of laths.
**la·ti·ces** (lā'tĭ-sēz', lăt'ĭ-) *n. var. pl. of* LATEX.
**la·tic·i·fer** (lā-tĭs'ə-fər) *n.* A plant duct containing latex. **—lat'i·cif'er·ous** (lăt'ĭ-sĭf'ər-əs) *adj.*
**lat·i·fun·di·um** (lăt'ə-fŭn'dē-əm) *n., pl.* **-di·a** (-dē-ə) [Lat. : *latus,* broad + *fundus,* estate.] A great landed estate, esp. of the ancient Romans.
**Lat·in** (lăt'n) *adj.* [ME < OFr. < Lat. *Latinus* < *Latium,* an ancient country in Italy.] **1.** Of or relating to Latium, its people, or its culture. **2.** Of or relating to ancient Rome, its people, or its culture. **3.** Of, relating to, or composed in the language of ancient Rome and Latium. **4.** Of or relating to those countries or peoples using Romance languages, esp. the countries of Latin America. **5.** Of or relating to the Roman Catholic Church. —*n.* **1.** The Italic language of ancient Latium and Rome that overspread western Europe and until modern times was the dominant language of church, school, and state. **2.** A native or resident of ancient Latium. **3.** A member of a Latin people, esp. of Latin America. **4.** A Roman Catholic.
**Latin alphabet** *n.* The Roman alphabet adopted from the Greek by way of the Etruscan alphabet, consisting of 23 letters on which the modern western European alphabets are founded.
**Lat·in-A·mer·i·can** (lăt'n-ə-mĕr'ĭ-kən) *adj.* Of, relating to, or designating Western Hemisphere nations south of the United States that have Spanish, Portuguese, or French as their official languages. **—Latin American** *n.*
**Lat·in·ate** (lăt'n-āt') *adj.* Of, relating to, or derived from Latin.
**Latin Church** *n.* The Roman Catholic Church.
**Latin cross** *n.* A cross having the horizontal bar shorter than the vertical bar.
**Lat·in·ism** (lăt'n-ĭz'əm) *n.* An idiom, structure, or word derived from or imitative of Latin.
**Lat·in·ist** (lăt'n-ĭst) *n.* A Latin scholar.
**La·tin·i·ty** (lə-tĭn'ĭ-tē) *n.* The way that Latin is spoken or written.
**Lat·in·ize** (lăt'n-īz') *v.* **-ized, -iz·ing, -iz·es.** —*vt.* **1. a.** To translate into Latin. **b.** To transliterate into the characters of the Latin alphabet : ROMANIZE. **2.** To cause to adopt or acquire Latin characteristics or customs. **3.** To cause to follow or resemble the Roman Catholic Church in dogma or practices. —*vi.* To use Latinisms. **—Lat'in·i·za'tion** *n.* **—Lat'in·iz'er** *n.*
**La·ti·no** (lə-tē'nō, lă-) *n.* A native or resident of Latin America.
**lat·ish** (lā'tĭsh) *adj. & adv. Informal.* Rather late.
**lat·i·tude** (lăt'ĭ-tōōd', -tyōōd') *n.* [ME < OFr. < Lat. *latitudo* < *latus,* wide.] **1.** Breadth : range. **2.** Freedom from the usual restraints, limitations, or regulations. **3. a.** The angular distance north or south of the equator, measured in degrees along a meridian, as on a map or globe. **b.** A region of the earth regarded in relation to its distance from the equator <temperate *latitudes*> **4.** *Astron.* The angular distance of a celestial body north or south of the ecliptic. **—lat'i·tu'di·nal** (-tōōd'n-əl, -tyōōd'-) *adj.* **—lat'i·tu'di·nal·ly** *adv.*
**lat·i·tu·di·nar·i·an** (lăt'ĭ-tōōd'n-âr'ē-ən, -tyōōd'-) *adj.* [Lat. *latitudo, latitudin-,* latitude + -ARIAN.] Encouraging freedom of thought and behavior, esp. in religion. **—lat'i·tu'di·nar'i·an** *n.* **—lat'i·tu'di·nar'i·an·ism** *n.*
**la·trine** (lə-trēn') *n.* [Fr. < Lat. *latrina* < *lavatrina* < *lavare,* to wash.] A communal toilet usu. in a barracks.
**-latry** *suff.* [< Gk. *latreia,* service, worship.] Worship <*bibliolatry*>
**lat·ten** (lăt'n) *n.* [ME *latton* < OFr. *laton.*] **1.** An alloy once made of or made to resemble brass, hammered thin and used in the manufacture of church vessels. **2.** A thin sheet of metal, esp. of tin.
**lat·ter** (lăt'ər) *adj.* [ME < OE *lattre.*] **1.** Indicating the second of two persons or things mentioned. **2.** Further advanced in time or sequence : LATER <*latter* part of the week> **3.** Closer to the end <the *latter* part of the game> —*n.* The second of two persons or things mentioned. ***usage:*** As used in contrast to *former, latter* refers to the second of only two, as in *The chairperson and the president attended the meeting; the latter presented a report.* In a sentence such as *The dean, the chancellor, and the provost will serve on the committee,* refer to the provost as *the last, the last of these,* or repeat the name or designation. **—lat'ter·ly** *adv.*
**lat·ter-day** (lăt'ər-dā') *adj.* Of present or recent times : MODERN.
**Latter-day Saint** *n.* A Mormon.
**lat·tice** (lăt'ĭs) *n.* [ME *latice* < OFr. *latiz.*] **1. a.** An open framework of interwoven strips that form regular, patterned spaces. **b.** A structure, as a screen, window, or gate, made of such a framework. **2.** Something, as a decorative motif or heraldic bearing, that resembles an open patterned framework. **3.** *Physics.* A regular, periodic configuration of points, particles, or objects throughout an area or space, esp. the arrangement of ions or molecules in a crystalline solid. —*vt.* **-ticed, -tic·ing, -tic·es.** To construct or furnish with a lattice or latticework. **—lat'ticed** *adj.*
**lat·tice·work** (lăt'ĭs-wûrk') *n.* **1.** An object, a structure, or material resembling a lattice. **2.** A structure made of lattices.
**Lat·vi·an** (lăt'vē-ən) *n.* **1.** A native or resident of Latvia. **2.** The Baltic language of the Latvians. **—Lat'vi·an** *adj.*
**laud** (lôd) *vt.* **laud·ed, laud·ing, lauds.** [Lat. *laudare* < *laus,* praise.] To give praise to : GLORIFY. —*n.* **1.** Glorification : praise. **2.** A hymn or song of praise. **3.** *often* **Lauds.** *(sing. or pl. in number).* The service of prayers following the matins and forming with them the first of the seven canonical hours. **—laud'er** *n.*
**laud·a·ble** (lô'də-bəl) *adj.* Deserving praise. **—laud'a·bil'i·ty, laud'a·ble·ness** *n.* **—laud'a·bly** *adv.*
**lau·da·num** (lôd'n-əm) *n.* [NLat.] A tincture of opium.
**lau·da·tion** (lô-dā'shən) *n.* The act of lauding : PRAISE.
**laud·a·tive** (lô'də-tĭv) *adj.* Laudatory.
**laud·a·to·ry** (lô'də-tôr'ē, -tōr'ē) *adj.* [LLat. *laudatorius* < Lat. *laudare,* to laud. —see LAUD.] Of, pertaining to, or giving praise.
**laugh** (lăf, läf) *v.* **laughed, laugh·ing, laughs.** [ME *laughen* < OE *hliehan.*] —*vi.* **1.** To express esp. mirth or derision usu. by a series of inarticulate sounds, with the mouth open in a wide smile. **2.** To show amusement <*laughed* at their silly mishaps> **3.** To feel derision <*laughed* at their hollow hopes> **4.** To produce sounds like laughter. —*vt.* **1.** To effect by laughter <*laughed* them out of the meeting> **2.** To express with a laugh. **—laugh away** (or **off**). To treat as ridiculously or laughably trivial. —*n.* **1. a.** The act of laughing. **b.** The sound of laughing. **2.** *Informal.* Something that is amusing, improbable, or ridiculous <What a *laugh!*> **—laugh'er** *n.* **—laugh'ing·ly** *adv.*
**laugh·a·ble** (lăf'ə-bəl, läf'-) *adj.* Causing or deserving laughter. **—laugh'a·ble·ness** *n.* **—laugh'a·bly** *adv.*
☆ ***syns:*** LAUGHABLE, COMIC, COMICAL, FARCICAL, FUNNY, LAUGHING, LUDICROUS, RIDICULOUS, RISIBLE *adj. core meaning* : causing or deserving laughter <a *laughable* economic proposal>
**laughing gas** *n.* Nitrous oxide.
**laughing jackass** *n.* The kookaburra.
**laugh·ing·stock** (lăf'ĭng-stŏk', läf'-) *n.* An object of ridicule.
**laugh·ter** (lăf'tər, läf'-) *n.* [ME < OE *hleahtor.*] **1.** The act of laughing. **2.** The sound produced by laughing. **3.** *Archaic.* A cause or subject for merriment.
**launce** (lăns, läns, lôns) *n.* [Perh. alteration of LANCE.] The sand lance.

**launch¹** (lônch, länch) *v.* **launched, launch·ing, launch·es.** [ME *launchen* < ONFr. *lancher*, var. of OFr. *lancier* < LLat. *lanceare*, to wield a lance < *lancea*, lance.] —*vt.* **1.** To discharge with force : PROPEL <*launch* a rocket> **2.** To put or lower (a boat) into the water, esp. for the first time. **3.** To put into action : INITIATE. **4.** To give a start to <*launch* a career> —*vi.* **1.** To rush or spring forward : PLUNGE. **2.** To make a beginning : COMMENCE. **3.** To move out to sea. **—launch** *n.*

**launch²** (lônch, länch) *n.* [Sp. and Port. *lancha*, both prob. of Malay orig.] **1.** A large ship's boat. **2.** A large open motorboat.

**launch·er** (lôn′chər, län′-) *n.* One that launches, as: **a.** A device for firing grenades. **b.** A device for firing rockets.

**launch pad** *also* **launching pad** *n.* The base or platform from which a rocket or space vehicle is launched.

**launch vehicle** *n.* BOOSTER 4b.

**laun·der** (lôn′dər, län′-) *v.* **-dered, -der·ing, -ders.** [< obs. *launder*, launderer < ME, alteration of *lavender* < OFr. *lavandier* < Lat. *lavanda*, things to be washed < *lavare*, to wash.] —*vt.* **1. a.** To wash (e.g., clothes). **b.** To wash and iron <*laundered* the week's linen> **2.** To clean up as if by laundering. **3.** To channel through an intermediate party in order to conceal the source <*laundered* the funds in foreign banks> —*vi.* **1.** To undergo washing in a specified way <Wool does not *launder* well.> **2.** To wash or wash and iron clothes or linens. —*n.* A wooden trough for water, used for washing ore. **—laun′der·er** *n.*

**laun·der·ette** (lôn′də-rĕt′, län′-) *n.* [Orig. a trademark.] A self-service laundry.

**laun·dress** (lôn′drĭs, län′-) *n.* A woman employed to wash and iron clothes or linens.

**Laun·dro·mat** (lôn′drə-măt′, län′-). A trademark for a commercial establishment equipped with washing machines and dryers, usu. coin-operated and self-service.

**laun·dry** (lôn′drē, län′-) *n., pl.* **-dries.** [< obs. *launder*, launderer. —see LAUNDER.] **1.** Soiled or laundered clothes and linens : WASH. **2.** A place where laundering is done.

**lau·re·ate** (lôr′ē-ĭt, lŏr′-) *adj.* [ME < Lat. *laureatus* < *laurea*, crown of laurel < *laureus*, of laurel < *laurus*, laurel.] **1.** Worthy of honor or distinction. **2.** Crowned or decked with laurel as a mark of honor. **3.** *Archaic.* Made of laurel sprigs, as a wreath or crown. —*n.* **1.** One honored for achievements esp. in the arts or sciences. **2.** A poet laureate. **—lau′re·ate·ship′** *n.*

**lau·rel** (lôr′əl, lŏr′-) *n.* [ME < Lat. *laureola*, dim. of *laurea*, laurel tree. —see LAUREATE.] **1.** A native Mediterranean shrub or tree, *Laurus nobilis*, with aromatic evergreen leaves and small blackish berries. **2.** A shrub or tree, as the mountain laurel, related to or resembling the laurel. **3.** *often* **laurels. a.** A wreath of laurel bestowed as a mark of honor upon poets, heroes, and victors in athletic contests during ancient times. **b.** Honor and glory won for achievement. —*vt.* **-reled, -rel·ing, -rels** *or* **-relled, -rel·ling, -rels.** To crown with laurel.

**Lau·ren·tian** (lô-rĕn′shən) *adj.* **1.** Of, relating to, or situated in the vicinity of the St. Lawrence River. **2.** *Geol.* Of or pertaining to the gneissic granite of the early Precambrian or Archeozoic.

**lau·ric acid** (lôr′ĭk, lŏr′-) *n.* [Lat. *laurus*, laurel + -IC.] A fatty acid, $C_{12}H_{24}O_2$, derived chiefly from coconut oil and used for making soaps, cosmetics, insecticides, and alkyd resins.

**lau·ryl alcohol** (lôr′əl, lŏr′-) *n.* [LAUR(EL) + -YL + ALCOHOL.] A colorless solid alcohol, $C_{12}H_{26}O$, used in synthetic detergents.

**la·va** (lä′və, lăv′ə) *n.* [Ital. < Lat. *labes*, fall.] **1.** Molten rock that issues from a volcano or a fissure in the earth's surface. **2.** The rock formed by the cooling and solidifying of lava.

**la·va·bo** (lə-vā′bō, -vä′bō) *n., pl.* **-boes.** [Lat., I shall wash < *lavare*, to wash.] **1.** *often* **Lavabo.** The ceremonial washing of the hands and recitation from the Psalms by the celebrant before the Eucharist in the Roman Catholic and Anglican churches. **2.** A washbowl and water tank with a spout that are fastened to a wall.

**lav·age** (lăv′ĭj, lä-väzh′) *n.* [Fr. < OFr. < *laver*, to wash < Lat. *lavare*.] *Med.* A washing, esp. of a hollow organ, as the stomach or lower bowel, with repeated injections of water.

**la·va-la·va** (lä′və-lä′və) *n.* [Samoan.] A draped kiltlike garment of cotton print worn by Polynesians and esp. Samoans.

**lav·a·liere** (lăv′ə-lîr′) *also* **la·val·lière** (lä′vä-lyâr′) *n.* [Fr. *lavallière*, after Louise de *la Vallière* (1644–1710).] A pendant worn on a chain around the neck.

**la·va·tion** (lă-vā′shən, lā-) *n.* [Lat. *lavatio* < *lavare*, to wash.] An act or instance of washing : CLEANSING.

**lav·a·to·ry** (lăv′ə-tôr′ē, -tōr′ē) *n., pl.* **-ries.** [ME *lavatorie*, piscina < Med. Lat. *lavatorium* < Lat. *lavare*, to wash.] **1.** A room with washing and often toilet facilities. **2.** A basin or bowl for washing, esp. one permanently installed with running water. **3.** A toilet.

**lave** (lāv) *v.* **laved, lav·ing, laves.** [ME *laven* < OE *lafian* < Lat. *lavare*.] —*vt.* **1.** To wash : bathe. **2. a.** To lap or wash against. **b.** To flow along or against <"The quiet and the cool *laved* her" —Edna Ferber> —*vi.* To bathe oneself.

**lav·en·der** (lăv′ən-dər) *n.* [ME < AN *lavendre* < Med. Lat. *lavendula*.] **1. a.** An aromatic Old World plant of the genus *Lavandula*, esp. *L. officinalis*, bearing small purplish flower clusters and yielding an oil used in perfumery. **b.** The fragrant dried leaves, stems, and flowers of the lavender. **2.** A pale to light or moderate purple to very light or very pale violet. **—lav′en·der** *adj.*

**la·ver¹** (lā′vər) *n.* [ME, water pitcher < OFr. *laveoir*.] **1.** A large basin used in ancient times by a Jewish priest for ablutions before making a sacrificial offering. **2.** *Archaic.* A vessel, stone basin, or trough used for washing.

**la·ver²** (lā′vər) *n.* [NLat. < Lat., a water plant.] An edible seaweed of the genus *Porphyra*.

**lav·er·ock** (lăv′ər-ək) *n.* [ME < OE *lāwerce*.] *Scot.* A skylark.

**lav·ish** (lăv′ĭsh) *adj.* [< ME *lavas* < OFr. *lavasse*, downpour < *laver*, to wash < Lat. *lavare*.] **1.** Spending or giving with liberality : PRODIGAL. **2.** Marked by or produced with extravagance and profusion <a *lavish* wedding supper> —*vt.* **-ished, -ish·ing, -ish·es.** To give or pour forth unstintingly. **—lav′ish·er** *n.* **—lav′ish·ly** *adv.* **—lav′ish·ness** *n.*

**†law** (lô) *n.* [ME < OE *lagu*.] **1. a.** A rule established by authority, society, or custom. **b.** The body or system of such rules. **c.** The authority or control imposed by such a system of rules. **2.** Common law. **3. a.** The actions or processes by which the laws of a society are enforced and through which redress for grievances is obtained. **b.** An agency of the law. **c.** An agent of the law. **4. a.** The science and study of law : JURISPRUDENCE. **b.** Knowledge of law. **c.** The profession of an attorney. **5. Law.** The body of precepts or principles held to express the divine will, esp. as revealed in the Bible <the *Law* of Moses> **6.** A rule or principle to be obeyed. **7.** A rule or procedural principle applicable in a particular domain <the *law* of the sea> **8.** A code of principles and regulations. **9.** A formulation describing a relationship that is presumed to hold between or among phenomena for all cases in which the specified conditions are met <Boyle's *law*> **10.** *Math.* A general principle or rule that holds in all cases to which it applies. —*vi. Regional.* **lawed, law·ing, laws.** To go to law : LITIGATE.

☆ ***syns:*** LAW, AXIOM, FUNDAMENTAL, PRINCIPLE, THEOREM, UNIVERSAL *n. core meaning* : a broad and basic rule or truth <the *laws* of physics>

**law-a·bid·ing** (lô′ə-bī′dĭng) *adj.* Abiding by the law.

**law·break·er** (lô′brā′kər) *n.* One that breaks the law.

**law·ful** (lô′fəl) *adj.* **1.** Allowed by law <*lawful* acts> **2.** Established, sanctioned, or recognized by the law <the *lawful* owner> **3.** Law-abiding. **—law′ful·ly** *adv.* **—law′ful·ness** *n.*

**law·giv·er** (lô′gĭv′ər) *n.* **1.** One who gives a code of laws to a people. **2.** A lawmaker.

**law·less** (lô′lĭs) *adj.* **1.** Unrestrained by law <a *lawless* frontier society> **2.** Heedless of or contrary to the law : DISOBEDIENT <*lawless* conduct> **3.** Not governed by law <the *lawless* areas of some inner cities> **—law′less·ly** *adv.* **—law′less·ness** *n.*

**law·mak·er** (lô′mā′kər) *n.* One who makes laws : LEGISLATOR. **—law′mak′ing** *n.*

**law merchant** *n., pl.* **laws merchant.** Rules and regulations applied to trade and commerce, derived from the customs of merchants in the past.

**lawn¹** (lôn) *n.* [ME *laund*, glade < OFr. *lande*, ult. of Germanic orig.] A usu. closely mown plot or area of grass or similar plants. **—lawn′y** *adj.*

▲ word history: The word *lawn¹*, which now denotes a carefully kept ornamental plot of closely mown grass, originated as a variant spelling of *laund*, a word now obsolete that meant "a woodland glade." *Laund* was borrowed from Old French *lande*. It is likely that *lande* entered Old French from a Celtic source, but the ultimate origin of *lande* is Germanic, and the word is cognate with—that is, has the same ancestor as—English *land*.

**lawn²** (lôn) *n.* [ME *laun*, after *Laon*, France.] A very fine, thin fabric of linen or cotton. **—lawn′y** *adj.*

**lawn mower** *n.* A hand- or power-operated machine for cutting grass or lawns.

**lawn tennis** *n.* Tennis played on a grass court.

**law of large numbers** *n.* Bernoulli's law.

**Law of Moses** *n.* The Pentateuch.

**law of nations** *n.* International law.

**law of parsimony** *n.* Ockham's razor.

**law·ren·ci·um** (lô-rĕn′sē-əm, lō-) *n.* [After Ernest O. *Lawrence* (1901–1958).] *Symbol* **Lw** A synthetic radioactive element having a single isotope, Lw 257; atomic number 103.

**law·suit** (lô′sōōt′) *n.* A case brought before a court of law.

**law·yer** (lô′yər) *n.* [ME *lauier* < *law*, law.] One whose profession is to give legal assistance and advice to clients and represent them in court or in other legal matters. **—law′yer·ly** *adj.*

**lax** (lăks) *adj.* **-er, -est.** [ME < Lat. *laxus*, loose.] **1.** Lacking in rigor, strictness, or firmness. **2.** Not taut, firm, or compact : SLACK. **3.** Loose and not easily retained or controlled. —Used of bowel movements. **4.** Pronounced with the muscles of the tongue and jaw partially relaxed, as the vowel *e* in *let*. **—lax′i·ty** (lăk′sĭ-tē) *n.* **—lax′ly** *adv.* **—lax′ness** *n.*

**lax·a·tive** (lăk′sə-tĭv) *n.* [ME < OFr. *laxatif* < Med. Lat. *laxativus,* preventing constipation < Lat. *laxare,* to relax < *laxus,* loose.] *Med.* A drug that stimulates evacuation of the bowels. *—adj.* **1.** Stimulating evacuation of the bowels. **2.** Causing looseness or relaxation, esp. of the bowels.

**lay[1]** (lā) *v.* **laid** (lād), **lay·ing, lays.** [ME *laien* < OE *lecgan.*] *—vt.* **1.** To cause to lie down <*lay* the baby in its bassinet> **2. a.** To bring to or place in a particular state or position. **b.** To bury. **3.** To put or set down : DEPOSIT. **4.** To produce and deposit <*lay* eggs> **5.** To cause to subside. **6.** To put up to or against <*lay* a finger to the nose> **7.** To put forward as a reproach or accusation <*laid* the blame on me> **8.** To put in order or readiness for use <*lay* places for six at dinner> **9.** To devise : make <*lay* plans> **10.** To spread over a surface <*lay* rail ties on a cinder bed> **11.** To place or give (importance) <*lay* emphasis on deportment> **12.** To impose as a burden or punishment <*lay* new taxes on the people> **13.** To present for examination <*lay* a case before a court> **14.** To put forward as a demand or assertion <*laid* claim to the treasure> **15.** To place (a bet) : WAGER. **16.** To aim (a gun or cannon). **17. a.** To place together (strands) to be twisted into rope. **b.** To make in this manner <*lay* up hawsers> *—vi.* **1.** To produce and deposit eggs. **2.** To bet : wager. **3.** *Nonstandard.* LIE[1] 1. **4.** To engage energetically in an action. **—lay aside. 1.** To give up : ABANDON <*lay aside* hope of success> **2.** To put aside for the future : SAVE. **—lay away. 1.** To reserve for the future : SAVE. **2.** To put aside and hold for future delivery. **—lay by.** To save. **—lay down. 1.** To store for the future. **2.** To specify as a guide or rule. **—lay in.** To store for future use. **—lay into.** To scold harshly. **—lay off. 1.** To terminate the employment of (a worker), esp. temporarily. **2.** To mark off. **3.** To give up : QUIT. **4.** To refrain from criticizing or annoying. **5.** To stop : cease. **—lay out. 1.** To make a detailed plan for. **2.** *Informal.* To speak. **3.** To prepare and clothe (a corpse) for burial. **4.** *Informal.* To knock to the ground or render unconscious by a blow. **—lay over.** To make a stopover in the course of a trip. **—lay to.** *Naut.* **1.** To bring (a ship) to a stop in open water. **2.** To remain stationary and face into the wind. **—lay up. 1.** To stock for future use. **2.** *Informal.* To confine with an injury or illness. **3.** To put (a ship) in dock, as for repairs. *—n.* **1. a.** The direction the strands of a rope or cable are twisted in <a left *lay*> **b.** The amount of such twist. **2.** The state of one that lays eggs <a hen coming into *lay*> **3. a.** *Chiefly Brit.* A line of activity. **b.** An occupation. **—lay down the law.** To assert positively and often arrogantly or harshly. **—lay it on thick. 1.** To exaggerate : overstate. **2.** To flatter effusively. **—lay of the land.** Nature, arrangement, or disposition.

**lay[2]** (lā) *adj.* [ME < OFr. *lai* < LLat. *laicus* < Gk. *laikos* < *laos,* the people.] **1.** Of or relating to the laity : SECULAR. **2.** Not of or coming from a particular profession <*lay* knowledge>

**lay[3]** (lā) *n.* [ME < OFr. *lai.*] A ballad.

**lay[4]** (lā) *v. p.t. of* LIE[1].

**lay·a·bout** (lā′ə-bout′) *n. Chiefly Brit.* A lazybones.

**lay day** *n.* [Prob. < LAY[1].] **1.** One of a specific number of days in port allowed the lessee of a ship without charge. **2.** A day of delay spent in port.

**lay·er** (lā′ər) *n.* [ME *leier,* one who lays stones < *laien,* to lay. —see LAY[1].] **1.** A single thickness, coating, or stratum spread out or covering a surface. **2.** One that lays, esp. a hen. **3.** A stem covered with soil for rooting while still part of a living plant. *—v.* **-ered, -er·ing, -ers.** *—vt.* To propagate (a plant) by means of a layer. *—vi.* **1.** To separate or split into layers. **2.** To take root as a result of layering.

**lay·er·ing** (lā′ər-ĭng) *also* **lay·er·age** (-ĭj) *n.* The process of rooting branches, twigs, or stems still attached to a parent plant, as by placing a specially treated part in moist soil.

**lay·ette** (lā-ĕt′) *n.* [Fr. < OFr., dim. of *laie,* box < MDu. *laege.*] Clothing and other equipment for a newborn child.

**lay figure** *n.* **1.** MANNEQUIN 2. **2.** A person likened to a puppet or dummy.

**lay·man** (lā′mən) *n.* **1.** One who is not a member of the clergy. **2.** One who is not a member of a particular profession or specialty.

**lay·off** (lā′ôf′, -ŏf′) *n.* **1.** Dismissal or suspension of employees. **2.** A period of temporary rest or inactivity.

**lay·out** (lā′out′) *n.* **1.** The act of planning or laying out. **2.** The arrangement or plan of something laid out <the *layout* of the steel mill> **3. a.** The spread and juxtaposition of printed matter, as of a newspaper, book, or magazine page. **b.** A dummy, sketch, or paste-up for matter to be printed. **4.** An establishment or quarters for a specific purpose <a restaurant in a new *layout* near the dock>

**lay·o·ver** (lā′ō′vər) *n.* A usu. brief stop or break in a journey.

**lay·per·son** (lā′pûr′sən) *n.* A layman or laywoman.

**lay reader** *n.* A layperson in the Anglican, Roman Catholic, or Eastern Orthodox church authorized to read parts of the service.

**lay-up** (lā′ŭp′) *n. Basketball.* A usu. one-handed, banked shot made close to the basket after driving in.

**lay·wom·an** (lā′wo͝om′ən) *n.* A woman who is not a member of the clergy.

o͞o **boot** ou **out** th **thin** *th* **this** ŭ **cut** ûr **urge** y **young**
yo͞o **abuse** zh **vision** ə **about,** it**e**m, edib**l**e, gall**o**p, circ**u**s

**la·zar** (lā′zər, lăz′ər) *n.* [ME < Med. Lat. *lazarus* < LLat. *Lazarus,* Lazarus, a beggar in the New Testament < Gk. *Lazaros.*] *Archaic.* One afflicted with a loathsome disease, esp. a leper.

**laz·a·ret·to** (lăz′ə-rĕt′ō) *also* **laz·a·ret** *or* **laz·a·rette** (lăz′ə-rĕt′) *n., pl.* **-tos.** [Ital. < *lazzaro,* lazar < Med. Lat. *lazarus.* —see LAZAR.] **1.** A hospital treating contagious diseases. **2.** A building or ship used as a quarantine station. **3.** *often* **lazaret.** A storage space between the decks of a ship.

**laze** (lāz) *v.* **lazed, laz·ing, laz·es.** [Back-formation < LAZY.] *—vi.* To be lazy : LOAF. *—vt.* To pass (time) in loafing. *—n.* A lazy person.

**laz·u·li** (lăz′yə-lē, lăzh′ə-) *n.* Lapis lazuli.

**laz·u·lite** (lăz′yo͞o-līt′, lăz′ə-, lăzh′ə-) *n.* [Med. Lat. *lazulum,* lapis lazuli + -ITE.] A rather rare, light to deep blue mineral, essentially $(Mg, Fe)Al_2(PO_4)_2(OH)_2$.

**laz·u·rite** (lăz′yo͞o-rīt′, lăz′ə-, lăzh′ə-) *n.* [G. *Lazurit* < Med. Lat. *lazur* < Ar. *lāzaward,* lapis lazuli. —see LAPIS LAZULI.] A rather rare, blue, violet-blue, or greenish-blue mineral, $Na_{4\text{-}5}Al_3Si_3O_{12}S$.

**la·zy** (lā′zē) *adj.* **-zi·er, -zi·est.** [Prob. of LG orig.] **1.** Resistant to work or exertion. **2.** Sluggish : slow-moving < *lazy* flow of the muddy river> **3.** Conducive to idleness or indolence <a *lazy* July afternoon> **4.** Depicted as reclining or lying on its side. —Used of a livestock brand. **—la′zi·ly** *adv.* **—la′zi·ness** *n.*

☆ ***syns:*** LAZY, IDLE, INDOLENT, SHIFTLESS, SLOTHFUL, TRIFLING *adj. core meaning* : resisting exertion, work, or other activity <a *lazy* student> ***ant:*** industrious

**la·zy·bones** (lā′zē-bōnz′) *pl.n. (sing. or pl. in number). Slang.* A lazy person.

**lazy eye** *n.* Amblyopia.

**lazy Susan** *n.* A revolving tray for food or condiments.

**lazy tongs** *pl.n.* Tongs with a jointed extensible framework operated by scissorslike handles, for grasping an object at a distance.

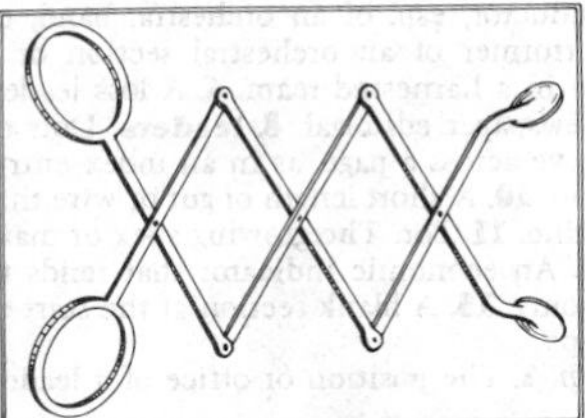

lazy tongs

**LCD** (ĕl′sē-dē′) *n.* [L(IQUID) C(RYSTAL) D(ISPLAY).] A digital display consisting of a liquid crystal material between sheets of glass that becomes readable in the presence of an applied voltage.

**L-do·pa** (ĕl-dō′pə) *n.* [L(EVOROTATORY) + DOPA.] A drug for treating Parkinson's disease.

**lea** (lē, lā) *also* **ley** (lā, lē) *n.* [ME < OE *lēah.*] Meadow : grassland.

**leach** (lēch) *v.* **leached, leach·ing, leach·es.** [Orig. unknown.] *—vt.* **1.** To remove soluble constituents from by the action of a percolating liquid. **2.** To remove from a substance by the action of a percolating liquid. *—vi.* **1.** To be dissolved and washed out by a percolating liquid. **2.** To lose or yield soluble matter to a percolating liquid. *—n.* **1.** An act or process of leaching. **2.** A porous, perforated, or sievelike receptacle that holds material to be leached. **3. a.** The substance through which a liquid is leached. **b.** The solution thus leached. **—leach′a·bil′i·ty** *n.* **—leach′a·ble** *adj.* **—leach′er** *n.*

**lead[1]** (lēd) *v.* **led** (lĕd), **lead·ing, leads.** [ME *leaden* < OE *lædan.*] *—vt.* **1.** To show the way to by going in advance <*lead* children to a playground> **2.** To guide or direct in a course <*lead* elephants> **3.** To serve as a route for <The new road *led* us through the mountains.> **4.** To guide the action or opinion of <*led* us to expect good news> **5.** To direct the performance or activities of <*lead* a scout troop> **6.** To play a principal or guiding role <*lead* the debate> **7.** To go or be at the head of <*led* the jungle patrol> **8.** To be ahead of <*led* the other contestants> **9.** To pass or go through : LIVE <*leading* the life you deserve> **10.** To begin or open with against an opponent <*led* hearts> **11.** To aim in front of (a moving target). *—vi.* **1.** To be first : be ahead. **2.** To go first as a guide. **3.** To act as commander, director, or guide. **4.** To afford a passage, course, or route. **5.** To tend toward a given goal or result. **6.** To make the first play, as in a game or contest. **7.** *Baseball.* To advance a few paces away from one's base toward the next while the pitcher is delivering the ball. —Used of a base runner. **—lead off.** *Baseball.* To be the first batter in an inning. **—lead on. 1.** To lure along a specific course. **2.** To make a gradual or indirect approach to a subject. *—n.* **1.** The foremost position. **2.** Margin of advantage or superiority. **3.** Information of possible use in a search : CLUE. **4.** Ability to lead : LEADERSHIP. **5.** An example : precedent <Your *lead* helped us.> **6. a.** The principal role in a dramatic production. **b.** the performer playing a principal role. **7. a.** The introductory part of a news story. **b.** An important, usu. prominently displayed news story. **8. a.** The first play in a game. **b.** The prerogative or turn to make the first play. **c.** A card so played. **9.** *Baseball.* A position taken by a base runner away from one base toward the next. **10.** LEASH 1. **11. a.** A deposit of gold ore found in an old riverbed. **b.** LODE 1. **12.** *Elect.* A conductor by which one circuit element is electrically connected to another or to the circuit.

**lead²** (lĕd) *n.* [ME < OE *lēad.*] **1.** *Symbol* **Pb** A soft, bluish-white, dense metallic element used in solder and type metal, bullets, radiation shielding, and paints; atomic number 82; atomic weight 207.19. **2.** A plumb bob suspended by a line for making soundings. **3. leads.** *Chiefly Brit.* A flat roof covered with sheets of lead. **4.** Bullets from or for firearms **:** SHOT. **5. leads.** Lead strips for holding the panes of a window. **6.** A thin strip of metal for separating lines of type. **7. a.** Any of various often graphitic compositions used as the writing substance in pencils. **b.** A thin stick of such material. —*vt.* **lead·ed, lead·ing, leads. 1.** To cover, line, weight, fill, or treat with lead. **2.** To provide space between (lines of type) with leads. **3.** To secure (window glass) with leads.

**lead acetate** (lĕd) *n.* A poisonous white crystalline compound, $Pb(C_2H_3O_2)_2 \cdot 3H_2O$, used for waterproofing and in dyes and varnishes.

**lead arsenate** (lĕd) *n.* A poisonous white crystalline compound, $Pb_3(AsO_4)_2$, used in insecticides and herbicides.

**lead carbonate** (lĕd) *n.* A poisonous white amorphous powder, $PbCO_3$, used as a paint pigment.

**lead chromate** (lĕd) *n.* A poisonous yellow crystalline compound, $PbCrO_4$, used as a paint pigment.

**lead colic** (lĕd) *n.* Painter's colic.

**lead·en** (lĕd'n) *adj.* **1.** Made of or containing lead. **2. a.** Heavy and inert. **b.** Sluggish **:** listless. **3.** Lacking sparkle or liveliness. **4.** Low in spirits **:** DEPRESSED <left home with a *leaden* heart> **5.** Dull, dark gray <*leaden* skies> **—lead'en·ly** *adv.* **—lead'en·ness** *n.*

**lead·er** (lē'dər) *n.* [ME *ledere* < OE *lǣdere.*] **1.** One that leads or guides. **2.** One in charge or in command of others. **3. a.** The head of a political organization. **b.** One who has power or influence, esp. of a political nature. **4. a.** A conductor, esp. of an orchestra, band, or chorus. **b.** The principal performer of an orchestral section or a group. **5.** The foremost horse in a harnessed team. **6.** A loss leader. **7.** *Chiefly Brit.* The main newspaper editorial. **8. leaders.** Dots or dashes in a row leading the eye across a page, as in an index entry. **9.** A pipe for conducting liquid. **10.** A short length of gut or wire that attaches a hook to a fishing line. **11.** *Bot.* The growing apex or main shoot of a shrub or tree. **12.** An economic indicator that tends to foretell a change in the economy. **13.** A blank section at the start of a reel of film or recording tape.

**lead·er·ship** (lē'dər-shĭp') *n.* **1.** The position or office of a leader. **2.** Capacity or ability to lead.

**lead glass** *n.* Flint glass.

**lead-in** (lēd'ĭn') *n.* **1.** An introduction. **2.** A program, as on television, that precedes another. **3.** The part of an antenna or aerial that leads to an electronic transmitter or receiver.

**lead·ing¹** (lē'dĭng) *adj.* **1.** Having a position in the lead **:** FOREMOST <the *leading* contender> **2.** Playing a lead in a theatrical production. **3.** Formulated so as to direct or control a response <a *leading* question> **—lead'ing·ly** *adv.*

**lead·ing²** (lĕd'ĭng) *n.* **1.** A border or rim of lead, as around a windowpane. **2.** The spacing between lines of printing type.

**lead·ing edge** (lē'dĭng) *n.* **1.** The edge of a sail that faces the wind. **2.** The front edge of an airplane propeller blade or wing.

**lead·ing tone** (lē'dĭng) *n. Mus.* The seventh tone or degree of a scale, a half tone below the tonic **:** SUBTONIC.

**lead line** (lĕd) *n.* A sounding line.

**lead monoxide** (lĕd) *n.* Litharge.

**lead·off** (lēd'ôf', -ŏf') *n.* **1.** An opening move or play. **2.** One that leads off.

**lead pencil** (lĕd) *n.* A pencil that uses graphite as its marking substance.

**lead·plant** (lĕd'plănt') *n.* A shrub, *Amorpha canescens* of central North America, bearing leaves covered with whitish hairs.

**lead poisoning** (lĕd) *n.* Acute or chronic poisoning by lead or any of its salts, with the acute form causing severe gastroenteritis, and the chronic form anemia, abdominal pain, constipation, partial paralysis, and convulsions.

**lead tetraethyl** (lĕd) *n.* Tetraethyl lead.

**lead-time** (lēd'tīm') *n.* The time between the decision to start a project and the completion of the work.

**lead·wort** (lĕd'wûrt', -wôrt') *n.* **1.** A chiefly tropical plant of the genus *Plumbago*, with variously colored flower clusters. **2.** A plant similar to the leadwort.

**leaf** (lēf) *n., pl.* **leaves** (lēvz) [ME < OE *lēaf.*] **1.** A usu. green, flattened structure of vascular plants, composed of a bladelike expansion attached to a stem and functioning as a principal organ of photosynthesis and transpiration. **2.** A leaflike organ or structure. **3.** Foliage. **4.** The leaves of a plant used or processed for a specific purpose <tobacco *leaf*> **5.** One of the folded sheets, as bound in a volume, each side of which forms a page. **6.** A very thin sheet of material <silver *leaf*> **7.** A removable or hinged section for a table top. **8.** A hinged or otherwise movable section of a folding door, blind, or gate. **9.** One of several metal strips forming a leaf spring. —*v.* **leafed, leaf·ing, leafs.** —*vi.* **1.** To put forth foliage. **2.** To turn through pages <*leafed* through the encyclopedia> —*vt.* To turn through the pages of. **—leaf'less** *adj.* **—leaf'like** *adj.*

**leaf·age** (lē'fĭj) *n.* Foliage.

**leaf·hop·per** (lēf'hŏp'ər) *n.* Any of numerous insects of the family Cicadellidae that suck juices from plants.

**leaf insect** *n.* Any of various chiefly Asiatic insects of the genus *Phyllium* and related genera that resemble leaves.

**leaf·let** (lē'flĭt) *n.* **1.** A segment of a compound leaf. **2.** A small leaf or leaflike part. **3.** A printed, usu. folded handbill or flier, as an advertising circular. —*v.* **-let·ed, -let·ing, -lets** *or* **-let·ted, -let·ting, -lets.** —*vi.* To hand out leaflets. —*vt.* To hand out leaflets to.

**leaf miner** *n.* Any of numerous small flies and moths that in the larval state dig into and feed on leaf tissue.

**leaf mold** *n.* Humus or compost made up of organic matter, as decomposed leaves.

**leaf spot** *n.* Any of various plant diseases causing distinct necrotic areas on the leaves.

**leaf spring** *n.* A composite spring, used esp. in automotive suspensions, consisting of several layers of flexible metallic strips joined to function as a unit.

**leaf·stalk** (lēf'stôk') *n.* The stalk by which a leaf is attached to a stem **:** PETIOLE.

**leaf·y** (lē'fē) *adj.* **-i·er, -i·est. 1.** Having or covered with leaves. **2.** Comprised of leaves. **3.** Similar to a leaf. **—leaf'i·ness** *n.*

**league¹** (lēg) *n.* [ME *ligg* < OFr. *ligue* < Ital. *liga* < *legare*, to bind < Lat. *ligare.*] **1.** An association of states, organizations, or individuals for common action **:** ALLIANCE. **2.** An association of sports teams or clubs that compete chiefly among themselves. **3.** *Informal.* A class of competition <not in the same *league* with me> —*v.* **leagued, leagu·ing, leagues.** —*vi.* To come together in or as if in a league. —*vt.* To bring together in or as if in a league.

**league²** (lēg) *n.* [ME *leuge* < Med. Lat. *leuga*, a measure of distance, of Celt. orig.] **1. a.** A unit of distance equal to 3 statute miles or approx. 4.83 kilometers. **b.** Any of various other units of about the same length. **2.** A square league.

**lea·guer¹** (lē'gər) *n.* [Du.] **1.** A siege. **2.** The camp esp. of a besieging army. —*vt.* **-guered, -guer·ing, -guers.** *Archaic.* To besiege.

**leagu·er²** (lē'gər) *n.* One that belongs to a league.

**leak** (lēk) *v.* **leaked, leak·ing, leaks.** [ME *leke*, prob. of MLG orig.] —*vi.* **1.** To allow the passage or escape of something through a breach or flaw. **2.** To escape or pass through a breach or flaw. **3.** To become publicly known through a breach of secrecy <The true facts and figures *leaked* out.> —*vt.* **1.** To allow (a substance) to escape or pass through a breach or flaw. **2.** To disclose (information) without authorization or official sanction <*leaked* the story to the press> —*n.* **1.** A crack or opening that permits something to escape from or enter a container or conduit. **2. a.** Loss of electric current due to faulty insulation. **b.** The path or place at which this loss takes place. **3.** An unauthorized disclosure of confidential information <"Sometimes we can't respond to stories based on *leaks*" —Ronald Reagan> **4.** An act or instance of leaking. **—leak'er** *n.*

**leak·age** (lē'kĭj) *n.* **1.** LEAK 4. **2.** Something that escapes by leaking. **3.** An amount lost as the result of leaking.

**leak·y** (lē'kē) *adj.* **-i·er, -i·est.** Allowing leakage.

**lean¹** (lēn) *v.* **leaned** *or* **leant** (lĕnt), **lean·ing, leans.** [ME *lenen* < OE *hleonian.*] —*vi.* **1.** To bend or slant away from the vertical. **2.** To incline the weight of the body so as to be supported <*leaning* against the porch pillar> **3.** To rely for assistance or support <*leaned* on each other for courage> **4.** To have a tendency or preference <They all *lean* toward the comedy programs on television.> **5.** *Informal.* To exert pressure <The supervisor is *leaning* on us.> —*vt.* **1.** To set or place so as to be resting or supported. **2.** To cause to incline. —*n.* A tilt or inclination away from the vertical.

**lean²** (lēn) *adj.* **-er, -est.** [ME *lene* < OE *hlǣne.*] **1.** Not fleshy or fat **:** THIN. **2.** Containing little or no fat <*lean* beef> **3. a.** Not productive or prosperous <*lean* years> **b.** Severely curtailed or reduced <a *lean* bank account> **4. a.** Lacking mineral value <*lean* ore> **b.** Lacking in combustible material <a *lean* fuel mixture> —*n.* Meat with little or no fat. **—lean'ly** *adv.* **—lean'ness** *n.*

**Le·an·der** (lē-ăn'dər) *n.* [Lat. < Gk. *Leandros.*] *Gk. Myth.* A youth who loved Hero and drowned during one of his nightly swims across the Hellespont to be with her.

**lean·ing** (lē'nĭng) *n.* A tendency **:** inclination <a *leaning* toward laziness>

**leant** (lĕnt) *v. var. p.t. & p.p. of* LEAN¹.

**lean-to** (lēn'to͞o') *n., pl.* **-tos. 1.** A structure with a single-pitch roof attached to the side of a building. **2.** A shelter or shed having a roof with only one slope.

**leap** (lēp) *v.* **leaped** *or* **leapt** (lĕpt, lēpt), **leap·ing, leaps.** [ME *lepen* < OE *hlēapan.*] —*vi.* **1.** To jump off the ground with a spring of the legs. **2.** To move quickly, abruptly, or impulsively <*leaps* at every opportunity for new experience> —*vt.* To jump over <tried to *leap* the ditch> —*n.* **1.** An act of leaping. **2.** The distance cleared in a leap. **3.** An abrupt or precipitous passage, shift, or transition <*leap* from obscurity to fame> **—leap′er** *n.*

**leap·frog** (lēp′frôg′, -frŏg′) *n.* A game in which one player leaps over another who is kneeling or bending over. *—v.* **-frogged, -frog·ging, -frogs.** *—vt.* **1.** To jump over in or as if in leapfrog. **2.** To advance (two military units) by engaging one with the enemy while moving the other to a position forward of the first unit. *—vi.* To advance in or as if in leapfrog.

**leapfrog test** *n. Computer Sci.* A method of checking the internal operations of a computer by performing arithmetic or logical operations on one section of storage, transferring the new data to another section, repeating the operations, and crosschecking the results.

**leapt** (lĕpt, lēpt) *v. var. p.t. & p.p. of* LEAP.

**leap year** *n.* **1.** A year in the Gregorian calendar having 366 days, with the extra day, Feb. 29, intercalated to compensate for the quarter-day difference between an ordinary year and the astronomical year. **2.** An intercalary year in a calendar.

**learn** (lûrn) *v.* **learned** *or* **learnt** (lûrnt), **learn·ing, learns.** [ME *lernen* < OE *leornian.*] *—vt.* **1.** To gain knowledge, comprehension, or mastery of through study or experience. **2.** To fix in the mind : MEMORIZE <*learned* the alphabet when four years old> **3.** To acquire through experience <*learned* humility> **4.** To become informed of : FIND OUT. **5.** *Nonstandard.* To cause to acquire knowledge : TEACH. **6.** *Obs.* To give information of. *—vi.* To gain knowledge, comprehension, or skill. **—learn′er** *n.*

**learn·ed** (lûr′nĭd) *adj.* **1.** Having or demonstrating profound knowledge or scholarship : ERUDITE. **2.** Directed toward scholars <a *learned* society>

☆ ***syns:*** LEARNED, ERUDITE, SCHOLARLY, SCHOLASTIC, WISE *adj. core meaning* : having or demonstrating profound knowledge and scholarship <a *learned* professor>

▲ word history: The adjective *learned* is the same word as the past participle of the verb *learn.* Since Old English times *learn* has always meant "to gain knowledge," but in Middle English times it also meant "to teach," although this sense is no longer current in standard Modern English. In Middle English the past participle *learned* had the sense "taught," and the word has survived as an adjective, especially with the meanings "educated" and "erudite."

**learn·ing** (lûr′nĭng) *n.* **1.** Education : instruction. **2.** Acquired wisdom, knowledge, or skill.

**learning disability** *n.* A condition thought to be associated with neurological dysfunction that is marked by inability to master a skill, as reading or numerical calculation.

**learn·ing-dis·a·bled** (lûr′nĭng-dĭs-ā′bəld) *adj.* Having a learning disability <*learning-disabled* children>

**learnt** (lûrnt) *v. var. p.t. & p.p. of* LEARN.

**lease** (lēs) *n.* [ME *les* < AN < *lesser,* to lease < OFr. *laissier,* to let go < Lat. *laxare,* to loosen < *laxus,* loose.] **1.** A contract granting occupation or use of property during a certain period in exchange for a specified rent. **2.** The term or duration of a lease. **3.** Property occupied or used under the terms of a lease. *—vt.* **leased, leas·ing, leas·es. 1.** To grant occupation or use of under the terms of a lease. **2.** To occupy or use under a lease. **—new lease on life.** A chance to continue under improved circumstances. **—leas′a·ble** *adj.*

**lease·hold** (lēs′hōld′) *n.* **1.** Possession by lease. **2.** Property held by lease. **—lease′hold′er** *n.*

**leash** (lēsh) *n.* [ME *lesh* < OFr. *laisse* < *laissier,* to let go. —see LEASE.] **1.** A chain, rope, or strap fastened to the collar or harness of an animal and used for leading it or holding it in check. **2.** Control : restraint <kept our laughter in *leash*> **3.** A set of three animals, as hounds. **4.** A set of three. *—vt.* **leashed, leash·ing, leash·es.** To restrain with or as if with a leash.

**leash law** *n.* An ordinance requiring that dogs be kept on a leash when not confined on their owners' property.

**leas·ing** (lē′sĭng) *n.* [ME *lesing* < OE *lēasung* < *lēasian,* to lie < *lēas,* false.] *Archaic.* **1.** The act of lying. **2.** A lie : falsehood.

**least** (lēst) *adj.* [ME < OE *lǣst.*] **1.** Lowest in rank or importance. **2.** Smallest in degree or magnitude. *—adv.* In or to the smallest degree. *—n.* One that is least. **—at least. 1.** According to the lowest possible assessment. **2.** In any event <You might *at least* let us know.> **—in the least.** At all <We won't object *in the least.*>

**least common denominator** *n. Math.* Lowest common denominator.

**least common multiple** *n.* The least quantity exactly divisible by each of two or more designated quantities; e.g., 12 is the least common multiple of 2, 3, 4, and 6.

**least flycatcher** *n.* A small grayish North American bird, *Empidonax minimus.*

**least squares** *pl.n.* A method of determining the line or curve that best fits a relation between two experimental sets of data, using the criterion that the sums of the squares of deviations of experimental points from curve ordinates be a minimum.

**†least·ways** (lēst′wāz′) *adv. Regional.* At least.

**least·wise** (lēst′wīz′) *adv. Informal.* At least : anyway.

**leath·er** (lĕth′ər) *n.* [ME *lether* < OE *leðer.*] **1.** The tanned or dressed hide of an animal, usu. with the hair having been removed. **2.** An article or part of leather, as a boot or strap. *—vt.* **-ered, -er·ing, -ers. 1.** To cover wholly or in part with leather. **2.** *Informal.* To beat with a leather strap.

**leath·er·back** (lĕth′ər-băk′) *n.* A large tropical marine turtle, *Dermochelys coriacea,* with a leathery, longitudinally ridged carapace.

**leath·er·ette** (lĕth′ə-rĕt′) *n.* [Orig. a trademark.] Imitation leather.

**leath·er·jack·et** (lĕth′ər-jăk′ĭt) *n.* A fish, *Oligoplites saurus* of Atlantic and Pacific waters, with tough, leathery skin and venomous spines on the anal fin.

**leath·ern** (lĕth′ərn) *adj. Archaic.* Made of, covered with, or resembling leather.

**leath·er·neck** (lĕth′ər-nĕk′) *n.* [From the leather neckband that was once part of the uniform.] *Slang.* A U.S. Marine.

**leath·er·wood** (lĕth′ər-wo͝od′) *n.* **1.** A shrub, *Dirca palustris* of eastern North America, with tough, pliable bark and small yellow flowers. **2.** The titi.

**leath·er·y** (lĕth′ə-rē) *adj.* Having the texture or appearance of leather <a *leathery* complexion>

**leave**[1] (lēv) *v.* **left** (lĕft), **leav·ing, leaves.** [ME *leven* < OE *lǣfan.*] *—vt.* **1.** To go away from or out of. **2.** To go without removing or taking <*left* the dishes in the sink> **3.** To have as a result, consequence, or remainder <The fire *left* only charred ashes.> **4.** To cause or allow to be or remain in a given condition <*left* the windows open> **5.** To have remaining after death <*left* a collection of rare books> **6.** To bequeath. **7.** To give over to another to control or act on <*left* all the arrangements to them> **8. a.** To forsake : abandon <*leave* one's children> **b.** To remove oneself from participation in or association with <*left* college a month after opening> **9.** To give or deposit, as for use or information, upon one's departure or in one's absence <*left* a package for you><*leave* a formal note at the embassy> **10.** *Nonstandard.* LET[1] 1. ***usage:*** It is entirely acceptable to use *leave alone* instead of *let alone* when the sense intended is "to allow to be or remain in a specified state," as in *If you leave me alone, I can finish the project on time.* But *leave* is not an acceptable substitute for *let* in the sense of "to allow or permit." Only *let* is acceptable in a sentence such as *Let me go.* *—vi.* To depart : go. **—leave off. 1.** To cease : stop. **2.** To stop using or doing.

**leave**[2] (lēv) *n.* [ME < OE *lēaf.*] **1.** Permission. **2. a.** Official permission to be absent from work or duty, esp. that granted to military personnel. **b.** The absence granted by such permission. **3.** Formal or verbal farewell <took *leave* of them and began our journey>

**leave**[3] (lēv) *vi.* **leaved, leav·ing, leaves.** [ME *leven* < *leaf,* leaf.] To put forth foliage : LEAF.

**leaved** (lēvd) *adj.* **1.** Having or bearing a leaf or leaves. **2.** Having a specified number or kind of leaves <three-*leaved*><serrate-*leaved*>

**leav·en** (lĕv′ən) *n.* [ME < OFr. *levain* < VLat. **levamen* < Lat. *levare,* to raise.] **1.** A substance, as yeast or cream of tartar, added to batters and doughs to produce fermentation. **2.** An element or influence that works subtly to enliven or lighten a whole. *—vt.* **-ened, -en·ing, -ens. 1.** To add leavening to. **2.** To produce fermentation in. **3.** To pervade with an enlivening or lightening influence.

**leav·en·ing** (lĕv′ə-nĭng) *n.* A fermentation-producing agent.

**leaves** (lēvz) *n. pl. of* LEAF.

**leave-tak·ing** (lēv′tā′kĭng) *n.* A farewell or departure.

**leav·ings** (lē′vĭngz) *pl.n.* Scraps : remains.

**Leb·a·nese** (lĕb′ə-nēz′, -nēs′) *adj.* Of or relating to Lebanon, its people, or their culture. *—n., pl.* **Lebanese.** A native or resident of Lebanon.

**le·bens·raum** (lā′bəns-roum′) *n.* [G.] Additional territory considered by a nation to be necessary for its continued existence or economic well-being.

**lech**[1] (lĕch) *n.* [Short for LECHER.] *Slang.* **1.** A lecher. **2.** A lecherous desire. *—vi.* **leched, lech·ing, lech·es.** To behave lecherously.

**lech**[2] (lĕch) *n. var. of* LETCH.

**lech·er** (lĕch′ər) *n.* [ME < OFr. < *lechier,* to live in debauchery, of Germanic orig.] A lecherous man.

**lech·er·ous** (lĕch′ər-əs) *adj.* Given to, marked by, or arousing lechery. **—lech′er·ous·ly** *adv.* **—lech′er·ous·ness** *n.*

**lech·er·y** (lĕch′ə-rē) *n., pl.* **-ies. 1.** Excessive indulgence in sexual activity. **2.** Lasciviousness : prurience.

**lec·i·thin** (lĕs′ə-thĭn) *n.* [Gk. *lekithos,* egg yolk + -IN.] Any of a group of phosphatides found in all plant and animal tissues, produced commercially from egg yolks, soybeans, and corn and used in the processing of foods, pharmaceuticals, cosmetics, paints and inks, and rubber and plastics.

**lec·i·thin·ase** (lĕs′ə-thə-nās′, -nāz′) *n.* An enzyme that hydrolyzes lecithin.

**lec·tern** (lĕk′tərn) *n.* [ME *lectorn* < OFr. *lettrun* < Med. Lat. *lecternum* < Lat. *legere,* to read.] **1.** A reading desk with a slanted top holding the books from which Scriptural passages are read during church services. **2.** A stand that supports a speaker's notes or books.

**lec·tion** (lĕk′shən) *n.* [Lat. *lectio,* a reading < *legere,* to read.] **1.** A variant reading or transcription of a text or copy. **2.** A scriptural reading that is a part of a church service.

**lec·tion·ar·y** (lĕk′shə-nĕr′ē) *n., pl.* **-ies.** [LLat. *lectionarium* < Lat. *lectio*, a reading. —see LECTION.] A book containing lections or a list of lections to be read at church services.

**lec·tor** (lĕk′tər) *n.* [ME < LLat. < Lat., reader < *legere*, to read.] **1.** One who reads aloud Scriptural passages in a church service. **2.** A public lecturer or reader in certain universities.

**lec·ture** (lĕk′chər) *n.* [ME, a reading < OFr. < Med. Lat. *lectura* < *legere*, to read.] **1.** An exposition of a given subject before an audience, esp. for instruction. **2.** A sober, often lengthy admonition or reproof. —*v.* **-tured, -tur·ing, -tures.** —*vi.* To deliver a lecture. —*vt.* **1.** To deliver a lecture to. **2.** To admonish or reprove soberly and often at length. **—lec′tur·er** *n.*

**led** (lĕd) *v. p.t. & p.p. of* LEAD[1].

**LED** (ĕl′ē-dē′, lĕd) *n.* [L(IGHT) E(MITTING) D(IODE).] A semiconductor diode that converts applied voltage to light and is used in digital displays, as of a calculator.

**Le·da** (lē′də) *n.* [Lat. < Gk. *Lēda.*] *Gk. Myth.* A Spartan queen and the mother, by Zeus in the form of a swan, of Helen and Pollux and by her husband of Castor and Clytemnestra.

**le·der·ho·sen** (lā′dər-hō′zən) *pl.n.* [G. < MHG *lederhose* : *leder*, leather + *hose*, trousers.] Leather shorts worn by men and boys, esp. in Bavaria.

**ledge** (lĕj) *n.* [ME, piece of wood used for bracing, prob. < *leggen*, to lay < OE *lecgan.*] **1.** A narrow shelf projecting from a wall. **2.** A cut or projection forming a shelf on a cliff or rock wall. **3.** An underwater ridge or rock shelf. **4.** A vein of rock-bearing ore. **—ledg′y** *adj.*

**ledg·er** (lĕj′ər) *n.* [ME *legger*, breviary, prob. < *leggen*, to lay. —see LEDGE.] **1. a.** A book for posting the monetary transactions of a business in the form of debits and credits. **b.** A book to which is transferred the record of accounts as final entry from original postings. **2.** A slab of stone laid flat over a grave. **3.** A horizontal timber in a scaffold, attached to the uprights and supporting the putlogs.

**ledger board** *n.* The top railing of a fence or balustrade.

**ledger line** *n. Mus.* A short line above or below a staff to accommodate notes higher or lower than the staff's range.

**lee** (lē) *n.* [ME *le* < OE *hleo*, shelter.] **1.** The side away from the direction from which the wind blows. **2.** Shelter : cover. —*adj.* Of or relating to the side sheltered from the wind <the *lee* side of a ship>

**lee·board** (lē′bôrd′, -bōrd′) *n.* One of a pair of movable boards or plates attached to the hull of sailboats to prevent slippage downwind.

**leech[1]** (lēch) *n.* [ME *leche*, physician, leech < OE *lǣce.*] **1.** Any of various chiefly aquatic bloodsucking or carnivorous annelid worms of the class Hirudinea, of which one species, *Hirudo medicinalis*, was once used by physicians to bleed their patients. **2.** One who preys on or clings to another : PARASITE. **3.** *Archaic.* A physician. —*v.* **leeched, leech·ing, leech·es.** —*vt.* To bleed with leeches. —*vi.* To attach oneself to another like a leech.

**leech[2]** (lēch) *n.* [ME *leche*, prob. < MLG *līk*, leech line.] **1.** Either vertical edge of a square sail. **2.** The after edge of a fore-and-aft sail.

**leek** (lēk) *n.* [ME *lek* < OE *lēac.*] A plant, *Allium porrum*, related to the onion and with a white, slender bulb and dark-green leaves.

**leer** (lîr) *vi.* **leered, leer·ing, leers.** [Prob. < obs. *leer*, cheek < ME *ler* < OE *hlēor.*] To look obliquely or suggestively, as with lustful interest, malicious intent, or insidious triumph. —*n.* A sly, suggestive, or cunning look.

**leer·y** (lîr′ē) *adj.* **-i·er, -i·est.** *Informal.* Suspicious or distrustful : WARY. **—leer′i·ly** *adv.* **—leer′i·ness** *n.*

**lees** (lēz) *pl.n.* [Pl. of obs. *lee*, sediment < ME *lie* < OFr. < Med. Lat. *lia.*] Sediment settling during fermentation, esp. in wine : DREGS.

**lee shore** *n.* A shore toward which the wind is blowing and toward which a ship is likely to be driven.

**lee·ward** (lē′wərd, lo͞o′ərd) *adj.* Situated on or moving toward the side toward which the wind is blowing. —*n.* The lee side or quarter. **—lee′ward** *adv.*

**lee·way** (lē′wā′) *n.* **1.** The drift of a ship or aircraft to leeward of true course. **2.** A margin of freedom or variation, as of activity, time, or expenditure : LATITUDE.

**left[1]** (lĕft) *adj.* [ME.] **1. a.** Of, designating, belonging to, or situated on the side of the body to the north when the subject is facing east. **b.** Of, designating, directed toward, or situated on the left side. **2.** *often* **Left.** Of or belonging to the political or intellectual Left. —*n.* **1. a.** The direction or position on the left side. **b.** The left side. **c.** The left hand. **d.** A turn in the direction of the left hand or side. **2.** *often* **Left. a.** Those who advocate the adoption of sometimes extreme measures to achieve the equality, freedom, and well-being of the citizens of a state. **b.** The opinion of those advocating such measures. **—left** *adv.*

**left[2]** (lĕft) *v. p.t. & p.p. of* LEAVE[1].

**left field** *n.* **1.** *Baseball.* **a.** The third of the outfield that is to the left looking from home plate. **b.** The position played by the left fielder. **2.** A position far from the center or mainstream, as of opinion or reason <out in *left field*>

**left fielder** *n. Baseball.* The player who defends left field.

**left-hand** (lĕft′hănd′) *adj.* **1.** Of, relating to, or located on the left. **2.** Intended for the left hand or use by a left-handed person.

**left-hand·ed** (lĕft′hăn′dĭd) *adj.* **1.** Having more dexterity in or using the left hand more easily than the right. **2.** Performed with the left hand. **3.** Designed for wear on or use by the left hand. **4.** Maladroit : clumsy. **5.** Of doubtful sincerity : DUBIOUS <*left-handed* compliments> **6.** Of, relating to, or born of a morganatic marriage. **7.** Turning or spiraling from right to left : COUNTERCLOCKWISE. **—left′-hand′ed, left′-hand′ed·ly** *adv.* **—left′-hand′ed·ness** *n.*

**left-hand·er** (lĕft′hăn′dər) *n.* One who is left-handed.

**left·ism** *also* **Left·ism** (lĕf′tĭz′əm) *n.* The ideology of the Left. **—left′ist** *n. & adj.*

**left·o·ver** (lĕft′ō′vər) *adj.* Being unused or uneaten. —*n. often* **leftovers. 1.** An unused portion or remnant, esp. of food. **2.** A dish made out of leftovers.

**left wing** *also* **Left Wing** *n.* A leftist faction of a group. **—left′-wing′** *adj.* **—left′-wing′er** *n.*

**left·y** (lĕf′tē) *n., pl.* **-ies.** *Slang.* **1.** A left-handed person. **2.** An advocate or member of the Left.

**leg** (lĕg) *n.* [ME < ON *leggr.*] **1. a.** A limb or appendage of an animal, used for locomotion or support. **b.** The hind or lower limb in humans and primates. **c.** The part of the vertebrate limb between the knee and foot. **d.** The back part of the hindquarter of a food animal. **2.** A supporting part similar to a leg in shape or function. **3.** One of the branches of a jointed or forked object. **4.** The part of a garment, esp. of a pair of pants, that covers all or part of the leg. **5.** Either side of a right triangle that is not the hypotenuse. **6.** A stage of a journey or course, esp.: **a.** The distance traveled by a sailing vessel on a single tack. **b.** The part of an air route or flight pattern that is between two successive stops, positions, or changes in direction. —*vi.* **legged, leg·ging, legs.** *Informal.* To go on foot. **—a leg to stand on.** A logical or justifiable basis for defense : SUPPORT <The prosecution left the defendant without *a leg to stand on.*> **—a leg up.** An act or instance of assisting : BOOST. **—on (one's) last legs.** At the end of one's health or strength. **—pull (someone's) leg.** *Informal.* To tease or make fun of.

**leg·a·cy** (lĕg′ə-sē) *n., pl.* **-cies.** [ME *legat* < OFr. < Med. Lat. *legantia* < Lat. *legare*, to bequeath as a legacy.] **1.** Money or property bequeathed by a will. **2.** Something handed down from an ancestor or from the past <a strong *legacy* of personal freedom>

**le·gal** (lē′gəl) *adj.* [ME < OFr. < Lat. *legalis* < *lex*, law.] **1.** Of, pertaining to, or concerned with law <*legal* papers> **2. a.** Authorized by or based on law <a *legal* case> **b.** Established by law : STATUTORY <the *legal* authority> **3.** Permitted by or in conformity with the law. **4.** Enforced or recognized by law rather than by equity. **5.** Created by the law <a *legal* defense> **6.** Applicable to or typical of lawyers or their profession. **—le′gal·ly** *adv.*

☆ ***syns:*** LEGAL, LAWFUL, LEGIT, LEGITIMATE *adj. core meaning*: within, allowed by, or sanctioned by the law <*legal* entry><a *legal* marriage> ***ant:*** illegal

**legal age** *n.* The age at which a person may by law assume the rights and responsibilities of an adult.

**legal cap** *n.* White, often ruled writing paper measuring 8½ by 13 to 16 inches.

**le·gal·ese** (lē′gə-lēz′, -lēs′) *n.* The specialized vocabulary of the legal profession.

**legal holiday** *n.* A holiday established by law and marked by a ban or limit on official business or work.

**le·gal·ism** (lē′gə-lĭz′əm) *n.* Strict, literal adherence to the law. **—le′gal·ist** *n.* **—le′gal·is′tic** *adj.* **—le′gal·is′ti·cal·ly** *adv.*

**le·gal·i·ty** (lē-găl′ĭ-tē) *n., pl.* **-ties. 1.** The quality or state of being legal. **2.** Observance of or adherence to the law. **3. legalities.** Requirements enjoined by law.

**le·gal·ize** (lē′gə-līz′) *vt.* **-ized, -iz·ing, -iz·es.** To make legal. **—le′gal·i·za′tion** *n.*

**legal pad** *n.* A ruled pad of writing paper measuring 8½ by 14 inches.

**legal reserve** *n.* The amount of money that a bank or insurance company is required by law to set aside as security.

**legal tender** *n.* Currency that may be offered to pay a debt and that a creditor must accept.

**leg·ate** (lĕg′ĭt) *n.* [ME < OFr. *legat* < Med. Lat. *legatus* < Lat. *legare*, to send as ambassador.] An official emissary, esp. an official representative of the pope. **—leg′ate·ship′** *n.*

**leg·a·tee** (lĕg′ə-tē′) *n.* The inheritor of a legacy.

**le·ga·tion** (lĭ-gā′shən) *n.* [ME *legacious* < OFr. < Lat. *legatio* < *legare*, to send as an ambassador.] **1.** The sending of a legate to a foreign country. **2. a.** A diplomatic mission ranking below an embassy. **b.** The legate and staff of such a mission. **c.** The premises occupied by a legation.

**le·ga·to** (lĭ-gä′tō) [Ital. < *legare*, to bind < Lat. *ligare.*] *Mus.* —*adv. & adj.* In a smooth, even style. —Used as a direction. —*n., pl.* **-tos.** A smooth, even style, performance, or passage.

**le·ga·tor** (lĭ-gā′tər) *n.* [Lat. < *legare*, to bequeath.] One who makes a will : TESTATOR.

**leg·end** (lĕj'ənd) *n.* [ME < OFr. *legende* < Med. Lat. *legenda* < Lat. neuter pl. gerundive of *legere,* to read.] **1. a.** An unverified popular story handed down from earlier times. **b.** A collection or body of such stories. **c.** A romanticized or popularized story of modern times. **2.** One who achieves legendary fame. **3. a.** An inscription or title on an object, as a coin. **b.** An explanatory caption accompanying a map, chart, or illustration.

▲ word history: The word *legend* is derived from Latin *legenda,* a participial form derived from the verb *legere,* "to read." In Medieval Latin *legenda* was used to mean "something to be read," especially the narrative of a saint's life. These biographies were considered to be important as both historical records and moral examples. The word *legenda* was borrowed into English as *legend* in the 14th century. It is likely that the utterly incredible exploits and events recounted in some legends led to the development of the sense "unverified popular tale; myth" for the English word.

**leg·en·dar·y** (lĕj'ən-dĕr'ē) *adj.* Of, comprising, based on, or of the nature of a legend.

**leg·er·de·main** (lĕj'ər-də-mān') *n.* [ME < OFr. *leger de main.*] **1.** Sleight of hand. **2.** Trickery or deception : HOCUS-POCUS.

**le·ges** (lē'jēz') *n. pl. of* LEX.

**leg·gings** (lĕg'ĭngz) *pl.n.* A leg covering of a strong material, as leather or canvas.

**leg·gy** (lĕg'ē) *adj.* **-gi·er, -gi·est. 1.** Having unusually long legs <a *leggy* colt> **2.** *Informal.* Having attractively long and slender legs. **3.** Having long, spindly, often leafless stems. **—leg'gi·ness** *n.*

**leg·horn** (lĕg'hôrn', -ərn) *n.* [After *Leghorn,* Italy.] **1. a.** The dried and bleached straw of an Italian wheat. **b.** A plaited fabric made from this straw. **c.** A hat made from this fabric. **2.** *often* **Leghorn.** A domestic fowl orig. bred in Mediterranean regions, noted for prolific egg production.

**leg·i·ble** (lĕj'ə-bəl) *adj.* [ME < LLat. *legibilis* < Lat. *legere,* to read.] Capable of being deciphered or read. **—leg'i·bil'i·ty, leg'i·ble·ness** *n.* **—leg'i·bly** *adv.*

**le·gion** (lē'jən) *n.* [ME *legioun* < OFr. *legion* < Lat. *legio* < *legere,* to gather.] **1.** The major unit of the Roman army, made up of 3,000 to 6,000 infantry troops and 100 to 200 cavalrymen. **2.** A large number : MULTITUDE. **3.** *often* **Legion.** A national organization of former armed forces personnel. *—adj.* Constituting a large number : MULTITUDINOUS <Their wrongs were *legion.*>

**le·gion·ar·y** (lē'jə-nĕr'ē) *adj.* Of, pertaining to, or constituting a legion. *—n., pl.* **-ies.** A soldier of a legion.

**legionary ant** *n.* An army ant.

**le·gion·naire** (lē'jə-nâr') *n.* [Fr. *légionnaire* < *legion,* legion < OFr. —see LEGION.] A member of a legion.

**Legionnaires' disease** *n.* [So called because it was first recognized when an outbreak occurred during the American Legion Convention in 1976.] A severe disease caused by the bacterium *Legionella premophilia* and marked by pneumonia, a dry cough, and muscular pain.

**Legion of Honor** *n.* A high French civilian and military decoration, instituted in 1802.

**Legion of Merit** *n.* A U.S. military decoration awarded for exceptionally meritorious conduct.

**leg·is·late** (lĕj'ĭ-slāt') *v.* **-lat·ed, -lat·ing, -lates.** [Back-formation < LEGISLATOR.] *—vi.* To pass a law or laws. *—vt.* To bring about by or as if by legislation.

**leg·is·la·tion** (lĕj'ĭ-slā'shən) *n.* **1.** The act or process of legislating : LAWMAKING. **2.** A proposed or enacted law or group of laws.

**leg·is·la·tive** (lĕj'ĭ-slā'tĭv) *adj.* **1.** Of or pertaining to legislation. **2.** Decided by or resulting from legislation. **3.** Empowered to create laws. **4.** Of or pertaining to a legislature. *—n.* A government's legislative body : LEGISLATURE. **—leg'is·la'tive·ly** *adv.*

**leg·is·la·tor** (lĕj'ĭ-slā'tər) *n.* [Lat. : *lex,* law + *lator,* proposer < *latus,* p.part. of *ferre,* to propose.] One who creates or enacts laws, esp. a member of a legislative body.

**leg·is·la·ture** (lĕj'ĭ-slā'chər) *n.* An officially selected body of persons with the power and responsibility to make laws for a nation or state.

**le·gist** (lē'jĭst) *n.* [ME *legiste* < OFr. < Med.Lat. *legista* < Lat. *lex,* law.] A specialist in law.

**le·git** (lə-jĭt') *adj. Slang.* Legitimate.

**le·git·i·mate** (lə-jĭt'ə-mĭt) *adj.* [ME, born in wedlock < OFr. *legitimer,* to be legitimate < Med. Lat. *legitimare* < Lat. *legitimus,* legitimate < *lex,* law.] **1.** Being in compliance with the law : LAWFUL <a *legitimate* business> **2.** Being in accordance with established or accepted patterns and standards <*legitimate* sales promotion> **3.** Based on logical reasoning : REASONABLE <a *legitimate* conclusion> **4.** Genuine : authentic <had a *legitimate* grievance> **5.** Born of legally married parents. **6.** Of, pertaining to, or ruling by hereditary right. **7.** Of or relating to professional theater that excludes burlesque, vaudeville, and some forms of musical comedy. *—vt.* (-māt') **-mat·ed, -mat·ing, -mates. 1.** To justify as legitimate : AUTHORIZE. **2.** To make, establish as, or declare to be legitimate. **—le·git'i·ma·cy** (-mə-sē) *n.* **—le·git'i·mate·ly** *adv.* **—le·git'i·ma'tion** *n.*

**le·git·i·ma·tize** (lə-jĭt'ə-mə-tīz') *vt.* **-tized, -tiz·ing, -tiz·es.** To legitimate.

**le·git·i·mist** (lə-jĭt'ə-mĭst) *n.* One who believes in or supports rule by hereditary right. **—le·git'i·mism** *n.*

**le·git·i·mize** (lə-jĭt'ə-mīz') *vt.* **-mized, -miz·ing, -miz·es.** To legitimate. **—le·git'i·mi·za'tion** *n.* **—le·git'i·miz'er** *n.*

**leg-of-mut·ton** (lĕg'ə-mŭt'n, lĕg'əv-) *adj.* Tapering sharply from one large end to a smaller end or point, as a sleeve.

**leg·room** (lĕg'rōōm', -rŏŏm') *n.* Room in which to stretch the legs while seated.

**leg·ume** (lĕg'yōōm', lə-gyōōm') *n.* [Fr. *légume* < Lat. *legumen,* bean.] **1. a.** A pod, as that of a pea or bean, that splits into two valves with the seeds attached to the lower edge of one of the valves. **b.** Such a pod or seed used as food. **2.** A plant of the family Leguminosae, typically bearing legumes.

**le·gu·mi·nous** (lə-gyōō'mə-nəs) *adj.* [NLat. *Leguminosae,* family name < Lat. *lugumen,* bean.] **1.** Of, belonging to, or typical of the family Leguminosae, which includes peas, beans, clover, alfalfa, and other plants. **2.** Resembling or having the nature of a legume.

**leg·work** (lĕg'wûrk') *n. Informal.* A task, as collecting information, that involves walking or traveling about.

**le·hu·a** (lā-hōō'ə) *n.* [Hawaiian.] A tree, *Metrosideros collina* of Hawaii and other Pacific islands, bearing showy red flowers.

**lei**[1] (lā, lā'ē) *n., pl.* **leis.** [Hawaiian.] A garland of flowers, esp. one worn around the neck.

**lei**[2] (lā) *n. pl. of* LEU.

**Leices·ter** (lĕs'tər) *n.* A sheep orig. bred in Leicestershire, England, producing long fine wool.

**leish·man·i·a·sis** (lēsh'mə-nī'ə-sĭs) *n.* [< NLat. *Leishmania,* genus of protozoans, after Sir William B. *Leishman* (1865–1926).] **1.** An infection with flagellate protozoans of the genus *Leishmania.* **2.** A disease, as kala-azar or either of two clinically distinct ulcerative skin diseases, caused by protozoans of the genus *Leishmania.*

**leis·ter** (lē'stər) *n.* [Of Scand. orig.] A three-pronged fishing spear. *—vt.* **-tered, -ter·ing, -ters.** To spear (a fish) with a leister.

**lei·sure** (lē'zhər, lĕzh'ər) *n.* [ME < OFr. *leisir* < Lat. *licēre,* to be permitted.] Freedom from time-consuming duties, responsibilities, or activities. **—at (one's) leisure.** When one has spare time : at one's convenience <completed the project at their *leisure*>

**lei·sured** (lē'zhərd, lĕzh'ərd) *adj.* Marked by leisure.

**lei·sure·ly** (lē'zhər-lē, lĕzh'ər-) *adj.* Without haste : UNHURRIED. *—adv.* In an unhurried way. **—lei'sure·li·ness** *n.*

**leisure suit** *n.* A suit for informal wear having a shirtlike jacket and matching slacks.

**leit·mo·tif** *also* **leit·mo·tiv** (līt'mō-tēf') *n.* [G. *Leitmotiv* : *leiten,* to lead (< OHG) + *motiv,* motive (< Fr. *motif*).] **1.** A melodic passage or phrase, esp. in Wagnerian opera, associated with a specific character, thing, or element. **2.** A dominant and recurring theme.

**lek** (lĕk) *n.* [Albanian.] —See table at CURRENCY.

**LEM** (lĕm) *n.* A lunar excursion module.

**lem·an** (lĕm'ən, lē'mən) *n.* [ME : *leof,* dear (< OE *lēof*) + *man,* man (< OE).] *Archaic.* **1.** A lover. **2.** A mistress.

**lem·ma**[1] (lĕm'ə) *n., pl.* **-mas** *or* **-ma·ta** (-ə-tə) [Lat. < Gk. *lēmma* < *lambanein,* to take.] **1.** A subsidiary proposition assumed to be valid and used to demonstrate a principal proposition. **2.** A theme, argument, or subject indicated in a title. **3.** A glossed word or phrase in a glossary or other listing.

**lem·ma**[2] (lĕm'ə) *n.* [Gk., husk < *lepein,* to peel.] *Bot.* The outer, lower bract enclosing the flower in a grass spikelet.

**lem·ming** (lĕm'ĭng) *n.* [Norw.] A rodent of the genus *Lemmus* or related genera of northern regions, as the European species *L. lemmus,* noted for its mass migrations into the sea.

**lem·nis·cus** (lĕm-nĭs'kəs) *n., pl.* **-nis·ci** (-nĭs'ī', -nĭs'kī', -nĭs'kē) [NLat. < Lat., ribbon < Gk. *lēmniskos.*] A bundle of brain nerve fibers.

**lem·on** (lĕm'ən) *n.* [ME *limon* < OFr. < Ar. *laymūn* < Pers. *līmūn.*] **1. a.** A spiny Asian evergreen tree, *Citrus limonia,* widely cultivated for its yellow egg-shaped fruit. **b.** The fruit of the lemon, with an aromatic rind and acidic, juicy pulp. **2.** Lemon yellow. **3.** *Informal.* One that is or proves to be defective or unsatisfactory : WASHOUT. *—adj.* Brilliant, vivid yellow to greenish yellow.

**lem·on·ade** (lĕm'ə-nād') *n.* A drink made of lemon juice, water, and sugar.

**lemon balm** *n.* BALM 3a.

**lem·on·grass** (lĕm'ən-grăs') *n.* A tropical grass of the genus *Cymbopogon,* esp. *C. citratus,* yielding an aromatic oil used in perfumery and as a flavoring.

**lemon verbena** *n.* An aromatic South American plant, *Lippia citriodora,* widely cultivated for its fragrant foliage and flowers.

**lem·on·y** (lĕm'ə-nē) *adj.* Having the characteristic odor, flavor, or color of lemons.

**lemon yellow** *n.* A brilliant, vivid yellow to greenish yellow.

**lem·pi·ra** (lĕm-pîr'ə) *n.* [Am. Sp., after *Lempira,* a 16th-cent. Indian leader who resisted the Spanish.] —See table at CURRENCY.

**le·mur** (lē′mər) *n.* [Lat. *lemures,* lemures.] Any of several arboreal primates chiefly of the family Lemuridae of Madagascar and adjacent islands, with large eyes, soft fur, and a long tail.

**lemur**
*Approximately 3 feet long including tail*

**lem·u·res** (lĕm′ə-rās′, lĕm′yə-rēz′) *pl.n.* [Lat.] The spirits of the dead regarded in ancient Rome as frightening specters.
**Len·a·pe** (lĕn′ə-pē) *also* **Len·i-Len·a·pe** *or* **Len·ni-Len·a·pe** (lĕn′ē-) *n.* DELAWARE¹. **—Len′a·pe** *adj.*
**lend** (lĕnd) *v.* **lent** (lĕnt), **lend·ing, lends.** [ME *lenden* < OE *lǣn* < *lǣn,* loan.] *—vt.* **1. a.** To give or allow the use of (something) temporarily on the condition that it or its equivalent will be returned. **b.** To provide (money) temporarily on the condition that the amount borrowed be returned, usu. with an interest fee. **2.** To contribute : impart <Cobwebs in corners *lent* a feeling of neglect to the room.> **3.** To put at another's service or use <*lend* a helping hand> *—vi.* To make a loan. **—lend (oneself) to.** To be suitable for <a subject that *lends* itself *to* controversy> **—lend′er** *n.*
☆ ***syns:*** LEND, ADVANCE, LOAN *v. core meaning* : to supply (money), esp. on credit <refused to *lend* them any more cash>
**lending library** *n.* A library from which books may be borrowed or rented for a minimal fee.
**length** (lĕngkth, lĕngth) *n.* [ME < OE *lengðu.*] **1.** The quality, state, or fact of being long. **2.** Measurement of the extent of an object along its greatest dimension <the *length* of the room> **3.** A piece of material, often of a standard size, normally measured along its greatest dimension <a *length* of cable> **4.** A measure used as a unit to estimate distances <an arm's *length*> **5.** The distance or extent from beginning to end as measured in space. **6.** The amount of time between specified moments : DURATION <the *length* of a movie> **7.** An extent to which an action or policy is carried <went to great *lengths* to make us comfortable> **8. a.** The quantity of a vowel. **b.** The quantity of a syllable. **—at length. 1.** After some time : EVENTUALLY. **2.** For a considerable time.
**length·en** (lĕngk′thən, lĕng′-) *vt. & vi.* **-ened, -en·ing, -ens.** To make or become longer. **—length′en·er** *n.*
**length·wise** (lĕngkth′wīz′, lĕngth′-) *adv. & adj.* Of or along the direction of length : LONGITUDINALLY.
**length·y** (lĕngk′thē, lĕng′-) *adj.* **-i·er, -i·est. 1.** Of considerable length, esp. in time. **2.** Excessively long : PROLONGED. **—length′i·ly** *adv.* **—length′i·ness** *n.*
**le·ni·ence** (lē′nē-əns, lēn′yəns) *n.* Leniency.
**le·ni·en·cy** (lē′nē-ən-sē, lēn′yən-) *n., pl.* **-cies. 1.** The quality or condition of being lenient. **2.** A lenient act.
**le·ni·ent** (lē′nē-ənt, lēn′yənt) *adj.* [Lat. *leniens, lenient-,* pr.part. of *lenire,* to pacify < *lenis,* soft.] **1.** Inclined not to be harsh : MERCIFUL. **2.** Not austere or strict : GENEROUS <*lenient* regulations> **—le′ni·ent·ly** *adv.*
**Len·i-Len·a·pe** (lĕn′ē-lĕn′ə-pē) *n. var. of* LENAPE.
**Len·in·ism** (lĕn′ə-nĭz′əm) *n.* The theory and practice of proletarian revolution as developed by Lenin. **—Len′in·ist** *n.*
**le·nis** (lē′nĭs, lā′-) *adj.* [Lat., soft.] Articulated with little or no aspiration, as the consonants *b* and *d* compared with *p* and *t.*
**len·i·tive** (lĕn′ĭ-tĭv) *adj.* [OFr. *lenitif* < Med. Lat. *lenitivus* < Lat. *lenire,* to soothe < *lenis,* soft.] Capable of easing pain or discomfort. *—n.* A lenitive medicine. **—len′i·tive·ly** *adv.*
**len·i·ty** (lĕn′ĭ-tē) *n.* [Lat. *lenitas* < *lenis,* soft.] LENIENCY 1.
**Len·ni-Len·a·pe** (lĕn′ē-lĕn′ə-pē) *n. var. of* LENAPE.
**lens** (lĕnz) *n.* [NLat. < Lat., lentil (from a double convex lens's resemblance to a lentil).] **1. a.** A carefully ground or molded piece of glass, plastic, or other transparent material with opposite surfaces either or both of which are curved, by means of which light rays are refracted so that they converge or diverge to form an image. **b.** A combination of two or more such pieces, occas. with other optical devices such as prisms, for forming an image for viewing or photographing. **2.** A device that causes radiation other than light to converge or diverge by an action analogous to that of an optical lens. **3.** A transparent, biconvex body of the eye between the iris and the vitreous humor that focuses light rays entering through the pupil to form an image on the retina.
**lent** (lĕnt) *v. p.t. & p.p. of* LEND.
**Lent** (lĕnt) *n.* [ME *lente,* spring, Lent < OE *lencten.*] The 40 weekdays from Ash Wednesday until Easter observed by Christians as a season of fasting and penitence.
▲ word history: The word *Lent* is derived from the Old English word *lencten,* which denoted both the season of spring and the ecclesiastical season of Lent. *Lencten* is ultimately derived from the same root as the adjective *long.* The meaning "spring" probably arose because spring is the time of year when the days grow longer. The ecclesiastical sense developed from the fact that Lent partially coincides with spring.
**Lent·en** (lĕn′tən) *adj.* **1.** Of or relating to Lent. **2.** Typical of or appropriate to Lent : SOMBER.
**len·ti·cel** (lĕn′tĭ-sĕl′) *n.* [NLat. *lenticella,* dim. of Lat. *lens,* lentil.] One of the small pores on the surface of the stems of woody plants that allows the passage of gases to and from the interior tissue. **—len′ti·cel′late** (-sĕl′ĭt) *adj.*
**len·tic·u·lar** (lĕn-tĭk′yə-lər) *adj.* [Lat. *lenticularis,* like a lentil < *lenticula,* lentil, dim. of *lens,* lentil.] **1.** Shaped like a biconvex lens. **2.** Of or relating to a lens.
**len·ti·go** (lĕn-tī′gō) *n., pl.* **-tig·i·nes** (-tĭj′ə-nēz′) [Lat. < *lens,* lentil.] **1.** A freckle. **2.** A nevus. **—len·tig′i·nous** (-tĭj′ə-nəs), **len·tig′i·nose′** (-nōs′) *adj.*
**len·til** (lĕn′təl) *n.* [ME < OFr. *lentille* < Lat. *lenticula,* dim. of *lens,* lentil.] **1.** An Old World leguminous plant, *Lens esculenta,* bearing pods that contain edible seeds. **2.** The round, flattened seed of the lentil.
**len·tisk** (lĕn′tĭsk′) *n.* [ME *lentiske* < Lat. *lentiscus.*] The Mediterranean mastic tree.
**len·to** (lĕn′tō) [Ital. < Lat. *lentus,* slow.] *Mus. —adv. & adj.* Slowly. —Used as a direction. *—n., pl.* **-tos.** A lento movement or passage.
**Le·o** (lē′ō) *n.* [Lat.] **1.** A constellation in the Northern Hemisphere. **2. a.** The fifth sign of the zodiac. **b.** One born under this sign.
**Leo Minor** *n.* A constellation in the Northern Hemisphere.
**le·one** (lē-ōn′) *n.* [After Sierra *Leone.*] —See table at CURRENCY.
**Le·o·nid** (lē′ə-nĭd) *n., pl.* **Le·o·nids** *or* **Le·on·i·des** (lē-ŏn′ĭ-dēz′) [< Lat. *Leo, Leon-,* the constellation Leo.] A falling star of the meteor shower that annually recurs in mid-Nov.
**le·o·nine** (lē′ə-nīn′) *adj.* [ME < OFr. < Lat. *leoninus* < *leo,* lion.] Of, relating to, or typical of a lion.
**leop·ard** (lĕp′ərd) *n.* [ME < OFr. < LLat. *leopardus* < LGk. *leopardos* : Gk. *leōn,* lion + Gk. *pardos,* pard.] **1. a.** A large feline mammal, *Panthera pardus* of Africa and Asia, with a tawny coat with dark rosettelike markings and also a black color phase. **b.** A feline, as the cheetah or the snow leopard. **c.** The pelt or fur of a leopard. **2.** *Heraldry.* A lion in side view, with one forepaw raised and the head facing the observer.
**leop·ard·ess** (lĕp′ər-dĭs) *n.* A female leopard.
**leopard lily** *n.* A tall plant, *Lilium pardalinum* of the western United States, bearing orange-red, dark-spotted flowers.
**leopard moth** *n.* A moth, *Zeuzera pyrina,* with spotted wings and larvae that damage trees by boring into the wood.
**leop·ard's-bane** (lĕp′ərdz-bān′) *n.* **1.** A plant of the widely cultivated genus *Doronicum,* with rayed yellow flowers. **2.** Any of several plants similar to or resembling a leopard's-bane.
**le·o·tard** (lē′ə-tärd′) *n.* [After Jules *Léotard* (1830–1870).] **1.** *often* **leotards.** A snugly fitting, elastic one-piece garment that covers the torso, worn esp. by dancers or acrobats. **2. leotards.** Tights.
**Lep·cha** (lĕp′chə) *n., pl.* **Lepcha** *or* **-chas. 1.** Any of a Mongoloid people living in Sikkim, India. **2.** The Tibeto-Burman language of the Lepcha.
**lep·er** (lĕp′ər) *n.* [ME < *lepre,* leprosy < OFr. < LLat. *lepra* < Gk. < *lepros,* scaly < *lepos, lepis,* scale.] **1.** A person suffering from leprosy. **2.** A person avoided by others.
**lepido-** *pref.* [< Gk. *lepis, lepid-,* scale.] Scale : flake <*lepidopteran*>
**le·pid·o·lite** (lĭ-pĭd′l-īt′) *n.* [G. *Lepidolith* : *Lepido-,* lepido- + *-lith,* -lith.] A lilac or pink to gray mica, $K_2Li_3Al_4Si_7O_{21}(OH, F)_3$, used as lithium ore and in ceramic production.
**lep·i·dop·ter·a** (lĕp′ĭ-dŏp′tər-ə) *n. pl. of* LEPIDOPTERON.
**lep·i·dop·ter·an** (lĕp′ĭ-dŏp′tər-ən) *n.* [NLat. *Lepidoptera,* order name : LEPIDO- + *pteron,* wing.] A lepidopterous insect.
**lep·i·dop·ter·ist** (lĕp′ĭ-dŏp′tər-ĭst) *n.* An entomologist specializing in the study of moths and butterflies.
**lep·i·dop·ter·on** (lĕp′ĭ-dŏp′tə-rŏn′, -tər-ən) *n., pl.* **-ter·a** (-tər-ə). A lepidopteran.
**lep·i·dop·ter·ous** (lĕp′ĭ-dŏp′tər-əs) *adj.* Of or belonging to the order Lepidoptera, which includes insects having four wings covered with small scales, as the butterflies and moths.
**lep·i·dote** (lĕp′ĭ-dōt′) *adj.* [Gk. *lepidōtos* < *lepis,* scale.] Covered with small scurfy scales.
**lep·o·rine** (lĕp′ə-rīn′, -ər-ĭn) *adj.* [Lat. *leporinus* < *lepus,* hare.] Of or typical of rabbits or hares.
**lep·re·chaun** (lĕp′rĭ-kôn′, -kŏn′) *n.* [Ir. Gael. *lupracān* < MIr. *luchrupán* < OIr. *luchorpán* : *lū,* small + *corp,* body < Lat. *corpus.*] An elf in Irish folklore who can reveal hidden treasure to the one who catches him.

**lep·ro·sar·i·um** (lĕp′rə-sâr′ē-əm) *n., pl.* **-i·ums** *or* **-i·a** (-ē-ə) [Med. Lat. < LLat. *leprosus,* leprous. —see LEPROUS.] A hospital for treating lepers.

**lep·rose** (lĕp′rōs′) *adj.* [LLat. *leprosus.* —see LEPROUS.] Scurfy or scaly : LEPROUS.

**lep·ro·sy** (lĕp′rə-sē) *n.* [< LEPROUS.] A chronic, infectious, granulomatous disease occurring almost exclusively in tropical and subtropical regions, caused by a bacillus, *Mycobacterium leprae,* and ranging in severity from noncontagious and spontaneously remitting forms to contagious, malignant forms with progressive anesthesia, paralysis, ulceration, nutritive disturbances, gangrene, and mutilation. **—lep·rot′ic** (lĕ-prŏt′ĭk) *adj.*

**lep·rous** (lĕp′rəs) *adj.* [ME < OFr. *lepros* < LLat. *leprosus* < *lepra,* leprosy. —see LEPER.] **1.** Having leprosy. **2.** Of, pertaining to, or resembling leprosy. **3.** *Biol.* Having or consisting of loose, scurfy scales.

**-lepsy** *suff.* [Gk. *-lēpsia* < *lēpsis,* seizure < *lambanein,* to take.] Fit : seizure <narco*lepsy*>

**lept-** *pref. var. of* LEPTO-.

**lep·ta** (lĕp′tə) *n. pl. of* LEPTON[1].

**lepto-** *or* **lept-** *pref.* [< Gk. *leptos,* fine, thin < *lepein,* to peel.] Slender : thin : fine <*lepto*cephalus>

**lep·to·ceph·a·lus** (lĕp′tə-sĕf′ə-ləs) *n., pl.* **-li** (-lī′). One of the slender, transparent larvae of eels and certain other fishes.

**lep·ton[1]** (lĕp′tŏn′) *n., pl.* **-ta** (-tə) [Mod. Gk. < Gk., small coin < *leptos,* fine, small < *lepein,* to peel.] —See table at CURRENCY.

**lep·ton[2]** (lĕp′tŏn′) *n.* Any of a family of subatomic particles including the electron, the muon, and their associated neutrinos, all having spin equal to 1/2 and masses less than those of the mesons. **—lep·ton′ic** (-tŏn′ĭk) *adj.*

**lepton number** *n.* A number calculated by subtracting the number of antileptons from the number of leptons in a system of elementary particles.

**lep·to·some** (lĕp′tə-sōm′) *n.* [G. *Leptosom* : Gk. *leptos,* slender + Gk. *sōma,* body.] A person with a slender, thin, or frail body. **—lep′·to·so·mat′ic** (-sō-măt′ĭk) *adj.*

**Le·pus** (lē′pəs) *n.* [Lat. *lepus,* hare.] A constellation in the Southern Hemisphere.

**Les·bi·an** (lĕz′bē-ən) *n.* **1.** A native or resident of Lesbos. **2.** The ancient Greek dialect of Lesbos. **—Les′bi·an** *adj.*

**lese maj·es·ty** *also* **lèse ma·jes·té** (lēz′ măj′ĭ-stē) *n.* [OFr. *lese majeste* < Lat. *laesa majestas* : *laesa,* p.part. of *laedere,* to injure + *majestas,* majesty. —see MAJESTY.] **1.** An offense or crime against the ruler or supreme power of a state. **2.** An affront to another's dignity.

**le·sion** (lē′zhən) *n.* [ME *lesioun* < OFr. *lesion* < Lat. *laesio* < *laedere,* to injure.] **1.** A wound or injury. **2.** A circumscribed pathological tissue alteration. **3.** A point or patch of a skin disease.

**les·pe·de·za** (lĕs′pĭ-dē′zə) *n.* [NLat. *Lespedeza,* genus name, after V.M. *Lespedez* (fl. 1785), Spanish governor of East Florida.] A plant of the genus *Lespedeza,* as the bush clover.

**less** (lĕs) *adj.* [ME *lesse* < OE *lǣssa.*] **1.** Not as great in amount or quantity <*less* money to spend> **2.** Lower in importance, esteem, or rank <no *less* a person than the governor> **3.** Consisting of a smaller number <*less* than ten> *—prep.* Minus : subtracting <Six *less* two is four.> *—adv.* To a smaller extent, degree, or frequency <*less* enthusiastic> *—n.* A smaller amount <expected *less* than I received> *—pron.* Fewer things or persons <Many things begin badly; *less* end well.> **—less than.** Not at all <a *less than* satisfactory conclusion> **—much (or still) less.** Certainly not <We're not looking for trouble, *much less* a fight.>

**-less** *suff.* [ME *-lesse* < OE *-lēas* < *lēas,* without.] **1.** Without : lacking <blame*less*> **2.** Unable to act or be acted upon in a specified way <daunt*less*>

**les·see** (lĕ-sē′) *n.* [ME < AN < OFr. *lesser,* to lease.] A holder of a lease.

**less·en** (lĕs′ən) *v.* **-ened, -en·ing, -ens.** [ME *lessenen* < *lessen,* to lessen < *lesse,* less.] *—vt.* **1.** To cause to decrease. **2.** To make little of : BELITTLE. *—vi.* To become less.

**less·er** (lĕs′ər) *adj.* [ME < *lesse,* less.] **1.** Smaller in amount, value, or importance, esp. in a comparison between two things <a *lesser* benefit> **2.** Of a smaller size than other similar forms <the *lesser* anteater>

**lesser celandine** *n.* A plant, *Ranunculus ficaria,* with heart-shaped leaves and yellow flowers.

**les·son** (lĕs′ən) *n.* [ME *lessoun* < OFr. *leson* < Lat. *lectio,* a reading. —see LECTION.] **1.** Something to be learned. **2. a.** A period of instruction : CLASS. **b.** An assignment or exercise in which something is to be learned. **c.** An act or instance of instructing : TEACHING. **3. a.** An experience, example, or observation that imparts wisdom or beneficial new knowledge. **b.** The knowledge or wisdom acquired. **4.** A reprimand. **5.** A reading from the Bible or other sacred writing as part of a religious service. *—vt.* **-soned, -son·ing, -sons. 1.** To teach a lesson to : INSTRUCT. **2.** To rebuke.

**les·sor** (lĕs′ôr′, lĕ-sôr′) *n.* [ME *lessour* < AN < OFr. *lesser,* to lease.] One who lets property under a lease.

**lest** (lĕst) *conj.* [ME < OE *ðȳ lǣs ðe.*] For fear that <whispered *lest* the visitors should hear them>

**let[1]** (lĕt) *v.* **let, let·ting, lets.** [ME *leten* < OE *lǣtan.*] *—vt.* **1.** To grant permission to : ALLOW. **2.** To cause to : MAKE <*Let* the words be printed.> **3.** Used to express a command, request, or proposal <*Let's* leave promptly at dawn.><*Let* y equal z.> ***usage:*** In informal speech *let's* has come increasingly into use as the mere indicator of a suggestion, so that one hears sentences such as *Let's us go* and *Don't let's us lose our heads.* These usages should be avoided in formal writing. **4.** To allow to enter, advance, or leave <*let* the cat out> **5.** To release from or as if from confinement <*let* the air out of the tires> **6.** To rent or lease <*let* an apartment> **7.** To award, esp. after bids have been submitted <*let* the contract to another company> *—vi.* **1.** To become rented or leased. **2.** To be or become assigned, as to a contractor. **—let down. 1.** To withdraw support from : FORSAKE. **2.** To fail to meet the expectations of : DISAPPOINT. **—let on. 1.** To allow it to be known <Don't *let on* that you expected to see us.> **2.** To pretend. **—let up. 1.** To make less : DIMINISH. **2.** To stop : cease. **3.** To be or become more lenient.

**let[2]** (lĕt) *n.* [ME < *letten,* to hinder < OE *lettan.*] **1.** Something that hinders : OBSTACLE <without *let* or hindrance> **2.** A stroke in tennis and other net games that is invalid and must be repeated. *—vt.* **let·ted** *or* **let, let·ting, lets.** *Archaic.* To hinder or obstruct.

**-let** *suff.* [ME *-lette* < OFr. *-elet.*] **1.** Small one <crater*let*> **2.** Something worn on <arm*let*>

**letch** *also* **lech** (lĕch) *n.* [Perh. back-formation < obs. *letcher,* var. of LECHER.] A strong, esp. sexual craving or desire.

**let·down** (lĕt′doun′) *n.* **1.** A decrease, decline, or relaxation, as of effort or energy. **2.** A disappointment. **3.** The descent of an aircraft preparing to land.

**le·thal** (lē′thəl) *adj.* [Lat. *lethalis* < *lethum,* death.] **1.** Capable of causing death. **2.** Of, relating to, or causing death. **3.** Extremely harmful : DEVASTATING <a *lethal* attack on one's integrity> **—le·thal′·i·ty** (lē-thăl′ĭ-tē) *n.* **—le′thal·ly** *adv.*

**le·thar·gic** (lə-thär′jĭk) *adj.* Of, causing, or marked by lethargy. **—le·thar′gi·cal·ly** *adv.*

**leth·ar·gy** (lĕth′ər-jē) *n., pl.* **-gies.** [ME *letargie* < OFr. *letargia* < Med. Lat. *litargia* < Lat. *lethargia* < Gk. *lēthargia* < *lethargos,* forgetful < *lēthē,* forgetfulness.] **1.** Drowsy or sluggish indifference : APATHY. **2.** A state of unconsciousness resembling deep sleep.

**le·the** (lē′thē) *n.* [Gk. *lēthē.*] **1. Lethe.** *Gk. Myth.* The river of forgetfulness in Hades. **2.** Loss of memory. **—le′the·an** *adj.*

**Le·to** (lē′tō′) *n.* [Gk. *Lētō.*] *Gk. Myth.* A consort of Zeus and the mother of Apollo and Artemis.

**let's** (lĕts). Let us.

**Lett** (lĕt) *n.* [G. *Lette* < Latvian *Latvi.*] A native or resident of Latvia : LATVIAN.

**let·ter** (lĕt′ər) *n.* [ME < OFr. *lettre* < Lat. *littera.*] **1.** A written symbol or character representing a speech sound and being a component of an alphabet. **2.** A written or printed communication sent to a recipient. **3.** *often* **letters.** A certified document granting rights to its bearer. **4.** Literal meaning <followed instructions to the *letter*> **5. letters.** *(sing. in number).* Literary culture or learning. **6. a.** A piece of type that prints a single character. **b.** A particular style of type. **c.** The characters in one style of type. **7.** An emblem in the shape of the initial of a school awarded to athletes. *—v.* **-tered, -ter·ing, -ters.** *—vt.* **1.** To write letters on. **2.** To write in letters. *—vi.* **1.** To write or form letters. **2.** To earn an athletic letter <*lettered* in football and basketball> **—let′ter·er** *n.*

☆ ***syns:*** **1.** LETTER, EPISTLE, MISSIVE, NOTE *n. core meaning* : a written communication directed to another <wrote a *letter* to the President> **2.** LETTER, LITERALITY, LITERALNESS *n. core meaning* : literal meaning <the *letter* of the law>

**letter bomb** *n.* An explosive mailed to its intended victim in an envelope.

**let·ter·box** (lĕt′ər-bŏks′) *n.* A mailbox.

**letter carrier** *n.* A mail carrier.

**let·tered** (lĕt′ərd) *adj.* **1. a.** Educated to read and write : LITERATE. **b.** Learned : erudite. **2.** Of or pertaining to literacy or learning. **3.** Inscribed or marked with or as if with letters.

**let·ter·form** (lĕt′ər-fôrm′) *n.* The development or design of the shape of an alphabet letter.

**let·ter·head** (lĕt′ər-hĕd′) *n.* **1.** The heading at the top of a sheet of letter paper, usu. a name and address. **2.** Stationery printed with a letterhead.

**letter of credence** *n.* An official document conveying the credentials of a diplomatic envoy to a foreign government.

**letter of credit** *n.* A letter issued by a bank authorizing the bearer to draw a stated amount of money from the issuing bank, its branches, or other associated banks or agencies.

**let·ter-per·fect** (lĕt′ər-pûr′fĭkt) *adj.* Absolutely correct.

**let·ter·press** (lĕt′ər-prĕs′) *n.* **1. a.** The process of printing from a raised inked surface. **b.** Something printed in this way. **2.** The text, as of a book, distinct from illustrations or other ornamentation.

**let·ter-qual·i·ty** (lĕt′ər-kwŏl′ĭ-tē) *adj.* Designating printed computer data that appears to have been typewritten.

**letters of administration** *pl.n.* A legal document entrusting an individual with the administration of the estate of a decedent.

**letters of marque** (märk) *pl.n.* [ME *lettres of marque* < OFr. *marque,* reprisal, of Germanic orig.] **1.** A document issued by a nation permitting a private citizen to seize citizens or goods of another nation. **2.** A document issued by a nation permitting a private citizen to equip a ship with arms in order to attack enemy ships.

**letters patent** *pl.n.* A government document granting a patentee exclusive right to the enjoyment or possession of an invention.

**letters testamentary** *pl.n.* A document issued by a probate court or officer informing an executor of a will of his or her appointment and empowering him or her to discharge the responsibilities.

**Let·tish** (lĕt′ĭsh) *adj.* Of or relating to the Latvians or their language. —*n.* LATVIAN 2.

**let·tuce** (lĕt′əs) *n.* [ME *lettuse* < OFr. *letues,* pl. of *laitue* < Lat. *lactuca* < *lac,* milk (from its milky juice).] **1. a.** A plant of the genus *Lactuca,* esp. *L. sativa,* cultivated for its edible leaves. **b.** The leaves of *L. sativa,* eaten as salad. **2.** *Slang.* Paper money.

**let·up** (lĕt′ŭp′) *n.* **1.** Reduction in force, pace, or intensity : SLOWDOWN. **2.** A temporary cessation : PAUSE.

**le·u** (lĕ′o͞o) *n., pl.* **lei** (lā) [Rum. < Lat. *leo,* lion.] —See table at CURRENCY.

**leuc-** *pref. var. of* LEUKO-.

**leu·cine** (lo͞o′sēn′) *n.* [LEUC(O)- + -INE.] An essential amino acid, $C_6H_{13}NO_2$, derived from the hydrolysis of protein by pancreatic enzymes and used as a nutrient.

**leu·cite** (lo͞o′sīt′) *n.* [G. *Leucit* : *leuc-,* leuko- + *-it,* -ite.] A white or gray mineral, essentially $KAl(SiO_3)_2$. **—leu·cit′ic** (-sĭt′ĭk) *adj.*

**leu·co-** *pref. var. of* LEUKO-.

**leu·co·plast** (lo͞o′kə-plăst′) *also* **leu·co·plas·tid** (lo͞o′kə-plăs′tĭd) *n.* [LEUCO- + PLAST(ID).] A colorless plastid in the cytoplasm of plant cells around which starch collects.

**leuk-** *pref. var. of* LEUKO-.

**leu·ke·mi·a** (lo͞o-kē′mē-ə) *n.* Any of a group of usu. fatal diseases of the reticuloendothelial system involving uncontrolled proliferation of leukocytes. **—leu·ke′mic** (-kē′mĭk) *adj. & n.*

**leuko-** *or* **leuk-** *also* **leuco-** *or* **leuc-** *pref.* [NLat. < Gk. *leukos,* clear, white.] **1.** White : colorless <*leuko*derma> **2.** Leukocyte <*leuko*penia>

**leu·ko·cyte** *also* **leu·co·cyte** (lo͞o′kə-sīt′) *n.* Any of the white or colorless nucleated cells occurring in blood. **—leu′ko·cyt′ic** (-sĭt′ĭk) *adj.* **—leu′ko·cy′toid′** *adj.*

**leu·ko·cy·to·sis** *also* **leu·co·cy·to·sis** (lo͞o′kə-sī-tō′sĭs) *n., pl.* **-ses** (-sēz′). A marked increase in the number of leukocytes in the blood. **—leu′ko·cy·tot′ic** (-tŏt′ĭk) *adj.*

**leu·ko·der·ma** *also* **leu·co·der·ma** (lo͞o′kə-dûr′mə) *n.* Partial or total lack of skin pigmentation. **—leu′ko·der′mal, leu′ko·der′mic** *adj.*

**leu·ko·ma** *also* **leu·co·ma** (lo͞o-kō′mə) *n.* [Gk. *leukōma* < *leukoun,* to make white < *leukos,* white.] A dense, white opacity of the cornea of the eye.

**leu·ko·pe·ni·a** *also* **leu·co·pe·ni·a** (lo͞o′kə-pē′nē-ə) *n.* An abnormally low number of leukocytes in the circulating blood. **—leu′ko·pe′nic** (-pē′nĭk) *adj.*

**leu·ko·poi·e·sis** *also* **leu·co·poi·e·sis** (lo͞o′kə-poi-ē′sĭs) *n.* Formation and growth of leukocytes. **—leu′ko·poi·et′ic** (-ĕt′ĭk) *adj.*

**leu·kor·rhe·a** *also* **leu·cor·rhe·a** (lo͞o′kə-rē′ə) *n.* Vaginal discharge containing mucus and pus. **—leu′kor·rhe′al** (-rē′əl) *adj.*

**lev** (lĕf) *n., pl.* **lev·a** (lĕv′ə) [Bulgarian.] —See table at CURRENCY.

**lev-** *pref. var. of* LEVO-.

**lev·a** (lĕv′ə) *n. pl. of* LEV.

**Lev·al·loi·si·an** (lĕv′ə-loi′zē-ən) *adj.* [After *Levallois*-Perret, a district of France.] Of or pertaining to a western European stage in lower Paleolithic culture, known from the method of striking off flake tools from pieces of flint.

**lev·al·lor·phan** (lĕv′ə-lôr′făn′, -fən) *n.* [LEV(OROTATORY) + AL-L(IUM) + (M)ORPH(INE) + -AN.] A morphine-related drug for counteracting morphine poisoning.

**le·vant** (lə-vănt′) *n.* [After *Levant,* countries bordering the eastern Mediterranean.] A heavy coarse-grained morocco leather often used in bookbinding.

**le·vant·er** (lə-văn′tər) *n.* **1.** A strong easterly wind of the Mediterranean area. **2. Levanter.** A native of the Levant.

**le·va·tor** (lə-vā′tər) *n., pl.* **lev·a·to·res** (lĕv′ə-tôr′ēz, -tōr′-) [< Lat. *levare, levat-,* to raise. —see LEVER.] **1.** *Anat.* A muscle that raises a bodily part. **2.** A surgical instrument for lifting the depressed part of a fractured skull.

**lev·ee¹** (lĕv′ē) *n.* [Fr. *levée* < OFr. *levee* < *lever,* to raise. —see LEVER.] **1.** An embankment, as of earth or concrete, used to prevent a river from overflowing. **2.** A small ridge bordering an irrigated field. **3.** A landing place on a river : PIER.

**lev·ee²** (lĕv′ē, lə-vē′, -vā′) *n.* [Fr. *levé,* var. of *lever,* a rising < *lever,* to rise < OFr. —see LEVER.] **1.** A reception once held by a monarch or other high-ranking personage upon arising from bed. **2.** A formal reception, as at a court.

**lev·el** (lĕv′əl) *n.* [ME, an instrument to check that a surface is horizontal < OFr. *livel* < Lat. *libella,* dim. of *libra,* balance.] **1.** Relative rank or position <a low *level* of accomplishment> **2.** A natural or proper position, place, or stage <Water seeks its *level.*> **3.** Position along a vertical axis : height or depth <a counter at chest *level*> **4. a.** A horizontal line or plane at right angles to the plumb. **b.** The height or position of such a line or plane. **5.** A flat, horizontal surface. **6.** A land area of uniform elevation. **7. a.** An instrument for determining whether a surface is horizontal, consisting of an encased, liquid-filled tube containing an air bubble that moves to a center window when the instrument is on a horizontal plane. **b.** Such a device combined with a telescope and used in surveying. **c.** Computation of the difference in elevation between two points by using such a device. **8.** *Computer Sci.* A bit, element, channel, or row of information. —*adj.* **1.** Having a flat, smooth surface. **2.** On a horizontal plane. **3.** Being at the same height or position as another : EVEN. **4.** Being at or having the same degree of rank, standing, or advantage as another : EQUAL. **5.** Without abrupt variations : STEADY <addressed the crowd in a *level* voice> **6.** Rational and measured <made a *level* evaluation of the situation> —*v.* **-eled, -el·ing, -els** *or* **-elled, -el·ling, -els.** —*vt.* **1.** To make horizontal, flat, or even <*leveled* the hillside with a bulldozer> **2.** To tear down : RAZE. **3.** To knock down with or as if with a blow. **4.** To place on the same level : EQUALIZE. **5.** To aim along a horizontal plane <*leveled* rifles at the oncoming deer> **6.** To direct emphatically toward someone <*leveled* accusations of perjury> **7.** To measure the various elevations of (a tract of land) with a level. —*vi.* **1.** To render persons or things equal : EQUALIZE. **2.** To aim a weapon horizontally. **3.** *Informal.* To be open and frank. **—level off. 1.** To move toward consistency or stability. **2.** To maneuver an aircraft into a line of flight parallel to the surface of the earth after losing or gaining altitude. —*adv.* Along a flat or even line or plane. **—(one's) level best.** The very best that one can do. **—on the level.** *Informal.* Without deception : HONEST. **—lev′el·ly** *adv.* **—lev′el·ness** *n.*

☆ ***syns:*** LEVEL, FLAT, FLUSH, PLANE, SMOOTH, STRAIGHT *adj. core meaning* : having no irregularities, roughness, or indentations <*level* boards>

**level compensator** *n.* An automatic gain control device used in the receiving equipment of telegraphic circuits.

**level crossing** *n. Chiefly Brit.* A grade crossing.

**lev·el·er** *also* **lev·el·ler** (lĕv′ə-lər) *n.* **1.** One that levels. **2. a.** One who urges abolition of social inequities. **b. Leveller.** A member of an English radical political movement in the 1640's that advocated universal male suffrage, parliamentary democracy, and religious tolerance.

**lev·el·head·ed** (lĕv′əl-hĕd′ĭd) *adj.* Sensible and self-composed. **—lev′el·head′ed·ness** *n.*

**leveling rod** *n.* A graduated pole with a movable marker, used with a surveyor's level for measuring differences in elevation.

**level of significance** *n.* The probability of a false rejection of the null hypothesis in a statistical test.

**lev·er** (lĕv′ər, lē′vər) *n.* [ME < OFr. *levier* < *lever,* to raise < Lat. *levare* < *levis,* light.] **1.** A simple machine consisting of a rigid body, usu. a metal bar, pivoted on a fixed fulcrum. **2.** A projecting handle for operating or adjusting a mechanism. **3.** A means of accomplishing : TOOL <used knowledge as a *lever* to gain advantage> —*vt.* **-ered, -er·ing, -ers.** To move or lift with or as if with a lever.

**lev·er·age** (lĕv′ər-ĭj, lē′vər-) *n.* **1. a.** The action of a lever. **b.** The mechanical advantage of a lever. **2.** Power to act effectively <Having knowledge and ability gives a person *leverage.*> **3.** Use of credit in order to improve one's speculative capacity. —*vt.* **-aged, -ag·ing, -ag·es. 1. a.** To provide (a company) with leverage. **b.** To supplement (e.g., money) with leverage. **2.** *Informal.* To affect, as if by leverage <a lifestyle *leveraged* by business responsibilities>

**lev·er·et** (lĕv′ər-ĭt) *n.* [ME < AN, dim. of *levre,* hare, var. of OFr. *lievre* < Lat. *lepus.*] A young hare, esp. one less than a year old.

**Le·vi** (lē′vī′) *n.* [Heb.] A tribe of Israel descended from Levi, a son of Jacob and Leah.

**lev·i·a·ble** (lĕv′ē-ə-bəl) *adj.* Capable of being levied or levied on.

**le·vi·a·than** (lə-vī′ə-thən) *n.* [ME < LLat. < Heb. *libhyāthān.*] **1.** A sea monster mentioned in the Old Testament. **2.** A very large animal. **3.** Something exceptionally large and awesome.

**lev·i·gate** (lĕv′ĭ-gāt′) *vt.* **-gat·ed, -gat·ing, -gates.** [Lat. *levigare, levigat-* : *levis,* smooth + *agere,* to make.] **1.** To make into a smooth, fine powder, as by grinding when moist. **2.** To suspend in a liquid. **3.** To polish. —*adj.* (-gāt′, -gĭt). Smooth. **—lev′i·ga′tion** *n.*

**lev·in** (lĕv′ĭn) *n.* [ME.] *Archaic.* Lightning.

**lev·i·rate** (lĕv′ər-ĭt, -ə-rāt′, lē′vər-ĭt, -və-rāt′) *n.* [< Lat. *levir,* husband's brother.] Marriage to the widow of one's brother, as required by ancient Hebrew law. **—lev′i·rat′ic** (-răt′ĭk), **lev′i·rat′i·cal** *adj.*

**Le·vi's** (lē′vīz′). A trademark for close-fitting trousers of heavy denim.

**lev·i·tate** (lĕv′ĭ-tāt′) *vi. & vt.* **-tat·ed, -tat·ing, -tates.** [< LEVITY.] To rise or cause to rise into the air and float in apparent defiance of gravity. **—lev′i·ta′tion** *n.* **—lev′i·ta′tion·al** *adj.* **—lev′i·ta′tor** *n.*

**Le·vite** (lē'vīt') *n.* [ME < LLat. *Levites* < Gk. *Leuitēs* < *Leui,* Levi < Heb. *Lēwī.*] One of the tribe of Levi, assistants to the Temple priests.

**Le·vit·i·cal** (lə-vĭt'ĭ-kəl) *also* **Le·vit·ic** (-vĭt'ĭk) *adj.* **1.** Of or pertaining to the Levites. **2.** Of or pertaining to Leviticus.

**Le·vit·i·cus** (lə-vĭt'ĭ-kəs) *n.* —See table at BIBLE.

**lev·i·ty** (lĕv'ĭ-tē) *n., pl.* **-ties.** [Lat. *levitas* < *levis,* light.] **1.** The quality or state of being light : BUOYANCY. **2.** A light manner or attitude, esp. when inappropriate : FRIVOLITY. **3.** Inconstancy : changeableness.

**le·vo** (lē'vō) *adj.* Levorotatory.

**levo-** or **lev-** *pref.* [Fr. *lévo-* < Lat. *laevus,* left.] **1.** To the left <*levorotatory*> **2.** Levorotatory <*levulose*>

**le·vo·ro·ta·ry** (lē'və-rō'tə-rē) *adj. var. of* LEVOROTATORY.

**le·vo·ro·ta·tion** (lē'və-rō-tā'shən) *n.* A counterclockwise rotation, esp. of the plane of polarized light.

**le·vo·ro·ta·to·ry** (lē'və-rō'tə-tôr'ē, -tōr'ē) *also* **le·vo·ro·ta·ry** (-rō'tə-rē) *adj.* **1.** Turning or rotating the plane of polarization of light to the left or counterclockwise. **2.** *Chem.* Of or relating to a chemical solution that rotates the plane of polarized light to the left or counterclockwise.

**lev·u·lose** (lĕv'yə-lōs', -lōz') *n.* [LEV- + *-ul* (of unknown orig.) + -OSE.] Fructose.

**lev·y** (lĕv'ē) *v.* **-ied, -y·ing, -ies.** [< ME *leve,* levy, tax < OFr. *levee* < *lever,* to raise. —see LEVER.] —*vt.* **1.** To impose or collect (e.g., a tax). **2.** To draft into military service. **3.** To declare and carry on (a war). —*vi.* To confiscate property, esp. in accordance with a legal judgment. —*n., pl.* **-ies. 1.** The act or process of levying. **2.** Money, property, or troops levied. **—lev'i·er** *n.*

**lewd** (lo͞od) *adj.* **-er, -est.** [ME *leued,* unlearned, lascivious < OE *lǣwede,* lay.] **1. a.** Preoccupied with sex : LUSTFUL. **b.** Indecent : obscene. **2.** *Obs.* Wicked. **—lewd'ly** *adv.* **—lewd'ness** *n.*

▲ word history: A thousand years ago it was no disgrace to be a *lewd* person. The word *lewd,* from Old English *lǣwede,* originally meant "lay, not belonging to the clergy." Each subsequent sense of *lewd* that developed, however, had a worse connotation than the one preceding. Such a pattern of sense development is called *pejoration.* During Middle English times the word *lewd* ran the gamut of senses from "lay" through "unlearned," "low-class," "ignorant; ill-mannered," and "wicked" to "lascivious." The last sense is the only one that survives in Modern English.

**lew·is** (lo͞o'ĭs) *n.* [Perh. from the name *Lewis.*] A dovetailed iron tenon made of several parts and designed to fit into a dovetail mortise in a large stone so that it can be lifted by a hoisting apparatus.

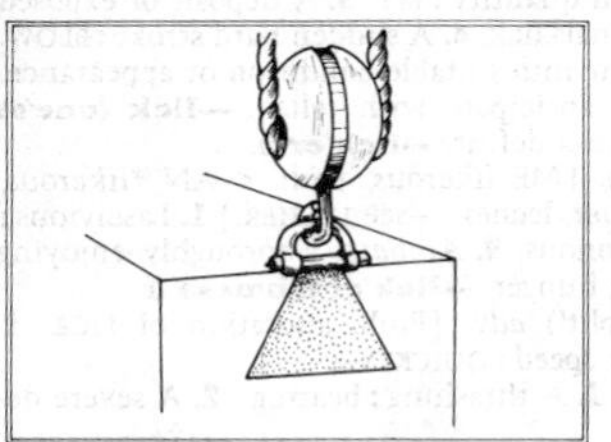
lewis

**lew·is·ite** (lo͞o'ĭ-sīt') *n.* [After Winford Lee *Lewis* (1878–1943).] An oily, colorless to violet or brown liquid, $C_2H_2AsCl_3$, for making a highly toxic military gas.

**lex** (lĕks) *n., pl.* **le·ges** (lē'jēz') [Lat.] Law.

**lex·eme** (lĕk'sēm') *n.* A meaningful speech form that is one of the vocabulary items of a language.

**lex·i·cal** (lĕk'sĭ-kəl) *adj.* [LEXIC(ON) + -AL.] **1.** Of or pertaining to the vocabulary, words, or morphemes of a language. **2.** Pertaining to lexicography or a lexicon. **—lex'i·cal'i·ty** (-kăl'ĭ-tē) *n.* **—lex'i·cal·ly** *adv.*

**lex·i·cog·ra·phy** (lĕk'sĭ-kŏg'rə-fē) *n.* Compilation of a dictionary. **—lex'i·cog'ra·pher** *n.* **—lex'i·co·graph'ic** (-kə-grăf'ĭk), **lex'i·co·graph'i·cal** *adj.* **—lex'i·co·graph'i·cal·ly** *adv.*

**lex·i·col·o·gy** (lĕk'sĭ-kŏl'ə-jē) *n.* A branch of linguistics that deals with the lexical components of language. **—lex'i·co·log'i·cal** (-kə-lŏj'ĭ-kəl) *adj.* **—lex'i·co·log'i·cal·ly** *adv.* **—lex'i·col'o·gist** *n.*

**lex·i·con** (lĕk'sĭ-kŏn') *n.* [LGk. *lexikon* < Gk. *lexikos,* of words < *lexis,* word < *legein,* to speak.] **1.** A dictionary. **2.** A stock of terms used in a particular subject, style, or profession : VOCABULARY. **3.** The morphemes of a language.

**ley** (lā, lē) *n. var. of* LEA.

**Ley·den jar** (līd'n) *n.* [After *Leyden,* Netherlands.] An early capacitor consisting of a glass jar lined inside and out with tinfoil, with a conducting rod connected to the inner foil lining passing out of the jar through an insulated stopper.

**Lha·sa ap·so** (lä'sə ăp'sō, lăs'ə) *n.* [After *Lhasa,* Tibet + Tibetan *apso,* Lhasa apso.] A small dog orig. bred in Tibet, having a long straight coat.

**li** (lē) *n., pl.* **li.** [Chin. (Mandarin) *li*3.] A Chinese measure of distance measuring approx. 1/3 of a mile or 0.52 kilometers.

**Li** *symbol for* LITHIUM.

**li·a·bil·i·ty** (lī'ə-bĭl'ĭ-tē) *n., pl.* **-ties. 1.** The state of being liable. **2. a.** Something for which one is liable : DEBT. **b. liabilities.** The financial obligations entered in the balance sheet of a business. **3.** Something that holds one back : HANDICAP. **4.** Likelihood.

**li·a·ble** (lī'ə-bəl) *adj.* [Prob. AN **liable* < OFr. *lier,* to bind < Lat. *ligare.*] **1.** Legally obligated : RESPONSIBLE <*liable* for jury duty> **2.** Subject : susceptible <*liable* to prosecution> **3.** Likely : apt <They are *liable* to return today.>

**li·ai·son** (lē'ā-zŏn', lē-ā'-) *n.* [Fr. < OFr. < *lier,* to bind. —see LIABLE.] **1. a.** Communication between different units or groups of an organization. **b.** A channel or means of communication <served as *liaison* between the forward echelon and the colonel's command post> **2. a.** A close relationship, connection, or link. **b.** An adulterous relationship : AFFAIR. **3.** The pronunciation of the usu. silent final consonant of a word when followed by a word starting with a vowel.

**li·an·a** (lē-ä'nə, -ăn'ə) *also* **li·ane** (-än', -ăn') *n.* [Fr. *liane.*] Any of various high-climbing, usu. woody vines common in the tropics.

**li·ang** (lē-äng') *n., pl.* **liang** or **-angs.** [Chin. (Mandarin) *liang*3.] A former Chinese unit of weight equivalent to 1/16 of the catty, approx. 1 1/3 ounces or 37.24 grams.

**li·ar** (lī'ər) *n.* One who tells lies.

**lib** (lĭb) *n. Informal.* LIBERATION 2.

**li·ba·tion** (lī-bā'shən) *n.* [ME *libacioun* < Lat. *libatio* < *libare,* to pour out as an offering.] **1. a.** The pouring of a liquid offering as a religious ritual. **b.** The liquid poured. **2.** *Informal.* A drink, esp. an intoxicating beverage. **—li·ba'tion·ar'y** (-shə-nĕr'ē) *adj.*

**lib·ber** (lĭb'ər) *n. Informal.* A proponent of liberation <a women's *libber*>

**li·bec·cio** (lĭ-bĕch'ē-ō', -bĕch'ō) *n.* [Ital. < Lat. *Libs,* the southwest wind < Gk. *Lips.*] A southwest wind in Italy.

**li·bel** (lī'bəl) *n.* [ME, litigant's written complaint < OFr. *libelle* < Lat. *libellus,* petition, dim. of *liber,* book.] **1.** *Law.* **a.** A written, printed, or pictorial statement that defames one's character or reputation or exposes one to public ridicule. **b.** The act of presenting such a statement to the public. **2.** The written claims presented by a plaintiff in an action at admiralty law or to an ecclesiastical court. —*vt.* **-beled, -bel·ing, -bels** or **-belled, -bel·ling, -bels.** To make or publish a libel about. **—li'bel·er, li'bel·ist** *n.*

☆ ***syns:*** LIBEL, ASPERSION, CALUMNY, DEFAMATION, DETRACTION, SCANDAL, SLANDER *n. core meaning* : the expression of injurious, malicious statements about someone <Statements about the executive's private life were pure *libel.*>

**li·bel·ant** *also* **li·bel·lant** (lī'bə-lənt) *n.* The plaintiff in a case of admiralty or ecclesiastical libel.

**li·bel·ee** *also* **li·bel·lee** (lī'bə-lē') *n.* The defendant in a case of admiralty or ecclesiastical libel.

**li·bel·ous** *also* **li·bel·lous** (lī'bə-ləs) *adj.* Constituting or involving a libel : DEFAMATORY. **—li'bel·ous·ly** *adv.*

**lib·er·al** (lĭb'ər-əl, lĭb'rəl) *adj.* [ME, generous < OFr. < Lat. *liberalis* < *liber,* free.] **1.** Holding, expressing, or following political views or policies that support civil liberties, democratic reforms, and use of governmental power to promote social progress. **2.** Holding, expressing, or following views or policies that support the freedom of individuals to act or express themselves as they choose. **3. Liberal.** Of, designating, or belonging to a political party that advocates liberal social or political views, esp. in the United States, Great Britain, and Canada. **4.** Of, pertaining to, or characteristic of representational forms of government. **5.** Tolerant of the ideas or behavior of others : BROAD-MINDED. **6. a.** Tending to give freely : GENEROUS <a *liberal* donor> **b.** Generous in amount : AMPLE <a *liberal* portion> **7.** Approximate or loose : not literal <a *liberal* interpretation> **8.** *Obs.* **a.** Permissible or suitable for a freeborn man. **b.** Morally unrestrained : LICENTIOUS. **9.** Of, pertaining to, or based on the liberal arts. —*n.* **1.** A person having or advocating liberal ideas or opinions. **2. Liberal.** A member of a Liberal political party. **—lib'er·al·ly** *adv.* **—lib'er·al·ness** *n.*

▲ word history: The Latin adjective *liberalis,* from which *liberal* is derived, is formed from the adjective *liber,* which meant "free," especially in the sense "freeborn, not a slave." Many senses of *liberalis,* and therefore of *liberal,* reflect this derivation. The sense "generous" denotes an attribute thought to be characteristic of a freeman, who was of a relatively high social status. The Latin word *liberalis* was extended to mean "noble, gentlemanly," in general. Although this sense is not recorded for English *liberal,* it survives in the phrase *liberal arts,* which originally denoted those branches of learning that were suitable for persons of high social rank.

**liberal arts** *pl.n.* Academic disciplines, as languages, history, and philosophy, that provide information of general cultural concern.

**lib·er·al·ism** (lĭb'ər-ə-lĭz'əm, lĭb'rə-) *n.* **1.** The quality or state of being liberal. **2.** Liberal views and policies, esp. in regard to political or social questions. **3.** A liberalizing movement within Protestantism. **—lib'er·al·ist** (-lĭst) *n.* **—lib'er·al·is'tic** (-lĭs'tĭk) *adj.*

**lib·er·al·i·ty** (lĭb'ə-răl'ĭ-tē) *n., pl.* **-ties. 1.** LIBERALISM 1. **2.** An instance of being liberal.

**lib·er·al·ize** (lĭb'ər-ə-līz', lĭb'rə-) *vt. & vi.* **-ized, -iz·ing, -iz·es.** To make or become liberal. **—lib'er·al·i·za'tion** *n.* **—lib'er·al·iz'er** *n.*

**lib·er·ate** (lĭb'ə-rāt') *vt.* **-at·ed, -at·ing, -ates.** [Lat. *liberare, liberat-* < *liber,* free.] **1.** To set free, as from oppression, confinement, or foreign control. **2.** *Chem.* To release from combination, as a gas. **3.** *Slang.* To obtain by illegal means, as by looting <handsome Samurai swords *liberated* from the Japanese> **—lib'er·a'tor** *n.*

**lib·er·a·tion** (lĭb'ə-rā'shən) *n.* **1.** The act of liberating or state of being liberated. **2.** The act or process of striving to achieve equal status and rights <women's *liberation*> **—lib'er·a'tion·ist** *n.*

**lib·er·tar·i·an** (lĭb'ər-târ'ē-ən) *n.* [< LIBERTY.] **1.** A believer in freedom of thought and action. **2.** A believer in free will. **—lib'er·tar'i·an·ism** *n.*

**lib·er·tin·age** (lĭb'ər-tē'nĭj) *n.* Libertinism.

**lib·er·tine** (lĭb'ər-tēn') *n.* [ME, freedman < Lat. *libertinus* < *libertus* < *liberare,* to liberate.] **1.** One who acts without moral restraint : RAKE. **2.** One who defies established religious precepts. *—adj.* Morally unrestrained : DISSOLUTE.

**lib·er·tin·ism** (lĭb'ər-tē-nĭz'əm) *n.* **1.** The quality or state of being libertine. **2.** The behavior of a libertine : PROMISCUITY.

**lib·er·ty** (lĭb'ər-tē) *n., pl.* **-ties.** [ME *liberte* < OFr. < Lat. *libertas* < *liber,* free.] **1. a.** The state of being free from control or restriction. **b.** The right to act, believe, or express oneself as one chooses. **c.** The state of being free from confinement, servitude, or forced labor. **2.** Freedom from undue or unjust governmental control. **3.** A right to engage in certain actions without interference or control <the *liberties* protected by the Bill of Rights> **4.** *often* **liberties. a.** A breach or overstepping of social convention or propriety. **b.** A statement, attitude, or act not warranted by actualities or conditions <a drama that takes *liberties* with the actual facts> **c.** An unwarranted risk : CHANCE <took reckless *liberties* on the racetrack> **5.** A period, usu. short, during which a sailor is authorized to go ashore. **—at liberty.** Not in confinement or under constraint : FREE.

☆ ***syns:*** LIBERTY, FREEDOM, LATITUDE, LEEWAY, LICENSE *n. core meaning* : departure from normal rules or procedures <took too much *liberty* with company guidelines>

**liberty cap** *n.* A brimless conical cap fitting snugly around the head that was used as a symbol of liberty by the French revolutionaries and was also worn in the United States before 1800.

**li·bid·i·nous** (lĭ-bĭd'n-əs) *adj.* [ME < Lat. *libidinosus* < *libido,* lust.] Lascivious. **—li·bid'i·nous·ly** *adv.* **—li·bid'i·nous·ness** *n.*

**li·bi·do** (lĭ-bē'dō, -bī'-) *n., pl.* **-dos.** [Lat., desire.] **1.** The psychic and emotional energy associated with instinctual biological drives. **2. a.** Sexual desire. **b.** Manifestation of the sexual drive. **—li·bid'i·nal** (-bĭd'n-əl) *adj.* **—li·bid'i·nal·ly** *adv.*

**Li·bra** (lī'brə, lē'-) *n.* [Lat. < *libra,* balance.] **1.** A constellation in the Southern Hemisphere. **2. a.** The seventh sign of the zodiac. **b.** One born under this sign. **3. libra** *pl.* **-brae** (-brē'). An ancient Roman unit of weight corresponding to a pound and equivalent to approx. 12 ounces.

**li·brar·i·an** (lī-brâr'ē-ən) *n.* **1.** A specialist in library work. **2.** *Computer Sci.* A program that originates, stores, and distributes the programs that make up an operating system. **—li·brar'i·an·ship'** *n.*

**li·brar·y** (lī'brĕr'ē) *n., pl.* **-ies.** [ME *librarie* < AN < Lat. *libraria,* bookseller's shop < *liber,* book.] **1. a.** A place in which literary and artistic materials, as books, periodicals, newspapers, pamphlets, and prints, are kept for reference or reading. **b.** A collection of such materials, esp. when systematically arranged for reference. **c.** An institution or foundation maintaining such a collection. **2.** A commercial establishment that lends books for a fee. **3.** A set or series of books issued by a publisher. **4.** A collection of recorded data organized for easy use.

**library science** *n.* The principles, practice, or study of library administration and care.

**library tape** *n. Computer Sci.* **1.** A magnetic tape stored separately from the computer from which it was generated. **2.** A magnetic tape containing a listing of library tapes.

**li·bra·tion** (lī-brā'shən) *n.* [Lat. *libratio,* oscillation < *librare,* to balance < *libra,* balance.] A very slow real or apparent oscillation of a satellite as seen from the larger celestial body around which it revolves. **—li·bra'tion·al** *adj.* **—li'bra·to'ry** (-brə-tôr'ē, -tōr'ē) *adj.*

**li·bret·tist** (lĭ-brĕt'ĭst) *n.* The writer of a libretto.

**li·bret·to** (lĭ-brĕt'ō) *n., pl.* **-bret·tos** *or* **-bret·ti** (-brĕt'ē) [Ital., dim. of *libro,* book < Lat. *liber.*] **1.** The text of an opera or other dramatic musical work. **2.** A book containing a libretto.

**Lib·y·an** (lĭb'ē-ən) *adj.* Of or pertaining to Libya, its people, or its language. *—n.* **1.** A native or resident of Libya. **2.** A Berber language of ancient northern Africa.

**lice** (līs) *n. pl. of* LOUSE 1.

**li·cence** (lī'səns) *n. Chiefly Brit. var. of* LICENSE.

**li·cense** (lī'səns) *n.* [ME *licence* < OFr. < Lat. *licentia* < *licēre,* to be permitted.] **1. a.** Official or legal permission to do or own a specified thing. **b.** Proof of permission granted, as in the form of a document <a driver's *license*> **2.** Deviation from normal rules, practices, or methods in order to achieve a particular effect <artistic *license*> **3.** Latitude of action, esp. in behavior or speech. **4.** Lack of due restraint : excessive freedom. **5.** Disregard for the standards of proper behavior : LICENTIOUSNESS. *—vt.* **-censed, -cens·ing, -cens·es. 1.** To grant permission to or for. **2.** To grant a license to or for. **—li'cens·a·ble** *adj.* **—li'cens·er, li'cen·sor'** *n.*

**licensed practical nurse** *n.* A nurse who has completed a practical nursing program and who is licensed by a state to provide routine patient care under the direction of a registered nurse or a physician.

**licensed vocational nurse** *n.* A licensed practical nurse who is licensed to practice in California or Texas.

**li·cens·ee** (lī'sən-sē') *n.* One to whom a license is granted.

**li·cen·ti·ate** (lī-sĕn'shē-ĭt) *n.* [Med. Lat. *licentiatus* < *licentiare,* to allow < Lat. *licentia,* freedom. —see LICENSE.] **1.** A person who is licensed by an authorized body to practice a specified profession. **2. a.** A degree from certain European universities ranking just below that of a doctor. **b.** One holding a licentiate.

**li·cen·tious** (lī-sĕn'shəs) *adj.* [Lat. *licentiosus* < *licentia,* freedom. —see LICENSE.] **1.** Morally undisciplined or sexually unrestrained. **2.** Having no regard for accepted rules or standards. **—li·cen'tious·ly** *adv.* **—li·cen'tious·ness** *n.*

**li·chee** (lē'chē) *n. var. of* LITCHI.

**li·chen** (lī'kən) *n.* [Lat. < Gk. *leikhēn.*] **1.** Any of numerous plants consisting of a fungus, usu. of the class Ascomycetes, in close combination with certain of the green or blue-green algae, typically forming a crustlike, scaly, or branching growth on rocks or tree trunks. **2.** *Pathol.* A skin eruption occurring primarily in lichenlike patches. *—vt.* **-chened, -chen·ing, -chens.** To cover with lichens. **—li'chen·ous** *adj.*

**li·chen·in** (lī'kə-nĭn) *n.* A white, starchlike, gelatinous compound, $C_6H_{10}O_5$, obtained from Iceland moss.

**lich gate** *also* **lych gate** (lĭch) *n.* [ME *lich,* corpse (< OE *lic*) + GATE.] A roofed gateway to a churchyard used orig. as a resting place for a bier before burial.

**lic·it** (lĭs'ĭt) *adj.* [ME < Lat. *licēre,* to be permitted.] Allowed by law : LEGAL. **—lic'it·ly** *adv.* **—lic'it·ness** *n.*

**lick** (lĭk) *v.* **licked, lick·ing, licks.** [ME *licken* < OE *liccian.*] *—vt.* **1.** To pass the tongue along or over. **2.** To lap up. **3.** To flicker at or move over like a tongue <The waves *licked* at the beach.> **4.** *Slang.* To punish with a beating : THRASH. **5.** *Slang.* To get the better of : DEFEAT <*licked* their drinking problem> *—vi.* To pass or move quickly <flames *licking* at the old barn> *—n.* **1.** An act or process of licking. **2.** A small quantity : BIT. **3.** A deposit of exposed natural salt that passing animals lick. **4.** A sudden hard stroke : BLOW. **—lick into shape.** To bring into suitable condition or appearance. **—lick (one's) chops.** To anticipate with relish. **—lick (one's) wounds.** To recuperate after a defeat. **—lick'er** *n.*

**lick·er·ish** (lĭk'ər-ĭsh) *adj.* [ME *likerous,* perh. < AN **likerous,* var. of OFr. *lechereus* < *lecher,* lecher. —see LECHER.] **1.** Lascivious : lecherous. **2.** Greedy : gluttonous. **3.** *Archaic.* Thoroughly enjoying good food. **4.** *Obs.* Arousing hunger. **—lick'er·ish·ness** *n.*

**lick·e·ty-split** (lĭk'ĭ-tē-splĭt') *adv.* [Prob. alteration of LICK + SPLIT.] *Informal.* With great speed : QUICKLY.

**lick·ing** (lĭk'ĭng) *n. Slang.* **1.** A thrashing : beating. **2.** A severe defeat or loss.

**lick·spit·tle** (lĭk'spĭt'l) *n.* A fawning underling : TOADY.

**lic·o·rice** (lĭk'ər-ĭs, -ĭsh) *n.* [ME < AN < LLat. *liquiritia,* alteration of Lat. *glycyrrhiza,* root of licorice < Gk. *glukurrhiza* : *glukus,* sweet + *rhiza,* root.] **1. a.** A Mediterranean plant, *Glycyrrhiza glabra,* with blue flowers and a sweet, distinctively flavored root. **b.** The root of the licorice, used as a flavoring in candy, liquors, tobacco, and medicines. **c.** A confection made from or flavored with the licorice root. **2.** A plant resembling licorice.

**lic·tor** (lĭk'tər) *n.* [ME *littoures,* lictors < Lat. *lictores.*] A Roman functionary who carried fasces when attending a magistrate in public appearances.

**lid** (lĭd) *n.* [ME < OE *hlid.*] **1.** A hinged or removable cover for a hollow receptacle. **2.** An eyelid. **3.** *Biol.* A flaplike covering, as an operculum. **4.** A curb or restraint <put a *lid* on federal government spending> **5.** *Slang.* A hat. **6.** *Slang.* An ounce, or 28 grams, of marijuana. *—vt.* **lid·ded, lid·ding, lids.** To cover with or as if with a lid. **—flip (one's) lid.** *Slang.* To lose one's composure or sanity.

**lid·less** (lĭd'lĭs) *adj.* **1.** Without a lid. **2.** *Archaic.* Sleepless.

**lie**[1] (lī) *vi.* **lay** (lā), **lain** (lān), **ly·ing, lies.** [ME *lien* < OE *licgan.*] **1.** To be or place oneself in a flat or horizontal position : RECLINE <*lay* on the couch to sleep> **2.** To be placed on or supported by a usu. horizontal surface <piles of books *lying* on the shelf> **3.** To be or remain in a specific condition <Spilled sugar *lay* on the floor for two days.> **4.** To be inherent <The difficulty *lies* in their obstinacy.> **5.** To occupy a place or position <The dump *lies* down the road.> **6.** To extend <Our land *lies* along both banks of the river.> **7.** *Archaic.* To stay for a night or short while. **8.** *Law.* To be admissible or maintainable. **—lie down. 1.** To submit passively to defeat, disappointment, or insults <took the rejection *lying down*> **2.** To fail to do or perform <*lying down* on the job> **—lie in.** To be in confinement for childbirth. **—lie to.** *Naut.* To remain stationary while facing the wind. **—lie with. 1.** To be decided by, dependent on, or up to <The decision *lies with* all of us.> **2.** *Archaic.* To have sexual intercourse with. *—n.* **1.** The way or position in which something is situated <the *lie* of the land> **2.** A haunt or hiding place of an animal. **3.** The position of a golf ball that has come to a stop. **—lie low. 1.** To keep oneself or one's plans hidden. **2.** To bide one's time but remain ready for action.

**lie[2]** (lī) *n.* [ME < OE *lyge.*] **1.** A false statement purposely put forward as truth : FALSEHOOD. **2.** Something meant to deceive or give a wrong impression. *—v.* **lied, ly·ing, lies.** *—vi.* **1.** To present false information with the purpose of deceiving. **2.** To convey a false image or impression <Numbers sometimes *lie.*> *—vt.* To cause to be in a particular condition or affect in a specific way by telling lies <*lied* themselves into one difficulty after another> **—give the lie to. 1.** To accuse of lying. **2.** To prove to be untrue.

☆ ***syns:*** LIE, CANARD, FALSEHOOD, FALSITY, FIB, FICTION, INVERACITY, PREVARICATION, STORY, TALE, UNTRUTH *n. core meaning* : an untrue declaration <spread *lies* about the star's personal life>

**lied** (lēt) *n., pl.* **lie·der** (lē′dər) [G. < OHG *liod.*] A German art song.

**Lie·der·kranz** (lē′dər-kränts′, -krănts′). A trademark for a soft cheese resembling a mild Limburger.

**lie detector** *n.* A polygraph used to detect lying.

**lief** (lēf) *adv.* [ME < OE *lēof,* dear.] Willingly : readily <I would as *lief* go now as later.> *—adj. Archaic.* **1.** Beloved : dear. **2.** Ready : willing.

**liege** (lēj) *n.* [ME *lege* < OFr. *lige* < LLat. *leticus* < *letus,* serf, of Germanic orig.] **1.** A feudal sovereign or lord. **2.** A feudal vassal or subject owing allegiance and services to a sovereign or lord. **3.** A monarch's loyal subject. *—adj.* **1. a.** Entitled to the allegiance and services of vassals or subjects <a *liege* lord> **b.** Bound to give allegiance and service to a lord or monarch. **2.** Faithful : loyal.

**liege·man** (lēj′mən) *n.* **1.** A feudal vassal or subject. **2.** A loyal supporter, follower, or subject.

**lien** (lēn, lē′ən) *n.* [OFr. < Lat. *ligamen,* bond < *ligare,* to bind.] *Law.* The right to take and sell or hold the property of a debtor as security or payment for a debt.

**li·e·nal** (lī-ē′nəl) *adj.* [Lat. *lien,* spleen + -AL.] Of or relating to the spleen.

**li·erne** (lē-ûrn′) *n.* [Fr.] A reinforcing rib that connects the intersections and bosses of the primary ribs in Gothic vaulting.

**lieu** (lo͞o) *n.* [OFr. < Lat. *locus.*] *Archaic.* Place : stead. **—in lieu of.** In place of : INSTEAD OF.

**lieu·ten·ant** (lo͞o-tĕn′ənt) *n.* [ME, deputy < OFr. : *lieu,* lieu + *tenir,* to hold < Lat. *tenēre.*] **1. a.** An officer in the U.S. Army, Air Force, or Marine Corps ranking below a captain. **b.** An officer in the U.S. Navy ranking above an ensign and below a lieutenant commander. **2.** (lĕf-tĕn′ənt). A commissioned officer in the British and Canadian navies ranking just below a lieutenant commander. **3.** An officer in a fire or police department ranking below a captain. **4.** One who acts in place of a superior : DEPUTY. **—lieu·ten′an·cy** *n.*

**lieutenant colonel** *n.* An officer in the U.S. Army, Air Force, or Marine Corps ranking above a major and below a colonel.

**lieutenant commander** *n.* An officer in the U.S. Navy or Coast Guard ranking above a lieutenant and below a commander.

**lieutenant general** *n.* An officer in the U.S. Army, Air Force, or Marine Corps ranking above a major general and below a general.

**lieutenant governor** *n.* **1.** An elected official ranking just below the governor of a U.S. state. **2.** The nonelective chief of government of a Canadian province. **—lieutenant governorship** *n.*

**life** (līf) *n., pl.* **lives** (līvz) [ME < OE *līf.*] **1.** The property or quality distinguishing living organisms from dead organisms and inanimate matter, manifested in functions such as growth, metabolism, response to stimuli, and reproduction. **2.** Living organisms as a whole <marine *life*><plant *life*> **3.** A living being, esp. a person <many *lives* lost in the hurricane> **4.** The physical, mental, and spiritual experiences that constitute a person's existence. **5. a.** The interval of time between birth and death : LIFETIME <spend their *lives* in hard work> **b.** A particular segment of this period <my adolescent *life*> **c.** The period from an occurrence until death <imprisoned for *life*> **6.** The time for which something functions or exists <the useful *life* of a television set> **7.** A spiritual state regarded as a transcending of death. **8.** An account of a person's life : BIOGRAPHY. **9.** Human existence or activity <real *life*><everyday *life*> **10. a.** A manner of living <wanted the good *life* for their children> **b.** The activities and interests of a particular area or realm <musical *life* in San Francisco> **11.** A source of vitality : animating force <You think you're the *life* of the party!> **12.** Liveliness or vitality : ANIMATION <an individual brimming with *life*> **13.** Something that actually exists regarded as a subject for an artist <a series of oils painted from *life*> **14.** Actual environment or reality : NATURE. **15. Life.** *Christian Science.* GOD 1c. **—for dear life.** Desperately or urgently. **—for the life of (one).** *Informal.* Though trying hard <*For the life of* me I couldn't solve the problem.>

☆ ***syns:*** LIFE, DURATION, EXISTENCE, LIFETIME, TERM *n. core meaning* : the period during which one exists <a warranty good for the *life* of the car><the poet's brief, stormy *life*>

**life belt** *n.* A life preserver worn like a belt.

**life·blood** (līf′blŭd′) *n.* **1.** Blood considered as essential for life. **2.** A vital or indispensable part <dedicated employees that are the *lifeblood* of the company>

**life·boat** (līf′bōt′) *n.* **1.** A boat carried aboard a ship for use in case the ship must be abandoned. **2.** A boat for rescue service.

**life buoy** *n.* BUOY 2.

**life cycle** *n.* **1.** The course of developmental changes through which an organism passes from its inception as a fertilized zygote to the mature state in which another zygote may be produced. **2.** A progression through a series of differing stages of development, as in insect metamorphosis.

**life expectancy** *n.* The years that a person is expected to live as ascertained by statistics.

**life·guard** (līf′gärd′) *n.* An expert swimmer hired to safeguard other swimmers, as at a beach or pool.

**life history** *n.* **1.** The history of changes undergone by an organism from inception or conception to death. **2.** Developmental history of an individual or group in society.

**life insurance** *n.* Insurance that guarantees a specific sum of money to a designated beneficiary upon the death of the insured.

**life jacket** *n.* A life preserver in the form of a vest or jacket.

**life·less** (līf′lĭs) *adj.* **1.** Having no life : INANIMATE. **2.** Uninhabited by living beings. **3.** Having lost life : DEAD. **4.** Lacking animation or vitality : DULL. **—life′less·ly** *adv.* **—life′less·ness** *n.*

**life·like** (līf′līk′) *adj.* Representing real life accurately <a *lifelike* portrait> **—life′like′ness** *n.*

**life line** *n.* **1. a.** An anchored line thrown as a support to a falling or drowning victim. **b.** A line shot to a ship in distress. **c.** A line for raising and lowering deep-sea divers. **2.** A means or route for transporting necessary supplies. **3.** A source of salvation in a crisis.

**life·long** (līf′lông′, -lŏng′) *adj.* Continuing or not changing for a lifetime.

**life preserver** *n.* A buoyant device, usu. shaped like a ring, belt, or jacket, designed to keep a person afloat in the water.

**lif·er** (lī′fər) *n. Slang.* **1.** A prisoner serving a life sentence. **2.** A career military officer or enlisted person. **3.** A right-to-lifer.

**life raft** *n.* A raft usu. of wood or inflatable material used by people who have been forced into water by an emergency.

**life·sav·er** (līf′sā′vər) *n.* **1.** One that saves a life. **2.** A lifeguard. **3.** A life preserver shaped like a ring. **4.** One that provides help in a crisis or emergency. **—life′sav′ing** *n.*

**life-size** (līf′sīz′) *also* **life-sized** (-sīzd′) *adj.* Being of the same size as an original <a *life-size* painting>

**life span** *n.* The period of time during which an organism remains alive under optimum or normal conditions.

**life·style** *also* **life-style** *or* **life style** (līf′stīl′) *n.* A way of life or style of living that reflects the values and attitudes of an individual or group.

**life-sup·port system** (līf′sə-pôrt′, -pōrt′) *n.* **1.** The equipment that provides a viable environment for crew or passengers aboard a spacecraft. **2.** Hospital equipment that makes possible the survival of a patient who might be unable to continue living independently.

**life·time** (līf′tīm′) *n.* **1.** The period of time during which an individual is alive. **2.** The period of time during which an object, property, process, or phenomenon exists or functions.

**life·way** (līf′wā′) *n.* A lifestyle.

**life·work** (līf′wûrk′) *n.* The chief work of a person's lifetime.

**lift** (lĭft) *v.* **lift·ed, lift·ing, lifts.** [ME *liften* < ON *lypta.*] *—vt.* **1.** To carry or direct from a lower to a higher position : RAISE <*lift* their faces><*lifted* the baby into the chair> **2. a.** To revoke by taking back : RESCIND <*lift* restrictions> **b.** To bring an end to (e.g., a siege) by removing forces. **c.** To cease (artillery fire) in an area. **3.** To raise in condition, rank, or esteem. **4.** To remove (plants) from the ground for transplanting. **5.** To sound or project in loud, clear tones <*lifted* our voices in song> **6.** *Informal.* To steal : pilfer <*lifted* my wallet> **7.** *Informal.* To copy from something already published : PLAGIARIZE. **8.** To clear or pay off (e.g., a debt). **9.** To perform cosmetic surgery on (the face), esp. in order to remove wrinkles. **10.** To hit (a golf ball) very high into the air. *—vi.* **1.** To rise : ascend. **2.** To disperse or disappear by or as if by rising <By noon the fog had *lifted.*> **3.** To become elevated : SOAR <Their hopes *lifted* when rescue came.> **—lift off.** To begin flight, as a spacecraft. *—n.* **1. a.** The act or process of lifting. **b.** The act or process of rising. **2.** Power or force available for raising <the *lift* of a derrick> **3.** An amount or weight lifted or capable of being lifted at one time : LOAD. **4. a.** The height or extent to which something is raised. **b.** The space or distance through which something is raised. **5.** An elevation or rise in the level of the ground. **6.** Spiritual elevation. **7.** A raised, high, or erect position, as of a part of the body <the *lift* of the model's chin> **8.** A machine or device designed to pick up, raise, or carry something. **9.** One of the layers of leather, rubber, or other material making up the heel of a shoe. **10.** *Chiefly Brit.* An elevator. **11.** A ride in a vehicle given to help someone reach a destination. **12.** Assistance : help. **13.** A set of pumps used in a mine. **14.** The component of the total aerodynamic force acting on an airfoil or on an entire aircraft or winged missile perpendicular to the relative wind and normally exerted in an upward direction, opposing the pull of gravity. **—lift′a·ble** *adj.* **—lift′er** *n.*

**lift·off** (lĭft′ôf′, -ŏf′) *n.* The initial movement or instant when a rocket or other aircraft commences flight.

**lig·a·ment** (lĭg′ə-mənt) *n.* [ME < Lat. *ligamentum,* bond < *ligare,* to bind.] **1.** *Anat.* A band or sheet of tough fibrous tissue joining two or more bones or cartilages or supporting an organ, fascia, or muscle. **2.** A connecting tie or bond. **—lig′a·ment′al** (-mĕn′tl), **lig′a·men′ta·ry** (-mĕn′tə-rē, -mĕn′trē), **lig′a·men′tous** *adj.*

**li·gan** (lī′gən) *n. var. of* LAGAN.

**li·gase** (lī′gās′) *n.* [Lat. *ligare,* to bind + -ASE.] An enzyme that catalyzes molecule linkage, gen. splitting off a pyrophosphate group from ATP concurrently.

**li·gate** (lī′gāt′) *vt.* **-gat·ed, -gat·ing, -gates.** [Lat. *ligare, ligat-.*] To tie or bind with a ligature.

**li·ga·tion** (lī-gā′shən) *n.* **1.** An act of binding or the state of being bound. **2.** Something that binds : LIGATURE.

**lig·a·ture** (lĭg′ə-cho͝or′, -chər) *n.* [ME < Lat. *ligatura* < *ligare,* to bind.] **1.** The act of binding or tying. **2. a.** A cord, wire, or bandage used for binding or tying. **b.** A surgical thread, wire, or cord for closing vessels or tying off ducts. **3.** Something that unites : BOND. **4.** A type or character, as æ, combining two or more letters. **5.** *Mus.* **a.** A group of notes intended to be sung or played as one phrase. **b.** A curved line indicating such a phrase : SLUR. *—vt.* **-tured, -tur·ing, -tures.** To ligate.

**light¹** (līt) *n.* [ME < OE *lēoht.*] **1.** *Physics.* **a.** Electromagnetic radiation that has a wavelength in the range from about 3,900 to about 7,700 angstroms and that may be perceived by the unaided, normal human eye. **b.** Electromagnetic radiation. **2.** The sensation of perceiving light <a brilliant *light* that dazzled them> **3. a.** A source of light, esp. an electric lamp <one *light* that was burning through the night> **b.** The illumination derived from such a source. **4.** A mechanical device that uses light as a warning or signal, esp. a beacon or a traffic signal. **5. a.** DAYLIGHT 1. **b.** DAWN 1. **6.** Something, as a window, that admits light. **7.** A source of fire, as a match or cigarette lighter. **8. a.** Spiritual awareness : ILLUMINATION. **b.** Something that enlightens or provides information <threw some *light* on the subject> **9.** Public attention : general knowledge <More misconduct came to *light.*> **10.** A way of looking at or considering a matter : ASPECT <viewed the problem in a different *light*> **11.** *Archaic.* Eyesight. **12. lights.** One's individual opinions, choices, or standards <behave according to their own *lights*> **13. a.** Visible light necessary for seeing <inadequate *light* to work by> **b.** A given kind of illumination <good *light* for reading> **14.** LUMINARY 3. **15.** An expression of the eyes. **16. Light.** The guiding spirit or divine presence in each person in Quaker doctrine. **17.** Representation of light in art. *—v.* **light·ed** *or* **lit** (lĭt), **light·ing, lights.** *—vt.* **1.** To set on fire : IGNITE. **2.** To cause to give out light <*lit* the candles> **3.** To provide, cover, or fill with light : ILLUMINATE <beacons *lighting* the night> **4.** To signal, direct, or guide with or as if with lights. **5.** To enliven : animate <a smile that *lit* the child's face> *—vi.* To start to burn. *—adj.* **-er, -est. 1.** Having a greater rather than lesser degree of lightness. —Used of a color. **2.** Characterized by or filled with light : BRIGHT <a *light* and colorful room> **3.** Not dark in color : FAIR <*light* complexion> **4.** Served with milk or cream <*light* coffee> **—in (the) light of.** In consideration of : in relationship to. **—see the light.** To comprehend something initially.

**light²** (līt) *adj.* **-er, -est.** [ME < OE *lēoht.*] **1. a.** Of relatively little weight : not heavy. **b.** Of relatively little weight for its size or bulk <Titanium is a *light* metal.> **c.** Of less than the standard, correct, or legal weight <a *light* pound> **2.** Exerting little impact or force : GENTLE <a *light* tap> **3. a.** Of little quantity : SCANTY <a *light* rainfall> **b.** Consuming or using relatively moderate amounts : ABSTEMIOUS <a *light* eater><a *light* smoker> **4.** Requiring little effort or exertion <*light* work> **5.** Having little importance : INSIGNIFICANT <*light* prattle> **6.** Intended mainly as entertainment : not profound <a *light* novel> **7.** Free from worries or troubles : BLITHE. **8.** Marked by frivolity. **9.** Liable to change : FICKLE. **10.** Mildly dizzy. **11.** Lacking in sexual discrimination : WANTON. **12.** Moving easily and quickly : NIMBLE <*light* on their feet> **13.** Designed for ease and quickness of movement <a *light* sailboat> **14.** Carrying little weight. **15.** Carrying little equipment or arms <*light* infantry> **16.** Requiring relatively little equipment and utilizing relatively simple processes <*light* manufacturing> **17.** Easily awakened or disturbed <a *light* sleeper> **18. a.** Easily digested <a *light* breakfast> **b.** Having a flaky or spongy texture <*light* pastry> **19.** Having a loose, porous consistency <*light* earth> **20.** Containing a relatively small amount of alcohol <a *light* wine> **21.** Designating a syllable or vowel uttered with little or no stress. *—adv.* **-er, -est. 1.** Lightly. **2.** With little weight and few burdens <traveled *light*> *—vi.* **light·ed** *or* **lit** (lĭt), **light·ing, lights. 1.** To get down, as from a horse : DISMOUNT. **2.** To come to rest : ALIGHT. **3.** To come upon one unexpectedly <Disaster *lighted* upon us.> **4.** To come upon by accident. **—light into.** To attack physically or verbally : ASSAIL. **—light out.** To leave hastily : RUN OFF.

**light adaptation** *n.* The process in which the eye adapts to increased illumination. **—light′-a·dapt′ed** (līt′ə-dăp′tĭd) *adj.*

**light·en¹** (līt′n) *v.* **-ened, -en·ing, -ens.** *—vt.* **1. a.** To make light or lighter : ILLUMINATE. **b.** To make (a color) lighter. **2.** *Archaic.* To enlighten. *—vi.* **1.** To become lighter : BRIGHTEN. **2.** To be luminous : SHINE. **3.** To give off flashes of lightning.

**light·en²** (līt′n) *v.* **-ened, -en·ing, -ens.** *—vt.* **1.** To make less heavy. **2.** To diminish the trouble, oppressiveness, or severity of. **3.** To relieve of worries or cares : GLADDEN. *—vi.* **1.** To become lighter. **2.** To become less troublesome, oppressive, or severe. **3.** To become cheerful.

**light·er¹** (lī′tər) *n.* **1.** One that ignites something. **2.** A mechanical device for lighting a cigarette, cigar, or pipe.

**light·er²** (lī′tər) *n.* [ME < *lighten,* to make less heavy < OE *līhtan.*] A large barge, esp. one used to deliver or unload goods to or from a cargo ship. *—vt.* **-ered, -er·ing, -ers.** To convey (cargo) in a lighter.

**light·er·age** (lī′tər-ĭj) *n.* **1.** Transportation of goods on a lighter. **2.** The fee charged for lightering.

**light·er-than-air** (lī′tər-*th*ən-âr′) *adj.* Weighing less than that of the air displaced. —Used of certain aircraft.

**light·face** (līt′fās′) *n.* A typeface or font of characters with relatively thin, light lines. **—light′faced′** *adj.*

**light-fin·gered** (līt′fĭng′gərd) *adj.* **1.** Having quick and nimble fingers : DEXTEROUS. **2.** Skilled at petty thievery. **—light′-fin′gered·ness** *n.*

**light-foot·ed** (līt′fo͝ot′ĭd) *also* **light·foot** (-fo͝ot′) *adj.* Treading with light and nimble ease. **—light′-foot′ed·ly** *adv.* **—light′-foot′ed·ness** *n.*

**light-hand·ed** (līt′hăn′dĭd) *adj.* Having a light, delicate touch. **—light′-hand′ed·ly** *adv.* **—light′-hand′ed·ness** *n.*

**light·head·ed** (līt′hĕd′ĭd) *adj.* **1.** Delirious, giddy, or faint. **2.** Frivolous. **—light′head′ed·ly** *adv.* **—light′head′ed·ness** *n.*

**light·heart·ed** (līt′här′tĭd) *adj.* Free from trouble or care : CHEERFUL. **—light′heart′ed·ly** *adv.* **—light′heart′ed·ness** *n.*

**light heavyweight** *n.* A boxer or wrestler weighing between 161 and 175 pounds.

**light·house** (līt′hous′) *n.* A tall structure topped by a powerful light used as a signal or beacon to aid marine navigation.

**light·ing** (lī′tĭng) *n.* **1.** The state of being lighted. **2. a.** The method or equipment for providing artificial illumination. **b.** The illumination so provided. **3.** The act or process of igniting.

**light·ly** (līt′lē) *adv.* **1.** With little weight or force : GENTLY. **2.** To a slight extent or amount <apply powder *lightly*> **3. a.** With little difficulty : EASILY. **b.** With agility and grace : NIMBLY. **4. a.** In a carefree manner : CHEERFULLY <took their bad luck *lightly*> **b.** Without due care or consideration <looked at the problem too *lightly*>

**light meter** *n.* An exposure meter.

**light-mind·ed** (līt′mīn′dĭd) *adj.* LIGHTHEADED 2. **—light′-mind′ed·ly** *adv.* **—light′-mind′ed·ness** *n.*

**light·ness¹** (līt′nĭs) *n.* **1.** The quality or state of being illuminated. **2.** The dimension of the color of an object by which the object appears to reflect or transmit more or less of the incident light, varying from black to white for surface colors and from black to colorless for transparent volume colors.

**light·ness²** (līt′nĭs) *n.* **1.** The quality or state of having little weight or force. **2.** Ease or quickness of movement : AGILITY. **3.** Ease or cheerfulness in manner or style. **4.** Freedom from trouble or worry. **5.** Lack of appropriate seriousness : LEVITY. **6.** Delicacy or subtlety in workmanship, performance, or effect.

**light·ning** (līt′nĭng) *n.* [ME < *lightenen,* to illuminate < *lighten* < *light,* illumination.] **1. a.** A large-scale high-tension natural electric discharge in the atmosphere. **b.** The visible flash of light accompanying such a discharge. **2.** *Informal.* A sudden, usu. improbable stroke of fortune. *—vi.* **-ninged** (-nĭngd), **-ning, -nings.** To discharge a flash of lightning. *—adj.* Moving with great suddenness or speed.

**lightning arrester** *n.* A protective device for electrical equipment that reduces excessive voltage caused by lightning to a safe level by grounding the discharge.

**lightning bug** *n.* A firefly.

**lightning rod** *n.* A grounded metal rod placed high on a structure to prevent damage by conducting lightning to the ground.

**light opera** *n.* An operetta.

**light pen** *n.* *Computer Sci.* A small photosensitive device connected to a computer and moved by hand over an output display in order to manipulate information in the computer.

**lights** (līts) *pl.n.* [ME *lightes* < *light,* light in weight.] The lungs, esp. the lungs of an animal used for food.

**light·ship** (līt′shĭp′) *n.* A ship with a powerful light or horn that is anchored in dangerous waters to warn other vessels.

---

ă **pat** ā **pay** âr **care** ä **father** ĕ **pet** ē **be** hw **which** ĭ **pit**
ī **tie** îr **pier** ŏ **pot** ō **toe** ô **paw, for** oi **noise** o͝o **took**

**light show** *n.* A display of colored lights in kaleidoscopic patterns, often accompanied by film loops and slides.

**light·some[1]** (līt'səm) *adj.* **1.** Providing light : LUMINOUS. **2.** Bright. **—light'some·ly** *adv.* **—light'some·ness** *n.*

**light·some[2]** (līt'səm) *adj.* **1.** Nimble or graceful. **2.** Free from worry or care. **3.** Frivolous. **—light'some·ly** *adv.* **—light'some·ness** *n.*

**lights out** *n.* **1.** A signal to extinguish lights for the night. **2.** Bedtime.

**light-struck** (līt'strŭk') *adj.* Fogged by accidental exposure to light. —Used of photosensitive materials.

**light stylus** *n.* A light pen.

**light water** *n.* Ordinary water.

**light·weight** (līt'wāt') *n.* **1.** One that weighs relatively little. **2.** A boxer or wrestler weighing between 127 and 135 pounds. **3.** A person of little ability, intelligence, influence, or importance.

**light·wood** (līt'wo͝od') *n.* Dry, easily ignited, often resinous wood, used esp. for kindling.

**light-year** *also* **light year** (līt'yîr') *n.* The distance that light travels in a vacuum for a period of one year, approx. 5.878 trillion (5.878 × $10^{12}$) miles or 9.46 trillion kilometers.

**lign-** *pref. var. of* LIGNI-.

**lig·ne·ous** (lĭg'nē-əs) *adj.* [Lat. *ligneus* < *lignum*, wood.] Composed of or having the appearance or texture of wood : WOODY.

**ligni-** *or* **ligno-** *or* **lign-** *pref.* [< Lat. *lignum*, wood.] Wood <*lignocellulose*>

**lig·ni·fy** (lĭg'nə-fī') *v.* **-fied, -fy·ing, -fies.** [Fr. *lignifier* < Lat. *lignum*, wood.] *—vi.* To form or turn into wood through the formation and deposit of lignin in cell walls. *—vt.* To make woody or woodlike by the deposit of lignin. **—lig'ni·fi·ca'tion** *n.*

**lig·nin** (lĭg'nĭn) *n.* A polymer functioning as a natural binder and support for the cellulose fibers of woody plants, and the chief non-carbohydrate constituent of wood.

**lig·nite** (lĭg'nīt') *n.* [Fr. < Lat. *lignum*, wood.] A low-grade, brownish-black coal. **—lig·nit'ic** (-nĭt'ĭk) *adj.*

**ligno-** *pref. var. of* LIGNI-.

**lig·no·cel·lu·lose** (lĭg'nō-sĕl'yə-lōs') *n.* A combination of lignin and cellulose that strengthens woody cells.

**lig·num vi·tae** (lĭg'nəm vī'tē) *n., pl.* **lignum vi·taes.** [Lat. *lignum*, wood + Lat. *vitae*, genitive of *vita*, life.] **1. a.** A tropical American tree, *Guaiacum officinale* or *G. sanctum*, with evergreen leaves and heavy, durable, resinous wood. **b.** The wood of the lignum vitae. **2.** A tree similar to or resembling the lignum vitae.

**lig·ro·in** (lĭg'rō-ən) *n.* [Orig. unknown.] A volatile, flammable fraction of petroleum, obtained by distillation and used as a solvent.

**lig·u·la** (lĭg'yə-lə) *n., pl.* **-lae** (-lē') *or* **-las.** [NLat., ligule. —see LIGULE.] A strap-shaped structure, esp. a mouth part occurring in certain insects.

**lig·u·late** (lĭg'yə-lĭt, -lāt') *adj.* **1.** Strap-shaped. **2.** Having a ligule.

**lig·ule** (lĭg'yo͞ol) *n.* [NLat. *ligula* < Lat., dim. of *lingua*, tongue.] A straplike structure, as a ray flower of a daisy or a sheathlike organ at the base of a grass leaf.

**lig·ure** (lĭg'yo͝or') *n.* [ME < LLat. *ligurius* < Gk. *ligurion.*] A precious stone of ancient Israel.

**lik·a·ble** *also* **like·a·ble** (lī'kə-bəl) *adj.* Attractive : pleasing. **—lik'a·ble·ness, like'a·ble·ness** *n.*

**†like[1]** (līk) *v.* **liked, lik·ing, likes.** [ME *liken* < OE *līcian.*] *—vt.* **1.** To find attractive or pleasant : ENJOY. **2.** To want to have <would *like* to go> **3.** To feel about : REGARD <How do you *like* their nerve!> **4.** *Regional.* To be pleasing to. *—vi.* **1.** To have an inclination or preference <If you *like*, we can remain here.> **2.** *Scot.* To be pleased. *—n.* Something liked : PREFERENCE <They had more *likes* than dislikes.>

☆ ***syns:*** LIKE, DIG, ENJOY, RELISH, SAVOR *n. core meaning* : to receive pleasure from <*likes* fine wine> ***ant:*** dislike

**like[2]** (līk) *prep.* [ME < *like*, similar (< OE *gelīc*), and *like*, similarly (< OE *gelīce*).] **1.** Possessing the characteristics of : similar to. **2.** In the typical manner of <It's not *like* you to take offense.> **3.** Inclined or disposed to <felt *like* quitting> **4.** As if the probability exists for <looks *like* a bad year for car sales> **5.** Such as <saved things *like* old clippings and string> *—adj.* **1.** Having the same or almost the same characteristics : SIMILAR <for this and *like* purposes> **2.** Alike <They are as *like* as two peas in a pod.> *—adv.* **1.** Probably : likely <*Like* as not you'll need more cash.> **2.** *Nonstandard.* —Used for emphasis <*Like* you know what I mean.> *—n.* One similar to another <preferred symphonic music, opera, and the *like*> *—conj.* **1.** In the same way that : AS <To play the violin *like* they do requires great talent.> **2.** As if <It looks *like* we'll leave on time.> **3.** —Used as an intensifier <worked *like* mad> ***usage:*** The best writers since Shakespeare's day have used *like* as a conjunction, but the usage has drawn such heavy fire from purists in recent times that it is best avoided. Thus, one should say *I answered as* (not *like*) *they did.* There can be no objection to the use of *like* when no verb follows, as in *You take to politics like a duck to water.* **—the like (or likes) of.** Such a one as <I don't care for *the like of* that kind of politician.>

☆ ***syns:*** LIKE, ALIKE, ANALOGOUS, COMPARABLE, CORRESPONDING, EQUIVALENT, PARALLEL, UNIFORM *adj. core meaning* : having the same or almost the same characteristics <coworkers of *like* opinions> ***ant:*** different, unlike

**like[3]** (līk) *also* **liked** (līkt) *aux.v.* [ME *liken*, to compare < *like*, similar < OE *gelīc.*] *Nonstandard.* To be just on the point of : be or come near to <I *like* to have killed them.>

**-like** *suff.* [< LIKE[2].] Resembling or typical of <lady*like*>

**like·a·ble** (lī'kə-bəl) *adj. var. of* LIKABLE.

**liked** (līkt) *aux.v. var. of* LIKE[3].

**like·li·hood** (līk'lē-ho͝od') *n.* **1.** The state of being likely or probable : PROBABILITY. **2.** Something probable.

**like·ly** (līk'lē) *adj.* **-li·er, -li·est.** [ME < both OE *gelīclic* (< *gelīc*, similar) and ON *līkligr* (< *līkr*, similar).] **1.** Possessing or displaying the characteristics or qualities that make something probable <They are *likely* to arrive this morning.> **2.** Within the realm of credibility : PLAUSIBLE <a *likely* result> **3.** Apparently appropriate : SUITABLE <a *likely* spot> **4.** Apt to yield a desired outcome : PROMISING <a *likely* subject for more research> **5.** Attractive. *—adv.* Probably.

☆ ***syns:*** LIKELY, PROBABLY *adj. core meaning* : showing a probability of happening or of being true <*likely* to rain at any moment> Both LIKELY and PROBABLY describe what seems to be true, but is not certain <a *likely* excuse><a *probable* explanation> ***ant:*** unlikely

**like-mind·ed** (līk'mīn'dĭd) *adj.* Having the same turn of mind as another.

**lik·en** (lī'kən) *vt.* **-ened, -en·ing, -ens.** [ME *liknen* < *liken*, to compare. —see LIKE[3].] To see, mention, or represent as similar : COMPARE.

**like·ness** (līk'nĭs) *n.* **1.** Resemblance to another. **2.** An imitative appearance : SEMBLANCE. **3.** An artistic representation : IMAGE.

**like·wise** (līk'wīz') *adv.* **1.** In the same way : SIMILARLY. **2.** As well : ALSO. ***usage:*** Since *likewise* is not a conjunction, it cannot take the place of *and.* Thus, in such sentences as *I risked my fortune and likewise my honor*, the word *and* should not be omitted.

**lik·ing** (lī'kĭng) *n.* **1.** Attraction : fondness. **2.** Preference or taste.

**li·ku·ta** (lē-ko͞o'tä) *n., pl.* **ma·ku·ta** (mä-ko͞o'tä) [Native word in Zaire.] —See table at CURRENCY.

**li·lac** (lī'lək, -lŏk', -lăk') *n.* [Obs. Fr. < Sp. < Ar. *līlak* < Pers., var. of *nīlak* < *nīl*, blue.] **1.** A shrub of the genus *Syringa*, esp. *S. vulgaris*, cultivated for its fragrant purplish or white flower clusters. **2.** A pale to light or moderate purple.

**li·lan·ge·ni** (lĭ-läng'gĕ-nē) *n.* [Native word in Swaziland.] —See table at CURRENCY.

**Lil·ith** (lĭl'ĭth) *n.* [Heb. *līlīth.*] **1.** An evil female spirit in ancient Semitic legend who was believed to haunt deserted places and attack children. **2.** Adam's first wife who was believed in Hebrew folklore to have existed before Eve was created.

**Lil·li·pu·tian** *also* **lil·li·pu·tian** (lĭl'ə-pyo͞o'shən) *n.* [After the *Lilliputians*, a people in *Gulliver's Travels* by Jonathan Swift (1667–1745).] A very small being or person. *—adj.* **1.** Tiny : diminutive. **2.** Petty : trivial.

**lilt** (lĭlt) *n.* [< ME *lilten*, to sound an alarm.] **1.** A light, happy song or tune. **2.** A lively or cheerful manner of speaking in which the pitch of the voice varies pleasantly. **3.** A resilient or light manner of moving or walking. *—v.* **lilt·ed, lilt·ing, lilts.** *—vt.* To say, sing, or play cheerfully and rhythmically. *—vi.* **1.** To speak, sing, or play with liveliness or rhythm. **2.** To move lightly and buoyantly.

**lil·y** (lĭl'ē) *n., pl.* **-ies.** [ME *lilie* < OE < Lat. *lilium.*] **1.** A plant of the genus *Lilium*, with variously colored, often trumpet-shaped flowers. **2. a.** A similar or related plant, as the day lily or the water lily. **b.** The flower of such a plant.

**lil·y-liv·ered** (lĭl'ē-lĭv'ərd) *adj.* Timid : cowardly.

**lily of the valley** *n., pl.* **lilies of the valley.** A widely cultivated plant, *Convallaria majalis*, bearing fragrant, bell-shaped white flowers on a leafless raceme.

**lily pad** *n.* A floating leaf of a water lily.

**lil·y-white** (lĭl'ē-hwīt', -wīt') *adj.* **1.** White as a lily. **2.** Beyond reproach : BLAMELESS. **3.** *Informal.* Excluding or seeking to exclude blacks, esp. from politics.

**li·ma bean** (lī'mə) *n.* [After *Lima*, Peru.] **1.** Any of several varieties of a tropical American plant, *Phaseolus limensis*, bearing flat pods with large light-green edible seeds. **2.** The seed of the lima bean.

**lim·a·cine** (lĭm'ə-sēn', lī'mə-) *adj.* [< Lat. *limax*, *limac-*, slug < *limus*, slime.] Of, relating to, or like a slug.

**limb[1]** (lĭm) *n.* [ME *lim* < OE.] **1.** A larger branch of a tree. **2.** An animal's jointed appendage, used for locomotion or grasping, as an arm, leg, wing, or flipper. **3.** A projecting part : EXTENSION. **4.** One regarded as an extension, member, or representative of a larger body or group. **5.** *Informal.* An impish child. *—vt.* **limbed, limb·ing, limbs.** To dismember. **—out on a limb.** *Informal.* In a difficult or vulnerable situation.

**limb[2]** (lĭm) *n.* [Lat. *limbus*, border.] **1.** *Astron.* The circumferential edge of the apparent disk of a celestial body. **2.** The edge of a graduated arc or circle of an instrument for measuring angles. **3.** *Bot.* The expanded tip of a petal or the expanded upper part of a corolla.

**lim·bate** (lĭm′bāt′) *adj.* [LLat. *limbatus,* bordered < Lat. *limbus,* border.] *Bot.* Having an edge or margin of a different color.
**lim·ber[1]** (lĭm′bər) *adj.* [Orig. unknown.] **1.** Bending or flexing readily : PLIABLE. **2.** Capable of bending or moving easily : SUPPLE. —*v.* **-bered, -ber·ing, -bers.** —*vt.* To make limber <*limbered* up my muscles> —*vi.* To make oneself limber <swimmers *limbering* up before a meet> **—lim′ber·ly** *adv.* **—lim′ber·ness** *n.*
**lim·ber[2]** (lĭm′bər) *n.* [ME *limour,* the shaft of a cart.] A two-wheeled, horse-drawn vehicle for towing a field gun.
**lim·bers** (lĭm′bərz) *pl.n.* [Alteration of Fr. *lumière* < OFr. *lumiere,* opening < LLat. *luminare,* light. —see LUMINARY.] Gutters or channels on each side of a ship's keelson that drain bilge water into the pump well.
**lim·bi** (lĭm′bī′) *n. pl. of* LIMBUS.
**lim·bo[1]** (lĭm′bō) *n., pl.* **-bos.** [ME < Med. Lat. *limbus* < Lat., border.] **1.** *often* **Limbo.** The abode of souls kept from Heaven through circumstance, as lack of baptism. **2.** A state of oblivion or neglect <kept the applicant in *limbo* for months> **3.** A state or place of confinement. **4.** An intermediate place or state.
**lim·bo[2]** (lĭm′bō) *n.* [Native word in the West Indies.] A West Indian dance in which the dancers bend over backward and pass repeatedly under a pole that is lowered slightly each time.
**Lim·burg·er** (lĭm′bûr′gər) *n.* [After *Limburg,* Belgium.] A soft white cheese with a pronounced flavor and odor.
**lim·bus** (lĭm′bəs) *n., pl.* **-bi** (-bī′) [Lat., border.] *Biol.* A distinctive border or edge.
**lime[1]** (līm) *n.* [Fr. < Prov. *limo* < Ar. *līmah.*] **1.** A spiny tree native to Asia, *Citrus aurantifolia,* with evergreen leaves, fragrant white flowers, and edible fruit. **2.** The egg-shaped fruit of the lime tree, with a green rind and acid juice used as flavoring.
**lime[2]** (līm) *n.* [Alteration of obs. *line,* linden < ME *lind* < OE.] Any of several Old World linden trees.
**lime[3]** (līm) *n.* [ME < OE *līm.*] **1. a.** Calcium oxide. **b.** Any of various mineral and industrial forms of calcium oxide differing chiefly in water content and percentage of such constituents as silica, alumina, and iron. **2.** BIRDLIME 1. —*vt.* **limed, lim·ing, limes. 1.** To treat with lime. **2.** To smear with birdlime. **3.** To catch or snare with or as if with birdlime. **—lim′y** *adj.*
**lime·ade** (lī-mād′) *n.* A sweetened drink of lime juice and water.
**lime·kiln** (līm′kĭl′, -kĭln′) *n.* A furnace for reducing naturally occurring forms of calcium carbonate to lime.
**lime·light** (līm′līt′) *n.* **1. a.** A stage light in which lime is heated to incandescence producing brilliant illumination. **b.** The brilliant light so produced. **2.** A focus of public attention.
**li·men** (lī′mən) *n., pl.* **li·mens** *or* **lim·i·na** (lĭm′ə-nə) [Lat., threshold.] The threshold of a physiological or psychological response. **—lim′i·nal** (lĭm′ə-nəl) *adj.*
**lim·er·ick** (lĭm′ər-ĭk) *n.* [After *Limerick,* a county of the Republic of Ireland.] A humorous or nonsensical verse of five anapestic lines usu. with the rhyme scheme *aabba.*
**lime·stone** (līm′stōn′) *n.* A shalelike or sandy sedimentary rock, chiefly $CaCO_3$, containing variable quantities of magnesium carbonate and quartz, used as a building stone and in manufacturing lime, carbon dioxide, and cement.
**lime-twig** (līm′twĭg′) *n.* **1.** A twig covered with birdlime to catch birds. **2.** A snare.
**lime·wa·ter** (līm′wô′tər, -wŏt′ər) *n.* A clear colorless alkaline aqueous solution of calcium hydroxide, used in skin preparations, as calamine lotion, and occas. as an antacid.
**lim·ey** (lī′mē) *n., pl.* **-eys.** *Slang.* **1.** A British sailor. **2.** An Englishman.
**li·mic·o·line** (lī-mĭk′ə-līn′, -lĭn) *adj.* [< NLat. *Limicolae,* former order name : Lat. *limus,* mud + *colere,* to inhabit.] Of or relating to shore birds of the suborder Charadrii, as the sandpiper.
**lim·it** (lĭm′ĭt) *n.* [ME *limite* < OFr. < Lat. *limes.*] **1.** The point, edge, or line beyond which something cannot or may not proceed. **2. limits.** The boundary surrounding a specific area : BOUNDS <within the park *limits*> **3.** A confining element : RESTRICTION. **4.** The greatest number or amount allowed. **5.** The largest amount which may be bet at one time in games of chance. **6.** *Math.* A number or point $k$ that is approached by a function $f(x)$ as x approaches $a$ if, for every positive number $\varepsilon$, there exists a number $\delta$ such that $|f(x)-k|<\varepsilon$ if $0<|x-a|<\delta$. **7.** *Informal.* One that approaches or exceeds certain limits, as of acceptability or credibility. —*vt.* **-it·ed, -it·ing, -its.** To restrict or confine within limits. **—lim′it·a·ble** *adj.* **—lim′it·er** *n.*
**lim·i·tar·y** (lĭm′ĭ-tĕr′ē) *adj. Archaic.* **1. a.** Of or pertaining to a limit or boundary. **b.** Restrictive : limiting. **2.** Limited.
**lim·i·ta·tion** (lĭm′ĭ-tā′shən) *n.* **1.** The act of limiting or state of being limited. **2.** LIMIT 3. **3.** *Law.* A limited period during which, by statute, an action may be brought.
**lim·it·ed** (lĭm′ĭ-tĭd) *adj.* **1.** Confined or restricted within certain limits <has only *limited* use> **2.** Not attaining the highest goals or achievement <a *limited* success> **3.** Having governmental or ruling powers restricted by enforceable limitations, as a constitution or legislative body. **4.** Indicating transportation facilities, as trains or buses, that make few stops and carry relatively few passengers. —*n.* A limited train or bus. **—lim′i·ted·ly** *adv.* **—lim′i·ted·ness** *n.*
**limited edition** *n.* An edition, as of a book, limited to a specified number of copies.
**limited war** *n.* A war whose objective is of smaller scope than the total defeat of the enemy.
**lim·i·ting** (lĭm′ĭ-tĭng) *adj.* **1.** Functioning or serving as a limit. **2.** Restricting the range of application of the noun modified.
**lim·it·less** (lĭm′ĭt-lĭs) *adj.* **1.** Having no limit. **2.** Unrestricted or unconfined. **—lim′it·less·ly** *adv.* **—lim′it·less·ness** *n.*
**limit point** *n. Math.* LIMIT 6.
**limn** (lĭm) *vt.* **limned, limn·ing** (lĭm′nĭng), **limns.** [ME *limnen,* to illuminate (a manuscript), prob. < *limnour,* an illuminator < AN *lymnour* < OFr. *enluminer,* to illuminate (a manuscript) < Lat. *illuminare,* to adorn < *lumen,* light.] **1.** To describe. **2.** To depict by drawing or painting. **—limn′er** (lĭm′nər) *n.*
**lim·net·ic** (lĭm-nĕt′ĭk) *adj.* [< Gk. *limnē,* lake.] Of, living, or occurring in the deeper, open waters of ponds or lakes.
**lim·nol·o·gy** (lĭm-nŏl′ə-jē) *n.* [Gk. *limnē,* lake + -LOGY.] Scientific study of the life and phenomena of lakes, ponds, and streams. **—lim′no·log′i·cal** (-nə-lŏj′ĭ-kəl) *adj.* **—lim′no·log′i·cal·ly** *adv.* **—lim·nol′o·gist** *n.*
**lim·o** (lĭm′ō) *n., pl.* **lim·os.** A limousine.
**lim·o·nene** (lĭm′ə-nēn′) *n.* [Fr. *limonène* < obs. Fr. *limon,* lemon < OFr. —see LEMON.] A liquid, $C_{10}H_{16}$, with a lemonlike fragrance, used in manufacturing resins and as a solvent, wetting agent, and dispersing agent.
**li·mo·nite** (lī′mə-nīt′) *n.* [G. *Limonit* < Gk. *leimōn,* meadow.] A yellowish-brown to black natural iron oxide, essentially $FeO(OH)\cdot nH_2O$, used as an ore of iron. **—li′mo·nit′ic** (-nĭt′ĭk) *adj.*
**lim·ou·sine** (lĭm′ə-zēn′, lĭm′ə-zēn′) *n.* [After *Limousin,* a region of France.] A large passenger vehicle, esp. a luxurious car usu. driven by a chauffeur.
**limp** (lĭmp) *vi.* **limped, limp·ing, limps.** [Prob. < obs. *limphalt,* lame.] **1.** To walk lamely, esp. with unevenness, as if favoring one leg. **2.** To move or proceed haltingly or unsteadily. —*n.* An irregular, jerky, or awkward gait. —*adj.* **-er, -est. 1.** Lacking or having lost rigidity, stiffness, or the ability to support itself <a *limp* stem> **2.** Lacking strength or firmness : WEAK <a *limp* wrist> **—limp′ly** *adv.* **—limp′ness** *n.*
☆ ***syns:*** LIMP, FLABBY, FLACCID, FLOPPY *adj. core meaning* **:** lacking stiffness or firmness <*limp* lettuce>
**lim·pet** (lĭm′pĭt) *n.* [Poss. < ME *lempet.*] **1.** Any of numerous marine gastropod mollusks, as of the families Acmaeidae and Patellidae, having a tent-shaped shell and adhering to rocks of tidal areas. **2.** One who clings persistently. **3.** An explosive designed to cling to the hull of a ship or the underside of a motor vehicle and to detonate on contact or by signal.

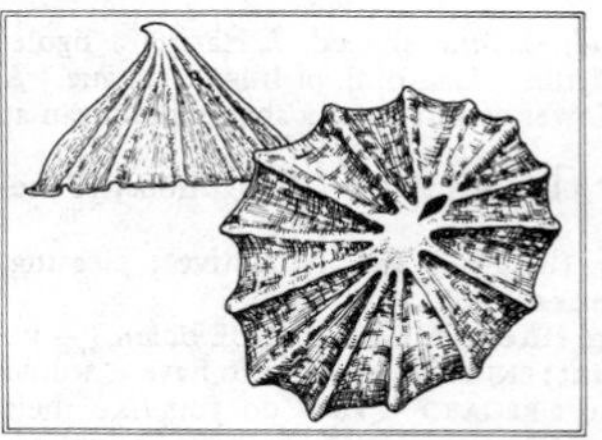
**limpet**

**lim·pid** (lĭm′pĭd) *adj.* [Fr. *limpide* < Lat. *limpidus.*] **1.** Transparently clear : PELLUCID <a *limpid* pond> **2.** Easily intelligible <writes with *limpid* prose> **3.** Calm and untroubled : SERENE. **—lim·pid′i·ty, lim′pid·ness** *n.* **—lim′pid·ly** *adv.*
**limp·kin** (lĭmp′kĭn) *n.* A brownish wading bird, *Aramus guarauna* of warm swampy regions of the New World, with a distinctive wailing call.
**lin·ac** (lĭn′ăk′) *n.* [LIN(EAR) AC(CELERATOR).] A linear accelerator.
**lin·age** *also* **line·age** (lī′nĭj) *n.* **1.** The number of lines of printed or written material. **2.** Payment for written work at a designated amount per line.
**lin·al·o·ol** (lĭ-năl′ō-ôl′, -ōl′) *n.* [Sp. *linaloe,* fragrant wood of a Mexican tree (< LLat. *lignum aloes,* wood of the aloe) + -OL[2].] A colorless, fragrant liquid, $C_{10}H_{18}O$, distilled from the oils of certain plants, as rosewood or bergamot, and used in perfumery.
**linch·pin** (lĭnch′pĭn′) *n.* [ME *linspin* : *lins,* linchpin (< OE *lynis*) + *pin,* pin (< OE *pinn*).] **1.** A locking pin inserted in the end of a shaft, as in an axle, to prevent a wheel from slipping off. **2.** A central and cohesive element <reduced spending that is the *linchpin* of our economic program>
**Lin·coln** (lĭng′kən) *n.* A sheep with long wool, orig. bred in Lincoln, England.
**Lin·coln·esque** (lĭng′kə-nĕsk′) *adj.* Of, relating to, or resembling Abraham Lincoln.

ă **pat** ā **pay** âr **care** ä **father** ĕ **pet** ē **be** hw **which** ĭ **pit**
ī **tie** îr **pier** ŏ **pot** ō **toe** ô **paw, for** oi **noise** ŏŏ **took**

**lin·den** (lĭn′dən) *n.* [ME, made of linden wood < OE < *lind,* linden.] A shade tree of the genus *Tilia,* with heart-shaped leaves and fragrant yellowish flowers.

**line**[1] (līn) *n.* [ME < OE, cord, and OFr. *ligne,* line, both < Lat. *linea,* string < *linum,* thread.] **1. a.** The locus of a point having one degree of freedom : CURVE. **b.** A set of points (*x, y*) that satisfy the linear equation $ax + by + c = 0$, where *a* and *b* are not both zero. **2. a.** A thin continuous mark, as that made by a pencil, pen, or brush applied to a surface. **b.** A similar mark scratched or cut into a surface. **c.** A crease in the skin, esp. on the face : WRINKLE. **3.** A real or imaginary straight line positioned in relation to fixed points of reference. **4. a.** A border or boundary <the state *line*> **b.** A constraint or limit <drew the *line* at nuclear intervention> **c.** A demarcation. **d.** A contour : outline. **5. a.** A mark for defining a shape or representing a contour. **b.** Any of the marks that make up the formal design of a picture. **6. a.** A cable, rope, string, cord, or wire. **b.** A rope used aboard a ship. **c.** A fishing line. **d.** A clothesline. **7.** A pipe or system of pipes for conveying a fluid. **8.** An electric-power transmission cable. **9. a.** A wire or system of wires linking telephone or telegraph systems. **b.** A functioning or open telephone connection. **10. a.** A passenger or cargo system of transportation, usu. over a particular route. **b.** A company managing or owning such a system. **11.** A railway track or system of tracks. **12.** A course of progress or movement : ROUTE <the *line* of travel> **13.** A general method, manner, or course of procedure. **14.** An official or prescribed policy <the government's *line*> **15.** A condition of agreement : ALIGNMENT. **16. a.** One's trade, occupation, or field of interest. **b.** Range of competency. **17.** *Archaic.* One's lot or position in life. **18.** Merchandise or services of a similar or related nature <Fabrics were not in the shop's *line.*> **19.** A group arranged in a row or series. **20. a.** A chronological series of persons or things that follow each other. **b.** Lineage or ancestry. **c.** A strain, as of livestock or plants, developed by selective breeding. **21. a.** A sequence of related things that leads to a particular ending <a *line* of thought> **b.** An ordered system of operations that makes possible sequential assembly or manufacture of goods at all or various production stages. **22. a.** A horizontal row of printed or written words or symbols. **b.** A unit of verse having a specific number of metrical feet typical of the verse. **23.** A brief letter : NOTE. **24.** *often* **lines.** The speeches and dialogue of a theatrical production. **25.** *Informal.* A calculated or glib way of speaking, usu. to obtain an undeclared end <They rattled off quite a *line* asking for help.> **26. lines.** *Chiefly Brit.* A marriage certificate. **27.** A horizontal demarcation separating categories of points scored in bridge. **28. a.** A source of information. **b.** The information itself. **29.** *Mus.* One of the five parallel marks making up a staff. **30. a.** A military formation in which elements, as troops, tanks, or ships, are arranged abreast of each other. **b.** The battle area nearest to the enemy. **c.** The troops in this area. **d.** Combatant troops. **e.** The officers in direct command of warships. **f.** A trench or bulwark. **g.** An extended system of such fortifications. **31. a.** A foul line in a sport. **b.** A real or imaginary mark demarcating a specified section of a playing area or field. **c.** A real or imaginary mark or point at which a race starts or finishes. **32.** *Football.* **a.** A line of scrimmage. **b.** The linemen. —*v.* **lined, lin·ing, lines.** —*vt.* **1.** To mark or incise with a line or lines. **2.** To represent with lines. **3.** To place in a row or series. **4.** To form a bordering line along. **5.** *Baseball.* To hit (a ball) sharply in a usu. straight line. —*vi. Baseball.* To hit a line drive. **—between the lines.** By inference : INDIRECTLY. **—down the line. 1.** All the way : COMPLETELY. **2.** At a point or end in the future. **—in line for.** Next in order for <*in line for* a promotion> **—line up. 1.** To arrange in a line : ALIGN. **2.** To organize and make ready <*lined up* notes against the measure> **—on the line.** *Informal.* **1.** Ready or available for immediate payment. **2.** In jeopardy <put my future *on the line*> **—out of line.** Uncalled-for : improper.

☆ **syns: 1.** LINE, COLUMN, FILE, QUEUE, RANK, ROW, STRING *n. core meaning* : a group of people or things arranged in a straight configuration <a long *line* at the bank> **2.** LINE, POLICY, PROCEDURE, PROGRAM *n. core meaning* : an official or prescribed plan or course of action <refused to follow the party *line*>

**line**[2] (līn) *vt.* **lined, lin·ing, lines.** [ME *linen* < *line,* flax < OE *līn* < Lat. *linum.*] **1.** To fit a covering to the inside surface of. **2.** To cover the inner surface of <Insulation *lined* the house's walls.> **3.** To fill generously, as with money or food.

**lin·e·age**[1] (lĭn′ē-ĭj) *n.* [ME < OFr. *lignage* < *ligne,* line. —see LINE[1].] **1. a.** Direct descent from an ancestor : ANCESTRY. **b.** Derivation. **2.** The descendants of a common ancestor regarded as the founder of the line.

**line·age**[2] (lī′nĭj) *n. var. of* LINAGE.

**lin·e·al** (lĭn′ē-əl) *adj.* [ME < AN *lineale* < Med. Lat. *linealis* < LLat. < Lat. *linea,* line < *linum,* thread.] **1.** Belonging to or being in the direct line of descent from an ancestor. **2.** Derived from or pertaining to a specific line of descent. **3.** Linear. **—lin′e·al·ly** *adv.*

**lin·e·a·ment** (lĭn′ē-ə-mənt) *n.* [ME *liniament* < Lat. *lineamentum* < *linea,* line < *linum,* thread.] **1.** A distinctive shape, contour, or line, esp. of the face. **2.** *often* **lineaments.** A characteristic or definitive feature.

**lin·e·ar** (lĭn′ē-ər) *adj.* [Lat. *linearis* < *linea,* line < *linum,* thread.] **1.** Of, pertaining to, or resembling a line : STRAIGHT. **2. a.** In, of, describing, described by, or related to a straight line. **b.** Having but one dimension. **3.** Marked by, composed of, or emphasizing drawn lines rather than painterly effects. **4.** *Bot.* Narrow and elongated <a *linear* petal> **—lin′e·ar·ly** *adv.*

**Linear A** *n.* An undeciphered writing system used on Crete from the 18th to the 15th cent. B.C.

**linear accelerator** *n.* An electron, proton, or heavy-ion accelerator in which the paths of the accelerated particles are essentially straight lines rather than spirals or circles.

**linear algebra** *n.* **1.** A branch of mathematics dealing with the theory of systems of linear equations, matrices, vector spaces, determinants, and linear transformations. **2.** A mathematical ring and vector space with scalars from an associated field, the multiplication of which is of the form $(aA)\ (bB) = (ab)\ (AB)$, where *a* and *b* are scalars and *A* and *B* are vectors.

**linear al·kyl·ate sulfonate** (ăl′kə-lāt′) *n.* A biodegradable surfactant that is a salt of sulfonic acid, used in detergents.

**Linear B** *n.* A syllabic script in Mycenaean Greek documents of Crete and Pylos from the 14th to the 12th cent. B.C.

**linear combination** *n.* A mathematical expression of first order, composed of the sums and differences of elements with non-zero coefficients.

**linear dependence** *n.* The property of a mathematical set, with its coefficients taken from another, of having at least one linear combination equal to zero when at least one of the coefficients is not equal to zero.

**linear equation** *n.* An algebraic equation, as $x + y + 5 = 0$, in which the highest degree term in the variable or variables is of the first degree.

**linear independence** *n.* The property of a mathematical set, with its coefficients taken from another, of having no linear combinations equal to zero unless all of the coefficients are equal to zero.

**lin·e·ar·ize** (lĭn′ē-ə-rīz′) *vt.* **-ized, -iz·ing, -iz·es.** To project or put in linear form. **—lin′e·ar·i·za′tion** *n.*

**linear measure** *n.* **1.** Measurement of length. **2.** A unit or system of units for measuring length.

**linear momentum** *n.* MOMENTUM 1.

**linear perspective** *n.* A form of perspective in painting or drawing in which parallel lines converge so as to give the illusion of depth and distance.

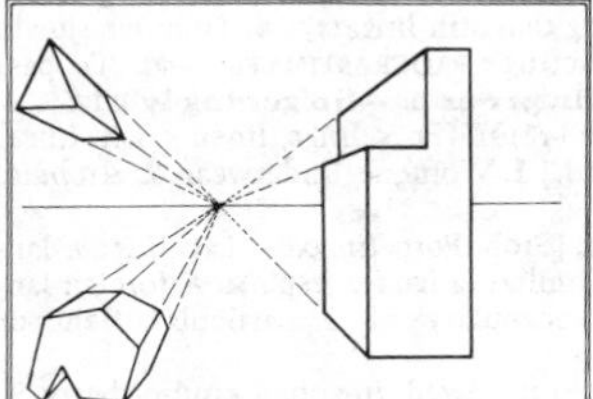

**linear perspective**

**lin·e·a·tion** (lĭn′ē-ā′shən) *n.* **1.** The act of marking or outlining with lines. **2.** An outline. **3.** An arrangement of lines.

**line·back·er** (līn′băk′ər) *n. Football.* One of the defensive players forming a second line of defense behind the ends and tackles. **—line′back′ing** *n.*

**line breeding** *n.* Selective breeding to perpetuate certain qualities or characteristics in a strain of livestock.

**line cut** *n.* A letterpress printing plate made from a line drawing by a photoengraving process.

**line drawing** *n.* A drawing made with lines only, esp. one used as copy for a line cut.

**line drive** *n. Baseball.* A hard-hit ball whose path approximates a straight line nearly parallel with the ground.

**line engraving** *n.* **1. a.** A metal plate, used in intaglio printing, on whose surface design lines have been hand engraved. **b.** The process of making such an engraving. **c.** A print made from such an engraving. **2.** A line cut.

**line·man** (līn′mən) *n.* **1.** One employed to install or repair telephone, telegraph, or electric power lines. **2.** *Football.* A player on the forward line.

**lin·en** (lĭn′ən) *n.* [ME < *linen,* of cloth < OE *līnen* < Lat. *linum,* thread.] **1. a.** Thread made from fibers of the flax plant. **b.** Cloth woven from this thread. **2.** Articles or garments made from linen or similar material. **3.** Paper made from flax fibers or given a linenlike luster. —*adj.* **1.** Made of linen or flax. **2.** Like linen.

**line of credit** *n.* CREDIT LINE 2.

**line of force** *n.* A theoretical line in a field of force, any tangent to which gives the direction of the field at the point of tangency.

**line of scrimmage** *n. Football.* An imaginary line across the field on which the ball rests and at which the teams line up for a new play.
**line of sight** *n.* **1.** An imaginary line from the eye to the object being looked at. **2.** An unobstructed path between electronic sending and receiving antennas.
**lin·e·o·late** (lĭn′ē-ə-lāt′) *adj.* [NLat. *lineolatus* < Lat. *lineola,* little line, dim. of *linea,* line < *linum,* thread.] Marked with fine lines.
**line printer** *n.* A high-speed printing device, used chiefly in data processing, that prints an entire line of type as a unit rather than printing each character individually.
**lin·er¹** (lī′nər) *n.* **1.** One that draws or makes lines. **2.** A commercial ship or aircraft, esp. one carrying passengers on a regular route. **3.** *Baseball.* A line drive.
**lin·er²** (lī′nər) *n.* **1.** One who makes or puts in linings. **2.** Something used as a lining.
**line score** *n. Baseball.* An inning-by-inning record of the runs, hits, and errors of a game.
**lines·man** (līnz′mən) *n.* **1. a.** *Football.* An official who marks the downs and the position of the ball and watches for certain violations from the sidelines. **b.** An official in various court games whose chief duty is to call shots that fall out of bounds. **2.** LINEMAN 1.
**line spectrum** *n.* A spectrum composed of a set of discrete, rather narrow lines.
**line squall** *n. Naut.* A squall occurring along a narrow band of thunderstorms.
**line storm** *n.* An equinoctial storm.
**line-up** *also* **line·up** (līn′ŭp′) *n.* **1.** A line of persons formed for inspection or identification. **2. a.** The players of a team chosen to start a game. **b.** A list of such players. **3.** A group of persons or things arrayed or enlisted for a specific purpose.
**ling¹** (lĭng) *n., pl.* **ling** *or* **lings.** [ME.] One of various marine food fishes related to or resembling the cod.
**ling²** (lĭng) *n.* [ME < ON *lyng.*] HEATHER 1.
**-ling¹** *suff.* [ME < OE.] **1.** One connected with <world*ling*> **2.** One having a specified quality <under*ling*> **3.** One that is young, small, or inferior <duck*ling*>
**-ling²** *suff.* [ME < OE.] In a given direction, manner, or condition <dark*ling*>
**lin·gam** (lĭng′gəm) *also* **lin·ga** (-gə) *n.* [Skt. *liṅgam,* penis.] A stylized phallus worshiped as a symbol of the Hindu god Shiva.
**ling·cod** (lĭng′kŏd′) *n., pl.* **lingcod** *or* **-cods.** A food fish, *Ophiodon elongatus* of northern Pacific waters.
**lin·ger** (lĭng′gər) *v.* **-gered, -ger·ing, -gers.** [ME *lengeren* < *lenger,* longer < OE *lengra.*] *—vi.* **1.** To delay in quitting or leaving something : TARRY. **2.** To remain very close to death for some time before dying. **3.** To persist <a feeling that still *lingers*> **4.** To move slowly : AMBLE. **5.** To be tardy in acting : PROCRASTINATE. *—vt.* To pass (time) in a leisurely way. **—lin′ger·er** *n.* **—lin′ger·ing·ly** *adv.*
**lin·ge·rie** (län′zhə-rā′, län′zhə-rē) *n.* [Fr. < *linge,* linen < Lat. *linea,* made of linen < *linum,* thread.] **1.** Women's underwear. **2.** *Archaic.* Linen articles, esp. garments.
**lin·go** (lĭng′gō) *n., pl.* **-goes.** [Prob. Port. *lingoa* < Lat. *lingua,* language.] Unintelligible or unfamiliar language, esp.: **a.** A foreign language. **b.** The specialized vocabulary of a particular field or discipline <computer *lingo*>
**lin·gon·ber·ry** (lĭng′ən-bĕr′ē) *n.* [Swed. *lingon,* a kind of berry + BERRY.] COWBERRY 2.
**lin·gua** (lĭng′gwə) *n., pl.* **-guae** (-gwē′) [Lat.] A tongue or tonguelike organ.
**lingua fran·ca** (frăng′kə) *n., pl.* **lingua fran·cas** (-kəz) *also* **lin·guae fran·cae** (lĭng′gwē frăng′kē) [Ital.] **1.** A mixture of Italian with French, Spanish, Arabic, Greek, and Turkish, spoken in the Mediterranean area, esp. in the Levant. **2.** A language used as a medium of communication between peoples who speak different languages. **3.** Something similar to a common language.
**lin·gual** (lĭng′gwəl) *adj.* **1.** Of, relating to, or like the tongue or a tonguelike organ. **2.** Pronounced with the tongue in conjunction with other organs of speech. **3.** Linguistic. *—n.* A sound articulated with the tongue in conjunction with other organs of speech, as the sounds (t), (l), and (n). **—lin′gual·ly** *adv.*
**lin·gui·ne** *also* **lin·gui·ni** (lĭng-gwē′nē) *n.* [Ital. < *lingua,* tongue < Lat.] *(sing. in number).* Pasta in long, flat, thin strands.
**lin·guist** (lĭng′gwĭst) *n.* [Lat. *lingua,* language + -IST.] **1.** A fluent speaker of several languages. **2.** A specialist in linguistics.
**lin·guis·tic** (lĭng-gwĭs′tĭk) *adj.* Of or relating to language or linguistics. **—lin·guis′ti·cal·ly** *adv.*
**linguistic atlas** *n.* A set of maps recording the geographic distribution of speech variations.
**linguistic form** *n.* A meaningful unit of speech, as an affix, word, phrase, or sentence.
**linguistic geography** *n.* Study of regional speech variations. **—linguistic geographer** *n.*
**lin·guis·tics** (lĭng-gwĭs′tĭks) *n. (sing. in number).* Study of the nature and structure of speech.
**lin·gu·late** (lĭng′gyə-lāt′) *adj.* [Lat. *lingulatus* < *lingula,* little tongue, dim. of *lingua.*] Shaped like a tongue.
**lin·i·ment** (lĭn′ə-mənt) *n.* [ME < LLat. *linimentum* < Lat. *linere,* to rub over.] A medicinal fluid applied to the skin as an anodyne or counterirritant.
**li·nin** (lī′nĭn) *n.* [Lat. *linum,* thread.] The filamentous, achromatic material in a cell nucleus that interconnects the chromatin granules.
**lin·ing** (lī′nĭng) *n.* **1. a.** An interior coating or covering. **b.** Material used for such coating or covering. **2.** Application of a lining.
**link¹** (lĭngk) *n.* [ME *linke,* of Scand. orig.] **1.** One of the rings or loops forming a chain. **2.** Something resembling a chain link in its physical arrangement or its connecting function, esp.: **a.** One of several sausages strung together. **b.** A unit in a transportation or communications system. **c.** A single connecting element. **3.** A cuff link. **4.** A unit of length used in surveying, equal to 0.01 chain, 7.92 inches, or approx. 20.12 centimeters. **5.** A lever or rod transmitting motion in a machine. **6.** *Computer Sci.* An identifying term attached to an element in a system to facilitate connection to other identified elements. *—vt. & vi.* **linked, link·ing, links.** To connect or become connected with or as if with links. **—link′er** *n.*
**link²** (lĭngk) *n.* [Poss. < Med. Lat. *linchinus,* candle < Lat. *lynchnus* < Gk. *lukhnos.*] A torch once used for lighting one's way in the streets.
**link·age** (lĭng′kĭj) *n.* **1.** An act of linking or the state of being linked. **2.** A system of interconnected machine parts, as rods, springs, and pivots, for transmitting power or motion. **3.** A measure of the induced voltage in a circuit caused by a magnetic flux and equal to the flux times the number of turns in the coil surrounding it. **4.** *Genetics.* A relationship between two or more nonallelic genes occupying the same chromosome that causes them to have closely associated inherited effects. **5.** A diplomatic negotiating strategy holding that progress on one issue is an essential element for progress on other issues <"We saw *linkage* . . . as synonymous with an overall strategic and geopolitical view" —Henry Kissinger>
**linked** (lĭngkt) *adj.* **1.** Connected by or as if by links. **2.** *Genetics.* Exhibiting linkage. **3.** *Computer Sci.* Provided with links.
**linking verb** *n.* COPULA 1.
**links** (lĭngks) *pl.n.* [ME < OE *hlincas,* pl. of *hlinc,* ridge.] **1.** A golf course. **2.** *Scot.* Sandy undulating ground usu. on a seashore.
**link·up** (lĭngk′ŭp′) *n.* **1.** An instance of meeting or contact, as of two spacecraft. **2. a.** Something serving to join or link. **b.** A functional unit derived from the linking up of separate elements.
**linn** (lĭn) *n.* [Sc. Gael. *linne.*] *Scot.* **1.** A waterfall. **2.** A steep ravine.
**Lin·nae·an** *also* **Lin·ne·an** (lĭ-nē′ən) *adj.* Of or relating to Linnaeus or his system of taxonomic classification and nomenclature.
**lin·net** (lĭn′ĭt) *n.* [OFr. *linette* < *lin,* flax < Lat. *linum.*] A small Old World songbird, *Acanthis cannabina,* with brownish plumage.
**lin·o·le·ic acid** (lĭn′ə-lē′ĭk) *n.* [Gk. *linon,* flax + OLEIC ACID.] A colorless to straw-colored liquid, $C_{18}H_{32}O_2$, an important component of drying oils and an essential fatty acid in the human diet.
**lin·o·len·ic acid** (lĭn′ə-lĕn′ĭk) *n.* [Alteration of LINOLEIC ACID.] A colorless liquid, $C_{18}H_{30}O_2$, an important component of natural drying oils and an essential fatty acid in the human diet.
**li·no·le·um** (lĭ-nō′lē-əm) *n.* [Orig. a trademark.] A durable material made in sheets by pressing a mixture of heated linseed oil, rosin, powdered cork, and pigments onto a burlap or canvas backing, used chiefly as a floor covering.
**Li·no·type** (lī′nə-tīp′). A trademark for a machine that sets type on a metal slug, operated by a keyboard.
**lin·sang** (lĭn′săng′) *n.* [Malay.] An Asian or African carnivorous mammal of the genera *Poiana* or *Prionodon,* with a spotted coat and a long banded tail.
**lin·seed** (lĭn′sēd′) *n.* [ME *linsed* < OE *līnsǣd* : *līn,* flax (< Lat. *linum*) + *sǣd,* seed.] The seed of flax, esp. when used as the source of linseed oil : FLAXSEED.
**linseed oil** *n.* A yellowish oil extracted from flaxseeds used as a drying oil in varnishes and paints and in printing inks, linoleum, and synthetic resins.
**lin·sey-wool·sey** (lĭn′zē-wo͝ol′zē) *n., pl.* **-seys.** [ME *linsiwolsie.*] A rough linen or cotton fabric woven with wool.
**lin·stock** (lĭn′stŏk′) *n.* [Du. *lontstok* : *lont,* match + *stok,* stick.] A long forked stick once used to hold a lighted match to fire a cannon.
**lint** (lĭnt) *n.* [ME < Med. Lat. *linteum* < Lat., linen cloth < *linum,* flax.] **1.** Clinging bits of fluff and fiber : FUZZ. **2.** Downy material obtained by scraping linen cloth and used for dressing wounds. **3.** The mass of soft fibers surrounding the seeds of unginned cotton.
**lin·tel** (lĭn′tl) *n.* [ME < OFr. < Lat. *limitaris,* of a threshold < *limes,* boundary.] The horizontal beam forming the upper member of a door or window frame and supporting part of the structure above it.
**lint·er** (lĭn′tər) *n.* **1. linters.** The short fibers clinging to cotton seeds after the first ginning. **2.** A machine for removing linters from cotton seeds.
**lint·white** (lĭnt′hwīt′, -wīt′) *n.* [ME *linkwhitte,* alteration of OE *līnetwige.*] A linnet.

**lin·u·ron** (lĭn'yə-rŏn') *n.* [Orig. unknown.] A herbicide, $C_9H_{10}O_2Cl_2N_2$, for selectively killing weeds.

**li·on** (lī'ən) *n.* [ME < OFr. < Lat. *leo* < Gk. *leōn.*] **1.** A large carnivorous feline mammal, *Panthera leo* of Africa and India, with a short tawny coat and a long heavy mane in the male. **2.** A large wildcat, esp. the cougar. **3.** A person felt to resemble a lion, as in ferocity or bravery. **4.** A person of extraordinary importance or prestige. **5. Lion.** LEO 2. **6. Lion.** A member of an international service club of business and professional people. **—lion's share.** The greatest part of a whole.

**li·on·ess** (lī'ə-nĭs) *n.* A female lion.

**li·on·heart·ed** (lī'ən-här'tĭd) *adj.* Extremely courageous.

**li·on·ize** (lī'ə-nīz') *vt.* **-ized, -iz·ing, -iz·es.** To treat (a person) as a celebrity. **—li'on·iz'er** *n.*

**lip** (lĭp) *n.* [ME < OE *lippa.*] **1.** *Anat.* Either of two fleshy, muscular folds that together surround the opening of the mouth. **2.** A structure or part encircling or bounding an orifice, esp.: **a.** *Anat.* A labium. **b.** The margin of flesh around a wound. **c.** Either of the margins of the aperture of a gastropod shell. **d.** The rim, as of a bell, vessel, or crater. **3.** *Bot.* One of the protruding divisions of an irregular corolla or calyx. **4.** The tip of a pouring spout. **5.** *Slang.* Back talk. —*vt.* **lipped, lip·ping, lips. 1. a.** To touch the lips to. **b.** To kiss. **2.** To utter. **3.** To splash or lap against. **4.** To hit a golf ball so that it stops just at the edge of (the hole). —*adj.* Uttered or formed with the help of the lips : LABIAL.

**lip–** *pref. var. of* LIPO-.

**lip·ase** (lĭp'ās', lī'pās') *n.* An enzyme that hydrolyzes fats to form fatty acids and glycerol.

**lip-gloss** (lĭp'glôs', -glŏs') *n.* A cosmetic giving shine to the lips.

**lip·id** (lĭp'ĭd, lī'pĭd) *also* **lip·ide** (lĭp'īd', lī'pīd') *n.* One of numerous fats and fatlike materials that are gen. insoluble in water but soluble in common organic solvents, are related to the fatty acid esters, and together with carbohydrates and proteins constitute the principal structural material of living cells. **—lip·id'ic** *adj.*

**lipo–** or **lip–** *pref.* [NLat. < Gk. *lipos,* fat.] Fat : fatty : fatty tissue <*lipolysis*>

**lip·oid** (lĭp'oid', lī'poid') *also* **li·poi·dal** (lĭ-poid'l, lī-) *adj.* Like fat : FATTY. **—lip'oid'** *n.*

**li·pol·y·sis** (lĭ-pŏl'ĭ-sĭs, lī-) *n.* Hydrolysis of fat.

**li·po·ma** (lĭ-pō'mə, lī-) *n., pl.* **-ma·ta** (-mə-tə) *or* **-mas.** A benign tumor of chiefly fatty cells. **—li·pom'a·tous** (-pŏm'ə-təs) *adj.*

**lip·o·pro·tein** (lĭp'ō-prō'tēn', -tē-ĭn, lī'pō-) *n.* A conjugated protein composed of a simple protein combined with a lipid group.

**lip·o·trop·ic** (lĭp'ō-trŏp'ĭk, -trō'pĭk, lī'pō-) *adj.* That prevents abnormal or excessive accumulation of fat in the liver. **—li·pot'ro·py** (lĭ-pŏt'rə-pē, lī-), **li·pot'ro·pism** *n.*

**lip-read** (lĭp'rēd') *v.* **-read** (-rĕd'), **-read·ing, -reads.** —*vt.* To interpret (another's utterance) by lip reading. —*vi.* To use lip reading.

**lip reading** *n.* A technique for understanding unheard speech by interpreting lip and facial movements. **—lip reader** *n.*

**lip service** *n.* Insincere respect or agreement.

**lip·stick** (lĭp'stĭk') *n.* A small stick of waxy or pastelike lip coloring usu. enclosed in a cylindrical case.

**lip-synch** (lĭp'sĭngk') *v.* **-synched, -synch·ing, -synchs.** [LIP + SYNCH(RONIZE).] —*vi.* To move the lips in synchronization with recorded sound. —*vt.* To move the lips in synchronization with.

**Lip·tau·er** (lĭp'tou'ər) *n.* [G., after *Liptau* (Liptow), Hungary.] **1.** A soft Hungarian cheese. **2. a.** A cheese spread made of Liptauer and seasonings. **b.** A cheese spread made with cream cheese or cottage cheese.

**li·quate** (lī'kwāt') *vt.* **-quat·ed, -quat·ing, -quates.** [Lat. *liquare, liquat-,* to melt.] To separate (the metals in an alloy) by melting some constituents while leaving others solid. **—li·qua'tion** *n.*

**liq·ue·fy** *also* **liq·ui·fy** (lĭk'wə-fī') *v.* **-fied, -fy·ing, -fies.** [OFr. *liquefier* < Lat. *liquefacere* : *liquēre,* to be liquid + *facere,* to make.] —*vt.* To cause to become liquid, esp.: **a.** To melt (a solid) by heating. **b.** To condense (a gas) by cooling. —*vi.* To become liquid. **—liq'ue·fac'tion** (-făk'shən) *n.* **—liq'ue·fi'er** *n.*

**li·ques·cent** (lĭ-kwĕs'ənt) *adj.* [Lat. *liquescens, liquescent-,* pr. part. of *liquescere,* to become liquid < *liquēre,* to be liquid.] Becoming or tending to become liquid : MELTING. **—li·ques'cence, li·ques'cen·cy** *n.*

**li·queur** (lĭ-kûr', -kyŏŏr') *n.* [Fr. < OFr. *liquor,* a liquid. —see LIQUOR.] A usu. sweet alcoholic beverage made with various flavorings : CORDIAL.

**liq·uid** (lĭk'wĭd) *n.* [< ME, of a liquid < OFr. *liquide* < Lat. *liquidus* < *liquēre,* to be liquid.] **1. a.** The state of matter in which a substance exhibits readiness to flow, little or no tendency to disperse, and relatively high incompressibility. **b.** Matter or a specific body of matter in this state. **2.** The sounds of *l* and *r,* which are nonfrictional and similar to vowels. —*adj.* **1.** Of or being a liquid. **2.** Liquefied, esp.: **a.** Melted by heating <*liquid* butter> **b.** Condensed by cooling <*liquid* nitrogen> **3.** Shining and clear <captivating *liquid* blue eyes> **4. a.** Smooth and clear <*liquid* verse> **b.** Articulated without friction and capable of being prolonged like a vowel, as the speech sounds *l* and *r.* **5.** Free-flowing. **6.** Easily converted into cash <*liquid* assets> **—li·quid'i·ty** (lĭ-kwĭd'ĭ-tē) *n.* **—liq'uid·ly** *adv.*

**liquid air** *n.* Air in the liquid state, condensed from the gas by cooling and occas. by pressure.

**liq·uid·am·bar** (lĭk'wĭ-dăm'bər) *n.* [NLat. *Liquidambar,* genus name : Lat. *liquidus,* liquid + Med. Lat. *ambar,* amber.] A tree of the genus *Liquidambar,* as the sweet gum.

**liq·ui·date** (lĭk'wĭ-dāt') *v.* **-dat·ed, -dat·ing, -dates.** [LLat. *liquidare, liquidat-,* to melt < Lat. *liquidus,* liquid.] —*vt.* **1. a.** To pay off or settle (e.g., a debt or claim). **b.** To settle the affairs of (e.g., a business) by ascertaining the liabilities and applying the assets to their discharge. **c.** To convert (assets) into cash. **2.** To terminate : abolish. **3.** To put to death : KILL. —*vi.* To enter into or be in the process of financial liquidation.

**liquid crystal** *n.* One of various liquids in which the atoms or molecules are regularly arrayed in either one dimension or two dimensions, the order giving rise to optical properties, as anisotropic scattering, associated with the crystals.

**liquid measure** *n.* **1.** A unit or system of units of liquid capacity. **2.** A measure for liquids.

**liq·ui·fy** (lĭk'wə-fī') *v. var. of* LIQUEFY.

**liq·uor** (lĭk'ər) *n.* [ME, a liquid < OFr. < Lat. < *liquēre,* to be liquid.] **1.** An alcoholic beverage made by distillation. **2.** A liquid substance, as broth, produced in cooking. **3.** An aqueous solution of a nonvolatile substance. **4.** An industrial suspension, solution, or emulsion. —*vt.* **-uored, -uor·ing, -uors. 1.** *Slang.* To cause to become drunk with alcoholic liquor <got all *liquored* up at the wedding> **2. a.** To treat (leather) with grease. **b.** To steep (e.g., malt).

**li·quo·rice** (lĭk'ər-ĭs, -ĭsh) *n. Chiefly Brit. var. of* LICORICE.

**li·ra** (lîr'ə, lē'rə) *n., pl.* **li·re** (lîr'ā, lē'rā) *or* **li·ras.** [Ital. < Lat. *libra,* a unit of weight.] —See table at CURRENCY.

**lir·i·pipe** (lîr'ə-pīp') *n.* [Med. Lat. *liripipium.*] A long cord or scarf attached to and hanging from a hood.

**li·sen·te** (lē-sĕn'tā) *n., pl.* **lisente.** [Sotho < E. CENT.] —See table at CURRENCY.

**lisle** (līl) *n.* [After *Lisle* (Lille), France.] **1.** A smooth, tightly twisted thread usu. spun from long-stapled cotton. **2.** Fabric knitted of lisle.

**lisp** (lĭsp) *n.* [ME *lispen* < OE *wlispian.*] **1.** A speech defect or mannerism typified by failure to produce normal sibilants, esp. by substitution of the sounds (th) and (*th*) for the sibilants (s) and (z). **2.** The sound of a lisp. —*v.* **lisped, lisp·ing, lisps.** —*vi.* **1.** To speak with a lisp. **2.** To speak falteringly, as a child might. —*vt.* To pronounce with a lisp. **—lisp'er** *n.*

**lis·some** *also* **lis·som** (lĭs'əm) *adj.* [Alteration of LITHESOME.] **1.** Easily flexed : SUPPLE. **2.** Capable of moving with ease : NIMBLE. **—lis'some·ly** *adv.* **—lis'some·ness** *n.*

**list[1]** (lĭst) *n.* [Fr. *liste* < OItal. *lista,* of Germanic orig.] An item-by-item series of words or numbers, as the names of persons or things, written or printed one after the other <a guest *list*><a shopping *list*> —*v.* **list·ed, list·ing, lists.** —*vt.* **1.** To make a list of : ITEMIZE. **2.** To put on a list : REGISTER. **3.** *Archaic.* To recruit. **4.** To put (oneself) in a specific category <*list* oneself as a writer> —*vi.* **1.** *Archaic.* To enlist in the armed forces. **2.** To have a specified list price <a watch that *lists* for $12 over wholesale>

**list[2]** (lĭst) *n.* [ME < OE *līste.*] **1. a.** A narrow strip, esp. of wood. **b.** A listel. **c.** A selvage or border of cloth. **2.** A band or stripe of color. **3.** *Obs.* A boundary : border. **4. lists. a.** An arena for tournaments, esp. jousting. **b.** A place of combat. **c.** An area of controversy. **5.** A ridge thrown up between two furrows by a lister in plowing. —*vt.* **list·ed, list·ing, lists. 1.** To line, cover, or edge with list. **2.** To cut a thin strip from the edge of. **3.** To furrow or plant (land) with a lister.

**list[3]** (lĭst) *n.* [Orig. unknown.] An inclination to one side, as of a ship : TILT. —*vi.* & *vt.* **list·ed, list·ing, lists.** To lean or cause to lean to the side : HEEL.

**list[4]** (lĭst) *vi.* & *vt.* **list·ed, list·ing, lists.** [ME *listen* < OE *hlystan.*] *Archaic.* To listen or listen to.

**list[5]** (lĭst) [ME *listen,* to desire < OE *lystan.*] *Archaic.* —*v.* **list·ed, list·ing, lists.** —*vt.* To be pleasing to : SATISFY. —*vi.* To be disposed : CHOOSE. —*n.* A desire or inclination.

**lis·tel** (lĭs'təl) *n.* [OFr. < OItal. *listello,* dim. of *lista,* border, of Germanic orig.] A narrow architectural border : FILLET.

**lis·ten** (lĭs'ən) *vi.* **-tened, -ten·ing, -tens.** [ME *listenen* < *listen,* to listen < *hlystan.*] **1.** To try to hear something. **2.** To pay attention : HEED. **—listen in. 1.** To tune in and listen to a broadcast. **2.** To listen to a conversation between others : EAVESDROP. **—lis'ten·er** *n.*

**list·er** (lĭs'tər) *n.* [Ult. < LIST[2].] A plow equipped with a double moldboard that turns up the soil on each side of the furrow, often with an attached drill for seed planting.

**list·ing** (lĭs'tĭng) *n.* **1.** An act or instance of making or entering in a list. **2.** An entry in a directory or list. **3.** A list.

**list·less** (lĭst'lĭs) *adj.* [ME *listles,* prob. < *list,* ability < OE.] Lacking energy or an inclination toward any effort : LETHARGIC. **—list'less·ly** *adv.* **—list'less·ness** *n.*

**list price** *n.* A basic advertised or published price, often subject to discount.

**lit[1]** (lĭt) *v. var. p.t. & p.p. of* LIGHT[1].

**lit[2]** (lĭt) *v. var. p.t. & p.p. of* LIGHT[2].

**lit·a·ny** (lĭt'n-ē) *n., pl.* **-nies.** [ME *letanie* < OFr. < Med. Lat. *letania* < LLat. *litania* < LGk. *litaneia* < Gk., entreaty < *litanuein,* to entreat < *litanos,* entreating < *litē,* supplication.] **1.** A liturgical prayer composed of phrases recited by a leader alternating with responses by a congregation. **2.** A repetitive or incantatory recital <a *litany* of compliments>

**li·tchi** *also* **li·chee** *or* **ly·chee** (lē'chē) *n.* [Chin. (Mandarin) *li[4] zhi[1].*] **1.** A Chinese tree, *Litchi chinensis,* bearing edible fruit. **2.** *also* **litchi nut.** The fruit of the litchi tree.

**-lite** *suff.* [Fr., alteration of *-lithe* < Gk. *lithos,* stone.] Stone : mineral : fossil <copro*lite*>

**li·ter** (lē'tər) *n.* [Fr. *litre* < obs. *litron,* measure of capacity < Med. Lat. *litra* < Gk., unit of weight.] A metric unit of volume equal to a cubic decimeter, approx. 1.056 liquid quarts or 0.908 dry quart.

**lit·er·al** (lĭt'ər-əl) *adj.* [ME < OFr. < LLat. *litteralis,* of letters < Lat. *littera,* letter.] **1.** In accordance with, conforming to, or upholding the primary or exact meaning of a word or words. **2.** Word for word. **3.** Concerned primarily with facts : PROSAIC <a *literal* person> **4.** Avoiding metaphor, exaggeration, or embellishment : PLAIN <a *literal* report> **5.** Consisting of, using, or expressed by letters <*literal* notation> *—n. Computer Sci.* A symbol or letter in data that represents itself. **—lit'er·al·ness** *n.*

☆ ***syns:*** LITERAL, VERBATIM *adj. core meaning* : word for word <a *literal* translation>

**lit·er·al·ism** (lĭt'ər-ə-lĭz'əm) *n.* **1.** Adherence to the explicit sense of a given doctrine or text. **2.** Literal portrayal : REALISM. **—lit'er·al·ist** *n.* **—lit'er·al·is'tic** *adj.*

**lit·er·al·ize** (lĭt'ər-ə-līz') *vt.* **-ized, -iz·ing, -iz·es.** To make literal.

**lit·er·al·ly** (lĭt'ər-ə-lē) *adv.* **1.** In a strict or literal sense. **2.** Really : actually <"There are people in the world who *literally* do not know how to boil water" —Craig Claiborne>

**lit·er·ar·y** (lĭt'ə-rĕr'ē) *adj.* [Lat. *litterarius* < *littera,* letter.] **1.** Of, pertaining to, or dealing with literature. **2.** Suited to literature rather than everyday writing or speech. **3.** Versed in or fond of literature or learning. **4.** Of or pertaining to writers or the profession of literature <*literary* cliques> **5.** Pedantic : bookish. **—lit'er·ar'i·ly** (-râr'ə-lē) *adv.* **—lit'er·ar'i·ness** *n.*

**lit·er·ate** (lĭt'ər-ĭt) *adj.* [ME *litterate* < Lat. *litteratus* < *littera,* letter.] **1.** Having the ability to read and write. **2.** Cultured : educated. **3.** Familiar with literature : WELL-READ. **4.** Well-written : polished <a *literate* analysis> *—n.* **1.** One who can read and write. **2.** An educated person. **—lit'er·a·cy** (-ə-sē) *n.* **—lit'er·ate·ly** *adv.* **—lit'er·ate·ness** *n.*

**lit·er·a·ti** (lĭt'ə-rä'tē) *pl.n.* [Ital. < Lat. *litteratus,* literate < *littera,* letter.] The literary intelligentsia.

**lit·er·a·tim** (lĭt'ə-rā'tĭm, -rä'-) *adv.* [Med. Lat. < Lat. *littera,* letter.] Literally.

**lit·er·a·ture** (lĭt'ər-ə-chŏor', -chər) *n.* [ME, book learning < OFr. < LLat. *litteratura* < *litterae,* literature < *littera,* letter.] **1.** A body of writings in prose or verse. **2.** Imaginative or creative writing, esp. of recognized artistic value : BELLES-LETTRES. **3.** The art or occupation of a literary writer. **4.** The body of written work produced by scholars in a given field <scientific *literature*> **5.** Printed material, as for a political or sales campaign. **6.** The aggregate of musical compositions, esp. for a specific instrument or ensemble.

**literature search** *n.* A systematic search for and investigation of published material relating to a given subject.

**lith-** *pref. var. of* LITHO-.

**-lith** *suff.* [< Gk. *lithos,* stone.] **1.** Rock : stone <xeno*lith*> **2.** Stone implement or structure <mega*lith*> **3.** Mineral concretion : calculus <cysto*lith*>

**lith·arge** (lĭth'ärj', lĭ-thärj') *n.* [ME *litarge* < OFr. < Lat. *lithargyrus* < Gk. *litharguros* : *lithos,* stone + *arguros,* silver.] A yellow lead oxide, PbO, used as a pigment and in glass and storage batteries.

**lithe** (līth) *adj.* [ME < OE *līðe.*] **1.** Readily flexed : SUPPLE. **2.** Effortlessly graceful <a *lithe* dancer> **—lithe'ly** *adv.* **—lithe'ness** *n.*

**lithe·some** (līth'səm) *adj.* LITHE 2.

**lith·i·a** (lĭth'ē-ə) *n.* [NLat. < Gk. *lithos,* stone.] Lithium oxide.

**li·thi·a·sis** (lĭ-thī'ə-sĭs) *n., pl.* **-ses** (-sēz'). *Pathol.* Formation of calculi in the body.

**lithia water** *n.* Mineral water containing some lithium salts.

**lith·ic** (lĭth'ĭk) *adj.* **1.** Of or relating to stone. **2.** Of or relating to lithium.

**-lithic** *suff.* [< LITHIC.] Pertaining to or typical of a given stage in the use of stone by human beings <Eo*lithic*>

**lith·i·um** (lĭth'ē-əm) *n. Symbol* **Li** A soft, silvery, highly reactive metallic element, used as a heat transfer medium, in thermonuclear weapons, and in alloys; atomic number 3; atomic weight 6.939.

**lithium oxide** *n.* A strongly alkaline white powder, $Li_2O$, used in glass and ceramics.

**litho-** *or* **lith-** *pref.* [Lat. < Gk. < *lithos,* stone.] **1.** Stone <*litho*sphere> **2.** Lithium <*lithic*>

**lith·o·graph** (lĭth'ə-grăf') *n.* [Back-formation < LITHOGRAPHY.] A print produced by lithography. *—vt.* **-graphed, -graph·ing, -graphs.** To produce by lithography. **—li·thog'raph·er** (lĭ-thŏg'rə-fər) *n.* **—lith'o·graph'ic, lith'o·graph'i·cal** *adj.* **—lith'o·graph'i·cal·ly** *adv.*

**li·thog·ra·phy** (lĭ-thŏg'rə-fē) *n.* [G. *Lithographie* : *litho-,* litho- + *-graphie,* -graphy.] A printing process in which the image configuration to be printed is rendered on a flat surface, as on stone or now chiefly on sheet zinc or aluminum, and treated so that only those areas to be printed will retain ink.

**li·thol·o·gy** (lĭ-thŏl'ə-jē) *n.* **1.** Gross physical character of a rock. **2.** Microscopic study, description, and classification of rocks. **—lith'o·log'ic** (lĭth'ə-lŏj'ĭk), **lith'o·log'i·cal** *adj.* **—lith'o·log'i·cal·ly** *adv.* **—li·thol'o·gist** *n.*

**lith·o·phyte** (lĭth'ə-fīt') *n.* **1.** A plant growing on a rocky surface. **2.** An organism, as coral, with a stony structure. **—lith'o·phyt'ic** (-fĭt'ĭk) *adj.*

**lith·o·pone** (lĭth'ə-pōn') *n.* [LITHO- + Gk. *ponos,* product.] A white pigment consisting of a mixture of zinc oxide, zinc sulfide, and barium sulfate.

**lith·o·sphere** (lĭth'ə-sfîr') *n.* **1.** The solid part of the earth. **2.** The earth's rocky crust.

**lith·o·stra·tig·ra·phy** (lĭth'ō-strə-tĭg'rə-fē) *n.* **1.** Stratigraphy based on the physical and petrographic properties of rocks. **2.** Interpretation of the physical characters of sedimentary rocks. **—lith'o·strat'i·graph'ic** (-străt'ĭ-grăf'ĭk) *adj.*

**li·thot·o·my** (lĭ-thŏt'ə-mē) *n., pl.* **-mies.** [LLat. *lithotomia* < Gk. : *lithos,* stone + *temnien,* to cut.] Surgery to remove calculi.

**li·thot·ri·ty** (lĭ-thŏt'rĭ-tē) *n., pl.* **-ties.** [LITHO- + alteration of Gk. *thruptein,* to crush.] Surgical pulverization of calculi in the urethra or bladder.

**Lith·u·a·ni·an** (lĭth'ōō-ā'nē-ən) *n.* **1.** A native or inhabitant of Lithuania. **2.** The Baltic language of the Lithuanians. **—Lith'u·a'ni·an** *adj.*

**lit·i·gant** (lĭt'ĭ-gənt) *n.* [Lat. *litigans, litigant-,* a disputant < pr.part. of *litigare,* to litigate. —see LITIGATE.] A party engaged in a lawsuit. *—adj.* Engaged in a lawsuit.

**lit·i·gate** (lĭt'ĭ-gāt') *vt. & vi.* **-gat·ed, -gat·ing, -gates.** [Lat. *litigare, litigat-* : *lis,* lawsuit + *agere,* to drive.] To subject to or engage in legal proceedings. **—lit'i·ga·ble** (-gə-bəl) *adj.* **—lit'i·ga'tion** *n.* **—lit'i·ga'tor** *n.*

**li·ti·gious** (lĭ-tĭj'əs) *adj.* **1.** Of, relating to, or marked by litigation. **2.** Tending to litigate. **—li·ti'gious·ly** *adv.* **—li·ti'gious·ness** *n.*

**lit·mus** (lĭt'məs) *n.* [Of Scand. orig.] A blue, amorphous powder derived from certain lichens that changes to red with increasing acidity and to blue with increasing alkalinity.

**litmus paper** *n.* An unsized white paper colored with litmus and used as an acid-base indicator.

**litmus test** *n.* **1.** A test for chemical acidity using litmus paper, which turns red in acid solutions and blue in alkaline solutions. **2.** A test that uses a single indicator to prompt a decision <"The word 'hopefully' has become the *litmus test* to determine whether one is a language snob or a language slob" —William Safire>

**li·to·tes** (lī'tə-tēz', lĭt'ə-) *n., pl.* **litotes.** [Gk. *litotēs* < *litos,* plain.] A figure of speech expressing an affirmative by the negation of its opposite, as in *no small feat.*

**li·tre** (lē'tər) *n. Chiefly Brit. var. of* LITER.

**lit·ter** (lĭt'ər) *n.* [ME < AN *littere* < Med. Lat. *litera* < Lat. *lectus,* bed.] **1.** A conveyance consisting typically of an enclosed couch mounted on shafts, used to carry a single passenger. **2.** A stretcher for the wounded or sick. **3.** Material, as straw, used as bedding for animals. **4.** The young produced at one birth by a multiparous mammal. **5.** A disorderly accumulation of objects, esp. carelessly discarded trash. **6.** The uppermost layer of the forest floor consisting mainly of decaying organic matter. *—v.* **-tered, -ter·ing, -ters.** *—vt.* **1.** To give birth to (a litter). **2.** To make untidy by discarding trash carelessly. **3.** To scatter about. **4.** To supply (animals) with bedding litter. *—vi.* **1.** To give birth to a litter. **2.** To scatter litter. **—lit'ter·er** *n.*

**lit·té·ra·teur** *also* **lit·ter·a·teur** (lĭt'ər-ə-tûr', lĭt'rə-) *n.* [Fr. < Lat. *litterator,* critic < *littera,* letter.] A man of letters.

**lit·ter·bag** (lĭt'ər-băg') *n.* A bag used, as in a motor vehicle, for temporary disposal of trash.

**lit·ter·bug** (lĭt'ər-bŭg') *n. Informal.* One who litters public areas with trash.

**lit·tle** (lĭt'l) *adj.* **lit·tler** *or* **less** (lĕs), **lit·tlest** *or* **least** (lēst) [ME < OE *lȳtel.*] **1.** Small in size <a thin *little* person> **2.** Short in duration or extent : BRIEF <*little* time> **3.** Small in degree or quantity <*little* funds> **4.** Not important : TRIVIAL. **5.** Narrow : petty <snide *little* cracks> **6.** Lacking power or influence <the *little* countries of the world> **7.** Being at an early stage of growth : YOUNG <a *little* plant> **8.** Captivating : endearing <a cute *little* thing> *—adv.* **less, least. 1.** Not much : SCARCELY <I sleep *little.*> **2.** Not at all : not in the least <They *little* expected such snow.> *—n.* **1.** A small amount or quantity <There was a *little* left.> **2.** Something much less than all <I know *little* of the building's history.> **3.** A short time or distance <a *little* down the path><a *little* past noon> **—little by little.** By small degrees or increments : GRADUALLY. **—lit'tle·ness** *n.*

☆ ***syns:*** **1.** LITTLE, BANTAM, PETITE, SMALL *adj. core meaning* :

notably below average in amount, size, or scope <a *little* car> **ant:** big **2.** LITTLE, INCONSEQUENTIAL, INSIGNIFICANT, TRIVIAL, UNIMPORTANT *adj. core meaning* **:** not of great importance <wasn't concerned with *little* things>

**little auk** *n.* The dovekie.

**Little Bear** *n.* Ursa Minor.

**Little Dipper** *n.* The seven brightest stars in Ursa Minor.

**lit·tle·neck** (lĭt'l-nĕk') *n.* [After *Littleneck* Bay, New York.] The quahog clam when small and suitable for eating raw.

**little owl** *n.* A small Old World owl, *Athene noctua,* with streaked brownish plumage.

**little slam** *n.* The winning of all but one of the tricks in one hand of bridge.

**little theater** *n.* A small theater usu. for a collegiate, experimental, or community drama group.

**little toe** *n.* The outermost and smallest toe of the human foot.

**lit·to·ral** (lĭt'ər-əl) *adj.* [Lat. *litoralis* < *litus,* shore.] Of, relating to, or existing on a shore. —*n.* A shore or coastal region.

**li·tur·gi·cal** (lĭ-tûr'jĭ-kəl) *also* **li·tur·gic** (-tûr'jĭk) *adj.* **1.** Of, pertaining to, or characteristic of liturgy. **2.** Used in or using liturgy. **—li·tur'gi·cal·ly** *adv.*

**li·tur·gics** (lĭ-tûr'jĭks) *n. (sing. in number).* Study of liturgies **:** LITURGIOLOGY.

**li·tur·gi·ol·o·gy** (lĭ-tûr'jē-ŏl'ə-jē) *n.* Liturgics.

**li·tur·gi·ol·o·gist** (lĭ-tûr'jē-ŏl'ə-jĭst) *n.* LITURGIST 2.

**lit·ur·gist** (lĭt'ər-jĭst) *n.* **1.** One using or advocating the use of liturgical forms. **2.** A scholar in liturgics.

**lit·ur·gy** (lĭt'ər-jē) *n., pl.* **-gies.** [LLat. *liturgia* < Gk. *leitourgia,* public service < *leitourgos,* public servant : *leos,* people + *ergon,* work.] **1.** The rite of the Eucharist. **2.** The prescribed form for a public religious service.

**liv·a·ble** *also* **live·a·ble** (lĭv'ə-bəl) *adj.* **1.** Fit to live in **:** HABITABLE. **2.** Endurable **:** bearable. **—liv'a·ble·ness** *n.*

**live¹** (lĭv) *v.* **lived, liv·ing, lives.** [ME *liven* < OE *libban.*] —*vi.* **1.** To be alive **:** EXIST. **2.** To continue to be alive. **3.** To support oneself **:** SUBSIST <*lived* on income from investments> **4.** To reside **:** DWELL <*lives* in a log cabin> **5.** To conduct one's life in a given manner <*lived* extravagantly> **6.** To pursue a positive, satisfying existence. **7.** To remain in human memory <a hero who *lives* in the minds of us all> —*vt.* **1.** To spend or pass (one's life). **2.** To go through **:** EXPERIENCE. **3.** To embody in one's manner of existence <*lived* their religious beliefs> **—live down.** To reduce or overcome the shame of (e.g., a misdeed) over time. **—live in.** To reside in one's place of employment. **—live it up.** *Informal.* To engage in as much pleasure as possible. **—live out.** To reside outside one's place of employment. **—live up to. 1.** To live in accordance with <*lived up to* their standards> **2.** To measure up to <*live up to* expectations> **3.** To carry out **:** FULFILL <*lived up to* the bargain> **—live with.** To put up with <had to *live with* a bad situation>

**live²** (līv) *adj.* [Short for ALIVE.] **1.** Having life **:** LIVING. **2.** Being of current interest <a *live* subject> **3.** Burning **:** glowing <*live* embers> **4.** Not yet exploded but capable of being fired <a *live* land mine> **5.** Carrying an electric current or energized with electricity. **6.** Not quarried **:** NATIVE <*live* marble> **7.** Broadcast while being performed <a *live* television newscast> **8.** Not yet typeset <*live* copy> **9.** Being or capable of being in play <a *live* ball> —*adv.* At, during, or from a live production <a show originating *live* from Las Vegas>

**live·a·ble** (lĭv'ə-bəl) *adj. var. of* LIVABLE.

**live-bear·er** (līv'bâr'ər) *n.* An ovoviviparous fish, as a guppy. **—live'-bear'ing** *adj.*

**live-for·ev·er** (lĭv'fər-ĕv'ər) *n.* The orpine.

**live-in** (lĭv'ĭn') *adj.* **1.** Residing in the place where one is employed <a *live-in* housekeeper> **2.** Residing together with another, esp. in sexual intimacy <a *live-in* lover>

**live·li·hood** (līv'lē-hŏŏd') *n.* [ME *liflode* < OE *līflād* : *līf,* life + *lād,* course.] Means of support **:** SUBSISTENCE.

**live·long** (lĭv'lông', -lŏng') *adj.* [ME *lefe long* : *lefe,* dear (< OE *lēof*) + *long,* long.] Complete **:** entire <complained the *livelong* day>

**live·ly** (līv'lē) *adj.* **-li·er, -li·est.** [ME *lifli* < OE *līflīc* < *līf,* life.] **1.** Full of energy or activity **:** VIGOROUS <a *lively* child> **2.** Full of spirit **:** ANIMATED <a *lively* rhythm> **3.** Marked by animated intelligence **:** BRIGHT <a *lively* discussion> **4.** Sparkling **:** effervescent. **5.** Keen **:** brisk <*lively* trade between nations> **6.** Bouncing readily on impact **:** RESILIENT. —*adv.* In a vigorous, spirited, or energetic manner. **—live'li·ly** *adv.* **—live'li·ness** *n.*

**li·ven** (lī'vən) *vi. & vt.* **-vened, -ven·ing, -vens.** To become or cause to become lively or livelier.

**live oak** (līv) *n.* An evergreen American oak, as *Quercus virginiana* of the southeastern United States or *Q. agrifolia* of southwestern North America.

**liv·er¹** (lĭv'ər) *n.* [ME < OE *lifer.*] **1.** *Anat.* A large compound tubular vertebrate gland that secretes bile and acts in formation of blood and in metabolism of carbohydrates, fats, proteins, minerals, and vitamins. **2.** An organ in invertebrates similar to the liver. **3.** The liver of an animal used as food.

**liv·er²** (lĭv'ər) *n.* One who has a specified lifestyle <a high *liver*>

**liver extract** *n.* A dry, brownish powder containing the soluble thermolabile fraction of mammalian livers, capable of increasing the number of red blood corpuscles of persons afflicted with pernicious anemia.

**liver fluke** *n.* **1.** A parasitic trematode worm, as *Fasciola hepatica* or *Opisthorchis sinensis,* infesting the liver of various animals, including human beings. **2.** Infestation with liver flukes.

**liv·er·ied** (lĭv'ə-rēd, lĭv'rēd) *adj.* Wearing livery.

**liv·er·ish** (lĭv'ər-ĭsh) *adj.* **1.** Resembling liver, esp. in color. **2.** Having a liver disorder **:** BILIOUS. **3.** Having an irritable disposition **:** PEEVISH. **—liv'er·ish·ness** *n.*

**liv·er·leaf** (lĭv'ər-lēf') *n.* Hepatica.

**liver starch** *n.* Glycogen.

**liv·er·wort** (lĭv'ər-wûrt', -wôrt') *n.* **1.** One of numerous green nonflowering plants of the class Hepaticae within the division Bryophyta. **2.** Hepatica.

**liv·er·wurst** (lĭv'ər-wûrst', -wŏŏrst') *n.* [Partial trans. of G. *Leberwurst* : *Leber,* liver + *Wurst,* sausage.] A sausage of ground liver.

**liv·er·y** (lĭv'ə-rē, lĭv'rē) *n., pl.* **-ies.** [ME < OFr. *livree,* delivery < *livrer,* to deliver < Lat. *liberare,* to free < *liber,* free.] **1.** The costume or insignia worn by a feudal lord's retainers. **2.** A distinctive uniform worn by the male servants of a household. **3.** The distinctive dress worn by the members of a particular group. **4. a.** The care and boarding of horses for a fee. **b.** The renting out of horses and carriages. **c.** A livery stable. **5.** *Law.* The official delivery of property, esp. land, to a new owner.

**liv·er·y·man** (lĭv'ə-rē-mən, lĭv'rē-) *n.* A keeper or employee of a livery stable.

**livery stable** *n.* A stable that boards horses and has horses and carriages for hire.

**lives** (līvz) *n. pl. of* LIFE.

**live steam** *n.* Steam coming from a boiler at full pressure.

**live·stock** (līv'stŏk') *n.* Domestic animals, as cattle or horses, raised for home use or profit.

**live wire** (līv) *n.* **1.** A wire carrying electric current. **2.** *Informal.* An alert, vivacious, or dynamic person.

**liv·id** (lĭv'ĭd) *adj.* [Fr. *livide* < Lat. *lividus* < *livēre,* to be livid.] **1.** Discolored, as from a bruise **:** BLACK-AND-BLUE. **2.** Pallid or ashen. **3.** Very angry **:** FURIOUS. **—li·vid'i·ty** (lĭ-vĭd'ĭ-tē), **liv'id·ness** *n.* **—liv'id·ly** *adv.*

**liv·ing** (lĭv'ĭng) *adj.* **1.** Having life **:** ALIVE <famous *living* artists> **2.** Being in active use or function <a *living* language> **3.** Of or pertaining to persons who are alive. **4.** Suited for daily life <the *living* area> **5.** Full of interest and vitality <made math a *living* subject> **6.** True to life **:** REAL <the photograph's *living* colors> **7.** *Informal.* Definite **:** absolute <a *living* dream> —*n.* **1.** The action or condition of maintaining life <the rising cost of *living*> **2.** A manner or style of life <easy *living*> **3.** A means of maintaining life <made their *living* by farming> **4.** *Chiefly Brit.* A church benefice and the revenue attached to it.

☆ **syns:** LIVING, KEEP, LIVELIHOOD, MAINTENANCE, SUBSISTENCE, SUPPORT, SUSTENANCE *n. core meaning* **:** the means needed to support life <had to work for a *living*>

**living room** *n.* A room in a residence intended for general use.

**living wage** *n.* A wage sufficient to provide minimally satisfactory living conditions.

**living will** *n.* A will in which the signer, in the event of a terminal illness, requests to be allowed to die rather than be kept alive by medical life-support systems.

**li·vre** (lē'vər, lē'vrə) *n.* [Fr. < Lat. *libra,* a unit of weight.] A former French monetary unit orig. worth one pound of silver.

**lix·iv·i·ate** (lĭk-sĭv'ē-āt') *vt.* **-at·ed, -at·ing, -ates.** [< LLat. *lixivium,* lye < Lat. *lixivius,* of lye < *lixa,* lye.] To percolate or wash the soluble matter from. **—lix·iv'i·a'tion** *n.*

**liz·ard** (lĭz'ərd) *n.* [ME < OFr. *lisarde* < Lat. *lacerta.*] **1.** One of numerous reptiles of the suborder Sauria, typically with an elongated scaly body, four legs, and a tapering tail. **2.** Leather made from the skin of a lizard.

**lizard fish** *n.* Any of various bottom-dwelling fishes of the family Synodontidae, of warm seas, with a lizardlike head.

**lizard fish**
*12–20 inches long*

**-'ll.** Shall : will <We'*ll* arrive later.><I'*ll* deal with them!>

**lla·ma** (lä'mə) *n.* [Sp. < Quechua.] A South American ruminant mammal, *Lama peruana,* related to the camel, raised for its soft fleecy wool and used as a beast of burden.

**lla·no** (lä'nō, lăn'ō) *n., pl.* **-nos.** [Sp., plain < Lat. *planum* < *planus,* level.] A large, grassy, almost treeless plain, esp. in Latin America.

**lo** (lō) *interj.* [ME < OE *lā.*] —Used to show suprise or call attention to something or someone.

**loach** (lōch) *n.* [ME *loche* < OFr.] Any of various Old World freshwater fishes of the family Cobitidae, with barbels around the mouth.

**load** (lōd) *n.* [ME *lode* < OE *lād.*] **1. a.** A mass or weight lifted, carried, or supported. **b.** The overall force to which a structure is subjected in supporting a weight or mass or in resisting externally applied forces. **2. a.** Something carried, as by a vehicle, person, or animal <a *load* of cordwood> **b.** The quantity of such material <a truck with a full *load* of watermelons> **3. a.** The share of work allocated to or required of a machine, individual, group, or organization. **b.** The demand for performance or services made on a system or machine. **4.** The amount that can be loaded into a device at one time. **5.** A single charge of ammunition for a firearm. **6. a.** Mental stress regarded as a burden <took a *load* off my mind> **b.** A responsibility regarded as an oppressive weight. **7.** The external mechanical resistance against which a machine acts. **8. a.** Power output, as of a power plant. **b.** A device to which power is delivered. **9.** The sales charge added to the price of a share in a mutual fund. **10.** *often* **loads.** *Informal.* A great amount or number <*loads* of fun> —*v.* **load·ed, load·ing, loads.** —*vt.* **1. a.** To place (a load) in or on a structure, device, or conveyance <*loading* ore onto a train> **b.** To place or put in or on (e.g., a conveyance) <*load* a truck> **2.** To provide or fill nearly to overflowing <*loaded* the cabinet with supplies> **3.** To weigh down : BURDEN <*loaded* with guilt> **4.** To charge (a firearm) with ammunition. **5. a.** To insert into a device or apparatus <*loaded* film into the camera> **b.** To insert something, as film, into. **6.** To make (dice) heavier on one side by adding weight. **7.** To distort (e.g., a question) so as to elicit a desired response. **8.** To dilute or adulterate. **9.** To raise the power demand in (an electrical circuit), as by adding resistance. **10.** To increase (e.g., a mutual-fund share price) by adding expenses or sale costs. —*vi.* **1.** To receive a load. **2.** To charge a firearm with ammunition. **3.** To place or put a load in or on a structure, device, or conveyance. **—get a load of.** *Slang.* To take notice of <*Get a load of* that swimsuit.> **—load'er** *n.*

**load·ed** (lō'dĭd) *adj.* **1.** Intended to trap or trick <a *loaded* question> **2.** *Slang.* Drunk. **3.** *Slang.* Rich : wealthy.

**load·er** (lō'dər) *n. Computer Sci.* A program that transfers data from off-line memory by means of an input or storage device.

**load·ing** (lō'dĭng) *n.* **1.** A weight, stress, or burden. **2.** An additive : filler. **3.** An addition to an insurance premium to cover extra costs. **4.** *Elect.* Addition of inductance to a transmission line to improve its transmission characteristics.

**loading program** *n. Computer Sci.* A sequence of instructions that starts the processing of a program entered by means of an automatic input device.

**load line** *n.* A Plimsoll mark.

**load·star** (lōd'stär') *n. var. of* LODESTAR.

**load·stone** (lōd'stōn') *n. var. of* LODESTONE.

**loaf**[1] (lōf) *n., pl.* **loaves** (lōvz) [ME *lof* < OE *hlāf.*] **1.** A shaped mass of bread baked in one piece. **2.** A shaped mass of food <pork *loaf*>

**loaf**[2] (lōf) *vi.* **loafed, loaf·ing, loafs.** [Orig. unknown.] To spend time aimlessly : IDLE. **—loaf'er** *n.*

**Loaf·er** (lō'fər). A trademark for a low leather step-in shoe with an upper resembling a moccasin but with a broad, flat heel.

**loam** (lōm) *n.* [ME *lome,* clay < OE *lām.*] **1.** Soil consisting chiefly of sand, clay, silt, and organic matter. **2.** A mixture of moist clay and sand, together with straw, used esp. in making bricks and foundry molds. —*vt.* **loamed, loam·ing, loams.** To fill or cover with loam. **—loam'y** *adj.*

**loan** (lōn) *n.* [ME *lone* < ON *lān.*] **1. a.** A sum of money lent at interest. **b.** Something lent for temporary use. **2.** A temporary transfer to a duty or place away from a regular job. —*vt.* **loaned, loan·ing, loans.** To lend. ***usage:*** *Loan,* which has long been established as a verb, is equally as acceptable as *lend* in all contexts except in such figurative expressions as *lend an ear* and *distance lends enchantment.*

**loan shark** *n. Informal.* A usurer, esp. one who is financed and supported by gangsters. **—loan'shark'ing** *n.*

**loan translation** *n.* A form of borrowing from one language to another whereby the semantic components of a given term are literally translated into their equivalents in the borrowing language, as *superman* from German *Ubermensch.*

**loan-word** *also* **loan·word** (lōn'wûrd') *n.* A word adopted from another language and at least partly naturalized, as *hors d'oeuvre.*

**loath** *also* **loth** (lōth, lō*th*) *adj.* [ME *loth* < OE *lāð,* displeasing.] Reluctant or unwilling : DISINCLINED <was *loath* to go>

**loathe** (lō*th*) *vt.* **loathed, loath·ing, loathes.** [ME *lothen* < OE *lāthian.*] To regard with great dislike and hostility : ABHOR. **—loath'er** *n.* **—loath'ing·ly** *adv.*

**loath·some** (lō*th*'səm, lōth'-) *adj.* [ME *lothsome* < *loth* hate, and *loth,* hateful. —see LOATH.] Arousing loathing : DETESTABLE. **—loath'some·ly** *adv.* **—loath'some·ness** *n.*

**loaves** (lōvz) *n. pl. of* LOAF[1].

**lob** (lŏb) *v.* **lobbed, lob·bing, lobs.** [Prob. of LG orig.] —*vt.* To throw, hit, or propel in a high arc. —*vi.* **1.** To hit a ball in a high arc. **2.** To move clumsily or heavily. —*n.* **1.** A ball thrown, hit, or propelled in a high arc. **2.** *Chiefly Brit.* A clumsy dull person : LOUT. **—lob'ber** *n.*

**lo·bar** (lō'bər, -bär') *adj.* Of or relating to a lobe.

**lo·bate** (lō'bāt') *also* **lo·bat·ed** (-bā'tĭd) *adj.* **1.** Having lobes. **2.** Resembling a lobe. **—lo'bate'ly** *adv.*

**lo·ba·tion** (lō-bā'shən) *n.* **1.** The state of being lobed. **2.** A lobe or part resembling a lobe.

**lob·by** (lŏb'ē) *n., pl.* **-bies.** [Med.Lat. *lobium,* monastic cloister, of Germanic orig.] **1.** A foyer, hall, or waiting room at or near the entrance to a building as a hotel or theater. **2.** A public room next to the assembly chamber of a legislative body. **3.** A group of persons trying to influence legislators, esp. in favor of a special interest. —*v.* **-bied, -by·ing, -bies.** —*vi.* To try to influence legislators, esp. in favor of a special interest. —*vt.* **1.** To try to influence legislators to pass (legislation). **2.** To try to influence (an official) to take a desired action. **—lob'by·er** *n.* **—lob'by·ism** *n.* **—lob'by·ist** *n.*

**lobe** (lōb) *n.* [OFr. < LLat. *lobus* < Gk. *lobos.*] **1.** A rounded projection, esp. a rounded, projecting anatomical part, as the fatty lobule of the auricle of the ear. **2.** A subdivision of an organ or part bounded by fissures, connective tissue, or other structural boundaries.

**lo·bec·to·my** (lō-bĕk'tə-mē) *n., pl.* **-mies.** Surgical excision of a lobe.

**lobed** (lōbd) *adj.* Having lobes <*lobed* leaves>

**lobe-fin** (lōb'fĭn') *n.* One of various mostly extinct bony fishes of the subclass Sarcopterygii, with the coelacanth being a living example. **—lobe'finned'** *adj.*

**lo·be·li·a** (lō-bē'lē-ə, -bēl'yə) *n.* [NLat. *Lobelia,* genus name, after Matthias de Lobel (1538–1616).] A plant of the genus *Lobelia,* with variously colored terminal flower clusters.

**†lob·lol·ly** (lŏb'lŏl'ē) *n., pl.* **-lies.** [Perh. dial. *lob,* to bubble + *lolly,* broth.] *Regional.* **1.** A mudhole : MIRE. **2.** A lout.

**loblolly pine** *n.* A pine, *Pinus taeda* of the southeastern United States, with strong wood used for lumber.

**†lo·bo** (lō'bō) *n., pl.* **-bos.** [Sp., wolf < Lat. *lupus.*] *Western U.S.* The gray or timber wolf, *Canis lupus.*

**lo·bot·o·my** (lō-bŏt'ə-mē) *n., pl.* **-mies. 1.** Surgical division of one or more cerebral nerve tracts. **2.** Surgical incision into a lobe.

**lob·scouse** (lŏb'skous') *n.* [Perh. dial. *lob,* to bubble + *scouse,* broth.] A sailor's stew made of vegetables, meat, and hardtack.

**lob·ster** (lŏb'stər) *n.* [ME *lobstere* < OE *loppestre,* perh. < Lat. *locusta.*] **1.** A relatively large marine crustacean of the genus *Homarus,* with five pairs of legs, the first pair modified into large claws. **2.** A crustacean, as the spiny lobster, related to the true lobster. **3.** The flesh of a lobster used as food.

**lobster pot** *n.* A slatted cage with an opening covered by a funnel-shaped net used to trap lobsters.

**lobster ther·mi·dor** (thûr'mĭ-dôr') *n.* [After *Thermidor,* the 11th month of the calendar used during the French Revolution.] A dish of cooked lobster meat mixed with a cream sauce, put into a lobster shell, sprinkled with cheese, and browned.

**lob·u·late** (lŏb'yə-lāt') *also* **lob·u·lat·ed** (-lā'tĭd) *adj.* Having or consisting of lobules. **—lob'u·la'tion** *n.*

**lob·ule** (lŏb'yōōl) *n.* [NLat. *lobulus,* dim. of LLat. *lobus,* lobe.] **1.** A small lobe. **2.** A subdivision or section of a lobe. **—lob'u·lar** (-yə-lər), **lob'u·lose'** (-yə-lōs') *adj.* **—lob'u·lar·ly** *adv.*

**lob·worm** (lŏb'wûrm') *n.* [Dial. *lob,* lout + WORM.] A lugworm.

**lo·cal** (lō'kəl) *adj.* [ME < LLat. *localis* < Lat. *locus,* place.] **1.** Of or relating to a specific place. **2.** Relating to, existing in, or serving a specific locality <*local* law enforcement> **3. a.** Not broad or general in scope. **b.** Narrow-minded. **4.** Affecting a limited part of the body <a *local* inflammation> **5.** Making all possible or scheduled stops on a route <a *local* bus> —*n.* **1.** A local public conveyance. **2.** A local branch or chapter of an organization, esp. of a labor union. **3.** A person from a specific locality. **—lo'cal·ly** *adv.*

**local anesthetic** *n.* An anesthetic that acts on and around the point where it is infused or applied.

**local color** *n.* The flavor or interest of a locality and its residents imparted by the sights and customs peculiar to it.

**lo·cale** (lō-kăl') *n.* [Fr. *local* < *local,* local < OFr. < LLat. *localis.* —see LOCAL.] **1.** A locality, esp. with reference to a particular characteristic or event. **2.** The setting or scene, as of a novel.

**lo·cal·ism** (lō'kə-lĭz'əm) *n.* **1. a.** A local idiom. **b.** A local custom or mannerism. **2.** Partiality to local customs and interests.

**lo·cal·i·ty** (lō-kăl'ĭ-tē) *n., pl.* **-ties.** [Fr. *localité* < LLat. *localitas* < *localis,* local. —see LOCAL.] **1.** A specific place, neighborhood, or district. **2.** The quality or fact of having position in space.

**lo·cal·ize** (lō'kə-līz') *v.* **-ized, -iz·ing, -iz·es.** —*vt.* **1.** To make local. **2.** To restrict or confine to a specific locality. **3.** To attribute to a

ă **pat** ā **pay** âr **care** ä **father** ĕ **pet** ē **be** hw **which** ĭ **pit** ī **tie** îr **pier** ŏ **pot** ō **toe** ô **paw, for** oi **noise** ŏŏ **took**

specific locality. —*vi.* To become local, esp. to become fixed in one part or area. **—lo'cal·i·za'tion** *n.*

**local option** *n.* Option granted usu. by a state government to a local government on such issues as the sale of alcoholic beverages.

**lo·cate** (lō'kāt', lō-kāt') *v.* **-cat·ed, -cat·ing, -cates.** [Lat. *locare, locat-*, to place < *locus,* place.] —*vt.* **1.** To determine or specify the position and boundaries of <*locate* Chicago on the map> **2.** To find by search, experimentation, or examination <*locate* the source of trouble> **3.** To situate, station, or store <*locate* an agent in Buffalo> —*vi.* To become established : SETTLE. **—lo'cat'a·ble** *adj.* **—lo'cat'er** *n.*

**lo·ca·tion** (lō-kā'shən) *n.* [Lat. *locatio,* a placing < *locare,* to place. —see LOCATE.] **1.** The act or process of locating. **2.** A place where something is or might be located : SITE. **3.** A site away from a motion-picture or television studio at which a scene is shot. **4.** A surveyed and marked off tract of land. **—lo·ca'tion·al** *adj.*

**loc·a·tive** (lŏk'ə-tĭv) *adj.* [Lat. *locus,* place + (VOC)ATIVE.] Being a noun case in certain Indo-European languages, as Sanskrit, denoting place or the place where. —*n.* **1.** The locative case. **2.** A term in the locative case.

**lo·ca·tor** (lō'kā'tər) *n.* One that locates, as a person who fixes the boundaries of a mining claim.

**loch** (lŏKH, lŏk) *n.* [ME *louch* < Sc. Gael. *loch.*] *Scot.* **1.** A lake. **2.** A sea arm similar to a fjord.

**lo·chi·a** (lō'kē-ə, lŏk'ē-ə) *pl.n.* [Gk. *lokhia* < *lokhios,* of childbirth < *lokhos,* childbirth.] Normal discharge of blood, tissue, and mucus from the vagina after childbirth. **—lo'chi·al** *adj.*

**lo·ci** (lō'sī', -kī') *n. pl. of* LOCUS.

**lock**[1] (lŏk) *n.* [ME < OE *loc,* bolt, bar, and *loca,* enclosure.] **1.** A device used, as on a door, to close, hold, or secure and operated by various means, as a key or combination. **2.** A section of a waterway, closed off with gates, used to raise or lower vessels from level to level. **3.** A mechanism in a firearm for exploding the charge. **4.** An interlocking or entanglement of elements or parts. **5.** A wrestling hold secured on a part of an opponent's body. —*v.* **locked, lock·ing, locks.** —*vt.* **1. a.** To secure with a lock, as against entry <*lock* the door> **b.** To make secure by or as if by locking <*locked* up the store> **2. a.** To safeguard or confine by or as if by means of a lock <*locking* the cat in for the night> **b.** To put and keep in a specific state or situation <*locked* into a boring job> **3.** To interlock and engage securely so as to be immobile. **4.** To link or clasp firmly : INTERTWINE. **5.** To contend in battle or struggle <*locked* in combat> **6. a.** To equip (a waterway) with locks. **b.** To pass (a vessel) through a lock. **7. a.** To secure (letterpress type) in a chase or press bed by tightening the quoins. **b.** To fasten (a curved plate) to the cylinder of a rotary press. **8.** To invest (funds) in such a way that they are not easily convertible to cash. **9.** *Computer Sci.* To end the processing of (a magnetic tape or disk) in such a way as to deny access to its contents. —*vi.* **1.** To become fastened by or as if by a lock. **2.** To become entangled : INTERLOCK. **3.** To become rigid or fixed. **4.** To flow or pass through a waterway lock. **—lock horns.** To become involved in conflict. **—lock out.** To withhold work from (employees) during a lockout. **—lock, stock, and barrel.** Together with everything : ENTIRELY.

**lock**[2] (lŏk) *n.* [ME *locke* < OE *locc, loca.*] **1. a.** A strand or curl of hair : TRESS. **b. locks.** The hair of the head. **2.** A small wisp or strand, as of wool or cotton.

**lock·age** (lŏk'ĭj) *n.* **1.** Passage of a ship through a lock. **2.** The toll for using a lock. **3.** A system of locks.

**lock·er** (lŏk'ər) *n.* **1.** One that locks. **2.** An enclosure that can be locked, esp. one at a public place for the safekeeping of items. **3.** A flat trunk for storage. **4.** A heavily insulated refrigerated room or compartment for storing frozen foods.

**locker room** *n.* **1.** A room, as in a gymnasium, with lockers in which clothing and equipment can be stored. **2.** A room for changing one's clothes, as at a public swimming pool.

**lock·er-room** (lŏk'ər-rōōm', -rŏŏm') *adj.* Of, pertaining to, or appropriate for use in a locker room <bought some *locker-room* equipment><*locker-room* language>

**lock·et** (lŏk'ĭt) *n.* [OFr. *locquet,* latch < *loc,* lock, of Germanic orig.] A small ornamental case for a keepsake or picture, usu. worn as a pendant.

**lock·jaw** (lŏk'jô') *n.* **1.** TETANUS 1. **2.** A symptom of tetanus, in which the jaw is locked closed because of a tonic spasm of the muscles of mastication.

**lock·nut** *also* **lock nut** (lŏk'nŭt') *n.* **1.** A usu. thin nut screwed down on another nut to keep it from loosening. **2.** A self-locking nut.

**lock·out** (lŏk'out') *n.* Withholding of work from employees and shutdown of a plant by an employer during a labor dispute.

**lock·smith** (lŏk'smĭth') *n.* One that repairs or makes locks.

**lock step** *n.* **1.** A way of marching in which the marchers follow each other as closely as possible. **2.** A standardized procedure closely and mindlessly followed.

**lock stitch** *n.* A stitch made on a sewing machine by the interlocking of the upper thread and the bobbin thread.

**lock·up** (lŏk'ŭp') *n.* **1.** *Informal.* A jail, esp. one in which accused persons are held prior to court hearings. **2.** An act of locking or the state of being locked.

**lo·co** (lō'kō) *n., pl.* **-cos.** [Sp., crazy.] **1.** Locoweed. **2.** Loco disease. —*vt.* **-coed, -co·ing, -cos. 1.** To poison with locoweed. **2.** *Slang.* To make insane : CRAZE. —*adj. Slang.* Crazy : insane.

**loco disease** *n.* A disease of livestock caused by eating locoweed and marked by lack of coordination, dullness, and partial paralysis.

**lo·co·ism** (lō'kō-ĭz'əm) *n.* Loco disease.

**lo·co·mo·tion** (lō'kə-mō'shən) *n.* [Lat. *locus,* place + MOTION.] **1.** The act of moving or capability of moving from place to place. **2.** Travel.

**lo·co·mo·tive** (lō'kə-mō'tĭv) *n.* [Lat. *locus,* place + LLat. *motivus,* moving.] A self-propelled vehicle, usu. diesel or electric, that travels on rails and moves railroad cars. —*adj.* **1.** Of or involved in locomotion. **2.** Of, relating to, or being a locomotive. **3.** Able to move independently from place to place. **4.** Of or relating to travel.

**lo·co·mo·tor** (lō'kə-mō'tər) *adj.* [Lat. *locus,* place + Lat. *motor,* mover < *movere,* to move.] LOCOMOTIVE 1.

**locomotor ataxia** *n.* Tabes dorsalis.

**lo·co·weed** (lō'kō-wēd') *n.* A plant of the genera *Oxytropis* or *Astragalus* of the western and central United States, causing severe poisoning when eaten by livestock.

**loc·u·lar** (lŏk'yə-lər) *also* **loc·u·late** (-lāt', -lĭt) *or* **loc·u·lat·ed** (-lā'tĭd) *adj.* [LOCUL(US) + -AR.] Having, formed of, or divided into small cells or cavities. **—loc'u·la'tion** *n.*

**loc·ule** (lŏk'yōōl) *n.* [NLat. < Lat. *loculus,* little place. —see LOCULUS.] A small cavity or compartment within an organ or part, as a plant ovary.

**loc·u·lus** (lŏk'yə-ləs) *n., pl.* **-li** (-lī') [NLat. < Lat., little place, dim. of *locus,* place.] A locule.

**lo·cum te·nens** (lō'kəm' tē'nĕnz', -nənz) *n.* [Med. Lat., one holding a place.] *Chiefly Brit.* A person, esp. a physician or member of the clergy, who substitutes for another.

**lo·cus** (lō'kəs) *n., pl.* **-ci** (-sī', -kī') [Lat.] **1.** A locality : place. **2.** The configuration or set of all points satisfying specified geometric conditions. **3.** The position that a gene occupies on a chromosome.

**locus clas·si·cus** (klăs'ĭ-kəs) *n., pl.* **loci clas·si·ci** (klăs'ĭ-sī', -kī') [NLat.] A passage from a standard or classic work that is cited as an illustrative example.

**lo·cust** (lō'kəst) *n.* [ME < Lat. *locusta.*] **1.** Any of numerous grasshoppers of the family Locustidae, often traveling in swarms and causing damage to vegetation. **2.** A cicada. **3. a.** A North American tree, *Robinia pseudoacacia,* with compound leaves, fragrant white flower clusters, and hard, durable wood. **b.** A similar or related tree, as the honey locust or the carob. **4.** The wood of a locust tree.

**lo·cu·tion** (lō-kyōō'shən) *n.* [ME *locucion* < Lat. *locutio* < *loqui,* to speak.] **1.** A particular phrase, word, or expression considered from a stylistic point of view. **2.** Style of speaking : PHRASEOLOGY.

**lode** (lōd) *n.* [ME *lode,* way < OE *lād.*] **1. a.** A rock fissure filled with a metalliferous ore. **b.** A vein of mineral ore deposited between clearly demarcated nonmetallic rock layers. **2.** A rich supply or source.

**lode·star** *also* **load·star** (lōd'stär') *n.* [ME *lodesterre* : *lode,* way (< OE *lād*) + *sterre,* star < OE *steorra.*] **1.** A star, esp. the North Star, used as a point of reference. **2.** A guiding interest or principle.

**lode·stone** *also* **load·stone** (lōd'stōn') *n.* [Obs. *lode,* way (< ME < OE *lād*) + STONE.] **1.** A magnetized piece of magnetite. **2.** A strong attraction.

**lodge** (lŏj) *n.* [ME < OFr. *loge,* of Germanic orig.] **1. a.** A cottage or cabin used as a temporary, often seasonal, residence <a hunting *lodge*> **b.** A small house on the grounds of an estate orig. used by a gatekeeper or caretaker. **c.** An inn. **2. a.** A North American Indian living unit, as a wigwam, hogan, or long house. **b.** The group of Indians living in such a unit. **3. a.** A local chapter of a fraternal organization. **b.** The meeting hall of such a chapter. **c.** The members of such a chapter. **4.** The den of an animal, as the dome-shaped structure built by beavers. —*v.* **lodged, lodg·ing, lodg·es.** —*vt.* **1. a.** To provide with quarters temporarily, esp. for sleeping. **b.** To rent a room to. **c.** To establish or place in quarters <*lodging* troops in homes> **2.** To serve as a repository for : HARBOR. **3.** To leave, place, or deposit, as for safety. **4.** To force, fix, or implant. **5.** To register (a charge) in court or with an appropriate party : FILE. **6.** To vest (e.g., authority). **7.** *Archaic.* To beat (crops) down flat. —*vi.* **1. a.** To live in a place temporarily. **b.** To rent accommodations, esp. for sleeping. **2.** To be or become embedded.

**lodge·ment** (lŏj'mənt) *n. var. of* LODGMENT.

**lodge·pole pine** (lŏj'pōl') *n.* A pine, *Pinus contorta* of western North America, having light wood used in construction.

**lodg·er** (lŏj'ər) *n.* One that lodges, esp. one who rents and lives in a furnished room.

**lodg·ing** (lŏj'ĭng) *n.* **1.** A place to live. **2.** *often* **lodgings.** Sleeping accommodations. **3. lodgings.** Rented rooms.

**lodg·ment** *also* **lodge·ment** (lŏj'mənt) *n.* **1.** The act of lodging or state of being lodged. **2.** A place for lodging. **3.** An accumulation or deposit. **4.** A military foothold gained in enemy territory.

**lod·i·cule** (lŏd′ĭ-kyo͞ol′) *n.* [Lat. *lodicula*, small blanket, dim. of *lodix*, blanket.] One of the small scales at the base of each flower in grasses.

**lo·ess** (lō′əs, lĕs, lŭs) *n.* [G. *Löss* < dial. G. *Lösch* < *lösch*, loose.] A buff to gray, fine-grained, calcareous clay or silt, held to be a deposit of wind-blown dust. **—lo·es′si·al** (lō-ĕs′ē-əl, lĕs′ē-əl) *adj.*

**loft** (lôft, lŏft) *n.* [ME, upstairs room < ON *lopt*.] **1.** A large, usu. unpartitioned floor over an industrial or commercial space. **2.** An open space under a roof : ATTIC. **3.** A balcony or gallery, as in a church. **4.** A hayloft. **5. a.** A coop for keeping pigeons. **b.** A flock of pigeons kept in a loft. **6.** A large room where full-scale plans of a vessel are laid out or where rigging is assembled <a sail *loft*> **7. a.** The backward slant of the face of a golf club head, designed to drive the ball in a high arc. **b.** A golf stroke that drives the ball in such an arc. **c.** The upward course of a ball driven in such an arc. —*vt.* **loft·ed, loft·ing, lofts. 1.** To place, keep, or store in a loft. **2.** To propel in a high arc. **3.** To lay out a drawing of (e.g., the parts of a ship's engine).

**loft·y** (lôf′tē, lŏf′-) *adj.* **-i·er, -i·est.** [ME, noble < *loft*, upstairs room. —see LOFT.] **1.** Of impressive height. **2.** Elevated in character : AUGUST. **3.** Haughtily proud. **—loft′i·ly** *adv.* **—loft′i·ness** *n.*

**log**[1] (lôg, lŏg) *n.* [ME.] **1. a.** A usu. large trunk of a fallen tree. **b.** A long thick section of trimmed but unhewn timber. **2.** A device trailed from a ship to determine its speed. **3. a.** A record of speed, progress, and important events, kept on a ship or aircraft. **b.** The book in which such a record is kept. **c.** A similar record or journal. —*v.* **logged, log·ging, logs.** —*vt.* **1. a.** To cut down the timber of (a section of land). **b.** To cut (trees) into logs. **2. a.** To enter in a ship's or aircraft's log. **b.** To travel (a specified distance, time, or speed). **3.** To spend (time) <*logged* 25 years with the same firm> —*vi.* To engage in logging.

**log**[2] (lôg, lŏg) *n.* Logarithm.

**log-** *pref. var. of* LOGO-.

**-log** *suff. var. of* -LOGUE.

**lo·gan·ber·ry** (lō′gən-bĕr′ē) *n.* [After James H. *Logan* (1841–1928).] **1.** A trailing prickly plant, *Rubus loganobaccus*, cultivated for its edible acidic fruit. **2.** The red fruit of the loganberry.

**log·a·rithm** (lô′gə-rĭth′əm, lŏg′ə-) *n.* [NLat. *logarithmus* : Gk. *logos*, reason + *arithmos*, number.] The exponent indicating the power to which a fixed number, the base, must be raised to produce a given number; e.g., if $n^x = a$, the logarithm of *a*, with *n* as the base, is *x*; symbolically, $\log_n a = x$. **—log′a·rith′mic** (-rĭth′mĭk), **log′a·rith′mi·cal** *adj.* **—log′a·rith′mi·cal·ly** *adv.*

**log·book** (lôg′bo͝ok′, lŏg′-) *n.* The official record book of a ship or aircraft.

**loge** (lōzh) *n.* [Fr. < OFr., lodge.] **1.** A small compartment, esp. a box in a theater. **2.** The front rows of a theater's mezzanine.

**log·ger** (lô′gər, lŏg′ər) *n.* **1.** A lumberjack. **2.** A machine for loading or hauling logs.

**log·ger·head** (lô′gər-hĕd′, lŏg′ər-) *n.* [Prob. dial. *logger*, wooden block (< LOG) + HEAD.] **1.** A marine turtle, *Caretta caretta*, with a large beaked head. **2.** An iron tool having a long handle with a bulbous end, used when heated to melt tar or warm liquids. **3.** A post on a whaleboat used to help secure a rope holding a harpooned whale. **4. a.** *Informal.* A blockhead : dolt. **b.** A disproportionately large head. **—at loggerheads.** In a dispute : at odds.

**loggerhead shrike** *n.* A North American bird, *Lanius ludovicianus*, with gray and white plumage and a hooked beak.

**log·gi·a** (lô′jē-ə, lŏj′ē-ə) *n.* [Ital. < Fr. *loge*. —see LOGE.] **1.** A roofed but open gallery or arcade along the front or side of a building, often at an upper level. **2.** An open balcony in a theater.

**log·ging** (lô′gĭng, lŏg′ĭng) *n.* The work or business of felling trees and conveying the logs to a mill.

**lo·gi·a** (lŏ′gē-ä′) *n. pl. of* LOGION.

**log·ic** (lŏj′ĭk) *n.* [ME < OFr. *logique* < Lat. *logica* < Gk. *logikē* < *logikos*, of reason < *logos*, reason.] **1.** Study of the principles of reasoning, esp. of the structure of propositions as distinguished from their content and of method and validity in deductive reasoning. **2. a.** A system of reasoning. **b.** A mode of reasoning. **c.** The formal guiding principles of a school, discipline, or science. **3.** Valid reasoning. **4.** The relationship of element to element to whole in a set of objects, principles, individuals, or events. **5.** *Computer Sci.* **a.** Computer circuitry. **b.** Graphic representation of computer circuitry.

**log·i·cal** (lŏj′ĭ-kəl) *adj.* **1.** Of, relating to, in accord with, or of the nature of logic. **2.** Displaying consistency in reasoning. **3.** Reasonable on the basis of previous events or statements <a *logical* conclusion> **4.** Able to reason clearly. **—log′i·cal′i·ty** (-kăl′ĭ-tē), **log′i·cal·ness** *n.* **—log′i·cal·ly** *adv.*

☆ **syns: 1.** LOGICAL, CONSEQUENT, INTELLIGENT, RATIONAL, REASONABLE, SENSIBLE *adj. core meaning* : consistent with reason and intellect <a *logical* explanation of the phenomena> **2.** LOGICAL, ANALYTICAL, RATIOCINATIVE, RATIONAL *adj. core meaning* : able to reason validly <a *logical* mind>

**logical circuit** *n. Computer Sci.* A switching circuit that performs an arithmetic logic function.

**logical positivism** *n.* A philosophy asserting the primacy of observation in assessing the truth of statements of fact and holding that subjective and metaphysical arguments not based on observable data are meaningless, meaningful statements being either a priori and analytic or a posteriori and synthetic.

**logic circuit** *n.* A logical circuit.

**lo·gi·cian** (lō-jĭsh′ən) *n.* A practitioner of a system of logic.

**lo·gi·on** (lō′gē-ŏn′) *n., pl.* **-gi·a** (-gē-ä′) [Gk., oracle < *logos*, word.] One of the sayings of Jesus not recorded in the Gospels but thought to have belonged to the same source material from which they were compiled.

**lo·gis·tic** (lō-jĭs′tĭk) *also* **lo·gis·ti·cal** (-tĭ-kəl) *adj.* [Gk. *logistikos*, of calculation < *logistēs*, calculator < *logizein*, to calculate < *logos*, reckoning.] **1.** Of or relating to symbolic logic. **2.** Of or relating to logistics. **—lo′gis·ti′cian** (-jĭ-stĭsh′ən) *n.*

**lo·gis·tics** (lō-jĭs′tĭks, lə-) *n. (sing. or pl. in number).* The procurement, maintenance, distribution, and replacement of personnel and materiel.

**log·jam** (lôg′jăm′, lŏg′-) *n.* **1.** An immovable mass of floating logs crowded together. **2.** A deadlock, as in negotiations.

**log·nor·mal** (lôg-nôr′məl, lŏg-) *adj.* Of, pertaining to, or being a logarithmic function with a normal distribution. **—log′nor·mal′i·ty** (-măl′ĭ-tē) *n.* **—log·nor′mal·ly** *adv.*

**lo·go** (lō′gō′, lô′gō′, lŏg′ō′) *n., pl.* **-gos.** LOGOTYPE 2.

**logo-** *or* **log-** *pref.* [Gk. < *logos*, word, speech.] Word : speech <*logo*gram>

**log·o·gram** (lô′gə-grăm′, lŏg′ə-) *n.* A symbol or letter representing an entire word, as ¢ for cents or £ for pounds. **—log′o·gram·mat′ic** (-grə-măt′ĭk) *adj.* **—log′o·gram·mat′i·cal·ly** *adv.*

**log·o·graph** (lô′gə-grăf′, lŏg′ə-) *n.* A logogram. **—log′o·graph′ic** *adj.* **—log′o·graph′i·cal·ly** *adv.*

**lo·gog·ra·phy** (lō-gŏg′rə-fē) *n.* The use of logotypes in printing and design.

**log·o·griph** (lô′gə-grĭf′, lŏg′ə-) *n.* [Fr. *logogriphe* : Gk. *logos*, word + Gk. *griphos*, fishing basket.] A word puzzle, as an anagram.

**Log·os** (lŏg′ŏs′, lō′gŏs′) *n.* [Gk., reason.] **1.** Cosmic reason, affirmed in ancient Greek philosophy as the source of world order and intelligibility. **2.** The self-revealing thought and will of God, as explicated in the Gospel of John, often associated with the second person of the Trinity.

**lo·go·type** (lô′gə-tīp′, lŏg′ə-) *n.* **1.** A single piece of type bearing two or more usu. separate elements. **2.** The name, trademark, or symbol of a company or publication, borne on one printing plate or piece of type.

**log·roll** (lôg′rōl′, lŏg′-) *v.* **-rolled, -roll·ing, -rolls.** —*vt.* To work toward the passage of (legislation) by logrolling. —*vi.* To engage in political logrolling.

**log·roll·ing** (lôg′rō′lĭng, lŏg′-) *n.* **1.** Birling. **2.** Exchange of political favors, esp. the trading of votes or influence among legislators to achieve passage of mutually advantageous projects. **—log′roll′er** *n.*

**-logue** *or* **-log** *suff.* [Gk. *-logos* < *legein*, to speak.] Speech : discourse <trave*logue*>

**log·wood** (lôg′wo͝od′, lŏg′-) *n.* **1.** A tropical American tree, *Haematoxylon campechianum*, with dark heartwood from which a dyestuff is obtained. **2. a.** The heartwood of the logwood. **b.** The blackish or brownish dye obtained from this heartwood.

**lo·gy** (lō′gē) *adj.* **-gi·er, -gi·est.** [Perh. < Du. *log*, heavy.] Marked by lethargy : SLUGGISH.

**-logy** *suff.* [ME *-logie* < OFr. < Lat. *-logia* < Gk. < *logos*, word, speech.] **1.** Discourse : expression <phraseo*logy*> **2.** Science : theory : study <dermato*logy*>

**loin** (loin) *n.* [ME *loine* < OFr. *loigne* < Lat. *lumbus*.] **1.** *Anat.* The part of the side and back between the ribs and the pelvis. **2.** A cut of meat taken from the loin of an animal. **3. loins. a.** The area of the groin or thighs. **b.** The reproductive organs.

**loin·cloth** (loin′klôth′, -klŏth′) *n.* A strip of cloth worn around the loins.

**loi·ter** (loi′tər) *vi.* **-tered, -ter·ing, -ters.** [ME *loiteren*.] **1.** To stand idly about. **2.** To proceed slowly or with frequent stops. **3.** To dawdle <*loitered* over their work> **—loi′ter·er** *n.*

**Lo·ki** (lō′kē) *n.* [ON.] *Norse Myth.* A god who created discord esp. among his fellow gods.

**loll** (lŏl) *v.* **lolled, loll·ing, lolls.** [ME *lollen* < MDu., to doze.] —*vi.* **1.** To stand, move, or recline in a relaxed or lazy way. **2.** To hang laxly. —*vt.* To allow to hang laxly. —*n. Archaic.* An act or attitude of lolling. **—loll′er** *n.* **—loll′ing·ly** *adv.*

**lol·la·pa·loo·za** (lŏl′ə-pə-lo͞o′zə) *n.* [Orig. unknown.] *Slang.* Something extraordinary of its kind.

**Lol·lard** (lŏl′ərd) *n.* [ME < MDu. *Lollaert*, heretic.] One of a sect of 14th- and 15th-cent. reformers who were followers of John Wycliffe.

**lol·li·pop** *also* **lol·ly·pop** (lŏl′ē-pŏp′) *n.* [Perh. dial. *lolly*, tongue + POP[1].] A piece of hard candy on the end of a stick.

**lol·lop** (lŏl′əp) *vi.* **-loped, -lop·ing, -lops.** [LOLL + (GALL)OP.] *Chiefly Brit.* **1.** To lounge about : LOLL. **2.** To move with a bobbing motion.

**lol·ly** (lŏl′ē) *n., pl.* **-lies.** [Short for LOLLIPOP.] *Chiefly Brit.* **1. a.** Popsicle. **b.** Lollipop. **2.** Money.

**lol·ly·gag** (lŏl′ĭ-găg′) *also* **lal·ly·gag** (lăl′ē-) *vi.* **-gagged, -gag·ging, -gags.** [Orig. unknown.] To waste time by fooling around.

**lol·ly·pop** (lŏl′ē-pŏp′) *n. var. of* LOLLIPOP.

**Lom·bard** (lŏm′bərd, -bärd′, lŭm′-) *n.* [ME < OFr. *lombart* < Ital. *lombardo* < Lat. *Langobardus.*] **1.** One of a Germanic people that invaded northern Italy in A.D. 568 and established a kingdom in the Po Valley. **2.** A native of Lombardy. **3.** *Archaic.* A banker or pawnbroker. **—Lom·bar′dic** (lŏm-bär′dĭk, lŭm-) *adj.*

**Lom·bar·dy poplar** (lŏm′bər-dē, lŭm′-) *n.* A tree, *Populus nigra italica,* with upward-pointing branches forming a slender columnar outline.

**lo·ment** (lō′mĕnt′) *n.* [NLat. *lomentum* < Lat., skin conditioner made of bean meal < *lavare,* to wash.] A pod, as of the tick trefoil or similar leguminous plants, with a series of constrictions separating the individual seeds.

**Lon·don broil** (lŭn′dən) *n.* A boneless cut of meat, as flank steak, that is usu. marinated, broiled, and cut into thin slices.

**lone** (lōn) *adj.* [ME, short for *alone.* —see ALONE.] **1. a.** Without companionship : ISOLATED. **b.** Preferring solitude. **2.** Being the only one of its kind : SOLE <the *lone* judge in the county> **3.** Located or standing by itself.

**lone·ly** (lōn′lē) *adj.* **-li·er, -li·est. 1. a.** Without companions : LONE. **b.** Marked by aloneness : SOLITARY <a *lonely* life> **2.** Unfrequented by people : DESOLATE <a *lonely* beach> **3. a.** Dejected by being alone. **b.** Producing such dejection <the *loneliest* holiday of the year> **—lone′li·ly** *adv.* **—lone′li·ness** *n.*

**lone·ly-hearts** (lōn′lē-härts′) *adj.* Of or pertaining to lonely people who are looking for companions or marriage partners <a *lonely-hearts* column in the newspaper>

**lon·er** (lō′nər) *n. Informal.* One who avoids the company of others.

**lone·some** (lōn′səm) *adj.* **1. a.** Dejected, due to lack of companionship. **b.** Producing the sense of loneliness <a *lonesome* voyage> **2.** Deserted : unfrequented <a *lonesome* meadow> **3.** Lone. *—n. Informal.* Self <went all by my *lonesome*>

**long**[1] (lông, lŏng) *adj.* **-er, -est.** [ME < OE *lang.*] **1. a.** Having great length. **b.** Tall. **2.** Of relatively great duration <a *long* time> **3.** Of a specified linear duration or extent <a foot *long*> <a minute *long*> **4.** Extending beyond a standard or average <a *long* tournament> **5.** Tediously protracted : LENGTHY <a *long* sermon> **6.** Concerned with distant issues : FAR-REACHING <a *long* view> **7.** Involving substantial chance : RISKY <*long* odds> **8.** Having an excess or an abundance of <*long* on optimism> **9.** Having a large holding of a security or commodity in expectation of a rise in price <*long* on wheat> **10. a.** Having a comparatively protracted phonetic duration. **b.** Of, relating to, or being a vowel sound in English, as the sound of *a* in *mate,* historically descended from a long vowel. **11. a.** Of relatively great duration. —Used of a syllable in prosody. **b.** Stressed or accented. *—adv.* **1.** During or for an extended period of time <a raise that was *long* due> **2.** Far <worked *long* into the night> **3.** For or throughout a specified period <swam all day *long*> **4.** At a time distant from that referred to <*long* before human life> *—n.* **1.** A long time. **2.** A long vowel, syllable, or consonant. **3.** One who acquires large holdings of a security expected to rise in price. **4. a.** A garment size for a tall person. **b. longs.** Full-length trousers. **—before long.** Soon. **—the long and the short.** The gist : substance <*The long and the short* of it is that they lost.>

☆ ***syns:*** LONG, ELONGATE, EXTENDED, LENGTHY *adj. core meaning* : having great physical length <a *long* distance> ***ant:*** short

**long**[2] (lông, lŏng) *vi.* **longed, long·ing, longs.** [ME *longen* < OE *langian.*] To desire or yearn greatly <*longed* to be free>

**lon·ga·nim·i·ty** (lŏng′gə-nĭm′ĭ-tē) *n.* [ME *longanimite* < OFr. < LLat. *longanimitas* < *longanimis,* patient : Lat. *longus,* long + Lat. *animus,* soul.] Equanimity in spite of adversity and suffering.

**long·boat** (lông′bōt′, lŏng′-) *n.* The longest boat carried by a sailing ship, esp. by a merchantman.

**long·bow** (lông′bō′, lŏng′-) *n.* **1.** A wooden bow approx. five to six feet long. **2.** A medieval English bow, occas. over six feet long.

**long distance** *n.* **1.** An operator or system that places long-distance telephone calls. **2.** A long-distance telephone call.

**long-dis·tance** (lông′dĭs′təns, lŏng′-) *adj.* **1.** Located at a long distance or far away. **2.** Covering a long distance. **3.** Of or involving telephone communications to a distant station. *—adv.* By long-distance telephone.

**long division** *n.* A process of division in arithmetic, usu. used when the divisor has more than one digit, in which the remainders leading to succeeding steps of the procedure are recorded in a determinate pattern.

**long dozen** *n.* A group of 13 : BAKER'S DOZEN.

**long-drawn-out** (lông′drôn′out, lŏng′-) *adj.* Extended to a great length : PROLONGED.

**lon·ge·ron** (lŏn′jər-ən) *n.* [Fr. < *longer,* to go along < LLat. *longare,* to lengthen < Lat. *longus,* long.] A structural member running from front to rear of an aircraft's fuselage.

**lon·gev·i·ty** (lŏn-jĕv′ĭ-tē) *n.* [LLat. *longaevitas* < Lat. *longaevus,* ancient : *longus,* long + *aevum,* age.] **1. a.** A long duration of life. **b.** Length of life. **2.** Long continuance or duration, as in an occupation. **—lon·ge′vous** (-jē′vəs) *adj.*

**long face** *n.* An unhappy or sullen facial expression.

**long green** *n. Slang.* Paper money.

**long·hair** (lông′hâr′, lŏng′-) *n. Informal.* **1.** One dedicated to the arts, esp. to classical music. **2.** One whose taste in the arts is thought to be overrefined. **3.** One who has long hair, esp. a hippie. **—long′-hair′, long′haired′** *adj.*

**long·hand** (lông′hănd′, lŏng′-) *n.* Cursive writing.

**long·head** (lông′hĕd′, lŏng′-) *n.* **1.** A head having a cephalic index of 75.9 or less. **2.** A person having a longhead.

**long·head·ed** *also* **long-head·ed** (lông′hĕd′ĭd, lŏng′-) *adj.* **1.** Dolichocephalic. **2.** Wise or foresighted.

**long·horn** (lông′hôrn′, lŏng′-) *n.* One of a breed of long-horned cattle bred at one time in the southwestern United States.

**long-horned beetle** (lông′hôrnd′, lŏng′-) *n.* Any of various beetles of the family Cerambycidae, having long legs and long antennae.

**long house** *n.* A long wooden dwelling, esp. of the Iroquois.

**lon·gi·corn** (lŏn′jĭ-kôrn′) *n.* [< NLat. *Longicornia,* former group name : Lat. *longus,* long + Lat. *cornu,* horn.] A long-horned beetle. *—adj.* **1.** Having long antennae. **2.** Of or belonging to the family Cerambycidae.

**long·ing** (lông′ĭng, lŏng′-) *n.* A strong persistent craving or desire, esp. one that is unfulfillable. **—long′ing·ly** *adv.*

**long·ish** (lông′ĭsh, lŏng′-) *adj.* Rather long.

**lon·gi·tude** (lŏn′jĭ-tōōd′, -tyōōd′) *n.* [ME, length < OFr. < Lat. *longitudo* < *longus,* long.] **1.** The angular distance on the earth or on a globe or map, east or west of the prime meridian at Greenwich, England, to the point on the earth's surface for which the longitude is being determined, expressed either in degrees or in hours, minutes, and seconds. **2.** The angular distance, measured in degrees eastward along the ecliptic from the vernal equinox to the great circle passing through the pole of the ecliptic and the celestial point being measured.

**lon·gi·tu·di·nal** (lŏn′jĭ-tōōd′n-əl, -tyōōd′-) *adj.* **1.** Of or relating to length. **2.** Running or placed lengthwise. **3.** Of or relating to longitude. **—lon′gi·tu′di·nal·ly** *adv.*

**long johns** *pl.n. Informal.* Long underwear.

**long jump** *n.* A jump in track and field events for distance, made usu. from a running start.

**long·leaf pine** (lông′lēf′, lŏng′-) *n.* An evergreen tree, *Pinus australis* of the southeastern United States, with long needles and heavy, tough resinous wood, valued as timber and a source of turpentine.

**long-lived** (lông′līvd′, -lĭvd′, lŏng′-) *adj.* **1.** Having a long life. **2.** Persistent <a *long-lived* belief> **—long′-lived′ness** *n.*

**long measure** *n.* Linear measure.

**Lon·go·bard** (lŏng′gō-bärd′) *n., pl.* **-bards** *or* **-bar·di** (-bär′dē) [Lat. *Longobardus.*] LOMBARD 1. **—Lon′go·bar′dic** *adj.*

**long-play·ing** (lông′plā′ĭng, lŏng′-) *adj.* Pertaining to or being a microgroove phonograph record, esp. one turning at 33⅓ revolutions per minute.

**long purples** *n.* Purple loosestrife.

**long-range** (lông′rānj′, lŏng′-) *adj.* **1.** Requiring or involving an extended time span <*long-range* plan> **2.** Of, appropriate for, or covering long distances <a *long-range* bomber>

**long·shore·man** (lông′shôr′mən, -shōr′mən, lŏng′-) *n.* A dock worker who loads and unloads ships.

**long shot** *n.* **1.** An entry, as in a dog race, with little chance of winning. **2. a.** A bet made at and against great odds. **b.** A risky venture paying off handsomely if successful. **—by a long shot.** By a great amount.

**long·sight·ed** (lông′sī′tĭd, lŏng′-) *adj.* Farsighted. **—long′sight′ed·ness** *n.*

**long·some** (lông′səm, lŏng′-) *adj.* Wearingly long.

**long·spur** (lông′spûr′, lŏng′-) *n.* A bird of the genera *Calcarius* or *Rhyncophanes* of northern regions, with brownish plumage and long-clawed hind toes.

**long-stand·ing** (lông′stăn′dĭng, lŏng′-) *adj.* Being of long duration.

**long-suf·fer·ing** (lông′sŭf′ər-ĭng, lŏng′-) *adj.* Patiently bearing difficulties or wrongs. *—n. also* **long-suf·fer·ance** (-əns). Patient endurance. **—long′-suf′fer·ing·ly** *adv.*

**long suit** *n.* **1.** A suit in certain card games containing more cards than any of the other suits in a hand. **2.** One's strongest personal asset or quality : FORTE.

**long-term** (lông′tûrm′, lŏng′-) *adj.* Involving, maturing, or being in effect after a number of years <a *long-term* commitment>

**long-time** (lông′tīm′, lŏng′-) *adj.* Long-standing.

**long ton** *n.* TON 1a.

**long-wind·ed** (lông′wĭn′dĭd, lŏng′-) *adj.* **1.** Tiresomely verbose <a *long-winded* politician> **2.** Not subject to quick loss of breath <a *long-winded* runner> **—long′-wind′ed·ly** *adv.* **—long′-wind′ed·ness** *n.*

**long·wise** (lông′wīz′, lŏng′-) *adv.* Lengthwise.

**loo¹** (lo͞o) *n., pl.* **loos.** [Short for obs. *lanterloo* < Fr. *lanturlu.*] A card game in which each player contributes stakes to a pool.
**loo²** (lo͞o) *n., pl.* **loos.** [Perh. < Fr. *lieux d'aisances.*] *Chiefly Brit.* A toilet.
**loo·by** (lo͞o'bē) *n., pl.* **-bies.** [ME *loby.*] A big, clumsy fellow : OAF.
**loo·fa** *or* **loo·fah** (lo͞o'fə) *n.* [Ar. *lūfah.*] **1. a.** An Old World tropical vine of the genus *Luffa.* **b.** The fruit of the loofa, having a fibrous, spongelike interior. **2.** The dried, fibrous part of the loofa fruit, used as a sponge or filter.

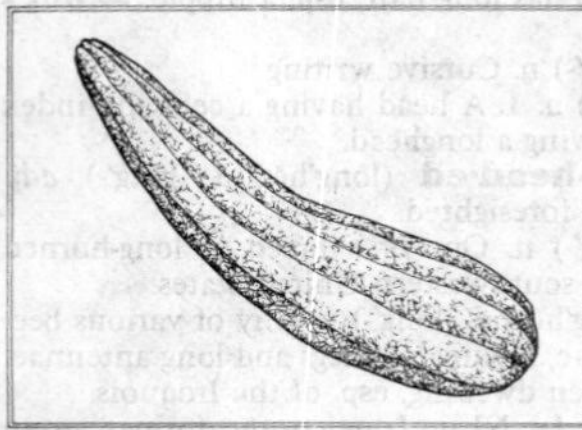
**loofa**

**look** (lo͝ok) *v.* **looked, look·ing, looks.** [ME *loken* < OE *lōcian.*] —*vi.* **1.** To use one's eyes in seeing : EXAMINE. **2. a.** To turn one's glance <*looked* to the side> **b.** To turn one's attention <*looked* to me for help> **3.** To appear or seem to be <*looked* happy> **4.** To face in a given direction <a cabin that *looks* on the river> —*vt.* **1.** To turn one's eyes on. **2.** To convey by one's expression <They *looked* daggers at me.> **3.** To have an appearance in accord with <*look* one's age> **—look after.** To take care of. **—look down on** To regard with scorn. **—look for.** To expect. **—look up. 1.** To search for and find, as in a reference book. **2.** *Informal.* To improve <Business is *looking up.*> **—look up to.** To admire. —*n.* **1. a.** An act of looking. **b.** A glance or gaze. **2.** Aspect or appearance. **3. looks.** Physical appearance, esp. when pleasing.
**look-a·like** (lo͝ok'ə-līk') *n.* One closely resembling another.
**look·down** (lo͝ok'doun') *n.* A marine fish, *Selene vomer* of Atlantic waters, with a steep frontal profile.
**look·er** (lo͝ok'ər) *n.* **1.** One who looks. **2.** *Slang.* An extremely attractive person.
**look·er-on** (lo͝ok'ər-ŏn', -ôn') *n., pl.* **look·ers-on.** A spectator.
**look-in** (lo͝ok'ĭn') *n.* **1.** A brief visit. **2.** A quick glance.
**looking glass** *n.* A mirror.
**look·out** (lo͝ok'out') *n.* **1.** An act of observing or keeping watch. **2.** A high place or structure with a broad view for observation. **3.** One who keeps watch. **4.** Outlook : view. **5.** A matter of worry or concern.
**look-see** (lo͝ok'sē') *n. Informal.* A quick survey.
**look-up** (lo͝ok'ŭp') *n. Computer Sci.* A procedure in which a table of values stored in a computer is searched for a specified value.
**loom¹** (lo͞om) *vi.* **loomed, loom·ing, looms.** [Orig. unknown.] **1.** To come into view as a massive, indistinct, or distorted image. **2.** To appear to the mind in an exaggerated and hostile form. **3.** To seem imminent : IMPEND. —*n.* A distorted, threatening appearance of something, as through fog or dimness.
**loom²** (lo͞om) *n.* [ME *lome* < OE *gelōma,* tool.] A device or machine from which cloth is made by interweaving yarn or thread at right angles.
**loon¹** (lo͞on) *n.* [Of Scand. orig.] A diving bird of the genus *Gavia* of northern regions, having a laughlike cry.
**loon²** (lo͞on) *n.* [ME *louen,* rogue.] **1.** A crazy or simple-minded person. **2.** An idler.
**loon·y** *also* **lun·y** (lo͞o'nē) [Shortening and alteration of LUNATIC.] *Informal.* —*adj.* **-i·er, -i·est. 1.** Extremely silly or foolish. **2.** Crazy : insane. —*n., pl.* **-ies.** LOON² 1.
**loony bin** *n. Informal.* An insane asylum.
**loop¹** (lo͞op) *n.* [ME *loupe.*] **1. a.** A length of line folded over and joined at the ends. **b.** The opening formed by such a doubled line. **2.** Something having a roughly oval, closed, or nearly closed turn or figure. **3.** *Elect.* A closed circuit. **4.** *Computer Sci.* A sequence of instructions that repeats until a terminal condition prevails. **5.** A flight maneuver in which an aircraft flies a circular path in a vertical plane with the lateral axis of the aircraft remaining horizontal. **6.** LEAGUE¹ 2. —*v.* **looped, loop·ing, loops.** —*vt.* **1.** To form into a loop. **2.** To join, fasten, or encircle with a loop or loops. **3.** To fly (an aircraft) in a loop. **4.** To move in a loop or arc. **5.** *Elect.* To join (conductors) so as to complete a circuit. —*vi.* **1.** To form a loop. **2.** To move in a loop. **3.** To make a loop in an aircraft.
**loop²** (lo͞op) *n.* [ME *loupe* < Med. Lat. *loupa,* of Germanic orig.] *Archaic.* LOOPHOLE 1.
**loop·er** (lo͞o'pər) *n.* **1.** One that makes loops. **2.** A measuring worm.
**loop·hole** (lo͞op'hōl') *n.* **1.** A small hole or slit in a wall, esp. one through which small arms may be fired. **2.** A way of escaping a difficulty, esp. an ambiguity or omission, as in the wording of a contract or law, that provides a means of evasion.
**loose** (lo͞os) *adj.* **loos·er, loos·est.** [ME *louse* < ON *lauss.*] **1.** Not fastened or restrained <*loose* tiles in the floor> **2.** Not taut or drawn up tightly : SLACK. **3.** Free from imprisonment or confinement : UNSHACKLED. **4.** Not tight-fitting. **5.** Not bound, bundled, stapled, or gathered together. **6.** Not compact or close in arrangement or structure. **7.** Not fast <a *loose* color> **8.** Lacking a sense of responsibility or restraint : IDLE <*loose* rumors> **9.** Lacking conventional moral restraint in sexual behavior. **10.** Not literal or precise <a *loose* interpretation> —*adv.* **1.** In a loose way. **2.** *Slang.* In a calm or unruffled condition <stay *loose*> —*v.* **loosed, loos·ing, loos·es.** —*vt.* **1.** To let loose : RELEASE. **2.** To make loose : UNDO. **3.** To cast loose : DETACH. **4.** To let fly : DISCHARGE. **5.** To release pressure on : EASE. **6.** To make less rigid : RELAX. —*vi.* **1.** To become loose. **2.** To discharge a missile : FIRE. **—loose'ly** *adv.* **—loose'ness** *n.*
☆ ***syns:*** LOOSE, LAX, RELAXED, SLACK *adj. core meaning* : not tightly bound to something else <a *loose* anchor line>
**loose-joint·ed** (lo͞os'join'tĭd) *adj.* **1.** Having freely articulated joints. **2.** Nimble or agile. **—loose'-joint'ed·ness** *n.*
**loose-leaf** (lo͞os'lēf') *adj.* Having leaves that can be easily removed, replaced, or rearranged.
**loos·en** (lo͞o'sən) *v.* **-ened, -en·ing, -ens.** [ME *lousnen* < *lousen* < *louse,* loose.] —*vt.* **1.** To unbind or make looser. **2.** To free from restraint, pressure, or strictness. **3.** To free (the bowels) from constipation. —*vi.* To become loose or looser.
**loose·strife** (lo͞os'strīf') *n.* [Intended as transl. of Gk. *lusimakheion* (interpreted as *lusis,* loosening + *machē,* battle), from the name *Lusimakhos.*] **1.** A plant of the genus *Lysimachia,* having usu. yellow flowers. **2.** A plant of the genus *Lythrum.*
**loot** (lo͞ot) *n.* [Hindi *lūṭ* < Skt. *lotram,* plunder.] **1.** Valuables plundered in time of war : SPOILS. **2. a.** Stolen goods. **b.** *Informal.* Goods illegally obtained, as by bribery. **3.** *Slang.* Money. —*v.* **loot·ed, loot·ing, loots.** —*vt.* **1.** To plunder : steal. **2.** To take as spoils. —*vi.* To engage in plunder. **—loot'er** *n.*
☆ ***syns:*** LOOT, PLUNDER, RANSACK *v. core meaning* : to rob on a large scale <*looted* the stores during the riot>
**lop¹** (lŏp) *vt.* **lopped, lop·ping, lops.** [Perh. < ME *loppe,* small branches < Med. Lat. *loppa.*] **1.** To cut off (a part) from : TRIM. **2.** To cut off from a tree or shrub. **3.** To eliminate as superfluous or undesirable. **—lop'per** *n.*
**lop²** (lŏp) *vi. & vt.* **lopped, lop·ping, lops.** [Orig. unknown.] To hang or let hang loosely : DROOP.
**lope** (lōp) *vi.* **loped, lop·ing, lopes.** [ME *lopen* < ON *hlaupa.*] To run or ride with a steady, easy gait. **—lope** *n.* **—lop'er** *n.*
**lop-eared** (lŏp'îrd') *adj.* Having bent or drooping ears.
**lop·py** (lŏp'ē) *adj.* **-pi·er, -pi·est.** Hanging limp : PENDULOUS.
**lop·sid·ed** (lŏp'sī'dĭd) *adj.* **1.** Larger, heavier, or higher on one side than on the other. **2.** Sagging or leaning to one side. **—lop'sid'ed·ly** *adv.* **—lop'sid'ed·ness** *n.*
**lo·qua·cious** (lō-kwā'shəs) *adj.* [Lat. *loquax, loquac-,* loquacious < *loqui,* to speak.] Extremely talkative : GABBY. **—lo·qua'cious·ly** *adv.* **—lo·qua'cious·ness, lo·quac'i·ty** (lō-kwăs'ĭ-tē) *n.*
**lo·quat** (lō'kwŏt', -kwăt') *n.* [Cantonese *lō kwat.*] **1.** A small tree native to eastern Asia, *Eriobotrya japonica,* having fragrant white flowers and yellow pear-shaped fruit. **2.** The fruit of the loquat.
**lo·ran** (lôr'ăn', lōr'-) *n.* [LO(NG-)RA(NGE) N(AVIGATION).] A long-range navigational system based on pulsed radio signals from two or more pairs of ground stations of known position, used by a navigator to establish the geographic position of an aircraft or ship
**lord** (lôrd) *n.* [ME < OE *hlāford* : *hlāf,* bread + *weard,* guardian.] **1.** A man of high rank in a feudal society or in one that retains feudal forms and institutions, esp.: **a.** A king. **b.** A territorial magnate. **c.** The proprietor of a manor. **2. Lord.** *Chiefly Brit.* A general masculine title of nobility or rank, used: **a.** Semiformally for any peer other than a duke. **b.** As the usual style for a baron. **c.** As a courtesy title for a younger son of a duke or marquis. **d.** As part of the titles of certain high officials and dignitaries. **e.** As a nominal title for a bishop. **3. a. Lord.** GOD 1a, b. **b.** *Archaic.* The head of a household. **c.** *Archaic.* A husband. **d.** A man of renowned power. **e.** A man who has mastery in a given activity or field. —*vi.* **lord·ed, lord·ing, lords.** To play the lord : DOMINEER <*lording* it over the strangers>
▲ word history: The actual as well as the symbolic importance of bread as a basic foodstuff is exhibited by the word *lord. Lord* is derived from a compound formed in very early Old English times from the words *hlāf,* "bread," and *weard,* "ward, guardian." *Lord,* therefore, literally means "guardian of the bread." Since such a position would be the dominant one in a household, *lord* came to denote a man of authority and rank in society at large.
**Lord Chancellor** *n., pl.* **Lords Chancellor.** The presiding officer of the House of Lords.
**lord·ing** (lôr'dĭng) *n.* **1.** *Archaic.* Lord : sir. **2.** *Obs.* A lordling.
**lord·ling** (lôrd'lĭng) *n.* An insignificant lord.
**lord·ly** (lôrd'lē) *adj.* **-li·er, -li·est. 1.** Of, relating to, or typical of a

ă **pat** ā **pay** âr **care** ä **father** ĕ **pet** ē **be** hw **which** ĭ **pit**
ī **tie** îr **pier** ŏ **pot** ō **toe** ô **paw, for** oi **noise** o͝o **took**

lord. **2.** Very refined and noble in nature. **3.** Pretentiously arrogant and overbearing. —*adv.* In a lordly way. **—lord'li·ness** *n.*

**Lord of Misrule** *n.* A master of traditional Christmas revelry in 15th- and 16th-cent. England.

**lor·do·sis** (lôr-dō'sĭs) *n.* [NLat. < Gk. *lordōsis* < *lordos*, bent backward.] Abnormal forward curvature of the spine in the lumbar region. **—lor·dot'ic** (-dŏt'ĭk) *adj.*

**Lord's Day** or **Lord's day** *n.* Sabbath : Sunday.

**lord·ship** (lôrd'shĭp') *n.* **1.** *often* **Lordship.** —Used as a form of address in speaking or referring to a man holding the title of lord. **2.** The authority or position of a lord. **3.** A feudal lord's territorial fief.

**Lord's Prayer** *n.* The prayer taught by Jesus to His disciples.

**Lord's Supper** *n.* **1.** The Last Supper. **2.** COMMUNION 4a.

**Lord's Table** *n.* The Communion table.

**lore**[1] (lôr, lōr) *n.* [ME < OE *lār.*] **1.** Accumulated tradition, fact, or belief about a subject. **2.** Knowledge gained through education or experience. **3.** *Archaic.* Something taught or learned.

**lore**[2] (lôr, lōr) *n.* [NLat. *lorum* < Lat., thong.] The area between a bird's eye and the base of the bill.

**Lo·re·lei** (lôr'ə-lī', lō'rə-lī') *n.* [G.] A legendary Germanic siren whose alluring singing enticed sailors to shipwreck.

**Lo·rentz contraction** (lō'rĕnts) *also* **Lo·rentz-Fitz·ger·ald contraction** (lō'rĕnts-fĭts-jĕr'əld) *n.* [After Hendrik A. *Lorentz* (1853–1928).] The contraction in length of a moving body, as measured by an observer at rest with respect to the body, by the factor $(1-v^2/c^2)^{1/2}$, where *v* is the relative speed of the moving body and *c* the speed of light.

**lor·gnette** (lôrn-yĕt') *n.* [Fr. < *lorgner*, to leer at < OFr. < *lorgne*, squinting.] Eyeglasses or opera glasses with a handle.

**lo·ri·ca** (lô-rī'kə, lō-) *n.*, *pl.* **-cae** (-sē') [Lat., leather cuirass < *lorum*, thong.] A protective external case or shell, as of a rotifer, diatom, or protozoan. **—lor'i·cate'** (lôr'ĭ-kāt', lŏr'-), **lor'i·ca'ted** (-kā'tĭd) *adj.*

**lor·i·keet** (lôr'ĭ-kēt', lŏr'-) *n.* [LOR(Y) + (PARA)KEET.] Any of several small Australasian parrots of the subfamily Loriinae.

**lo·ris** (lôr'ĭs, lōr'-) *n.* [Fr., poss < obs. Du. *loeris*, clown.] A small, nocturnal, arboreal primate of the genera *Loris* or *Nycticebus* of tropical Asia, with dense fur, large eyes, and a vestigial tail.

**lorn** (lôrn) *adj.* [ME, p.part. of *lesen*, to suffer a loss < OE *leosan.*] Forsaken : forlorn.

**lor·ry** (lôr'ē, lŏr'ē) *n.*, *pl.* **-ries.** [Orig. unknown.] **1.** A low, horse-drawn four-wheeled wagon. **2.** *Chiefly Brit.* A motor truck. **3.** A flatbed freight car that moves on rails.

**lo·ry** (lôr'ē, lōr'ē) *n.*, *pl.* **-ries.** [Malay *luri.*] Any of various brightly colored Australasian parrots of the subfamily Loriinae, having a tongue with a brushlike tip.

**lose** (lo͞oz) *v.* **lost** (lôst, lŏst), **los·ing, los·es.** [ME *losen* < OE *lōsian* < *los*, loss.] —*vt.* **1. a.** To be unable to find. **b.** To incur the deprivation of, as by negligence or accident. **2.** To be unable to maintain, sustain, or keep <*lost* their supporters> **3. a.** To be deprived of <*lost* everything in the flood> **b.** To be deprived of through death <*lost* both parents> **4.** To fail to win <*lose* the contest> **5.** To fail to take advantage of or use <*lose* an opportunity> **6.** To fail to see, hear, or comprehend. **7.** To remove (oneself), as from everyday reality into a fantasy world. **8.** To rid oneself of <*lose* weight> **9.** To wander or stray from <*lose* one's way> **10.** To make (oneself) disappear or fade from view, as in a crowd. **11.** To outdistance or elude <*lost* the trackers> **12.** To cause or result in the loss of <a mistake that *lost* the contract> **13.** To cause to be destroyed <Both ships were *lost* in the storm.> —*vi.* **1.** To suffer loss. **2.** To be defeated. **3.** To run slow. —Used of a timepiece. **—lose out.** To fail to receive an expected gain. **—lose out on.** To miss (e.g., an opportunity).

☆ **syns:** LOSE, MISLAY, MISPLACE *v. core meaning* : to be unable to find <*lost* my gloves>

**lo·sel** (lō'zəl, lo͞o'-, lŏz'əl) *n.* [ME < *lesen*, to suffer a loss. —see LORN.] One that is worthless.

**los·er** (lo͞o'zər) *n.* **1.** One that loses. **2. a.** One who fails consistently, esp. because of incompetence. **b.** Something destined to fail.

**loss** (lôs, lŏs) *n.* [ME *los*, perh. back-formation < *lost*, p.part. of *losen*, to lose.] **1.** An act or instance of losing. **2.** One that is lost. **3.** Injury or suffering caused by losing or by being lost. **4. losses. a.** Casualties. **b.** Setbacks. **5.** Destruction. **6.** *Elect.* The power decrease in a circuit, circuit element, or device caused by resistance. **7.** The amount of a claim on an insurer by an insured. **—at a loss.** Perplexed : uncertain.

**loss leader** *n.* An article offered by a retail store at cost or less than at cost to attract customers.

**loss ratio** *n.* The ratio between the premiums paid to an insurance company and the claims settled by the company.

**lost** (lôst, lŏst) *adj.* [P.part. of LOSE.] **1.** Unable to find one's way. **2. a.** No longer in one's possession. **b.** No longer practiced or known. **3.** Unable to act, function, or make progress. **4.** Spiritually or physically destroyed. **5.** Completely absorbed : RAPT <stood by the window, *lost* in thought>

**lot** (lŏt) *n.* [ME < OE *hlot.*] **1.** An object used to make a determination or choice by chance. **2. a.** The use of lots for selection. **b.** The selections made. **3.** Something that befalls an individual as a result of determination by lot. **4.** One's fate in life : FORTUNE. **5.** A number of associated things or people. **6.** Kind or sort. **7.** Miscellaneous articles sold as one unit. **8.** A large amount, extent, or number <has a *lot* of money><made *lots* of enemies> **9. a.** A piece of land. **b.** A piece of land with designated boundaries. **c.** A motion-picture studio. —*vt.* **lot·ted, lot·ting, lots. 1.** To apportion by lots : ALLOT. **2.** To divide (land) into lots.

**lo·tah** *also* **lo·ta** (lō'tə) *n.* [Hindi *loṭā.*] A rounded brass or copper container used in India for carrying water or storing food.

**loth** (lōth, lōth) *adj. var. of* LOATH.

**Lo·thar·i·o** (lō-thâr'ē-ō) *n.*, *pl.* **-os.** [After *Lothario*, a character in *The Fair Penitent*, a play by Nicholas Rowe (1674–1718).] A seducer.

**lo·ti** (lō'tē) *n.*, *pl.* **loti.** [Sotho < *Loti*, a range of mountains in Lesotho.] —See table at CURRENCY.

**lo·tic** (lō'tĭk) *adj.* [< Lat. *lotus*, p.part. of *lavere*, to wash.] Of, relating to, or living in moving water.

**lo·tion** (lō'shən) *n.* [ME *locion* < Lat. *lotio*, a washing < *lavere*, to wash.] **1.** A medicated liquid for external application. **2.** An externally applied cosmetic liquid <hand *lotion*>

**lo·tos** (lō'təs) *n. var. of* LOTUS.

**lot·ter·y** (lŏt'ə-rē) *n.*, *pl.* **-ies.** [Prob. < Du. *loterije* < *lot*, lot.] **1.** A contest in which tokens are distributed or sold, the winning token or tokens being secretly predetermined or ultimately selected in a chance drawing. **2.** An event or activity regarded as having an outcome depending on chance.

**lot·to** (lŏt'ō) *n.* [Ital. < Fr. *lot*, lot, of Germanic orig.] A game of chance similar to bingo.

**lo·tus** *also* **lo·tos** (lō'təs) *n.* [Lat., name of several plants < Gk. *lōtos.*] **1. a.** An aquatic plant native to southern Asia, *Nelumbo nucifera*, with large leaves, fragrant pinkish flowers, and a broad rounded perforated seed pod. **b.** Any of various plants similar or related to the lotus, as certain water lilies. **2.** A representation of a lotus plant in classical, usu. Egyptian, architecture, sculpture, and art. **3.** A leguminous plant of the genus *Lotus*. **4. a.** A small tree or shrub, *Zizyphus lotus* of the Mediterranean region, whose fruit is said to be that eaten by the lotus-eaters. **b.** The fruit of this tree.

**lo·tus-eat·er** (lō'təs-ē'tər) *n.* One of a North African people described in the Homeric *Odyssey* who fed on the lotus and hence lived in drugged indolence.

**loud** (loud) *adj.* **-er, -est.** [ME < OE *hlūd.*] **1.** Marked by high volume and intensity of sound. **2.** Producing a sound of high volume and intensity. **3.** Clamorous and insistent. **4. a.** Having offensively bright colors <a *loud* necktie> **b.** Having an offensively strong odor. **c.** Offensive in manner. **—loud, loud'ly** *adv.* **—loud'ness** *n.*

**loud·en** (loud'n) *vt.* & *vi.* **-ened, -en·ing, -ens.** To make or become louder.

**loud·mouth** (loud'mouth') *n.* One given to loud and offensive talk. **—loud'mouthed'** (-mouthd', -moutht') *adj.*

**loud·speak·er** (loud'spē'kər) *n.* A device that converts electric signals to audible sound.

**lough** (lŏKH, lŏk) *n.* [ME < OE *luh*, of Celt. orig.] *Ir.* **1.** A lake. **2.** A bay or inlet of the sea.

**lou·is d'or** (lo͞o'ē dôr') *also* **lou·is** (lo͞o'ē) *n.* [Fr. : *Louis*, Louis XIII + *d'or*, of gold.] **1.** A gold French coin in circulation from 1640 until the Revolution. **2.** A post-Revolutionary French 20-franc gold coin.

**Lou·i·si·an·a French** (lo͞o-ē'zē-ăn'ə) *n.* French as spoken by the descendants of the original French settlers of Louisiana.

**Lou·is Qua·torze** (lo͞o'ē kă-tôrz') *adj.* [Fr., Louis XIV.] Of, relating to, or typical of the baroque style in architecture, furniture, and decoration of the reign of Louis XIV.

**Lou·is Quinze** (lo͞o'ē kăNZ') *adj.* [Fr., Louis XV.] Of, relating to, or typical of the rococo style in architecture, furniture, and decoration of the reign of Louis XV.

**Lou·is Seize** (lo͞o'ē sĕz') *adj.* [Fr., Louis XVI.] Of, relating to, or typical of the neoclassic style in architecture, furniture, and decoration of the reign of Louis XVI.

**Lou·is Treize** (lo͞o'ē trĕz') *adj.* [Fr., Louis XIII.] Of, relating to, or typical of the heavy late-Renaissance style in architecture, furniture, and decoration of the reign of Louis XIII.

**lounge** (lounj) *v.* **lounged, loung·ing, loung·es.** [Orig. unknown.] —*vi.* **1.** To move or act in a relaxed, lazy manner : LOLL. **2.** To pass time idly. —*vt.* To pass (time) in lounging. —*n.* **1.** A public waiting room, as in a hotel or air terminal, often with smoking or lavatory facilities. **2. a.** A living room. **b.** A lobby. **c.** A bar where alcoholic beverages are served. **3.** A long couch, esp. one with no back and a headrest at one end. **—loung'er** *n.*

**loupe** (lo͞op) *n.* [Fr.] A small magnifying glass usu. set in an eyepiece and used chiefly by jewelers and watchmakers.

**loup-ga·rou** (lo͞o'gə-ro͞o', -gä-) *n.*, *pl.* **loups-ga·rous** (lo͞o'gə-ro͞oz', -gä-ro͞o') [Fr. : *loup*, wolf (< Lat. *lupus*) + *garou*, werewolf, of Germanic orig.] A werewolf.

**loup·ing ill** (lou'pĭng, lō'-) *n.* [< obs. *loup*, to leap < ME *loupen.*] TREMBLE 3a.

**lour** (lou'ər) *v.* & *n. var. of* LOWER[1].

**louse** (lous) *n.* [ME < OE *lūs.*] **1.** *pl.* **lice** (līs). One of numerous small, flat-bodied, wingless, biting or sucking insects of the order Anoplura, many of which are external parasites on various animals, including human beings. **2.** *pl.* **lous·es.** *Slang.* A mean or contemptible person. —*vt.* **loused, lous·ing, lous·es.** *Slang.* To bungle <*loused* up everything>

**louse·wort** (lous'wûrt', -wôrt') *n.* A plant of the genus *Pedicularis,* with irregular, variously colored flower clusters.

**lous·y** (lou'zē) *adj.* **-i·er, -i·est. 1.** Infested with lice. **2.** Extremely despicable : NASTY. **3.** Very painful or unpleasant <a *lousy* toothache> **4.** Inferior or worthless <a *lousy* car> **5.** *Slang.* Abundantly supplied <*lousy* with money> **—lous'i·ly** *adv.* **—lous'i·ness** *n.*

**lout**[1] (lout) *n.* [Poss. < LOUT[2].] An awkward, stupid person : OAF.

**lout**[2] (lout) *vi.* **lout·ed, lout·ing, louts.** [ME *louten* < OE *lūtan.*] *Archaic.* **1.** To curtsy or bow. **2.** To stoop or bend.

**lout·ish** (lou'tĭsh) *adj.* Clumsy or oafish. **—lout'ish·ly** *adv.* **—lout'ish·ness** *n.*

**lou·ver** *also* **lou·vre** (lo͞o'vər) *n.* [ME *luver,* hole in the roof < OFr. < Med. Lat. *luvarium.*] **1. a.** A framed opening in a wall fitted with movable or fixed slanted slats. **b.** One of the slats used in a louver. **2.** A lantern-shaped cupola on the roof of many medieval buildings to admit air and provide for the escape of smoke. **3.** A slatted opening for ventilation, as on a car hood. **—lou'vered** *adj.*

**lov·a·ble** *also* **love·a·ble** (lŭv'ə-bəl) *adj.* Easy to love. **—lov'a·bil'i·ty, lov'a·ble·ness** *n.* **—lov'a·bly** *adv.*

**lov·age** (lŭv'ĭj) *n.* [ME < OFr. *luvesche* < LLat. *levisticum,* alteration of Lat. *ligusticum* < *ligusticus,* Ligurian.] A plant, *Levisticum officinale,* with small aromatic seeds used as seasoning.

**love** (lŭv) *n.* [ME < OE *lufu.*] **1. a.** An intense affection for another person based on personal or familial ties. **b.** A strong affection for or attachment to another person based on regard or shared experiences or interests. **2.** An expression of one's affection <send them my *love*> **3. a.** An intense attraction to another person based mainly on sexual desire. **b.** The deep tenderness, affection, and concern felt for a person with whom one has or wishes to have a relationship based on sexual attraction. **c.** The object of such an attraction : BELOVED. **4. a.** Intense sexual passion. **b.** Sexual intercourse. **c.** LOVE AFFAIR 1. **5.** An intense emotional attachment, as for a pet or treasured object. **6. a.** A strong enthusiasm <a *love* of art> **b.** The object of such an enthusiasm <Fishing is my greatest *love.*> **7. Love.** *Myth.* Eros or Cupid, the god of love. **8. a.** God's mercy and benevolence toward humans. **b.** Humankind's devotion to or adoration of God. **c.** A feeling of kindness or brotherhood toward others. **9. Love.** *Christian Science.* GOD 1c. **10.** A zero score in tennis. —*v.* **loved, lov·ing, loves.** —*vt.* **1.** To feel love for. **2. a.** To caress or embrace. **b.** To have sexual intercourse with. **3.** To desire enthusiastically <*loves* to read> **4.** To thrive on : NEED <a plant that *loves* humidity> —*vi.* To feel affection or experience sexual desire for another. **—fall in love.** To become enamored of or sexually attracted to another. **—for love or money.** Under any circumstances <wouldn't do that *for love or money*>

☆ ***syns:*** LOVE, ADORE, CHERISH *v. core meaning* : to feel love or strong affection for <*loved* their children><*love* one's spouse> LOVE is the most neutral. ADORE stresses great devotion <*adore* one's parents>; used in an informal sense, it merely means to like very much <*adored* skiing> CHERISH emphasizes tender care <*cherished* the baby as if it were their own><*cherished* their antiques> ***ant:*** hate

**love·a·ble** (lŭv'ə-bəl) *adj. var. of* LOVABLE.

**love affair** *n.* **1.** An intimate sexual relationship or episode between lovers. **2.** A strong enthusiasm <America's continuing *love affair* with television>

**love apple** *n. Archaic.* A tomato.

**love·bird** (lŭv'bûrd') *n.* Any of various small Old World parrots, chiefly of the genus *Agapornis,* often caged as a pet.

**love child** *n.* An illegitimate child.

**love feast** *n.* **1. a.** A meal eaten with others as a symbol of love among early Christians. **b.** A similar symbolic meal among certain modern Christian sects. **2.** An assembly intended to promote good will among the participants.

**love-in-a-mist** (lŭv'ĭn-ə-mĭst') *n.* A native European plant, *Nigella damascena,* with blue or whitish flowers surrounded by many threadlike bracts.

**love knot** *n.* A stylized knot regarded as a symbol of love.

**love·less** (lŭv'lĭs) *adj.* **1.** Marked by an absence of love <a *loveless* marriage> **2.** Feeling no love. **3.** Receiving no love.

**love-lies-bleed·ing** (lŭv'līz-blē'dĭng) *n.* A tropical plant, *Amaranthus caudatus,* with small red flower clusters.

**love·lock** (lŭv'lŏk') *n.* A lock of hair, often tied with ribbon, worn by 17th- and 18th-cent. courtiers.

**love·lorn** (lŭv'lôrn') *adj.* Bereft of love or one's lover.

**love·ly** (lŭv'lē) *adj.* **-li·er, -li·est. 1.** Full of love : LOVING. **2.** Inspiring love or affection. **3.** Having attractive or pleasing qualities. **4.** Enjoyable : delightful. —*n., pl.* **-lies.** *Informal.* **1.** A beautiful person, esp. a woman. **2.** A lovely object. **—love'li·ness** *n.* **—love'ly** *adv.*

**love·mak·ing** (lŭv'mā'kĭng) *n.* **1.** Sexual activity, esp. sexual intercourse. **2.** Courtship.

**lov·er** (lŭv'ər) *n.* **1.** One who loves another, esp. one who feels sexual love. **2. lovers.** A couple in love with one other. **3. a.** A paramour. **b.** A sexual partner. **4.** A devotee or admirer : FAN. **—lov'er·ly** *adj. & adv.*

**lovers' knot** *n.* A true lovers' knot.

**love seat** *n.* A small sofa that seats two people.

**love·sick** (lŭv'sĭk') *adj.* **1.** So deeply affected by love as to be unable to act normally. **2.** Exhibiting a lover's longing. **—love'sick'ness** *n.*

**lov·ing cup** (lŭv'ĭng) *n.* **1.** A large ornamental wine vessel, usu. with two or more handles. **2.** A loving cup given as an award or trophy.

**lov·ing-kind·ness** (lŭv'ĭng-kīnd'nĭs) *n.* Tender and benevolent affection.

**low**[1] (lō) *adj.* **-er, -est.** [ME *loue* < ON *lāgr.*] **1. a.** Having relatively little height. **b.** Rising only slightly above surrounding surfaces. **c.** Placed or situated below normal height <a *low* railing> **d.** Situated below the surrounding surfaces <rain collecting in *low* spots> **e.** *Archaic.* Dead and buried. **f.** Having a low neckline : DÉCOLLETÉ. **2.** Near or at the horizon <The moon was *low* in the sky.> **3.** Sounded with all or part of the tongue depressed. —Used of a vowel <The vowel (ä) in "large" is *low.*> **4.** Of less than average or usual depth : SHALLOW <The lake is *low.*> **5. a.** Humble in character or status. **b.** Of relatively simple structure in the scale of living organisms. **6.** Morally blameworthy <a *low* deal> **7. a.** Lacking vigor or strength : WEAK. **b.** Emotionally or mentally depressed. **8. a.** Below average in degree, intensity, or amount <*low* temperatures> **b.** Below a standard or average <*low* pay><a *low* level of quality> **c.** Relating to or designating latitudes nearest to the equator. **d.** Of relatively small price. **e.** Not advancing or flourishing. **9. a.** *Mus.* Being a sound produced by a relatively small frequency of vibrations <a *low* note> **b.** Not loud : SUBDUED. **10.** Being almost without money <Funds are *low.*> **11.** Not adequately provided with or equipped for : SHORT <*low* on supplies> **12.** Depreciatory : disparaging <a *low* opinion of their lifestyle> **13.** Reduced in wealth or health. —*adv.* **1. a.** In a low level, position, or space. **b.** In a low rank or condition : HUMBLY. **2.** In or to a reduced, humbled, or degraded condition <brought *low* by financial reverses> **3.** Softly : quietly <speak *low*> **4.** With a deep pitch. **5.** At a small price <bought *low,* sold high> —*n.* **1.** A low position, level, or degree. **2.** A region of depressed barometric pressure. **3.** The gear configuration or setting that produces the lowest range of output speeds, as in an automotive transmission. **—low'ness** *n.*

**low**[2] (lō) *n.* [ME *lowen* < OE *hlōwan.*] The sound uttered by cattle : MOO. —*vi.* **lowed, low·ing, lows.** To utter a low : MOO.

**low beam** *n.* A low-intensity vehicular headlight.

**low·born** (lō'bôrn') *adj.* Being of humble birth.

**low·boy** (lō'boi') *n.* A low tablelike chest of drawers.

**low·bred** (lō'brĕd') *adj.* Vulgar : coarse.

**low·brow** (lō'brou') *n. Informal.* One having uncultivated tastes. —*adj. also* **low·browed** (-broud'). Uncultivated.

**Low-Church** (lō'chûrch') *adj.* Of or pertaining to a faction in the Anglican Church opposing excessive ritualism and favoring a more evangelical doctrine. **—Low'-Church'man** *n.*

**low comedy** *n.* Comedy marked by burlesque, slapstick, and horseplay.

**low·down** (lō'doun') *n. Slang.* The whole truth.

**low-down** (lō'doun') *adj.* **1.** Despicable. **2.** Emotionally depressed.

**low·er**[1] *also* **lour** (lou'ər) [ME *louren.*] —*vi.* **-ered, -er·ing, -ers** *also* **loured, lour·ing, lours. 1.** To look sullen, angry, or threatening : SCOWL. **2.** To appear threatening or dark, as the sky. —*n.* **1.** A sullen, threatening, or angry look. **2.** An ominous appearance, as of thunderheads. **—low'er·ing·ly** *adv.*

**low·er**[2] (lō'ər) *adj.* **1.** Being below another in position, rank, or authority. **2.** Being below a comparable or similar thing <a *lower* shelf> **3.** Being an earlier division of a specified geologic or archaeological period. **4.** Denoting the larger and usu. more representative house of a bicameral legislature. —*v.* **-ered, -er·ing, -ers.** —*vt.* **1.** To bring, let, or move down to a lower level. **2.** To reduce in value, quality, or degree. **3.** To weaken : undermine <A cold *lowers* one's resistance.> **4.** To reduce in respect or standing. —*vi.* **1.** To move down. **2.** To become less : DIMINISH.

☆ ***syns:*** LOWER, DEPRESS, DROP, LET (down) *v. core meaning* : to cause to descend <*lowered* the shade><*lowered* the flag> ***ant:*** raise

**lower bound** *n. Math.* A number that is not greater than any number in a set.

**Lower Carboniferous** *n.* Mississippian.

**low·er-case** (lō'ər-kās') *adj.* Of or relating to small letters as distinguished from capitals. —*vt.* **-cased, -cas·ing, -cas·es.** To set (type) in lower case. **—lower case** *n.*

**lower class** *n.* The socioeconomic class or classes of lower than middle rank. **—low'er-class'** *adj.*

**low·er·class·man** (lō'ər-klăs'mən) *n.* An underclassman.

**Lower Cretaceous** *n.* Comanchean.
**lower criticism** *n.* Textual criticism and verbal examination of a written work, esp. of the Bible.
**low·er·most** (lō'ər-mōst') *adj.* Being lowest.
**lower world** *n.* The abode of the dead, held by the ancients to be under the earth's surface.
**low·er·y** (lou'ə-rē) *adj.* Overcast : threatening.
**lowest common denominator** *n.* The least common multiple of the denominators of a set of fractions.
**lowest common multiple** *n.* Least common multiple.
**lowest terms** *pl.n.* The numerator and denominator of a fraction that have had all common factors but 1 factored out and canceled.
**low frequency** *n.* A radio frequency in the range from 30 to 300 kilocycles per second.
**Low German** *n.* **1.** The German dialects of northern Germany, esp. as used since the beginning of the modern period. **2.** The continental West Germanic languages except High German.
▲ word history: The name *Low German* for the northern German dialects is a translation of German *Plattdeutsch,* literally "flat German." The term refers to the topography of the area where these dialects are spoken: northern Germany, bounded by the North and Baltic seas, is low-lying and flat country. Toward the south the terrain becomes progressively more mountainous, culminating in the Alps. The dialects of the more elevated country are consequently called *High German.*
**low-key** (lō'kē') *also* **low-keyed** (lō'kēd') *adj.* **1.** Low in intensity : SUBDUED. **2.** Having or producing uniformly dark tones with little contrast.
**low·land** (lō'lənd) *n.* Land that is low in relation to the surrounding countryside.
**low·land·er** (lō'lən-dər) *n.* **1.** A native or inhabitant of a lowland. **2. Lowlander.** An inhabitant of the Scottish Lowlands.
**low-lev·el** (lō'lĕv'əl) *adj.* **1.** Relating to or being of low rank or importance. **2.** Located in or occurring at a low level.
**low·life** (lō'līf') *n.* One of low social status or moral character.
**low·ly** (lō'lē) *adj.* **-li·er, -li·est. 1.** Having or suited for a low position or rank. **2.** Humble : meek. **3.** Plain or prosaic. —*adv.* **1.** In a low condition, manner, or position. **2.** In a meek or humble manner. **3.** Low in sound. **—low'li·ness** *n.*
**Low Mass** *n.* A nonelaborate Mass recited rather than sung by the priest.
**low-mind·ed** (lō'mīn'dĭd) *adj.* Showing a crude, vulgar character. **—low'-mind'ed·ly** *adv.* **—low'-mind'ed·ness** *n.*
**low-necked** (lō'nĕkt') *also* **low-neck** (-nĕk') *adj.* Having a low-cut neckline : DÉCOLLETÉ.
**low-pitched** (lō'pĭcht') *adj.* **1.** Low in tone or tonal range. **2.** Having a moderate slope.
**low-pres·sure** (lō'prĕsh'ər) *adj.* **1.** Having, working under, or exerting little pressure. **2.** Calm : easygoing.
**low profile** *n.* Behavior or activity carried out inconspicuously.
**low relief** *n.* Sculptural relief that projects very little from the background.
**low-rise** (lō'rīz') *adj.* Being one or two stories high and having no elevators <*low-rise* condominiums>
**low-spir·it·ed** (lō'spĭr'ĭ-tĭd) *adj.* In low spirits : DEPRESSED. **—low'-spir'it·ed·ly** *adv.* **—low'-spir'it·ed·ness** *n.*
**Low Sunday** *n.* The Sunday after Easter.
**low-ten·sion** (lō'tĕn'shən) *adj.* **1.** Of or at low potential or voltage. **2.** Operating at low voltage.
**low-test** (lō'tĕst') *adj.* Having low volatility and a high boiling point. —Used of gasoline.
**low tide** *n.* **1.** The tide at its lowest ebb. **2.** The time at which low tide occurs.
**low water** *n.* **1.** The lowest level of water in a body of water, as a lake, river, or reservoir. **2.** Low tide.
**lox¹** (lŏks) *n.* [Yiddish *laks* < MHG *lahs,* salmon < OHG.] Smoked salmon.
**lox²** (lŏks) *n.* [L(IQUID) OX(YGEN).] Liquid oxygen, esp. as a rocket fuel oxidizer.
**lox·o·drom·ic** (lŏk'sə-drŏm'ĭk) *also* **lox·o·drom·i·cal** (-ĭ-kəl) *adj.* [Gk. *loxos,* slanting + Gk. *dromos,* course.] *Naut.* Relating to sailing on a rhumb line. **—lox'o·drom'i·cal·ly** *adv.*
**loxodromic curve** *n.* A rhumb line.
**loy·al** (loi'əl) *adj.* [Fr. < OFr. *loial* < Lat. *legalis,* legal < *lex,* law.] **1.** Firm in allegiance to one's government, homeland, or sovereign. **2.** Faithful to a person, custom, or ideal. **3.** Of or professing loyalty. **—loy'al·ly** *adv.*
**loy·al·ist** (loi'ə-lĭst) *n.* **1.** One who maintains loyalty to a lawful government, political party, or sovereign, esp. during war or revolutionary change. **2. Loyalist.** A Tory. **—loy'al·ism** *n.*
**loy·al·ty** (loi'əl-tē) *n., pl.* **-ties. 1.** The quality or state of being loyal. **2. loyalties.** Feelings of devoted attachment and affection.
**loz·enge** (lŏz'ĭnj) *n.* [ME *losenge* < OFr.] **1.** A four-sided planar figure with a diamondlike shape. **2.** A lozenge-shaped, medicated candy.
**LP** (ĕl'pē') *n.* [Orig. a trademark.] A long-playing record.
**LSD** (ĕl'ĕs-dē') *n.* Lysergic acid diethylamide.
**Lu** *symbol for* LUTETIUM.
**lu·au** (lo͞o-ou', lo͞o'ou') *n.* [Hawaiian *lu'au.*] A Hawaiian feast.
**lub·ber** (lŭb'ər) *n.* [ME *lobur,* lazy lout.] **1.** An awkward fellow. **2.** An inexperienced sailor.
**lubber line** *also* **lubber's line** *n.* A mark or line on a compass or cathode-ray indicator that represents the heading of a ship or aircraft.
**lubber's hole** *n.* A hole through the platform surrounding the upper part of a ship's mast, through which one may climb to go aloft.
**lu·bri·cant** (lo͞o'brĭ-kənt) *n.* A usu. oily substance, as grease, that reduces friction, heat, and wear when applied as a surface coating to moving parts or between solid surfaces. **—lu'bri·cant** *adj.*
**lu·bri·cate** (lo͞o'brĭ-kāt') *v.* **-cat·ed, -cat·ing, -cates.** [Lat. *lubricare, lubricat-* < *lubricus,* slippery.] —*vt.* **1.** To apply a lubricant to. **2.** To make smooth or slippery. —*vi.* To act as a lubricant. **—lu'bri·ca'tion** *n.* **—lu'bri·ca'tive** *adj.*
**lu·bri·ca·tor** (lo͞o'brĭ-kā'tər) *n.* **1.** One who lubricates. **2.** A lubricant. **3.** A device for applying a lubricant.
**lu·bri·cious** (lo͞o-brĭsh'əs) *also* **lu·bri·cous** (-brĭk'əs) *adj.* [Lat. *lubricus,* slippery.] **1.** Having a smooth or slippery quality. **2. a.** Marked by lewdness : WANTON. **b.** Sexually arousing : SALACIOUS. **3.** Characterized by shiftiness or trickery. **—lu·bri'cious·ly** *adv.* **—lu·bri'cious·ness** *n.*
**lu·bric·i·ty** (lo͞o-brĭs'ĭ-tē) *n.* [LLat. *lubricitas,* slipperiness < Lat. *lubricus,* slippery.] The quality or condition of being lubricious.
**lu·carne** (lo͞o-kärn') *n.* [OFr. *lucane,* poss. of Germanic orig.] A dormer window.
**lu·cent** (lo͞o'sənt) *adj.* [Lat. *lucens, lucent-,* pr.part. of *lucēre,* to shine.] **1.** Giving off light : LUMINOUS. **2.** Clear : translucent. **—lu'cen·cy** *n.* **—lu'cent·ly** *adv.*
**lu·cerne** (lo͞o-sûrn') *n.* [Fr. *luzerne* < Prov. *luzerno.*] *Chiefly Brit.* Alfalfa.
**lu·ces** (lo͞o'sēz') *n. var. pl. of* LUX.
**lu·cid** (lo͞o'sĭd) *adj.* [Lat. *lucidus* < *lux,* light < *lucēre,* to shine.] **1.** Easily understood. **2.** Mentally sound : SANE. **3.** Translucent. **—lu·cid'i·ty** (-sĭd'ĭ-tē), **lu'cid·ness** *n.* **—lu'cid·ly** *adv.*
**Lu·ci·fer** (lo͞o'sə-fər) *n.* [ME < Lat., morning star < *lucifer,* light-bearing : *lux,* light + *ferre,* to bring.] **1.** The archangel who was expelled from heaven : SATAN. **2.** The planet Venus in its appearance as the morning star. **3. lucifer.** A friction match.
**lu·cif·er·ase** (lo͞o-sĭf'ə-rās') *n.* An enzyme that catalyzes the oxidation of luciferin.
**lu·cif·er·in** (lo͞o-sĭf'ər-ĭn) *n.* [Lat. *lucifer,* light-bringing + -IN.] A pigment in bioluminescent animals, as fireflies or certain marine crustaceans, that produces an almost heatless, bluish-green light when oxidized.
**Lu·ci·na** (lo͞o-sī'nə) *n.* [Lat., goddess of childbirth < *lucinus,* light-bringing < *lux,* light < *lucēre,* to shine.] *Archaic.* A midwife.
**Lu·cite** (lo͞o'sīt'). A trademark for a transparent thermoplastic acrylic resin.
**luck** (lŭk) *n.* [ME *lucke* < MDu. *luc.*] **1.** The chance happening of adverse or fortunate events. **2.** Good fortune or prosperity : SUCCESS. —*vi.* **lucked, luck·ing, lucks.** To gain success or something desirable by chance <*lucked* on to a good job> **—luck'less** *adj.*
**luck·y** (lŭk'ē) *adj.* **-i·er, -i·est. 1.** Having good luck. **2.** Occurring by chance : FORTUITOUS. **3.** Believed to bring good luck <a *lucky* charm> **—luck'i·ly** (-lē) *adv.* **—luck'i·ness** *n.*
**lu·cra·tive** (lo͞o'krə-tĭv) *adj.* [ME *lucratif* < OFr. < Lat. *lucrativus* < *lucrari,* to profit < *lucrum,* profit.] Producing wealth or profits.
**lu·cre** (lo͞o'kər) *n.* [ME < Lat. *lucrum.*] Money or profits.
**lu·cu·brate** (lo͞o'kyo͞o-brāt') *vi.* **-brat·ed, -brat·ing, -brates.** [Lat. *lucubrare, lucubrat-,* to work at night by lamplight < *lux,* light < *lucēre,* to shine.] To write in a scholarly way.
**lu·cu·bra·tion** (lo͞o'kyo͞o-brā'shən) *n.* **1.** Laborious writing or study. **2.** Pedantry in writing or speech.
**lu·cu·lent** (lo͞o'kyo͞o-lənt) *adj.* [ME, shiny < Lat. *luculentus* < *lux,* light < *lucēre,* to shine.] Clearly understood : LUCID.
**Lu·cul·lan** (lo͞o-kŭl'ən) *adj.* [After Lucius Licinius *Lucullus* (110?–57? B.C.), Roman patron of arts.] Lavish : luxurious <a *Lucullan* banquet>
**Lud·dite** (lŭd'īt') *n.* [After Ned *Ludd,* a legendary leader.] One of a group of British workers who between 1811 and 1816 rioted and destroyed laborsaving textile machinery in the belief that such machinery would diminish employment.
**lu·di·crous** (lo͞o'dĭ-krəs) *adj.* [Lat. *ludicrus,* sportive < *ludus,* game < *ludere,* to play.] So obviously absurd or incongruous as to be laughable. **—lu'di·crous·ly** *adv.* **—lu'di·crous·ness** *n.*
**lu·es** (lo͞o'ēz) *n., pl.* **lues.** [NLat. < Lat., plague.] Syphilis. **—lu·et'ic** (-ĕt'ĭk) *adj.* **—lu·et'i·cal·ly** *adv.*
**luff** (lŭf) [ME *loffe,* spar for holding a square sail on a windward tack < OFr. *lof,* prob. of Germanic orig.] *Naut.* —*n.* **1.** The act of sailing closer into the wind. **2.** The forward side of a fore-and-aft sail.

3. The fullest part of the bow of a ship. —*vi.* **luffed, luff·ing, luffs.** **1.** To steer a sailing vessel nearer into the wind. **2.** To flap while losing wind. —Used of a sail.

**lug¹** (lŭg) *n.* [ME *lugge,* earflap.] **1.** A projection or handle on a vessel or machine shaped like an ear, used as a hold or support. **2.** A nut, esp. one closed at one end to serve as a cap. **3.** A loop, usu. of leather, at the side of the saddle of a harness rig through which one of the shafts passes. **4.** A brass or copper fitting to which electrical wires can be connected, as by soldering. **5.** *Slang.* A clumsy fool : DOLT.

**lug²** (lŭg) *v.* **lugged, lug·ging, lugs.** [ME *luggen,* of Scand. orig.] —*vt.* **1.** To haul or drag (something) laboriously. **2.** To drag or pull with short jerks. —*vi.* **1.** To pull with difficulty : TUG. **2.** To move by jerks or as if under a heavy burden. —*n.* **1.** *Archaic.* An act of lugging. **2.** *Archaic.* Something lugged. **3.** A lugsail. **4.** *Slang.* An extortion of money <put the *lug* on>

**luge** (lo͞ozh) *n.* [Fr.] **1.** A sled similar to a toboggan that is ridden in a supine position. **2.** A competition in which luges race against a clock.

**luge**

**lug·gage** (lŭg'ĭj) *n.* [Probably LUG² + (BAG)GAGE.] **1.** Containers for a traveler's belongings. **2.** The cases and belongings of a traveler.

**lug·ger** (lŭg'ər) *n.* [< LUGSAIL.] A small fishing or sailing boat equipped with one or more lugsails.

**lug·sail** (lŭg'səl) *n.* [Perh. < LUG¹.] *Naut.* A quadrilateral sail lacking a boom and having the foot larger than the head, bent to a yard hanging obliquely on the mast.

**lu·gu·bri·ous** (lo͝o-go͞o'brē-əs, lo͝o-gyo͞o'-) *adj.* [Lat. *lugubris* < *lugēre,* to mourn.] Mournful, esp. to a ludicrous or exaggerated degree. **—lu·gu'bri·ous·ly** *adv.* **—lu·gu'bri·ous·ness** *n.*

**lug·worm** (lŭg'wûrm') *n.* [Orig. unknown.] A segmented burrowing marine worm of the genus *Arenicola,* esp. *A. marina,* often used as bait.

**Luke** (lo͝ok) *n.* —See table at BIBLE.

**luke·warm** (lo͞ok'wôrm') *adj.* [ME *leukwarm* : *leuk,* lukewarm + *warm,* warm.] **1.** Mildly warm : TEPID. **2.** Lacking enthusiasm : HALF-HEARTED. **—luke'warm'ly** *adv.* **—luke'warm'ness** *n.*

**lull** (lŭl) *v.* **lulled, lull·ing, lulls.** [ME *lullen.*] —*vt.* **1.** To cause to sleep or rest : SOOTHE. **2.** To deceive into trustfulness. —*vi.* To become calm. —*n.* **1.** A relatively calm interval during a storm. **2.** An interval of lessened activity <a *lull* in sales>

**lull·a·by** (lŭl'ə-bī') *n., pl.* **-bies.** [Obs. *lulla,* interjection used in lullabies (< ME *lullai* < *lullen,* to lull) + (BYE)-BY(E).] A soothing song intended to lull a child to sleep. —*vt.* **-bied, -by·ing, -bies.** To quiet with or as if with a lullaby.

**lu·lu** (lo͞o'lo͞o) *n.* [Perh. from *Lulu,* a nickname for *Louise.*] *Slang.* Something remarkable : LOLLAPALOOZA.

**lum·ba·go** (lŭm-bā'gō) *n.* [Lat. < *lumbus,* loin.] A painful inflammatory rheumatism of the lumbar tendons and muscles.

**lum·bar** (lŭm'bər, -bär') *adj.* [NLat. *lumbaris* < Lat. *lumbus,* loin.] Of or located in the part of the back and sides between the lowest ribs and the pelvis. —*n.* A lumbar nerve, artery, vertebra, or part.

**lum·ber¹** (lŭm'bər) *n.* [Perh. < LUMBER².] **1.** Timber sawed into standardized structural members, as boards or planks. **2.** *Chiefly Brit.* Miscellaneous stored items. **3.** Something cumbersome or useless. —*v.* **-bered, -ber·ing, -bers.** —*vt.* **1. a.** To cut down (trees) and prepare as marketable timber. **b.** To cut down the timber of. **2.** *Chiefly Brit.* To clutter with or as if with unused items. —*vi.* To cut and prepare timber for the market. **—lum'ber** *adj.* **—lum'ber·man, lum'ber·er** *n.*

**lum·ber²** (lŭm'bər) *vi.* **-bered, -ber·ing, -bers.** [ME *lomeren.*] **1.** To move or walk heavily and clumsily. **2.** To move with a rumbling noise. **—lum'ber·er** *n.*

**lum·ber·jack** (lŭm'bər-jăk') *n.* **1.** One who fells trees and transports the timber to a mill. **2.** A short warm outer jacket.

**lum·ber·yard** (lŭm'bər-yärd') *n.* A commercial yard from which lumber and other building materials are sold.

**lum·bri·coid** (lŭm'brĭ-koid') *n.* [Lat. *lumbricus,* earthworm + -OID.] A parasitic roundworm, *Ascaris lumbricoides,* that infests the human intestine. —*adj.* Resembling an earthworm.

**lu·men** (lo͞o'mən) *n., pl.* **-mens** or **-mi·na** (-mə-nə) [NLat. < Lat., an opening.] **1.** *Anat.* The inner open space of a tubular organ, as of a blood vessel or an intestine. **2.** *Physics.* The unit of luminous flux in the International System, equal to the luminous flux emitted in a solid angle of one steradian by a uniform point source having an intensity of one candle. **—lu'men·al, lu'min·al** *adj.*

**lu·mi·nance** (lo͞o'mə-nəns) *n.* **1.** The quality or state of being luminous. **2.** *Physics.* The luminous intensity per unit projected area of a given surface viewed from a given direction.

**lu·mi·nar·y** (lo͞o'mə-nĕr'ē) *n., pl.* **-ies.** [ME < OFr. *luminarie* and Med. Lat. *luminarium* < LLat. *luminare* < Lat. *lumen,* light.] **1.** An object, as a celestial body, that gives light. **2.** A source of spiritual or intellectual light. **3.** A notable person. **—lu'mi·nar'y** *adj.*

**lu·mi·nesce** (lo͞o'mə-nĕs') *vi.* **-nesced, -nesc·ing, -nesc·es.** [Back-formation < LUMINESCENCE.] To be or become luminescent.

**lu·mi·nes·cence** (lo͞o'mə-nĕs'əns) *n.* [Lat. *lumen, lumin-,* light + -ESCENCE.] **1.** Emission of light, as in phosphorescence, fluorescence, and bioluminescence, by processes that derive energy from essentially nonthermal sources, as chemical, biochemical, or crystallographic changes, motion of subatomic particles, or excitation of an atomic system by radiation, esp. such emission distinguished from incandescence. **2.** Light emitted by luminescence.

**lu·mi·nes·cent** (lo͞o'mə-nĕs'ənt) *adj.* Capable of, showing, or suitable for the emission of luminescence.

**lu·mi·nif·er·ous** (lo͞o'mə-nĭf'ər-əs) *adj.* [Lat. *lumen, lumin-,* light + -FEROUS.] Generating, transmitting, or yielding light.

**lu·mi·nos·i·ty** (lo͞o'mə-nŏs'ĭ-tē) *n.* **1.** LUMINANCE 1. **2.** Something luminous. **3.** *also* **luminosity factor.** The ratio of luminous flux at a specific wavelength to the radiant flux at the same wavelength.

**lu·mi·nous** (lo͞o'mə-nəs) *adj.* [ME < OFr. *lumineux* < Lat. *luminosus* < *lumen,* light.] **1.** Emitting light, esp. self-generated light. **2.** Full of light : ILLUMINATED. **3. a.** Easily understood : LUCID. **b.** Enlightened and intelligent. **—lu'mi·nous·ly** *adv.* **—lu'mi·nous·ness** *n.*

**luminous efficiency** *n.* The ratio of the total luminous flux to the total radiant flux of an emitting source.

**luminous energy** *n.* Radiant energy of electromagnetic waves in the visible portion of the electromagnetic spectrum.

**luminous flux** *n.* The rate of flow of light per unit time, esp. the flux of visible light expressed in lumens.

**luminous intensity** *n.* The luminous flux density per solid angle as measured in a given direction relative to the emitting source.

**luminous paint** *n.* A paint containing a phosphorescent or fluorescent substance that makes it glow in the dark.

**lum·mox** (lŭm'əks) *n.* [Orig. unknown.] An ungainly person.

**lump¹** (lŭmp) *n.* [ME.] **1.** An irregularly shaped piece or mass. **2.** A small cube, as of sugar. **3.** *Pathol.* A swelling or small palpable mass. **4.** An aggregate : collection. **5.** A lummox. **6. lumps. a.** Punishment in the form of beatings <took a lot of *lumps* growing up small> **b.** One's just deserts : COMEUPPANCE. —*adj.* **1.** Formed into lumps. **2.** Undivided : unbroken <one *lump* sum> —*v.* **lumped, lump·ing, lumps.** —*vt.* **1.** To put together in a single group without discrimination. **2.** To make into lumps. —*vi.* **1.** To become lumpy. **2.** To move heavily or awkwardly.

**lump²** (lŭmp) *vt.* **lumped, lump·ing, lumps.** [Orig. unknown.] *Informal.* To put up with : TOLERATE <like it or *lump* it>

**lum·pen** (lŭm'pən, lo͝om'-) *adj.* [< G. *Lumpenproletariat,* the lowest section of the proletariat < *Lump,* contemptible person < *Lumpen,* rags.] Of or relating to dispossessed, often displaced individuals who have been cut off from the socioeconomic class with which they would ordinarily be identified <*lumpen* intellectuals>

**lump·er** (lŭm'pər) *n.* [< LUMP¹.] A laborer employed to load and unload ships.

**lump·fish** (lŭmp'fĭsh') *n., pl.* **lumpfish** or **-fish·es.** [Obs. *lump,* lumpfish + FISH.] A fish, *Cyclopterus lumpus* of Atlantic waters, with a body covered with tuberous excrescences.

**lump·ish** (lŭm'pĭsh) *adj.* **1.** Dull : stupid. **2.** Awkward : cumbersome. **—lump'ish·ly** *adv.* **—lump'ish·ness** *n.*

**lump·y** (lŭm'pē) *adj.* **-i·er, -i·est.** **1.** Covered or filled with lumps. **2.** Thickset or cumbersome. **3.** Marked by short, jumbled waves, as a tidal rip.

**lumpy jaw** *n.* Actinomycosis.

**Lu·na** (lo͞o'nə) *n.* [Lat. < *luna,* moon.] **1.** *Myth.* The goddess of the moon. **2. luna.** An alchemical designation for silver.

**lu·na·cy** (lo͞o'nə-sē) *n., pl.* **-cies.** [< LUNATIC.] **1.** Intermittent mental derangement. **2.** Insanity. **3. a.** Wild and irresponsible foolishness : FOLLY. **b.** A foolish act.

**luna moth** *n.* [NLat. *luna,* specific epithet of *Actias luna* < Lat., moon.] A large, pale-green North American moth, *Actias luna,* with a long projection on each hind wing.

**lu·nar** (lo͞o'nər) *adj.* [Lat. *lunaris* < *luna,* moon.] **1.** Of, involving, caused by, or affecting the moon. **2.** Measured by the revolution of the moon. **3.** Of or pertaining to silver.

**lunar caustic** *n.* Silver nitrate used in cauterization.

**lunar excursion module** *also* **lunar module** *n.* A spacecraft designed to transport astronauts from a command module orbiting the moon to the lunar surface and back.

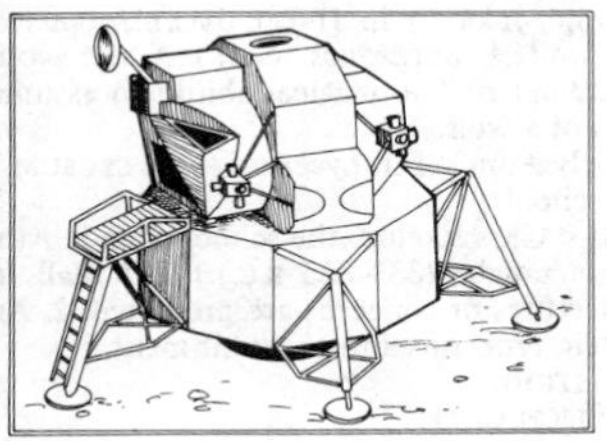

**lunar excursion module**

**lunar month** *n.* MONTH 5.
**lu·nar·naut** (lo͞o'nər-nôt') *n.* [LUNAR + (ASTRO)NAUT.] An astronaut who explores the moon.
**lunar year** *n.* An interval of 12 lunar months.
**lu·nate** (lo͞o'nāt') *also* **lu·nat·ed** (-nā'tĭd) *adj.* [Lat. *lunatus* < *lunare,* to bend like a crescent < *luna,* moon.] Crescent-shaped.
**lunate bone** *n.* The second of three bones forming the upper row of bones in the wrist.
**lu·na·tic** (lo͞o'nə-tĭk) *adj.* [ME *lunatik* < OFr. *lunatique* < Lat. *lunaticus* < *luna,* moon.] **1.** Suffering from lunacy : INSANE. **2.** Of or for the insane <a *lunatic* asylum> **3.** Wildly foolish : SILLY <a *lunatic* scheme to get rich quick> **—lu'na·tic** *n.*
**lunatic fringe** *n.* The extremist, fanatical, or irrational members of a group, esp. such members of a religious or political group.
**lu·na·tion** (lo͞o-nā'shən) *n.* [ME *lunacioun* < Med. Lat. *lunatio* < Lat. *luna.*] The elapsed time between two successive new moons, averaging 29 days, 12 hours, 44 minutes, 28 seconds.
**lunch** (lŭnch) *n.* [Perh. < Sp. *lonja,* slice < OFr. *longe.*] **1.** A midday meal. **2.** Food for a midday meal. **—lunch** *v.* **(lunched, lunch·ing, lunch·es).** **—lunch'er** *n.*
**lunch·eon** (lŭn'chən) *n.* [Prob. < LUNCH.] **1.** LUNCH 1. **2.** An afternoon party at which a light meal is served.
**lunch·eon·ette** (lŭn'chə-nĕt') *n.* A small restaurant featuring inexpensive meals.
**lunch·room** (lŭnch'ro͞om', -ro͝om') *n.* **1.** A luncheonette. **2.** A room in a facility, as a school, where lunches may be purchased or those brought from home may be eaten.
**lune** (lo͞on) *n.* [Lat. *luna,* moon.] A section of a sphere enclosed between two semicircles with their common end points at opposite poles.
**lu·nette** (lo͞o-nĕt') *n.* [Fr. < OFr., moon-shaped object < *lune,* moon < Lat. *luna,* moon.] **1. a.** A small, crescent-shaped or circular opening in a vaulted roof. **b.** A crescent-shaped or semicircular space, usu. over a window or door, that may contain another window, a sculpture, or a mural. **2.** A military fortification with two projecting faces and two parallel flanks.
**lung** (lŭng) *n.* [ME *lunge* < OE *lungen.*] **1.** Either of two spongy, saclike thoracic organs in most vertebrates, functioning to remove carbon dioxide from the blood and provide it with oxygen. **2.** An invertebrate structure similar to the lung, as in terrestrial snails.
**lunge** (lŭnj) *n.* [Alteration of obs. *allonge* < Fr. *alongier,* to lunge : *à,* to + *long,* long.] **1.** A sudden thrust or pass, as with a sword. **2.** A sudden forward movement : PLUNGE. *—v.* **lunged, lung·ing, lung·es.** *—vi.* **1.** To make a pass or thrust. **2.** To move with a lunge. *—vt.* To cause (someone) to lunge.
**lung·er¹** (lŭn'jər) *n.* One that lunges.
**lung·er²** (lŭng'ər) *n. Informal.* One afflicted with tuberculosis of the lungs.
**lung·fish** (lŭng'fĭsh') *n., pl.* **lungfish** or **-fish·es.** Any of several elongated tropical freshwater fishes of the order Dipnoi or Dipneusti, having lungs as well as gills, and in certain species constructing a mucus-lined mud covering in which to withstand an extended drought.
**lung·wort** (lŭng'wûrt', -wôrt') *n.* **1.** A plant of the genus *Mertensia,* with tubular, usu. blue flower clusters. **2.** A native European plant of the genus *Pulmonaria,* with long-stalked leaves and coiled blue or purple flower clusters, once used to treat respiratory disorders.
**lu·ni·so·lar** (lo͞o'nĭ-sō'lər) *adj.* [Lat. *luna,* moon + SOLAR.] Of or caused by both the sun and the moon.
**lu·ni·ti·dal** (lo͞o'nĭ-tīd'l) *adj.* [Lat. *luna,* moon + TIDAL.] Of or relating to tidal phenomena caused by the moon.
**lunitidal interval** *n.* The elapsed time between the moon's transit of a particular meridian and the next high tide at that meridian.
**lunk·er** (lŭng'kər) *n.* [Orig. unknown.] *Informal.* One unusually large of its kind, esp. a game fish.
**lunk·head** (lŭngk'hĕd') *n.* [Prob. alteration of LUMP + HEAD.] *Slang.* A stupid person : DOLT. **—lunk'head'ed** *adj.*
**lu·nu·la** (lo͞on'yə-lə) *n., pl.* **-lae** (-lē') [Lat. < *luna,* moon.] A small crescent-shaped marking or structure.
**lu·nu·lar** (lo͞on'yə-lər) *adj.* Crescent-shaped.
**lu·nu·late** (lo͞on'yə-lāt', -lĭt) *also* **lu·nu·lat·ed** (-lā'tĭd) *adj.* **1.** Small and lunular. **2.** Having crescent-shaped markings.
**lu·nule** (lo͞on'yo͞ol) *n.* A lunula.
**lun·y** (lo͞o'nē) *adj. var. of* LOONY.
**Lu·per·ca·li·a** (lo͞o'pər-kā'lē-ə) *n.* [Lat. < *Lupercus,* Roman god of flocks.] An ancient Roman fertility festival celebrated on Feb. 15 in honor of the pastoral god Lupercus. **—Lu'per·ca'li·an** *adj.*
**lu·pine¹** *also* **lu·pin** (lo͞o'pən) *n.* [ME < Lat. *lupinum* < *lupinus,* wolflike < *lupus,* wolf.] A plant of the genus *Lupinus,* with stalks of variously colored flower clusters.
**lu·pine²** (lo͞o'pīn') *adj.* [Lat. *lupinus* < *lupus,* wolf.] **1.** Wolflike. **2.** Fierce : RAPACIOUS.
**lu·pu·lin** (lo͞op'yə-lən) *n.* [NLat. *lupulus,* specific epithet for a hop species (< Lat. *lupus,* a hop plant) + -IN.] Minute yellowish-brown hairs from the strobiles of the hop plant, once used as a sedative.
**lu·pus** (lo͞o'pəs) *n.* [Med. Lat. < Lat., wolf.] A disease of the skin and mucous membranes causing disfiguring lesions, esp.: **a.** Lupus vulgaris, marked by ulcerating, nodular facial lesions, esp. around the nose and ears. **b.** Lupus erythematosus, marked by eruption of atrophic scarred lesions with chronically inflamed margins.
**lurch¹** (lûrch) *vi.* **lurched, lurch·ing, lurch·es.** [Orig. unknown.] **1.** To stagger. **2.** To pitch or roll erratically or suddenly, as a ship during a storm. *—n.* **1.** A staggering or tottering gait or movement. **2.** An abrupt pitching or rolling.
**lurch²** (lûrch) *n.* [OFr. *lourche,* a kind of game < *lourche,* soundly defeated.] The losing position of a cribbage player who scores 30 points or less to the winner's 61. **—in the lurch.** In a difficult position.
**lurch·er** (lûr'chər) *n.* [< obs. *lurch,* to lurk < ME *lorchen,* perh. < *lurken.* —see LURK.] **1.** *Archaic.* A sneak thief. **2.** *Chiefly Brit.* A crossbred dog used by poachers.
**lure** (lo͝or) *n.* [ME < OFr., of Germanic orig.] **1. a.** Something tempting or attracting : ENTICEMENT. **b.** An appeal or attraction. **2.** A decoy used in catching animals, esp. an artificial bait used for catching fish. **3.** A bunch of feathers attached to a long cord, used in falconry to recall the hawk. *—vt.* **lured, lur·ing, lures. 1.** To attract by temptation or wiles : ENTICE. **2.** To recall (a falcon) with a lure.
**lu·rid** (lo͝or'ĭd) *adj.* [Lat. *luridus,* pale < *luror,* paleness.] **1. a.** Causing shock or horror : GHASTLY. **b.** Emphasizing violence or sensationalism <*lurid* headlines> **2.** Glowing or shining with the glare of fire through a haze. **3.** Pallid or sallow. **—lu'rid·ly** *adv.* **—lu'rid·ness** *n.*
**lurk** (lûrk) *vi.* **lurked, lurk·ing, lurks.** [ME *lurken,* of Scand. orig.] **1.** To lie in wait, as in ambush. **2.** To move stealthily : SNEAK. **3.** To exist unsuspected or unobserved <dangers *lurking* everywhere>
**lus·cious** (lŭsh'əs) *adj.* [ME *lucius,* short for *delicious,* delicious. —see DELICIOUS.] **1.** Sweet and pleasant to smell or taste <a *luscious* dessert> **2.** Having strong sensory appeal : SEDUCTIVE. **3.** *Archaic.* Excessively sweet : CLOYING.
**lush¹** (lŭsh) *adj.* **-er, -est.** [ME *lusch,* soft.] **1. a.** Having or marked by luxuriant vegetation. **b.** Marked by abundance : PLENTIFUL. **c.** Very productive : THRIVING. **2. a.** Sumptuous : opulent. **b.** Delicious : savory. **c.** Voluptuous : sensual. **3.** Excessively elaborate : EXTRAVAGANT <*lush* verse> **—lush'ly** *adv.* **—lush'ness** *n.*
**lush²** (lŭsh) [Orig. unknown.] *Slang.* *—n.* **1.** A drunkard. **2.** Intoxicating liquor. *—vi.* **lushed, lush·ing, lush·es.** To drink liquor to excess : GUZZLE.
**lust** (lŭst) *n.* [ME < OE, desire.] **1.** Intense or unrestrained sexual desire. **2.** An overwhelming craving <a *lust* for money> **3.** *Obs.* Pleasure : relish. *—vi.* **lust·ed, lust·ing, lusts.** To have an intense or obsessive desire, esp. sexual desire.

▲ word history: The noun *lust* preserves the same form it had in Old English but its meaning is now quite different. It originally was a word of neutral connotations, meaning simply "pleasure." *Lust* is related to the now archaic verb *list,* meaning "to wish to, to be inclined to." In theological usage Old English *lust* was used to refer to pleasures and desires that were considered sinful, especially sexual desire. In this context *lust* was a term of opprobrium and reproach. This disapproval has carried over to the most recent sense of *lust,* "an overwhelming desire or craving." The meaning "pleasure" is now obsolete.

**lus·ter** (lŭs'tər) *n.* [OFr. *lustre* < OItal. *lustro* < *lustrare,* to make bright < Lat. < *lustrum,* purification < *luere,* to set free.] **1.** Soft reflected light : SHEEN. **2.** Radiance or brilliance of light : BRIGHTNESS. **3.** Glorious or radiant quality : SPLENDOR. **4.** A glass pendant, esp. on a chandelier. **5.** A decorative object, as a chandelier with glass pendants. **6.** A fabric, as alpaca, with a glossy surface. **7.** The appearance of a mineral surface judged by its brilliance and ability to reflect light. *—vt. & vi.* **-tered, -ter·ing, -ters.** To give a luster to or be or become lustrous.
**lus·ter·ware** (lŭs'tər-wâr') *n.* Pottery with a metallic sheen.
**lust·ful** (lŭst'fəl) *adj.* Filled with lust : LASCIVIOUS. **—lust'ful·ly** *adv.* **—lust'ful·ness** *n.*
**lus·tral** (lŭs'trəl) *adj.* [Lat. *lustralis* < *lustrium,* lustrum. —see LUSTER.] **1.** Of, relating to, or used in a purification rite. **2.** Relating to a lustrum.

**lus·trate** (lŭs′trāt′) *vt.* **-trat·ed, -trat·ing, -trates.** [Lat. *lustrare, lustrat-,* to purify, make bright. —see LUSTER.] To purify by ceremony. **—lus·tra′tion** *n.* **—lus′tra·tive** (-trə-tĭv) *adj.*

**lus·tre** (lus′tər) *n.* & *v. Chiefly Brit. var. of* LUSTER.

**lus·trous** (lŭs′trəs) *adj.* **1.** Having a glow or sheen. **2.** Gleaming with or as if with brilliant light : RADIANT. **—lus′trous·ly** *adv.* **—lus′trous·ness** *n.*

**lus·trum** (lŭs′trəm) *n.* [Lat. —see LUSTER.] **1.** Ceremonial purification of the entire ancient Roman population after the census every five years. **2.** A five-year period.

**lust·y** (lŭs′tē) *adj.* **-i·er, -i·est. 1.** Full of vitality : VIGOROUS. **2.** Powerful. **3.** Lustful. **4.** *Archaic.* Merry : joyous. **—lust′i·ly** *adv.* **—lust′i·ness** *n.*

**lu·sus na·tu·rae** (lo͞o′səs nə-to͝or′ē, -tyo͝or′ē) *n.* [Lat.] A freak of nature.

**lu·ta·nist** (lo͞ot′n-ĭst) *n.* [Med. Lat. *lutanista* < *lutana,* lute, poss. < OFr. *lut.* —see LUTE[1].] LUTIST 2.

**lute[1]** (lo͞ot) *n.* [ME < OFr. *lut* < Ar. *al-'ud.*] A musical stringed instrument with a body shaped like half a pear and usu. a bent neck with a fretted fingerboard with pegs for tuning.

**lute[2]** (lo͞ot) *n.* [ME < OFr. *lut* < Lat. *lutum,* potter's clay.] A substance, as dried cement or clay, used to pack and seal joints or to coat a porous surface to make it impervious.

**lu·te·al** (lo͞o′tē-əl) *adj.* Of or pertaining to the corpus luteum.

**lu·te·ci·um** (lo͞o-tē′shē-əm) *n. var. of* LUTETIUM.

**lu·te·in** (lo͞o′tē-ĭn, -tēn′) *n.* [Lat. *luteum* (< *luteus,* yellow < *lutum,* yellowweed) + -IN.] A yellow pigment isolated from the corpus luteum and found in body fats and egg yolk.

**lu·te·ous** (lo͞o′tē-əs) *adj.* [Lat. *luteus,* yellow. —see LUTEIN.] Of a light or moderate greenish yellow.

**lu·te·ti·um** *also* **lu·te·ci·um** (lo͞o-tē′shē-əm) *n.* [Lat. *Lutetia,* ancient name of Paris, France + -IUM.] *Symbol* **Lu** A silvery-white rare-earth element used in nuclear technology; atomic number 71; atomic weight 174.97.

**Lu·ther·an** (lo͞o′thər-ən) *adj.* **1.** Of or pertaining to Martin Luther or his religious teachings and esp. to the doctrine of justification by faith alone. **2.** Of or pertaining to the branch of the Protestant Church adhering to the views of Martin Luther. **—Lu′ther·an** *n.* **—Lu′ther·an·ism** *n.*

**lu·tist** (lo͞o′tĭst) *n.* **1.** A maker of lutes. **2.** A lute player.

**lux** (lŭks) *n., pl.* **lux·es** or **lu·ces** (lo͞o′sēz′) [Lat., light < *lucēre,* to shine.] The International System unit of illumination, equal to one lumen per square meter.

**lux·ate** (lŭk′sāt′) *vt.* **-at·ed, -at·ing, -ates.** [Lat. *luxare, luxat-* < *luxus,* dislocated.] To put out of joint : DISLOCATE. **—lux·a′tion** *n.*

**luxe** (lo͝oks, lŭks) *n.* [Fr., luxury < Lat. *luxus.*] Elegant sumptuousness : LUXURY.

**lux·u·ri·ant** (lŭg-zho͝or′ē-ənt, lŭk-sho͝or′-) *adj.* [Lat. *luxurians, luxuriant-,* pr.part. of *luxuriare,* to be luxuriant. —see LUXURIATE.] **1. a.** Marked by profuse or rich growth. **b.** Producing or yielding in abundance. **2.** Excessively elaborate : FLORID. **3.** Marked by or showing luxury : LUXURIOUS. **—lux·u′ri·ance** *n.* **—lux·u′ri·ant·ly** *adv.*

**lux·u·ri·ate** (lŭg-zho͝or′ē-āt′, lŭk-sho͝or′-) *vi.* **-at·ed, -at·ing, -ates.** [Lat. *luxuriare, luxuriat-,* to be luxuriant < *luxuria,* luxury.] **1.** To take luxurious pleasure : indulge oneself. **2.** To proliferate. **3.** To grow profusely : THRIVE.

**lux·u·ri·ous** (lŭg-zho͝or′ē-əs, lŭk-sho͝or′-) *adj.* **1.** Fond of or inclined to luxury. **2.** Of or pertaining to self-indulgent or unrestrained gratification of the senses. **3. a.** Of the most expensive or choice variety. **b.** Marked by extravagant, ostentatious magnificence.

☆ ***syns:*** LUXURIOUS, LAVISH, LUSH, LUXURIANT, OPULENT, PALATIAL, PLUSH, RICH *adj. core meaning* : marked by extravagant, ostentatious magnificence <a *luxurious* royal yacht>

**lux·u·ry** (lŭg′zhə-rē, lŭk′shə-) *n., pl.* **-ries.** [ME *luxurie,* lust < OFr. < Lat. *luxuria* < *luxus,* luxury.] **1.** Something not essential but conducive to comfort or pleasure. **2.** Something expensive or hard to obtain. **3.** Sumptuous living or surroundings <lives in *luxury*>

**Lw** *symbol for* LAWRENCIUM.

**lwei** (lwā) *n., pl.* **lwei.** [Of Bantu orig.] —See table at CURRENCY.

**-ly[1]** *suff.* [ME *-li* < OE *-līc.*] **1.** Like : resembling : having the characteristics of <sister*ly*> **2.** Recurring at a specified interval of time <hour*ly*>

▲ word history: It is no coincidence that the adjectival suffix *-ly*[1] means "like" because *-ly*[1] and *like*[2] are ultimately derived from the same form. The prehistoric Germanic root *līko-* is the source of both Modern English forms. *Līko-* meant basically "form, shape, body," and the original adjectival suffix, represented in Old English as *-līc,* meant "having the form of." The same Germanic root also appeared in Old English as the noun *līc,* which meant "body," either living or dead. The noun survives in Modern English only in the compound *lich gate,* which denotes the gateway to a churchyard burial ground.

**-ly[2]** *suff.* [ME *-li* < OE *-līce* < *-līc,* adj. suffix.] **1.** In a specified manner : in the manner of <gradual*ly*> **2.** At a specified interval of time <week*ly*> **3.** With respect to <part*ly*>

**ly·ase** (lī′ās′) *n.* [Gk. *luein,* to loosen + -ASE.] An enzyme that catalyzes the formation of double bonds by removing chemical groups from a substrate without hydrolysis or that adds chemical groups at double bonds.

**ly·can·thrope** (lī′kən-thrōp′, lī-kăn′-) *n.* [NLat. *lycanthropus* < Gk. *lukanthrōpos* : *lukos,* wolf + *anthrōpos,* man.] A werewolf.

**ly·can·thro·py** (lī-kăn′thrə-pē) *n.* The magical ability to assume the form and characteristics of a wolf.

**ly·cée** (lē-sā′) *n.* [Fr. < OFr., lyceum < Lat. *Lyceum.* —see LYCEUM.] A French public secondary school.

**ly·ce·um** (lī-sē′əm) *n.* [Lat. < Gk. *Lukeion,* the school outside Athens, Greece, where Aristotle taught (335–323 B.C.).] **1.** A hall in which public programs, as lectures or concerts, are presented. **2.** An organization sponsoring public programs and entertainment.

**ly·chee** (lē′chē) *n. var. of* LITCHI.

**lych gate** (lĭch) *n. var. of* LICH GATE.

**lych·nis** (lĭk′nĭs) *n.* [NLat. *Lychnis,* genus name < Lat. *lychnis,* a red flower < Gk. *lukhnis* < *lukhinos,* lamp.] A plant of the genus *Lychnis,* as the campion.

**Ly·ci·an** (lĭsh′ē-ən, lĭsh′ən) *n.* **1.** A native or inhabitant of ancient Lycia. **2.** The Anatolian language of the Lycians. *—adj.* Of or relating to the Lycians or their language.

**ly·co·po·di·um** (lī′kə-pō′dē-əm) *n.* [NLat. *Lycopodium,* genus name : Gk. *lukos,* wolf + Gk. *pous,* foot.] **1.** A plant of the genus *Lycopodium,* as the club moss. **2.** The yellowish powdery spores of a club moss, esp. *Lycopodium clavatum,* used in fireworks and explosives and as a covering for pills.

**lyd·dite** (lĭd′īt′) *n.* [After Lydd, England.] An explosive consisting primarily of picric acid.

**Lyd·i·an** (lĭd′ē-ən) *n.* **1.** One of a people of ancient Lydia. **2.** The Anatolian language of the Lydians. *—adj.* Of or relating to the Lydians or their language.

**lye** (lī) *n.* [ME *lie* < OE *lēag.*] **1.** The liquid obtained by leaching wood ashes. **2.** Potassium hydroxide. **3.** Sodium hydroxide.

**ly·ing[1]** (lī′ĭng) *adj.* [Pr.part. of LIE[1].] Reclining.

**ly·ing[2]** (lī′ĭng) *adj.* [Pr.part. of LIE[2].] Untruthful.

**ly·ing-in** (lī′ĭng-ĭn′) *n., pl.* **ly·ings-in** or **ly·ing-ins.** Confinement of a woman in childbirth.

**lymph** (lĭmf) *n.* [Lat. *lympha,* water, prob. < Gk. *numphē,* water spirit.] **1.** A clear, transparent, watery, occas. faintly yellowish liquid that contains white blood cells and some red blood cells, travels through the lymphatic system to return to the venous blood stream through the thoracic duct, and acts to remove bacteria and certain proteins from the tissues, transport fat from the intestines, and supply lymphocytes to the blood. **2.** *Archaic.* A spring or stream of pure, clear water.

**lymph-** *pref. var. of* LYMPHO-.

**lym·phad·e·ni·tis** (lĭm-făd′n-ī′tĭs, lĭm′fə-də-nī′-) *n.* [LYMPH + Gk. *adēn,* gland + -ITIS.] Inflammation of the lymph nodes.

**lym·phat·ic** (lĭm-făt′ĭk) *adj.* [NLat. *lymphaticus* < *lympha,* lymph.] **1.** Of or pertaining to lymph, a lymph vessel, or a lymph node. **2.** Lacking vitality or energy : LETHARGIC. *—n.* A vessel that conveys lymph.

**lymphatic system** *n.* The interconnected system of spaces and vessels between tissues and organs by which lymph is circulated throughout the body.

**lym·pha·tism** (lĭm′fə-tĭz′əm) *n.* A pathological condition of infancy and childhood marked by hyperplasia of the lymphatic structures, spleen, and bone marrow.

**lym·pha·ti·tis** (lĭm′fə-tī′tĭs) *n.* [LYMPHAT(IC) + -ITIS.] Inflammation of lymph vessels or nodes.

**lymph follicle** *n.* One of the round masses of lymphocytes in the cortex of a lymph node.

**lymph gland** *n.* A lymph node.

**lymph node** *n.* Any of numerous oval or round bodies, located along the lymphatic vessels, that supply lymphocytes to the circulatory system and remove bacteria and foreign particles from the lymph.

**lymph nodule** *n.* A lymph follicle.

**lympho-** or **lymph-** *pref.* [< LYMPH.] Lymphatic system : LYMPH <*lympho*cyte>

**lym·pho·blast** (lĭm′fə-blăst′) *n.* An immature lymphocyte.

**lym·pho·cyte** (lĭm′fə-sīt′) *n.* A white blood cell formed in lymphoid tissue, as in the lymph nodes, spleen, thymus, and tonsils, constituting 22–28% of all leukocytes in the normal adult human's blood. **—lym′pho·cyt′ic** (-sĭt′ĭk) *adj.*

**lym·pho·cy·to·sis** (lĭm′fō-sī-tō′sĭs) *n.* A leukocytosis in which the number of lymphocytes is increased markedly. **—lym′pho·cy·tot′ic** (-tŏt′ĭk) *adj.*

**lym·phoid** (lĭm′foid′) *adj.* Of or relating to lymph, lymphatic tissue, or the lymphatic system.

**lym·pho·ma** (lĭm-fō′mə) *n., pl.* **-ma·ta** (-mə-tə) *or* **-mas.** Any of various abnormally proliferative diseases of lymphoid tissue. **—lym·pho′ma·toid′, lym·phom′a·tous** (-fŏm′ə-təs) *adj.*

**lym·pho·poi·e·sis** (lĭm′fō-poi-ē′sĭs) *n., pl.* **-ses** (-sēz′). Formation of lymphocytes. **—lym′pho·poi·et′ic** (-ĕt′ĭk) *adj.*

**lyn·ce·an** (lĭn-sē′ən) *adj.* [Lat. *lynceus* < Gk. *Lunkeios,* pertaining to Lynceus, an Argonaut noted for his excellent sight.] Sharp-sighted.

**lynch** (lĭnch) *vt.* **lynched, lynch·ing, lynch·es.** [< LYNCH LAW.] To execute without due process of law, esp. to hang.
**lynch law** *n.* [Perh. after Charles *Lynch* (1736–1796).] Punishment of criminal suspects without due process of law.
**lynx** (lĭngks) *n.* [Lat. < Gk. *lunx.*] A wild cat of the genus *Lynx,* esp. *L. canadensis* of northern North America or *L. lynx* of Eurasia, with thick soft fur, a short tail, and tufted ears.
**lynx-eyed** (lĭngks'īd') *adj.* Having keen vision : SHARP-SIGHTED.
**lyo-** *pref.* [< Gk. *luein,* to loosen, dissolve.] Dispersion : dissolution <*lyo*philic>
**ly·on·naise** (lī'ə-nāz', lē'-) *adj.* [< Fr. *à la Lyonnaise,* in the manner of Lyon < *Lyon,* Lyon, France.] Cooked with onions <*lyonnaise* potatoes><potatoes *lyonnaise*>
**ly·o·phil·ic** (lī'ə-fĭl'ĭk) *adj.* Of, pertaining to, or exhibiting a strong affinity between the dispersed phase and the dispersing medium of a colloid.
**ly·oph·i·liz·er** (lī-ŏf'ə-lī'zər) *n.* [LYOPHIL(IC) + -IZE.] A device for freeze-drying.
**ly·o·pho·bic** (lī'ə-fō'bĭk) *adj.* Of, pertaining to, or exhibiting a lack of strong affinity between the dispersed phase and the dispersing medium of a colloid.
**Ly·ra** (lī'rə) *n.* [Lat. *lyra,* lyre.] A constellation in the Northern Hemisphere.
**ly·rate** (lī'rāt', -rĭt) *adj.* Shaped or formed like a lyre.
**lyre** (līr) *n.* [ME *lire* < OFr. < Lat. *lyra* < Gk. *lura.*] A stringed instrument of the harp family used to accompany a reader or singer of poetry, esp. in ancient Greece.
**lyre·bird** (līr'bûrd') *n.* Either of two Australian birds of the genus *Menura,* of which the male has a long tail spread during courtship in a lyre-shaped display.
**lyr·ic** (lĭr'ĭk) *adj.* [OFr. *lyrique,* of a lyre < Lat. *lyricus* < Gk. *lurikos* < *lura,* lyre.] **1. a.** Of or relating to a category of poetic literature representational of music in its sound patterns and gen. characterized by sensuality and subjectivity of expression. **b.** Relating to or constituting a poem in this category, as a sonnet or ode. **c.** Being a poet of lyric verse. **2.** Extremely enthusiastic : EXUBERANT. **3. a.** Of or relating to the harp or lyre. **b.** Suitable for accompaniment by the lyre. **4.** Having a singing voice of modest range and light volume. **5.** LYRICAL 1. —*n.* **1.** A lyric poem or poet. **2.** *often* **lyrics.** The words of a song.
**lyr·i·cal** (lĭr'ĭ-kəl) *adj.* **1.** Expressing emotion or feeling, esp. deep personal emotion, in a direct manner. **2.** Lyric. **—lyr'i·cal·ly** *adv.*
**lyr·i·cism** (lĭr'ĭ-sĭz'əm) *n.* **1. a.** The quality or character of subjectivity and sensuality of expression, esp. in the arts. **b.** Melodiousness. **2.** An intense outpouring of exuberant emotion.
**lyr·i·cist** (lĭr'ĭ-sĭst) *n.* A writer of song lyrics.
**lyr·ism** (lĭr'ĭz'əm) *n.* [Fr. *lyrisme* < Gk. *lurismos,* played on the lyre < *lura,* lyre.] Lyricism.
**lyr·ist** (lĭr'ĭst) *n.* [Lat. *lyristes,* lyre player < Gk. *luristēs* < *lura,* lyre.] **1.** A lyricist. **2.** (līr'ĭst). A player of a lyre.
**lys-** *pref. var. of* LYSO-.
**lyse** (līs, līz) *vi.* & *vt.* **lysed, lys·ing, lys·es.** [< LYSIS.] To undergo or cause to undergo lysis.
**Ly·sen·ko·ism** (lī-sĕng'kō-ĭz'əm) *n.* The biological doctrine of Trofim Lysenko that maintains the possibility of inheriting environmentally acquired characteristics.
**ly·ser·gic acid** (lĭ-sûr'jĭk, lī-) *n.* [LYS(O)- + ERG(OT) + -IC + ACID.] A crystalline alkaloid, $C_{16}H_{16}N_2O_2$, derived from ergot and used in medical research.
**lysergic acid di·eth·yl·am·ide** (dī'ĕth-əl-ăm'īd') *n.* A hallucinogenic drug, $C_{20}H_{25}N_3O$, derived from lysergic acid.
**ly·ses** (lī'sēz') *n. pl. of* LYSIS.
**lysi-** *pref. var. of* LYSO-.
**ly·sin** (lī'sĭn) *n.* A specific antibody that destroys blood cells, tissues, or microorganisms.
**ly·sine** (lī'sēn', -sĭn) *n.* An essential, crystalline amino acid, $C_6H_{14}N_2O_2$, used in nutrition studies and culture media and to fortify foods and feeds.
**ly·sis** (lī'sĭs) *n., pl.* **-ses** (-sēz') [NLat. < Gk. *lusis,* a loosening < *luein,* to loosen.] **1.** *Biochem.* Dissolution or destruction of red blood cells, bacteria, or other antigens by a specific lysin. **2.** *Med.* Gradual subsiding of the symptoms of an acute disease.
**-lysis** *suff.* [NLat. < Gk. *lusis,* a loosening < *luein,* to unbind.] Decomposition : dissolving : disintegration <hydro*lysis*>
**lyso-** *or* **lysi-** *or* **lys-** *pref.* [< Gk. *lusis,* a loosening < *luein,* to unbind.] **1.** Lysis <*lys*in> **2.** Lysin <*lyso*genesis>
**ly·so·gen·e·sis** (lī'sō-jĕn'ĭ-sĭs) *n.* Production of lysins.
**Ly·sol** (lī'sôl', -sōl', -sŏl'). A trademark for a liquid antiseptic and disinfectant.
**ly·so·zyme** (lī'sə-zīm') *n.* A naturally-occurring enzyme in tears, capable of destroying the cell walls of certain bacteria and thus acting as a mild antiseptic.
**-lyte** *suff.* [< Gk. *lutos,* soluble < *luein,* to unbind.] A substance that can be decomposed by a specified process <electro*lyte*>
**lyt·ic** (lĭt'ĭk) *adj.* [Gk. *lutikos,* able to loose < *luein,* to loosen.] **1.** Of, relating to, or causing lysis. **2.** Of or relating to a lysin.
**-lytic** *suff.* [< Gk. *lutikos,* able to loosen < *luein,* to unbind.] Of, relating to, or bringing about a specified kind of decomposition <cellulo*lytic*>
**lyt·ta** (lĭt'ə) *n., pl.* **lyt·tae** (lĭt'ē') [Lat. < Gk. *lutta, lussa,* madness.] A thin cartilaginous strip found on the underside of the tongues of some carnivorous mammals, as dogs.
**-lyze** *suff.* [< LYSIS.] To cause or undergo lysis <pyro*lyze*>

# Mm

**m** *or* **M** (ĕm) *n., pl.* **m's** *or* **M's. 1.** The 13th letter of the English alphabet. **2.** A speech sound represented by the letter *m.* **3.** The 13th in a series. **4.** Something shaped like the letter M. **5. M** The Roman numeral for 1,000. **6.** An em in printing.
**-'m.** Am <I'*m* leaving soon.>
**ma** (mä, mô) *n.* [Short for MAMA.] *Informal.* Mother.
**ma'am** (măm) *n.* Madam.
**mac** (măk) *n. Chiefly Brit.* A raincoat : mackintosh.
**Mac** (măk) *n.* [Prob. < *Mac-,* a common prefix in Scottish and Irish surnames.] *Informal.* A fellow. —Used to address a man whose name is unknown <Move your car, *Mac.*>
**ma·ca·bre** (mə-kä'brə, -bər) *adj.* [Fr. < (*danse*) *macabre* (dance of) death < OFr. (*danse de*) *Macabre.*] **1.** Suggestive of the horror of death : GRUESOME <a *macabre* story of revenge> **2.** Comprising or including a depiction of death. **—ma·ca'bre·ly** *adv.*
**mac·ad·am** (mə-kăd'əm) *n.* [After John L. *McAdam* (1756–1836).] A road pavement of layers of compacted broken stone, now usu. bound with tar or asphalt.
**mac·a·da·mi·a nut** (măk'ə-dā'mē-ə) *n.* [NLat. *Macadamia,* genus name, after John *Macadam* (1827–1865).] The round, hard-shelled, edible nut of an Australian tree, *Macadamia ternifolia,* now cultivated in Hawaii.
**mac·ad·am·ize** (mə-kăd'ə-mīz') *vt.* **-ized, -iz·ing, -iz·es.** To construct or pave (a road) with macadam. **—mac·ad'am·i·za'tion** *n.* **—mac·ad'am·iz'er** *n.*
**ma·caque** (mə-kăk', -käk') *n.* [Fr. < Port. *macaco,* poss. < Kongo.] A short-tailed monkey of the genus *Macaca* of southeastern Asia, Japan, Gibraltar, and northern Africa.
**mac·a·ro·ni** (măk'ə-rō'nē) *n.* [Ital. *maccheroni,* pl. of *maccherone* < dial. Ital. *maccarone,* macaroni.] **1.** *pl.* **macaroni.** A pasta or paste of wheat flour pressed into hollow tubes or other shapes, dried, and prepared for eating by boiling. **2.** *pl.* **-ni** *or* **-nies. a.** A member of a class of young Englishmen who traveled much in Europe in the late 18th and early 19th cent. and who affected foreign fashions, customs, and manners. **b.** A fashionable fop.
**mac·a·ron·ic** (măk'ə-rŏn'ĭk) *adj.* [NLat. *macaronicus* < dial. Ital. *maccarone,* macaroni.] **1.** Of or relating to a literary composition containing a mixture of vernacular words with Latin words or with non-Latin words given Latin terminations. **2.** Of or involving a mixture of two or more languages. **—mac'a·ron'ic** *n.*

▲ word history: The adjective *macaronic* is derived from the same Italian word as the English noun *macaroni.* Macaronic compositions are written in a mixture of Latin and the vernacular language, usually with a comic or satiric effect. The name was invented by a 16th-century Italian poet who compared such compositions with macaroni, since both were mixtures of various ingredients.

**mac·a·roon** (măk'ə-ro͞on') *n.* [Fr. *macaron* < dial. Ital. *maccarone,* small cake, macaroni.] A chewy cookie made with sugar, egg whites, and almond paste or coconut.

**ma·caw** (mə-kô′) *n.* [Port. *macaú.*] Any of various tropical American parrots of the genera *Ara* and *Anodorhynchus,* including the largest parrots, typified by long saber-shaped tails, powerful curved bills, and usu. brilliant plumage.

**Mac·ca·bees** (măk′ə-bēz′) *pl.n.* *(sing. in number).* —See table at BIBLE. **—Mac′ca·be′an** *adj.*

**mac·ca·boy** (măk′ə-boi′) *n.* [Fr. *macouba,* after *Macouba,* a district of Martinique.] A perfumed snuff made in Martinique.

**mace**[1] (mās) *n.* [ME < OFr.] **1.** A heavy medieval war club with a flanged or spiked metal head, for crushing armor. **2.** A ceremonial staff carried or exhibited as the symbol of authority of a legislative body. **3.** A macebearer.

**mace**[1]
*Two types of maces*

**mace**[2] (mās) *n.* [ME < OFr. < Med. Lat. *macis* < Gk. *makir,* an Indian spice.] An aromatic spice made from the dried, waxy covering that partly encloses the nutmeg kernel.

**Mace** (mās). An alternate trademark for Chemical Mace.

**mace·bear·er** (mās′bâr′ər) *n.* An official who carries a mace.

**mac·é·doine** (măs′ə-dwän′) *n.* [Fr. < *Macédoine,* Macedonian (perh. from the variety of races in Macedonia).] **1.** A mixture of finely cut fruits or vegetables, occas. jellied, served as a salad, dessert, or appetizer. **2.** A mixture : medley.

**Mac·e·do·ni·an** (măs′ĭ-dō′nē-ən) *adj.* Of or relating to ancient or modern Macedonia or the people or languages of these regions. —*n.* **1.** A native or resident of ancient or modern Macedonia. **2.** The language of ancient Macedonia, of uncertain linguistic affiliation but having features regarded as Indo-European. **3.** The Slavic language of modern Macedonia.

**mac·er·ate** (măs′ə-rāt′) *v.* **-at·ed, -at·ing, -ates.** [Lat. *macerare, macerat-,* to macerate.] —*vt.* **1.** To make soft by soaking in a liquid. **2.** To separate into constituents by soaking. **3.** To cause to become thin, usu. by starvation. —*vi.* To become macerated. **—mac′er·a′tion** *n.* **—mac′er·at′er, mac′er·a′tor** *n.*

**Mach** *also* **mach** (mäk) *n.* Mach number.

**ma·chet·e** (mə-shĕt′ē, -chĕt′ē) *n.* [Sp., dim. of *macho,* mace.] A large heavy knife with a broad blade, used as a weapon and for cutting vegetation.

**Mach·i·a·vel·li·an** (măk′ē-ə-vĕl′ē-ən) *adj.* **1.** Of or relating to Machiavelli or Machiavellianism. **2.** Suggestive of or marked by the principles of expediency, deceit, and cunning imputed to Machiavelli. **—Mach′i·a·vel′li·an, Mach′i·a·vel′list** (-vĕl′ĭst) *n.*

**Mach·i·a·vel·li·an·ism** (măk′ē-ə-vĕl′ē-ə-nĭz′əm) *also* **Mach·i·a·vel·lism** (-vĕl′ĭz′əm) *n.* The political doctrine of Machiavelli, which denies the relevance of morality in political affairs and holds that craft and deceit are justified in pursuing and retaining political power.

**ma·chic·o·late** (mə-chĭk′ə-lāt′) *vt.* **-lat·ed, -lat·ing, -lates.** [Med. Lat. *machicolare, machicolat-* < OFr. *machicoller* < *machicoleis,* machicolation : *macher,* to crush + *col,* neck < Lat. *collum.*] To provide with machicolations.

**ma·chic·o·la·tion** (mə-chĭk′ə-lā′shən) *n.* **1. a.** A projecting gallery at the top of a castle wall, supported by a row of corbeled arches, with openings in the floor through which stones and boiling liquids could be dropped on attackers. **b.** One of these openings. **2.** A row of small corbeled arches used as an ornamental architectural feature.

**mach·i·nate** (măk′ə-nāt′, măsh′-) *v.* **-nat·ed, -nat·ing, -nates.** [Lat. *machinari, machinat-,* to design < *machina,* device. —see MACHINE.] —*vt.* To concoct (a plot). —*vi.* To plot. **—mach′i·na′tion** (-nā′shən, măsh′-) *n.* **—mach′i·na′tor** *n.*

**ma·chine** (mə-shēn′) *n.* [Fr. < OFr. < Lat. *machina* < Gk. *makhana, mēkhanē* < *mēkhos,* means.] **1. a.** A system, usu. of rigid bodies, constructed and connected to change, transmit, and direct applied forces in a predetermined way to accomplish a particular objective, as performance of useful work. **b.** A simple device, as a lever, pulley, or inclined plane, that changes the direction or magnitude, or both, of an applied force. **2.** A device or system along with its source of power and auxiliary equipment, as a motor vehicle or a jackhammer. **3.** A system or device, as a computer, that performs or helps in performing a human task. **4. a.** An intricate natural system or organism, as the human body. **b.** A functional unit of such a system, as the heart or lungs. **5.** One who acts in a mechanical or rigid way. **6.** An organized group under the control of one or more leaders <a political *machine*> **7.** Deus ex machina. —*vt.* & *vi.* **-chined, -chin·ing, -chines.** —*vt.* To cut, shape, or finish or be cut, shaped, or finished by machine. **—ma·chin′a·ble** *adj.*

**machine finish** *n.* Mill finish.

**machine gun** *n.* An automatic gun, often mounted, that fires repeatedly and rapidly.

**ma·chine-gun** (mə-shēn′gŭn′) *vt.* **-gunned, -gun·ning, -guns.** To fire at or kill with a machine gun.

**machine language** *n.* *Computer Sci.* Any of various systems of symbols for coding input data.

**ma·chine-read·a·ble** (mə-shēn′rē′də-bəl) *adj.* *Computer Sci.* Capable of being read or used by a computer.

**ma·chin·er·y** (mə-shē′nə-rē, -shēn′rē) *n., pl.* **-ies. 1.** Machines or machine parts in general. **2.** The working parts of a machine. **3.** A system of related elements that operates in a definable way <legislative *machinery*> **4. a.** A device or means for achieving a result. **b.** A literary device for effecting a calculated result, as a happy ending.

**machine shop** *n.* A workshop where power tools are used for making, finishing, or repairing machines or machine parts.

**machine tool** *n.* A power tool for machining.

**machine translation** *n.* Automatic translation, as by computer, from one language to another.

**ma·chin·ist** (mə-shē′nĭst) *n.* **1.** A skilled machine tool operator. **2.** One who makes, operates, or repairs machines. **3.** A warrant officer who assists the engineering officer in the engine room of a naval vessel. **4.** *Archaic.* One in charge of stage machinery.

**ma·chis·mo** (mä-chēz′mō) *n.* [Sp. < *macho,* male. —see MACHO.] An exaggerated sense of masculinity stressing attributes such as courage, virility, aggressiveness, and domination of women.

**Mach·me·ter** (mäk′mē′tər) *n.* An aircraft instrument that indicates speed in Mach numbers.

**Mach number** *n.* [After Ernst *Mach* (1836–1916).] The ratio of the speed of an object to the speed of sound in the surrounding medium.

**ma·cho** (mä′chō′) *adj.* [Sp., male < Lat. *masculus.*] Characterized by machismo. —*n., pl.* **-chos. 1.** Machismo. **2.** A man characterized by machismo.

**mac·in·tosh** (măk′ĭn-tŏsh′) *n.* var. of MACKINTOSH.

**mack·er·el** (măk′ər-əl, măk′rəl) *n., pl.* **mackerel** *or* **-els.** [ME *makerel* < OFr. *maquerel.*] **1.** A marine fish of the family Scombridae, esp. the Atlantic mackerel, *Scomber scombrus,* an important food fish with dark wavy bars on the back and a silvery belly. **2.** A smaller fish of the suborder Scombroidea, as the Spanish mackerel. **3.** Any of various fishes similar to the mackerel.

**mackerel sky** *n.* A cirrocumulus or altocumulus cloud formation resembling the bars on a mackerel's back.

**mack·i·naw** (măk′ə-nô′) *n.* [After *Mackinaw* City, Michigan.] **1.** A short double-breasted coat of thick, usu. plaid woolen material. **2.** The cloth from which a mackinaw coat is made, usu. of wool and often with a heavy nap. **3.** A flat-bottomed boat with a pointed bow and square stern, once common on the upper Great Lakes.

**Mackinaw blanket** *n.* A thick blanket in solid colors or stripes, once used in northern and western North America by Indians, traders, and trappers.

**Mackinaw trout** *n.* The lake trout.

**mack·in·tosh** *also* **mac·in·tosh** (măk′ĭn-tŏsh′) *n.* [After Charles *Macintosh* (1766–1843), its inventor.] *Chiefly Brit.* **1.** A raincoat. **2.** A lightweight waterproof fabric orig. made of rubberized cotton.

**mack·le** (măk′əl) *also* **mac·ule** (măk′yo͞ol) [Fr. *macule* < Lat. *macula,* spot.] —*n.* A spot, esp. a blurred or double impression made by a slipping of printing type or wrinkle in the paper. —*v.* **-led, -ling, -les** *also* **-uled, -ul·ing, -ules.** —*vt.* To blur or double (a printed impression). —*vi.* To become blurred.

**mac·le** (măk′əl) *n.* [Fr. < OFr., lozenge < Lat. *macula,* mesh.] **1.** Chiastolite. **2.** TWIN 4. **3.** A spot or discoloration in a mineral.

**macr-** *pref. var. of* MACRO-.

**mac·ra·mé** (măk′rə-mā′) *n.* [Fr. < Ital. *macramè* < Turk. *makrama,* towel < Ar. *miqramah,* striped cloth.] Coarse lacework made by weaving and knotting cords into a pattern.

**mac·ren·ceph·a·ly** (măk′rĕn-sĕf′ə-lē) *also* **mac·ren·ce·pha·li·a** (-sə-fā′lē-ə) *n.* Abnormal enlargement of the brain.

**mac·ro** (măk′rō′) *n.* [Short for MACROINSTRUCTION.] *Computer Sci.* An instruction in assembly language that is carried out by a sequence of instructions in machine language.

**macro-** *or* **macr-** *pref.* [< Gk. *makros,* large.] **1.** Large <*macro*nucleus> **2.** Long <*macro*biosis> **3.** Inclusive <*macro*instruction>

**mac·ro·bi·o·sis** (măk′rō-bī-ō′sĭs) *n.* [LGk. *makrobiōsis* : Gk. *makros,* long + Gk. *biōsis,* way of life.] Longevity.

**mac·ro·bi·o·ta** (măk′rō-bī-ō′tə) *n.* Macroscopic plant and animal life indigenous to a particular region.

**mac·ro·bi·ot·ics** (măk′rō-bī-ŏt′ĭks) *n.* *(sing. in number).* **1.** The theory or practice of promoting longevity. **2.** A method professing promotion of longevity, chiefly by means of diet consisting mainly of whole grains, vegetables, and fish. **—mac'ro·bi·ot'ic** *adj.*

**mac·ro·ceph·a·ly** (măk'rō-sĕf'ə-lē) *also* **mac·ro·ce·pha·li·a** (-sə-fā'lē-ə) *n.* Abnormally large cranial capacity. **—mac'ro·ce·phal'ic** (-sə-făl'ĭk), **mac'ro·ceph'a·lous** *adj.*

**mac·ro·cli·mate** (măk'rō-klī'mĭt) *n. Meteorol.* The climate of a large geographic area. **—mac'ro·cli·mat'ic** *adj.*

**mac·ro·code** (măk'rə-kōd') *n.* **1.** A coding system that assembles sets of computer instructions. **2.** A single code representing a set of computer instructions.

**mac·ro·con·sum·er** (măk'rō-kən-so͞o'mər) *n. Ecol.* A heterotrophic organism that ingests other organisms or particulate organic matter.

**mac·ro·cosm** (măk'rə-kŏz'əm) *n.* [Fr. *macrocosme* < Med. Lat. *macrocosmus* : Gk. *makros,* large + Gk. *kosmos,* world.] **1.** The entire world : UNIVERSE. **2.** A system reflecting on a large scale one of its component parts or systems. **—mac'ro·cos'mic** *adj.* **—mac'ro·cos'mic·al·ly** *adv.*

**mac·ro·cyte** (măk'rō-sīt') *n.* An abnormally large red blood cell occurring in some forms of anemia. **—mac'ro·cyt'ic** (-sĭt'ĭk) *adj.*

**mac·ro·cy·to·sis** (măk'rō-sī-tō'sĭs) *n.* A condition in which the blood contains macrocytes. **—mac'ro·cy·tot'ic** (-tŏt'ĭk) *adj.*

**mac·ro·ec·o·nom·ics** (măk'rō-ĕk'ə-nŏm'ĭks, -ē'kə-) *n. (sing. in number).* The study of the overall aspects of a national economy, as income, output, and the interrelationship among diverse economic sectors. **—mac'ro·ec'o·nom'ic** *adj.*

**mac·ro·ev·o·lu·tion** (măk'rō-ĕv'ə-lo͞o'shən, -ē'və-) *n.* Evolution involving whole species or large groups of organisms. **—mac'ro·ev'o·lu'tion·ar'y** *adj.*

**mac·ro·ga·mete** (măk'rō-gə-mēt', -găm'ēt') *n.* The larger of two conjugating cells, usu. female, in protozoans.

**mac·ro·glob·u·lin** (măk'rō-glŏb'yə-lĭn) *n.* A globulin with high molecular weight.

**mac·ro·glob·u·lin·e·mi·a** (măk'rō-glŏb'yə-lə-nē'mē-ə) *n.* Presence of an abnormally large number of macroglobulins in the blood serum.

**mac·ro·graph** (măk'rō-grăf') *n.* A representation of an object at least as large as the object.

**ma·crog·ra·phy** (mə-krŏg'rə-fē) *n.* **1.** Examination of objects with the naked eye. **2.** Abnormally large handwriting, occas. symptomatic of a nervous disorder.

**mac·ro·in·struc·tion** (măk'rō-ĭn-strŭk'shən) *n.* A macro.

**mac·ro·mere** (măk'rə-mîr') *n.* A large blastomere.

**mac·ro·mol·e·cule** (măk'rō-mŏl'ĭ-kyo͞ol') *n.* A very large molecule, as a polymer or protein, made up of many smaller structural units. **—mac'ro·mo·lec'u·lar** *adj.*

**ma·cron** (mā'krŏn', -krən) *n.* [Gk. *makron* < *makros,* long.] **1.** A diacritical mark over a vowel indicating a long sound or phonetic value in pronunciation, as (ā) in the word *make.* **2.** The horizontal mark (—) indicating a stressed or long syllable in a foot of verse.

**mac·ro·nu·cle·us** (măk'rō-no͞o'klē-əs, -nyo͞o'-) *n., pl.* **-cle·i** (-klē-ī'). A large trophic nonreproductive nucleus in the cells of ciliated protozoans.

**mac·ro·nu·tri·ent** (măk'rō-no͞o'trē-ənt, -nyo͞o'-) *n.* An element, as carbon, hydrogen, oxygen, or nitrogen, needed in large amounts for plant growth and development.

**mac·ro·phage** (măk'rə-fāj') *n.* A large phagocytic cell of the reticuloendothelial system. **—mac'ro·phag'ic** *adj.*

**mac·ro·phys·ics** (măk'rō-fĭz'ĭks) *n. (sing. in number).* The physics of macroscopic phenomena.

**mac·ro·phyte** (măk'rə-fīt') *n.* A macroscopic plant in an aquatic environment. **—mac'ro·phyt'ic** (-fĭt'ĭk) *adj.*

**ma·crop·ter·ous** (mə-krŏp'tər-əs) *adj.* [Gk. *makropteros* : *makros,* large + *pteron,* wing.] Having unusually large fins or wings.

**mac·ro·scop·ic** (măk'rə-skŏp'ĭk) *also* **mac·ro·scop·i·cal** (-ĭ-kəl) *adj.* [MACRO- + -SCOP(Y) + -IC.] **1.** Large enough to be examined without instrumentation, esp. by the naked eye. **2.** Relating to observations made without magnifying instruments, esp. by the naked eye. **—mac'ro·scop'i·cal·ly** *adv.*

**mac·ro·spo·ran·gi·um** (măk'rō-spə-răn'jē-əm) *n., pl.* **-gi·a** (-jē-ə). A megasporangium.

**mac·ro·spore** (măk'rə-spôr', -spōr') *n. Bot.* MEGASPORE 2.

**mac·u·la** (măk'yə-lə) *also* **mac·ule** (-yo͞ol) *n., pl.* **-lae** (-lē') or **-las** *also* **-ules.** [Lat., spot.] **1.** A spot, stain, or pit, esp. a skin discoloration caused by excess or lack of pigment : BLEMISH. **2.** A sunspot. **—mac'u·lar** (-lər) *adj.*

**macula lu·te·a** (lo͞o'tē-ə) *n., pl.* **maculae lu·te·ae** (lo͞o'tē-ē') [NLat. : Lat. *macula,* spot + Lat. *lutea,* yellow.] An area in the eye near the center of the retina at which visual perception is most acute.

**mac·u·late** (măk'yə-lāt') *vt.* **-lat·ed, -lat·ing, -lates.** [ME *maculaten* < Lat. *maculare* < *macula,* spot.] To spot, blemish, or pollute. —*adj.* (măk'yə-lĭt). **1.** Spotted : blotched. **2.** Impure : stained.

**mac·u·la·tion** (măk'yə-lā'shən) *n.* **1.** The act of spotting or staining or state of being spotted or stained. **2.** The spotted markings of a plant or animal, as the spots of the leopard.

**mac·ule** (măk'yo͞ol) *v.* [ME *maculen* < OFr. *maculer* < *macule,* spot < Lat. *macula.*] *var. of* MACKLE. —*n.* **1.** *var. of* MACKLE. **2.** *var. of* MACULA.

**mad** (măd) *adj.* **mad·der, mad·dest.** [ME < OE *gemǣdde* < *ge*mǣdan,* to madden < *gemǣd,* mad.] **1.** Afflicted with a mental disorder : INSANE. **2.** Temporarily or apparently deranged by violent sensations, emotions, or ideas <*mad* with rage> **3.** *Informal.* Feeling or displaying strong liking or enthusiasm <*mad* about antique cars> **4.** *Informal.* Angry : resentful. **5.** Lacking restraint or reason : FOOLISH <a *mad* whim> **6.** Marked by extreme excitement, confusion, or agitation : FRANTIC <a *mad* dash to escape the rain> **7.** Boisterously gay : HILARIOUS <had a *mad* time at Mardi gras> **8.** Affected by rabies : RABID. —*vt.* & *vi.* **mad·ded, mad·ding, mads.** To madden. **—like mad.** *Slang.* Wildly : impetuously. **—mad'ly** *adv.* **—mad'ness** *n.*

**Mad·a·gas·car periwinkle** (măd'ə-găs'kər) *n.* A plant native to Madagascar, *Vinca rosea,* bearing pink or white flowers.

**Mad·am** (măd'əm) *n.* [ME < OFr. *ma dame* : *ma,* my + *dame,* lady.] **1.** *pl.* **Mes·dames** (mā-däm'). **a.** —Used as a courtesy title in addressing a woman. **b.** —Used at one time as a courtesy title before a given name but now used only before a surname or a title indicating rank or office <*Madam* Justice> **2. madam.** The mistress of a household. **3. madam.** A woman manager of a brothel.

**Ma·dame** (mə-dăm', măd'əm) *n., pl.* **Mes·dames** (mā-däm') [Fr. < OFr. *ma dame* : *ma,* my + *dame,* lady.] —Used as a French courtesy title for a married woman.

**mad·cap** (măd'kăp') *n.* A rash, impulsive person. —*adj.* Impulsive : wild <*madcap* antics>

**mad·den** (măd'n) *v.* **-dened, -den·ing, -dens.** —*vt.* **1.** To make mad : drive insane. **2.** To make angry : IRRITATE. —*vi.* To become furious.

**mad·den·ing** (măd'n-ĭng) *adj.* **1.** Tending to make mad. **2.** Tending to irritate or anger. **—mad'den·ing·ly** *adv.*

**mad·der**[1] (măd'ər) *n.* [ME < OE *mædere.*] **1. a.** A plant of the genus *Rubia,* esp. a Eurasian species, *R. tinctoria,* with small yellow flowers and a red fleshy root. **b.** The root of this plant, once an important source of dye. **2.** A red dye derived from the madder root. **3.** A medium to strong red or reddish orange.

**mad·der**[2] (măd'ər) *adj. compar. of* MAD.

**mad·ding** (măd'ĭng) *adj. Archaic.* Frenzied <"Far from the *madding* crowd's ignoble strife" —Thomas Gray>

**mad-dog skullcap** (măd'dôg', -dŏg') *n.* A North American plant, *Scutellaria lateriflora,* bearing two-lipped blue or white flowers in one-sided clusters.

**made** (mād) *adj.* [< p.part. of MAKE.] **1.** Produced or manufactured by constructing, shaping, or forming. **2.** Produced or created artificially. **3.** Contrived : invented. **4.** Assured of success. **—made for.** Perfectly suited for <a day *made for* hiking>

**Ma·dei·ra** (mə-dîr'ə) *n.* A fortified dessert wine, esp. from the island of Madeira.

**Madeira vine** *n.* A tropical American vine, *Boussingaultia baselloides,* bearing small, white, fragrant flowers.

**mad·e·leine** (măd'ə-lĕn') *n.* [After *Madeleine* Paulmier, a 19th-cent. French pastry cook.] A small rich cake, baked in a shell-shaped mold.

**Mad·e·moi·selle** (măd'ə-mə-zĕl', mäd-mwä-zĕl') *n., pl.* **Mad·e·moi·selles** (-zĕlz) or **Mes·de·moi·selles** (mād'mwä-zĕl') [Fr. < OFr. *ma demoiselle* : *ma,* my + *demoiselle,* young lady.] **1.** —Used as a French courtesy title for a young girl or unmarried woman. **2. mademoiselle.** A French governess. **3. mademoiselle.** A marine fish of the genus *Bairdiella,* esp. *B. chrysura* of the U.S. Atlantic and Gulf coasts.

**made-to-or·der** (mād'to͞o-ôr'dər) *adj.* **1.** Made according to particular requirements or instructions : CUSTOM-MADE. **2.** Highly suitable <a resort *made-to-order* for tourists>

**made-up** (mād'ŭp') *adj.* **1.** Invented : fabricated <a *made-up* alibi> **2.** Adorned or changed by the application of cosmetics or make-up <a *made-up* clown> **3. a.** Finished : complete <a *made-up* bed> **b.** Put together : ARRANGED <a *made-up* newspaper page>

**mad·house** (măd'hous') *n.* **1.** An asylum for the mentally ill. **2.** *Informal.* A place of great confusion and disorder.

**Mad·i·son Avenue** (măd'ĭ-sən) *n.* [After *Madison Avenue,* New York, New York, the center of American advertising.] The American advertising industry.

**mad·man** (măd'măn', -mən) *n.* One who is or appears mentally ill.

**Ma·don·na** (mə-dŏn'ə) *n.* [Ital. : *ma,* my (< Lat. *mea*) + *donna,* lady (< Lat. *domina*).] **1.** The Virgin Mary. **2.** *Obs.* A married woman in Italy.

**Madonna lily** *n.* A native Eurasian plant, *Lilium candidum,* with white trumpet-shaped flowers.

**ma·dras** (măd'rəs, mə-drăs', -dräs') *n.* [After *Madras,* India.] **1.** A fine-textured cotton cloth, usu. with a plaid, striped, or checked pattern. **2.** A gen. striped silk cloth. **3. a.** A light cotton cloth used for drapery. **b.** A similar cloth of rayon. **4.** A large handkerchief of brightly colored cotton or silk, often worn as a turban.

**mad·re·pore** (măd′rə-pôr′, -pōr′) *n.* [Fr. < Ital. *madrepora* : *madre,* mother (< Lat. *mater*) + *porus,* tufa (< Gk. *pōros*).] Any of various corals of the genus *Madrepora,* including the tropical reef builders. **—mad′re·por′ic** *adj.*

**mad·re·por·ite** (măd′rə-pôr′īt, -pōr′-) *n.* [So called because the perforations resemble those of the madrepore.] A perforated structure found in most echinoderms that forms the intake for their water-vascular system.

**mad·ri·gal** (măd′rĭ-gəl) *n.* [Ital. *madrigale* < Med. Lat. *matricalis,* simple < *matrix,* womb < *mater,* mother.] **1.** An unaccompanied vocal composition for two or three voices in simple harmony, developed in Italy in the late 13th and early 14th cent. **2.** A polyphonic part song, developed in Italy in the 16th cent. and very popular in England in the 16th and early 17th cent., usu. unaccompanied and featuring parts for four to six voices. **3.** A lyric poem with a pastoral, idyllic, or amatory subject, developed from the lyrics of the 13th-cent. Italian madrigal. **4.** A part song, esp. a glee. **—mad′ri·gal·ist** *n.*

**ma·dri·lène** *also* **ma·dri·lene** (măd′rĭ-lĕn′) *n.* [Fr. (*consommé*) *madrilène,* Madrid (consommé).] A consommé flavored with tomato, gen. served chilled.

**ma·dro·ña** (mə-drō′nyə) *also* **ma·dro·ño** (mə-drō′nyō) *or* **ma·dro·ne** (mə-drō′nə) *n.* [Sp. *madroño.*] A tree, *Arbutus menziesi* of western North America, with glossy evergreen leaves, white flowers, and red-orange fruit.

**mad tom** *n.* A small freshwater North American catfish of the genus *Noturus,* common in the east-central United States, having poisonous spines.

**ma·du·ro** (mə-dŏŏr′ō) *n., pl.* **-ros.** [Sp. < *maduro,* mature < Lat. *maturus.*] A strong-flavored, dark-colored cigar. **—ma·du′ro** *adj.*

**mad·wom·an** (măd′wŏŏm′ən) *n.* A woman who is or appears to be mentally ill.

**mad·wort** (măd′wûrt′, -wôrt′) *n.* **1.** A low-growing, native Eurasian plant, *Asperugo procumbens,* with rough stems and small blue flowers. **2.** ALYSSUM 1.

**Mae·ce·nas** (mĭ-sē′nəs, mī-) *n.* [After Gaius *Maecenas* (70–8 B.C.).] A patron, esp. one generous to artists.

**mael·strom** (māl′strəm) *n.* [Obs. Du. : *malen,* to whirl + *stroom,* stream.] **1.** An extraordinarily large or violent whirlpool. **2.** A situation resembling a whirlpool in violence or turbulence <caught in the *maelstrom* of political scandal>

**mae·nad** (mē′năd′) *n.* [Lat. *maenas, maenad-,* maenad < Gk. *mainas* < *mainesthai,* to be mad.] **1.** *Gk. Myth.* A woman member of the orgiastic cult of Dionysus. **2.** A frenzied woman.

**ma·es·to·so** (mä′ĕs-tō′sō, -zō) *adj. & adv.* [Ital. < *maestà,* majesty < Lat. *majestas.*] *Mus.* In a majestic and dignified manner. —Used as a direction.

**maes·tro** (mīs′trō) *n., pl.* **-tros** *or* **-tri** (-trē) [Ital. < Lat. *magister,* master.] A master in an art, esp. a composer or conductor.

**Mae West** (mā′ wĕst′) *n.* [After *Mae West* (1892–1980), so called from its resemblance to her curvaceous torso.] An inflatable vestlike life preserver.

**Ma·fi·a** (mä′fē-ə) *n.* [Ital. < dial. Ital. *mafia,* bluster, boldness, prob. < Ar. *mahyah,* boasting.] **1.** A secret terrorist organization in Sicily, operating since the early 19th cent. in opposition to legal authority. **2.** An alleged international criminal organization believed active, esp. in Italy and the United States, since the late 19th cent. **3. mafia.** An exclusive group allegedly exercising control over a particular field, esp. politics <the Georgia *mafia*>

**Ma·fi·o·so** (mä′fē-ō′sō) *n., pl.* **-si** (-sē) [Ital. < *mafia,* mafia.] A member of the Mafia.

**mag** (măg) *n. Slang.* MAGAZINE 2.

**mag·a·zine** (măg′ə-zēn′, măg′ə-zēn′) *n.* [OFr. *magazin* < OItal. *magazzino* < Ar. *makhāzin,* pl. of *makhzan,* storehouse < *khazana,* to store.] **1. a.** A place for storage, esp. of ammunition. **b.** The contents of a storehouse, esp. a stock of ammunition. **2.** A periodical containing a collection of articles, stories, pictures, or other features. **3. a.** A compartment in some types of firearms for holding cartridges that are fed into the firing chamber. **b.** A compartment in a camera for holding rolls or cartridges of film that are fed through the exposure mechanism. **c.** Any of various compartments attached to machines for storing or supplying necessary material.

▲ word history: The use of *magazine* to mean "a periodical publication" is a specialized development of the original, more general sense "storehouse." *Magazine* was at one time used in book titles to mean a storehouse of information on a special topic, equivalent to the use of *encyclopedia* today. The word was also used in titles of periodical publications that contained a storehouse of miscellaneous literary works, articles on various topics, and other features. From the latter use the word *magazine* became a generic term for all such publications.

**mag·da·len** (mag′də-lən) *also* **mag·da·lene** (-lēn′) *n.* [After *Mary Magdalene,* traditionally identified with the woman taken in adultery in the New Testament.] **1.** A reformed prostitute. **2.** A reformatory for prostitutes.

**Mag·da·le·ni·an** (măg′də-lē′nē-ən) *adj.* [Fr. *magdalénien,* after La *Madeleine,* a prehistoric site in Dordogne, France.] Of or relating to the last upper Paleolithic culture of Europe, succeeding the Aurignacian.

**Mag·el·lan·ic cloud** (măj′ə-lăn′ĭk) *n.* [After Ferdinand *Magellan* (1480?–1521).] Either of two small galaxies, the closest to the Milky Way, faintly visible near the south celestial pole.

**Ma·gen Da·vid** *also* **Mo·gen Da·vid** (mô′gən dô′vĭd) *n.* [Heb. *māgen Dāwid.*] A six-pointed star, formed by placing two triangles together, one upon the other or interlaced, that is a symbol of Judaism and Israel.

**ma·gen·ta** (mə-jĕn′tə) *n.* [After *Magenta,* Italy.] **1.** Fuchsin. **2.** A moderate to vivid purplish red or dark to strong reddish purple.

**mag·got** (măg′ət) *n.* [ME *magot,* prob. of Scand. orig.] **1.** The legless soft-bodied larva of any of various insects of the order Diptera, esp. of the housefly and bluebottle fly, usu. found in decaying matter or as a parasite. **2.** A whim : notion. **—mag′got·y** *adj.*

**ma·gi** (mā′jī′) *n. pl. of* MAGUS.

**mag·ic** (măj′ĭk) *n.* [ME *magik* < OFr. *magique* < LLat. *magice* < Gk. *magikē* < *magos,* magician. —see MAGUS.] **1.** The art that purports to control or forecast natural events, effects, or forces by invoking the supernatural. **2. a.** The practice of attempting to produce supernatural effects or control events in nature through the use of charms, spells, or rituals. **b.** The charms, spells, and rituals so used. **3.** Exercise of sleight of hand or conjuring for entertainment. **4.** A mysterious quality of enchantment <the *magic* of exotic lands>

☆ ***syns:*** MAGIC, CONJURATION, CONJURY, SORCERY, SORTILEGE, THAUMATURGY, THEURGY, WITCHCRAFT, WITCHING, WIZARDRY *n. core meaning* : the use of supernatural powers to influence or predict events <believed in the power of *magic*>

**mag·i·cal** (măj′ĭ-kəl) *adj.* **1.** Relating to or created by magic. **2.** Bewitching : enchanting. **—mag′i·cal·ly** *adv.*

**ma·gi·cian** (mə-jĭsh′ən) *n.* **1.** A wizard : sorcerer. **2.** One who performs magic for entertainment. **3.** One whose skill or art appears to be magical <a *magician* with the camera>

**magic lantern** *n.* An optical device once used for projecting the enlarged image of a picture.

**magic number** *n. Physics.* Any of the numbers 2, 6, 8, 14, 20, 28, 50, 82, 126, that represent the number of neutrons or protons in strongly bound, exceptionally stable, and abundant atomic nuclei.

**mag·is·te·ri·al** (măj′ĭ-stîr′ē-əl) *adj.* [LLat. *magisterius* < Lat. *magister,* master.] **1.** Of, relating to, or typical of a master or teacher : AUTHORITATIVE. **2.** Overbearing : dogmatic. **3.** Of or relating to a magistrate or his or her official functions. **—mag′is·te′ri·al·ly** *adv.*

**mag·is·te·ri·um** (măj′ĭ-stîr′ē-əm) *n.* [Lat., teaching authority.] The authority to teach religious doctrine, esp. as claimed by the Roman Catholic Church.

**mag·is·tra·cy** (măj′ĭs-trə-sē) *also* **mag·is·tra·ture** (-trə-chŏŏr′) *n., pl.* **-cies** *also* **-tures. 1.** The position, function, or term of office of a magistrate. **2.** A body of magistrates. **3.** The district under a magistrate's jurisdiction.

**mag·is·tral** (măj′ĭ-strəl) *adj.* [LLat. *magistralis,* belonging to a master < *magister,* master.] **1.** Magisterial. **2.** Prepared as directed by a physician's prescription. —Used of medications. **3.** Principal : main <the *magistral* line of defense>

**mag·is·trate** (măj′ĭ-strāt′, -strĭt) *n.* [Lat. *magistratus* < *magister,* master.] **1.** A civil officer empowered to administer and enforce law. **2.** A minor official with limited judicial authority, as a justice of the peace or the judge of a police court.

**mag·is·tra·ture** (măj′ĭ-strā′chər, -strə-chŏŏr′) *n.* Magistracy.

**Ma·gle·mo·si·an** (mä′glə-mō′zē-ən) *adj.* [After *Maglemose,* Denmark.] Of or relating to a northern European forest culture of the Mesolithic age.

**mag·ma** (măg′mə, mäg′-) *n., pl.* **-ma·ta** (-mä′tə) *or* **-mas.** [ME < Lat. < Gk., unguent < *massein,* to knead.] **1.** A mixture of finely divided solids with sufficient liquid to produce a pasty mass. **2.** *Geol.* Molten matter beneath the earth's crust, from which igneous rock is formed by cooling. **3.** A pharmacological particulate suspension in a liquid, as milk of magnesia. **4.** The residue of fruits after the juice has been expressed : POMACE. **—mag·mat′ic** (-măt′ĭk) *adj.*

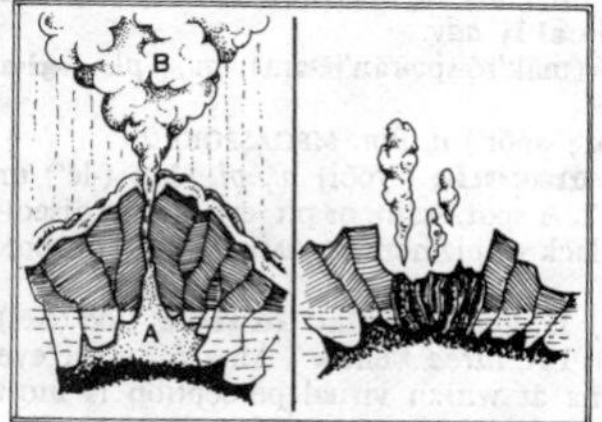


**magma**
*An erupting volcano showing A. magma deposit and B. lava flow*

**Mag·na Char·ta** *or* **Mag·na Car·ta** (măg′nə kär′tə) *n.* [Med. Lat. : Lat. *magna,* great + *charta,* charter.] **1.** The charter of English political and civil liberties granted by King John at Runnymede on

Jun. 15, 1215. **2.** A document or piece of legislation guaranteeing basic rights.
**mag·na cum lau·de** (măg′nə ko͝om lou′də) *adv.* [Lat.] With high honors <graduated *magna cum laude*>
**mag·na·nim·i·ty** (măg′nə-nĭm′ĭ-tē) *n., pl.* **-ties. 1.** The quality of being magnanimous. **2.** A magnanimous act.
**mag·nan·i·mous** (măg-năn′ə-məs) *adj.* [Lat. *magnanimus* : *magnus,* great + *animus,* soul.] Noble of heart and mind, esp. generous in forgiving : UNSELFISH. **—mag·nan′i·mous·ly** *adv.* **—mag·nan′i·mous·ness** *n.*
**mag·nate** (măg′nāt′, -nĭt) *n.* [ME *magnat* < LLat. *magnatus* < Lat. *magnus,* great.] A powerful or influential person, esp. in business or industry : TYCOON <an oil *magnate*>
**mag·ne·sia** (măg-nē′zhə, -shə) *n.* [Med. Lat., mineral ingredient of the philosopher's stone < Gk. *Magnēsia (lithos),* Magnesian (stone) < *Magnēs,* of Magnesia, an ancient city in Asia Minor.] Magnesium oxide, esp. when processed for purity. **—mag·ne′sian** *adj.*
**mag·ne·site** (măg′nə-sīt′) *n.* **1.** A white, yellowish, or brown, usu. crystalline mineral of magnesium carbonate, $MgCO_3$, used in manufacturing magnesium oxide and carbon dioxide. **2.** Any of several grades of magnesium oxide obtained from magnesite.
**mag·ne·si·um** (măg-nē′zē-əm, -zhəm) *n.* [< MAGNESIA.] *Symbol* **Mg** A light, silvery, moderately hard metallic element used in structural alloys, pyrotechnics, flash photography, and incendiary bombs; atomic number 12; atomic weight 24.312.
**magnesium carbonate** *n.* A very light, odorless, white powdery compound, $MgCO_3$, used in products such as inks, glass, dentifrices, and cosmetics.
**magnesium hydroxide** *n.* A white powder, $Mg(OH)_2$, used as an antacid and laxative.
**magnesium oxide** *n.* A white powdery compound, MgO, having a high melting point (2,800°C) and used in high-temperature refractories, electric insulation, food packaging, and semiconductors.
**magnesium sulfate** *n.* A colorless crystalline compound, $MgSO_4$, used in fireproofing, ceramics, matches, explosives, and fertilizers.
**mag·net** (măg′nĭt) *n.* [ME < Lat. < Gk. *magnēs* < *Magnēs lithos,* Magnesian stone < *Magnēs,* of Magnesia, an ancient city in Asia Minor.] **1.** A body that attracts certain materials, as iron, by virtue of a surrounding field of force created by the motion of its atomic electrons and the alignment of its atoms. **2.** An electromagnet. **3.** One that attracts.
**magnet-** *pref. var. of* MAGNETO-.
**mag·net·ic** (măg-nĕt′ĭk) *adj.* **1.** Of or pertaining to magnetism or magnets. **2.** Having the properties of a magnet. **3.** Pertaining to the magnetic poles of the earth. **4.** Capable of being magnetized or of being attracted by a magnet. **5.** Operating by means of magnetism. **6.** Having an unusual ability or power to attract <a *magnetic* temperament> **—mag·net′i·cal·ly** *adv.*
**magnetic bottle** *n.* A magnetic field for confining plasma, as during nuclear fusion.
**magnetic bubble** *n.* A small, stable, cylindrical region of magnetization in thin film or material that can be manipulated by an external magnetic field and used for representing data in a computer memory.
**magnetic compass** *n.* An instrument using a magnetic needle to show direction relative to the earth's magnetic field.
**magnetic core** *n.* CORE 4.
**magnetic declination** *n.* The angle between the geographic meridian and the local magnetic meridian, in navigation indicated as degrees plus (+) to the east, or degrees minus (−) to the west, of the geographic meridian.
**magnetic dip** *n.* The angle that the earth's magnetic field makes with the horizontal plane at any given location.
**magnetic equator** *n.* A line joining all points on the earth's surface where there is no magnetic dip.
**magnetic field** *n.* A condition in a region of space, as that around a magnet or an electric current, marked by the existence of a detectable magnetic force at every point in the region.
**magnetic field strength** *n.* **1.** Magnetic intensity. **2.** MAGNETIC INDUCTION 1.
**magnetic flux** *n.* The total number of magnetic lines of force passing through a bounded area in a magnetic field.
**magnetic flux density** *n.* MAGNETIC INDUCTION 1.
**magnetic force** *n.* **1.** The force on a magnetic pole in a magnetic field. **2.** The force on an electrically charged particle or electric current in a magnetic field.
**magnetic head** *n.* A device, as in a tape recorder, that converts electric impulses into variations in the magnetism of a surface for storage and subsequent retrieval.
**magnetic hysteresis** *n.* Failure of the magnetization in a body to return to its original value when the external field is reduced.
**magnetic inclination** *n.* Magnetic dip.
**magnetic induction** *n.* **1.** A vector quantity that specifies the direction and magnitude of magnetic force at every point in a magnetic field. **2.** Temporary conversion of certain materials, as a piece of iron, into a magnet by a magnetic field.
**magnetic intensity** *n.* That part of a magnetic field related solely to external currents as a cause, without reference to the presence of matter.
**magnetic line of force** *n.* A curve whose tangent at any point is along the direction of magnetic force at that point; the number of lines of force per unit area in the neighborhood of a point is proportional to the magnetic induction at that point.
**magnetic meridian** *n.* A meridian passing through the earth's magnetic poles.
**magnetic mine** *n.* A marine mine detonated by a mechanism that responds to a mass of magnetic material, as the steel hull of a ship.
**magnetic moment** *n.* The ratio of the maximum torque exerted on a magnet or electric current loop in a magnetic field to the magnetic induction of the field.
**magnetic needle** *n.* A needle-shaped bar magnet usu. suspended on a low-friction mounting and used in instruments, as a compass, for indicating alignment of a local magnetic field.
**magnetic north** *n.* The direction of the earth's magnetic pole, to which the north-seeking pole of a magnetic needle points when free from local magnetic influence.
**magnetic permeability** *n.* A measure of the ability of a medium to modify a magnetic field, equal to the ratio of magnetic induction to magnetic intensity.
**magnetic pickup** *n.* A phonograph pickup that utilizes a coil in a magnetic field for receiving vibrations from the stylus and converting them into electric impulses.
**magnetic pole** *n.* **1.** Either of two limited regions in a magnet at which the magnet's field is most intense, each of which is designated by the approximate geographic direction to which it is attracted. **2.** Either of two variable points on the earth, close to but not coinciding with the geographic poles, where the earth's magnetic field is most intense.
**magnetic pole strength** *n.* A measure of the effectiveness of a magnet, equal to the quotient of the magnetic moment by the length of the magnet.
**magnetic pyrites** *n.* Pyrrhotite.
**magnetic recording** *n.* **1.** The recording of a signal, as sound or computer instructions, in the form of a magnetic pattern on a magnetizable surface for storage and subsequent retrieval. **2.** A surface containing a magnetic recording.
**magnetic storm** *n.* A severe but transitory fluctuation in the earth's magnetic field believed to be produced by currents of charged particles and gamma rays, resulting from abnormal solar activity.
**magnetic susceptibility** *n.* The ratio of the magnetic permeability of a medium to that of a vacuum, minus one; it is positive for a paramagnetic or ferromagnetic medium, negative for a diamagnetic medium.
**magnetic tape** *n.* Plastic tape coated with iron oxide for use in magnetic recording.
**magnetic variation** *n.* Magnetic declination.
**mag·net·ism** (măg′nĭ-tĭz′əm) *n.* **1.** The class of phenomena exhibited by the field of force created by a magnet or electric current. **2.** Study of magnets and their effects. **3.** The force exerted by a magnetic field. **4.** Magnetic flux. **5.** Power to fascinate, attract, or influence <the *magnetism* of rock stars> **6.** Animal magnetism.
**mag·net·ite** (măg′nĭ-tīt′) *n.* The mineral form of black iron oxide, $Fe_3O_4$, often occurring with titanium or magnesium and an important ore of iron.
**mag·net·i·za·tion** (măg′nĭ-tĭ-zā′shən) *n.* **1.** The process of making a substance temporarily or permanently magnetic. **2.** The magnetic moment per unit volume induced in a body by an external field. **3.** The property of being magnetic.
**mag·net·ize** (măg′nĭ-tīz′) *vt.* **-ized, -iz·ing, -iz·es. 1.** To make magnetic. **2.** To fascinate, attract, or influence <a speaker who *magnetized* the audience> **—mag′net·iz′a·ble** *adj.* **—mag′net·iz′er** *n.*
**magnetizing force** *n.* The magnetic intensity at any point in a substance capable of being magnetized.
**mag·ne·to** (măg-nē′tō) *n., pl.* **-tos.** [Short for *magnetoelectric machine.*] A small generator of alternating current with permanent magnets, used in the ignition systems of some internal-combustion engines.
**magneto-** *or* **magnet-** *pref.* [< MAGNET.] **1.** Magnetism : magnetic <*magneto*chemistry> **2.** Magnetic field <*magneto*meter>
**mag·ne·to·e·lec·tric** (măg-nē′tō-ĭ-lĕk′trĭk, -ē′lĕk′-) *adj.* Of or relating to magnetically produced electricity. **—mag·ne′to·e·lec·tric′i·ty** *n.*
**mag·ne·to·flu·id·dy·nam·ics** (măg-nē′tō-flo͞o′ĭd-dī-năm′ĭks, -nĕt′ō-) *n.* (*sing. or pl. in number*). Magnetohydrodynamics. **—mag·ne′to·flu′id·dy·nam′ic** *adj.*
**mag·ne·to·gas·dy·nam·ics** (măg-nē′tō-găs′dī-năm′ĭks) *n.* (*sing. or pl. in number*). Magnetohydrodynamics. **—mag·ne′to·gas′dy·nam′ic** *adj.*
**mag·ne·to·graph** (măg-nē′tō-grăf′) *n.* A magnetometer equipped for recording, as by photography.

**mag·ne·to·hy·dro·dy·nam·ics** (măg-nē′tō-hī′drō-dī-năm′ĭks) *n. (sing. or pl. in number).* Study of electrically conducting fluids, as molten metal or plasma, in electric and magnetic fields. **—mag·ne′to·hy′dro·dy·nam′ic** *adj.*

**mag·ne·tom·e·ter** (măg′nĭ-tŏm′ĭ-tər) *n.* An instrument for comparing the intensity and direction of magnetic fields. **—mag·ne′to·met′ric** (măg-nē′tō-mĕt′rĭk) *adj.* **—mag′ne·tom′e·try** *n.*

**mag·ne·to·mo·tive force** (măg-nē′tō-mō′tĭv) *n.* **1.** The agency that produces magnetic flux in a magnetic circuit. **2.** The strength of a magnetic flux-producing agency, equal to the work required to carry a hypothetical isolated magnetic pole of unit strength completely around the circuit.

**mag·ne·ton** (măg′nĭ-tŏn′) *n.* A unit of the magnetic moment of a subatomic particle, equal to $eh/4\pi mc$, where *e* is the particle's electric charge, *m* its mass, *h* Planck's constant, and *c* the speed of light, esp.: **a.** The Bohr magneton, calculated using the mass and charge of the electron. **b.** The nuclear magneton, calculated using the mass of the nucleon.

**mag·ne·to·plas·ma·dy·nam·ics** (măg-nē′tō-plăz′mə-dī-năm′ĭks, -nĕt′ō-) *n. (sing. or pl. in number).* Magnetohydrodynamics. **—mag·ne′to·plas′ma·dy·nam′ic** *adj.*

**mag·ne·to·sphere** (măg-nē′tō-sfîr′) *n.* An asymmetric region surrounding the earth, extending from approx. 400 to several thousand miles above the surface, in which charged particles are trapped and their behavior is dominated by the earth's magnetic field.

**mag·ne·to·stric·tion** (măg-nē′tō-strĭk′shən) *n.* Deformation of a ferromagnetic material subjected to a magnetic field.

**mag·ne·tron** (măg′nĭ-trŏn′) *n.* [MAGNE(T) + -TRON.] A thermionic tube in which control of the electron beam by electromagnetic fields generates high-power microwaves.

**mag·nif·ic** (măg-nĭf′ĭk) *adj.* [ME *magnifique* < OFr. < Lat. *magnificus* : *magnus,* great + *facere,* to make.] **1.** MAGNIFICENT 2. **2.** Impressive in size or extent : LARGE. **3. a.** Exalted. **b.** Grandiloquent : pompous <*magnific* rhetoric> **—mag·nif′i·cal·ly** *adv.*

**Mag·nif·i·cat** (măg-nĭf′ĭ-kăt′) *n.* [Lat., it magnifies.] **1. a.** The canticle beginning *Magnificat anima mea Dominum,* "My soul doth magnify the Lord." **b.** A musical setting of this text. **2. magnificat.** A hymn or song of praise.

**mag·ni·fi·ca·tion** (măg′nə-fĭ-kā′shən) *n.* **1. a.** The act of magnifying or state of being magnified. **b.** Enlargement in size, as of an optical image. **c.** Something magnified. **2.** The ratio of an optical image size to object size.

**mag·nif·i·cence** (măg-nĭf′ĭ-səns) *n.* **1.** Lavishness or greatness of surroundings : SPLENDOR. **2.** Imposing or grand beauty <the *magnificence* of the mountain scenery>

**mag·nif·i·cent** (măg-nĭf′ĭ-sənt) *adj.* [Lat. *magnificens, magnificent-,* var. of *magnificus.* —see MAGNIFIC.] **1.** Splendid in appearance : GRAND <a *magnificent* civic center> **2.** Grand or noble in thought or deed : EXALTED. **3.** Outstanding of its kind : SUPERLATIVE <performed a *magnificent* pas de deux> **—mag·nif′i·cent·ly** *adv.*

**mag·nif·i·co** (măg-nĭf′ĭ-kō′) *n., pl.* **-coes.** [Ital. < *magnifico,* magnificent < Lat. *magnificus.* —see MAGNIFIC.] **1.** A nobleman of the Venetian Republic. **2.** A person of distinguished rank, importance, or appearance.

**mag·ni·fi·er** (măg′nə-fī′ər) *n.* **1. a.** A magnifying glass. **b.** A system of optical components that magnifies. **2.** One that magnifies.

**mag·ni·fy** (măg′nə-fī′) *v.* **-fied, -fy·ing, -fies.** [ME *magnifien* < OFr. *magnifier* < Lat. *magnificare* < *magnificus,* magnificent. —see MAGNIFIC.] —*vt.* **1.** To make greater in size : ENLARGE. **2.** To cause to appear greater or seem more important : EXAGGERATE <They greatly *magnified* their unimportant contributions.> **3.** To increase the apparent size of, esp. by means of a lens. **4.** To praise or glorify. —*vi.* To increase or have the power to increase the size or volume of an image or sound.

**magnifying glass** *n.* A converging lens that enlarges the image of an object.

**mag·nil·o·quent** (măg-nĭl′ə-kwənt) *adj.* [Lat. *magniloquus* : *magnus,* great + *loqui,* to speak.] Lofty and extravagant in speech : SONOROUS. **—mag·nil′o·quence** *n.* **—mag·nil′o·quent·ly** *adv.*

**mag·ni·tude** (măg′nĭ-to͞od′, -tyo͞od′) *n.* [ME, great size < Lat. *magnitudo* < *magnus,* great.] **1. a.** Greatness of position or rank. **b.** Greatness in extent or size. **c.** Greatness in influence or significance. **2.** *Astron.* Relative brightness of a celestial body designated on a numerical scale, orig. integers from 1 (brightest) through 6 (faintest visible), now extended to include negative integers, integers above 6, and decimals, with the scale rule such that a decrease of 1 unit represents an increase in apparent brightness by a factor of 2.512. **3.** *Math.* **a.** A number assigned to a member of a set to form the basis of comparison with other members of the same set. **b.** A property that can be quantitatively described, as the volume of a sphere or length of a vector.

**mag·no·lia** (măg-nōl′yə) *n.* [NLat. *Magnolia,* genus name, after Pierre *Magnol* (1638–1715).] **1.** Any of various deciduous or evergreen trees and shrubs of the genus *Magnolia* of the Western Hemisphere and Asia, many of which are cultivated for their showy white, pink, purple, or yellow flowers. **2.** The flower of the magnolia.

**magnolia warbler** *n.* A black-and-yellow, ground-nesting songbird, *Dendroica magnolia* of North America.

**mag·num** (măg′nəm) *n.* [Lat., neuter of *magnus,* great.] **1.** A bottle for wine or liquor, with a capacity of approx. two fifths of a gallon. **2.** The amount of liquid in a magnum.

**mag·num o·pus** (măg′nəm ō′pəs) *n.* [Lat.] **1.** A great work, esp. a literary or artistic masterpiece. **2.** An artist's, writer's, or composer's greatest single work.

**mag·nus hitch** (măg′nəs) *n.* [Orig. unknown.] A clove hitch with one extra turn.

**mag·pie** (măg′pī′) *n.* [*Mag,* a nickname for *Margaret* + PIE[2].] **1.** Any of various birds of the family Corvidae, found worldwide, with a long graduated tail and black, blue, or green coloring and white markings, noted for their chattering call. **2.** Any of various birds similar to the magpie. **3.** Any of several piping crows and bell magpies of the family Cracticidae of Australia. **4.** One who chatters.

**ma·guey** (mə-gā′) *n.* [Sp., of Cariban orig.] **1.** A tropical American plant of the genus *Agave.* **2.** A plant of the genus *Furcraea.* **3.** The fiber obtained from the maguey.

**ma·gus** (mā′gəs) *n., pl.* **ma·gi** (-jī′) [ME < Lat. *magus,* sorcerer < Gk. *magos* < Pers. *maguš.*] **1.** A member of the Zoroastrian priestly caste of the Medes and Persians. **2. Magus.** One of the three wise men from the East who traveled to Bethlehem with gifts for the infant Jesus. **3.** A sorcerer : magician. **—ma′gi·an** (-jē-ən) *n.*

**Mag·yar** (măg′yär′, mäg′-, mŭd′-) *n.* [Hung.] **1.** A member of the principal ethnic group of Hungary. **2.** The Finno-Ugric language of the Magyars that is the official language of Hungary. **—Mag′yar** *adj.*

**ma·ha·ra·jah** *or* **ma·ha·ra·ja** (mä′hə-rä′jə, -zhə) *n.* [Hindi *mahārājā* < Skt. : *mahā-,* great + *rājā,* king.] A king or prince in India ranking above a raja, esp. the ruler of one of the former native states.

**ma·ha·ra·ni** *or* **ma·ha·ra·nee** (mä′hə-rä′nē) *n.* [Hindi *mahārānī* < Skt. *mahārājñī* : *mahā-,* great + *rājñī,* queen.] **1.** The wife of a maharajah. **2.** A princess in India ranking above a rani, esp. the ruler of one of the former native states.

**ma·ha·ri·shi** (mä′hə-rē′shē, mə-här′ə-shē) *n.* [Skt. *mahārṣi* : *mahat-,* great + *ṛṣi,* sage.] A Hindu teacher of mysticism and spiritual knowledge.

**ma·hat·ma** (mə-hät′mə, -hăt′-) *n.* [Skt. *mahātman* : *mahā-,* great + *ātman,* self, soul.] **1.** One of a class of persons venerated in India and Tibet for great knowledge and love of humanity. **2. Mahatma.** A Hindu title of respect for a man famed for spirituality and high-mindedness.

**Ma·ha·ya·na** (mä′hə-yä′nə) *n.* [Skt. *Mahāyānam* : *mahā-,* great + *yānam,* vehicle.] A major school of Buddhism, active in Japan, Korea, Nepal, Tibet, Mongolia, and China, that emphasizes lack of concern for oneself and compassion for universal suffering. **—Ma′ha·ya′nist** *n.* **—Ma′ha·ya·nis′tic** *adj.*

**Mah·di** (mä′dē) *n.* [Ar. *mahdīy* < *madā,* he led in the right way.] **1.** The Islamic messiah who, it is believed, will appear at the world's end and establish a reign of peace and righteousness. **2.** An Islamic leader assuming a messiah's role. **—Mah′dism** *n.* **—Mah′dist** *n.*

**Ma·hi·can** (mə-hē′kən) *also* **Mo·hi·can** (mō-, mə-) *n., pl.* **Mahican** *or* **-cans** *also* **Mohican** *or* **-cans. 1. a.** A tribe or confederacy of Indians once living between the upper Hudson River Valley and Lake Champlain. **b.** A member of this tribe or confederacy. **2.** The Algonquian language of the Mahican.

**mah·jong** *also* **mah·jongg** (mä′zhŏng′, -zhông′) *n.* [Chin. (Mandarin) *ma*[2] *jiang*[4].] A game of Chinese origin usu. played by four persons using tiles similar to dominoes and bearing various designs, which are drawn and discarded until one player wins with a hand of four combinations of three tiles each and a pair of matching tiles.

**mahl·stick** (môl′stĭk′) *n. var. of* MAULSTICK.

**ma·hog·a·ny** (mə-hŏg′ə-nē) *n., pl.* **-nies.** [Orig. unknown.] **1. a.** A tropical American tree of the genus *Swietenia,* prized for its hard, reddish-brown wood. **b.** The wood of the mahogany, esp. that of *S. mahogani,* used for making furniture. **2. a.** Any of several trees with wood similar to true mahogany. **b.** The wood of such trees. **3.** A moderate reddish brown.

**ma·hout** (mə-hout′) *n.* [Hindi *mahāut* < Skt. *mahāmātraḥ,* an honorific title: *mahā-,* great + *mātram,* measure, size.] The keeper and driver of an elephant.

**Mah·ra·ti** (mə-rä′tē) *n. var. of* MARATHI.

**Mah·rat·ta** (mə-rä′tə) *n. var. of* MARATHA.

**Mah·rat·ti** (mə-rä′tē) *n. var. of* MARATHI.

**ma·huang** (mä-hwäng′) *n.* [Chin. (Mandarin) *ma*[2] *huang*[2] : *ma*[2], hemp + *huang*[2], yellow.] An Asiatic plant or shrub of the genus *Ephedra,* esp. *E. sinica,* from which ephedrine is extracted.

**mah·zor** (mäKH′zôr′, -zər) *n., pl.* **mah·zor·im** (mäKH-zôr′ĭm) *or* **-zors.** [Heb. *maḥzor.*] The Hebrew prayer book containing rituals prescribed for holidays.

**Mai·a** (mā′ə, mī′ə) *n.* [Lat. < Gk. < *maia,* mother.] **1.** *Gk. Myth.* A goddess, the eldest of the Pleiades. **2.** The brightest star in the Pleiades.

**maid** (mād) *n.* [ME < OE *mægden.*] **1. a.** A girl or an unmarried woman. **b.** A virgin. **2.** A woman servant.

**maid·en** (mād′n) *n.* [ME < OE *mægden.*] **1. a.** An unmarried girl or woman. **b.** A virgin. **2.** A machine similar to the guillotine, for beheading criminals in 16th- and 17th-cent. Scotland. **3.** A racehorse that has never won a race. **4.** A maiden over in cricket. **5.** A one-year-old woody plant. *—adj.* **1.** Of, relating to, or suitable for a maiden. **2.** Being an unmarried woman. **3.** Inexperienced : untried. **4.** Designating a racehorse that has never won a race. **5.** First or earliest <a *maiden* flight>

**maid·en·hair** *or* **maid·en·hair fern** (mād′n-hâr′) *n.* [From the fineness of its stems.] A fern of the genus *Adiantum,* with dark stems and light-green feathery fronds having fan-shaped leaflets.

**maidenhair tree** *n.* The ginkgo.

**maid·en·head** (măd′n-hĕd′) *n.* [ME *maidenhed* : *maiden* (< OE *mægden*) + *-hed,* -hood < OE *-hād.*] The quality or state of being a maiden : VIRGINITY.

**maid·en·hood** (mād′n-hŏŏd′) *n.* The condition or time of being a maiden.

**maid·en·ly** (mād′n-lē) *adj.* Relating to or befitting for a maiden. **—maid′en·li·ness** *n.*

**maiden name** *n.* A woman's family name before marriage.

**maiden over** *n.* An over in cricket during which no runs are scored.

**maid·hood** (măd′hŏŏd′) *n.* Maidenhood.

**maid in waiting** *n., pl.* **maids in waiting.** An unmarried woman attending a queen or princess.

**maid of honor** *n., pl.* **maids of honor. 1.** A noblewoman who attends a queen or princess. **2.** The chief unmarried woman attendant of a bride.

**maid·ser·vant** (mād′sûr′vənt) *n.* A woman servant.

**Mai·du** (mī′dōō) *n., pl.* **Maidu** *or* **-dus.** [Maidu.] **1. a.** An Indian tribe once living in the Sacramento Valley area of California. **b.** A member of this tribe. **2.** The Penutian language of the Maidu. **—Mai′du** *adj.*

**ma·ieu·tic** (mā-yōō′tĭk, mī-) *also* **ma·ieu·ti·cal** (-tĭ-kəl) *adj.* [Gk. *maieutikos* < *maieuesthai,* to act as midwife < *maia,* midwife.] Relating to that aspect of the Socratic method that aids an individual in the formulation of latent concepts.

**mail**[1] (māl) *n.* [ME *male,* bag < OFr., of Germanic orig.] **1. a.** Materials handled in a postal system, as letters and packages. **b.** Postal material for a specific recipient. **c.** Material processed for distribution from a post office at a specified time <the morning *mail*> **2.** *often* **mails.** A system for transporting postal materials, as letters and packages. **3.** A vehicle for transporting mail. *—v.* **mailed, mail·ing, mails.** *—vt.* To send by mail. *—vi.* To send letters and other postal material by mail. **—mail′a·bil′i·ty** *n.* **—mail′a·ble** *adj.*

▲ word history: The word *mail*[1], which denotes the material handled by the post office, is a survivor of the days when the few letters and dispatches that were exchanged were carried by horsemen in their traveling bags. In Middle English times the word *mail* meant simply "bag," especially one used by a traveler for provisions. Such bags were used to carry letters, and the word *mail* eventually came to designate the contents rather than the container.

**mail**[2] (māl) *n.* [ME < OFr. *maile* < Lat. *macula,* mesh.] **1.** Flexible armor made of small overlapping metal rings, loops of chain, or scales. **2.** The protective shell or covering of certain animals, as the turtle. **3.** The full-grown breast feathers of a hawk. *—vt.* **mailed, mail·ing, mails.** To cover or armor with mail.

**mail**[3] (māl) *n.* [ME *maile* < OE *māl,* agreement < ON.] *Chiefly Scot.* Rent : payment.

**mail·bag** (māl′băg′) *n.* **1.** A large canvas sack for transporting mail. **2.** A bag hung over the shoulder, used by letter carriers for carrying mail.

**mail·box** (māl′bŏks′) *n.* **1.** A public container where outgoing mail is deposited for collection. **2.** A private box for incoming mail.

**mail call** *n.* Distribution of mail to members of a military unit.

**mail carrier** *n.* **1.** A vehicle or other device for transporting mail. **2.** One who carries and delivers mail.

**mail drop** *n.* **1.** A receptacle or slot for mail delivery. **2.** An address at which one receives mail but does not reside.

**mailed** (māld) *adj.* **1.** Covered with or made of plates of mail. **2.** Having a hard covering of scales, spines, or horny plate, as an armadillo.

**mailed fist** *n.* The threat of military force.

**mail·er** (mā′lər) *n.* **1.** A user of the mails. **2. a.** One who addresses, stamps, or otherwise prepares mail. **b.** A mailing machine. **3.** A ship that carries mail. **4.** A container, as a cardboard tube, for holding material to be mailed. **5.** An advertising leaflet included with a letter.

**Mail·gram** (māl′grăm′). A trademark for a telegram delivered by the postal service.

**mail·ing** (mā′lĭng) *n.* **1.** Something sent by mail. **2.** A batch of mail sent at one time by a mailer.

**mailing machine** *n.* Any of various machines that stamp, address, or seal material for mailing.

**mail·lot** (mä-yō′) *n.* [Fr. < OFr., swaddling clothes < *maille,* band of cloth < Lat. *macula,* mesh.] **1.** A coarsely knitted, stretchable jersey fabric. **2. a.** A pair of tights or a gymnastic suit made of maillot. **b.** A bathing suit of maillot, esp. a usu. one-piece suit for women.

**mail·man** (māl′măn′, -mən) *n.* MAIL CARRIER 2.

**mail order** *n.* A request for services or goods received and often filled through the mail.

**mail-or·der house** (māl′ôr′dər) *n.* A business establishment mainly organized to promote, receive, and fill orders for merchandise or services through the mail.

**maim** (mām) *vt.* **maimed, maim·ing, maims.** [ME *maimen* < OFr. *mahaignier.*] **1.** To disable or disfigure, usu. by depriving of the use of a limb or bodily member : MUTILATE. **2.** To make imperfect or defective : IMPAIR. **—maim′er** *n.*

**main**[1] (mān) *adj.* [ME < OE *mægen.*] **1.** Most important : PRINCIPAL. **2.** Exerted to the utmost : SHEER <by *main* force> **3.** *Obs.* Of or relating to a continuous area or stretch, as of land or water. **4.** *Naut.* Connected to or located near the mainmast <a *main* skysail> *—n.* **1.** The principal or most important part or point <children who are, in the *main,* well-behaved> **2.** The principal pipe or conduit in a system for conveying a utility, as water, gas, or oil. **3.** Physical power <might and *main*> **4.** The mainland. **5.** The open ocean. **6.** *Naut.* **a.** The mainsail. **b.** The mainmast. **—main′ly** *adv.*

**main**[2] (mān) *n.* [Prob. < MAIN[1].] **1.** A number greater than four but not exceeding nine in the game of hazard that is called by the caster before the throw of the dice. **2.** An off number of matches in a series of cockfights.

**main chance** *n.* One's best opportunity.

**main clause** *n.* The principal clause in a complex sentence.

**main deck** *n.* The principal deck of a ship.

**main drag** *n. Slang.* The principal thoroughfare of a city or town.

**main·frame** (mān′frām′) *n.* The central processing unit of a computer exclusive of peripheral and remote devices.

**main·land** (mān′lănd′, -lənd) *n.* The principal land mass of a continent.

**main·line** (mān′līn′) *vi.* **-lined, -lin·ing, -lines.** *Slang.* To inject narcotics directly into a major vein. **—main′lin′er** *n.*

**main line** *n.* **1.** A principal section of a railroad line. **2.** *Slang.* **a.** A principal and easily accessible vein, usu. in the arm or leg, into which narcotics can be injected. **b.** The injection of a narcotic into an easily accessible vein.

**main·mast** (mān′məst, -măst′) *n. Naut.* **1.** The principal mast of a vessel. **2.** The taller mast, forward or aft, of a two-masted sailing vessel. **3.** The second mast aft of a sailing ship with at least three masts.

**main roy·al·mast** (roi′əl-məst, -măst′) *n. Naut.* The section of the mainmast of a square-rigged vessel above the main topgallantmast.

**main·sail** (mān′səl, -sāl′) *n. Naut.* **1.** The principal sail of a vessel. **2.** A quadrilateral or triangular sail set from the after part of the mainmast on a fore-and-aft rigged vessel. **3.** A square sail set from the main yard on a square-rigged vessel.

**main sequence** *n.* A major grouping of stars, containing the sun and 90% of the known stars in the vicinity of the sun, typified by an approx. uniform average increase of luminosity with surface temperature as shown by a single band on the Hertzsprung-Russell diagram.

**main·sheet** (mān′shēt′) *n. Naut.* The rope that controls the angle at which the mainsail is trimmed and set.

**main·spring** (mān′sprĭng′) *n.* **1.** The principal spring in a mechanical device, esp. in a watch or clock, that drives the mechanism by uncoiling. **2.** A motivating force.

**main·stay** (mān′stā′) *n.* **1.** *Naut.* A strong rope that steadies and supports the mainmast of a sailing vessel. **2.** A principal support <Tidal flow is a *mainstay* of marine life.>

**main stem** *n.* **1.** MAIN STREET 1. **2.** MAIN LINE 1.

**main·stream** (mān′strēm′) *n.* The prevailing current or direction of a movement or influence <"Ireland is moving in the *mainstream* of current world events" —John F. Kennedy> *—adj.* Having, influenced by, or in agreement with the prevailing attitudes and values of a society or group <*mainstream* opinion> *—vt.* **-streamed, -stream·ing, -streams.** To place (a handicapped student) in regular school classes.

**main street** *n.* [After *Main Street,* a novel by Sinclair Lewis (1885–1951).] **1.** The principal street of a city or town, esp. a small one. **2. Main Street.** The culture of smug, materialistic, and provincial small towns.

**main·tain** (mān-tān′) *vt.* **-tained, -tain·ing, -tains.** [ME *maintainen* < OFr. *maintenir* < Med. Lat. *manutenēre* < Lat. *manu tenēre,* to hold in the hand.] **1.** To continue : carry on <*maintain* a sound economic policy> **2.** To preserve or keep in a given existing condition, as of efficiency or good repair <*maintain* bridges> **3. a.** To provide for <*maintain* an aged relative> **b.** To keep in existence : SUSTAIN <*maintaining* life in a spacecraft environment> **4.** To defend, as against danger or attack <"Perhaps the Germans could not *maintain* the corridor" —Winston Churchill> **5.** To declare to be true : AFFIRM <They *maintained* the accuracy of the report.> **—main·tain′a·ble** *adj.* **—main·tain′er** *n.*

**main·te·nance** (mān′tə-nəns) *n.* [ME < OFr. < *maintenir,* to maintain. —see MAINTAIN.] **1.** The act of maintaining or state of being maintained. **2.** The work of keeping something in suitable condition. **3.** A means of maintaining or supporting <wages that do not

guarantee *maintenance*> **4.** *Law.* Unlawful interference in a lawsuit by aiding either party with the wherewithal to carry it on.

**main·top** (mān′tŏp′) *n. Naut.* A platform at the head of the mainmast on a square-rigged vessel.

**main topgallant** *n. Naut.* A sail or yard set from the topgallant section of a mainmast.

**main top·gal·lant·mast** (tə-găl′ənt-məst, tŏp-) *n. Naut.* The section of the mainmast next above the main topmast on a square-rigged vessel.

**main topmast** *n. Naut.* The section of the mainmast on a square-rigged sailing vessel between the lower mast and the main topgallantmast.

**main topsail** *n. Naut.* The sail set above the mainsail.

**main yard** *n. Naut.* The lower yard on a mainmast.

**mai tai** (mī′ tī′) *n.* [Tahitian *maitai*, good.] A cocktail consisting of rum, curaçao, and fruit juices.

**maî·tre d′** (mā′trə dē′, mā′tər) *n., pl.* **maî·tre d′s** (dēz′). *Informal.* MAÎTRE D'HÔTEL 1, 2.

**maî·tre d'hô·tel** (mā′trə dō-tĕl′) *n., pl.* **maî·tres d'hô·tel** (mā′trə dō-tĕl′) [Fr. *maître*, master + *de*, of + *hôtel*, house.] **1.** A headwaiter. **2.** A major-domo. **3.** A sauce of melted butter, chopped parsley, lemon juice, salt, and pepper.

**maize** (māz) *n.* [Sp. *maíz* < Carib *mahiz.*] **1.** CORN¹ 1. **2.** A light yellow to moderate orange yellow.

**ma·jes·tic** (mə-jĕs′tĭk) *also* **ma·jes·ti·cal** (-tĭ-kəl) *adj.* Having or showing stateliness or great dignity. **—ma·jes′ti·cal·ly** *adv.*

**maj·es·ty** (măj′ĭ-stē) *n., pl.* **-ties.** [ME *majeste* < OFr. < Lat. *majestas* < *major*, greater. —see MAJOR.] **1. a.** A sovereign's greatness and dignity. **b.** The sovereignty and power of God. **2.** Supreme power or authority <the *majesty* of the law> **3. a.** A royal personage. **b. Majesty.**—Used as a title in speaking of or to a sovereign <Your *Majesty*> **4. a.** Royal dignity of bearing : GRANDEUR. **b.** Stateliness, magnificence, or splendor, as of appearance, style, or character <the *majesty* of the Taj Mahal>

**ma·jol·i·ca** (mə-jŏl′ĭ-kə, -yŏl′-) *n.* [OItal. *maiolica* < *Majolica*, Majorca.] **1.** Tin-glazed earthenware often richly decorated and colored, esp. an earthenware of this type made in Italy. **2.** A modern pottery copying majolica.

**ma·jor** (mā′jər) *adj.* [ME *majour* < Lat. *major*, comp. of *magnus*, great.] **1.** Greater in importance, rank, or stature <a *major* breakthrough in cancer research> **2.** Demanding great attention or concern : SERIOUS <a *major* accident> **3.** *Law.* Having attained full legal age. **4.** Of a greater number, quantity, or extent. **5.** Designating the principal field of academic specialization selected by college or university students. **6.** More inclusive in scope. **7.** *Mus.* **a.** Designating a scale or mode with half steps between the third and fourth and the seventh and eighth degrees. **b.** Equivalent to the distance between the tonic note and the second or third or sixth or seventh degrees of a major scale or mode <a *major* interval> **c.** Based on a major scale <*major* key> *—n.* **1. a.** An officer in the U.S. Army, Air Force, or Marine Corps ranking above a captain and below a lieutenant colonel. **b.** An officer of similar rank in other military or paramilitary organizations. **2.** *Law.* One of full legal age. **3. a.** A field or subject selected as an academic specialization. **b.** A student specializing in such a field <a geology *major*> **4.** *Logic.* A major premise or major term. **5.** *Mus.* A major scale, key, interval, or mode. **6. majors.** The major leagues. *—vi.* **-jored, -jor·ing, -jors.** To pursue academic studies in a major field.

**major axis** *n.* **1.** The line intersecting an ellipse and passing through both of its focuses. **2.** The longest axis of an ellipsoid.

**ma·jor-do·mo** (mā′jər-dō′mō) *n., pl.* **-mos.** [Ital. *maggiordomo* or Sp. *mayordomo*, both < Med. Lat. *major domus* : *major*, chief + Lat. *domus*, of the house.] **1.** The chief steward or butler in the household of a sovereign or noble. **2.** A butler or steward.

**ma·jor·ette** (mā′jə-rĕt′) *n.* A drum majorette.

**major gene** *n.* Oligogene.

**major general** *n.* A U.S. Army, Air Force, or Marine Corps officer ranking above a brigadier general and below a lieutenant general.

**ma·jor·i·tar·i·an·ism** (mə-jôr′ĭ-târ′ē-ə-nĭz′əm, -jŏr′-) *n.* Belief in or the practice of decision-making in an organized group by a numerical majority of its members. **—ma·jor′i·tar′i·an** *n.*

**ma·jor·i·ty** (mə-jôr′ĭ-tē, -jŏr′-) *n., pl.* **-ties.** [OFr. *majorite* < Med. Lat. *majoritas* < Lat. *major*, greater. —see MAJOR.] **1.** The greater part or number. **2. a.** A number more than half of the total number of a given group. **b.** The number of election votes above the total number of all other votes cast. **3.** The legal age status at which full personal and civil rights may be exercised. **4.** The political party, group, or faction having the greatest power by virtue of its bigger representation or electoral strength. **5.** The military rank, commission, or office of a major. **6.** *Obs.* The fact or state of being greater : SUPERIORITY. ***usage:*** When *majority* refers to a particular number of votes, it takes a singular verb, as in *That candidate's majority was five votes.* When referring to a group that constitutes a majority, depending on whether the group is considered as a whole or as a set of individuals, *majority* may take either a singular or plural verb. So we say *The majority elects the candidate it wants,* since the action is accomplished by the group as a whole; but *The majority of the voters live in the city,* since that action is performed by each voter individually.

**majority leader** *n.* The leader of the majority party in a legislative body.

**majority rule** *n.* A political doctrine by which a numerical majority, usu. one more than 50% of a given group, holds the power to make decisions binding on all.

**major league** *n.* **1.** Either of the two chief groups of U.S. professional baseball teams. **2.** A league of principal importance in other professional sports, as basketball, football, or ice hockey.

**ma·jor-league** (mā′jər-lēg′) *adj.* In a leading position <Computer software is now a *major-league* business.>

**ma·jor-med·i·cal** (mā′jər-mĕd′ĭ-kəl) *adj.* Of, pertaining to, or being a kind of insurance plan that covers most of the medical bills of major illnesses.

**major orders** *pl.n.* Holy orders.

**major party** *n.* A political party able to gain control of a government with comparative regularity.

**major premise** *n.* The premise containing the major term in a syllogism.

**Major Prophets** *pl.n.* The Hebrew prophets Isaiah, Jeremiah, and Ezekiel.

**major scale** *n. Mus.* A diatonic scale with half steps between the third and fourth and the seventh and eighth tones.

**major suit** *n.* A bridge suit of superior scoring value, either hearts or spades.

**major term** *n.* A term of a syllogism that forms the predicate of the conclusion and the subject or predicate of the major premise.

**ma·jus·cule** (mə-jŭs′kyōōl, măj′ə-skyōōl′) *n.* [Fr. < Lat. *majusculus*, somewhat great, dim. of *major*, greater. —see MAJOR.] **1.** A large letter, either capital or uncial, used in writing or printing. **2.** Writing that uses majuscule letters. **—ma·jus′cule, ma·jus′cu·lar** (mə-jŭs′kyə-lər) *adj.*

**mak·ar** (mä′kər, mā′-) *n.* [ME *maker*, maker.] *Chiefly Scot.* A poet.

**make** (māk) *v.* [ME *maken* < OE *macian.*] **made** (mād), **mak·ing, makes.** *—vt.* **1. a.** To cause to exist or happen : CREATE. **b.** To bring into existence by forming or modifying materials <*make* a shirt> **c.** To create by putting together component parts <*make* a cake> **d.** To form by assembling individuals or constituents <*make* a quartet> **2. a.** To cause to be or become <*made* our feelings known> **b.** To cause to assume a specified function or role <*made* you a sergeant> **3.** To cause to be experienced <*made* trouble for us> **4. a.** To construct or formulate, esp. by the use of mental or imaginative power <*make* new plans> **b.** To compose <*make* rhymes> **5. a.** To prepare : fix <*make* dinner> **b.** To get ready or set in order for use <*make* camp> **c.** To gather and light the materials for (a fire). **6. a.** To act so as to engage in or carry out. —Used with a noun object indicating the nature of the action <*make* war> **b.** To perform by moving the body or a part of the body <*make* a turn to the left> **c.** To achieve or produce by action or effort <*make* work for the unemployed> **7. a.** To institute or establish : ENACT <*make* new laws> **b.** To draw up and execute in a suitable form <*make* a lease> **8. a.** To reach by traveling <*made* Paris in three days> **b.** To accomplish or complete by traveling across or over <*make* the rounds> **9. a.** To attain : reach <*make* it to the top of the mountain> **b.** To attain the rank or position of <*made* captain> **c.** To acquire a place in or on <*made* the soccer team> **10. a.** To acquire (e.g., money), as by work. **b.** To gain through behavior or effort <*make* friends> **c.** To achieve or score in a sport or game. **11.** To be adequate guarantee for the success of <The move could *make* your career.> **12.** To compel to act <*made* them go> **13. a.** To be capable of conversion into, esp. by a process of fabrication or manufacture <Birch *makes* handsome furniture.> **b.** To be capable of growing or developing into <*made* a good lawyer> **14. a.** To draw a conclusion as to the nature or significance of <unable to *make* heads or tails of it> **b.** To calculate, conjecture, or estimate to be. **c.** To consider as being <weren't the geniuses some people *made* them> **15. a.** To amount to <*makes* no difference> **b.** To bring up to the sum of <This *makes* the only time I've failed.> **c.** To constitute the essential being or nature of. **16.** To succeed in reaching and boarding : CATCH <*make* the plane> **17.** *Slang.* To succeed in having sexual intercourse with. **18.** To appear to begin (an action). *—vi.* **1.** To act : behave. **2.** To begin or appear to begin an action <*made* as if to go> **3.** To proceed <The dog got loose, and we *made* after it.> **4.** To have a particular effect <imaginative food preparation that *makes* for good eating> **5.** To undergo fabrication or manufacture <This material *makes* up nicely.> **—make off. 1.** To depart with haste. **2.** To snatch : steal <*made off* with all the best furniture> **—make out. 1.** To discern or see, esp. with difficulty <could barely *make out* the highway in the pouring rain> **2.** To understand : comprehend <couldn't *make out* the handwriting> **3.** To fill in (e.g., a document) <*made out* an insurance application> **4.** *Informal.* To prove or imply <This *makes* me *out* a fool.> **5.** *Informal.* To pretend to be true <*made out* that the loafer was a hard worker> **6.** *Slang.* **a.** To neck : pet. **b.** To have sexual intercourse. **7.** *Slang.* To get along in a given way : FARE <*made out* poorly

ă **pat** ā **pay** âr **care** ä **father** ĕ **pet** ē **be** hw **which** ĭ **pit**
ī **tie** îr **pier** ŏ **pot** ō **toe** ô **paw, for** oi **noise** ŏŏ **took**

in politics> **—make over. 1.** To redo : renovate. **2.** To change or transfer the ownership of, usu. by way of a legal document. **—make up. 1.** To construct, create, or shape by collecting and fitting components, parts, or materials together. **2.** To apply cosmetics to the face. **3.** To construct falsely or fictionally : FABRICATE <*made up* an alibi> **4.** To come to a decision <*made up* my mind to go> **5. a.** To offset a deficit <*make up* the shortage> **b.** To compensate for a mistake, offense, or omission <*make up* for their crimes> **6.** To resolve a quarrel <kissed and *made up*> **7.** To make ingratiating overtures to someone <*made up* to the company president> **8.** To take (e.g., a test) again or at a later time because of a previous absence or failure. **9.** To arrange material into (columns or pages) for printing <*made up* the editorial page> *—n.* **1.** An act or process of making. **2.** The manner or style in which an object is made <The *make* of this coat flatters you.> **3. a.** A manufacturing style. **b.** A distinctive line of manufactured goods, identified by the maker's name or the registered trademark <a famous *make* of shoe> **4.** The physical or moral nature of a person. **5.** The amount produced, esp. the output of a factory. **—make a face.** To grimace. **—make a go of.** To achieve success in <*made a go of* the new enterprise> **—make a mountain out of a molehill.** To attach too much significance to a trivial issue. **—make away with. 1.** To make off with. **2.** To use up : CONSUME. **—make believe.** To pretend. **—make bold.** To venture. **—make book.** To accept bets on a race, game, or contest. **—make do.** To manage to get along with whatever is available. **—make ends meet.** To manage carefully so as to make one's means adequate for one's needs. **—make eyes.** To ogle. **—make fun of.** To mock : ridicule. **—make good. 1. a.** To carry out successfully : ACHIEVE <*made good* their career plans> **b.** To carry out (e.g., a promise) <*made good* our vow> **2.** To repay : indemnify <*make good* one's obligations> **3.** To succeed <*make good* in the leading role> **—make hay.** To take advantage of an opportunity <Speculators *made hay* of the drop in land prices.> **—make it.** *Informal.* To be successful <finally *made it* as an actor> **—make light of.** To treat as unimportant <*made light of* my illness> **—make love. 1.** To court : woo. **2.** To embrace and caress : PET. **3.** To engage in sexual intercourse. **—make much of.** To treat as of great importance. **—make no bones about.** To be forthright, unequivocal, and sure <*made no bones about* my outrage> **—make public.** To disclose to public knowledge. **—make sail. 1.** To begin a voyage. **2.** To set sail. **—make the grade.** To measure up to a particular standard. **—make the most of.** To use to the most advantage. **—make the scene.** *Slang.* To put in an appearance <*made the scene* at the luncheon> **—make time. 1.** To move fast. **2.** *Slang.* To make progress toward winning, esp. romantic or sexual favors. **—make tracks.** *Slang.* To move rapidly. **—make waves.** *Slang.* To cause a controversy or disturbance. **—make way. 1.** To give room for passage. **2.** To make progress. **—make with.** *Slang.* To perform : produce <started *making with* the hard work> **—on the make.** Eagerly and often offensively striving for financial or social improvement.

☆ ***syns:*** MAKE, ASSEMBLE, BUILD, CONSTRUCT, FABRICATE, FASHION, MANUFACTURE, PRODUCE, SHAPE *v. core meaning* : to create by forming, combining, or altering materials <*make* a sandwich><*make* a house from logs>

**make-be·lieve** (māk′bĭ-lēv′) *n.* Fanciful or playful pretense. **—make′be·lieve′** *adj.*

**make·fast** (māk′făst′) *n. Naut.* An object, as a buoy, post, or pile, to which a boat is moored.

**mak·er** (mā′kər) *n.* **1.** One that makes. **2.** *Law.* A signer of a promissory note. **3. Maker.** GOD 1a, b. **4.** *Archaic.* A poet.

**make-read·y** (māk′rĕd′ē) *n.* Preparation of a form for printing by adjusting and leveling the plates to ensure a clear impression.

**make·shift** (māk′shĭft′) *n.* A temporary or expedient substitute. **—make′shift′** *adj.*

**make-up** *also* **make·up** (māk′ŭp′) *n.* **1.** The way in which something is arranged or composed : CONSTRUCTION. **2.** Arrangement or composition, as of printing type or illustrations, in a book or on a page. **3.** The qualities or temperament constituting a personality : DISPOSITION. **4.** Cosmetics applied esp. to the face. **5.** The cosmetics that a performer uses in portraying a role. **6.** A special examination for a student who has missed or has failed a previous examination.

**make·weight** (māk′wāt′) *n.* **1.** Something added on a scale so as to meet a required weight. **2.** A counterweight.

**make-work** (māk′wûrk′) *n.* Busywork.

**ma·ki·mo·no** (mä′kĭ-mō′nō) *n.* [J., scroll : *maki*, roll + *mono*, thing.] A horizontal Japanese scroll with pictures or calligraphy.

**mak·ing** (mā′kĭng) *n.* **1.** The act of one that makes or the process of being made. **2.** The means or a process followed in gaining success. **3. a.** Something made. **b.** An amount or quantity of something made at one time. **4.** *often* **makings.** The materials or substances needed for doing or making something <We have the *makings* for success in this business.> **5. makings.** *Slang.* The tobacco and paper for rolling a cigarette.

**ma·ko shark** (mä′kō) *n.* [Maori *mako.*] Either of two sharks of the genus *Isurus*, characterized by a large heavy body and an almost symmetric tail.

**ma·ku·ta** (mä-kōō′tä) *n. pl. of* LIKUTA.

**mal-** *pref.* [ME < OFr. < Lat. < *male*, badly and *malus*, bad.] **1.** Bad : badly <*mal*administer> **2.** Abnormal : abnormally <*mal*formation><*mal*absorption>

**mal·ab·sorp·tion** (măl′əb-sôrp′shən, -zôrp′-) *n.* Defective or inadequate absorption of nutrients from the intestinal tract.

**Ma·lac·ca** (mə-lăk′ə) *n.* [After *Malacca*, Malaysia.] The stem of the rattan palm of Asia, used for walking sticks.

**Mal·a·chi** (măl′ə-kī′) *n.* [Heb. *Mal'ākhī.*] —See table at BIBLE.

**mal·a·chite** (măl′ə-kīt′) *n.* [ME *melochite* < Lat. *molochites* < Gk. *molokhitis* < *molokhē*, mallow.] A green to nearly black mineral carbonate of copper, $CuCO_3 \cdot Cu(OH)_2$, used as a source of copper and for ornamental stoneware.

**mal·a·col·o·gy** (măl′ə-kŏl′ə-jē) *n.* [Fr. *malacologie*, contraction of *malacozoologie* < NLat. *Malacozoa*, a classification that includes mollusks : Gk. *malakos*, soft + Gk. *zōia*, animal.] Scientific study of mollusks.

**mal·ad·ap·ta·tion** (măl′ăd-ăp-tā′shən) *n.* Faulty or inadequate adaptation.

**mal·a·dapt·ed** (măl′ə-dăp′tĭd) *adj.* Not suited, as to a certain situation or function.

**mal·a·dap·tive** (măl′ə-dăp′tĭv) *adj.* **1.** Characterized by faulty or inadequate adaptation. **2.** Not aiding or furthering adaptation.

**mal·ad·just·ment** (măl′ə-jŭst′mənt) *n.* **1.** Faulty adjustment, as in a machine. **2.** *Psychol.* Inability to adjust personality needs to the demands of the environment. **3.** Imbalance in economic and social relations, as between city and country or supply and demand. **—mal′ad·just′ed** *adj.*

**mal·ad·min·is·ter** (măl′əd-mĭn′ĭ-stər) *vt.* **-tered, -ter·ing, -ters.** To administer dishonestly or inefficiently. **—mal′ad·min′is·tra′tion** *n.*

**mal·a·droit** (măl′ə-droit′) *adj.* [Fr. : *mal-*, mal- + *adroit*, adroit.] **1.** Marked by a lack of dexterity : CLUMSY. **2.** Marked by a lack of perception : TACTLESS. **—mal′a·droit′ly** *adv.*

**mal·a·dy** (măl′ə-dē) *n., pl.* **-dies.** [ME *maladie* < OFr. < *malade*, sick < Lat. *male habitus*, in poor condition.] **1.** A disease, disorder, or ailment. **2.** An unwholesome condition <the *malady* of urban blight>

**ma·la fi·de** (mā′lə fī′dē) *adv. & adj.* [Lat.] In or with bad faith.

**Mal·a·ga** (măl′ə-gə) *n.* A sweet wine orig. from Málaga, Spain.

**Mal·a·gas·y** (măl′ə-găs′ē) *n., pl.* **Malagasy** *or* **-gas·ies. 1.** A native or resident of Madagascar. **2.** The Austronesian language of the Malagasy. **—Mal′a·gas′y** *adj.*

**ma·la·gue·ña** (mä′lə-gā′nyə) *n.* [Sp., fem. of *malagueño*, of Málaga, after *Málaga*, Spain.] **1.** A dance native to Málaga, Spain, and a variety of the fandango. **2.** A folk tune native to Málaga that is similar to the fandango.

**mal·aise** (mă-lāz′, -lĕz′) *n.* [Fr. < OFr. : *mal-*, mal- + *aise*, ease.] A vague feeling of depression or illness.

**mal·a·mute** *or* **mal·e·mute** (măl′ə-myōōt′) *n.* [*Malemute*, an Alaskan Eskimo people.] A powerful dog orig. bred in Alaska as a sled dog, having a thick gray, black, or white coat.

**mal·a·pert** (măl′ə-pûrt′) *adj.* [ME < OFr. : *mal-*, mal- + *apert*, clever, alteration of Lat. *expertus.*] Saucy in manner or speech. *—n.* A saucy person. **—mal′a·pert′ly** *adv.* **—mal′a·pert′ness** *n.*

**mal·ap·por·tioned** (măl′ə-pôr′shənd, -pōr′-) *adj.* Marked by an unfair or inappropriate proportional distribution of legislative representatives. **—mal′ap·por′tion·ment** *n.*

**mal·a·prop·ism** (măl′ə-prŏp-ĭz′əm) *n.* [After Mrs. *Malaprop*, a character in *The Rivals*, a play by Richard B. Sheridan (1751–1816).] Ludicrous misuse of a word. **—mal′a·prop′i·an** *adj.*

**mal·a·pro·pos** (măl′ăp-rə-pō′) *adj.* [Fr. *mal à propos.*] Out of place : INAPPROPRIATE. **—mal′a·pro·pos′** *adv.*

**ma·lar** (mā′lər, -lär′) *adj.* [Lat. *malaris* < *mala*, cheekbone.] Of or relating to the cheekbone or the cheek. *—n.* The zygomatic bone in the cheek.

**ma·lar·i·a** (mə-lâr′ē-ə) *n.* [Ital. *mal'aria* < *mala aria*, bad air.] **1.** An infectious disease transmitted by the bite of the infected female anopheles mosquito and marked by cycles of chills, fever, and sweating. **2.** *Archaic.* Bad or foul air. **—ma·lar′i·al, ma·lar′i·an, ma·lar′i·ous** *adj.*

**ma·lar·i·ol·o·gy** (mə-lâr′ē-ŏl′ə-jē) *n.* Study of malaria. **—ma·lar′i·ol′o·gist** *n.*

**ma·lar·key** *also* **ma·lar·ky** (mə-lär′kē) *n.* [Orig. unknown.] *Slang.* Nonsense.

**mal·as·sim·i·la·tion** (măl′ə-sĭm′ə-lā′shən) *n.* Incomplete assimilation of injested food.

**mal·ate** (măl′āt′, mā′lāt′) *n.* [MAL(IC ACID) + -ATE[2].] A salt or ester of malic acid.

**Mal·a·thi·on** (măl′ə-thī′ŏn′). A trademark for the organic compound, $C_{10}H_{19}O_6PS_2$, used as an insecticide.

**Ma·lay** (mā′lā′, mə-lā′) *n.* [Obs. Du. *Malayo* < Malay *Mĕlayu.*] **1.** One of a people living in the Malay Peninsula, other parts of Malaysia, Indonesia, and some adjacent areas. **2.** The Austronesian language of the Malays. **3.** A fowl having red and black plumage, domesticated in Asia. —*adj.* **1.** Of, relating to, or typical of the Malays or their language. **2.** Of or relating to Malaya or Malaysia. **—Ma·lay'an** (mə-lā'ən) *adj.* & *n.*

**Mal·a·ya·lam** (măl'ə-yä'ləm) *n.* A Dravidian language of the Malabar coast in southwestern India.

**Ma·lay·o-Pol·y·ne·sian** (mə-lā'ō-pŏl'ə-nē'zhən, -shən) *n.* Austronesian. **—Ma·lay'o-Pol'y·ne'sian** *adj.*

**mal·con·tent** (măl'kən-tĕnt') *adj.* [OFr. : *mal-*, mal- + *content,* content.] Discontented with existing circumstances. —*n.* A discontented person.

**mal de mer** (măl' də mâr') *n.* [Fr.] Seasickness.

**mal·dis·tri·bu·tion** (măl'dĭs-trə-byōō'shən) *n.* Faulty distribution or apportionment over an area or among a group.

**male** (māl) *adj.* [ME < OFr. < Lat. *masculus,* dim. of *mas,* male.] **1.** Of, relating to, or designating the sex that has organs to produce spermatozoa for fertilizing ova. **2.** Of or typical of the male sex : MASCULINE. **3.** Manly : virile. **4.** Made up of men or boys or both <a *male* choir> **5.** *Bot.* **a.** Relating to or designating organs, as stamens or anthers, that are capable of fertilizing female organs. **b.** Bearing stamens but not pistils : STAMINATE <*male* flowers> **6.** Designating an object, as an electric plug, designed for insertion into a fitted bore or socket. —*n.* **1.** An individual of the sex that begets young by fertilizing ova. **2.** A plant bearing only staminate flowers. **—male'ness** *n.*

**ma·le·ate** (mā'lē-āt', mə-lē'ət) *n.* [MALE(IC ACID) + -ATE[2].] A salt or ester of maleic acid.

**Mal·e·cite** (măl'ĭ-sīt') *also* **Mal·i·seet** (-sēt') *n., pl.* **Malecite** *or* **-cites** *also* **Maliseet** *or* **-seets.** [Prob. < Micmac *Maliisit.*] **1. a.** A tribe of Indians once inhabiting New Brunswick and northeastern Maine. **b.** A member of this tribe. **2.** The Algonquian language of the Malecite.

**mal·e·dict** (măl'ĭ-dĭkt') *adj.* [ME *maladicte* < Lat. *maledictus* < *maledicere,* to curse : *male,* ill (< *malus,* bad) + *dicere,* to speak.] That is accursed. —*vt.* **-dict·ed, -dict·ing, -dicts.** To invoke against. **—mal'e·dic'·tion** (-dĭk'shən) *n.*

**mal·e·fac·tor** (măl'ə-făk'tər) *n.* [ME *malefactour* < Lat. *malefactor* < *malefacere,* to do wrong : *male,* ill (< *malus,* bad) + *facere,* to do.] **1.** One who has committed a crime : CRIMINAL. **2.** An evildoer. **—mal'e·fac'tion** (-făk'shən) *n.*

**male fern** *n.* A fern, *Dryopteris filix-mas,* that yields a drug for treating tapeworm infestation.

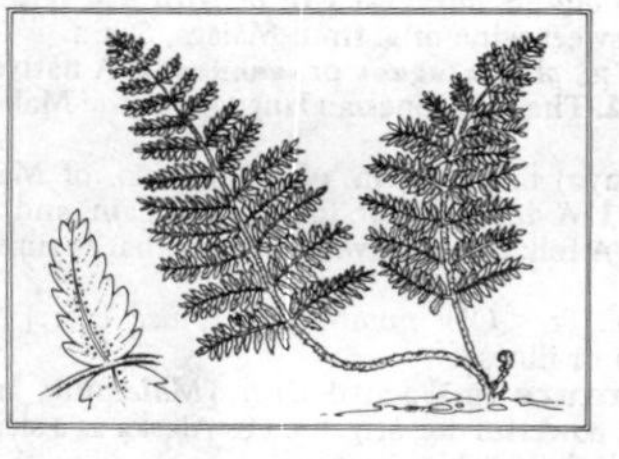

**male fern**

**ma·lef·ic** (mə-lĕf'ĭk) *adj.* [Lat. *maleficus* : *male,* ill (< *malus,* bad) + *facere,* to make.] **1.** Having a malignant influence : SINISTER. **2.** Malicious : evil.

**ma·lef·i·cence** (mə-lĕf'ĭ-səns) *n.* [Lat. *maleficentia* < *maleficus,* malefic.] **1.** Evil or harm : MISCHIEF. **2.** The quality or state of being evil or malignant. **—ma·lef'i·cent** *adj.*

**ma·le·ic acid** (mə-lē'ĭk) *n.* [Fr. *acide maléique,* alteration of *acide malique,* malic acid.] A colorless crystalline acid, $C_4H_4O_4$, used in synthesis of resins and as an oil and fat preservative.

**mal·e·mute** (măl'ə-myōōt') *n. var. of* MALAMUTE.

**mal·en·ten·du** (măl'ŏn-tŏn-dōō') *n.* [Fr. < *mal entendu,* misunderstood.] A misunderstanding.

**ma·lev·o·lence** (mə-lĕv'ə-ləns) *n.* **1.** The quality or condition of being malevolent. **2.** Ill will : MALICE.

**ma·lev·o·lent** (mə-lĕv'ə-lənt) *adj.* [Lat. *malevolens, malevolent-,* malevolent : *male,* ill (< *malus,* bad) + *volens,* wishing (< *velle,* to wish).] **1.** Having or showing ill will or spite. **2.** Having an evil influence. **—ma·lev'o·lent·ly** *adv.*

☆ ***syns:*** MALEVOLENT, BITCHY, DESPITEFUL, EVIL, MALICIOUS, MALIGN, MALIGNANT, MEAN, NASTY, POISONOUS, SPITEFUL, VENOMOUS, VICIOUS, WICKED *adj. core meaning* : marked by ill will or spite <held a *malevolent* attitude toward all strangers> ***ant:*** benevolent

**mal·fea·sance** (măl-fē'zəns) *n.* [MAL- + obs. *feasance,* doing < OFr. *faisance* < Med. Lat. *facientia* < Lat. *facere,* to do.] *Law.* Misconduct, esp. by a public official. **—mal·fea'sant** *adj.* & *n.*

**mal·for·ma·tion** (măl'fôr-mā'shən) *n.* **1.** The state of being malformed. **2.** An abnormal structure or form.

**mal·formed** (măl-fôrmd') *adj.* Abnormally or imperfectly formed.

**mal·func·tion** (măl-fŭngk'shən) *vi.* **-tioned, -tion·ing, -tions.** **1.** To fail to function. **2.** To function abnormally or imperfectly. **—mal·func'tion** *n.*

**mal·ic acid** (măl'ĭk, mā'lĭk) *n.* [Fr. *acide malique* < Lat. *malum,* apple < Gk. *mēlon, malon.*] A colorless crystalline compound, $COOHCH_2CH(OH)COOH$, naturally occurring in a wide variety of unripe fruit, including apples, cherries, and tomatoes, and used as a flavoring and in the aging of wine.

**mal·ice** (măl'ĭs) *n.* [ME < OFr. < Lat. *malitia* < *malus,* bad.] **1.** A desire to harm others or to see others suffer. **2.** *Law.* Intent, without just cause or reason, to commit an unlawful act injurious to another or others.

**ma·li·cious** (mə-lĭsh'əs) *adj.* Resulting from, inclined to, or marked by malice <*malicious* words> **—ma·li'cious·ly** *adv.* **—ma·li'cious·ness** *n.*

**ma·lign** (mə-līn') *vt.* **-ligned, -lign·ing, -ligns.** [ME *malignen,* to attack < OFr. *malignier* < LLat. *malignari* < Lat. *malignus,* malign.] To speak evil of. —*adj.* **1.** Evil in nature, intent, or influence. **2.** Strongly suggestive of evil, menace, or harm. **—ma·lign'er** *n.* **—ma·lign'ly** *adv.*

☆ ***syns:*** MALIGN, BALEFUL, SINISTER *adj. core meaning* : strongly suggestive of evil, menace, or harm <the hangman, faceless and *malign*> ***ant:*** benign

**ma·lig·nan·cy** (mə-lĭg'nən-sē) *also* **ma·lig·nance** (-nəns) *n., pl.* **-nan·cies** *also* **-nanc·es.** **1.** The quality or state of being malignant. **2.** A malignant tumor.

**ma·lig·nant** (mə-lĭg'nənt) *adj.* **1.** Showing great malevolence : EVIL. **2.** Highly injurious : PERNICIOUS. **3.** *Pathol.* **a.** Designating an abnormal growth that tends to metastasize. **b.** Life-threatening : virulent <a *malignant* disease> **—ma·lig'nant·ly** *adv.*

**ma·lig·ni·ty** (mə-lĭg'nĭ-tē) *n., pl.* **-ties. 1. a.** Intense ill will or hatred. **b.** An act or feeling of great malice. **2.** The quality or state of being highly evil or injurious.

**ma·li·hi·ni** (mä'lĭ-hē'nē) *n.* [Hawaiian.] A newcomer, foreigner, or stranger among the people of Hawaii.

**ma·lines** (mə-lēn') *n.* [Fr. < *Malines,* Mechlin, Belgium.] **1.** *also* **ma·line** (mə-lēn'). Thin stiff veiling woven in a hexagonal pattern. **2.** Mechlin.

**ma·lin·ger** (mə-lĭng'gər) *vi.* **-gered, -ger·ing, -gers.** [< Fr. *malingre,* sickly.] To pretend to be ill or injured in order to avoid responsibilities or work. **—ma·lin'ger·er** *n.*

**Ma·lin·ke** (mə-lĭng'kē) *n., pl.* **Malinke** *or* **-kes. 1. a.** A western African people related to the Mandingos. **b.** A member of this people. **2.** The language of the Malinke.

**Mal·i·seet** (măl'ĭ-sēt') *n. var. of* MALECITE.

**mal·i·son** (măl'ĭ-sən, -zən) *n.* [ME < OFr. *maleicon* < Lat. *maledictio* < *maledicere,* to speak ill. —see MALEDICT.] A curse.

**mall[1]** (môl, măl) *n.* [After *The Mall,* London, England, orig. a pallmall alley.] **1.** A shady public walk or promenade. **2. a.** A street lined with shops and closed to vehicles. **b.** A shopping center. **c.** A large building or group of buildings with shops, businesses, and restaurants usu. accessible by common passageways. **3.** A median strip dividing a road or highway.

**mall[2]** (môl) *n.* & *v. var. of* MAUL.

**mal·lard** (măl'ərd) *n., pl.* **mallard** *or* **-lards.** [ME < OFr. *malarde,* poss. of Germanic orig.] A wild duck, *Anas platyrhynchos,* of which the male has a green head and neck.

**mal·le·a·ble** (măl'ē-ə-bəl) *adj.* [ME < OFr. < Med. Lat. *malleabilis* < *malleare,* to hammer < Lat. *malleus,* hammer.] **1.** Capable of being shaped or formed, as by pressure or hammering. **2.** Capable of being altered or influenced : TRACTABLE <a *malleable* mind> **—mal'le·a·bil'i·ty, mal'le·a·ble·ness** *n.* **—mal'le·a·bly** *adv.*

☆ ***syns:*** MALLEABLE, DUCTILE, FLEXIBLE, FLEXUOUS, MOLDABLE, PLASTIC, PLIABLE, PLIANT, SUPPLE, WORKABLE *adj. core meaning* : capable of being shaped, bent, or drawn out, as by hammering or pressure <Gold and copper are *malleable* metals.> ***ant:*** inflexible, rigid

**mal·lee** (măl'ē) *n.* [Native word in Australia.] **1.** Any of several low, scrubby evergreen trees of the genus *Eucalyptus* of western Australia. **2.** A thicket or growth of mallee.

**mal·le·muck** (măl'ə-mŭk') *n.* [Du. *mallemok,* fulmar : *mal,* silly + *mok,* gull.] A sea bird, as the fulmar, albatross, or shearwater.

**mal·let** (măl'ĭt) *n.* [ME < OFr. *maillet,* dim. of *mail,* maul. —see MAUL.] **1. a.** A short-handled hammer, usu. with a cylindrical head of wood, used mainly to drive a chisel or wedge. **b.** A tool with a large head for striking a surface without damaging it. **2.** A long-handled implement for striking a ball, as in croquet and polo. **3.** A light hammer with a rounded, usu. padded head for striking a percussion instrument.

**mal·le·us** (măl'ē-əs) *n., pl.* **mal·le·i** (măl'ē-ī') [NLat. < Lat., hammer.] *Anat.* The largest of three small bones in the middle ear.

**mal·low** (măl'ō) *n.* [ME *malow* < OFr. *malve* < Lat. *malva.*] **1.** A plant of the widely distributed genus *Malva,* bearing pink or white flowers. **2.** A plant related to the mallow, as the rose mallow.

**malm** (mäm) *n.* [ME, chalky soil < OE *mealm.*] **1. a.** A soft, easily crumbled limestone. **b.** Loam formed by the disintegration of soft limestone. **2.** A mixture of chalk and clay for making bricks.

**malm·sey** (mäm'zē) *n., pl.* **-seys.** [ME < Med. Lat. *malmasia* < *Monembasia,* Monemvasia, Greece.] A sweet fortified white wine orig. made in Greece.

**mal·nour·ished** (măl-nûr′ĭsht) *adj.* Suffering from insufficient food or improper nutrition.

**mal·nu·tri·tion** (măl′nōō-trĭsh′ən, -nyōō-) *n.* Poor nutrition resulting from an insufficient or poorly balanced diet or from defective digestion or defective assimilation of foods.

**mal·oc·clu·sion** (măl′ə-klōō′zhən) *n.* Faulty closure of teeth.

**mal·o·dor·ous** (măl-ō′dər-əs) *adj.* Having a foul odor : SMELLY. **—mal·o′dor** *n.* **—mal·o′dor·ous·ly** *adv.* **—mal·o′dor·ous·ness** *n.*

**ma·lo·nic acid** (mə-lō′nĭk, -lŏn′ĭk) *n.* [Fr. *acide malonique*, alteration of *acide malique*, malic acid.] A colorless crystalline acid, $C_3H_4O_4$, obtained from malic acid and used in making barbiturates.

**Mal·pigh·i·an corpuscle** (măl-pĭg′ē-ən) *n.* [After Marcello *Malpighi* (1628–1694).] *Anat.* **1.** *also* **Malpighian body.** A mass of arterial capillaries enclosed in a capsule and attached to a tubule in the kidney. **2.** A lymph nodule surrounding the smaller arteries in the spleen.

**Malpighian layer** *n.* [After Marcello *Malpighi* (1628–1694).] *Anat.* The deepest layer of the epidermis from which the outer layers develop.

**Malpighian tube** *n.* [After Marcello *Malpighi* (1628–1694).] One of the excretory tubes leading from the digestive tract in insects.

**mal·po·si·tion** (măl′pə-zĭsh′ən) *n.* Abnormal positioning, esp. of a fetus.

**mal·prac·tice** (măl-prăk′tĭs) *n.* **1.** Improper or negligent treatment of a patient by medical care providers, as physicians or nurses, resulting in injury or damage to the patient. **2.** Improper or unethical conduct by the holder of an official or professional position <legal *malpractice*> **3.** An act or instance of improper practice. **—mal′-prac·ti′tion·er** (-tĭsh′ə-nər) *n.*

**malt** (môlt) *n.* [ME < OE *mealt.*] **1.** Grain, usu. barley, that has been allowed to sprout, used chiefly in brewing and distilling. **2.** An alcoholic beverage brewed from malt. **3.** MALTED MILK 2. *—v.* **malt·ed, malt·ing, malts.** *—vt.* **1.** To process (grain) into malt. **2.** To treat or to mix with malt or a malt extract. *—vi.* To become malt.

**Mal·ta fever** (môl′tə) *n.* Undulant fever.

**mal·tase** (môl′tās′, -tāz′) *n.* An enzyme that hydrolyzes maltose to glucose.

**malt·ed milk** (môl′tĭd) *n.* **1.** A soluble powder made of dried milk, malted barley, and wheat flour. **2.** *also* **malt** or **malted.** A beverage made by mixing malted milk powder in milk and adding ice cream and flavoring.

**Mal·tese** (môl-tēz′, -tēs′) *adj.* **1.** Of or relating to Malta, its inhabitants, or the language spoken in Malta. **2.** Of or relating to the Knights of Malta. *—n., pl.* **Maltese. 1.** A native or resident of Malta. **2.** The Semitic language of the Maltese. **3.** A small dog of an ancient breed, with long, silky white hair. **4.** A Maltese cat.

**Maltese cat** *n.* A domestic cat with short, silky, bluish-gray hair.

**Maltese cross** *n.* A cross in the form of four arrowheads placed with their points toward the center of a circle.

**mal·tha** (măl′thə) *n.* [Lat. < Gk., a mixture of wax and pitch.] A black, viscous natural bitumen.

**Mal·thu·sian** (măl-thōō′zhən, môl-) *adj.* Of or relating to the theory of Thomas Malthus that population tends to increase more rapidly than food supply, with inevitably disastrous results unless the increase in population can be controlled. **—Mal·thu′sian** *n.* **—Mal·thu′sian·ism** *n.*

**malt liquor** *n.* Fermented beer or ale made with malt.

**mal·tose** (môl′tōs′, -tōz′) *n.* [Fr. < E. MALT.] A sugar, $C_{12}H_{22}O_{11} \cdot H_2O$.

**mal·treat** (măl-trēt′) *vt.* **-treat·ed, -treat·ing, -treats.** [Fr. *maltraiter* : *mal-*, mal- + *traiter*, to treat.] To treat roughly or cruelly. **—mal·treat′ment** *n.*

**malt sugar** *n.* Maltose.

**mal·va·si·a** (măl′və-zē′ə) *n.* [Ital., after *Monemvasia*, Greece.] **1.** A grape from which malmsey is made. **2.** Malmsey.

**mal·ver·sa·tion** (măl′vər-sā′shən) *n.* [OFr. < *malverser*, to misbehave < Lat. *male versari.*] Misconduct in public office.

**mal·voi·sie** (măl′vwə-zē′) *n.* [ME *malvesie* < OFr., after *Monemvasia*, Greece.] Malmsey.

**ma·ma** *also* **mam·ma** (mä′mə) *n.* [Of baby-talk orig.] **1.** (*also* mə-mä′). Mother. **2.** *Slang.* **a.** A woman. **b.** A wife.

**mam·ba** (mäm′bə) *n.* [Zulu *í-mâmbá.*] A venomous snake of the genus *Dendraspis* of tropical Africa, esp. *D. angusticeps*, a black or green tree snake.

**mam·bo** (mäm′bō) *n., pl.* **-bos.** [Sp. (Cuba) < *mamboo*, percussion instrument.] **1.** A dance of Latin-American origin, resembling the rumba. **2.** The syncopated music for the mambo, in 4/4 time. *—vi.* **-boed, -bo·ing, -bos.** To dance the mambo.

**Mam·e·luke** (măm′ə-lōōk′) *n.* [Ar. *mamlūk*, Mameluke, slave.] A member of a former military caste, orig. composed of slaves from Turkey, that held the Egyptian throne from about 1250 until 1517 and remained powerful until 1811.

**mam·ma**[1] (mä′mə) *n. var. of* MAMA.

**mam·ma**[2] (măm′ə) *n., pl.* **mam·mae** (măm′ē) [Lat.] An organ of female mammals containing lactation glands. **—mam′mate′** *adj.*

**mam·mal** (măm′əl) *n.* [< LLat. *mammalis*, of the breast < Lat. *mamma*, breast.] A member of the class Mammalia. **—mam·ma′li·an** (mə-mā′lē-ən) *adj.* & *n.*

**Mam·ma·li·a** (mə-mā′lē-ə) *pl.n.* [NLat. < LLat. *mammalis*, of the breast < *mamma*, breast.] A class of vertebrate animals of more than 15,000 species, including humans, distinguished by self-regulating body temperature, hair, and, in the females, mammae.

**mam·mal·o·gy** (mə-măl′ə-jē, -mŏl′-) *n.* [MAMMA(L) + -LOGY.] The branch of zoology dealing with the study of Mammalia. **—mam′ma·log′i·cal** (măm′ə-lŏj′ĭ-kəl) *adj.* **—mam·mal′o·gist** *n.*

**mam·ma·plas·ty** (măm′ə-plăs′tē) *n., pl.* **-ties.** Plastic surgery of the breast.

**mam·ma·ry** (măm′ə-rē) *adj.* Of or relating to a breast or mamma.

**mammary gland** *n.* A milk-producing organ in female mammals, consisting of clusters of alveoli or small cavities with ducts terminating in a nipple.

**mam·mif·er·ous** (mə-mĭf′ər-əs) *adj.* [Fr. *mammifère* : Lat. *mamma*, breast + *-fère*, -fer.] Having mammary glands.

**mam·mil·la** (mə-mĭl′ə) *n., pl.* **-mil·lae** (-mĭl′ē) [Lat., dim of *mamma*, breast.] **1.** A nipple : teat. **2.** A nipple-shaped protuberance. **—mam′mil·lar′y** (măm′ə-lĕr′ē) *adj.*

**mam·mil·late** (măm′ə-lāt′) *also* **mam·mil·lat·ed** (-lā′tĭd) *adj.* **1.** Having nipples or mammillae. **2.** Shaped like a nipple or mammilla. **—mam′mil·la′tion** *n.*

**mam·mo·gram** (măm′ə-grăm′) *n.* [MAMM(A)[2] + -O- + -GRAM.] An x-ray photograph or radiograph of the breast.

**mam·mog·ra·phy** (mə-mŏg′rə-fē) *n.* [MAMM(A)[2] + -O- + -GRAPHY.] Examination of the breast by x-rays in order to detect tumors before they can be felt by palpation.

**Mam·mon** (măm′ən) *n.* [ME < LLat. *mammon* < Gk. *mamōnas* < Aram. *māmōnā*, riches.] **1.** Riches, greed, and worldly gain personified as a false god in the New Testament. **2.** *often* **mammon.** Riches considered as an evil influence of object or worship.

**mam·moth** (măm′əth) *n.* [Obs. R. *mammot′.*] **1.** An extinct elephant of the genus *Mammuthus*, once found throughout the Northern Hemisphere. **2.** Something of great size. *—adj.* Enormous.

**man** (măn) *n., pl.* **men** (mĕn) [ME < OE.] **1.** An adult male human being. **2.** A human being, regardless of sex. ***usage:*** The use of *man* in its primary sense of "an adult male human being" makes this term unacceptable to many when the sense intended is "a human being, regardless of sex." Consequently, such terms as *men and women, human beings*, and *persons* have come to be used as substitutes for *man* in the latter sense. **3.** The human race : MANKIND. **4.** *Zool.* A member of the genus *Homo*, family Hominidae, order Primates, class Mammalia, characterized by erect posture and an opposable thumb, esp. a member of the only extant species, *Homo sapiens*, distinguished by the ability to communicate by means of organized speech and to record information in a variety of symbolic systems. **5.** A being composed of a body and a soul or spirit in Christianity and Judaism. **6.** *Informal.* **a.** A husband. **b.** A lover or sweetheart. **7. men. a.** Male members of the labor force in an industry or factory. **b.** Enlisted servicemen. **8.** A male servant or subordinate. **9.** *Informal.* Fellow. —Used as a term of address. **10.** One who swore allegiance to a lord in the Middle Ages : VASSAL. **11.** Any of the pieces used in a board game, as chess or checkers. **12.** *Naut.* A ship <merchant*man*><*man*-of-war> **13. the Man.** *Slang.* **a.** A policeman. **b.** A white man. *—vt.* **manned, man·ning, mans. 1.** To supply or furnish with men for defense, support, or service <*manning* the lines> **2.** To be stationed at in order to defend, care for, or operate <*man* the guns> *—interj.* —Used as an expletive to indicate intense feeling <*Man!* It's really wild out there!> **—one's own man.** Independent in opinion and conduct. **—to a man.** Without exception <They objected to the proposal *to a man.*>

**ma·na** (mä′nə) *n.* [Maori.] **1.** An impersonal supernatural force believed to be inherent in a person, god, or sacred object. **2.** Power : authority.

**man about town** *n., pl.* **men about town.** A worldly and socially knowledgeable man who spends much time in fashionable places.

**man·a·cle** (măn′ə-kəl) *n.* [ME < OFr. *manicle* < Lat. *manicula*, little hand, dim. of *manus*, hand.] **1.** A device for confining the hands, usu. consisting of two metal rings that are fastened about the wrists and joined by a metal chain : HANDCUFF. **2.** A fetter : restraint. *—vt.* **-cled, -cling, -cles.** To confine or restrain with or as if with manacles : FETTER.

**man·age** (măn′ĭj) *v.* **-aged, -ag·ing, -ag·es.** [Ital. *maneggiare* < VLat. **manidiare* < Lat. *manus*, hand.] *—vt.* **1.** To direct or control the use of. **2. a.** To exert control over. **b.** To make submissive to one's authority, discipline, or persuasion. **3.** To direct or administer (e.g., a business). **4.** To arrange or contrive <*managed* to get better tickets> *—vi.* **1.** To direct, supervise, or carry on business or other affairs. **2.** To carry on : get along <I'll *manage* somehow.>

**man·age·a·ble** *adj.* Capable of being managed or controlled <a *manageable* crisis><a *manageable* horse> **—man′age·a·bil′i·ty, man′age·a·ble·ness** *n.* **—man′age·a·bly** *adv.*

**managed currency** *n.* A monetary system in which a governmental agency or central bank, rather than the gold standard, regulates the money supply and its purchasing power.
**managed news** *n.* News generated and circulated by a government so as to slant it favorably toward the government's position or interests.
**man·age·ment** (măn'ĭj-mənt) *n.* **1.** The act, manner, or practice of managing, supervising, or controlling. **2.** The persons who manage a business establishment, organization, or institution. **3.** Executive skill <good editorial *management*>
**man·ag·er** (măn'ĭ-jər) *n.* **1.** One who manages a business or enterprise. **2.** One in charge of the business affairs of an entertainer. **3. a.** One in charge of the training and performance of an athlete or team. **b.** A student in charge of the equipment and records of a school or college team. **—man'ag·er·ship'** *n.*
**man·a·ge·ri·al** (măn'ĭ-jîr'ē-əl) *adj.* Of, relating to, or typical of a manager or management. **—man'a·ge'ri·al·ly** *adv.*
**managing editor** *n.* An executive who supervises editorial work.
**man·a·kin** (măn'ə-kĭn) *n.* [Alteration of MANIKIN.] Any of various small colorful birds of the family Pipridae, indigenous to Central and South American forests.
**ma·ña·na** (mä-nyä'nə) *adv.* [Sp. < *(cras) mañana,* early (tomorrow).] **1.** Tomorrow. **2.** At an unspecified future time. *—n.* An indefinite time in the future.
**Ma·nas·seh** (mə-năs'ə) *n.* [Heb. *Mĕnashsheh.*] A tribe of Israel descended from Manasseh, the elder son of Joseph.
**man-at-arms** (măn'ət-ärmz') *n., pl.* **men-at-arms.** A soldier, esp. a medieval cavalryman supplied with heavy arms.
**man·a·tee** (măn'ə-tē') *n.* [Sp. *manatí* < Carib, breast.] An aquatic mammal of the genus *Trichechus,* found in tropical Atlantic coastal waters.

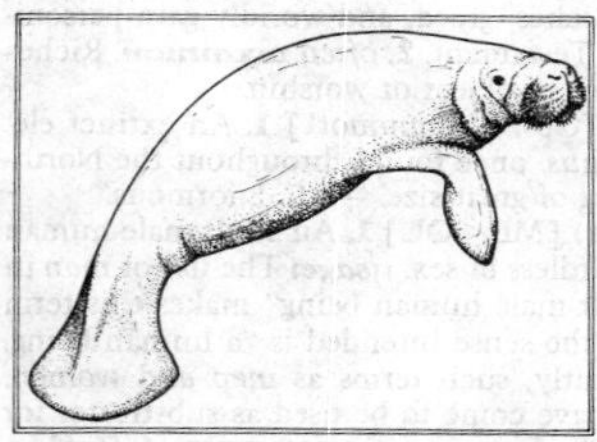
**manatee**
*8–15 feet long*

**Man·ches·ter terrier** (măn'chĕs'tər, -chĭ-stər) *n.* A short-haired black-and-tan dog orig. bred in Manchester, England.
**man·chi·neel** (măn'chĭ-nēl') *n.* [Fr. *mancenille* < Sp. *manzanilla,* dim. of *manzana,* apple.] A tropical American tree, *Hippomane mancinella,* bearing a poisonous fruit and a poisonous milky sap that produces skin blisters on contact.
**Man·chu** (măn'cho͞o, măn-cho͞o') *n., pl.* **Manchu** or **-chus.** [Manchu.] **1.** One of a nomadic Mongoloid people native to Manchuria, who conquered China in 1644 and established a dynasty that was overthrown in 1911. **2.** The Tungusic language of the Manchu. **—Man'chu** *adj.*
**-mancy** *suff.* [ME < OFr. *-mancie* < LLat. *-mantia* < Gk. *manteia* < *manteuesthai,* to prophesy < *mantis,* prophet.] Divination <biblio*mancy*>
**Man·dae·an** (măn-dē'ən) *n. var. of* MANDEAN.
**man·da·la** (mŭn'də-lə) *n.* [Skt. *maṇḍalam,* circle, prob. < Tamil *muṭalai.*] A design symbolizing the universe. **—man'dal·ic** *adj.*
**man·da·mus** (măn-dā'məs) *n.* [Lat., we order < *mandare,* to order.] *Law.* A writ issued by a superior court ordering a public official or body or a lower court to perform a specified duty. *—vt.* **-mused, -mus·ing, -mus·es.** To serve with a mandamus.
**Man·dan** (măn'dăn') *n., pl.* **Mandan** or **-dans. 1. a.** A tribe of Indians that lived in the Missouri River valley in North Dakota. **b.** A member of this tribe. **2.** The Siouan language of the Mandan.
**man·da·rin** (măn'də-rĭn) *n.* [Port. < Malay *mĕntĕri* < Skt. *mantrī,* counselor < *mantraḥ,* counsel.] **1.** A member of one of the nine ranks of high public officials in imperial China. **2. a.** A civil servant **:** BUREAUCRAT. **b.** An influential or highly placed person, esp. in intellectual or political circles. **3. Mandarin.** The standard vernacular language of China, based on the principal dialect spoken in and around Beijing. *—adj.* **1.** Of or like a mandarin. **2.** Marked by elaborate and refined language or literary style.
**mandarin collar** *n.* A narrow upright collar usu. divided in front.
**mandarin duck** *n.* A waterfowl, *Aix galericulata* of Asia, with brightly colored plumage and a crested head.
**mandarin orange** *n.* [Fr. *mandarine* < Sp. *mandarina* < *mandarin,* mandarin < Port. (from the color of a mandarin's robe). —see MANDARIN.] **1.** A small Chinese orange tree, *Citrus reticulata,* of the rue family. **2.** The small loose-skinned fruit of the mandarin orange.
**man·date** (măn'dāt') *n.* [Lat. *mandatum* < *mandare,* to order.] **1.** An authoritative instruction or command. **2.** The wishes of a political electorate, expressed by election results to its representatives in government. **3. a.** A commission from the League of Nations authorizing a nation to administer a territory. **b.** A region under such administration. **4.** *Law.* **a.** An order issued by a superior law court to a lower court. **b.** A contract by which one party agrees to perform services for another without payment. *—vt.* **-dat·ed, -dat·ing, -dates. 1.** To assign (e.g., a territory) to a specified nation under a mandate. **2.** To make mandatory **:** REQUIRE <*mandated* deregulation of gas prices> **—man'da'tor** *n.*
**man·da·to·ry** (măn'də-tôr'ē, -tōr'ē) *adj.* **1.** Of, relating to, having the nature of, or containing a mandate. **2.** Required by or as if by mandate **:** OBLIGATORY. **3.** Holding a mandate over a region.
**man-day** (măn'dā') *n.* An industrial unit of production equal to the work of one person in one day.
**Man·de** (män'dā') *n., pl.* **Mande** or **-des.** [Mandingo.] **1.** A people of West Africa in the upper Niger valley. **2.** A branch of the Niger-Congo language family.
**Man·de·an** *also* **Man·dae·an** (măn-dē'ən) *n.* [Mandean *mandaya,* having knowledge < *manda,* knowledge.] **1.** A member of an ancient Gnostic sect of Mesopotamia. **2.** A form of Aramaic used by the Mandeans. **—Man·de'an** *adj.*
**man·di·ble** (măn'də-bəl) *n.* [OFr. < LLat. *mandibula* < *mandere,* to chew.] A jaw or a part of a jaw, esp.: **a.** The lower jaw in vertebrates. **b.** Either the lower or upper part of a bird's beak. **c.** Any of various insect mouth parts. **—man·dib'u·lar** (măn-dĭb'yə-lər) *adj.*
**man·dib·u·late** (măn-dĭb'yə-lĭt, -lāt') *n.* An animal with mandibles. **—man·dib'u·late'** *adj.*
**Man·din·go** (măn-dĭng'gō) *n., pl.* **-gos** or **-goes.** [Mandingo.] **1.** A member of any of various Negroid peoples of the upper Niger River valley of western Africa. **2.** Any of the languages of the Mandingos.
**man·do·lin** (măn'də-lĭn', măn'də-lĭn') *n.* [Fr. *mandoline* < Ital. *mandolino,* dim. of *mandola,* lute < Fr. *mandore* < LLat. *pandura,* three-string lute < Gk. *pandoura.*] A musical instrument with a usu. pear-shaped body and a fretted neck over which several pairs of strings are stretched. **—man'do·lin'ist** *n.*
**man·drag·o·ra** (măn-drăg'ə-rə) *n.* [ME < Lat. *mandragoras* < Gk.] The mandrake.
**man·drake** (măn'drāk') *n.* [ME < MDu. *mandrage* and OE *mandragora,* both < Lat. *mandragoras* < Gk.] **1. a.** A Eurasian plant, *Mandragora officinarum,* bearing purplish flowers and a branched root thought to look like the human body. **b.** The root of the mandrake, once used for a narcotic preparation. **2.** MAY APPLE 1.
**man·drel** or **man·dril** (măn'drəl) *n.* [Prob. alteration of Fr. *mandrin,* lathe.] **1.** A spindle or axle for securing or supporting material being machined. **2.** A metal core around which material, as wood, can be cast and shaped. **3.** A shaft on which a working tool is mounted, as in dental drills.
**man·drill** (măn'drəl) *n.* [MAN + DRILL⁴.] A large fierce baboon, *Mandrillus sphinx* of western Africa, with a beard, crest, and mane, tawny-greenish body hair with yellowish hair on the lower part, and brilliant blue, purple, and scarlet facial markings in the adult male.
**mane** (mān) *n.* [ME < OE *manu.*] **1. a.** The long hair on the top and sides of the neck of mammals such as the horse and the male lion. **b.** The feathers on the back of the neck and head of some pigeons. **2.** A long thick growth of human hair.
**man-eat·er** (măn'ē'tər) *n.* **1.** An animal that feeds on human flesh. **2.** A cannibal.
**ma·nège** *also* **ma·nege** (mă-nĕzh') *n.* [Fr. < Ital. *maneggio* < *maneggiare,* to manage. —see MANAGE.] **1.** The art and practice of training a horse in classical riding maneuvers and exercises. **2.** A riding academy.
**ma·nes** or **Ma·nes** (mā'nēz', mä'nās') *pl.n.* [Lat. < *manus,* good.] **1.** The spirits of the dead, esp. ancestors, deified as minor gods in ancient Rome. **2.** A revered spirit of one who has died.
**ma·neu·ver** (mə-no͞o'vər, -nyo͞o'-) *n.* [Fr. *manoeuvre* < OFr. *maneuvre,* manual work < Med. Lat. *manuopera* < Lat. *manu operari,* to work by hand.] **1. a.** A tactical or strategic military movement. **b.** *often* **maneuvers.** A large-scale military training exercise simulating combat. **2. a.** A physical movement or way of doing something requiring dexterity and skill. **b.** A controlled change in flight path of an aircraft, rocket, or space vehicle. **3.** A calculated and skillful movement, act, or stratagem. *—v.* **-vered, -ver·ing, -vers.** *—vi.* **1.** To carry out a military maneuver. **2.** To make a change or a series of changes in position for a purpose. **3.** To change tactics. **4.** To try to effect an end by planning or scheming. *—vt.* **1.** To alter the tactical placement of (troops or warships). **2.** To manipulate into a desired position or toward a predetermined goal. **—ma·neu'ver·a·bil'i·ty** *n.* **—ma·neu'ver·a·ble** *adj.* **—ma·neu'ver·er** *n.*
**man Friday** *n.* [After *Man Friday,* a character in *Robinson Crusoe,* a novel by Daniel Defoe (1660–1731).] A devoted man servant, aide, or employee.
**man·ful** (măn'fəl) *adj.* **1.** Courageous **:** resolute. **2.** Manly. **—man'·ful·ly** *adv.* **—man'ful·ness** *n.*
**man·ga·bey** (măng'gə-bā', -bē') *n.* [After *Mangaby,* a region of

Madagascar.] A monkey of the genus *Cercocebus* of equatorial Africa, with a long tail and a relatively long muzzle.

**mangan-** *pref. var. of* MANGANO-.

**man·ga·nate** (măng'gə-nāt') *n.* A salt containing manganese in its anion, esp. a salt containing the $MnO_4$ radical.

**man·ga·nese** (măng'gə-nēz', -nēs') *n.* [Fr. *manganèse* < Ital. *manganese* < Med. Lat. *magnesia.* —see MAGNESIA.] *Symbol* **Mn** A gray-white, brittle metallic element, alloyed with steel to increase such properties as strength, hardness, and wear resistance; atomic number 25; atomic weight 54.9380. **—man'ga·ne'sian** *adj.*

**manganese dioxide** *n.* A black crystalline compound, $MnO_2$, used as a depolarizer of dry-cell batteries and in textile dyeing.

**manganese spar** *n.* **1.** Rhodonite. **2.** Rhodochrosite.

**man·gan·ic** (măn-găn'ĭk, măng-) *adj.* Relating to trivalent manganese or a compound containing trivalent manganese.

**man·ga·nite** (măng'gə-nīt') *n.* A steel-gray to black mineral form of manganese oxide, $Mn_2O_3 \cdot H_2O$, found in North America and Europe, an important ore of manganese.

**mangano-** or **mangan-** *pref.* [< G. *Mangan* < Fr. *manganèse.*] Manganese <*manganite*>

**man·ga·nous** (măng'gə-nəs) *adj.* Relating to bivalent manganese or compound containing bivalent manganese.

**mange** (mānj) *n.* [ME *manjeue* < OFr. *manjue* < *mangier,* to eat. —see MANGER.] A contagious skin disease of mammals, occas. including humans, caused by parasitic mites and symptomized by itching and loss of hair.

**man·gel-wur·zel** (măng'gəl-wûr'zəl) *n.* [G. *Mangoldwurzel* : *Mangold,* beet + *Wurzel,* root.] A common beet with a large yellowish root, used chiefly as cattle feed.

**man·ger** (mān'jər) *n.* [ME < OFr. *mangeoire* < *mangier,* to eat < Lat. *manducare* < *manduco,* glutton < *mandere,* to chew.] A trough holding livestock feed.

**man·gle¹** (măng'gəl) *vt.* **-gled, -gling, -gles.** [ME *manglen* < AN *mangler.*] **1.** To disfigure or mutilate by battering, hacking, cutting, or tearing. **2.** To ruin or spoil through ineptitude or ignorance : BOTCH <*mangle* a piano piece> **—man'gler** *n.*

**man·gle²** (măng'gəl) *n.* [Du. < G. < MHG, dim. of *mange,* mangonel < LLat. *manganum,* catapult. —see MANGONEL.] **1.** A laundry machine for pressing fabrics. **2.** *Chiefly Brit.* A clothes wringer. —*vt.* **-gled, -gling, -gles.** To smooth or press with a mangle.

**man·go** (măng'gō) *n., pl.* **-goes** or **-gos.** [Port. *manga,* fruit of the mango tree < Malay *manggā* < Tamil *mānkāy* : *mān,* mango tree + *kāy,* fruit.] **1. a.** A tropical Asian evergreen tree, *Mangifera indica,* widely cultivated for its edible fruit. **b.** The ovoid fruit of this tree, with a smooth rind and sweet, juicy, yellow-orange flesh. **2.** A pickle, esp. a pickled stuffed sweet pepper.

**man·go·nel** (măng'gə-nĕl') *n.* [ME < OFr. < Med. Lat. *mangonellus,* dim. of LLat. *manganum,* catapult < Gk. *manganon,* war machine.] A war machine used during the Middle Ages for hurling missiles, as stones.

**man·go·steen** (măng'gə-stēn') *n.* [Malay *mangustan.*] **1.** A tropical tree, *Garcinia mangostana,* with thick, leathery leaves and edible fruit. **2.** The fruit of the mangosteen tree, with a hard rind and segmented, sweet, juicy pulp.

**man·grove** (măn'grōv', măng'-) *n.* [Prob. Port. *mangue,* mangrove (< Taino *mangle*) + GROVE.] **1.** A tropical evergreen tree or shrub of the genus *Rhizophora,* having stiltlike roots and stems and forming dense thickets along tidal shores. **2.** A tree or shrub similar to the mangrove, esp. one of the genus *Avicennia.*

**mang·y** (mān'jē) *adj.* **-i·er, -i·est. 1.** Affected with or caused by mange. **2.** Having many bare spots : SHABBY <a *mangy* bearskin rug> **—mang'i·ly** *adv.* **—mang'i·ness** *n.*

**man·han·dle** (măn'hăn'dəl) *vt.* **-dled, -dling, -dles. 1.** To handle roughly. **2.** To move by human strength.

**Man·hat·tan** (măn-hăt'n, mən-) *n.* [After *Manhattan,* a borough of New York City.] A cocktail made of sweet vermouth and whiskey, often with a dash of bitters and a maraschino cherry.

**man·hole** (măn'hōl') *n.* A hole through which one may enter a sewer, boiler, pipe, conduit, or drain.

**man·hood** (măn'ho͝od') *n.* **1.** The state or condition of being an adult male. **2.** The composite of qualities, as courage, determination, and vigor, often attributed to an adult male. **3.** Men as a group. **4.** The state of being part of or endowed with humanity.

**man-hour** (măn'our') *n.* An industrial unit of production equal to the work of one person in one hour.

**man·hunt** (măn'hŭnt') *n.* An extensive organized search, as for a criminal fugitive or missing person.

**ma·ni·a** (mā'nē-ə, mān'yə) *n.* [ME, madness < LLat. < Gk.] **1.** An intense or unreasonable desire or enthusiasm : CRAZE. **2.** A manifestation of manic-depressive psychosis, symptomized by profuse and quickly changing ideas, exaggerated gaiety, and physical overactivity. **3.** Violent abnormal behavior. **—man'ic** (măn'ĭk) *adj.*

**-mania** *suff.* [< MANIA.] An exaggerated or unreasonable desire or enthusiasm for <balleto*mania*>

**ma·ni·ac** (mā'nē-ăk') *n.* [< LLat. *maniacus,* maniacal < Gk. *maniakos* < *mania,* madness.] **1.** An insane person. **2.** One who has an excessive enthusiasm or desire for something <a golf *maniac*>

**ma·ni·a·cal** (mə-nī'ə-kəl) *also* **ma·ni·ac** (mā'nē-ăk') *adj.* **1.** Insane <a *maniacal* arsonist> **2.** Marked by excessive enthusiasm <a *maniacal* passion for sailing> **—ma·ni'a·cal·ly** *adv.*

**man·ic-de·pres·sive** (măn'ĭk-dĭ-prĕs'ĭv) *adj. Psychiat.* Characterizing a psychosis in which periods of manic excitation alternate with melancholic depression. **—man'ic-de·pres'sive** *n.*

**Man·i·chae·an** or **Man·i·che·an** (măn'ĭ-kē'ən) *also* **Man·i·chee** (măn'ĭ-kē') [ME *Maniche* < LLat. *Manichaeus* < LGk. *Manikhaios* < *Manikhaios,* Manes, the founder of the philosophy.] —*n.* A believer in Manichaeism. —*adj.* Of or relating to Manichaeism.

**Man·i·chae·ism** (măn'ĭ-kē'ĭz'əm) *also* **Man·i·chae·an·ism** (-kē'ə-nĭz'əm) *n.* **1.** The syncretic dualistic religious philosophy taught by the Persian prophet Manes about the 3rd cent. A.D., combining elements of Zoroastrian, Christian, and Gnostic thought. **2.** A dualistic philosophy similar to Manichaeism, esp. one regarded as a heresy by the Roman Catholic Church.

**Man·i·che·an** (măn'ĭ-kē'ən) or **Man·i·chee** (măn'ĭ-kē') *n.* & *adj. vars. of* MANICHAEAN.

**man·i·cot·ti** (măn'ĭ-kŏt'ē) *n., pl.* **manicotti.** [Ital., pl. of *manicotto,* muff < *manica,* sleeve < Lat. < *manus,* hand.] An Italian dish of pasta filled with chopped ham and ricotta cheese, usu. served hot with a tomato sauce.

**man·i·cure** (măn'ĭ-kyo͝or') *n.* [Fr. : Lat. *manus,* hand + Lat. *cura,* care.] Treatment of the hands and fingernails, including shaping, cleaning, and polishing of the nails. —*vt.* **-cured, -cur·ing, -cures. 1.** To care for (the fingernails) by shaping, cleaning, and polishing. **2.** To clip or trim evenly and closely <*manicure* the grounds of the estate> **—man'i·cur'ist** *n.*

**man·i·fest** (măn'ə-fĕst') *adj.* [ME < Lat. *manifestus.*] Clearly apparent to the sight or understanding : OBVIOUS. —*vt.* **-fest·ed, -fest·ing, -fests. 1.** To show or demonstrate plainly : REVEAL. **2.** To be evidence of : PROVE. **3. a.** To record in a ship's manifest. **b.** To display or present a manifest of (cargo). —*n.* **1.** A list of cargo or passengers. **2.** A list of railroad cars according to owner and location. **—man'i·fest'ly** *adv.*

**man·i·fes·ta·tion** (măn'ə-fĕ-stā'shən) *n.* **1. a.** The act of manifesting or state of being manifested. **b.** Demonstration of the existence, reality, or presence of a person, object, or quality <a *manifestation* of their best intentions> **c.** One of the forms in which one, as a divine being, an individual, or an idea, is revealed. **2.** A public, usu. political, demonstration.

**Manifest Destiny** *n.* **1.** A future event regarded as unavoidable. **2.** A U.S. policy during the 19th and early 20th cent. of imperialistic expansion defended as necessary or benevolent.

**man·i·fes·to** (măn'ə-fĕs'tō) *n., pl.* **-toes** or **-tos.** [Ital. < *manifesto,* clear < Lat. *manifestus.*] A public declaration of intentions or principles, esp. of a political nature. —*vi.* **-toed, -to·ing, -toes.** To issue a manifesto.

**man·i·fold** (măn'ə-fōld') *adj.* [ME < OE *manigfeald* : *manig,* many + *-feald,* -fold.] **1.** Of many kinds : MULTIPLE <the *manifold* problems of life> **2.** Having many features or forms <*manifold* talent> **3.** Composed of or operating several of one kind. —*n.* **1.** A whole made up of diverse elements. **2.** One of numerous copies. **3.** A pipe with several apertures for making multiple connections. **4.** *Math.* A set of elements sharing a number of usu. topologic properties, as orientability, differentiability, and dimensionality. —*vt.* **-fold·ed, -fold·ing, -folds. 1.** To make numerous copies of. **2.** To make manifold : MULTIPLY. **—man'i·fold'ly** *adv.* **—man'i·fold'ness** *n.*

**man·i·kin** or **man·ni·kin** (măn'ĭ-kĭn) *n.* [Du. *mannekijn* < MDu., dim. of *man,* man.] **1.** A dwarf. **2.** A mannequin.

**ma·nil·a** or **ma·nil·la** (mə-nĭl'ə) *n. often* **Manila** or **Manilla. 1.** A cheroot made in Manila. **2.** Manila hemp. **3.** Manila paper. **4.** A light yellow brown.

**Manila hemp** *n.* The fiber of the abaca, used for making rope, cordage, and paper.

**Manila paper** *n.* Strong, usu. buff paper or thin cardboard with a smooth finish, made from Manila hemp or wood fibers similar to it.

**ma·nil·la** (mə-nĭl'ə) *n. var. of* MANILA.

**man in the street** *n.* An average citizen.

**man·i·oc** (măn'ē-ŏk') *also* **man·i·o·ca** (măn'ē-ō'kə) *n.* [Fr., of Tupian orig.] CASSAVA 1.

**man·i·ple** (măn'ə-pəl) *n.* [ME < OFr. < Lat. *manipulus,* handful.] **1.** An ornamental silk band hung as an ecclesiastical vestment on the left arm near the wrist. **2.** A subdivision of an ancient Roman legion, having 60 or 120 men.

**ma·nip·u·lar** (mə-nĭp'yə-lər) *adj.* **1.** Of or pertaining to an ancient Roman maniple. **2.** Of or pertaining to manipulation. —*n.* A Roman soldier in a maniple.

**ma·nip·u·late** (mə-nĭp'yə-lāt') *vt.* **-lat·ed, -lat·ing, -lates.** [Back-formation < MANIPULATION.] **1.** To control or operate by skilled use of the hands : HANDLE. **2.** To manage or influence shrewdly or deviously <*manipulated* the news> **3.** To falsify or tamper with (financial records) for personal gain. **—ma·nip'u·la·bil'i·ty** *n.* **—ma·nip'u·la·ble** (-lə-bəl) *adj.* **—ma·nip'u·la'tive, ma·nip'u·la·to'ry** (-lə-tôr'ē, -tōr'ē) *adj.* **—ma·nip'u·la'tor** *n.*

**ma·nip·u·la·tion** (mə-nĭp′yə-lā′shən) *n.* [Fr. < *manipule,* handful < Lat. *manipulus* < *manus,* hand.] **1.** The act of manipulating or state of being manipulated. **2.** Shrewd, often devious management, esp. for one's own advantage.

**man·i·tou** *or* **man·i·tu** (măn′ĭ-to͞o′) *also* **man·i·to** (-tō′) *n.* [Fr. < Ojibwa *manitoo.*] **1.** A deified spirit or force of nature, either good or bad, in the religion of the Algonquian Indians. **2.** A representation or image of a manitou.

**man·kind** (măn′kīnd′) *n.* **1.** (*also* măn′kīnd′). The human race. **2.** Men as opposed to women.

**man·like** (măn′līk′) *adj.* **1.** Resembling a man. **2.** Belonging to or appropriate for a man.

**man·ly** (măn′lē) *adj.* **-li·er, -li·est. 1.** Having qualities traditionally attributed to a man. **2.** MANLIKE 2. —*adv.* In a manly way. **—man′li·ness** *n.*

**man-made** (măn′mād′) *adj.* Made by human beings.

**man·na** (măn′ə) *n.* [ME < OE < LLat. < Gk. < Aram. *mannā* < Heb. *mān.*] **1.** The food miraculously provided for the Israelites in the wilderness on the flight from Egypt. **2.** Spiritual nourishment from a divine source. **3.** Something valuable that one receives unexpectedly. **4.** The dried exudate of certain plants, esp. that of a Eurasian ash tree, *Fraxinus ornus,* once used as a laxative.

**man·nan** (măn′ăn′, -ən) *n.* [MANN(OSE) + -AN.] Any of a group of polysaccharides that are polymers of mannose.

**manned** (mănd) *adj.* Conveying or operated by a human being <a *manned* space station>

**man·ne·quin** (măn′ĭ-kĭn) *n.* [Fr. < MDu. *mannekijn,* manikin. —see MANIKIN.] **1.** A life-size complete or partial representation of the human body, used to fit or display clothes : DUMMY. **2.** A jointed model of the human body used by artists, esp. to demonstrate the arrangement of drapery. **3.** One who models clothes.

**man·ner** (măn′ər) *n.* [ME < OFr. *maniere* < VLat. **manuaria* < Lat. *manuarius,* of the hand < Lat. *manus,* hand.] **1.** A way of doing something or the way in which a thing is done or takes place. **2.** One's bearing or behavior <a dignified *manner*> **3. manners. a.** Socially correct behavior. **b.** The prevailing systems or modes of social conduct of a specific society, period, or group, esp. as the subject of a literary work. **4.** Practice, style, execution, or method in the arts <the artist's early *manner*> **5.** Kind : sort <collected all *manner* of early American antiques>

☆ ***syns:*** MANNERS, DECORUM, ETIQUETTE, PROPRIETIES *n. core meaning* : socially correct behavior <Mind your *manners* at the company party.>

**man·nered** (măn′ərd) *adj.* **1.** Having manners of a specific kind <an ill-*mannered* boor> **2.** Affected or artificial <*mannered* writing> **3.** Of, relating to, or displaying mannerisms.

**man·ner·ism** (măn′ə-rĭz′əm) *n.* **1.** A distinctive behavioral trait : IDIOSYNCRASY. **2.** Exaggerated or affected style or habit, as in dress, speech, or art. **3. Mannerism.** A late 16th-cent. artistic style marked by distortion, as of scale and perspective. **—man′ner·ist** *n.*

**man·ner·ly** (măn′ər-lē) *adj.* Having good manners : POLITE. **—man′ner·li·ness** *n.* **—man′ner·ly** *adv.*

**man·ni·kin** (măn′ĭ-kĭn) *n. var. of* MANIKIN.

**man·nish** (măn′ĭsh) *adj.* **1.** Of or befitting a man. **2.** Resembling a man. **—man′nish·ly** *adv.* **—man′nish·ness** *n.*

**man·nite** (măn′īt′) *n.* [Fr. < *manna,* manna < LLat.] Mannitol.

**man·ni·tol** (măn′ĭ-tôl′, -tōl′) *n.* [MANN(A) + -IT(E) + -OL.] An alcohol, $C_6H_8(OH)_6$, used as a nutrient and dietary supplement and as the basis of dietetic sweets.

**man·nose** (măn′ōs′) *n.* [MANN(A) + -OSE.] A monosaccharide, $C_6H_{12}O_6$, derived from the oxidation of mannitol.

**ma·noeu·vre** (mə-no͞o′vər, -nyo͞o′-) *n.* & *v. Chiefly Brit. var. of* MANEUVER.

**man of God** *n.* A clergyman.

**man of letters** *n.* A man involved in scholarly or literary pursuits and activities.

**man of the cloth** *n.* A clergyman.

**man of the house** *n.* The male head of a household.

**man of the world** *n.* A sophisticated man.

**man-of-war** (măn′ə-wôr′) *n., pl.* **men-of-war** (mĕn′-). **1.** A warship. **2.** The Portuguese man-of-war.

**ma·nom·e·ter** (mă-nŏm′ĭ-tər) *n.* [Fr. *manomètre* : Gk. *manos,* sparse + Fr. *-mètre,* -meter.] **1.** An instrument for measuring the pressure of liquids and gases. **2.** A sphygmomanometer. **—man′o·met′ric** (măn′ə-mĕt′rĭk), **man′o·met′ri·cal** *adj.* **—man′o·met′ri·cal·ly** *adv.* **—ma·nom′e·try** *n.*

**man·or** (măn′ər) *n.* [ME < AN *maner* < OFr. *maneir,* to dwell < Lat. *manēre.*] **1. a.** The district over which a medieval lord had domain in western Europe. **b.** The lord's residence in such a district. **2.** A landed estate. **3.** The main house on an estate : MANSION. **4.** A tract of land with hereditary rights granted to the proprietor by royal charter in some North American colonies. **—ma·no′ri·al** (mə-nôr′ē-əl, -nōr′-) *adj.*

**man-o'-war bird** (măn′ə-wôr′) *n.* Frigate bird.

**man·pow·er** (măn′pou′ər) *n.* **1.** Human physical strength. **2.** Power in terms of the workers available to a particular group or needed for a specific task.

**man·qué** (mäɴ-kā′) *adj.* [Fr. < *manquer,* to fail < VLat. **mancare* < Lat. *mancus,* maimed.] Frustrated : unfulfilled <a writer *manqué*>

**man·rope** (măn′rōp′) *n. Naut.* A rope handrail on a ladder or gangplank.

**man·sard** (măn′särd′) *n.* [Fr. *mansarde,* after François *Mansart* (1598–1666).] **1.** A roof with two slopes on all four sides, the lower slope being nearly vertical and the upper nearly horizontal. **2.** The upper story formed by the lower slope of a mansard roof.

**manse** (măns) *n.* [ME *manss,* a dwelling < Med. Lat. *mansa* < Lat. *manēre,* to dwell.] **1. a.** *Chiefly Scot.* A clergyman's house and land. **b.** A Presbyterian minister's house. **2.** *Archaic.* A mansion.

**man·ser·vant** (măn′sûr′vənt) *n., pl.* **men·ser·vants.** A male servant, esp. a valet.

**man·sion** (măn′shən) *n.* [ME *mansioun,* a dwelling < OFr. *mansion* < Lat. *mansio* < *manēre,* to dwell.] **1.** A large stately house. **2.** A manor house. **3.** *Archaic.* An abode. **4. a.** HOUSE 8b. **b.** One of the 28 divisions of the moon's monthly path.

**man-sized** (măn′sīzd′) *also* **man-size** (-sīz′) *adj. Informal.* Quite large <a *man-sized* sandwich>

**man·slaugh·ter** (măn′slô′tər) *n.* **1.** The unpremeditated taking of human life. **2.** *Law.* The unlawful killing of one human being by another without express or implied intent to do injury.

**man·slay·er** (măn′slā′ər) *n.* One that kills a human being.

**man·sue·tude** (măn′swĭ-to͞od′, -tyo͞od′) *n.* [ME < Lat. *mansuetudo* < *mansuescere,* to tame : *manus,* hand + *suescere,* to accustom.] Gentleness : mildness.

**man·ta** (măn′tə) *n.* [Sp., cape < Lat. *mantum.*] **1.** A rough-textured cotton fabric or blanket made and used in Latin America and the southwestern United States. **2.** *also* **manta ray.** Any of several fishes of the family Mobulidae, having large, very flat bodies with winglike pectoral fins.

**man·teau** (măn-tō′) *n., pl.* **-teaus** (-tōz′) *or* **-teaux** (-tō′) [Fr. < OFr. *mantel.* —see MANTLE.] A loose cloak.

**man·tel** *also* **man·tle** (măn′tl) *n.* [Var. of MANTLE.] **1.** An ornamental facing surrounding a fireplace. **2.** The shelf over a fireplace.

**man·tel·et** (măn′tl-ĭt, mănt′lĭt) *n.* [ME < OFr., dim. of *mantel,* mantle. —see MANTLE.] **1.** A short cape. **2.** *also* **mant·let** (mănt′lĭt). A mobile screen or shield once used to protect besieging soldiers.

**man·tel·let·ta** (măn′tə-lĕt′ə) *n.* [Ital. < OFr. *mantelet,* mantelet.] A knee-length sleeveless vestment worn by Roman Catholic prelates.

**man·tel·piece** *also* **man·tle·piece** (măn′tl-pēs′) *n.* MANTEL 2.

**man·tel·tree** (măn′tl-trē′) *n.* [ME.] A beam, stone, or arch that functions as a lintel on a fireplace.

**man·tes** (măn′tēz′) *n. var. pl. of* MANTIS.

**man·tic** (măn′tĭk) *adj.* [Gk. *mantikos* < *mantis,* prophet.] Of, relating to, or having the power of divination : PROPHETIC.

**man·ti·core** (măn′tĭ-kôr′, -kōr′) *n.* [ME < Lat. *mantichōra* < Gk. *mantikhōras.*] A fabulous monster with the head of a man, body of a lion, and tail of a scorpion or dragon.

**man·tid** (măn′tĭd) *n.* [< NLat. *Mantidae,* family name < *Mantis,* mantis genus < Gk. *mantis,* prophet.] Mantis.

**man·til·la** (măn-tē′yə, -tĭl′ə) *n.* [Sp., dim. of *manta,* cape. —see MANTA.] **1.** A usu. lace scarf worn over the head and shoulders, often over a high comb, by Spanish and Latin-American women. **2.** A short cloak.

**man·tis** (măn′tĭs) *n., pl.* **-tis·es** *or* **-tes** (-tēz′) [NLat. *Mantis,* genus name < Gk. *mantis,* prophet.] Any of various chiefly tropical insects of the family Mantidae, including a few Temperate Zone species, usu. pale-green and with two pairs of walking legs and powerful forelimbs often folded in a praying position.

**mantis crab** *n.* The squilla.

**man·tis·sa** (măn-tĭs′ə) *n.* [Lat., makeweight.] *Math.* The decimal part of a common logarithm when the logarithm is written as the sum of an integer and a decimal.

**mantis shrimp** *n.* The squilla.

**man·tle** (măn′tl) *n.* [ME < OE *mentel* and OFr. *mantel,* both < Lat. *mantellum.*] **1.** A loose sleeveless coat : CLOAK. **2.** Something that covers, envelops, or conceals <a *mantle* of dust> **3.** *var. of* MANTEL. **4.** The outer covering of a wall. **5.** A zone of hot gases around a flame. **6.** A device in gas lamps consisting of a sheath of threads that gives off brilliant illumination when heated by the flame. **7.** *Anat.* The cerebral cortex. **8.** *Geol.* The layer of the earth between the crust and the core. **9.** The outer wall and casing of a blast furnace above the hearth. **10.** The wings, shoulder feathers, and back of a bird when colored differently from the rest of the body. **11.** *Zool.* A membrane between the body and the shell of a mollusk or a brachiopod. —*v.* **-tled, -tling, -tles.** —*vt.* To cover with or as if with a mantle : CONCEAL. —*vi.* **1.** To spread or become extended over a surface. **2.** To become covered with a coating, as scum on the surface of a liquid. **3.** To be or become covered with blushes or colors, as the face.

**man·tle·piece** (măn′tl-pēs′) *n. var. of* MANTELPIECE.

**mantle rock** *n.* Regolith.

**mant·let** (mănt′lĭt) *n. var. of* MANTELET 2.

**man-to-man** (măn′tə-măn′) *adj.* **1.** Forthright and honest <a

*man-to-man* talk> **2.** Of, relating to, or being a system of defense in which a defensive player guards a specific offensive player.

**Man·toux test** (măn'to͞o', män-to͞o') *n.* [After Charles *Mantoux* (1877–1947).] An intracutaneous test for tuberculin sensitivity that indicates past or present tuberculous infection.

**man·tra** (măn'trə, mŭn'-) *n.* [Skt. *mantraḥ.*] A sacred Hindu formula believed to embody the divinity invoked and to possess magical power.

**man·tu·a** (măn'cho͞o-ə, -to͞o-ə) *n.* [Alteration of MANTEAU.] A loose gown, open in front to reveal an underskirt, worn in the 17th and 18th cent.

**man·u·al** (măn'yo͞o-əl) *adj.* [ME < OFr. *manuel* < Lat. *manualis* < *manus*, hand.] **1. a.** Of, relating to, or done by the hands <*manual* skill> **b.** Used by or operated with the hands <a *manual* gearshift> **c.** Employing human rather than mechanical energy <*manual* work> **2.** Of, relating to, or resembling a manual or guidebook. —*n.* **1.** A small reference book, esp. one providing instructions. **2.** An organ keyboard played with the hands. **3.** Prescribed movements in the handling of a weapon, esp. a rifle <the military *manual* of arms> **—man'u·al·ly** *adv.*

**manual alphabet** *n.* An alphabet of hand signals used by deaf-mutes for communication.

**manual training** *n.* A course of training for developing manual dexterity in practical arts, as woodworking or handicrafts.

**ma·nu·bri·um** (mə-no͞o'brē-əm, -nyo͞o'-) *n., pl.* **-bri·a** (-brē-ə) [NLat. < Lat., handle < *manus*, hand.] **1.** The upper portion of the breastbone or sternum. **2.** The handle-shaped projection of the malleus in the ear.

**man·u·fac·to·ry** (măn'yə-făk'tə-rē) *n., pl.* **-ries.** [MANUFAC-T(URE) + -ORY.] A factory.

**man·u·fac·ture** (măn'yə-făk'chər) *v.* **-tured, -tur·ing, -tures.** [< OFr., a making by hand < LLat. *manufactus* : Lat. *manus*, hand + Lat. *facere*, to make.] —*vt.* **1. a.** To make or process (a raw material) into a finished product, esp. by a large-scale industrial operation. **b.** To make or process (a product), esp. with industrial machines. **2.** To produce, create, or turn out in a mechanical way <". . . books seem to have been *manufactured* rather than composed" —Dwight Macdonald> **3.** To concoct : fabricate <*manufacture* an alibi> —*vi.* To make or process goods, esp. in large quantities, by industrial machinery. —*n.* **1.** The act, craft, or process of manufacturing. **2.** A manufactured product. **3.** An industry. **—man'u·fac'tur·a·ble** *adj.* **—man'u·fac'tur·er** *n.*

**manufactured gas** *n.* A gaseous fuel made from soft coal or various petroleum products.

**man·u·mit** (măn'yə-mĭt') *vt.* **-mit·ted, -mit·ting, -mits.** [ME *manumitten* < OFr. *manumitter* < Lat. *manumittere* : *manus*, hand + *mittere*, to send from.] To free from bondage : EMANCIPATE. **—man'u·mis'sion** (-mĭsh'ən) *n.*

**ma·nure** (mə-no͝or', -nyo͝or') *n.* [< ME *manuren*, to cultivate land < AN *meinourer* < Med. Lat. *manuoperari* : Lat. *manus*, hand + Lat. *operari*, to work.] Material for fertilizing soil, as animal dung or compost. —*vt.* **-nured, -nur·ing, -nures.** To apply manure to. **—ma·nur'er** *n.*

**ma·nus** (mā'nəs, mä'-) *n., pl.* **manus.** [Lat., hand.] The end of the vertebrate forelimb, as the hand, claw, or hoof.

**man·u·script** (măn'yə-skrĭpt') *n.* [< Med. Lat. *manuscriptus*, handwritten : Lat. *manus*, hand + Lat. *scriptus*, p.part. of *scribere*, to write.] **1.** A composition, as a book or document, written by hand. **2.** A typewritten or handwritten version of a work, as a book, article, or document, esp. the author's own copy, prepared and submitted for publication. **3.** Handwriting.

**man·wise** (măn'wīz') *adv.* In a way typical of humankind.

**Manx** (măngks) *adj.* Of or relating to the Isle of Man or the Manx language. —*n., pl.* **Manx. 1.** A native or resident of the Isle of Man. **2.** The nearly extinct Goidelic language of the Manx. **3.** A Manx cat.

**Manx cat** or **manx cat** *n.* A breed of domestic cat with short hair usu. of solid color, and an internal vestigial tail.

**man·y** (mĕn'ē) *adj.* **more** (môr, mōr), **most** (mōst) [ME < OE *manig.*] **1.** Being one of a large indefinite number : NUMEROUS <*many* a time> **2.** Amounting to or consisting of a large indefinite number. —*n. (pl. in number).* **1.** A large, indefinite number of persons or things <*Many* of the players had injuries.> **2.** *often* **the many.** The great body of the people : the masses <"The *many* fail; the one succeeds" —Tennyson> —*pron. (pl. in number).* A large number of persons or things <"*Many* are called, but few are chosen" —Matthew 22:14>

☆ **syns:** MANY, LEGION, MULTITUDINOUS, MYRIAD, NUMEROUS, VOLUMINOUS *adj. core meaning* : amounting to or consisting of a large, indefinite number <*many* stars in the night sky><*many* papers to read> **ant:** few

**man·y·fold** (mĕn'ē-fōld') *adv.* By many times.

**man·y·plies** (mĕn'ĭ-plīz') *n.* The omasum.

**Man·za·ni·lla** (măn'zə-nē'yə, -nĭl'ə) *n.* [Sp., dim. of *manzana*, apple.] A pale dry Spanish sherry.

**man·za·ni·ta** (măn'zə-nē'tə) *n.* [Sp., dim. of *manzana*, apple.] An evergreen shrub of the genus *Arctostaphylos* of the North American Pacific coast, esp. *A. manzanita*, bearing white or pink flower clusters.

**Mao·ism** (mou'ĭz'əm) *n.* Marxism-Leninism developed in China chiefly by Mao Zedong. **—Mao'ist** *n.*

**Mao·ri** (mou'rē) *n., pl.* **Maori** *or* **-ris.** [Maori.] **1.** A member of the aboriginal people of New Zealand, of Polynesian-Melanesian descent. **2.** The Austronesian language of the Maori. **—Mao'ri** *adj.*

**map** (măp) *n.* [Med. Lat. *mappa* < Lat., napkin.] **1.** A usu. plane surface representation of a region of the earth or sky. **2.** Something resembling a map in clarity of representation. **3.** *Slang.* The face. —*vt.* **mapped, map·ping, maps. 1.** To make a map of. **2.** To explore or make a survey of (a region) to make a map. **3.** To plan or delineate, esp. in detail : ARRANGE <*mapping* out a new sales campaign> **4.** *Math.* To establish a mapping of (a set or aggregate). **—put on the map.** To make famous. **—wipe off the map.** To destroy completely : ANNIHILATE. **—map'per** *n.*

**ma·ple** (mā'pəl) *n.* [ME < OE *mapul.*] **1.** A usu. tall, deciduous tree or shrub of the genus *Acer* of the North Temperate Zone, with lobed leaves and winged seeds borne in pairs. **2.** The wood of a maple, esp. the hard close-grained wood of the sugar maple, used for furniture and flooring. **3.** The flavor of the concentrated sap of the sugar maple.

**maple sugar** *n.* A sugar made by boiling down maple syrup.

**maple syrup** *n.* **1.** A sweet syrup made from the sap of the sugar maple. **2.** Syrup made from various sugars and flavored with maple syrup or artificial maple flavoring.

**map·mak·er** (măp'mā'kər) *n.* A maker of maps : CARTOGRAPHER. **—map'mak'ing** *n.*

**map·ping** (măp'ĭng) *n. Math.* A rule of correspondence established between two mathematical sets that associates each member of the first set with a single member of the second.

**ma·qui** (mä'kē) *n.* [Sp., of Araucanian orig.] **1.** An evergreen shrub, *Aristotelia macqui* of Chile, bearing purple berries. **2.** A Chilean wine made from maqui berries.

**ma·quil·lage** (mä'kē-äzh') *n.* [Fr.] MAKE-UP 4.

**ma·quis** (mä-kē') *n., pl.* **maquis.** [Fr. < Ital. *macchie*, pl. of *macchia*, thicket < Lat. *macula*, spot.] **1.** A dense growth of small trees and shrubs in the Mediterranean area. **2.** *often* **Maquis. a.** The French resistance organization that fought against German occupation forces during World War II. **b.** A member of this group.

**mar** (mär) *vt.* **marred, mar·ring, mars.** [ME *maren* < OE *mierran.*] **1.** To deface or damage. **2.** To spoil the quality of. —*n.* A disfiguring mark : BLEMISH.

**mar·a·bou** *also* **mar·a·bout** (măr'ə-bo͞o') *n.* [Fr. *marabout*, Moslem hermit, marabout. —see MARABOUT[1].] **1.** A large Old World stork of the genus *Leptoptilus*, with a soft down used for trimming women's garments. **2. a.** The down of the marabou or an imitation made from other bird feathers. **b.** A garment or part of a garment trimmed with the down of the marabou or an imitation of it. **3. a.** A raw silk that can be dyed without being separated from the gum. **b.** A fabric or a garment made from such silk.

**mar·a·bout**[1] (măr'ə-bo͞o', -bo͞ot') *n.* [Fr. *marabout* < Port. *marabuto* < Ar. *murābit.*] **1.** A Moslem saint or hermit, esp. in northern Africa. **2.** The tomb of a marabout or a memorial shrine.

**mar·a·bout**[2] (măr'ə-bo͞o') *n. var. of* MARABOU.

**ma·ra·ca** (mə-rä'kə) *n.* [Port. *maracá* < Tupi.] A percussion instrument consisting of a hollow-gourd rattle containing beans or pebbles, often played in pairs.

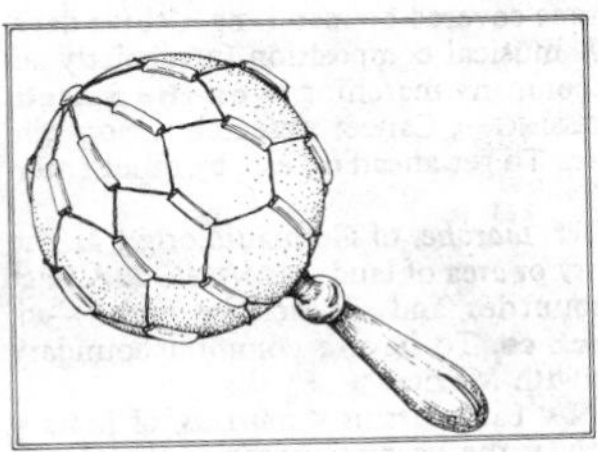

**maraca**

**ma·ran·ta** (mə-răn'tə) *n.* [NLat. *Maranta*, genus name, after Bartolommeo *Maranta* (d. 1571).] **1.** A plant of the tropical American genus *Maranta*, one species of which yields arrowroot. **2.** A starch made from arrowroot.

**ma·ras·ca** (mə-răs'kə) *n.* [Ital.] A European cherry tree, *Prunus cerasus marasca*, bearing bitter red fruit from which maraschino is made.

**mar·a·schi·no** (măr'ə-skē'nō, -shē'-) *n.* [Ital. < *marasca*, marasca.] A cordial made from the crushed pits and fermented juice of the marasca cherry.

**maraschino cherry** *n.* A maraschino-flavored preserved cherry.

**ma·ras·mus** (mə-răz'məs) *n.* [LLat. < Gk. *marasmos* < *marainein*, to waste away.] A wasting away of the body, associated with inadequate or inadequately assimilated food. **—ma·ras'mic** *adj.*

**Ma·ra·tha** *also* **Mah·rat·ta** (mə-rä'tə) *n., pl.* **Maratha** *or* **-thas** *also* **Mahratta** *or* **-tas.** [Marathi.] **1.** A Scytho-Dravidian people of southwestern India. **2.** A member of the Maratha people.

**Ma·ra·thi** *also* **Mah·ra·ti** *or* **Mah·rat·ti** (mə-rä'tē) *n.* [Marathi < *Maratha,* Maratha.] The principal Indic language of the state of Maharashtra in western India.

**mar·a·thon** (măr'ə-thŏn') *n.* [After *Marathon,* Greece (so called because a messenger ran from there to Athens to announce a victory over the Persians in 490 B.C.).] **1.** A cross-country footrace of 26 miles, 385 yards or approx. 42 kilometers. **2.** A long-distance race <a swimming *marathon*> **3.** A contest of endurance <a bicycle *marathon*> **4.** A task or action that requires prolonged effort or endurance <a TV fund-raising *marathon*>

**ma·raud** (mə-rôd') *v.* **-raud·ed, -raud·ing, -rauds.** [Fr. *marauder* < *maraud,* vagabond.] *—vi.* To raid for plunder : PILLAGE. *—vt.* To invade for loot. **—ma·raud'er** *n.*

**mar·ble** (mär'bəl) *n.* [ME < OFr. *marbre* < Lat. *marmor* < Gk. *marmaros.*] **1.** A metamorphic rock, mainly calcium carbonate, $CaCO_3$, often irregularly colored by impurities, used for architectural and ornamental purposes. **2.** A piece of marble. **3.** A sculpture of marble <the Elgin *marbles*> **4.** A small hard ball, usu. of glass, used in children's games. **5. marbles.** *(sing. in number).* Any of various games played with marbles. **6. marbles.** *Slang.* Common sense : sanity <must have lost their *marbles*> *—vt.* **-bled, -bling, -bles. 1.** To streak and mottle with veins and colors that imitate marble <*marbling* book edges> **2.** To intermix with flecks or thin strips of fat <*marble* a steak> *—adj.* Resembling marble in consistency, texture, venation, color, or coldness. **—mar'bled, mar'bly** *adj.*

**marble cake** *n.* A cake with a mottled or streaked appearance achieved by mixing light and dark batter.

**mar·ble·ize** (mär'bə-līz') *vt.* **-ized, -iz·ing, -iz·es.** To marble.

**mar·ble·wood** (mär'bəl-wo͝od') *n.* An Asian tree, *Diospyros kurzii,* having mottled gray wood used in cabinetwork.

**marc** (märk) *n.* [Fr. < OFr. *marchier,* to trample.] **1.** The pulpy residue left after the juice has been pressed from grapes, apples, or other fruits. **2.** Brandy distilled from grape or apple residue.

**mar·ca·site** (mär'kə-sīt', -zīt') *n.* [ME < Med. Lat. *marcasita* < Ar. *marqashītā* < Pers.] **1.** A mineral of iron disulfide, $FeS_2$, with the same composition as pyrite but differing in crystalline structure. **2.** An ornament of pyrite, polished steel, or white metal. **3.** A mineral resembling iron disulfide. **—mar'ca·sit'i·cal** (-sĭt'ĭ-kəl) *adj.*

**mar·cel** (mär-sĕl') *n.* [After *Marcel* Grateau (1852–1936).] A hair style popular in the past, characterized by deep, regular waves made by a heated curling iron. *—vt.* **-celled, -cel·ling, -cels.** To style (the hair) in a marcel.

**mar·ces·cent** (mär-sĕs'ənt) *adj.* [Lat. *marcescens, marcescent-,* becoming withered < *marcescere,* inchoative of *marcēre,* to wither.] Withering but not dropping off, as a blossom that persists on a twig after flowering.

**march¹** (märch) *v.* **marched, march·ing, march·es.** [ME *marchen* < OFr. *marchier,* prob. of Germanic orig.] *—vi.* **1. a.** To walk in a formal military manner at a steady rate with measured steps. **b.** To begin to move in such a manner <The band will *march* in the parade.> **2.** To proceed or advance with steady movement. **3.** To stand arranged in an orderly way <lindens *marching* up the hill> *—vt.* **1.** To cause to march <*march* troops in review> **2.** To traverse by marching <They *marched* across the mountains in a day.> *—n.* **1.** The act of marching, esp.: **a.** The steady forward movement of a body of troops. **b.** A long tiring journey on foot. **2.** Forward movement : PROGRESSION <the *march* of history> **3.** A regulated pace <quick *march*> **4.** The distance covered by marching <three days' *march* from here> **5.** *Mus.* A musical composition in regularly accented usu. duple meter to accompany marching. **—on the march.** Advancing steadily : PROGRESSING <Cancer research is *on the march.*> **—steal a march on.** To get ahead of, esp. by quiet enterprise or secretive methods.

**march²** (märch) *n.* [ME < OFr. *marche,* of Germanic orig.] **1.** The border or boundary of a country or area of land : FRONTIER. **2.** A tract of land bordering on two countries and claimed by both. *—vi.* **marched, march·ing, march·es.** To have a common boundary <The United States *marches* with Mexico.>

**March** (märch) *n.* [ME < AN < Lat. *martius* < *martius,* of Mars < *Mars,* Mars.] The third month of the year according to the Gregorian calendar. —See table at CALENDAR.

**march·er** (mär'chər) *n.* One who marches, esp. for a specific cause <a civil-rights *marcher*>

**marching orders** *pl.n.* Orders to move on or depart.

**mar·chio·ness** (mär'shə-nĭs, mär'shə-nĕs') *n., pl.* **-ness·es.** [Med. Lat. *marchionissa,* fem. of *marchio,* marquis < *marca,* boundary, of Germanic orig.] **1.** The wife or widow of a marquis. **2.** A woman holding the title of marquis in her own right.

**march·land** (märch'lănd') *n.* MARCH².

**march·pane** (märch'pān') *n.* Marzipan.

**Mar·cion·ism** (mär'shə-nĭz'əm) *n.* [After *Marcion* (ca. 2nd cent. A.D.), its founder.] A Gnostic movement of the 2nd and 3rd cent. A.D. that rejected the Old Testament.

**Mar·co·ni rig** (mär-kō'nē) *n.* [After Guglielmo *Marconi* (1874–1937), from its resemblance to the early antennae used by him for his wireless telegraphy.] A Bermuda rig.

**Mar·di gras** (mär'dē grä') *n.* [Fr.] Shrove Tuesday, celebrated by carnivals, masquerade balls, and parades of costumed merrymakers.

**Mar·duk** (mär'do͝ok) *n.* [Babylonian.] *Myth.* The chief Babylonian god.

**mare¹** (mâr) *n.* [ME < OE *mere.*] A female horse or the female of other equine species.

**ma·re²** (mä'rā) *n., pl.* **-ri·a** (-rē-ə) [NLat. < Lat. *mare,* sea.] *Astron.* Any of the large dark areas on the moon or Mars.

**ma·re clau·sum** (mä'rā klou'səm, klô'-) *n.* [NLat : Lat. *mare,* sea + Lat. *clausum,* closed.] A navigable body of water, as a sea, under the jurisdiction of one nation but closed to all others.

**ma·re li·be·rum** (mä'rā lē'bə-ro͞om') *n.* [NLat. : Lat. *mare,* sea + Lat. *liberum,* free.] A navigable body of water, as a sea, open to navigation by vessels of all nations.

**Ma·ren·go** (mə-rĕng'gō) *adj.* [After *Marengo,* Italy, prob. from the chicken dish served to Napoleon following his victory here in 1800 over the Austrians.] Browned in oil and sautéed in a sauce of tomatoes, mushrooms, garlic or onion, and white wine <chicken *Marengo*><veal *Marengo*>

**ma·re nos·trum** (mä'rā nō'strəm) *n.* [Lat., our sea.] A navigable body of water, as a sea, under the jurisdiction of one nation or shared by two or more nations.

**mare's nest** *n., pl.* **mare's nests** *or* **mares' nests. 1.** A hoax : fraud. **2.** A very complex situation.

**mare's-tail** (mârz'tāl') *n.* An aquatic plant, *Hippuris vulgaris* of the North Temperate Zone, bearing minute flowers and whorls of tapering leaves.

**mar·gar·ic** (mär-găr'ĭk) *adj.* [< Gk. *margaron,* pearl.] Like a pearl or pearls : PEARLY.

**margaric acid** *n.* [Fr. *margarique* < Gk. *margaron,* pearl.] A synthetic crystalline fatty acid, $C_{17}H_{34}O_2$.

**mar·ga·rine** *also* **mar·ga·rin** (mär'jər-ĭn) *n.* [Fr. < *margarique,* margaric acid.] A fatty solid butter substitute consisting of a blend of hydrogenated vegetable oils mixed with other ingredients, as emulsifiers, vitamins, and coloring matter.

▲ word history: Margarine, a relatively new substance, is the product of modern chemistry. Both the substance and its name originated in France in the 19th century. The substance was originally made from a combination of animal fats, one of which was called *margaric acid* (*acide margarique* in French). The adjective *margaric* or *margarique* is derived from Greek *margaron,* "pearl," and was applied to the compound in question because its crystals had a pearly sheen. Margarine is naturally white; it is colored yellow in imitation of the natural yellow color of butter.

**mar·ga·ri·ta** (mär'gə-rē'tə) *n.* [Sp., from the name *Margarita,* Margaret.] A cocktail of tequila and lemon or lime juice, usu. served with salt encrusted on the rim of the glass.

**mar·ga·rite¹** (mär'gə-rīt') *n.* [G. *Margarit* < Gk. *margaritēs,* pearl < *margaron.*] A mineral, $CaAl_2(Si_2Al_2)O_{10}(OH)_2$, formed in sheets of monoclinic crystals, with a pearly, translucent luster.

**mar·ga·rite²** (mär'gə-rīt') *n.* [ME < OFr. < Lat. *margarita* < Gk. *margaritēs* < *margaron.*] **1.** A rock formation resembling beads. **2.** *Archaic.* A pearl.

**mar·gay** (mär'gā', mär-gā') *n.* [Sp. (South America) < Tupi *marakaya.*] A spotted wildcat, *Felis weidii,* found from Texas to Brazil and resembling a small, long-tailed ocelot.

**mar·gin** (mär'jĭn) *n.* [ME < Lat. *margo.*] **1.** An edge and the area adjacent to it : BORDER. **2.** The blank space bordering the printed or written area on a page. **3.** A limit of a state or process <the *margin* of truth> **4.** An amount allowed beyond the necessary <a *margin* of error> **5.** A measure, quantity, or degree of difference <victory by a handsome *margin*> **6. a.** The minimum return that an enterprise may earn and still pay for itself. **b.** The difference between the cost and the selling price of securities or commodities. **c.** The difference between the market value of collateral and the face value of a loan. **7.** An amount in money, or represented by securities, deposited by a customer with a broker as a provision against loss on transactions made on account. **8.** *Bot.* The border of a leaf. **9.** *Biol.* The boundary area of an insect's wing. *—vt.* **-gined, -gin·ing, -gins. 1.** To provide with a margin. **2.** To be a margin to : BORDER. **3.** To enter or inscribe in the margin of a page. **4. a.** To add margin to <*margin* up a brokerage account> **b.** To deposit margin for <*margin* a transaction> **c.** To buy or hold (securities) by depositing or adding to a margin.

**mar·gin·al** (mär'jə-nəl) *adj.* [Med. Lat. *marginalis* < Lat. *margo,* margin.] **1.** Of, relating to, or comprising a margin <the *marginal* strip of farmland> **2.** Geographically adjacent <states *marginal* to Virginia> **3.** Written or printed in the margin of a book <*marginal* comments> **4.** Barely within a lower standard or limit of quality <*marginal* reading aptitude> **5. a.** Designating enterprises that produce goods or are capable of producing goods at a rate that barely covers production costs. **b.** Relating to commodities thus manufactured and sold. **6.** *Psychol.* Relating to or located at the fringe of consciousness. **—mar'gin·al'i·ty** *n.* **—mar'gin·al·ly** *adv.*

**mar·gi·na·li·a** (mär'jə-nā'lē-ə) *pl.n.* [NLat. < Med. Lat. *marginalis,* marginal. —see MARGINAL.] Notes in a book margin.

**mar·gin·ate** (mär'jə-nāt') *vt.* **-at·ed, -at·ing, -ates.** To provide with margins or a margin. *—adj.* (mär'jə-nĭt, -nāt') *also* **margin·at·ed** (-nā'tĭd). *Biol.* With a border or edge of distinctive pattern or color. **—mar'gin·a'tion** *n.*

**mar·gra·vate** (mär'grə-vāt') *n. var. of* MARGRAVIATE.

**mar·grave** (mär'grāv') *n.* [MDu. *markgrave* : *mark,* march, border, + *grave,* count.] **1.** The medieval lord or military governor of a German border province. **2.** A hereditary title of certain princes in the Holy Roman Empire.

**mar·gra·vi·ate** (mär-grā'vē-ĭt, -āt') *also* **mar·gra·vate** (mär'grə-vāt') *n.* The territory ruled by a margrave.

**mar·gra·vine** (mär'grə-vēn') *n.* [MDu. *markgravin* < *markgrave,* margrave. —see MARGRAVE.] The wife or widow of a margrave.

**mar·gue·rite** (mär'gə-rēt', -gyə-) *n.* [Fr. < OFr. *margarite,* daisy < Lat. *margarita,* pearl < Gk. *margaritēs* < *margaron.*] **1.** A plant native to the Canary Islands, *Chrysanthemum frutescens,* with white or pale-yellow flowers similar to those of the common American daisy. **2.** A plant similar to or resembling the marguerite.

**ma·ri·a** (mä'rē-ə) *n. pl. of* MARE[2].

**ma·ri·a·chi** (mä'rē-ä'chē) *n.* [Mex. Sp.] **1.** A Mexican street band. **2. a.** The music performed by a mariachi. **b.** A musician playing in a mariachi.

**Mar·i·an** (mâr'ē-ən, măr'-) *n.* **1.** A devotee of the Virgin Mary. **2. a.** A supporter of Queen Mary I of England. **b.** A supporter of Mary Queen of Scots. *—adj.* Of or relating to the Virgin Mary, Queen Mary I of England, or Mary Queen of Scots.

**ma·ric·o·lous** (mə-rĭk'ə-ləs) *adj.* [Lat. *mare, mari-,* sea + -COLOUS.] Living in the sea.

**mar·i·cul·ture** (măr'ĭ-kŭl'chər) *n.* [Lat. *mare, mari-,* sea + CULTURE.] Cultivation of marine organisms in their natural habitats.

**mar·i·gold** (măr'ĭ-gōld', mâr'-) *n.* [ME : *Mary,* Mary + *gold,* gold.] **1.** A plant of the genus *Tagetes,* native to tropical America and widely cultivated for its bright orange or yellow flowers. **2.** A plant with flowers similar to those of the marigold, as the corn marigold or marsh marigold.

**mar·i·jua·na** *or* **mar·i·hua·na** (măr'ə-wä'nə) *n.* [Mex. Sp. *marihuana.*] **1.** Hemp. **2.** The dried flower clusters and leaves of the hemp plant, esp. when taken to induce euphoria.

**ma·rim·ba** (mə-rĭm'bə) *n.* [Kimbundu.] A large xylophone with resonators.

**ma·ri·na** (mə-rē'nə) *n.* [Ital., seashore < *marino,* belonging to the sea < Lat. *marinus.* —see MARINE.] A boat basin with facilities for small boats, as docks, moorings, and supplies.

**mar·i·nade** (măr'ə-nād') *n.* [Fr. < Sp. *marinada* < *marinar,* to marinate < *marino,* belonging to the sea. —see MARINA.] A liquid mixture, usu. of vinegar or wine and oil with various herbs and spices, in which to soak meat, fowl, and fish before cooking. *—vt.* **-nad·ed, -nad·ing, -nades.** To marinate.

**ma·ri·na·ra** (măr'ə-năr'ə, mär'ə-när'ə) *adj.* [Ital. *(alla) marinara,* in sailor style < *marinaro,* of sailors < *marino,* marine < Lat. *marinus.*] Being or served with a sauce made of tomatoes, onions, garlic, and spices <veal *marinara*>

**mar·i·nate** (măr'ə-nāt') *vt.* **-nat·ed, -nat·ing, -nates.** [Alteration of MARINADE.] To soak (e.g., meat) in a marinade.

**ma·rine** (mə-rēn') *adj.* [ME < OFr. < Lat. *marinus* < *mare,* sea.] **1. a.** Of or relating to the sea <*marine* research> **b.** Native to or formed by the sea <*marine* organisms> **2.** Of or pertaining to shipping or maritime affairs. **3.** Of or relating to sea navigation : NAUTICAL <buoys and other *marine* aids> **4.** Designating or relating to troops that serve at sea as well as on land, specif. the U.S. Marine Corps. *—n.* **1.** The naval or mercantile ships or shipping fleet of a country. **2. a.** A soldier serving on a ship or at a naval installation. **b. Marine.** A member of the U.S. Marine Corps. **3.** The governmental department in charge of naval affairs in some nations. **4.** A photograph or painting of the sea.

**Marine Corps** *n.* A branch of the U.S. armed forces composed mainly of amphibious troops under the authority of the Secretary of the Navy.

**mar·i·ner** (măr'ə-nər) *n.* [ME < AN < OFr. *marine,* marine.] One who navigates a ship : SAILOR.

**Mar·i·ol·a·try** (mâr'ē-ŏl'ə-trē) *n.* Idolatrous or excessive worship of the Virgin Mary.

**Mar·i·ol·o·gy** *also* **Mar·y·ol·o·gy** (mâr'ē-ŏl'ə-jē) *n.* The body of belief pertaining to the Virgin Mary.

**mar·i·o·nette** (măr'ē-ə-nĕt') *n.* [Fr., from the name *Marion,* Marion.] A jointed puppet manipulated by wires or strings attached to its limbs.

**mar·i·po·sa lily** *also* **mar·i·po·sa tulip** (măr'ə-pō'zə, -sə) *n.* [Sp. *mariposa,* butterfly : *Maria,* Mary + *posar,* to perch < LLat. *pausare,* to pause < Lat. *pausa,* pause.] A bulbous plant of the genus *Calochortus* of the southwestern United States and Mexico, with variously colored, tuliplike flowers.

**Mar·ist** (mâr'ĭst, măr'-) *n.* **1.** A member of the Society of Mary, a congregation of Roman Catholic missionary priests founded in 1824. **2.** A member of the Little Brothers of Mary, a Roman Catholic teaching order founded in 1817.

**mar·i·tal** (măr'ĭ-təl) *adj.* [Lat. *maritalis* < *maritus,* married.] **1.** Of or relating to marriage <*marital* problems> **2.** Of or relating to a husband. **—mar'i·tal·ly** *adv.*

**mar·i·time** (măr'ĭ-tīm') *adj.* [Lat. *maritimus* < *mare,* sea.] **1.** Located on or close to the sea. **2.** Of or concerned with shipping or navigation. **3.** Of or suggesting a mariner.

**mar·jo·ram** (mär'jər-əm) *n.* [ME *majorane* < OFr. < Med. Lat *majorana.*] **1.** An aromatic plant, *Majorana hortensis,* with small purplish-white flowers and leaves used as seasoning. **2.** A plant, *Origanum vulgare,* similar to the marjoram, with spikes of pinkish flowers and leaves used in cooking.

**mark[1]** (märk) *n.* [ME < OE *mearc.*] **1.** A visible trace or impression, as a spot, dent, or line. **2.** A cross or other sign used instead of a signature. **3.** A punctuation mark. **4. a.** A number, letter, or symbol indicating grades of scholastic achievement. **b.** *often* **marks.** An appraisal <got high *marks* for leadership> **5. a.** An inscription, name, stamp, label, or seal placed on an article to signify ownership, quality, manufacture, or origin. **b.** A notch in an animal's ear or hide indicating ownership. **6.** *Naut.* **a.** A knot or piece of material placed at various measured lengths on a lead line to indicate the depth of the water. **b.** A Plimsoll mark. **7. a.** A visible indication of a quality, property, or feature <The old castle now bears no *marks* of a fortress.> **b.** A visible sign or symbol, as a badge or brand adopted by or imposed on a person. **c. Mark.** A mode, brand, size, or quality of a product <This rifle is the *Mark* II model.> **8.** A recognized standard of quality : NORM <a production that is not up to the *mark*> **9.** Quality : importance <"A fellow of no *mark* nor likelihood"—Shakespeare> **10.** Notice : attention <an accomplishment worthy of *mark*> **11.** A target. **12.** Something that one wishes to achieve : GOAL. **13.** An object or point serving as a guide. **14.** *Slang.* The intended victim of a swindler : DUPE <an easy *mark* for con artists> **15.** The place from which racers begin and occas. end their contest. **16.** A point reached or gained <the halfway *mark* in the book> **17.** A record <set a new *mark* in the pole vault> **18.** A spare or strike in bowling. **19.** A stationary ball in bowls : JACK. **20.** A boundary between countries. **21.** A tract of land held in common by a medieval English or German community. **22.** The numerical value assigned for computational convenience to a statistical observation falling within one of a number of intervals. *—v.* **marked, marking, marks.** *—vt.* **1.** To make a visible impression on, as with a spot, line, or dent. **2.** To form, make, or depict by making a visible impression, as with a spot, line, or dent. **3. a.** To indicate or distinguish by making a visible impression <*marked* the map to show the meeting place> **b.** To distinguish, indicate, or characterize <the bad taste that *marks* many paperback romances> **4.** To set off or separate by or as if by a mark. **5.** To attach price tags, maker's labels, or other identification to (articles for sale). **6.** To grade and correct (scholastic work) by evaluating with a scale of letters or numbers. **7. a.** To give attention to : NOTICE. **b.** To take note of in writing. **8.** To consider : study. **9.** To keep (score) in various games. *—vi.* **1.** To make a visible impression <This crayon will not *mark* on slick paper.> **2.** To receive a visible impression <The paper *marks* easily.> **3.** To pay attention : NOTICE. **4.** To keep score. **5.** To determine scholastic grades <That teacher *marks* leniently.> **—mark down.** To mark for sale at a lower price. **—mark up.** To mark for sale at a higher price.

**mark[2]** (märk) *n.* [ME, unit of weight < OE *marc.*] **1.** A former English and Scottish monetary unit equal to 13 shillings and 4 pence. **2.** Any of several former European units of weight equal to approx. eight ounces, used esp. for weighing gold and silver. **3.** —See table at CURRENCY.

**Mark** (märk) *n.* **1.** —See table at BIBLE. **2.** A king of Cornwall, the husband of Iseult and uncle of Tristan in Arthurian legend.

**mark·down** (märk'doun') *n.* **1.** A price reduction. **2.** The amount by which a price is reduced.

**marked** (märkt) *adj.* **1.** Bearing a mark or marks. **2.** Having a cleary defined character : NOTICEABLE <*marked* temperature changes in the ocean> **3.** Singled out, esp. for a dire fate. **—mark'ed·ly** (mär'kĭd-lē) *adv.*

**mark·er** (mär'kər) *n.* **1.** One that marks or distinguishes, as a bookmark, tombstone, or milestone. **2.** An implement used for marking or writing. **3.** One who marks objects, esp. for industrial purposes. **4.** One who grades scholastic papers. **5.** A device, as a line, stake, or flag, on a playing field that shows the playing or scoring position. **6. a.** One that keeps score in various games. **b.** A score in a game. **7.** *Slang.* A written, signed promissory note : IOU.

**mar·ket** (mär'kĭt) *n.* [ME < OE < ONFr. < Lat. *mercatus* < *mercari,* to buy < *merx,* merchandise.] **1.** A public gathering held for buying and selling merchandise. **2.** A place where goods are offered for sale. **3.** A store or shop that sells a particular type of merchandise <a meat *market*> **4. a.** The business of buying and selling a specified commodity. **b.** A market price. **c.** A geographic region regarded as a place for sales <seafood for the domestic *market*> **d.** A subdivision of a population considered as buyers <the college *market*> **5.** Opportunity to buy or sell : demand for merchandise <a new *market* for home appliances> **6. a.** An exchange for buying and selling stocks or commodities <securities traded on the Chicago *market*> **b. the market.** The entire enterprise of buying and selling commodities and securities. —*v.* **-ket·ed, -ket·ing, -kets.** —*vt.* **1.** To offer for sale. **2.** To sell. —*vi.* **1.** To deal in a market. **2.** To buy household supplies <*marketed* for the week's groceries> **—in the market.** Interested in buying. **—on the market.** For sale. **—mar'ket·er** *n.*

**mar·ket·a·ble** (mär'kĭ-tə-bəl) *adj.* **1.** Fit to be offered for sale. **2.** Salable <a *marketable* item><a *marketable* idea> **3.** Having to do with buying or selling. **—mar'ket·a·bil'i·ty** *n.*

**market basket** *n.* **1.** A grocery cart. **2.** A selection of foods required for a statistical household of 3.2 persons or for a family of four, considered in terms of its fluctuating cost.

**mar·ket·ing** (mär'kĭ-tĭng) *n.* **1.** The act or process of buying and selling in a market. **2.** The commercial functions involved in transferring goods from producer to consumer.

**market order** *n.* An order to buy or sell stocks or commodities at the prevailing market price.

**mar·ket·place** *also* **market place** (mär'kĭt-plās') *n.* **1.** A place in which a market is set up. **2.** The business world. **3.** The arena in which works, opinions, or ideas are debated and exchanged.

**market price** *n.* The prevailing price at which merchandise, securities, or commodities are sold.

**market value** *n.* The amount that a seller may expect to receive in the open market for merchandise, services, or securities.

**mark·ing** (mär'kĭng) *n.* **1.** A mark or marks. **2.** The characteristic coloration of an animal or plant.

**mark·ka** (mär'kä') *n., pl.* **-kaa** (-kä') [Finn. < Swed. *mark*, unit of value.] —See table at CURRENCY.

**marks·man** (märks'mən) *n.* **1.** One skilled at firing a gun or other weapon. **2.** A U.S. military classification for the lowest of three ratings of rifle proficiency. **—marks'man·ship'** *n.*

**marks·wo·man** (märks'wŏŏm'ən) *n.* A woman skilled at firing a gun or other weapon.

**mark·up** (märk'ŭp') *n.* **1.** A raise in price. **2.** An amount added to a cost price in calculating a selling price, esp. an amount that takes into account overhead and profit.

**marl** (märl) *n.* [ME < OFr. < LLat. *margila*, dim. of Lat. *marga*, marl.] A mixture of clays, carbonates of calcium and magnesium, and remnants of shells, forming a loam used as fertilizer. —*vt.* **marled, marl·ing, marls.** To fertilize with marl.

**mar·lin** (mär'lĭn) *n.* [Short for MARLINESPIKE (from the pointed shape of its snout).] A large game fish of the genus *Makaira* of the Atlantic and Pacific oceans.

**mar·line** (mär'lĭn) *also* **mar·ling** (-lĭng) *n.* [MDu. *marlijn* : *marren*, to tie + *lijn*, line < Lat. *linea*.] *Naut.* A light rope made of two loosely twisted strands.

**mar·line·spike** (mär'lĭn-spīk') *also* **mar·ling·spike** (-lĭng-spīk') *n. Naut.* A pointed metal spike for separating strands of rope in splicing.

**mar·lite** (mär'līt') *n.* A marl containing 25–75% clay, the remainder being calcium carbonate, that is resistant to decomposition in air. **—mar·lit'ic** (-lĭt'ĭk) *adj.*

**marl·stone** (märl'stōn') *n.* Marlite.

**mar·ma·lade** (mär'mə-lād') *n.* [Fr. *marmelade* < Port. *mamelada* < *marmelo*, quince < Lat. *melimelum*, a kind of sweet apple < Gk. *melimēlon* : *meli*, honey + *mēlon*, apple.] A preserve made from the pulp and rind of fruits, esp. citrus fruits.

**marmalade plum** *n.* A tree, *Calocarpum zapota* of the American tropics, bearing edible fruit.

**mar·mo·re·al** (mär-môr'ē-əl, -mōr'-) *also* **mar·mo·re·an** (-ē-ən) *adj.* [Lat. *marmoreus* < *marmor*, marble.] Like marble, as in smoothness, whiteness, or hardness.

**mar·mo·set** (mär'mə-sĕt', -zĕt') *n.* [ME < OFr. < *marmouser*, to murmur.] A small monkey of the genera *Callithrix*, *Cebuella*, *Saguinus*, or *Leontideus* of tropical American forests, with soft, dense fur, tufted ears, and a long tail.

**mar·mot** (mär'mət) *n.* [Fr. *marmotte.*] A stocky, coarse-furred rodent of the genus *Marmota*, found throughout the Northern Hemisphere, having short legs and ears and bushy tails.

**marmot**
*20–28 inches long*

**Mar·o·nite** (măr'ə-nīt') *n.* [Med. Lat. *maronita* < *Maro*, Syrian monk of the 5th cent. A.D.] A member of a Christian Uniat Church, chiefly of Lebanon, whose liturgy is conducted in Syriac. **—Mar'o·nite'** *adj.*

**ma·roon¹** (mə-rōōn') *vt.* **-rooned, -roon·ing, -roons.** [Fr., *marron*, fugitive slave < Am. Sp. *cimarrón*, poss. < *cima*, of a mountain.] **1.** To put (a person) ashore on a deserted island or isolated coast. **2.** To abandon or isolate (a person) with small chance of escape or rescue. —*n.* **1. a.** A fugitive Negro slave in the West Indies in the 17th and 18th cent. **b.** A descendant of such a slave. **2.** A marooned person.

**ma·roon²** (mə-rōōn') *n.* [Fr. *marron* < Ital. *marrone.*] A dark reddish brown to dark purplish red.

**mar·plot** (mär'plŏt') *n.* [After *Marplot*, a character in *The Busy Body*, a play by Susanna Centlivre (1667?–1723).] A stupid, officious meddler whose interference spoils the success of an undertaking.

**mar·quee** (mär-kē') *n.* [Fr. *marquise*, marquee, marquise.] **1.** A large open-sided tent, used chiefly for outdoor entertainment. **2.** A rooflike structure, often bearing a signboard, projecting over an entrance to a theater or other building.

**Mar·que·san** (mär-kā'zən, -sən) *n.* **1.** An inhabitant of the Marquesas Islands. **2.** The Austronesian language of the Marquesans. —*adj.* Of or relating to the Marquesas Islands, their inhabitants, or their language.

**mar·quess** (mär'kwĭs) *n. Chiefly Brit. var. of* MARQUIS.

**mar·que·try** *also* **mar·que·terie** (mär'kĭ-trē) *n., pl.* **-tries.** [OFr. *marqueterie* < *marqueter*, to checker < *marque*, mark, of Germanic orig.] An inlay, as of wood or ivory, used chiefly for decorating furniture.

**mar·quis** (mär'kwĭs, mär-kē') *n., pl.* **mar·quis·es** (mär'kwĭ-sĭz) or **mar·quis** (mär-kēz') [ME < OFr. < *marche*, boundary, of Germanic orig.] A nobleman ranking above an earl or count and below a duke.

**mar·quis·ate** (mär'kwĭ-zĭt, -sĭt) *n.* A marquis' rank or territory.

**mar·quise** (mär-kēz') *n.* [Fr., fem. of *marquis*, marquis.] **1.** A marchioness. **2.** A marquee. **3. a.** A ring set with a pointed oval stone or pointed oval stones in a cluster. **b.** A gemstone, as a diamond, cut in a pointed oval shape.

**mar·qui·sette** (mär'kĭ-zĕt', -kwĭ-) *n.* [MARQUISE + -ETTE.] A sheer fabric of cotton, rayon, silk, or nylon, used for clothing, curtains, and mosquito nets.

**mar·ram** (măr'əm) *n.* [Of Scand. orig.] A beach grass, *Ammophila arenaria*, planted to stabilize shifting dunes.

**mar·riage** (măr'ĭj) *n.* [ME *mariage* < OFr. < *marier*, to marry. —see MARRY.] **1. a.** The state of being married : WEDLOCK. **b.** Legal union of a man and woman as husband and wife. **2.** The act of marrying or the ceremony of being married. **3.** Close union <a *marriage* of minds> **4.** The combination of the king and queen of the same suit, as in pinochle.

☆ ***syns:*** MARRIAGE, NUPTIALS, WEDDING *n. core meaning* : the act or ceremony of being married <civil and religious *marriages*>

**mar·riage·a·ble** (măr'ĭ-jə-bəl) *adj.* Suitable or eligible for marriage. **—mar'riage·a·bil'i·ty, mar'riage·a·ble·ness** *n.*

**marriage of convenience** *n.* A marriage or joint undertaking arranged for political, economic, or social benefit rather than from personal attachment.

**mar·ried** (măr'ēd) *adj.* **1. a.** Having a spouse <Are you *married*?> **b.** United in matrimony <a *married* couple> **2.** Of or relating to marriage <*married* bliss> —*n.* A married person <a line of inexpensive furniture for young *marrieds*>

**mar·ron** (mă-rôN') *n.* [Fr. —see MAROON².] SPANISH CHESTNUT 2.

**mar·row** (măr'ō) *n.* [ME *marow* < OE *mearg.*] **1.** The soft material that fills bone cavities, consisting of maturing blood cells and fat cells along with supporting connective tissue and many blood vessels. **2. a.** Spinal marrow. **b.** The spinal cord. **3. a.** The inmost, choicest, or most essential part : PITH. **b.** Strength or vigor : VITALITY. **4.** *Chiefly Brit.* The vegetable marrow.

**mar·row·bone** (măr'ō-bōn') *n.* A bone for flavoring soup.

**marrow squash** *n.* The vegetable marrow.

**mar·ry¹** (măr'ē) *v.* **-ried, -ry·ing, -ries.** [ME *marien* < OFr. *marier* < Lat. *maritare* < *maritus*, married.] —*vt.* **1. a.** To become united with in matrimony. **b.** To take as a husband or wife. **c.** To give in marriage. **2.** To perform a marriage ceremony for. **3.** To obtain by marriage <*marry* power> **4.** *Naut.* To join (two ropes) end to end by interweaving their strands. **5.** To unite in a close, usu. permanent way. —*vi.* **1.** To take a husband or wife : WED. **2.** To enter into a close relationship : UNITE.

**mar·ry²** (măr'ē) *interj.* [ME *Marie*, Mary, the mother of Jesus.] *Archaic.* —Used as an exclamation of surprise or emphasis.

**Mars** (märz) *n.* [ME < Lat.] **1.** *Rom. Myth.* The god of war. **2.** The fourth planet from the sun, having a sidereal period of revolution about the sun of 687 days at a mean distance of 141.6 million miles or 227.8 million kilometers, a mean radius of approx. 2,090 miles or 3,363 kilometers, and a mass approx. 0.15 that of Earth.

**mar·seille** (mär-sāl') *also* **mar·seilles** (-sālz') *n.* [After *Mar-*

ă pat ā pay âr care ä father ĕ pet ē be hw which ĭ pit
ī tie îr pier ŏ pot ō toe ô paw, for oi noise ŏŏ took

*seilles,* France.] A heavy cotton fabric with a raised pattern of figures or stripes.

**marsh** (märsh) *n.* [ME < OE *mersc.*] An area of low-lying wet land.

**mar·shal** (mär′shəl) *n.* [ME < OFr. *mareschal,* of Germanic orig.] **1.** A military officer of the highest rank in some countries. **2. a.** A U.S. federal officer who carries out court orders. **b.** A U.S. city officer who carries out court orders. **c.** The head of a U.S. fire or police department. **3.** One in charge of a parade or ceremony. **4.** A high official in a royal court, esp. one aiding the sovereign in military affairs. —*v.* **-shaled, -shal·ing, -shals** *also* **-shalled, -shal·ling, -shals.** —*vt.* **1.** To arrange (e.g., troops) in line for a military formation. **2.** To arrange in methodical order <*marshal* points for debate> **3.** To enlist and organize <*marshaled* our allies against the enemy> **4.** To guide (a person) ceremoniously : USHER. —*vi.* To take up positions in or as if in a military formation. **—mar′shal·cy, mar′shal·ship′** *n.*

**marsh elder** *n.* A coarse shrub of the genus *Iva* of eastern and central North America, often growing in salt marshes.

**marsh gas** *n.* Methane.

**marsh hawk** *n.* A hawk, *Circus cyaneus,* found in North American and Eurasian marshy areas.

**marsh hen** *n.* Any of various marsh birds of the family Rallidae, including the gallinules, coots, and rails.

**marsh·land** (märsh′lănd′) *n.* A marshy tract of land.

**marsh·mal·low** (märsh′mĕl′ō, -măl′ō) *n.* **1.** A confection of sweetened paste, once made from the root of the marsh mallow. **2.** A confection made of corn syrup, gelatin, sugar, and starch and dusted with powdered sugar. **3.** *Slang.* A timid or cowardly person.

**marsh mallow** *n.* [ME *mershmalwe* < OE *merscmealwe* : *mersc,* marsh + *mealwe,* mallow.] A plant, *Althaea officinalis,* native to Europe and naturalized in marshes of eastern North America, with pink flowers and a mucilaginous root used as a demulcent and in confectionery.

**marsh marigold** *n.* A plant of the genus *Caltha,* esp. *C. palustris,* growing in swampy places and bearing bright yellow flowers.

**marsh·y** (mär′shē) *adj.* **-i·er, -i·est.** Of, resembling, or marked by a marsh or marshes : BOGGY. **—marsh′i·ness** *n.*

**mar·su·pi·a** (mär-sōō′pē-ə) *n. pl. of* MARSUPIUM.

**mar·su·pi·al** (mär-sōō′pē-əl) *n.* [NLat. *marsupialis* < *marsupium,* marsupium. —see MARSUPIUM.] A mammal of the order Marsupialia, including the kangaroo, opossum, bandicoot, and wombat, found chiefly in Australia and South and Central America. **—mar·su′pi·al** *adj.*

**mar·su·pi·um** (mär-sōō′pē-əm) *n., pl.* **-pi·a** (-pē-ə) [NLat., pouch < Gk. *marsupion, marsipion,* dim. of *marsippos,* purse.] **1.** An external abdominal pouch in female marsupials containing mammary glands and sheltering the young until fully developed. **2.** A temporary egg pouch in various animals.

**mart** (märt) *n.* [ME < MDu. < ONFr. *market,* market. —see MARKET.] **1.** A trading center : MARKET. **2.** *Archaic.* A fair.

**mar·ta·gon** (mär′tə-gən) *n.* [ME < OFr. < OSp. < Turk. *martagān,* a kind of turban.] A lily, *Lilium martagon,* bearing pinkish-purple spotted flowers.

**Mar·tel·lo tower** (mär-tĕl′ō) *n.* [After Cape *Mortella,* Corsica.] A small circular fort, once used for coastal defense in Europe.

**mar·ten** (mär′tn) *n., pl.* **marten** or **-tens.** [ME < OFr. *martrine* < Med. Lat. *martrina,* of Germanic orig.] **1.** A carnivore of the genus *Martes* of northern wooded areas. **2.** The fur of a marten.

**mar·ten·site** (mär′tn-zīt′) *n.* [After Adolf *Martens* (1850–1914).] A solid solution of iron and up to 1% carbon, the chief constituent of hardened carbon tool steels. **—mar′ten·sit′ic** (-zĭt′ĭk) *adj.*

**mar·tial** (mär′shəl) *adj.* [ME < Lat. *martialis* < *Mars,* Mars.] **1.** Of, relating to, or like war. **2.** Relating to or connected with the armed forces or the military profession. **3.** Typical of or suitable for a warrior. **—mar′tial·ism** *n.* **—mar′tial·ist** *n.* **—mar′tial·ly** *adv.*

**martial art** *n.* Any of several Oriental arts of self-defense or combat, as karate, judo, or tae kwon do, usu. practiced as sport.

**martial law** *n.* Temporary rule by military authorities imposed on a civilian population esp. in time of war or when civil authority is unable to preserve public safety.

**Mar·tian** (mär′shən) *adj.* [ME < Lat. *martius* < *Mars,* Mars.] Of or relating to the planet Mars. —*n.* A hypothetical inhabitant of the planet Mars, esp. as a stock fictional character.

**mar·tin** (mär′tn) *n.* [OFr., prob. < the name *Martin,* Martin.] A bird, as the house martin or purple martin, that resembles and is closely related to the swallows.

**mar·ti·net** (mär′tn-ĕt′) *n.* [After Jean *Martinet* (d. 1672), a French army officer.] **1.** A rigid military disciplinarian. **2.** One who demands absolute adherence to rules.

**mar·tin·gale** (mär′tn-găl′) *also* **mar·tin·gal** (-găl′) *n.* [OFr.] **1.** The strap of a horse's harness connecting the girth to the nose band and designed to prevent the animal from throwing back its head. **2.** *Naut.* Any of several parts of standing rigging that strengthen the bowsprit and jib boom against the force of the head stays. **3.** A method of gambling in which one doubles the stakes after each loss. **4.** A loose half belt or strap on the back of a garment.

**mar·ti·ni** (mär-tē′nē) *n., pl.* **-nis.** [Orig. unknown.] A cocktail of gin or vodka and dry vermouth.

**Martin Luther King Day** *n.* Jan. 15, Martin Luther King, Jr.'s birthday, observed on the third Monday in Jan. as a legal holiday.

**Mar·tin·mas** (mär′tn-məs) *n.* [ME *martinmesse* : *Martin,* Martin + *messe,* Mass.] A Christian festival celebrated annually on Saint Martin's Day, Nov. 11.

**mart·let** (märt′lĭt) *n.* [OFr. *martelet,* prob. alteration of *martinet,* dim. of *martin,* martin.] **1.** The martin. **2.** *Heraldry.* A representation of a bird without feet, used as a crest or bearing to indicate a fourth son.

**mar·tyr** (mär′tər) *n.* [ME < OE < LLat. < Gk. *murtus,* witness.] **1.** One who chooses to die rather than renounce religious principles. **2.** One who suffers much or makes great sacrifices in order to advance a belief, cause, or principle. **3.** One who endures great suffering. **4.** One who makes a great show of suffering in order to arouse sympathy. —*vt.* **-tyred, -tyr·ing, -tyrs. 1.** To make a martyr of (a person). **2.** To inflict great pain on : TORMENT.

**mar·tyr·dom** (mär′tər-dəm) *n.* **1. a.** The state of being a martyr. **b.** The suffering of death by a martyr. **2.** Extreme suffering.

**mar·tyr·ize** (mär′tə-rīz′) *vt.* **-ized, -iz·ing, -iz·es.** To martyr.

**mar·tyr·ol·o·gy** (mär′tə-rŏl′ə-jē) *n., pl.* **-gies. 1.** A list or catalogue of religious martyrs. **2. a.** The history of religious, esp. Christian, martyrs. **b.** A history or account of martyrs. **—mar′tyr·ol′o·gist** *n.*

**mar·vel** (mär′vəl) *n.* [ME *marvail* < OFr. *merveille* < Lat. *marabilis,* wonderful < *mirari,* to wonder.] **1.** One that evokes surprise, admiration, or wonder. **2.** Strong surprise : ASTONISHMENT. —*v.* **-veled, -vel·ing, -vels** *also* **-velled, -vel·ling, -vels.** —*vi.* To become filled with wonder or astonishment <*marveled* at such keyboard virtuosity> —*vt.* To wonder at or about <I *marvel* the fact that everyone was rescued.>

☆ ***syns:*** MARVEL, MIRACLE, PHENOMENON, PRODIGY, SENSATION, STUNNER, WONDER, WONDERMENT *n. core meaning* : one that evokes great surprise and admiration <scientific *marvels*><a person who is a *marvel* of efficiency>

**mar·vel·ous** *also* **mar·vel·lous** (mär′və-ləs) *adj.* **1.** Causing astonishment or wonder <a *marvelous* tennis match> **2.** Supernatural : miraculous. **3.** Of the highest or best kind or quality : FIRST-RATE. **—mar′vel·ous·ly** *adv.* **—mar′vel·ous·ness** *n.*

**Marx·i·an** (märk′sē-ən) *n.* A student, advocate, or practitioner of Karl Marx's philosophical or socioeconomic concepts as a means of interpretation and analysis, as in political economy or historical or literary criticism. **—Marx′i·an** *adj.* **—Marx′i·an·ism** *n.*

**Marx·ism** (märk′sĭz′əm) *n.* The economic and political ideas of Karl Marx and Friedrich Engels as developed into a system of thought that gives a primary role to class struggle in leading society from capitalistic bourgeois democracy to a socialist society and thence to communism.

**Marx·ism-Len·in·ism** (märk′sĭz′əm-lĕn′ĭ-nĭz′əm) *n.* Expansion of Marxism to include Lenin's concept of imperialism as the final form of capitalism and a shift in the focus of struggle from developed to underdeveloped countries.

**Marx·ist** (märk′sĭst) *n.* A believer in or follower of the ideas of Marx and Engels, esp. a militant Communist.

**Mary Jane** (mâr′ē jān′) *n.* [By folk ety. < MARIJUANA.] *Slang.* Marijuana.

**Mar·y·ol·o·gy** (mâr′ē-ol′ə-jē) *n. var. of* MARIOLOGY.

**mar·zi·pan** (mär′zə-păn′, märt′sə-păn′) *n.* [G. < Ital. *marzapane,* marzipan, a medieval coin < Ar. *mawthabān,* a medieval coin.] A confection of ground almonds, egg whites, and sugar, molded into decorative forms.

**Ma·sai** (mä-sī′) *n., pl.* **Masai** or **-sais. 1.** A member of a people of Kenya and parts of Tanzania. **2.** The Nilotic language of the Masai. **—Ma·sai′** *adj.*

**mas·car·a** (mă-skăr′ə) *n.* [Sp. *máscara* < Ital. *maschera,* mask. —see MASK.] A cosmetic for darkening the eyelashes and eyebrows.

**mas·cot** (măs′kŏt′, -kət) *n.* [Fr. *mascotte* < Prov. *mascoto* < *masca,* witch < LLat.] One that is thought to bring good luck, esp. one kept as the symbol of an athletic team.

**mas·cu·line** (măs′kyə-lĭn) *adj.* [ME < OFr. *masculin* < Lat. *masculinus* < *masculus,* male.] **1.** Of or relating to men or boys : MALE. **2.** Suggestive or typical of a man. **3.** Of, designating, or constituting the gender of words or grammatical forms denoting or referring normally to males <a *masculine* ending> **4.** Of or with a stressed terminal syllable or syllables <the *masculine* rhyme of "annoy, enjoy"> **5.** *Mus.* Ending on an accented beat <a *masculine* cadence> —*n.* **1.** The masculine gender. **2.** A word or word form of the masculine gender. **3.** A male. **—mas′cu·line·ly** *adv.* **—mas′cu·lin′i·ty, mas′cu·line·ness** *n.*

**mas·cu·lin·ize** (măs′kyə-lə-nīz′) *vt.* **-ized, -iz·ing, -iz·es. 1.** To give a masculine character or appearance to. **2.** To cause to assume masculine characteristics.

**ma·ser** (mā′zər) *n.* [Acronym for *microwave amplification by stimulated emission of radiation.*] Any of several devices that convert incident electromagnetic radiation from a wide range of frequencies to one or more discrete frequencies of highly amplified and coherent microwave radiation.

**mash** (măsh) *n.* [ME < OE *max.*] **1.** A fermentable starchy mixture from which spirits or alcohol can be distilled. **2.** A mixture of ground grain and nutrients fed to fowl and livestock. **3.** A soft pulpy mixture or mass. **4.** A grinding or crushing. *—vt.* **mashed, mash·ing, mash·es. 1.** To convert (malt or grain) into mash. **2.** To convert into a soft pulpy mixture <*mash* carrots> **3.** To grind or crush. **4.** *Slang.* To flirt with or approach aggressively.

**mash·er** (măsh′ər) *n.* **1.** A kitchen utensil for mashing foods. **2.** *Slang.* A man who attempts to force his attentions on a woman.

**mash·ie** *also* **mash·y** (măsh′ē) *n., pl.* **-ies.** [Orig. unknown.] A golf club of medium loft.

**mas·jid** (mŭs′jĭd) *n.* [Ar.] A mosque.

**mask** (măsk) *n.* [Fr. *masque* < Ital. *maschera,* poss. < Ar. *maskharah,* buffoon.] **1.** A covering to conceal one's face, as: **a.** A cloth covering with openings for the eyes, entirely or partly concealing the face, and worn esp. at a masquerade ball. **b.** A representation of a grotesque face. **c.** A facial covering worn for ritual. **d.** A figure of a head worn by actors in Greek and Roman drama to identify a trait. **2.** A protective covering for the face or head. **3.** A gas mask. **4.** A representation of a face or head: **a.** A mold of a person's face, often made after death. **b.** An often grotesque representation of a head and face, used for ornamentation. **5.** The face or facial markings of certain animals, as a fox or dog. **6.** A face having a blank, fixed, or enigmatic expression. **7.** A disguise or concealment <hid behind a *mask* of indifference> **8.** A natural or artificial feature of terrain concealing and protecting military forces or installations. **9. a.** An opaque pattern or border between a source of light and a photosensitive surface that prevents exposure of specified areas of the surface. **b.** The translucent border framing a television picture tube and screen. **10.** A cosmetic preparation that cleans and tightens the skin of the face. **11.** *var. of* MASQUE. **12.** One wearing a mask. *—v.* **masked, mask·ing, masks.** *—vt.* **1.** To cover with a decorative or protective mask. **2.** To disguise by making indistinct or blurred to the senses <incense *masking* unpleasant odors> **3.** To cover up for concealment or protection. **4.** To block the view of <Camouflage *masked* the gun battery.> **5.** To cover (a part of a photographic film) by applying an opaque border. **6.** *Chem.* To inhibit (a compound or radical) with a reagent more active in a specific reaction. *—vi.* **1.** To put on a mask, esp. for a masquerade ball. **2.** To conceal one's true personality, character, or intentions.

**masked** (măskt) *adj.* **1.** Wearing a mask. **2.** Disguised : concealed. **3.** Latent or hidden, as a symptom or disease. **4.** *Bot.* Personate. **5.** *Zool.* Having masklike markings on the head or face.

**masked ball** *n.* A ball at which masks and costumes are required.

**mas·keg** (măs′kĕg′) *n. var. of* MUSKEG.

**mask·er** *also* **mas·quer** (măs′kər) *n.* A participant in a masquerade or masque.

**mask·ing** (măs′kĭng) *n.* **1.** *Psychol.* Concealment or screening of one sensory process by another. **2.** Theatrical scenery for concealing a part of the stage from the audience. **3.** *Computer Sci.* The process of selecting bits from a storage unit by using an instruction that eliminates the other bits in the unit.

**masking paper** *n.* Paper for covering and protecting a surface not to be painted.

**masking tape** *n.* An adhesive tape for covering and protecting a surface not to be painted.

**mas·o·chism** (măs′ə-kĭz′əm) *n.* [After Leopold von Sacher-*Masoch* (1836–1895), Austrian novelist.] **1. a.** Derivation of pleasure from being offended, dominated, or mistreated. **b.** The tendency to seek such mistreatment. **2.** Direction of destructive tendencies inward or upon oneself. **—mas′o·chist** *n.* **—mas′o·chis′tic** *adj.* **—mas′o·chis′ti·cal·ly** *adv.*

**ma·son** (mā′sən) *n.* [ME < OFr., of Germanic orig.] **1.** One who builds or works with stone or brick. **2.** A stonecutter. **3. Mason.** FREEMASON 2. *—vt.* **-soned, -son·ing, -sons.** To build of or strengthen with masonry.

**mason bee** *n.* Any of various solitary bees of the genus *Anthidium,* found worldwide, that build clay nests.

**Ma·son·ic** (mə-sŏn′ĭk) *adj.* Of or relating to Freemasons or Freemasonry.

**Ma·son·ite** (mā′sə-nīt′). A trademark for a type of fiberboard used for insulation, paneling, or partitions.

**Ma·son jar** (mā′sən) *n.* [After John L. *Mason* (1832–1902), its inventor.] A wide-mouthed glass jar with a screw top, used for home canning and preserving.

**ma·son·ry** (mā′sən-rē) *n., pl.* **-ries. 1.** The trade of a mason. **2.** Work done by a mason. **3.** Stonework or brickwork. **4. Masonry.** FREEMASONRY 2.

**masonry cement** *n.* A kind of cement esp. prepared for use in the mortar of brick and block masonry.

**Ma·so·ra** *also* **Ma·so·rah** (mə-sôr′ə, -sōr′ə) *n.* [Heb. *māsōrāh* < Heb. *māsar,* to hand over.] **1.** The body of tradition pertaining to correct textual reading of the Old Testament. **2.** The critical notes embodying the Masora, made by Jewish scholars before the 10th cent. A.D. **—Mas′o·ret′ic** (măs′ə-rĕt′ĭk) *adj.*

**masque** *also* **mask** (măsk) *n.* [Fr. < Ital. *maschera,* mask.] **1.** A popular 16th- and early 17th-cent. English entertainment, usu. based on a mythological or allegorical theme. **2.** A verse composition written for a masque production. **3.** MASQUERADE 1.

**mas·quer** (măs′kər) *n. var. of* MASKER.

**mas·quer·ade** (măs′kə-rād′) *n.* [OFr. *mascarade* < OSp. *mascarada,* poss. < Ar. *maskharah,* buffoon.] **1. a.** A costume party at which masks are worn. **b.** A costume for such a party or ball. **2. a.** A disguise or false outward show : PRETENSE <a *masquerade* of courage> **b.** An involved scheme : CHARADE. *—vi.* **-ad·ed, -ad·ing, -ades. 1.** To wear a mask or disguise, as at a masquerade. **2.** To go about as if in disguise <The secret agent *masqueraded* as a sailor.> **—mas′quer·ad′er**

**mass** (măs) *n.* [ME *masse* < OFr. < Lat. *massa* < Gk. *maza.*] **1.** A unified body of matter without specific shape. **2.** A grouping of individual parts or elements composing a unified body of unspecified size or quantity <the *mass* of undergraduates> **3.** A large, nonspecific amount or number <a *mass* of insect bites> **4.** The major part : MAJORITY <the *mass* of the continent> **5.** Physical volume or bulk of a solid body. **6.** *Physics.* The measure of a body's resistance to acceleration; the mass of a body is different from but proportional to its weight, is independent of the body's position but dependent on its motion with respect to other bodies, and may be expressed in mass units, as kilograms or slugs, or corresponding energy units, by means of the mass-energy relationship of the special theory of relativity. **7.** An area of unified light, shade, or color in a painting. **8.** A thick pasty mixture of drugs for forming pills. **9.** A shapeless mineral deposit. **10. the masses.** The common people. *—vt. & vi.* **massed, mass·ing, mass·es.** To gather or form into a mass. *—adj.* **1.** Of, relating to, typical of, or attended by a large number of people <*mass* entertainment> **2.** Directed at or reaching a large number of people <*mass* media><*mass* transportation> **3.** Involving great numbers or large amounts : LARGE-SCALE <a *mass* movement> **4.** Total : complete <The *mass* effect is unpleasant.>

**Mass** *also* **mass** (măs) *n.* [ME *masse* < OE *mæsse* < LLat. *missa* < Lat. *mittere,* to send.] **1. a.** Celebration of the Eucharist in Roman Catholic and some Protestant churches. **b.** A service including this celebration. **2.** A musical setting of certain parts of the Mass.

**Mas·sa·chu·set** *also* **Mas·sa·chu·sett** (măs′ə-chōō′sĭt, -zĭt) *n.* [Massachuset.] **1. a.** A large tribe of Indians who lived on or close to Massachusetts Bay. **b.** A member of this tribe. **2.** The Algonquian language of the Massachuset.

**mas·sa·cre** (măs′ə-kər) *n.* [OFr.] **1.** The savage killing of many victims. **2.** Slaughter of many animals. **3.** *Informal.* A severe defeat, as in sports. *—vt.* **-cred** (-kərd), **-cring** (-krĭng, -kər-ĭng), **-cres. 1.** To kill wantonly and indiscriminately : SLAUGHTER. **2.** *Informal.* To defeat decisively. **—mas′sa·crer** (-kər-ər, -krər) *n.*

☆ ***syns:*** MASSACRE, BLOOD BATH, BLOODLETTING, BLOODSHED, BUTCHERY, CARNAGE, POGROM, SLAUGHTER *n. core meaning* : the savage killing of many victims <troops involved in *massacres* of innocent civilians>

**mas·sage** (mə-säzh′, -säj′) *n.* [Fr. < *masser,* to massage < Ar. *massa,* he touched.] The kneading or rubbing of parts of the body to help circulation or relax muscles. *—vt.* **-saged, -sag·ing, -sag·es. 1.** To give a massage to. **2.** To treat by means of a massage. **3.** *Computer Sci.* To manipulate (input data) to produce output in a desired format.

**mas·sa·sau·ga** (măs′ə-sô′gə) *n.* [After the *Missisauga* River, Ontario, Canada.] A brown and white venomous North American rattlesnake, *Sistrurus catenatus.*

**mass·cult** (măs′kŭlt′) *n.* [MASS + CULT(URE).] Culture at the level of the masses.

**mass defect** *n.* The amount by which the mass of an atomic nucleus is less than the sum of the masses of its constituent particles.

**mass deficiency** *n.* The mass defect.

**mas·sé** (mă-sā′) *n.* [Fr. < *masser,* to cue < *masse,* cue.] A stroke in billiards made by hitting the cue ball on its side with the cue held almost vertical, so that the cue ball will curve around one ball before hitting another.

**mass-en·er·gy equivalence** (măs′ĕn′ər-jē) *n.* The physical principle that a measured quantity of energy is equivalent to a measured quantity of mass, expressed by Einstein's equation, $E = mc^2$, where *E* represents energy, *m* the equivalent mass, and *c* the speed of light.

**mas·se·ter** (mə-sē′tər, mă-) *n.* [NLat. < Gk. *masētēr* < *masasthai,* to chew.] A large masticatory muscle that raises the lower jaw.

**mas·seur** (mă-sûr′, mə-) *n.* [Fr. < *masser,* to massage.] A man who gives massages professionally.

**mas·seuse** (mă-soez′) *n.* [Fr., fem. of *masseur.*] A woman who gives massages professionally.

**mas·si·cot** (măs′ĭ-kŏt′, -kō′) *n.* [ME *masticot.*] The yellow crystalline mineral form of lead monoxide, PbO.

**mas·sif** (mă-sēf′) *n.* [Fr. < *massif,* massive < OFr. —see MASSIVE.] A large mountain mass or compact group of connected mountains forming an independent portion of a range.

**mas·sive** (măs′ĭv) *adj.* [ME *massif* < OFr. < *masse,* mass. —see MASS[1].] **1.** Consisting of or making up a large mass : BULKY <a *massive* highboy> **2.** Unusually large or impressive <a *massive* structure> **3.** Large or imposing in quantity, scope, degree, intensity, or scale <a *massive* project> **4.** *Med.* Large in comparison with the usual amount <a *massive* injection> **5.** *Pathol.* Severely affecting a widespread area of bodily tissue <a *massive* infection requiring antibiotics> **6.** Lacking crystalline structure : AMORPHOUS. **—mas′sive·ly** *adv.* **—mas′sive·ness** *n.*

**mass·less** (măs′lĭs) *adj.* Lacking mass.

**mass medium** *n., pl.* **mass media.** A means of public communication reaching a large audience.

**mass noun** *n.* A noun, as *flour, coal,* or *dishonesty,* typically denoting a concept or substance that in English is preceded in the singular by modifiers such as *some* or *much* rather than *a* or *an.*

**mass number** *n.* The total number of neutrons and protons in an atomic nucleus.

**mass-pro·duce** (măs′prə-do͞os′, -dyo͞os′) *vt.* **-duced, -duc·ing, -duc·es.** To manufacture in large amounts.

**mass production** *n.* The manufacture of goods in large amounts.

**mass spectrograph** *n.* An instrument for separating charged particles from a prepared beam by means of an electromagnetic field and for photographing the resulting distribution or spectrum of masses.

**mass·y** (măs′ē) *adj.* **-i·er, -i·est.** Having great mass or bulk.

**mast**[1] (măst) *n.* [ME < OE *mæst.*] **1.** A tall vertical spar that rises from the keel of a sailing vessel to support the sails and running rigging. **2.** A vertical pole.

**mast**[2] (măst) *n.* [ME < OE *mæst.*] The nuts of forest trees accumulated on the ground, used esp. to feed swine.

**mast-** *pref. var. of* MASTO-.

**mas·ta·ba** *also* **mas·ta·bah** (măs′tə-bə) *n.* [Ar. *maṣṭabah,* stone bench.] An ancient Egyptian tomb with a rectangular base, sloping sides, and a flat roof.

**mast cell** *n.* [Partial transl. of G. *Mast Zelle* : *Mast,* food + *Zelle,* cell.] A cell found principally in connective tissue that contains numerous basophilic granules of heparin.

**mas·tec·to·my** (măs-tĕk′tə-mē) *n., pl.* **-mies.** Surgical removal of a breast.

**mas·ter** (măs′tər) *n.* [ME < OE *magister* and OFr. *maistre,* both < Lat. *magister.*] **1.** One with control over the action of another or others. **2.** The captain of a merchant ship. **3. a.** One who employs an apprentice. **b.** An employer. **4.** The owner of a slave or an animal. **5.** The head of a household. **6.** One who has control over something : POSSESSOR <the *master* of a great cotton plantation> **7.** One who defeats another : VICTOR. **8.** A teacher, schoolmaster, or tutor. **9. a.** One whose teachings or doctrines are accepted by followers. **b. Master.** Christ. **10.** A learned person : SCHOLAR. **11. a.** A university or college degree signifying completion of at least one year of prescribed study beyond the bachelor's degree. **b.** One holding such a degree. **12.** An artist or performer of great and exemplary skill. **13.** A worker qualified to teach apprentices. **14.** An old master. **15.** An expert <a *master* of deception> **16.** A title once used for a man holding a naval office ranking next below a lieutenant on a warship. **17.** The title of the head or presiding officer of certain societies, clubs, orders, or institutions. **18.** *Chiefly Brit.* The title of any of various law court officers. **19. Master.** The title of any of various officers with specified duties involving the management of the British royal household. **20. Master.** A person who owns a pack of hounds or is a chief officer of a hunt. **21.** *Archaic.* A form of address for a man : MISTER. **22. Master.** —Used as a courtesy title with the name of a boy not old enough to be addressed as Mister. **23.** An original from which copies can be made. —*adj.* **1.** Of, relating to, or typical of a master. **2.** Being the chief or leading force. **3.** Being something specified in a superlative degree <a *master* liar> **4.** Highly skilled : EXPERT <a *master* mechanic> **5.** Controlling all other parts of a mechanism <a *master* valve> **6.** Being an original from which copies are made. —*vt.* **-tered, -ter·ing, -ters. 1.** To act as or be the master of. **2.** To make oneself a master of <*mastered* the game> **3.** To overcome or defeat, as an addiction. **4.** To break or tame (a person or animal) : SUBJUGATE. **5.** To season or age (dyed goods). **—mas′ter·dom** *n.*

**mas·ter-at-arms** (măs′tər-ət-ärmz′) *n., pl.* **mas·ters-at-arms.** A petty officer assigned to keep order on board a warship.

**master bedroom** *n.* A main bedroom.

**mas·ter·ful** (măs′tər-fəl) *adj.* **1.** Given to playing the master : IMPERIOUS. **2.** Fit to command. **3.** Revealing mastery <*masterful* treatment of the subject> **4.** Expert <*masterful* piano technique> **—mas′ter·ful·ly** *adv.* **—mas′ter·ful·ness** *n.*

**master key** *n.* A key that opens several different locks whose keys are not the same.

**mas·ter·ly** (măs′tər-lē) *adj.* Having or displaying the skill or knowledge of a master. **—mas′ter·ly** *adv.* **—mas′ter·li·ness** *n.*

**master mariner** *n.* MASTER 2.

**master mason** *n.* **1.** An expert mason. **2. Master Mason.** The third degree of Freemasonry.

**mas·ter·mind** (măs′tər-mīnd′) *n.* A highly intelligent person, esp. one who plans and directs a complex project. —*vt.* **-mind·ed, -mind·ing, -minds.** To direct, plan, or supervise (e.g., a project).

**master of ceremonies** *n.* **1.** The host at a formal event, making the welcoming speech and introducing other speakers. **2.** A performer who leads a program of varied entertainment by introducing other performers to the audience.

**mas·ter·piece** (măs′tər-pēs′) *n.* [Prob. transl. of Du. *meesterstuk* or G. *Meisterstück.*] **1.** An outstanding work of art or craft. **2.** The greatest single work, as of an artist or composer. **3.** Something superlative <a *masterpiece* of deceit>
☆ ***syns:*** MASTERPIECE, CHEF-D'OEUVRE, MAGNUM OPUS, MASTERWORK, TOUR DE FORCE *n. core meaning* : an outstanding and ingenious work <The writer's first novel was a *masterpiece.*>

**master plan** *n.* A plan giving complete instruction or guidance.

**master race** *n.* A people who regard themselves as superior to other races and therefore suited to rule over them.

**master sergeant** *n.* A noncommissioned officer of the next to highest rating in the U.S. Army, Air Force, and Marine Corps.

**mas·ter·ship** (măs′tər-shĭp′) *n.* **1.** The office, function, or authority of a master. **2.** The skill of a master.

**mas·ter·sing·er** (măs′tər-sĭng′ər) *n.* A Meistersinger.

**mas·ter·stroke** (măs′tər-strōk′) *n.* A masterly achievement or action <a *masterstroke* of legislative bargaining>

**mas·ter·work** (măs′tər-wûrk′) *n.* A masterpiece.

**mas·ter·y** (măs′tə-rē) *n., pl.* **-ies. 1.** Consummate skill or knowledge. **2.** The status of master or ruler : CONTROL <*mastery* of the skies> **3.** Full command of a subject of study <a student's *mastery* of algebra>

**mast·head** (măst′hĕd′) *n.* **1.** The top of a ship's mast. **2.** The listing in a publication, as a newspaper or magazine, of information about its staff and operation.

**mas·tic** (măs′tĭk) *n.* [ME < OFr. < Lat. *mastiche* < Gk. *mastikhē.*] **1.** The mastic tree. **2.** The aromatic resin of the mastic tree, used in varnishes and lacquers and as an astringent. **3.** A Near Eastern liquor with mastic resin and aniseed flavoring. **4.** A pastelike cement, esp. one made with powdered lime or brick and tar.

**mas·ti·cate** (măs′tĭ-kāt′) *vt.* **-cat·ed, -cat·ing, -cates.** [LLat. *masticare, masticat-,* to masticate < Gk. *mastikhan,* to grind the teeth.] **1.** To chew. **2.** To grind and knead. **—mas′ti·ca′tion** *n.* **—mas′ti·ca′tor** *n.*

**mas·ti·ca·to·ry** (măs′tĭ-kə-tôr′ē, -tōr′ē) *adj.* **1.** Of, relating to, or used in mastication. **2.** Adapted for chewing. —*n., pl.* **-ries.** A substance chewed to increase salivation.

**mastic tree** *n.* **1.** A small evergreen tree, *Pistacia lentiscus* of the Mediterranean region. **2.** The pepper tree.

**mas·tiff** (măs′tĭf) *n.* [ME *mastif* < OFr. *mastin* < Lat. *mansuetus,* tamed : *manus,* hand + *suescere,* to accustom.] A large dog prob. bred orig. in ancient Asia, with a short fawn-colored coat.

**mastiff**
*30 inches high at shoulder*

**mastiff bat** *n.* Any of various tropical bats of the family Molossidae, with narrow wings and brown, gray, or black fur.

**mas·ti·goph·o·ran** (măs′tĭ-gŏf′ə-rən) *also* **mas·tig·o·phore** (mă-stĭg′ə-fôr′, -fōr′) *n.* [NLat. *Mastigophora,* class name : Gk. *mastix,* whip + Gk. *pherein,* to bear.] A member of the class Mastigophora, which includes protozoans with one or more flagella. **—mas′ti·goph′o·ran** *adj.*

**mas·ti·tis** (mă-stī′tĭs) *n.* Inflammation of the breast or udder. **—mas·tit′ic** (-tĭt′ĭk) *adj.*

**masto-** *or* **mast-** *pref.* [NLat. < Gk. *mastos,* breast.] Breast : mammary gland : nipple <*mastectomy*>

**mas·to·don** (măs′tə-dŏn′) *n.* [NLat. *Mastodon,* genus name : Gk. *mastos,* nipple + Gk. *odous,* tooth.] An extinct mammal of the genus *Mammut,* occas. called *Mastodon,* resembling the elephant.

**mas·to·dont** (măs′tə-dŏnt′) *adj.* **1.** Having teeth like those of a mastodon. **2.** Of, relating to, or typical of mastodons.

**mas·toid** (măs′toid′) *n.* **1.** The mastoid process. **2.** Mastoiditis. —*adj.* **1.** Relating to the mastoid process. **2.** Shaped like a breast or nipple.

**mastoid bone** *n.* The mastoid process.

**mastoid cell** *n.* One of the small air-filled spaces in the mastoid process.

**mas·toid·ec·to·my** (măs′toid-ĕk′tə-mē) *n., pl.* **-mies.** Surgical removal of part or all of the mastoid process.
**mas·toid·i·tis** (măs′toid-ī′tĭs) *n.* Inflammation of part or all of the mastoid process.
**mastoid process** *n.* The rear portion of the temporal bone on each side of the head behind the ear in humans and many other vertebrates.
**mat¹** (măt) *n.* [ME < OE *matta* < LLat.] **1.** A flat piece of fabric or other material used for wiping one's shoes or feet, or in various other forms as a floor covering. **2.** A small flat piece of decorated material placed under an object, as a lamp or dish of food. **3.** A floor pad to protect athletes, as in wrestling or gymnastics. **4.** A densely woven or thickly tangled mass <a *mat* of long fur> **5.** The solid part of a lace design. **6.** A heavy, woven net of rope or wire cable placed over a blasting site to keep debris from scattering. *—v.* **mat·ted, mat·ting, mats.** *—vt.* **1.** To cover, protect, or decorate with a mat or mats. **2.** To cover with or interweave into a thick mass <weeds *matting* the driveway> *—vi.* To be interwoven into a thick mass.
**mat²** (măt) *n.* [< Fr., dull < OFr., helpless. —see MATE².] **1.** A border of cardboard or similar material placed around a picture as a frame or to serve as a contrast between the picture and the frame. **2.** *also* **matte. a.** A dull, often rough finish, as on glass, metal, or paper. **b.** A special tool for producing such a surface or finish. **3.** MATRIX 10b. *—vt.* **mat·ted, mat·ting, mats. 1.** To put a mat around (a picture). **2.** To produce a dull finish on. *—adj. also* **matte.** Having a dull finish.
**Mat·a·be·le** (măt′ə-bē′lē) *n., pl.* **Matabele** *or* **-les. 1. a.** A Zulu tribe driven out of the Transvaal by the Boers in 1837. **b.** A member of this tribe. **2.** The Bantu language of the Matabele.
**mat·a·dor** (măt′ə-dôr′) *n.* [Sp. < Lat. *mactare,* to sacrifice < *mactus,* sacred.] **1.** A bullfighter who performs the final passes and kills the bull. **2.** One of the highest trumps in certain card games.
**match¹** (măch) *n.* [ME *macch* < OE *gemæcca,* companion.] **1. a.** One exactly like another : COUNTERPART. **b.** One like another in one or more specified qualities. **2. a.** One that closely harmonizes with or resembles another. **b.** A pair made up of two that harmonize with or resemble each other <The textures were a close *match.*> **3.** One equal in qualities or able to compete with another of the same class or type. **4. a.** An athletic contest or game in which two or more contestants or teams oppose and vie with one other <a tennis *match*> **b.** A race between horses belonging to two different owners who have set the conditions and terms of the race. **c.** A tennis match decided on the basis of victory in a specified number of sets, usu. two out of three or three out of five. **5.** A marriage <a perfect *match*> **6.** One viewed as a prospective marriage partner. *—v.* **matched, match·ing, match·es.** *—vt.* **1. a.** To be exactly like : correspond precisely. **b.** To be like with respect to specified qualities. **2.** To resemble or harmonize with <The lipstick *matches* the nail polish.> **3.** To adapt so that a balanced or harmonious result is achieved <"Let poets *match* their subject to their strength" —Earl Roscommon> **4.** To fit together or cause to fit together, esp. to cut (boards) with a tongue and groove. **5.** To join or give in marriage. **6.** To place in opposition or competition : PIT <*match* wits> **7.** To provide with an adversary or competitor. **8.** To do as well as or better than in competition : EQUAL. **9.** To set in comparison : COMPARE <beauty that could never be *matched*> **10.** To provide funds so as to equal or complement <The company will *match* all private gifts to the orchestra.> **11.** To flip or toss (coins) and compare the sides that land face up. **12.** To couple (electric circuits) using a transformer. *—vi.* To be a close counterpart : CORRESPOND. **—match′a·bil′i·ty** *n.* **—match′a·ble** *adj.* **—match′er** *n.*
**match²** (măch) *n.* [ME *matche,* lamp wick < OFr. *meche* < Med. Lat. *myxa* < Lat., a lamp's nozzle < Gk. *muxa,* lamp wick.] **1.** A narrow strip of wood, cardboard, or wax coated on one end with a compound easily ignitable by friction. **2.** An easily ignited cord or wick, once used to detonate powder charges or fire cannons and muzzle-loading firearms.
**match·a·ble** (măch′ə-bəl) *adj.* Capable of being matched. **—match′a·bil′i·ty** *n.*
**match·board** (măch′bôrd′, -bōrd′) *n.* A board cut with a tongue on one side and a matching groove on the other to fit with other boards identically cut.
**match·book** (măch′bo͝ok′) *n.* A small folder containing matches.
**match·box** (măch′bŏks′) *n.* A box for holding or storing matches.
**matching funds** *pl.n.* Sums of money contributed by an individual or institution, as a corporation or government agency, to match funds raised by the recipient, usually for a specified project.
**matching grant** *n.* Matching funds.
**match·less** (măch′lĭs) *adj.* Without match or equal : PEERLESS <*matchless* ability> **—match′less·ly** *adv.* **—match′less·ness** *n.*
**match·lock** (măch′lŏk′) *n.* **1.** A gunlock in which a match ignites powder. **2.** A musket with a matchlock.
**match·mak·er** (măch′mā′kər) *n.* **1.** One who habitually tries to arrange marriages. **2.** An arranger of athletic competitions.
**match play** *n.* A method of scoring golf games by counting only the number of holes won by each side instead of the number of strokes taken.
**match point** *n.* The final point required to win a sports match, esp. in tennis.
**match·wood** (măch′wo͝od′) *n.* Pieces or splinters of wood suitable esp. for making matches.
**mate¹** (māt) *n.* [ME < MLG.] **1.** One of a matched pair. **2.** A spouse. **3. a.** One of a conjugal pair of animals or birds. **b.** One of a pair of animals brought together for breeding. **4. a.** A close associate. **b.** A good friend or close companion. **5.** A deck officer on a merchant ship ranking below the master. **6.** A U.S. Navy petty officer who is an assistant to a warrant officer. *—v.* **mat·ed, mat·ing, mates.** *—vt.* **1.** To join closely : PAIR. **2.** To unite in marriage. **3.** To pair (animals) for breeding. *—vi.* **1.** To become joined in marriage. **2.** To become mated : BREED.
**mate²** (māt) *n.* [ME < OFr. *mat,* checkmated < Pers. *māt,* helpless.] A checkmate. *—vt. & vi.* **mat·ed, mat·ing, mates.** To checkmate or achieve a checkmate.
**ma·té** (mä′tā′) *n.* [Am. Sp. < Quechua *mate,* beverage.] **1.** An evergreen tree, *Ilex paraguayensis,* widely cultivated in South America, its indigenous habitat. **2.** A mildly stimulant beverage, popular in South America, made from the dried leaves of the maté.
**mat·e·lote** (măt′ə-lōt′) *also* **mat·e·lotte** (-lŏt′) *n.* [Fr. < *matelot,* sailor < OFr. *matenot,* prob. < MDu. *mattenoot* : *matte,* bed + *noot,* fellow.] **1.** A wine sauce for fish. **2.** Fish stewed in a wine sauce.
**ma·ter** (mā′tər) *n.* [Lat.] *Chiefly Brit.* Mother.
**ma·ter·fa·mil·i·as** (mā′tər-fə-mĭl′ē-əs) *n.* [Lat. : *mater,* mother + *familia,* family.] The mother of a family.
**ma·te·ri·al** (mə-tîr′ē-əl) *n.* [ME < *material,* consisting of matter < LLat. *materialis* < Lat. *materia,* matter.] **1.** The substance or substances out of which a thing is or can be made. **2.** Something, as an idea or sketch, to be refined and made or incorporated into a finished effort <*material* for a television series> **3. materials.** Tools or apparatus for performing a given task <painting *materials*> **4.** Yard goods or cloth. *—adj.* **1.** Composed of or relating to physical substances. **2.** Of, relating to, or affecting physical well-being <*material* necessities> **3.** Of or concerned with the physical as opposed to the intellectual or spiritual. **4.** Substantial : noticeable <a *material* advance> **5.** Of importance to a case : RELEVANT <discussed only the most *material* concerns> **6.** *Philos.* Of or relating to the matter of reasoning, rather than the form. **—ma·te′ri·al·ness** *n.*
**ma·te·ri·al·ism** (mə-tîr′ē-ə-lĭz′əm) *n.* **1.** *Philos.* The theory that physical matter in its movements and modifications is the only reality and that everything in the universe, including thought, feeling, mind, and will, is explainable in terms of physical laws. **2.** The theory that physical well-being and worldly possessions constitute the highest value and greatest good in life. **3.** A great or excessive regard for worldly concerns. **—ma·te′ri·al·ist** *n.* **—ma·te′ri·al·is′tic** *adj.* **—ma·te′ri·al·is′ti·cal·ly** *adv.*
**ma·te·ri·al·i·ty** (mə-tîr′ē-ăl′ĭ-tē) *n., pl.* **-ties. 1.** The quality or state of being material. **2.** Physical substance : MATTER.
**ma·te·ri·al·ize** (mə-tîr′ē-ə-līz′) *v.* **-ized, -iz·ing, -iz·es.** *—vt.* To cause to become real or actual <By building the boat, I *materialized* my dream.> *—vi.* **1.** To assume material or effective form <Support on the picket line did not *materialize.*> **2.** To take physical shape or form. **3.** To appear <"A smoldering cigarette suggested the occupant was in the vicinity, but . . . no one *materialized*" —Robin Cook> **—ma·te′ri·al·i·za′tion** *n.* **—ma·te′ri·al·iz′er** *n.*
**ma·te·ri·al·ly** (mə-tîr′ē-ə-lē) *adv.* **1.** With regard to matter as opposed to form. **2.** To a significant degree or extent : IMPORTANTLY. **3.** With regard to the physical world.
**ma·te·ri·als-in·ten·sive** (mə-tîr′ē-əlz-ĭn-tĕn′sĭv) *adj.* Requiring great use of or large expenditures on materials <new, less *materials-intensive* technologies>
**ma·te·ri·a med·i·ca** (mə-tîr′ē-ə mĕd′ĭ-kə) *n.* [NLat. : Lat. *materia,* material + Lat. *medica,* medical.] *Med.* **1.** Study of remedies and their sources, preparation, and use. **2. a.** Substances used as or in preparing remedies : MEDICINES. **b.** A specific medication.
**ma·te·ri·el** *or* **ma·té·ri·el** (mə-tîr′ē-ĕl′) *n.* [Fr. < *matériel,* consisting of matter < OFr. *materiel* < LLat. *materialis.* —see MATERIAL.] **1.** The equipment and supplies, as guns and ammunition, of a military force. **2.** An organization's equipment and supplies.
**ma·ter·nal** (mə-tûr′nəl) *adj.* [ME < OFr. *maternel* < Lat. *maternus* < *mater,* mother.] **1.** Pertaining to or typical of a mother or motherhood : MOTHERLY <*maternal* love> **2.** Inherited from one's mother <a *maternal* characteristic> **3.** Related through one's mother <my *maternal* grandparents> **—ma·ter′nal·ly** *adv.*
**ma·ter·ni·ty** (mə-tûr′nĭ-tē) *n.* [Fr. *maternité* < Med. Lat. *maternitas* < Lat. *maternus,* maternal < *mater,* mother.] **1.** The state of being a mother : MOTHERHOOD. **2.** The feelings or characteristics associated with being a mother : MOTHERLINESS. **3.** A section of a hospital for the care of mothers before and after childbirth and for the care of their babies. *—adj.* Relating to or designed for pregnancy <*maternity* leave><a *maternity* dress>
**mat·ey** (mā′tē) *adj. Chiefly Brit.* Friendly : sociable.

**math** (măth) *n.* Mathematics.

**math·e·mat·i·cal** (măth'ə-măt'ĭ-kəl) *also* **math·e·mat·ic** (-ĭk) *adj.* [ME < Med. Lat. *mathematicus* < Lat. < Gk. *mathēmatikos* < *mathēma*, science < *manthanein*, to learn.] **1.** Of or relating to mathematics. **2.** Precise : exact. **3.** Absolute : certain. **—math'e·mat'i·cal·ly** *adv.*

**mathematical induction** *n.* INDUCTION 3b.

**mathematical logic** *n.* Symbolic logic.

**math·e·ma·ti·cian** (măth'ə-mə-tĭsh'ən) *n.* A specialist in mathematics.

**math·e·mat·ics** (măth'ə-măt'ĭks) *n.* [ME *mathematik* < OFr. *matimatique* < Lat. *mathematica* < Gk. *mathēmatika* < *mathēmatikos*, mathematical. —see MATHEMATICAL.] (*sing. in number*). The study of numbers and their form, arrangement, and associated relationships, using rigorously defined literal, numerical, and operational symbols.

**maths** (măths) *n. Chiefly Brit.* Mathematics.

**ma·til·i·ja poppy** (mə-tĭl'ē-hä') *n.* [After *Matilija* Canyon, California.] A shrubby plant, *Romneya coulteri* of California and Mexico, with very large solitary white flowers.

**mat·in** (măt'n) *also* **mat·in·al** (-əl) *adj.* [< MATINS.] Of or relating to matins or to the early part of the day.

**mat·i·nee** *or* **mat·i·née** (măt'n-ā') *n.* [Fr. *matinée* < *matin*, morning < Lat. *matutinus*, of the morning. —see MATINS.] A daytime dramatic or musical performance.

**mat·ins** (măt'nz) *n.* [ME *matines* < OFr. < Med. Lat. (*vigiliae*) *matutinae*, morning (vigils) < Lat. *matutinus*, of the morning < *Matuta*, goddess of dawn.] (*sing. or pl. in number*). **1.** *Rom. Cath. Ch.* The office, together with lauds, constituting the first of the seven canonical hours. **2.** *often* **Matins.** Morning Prayer.

**matri-** *or* **matro-** *or* **matr-** *pref.* [Lat. < *mater*, mother.] Mother <*matri*lineal>

**ma·tri·arch** (mā'trē-ärk') *n.* **1.** A woman who rules a family, clan, or tribe. **2.** A woman who dominates a group or activity <the *matriarch* of Parisian literary circles> **—ma'tri·ar'chal** (-är'kəl), **ma'tri·ar'chic** (-är'kĭk) *adj.* **—ma'tri·ar'chal·ism** *n.*

**ma·tri·ar·chate** (mā'trē-är'kĭt, -kāt') *n.* **1. a.** A society, tribe, or state in which women hold the dominant authority. **b.** The authority held by matriarchs in such a society. **2.** A hypothetical stage in the evolution of society in which matriarchs wield authority.

**ma·tri·ar·chy** (mā'trē-är'kē) *n., pl.* **-chies. 1.** A social system in which descent is traced through the mother's side of the family. **2.** A matriarchate.

**ma·tri·ces** (mā'trĭ-sēz', măt'rĭ-) *n. var. pl. of* MATRIX.

**mat·ri·cide** (măt'rə-sīd') *n.* [Lat. *matricidium* : *mater*, mother + *caedere*, to kill.] **1.** Murder of one's own mother. **2.** One who murders one's own mother. **—mat'ri·ci'dal** (-sīd'l) *adj.*

**mat·ri·cli·nous** (măt'rĭ-klī'nəs) *also* **mat·ro·cli·nous** (măt'rə-) *adj.* [MATRI- + Gk. *klinein*, to lean.] Having predominantly maternal hereditary traits.

**ma·tric·u·late** (mə-trĭk'yə-lāt') *vt.* & *vi.* **-lat·ed, -lat·ing, -lates.** [Med. Lat. *matriculare, matriculat-*, to matriculate < LLat. *matricula*, list < *matrix*, list. —see MATRIX.] To admit or be admitted into a group, esp. a college or university : ENROLL. **—ma·tric'u·lant, ma·tric'u·late'** *n.* **—ma·tric'u·la'tion** *n.*

**mat·ri·lin·e·al** (măt'rə-lĭn'ē-əl) *adj.* Pertaining to, based on, or tracing ancestral descent through the maternal line. **—mat'ri·lin'e·al·ly** *adv.*

**mat·ri·lo·cal** (măt'rə-lō'kəl) *adj.* Pertaining to the home territory of a wife's kin group or clan in primitive societies. **—mat'ri·lo'cal·ly** *adv.*

**mat·ri·mo·ny** (măt'rə-mō'nē) *n., pl.* **-nies.** [ME < OFr. *matrimonie* < Lat. *matrimonium* < *mater*, mother.] MARRIAGE 1, 2. **—mat'ri·mo'ni·al** *adj.* **—mat'ri·mo'ni·al·ly** *adv.*

**matrimony vine** *n.* Any of various often thorny shrubs of the genus *Lycium*, species of which are cultivated for their purplish flowers and brightly colored berries.

**ma·trix** (mā'trĭks) *n., pl.* **ma·tri·ces** (mā'trĭ-sēz', măt'rĭ-) *or* **ma·trix·es.** [LLat. < Lat., breeding animal < *mater*, mother.] **1.** A situation or surrounding substance within which something originates, develops, or is contained. **2.** The womb. **3.** *Anat.* The formative cells of a tooth or fingernail. **4.** *Geol.* **a.** The solid matter in which a fossil or crystal is embedded. **b.** The impression left in a rock when an object, as a gemstone, has been removed. **5.** A die or mold. **6.** The principal metal in an alloy, as the iron in steel. **7.** A binding substance, as cement in concrete. **8.** *Math.* A rectangular array of algebraic or numerical quantities treated as an algebraic entity. **9.** *Computer Sci.* The network of intersections between input and output leads in a computer, functioning as an encoder or decoder. **10. a.** A metal plate used for casting typefaces. **b.** A mold used in stereotyping and designed to receive positive impressions of type or illustrations from which metal plates can be cast.

**matro-** *pref. var. of* MATRI-.

**mat·ro·cli·nous** (măt'rə-klī'nəs) *adj. var. of* MATRICLINOUS.

**ma·tron** (mā'trən) *n.* [ME *matrone* < OFr. < Lat. *matrona* < *mater*, mother.] **1.** A married woman, esp. a mature mother with established dignity and social position. **2.** A woman supervisor in a public institution, as a hospital, school, or prison. **—ma'tron·al** *adj.* **—ma'tron·li·ness** *n.* **—ma'tron·ly** *adj.* & *adv.*

**matron of honor** *n., pl.* **matrons of honor.** A married woman serving as chief attendant of a bride at a wedding.

**mat·ro·nym·ic** (măt'rə-nĭm'ĭk) *adj.* & *n. var. of* METRONYMIC.

**matte**[1] (măt) *n. var. of* MAT[2] 2. *—adj. var. of* MAT[2].

**matte**[2] (măt) *n.* [Fr.] A mixture of a metal with its oxides and sulfides, produced by smelting certain sulfide ores.

**mat·ted** (măt'ĭd) *adj.* **1.** Made from or covered with mats. **2.** Tangled in a dense mass <a *matted* beard>

**mat·ter** (măt'ər) *n.* [ME < OFr. *matere* < Lat. *materia* < *mater*, mother.] **1. a.** Something that occupies space and can be perceived by one or more senses. **b.** *Physics.* An entity displaying inertia and gravitation when at rest as well as when in motion. **2.** A specific type of substance <organic *matter*> **3.** Waste or discharge from a living organism, as pus or feces. **4.** The actual substance of thought or expression as opposed to the way in which it is stated or conveyed. **5.** *Philos.* In Aristotelian and Scholastic use, that which is in itself formless and undifferentiated and which, as the subject of development and change, receives form and becomes substance and experience. **6.** *Christian Science.* That which is postulated by the mortal mind, regarded as illusion and as the opposite of substance or God. **7.** The subject of concern, feeling, or action <diplomatic *matters*><a private *matter*> **8.** Trouble or difficulty <Is something the *matter*?> **9.** An approximated quantity, amount, or extent <a *matter* of several additional weeks> **10.** Something printed or otherwise set down in writing. **11.** Something sent by mail. **12. a.** Composed printing type. **b.** Material to be set in type. *—vi.* **-tered, -ter·ing, -ters.** To be of importance <It really *matters* a lot.> **—for that matter.** So far as that is concerned. **—no matter.** Regardless of <wanted it *no matter* what the cost>

**matter of course** *n.* A logical or natural outcome.

**mat·ter-of-fact** (măt'ər-əv-făkt') *adj.* Relating to or adhering to facts : LITERAL <a *matter-of-fact* account> **—mat'ter-of-fact'ly** *adv.* **—mat'ter-of-fact'ness** *n.*

**Mat·thew** (măth'yo͞o) *n.* —See table at BIBLE.

**mat·ting**[1] (măt'ĭng) *n.* **1.** Coarsely woven fabric for covering floors. **2.** Mat making.

**mat·ting**[2] (măt'ĭng) *n.* **1.** A dull finish or surface. **2.** The process of dulling a surface, as of metal. **3.** A border or mat used for framing a picture.

**mat·tins** (măt'nz) *n. Chiefly Brit. var. of* MATINS.

**mat·tock** (măt'ək) *n.* [ME < OE *mattuc.*] A digging tool with a blade at right angles to the handle, used with a downward motion.

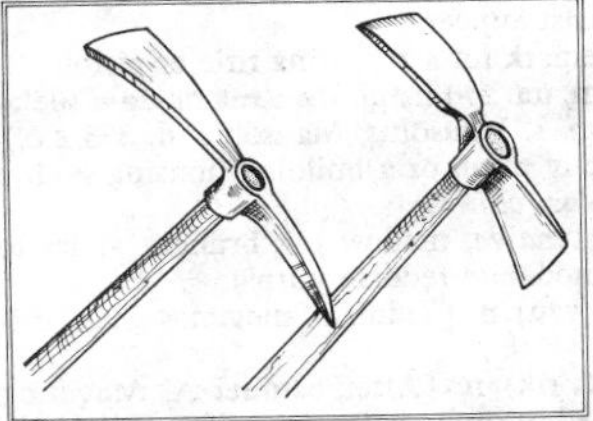

**mattock**

**mat·tress** (măt'rĭs) *n.* [ME *mattresse* < OFr. *materas* < OItal. *materasso* < Ar. *maṭraḥ*, place where something is thrown < *ṭaraḥ*, he threw.] **1. a.** A rectangular pad of heavy cloth filled with soft material, used as or on a bed. **b.** An airtight inflatable pad used as or on a bed or as a cushion. **2.** A closely woven mat of brush and poles for protecting an embankment, dike, or dam from erosion.

**mat·u·rate** (măch'ə-rāt') *vi.* & *vt.* **-rat·ed, -rat·ing, -rates.** [Lat. *maturare, maturat-*, to maturate < *maturus*, mature.] **1.** To mature or ripen. **2.** *Archaic.* To suppurate or cause to suppurate. **—mat'u·ra'tive** *adj.*

**mat·u·ra·tion** (măch'ə-rā'shən) *n.* **1.** The process of becoming mature. **2.** *Archaic.* SUPPURATION 1. **3.** *Biol.* **a.** Gametogenesis. **b.** Final differentiation processes, as the final ripening of a seed.

**maturation division** *n.* MEIOSIS 1.

**ma·ture** (mə-tyo͝or', -to͝or', -cho͝or') *adj.* **-tur·er, -tur·est.** [ME < Lat. *maturus.*] **1. a.** Having reached full natural growth or development <a *mature* cell> **b.** Fully developed : RIPE <a *mature* melon> **2.** Of, relating to, or characteristic of full mental or physical development. **3.** Worked out fully by the mind : CONSIDERED <a *mature* decision> **4.** Having reached the limit of its time : DUE <a *mature* certificate of deposit> **5.** *Geol.* Being a landscape in which hills and valleys predominate over flat areas due to erosion. *—v.* **-tured, -tur·ing, -tures.** *—vt.* **1.** To bring to full development : RIPEN. **2.** To work out completely in the mind. *—vi.* **1.** To evolve toward full development <The child's judgment *matures* with age.> **2.** To become due.

—Used of financial obligations, as notes and bonds. **—ma·ture'ly** *adv.* **—ma·ture'ness** *n.*

☆ ***syns:*** MATURE, ADULT, BIG, DEVELOPED, FULL-BLOWN, FULL-FLEDGED, FULL-GROWN, GROWN, GROWN-UP, RIPE, RIPENED *adj. core meaning* : having reached full growth and development <*mature* fruit><*mature* eagles> ***ant:*** immature

**ma·tur·i·ty** (mə-tyŏŏr'ĭ-tē, -tŏŏr'-, -chŏŏr'-) *n., pl.* **-ties.** [ME *maturite* < OFr. < Lat. *maturitas* < *maturus,* mature.] **1.** The quality or state of being mature. **2. a.** The time at which an obligation, as a note or bond, is due. **b.** The state of a note or bond being due.

**ma·tu·ti·nal** (mə-tōōt'n-əl, -tyōōt'-, măch'ōō-tī'nəl) *adj.* [LLat. *matutinalis* < Lat. *matutinus.* —see MATINS.] Of, relating to, or occurring in the morning : EARLY. **—ma·tu'ti·nal·ly** *adv.*

**mat·zo** (mät'sə, -sō) *n., pl.* **-zoth** (-sōth', -sōt', -sōs') *or* **-zos** (-səz, -səs, -sōz') *or* **-zot** (-sōt') [Yiddish *matse* < Heb. *maṣṣah.*] A flat, brittle piece of unleavened bread, eaten esp. during the Passover.

**matzo ball** *n.* A small dumpling from matzo meal.

**maud·lin** (môd'lĭn) *adj.* [Alteration of Mary *Magdalene,* who was frequently depicted as a tearful penitent.] Effusively or tearfully sentimental. **—maud'lin·ly** *adv.* **—maud'lin·ness** *n.*

▲ word history: The word *maudlin* is the regular English development of the second name of Mary Magdalene, a woman mentioned by name in the gospel of Luke. She has also been identified with certain anonymous women in the New Testament, especially the sinful woman who washed Jesus' feet with her tears (Luke 7). From these and other fragments of Scripture evolved a popular legend of Mary Magdalene as a reformed prostitute who became one of Jesus' most devoted and favored female disciples. She was frequently depicted in art as weeping copiously for her sins, and it is this attribute of hers that gave rise to the current sense of the adjective *maudlin.*

**maul** *also* **mall** (môl) [ME < OFr. *mail* < Lat. *malleus.*] *—n.* A heavy long-handled hammer for driving stakes, piles, or wedges. *—vt.* **mauled, maul·ing, mauls** *also* **malled, mall·ing, malls. 1.** To split (wood) with a maul and wedge. **2.** To handle fiercely by slashing, tearing, and bruising <A lion *mauled* the hunter.> **3.** To injure by or as if by beating. **—maul'er** *n.*

**maul·stick** *also* **mahl·stick** (môl'stĭk') *n.* [Partial transl. of Du. *maalstok* : obs. Du. *malen,* to point + *stok,* stick.] A long wooden stick used by painters for supporting the hand holding the brush.

**maund** (mônd) *n.* [Hindi *mān.*] A unit of weight varying in different countries of Asia from 11.2 to 37.358 kilograms or 24.7 to 82.286 pounds avoirdupois, the latter being the official maund in India.

**maun·der** (môn'dər, män'-) *vi.* **-dered, -der·ing, -ders.** [Prob. imit.] **1.** To talk incoherently or aimlessly. **2.** To move or act aimlessly or vaguely : WANDER.

**Maun·dy Thursday** (môn'dē, män'-) *n.* [ME *maunde,* ceremony of washing the feet of the poor on Maundy Thursday < OFr. *mande* < Lat. *mandatum,* mandate. —see MANDATE.] The Thursday before Easter, commemorating the Last Supper.

**Mau·ser** (mou'zər). A trademark for a repeating rifle or pistol.

**mau·so·le·um** (mô'sə-lē'əm, mô'zə-) *n., pl.* **-le·ums** *or* **-le·a** (-lē'ə) [ME < Lat. < Gk. *mausoleion,* < *Mausōlus,* Mausolus (d. 353 B.C.), satrap of Caria.] A large stately tomb or a building housing such a tomb or tombs. **—mau'so·le'an** *adj.*

**mauve** (mōv) *n.* [Fr. < Lat. *malva,* mallow.] A brilliant violet to strong or brilliant purple to moderate reddish purple.

**ma·ven** *also* **ma·vin** (mā'vən) *n.* [Yiddish *meyvn* < Heb. *mēbhīn.*] An expert.

**mav·er·ick** (măv'ər-ĭk, măv'rĭk) *n.* [After Samuel A. *Maverick* (1803–1870).] **1.** An unbranded or orphaned range calf or colt, traditionally regarded as the property of the first person who brands it. **2.** A horse or steer that has escaped from a herd. **3.** An independently minded person who refuses to abide by the dictates of or resists adherence to a particular group.

**ma·vie** (mā'vē) *n. var. of* MAVIS.

**ma·vin** (mā'vən) *n. var. of* MAVEN.

**ma·vis** (mā'vĭs) *also* **ma·vie** (-vē) *n.* [ME < OFr. *mauvis.*] The song thrush.

**ma·vour·neen** *also* **ma·vour·nin** (mə-vŏŏr'nēn') *n.* [Ir. Gael. *mo mhuirnín.*] *Ir.* My darling.

**maw** (mô) *n.* [ME < OE *maga.*] **1.** The stomach, mouth, or gullet of a voracious animal. **2.** An opening that gapes as if with voracious appetite <civilized peoples swept into the *maw* of Nazi Germany and destroyed>

**mawk·ish** (mô'kĭsh) *adj.* [< ME *mawke,* maggot, var. of *magot.*] **1.** Excessively and objectionably sentimental. **2.** Sickening or insipid in taste. **—mawk'ish·ly** *adv.* **—mawk'ish·ness** *n.*

**max·i** (măk'sē) *n.* [< MAXIMUM.] An ankle- or floor-length garment, as a skirt or coat.

**max·il·la** (măk-sĭl'ə) *n., pl.* **max·il·lae** (măk-sĭl'ē) *or* **max·il·las.** [Lat., jaw bone.] **1.** *Anat.* One of a pair of bones forming the upper jaw. **2.** *Zool.* Either of two laterally moving appendages behind the mandibles in insects and most other arthropods.

**max·il·lar·y** (măk'sə-lĕr'ē) *n., pl.* **-ies.** A jaw or jawbone. **—max'il·lar'y** *adj.*

**max·il·li·ped** (măk-sĭl'ə-pĕd') *n.* [MAXILL(A) + -PED.] *Zool.* One of the three pairs of crustacean head appendages located just behind the maxillae.

**max·il·lo·fa·cial** (măk-sĭl'ō-fā'shəl) *adj.* [MAXILL(A) + FACIAL.] Relating to or involving the maxilla and the face.

**max·im** (măk'sĭm) *n.* [ME *maxime* < OFr. < Med. Lat. *maxima* < Lat. *maximus,* greatest < *magnus,* great.] A concise formulation of a fundamental principle or rule of conduct : SAYING.

**max·i·ma** (măk'sə-mə) *n. var. pl. of* MAXIMUM.

**max·i·mal** (măk'sə-məl) *adj.* **1.** Of, relating to, or consisting of a maximum. **2.** Being the greatest or highest possible. **3.** *Math.* Designating the maximal of an ordered set. *—n. Math.* An element in an ordered set that is followed by no other. **—max'i·mal·ly** *adv.*

**max·i·mal·ist** (măk'sə-mə-lĭst) *n.* [Fr. *maximaliste,* prob. < E. *maximal.*] An advocate of direct revolutionary action to secure social and political gains.

**max·i·mize** (măk'sə-mīz') *vt.* **-mized, -miz·ing, -miz·es. 1.** To make as great as possible. **2.** To assign the highest possible importance to. **3.** *Math.* To find a maximum value of (a function). **—max'i·mi·za'tion** *n.* **—max'i·miz'er** *n.*

**max·i·mum** (măk'sə-məm) *n., pl.* **-mums** *or* **-ma** (-mə) [< Lat., neuter of *maximus,* greatest. —see MAXIM.] **1. a.** The greatest possible quantity, degree, or number. **b.** The time or period during which the highest point or degree is attained. **2.** An upper limit delineated by law or other authority. **3.** *Astron.* **a.** The moment when a variable star is most brilliant. **b.** The magnitude of the star at such a moment. **4.** *Math.* **a.** The value of a function that is not exceeded by neighboring values. **b.** The greatest value assumed by a function within some subset of its domain of definition. **c.** The largest number in a set. *—adj.* **1.** Having or being the greatest quantity or the highest degree that has been or can be attained. **2.** Of, relating to, or constituting a maximum <a *maximum* number in a series>

**ma·xixe** (mə-shēsh', -shē'shə) *n.* [Port. (Brazil).] A Brazilian dance resembling the two-step.

**max·well** (măks'wĕl', -wəl) *n.* [After James *Maxwell* (1831–1879).] A unit of magnetic flux in the centimeter-gram-second electromagnetic system, equal to the flux perpendicularly intersecting an area of one square centimeter in a region where the magnetic induction is one gauss.

**may** (mā) *aux.v. p.t.* **might** (mīt), *present* **may** *for sing. & pl.* [ME, to be able < OE *mæg,* 1st and 3rd person singular indicative of *magan,* to be strong, be able.] **1.** To be allowed or permitted to <You *may* borrow the car today.> **2.** —Used to indicate a certain measure of likelihood or possibility <It *may* snow later today.> **3.** To be obliged : MUST. —Used in statutes, deeds, and other legal documents. **4.** —Used to express a desire or fervent wish <*May* I do as well!> **5.** —Used to express contingency, purpose, or result, in clauses introduced by *that* or *so that* <explaining nuclear energy so that the average person *may* understand it>

**May** (mā) *n.* [ME < OFr. *Mai* < Lat. *Maius* < *Maia,* an Italic goddess.] **1.** The fifth month of the year according to the Gregorian calendar. —See table at CALENDAR. **2.** The springtime of life : YOUTH. **3.** Celebration of May Day. **4. may.** *Chiefly Brit.* The hawthorn's blossoms.

**ma·ya** (mä'yə) *n.* [Skt. *māyā.*] **1.** The origin of the world in Hinduism. **2.** The illusory appearance of the world.

**Ma·ya** (mä'yə) *n., pl.* **Maya** *or* **-yas.** [Sp.] **1.** A member of an Indian people of southern Mexico and Central America whose civilization reached its height around A.D. 1000. **2.** The Mayan language of the Maya. **—Ma'ya** *adj.*

**Ma·yan** (mä'yən) *adj.* Of or relating to the Mayas, their culture, their language, or the language group to which it belongs. *—n.* **1.** A Maya. **2.** A linguistic stock of Central America that includes Maya and Yucatec.

**May apple** *n.* **1.** A plant, *Podophyllum peltatum* of eastern North America, bearing a single nodding white flower, oval yellow fruit, the pulp of which is edible when ripe, and poisonous roots, leaves, and seeds. **2.** The fruit of the May apple.

**may·be** (mā'bē) *adv.* Perhaps : possibly.

**May beetle** *n.* The June beetle.

**may·day** (mā'dā') *n.* [Alteration of Fr. *m'aidez,* Help me!] An international radio-telephone signal word used by aircraft and ships in distress.

**May Day** *n.* **1.** The first day of May, marked by the celebration of spring. **2.** May 1, regarded in a number of places as an international holiday to honor labor organizations.

**may·est** (mā'ĭst) *or* **mayst** (māst) *aux.v. Archaic. 2nd person sing. present tense of* MAY¹.

**may·flow·er** (mā'flou'ər) *n.* **1.** Any of a wide variety of plants that blossom in May. **2.** The trailing arbutus.

**may·fly** (mā'flī') *n.* Any of various fragile, winged insects of the order Ephemeroptera that develop from aquatic nymphs and live in the adult stage for only a few hours.

**may·hap** (mā'hăp', mā-hăp') *adv.* [< the phrase *it may hap.*] Perhaps : perchance.

**may·hem** (mā'hĕm', mā'əm) *n.* [ME < AN *mahem* < OFr. *mahaignier*, to maim.] **1.** *Law.* The willful maiming or crippling of a person. **2.** Infliction of violent injury on a person or thing : wanton destruction. **3.** Violent disorder : HAVOC.

**may·n't** (mā'ənt, mānt). May not.

**may·o** (mā'ō) *n. Slang.* Mayonnaise.

**may·on·naise** (mā'ə-nāz', mā'ə-nāz') *n.* [Fr.] A dressing made of beaten raw egg yolk, oil, lemon juice or vinegar, and seasonings.

**may·or** (mā'ər, mâr) *n.* [ME *maire* < OFr. < Med. Lat. *major* < Lat., greater. —see MAJOR.] The chief magistrate of a city, town, borough, or municipal corporation. **—may'or·al** *adj.* **—may'or·ship'** *n.*

**may·or·al·ty** (mā'ər-əl-tē, mâr'əl-) *n., pl.* **-ties.** [ME *mairalte* < AN < OFr. *maire*, mayor. —see MAYOR.] **1.** The office of a mayor. **2.** A mayor's term of office.

**may·or·ess** (mā'ər-ĭs, mâr'ĭs) *n.* **1.** A woman who is the chief magistrate of a city, town, borough, or municipal corporation. **2.** The wife of a mayor.

**May·pole** *also* **may·pole** (mā'pōl') *n.* A pole hung with streamers that May Day merrymakers hold while dancing.

**may·pop** (mā'pŏp') *n.* [Alteration of *maycock* < Powhatan *mahcawq.*] **1.** A vine, *Passiflora incarnata* of the southeastern United States, bearing purple and white flowers and edible yellow fruit. **2.** The fruit of the maypop.

**mayst** (māst) *aux.v. var. of* MAYEST.

**may tree** *n. Chiefly Brit.* The hawthorn.

**may·weed** (mā'wēd') *n.* A widespread weed, *Anthemis cotula*, with rank-smelling leaves and white flowers.

**May wine** *n.* [Transl. of G. *Maiwein.*] **1.** A still white wine flavored with woodruff and often containing slices of orange or pineapple. **2.** A punch of champagne, claret, and Moselle or Rhine wine, flavored with woodruff.

**ma·zae·di·um** (mə-zē'dē-əm) *n., pl.* **-di·a** (-dē-ə) [NLat. : Gk. *maza*, lump + Lat. *aedes*, house.] A fruiting body of some lichens in which the spores lie freely in a powdery mass enclosed in a peridium.

**ma·zal tov** (mä'zəl tôf') *interj. var. of* MAZEL TOV.

**Maz·da·ism** *also* **Maz·de·ism** (măz'də-ĭz'əm) *n.* [< Avestan *mazda*, the good principle.] Zoroastrianism.

**†maze** (māz) *n.* [ME *mase*, maze, confusion < *masen*, to confuse < OE *āmasian*, to confound.] **1. a.** An intricate, usu. confusing, network of walled or hedged pathways : LABYRINTH. **b.** A physical situation in which it is easy to get lost. **c.** Any of various networks of pathways, some blind and some leading to a goal, used experimentally for investigation of learning in animals. **2.** A graphic puzzle whose solution is an uninterrupted path through an intricate pattern of line segments from a starting point to a goal. **3.** Something composed of many confused or conflicting elements : TANGLE <a *maze* of bureaucratic red tape> —*vt.* **mazed, maz·ing, maz·es. 1.** *Chiefly Regional.* To stupefy : daze. **2.** To bewilder. **—maz'y** *adj.*

**ma·zel tov** *also* **ma·zal tov** (mä'zəl tôf') *interj.* [Heb. *mazāl ṭob* : *mazāl*, luck + *ṭob*, good.] Congratulations.

**ma·zer** (mā'zər) *n.* [ME < OFr. *mazre*, of Germanic orig.] A large drinking bowl or goblet made of metal or hard wood.

**ma·zour·ka** (mə-zûr'kə, -zoor'-) *n. var. of* MAZURKA.

**ma·zu·ma** (mə-zoo'mə) *n.* [Yiddish *mazumen* < Heb. *mzumān*, fixed.] *Slang.* Money : cash.

**ma·zur·ka** *also* **ma·zour·ka** (mə-zûr'kə, -zoor'-) *n.* [Pol., accusative of *mazurek*, dim. of *mazur*, someone from Mazovia, a province of Poland.] **1.** A lively Polish dance similar to the polka, frequently adopted as a ballet form. **2.** Music for a mazurka, written in 3/4 or 3/8 time with a heavily accented second beat.

**maz·zard** (măz'ərd) *n.* [Orig. unknown.] A wild sweet cherry, *Prunus avium*, often used as grafting stock.

**mbi·ra** (ĕm-bîr'ə, əm-) *n.* [Of Bantu orig.] A thumb piano.

**MC** (ĕm'sē') *n.* A master of ceremonies.

**Mc·Car·thy·ism** (mə-kär'thē-ĭz'əm) *n.* [After Joseph R. *McCarthy* (1909–1957).] **1.** The political practice of publicizing accusations of subversion or disloyalty with insufficient regard to evidence. **2.** Use of unfair investigative and accusatory methods in order to suppress opposition. **—Mc·Car'thy·ist** *n.*

**Mc·Coy** (mə-koi') *n.* [After Kid *McCoy* (Norman Selby, 1873–1940), American boxer.] *Slang.* The authentic thing or quality <the real *McCoy*>

**Mc·In·tosh** (măk'ĭn-tŏsh') *n.* [After John *McIntosh* (fl. 1796), its first cultivator.] A red eating apple, grown commercially in the northern United States.

**Md** *symbol for* MENDELEVIUM.

**M-day** (ĕm'dā') *n.* The day on which national mobilization for war is ordered.

**me** (mē) *pron.* [ME < OE *mē.*] *The objective case of* I. —Used: **1.** As the direct object of a verb <They helped *me.*> **2.** As the indirect object of a verb <They offered *me* money.> **3.** As the object of a preposition <This memorandum is addressed to *me.*>

**me·a cul·pa** (mā'ə kŭl'pə) *n.* [Lat., through my fault.] Acknowledgment of a personal fault or error.

**mead[1]** (mēd) *n.* [ME < OE *medu.*] An alcoholic beverage made from fermented honey and water.

**mead[2]** (mēd) *n.* [ME *meade* < OE *mǣd.*] *Archaic.* A meadow.

**mead·ow** (mĕd'ō) *n.* [ME *medoue* < OE *mǣdwe* < *mǣd*, meadow.] A tract of grassland. **—mead'ow·y** *adj.*

▲ word history: *Mead*[2] and *meadow*, both meaning "grassland," are descended from the same Old English word, *mǣd*. The form *mead* comes from the nominative singular form of the Old English word, which was *mǣd*. The form *meadow* comes from the inflected forms of *mǣd*, all of which contained a *w* in the suffix. *Mǣdwe* was the form of the genitive, dative, and accusative singular. A new nominative, *mǣdwe*, later *meadow*, arose to fit the pattern of the inflected forms.

**meadow beauty** *n.* A plant of the genus *Rhexia* of eastern North America, growing in wet ground and bearing purple flowers.

**meadow fescue** *n.* A grass, *Festuca eliator*, used for hay.

**meadow hen** *n.* A bird of the family Rallidae, esp. a rail or coot.

**mead·ow·land** (mĕd'ō-lănd') *n.* A meadow.

**mead·ow·lark** (mĕd'ō-lärk') *n.* A bird of the genus *Sturnella* of North America, two species of which, *S. magna*, the eastern meadowlark, and *S. neglecta*, the western meadowlark, are noted for their song.

**meadow mouse** *n.* The field mouse.

**meadow mushroom** *n.* An edible mushroom, *Agaricus campestris*, that flourishes in moist soil and is cultivated for human consumption.

**meadow nematode** *n.* Any of various nematodes of the genus *Pratylenchus*, parasitic on plant roots.

**meadow rue** *n.* A plant of the genus *Thalictrum*, bearing small white, yellowish, or purplish flower clusters.

**meadow saffron** *n.* The autumn crocus.

**mead·ow·sweet** (mĕd'ō-swēt') *n.* A plant of the genus *Spiraea*, esp. *S. alba* or *S. latifolia* of eastern North America, bearing pyramidal flower clusters.

**mea·ger** *also* **mea·gre** (mē'gər) *adj.* [ME *megre* < OFr. < Lat. *macer.*] **1.** Having little flesh : LEAN. **2.** Conspicuously deficient in quantity, fullness, or extent <*meager* income> **3.** Deficient in richness, fertility, or vigor <the *meager* crops produced on worn-out land> **—mea'ger·ly** *adv.* **—mea'ger·ness** *n.*

☆ **syns:** MEAGER, EXIGUOUS, MEASLY, POOR, PUNY, SCANT, SCANTY, SKIMPY, SPARE, SPARSE, STINGY *adj. core meaning* : conspicuously deficient in quantity, fullness, or extent <eked out a *meager* existence><a *meager* increase in salary> ***ant:*** ample

**meal[1]** (mēl) *n.* [ME *meale* < OE *melu.*] **1.** The edible seed or other edible part of a coarsely ground grain. **2.** A granular substance made by grinding.

**meal[2]** (mēl) *n.* [ME < OE *mǣl.*] **1.** The food served and eaten in one sitting. **2.** A customary time or occasion of eating food.

▲ word history: The usual current sense of *meal*[2], "food eaten at one sitting," has existed since the 13th century but it was actually an extension of the Old English sense, "fixed time for eating." *Meal*[2], descended from an Indo-European root that meant basically "measure" (a meaning recorded for *meal* in medieval times), is related to Latin *metiri*, "to measure," and Greek *metron*, "measure, poetic meter."

**meal·ie** (mē'lē) *n.* [Afr. *milie* < Port. *milho*, millet < Lat. *milium.*] *So. Afr.* **1.** An ear of corn. **2. mealies.** Corn : maize.

**meal ticket** *n.* **1.** A card or ticket entitling the holder to a meal or meals. **2.** *Slang.* One relied on as a source of financial support.

**meal·time** (mēl'tīm') *n.* The usual time for eating a meal.

**meal·worm** (mēl'wûrm') *n.* The larva of any of several beetles of the genus *Tenebrio* that infest flour and other grain products and are raised for bird feed.

**meal·y** (mē'lē) *adj.* **-i·er, -i·est. 1.** Resembling meal in texture or consistency : GRANULAR <*mealy* potatoes> **2. a.** Made of or containing meal. **b.** Sprinkled or covered with meal or a similar granular substance. **3.** Flecked with spots : MOTTLED. **4.** Lacking healthy coloring : PALE. **5.** Mealy-mouthed. **—meal'i·ness** *n.*

**meal·y·bug** (mē'lē-bŭg') *n.* [So called because it is covered with a white powdery substance.] Any of various insects of the genus *Pseudococcus*, some of which are harmful to plants, esp. citrus trees.

**meal·y-mouthed** (mē'lē-mouthd', -moutht') *adj.* Unwilling to state facts or opinions simply and directly : EVASIVE.

**mean[1]** (mēn) *v.* **meant** (mĕnt), **mean·ing, means.** [ME *menen* < OE *mǣnan*, to tell of.] —*vt.* **1. a.** To be defined or described as : DENOTE <The word "cat" *means* a certain species of mammal.> **b.** To convey the same sense as <The Spanish word "ventana" *means* "window."> **c.** To act as a symbol of : SIGNIFY <In this poem, the rose *means* love.> **2.** To intend to indicate or convey <What do they *mean* by those gestures?> **3.** To have as a purpose or intention : INTEND <I didn't *mean* to make you angry.> **4.** To design or intend for a certain purpose or end <a truck *meant* for light hauling> **5.** To have as a consequence : BRING ABOUT <Freezing rain *means* cold.> **6.** To be attended by or associated with <A rainbow *means* the storm is over.> —*vi.* **1.** To be of a specified significance or importance : MATTER <Wealth *meant* little to them.> **2.** To have intentions of a specified kind <*means* well but lacks tact> **—mean business.** *Informal.* To be in earnest.

oo **boot** ou **out** th **thin** *th* **this** ŭ **cut** ûr **urge** y **young**
yoo ab**u**se zh vi**s**ion ə **a**bout, it**e**m, ed**i**ble, gall**o**p, circ**u**s

**mean²** (mēn) *adj.* **-er, -est.** [ME < OE *gemǣne,* common.] **1.** Low in quality or grade : INFERIOR. **2.** Low in social status. **3.** Poor or common in appearance : SHABBY <a *mean* cabin> **4.** Ignoble : base <a *mean* purpose> **5.** Low in amount or value : PALTRY. **6.** Miserly : stingy. **7. a.** Lacking elevating human qualities, as kindness and good will. **b.** Reluctant to oblige or accommodate. **c.** Displaying malice : MALICIOUS. **8.** *Informal.* Ill-tempered. **9.** *Slang.* **a.** Hard to cope with : DIFFICULT <a *mean* road to travel in winter> **b.** Hard to defeat <plays a *mean* game of tennis> **—mean'ly** *adv.*

**mean³** (mēn) *n.* [ME, middle < OFr. < Lat. *medianus* < *medius.*] **1.** The middle point between two extremes. **2.** The avoidance of extremes of behavior : MODERATION <"Every virtue, as we were taught in youth, is a *mean* between two extremes" —Max Beerbohm> **3.** *Math.* **a.** A number that represents a set of numbers in any of several ways determined by a rule involving all members of the set : AVERAGE. **b.** The arithmetic mean. **4.** *Logic.* The middle term in a syllogism. **5. means.** A course of action, method, or instrument by which an act can be accomplished or an end achieved <suggested the end justified the *means*> **6. means.** Wealth, as money or property. ***usage:*** In the sense of "financial resources," *means* takes a plural verb, as in *Our means were more than adequate.* In the sense of "a way to an end," it may take a singular or plural verb; the choice of a modifier such as *any* or *all* gen. determines the number of the verb, as in: *Every means available was considered; There are several means at our disposal.* *—adj.* **1.** Occupying a middle position between two extremes. **2.** Intermediate in extent, size, quality, degree, or time : MEDIUM. **—by all means.** Without fail : CERTAINLY. **—by any means.** In any way possible : in any case <not *by any means* an easy job> **—by means of.** With the use of : owing to <succeeded *by means of* patience> **—by no means.** In no sense : certainly not.

**mean calorie** *n.* CALORIE 2.

**me·an·der** (mē-ăn'dər) *vi.* **-dered, -der·ing, -ders.** [< Lat. *maeander,* circuitous windings < Gk. *maiandros,* after *Maeander,* a river in Turkey noted for its winding course.] **1.** To follow a turning and winding course <streams *meandering* through fields> **2.** To wander aimlessly without fixed direction *—n.* **1. meanders.** Windings or turns, as of a stream or path. **2.** *often* **meanders.** A circuitous excursion or journey : RAMBLE. **3.** The Greek artistic and architectural fret or key pattern. **—me·an'der·er** *n.* **—me·an'der·ing·ly** *adv.* **—me·an'drous** (-drəs) *adj.*

**mean deviation** *n.* The arithmetic mean of the absolute values of deviations from the arithmetic mean, or from the median, in a statistical distribution.

**mean·ing** (mē'nĭng) *n.* **1.** Something signified by a word. **2.** Something one wishes to convey, esp. by language. **3.** Something interpreted to be the intent, goal, or end. **4.** Something regarded as the inner significance <"But who can comprehend the *meaning* of the voice of the city?" —O. Henry> **5.** Functional value : EFFICACY <ceremonies now empty of all *meaning*> *—adj.* **1.** MEANINGFUL 2. **2.** Intentioned or disposed in a specified manner <a well-*meaning* person>

☆ ***syns:*** MEANING, ACCEPTATION, IMPORT, INTENT, MESSAGE, PURPORT, SENSE, SIGNIFICANCE, SIGNIFICATION, VALUE *n. core meaning* : what is signified by a term <synonyms—words with the same *meaning*>

**mean·ing·ful** (mē'nĭng-fəl) *adj.* **1.** Having meaning, function, or purpose : SIGNIFICANT. **2.** Full of meaning : EXPRESSIVE <a *meaningful* glance> **—mean'ing·ful·ly** *adv.* **—mean'ing·ful·ness** *n.*

**mean·ing·less** (mē'nĭng-lĭs) *adj.* Having no significance or meaning : SENSELESS. **—mean'ing·less·ly** *adv.* **—mean'ing·less·ness** *n.*

**mean·ly** (mēn'lē) *adv.* In a mean, poor, or base manner.

**mean·ness** (mēn'nĭs) *n.* **1.** The state of being inferior in quality, value, or character : COMMONNESS. **2.** Stinginess : selfishness. **3.** A malicious or spiteful act.

**mean solar day** *n.* The period of time between two successive transits of the mean sun : the standard for the 24-hour day, measured from midnight to midnight.

**mean square** *n.* The arithmetic mean of the squares of a set of numbers.

**mean sun** *n.* A hypothetical sun defined as moving at a uniform rate along the celestial equator so that it completes its orbit in the same period as the apparent sun, used to compute the mean solar day.

**meant** (mĕnt) *v. p.t. & p.p. of* MEAN¹.

**mean·time** (mēn'tīm') *n.* The time between one occurrence and another : INTERVAL. *—adv.* During a period of intervening time : MEANWHILE <"*Meantime,* let wonder seem familiar" —Shakespeare> ***usage:*** *Meantime* serves chiefly as a noun: *In the meantime we played cards.* As an adverb, *meanwhile* is more common: *Meanwhile we played cards.*

**mean time** *n.* Time measured with reference to the mean sun, giving equal 24-hour days throughout the year.

**mean·while** (mēn'hwīl', -wīl') *n.* The intervening time. *—adv.* **1.** During or in the intervening time. **2.** At the same time <The jury is out; *meanwhile,* we must wait.>

**mea·sles** (mē'zəlz) *n.* [ME *maseles,* pl. of *masel,* measles-spot, of MLG orig.] *(used with a sing. verb).* **1. a.** An acute contagious virus disease, usu. occurring in childhood and characterized by eruption of red spots. **b.** One of several diseases displaying similar but milder symptoms, esp. German measles. **2.** A disease of cattle and swine, caused by tapeworm larvae. **3.** A plant disease usu. caused by fungi, and producing minute spots on stems and leaves.

**mea·sly** (mēz'lē) *adj.* **-sli·er, -sli·est. 1.** Infected or spotted with measles : MEASLED. **2.** *Slang.* Odiously small : MEAGER.

**meas·ur·a·ble** (mĕzh'ər-ə-bəl) *adj.* **1.** Able to be measured. **2.** Of distinguished importance : SIGNIFICANT. **3.** Not so great as to escape all comparison or measure : MODERATE. **—meas'ur·a·bil'i·ty** *n.* **—meas'ur·a·bly** *adv.*

**meas·ure** (mĕzh'ər) *n.* [ME < OFr. *mesure* < Lat. *mensura* < *metiri,* to measure.] **1.** Dimensions, capacity, or quantity as determined by measuring. **2.** A reference sample or standard used for the quantitative comparison of properties. **3.** A unit specified by a scale, as an inch, or by variable conditions, as a day's march. **4.** A system of measurement, as the metric system. **5.** A device, as a marked tape or a graduated container, used for measuring. **6.** An act of measurement. **7.** A basis of comparison : CRITERION <"the final *measure* of the worth of a society" —Joseph Wood Krutch> **8.** The degree or extent of something. **9.** A fitting amount <a *measure* of appreciation> **10.** A limited degree or amount. **11.** Limit : bounds <wealth knowing no *measure*> **12.** Appropriate restraint : MODERATION <discipline in *measure*> **13.** *often* **measures.** An action taken as a means to an end : EXPEDIENT. **14.** A legislative bill or enactment. **15.** Poetic meter. **16.** *Mus.* The metric unit between two bars on the staff : BAR. *—v.* **-ured, -ur·ing, -ures.** *—vt.* **1.** To determine the dimensions, quantity, or capacity of. **2.** To mark, establish, or lay out dimensions for by measuring. **3.** To estimate by comparison or evaluation <"I gave them an account . . . of the situation as far as I could *measure* it" —Winston Churchill> **4.** To bring into opposition <*measured* our power with the adversary> **5.** To mark off, usu. with reference to a given unit of measurement <*measure* out a cup of sugar> **6.** To serve as a measure of <An altimeter *measures* altitude.> **7.** To distribute or allot as if by measuring : METE <The judge *measured* out severe sentences.> **8.** To choose or consider with care : WEIGH <*measure* one's words> **9.** *Archaic.* To travel over <"We must *measure* much ground today" —Shakespeare> *—vi.* **1.** To have a measurement of <The kitchen *measures* 10 by 12 feet.> **2.** To allow of measurement. **—beyond measure. 1.** In excess. **2.** Without limit. **—for good measure.** In addition to the amount required. **—in a (**or **some) measure.** To a degree <The proposal was *in some measure* faulty.> **—measure up. 1.** To be the equal of. **2.** To have the necessary qualifications. **—meas'ur·er** *n.*

**meas·ured** (mĕzh'ərd) *adj.* **1.** Determined by measurement <a precise, *measured* distance> **2.** Regular, as in number and rhythm <"A clock struck slowly in the house with a *measured,* solemn chime" —Thomas Wolfe> **3.** Careful : restrained. **4.** Calculated <with *measured* sarcasm> **5.** Slow and stately. **6.** Metrical. **7.** *Mus.* Mensural. **8.** Limited <a *measured* capacity for creativity> **—meas'ured·ly** *adv.* **—meas'ured·ness** *n.*

**meas·ure·less** (mĕzh'ər-lĭs) *adj.* Without limits : INFINITE. **—meas'ure·less·ly** *adv.*

**meas·ure·ment** (mĕzh'ər-mənt) *n.* **1.** The act of measuring or process of being measured. **2.** A system of measuring. **3.** The dimension, capacity, or quantity determined by measuring.

**measuring worm** *n.* A geometrid caterpillar that moves in alternate contractions and expansions suggestive of measuring.

**meat** (mēt) *n.* [ME *mete* < OE, food.] **1.** The edible flesh of mammals. **2.** An edible, fleshy, inner part <lobster *meat*> **3.** The edible portions of eggs, nuts, or fruits. **4.** The essence or central part <the *meat* of the essay> **5.** *Slang.* Something one enjoys or excels in : FORTE. **6.** Something eaten for nourishment : FOOD.

▲ word history: The word *meat* is an example of a word whose signification has become narrower in the course of its development. In Old English times the word denoted food of any kind, but chiefly solid food in contrast to liquids. This is the sense of *meat* in such compounds as *sweetmeat,* "a piece of candy," and *nutmeat,* "the edible part of a nut." In later medieval times *meat* came to signify animal flesh in contrast to fish, and at times in contrast to poultry as well. In very recent times *meat* has occasionally been restricted to denoting a particular kind of animal flesh, such as pork.

**meat·ball** (mēt'bôl') *n.* **1.** A small ball of cooked, often spiced ground meat. **2.** *Slang.* A stupid, awkward, or dull person.

**meat·less** (mēt'lĭs) *adj.* **1.** Lacking meat or food. **2.** Being or relating to a time when meat is not to be eaten, as for religious reasons.

**meat loaf** *n.* A usu. loaf-shaped baked dish of seasoned ground meat.

**me·a·tus** (mē-ā'təs) *n., pl.* **-tus·es** or **meatus.** [Lat., passage < *meare,* to pass.] A body canal or passage, as the opening of the ear or the urethral canal.

**meat·y** (mē'tē) *adj.* **-i·er, -i·est. 1. a.** Of or relating to meat. **b.** Having the smell or flavor of meat. **c.** Full of or containing meat. **2.** Heavily fleshed. **3.** Amply thought-provoking <a *meaty* subject for debate> **—meat'i·ness** *n.*

**mec·a·myl·a·mine** (mĕk'ə-mĭl'ə-mēn') *n.* [Orig. a trademark.] A drug, $C_{11}H_{21}N \cdot HCl$, administered orally to reduce highly elevated blood pressure.
**mec·ca** (mĕk'ə) *n.* [After *Mecca*, Saudi Arabia, from its being a place of pilgrimage.] **1. a.** A place held to be the center of an activity or interest. **b.** A goal to which adherents of a religious faith or practice fervently aspire. **2.** A place visited by many people <a tourist *mecca*>
**mechan-** *pref. var. of* MECHANO-.
**me·chan·ic** (mĭ-kăn'ĭk) *n.* [< ME, mechanical < OFr. *mecanique* < Lat. *mechanicus* < Gk. *mēkhanikos* < *mēkhanē*, machine < *mēkhos*, means.] A worker skilled in using, making, or repairing machines and tools. **—me·chan'ic** *adj.*
**me·chan·i·cal** (mĭ-kăn'ĭ-kəl) *adj.* [ME < *mechanic*, mechanical. —see MECHANIC.] **1.** Of or relating to machines or tools. **2.** Operated or produced by a machine. **3.** Of, relating to, or governed by mechanics. **4.** Performing or acting like a machine : AUTOMATIC <an insincere, *mechanical* greeting> **5.** Relating to, produced by, or dominated by physical forces. **6.** Interpreting and explaining the phenomena of the universe by referring to causally determined material forces : MECHANISTIC. **7.** Of or relating to manual labor, its tools, and its skills. *—n.* A layout of type proofs, artwork, or both, exactly positioned and prepared for making a printing plate, as an offset. **—me·chan'i·cal·ly** *adv.* **—me·chan'i·cal·ness** *n.*
**mechanical advantage** *n.* The ratio of the output force of a machine to the input force.
**mechanical drawing** *n.* **1.** Drafting. **2.** A drawing, as an architect's plans, that enables measurements to be interpreted.
**mechanical engineering** *n.* The branch of engineering that encompasses the generation and application of heat and mechanical power and the production, design, and use of machines and tools. **—mechanical engineer** *n.*
**mech·a·ni·cian** (mĕk'ə-nĭsh'ən) *n.* One who uses, makes, or repairs machines and tools.
**me·chan·ics** (mĭ-kăn'ĭks) *n. (sing. or pl. in number).* **1.** Analysis of the action of forces on matter or material systems. **2.** Design, operation, construction, and application of machinery or mechanical structures. **3.** The technical and functional aspects of an activity <the *mechanics* of baseball>
**mech·a·nism** (mĕk'ə-nĭz'əm) *n.* [LLat. *mechanisma* < Gk. *mēkhanē*, machine. —see MECHANIC.] **1. a.** A mechanical device : MACHINE. **b.** Arrangement of machine parts. **2.** A system of parts that interact or operate like those of a machine <the *mechanism* of the brain> **3.** A physical instrument or process by which something is done or originates <"The *mechanism* of oral learning is largely that of continuous repetition" —T.G.E. Powell> **4.** *Psychol.* **a.** Automatic and consistent response of an organism to various stimuli. **b.** A habitual manner of acting to achieve an end. **5.** *Psychoanal.* A usu. unconscious mental and emotional pattern that dominates behavior. **6.** *Chem.* The sequence of steps in a chemical reaction. **7.** *Philos.* The doctrine that all natural phenomena are explainable by material causes and mechanical principles.
**mech·a·nist** (mĕk'ə-nĭst) *n.* **1.** One who adheres to the philosophical doctrine of mechanism. **2.** A mechanician.
**mech·a·nis·tic** (mĕk'ə-nĭs'tĭk) *adj.* **1.** Mechanically determined. **2.** *Philos.* Of or relating to mechanism, esp. tending to explain phenomena only by reference to physical or biological causes. **3.** Mechanical. **—mech'a·nis'ti·cal·ly** *adv.*
**mech·a·nize** (mĕk'ə-nīz') *vt.* **-nized, -niz·ing, -niz·es. 1.** To equip with machinery. **2.** To equip (a military unit) with motor vehicles, as tanks and trucks. **3.** To make automatic or unspontaneous. **4.** To produce by or as if by machines. **—mech'a·ni·za'tion** *n.* **—mech'a·niz'er** *n.*
**mechano-** *or* **mechan-** *pref.* [ME *mechan-* < Lat. < Gk. *mēkhan-* < *mēkhanē*, machine.] **1.** Machine : machinery <*mechanize*> **2.** Mechanical <*mechano*therapy>
**mech·a·no·chem·i·cal coupling** (mĕk'ə-nō-kĕm'ĭ-kəl) *n.* Reversible conversion of chemical energy into mechanical work.
**mech·a·no·re·cep·tor** (mĕk'ə-nō-rĭ-sĕp'tər) *n.* A receptor that responds to mechanical stimuli, as tension and pressure. **—mech'a·no·re·cep'tion** *n.* **—mech'a·no·re·cep'tive** *adj.*
**mech·a·no·ther·a·py** (mĕk'ə-nō-thĕr'ə-pē) *n.* Medical treatment by mechanical methods, as massage. **—mech'an·o·ther'a·pist** *n.*
**Mech·lin** (mĕk'lĭn) *n.* [After *Mechlin*, Belgium.] Lace in which the pattern details are defined by a flat thread.
**me·co·ni·um** (mĭ-kō'nē-əm) *n.* [Lat. < Gk. *mēkōneion* < *mēkōn*, poppy.] Excrement in the fetal intestinal tract discharged at birth.
**me·cop·ter·an** (mĭ-kŏp'tər-ən) *n.* [< NLat. *Mecoptera*, order name : Gk. *mēkos*, length + Gk. *pteron*, wing.] One of various carnivorous insects of the order Mecoptera, distinguished by an elongated head similar to a beak with chewing mouthparts at the tip.
**me·da·ka** (mĭ-dä'kə) *n.* [J., killifish.] **1.** The Japanese rice fish, *Oryzias latipes*, used in biological research. **2.** A fish of the Asiatic and Indo-Malayan genus *Oryzias*.

**med·al** (mĕd'l) *n.* [Fr. *médaille* < OItal. *medaglia*, coin worth half a denarius, medal < VLat. **medalis* < LLat. *medialis*, middle < Lat. *medius*.] **1.** A flat piece of metal stamped with a commemorative design or inscription, often presented as an award. **2.** A piece of metal stamped with a religious symbol, used as an object of worship or commemoration.
**Medal for Merit** *n.* A decoration awarded by the United States to civilians for outstanding services in war or peace.
**med·al·ist** (mĕd'l-ĭst) *n.* **1.** One who makes, designs, or collects medals. **2.** A recipient of a medal.
**me·dal·lion** (mĭ-dăl'yən) *n.* [Fr. *médaillon* < Ital. *medaglione*, aug. of *medaglia*, medal < OItal. —see MEDAL.] **1.** A large medal. **2.** One of various large ancient Greek coins. **3.** Something like a large medal.
**med·al·list** (mĕd'l-ĭst) *n. Chiefly Brit. var. of* MEDALIST.
**Medal of Freedom** *n.* A decoration awarded by the United States to civilians for outstanding achievement in various fields of endeavor.
**Medal of Honor** *n.* The highest U.S. military decoration, awarded by Congress to military personnel for gallantry and bravery beyond the call of duty in action against an enemy.
**med·dle** (mĕd'l) *vi.* **-dled, -dling, -dles.** [ME *medlen* < OFr. *medler*, var. of *mesler* < VLat. **misculare*, freq. of Lat. *miscēre*, to mix.] **1.** To intrude in other people's business or affairs : INTERFERE. **2.** To handle something ignorantly or idly : TAMPER. **—med'dler** (mĕd'lər, mĕd'l-ər) *n.*
**med·dle·some** (mĕd'l-səm) *adj.* Inclined to meddle. **—med'dle·some·ly** *adv.* **—med'dle·some·ness** *n.*
**Me·de·a** (mĭ-dē'ə) *n.* [Lat. < Gk. *Mēdeia*.] *Gk. Myth.* A princess and sorceress of Colchis who helped Jason obtain the Golden Fleece.
**Med·fly** *also* **med·fly** (mĕd'flī') *n.* The Mediterranean fruit fly.
**me·di·a¹** (mē'dē-ə) *n. var. pl. of* MEDIUM.
**me·di·a²** (mē'dē-ə) *n.* MEDIAL 1.
**me·di·a·cy** (mē'dē-ə-sē) *n.* **1.** The quality or state of being mediate. **2.** Mediation.
**me·di·ae·val** (mē'dē-ē'vəl, mĕd-ē'-) *adj. var. of* MEDIEVAL.
**me·di·ae·val·ism** (mē'dē-ē'və-lĭz'əm, mĕd-ē'-) *n. var. of* MEDIEVALISM.
**me·di·ae·val·ist** (mē'dē-ē'və-lĭst, mĕd-ē'-) *n. var. of* MEDIEVALIST.
**media event** *n.* An occasion that is orchestrated and publicized so as to achieve wide coverage by the print and electronic media <turned an otherwise dull press conference into a *media event*>
**me·di·al** (mē'dē-əl) *adj.* [LLat. *medialis* < Lat. *medius*, middle.] **1.** Relating to, situated in, or extending toward the middle : MEDIAN. **2.** Being a sound, syllable, or letter occurring between the initial and final positions in a word or morpheme. **3.** Being or relating to a mathematical mean or average. **4.** Ordinary : average. *—n.* **1.** A voiced stop, as *b, d,* or *g.* **2.** An element, as a sound, letter, or form of a letter, used in the middle of a word. **—me'di·al·ly** *adv.*
**me·di·an** (mē'dē-ən) *adj.* [Lat. *medianus* < *medius*, middle.] **1.** Relating to, situated in, or directed toward the middle : MEDIAL. **2.** *Anat.* & *Zool.* Of, relating to, or lying in the plane that divides a bilaterally symmetric animal into right and left halves : MESIAL. **3.** *Statistics.* Pertaining to or constituting the middle value in a distribution. *—n.* **1.** A median point, line, plane, or part. **2.** *Statistics.* The middle value in a distribution, above and below which lie an equal number of values. **3.** *Math.* **a.** A line that joins a vertex of a triangle to the midpoint of the opposite side. **b.** The line that joins the midpoints of the nonparallel sides of a trapezoid. **—me'di·an·ly** *adv.*
**median plane** *n.* A plane dividing a bilaterally symmetric animal into right and left halves.
**median point** *n.* The intersection of the medians of a triangle.
**median strip** *n.* The dividing area, either landscaped or paved, between opposing highway traffic lanes.
**me·di·ant** (mē'dē-ənt) *n. Mus.* The third tone in a diatonic scale between the tonic and the dominant and related harmonically to them.
**me·di·as·ti·na** (mē'dē-ə-stī'nə) *n. pl. of* MEDIASTINUM.
**me·di·as·ti·ni·tis** (mē'dē-ăs'tə-nī'tĭs) *n.* Inflammation of the mediastinum.
**me·di·as·ti·num** (mē'dē-ə-stī'nəm) *n., pl.* **-na** (-nə) [NLat. < Med. Lat. *mediastinus*, medial < Lat., drudge < *medius*, middle.] The septum that divides the pleural sacs in mammals, containing all the thoracic viscera except the lungs. **—me'di·as·ti'nal** *adj.*
**me·di·ate** (mē'dē-āt') *v.* **-at·ed, -at·ing, -ates.** [LLat. *mediare, mediat-*, to be in the middle < Lat. *medius*, middle.] *—vt.* **1.** To settle or resolve (differences) by acting as an intermediary between two or more opposing parties. **2.** To bring about (e.g., a settlement) by action as an intermediary. **3.** To transmit or convey as an intermediary agent or mechanism. *—vi.* **1.** To intervene between two or more disputing parties in order to effect a settlement, agreement, or compromise. **2.** To reconcile differences. *—adj.* (mē'dē-ĭt). Acting through, involving, or dependent on an intervening agency. **—me'di·ate·ly** (-ĭt-lē) *adv.*
**me·di·a·tion** (mē'dē-ā'shən) *n.* **1.** The act of mediating or state of being mediated. **2.** *Law.* An attempt to effect a peaceful settlement or compromise between disputing nations through the benevolent intervention of a neutral power. **—me'di·a'tive, me'di·a·to'ry** *adj.*

**me·di·a·tize** (mē′dē-ə-tīz′) *vt.* **-tized, -tiz·ing, -tiz·es.** [G. *mediatisieren* < *mediat*, mediate < LLat. *mediare*, to be in the middle. —see MEDIATE.] To annex (a lesser state) to a greater state as a means of permitting the ruler of the lesser power to retain the title and part of the former authority of his or her office. **—me′di·a·ti·za′tion** *n.*

**me·di·a·tor** (mē′dē-ā′tər) *n.* One that mediates, esp. one who serves as an intermediary to reconcile differences.

**med·ic**[1] *or* **med·ick** (mĕd′ĭk) *n.* [ME *medike* < Lat. *Medica* < Gk. *Mēdikē* < *Mēdikos*, of Media, an ancient country of southwestern Asia < *Mēdos*, an inhabitant of Media < OPers. *mada.*] An Old World plant of the genus *Medicago*, with small, usu. yellow flower clusters and compound leaves having three leaflets.

**med·ic**[2] (mĕd′ĭk) *n.* [Lat. *medicus.* —see MEDICAL.] *Informal.* **1.** A surgeon or physician. **2.** A medical student or intern. **3.** A military medical corpsman.

**med·i·ca·ble** (mĕd′ĭ-kə-bəl) *adj.* Potentially responsive to treatment with medicine : CURABLE.

**Med·i·caid** *also* **med·i·caid** (mĕd′ĭ-kād′) *n.* [MEDIC(AL) + AID.] A program jointly funded by the states and the federal government, that provides medical aid for people who are unable to finance their own medical expenses.

**med·i·cal** (mĕd′ĭ-kəl) *adj.* [Fr. *médical* < LLat. *medicalis* < Lat. *medicus*, physician < *medēri*, to heal.] **1.** Of or relating to the study or practice of medicine. **2.** Requiring medical as distinct from surgical treatment.

**medical examiner** *n.* **1.** A physician officially authorized by a governmental unit, as a city or county, to ascertain causes of deaths, esp. those occurring under unnatural circumstances. **2.** A physician who examines applicants for life insurance.

**medical law** *n.* A branch of law concerned with the legal regulation of medicine and medical practice.

**me·dic·a·ment** (mĭ-dĭk′ə-mənt, mĕd′ĭ-kə-) *n.* [Lat. *medicamentum* < *medicari*, to medicate.] An agent that promotes recovery from injury or ailment : MEDICINE.

**Med·i·care** *also* **med·i·care** (mĕd′ĭ-kâr′) *n.* [MEDI(CAL) + CARE.] A program under the Social Security Administration that provides medical care for the elderly.

**med·i·cate** (mĕd′ĭ-kāt′) *vt.* **-cat·ed, -cat·ing, -cates.** [Lat. *medicari, medicat-* < *medicus*, doctor < *medēri*, to heal.] **1.** To treat medicinally. **2.** To permeate or tincture with a medicinal substance. **—med′i·ca′tive** *adj.*

**med·i·ca·tion** (mĕd′ə-kā′shən) *n.* **1.** A medicine. **2.** The act or process of being medicated. **3.** Administration of medicine to a patient.

**me·dic·i·nal** (mə-dĭs′ə-nəl) *adj.* Relating to or having the properties of medicine. **—me·dic′i·nal·ly** *adv.*

**med·i·cine** (mĕd′ĭ-sĭn) *n.* [ME < Lat. *medicina* < *medicus*, physician. —see MEDICAL.] **1. a.** The science of diagnosing, treating, or preventing disease or damage to the body or mind. **b.** The branch of this science encompassing nonsurgical management of disease or physical and mental impairment by drugs, diet, and exercise. **2.** The practice of medicine. **3.** An agent, as a drug, for treating injury or disease. **4.** Something believed by North American Indians to control natural or supernatural powers and serve as a preventive or remedy.

**medicine ball** *n.* A large heavy ball used for exercising.

**medicine dance** *n.* A ritual dance performed by some Plains Indians to obtain supernatural assistance.

**medicine lodge** *n.* A large wooden structure used by some North American Indian tribes for ritualistic ceremonies.

**medicine man** *n.* **1.** One believed among preliterate peoples to hold supernatural powers, as for healing or invoking spirits : SHAMAN. **2.** A hawker of potions and brews among the audience in a medicine show.

**medicine show** *n.* A traveling show, popular esp. in 19th-cent. America, that offered varied entertainment, between the acts of which nostrums were peddled.

**med·ick** (mĕd′ĭk) *n. var. of* MEDIC[1].

**med·i·co** (mĕd′ĭ-kō′) *n., pl.* **-cos.** [Ital. and Sp., both < Lat. *medicus.* —see MEDICAL.] *Informal.* **1.** A physician. **2.** A medical student.

**med·i·co·le·gal** (mĕd′ĭ-kō-lē′gəl) *adj.* [MEDIC(AL) + LEGAL.] Of or relating to both medicine and law.

**me·di·e·val** *also* **me·di·ae·val** (mē′dē-ē′vəl, mĕd′ē-) *adj.* [Lat. *medius*, middle + Lat. *aevum*, age + -AL.] Relating or belonging to the Middle Ages. **—me′di·e′val·ly** *adv.*

**Medieval Greek** *n.* Greek as used from about 800 to 1500.

**me·di·e·val·ism** *also* **me·di·ae·val·ism** (mē′dē-ē′və-lĭz′əm, mĕd′ē-) *n.* **1.** The body of customs, beliefs, or practices of the Middle Ages. **2.** Devotion to or acceptance of the ideas of the Middle Ages. **3.** Scholarly study of the Middle Ages.

**me·di·e·val·ist** *also* **me·di·ae·val·ist** (mē′dē-ē′və-lĭst, mĕd-ē′-) *n.* **1.** A specialist in medieval studies. **2.** A connoisseur of medieval culture.

**Medieval Latin** *n.* Latin as used from about 700 to 1500.

**me·di·o·cre** (mē′dē-ō′kər) *adj.* [Lat. *mediocris* < *medius*, middle.] Moderate to low in quality : AVERAGE.

**me·di·oc·ri·ty** (mē′dē-ŏk′rĭ-tē) *n., pl.* **-ties. 1.** The quality or state of being mediocre. **2.** Mediocre achievement, ability, or performance. **3.** One who displays mediocre qualities.

**med·i·tate** (mĕd′ə-tāt′) *v.* **-tat·ed, -tat·ing, -tates.** [Lat. *meditari, meditat-*, to meditate.] —*vt.* **1.** To reflect on : PONDER. **2.** To plan or intend in the mind <*meditated* a crime> —*vi.* To engage in contemplation. **—med′i·ta′tor** *n.*

**med·i·ta·tion** (mĕd′ĭ-tā′shən) *n.* **1. a.** The act or process of meditating. **b.** A devotional exercise of contemplation. **2.** A contemplative discourse, usu. on a philosophical or religious subject.

**med·i·ta·tive** (mĕd′ĭ-tā′tĭv) *adj.* Marked by or expressing meditation. **—med′i·ta′tive·ly** *adv.* **—med′i·ta′tive·ness** *n.*

**med·i·ter·ra·ne·an** (mĕd′ĭ-tə-rā′nē-ən, -rān′yən) *adj.* [Lat. *mediterraneus* : *medius*, middle + *terra*, land.] **1.** Surrounded nearly or completely by dry land. —Used of large bodies of water, as lakes or seas. **2. Mediterranean.** Of, relating to, or typical of the Mediterranean Sea or the countries that border on it.

**Mediterranean fever** *n.* Undulant fever.

**Mediterranean flour moth** *n.* A small pale-gray moth, *Ephestia kuehniella*, now found worldwide, the larvae of which destroy flour, whole grains, and beehive pollen.

**Mediterranean fruit fly** *n.* A black and white two-winged fly, *Ceratitis capitata*, that attacks citrus and other fruits.

**me·di·um** (mē′dē-əm) *n., pl.* **-di·a** (-dē-ə) *or* **-di·ums.** [Lat. < *medius*, middle.] **1.** Something, as an intermediate course of action, that occupies a position or has a condition midway between extremes. **2.** An intervening substance through which something is transmitted or carried on, as an agency for transmitting energy. **3.** An agency by which something is conveyed, accomplished, or transferred <money as a *medium* of exchange> **4.** *pl.* **media.** A means of mass communication, as magazines, newspapers, or television. **5.** *pl.* **mediums.** One thought to have the power to communicate with the spirits of the dead. **6.** *pl.* **media. a.** A surrounding environment in which something thrives and functions. **b.** The substance in which a specific organism lives and thrives. **c.** A substance in which bacteria are cultivated for scientific purposes. **7. a.** A specific type of artistic technique or means of expression as determined by the materials used or the creative methods employed. **b.** The materials used in a specific artistic technique. **8.** A solvent for thinning paint. **9.** *Chem.* A filtering substance, as filter paper. **10.** A size of paper, usu. 18 × 23 inches or 17½ × 22 inches. —*adj.* **1.** Occurring or being between two amounts, degrees, or quantities : INTERMEDIATE <wears a *medium* shirt> **2.** Average <a *medium*-grade ore>

**medium of exchange** *n.* Something, as money, commonly used in a specific area or among a certain group of people.

**med·lar** (mĕd′lər) *n.* [ME *medler* < OFr. < **medle*, var. of *mesle*, fruit of the medlar < Lat. *mespila* < Gk. *mespilē.*] **1.** A tree, *Mespilus germanica*, cultivated for its fruit. **2.** The fruit of the medlar.

**med·ley** (mĕd′lē) *n., pl.* **-leys.** [ME *medlee* < OFr., var. of *meslee* < VLat. **misculare.* —see MEDDLE.] **1.** A jumbled assortment. **2.** A piece of music arranged from a series of melodies from various other sources.

**Mé·doc** (mā-dôk′, -dŏk′) *n.* [After *Médoc*, a region of France.] A red Bordeaux wine.

**me·dul·la** (mə-dŭl′ə) *n., pl.* **-las** *or* **-lae** (-ē) [Lat.] **1.** *Anat.* The inner core of certain vertebrate body structures, as the marrow of bone. **2.** The medulla oblongata. **3.** *Bot.* The pith in some plant stems. **—me·dul′lar, med′ul·lar′y** (mĕd′ə-lĕr′ē, mə-dŭl′ə-rē) *adj.*

**medulla ob·lon·ga·ta** (ŏb′lông-gä′tə) *n., pl.* **medulla ob·lon·ga·tas** *or* **medullae ob·lon·ga·tae** (-gä′tē). [NLat., oblong medulla.] The nerve tissue at the base of the brain that controls respiration, circulation, and various other bodily functions.

**medullary sheath** *n.* **1.** *Anat.* Myelin. **2.** *Bot.* A layer of thick-walled cells surrounding the pith in the stems of plants.

**med·ul·lat·ed** (mĕd′ə-lā′tĭd) *adj.* [LLat. *medullatus*, having marrow < Lat. *medulla*, medulla.] Myelinated.

**med·ul·li·za·tion** (mĕd′ə-lĭ-zā′shən) *n.* Replacement of bone tissue by marrow, as in inflammatory bone disease.

**Me·du·sa** (mə-dōō′sə, -zə, -dyōō′-) *n., pl.* **-sas** *or* **-sae** (-sē, -zē) [Lat. < Gk. *Medousa.*] **1.** *Gk. Myth.* One of the three Gorgons. **2. medusa.** The tentacled, usu. bell-shaped, free-swimming sexual stage in the life cycle of a coelenterate of the class Scyphozoa or Hydrozoa.

**me·du·soid** (mə-dōō′soid′, -zoid′, -dyōō′-) *n.* **1.** A shape resembling a jellyfish. **2.** A jellyfish. **—me·du′soid′** *adj.*

**meed** (mēd) *n.* [ME *mede* < OE *mēd.*] **1.** *Archaic.* A merited gift or wage. **2.** Fitting recompense.

**meek** (mēk) *adj.* **-er, -est.** [ME *meke* < ON *mjūkr*, soft.] **1.** Exhibiting humility and patience : GENTLE. **2.** Easily imposed on : SUBMISSIVE. **—meek′ly** *adv.*

**meer·schaum** (mîr′shəm, -shôm′) *n.* [G. : *Meer*, sea + *Schaum*, foam.] **1.** A tough, compact, usu. white mineral of hydrous magnesium silicate, $H_4Mg_2Si_3O_{10}$, found in the Mediterranean area and used in fashioning tobacco pipes and as a building stone. **2.** A tobacco pipe with a bowl of meerschaum.

**meerschaum**
*A meerschaum pipe*

**meet¹** (mēt) *v.* **met** (mĕt), **meet·ing, meets.** [ME *meten* < OE *mētan.*] —*vt.* **1.** To come upon by arrangement or chance. **2.** To be present at the arrival of <*met* the bus> **3.** To be introduced to. **4.** To come into conjunction with : JOIN <where the land *meets* the sea> **5.** To come into the company or presence of, as for a conference. **6.** To come to the notice of (the senses). **7.** To experience : undergo <*met* great wealth with humility> **8.** To deal with : OPPOSE <"We have *met* the enemy and they are ours" —Oliver Hazard Perry> **9.** To cope with <*met* each and every obligation> **10.** To come into accord with the wishes, views, or opinions of <The company *met* us only on some points.> **11.** To satisfy (e.g., a need) : FULFILL <*met* all the conditions> **12.** To pay : settle <money to *meet* debts> —*vi.* **1.** To come together <Let's *meet* for lunch.> **2.** To come into conjunction. **3.** To come together as opponents : CONTEND. **4.** To become introduced. **5.** To assemble. **6.** To experience or undergo <The clean-air bill *met* with approval.> **7.** To occur together, esp. in one person or entity : UNITE. —*n.* A contest or meeting, esp. an athletic competition. **—meet (one) halfway.** To compromise.

**meet²** (mēt) *adj.* [ME *mete* < OE *gemǣte.*] Fitting : proper <It is *meet* that we give thanks.> **—meet'ly** *adv.*

**meet·ing** (mē'tĭng) *n.* **1.** The act or process of coming together : ENCOUNTER. **2.** An assembly of people. **—meeting of the minds.** Agreement : concord.

**meet·ing·house** (mē'tĭng-hous') *n.* A building for public meetings and esp. for Quaker religious services.

**mef·e·nam·ic acid** (mĕf'ə-năm'ĭk) *n.* [(DI)ME(THYL) + *fen* (alteration of PHENYL) + AM(INOBENZO)IC ACID.] A crystalline compound, $C_{15}H_{15}NO_2$, used as an anti-inflammatory drug and as an analgesic.

**mega-** *pref.* [Gk. < *megas,* great.] **1.** Large <*mega*dose> **2.** One million ($10^6$) <*mega*hertz>

**meg·a·buck** (mĕg'ə-bŭk') *n. Slang.* One million dollars.

**meg·a·ceph·a·ly** (mĕg'ə-sĕf'ə-lē) *n.* Macrocephaly. **—meg'a·ce·phal'ic** (-sə-făl'ĭk), **meg'a·ceph'a·lous** (-sĕf'ə-ləs) *adj.*

**meg·a·cy·cle** (mĕg'ə-sī'kəl) *n. Physics.* **1.** One million cycles. **2.** Megahertz.

**meg·a·death** (mĕg'ə-dĕth') *n.* One million deaths. —Used as a unit in reference to nuclear warfare.

**Me·gae·ra** (mə-jîr'ə) *n.* [Lat. < Gk. *Megaira.*] *Gk. Myth.* One of the Furies.

**meg·a·gam·ete** (mĕg'ə-găm'ēt', -gə-mēt') *n.* Macrogamete.

**meg·a·ga·me·to·phyte** (mĕg'ə-gə-mē'tə-fīt') *n.* The female gametophyte arising from a megaspore of a heterosporous plant.

**meg·a·hertz** (mĕg'ə-hûrts') *n., pl.* **megahertz.** *Physics.* One million cycles per second, used esp. as a radio-frequency unit.

**meg·a·kar·y·o·cyte** (mĕg'ə-kăr'ē-ō-sīt', -ə-sīt') *n.* A large bone marrow cell with a lobulate nucleus, the precursor to blood platelets.

**megal-** *pref. var. of* MEGALO-.

**meg·a·lith** (mĕg'ə-lĭth') *n.* A large stone used in various prehistoric monuments. **—meg'a·lith'ic** *adj.*

**megalo-** or **megal-** *pref.* [Gk. < *megas,* great.] Of exaggerated size or greatness : LARGE <*megalo*cephaly>

**meg·a·lo·blast** (mĕg'ə-lō-blăst') *n.* A large nucleated erythroblast found in the blood in pernicious anemia. **—meg'a·lo·blas'tic** *adj.*

**meg·a·lo·car·di·a** (mĕg'ə-lō-kär'dē-ə) *n.* [MEGALO- + Gk. *kardia,* heart.] Enlargement of the heart.

**meg·a·lo·ceph·a·ly** (mĕg'ə-lō-sĕf'ə-lē) *n.* Macrocephaly. **—meg'a·lo·ce·phal'ic** (-sə-făl'ĭk), **meg'a·lo·ceph'a·lous** (-sĕf'ə-ləs) *adj.*

**meg·a·lo·ma·ni·a** (mĕg'ə-lō-mā'nē-ə, -mān'yə) *n.* A psychopathological condition in which fantasies of power or wealth predominate. **—meg'a·lo·ma'ni·ac'** *n.* **—meg'a·lo·ma·ni'a·cal** (-mə-nī'ə-kəl), **meg'a·lo·man'ic** (-măn'ĭk) *adj.*

**meg·a·lop·o·lis** (mĕg'ə-lŏp'ə-lĭs) *also* **me·gap·o·lis** (mə-găp'ə-lĭs, mĕ-) *n.* [MEGALO- + Gk. *polis,* city.] A region composed of several large cities and their suburbs in sufficiently close proximity to be considered a single urban complex. **—meg'a·lop·o·lis'tic** *adj.* **—meg'a·lo·pol'i·tan** (-lō-pŏl'ĭ-tən) *adj.*

**meg·a·lo·saur** (mĕg'ə-lə-sôr') *n.* [NLat. *Megalosaurus,* genus name : MEGALO- + Gk. *sauros,* lizard.] An extinct gigantic carnivorous dinosaur of the Jurassic period. **—meg'a·lo·sau'ri·an** *n.* & *adj.*

**meg·a·phone** (mĕg'ə-fōn') *n.* A funnel-shaped device used to direct and amplify the voice. —*vt.* & *vi.* **-phoned, -phon·ing, -phones.** To speak or transmit through or as if through a megaphone. **—meg'a·phon'ic** (-fŏn'ĭk) *adj.* **—meg'a·phon'i·cal·ly** *adv.*

**meg·a·pode** (mĕg'ə-pōd') *n.* [< NLat. *Megapodiidae,* family name : MEGA- + Gk. *pous,* foot.] One of various birds of the family Megapodiidae, found in Australia and many South Pacific islands.

**me·gap·o·lis** (mə-găp'ə-lĭs, mĕ-) *n. var. of* MEGALOPOLIS.

**meg·a·scop·ic** (mĕg'ə-skŏp'ĭk) *adj.* MACROSCOPIC 2. **—meg'a·scop'i·cal·ly** *adv.*

**meg·a·spo·ran·gi·um** (mĕg'ə-spə-răn'jē-əm) *n., pl.* **-gi·a** (-jē-ə). A structure enclosing a megaspore.

**meg·a·spore** (mĕg'ə-spôr', -spōr') *n.* **1.** The larger of two types of spores formed by heterosporous plants, as ferns, giving rise to the female gametophyte. **2.** A spore that forms the embryo sac in seed plants. **—meg'a·spor'ic** *adj.*

**meg·a·spo·ro·gen·e·sis** (mĕg'ə-spôr'ə-jĕn'ĭ-sĭs, -spōr'-) *n.* Production of megaspores.

**meg·a·spo·ro·phyll** (mĕg'ə-spôr'ə-fĭl', -spōr'-) *n.* A leaflike structure producing megasporangia.

**meg·a·struc·ture** (mĕg'ə-strŭk'chər) *n.* A huge, tall building.

**meg·a·there** (mĕg'ə-thîr') *n.* [< NLat. *Megatheriidae,* family name : MEGA- + Gk. *thēr,* beast.] A member of the extinct family Megatheriidae, composed of large ground sloths of the Miocene and Pleistocene epochs. **—meg'a·the'ri·an** (-ē-ən) *adj.*

**meg·a·ton** (mĕg'ə-tŭn') *n.* A unit of explosive force equal to one million tons of TNT. **—meg'a·ton'nage** *n.*

**Me·gil·lah** (mə-gĭl'ə) *n.* [Heb. *mĕgillāh* < *gālal,* to roll.] **1.** The Judaic scroll containing the Biblical narrative of the Book of Esther, traditionally read in synagogues to celebrate the festival of Purim. **2. megillah.** *Slang.* A tediously detailed story or account.

**Me·grez** (mē'grĕz') *n.* [Short for Ar. *maghriz adh-Dhanab ad-Dubb al-Akbar,* the root of the tail of the greater bear.] A star in the Big Dipper.

**me·grim** (mē'grĭm) *n.* [ME *migrem* < OFr. *migraine,* migraine. —see MIGRAINE.] **1.** A migraine. **2.** *often* **megrims.** A whim or fancy. **3. megrims.** Low spirits : DEPRESSION.

**mei·o·sis** (mī-ō'sĭs) *n., pl.* **-ses** (-sēz') [NLat. < Gk. *meiōsis,* diminution < *meioun,* to diminish < *meiōn,* less.] **1.** *Biol.* Cell division in sexually reproducing organisms that reduces the number of chromosomes in reproductive cells, leading to production of gametes in animals and spores in plants. **2.** Rhetorical understatement. **—mei·ot'ic** (-ŏt'ĭk) *adj.* **—mei·ot'i·cal·ly** *adv.*

**Meis·sen** (mī'sən) *n.* A delicate porcelain ware made in Meissen, Germany.

**Meis·ter·sing·er** (mīs'tər-sĭng'ər) *n., pl.* **Meistersinger** or **-ers.** [G. < MHG : *meister,* master (< OHG *meistar* < Lat. *magister*) + *singer,* singer (< *singen,* to sing < OHG *singan*).] A member of one of the guilds organized in the principal cities of Germany in the 14th, 15th, and 16th cent. for the purpose of establishing competitive standards for the composition and performance of music and poetry.

**mel·a·mine** (mĕl'ə-mēn') *n.* [G. *Melamin.*] A white crystalline compound, $C_3H_6N_6$, used to make melamine resins.

**melamine resin** *n.* A thermosetting resin used for molded products, adhesives, and surface coatings.

**melan-** *pref. var. of* MELANO-.

**mel·an·cho·li·a** (mĕl'ən-kō'lē-ə) *n.* [NLat. < LLat., melancholy.] A mental disorder marked by feelings of dejection and usu. by withdrawal, often a phase of manic-depressive psychosis. **—mel'an·cho'li·ac** (-lē-ăk') *adj.* & *n.*

**mel·an·chol·ic** (mĕl'ən-kŏl'ĭk) *adj.* **1.** Afflicted with or subject to melancholy. **2.** Of or relating to melancholia. **—mel'an·chol'ic** *n.* **—mel'an·chol'i·cal·ly** *adv.*

**mel·an·chol·y** (mĕl'ən-kŏl'ē) *n.* [ME *melancolie* < OFr. < Med. Lat. *melencholia* < LLat. *melancholia* < Gk. *melankholia* : *melas,* black + *kholē,* bile.] **1.** Sadness or depression of the spirits : GLOOM. **2.** Pensive contemplation. **3. a.** Black bile. **b.** An emotional state characterized by depression and outbreaks of violent anger, once believed to arise from an excess of black bile. —*adj.* **1.** Sad : depressed. **2.** Tending to promote gloom or sadness <a *melancholy* ballad> **3.** Pensive. **—mel'an·chol'i·ly** *adv.* **—mel'an·chol'i·ness** *n.*

**Mel·a·ne·sian** (mĕl'ə-nē'zhən, -shən) *adj.* Of or relating to Melanesia, its people, or their languages. —*n.* **1.** A native inhabitant of Melanesia. **2.** A subfamily of the Austronesian languages that includes the languages of Melanesia.

**mé·lange** *also* **me·lange** (mā-länzh') *n.* [Fr. < OFr. < *mesler,* to mix. —see MEDDLE.] A mixture.

**me·la·ni·an** (mə-lā'nē-ən) *adj.* Relating to dark-brown or black pigmentation.

**me·lan·ic** (mə-lăn'ĭk) *adj.* **1.** Of, relating to, or exhibiting melanism. **2.** Afflicted with melanosis.

**mel·a·nin** (mĕl'ə-nĭn) *n.* A dark-brown or black pigment of plants and animals, as that found in the skin, retina, and hair of human beings.

**mel·a·nism** (mĕl'ə-nĭz'əm) *n.* **1.** Melanosis. **2.** Darkness of the skin, hair, or eyes due to high pigmentation. **—mel'a·nis'tic** *adj.*

**mel·a·nite** (mĕl'ə-nīt') *n.* [G. *Melanit* < Gk. *melas,* black.] A black variety of garnet. **—mel'a·nit'ic** (-nĭt'ĭk) *adj.*

**mel·a·no-** or **melan-** *pref.* [NLat. < Gk. < *melas,* black.] Black : dark <*melanin*>
**mel·a·no·blast** (mĕl′ə-nō-blăst′, mə-lăn′ə-) *n.* A precursor cell of a melanocyte or melanophore. **—mel′a·no·blas′tic** *adj.*
**mel·a·no·blas·to·ma** (mĕl′ə-nō-blă-stō′mə, mə-lăn′ə-) *n.* A malignant tumor composed chiefly of melanoblasts.
**mel·a·noch·ro·i** (mĕl′ə-nŏk′rō-ī′, -nŏk′roi′) *pl.n.* [NLat. : Gk. *melas,* dark + Gk. *ōkhros,* pale.] Caucasian peoples with dark hair and light complexions. **—mel′a·no·chro′ic** (-nō-krō′ĭk), **mel′a·noch′·roid′** (-nŏk′roid′) *adj.*
**mel·a·no·cyte** (mĕl′ə-nō-sīt′) *n.* An epidermal cell capable of synthesizing the black pigment melanin and responsible for color variations in the skin of humans and many animals.
**mel·a·no·cyte-stim·u·lat·ing hormone** (mĕl′ə-nō-sīt′stĭm′yə-lā′tĭng, mə-lăn′ə-) *n.* A hormone secreted by the pituitary gland that stimulates melanocytes or melanophores to disperse melanin.
**mel·a·noid** (mĕl′ə-noid′) *adj.* **1.** Black-pigmented. **2.** MELANIC 2. **—mel′a·noid′** *n.*
**mel·a·no·ma** (mĕl′ə-nō′mə) *n., pl.* **-mas** *or* **-ma·ta** (-mə-tə). A dark-pigmented malignant mole or tumor.
**mel·a·no·phore** (mĕl′ə-nə-fôr′, -fōr′, mə-lăn′ə-) *n.* A chromatophore containing melanin.
**mel·a·no·sis** (mĕl′ə-nō′sĭs) *n.* Abnormally dark pigmentation of the skin or other tissues, resulting from sunburn and various dermatoses. **—mel′a·not′ic** (-nŏt′ĭk) *adj.*
**mel·a·nous** (mĕl′ə-nəs) *adj.* Having a dark or black complexion and black hair. **—mel′a·nos′i·ty** (-nŏs′ĭ-tē) *n.*
**mel·a·phyre** (mĕl′ə-fīr′) *n.* [Fr. *mélaphyre* : Gk. *melas,* black + F. *porphyre,* porphyry < Med. Lat. *porphyrium.* —see PORPHYRY.] A dark igneous porphyry embedded with feldspar crystals.
**Mel·ba toast** (mĕl′bə) *n.* [After Nellie *Melba* (1861–1931).] Very thinly sliced crisp toast.
**Mel·chiz·e·dek** (mĕl-kĭz′ə-dĕk′) *n.* [Gk. *Melkhisedek* < Heb. *Malkī-ṣedheq.*] *Mormon Ch.* The higher order of priesthood.
**meld¹** (mĕld) *v.* **meld·ed, meld·ing, melds.** [G. *melden,* to announce < OHG *meldōn.*] *—vt.* To display or declare (a card or combination of cards in a hand) for inclusion in one's score in various card games, as pinochle. *—vi.* To present a meld. *—n.* A combination of cards to be declared for a score.
**meld²** (mĕld) *vt. & vi.* **meld·ed, meld·ing, melds.** [Blend of MELT and WELD.] To cause to merge or become merged.
**me·lee** (mā′lā′, mā-lā′) *also* **mê·lée** (mĕ-lā′) *n.* [Fr. *mêlée* < OFr. *meslee,* medley. —see MEDLEY.] **1.** Confused, hand-to-hand fighting : FREE-FOR-ALL. **2.** A confused and tumultuous mingling, as of a crowd.
**mel·i·lot** (mĕl′ə-lŏt′) *n.* [ME *melilote* < OFr. < Lat. *melilotos* < Gk. *melilōtos* : *meli,* honey + *lōtos,* lotus.] An Old World plant of the genus *Melilotus,* with compound leaves and narrow, small, fragrant white or yellow flower clusters.
**mel·i·nite** (mĕl′ə-nīt′) *n.* [Fr. *mélinite* < Gk. *mēlinos,* quince-yellow < *mēlon,* quince.] A high explosive made with picric acid.
**mel·io·rate** (mĕl′yə-rāt′, mē′lē-ə-) *vt. & vi.* **-rat·ed, -rat·ing, -rates.** [LLat. *meliorare, meliorat-* < *melior,* better.] To make or become better : IMPROVE. **—mel′io·ra·ble** (-rə-bəl) *adj.* **—mel′io·ra′·tive** *adj. & n.* **—mel′io·ra′tor** *n.*
**mel·io·ra·tion** (mĕl′yə-rā′shən, mē′lē-ə-) *n.* **1. a.** The act or process of improving or the state of being improved. **b.** An improvement. **2.** The linguistic process by which a word grows gradually more elevated in meaning or more positive in connotation.
**mel·io·rism** (mĕl′yə-rĭz′əm, mē′lē-ə-) *n.* [Lat. *melior,* better + -ISM.] The belief that society tends toward improvement and that human effort can further its improvement. **—mel′io·rist** *n.* **—mel′·io·ris′tic** *adj.*
**me·lis·ma** (mə-lĭz′mə) *n., pl.* **-ma·ta** (-mə-tə) *or* **-mas.** [NLat. < Gk., song < *melizein,* to sing < *melos,* song.] *Mus.* A passage sung to one syllable of text, as in Gregorian chant. **—mel′is·mat′ic** (mĕl′ĭz-măt′ĭk) *adj.*
**mel·lif·er·ous** (mə-lĭf′ər-əs) *also* **mel·lif·ic** (mə-lĭf′ĭk) *adj.* [< Lat. *mellifer* : *mel,* honey + *ferre,* to bear.] Forming or bearing honey.
**mel·lif·lu·ent** (mə-lĭf′lo͞o-ənt) *adj.* Mellifluous. **—mel·lif′lu·ent·ly** *adv.*
**mel·lif·lu·ous** (mə-lĭf′lo͞o-əs) *adj.* [Lat. *mellifluus* : *mel,* honey + *fluere,* to flow.] **1.** Flowing with sweetness or honey. **2.** Smooth and sweet : HONEYED <*mellifluous* singing> **—mel·lif′lu·ous·ly** *adv.* **—mel·lif′lu·ous·ness** *n.*
**mel·lo·phone** (mĕl′ō-fōn′) *n.* [MELLO(W) + -PHONE.] A brass musical wind instrument occas. used as a substitute for the French horn.
**mel·low** (mĕl′ō) *adj.* **-er, -est.** [ME *melowe.*] **1. a.** Soft, juicy, sweet, and full-flavored because of ripeness. **b.** Suggesting sweetness, softness, full flavor, or juiciness. **2.** Rich and soft in quality <a *mellow* tone> **3.** Having the wisdom, dignity, or gentleness that is often characteristic of maturity. **4.** Relaxed and at ease : GENIAL. **5. a.** Slightly and pleasantly intoxicated. **b.** Pleasantly high from a drug, esp. from smoking marijuana. **6.** Soft and loamy. —Used of soil. *—vt. & vi.* **-lowed, -low·ing, -lows.** To make or become mellow. **—mel′low·ly** *adv.* **—mel′low·ness** *n.*
**me·lo·de·on** (mə-lō′dē-ən) *n.* [G. *Melodion* < *Melodie,* melody < OFr. —see MELODY.] A small reed organ.
**me·lod·ic** (mə-lŏd′ĭk) *adj.* Of, relating to, or containing melody. **—me·lod′i·cal·ly** *adv.*
**me·lo·di·ous** (mə-lō′dē-əs) *adj.* **1.** Of, relating to, or containing a melody. **2.** Agreeable to the ear. **—me·lo′di·ous·ly** *adv.* **—me·lo′di·ous·ness** *n.*
**mel·o·dize** (mĕl′ə-dīz′) *v.* **-dized, -diz·ing, -diz·es.** *—vt.* **1.** To write a melody for (a song lyric). **2.** To make melodious. *—vi.* To compose a melody. **—mel′o·diz′er, mel′o·dist** *n.*
**mel·o·dra·ma** (mĕl′ə-drä′mə, -drăm′ə) *n.* [Fr. *mélodrame* : Gk. *melos,* song + Fr. *drame,* drama < LLat. *drama.* —see DRAMA.] **1. a.** A dramatic presentation marked by heavy use of suspense, sensational episodes, romantic sentiment, and a conventionally happy ending. **b.** The dramatic genre including such works. **2.** Behavior or occurrences having melodramatic characteristics.
**mel·o·dra·mat·ic** (mĕl′ə-drə-măt′ĭk) *adj.* **1.** Having the excitement and emotional appeal of melodrama. **2.** Highly emotional or sentimental : HISTRIONIC. **—mel′o·dra·mat′i·cal·ly** *adv.*
**mel·o·dra·mat·ics** (mĕl′ə-drə-măt′ĭks) *n. (sing. or pl. in number).* **1.** Melodramatic theatrical performance. **2.** Melodramatic actions.
**mel·o·dy** (mĕl′ə-dē) *n., pl.* **-dies.** [ME *melodie* < OFr. < LLat. *melodia* < Gk. *melōidia,* choral song : *melos,* tune + *aoidein,* to sing.] **1.** A pleasing succession or arrangement of sounds. **2.** Musical quality, as in lyric poetry. **3.** *Mus.* **a.** A rhythmic sequence of single tones organized so as to make up a particular musical phrase or idea. **b.** The structure of music with respect to the arrangement of single notes in succession. **c.** The leading part or the air in a harmonic composition. **4.** A poem appropriate for setting to music or singing.
**mel·oid** (mĕl′oid′, mĕl′ō-ĭd) *n.* [< NLat. *Meloidae,* family name < *Meloe,* beetle genus.] A blister beetle. **—mel′oid** *adj.*
**mel·on** (mĕl′ən) *n.* [ME < LLat. *melo,* short for Lat. *melopepon* < Gk. *mēlopepōn* : *melon,* apple + *pepōn,* gourd.] **1.** A variety of either of two related vines, *Cucumis melo* or *Citrullus vulgaris,* widely cultivated for its edible fruit. **2.** The fruit of a melon vine, with a hard rind and juicy flesh.
**Mel·pom·e·ne** (mĕl-pŏm′ə-nē′) *n.* [Lat. < Gk. *Melpomenē* < *melpesthai,* to sing.] *Gk. Myth.* The Muse of tragedy.
**melt** (mĕlt) *v.* **melt·ed, melt·ing, melts.** [ME *melten* < OE *meltan.*] *—vi.* **1.** To be transformed from a solid to a liquid state by application of heat, pressure, or both. **2.** To dissolve. **3.** To disappear or vanish gradually as if by dissolving <inhibitions *melting* away> **4.** To pass or merge imperceptibly into something else <*melted* into the crowd> **5.** To become softened in feeling. **6.** *Obs.* To be crushed or overcome, as by grief, dismay, or fear. *—vt.* **1.** To reduce from a solid to a liquid state by application of heat, pressure, or both. **2.** To dissolve <*melt* honey in hot tea> **3.** To cause to disappear gradually : DISPERSE. **4.** To cause (units) to blend. **5.** To make gentle or tender : SOFTEN. *—n.* **1.** A melted solid. **2.** The state of being melted. **3. a.** The act or process of melting. **b.** The quantity melted during a single operation or at one time. **—melt′a·bil′i·ty** *n.* **—melt′a·ble** *adj.* **—melt′er** *n.*
**melt·age** (mĕl′tĭj) *n.* **1.** The quantity or substance produced by a melting process. **2.** The act or process of melting.
**melt·down** (mĕlt′doun′) *n.* The melting of a nuclear reactor core.
**melting point** *n.* **1.** The temperature at which a solid becomes a liquid at standard atmospheric pressure. **2.** The temperature at which a solid and its liquid are in equilibrium, at any fixed pressure.
**melting pot** *n.* **1.** A container for melting something. **2.** A place where immigrants of different races or cultures form an integrated society.
**mel·ton** (mĕl′tən) *n.* [After *Melton* Mowbray, England.] A heavy woolen cloth used chiefly for making overcoats and hunting jackets.
**mem** (mĕm) *n.* [Heb.] The 13th letter of the Hebrew alphabet. —See table at ALPHABET.
**mem·ber** (mĕm′bər) *n.* [ME < OFr. *membre* < Lat. *membrum.*] **1.** A distinct part of a whole, esp.: **a.** A syntactic unit of a sentence : CLAUSE. **b.** A proposition of a syllogism. **c.** An element in a mathematical set. **2.** A part or organ of a human or animal body, as: **a.** A limb, as an arm or leg. **b.** The penis. **3.** A part of a plant. **4.** One who belongs to a group or organization. **5.** *Math.* The expression on either side of an equality sign.
**mem·ber·ship** (mĕm′bər-shĭp′) *n.* **1.** The state of being a member. **2.** The total number of members in a group.
**mem·brane** (mĕm′brān′) *n.* [Lat. *membrana,* skin < *membrum,* member.] **1.** *Biol.* A thin pliable layer of tissue covering surfaces or separating or connecting regions, structures, or organs of an animal or plant. **2.** A piece of parchment. **3.** *Chem.* A thin sheet of natural or synthetic material permeable to substances in solution. **—mem′·bra·nal** (-brə-nəl) *adj.*
**membrane bone** *n.* A bone formed directly in the connective tissue, as some cranial bones.
**mem·bra·nous** (mĕm′brə-nəs) *adj.* **1.** Made of or similar to a membrane. **2.** *Pathol.* Marked by membrane formation.
**membranous labyrinth** *n.* The soft-tissue sensory structures of the inner ear.

**me·men·to** (mə-mĕn′tō) *n., pl.* **-tos** or **-toes.** [ME < Lat. *memento,* imper. of *meminisse,* to remember.] A keepsake : souvenir.
**me·men·to mo·ri** (mə-mĕn′tō môr′ē) *n.* [Lat., remember that you must die.] **1.** A reminder of death or mortality, esp. a death's-head. **2.** A reminder of human errors or failures.
**Mem·non** (mĕm′nŏn′) *n.* [Gk. *Memnōn.*] *Gk. Myth.* An Ethiopian king killed by Achilles and made immortal by Zeus.
**mem·o** (mĕm′ō) *n., pl.* **-os.** A memorandum.
**mem·oir** (mĕm′wär′, -wôr′) *n.* [Fr. *mémoire* < OFr. *memoire,* memory.] **1.** An account of the personal experiences of an author. **2.** *often* **memoirs.** An autobiography. **3.** A biography or biographical sketch. **4.** A report, esp. on a scholarly or scientific subject. **5. memoirs.** The report of the proceedings of a learned society.
**mem·o·ra·bil·i·a** (mĕm′ər-ə-bĭl′ē-ə, -bĭl′yə) *pl.n.* [Lat. < *memorabilis,* memorable.] Notable things worthy of remembrance.
**mem·o·ra·ble** (mĕm′ər-ə-bəl) *adj.* [ME < Lat. *memorabilis* < *memorare,* to remember < *memor,* mindful.] Worth being remembered : REMARKABLE <a *memorable* occasion> **—mem′o·ra·bil′i·ty, mem′o·ra·ble·ness** *n.* **—mem′o·ra·bly** *adv.*
**mem·o·ran·dum** (mĕm′ə-răn′dəm) *n., pl.* **-dums** or **-da** (-də) [ME < Lat. *memorandus,* to be remembered < *memorare,* to remember. —see MEMORABLE.] **1.** A short informal note written as a reminder. **2.** A written record or communication, as in a business office. **3.** *Law.* A short written statement outlining the terms of an agreement, transaction, or contract. **4.** A business statement made by a consignor about a shipment of goods that may be returned. **5.** A brief unsigned diplomatic communication.
**me·mo·ri·al** (mə-môr′ē-əl, -mōr′-) *n.* [ME < LLat. *memoriale* < *memorialis,* belonging to memory < *memoria,* memory. —see MEMORY.] **1.** Something, as a monument or a holiday, designed or established to preserve the memory of a person or event. **2.** A petition or written statement of facts presented to a legislative body or an executive. *—adj.* **1.** Serving as a remembrance of a person or event. **2.** Of, relating to, or in memory. **—me·mo′ri·al·ly** *adv.*
**Memorial Day** *n.* May 30, a U.S. holiday observed on the last Monday in May in honor of servicemen killed in war.
**me·mo·ri·al·ist** (mə-môr′ē-ə-lĭst, -mōr′-) *n.* **1.** A writer of memoirs. **2.** One who signs or writes a memorial.
**me·mo·ri·al·ize** (mə-môr′ē-ə-līz′, mə-mōr′-) *vt.* **-ized, -iz·ing, -iz·es. 1.** To commemorate. **2.** To present a memorial to : PETITION. **—me·mo′ri·al·i·za′tion** *n.* **—me·mo′ri·al·iz′er** *n.*
**mem·o·rize** (mĕm′ə-rīz′) *vt.* **-rized, -riz·ing, -riz·es.** To commit to memory. **—mem′o·riz′a·ble** *adj.* **—mem′o·ri·za′tion** *n.* **—mem′o·riz′er** *n.*
**mem·o·ry** (mĕm′ə-rē) *n., pl.* **-ries.** [ME *memorie* < OFr. *memoire* < Lat. *memoria* < *memor,* mindful.] **1.** The mental faculty of retaining and recalling past experience. **2.** The act or an instance of remembrance : RECOLLECTION. **3.** All that a person can remember. **4.** Something remembered. **5.** The fact of being remembered : REMEMBRANCE. **6.** The time period covered by the remembrance or recollection of a person or group of persons <within the *memory* of humankind> **7.** *Biol.* Persistent modification of behavior resulting from the organism's experience. **8.** *Computer Sci.* **a.** A unit of a computer that stores data for retrieval. **b.** Capacity for storing information. **9.** *Statistics.* The set of past events affecting a given event in a stochastic process.
**memory engram** *n.* Engram.
**mem·sa·hib** (mĕm′sä′ĭb) *n.* [MA'AM + SAHIB.] —Used as a title of respect for a white European woman in colonial India.
**men** (mĕn) *n. pl. of* MAN.
**men-** *pref. var. of* MENO-.
**men·ace** (mĕn′ĭs) *n.* [ME < OFr. < Lat. *minacia* < *minax,* threatening < *minari,* to threaten.] **1. a.** A threat <the *menace* of war> **b.** An act of threatening. **2.** A vexatious or annoying person. *—v.* **-aced, -ac·ing, -ac·es.** *—vt.* **1.** To threaten. **2.** To be a threat to. *—vi.* To make threats. **—men′ac·er** *n.* **—men′ac·ing·ly** *adv.*
**men·a·di·one** (mĕn′ə-dī′ōn′) *n.* [ME(THYL) + NA(PHTHA) + DI-[1] + -ONE.] A yellow crystalline powder, $C_{10}H_5CH_3O_2$, with physiological effects similar to vitamin K, used as a medicine and fungicide.
**mé·nage** (mā-näzh′) *n.* [Fr. < OFr. *mesnage,* dwelling < Lat. *mansio.* —see MANSION.] **1.** Persons living together as a unit : HOUSEHOLD. **2.** Household management.
**ménage à trois** (ä trwä) *n.* [Fr., household for three.] Cohabitation by three people, as a married couple and a lover.
**me·nag·er·ie** (mə-năj′ə-rē, mə-năzh′-) *n.* [Fr. *ménagerie* < *ménage,* ménage.] **1.** A collection of live wild animals on exhibition. **2.** The enclosure in which wild animals are kept.
**me·nar·che** (mə-när′kē) *n.* [NLat. : MEN(O)- + Gk. *arkhē,* beginning.] The first occurrence of menstruation. **—me·nar′che·al** *adj.*
**men·a·zon** (mĕn′ə-zŏn′) *n.* [(DI)ME(THYL) + (DIAMI)N(E) + (TRI)AZ(INE) + (THI)ON(ATE).] A colorless crystalline compound, $C_6H_{12}N_5O_2PS_2$, used as an insecticide.
**mend** (mĕnd) *v.* **mend·ed, mend·ing, mends.** [ME *menden* < *amenden,* to amend. —see AMEND.] *—vt.* **1.** To make right or correct : REPAIR. **2.** To improve or reform. *—vi.* **1. a.** To improve in health. **b.** To heal. **2.** To correct errors : set right. *—n.* **1.** The act of mending. **2.** A mended place. **—on the mend.** Improving, esp. in health. **—mend′a·ble** *adj.* **—mend′er** *n.*
**men·da·cious** (mĕn-dā′shəs) *adj.* [Lat. *mendax, mendac-,* mendacious.] **1.** Lying : untruthful <a *mendacious* person> **2.** Not true : FALSE <a *mendacious* rumor> **—men·da′cious·ly** *adv.*
**men·dac·i·ty** (mĕn-dăs′ĭ-tē) *n.* **1.** The state of being mendacious. **2.** A lie : falsehood.
**men·de·le·vi·um** (mĕn′də-lē′vē-əm) *n.* [After Dmitri *Mendeleev* (1834–1907).] *Symbol* **Md** A radioactive transuranium element of the actinide series; atomic number 101; mass numbers 255 and 256.
**Men·de·li·an** (mĕn-dē′lē-ən, -dēl′yən) *adj.* Of or relating to Gregor Mendel or his theories of genetics.
**Men·del·ism** (mĕn′dl-ĭz′əm) *also* **Men·de·li·an·ism** (mĕn-dē′lē-ə-nĭz′əm) *n.* The theoretical principles of heredity formulated by Gregor Mendel.
**Men·del's law** (mĕn′dlz) *n.* The principles of heredity of sexually reproducing organisms formulated by Gregor Mendel, now usu. summarized in three laws: **a.** Law of Segregation: Certain paired characteristics, one from each parent, do not blend with or alter each other in the offspring, thus accounting for contrasting traits in successive generations. **b.** Law of Independent Combination: The genes determining such pairs of traits combine in the offspring according to the statistics of chance. **c.** Law of Dominance: If one of a pair of genes is dominant and the other recessive, the recessive trait may appear in an offspring only if both genes of its pair are recessive.
**men·di·cant** (mĕn′dĭ-kənt) *adj.* [Lat. *mendicans, mendicant-,* pr. part. of *mendicare,* to beg < *mendicus,* beggar < *mendum,* fault.] Depending on begging for a living. *—n.* **1.** A beggar. **2.** A friar of a mendicant order. **—men′di·can·cy, men·dic′i·ty** (-dĭs′ĭ-tē) *n.*
**Men·e·la·us** (mĕn′ə-lā′əs) *n.* [Lat. < Gk. *Menelaos.*] *Gk. Myth.* The king of Sparta, husband of Helen and brother of Agamemnon.
**men·folk** (mĕn′fōk′) *or* **men·folks** (-fōks′) *pl.n.* **1.** Men as a group. **2.** The male members of a family or community.
**men·ha·den** (mĕn-hād′n) *n., pl.* **menhaden** or **-dens.** [Narraganset *munnawhatteaug.*] An inedible fish, *Brevoortia tyrannus* of American Atlantic and Gulf waters, used as a source of fish oil, fish meal, fertilizer, and bait.
**men·hir** (mĕn′hîr′) *n.* [Fr. < Breton : *men,* stone + *hir,* long.] A prehistoric monument consisting of a single tall upright megalith.
**me·ni·al** (mē′nē-əl, mēn′yəl) *adj.* [ME, belonging to a household < *meine,* household servants < AN < VLat. **mansionata* < Lat. *mansio,* house. —see MANSION.] **1.** Of, relating to, or appropriate for a servant. **2.** Of or relating to work or a job regarded as servile. *—n.* A servant, esp. a domestic servant. **—me′ni·al·ly** *adv.*
**mening-** *pref. var. of* MENINGO-.
**me·nin·ge·al** (mə-nĭn′jē-əl) *adj.* Of, pertaining to, or concerned with a meninx or meninges.
**me·nin·ges** (mə-nĭn′jēz) *n. pl. of* MENINX.
**meningi-** *pref. var. of* MENINGO-.
**men·in·gi·tis** (mĕn′ĭn-jī′tĭs) *n.* Inflammation of any or all of the meninges of the brain and spinal cord, usu. caused by a bacterial infection. **—men′in·git′ic** (-jĭt′ĭk) *adj.*
**meningo-** or **meningi-** or **mening-** *pref.* [NLat. < *meninx,* meninx.] Meninges <*meningo*coccus>
**me·nin·go·coc·cus** (mə-nĭng′gə-kŏk′əs, -nĭn′jə-) *n., pl.* **-coc·ci** (-kŏk′sī). A bacterium, *Neisseria meningitidis,* the cause of epidemic cerebrospinal meningitis. **—me·nin′go·coc′cal, me·nin′go·coc′cic** (-kŏk′sĭk) *adj.*
**me·nin·go·en·ceph·a·li·tis** (mə-nĭng′gō-ĕn-sĕf′ə-lī′tĭs) *n.* Inflammation of the brain and meninges. **—me·nin′go·en·ceph′a·lit′ic** (-lĭt′ĭk) *adj.*
**me·ninx** (mē′nĭngks) *n., pl.* **me·nin·ges** (mə-nĭn′jēz) [NLat. < Gk. *mēninx,* membrane.] One of the membranes enclosing the vertebrate brain and spinal cord.
**me·nis·cus** (mə-nĭs′kəs) *n., pl.* **-nis·ci** (-nĭs′ī′) or **-nis·cus·es.** [NLat. < Gk. *mēniskos,* dim. of *mēnē,* moon.] **1.** A crescent-shaped body. **2.** A concavo-convex lens. **3.** The curved upper surface of a nonturbulent liquid in a container that is concave if the liquid wets the container walls and convex if it does not. **4.** *Anat.* A cartilage disk that cushions the ends of bones meeting in a joint. **—me·nis′cal** (-kəl), **me·nis′cate′** (-kāt′), **me·nis′coid′** (-koid′), **men′is·coi′dal** (mĕn′ĭs-koid′l) *adj.*
**Men·non·ite** (mĕn′ə-nīt′) *n.* [G. *Mennonit,* after *Menno* Simons (1492–1559).] A member of an Evangelical Protestant Christian sect opposed to taking oaths, holding public office, or serving in the armed forces.
**meno-** or **men-** *pref.* [NLat. < Gk. *meis, mēn-,* month.] **1.** Menstruation <*men*arche> **2.** Menses <*meno*rrhagia>
**men·o·pause** (mĕn′ə-pôz′) *n.* [Fr. : Gk. *meis, mēn-,* moon + Fr. *pause,* pause < Lat. *pausa.* —see PAUSE.] The period of cessation of menstruation, occurring usu. between the ages of 45 and 50. **—men′o·paus′al** *adj.*
**Me·no·rah** (mə-nôr′ə, -nōr′ə) *n.* [Heb. *mənōrāh.*] **1.** A ceremonial seven-branched candelabrum of the Jewish Temple symbolizing the seven days of the Creation. **2.** A nine-branched candelabrum used in celebrating Chanukah.

**men·or·rha·gi·a** (mĕn′ə-rā′jē-ə) *n. Pathol.* Abnormally heavy menstrual flow. **—men′or·rha′gic** (-jĭk) *adj.*

**Men·sa** (mĕn′sə) *n.* [Lat. *mensa,* table.] A constellation in the Southern Hemisphere.

**men·sal** (mĕn′səl) *adj.* [LLat. *mensalis* < Lat. *mensa,* table.] Of, relating to, or used at table.

**mensch** (mĕnsh) *n.* [Yiddish < MHG, man < OHG *mennisco.*] *Informal.* One having admirable characteristics, as integrity and compassion.

**men·ses** (mĕn′sēz′) *pl.n.* [Lat., pl. of *mensis,* month.] *(sing. or pl. in number). Physiol.* Blood and dead cell debris discharged from the uterus through the vagina by adult women at approx. monthly intervals between puberty and menopause.

**Men·she·vik** (mĕn′shə-vĭk) *n., pl.* **Men·she·viks** *or* **Men·she·vi·ki** (mĕn′shə-vē′kē) [R. *men′shevik* < *men′she,* less.] A member of the liberal minority faction of the Russian Social Democratic Party that struggled against the Bolsheviks before and during the Russian Revolution. **—Men′she·vism** *n.* **—Men′she·vist** *n.*

**men's room** *n.* A restroom for men.

**men·stru·a** (mĕn′strōō-ə) *n. var. pl. of* MENSTRUUM.

**men·stru·al** (mĕn′strōō-əl) *also* **men·stru·ous** (mĕn′strōō-əs) *adj.* [ME < Lat. *menstrualis* < *menstruus,* menstrual < *mensis,* month.] **1.** *Physiol.* Of or relating to menstruation. **2. a.** Occurring monthly. **b.** Of a monthly duration.

**men·stru·ate** (mĕn′strōō-āt′) *vi.* **-at·ed, -at·ing, -ates.** [Lat. *menstruare, menstruat-* < *menstruus,* menstrual.] *Physiol.* To undergo menstruation.

**men·stru·a·tion** (mĕn′strōō-ā′shən) *n.* The process or an instance of discharging the menses.

**men·stru·ous** (mĕn′strōō-əs) *adj. var. of* MENSTRUAL.

**men·stru·um** (mĕn′strōō-əm) *n., pl.* **-stru·ums** *or* **-stru·a** (-strōō-ə) [ME, menstruation < Med. Lat. < Lat. *menstruus,* menstrual.] A solvent, esp. one used in extracting and preparing drugs.

**men·su·ra·ble** (mĕn′sər-ə-bəl, -shər-) *adj.* **1.** Capable of being measured. **2.** MENSURAL 2. **—men′su·ra·bil′i·ty, men′su·ra·ble·ness** *n.*

**men·su·ral** (mĕn′sər-əl, -shər-) *adj.* [Lat. *mensuralis* < *mensura,* measure. —see MEASURE.] **1.** Of or relating to measure. **2.** *Mus.* Having notes of fixed rhythmic value.

**men·su·ra·tion** (mĕn′sə-rā′shən, -shə-) *n.* **1.** The act, process, or art of measuring. **2.** Measurement of geometric quantities. **—men′su·ra′tive** *adj.*

**mens·wear** (mĕnz′wâr′) *n.* Garments for men.

**-ment** *suff.* [ME < OFr. < Lat. *-mentum,* n. suffix.] **1.** Action : process <appease*ment*> **2.** Result of an action or process <advance*ment*> **3.** Means, instrument, or agent of an action or process <adorn*ment*>

**men·tal** (mĕn′tl) *adj.* [ME < Lat. *mentalis* < *mens,* mind.] **1.** Of or pertaining to the mind : INTELLECTUAL. **2.** Executed or performed by the mind. **3.** Of, pertaining to, or affected by mental disorder. **4.** Intended for treatment of people affected with mental disorder <a *mental* ward> **5.** Of or pertaining to telepathy. **—men′tal·ly** *adv.*

**mental age** *n.* A measure of mental development as determined by intelligence tests, gen. restricted to children and expressed as the age at which that level is average.

**mental deficiency** *n.* Subnormal intellectual development, either congenital or induced by disease or brain injury, characterized broadly by deficiencies ranging from impaired learning ability to social incompetence.

**men·tal·i·ty** (mĕn-tăl′ĭ-tē) *n., pl.* **-ties. 1.** The sum of a person's intellectual capabilities. **2.** A way of thinking : VIEWPOINT.

**mental retardation** *n.* Mental deficiency.

**mental telepathy** *n.* Telepathy.

**men·thol** (mĕn′thôl′) *n.* [G. : Lat. *mentha,* mint + *-ol,* -ol.] A white, crystalline, organic compound, $CH_3C_6H_9(C_3H_7)OH$, obtained from peppermint oil or synthesized and used in perfumery, and as a flavoring and mild anesthetic. **—men′tho·lat′ed** *adj.*

**men·tion** (mĕn′shən) *vt.* **-tioned, -tion·ing, -tions.** [< ME, mention < OFr. < Lat. *mentio* < *mens,* mind.] To refer to or cite incidentally. *—n.* **1. a.** The act of casually or briefly referring to something. **b.** An incidental reference. **2.** Honorable mention. **—men′tion·a·ble** *adj.*

**Men·tor** (mĕn′tôr′, -tər) *n.* [Lat. < Gk. *Mentōr.*] **1.** *Gk. Myth.* Odysseus' trusted counselor, under whose disguise Athena became the guardian and teacher of Telemachus. **2. mentor.** A wise and trusted teacher or counselor.

**men·u** (mĕn′yōō, mā′nyōō) *n.* [Fr. < *menu,* small < Lat. *minūtus.* —see MINUTE².] **1.** A list of the food and drink available or to be served for a meal. **2.** The food and drink available or served at a meal.

**me·ow** (mē-ou′) *n.* [Imit.] **1.** The cry of a cat. **2.** A spiteful comment. *—vi.* **-owed, -ow·ing, -ows.** To make or utter a meow.

**me·per·i·dine hydrochloride** (mə-pĕr′ĭ-dēn′) *n.* [ME(THYL) + (PI)PERIDINE.] An organic compound, $C_{15}H_{21}NO_2 \cdot HCl$, used as an analgesic and sedative.

**Meph·i·stoph·e·les** (mĕf′ĭ-stŏf′ə-lēz′) *n.* [G.] The devil in the Faust legend to whom Faust sold his soul. **—Me·phis′to·phe′le·an, Me·phis′to·phe′li·an** (mə-fĭs′tō-fē′lē-ən, -fēl′yən, mĕf′ĭ-stō-) *adj.*

**me·phi·tis** (mə-fī′tĭs) *n.* [Lat.] **1.** An offensive smell : STENCH. **2.** A foul-smelling poisonous gas emitted from the earth. **—me·phit′ic** (-fĭt′ĭk), **me·phit′i·cal** *adj.* **—me·phit′i·cal·ly** *adv.*

**mep·ro·bam·ate** (mĕp′rō-băm′āt′, mĕ-prō′bə-māt′) *n.* [ME(THYL) + PRO(PYL) + (DICAR)BAMATE.] A bitter white powder, $CH_3(C_3H_7)C(CH_2OOCNH_2)_2$, used as a tranquilizer.

**-mer** *suff. var. of* -MERE.

**mer·bro·min** (mər-brō′mĭn) *n.* [MER(CURIC) + BROM(O)– + –IN.] A green crystalline organic compound, $C_{20}H_8Br_2HgNa_2O_6$, forming a red aqueous solution, used as a germicide and antiseptic.

**mer·can·tile** (mûr′kən-tēl′, -tīl′, -tĭl) *adj.* [Fr. < Ital. < *mercante,* merchant < Lat. *mercans* < *mercari,* to trade.] **1.** Of or relating to merchants or trade. **2.** Of or relating to mercantilism.

**mer·can·til·ism** (mûr′kən-tē-lĭz′əm, -tĭ-) *n.* [MERCANTIL(E) + -ISM.] **1.** The theory and system of political economy prevailing in Europe after the decline of feudalism based on national policies of amassing bullion, establishing colonies and a merchant marine, and developing industry to achieve a favorable balance of trade. **2.** Commercialism. **—mer′can·til′ist** *n.* **—mer′can·til·is′tic** *adj.*

**mer·cap·tan** (mər-kăp′tăn′) *n.* [G. < Dan. < Med. Lat. *mercurium captans,* (one) seizing mercury.] A sulfur-containing organic compound with the general formula RSH, R being any radical, e.g., ethyl mercaptan, $C_2H_5SH$.

**Mer·ca·tor projection** (mər-kā′tər) *n.* [After Gerhardus *Mercator* (1512–1594).] A map projection in which the meridians and parallels of latitude appear as lines crossing at right angles and in which areas appear greater farther from the equator.

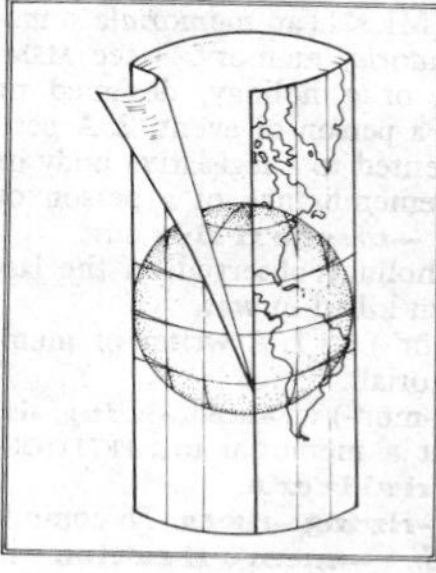

**Mercator projection**

**mer·ce·nar·y** (mûr′sə-nĕr′ē) *adj.* [ME, a mercenary < Lat. *mercenarius* < *merces,* wages.] **1.** Motivated solely by a desire for monetary or material gain. **2.** Hired for service in a foreign army. *—n., pl.* **-ies. 1.** One who serves merely for monetary gain : HIRELING. **2.** A professional soldier hired by a foreign country. **—mer′ce·nar′i·ly** *adv.* **—mer′ce·nar′i·ness** *n.*

**mer·cer** (mûr′sər) *n.* [ME < OFr. *mercier* < Lat. *merx,* merchandise.] *Chiefly British.* A textile dealer.

**mer·cer·ize** (mûr′sə-rīz′) *vt.* **-ized, -iz·ing, -iz·es.** [After John *Mercer* (1791–1866), its inventor.] To treat (cotton thread) with sodium hydroxide so as to shrink the fiber and increase its color absorption and luster.

**mer·chan·dise** (mûr′chən-dīz′, -dīs′) *n.* [ME *merchaundise* < OFr. *marcheandise* < *marcheant,* merchant.] Goods or commodities that may be bought or sold. *—v.* (mûr′chən-dīz′) **-dised, -dis·ing, -dis·es.** *—vt.* **1.** To buy and sell (goods). **2.** To promote the sale of, as by advertising. *—vi.* To trade commercially. **—mer′chan·dis′er** *n.*

**mer·chant** (mûr′chənt) *n.* [ME < OFr. *marcheant* < VLat. **mercatans* < Lat. *mercari,* to trade.] **1.** One whose occupation is the buying and selling of goods for profit. **2.** One who runs a retail business : SHOPKEEPER. *—adj.* **1.** Of or relating to a merchant, merchandise, or commercial trade. **2.** Of or relating to the merchant marine.

**mer·chant·a·ble** (mûr′chənt-ə-bəl) *adj.* Appropriate for buying and selling : MARKETABLE.

**mer·chant·man** (mûr′chənt-mən) *n.* **1.** A commercial ship. **2.** *Archaic.* A merchant.

**merchant marine** *n.* **1.** A nation's commercial ships. **2.** The personnel of merchant ships.

**merchant ship** *n.* MERCHANTMAN 1.

**Mer·ci·an** (mûr′shē-ən, -shən) *n.* **1.** A native or resident of Mercia. **2.** The Old English dialect of Mercia. **—Mer′ci·an** *adj.*

**mer·ci·ful** (mûr′sĭ-fəl) *adj.* Full of mercy. **—mer′ci·ful·ly** *adv.* **—mer′ci·ful·ness** *n.*

**mer·ci·less** (mûr′sĭ-lĭs) *adj.* Having or displaying no mercy. **—mer′ci·less·ly** *adv.* **—mer′ci·less·ness** *n.*

☆ ***syns:*** MERCILESS, PITILESS, REMORSELESS, UNMERCIFUL *adj. core meaning* : having or displaying no mercy <continued the *merciless* bombardment> ***ant:*** merciful

**mercur-** *pref. var. of* MERCURO-.

**mer·cu·rate** (mûr′kyə-rāt′) *vt.* **-rat·ed, -rat·ing, -rates.** To treat or combine with mercury or a mercury compound. **—mer′cu·ra′tion** *n.*

**mer·cu·ri·al** (mər-kyŏŏr′ē-əl) *n.* [Lat. *mercurialis* < *Mercurius,* Mercury.] A chemical or pharmaceutical containing mercury. —*adj.* **1.** *often* **Mercurial.** Of or relating to the Roman god Mercury or the planet Mercury. **2.** Having the characteristics of eloquence, swiftness, shrewdness, and thievishness attributed to the god Mercury. **3.** Containing or caused by the action of the element mercury. **4.** Being rapidly changeable in character <a *mercurial* disposition> **—mer·cu′ri·al·ly** *adv.*

**mer·cu·ri·al·ism** (mər-kyŏŏr′ē-ə-lĭz′əm) *n.* Poisoning caused by mercury or its compounds.

**mer·cu·ric** (mər-kyŏŏr′ĭk) *adj. Chem.* Relating to or containing bivalent mercury.

**mercuric chloride** *n.* A poisonous white crystalline compound, $HgCl_2$, used as an antiseptic and disinfectant, in insecticides, preservatives, and batteries, and in metallurgy and photography.

**mercuric sulfide** *n.* A poisonous compound, HgS, having two forms: **a.** Black mercuric sulfide, a black powder obtained from mercury salts or by the reaction of mercury with sulfur, used as a pigment. **b.** Red mercuric sulfide, a bright scarlet powder derived from heating mercury with sulfur, used as a pigment.

**mercuro-** *or* **mercur-** *pref.* [< MERCURY.] Mercury <*mercurous*>

**Mer·cu·ro·chrome** (mər-kyŏŏr′ə-krōm′). A trademark for a solution of merbromin, used as an antiseptic.

**mer·cu·rous** (mər-kyŏŏr′əs, mûr′kyər-əs) *adj. Chem.* Relating to or containing monovalent mercury.

**mercurous chloride** *n.* Calomel.

**Mer·cu·ry** (mûr′kyə-rē) *n.* [ME *Mercurie* < OFr. < Lat. *Mercurius.*] **1.** *Rom. Myth.* A god that served as messenger to the other gods and was himself the god of travel, commerce, and thievery. **2. mercury.** *Symbol* **Hg** A silvery-white poisonous metallic element used in thermometers, barometers, vapor lamps, and batteries and in the preparation of chemical pesticides; atomic number 80; atomic weight 200.59. **3. mercury.** Temperature <The *mercury* is rising.> **4. mercury.** A weedy plant of the genera *Mercurialis* or *Acalypha.* **5.** The smallest of the planets and the one nearest the sun, having a sidereal period of revolution about the sun of 88.0 days at a mean distance of 36.2 million miles or 58.3 million kilometers, a mean radius of approx. 1,500 miles or 2,414 kilometers, and a mass approx. 0.05 that of Earth.

**mer·cu·ry-va·por lamp** (mûr′kyə-rē-vā′pər) *n.* A lamp in which ultraviolet and yellowish-green to blue visible light is produced by an electric discharge through mercury vapor.

**mer·cy** (mûr′sē) *n., pl.* **-cies.** [ME < OFr. *merci* < Med. Lat. *merces* < Lat., reward.] **1.** Kind and compassionate treatment : CLEMENCY. **2.** A disposition to be forgiving and kind. **3.** A fortunate occurrence <a *mercy* we weren't killed> **4.** Alleviation of distress : RELIEF.

**mercy killing** *n.* Euthanasia.

**mercy seat** *n.* **1.** The golden covering of the ark of the covenant, held to be the resting place of God. **2.** The throne of God.

**mere¹** (mîr) *adj. superl.* **mer·est.** [ME, absolute < OFr. *mer* < Lat. *merus,* pure.] **1.** Being nothing more than what is specified <a *mere* whim> **2.** Unadulterated : pure.

**mere²** (mîr) *n.* [ME < OE.] A small pond, lake, or marsh.

**mere³** (mîr) *n.* [ME < OE *mǣre.*] *Archaic.* A boundary.

**-mere** *or* **-mer** *suff.* [Fr. *-mere* < Gk. *meros,* part.] Part : segment <blasto*mere*>

**mere·ly** (mîr′-lē) *adv.* Nothing more than.

**mer·e·tri·cious** (mĕr′ĭ-trĭsh′əs) *adj.* [Lat. *meritricius* < *meretrix,* prostitute < *merēre,* to earn money.] **1.** Relating to or resembling a prostitute. **2. a.** Attractive in a tawdry manner <*meretricious* costume jewelry> **b.** Based on pretense : INSINCERE <*meretricious* remarks> **—mer′e·tri′cious·ly** *adv.* **—mer′e·tri′cious·ness** *n.*

**mer·gan·ser** (mər-găn′sər) *n.* [NLat. : Lat. *mergus,* diver (< *mergere,* to plunge) + Lat. *anser,* goose.] A fish-eating duck of the genus *Mergus,* with a slim hooked bill.

**merge** (mûrj) *v.* **merged, merg·ing, merg·es.** [Lat. *mergere,* to plunge.] —*vt.* **1.** To cause to be united or gradually absorbed in stages. **2.** To combine, as sets of data. —*vi.* To blend together gradually. **—mer′gence** *n.*

**merg·er** (mûr′jər) *n.* **1.** The union of two or more commercial interests or corporations. **2.** *Law.* Absorption of a lesser estate, liability, action, right, or offense into a greater one.

**me·rid·i·an** (mə-rĭd′ē-ən) *n.* [ME < OFr. *meridiane,* midday < Lat. *meridianus,* of midday < *meridies,* midday : *medius,* middle + *dies,* day.] **1. a.** A great circle on the earth's surface passing through both geophysical poles. **b.** Either half of such a great circle lying between the poles. **2.** *Astron.* A great circle passing through the two poles of the celestial sphere and the observer's zenith. **3.** *Math.* **a.** A curve on a surface of revolution, formed by the intersection of a plane containing the axis of revolution with the surface. **b.** A plane section of a surface of revolution containing the axis of revolution. **4.** The highest point or stage of development : ZENITH <a leader at the *meridian* of power> **5.** *Archaic.* Noon. **—me·rid′i·an** *adj.*

**me·rid·i·o·nal** (mə-rĭd′ē-ə-nəl) *adj.* [ME, pertaining to the sun's position at noon < OFr. *meridionel* < LLat. *meridionalis* < Lat. *meridianus,* of midday. —see MERIDIAN.] **1.** Of or relating to a meridian. **2.** Typical of southern areas or people. **3.** Situated in the south : SOUTHERLY. —*n.* A resident of a southern region, esp. of France.

**me·ringue** (mə-răng′) *n.* [Fr. *méringue.*] **1.** A dessert topping made of beaten and baked egg whites. **2.** A small pastry shell or cake of meringue, often containing nutmeats or fruit.

**me·ri·no** (mə-rē′nō) *n., pl.* **-nos.** [Sp.] **1. a.** A sheep orig. bred in Spain. **b.** The fine wool of the merino. **2.** A soft lightweight fabric made orig. of merino wool but now of any fine wool. **3. a.** A fine wool and cotton yarn used for knitted clothing and hosiery. **b.** A fabric knitted from merino yarn.

**mer·i·stem** (mĕr′ĭ-stĕm′) *n.* [< Gk. *meristos,* divided < *merizein,* to divide < *meris,* division.] The growing point or area of rapidly dividing cells at the tip of a root, stem, or branch. **—mer′i·ste·mat′ic** *adj.* **—mer′is·te·mat′i·cal·ly** *adv.*

**me·ris·tic** (mə-rĭs′tĭk) *adj.* [Gk. *meristos,* divided. —see MERISTEM.] **1.** Composed of segments <*meristic* worms> **2.** Modified by changes in the number or placement of entire body parts. **—me·ris′ti·cal·ly** *adv.*

**mer·it** (mĕr′ĭt) *n.* [ME < OFr. *merite,* reward or punishment < Lat. *meritum* < *merēre,* to deserve.] **1.** A usu. high level of superiority : WORTH. **2.** An aspect of a person's character or actions deserving approval or disapproval. **3.** Spiritual credit granted for good works. **4. merits.** *Law.* **a.** A party's strict legal rights, excluding jurisdictional or technical aspects. **b.** The factual substance of a case as distinguished from its form and procedural aspects. **5.** The intrinsic right or wrong or actual facts of a matter. —*v.* **-it·ed, -it·ing, -its.** —*vt.* **1.** To earn : deserve. **2.** To have the right to claim (a divine reward). —*vi.* To gain spiritual merit. **—mer′it·ed·ly** *adv.*

☆ ***syns:*** MERIT, CALIBER, QUALITY, STATURE, VALUE, VIRTUE, WORTH *n. core meaning* : a usu. high level of superiority <a book of some *merit*>

**mer·i·toc·ra·cy** (mĕr′ĭ-tŏk′rə-sē) *n., pl.* **-cies. 1.** A system in which advancement is based on achievement or ability. **2.** Leadership by talented achievers. **—mer′it·o·crat′** (-tə-krăt′) *n.* **—mer′it·o·crat′ic** *adj.*

**mer·i·to·ri·ous** (mĕr′ĭ-tôr′ē-əs, -tōr′-) *adj.* [Lat. *meritorius,* earning money < *merēre,* to earn.] Deserving praise or reward : ADMIRABLE. **—mer′i·to′ri·ous·ly** *adv.*

**merit system** *n.* A system of promoting and appointing civil service personnel on the basis of merit, determined by competitive examinations.

**merle** *also* **merl** (mûrl) *n.* [ME < OFr. < Lat. *merulus.*] An Old World songbird : BLACKBIRD.

**mer·lin** (mûr′lĭn) *n.* [ME < AN *merilun,* of Germanic orig.] The pigeon hawk.

**Mer·lin** (mûr′lĭn) *n.* A magician and prophet who was a counselor to King Arthur.

**mer·lon** (mûr′lən) *n.* [Fr. < Ital. *merlone,* aug. of *merlo,* battlement < Med. Lat. *merulus* < Lat., merle.] The solid section of a crenelated wall between two open spaces.

**mer·maid** (mûr′mād′) *n.* [ME : *mere,* sea + *maid,* maid.] A legendary sea creature with the head and upper body of a woman and the tail of a fish.

**mer·man** (mûr′măn′, -mən) *n.* [MER(MAID) + MAN.] A legendary sea creature with the head and upper body of a man and the tail of a fish.

**mero-** *pref.* [NLat. < Gk. *meros,* part.] **1.** Part : segment <*mero*zoite> **2.** Partial : partially <*mero*pia>

**mer·o·blas·tic** (mĕr′ə-blăs′tĭk) *adj.* Undergoing partial cleavage. —Used of an egg with a large yolk. **—mer′o·blas′ti·cal·ly** *adv.*

**mer·o·crine** (mĕr′ə-krĭn, -krīn′, -krēn′) *adj.* [MERO- + Gk *krinein,* to separate.] Of or relating to a gland whose cells remain intact during secretion.

**mer·o·my·o·sin** (mĕr′ə-mī′ə-sĭn) *n.* Either of two protein subunits of a myosin molecule.

**Mer·o·pe** (mĕr′ə-pē′) *n.* [Gk. *Meropē.*] **1.** *Gk. Myth.* One of the Pleiades, who hid her face in shame after marrying a mortal. **2.** The seventh star in the Pleiades cluster and the only one invisible to the naked eye.

**me·ro·pi·a** (mə-rō′pē-ə) *n.* Partial blindness. **—me·ro′pic** (-rō′pĭk, -rŏp′ĭk) *adj.*

**-merous** *suff.* [NLat. *-merus* < Gk. *meros,* part.] Having a given kind or number of parts <aniso*merous*>

**Mer·o·vin·gi·an** (mĕr′ə-vĭn′jē-ən, -jən) *adj.* [Fr. *mérovingien* < Med. Lat. *Merovingi,* Merovingians < *Merovaeus,* Mérovée (d. 458), second Frankish king.] Of or relating to the first dynasty of Frankish kings that ruled over Gaul from about A.D. 500 until A.D. 751. **—Mer′o·vin′gi·an** *n.*

**mer·o·zo·ite** (mĕr′ə-zō′īt′) *n.* A cell produced by fission of a sporozoan.

**mer·ri·ment** (mĕr′ĭ-mənt) *n.* Gaiety : merrymaking.

**mer·ry** (mĕr′ē) *adj.* **-ri·er, -ri·est.** [ME *merri* < OE *myrige.*] **1.** Full of high-spirited gaiety : JOLLY. **2.** Characterized by or offering fun and festivity. **3.** Delightful : entertaining. **4.** Brisk. **—mer′ri·ly** *adv.* **—mer′ri·ness** *n.*

☆ ***syns:*** MERRY, FESTIVE, GALA, GLAD, GLADSOME, HAPPY, JOYFUL, JOYOUS *adj. core meaning* : marked by celebrations <a *merry* holiday season>

**mer·ry-an·drew** (mĕr′ē-ăn′dro͞o) *n.* [MERRY + the name *Andrew.*] A clown : buffoon.

**mer·ry-bells** (mĕr′ē-bĕlz′) *n. (sing. or pl. in number).* The bellwort.

**mer·ry-go-round** (mĕr′ē-gō-round′) *n.* **1.** An amusement park ride with seats, often in the form of horses, fitted on a revolving circular platform. **2.** Playground equipment consisting of a small circular platform that revolves when pushed or pedaled. **3.** A whirl or swift round <a *merry-go-round* of appointments>

**mer·ry·mak·ing** (mĕr′ē-mā′kĭng) *n.* **1.** Participation in a revel. **2.** A festivity : revelry. **—mer′ry·mak′er** *n.*

**Mer·thi·o·late** (mər-thī′ə-lāt′). A trademark for thimerosal.

**mes-** *pref. var. of* MESO-.

**me·sa** (mā′sə) *n.* [Sp. < OSp. < Lat. *mensa,* table.] A flat-topped elevation with one or more clifflike sides.

**mé·sal·li·ance** (mā′ză-lyäɴs′, mā-zăl′ē-əns) *n.* [Fr. *més-*, mis- + *alliance,* alliance.] A marriage with one of inferior social position.

**mes·cal** (mĕs-kăl′) *n.* [Sp. < Nahuatl *mexcalli,* mescal liquor.] **1.** A spineless globe-shaped cactus, *Lophophora williamsii* of Mexico and the southwestern United States, with buttonlike tubercles that are dried and chewed as a drug by certain Indian tribes. **2.** A Mexican liquor distilled from the fermented juice of certain species of agave. **3.** MAGUEY 1.

**mes·ca·line** (mĕs′kə-lēn′, -lən) *n.* An alkaloid drug, $(CH_3O)_3C_6H_2(CH_2CH_2NH_2)$, that produces hallucinations.

**Mes·dames** (mā-däm′) *n.* **1.** *pl. of* MADAM 1. **2.** *pl. of* MADAME.

**Mes·de·moi·selles** (măd′mwä-zĕl′) *n. pl. of* MADEMOISELLE.

**mes·en·ceph·a·lon** (mĕz′ĕn-sĕf′ə-lŏn′, mĕs′-) *also* **mes·o·ceph·a·lon** (mĕz′ō-, mĕs′ō-) *n.* The middle section of the embryonic brain. **—mes′en·ce·phal′ic** (-sə-făl′ĭk) *adj.*

**mes·en·chyme** (mĕz′ən-kīm′, mĕs′-) *also* **mes·en·chy·ma** (mĕz-ĕng′kĭ-mə, mĕs-) *n.* [G. *Mesenchym* : *mes-*, meso- + *-enchym,* -enchyma.] The part of the embryonic mesoderm from which develop connective tissue, skeletal tissue, and the circulatory and lymphatic systems. **—mes′en′chy·mal, mes′en·chym′a·tous** (-kĭm′ə-təs, -kī′mē-) *adj.*

**mes·en·ter·i·tis** (mĕz-ĕn′tə-rī′tĭs, mĕs-) *n.* Inflammation of the mesentery.

**mes·en·ter·i·um** (mĕz′ən-tîr′ē-əm, mĕs′-) *n., pl.* **-i·a** (-ē-ə) [NLat. —see MESENTERY.] Mesentery.

**mes·en·ter·on** (mĕz-ĕn′tə-rŏn′, mĕs-) *n.* **1.** The midgut. **2.** The middle section of the gastrovascular cavity in anthozoans. **—mes·en′ter·on′ic** *adj.*

**mes·en·ter·y** (mĕz′ən-tĕr′ē, mĕs′-) *n., pl.* **-ies.** [NLat. *mesenterium* < Gk. *mesenterion* : *mesos,* middle + *enteron,* enteron.] One of several peritoneal folds that connect the intestines to the dorsal abdominal wall. **—mes′en·ter′ic** *adj.*

**mesh** (mĕsh) *n.* [Prob. < MDu. *maesche.*] **1. a.** One of the open spaces in a thread, cord, or wire network. **b.** *often* **meshes.** The threads, cords, or wires surrounding these spaces. **2.** A net or network. **3.** A fabric with an open network of interlacing threads. **4.** Something that entraps or snares <the *meshes* of politics> **5. a.** The engagement of gear teeth. **b.** The state of being so engaged. —*v.* **meshed, mesh·ing, mesh·es.** —*vt.* **1.** To ensnare or entangle. **2.** To cause (gear teeth) to become engaged. **3.** To cause to work closely together. —*vi.* **1.** To be or become entangled. **2.** To be or become engaged, as gear teeth. **3. a.** To coordinate or fit harmoniously. **b.** To accord with another : HARMONIZE. **—mesh′y** *adj.*

**mesh·work** (mĕsh′wûrk′) *n.* Meshes : network.

**me·si·al** (mē′zē-əl, -zhəl) *adj.* Of, in, near, or toward the middle. **—me′si·al·ly** *adv.*

**me·sit·y·lene** (mə-sĭt′l-ēn′) *n.* [MESITYL(OXIDE) + -ENE.] A hydrocarbon, $(CH_3)_3C_6H_3$, occurring in petroleum and coal tar and synthesized from acetone.

**mes·i·tyl oxide** (mĕs′ĭ-tĭl) *n.* [< Gk. *mesitēs,* mediator < *mesos,* middle.] An oily liquid, $(CH_3)_2C{:}CHCOCH_3$, obtained from acetones and used as a solvent and insect repellent.

**mes·mer·ism** (mĕz′mə-rĭz′əm, mĕs′-) *n.* [After Franz *Mesmer* (1734–1815).] **1.** Hypnotic induction held to involve animal magnetism. **2.** Hypnotism. **3.** Hypnotic appeal. **—mes·mer′ic** (-mĕr′ĭk) *adj.* **—mes·mer′i·cal·ly** *adv.* **—mes′mer·ist** *n.*

**mes·mer·ize** (mĕz′mə-rīz′, mĕs′-) *vt.* **-ized, -iz·ing, -iz·es. 1.** To hypnotize. **2.** To enthrall <The ballerina *mesmerized* the audience.> **—mes′mer·iz′er** *n.*

**mesne** (mēn) *adj.* [OFr. < AN *meen* < OFr. *meien,* middle < Lat. *medianus* < *medius.*] Intervening : intermediate.

**mesne lord** *n.* A feudal lord ranking between a superior lord and his own tenants or vassals.

**meso-** *or* **mes-** *pref.* [Gk. < *mesos,* middle.] **1.** In the middle : MIDDLE <*mesoderm*> **2.** Intermediate <*mesomorph*>

**mes·o·blast** (mĕz′ə-blăst′, mĕs′-) *n.* The middle germinal embryonic layer. **—mes′o·blas′tic** *adj.*

**mes·o·carp** (mĕz′ə-kärp′, mĕs′-) *n. Bot.* The middle, usu. fleshy layer of a pericarp.

**mes·o·ceph·a·lon** (mĕz′ō-sĕf′ə-lŏn′, mĕs′ō-) *n. var. of* MESENCEPHALON.

**mes·o·derm** (mĕz′ə-dûrm′, mĕs′-) *n.* The embryonic germ layer, lying between the ectoderm and the endoderm, from which develop connective tissue, muscles, and the urogenital and vascular systems. **—mes′o·der′mal, mes′o·der′mic** *adj.*

**mes·o·gle·a** *also* **mes·o·gloe·a** (mĕz′ə-glē′ə, mĕs′-, mē′zə-, -sə-) *n.* [MESO- + Gk. *gloia,* glue.] A gelatinous substance between the ectoderm and endoderm in some coelenterates and ctenophores. **—mes′o·gle′al** *adj.*

**Mes·o·lith·ic** (mĕz′ə-lĭth′ĭk, mĕs′-) *adj.* Designating the cultural period between the Paleolithic and Neolithic ages, marked by the appearance of the bow and of cutting tools. —*n.* The Mesolithic age.

**mes·o·morph** (mĕz′ə-môrf′, mĕs′-) *n.* A human body characterized by powerful musculature and a predominantly bony framework.

**mes·o·mor·phic** (mĕz′ə-môr′fĭk, mĕs′-) *adj.* **1.** *also* **mes·o·mor·phous** (-môr′fəs). Of, relating to, or existing in a state of matter intermediate between liquid and crystal. **2.** Of or relating to a mesomorph. **—mes′o·mor′phism, mes′o·mor′phy** *n.*

**mes·on** (mĕz′ŏn′, mē′zŏn′, mĕs′ŏn′, mē′sŏn′) *n.* A subatomic particle having integral spins and masses gen. intermediate between a lepton and baryon. **—me·son′ic** (mĕ-zŏn′ĭk, -sŏn′ĭk, mē-) *adj.*

**mes·o·neph·ros** (mĕz′ə-nĕf′rəs, -rŏs′, mĕs′-) *n.* [MESO- + Gk. *nephros,* kidney.] The midpart of the embryonic excretory system in vertebrates that becomes the functioning kidney in fish and amphibians. **—mes′o·neph′ric** *adj.*

**mes·o·pause** (mĕz′ə-pôz′, mĕs′-) *n.* The atmospheric zone approx. 50 miles or 80 kilometers above the earth, forming the upper limit of the mesosphere.

**mes·o·phyll** (mĕz′ə-fĭl′, mĕs′-) *n.* The soft tissue between the epidermal layers of a leaf. **—mes′o·phyl′lic, mes′o·phyl′lous** *adj.*

**mes·o·phyte** (mĕz′ə-fīt′, mĕs′-) *n.* A land plant that grows in a moderately moist environment. **—mes′o·phyt′ic** (-fĭt′ĭk) *adj.*

**mes·o·some** (mĕz′ə-sōm′, mĕs′-, mē′zə-, -sə-) *n.* A convoluted invagination of the cytoplasmic membrane in some bacterial cells.

**mes·o·sphere** (mĕz′ə-sfîr′, mĕs′-) *n.* The part of the atmosphere from approx. 20 to 50 miles or 30 to 80 kilometers above the earth, marked by a temperature range that decreases from 50°F to −130°F with increasing altitude. **—mes′o·spher′ic** (-sfîr′ĭk, -sfĕr′-) *adj.*

**mes·o·the·li·o·ma** (mĕz′ə-thē′lē-ō′mə, mĕs′-, mē′zə-, -sə-) *n., pl.* **-ma·ta** (-mə-tə) *or* **-mas.** A tumor derived from mesothelium.

**mes·o·the·li·um** (mĕz′ə-thē′lē-əm, mĕs′-) *n., pl.* **-li·a** (-lē-ə) [MESO- + (EPI)THELIUM.] **1.** The layer of flat cells lining the embryonic body cavity. **2.** A layer of epithelial squamous cells lining the peritoneum, pericardium, and pleura. **—mes′o·the′li·al** *adj.*

**mes·o·tho·rax** (mĕz′ə-thôr′ăks′, -thōr′-, mĕs′-) *n., pl.* **-rax·es** *or* **-ra·ces** (-ə-sēz′). The midsection of an insect's thoracic region, bearing the middle legs and the rear wings. **—mes′o·tho·rac′ic** *adj.*

**mes·o·tho·ri·um** (mĕz′ə-thôr′ē-əm, -thōr′-, mĕs′-) *n.* A decay product of thorium, as: **a.** MESOTHORIUM I. **b.** MESOTHORIUM II.

**mesothorium I** *n.* An isotope of radium.

**mesothorium II** *n.* An isotope of actinium.

**Mes·o·zo·ic** (mĕz′ə-zō′ĭk, mĕs′-) *adj.* Of, belonging to, or designating the third era of geologic time, including the Cretaceous, Jurassic, and Triassic periods and marked by the predominance of reptilian life forms. —*n.* The Mesozoic era.

**mes·quite** (mĕ-skēt′, mə-) *n.* [Sp. < Nahuatl *mizquitl.*] A shrub or small tree of the genus *Prosopis,* esp. *P. juliflora* of the southwestern United States and Mexico, whose pods are used as forage.

**mess** (mĕs) *n.* [ME *mes,* course of a meal < OFr. < Lat. *missus* < *mittere,* to place.] **1.** A disorderly accumulation of objects. **2.** A cluttered, untidy, usu. dirty condition. **3. a.** A confusing, disturbing, and troublesome state : MIX-UP. **b.** Senseless confusion and discontinuity : CHAOS. **4.** One who looks unkempt. **5. a.** *Archaic.* An amount of food for a meal, course, or dish. **b.** A serving of soft semiliquid food. **c.** A number or amount acquired <caught a *mess* of trout> **6. a.** A group of persons, usu. in the military, who regularly eat meals together. **b.** A mess hall. **c.** A meal eaten in a mess hall. —*v.* **messed, mess·ing, mess·es.** —*vt.* **1.** To make disorderly and dirty <*messed* up the kitchen> **2.** To mismanage : botch <*messed* up the exam> **3.** *Slang.* To be rough with : MANHANDLE. —*vi.* **1.** To cause or make a mess. **2.** To interfere : MEDDLE. **3.** To take a meal in a military mess. **—mess around.** *Informal.* **1.** To occupy time by tinkering or puttering. **2.** To associate with.

▲ word history: The word *mess* has been in the English language since the 14th century, but its meanings "a disorderly jumble" and "an untidy condition" did not develop until the 19th century. *Mess* originally meant "a portion of food," and in this sense it was used without disparagement for 500 years. In the 19th century *mess* was also used to refer to an unpalatable mixture of food, and it is likely that this use gave rise to the modern meanings, which have been extended beyond reference to food. In addition to "a portion of food," *mess* in medieval times also meant "a group of persons who usually eat together." This sense has been preserved in such expressions as *mess hall* and *officers' mess.*

**mes·sage** (mĕs′ĭj) *n.* [ME < OFr. < VLat. **missaticum* < Lat. *mittere,* to send.] **1.** A communication transmitted by various means, as speech or writing, from one person or group to another. **2.** An often formal statement made or read before a gathering. **3.** A basic theme or idea : SIGNIFICANCE <I get your *message.*>

**mes·sa·line** (mĕs′ə-lēn′) *n.* [Fr.] A soft, lightweight, shiny silk cloth with a twilled or satin weave.

**Mes·sei·gneurs** (mā-sĕ-nyœr′) *n. pl. of* MONSEIGNEUR.

**mes·sen·ger** (mĕs′ən-jər) *n.* [ME *messanger* < OFr. < *message,* message.] **1.** One who is charged with transmitting messages or performing errands, as: **a.** One employed to carry letters, telegrams, or parcels. **b.** A military or official courier. **c.** An envoy. **d.** A prophet. **2.** A bearer of news. **3.** *Archaic.* A forerunner : harbinger. **4.** *Naut.* A chain or rope for hauling in a cable.

**messenger ribonucleic acid** *n.* Messenger RNA.

**messenger RNA** *n.* A ribonucleic acid that carries the genetic information required for protein synthesis in cells.

**mess hall** *n.* **1.** A hall in which mess is served. **2.** A building used for serving and eating meals.

**Mes·si·ah** (mə-sī′ə) *n.* [Aram. *mĕshīḥa* or Heb. *māshiaḥ.*] **1. a.** *also* **Mes·si·as** (mə-sī′əs). The anticipated deliverer and king of the Jews. **b.** Jesus Christ. **2. messiah.** A liberator or deliverer.

**mes·si·an·ic** *also* **Mes·si·an·ic** (mĕs′ē-ăn′ĭk) *adj.* [NLat. **messianicus* < LLat. *messias,* messiah < Gk. < Aram. *mĕshīḥa.*] **1.** Of or relating to a messiah. **2.** Invoking the aura of a messiah.

**mes·si·a·nism** (mĕs′ē-ə-nĭz′əm, mə-sī′-) *n.* **1.** Belief in a messiah. **2.** Belief that a cause is totally right.

**Mes·si·as** (mə-sī′əs) *n. var. of* MESSIAH 1a.

**Mes·sieurs** (mĕs′ərz, mā-syœ′) *n. pl. of* MONSIEUR.

**mess jacket** *n.* A man's fitted waist-length uniform jacket, worn by some military officers on formal occasions.

**mess kit** *n.* A compactly arranged kit of cooking and eating utensils, used by campers and soldiers.

**mess·mate** (mĕs′māt′) *n.* One with whom one eats regularly, as in a military mess.

**mes·suage** (mĕs′wĭj) *n.* [ME < AN, prob. alteration of OFr. *mesnage.* —see MÉNAGE.] *Law.* A dwelling house with its outbuildings and adjoining lands.

**mess·y** (mĕs′ē) *adj.* **-i·er, -i·est. 1.** Marked by dirt and disorder. **2.** Lacking precision and neatness <a *messy* book report> **3.** Unpleasantly difficult to resolve or settle <a *messy* divorce case> **—mess′i·ly** *adv.* **—mess′i·ness** *n.*

☆ ***syns:*** MESSY, DISHEVELED, ILL-KEMPT, MUSSY, SLIPSHOD, SLOPPY, SLOVENLY, UNKEMPT, UNTIDY *adj. core meaning* : marked by an absence of cleanliness and order <a *messy* kitchen> ***ant:*** neat

**mes·ti·za** (mĕs-tē′zə) *n.* [Sp., fem. of *mestizo.* —see MESTIZO.] A woman of mixed European and American Indian ancestry.

**mes·ti·zo** (mĕs-tē′zō) *n., pl.* **-zos** or **-zoes.** [Sp. < *mestizo,* mixed < OSp. < Lat. *miscēre,* to mix.] A man of mixed European and American Indian ancestry.

**met** (mĕt) *v. p.t. & p.p. of* MEET[1].

**met-** *pref. var. of* META-.

**met·a** (mĕt′ə) *adj.* [< META-.] *Chem.* **1.** Of, relating to, or designating positions in a benzene ring separated by one carbon atom. **2.** Designating closely related, esp. isomeric, compounds.

**meta-** or **met-** *pref.* [Gk. < *meta,* beside, after.] **1. a.** Later in time <*met*estrus> **b.** At a later stage of development <*meta*nephros> **2.** Situated behind <*meta*carpus> **3. a.** Change : transformation <*meta*chromatism> **b.** Alternation <*meta*genesis> **4. a.** Beyond : transcending : more comprehensive <*meta*linguistics> **b.** At a higher stage of development <*meta*zoan> **5.** Having undergone metamorphosis <*meta*somatic> **6.** Derivative or related chemical substance <*meta*protein>

**met·a·bol·ic** (mĕt′ə-bŏl′ĭk) *adj.* [Gk. *metabolikos,* changeable < *metabolē,* change. —see METABOLISM.] **1.** *Biol.* Of, relating to, or exhibiting metabolism. **2.** *Zool.* Of, relating to, or undergoing metamorphosis. **—met′a·bol′i·cal·ly** *adv.*

**me·tab·o·lism** (mə-tăb′ə-lĭz′əm) *n.* [Gk. *metabolē,* change < *metaballein,* to change : *meta,* change + *ballein,* to throw.] *Biol.* **1.** The complex of chemical and physical processes involved in the maintenance of life. **2.** The functioning of a specific substance within the living body <water *metabolism*><iodine *metabolism*>

**me·tab·o·lite** (mə-tăb′ə-līt′) *n.* [METABOL(ISM) + -ITE.] One of various organic compounds produced by metabolism.

**me·tab·o·lize** (mə-tăb′ə-līz′) *v.* **-lized, -liz·ing, -liz·es.** *—vt.* To subject to metabolism. *—vi.* To undergo change by metabolism.

**met·a·car·pal** (mĕt′ə-kär′pəl) *adj.* Relating to the metacarpus. *—n.* A bone of the metacarpus.

**met·a·car·pus** (mĕt′ə-kär′pəs) *n.* The part of the hand or forefoot that includes the five bones between the phalanges and the carpus.

**met·a·cen·ter** (mĕt′ə-sĕn′tər) *n.* The intersection of the verticals through the center of buoyancy of a floating body when in equilibrium and when tilted. **—met′a·cen′tric** (-sĕn′trĭk) *adj.*

**met·a·chro·ma·tism** (mĕt′ə-krō′mə-tĭz′əm) *n.* A change in color caused by variation of the physical conditions to which a body is subjected, as in heating. **—met′a·chro·mat′ic** (-măt′ĭk) *adj.*

**met·a·eth·ics** (mĕt′ə-ĕth′ĭks) *n. (sing. in number).* The study of ethical terms, arguments, and judgments. **—met′a·eth′i·cal** *adj.*

**met·a·gal·ax·y** (mĕt′ə-găl′ək-sē) *n., pl.* **-ies.** The total physical universe including all galaxies.

**met·age** (mē′tĭj) *n.* **1.** Official measurement of weight or contents, as of trucks using state roads. **2.** The fee charged for metage.

**met·a·gen·e·sis** (mĕt′ə-jĕn′ĭ-sĭs) *n. Biol.* The occurrence in certain organisms of alternating sexual (gametophyte) and asexual (sporophyte) reproductive cycles. **—met′a·ge·net′ic** (-jə-nĕt′ĭk) *adj.*

**me·tag·na·thous** (mə-tăg′nə-thəs) *adj.* Having a beak in which the tips of the mandibles cross. —Used of birds.

**met·al** (mĕt′l) *n.* [ME < OFr. < Lat. *metallum* < Gk. *metallon.*] **1.** One of a category of electropositive elements that are usu. whitish, lustrous, and, in the transition metals, typically ductile and malleable with high tensile strength. **2.** An alloy of two or more metallic elements. **3.** An object made of metal. **4.** Basic character : METTLE. **5.** Molten glass, esp. when used in glassmaking. **6.** Molten cast iron. **7.** Printing type made of metal.

**met·a·lin·guis·tics** (mĕt′ə-lĭng-gwĭs′tĭks) *n. (sing. in number).* The study of the interrelationship between language and other cultural behavioral phenomena.

**metall-** or **metalli-** *pref. var. of* METALLO-.

**me·tal·lic** (mə-tăl′ĭk) *adj.* **1.** Of, relating to, or having the characteristics of a metal. **2.** Containing a metal <a *metallic* powder> **3.** Having a quality resembling metal, as: **a.** Having iridescence <*metallic* fabric> **b.** Acrid <a *metallic* taste> **4.** Harshly resonant <a *metallic* voice> **—me·tal′li·cal·ly** *adv.*

**metallic bond** *n.* The chemical bond typical of metals, produced by the sharing of valence electrons between atoms in a usu. stable crystalline structure.

**metallic soap** *n.* A soft waxlike organic compound composed of a metal oxide and a fatty acid, used as a drier or lubricant.

**met·al·lif·er·ous** (mĕt′l-ĭf′ər-əs) *adj.* [Lat. *metallifer* : *metallum,* metal + *ferre,* to bear.] Containing metal.

**met·al·line** (mĕt′l-īn, -ĭn′) *adj.* **1.** Of, resembling, or having the properties of a metal. **2.** Containing metal ions.

**metallo-** or **metalli-** or **metall-** *pref.* [< Lat. *metallum,* metal.] Metal <*metallo*therapy>

**met·al·log·ra·phy** (mĕt′l-ŏg′rə-fē) *n.* Study of the structure of metals and their compounds, esp. with a microscope. **—met′al·log′ra·pher** *n.* **—me·tal′lo·graph′ic** (mə-tăl′ə-grăf′ĭk) *adj.* **—me·tal′lo·graph′i·cal·ly** *adv.*

**met·al·loid** (mĕt′l-oid′) *n.* A nonmetallic element, as arsenic, that has some of the chemical properties of a metal, or one, as carbon, that can form an alloy with metals. *—adj. also* **met·al·loi·dal** (mĕt′l-oid′l). **1.** Pertaining to or having the properties of a metalloid. **2.** Resembling a metal in appearance.

**me·tal·lo·ther·a·py** (mə-tăl′ō-thĕr′ə-pē) *n. Med.* Use of metals or metal compounds in treating disease.

**met·al·lur·gy** (mĕt′l-ûr′jē) *n.* [NLat. *metallurgia* < Gk. *metallourgos,* miner : *metallon,* a mine + *ergon,* work.] **1.** The science or procedures of extracting metals from their ores, of purifying metals, and of creating useful items from metals. **2.** Knowledge and study of metals and their properties in bulk and at the atomic level. **—met′al·lur′gic, met′al·lur′gi·cal** *adj.* **—met′al·lur′gi·cal·ly** *adv.* **—met′al·lur′gist** *n.*

**met·al·work** (mĕt′l-wûrk′) *n.* Work done in metal.

**met·al·work·ing** (mĕt′l-wûr′kĭng) *n.* The act or process of shaping objects from metal. **—met′al·work′er** *n.*

**met·a·math·e·mat·ics** (mĕt′ə-măth′ə-măt′ĭks) *n. (sing. in number).* The study of the principles, conceptual elements, consistency, and other aspects of logical systems, esp. of mathematical systems. **—met′a·math′e·mat′i·cal** *adj.* **—met′a·math′e·ma·ti′cian** *n.*

**met·a·mere** (mĕt′ə-mîr′) *n. Zool.* One of a series of homologous body segments, as in worms and lobsters. **—met′a·mer′ic** (-mĕr′ĭk, -mîr′-) *adj.* **—met′a·mer′i·cal·ly** *adv.*

**me·tam·er·ism** (mə-tăm′ə-rĭz′əm) *n.* The condition of having the body divided into metameres.

**met·a·mor·phic** (mĕt′ə-môr′fĭk) *also* **met·a·mor·phous** (-fəs) *adj.* [METAMORPH(OSIS) + -IC.] **1.** Of or relating to metamorphosis. **2.** *Geol.* Characteristic of, relating to, or changed by metamorphism.

**met·a·mor·phism** (mĕt′ə-môr′fĭz′əm) *n.* [METAMORPH(OSIS) + -ISM.] *Geol.* An alteration in texture, composition, or structure of rock masses caused by great heat or pressure.

**met·a·mor·phose** (mĕt′ə-môr′fōz′, -fōs′) *v.* **-phosed, -phos·ing, -phos·es.** [OFr. *metamorphoser* < *metamorphose,* metamorphosis < Lat. *metamorphosis.* —see METAMORPHOSIS.] *—vt.* **1.** To change from one form to another, as by sorcery. **2.** To subject to metamorphosis or metamorphism. *—vi.* To be transformed by or as if by metamorphosis or metamorphism.

**met·a·mor·pho·sis** (mĕt′ə-môr′fə-sĭs) *n., pl.* **-ses** (-sēz′) [Lat. < Gk. *metamorphōsis* < *metamorphoun,* to transform : *meta,* change

+ *morphē,* form.] **1.** A transformation, as by supernatural means. **2.** A marked alteration in appearance, condition, character, or function. **3.** *Biol.* Change in the structure and habits of an animal during normal growth, usu. in the postembryonic stage. **4.** *Pathol.* Transformation of one kind of tissue into another, esp. degeneration.

**met·a·mor·phous** (mĕt′ə-môr′fəs) *adj. var. of* METAMORPHIC.

**met·a·neph·ros** (mĕt′ə-nĕf′rŏs′) *n.* [META- + Gk. *nephros,* kidney.] The embryonic vertebrate kidney, in its third stage, which becomes the adult kidney.

**met·a·phase** (mĕt′ə-fāz′) *n.* The stage of mitosis during which the chromosomes are aligned along the equator of the mitotic spindle.

**met·a·phor** (mĕt′ə-fôr′, -fər) *n.* [OFr. *metaphore* < Lat. *metaphora* < Gk. < *metapherein,* to transfer : *meta,* change + *pherein,* to bear.] A figure of speech in which a term is transferred from the object it ordinarily designates to an object it may designate only by implicit comparison or analogy, as in the phrase *evening of life.* **—met′a·phor′ic, met′a·phor′i·cal** *adj.* **—met′a·phor′i·cal·ly** *adv.*

**met·a·phos·phate** (mĕt′ə-fŏs′fāt′) *n.* The inorganic anion $PO_3^-$, or a compound containing it.

**met·a·phos·phor·ic acid** (mĕt′ə-fŏs-fôr′ĭk, -fŏr′-) *n.* An inorganic compound, $HPO_3$, used as a dehydrating agent and in dental cements.

**met·a·phrase** (mĕt′ə-frāz′) *n.* [NLat. *metaphrasis* < Gk. < *metaphrazein,* to translate : *meta,* change + *phrazein,* to tell.] A word-for-word translation. —*vt.* **-phrased, -phras·ing, -phras·es.** To manipulate the wording of (a text), esp. as a means of subtly altering the sense. **—met′a·phras′tic** (-frăs′tĭk) *adj.*

**met·a·phrast** (mĕt′ə-frăst′) *n.* [Med. Gk. *metaphrastēs* < Gk. *metaphrazein,* to translate. —see METAPHRASE.] One who renders a text into a different form, as by recasting prose in verse.

**met·a·phys·i·cal** (mĕt′ə-fĭz′ĭ-kəl) *adj.* [ME *metaphisicalle* < Med. Lat. *metaphysicalis* < *metaphysica,* metaphysics.] **1.** Of or relating to metaphysics. **2.** Based on abstract or speculative reasoning. **3.** Excessively subtle : ABSTRUSE. **4. a.** Immaterial : incorporeal. **b.** Supernatural. **5.** *often* **Metaphysical.** Of or designating a kind of 17th-cent. poetry marked by complex imagery and elaborate conceits. **—met′a·phys′i·cal·ly** *adv.*

**met·a·phy·si·cian** (mĕt′ə-fə-zĭsh′ən) *n.* One who specializes or is skilled in metaphysics.

**met·a·phys·ics** (mĕt′ə-fĭz′ĭks) *n.* [Med. Lat. *metaphysica* < Gk. *Ta meta ta phusika,* the things after the physics, the title of Aristotle's treatise on first principles (so called because it followed his work on physics).] *(sing. in number).* **1.** The branch of philosophy that systematically investigates the nature of first principles and problems of ultimate reality, including ontology and often cosmology. **2.** Critical or speculative philosophy.

**met·a·pla·sia** (mĕt′ə-plā′zhə, -zhē-ə) *n.* **1.** Metamorphosis of tissue from one type to another, as in ossification. **2.** The change of cells from a normal to an abnormal state. **—met′a·plas′tic** (-plăs′tĭk) *adj.*

**met·a·plasm** (mĕt′ə-plăz′əm) *n.* [Lat. *metaplasmus,* transformation < Gk. *metaplasmos* < *metaplassein,* to remold : *meta,* change + *plassein,* to mold.] **1.** Alteration of a word by adding, deleting, or transposing letters or syllables. **2.** *Biol.* Inert material in the protoplasm of a cell, as an egg yolk. **—met′a·plas′mic** (-plăz′mĭk) *adj.*

**met·a·pro·tein** (mĕt′ə-prō′tēn′, -prō′tē-ĭn) *n.* Any of various organic compounds resulting from a reaction between an acid or alkali and a protein that are soluble in weak acids or alkalis and insoluble in neutral solutions.

**met·a·psy·chol·o·gy** (mĕt′ə-sī-kŏl′ə-jē) *n.* Philosophical speculation on the structure, origin, and function of the mind and the relationship between the mind and objective reality.

**met·a·so·ma·tism** (mĕt′ə-sō′mə-tĭz′əm) *also* **met·a·so·ma·to·sis** (-tō′sĭs) *n. Geol.* Metamorphism in which chemical as well as physical changes occur as a result of reaction with external material. **—met′a·so·mat′ic** (-măt′ĭk) *adj.* **—met′a·so·mat′i·cal·ly** *adv.*

**met·a·sta·ble** (mĕt′ə-stā′bəl) *adj.* Designating a relatively unstable, transient, but significant state of a physical or chemical system, as of a supersaturated solution or an energetically excited atom. **—met′a·sta·bil′i·ty** (-stə-bĭl′ĭ-tē) *n.*

**me·tas·ta·sis** (mə-tăs′tə-sĭs) *n., pl.* **-ses** (-sēz′) [NLat. < LLat., transition < Gk. < *methistanai,* to change : *meta,* change + *histanai,* to cause to stand.] **1.** *Pathol.* Transmission of disease from an original site to one or more sites elsewhere in the body, as in cancer. **2.** A sudden transition from one point to another in rhetoric. **3.** Paramorphism. **—met′a·stat′ic** (mĕt′ə-stăt′ĭk) *adj.*

**me·tas·ta·size** (mə-tăs′tə-sīz′) *vi.* **-sized, -siz·ing, -siz·es.** To be transmitted, transformed, or transferred by or as if by metastasis.

**met·a·tar·sal** (mĕt′ə-tär′səl) *adj.* Of or relating to the metatarsus. —*n.* One of the bones of the metatarsus. **—met′a·tar′sal·ly** *adv.*

**met·a·tar·sus** (mĕt′ə-tär′səs) *n., pl.* **-si** (-sī′). **1.** The mid part of the human foot that forms the instep, made up of the five bones between the toes and the tarsus. **2.** A part of the hind foot in four-legged animals or of the foot in birds analogous to the metatarsus.

**me·tath·e·sis** (mə-tăth′ĭ-sĭs) *n., pl.* **-ses** (-sēz′) [Gk. < *metatithenai,* to transpose : *meta,* change + *tithenai,* to place.] **1.** Transposition within a word of letters, sounds, or syllables, as in the change from Old English *brid* to modern English *bird* or the confusion of *revelant* for *relevant.* **2.** *Chem.* Double decomposition. **—met′a·thet′ic** (mĕt′ə-thĕt′ĭk), **met′a·thet′i·cal** *adj.* **—met′a·thet′i·cal·ly** *adv.*

**me·tath·e·size** (mə-tăth′ĭ-sīz′) *vt.* & *vi.* **-sized, -siz·ing, -siz·es.** To subject to or undergo metathesis.

**met·a·tho·rax** (mĕt′ə-thôr′ăks′, -thōr′-) *n., pl.* **-rax·es** or **-ra·ces** (-ə-sēz′). The hindmost of the three thoracic segments of an insect. **—me′ta·tho·rac′ic** (-thə-răs′ĭk) *adj.*

**met·a·zo·an** (mĕt′ə-zō′ən) *n.* [NLat. *Metazoa,* a subdivision of the animal kingdom : META- + Gk. *zōia,* pl. of *zōion,* animal.] A member of one of two divisions of the animal kingdom, the Metazoa, which includes all animals more complex than the one-celled protozoan. **—met′a·zo′al, met′a·zo′an, met′a·zo′ic** *adj.*

**mete**[1] (mēt) *vt.* **met·ed, met·ing, metes.** [ME *meten* < OE *metan.*] **1.** To distribute by or as if by measure : dole out <a judge who *meted* out stiff sentences> **2.** *Archaic.* To measure.

**mete**[2] (mēt) *n.* [AN < Lat. *meta.*] A boundary.

**me·tem·psy·cho·sis** (mə-tĕm′sĭ-kō′sĭs, mĕt′əm-sī-) *n., pl.* **-ses** (-sēz′) [Gk. *metempsukhōsis* : *meta,* change + *empsukhos,* animate (*en,* in + *psukhē,* soul).] Transmigration of souls.

**met·en·ceph·a·lon** (mĕt′ĕn-sĕf′ə-lŏn′) *n., pl.* **-la** (-lə). The portion of the embryonic hindbrain from which the cerebellum and the pons develop. **—met′en·ce·phal′ic** (-sə-făl′ĭk) *adj.*

**me·te·or** (mē′tē-ər, -ôr′) *n.* [ME < OFr. *meteore* < Med. Lat. *meteorum* < Gk. *meteōron,* astronomical phenomenon < *meteōros,* high in the air.] **1.** The incandescent trail or streak that appears in the sky when a meteoroid is made luminous by friction with the earth's atmosphere. **2.** A meteoroid. **3.** An atmospheric phenomenon, as a rainbow, lightning, or snow.

**me·te·or·ic** (mē′tē-ôr′ĭk, -ŏr′-) *adj.* **1. a.** Of or relating to a meteor. **b.** Of, relating to, or formed by a meteoroid. **2.** Resembling a meteor in swiftness and brilliance <a *meteoric* career> **3.** Of or relating to the earth's atmosphere. **—me′te·or′i·cal·ly** *adv.*

**me·te·or·ite** (mē′tē-ə-rīt′) *n.* The metallic or stony material of a meteoroid that survives passage through the atmosphere and reaches the earth's surface. **—me′te·or·it′ic, me′te·or·it′i·cal** *adj.*

**me·te·or·o·graph** (mē′tē-ôr′ə-grăf′, -ŏr′-) *n.* An instrument that records simultaneously several meteorological conditions, as temperature, moisture, and barometric pressure.

**me·te·or·oid** (mē′tē-ə-roid′) *n.* Any of numerous celestial bodies, ranging in size from minute specks of dust to asteroids weighing thousands of tons, that appear as meteors when entering the earth's atmosphere.

**me·te·or·ol·o·gy** (mē′tē-ə-rŏl′ə-jē) *n.* [Gk. *meteōrologia,* discussion of astronomical phenomena : *meteōron,* astronomical phenomenon + *logos,* speech.] The science concerned with atmospheric phenomena, esp. weather and weather conditions. **—me′te·or·o·log′i·cal** (-ər-ə-lŏj′ĭ-kəl), **me′te·or·o·log′ic** *adj.* **—me′te·or·o·log′i·cal·ly** *adv.* **—me′te·or·ol′o·gist** *n.*

**meteor shower** *n.* A group of meteors appearing together and having an apparently common origin.

**me·ter**[1] (mē′tər) *n.* [ME < OFr. *metre* < Lat. *metrum* < Gk. *metron.*] **1. a.** The measured rhythm typical of verse. **b.** A specified rhythmic pattern of verse, usu. determined by the number and kinds of metrical units in a typical line <iambic *meter*> **2. a.** The division of music into measures or bars. **b.** A particular musical rhythm determined by the number of beats and the time value assigned to each note in a measure.

**me·ter**[2] (mē′tər) *n.* [Fr. *mètre* < Gk. *metron,* measure.] The fundamental unit of length in the metric system, equivalent to 39.37 inches, which was defined in 1790 as one ten-millionth ($10^{-7}$) of the earth's quadrant passing through Paris but was redefined in 1960 as the length equal to 1,650,763.73 wavelengths in a vacuum of the orange-red radiation of krypton 86.

**me·ter**[3] (mē′tər) *n.* [< -METER.] **1.** A device designed to measure time, speed, distance, or intensity or indicate and regulate or record the volume or amount of something, as a flow of gas or an electric current. **2.** A postage meter. **3.** A parking meter. —*vt.* **-tered, -ter·ing, -ters. 1. a.** To measure with a metering device. **b.** To supply or regulate by means of a metering device. **2.** To imprint with postage by means of a postage meter.

**-meter** *suff.* [Fr. *-mètre* < Gk. *metron,* measure.] Measuring device <anemo*meter*>

**me·ter-kil·o·gram-sec·ond system** (mē′tər-kĭl′ə-grăm-sĕk′ənd) *n.* A coherent system of units for mechanics, using the meter, kilogram, and second as basic units of length, mass, and time.

**meter maid** *n.* A woman member of a police traffic department who writes parking tickets.

**me·tes·trus** (mĕ-tĕs′trəs) *n.* The period of sexual inactivity that follows estrus. **—me·tes′trous** (-trəs) *adj.*

**meth-** *pref.* [< METHYL.] Methyl <*meth*ane>

**meth·ac·ry·late** (mĕth-ăk′rə-lāt′) *n.* [METH- + ACRYL(IC) + -ATE.] **1.** An ester of methacrylic acid, $CH_2{:}C(CH_3)COOR$, R being an organic radical, used to make plastics. **2.** A resin derived from methacrylic acid.

**meth·a·cryl·ic acid** (mĕth'ə-krĭl'ĭk) *n.* A colorless liquid, $CH_2{:}C(CH_3)COOH$, used in making resins and plastics.

**meth·a·done hydrochloride** (mĕth'ə-dōn') *n.* [(DI)METH(YL) + A(MINO) + D(IPHENYL) + *(heptan)one,* a ketone.] An organic compound, $C_{21}H_{27}NO{\cdot}HCl$, used as an analgesic and in treating heroin addiction.

**meth·am·phet·a·mine** (mĕth'ăm-fĕt'ə-mēn', -mĭn) *n.* An amine derivative, $C_{10}H_{15}N$, of amphetamine used in the form of its crystalline hydrochloride as a stimulant.

**meth·ane** (mĕth'ān') *n.* An odorless, colorless, flammable gas, $CH_4$, the major constituent of natural gas, used as a fuel and an important source of hydrogen and in a wide variety of organic compounds.

**methane series** *n.* Paraffin series.

**meth·a·nol** (mĕth'ə-nôl', -nōl') *n.* A colorless flammable liquid, $CH_2(OCH_3)_2$, used as an antifreeze, fuel, general solvent, and denaturant for ethyl alcohol.

**meth·a·qua·lone** (mĕth'ə-kwā'lōn') *n.* [Blend of METH- and *quinazoline,* a derivative of quinoline.] A habit-forming sedative and hypnotic drug, $C_{16}H_{14}N_2O$.

**Meth·e·drine** (mĕth'ĭ-drēn', -drĭn). A trademark for methamphetamine.

**me·theg·lin** (mə-thĕg'lĭn) *n.* [Welsh *meddyglyn* : *meddyg,* medicinal (< Lat. *medicus* < *mederi,* to heal) + *llyn,* a liquid.] A beverage usu. made of fermented honey and water : MEAD.

**met·he·mo·glo·bin** (mĕt-hē'mə-glō'bĭn) *n.* [MET(A)- + HEMOGLOBIN.] A brownish-red crystalline organic compound formed by oxidation of hemoglobin and found in the blood after poisoning by substances such as chlorates, nitrates, or ferricyanides.

**me·the·na·mine** (mə-thē'nə-mēn', -mĭn) *n.* [METH- + -EN(E) + AMINE.] An organic compound, $(CH_2)_6N_4$, used as a urinary tract antiseptic and in rubber vulcanizing.

**meth·i·cil·lin** (mĕth'ĭ-sĭl'ĭn) *n.* [METH- + (PEN)ICILLIN.] A synthetic penicillin resistant to penicillinase.

**me·thinks** (mĭ-thĭngks') *v. p.t.* **me·thought** (-thôt') [ME *me thinketh* < OE *mē þyncð* : *mē,* to me + *þyncð,* it seems.] *Archaic.* It seems to me.

▲ word history: The archaic form *methinks* is strictly speaking not one word but two: the pronoun *me* and the obsolete verb *think,* meaning "to seem." The obsolete verb is related to, but not the same as, the current verb *think,* meaning "to have in mind." *Methinks* is an impersonal construction, that is, one without an expressed subject. In *methinks, me* is a dative form of the pronoun *I* functioning as the indirect object of *think,* and *thinks* is the third person singular form of the verb. The construction *methinks* means "it seems to me," not "I think."

**me·thi·o·nine** (mə-thī'ə-nēn') *n.* [ME(TH)- + THION- + -INE².] An organic compound, $C_5H_{11}NO_2S$, derived from protein and used as a dietary supplement and in pharmaceuticals.

**meth·od** (mĕth'əd) *n.* [Fr. *méthode* < Lat. *methodus* < Gk. *methodos* : *meta,* after + *hodos,* journey.] **1.** A manner or means of procedure, esp. a systematic and regular way of accomplishing a given task. **2.** Orderly and planned arrangement. **3.** The procedures and techniques characteristic of a particular discipline or field of knowledge. **4. Method.** A system of acting in which a theatrical performer recalls emotions and reactions from past experience and utilizes them in the role being played.

☆ ***syns:*** METHOD, FASHION, MANNER, MODE, SYSTEM, WAY *n. core meaning* : the procedures or plans followed to accomplish a given task <practical *methods* of pollution control> METHOD often suggests regularity of procedure; it emphasizes detailed, logically ordered plans <three *methods* of purifying water> SYSTEM stresses order and regularity affecting all parts and details of a procedure <a *system* for improving production> MANNER, FASHION, and MODE refer more to individual and distinctive procedure, as that dictated by preference, tradition, or custom <taught in an innovative *manner*><sings in an interesting *fashion*><an unusual *mode* of painting> WAY is the most neutral and general of these terms and is often an inclusive synonym for them <found a better *way* of solving our problems>

**me·thod·i·cal** (mə-thŏd'ĭ-kəl) *also* **me·thod·ic** (-ĭk) *adj.* **1.** Arranged or proceeding in a systematic order. **2.** Marked by ordered and systematic behavior. **—me·thod'i·cal·ly** *adv.* **—me·thod'i·cal·ness** *n.*

**Meth·od·ism** (mĕth'ə-dĭz'əm) *n.* **1.** The beliefs, worship, and system of organization of the Methodists. **2. methodism.** Emphasis on systematic procedure.

**Meth·od·ist** (mĕth'ə-dĭst) *n.* **1.** A member of a Protestant Christian denomination whose theology developed from the teachings of John and Charles Wesley and others in England in the early 18th cent., is marked by an emphasis on the doctrines of free grace and individual responsibility. **2. methodist.** One who insists on or emphasizes systematic methods. **—Meth'od·is'tic** *adj.*

**meth·od·ize** (mĕth'ə-dīz') *vt.* **-ized, -iz·ing, -iz·es.** To reduce to or organize according to a method : SYSTEMATIZE. **—meth'od·i·za'tion** *n.* **—meth'od·iz'er** *n.*

**meth·od·ol·o·gy** (mĕth'ə-dŏl'ə-jē) *n., pl.* **-gies. 1.** The system of principles, procedures, and practices applied to a particular branch of knowledge. **2.** The branch of logic dealing with the general principles of the formation of knowledge. **—meth'od·o·log'i·cal** (mĕth'ə-də-lŏj'ĭ-kəl) *adj.* **—meth'od·o·log'i·cal·ly** *adv.* **—meth'od·ol'o·gist** (-jĭst) *n.*

**meth·o·trex·ate** (mĕth'ə-trĕk'sāt') *n.* [METH- + *trex-* (of unknown orig.) + -ATE.] A toxic antimetabolite, $C_{20}H_{22}N_8O_5$, used in treating cancer.

**me·thought** (mĭ-thôt') *v. Archaic. p.t. of* METHINKS.

**me·thox·y·chlor** (mə-thŏk'sĭ-klôr', -klōr') *n.* [METH- + OXY- + (TRI)CHLOR(ETHANE).] A white crystalline compound, $C_{16}Cl_3H_{15}O_2$, used as an insecticide.

**Me·thu·se·lah** (mə-thōō'zə-lə) *n.* [Heb. *Methūshelāh.*] **1.** A Biblical patriarch said to have lived 969 years. **2. methuselah.** A large wine bottle holding approx. 6½ quarts.

**meth·yl** (mĕth'əl) *n.* [Fr. *méthyle,* back-formation < *méthylène,* methylene.] The univalent organic radical $CH_3$, derived from methane and occurring in many important organic compounds. **—me·thyl'ic** (mə-thĭl'ĭk) *adj.*

**methyl acetate** *n.* An organic compound, $C_3H_6O_2$, used as a paint remover and general solvent and in making perfumes.

**meth·yl·al** (mĕth'ə-lăl') *n.* A colorless flammable liquid, $CH_2(OCH_3)_2$, used to make perfumes, adhesives, and protective coatings.

**methyl alcohol** *n.* Methanol.

**meth·yl·a·mine** (mĕth'ə-lə-mēn', -lăm'ēn', mə-thĭl'ə-mēn') *n.* A flammable gas, $CH_3NH_2$, produced by decomposition of organic matter and synthesized for use as a solvent and in making dyes and insecticides.

**meth·yl·ase** (mĕth'ə-lās', -lāz') *n.* An enzyme that catalyzes a methylation reaction.

**meth·yl·ate** (mĕth'ə-lāt') *n.* An organic compound in which the hydrogen of the hydroxyl group (OH) of methyl alcohol is replaced by a metal. *—vt.* **-at·ed, -at·ing, -ates. 1.** To combine or mix with methyl alcohol. **2.** To combine with the methyl radical. **—meth'yl·a'tion** *n.* **—meth'yl·a'tor** *n.*

**methylated spirit** *n. often* **methylated spirits.** A denatured alcohol composed of a mixture of ethyl alcohol and methyl alcohol.

**methyl bromide** *n.* A toxic gas, $CH_3Br$, used as a fumigant.

**methyl chloride** *n.* An explosive gas, $CH_3Cl$, used in organic synthesis and polymerization and as a refrigerant and anesthetic.

**meth·yl·ene** (mĕth'ə-lēn') *n.* [Fr. *méthylène* : Gk. *methu,* wine + Gk. *hulē,* wood + *-ène,* -ene.] A bivalent organic radical, $CH_2$, a component of unsaturated hydrocarbons.

**methylene blue** *n.* An organic compound, $C_{16}H_{18}N_3SCl{\cdot}3H_2O$, whose dark-green crystals or powder forms a deep-blue solution when dissolved in water, used as a biological stain and antidote for cyanide poisoning.

**methyl ethyl ketone** *n. Chem.* Butanone.

**methyl methacrylate** *n.* A colorless liquid, $CH_2{:}C(CH_3)COOCH_3$, used as a monomer in plastics.

**meth·yl·naph·tha·lene** (mĕth'əl-năf'thə-lēn', -năp'thə-) *n.* An organic compound, $C_{10}H_7CH_3$, obtained from coal tar in two isomeric forms, one a liquid, the other a solid.

**met·i·cal** (mĕt'ĭ-kəl) *n.* —See table at CURRENCY.

**me·tic·u·lous** (mə-tĭk'yə-ləs) *adj.* [Lat. *meticulosus,* timid < *metus,* fear.] **1.** Very careful : PRECISE. **2.** Excessively concerned with details : NIT-PICKING. **—me·tic'u·los'i·ty** (-lŏs'ĭ-tē), **me·tic'u·lous·ness** *n.* **—me·tic'u·lous·ly** *adv.*

**mé·tier** (mā-tyā') *n.* [Fr. < OFr. *mestier* < Lat. *ministerium,* occupation.] **1.** An occupation, trade, or profession. **2.** Work or activity for which one is especially suited : LONG SUIT.

**mé·tis** (mā-tēs') *n., pl.* **métis.** [Canadian Fr. < OFr. *metis,* of mixed race < Lat. *miscēre,* to mix.] **1.** One of mixed Indian and French-Canadian ancestry. **2.** A crossbred animal.

**Me·ton·ic cycle** (mə-tŏn'ĭk) *n.* [After *Meton,* Athenian astronomer of the 5th cent. B.C., its discoverer.] A period of 235 lunar months, or about 19 Julian years, at the end of which the phases of the moon recur in the same order and on the same days as in the preceding cycle.

**met·o·nym** (mĕt'ə-nĭm') *n.* A word used in metonymy.

**me·ton·y·my** (mə-tŏn'ə-mē) *n., pl.* **-mies.** [LLat. *metonymia* < Gk. *metōnumia* : *meta-,* changing + *onoma,* name.] A figure of speech in which an attribute or commonly associated feature is used to name or designate something, as in *The pen is mightier than the sword.* **—met'o·nym'ic** (mĕt'ə-nĭm'ĭk), **met'o·nym'i·cal** (-ĭ-kəl) *adj.* **—met'o·nym'i·cal·ly** *adv.*

**me-too** (mē'tōō') *adj. Informal.* Advocating practices or principles closely similar to and copied from those of a rival. **—me'-too'er** *n.*

**met·o·pe** (mĕt'ə-pē) *n.* [Lat. *metopa* < Gk. *metopē* : *meta,* between + *opē,* opening.] The space between two triglyphs on a Doric frieze.

**me·top·ic** (mə-tŏp'ĭk) *adj.* [Gk. *metōpikos* < *metōpon,* forehead : *meta,* between + *ōps,* eye.] Of or relating to the forehead.

**met·o·pon hydrochloride** (mĕt'ə-pŏn') *n.* [E. *metopon,* a morphine derivative + HYDROCHLORIDE.] A narcotic drug, $C_{18}H_{21}NO_3{\cdot}HCl$, derived from morphine.

**metr-** *pref. var. of* METRO-.
**Met·ra·zol** (mĕt′rə-zôl′, -zōl′). A trademark for pentylenetetrazol.
**me·tre**[1] (mē′tər) *n. Chiefly Brit. var. of* METER[1].
**me·tre**[2] (mē′tər) *n. Chiefly Brit. var. of* METER[2].
**met·ric**[1] (mĕt′rĭk) *adj.* [Fr. *métrique* < *mètre,* meter < Gk. *metron,* measure.] Designating, relating to, or using the metric system.
**met·ric**[2] (mĕt′rĭk) *n.* **1.** A standard of measurement. **2.** *Math.* A geometric function defined for a coordinate system such that the distance between any two points in that system may be determined from their coordinates.
**met·ric**[3] (mĕt′rĭk) *n.* [Gk. *metrikē (tekhnē),* (the art) of meter.] Metrics.
**met·ri·cal** (mĕt′rĭ-kəl) *adj.* [Lat. *metricus* < Gk. *metrikos* < *metron,* meter.] **1.** Of, relating to, or composed in rhythmic meter. **2.** Of or relating to measurement. **—met′ri·cal·ly** *adv.*
**met·ri·ca·tion** (mĕt′rĭ-kā′shən) *n.* Conversion to the metric system of weights and measures : METRIFICATION.
**metric centner** *n.* A unit of mass equal to 100 kilograms.
**metric hundredweight** *n.* A unit of mass equal to 50 kilograms.
**met·rics** (mĕt′rĭks) *n. (sing. in number).* The branch of prosody dealing with measure and metrical structures.
**-metrics** *suff.* [< METRIC[2].] The application of statistics and mathematical analysis to a specified field of study <econo*metrics*>
**metric system** *n.* A decimal system of weights and measures based on the meter as a unit length and the kilogram as a unit mass.
**metric ton** *n.* A unit of mass equal to 1,000 kilograms.
**met·ri·fy** (mĕt′rə-fī′) *vt. & vi.* **-fied, -fy·ing, -fies.** [OFr. *metrifier* < Med. Lat. *metrificare* : Lat. *metrum,* measure (< Gk. *metron*) + Lat. *facere,* to make.] **1.** To compose in or put into rhythmic meters. **2.** To convert to or adopt the metric system. **—met′ri·fi·ca′tion** *n.*
**me·tri·tis** (mə-trī′tĭs) *n.* Inflammation of the uterus.
**met·ro** (mĕt′rō) *n., pl.* **-ros.** [Fr., short for *(chemin de fer) métropolitain,* metropolitan (railway).] A subway system.
**metro-** *or* **metr-** *pref.* [NLat. < Gk. *mētro-* < *mētra,* uterus < *mētēr,* mother.] Uterus <*metritis*>
**me·trol·o·gy** (mĕ-trŏl′ə-jē) *n., pl.* **-gies.** [Fr. *métrologie* < Gk. *metrologia,* theory of ratios : *metron,* measure + *logos,* reckoning.] **1.** The science that deals with measurement. **2.** A system of measurement. **—met′ro·log′i·cal** (mĕt′rə-lŏj′ĭ-kəl) *adj.* **—met′ro·log′i·cal·ly** *adv.* **—me·trol′o·gist** *n.*
**met·ro·nome** (mĕt′rə-nōm′) *n.* [Gk. *metron,* measure + *nomos,* rule.] A device to mark time at a steady beat in adjustable intervals. **—met′ro·nom′ic** (-nŏm′ĭk) *adj.* **—met′ro·nom′i·cal·ly** *adv.*
**me·tro·nym·ic** (mē′trə-nĭm′ĭk, mĕt′rə-) *also* **ma·tro·nym·ic** (măt′-) *adj.* [Gk. *mētēr,* mother + Gk. *onuma,* name + -IC.] Of, relating to, or derived from the name of one's mother or female ancestor. —*n.* A metronymic name.
**me·trop·o·lis** (mə-trŏp′ə-lĭs) *n.* [LLat. < Gk. *mētropolis,* mother city : *mētēr,* mother + *polis,* city.] **1.** A major city. **2.** A city regarded as the center of a specific activity <a great entertainment *metropolis*> **3.** The chief see of a metropolitan bishop, esp. the main diocese of a specific ecclesiastical province. **4.** The mother city of an ancient Greek colony or state. **5.** *Zool.* A region in which a particular kind of organism lives and thrives.
**met·ro·pol·i·tan** (mĕt′rə-pŏl′ĭ-tən) *adj.* [ME < LLat. *metropolitanus* < Gk. *mētropolitēs,* citizen of a metropolis < *mētropolis,* mother city. —see METROPOLIS.] **1. a.** Of, relating to, or typical of a metropolis. **b.** Making up a metropolis. **2.** Relating to or comprising the home territory of a sovereign state. **3.** Of or relating to a metropolitan. —*n.* **1. a.** An archbishop who has authority over bishops in the Roman Catholic and other episcopal churches. **b.** A bishop ranking next below the patriarch who serves as the head of an ecclesiastical province in the Eastern Orthodox Church. **2.** A resident of a metropolis, esp. one who displays big-city attitudes, characteristics, and values.
**me·tror·rha·gi·a** (mē′trə-rā′jē-ə, -jə) *n.* An abnormal uterine hemorrhage, esp. between menstrual flows. **—me′tror·rha′gic** *adj.*
**-metry** *suff.* [ME *-metrie* < OFr. < Lat. *-metria* < Gk. < *metron,* measure.] Process or science of measuring <iso*metry*>
**met·tle** (mĕt′l) *n.* [Alteration of METAL.] **1.** Inherent quality of character and temperament. **2.** Fortitude and courage : SPIRIT. **—on (one's) mettle.** Prepared to put one's spirit or courage to the test.
**met·tle·some** (mĕt′l-səm) *adj.* Full of courage : HIGH-SPIRITED.
**mew**[1] (myōō) *n.* [ME < OFr. *mue* < *muer,* to molt < Lat. *mutare,* to change.] **1.** A cage for hawks, esp. when molting. **2.** A secret place : HIDEAWAY. **3. mews.** A small street behind a residential street that contains small apartments. —*v.* **mewed, mew·ing, mews.** —*vt.* To confine in or as if in a cage. —*vi.* To molt. —Used of a hawk.
**mew**[2] (myōō) *vi.* **mewed, mew·ing, mews.** [ME *mewen.*] To make the high-pitched, crying sound of a cat : MEOW. **—mew** *n.*
**mew**[3] (myōō) *n.* [ME < OE *mæw.*] A sea bird, *Larus canus,* one of the gulls found in northern Eurasia and western North America.
**mewl** (myōōl) *vi.* **mewled, mewl·ing, mewls.** [Imit.] To cry weakly.
**Mex·i·can** (mĕk′sĭ-kən) *adj.* Of or relating to Mexico or to its inhabitants, their language, or their culture. —*n.* A native or resident of Mexico.
**Mexican hairless** *n.* One of a breed of small dogs with a smooth hairless body except for tufts on the head and tail.

**Mexican hairless**
*16–20 inches high at shoulder*

**Mexican Spanish** *n.* The Spanish language as used in Mexico.
**me·ze·re·on** (mə-zîr′ē-ən) *n.* [ME *mizerion* < Med. Lat. *mezereon* < Ar. *māzaryūn.*] **1.** A native Eurasian shrub, *Daphne mezereum,* having fragrant lilac-purple flowers and small scarlet fruit. **2.** MEZEREUM 2.
**me·ze·re·um** (mə-zîr′ē-əm) *n.* [Alteration of MEZEREON.] **1.** MEZEREON 1. **2.** The dried bark of certain shrubs of the genus *Daphne,* that was once used externally as a vesicant and internally for arthritis.
**me·zu·zah** *also* **me·zu·za** (mə-zōōz′ə) *n.* [Heb. *mzūzāh,* doorpost.] A small piece of parchment inscribed with the Biblical passages Deuteronomy 6:4–9 and 11:13–21 and marked with the word "Shaddai," a name of the Almighty, that is rolled up in a container and affixed to a door frame as a sign that a Jewish family lives within.
**mez·za·nine** (mĕz′ə-nēn′, mĕz′ə-nēn′) *n.* [Fr. < Ital. *mezzanino* < *mezzano,* middle < Lat. *medianus,* in the middle < *medius,* middle.] **1.** A partial story between two main stories of a building. **2.** The lowest balcony in a theater or the first few rows of that balcony.
**mez·zo** (mĕt′sō, mĕd′zō, mĕz′ō) *n., pl.* **-zos.** A mezzo-soprano.
**mezzo for·te** (fôr′tā′) *adj. & adv.* [Ital.] *Mus.* Moderately loud.
**mezzo pi·a·no** (pē-ä′nō) *adj. & adv.* [Ital.] *Mus.* Moderately soft.
**mez·zo-re·lie·vo** (mĕt′sō-rĭ-lē′vō, -rĕl-yä′vō, mĕd′-, mĕz′-) *n., pl.* **-vos.** [Ital. *mezzorilievo* : *mezzo,* half (< Lat. *medius*) + *rilievo,* relief < *rilevare,* to raise < Lat. *relevare,* to raise, relieve. —see RELIEVE.] Sculptural relief having modeled forms that project approx. halfway from the background.
**mez·zo-so·pran·o** (mĕt′sō-sə-prăn′ō, -prä′nō, mĕd′-, mĕz′-) *n., pl.* **-os.** [Ital. : *mezzo,* half (< Lat. *medius*) + *soprano,* soprano. —see SOPRANO.] **1. a.** A voice with a range between soprano and contralto. **b.** A vocal part for a voice with such a range. **2.** A woman with a mezzo-soprano voice.
**mez·zo·tint** (mĕt′sō-tĭnt′, mĕd′-) *n.* [Ital. *mezzotinto* : *mezzo,* half (< Lat. *medius*) + *tinto,* tint < Lat. *tingere,* to dye.] **1.** A method of engraving a steel or copper plate by burnishing and scraping areas to produce effects of shadow and light. **2.** A print made from a plate engraved by mezzotint.
**Mg** *symbol for* MAGNESIUM.
**mho** (mō) *n., pl.* **mhos.** [Backward spelling of OHM.] A unit of conductance reciprocal to the ohm.
**mi** (mē) *n.* [Med. Lat. —see GAMUT.] *Mus.* The third tone of the diatonic scale in solfeggio.
**MIA** (ĕm′ī-ā′) *n.* [M(ISSING) I(N) A(CTION).] A serviceman who is reported missing following a combat mission and whose death can be neither confirmed nor denied.
**Mi·am·i** (mī-ăm′ē, -ăm′ə) *n., pl.* **Miami** *or* **-is.** A member of an Algonquian tribe of Indians once living in Ohio, Indiana, Illinois, and Wisconsin.
**mi·as·ma** (mī-ăz′mə, mē-) *n., pl.* **-mas** *or* **-ma·ta** (-mə-tə) [Gk. < *miainein,* to pollute.] **1. a.** A poisonous atmosphere once believed to rise from swamps and putrid matter and cause disease. **b.** A thick vaporous atmosphere or emanation <a *miasma* of cigar smoke> **2.** A harmful influence or atmosphere <a *miasma* of evil> **—mi·as′mal, mi′as·mat′ic** (-măt′ĭk), **mi·as′mic** (-mĭk) *adj.*
**mi·ca** (mī′kə) *n.* [NLat. < Lat., grain.] Any of a group of chemically and physically related mineral silicates, common in igneous and metamorphic rocks, that contain hydroxyl, alkali, and aluminum silicate groups and can be split into flexible sheets used in insulation. **—mi·ca′ceous** (-kā′shəs) *adj.*
**Mi·cah** (mī′kə) *n.* [Heb. *Mīkhāh.*] —See table at BIBLE.
**mice** (mīs) *n. pl. of* MOUSE.
**mi·celle** (mī-sĕl′) *n.* [NLat. *micella* < Lat. *mica,* grain.] **1.** A submicroscopic aggregation of molecules, as a droplet in a colloidal system. **2.** A colloidal organic particle found in coal. **3.** A coherent strand or structure in synthetic or natural fibers. **4.** A submicroscopic structural unit of protoplasm. **—mi·cel′lar** (mī-sĕl′ər) *adj.*

**Mi·chael** (mī′kəl) *n.* [Heb. *Mīkā′ēl.*] The guardian archangel of the Jews in the Old Testament.
**Mich·ael·mas** (mĭk′əl-məs) *n.* [ME *mychelmesse* < OE *Michaeles mæsse,* Michael's mass.] A church festival celebrated on Sept. 29 in honor of the archangel Michael.
**Michaelmas daisy** *n.* Any of several hybrid asters derived primarily from North American species such as the New England aster and New York aster.
**Mick·ey Finn** (mĭk′ē fĭn′) *n.* [Orig. unknown.] *Slang.* An alcoholic beverage surreptitiously doctored with a laxative or drug.
**mick·le** (mĭk′əl) [ME *mikell* < OE *micel* and ON *mikill.*] *Scot.* —*adj.* Great. —*adv.* Greatly.
**Mic·mac** (mĭk′măk′) *n., pl.* **Micmac** or **-macs.** [Micmac *miigemuaġ.*] **1. a.** A tribe of North American Indians once living in what is now Nova Scotia and New Brunswick. **b.** A member of this tribe. **2.** The Algonquian language of the Micmac.
**micr-** *pref. var. of* MICRO-.
**mi·cra** (mī′krə) *n. var. pl. of* MICRON.
**micro-** or **micr-** *pref.* [ME < Lat. < Gk. *mikro-* < *mikros,* small.] **1. a.** Small <*micro*circuit> **b.** Abnormally small <*micro*cephaly> **c.** Requiring or involving microscopy <*micro*surgery> **2.** One millionth ($10^{-6}$) <*micro*calorie>
**mi·cro·a·nal·y·sis** (mī′krō-ə-năl′ĭ-sĭs) *n.* Chemical analysis of amounts weighing a milligram or less. **—mi′cro·an′a·lyst** (-ăn′ə-lĭst) *n.* **—mi′cro·an′a·lyt′ic** (-ăn′ə-lĭt′ĭk), **mi′cro·an′a·lyt′i·cal** *adj.*
**mi·cro·a·nat·o·my** (mī′krō-ə-năt′ə-mē) *n.* Histology. **—mi′cro·an′a·tom′i·cal** (-ăn′ə-tŏm′ĭ-kəl) *adj.*
**mi·crobe** (mī′krōb′) *n.* [MICRO- + Gk. *bios,* life.] A minute life form, esp. a disease-causing microorganism. **—mi·cro′bi·al** (mī-krō′bē-əl), **mi·cro′bic** (-krō′bĭk) *adj.*
**mi·cro·bi·cide** (mī-krō′bĭ-sīd′) *n.* An agent that kills microbes. **—mi·cro′bi·cid′al** (-sīd′l) *adj.*
**mi·cro·bi·ol·o·gy** (mī′krō-bī-ŏl′ə-jē) *n.* The science that deals with microorganisms and esp. with their effects on other forms of life. **—mi′cro·bi′o·log′i·cal** (-bī′ə-lŏj′ĭ-kəl), **mi′cro·bi′o·log′ic** *adj.* **—mi′cro·bi′o·log′i·cal·ly** *adv.* **—mi′cro·bi·ol′o·gist** *n.*
**mi·cro·bus** (mī′krō-bŭs′) *n.* A station wagon shaped like a small motorbus.
**mi·cro·ceph·a·ly** (mī′krō-sĕf′ə-lē) *n.* Abnormal smallness of the head, often associated with pathological mental conditions. **—mi′cro·ce·phal′ic** (-sə-făl′ĭk) *n.* & *adj.* **—mi′cro·ceph′a·lous** (-sĕf′ə-ləs) *adj.*
**mi·cro·chem·is·try** (mī′krō-kĕm′ĭ-strē) *n.* Chemistry that deals with minute quantities of materials weighing one milligram or less. **—mi′cro·chem′i·cal** (-ĭ-kəl) *adj.* **—mi′cro·chem′ist** *n.*
**mi·cro·chip** (mī′krō-chĭp′) *n.* An integrated circuit.
**mi·cro·cir·cuit** (mī′krō-sûr′kĭt) *n.* An electric circuit of miniaturized components. **—mi′cro·cir′cuit·ry** (-kĭ-trē) *n.*
**mi·cro·cli·mate** (mī′krō-klī′mĭt) *n.* The climate of a specific place within a given area. **—mi′cro·cli·mat′ic** (-măt′ĭk) *adj.* **—mi′cro·cli′ma·to·log′ic** (-mə-tə-lŏj′ĭk), **mi′cro·cli′ma·to·log′i·cal** *adj.* **—mi′cro·cli′ma·tol′o·gy** (-mə-tŏl′ə-jē) *n.*
**mi·cro·cline** (mī′krō-klīn′) *n.* [G. *Mikroklin* : Gk. *mikros,* small + Gk. *klinein,* to lean.] A mineral of the feldspar group, chiefly $KAlSi_3O_8$, used to make pottery.
**mi·cro·coc·cus** (mī′krō-kŏk′əs) *n., pl.* **-coc·ci** (-kŏk′sī′, -kŏk′ī′) [NLat. *Micrococcus,* genus name : MICRO- + *coccus,* coccus.] A spherical bacterium of several species of the genus *Micrococcus,* found in irregular clusters. **—mi′cro·coc′cal** (-kŏk′əl) *adj.*
**mi·cro·com·put·er** (mī′krō-kəm-pyōō′tər) *n.* A very small computer built around a microprocessor.
**mi·cro·cop·y** (mī′krō-kŏp′ē) *n.* A greatly reduced photographic copy, usu. reproduced by projection.
**mi·cro·cosm** (mī′krə-kŏz′əm) *n.* [ME *microcosme* < Med. Lat. *microscosmus* < Gk. *mikros kosmos,* small world.] A diminutive, representative system analogous to a larger system in make-up, development, or configuration. **—mi′cro·cos′mic** (-kŏz′mĭk), **mi′cro·cos′mi·cal** (-mĭ-kəl) *adj.* **—mi′cro·cos′mi·cal·ly** *adv.*
**mi·cro·crys·tal·line** (mī′krō-krĭs′tə-lĭn) *adj.* Having a crystalline structure visible only under the microscope. **—mi′cro·crys′tal** *n.*
**mi·cro·cyte** (mī′krə-sīt′) *n.* [MICRO- + (ERYTHRO)CYTE.] An abnormally small red blood cell, less than five microns in diameter. **—mi′cro·cyt′ic** (-sĭt′ĭk) *adj.*
**mi·cro·dot** (mī′krō-dŏt′) *n.* A photographic reproduction of written or printed material reduced to the size of a dot, esp. for secret transmittal.
**mi·cro·e·lec·trode** (mī′krō-ĭ-lĕk′trōd′) *n.* A very small electrode often used in the study of the electrical characteristics of living cells and tissues.
**mi·cro·e·lec·tron·ics** (mī′krō-ĭ-lĕk-trŏn′ĭks) *n.* (*sing. in number*). The branch of electronics dealing with miniature components. **—mi′cro·e·lec·tron′ic** *adj.*
**mi·cro·en·vi·ron·ment** (mī′krō-ĕn-vī′rən-mənt) *n.* A microhabitat.

**mi·cro·fiche** (mī′krō-fēsh′) *n., pl.* **microfiche** or **-fich·es.** [Fr. : *micro-,* small (< Gk. *mikros*) + *fiche,* card < OFr., peg < *fichier,* to drive in < Lat. *figere.*] A sheet of microfilm containing a large number of pages, as of printed text, in reduced form.
**mi·cro·fi·lar·i·a** (mī′krə-fə-lâr′ē-ə) *n.* A slender larval filaria. **—mi′cro·fi·lar′i·al** (-ē-əl) *adj.*
**mi·cro·film** (mī′krə-fĭlm′) *n.* **1.** A film on which printed materials are photographed greatly reduced in size. **2.** A reproduction on microfilm. —*vt.* **-filmed, -film·ing, -films.** To reproduce (e.g., documents) on microfilm.
**mi·cro·form** (mī′krə-fôrm′) *n.* An arrangement of images reduced in size, as on microfilm.
**mi·cro·ga·mete** (mī′krō-gə-mēt′, -găm′ēt′) *n.* The smaller of a pair of conjugating gametes, the male gamete.
**mi·cro·ga·me·to·cyte** (mī′krō-gə-mē′tə-sīt′) *n.* A gametocyte that generates microgametes.
**mi·crog·ra·phy** (mī-krŏg′rə-fē) *n.* Study or graphic representation of microscopic objects. **—mi′cro·graph′ic** (mī′krə-grăf′ĭk) *adj.*
**mi·cro·groove** (mī′krō-grōōv′) *n.* [Orig. a trademark.] A long-playing phonograph record.
**mi·cro·hab·i·tat** (mī′krō-hăb′ĭ-tăt′) *n.* The smallest unit of a habitat, as a clump of grass or space between rocks.
**mi·cro·ma·nip·u·la·tor** (mī′krō-mə-nĭp′yə-lā′tər) *n.* A device for manipulating minute instruments and needles under a microscope in order to perform microsurgery. **—mi′cro·ma·nip′u·la′tion** *n.*
**mi·cro·mere** (mī′krō-mîr′) *n.* [MICRO- + (BLASTO)MERE.] A tiny blastomere.
**mi·cro·me·te·or·ite** (mī′krō-mē′tē-ə-rīt′) *n.* A micrometeoroid, esp. one found on the earth or the moon.
**mi·cro·me·te·or·oid** (mī′krō-mē′tē-ə-roid′) *n.* One of numerous relatively small meteoroids distinguished by increasing occurrence as meteors with decreasing meteoric mass.
**mi·cro·me·te·or·ol·o·gy** (mī′krō-mē′tē-ə-rŏl′ə-jē) *n.* The study of meteorologic conditions in a small region, usu. a shallow layer up to a few hundred feet above ground in which temperature and humidity extremes are found. **—mi′cro·me′te·or′o·log′i·cal** (-ôr′ə-lŏj′ĭ-kəl, -ər-ə-) *adj.* **—mi′cro·me′te·or·ol′o·gist** *n.*
**mi·crom·e·ter** (mī-krŏm′ĭ-tər) *n.* [Fr. *micromètre* : *micro-,* small (< Gk. *mikros*) + *mètre,* meter < Gk. *metron,* measure.] A device for measuring minute distances, esp. one based on the rotation of a finely threaded screw, as in relation to a microscope.
**mi·crom·e·try** (mī-krŏm′ĭ-trē) *n.* Measurement with a micrometer. **—mi′cro·met′ric** (mī′krō-mĕt′rĭk), **mi′cro·met′ri·cal** *adj.* **—mi′cro·met′ri·cal·ly** *adv.*
**mi·cron** *also* **mi·kron** (mī′krŏn′) *n., pl.* **-crons** or **-cra** (-krə) *also* **-krons** or **-kra.** [Gk., neuter of *mikros,* small.] A unit of length equal to one millionth ($10^{-6}$) of a meter.
**Mi·cro·ne·sian** (mī′krə-nē′zhən, -shən) *adj.* Of or relating to Micronesia, its inhabitants, their languages, or their culture. —*n.* **1.** A native or resident of Micronesia. **2.** A subfamily of the Austronesian language family that includes the languages of Micronesia.
**mi·cron·ize** (mī′krə-nīz′) *vt.* **-ized, -iz·ing, -iz·es.** To reduce to particles that are only a few microns in diameter.
**mi·cro·nu·cle·us** (mī′krō-nōō′klē-əs, -nyōō′-) *n., pl.* **-cle·i** (-klē-ī′) or **-cle·us·es.** The smaller nuclear mass in ciliated and suctorial protozoans as opposed to the macronucleus.
**mi·cro·nu·tri·ent** (mī′krō-nōō′trē-ənt, -nyōō′-) *n.* A substance that in minute amounts is essential to life.
**mi·cro·or·gan·ism** (mī′krō-ôr′gə-nĭz′əm) *n.* An animal or plant of microscopic size, esp. a bacterium or protozoan.
**mi·cro·pa·le·on·tol·o·gy** (mī′krō-pā′lē-ŏn-tŏl′ə-jē, -ən-) *n.* The study of microscopic fossils. **—mi′cro·pa′le·on′to·log′ic** (-tə-lŏj′ĭk), **mi′cro·pa′le·on′to·log′i·cal** *adj.* **—mi′cro·pa′le·on·tol′o·gist** *n.*
**mi·cro·phage** (mī′krə-fāj′) *n.* A small phagocyte.
**mi·cro·phone** (mī′krə-fōn′) *n.* An instrument that converts acoustical waves into an electric current, usu. fed into an amplifier, recorder, or broadcast transmitter. **—mi′cro·phon′ic** (-fŏn′ĭk) *adj.*
**mi·cro·pho·to·graph** (mī′krō-fō′tə-grăf′) *n.* **1.** A photograph requiring magnification for viewing. **2.** A photograph on microfilm. **3.** A photomicrograph. **—mi′cro·pho′to·graph′** *v.* **(-graphed, -graph·ing, -graphs). —mi′cro·pho·tog′ra·pher** (-fə-tŏg′rə-fər) *n.* **—mi′cro·pho′to·graph′ic** (-grăf′ĭk) *adj.* **—mi′cro·pho·tog′ra·phy** (-rə-fē) *n.*
**mi·cro·phys·ics** (mī′krō-fĭz′ĭks) *n.* (*sing. in number*). The physics of molecular, atomic, nuclear, and subnuclear systems. **—mi′cro·phys′i·cal** (-ĭ-kəl) *adj.* **—mi′cro·phys′i·cal·ly** *adv.* **—mi′cro·phys′i·cist** (-ĭ-sĭst) *n.*
**mi·cro·phyte** (mī′krə-fīt′) *n.* A plant of microscopic size. **—mi′cro·phyt′ic** (-fĭt′ĭk) *adj.*
**mi·cro·print** (mī′krə-prĭnt′) *n.* The positive or printed reproduction of a microphotograph.
**mi·cro·proc·es·sor** (mī′krō-prŏs′ĕs-ər) *n.* *Computer Sci.* A semiconductor central processing unit usu. contained on a single integrated circuit chip.
**mi·cro·pyle** (mī′krə-pīl′) *n.* [MICRO- + Gk. *pulē,* gate.] **1.** *Bot.* A minute opening in the ovule of a plant through which the pollen tube usu. enters. **2.** *Zool.* A pore in the membrane of the ova of some animals through which the spermatozoon enters.

**mi·cro·scope** (mīʹkrə-skōpʹ) *n.* [NLat. *microscopium* : MICRO- + -*scopium*, -scope.] An optical instrument that uses a combination of lenses to produce magnified images of small objects, esp. of objects too small to be seen by the unaided eye.

**mi·cro·scop·ic** (mīʹkrə-skŏpʹĭk) *also* **mi·cro·scop·i·cal** (-ĭ-kəl) *adj.* **1.** Too small to be seen by the unaided eye but large enough to be studied under a microscope. **2.** Exceptionally small : MINUTE. **3.** Marked by or done with extreme attention to detail. **4.** Of, relating to, or concerned with a microscope. **5.** Suggesting or resembling a microscope in having the ability to observe very small objects. **—miʹcro·scopʹi·cal·ly** *adv.*

**Mi·cro·sco·pi·um** (mīʹkrə-skōʹpē-əm) *n.* [NLat., microscope.] A constellation in the Southern Hemisphere.

**mi·cros·co·py** (mī-krŏsʹkə-pē) *n.* **1.** The study or use of microscopes. **2.** Investigation using a microscope. **—mi·crosʹco·pist** *n.*

**mi·cro·seism** (mīʹkrə-sīʹzəm) *n.* A faint, recurrent tremor of the earth's crust. **—miʹcro·seisʹmic** (-sīzʹmĭk, -sīsʹ-) *adj.*

**mi·cro·some** (mīʹkrə-sōmʹ) *n.* [G. *Mikrosom* : Gk. *mikros,* small + Gk. *sōma,* body.] A ribosome. **—miʹcro·soʹmi·al** (-sōʹmē-əl), **miʹcro·soʹmic** (-sōʹmĭk) *adj.*

**mi·cro·spo·ran·gi·um** (mīʹkrō-spə-rănʹjē-əm) *n., pl.* **-gi·a** (-jē-ə). A structure or receptacle in which microspores are formed. **—miʹcro·spo·ranʹgi·ate** (-jē-ĭt) *adj.*

**mi·cro·spore** (mīʹkrə-spôrʹ, -spōrʹ) *n.* **1.** *Bot.* The smaller of two types of spores produced by heterosporous plants, as ferns, giving rise to the male gametophyte. **2.** The smaller of two spores formed by Radiolaria and certain other protozoans. **—miʹcro·spoʹric, miʹcro·spoʹrous** (mīʹkrə-spôrʹəs, -spōrʹ-, mī-krŏsʹpər-əs) *adj.*

**mi·cro·spo·ro·cyte** (mīʹkrə-spôrʹə-sītʹ, -spōrʹ-) *n.* A cell that generates microspores.

**mi·cro·spo·ro·gen·e·sis** (mīʹkrə-spôrʹə-jĕnʹĭ-sĭs, -spōrʹ-) *n.* Production or formation of microspores.

**mi·cro·spo·ro·phyll** (mīʹkrə-spôrʹə-fĭlʹ, -spōrʹ-) *n.* A structure that produces microsporangia.

**mi·cro·state** (mīʹkrō-stātʹ) *n.* An independent country having a small area and population.

**mi·cro·sur·ger·y** (mīʹkrō-sûrʹjə-rē) *n.* Surgery on minute living structures or cells by means of a micromanipulator. **—miʹcro·surʹgi·cal** (-jĭ-kəl) *adj.*

**mi·cro·teach·ing** (mīʹkrə-tēʹchĭng) *n.* A method of practice teaching in which a videotape of a small segment of a student's classroom teaching is made and later evaluated.

**mi·cro·tome** (mīʹkrə-tōmʹ) *n.* An instrument used to cut tissue into thin sections for microscopic examination.

**mi·crot·o·my** (mī-krŏtʹə-mē) *n.* Preparation of specimens with a microtome. **—miʹcro·tomʹic** (mīʹkrə-tŏmʹĭk) *adj.*

**mi·cro·tone** (mīʹkrə-tōnʹ) *n. Mus.* An interval smaller than a half tone. **—miʹcro·tonʹal** (-tōʹnəl) *adj.* **—miʹcro·to·nalʹi·ty** (-tō-nălʹĭ-tē) *n.* **—miʹcro·tonʹal·ly** *adv.*

**mi·cro·vil·lus** (mīʹkrō-vĭlʹəs) *n., pl.* **-vil·li** (-vĭlʹīʹ). Microscopic projections from the surface of a cell. **—miʹcro·vilʹlar** *adj.*

**mi·cro·wave** (mīʹkrə-wāvʹ, -krō-) *n.* An electromagnetic wave with a wavelength ranging approx. from one millimeter to one meter, the region between infrared and short-wave radio wavelengths.

**microwave oven** *n.* An oven in which microwaves heat and cook food.

**mic·tu·rate** (mĭkʹchə-rātʹ, mĭkʹtə-) *vi.* **-rat·ed, -rat·ing, -rates.** [< Lat. *micturire,* to want to urinate < *mingere,* to urinate.] To urinate. **—micʹtu·riʹtion** (-rĭshʹən) *n.*

**mid**[1] (mĭd) *adj.* [ME < OE *midd.*] **1.** Middle : central. **2.** Being the part in the middle or center <a ship lost in the *mid* Atlantic> **3.** Produced with the tongue in a position approx. intermediate between high and low. —Used of vowel sounds.

**mid**[2] (mĭd) *prep.* Amid <*mid* smoke and flame>

**mid-** *pref.* [< MID[1].] Middle <*mid*summer>

**mid·air** (mĭdʹârʹ) *n.* A point or region in the middle of the air : SPACE. **—midʹairʹ** *adj.*

**Mi·das** (mīʹdəs) *n.* [Lat. < Gk.] The fabled king of Phrygia to whom Dionysus gave the power of turning to gold all that he touched.

**mid·brain** (mĭdʹbrānʹ) *n.* **1.** The mesencephalon. **2.** The parts of the brain that develop from the midbrain region.

**mid·course** (mĭdʹkôrsʹ, -kōrsʹ) *n. Aerospace.* The part of a missile's flight between burnout and re-entry, during which corrective maneuvers are made.

**mid·cult** (mĭdʹkŭltʹ) *n.* [MID(DLEBROW) + CULT(URE).] A form of intellectual and artistic culture that has qualities of both high culture and mass culture without being either.

**mid·day** (mĭdʹdāʹ) *n.* [ME < OE *middæg* : *midd,* mid + *dæg,* day.] The middle of the day : NOON.

**†mid·den** (mĭdʹn) *n.* [ME *myddung,* prob. of ON orig.] **1.** A dunghill or refuse heap, esp. of a primitive habitation. **2.** A kitchen midden.

**mid·dle** (mĭdʹl) *adj.* [ME *middel* < OE.] **1.** Being equally distant from extremes or limits : CENTRAL <the *middle* car in the line> **2.** Being at neither one extreme nor the other. **3. a.** Intervening between an earlier and a later period of time : part of a sequence or series. **b. Middle.** *Geol.* Designating a division between an earlier and a later division <the *Middle* Paleozoic> **4. Middle.** Designating a stage in the development of a language or literature between earlier and later stages. **5.** *Logic.* Designating a term that appears in both premises of a syllogism but not in the conclusion. **6.** Of a verb form or voice in which the subject both performs and is affected by the action specified. *—n.* **1.** An area or point equidistant between extremes : CENTER <the *middle* of the room> **2.** Something intermediate between extremes : MEAN. **3.** The interior portion. **4.** The middle part of the human body : WAIST. **5.** *Logic.* A middle term. *—vt.* **-dled, -dling, -dles. 1.** To place in the middle. **2.** *Naut.* To fold in the middle.

☆ ***syns:*** MIDDLE, CENTRAL, INTERMEDIATE, MEAN, MEDIAL, MEDIAN, MID, MIDDLE-OF-THE-ROAD, MIDWAY *adj. core meaning* : not extreme <took a *middle* position in the argument>

**middle age** *n.* The time of human life from about 40 to 60.

**mid·dle-aged** (mĭdʹl-ājdʹ) *adj.* Of or relating to middle age.

**Middle Ages** *pl.n.* The period in European history between antiquity and the Renaissance, often dated from A.D. 476 to 1453.

**Middle America** *n.* **1.** That part of the U.S. middle class thought of as being average in income and education and conservative in values and attitudes. **2.** The American heartland thought of as being made up of small towns, small cities, and suburbs. **—Middle American** *n.*

**mid·dle·brow** (mĭdʹl-brouʹ) *n. Informal.* A moderately cultured person. **—midʹdle·browʹ** *adj.*

**middle C** *n. Mus.* The tone represented by the first ledger line below the treble clef or the first ledger line above the bass clef.

**middle class** *n.* The members of society occupying a socioeconomic position intermediate between the laboring classes and the wealthy. **—midʹdle-classʹ** *adj.*

**middle distance** *n.* **1.** The area between the foreground and background in a drawing, painting, or photograph. **2.** A division of competition in racing with events usu. ranging from 400 meters to 1,500 meters or 440 yards to 1 mile.

**Middle Dutch** *n.* The Dutch language from the middle of the 12th through the 15th cent.

**middle ear** *n.* The space between the tympanic membrane and the internal ear, containing the auditory ossicles that convey vibrations to the auditory tube.

**Middle English** *n.* The English language from about 1100 to 1500.

**middle ground** *n.* **1.** MIDDLE DISTANCE 1. **2.** A point of view midway between extremes.

**Middle High German** *n.* High German from the 11th through the 15th cent.

**Middle Irish** *n.* Irish from the 10th through the 13th cent.

**Middle Low German** *n.* Low German from the middle of the 13th through the 15th cent.

**mid·dle·man** (mĭdʹl-mănʹ) *n.* **1.** A trader who buys from producers and sells to retailers or consumers. **2.** An intermediary.

**middle management** *n.* A group of persons occupying an intermediate managerial position below the level of upper executives. **—middle manager** *n.*

**mid·dle·most** (mĭdʹl-mōstʹ) *adj.* [ME *middelmast* : *middel,* middle + -*mast,* -most.] Midmost.

**mid·dle-of-the-road** (mĭdʹl-əv-thə-rōdʹ) *adj.* **1.** Following a course of action midway between extremes, esp. being neither liberal nor conservative in politics. **2.** Of, pertaining to, or being a type of popular music that appeals to a wide audience.

**middle school** *n.* A school that usu. includes grades five through eight.

**middle term** *n.* The term in a syllogism presented in both premises but not appearing in the conclusion.

**mid·dle·weight** (mĭdʹl-wātʹ) *n.* A boxer or wrestler weighing between 147 and 160 pounds.

**Middle Welsh** *n.* Welsh from the 12th through the 15th cent.

**†mid·dling** (mĭdʹlĭng, -lĭn) *adj.* [ME *mydlyn* : *mid,* mid + -*ling,* small.] **1.** Of medium size, position, or quality. **2.** Mediocre. *—n.* **1.** *often* **middlings.** *Chiefly Southeastern U.S.* Pork or bacon cut from between the ham and shoulder of a pig. **2.** Any of various products intermediate in quality, price, size, or grade. **3. middlings** *(sing. or pl. in number).* Coarsely ground wheat mixed with bran. *—adv. Informal.* Fairly : moderately. **—midʹdling·ly** *adv.*

**mid·dy** (mĭdʹē) *n., pl.* **-dies. 1.** *Informal.* A midshipman. **2.** A middy blouse.

**middy blouse** *n.* A woman's or child's loose blouse with a sailor collar.

**mid·field** (mĭdʹfēldʹ) *n.* **1.** The section of a playing field midway between goals. **2.** Players on a team whose usual position is in the midfield. **—midʹfieldʹer** *n.*

**Mid·gard** (mĭdʹgärdʹ) *n.* [ON *Miðgarðr.*] *Norse Myth.* The part of the world inhabited by humans, imagined as a fortress built by the gods around the middle region of the universe and encircled by an enormous serpent.

**midge** (mĭj) *n.* [ME *migge* < OE *mycg.*] **1.** A gnatlike fly of the family Chironomidae, found worldwide. **2.** A small person.

**midg·et** (mĭjʹĭt) *n.* [Dim. of MIDGE.] **1.** An extremely small person who is otherwise normally proportioned. **2.** A small or miniature version. **3.** A class of very small objects, as very small sailboats, submarines, or cars. *—adj.* **1.** Miniature : diminutive. **2.** Belonging to a type or class much smaller than what is considered standard.

**mid·gut** (mĭd′gŭt′) *n.* The middle section of the embryonic digestive tract from which the ileum and the jejunum develop.

**mid·i** (mĭd′ē) *n.* [< MIDDLE.] A mid-calf length skirt or coat.

**Mid·i·an·ite** (mĭd′ē-ə-nīt′) *n.* One of the ancient Arabian tribe of Midian. **—Mid′i·an·ite′** *adj.*

**mid·i·ron** (mĭd′ī′ərn) *n.* An iron golf club with more loft than a driver and less than a mashie, used for medium fairway shots and long approach shots.

**mid·land** (mĭd′lənd) *n.* The interior or middle part of a specific region or country. **—mid′land** *adj.*

**mid·line** (mĭd′līn′) *n.* A medial line, esp. the medial line or plane of the body.

**mid·most** (mĭd′mōst′) *adj.* [ME *midmest* < OE : *midd,* mid + *-mest,* -most.] **1.** Located in the exact middle : MIDDLEMOST. **2.** Located nearest the middle. **—mid′most′** *adv.*

**mid·night** (mĭd′nīt′) *n.* [ME < OE *midniht* : *midd,* mid + *niht,* night.] **1.** The middle of the night. **2. a.** Intense gloom or darkness. **b.** A period of darkness and gloom. **—burn the midnight oil.** To work or study very late at night.

**midnight sun** *n.* The sun as seen at midnight during the summer within the Arctic or Antarctic Circle.

**mid·point** (mĭd′point′) *n.* **1.** The point of a line segment or curvilinear arc that divides it into two parts of the same length. **2.** A position midway between two extremes.

**Mid·rash** (mĭd′răsh′) *n., pl.* **Mid·rash·im** (mĭd-rä′shĭm) [Heb. *midhrāsh,* commentary.] One of a group of Jewish commentaries on the Hebrew Scriptures written between A.D. 400 and 1200.

**mid·rib** (mĭd′rĭb′) *n.* The central or principal vein of a leaf.

**mid·riff** (mĭd′rĭf′) *n.* [ME *midrif* < OE *midhrif* : *midd,* mid + *hrif,* belly.] **1.** DIAPHRAGM 1. **2.** The middle outer part of the front of the human body extending from just below the breast to the waistline. **—mid′riff** *adj.*

**mid·sec·tion** (mĭd′sĕk′shən) *n.* A middle section, esp. the midriff of the human body.

**mid·ship** (mĭd′shĭp′) *adj.* Of, relating to, or located in the middle of a ship.

**mid·ship·man** (mĭd′shĭp′mən, mĭd-shĭp′mən) *n.* **1.** A student training to be a commissioned naval officer, esp. a student at a naval academy. **2.** A fish of the genus *Porichthys,* with several rows of light-producing organs along its body.

**mid·ships** (mĭd′shĭps′) *adv.* [Short for AMIDSHIPS.] *Naut.* **1.** Amidships. **2.** In the center position. —Used of the helm.

**midst** (mĭdst, mĭtst) *n.* [ME *middest,* alteration of *middes* < *midde,* in the middle < OE, middle.] **1.** The middle position or part : CENTER <in the *midst* of the forest> **2.** A position of proximity to other individuals or members <a foreigner in our *midst*> **3.** The condition of being surrounded by or beset by something <in the *midst* of cooking dinner> **4.** A time period about the middle of a continuing condition or act <in the *midst* of their careers> *—prep.* Among.

**mid·stream** (mĭd′strēm′) *n.* **1.** The middle part of a stream. **2.** The middle part of a course or career <the *midstream* of life>

**mid·sum·mer** (mĭd′sŭm′ər) *n.* **1.** The middle of the summer. **2.** The summer solstice, about Jun. 21.

**Midsummer Day** *n.* The feast of the birth of John the Baptist, celebrated Jun. 24.

**mid·term** (mĭd′tûrm′) *n.* **1.** The middle of an academic term or a political term of office. **2. a.** An examination given in the middle of a school term. **b. midterms.** A series of such examinations. **—mid′term′** *adj.*

**mid·town** (mĭd′toun′) *n.* A central portion of a city.

**mid-Vic·to·ri·an** (mĭd′vĭk-tôr′ē-ən, -tōr′-) *adj.* Relating to, occurring in, or typical of the middle period of the reign of Queen Victoria in Great Britain (1837–1901), a period known for rigid social standards. *—n.* **1.** A person living in the mid-Victorian period. **2.** A person having mid-Victorian ideas.

**mid·way** (mĭd′wā′) *n.* **1.** The area of a fair, carnival, circus, or exposition where amusements and concessions are located. **2.** *Obs.* **a.** The middle of a way or distance. **b.** A middle way or course of action or thought. *—adv.* In the middle of a way or distance : HALFWAY. **—mid′way′** *adj.*

**mid·week** (mĭd′wēk′) *n.* The middle of the week. **—mid′week′ly** *adj. & adv.*

**mid·wife** (mĭd′wīf′) *n.* [ME *midwif* : *mid,* with (< OE) + *wif,* woman < OE *wīf.*] **1.** A woman who assists women in childbirth. **2.** One who helps in bringing something about. *—vt.* **-wifed** or **-wived** (-wīvd′), **-wif·ing** or **-wiv·ing** (-wī′vĭng), **-wifes** or **-wives** (-wīvz′). To assist in bringing forth or about.

**mid·wife·ry** (mĭd′wīf′rē, -wī′fə-rē, mĭd-wĭf′ə-rē) *n.* The techniques and practice of a midwife.

**mid·win·ter** (mĭd′wĭn′tər) *n.* **1.** The middle of the winter. **2.** The period of the winter solstice, about Dec. 22.

**mid·year** (mĭd′yîr′) *n.* **1.** The middle of the calendar or academic year. **2. a.** An examination given in the middle of the school year. **b. midyears.** A series of such examinations. **—mid′year′** *adj.*

**mien** (mēn) *n.* [< DEMEAN.] **1.** Bearing or manner : EXPRESSION <walked with a dignified *mien*> **2.** An aspect or appearance.

**miff** (mĭf) *n.* [Orig. unknown.] **1.** An irritable, bad-tempered mood : HUFF. **2.** A trivial argument : TIFF. *—vt.* **miffed, miff·ing, miffs.** To cause to become offended or annoyed.

**mif·fy** (mĭf′ē) *adj.* **-fi·er, -fi·est.** **1.** *Informal.* Overly sensitive : TOUCHY. **2.** *Bot.* Difficult to raise except under perfect conditions. —Used of certain plants. **—mif′fi·ness** *n.*

**might[1]** (mīt) *n.* [ME < OE *miht.*] **1.** Extraordinary force, power, or influence <the *might* of mechanized armies> **2.** Physical or bodily strength. **3.** Ability to do something.

**might[2]** (mīt) *aux.v.* [ME < OE *meahte,* 1st and 3rd person p. indicative of *magan,* to be able.] **1.** —Used to indicate: **a.** A condition or state contrary to fact <You *might* help if you knew the facts.> **b.** A possibility or probability weaker than *may* <We *might* discover the fountain of youth.> **2.** —Used to express possibility or probability or permission in the past. **3.** —Used to express a higher degree of deference or politeness than *may* <*Might* I express my concern?> ***usage:*** In certain dialects of English *might* occurs in construction with *could* and, less frequently, with *should,* as in *We might could park here.* Such constructions, however, are unfamiliar to most speakers of American English and should be avoided in formal writing or speech.

**might·y** (mī′tē) *adj.* **-i·er, -i·est.** **1.** Having or showing great power, strength, skill, or force <a *mighty* speaker><a *mighty* wind> **2.** Imposing or awesome in size, degree, or extent <a *mighty* stone barricade> *—adv. Informal.* To a great degree : EXTREMELY. **—might′i·ly** *adv.* **—might′i·ness** *n.*

**mi·gnon·ette** (mĭn′yə-nĕt′) *n.* [Fr., fem. of obs. *mignonnet,* dim. of *mignon,* dainty.] A Mediterranean plant, *Reseda odorata,* widely cultivated for its fragrant greenish flower clusters.

**mi·graine** (mī′grān′) *n.* [ME < OFr. < LLat. *hemicrania,* pain in one side of the head < Gk. *hēmikrania* : *hemi-,* half + *kranion,* head.] Severe, recurrent headache, usu. affecting only one side of the head, marked by sharp pain and often nausea. **—mi′grain·ous** *adj.*

**mi·grant** (mī′grənt) *n.* [Lat. *migrans, migrant-,* pr.part. of *migrare,* to migrate.] **1.** One that moves from one region to another by chance, instinct, or plan. **2.** An itinerant worker. *—adj.* Migratory.

**mi·grate** (mī′grāt′) *vi.* **-grat·ed, -grat·ing, -grates.** [Lat. *migrare, migrat-.*] **1.** To move from one country or area and settle in another. **2.** To change location periodically, esp. to move seasonally from one region to another. ***usage:*** *Migrate,* which can be followed by *from* or *to,* is used with reference to both a place of departure and a destination. *Emigrate* is used with reference to movement away from a place of departure, as in *My ancestors emigrated from France during the Reign of Terror. Immigrate* refers to movement toward a destination, as in *Many people immigrated to this country to find a better way of life.* **—mi′gra′tor** *n.*

**mi·gra·tion** (mī-grā′shən) *n.* **1.** The act or process of migrating. **2.** A group migrating together. **3.** *Chem.* **a.** Movement of one or more atoms from one position in a molecule to another. **b.** Movement of ions toward one electrode or the other during electrolysis. **—mi·gra′tion·al** *adj.*

**mi·gra·to·ry** (mī′grə-tôr′ē, -tōr′ē) *adj.* **1.** Marked by migration <*migratory* hawks> **2.** Of or pertaining to a migration. **3.** Wandering : nomadic.

**mi·ka·do** (mĭ-kä′dō) *n., pl.* **-dos.** [J. : *mi,* honorific prefix + *kado,* gate.] A Japanese emperor.

**mike** (mīk) *n. Informal.* A microphone.

**mi·kron** (mī′krŏn′) *n. var. of* MICRON.

**mil[1]** (mĭl) *n.* [< Lat. *mille,* thousand.] **1.** A unit of length equal to one thousandth ($10^{-3}$) of an inch or .0254 millimeter, used chiefly to specify the diameter of wire. **2.** A milliliter or one cubic centimeter. **3.** A unit of angular measurement used in artillery and equal to 1/6400 of a complete revolution.

**mil[2]** (mĭl) *n.* [< Lat. *mille,* thousand.] —See table at CURRENCY.

**mi·la·dy** (mĭ-lā′dē) *n.* [Fr. < E. *my lady.*] **1.** —Used as a title of respect for an English noblewoman or gentlewoman. **2.** A chic or fashionable woman.

**mil·age** (mī′lĭj) *n. var. of* MILEAGE.

**milch** (mĭlch) *adj.* [ME *milche* < OE *-milce.*] Giving milk <*milch* cows>

**mil·chig** (mĭl′KHĭk) *adj.* [Yiddish < *milch,* milk < MHG < OHG *miluh.*] Derived from or made of milk or dairy products.

**mild** (mīld) *adj.* **-er, -est.** [ME < OE *milde.*] **1.** Gentle or kind. **2. a.** Moderate in type, degree, effect, or force <a *mild* beer><a *mild* tranquilizer> **b.** Free from extremes, esp. in temperature <a *mild* winter> **3.** Not very severe <a *mild* infection> **4.** Rather easily shaped, molded, or worked : MALLEABLE <a *mild* alloy> **—mild′ly** *adv.* **—mild′ness** *n.*

☆ ***syns:*** MILD, MODERATE, TEMPERATE *adj. core meaning* : free from extremes in temperature <a *mild* climate> ***ant:*** harsh

**mil·dew** (mĭl′do͞o′, -dyo͞o′) *n.* [ME < OE *mildēaw.*] **1.** A plant disease in which a fungus forms a superficial growth on the plant. **2.** A superficial coating or discoloring of organic materials, as paint, paper, cloth, or leather, caused by fungi, esp. under damp conditions. —*vt.* & *vi.* **-dewed, -dew·ing, -dews.** To affect or become affected with mildew. **—mil′dew′y** *adj.*

**mile** (mīl) *n.* [ME < OE *mīl* < Lat. *milia,* miles, pl. of *mille (passuum),* thousand (paces).] **1.** A unit of length equal to 5,280 feet, 1,760 yards, or 1,609.34 meters. **2.** A nautical mile. **3.** An air mile. **4.** A race of a mile. **5.** A relatively long distance.

**mile·age** *also* **mil·age** (mī′lĭj) *n.* **1.** Total extent, length, or distance measured or expressed in miles. **2.** Total miles covered in a given time. **3.** The amount of use, service, or wear estimated by miles used or traveled <got a lot of *mileage* from these tires> **4.** The number of miles traveled by a motor vehicle on a certain quantity of fuel. **5.** An allowance for travel expenses established at a specified rate per mile. **6.** Expense per mile, as for the use of a car. **7.** *Informal.* The amount of service something has yielded or may yield in the future : USEFULNESS.

**mile·post** (mīl′pōst′) *n.* A post set up to indicate distance in miles, as along a highway.

**mil·er** (mī′lər) *n.* One trained to race a mile.

**mi·les glo·ri·o·sus** (mē′lās glôr′ē-ō′səs, glōr′-) *n., pl.* **mi·li·tes glo·ri·o·si** (mē′lĭ-tās glôr′ē-ō′sē, glōr′-) [Lat., after *Miles Gloriosus,* a comedy by Plautus.] A boastful swaggering soldier, esp. as a stock character in comedy.

**Mi·le·sian[1]** (mī-lē′zhən, -shən) *adj.* Of or relating to Miletus or its inhabitants. —*n.* A native or resident of Miletus.

**Mi·le·sian[2]** (mī-lē′zhən, -shən) *n.* [After *Milesius,* legendary ancestor of the Irish people.] A native of Ireland : IRISHMAN. —*adj.* Of or relating to Ireland or its people : IRISH.

**mile·stone** (mīl′stōn′) *n.* **1.** A stone marker on a roadside that indicates the distance in miles from a given point. **2.** A significant event in one's career or history.

**mil·foil** (mĭl′foil′) *n.* [ME < OFr. < Lat. *millefolium* : *mille,* thousand + *folium,* leaf.] **1.** The yarrow. **2.** Water milfoil.

**mil·i·a** (mĭl′ē-ə) *n. pl. of* MILIUM.

**mil·i·ar·i·a** (mĭl′ē-âr′ē-ə) *n.* [NLat. *(febris) miliaria,* miliary (fever).] A skin disease caused by an inflammation of the sweat glands and marked by blebs, redness, and a burning or prickling sensation. **—mil′i·ar′i·al** *adj.*

**mil·i·ar·y** (mĭl′ē-ĕr′ē) *adj.* [Lat. *miliarius,* of millet < *milium,* millet.] **1.** Designating a lesion or growth approx. one-eighth inch in diameter. **2.** Designating a disease characterized by small skin lesions that look like millet seeds.

**miliary tuberculosis** *n.* Acute tuberculosis marked by very small tubercles in various bodily organs, caused by the spread of tubercle bacilli through the blood stream.

**mi·lieu** (mēl-yœ′) *n.* [Fr. < OFr., center : *mi,* middle (< Lat. *medius*) + *lieu,* place < Lat. *locus.*] Surroundings : environment.

**mil·i·tant** (mĭl′ĭ-tənt) *adj.* [ME < OFr. < Lat. *militans,* pr.part. of *militare,* to serve as a soldier < *miles,* soldier.] **1.** Fighting or warring. **2.** Combative, esp. in the service of a cause <a *militant* civil-rights activist> **—mil′i·tan·cy** *n.* **—mil′i·tant** *n.* **—mil′i·tant·ly** *adv.*

**mil·i·ta·rism** (mĭl′ĭ-tə-rĭz′əm) *n.* **1.** Exaltation of the ideals of a professional military class. **2.** Predominance of the military in the administration or policy of a state. **3.** A policy in which military preparedness is of primary importance. **—mil′i·ta·rist** *n.* **—mil′i·ta·ris′tic** (-rĭs′tĭk) *adj.* **—mil′i·ta·ris′ti·cal·ly** *adv.*

**mil·i·ta·rize** (mĭl′ĭ-tə-rīz′) *vt.* **-rized, -riz·ing, -riz·es. 1.** To equip or train for war. **2.** To imbue with militarism. **3.** To adopt for use by or in the military. **—mil′i·ta·ri·za′tion** *n.*

**mil·i·tar·y** (mĭl′ĭ-tĕr′ē) *adj.* [Fr. *militaire* < Lat. *militaris* < *miles,* soldier.] **1.** Of, relating to, or typical of soldiers or the armed forces <a *military* haircut><*military* uniforms> **2.** Performed or supported by the armed forces. **3.** Of or relating to war <*military* maneuvers> —*n.* Armed forces <government by the *military*> **—mil′i·tar′i·ly** (-târ′ə-lē) *adv.*

**military attaché** *n.* An officer of the armed forces assigned to the official staff of an ambassador, consul general, or minister to a foreign country.

**military intelligence** *n.* **1.** Information important for its military value. **2.** The branch of the military that procures, analyzes, and uses information of strategic and tactical value.

**military law** *n.* Regulations and rules relating to the discipline and administration of the armed forces.

**military police** *n.* A branch of an army assigned to perform police duties. **—military policeman** *n.*

**military science** *n.* The tactical principles of warfare.

**mil·i·tate** (mĭl′ĭ-tāt′) *vi.* **-tat·ed, -tat·ing, -tates.** [Lat. *militare, militat-,* to serve as a soldier < *miles,* soldier.] To have force or influence <Experience *militated* against taking strong action.>

**mi·li·tia** (mə-lĭsh′ə) *n.* [Lat., warfare < *miles,* soldier.] **1.** A military force that is not part of a regular army and is subject to call for service in an emergency. **2.** The whole body of physically fit male civilians eligible by law for military service.

**mi·li·tia·man** (mə-lĭsh′ə-mən) *n.* A member of a militia.

**mil·i·um** (mĭl′ē-əm) *n., pl.* **-i·a** (-ē-ə) [ME < Lat., millet.] A small, hard, white or yellowish mass just below the surface of the skin, caused by retention of the secretion of a sebaceous gland.

**milk** (mĭlk) *n.* [ME < OE *milc.*] **1. a.** A whitish liquid produced by the mammary glands of all mature female mammals after they have given birth and used for feeding their young. **b.** The milk of animals, esp. cows, used as food by humans. **2.** A liquid similar to milk in appearance, as coconut milk, milkweed sap, or plant latex. **3.** Any of various medicinal emulsions. —*v.* **milked, milk·ing, milks.** —*vt.* **1.** To draw milk from the teat or udder of (a female mammal). **2.** To press out, drain off, or remove by or as if by milking <*milk* a cobra for its venom> **3.** To draw out or extract something from as if by milking. **4.** To use for one's own gain : EXPLOIT <officials *milking* the state's revenues> —*vi.* **1.** To yield or supply milk. **2.** To draw milk from a female mammal. **—milk′er** *n.*

**milk adder** *n.* The milk snake.

**milk chocolate** *n.* Sweetened chocolate made with milk.

**milk fever** *n.* **1.** A mild fever, usu. occurring at the beginning of lactation, associated with infection following childbirth. **2.** A disease affecting cows and occas. sheep or goats, esp. after giving birth.

**milk·fish** (mĭlk′fĭsh′) *n., pl.* **milkfish** *or* **-fish·es.** [From its color.] A large fish, *Chanos chanos* of the South Pacific and Indian oceans, widely used for food.

**milk glass** *n.* An opaque or translucent whitish glass.

**milk leg** *n.* A painful swelling of the leg that occurs in women after childbirth, caused by inflammation of the femoral veins.

**milk·maid** (mĭlk′mād′) *n.* A girl or woman who milks cows.

**milk·man** (mĭlk′măn′) *n.* A man who sells or delivers milk.

**milk of magnesia** *n.* A liquid suspension of magnesium hydroxide, $Mg(OH)_2$, used as an antacid and laxative.

**milk run** *n.* [From its being as uneventful and routine as the daily delivery of milk.] *Slang.* A routine aerial mission or flight.

**milk shake** *n.* A drink made of milk, flavoring, and usu. ice cream, shaken or blended until foamy.

**milk sickness** *n.* **1.** An acute disease marked by trembling, vomiting, and severe intestinal pain caused by eating the dairy products or flesh of cattle poisoned by eating white snakeroot. **2.** TREMBLES 3b.

**milk snake** *n.* A nonvenomous grayish or tan snake, *Lampropeltis doliata* or *L. triangulum* of the northeastern United States.

**milk·sop** (mĭlk′sŏp′) *n.* A man lacking courage and virility : WEAKLING. **—milk′sop′py** *adj.*

**milk sugar** *n.* Lactose.

**milk tooth** *n.* One of the temporary first teeth of a mammal.

**milk vetch** *n.* [From the belief that it increases the milk yield of goats.] A plant of the genus *Astragalus,* with compound leaves and purple, white, or yellowish flower clusters.

**milk·weed** (mĭlk′wēd′) *n.* **1.** A plant of the genus *Asclepias,* most of which have milky juice, esp. *A. syriaca,* the common milkweed of eastern North America, which has fragrant, dull purple flower clusters and pointed pods that split open to release seeds with downy tufts. **2.** Any of various plants with milklike juice.

**milkweed butterfly** *n.* MONARCH 5.

**milk·wort** (mĭlk′wûrt′, -wôrt′) *n.* [From the belief that it increases human lactation.] A plant of the genus *Polygala,* with variously colored, usu. small flowers.

**milk·y** (mĭl′kē) *adj.* **-i·er, -i·est. 1.** Like milk in color or consistency. **2.** Filled with, consisting of, or yielding milk or a fluid resembling milk. **3.** Subdued : mild. **—milk′i·ness** *n.*

**milky disease** *n.* A bacterial disease of Japanese beetle larvae and other scarabaeid grubs that eventually turns the blood of the grub a milky white color.

**Milky Way** *n.* [ME, transl. of Lat. *via lactea.*] The galaxy in which the solar system is located, visible at night as a luminous band.

**mill[1]** (mĭl) *n.* [ME *mille* < OE *mylen* < LLat. *molina* < *molinus,* of a mill < Lat. *mola,* millstone.] **1.** A building equipped with machinery for grinding grain into flour or meal. **2.** A mechanism or device, as rotating millstones, that grinds grain. **3.** A machine or device that grinds, crushes, or presses a solid or coarse substance into a pulp or minute grains. **4.** A machine that releases the juice of fruits and vegetables by pressing or grinding. **5. a.** A machine, as one for stamping coins, that produces something by the repetition of a simple process. **b.** Any of various machines for shaping, polishing, cutting, or dressing metal surfaces. **6. a.** A building or group of buildings equipped with machinery for processing raw materials into finished or industrial products <a paper *mill*> **b.** A building or group of buildings having machinery for manufacture : FACTORY. **7.** A process, agency, or institution that operates in a routine way or turns out products in the manner of a factory <a diploma *mill*> **8.** A slow or laborious process. **9.** A roller of hardened steel bearing a raised design, used for making a die or printing plate by pressure. —*v.* **milled, mill·ing, mills.** —*vt.* **1.** To pulverize, grind, or break down into smaller particles in a mill. **2.** To transform or process mechanically in a mill. **3.** To shape, dress, polish, or finish in a mill or with a milling tool. **4. a.** To produce a ridge around the edge of (a coin).

ă **pat** ā **pay** âr **care** ä **father** ĕ **pet** ē **be** hw **which** ĭ **pit** ī **tie** îr **pier** ŏ **pot** ō **toe** ô **paw, for** oi **noise** o͝o **took**

**b.** To groove or flute the rim of (a coin). **5.** To stir or agitate until foamy. —*vi.* **1.** To move around in churning confusion <a mob *milling* through the streets> **2.** *Slang.* To fight with the fists : BOX. **3.** To undergo milling.

**mill[2]** (mĭl) *n.* [< Lat. *mille,* thousand.] A monetary unit equal to 1/1000 of a U.S. dollar or 1/10 of a cent.

**mill·board** (mĭl'bôrd', -bōrd') *n.* [Alteration of *milled board.*] A stiff heavy paperboard used chiefly for book covers.

**mill·dam** (mĭl'dăm') *n.* A dam constructed across a stream to raise the water level so the overflow will have sufficient power to turn a mill wheel.

**mil·le·nar·i·an** (mĭl'ə-nâr'ē-ən) *adj.* [LLat. *millenarius,* millenary.] **1.** Of or relating to a thousand, esp. to a thousand years. **2.** Of, relating to, or believing in the doctrine of the millennium. —*n.* One who believes the millennium will occur. **—mil'le·nar'i·an·ism** *n.*

**mil·le·nar·y** (mĭl'ə-nĕr'ē, mə-lĕn'ə-rē) *adj.* [LLat. *millenarius* < *milleni,* a thousand each < *mille,* thousand.] Millenarian. —*n., pl.* **-ies. 1.** A total or sum of one thousand, esp. a thousand years. **2.** A millenarian.

**mil·len·ni·um** (mə-lĕn'ē-əm) *n., pl.* **-len·ni·ums** or **-len·ni·a** (-lĕn'ē-ə) [NLat. : Lat. *mille,* thousand + Lat. *annus,* year.] **1.** A span of one thousand years. **2.** A thousand-year period of holiness during which Christ is to rule on earth. **3.** A hoped-for period of happiness, peace, prosperity, and justice. **—mil·len'ni·al** *adj.* **—mil·len'ni·al·ism** *n.* **—mil·len'ni·al·ist** *n.* **—mil·len'ni·al·ly** *adv.*

**mil·le·pede** (mĭl'ə-pēd') *n. var. of* MILLIPEDE.

**mil·le·pore** (mĭl'ə-pôr', -pōr') *n.* [NLat. *Millepora,* genus name : Lat. *mille,* thousand + *porus,* pore < Gk. *poros.*] A reef-building hydrocoral of the genus *Millepora* of tropical marine waters, producing white or yellowish calcareous formations and resembling the true corals.

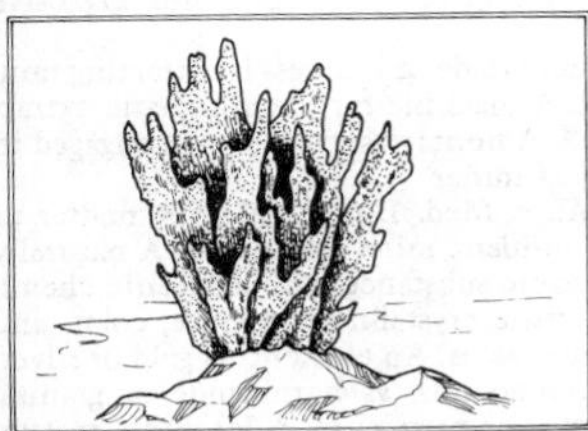

**millepore**

**mill·er** (mĭl'ər) *n.* **1.** One who works in, operates, or owns a mill, esp. a grain mill. **2.** A milling machine. **3.** Any of various moths with wings and bodies covered with a powdery substance.

**mil·ler·ite** (mĭl'ə-rīt') *n.* [G. *Millerit,* after William Hallowes *Miller* (1801–1880).] A mineral of nickel sulfide, NiS, occurring mainly in long slender crystals and used as a nickel ore.

**miller's thumb** *n.* [ME *millarys thowmbe.*] A freshwater fish of the genus *Cottus* of Europe and North America, with a spiny head and fins.

**mil·les·i·mal** (mə-lĕs'ə-məl) *adj.* [Lat. *millesimus* < *mille,* thousand.] **1.** Thousandth. **2.** Consisting of a thousandth. **3.** Relating to thousandths. —*n.* A thousandth. **—mil·les'i·mal·ly** *adv.*

**mil·let** (mĭl'ĭt) *n.* [ME *milet* < OFr. < *mil,* millet < Lat. *milium.*] **1. a.** A grass, *Panicum miliaceum,* cultivated in Eurasia for its seed and in North America for hay. **b.** The white seeds of the millet used as a food grain in the Old World. **2.** Any of various grasses related to or resembling the millet.

**milli-** *pref.* [Fr. < Lat. *milli-* < *mille,* thousand.] One thousandth ($10^{-3}$) <*milli*second>

**mil·liard** (mĭl'yərd, -yärd', mĭl'ē-ärd') *n.* [Fr. < OFr. *milliart* < *milion,* million.] *Chiefly Brit.* BILLION 1.

**mil·li·ar·y** (mĭl'ē-ĕr'ē) *adj.* [Lat. *milliarius,* consisting of a thousand < *mille (passuum),* thousand (paces), mile.] Relating to or marking the distance of an ancient Roman mile, which equaled 1,000 paces.

**mil·lieme** (mēl-yĕm', mē-yĕm') *n.* [Prob. < Fr. *millième,* thousandth < *mille,* thousand < OFr. < Lat.] —See table at CURRENCY.

**mil·li·gram** (mĭl'ĭ-grăm') *n.* A unit of mass or weight equal to one thousandth ($10^{-3}$) of a gram.

**mil·li·li·ter** (mĭl'ə-lē'tər) *n.* A unit of volume equal to one thousandth of a liter, or 0.001 liquid quart or 0.0009 dry quart.

**mil·li·me·ter** (mĭl'ə-mē'tər) *n.* A unit of length equal to one thousandth ($10^{-3}$) of a meter, or 0.0394 inch.

**mil·line** (mĭl'līn') *n.* [MIL(LION) + LINE.] **1.** A unit of advertising copy equal to one agate line one column wide printed in one million copies of a publication. **2.** The cost of a unit of advertising copy.

**mil·li·ner** (mĭl'ə-nər) *n.* [Alteration of obs. *Milaner,* native of Milan, Italy, from the importation of fashion accessories from Milan.] One who makes, designs, trims, or sells women's hats.

**mil·li·ner·y** (mĭl'ə-nĕr'ē) *n.* **1.** Articles, esp. women's hats, sold by a milliner. **2.** The profession or business of a milliner.

**†mill·ing** (mĭl'ĭng) *n.* **1.** The act or process of grinding, esp. grain into meal or flour. **2.** The operation of shaping, cutting, finishing, or working products manufactured in a mill. **3.** The ridges cut on the edges of coins. **4.** *Western U.S.* The process of halting a cattle stampede by turning the lead animals in a wide arc so that they form the center of a gradually tightening spiral.

**mil·lion** (mĭl'yən) *n., pl.* **million** or **-lions.** [ME < OFr. *milion* < OItal. *milione,* aug. of *mille,* thousand < Lat.] **1.** The cardinal number equal to 1,000 × 1,000 or $10^6$. **2.** A million monetary units, as dollars <made a *million* in the computer business> **3.** *often* **millions.** A large, unspecified number <*millions* of people at the beach> **4.** The common people. **—mil'lion** *adj.*

**mil·lion·aire** (mĭl'yə-nâr') *n.* [Fr. *millionnaire* < *million,* million < OFr. *milion.* —see MILLION.] One whose wealth amounts to a million or more dollars or its equivalent in another currency.

**mil·lionth** (mĭl'yənth) *n.* **1.** The ordinal number matching the number one million in a series. **2.** One of a million equal parts. **—mil'lionth** *adj.* & *adv.*

**mil·li·pede** or **mil·le·pede** (mĭl'ə-pēd') *n.* [Lat. *millepeda,* a kind of insect : *mille,* thousand + *pes,* foot.] A crawling herbivorous arthropod of the class Diplopoda, found worldwide and having a wormlike body with legs attached in double pairs to most of its body segments.

**mil·li·sec·ond** (mĭl'ĭ-sĕk'ənd) *n.* One thousandth of a second.

**mill·pond** (mĭl'pŏnd') *n.* A pond formed by a milldam.

**mill·race** (mĭl'rās') *n.* **1.** The swift flow of water that drives a mill wheel. **2.** The channel for the water that drives a mill wheel.

**mill·run** (mĭl'rŭn') *n.* **1.** A millrace. **2.** The output of a sawmill. **3. a.** A test of the mineral quality or content of a rock or ore by the process of milling. **b.** The mineral yielded by a millrun.

**mill-run** (mĭl'rŭn') *adj.* Unsorted and uninspected, as a product is when it leaves a mill <*mill-run* yarn>

**mill·stone** (mĭl'stōn') *n.* **1.** One of a pair of circular stones used in a mill for grinding grain. **2.** A heavy burden.

**mill·stream** (mĭl'strēm') *n.* MILLRACE 1.

**mill·wright** (mĭl'rīt') *n.* One that designs, builds, or repairs mills or mill machinery.

**mil·neb** (mĭl'nĕb') *n.* [Orig. unknown.] A white crystalline compound, $C_{12}H_{22}N_4S_4$, used as an agricultural fungicide.

**mi·lo** (mī'lō) *n., pl.* **-los.** [Poss. < Afr. *mealie,* corn, prob. < Port. *milho* < Lat. *milium,* grain.] An early-growing, usu. drought-resistant grain sorghum resembling millet.

**mi·lord** (mĭ-lôrd') *n.* [Fr. < E. *my lord.*] —Used as a title of respect for an English nobleman or gentleman.

**milque·toast** (mĭlk'tōst') *n.* [After Caspar *Milquetoast,* a comic-strip character created by Harold T. Webster (1885–1952).] A timid, meek person.

**milt** (mĭlt) *n.* [Prob. < MDu. *milte.*] **1. a.** Fish sperm, including the seminal fluid. **b.** The reproductive glands of male fishes when filled with this fluid. **2.** *Zool.* SPLEEN 2. —*vt.* **milt·ed, milt·ing, milts.** To fertilize (fish roe) with milt.

**milt·er** (mĭl'tər) *n.* A male fish ready to breed.

**mime** (mīm) *n.* [Lat. *mimus* < Gk. *mimos.*] **1. a.** A form of ancient Greek and Roman drama in which realistic characters and situations were farcically portrayed and actual persons mimicked on the stage. **b.** A performance of or dialogue for such a drama. **c.** A performer in such a drama. **2.** A modern performer who specializes in comic mimicry. **3. a.** PANTOMIME 4. **b.** A performance of pantomime. **c.** A performer skilled in pantomime. —*v.* **mimed, mim·ing, mimes.** —*vt.* **1.** To ridicule by imitation. **2.** To act out with gestures and body movements. —*vi.* **1.** To act as a mimic. **2.** To portray characters and situations by wordless gesture and body movement. **—mim'er** *n.*

**mim·e·o** (mĭm'ē-ō') *n., pl.* **-os.** A mimeographed publication.

**mim·e·o·graph** (mĭm'ē-ə-grăf') *n.* [Orig. a trademark.] **1.** A duplicator that makes copies of written, drawn, or typed material from a stencil fitted around an inked drum. **2.** A copy made by a mimeograph. —*v.* **-graphed, -graph·ing, -graphs.** —*vt.* To make (copies) on a mimeograph. —*vi.* To use a mimeograph.

**mi·me·sis** (mĭ-mē'sĭs, mī-) *n.* [Gk. *mimēsis* < *mimeisthai,* to imitate < *mimos,* imitator.] **1.** Imitation or representation of nature, esp. in literature and art. **2.** *Biol.* MIMICRY 2. **3.** *Med.* The appearance, often due to hysteria, of symptoms of a disease not actually present.

**mi·met·ic** (mĭ-mĕt'ĭk, mī-) *adj.* [Gk. *mimētikos* < *mimeisthai,* to imitate < *mimos,* imitator.] **1.** Relating to, typical of, or displaying mimicry. **2. a.** Of or relating to an imitation. **b.** Using imitative means of representation. **—mi·met'i·cal·ly** *adv.*

**mim·ic** (mĭm'ĭk) *vt.* **-icked, -ick·ing, -ics.** [Lat. *mimicus* < Gk. *mimikos* < *mimos,* imitator.] **1.** To imitate closely, esp. in expression, speech, and gesture : APE. **2.** To imitate so as to ridicule : MOCK <*mimicked* the President> **3.** To resemble closely : SIMULATE. **4.** To take on the appearance of. —*n.* **1.** One who imitates, esp.: **a.** A mime. **b.** A person who copies or mimics others, as for amusement. **2.** A copy or imitation. —*adj.* **1.** Relating to, acting as, resembling, or typical of a mimic or mimicry. **2.** Artificial : mock <a *mimic* conflict> **—mim'ick·er** *n.*

**mim·ic·ry** (mĭm'ĭ-krē) *n., pl.* **-ries. 1. a.** The act, practice, or art of mimicking. **b.** An instance of mimicking. **2.** *Biol.* The resemblance, through natural selection, of one organism to another or to a natural object, as an aid in concealment.

**Mi·mir** (mē'mîr') *n.* [ON.] *Norse Myth.* A giant who dwelt by the roots of Yggdrasil, where he guarded the well of wisdom.

**mi·mo·sa** (mĭ-mō'sə, -zə) *n.* [NLat. *Mimosa,* genus name < Lat. *mimus,* mime < Gk. *mimos.*] **1.** Any of various mostly tropical plants, shrubs, and trees of the genus *Mimosa,* with ball-like flower clusters and compound leaves that are often sensitive to touch or light. **2.** The silk tree.

**mi·na¹** (mī'nə) *n., pl.* **-nas** *or* **-nae** (-nē) [Lat. < Gk. *mna* < Akkadian *manru,* a unit of weight.] A unit of weight or money used in ancient Greece and Asia.

**mi·na²** (mī'nə) *n. var. of* MYNA.

**mi·na·cious** (mĭ-nā'shəs) *adj.* [Lat. *minax, minac-* < *minari,* to threaten < *minae,* threats.] Threatening or menacing. **—mi·na'cious·ly** *adv.* **—mi·na'cious·ness, mi·nac'i·ty** (mĭ-năs'ĭ-tē) *n.*

**min·a·ret** (mĭn'ə-rĕt') *n.* [Fr. < Turk. *minārat* < Ar. *manārat.*] A tall slender tower on a mosque with one or more projecting balconies from which a muezzin summons the people to prayer.

**min·a·to·ry** (mĭn'ə-tôr'ē, -tōr'ē) *also* **min·a·to·ri·al** (mĭn'ə-tôr'ē-əl, -tōr'-) *adj.* [Fr. *minatoire* < LLat. *minatorius* < Lat. *minari,* to threaten < *minae,* threats.] Minacious. **—min'a·to'ri·ly** *adv.*

**mince** (mĭns) *v.* **minced, minc·ing, minc·es.** [ME *mincen* < OFr. *mincier* < VLat. **minutiare* < LLat. *minutia,* smallness. —see MINUTIA.] —*vt.* **1.** To cut or chop into very small pieces. **2.** To pronounce affectedly. **3.** To restrain or moderate for the sake of decorum and politeness : EUPHEMIZE <Let's not *mince* words: they're thieves.> —*vi.* **1.** To walk with very short steps or with excessive primness. **2.** To speak in an affected manner. —*n.* Finely chopped food : MINCEMEAT. **—minc'er** *n.*

**mince·meat** (mĭns'mēt') *n.* **1.** Finely chopped meat. **2.** A mixture of finely chopped apples, spices, suet, and sometimes meat, used esp. as a pie filling. **—make mincemeat of.** *Slang.* To destroy totally as if by cutting into little pieces.

**mince pie** *n.* A pie filled with mincemeat.

**minc·ing** (mĭn'sĭng) *adj.* Affectedly refined. **—minc'ing·ly** *adv.*

**†mind** (mīnd) *n.* [ME *minde* < OE *gemynd.*] **1.** The human consciousness that originates in the brain and is manifested esp. in thought, memory, perception, feeling, will, or imagination. **2.** All of the conscious and unconscious processes of the brain and central nervous system that direct the mental and physical behavior of a sentient organism. **3.** The principle of intelligence. **4.** The faculty of reasoning, thinking, and applying knowledge. **5.** One of superior mental ability : BRAIN <the greatest *mind* in government> **6. a.** Individual consciousness, memory, or recollection <I'll keep your suggestion in *mind.*> **b.** An individual or a group that embodies certain mental qualities <the corporate *mind*> **c.** The thought processes characteristic of an individual or a group : PSYCHOLOGY <the military *mind*> **7.** Opinion or sentiment <changed my *mind*> **8.** A desire for a thing or activity <I have a *mind* to go to a concert.> **9.** Focus of thought : ATTENTION. **10.** A healthy mental state : SANITY <going out of my *mind*> **11. Mind.** *Christian Science.* The Deity regarded as the perfect intelligence ruling over all of divine creation. —*v.* **mind·ed, mind·ing, minds.** —*vt.* **1.** *Chiefly Regional.* **a.** To put (a person) in mind of something : REMIND. **b.** To bring (an object or idea) to mind : REMEMBER. **2. a.** To become aware of : NOTICE. **b.** *Chiefly Regional.* To have in mind as a purpose or goal : INTEND. **3.** To heed in order to obey < *Mind* your parents.> **4.** To attend to <*Mind* what you're told.> **5.** To be careful about <*Mind* the high step.> **6.** To object to : DISLIKE. **7.** To take care or charge of : look after. **8.** To care or be concerned about. —*vi.* **1.** To take notice : HEED. **2.** To behave obediently. **3.** To be concerned : CARE <"Not *minding* about bad food has become a national obsession" —*Times Literary Supplement*> **4.** To be careful or cautious. **—make up (one's) mind.** To come to a decision or reach an opinion. **—piece of (one's) mind.** *Informal.* A strongly worded rebuke. **—put (one) in mind.** *Informal.* To fill one with memories : REMIND <The movie *put* me *in mind* of my childhood.>

**mind-blow·ing** (mīnd'blō'ĭng) *adj. Slang.* **1.** Producing hallucinatory effects <a *mind-blowing* drug> **2.** Strongly affecting the mind or emotions <a *mind-blowing* war story> **—mind'blow'er** *n.*

**mind-bog·gling** (mīnd'bŏg'lĭng) *adj. Informal.* Overwhelming, as with wonder or perplexity <a *mind-boggling* task>

**mind·ed** (mīn'dĭd) *adj.* **1.** Disposed : inclined. **2.** Having a specified kind of mind <broad-*minded*>

**mind-ex·pand·ing** (mīnd'ĭk-spăn'dĭng) *adj.* Psychedelic <*mind-expanding* drugs>

**mind·ful** (mīnd'fəl) *adj.* Attentive : heedful <*mindful* of my duties> **—mind'ful·ly** *adv.* **—mind'ful·ness** *n.*

**mind·less** (mīnd'lĭs) *adj.* **1. a.** Lacking good sense or intelligence : FOOLISH. **b.** Devoid of intelligent purpose, direction, or meaning <*mindless* violence> **2.** Giving or showing little care or attention : HEEDLESS <*mindless* of any hazard> **—mind'less·ly** *adv.* **—mind'less·ness** *n.*

**mind reading** *n.* **1.** The act of guessing what someone else is thinking by observing facial expressions and other signs. **2.** Telepathy. **—mind reader** *n.*

**mind·set** (mīnd'sĕt') *n.* A mental disposition or attitude that predetermines one's responses to and interpretations of situations.

**mind's eye** *n.* Inherent ability to imagine or remember scenes.

**mine¹** (mīn) *n.* [ME < OFr.] **1. a.** An underground or in-ground excavation from which minerals or ore can be extracted. **b.** The site of such an excavation, with its buildings, elevator shafts, and equipment. **2.** A deposit of minerals or ore in or on the surface of the earth. **3.** An abundant supply or source of something valuable <The museum curator is a *mine* of facts about art.> **4. a.** A tunnel dug beneath an enemy position to gain an avenue of attack or to lay explosives. **b.** An explosive device for destroying enemy personnel, fortifications, or equipment, often in a concealed position and designed to be detonated by contact, by a time fuse, or electronically. **5.** An insect's burrow, tunnel, or gallery. —*v.* **mined, min·ing, mines.** —*vt.* **1. a.** To extract (minerals or ore) from the earth. **b.** To dig a mine in (the earth) to obtain minerals or ore. **2. a.** To tunnel beneath (the earth or a surface feature). **b.** To make (a tunnel) by digging. **3.** To lay explosive mines in or under <*mined* the entrance to the port> **4.** To attack, damage, or destroy by underhand means : SUBVERT. **5.** To delve into and make use of : EXPLOIT <*mine* the old land records for genealogical data> —*vi.* **1. a.** To excavate the earth in order to extract minerals or ore. **b.** To work in a mine. **2.** To dig a tunnel under the earth, esp. beneath an enemy position. **3.** To lay explosive mines.

**mine²** (mīn) *pron.* [ME *min* < OE *mīn.*] *(sing. or pl. in number).* That or those belonging to me <The red socks are *mine.*><If you can't find your jacket, take *mine.*>

**mine detector** *n.* An electromagnetic device for locating explosive mines. **—mine detection** *n.*

**mine·field** (mīn'fēld') *n.* An area laid with explosive mines.

**mine·lay·er** (mīn'lā'ər) *n.* A ship equipped for laying explosive underwater mines.

**min·er** (mī'nər) *n.* **1.** One whose trade or business is extracting minerals or ore from the earth. **2.** A machine for the automatic extraction of minerals, esp. of coal. **3.** A military service person engaged in laying explosive mines. **4.** A leaf miner.

**min·er·al** (mĭn'ər-əl) *n.* [ME < Med. Lat. *minerale* < neuter of *mineralis,* of minerals < OFr. *miniere,* mine < *mine.*] **1.** A naturally occurring, homogeneous inorganic substance with a specific chemical composition and characteristic crystalline structure, color, and hardness. **2.** A natural substance, as: **a.** An element, as gold or silver. **b.** A mixture of inorganic compounds, as hornblende or granite. **c.** An organic derivative, as coal or petroleum. **3.** Inorganic matter. **4.** An ore. **5. minerals.** *Chiefly Brit.* Mineral water. —*adj.* **1.** Of or relating to minerals <a *mineral* discovery> **2.** Impregnated with minerals.

**min·er·al·ize** (mĭn'ər-ə-līz') *v.* **-ized, -iz·ing, -iz·es.** —*vt.* **1.** To convert to a mineral substance. **2.** To transform (a metal) into a mineral by oxidation. **3.** To impregnate with minerals. —*vi.* To develop or hasten mineral formation. **—min'er·al·i·za'tion** *n.* **—min'er·al·iz'er** *n.*

**mineral kingdom** *n.* The group of substances and objects composed only of inorganic matter.

**min·er·al·o·cor·ti·coid** (mĭn'ər-ə-lō-kôr'tĭ-koid') *n.* Any of a group of steroid hormones secreted by the adrenal cortex that regulate the concentrations of electrolytes, as sodium and potassium, in the extracellular fluids.

**min·er·al·o·gy** (mĭn'ə-rŏl'ə-jē, -răl'-) *n.* [MINERA(L) + -LOGY.] Study of minerals, including their identification, distribution, and properties. **—min'er·a·log'i·cal** (-ər-ə-lŏj'ĭ-kəl) *adj.* **—min'er·a·log'i·cal·ly** *adv.* **—min'er·al'o·gist** *n.*

**mineral oil** *n.* **1.** A light hydrocarbon oil, esp. a distillate of petroleum. **2.** A refined petroleum oil used medicinally as a laxative.

**mineral tar** *n.* Maltha.

**mineral water** *n.* Naturally occurring or prepared water containing dissolved minerals or gases.

**mineral wax** *n.* Ozocerite.

**mineral wool** *n.* An inorganic fibrous material produced by steam blasting and by cooling molten silicate or a similar substance and used as an insulator and filtering medium.

**miner's lettuce** *n.* Winter purslane.

**Mi·ner·va** (mĭ-nûr'və) *n.* [Lat.] *Rom. Myth.* The goddess of wisdom, invention, the arts, and martial prowess.

**min·e·stro·ne** (mĭn'ĭ-strō'nē) *n.* [Ital., aug. of *minestra,* soup < *minestrare,* to serve < Lat. *ministrare* < *minister,* servant.] A soup of Italian origin containing vegetables, vermicelli, and herbs in a meat or vegetable broth.

**mine sweeper** *n.* A ship equipped for destroying, removing, or neutralizing explosive marine mines.

**min·gle** (mĭng'gəl) *v.* **-gled, -gling, -gles.** [ME *menglen,* freq. of *mengen,* to mix < OE *mengan.*] —*vt.* **1.** To mix or bring together in close association : COMBINE. **2.** To mix so that the components become united : MERGE. —*vi.* **1.** To be or become mixed or united. **2.** To join or take part with others <*mingled* with the spectators> **—min′gler** *n.*

**min·gy** (mĭn′jē) *adj.* **-gi·er, -gi·est.** [Perh. a blend of MEAN and STINGY.] *Informal.* **1.** Small in quantity <a *mingy* amount> **2.** Mean and stingy <a *mingy* individual>

**min·i** (mĭn′ē) *n.* **1.** Something much smaller or shorter than other members of its class. **2.** A miniskirt.

**mini-** *pref.* [< MINIATURE.] Small : miniature <*minicar*>

**min·i·a·ture** (mĭn′ē-ə-cho͝or′, mĭn′ə-, -chər) *n.* [Ital. *miniatura,* illumination of manuscripts < *miniare,* to illuminate < Lat., to color red < *minium,* red lead.] **1. a.** A copy or model that reproduces or represents something in a greatly reduced size. **b.** Something small of its class. **2. a.** A small painting done with great detail, often on a surface such as ivory or vellum. **b.** A small portrait, picture, or decorative letter on an illuminated manuscript. **c.** The art of painting miniatures. —*adj.* Greatly reduced in size or scale : TINY <*miniature* houses>

▲ word history: The idea of smallness was not originally part of the meaning of *miniature. Miniature* is derived from Latin *minium,* "red lead," a compound of lead used as a pigment. In medieval times chapter headings and other important divisions of a text were distinguished by being written in red, while the rest of the book was written in black. The Latin verb *miniare,* derived from *minium,* meant "to color red." *Miniatura,* the future participle, denoted the process of writing in red. Sections of a manuscript were also marked off with large ornate initial capital letters, which were often decorated with small paintings. *Miniatura* was used to denote these paintings as well. Since the paintings were necessarily very small, *miniatura* came to denote a small painting or a small object of any kind.

**miniature golf** *n.* A simplified version of golf played on a miniature course.

**min·i·a·tur·ize** (mĭn′ē-ə-chə-rīz′, mĭn′ə-) *vt.* **-ized, -iz·ing, -iz·es.** To plan or make in a very small size. **—min′i·a·tur′i·za′tion** *n.*

**min·i·bike** (mĭn′ē-bīk′) *n.* [Orig. a trademark.] A small motorbike with a low frame, small wheels, and elevated handlebars.

**min·i·bus** (mĭn′ē-bŭs′) *n.* A small bus.

**min·i·cab** (mĭn′ē-kăb′) *n.* A minicar used as a taxicab, esp. in Great Britain.

**min·i·car** (mĭn′ē-kär′) *n.* A subcompact car.

**min·i·com·put·er** (mĭn′ē-kəm-pyo͞o′tər) *n.* A small computer with more memory and higher execution speed than a microcomputer.

**min·ié ball** (mĭn′ē, mĭn′ē-ā′) *n.* [After Claude Etienne *Minié* (1814–1879), its inventor.] A 19th-cent. conical rifle bullet with a hollow base that expanded when fired.

**min·i·fy** (mĭn′ə-fī′) *vt.* **-fied, -fy·ing, -fies.** [MIN(IMUM) + (MAGN)IFY.] To make smaller or less significant : REDUCE.

**min·i·kin** (mĭn′ĭ-kĭn) *n.* [MDu. *minneken,* darling, dim. of *minne,* love.] *Archaic.* A small, delicate creature.

**min·im** (mĭn′əm) *n.* [< Med. Lat. *minimus,* least < Lat.] **1.** A unit of fluid measure: **a.** In the United States, 1/60 of a fluid dram or 0.00376 cubic inches. **b.** In Great Britain, 1/20 of a scruple or 0.00361 cubic inches. **2.** *Mus.* A half note. **3.** An insignificantly small portion or thing : JOT. **4.** A downward vertical stroke in handwriting.

**min·i·ma** (mĭn′ə-mə) *n. var. pl. of* MINIMUM.

**min·i·mal** (mĭn′ə-məl) *adj.* **1.** Smallest in degree or amount. **2.** Of, pertaining to, or being minimal art. —*n. Math.* A number that precedes all others in an ordered set. **—min′i·mal′i·ty** (-măl′ĭ-tē) *n.* **—min′i·mal·ly** *adv.*

**minimal art** *n.* Nonrepresentational art that consists mainly of basic geometric shapes. **—minimal artist** *n.*

**min·i·mal·ism** (mĭn′ə-mə-lĭz′əm) *n.* Minimal art. **—min′i·mal·ist** *n.*

**min·i·max** (mĭn′ə-măks′) *adj.* [MINI(MUM) + MAX(IMUM).] *Math.* Of or relating to the strategic principle in game theory by which a player selects the strategy to minimize an opponent's greatest possible gain and maximize his or her own.

**min·i·mize** (mĭn′ə-mīz′) *vt.* **-mized, -miz·ing, -miz·es.** [< MINIMUM.] **1.** To reduce to the smallest possible amount, size, extent, or degree. **2.** To represent as having the least degree of importance or value : DEPRECATE <*minimized* the value of their contribution> **—min′i·mi·za′tion** *n.* **—min′i·miz′er** *n.*

**min·i·mum** (mĭn′ə-məm) *n., pl.* **-mums** *or* **-ma** (-mə) [Lat., neuter of *minimus,* least.] **1.** The least possible quantity or degree. **2.** The lowest amount or degree reached or recorded. **3.** *Math.* **a.** A number not greater than any other in a finite set of numbers. **b.** A value of a function that is exceeded for any sufficiently small increase or decrease in the function's variables. —*adj.* Of, consisting of, or representing the lowest possible amount or degree allowable or attainable.

**minimum wage** *n.* **1.** The lowest wage, set by law or contract, that may be paid for a specified job. **2.** A living wage.

**min·ion** (mĭn′yən) *n.* [Fr. *mignon,* darling < OFr. *mignot.*] **1.** One who is esteemed or favored. **2. a.** An obsequious follower : SYCOPHANT. **b.** A subordinate official.

**min·i·se·ries** (mĭn′ē-sîr′ēz) *n.* **1.** A sequence of episodes making up a televised dramatic production. **2.** A short series of performances or athletic contests.

**min·i·ski** (mĭn′ē-skē′) *n.* A short ski used by beginners.

**min·i·skirt** (mĭn′ē-skûrt′) *n.* An extremely short skirt. **—min′i·skirt′ed** *adj.*

**min·i·state** (mĭn′ē-stāt′) *n.* A microstate.

**min·is·ter** (mĭn′ĭ-stər) *n.* [ME *ministre* < OFr. < Lat. *minister.*] **1.** One serving as an agent for another by carrying out specified functions or orders. **2. a.** One authorized to perform religious functions in a church, esp. a Protestant church. **b.** *Rom. Cath. Ch.* The superior in certain orders. **3.** A high officer of state appointed to head an executive or administrative department of government. **4.** A representative of one's government in diplomatic dealings with other governments, usu. ranking next below an ambassador. —*v.* **-tered, -ter·ing, -ters.** —*vi.* **1.** To attend to the needs and wants of others. **2.** To perform the functions of a member of the clergy. —*vt.* To administer : dispense.

**min·is·te·ri·al** (mĭn′ĭ-stîr′ē-əl) *adj.* **1.** Of, relating to, or typical of a religious minister or the ministry. **2.** Of or relating to administrative and executive duties and functions of government. **3.** *Law.* Of or designating a mandatory act or duty admitting of no personal judgment or discretion in its performance. **4.** Functioning as an agent : INSTRUMENTAL. **—min′is·te′ri·al·ly** *adv.*

**minister plenipotentiary** *n.* A diplomatic representative with full authority to speak and act for a government : PLENIPOTENTIARY.

**minister resident** *n.* A diplomatic agent ranking below a minister plenipotentiary.

**min·is·trant** (mĭn′ĭ-strənt) *adj.* [< Lat. *ministrans, ministrant-,* pr.part. of *ministrare,* to serve < *minister,* servant.] Serving as a minister. —*n.* One who ministers.

**min·is·tra·tion** (mĭn′ĭ-strā′shən) *n.* [Lat. *ministratio* < *ministrare,* to serve < *minister,* servant.] **1.** The act or process of serving or aiding. **2.** The act of performing the duties of a member of the clergy. **—min′is·tra′tive** *adj.*

**min·is·try** (mĭn′ĭ-strē) *n., pl.* **-tries.** [ME *ministerie* < Lat. *ministerium* < *minister,* minister.] **1. a.** An act of serving : MINISTRATION. **b.** One that serves as a means : INSTRUMENTALITY. **2. a.** The profession, services, and duties of a minister of religion. **b.** The clergy. **c.** The period of service of a minister of religion. **3. a.** A governmental department presided over by a minister. **b.** The building housing such a department. **c.** The duties, functions, or term of a governmental minister and his or her staff. **d.** *often* **Ministry.** Governmental ministers as a group.

**min·i·track** (mĭn′ē-trăk′) *n.* An electronic measuring system for following the course of satellites and rockets and correlating radio signals received by a network of ground stations.

**min·i·um** (mĭn′ē-əm) *n.* [Lat.] Red lead.

**min·i·ver** (mĭn′ə-vər) *n.* [ME *meniver* < OFr. *menu vair,* small vair.] A white or light-gray fur used to trim medieval robes or current-day robes of state.

**mink** (mĭngk) *n., pl.* **mink** *or* **minks.** [ME *mynk.*] **1.** A semiaquatic carnivore of the genus *Mustela,* esp. *M. vison* of North America, similar to the weasel and having short ears, a pointed snout, short legs, and partly webbed toes. **2. a.** The soft, thick, lustrous fur of the mink. **b.** A garment made of mink.

**Min·nan** (mĭ-nän′) *n.* [Chin. (Mandarin) *min*$^{3}$ *nan*$^{2}$ : *min*$^{3}$, Fujian Province + *nan*$^{2}$, south.] The dialect of Chinese spoken in Taiwan.

**min·ne·sing·er** (mĭn′ĭ-sĭng′ər, mĭn′ə-zĭng′ər) *n.* [G. < MHG : *minne,* love (< OHG *minna*) + *singer,* singer < *singen,* to sing < OHG *singan.*] Any of the German lyric poets and singers who flourished from the 12th to the 14th cent.

**min·now** (mĭn′ō) *n., pl.* **minnow** *or* **-nows.** [ME *meneu.*] **1.** Any of a large number of small freshwater fishes of the family Cyprinidae, widely used as bait. **2.** A small silver-colored fish.

**Mi·no·an** (mĭ-nō′ən) *adj.* [Lat. *Minous,* of Minos < Gk. *Minōios* < *Minōs,* Minos.] Of or relating to the advanced Bronze Age culture that flourished in Crete from about 3000 to 1100 B.C.

**mi·nor** (mī′nər) *adj.* [ME < Lat.] **1.** Lesser or smaller in amount, extent, or size. **2.** Lesser in importance, rank, or stature <a *minor* poet> **3.** Lesser in seriousness or danger <a *minor* accident> **4.** *Law.* Not yet of legal age. **5.** *Chiefly Brit.* Designating the junior or younger of two pupils with the same surname <Jones *minor*> **6.** Designating or pertaining to a secondary area of academic specialization. **7.** *Logic.* Dealing with a more restricted category. **8.** *Mus.* **a.** Designating a minor scale. **b.** Less in distance by a half step than the corresponding major interval. **c.** Based on a minor scale <a *minor* key> —*n.* **1.** One that is lesser in comparison to others of the same class. **2.** *Law.* One who has not reached full legal age. **3. a.** A secondary area of specialized academic study. **b.** One studying a minor <a geology *minor*> **4.** *Logic.* **a.** A minor premise. **b.** A minor term. **5.** *Mus.* A minor key, scale, or interval. **6. minors.** The minor leagues of a sport. —*vi.* **-nored, -nor·ing, -nors.** To pursue academic studies in a minor field.

☆ ***syns:*** MINOR, INFERIOR, LOW, LOWER, MINOR-LEAGUE, PETTY, SECONDARY, SMALL, SMALLTIME, UNDER *adj. core meaning* **:** of subordinate standing or importance <*minor* government officials> ***ant:*** major

**Mi·nor·ca** (mĭ-nôr′kə) *n.* [After *Minorca,* an island in the Mediterranean.] A domestic fowl orig. bred in the Mediterranean region, with white or black plumage.

**Mi·nor·ite** (mī′nə-rīt′) *n.* [< Med. Lat. *(Fratres) Minores,* minor (friars).] A Franciscan friar.

**mi·nor·i·ty** (mə-nôr′ĭ-tē, -nŏr′-, mī-) *n., pl.* **-ties.** [Fr. *minorité* < Med. Lat. *minoritas* < Lat. *minor,* smaller.] **1.** The smaller in number of two groups making a whole. **2. a.** A racial, religious, political, national, or other group thought to be different from the larger group of which it is part. **b.** A member of a minority group. **3.** The state or period of being below legal age.

**minority leader** *n.* The head of the minority party in a legislative body <the Senate *minority leader*>

**minor league** *n.* A league of professional sports clubs, esp. baseball teams, not in the major leagues.

**mi·nor-league** (mī′nər-lēg′) *adj.* **1.** Relating or belonging to a minor sports league. **2.** Subordinate in position or importance <a *minor-league* author> **—mi′nor-leagu′er** *n.*

**minor orders** *pl.n. Rom. Cath. Ch.* The orders of acolyte, exorcist, reader or lector, and doorkeeper.

**minor premise** *n. Logic.* The premise in a syllogism containing the minor term, which will form the subject of the conclusion.

**Minor Prophets** *pl.n.* The Hebrew prophets Hosea, Joel, Amos, Obadiah, Jonah, Micah, Nahum, Habakkuk, Zephaniah, Haggai, Zechariah, and Malachi.

**minor scale** *n. Mus.* A diatonic scale with a minor third between the first and third tones and with several forms having different intervals above the fifth.

**minor suit** *n.* The suit of clubs or of diamonds in bridge.

**minor term** *n. Logic.* The term in a syllogism that is stated in the minor premise and forms the subject of the conclusion.

**Mi·nos** (mī′nəs, -nŏs′) *n.* [Lat. < Gk. *Minōs.*] *Gk. Myth.* A king of Crete, the son of Zeus and Europa.

**Min·o·taur** (mĭn′ə-tôr′, mī′nə-) *n.* [Lat. *Minotaurus* < Gk. *Minōtauros* : *Minōs,* Minos + *tauros,* bull.] *Gk. Myth.* The son of Pasiphaë by a sacred bull, half man and half bull in body, slain by Theseus.

**min·ster** (mĭn′stər) *n.* [ME *mynster* < OE *mynster* < LLat. *monasterium,* monastery. —see MONASTERY.] *Chiefly Brit.* **1.** A monastery church. **2.** The title of certain large cathedrals <York *Minster*>

**min·strel** (mĭn′strəl) *n.* [ME *ministral* < OFr. *menestral* < LLat. *ministerialis,* official in the imperial household < Lat. *ministerium,* ministry < *minister,* minister.] **1.** A medieval musician who sang and recited poetry, usu. to the accompaniment of a harp, while traveling from place to place. **2. a.** A lyric poet. **b.** A musician. **3. a.** A performer in a minstrel show. **b.** A performance of a minstrel show.

**minstrel show** *n.* A variety show in which performers sing, dance, tell jokes, and perform comic skits.

**min·strel·sy** (mĭn′strəl-sē) *n., pl.* **-sies. 1.** The art or profession of a minstrel. **2.** A troupe of minstrels. **3.** Ballads and lyrics that minstrels sing.

**mint**[1] (mĭnt) *n.* [ME *mynt* < OE *mynet,* money < Lat. *moneta.* —see MONEY.] **1.** A place where the coins of a country are manufactured under government authority. **2.** An abundant amount, esp. of money. **3.** A source <a *mint* of valuable advice> *—vt.* **mint·ed, mint·ing, mints. 1.** To produce (money) by stamping metal **:** COIN. **2.** To invent or fabricate <a speech *minted* for the ceremony> *—adj.* Undamaged as if freshly minted <lawn furniture still in *mint* condition>

**mint**[2] (mĭnt) *n.* [ME *minte* < OE < Lat. *menta* < Gk. *minthē.*] **1.** A plant of the genus *Mentha,* with aromatic foliage and two-lipped flowers, often cultivated for its oil and used for flavoring. **2.** A plant, as the mountain mint or stone mint, resembling the mint. **3.** A candy with mint flavor. **—mint′y** *adj.*

**mint·age** (mĭn′tĭj) *n.* **1.** The act or process of minting coins. **2.** Coins produced in a mint. **3.** The fee paid to a mint by a government. **4.** The impression stamped on a coin.

**mint julep** *n.* A tall frosted drink made of bourbon whiskey, sugar, crushed mint leaves, and ice.

**min·u·end** (mĭn′yo͞o-ĕnd′) *n.* [Lat. *minuendum,* thing to be diminished, neuter gerund. of *minuere,* to lessen.] *Math.* The quantity from which another quantity, the subtrahend, is to be subtracted.

**min·u·et** (mĭn′yo͞o-ĕt′) *n.* [Fr. *menuet* < OFr. *menu,* small < Lat. *minutus* < *minuere,* to lessen.] **1.** A slow, stately pattern dance in 3/4 time for groups of couples, originated in 17th-cent. France. **2.** The music for or in the rhythm of the minuet.

**mi·nus** (mī′nəs) *prep.* [ME < Lat. *minus,* less.] **1.** *Math.* Reduced by the subtraction of **:** LESS. **2.** *Informal.* Lacking **:** without <arrived in class *minus* their notebooks> *—adj.* **1.** *Math.* Negative or on the negative part of a scale <a *minus* value><*minus* 15°> **2.** Designating one subdivision of a grade less than <a grade of C *minus*> *—n.* **1.** The minus sign (–). **2.** A negative quantity. **3.** A deficiency or defect.

**min·us·cule** (mĭn′ə-skyo͞ol′, mĭ-nŭs′kyo͞ol′) *n.* [Fr. < Lat. *minusculus,* very small, dim. of *minor,* smaller.] **1.** A small cursive script developed from uncial between the 7th and 9th cent. A.D. and used in medieval manuscripts. **2.** A letter written in minuscule. **3.** A lower-case letter. *—adj.* **1.** Of, relating to, or written in minuscule. **2.** Extremely small **:** TINY. **—mi·nus′cu·lar** (mĭ-nŭs′kyə-lər) *adj.*

**minus sign** *n. Math.* A symbol (–) used to indicate subtraction or a negative quantity.

**min·ute**[1] (mĭn′ĭt) *n.* [ME < OFr. < LLat. *minuta* < Lat. *minutus,* small < *minuere,* to lessen.] **1. a.** A unit of time equal to one sixtieth of an hour or 60 seconds. **b.** A unit of angular measurement equal to one sixtieth of a degree or 60 seconds. **2.** A short interval of time <They'll be here in a *minute.*> **3.** A specific point in time. **4.** A note or summary covering points to be remembered **:** MEMORANDUM. **5. minutes.** An official record of proceedings at the meeting of an organization. *—vt.* **-ut·ed, -ut·ing, -utes. 1.** To record in a memorandum or other notation. **2.** To record in the minutes of a meeting.

**mi·nute**[2] (mī-no͞ot′, -nyo͞ot′, mĭ-) *adj.* [Lat. *minutus* < *minuere,* to lessen.] **1.** Exceptionally small **:** TINY. **2.** Insignificant. **3.** Marked by close examination. **—mi·nute′ness** *n.*

▲ word history: The noun *minute*[1] and the adjective *minute*[2] are both descended from Latin *minutus,* "small." The adjective is a direct borrowing of the Latin word. The noun entered English through French and has a more complex history. The noun *minute* is derived from Latin *minuta,* the feminine singular form of *minutus,* because the adjective occurred in the phrase *pars minuta prima,* "first small part," which specifically denoted a unit in a system of fractions with denominators that were powers of 60. In Medieval Latin *minuta* was used as a noun to mean "one sixtieth of a unit." Units such as the degree of the circle and the hour were, and still are, divided into 60 parts.

**minute hand** *n.* The long hand on a clock or watch indicating the minutes.

**min·ute·ly**[1] (mĭn′ĭt-lē) *adj. Archaic.* On a minute-by-minute basis.

**mi·nute·ly**[2] (mī-no͞ot′lē, -nyo͞ot′-, mĭ-) *adv.* **1.** With attention to minutiae. **2.** On a very small scale.

**min·ute·man** (mĭn′ĭt-măn′) *n.* An armed civilian ready to fight on a minute's notice just before and during the Revolutionary War.

**minute of arc** *n.* MINUTE[1] 1b.

**mi·nu·ti·a** (mĭ-no͞o′shē-ə, -shə, -nyo͞o′-) *n., pl.* **-ti·ae** (-shē-ē′) [Lat., smallness < *minutus,* small < *minuere,* to lessen.] *often* **minutiae.** A small, trivial detail <the *minutiae* of family life>

**minx** (mĭngks) *n.* [Orig. unknown.] **1.** A pert, impudent, or flirtatious young girl. **2.** *Obs.* A promiscuous woman.

**Mi·o·cene** (mī′ə-sēn′) *adj.* [Gk. *meiōn,* less + -CENE.] Of, belonging to, or typical of the geologic time, rock series, and sedimentary deposits of the fourth epoch of the Tertiary period, marked by the appearance of primitive apes, whales, and grazing animals. *—n.* **1.** The Miocene epoch. **2.** The deposits of the Miocene epoch.

**mi·o·sis** *also* **my·o·sis** (mī-ō′sĭs) *n., pl.* **-ses** (-sēz′) [Gk. *muein,* to close the eyes + -OSIS.] Excessive contraction of the pupil of the eye.

**mi·ot·ic** (mī-ŏt′ĭk) *n.* [< MIOSIS.] An agent causing contraction of the pupil of the eye. **—mi·ot′ic** *adj.*

**mir** (mîr) *n.* [R.] A prerevolutionary Russian peasant commune.

**mi·ra·bi·le dic·tu** (mĭ-rä′bĭ-lē dĭk′to͞o) *adv.* [Lat.] Wonderful to relate.

**mir·a·cle** (mĭr′ə-kəl) *n.* [ME < OFr. < Lat. *miraculum* < *mirari,* to wonder at.] **1.** An event that seems impossible to explain by natural laws and so is regarded as supernatural in origin or as an act of God. **2.** One that excites admiring awe. **3.** A miracle play.

**miracle play** *n.* A religious drama of the Middle Ages representing events in the lives of miracle-working saints and martyrs.

**mi·rac·u·lous** (mĭ-răk′yə-ləs) *adj.* [OFr. *miraculeux* < Med. Lat. *miraculosus* < Lat. *miraculum,* miracle.] **1.** Of the nature of a miracle. **2.** Caused by or as if by a miracle. **3.** Having the power to work miracles. **—mi·rac′u·lous·ly** *adv.* **—mi·rac′u·lous·ness** *n.*

**mi·rage** (mĭ-räzh′) *n.* [Fr. < *mirer,* to look at < Lat. *mirari,* to wonder at.] **1.** An optical phenomenon creating the illusion of water, often with inverted reflections of distant objects, resulting from light distortion by alternate layers of hot and cool air. **2.** Something insubstantial or illusory.

**mire** (mīr) *n.* [ME < ON *mýrr.*] **1.** An area of wet, soggy, and muddy ground **:** BOG. **2.** Deep slimy soil or mud. *—v.* **mired, mir·ing, mires.** *—vt.* **1.** To cause to sink or become stuck in mire. **2.** To soil with mud. **3.** To trap or entangle as if in mire. *—vi.* To sink or become stuck in or as if in a mire.

**mi·rex** (mī′rĕks′) *n.* [Perh. (PIS)MIR(E) + EX(TERMINATE).] An insecticide, $C_{10}Cl_{12}$, used against ants.

**mirk** (mûrk) *n. & adj. var. of* MURK.

**mirk·y** (mûr′kē) *adj. var. of* MURKY.

**mir·ror** (mĭr′ər) *n.* [ME *mirour* < OFr. < *mirer,* to look at < Lat. *mirari,* to wonder at.] **1.** A surface able to reflect enough undiffused light to form a virtual image of an object placed before it. **2.** Something that faithfully reflects or gives a true picture of something else.

ă **pat** ā **pay** âr **care** ä **father** ĕ **pet** ē **be** hw **which** ĭ **pit**
ī **tie** îr **pier** ŏ **pot** ō **toe** ô **paw, for** oi **noise** o͞o **took**

3. Something worthy of imitation. —*vt.* **-rored, -ror·ing, -rors.** To reflect in or as if in a mirror.

**mirth** (mûrth) *n.* [ME < OE *myrgð.*] Gaiety and gladness, esp. when expressed by laughter.

**mirth·ful** (mûrth'fəl) *adj.* **1.** Full of mirth. **2.** Marked by or expressing mirth. **—mirth'ful·ly** *adv.* **—mirth'ful·ness** *n.*

**mirth·less** (mûrth'lĭs) *adj.* Devoid of mirth. **—mirth'less·ly** *adv.* **—mirth'less·ness** *n.*

**MIRV** (mûrv) *n.* [M(ULTIPLE) I(NDEPENDENTLY TARGETED) R(E-ENTRY) V(EHICLES).] **1.** An offensive ballistic-missile system in which a number of warheads aimed at independent targets can be launched by a single booster rocket. **2.** Any of the warheads of a MIRV.

**mir·y** (mīr'ē) *adj.* **-i·er, -i·est. 1.** Full of or resembling mire : SWAMPY. **2.** Smeared with mire : MUDDY. **—mir'i·ness** *n.*

**mis-¹** *pref.* [Partly < ME *mis-* (< OE), and partly < ME *mes-* < OFr. < Lat. *minus,* less.] **1.** Bad : badly : wrong : wrongly <*misconduct*> **2.** Failure : lack <*misfire*> **3.** —Used as an intensive <*misdoubt*>

**mis-²** *pref. var. of* MISO-.

**mis·ad·ven·ture** (mĭs'əd-vĕn'chər) *also* **mis·ven·ture** (mĭs-vĕn'-) *n.* [ME *misaventure* < OFr. *mesaventure* < *mesavenir,* to result in misfortune : *mes-,* badly (< Lat. *minus,* less) + *avenir,* to turn out < Lat. *advenire,* to come to (*ad-,* to + *venire,* to come).] Great misfortune.

**mis·ad·vise** (mĭs'əd-vīz') *vt.* **-vised, -vis·ing, -vis·es.** To advise wrongly.

**mis·a·ligned** (mĭs'ə-līnd') *adj.* Wrongly aligned. **—mis'a·lign'ment** *n.*

**mis·al·li·ance** (mĭs'ə-lī'əns) *n.* [Fr. *mésalliance* : *més-,* bad (< Lat. *minus,* less) + *alliance,* alliance < OFr. *aliance.* —see ALLIANCE.] **1.** An unsuitable alliance. **2.** An unsuitable marriage.

**mis·al·lo·ca·tion** (mĭs-ăl'ə-kā'shən) *n.* Faulty or inappropriate allocation <a gross *misallocation* of government funds>

**mis·al·ly** (mĭs'ə-lī') *vt.* **-lied, -ly·ing, -lies.** To ally badly.

**mis·an·thrope** (mĭs'ən-thrōp', mĭz'-) *also* **mis·an·thro·pist** (mĭs-ăn'thrə-pĭst) *n.* [Fr. < Gk. *misanthrōpos,* hating mankind : *misein,* to hate + *anthrōpos,* man.] One who detests or distrusts humanity.

**mis·an·throp·ic** (mĭs'ən-thrŏp'ĭk, mĭz'-) *adj.* **1.** Of or typical of a misanthrope. **2.** Marked by detestation or distrust of humanity. **—mis'an·throp'i·cal·ly** *adv.*

**mis·an·thro·py** (mĭs-ăn'thrə-pē, mĭz-) *n.* Hatred of humanity.

**mis·ap·ply** (mĭs'ə-plī') *vt.* **-plied, -ply·ing, -plies.** To use or apply badly or wrongly. **—mis·ap'pli·ca'tion** (-ăp'lĭ-kā'shən) *n.*

**mis·ap·pre·hend** (mĭs-ăp'rĭ-hĕnd') *vt.* **-hend·ed, -hend·ing, -hends.** To fail to interpret correctly : MISUNDERSTAND. **—mis·ap'pre·hen'sion** (-hĕn'shən) *n.*

**mis·ap·pro·pri·ate** (mĭs'ə-prō'prē-āt') *vt.* **-at·ed, -at·ing, -ates. 1. a.** To appropriate wrongly <*misappropriating* the philosophy of Thoreau> **b.** To appropriate dishonestly for one's own use : EMBEZZLE. **2.** To use illegally. **—mis'ap·pro'pri·a'tion** *n.*

**mis·be·come** (mĭs'bĭ-kŭm') *vt.* **-came** (-kām'), **-come, -com·ing, -comes.** To be unsuitable or inappropriate for.

**mis·be·got·ten** (mĭs'bĭ-gŏt'n) *adj.* **1.** Illegally or abnormally begotten, esp. illegitimate. **2.** Having an improper basis or origin : ILL-CONCEIVED <*misbegotten* economic theories>

**mis·be·have** (mĭs'bĭ-hāv') *vi.* **-haved, -hav·ing, -haves.** To behave badly. **—mis'be·hav'er** *n.* **—mis'be·hav'ior** (-hāv'yər) *n.*

**mis·be·lief** (mĭs'bĭ-lēf') *n.* **1.** A wrong or faulty belief. **2.** A heretical or unorthodox religious belief.

**mis·cal·cu·late** (mĭs-kăl'kyə-lāt') *vt.* & *vi.* **-lat·ed, -lat·ing, -lates.** To calculate wrongly. **—mis·cal'cu·la'tion** *n.*

**mis·call** (mĭs-kôl') *vt.* **-called, -call·ing, -calls.** To misname.

**mis·car·riage** (mĭs-kăr'ĭj) *n.* **1. a.** Bad administration : MISMANAGEMENT <an egregious *miscarriage* of justice> **b.** Failure to attain the right or desired end <the *miscarriage* of our fondest hopes> **2.** Premature expulsion of a nonviable fetus from the uterus.

**mis·car·ry** (mĭs-kăr'ē) *vi.* **-ried, -ry·ing, -ries. 1.** To fail to reach the proper conclusion : go wrong <Their scheme *miscarried.*> **2.** To bring forth a fetus prematurely : ABORT.

**mis·cast** (mĭs-kăst') *vt.* **-cast, -cast·ing, -casts. 1.** To cast in an unsuitable role. **2.** To cast (a role or a theatrical production) inappropriately.

**mis·ceg·e·na·tion** (mĭ-sĕj'ə-nā'shən, mĭs'ĭ-jə-) *n.* [Lat. *miscēre,* to mix + *genus,* race.] The interbreeding of what are presumed to be distinct human races, esp. marriage or cohabitation between white and nonwhite persons. **—mis·ceg'e·na'tion·al** *adj.*

**mis·cel·la·ne·a** (mĭs'ə-lā'nē-ə) *pl.n.* [Lat. < neuter pl. of *miscellaneus,* miscellaneous.] A conglomeration of miscellaneous items.

**mis·cel·la·ne·ous** (mĭs'ə-lā'nē-əs) *adj.* [Lat. *miscellaneus* < *miscēre,* to mix.] **1.** Composed of a variety of parts or ingredients. **2.** Having a variety of characteristics, abilities, or appearances. **3.** Concerned with diverse subjects or aspects. **—mis'cel·la'ne·ous·ly** *adv.* **—mis'cel·la'ne·ous·ness** *n.*

**mis·cel·la·ny** (mĭs'ə-lā'nē) *n., pl.* **-nies.** [Lat. *miscellanea,* miscellanea.] **1.** A collection of various items, parts, or ingredients, esp. one made up of diverse literary works. **2. miscellanies.** A publication containing a variety of literary works.

**mis·chance** (mĭs-chăns') *n.* [ME *mischaunce* < OFr. *meschance* : *mes-,* bad (< Lat. *minus,* less) + *chance,* chance. —see CHANCE.] **1.** An unfortunate occurrence. **2.** Bad luck.

**mis·chief** (mĭs'chĭf) *n.* [ME *mischef* < OFr. *meschef,* misfortune : *mes-,* badly (< Lat. *minus,* less) + *chef,* head, end < Lat. *caput.*] **1.** Behavior causing discomfiture or annoyance in another. **2.** An inclination or tendency to play pranks or cause embarrassment. **3.** The cause of minor trouble or disturbance <The child was a *mischief* all day.> **4.** Damage, destruction, or injury caused by a specific person or thing <The flooded basement was the *mischief* of vandals.> **5.** The quality or state of being mischievous <children with *mischief* in their hearts>

**mis·chie·vous** (mĭs'chə-vəs) *adj.* [ME *meschevous* < AN < OFr. *meschef,* misfortune. —see MISCHIEF.] **1.** Causing mischief. **2.** Teasing : playful. **3.** Troublesome : irritating <a *mischievous* act> **4.** Causing harm, injury, or damage <*mischievous* gossip> **—mis'chie·vous·ly** *adv.* **—mis'chie·vous·ness** *n.*

**mis·ci·ble** (mĭs'ə-bəl) *adj.* [Med. Lat. *miscibilis* < Lat. *miscēre,* to mix.] *Chem.* Capable of being mixed. **—mis'ci·bil'i·ty** *n.*

**mis·clas·si·fy** (mĭs-klăs'ə-fī') *vt.* **-fied, -fy·ing, -fies.** To classify incorrectly. **—mis·clas'si·fi·ca'tion** *n.*

**mis·con·ceive** (mĭs'kən-sēv') *vt.* **-ceived, -ceiv·ing, -ceives.** To mistake the meaning of. **—mis'con·ceiv'er** *n.* **—mis'con·cep'tion** (-sĕp'shən) *n.*

**mis·con·duct** (mĭs-kŏn'dŭkt) *n.* **1. a.** Behavior not in conformity with prevailing standards or laws : IMPROPRIETY. **b.** Adultery. **2.** Dishonest or bad management, esp. by persons entrusted or engaged to act on behalf of another. **3.** Malfeasance, esp. by government officials. **—mis'con·duct'** (mĭs'kən-dŭkt') *v.* **(-duct·ed, -duct·ing, -ducts).**

**mis·con·struc·tion** (mĭs'kən-strŭk'shən) *n.* **1.** An inaccurate explanation, interpretation, or report : MISUNDERSTANDING. **2.** A faulty construction, esp. of a sentence or clause.

**mis·con·strue** (mĭs'kən-strōō') *vt.* **-strued, -stru·ing, -strues.** To mistake the meaning of : MISINTERPRET.

**mis·count** (mĭs-kount') *vt.* & *vi.* **-count·ed, -count·ing, -counts.** To count incorrectly : MISCALCULATE. —*n.* (mĭs'kount'). An inaccurate count.

**mis·cre·ant** (mĭs'krē-ənt) *n.* [ME *miscreaunt,* heretic < OFr. *mescreant,* pr.part. of *mescroire,* to disbelieve : *mes-* (reversal < Lat. *minus,* less) + *croire,* to believe < Lat. *credere.*] **1.** One who behaves badly or criminally. **2.** An infidel. **—mis'cre·ant** *adj.*

**mis·cre·ate** (mĭs'krē-āt') *vt.* **-at·ed, -at·ing, -ates.** To construct or shape badly. —*adj.* (mĭs'krē-ĭt, -āt'). Formed unnaturally : DEFORMED. **—mis'cre·a'tion** *n.*

**mis·cue** (mĭs-kyōō') *n.* **1.** A billiards stroke that misses or just brushes the ball because of the cue slipping. **2.** A blunder. —*vi.* **-cued, -cu·ing, -cues. 1.** To make a miscue. **2.** To miss a stage cue.

**mis·deal** (mĭs-dēl') *v.* **-dealt** (-dĕlt'), **-deal·ing, -deals.** —*vt.* To deal (playing cards) improperly. —*vi.* To deal playing cards improperly. **—mis'deal'** *n.* **—mis·deal'er** *n.*

**mis·deed** (mĭs-dēd') *n.* An illegal or wicked deed.

**mis·de·mean·ant** (mĭs'dĭ-mē'nənt) *n. Law.* One guilty of or convicted and sentenced for a misdemeanor.

**mis·de·mean·or** (mĭs'dĭ-mē'nər) *n.* **1.** A misdeed. **2.** *Law.* An offense less serious than a felony.

**mis·di·ag·nose** (mĭs-dī'əg-nōs', -nōz') *vt.* **-nosed, -nos·ing, -nos·es.** To diagnose incorrectly. **—mis·di'ag·no'sis** (-nō'sĭs) *n.*

**mis·di·rect** (mĭs'dĭ-rĕkt', -dī-) *vt.* **-rect·ed, -rect·ing, -rects. 1.** To instruct incorrectly. **2.** To put a wrong address on.

**mis·di·rec·tion** (mĭs'dĭ-rĕk'shən, -dī-) *n.* **1.** Inaccurate or wrong instructions or guidance. **2.** *Law.* An error made by a judge in charging a jury.

**mis·do** (mĭs-dōō') *vt.* **-did** (-dĭd'), **-done** (-dŭn'), **-do·ing, -does** (-dŭz'). To do wrongly or ineptly. **—mis·do'er** *n.* **—mis·do'ing** *n.*

**mis·doubt** (mĭs-dout') *vt.* **-doubt·ed, -doubt·ing, -doubts.** To feel wary of : SUSPECT.

**mise en scène** (mēz' äN sĕn') *n.* [Fr., putting on stage.] **1.** The arrangement of performers and properties on a stage for a theatrical production. **2. a.** A stage setting. **b.** A physical environment.

**mi·ser** (mī'zər) *n.* [< Lat., wretched.] **1.** One who deprives oneself of all but the barest essentials in order to hoard money. **2.** One who is greedy or avaricious.

**mis·er·a·ble** (mĭz'ər-ə-bəl, mĭz'rə-bəl) *adj.* [ME < OFr. < Lat. *miserabilis,* pitiable < *miserari,* to pity < *miser,* wretched.] **1.** Very uncomfortable or unhappy : WRETCHED. **2.** Causing or accompanied by wretchedness <a *miserable* place in the winter> **3.** Mean : shameful <a *miserable* lack of concern for the children> **4.** Wretchedly inadequate <*miserable* living quarters> **5.** Of poor quality : INFERIOR. **—mis'er·a·ble·ness** *n.* **—mis'er·a·bly** *adv.*

**mis·e·re·re** (mĭz'ə-râr'ē, -rîr'ē) *n.* [Lat., have mercy, the first word of the psalm.] **1. Miserere.** The 51st Psalm. **2.** A vocal lament or complaint. **3.** MISERICORD 2.

**mis·er·i·cord** or **mis·er·i·corde** (mĭz′ər-ĭ-kôrd′, mĭ-zĕr′ĭ-kôrd′) *n.* [ME, pity < OFr. < Lat. *misericordia* < *misericors,* merciful : *miserēre,* to feel pity + *cors,* heart.] **1. a.** Modification of monastic rules, as an exemption from fasting. **b.** The room in a monastery used by monks granted such an exemption. **2.** A bracket attached to the underside of a hinged seat in a church stall against which a standing person may lean when the seat is turned up. **3.** A narrow dagger used in medieval times to deliver the death stroke to a seriously wounded knight.

**mi·ser·ly** (mī′zər-lē) *adj.* Characteristic of a miser, esp. tending to hoard money or possessions. **—mi′ser·li·ness** *n.*

**mis·er·y** (mĭz′ə-rē) *n., pl.* **-ies.** [ME *miserie* < OFr. < Lat. *miseria* < *miser,* wretched.] **1. a.** Suffering and want resulting from physical conditions or extreme poverty. **b.** Mental or emotional unhappiness or distress. **2.** A cause of suffering. **3.** *Informal.* A physical ailment.

**mis·es·teem** (mĭs′ə-stēm′) *vt.* **-teemed, -teem·ing, -teems.** To fail to regard with deserved esteem : DISRESPECT.

**mis·es·ti·mate** (mĭs-ĕs′tə-māt′) *vt.* **-mat·ed, -mat·ing, -mates.** To estimate incorrectly. **—mis·es′ti·ma′tion** *n.*

**mis·fea·sance** (mĭs-fē′zəns) *n.* [OFr. *mesfaisance* < *mesfaire,* to do wrong : *mes-,* wrongly (< Lat. *minus*) + *faire,* to do < Lat. *facere.*] *Law.* Improper and unlawful execution of an act that in itself is lawful and proper.

**mis·fea·sor** (mĭs-fē′zər) *n.* [OFr. *mesfesour* < *mesfaire,* to do wrong. —see MISFEASANCE.] *Law.* One guilty of misfeasance.

**mis·file** (mĭs-fīl′) *vt.* **-filed, -fil·ing, -files.** To file incorrectly.

**mis·fire** (mĭs-fīr′) *vi.* **-fired, -fir·ing, -fires. 1. a.** To fail to ignite when expected <The engine *misfired.*> **b.** To fail to fire when expected <The pistol *misfired.*> **2.** To fail to achieve an anticipated result <a plan that *misfired*> **—mis′fire′** (mĭs′fīr′, mĭs-fīr′) *n.*

**mis·fit** (mĭs′fĭt′, mĭs-fĭt′) *n.* **1.** Something of the wrong shape or size for its purpose. **2.** One who is maladjusted or disturbingly different from those with whom he or she associates.

**mis·for·tune** (mĭs-fôr′chən) *n.* **1. a.** Ill luck : bad fortune. **b.** The condition arising from ill luck or bad fortune. **2.** An unexpected and distressing occurrence.

**mis·give** (mĭs-gĭv′) *v.* **-gave** (-gāv′), **-giv·en** (-gĭv′ən), **-giv·ing, -gives.** [MIS- + GIVE, to suggest (obs.).] —*vt.* To arouse suspicion or apprehension in. —*vi.* To be suspicious or apprehensive.

**mis·giv·ing** (mĭs-gĭv′ĭng) *n.* Uncertainty or apprehension <had *misgivings* about such a long automobile trip>

**mis·gov·ern** (mĭs-gŭv′ərn) *vt.* **-erned, -ern·ing, -erns.** To govern badly or inefficiently. **—mis·gov′ern·ment** *n.*

**mis·guide** (mĭs-gīd′) *vt.* **-guid·ed, -guid·ing, -guides.** To lead astray. **—mis·guid′ance** *n.* **—mis·guid′ed·ly** (-gī′dĭd-lē) *adv.* **—mis·guid′er** *n.*

**mis·han·dle** (mĭs-hăn′dl) *vt.* **-dled, -dling, -dles. 1.** To deal with inefficiently or clumsily. **2.** To treat roughly : MALTREAT.

**mis·hap** (mĭs′hăp′, mĭs-hăp′) *n.* **1.** MISFORTUNE 1a. **2.** An unfortunate accident.

**mis·hear** (mĭs-hîr′) *vt.* **-heard** (-hûrd′), **-hear·ing, -hears.** To hear incorrectly : MISUNDERSTAND.

**mish·mash** (mĭsh′măsh′, -mäsh′) *n.* [Redup. of MASH.] A collection of unrelated things : HODGEPODGE.

**Mish·nah** *also* **Mish·na** (mĭsh′nə) *n.* [Heb. *mishnāh,* instruction < *shānāh,* he repeated.] **1.** The first section of the Talmud, a compilation of early oral interpretations of the Scriptures dating from about A.D. 200. **2.** A paragraph from the Mishnah. **3.** The teaching of a rabbi or other noted authority on Jewish laws. **—Mish·na′ic** (mĭsh-nā′ĭk) *adj.*

**mis·i·den·ti·fy** (mĭs′ī-dĕn′tə-fī′) *vt.* **-fied, -fy·ing, -fies.** To identify incorrectly. **—mis′i·den′ti·fi·ca′tion** *n.*

**mis·im·pres·sion** (mĭs′ĭm-prĕsh′ən) *n.* An incorrect impression.

**mis·in·form** (mĭs′ĭn-fôrm′) *vt.* **-formed, -form·ing, -forms.** To give inaccurate information to. **—mis′in·form′ant** (-fôr′mənt), **mis′in·form′er** *n.* **—mis·in′for·ma′tion** *n.*

**mis·in·ter·pret** (mĭs′ĭn-tûr′prĭt) *vt.* **-pret·ed, -pret·ing, -prets. 1.** To explain inaccurately. **2.** To interpret incorrectly. **—mis′in·ter′pre·ta′tion** *n.* **—mis′in·ter′pret·er** *n.*

**mis·join·der** (mĭs-join′dər) *n.* *Law.* Improper joining of different causes of action or different parties to a lawsuit.

**mis·judge** (mĭs-jŭj′) *vt. & vi.* **-judged, -judg·ing, -judg·es.** To judge incorrectly. **—mis·judg′ment** *n.*

**mis·la·bel** (mĭs-lā′bəl) *vt.* **-beled, -bel·ing, -bels** *also* **-belled, -bel·ling, -bels.** To label inaccurately.

**mis·lay** (mĭs-lā′) *vt.* **-laid** (-lād′), **-lay·ing, -lays. 1.** To put in a place that is afterward forgotten <*mislaid* my keys> **2.** To place or put down incorrectly <*mislaid* the new carpeting> **—mis·lay′er** *n.*

**mis·lead** (mĭs-lēd′) *vt.* **-led** (-lĕd′), **-lead·ing, -leads. 1.** To lead in the wrong direction. **2.** To lead into error or wrongdoing : DECEIVE.

**mis·lead·ing** (mĭs-lē′dĭng) *adj.* Tending to mislead : DECEPTIVE <*misleading* statements> **—mis·lead′ing·ly** *adv.*

**mis·like** (mĭs-līk′) *vt.* **-liked, -lik·ing, -likes.** [ME *misliken* < OE *mislīcian* : *mis-,* ill + *līcian,* to please.] **1.** *Archaic.* To be displeasing to. **2.** To disapprove of : DISLIKE. —*n.* Disapproval : dislike.

**mis·man·age** (mĭs-măn′ĭj) *vt.* **-aged, -ag·ing, -ag·es.** To manage badly. **—mis·man′age·ment** *n.*

**mis·mar·riage** (mĭs-măr′ĭj) *n.* An unsuitable marriage.

**mis·match** (mĭs-măch′) *vt.* **-matched, -match·ing, -match·es.** To match unsuitably or inaccurately. **—mis·match′** (mĭs-măch′, mĭs′măch′) *n.*

**mis·mate** (mĭs-māt′) *vt.* **-mat·ed, -mat·ing, -mates.** To mate or match unsuitably.

**mis·name** (mĭs-nām′) *vt.* **-named, -nam·ing, -names.** To call by a wrong name.

**mis·no·mer** (mĭs-nō′mər) *n.* [ME < OFr. *mesnommer,* to misname : *mes-,* wrongly (< Lat. *minus,* less) + *nommer,* to name < Lat. *nominare* < *nomen,* name.] **1.** An error in naming a person or place. **2.** A name unsuitably or wrongly applied.

**mi·so** (mē′sō) *n.* [J.] A thick fermented paste made by grinding together cooked soybeans, rice, and salt, used esp. in making soups.

**miso-** or **mis-** *pref.* [Gk. < *misein,* to hate.] Hatred <*miso*gamy>

**mi·sog·a·my** (mĭ-sŏg′ə-mē) *n.* Hatred of marriage. **—mi·sog′a·mist** *n.*

**mi·sog·y·ny** (mĭ-sŏj′ə-nē) *n.* [Gk. *misogunia* : *misein,* to hate + *gunē,* women.] Hatred of women. **—mis′o·gyn′ic** (mĭs′ə-jĭn′ĭk, -gī′nĭk) *adj.* **—mi·sog′y·nist** *n.* **—mi·sog′y·nis′tic** (-sŏj′ə-nĭs′tĭk), **mi·sog′y·nous** (-sŏj′ə-nəs) *adj.*

**mi·sol·o·gy** (mĭ-sŏl′ə-jē) *n.* Hatred of reason, argument, or enlightenment. **—mi·sol′o·gist** *n.*

**mis·o·ne·ism** (mĭs′ə-nē′ĭz′əm) *n.* [Ital. *misoneismo* : Gk. *misein,* to hate + Gk. *neos,* new.] Hatred of change or innovation. **—mis′o·ne′ist** *n.*

**mis·o·ri·ent** (mĭs-ôr′ē-ənt, -ĕnt′, -ōr′-) *vt.* **-ent·ed, -ent·ing, -ents.** To orient incorrectly or inappropriately. **—mis·o′ri·en·ta′tion** *n.*

**mis·per·ceive** (mĭs′pər-sēv′) *vt.* **-ceived, -ceiv·ing, -ceives.** To misunderstand.

**mis·place** (mĭs-plās′) *vt.* **-placed, -plac·ing, -plac·es. 1. a.** To put in a wrong place. **b.** MISLAY 1. **2.** To bestow (e.g., confidence) on an improper, unsuitable, or unworthy object. **—mis·place′ment** *n.*

**mis·play** (mĭs′plā′, mĭs-plā′) *n.* A mistaken action in a game. —*vt.* (mĭs-plā′) **-played, -play·ing, -plays.** To make a misplay of.

**mis·print** (mĭs-prĭnt′) *vt.* **-print·ed, -print·ing, -prints.** To print incorrectly. —*n.* (mĭs′prĭnt′, mĭs-prĭnt′). A printing error.

**mis·pri·sion** (mĭs-prĭzh′ən) *n.* [ME < OFr. *mesprison* < *mesprendre,* to make a mistake : *mes-,* wrongly (< Lat. *minus,* less) + *prendre,* to take < Lat. *prehendere.*] *Law.* **1.** Maladministration of public office. **2.** Neglect in preventing or reporting a crime.

**mis·prize** (mĭs-prīz′) *vt.* **-prized, -priz·ing, -priz·es. 1.** To undervalue. **2.** To scorn : despise.

**mis·pro·nounce** (mĭs′prə-nouns′) *vt. & vi.* **-nounced, -nounc·ing, -nounc·es.** To pronounce incorrectly or make an incorrect pronunciation. **—mis′pro·nun′ci·a′tion** (-nŭn′sē-ā′shən) *n.*

**mis·quote** (mĭs-kwōt′) *vt.* **-quot·ed, -quot·ing, -quotes.** To quote incorrectly. **—mis′quo·ta′tion** (-kwō-tā′shən) *n.*

**mis·read** (mĭs-rēd′) *vt.* **-read** (-rĕd′), **-read·ing, -reads. 1.** To read inaccurately. **2.** MISINTERPRET 2.

**mis·reck·on** (mĭs-rĕk′ən) *vt. & vi.* **-oned, -on·ing, -ons.** To miscalculate.

**mis·re·mem·ber** (mĭs′rĭ-mĕm′bər) *vt.* **-bered, -ber·ing, -bers.** To recollect incorrectly.

**mis·re·port** (mĭs′rĭ-pôrt′, -pōrt′) *vt.* **-port·ed, -port·ing, -ports.** To report falsely or mistakenly. —*n.* A wrong or inaccurate report. **—mis′re·port′er** *n.*

**mis·rep·re·sent** (mĭs-rĕp′rĭ-zĕnt′) *vt.* **-sent·ed, -sent·ing, -sents. 1.** To give a misleading or incorrect representation of. **2.** To serve dishonestly or incorrectly as an official representative of. **—mis·rep′re·sen·ta′tion** *n.* **—mis·rep′re·sen′ta·tive** (-zĕn′tə-tĭv) *adj.* **—mis·rep′re·sent′er** *n.*

**mis·rule** (mĭs-ro͞ol′) *vt.* **-ruled, -rul·ing, -rules.** To rule wrongly, unwisely, or unjustly : MISGOVERN. —*n.* **1.** Misgovernment. **2.** Lawless confusion : DISORDER.

**miss¹** (mĭs) *v.* **missed, miss·ing, miss·es.** [ME *missen* < OE *missan.*] —*vt.* **1.** To fail to hit, reach, catch, meet, or otherwise make contact with (a specific object). **2.** To fail to perceive, comprehend, or otherwise experience <*missed* the message in the play> **3.** To fail to accomplish, achieve, or attain (a goal). **4.** To fail to attend or perform <*missed* the first game of the season> **5. a.** To leave out : OMIT. **b.** To let slip <*miss* a big opportunity> **6.** To escape or avoid. **7.** To discover the absence or loss of. **8.** To feel the lack or loss of <*missed* friends from college> —*vi.* **1.** To fail to hit or make contact with something <swung at the pitch and *missed*> **2. a.** To be unsuccessful. **b.** To lose a benefit or opportunity <*missed* out on the pay raise> —*n.* A failure to hit, succeed, or find.

**miss²** (mĭs) *n.* [Short for MISTRESS.] **1. Miss.** —Used as a courtesy title before the name of an unmarried woman <*Miss* Watson><*Miss* Lee Watson> **2.** —Used as a form of address without a name in speaking to a usu. young woman <May I help you, *miss?*> **3.** A young unmarried woman.

**mis·sa can·ta·ta** (mĭs′ə) *n.* [NLat., sung mass.] A High Mass.

**mis·sal** (mĭs′əl) *n.* [ME *messel* < Med. Lat. *missale* < neuter of *missalis,* of the Mass < LLat. *missa,* Mass < Lat. *mittere,* to send.] **1.** *Rom. Cath. Ch.* A book containing all the prayers and responses for celebration of the Mass. **2.** A prayer book.

**mis·sel thrush** *also* **mis·tle thrush** (mĭs′əl) *n.* [< obs. *missel,* mistletoe < ME *mistel* < OE.] A European thrush, *Turdus viscivorus,* that eats berries, chiefly of the mistletoe.

**missel thrush**
*10½ inches long*

**mis·shape** (mĭs-shāp′) *vt.* **-shaped** *or* **-shap·en** (-shā′pən), **-shap·ing, -shapes.** To shape badly : DEFORM. **—mis·shap′en·ly** *adv.* **—mis·shap′er** *n.*

**mis·sile** (mĭs′əl, -īl′) *n.* [Lat. < neuter of *missilis,* able to be thrown < *mittere,* to let go.] **1.** An object fired, thrown, dropped, or otherwise projected at a target : PROJECTILE. **2.** A guided missile. **3.** A ballistic missile.

**mis·sile·man** (mĭs′əl-mən) *n.* One who designs, constructs, or launches guided missiles.

**mis·sile·ry** *also* **mis·sil·ry** (mĭs′əl-rē) *n.* **1.** The science of making and deploying guided or ballistic missiles. **2.** Missiles in general.

**miss·ing** (mĭs′ĭng) *adj.* **1.** Not present : LOST. **2.** Lacking <a new jacket with a button *missing*>

**missing link** *n.* **1.** A theoretical primate postulated to bridge the evolutionary gap between the anthropoid apes and human beings. **2.** Something lacking but necessary for completion of a series.

**mis·sion** (mĭsh′ən) *n.* [Fr. < Lat. *missio* < *mittere,* to send.] **1. a.** A body of persons sent to negotiate or establish relations with a foreign country. **b.** The business with which such a body of persons is charged. **2. a.** A body of persons sent to do religious work in a foreign land. **b.** An establishment of missionaries abroad. **c.** The district assigned to a missionary. **d. missions.** Missionary work. **e.** A missionary building or compound. **f.** An organization for carrying on missionary work in a territory. **3.** A permanent diplomatic office in a foreign country. **4.** A combat operation assigned to an individual or unit. **5.** A welfare or educational organization set up for the poor. **6.** A church or congregation without its own priest. **7.** A series of special religious services for purposes of proselytizing. **8.** A self-imposed duty. *—vt.* **-sioned, -sion·ing, -sions. 1.** To send on a mission. **2.** To establish a mission among or in. **—mis′sion·er** *n.*

☆ **syns:** MISSION, COMMISSION, ERRAND, OFFICE, TASK *n. core meaning* : an assignment one is sent to carry out <a *mission* of great urgency>

**mis·sion·ar·y** (mĭsh′ə-nĕr′ē) *n., pl.* **-ies. 1.** One sent on a mission, esp. one sent to do religious or charitable work in a territory or foreign country. **2.** A propagandist. *—adj.* **1.** Of or relating to missions or missionaries. **2.** Engaged in the activities of a mission or missionary. **3.** Tending to propagandize or use insistent persuasion <*missionary* zeal>

**mis·sis** *or* **mis·sus** (mĭs′ĭz, -ĭs) *n.* [Alteration of MISTRESS.] *Informal.* **1.** The mistress of a household. **2.** One's wife.

**Mis·sis·sip·pi·an** (mĭs′ĭ-sĭp′ē-ən) *adj.* **1.** Of, belonging to, or designating the geologic time, system of rocks, and sedimentary deposits of the fifth period of the Paleozoic era, marked by the submergence of extensive land areas beneath shallow seas. **2.** Of or concerned with the state of Mississippi. *—n.* **1.** *Geol.* The Mississippian period. **2.** A native or resident of Mississippi.

**mis·sive** (mĭs′ĭv) *n.* [< ME *(letter) missive,* (letter) sent (by superior authority) < Med. Lat. *(littere) missive* < neuter of *missivus,* sent < Lat. *mittere,* to send.] A message or letter.

**Mis·sou·ri** (mĭ-zo͝or′ē) *n., pl.* **Missouri** *or* **-ris. 1. a.** A tribe of North American Indians formerly living in what is now northern Missouri. **b.** A member of this tribe. **2.** The Siouan language of the Missouri.

**mis·spell** (mĭs-spĕl′) *vt.* **-spelled** *or* **-spelt** (-spĕlt′), **-spell·ing, -spells.** To spell incorrectly. **—mis·spell′ing** *n.*

**mis·spend** (mĭs-spĕnd′) *vt.* **-spent** (-spĕnt′), **-spend·ing, -spends.** To spend extravagantly or improperly : SQUANDER.

**mis·state** (mĭs-stāt′) *vt.* **-stat·ed, -stat·ing, -states.** To state falsely or incorrectly. **—mis·state′ment** *n.*

**mis·step** (mĭs-stĕp′) *n.* **1.** An awkward or misplaced step. **2.** An instance of improper or wrong conduct : BLUNDER.

**mis·sus** (mĭs′ĭz, -ĭs) *n. var. of* MISSIS.

**miss·y** (mĭs′ē) *n., pl.* **-ies.** *Informal.* MISS² 3.

**mist** (mĭst) *n.* [ME < OE.] **1.** A mass of fine droplets of water in the atmosphere near or in contact with the earth. **2.** Water vapor condensed on and clouding the appearance of a surface. **3.** Fine drops of a liquid, as perfume, sprayed into the air. **4.** A colloidal suspension of a liquid in a gas. **5.** A dimming or concealing element or factor. **6.** Something that produces or gives the impression of dimness or obscurity <the *mists* of the past> *—v.* **mist·ed, mist·ing, mists.** *—vi.* **1.** To be or become obscured or blurred by or as if by mist. **2.** To rain in a fine shower. *—vt.* To conceal or veil as if with mist.

**mis·tak·a·ble** (mĭ-stā′kə-bəl) *adj.* Capable of being misunderstood or mistaken. **—mis·tak′a·bly** *adv.*

**mis·take** (mĭ-stāk′) *n.* [< ME *mistaken,* to misunderstand < ON *mistaka,* to take in error : *mis-,* wrongly + *taka,* to take.] **1.** An error : fault. **2.** A misconception : misunderstanding. *—v.* **-took** (-sto͝ok′), **-tak·en, -tak·ing, -takes.** *—vt.* **1.** To understand wrongly : MISINTERPRET <*mistook* my purpose> **2.** To recognize or identify incorrectly <*mistook* me for you> *—vi.* To make a mistake : ERR. **—mis·tak′er** *n.*

**mis·tak·en** (mĭ-stā′kən) *adj.* **1.** Incorrect in opinion, understanding, or perception. **2.** Based on error : WRONG <a *mistaken* approach to solving the problem> **—mis·tak′en·ly** *adv.*

**Mis·ter** (mĭs′tər) *n.* [Alteration of MASTER.] **1.** —Used as a courtesy title when speaking to or of a man, usu. written in its abbreviated form and placed before a man's surname or title of office <*Mr.* Davis><*Mr.* Mayor> **2.** —Used as a form of address for certain U.S. military personnel, as petty officers or warrant officers. **3. mister.** *Informal.* —Used as a form of address without a name in speaking to a man <May I help you, *mister?*>

**mist·flow·er** (mĭst′flou′ər) *n.* A plant, *Eupatorium coelestinum* of southeastern North America, bearing small blue flower clusters.

**mis·tle thrush** (mĭs′əl) *n. var. of* MISSEL THRUSH.

**mis·tle·toe** (mĭs′əl-tō′) *n.* [ME *mistelto* < OE *misteltān* : *mistel,* mistletoe + *tān,* twig.] **1.** A Eurasian parasitic shrub, *Viscum album,* with leathery evergreen leaves and waxy white berries. **2.** Any of various American parasitic shrubs, as *Phoradendron flavescens* of eastern North America. **3.** A mistletoe sprig, often used as a Christmas decoration.

**mistletoe cactus** *n.* A leafless epiphytic tropical American cactus, *Rhipsalis cassytha.*

**mis·took** (mĭ-sto͝ok′) *v. p.t. of* MISTAKE.

**mis·tral** (mĭs′trəl, mĭ-sträl′) *n.* [Fr. < Prov. < Lat. *magistralis,* of a master < *magister,* master.] A cold, dry northerly wind that blows in squalls through the Rhone Valley and nearby areas toward the Mediterranean coast of southern France.

**mis·treat** (mĭs-trēt′) *vt.* **-treat·ed, -treat·ing, -treats.** To handle or treat roughly. **—mis·treat′ment** *n.*

**mis·tress** (mĭs′trĭs) *n.* [ME *mistres* < OFr. *maistresse* < *maistre,* master < Lat. *magister.*] **1.** A woman in a position of authority, control, or ownership, as the head of a household. **2.** A woman owning an animal or formerly a slave. **3.** A woman who has ultimate control over something <the *mistress* of our affections> **4. a.** A nation or country with supremacy over others. **b.** Something that directs or reigns and is personified as female. **5.** A woman who has mastered a skill. **6.** A woman who has a continuing sexual relationship with a man to whom she is not married, esp. one who receives financial support from the man. **7. Mistress.** —Used formerly as a courtesy title when speaking to or of a woman. **8.** *Chiefly Brit.* A woman teacher.

**mis·tri·al** (mĭs-trī′əl, -trīl′) *n. Law.* **1.** A trial that becomes invalid because of basic error in procedure. **2.** An inconclusive trial, as one in which the jurors cannot reach a verdict.

**mis·trust** (mĭs-trŭst′) *n.* Lack of trust. *—v.* **-trust·ed, -trust·ing, -trusts.** *—vt.* To regard without confidence. *—vi.* To be wary or doubtful. **—mis·trust′ful** *adj.* **—mis·trust′ful·ly** *adv.* **—mis·trust′ful·ness** *n.*

**mist·y** (mĭs′tē) *adj.* **-i·er, -i·est. 1.** Made up of, filled with, or resembling mist <a *misty* fog> **2.** Obscured or clouded by or as if by mist. **3.** Lacking clarity : VAGUE. **—mist′i·ly** *adv.* **—mist′i·ness** *n.*

**mist·y-eyed** (mĭs′tē-īd′) *adj.* **1.** With eyes blurred as if by mist. **2.** Sentimental or dreamy.

**mis·un·der·stand** (mĭs-ŭn′dər-stănd′) *vt.* **-stood** (-sto͝od′), **-stand·ing, -stands.** To understand incorrectly.

**mis·un·der·stand·ing** (mĭs-ŭn′dər-stăn′dĭng) *n.* **1.** Failure to understand correctly. **2.** A disagreement : quarrel.

**mis·us·age** (mĭs-yo͞o′sĭj, -zĭj) *n.* **1.** Abusive treatment. **2.** Improper application, as of words.

**mis·use** (mĭs-yo͞os′) *n.* Improper or incorrect use : MISAPPLICATION. *—vt.* (mĭs-yo͞oz′) **-used, -us·ing, -us·es. 1.** To use incorrectly. **2.** To mistreat or abuse.

**mis·val·ue** (mĭs-văl′yo͞o) *vt.* **-ued, -u·ing, -ues.** To value or estimate incorrectly.

**mis·ven·ture** (mĭs-vĕn′chər) *n. var. of* MISADVENTURE.

**mis·word** (mĭs-wûrd′) *vt.* **-word·ed, -word·ing, -words.** To express incorrectly.

**mite¹** (mīt) *n.* [ME < OE *mīte.*] Any of various small, often parasitic arachnids.

**mite[2]** (mīt) *n.* [ME < MDu., a small Flemish coin.] **1. a.** A very small amount of money or contribution. **b.** A widow's mite. **2.** A coin of little value, esp. an obsolete British coin worth half a farthing. **3.** A very small object or creature.

**mi·ter** (mī′tər) *n.* [ME *mitre* < OFr. < Lat. *mitra,* headband < Gk.] **1.** A tall pointed hat with peaks in front and back, worn by bishops and certain other ecclesiastics. **2. a.** A thong for binding the hair, worn by ancient Greek women. **b.** The ceremonial headdress of ancient Jewish high priests. **3.** A covering or top of a chimney allowing the release of smoke while keeping out rain and debris. **4. a.** A miter joint. **b.** The edge of a piece of material that has been beveled preparatory to making a miter joint. **c.** A miter square. —*v.* **-tered, -ter·ing, -ters.** —*vt.* **1.** To bestow a miter upon. **2.** To make join with a miter joint. —*vi.* To meet in a miter joint.

**miter box** *n.* **1.** A box open at the ends, with slotted sides that guide a saw in cutting miter joints. **2.** A device for handsaws that may be set to guide cuts in lumber at various degrees.

**miter joint** *n.* A joint made by beveling each of two surfaces to be joined, usu. at a 45° angle, to form a 90° corner.

**miter square** *n.* A carpenter's square with a blade that is set at a 45° angle or is adjustable.

**mi·ter·wort** (mī′tər-wûrt′, -wôrt′) *n.* A North American plant of the genus *Mitella,* with heart-shaped leaves and small white flower clusters.

**Mith·ra·ism** (mĭth′rə-ĭz′əm, -rā-) *n.* A Persian religious cult that flourished in the late Roman Empire, rivaling Christianity. **—Mith·ra′ic** (-rā′ĭk) *adj.* **—Mith·ra′ist** (-rā′ĭst) *n.*

**Mith·ras** (mĭth′rəs) *n.* [Lat. < Gk. < OPers. *mithra.*] *Myth.* The Persian god of light and guardian against evil.

**mith·ri·date** (mĭth′rĭ-dāt′) *n.* [After *Mithridates* (132?–63 B.C.), who is said to have acquired tolerance for poison.] A substance regarded as an antidote against poison.

**mith·ri·da·tism** (mĭth′rĭ-dā′tĭz′əm) *n.* Tolerance for a poison acquired by taking ever larger doses. **—mith′ri·dat′ic** (-dăt′ĭk) *adj.*

**mit·i·gate** (mĭt′ĭ-gāt′) *vt.* & *vi.* **-gat·ed, -gat·ing, -gates.** [ME *mitigaten* < Lat. *mitigare* < *mitis,* soft.] To make or become less severe or intense : MODERATE. **—mit′i·ga·ble** (-gə-bəl) *adj.* **—mit′i·ga′tion** *n.* **—mit′i·ga′tive, mit′i·ga·to′ry** (-gə-tôr′ē, -tōr′ē) *adj.* **—mit′i·ga′tor** *n.*

**mi·to·chon·dri·on** (mī′tə-kŏn′drē-ən) *n., pl.* **-dri·a** (-drē-ə) [NLat. : Gk. *mitos,* thread + Gk. *khondrion,* dim. of *khondros,* grain.] *Biol.* A microscopic body occurring in the cells of nearly all living organisms and containing enzymes responsible for the conversion of food to usable energy. **—mi′to·chon′dri·al** (-drē-əl) *adj.*

**mi·to·gen** (mī′tə-jən) *n.* [MITO(SIS) + -GEN.] A mitosis-inducing agent. **—mi′to·gen′ic** (mī′tə-jĕn′ĭk, mĭt′ə-) *adj.* **—mi′to·ge·nic′i·ty** (-jə-nĭs′ĭ-tē) *n.*

**mi·to·my·cin** (mī′tə-mī′sĭn) *n.* [MITO(SIS) + -MYCIN.] A complex of antibiotics produced by the bacterium *Streptomyces caespitosus* that is occas. used in cancer chemotherapy.

**mi·to·sis** (mī-tō′sĭs) *n., pl.* **-ses** (-sēz′) [Gk. *mitos,* thread + -OSIS.] *Biol.* **1.** Sequential differentiation and segregation of replicated chromosomes in a cell nucleus that precedes complete cell division. **2.** The entire sequence of processes in cell division in which the diploid number of chromosomes is retained in both daughter cells. **—mi·tot′ic** (mī-tŏt′ĭk) *adj.* **—mi·tot′i·cal·ly** *adv.*

**mi·tral** (mī′trəl) *adj.* [Fr. < Lat. *mitra,* miter.] **1.** Relating to or resembling a miter. **2.** Relating to a mitral valve.

**mitral valve** *n.* The cardiac valve between the left auricle and the left ventricle regulating blood flow from the auricle to the ventricle.

**mi·tre** (mī′tər) *n.* & *v. Chiefly Brit. var. of* MITER.

**mitt** (mĭt) *n.* [Short for MITTEN.] **1.** A woman's glove extending over the hand but only partially covering the fingers. **2.** A mitten. **3.** A baseball glove, esp. a large leather padded glove used by catchers and first basemen. **4.** *Slang.* A hand : fist.

**mit·ten** (mĭt′n) *n.* [ME *mytan* < OFr. *mitaine,* prob. < Lat. *medietas,* half < *medius,* middle.] A covering for the hand that encases the thumb separately and the four fingers together.

**mit·ti·mus** (mĭt′ə-məs) *n., pl.* **-mus·es.** [Lat., we send (the first word of the writ).] *Law.* A writ instructing a jailer to hold a prisoner.

**mitz·vah** (mĭts′və) *n., pl.* **-voth** (-vōth′, -vōt′) or **-vahs.** [Heb. *miṣwāh.*] **1. a.** A commandment of the Jewish law. **b.** Fulfillment of such a commandment. **2.** A worthy deed.

**mix** (mĭks) *v.* **mixed, mix·ing, mix·es.** [Back-formation < obs. *mixte,* mixed < ME < OFr. < Lat. *mixtus,* p.part. of *miscēre,* to mix.] —*vt.* **1. a.** To combine or blend into one mass or mixture, rendering the constituent parts indistinguishable. **b.** To create or form by adding ingredients together. **c.** To add (an ingredient or element) to another. **2.** To combine or join <*mix* happiness with disappointment> **3.** To bring into social contact <*mix* young and old at a party> **4.** To crossbreed. —*vi.* **1. a.** To become mixed or blended together. **b.** To be capable of being blended together <Eggs *mix* with milk.> **2.** To associate socially or get along with others <*mix* with the crowd> **3.** To be crossbred. **—mix up. 1.** To confuse : confound. **2.** To involve <I got *mixed up* in a power struggle.> —*n.* **1.** An act of mixing. **2.** A mixture, esp. of ingredients packaged and sold commercially <muffin *mix*> **3.** A tape recording or a phonograph record produced by adjusting and combining sounds. **—mix it up.** *Slang.* To fight. **—mix′a·ble** *adj.*

**mixed** (mĭkst) *adj.* **1.** Blended together into a single unit or mass : INTERMINGLED. **2.** Composed of a variety of differing, sometimes conflicting entities <*mixed* reactions> **3.** Composed of people of different sexes, races, religions, or social classes.

**mixed bag** *n.* A collection of dissimilar things : ASSORTMENT.

**mixed drink** *n.* A drink of one or more kinds of liquor combined with other ingredients, usu. shaken or stirred before serving.

**mixed grill** *n.* A dish consisting of a variety of broiled meats and vegetables, typically including a lamb chop.

**mixed marriage** *n.* Marriage between persons of different races or religions.

**mixed-me·di·a** (mĭkst′mē′dē-ə) *adj.* Multimedia.

**mixed metaphor** *n.* A succession of metaphors that produce an incongruous and ludicrous effect, as *soaring ambitions that were stalled by a rock of resistance.*

**mixed nerve** *n.* A nerve having both sensory and motor fibers.

**mixed number** *n.* A number, as 7¼, equal to the sum of an integer and a fraction.

**mixed-up** (mĭkst′ŭp′) *adj. Informal.* Muddled : confused.

**mix·er** (mĭk′sər) *n.* **1.** One that mixes. **2.** One who is sociable. **3.** An informal dance or party that gives members of a group a chance to become acquainted with each other. **4.** A device that blends or mixes substances or ingredients, esp. by mechanical agitation. **5.** A beverage, as soda water or ginger ale, for diluting alcoholic drinks.

**mix·ol·o·gy** (mĭk-sŏl′ə-jē) *n.* The study or skill of preparing mixed drinks. **—mix·ol′o·gist** *n.*

**mixt** (mĭkst) *v. Archaic. var. p.t.* & *p.p. of* MIX.

**mix·ture** (mĭks′chər) *n.* [Fr. < Lat. *mixtura* < *miscēre,* to mix.] **1.** Something made by mixing. **2.** Something composed of diverse elements <a *mixture* of rage and frustration> **3.** A fabric consisting of different kinds of thread or yarn. **4.** The act or process of mixing or state of being mixed. **5.** *Chem.* A composition of two or more substances not chemically bound to each other.

**mix-up** (mĭks′ŭp′) *n.* **1.** Muddle : confusion. **2.** *Informal.* A fight.

**Mi·zar** (mī′zär′) *n.* [Ar. *mi'zar,* Mizar, veil.] The star at the crook of the handle of the Big Dipper.

**miz·zen** or **miz·en** (mĭz′ən) *n.* [ME *meson* < OFr. *misaine,* prob. < OItal. *mezzana* < *mezzano,* middle < Lat. *medianus* < *medius,* half.] *Naut.* **1.** A fore-and-aft sail set on the mizzenmast. **2.** A mizzenmast. **—miz′zen** *adj.*

**miz·zen·mast** or **miz·en·mast** (mĭz′ən-məst, -măst′) *n. Naut.* **1.** The third mast aft on sailing ships carrying three or more masts. **2.** JIGGER MAST 1.

**miz·zle[1]** (mĭz′əl) *vi.* **-zled, -zling, -zles.** [ME *misellen.*] To rain in fine, mistlike droplets. **—miz′zle** *n.* **—miz′zly** *adv.*

**miz·zle[2]** (mĭz′əl) *vi.* **-zled, -zling, -zles.** [Orig. unknown.] *Chiefly Brit.* To make a sudden departure.

**Mn** *symbol for* MANGANESE.

**mne·mon·ic** (nĭ-mŏn′ĭk) *adj.* [Gk. *mnēmonikos,* of memory < *mnēmōn,* mindful < *mnasthai,* to remember.] Pertaining to, aiding, or intended to aid the memory. —*n.* A device, as a formula or rhyme, used as an aid in remembering. **—mne·mon′i·cal·ly** *adv.*

**mne·mon·ics** (nĭ-mŏn′ĭks) *n. (sing. in number).* A system to enhance or develop the memory.

**Mne·mos·y·ne** (nĭ-mŏs′ə-nē, -mŏz′-) *n.* [Lat. < Gk. *Mnēmosunē* < *mnasthai,* to remember.] *Gk. Myth.* The goddess of memory.

**-mo** *suff.* [< DUODECIMO.] —Used after numerals to indicate the number of leaves that results from folding a sheet of paper <twelvemo>

**Mo** *symbol for* MOLYBDENUM.

**mo·a** (mō′ə) *n.* [Maori.] Any of various large, long-necked, flightless birds of the order Dinorthiformes, native to New Zealand and extinct for more than a hundred years.

**Mo·ab·ite** (mō′ə-bīt′) *n.* **1.** A descendant of Moab, the son of Lot in the Old Testament. **2.** An inhabitant of Moab. **—Mo′a·bite′** *adj.*

**moan** (mōn) *n.* [ME *mone,* complaint.] **1. a.** A low, drawn out, mournful sound, usu. indicating sorrow or pain. **b.** A sound similar to a moan. **2.** Lamentation. —*v.* **moaned, moan·ing, moans.** —*vi.* **1.** To utter a moan or make a sound like a moan. **2.** To complain, lament, or grieve. —*vt.* **1.** To bewail. **2.** To utter with a moan.

**moat** (mōt) *n.* [ME *mote* < OFr., mound.] A wide, deep ditch, usu. filled with water, around a medieval town, fortress, or castle for protection against assault. —*vt.* **moat·ed, moat·ing, moats.** To surround with or as if with a moat.

**mob** (mŏb) *n.* [Short for obs. *mobile* < Lat. *mobile (vulgus),* fickle (crowd).] **1.** A large disorderly crowd. **2.** The common people. **3.** *Informal.* An organized gang of criminals. —*vt.* **mobbed, mob·bing, mobs. 1.** To crowd around and jostle or annoy, esp. in anger or excessive enthusiasm <Dozens of shouting fans *mobbed* the actor.>

**2.** To crowd into <Spectators *mobbed* the arena.> **3.** To attack violently, usu. in a disorderly crowd.

▲ word history: Every age has its linguistic fads and ephemeral coinages; occasionally some of them survive and become part of the standard vocabulary. In the 17th and 18th centuries the abbreviation of long words or phrases to one or two syllables had a vogue that was much deplored by the self-appointed literary watchdogs of the day. *Mob* is one such abbreviation that caught on. It is short for *mobile,* which was used in the early 17th century to denote "the masses." In this usage *mobile* was itself a shortening of the Latin phrase *mobile vulgus,* "the excitable populace." The note of contempt originally inherent in English *mob* was also found in the Latin phrase.

**mob·cap** (mŏb′kăp′) *n.* [Obs. *mob,* mobcap (poss. < obs. Du. *mop*) + CAP.] A large high cap trimmed with ribbons and frills, worn by 18th- and early 19th-cent. women.

**mo·bile** (mō′bəl, -bēl′, -bīl′) *adj.* [OFr. < Lat. *mobilis* < *movēre,* to move.] **1.** Capable of moving or of being moved from one place to another. **2.** Moving rapidly from one state to another <a *mobile,* beautiful face> **3. a.** Marked by the easy intermixing of different social groups <a *mobile* community> **b.** Having the possibility of relatively easy movement from one social class or level to another <an upwardly *mobile* generation> **4.** Flowing freely <a *mobile* liquid> *—n.* (mō′bēl′). A sculpture composed of parts that move, esp. in response to air currents. **—mo·bil′i·ty** (-bĭl′ĭ-tē) *n.*

**mobile home** *n.* A house trailer serving as a permanent home and usu. hooked up to utilities.

**mo·bi·lize** (mō′bə-līz′) *v.* **-lized, -liz·ing, -liz·es.** [Fr. *mobiliser* < *mobile,* mobile < OFr.] *—vt.* **1.** To make mobile. **2. a.** To assemble, prepare, or put into operation for war or another emergency. **b.** To coordinate or assemble for a particular purpose <*mobilized* the church's financial resources> *—vi.* To become prepared for war or other emergency. **—mo′bi·li·za′tion** *n.*

☆ ***syns:*** MOBILIZE, MARSHAL, MUSTER, ORGANIZE, RALLY *v. core meaning* : to assemble, prepare, or put into operation, as for war or a similar emergency <*mobilized* a rescue crew>

**Mö·bi·us strip** (mœ′bē-əs) *n.* [After August *Möbius* (1790–1868).] A one-sided surface that can be formed from a rectangular strip by twisting one end 180° and fastening it to the other end.

**mob·oc·ra·cy** (mŏb-ŏk′rə-sē) *n., pl.* **-cies. 1.** Political control by a mob. **2.** A mob as the source of political control. **—mob′o·crat′** (mŏb′ə-krăt′) *n.* **—mob′o·crat′ic** (-krăt′ĭk), **mob′o·crat′i·cal** *adj.*

**mob·ster** (mŏb′stər) *n.* A member of a criminal gang.

**moc·ca·sin** (mŏk′ə-sĭn) *n.* [Natick *mohkussin.*] **1.** A soft leather slipper worn by American Indians. **2.** A shoe or slipper similar to an Indian moccasin. **3.** The water moccasin.

**moccasin flower** *n.* A North American orchid of the genus *Cypripedium,* esp. *C. acaule* of eastern North America, bearing a solitary flower with a pouchlike pink lip.

**mo·cha** (mō′kə) *n.* [After *Mocha,* a port of Yemen, from which it was orig. exported.] **1.** A rich pungent Arabian coffee. **2.** A fine grade of coffee. **3.** A flavoring made of coffee often mixed with chocolate. **4.** A soft thin glove leather usu. of goatskin. **5.** A dark olive brown.

**mock** (mŏk) *v.* **mocked, mock·ing, mocks.** [ME *mokken* < OFr. *mocquer.*] *—vt.* **1.** To treat with contempt or ridicule : DERIDE. **2. a.** To mimic, as in sport or derision. **b.** To imitate : counterfeit. **3.** To frustrate the hopes of. *—vi.* To express scorn or ridicule <They *mocked* at your suggestion.> *—n.* **1. a.** An act of mocking. **b.** Derision : mockery. **2.** Something worthy of derision. **3.** An imitation or counterfeit. *—adv.* In an insincere way. *—adj.* Simulated : sham <a *mock* battle> **—mock′er** *n.* **—mock′ing·ly** *adv.*

**mock·er·y** (mŏk′ə-rē) *n., pl.* **-ies. 1.** Scornful derision or ridicule. **2.** A specific action of derision or ridicule. **3.** An object of scorn or ridicule. **4.** A false, derisive, or impudent imitation. **5.** Something ludicrously unsuitable or futile.

☆ ***syns:*** MOCKERY, BURLESQUE, CARICATURE, FARCE, MOCK, PARODY, SHAM, TRAVESTY *n. core meaning* : a false, derisive, or impudent imitation <The trial was a *mockery* of justice.>

**mock-he·ro·ic** (mŏk′hĭ-rō′ĭk) *n., pl.* **mock-he·ro·ics.** A satirical imitation or burlesque of the heroic manner or style. **—mock′-he·ro′ic** *adj.* **—mock′-he·ro′i·cal·ly** *adv.*

**mock·ing·bird** (mŏk′ĭng-bûrd′) *n.* A species of New World birds of the family Mimidae, esp. *Mimus polyglottos,* a gray and white bird of the southern United States.

**mock moon** *n.* A paraselene.

**mock orange** *n.* **1.** A deciduous shrub of the genus *Philadelphus,* with white, usu. fragrant flowers. **2.** A shrub or tree with flowers or fruit resembling those of the orange.

**mock turtle soup** *n.* Soup made from calf's head or veal and spiced to taste like green turtle soup.

**mock·up** *also* **mock-up** (mŏk′ŭp′) *n.* A usu. full-sized scale model of a structure, used for demonstration, study, or testing.

**mod** (mŏd) *n.* [After *the Mods,* name of several gangs of English youths, short for MODERN.] An unconventional and modern style of fashionable dress originating in England. *—adj.* **1.** In or typical of the mod style. **2.** Stylishly up-to-date, esp. in dress.

**mo·dal** (mōd′l) *adj.* [Med. Lat. *modalis* < Lat. *modus,* measure.] **1.** Of, relating to, or typical of a mode. **2.** Of, relating to, or expressing the mood of a verb. **3.** *Mus.* Of, relating to, typical of, or written in any of the modes typical of medieval church music. **4.** *Philos.* Of or relating to mode as opposed to substance. **5.** *Logic.* Expressing or marked by modality. **6.** *Statistics.* Of or relating to a statistical mode. **—mo′dal·ly** *adv.*

**modal auxiliary** *n.* One of a set of English verbs, including *can, may, must, ought, shall, should, will,* and *would,* that express mood or tense when used with other verbs.

**mo·dal·i·ty** (mō-dăl′ĭ-tē) *n., pl.* **-ties. 1.** The quality, state, or fact of being modal. **2.** A tendency to conform to a general type or pattern. **3.** *Logic.* The classification of propositions on the basis of whether they assert or deny the possibility, impossibility, contingency, or necessity of their content. **4.** *Med.* **a.** A method of therapy, usu. physical, as massage. **b.** An apparatus for such a therapy.

**mode** (mōd) *n.* [ME, tune < Lat. *modus,* tune, manner.] **1. a.** Manner, way, or method of doing or acting <ancient *modes* of communication> **b.** A particular form, variety, or way <a *mode* of behavior> **c.** A given condition of functioning : STATUS <a spacecraft in its re-entry *mode*> **2.** The current or customary fashion or style. **3.** *Mus.* **a.** Any of certain arrangements of the diatonic tones of an octave. **b.** One of several patterned arrangements characteristic of classical Greek and medieval church music. **4.** *Philos.* The particular manner or form in which an underlying substance, or some permanent attribute or aspect of it, is manifested. **5.** *Logic.* **a.** The arrangement or order of the propositions in a syllogism according to both quality and quantity. **b.** The modality of a proposition. **6.** *Statistics.* The value or item occurring most frequently in a series of observations or statistical data. **7.** *Geol.* The mineral composition of a specific sample of igneous rock. **8.** *Physics.* Any of numerous patterns of wave motion, as of acoustic waves.

**mod·el** (mŏd′l) *n.* [OFr. *modelle* < OItal. *modello* < Lat. *modulus,* dim. of *modus,* measure.] **1.** A small object, usu. built to scale, that represents another, often larger object. **2.** A preliminary pattern serving as the plan from which an item not yet constructed will be produced. **3.** A tentative description of a theory or system that accounts for all of its known properties. **4.** A design or style of an item. **5.** An example to be imitated or compared <a *model* of politeness> **6.** The subject for an artist or photographer. **7.** One whose job is to display clothes or other merchandise. *—v.* **-eled, -el·ing, -els** *also* **-elled, -el·ling, -els.** *—vt.* **1.** To make or build a model of. **2.** To plan or fashion according to a model. **3.** To make by shaping a plastic substance. **4.** To make conform to a selected standard <They *modeled* their actions on ours.> **5.** To display by wearing or posing. **6.** To give a three-dimensional appearance to, as by shading in painting and drawing. *—vi.* **1.** To make a model. **2.** To serve as a model. *—adj.* **1.** Serving as or used as a model. **2.** Worthy of imitation <a *model* couple> **—mod′el·er** *n.*

**mo·dem** (mō′dĕm′) *n.* [MO(DULATOR) + DEM(ODULATOR).] A device that converts data from one form into another, as from one usable in data processing to one usable in telephonic transmission.

**mod·er·ate** (mŏd′ər-ĭt) *adj.* [Lat. *moderatus,* p.part. of *moderare,* to reduce.] **1.** Within limits : REASONABLE <a *moderate* fee> **2.** Not violent or harsh : MILD <*moderate* weather> **3.** Of average or medium quantity, quality, or extent. **4.** Opposed to extreme or radical views or measures, esp. in politics and religion. *—n.* One having moderate views or opinions. *—v.* (mŏd′ə-rāt′) **-at·ed, -at·ing, -ates.** *—vt.* **1.** To make less violent, severe, or extreme. **2.** To preside over. *—vi.* **1.** To become less violent, severe, or extreme : ABATE. **2.** To act as a moderator. **—mod′er·ate·ly** *adv.* **—mod′er·a′tion** *n.*

**mod·e·ra·to** (mŏd′ə-rä′tō) *adv.* & *adj.* [Ital. < Lat. *moderatus,* moderate.] *Mus.* In moderate tempo : slower than allegretto but faster than andante. —Used as a direction. **—mod′e·ra′to** *n.*

**mod·er·a·tor** (mŏd′ə-rā′tər) *n.* **1.** One that moderates. **2.** The officer who presides over a synod or general assembly of the Presbyterian Church. **3.** A substance, as water or graphite, used in a nuclear reactor to decrease the speed of fast neutrons and increase the likelihood of fission.

**mod·ern** (mŏd′ərn) *adj.* [OFr. *moderne* < LLat. *modernus* < Lat. *modo,* just now < *modus,* measure.] **1.** Of, relating to, or typical of the present or recent times. **2.** Of or relating to advanced style, technique, or technology. *—n.* **1.** One who lives in modern times. **2.** One who has modern ideas, standards, or beliefs. **3.** A typeface marked by strongly contrasted heavy and thin parts. **—mo·dern′i·ty** (mŏ-dûr′nĭ-tē), **mod′ern·ness** *n.* **—mod′ern·ly** *adv.*

☆ ***syns:*** MODERN, CURRENT, LATTER-DAY, PRESENT-DAY *adj. core meaning* : of or relating to the present or recent times <*modern* medicine><*modern* astronomy>

**Modern English** *n.* English since approx. 1500.

**Modern Greek** *n.* Greek since the early 16th cent.

**Modern Hebrew** *n.* New Hebrew.

**mod·ern·ism** (mŏd′ər-nĭz′əm) *n.* **1.** Modern thought, character, or practice. **2.** Something, as a peculiarity of usage or style, typical of modern times. **3.** *often* **Modernism.** The name given in Christian churches to movements that try to adapt church teachings to modern revolutions in science and philosophy. **4.** The practice and theory of modern art. **—mod'ern·ist** *n.* **—mod'ern·is'tic** *adj.*

**mod·ern·ize** (mŏd'ər-nīz') *v.* **-ized, -iz·ing, -iz·es.** —*vt.* To make modern in appearance or style. —*vi.* To accept or adopt modern ways, ideas, or styles. **—mod'ern·i·za'tion** *n.*

☆ ***syns:*** MODERNIZE, REFURBISH, REJUVENATE, RENOVATE, RESTORE, REVAMP, UPDATE *v. core meaning*: to make modern in appearance or style <*modernized* the old factory>

**modern pentathlon** *n.* A pentathlon.

**mod·est** (mŏd'ĭst) *adj.* [OFr. *modeste* < Lat. *modestus.*] **1.** Having or displaying a moderate estimation of one's own talents, abilities, and value. **2.** Shy and retiring in disposition. **3.** Observing conventional proprieties in speech, behavior, or dress: DECOROUS. **4.** Quiet and humble in appearance: UNPRETENTIOUS <a *modest* dwelling> **5.** Not extreme: MODERATE <a *modest* price> **—mod'est·ly** *adv.*

☆ ***syns:*** MODEST, BACKWARD, BASHFUL, COY, DEMURE, DIFFIDENT, RETIRING, SELF-EFFACING, SHY, TIMID *adj. core meaning*: reticent or reserved in manner <too *modest* to ask for a dance>

**mod·es·ty** (mŏd'ĭ-stē) *n., pl.* **-ties. 1.** The quality or state of being modest. **2.** Reserve or propriety in speech, dress, or behavior. **3.** Lack of pretentiousness.

**mod·i·cum** (mŏd'ĭ-kəm) *n., pl.* **-cums** *or* **-ca** (-kə) [ME < Lat., neuter of *modicus,* moderate < *modus,* measure.] A very small amount: JOT.

**mod·i·fi·ca·tion** (mŏd'ə-fĭ-kā'shən) *n.* **1.** The act of modifying or state of being modified. **2.** A small alteration, adjustment, or limitation. **3.** *Biol.* A physical change in an organism because of environment or activity. **4. a.** A change in a word as it passes from one language to another. **b.** The linguistic change of a morpheme from one construction to another. **—mod'i·fi·ca'tor** *n.* **—mod'i·fi·ca'to·ry** (-kā'tə-rē), **mod'i·fi·ca'tive** (-kā'tĭv) *adj.*

**modified American plan** *n.* A system of hotel management in which guests are charged a fixed daily or weekly rate for room, breakfast, and lunch or dinner.

**mod·i·fi·er** (mŏd'ə-fī'ər) *n.* One that modifies, esp. a word, phrase, or clause that qualifies the sense of another word or word group.

**mod·i·fy** (mŏd'ə-fī') *v.* **-fied, -fy·ing, -fies.** [ME *modifien,* to limit < OFr. *modifier* < Lat. *modificare* : *modus,* measure + *facere,* to make.] —*vt.* **1.** To change in form or character: ALTER. **2.** To make less extreme, severe, or strong. **3.** To qualify or limit the meaning of; e.g., *chilly* modifies *wind* in *a chilly wind.* **4.** To change (a vowel) by umlaut. —*vi.* To be or become modified. **—mod'i·fi'a·ble** *adj.*

**mo·dil·lion** (mō-dĭl'yən) *n.* [Ital. *modiglione* < VLat. **mutillio* < Lat. *mutulus.*] An ornamental bracket used in series under the cornice of the Corinthian, Composite, or Ionic orders.

**mo·di·o·lus** (mō-dī'ə-ləs) *n., pl.* **-li** (-lī') [NLat. < Lat., dim. of *modius,* a measure for grain.] The central, conical, bony shaft of the cochlea.

**mod·ish** (mō'dĭsh) *adj.* [< MODE.] Conforming to the prevailing or current fashion: STYLISH. **—mod'ish·ly** *adv.* **—mod'ish·ness** *n.*

**mo·diste** (mō-dēst') *n.* [Fr. < *mode,* fashion < Lat. *modus.*] One who makes, designs, or deals in women's fashions.

**mod·u·lar** (mŏj'ə-lər) *adj.* **1.** Of, pertaining to, or based on a module. **2.** Designed with standardized units or dimensions for flexible use <*modular* living room furniture> **—mod'u·lar'i·ty** (-lăr'ĭ-tē) *n.* **—mod'u·lar·ly** *adv.*

**mod·u·lar·ized** (mŏj'ə-lə-rīzd', mŏd'yə-) *adj.* Having or made up of modules.

**mod·u·late** (mŏj'ə-lāt') *v.* **-lat·ed, -lat·ing, -lates.** [Lat. *modulari, modulat-* < *modulus,* dim. of *modus,* measure.] —*vt.* **1.** To adjust or adapt to a certain proportion: TEMPER. **2.** To alter the pitch, intensity, or tone of. **3.** *Electron.* To vary the frequency, amplitude, or phase of (a carrier wave). —*vi.* **1.** To pass from one key or tonality to another by means of a regular melodic chord or progression. **2.** To sing or play with modulation. **—mod'u·la·bil'i·ty** *n.* **—mod'u·la'tive, mod'u·la·to'ry** (-lə-tôr'ē, -tōr'ē) *adj.*

**mod·u·la·tion** (mŏj'ə-lā'shən) *n.* **1.** The act or process of modulating or the state of being modulated. **2.** *Mus.* A passing from one tonality to another by means of a regular melodic or chord progression. **3.** A change in loudness or pitch of the voice, esp. the use of a particular intonation or inflection of the voice to convey meaning. **4.** *Electron.* The variation of a property of an electromagnetic wave or signal, as its amplitude, frequency, or phase, in a manner determined by another wave or signal.

**mod·u·la·tor** (mŏj'ə-lā'tər) *n.* **1.** One that modulates. **2.** *Electron.* A device or electric circuit for modulating a carrier wave. **3.** *Anat.* A nerve fiber in the retina of the eye, related to color discrimination.

**mod·ule** (mŏj'ōōl) *n.* [Lat. *modulus,* dim. of *modus,* measure.] **1.** A standard or unit of measurement. **2. a.** The part of an architectural construction used as a standard to which the rest is proportioned. **b.** A uniform structural component used repeatedly in a building. **c.** A standardized unit designed for use with others of its kind. **3.** *Electron.* A self-contained assembly of electronic components and circuitry, as a stage in a computer. **4.** A self-contained unit of a spacecraft that performs a specific task or class of tasks for supporting the major function of the craft. **5.** A unit of instruction covering a single topic or a small section of a broad topic.

**mod·u·lus** (mŏj'ə-ləs) *n., pl.* **-li** (-lī') [NLat. < Lat., dim. of *modus,* measure.] **1.** *Physics.* A constant or coefficient that expresses the degree to which a substance possesses some property. **2.** *Math.* **a.** The absolute value of a complex number. **b.** A number or quantity that produces the same remainder when divided into each of two quantities. **c.** The number by which a logarithm in one system must be multiplied to obtain the corresponding logarithm in another system. **3.** A standard: norm. **—mod'u·lar** (-lər) *adj.*

**mo·dus op·er·an·di** (mō'dəs ŏp'ə-răn'dē, -dī') *n.* [Lat.] A method of operating.

**modus vi·ven·di** (vĭ-vĕn'dē, -dī') *n.* [Lat.] **1.** A way of living. **2.** A temporary agreement between disputants pending a final settlement.

**mo·fette** *also* **mof·fette** (mō-fĕt') *n.* [Fr., gaseous exhalation < Ital. *moffetta,* prob. of Germanic orig.] **1.** An opening in the earth from which carbon dioxide and other gases escape, usu. marking the final phase of volcanic activity. **2.** The gases escaping from a mofette.

**Mo·gen Da·vid** (mô'gən dô'vĭd) *n. var. of* MAGEN DAVID.

**Mo·ghul** (mōō-gŭl') *n. var. of* MOGUL 1.

**mo·gul** (mō'gəl) *n.* [Prob. of Scand. orig.] A bump on a ski slope.

**Mo·gul** (mō'gəl, mō-gŭl') *n.* [Pers. *Mughul* < Mongolian *Mongul.*] **1.** *also* **Mo·ghul** (mōō-gŭl'). **a.** One of the followers of Baber who conquered India in 1526 and founded a Moslem empire that formally lasted until 1857. **b.** A descendant of a follower of Baber. **2.** A Mongol or Mongolian. **3. mogul.** A very rich or powerful person: MAGNATE. **—Mo'gul** *adj.*

**mo·hair** (mō'hâr') *n.* [Alteration of obs. *mocayare* < Ital. *moccaiaro* < Ar. *mukhayyar.*] **1. a.** The hair of the Angora goat. **b.** A shiny, heavy, woolly fabric made of mohair, often with a mixture of cotton. **2.** An upholstery fabric with mohair pile.

**Mo·ham·med·an** (mō-hăm'ĭ-dən) *also* **Mu·ham·mad·an** *or* **Mu·ham·med·an** (mōō-) *adj.* Of or relating to Mohammed or Islam: MOSLEM. —*n.* A follower of Mohammed or Islam: MOSLEM.

**Mo·ham·med·an·ism** (mō-hăm'ĭ-də-nĭz'əm) *also* **Mu·ham·mad·an·ism** (mōō-) *n.* Islam.

**Mo·har·ram** (mō-hăr'əm) *n. var. of* MUHARRAM.

**Mo·ha·ve** *also* **Mo·ja·ve** (mō-hä'vē) *n., pl.* **Mohave** *or* **-ves** *also* **Mojave** *or* **-ves.** [Mohave *hamokhava,* three peaks.] **1. a.** A tribe of Indians once living along the Gila and Colorado rivers. **b.** A member of this tribe. **2.** The Yuman language of the Mohave. **—Mo·ha've** *adj.*

**Mo·hawk** (mō'hôk') *n., pl.* **Mohawk** *or* **-hawks.** [Narraganset *mohowaugsuck.*] **1. a.** A tribe of Indians once occupying the area from the Mohawk River to the St. Lawrence. **b.** A member of this tribe. **2.** The Iroquoian language of the Mohawk. **—Mo'hawk'** *adj.*

**Mo·he·gan** (mō-hē'gən) *n., pl.* **Mohegan** *or* **-gans.** [Prob. of Algonquian orig.] **1. a.** A tribe of Indians once living in the area around the Thames River, Connecticut. **b.** A member of this tribe. **2.** The Algonquian language of the Mohegan. **—Mo·he'gan** *adj.*

**Mo·hi·can** (mō-hē'kən, mə-) *n. var. of* MAHICAN.

**Mo·ho** (mō'hō') *n.* The Mohorovičić discontinuity.

**Mo·hock** (mō'hŏk') *n.* [Alteration of MOHAWK.] One of a band of young aristocrats who vandalized London in the early 18th cent. **—Mo'hock·ism** *n.*

**Mo·ho·ro·vi·čić discontinuity** (mō'hə-rō'və-chĭch) *n.* [After Andrija *Mohorovičić* (1857–1936).] The boundary between the earth's crust and the subjacent mantle rock, ranging in depth from 6 to 8 miles or 9.65 to 12.87 kilometers under ocean basins and from 20 to 25 miles or 32.18 to 40.23 kilometers under continents.

**Mohs scale** (mōz) *n.* [After Friedrich *Mohs* (1773–1839).] A scale for determining the relative hardness of a mineral according to its resistance to scratching by one of the following minerals, arranged in order of increasing hardness: 1. talc; 2. gypsum; 3. calcite; 4. fluorite; 5. apatite; 6. feldspar; 7. vitreous silica; 8. quartz; 9. topaz; 10. garnet; 11. fused zirconia; 12. fused alumina; 13. silicon carbide; 14. boron carbide; 15. diamond.

**mo·hur** (mō'ər, mə-hōōr') *n.* [Hindi *muhar* < Pers. *muhr,* seal.] A gold coin formerly in circulation in India.

**moi·dore** (moi'dôr', -dōr', moi-dôr', -dōr') *n.* [Alteration of Port. *moeda de ouro,* coin of gold.] A former Portuguese or Brazilian gold coin.

**moi·e·ty** (moi'ĭ-tē) *n., pl.* **-ties.** [ME *moite* < OFr. < LLat. *medietas,* half < Lat. *medius,* middle.] **1.** A half. **2.** A part, portion, or share. **3.** Either of two basic units that make up a tribe on the basis of unilateral descent.

**moil** (moil) *vi.* **moiled, moil·ing, moils.** [ME *moillen,* to wet < OFr. *moillier* < Lat. *mollis,* soft.] **1.** To toil: slave. **2.** To churn about. —*n.* **1.** Drudgery: toil. **2.** Turmoil: confusion. **—moil'er** *n.*

**moi·ré** (mwä-rā') *also* **moire** (mwär, mwä-rā') *n.* [Fr. < *moirer,* to water.] **1.** Cloth, esp. silk, with a watered or wavy pattern. **2.** A watered pattern created on cloth by engraved rollers. **—moi·ré'** *adj.*

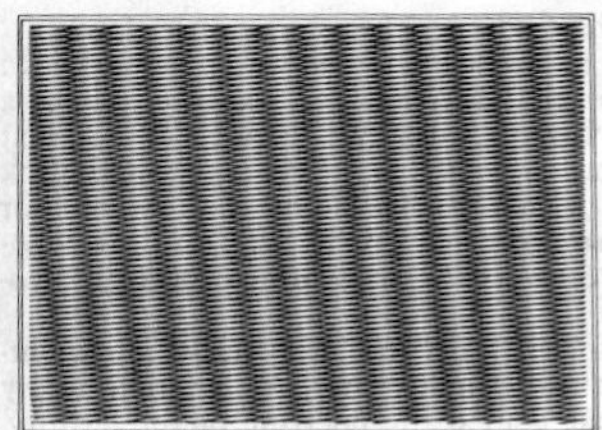
moiré

**moist** (moist) *adj.* **-er, -est.** [ME *moiste* < OFr. < Lat. *mucidus,* moldy < *mucus,* mucus.] **1.** Slightly wet. **2.** Filled with moisture. **3.** Tearful. **—moist'ly** *adv.* **—moist'ness** *n.*
**mois·ten** (moi'sən) *vt.* & *vi.* **-tened, -ten·ing, -tens.** To make or become moist. **—mois'ten·er** *n.*
**mois·ture** (mois'chər) *n.* [ME < OFr. *moistour* < *moiste,* moist.] Diffuse wetness that can be felt as vapor in the atmosphere or as condensed liquid on the surfaces of objects : DAMPNESS.
**mois·tur·ize** (mois'chə-rīz') *vt.* **-ized, -iz·ing, -iz·es.** To add moisture to (e.g., skin). **—mois'tur·iz'er** *n.*
**mo·jar·ra** (mō-här'ə) *n., pl.* **mojarra** or **-ras.** [Am. Sp. < Sp., a kind of fish found off the coast of Spain < Ar. *muḥarrab,* pointed < *ḥarrab,* he sharpened.] **1.** Any of several species of American marine fishes of the family Gerridae, with protrusile mouths. **2.** Any of several tropical American freshwater fishes of the family Cichlidae.
**Mo·ja·ve** (mō-hä'vē) *n. var. of* MOHAVE.
**moke** (mōk) *n.* [Orig. unknown.] **1.** *Chiefly Brit.* A donkey. **2.** *Austral.* An old, broken-down horse.
**mol** (mōl) *n. var. of* MOLE⁵.
**mo·lal** (mō'ləl) *adj.* [< MOLE (gram molecule).] Of or designating a solution containing one mole of solute in 1,000 grams of solvent, usu. water.
**mo·lal·i·ty** (mō-lăl'ĭ-tē) *n., pl.* **-ties.** The molal concentration of a solute, usu. expressed as the number of moles of solute per 1,000 grams of solvent.
**mo·lar¹** (mō'lər) *adj.* [< MOLE⁵.] **1.** *Physics.* Of or relating to a body of matter as a whole, perceived apart from molecular or atomic properties. **2.** *Chem.* **a.** Containing one mole of a substance. **b.** Relating to or designating a solution that contains one mole of solute per liter of solution.
**mo·lar²** (mō'lər) *n.* [Lat. *molaris* < *mola,* millstone.] A tooth with a broad crown for grinding food, situated behind the bicuspids. *—adj.* **1.** Of or relating to the molar teeth. **2.** Capable of grinding.
**mo·lar·i·ty** (mō-lăr'ĭ-tē) *n., pl.* **-ties.** The molar concentration of a solute, usu. expressed as the number of moles of solute per liter of solution.
**mo·las·ses** (mə-lăs'ĭz) *n.* [Port. *melaço* < LLat. *mellaceum,* must < Lat. *mel,* honey.] A dark, thick syrup produced in refining sugar.
**mold¹** (mōld) *n.* [ME *molde* < OFr. *modle* < Lat. *modulus,* dim. of *modus,* measure.] **1.** A hollow form or matrix for shaping a fluid or plastic substance. **2.** A frame or model around or on which an object is formed or shaped. **3.** An object made in or shaped on a mold. **4.** The pattern of a mold. **5.** General shape or form <the oval *mold* of a face> **6.** Distinctive character or type <in the *mold* of their forefathers> **7.** MOLDING 2. *—vt.* **mold·ed, mold·ing, molds. 1.** To shape in or on a mold. **2. a.** To form into a particular shape. **b.** To guide or determine the growth or development of : INFLUENCE. **3.** To fit closely. —Used of clothes. **4.** To make a mold of or from, before casting. **—mold'a·ble** *adj.* **—mold'er** *n.*
**mold²** (mōld) *n.* [ME *molde.*] **1.** A fungous growth often producing disintegration of organic matter. **2.** A fungus that causes mold. *—vi.* **mold·ed, mold·ing, molds.** To become moldy.
**mold³** (mōld) *n.* [ME < OE *molde.*] **1.** Loose, friable soil, rich in humus and fit for planting. **2.** *Chiefly Brit. Regional.* **a.** The earth : the ground. **b.** The earth of the grave. **3.** *Obs.* Earth as the substance of the human body.
**mold·board** (mōld'bôrd', -bōrd') *n.* [MOLD³ + BOARD.] The curved plate of a plow that turns over the furrow slice.
**mold·er** (mōl'dər) *v.* **-ered, -er·ing, -ers.** [Prob. of Scand. orig.] *—vi.* To become dust gradually, as by natural decay : CRUMBLE. *—vt.* To cause to decay or crumble.
**mold·ing** (mōl'dĭng) *n.* **1.** Something molded. **2.** A strip for decorating a surface in architecture.
**mold·y** (mōl'dē) *adj.* **-i·er, -i·est. 1.** Covered with or containing mold. **2.** Musty or stale, as from age or decay. **—mold'i·ness** *n.*
**mole¹** (mōl) *n.* [ME < OE *māl.*] A small congenital growth on the human skin, usu. slightly raised and dark and occas. hairy.
**mole²** (mōl) *n.* [ME *molle,* poss. < MLG *mol.*] **1.** Any of various small, insectivorous, burrowing mammals with thickset bodies bearing silky light-brown to dark-gray fur, rudimentary eyes, tough muzzles, and strong forefeet for digging and usu. living underground. **2.** MOLESKIN 1. **3.** A dark gray.
**mole³** (mōl) *n.* [Fr. *môle* < LGk. *mōlos* < Lat. *moles,* mass.] **1.** A massive stone wall used esp. as a breakwater or to enclose an anchorage or harbor. **2.** The anchorage or harbor enclosed by a mole.
**mole⁴** (mōl) *n.* [Fr. *môle* < Lat. *mola,* millstone.] A uterine mass caused by degeneration or abortive development of an ovum.
**mole⁵** *also* **mol** (mōl) *n.* [G. *Mol,* short for *Molekulargewicht,* molecular weight.] The amount of a substance with a weight in grams numerically equal to the molecular weight of the substance.
**mo·lec·u·lar** (mə-lĕk'yə-lər) *adj.* Relating to, consisting of, caused by, or existing between molecules. **—mo·lec'u·lar'i·ty** (-yə-lăr'ĭ-tē) *n.* **—mo·lec'u·lar·ly** *adv.*
**molecular beam** *n. Physics.* A highly collimated, internally collisionless stream of molecules for studying electromagnetic phenomena as the stream traverses an evacuated chamber.
**molecular biology** *n.* The branch of biology concerned with the structure and development of biological systems as analyzed in terms of the physics and chemistry of their molecular constituents. **—molecular biologist** *n.*
**molecular film** *n.* A surface film of thickness comparable to that of a single molecule.
**molecular formula** *n.* A chemical formula showing the number of atoms of each element in a molecule of a compound.
**molecular weight** *n.* The sum of the atomic weights of a molecule's constituent atoms.
**mol·e·cule** (mŏl'ĭ-kyo͞ol') *n.* [Fr. *molécule* < NLat. *molecula,* dim. of Lat. *moles,* mass.] **1.** A stable configuration of atomic nuclei and electrons bound together by electrostatic and electromagnetic forces, and the simplest structural unit displaying the characteristic physical and chemical properties of a compound. **2.** A small particle : BIT.
**mole·hill** (mōl'hĭl') *n.* A small mound of loose earth thrown up by a burrowing mole.
**mole·skin** (mōl'skĭn') *n.* **1.** The short, soft, silky fur of the mole. **2. a.** A heavy-napped cotton twill fabric. **b. moleskins.** Clothing, esp. trousers, of moleskin fabric.
**mo·lest** (mə-lĕst') *vt.* **-lest·ed, -lest·ing, -lests.** [ME *molesten* < OFr. *molester* < Lat. *molestare* < *molestus,* troublesome.] **1.** To disturb or interfere with : ANNOY. **2.** To accost and harass sexually. **—mo'les·ta'tion** (mō'lĕs-tā'shən) *n.* **—mo·lest'er** *n.*
**moll** (mŏl) *n.* [< *Moll,* nickname for *Mary.*] *Slang.* **1.** A woman companion of a gangster. **2.** A prostitute.
**mol·lie** (mŏl'ē) *n. var. of* MOLLY.
**mol·li·fy** (mŏl'ə-fī') *vt.* **-fied, -fy·ing, -fies.** [ME *mollifien* < OFr. *mollifier* < Lat. *mollificare* : *mollis,* soft + *facere,* to make.] **1.** To allay the anger of : PLACATE. **2.** To lessen in intensity : TEMPER. **3.** To reduce the rigidity of : SOFTEN. **—mol'li·fi'a·ble** *adj.* **—mol'li·fi·ca'tion** *n.* **—mol'li·fi'er** *n.* **—mol'li·fy'ing·ly** *adv.*
**mol·lusc** (mŏl'əsk) *n. var. of* MOLLUSK.
**mol·lus·can** *also* **mol·lus·kan** (mə-lŭs'kən) *adj.* Of or pertaining to the mollusks. *—n.* A mollusk.
**mol·lus·ci·cide** (mə-lŭs'kĭ-sīd') *n.* [Lat. *molluscus,* soft (< *mollis*) + -CIDE.] A mollusk-killing agent. **—mol·lus'ci·cid'al** *adj.*
**mol·lusk** *also* **mol·lusc** (mŏl'əsk) *n.* [Fr. *mollusque* < NLat. *Mollusca,* phylum name < Lat. *molluscus,* soft < *mollis.*] Any of various members of the phylum Mollusca, largely marine invertebrates, including the edible shellfish and some 100,000 other species. **—mol·lus'cous** (mə-lŭs'kəs) *adj.*
**mol·lus·kan** (mə-lŭs'kən) *adj. var. of* MOLLUSCAN.
**mol·ly** *also* **mol·lie** (mŏl'ē) *n., pl.* **-lies.** [< NLat. *Mollienisia,* genus name, after Comte François N. *Mollien* (1758–1850).] A tropical or subtropical fish of the genus *Mollienisia,* often seen in aquariums.
**mol·ly·cod·dle** (mŏl'ē-kŏd'l) *n.* [Brit. slang *molly,* milksop (< the name *Molly*) + CODDLE.] A pampered and protected person. *—vt.* **-dled, -dling, -dles.** To be overprotective and indulgent toward. **—mol'ly·cod'dler** *n.*
**Mo·loch** (mō'lŏk', mŏl'ək) *n.* [LLat. *Moloch* < Gk. *Molokh* < Heb. *Molekh.*] The god of the Ammonites and Phoenicians to whom children were sacrificed in the Old Testament.
**Mo·lo·tov cocktail** (mŏl'ə-tôf', môl'-, mō'lə-) *n.* [After Vyacheslav M. *Molotov* (1890–1981).] A makeshift incendiary bomb made of a breakable container full of flammable liquid and equipped with a rag wick.
**molt** (mōlt) *v.* **molt·ed, molt·ing, molts.** [ME *mouten* < OE *mūtian* < Lat. *mutare,* to change.] *—vi.* To shed an outer covering, as feathers, cuticle, or skin, which is replaced periodically by a new growth. *—vt.* To shed by the process of molting. *—n.* **1.** An act or process of molting. **2.** The material cast off during molting.
**mol·ten** (mōl'tən) *v. Archaic. var. p.p. of* MELT. *—adj.* **1.** Made liquid by heat : MELTED. **2.** Made by melting and casting <a *molten* image> **3.** Brilliantly glowing.
**mol·to** (mōl'tō) *adv.* [Ital. < Lat. *multum.*] *Mus.* Very : much. —Used as a direction.
**mo·ly** (mō'lē) *n., pl.* **-lies.** [Lat. < Gk. *mōlu.*] A mythical magic herb with black roots and white flowers.
**mo·lyb·de·nite** (mə-lĭb'də-nīt') *n.* [MOLYBDEN(UM) + -ITE.] A

mineral form of molybdenum sulfide, $MoS_2$, the chief ore of molybdenum.

**mo·lyb·de·num** (mə-lĭb′də-nəm) *n.* [NLat. < Lat. *molybdaena,* galena < Gk. *molubdaina* < *molubdos,* lead.] *Symbol* **Mo** A hard, gray, metallic element used to toughen alloy steels; atomic number 42; atomic weight 95.94.

**mo·lyb·dic** (mə-lĭb′dĭk) *adj.* Designating molybdenum or a compound containing molybdenum in its higher valences.

**mo·lyb·dous** (mə-lĭb′dəs) *adj.* Designating molybdenum or a compound containing molybdenum in its lower valences.

**mom** (mŏm) *n.* [Shortening and alteration of MAMMA.] *Informal.* MOTHER[1] 1.

**mom-and-pop** (mŏm′ən-pŏp′) *adj.* Of or designating a small business run by the owners <a *mom-and-pop* delicatessen>

**mo·ment** (mō′mənt) *n.* [ME < OFr. < Lat. *momentum* < *movēre,* to move.] **1.** A brief indefinite interval of time. **2.** A specific point in time, esp. the present <I am eating at the *moment.*> **3.** A particular period of importance, significance, or excellence. **4.** Outstanding significance or value : IMPORTANCE <a scientific discovery of great *moment*> **5.** *Philos.* **a.** An essential or constituent element. **b.** A phase or aspect of a thing. **6.** *Physics.* **a.** The product of a quantity and its perpendicular distance from a reference point. **b.** Rotation produced in a body when a force is applied : TORQUE. **7.** *Statistics.* The expected value of a positive integral power of a random variable.

☆ ***syns:*** MOMENT, FLASH, INSTANT, JIFFY, MINUTE, SECOND, TRICE *n. core meaning* : a brief but usu. not insignificant period <had to wait a *moment*><a great *moment* in history> MINUTE, used strictly, is specific; it is also interchangeable with MOMENT <had to wait a *minute*><a critical *minute* in the testimony> An INSTANT or a FLASH is a period of time almost too brief to detect; each word implies haste and usu. urgency <must have the papers this *instant*><came in a *flash*> SECOND may be used specifically or loosely as the equivalent of INSTANT <held the lever down for a *second*><need to see you this *second*> TRICE, a literary term, and JIFFY are usu. preceded by *in a* <in a *trice*><in a *jiffy*>; they are imprecise but approx. equal in duration to INSTANT.

**mo·men·ta** (mō-mĕn′tə) *n. pl. of* MOMENTUM.

**mo·men·tar·i·ly** (mō′mən-târ′ə-lē) *adv.* **1.** For only a moment. **2.** Very soon. **3.** From moment to moment.

**mo·men·tar·y** (mō′mən-tĕr′ē) *adj.* [Lat. *momentarius* < *momentum,* moment < *movēre,* to move.] **1.** Lasting only a short time. **2.** Occurring or present at every moment <in *momentary* dread of falling> **3.** Short-lived. —Used of a living organism. **—mo′men·tar′i·ness** *n.*

**moment of inertia** *n.* **1.** A measure of a body's resistance to angular acceleration, equal to: **a.** The product of the mass of a particle and the square of its distance from a reference. **b.** The sum of the products of each mass element of a body multiplied by the square of its distance from an axis. **2.** The sum of the products of each element of an area multiplied by the square of its distance from a coplanar axis.

**moment of momentum** *n.* Angular momentum.

**moment of truth** *n.* **1.** The final kill in a bullfight. **2.** A decisive or crucial point.

**mo·men·tous** (mō-mĕn′təs) *adj.* Of utmost importance or outstanding significance. **—mo·men′tous·ly** *adv.* **—mo·men′tous·ness** *n.*

**mo·men·tum** (mō-mĕn′təm) *n., pl.* **-ta** (-tə) or **-tums.** [Lat., movement < *movēre,* to move.] **1.** *Physics.* The product of a body's mass and linear velocity. **2.** Impetus in human affairs or actions <a career that gained *momentum*>

**Mo·mus** (mō′məs) *n.* [Gk. *Mōmos* < *mōmos,* blame.] *Gk. Myth.* The god of blame and ridicule.

**mon** (mŏn) *n. Scot.* A man.

**Mon** (mōn) *n.* **1. a.** The principal native people of the Pegu region in Burma. **b.** A member of this people. **2.** The Mon-Khmer language of the Mon.

**mon-** *pref. var. of* MONO-.

**mon·a·chism** (mŏn′ə-kĭz′əm) *n.* [ME < Med. Lat. *monachismus* < LGk. *monakhismos* < *monakhos,* monk. —see MONK.] Monasticism.

**mo·nad** (mō′năd′) *n.* [LLat. *monas, monad-,* unit < Gk. < *monos,* single.] **1.** An impenetrable and indivisible unit of substance viewed as the basic constituent element of physical reality in the philosophy of Leibnitz. **2.** *Biol.* A single-celled microscopic organism, esp. a flagellate protozoan. **3.** *Chem.* An atom or radical with a valence of 1. **—mo·nad′ic, mo·nad′i·cal** *adj.* **—mo·nad′i·cal·ly** *adv.*

**mon·a·del·phous** (mŏn′ə-dĕl′fəs, mō′nə-) *adj. Bot.* **1.** United by the filaments into a single tubelike group. —Used of stamens. **2.** Having monadelphous stamens.

**mon·a·des** (mŏn′ə-dēz′) *n. pl. of* MONAS.

**mo·nad·nock** (mə-năd′nŏk′) *n.* [After Mt. Monadnock in New Hampshire.] A mountain or rocky mass that has withstood erosion and rises alone in a plain or peneplain.

**mo·nan·drous** (mə-năn′drəs) *adj.* **1.** Having a single stamen. **2.** Having flowers bearing a single stamen. **3.** Of or pertaining to the practice of monandry.

**mo·nan·dry** (mə-năn′drē) *n.* **1.** The practice of having one husband at a time. **2.** *Bot.* The state of being monandrous.

**mo·nan·thous** (mə-năn′thəs) *adj. Bot.* Bearing a single flower.

**mon·arch** (mŏn′ərk, -ärk′) *n.* [LLat. *monarcha* < Gk. *monarkhēs* : *monos,* single + *arkhein,* to rule.] **1.** A sole and absolute ruler of a state. **2.** A sovereign. **3.** One that presides over or rules. **4.** One that outdoes others in pre-eminence or power. **5.** A large orange and black butterfly, *Danaus plexippus,* with a wingspread of up to 4 inches or approx. 10 centimeters. **—mo·nar′chal** (mə-när′kəl), **mo·nar′chi·al, mo·nar′chic, mo·nar′chi·cal** *adj.* **—mo·nar′chal·ly, mo·nar′chi·cal·ly** *adv.*

**mo·nar·chi·an·ism** (mə-när′kē-ə-nĭz′əm) *n.* [< LLat. *monarchiani,* the monarchians < *monarchia,* monarchy.] A Christian heresy of the 2nd and 3rd cent. that denied the doctrine of the Trinity. **—mo·nar′chi·an** (mə-när′kē-ən) *n.*

**mon·ar·chism** (mŏn′ər-kĭz′əm, -är′-) *n.* **1.** The principles of monarchy. **2.** Belief in or advocacy of monarchy. **—mon′ar·chist** (-kĭst) *n.* **—mon′ar·chis′tic** *adj.*

**mon·ar·chy** (mŏn′ər-kē, -är′-) *n., pl.* **-chies.** [ME *monarchie* < OFr. < LLat. *monarchia* < Gk. *monarkhia* < *monarkhēs,* monarch.] **1.** Government by a monarch. **2.** A state ruled by a monarch.

**mo·nar·da** (mə-när′də) *n.* [NLat. *Monarda,* genus name, after Nicolas *Monardes* (d. 1588).] An aromatic plant of the genus *Monarda,* as the Oswego tea.

**mo·nas** (mō′năs′) *n., pl.* **mon·a·des** (mŏn′ə-dēz′) [LLat. < Gk. < *monos,* single.] *Biol.* MONAD 2.

**mon·as·ter·y** (mŏn′ə-stĕr′ē) *n., pl.* **-ies.** [ME *monasterie* < LLat. *monasterium* < LGk. *monastērion* < Gk. *monazein,* to live alone < *monos,* alone.] **1.** The dwelling place of a community of religious persons, esp. friars. **2.** The community of friars residing in a monastery. **—mon′as·te′ri·al** (mŏn′ə-stîr′ē-əl, -stĕr′-) *adj.*

**mo·nas·tic** (mə-năs′tĭk) *also* **mo·nas·ti·cal** (-tĭ-kəl) *adj.* [LLat. *monasticus* < LGk. *monastikos* < Gk. *monazein,* to live alone < *monos,* alone.] Of, relating to, typical of, or similar to monasteries or persons living in religious or contemplative seclusion. —*n.* A friar. **—mo·nas′ti·cal·ly** *adv.*

**mo·nas·ti·cism** (mə-năs′tĭ-sĭz′əm) *n.* The monastic life or system.

**mon·a·tom·ic** (mŏn′ə-tŏm′ĭk) *adj.* **1.** Occurring as single atoms, as helium. **2.** Having one replaceable atom or radical. **3.** Univalent. **—mon′a·tom′ic·al·ly** *adv.*

**mon·au·ral** (mŏn-ôr′əl) *adj.* **1.** Designating sound reception by one ear. **2.** Pertaining to a system of transmitting, recording, or reproducing sound whereby one or more sources are channeled into a single carrier.

**mon·ax·i·al** (mŏn-ăk′sē-əl) *adj.* Uniaxial.

**mon·ax·on** (mŏn-ăk′sŏn′) *n.* [MON(O)- + Gk. *axōn,* axis.] *Zool.* A straight spicule in sponges.

**mon·a·zite** (mŏn′ə-zīt′) *n.* [G. *Monazit* < Gk. *monazein,* to live alone < *monos,* alone.] A yellow or reddish-brown mineral phosphate of rare-earth metals, chiefly cerium and lanthanum, usu. together with thorium.

**Mon·day** (mŭn′dē, -dā′) *n.* [ME < OE *Mōnandæg,* transl. of Lat. *dies lunae,* day of the moon.] The second day of the week, occurring after Sunday and before Tuesday.

**mo·ne·cious** (mə-nē′shəs) *adj. var. of* MONOECIOUS.

**Mo·né·gasque** (mô-nā-găsk′) *n.* [Fr. < Prov. *mounegasc* < *Mounegue,* Monaco.] A citizen of Monaco. **—Mo·né·gasque′** *adj.*

**Mo·nel** (mō-nĕl′). A trademark for an alloy of nickel, copper, iron, and manganese.

**mon·es·trous** (mŏn-ĕs′trəs) *adj.* Having one estrous cycle per year.

**mon·e·tar·ism** (mŏn′ĭ-tə-rĭz′əm) *n.* The theory or policy of regulating the economy, esp. with regard to inflation, by increasing or decreasing the amount of money in circulation. **—mon′e·tar·ist** *n.*

**mon·e·tar·y** (mŏn′ĭ-tĕr′ē, mŭn′-) *adj.* [LLat. *monetarius* < Lat. *moneta,* money.] Of or relating to money or its means of circulation. **—mon′e·tar′i·ly** *adv.*

**mon·e·tize** (mŏn′ĭ-tīz′, mŭn′-) *vt.* **-tized, -tiz·ing, -tiz·es.** [< Lat. *moneta,* money.] **1.** To establish as legal tender. **2.** To coin (money). **—mon′e·ti·za′tion** *n.*

**mon·ey** (mŭn′ē) *n., pl.* **mon·eys** or **mon·ies.** [ME < OFr. *moneie* < Lat. *moneta* < *Moneta,* epithet of Juno, whose temple in Rome housed the mint.] **1.** A commodity, as silver or gold, that is legally established as an exchangeable equivalent of all other commodities and is used as a measure of their comparative values on the market. **2.** The official currency, coins, and negotiable paper notes issued by a government. **3.** Property and assets considered in terms of monetary value : WEALTH. **4.** Pecuniary profit or loss <lost *money* on the sale of the stock> **5.** An indefinite amount of currency <*money* for household expenses> **6.** *often* **moneys** or **monies.** Sums of money collected <highway toll *moneys*> **—on the money.** Exact : precise.

▲ word history: Money may be the root of all evil, but the root of the word *money* is a divine epithet. One of the titles of the Roman goddess Juno was *Moneta,* a name whose exact meaning is not known. *Moneta* is probably derived from the same root as the verb *monēre,* "to bring to mind, remind." The temple of Juno Moneta was the place where Roman money was coined, and the name *Moneta* became a word denoting "the mint" and by extension "coined money." *Moneta* developed into Old French *moneie* with the sense "coined money," which was borrowed into English in the 14th century. The Latin term *moneta* meaning "mint, a place where money is coined," was borrowed into the Germanic languages before Old English times and descended into Old English as *mynet,* whose modern form is *mint*[1].

**mon·ey·bag** (mŭn'ē-băg') *n.* **1.** A bag for holding money. **2. moneybags.** *(sing. or pl. in number).* Wealth. **3. moneybags.** *(sing. in number).* A rich and greedy person.

**mon·ey·chang·er** (mŭn'ē-chān'jər) *n.* **1.** One who exchanges money, as from one currency to another. **2.** A machine that holds and dispenses coins.

**mon·eyed** *also* **mon·ied** (mŭn'ēd) *adj.* **1.** Having a lot of money : RICH <the *moneyed* families> **2.** Representing or arising from the possession of money <*moneyed* interests>

**mon·ey·er** (mŭn'ē-ər) *n.* [ME *monyer* < OFr. *monier* < LLat. *monetarius* < *moneta,* money.] One authorized to coin or mint money.

**mon·ey·lend·er** (mŭn'ē-lĕn'dər) *n.* One in the business of lending money at an interest rate.

**mon·ey·mak·ing** (mŭn'ē-mā'kĭng) *n.* Acquisition of money or other wealth. *—adj.* **1.** Engaged in acquiring wealth. **2.** Actually or potentially profitable. **—mon'ey·mak'er** *n.*

**money market** *n.* Trade in negotiable instruments, as U.S. Treasury securities or certificates of deposit. **—mon'ey-mar'ket** *adj.*

**money of account** *n.* Any of the various monetary units in which accounts are kept, which may or may not correspond to actual current denominations.

**money order** *n.* An order for the payment of a stated sum of money, usu. issued and payable at a bank or post office.

**mon·ey·wort** (mŭn'ē-wûrt', -wôrt') *n.* [From the round shape of its leaves.] A creeping plant, *Lysimachia nummularia* of Europe and eastern North America, with rounded, opposite leaves and yellow flowers.

**mon·ger** (mŭng'gər, mŏng'-) *n.* [ME *mongere* < OE *mangere* < Lat. *mango,* of Gk. orig.] **1.** A dealer in a particular commodity <iron*monger*> **2.** One promoting something undesirable <rumor*monger*><war*monger*> *—vt.* **-gered, -ger·ing, -gers.** To peddle.

**mon·go** (mŏng'gō) *n., pl.* **mongo.** [Mongolian.] —See table at CURRENCY.

**Mon·gol** (mŏng'gəl, -gōl') *n.* [Mongol.] **1.** A member of one of the nomadic tribes of Mongolia. **2.** A native of Mongolia. **3.** MONGOLIAN 3. **4.** A member of the Mongoloid ethnic group. **—Mon'gol** *adj.*

**Mon·go·li·an** (mŏng-gō'lē-ən, -gōl'yən, mŏn-) *adj.* **1.** Of, relating to, or constituting the Mongolian People's Republic, the Mongols, or Mongolian. **2. mongolian.** MONGOLOID 2. *—n.* **1.** A native or inhabitant of Mongolia. **2.** A member of the Mongoloid race. **3.** Any of the Mongolic languages of Mongolia.

**Mon·gol·ic** (mŏng-gŏl'ĭk, mŏn-) *n.* A subfamily of the Altaic language family that includes Mongolian. *—adj.* **1.** Of or relating to the Mongoloid ethnic division. **2.** Of or relating to the subfamily of Altaic languages spoken in Mongolia.

**mon·gol·ism** *also* **Mon·gol·ism** (mŏng'gə-lĭz'əm, mŏn'-) *n.* [From the resemblance of the features of an affected person to those of Mongolians.] Down's syndrome.

**Mon·gol·oid** (mŏng'gə-loid', mŏn'-) *adj.* **1.** Of, relating to, or designating a major ethnic division of the human species whose members have yellowish-brown to white pigmentation, coarse straight black hair, dark eyes with decided epicanthic folds, and prominent cheekbones. **2.** Characteristic of or like a Mongol. **3. mongoloid.** Characterized by, affected with, or relating to Down's syndrome. *—n.* **1.** A member of the Mongoloid ethnic division of the human species. **2. mongoloid.** A person afflicted with Down's syndrome.

**mon·goose** (mŏng'go͞os', mŏn'-) *n., pl.* **-goos·es.** [Marathi *mangūs,* of Dravidian orig.] An Old World carnivorous mammal of the genus *Herpestes* or related genera, with a slender body and a long tail and well-known for its ability to kill venomous snakes.

**mon·grel** (mŭng'grəl, mŏng'-) *n.* [Prob. < ME *mong,* mixture < OE *(ge)mang.*] **1.** An animal or plant, esp. a dog, resulting from various interbreedings. **2.** A cross between two different breeds. *—adj.* Of mixed origin or character. **—mon·grel'ism** *n.* **—mon'grel·ly** *adv.*

**mon·grel·ize** (mŭng'grə-līz', mŏng'-) *vt.* **-ized, -iz·ing, -iz·es.** To make mongrel in race, nature, or character. **—mon'grel·i·za'tion** *n.*

**mon·ick·er** (mŏn'ĭ-kər) *n. var. of* MONIKER.

**mon·ied** (mŭn'ēd) *adj. var. of* MONEYED.

**mon·ies** (mŭn'ēz') *n. var. pl. of* MONEY.

**mon·i·ker** or **mon·ick·er** (mŏn'ĭ-kər) *n.* [Orig. unknown.] *Slang.* A personal name or nickname.

**mo·nil·i·a·sis** (mō'nə-lī'ə-sĭs, mŏn'ə-) *n.* [NLat. *Monilia,* fungus genus + -IASIS.] Candidiasis.

**mo·nil·i·form** (mō-nĭl'ə-fôrm') *adj.* [Lat. *monile,* necklace + -FORM.] Similar to a string of beads, as various plant roots, the antennae of certain insects, and the nuclei of some members of the Ciliata. **—mo·nil'i·form'ly** *adv.*

**mon·ish** (mŏn'ĭsh) *vt.* **-ished, -ish·ing, -ish·es.** [ME *monesen* < OFr. *monester* < VLat. **monestare* < Lat. *monēre,* to warn.] To admonish.

**mo·nism** (mō'nĭz'əm, mŏn'ĭz'əm) *n. Philos.* A metaphysical system in which reality is viewed as a unified whole. **—mo'nist** *n.* **—mo·nis'tic** (mō-nĭs'tĭk, mŏ-) *adj.* **—mo·nis'ti·cal·ly** *adv.*

**mo·ni·tion** (mō-nĭsh'ən, mə-) *n.* [ME *monicion* < OFr. *monition* < Lat. *monitio* < *monēre,* to warn.] **1.** A warning of approaching danger. **2. a.** An admonition. **b.** Advice : counsel. **3.** A formal order from an ecclesiastical court or bishop to refrain from a specified offense.

**mon·i·tor** (mŏn'ĭ-tər) *n.* [Lat. < *monēre,* to warn.] **1.** One that admonishes, cautions, or reminds. **2.** A pupil who aids a teacher with routine duties. **3. a.** A device for recording or controlling a process or activity. **b.** A screen for viewing or checking the picture being picked up by a television camera. **4.** An articulated device holding a rotating nozzle, used in mining and fire fighting. **5.** A heavily ironclad 19th-cent. warship with a flat low deck and one or more gun turrets. **6.** Any of various tropical carnivorous lizards of the genus *Varanus,* from several inches to ten feet in length. *—v.* **-tored, -tor·ing, -tors.** *—vt.* **1.** To check (the transmission quality of a signal) by a receiver. **2.** To test for radiation intensity. **3.** To track by or as if by an electronic device. **4.** To check by means of a receiver for significant content. **5.** To check systematically or scrutinize for the purpose of collecting specified categories of data. **6.** To keep watch over : SUPERVISE <*monitor* the history test> **7.** To direct as a monitor. *—vi.* To act as a monitor.

**mon·i·to·ri·al** (mŏn'ĭ-tôr'ē-əl, -tōr'-) *adj.* Of, relating to, or performed by monitors. **—mon'i·to'ri·al·ly** *adv.*

**mon·i·to·ry** (mŏn'ĭ-tôr'ē, -tōr'ē) *adj.* [Lat. *monitorius* < *monitor,* monitor.] Conveying an admonition or warning <a *monitory* frown> *—n., pl.* **-ries.** A letter of admonition.

**monk** (mŭngk) *n.* [ME *munk* < OE *munuc* < LLat. *monachus* < LGk. *monakhos* < Gk., single < *monos.*] A member of a religious brotherhood residing in a monastery and devoted to a discipline prescribed by his order.

**mon·key** (mŭng'kē) *n., pl.* **-keys.** [Prob. of LG orig.] **1.** A member of the order Primates, excluding human beings; specif., most long-tailed primates, including the Old and New World monkeys and the marmosets, and usu. excluding the anthropoid apes and the lemurs, lorises, tree shrews, and tarsiers. **2.** A mischievous, playful young person. **3.** The iron block of a pile driver. **4.** *Slang.* One who is mocked, duped, or made to look like a fool. **5.** *Slang.* Drug addiction, regarded as a burden <have a *monkey* on one's back> *—v.* **-keyed, -key·ing, -keys.** *—vi. Informal.* **1.** To play, fiddle, or tamper with something idly. **2.** To behave in a mischievous or apish way. *—vt.* To imitate : mimic.

**monkey bread** *n.* The fruit of the baobab.

**monkey business** *n. Slang.* Silly, mischievous, or deceitful acts.

**mon·key-flow·er** (mŭng'kē-flou'ər) *n.* A plant of the genus *Mimulus,* bearing variously colored, two-lipped flowers.

**monkey jacket** *n.* **1.** A short, tight-fitting jacket, once worn by sailors. **2.** *Slang.* A mess jacket.

**monkey pot** *n.* **1. a.** The large urn-shaped lidded pod of tropical trees of the genus *Lecythis.* **b.** A tree bearing this type of pod. **2.** A cylindrical or barrel-shaped melting pot used in making flint glass.

**monkey puzzle** *n.* An evergreen tree native to Chile, *Araucaria araucana,* having intricately ramifying branches covered with stiff prickle-tipped leaves.

**mon·key·shine** (mŭng'kē-shīn') *n. often* **monkeyshines.** *Slang.* A playful trick : PRANK.

**monkey wrench** *n.* [Orig. unknown.] **1.** A hand tool with adjustable jaws for turning nuts of varying sizes. **2.** *Informal.* A disrupting element <threw a *monkey wrench* into the preparations>

**monk·fish** (mŭngk'fĭsh') *n., pl.* **monkfish** or **-fish·es.** [From the cowled appearance of its head.] The goosefish.

**Mon-Khmer** (mōn'kmĕr') *n.* A subfamily of the Austro-Asiatic language family that includes Mon, Khmer and other languages of Southeast Asia.

**monk·hood** (mŭngk'ho͝od') *n.* **1.** The state or profession of a monk : MONASTICISM. **2.** Monks as a group.

**monk·ish** (mŭng'kĭsh) *adj.* **1.** Of, pertaining to, or typical of monks or monasticism. **2.** Of or tending toward self-denial.

**monk's cloth** *n.* A heavy cotton cloth in a coarse basket weave.

**monks·hood** (mŭngks'ho͝od') *n.* A usu. poisonous plant of the genus *Aconitum,* with hooded flowers.

**mon·o**[1] (mŏn'ō) *n.* MONONUCLEOSIS 2.

**mon·o**[2] (mŏn'ō) *adj.* MONAURAL 2.

**mono-** or **mon-** *pref.* [ME < OFr. < Lat. < Gk. < *monos,* single, alone.] **1.** One : single : alone <*mono*morphic> **2.** Containing a single atom, radical, or group <*mono*basic> **3.** Monomolecular : monatomic <*mono*layer>

**mon·o·ac·id** (mŏn'ō-ăs'ĭd) *also* **mon·o·a·cid·ic** (-ə-sĭd'ĭk) *adj.* Having only one hydroxyl group to react with acids. *—n.* **monoacid.** An acid having one replaceable hydrogen atom.

**mon·o·am·ine** (mŏn'ō-ăm'ēn, -ə-mēn') *n.* An amine compound with one amino group.

**monoamine oxidase** *n.* An enzyme that acts as a catalyst in the oxidative deamination of monoamines.

**mon·o·ba·sic** (mŏn′ə-bā′sĭk) *adj.* **1.** Monoprotic. **2.** Having only one metal ion or positive radical.

**mon·o·carp** (mŏn′ə-kärp′) *n.* A monocarpic plant.

**mon·o·car·pel·lar·y** (mŏn′ə-kär′pə-lĕr′ē) *adj.* Consisting of only one carpel.

**mon·o·car·pic** (mŏn′ə-kär′pĭk) *also* **mon·o·car·pous** (-kär′pəs) *adj.* Flowering and bearing fruit only once.

**Mo·noc·er·os** (mə-nŏs′ər-əs) *n.* [ME, unicorn < OFr. < Lat. < Gk. *monokerōs,* having one horn : *monos,* one + *keras,* horn.] **1.** A constellation in the Southern Hemisphere. **2. monoceros.** *Obs.* **a.** A one-horned fish, as the swordfish. **b.** A unicorn.

**mon·o·cha·si·um** (mŏn′ə-kā′zē-əm, -zhē-əm) *n., pl.* **-si·a** (-zē-ə, -zhē-ə) [MONO- + (DI)CHASIUM.] A cyme with a single main stem. **—mon′o·cha′si·al** *adj.*

**mon·o·chord** (mŏn′ə-kôrd′) *n.* [ME *monocorde* < OFr. < Med. Lat. *monochordum* < Gk. *monokhordon* : *monos,* one + *khordē,* string.] An acoustical instrument consisting of a sounding box with one string and a movable bridge, for studying musical tones.

**mon·o·chro·mat·ic** (mŏn′ə-krō-măt′ĭk) *also* **mon·o·chro·ic** (-krō′ĭk) *adj.* [Lat. *monochromatos* < Gk. *monokhrōmatos* : *monos,* one + *khrōma,* color.] **1.** Of a single color. **2.** Having or producing light of only one wavelength. **—mon′o·chro·mat′i·cal·ly** *adv.* **—mon′o·chro′ma·tic′i·ty** (-mə-tĭs′ĭ-tē) *n.*

**mon·o·chrome** (mŏn′ə-krōm′) *n.* [Med. Lat. *monochroma* < Gk. *monokhrōmos,* of one color : *monos,* one + *khrōma,* color.] **1.** A painting in different shades of one color. **2.** The technique of executing monochrome paintings. **—mon′o·chro′mic** (-krō′mĭk) *adj.*

**mon·o·cle** (mŏn′ə-kəl) *n.* [Fr. < LLat. *monoculus,* having one eye : Gk. *monos,* one + Lat. *oculus,* eye.] An eyeglass for one eye. **—mon′o·cled** (-kəld) *adj.*

**mon·o·cline** (mŏn′ə-klīn′) *n.* A geologic formation in which all strata are inclined in the same direction. **—mon′o·cli′nal** *adj.*

**mon·o·clin·ic** (mŏn′ə-klĭn′ĭk) *adj.* Of or relating to three unequal crystal axes, two of which intersect obliquely and are perpendicular to the third.

**mon·o·cli·nous** (mŏn′ə-klī′nəs) *adj.* [NLat. *monoclinus* : Gk. *monos,* one + Gk. *klinē,* couch.] Bearing pistils and stamens in the same flower.

**mon·o·coque** (mŏn′ə-kŏk′, -kōk′) *n.* [Fr. : Gk. *monos,* one + *coque,* shell < Lat. *coccum,* berry < Gk. *kokkos.*] A metal structure, as of an aircraft, in which the covering absorbs much of the stress to which the body is subjected.

**mon·o·cot·y·le·don** (mŏn′ə-kŏt′l-ēd′n) *also* **mon·o·cot** (mŏn′ə-kŏt′) *n.* [NLat. *Monocotyledones,* class name : MONO- + Lat. *cotyledon,* navelwort. —see COTYLEDON.] Any of various plants of the Monocotyledonae, one of the two major divisions of angiosperms, marked by a single embryonic seed leaf appearing at germination. **—mon′o·cot′y·le′don·ous** *adj.*

**mo·noc·ra·cy** (mō-nŏk′rə-sē, mə-) *n.* Government or rule by a single person : AUTOCRACY. **—mon′o·crat′** (mŏn′ə-krăt′) *n.* **—mon′o·crat′ic** *adj.*

**mo·noc·u·lar** (mŏ-nŏk′yə-lər, mə-) *adj.* [LLat. *monoculus,* having one eye. —see MONOCLE.] **1.** Having or relating to one eye. **2.** Adapted for the use of only one eye. **—mo·noc′u·lar·ly** *adv.*

**mon·o·cy·cle** (mŏn′ə-sī′kəl) *n.* A unicycle.

**mon·o·cyte** (mŏn′ə-sīt′) *n.* A large white blood corpuscle, with a pale, oval nucleus and more protoplasm than a lymphocyte. **—mon′o·cyt′ic** (-sĭt′ĭk), **mon′o·cy′toid′** (-sī′toid′) *adj.*

**mon·o·dac·tyl** (mŏn′ə-dăk′tl) *n.* [< Gk. *monodaktulos,* having one toe : *monos,* one + *daktulos,* toe.] An animal with only one claw on each extremity. **—mon′o·dac′ty·lous** *adj.*

**mon·o·dra·ma** (mŏn′ə-drä′mə, -drăm′ə) *n.* A dramatic composition for one performer. **—mon′o·dra·mat′ic** (-drə-măt′ĭk) *adj.*

**mon·o·dy** (mŏn′ə-dē) *n., pl.* **-dies.** [LLat. *monodia* < Gk. *monōidia* : *monos,* one + *oidē,* song.] **1.** An ode for one performer, as in Greek drama. **2.** An elegiac verse expressing personal lament. **3.** *Mus.* **a.** A style of composition with one melodic line or vocal part predominating. **b.** A composition in this style. **—mo·nod′ic** (mə-nŏd′ĭk), **mo·nod′i·cal** *adj.* **—mo·nod′i·cal·ly** *adv.* **—mon′o·dist** *n.*

**mo·noe·cious** *also* **mo·ne·cious** (mə-nē′shəs) *adj.* [< NLat. *Monoecia,* class name : Gk. *monos,* one + Gk. *oikia,* house.] **1.** *Bot.* Bearing male and female reproductive organs in separate flowers on the same plant. **2.** *Zool.* Hermaphroditic. **—mo·noe′cious·ly** *adv.*

**mon·o·es·ter** (mŏn′ō-ĕs′tər) *n.* An ester with one ester group.

**mo·nog·a·my** (mə-nŏg′ə-mē) *n.* **1.** The custom or state of being married to one person at a time. **2.** The state of having one mate for life. **—mo·nog′a·mist** *n.* **—mo·nog′a·mous** *adj.* **—mo·nog′a·mous·ly** *adv.*

**mon·o·gen·e·sis** (mŏn′ə-jĕn′ĭ-sĭs) *n.* **1.** The theory that all living organisms are descended from a single cell. **2.** Asexual reproduction, as by sporulation. **3.** Development of an ovum into an organism resembling the parent, without metamorphosis. **—mo·nog′e·nous** (mə-nŏj′ə-nəs) *adj.*

**mon·o·ge·net·ic** (mŏn′ə-jə-nĕt′ĭk) *adj.* **1.** Relating to or showing monogenesis. **2.** Asexual. **3.** Arising from a single formation process, as a mountain range.

**mon·o·gen·ic** (mŏn′ə-jĕn′ĭk) *adj.* **1.** Having a common or single origin, as igneous rocks composed of a single mineral. **2. a.** Of or relating to monogenesis : MONOGENETIC. **b.** Of or relating to monogenism. **3.** Of or regulated by one gene or one of a pair of allelic genes. **4.** Producing offspring mostly of one sex. **—mon′o·gen′i·cal·ly** *adv.*

**mo·nog·e·nism** (mə-nŏj′ə-nĭz′əm) *n.* The theory that humankind has descended from a single pair of ancestors. **—mo·nog′e·nist** *n.* **—mo·nog′e·nis′tic** *adj.*

**mon·o·gram** (mŏn′ə-grăm′) *n.* [LLat. *monogramma* < Gk. *monogrammon* : *monos,* single + *gramma,* letter < *graphein,* to write.] A design of one or more letters, usu. the initials of a name. —*vt.* **-grammed, -gram·ming, -grams** *also* **-gramed, gram·ing, -grams.** To mark with a monogram. **—mon′o·gram·mat′ic** (-grə-măt′ĭk) *adj.*

**mon·o·graph** (mŏn′ə-grăf′) *n.* A scholarly book, article, or pamphlet on a specific, usu. limited subject. **—mon′o·graph′** *v.* **(-graphed, -graph·ing, -graphs). —mo·nog′ra·pher** (mə-nŏg′rə-fər) *n.* **—mon′o·graph′ic** *adj.* **—mon′o·graph′i·cal·ly** *adv.*

**mo·nog·y·ny** (mə-nŏj′ə-nē) *n.* The practice or state of having only one wife at a time. **—mo·nog′y·nist** *n.* **—mo·nog′y·nous** *adj.*

**mon·o·hy·brid** (mŏn′ō-hī′brĭd) *n.* Hybrid offspring of parents differing in a single characteristic or genetic factor.

**mon·o·hy·drate** (mŏn′ō-hī′drāt′) *n.* A compound of singly hydrated molecules. **—mon′o·hy′drat′ed** *adj.*

**mon·o·hy·dric** (mŏn′ō-hī′drĭk) *adj.* Containing one replaceable hydroxyl radical.

**mo·noi·cous** (mə-noi′kəs) *adj.* [Alteration of MONOECIOUS.] Bearing archegonia and antheridia on the same plant : BISEXUAL.

**mon·o·lay·er** (mŏn′ō-lā′ər) *n.* A stratum or film of a compound one molecule thick.

**mon·o·lin·gual** (mŏn′ə-lĭng′gwəl) *adj.* Knowing or using only one language. **—mon′o·lin′gual** *n.*

**mon·o·lith** (mŏn′ə-lĭth′) *n.* [Fr. *monolithe* < Gk. *monolithos,* consisting of a single stone : *monos,* one + *lithos,* stone.] **1.** A large stone block, esp. one used in sculpture or architecture. **2.** A large organization, as a corporation, that functions as a powerful unit.

**mon·o·lith·ic** (mŏn′ə-lĭth′ĭk) *adj.* **1.** Composed of a monolith. **2.** Massive, solid, and uniform. **—mon′o·lith′i·cal·ly** *adv.*

**mon·o·logue** *also* **mon·o·log** (mŏn′ə-lôg′, -lŏg′) *n.* [Fr. : Gk. *monos,* one + Gk. *logos,* speech.] **1.** A long speech by one person, often monopolizing conversation. **2. a.** A dramatic soliloquy. **b.** A literary composition in the form of a soliloquy. **3.** A continuous series of comic stories or jokes delivered by a single comedian. **—mon′o·log′ic** (mŏn′ə-lŏj′ĭk), **mon′o·log′i·cal** *adj.* **—mo·nol′o·gist** (mə-nŏl′ə-jĭst, mŏn′ə-lôg′ĭst, -lŏg′-) *n.*

**mon·o·ma·ni·a** (mŏn′ə-mā′nē-ə, -mān′yə) *n.* **1.** Pathological obsession with a single idea. **2.** Intent concentration on or exaggerated enthusiasm for a subject or an idea. **—mon′o·ma′ni·ac′** (-mā′nē-ăk′) *n.* **—mon′o·ma·ni′a·cal** (-mə-nī′ə-kəl) *adj.*

**mon·o·mer** (mŏn′ə-mər) *n.* A molecule that can be chemically bound as a unit of a polymer. **—mon′o·mer′ic** (-mĕr′ĭk) *adj.*

**mon·o·me·tal·lic** (mŏn′ō-mə-tăl′ĭk) *adj.* **1.** Having or containing one metal. **2.** Relating to monometallism.

**mon·o·met·al·lism** *also* **mon·o·met·al·ism** (mŏn′ō-mĕt′l-ĭz′əm) *n.* The economic theory or practice of having only one metal as a standard of money. **—mon′o·met′al·list** *n.*

**mo·no·mi·al** (mŏ-nō′mē-əl, mə-) *n.* [MON(O)- + (BIN)OMIAL.] **1.** *Math.* An algebraic expression having only one term. **2.** *Biol.* A taxonomic name composed of one word. **—mo·no′mi·al** *adj.*

**mon·o·mo·lec·u·lar** (mŏn′ō-mə-lĕk′yə-lər) *adj.* **1.** Of or relating to a single molecule. **2.** Of or having a layer one molecule thick. **—mon′o·mo·lec′u·lar·ly** *adv.*

**mon·o·mor·phic** (mŏn′ō-môr′fĭk) *also* **mon·o·mor·phous** (-fəs) *adj.* **1.** *Chem.* Having but one form, as one crystal form. **2.** *Zool.* Having a basic structure remaining unchanged through a series of developmental changes. **—mon′o·mor′phism** *n.*

**mon·o·nu·cle·ar** (mŏn′ō-nōō′klē-ər, -nyōō′-) *adj.* Having a single nucleus.

**mon·o·nu·cle·o·sis** (mŏn′ō-nōō′klē-ō′sĭs, -nyōō-) *n.* [MONO- + NUCLE(US) + -OSIS.] **1.** Presence of an abnormally large number of leucocytes with single nuclei in the blood stream. **2.** An acute, infectious, viral disease chiefly affecting lymphoid tissues.

**mon·o·nu·cle·o·tide** (mŏn′ō-nōō′klē-ə-tīd′, -nyōō′-) *n.* A nucleotide containing one molecule each of a phosphoric acid, a pentose, and either a purine or pyrimidine base.

**mon·o·pet·al·ous** (mŏn′ə-pĕt′l-əs) *adj.* Having petals united to form one corolla : GAMOPETALOUS.

**mo·noph·a·gous** (mə-nŏf′ə-gəs) *adj.* Eating but one type of food. **—mo·noph′a·gy** (-ə-jē) *n.*

**mon·o·pho·bi·a** (mŏn′ō-fō′bē-ə) *n.* Abnormal fear of solitude. **—mon′o·pho′bic** (-fō′bĭk) *adj.*

**mon·o·phon·ic** (mŏn′ə-fŏn′ĭk) *adj.* **1.** *Mus.* Having a single melodic line. **2.** MONAURAL 2. **—mon′o·phon′i·cal·ly** *adv.*

**mo·noph·o·ny** (mə-nŏf′ə-nē) *n.* Music with but one melodic line.

**mon·oph·thong** (mŏn′əf-thông′, -thŏng′) *n.* [LGk. *monophthongos* : *monos*, single + *phthongos*, sound.] **1.** A single vowel sound made with the supraglottal speech organs in fixed position. **2.** Two written vowels representing a single sound; e.g., *oa* in *boat* is a monophthong. **—mon′oph·thon′gal** (-thông′gəl, -thŏng′-) *adj.*

**mon·o·phy·let·ic** (mŏn′ō-fī-lĕt′ĭk) *adj.* **1.** Of or concerning a single phylum of plants or animals. **2.** Descended or derived from one stock or source. **—mon′o·phy·let′ic·al·ly** *adv.*

**Mo·noph·y·site** (mə-nŏf′ə-sīt′) *n.* [Med. Lat. *monophysita* < Med. Gk. *monophusitēs* : *monos*, single + *phusis*, nature.] An adherent of the doctrine that in the person of Christ there was but a single divine nature. **—Mo·noph′y·sit′ic** (-sĭt′ĭk) *adj.* **—Mo·noph′y·sit′ism** *n.*

**mon·o·plane** (mŏn′ə-plān′) *n.* An aircraft with a single pair of wings.

**mon·o·ple·gi·a** (mŏn′ə-plē′jē-ə, -plē′jə) *n.* Paralysis of one limb or body part, as one side of the face. **—mon′o·ple′gic** (-plē′jĭk) *adj.*

**mon·o·ploid** (mŏn′ə-ploid′) *adj.* Having a haploid set of chromosomes. —*n.* A monoploid individual.

**mon·o·pode** (mŏn′ə-pōd′) *n.* [LLat. *monopodius*, one-footed. —see MONOPODIUM.] **1.** A creature with but one foot, specif. a member of a fabled people in Africa. **2.** *Bot.* Monopodium.

**mon·o·po·di·um** (mŏn′ə-pō′dē-əm) *n., pl.* **-dia** (-dē-ə) [NLat. < LLat. *monopodius*, one-footed < Gk. *monopous* : *monos*, one + *pous*, foot.] A main axis of a plant, as the trunk of certain conifers, that follows a single line of growth, giving off lateral branches. **—mon′o·po′di·al** *adj.*

**mo·nop·o·lize** (mə-nŏp′ə-līz′) *vt.* **-lized, -liz·ing, -liz·es. 1.** To acquire or maintain a monopoly of. **2.** To dominate by excluding others. **—mo·nop′o·li·za′tion** *n.* **—mo·nop′o·liz′er** *n.*

**mo·nop·o·ly** (mə-nŏp′ə-lē) *n., pl.* **-lies.** [Lat. *monopolium* < Gk. *monopōlion* : *monos*, one + *pōlein*, to sell.] **1.** Exclusive control by one group of the means of producing or selling a commodity or service. **2.** *Law.* A right granted by a government giving exclusive control over a specified commercial activity to a single party. **3. a.** A company or group having exclusive control over a commercial activity. **b.** A commodity or service so controlled. **4.** Exclusive possession or control. **—mo·nop′o·lism** *n.* **—mo·nop′o·list** *n.* **—mo·nop′o·lis′tic** *adj.* **—mo·nop′o·lis′ti·cal·ly** *adv.*

**mon·o·pro·pel·lant** (mŏn′ō-prə-pĕl′ənt) *n.* A rocket propellant that combines fuel and oxidizer, as a mixture of hydrogen peroxide and alcohol.

**mon·o·pro·tic** (mŏn′ə-prō′tĭk) *adj.* [MONO- + PROT(ON) + -IC.] Having only one hydrogen ion to donate to a base in an acid-base reaction.

**mo·nop·so·ny** (mə-nŏp′sə-nē) *n., pl.* **-nies.** [MON(O)- + Gk. *opsōnia*, purchase of food < *opsōnein*, to buy food.] A situation in which only one buyer seeks the product or service of several sellers.

**mon·o·rail** (mŏn′ə-rāl′) *n.* **1.** One rail on which a vehicle or train of cars travels. **2.** A railway system using a monorail.

**mon·o·sac·cha·ride** (mŏn′ə-săk′ə-rīd′, -rĭd) *n.* A simple sugar that cannot be decomposed by hydrolysis, esp. one of the hexoses, having the general formula $C_6H_{12}O_6$.

**mon·o·sep·al·ous** (mŏn′ə-sĕp′ə-ləs) *adj.* Having sepals united to form a single calyx : GAMOSEPALOUS.

**mon·o·sex·u·al** (mŏn′ə-sĕk′shōō-əl) *adj.* Composed of or intended for individuals of one sex <*monosexual* boarding schools>

**mon·o·so·di·um glu·ta·mate** (mŏn′ə-sō′dē-əm glōō′tə-māt′) *n.* Sodium glutamate.

**mon·o·some** (mŏn′ə-sōm′) *n.* **1.** A cell lacking one or more chromosomes. **2.** An unpaired X chromosome. **3.** A single ribosome. **—mon′o·so′mic** (-sō′mĭk-) *adj.*

**mon·o·sper·mous** (mŏn′ə-spûr′məs) *also* **mon·o·sper·mal** (-məl) *adj.* Having a single seed.

**mon·o·stome** (mŏn′ə-stōm′) *also* **mo·nos·to·mous** (mə-nŏs′tə-məs) *adj.* Having one oral sucker only, as certain flatworms. —*n.* **monostome.** A trematode worm.

**mon·o·sty·lous** (mŏn′ə-stī′ləs) *adj. Bot.* Having one style.

**mon·o·syl·lab·ic** (mŏn′ə-sĭ-lăb′ĭk) *adj.* **1.** Having only one syllable. **2.** Characterized by or made up of monosyllables. **—mon′o·syl·lab′ic·al·ly** *adv.*

**mon·o·syl·la·ble** (mŏn′ə-sĭl′ə-bəl) *n.* [LLat. *monosyllabum* < Gk. *monosullabon* : *monos*, one + *sullabē*, syllable. —see SYLLABLE.] A one-syllable word or utterance.

**mon·o·syn·ap·tic** (mŏn′ō-sə-năp′tĭk) *adj.* Having one neural synapse. **—mon′o·syn·ap′ti·cal·ly** *adv.*

**mon·o·the·ism** (mŏn′ə-thē-ĭz′əm) *n.* The doctrine or belief that there is only one God. **—mon′o·the′ist** *n.* **—mon′o·the·is′tic** *adj.* **—mon′o·the·is′ti·cal·ly** *adv.*

**mon·o·the·mat·ic** (mŏn′ō-thē-măt′ĭk) *adj. Mus.* Having a single theme.

**mon·o·tint** (mŏn′ə-tĭnt′) *n.* MONOCHROME 1.

**mon·o·tone** (mŏn′ə-tōn′) *n.* [< Gk. *monotonos*, monotonous.] **1.** A succession of sounds, syllables, or words uttered in a single tone of voice. **2.** *Mus.* **a.** A single tone repeated with different words or time values, as in plainsong. **b.** A chant in a single tone. **3.** Repetition or dull sameness in sound, style, manner, or color. —*adj.* **1.** Of, relating to, or typical of sounds emitted at a single pitch. **2.** Of or having a single color. **3.** *also* **mon·o·ton·ic** (mŏn′ə-tŏn′ĭk). *Math.* Designating sequences whose successive members either consistently increase or decrease but do not oscillate in relative value. **—mon′o·ton′ic** (-tŏn′ĭk) *adj.* **—mon′o·ton′i·cal·ly** *adv.*

**mo·not·o·nous** (mə-nŏt′n-əs) *adj.* [Gk. *monotonos* : *monos*, one + *tonos*, tone.] **1.** Spoken or sounded in an unvarying tone. **2.** Repetitiously dull. **—mo·not′o·nous·ly** *adv.* **—mo·not′o·nous·ness** *n.*

**mo·not·o·ny** (mə-nŏt′n-ē) *n.* [Gk. *monotonia* < *monotonos*, monotonous.] **1.** Uniformity or lack of variation in pitch, intonation, or inflection. **2.** Tedious sameness.

**mon·o·treme** (mŏn′ə-trēm′) *n.* [< NLat. *Monotremata*, order name : Gk. *monos*, one + Gk. *trēma*, hole < *tetrainein*, to perforate.] A member of the Monotremata, an order of egg-laying mammals restricted to Australia and New Guinea and including the platypus and the echidna. **—mon′o·trem′a·tous** (-trĕm′ə-təs) *adj.*

**mo·not·ri·chous** (mə-nŏt′rĭ-kəs) *also* **mon·o·trich·ic** (mŏn′ə-trĭk′ĭk) *adj.* Having one flagellum at only one pole or end, as certain bacteria.

**mon·o·tro·phic** (mŏn′ə-trō′fĭk) *adj.* Needing but one kind of food : MONOPHAGOUS.

**mon·o·type** (mŏn′ə-tīp′) *n. Biol.* The sole member of its group, as a species that also constitutes a genus. **—mon′o·typ′ic** (-tĭp′ĭk) *adj.*

**Mon·o·type** (mŏn′ə-tīp′). A trademark for a typesetting machine operated from a keyboard that activates a unit that casts individual letters from matrices and assembles them.

**mon·o·va·lent** (mŏn′ə-vā′lənt) *adj.* **1.** *Chem.* Possessing a valence of 1 : UNIVALENT. **2.** *Pathol.* Able to resist a specific pathogen because the proper antibodies or antigens are present. **—mon′o·va′lence, mon′o·va′len·cy** *n.*

**mon·ox·ide** (mə-nŏk′sīd′) *n.* An oxide with each molecule containing one oxygen atom.

**mon·o·zy·got·ic** (mŏn′ō-zī-gŏt′ĭk) *adj.* Derived from a single fertilized ovum. —Usu. used of identical twins.

**Monroe Doctrine** *n.* [After James *Monroe* (1758–1831).] The U.S. policy of opposition to outside interference in the Americas.

**mons** (mŏnz) *n., pl.* **mon·tes** (mŏn′tēz) [Lat., mountain.] A bodily protuberance, esp. that formed by the pubic bones.

**Mon·sei·gneur** (môN-sĕ-nyœr′) *n., pl.* **Mes·sei·gneurs** (mā-sĕ-nyœr′) [Fr. < OFr. : *mon*, my (< Lat. *meum*) + *seigneur*, sir, prob. < Lat. *senior*, older, comp. of *senex*, old.] A French title of honor or respect given to princes and prelates.

**Mon·sieur** (mə-syœ′) *n., pl.* **Mes·sieurs** (mĕs′ərz, mā-syœ′) [Fr. < OFr. : *mon*, my (< Lat. *meum*) + *sieur*, sir < *seigneur*, prob. < Lat. *senior*, older, comp. of *senex*, old.] —Used as a courtesy title with the name of a Frenchman.

**Mon·si·gnor** *also* **mon·si·gnor** (mŏn-sēn′yər) *n., pl.* **-gnors.** [Ital. < Fr. *Monseigneur*. —see MONSEIGNEUR.] *Rom. Cath. Ch.* A prelate with a rank or title, as of chamberlain, that is usu. conferred by the Pope. —Used as a title with a surname or with a first name and a surname. **—Mon′si·gnor′i·al** *adj.*

**mon·soon** (mŏn-sōōn′) *n.* [Obs. Du. *monssoen* < Port. *monção* < Ar. *mausim*, season.] A wind system that affects large climatic regions and reverses direction seasonally, esp. the Asiatic monsoon that produces dry and wet seasons in India and southern Asia. **—mon·soon′al** *adj.*

**mon·ster** (mŏn′stər) *n.* [ME < OFr. *monstre* < Lat. *monstrum*, portent, monster < *monēre*, to warn.] **1.** A creature with a frightening or bizarre shape or appearance. **2.** An animal or plant with structural defects or deformities. **3.** *Pathol.* A grotesquely abnormal fetus or infant. **4.** A very large animal, plant, or object. **5.** One arousing disgust or horror.

**mon·strance** (mŏn′strəns) *n.* [ME < OFr. < Med. Lat. *monstrantia* < Lat. *monstrare*, to show < *monstrum*, portent, monster.] *Rom. Cath. Ch.* A receptacle that holds the Host.

**mon·strous** (mŏn′strəs) *adj.* [ME *monstrows* < OFr. *monstreux* < Lat. *monstruosus* < *monstrum*, monster.] **1.** Deviating greatly from the norm in structure or appearance : ABNORMAL. **2.** Unusually large : ENORMOUS. **3.** Hideous : shocking <"a *monstrous* tyranny, never surpassed" —Winston Churchill> **—mon′strous·ly** *adv.* **—mon·stros′i·ty** (-strŏs′ĭ-tē) *n.* **—mon′strous·ness** *n.*

**mons ve·ne·ris** (vĕn′ər-ĭs) *n.* [N Lat., eminence of Venus.] The female mons.

**mon·tage** (mŏn-täzh′, môN-) *n.* [Fr. < *monter*, to mount < OFr. —see MOUNT[1].] **1. a.** The art, style, or process of making a pictorial composition by closely arranging or superimposing many pictures or designs. **b.** A picture so made. **2. a.** A rapid sequence of related short scenes or images used as a motion-picture or television technique for underscoring a theme. **b.** A sequence using this effect. **3.** A mixture of miscellaneous elements : JUMBLE. **—mon·tage′** *v.* **(-taged, -tag·ing, -tag·es).**

**Mon·ta·gnard** (mŏn′tən-yärd′) *n.* [Fr., mountaineer < *montagne*, mountain. —see MOUNTAIN.] A member of a tribal people inhabiting a mountainous region of southern Vietnam near the border of Cambodia.

**mon·tane** (mŏn-tān′, mŏn′tān′) *adj.* [Lat. *montanus* < *mons,* mountain.] Of, growing in, or inhabiting mountain areas.

**mon·tan wax** (mŏn′tən, -tăn′) *n.* [< Lat. *montanus,* montane.] A hard, white wax extracted from lignite and used to make polishes, candles, and insulators.

**mon·te** (mŏn′tē) *n.* [Sp., *monte,* mountain < Lat. *mons,* mountain.] A card game in which a player bets that one of two cards will be matched by the dealer before the other one.

**mon·te·ro** (mŏn-târ′ō) *n., pl.* **-ros.** [Sp., hunter < *monte,* mountain < Lat. *mons.*] A huntsman's cap with side flaps.

**mon·tes** (mŏn′tēz) *n. pl. of* MONS.

**Mon·tes·so·ri·an** (mŏn′tĭ-sôr′ē-ən, -sōr′-) *adj.* Of or pertaining to the Montessori method.

**Mon·tes·so·ri method** (mŏn′tĭ-sôr′ē, -sōr′ē) *n.* [After Maria *Montessori* (1870–1952).] A method of instructing young children that emphasizes development of a child's initiative.

**month** (mŭnth) *n.* [ME *moneth* < OE *mōnað.*] **1.** One of the 12 divisions of a year according to the Gregorian calendar. **2.** A period extending from a date in one calendar month to the corresponding date the next month. **3. a.** Four weeks. **b.** Thirty days. **4.** The average period of revolution of the moon around the earth determined by using a fixed star as a reference point and equal to 27 days 7 hours 43 minutes. **5.** The average time between successive new, or full, moons that is equal to 29 days 12 hours 44 minutes. **6.** One twelfth of a tropical year, totaling 30 days 10 hours 29 minutes 3.8 seconds. **—month of Sundays.** *Informal.* An indefinitely long period.

**month·ly** (mŭnth′lē) *adj.* **1.** Occurring, appearing, or coming due every month. **2.** Continuing or lasting for a month. *—adv.* Once a month : every month. *—n., pl.* **-lies. 1.** A periodical published once each month. **2. monthlies.** *Informal.* The menses.

**mon·ti·cule** (mŏn′tĭ-kyōōl′) *n.* [Fr. < LLat. *monticulus,* dim. of Lat. *mons,* mountain.] A secondary volcanic cone of a volcano.

**mon·u·ment** (mŏn′yə-mənt) *n.* [ME < Lat. *monumentum,* memorial < *monēre,* to remind.] **1.** A structure, as a building or sculpture, erected as a memorial. **2.** An inscribed stone or other marker at a grave : TOMBSTONE. **3.** Something venerated for its aesthetic or historic significance. **4.** A national monument. **5. a.** An enduring and outstanding achievement. **b.** An exceptional example <a *monument* of ignorance> **6.** A boundary marker, as a stone or post. **7.** A written document, esp. a legal one.

**mon·u·men·tal** (mŏn′yə-mĕn′tl) *adj.* **1.** Of, similar to, or serving as a monument. **2.** Impressively large, sturdy, and enduring. **3.** Of outstanding significance <a *monumental* achievement in American painting> **4.** Astounding <*monumental* selfishness> **—mon′u·men·tal′i·ty** *n.* **—mon′u·men′tal·ly** *adv.*

**mon·u·men·tal·ize** (mŏn′yə-mĕn′tl-īz′) *vt.* **-ized, -iz·ing, -iz·es.** To memorialize with a monument.

**mon·u·ron** (mŏn′yə-rŏn′) *n.* [MON(O)- + UR(EA) + -ON.] A crystalline compound, $C_9H_{11}ClN_2O$, used as a herbicide for grasses and broadleaf weeds.

**mon·zo·nite** (mŏn-zō′nīt′, mŏn′zə-nīt′) *n.* [Fr., after Mt. *Monzoni* in northeastern Italy, where it was discovered.] An igneous rock composed mainly of plagioclase and orthoclase, with small amounts of other minerals. **—mon′zo·nit′ic** *adj.*

**moo** (mōō) *vi.* **mooed, moo·ing, moos.** [Imit.] To emit the deep bellowing sound made by a cow. **—moo** *n.*

**mooch** (mōōch) *v.* **mooched, mooch·ing, mooch·es.** [ME *mowchen* < OFr. *muchier.*] *Slang.* *—vt.* **1.** To obtain (something) free of charge, as by cajolery. **2.** To steal or filch. *—vi.* **1.** To live off others : SPONGE. **2.** To wander around aimlessly. **3.** To skulk about. **—mooch′er** *n.*

**mood[1]** (mōōd) *n.* [ME *mod* < OE *mōd.*] **1.** A temporary state of mind or feeling, as indicated by the trend of one's thoughts <a somber *mood*> **2.** A pervading impression on the feelings of an observer <the melancholy *mood* of the poem> **3.** Angry or sulking behavior. **4.** Inclination : disposition.

**mood[2]** (mōōd) *n.* [Alteration of MODE.] **1.** A verb form or set of verb forms inflected to indicate the manner in which the action or state expressed by a verb is viewed concerning functions such as factuality, possibility, or command. **2.** *Logic.* Arrangement or form of a proposition.

**mood·y** (mōō′dē) *adj.* **-i·er, -i·est. 1.** Given to changeable emotional states, esp. of gloom. **2.** Gloomy : uneasy <a *moody* glance> **—mood′i·ly** *adv.* **—mood′i·ness** *n.*

☆ ***syns:*** MOODY, TEMPERAMENTAL *adj. core meaning* : given to changeable emotional states, as of anger or gloom <a *moody* poet>

**moo goo gai pan** (mōō′ gōō′ gī′ păn′) *n.* [Cantonese, corresponding to Mandarin *mu*[2] *gu*[2] *ji*[1] *pian*[1] : *mu*[2] *gu*[2], mushroom + *ji*[1], chicken + *pian*[1], slice.] A Cantonese dish of mushrooms, chicken, vegetables, and spices steamed together.

**moo·la** or **moo·lah** (mōō′lə) *n.* [Orig. unknown.] *Slang.* Money.

**moon** (mōōn) *n.* [ME *moone* < OE *mōna.*] **1.** The natural satellite of the earth, varying in distance from the earth between 221,600 miles or 356,000 kilometers and 252,950 miles or 406,997 kilometers, having a mean diameter of 2,160 miles or 3,475 kilometers, a mass approx. 1/80 that of the earth, and an average period of revolution around the earth of 29 days 12 hours 44 minutes. **2.** A natural satellite revolving around a planet. **3.** The moon as it appears at a particular time in its cycle of phases <the new *moon*><a crescent *moon*> **4.** A month, esp. a lunar month. **5.** A disk, globe, or crescent similar to the moon. **6.** Moonlight. *—vi.* **mooned, moon·ing, moons. 1.** To wander about or pass time aimlessly and languidly. **2.** To display infatuation. **—moon′like′** *adj.*

**moon·beam** (mōōn′bēm′) *n.* A ray of moonlight.

**moon·blind** (mōōn′blīnd′) *adj.* Suffering from moon blindness.

**moon blindness** *n.* **1.** Recurrent inflammation of a horse's eyes, often leading to eventual blindness. **2.** Night blindness.

**moon·calf** (mōōn′kăf′, -käf′) *n.* **1.** A fool. **2.** A freak.

**moon dog** *n.* A paraselene.

**moon·eye** (mōōn′ī′) *n.* **1.** A silvery freshwater fish, *Hiodon tergisus* of northern North America. **2.** MOON BLINDNESS 1.

**moon-faced** (mōōn′fāst′) *adj.* Having a round face.

**moon·fish** (mōōn′fĭsh′) *n., pl.* **moonfish** or **-fish·es. 1.** Either of two marine fishes of the family Carangidae of warm American coastal waters, with short, compressed, silver to yellowish bodies. **2.** The opah.

**moon·flow·er** (mōōn′flou′ər) *n.* Any of several night-blooming vines related to the morning-glories.

**Moon·ie** (mōō′nē) *n.* A member of the Unification Church established and headed by Sun Myung Moon.

**moon·light** (mōōn′līt′) *n.* Light reflected from the moon's surface. *—vi.* **-light·ed, -light·ing, -lights.** *Informal.* To work at another job, often at night, in addition to a full-time job. **—moon′light′er** *n.*

**moon·lit** (mōōn′lĭt′) *adj.* That is lighted by the moon.

**moon·scape** (mōōn′skāp′) *n.* **1.** A view or picture of the surface of the moon. **2.** A desolate landscape.

**moon·seed** (mōōn′sēd′) *n.* A climbing vine of the genus *Menispermum* or related genera, bearing red or blackish fruit with crescent-shaped or ring-shaped seeds.

**moon shell** *n.* Any of various marine gastropod mollusks of the family Naticidae, with smooth spherical shells.

**moon·shine** (mōōn′shīn′) *n.* **1.** Moonlight. **2.** *Informal.* Foolish or nonsensical talk, thought, or action. **3.** *Slang.* Illegally distilled whiskey. *—v.* **-shined, -shin·ing, -shines.** *—vt.* To distill (liquor) illegally. *—vi.* To operate an illegal still. **—moon′shin′er** *n.*

**moon·shot** (mōōn′shŏt′) *n.* Launch of a spacecraft on a mission to the moon.

**moon·stone** (mōōn′stōn′) *n.* A feldspar prized as a gem for its pearly translucence.

**moon·struck** (mōōn′strŭk′) *also* **moon·strick·en** (-strĭk′ən) *adj.* [From the belief that the moon caused insanity.] **1.** Afflicted with insanity. **2.** Distracted or dazed with romantic sentiment.

**moon·walk** (mōōn′wôk′) *n.* An exploratory walk on the moon by an astronaut. **—moon′-walk′** *v.* **(-walked, -walk·ing, -walks). —moon′walk·er** *n.*

**moon·ward** (mōōn′wərd) *adv.* Toward the moon.

**moon·wort** (mōōn′wûrt′, -wôrt′) *n.* A grape fern.

**moon·y** (mōō′nē) *adj.* **-i·er, -i·est. 1. a.** Like the moon. **b.** Like moonlight. **2.** Moonlit. **3.** Given to reverie : DREAMY.

**moor[1]** (mŏŏr) *v.* **moored, moor·ing, moors.** [ME *moren* < MLG *mōręn.*] *—vt.* **1.** To secure or make fast (e.g., a vessel) by cables, anchors, or lines. **2.** To fix in place : SECURE. *—vi.* **1.** To secure a vessel or aircraft. **2.** To be secured, as a vessel.

**moor[2]** (mŏŏr) *n.* [ME *mor* < OE *mōr.*] A broad tract of open land, often high but poorly drained, with patches of heath and peat bogs.

**Moor** (mŏŏr) *n.* [ME *More* < OFr. < Lat. *Maurus.*] **1.** One of a Moslem people of mixed Berber and Arab descent, now living chiefly in northern Africa. **2.** One of the Saracens who invaded Spain in the 8th cent. A.D. **—Moor′ish** *adj.*

**moor·age** (mŏŏr′ĭj) *n.* **1.** A place where a ship may be moored. **2.** The act of mooring or state of being moored. **3.** A charge for the use of mooring facilities.

**moor·hen** (mŏŏr′hĕn′) *n.* *Chiefly Brit.* A widely distributed gallinule, *Gallinula chloropus.*

**moor·ing** (mŏŏr′ĭng) *n.* **1.** Equipment, as anchors or chains, for making fast a vessel or aircraft. **2.** The act of securing a vessel or aircraft. **3.** A place at which a vessel or aircraft can be moored. **4.** *often* **moorings.** Elements supplying security or stability <lost my psychological *moorings* when the truth was revealed>

**moor·wort** (mŏŏr′wûrt′, -wôrt′) *n.* The bog rosemary.

**moose** (mōōs) *n., pl.* **moose.** [Natick *moos.*] A hoofed mammal of the deer family, *Alces alces* or *A. americana,* dwelling in forests of northern North America and in Eurasia, with a broad pendulous muzzle and large flat antlers in the male.

**moose·bird** (mōōs′bûrd′) *n.* The Canada jay.

**moose·wood** (mōōs′wŏŏd′) *n.* A slender maple, *Acer pennsylvanicum* of eastern North America, having smooth bark with vertical whitish or greenish stripes.

**moot** (mōōt) *n.* [ME < OE *mōt.*] **1.** An ancient English meeting, esp. a representative meeting of the freemen of a shire. **2.** A hypothetical case argued as an exercise by law students. *—vt.* **moot·ed, moot·ing, moots. 1. a.** To bring up as a subject for debate or discussion.

**b.** To debate or discuss. **2.** To plead or argue (a case) in a moot court. —*adj.* **1.** Subject to debate : ARGUABLE <a *moot* point> **2. a.** *Law.* Lacking legal significance, through having been previously decided or settled. **b.** Of no practical importance : ACADEMIC.

**moot court** *n.* A mock court where hypothetical cases are tried for the training of law students.

**mop** (mŏp) *n.* [ME *mappe.*] **1.** A cleaning implement made of absorbent material fastened to a handle and used for dusting, washing, and drying floors. **2.** A loosely tangled bunch or mass <a *mop* of curls> —*vt.* **mopped, mop·ping, mops.** To wash or wipe with or as if with a mop. **—mop up. 1.** To clear (an area) of enemy troops remaining after a victory. **2.** *Informal.* To finish a task or action. **—mop'per** *n.*

**mope** (mōp) *vi.* **moped, mop·ing, mopes.** [Orig. unknown.] **1. a.** To be dejected or gloomy. **b.** To sulk or brood. **2.** To move in a leisurely or aimless way : DAWDLE. —*n.* **1.** One given to dejected or gloomy moods. **2. mopes.** Low spirits. **—mop'er** *n.* **—mop'ing·ly** *adv.* **—mop'ish** *adj.* **—mop'ish·ly** *adv.*

**mo·ped** (mō'pĕd') *n.* [MO(TOR) + PED(AL).] A motor-driven two-wheeled vehicle that can also be pedaled.

**mop·pet** (mŏp'ĭt) *n.* [< obs. *mop,* fool, child < ME.] A young child.

**mop-up** (mŏp'ŭp') *n.* A finishing action.

**mo·quette** (mō-kĕt') *n.* [Fr., alteration of *moucade.*] **1.** A heavy fabric with a thick nap, used for upholstery. **2.** Carpet with a deep tufted pile.

**mo·ra** (môr'ə, mōr'ə) *n., pl.* **mo·rae** (môr'ē, mōr'ē) *or* **mo·ras.** [< Lat., pause.] The unit of metrical time in quantitative verse equal to the short syllable.

**mo·raine** (mə-rān') *n.* [Fr.] Debris, as boulders or stones, deposited by a glacier. **—mo·rain'al, mo·rain'ic** *adj.*

**mor·al** (môr'əl, mŏr'-) *adj.* [ME < OFr. < Lat. *moralis* < *mos,* custom.] **1.** Of or concerned with the principles of right and wrong in relation to human action and character. **2.** Teaching or displaying good or correct character and behavior <a *moral* lesson> **3.** In accord with standards of what is right or just in behavior : VIRTUOUS. **4.** Arising from conscience or the sense of right and wrong <a *moral* duty> **5.** Having psychological rather than physical or tangible effects <a *moral* success> **6.** Based on strong likelihood or conviction rather than on solid evidence <a *moral* certainty> —*n.* **1.** The lesson or principle contained in or taught by a fable, story, or event. **2.** A concisely expressed precept or general truth : MAXIM. **3. morals.** Rules or habits of conduct, esp. sexual conduct, with regard to standards of right and wrong. **—mor'al·ly** *adv.*

**mo·rale** (mə-răl') *n.* [Fr. < fem. of *moral,* moral < OFr. —see MORAL.] A strong sense of enthusiasm and dedication to a commonly shared goal that unifies a group.

☆ ***syns:*** MORALE, ESPRIT, ESPRIT DE CORPS *n. core meaning* : a strong sense of enthusiasm and dedication to a common goal that unites a group <The defeats never broke the team's *morale.*>

**moral hazard** *n.* A risk to an insurance company arising from doubt about the integrity of the insured.

**mor·al·ism** (môr'ə-lĭz'əm, mŏr'-) *n.* **1.** A conventional moral maxim or attitude. **2.** The act or practice of moralizing. **3.** An often excessive concern for morality.

**mor·al·ist** (môr'ə-lĭst, mŏr'-) *n.* **1.** A teacher or student of morals and moral problems. **2.** One who follows a system of moral principles. **3.** One unduly concerned with the morals of others.

**mor·al·is·tic** (môr'ə-lĭs'tĭk, mŏr'-) *adj.* **1.** Marked by or showing an interest in or concern with morality. **2.** Marked by a narrow-minded morality. **—mor'al·is'ti·cal·ly** *adv.*

**mo·ral·i·ty** (mə-răl'ĭ-tē, mô-) *n., pl.* **-ties. 1.** The quality of being in accord with standards of good or right conduct. **2.** A system of ideas of right and wrong conduct. **3.** Virtuous conduct. **4.** A rule or lesson in moral conduct.

**morality play** *n.* An allegorical play of the 15th and 16th cent. in which the characters personify virtues and vices.

**mor·al·ize** (môr'ə-līz', mŏr'-) *v.* **-ized, -iz·ing, -iz·es.** —*vt.* **1.** To explain or interpret the moral meaning of. **2.** To improve the morals of : REFORM. —*vi.* To think about or discuss moral or ethical issues. **—mor'al·i·za'tion** *n.* **—mor'al·iz'er** *n.*

**moral philosophy** *n.* Ethics.

**mo·rass** (mə-răs', mô-) *n.* [Du. *moeras* < OFr. *maresc,* prob. of Germanic orig.] **1.** Low-lying, soggy ground : BOG. **2.** Something that hinders, engulfs, or overwhelms.

**mor·a·to·ri·um** (môr'ə-tôr'ē-əm, -tōr'-, mŏr'-) *n., pl.* **-to·ri·ums** *or* **-to·ri·a** (-tôr'ē-ə, -tōr'ē-ə) [< LLat., neuter of *moratorius,* delaying. —see MORATORY.] **1.** *Law.* An authorization to a debtor, as a bank or nation, permitting temporary suspension of payments. **2.** A suspension of action.

**mor·a·to·ry** (môr'ə-tôr'ē, -tōr'ē, mŏr'-) *adj.* [Fr. *moratoire* < LLat. *moratorius,* delaying < Lat. *morari,* to delay < *mora,* delay.] Authorizing delay in payment.

**Mo·ra·vi·an** (mə-rā'vē-ən) *n.* **1.** A native or inhabitant of Moravia. **2.** The Czech dialects of Moravia. **3.** A member of a Protestant denomination founded in Saxony in 1722 by Hussite emigrants from Moravia. **—Mo·ra'vi·an** *adj.*

**mo·ray** (môr'ā, mə-rā') *n.* [Port. *moreia* < Lat. *murena* < Gk. *muraina.*] Any of various often voracious marine eels of the family Muraenidae of chiefly tropical coastal waters.

**mor·bid** (môr'bĭd) *adj.* [Lat. *morbidus,* diseased < *morbus,* disease.] **1. a.** Of, pertaining to, or resulting from disease. **b.** Psychologically unhealthy <a *morbid* fear of elevators> **2.** Subject to or excessively concerned with unwholesome matters. **3.** Grisly : gruesome. **—mor·bid'i·ty, mor'bid·ness** *n.* **—mor'bid·ly** *adv.*

☆ ***syns:*** MORBID, MACABRE, SICK, SICKLY, UNHEALTHY, UNWHOLESOME *adj. core meaning* : susceptible to or marked by preoccupation with unwholesome matters <a *morbid* interest in torture>

**mor·da·cious** (môr-dā'shəs) *adj.* [Lat. *mordax, mordac-* < *mordēre,* to bite.] **1.** Prone to biting. **2.** Sarcastic : caustic. **—mor·da'cious·ly** *adv.* **—mor·dac'i·ty** (-dăs'ĭ-tē) *n.*

**mor·dant** (môr'dnt) *adj.* [Fr. < OFr. < pr.part. of *mordre,* to bite < Lat. *mordēre,* to bite.] **1. a.** Bitingly sarcastic. **b.** Trenchant and incisive. **2.** Bitingly painful. **3.** Suitable for fixing colors in dyeing. —*n.* **1.** A reagent, as tannic acid, for fixing coloring matter in textiles, leather, or other material. **2.** A corrosive substance, as an acid, used in etching. —*vt.* **-dant·ed, -dant·ing, -dants.** To treat with a mordant. **—mor'dan·cy** *n.* **—mor'dant·ly** *adv.*

**mor·dent** (môr'dnt, môr-dĕnt') *n.* [G. < Ital. *mordente* < *mordere,* to bite < Lat. *mordēre.*] *Mus.* A melodic ornament in which a principal note is rapidly alternated with a note a half or full step below.

**more** (môr, mōr) *adj.* [ME < OE *māra.*] **1. a.** Greater in number. **b.** Greater in size, amount, extent, or degree. **2.** Extra : additional <We need *more* heat.> —*n.* An additional or greater quantity, number, degree, or amount <*More* of them are expected.> —*pron.* **1.** Something greater or better <*more* for your money> **2.** *(pl. in number).* A greater number of persons or things <hired six workers but *more* will be needed> —*adv.* **1.** To or in a greater extent or degree. —Used to form the comparative of many adjectives and adverbs <*more* eager><*more* readily> **2.** In addition. **—more and more.** To a growing extent or degree <I dislike them *more and more.*> **—more or less. 1.** About : approximately. **2.** To an indefinite degree.

**mo·reen** (mə-rēn', mô-) *n.* [Poss. < MOIRE.] A sturdy ribbed fabric of wool, cotton, or wool and cotton often with an embossed finish, used for upholstery and clothing.

**mo·rel** (mə-rĕl', mô-) *n.* [Fr. *morille* < OFr., prob. of Germanic orig.] An edible mushroom of the genus *Morchella* or related genera, marked by a brownish spongelike cap.

**mo·rel·lo** (mə-rĕl'ō) *n., pl.* **-los.** [Prob. < Ital. *amarello* < Med. Lat. *amarellum,* dim. of Lat. *amarus,* bitter.] A variety of the sour cherry, *Prunus cerasus austera,* bearing fruit with dark-red skin.

**more·o·ver** (môr-ō'vər, mōr-, môr'ō'vər, mōr'-) *adv.* Beyond what has already been stated : BESIDES.

**mo·res** (môr'āz', -ēz, mōr'-) *pl.n.* [Lat., pl. of *mos,* custom.] **1.** The approved traditional customs and usages of a particular social group. **2.** Moral attitudes. **3.** Ways : manners.

**Mo·resque** (mô-rĕsk', mə-) *adj.* [Fr.< Sp. *Morisco,* Morisco.] Typical of Moorish art and architecture. —*n.* An ornament or decoration in Moorish style.

**Mor·gan** (môr'gən) *n.* [After Justin *Morgan* (1747–1798).] A saddle and trotting horse orig. bred in America.

**mor·ga·nat·ic** (môr'gə-năt'ĭk) *adj.* [NLat. *morganaticus* < Med. Lat. *matrimonium ad morganaticam,* marriage for the morning-gift.] Of or relating to a legal marriage between one of royal or noble birth and a partner of lower rank, in which no titles or estates of the royal or noble partner may be claimed by the partner of inferior rank nor by any of the offspring of the marriage. **—mor'ga·nat'i·cal·ly** *adv.*

**mor·gan·ite** (môr'gə-nīt') *n.* [After John Pierpont *Morgan* (1837–1913).] A rosy-pink silicate of beryllium and aluminum valued as a semiprecious gem.

**Mor·gan le Fay** (môr'gən lə fā') *n.* [OFr. *Morgain la fee,* Morgan the fairy.] The sorceress sister and enemy of King Arthur according to Arthurian legend.

**mor·gen** (môr'gən) *n., pl.* **morgen** *or* **-gens.** [Du. < MDu. *morghen,* morning.] A Dutch and South African unit of land area equal to 2.116 acres.

**morgue** (môrg) *n.* [Fr.] **1.** A place in which the bodies of persons found dead are kept until identified and claimed or until burial arrangements have been made. **2.** A newspaper or magazine reference file.

**mor·i·bund** (môr'ə-bŭnd', mŏr'-) *adj.* [Lat. *moribundus* < *mori,* to die.] At or near the point of death. **—mor'i·bun'di·ty** *n.* **—mor'i·bund'ly** *adv.*

**mo·ri·on**[1] (môr'ē-ŏn', mōr'-) *n.* [Fr. < Sp. *morrion* < *morro,* round object.] A crested metal helmet with a curved peak in front and back, worn by 16th- and 17th-cent. soldiers.

**mo·ri·on**[2] (môr'ē-ŏn', mōr'-) *n.* [Alteration of Lat. *mormorion.*] An often nearly black smoky quartz.

**Mo·ris·co** (mə-rĭs'kō) *n., pl.* **-cos** *or* **-coes.** [Sp. < *Moro,* Moor < Lat. *Maurus.*] A Spanish Moor. **—Mo·ris'co** *adj.*

**Mor·mon** (môr'mən) *n.* **1.** *Mormon Ch.* A prophet, warrior, and historian revealed to Joseph Smith as the author of a sacred history of the Americas which Smith translated as the Book of Mormon. **2.** A member of the Church of Jesus Christ of Latter-day Saints, founded by Joseph Smith in 1830. **—Mor'mon·ism** *n.*

**morn** (môrn) *n.* [ME < OE *morgen.*] **1.** The morning. **2.** The dawn.

**morn·ing** (môr'nĭng) *n.* [ME < *morn,* morn < OE *morgen.*] **1.** The early or first part of the day, stretching from midnight to noon or from sunrise to noon. **2.** The dawn. **3.** The first or early part.

**morn·ing-glo·ry** (môr'nĭng-glôr'ē, -glōr'ē) *n.* Any of various usu. twining vines of the genus *Ipomoea,* bearing funnel-shaped, variously colored flowers that close late in the day.

**Morning Prayer** *n.* The morning liturgy in the Anglican Church.

**morn·ings** (môr'nĭngz) *adv.* Regularly in the morning.

**morning sickness** *n.* Nausea and vomiting, often upon rising in the morning, esp. during the early stages of pregnancy.

**morning star** *n.* A planet, esp. Venus, seen in the east just before or at sunrise.

**Mo·ro** (môr'ō, mōr'ō) *n., pl.* **-ros.** [Sp. < Lat. *Maurus,* Moor.] **1.** A member of the Moslem Malay tribes of the southern Philippines. **2.** Any of the Austronesian languages of the Moro. **—Mo'ro** *adj.*

**mo·roc·co** (mə-rŏk'ō) *n., pl.* **-cos.** [After *Morocco,* where it was orig. made.] A soft fine leather of goatskin tanned with sumac.

**mo·ron** (môr'ŏn', mōr'-) *n.* [< Gk. *mōron,* neuter of *mōros,* stupid.] **1.** A mentally retarded person with a mental age between 7 and 12 years or an intelligence quotient between 50 and 75. **2.** A particularly stupid person. **—mo·ron'ic** (mə-rŏn'ĭk, mô-) *adj.* **—mo·ron'i·cal·ly** *adv.* **—mo'ron'ism, mo·ron'i·ty** (mə-rŏn'ĭ-tē, mô-) *n.*

**mo·rose** (mə-rōs', mô-) *adj.* [Lat. *morosus,* peevish < *mos,* manner.] **1.** Sullenly melancholy. **2.** Marked by or showing gloom. **—mo·rose'ly** *adv.* **—mo·rose'ness** *n.*

**morph** (môrf) *n.* **1.** An allomorph. **2.** A phoneme or sequence of phonemes assumed to be an allomorph although its assignment to a particular morpheme has not been established.

**morph-** *pref. var. of* MORPHO-.

**-morph** *suff.* [Gk. *-morphos* < *morphē,* shape.] **1.** Form : shape : structure <endo*morph*> **2.** Morpheme <allo*morph*>

**mor·phal·lax·is** (môr'fə-lăk'sĭs) *n.* [NLat. : MORPH(O)- + Gk. *allaxis,* exchange < *allassein,* to exchange < *allos,* other.] *Biol.* Regeneration of a part or transformation of one part into another by means of structural reorganization with only limited production of new cells, a process observed chiefly in invertebrate organisms, as certain lobsters.

**mor·pheme** (môr'fēm') *n.* [Fr. *morphème* < Gk. *morphē,* form.] A meaningful linguistic unit consisting of a word, as *man,* or a word element, as *-ed* of *walked,* that cannot be divided into smaller meaningful parts. **—mor·phem'ic** *adj.* **—mor·phem'i·cal·ly** *adv.*

**mor·phem·ics** (môr-fē'mĭks) *n. (sing. in number).* **1.** The study of morphemes as a branch of linguistic analysis. **2.** Morphemic structure of a language.

**Mor·phe·us** (môr'fē-əs, -fyōōs') *n.* [Lat.] The god of dreams in Ovid's *Metamorphoses.* **—Mor'phe·an** (-fē-ən) *adj.*

**mor·phi·a** (môr'fē-ə) *n.* [NLat. < Lat. *Morpheus,* Morpheus.] Morphine.

**mor·phic** (môr'fĭk) *adj.* Relating to form : MORPHOLOGICAL. **—mor'phi·cal·ly** *adv.*

**-morphic** *suff.* Having a specified form <geo*morphic*>

**mor·phine** (môr'fēn') *n.* [Fr. < *Morphée,* Morpheus < Lat. *Morpheus.*] An organic compound, $C_{17}H_{19}NO_3$, derived from opium, the soluble salts of which are used in human and veterinary medicine as a sedative or light anesthetic.

**mor·phin·ism** (môr'fē-nĭz'əm, môr'fə-) *n.* **1.** Morphine addiction. **2.** Poisoning produced by immoderate dosage of morphine.

**-morphism** *suff.* The state or quality of having a specified form <homo*morphism*>

**morpho-** or **morph-** *pref.* [G. < Gk. < *morphē,* shape.] **1.** Form : shape : structure <*morpho*genesis> **2.** Morpheme <*morpho*phonemics>

**mor·pho·gen·e·sis** (môr'fō-jĕn'ĭ-sĭs) *n.* Evolutionary or embryological development of the structure of an organism or part. **—mor'pho·ge·net'ic** (-jə-nĕt'ĭk), **mor'pho·gen'ic** *adj.* **—mor'pho·ge·net'i·cal·ly** *adv.*

**mor·phol·o·gy** (môr-fŏl'ə-jē) *n.* [G. *Morphologie* : Gk. *morphē,* shape + Gk. *logos,* word, discussion.] **1.** Study of the structure and form of living organisms. **2.** An organism's structure and form, excluding its functions. **3.** *Geol.* Geomorphology. **4.** Study of word formation in a language, including inflection, derivation, and the formation of compounds. **—mor'pho·log'i·cal** (-fə-lŏj'ĭ-kəl), **mor'pho·log'ic** *adj.* **—mor'pho·log'i·cal·ly** *adv.* **—mor·phol'o·gist** *n.*

**mor·pho·pho·ne·mics** (môr'fō-fə-nē'mĭks) *n. (sing. in number).* **1.** Linguistic structure in terms of the phonological patterning of morphemes, as through variations, such as the addition, loss, or substitution of phonemes, including stress shifts, which determine the different shapes of morphemically related words. **2.** Study of the morphophonemics of a language. **—mor'pho·pho·ne'mic** *adj.*

**mor·pho·sis** (môr-fō'sĭs) *n., pl.* **-ses** (-sēz') [Gk. *morphōsis,* process of forming < *morphoun,* to form < *morphē,* form.] The manner in which an organism or one of its parts changes form or the manner or order of its development.

**-morphous** *suff.* Having a specified form <hetero*morphous*>

**mor·ris** (môr'ĭs, mŏr'-) *n.* [ME *moreys (dance),* morris (dance) < *moreys,* Moorish < *More,* Moor. —see MOOR.] An English folk dance in which costumed dancers act out a story.

**Mor·ris chair** (môr'ĭs, mŏr'-) *n.* [After William *Morris* (1834–1896).] A large easy chair with arms, an adjustable back, and removable cushions.

**mor·row** (môr'ō, mŏr'ō) *n.* [ME *morow* < OE *morgen.*] **1.** The day after a specified day. **2.** The time just following a particular event. **3.** *Archaic.* The morning.

**Morse code** (môrs) *n.* [After Samuel F. B. *Morse* (1791–1872).] One of the codes in which short and long elements, esp. in the form of sounds or flashes of light, represent letters of the alphabet and numbers.

### MORSE CODE

| | | | | | | | |
|---|---|---|---|---|---|---|---|
| A | ·– | N | –· | Á | ·––·– | 8 | –––·· |
| B | –··· | O | ––– | Ä | ·–·– | 9 | ––––· |
| C | –·–· | P | ·––· | É | ··–·· | 0 | ––––– |
| D | –·· | Q | ––·– | Ñ | ––·–– | ,(comma) | ––··–– |
| E | · | R | ·–· | Ö | –––· | .(period) | ·–·–·– |
| F | ··–· | S | ··· | Ü | ··–– | ? | ··––·· |
| G | ––· | T | – | 1 | ·–––– | ; | –·–·–· |
| H | ···· | U | ··– | 2 | ··––– | : | –––··· |
| I | ·· | V | ···– | 3 | ···–– | / | –··–· |
| J | ·––– | W | ·–– | 4 | ····– | -(hyphen) | –····– |
| K | –·– | X | –··– | 5 | ····· | apostrophe | ·––––· |
| L | ·–·· | Y | –·–– | 6 | –···· | parenthesis | –·––·– |
| M | –– | Z | ––·· | 7 | ––··· | underline | ··––·– |

**mor·sel** (môr'səl) *n.* [ME < OFr., dim. of *mors,* bite < Lat. *morsus* < *mordēre,* to bite.] **1.** A small piece or bite of food. **2.** A small piece or amount. **3.** A tasty delicacy. **4.** One that is extremely pleasing.

**mort¹** (môrt) *n.* [ME, death < OFr. < Lat. *mors.*] The note blown on a hunting horn to announce the killing of a deer.

**mort²** (môrt) *n.* [Poss. < MORTAL.] A great number or quantity.

**†mor·tal** (môr'tl) *adj.* [ME < OFr. < Lat. *mortalis* < *mors,* death.] **1.** Liable or subject to death. **2.** Of or relating to human beings. **3.** Of, relating to, or associated with death <*mortal* pain> **4.** Causing death : FATAL <a *mortal* wound> **5.** Implacable : unrelenting <a *mortal* enemy> **6.** Of great intensity or severity : DIRE <in *mortal* fear> **7.** Conceivable <an item of no *mortal* use> **8.** —Used as an intensifier <kept us waiting for six *mortal* hours> —*n.* A human being. —*adv. Regional.* Extremely : very. **—mor'tal·ly** *adv.*

**mor·tal·i·ty** (môr-tăl'ĭ-tē) *n., pl.* **-ties. 1.** The quality or state of being mortal. **2.** Frequency of number of deaths in proportion to a population : DEATH RATE. **3.** *Archaic.* Death. **4.** *Archaic.* Deadliness.

**mortality table** *n.* An actuarial table based on death statistics over a certain number of years.

**mortal sin** *n. Rom. Cath. Ch.* A serious and deliberate sin, as murder or suicide, that robs the soul of sanctifying grace.

**mor·tar** (môr'tər) *n.* [ME *morter* < OE *mortere* and OFr. *mortier,* both < Lat. *mortarium.*] **1.** A vessel in which substances are crushed or ground with a pestle. **2.** A machine for grinding or blending materials. **3. a.** *also* **trench mortar.** A muzzle-loading cannon for firing shells at low velocities, short ranges, and high trajectories. **b.** A device for shooting life lines over a stretch of water. **4.** A mix of cement or lime with sand and water used in building. —*vt.* **-tared, -tar·ing, -tars.** To plaster or join with mortar.

**mor·tar·board** (môr'tər-bôrd', -bōrd') *n.* **1.** A square board with a handle for holding and carrying mortar. **2.** An academic cap topped by a flat square.

**mort·gage** (môr'gĭj) *n.* [ME *morgage* < OFr. : *mort,* dead (< Lat. *mortuus* < *mors,* death) + *gage,* pledge, of Germanic orig.] **1.** A temporary and conditional pledge of property to a creditor as security for the performance of an obligation or repaying of a debt. **2.** A contract or deed defining the terms of a mortgage. **3.** The claim of the mortgagee upon mortgaged property. —*vt.* **-gaged, -gag·ing, -gag·es. 1.** To pledge or convey with a mortgage. **2.** To make subject to a pledge or claim.

**mort·ga·gee** (môr'gĭ-jē') *n.* A mortgage holder.

**mort·ga·gor** (môr'gĭ-jôr', môr'gĭ-jər) *also* **mort·gag·er** (môr'gĭ-jər) *n.* One who mortgages one's property.

**mor·tice** (môr'tĭs) *n. & v. var. of* MORTISE.

**mor·ti·cian** (môr-tĭsh'ən) *n.* [MORT(UARY) + -ICIAN.] A funeral director : UNDERTAKER.

**mor·ti·fi·ca·tion** (môr'tə-fĭ-kā'shən) *n.* **1.** A feeling of shame, humiliation, or wounded pride. **2.** The mortifying of the body and appetites. **3.** Death or decay of one part of a living body : GANGRENE.
**mor·ti·fy** (môr'tə-fī') *v.* **-fied, -fy·ing, -fies.** [ME *mortifien* < OFr. *mortifier* < LLat. *mortificare,* to kill : *mors,* death + *facere,* to make.] —*vt.* **1.** To cause to experience shame, humiliation, or wounded pride : HUMILIATE. **2.** To discipline (one's body and physical appetites) by austerity and self-denial. —*vi.* **1.** To practice ascetic discipline or self-denial of the body and its appetites. **2.** To become gangrenous or necrotic. **—mor'ti·fy'ing** *adj.*
**mor·tise** *also* **mor·tice** (môr'tĭs) [ME *mortays* < OFr. *mortoise.*] —*n.* **1.** A usu. rectangular cavity in a piece of material, as wood or stone, for receiving a tenon. **2.** A hole cut in a printing plate for inserting type. —*vt.* **-tised, -tis·ing, -tis·es** *also* **-ticed, -tic·ing, -tic·es. 1.** To join or fasten securely, as with a mortise and tenon. **2.** To cut or make a mortise in. **3. a.** To cut a hole in (a printing plate) for inserting type. **b.** To cut such a hole and insert (type).

**mortise**

**mort·main** (môrt'mān') *n.* [ME *mortemayne* < OFr. *mortemain* : *morte,* fem. of *mort,* dead (< Lat. *mortuus* < *mors,* death) + *main,* hand < Lat. *manus.*] **1.** *Law.* Perpetual ownership of real estate by institutions, as churches, that cannot sell or transfer them. **2.** The often heavy weight of the past upon the present.
**mor·tu·ar·y** (môr'cho͞o-ĕr'ē) *n., pl.* **-ies.** [ME *mortuarie* < LLat. *mortuarium* < *mortuarius,* of burial < Lat. *mortuus,* dead < *mors,* death.] A place, esp. a funeral home, for keeping dead bodies before burial or cremation.
**mor·u·la** (môr'yə-lə- môr'ə-) *n., pl.* **-lae** (-lē') [NLat. < Lat. *morum,* mulberry.] **1.** The spherical embryonic mass of blastomeres formed before complete blastulation. **2.** A spherical mass of developing male gametes found esp. in certain annelid worms. **—mor'u·lar** (-lər) *adj.* **—mor'u·la'tion** *n.*
**mo·sa·ic** (mō-zā'ĭk) *n.* [ME *musycke* < OFr. *mosaique* < OItal. *mosaico* < Med. Lat. *musaicus* < Gk. *mouseios,* of the Muses < *Mousa,* Muse.] **1.** A decorative design or picture made by setting small colored pieces, as tile, in mortar. **2.** Something resembling a mosaic <a *mosaic* of ideas> **3.** A virus disease of plants, causing light and dark areas in the leaves, which often shrivel and become dwarfed. **4.** Overlapping photographs, usu. aerial, assembled into a composite picture. **5.** A photosensitive surface in the iconoscope of a television camera. —*vt.* **-icked, -ick·ing, -ics. 1.** To make a mosaic. **2.** To adorn with or as if with mosaic. **—mo·sa'i·cist** (mō-zā'ĭ-sĭst) *n.*
**Mo·sa·ic** (mō-zā'ĭk) *adj.* Of or relating to Moses or the laws and writings ascribed to him.
**mosaic gold** *n.* ORMOLU 1.
**mo·sa·i·cism** (mō-zā'ĭ-sĭz'əm) *n.* The condition in which tissues of genetically different types are present in the same organism.
**Mosaic Law** *n.* The Pentateuch.
**mos·cha·tel** (mŏs'kə-tĕl', mŏs'kə-tĕl') *n.* [Fr. *moscatelle* < Ital. *moscatella* < *moscato,* musk < LLat. *muscus.* —see MUSK.] A plant, *Adoxa moschatellina* of northern regions, bearing greenish-white, musk-scented flowers.
**Mo·selle** (mō-zĕl') *n.* A light, dry white wine made in the valley of the Moselle River.
**mo·sey** (mō'zē) *vi.* **-seyed, -sey·ing, -seys.** [Orig. unknown.] *Informal.* **1.** To move in a leisurely, idle fashion. **2.** To move along.
**mo·shav** (mō-shäv') *n., pl.* **mo·sha·vim** (mō'shə-vēm') [Mod. Heb. *mōshābh* < Heb., dwelling.] An Israeli cooperative settlement composed of small farms.
**Mos·lem** (mŏz'ləm, mŏs'-) *n.* [Ar. *muslim.*] A believer in or adherent of Islam. **—Mos'lem** *adj.*
**mosque** (mŏsk) *n.* [Fr. *mosquée* < OItal. *moschea* < Ar. *masjid* < *sajada,* to worship.] A Moslem house of worship.
**mos·qui·to** (mə-skē'tō) *n., pl.* **-toes** *or* **-tos.** [Sp. < dim. of *mosca,* fly < Lat. *musca.*] Any of various winged insects of the family Culicidae, in which the female of most species has a long proboscis for sucking blood and some species of which are vectors of diseases such as malaria and yellow fever.
**Mosquito** *n., pl.* **-to** *or* **-tos. 1.** A South American people living in Nicaragua and Honduras. **2.** The language of the Mosquito.
**mosquito boat** *n. Chiefly Brit.* A PT boat.
**mosquito hawk** *n.* NIGHTHAWK 1.
**mosquito net** *n.* A fine net or screen for keeping out mosquitoes.
**moss** (môs, mŏs) *n.* [ME < OE *mos.*] **1. a.** Any of various green, usu. small plants of the class Musci within the division Bryophyta. **b.** A patch or covering of such plants. **2.** A plant similar to moss in appearance or growth, as the club moss, Irish moss, or Spanish moss. **3.** *Chiefly Scot.* A peat bog or moor. **—moss'like'** *adj.*
**moss·back** (môs'băk', mŏs'-) *n.* **1.** An old shellfish or turtle with algae growing on its back. **2.** *Slang.* A very conservative or old-fashioned person. **—moss'backed'** *adj.*
**Möss·bau·er effect** (môs'bou'ər) *n.* [After Rudolf *Mössbauer* (b. 1929).] The recoilless fluorescence of gamma rays with an extremely narrow frequency range from atomic nuclei bound in solids.
**moss·bunk·er** (môs'bŭng'kər, mŏs'-) *n.* [Du. *marsbanker.*] The menhaden.
**moss campion** *n.* A low-growing plant, *Silene acaulis* of cool regions, bearing purplish-red flowers and forming dense, cushionlike mats.
**moss green** *n.* A moderate yellow green to grayish or moderate olive or dark yellowish green. **—moss'-green'** *adj.*
**moss·grown** (môs'grōn', mŏs'-) *adj.* **1.** Overgrown with moss. **2.** Old-fashioned.
**mos·so** (môs'sō) *adv.* [Ital. < *muovere,* to move < Lat. *movēre.*] *Mus.* With motion or animation. —Used as a direction.
**moss pink** *n.* A low-growing plant, *Phlox subulata,* forming dense mosslike mats and widely cultivated for its abundant pink or white flowers.
**moss rose** *n.* A variety of rose, *Rosa centifolia muscosa,* bearing fragrant pink flowers with a mossy flower stalk and calyx.
**moss-troop·er** (môs'tro͞o'pər, mŏs'-) *n.* **1.** One of a band of raiders roving through the bogs on the borders of England and Scotland during the 17th cent. **2.** A plunderer : marauder.
**moss·y** (mô'sē, mŏs'ē) *adj.* **-i·er, -i·est. 1.** Covered with moss. **2.** Like moss. **3.** Old-fashioned. **—moss'i·ness** *n.*
**most** (mōst) *adj.* [ME < OE *mæst.*] **1.** Greatest in number, quantity, size, or degree. **2.** The greatest part of <*most* politicians> —*n.* The greatest amount <had the *most* to lose and the least to gain> —*pron. (sing. or pl. in number).* The greatest part <*Most* of the horses roamed freely.><*Most* of the house was damp.> —*adv.* **1.** In the highest degree, quantity, or extent. —Used with many adjectives and adverbs to form the superlative degree <*most* reliable><*most* angrily> **2.** Very <a *most* satisfactory evening of entertainment> **3.** *Informal.* Almost <We'll be there *most* any time.> **—at (the) most.** At the maximum <saw them for a day *at the most*> **—for the most part.** In most cases : USUALLY.
**-most** *suff.* [ME, alteration of *-mest,* as in *formest,* foremost.] **1.** Most <inner*most*> **2.** Nearest to <aft*most*>
**mos·tac·cio·li** (mô-stä'chə-lē') *n.* [Ital. < *mostaccio,* mustache.] Pasta shaped like a short tube with slanted ends.
**most·ly** (mōst'lē) *adv.* For the greatest part : almost completely.
**mot** (mō) *n.* [Fr. < OFr., word, prob. < Lat. *muttum,* grunt < *muttire,* to mutter.] A witty or penetrating remark.
**mote**[1] (mōt) *n.* [ME *mot* < OE.] A tiny particle : SPECK.
**mote**[2] (mōt) *aux.v.* [ME *moten* < OE *mōtan.*] *Archaic.* May : might.
**mo·tel** (mō-tĕl') *n.* [Blend of MOTOR and HOTEL.] A hotel for motorists, usu. with rooms adjoining a parking area.
**mo·tet** (mō-tĕt') *n.* [ME < OFr. < *mot,* word. —see MOT.] A polyphonic musical composition based on a sacred text and usu. sung without accompaniment.
**moth** (môth, mŏth) *n., pl.* **moths** (môthz, mŏthz, môths, mŏths) [ME *motthe* < OE *moððe.*] **1.** Any of numerous insects of the order Lepidoptera, gen. distinguished from butterflies by their nocturnal activity, hairlike or feathery antennae, and stout bodies. **2.** The clothes moth.
**moth·ball** (môth'bôl', mŏth'-) *n.* **1.** A marble-sized ball, orig. of camphor but now of naphthalene, placed with clothes to keep moths away. **2. mothballs.** Long-term storage <aircraft in *mothballs*>
**moth-ball** (môth'bôl', mŏth'-) *vt.* **-balled, -ball·ing, -balls.** To deactivate (e.g., a ship) and put into long-term storage.
**moth-eat·en** (môth'ēt'n, mŏth'-) *adj.* **1.** Eaten away by moths. **2.** Old and timeworn <a *moth-eaten* maxim> **3.** Shabby <a *moth-eaten* old house>
**moth·er**[1] (mŭth'ər) *n.* [ME *moder* < OE *mōdor.*] **1.** A female parent. **2.** A woman holding an authoritative or responsible position similar to that of a mother. **3.** A creative source : ORIGIN <Religious rites were the *mother* of the dramatic arts.> **4.** An elderly or old woman. **5.** Qualities attributed to a mother, as capacity to love. —*vt.* **-ered, -er·ing, -ers. 1.** To give birth to : be the mother of. **2.** To create : produce. **3.** To watch over, nourish, and protect. **—moth'er·hood'** *n.* **—moth'er·less** *adj.* **—moth'er·less·ness** *n.*
**moth·er**[2] (mŭth'ər) *n.* [Poss. < MOTHER[1].] A stringy slime of yeast cells and bacteria that forms on the surface of fermenting liquids and is added to wine or cider to start production of vinegar.
**Mother Car·ey's chicken** (kâr'ēz) *n.* [Orig. unknown.] A petrel, esp. the storm petrel.
**mother cell** *n.* A cell generating other cells.
**moth·er·house** (mŭth'ər-hous') *n.* **1.** The convent in which the

mother superior of a religious community resides. **2.** The original convent of a religious community.
**Mother Hub·bard** (hŭb′ərd) *n.* A loose beltless dress.
**moth·er-in-law** (mŭth′ər-ĭn-lô′) *n., pl.* **moth·ers-in-law.** The mother of one's wife or husband.
**moth·er·land** (mŭth′ər-lănd′) *n.* **1.** The land or country of one's birth. **2.** The native land of one's ancestors. **3.** A country regarded as the place of origin, as of a movement.
**moth·er·ly** (mŭth′ər-lē) *adj.* Of, suitable for, resembling, or typical of a mother : MATERNAL. **—moth′er·li·ness** *n.*
**moth·er-na·ked** (mŭth′ər-nā′kĭd) *adj.* Entirely naked.
**moth·er-of-pearl** (mŭth′ər-əv-pûrl′) *n.* The pearly internal layer of certain mollusk shells, used for making decorative objects. **—moth′er-of-pearl′** *adj.*
**Mother's Day** *n.* A day for honoring mothers and motherhood, observed annually on the second Sunday in May.
**mother superior** *n.* A woman in charge of a religious community of women.
**mother tongue** *n.* **1.** One's native language. **2.** A language from which another derives.
**mother wit** *n.* Common sense or innate intelligence.
**moth·er·wort** (mŭth′ər-wûrt′, -wôrt′) *n.* [ME *moderwort* : *moder*, mother + *wort*, wort < OE *wyrt*, plant.] A plant of the genus *Leonurus*, esp. *L. cardiaca*, a weed with small purple or pink flower clusters.
**moth·proof** (môth′pro͞of′, mŏth′-) *adj.* Resistant to damage by moths. *—vt.* **-proofed, -proof·ing, -proofs.** To make resistant to damage by moths. **—moth′proof′er** *n.*
**moth·y** (mô′thē, mŏth′ē) *adj.* **-i·er, -i·est.** Infested with moths.
**mo·tif** (mō-tēf′) *also* **mo·tive** (mō′tĭv, mō-tēv′) *n.* [Fr. < OFr., motive. —see MOTIVE.] **1. a.** A recurrent thematic element in an artistic or literary work. **b.** A dominant theme. **2.** A brief important phrase in a musical composition. **3.** A recurring architectural or decorative design.
**mo·tile** (mōt′l, mō′tīl′) *adj.* [MOT(ION) + -ILE¹.] Moving or capable of moving spontaneously, as certain spores and microorganisms. *—n. Psychol.* One whose mental imagery mainly consists of his or her own bodily motion. **—mo·til′i·ty** (mō-tĭl′ĭ-tē) *n.*
**mo·tion** (mō′shən) *n.* [ME *mocioun* < OFr. *motion* < Lat. *motio* < *movēre*, to move.] **1.** The act or process of changing position. **2.** A significant or expressive shift in the position of the body or a part of the body : GESTURE. **3.** Active functioning. **4.** Ability or power to move. **5.** An inner impulse or inclination. **6.** *Mus.* Melodic ascent and descent of pitch. **7.** *Law.* Application to a court for a ruling. **8.** A formal proposal put to the vote under parliamentary procedures. *—v.* **-tioned, -tion·ing, -tions.** *—vt.* To direct by making a gesture. *—vi.* To signal by making a gesture.
**mo·tion·less** (mō′shən-lĭs) *adj.* Not moving. **—mo′tion·less·ly** *adv.* **—mo′tion·less·ness** *n.*
**motion picture** *n.* **1.** A series of filmed images watched in rapid enough succession to create the illusion of motion and continuity. **2.** A connected narrative told through motion pictures.
**motion sickness** *n.* Sickness caused by motion, as in travel by aircraft, car, ship, or other vehicle and marked by nausea, vomiting, and often dizziness.
**motion study** *n.* A time study.
**mo·ti·vate** (mō′tə-vāt′) *vt.* **-vat·ed, -vat·ing, -vates.** To provide with an incentive or motive : IMPEL. **—mo′ti·va′tion** (-vā′shən) *n.* **—mo′ti·va′tion·al** *adj.* **—mo′ti·va′tion·al·ly** *adv.*
**motivational research** *n.* The use of specific techniques borrowed from psychology and sociology, esp. by advertisers and marketers, to evaluate consumer attitudes toward products and services.
**mo·tive** (mō′tĭv) *n.* [ME < OFr. *motif* < *motive*, causing motion < LLat. *motivus* < Lat. *movēre*, to move.] **1.** An impulse, as an emotion, desire, or physiological need, acting as incitement to action. **2.** (mō′tĭv, mō-tēv′). *var. of* MOTIF. *—adj.* **1.** Causing or able to cause motion. **2.** Of, relating to, or constituting a motive. *—vt.* **-tived, -tiv·ing, -tives.** To motivate. **—mo·tiv′i·ty** *n.*
**mot juste** (mō zhüst′) *n., pl.* **mots justes** (mō zhüst′) [Fr. : *mot*, word + *juste*, right.] The most appropriate word or expression.
**mot·ley** (mŏt′lē) *adj.* [ME, poss. < *mot*, speck < OE.] **1.** With diverse components of great variety : HETEROGENEOUS. **2.** Displaying or having many colors. *—n.* **1.** A multicolored costume, esp. one worn by a clown or court jester. **2.** A heterogeneous mixture.
**mot·mot** (mŏt′mŏt′) *n.* [Am. Sp. *mot-mot.*] A tropical American bird of the family Momotidae, usu. with green and blue plumage.
**mo·to·cross** (mō′tō-krôs′, -krŏs′) *n.* [MOTO(R) + CROSS(-COUNTRY).] A cross-country motorcycle race over a course of rough terrain, as steep hills and hairpin curves.
**mo·to·neu·ron** (mō′tə-no͝or′ŏn′, -nyo͝or′-) *n.* [MOTO(R) + NEURON.] A neuron that stimulates motion.
**mo·tor** (mō′tər) *n.* [Lat. < *movēre*, to move.] **1.** A device, as a machine or engine, that imparts or generates motion. **2.** A device that converts a form of energy into mechanical energy, esp. an internal-combustion engine or an arrangement of coils and magnets that converts electric current into mechanical power. **3.** A motorized conveyance, esp. a car. *—adj.* **1.** Causing or producing motion <*motor* power> **2.** Of, pertaining to, or indicating nerves carrying impulses from the nerve centers to the muscles. **3.** Of or relating to movements of the muscles <*motor* coordination> *—v.* **-tored, -tor·ing, -tors.** *—vi.* To drive or travel in a motor vehicle. *—vt.* To carry by motor vehicle.
**mo·tor·bike** (mō′tər-bīk′) *n.* **1.** A lightweight motorcycle. **2.** A pedal bicycle with a motor attached.
**mo·tor·boat** (mō′tər-bōt′) *n.* A boat propelled by an internal-combustion engine.
**mo·tor·bus** (mō′tər-bŭs′) *n., pl.* **-bus·es** *or* **-bus·ses.** A passenger bus powered by a motor.
**mo·tor·cade** (mō′tər-kād′) *n.* A motor vehicle procession.
**mo·tor·car** (mō′tər-kär′) *n.* An automobile.
**motor court** *n.* A motel.
**mo·tor·cy·cle** (mō′tər-sī′kəl) *n.* A vehicle with two wheels in tandem, propelled by an internal-combustion engine and sometimes having a sidecar with a third wheel. *—vi.* **-cled, -cling, -cles.** To drive or ride on a motorcycle. **—mo′tor·cy′clist** *n.*
**motor drive** *n.* A system consisting of an electric motor and accessory parts, for powering machinery.
**motor home** *n.* A motor vehicle built on a truck or bus chassis and equipped to serve as self-contained living quarters for recreational travel.
**motor inn** *or* **motor hotel** *n.* An urban motel usu. with several stories and space for guest parking.
**mo·tor·ist** (mō′tər-ĭst) *n.* A driver or traveler in a motor vehicle.
**mo·tor·ize** (mō′tə-rīz′) *vt.* **-ized, -iz·ing, -iz·es. 1.** To equip with a motor. **2.** To supply with motor-driven vehicles. **3.** To provide with cars. **—mo′tor·i·za′tion** *n.*
**motor lodge** *n.* A motel.
**mo·tor·man** (mō′tər-mən) *n.* A driver of an electrically powered streetcar, locomotive, or subway train.
**motor neuron** *n.* Motoneuron.
**motor pool** *n.* A centrally controlled group of motor vehicles intended for the use of personnel, as of a governmental agency or military installation.
**motor scooter** *n.* A two-wheeled vehicle with small wheels and a low-powered gasoline engine geared to the rear wheel.
**motor vehicle** *n.* A self-propelled wheeled conveyance not running on rails.
**mo·tor·way** (mō′tər-wā′) *n. Chiefly Brit.* A superhighway.
**†motte** *also* **mott** (mŏt) *n.* [Mex. Sp. *mata* < Sp., shrub, prob. < LLat. *matta*, mat.] *Western U.S.* A small stand of trees on a prairie.
**mot·tle** (mŏt′l) *vt.* **-tled, -tling, -tles.** [Prob. back-formation < MOTLEY.] To cover (a surface) with streaks or spots of different colors or shades. *—n.* **1.** A spot or blotch of color. **2.** A variegated pattern, as on marble. **—mot′tler** *n.*
**mot·to** (mŏt′ō) *n., pl.* **-toes** *or* **-tos.** [Ital., motto, word < Lat. *muttum*, grunt < *muttire*, to mutter.] **1.** A brief statement expressing a principle, goal, or ideal. **2.** A sentence, phrase, or word of suitable character inscribed on or fastened to an object.
**mouch** (mo͞och) *v. Chiefly Brit. var. of* MOOCH.
**moue** (mo͞o) *n.* [Fr. < OFr., of Germanic orig.] A pout : grimace.
**mou·flon** *also* **mouf·flon** (mo͞of′lŏn′) *n., pl.* **mouflon** *or* **-flons** *also* **moufflon** *or* **-flons.** [Fr. < dial. Ital. *muvrone* < LLat. *mufro.*] A wild sheep, *Ovis musimon* of Sardinia and Corsica.
**mouil·lé** (mo͞o-yā′) *adj.* [Fr. < *mouiller*, to moisten, palatalize < OFr. *moullier*, to soften by soaking < VLat. **molliare* < Lat. *mollis*, soft.] Pronounced palatally, as the *ll* in the French word *mouillé.*
**mou·jik** (mo͞o-zhēk′, -zhĭk′) *n. var. of* MUZHIK.
**mou·lage** (mo͞o-läzh′) *n.* [Fr. < OFr. < *mouler*, to mold < *modle*, mold. —see MOLD¹.] **1.** Construction of a mold of a mark, as a footprint, for evidence in a criminal investigation. **2.** A mold used in a criminal investigation.
**mould** (mōld) *n. & v. Chiefly Brit. var. of* MOLD.
**moul·der** (mōl′dər) *v. Chiefly Brit. var. of* MOLDER.
**mould·ing** (mōl′dĭng) *n. Chiefly Brit. var. of* MOLDING.
**mould·y** (mōl′dē) *adj. Chiefly Brit. var. of* MOLDY.
**mou·lin** (mo͞o-lăN′) *n.* [Fr., moulin, mill < OFr., mill < LLat. *molina.* —see MILL¹.] A vertical shaft in a glacier, kept open by falling water and rock debris.
**moult** (mōlt) *v. & n. Chiefly Brit. var. of* MOLT.
**mound** (mound) *n.* [Orig. unknown.] **1.** A pile of earth, gravel, sand, rocks, or debris heaped for protection or concealment. **2.** A natural elevation, as a small hill. **3.** A raised mass <a *mound* of hay> **4.** *Baseball.* The slightly elevated pitcher's area in the center of the diamond. *—vt.* **mound·ed, mound·ing, mounds. 1.** To fortify or conceal with a mound. **2.** To heap in a mound.
**Mound Builder** *n.* A member of a prehistoric North American Indian tribe who built burial and effigy mounds, principally in the Mississippi valley.
**mount¹** (mount) *v.* **mount·ed, mount·ing, mounts.** [ME *mounten* < OFr. *monter* < VLat. **montare* < Lat. *mons*, mountain.] *—vt.* **1.** To ascend or climb. **2.** To place oneself on <*mount* a camel> **3.** To get up on in order to copulate. —Used of males. **4.** To provide with a riding horse. **5. a.** To secure firmly to a support. **b.** To place

or fix on or in a secure place for display, study, or use. **6.** To equip with scenery, costumes, and other accessories <*mount* an elaborate musical show> **7. a.** To set (guns) in position. **b.** To plan and start to carry out <*mount* an offensive> **c.** To be equipped with <The battleship *mounted* many heavy guns.> **d.** To post (a guard) <*mount* patrols> *—vi.* **1.** To go or move upward. **2.** To get or climb up on a vehicle or horse. **3.** To increase, as in amount, degree, extent, intensity, or number. *—n.* **1. a.** An animal or vehicle on which to ride. **b.** The chance to ride a horse in a race. **2.** An object to which another is affixed or on which another is placed for convenience, display, or use, esp.: **a.** A glass slide for use with a microscope. **b.** A setting for a jewel. **c.** An undercarriage or stand on which a device rests while in use. **—mount'a·ble** *adj.* **—mount'er** *n.*

**mount²** (mount) *n.* [ME *mont* < OE *munt* < Lat. *mons.*] **1.** A mountain or hill <*Mount* Katahdin> **2.** Any of the seven fleshy cushions around the edges of the palm of the hand in palmistry.

**moun·tain** (moun'tən) *n.* [ME < OFr. *montaigne* < VLat. ***montanea** < Lat. *montanus,* of a mountain < *mons,* mountain.] **1.** A natural elevation of the earth's surface with considerable mass, gen. steep sides, and a height exceeding that of a hill. **2. a.** A large heap <a *mountain* of fallen leaves> **b.** A huge quantity <a *mountain* of problems> **—moun'tain·y** *adj.*

**mountain ash** *n.* A deciduous tree of the genus *Sorbus,* as the rowan, bearing small white flowers and bright orange-red berries in clusters.

**mountain cat** *n.* The mountain lion.

**mountain cranberry** *n.* The cowberry.

**moun·tain·eer** (moun'tə-nîr') *n.* **1.** A native or inhabitant of a mountainous area. **2.** One who climbs mountains for sport. *—vi.* **-eered, -eer·ing, -eers.** To climb mountains for sport.

**mountain goat** *n.* A hoofed mammal, *Oreamnos americanus* of the northwestern North American mountains, with short, curved black horns and yellowish-white hair and beard.

**mountain laurel** *n.* An evergreen shrub, *Kalmia latifolia* of eastern North America, with poisonous leathery leaves and pink or white flower clusters.

**mountain lion** *n.* A large powerful wild cat, *Felis concolor* of mountainous regions of the Western Hemisphere, with an unmarked tawny body.

**moun·tain·ous** (moun'tə-nəs) *adj.* **1.** Of or relating to a region with many mountains. **2.** Of imposing size or height.

**mountain range** *n.* A series of mountain ridges alike in form, direction, and origin.

**mountain sheep** *n.* **1.** The bighorn. **2.** A wild sheep native to a mountainous area.

**mountain sickness** *n.* Sickness caused by insufficient oxygen at high altitudes and marked by shortness of breath and nausea.

**moun·tain·side** (moun'tən-sīd') *n.* A sloping side of a mountain.

**Mountain Standard Time** *n.* Time at the 105th meridian west of Greenwich, England, and in the seventh time zone based on it in North America.

**moun·tain·top** (moun'tən-tŏp') *n.* The top of a mountain.

**moun·te·bank** (moun'tə-băngk') *n.* [Ital. *montambanco* < the phrase *monta im banco,* he gets up onto the bench.] **1.** A roving hawker of nostrums who attracts customers with stories, jokes, or tricks. **2.** An imposter or trickster. *—vi.* **-banked, -bank·ing, -banks.** To act as a mountebank.

**Mount·ie** (moun'tē) *also* **Mount·y** *n., pl.* **-ies.** *Informal.* A member of the Royal Canadian Mounted Police.

**mount·ing** (moun'tĭng) *n.* Something that provides a backing or appropriate setting <a *mounting* for a cameo>

**Mount·y** (moun'tē) *n. var. of* MOUNTIE.

**mourn** (môrn, mōrn) *v.* **mourned, mourn·ing, mourns.** [ME *mournen* < OE *murnan.*] *—vi.* **1.** To express or feel grief or sorrow. **2.** To express public grief for a death by conventional signs. *—vt.* To feel or express sorrow for. **—mourn'er** *n.*

**mourn·ful** (môrn'fəl, mōrn'-) *adj.* **1.** Feeling or showing grief. **2.** Arousing or suggesting sadness <the *mournful* cry of a bird> **—mourn'ful·ly** *adv.* **—mourn'ful·ness** *n.*

**mourn·ing** (môr'nĭng, mōr'-) *n.* **1.** The acts or expressions of one who has endured a bereavement. **2.** The symbols or conventional outward signs of grief for the dead, as a black tie or armband or black clothing. **3.** The period during which a death is mourned. **—mourn'ing·ly** *adv.*

**mourning cloak** *n.* A butterfly, *Nymphalis antiopa* of Europe and North America, with purplish-brown wings having a broad yellow border.

**mourning dove** *n.* A wild dove, *Zenaidura macroura* of North America, known for its sorrowful call.

**mourning warbler** *n.* A warbler, *Oporornis philadelphia* of eastern North America, known for its plaintive song.

**mouse** (mous) *n., pl.* **mice** (mīs) [ME < OE *mūs.*] **1. a.** A small rodent of the family Muridae or Cricetidae, as the common house mouse, *Mus musculus,* with a long naked or nearly hairless tail. **b.** An animal resembling or related to the mouse, as the jumping mouse or pocket mouse. **2.** *Informal.* A timid or cowardly person. **3.** *Slang.* BLACK EYE 1. *—vi.* (mouz) **moused, mous·ing, mous·es.** **1.** To hunt, stalk, or catch mice. **2.** To search stealthily for something : PROWL.

▲ word history: Until about 500 years ago the word *mouse* had remained unchanged from the days, about 6,000 years ago, when the parent Indo-European language was spoken. The Indo-European ancestor of mouse was *mūs–,* a form that survived almost unchanged in many of the descendent languages: Latin *mūs,* Greek *mūs,* Sanskrit *mūs–,* Slavic *myši,* Germanic *mūs.* The original vowel of Indo-European *mūs–* was probably pronounced like the vowel of English *loom,* and the long *u* of Old English *mūs* had a similar sound. Beginning around the 15th century all the long vowels in English underwent a radical change. The long *u* in words inherited from Old English became the diphthong that now occurs in the words *mouse, house, cow, now,* and *how.* The other long vowels also changed at the same time in such a way that they remained distinct from each other although none preserved its original sound. This wholesale and systematic change, called by philologists the "Great Vowel Shift," has caused many inconsistencies in English spelling.

**mouse deer** *n.* A chevrotain.

**mous·er** (mou'zər) *n.* An animal that catches mice.

**mouse·tail** (mous'tāl') *n.* A plant of the genus *Myosurus,* esp. *M. minimus,* bearing a taillike flower spike.

**mouse·trap** (mous'trăp') *n.* A trap for catching mice.

**mous·ey** (mou'sē, -zē) *adj. var. of* MOUSY.

**mous·ing** (mou'zĭng) *n. Naut.* A binding or metal shackle around the point and shank of a hook to keep it from slipping from an eye.

**mous·sa·ka** (mo͞o-sä'kə, mo͞o'sä-kä') *n.* [Mod. Gk. *moussakas.*] A Greek dish made of layers of ground beef or lamb and sliced eggplant topped with a cheese sauce.

**mousse** (mo͞os) *n.* [Fr., mousse, foam < Lat. *mulsa,* honeywater < *mulsus,* p.part. of *mulcēre,* to delight.] **1.** A chilled dessert made with flavored or whipped cream and gelatin. **2.** A molded dish made from a purée of meat, fish, or shellfish with whipped cream.

**mousse·line** (mo͞os-lēn') *n.* [Fr., muslin. —see MUSLIN.] A fine cotton fabric orig. made in Mosul, Iraq.

**mous·tache** (mŭs'tăsh', mə-stăsh') *n. var. of* MUSTACHE.

**Mous·te·ri·an** (mo͞o-stîr'ē-ən) *adj.* [Fr. *moustérien* < *Le Moustier,* a cave in southwestern France where archaeological finds were made.] Indicating or belonging to a Middle Paleolithic culture following the Acheulian.

**mous·y** *also* **mous·ey** (mou'sē, -zē) *adj.* **-i·er, -i·est.** **1.** Quiet : stealthy. **2. a.** Colorless or grayish brown in color. **b.** Shy.

**mouth** (mouth) *n., pl.* **mouths** (mouthz) [ME < OE *mūð.*] **1.** *Anat.* **a.** The bodily opening through which an animal ingests food. **b.** The system of related organs including the lips, teeth, tongue, and associated parts, with which food is chewed and swallowed and sounds and speech are articulated. **2.** The part of the lips that can be seen on the human face. **3.** A person regarded as a consumer of food <had 12 *mouths* to feed> **4.** A pout, grimace, or similar expression <making *mouths* at each other> **5. a.** The faculty of speech. **b.** A tendency to talk too much or unwisely <a big *mouth*> **c.** Rude or foul language <Watch your *mouth.*> **6.** MOUTHPIECE 3. **7.** A natural opening, as the part of a stream that empties into a larger body of water or the entrance to a harbor, canyon, valley, or cave. **8.** An opening in a container. **9.** An opening in tools and devices whose function is to grip or hold. **10. a.** An opening in the pipe of an organ. **b.** The opening in the mouthpiece of a flute across which the player blows. *—v.* (mou*th*) **mouthed, mouth·ing, mouths.** *—vt.* **1.** To speak : pronounce. **2. a.** To utter in a bombastic style <always *mouthing* dull platitudes> **b.** To utter automatically, without conviction or comprehension <*mouthing* pious thoughts> **3.** To form or articulate words soundlessly <*mouthed* the song in time with the chorus> **4.** To put, take, or move around in the mouth. *—vi.* **1.** To orate affectedly : DECLAIM. **2.** To grimace. **—mouth off.** *Slang.* To complain, criticize, or brag loudly and indiscreetly.

**mouth·breed·er** (mouth'brē'dər) *n.* Any of various unrelated fishes that carry their eggs and young in their mouths.

**mouth·ful** (mouth'fo͝ol') *n.* **1.** The amount of food or other material that can be placed or kept in the mouth at one time. **2.** A small amount to be eaten or tasted. **3.** An utterance that is complicated or difficult to pronounce. **4.** An important or discerning comment.

**mouth organ** *n.* **1.** HARMONICA 1. **2.** A panpipe.

**mouth·piece** (mouth'pēs') *n.* **1.** A part, as of a musical instrument or a telephone, that functions in or near the mouth. **2.** A protective rubber device that boxers wear over the teeth. **3.** *Informal.* A spokesperson. **4.** *Slang.* A defense lawyer.

**mouth-to-mouth** (mouth'tə-mouth') *adj.* Of, pertaining to, or being a method of artificial resuscitation in which the rescuer places his or her mouth directly over the victim's and forces air into the victim's lungs every few seconds.

**mouth·wash** (mouth'wŏsh', -wôsh') *n.* An antiseptic, usu. flavored liquid for cleaning the mouth and freshening breath.

**mouth·y** (mou'thē, -*th*ē) *adj.* **-i·er, -i·est.** Given to ranting : BOMBASTIC. **—mouth'i·ness** *n.*

**mou·ton** (mo͞o'tŏn') *n.* [Fr., sheep < OFr., of Celt. orig.] Sheepskin sheared and processed to look like beaver or seal.
**mou·ton·née** (mo͞o'tə-nā') *also* **mou·ton·néed** (-nād') *adj.* [Short for Fr. *roche moutonnée* : *roche*, rock + *moutonnée*, fleecy.] *Geol.* Rounded by glacial action to a shape suggestive of a sheep's back, as a rock formation.
**mov·a·ble** *also* **move·a·ble** (mo͞o'və-bəl) *—adj.* **1.** Capable of being moved. **2.** Varying in date from year to year <a *movable* holiday> **3.** *Law.* Of or relating to personal property that can be moved. *—n.* **1.** *often* **movables.** Something that can be moved, esp. furniture, as opposed to permanent fixtures. **2.** *Law.* Personal property. **—mov'a·bil'i·ty, mov'a·ble·ness** *n.* **—mov'a·bly** *adv.*
**move** (mo͞ov) *v.* **moved, mov·ing, moves.** [ME *moven* < AN *mover* < Lat. *movēre.*] *—vi.* **1.** To change from one position to another <*moved* away from the door> **2.** To advance step by step : go forward <events *moved* slowly> **3.** To follow a particular course <The earth *moves* in orbit around the sun.> **4.** To progress toward a given state <*moving* up in the world> **5.** To settle in a new place of business or residence : RELOCATE. **6.** To change hands commercially <Apartment leases *move* fast in the fall.> **7.** To change posture or position : STIR <*moved* in their chairs> **8.** To be stirred <flowers *moving* in the breeze> **9.** To be put into motion or turn in a prescribed fashion. —Used of machinery. **10.** To hum with activity. **11.** To start an action. **12.** To live or be active in a specific environment <*move* in theater circles> **13.** To make a formal motion in parliamentary procedure <*move* for adoption of the resolution> *—vt.* **1. a.** To change the position or place of <*move* our business><*move* my legs> **b.** To change from one position to another in a board game <*move* a knight> **2.** To dissuade from a firm opinion. **3.** To cause to take action : ROUSE <Resentment *moved* them to seek revenge.> **4. a.** To set or keep in motion. **b.** To cause to function. **c.** To cause to advance or progress. **5.** To set astir : SHAKE <The wind *moved* the leaves.> **6. a.** To arouse or stir the emotions of <The plight of the refugees *moved* us greatly.> **b.** To excite or provoke to the manifestation of a feeling or emotion <We were *moved* to tears.> **7.** To propose or request in formal parliamentary procedure <*moved* adjournment> **8.** To cause to change hands commercially <easy to *move* calculators> **9.** To cause (the bowels) to evacuate. **—move in.** To occupy a residence or place of business. **—move in on. 1.** To make advances toward. **2.** To try to get control of. *—n.* **1.** An act of moving. **2.** A change of residence or place of business. **3. a.** An act of shifting a piece from one position to another in board games. **b.** The prescribed manner in which a piece may be maneuvered. **c.** A player's turn to maneuver a piece. **4.** One of a series of calculated actions embarked on to gain an end. **—get a move on.** *Informal.* To get started : get going. **—on the move. 1.** In the process of moving about. **2.** Making progress : ADVANCING <a career *on the move*>
**move·a·ble** (mo͞o'və-bəl) *adj.* & *n. var. of* MOVABLE.
**move·ment** (mo͞ov'mənt) *n.* [ME < OFr. < Med. Lat. *movimentum* < Lat. *movēre*, to move.] **1. a.** An act of moving. **b.** A strategic or tactical shift in the location of military troops, ships, or aircraft : MANEUVER. **2. a.** The activities of a group of people to achieve a specific goal <the peace *movement*> **b.** A tendency or trend. **3.** Activity, esp. in business or commerce. **4. a.** Evacuation of the bowels. **b.** The waste matter so evacuated. **5.** The quality that manifests the illusion or effect of motion in the fine arts. **6.** Progression of events in the development of a literary plot. **7.** The metrical or rhythmical structure of a poetic composition. **8.** *Mus.* A self-contained component section of a composition. **9.** A mechanism that produces or transmits motion, as the works of a watch.
**mov·er** (mo͞o'vər) *n.* **1.** One that moves. **2.** One whose occupation is transporting furnishings.
**mov·ie** (mo͞o'vē) *n.* [Short for MOVING PICTURE.] *Informal.* **1.** A motion picture. **2.** A theater that shows motion pictures. **3. movies. a.** A showing of a motion picture. **b.** The motion picture industry.
**mov·ie·dom** (mo͞o'vē-dəm) *n.* Filmdom.
**mov·ie·go·er** (mo͞o'vē-gō'ər) *n.* A filmgoer.
**mov·ie·mak·er** (mo͞o've-mā'kər) *n.* One who makes movies.
**mov·ing** (mo͞o'vĭng) *adj.* **1.** Changing or able to change position. **2.** Of or relating to a change of residence <a *moving* van> **3.** Causing or producing motion or action. **4.** Deeply affecting the emotions. **—mov'ing·ly** *adv.*

☆ ***syns:*** **1.** MOVING, MOBILE, MOVABLE *adj. core meaning* : capable of moving or being moved from place to place <a *moving* automobile><*moving* targets> **2.** MOVING, POIGNANT, STIRRING, TOUCHING *adj. core meaning* : eliciting a deep emotional response <a *moving* farewell speech>

**moving picture** *n.* A motion picture.
**mow¹** (mou) *n.* [ME, stack of hay < OE *mūga.*] **1.** A place for storing grain or hay. **2.** Feed stored, esp. in a barn.
**mow²** (mō) *vt.* **mowed, mowed** *or* **mown** (mōn), **mow·ing, mows.** [ME *mowen* < OE *māwan.*] **1.** To cut down with a scythe or mechanical device. **2.** To cut (growth) from <*mow* the meadow> **—mow down.** To destroy in great numbers, as in battle. **—mow'er** (mō'ər) *n.*
**mox·ie** (mŏk'sē) *n.* [< MOXIE.] **1.** The ability to face problems with spirit : PLUCK. **2.** Energy or pep.
**Mox·ie** (mŏk'sē). A trademark for a soft drink.
**Moz·ar·ab** (mō-zăr'əb) *n.* [Sp. *Mozárabe* < Ar. *musta'rib*, would-be Arab < *'arab*, Arab.] One of a group of Spanish Christians who practiced a modified form of their religion under the Moslems. **—Moz·ar'a·bic** *adj.*
**mo·zet·ta** (mō-zĕt'ə, mōt-sĕt'ə) *n.* [Ital. < *almozetta* < Med. Lat. *almutia.*] *Rom. Cath. Ch.* A short hooded cape worn by bishops.
†**mo·zo** (mō'zō) *n., pl.* **-zos.** [Sp., boy < OSp. *moço.*] *Western U.S.* A man who helps with a pack train or acts as a porter.
**moz·za·rel·la** (mŏt'sə-rĕl'ə, mŏt'-) *n.* [Ital., dim. of *mozza*, a kind of cheese < *mozzare*, to cut off.] A soft white Italian curd cheese.
**Mpon·gwe** (əm-pŏng'wā) *n.* A Bantu language used in the area around the estuary of the Gabon river.
**Mr.** (mĭs'tər) *n., pl.* **Messrs.** [Abbr. of MISTER.] —Used as a courtesy title with a man's name <*Mr.* Roe><*Mr.* James Roe>
**Mrs.** (mĭs'ĭz) *n., pl.* **Mmes.** [Abbr. of MISTRESS.] —Used as a courtesy title with a married woman's name <*Mrs.* Smith><*Mrs.* John Smith>
**Mrs. Grun·dy** (grŭn'dē) *n.* [After *Mrs. Grundy*, character alluded to in the play *Speed the Plough* by Thomas Morton (1764–1838).] A highly conventional, priggish person.
**Ms.** *or* **Ms** (mĭz) *n., pl.* **Mses.** *or* **Mss.** [Blend of MISS and MRS.] —Used as a courtesy title with a woman's name when her marital status is unknown or irrelevant <*Ms.* Doe><*Ms.* Jane Doe> ***usage:*** *Ms.*, the equivalent of *Mr.*, is now widely used in business, professional, and social situations.
**mu** (myo͞o, mo͞o) *n.* [Gk., of Phoenician orig.; akin to Hebrew *mēm.*] The 12th letter in the Greek alphabet. —See table at ALPHABET.
**muc-** *pref. var. of* MUCO-.
**much** (mŭch) *adj.* **more** (môr, mōr), **most** (mōst) [ME *muche* < *muchel* < OE *mycel.*] Great in quantity, degree, or extent <*much* heat> *—n.* **1.** A large quantity or amount. **2.** One that is important or remarkable <As a swimmer, you certainly aren't *much.*> *—adv.* **more, most. 1.** To a great degree or extent <*much* discussed> **2.** Just about : ALMOST <*much* the same> **—much'ness** (-nĭs) *n.*
**much as** *conj.* However much.
**muci-** *pref. var. of* MUCO-.
**mu·cic acid** (myo͞o'sĭk) *n.* [< MUCUS.] An organic acid, HOOC-$(CHOH)_4$COOH, often derived from milk sugar.
**mu·cif·er·ous** (myo͞o-sĭf'ər-əs) *adj.* Secreting or producing mucus.
**mu·ci·lage** (myo͞o'sə-lĭj) *n.* [ME *muscilage* < OFr. *mucilage* < LLat. *mucilago*, musty juice < Lat. *mucus*, mucus.] **1.** A sticky adhesive substance. **2.** A gummy substance extracted from certain plants. **—mu'ci·lag'i·nous** (-lăj'ə-nəs) *adj.*
**mu·cin** (myo͞o'sĭn) *n.* Any of a group of organic compounds produced by mucous membranes. **—mu'cin·ous** *adj.*
**muck** (mŭk) *n.* [ME *muk.*] **1.** A moist sticky mixture, esp. of mud and filth. **2.** Moist animal dung mixed with decayed vegetable matter for use as a fertilizer : MANURE. **3.** Dark, fertile soil containing putrid vegetable matter. **4.** Something disgusting or filthy. **5.** Earth, rocks, or clay excavated in mining. *—vt.* **mucked, muck·ing, mucks. 1.** To fertilize with compost or manure. **2.** *Informal.* To soil or make dirty with or as if with muck. **3.** To remove muck or dirt from (e.g., a mine). **—muck about.** *Chiefly Brit. Informal.* To spend time idly : PUTTER. **—muck up.** *Informal.* To bungle : botch. **—muck'i·ly** *adv.* **—muck'y** *adj.*
**muck·a·muck** (mŭk'ə-mŭk') *n.* [Short for HIGH MUCKAMUCK.] *Slang.* An important person.
**muck·rake** (mŭk'rāk') *vi.* **-raked, -rak·ing, -rakes.** To search for and expose corruption. **—muck'rak'er** *n.*
**muco-** *or* **muci-** *or* **muc-** *pref.* [< Lat. *mucus*, mucus.] **1.** Mucus <*mucoprotein*> **2.** Mucous membrane <*mucin*> **3.** Mucin <*mucoid*>
**mu·co·cu·ta·ne·ous** (myo͞o'kō-kyo͞o-tā'nē-əs) *adj.* Relating to the skin and a mucous membrane.
**mu·coid** (myo͞o'koid') *n.* Any of a group of organic compounds similar to the mucins and found in connective tissue. **—mu'coid, mu·coi'dal** (myo͞o-koid'l) *adj.*
**mu·co·lyt·ic** (myo͞o'kə-lĭt'ĭk) *adj.* Breaking down or hydrolyzing mucin or mucopolysaccharides.
**mu·co·pol·y·sac·cha·ride** (myo͞o'kō-pŏl'ē-săk'ə-rīd') *n.* Any of the polysaccharides that form chemical bonds with water to produce mucilaginous and lubricating fluids.
**mu·co·pro·tein** (myo͞o'kō-prō'tēn', -prō'tē-ĭn) *n.* Any of a group of organic compounds, as the mucins, that contain proteins and mucopolysaccharides.
**mu·co·pu·ru·lent** (myo͞o'kō-pyo͝or'ə-lənt, -yə-lənt) *adj.* Containing mucus and pus.
**mu·co·sa** (myo͞o-kō'sə) *n., pl.* **-sae** (-sē) *or* **-sas.** [NLat. < Lat., fem. of *mucosus*, mucous.] A mucous membrane. **—mu·co'sal** *adj.*
**mu·cous** (myo͞o'kəs) *also* **mu·cose** (-kōs') *adj.* [Lat. *mucosus* < *mucus*, mucus.] **1.** Producing or secreting mucus. **2.** Relating to, comprising, or similar to mucus.

**mucous membrane** *n.* The membrane lining all bodily channels that communicate with the air, as the respiratory and alimentary tracts, the glands of which secrete mucus.

**mu·cro** (myōō'krō) *n., pl.* **mu·cro·nes** (myōō-krō'nēz) [NLat. < Lat., sword point.] A sharp tip on plant and animal organs.

**mu·cro·nate** (myōō'krə-nāt') *adj.* [Lat. *mucronatus* < *mucro,* sword point.] *Biol.* Of or having a mucro. **—mu'cro·na'tion** *n.*

**mu·cus** (myōō'kəs) *n.* [Lat.] The viscous suspension of mucin, water, cells, and inorganic salts secreted as a protective, lubricant coating by glands in the mucous membrane.

**mud** (mŭd) *n.* [ME *mudde.*] **1.** Wet, sticky, soft earth. **2.** Slanderous or defamatory charges or comments. *—vt.* **mud·ded, mud·ding, muds.** To soil or bury with or as if with mud.

**mud cat** *n.* Any of several large American catfish of the Mississippi valley and southeastern U.S. streams.

**mud dauber** *n.* A wasp with long hind legs and a slender abdomen terminating in a bulb, the female of which lays eggs in paralyzed insect larvae that are then placed in a nest of mud.

**mud·der** (mŭd'ər) *n.* A racehorse that runs well on a wet or muddy track.

**mud·dle** (mŭd'l) *v.* **-dled, -dling, -dles.** [Poss. < obs. Du. *moddelen* < MDu. < *modde,* mud.] *—vt.* **1.** To make turbid: MUDDY. **2.** To mix confusedly: JUMBLE. **3.** To befuddle, as with alcohol. **4.** To mismanage: bungle. **5.** To stir or mix (a beverage) gently. *—vi.* To act or think in a confused manner. **—muddle through.** To push on to a successful conclusion in a disorganized way. **—mud'dle** *n.* **—mud'dler** *n.*

**mud·dle-head·ed** (mŭd'l-hĕd'ĭd) *adj.* **1.** Mentally confused. **2.** Inept: stupid. **—mud'dle-head'ed·ness** *n.*

**mud·dy** (mŭd'ē) *adj.* **-di·er, -di·est. 1.** Covered, full of, or spattered with mud. **2.** Not pure or clear, as a color or liquid. **3.** Dull in color <a *muddy* complexion> **4.** Confused, vague, or obscure, as in expression or meaning <a *muddy* prose style> *—vt.* **-died, -dy·ing, -dies. 1.** To make dirty or muddy. **2.** To make dull or cloudy. **3.** To make obscure or confused <*muddied* the issue> **—mud'di·ly** *adv.* **—mud'di·ness** *n.*

**mud eel** *n.* An amphibian found in swamps of the southeastern United States, *Siren lacertina,* resembling an eel and having only inconspicuous front legs.

**mud·fish** (mŭd'fĭsh') *n., pl.* **mudfish** *or* **-fish·es.** A fish living in mud or muddy water, as the bowfin.

**mud flat** *n.* Land under water at high tide but exposed at low tide.

**mud·guard** (mŭd'gärd') *n.* A shield over a vehicle's wheel.

**mud hen** *n.* A bird dwelling in marshy or coastal regions, as a coot or rail.

**mud minnow** *n.* A brown fish of the genus *Umbra,* found in muddy portions of North American and European lakes and ponds.

**mud puppy** *n.* An aquatic salamander of the genus *Necturus,* esp. *N. maculosus* of North America, with conspicuous clusters of external gills.

**mu·dra** (mə-drä') *n.* [Skt. *mudrā,* sign, token.] A code of body postures and hand movements used in East Indian classical dance to enact a story.

**mud·skip·per** (mŭd'skĭp'ər) *n.* Any of several species of fishes of the family Gobiidae, found in the Indo-Pacific region and parts of tropical Africa and noted for their maneuverability on land.

**mud·sling·er** (mŭd'slĭng'ər) *n.* One who makes malicious charges against an opponent. **—mud'sling'ing** *n.*

**mud snake** *n.* A burrowing snake, *Farancia abacura* of the southeastern United States, having black scales with reddish markings.

**mud·stone** (mŭd'stōn') *n.* A dark-gray, fine-grained shale that decomposes into mud when exposed to the atmosphere.

**mud turtle** *n.* A turtle of the genus *Kinosternon,* found in sluggish fresh waters throughout the Western Hemisphere.

**mud wasp** *n.* The potter wasp.

**Muen·ster** *or* **Mun·ster** (mŭn'stər, mōōn'-) *n.* [After *Munster,* city in northeastern France where it was orig. made.] A mild, semisoft, creamy, yellow fermented Alsatian cheese.

**mu·ez·zin** (myōō-ĕz'ĭn, mōō-) *n.* [Ar. *mu'adhdhin.*] The crier who calls the Moslem faithful to prayer five times a day.

**muff¹** (mŭf) *v.* **muffed, muff·ing, muffs.** [Orig. unknown.] *—vt.* **1.** To perform clumsily: BUNGLE. **2.** To fail to catch (the ball) in a sport. *—vi.* **1.** To perform an act clumsily: BUNGLE. **2.** To fail to catch a ball in a sport. **—muff** *n.*

**muff²** (mŭf) *n.* [Du. *mof* < MDu. *moffel* < Med. Lat. *muffla.*] A small cylindrical fur or cloth cover, open at both ends, in which the hands are kept warm.

**muf·fin** (mŭf'ĭn) *n.* [Poss. < LG *muffen,* pl. of *muffe,* cake.] A small, cup-shaped bread, often sweetened and usu. served hot.

**muf·fle¹** (mŭf'əl) *vt.* **-fled, -fling, -fles.** [ME *muflen,* poss. < OFr. *moufle,* glove < Med. Lat. *muffla.*] **1.** To wrap up or cover for warmth, protection, or secrecy. **2. a.** To wrap or pad so as to deaden the sound of. **b.** To deaden (a sound). **3.** To make vague or obscure <"his message was so *muffled* by learning and 'artiness'" —Walter Blair> *—n.* **1.** Something that muffles. **2.** A kiln or part of a kiln in which pottery can be fired without exposure to direct flame.

☆ ***syns:*** MUFFLE, DAMPEN, DEADEN, MUTE, STIFLE *v. core meaning:* to decrease or dull the sound of <The thick wall *muffled* the explosion.>

**muf·fle²** (mŭf'əl) *n.* [Fr. *mufle.*] The hairless snout of certain mammals.

**muf·fler** (mŭf'lər) *n.* **1.** A heavy scarf wrapped about the neck for warmth. **2.** A device that absorbs noise, esp. one used with an internal-combustion engine.

**muf·ti¹** (mŭf'tē, mōōf'-) *n., pl.* **-tis.** [Ar. *muftī* < *aftā,* give a decision.] A Moslem judge who interprets religious law.

**muf·ti²** (mŭf'tē) *n.* [Prob. < MUFTI¹.] Civilian dress, esp. such clothing when worn by one whose regular garb is a uniform.

**mug¹** (mŭg) *n.* [Orig. unknown.] **1.** A cylindrical drinking vessel often having a handle. **2.** The liquid in such a vessel.

**mug²** (mŭg) *n.* [Prob. < MUG¹.] **1.** A person's face. **2.** The human mouth, chin, and jaw. **3.** A photograph of the face, esp. one used by police for identification. **4.** A grimace. **5.** A thug: hoodlum. *—v.* **mugged, mug·ging, mugs.** *—vt.* **1.** To photograph (a person's face) for police files. **2.** To waylay and assault, usu. with intent to rob. *—vi.* To grimace, esp. to make distorted facial expressions.

**mug·ger¹** (mŭg'ər) *n.* [Hindi *magar* < Skt. *makaraḥ,* crocodile, of Dravidian orig.] A large crocodile, *Crocodilus palustris* of southwestern Asia, with an exceptionally broad wrinkled snout.

**mug·ger²** (mŭg'ər) *n.* A thug who waylays and assaults a victim, usu. with intent to rob.

**mug·ging** (mŭg'ĭng) *n.* Aggravated assault usu. perpetrated by surprise and with intent to rob.

**mug·gy** (mŭg'ē) *adj.* **-gi·er, -gi·est.** [< dial. *mug,* mist < ME *muggen,* to drizzle.] Hot and extremely humid. **—mug'gi·ness** *n.*

**mug·wump** (mŭg'wŭmp') *n.* [Natick *mugwomp,* captain.] **1.** A Republican who bolted the party in 1884, protesting the candidacy of James G. Blaine for the U.S. Presidency. **2.** One who acts independently, esp. in politics. **—mug'wump'er·y** *n.*

**Mu·ham·mad·an** *or* **Mu·ham·med·an** (mōō-hăm'ĭ-dən) *adj.* & *n. vars. of* MOHAMMEDAN.

**Mu·ham·mad·an·ism** (mōō-hăm'ə-də-nĭz'əm) *n. var. of* MOHAMMEDANISM.

**Mu·har·ram** (mōō-hăr'əm) *also* **Mo·har·ram** (mō-) *or* **Mu·har·rum** (mōō-) *n.* [Ar. *muḥarram,* p.part. of *ḥarrama,* he forbade.] **1.** The first month of the Moslem year. —See table at CALENDAR. **2.** A festival during the first ten days of Muharram.

**mu·jik** (mōō-zhēk', -zhĭk') *n. var. of* MUZHIK.

**muk·luk** (mŭk'lŭk') *n.* [Eskimo *muklok,* large seal.] **1.** A soft Eskimo boot made of sealskin or reindeer skin. **2.** A slipper similar to the mukluk.

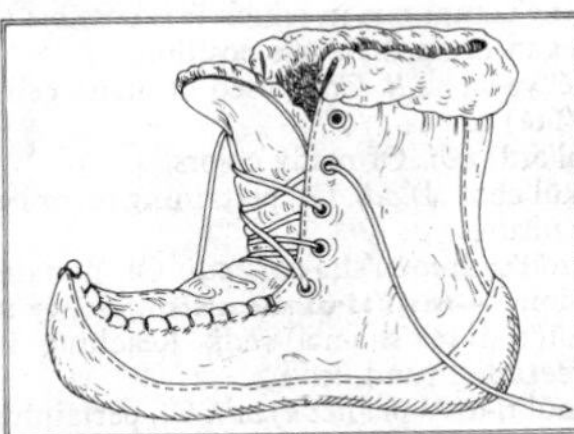
**mukluk**

**mu·lat·to** (mōō-lăt'ō, -lä'tō, myōō-) *n., pl.* **-tos** *or* **-toes.** [Sp. *mulato,* mulatto, young mule < *mulo,* mule < Lat. *mulus.*] **1.** One having one white and one black parent. **2.** One of mixed black and white ancestry.

**mul·ber·ry** (mŭl'bĕr'ē, -bə-rē) *n.* [ME *mulberrie* : OFr. *moure,* mulberry (< Lat. *morum*) + OE *berie,* berry.] **1.** A tree of the genus *Morus,* bearing edible fruit. **2.** The sweet berrylike fruit of the mulberry. **3.** A tree resembling or related to the mulberry. **4.** A grayish to dark purple.

**mulch** (mŭlch) *n.* [Poss. < dial. *melch,* soft < ME *melsch.*] A protective covering of various substances, esp. organic, placed on the earth around plants to retard weed growth and prevent moisture evaporation and freezing of roots. *—vt.* **mulched, mulch·ing, mulch·es.** To cover with a mulch.

**mulct** (mŭlkt) *n.* [Lat. *mulcta.*] A fine or similar penalty. *—vt.* **mulct·ed, mulct·ing, mulcts. 1.** To penalize by fining or demanding forfeiture. **2.** To acquire or take away from by trickery or deception.

**mule¹** (myōōl) *n.* [ME < OFr. *mul* < Lat. *mulus.*] **1.** A sterile hybrid of a male ass and a female horse. **2.** A sterile hybrid, as between a canary and other birds or between certain plants. **3.** *Informal.* A stubborn person. **4.** A type of spinning machine for making thread or yarn from fibers. **5.** A small, usu. electric tractor or locomotive for hauling short distances.

**mule²** (myōōl) *n.* [OFr., slipper < Lat. *mulleus (calceus),* red (shoe).] A heelless slipper.

**mule deer** *n.* A hoofed mammal, *Odocoileus hemionus* of western North America, with long ears and two-pronged antlers.

**mule·skin·ner** (myo͞ol′skĭn′ər) *n. Informal.* A driver of mules.

**mu·le·ta** (mo͞o-lā′tə, -lĕt′ə) *n.* [Sp., crutch < *mula,* she-mule < Lat., fem. of *mulus,* mule.] A short red cape, hung from a hollow staff, used by a matador for maneuvering the bull during the final passes before the kill.

**mu·le·teer** (myo͞o′lə-tîr′) *n.* [OFr. *muletier* < *mulet,* dim. of *mul,* mule.] A muleskinner.

**mu·ley** (myo͞o′lē, mo͝ol′ē, mo͞o′lē) *adj.* [Of Celt. orig.] Having no horns. —Used of cattle. —*n., pl.* **-leys.** A hornless animal.

**mu·li·eb·ri·ty** (myo͞o′lē-ĕb′rĭ-tē) *n.* [LLat. *muliebritas* < Lat. *muliebris,* womanly < *mulier,* woman.] **1.** The state of being a woman. **2.** The qualities held to be typical of women.

**mul·ish** (myo͞o′lĭsh) *adj.* Unreasonably obstinate : STUBBORN. **—mul′ish·ly** *adv.* **—mul′ish·ness** *n.*

**mull¹** (mŭl) *vt.* **mulled, mull·ing, mulls.** [Orig. unknown.] To heat and spice (e.g., wine).

**mull²** (mŭl) *v.* **mulled, mull·ing, mulls.** [ME *mullen,* to grind < *mul,* dust, prob. < MDu.] —*vt.* To go over mentally : DELIBERATE. —*vi.* To ponder over something <*mull* over the problem>

**mull³** (mŭl) *n.* [Short for obs. *mulmull* < Hindi *malmal* < Pers.] A soft, thin kind of muslin used in dresses and for trimmings.

**mul·lah** *also* **mul·la** (mŭl′ə, mo͝ol′ə) *n.* [Turk., Pers. and Urdu *mullā,* all < Ar. *mawlā,* master.] —Used as a title for a Moslem religious teacher or leader. **—mul′lah·ism** *n.*

**mul·lein** (mŭl′ən) *n.* [ME *moleyne* < OFr. *moleine.*] **1.** A plant of the genus *Verbascum,* esp. the native Eurasian species *V. thapsus,* a tall plant having leaves covered with dense woolly down and closely clustered yellow flowers. **2.** Cretan mullein.

**mullein pink** *n.* Rose campion.

**mul·ler** (mŭl′ər) *n.* [ME *molour,* prob. < *mullen,* to grind. —see MULL².] A device with a hard base, for grinding paints or drugs.

**mul·let** (mŭl′ĭt) *n., pl.* **mullet** or **-lets.** [ME *molet* < OFr. *mulet* < Lat. *mullus,* red mullet < Gk. *mullos.*] Any of various edible fishes of the family Mugilidae, found worldwide in tropical and temperate coastal waters and some freshwater streams.

**mul·li·gan stew** (mŭl′ĭ-gən) *n.* [Prob. from the name *Mulligan.*] A stew of various meats and vegetables.

**mul·li·ga·taw·ny** (mŭl′ĭ-gə-tô′nē) *n.* [Tamil *miḷagutaṇṇī.*] An East Indian meat soup strongly flavored with curry.

**mul·lion** (mŭl′yən) *n.* [Alteration of obs. *monial* < ME *moniel* < OFr. *moynel* < *moyen,* middle < Lat. *medianus* < *medius.*] A vertical strip separating windowpanes. **—mul′lioned** *adj.*

**multi-** *pref.* [ME < Lat. < *multus,* much.] **1.** Many : much : multiple <*multi*colored> **2. a.** More than one <*multi*parous> **b.** More than two <*multi*lateral>

**mul·ti·ad·dress** (mŭl′tē-ăd′rĕs′) *adj. Computer Sci.* Designating a storage system of data-processing computers in which it is possible to store instructions or quantities in more than one position.

**mul·ti·cel·lu·lar** (mŭl′tĭ-sĕl′yə-lər) *adj.* Composed of many cells. **—mul′ti·cel′lu·lar′i·ty** (-lăr′ĭ-tē) *n.*

**mul·ti·col·ored** (mŭl′tĭ-kŭl′ərd) *adj.* Of many colors.

**mul·ti·cul·tur·al** (mŭl′tĭ-kŭl′chər-əl) *adj.* Of, pertaining to, or designed for several individual cultures.

**mul·ti·di·men·sion·al** (mŭl′tĭ-dĭ-mĕn′shə-nəl) *adj.* Of, pertaining to, or with several dimensions. **—mul′ti·di·men′sion·al′i·ty** *n.*

**mul·ti·di·rec·tion·al** (mŭl′tĭ-dĭ-rĕk′shə-nəl) *adj.* Reaching in many directions <a *multidirectional* fund drive>

**mul·ti·dis·ci·pli·nar·y** (mŭl′tĭ-dĭs′ə-plə-nĕr′ē) *adj.* Of, pertaining to, or making use of several disciplines at once <*multidisciplinary* lesson materials>

**mul·ti·eth·nic** (mŭl′tē-ĕth′nĭk) *adj.* Of, relating to, or including a variety of ethnic groups <a *multiethnic* theatrical production>

**mul·ti·fac·et·ed** (mŭl′tə-făs′ĭ-tĭd, -tē-) *adj.* Having several individual facets <a *multifaceted* question>

**mul·ti·fam·i·ly** (mŭl′tĭ-făm′ə-lē) *adj.* Of, relating to, or intended for use by several individual families <a *multifamily* house>

**mul·ti·far·i·ous** (mŭl′tə-fâr′ē-əs) *adj.* [LLat. *multifarius* < *multus,* many.] Having great variety : DIVERSE. **—mul′ti·far′i·ous·ly** *adv.* **—mul′ti·far′i·ous·ness** *n.*

**mul·ti·fid** (mŭl′tə-fĭd′) *adj.* [Lat. *multifidus,* divided into many parts : *multus,* many + *findere,* to cleave.] *Biol.* Having many clefts forming lobes <*multifid* leaves>

**mul·ti·flo·ra rose** (mŭl′tə-flôr′ə, -flōr′ə) *n.* [NLat. *Rosa multiflora,* species name : Lat. *rosa,* rose + Lat. *multiflora,* fem. of *multiflorus,* multiflorous.] A climbing or sprawling Asian shrub, *Rosa multiflora,* bearing small, fragrant flower clusters.

**mul·ti·flo·rous** (mŭl′tə-flôr′əs, -flōr′əs) *adj.* [Lat. *multiflorus* : *multus,* many + *flos,* flower.] *Bot.* Bearing many flowers.

**mul·ti·foil** (mŭl′tə-foil′) *n.* A flat object or opening with scalloped edges or ornaments.

**mul·ti·fold** (mŭl′tə-fōld′) *adj.* Many times doubled : MANIFOLD.

**mul·ti·form** (mŭl′tə-fôrm′) *adj.* [Lat. *multiformis* : *multus,* many + *forma,* shape.] Occurring in or having many forms, shapes, or appearances. **—mul′ti·for′mi·ty** *n.*

**mul·ti·lane** (mŭl′tə-lān′, -tē-) *adj.* Having several lanes <a *multilane* expressway>

**mul·ti·lat·er·al** (mŭl′tĭ-lăt′ər-əl) *adj.* **1.** Having many sides. **2.** Involving more than two nations <*multilateral* arms agreements> **—mul′ti·lat′er·al·ly** *adv.*

**mul·ti·lay·ered** (mŭl′tə-lā′ərd) *adj.* Consisting of or involving several individual layers <a *multilayered* torte>

**mul·ti·lev·el** (mŭl′tə-lĕv′əl) *also* **mul·ti·lev·eled** (-lĕv′əld) *adj.* Having several levels <a *multilevel* parking garage>

**mul·ti·lin·gual** (mŭl′tə-lĭng′gwəl) *adj.* **1.** Of, including, or depicted in several languages <*multilingual* signs> **2.** Using or able to use several languages. **—mul′ti·lin′gual·ism** *n.*

**Mul·ti·lith** (mŭl′tĭ-lĭth′). A trademark for a small rotary offset press.

**mul·ti·me·di·a** (mŭl′tĭ-mē′dē-ə) *adj.* Including or involving the use of several media <a *multimedia* display> **—mul′ti·me′di·a** *n.*

**mul·ti·mil·lion·aire** (mŭl′tĭ-mĭl′yə-nâr′) *n.* One who has financial assets of many millions of dollars.

**mul·ti·na·tion·al** (mŭl′tĭ-năsh′ə-nəl, -năsh′nəl) *adj.* **1.** Having operations, subsidiaries, or investments in several countries <a *multinational* corporation> **2.** Of, in, or involving several or many countries <a *multinational* research project> —*n.* A multinational company or corporation. **—mul′ti·na′tion·al·ism** *n.*

**mul·ti·no·mi·al** (mŭl′tĭ-nō′mē-əl) *n.* [MULTI- + (BI)NOMIAL.] POLYNOMIAL 2b. **—mul′ti·no′mi·al** *adj.*

**multinomial theorem** *n. Math.* The theorem that establishes the rule for forming the terms of a polynomial expansion.

**mul·ti·nu·cle·ar** (mŭl′tē-no͞o′klē-ər, -nyo͞o′-) *adj.* Multinucleate.

**mul·ti·nu·cle·ate** (mŭl′tē-no͞o′klē-ət, -nyo͞o′-) *also* **mul·ti·nu·cle·at·ed** (-ā′tĭd) *adj.* Having more than two nuclei.

**mul·tip·a·rous** (mŭl-tĭp′ər-əs) *adj.* **1.** Having borne more than one child. **2.** Giving birth to more than one offspring at one time.

**mul·ti·par·tite** (mŭl′tĭ-pär′tīt′) *adj.* [Lat. *multipartitus* : *multus,* many + *partitus,* p.part. of *partire,* to divide < *pars,* part.] **1.** Having many parts. **2.** Multilateral.

**mul·ti·par·ty** (mŭl′tə-pär′tē) *adj.* Of, pertaining to, or involving more than two political parties.

**mul·ti·ped** (mŭl′tə-pĕd′) *also* **mul·ti·pede** (-pēd′) *adj.* [Lat. *multipedes* : *multus,* many + *pes,* foot.] Having many feet.

**mul·ti·ple** (mŭl′tə-pəl) *adj.* [Fr. < Lat. *multiplex,* multiplex.] Having, relating to, or consisting of more than one individual, element, component, or part : MANIFOLD. —*n. Math.* A quantity into which another quantity may be divided with zero remainder <The numbers 4, 6, and 12 are *multiples* of 2.>

**multiple allele** *n. Genetics.* A set of three or more alleles, or alternative states of a gene, only two of which can be present in a somatic cell at the same time.

**mul·ti·ple-choice** (mŭl′tə-pəl-chois′) *adj.* Posing various solutions from which the correct one is to be chosen <*multiple-choice* history tests>

**multiple factor** *n. Genetics.* A combination of two or more genes acting as a unit to produce a quantitative inheritance trait, as leaf shape or eye color.

**multiple fruit** *n.* A fruit, as a pineapple, in which the fruits of several flowers are united in a single structure.

**multiple myeloma** *n.* A malignant disease of bone marrow marked by the presence of myelomas in many bones of the body.

**multiple neuritis** *n.* Inflammation of more than one nerve at a time.

**multiple root** *n.* ROOT 9c.

**multiple sclerosis** *n.* A degenerative disease of the central nervous system in which hardening of tissue occurs throughout the brain or spinal cord or both.

**multiple star** *n.* Three or more stars, usu. with a common gravitational center, that appear as one to the naked eye.

**multiple store** *n. Chiefly Brit.* A chain store.

**mul·ti·plet** (mŭl′tə-plĕt′, -plĭt) *n.* [< MULTIPLE.] **1.** A spectral line having more than one component representing slight variations in energy states characteristic of an atom. **2.** Any of several classes or groupings of subatomic particles, as the nucleon, each member of which has the same set of quantum numbers except for electric charge.

**mul·ti·plex** (mŭl′tə-plĕks′) *adj.* [Lat. < *multus,* many.] **1.** Multiple : manifold. **2.** Indicating or being a simultaneous communication of two or more messages on the same wire or radio channel. —*v.* **-plexed, -plex·ing, -plex·es.** —*vi.* To send signals or messages in a multiplex system. —*vt.* To send simultaneously (more than one signal) on a single radio frequency.

**mul·ti·pli·a·ble** (mŭl′tə-plī′ə-bəl) *also* **mul·ti·plic·a·ble** (-plĭk′ə-bəl) *adj.* Capable of being multiplied.

**mul·ti·pli·cand** (mŭl′tə-plĭ-kănd′) *n.* [Lat. *multiplicandum,* neuter of *multiplicandus,* gerund. of *multiplicare,* to multiply. —see MULTIPLY.] The number that is or is to be multiplied by another.

**mul·tip·li·cate** (mŭl-tĭp′lĭ-kĭt) *adj.* [ME < Lat. *multiplicatus,* p. part. of *multiplicare,* to multiply. —see MULTIPLY.] **1.** Having more than one layer or fold, as some shells or leaves. **2.** Multiple.

ă **pat** ā **pay** âr **care** ä **father** ĕ **pet** ē **be** hw **which** ĭ **pit** ī **tie** îr **pier** ŏ **pot** ō **toe** ô **paw, for** oi **noise** o͝o **took**

**mul·ti·pli·ca·tion** (mŭl′tə-plĭ-kā′shən) *n.* **1. a.** The act or process of multiplying. **b.** The condition of being multiplied. **2.** Propagation of plants and animals. **3.** *Math.* **a.** The conjunction of two real numbers in which the number of times either is taken in summation is determined by the value of the other. **b.** Any of certain analogous operations conjoining expressions other than real numbers. **4.** An increase achieved by adding. **—mul′ti·pli·ca′tion·al** *adj.*

**multiplication sign** *n.* The sign (×) placed between multiplicand and multiplier or operand and operator, as in $a \times b$.

**multiplication table** *n.* A table listing the products of certain numbers multiplied together, esp. the numbers 1 to 12.

**mul·ti·pli·ca·tive** (mŭl′tə-plĭk′ə-tĭv, mŭl′tə-plĭ-kā′tĭv) *adj.* **1.** Likely to multiply or capable of multiplying or increasing. **2.** Of or involving multiplication. **—mul′ti·pli′ca·tive·ly** *adv.*

**multiplicative inverse** *n.* INVERSE 2a.

**mul·ti·plic·i·ty** (mŭl′tə-plĭs′ĭ-tē) *n., pl.* **-ties.** [Fr. *multiplicité* < LLat. *multiplicitas* < *multiplex,* multiplex. —see MULTIPLEX.] **1.** The state of being manifold. **2.** A large number <a *multiplicity* of ideas> **3.** The number of subatomic particles in a multiplet.

**mul·ti·pli·er** (mŭl′tə-plī′ər) *n.* **1.** One that multiplies. **2.** *Math.* The number by which the multiplicand is multiplied. **3.** *Physics.* A device, as a phototube, for enhancing or increasing an effect.

**mul·ti·ply** (mŭl′tə-plī′) *v.* **-plied, -ply·ing, -plies.** [ME *multiplien* < OFr. *multiplier* < Lat. *multiplicare* < *multiplex,* multiplex.] —*vt.* **1.** To increase the amount, number, or degree of. **2.** *Math.* To perform multiplication on. —*vi.* **1.** To become more in number, amount, or degree. **2.** To breed : propagate. **3.** *Math.* To perform multiplication.

**mul·ti·pur·pose** (mŭl′tə-pûr′pəs) *adj.* Designed or used for several purposes <*multipurpose* tents>

**mul·ti·ra·cial** (mŭl′tē-rā′shəl) *adj.* Made up of, involving, or acting on behalf of various races <a *multiracial* council>

**mul·ti·sense** (mŭl′tĭ-sĕns′) *adj.* Having multiple meanings.

**mul·ti·sen·so·ry** (mŭl′tĭ-sĕn′sə-rē) *adj.* Pertaining to or making use of several bodily senses <*multisensory* methods of teaching reading>

**mul·ti·stage** (mŭl′tĭ-stāj′) *adj.* Functioning by stages.

**multistage rocket** *n.* A rocket of at least two stages, each stage firing in succession.

**mul·ti·state** (mŭl′tĭ-stāt′) *adj.* Of, pertaining to, or involving several states <a *multistate* park project>

**mul·ti·sto·ry** (mŭl′tĭ-stôr′ē, -stōr′ē) *also* **mul·ti·sto·ried** (-stôr′ēd, -stōr′-) *adj.* Having several stories <a *multistory* motel>

**mul·ti·tude** (mŭl′tĭ-to͞od′, -tyo͞od′) *n.* [ME < OFr. < Lat. *multitudo* < *multus,* many.] **1.** The quality or condition of being numerous. **2.** A great, indefinite number. **3.** The masses : the populace.

**mul·ti·tu·di·nous** (mŭl′tĭ-to͞od′n-əs, -tyo͞od′-) *adj.* **1.** Very numerous. **2.** Having many parts. **3.** Crowded. **—mul′ti·tu′di·nous·ly** *adv.*

**mul·ti·va·lent** (mŭl′tĭ-vā′lənt, mŭl-tĭv′ə-lənt) *adj.* **1.** *Chem.* POLYVALENT 2b. **2.** Of or relating to homologous chromosomes in synapsis. **3.** Having various meanings or values. **—mul′ti·va′lence** *n.*

**mul·ti·ver·si·ty** (mŭl′tĭ-vûr′sĭ-tē) *n., pl.* **-ties.** [MULTI- + (UNI)VERSITY.] A university that has numerous constituent and affiliated institutions, as separate colleges, campuses, and research centers.

**mul·ti·vi·ta·min** (mŭl′tə-vī′tə-mĭn) *adj.* Containing many vitamins. **—mul′ti·vi′ta·min** *n.*

**mul·ture** (mŭl′chər) *n.* [ME *multyr* < OFr. *molture,* grinding < Lat. *molere,* to grind.] A fee paid for grinding grain at a mill.

**mum¹** (mŭm) *adj.* [ME.] Silent.

**mum²** (mŭm) *vi.* **mummed, mum·ming, mums.** [ME *mummen* < OFr. *momer.*] **1.** To act or play in a pantomime. **2.** To go merrymaking in a disguise or mask esp. during a festival.

**mum³** (mŭm) *n. Informal. Chiefly Brit.* MOTHER¹ 1.

**mum⁴** (mŭm) *n. Informal.* A chrysanthemum.

**mum·ble** (mŭm′bəl) *v.* **-bled, -bling, -bles.** [ME *momelen.*] —*vt.* **1.** To utter unclearly by lowering the voice or partially closing the mouth. **2.** To chew slowly or painfully without or as if without teeth. —*vi.* **1.** To speak words unclearly, as by lowering the voice or partially closing the mouth. **2.** To chew food slowly or painfully, as if without teeth. —*n.* A low, indistinct sound or utterance. **—mum′bler** *n.*

**mum·ble·ty-peg** (mŭm′bəl-tē-pĕg′, mŭm′blē-pĕg′) *also* **mum·ble-the-peg** (mŭm′bəl-*thə*-pĕg′) *n.* [< the phrase *mumble the peg,* from the fact that orig. the loser had to pull up with his teeth a peg driven into the ground.] A game in which the players throw a knife from different positions, the object being to stick the blade firmly in the ground.

**mum·bo jum·bo** (mŭm′bō jŭm′bō) *n.* [Prob. of Mandingo orig.] **1.** An object believed to have supernatural powers : FETISH. **2.** An obscure or complicated ritual. **3.** Language or ritualistic activity designed to confuse. **4.** Incomprehensible language.

**mu meson** *n. Physics.* The muon.

**mum·mer** (mŭm′ər) *n.* [ME *mummar* < OFr. *momeur* < *momer,* to pantomime.] **1. a.** One who acts or plays in a pantomime. **b.** An actor. **2.** A masked merrymaker esp. at a festival.

**mum·mer·y** (mŭm′ə-rē) *n., pl.* **-ies. 1.** A performance by mummers. **2.** A pretentious or hypocritical show or ceremony.

**mum·mi·fy** (mŭm′ə-fī′) *v.* **-fied, -fy·ing, -fies.** —*vt.* **1.** To make into a mummy by embalming and drying. **2.** To cause to shrivel and dry up. —*vi.* To shrivel or dry up like a mummy. **—mum′mi·fi·ca′tion** *n.*

**mum·my¹** (mŭm′ē) *n., pl.* **-mies.** [ME *mummie,* embalming ointment < OFr. *momie* < Med. Lat. *mumia* < Ar. *mūmiyā* < *mūm,* wax.] **1.** The body of a human being or animal embalmed after death in keeping with the practice of the ancient Egyptians. **2.** A withered or shrunken body resembling a mummy.

**mum·my²** (mŭm′ē) *n., pl.* **-mies.** [Alteration of MAMMY.] *Informal.* MOTHER¹ 1.

**mumps** (mŭmps) *pl.n.* [< pl. of dial. *mump,* grimace.] (*sing.* or *pl. in number*). An acute, inflammatory, contagious disease of the salivary glands, esp. the parotids, and occas. of the pancreas, ovaries, or testes, caused by a virus, *Rubula inflans.*

**munch** (mŭnch) *vt.* **munched, munch·ing, munch·es.** [ME *monchen.*] To chew with a crunching sound. **—munch′er** *n.*

**munch·ies** (mŭn′chēz) *pl.n. Slang.* **1.** Food for snacking. **2.** A craving for snack foods.

**mun·dane** (mŭn′dān′, mŭn′dān′) *adj.* [ME *mondeyne* < OFr. *mondain* < LLat. *mundanus* < Lat. *mundus,* world.] **1.** Of, pertaining to, or typical of this world. **2.** Characteristic of or concerned with the ordinary. **—mun·dane′ly** *adv.*

**mung bean** (mŭng) *n.* [Hindi *mūg* < Skt. *mudgaḥ.*] A bean, *Phaseolus aureus,* cultivated esp. as the source of bean sprouts.

**mun·go** (mŭng′gō) *n.* [Orig. unknown.] Low-quality reclaimed wool.

**mu·nic·i·pal** (myo͞o-nĭs′ə-pəl) *adj.* [Lat. *municipalis* < *municipium,* town < *municeps,* citizen : *munus,* public office + *capere,* to take.] **1. a.** Of, relating to, or typical of a municipality. **b.** Having local self-government. **2.** Of or relating to a nation's internal affairs as opposed to its foreign affairs. **—mu·nic′i·pal·ly** *adv.*

**mu·nic·i·pal·i·ty** (myo͞o-nĭs′ə-păl′ĭ-tē) *n., pl.* **-ties. 1.** A political unit, as a city or town, incorporated for local self-government. **2.** A body of officials appointed to manage the affairs of a municipality.

**mu·nic·i·pal·ize** (myo͞o-nĭs′ə-pə-līz′) *vt.* **-ized, -iz·ing, -iz·es. 1.** To place under municipal ownership. **2.** To make a municipality of. **—mu·nic′i·pal·i·za′tion** *n.*

**mu·nif·i·cent** (myo͞o-nĭf′ĭ-sənt) *adj.* [Lat. *munificens* < *munificus* : *munus,* gift + *facere,* to make.] **1.** Extremely liberal in giving. **2.** Showing great generosity <a *munificent* donation> **—mu·nif′i·cence** *n.* **—mu·nif′i·cent·ly** *adv.*

**mu·ni·ment** (myo͞o′nə-mənt) *n.* [ME < OFr. < Med. Lat. *munimentum* < Lat., defense < *munire,* to fortify.] **1. muniments.** *Law.* Documentary evidence by which one can defend a title to property or a claim to rights. **2.** *Archaic.* A means of defense or protection.

**mu·ni·tion** (myo͞o-nĭsh′ən) *n.* [OFr., fortification < Lat. *munitio* < *munire,* to defend.] *often* **munitions.** War materiel, esp. weapons and ammunition. —*vt.* **-tioned, -tion·ing, -tions.** To supply with munitions.

**Mun·ster** (mŭn′stər, mo͝on′-) *n. var. of* MUENSTER.

**munt·jac** *also* **munt·jak** (mŭnt′jăk′) *n.* [Malay *menjangan,* deer.] A small deer of the genus *Muntiacus* of southeastern Asia and the East Indies.

**mu·on** (myo͞o′ŏn′) *n. Physics.* A subatomic particle in the lepton family, having a mass 207 times that of the electron, a negative electric charge, and a mean lifetime of $2.2 \times 10^{-6}$ second.

**mu·ral** (myo͝or′əl) *n.* [OFr. < Lat. *muralis* < *murus,* wall.] A work of art, as a painting, applied directly to a ceiling or wall. —*adj.* **1.** Of, relating to, or like a wall. **2.** Applied to a wall. **—mu′ral·ist** *n.*

**mu·ram·ic acid** (myo͞o-răm′ĭk) *n.* [Lat. *murus,* wall + AM(IDE) + -IC.] An amino sugar, $C_9H_{17}NO_7$, occurring in the murein in the cell walls of bacteria and blue-green algae.

**mur·der** (mûr′dər) *n.* [ME *murther* < OE *morðor.*] **1.** The unlawful killing of one human being by another, esp. with premeditated malice. **2.** *Slang.* Something very difficult, uncomfortable, or dangerous. —*v.* **-dered, -der·ing, -ders.** —*vt.* **1.** To kill (a human being) unlawfully. **2.** To kill brutally or inhumanly. **3.** To put an end to : DESTROY. **4.** To spoil by ineptness : MUTILATE <*murdered* the piano selections> **5.** *Slang.* To defeat decisively. —*vi.* To commit murder. **—get away with murder.** *Informal.* To escape punishment for or detection of a blameworthy act. **—mur′der·er** *n.* **—mur′der·ess** *n.*

☆ **syns:** MURDER, BUMP OFF, DO IN, HIT, KILL, KNOCK OFF, LIQUIDATE, OFF, PUT AWAY, RUB OUT, SLAY, WASTE, WIPE OUT, ZAP *v. core meaning* : to take the life of (a person or persons) unlawfully and deliberately <*murdered* those who dared to disagree>

**mur·der·ous** (mûr′dər-əs) *adj.* **1.** Capable of, guilty of, or intending murder <*murderous* street ruffians> **2.** Typical of or giving rise to murder or bloodshed. **3.** *Informal.* Severe or overwhelming <a *murderous* budget problem> **—mur′der·ous·ly** *adv.* **—mur′der·ous·ness** *n.*

**mure** (myo͝or) *vt.* **mured, mur·ing, mures.** [ME *muren* < OFr. *murer* < LLat. *murare,* to wall in < Lat. *murus,* wall.] To immure.

**mu·re·in** (myŏŏr'ē-ĭn, myŏŏr'ēn') *n.* [MUR(AMIC ACID) + -EIN.] A peptidoglycan that contains muramic acid and occurs in procaryotic cell walls.

**mu·rex** (myŏŏr'ĕks') *n., pl.* **mu·ri·ces** (myŏŏr'ĭ-sēz') *or* **mu·rex·es.** [NLat. *Murex*, genus name < Lat. *murex*, purple fish.] A marine gastropod of the genus *Murex*, with rough spiny shells, common in warm seas and yielding a purple dye.

**mu·ri·cate** (myŏŏr'ĭ-kāt') *also* **mu·ri·cat·ed** (-kā'tĭd) *adj.* [Lat. *muricatus*, shaped like a murex < *murex*, murex.] Having a roughened surface because of many short spines.

**mu·ri·ces** (myŏŏr'ĭ-sēz') *n. var. pl. of* MUREX.

**mu·rine** (myŏŏr'īn') *adj.* [Lat. *murinus*, of mice < *mus*, mouse.] **1.** Of or relating to a member of the rodent family Muridae, including rats and mice. **2.** Caused, transmitted, or affected by rodents of the family Muridae <a *murine* plague> **—mu'rine'** *n.*

**murk** *also* **mirk** (mûrk) [ME *mirke*, poss. < OE *mirce.*] *—n.* Gloom : darkness. *—adj. Archaic.* Dark : gloomy.

**murk·y** *also* **mirk·y** (mûr'kē) *adj.* **-i·er, -i·est. 1.** Dark or gloomy. **2. a.** Heavy with smoke, mist, or fog : HAZY. **b.** Turbid with sediment <*murky* waters> **3. a.** Cloudy in color : DULL. **b.** Cloudy in mind : MUDDLED. **4.** Hard to understand : OBSCURE. **—murk'i·ly** *adv.* **—murk'i·ness** *n.*

**mur·mur** (mûr'mər) *n.* [ME *murmure* < OFr. < Lat. *murmur.*] **1.** A low, unclear, and continuous sound <the *murmur* of the rain> **2.** An indistinct complaint : MUTTER. **3.** A whispered utterance <a *murmur* of dismay> **4.** *Med.* An abnormal sound, usu. in the thoracic cavity, originating in the heart or lungs and detected by the ear or a device such as a stethoscope. *—v.* **-mured, -mur·ing, -murs.** *—vi.* **1.** To make a low, continuous, and unclear sound or sequence of sounds. **2.** To complain in low mumbling tones : GRUMBLE. *—vt.* To utter indistinctly. **—mur'mur·er** *n.* **—mur'mur·ing·ly** *adv.* **—mur'mur·ous** *adj.* **—mur'mur·ous·ly** *adv.*

☆ ***syns:*** MURMUR, MUMBLE, SIGH, SOUGH, SUSURRATION, WHISPER *n. core meaning* : a low, indistinct, often continuous sound <the *murmur* of distant voices>

**mur·phy** (mûr'fē) *n., pl.* **-phies.** [< *Murphy*, a common Irish name, from the fact that the potato was a staple Irish food.] *Slang.* A potato.

**Mur·phy bed** (mûr'fē) *n.* [After William *Murphy* (1876–1959).] A bed that folds or swings into a closet.

**mur·rain** (mûr'ĭn) *n.* [ME *moreyne* < OFr. *morine* < *morir*, to die < Lat. *mori.*] **1.** A malignant and highly infectious disease of domestic plants or animals, as potato blight or anthrax. **2.** A pestilence.

**murre** (mûr) *n., pl.* **murre** *or* **murres.** [Orig. unknown.] A sea bird of the genus *Uria* of north temperate and arctic regions.

**mur·rey** (mûr'ē) *n.* [ME < OFr. *more* < Med. Lat. *moratus* < Lat. *morum*, mulberry.] MULBERRY 4.

**mur·ther** (mûr'thər) *n. & v. Obs. var. of* MURDER.

**Mus·ca** (mŭs'kə) *n.* [< Lat. *musca*, fly.] A constellation in the Southern Hemisphere.

**Mus·ca·det** (mŭs'kə-dā') *n.* [Fr. < Prov., muscadet grape < *musc*, musk odor.] A dry white wine of French origin.

**mus·ca·dine** (mŭs'kə-dīn', -dĭn) *n.* [Alteration of MUSCATEL.] A woody vine, *Vitis rotundifolia* of the southeastern United States, bearing a musky grape used for making wine.

**mus·cae vo·li·tan·tes** (mŭs'ē vŏl'ĭ-tăn'tēz) *pl.n.* [Lat., flying flies.] Small motes and threads that seem to move about the field of vision, caused by cell fragments or other defects in the vitreous humor and the lens of the eye.

**mus·ca·rine** (mŭs'kə-rēn') *n.* [< NLat. *muscaria*, specific epithet of *Amanita muscaria*, fly agaric < Lat. *muscarius*, of flies < *musca*, fly.] A highly toxic organic compound, $C_9H_{21}O_3N$, related to the cholines, obtained from the red form of the mushroom *Amanita muscaria* and found in dead animal tissue. **—mus'ca·rin'ic** *adj.*

**mus·cat** (mŭs'kăt', -kət) *n.* [OFr. < OProv. < *musc*, musk < LLat. *muscus.*] **1.** Any of various sweet white grapes used for making wine or raisins. **2.** MUSCATEL 1.

**mus·ca·tel** (mŭs'kə-tĕl') *n.* [ME *muscadelle* < OFr. *muscadel* < OProv. *muscadel*, dim. of *muscat*, muscat.] **1.** A rich, sweet wine made from muscat grapes. **2.** A muscat grape or raisin.

**mus·cle** (mŭs'əl) *n.* [OFr. < Lat. *musculus*, dim. of *mus*, mouse.] **1.** A tissue made up of fibers that can contract and relax to effect bodily movement. **2.** A contractile organ composed of muscle tissue. **3.** Muscular strength <enough *muscle* to hurl the discus> **4.** Power or authority. **5.** *Slang.* A hired thug <beaten up by *muscle* from out of town> *—vi.* **-cled, -cling, -cles.** To make one's way by or as if by force <*muscled* into the conversation>

☆ ***syns:*** MUSCLE, CLOUT, FORCE, POWER, WEIGHT *n. core meaning* : effective means of influencing, compelling, or pushing <putting *muscle* into law enforcement>

**mus·cle-bound** (mŭs'əl-bound') *adj.* **1.** Having stiff, overdeveloped muscles, usu. because of too much exercise. **2.** Marked by inflexibility : RIGID.

**muscle fiber** *n.* An elongated, contractile cell with highly striated cytoplasm.

**muscle sugar** *n.* Inositol.

**mus·co·vite** (mŭs'kə-vīt') *n.* [< *Muscovy glass*, its former name.] A mineral, the most common form of mica, consisting essentially of hydrous potassium aluminum silicate, $KAl_2(AlSi_3O_{10})(OH)_2$, that ranges from colorless or pale yellow to gray and brown, has a pearly luster, and is used as an insulator.

**Mus·co·vite** (mŭs'kə-vīt') *n.* A native or resident of Moscow or of Muscovy. **—Mus'co·vite'** *adj.*

**Mus·co·vy duck** (mŭs'kə-vē, -kō'-) *n.* [After *Muscovy*, the principality of Moscow.] A waterfowl, *Cairina moschata*, found wild from Mexico to Brazil but widely domesticated for its tasty flesh.

**mus·cu·lar** (mŭs'kyə-lər) *adj.* [< Lat. *musculus*, muscle.] **1.** Relating to or consisting of muscle. **2.** Accomplished with the use of the muscles. **3.** Having well-developed muscles : BRAWNY. **4.** Of great strength. **—mus'cu·lar'i·ty** (-lăr'ĭ-tē) *n.* **—mus'cu·lar·ly** *adv.*

**muscular dystrophy** *n.* A chronic noncontagious disease of unknown etiology in which total incapacitation follows gradual but irreversible muscular deterioration.

**mus·cu·la·ture** (mŭs'kyə-lə-chŏŏr') *n.* [Fr. < Lat. *musculus*, muscle.] The system of muscles of an animal or a body part.

**mus·cu·lo·skel·e·tal** (mŭs'kyə-lō-skĕl'ĭ-tl) *adj.* [Lat. *musculus*, muscle + SKELETAL.] Of or involving the muscles and the skeleton.

**muse** (myōōz) *v.* **mused, mus·ing, mus·es.** [ME *musen* < OFr. *muser* < *mus*, snout < Med. Lat. *musum.*] *—vi.* To consider or meditate at length <*muse* on the frailties of humankind> *—vt.* To consider reflectively <*muse* the problem> *—n.* A state of deep reflection or meditation.

**Muse** (myōōz) *n.* [ME < OFr. < Lat. *Musa* < Gk. *Mousa.*] **1.** *Gk. Myth.* One of the nine daughters of Mnemosyne and Zeus, each of whom presided over a different art or science. **2. muse. a.** A guiding spirit. **b.** A source of inspiration. **3. muse.** A poet.

**mu·sette** (myōō-zĕt') *n.* [ME < OFr. < *muser*, to play the musette, muse.] **1.** A small French bagpipe having a soft sound. **2.** A small leather or canvas bag with a shoulder strap.

**mu·se·um** (myōō-zē'əm) *n.* [Lat. < Gk. *mouseion* < *mouseios*, of the Muses < *Mousa*, Muse.] An institution for the acquisition, preservation, study, and exhibition of works of artistic, historical, or scientific value.

**mush**[1] (mŭsh) *n.* [Prob. alteration of MASH.] **1.** A thick mixture of cornmeal boiled in liquid. **2.** Thick, soft, pulpy matter. **3.** *Informal.* **a.** Maudlin sentimentality. **b.** Foolish puerile amorousness.

**mush**[2] (mŭsh) *vi.* **mushed, mush·ing, mush·es.** [Canadian Fr. *mouche*, imper. of *moucher*, to hasten < Fr. *mouche*, fly < Lat. *musca.*] To travel with a dog sled. —Used often as a command to a team of dogs. *—n.* A journey by dog sled. **—mush'er** *n.*

**mush·room** (mŭsh'rōōm', -rŏŏm') *n.* [ME *muscheron* < OFr. *mousseron.*] **1.** Any of various fleshy fungi of the class Basidiomycetes, with an umbrella-shaped cap borne on a stalk. **2.** Something shaped like a mushroom. *—vi.* **-roomed, -room·ing, -rooms. 1.** To multiply, grow, or expand rapidly <New towns *mushroomed* in the postwar years.> **2.** To spread out, flatten, or swell into a shape like a mushroom. **—mush'room'like** *adj.*

**mush·y** (mŭsh'ē) *adj.* **-i·er, -i·est. 1.** Soft in consistency. **2.** *Informal.* **a.** Excessively sentimental : MAUDLIN. **b.** Foolishly amorous : INFATUATED. **—mush'i·ly** *adv.* **—mush'i·ness** *n.*

**mu·sic** (myōō'zĭk) *n.* [ME < OFr. *musique* < Lat. *musica* < Gk. *mousikē (tekhnē)*, (art) of the Muses < *mousikos*, of the Muses < *Mousa*, Muse.] **1.** The art of arranging tones in an orderly sequence so as to produce a unified and continuous composition. **2.** Vocal or instrumental sounds with rhythm, melody, and harmony. **3. a.** A musical composition. **b.** The written or printed score for a musical composition. **4.** A musical accompaniment. **5.** A particular category or kind of music <dance *music*><bluegrass *music*> **6.** An aesthetically pleasing or harmonious sound or mix of sounds.

▲ word history: In Greek the word *mousikē*, originally an adjective that meant "pertaining to the Muses," was used of any of the arts over which the Muses presided, including poetry, drama, and dance, as well as song and instrumental music. Music was so important to the Greeks, however, that *mousikē* was used to refer especially to that branch of the arts. The Romans borrowed the Greek word and by medieval times Latin *musica* referred only to the musical art.

**mu·si·cal** (myōō'zĭ-kəl) *adj.* **1.** Of, relating to, or capable of producing music <a *musical* instrument> **2.** Typical of or similar to music : MELODIOUS <a *musical* lilt in their speech> **3.** Set to or accompanied by music <a *musical* program> **4.** Fond of or skilled in music. *—n.* **1.** A musical comedy. **2.** *Archaic.* A musicale. **—mu'si·cal'i·ty** (-kăl'ĭ-tē) *n.* **—mu'si·cal·ly** *adv.*

**musical chairs** *pl.n. (sing. in number).* **1.** A game in which the players walk to music around a row of chairs having one chair fewer than the number of players and rush to sit down when the music stops. **2.** *Informal.* A rearrangement, as of the elements of a situation, with slight practical influence or importance.

**musical comedy** *n.* A play or motion picture combining dialogue with songs and dances.

**mu·si·cale** (myōō'zĭ-kăl') *n.* [Fr. < *(soirée) musicale*, musical (evening).] A program of music at a party or social gathering.

**music box** *n.* A box enclosing a mechanical device that produces music when activated by a clockwork.
**music drama** *n.* An opera in which the continuity is not broken by arias, recitatives, or ensembles, and in which the music reflects or embodies the dramatic action.
**music hall** *n.* **1.** An auditorium for musical performances. **2.** *Chiefly Brit.* **a.** A vaudeville theater. **b.** Vaudeville.
**mu·si·cian** (myo͞o-zĭsh'ən) *n.* [ME *musicien* < OFr. < Lat. *musica,* music.] **1.** A composer or performer of music. **2.** A performer esp. of instrumental music. **—mu·si'cian·ly** *adj.*
**mu·si·cian·ship** (myo͞o-zĭsh'ən-shĭp') *n.* Skill, insight, and artistry in the performance of music.
**music of the spheres** *n.* An inaudible harmony that Pythagoras thought to be produced by the movements of celestial bodies.
**mu·si·col·o·gy** (myo͞o'zĭ-kŏl'ə-jē) *n.* Historical and scientific study of music. **—mu'si·col'o·gist** *n.*
**mus·ing** (myo͞o'zĭng) *adj.* Absorbed in thought : ABSTRACTED. *—n.* Meditation : contemplation. **—mus'ing·ly** *adv.*
**mu·sique con·crète** (mo͞o-sēk' kŏn-krĕt') *n.* [Fr. : *musique,* music + *concrète,* concrete.] A musical composite of heterogeneous recorded sounds randomly modified and arranged.
**musk** (mŭsk) *n.* [ME *muske* < OFr. *musc* < LLat. *muscus* < Gk. *moskhos* < Pers. *muskh,* prob. < Skt. *muṣkaḥ,* testicle.] **1. a.** A greasy secretion with a strong odor, produced in a glandular sac under the skin of the abdomen of the male musk deer and used in perfumery. **b.** A similar secretion of certain other vertebrates, as the otter or civet. **c.** A synthetic chemical similar to natural musk in odor or use. **2. a.** The odor of musk. **b.** An odor similar to musk. **3.** The musk deer. **—musk'i·ness** *n.* **—musk'y** *adj.*
**musk beaver** *n.* The muskrat.
**musk deer** *n.* A small hornless deer, *Moschus moschiferus* of central and northeastern Asia, the male of which secretes musk.
**musk duck** *n.* **1.** The Muscovy duck. **2.** A waterfowl, *Biziura lobata* of Australia, the male of which has a leathery chin lobe and gives off a musky odor during the breeding season.
**mus·keg** (mŭs'kĕg') *also* **mas·keg** (măs'-) *n.* [Cree *maskeek.*] A swamp or bog formed by accumulated sphagnum moss, leaves, and decayed matter resembling peat.
**mus·kel·lunge** (mŭs'kə-lŭnj') *n., pl.* **muskellunge** *or* **-lung·es.** [Of Algonquian orig.] A large North American freshwater game fish, *Esox masquinongy,* similar to the pike.
**mus·ket** (mŭs'kĭt) *n.* [Fr. *mousquet* < Ital. *moschetto,* bolt for a crossbow, musket, dim. of *mosca,* fly < Lat. *musca.*] A smoothbore shoulder gun used from the late 16th through the 18th cent.
**mus·ket·eer** (mŭs'kĭ-tîr') *n.* [Fr. *mousquetaire* < *mousquet,* musket.] **1.** A soldier armed with a musket. **2.** A member of the 17th- and 18th-cent. French royal household bodyguard.
**mus·ket·ry** (mŭs'kĭ-trē) *n.* **1.** Muskets in general. **2.** Musket fire. **3.** The technique of using small arms.
**Mus·kho·ge·an** (mŭs-kō'gē-ən) *n.* A family of American Indian languages that includes Chickasaw, Choctaw, Creek, and Seminole.
**mus·kie** *or* **mus·ky** (mŭs'kē) *n., pl.***-kies.** The muskellunge.
**musk·mel·on** (mŭsk'mĕl'ən) *n.* **1.** A variety of the melon *Cucumis melo,* as the cantaloupe, having fruit with a netted rind and flesh with a musky aroma. **2.** The fruit of the muskmelon.
**musk ox** *n.* A large hoofed mammal, *Ovibos moschatus* of northern Canada and Greenland, that emits a musky odor and has horns and a long, shaggy, brown-to-black coat.
**musk·rat** (mŭs'krăt') *n., pl.* **muskrat** *or* **-rats. 1.** An aquatic rodent, *Ondatra zibethica* of North America, having a brown coat used as a fur. **2.** The fur of the muskrat.
**musk rose** *n.* A prickly shrub native to the Mediterranean region, *Rosa moschata,* grown for its musk-scented white flower clusters.
**musk turtle** *n.* A small freshwater turtle of the genus *Sternotherus* of eastern North America, having a musky odor.
**mus·ky** (mŭs'kē) *n. var. of* MUSKIE.
**Mus·lim** (mŭz'ləm, mo͝oz'-, mŭs'-, mo͝os'-) *n.* [Ar. *muslim,* active part. of *aslama,* he surrendered.] **1.** A Moslem. **2.** A Black Muslim. **—Mus'lim** *adj.*
**mus·lin** (mŭz'lĭn) *n.* [Fr. *mousseline* < Ital. *mussolina* < *Mussolo,* Mosul, Iraq < Ar. *Al-Mawṣil.*] **1.** A sturdy plain-weave cotton fabric, used esp. for sheets. **2.** A model, as of a garment, used as a pattern.
**mus·quash** (mŭs'kwŏsh', -kwôsh') *n.* [Of Algonquian orig.] The muskrat.
**†muss** (mŭs) *vt.* **mussed, muss·ing, muss·es.** [Prob. alteration of MESS.] To make messy or untidy : RUMPLE. *—n.* **1.** A mess : disorder. **2.** *Regional.* A squabble : row. **—muss'i·ly** *adv.* **—muss'i·ness** *n.* **—muss'y** *adj.*
**mus·sel** (mŭs'əl) *n.* [ME *muscle* < OE *muscelle.*] **1.** A marine bivalve mollusk, esp. the edible *Mytilus edulis,* with a blue-black shell. **2.** Any of several freshwater bivalve mollusks of the genera *Anodonta* and *Unio* of the central United States, whose shells provide mother-of-pearl.
**Mus·sul·man** (mŭs'əl-mən) *n., pl.* **-men** *or* **-mans.** [Turk. *musulmān,* prob. alteration of Ar. *muslim,* Muslim.] A Moslem.
**must**[1] (mŭst) *aux.v.* [ME *moste* < OE *mōste,* p.t. of *mōtan,* to be allowed.] **1.** To be required or obliged by law, morality, or custom <You *must* obey the speed limit.> **2.** To be compelled, as by a physical necessity or requirement <Animals *must* have food in order to live.> **3.** —Used to express a command or admonition <You *must* not do that again.><You simply *must* accept our invitation.> **4.** To be determined to <If you *must* talk, do it quietly.><I *must* have a signed contract before I can proceed.> **5.** —Used to indicate: **a.** Inevitability or certainty <Each day *must* end.> **b.** Logical probability or presumptive certainty <If their car is in the driveway, they *must* be at home.> *—vi. Archaic.* To be obliged or required to go <"I *must* from hence" —Shakespeare> *—n.* **1.** An absolute requirement <Punctuality is a *must.*> **2.** Something indispensable <A swimsuit is a *must* at the beach.>
**must**[2] (mŭst) *n.* [Prob. back-formation < MUSTY.] **1.** Staleness : mustiness. **2.** Musk.
**must**[3] (mŭst) *n.* [ME < OE < Lat. *mustum,* neuter of *mustus,* new.] Unfermented or fermenting juice esp. from grapes.
**mus·tache** *or* **mous·tache** (mŭs'tăsh', mə-stăsh') *n.* [OFr. *moustache* < Ital. *mustaccio* < Med. Gk. *moustakion* < Gk. *mustax,* mustache, upper lip.] **1.** The hair growing on the human upper lip, esp. when cultivated and groomed. **2.** Something similar to a mustache in position and appearance, esp.: **a.** Whiskers around an animal's mouth. **b.** Distinctive coloring or feathers near a bird's beak.
**mus·ta·chio** (mə-stăsh'ō, -stăsh'ē-ō', -stä'shō, -shē-ō') *n., pl.* **-chios.** [Ult. < Ital. *mustaccio,* mustache.] A mustache, esp. a luxuriant one.
**mus·tang** (mŭs'tăng') *n.* [Sp. *mesteño,* stray animal < OSp. *mesta,* association of cattle owners < Lat. *miscere,* to mix.] A wild horse of the North American plains, descended from Spanish horses.
**mus·tard** (mŭs'tərd) *n.* [ME < OFr. *moustarde* < Lat. *mustum,* must.] **1.** A plant of the genus *Brassica,* native to Eurasia, having four-petaled yellow flowers and slender pods, esp. *B. nigra* and *B. alba,* which are cultivated for their pungent seeds. **2. a.** Powdered mustard seeds used medicinally. **b.** A condiment made from powdered mustard seeds. **3.** A dark yellow to light olive brown.
**mustard gas** *n.* An oily volatile liquid, $(ClCH_2CH_2)_2S$, used in warfare as a gaseous blistering agent.
**mustard oil** *n.* An oil obtained from mustard seeds.
**mustard plaster** *n.* A medicinal plaster containing a mixture of powdered mustard, flour, and water.
**mus·te·line** (mŭs'tə-līn', -lĭn) *adj.* [Lat. *mustelinus,* of a weasel < *mustela,* weasel, prob. < *mus,* mouse.] Of or relating to fur-bearing mammals of the family Mustelidae, which includes the badger, mink, otter, and weasel.
**mus·ter** (mŭs'tər) *v.* **-tered, -ter·ing, -ters.** [ME *mustren* < OFr. *moustrer* < Lat. *monstrare,* to show < *monstrum,* portent < *monēre,* to warn.] *—vt.* **1. a.** To summon or assemble (troops). **b.** To convene or collect together. **2.** To gather or collect (e.g., courage). *—vi.* To gather or assemble. **—muster in.** To enlist (someone) in military service <was *mustered in* just after graduation> **—muster out.** To discharge (someone) from military service <was *mustered out* near the end of the war> *—n.* **1. a.** A gathering, esp. of troops, for service, review, inspection, or roll call. **b.** The persons assembled for such a gathering. **2.** A muster roll. **3.** An assembly : gathering <a *muster* of community leaders at a luncheon>
**muster roll** *n.* **1.** The official roll of military personnel. **2.** An inventory : roster.
**must·n't** (mŭs'ənt). Must not.
**must·y** (mŭs'tē) *adj.* **-i·er, -i·est.** [Alteration of obs. *moisty* < MOIST.] **1.** Having a moldy or stale odor or taste. **2. a.** Worn : trite. **b.** Antiquated. **—must'i·ly** *adv.* **—must'i·ness** *n.*
**mu·ta·ble** (myo͞o'tə-bəl) *adj.* [Lat. *mutabilis* < *mutare,* to change.] **1.** Capable of or subject to alteration or change. **2.** Prone to frequent change. **—mu'ta·bil'i·ty, mu'ta·ble·ness** *n.* **—mu'ta·bly** *adv.*
**mu·ta·gen** (myo͞o'tə-jən, -jĕn') *n.* [MUTA(TION) + -GEN.] An agent, as a radioactive element or ultraviolet light, that causes biological mutation. **—mu'ta·gen'ic** *adj.* **—mu'ta·gen'i·cal·ly** *adv.*
**mu·tant** (myo͞ot'nt) *n.* [Lat. *mutans, mutant-,* changing, pr.part. of *mutare,* to change.] An organism or individual differing from the parental strain or strains as a result of mutation. **—mu'tant** *adj.*
**mu·tate** (myo͞o'tāt', myo͞o-tāt') *vt. & vi.* **-tat·ed, -tat·ing, -tates.** [Lat. *mutare, mutat-.*] To cause to undergo or to undergo change by mutation. **—mu'ta'tive** (-tā'tĭv, -tə-tĭv) *adj.*
**mu·ta·tion** (myo͞o-tā'shən) *n.* [ME *mutacioun* < OFr. *mutation* < Lat. *mutatio* < *mutare,* to change.] **1.** The act or process of being altered or changed. **2.** An alteration, as in form, nature, or quality. **3.** *Biol.* **a.** An inheritable alteration of the genes or chromosomes of an organism. **b.** A mutant. **4.** The change caused in the sound of one vowel by its assimilation to another vowel, esp. an umlaut. **—mu·ta'tion·al** *adj.* **—mu·ta'tion·al·ly** *adv.*
**mu·ta·tis mu·tan·dis** (mo͞o-tä'tĭs mo͞o-tän'dĭs) *adv.* [Lat.] The necessary changes having been made.
**mutch·kin** (mŭch'kĭn) *n.* [ME *muchekyn* < MDu. *mudseken.*] *Scot.* A unit of liquid measure equal to 0.9 U.S. pint.
**mute** (myo͞ot) *adj.* **mut·er, mut·est.** [ME *muet* < OFr. < *mu* < Lat. *mutus.*] **1.** Refraining from producing speech or vocal sound.

**2. a.** Unable to speak. **b.** Unable to vocalize, as certain animals. **3.** Expressed without speech : UNSPOKEN <a *mute* question> **4.** *Law.* Refusing to plead when under arraignment <stand *mute*> **5. a.** Silent, as the *e* in *house.* **b.** Pronounced with a temporary stoppage of breath, as the sounds of *p* and *b* : PLOSIVE. —*n.* **1.** One incapable of speech. **2.** *Law.* A defendant who refuses to plead when under arraignment. **3.** One of various devices for softening or muffling the tone of a musical instrument. **4. a.** A silent or unpronounced letter. **b.** A plosive : stop. —*vt.* **mut·ed, mut·ing, mutes. 1.** To soften or muffle the sound of. **2.** To soften the tone, shade, color, or hue of. **—mute'ly** *adv.* **—mute'ness** *n.*

**mut·ed** (myōō'tĭd) *adj.* **1.** Produced by or provided with a mute. **2. a.** Softened : subdued. **b.** Muffled : indistinct. **—mut'ed·ly** *adv.*

**mu·ti·late** (myōōt'l-āt') *vt.* **-lat·ed, -lat·ing, -lates.** [Lat. *mutilare, mutilat-* < *mutilus,* maimed.] **1.** To cut off or destroy an essential part, as a limb. **2.** To render imperfect by cutting up or radically altering a part <censors who *mutilated* the news story> **—mu'ti·la'tion** *n.* **—mu'ti·la'tive** *adj.* **—mu'ti·la'tor** *n.*

**mu·ti·neer** (myōōt'n-îr') *n.* [Obs. Fr. *mutinier* < OFr. *mutin,* rebellious. —see MUTINY.] A participant in a mutiny.

**mu·ti·nous** (myōōt'n-əs) *adj.* [< obs. *mutine,* mutiny.] **1.** Relating to, engaged in, or disposed toward mutiny. **2.** Unruly : uncontrollable. **—mu'ti·nous·ly** *adv.* **—mu'ti·nous·ness** *n.*

**mu·ti·ny** (myōōt'n-ē) *n., pl.* **-nies.** [Obs. *mutine* < OFr. *mutin,* rebellious < *muete,* revolt < VLat. **movita* < Lat. *movēre,* to move.] Open rebellion against lawful authority, esp. rebellion of military enlisted personnel against officers. —*vi.* **-nied, -ny·ing, -nies.** To commit mutiny.

**mut·ism** (myōō'tĭz'əm) *n.* Inability to speak.

**mutt** (mŭt) *n.* [Short for MUTTONHEAD.] *Slang.* A mongrel dog.

**mut·ter** (mŭt'ər) *v.* **-tered, -ter·ing, -ters.** [ME *muttren.*] —*vi.* **1.** To speak low and indistinctly. **2.** To complain or grumble sullenly. —*vt.* To utter or say in low tones. **—mut'ter** *n.* **—mut'ter·er** *n.*

**mut·ton** (mŭt'n) *n.* [ME *motoun* < OFr. *moton,* of Celt. orig.] The meat of fully grown sheep.

**mut·ton·chops** (mŭt'n-chŏps) *pl.n.* Side whiskers shaped like chops of meat.

**mut·ton·fish** (mŭt'n-fĭsh') *n., pl.* **muttonfish** or **-fish·es.** The eelpout.

**mut·ton·head** (mŭt'n-hĕd') *n.* [From the stupidity of sheep.] *Slang.* A stupid person. **—mut'ton·head'ed** *adj.*

**mu·tu·al** (myōō'chōō-əl) *adj.* [ME *mutuall* < OFr. *mutuel* < Lat. *mutuus* < *mutare,* to change.] **1.** Having the same relationship each to the other <*mutual* enemies> **2.** Directed and received in equal amount <*mutual* admiration> **3.** Possessed in common <*mutual* avocations> **—mu'tu·al'i·ty** (-ăl'ĭ-tē) *n.* **—mu'tu·al·ly** *adv.*

**mutual fund** *n.* A company without fixed capitalization, freely buying and selling its own shares and using its capital to invest in other companies.

**mutual inductance** *n.* **1.** The ratio expressed by the flux linking one circuit with a neighboring circuit divided by the current in the neighboring circuit. **2.** The ratio expressed by the electromotive force induced in a circuit by a neighboring circuit divided by the corresponding change of current in the neighboring circuit.

**mutual insurance** *n.* An insurance system in which the insured persons become company members, each paying specified amounts into a common fund from which members are entitled to indemnification in case of loss.

**mu·tu·al·ism** (myōō'chōō-ə-lĭz'əm) *n.* An association, as parasitism or symbiosis, between two organisms.

**mu·tu·al·ize** (myōō'chōō-ə-līz') *vt.* **-ized, -iz·ing, -iz·es. 1.** To make mutual. **2.** To reorganize or set up (a corporation) so that the majority of common stock is owned by employees or customers. **—mu'tu·al·i·za'tion** *n.*

**muu·muu** (mōō'mōō') *n.* [Hawaiian *mu'u mu'u.*] A long loose dress that hangs free from the shoulders.

muumuu

**Mu·zak** (myōō'zăk'). A trademark for recorded background music transmitted by wire, as to businesses, on a subscription basis.

**mu·zhik** *also* **mou·jik** *or* **mu·jik** *or* **mu·zjik** (mōō-zhēk', -zhĭk') *n.* [R. < *muzh,* man.] A Russian peasant.

**muz·zle** (mŭz'əl) *n.* [ME *musell* < OFr. *musel* < LLat. *musum,* snout.] **1.** The forward, projecting part of the head of certain animals, including the nose and jaws. **2.** A leather or wire device that, when fitted over an animal's snout, prevents biting and eating. **3.** A restraint to free expression or movement. **4.** The forward discharging end of a firearm barrel. —*vt.* **-zled, -zling, -zles. 1.** To put a muzzle on (an animal). **2.** To restrain <The gag order effectively *muzzled* all press coverage.> **—muz'zler** *n.*

**muz·zy** (mŭz'ē) *adj.* **-zi·er, -zi·est.** [Orig. unknown.] *Informal.* **1.** Mentally confused : MUDDLED. **2.** Indistinct : blurred. **—muz'zi·ly** *adv.* **—muz'zi·ness** *n.*

**my** (mī) *adj.* [ME < OE *mīn.*] **1.** —Used attributively to indicate possession, agency, or reception of an action by the speaker <*my* hat><doing *my* chores><suffered *my* first defeat> **2.** —Used preceding various forms of affectionate, polite, or familiar address <*my* lady><*my* dear Dr. Smith><*my* good sir> **3.** —Used in various interjectional phrases <*My* word!><*My* gosh!> —*interj.* —Used to express surprise, pleasure, or dismay <Oh, *my!*><*My, my!*>

**my-** *pref. var. of* MYO-.

**my·al·gi·a** (mī-ăl'jē-ə, -jə) *n.* [MY(O)- + -ALGIA.] **1.** Muscular rheumatism. **2.** Muscular pain.

**my·as·the·ni·a** (mī'əs-thē'nē-ə) *n.* [MY(O)- + ASTHENIA.] Abnormal muscular weakness or fatigue. **—my'as·then'ic** (-thĕn'ĭk) *adj.*

**myc-** *pref. var. of* MYCO-.

**my·ce·li·um** (mī-sē'lē-əm) *n., pl.* **-li·a** (-lē-ə) [NLat. : MYC(O)- + Gk. *hēlos,* wart.] The vegetative part of a fungus, made up of a mass of branching, threadlike filaments called hyphae. **—my·ce'li·al** *adj.*

**My·ce·nae·an** (mī'sə-nē'ən) *adj.* Of, relating to, or designating the Aegean civilization that spread its influence from Mycenae to many parts of the Mediterranean region from about 1400 B.C. to 1150 B.C. **—My'ce·nae'an** *n.*

**-mycete** *suff.* [NLat. *-mycetes* < Gk. *mukētes,* pl. of *mukēs,* fungus.] Fungus <actino*mycete*>

**my·ce·to·ma** (mī'sĭ-tō'mə) *n., pl.* **-mas** or **-ma·ta** (-mə-tə) [Gk. *mukēs, mukēt-,* fungus + -OMA.] **1.** A chronic fungous infection usu. affecting the foot, characterized by nodules that discharge oily pus. **2.** A mycetoma nodule. **—my'ce·to'ma·tous** (-tō'mə-təs, -tŏm'ə-təs) *adj.*

**my·ce·to·zo·an** (mī-sē'tə-zō'ən) *n.* [< NLat. *Mycetozoa,* order name : Gk. *mukēs,* fungus + Gk. *zōia,* pl. of *zōion,* animal.] A slime mold. **—my·ce'to·zo'an** *adj.*

**-mycin** *suff.* [MYC(O)- + IN.] A substance derived from a fungus <neo*mycin*>

**myco-** or **myc-** *pref.* [NLat. < Gk. *mukēs,* fungus.] Fungus <*mycology*><*mycosis*>

**my·co·bac·te·ri·um** (mī'kō-băk-tîr'ē-əm) *n., pl.* **-te·ri·a** (-tîr'ē-ə). Any of various slender, rod-shaped bacteria of the genus *Mycobacterium,* which includes the bacterium that causes tuberculosis.

**my·col·o·gy** (mī-kŏl'ə-jē) *n.* **1.** The branch of botany concerned with fungi. **2.** The fungi native to a region. **—my'co·log'i·cal** (-kə-lŏj'ĭ-kəl), **my'co·log'ic** *adj.* **—my·col'o·gist** *n.*

**my·cor·rhi·za** or **my·co·rhi·za** (mī'kə-rī'zə) *n., pl.* **-zae** (-zē') or **-zas.** [MYCO- + Gk. *rhiza,* root.] *Bot.* The symbiotic association of the mycelium of a fungus with the roots of certain plants, as conifers, beeches, or orchids. **—my'cor·rhi'zal** *adj.*

**my·co·sis** (mī-kō'sĭs) *n., pl.* **-ses** (-sēz'). **1.** A fungous growth in the body. **2.** A disease caused by a fungous growth.

**my·dri·a·sis** (mĭ-drī'ə-sĭs) *n.* [Lat. < Gk. *mudriasis.*] Prolonged abnormal dilatation of the pupil of the eye due to disease or a drug.

**myd·ri·at·ic** (mĭd'rē-ăt'ĭk) *n.* [< MYDRIASIS.] A drug producing dilatation of the pupils. **—myd'ri·at'ic** *adj.*

**myel-** *pref. var. of* MYELO-.

**my·e·len·ceph·a·lon** (mī'ə-lĕn-sĕf'ə-lŏn') *n.* The rear part of the embryonic hindbrain from which the medulla oblongata develops. **—my'e·len·ce·phal'ic** (-sə-făl'ĭk) *adj.*

**my·e·lin** (mī'ə-lĭn) *also* **my·e·line** (-lĭn, -lēn') *n.* **1.** A fatty white material encasing some nerve fibers. **2.** One of several fatlike substances found in body tissues. **—my'e·lin'ic** *adj.*

**my·e·li·nat·ed** (mī'ə-lə-nā'tĭd) *adj.* Having a myelin sheath.

**my·e·li·ni·za·tion** (mī'ə-lə-nĭ-zā'shən) *also* **my·e·li·na·tion** (-nā'shən) *n.* The process of growing a myelin sheath or state of having a myelin sheath.

**my·e·li·tis** (mī'ə-lī'tĭs) *n.* Inflammation of bone marrow or the spinal column.

**myelo-** or **myel-** *pref.* [NLat. < Gk. *muelos,* marrow < *mus,* muscle.] Spinal cord : marrow <*myelitis*>

**my·e·loid** (mī'ə-loid') *adj.* **1.** Of, related to, or derived from bone marrow. **2.** Of or relating to the spinal cord.

**my·e·lo·ma** (mī'ə-lō'mə) *n., pl.* **-mas** or **-ma·ta** (-mə-tə). A malignant tumor of the bone marrow. **—my'e·lo'ma·toid'** *adj.*

**my·ia·sis** (mī'ə-sĭs, mī-ī'ə-sĭs) *n., pl.* **my·ia·ses** (mī'ə-sēz) [Gk. *mua,* fly + -IASIS.] Infestation of human tissue by fly maggots or flies or a disease resulting from it.

**My·lar** (mī'lär'). A trademark for a thin strong polyester film.

**my·lo·nite** (mī'lə-nīt') *n.* [Gk. *mulōn,* mill < *mulos* + -ITE.] Fine-grained laminated rock formed by the shifting of rock layers.

**my·na** or **my·nah** *also* **mi·na** (mī'nə) *n.* [Hindi *mainā* < Skt.

*madanaḥ,* passion, lust.] Any of various birds of the family Sturnidae of southeastern Asia, related to the starlings and having blue-black to dark-brown coloration and yellow bills, certain species of which are known for their mimicry of human speech.

**myn·heer** (mīn-hâr′, -hîr′) *n.* [Du. *mijnheer* : *mijn,* my (< MDu. *mijni*) + *heer,* lord < MDu.] **1.** *often* **Mynheer.** The Dutch title of courtesy and respect equivalent to the English *sir* or *Mr.* **2.** *Informal.* A Dutchman.

**myo-** or **my-** *pref.* [NLat. < Gk. *mus,* muscle.] Muscle <*myo*graph><*myo*genic>

**my·o·car·di·o·graph** (mī′ō-kär′dē-ə-grăf′) *n.* An instrument for graphically depicting the action of the heart muscle.

**my·o·car·di·tis** (mī′ō-kär-dī′tĭs) *n.* [MYOCARD(IUM) + -ITIS.] Inflammation of the myocardium.

**my·o·car·di·um** (mī′ō-kär′dē-əm) *n.* [NLat. : MYO- + Gk. *kardia,* heart.] Cardiac muscle tissue. **—my′o·car′di·al** *adj.*

**my·o·gen·ic** (mī′ə-jĕn′ĭk) *also* **my·o·ge·net·ic** (mī′ō-jə-nĕt′ĭk) *adj.* **1.** Generating muscle tissue. **2.** Of muscular origin.

**my·o·glo·bin** (mī′ə-glō′bĭn) *n.* The form of hemoglobin found in muscle fibers.

**my·o·graph** (mī′ə-grăf′) *n.* An instrument for graphically depicting muscular contractions.

**my·ol·o·gy** (mī-ŏl′ə-jē) *n.* Study of muscles. **—my′o·log′ic** (mī′ə-lŏj′ĭk) *adj.* **—my·ol′o·gist** *n.*

**my·o·ma** (mī-ō′mə) *n., pl.* **-mas** *or* **-ma·ta** (-mə-tə). A tumor composed of muscle tissue. **—my·o′ma·tous** (-ō′mə-təs, -ŏm′ə-) *adj.*

**my·ope** (mī′ōp′) *n.* [Fr. < LLat. *myops,* myopic < Gk. *muōps.* —see MYOPIA.] A person who has myopia.

**my·o·pi·a** (mī-ō′pē-ə) *n.* [Gk. *muōpia* < *muōps,* myopic : *muein,* to close the eyes + *ōps,* eye.] **1.** A visual defect in which distant objects appear blurred because their images are focused in front of the retina rather than on it : NEARSIGHTEDNESS. **2.** Lack of discernment or shortsightedness in thinking or planning <political *myopia*> **—my·op′ic** (-ŏp′ĭk, -ō′pĭk) *adj.* **—my·op′i·cal·ly** *adv.*

**my·o·sin** (mī′ə-sĭn) *n.* The commonest protein in muscle, that with actin, forms actomyosin.

**my·o·sis** (mī-ō′sĭs) *n. var. of* MIOSIS.

**my·o·so·tis** (mī′ə-sō′tĭs) *n.* [NLat. *Myosotis,* genus name < Lat. *myosotis,* a kind of plant < Gk. *muosotis* : *mus,* mouse + *ous* ear.] A plant of the genus *Myosotis,* as the forget-me-not.

**my·o·to·ni·a** (mī′ə-tō′nē-ə) *n.* Tonic spasm or temporary muscular rigidity. **—my′o·ton′ic** (-tŏn′ĭk) *adj.*

**myr·i·ad** (mîr′ē-əd) *adj.* [Gk. *murias,* ten thousand < *murios,* countless.] **1.** Constituting an extremely large, indefinite number : INNUMERABLE. **2.** Composed of numerous diverse elements or facets <the *myriad* responsibilities of a parent> *—n.* **1.** *Archaic.* Ten thousand. **2.** A great number.

**myr·i·a·pod** (mîr′ē-ə-pŏd′) *n.* An arthropod, as the centipede, having a segmented body and numerous legs. **—myr′i·ap′o·dan** (-ăp′ə-dən) *adj.* & *n.* **—myr′i·ap′o·dous** (-ăp′ə-dəs) *adj.*

**my·ris·tic acid** (mə-rĭs′tĭk, mī-) *n.* [Gk. *muristikos,* fragrant < *muron,* perfume.] An organic compound, $CH_3(CH_2)_{12}COOH$, occurring in animal and vegetable fats and used in cosmetics and flavors.

**myrmeco-** *pref.* [Gk. *murmeco-* < *murmēx,* ant.] Ant <*myrme*cophagous>

**myr·me·col·o·gy** (mûr′mĭ-kŏl′ə-jē) *n.* Study of ants. **—myr′me·co·log′i·cal** (-kə-lŏj′ĭ-kəl) *adj.* **—myr′me·col′o·gist** *n.*

**myr·me·coph·a·gous** (mûr′mĭ-kŏf′ə-gəs) *adj.* Feeding on ants.

**myr·me·co·phile** (mûr′mĭ-kə-fīl′) *n.* An organism that habitually shares the nest of an ant colony. **—myr′me·coph′i·lous** (-kŏf′ə-ləs) *adj.* **—myr′me·coph′i·ly** (-kŏf′ə-lē) *n.*

**Myr·mi·don** (mûr′mə-dŏn′, -dən) *n.* [< Lat. *Myrmidones,* Myrmidons < Gk. *Murmidones.*] **1.** One of a legendary Greek warrior people of ancient Thessaly who followed their king Achilles on the expedition against Troy. **2. myrmidon.** A loyal, unquestioning follower.

**my·rob·a·lan** (mī-rŏb′ə-lən, mə-) *n.* [OFr. *mirobalan* < Lat. *myrobalanum,* fruit of a myrobalan < Gk. *murobalanos* : *muron,* perfume + *balanos,* fruit shaped like an acorn.] **1.** A native Asian tree, *Prunus cerasifera,* yielding edible red or yellow fruit. **2.** The Indian almond. **3.** The fruit of a myrobalan.

**myrobalan**
*A detail of the leaves that grow only at the end of the branch*

**myrrh** (mûr) *n.* [ME *myrr* < OE *myrrha* < Lat. < Gk. *murrha,* prob. of Semitic orig.] **1.** An aromatic gum resin obtained from several trees and shrubs of the genus *Commiphora* of India, Arabia, and eastern Africa, used in perfume and incense. **2.** SWEET CICELY 2.

**myr·tle** (mûr′tl) *n.* [ME *mirtille* < OFr. < Med. Lat. *myrtillus,* dim. of Lat. *myrtus* < Gk. *murtos.*] **1.** An evergreen shrub or tree of the genus *Myrtus,* esp. *M. communis,* an aromatic shrub native to the Mediterranean region and western Asia, having pink or white flowers and blue-black berries. **2.** PERIWINKLE².

**my·self** (mī-sĕlf′) *pron.* [ME < OE *mē self.*] **1.** The one identical with me. —Used: **a.** Reflexively as the direct or indirect object of a verb or the object of a preposition <injured *myself*><give *myself* time to think><muttered to *myself*> **b.** For emphasis <I *myself* did it.> **c.** In an absolute construction <*Myself* in debt, I could offer no help.> **2.** My normal or healthy condition or state <I have not been *myself* lately.> ***usage:*** In informal speech, reflexive pronouns like *myself* often occur in compound phrases such as *my husband and myself.* Such a construction is best avoided in formal writing.

**my·so·phil·i·a** (mī′sō-fĭl′ē-ə) *n.* [Gk. *musos,* filth + -PHILIA.] Pathological interest in and attraction to excreta.

**my·so·pho·bi·a** (mī′sō-fō′bē-ə) *n.* [Gk. *musos,* filth + -PHOBIA.] Pathological fear of dirt or contamination.

**mys·ta·gogue** (mĭs′tə-gŏg′, -gôg′) *n.* [OFr. < Lat. *mystagogus* < Gk. *mustagōgos* : *mustēs,* an initiate + *agōgos,* leader < *agein,* to lead.] **1.** One who prepared candidates for initiation into a mystery cult. **2.** One who holds or spreads mystical doctrines. **—mys′ta·gog′ic** (-gŏj′ĭk) *adj.* **—mys′ta·go′gy** (-gō′jē) *n.*

**mys·te·ri·ous** (mĭ-stîr′ē-əs) *adj.* [OFr. *mysterieux* < *mystere,* mystery < Lat. *mysterium,* mystery.] **1.** Of, relating to, or being a mystery <*mysterious* and enduring truths> **2.** Simultaneously arousing and eluding the desire to comprehend or explain <a *mysterious* phone call> **—mys·te′ri·ous·ly** *adv.* **—mys·te′ri·ous·ness** *n.*

☆ ***syns:*** MYSTERIOUS, ARCANE, CABALISTIC, MYSTIC, MYSTICAL, MYSTIFYING, UNACCOUNTABLE, UNEXPLAINABLE, UNFATHOMABLE *adj. core meaning* : difficult to explain or understand <a cult leader who held a *mysterious* power over the congregation>

**mys·ter·y¹** (mĭs′tə-rē) *n., pl.* **-ies.** [ME < Lat. *mysterium* < Gk. *mustērion,* secret rite < *muein,* to be closed.] **1. a.** Something not fully understood or understandable. **b.** A mysterious quality or character <a country of *mystery* and charm> **c.** A fictional work dealing with a baffling crime. **2.** A specialized practice or skill of a particular profession or group. **3.** A religious truth comprehensible only by divine revelation. **4.** One of 15 incidents, as the Annuciation or the Ascension, serving as a subject for meditation during the recitation of the rosary. **5.** A Christian sacrament, esp. the Eucharist. **6. a.** A religious cult practicing secret rites to which only initiates were admitted. **b.** A secret rite of such a cult.

**mys·ter·y²** (mĭs′tə-rē) *n., pl.* **-ies.** [ME < LLat. *misterium,* alteration of Lat. *ministerium,* occupation < *minister,* servant.] **1.** *Archaic.* A trade or occupation. **2.** *Archaic.* A guild. **3.** A mystery play.

**mystery play** *n.* [< MYSTERY².] A medieval drama based on Scriptural events esp. in the life of Christ.

**mys·tic** (mĭs′tĭk) *adj.* [ME < Lat. *mysticus* < Gk. *mustikos* < *mustērion,* secret rite. —see MYSTERY¹.] **1.** Of or relating to religious mysteries or occult rites and practices. **2.** Of or relating to mysticism or mystics. **3.** Inspiring a sense of wonder and mystery. **4.** MYSTICAL 1. **5. a.** Mysterious : bizarre. **b.** Puzzling : obscure. *—n.* One who practices or believes in mysticism or a specified form of mysticism.

**mys·ti·cal** (mĭs′tĭ-kəl) *adj.* **1.** Of or having a spiritual reality or import unapparent to the intelligence or senses. **2.** Of, relating to, or stemming from direct communion with God or with ultimate reality. **3.** Of or founded on subjective experience <*mystical* religions> **4.** MYSTIC 1. **5.** Arcane : cryptic. **—mys′ti·cal·ly** *adv.* **—mys′ti·cal·ness** *n.*

**mys·ti·cete** (mĭs′tĭ-sēt′) *n.* [NLat. *mysticetus* < Gk. *mustikētos,* a kind of whale, prob. alteration of *ho mus to kētos,* the whale (called) the mouse.] Any of several whales with symmetric skulls, paired blowholes, and plates of whalebone instead of teeth. **—mys′ti·ce′tous** (-sē′təs) *adj.*

**mys·ti·cism** (mĭs′tĭ-sĭz′əm) *n.* **1. a.** A spiritual discipline aiming at direct union or communion with God or with ultimate reality through trancelike contemplation or deep meditation. **b.** The experience of such communion as described by mystics. **2.** A belief in the existence of realities beyond intellectual or perceptual apprehension that are central to being and directly accessible by subjective experience, as by intuition. **3.** Vague and unsupportable speculation.

**mys·ti·fy** (mĭs′tə-fī′) *vt.* **-fied, -fy·ing, -fies.** [Fr. *mystifier* < *mystère,* mystery < Lat. *mysterium.* —see MYSTERY¹.] **1.** To bewilder : perplex. **2.** To make obscure or mysterious. **—mys′ti·fi·ca′tion** (-fĭ-kā′shən) *n.* **—mys′ti·fi′er** *n.* **—mys′ti·fy′ing·ly** *adv.*

**mys·tique** (mĭ-stēk′) *n.* [Fr. < *mystique,* mystic < Lat. *mysticus.* —see MYSTIC.] **1.** A body of mystical beliefs and attitudes associated with a particular person, thing, or idea. **2.** The specialized expertise required for an activity or occupation.

**myth** (mĭth) *n.* [Lat. *mythos* < Gk. *muthos.*] **1. a.** A traditional story originating in a preliterate society, dealing with supernatural beings, ancestors, or heroes that serve as primordial types in a primitive view of the world. **b.** A body of such stories told among a given people : MYTHOLOGY. **c.** All such stories as a whole. **2.** A real or fictional story, recurring theme, or character type that appeals to the consciousness of a people by embodying its cultural ideals or by giving expression to deep, commonly felt emotions. **3.** A fiction or half-truth, esp. one that forms part of the ideology of a society <the *myth* of ethnic superiority> **4.** A fictitious story, person, or thing.

**myth·i·cal** (mĭth'ĭ-kəl) *also* **myth·ic** (-ĭk) *adj.* **1.** Having the nature of a myth. **2.** Of or existing in myth <the *mythical* winged horse> **3.** Illusory : fictitious. **—myth'i·cal·ly** *adv.*

**myth·i·cize** (mĭth'ĭ-sīz') *vt.* **-cized, -ciz·ing, -ciz·es. 1.** To turn (a person or event) into myth. **2.** To interpret as a myth or in terms of mythology.

**myth·mak·er** (mĭth'mā'kər) *n.* A creator of myths or mythical situations. **—myth'mak'ing** *n.*

**my·thog·ra·pher** (mĭ-thŏg'rə-fər) *n.* [Gk. *mythographos* : *muthos,* myth + *-graphos,* writer < *graphein,* to write.] A recorder or narrator of myths.

**my·thog·ra·phy** (mĭ-thŏg'rə-fē) *n., pl.* **-phies. 1.** Artistic representation of mythical subjects. **2.** A collection of myths with critical commentary.

**my·thoi** (mī'thoi', mĭth'oi') *n. pl. of* MYTHOS.

**myth·o·log·i·cal** (mĭth'ə-lŏj'ĭ-kəl) *also* **myth·o·log·ic** (-ĭk) *adj.* **1.** Of, relating to, or celebrated in mythology. **2.** Fabulous : imaginary. **—myth'o·log'i·cal·ly** *adv.*

**my·thol·o·gist** (mĭ-thŏl'ə-jĭst) *n.* A student of mythology.

**my·thol·o·gize** (mĭ-thŏl'ə-jīz') *v.* **-gized, -giz·ing, -giz·es.** *—vt.* To convert into myth. *—vi.* **1.** To construct or relate a myth. **2.** To write about or interpret myths or mythology. **—my·thol'o·giz'er** *n.*

**my·thol·o·gy** (mĭ-thŏl'ə-jē) *n., pl.* **-gies.** [Fr. *mythologie* < LLat. *mythologia* < Gk. *muthologia,* storytelling : *muthos,* story + *logos,* speech.] **1. a.** A body of myths about the history and origin of a people and their ancestors, deities, and heroes. **b.** A body of myths concerning an individual, event, or institution <"A new *mythology,* essential to the . . . American funeral rite, has grown up" —Jessica Mitford> **2.** The field of scholarship dealing with the systematic collection and study of myths.

**myth·o·ma·ni·a** (mĭth'ə-mā'nē-ə, -mān'yə) *n.* [Gk. *muthos,* story + -MANIA.] A compulsion to exaggerate, tell lies, or embellish the truth. **—myth'o·ma'ni·ac'** (-ăk') *n.*

**myth·o·poe·ic** *also* **myth·o·pe·ic** (mĭth'ə-pē'ĭk) *adj.* [< Gk. *muthopoios,* composer of fiction < *mythopoiein,* to relate a story : *muthos,* story + *poiein,* to make.] Productive of myths : MYTHMAKING. **—myth'o·poe'ia** (-pē'ə), **myth'o·po·e'sis** (-pō-ē'sĭs) *n.*

**my·thos** (mī'thŏs', mĭth'ŏs') *n., pl.* **my·thoi** (mī'thoi', mĭth'oi') [Gk. *muthos.*] **1.** Myth. **2.** Mythology. **3.** The pattern of basic values and attitudes of a people, characteristically transmitted through the arts. **4.** A deliberately fostered aura or cult.

**myx-** *pref. var. of* MYXO-.

**myx·e·de·ma** *or* **myx·oe·de·ma** (mĭk'sĭ-dē'mə) *n.* A disease caused by decreased activity of the thyroid gland in adults, and marked by dry skin, swellings around the lips and nose, mental deterioration, and a subnormal basal metabolic rate. **—myx'e·dem'a·tous** (-dĕm'ə-təs, -dē'mə-), **myx'e·dem'ic** (-dĕm'ĭk) *adj.*

**myxo-** *or* **myx-** *pref.* [NLat. < Gk. *muxa,* mucus, slime.] Mucus <*myxocyte*>

**myx·o·cyte** (mĭk'sə-sīt') *n.* A large cell found in mucous tissue.

**myx·oid** (mĭk'soid') *adj.* Containing mucus : MUCOID.

**myx·o·ma** (mĭk-sō'mə) *n., pl.* **-mas** *or* **-ma·ta** (-mə-tə). A benign tumor composed of connective tissue and mucous elements. **—myx·o'ma·tous** (-sō'mə-təs, -sŏm'ə-) *adj.*

**myx·o·ma·to·sis** (mĭk-sō'mə-tō'sĭs) *n., pl.* **-ses** (-sēz). **1.** *Pathol.* A condition marked by the growth of numerous myxomas. **2.** A highly infectious, usu. fatal, viral disease of rabbits marked by numerous skin tumors similar to myxomas.

**myx·o·my·cete** (mĭk'sō-mī-sēt', -mī'sēt') *n.* [NLat. *Myxomycetes,* class name : MYXO- + Gk. *mukētes,* pl. of *mukēs,* fungus.] A slime mold. **—myx'o·my·ce'tous** (-mī-sē'təs) *adj.*

# Nn

**n** *or* **N** (ĕn) *n., pl.* **n's** *or* **N's. 1.** The 14th letter of the English alphabet. **2.** A speech sound represented by the letter *n.* **3.** The 14th in a series. **4.** Something shaped like the letter N.

**N** *symbol for* NITROGEN.

**Na** [< NLat. *natrium* < NATRON.] *symbol for* SODIUM.

**nab** (năb) *vt.* **nabbed, nab·bing, nabs.** [Perh. var. of dial. *nap,* to seize.] *Slang.* **1.** To seize (a fugitive or wrongdoer) : ARREST. **2.** To grab : snatch. **—nab'ber** *n.*

**na·bob** (nā'bŏb') *n.* [Port. *nababo* < Hindi *nawwāb.* —see NAWAB.] **1.** A governor in India under the Mogul Empire. **2. a.** A prominent, wealthy man. **b.** A self-important person.

**na·celle** (nə-sĕl') *n.* [Fr., nacelle, small boat < LLat. *navicella,* dim. of Lat. *navis,* ship.] A separate streamlined enclosure on an aircraft for housing an engine or sheltering the crew or cargo.

**na·cho** (nä'chō') *n.* [Poss. < Sp., flat-nosed.] A small often triangular piece of tortilla topped with chili sauce or cheese and broiled.

**na·cre** (nā'kər) *n.* [Fr. < OItal. *nacarra,* nacre, drum < Ar. *naqqārah,* drum.] Mother-of-pearl. **—na'cred** (-kərd), **na'cre·ous** (-krē-əs) *adj.*

**Na-Den·e** *also* **Na-Dé·né** (nä-dĕn'ē) *n.* [Of Athapascan orig.] A phylum of North American Indian languages spoken in western North America from Alaska to Mexico.

**na·dir** (nā'dər, -dîr') *n.* [ME < Med. Lat. < Ar. *naḍīr assamt,* opposite the zenith.] **1.** A point on the celestial sphere diametrically opposite the zenith. **2.** The lowest point <the *nadir* of my career>

**nae** (nā) *adv. Scot.* **1.** No. **2.** Not.

**nag**[1] (năg) *v.* **nagged, nag·ging, nags.** [Prob. of Scand. orig.] *—vt.* **1.** To annoy by constant complaining, scolding, or urging. **2.** To torment persistently, as with anxiety. *—vi.* **1.** To complain, scold, or find fault constantly. **2.** To be a continuing source of worry or irritation. **—nag, nag'ger** *n.* **—nag'ging·ly** *adv.*

☆ ***syns:*** NAG, CARP (at), FUSS (at), HENPECK, PECK, PICK AT, PICK (on) *v. core meaning* : to scold or find fault (with) constantly <continually *nagging* them about their habits><always *nagging* about money>

**nag**[2] (năg) *n.* [ME.] **1.** A horse, esp.: **a.** An old or worn-out horse. **b.** *Slang.* A racehorse. **2.** *Archaic.* A small saddle horse or pony.

**na·ga·na** *also* **n'ga·na** (nə-gä'nə) *n.* [Zulu *u-nakane.*] An often fatal disease of African livestock transmitted by the bite of the tsetse or other flies.

**Na·hua·tl** (nä'wät'l) *n., pl.* **Nahuatl** *or* **-tls.** [Sp. < Nahuatl *Nahuatl,* sing. of *Nahua,* the Nahuatl people.] **1. a.** A group of Mexican and Central American Indian tribes, including the Aztecs. **b.** A member of any of these tribes. **2.** The Uto-Aztecan language of the Nahuatl.

**Na·hum** (nā'həm, nā'əm) *n.* [Heb. *Naḥūm.*] —See table at BIBLE.

**nai·ad** (nā'əd, -ăd', nī'-) *n., pl.* **-a·des** (-ə-dēz') *or* **-ads.** [Gk. *naias, naiad-.*] **1.** *Gk. Myth.* One of the nymphs living in and presiding over brooks, fountains, and springs. **2.** The aquatic nymph of an insect, as the mayfly. **3.** A freshwater mussel of the family Unionidae. **4.** An aquatic plant of the genus *Naias.*

**na·if** *or* **na·ïf** (nä-ēf') *adj. & n. vars. of* NAIVE.

**nail** (nāl) *n.* [ME < OE *nægl.*] **1.** A slim, pointed piece of metal used as a fastener and made to be hammered in. **2. a.** A fingernail or toenail. **b.** A talon or claw. **3.** Something resembling a nail. **4.** A former measure of length for cloth, equal to 1/16 yard. *—vt.* **nailed, nail·ing, nails. 1.** To join, fasten, or attach with or as if with a nail. **2.** To cover, enclose, or shut by fastening with nails <*nail* the door shut> **3.** To keep motionless or intent <Terror *nailed* me to my seat.> **4.** *Informal.* To stop and seize : CATCH. **5.** *Informal.* To detect and expose <*nail* a lie> **6.** *Informal.* To strike or bring down, esp. with an object shot or hurled <*nail* a pheasant in flight> **7.** *Baseball.* To put out (a base runner). **—nail down. 1.** To discover or establish decisively <*nailed down* the truth> **2.** To win or gain decisively. **—nail'er** *n.*

**nail·brush** (nāl'brŭsh') *n.* A small brush with firm bristles for washing the hands and cleaning the fingernails.

**nail file** *n.* A small flat file for shaping the fingernails.

**nail fold** *n.* A keratinous material overlapping the base of a fingernail or toenail as a circular fold : CUTICLE.
**nail polish** *n.* A clear or colored cosmetic lacquer applied to the fingernails or toenails.
**nail scissors** *pl.n.* *(sing. or pl. in number).* Small scissors with short curved blades for trimming and shaping fingernails or toenails.
**nain·sook** (nān′so͝ok′) *n.* [< Hindi *nainsukh,* pleasant : *nain,* eye (< Skt. *nayanam* < *nayati,* he leads) + *sukh,* pleasure (< Skt. *sukha-,* pleasant).] A soft light cotton material, often with a woven stripe.
**nai·ra** (nī′rə) *n.* [Alteration of *Nigeria.*] —See table at CURRENCY.
**na·ive** *or* **na·ïve** (nä-ēv′) *also* **na·if** *or* **na·ïf** (nä-ēf′) [Fr. *naïve,* fem. of *naïf* < OFr. < Lat. *nativus,* natural < *nasci,* to be born.] —*adj.* **1. a.** Lacking sophistication and worldliness : ARTLESS. **b.** Simple and credulous : INGENUOUS. **2.** Lacking critical ability or analytical insight. **3.** Not previously subjected to experiments <testing *naive* rabbits> —*n.* A naive person. **—na·ive′ly** *adv.* **—na·ive′ness** *n.*
**na·ive·té** *or* **na·ïve·té** (nä′ēv-tā′, nä-ē′vĭ-tā′) *n.* [Fr. < *naïf,* artless. —see NAIVE.] **1.** The quality or state of being naive. **2.** A naive statement or action.
**na·ive·ty** *or* **na·ïve·ty** (nä-ēv′tē, -ē′vĭ-tē) *n., pl.* **-ties.** Naiveté.
**na·ked** (nā′kĭd) *adj.* [ME < OE *nacod.*] **1.** Without clothing on the body : NUDE. **2.** Without covering, esp. the usual covering <a *naked* blade> **3.** Devoid of vegetation or foliage. **4.** Without addition, disguise, or embellishment <presented the *naked* facts> **5.** Devoid of a specified element or aspect <*naked* of all affectation> **6.** Defenseless : vulnerable. **7.** *Bot.* **a.** Not encased in ovaries <*naked* seeds> **b.** Unprotected by scales <*naked* buds> **c.** Lacking a perianth <*naked* flowers> **d.** Without leaves or pubescence <*naked* stalks> **8.** Lacking protective covering such as scales, feathers, fur, or a shell. **9.** *Law.* Unsupported or uncorroborated by authority, evidence, or proof. **—na′ked·ly** *adv.* **—na′ked·ness** *n.*
**naked eye** *n.* The eye unassisted by an optical instrument.
**naked option** *n.* A securities option sold by a trader who does not own the optioned stock.
**na·led** (nā′lĕd′) *n.* [Orig. unknown.] A nonpersistent chemical, $C_4H_7O_4PBr_2Cl_2$, used as an insecticide.
**nal·ox·one** (năl′ək-sōn′) *n.* [N(ORMAL) + AL(LYL) + (HY-DR)OX(Y) + -ONE.] A drug, $C_{19}H_{21}NO_4$, used in the form of its hydrochloride as an antagonist to narcotic drugs such as morphine.
**nam·by-pam·by** (năm′bē-păm′bē) *adj.* [After *Namby-Pamby,* a satire on the poetry of Ambrose Philips by Henry Carey (d. 1743).] **1.** Insipid and sentimental. **2.** Lacking decisiveness or vigor : WEAK. **—nam′by-pam′by** *n.*
**name** (nām) *n.* [ME < OE *nama.*] **1.** A word or words by which an entity is designated and set apart from others. **2.** A word or words used to evaluate or describe, often disparagingly <called me *names*> **3.** Representation or repute as opposed to reality <a democracy in *name* only> **4. a.** General reputation <a bad *name*> **b.** A distinguished reputation : RENOWN. **5.** *Informal.* One who is famous or outstanding <a big *name* in government> —*vt.* **named, nam·ing, names. 1.** To give a name to. **2.** To identify by name. **3.** To specify, mention, or cite by name. **4.** To call by an epithet <*named* them all crooks> **5.** To nominate or appoint to a specific office, duty, or honor. **6.** To specify : fix <*name* the date of our appointment> —*adj. Informal.* Well-known by a name <*name* brands> **—in the name of.** By the authority of. **—to (one's) name.** Belonging to one <not a penny *to my name*> **—nam′a·ble, name′a·ble** *adj.* **—nam′er** *n.*
☆ ***syns:*** NAME, APPELLATION, APPELLATIVE, COGNOMEN, DENOMINATION, DESIGNATION, HANDLE, MONIKER, TAG *n. core meaning* : the word or words by which one is called and identified <the product's generic *name*><the *name* of a street>
**name day** *n.* **1.** The feast day of the saint after whom one is named. **2.** The day on which one is baptized.
**name·less** (nām′lĭs) *adj.* **1.** Having no name. **2.** Unknown by name : OBSCURE. **3.** Not designated by name : ANONYMOUS <a *nameless* patron> **4.** Not capable of being described : INEXPRESSIBLE <*nameless* repulsion> **—name′less·ly** *adv.* **—name′less·ness** *n.*
**name·ly** (nām′lē) *adv.* That is to say : SPECIFICALLY.
**name·plate** (nām′plāt′) *n.* **1.** A plate, as on an office door, inscribed with a name. **2.** A brand of merchandise.
**name·sake** (nām′sāk′) *n.* One named after another.
**nan·a** (năn′ə, nä′nə) *n.* [Of baby-talk orig.] **1.** A nurse or nursemaid. **2.** A grandmother.
**NAND** (nănd) *n.* [N(OT) + AND.] A machine logic circuit that produces an output inverse to that of an AND circuit.
**nan·din** (năn′dĭn) *n.* [NLat. *Nandina,* genus name < J. *nanten,* southern sky < Chin. (Mandarin) *nan*[2] *tian*[1] : *nan*[2], south + *tian*[1], sky.] An evergreen Asiatic shrub, *Nandina domestica,* with compound leaves and small white flowers that grow in a branching cluster and are followed by bright red berries.
**nan·keen** (năn-kēn′) *n.* [After *Nanjing* (Nanking), China.] **1. a.** A sturdy yellow or buff cotton cloth. **b. nankeens.** Trousers made of nankeen. **2. Nankeen.** A Chinese porcelain with a blue-and-white pattern.
**nan·nie** (năn′ē) *n. Chiefly Brit. var. of* NANNY.
**nan·no·plank·ton** (năn′ə-plăngk′tən) *n. var. of* NANOPLANKTON.
**nan·ny** *also* **nan·nie** (năn′ē) *n., pl.* **-nies.** [Alteration of NANA.] *Chiefly Brit.* A children's nurse.
**nan·ny·ber·ry** (năn′ē-bĕr′ē) *n.* The sheepberry.
**nanny goat** *n.* [< *Nanny,* nickname for *Ann.*] A female goat.
**nano-** *pref.* [Lat. *nanus,* dwarf < Gk. *nanos.*] **1.** Extremely small <*nanoplankton*> **2.** One billionth ($10^{-9}$) <*nanosecond*>
**nan·o·gram** (năn′ə-grăm′) *n.* One billionth ($10^{-9}$) of a gram.
**nan·o·me·ter** (năn′ə-mē′tər) *n.* One billionth ($10^{-9}$) of a meter.
**na·no·plank·ton** (nā′nə-plăngk′tən, năn′ə-) *also* **nan·no·plank·ton** (năn′ə-plăngk′tən) *n.* Microscopic aquatic animal and plant organisms comprising the smallest of the plankton.
**nan·o·sec·ond** (năn′ə-sĕk′ənd) *n.* One billionth ($10^{-9}$) of a second.
**nap**[1] (năp) *n.* [< ME *nappen,* to doze < OE *hnappian.*] A brief sleep, often during the day. —*vi.* **napped, nap·ping, naps. 1.** To sleep for a brief period, often during the day : DOZE. **2.** To be unaware of imminent danger or trouble <The storm caught us *napping.*>
**nap**[2] (năp) *n.* [ME *noppe* < MDu.] A soft or fuzzy surface on fabric or leather. —*vt.* **napped, nap·ping, naps.** To form or raise a nap on (fabric or leather).
**na·palm** (nā′päm′) *n.* [E. *naphthenate,* salt of naphthenic acid (< NAPHTHENE) + PALM(ITATE).] **1. a.** An aluminum soap of various fatty acids that when mixed with gasoline makes a firm jelly used in flame throwers and incendiary bombs. **b.** This jelly used in flame throwers and bombs. **2.** An incendiary mixture made of polystyrene, benzene, and gasoline. —*vt.* **-palmed, -palm·ing, -palms.** To attack with napalm.
**nape** (nāp, năp) *n.* [ME.] The back of the neck.
**na·per·y** (nā′pə-rē) *n.* [ME *naperie* < OFr. < *nape,* tablecloth < Lat. *mappa,* napkin.] Household linen, esp. table linen.
**Naph·ta·li** (năf′tə-lī′) *n.* [Heb. *Naphtālī* < *niphtal,* he wrestled.] **1.** A son of Jacob in the Old Testament. **2.** A tribe of Israel descended from Naphtali.
**naph·tha** (năf′thə, năp′-) *n.* [Gk., liquid bitumen.] **1.** A colorless flammable liquid obtained from crude petroleum and used as a solvent and cleaning fluid and as a raw material for gasoline. **2.** Any of several volatile hydrocarbon liquids derived chiefly from coal tar and used as solvents. **3.** *Obs.* Petroleum.
**naph·tha·lene** (năf′thə-lēn′, năp′-) *also* **naph·tha·line** *or* **naph·tha·lin** (-lĭn) *n.* [NAPHTH(A) + AL(COHOL) + -ENE.] A white crystalline compound, $C_{10}H_8$, derived from coal tar or petroleum and used to manufacture dyes, explosives, moth repellents, and solvents. **—naph′tha·len′ic** (-lĕn′ĭk) *adj.*
**naph·thene** (năf′thēn′, năp′-) *n.* [NAPHTH(A) + -ENE.] Any of several cycloparaffin hydrocarbons with the general formula $C_nH_{2n}$, and found in various petroleums. **—naph·then′ic** (-thĕn′ĭk) *adj.*
**naph·thol** (năf′thôl′, -thōl′, năp′-) *also* **naph·tol** (-tôl′, -tōl′) *n.* [NAPHTH(ALENE) + -OL[2].] An organic compound, $C_{10}H_7OH$, occurring in two isomeric forms: **a.** alpha-naphthol, colorless or yellow prisms or powder used in organic synthesis, dyes, and perfumes. **b.** beta-naphthol, white lustrous leaflets or powder used in dyes and insecticides and in the manufacture of rubber.
**Na·pier·i·an logarithm** (nə-pîr′ē-ən, nā-) *n.* [After John *Napier* (1550–1617).] *Math.* A logarithm to the base *e* (=2.71828...).
**Na·pi·er's bones** (nā′pē-ərz) *n.* [After John *Napier* (1550–1617).] A data or logarithm table used for rapid multiplication.
**na·pi·form** (nā′pə-fôrm′) *adj.* [Lat. *napus,* turnip + -FORM.] Shaped like a turnip <*napiform* roots>
**nap·kin** (năp′kĭn) *n.* [ME, dim. of *nape,* tablecloth < OFr. < Lat. *mappa,* napkin.] **1.** A piece of cloth or absorbent paper used at table to protect the clothes or wipe the lips and fingers. **2.** A cloth or towel. **3.** *Chiefly Brit.* A diaper. **4.** A sanitary napkin.
**na·po·le·on** (nə-pō′lē-ən, -pōl′yən) *n.* [After *Napoleon* I (1769–1821).] **1.** A rectangular piece of pastry made with crisp flaky layers filled with custard cream. **2.** A 20-franc gold coin once in circulation in France.
**nappe** (năp) *n.* [Fr., sheet < OFr., tablecloth < Lat. *mappa,* napkin.] **1.** A sheet of water flowing over a barrier, as a dam. **2.** *Geol.* **a.** A recumbent anticline or fold of strata. **b.** A mass of rock moved from its original position by an anticline. **3.** *Math.* Either of the two parts into which a cone is divided by the vertex.
**nap·py**[1] (năp′ē) *adj.* **-pi·er, -pi·est. 1.** Having a nap : FUZZY. **2.** KINKY 1.
**nap·py**[2] (năp′ē) *n., pl.* **-pies.** [Prob. < dial. *nap,* bowl < ME < OE *hnæp.*] A shallow, round serving or cooking dish with a flat bottom and sloping sides.
**nap·py**[3] (năp′ē) *n., pl.* **-pies.** [Shortening and alteration of NAPKIN.] *Chiefly Brit.* An infant's diaper.
**narc** *or* **nark** (närk) *n.* [Short for *narcotics agent.*] *Slang.* A law enforcement officer who deals with narcotics violations.
**nar·cism** (när′sĭz′əm) *n. var. of* NARCISSISM.
**nar·cis·si** (när-sĭs′ī′, -sĭs′ē) *n. var. pl. of* NARCISSUS 2.
**nar·cis·sism** (när′sĭ-sĭz′əm) *also* **nar·cism** (när′sĭz′əm) *n.* [After NARCISSUS.] **1.** Excessive admiration or love of oneself. **2.** *Psychoanal.* Arrestment of development at or regression to the infantile

developmental stage in which one's own body is the object of erotic interest. **—nar'cis·sist** *n.* **—nar'cis·sis'tic** *adj.*

**Nar·cis·sus** (när-sĭs'əs) *n.* [Lat. < Gk. *Narkissos.*] **1.** *Gk. Myth.* A youth who pined away in love for his own image in a pool of water and was transformed into the flower that bears his name. **2. narcis·sus** *pl.* **-cis·sus·es** *or* **-cis·si** (-sĭs'ī', -sĭs'ē). Any of several widely cultivated plants of the genus *Narcissus,* with narrow grasslike leaves and usu. yellow or white flowers marked by a cup-shaped or trumpet-shaped central crown.

**narco-** *pref.* [Gk. *narko-* < *narkoun,* to numb < *narkē,* numbness.] **1.** Numbness : stupor : lethargy <*narcolepsy*> **2.** Narcotic drug <*narcoanalysis*>

**nar·co** (när'kō) *n., pl.* **-cos.** *Slang.* A narc.

**nar·co·a·nal·y·sis** (när'kō-ə-năl'ĭ-sĭs) *n.* Psychoanalysis conducted while the patient is in a drug-induced drowsy state. **—nar'co·an'a·lyt'ic** (-ăn'ə-lĭt'ĭk) *adj.*

**nar·co·lep·sy** (när'kə-lĕp'sē) *n.* A condition marked by sudden, uncontrollable, and usu. brief attacks of deep sleep. **—nar'co·lep'tic** (-lĕp'tĭk) *adj.* & *n.*

**nar·co·sis** (när-kō'sĭs) *n.* [Gk. *narkōsis,* a numbing < *narkoun,* to numb < *narkē,* numbness.] **1.** A deep, drug-induced unconsciousness. **2.** *Biol.* Immobility in an organism caused by chemicals such as carbon dioxide.

**nar·co·syn·the·sis** (när'kō-sĭn'thĭ-sĭs) *n.* Narcoanalysis directed toward making the patient recall suppressed memories and emotional traumas for later interpretation.

**nar·cot·ic** (när-kŏt'ĭk) *n.* [ME *narkotik* < Med. Lat. *narcoticum* < Gk. *narkōtikos,* numbing < *narkoun,* to numb < *narkē,* numbness.] **1.** A habit-forming drug that lessens sensibility, relieves pain, and produces sleep when administered therapeutically but that causes coma, convulsions, or stupor when administered in large doses. **2.** Something that soothes, numbs, or induces a dreamlike state. **3.** A narcotics addict. *—adj.* **1.** Inducing sleep or stupor. **2.** Of or relating to narcotics, their effects, or their use. **3.** Of or relating to a narcotics addict. **—nar·cot'i·cal·ly** *adv.*

**nar·co·tism** (när'kə-tĭz'əm) *n.* [NARCOT(IC) + -ISM.] **1.** Addiction to narcotics such as heroin, opium, or morphine. **2.** NARCOSIS 1.

**nar·co·tize** (när'kə-tīz') *vt.* **-tized, -tiz·ing, -tiz·es. 1.** To put under the influence of a narcotic. **2.** To put to sleep : LULL. **3.** To dull : deaden. **—nar'co·ti·za'tion** *n.*

**nard** (närd) *n.* [ME *narde* < OFr. < Lat. *nardus* < Gk. *nardos.*] **1.** SPIKENARD 1. **2.** A balm from spikenard. **3.** Any of several plants of the genus *Valeriana* or related plants whose aromatic roots have been used medicinally.

**nar·es** (nâr'ēz) *n. pl. of* NARIS.

**nar·ghi·le** *also* **nar·gi·leh** (när'gə-lē') *n.* [Fr. *narguilé* < Pers. *nārgīleh* < *nārgīl,* coconut.] A hookah.

**nar·is** (nâr'ĭs) *n., pl.* **-es** (-ēz) [Lat.] An opening in the vertebrate nasal cavity. **—nar'i·al** (-ē-əl) *adj.*

**nark**[1] (närk) *n. Slang. var. of* NARC.

**nark**[2] (närk) [Perh. < Romany *nāk,* nose.] *Chiefly Brit. —n.* A police informer. *—vi.* **narked, nark·ing, narks.** To be a nark.

**Nar·ra·gan·set** *also* **Nar·ra·gan·sett** (năr'ə-găn'sĭt) *n., pl.* **Narraganset** *or* **-sets** *also* **Narragansett** *or* **-setts.** [Alteration of Narraganset *Nanhigganeuck.*] **1. a.** A tribe of Indians that once lived in the area of Rhode Island. **b.** A member of this tribe. **c.** The Algonquian language of the Narraganset. **2.** A small sturdy saddle horse orig. bred. in Rhode Island. **—Nar'ra·gan'set** *adj.*

**nar·rate** (năr'āt', nă-rāt') *v.* **-rat·ed, -rat·ing, -rates.** [Lat. *narrare, narrat-* < *gnarus,* knowing.] *—vt.* To give an account of : RELATE. *—vi.* **1.** To give a description or account. **2.** To supply a running commentary for a performance, as of a film. **—nar·ra'tion** (nă-rā'shən, nə-) *n.* **—nar·ra'tion·al** *adj.* **—nar'ra'tor, nar'rat'er** *n.*

**nar·ra·tive** (năr'ə-tĭv) *n.* **1.** A narrated account : STORY. **2.** The act, technique, or process of narrating. **3.** *Computer Sci.* Data in a computer program that does not function in the program itself but is used by the programmer to identify and correct individual machine instructions. *—adj.* **1.** Consisting of or marked by the telling of a story <*narrative* verse> **2.** Of or relating to narration <*narrative* technique> **—nar'ra·tive·ly** *adv.*

**†nar·row** (năr'ō) *adj.* **-er, -est.** [ME *narwe* < OE *nearu.*] **1.** Small or slender in width, esp. compared with length. **2.** Limited in area or scope : CONFINED. **3.** Lacking flexibility : RIGID <*narrow* attitudes> **4.** Barely sufficient : CLOSE <won by a *narrow* margin> **5.** Painstakingly thorough or precise : METICULOUS. **6.** *Regional.* Miserly : stingy. **7.** TENSE[1] 4. *—v.* **-rowed, -row·ing, -rows.** *—vt.* **1.** To make narrower : CONTRACT. **2.** To restrict or limit <*narrowed* down the odds> *—vi.* To become narrower : CONTRACT. *—n.* **1.** A narrow part, as a narrow pass through mountains. **2. a. narrows** *(sing. or pl. in number).* A narrow body of water connecting two larger ones. **b.** A narrow part of a river or ocean current. **—nar'row·ly** *adv.* **—nar'row·ness** *n.*

**narrow gauge** *n.* **1.** A distance between the rails of a railroad track less than the standard width of 56½ inches or approx. 1.4 meters. **2.** A locomotive, car, or railway of a narrow gauge. **—nar'row-gauge', nar'row-gauged'** *adj.*

**nar·row-mind·ed** (năr'ō-mīn'dĭd) *adj.* Lacking tolerance, breadth of view, or sympathy : PETTY. **—nar'row-mind'ed·ly** *adv.* **—nar'row-mind'ed·ness** *n.*

**nar·thex** (när'thĕks') *n.* [LGk. *narthēx* < Gk., box, giant fennel.] **1.** A portico or lobby of an early Christian or Byzantine church or basilica. **2.** An entrance hall leading to a church nave.

**nar·whal** *also* **nar·wal** (när'wəl) *or* **nar·whale** (-hwāl', -wāl') *n.* [Norw. or Dan. *narhval* < ON *nāhvalr* : *nār,* corpse + *hvalr,* whale (so called from its whitish color).] An aquatic arctic mammal, *Monodon monoceros,* that has a spotted pelt and is characterized in the male by a long spirally twisted ivory tusk.

**narwhal**
*13–16 feet long*

**†nar·y** (nâr'ē) *adj.* [Alteration of *ne'er a.*] *Regional.* Not one : NO.

**na·sal** (nā'zəl) *adj.* [Med. Lat. *nasalis* < Lat. *nasus,* nose.] **1.** Of or relating to the nose. **2.** Uttered by lowering the soft palate and occluding the mouth so that most of the air passes through the nose, as in sounding *m, n,* and *ng.* **3.** Marked by or resembling a resonant sound produced through the nose <a *nasal* twang> *—n.* **1.** A nasal consonant. **2.** A nasal part or bone. **3.** The nosepiece of a helmet. **—na·sal'i·ty** (nā-zăl'ĭ-tē) *n.* **—na'sal·ly** *adv.*

**na·sal·ize** (nā'zə-līz') *vt.* & *vi.* **-ized, -iz·ing, -iz·es.** To make nasal or produce nasal sounds. **—na'sal·i·za'tion** *n.*

**nas·cent** (năs'ənt, nā'sənt) *adj.* [Lat. *nascens, nascent-,* pr.part. of *nasci,* to be born.] That is coming into existence. **—nas'cence, nas'cen·cy** *n.*

**nase·ber·ry** (nāz'bĕr'ē) *n.* [Sp. *néspera* < Lat. *mespila,* medlar < Gk. *mespilē.*] SAPODILLA 2.

**Nash·ville** (năsh'vĭl') *n.* [After *Nashville,* Tennessee.] **1.** Country music. **2.** The country music industry.

**naso-** *pref.* [NLat. < Lat. *nasus,* nose.] Nose <*nasopharynx*>

**na·so·fron·tal** (nā'zō-frŭn'təl) *adj.* Of or relating to the nasal and frontal bones.

**na·so·phar·ynx** (nā'zō-făr'ĭngks) *n.* The portion of the pharynx behind the nasal cavity and above the soft palate. **—na'so·pha·ryn'·ge·al** (-fə-rĭn'jē-əl, -jəl, -făr'ən-jē'əl) *adj.*

**nas·tic** (năs'tĭk) *adj.* [< Gk. *nastos,* pressed close < *nassein,* to press.] Of, relating to, or marked by a tendency in plants to grow or change according to internal cell pressures.

**nas·tur·tium** (nə-stûr'shəm, nă-) *n.* [Lat., a kind of cress : *nasus,* nose + **tortare,* freq. of *torquēre,* to twist (from its pungent smell).] **1.** Any of various plants of the genus *Tropaeolum,* having flowers with five broad petals that are usu. yellow, orange, or red. **2.** A brilliant orange yellow.

**nas·ty** (năs'tē) *adj.* **-ti·er, -ti·est.** [ME *nasti,* of Scand. orig.] **1. a.** Disgustingly dirty : FILTHY. **b.** Physically repellent : FOUL. **2.** Morally offensive : OBSCENE. **3.** Malicious : mean. **4.** Causing discomfort or trouble : UNPLEASANT <a *nasty* cold> **5.** Painful or dangerous : GRAVE <a *nasty* accident> **6.** Difficult to understand or handle : EXASPERATING. **—nas'ti·ly** *adv.* **—nas'ti·ness** *n.*

**-nasty** *suff.* [G. *nastie* < Gk. *nastos,* pressed down < *nassein,* to press.] Nastic response or change <*epinasty*>

**na·tal** (nāt'l) *adj.* [Lat. *natalis* < *nasci,* to be born.] **1.** Of, pertaining to, or accompanying birth <*natal* trauma> **2.** Of or relating to the time or place of one's birth.

**na·tal·i·ty** (nā-tăl'ĭ-tē, nə-) *n., pl.* **-ties.** Birthrate.

**Na·tal plum** (nə-tăl', -tāl') *n.* [After *Natal,* South Africa.] A South African shrub, *Carissa grandiflora,* having forked spines, white flowers, and an edible scarlet berry.

**na·tant** (nāt'nt) *adj.* [Lat. *natans, natant-,* pr.part. of *natare,* to swim.] Floating or swimming in the water.

**na·ta·tion** (nā-tā'shən, nă-) *n.* [Lat. *natatio* < *natare,* to swim.] The act or skill of swimming.

**na·ta·to·ri·a** (nā'tə-tôr'ē-ə, -tōr'-, năt'ə-) *n. var. pl. of* NATATORIUM.

**na·ta·to·ri·al** (nā'tə-tôr'ē-əl, -tōr'-, năt'ə-) *also* **na·ta·to·ry** (nā'-tə-tôr'ē, -tōr'ē, năt'ə-) *adj.* [LLat. *natatorius* < Lat. *natare,* to swim.] Of, relating to, or adapted for swimming.

**na·ta·to·ri·um** (nā'tə-tôr'ē-əm, -tōr'-, năt'ə-) *n., pl.* **-to·ri·ums** *or* **-to·ri·a** (-tôr'ē-ə, -tōr'-) [LLat. < Lat. *natare,* to swim.] An indoor swimming pool.

**na·ta·to·ry** (nā'tə-tôr'ē, -tōr'ē, năt'ə-) *adj. var. of* NATATORIAL.

**natch** (năch) *adv.* [Shortening and alteration of NATURALLY.] *Slang.* Of course : NATURALLY.

**Natch·ez** (năch′ĭz) *n., pl.* **Natchez.** [Fr.] **1. a.** A tribe of Indians once living in the area of Mississippi. **b.** A member of this tribe. **2.** The Muskhogean language of the Natchez. **—Natch′ez** *adj.*
**na·tes** (nā′tēz′) *pl.n.* [Lat., pl. of *natis,* buttock.] The buttocks.
**nathe·less** (năth′lĭs) *also* **nath·less** (năth′-) *adv.* [ME *nathles* < OE *nā ðē lǣs,* not less by that.] *Archaic.* Notwithstanding.
**Na·tick** (nā′tĭk) *n.* [Orig. unknown.] A dialect of Massachuset.
**na·tion** (nā′shən) *n.* [ME *nacioun* < OFr. *nacion* < Lat. *natio* < *nascī,* to be born.] **1.** A people who share common customs, origins, history, and often language : NATIONALITY. **2.** A rather large group of people organized under a single, usu. independent government : COUNTRY. **3.** The government of a sovereign state. **4. a.** A federation or tribe, esp. one of North American Indians. **b.** The territory occupied by such a federation or tribe. **—na′tion·hood′** *n.*
☆ ***syns:*** NATION, COUNTRY, LAND, STATE *n. core meaning* : an organized geopolitical unit <*nations* of the free world> NATION primarily signifies a political body—a group of human beings organized under a single government, without close regard for their origins <the new *nations* of Africa>; it also denotes the territory occupied by a political body <all across the *nation*> STATE even more specifically indicates governmental organization, gen. on a sovereign basis and in a well-defined area <the *state* of Israel> COUNTRY signifies the territory of one nation <the *country* of France> and is also used in the sense of *nation* <all the *countries* of the free world> LAND is a term for a region <the *land* of the bison and beaver>; it also can be used to mean *nation* <the highest elective office in the *land*>
**na·tion·al** (năsh′ə-nəl, năsh′nəl) *adj.* **1.** Of, relating to, or belonging to a nation as an organized whole. **2.** Of or pertaining to nationality. **3.** Characteristic of or peculiar to the people of a nation. **4.** Of or maintained by the government of a nation. **5.** In the interest of one's own nation. **6.** Devoted to one's own nation or its interests : NATIONALISTIC. *—n.* **1.** A citizen of a specific nation. **2.** *often* **nationals.** A tournament or contest involving participants from all parts of a nation. **—na′tion·al·ly** *adv.*
**national bank** *n.* **1.** A bank in a U.S. system of federally chartered privately owned banks, each required by law to be an investing member of its district Federal Reserve Bank and to be insured by the Federal Deposit Insurance Corporation. **2.** A bank usu. owned or controlled by a government.
**national debt** *n.* A national government's total financial obligations.
**national forest** *n.* A large expanse of forest protected by a government and harvestable only under controlled conditions.
**National Guard** *n.* The military reserve units controlled by each U.S. state, equipped by the federal government and subject to the call of either the federal or the state government.
**national income** *n.* The total net value of all goods and services produced within a nation over a specified time period, representing the sum of wages, profits, rents, interest, and pension payments to residents of the nation.
**na·tion·al·ism** (năsh′ə-nə-lĭz′əm, năsh′nə-) *n.* **1.** Devotion to the interests or culture of a nation. **2.** The belief that nations would benefit from acting independently rather than collectively, emphasizing national rather than international goals. **3.** Aspirations for national independence in a country under foreign domination. **—na′tion·al·ist** *n.* **—na′tion·al·is′tic** *adj.* **—na′tion·al·is′ti·cal·ly** *adv.*
**na·tion·al·i·ty** (năsh′ə-năl′ĭ-tē, năsh-năl′-) *n., pl.* **-ties. 1.** The status of belonging to a particular nation by origin, birth, or naturalization. **2.** A people having common traditions or origins and often constituting a nation. **3.** Existence as a politically autonomous entity. **4.** National character. **5.** Nationalism.
**na·tion·al·ize** (năsh′ə-nə-līz′, năsh′nə-) *vt.* **-ized, -iz·ing, -iz·es. 1.** To convert from private to governmental control and ownership <*nationalize* the railroads> **2.** To make national in character. **—na′tion·al·i·za′tion** *n.* **—na′tion·al·iz′er** *n.*
**national monument** *n.* A natural landmark or a structure or site of historic interest set aside by a national government and maintained for public enjoyment or study.
**national park** *n.* A tract of land declared public property by a national government so as to be preserved and developed for recreational or cultural purposes.
**national seashore** *n.* A federally protected and maintained seacoast recreational area.
**National Socialism** *n.* Nazism.
**nation-state** (nā′shən-stāt′) *n.* An autonomous state inhabited esp. by a predominantly homogeneous people.
**na·tion·wide** (nā′shən-wīd′) *adj.* Pervasive throughout a nation.
**na·tive** (nā′tĭv) *adj.* [ME *natif* < OFr. < Lat. *nativus* < *nasci,* to be born.] **1.** Existing in or belonging to one by nature : INNATE <*native* skill> **2.** Being such by birth or origin <a *native* German> **3.** Being one's own because of the place or circumstances of one's birth <our *native* country> **4.** Growing, living, or produced orig. in a certain place : INDIGENOUS <a plant *native* to Africa> **5.** Of, belonging to, or characteristic of the original inhabitants of a given place, esp. those of primitive culture. **6.** Occurring in nature pure or uncombined with other substances <*native* gold> **7.** Natural : unaffected. **8.** *Archaic.* Closely related, as by birth or race. *—n.* **1.** One born in or connected with a place by birth. **2.** An original inhabitant or lifelong resident. **3.** One belonging to a primitive people orig. occupying a country. **4.** Something, esp. an animal or plant, that originated in a given place. **—na′tive·ly** *adv.* **—na′tive·ness** *n.*
☆ ***syns:*** NATIVE, ENDEMIC, INDIGENOUS *adj. core meaning* : belonging to one esp. because of the place of one's birth <our *native* land> NATIVE indicates birth or immediate origin in a specific place <a *native* Italian> ENDEMIC stresses restriction to a localized area <revolutionary activity and guerrilla fighters *endemic* to the mountains> INDIGENOUS, similar to ENDEMIC, specifies that something or someone is of a kind orig. living or growing in a region rather than coming or being brought from another part of the world <the bison—*indigenous* to North America><the Ainu—*indigenous* to the northernmost islands of Japan>
**Native American** *n.* An American Indian. ***usage:*** In the United States many now prefer the designation *Native American* instead of *Indian* for the first inhabitants of the Western Hemisphere, although usage may vary according to tribe and region. In Canada and Alaska in particular, *American Indian* is still preferred as making a useful distinction from Eskimos.
**na·tive-born** (nā′tĭv-bôrn′) *adj.* Belonging to a place by birth.
**na·tiv·ism** (nā′tĭ-vĭz′əm) *n.* **1.** A sociopolitical policy favoring the interests of native inhabitants over those of immigrants. **2.** Reestablishment or perpetuation of native cultural traits, esp. in opposition to acculturation. **—na′tiv·ist** *n.* **—na′tiv·is′tic** *adj.*
**na·tiv·i·ty** (nə-tĭv′ə-tē, nā-) *n., pl.* **-ties.** [ME *nativite* < Lat. *nativitas* < *nativus,* born < *nasci,* to be born.] **1.** Birth, esp. the place, conditions, or circumstances of being born. **2. Nativity. a.** The birth of Jesus. **b.** A representation, as a painting, of Jesus' birth. **c.** Christmas. **3.** A horoscope for the time of one's birth.
**na·tro·lite** (nā′trə-līt′) *n.* [G. *Natrolith* : *natron,* natron (< Fr.) + *-lith,* -lite.] A colorless to white zeolite with composition $Na_2(Al_2Si_3O_{10})\cdot 2H_2O$.
**na·tron** (nā′trŏn′, -trən) *n.* [Fr. < Sp. *natrón* < Ar. *naṭrūn,* niter < Gk. *nitron,* of Semitic orig.] A mineral of hydrous sodium carbonate, $Na_2CO_3\cdot 10H_2O$, often found crystallized with other salts.
**nat·ter** (năt′ər) *vi.* **-tered, -ter·ing, -ters.** [Imit.] *Chiefly Brit.* To talk idly : CHATTER.
**nat·ty** (năt′ē) *adj.* **-ti·er, -ti·est.** [Perh. var. of obs. *netty* < *net,* elegant < Fr. < OFr. —see NEAT.] *Informal.* Neat and trim : DAPPER. **—nat′ti·ly** *adv.* **—nat′ti·ness** *n.*
**nat·u·ral** (năch′ər-əl, năch′rəl) *adj.* [ME < Lat. *naturalis* < *natura,* nature < *nasci,* to be born.] **1.** Present in or produced by nature. **2.** Of, relating to, or concerning nature. **3.** Conforming to the usual or ordinary course of nature <died from *natural* causes> **4. a.** Not acquired : INNATE <a *natural* eye for beauty> **b.** Having a particular character by nature <a *natural* comic> **5.** Free from pretension or artificiality : SPONTANEOUS. **6.** Not treated, altered, or disguised <*natural* flavoring> **7.** Faithfully representing life or nature. **8.** Expected and accepted <Success was a *natural* consequence of their efforts.> **9.** Established by moral conviciton or certainty. **10.** Being in a primitive or unregenerate state : UNCIVILIZED. **11. a.** Related by blood rather than adoption. **b.** Born of unwed parents : ILLEGITIMATE. **12.** *Math.* Of or relating to positive integers. **13.** *Mus.* **a.** Not sharped or flatted. **b.** Having no sharps or flats. **14.** Afro. *—n.* **1. a.** One having all the qualifications necessary for success <a *natural* for the position> **b.** One suited by nature for a certain function or purpose. **2.** A person with subnormal intelligence. **3.** *Mus.* **a.** The sign (♮) placed before a note to cancel a preceding sharp or flat. **b.** A note so affected. **4.** A yellowish gray to pale orange yellow. **5.** A combination that wins immediately in certain card and dice games. **6.** An Afro hair style. **—nat′u·ral·ness** *n.*
☆ ***syns:*** NATURAL, ORGANIC, UNADULTERATED *adj. core meaning* : produced by nature <*natural* foods>
**natural childbirth** *n.* Childbirth regarded as a natural process involving little stress or pain and requiring preparatory training and medical supervision but no anesthesia or surgical aid.
**natural food** *n.* Food containing no additives, as preservatives or artificial coloring or flavoring.
**natural gas** *n.* A mixture of hydrocarbon gases that occurs with petroleum deposits, chiefly methane together with varying quantities of ethane, butane, propane, and other gases, and used as a fuel and in manufacturing organic compounds.
**natural history** *n.* The study of natural objects and organisms and their evolution, origins, description, and interrelationships.
**nat·u·ral·ism** (năch′ər-ə-lĭz′əm, năch′rə-) *n.* **1.** Conformity to nature : realistic or factual representation, esp. in literature and art. **2.** *Philos.* The system of thought holding that all phenomena can be explained in terms of natural causes and laws without attributing supernatural significance to them. **3.** The doctrine that all religious truths are derived from nature and natural causes and not from revelation. **4.** Thought or conduct based on natural desires or instincts.
**nat·u·ral·ist** (năch′ər-ə-lĭst, năch′rə-) *n.* **1.** One versed in natural history, esp. in botany or zoology. **2.** An advocate of naturalism.

**nat·u·ral·is·tic** (năch′ər-ə-lĭs′tĭk, năch′rə-) *adj.* **1.** Imitating or producing the effect or appearance of nature. **2.** Of, relating to, or being in accordance with the doctrines of naturalism. **—nat′u·ral·is′ti·cal·ly** *adv.*

**nat·u·ral·ize** (năch′ər-ə-līz′, năch′rə-) *v.* **-ized, -iz·ing, -iz·es.** —*vt.* **1.** To grant full citizenship to (one of foreign birth). **2.** To adopt (something foreign) into general use. **3.** To acclimate or adapt (e.g., a plant) to life in a new environment. **4.** To cause to conform to nature. —*vi.* To become acclimated or naturalized : ADAPT. **—nat′u·ral·i·za′tion** *n.*

**natural language** *n.* A human written or spoken language.

**natural law** *n.* A law or body of laws that derives from nature and is thought to be binding on human actions apart from or in conjunction with laws established by human authority.

**natural logarithm** *n.* A Napierian logarithm.

**nat·u·ral·ly** (năch′ər-ə-lē, năch′rə-) *adv.* **1.** In a natural way. **2.** By nature : INNATELY. **3.** Without a doubt : POSITIVELY.

**natural number** *n. Math.* One of the set of positive whole numbers : positive integer.

**natural philosophy** *n.* The study of nature and the physical universe. **—natural philosopher** *n.*

**natural resource** *n.* A material source of wealth, as timber, fresh water, or a mineral deposit, occurring in a natural state.

**natural science** *n.* A science, as biology, chemistry, or physics, based chiefly on objective quantitative hypotheses.

**natural selection** *n.* The principle that individuals possessing characteristics advantageous for survival in a specific environment constitute an increasing proportion of their species in that environment with each succeeding generation.

**natural theology** *n.* A theology holding that knowledge of God may be acquired without recourse to revelation.

**na·ture** (nā′chər) *n.* [ME, essential properties of a thing < Lat. *natura* < *nasci,* to be born.] **1.** The material world and its phenomena. **2.** The processes and forces that produce and control all the phenomena of the material world. **3.** The world of living things and the outdoors <the serenity of *nature*> **4.** A primitive state of existence undefiled by civilization. **5.** The human condition or state unredeemed by grace. **6.** Kind : type <an object of that *nature*> **7.** Essential characteristics and qualities <the *nature* of the dilemma><the *nature* of the typical dictator> **8.** An individual's fundamental character or disposition : TEMPERAMENT <had a kind *nature*> **9.** The natural or real aspect of a person, place, or thing. **10.** Bodily functions and processes <the call of *nature*>

**nature trail** *n.* A trail, as through woods or by a seashore, usu. with natural features labeled esp. for study.

**na·tur·op·a·thy** (nā′chə-rŏp′ə-thē) *n.* [NATUR(E) + -PATHY.] A system of therapy that relies exclusively on natural remedies, as sunlight supplemented with diet and massage, to treat the sick. **—na′tur·o·path′** (nā′chər-ə-păth′, nə-chŏŏr′-) *n.* **—na·tur′o·path′ic** (nə-chŏŏr′ə-păth′ĭk) *adj.*

**Nau·ga·hyde** (nô′gə-hīd′). A trademark for vinyl-coated fabrics.

**naught** *also* **nought** (nôt) [ME *nauht* < OE *nāwiht* : *nā,* no + *wiht,* thing.] —*n.* **1.** Nothing <all for *naught*> **2.** The figure 0 : ZERO. —*adj.* Of no value : WORTHLESS. —*adv.* Not in the least.

**naugh·ty** (nô′tē) *adj.* **-ti·er, -ti·est.** [< ME *noughti,* wicked < *nought,* evil < OE *nāwiht,* nothing, evil. —see NAUGHT.] **1.** Unruly : mischievous. **2.** Indecent : improper <a *naughty* bikini> **3.** *Archaic.* Wicked : immoral. **—naugh′ti·ly** *adv.* **—naugh′ti·ness** *n.*

**nau·pli·us** (nô′plē-əs) *n., pl.* **-pli·i** (-plē-ī′) [Lat., a kind of shellfish < Gk. *nauplios.*] The microscopic free-swimming first stage of the larva of certain crustaceans.

**nau·se·a** (nô′zē-ə, -zhə, -sē-ə, -shə) *n.* [Lat. < Gk. *nausia,* seasickness < *naus,* ship.] **1.** A stomach disturbance marked by a feeling of the need to vomit. **2.** Strong aversion : DISGUST.

**nau·se·ate** (nô′zē-āt′, -zhē-, -sē-, -shē-) *vi. & vt.* **-at·ed, -at·ing, -ates.** [Lat. *nauseare, nauseat-* < *nausea,* nausea.] **1.** To feel or cause to feel nausea. **2.** To feel or cause to feel disgust or loathing.

**nau·seous** (nô′shəs, -zē-əs) *adj.* **1.** Causing nausea : SICKENING. **2.** Affected with nausea. ***usage:*** Although *nauseous* has developed a second sense meaning "affected with nausea," purists still prefer to restrict this sense to the term *nauseated* and to use *nauseous* only in the sense of "causing nausea." **—nau′seous·ly** *adv.*

**nautch** (nôch) *n.* [Hindi *nāc* < Prakrit *nacca,* dance < Skt. *nṛtyam* < *nṛyati,* he dances.] A dance form of northern India for a single girl dancer accompanied by several musicians and sometimes by a singer.

**nau·ti·cal** (nô′tĭ-kəl) *adj.* [Lat. *nauticus* < Gk. *nautikos* < *nautēs,* sailor < *naus,* ship.] Of, relating to, or characteristic of ships, shipping, sailors, or navigation on a body of water. **—nau′ti·cal·ly** *adv.*

**nautical mile** *n.* A unit of length used in sea and air navigation, based on the length of one minute of arc of a great circle, esp. an international and U.S. unit equal to 1,852 meters or about 6,076 feet.

**nau·ti·li** (nôt′l-ī′) *n. var. pl. of* NAUTILUS.

**nau·ti·loid** (nôt′l-oid′) *n.* [< NLat. *Nautiloidea,* subclass name < Lat. *nautilus,* nautilus.] A mollusk of the subclass Nautiloidea, including the nautiluses and numerous extinct species known only as fossils.

**nau·ti·lus** (nôt′l-əs) *n., pl.* **-ti·lus·es** *or* **-ti·li** (-l-ī′) [Lat. < Gk. *nautilos,* nautilus, sailor < *nautēs,* sailor < *naus,* ship.] **1.** A mollusk of the genus *Nautilus* found in the Indian and Pacific oceans and having a spiral shell with a series of air-filled chambers. **2.** The chambered nautilus.

**Nav·a·ho** *also* **Nav·a·jo** (năv′ə-hō′, nä′və-) *n., pl.* **Navaho** *or* **-hos** *or* **-hoes** *also* **Navajo** *or* **-jos** *or* **-joes.** [Mex. Sp. *(Apache de) Navajo,* (Apache of) Navaho < Tewa *Navahu,* the name of a Tewa pueblo.] **1. a.** A group of Indians occupying an extensive reservation in parts of New Mexico, Arizona, and Utah. **b.** A member of this group. **2.** The Athapascan language of the Navaho. **—Nav′a·ho′** *adj.*

**na·val** (nā′vəl) *adj.* [Lat. *navalis* < *navis,* ship.] **1.** Of or relating to ships or shipping. **2.** Of or relating to a navy. **3.** Having a navy.

**naval architect** *n.* An architect who designs ships.

**naval stores** *pl.n.* Products such as turpentine or pitch, orig. used to caulk the seams of wooden ships.

**nav·ar** (năv′är′) *n.* [NAV(IGATIONAL) + (RAD)AR.] A method of air navigation in which traffic in a pilot's vicinity is observed by ground radar and relayed to the pilot's radarscope.

**nave**[1] (nāv) *n.* [Med. Lat. *navis* < Lat., ship.] The central part of a church from the narthex to the chancel, flanked by aisles.

**nave**[2] (nāv) *n.* [ME < OE *nafa.*] The hub of a wheel.

**na·vel** (nā′vəl) *n.* [ME < OE *nafela.*] **1.** The mark on the mammalian abdomen where the umbilical cord was attached during gestation. **2.** A central point : MIDDLE.

**navel orange** *n.* A sweet, usu. seedless orange having at its apex a navellike formation enclosing an underdeveloped fruit.

**na·vel·wort** (nā′vəl-wûrt′, -wôrt′) *n.* [From the navellike depression on its leaves.] **1.** PENNYWORT b. **2.** A plant of the genus *Omphalodes,* with one-sided clusters of usu. blue flowers.

**na·vic·u·lar** (nə-vĭk′yə-lər) *n.* [LLat. *navicularis* < Lat. *navicula,* dim. of *navis,* ship.] **1.** A bone of the wrist shaped like a comma. **2.** The concave bone in front of the anklebone on the instep of the foot. —*adj.* Shaped like a boat.

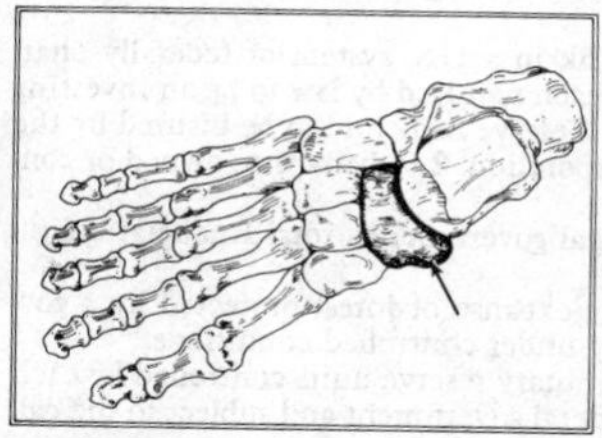

**navicular**

**nav·i·ga·ble** (năv′ĭ-gə-bəl) *adj.* **1.** Wide or deep enough for vehicular passage. **2.** Capable of being steered. —Used of vessels or aircraft. **—nav′i·ga·bil′i·ty, nav′i·ga·ble·ness** *n.* **—nav′i·ga·bly** *adv.*

**nav·i·gate** (năv′ĭ-gāt′) *v.* **-gat·ed, -gat·ing, -gates.** [Lat. *navigare, navigat-* : *navis,* ship + *agere,* to direct.] —*vt.* **1.** To record, plan, and control the position and course of (a ship or aircraft). **2.** To follow a planned course on, across, or through <*navigate* a river> —*vi.* **1.** To control the course of a ship or aircraft. **2.** To voyage over water in a boat or ship : SAIL. **3.** *Informal.* **a.** To make one's way. **b.** To walk <too drunk to *navigate*>

**nav·i·ga·tion** (năv′ĭ-gā′shən) *n.* **1.** The theory and practice of navigating, esp. the charting of a course for a ship or aircraft. **2.** Travel or traffic by vessels, esp. commercial shipping. **—nav′i·ga′tion·al** *adj.*

**nav·i·ga·tor** (năv′ĭ-gā′tər) *n.* **1.** One who navigates. **2.** A device that directs the course of an aircraft or missile.

**nav·vy** (năv′ē) *n., pl.* **-vies.** [Alteration of NAVIGATOR, canal laborer (obs.).] *Chiefly Brit.* A laborer, esp. one employed in construction or excavation projects.

**na·vy** (nā′vē) *n., pl.* **-vies.** [ME *navie* < OFr. < VLat. **navia* < Lat. *navis,* ship.] **1.** All of a nation's warships. **2.** *often* **Navy.** A nation's entire military organization for sea warfare and defense, including personnel, vessels, and shore establishments. **3.** A group of ships : FLEET. **4.** Navy blue.

**navy bean** *n.* [From its former use as a standard provision of the U.S. Navy.] One of several varieties of the kidney bean, cultivated for their nutritious white seeds.

**navy blue** *n.* [From the color of the British naval uniform.] A dark blue.

**Navy Cross** *n.* A decoration awarded by the U.S. Navy for exceptional heroism in action.

**navy yard** *n.* A dockyard for building, repairing, or docking naval ships.

**na·wab** (nə-wŏb′) *n.* [Hindi *nawwāb* < Ar. *nuwwāb,* pl. of *na'ib,* deputy.] **1.** A governor or ruler in India under the Mogul empire. **2.** NABOB 2.

**nay** (nā) *adv.* [ME < ON *nei* : *ne,* not + *ei,* ever.] **1.** No <All but four senators voted *nay.*> **2.** And moreover <was good-looking, *nay,* beautiful> —*n.* **1.** A refusal or denial. **2.** A negative vote or voter.

**Naz·a·rene** (năz′ə-rēn′, năz′ə-rēn′) *n.* [ME < LLat. *Nazarenus* < Gk. *Nazarēnos* < *Nazaret,* Nazareth.] **1. a.** A native or inhabitant of Nazareth. **b.** Jesus. **2.** A member of a sect of early Christians of Jewish origin who retained many of the prescribed Jewish observances. **3.** A member of an American Protestant denomination that follows many of the doctrines of early Methodism.

**Na·zi** (nät′sē, năt′-) *n., pl.* **-zis.** [G., contraction of *Nationalsozialist,* National Socialist.] **1.** A member of the National Socialist German Workers' Party, founded in Germany in 1919 and brought to power in 1933 under Adolf Hitler. **2.** *often* **nazi.** An advocate or adherent of policies characteristic of Nazism : FASCIST. **—Na′zi** *adj.* **—Na′zi·fi·ca′tion** (-sə-fĭ-kā′shən) *n.* **—Na′zi·fy′** (-sə-fī′) *v.* **(-fied, -fy·ing, -fies).**

**Na·zism** (nät′sĭz′əm, năt′-) *also* **Na·zi·ism** (-sē-ĭz′əm) *n.* The ideology and practice of the Nazis, esp. the policy of state control of the economy, racist nationalism, and national expansion.

**Nb** *symbol for* NIOBIUM.

**Nd** *symbol for* NEODYMIUM.

**Ne** *symbol for* NEON.

**Ne·an·der·thal** (nē-ăn′dər-thôl′, -tôl′, nā-än′dər-täl′) *n.* **1.** Neanderthal man. **2.** *Slang.* A boor. *—adj.* **1.** Of, relating to, or resembling a Neanderthal man. **2.** *Slang.* Crude : boorish. **—Ne·an′der·thal′oid′** (-thô′loid′, -tô′-, -tä′-) *adj.*

**Neanderthal man** *n.* [After *Neanderthal,* a valley near Düsseldorf, West Germany.] An extinct species or race of man, *Homo neanderthalensis,* living during the late Pleistocene epoch in the Old World and associated with Middle Paleolithic tools.

**ne·an·throp·ic** (nē′ən-thrŏp′ĭk) *adj.* [NE(O)- + ANTHROP(O)- + -IC.] Of or relating to members of the extant species *Homo sapiens* as compared with other, now extinct species of *Homo.*

**Ne·a·pol·i·tan** (nē′ə-pŏl′ĭ-tən) *adj.* Of, belonging to, or characteristic of Naples, Italy. *—n.* A resident or native of Naples, Italy.

**Neapolitan ice cream** *n.* Brick ice cream with layers of different colors and flavors.

**neap tide** (nēp) *n.* [ME **neep* < OE *nēp(flōd),* neap (tide).] A tide of lowest range, occurring when the sun and moon are in quadrature.

**near** (nîr) *adv.* **-er, -est.** [ME *ner* < OE *nēar,* comp. adv. of *nēah,* near.] **1.** To, at, or within a short distance or interval in space or time. **2.** Nearly : almost <*near* played out> **3.** With or in a close relationship. *—adj.* **-er, -est. 1.** Close in time, position, space, or degree <the *near* past> **2.** Closely related by kinship or association : INTIMATE. **3.** Failing or succeeding by a very small margin <a *near* bull's-eye> **4.** Closely corresponding to or resembling an original <a *near* copy> **5. a.** Closer of two or more. **b.** On the left side of a vehicle or draft team. **6.** Short and direct <the *nearest* route to the airport> **7.** Stingy : cheap. *—prep.* Close to <a motel *near* town> *—v.* **neared, near·ing, nears.** *—vt.* To come close or closer to. *—vi.* To draw near or nearer : APPROACH. **—near′ness** *n.*

☆ ***syns:*** NEAR, CLOSE, IMMEDIATE, NEARBY, NIGH *adj. core meaning* : not far from another in space, time, or relation <The airport was *near* the town.><They were *near* to me in age.>

**near beer** *n.* A malt liquor not containing enough alcohol to be considered an alcoholic beverage.

**near·by** (nîr′bī′) *adj. & adv.* Close at hand.

**Ne·arc·tic** (nē-ärk′tĭk, -är′tĭk) *adj.* [NE(O)- + ARCTIC.] Of or designating the zoogeographic region that includes the Arctic and Temperate areas of North America and Greenland.

**near·ly** (nîr′lē) *adv.* **1.** Almost but not quite. **2.** Closely : intimately <were *nearly* associated for years>

**near·sight·ed** (nîr′sī′tĭd) *adj.* Not able to see distant objects clearly : MYOPIC. **—near′sight′ed·ly** *adv.* **—near′sight′ed·ness** *n.*

**neat**[1] (nēt) *adj.* **-er, -est.** [OFr. *net* < Lat. *nitidus,* elegant < *nitēre,* to shine.] **1.** Clean and orderly : TIDY. **2.** Orderly and exact in procedure : SYSTEMATIC. **3.** Characterized by creativity and skill : ADROIT. **4.** Not diluted or mixed with other substances <*neat* liqueur> **5.** NET[2] 1. **6.** *Slang.* Wonderful : terrific <went to a *neat* picnic> *—adv.* Without dilution : STRAIGHT <takes whiskey *neat*> **—neat′ly** *adv.* **—neat′ness** *n.*

☆ ***syns:*** NEAT, ORDERLY, SHIPSHAPE, SNUG, SPICK-AND-SPAN, TIDY, TRIM *adj. core meaning* : in good order or clean condition <kept a *neat* yard> ***ant:*** disorderly, messy

▲ word history: The adjective *neat* is derived from Old French *net,* which had several meanings. A basic sense of the French word was "free from dirt; clean." Liquor that contained no impurities was called *neat,* a sense that survives in the meaning "undiluted." An amount of money that was not liable to any reduction was also considered *neat.* A variant form of Old French *net* had the same spelling and meanings but a different pronunciation. It survives in the English word *net,* meaning "remaining after all deductions." A *neat* profit and a *net* profit are in meaning and origin exactly the same.

**neat**[2] (nēt) *n., pl.* **neat.** [ME *net* < OE *nēat.*] *Archaic.* A domestic bovine animal.

**neat·en** (nēt′n) *vt.* **-ened, -en·ing, -ens.** To make orderly : TIDY.

**neath** or **′neath** (nēth) *prep.* Beneath <*neath* sun and sky>

**neat's-foot oil** (nēts′fo͝ot′) *n.* A light yellow oil obtained from the feet and shinbones of cattle, used chiefly to dress leather.

**neb** (nĕb) *n.* [ME < OE.] **1. a.** A bird's beak. **b.** A nose : snout. **2.** A projecting part, esp. a nib.

**neb·bish** (nĕb′ĭsh, -ĭKH) *n.* [Yiddish *nebech.*] One who is weak-willed, shy, and ineffectual.

**neb·u·la** (nĕb′yə-lə) *n., pl.* **-lae** (-lē′) *or* **-las.** [Lat., cloud.] **1.** *Astron.* **a.** A diffuse mass of interstellar dust or gas or both, visible as luminous patches or dark areas depending on the way the mass reflects or absorbs incident radiation. **b.** Such a mass that absorbs ultraviolet radiation from stars and re-emits it as visible light. **c.** Such a mass that absorbs all incident radiation without re-emission. **d.** Such a mass that reflects visible radiation. **e.** A galactic nebula. **2.** *Pathol.* **a.** A cloudy spot on the cornea. **b.** Cloudiness in urine. **3.** A liquid medication applied by spraying. **—neb′u·lar** *adj.*

**nebular hypothesis** *n.* A theory of the origin of the solar system according to which a rotating nebula cooled and contracted, throwing off rings of matter that contracted into the planets and their smaller moons, while the great mass of the condensing nebula became the sun.

**neb·u·lize** (nĕb′yə-līz′) *vt.* **-lized, -liz·ing, -liz·es.** [< NEBULA.] **1.** To convert (a liquid) to a fine spray : ATOMIZE. **2.** To treat with a medicated spray. **—neb′u·li·za′tion** *n.* **—neb′u·liz′er** *n.*

**neb·u·los·i·ty** (nĕb′yə-lŏs′ĭ-tē) *n., pl.* **-ties. 1.** The quality or condition of being nebulous. **2. a.** NEBULA 1. **b.** A mass of material constituting a nebula.

**neb·u·lous** (nĕb′yə-ləs) *adj.* [Lat. *nebulosus* < *nebula,* cloud.] **1.** Misty, cloudy, or hazy. **2.** Having indefinite form or limits : VAGUE <*nebulous* ideas> **3.** Of, relating to, or characteristic of a nebula. **—neb′u·lous·ly** *adv.* **—neb′u·lous·ness** *n.*

**nec·es·sar·i·ly** (nĕs′ĭ-sâr′ə-lē) *adv.* Of necessity : INEVITABLY.

**nec·es·sar·y** (nĕs′ĭ-sĕr′ē) *adj.* [ME *necessarie* < Lat. *necessarius* < *necesse.*] **1.** Absolutely required : INDISPENSABLE. **2.** Needed to bring about a certain effect or result <the *necessary* equipment> **3. a.** Unavoidably determined by prior conditions or circumstances : INEVITABLE <the *necessary* consequence of self-indulgence> **b.** Logically inevitable. **4.** Required by obligation, convention, or compulsion. *—n., pl.* **-ies.** NECESSITY 1b.

☆ ***syns:*** NECESSARY, ESSENTIAL, INDISPENSABLE, VITAL *adj. core meaning* : needed to achieve a result or fulfill a requirement <fresh vegetables *necessary* to good nutrition> NECESSARY stresses that which fulfills a requirement <fill out the *necessary* forms> VITAL refers to what is urgent and therefore crucially important <irrigation being *vital* to early civilization> INDISPENSABLE even more strongly connotes what cannot be ignored or omitted <oxygen—*indispensable* for human life>

**necessary condition** *n.* **1.** *Logic.* A proposition whose falsity assures the falsity of another proposition derived from it. **2.** A condition that is an essential antecedent of another.

**ne·ces·si·tar·i·an·ism** (nə-sĕs′ĭ-târ′ē-ə-nĭz′əm) *n.* The doctrine that events are inevitably determined by preceding causes. **—ne·ces′si·tar′i·an** *adj. & n.*

**ne·ces·si·tate** (nə-sĕs′ĭ-tāt′) *vt.* **-tat·ed, -tat·ing, -tates.** [Med. Lat. *necessitare, necessitat-* < Lat. *necessitas,* necessity < *necesse,* necessary.] **1.** To make necessary or unavoidable. **2.** To compel or require. **—ne·ces′si·ta′tion** *n.* **—ne·ces′si·ta′tive** *adj.*

**ne·ces·si·tous** (nə-sĕs′ĭ-təs) *adj.* [Fr. *nécessiteux* < OFr. *necessite,* necessity.] **1.** Needy : poor. **2.** Urgent. **—ne·ces′si·tous·ly** *adv.*

**ne·ces·si·ty** (nə-sĕs′ĭ-tē) *n., pl.* **-ties.** [ME *necessite* < OFr. < Lat. *necessitas* < *necesse,* necessary.] **1. a.** The quality or state of being necessary. **b.** Something necessary. **2. a.** Something dictated by invariable physical laws. **b.** The force exerted by circumstance. **3.** Neediness.

**neck** (nĕk) *n.* [ME *nekke* < OE *hnecca.*] **1.** The part of the body linking the head and trunk. **2.** The part of a garment around or near the neck. **3.** *Anat.* A rather narrow portion of a structure, as of a bone or organ, that joins its parts. **4.** A rather narrow elongation, projection, or connecting part <the *neck* of a bottle> **5.** The narrow part along which the strings of a stringed instrument extend to the pegs. **6.** *Geol.* Solidified lava filling the vent of an extinct volcano. **7.** The siphon of a bivalve mollusk, as a clam. **8.** A narrow margin <lost by a *neck*> *—v.* **necked, neck·ing, necks.** *—vi. Slang.* To kiss and caress : PET. *—vt.* To strangle or decapitate (a fowl). **—neck and neck.** Even in a competition. **—stick (one's) neck out.** To risk criticism, trouble, or harm unnecessarily.

**neck·er·chief** (nĕk′ər-chĭf, -chēf′) *n.* A kerchief worn around the neck.

**neck·ing** (nĕk′ĭng) *n.* **1.** A molding between the upper part of a column and the projecting part of the capital. **2.** The act or practice of kissing and petting.

**neck·lace** (nĕk′lĭs) *n.* A neck ornament.

**neck·line** (nĕk′līn′) *n.* The line formed by the edge of a garment at or near the neck.

**neck·tie** (nĕk′tī′) *n.* A long narrow band of fabric worn around the neck and tied in a knot or bow close to the throat.

**neck·wear** (nĕk′wâr′) *n.* Articles worn around the neck.

**necro-** *or* **necr-** *pref.* [NLat. < Gk. *nekros,* corpse.] **1.** Dead body : corpse <*necro*phagous> **2.** Death <*necro*biosis>

**nec·ro·bi·o·sis** (nĕk′rō-bī-ō′sĭs) *n.* Natural degeneration and death of cells and tissues as opposed to death from injury or disease and distinguished from death of the entire organism. **—nec′ro·bi·ot′ic** (-ŏt′ĭk) *adj.*

**ne·crol·o·gy** (nə-krŏl′ə-jē, nĕ-) *n., pl.* **-gies. 1.** A list of deceased people esp. of the near past. **2.** An obituary. **—nec′ro·log′ic** (nĕk′rə-lŏj′ĭk), **nec′ro·log′i·cal** *adj.* **—ne·crol′o·gist** *n.*

**nec·ro·man·cy** (nĕk′rə-măn′sē) *n.* [Alteration of ME *nigromancie* < OFr. *nigremancie* < Med. Lat. *nigromantia,* alteration of *necromantia* < Gk. *nekromanteia* : *nekros,* corpse + *manteia,* divination. —see -MANCY.] **1.** Communication with the spirits of the dead in order to predict the future. **2.** Black magic. **3.** Magical qualities. **—nec′ro·man′cer** *n.* **—nec′ro·man′tic** (-măn′tĭk) *adj.*

**nec·ro·pha·gia** (nĕk′rə-fā′jə) *also* **ne·croph·a·gy** (nə-krŏf′ə-jē) *n.* The act or practice of eating the flesh of corpses or carrion.

**ne·croph·a·gous** (nə-krŏf′ə-gəs, nĕ-) *adj.* Feeding on carrion or corpses.

**ne·croph·a·gy** (nə-krŏf′ə-jē) *n. var. of* NECROPHAGIA.

**ne·crop·o·lis** (nə-krŏp′ə-lĭs, nĕ-) *n., pl.* **-lis·es** *or* **-leis** (-lās′). A cemetery, esp. a large and elaborate one belonging to an ancient city.

**nec·rop·sy** (nĕk′rŏp′sē) *n., pl.* **-sies.** An autopsy. **—nec′rop′sy** *v.* **(-sied, -sy·ing, -sies).**

**ne·crose** (nĕ-krōs′, -krōz′, nĕk′rōs′, -rōz′) *vt.* & *vi.* **-crosed, -cros·ing, -cros·es.** [Back-formation < NECROSIS.] To affect or be affected with necrosis.

**ne·cro·sis** (nə-krō′sĭs, nĕ-) *n., pl.* **-ses** (-sēz′) [LLat. < Gk. *nekrōsis,* death < *nekroun,* to make dead < *nekros,* corpse.] Pathologic death of living plant or animal tissue. **—ne·crot′ic** (-krŏt′ĭk) *adj.*

**nec·ro·tize** (nĕk′rə-tīz′) *vt.* & *vi.* **-tized, -tiz·ing, -tiz·es.** To necrose.

**nec·tar** (nĕk′tər) *n.* [Lat. < Gk. *nektar.*] **1.** *Gk. & Rom. Myth.* The drink of the gods. **2.** A delicious or stimulating drink. **3.** A sweet liquid secreted by various flowers and gathered by bees for honey.

**nec·tar·ine** (nĕk′tə-rēn′) *n.* [< obs. *nectarine,* sweet as nectar < NECTAR.] A peach with a smooth waxy skin.

**nec·tar·iv·o·rous** (nĕk′tə-rĭv′ə-rəs) *adj.* Feeding on nectar, as certain insects do.

**nec·tar·ous** (nĕk′tə-rəs) *adj.* **1.** Of, pertaining to, or consisting of nectar. **2.** Resembling nectar, as in sweetness or fragrance.

**nec·ta·ry** (nĕk′tə-rē) *n., pl.* **-ries.** [NLat. *nectarium* < NECTAR.] **1.** A glandlike organ, usu. at the base of a flower, that secretes nectar. **2.** The part of a flower in which a nectary is contained. **—nec·tar′i·al** (-târ′ē-əl) *adj.*

**née** *also* **nee** (nā) *adj.* [Fr., fem. p.part. of *naître,* to be born < Lat. *nasci.*] Born. —Used to indicate the maiden name of a married woman <Mary Smith, *née* Jones>

**need** (nēd) *n.* [ME *nede* < OE *nēod.*] **1.** A lack of something necessary or desirable <plants in *need* of water><a *need* for love> **2.** Something necessary or wanted : REQUISITE <had modest *needs*> **3.** Necessity : obligation <no *need* for apologies> **4.** Poverty or misfortune <in dire *need*> *—v.* **need·ed, need·ing, needs.** *—aux.* To be under the necessity of or the obligation to <You *need* not answer.> *—vt.* To have need of : REQUIRE. *—vi.* **1.** To be in need or want. **2.** *Archaic.* To be necessary.
☆ **syns:** NEED, EXIGENCY, NECESSITY *n. core meaning* : a condition in which something necessary is required or wanted <patients in *need* of special care>

**need·ful** (nēd′fəl) *adj.* Necessary : essential. **—need′ful·ly** *adv.* **—need′ful·ness** *n.*

**nee·dle** (nēd′l) *n.* [ME *nedle* < OE *nædl.*] **1.** A small slender sewing implement, usu. of polished steel, with an eye at one end through which thread is passed and held. **2.** An implement similar to a needle, as one used in knitting. **3.** A small pointed stylus for transmitting vibrations from the grooves of a phonograph record. **4. a.** A slender pointer or indicator on a dial, scale, or similar part of a mechanical device. **b.** A magnetic needle. **5.** A hypodermic needle. **6.** A stiff narrow leaf, as of a conifer. **7.** A sharp fine projection, as a spine of a sea urchin or a crystal. **8.** A sharp, pointed instrument for engraving. *—v.* **-dled, -dling, -dles.** *—vt.* **1.** To pierce, prick, or stitch with or as if with a needle. **2.** *Informal.* To goad or provoke : TEASE. **3.** *Slang.* To increase the alcoholic content of (a beverage). *—vi.* To sew or do similar work with a needle. **—nee′dler** *n.*

**nee·dle·fish** (nēd′l-fĭsh′) *n., pl.* **needlefish** *or* **-fish·es. 1.** Any of several marine fishes of the family Belonidae, with slender bodies and narrow jaws. **2.** A fish with projecting jaws, as the pipefish.

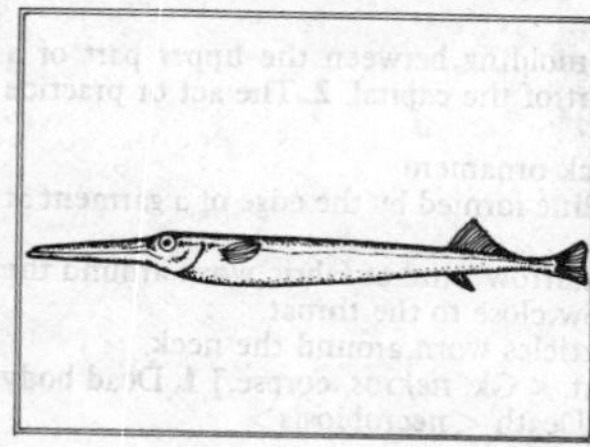

**needlefish**
*Up to 4 inches long*

**nee·dle·point** (nēd′l-point′) *n.* **1.** Decorative needlework on canvas, usu. in a diagonal stitch covering the entire surface of the material. **2.** A lace worked with a needle on paper patterns.

**need·less** (nēd′lĭs) *adj.* Not needed : UNNECESSARY. **—need′less·ly** *adv.* **—need′less·ness** *n.*

**needle valve** *n.* A valve having a slender point fitting into a conical seat, for accurately regulating the flow of a liquid or gas.

**nee·dle·wom·an** (nēd′l-wŏŏm′ən) *n.* A woman who does needlework.

**nee·dle·work** (nēd′l-wûrk′) *n.* Work done with a needle. **—nee′dle·work′er** *n.*

**need·n't** (nēd′nt). Need not.

**needs** (nēdz) *adv.* [ME *nedes* < *nede,* necessarily < OE *nēde.*] Of necessity : NECESSARILY <I must *needs* go.>

**need·y** (nē′dē) *adj.* **-i·er, -i·est.** Being in need <*needy* families> **—need′i·ness** *n.*

**ne'er** (nâr) *adv.* Never.

**ne'er-do-well** (nâr′dōō-wĕl′) *n.* An idle irresponsible person. **—ne′er′-do-well′** *adj.*

**ne·far·i·ous** (nə-fâr′ē-əs) *adj.* [Lat. *nefarius* < *nefas,* crime : *ne-,* not + *fas,* law.] Extremely infamous or wicked : EVIL. **—ne·far′i·ous·ly** *adv.* **—ne·far′i·ous·ness** *n.*

**ne·gate** (nĭ-gāt′) *vt.* **-gat·ed, -gat·ing, -gates.** [Lat. *negare, negat-,* to deny.] **1.** To make ineffective or null : INVALIDATE. **2.** To rule out : DENY. **3.** *Computer Sci.* To perform the machine logic operation NOT. **—ne·ga′tor, ne·gat′er** *n.*

**ne·ga·tion** (nĭ-gā′shən) *n.* **1.** The act or process of negating. **2.** A denial, contradiction, or negative statement. **3.** The opposite or absence of something regarded as positive, actual, or affirmative. **—ne·ga′tion·al** *adj.*

**neg·a·tive** (nĕg′ə-tĭv) *adj.* [LLat. *negatitivus* < *negare,* to deny.] **1.** Expressing, consisting of, or containing a negation, refusal, or denial <a *negative* reply> **2. a.** Lacking the quality of being positive or affirmative <*negative* thoughts> **b.** Indicating opposition or resistance <a *negative* response to a new gasoline tax> **3.** Not positive or constructive. **4.** Not indicative of the presence of microorganisms, disease, or a specific condition <*negative* blood tests> **5.** *Logic.* Denying agreement between a subject and its predicate. —Used of a proposition. **6.** *Math.* Relating to or denoting: **a.** A quantity less than zero. **b.** The sign (–). **c.** A quantity to be subtracted from another. **d.** A quantity, angle, number, velocity, or direction, in a sense opposite to another of the same magnitude indicated or understood to be positive. **7.** *Physics.* Relating to or denoting: **a.** Electric charge of the same sign as that of an electron, designated by the symbol (–). **b.** A body having an excess of electrons. **8.** *Chem.* Relating to or denoting an ion, the anion, that is attracted to a positive electrode. **9.** *Biol.* Indicating resistance to, opposition to, or motion away from a stimulus <a *negative* tropism> *—n.* **1.** A statement or act indicating or expressing a contradiction, refusal, or denial. **2.** The opposite of something positive. **3.** A word or part of a word, as *no, not,* or *non-,* that indicates negation. **4.** The side in a debate that opposes or contradicts the question being debated. **5. a.** An image in which the light areas of the object rendered appear dark and the dark areas appear light. **b.** Photographic material, as a film or plate, containing such an image. **6.** *Math.* A negative quantity. *—vt.* **-tived, -tiv·ing, -tives. 1.** To refuse to approve : VETO. **2.** To contradict : deny. **3.** To demonstrate to be false : DISPROVE. **4.** To counteract : neutralize. **—neg′a·tive·ly** *adv.* **—neg′a·tive·ness, neg′a·tiv′i·ty** (-tĭv′ĭ-tē) *n.*

**negative feedback** *n.* Feedback that reduces the output of a system, as the action of heat on a thermostat to limit the output of a furnace.

**negative income tax** *n.* Government payments to those whose income is below a specified level, proposed as an alternative to welfare.

**negative transfer** *n.* Interference with current learning or performance as a result of the transfer of previously learned responses.

**neg·a·tiv·ism** (nĕg′ə-tĭ-vĭz′əm) *n.* **1.** Habitual skepticism or resistance to the orders, suggestions, or instructions of others. **2.** Behavior marked by determined refusal to carry out the orders, suggestions, or instructions of others. **—neg′a·tiv·ist** *n.* **—neg′a·tiv·is′tic** *adj.*

**neg·a·tron** (nĕg′ə-trŏn′) *n.* [NEGA(TIVE) + (ELEC)TRON.] An electron.

**ne·glect** (nĭ-glĕkt′) *vt.* **-glect·ed, -glect·ing, -glects.** [Lat. *neglegere, neglect-* : *neg-,* not + *legere,* to choose.] **1.** To ignore : disregard. **2.** To fail to care for or give proper attention to. **3.** To fail to do or carry out, as through oversight or carelessness. *—n.* **1.** An act or instance of neglecting. **2.** The state of being neglected. **3.** Habitual lack of care. **—ne·glect′er** *n.*

**ne·glect·ful** (nĭ-glĕkt′fəl) *adj.* Marked by neglect : CARELESS. **—ne·glect′ful·ly** *adv.* **—ne·glect′ful·ness** *n.*

**neg·li·gee** *also* **neg·li·gée** *or* **neg·li·gé** (nĕg′lĭ-zhā′, nĕg′lĭ-zhā′) *n.* [Fr. < *négliger,* to neglect < Lat. *neglegere.*] **1.** A woman's loose dressing gown, often of soft sheer fabric. **2.** Informal or incomplete attire.

**neg·li·gence** (nĕg′lĭ-jəns) *n.* **1.** The quality or state of being negligent. **2.** A negligent act or failure to act. **3.** *Law.* Omission or neglect of reasonable care, precaution, or action.

**neg·li·gent** (nĕg′lĭ-jənt) *adj.* [ME < OFr. < Lat. *neglegens*, pr.part. of *neglegere*, to neglect.] **1.** Marked by or inclined to neglect, esp. habitually. **2.** Extremely heedless or casual. **—neg′li·gent·ly** *adv.*

☆ ***syns:*** NEGLIGENT, DERELICT, DISREGARDFUL, LAX, NEGLECTFUL, REMISS, SLACK *adj. core meaning:* lacking or exhibiting a lack of due care or concern <physicians accused of being *negligent* in their care of the patient>

**neg·li·gi·ble** (nĕg′lĭ-jə-bəl) *adj.* [< Lat. *neglegere*, to neglect.] So insignificant as to be unworthy of consideration <*negligible* risks> **—neg′li·gi·bil′i·ty, neg′li·gi·ble·ness** *n.* **—neg′li·gi·bly** *adv.*

**ne·go·tia·ble** (nĭ-gō′shə-bəl, -shē-ə-) *adj.* **1.** Capable of being negotiated. **2.** Capable of being legally transferred from one to another, either by delivery or by delivery and endorsement. **—ne·go′tia·bil′i·ty** *n.* **—ne·go′tia·bly** *adv.*

**ne·go·ti·ant** (nĭ-gō′shē-ənt, -shənt) *n.* One that negotiates.

**ne·go·ti·ate** (nĭ-gō′shē-āt′) *v.* **-at·ed, -at·ing, -ates.** [Lat. *negotiari, negotiat-*, to transact business < *negotium*, business : *neg-*, not + *otium*, leisure.] *—vi.* To confer with another so as to come to terms or reach an agreement. *—vt.* **1.** To settle or arrange by conferring or discussing <*negotiate* a new contract> **2. a.** To transfer title to or ownership of (e.g., notes) to another party in return for value received. **b.** To discount or sell. **3.** To succeed in going over, accomplishing, or coping with <*negotiate* a downhill curve> **—ne·go′ti·a′tion** (-ā′shən) *n.* **—ne·go′ti·a′tor** *n.* **—ne·go′tia·to′ry** (-shə-tôr′ē, -tōr′ē, -shē-ə-) *adj.*

**Ne·gril·lo** (nĭ-grĭl′ō, -grē′yō) *n.*, *pl.* **-los** or **-loes.** [Sp., dim. of *negro*, black person. —see NEGRO.] One of a group of Negroid peoples of Africa, including the Bushmen and the Pygmies, who are short in stature.

**Ne·gri·to** (nĭ-grē′tō) *n.*, *pl.* **-tos** or **-toes.** [Sp., dim. of *negro*, black person. —see NEGRO.] **1.** A Negrillo. **2.** One of various groups of Negroid people of short stature inhabiting parts of Malaysia, the Philippines, and southeastern Asia.

**ne·gri·tude** (nē′grĭ-to͞od′, -tyo͞od′, nĕg′rĭ-) *n.* [Fr. *négritude* < *nègre*, Negro < Sp. *negro*.] An ideological and aesthetic concept affirming the independent validity of Negro culture.

**Ne·gro** (nē′grō) *n.*, *pl.* **-groes.** [Sp. and Port. < *negro*, black < Lat. *niger*.] **1.** A member of the Negroid ethnic division of the human species, esp. one of various peoples of central and southern Africa. **2.** One of Negro descent. **—Ne′gro** *adj.*

**Ne·groid** (nē′groid′) *adj.* [NEGR(O)- + -OID.] **1.** Of, relating to, or designating a major ethnic division of the human species whose members are marked by brown to black pigmentation and often by tightly curled hair. **2.** Of or characteristic of Negroes. *—n.* A member of the Negroid ethnic division of the human species.

**ne·gus** (nē′gəs) *n.* [After Colonel Francis *Negus* (d. 1732).] A beverage of wine, hot water, lemon juice, sugar, and nutmeg.

**Ne·gus** (nē′gəs, nĭ-go͞os′) *n.* [Amharic *negūs* < Ethiopic *nĕgūśā*.] The title of the former emperor of Ethiopia.

**Ne·he·mi·ah** (nē′hə-mī′ə, nē′ə-) *n.* [Heb. *Nĕḥemyāh*.] —See table at BIBLE.

**neigh** (nā) *n.* [< ME *neighen*, to neigh < OE *hnægan*.] The long, high-pitched sound made by a horse. **—neigh** *v.* **(neighed, neigh·ing, neighs).**

**neigh·bor** (nā′bər) *n.* [ME *neighebor* < OE *nēahgebūr* : *neāh*, near + *gebūr*, dweller.] **1.** One who lives or is located near or next to another. **2.** A fellow human being. *—v.* **-bored, -bor·ing, -bors.** *—vt.* To lie close to or border directly on. *—vi.* To live or be located close by.

**neigh·bor·hood** (nā′bər-ho͝od′) *n.* **1.** A district or area with distinctive characteristics. **2.** The people who live in a particular district or area. **3.** *Informal.* Approximate range or amount <in the *neighborhood* of 1,000 miles> **4.** *Math.* The set of points surrounding a specified point, each of which is at a distance from the specified point less than an arbitrary bound.

**neigh·bor·ly** (nā′bər-lē) *adj.* Appropriate to, characteristic of, or exhibiting the feelings of a friendly neighbor <*neighborly* greetings across the fence> **—neigh′bor·li·ness** *n.*

**neigh·bour** (nā′bər) *n.* & *v. Chiefly Brit. var. of* NEIGHBOR.

**nei·ther** (nē′thər, nī′-) *adj.* [ME < OE *nāwðer* : *nā*, not + *hwæðer*, which of two.] Not one or the other <*Neither* glove fits well.> *—pron.* Not the one nor the other: not either one <*Neither* of them fits.> *—conj.* **1.** Not either. —Used with the correlative conjunction *nor* <*Neither* we nor they desire it.> **2.** Also not <If I can't do it, *neither* can you.> *—adv.* Similarly not <Just as you would not, so *neither* would they.> ***usage:*** According to the traditional rule, *neither* should be construed as singular when it occurs as the subject of a sentence, as in *Neither of the two is willing to go.* Accordingly, a pronoun with *neither* as an antecedent must also be singular, as in *Neither architect submitted his model in the competition.*

**nek·ton** (nĕk′tən, -tŏn′) *n.* [G. < Gk. *nēkton*, neuter of *nēktos*, swimming < *nēkhein*, to swim.] The total population of marine animal organisms that swim independently of currents, ranging in size from microscopic organisms to whales. **—nek·ton′ic** (-tŏn′ĭk) *adj.*

**nel·son** (nĕl′sən) *n.* [Perh. < the name *Nelson*.] A wrestling hold in which the user places an arm under the opponent's arm and applies pressure with the palm of the hand against the opponent's neck.

**nemato-** or **nemat-** *pref.* [NLat. < Gk. *nēma*, thread.] **1.** Thread : threadlike <*nemato*cyst> **2.** Nematode <*nemato*cide>

**nem·a·to·cyst** (nĕm′ə-tə-sĭst′, nĭ-măt′ə-) *n. Zool.* A minute stinging organ in various coelenterates, as jellyfish, that when stimulated ejects a coiled tube that chemically paralyzes its victim. **—nem′a·to·cys′tic** *adj.*

**nem·a·tode** (nĕm′ə-tōd′) *n.* [NLat. *Nematoda*, phylum name < Gk. *nemā*, thread.] Any of various worms of the phylum Nematoda, having unsegmented threadlike bodies, many of which, as the hookworm, are parasitic.

**Nem·bu·tal** (nĕm′byə-tôl′). A trademark for the drug pentobarbital sodium.

**ne·mer·te·an** (nĭ-mûr′tē-ən) *also* **nem·er·tine** (nĕm′ər-tīn′) or **nem·er·tin·e·an** (nĕm′ər-tĭn′ē-ən) *adj.* [< NLat. *Nemertea*, phylum name < *Nemertes*, genus name < Gk. *Nēmertēs*, name of a Nereid.] Of, relating to, or belonging to the phylum Nemertea, consisting chiefly of marine worms with soft cylindrical or flattened, usu. brightly colored bodies. **—ne·mer′te·an** *n.*

**nem·e·sis** (nĕm′ĭ-sĭs) *n.*, *pl.* **-ses** (-sēz′) [Gk. < *nemesis*, retribution < *nemein*, to deal out.] **1. Nemesis.** *Gk. Myth.* The goddess of retributive justice or vengeance. **2.** One that inflicts retribution or vengeance. **3.** An unbeatable rival. **4.** Retributive justice in its execution or outcome. **5.** A source of harm or destruction.

**ne·ne** (nā′nā) *n.* [Hawaiian *nēnē*.] A goose, *Branta sandvicensis* of the Hawaiian Islands, now very rare.

**neo-** *pref.* [Gk. < *neos*, new.] **1.** New : recent <*Neo*lithic>. **2. a.** New and different <*neo*impressionism> **b.** New and abnormal <*neo*plasm> **3.** New World <*Neo*tropical>

**ne·o·ars·phen·a·mine** (nē′ō-ärs-fĕn′ə-mēn′) *n.* A yellow powder, $C_{13}H_{13}As_2N_2NaO_4S$, containing arsenic, used chiefly in treating syphilis and yaws.

**Ne·o·cene** (nē′ə-sēn′) *n.* A division of the Tertiary period comprising the Miocene and Pliocene. **—Ne′o·cene′** *adj.*

**ne·o·clas·si·cism** (nē′ō-klăs′ĭ-sĭz′əm) *n.* A revival of classical aesthetics and forms, esp. in the arts. **—ne′o·clas′sic, ne′o·clas′si·cal** *adj.* **—ne′o·clas′si·cist** *n.*

**ne·o·col·on·i·al·ism** (nē′ō-kə-lō′nē-ə-lĭz′əm) *n.* **1.** Control of former colonies by colonial powers, esp. by economic means. **2.** Indirect political or economic influence upon other nations or peoples by a powerful nation. **—ne′o·co·lo′ni·al** *adj.* **—ne′o·co·lo′ni·al·ist** *n.*

**ne·o·cor·tex** (nē′ō-kôr′tĕks′) *n.* The dorsal region of the cerebral cortex. **—ne′o·cor′ti·cal** (-tĭ-kəl) *adj.*

**Ne·o-Dar·win·ism** (nē′ō-där′wə-nĭz′əm) *n.* The theory that the evolutionary development of animals and plants is determined chiefly by natural selection and that acquired characteristics cannot be inherited. **—Ne′o-Dar·win′i·an** (-där-wĭn′ē-ən) *adj.* **—Ne′o-Dar′win·ist** *n.*

**ne·o·dym·i·um** (nē′ō-dĭm′ē-əm) *n.* [NEO- + (DI)DYMIUM.] *Symbol* **Nd** A bright silvery rare-earth metallic element, used for coloring glass and in some lasers; atomic number 60; atomic weight 144.24.

**Ne·o-Freud·i·an** (nē′ō-froi′dē-ən) *adj. Psychoanal.* Of or pertaining to a theory based on Freudian philosophy but emphasizing the significance of cultural and social influences on personality development. **—Ne′o-Freud′i·an** *n.*

**Ne·o·gae·a** *also* **Ne·o·ge·a** (nē′ə-jē′ə) *n.* [NEO- + Gk. *gaia*, earth.] A region coextensive with the Neotropical region and considered one of the primary zoogeographic realms. **—Ne′o·gae′an** *adj.*

**ne·o·gen·e·sis** (nē′ō-jĕn′ĭ-sĭs) *n.* Tissue regeneration. **—ne′o·ge·net′ic** (-jə-nĕt′ĭk) *adj.*

**ne·o·im·pres·sion·ism** *also* **ne·o-im·pres·sion·ism** (nē′ō-ĭm-prĕsh′ə-nĭz′əm) *n.* A 19th-cent. artistic movement that sought to make impressionism more meticulous and formal, marked by juxtaposition of dots of primary colors to achieve brighter secondary colors, with the mixture of pure tones being left for the eye itself to complete. **—ne′o·im·pres′sion·ist** *n.* & *adj.*

**Ne·o-La·marck·ism** (nē′ō-lə-mär′kĭz′əm) *n.* The theory that acquired characteristics can be transmitted but that natural selection is also valid. **—Ne′o-La·marck′i·an** (-kē-ən) *adj.* & *n.*

**ne·o·lith** (nē′ə-lĭth′) *n.* [Back-formation < NEOLITHIC.] A Neolithic stone implement.

**Ne·o·lith·ic** (nē′ə-lĭth′ĭk) *adj.* Of or denoting the cultural period beginning around 10,000 B.C. in the Middle East and later elsewhere and marked by the invention of farming and the making of technically advanced stone implements. *—n.* The Neolithic Age.

**ne·ol·o·gism** (nē-ŏl′ə-jĭz′əm) *n.* **1.** A newly coined word, phrase, or expression. **2.** A meaningless word or phrase coined or used by a psychotic. **—ne·ol′o·gist** *n.* **—ne·ol′o·gis′tic, ne·ol′o·gis′ti·cal** *adj.*

**ne·ol·o·gy** (nē-ŏl′ə-jē) *n.*, *pl.* **-gies.** **1.** A neologism. **2.** Use of a newly coined word, phrase, or expression or of a new meaning for an already established word. **—ne′o·log′i·cal** (nē′ə-lŏj′ĭ-kəl) *adj.* **—ne′o·log′i·cal·ly** *adv.*

**ne·o·morph** (nē′ə-môrf′) *n.* A biological structure that has not evolved from a similar structure in an ancestor. **—ne′o·morph′ic** (-môr′fĭk) *adj.*

**ne·o·my·cin** (nē′ə-mī′sĭn) *n.* An antibiotic drug consisting of a group of organic complexes produced by the metabolism of bacteria.

**ne·on** (nē′ŏn′) *n.* [< Gk., neuter of *neos,* new.] *Symbol* **Ne** An inert gaseous element occurring in the atmosphere to the extent of 18 parts per million, used in display and television tubes; atomic number 10; atomic weight 20.183.

**ne·o·nate** (nē′ə-nāt′) *n.* [NEO- + Lat. *natus,* p.part. of *nasci,* to be born.] A newborn child. **—ne′o·na′tal** (nē′ō-nāt′l) *adj.*

**ne·o·na·tol·o·gy** (nē′ō-nā-tŏl′ə-jē) *n.* The medical study of the first 60 days of an infant's life. **—ne′o·na·tol′o·gist** *n.*

**neon tetra** *n.* A small tropical American freshwater fish, *Hyphessobrycon innesi,* with iridescent blue and red markings.

**ne·o·or·tho·dox·y** (nē′ō-ôr′thə-dŏk′sē) *n.* A 20th-cent. Protestant movement that opposes liberalism and aims to revive adherence to certain Reformation doctrines. **—ne′o-or′tho·dox′** *adj.*

**ne·o·phyte** (nē′ə-fīt′) *n.* [LLat. *neophytus* < Gk. *neophutos* : *neous,* new + *phuton,* plant < *phuein,* to bring forth.] **1.** A recent convert. **2. a.** A newly ordained Roman Catholic priest. **b.** A novice of a religious order. **3.** A beginner : novice.

**ne·o·plasm** (nē′ə-plăz′əm) *n.* An abnormal new growth of tissue in animals or plants : TUMOR. **—ne′o·plas′tic** *adj.*

**Ne·o-Pla·to·nism** *also* **Ne·o·pla·to·nism** (nē′ō-plāt′n-ĭz′əm) *n.* **1.** A philosophical system developed at Alexandria in the 3rd cent. A.D., based on a modified form of Platonism combined with elements of Oriental mysticism and some Judaic and Christian concepts and positing a single source from which all existence emanates and with which an individual soul can be mystically united. **2.** A revival of Neo-Platonism or a system derived from it, as in the Middle Ages. **—Ne′o-Pla·ton′ic** (-plə-tŏn′ĭk) *adj.* **—Ne′o-Pla′to·nist** *n.*

**ne·o·prene** (nē′ə-prēn′) *n.* [NEO- + PR(OPYL) + -ENE.] A synthetic rubber produced by polymerization of chloroprene and marked by its durability and resistance esp. to oil.

**Ne·op·tol·e·mus** (nē′ŏp-tŏl′ə-məs) *n. Gk. Myth.* A son of Achilles who killed Priam during the capture of Troy.

**Ne·o-Scho·las·ti·cism** (nē′ō-skə-lăs′tĭ-sĭz′əm) *n.* A movement that attempts to revive medieval Scholasticism by infusing it with modern concepts. **—Ne′o-Scho·las′tic** (-lăs′tĭk) *adj.*

**ne·o·ter·ic** (nē′ə-tĕr′ĭk) *adj.* [LLat. *neotericus* < Gk. *neōterikos,* modern < *neōteros,* younger, comp. of *neos,* new.] Of recent origin : MODERN <*neoteric* compositions for woodwinds and brass>

**Ne·o·trop·i·cal** (nē′ō-trŏp′ĭ-kəl) *adj.* Of or designating the zoogeographic region stretching southward from the tropic of Cancer and including southern Mexico, Central and South America, and the West Indies.

**Ne·o·zo·ic** (nē′ə-zō′ĭk) *adj.* Of, relating to, or designating the geologic period from the end of the Mesozoic era to the present.

**Nep·al·ese** (nĕp′ə-lēz′, -lēs′) *n., pl.* **Nepalese. 1.** A native or inhabitant of Nepal. **2.** The Indic language of Nepal. *—adj.* Of, relating to, or designating Nepal, its language, inhabitants, or culture.

**ne·pen·the** (nĭ-pĕn′thē) *n.* [Gk. *nēpenthes (pharmakon),* grief-banishing (drug) : *ne-,* not + *penthos,* grief.] **1.** A legendary drug of ancient times, used as a remedy for sorrow. **2.** Something inducing forgetfulness of grief or easing pain. **—ne·pen′the·an** (-thē-ən) *adj.*

**neph·e·line** (nĕf′ə-lēn′, -lĭn) *also* **neph·e·lite** (-līt′) *n.* [Fr. *néphéline* < Gk. *nephelē,* cloud (because it becomes cloudy when placed in nitric acid).] A mineral of sodium- or potassium-aluminum silicate, occurring worldwide in igneous rocks and used in making ceramics and enamels. **—neph′e·lin′ic** (-lĭn′ĭk) *adj.*

**neph·e·lin·ite** (nĕf′ə-lĭ-nīt′) *n.* An igneous rock consisting chiefly of pyroxene and nepheline.

**neph·e·lite** (nĕf′ə-līt′) *n. var. of* NEPHELINE.

**neph·e·lom·e·ter** (nĕf′ə-lŏm′ĭ-tər) *n.* [Gk. *nephelē,* cloud + -METER.] An apparatus for measuring the size and concentration of particles in a liquid by analysis of light transmitted through or reflected by the liquid. **—neph′e·lo·met′ric** (-lō-mĕt′rĭk) *adj.* **—neph′e·lom′e·try** *n.*

**neph·ew** (nĕf′yōō) *n.* [ME *neveu* < OFr. < Lat. *nepos.*] **1.** The son of one's brother or sister or of one's brother-in-law or sister-in-law. **2.** A son of a celibate ecclesiastic.

**ne·phol·o·gy** (nĕ-fŏl′ə-jē) *n.* [Gk. *nephos,* cloud + -LOGY.] The science of clouds. **—neph′o·log′i·cal** (nĕf′ə-lŏj′ĭ-kəl) *adj.*

**nephr-** *pref. var. of* NEPHRO-.

**ne·phrec·to·my** (nə-frĕk′tə-mē) *n., pl.* **-mies.** Surgical removal of a kidney.

**ne·phrid·i·um** (nə-frĭd′ē-əm) *n., pl.* **-i·a** (-ē-ə). **1.** An excretory organ in many invertebrates. **2.** The excretory organ of a vertebrate embryo from which the kidney develops. **—ne·phrid′i·al** *adj.*

**neph·rite** (nĕf′rīt′) *n.* [G. *Nephrit* < Gk. *nephros,* kidney, from the belief that it cured kidney diseases.] A white to dark green jade, chiefly a metasilicate of iron, calcium, and magnesium.

**ne·phrit·ic** (nə-frĭt′ĭk) *adj.* **1.** Of or relating to the kidneys. **2.** Of, relating to, or affected with nephritis.

**ne·phri·tis** (nə-frī′tĭs) *n.* [LLat. < Gk. < *nephros,* kidney.] Acute or chronic inflammation of the kidneys.

**nephro-** *or* **nephr-** *pref.* Kidney : kidneylike structure <*nephro*sis><*nephro*tomy>

**neph·ro·gen·ic** (nĕf′rə-jĕn′ĭk) *adj.* Nephrogenous.

**ne·phrog·e·nous** (nə-frŏj′ə-nəs) *adj.* **1.** Originating in the kidney. **2.** Capable of generating new kidney tissue.

**neph·rol·o·gy** (nə-frŏl′ə-jē) *n.* The science that deals with the kidneys, esp. their diseases or functions. **—ne·phrol′o·gist** *n.*

**ne·phro·sis** (nə-frō′sĭs) *n.* A kidney disease, esp. one that is marked by degenerative lesions of the renal tubules. **—ne·phrot′ic** (-frŏt′ĭk) *adj.*

**ne·phrot·o·my** (nə-frŏt′ə-mē) *n., pl.* **-mies.** Surgical incision into the kidney.

**ne plus ul·tra** (nē′ plŭs ŭl′trə, nā′ plōōs ōōl′trä) *n.* [Lat., (go) no more beyond (this point).] The extreme or highest point, esp. of excellence or achievement.

**nep·o·tism** (nĕp′ə-tĭz′əm) *n.* [Fr. *népotisme* < Ital. *nepotismo* < *nepote,* nephew < Lat. *nepos.*] Favoritism shown by persons in high office to relatives or close friends esp. in granting jobs. **—nep′o·tist** *n.* **—nep′o·tis′tic, nep′o·tis′ti·cal** *adj.*

**Nep·tune** (nĕp′tōōn′, -tyōōn′) *n.* [Lat. *Neptunus.*] **1. a.** *Rom. Myth.* The god of the sea. **b.** The ocean <"sat with me on *Neptune's* yellow sands" —Shakespeare> **2.** The eighth planet from the sun, having a sidereal period of revolution around the sun of 164.8 years at a mean distance of 2.8 billion miles or 4.5 billion kilometers, a mean radius of 14,000 miles or 22,500 kilometers, and a density 17.2 times that of earth. **—Nep·tu′ni·an** (-tōō′nē-ən, -tyōō′-) *adj.*

**nep·tu·ni·um** (nĕp-tōō′nē-əm, -tyōō′-) *n.* [After the planet *Neptune,* so called because Neptune is the next planet after Uranus and neptunium follows uranium in the periodic table.] *Symbol* **Np** A silvery, metallic, naturally radioactive element; atomic number 93; longest-lived isotope Np 237.

**nerd** *also* **nurd** (nûrd) *n.* [Prob. alteration of NUT.] *Slang.* One who is socially inept, foolish, or ineffectual.

**Ne·re·id** (nîr′ē-ĭd) *n.* [Gk. *Nēreis, Nēreid-* < *Nēreus,* Nereus.] **1.** *Gk. Myth.* Any of the sea nymphs, daughters of Nereus. **2.** The smaller of the two satellites of the planet Neptune.

**ne·re·is** (nîr′ē-ĭs) *n., pl.* **ne·re·i·des** (nə-rē′ĭ-dēz′) [NLat. *Nereis,* genus name < Lat., Nereid < Gk. —see NEREID.] Any of several marine worms of the genus *Nereis,* having a long flat segmented body and a pair of paddles on each segment.

**Ne·re·us** (nîr′ē-əs, nîr′ōōs′) *n.* [Gk. *Nēreus.*] *Gk. Myth.* A sea god, the father of the Nereids.

**ne·rit·ic** (nə-rĭt′ĭk) *adj.* [Perh. < Lat. *nerita,* sea snail < Gk. *nēritēs* < *Nēreus,* Nereus.] Of or relating to shoreline waters and deposits.

**ner·o·li oil** (nĕr′ə-lē) *n.* [Fr., after Anna Maria de la Trémoille, 17th-cent. princess of *Neroli.*] An essential oil distilled from orange flowers and used in perfumery.

**nerts** (nûrts) *interj.* [Alteration of NUTS.] *Slang.* —Used to express contempt, disgust or refusal.

**ner·vate** (nûr′vāt′) *adj. Bot.* Having veins. —Used of leaves.

**ner·va·tion** (nûr-vā′shən) *n.* Venation.

**nerve** (nûrv) *n.* [Lat. *nervus.*] **1.** Any of the bundles of fibers interconnecting the central nervous system and the organs or parts of the body, capable of transmitting both sensory stimuli and motor impulses from one part of the body to another. **2.** A muscle or tendon <strain every *nerve* to win> **3.** A sensitive point or subject. **4.** A source from which energy or dynamic action emanates. **5. a.** Patience : endurance. **b.** Forcefulness : fortitude. **c.** Courage. **d.** *Informal.* Audacity : gall. **6.** Nervous agitation caused by anxiety, fear, or stress : HYSTERIA. **7.** A vein in an insect's wing. **8.** The midrib and larger veins in a leaf. *—vt.* **nerved, nerv·ing, nerves.** To give strength or courage to.

**nerve cell** *n.* A neuron.

**nerve center** *n.* **1.** A group of nerve cells that perform a specific function. **2.** A source or focus of control or power.

**nerve fiber** *n.* A threadlike process that is part of a nerve.

**nerve impulse** *n.* The wavelike progression of chemical and electrical disturbance along a stimulated nerve fiber.

**nerve·less** (nûrv′lĭs) *adj.* **1. a.** Lacking energy or strength. **b.** Lacking courage. **2.** Showing calmness and poise : COOL. **—nerve′less·ly** *adv.* **—nerve′less·ness** *n.*

**nerve-rack·ing** *also* **nerve-wrack·ing** (nûrv′răk′ĭng) *adj.* Extremely irritating to the nerves.

**nerv·ous** (nûr′vəs) *adj.* [ME, containing nerves < Lat. *nervosus* < *nervus,* nerve.] **1. a.** Of or relating to the nerves or nervous system. **b.** Stemming from or affecting the nerves or nervous system <a *nervous* tic> **2.** Easily excited or distraught : JUMPY. **3.** Characterized by uneasiness : APPREHENSIVE. **4.** *Archaic.* Strong : vigorous. **—nerv′ous·ly** *adv.* **—nerv′ous·ness** *n.*

**nervous breakdown** *n.* **1.** Neurasthenia. **2.** A severe or incapacitating emotional disorder.

**nervous Nel·lie** *or* **nervous Nel·ly** (nĕl′ē) *n., pl.* **-lies.** *Informal.* One who worries excessively or is overly timid.

**nervous system** *n. Anat.* A coordinating mechanism in all multicellular animals except sponges that regulates internal body functions and responses to external stimuli; in vertebrates it consists of the brain, spinal cord, nerves, ganglia, and parts of receptor and effector organs.

**ner·vure** (nûr′vyər) *n.* [Fr. < Lat. *nervus,* sinew.] **1.** One of the vascular ridges that form the framework of a leaf. **2.** One of the thickened ribs of tissue that form the framework of an insect's wing.

**nerv·y** (nûr′vē) *adj.* **-i·er, -i·est. 1.** Insolently confident : BRAZEN. **2.** Showing or requiring fortitude, energy, or stamina. **3.** *Chiefly Brit.* NERVOUS 2. **4.** *Archaic.* Full of muscular force. **—nerv′i·ness** *n.*

**nes·cience** (nĕsh′əns, nĕsh′ē-əns, nēsh′-, nĕs′ē-əns, nē′sē-) *n.* [LLat. *nescientia* < *nesciens,* ignorant, pr.part. of *nescire,* to be ignorant : *ne-,* not + *scire,* to know.] Absence of knowledge or awareness : IGNORANCE. **—nes′cient** *adj.* & *n.*

**ness** (nĕs) *n.* [ME *nasse* < OE *naessa.*] A cape or headland.

**-ness** *suff.* [ME *-nes* < OE.] State : quality : condition : degree <bright*ness*>

**Nes·sel·rode** (nĕs′əl-rōd′) *n.* [After Count Karl von *Nesselrode* (1780–1862).] A mixture of cherries, chestnuts, candied fruits, and liqueur, used in puddings, ice cream, or pies.

**nest** (nĕst) *n.* [ME < OE.] **1. a.** A shelter made by a bird for holding its eggs and young. **b.** A structure or shelter in which fishes or insects deposit eggs or keep their young. **c.** A spot where young are reared : LAIR. **d.** A number of animals, as birds or insects, occupying a place where young are reared. **2.** A snug lodging. **3. a.** A place or environment that fosters rapid growth or development : HOTBED. **b.** The persons occupying or frequenting such a place or environment. **4.** A set of objects of graduated size that can be stacked together, each fitting within a larger one. **5.** *Computer Sci.* One subroutine or set of data contained sequentially within another. *—v.* **nest·ed, nest·ing, nests.** *—vi.* **1.** To occupy or build a nest. **2.** To hunt for birds' nests, esp. in order to collect the eggs. *—vt.* **1.** To place in or as if in a nest. **2.** To put snugly together or inside one another.

**nest egg** *n.* **1.** An artificial or natural egg placed in a nest to induce a bird to lay. **2.** A reserve of money.

**†nest·er** (nĕs′tər) *n.* **1.** One that nests. **2.** *Western U.S.* A squatter, homesteader, or farmer who settles in cattle-grazing territory.

**nes·tle** (nĕs′əl) *v.* **-tled, -tling, -tles.** [ME *nestlen* < OE *nestlian,* to make a nest < *nest,* nest.] *—vi.* **1. a.** To settle snugly and comfortably. **b.** To lie in a sheltered place <a boat that *nestled* in the harbor> **2.** To draw or press close, esp. in an affectionate way <The child *nestled* up to the teddy bear.> **3.** *Archaic.* To nest. *—vt.* To snuggle, rest, or press contentedly. **—nes′tler** *n.*

**nest·ling** (nĕst′lĭng) *n.* **1.** A bird too young to leave its nest. **2.** A young child.

**Nes·tor** (nĕs′tər, -tôr′) *n.* [Gk. *Nestōr.*] **1.** A hero celebrated for his age and for his wise advice in the Homeric poems. **2.** *often* **nestor.** A venerable, wise old man.

**Nes·to·ri·an** (nĕ-stôr′ē-ən, -stōr′-) *adj.* Designating a church of the East that adheres to the doctrines of Nestorius, which assert that Christ had two distinct natures, divine and human. **—Nes·to′ri·an** *n.* **—Nes·to′ri·an·ism** *n.*

**net¹** (nĕt) *n.* [ME < OE *nett.*] **1.** Openwork fabric made of cords, threads, or ropes woven or knotted together at regular intervals. **2.** Something made of net, esp.: **a.** A device for capturing animals, as fish or butterflies. **b.** A mesh for holding the hair in place. **3. a.** A barrier strung between two posts to divide a tennis, badminton, or volleyball court in half. **b.** A ball that is hit into a net. **c.** An ice hockey goal. **4.** A meshed network of figures, lines, or fibers. **5.** Something that ensnares. *—vt.* **net·ted, net·ting, nets. 1.** To catch in or as if in a net. **2.** To protect, cover, or surround with or as if with a net. **3.** To hit (a ball) into a net. **4.** To make into a net. **—net′ter** *n.*

**net²** (nĕt) *adj.* [ME < OFr., elegant. —see NEAT.] **1.** Remaining after all deductions and adjustments have been made <*net* profits><*net* volume> **2.** Ultimate : final <the *net* outcome> *—n.* A net amount, as of profit or weight. *—vt.* **net·ted, net·ting, nets. 1.** To bring in or yield as profit. **2.** To clear as profit.

**neth·er** (nĕth′ər) *adj.* [ME < OE *niðera* < *niðer,* down.] **1.** Located or thought to be located beneath the earth's surface <the *nether* regions> **2.** Located below or down <the *nether* end>

**neth·er·most** (nĕth′ər-mōst′) *adj.* Farthest down : LOWEST.

**neth·er·world** (nĕth′ər-wûrld′) *n.* **1.** The world of the dead : HADES. **2.** The underworld : HELL.

**net·keep·er** (nĕt′kē′pər) *n.* A goalkeeper.

**net·su·ke** (nĕt′sə-kē) *n.* [J.] A small Japanese toggle, usu. decorated with inlays or carving, used esp. to fasten a purse to a kimono sash.

netsuke

**net·ting** (nĕt′ĭng) *n.* **1.** An openwork fabric : NET. **2.** The act or process of making a net. **3.** The act or process of fishing with a net.

**net·tle** (nĕt′l) *n.* [ME < OE *netele.*] **1.** A plant of the genus *Urtica,* having toothed leaves covered with hairs that secrete a stinging fluid that affects the skin on contact. **2.** A stinging or prickly plant. *—vt.* **-tled, -tling, -tles. 1.** To sting with or as if with a nettle. **2.** To vex : irritate <was *nettled* by their criticism>

**nettle rash** *n.* Urticaria.

**net·tle·some** (nĕt′l-səm) *adj.* Annoying : vexatious.

**net ton** *n.* TON 1b.

**net·work** (nĕt′wûrk′) *n.* **1.** An openwork fabric or structure in which rope, thread, or wires cross at regular intervals. **2.** Something resembling a net <a *network* of spies><a *network* of highways> **3.** A chain of interconnected broadcasting stations, usu. sharing a large proportion of their programs <a TV *network*> **4.** A group or system of electric components and connecting circuitry designed to function in a specific way. *—v.* **-worked, -work·ing, -works.** *—vt.* **1.** To overlay with or as if with a network. **2.** To broadcast over a network. *—vi. Informal.* To make connections among people or groups of a like kind.

**Neuf·châ·tel** (no͞o′shə-tĕl′, nœ′shä-) *n.* [After *Neufchâtel,* France.] A soft white cheese made from skimmed or whole milk or cream.

**neume** *or* **neum** (no͞om, nyo͞om) *n.* [ME, series of notes sung on one syllable < Med. Lat. *pneuma* < Gk., breath.] A symbol used in notation of plainsong in the Middle Ages, surviving in transcriptions of Gregorian chants. **—neu·mat′ic** (no͞o-măt′ĭk, nyo͞o-) *adj.*

**neur-** *pref. var. of* NEURO-.

**neu·ral** (no͝or′əl, nyo͝or′-) *adj.* **1.** Of or relating to a nerve or the nervous system. **2.** Of, relating to, or located on the same side of the body as the spinal cord : DORSAL. **—neu′ral·ly** *adv.*

**neu·ral·gia** (no͞o-răl′jə, nyo͞o-) *n.* Paroxysmal pain along a nerve. **—neu·ral′gic** *adj.*

**neu·ras·the·ni·a** (no͝or′əs-thē′nē-ə, nyo͝or′-) *n.* A condition characterized by fatigue, loss of memory and energy, and feelings of inadequacy, once thought to result from nervous exhaustion. **—neu′ras·then′ic** (-thĕn′ĭk) *adj.* **—neu′ras·then′i·cal·ly** *adv.*

**neu·rec·to·my** (no͞o-rĕk′tə-mē, nyo͞o-) *n., pl.* **-mies.** Surgical removal of a nerve or part of a nerve.

**neu·ri·lem·ma** (no͝or′ə-lĕm′ə, nyo͝or′-) *n.* [NEUR- + Gk. *eilēma,* veil < *eilein,* to wind.] The outer covering of a nerve fiber. **—neu′ri·lem′mal** *adj.*

**neu·ris·tor** (no͞o-rĭs′tər, nyo͞o-) *n.* [NEUR(ON) + (TRANS)ISTOR.] An electronic device capable of relaying a signal without attenuation in velocity.

**neu·ri·tis** (no͞o-rī′tĭs, nyo͞o-) *n.* Inflammation of a nerve that causes pain, loss of reflexes, and muscular atrophy. **—neu·rit′ic** (-rĭt′ĭk) *adj.*

**neuro-** *or* **neur-** *pref.* [NLat. < Gk. *neuron,* tendon, nerve.] **1.** Nerve <*neuro*blast> **2.** Neural <*neuro*pathology>

**neu·ro·blast** (no͝or′ə-blăst′, nyo͝or′-) *n.* An embryonic cell from which a nerve cell develops.

**neu·ro·cyte** (no͝or′ə-sīt′, nyo͝or′-) *n.* A nerve cell and its processes.

**neu·rog·li·a** (no͞o-rŏg′lē-ə, nyo͞o-, no͞o′rə-glē′ə, nyo͞o′-, -glī′-) *n.* [NEURO- + Med. Gk. *glia,* glue.] The network of branched cells and fibers that supports the tissue of the central nervous system. **—neu·rog′li·al** *adj.*

**neu·rol·o·gy** (no͞o-rŏl′ə-jē, nyo͞o-) *n.* The medical science of the nervous system and its disorders. **—neu′ro·log′i·cal** (no͝or′ə-lŏj′ĭ-kəl, nyo͝or′-) *adj.* **—neu·rol′o·gist** *n.*

**neu·ro·ma** (no͞o-rō′mə, nyo͞o-) *n., pl.* **-mas** *or* **-ma·ta** (-mə-tə). A tumor composed of nerve tissue.

**neu·ron** (no͝or′ŏn′, nyo͝or′-) *also* **neu·rone** (-ōn′) *n.* [Gk., nerve.] Any of the cells of nerve tissue consisting of a nucleated portion and cytoplasmic extensions, the cell body, and the dendrites and axons. **—neu·ron′ic** *adj.* **—neu·ron′i·cal·ly** *adv.*

**neu·ro·path** (no͝or′ə-păth′, nyo͝or′-) *n.* One afflicted with or having a hereditary tendency toward neurosis or nervous disorders. **—neu′ro·path′ic, neu′ro·path′i·cal** *adj.* **—neu′ro·path′i·cal·ly** *adv.*

**neu·ro·pa·thol·o·gy** (no͝or′ō-pə-thŏl′ə-jē, nyo͝or′-) *n.* Study of diseases of the nervous system. **—neu′ro·path′o·log′ic** (-păth′ə-lŏj′ĭk), **neu′ro·path′o·log′i·cal** *adj.* **—neu′ro·pa·thol′o·gist** *n.*

**neu·rop·a·thy** (no͞o-rŏp′ə-thē, nyo͞o-) *n.* Disease or abnormality of the nervous system.

**neu·ro·psy·chi·a·try** (no͝or′ō-sī-kī′ə-trē, -sĭ-, nyo͝or′-) *n.* The integrated medical study of both psychiatric and neurological disorders. **—neu′ro·psy′chi·at′ric** (-sī′kē-ăt′rĭk) *adj.* **—neu′ro·psy·chi′a·trist** *n.*

**neu·rop·ter·an** (no͞o-rŏp′tər-ən, nyo͞o-) *n.* [< NLat. *Neuroptera,* order name : NEURO- + Gk. *pteron,* wing.] An insect of the order Neuroptera, as the ant lion or lacewing, having four net-veined wings. **—neu·rop′ter·an** *adj.* **—neu·rop′ter·ous** *adj.*

**neu·ro·sis** (no͞o-rō′sĭs, nyo͞o-) *n., pl.* **-ses** (-sēz′). Any of various functional disorders of the mind or emotions without obvious organic lesion or change and involving abnormal behavior symptoms, as anxiety or phobia.

**neu·ro·sur·ger·y** (no͝or′ō-sûr′jə-rē, nyo͝or′-) *n.* Surgery of the nervous system or of any of its parts. **—neu′ro·sur′geon** *n.* **—neu′ro·sur′gi·cal** *adj.*

**neu·rot·ic** (no͞o-rŏt′ĭk, nyo͞o-) *adj.* **1.** Of, relating to, or derived from a neurosis. **2.** Of, relating to, or afflicted with neurosis. **—neu·rot′ic** *n.* **—neu·rot′i·cal·ly** *adv.*

**neu·rot·o·my** (no͞o-rŏt′ə-mē, nyo͞o-) *n., pl.* **-mies.** Surgical cutting or stretching of a nerve, usu. so as to relieve pain.

**neu·ter** (no͞o′tər, nyo͞o′-) *adj.* [ME *neutre* < OFr. < Lat. *neuter,* neither : *ne-,* not + *uter,* either.] **1. a.** Neither masculine nor feminine in grammatical gender. **b.** Neither active nor passive. —Used of verbs. **2. a.** *Biol.* Having no sexual organs. **b.** *Bot.* Having no pistils or stamens : ASEXUAL. **c.** *Zool.* Sexually undeveloped. **3.** Taking no side : NEUTRAL. —*n.* **1. a.** The neuter grammatical gender. **b.** A neuter word. **2. a.** A castrated animal. **b.** A sexually undeveloped or imperfectly developed female insect : WORKER. **c.** *Bot.* A plant without stamens or pistils. **3.** A neutral person. —*vt.* **-tered, -ter·ing, -ters.** To castrate or spay.

**neu·tral** (no͞o′trəl, nyo͞o′-) *adj.* [Lat. *neutralis,* grammatically neuter < *neuter.* —see NEUTER.] **1.** Not allied with, supporting, or favoring either side in a dispute, war, or contest. **2.** Belonging to neither side nor party <on *neutral* land> **3.** Not one thing or the other : INDIFFERENT. **4.** Of no sex : NEUTER. **5.** *Chem.* **a.** Of or relating to a compound that is neither alkaline nor acidic. **b.** Of or relating to a solution in which the concentrations of positive and negative ions are equal. **6.** *Physics.* **a.** Of or relating to an object, particle, or system that has neither positive nor negative electric charge. **b.** Of or relating to a particle, object, or system that has a net electric charge of zero. **7.** Of or indicating a color, as gray, black, or white, that lacks hue : ACHROMATIC. **8.** Pronounced with the tongue in a relaxed middle position, as the *a* in *around.* —*n.* **1.** One that is neutral. **2.** A neutral color. **3.** A position in which a set of gears is disengaged so that power cannot be transmitted. **—neu′tral·ly** *adv.*

☆ ***syns:*** NEUTRAL, DETACHED, DISINTERESTED, DISPASSIONATE, IMPERSONAL *adj. core meaning* : not supporting or favoring either side, as in a dispute <held a *neutral* attitude through the entire discussion>

**neu·tral·ism** (no͞o′trə-lĭz′əm, nyo͞o′-) *n.* **1.** Neutrality. **2.** A political attitude of noninvolvement or nonalignment with conflicting alliances. **—neu′tral·ist** *n.* **—neu′tral·is′tic** *adj.*

**neu·tral·i·ty** (no͞o-trăl′ĭ-tē, nyo͞o-) *n.* The state or policy of being neutral, esp. nonparticipation in war.

**neu·tral·i·za·tion** (no͞o′trə-lĭ-zā′shən, nyo͞o′-) *n.* **1.** The act or process of neutralizing. **2.** *Chem.* A reaction between an acid and a base that yields a salt and water.

**neu·tral·ize** (no͞o′trə-līz′, nyo͞o′-) *vt.* **-ized, -iz·ing, -iz·es. 1.** To make neutral. **2.** To counterbalance the effect of so as to render ineffective. **3.** To make neutral and immune from use, invasion, or control by a warring nation. **4.** *Chem.* **a.** To make (a solution) chemically neutral. **b.** To cause (an acid or base) to undergo neutralization. **5.** *Med.* To counteract the effect of (a toxin or drug). **—neu′tral·iz′er** *n.*

**neutral spirits** *n. (sing. or pl. in number).* Ethyl alcohol distilled at or above 190 proof and frequently used in alcoholic beverage blends.

**neu·tri·no** (no͞o-trē′nō, nyo͞o-) *n., pl.* **-nos.** [Ital., dim. of *neutrone,* neutron < E. NEUTRON.] *Physics.* Either of two massless, electrically neutral, stable subatomic particles in the lepton family.

**neu·tron** (no͞o′trŏn′, nyo͞o′-) *n.* [NEUTR(AL) + -ON[1].] An electrically neutral subatomic particle in the baryon family, having a mass 1,839 times that of the electron, stable when bound in an atomic nucleus, and having a mean lifetime of approx. 16.6 minutes as a free particle.

**neutron star** *n.* A celestial body hypothesized to occur in a terminal stage of stellar evolution, consisting of a superdense mass essentially of neutrons, and having a powerful gravitational attraction from which only neutrinos and high-energy photons could escape, thus rendering the body invisible except to x-ray detection.

**neu·tro·phil** (no͞o′trə-fĭl′, nyo͞o′-) *also* **neu·tro·phile** (-fīl′) *or* **neu·tro·phil·ic** (no͞o′trə-fĭl′ĭk, nyo͞o′-) *adj.* [NEUTR(AL) + -PHIL(E).] Easily stained by neutral dyes. —Used of cells such as leucocytes. **—neu′tro·phil′, neu′tro·phile′** *n.*

**né·vé** (nā-vā′) *n.* [Fr., ult. < Lat. *nix,* snow.] **1.** The upper part of a glacier where the snow turns into ice. **2. a.** A snow field at the head of a glacier. **b.** The granular snow typically found in such a field.

**nev·er** (nĕv′ər) *adv.* [ME < OE *nǣfre* : *ne,* not + *ǣfre,* ever.] **1.** Not ever <I have *never* seen the movie.> **2.** Not at all <*Never* mind.><*Never* fear.>

**nev·er·more** (nĕv′ər-môr′, -mōr′) *adv.* Never again.

**nev·er-nev·er land** (nĕv′ər-nĕv′ər) *n.* An illusory, idyllic, or ideal place.

**nev·er·the·less** (nĕv′ər-thə-lĕs′) *adv.* Nonetheless <The project may fail, but we will try *nevertheless.*>

**ne·vus** (nē′vəs) *n., pl.* **-vi** (-vī′) [Lat. *naevus.*] A congenital growth or mark on the skin, as a birthmark. **—ne′void′** (-void′) *adj.*

**new** (no͞o, nyo͞o) *adj.* **-er, -est.** [ME *newe* < OE *nīwe.*] **1.** Having existed or been made for only a short time : RECENT <a *new* book> **2. a.** Not yet old : FRESH <a *new* coat of wax> **b.** Never used before <a *new* suit> **3.** Just discovered, found, or learned <*new* evidence> **4.** Unfamiliar : novel <customs that were *new* to me> **5.** Starting again in a cycle <the *new* lunar cycle> **6.** Recently acquired or obtained <*new* military strength> **7.** Additional <developed *new* sources of energy> **8.** Recently arrived or established in a position, place, or relationship <a *new* boss> **9.** Changed for the better : REJUVENATED <My vacation made a *new* person of me.> **10.** Coming after or taking the place of a previous one or ones <a *new* model> **11. a.** Modern : current <*new* dances> **b. New.** In the most recent period, form, or development <*New* Latin> **12.** Inexperienced : untrained <*new* at the job> —*adv.* Freshly : recently <the smell of *new*-mown grass> **—new′ness** *n.*

**new·born** (no͞o′bôrn′, nyo͞o′-) *adj.* **1.** Born very recently. **2.** Born anew. —*n.* A neonate.

**New·burg** *also* **New·burgh** (no͞o′bûrg′, nyo͞o′-) *adj.* [Orig. unknown.] Served in a rich sauce made of cream, butter, egg yolks, wine, and usu. nutmeg <crab *Newburg*>

**new·com·er** (no͞o′kŭm′ər, nyo͞o′-) *n.* A recent arrival.

**New Criticism** *n.* A method of literary analysis developed in the mid-20th cent. that stresses a close examination of the language, imagery, structure, and thematic tensions of the text and posits that the facts of an author's life or intentions for the work are irrelevant. **—New Critic** *n.*

**New Deal** *n.* **1.** The policies and programs for economic recovery and reform, relief, and social security introduced during the 1930's by President Franklin D. Roosevelt and his administration. **2.** The period during which the policies and programs of the New Deal were put into effect. **—New Dealer** *n.*

**new·el** (no͞o′əl, nyo͞o′-) *n.* [ME *novel* < OFr. < VLat. **nucale* < Lat. *nux,* nut.] **1.** The vertical support at the center of a circular staircase. **2.** A post that supports a handrail at the bottom or at the landing of a staircase.

**New English Bible** *n.* A British interdenominational translation of the Bible first published in complete form in 1970.

**new·fan·gled** (no͞o′făng′gəld, nyo͞o′-) *adj.* [< ME *neuefangel,* fond of novelty.] New : novel. **—new′fan′gled·ness** *n.*

**new-fash·ioned** (no͞o′făsh′ənd, nyo͞o′-) *adj.* **1.** Up-to-date. **2.** Created in a new fashion or form.

**New Federalism** *n.* A political and economic program of decentralization, whereby federally administered functions and tax revenues are shared with the 50 U.S. states. **—New Federalist** *n.*

**new·found** (no͞o′found′, nyo͞o′-) *adj.* Recently discovered.

**New·found·land** (no͞o′fən-lənd, nyo͞o′-) *n.* [After *Newfoundland,* Canada.] A large breed of dog with a broad head and square muzzle, a powerful body, and a dense, usu. black coat.

**New Greek** *n.* Modern Greek.

**New Hebrew** *n.* The Hebrew language used in Israel at the present time.

**New Jerusalem** *n.* **1.** The final resting place of souls redeemed by Christ. **2.** An ideal community on earth.

**New Journalism** *n.* Journalism marked by the reporter's subjective interpretations and personal involvement, often featuring fictional dramatized elements to emphasize that involvement. **—New Journalist** *n.*

**New Latin** *n.* Latin as used since about 1500.

**New Left** *n.* A U.S. radical political movement originating in the 1960's and marked by active advocacy of revolutionary changes, as in politics, government, education, and society. **—New Leftist** *n.*

**new·ly** (no͞o′lē, nyo͞o′-) *adv.* **1.** Lately : recently <a *newly* baked pie> **2.** Once more : ANEW <a *newly* decorated room> **3.** In a new or different way : FRESHLY <an old song *newly* arranged>

**new·ly-wed** *also* **new·ly·wed** (no͞o′lē-wĕd′, nyo͞o′-) *n.* A recently married person.

**new math** *n.* Mathematics taught in elementary and secondary schools that constructs mathematical relationships from set theory.

**new moon** *n.* **1.** The phase of the moon occurring when it passes

between the earth and the sun and is invisible or visible only as a narrow crescent at sunset. **2.** The crescent moon.
**news** (nōōz, nyōōz) *pl.n. (sing. in number).* **1.** Recent events and happenings, esp. those that are notable or unusual. **2. a.** Information about recent events of general interest, esp. as reported by the print and broadcast media. **b.** A presentation of such information. **3.** Newsworthy material.
**news agency** *n.* An organization that provides news coverage to subscribers, as newspapers or periodicals.
**news·boy** (nōōz'boi', nyōōz'-) *n.* One who delivers or sells newspapers to the public.
**news·break** (nōōz'brāk', nyōōz'-) *n.* An event worthy of reporting, as on television.
**news·cast** (nōōz'kăst', nyōōz'-) *n.* [NEWS + (BROAD)CAST.] A radio or television broadcast of events in the news. **—news'cast'er** *n.*
**news conference** *n.* A press conference.
**news·let·ter** (nōōz'lĕt'ər, nyōōz'-) *n.* A printed report giving news or information of interest to a special group.
**news·man** (nōōz'măn', -mən, nyōōz'-) *n.* One who gathers, reports, or edits news.
**news·mon·ger** (nōōz'mŭng'gər, -mŏng'-, nyōōz'-) *n.* One who gathers and repeats news, esp. a gossip.
**news·pa·per** (nōōz'pā'pər, nyōōz'-) *n.* **1.** A typically daily or weekly publication containing recent news, feature articles, editorials, and usu. advertising. **2.** Newsprint.
**news·pa·per·man** (nōōz'pā'pər-măn', nyōōz'-) *n.* One who owns or is employed by a newspaper.
**news·pa·per·wom·an** (nōōz'pā'pər-wŏŏm'ən, nyōōz'-) *n.* A woman who owns or is employed by a newspaper.
**new·speak** (nōō'spēk', nyōō'-) *n.* [< *Newspeak,* a language invented by George Orwell (1903–1950) in the novel *Nineteen Eighty-Four.*] Ambiguous and contradictory language used as propaganda.
**news·person** (nōōz'pûr'sən, nyōōz'-) *n.* A reporter or newscaster.
**news·print** (nōōz'prĭnt', nyōōz'-) *n.* Inexpensive paper made from wood pulp, used chiefly for printing newspapers.
**news·reel** (nōōz'rēl', nyōōz'-) *n.* A short film dealing with current news events.
**news release** *n.* HANDOUT 3.
**news·room** (nōōz'rōōm', -rŏŏm', nyōōz'-) *n.* A room, as in a newspaper office or a radio or television station, in which news is prepared for release.
**news·stand** (nōōz'stănd', nyōōz'-) *n.* A place at which newspapers and periodicals are sold.
**New Style** *n.* The method of reckoning the months and days of the year according to the Gregorian calendar.
**news·wom·an** (nōōz'wŏŏm'ən, nyōōz'-) *n.* A woman who gathers, reports, or edits news.
**news·wor·thy** (nōōz'wûr'*th*ē, nyōōz'-) *adj.* Worthy of reporting to the general public.
**news·y** (nōō'zē, nyōō'-) *adj.* **-i·er, -i·est.** *Informal.* Full of news.
**newt** (nōōt, nyōōt) *n.* [ME *a neute,* alteration of *an eute,* var. of *evete* < OE *efete.*] Any of several small semiaquatic salamanders of the genus *Triturus* and related genera.
**New Testament** *n.* The Gospels, Acts, Pauline and other Epistles, and the Book of Revelation, together viewed by Christians as forming the record of the new dispensation belonging to the Church. —See table at BIBLE.
**New Thought** *n.* A modern religious movement that emphasizes spiritual healing and the creative power of positive thought.
**new·ton** (nōōt'n, nyōōt'n) *n.* [After Sir Isaac *Newton* (1642–1727).] *Physics.* The unit of force in the meter-kilogram-second system that is needed to accelerate a mass of one kilogram one meter per second per second, equal to 100,000 dynes.
**New·to·ni·an** (nōō-tō'nē-ən, nyōō-) *adj.* Of, relating to, or in accordance with the work of Sir Isaac Newton, esp. in mechanics and gravitation.
**new town** *n.* A planned urban community designed for self-sufficiency and comprising housing, commercial, industrial, and recreational facilities.
**New Wave** *n.* **1.** A filmmaking movement marked by unconventional techniques, as abstraction and subjective symbolism, and often by experimental photography. **2.** A new movement in a particular area, as gourmet cooking. **3.** Rock music characterized by ensemble playing rather than lengthy solo passages and by lyrics that express anger or social alienation.
**New World** *n.* The Western Hemisphere.
**New Year** *n.* **1.** The first day or days of the calendar year. **2.** Rosh Hashanah.
**New Year's Day** *n.* Jan. 1, the first day of the year, celebrated as a holiday in many countries.
**New York aster** (yôrk) *n.* A wild aster, *Aster novi-belgi* of eastern North America, with pointed leaves and bluish-violet flowers.

New York aster

**next** (nĕkst) *adj.* [ME *nexte* < OE *nēahst,* superl. of *nēah,* near.] **1.** Closest or nearest in position or space : ADJACENT <the *next* house> **2.** Immediately following or preceding in time, order, or sequence <*next* Friday><the *next* governor> *—adv.* **1.** In the time, position, or order nearest or immediately following <They come *next.*> **2.** On the first subsequent occasion <when *next* I visit> *—prep.* Close to : NEAREST.
**next door** *adv.* To or in the adjacent building, house, room, or apartment. **—next'-door'** (nĕkst'dôr', -dōr') *adj.*
**next friend** *n. Law.* One appointed by or admitted to a court to sue as the representative of a minor or a person under legal disability.
**next of kin** *n.* **1.** The person or persons most closely related to another person. **2.** *Law.* **a.** A deceased person's closet relative. **b.** Those relatives entitled to the estate of a deceased person in accordance with the statutes of distribution.
**nex·us** (nĕk'səs) *n., pl.* **nexus** or **-us·es.** [Lat. < *nectere,* to bind.] **1.** A means of connection : LINK. **2.** A connected series or group.
**Nez Perce** (nĕz' pûrs', nĕs') *n., pl.* **Nez Perce** or **Nez Perc·es** (pûr'sĭz) [Canadian Fr.] **1. a.** A tribe of Indians once occupying much of the Pacific Northwest. **b.** A member of this tribe. **2.** The Sahaptin language of the Nez Perce.
**n'ga·na** (nə-gä'nə) *n. var. of* NAGANA.
**ngul·trum** (ĕn-gŭl'trəm, ĕng-) *n.* [Native word in Bhutan.] —See table at CURRENCY.
**ngwee** (ĕn-gwē') *n., pl.* **ngwee.** [Of Bantu orig.] —See table at CURRENCY.
**Ni** *symbol for* NICKEL.
**ni·a·cin** (nī'ə-sĭn) *n.* [NI(COTINIC) AC(ID) + -IN.] Nicotinic acid.
**nib** (nĭb) *n.* [Perh. alteration of NEB.] **1. a.** The point of a quill pen, esp. when sharpened. **b.** The point of a pen. **2.** A sharp tip or point. **3.** A bird's beak or bill.
**nib·ble** (nĭb'əl) *v.* **-bled, -bling, -bles.** [Orig. unknown.] *—vt.* **1.** To bite at gently and repeatedly. **2.** To eat with small quick bites or in small morsels. *—vi.* To take small or hesitant bites. *—n.* **1.** A small quantity, esp. of food. **2.** An act of nibbling.
**Ni·be·lung** (nē'bə-lŏŏng') *n.* [G. < MHG *Nibelungen.*] *Myth.* **1.** Any of a race of subterranean dwarfs whose hoard of riches and magic ring were taken from them by Siegfried. **2.** A follower of Siegfried. **3.** One of the Burgundian kings in the *Nibelungenlied.*
**Ni·be·lung·en·lied** (nē'bə-lŏŏng'ən-lēd') *n.* An early 13th-cent. Middle High German epic poem based on the legends of Siegfried and of the Burgundian kings.
**nib·lick** (nĭb'lĭk) *n.* [Orig. unknown.] A nine iron.
**nic·co·lite** (nĭk'ə-līt') *n.* [NLat. *niccolum,* nickel (perh. < Swed. *nickel*) + -ITE.] A nickel ore, essentially nickel arsenide, NiAs, found in America and Europe.
**nice** (nīs) *adj.* **nic·er, nic·est.** [ME, foolish < OFr. < Lat. *nescius,* ignorant < *nescire,* to be ignorant. —see NESCIENCE.] **1.** Enjoyable and agreeable : PLEASANT. **2.** Having an appealing or attractive appearance <*nice* clothing><a *nice* figure> **3.** Courteous and polite : CONSIDERATE <a *nice* offer> **4.** Of good character and reputation : RESPECTABLE. **5.** Excessively fussy : FASTIDIOUS. **6.** Showing or characterized by great precision and sensitive discernment : SUBTLE <a *nice* distinction> **7.** Executed with accuracy, delicacy, and skill <a *nice* bit of workmanship> **8.** —Used as an intensive with *and* <*nice* and warm> **9.** *Obs.* **a.** Wanton : profligate. **b.** Affectedly modest : COY. **—nice'ly** *adv.* **—nice'ness** *n.*

▲ word history: Since its adoption in the 13th century the word *nice* has developed from a term of abuse to a term of praise, a process called *melioration. Nice* is derived from Latin *nescius,* "ignorant," and was used in Middle English to mean "foolish; without sense." By the 15th century *nice* had acquired the sense "elegant" in conduct and dress, but not in a complimentary sense: *nice* meant "over-refined, overly delicate." This sense survives in the meanings "fastidious" and "precise, subtle." To the extent that delicacy, refinement, and precision have favorable connotations, *nice* developed corresponding senses.

**Ni·cene Creed** (nī'sēn', nī-sēn') *n.* A formal statement of doctrine of the Christian faith adopted at the Council of Nicaea in A.D. 325 and expanded in later councils.
**nice-nel·ly** (nīs'nĕl'ē) *adj.* [< the name *Nelly.*] **1.** PRIM[1] 1. **2.** Characterized by the use of euphemism.
**ni·ce·ty** (nī'sĭ-tē) *n., pl.* **-ties.** [ME *nicete,* exactitude, silliness < OFr., silliness < *nice,* silly. —see NICE.] **1.** The quality of showing or

requiring precise and careful treatment. **2.** Delicacy : fastidiousness. **3.** A fine point or subtle distinction. **4.** An elegant or refined feature : AMENITY <the *niceties* of formal dining>
**niche** (nĭch, nēsh) *n.* [Fr. < OFr. < *nichier*, to nest < VLat. **nidicare* < Lat. *nidus*, nest.] **1.** A recess in a wall, as for holding a statue. **2.** A hollow or crevice, as in rock. **3.** A situation or activity specially suited to one's character or abilities. **4.** *Ecol.* **a.** The set of functional relationships of an organism or population to the environment it occupies. **b.** The area within a habitat occupied by an organism. —*vt.* **niched, nich·ing, nich·es.** To place in a niche.
**nick** (nĭk) *n.* [ME *nik.*] A shallow cut, notch, or indentation on a surface. —*vt.* **nicked, nick·ing, nicks. 1. a.** To cut a nick or notch in. **b.** To graze and wound slightly <was *nicked* by the bullet> **2.** To cut short : CHECK. **3.** *Slang.* To cheat or overcharge. **—in the nick of time.** Just at the critical moment.
**nick·el** (nĭk'əl) *n.* [Swed., short for G. *Kupfernickel*, niccolite : *Kupfer*, copper (< OHG *kupfar* < LLat. *cuprum*) + *Nickel*, demon, from the deceptive copper color of the ore.] **1.** *Symbol* **Ni** A silvery, hard, ductile, metallic element used in alloys, in corrosion-resistant surfaces and batteries, and for electroplating; atomic number 28; atomic weight 58.71. **2.** A U.S. coin worth five cents, made of a nickel and copper alloy. —*vt.* **-eled, -el·ing, -els** *or* **-elled, -el·ling, -els.** To coat with a thin layer of nickel.
▲ word history: Copper has been known and worked since ancient times, but the metal nickel was isolated and identified only in the 18th century. Nickel is found in an ore that resembles copper ore in appearance; the German name for the ore is *Kupfernickel*, literally "copper demon." The ore was so called because it produces no copper but was mistaken for copper ore, just as "fool's gold" looks deceptively like gold. Axel von Cronstadt, the mineralogist who first isolated the metal nickel, took its name from the second element of *Kupfernickel*.
**nick·el-and-dime** (nĭk'əl-ən-dīm') *Informal.* —*adj.* **1.** Of, involving, or paying only a small amount of money <a *nickel-and-dime* job> **2.** Minor : small-time <*nickel-and-dime* gangsters> —*v.* **-dimed, -dim·ing, -dimes.** —*vi.* To spend money prudently, esp. out of necessity. —*vt.* To change for the worse, esp. to diminish or destroy gradually through persistent attention to petty financial details <*nickel-and-dimed* me out of business>
**nick·el·ic** (nĭ-kĕl'ĭk) *adj.* **1.** Of or containing nickel. **2.** Of or containing trivalent nickel, $Ni^{3+}$.
**nick·el·if·er·ous** (nĭk'ə-lĭf'ər-əs) *adj.* Yielding or containing nickel. —Used of ores.
**nick·el·o·de·on** (nĭk'ə-lō'dē-ən) *n.* [NICKEL + (MEL)ODEON.] **1.** An early movie house charging an admission price of five cents. **2. a.** A player piano. **b.** A juke box.
**nick·el·ous** (nĭk'ə-ləs) *adj.* **1.** Of or containing nickel. **2.** Of or containing bivalent nickel, $Ni^{2+}$.
**nickel silver** *n.* A silvery, hard, corrosion-resistant, malleable alloy of zinc, copper, and nickel.
**nick·er** (nĭk'ər) *vi.* **-ered, -er·ing, -ers.** [Perh. alteration of NEIGH.] To neigh softly. **—nick'er** *n.*
**nick·nack** (nĭk'năk') *n. var. of* KNICKKNACK.
**nick·name** (nĭk'nām') *n.* [ME *a nekename*, alteration of *an ekename* : *eke*, addition (< OE *ēaca*) + *name*, name < OE *nama*.] **1.** A descriptive name added to or replacing the actual name of a person, place, or thing. **2.** A shortened or familiar form of a proper name. —*vt.* **-named, -nam·ing, -names. 1.** To give a nickname to. **2.** To call by an incorrect name : MISNAME. **—nick'nam'er** *n.*
▲ word history: A *nickname* is literally an additional name. The Middle English form of the word *nickname* was *ekename*, from *eke*, "addition" (related to the verb *eke*), and *name*, "name." *Ekename* acquired an initial *n* from the indefinite article *an*, which frequently preceded it: *an ekename* came to be spelled and pronounced as if it were *a nekename*. In modern times the first syllable *neke-*, which was no longer recognized as an English word, was respelled *nick-*, and *nickname* has been the usual form ever since.
**ni·co·ti·an·a** (nĭ-kō'shē-ăn'ə, -ā'nə, -ä'nə) *n.* [NLat. (*herba*) *nicotiana*, (herb of) Nicot, after Jean *Nicot* (1530?–1600).] A flowering tobacco plant of the genus *Nicotiana* native to the Americas.
**nicotin-** *pref.* [< NICOTINE.] **1.** Nicotine <*nicotinic*> **2.** Nicotinic acid <*nicotinamide*>
**nic·o·tin·a·mide-ad·e·nine di·nu·cle·o·tide** (nĭk'ə-tē'nə-mīd-ăd'n-ēn' dī-nōō'klē-ə-tīd', -nyōō'-, -tĭn'ə-) *n.* An enzyme, $C_{21}H_{27}N_7O_{14}P_2$, occurring in most living cells and utilized alternately as an oxidizing or reducing agent.
**nicotinamide-adenine dinucleotide phosphate** *n.* An enzyme, $C_{21}H_{28}N_7O_{17}P_3$, occurring in most living cells and utilized similarly to nicotinamide-adenine dinucleotide but interacting with different metabolites.
**nic·o·tine** (nĭk'ə-tēn') *n.* [Fr. < NLat. *nicotiana*. —see NICOTIANA.] A poisonous alkaloid, $C_5H_4NC_4H_7NCH_3$, derived from the tobacco plant and used in medicine and as an insecticide.
**nic·o·tin·ic** (nĭk'ə-tĭn'ĭk, -tē'nĭk) *adj.* **1.** Of or relating to nicotine. **2.** Of or relating to nicotinic acid.
**nicotinic acid** *n.* [So called because it is often obtained by oxidizing nicotine.] A member of the vitamin B complex, $C_5H_4NCOOH$, occurring in living cells as an essential substance for growth and synthesized for use in treating pellagra.
**nic·o·tin·ism** (nĭk'ə-tē-nĭz'əm) *n.* Nicotine poisoning.
**nic·ti·tate** (nĭk'tĭ-tāt') *also* **nic·tate** (nĭk'tāt') *vi.* **-tat·ed, -tat·ing, -tates.** [Med. Lat. *nictitare, nictitat-*, freq. of Lat. *nictare*, to wink.] To wink. **—nic'ti·ta'tion** *n.*
**nictitating membrane** *also* **nictating membrane** *n.* An inner eyelid in birds, reptiles, and some mammals that helps to keep the eye clean.
**nid·der·ing** (nĭd'ər-ĭng) *n.* [Alteration of ME *nithing* < OE *nīðing* < ON *nīðingr*.] *Archaic.* A wretched coward.
**nide** (nīd) *n.* [Lat. *nidus*, nest.] A nest or brood of pheasants.
**ni·di** (nī'dī') *n. var. pl. of* NIDUS.
**nid·i·fy** (nĭd'ə-fī') *vi.* **-fied, -fy·ing, -fies.** [Lat. *nidificare* : *nidus*, nest + *facere*, to make.] To construct a nest. **—nid'i·fi·ca'tion** *n.*
**ni·dus** (nī'dəs) *n., pl.* **-dus·es** *or* **-di** (-dī') [Lat.] **1.** A nest, esp. one for insect or spider eggs. **2.** A cavity in which spores develop. **3.** *Pathol.* The seat of bacterial growth in a living organism.
**niece** (nēs) *n.* [ME *nece* < OFr. < VLat. **neptia* < Lat. *neptis*.] **1.** A daughter of one's brother, brother-in-law, sister, or sister-in-law. **2.** The daughter of a celibate ecclesiastic.
**ni·el·lo** (nē-ĕl'ō) *n., pl.* **-el·li** (-ĕl'ē) *or* **-el·los.** [Ital. < Med. Lat. *nigellum* < Lat. *nigellus*, dim. of *niger*, black.] **1.** Any of several black metallic alloys of sulfur with silver, copper, or lead, used to fill an incised design on the surface of another metal. **2.** A surface or object decorated with niello. **3.** The art or process of ornamenting metal surfaces with niello. —*vt.* **-loed, -lo·ing, -los.** To inlay or decorate with niello. **—ni·el'list** *n.*
**Nif·l·heim** (nĭv'əl-hām') *n.* [ON *niflheimr* : *nifl*, mist + *heimr*, home.] *Norse Myth.* The world of the dead.
**nif·ty** (nĭf'tē) [Orig. unknown.] *Slang.* —*adj.* **-ti·er, -ti·est.** Excellent : first-rate. —*n., pl.* **-ties.** Something regarded as nifty. **—nif'ti·ly** *adv.* **—nif'ti·ness** *n.*
**Ni·ger-Con·go** (nī'jər-kŏng'gō) *n.* A large language family of south, central, and west Africa that includes the Mande, Kwa, and Bantu groups.
**nig·gard** (nĭg'ərd) *n.* [ME *nigard*, of Scand. orig.] A stingy person : CHEAPSKATE. **—nig'gard** *adj.*
**nig·gard·ly** (nĭg'ərd-lē) *adj.* **1.** Not willing to spend, give, or share : CHEAP. **2.** Scanty : meager. **—nig'gard·li·ness** *n.* **—nig'gard·ly** *adv.*
**nig·gle** (nĭg'əl) *vi.* **-gled, -gling, -gles.** [Orig. unknown.] **1.** To be preoccupied with petty details. **2.** To find fault constantly and trivially : CARP. **—nig'gler** *n.*
**nig·gling** (nĭg'lĭng) *adj.* **1.** Inordinately concerned with details : FUSSY. **2.** Persistently nagging : PETTY. **3.** Showing or requiring close attention to details : EXACTING. **—nig'gling** *n.* **—nig'gling·ly** *adv.*
**†nigh** (nī) *adv.* **-er, -est.** [ME *neigh* < OE *nēah*.] **1.** Near in time, place, or relationship <as the holiday drew *nigh*> **2.** Nearly : almost <*nigh* onto two days> —*adj.* **-er, -est. 1.** Being near in time, place, or relationship : CLOSE. **2.** *Regional.* Direct : short. —*prep.* Not far from : NEAR. —*vt. & vi.* **nighed, nigh·ing, nighs.** To come near to or draw near.
**night** (nīt) *n.* [ME < OE *niht*.] **1.** The period between sunset and sunrise, esp. the hours of darkness. **2.** Nightfall. **3.** Darkness. **4.** A time or condition of gloom, obscurity, ignorance, or sadness <"In a real dark *night* of the soul it is always three o'clock in the morning" —F. Scott Fitzgerald>
**night-blind** (nīt'blīnd') *adj.* Suffering from night blindness.
**night blindness** *n.* Vision that is normal in daylight but abnormally weak when the light is poor.
**night-bloom·ing cereus** (nīt'blōō'mĭng) *n.* Any of several flowering cacti bearing large fragrant flowers that open at night.
**night·cap** (nīt'kăp') *n.* **1.** A cloth cap worn esp. in bed. **2.** *Informal.* A usu. alcoholic drink taken just before bedtime. **3.** *Slang.* The last event in a day's competition, esp. the second game in a baseball double-header.
**night·clothes** (nīt'klōz', -klōthz') *pl.n.* Clothes worn in bed.
**night·club** (nīt'klŭb') *n.* A commercial establishment that provides food, drink, and entertainment and stays open late at night.
**night crawler** *n.* An earthworm that crawls out from the ground at night.
**night·dress** (nīt'drĕs') *n.* **1.** A nightgown. **2.** Nightclothes.
**night·fall** (nīt'fôl') *n.* The approach of darkness : DUSK.
**night·glow** (nīt'glō') *n.* Airglow occurring at night.
**night·gown** (nīt'goun') *n.* A usu. loose garment worn in bed.
**night·hawk** (nīt'hôk') *n.* **1. a.** A chiefly nocturnal bird of the genus *Chordeiles*, with buff to black mottled feathers, esp. *C. minor* of North America. **b.** A related European bird, the nightjar. **2.** *Informal.* A night owl.
**night heron** *n.* A nocturnal heron of the genus *Nycticorax*, esp. the black-crowned heron, *N. nycticorax*.
**night·ie** *or* **night·y** (nī'tē) *n., pl.* **-ies.** [Shortening and alteration of NIGHTGOWN.] *Informal.* A nightgown for women and girls.

**night·in·gale** (nīt′n-gāl′, nī′tĭng-) *n.* [ME *nightegale* < OE *nihtegale* : *niht,* night + *galan,* to sing.] **1.** A European songbird, *Luscinia megarhynchos,* with brownish plumage, noted for its nocturnal song. **2.** Any of various nocturnal songbirds.
**night·jar** (nīt′jär′) *n.* A nocturnal bird of the family Caprimulgidae, esp. the common European nightjar, *Caprimulgus europaeus.*
**night jasmine** *n.* **1.** A shrub, *Nyctanthes arbortristis,* cultivated for its fragrant white flowers. **2.** A West Indian shrub, *Cestrum nocturnum,* bearing small greenish-white flowers that are very fragrant at night.
**night latch** *n.* A spring lock that can be opened from the inside by turning a knob but from the outside only with a key.
**night letter** *n.* A telegram sent at night at a reduced rate for delivery the next morning.
**night·life** (nīt′līf′) *n.* Social activities or entertainment available or pursued at night.
**night-light** (nīt′līt′) *n.* A usu. small light left on all night.
**night·long** (nīt′lông′, -lŏng′) *adj.* Lasting through the night. —*adv.* Through the night : all night.
**night·ly** (nīt′lē) *adj.* **1.** Of or taking place during the night : NOCTURNAL <*nightly* burglarizing> **2.** Happening or done every night <*nightly* insomnia> **—night′ly** *adv.*
**night·mare** (nīt′mâr′) *n.* [ME, a female demon that afflicts sleeping people : *night,* night + *mare,* goblin < OE.] **1.** A dream causing strong feelings of fear, horror, and distress. **2.** An intensely distressing experience or event. **3.** A demon or spirit once believed to plague sleeping people. **—night′mar′ish** *adj.* **—night′mar′ish·ly** *adv.* **—night′mar′ish·ness** *n.*
▲ word history: In Old and Middle English *mare* was a word denoting an evil spirit. Although the spirit was imagined to be female, the word *mare* is unrelated to the modern word *mare* meaning "female horse." *Nightmare* is a compound of *night* and the old word *mare;* it denoted an evil spirit thought to afflict sleeping persons by sitting on them and causing a feeling of suffocation. *Nightmare* was also used to denote both the feeling itself and the dreams that produced it.
**night owl** *n.* One who habitually stays up late at night.
**night·rid·er** (nīt′rī′dər) *n.* One of a band of mounted and usu. masked men in the southern United States who engaged in nocturnal terrorism for intimidation or revenge.
**nights** (nīts) *adv.* Regularly or habitually in the nighttime.
**night-scent·ed stock** (nīt′sĕn′tĭd) *n.* Evening stock.
**night school** *n.* A school that offers classes in the evening.
**night·shade** (nīt′shād′) *n.* [ME < OE *nihtscada.*] Any of several plants of the genus *Solanum,* many of them yielding a poisonous juice, esp. the deadly nightshade and the bittersweet.
**night·shirt** (nīt′shûrt′) *n.* A long shirt worn in bed.
**night·side** nīt′sīd′ *n.* The side of a heavenly body, as the moon, not in sunlight.
**night sight** *n.* An infrared sight on a weapon making vision possible at night.
**night soil** *n.* Human excrement collected for use as fertilizer.
**night·stand** (nīt′stănd′) *n.* A night table.
**night·stick** (nīt′stĭk′) *n.* A club carried by a police officer.
**night table** *n.* A small table or stand placed at a bedside.
**night·tide** (nīt′tīd′) *n.* Nighttime.
**night·time** (nīt′tīm′) *n.* The time between sunset and sunrise.
**night·walk·er** (nīt′wô′kər) *n.* **1.** One who walks around in the streets at night, esp. for illicit purposes. **2.** A night crawler.
**night·y** (nī′tē) *n. var. of* NIGHTIE.
**ni·gres·cence** (nī-grĕs′əns) *n.* [< Lat. *nigrescens, nigrescent-,* pr. part. of *nigrescere,* to become black < *niger,* black.] **1.** The process of becoming black or dark. **2.** Blackness or darkness, as of complexion. **—ni·gres′cent** *adj.*
**nig·ri·tude** (nĭg′rĭ-to͞od′, -tyo͞od′, nī′grĭ-) *n.* [Lat. *nigritudo* < *niger,* black.] Blackness.
**ni·gro·sine** (nī′grə-sēn′, -sĭn) *n.* [Lat. *niger, nigr-,* black + -OS(E) + -INE.] Any of a class of dyes, varying from blue to black, used in making inks and dyeing wood and textiles.
**ni·hil·ism** (nī′ə-lĭz′əm, nī′hə-, nē′-) *n.* [< Lat. *nihil,* nothing.] **1.** A doctrine that all values are worthless and that nothing is knowable or can be communicated. **2.** Rejection of all distinctions in moral value, constituting a willingness to refute all previous theories of morality. **3.** The belief that destruction of existing political or social institutions is necessary for future improvement. **4.** *also* **Nihilism.** The doctrine of a 19th-cent. Russian movement that advocated assassination and terrorism. **—ni′hil·ist** (-lĭst) *n.* **—ni′hil·is′tic** *adj.*
**ni·hil·i·ty** (nī-hĭl′ĭ-tē, nē-) *n.* Nonexistence.
**ni·hil ob·stat** (nī′hĭl ŏb′stät′, -stăt′, nē′-) *n.* [Lat., nothing hinders.] **1.** *Rom. Cath. Ch.* An attestation that a book contains nothing damaging to faith or morals. **2.** Official approval, esp. of an artistic work.
**-nik** *suff.* [Yiddish, of Slavic orig.] One characterized by or associated with <jazznik>

**Ni·ke** (nī′kē) *n.* [Gk. *Nikē.*] *Gk. Myth.* The goddess of victory.
**nil** (nĭl) *n.* [Lat., short for *nihil.*] Not anything : ZERO. **—nil** *adj.*
**Nile blue** (nīl) *n.* A light bluish green.
**Nile green** *n.* A moderate yellow green to vivid pale green.
**nill** (nĭl) *v.* **nilled, nill·ing, nills.** [ME *nilen* < OE *nyllan* : *ne,* not + *wyllan,* to wish.] *Obs.* —*vt.* Not to will. —*vi.* To be unwilling.
**Ni·lot·ic** (nī-lŏt′ĭk) *adj.* [Lat. *Niloticus* < Gk. *Neilotikos* < *Neilos,* Nile.] **1.** Of of relating to the Nile or the Nile Valley. **2.** Of or relating to a Negroid group of peoples in eastern Africa. —*n.* A large group of related African languages spoken in southern Sudan, Uganda, Kenya, and northern Tanzania.
**nil·po·tent** (nĭl-pŏt′nt, nĭl′pŏt′nt) *n.* An algebraic quantity that when raised to some power equals zero. **—nil·po′ten·cy** *n.*
**nim** (nĭm) *vt.* & *vi.* **nimmed, nim·ming, nims.** [ME *nimen,* to take < OE *niman.*] *Archaic.* To steal : pilfer.
**nim·bi** (nĭm′bī′) *n. var. pl. of* NIMBUS.
**nim·ble** (nĭm′bəl) *adj.* **-bler, -blest.** [ME *nemel,* prob. of OE orig.] **1.** Moving or acting with agility : DEFT <*nimble* feet> **2.** Quick and clever in thinking, apprehending, or responding <a *nimble* mind> **—nim′ble·ness** *n.* **—nim′bly** *adv.*
**nim·bo·stra·tus** (nĭm′bō-strā′təs, -străt′əs) *n.* [NIMB(US) + STRATUS.] A low, gray, often dark cloud that precipitates rain, snow, or sleet.
**nim·bus** (nĭm′bəs) *n., pl.* **-bi** (bī′) *or* **-bus·es.** [Lat., cloud.] **1.** A cloudy radiance thought to surround a deity when on earth. **2.** A radiant light that appears usu. in the form of a circle about the head in the representation of a saint or diety. **3.** An atmosphere or aura, as of mystery, that surrounds a person or thing. **4.** A uniformly gray rain cloud extending over the sky.
**ni·mi·e·ty** (nĭ-mī′ĭ-tē) *n.* [LLat. *nimietas* < Lat. *nimius,* excessive < *nimis,* excessively.] Excess : redundancy.
**nim·i·ny-pim·i·ny** (nĭm′ə-nē-pĭm′ə-nē) *adj.* [Perh. alteration of NAMBY-PAMBY.] Persnickety. **—nim′i·ny-pim′i·ny** *n.*
**Nim·rod** *also* **nim·rod** (nĭm′rŏd′) *n.* [After *Nimrod,* a hunter and king in the Old Testament.] A hunter.
**nin·com·poop** (nĭn′kəm-po͞op′, nĭng′-) *n.* [Orig. unknown.] A foolish or stupid person. **—nin′com·poop′er·y** *n.*
**nine** (nīn) *n.* [ME < OE *nigon* : akin to G. *neun,* Lat. *novem,* Gk. *ennea,* Skt. *nava.*] **1.** The cardinal number equal to 8 + 1. **2.** The ninth in a set or sequence. **3.** Something having nine parts, units, or members. **4.** A set of nine, esp.: **a.** A baseball team. **b. Nine.** The nine Muses. **—to the nines.** To the highest degree <dressed *to the nines*> **—nine** *adj.* & *pron.*
**nine·bark** (nīn′bärk′) *n.* [From the many layers in its bark.] A shrub, *Physocarpus opulifolius,* of eastern North America, with shredding or peeling bark and small white flower clusters.
**nine days' wonder** *n.* One that creates a brief stir.
**nine·fold** (nīn′fōld′) *adj.* **1.** Having nine parts. **2.** Nine times as much or as many. **—nine′fold** *adv.*
**nine iron** *n.* An iron-headed golf club with a face slanted at a greater angle than any other iron.
**nine·pin** (nīn′pĭn′) *n.* **1.** A wooden pin used in the game of ninepins. **2. ninepins.** *(sing. in number).* A bowling game in which nine pins are the target.
**nine·teen** (nīn-tēn′) *n.* [ME *nintene* < OE *nigontȳne* < *nigon,* nine.] **1.** The cardinal number equal to 18 + 1. **2.** The 19th in a set or sequence. **—nine·teen′** *adj.* & *pron.*
**nine·teenth** (nīn-tēnth′) *n.* **1.** The ordinal number matching the number 19 in a series. **2.** One of 19 equal parts. **—nine·teenth′** *adj.* & *adv.*
**nine·ti·eth** (nīn′tē-ĭth) *n.* **1.** The ordinal number matching the number 90 in a series. **2.** One of 90 equal parts. **—nine′ti·eth** *adj.* & *adv.*
**nine-to-fiv·er** (nīn′tə-fī′vər) *n.* One who works regular daytime hours, as in an office.
**nine·ty** (nīn′tē) *n., pl.* **-ties.** [ME *ninti* < OE *nigontig* : *nigon,* nine + *-tig,* -ty.] The cardinal number equal to 9 × 10. **—nine′ty** *adj.* & *pron.*
**nin·ny** (nĭn′ē) *n., pl.* **-nies.** [Perh. alteration of INNOCENT.] A fool.
**ni·non** (nē′nŏn′) *n.* [Prob. < Fr. *Ninon,* nickname for *Anne.*] A sheer fabric of silk, rayon, or nylon made in various open lacy patterns or tight smooth weaves.
**ninth** (nīnth) *n.* [ME *ninthe* < *nine,* nine < OE *nigon.*] **1.** The ordinal number matching the number nine in a series. **2.** One of nine equal parts. **3.** *Mus.* **a.** A harmonic or melodic interval of an octave and a second. **b.** The tone at the upper limit of such an interval. **c.** A chord having a root with its third, seventh, and ninth. **—ninth** *adj.* & *adv.*
**Ni·nus** (nī′nəs) *n.* [Lat. < Gk. *Ninos.*] *Myth.* The legendary founder of Nineveh and husband of Semiramis.
**Ni·o·be** (nī′ə-bē) *n.* [Gk. *Niobē.*] *Gk. Myth.* The daughter of Tantalus who turned to stone while lamenting the loss of her children.
**ni·o·bi·um** (nī-ō′bē-əm) *n.* [NLat., after *Niobe,* so called because it is obtained from tantalite. —see TANTALITE.] *Symbol* **Nb** A silvery, soft, ductile metallic element used in steel alloys, arc welding, and superconductivity research; atomic number 41; atomic weight 92.906.

▲ word history: Chemistry is a serious science, but it has its fanciful moments, which are revealed, for example, in the names of some of the elements. The word *niobium* is derived from *Niobe*, the name of a figure in Greek mythology. She was the daughter of Tantalus, who gave his name to the mineral ore tantalite. Because the element is extracted from tantalite, it was named after Niobe.

**nip[1]** (nĭp) *v.* **nipped, nip·ping, nips.** [ME *nippen* < MDu. *nipen.*] —*vt.* **1.** To seize and bite or pinch. **2.** To remove or sever, as a plant leaf, by or as if by nipping. **3.** To sting with the cold : CHILL. **4.** To halt the further growth or development of <*nipped* my cold in the bud> **5.** *Slang.* **a.** To snatch up hastily. **b.** To steal. —*vi. Chiefly Brit. Slang.* To move quickly : DART. —*n.* **1. a.** The act of nipping. **b.** A small sharp bite or pinch. **2.** A small portion or bit. **3. a.** A sharp stinging quality, as of frosty air. **b.** Intensely sharp cold. **4.** A cutting remark. **5.** A sharp biting flavor : TANG. **—nip and tuck.** Very close : neck and neck.

**nip[2]** (nĭp) *n.* [< *nip*, a small amount of spirits, prob. short for *nipperkin*, of Germanic orig.] A small amount of liquor. —*v.* **nipped, nip·ping, nips.** —*vt.* To drink (alcoholic liquor) in small amounts. —*vi.* To take a nip of alcoholic liquor. **—nip'per** *n.*

**ni·pa** (nē'pə) *n.* [NLat. < Malay *nipah.*] **1.** A large palm, *Nipa fruticans* of the Philippines and Australia, with long leaves used for thatching. **2.** An alcoholic liquor made from the sap of the nipa.

**nip·per** (nĭp'ər) *n.* **1.** *often* **nippers.** A tool, as pincers or pliers, used for grasping or nipping. **2.** A pincerlike part, as the large claw of a crab or lobster. **3.** *Chiefly Brit.* A small boy.

**nip·ping** (nĭp'ĭng) *adj.* Sharp and biting <a *nipping* wind> **—nip'ping·ly** *adv.*

**nip·ple** (nĭp'əl) *n.* [< obs. *neble*, dim of NEB.] **1.** The small conical protuberance near the center of the mammary gland containing the outlets of the milk ducts. **2. a.** The rubber cap on a bottle from which a baby nurses. **b.** A pacifier for a baby. **3.** A device resembling or functioning like a nipple, esp.: **a.** A regulated opening for discharging a liquid, as in a small stopcock. **b.** A pipe coupling threaded on both ends. **c.** A short length of pipe to which a nozzle can be attached. **4.** A natural projection, as a mountain crest.

**nip·ple·wort** (nĭp'əl-wûrt', -wôrt') *n.* [From its former use in folk medicine to cure breast tumors.] A plant, *Lapsana communis*, with a milky juice and small yellow flower heads.

**Nip·pon·ese** (nĭp'ə-nēz', -nēs') *adj.* & *n.* [< *Nippon* (Japan).] Japanese.

**nip·py** (nĭp'ē) *adj.* **-pi·er, -pi·est. 1.** Sharp or biting <*nippy* salad dressing> **2.** Bitingly cold. **—nip'pi·ly** *adv.* **—nip'pi·ness** *n.*

**nip-up** (nĭp'ŭp') *n.* An acrobatic leap from a supine to an upright position.

**nir·va·na** (nîr-vä'nə, nər-) *n.* [Skt. *nirvāṇam* : *nir-*, out, away + *vāti*, he blows.] **1. Nirvana. a.** The state of total blessedness, marked by release from the cycle of reincarnations and achieved through the extinction of the self in Buddhism. **b.** A similar state in which reunion with Brahma is attained through the suppression of individual existence. **2.** An ideal condition of perfect harmony and peace.

**Ni·san** (nĭs'ən, nē-sän') *n.* [Heb. *Nīsān* < Akkadian *Nissanu.*] The seventh month of the civil year and the first of the religious year in the Hebrew calender. —See table at CALENDAR.

**Ni·sei** (nē-sā', nē'sā') *n., pl.* **Nisei** *or* **-seis.** [J. : *ni*, second + *sei*, generation.] One born in America of Japanese immigrant parents.

**ni·si** (nī'sī') *adj.* [Lat. : *ne-*, not + *si*, if.] *Law.* Taking effect at a specified date unless cause is shown for nullification or modification <a ruling *nisi*><a divorce *nisi*>

**Nis·sen hut** (nĭs'ən) *n.* [After Lieut. Col. Peter N. *Nissen* (1871–1930).] A prefabricated hut of corrugated steel shaped like a half cylinder, used esp. as a military shelter.

**ni·sus** (nī'səs) *n., pl.* **nisus.** [Lat. < *niti*, to strive.] An endeavor to realize an objective.

**nit** (nĭt) *n.* [ME *nite* < OE *hnitu.*] **1.** The egg or young of a parasitic insect, as a louse. **2.** A unit of illuminative brightness equal to one candle per square meter, measured perpendicular to the rays of the source. **—nit'ty** *adj.*

**ni·ter** (nī'tər) *n.* [ME *nitre* < OFr. < Lat. *nitrum*, natron < Gk. *nitron*, of Semitic orig.] A colorless, white, or gray mineral of potassium nitrate, $KNO_3$, used to make gunpowder.

**nit-pick** (nĭt'pĭk') *vi.* **-picked, -pick·ing, -picks.** *Informal.* To be concerned with or critical of insignificant details. **—nit'-pick'er** *n.* **—nit'-pick'ing** *n.*

**nitr-** *pref. var. of* NITRO-.

**ni·trate** (nī'trāt', -trĭt) *n.* **1.** The radical $NO_3^-$ or a compound containing it, as a salt or ester of nitric acid. **2.** Sodium nitrate or potassium nitrate used as fertilizer. —*vt.* **-trat·ed, -trat·ing, -trates.** To treat with a nitrate or nitric acid, usu. to change an organic compound into a nitrate. **—ni·tra'tion** *n.* **—ni'tra'tor** *n.*

**ni·tre** (nī'tər) *n. Chiefly Brit. var. of* NITER.

**ni·tric** (nī'trĭk) *adj.* Of, derived from, or having nitrogen, esp. in a valence state higher than that in a comparable nitrous compound.

**nitric acid** *n.* A transparent, colorless to yellowish, fuming corrosive liquid, $HNO_3$, a highly reactive oxidizing agent used in industrial metallurgical processes and in making fertilizers, explosives, and rocket fuels.

**nitric oxide** *n.* A colorless, poisonous gas, NO, produced as an intermediate during the manufacture of nitric acid from atmospheric nitrogen or ammonia.

**ni·tride** (nī'trīd') *n.* A compound having nitrogen and another more electropositive element.

**ni·tri·fy** (nī'trə-fī') *vt.* **-fied, -fy·ing, -fies. 1.** To oxidize into nitric acid, nitrous acid, or any nitrate or nitrite, as by the action of nitrobacteria. **2.** To treat or combine with nitrogen or nitrogen-containing compounds. **—ni'tri·fi·ca'tion** *n.* **—ni'tri·fi'er** *n.*

**ni·trile** *also* **ni·tril** (nī'trəl) *n.* A compound having trivalent nitrogen, $N^{-3}$, in a cyanogen group.

**ni·trite** (nī'trīt') *n.* [NITR(O)- + -ITE.] A salt or ester of nitrous acid.

**nitro-** *or* **nitr-** *pref.* [NLat. < Lat. *nitrum*, natron < Gk. *nitron*, of Semitic orig.] **1.** Nitrate : niter <*nitro*bacterium> **2. a.** Nitrogen <*nitri*le> **b.** Containing the univalent group $NO_2$ <*nitro*methane>

**ni·tro·bac·te·ri·um** (nī'trō-băk-tîr'ē-əm) *n.* A soil bacterium that produces nitrification.

**ni·tro·ben·zene** (nī'trō-bĕn'zēn', -bĕn-zēn') *n.* A poisonous organic compound, $C_6H_5NO_2$, either an oily liquid or bright-yellow crystals, having the odor of almonds and used to make aniline, insulating compounds, and polishes.

**ni·tro·cel·lu·lose** (nī'trō-sĕl'yə-lōs', -lōz') *n.* A pulpy or cottonlike polymer derived from cellulose treated with sulfuric and nitric acids and used in making explosives, collodion, plastics, and solid monopropellants. **—ni'tro·cel'lu·los'ic** (-lō'sĭk, -zĭk) *adj.*

**ni·tro·chlo·ro·form** (nī'trō-klôr'ə-fôrm', -klōr'-) *n.* Chloropicrin.

**ni·tro·gen** (nī'trə-jən) *n. Symbol* **N** A nonmetallic element constituting nearly four fifths of the air by volume, occurring as a colorless, odorless, almost inert gas, in various minerals, and in all proteins; atomic number 7; atomic weight 14.0067. **—ni·trog'e·nous** (nī-trŏj'ə-nəs) *adj.*

**ni·trog·e·nase** (nī-trŏj'ə-nās', nī'trə-jə-) *n.* An enzyme utilized in nitrogen fixation to convert molecular nitrogen to ammonia.

**nitrogen balance** *n.* The difference between the amounts of nitrogen taken into and lost by the body or the soil.

**nitrogen cycle** *n.* **1.** The continuous cyclic progression of chemical reactions in which atmospheric nitrogen is compounded, dissolved in rain, deposited in the soil, assimilated and metabolized by bacteria and plants, and returned to the atmosphere by organic decomposition. **2.** Carbon-nitrogen cycle.

**nitrogen dioxide** *n.* A mildly poisonous brown gas, $NO_2$, often found in automobile exhaust fumes and smog and synthesized for use as a catalyst and oxidizing or nitrating agent.

**nitrogen fixation** *n.* **1.** Conversion of atmospheric nitrogen into nitrogenous compounds by natural agencies or industrial processes. **2.** Conversion by certain algae and soil bacteria of inorganic nitrogen compounds into organic compounds assimilable by plants. **—ni'tro·gen-fix'ing** (nī'trə-jən-fĭk'sĭng) *adj.*

**ni·tro·gen·ize** (nī-trŏj'ə-nīz', nī'trə-jə-nīz') *vt.* **-ized, -iz·ing, -iz·es.** To treat or combine with nitrogen.

**ni·tro·glyc·er·in** *also* **ni·tro·glyc·er·ine** (nī'trō-glĭs'ər-ĭn) *n.* A thick, pale-yellow liquid, $CH_2NO_3CHNO_3CH_2NO_3$, that is explosive on concussion or exposure to sudden heat and is used to make dynamite and blasting gelatin and in medicine as a vasodilator.

**ni·tro·hy·dro·chlo·ric acid** (nī'trō-hī'drə-klôr'ĭk) *n.* Aqua regia.

**ni·tro·meth·ane** (nī'trō-mĕth'ān') *n.* An oily, colorless liquid, $CH_3NO_2$, used in organic synthesis, in making dyes and resins, and as a rocket propellant.

**ni·tro·par·af·fin** (nī'trō-păr'ə-fĭn) *n.* Any of a group of organic compounds formed by replacing one or more of the hydrogen atoms of a paraffin hydrocarbon with the nitro group, $NO_2^-$, as in nitromethane $CH_3NO_2$.

**ni·tro·starch** (nī'trō-stärch') *n.* A highly explosive orange-colored powder, $C_{12}H_{12}(NO_2)_8O_{10}$, derived from starch and used as a demolition agent.

**ni·trous** (nī'trəs) *adj.* Of, derived from, or containing nitrogen, esp. in a valence state lower than that in a comparable nitric compound.

**nitrous acid** *n.* An unstable inorganic acid, $HNO_2$, existing only in solution.

**nitrous oxide** *n.* A colorless, sweet inorganic gas, $N_2O$, used as an anesthetic : LAUGHING GAS.

**nits-and-lice** (nĭts'ənd-līs') *n.* (*sing. or pl. in number*). A plant, *Hypericum drummondii* of the central United States, with narrow leaves and yellow flowers.

**nit·ty-grit·ty** (nĭt'ē-grĭt'ē, nĭt'ē-grĭt'ē) *n.* [Orig. unknown.] *Slang.* The most fundamental and specific details : CORE.

**nit·wit** (nĭt'wĭt') *n. Informal.* A silly or stupid person.

**ni·val** (nī'vəl) *adj.* [Lat. *nivalis* < *nix*, snow.] Of or growing in or under snow.

**niv·e·ous** (nĭv'ē-əs) *adj.* [Lat. *niveus* < *nix*, snow.] Like snow.

**nix[1]** (nĭks) *n.* [G. < MHG *nickes* < OHG *nihhus.*] *Ger. Myth.* A water sprite, usu. in human form or half-human and half-fish.

**nix[2]** (nĭks) [< G. *nichts,* nothing < MHG *nihtes* < *niht* < OHG *niwiht* : *ni-,* no + *wiht,* thing.] *Slang.* —*n.* Nothing. —*adv.* No. —*vt.* **nixed, nix·ing, nix·es.** To veto : forbid.

**nix·ie** *also* **nix·y** (nĭk'sē) *n., pl.* **-ies.** [< NIX[2].] *Slang.* A misaddressed piece of mail.

**Ni·zam** (nĭ-zäm', -zăm', nī-) *n.* [Hindi *nizām (-almulk),* governor (of the empire) < *niẓām,* government < Ar. *niẓām* < *naẓama,* he arranged.] The title of the former rulers of Hyderabad, India.

**no[1]** (nō) *adv.* [ME < OE *nā* : *ne,* not + *ā,* ever.] **1.** Not so. —Used to express refusal, denial, or disagreement <*No,* it is not raining.> **2.** Not at all. —Used with the comparative <*no* worse> <*no* less> **3.** Not <like it or *no*> —*n., pl.* **noes. 1.** A negative response : refusal or denial. **2.** A negative vote or voter.

**no[2]** (nō) *adj.* [ME < OE *nā* < *nān,* none.] **1.** Not any : not one <*No* books were damaged.> **2.** Not close to being : not at all <I am *no* expert.> ***usage:*** When *no* introduces a compound phrase, its elements should be connected with or instead of *nor,* as in *The applicant has no experience or interest in carpentry.*

**No[1]** *also* **Noh** (nō) *n., pl.* **No** *also* **Noh.** [J. *nō,* talent, ability < Chin. *neng[2].*] The classical drama of Japan, with music and dance performed by elaborately dressed actors in a highly stylized manner.

**No[2]** *symbol for* NOBELIUM.

**no-ac·count** (nō'ə-kount') *adj.* Good-for-nothing : worthless.

**No·a·chi·an** (nō-ā'kē-ən) *adj.* **1.** Of or pertaining to Noah or his time. **2.** Ancient : antiquated.

**nob[1]** (nŏb) *n.* [Perh. var. of KNOB.] *Slang.* The human head.

**nob[2]** (nŏb) *n.* [Orig. unknown.] *Slang.* One who has wealth or social position.

**nob·ble** (nŏb'əl) *vt.* **-bled, -bling, -bles.** [Orig. unknown.] *Chiefly Brit.* **1.** To disable (a racehorse), esp. by drugging. **2. a.** To win over to one's point of view. **b.** To outdo or get the better of by devious means. **c.** To filch or steal. **—nob'bler** *n.*

**nob·by** (nŏb'ē) *adj.* **-bi·er, -bi·est.** [< NOB[2].] Fashionable : smart.

**No·bel·ist** (nō-bĕl'ĭst) *n.* A recipient of a Nobel Prize.

**no·bel·i·um** (nō-bĕl'ē-əm) *n.* [After the *Nobel* Institute at Stockholm.] *Symbol* **No** A synthetic radioactive element produced in trace amounts; atomic number 102; longest-lived isotope No 254.

**No·bel Prize** (nō-bĕl') *n.* Any of the international prizes awarded annually by the Nobel Foundation for outstanding achievements in the fields of physics, chemistry, physiology or medicine, literature, and economics and for the promotion of world peace.

**no·bil·i·ar·y** (nō-bĭl'ē-ĕr'ē, -bĭl'yə-rē) *adj.* [Fr. *nobiliaire* < Lat. *nobilis,* noble.] Of or relating to the nobility.

**no·bil·i·ty** (nō-bĭl'ĭ-tē) *n., pl.* **-ties.** [ME *nobilite,* the quality of being noble < OFr. < Lat. *nobilitas* < *nobilis,* noble.] **1.** A class of persons set apart by high birth or rank : ARISTOCRACY. **2.** Noble rank or status. **3.** The quality or state of being noble.

**no·ble** (nō'bəl) *adj.* **-bler, -blest.** [ME < OFr. < Lat. *nobilis.*] **1.** Of, in, or belonging to the nobility. **2.** Having or displaying qualities of high moral character, as honor, generosity, or courage <a *noble* soul> **3. a.** Superior in nature or character : EXALTED. **b.** Grand and stately : MAJESTIC <a *noble* vista> **4.** Designating an esp. corrosion-resistant metal, as gold. **5.** *Chem.* Inactive or inert. —*n.* **1.** A member of the nobility. **2.** A former English gold coin worth six shillings and eight pence. **—no'ble·ness** *n.* **—no'bly** *adv.*

**no·ble·man** (nō'bəl-mən) *n.* A man of noble rank.

**no·blesse** (nō-blĕs') *n.* [ME < OFr. < *noble,* noble < Lat. *nobilis.*] **1.** Noble birth or condition. **2.** The aristocracy.

**noblesse o·blige** (ō-blēzh') *n.* [Fr. : *noblesse,* nobility + *oblige,* obligates.] Honorable and charitable behavior regarded as the responsibility of persons of noble birth or rank.

**no·ble·wom·an** (nō'bəl-wo͝om'ən) *n.* A woman of noble rank.

**no·bod·y** (nō'bŏd'ē, -bə-dē) *pron.* No person : NO ONE. —*n., pl.* **-ies.** One of no significance or influence.

**no·cent** (nō'sənt) *adj.* [ME *nocent,* guilty < Lat. *nocens,* pr.part. of *nocēre,* to harm.] Causing harm : INJURIOUS.

**nock** (nŏk) *n.* [ME *nokke.*] **1.** The groove at either end of a bow for holding the bowstring. **2.** The notch in the end of an arrow that fits on the bowstring. —*vt.* **nocked, nock·ing, nocks. 1.** To put a nock in (a bow or arrow). **2.** To fit (an arrow) to a bowstring.

**noct-** *pref. var. of* NOCTI-.

**noc·tam·bu·lism** (nŏk-tăm'byə-lĭz'əm) *also* **noc·tam·bu·la·tion** (-tăm'byə-lā'shən) *n.* [NOCTI- + Lat. *ambulare,* to walk + -ISM.] Somnambulism. **—noc·tam'bu·list** *n.*

**nocti-** *or* **noct-** *pref.* [NLat. < Lat. *nox,* night.] Night <*noctilucent*>

**noc·ti·lu·ca** (nŏk'tə-lo͞o'kə) *n.* [NLat. *Noctiluca,* genus name < Lat. *noctiluca,* lantern : *nox,* night + *lucēre,* to shine.] Any of various plantlike, bioluminescent marine organisms of the genus *Noctiluca* that when massed together make the sea phosphorescent.

**noc·ti·lu·cent** (nŏk'tə-lo͞o'sənt) *adj.* Luminous at night. —Used esp. of certain high clouds.

**noc·tu·id** (nŏk'cho͞o-ĭd) *n.* [< NLat. *Noctuidae,* family name < Lat. *noctua,* night owl.] Any of the night-flying moths of the family Noctuidae, whose larvae are destructive pests.

**noc·tule** (nŏk'cho͞ol') *n.* [Fr. < Ital. *nottola* < LLat. *noctula,* dim. of Lat. *noctua,* night owl < *nox,* night.] A large, reddish-brown, insectivorous bat of the genus *Nyctalus,* found in Eurasia, Indonesia, and the Philippines.

**noc·turn** (nŏk'tûrn') *n.* [ME *nocturne* < Med. Lat. *nocturna* < Lat. *nocturnus,* of the night < *nox,* night.] Any of the three canonical divisions of the office of matins.

**noc·tur·nal** (nŏk-tûr'nəl) *adj.* [LLat. *nocturnalis* < Lat. *nocturnus* < *nox,* night.] **1.** Of, relating to, or taking place during the night. **2.** *Bot.* Having flowers that open during the night. **3.** *Zool.* Active at night <*nocturnal* animals> **—noc·tur'nal·ly** *adv.*

**noc·turne** (nŏk'tûrn') *n.* [Fr. < *nocturne,* nocturnal < Lat. *nocturnus* < *nox,* night.] **1.** A painting of a night scene. **2.** A romantic musical composition designed to evoke thoughts and feelings of night.

**noc·u·ous** (nŏk'yo͞o-əs) *adj.* [Lat. *nocuus* < *nocēre,* to harm.] Harmful : noxious. **—noc'u·ous·ly** *adv.*

**nod** (nŏd) *v.* **nod·ded, nod·ding, nods.** [ME *nodden.*] —*vi.* **1.** To lower and raise the head quickly, as in agreement. **2.** To doze momentarily <*nod* off> **3.** To be momentarily inattentive or careless. **4.** To sway, sag, or bend downward, as flowers in the wind. —*vt.* **1.** To lower and raise (the head) quickly in agreement or acknowledgment. **2.** To express by lowering and raising the head <*nodded* approval> **3.** To summon, guide, or send by nodding the head <*nodded* us into the auditorium> —*n.* **1.** An act or instance of nodding. **2.** An indication of approval or agreement <got the *nod* to begin the project> **—nod'der** *n.*

**nod·al** (nōd'l) *adj.* Of, relating to, being, or located near or at a node. **—nod'al·ly** *adv.*

**nodding pogonia** *n.* A North American orchid, *Triphora trianthophora,* with pink or white flowers.

**nod·dle** (nŏd'l) *n.* [ME *nodel,* back of the head.] The human head.

**nod·dy** (nŏd'ē) *n., pl.* **-dies.** [Prob. < obs. *noddy,* foolish.] **1.** A fool : simpleton. **2.** A dark brown, white-headed tropical tern of the genus *Anous.*

**node** (nōd) *n.* [Lat. *nodus,* knot.] **1.** A protuberance or swelling. **2.** *Bot.* The often enlarged point on a stem where a leaf, bud, or other organ is attached : JOINT. **3.** *Physics.* A point or region of minimum or zero amplitude in a periodic system. **4.** *Math.* The point at which a continuous curve crosses itself. **5.** The intersection or terminating point of two or more lines or curves : VERTEX. **6.** *Astron.* **a.** Either of two diametrically opposite points at which the orbit of a planet intersects the ecliptic. **b.** Either of two points at which the orbit of a satellite intersects the orbital plane of a planet.

**node**
*Two examples of nodes:* (left) *buckwheat and* (right) *rye grass*

**no·di** (nō'dī') *n. pl. of* NODUS.

**nod·ule** (nŏj'o͞ol) *n.* [Lat. *nodulus,* dim. of *nodus,* knot.] **1.** A small node, as of body tissue. **2.** *Anat.* A localized swelling. **3.** *Bot.* A small knoblike outgrowth, as those found on the roots of most leguminous plants. **4.** A small lump of a mineral or mixture of minerals. **—nod'u·lar** (nŏj'ə-lər), **nod'u·lose'** (-lōs'), **nod'u·lous** (-ləs) *adj.*

**no·dus** (nō'dəs) *n., pl.* **-di** (-dī') [Lat., knot.] A complicated problem or situation : DIFFICULTY.

**No·ël** *also* **No·el** (nō-ĕl') *n.* [Fr. < OFr. *novel* < Lat. *natalis (dies),* (day) of birth < *nasci,* to be born.] **1.** Christmas. **2. noël** *also* **noel.** A Christmas carol.

**no·e·sis** (nō-ē'sĭs) *n.* [Gk. *noēsis,* understanding < *noein,* to perceive < *nous,* mind.] *Psychol.* The cognitive process : COGNITION.

**no·et·ic** (nō-ĕt'ĭk) *adj.* [Gk. *noētikos* < *noēsis,* understanding. —see NOESIS.] Of, pertaining to, arising from, or understood by the intellect <*noetic* truths>

**no-fault** (nō'fôlt') *adj.* **1.** Of or indicating a system of automotive insurance in which accident victims are compensated by their insurance companies without assignment of blame. **2.** Of or indicating a type of divorce in which blame is assigned to neither party.

**no-frills** (nō'frĭlz') *adj.* Devoid of special or extra features : BASIC <a *no-frills* airline>

**nog[1]** (nŏg) *n.* [Orig. unknown.] A wooden block built into a masonry wall to hold nails that support joinery structures.

**nog[2]** (nŏg) *n.* [Orig. unknown.] Eggnog.

**nog·gin** (nŏg′ĭn) *n.* [Orig. unknown.] **1.** A small cup or mug. **2.** A unit of liquid measure equal to one-quarter of a pint. **3.** *Slang.* The human head.

**no-go** (nō′gō′) *adj. Slang.* Not in a suitable condition for functioning or proceeding properly <a *no-go* space launch>

**no-good** (nō′gŏŏd′) *adj.* **1.** Having no worth, merit, or use. **2.** Despicable : vile. —*n.* One that is no-good.

**Noh** (nō) *n. var. of* NO[1].

**no-hit** (nō′hĭt′) *adj. Baseball.* Of, relating to, or being a game in which one pitcher allows the opposing team no hits and no runs. **—no-hit′ter** *n.*

**no·how** (nō′hou′) *adv. Nonstandard.* In no way : not at all.

**noil** (noil) *n.* [OFr. *noel* < Med. Lat. *nodellus,* dim. of Lat. *nodus,* knot.] A short fiber combed from long fibers during the preparation of textile yarns.

**noise** (noiz) *n.* [ME < OFr. < Lat. *nausea,* seasickness. —see NAUSEA.] **1. a.** Sound or a sound that is loud, disagreeable, or unwanted. **b.** Sound or a sound of any kind. **2.** A loud clamor or commotion. **3.** *Physics.* A usu. random and persistent disturbance that obscures or reduces the clarity or quality of a signal. **4.** *Computer Sci.* Irrelevant or meaningless data generated by a computer along with desired data. **5.** *Informal.* **a.** A protest or complaint. **b.** Rumor : talk. —*v.* **noised, nois·ing, nois·es.** —*vt.* To spread the rumor or report of. —*vi.* **1.** To talk volubly or loudly. **2.** To be noisy.

**noise·less** (noiz′lĭs) *adj.* Creating no noise : SILENT. **—noise′less·ly** *adv.* **—noise′less·ness** *n.*

**noise·mak·er** (noiz′mā′kər) *n.* One that makes noise, esp. a device for making noise at a party. **—noise′mak′ing** *n.*

**noise pollution** *n.* Annoying or harmful environmental noise.

**noi·some** (noi′səm) *adj.* [ME *noisom* : *noy,* harm (< *anoi,* annoyance < OFr. < Lat. *in odio,* hateful) + *-some,* -some.] **1.** Offensive to the point of arousing disgust : FOUL <a *noisome* smell> **2.** Harmful or dangerous <*noisome* industrial fumes> **—noi′some·ly** *adv.* **—noi′some·ness** *n.*

▲ word history: Neither in meaning nor in origin does *noisome* have any connection with the word *noise*; the similarities are purely coincidental. *Noisome* is a compound formed in Middle English from *noy,* an obsolete word meaning "harm" or "annoyance," and the suffix *-some,* which is still current in Modern English, where it is used to form adjectives with the general sense of "characterized by some quality." *Noy* is related to the modern verb *annoy,* whose source is the Latin phrase *in odium,* "hateful, odious."

**nois·y** (noi′zē) *adj.* **-i·er, -i·est. 1.** Making noise. **2.** Full of, marked by, or accompanied by noise. **—nois′i·ly** *adv.* **—nois′i·ness** *n.*

**no·li me tan·ge·re** (nō′lē mē tăn′jə-rē) *n.* [Lat., don't touch me.] A warning or prohibition against touching or meddling.

**nol·le pros·e·qui** (nŏl′ē prŏs′ĭ-kwī′) *n.* [Lat., to be unwilling to pursue.] *Law.* A declaration that the plaintiff in a civil case or the prosecutor in a criminal case will drop prosecution of all or part of a suit or indictment.

**no·lo** (nō′lō) *n.* Nolo contendere.

**no-load** (nō′lōd′) *adj.* Designating a mutual fund sold at net asset value without a sales commission.

**no·lo con·ten·de·re** (nō′lō kən-tĕn′də-rē) *n.* [Lat., I do not wish to contend.] *Law.* A plea made by the defendant in a criminal action that is equivalent to an admission of guilt and subjects the defendant to punishment but leaves open the possibility for him or her to deny the alleged facts in other proceedings.

**nol-pros** (nŏl′prŏs′) *vt.* **-prossed, -pros·sing, -pros·ses.** *Law.* To drop prosecution of by entering a nolle prosequi on the court records.

**no·ma** (nō′mə) *n.* [Lat. < Gk. *nomē* < *nemein,* to spread.] A severe, often gangrenous inflammation of the mouth, occurring esp. in a young child after a debilitating disease.

**no·mad** (nō′măd′) *n.* [Lat. *nomas, nomad-* < Gk. *nomas,* wandering around for pasture < *nemein,* to pasture.] **1.** A member of a group of people who have no permanent home and wander from place to place in search of food, water, and grazing land. **2.** One who roams about : WANDERER. **—no′mad′ism** *n.*

**no·mad·ic** (nō-măd′ĭk) *adj.* Wandering from place to place : ITINERANT. **—no·mad′i·cal·ly** *adv.*

☆ ***syns:*** NOMADIC, ITINERANT, PERIPATETIC, ROAMING, ROVING, VAGABOND, VAGRANT, WANDERING *adj. core meaning* : leading the life of a person without a fixed domicile <a *nomadic* tribe>

**no man's land** *n.* **1.** An unowned or unclaimed piece of land. **2.** Land under dispute by two opposing parties, esp. the field of battle between two opposing entrenched armies. **3.** An area of ambiguity.

**nom·bril** (nŏm′brəl) *n.* [OFr., nombril, navel, alteration of *ombril,* navel < Lat. *umbilicus.*] *Heraldry.* The point on an escutcheon between the fess point and the base point.

**nom de guerre** (nŏm′ də gâr′) *n.* [Fr., war name.] A pseudonym.

**nom de plume** (nŏm′ də plŏŏm′) *n.* [Fr., pen name.] A fictitious name adopted by a writer.

**nome** (nōm) *n.* [Gk. *nomos,* district, habitation.] **1.** A province of ancient Egypt. **2.** A province of modern Greece.

**no·men·cla·tor** (nō′mən-klā′tər) *n.* [Lat., a slave who accompanied his master to tell him the names of people he met : *nomen,* name + *calare,* to call.] One who assigns names, as in scientific classification.

**no·men·cla·to·ri·al** (nō′mən-klə-tôr′ē-əl, -tōr′-) *adj.* Of or relating to nomenclature.

**no·men·cla·ture** (nō′mən-klā′chər, nō-mĕn′klə-) *n.* [Lat. *nomenclatura* < *nomenclator,* nomenclator.] A system of names used in an art or science.

**nom·i·nal** (nŏm′ə-nəl) *adj.* [Lat. *nominalis* < *nomen,* name.] **1. a.** Of, like, relating to, or consisting of a name or names. **b.** Bearing one's name <*nominal* shares of stock> **2.** Existing in name only and not in reality. **3.** Small : trifling <a *nominal* fee> **4.** Of or pertaining to a noun or a word group functioning as a noun. **5.** According to plan <The pre-launch conditions were *nominal.*> —*n.* A word or group of words that functions as a noun. **—nom′i·nal·ly** *adv.*

**nom·i·nal·ism** (nŏm′ə-nə-lĭz′əm) *n. Philos.* The doctrine that abstract concepts, general terms, or universals are without objective reference and exist soley as names. **—nom′i·nal·ist** *n.* **—nom′i·nal·is′tic** *adj.*

**nominal value** *n.* The stated, par, or book value of a share of stock as opposed to the real or market value.

**nom·i·nate** (nŏm′ə-nāt′) *vt.* **-nat·ed, -nat·ing, -nates.** [Lat. *nominare, nominat-* < *nomen,* name.] **1.** To propose as a candidate, as for election. **2. a.** To designate or appoint to an office or responsibility. **b.** To propose for an honor. **—nom′i·na′tion** (-nā′shən) *n.* **—nom′i·na′tor** *n.*

**nom·i·na·tive** (nŏm′ə-nā′tĭv) *adj.* **1. a.** Appointed to office. **b.** Nominated as a candidate for office. **2.** Having or bearing a person's name <*nominative* shares of stock> **3.** (-nə-tĭv). Of, relating to, or belonging to a grammatical case that usu. indicates the subject of a verb. —*n.* (-nə-tĭv). The nominative case.

**nom·i·nee** (nŏm′ə-nē′) *n.* [NOMIN(ATE) + -EE.] One nominated.

**nom·o·gram** (nŏm′ə-grăm′, nō′mə-) *n.* [Gk. *nomos,* law + -GRAM.] A nomograph.

**nom·o·graph** (nŏm′ə-grăf′, nō′mə-) *n.* [Gk. *nomos,* law + -GRAPH.] **1.** A graph having three coplanar curves, usu. parallel straight lines, each graduated for a different variable so that a straight line cutting all three curves intersects the related values of each variable. **2.** A chart representing numerical relationships. **—nom′o·graph′ic** (-grăf′ĭk) *adj.* **—no·mog′ra·phy** (nō-mŏg′rə-fē) *n.*

**-nomy** *suff.* [Lat. *-nomia* < Gk. < *nomos,* law < *nemnein,* to arrange.] A system of laws governing or a body of knowledge about a specified field <*aeronomy*>

**non-** *pref.* [ME < OFr. < Lat. *non,* not.] Not <*noncombatant*>

**nona-** *pref.* [< Lat. *nonus,* ninth.] Ninth : nine <*nonagon*>

**non·ad·di·tive** (nŏn-ăd′ĭ-tĭv) *adj.* **1.** Having a numerical value different from the sum of the component parts. **2.** Of, relating to, or being a genic effect that is nonadditive. **—non′ad·di·tiv′i·ty** *n.*

**non·age** (nŏn′ĭj, nō′nĭj) *n.* [ME *nounage* < AN < OFr. *nonage* : *non-,* non- + *age,* age.] **1.** The period during which one is legally underage. **2.** A period of immaturity.

**non·a·ge·nar·i·an** (nŏn′ə-jə-nâr′ē-ən, nō′nə-) *adj.* [< Lat. *nonagenarius* < *nonageni,* ninety each < *nonagintia,* : *novem,* nine + *-ginta,* ten times.] Being 90 years old or between 90 and 100 years old. **—non′a·ge·nar′i·an** *n.*

**non·a·gon** (nŏn′ə-gŏn′, nō′nə-) *n.* A polygon with nine sides.

**non·a·ligned** (nŏn′ə-līnd′) *adj.* Not in alliance with any other nation or bloc : NEUTRAL. **—non′a·lign′ment** *n.*

**non·a·no·ic acid** (nŏn′ə-nō′ĭk) *n.* [< *nonane,* a paraffin : NONA- + -ANE (so called because it is ninth in the methane series).] Pelargonic acid.

**non·book** nŏn′bŏŏk′ *n.* A book that is usu. a collection, as of photos and news clippings.

**non·ca·lor·ic** (nŏn′kə-lôr′ĭk, -lōr′-) *adj.* Having few or no calories.

**non·can·di·date** (nŏn-kăn′dĭ-dāt′, -dĭt) *n. Slang.* One who is not a candidate, esp. for political office. **—non′can′di·da·cy** *n.*

**nonce** (nŏns) *n.* [< ME *for the nones,* for the occasion.] The present or particular occasion <Let's disregard it for the *nonce.*>

▲ word history: Many words that have come about by the addition or loss of an initial *n,* such as *apron* and *nickname,* have done so because of their conjunction with the indefinite article *an*: *a napron* became *an apron,* and *an ekename* became *a nickname.* The word *nonce* arose from a similar process, but the initial *n* is derived from the inflected demonstrative pronoun. The original Middle English phrase was *to* or *for than anes,* which literally meant "for that one (occasion)." The phrase was respelled as *to* or *for tha nanes* at a very early date. *Nanes* became *nones* by a regular process that changed long *a* to long *o*; *nones* was later respelled *nonce.* With the disappearance of inflected adjectives and pronouns *tha* was interpreted as the definite article *the.* The word *anes,* without the added *n,* survived to become the modern word *once.*

**nonce word** *n.* A word invented, occurring, or used only for a particular occasion.

**non'a·ban'don·ment** *n.*
**non'ab·di·ca'tion** *n.*
**non'ab·o·li'tion** *n.*
**non'a·bra'sive** *adj.*
**non'a·bridg'ment** *n.*
**non·ab'so·lute** *adj. & n.*
**non'ab·sorb'ent** *adj. & n.*
**non'ab·sorp'tion** *n.*
**non'ab·stain'er** *n.*
**non·ab'stract** *adj. & n.*
**non'a·bu'sive** *adj.*
**non'ac·a·dem'ic** *adj.*
**non'ac·cel'er·a'tion** *n.*
**non·ac'cent** *n.*
**non'ac·cept'ance** *n.*
**non'ac·ces'so·ry** *adj. & n.*
**non'ac·ci·den'tal** *adj. & n.*
**non'ac·com'mo·dat'ing** *adj.*
**non'ac·cord'** *n.*
**non'ac·cu·mu·la'tion** *n.*
**non·ac'id** *n.*
**non'ac·qui·es'cent** *adj.*
**non'ac·quit'tal** *n.*
**non·ac'tion** *n.*
**non·ac'tive** *adj.*
**non·ac'tu·al** *adj.*
**non'a·cute'** *adj.*
**non'a·dapt'a·ble** *adj.*
**non'ad·dic'tive** *adj.*
**non'a·dept'** *adj.*
**non'ad·her'ence** *n.*
**non'ad·he'sive** *adj.*
**non'ad·ja'cent** *adj.*
**non'ad·jec·ti'val** *adj.*
**non'ad·join'ing** *adj.*
**non'ad·just'a·ble** *adj.*
**non'ad·min'is·tra'tive** *adj.*
**non'ad·mis'sion** *n.*
**non'a·dult'** *adj. & n.*
**non'ad·vance'ment** *n.*
**non'ad·van·ta'geous** *adj.*
**non'ad·ven'tur·ous** *adj.*
**non'ad·ver'bi·al** *adj.*
**non'aes·thet'ic** *adj.*
**non'af·fec·ta'tion** *n.*
**non'af·fil'i·a'tion** *n.*
**non-Af'ri·can** *adj. & n.*
**non'al·co·hol'ic** *adj.*
**non'a·pol'o·get'ic** *adj.*
**non'ap·par'ent** *adj.*
**non'ap·pear'ance** *n.*
**non'ap·pre'ci·a'tion** *n.*
**non'ap·proach'a·ble** *adj.*
**non·a'que·ous** *adj.*
**non·ar'bi·trar'y** *adj.*
**non'ar·o·mat'ic** *adj.*
**non-A'sian** *n. & adj.*
**non'as·ser'tive** *adj.*
**non'as·sim'i·la'tion** *n.*
**non·ath'lete** *n.*
**non'a·tom'ic** *adj.*
**non'at·tach'ment** *n.*
**non'at·tain'a·ble** *adj.*
**non'at·ten'dance** *n.*
**non'at·trib'u·tive** *adj. & n.*
**non'au·to·mat'ic** *adj.*
**non'bac·te'ri·al** *adj.*
**non-Bap'tist** *n.*
**non·ba'sic** *adj.*
**non·beau'ty** *n.*
**non·be'ing** *n.*
**non'be·liev'er** *n.*
**non'be·nev'o·lent** *adj.*
**non-Bib'li·cal** *adj.*
**non-Brit'ish** *adj.*
**non·bus'y** *adj.*
**non·caf·feine'** *n.*
**non·cak'ing** *adj.*
**non'can'cer·ous** *adj.*
**non'ca·non'i·cal** *adj.*
**non'cap·i·tal·is'tic** *adj.*
**non'car·niv'o·rous** *adj.*
**non'cat·e·gor'i·cal** *adj.*
**non-Cath'o·lic** *adj. & n.*
**non'-Cau·ca'sian** *adj. & n.*
**non'ce·les'tial** *adj.*
**non·cel'lu·lar** *adj.*
**non-Celt'ic** *adj.*
**non'cere·mo'ni·al** *adj.*
**non'cer'ti·fied'** *adj.*
**non·chem'i·cal** *adj.*
**non'-Chi·nese'** *adj. & n.*
**non-Chris'tian** *adj. & n.*
**non·cit'i·zen** *n.*
**non·civ'i·lized'** *adj.*
**non'clas·si·fi·ca'tion** *n.*
**non·cler'i·cal** *adj.*
**non·clin'i·cal** *adj.*
**non'co·er'cive** *adj.*
**non'co·he'sive** *adj.*
**non'col·laps'i·ble** *adj.*
**non'col'le·giate** *adj.*
**non'com·bus'ti·ble** *adj. & n.*
**non'com·mer'cial** *adj. & n.*
**non'com·mu'ni·ca·ble** *adj.*
**non·com'mu·nist** *n. & adj.*
**non'com·pen·sa'tion** *n.*
**non·com'pe·tent** *adj.*
**non·com'pe·tent·ly** *adv.*
**non'com·pet'i·tive** *adj.*
**non'com·pet'i·tive·ness** *n.*
**non'com·ple'tion** *n.*
**non'com·pres'sion** *n.*
**non'com·pul'sion** *n.*
**non'con·cur'rence** *n.*
**non'con·cur'rent** *adj.*
**non'con·den·sa'tion** *n.*
**non'con·fi·den'tial** *adj.*
**non'con·fine'ment** *n.*
**non'con·flic'tive** *adj.*
**non'con·ges'tion** *n.*
**non'-Con·gres'sion·al** *adj.*
**non'con·nec'tive** *adj. & n.*
**non·con'no·ta'tive** *adj.*
**non·con'scious** *adj.*
**non'con·sec'u·tive** *adj.*
**non'con·sent'** *n.*
**non'con·ser'va·tive** *adj. & n.*
**non'con·sid'er·a'tion** *n.*
**non'con·spir'a·tor** *n.*
**non'con·sult'a·tive** *adj.*
**non'con·ta'gious** *adj.*
**non'con·tem'po·rar'y** *adj.*
**non'con·tin'u·ous** *adj.*
**non·con'tra·band'** *n. & adj.*
**non'con·tra·dic'tion** *n.*
**non'con·tri·bu'tion** *n.*
**non'con·trib'u·tor** *n.*
**non'con·tro·ver'sial** *adj.*
**non'con·ver'gence** *n.*
**non'con·ver'sion** *n.*
**non'con·vic'tion** *n.*
**non'co·or'di·na'tion** *n.*
**non·cor'po·rate** *adj.*
**non'cor·rec'tive** *adj. & n.*
**non·cor're·lat'ing** *adj.*



**non'cor·re·la'tion** *n.*
**non·cos'mic** *adj.*
**non'cre·a'tive** *adj.*
**non·cred'i·ble** *adj.*
**non·crim'i·nal** *adj. & n.*
**non·cul'pa·ble** *adj.*
**non·cul'ture** *n.*
**non·cu'mu·la'tive** *adj.*
**non·cur'rent** *adj.*
**non'cur·tail'ment** *n.*
**non·cus'tom·ar'y** *adj.*
**non'de·cay'ing** *adj.*
**non'de·cep'tive** *adj.*
**non'de·liv'er·y** *n.*
**non'dem·o·crat'ic** *adj.*
**non'de·part·men'tal** *adj.*
**non'de·par'ture** *n.*
**non'de·pen'dence** *n.*
**non'de·pos'i·tor** *n.*
**non'de·pre'ci·a'tion** *n.*
**non'de·riv'a·tive** *adj. & n.*
**non'de·rog'a·to'ry** *adj.*
**non'de·tach'a·ble** *adj.*
**non'det·ri·men'tal** *adj.*
**non'de·vel'op·ment** *n.*
**non'de·vi·a'tion** *n.*
**non'de·vo'tion·al** *adj.*
**non'di·a·lec'tal** *adj.*
**non'di·dac'tic** *adj.*
**non'dif·fer·en'ti·a'tion** *n.*
**non'di·gest'i·ble** *adj.*
**non'di·rec'tion·al** *adj.*
**non'dis·cern'ment** *n.*
**non·dis'ci·pli·nar'y** *adj.*
**non'dis·pos'al** *n.*
**non'dis·rup'tive** *adj.*
**non·dis'si·dence** *n.*
**non'dis·tri·bu'tion** *n.*
**non'di·ver'gence** *n.*
**non'di·ver'si·fi·ca'tion** *n.*
**non'di·vis'i·ble** *adj.*
**non·doc'tri·nal** *adj.*
**non'doc·u·men'ta·ry** *adj.*
**non'dog·mat'ic** *adj.*
**non'do·mes'tic** *adj. & n.*
**non'dra·mat'ic** *adj.*
**non·dry'ing** *adj.*
**non'ec·cle'si·as'ti·cal** *adj.*
**non'ec·lec'tic** *adj.*
**non'ec·o·nom'ic** *adj.*
**non·ed'i·ble** *adj. & n.*
**non'ed·i·to'ri·al** *adj.*
**non·ed'u·ca·ble** *adj.*
**non'ed·u·ca'tion·al** *adj.*
**non'ef·fec'tive** *adj.*
**non'ef·fer·ves'cent** *adj.*
**non'ef·fi'cien·cy** *n.*
**non'e·las'tic** *adj.*
**non'e·lect'** *n.*
**non'e·lec'tion** *n.*
**non'e·lec'tive** *adj.*
**non'e·lec'tric** *adj.*
**non'el·e·men'ta·ry** *adj.*
**non'e·lim'i·na'tion** *n.*
**non'e·mo'tion·al** *adj.*
**non'em·pir'i·cal** *adj.*
**non'en·cy'clo·pe'dic** *adj.*
**non'en·force'a·ble** *adj.*
**non-Eng'lish** *adj. & n.*
**non·en'ter·pris'ing** *adj.*
**non'-E·pis'co·pa'lian** *adj.*
**non·e'qual** *adj. & n.*
**non'e·quiv'a·lent** *adj. & n.*
**non'e·ra'sure** *n.*
**non'e·rot'ic** *adj.*
**non·er'u·dite'** *adj.*
**non'es·sen'tial** *adj. & n.*
**non'es·thet'ic** *adj.*
**non'e·ter'nal** *adj.*
**non·eth'i·cal** *adj.*
**non'eth·no·log'i·cal** *adj.*
**non'-Eu·ro·pe'an** *adj. & n.*
**non'e·vac'u·a'tion** *n.*
**non'e·va'sion** *n.*
**non'e·va'sive** *adj.*
**non'e·vic'tion** *n.*
**non'ev·o·lu'tion·ar'y** *adj.*
**non'ex·pan'sive** *adj.*
**non'ex·pend'a·ble** *adj.*
**non'ex·per'i·men'tal** *adj.*
**non'ex·plo'sive** *adj. & n.*
**non·ex'tant** *adj.*
**non'ex·ter'nal** *adj.*
**non'ex·tinct'** *adj.*
**non'ex·tra·di'tion** *n.*
**non'ex·tra'ne·ous** *adj.*
**non'fa·ce'tious** *adj.*
**non·fac'tu·al** *adj.*
**non·fan'ta·sy** *n.*
**non·fas'cist** *n. & adj.*
**non'fas·tid'i·ous** *adj.*
**non·fa'tal** *adj.*
**non·fat'ten·ing** *adj.*
**non·fed'er·al** *adj.*
**non·fer'tile** *adj.*
**non·fes'tive** *adj.*
**non·feu'dal** *adj.*
**non·fig'ur·a·tive** *adj.*
**non'fi·nan'cial** *adj.*
**non·fi'nite** *adj. & n.*
**non·fire'proof'** *adj.*
**non·for'mal** *adj.*
**non'for·tu'i·tous** *adj.*
**non'fra·ter'nal** *adj.*
**non·fraud'u·lent** *adj.*
**non·free'dom** *n.*
**non-French'** *adj. & n.*
**non·fre'quent** *adj.*
**non'ful·fill'ment** *n.*
**non·func'tion·al** *adj.*
**non'fun·da·men'tal** *adj.*
**non·gas'e·ous** *adj.*
**non'ge·lat'i·nous** *adj.*
**non'ge·net'ic** *adj.*
**non·gov'ern·ment'al** *adj.*
**non'gre·gar'i·ous** *adj.*
**non·hab'it·a·ble** *adj.*
**non'ha·bit'u·al** *adj.*
**non·haz'ard·ous** *adj.*
**non·hea'then** *n. & adj.*
**non'he·red'i·tar'y** *adj.*
**non'he·ret'i·cal** *adj.*
**non'his·tor'ic** *adj.*
**non'ho·mo·ge'ne·ous** *adj.*
**non·hos'tile** *adj.*
**non·hu'man** *adj.*
**non·hu'man·ness** *n.*
**non·hu'mor·ous** *adj.*
**non'i·den'ti·ty** *n.*
**non'id·i·o·mat'ic** *adj.*
**non'i·dol'a·trous** *adj.*
**non'im·ag'i·nar'y** *adj.*
**non·im'i·ta'tive** *adj.*
**non'im·mune'** *adj.*
**non'im·pair'ment** *n.*
**non'im·pe'ri·al** *adj.*
**non'im·prove'ment** *n.*
**non'im·pul'sive** *adj.*
**non'in·can·des'cent** *adj.*
**non'in·clu'sive** *adj.*
**non·in'crease** *n.*
**non'in·de·pend'ent** *adj.*
**non·in'dexed** *adj.*
**non'in·dict'ment** *n.*

**non·cha·lant** (nŏn′shə-länt′) *adj.* [Fr. < OFr. < *nonchaloir,* to be unconcerned : *non-,* non- + *chaloir,* to be concerned < Lat. *calere,* to be warm.] Casually unconcerned : coolly indifferent. **—non′cha·lance′** -läns′) *n.* **—non′cha·lant′ly** *adv.*

**non·chro·mo·som·al** (nŏn′krō-mə-sō′məl) *adj.* Not located on or involving a chromosome.

**non·com** (nŏn′kŏm′) *n. Informal.* A noncommissioned officer.

**non·com·bat·ant** (nŏn′kəm-băt′nt, -kŏm′bə-tənt) *n.* **1.** A member of the armed forces, as a chaplain, whose duties do not involve combat. **2.** A civilian in wartime, esp. one in a war zone.

**non·com·mis·sioned officer** (nŏn′kə-mĭsh′ənd) *n.* An enlisted member of the armed forces, as a sergeant, appointed to a rank conferring leadership over others.

**non·com·mit·tal** (nŏn′kə-mĭt′l) *adj.* Not revealing what one thinks or feels. **—non′com·mit′tal·ly** *adv.*

**non·com·pli·ance** (nŏn′kəm-plī′əns) *n.* Refusal or failure to comply. **—non′com·pli′ant** *adj.* & *n.*

**non com·pos men·tis** (nŏn kŏm′pəs mĕn′tĭs) *adj.* [Lat., not in control of the mind.] *Law.* Not of sound mind and hence not legally responsible.

**non·con·duc·tor** (nŏn′kən-dŭk′tər) *n.* A substance that conducts little or no electricity or heat.

**non·con·form·ist** (nŏn′kən-fôr′mĭst) *n.* **1.** One who does not conform to conventional beliefs, customs, or practices. **2.** *often* **Nonconformist.** One who does not belong to a national or established church, esp. to the Church of England.

**non·con·form·i·ty** (nŏn′kən-fôr′mĭ-tē) *n.* **1.** Refusal or failure to conform to conventional customs, beliefs, or practices. **2.** *often* **Nonconformity.** Refusal to accept or conform to the doctrines or practices of the Church of England.

**non·co·op·er·a·tion** (nŏn′kō-ŏp′ə-rā′shən) *n.* Refusal or failure to cooperate. **—non′co·op′er·a′tion·ist** *n.* **—non′co·op′er·a·tive** (-ŏp′ər-ə-tĭv, -ŏp′ə-rā′-) *adj.* **—non′co·op′er·a′tor** *n.*

**non·cred·it** (nŏn-krĕd′ĭt) *adj.* Of, relating to, or constituting an academic course that does not offer credit toward a degree.

**non·dair·y** (nŏn-dâr′ē) *adj.* Not made with dairy products.

**non·de·duct·i·ble** (nŏn′dĭ-dŭk′tə-bəl) *adj.* Not deductible, esp. for income-tax purposes.

**non·de·nom·i·na·tion·al** (nŏn′dĭ-nŏm′ə-nā′shə-nəl) *adj.* Not associated with or restricted to a particular religious denomination.

**non·de·script** (nŏn′dĭ-skrĭpt′) *adj.* [NON- + Lat. *descriptus,* p.part. of *describere,* to describe. —see DESCRIBE.] Lacking in distinctive qualities, character, or form. **—non′de·script′** *n.*

**non·de·struc·tive** (nŏn′dĭ-strŭk′tĭv) *adj.* Of, relating to, or being a process that does not result in the destruction of the material under investigation. **—non′de·struc′tive·ly** *adv.*

**non·di·rec·tive** (nŏn′dĭ-rĕk′tĭv, -dī-) *adj.* Of, relating to, or being a psychotherapeutic technique in which the therapist takes an unobtrusive role in order to encourage free expression by the patient.

**non·dis·crim·i·na·tion** (nŏn′dĭ-skrĭm′ə-nā′shən) *n.* **1.** Absence of discrimination. **2.** The policy or practice of refraining from discrimination. **—non′dis·crim′i·na·to′ry** (-nə-tôr′ē, -tōr′ē) *adj.*

**non·dis·junc·tion** (nŏn′dĭs-jŭngk′shən) *n. Biol.* Failure of paired chromosomes to separate during the process of mitosis. **—non′dis·junc′tion·al** *adj.*

**non·dis·tinc·tive** (nŏn′dĭs-tĭngk′tĭv) *adj.* Not serving to distinguish meaning.

**non·drink·er** (nŏn-drĭng′kər) *n.* One who does not drink alcoholic beverages.

**none** (nŭn) *pron.* [ME < OE *nān* : *ne,* no + *ān,* one.] **1.** Not one : NOBODY <*None* wanted to go.> **2.** Not any <*None* of my possessions survived the fire.> **3.** No part : not any <*none* of your concern> *—adv.* In no way : not at all <I was *none* too pleased.>

**non·e·go** (nŏn-ē′gō) *n.* [Transl. of G. *Nichtich* : *nicht,* not + *Ich,* I.] All that is not part of the ego or the conscious self.

**non·en·ti·ty** (nŏn-ĕn′tĭ-tē) *n., pl.* **-ties. 1.** Nonexistence. **2.** Something that does not exist or that exists only in the imagination. **3.** One of no importance or influence.

**nones** (nōnz) *pl.n.* [ME < Lat. *nonae,* fem. pl. of *nonus,* ninth.] **1.** The seventh day of Mar., May, Jul., or Oct. and the fifth day of the other months in the ancient Roman calendar. **2. a.** The fifth of the seven canonical hours. **b.** The time of day set aside for this prayer, usu. the ninth hour after sunrise.

**none·such** *also* **non·such** (nŭn′sŭch′) *n.* **1.** One without equal. **2.** The black medic. **—none′such′** *adj.*

**none·the·less** (nŭn′thə-lĕs′) *adv.* However : nevertheless.

**non-Eu·clid·e·an** (nŏn′yōō-klĭd′ē-ən) *adj.* Designating any of several modern geometries not based on the postulates of Euclid.

**non·e·vent** (nŏn′ĭ-vĕnt′) *n. Slang.* An anticipated event that does not take place or that proves anticlimactic.

**non·ex·ist·ence** (nŏn′ĭg-zĭs′təns) *n.* **1.** The condition of not existing. **2.** Something that does not exist. **—non′ex·ist′ent** *adj.*

**non·fat** (nŏn′făt′) *adj.* Lacking fat solids or having the fat content removed, as milk.

**non·fea·sance** (nŏn-fē′zəns) *n.* [NON- + obs. *feasance,* a doing < OFr. *faisance.* —see MALFEASANCE.] *Law.* Failure to perform either a legal requirement or an official duty.

**non′in·di·vid′u·al** *adj.*
**non′in·duc′tive** *adj.*
**non′in·dus′tri·al** *adj.*
**non′in·fal′li·ble** *adj.*
**non′in·flam′ma·ble** *adj.*
**non′in·flec′tion·al** *adj.*
**non′in·form′a·tive** *adj.*
**non·in′ju·ry** *n.*
**non′in·quir′ing** *adj.*
**non′in·struc′tion·al** *adj.*
**non′in·stru·men′tal** *adj.*
**non′in·tel·lec′tu·al** *adj.* & *n.*
**non′in·tel′li·gent** *adj.*
**non′in·ter·change′a·ble** *adj.*
**non′in·tox′i·cant** *adj.*
**non·ir′ri·tant** *adj.* & *n.*
**non-Jew′** *n.*
**non′ju·di′cial** *adj.*
**non·ko′sher** *adj.*
**non-Lat′in** *adj.* & *n.*
**non·le′gal** *adj.*
**non·le′thal** *adj.*
**non′lib·er·a′tion** *n.*
**non·life′** *n.*
**non′lim·i·ta′tion** *n.*
**non′liq·ui·da′tion** *n.*
**non·list′ing** *adj.*
**non·lit′er·a·cy** *n.*
**non·lit′er·ar′y** *adj.*
**non′li·tur′gi·cal** *adj.*
**non·liv′ing** *adj.* & *n.*
**non·lo′cal** *adj.* & *n.*
**non·lov′ing** *adj.*
**non·lu′cid** *adj.*
**non′lu·mi·nes′cent** *adj.*
**non′mag·net′ic** *adj.*
**non·main′te·nance** *n.*
**non′ma·lig′nant** *adj.*
**non·mal′le·a·ble** *adj.*
**non·mar′i·tal** *adj.*
**non·mar′i·time′** *adj.*
**non·mar′ry·ing** *adj.*
**non·mar′tial** *adj.*
**non′ma·te′ri·al** *adj.*
**non′ma·ter′nal** *adj.*
**non′me·chan′i·cal** *adj.*
**non·med′i·cal** *adj.*
**non′me·lo′di·ous** *adj.*
**non·mem′ber** *n.*
**non′-Meth′od·ist** *adj.*
**non′met·ro·pol′i·tan** *adj.*
**non·mi′gra·to′ry** *adj.*
**non·mil′i·tant** *adj.* & *n.*
**non·min′er·al** *n.* & *adj.*
**non′min·is·te′ri·al** *adj.*
**non·mo′bile** *adj.*
**non·mor′tal** *adj.* & *n.*
**non-Mos′lem** *adj.* & *n.*
**non′mo·tile** *adj.*
**non′mu·nic′i·pal** *adj.*
**non·mus′cu·lar** *adj.*
**non·mys′ti·cal** *adj.*
**non·myth′i·cal** *adj.*
**non·na′tion·al** *adj.* & *n.*
**non·nau′ti·cal** *adj.*
**non·na′val** *adj.*
**non′ne·go′tia·ble** *adj.*
**non·neu′tral** *adj.* & *n.*
**non·nu′tri·ent** *n.* & *adj.*
**non′o·be′di·ence** *n.*
**non′o·blig′a·to′ry** *adj.*
**non′ob·ser′vance** *n.*
**non′ob·struc′tive** *adj.*
**non′oc·cu·pa′tion** *n.*
**non′oc·cur′rence** *n.*
**non·o′dor·ous** *adj.*
**non′of·fen′sive** *adj.*
**non′of·fi′cial** *adj.*
**non·op′er·a′tive** *adj.*
**non·op′tion·al** *adj.*
**non′or·gan′ic** *adj.*
**non·own′er** *n.*
**non′pa·cif′ic** *adj.*
**non·pa′gan** *n.* & *adj.*
**non·pa′pal** *adj.*
**non·par′** *adj.*
**non·par′al·lel′** *adj.* & *n.*
**non′par·a·sit′ic** *adj.*
**non′pa·ren′tal** *adj.*
**non′pa·rish′ion·er** *n.*
**non′par·lia·men′ta·ry** *adj.*
**non′pa·ro′chi·al** *adj.*
**non·par′tial** *adj.*
**non′par·tic′i·pant** *n.*
**non′par·tic′i·pat′ing** *adj.*
**non′par·tic′i·pa′tion** *n.*
**non′pa·ter′nal** *adj.*
**non′path·o·gen′ic** *adj.*
**non·pay′ment** *n.*
**non′per·cep′tu·al** *adj.*
**non′per·form′ance** *n.*
**non′pe·ri·od′i·cal** *adj.* & *n.*
**non·per′ish·a·ble** *adj.* & *n.*
**non·per′ma·nent** *adj.*
**non·per′me·a·ble** *adj.*
**non′per·pet′u·al** *adj.*
**non′phil·an·throp′ic** *adj.*
**non·phys′i·cal** *adj.*
**non′phys·i·o·log′i·cal** *adj.*
**non·plan′e·tar′y** *adj.*
**non′po·et′ic** *adj.*
**non·poi′son·ous** *adj.*
**non′po·lit′i·cal** *adj.*
**non·po′rous** *adj.*
**non′pred′a·to′ry** *adj.*
**non′pre·dict′a·ble** *adj.*
**non′pref·er·en′tial** *adj.*
**non′pre·scrip′tion** *adj.*
**non′pres·i·den′tial** *adj.*
**non′prev′a·lent** *adj.*
**non·priest′ly** *adj.*
**non′pro·duc′er** *n.*
**non′pro·fi′cien·cy** *n.*
**non′pro·gres′sive** *adj.* & *n.*
**non′pro·lif′ic** *adj.*
**non′pro·por′tion·al** *adj.*
**non′pro·pri′e·ty** *n.*
**non′pro·tec′tion** *n.*
**non-Prot′es·tant** *n.* & *adj.*
**non·psy′chic** *adj.* & *n.*
**non·pub′lic** *adj.*
**non·ra′cial** *adj.*

ă **pat** ā **pay** âr **care** ä **father** ĕ **pet** ē **be** hw **which** ĭ **pit**
ī **tie** îr **pier** ŏ **pot** ō **toe** ô **paw, for** oi **noise** ŏŏ **took**

**non·fer·rous** (nŏn-fĕr′əs) *adj.* **1.** Not containing or made of iron. **2.** Of or relating to metals other than iron.

**non·fic·tion** (nŏn-fĭk′shən) *n.* Literary works other than fiction. **—non·fic′tion·al** *adj.*

**non·flam·ma·ble** (nŏn-flăm′ə-bəl) *adj.* Not flammable, esp. not readily ignited.

**non·food** (nŏn′fo͞od′) *adj. Slang.* Of, pertaining to, or being something inedible, as certain articles sold in a supermarket.

**non·grad·ed** (nŏn-grā′dĭd) *adj.* Being without grade levels, as an elementary school.

**non·he·ro** (nŏn-hîr′ō) *n.* An antihero. **—non′he·ro′ic** *adj.*

**non·i·den·ti·cal** (nŏn′ī-dĕn′tĭ-kəl) *adj.* **1.** Not the same : DIFFERENT. **2.** FRATERNAL 3.

**no·nil·lion** (nō-nĭl′yən) *n.* [Fr. < OFr. : Lat. *nonus,* nine + Fr. *million,* million < OFr. *milion.* —see MILLION.] **1.** The cardinal number equal to $10^{30}$. **2.** *Chiefly Brit.* The cardinal number equal to $10^{54}$. **—no·nil′lion** *adj.*

**no·nil·lionth** (nō-nĭl′yənth) *n.* The ordinal number nonillion in a series. **—no·nil′lionth** *adj.* & *adv.*

**non·in·duc·tive** (nŏn′ĭn-dŭk′tĭv) *adj.* Having low inductance.

**non·in·ter·ven·tion** (nŏn′ĭn-tər-vĕn′shən) *n.* Refusal or failure to intervene, esp. in the affairs of other nations. **—non′in·ter·ven′tion·ist** *n.*

**non·in·volve·ment** (nŏn′ĭn-vŏlv′mənt) *n.* **1.** Lack of emotional involvement. **2.** Nonintervention.

**non·join·der** (nŏn-join′dər) *n. Law.* The omission of a party, plaintiff, defendant, or cause of action that should have been included as a part of an action or suit.

**non·judg·men·tal** (nŏn′jŭj-mĕn′tl) *adj.* Refraining from judgment, esp. one based on personal standards.

**non·ju·ror** (nŏn-jo͝or′ər, -ôr′) *n.* [NON- + JUROR, one who takes an oath (obs.).] **1.** One who refuses to take an oath, as of allegiance. **2. Nonjuror.** An Anglican clergyman who refused to swear allegiance to William and Mary in 1689.

**non·lin·e·ar** (nŏn-lĭn′ē-ər) *adj.* **1.** Not being in a straight line. **2.** Occurring as a result of a nonadditive operation.

**non·lit·er·ate** (nŏn-lĭt′ər-ĭt) *adj.* Having no written language <*nonliterate* cultures> **—non·lit′er·ate** *n.*

**non·met·al** (nŏn-mĕt′l) *n. Chem.* Any of a number of elements, as oxygen or sulfur, that gen. occur as negatively charged ions or radicals, form oxides that produce acids, and are poor conductors of heat and electricity when solid.

**non·me·tal·lic** (nŏn′mə-tăl′ĭk) *adj.* **1.** Not metallic. **2.** *Chem.* Of, relating to, or being a nonmetal.

**non·mor·al** (nŏn-môr′əl, -mŏr′-) *adj.* Unrelated to moral or ethical considerations.

**non·neg·a·tive** (nŏn-nĕg′ə-tĭv) *adj.* Of, relating to, or being a quantity that is either positive or zero.

**non·nu·cle·ar** (nŏn-no͞o′klē-ər, -nyo͞o′-) *adj.* **1.** Not causing, involving, or operated by nuclear energy **2. a.** Not possessing nuclear weapons. **b.** Not involving nuclear weapons <a *nonnuclear* war>

**no-no** (nō′nō′) *n. Informal.* **1.** Something unacceptable or not permissible. **2.** An embarassing blunder.

**non·ob·jec·tive** (nŏn′əb-jĕk′tĭv) *adj.* Designating or being a style of graphic art in which natural objects are not depicted realistically.

**non ob·stan·te** (nŏn′ əb-stăn′tē, nŏn′-) *prep.* [Lat.] Notwithstanding.

**no-non·sense** (nō-nŏn′sĕns′, -səns) *adj.* Practical : serious.

**non·pa·reil** (nŏn′pə-rĕl′) *adj.* [ME *nonparail* < OFr. *nonpareil* : *non-,* non- + *pareil,* equal < VLat. **pariculus,* dim. of Lat. *par,* equal.] Having no equal : PEERLESS. —*n.* **1.** One without equal : PARAGON. **2.** The painted bunting. **3.** A small, flat chocolate drop covered with white pellets of sugar.

**non·par·ti·san** (nŏn-pär′tĭ-zən, -sən) *adj.* **1.** Not partisan. **2.** Not based on, influenced by, associated with, or supporting the policies or interests of a political party. **—non·par′ti·san** *n.* **—non·par′ti·san·ship′** *n.*

**non·per·sis·tent** (nŏn′pər-sĭs′tənt) *adj.* Of or relating to a chemical or biological agent that has a short existence under natural conditions. **—non′per·sist′ence** *n.*

**non·per·son** (nŏn-pûr′sən) *n.* **1.** One whose existence is expunged from public attention and memory, esp. by the government and usu. for reasons of ideological or political deviation. **2.** One who has no legal or social standing.

**non·plus** (nŏn-plŭs′) *n.* [< Lat. *non plus,* no more.] Perplexity, confusion, or bewilderment. —*vt.* **-plused, -plus·ing, -plus·es** *also* **-plussed, -plus·sing, -plus·ses.** To bewilder : perplex.

**non·pol·lut·ing** (nŏn′pə-lo͞o′tĭng) *adj.* Not causing pollution.

**non pos·su·mus** (nŏn pŏs′ə-məs, nōn) *n.* [Lat., we cannot.] A statement indicating inability to do a particular thing.

**non·pro·duc·tive** (nŏn′prə-dŭk′tĭv) *adj.* **1.** Not yielding or producing <*nonproductive* soil> **2.** Not involved in the direct production of goods, as clerical personnel. **3.** Not producing phlegm : DRY <a *nonproductive* cough> **—non′pro·duc′tive·ly** *adv.*

**non·pro·fes·sion·al** (nŏn′prə-fĕsh′ə-nəl) *n.* One who is not a professional. —*adj.* Not professional. **—non′pro·fes′sion·al·ly** *adv.*

**non·prof·it** (nŏn-prŏf′ĭt) *adj.* Not seeking profit.

**non·pro·lif·er·a·tion** (nŏn′prə-lĭf′ə-rā′shən) *adj.* Of, relating to,

**non·rad′i·cal** *adj.* & *n.*
**non·ra′tion·al** *adj.*
**non′re·ac′tive** *adj.*
**non′re·ceiv′a·ble** *adj.* & *n.*
**non′re·cip′ro·cal** *adj.* & *n.*
**non′rec·og·ni′tion** *n.*
**non′rec·tan′gu·lar** *adj.*
**non′re·cur′rent** *adj.*
**non′re·deem′a·ble** *adj.*
**non′re·gen′er·a′tive** *adj.*
**non′reg·i·men′tal** *adj.*
**non′reg·is·tra′tion** *n.*
**non′re·li′ance** *n.*
**non′re·lig′ious** *adj.*
**non′re·mu′ner·a′tive** *adj.*
**non′re·new′a·ble** *adj.*
**non′re·pay′a·ble** *adj.*
**non′re·pen′tant** *adj.*
**non′re·pet′i·tive** *adj.*
**non′rep·re·hen′si·bly** *adv.*
**non′rep·re·sen·ta′tion** *n.*
**non′rep·re·sen′ta·tive** *adj.*
**non′re·sem′blance** *n.*
**non′res·i·den′tial** *adj.*
**non′re·sid′u·al** *adj.*
**non′re·solv′a·ble** *adj.*
**non′re·ten′tion** *n.*
**non′re·ten′tive** *adj.*
**non′re·tir′ing** *adj.*
**non′re·vers′i·ble** *adj.*
**non′re·volt′ing** *adj.*
**non′rhe·tor′i·cal** *adj.*
**non·ri′val** *n.*
**non-Ro′man** *adj.* & *n.*
**non·ru′ral** *adj.*
**non′sac·ra·men′tal** *adj.*
**non·sa′cred** *adj.*
**non·sal′a·ble** *adj.*
**non·sal′a·ried** *adj.*
**non′scho·las′tic** *adj.*
**non′sci·en·tif′ic** *adj.*
**non·scor′ing** *adj.*
**non·sea′son·al** *adj.*
**non′sec·re·tar′i·al** *adj.*
**non·sec′u·lar** *adj.*
**non′seg·men′tal** *adj.*
**non′se·lec′tive** *adj.*
**non-Sem′ite** *n.*
**non′sen·a·to′ri·al** *adj.*
**non·sen′si·tive** *adj.*
**non·sen′su·al** *adj.*
**non·ser′vile** *adj.*
**non·sex′u·al** *adj.*
**non·slip′per·y** *adj.*
**non·smok′er** *n.*
**non·smok′ing** *adj.*
**non·so′cial** *adj.*
**non·so′cial·ist** *n.*
**non·sol′vent** *n.*
**non·spar′ing** *adj.*
**non·spar′kling** *adj.*
**non′spe·cif′ic** *adj.*
**non′spec·tac′u·lar** *adj.*
**non·spec′u·la′tive** *adj.*
**non·spher′i·cal** *adj.*
**non·spir′i·tu·al** *adj.* & *n.*
**non′spon·ta′ne·ous** *adj.*
**non·sport′ing** *adj.*
**non·stand′ard·ized′** *adj.*
**non·sta′tion·ar′y** *adj.*
**non·stat′u·to′ry** *adj.*
**non′stim·u·la′tion** *n.*
**non′stra·te′gic** *adj.*
**non·stretch′a·ble** *adj.*
**non·stri′at·ed** *adj.*
**non·strik′er** *n.*
**non·struc′tur·al** *adj.*
**non′sub·mis′sive** *adj.*
**non′sub·scrib′er** *n.*
**non′suc·cess′** *n.*
**non′sup·port′er** *n.*
**non·sur′gi·cal** *adj.*
**non′sus·tain′ing** *adj.*
**non′sym·pa·thet′ic** *adj.*
**non·sym′pa·thy** *n.*
**non′sym·phon′ic** *adj.*
**non′symp·to·mat′ic** *adj.*
**non·syn′chro·nous** *adj.*
**non′syn·tac′tic** *adj.*
**non′sys·tem·at′ic** *adj.*
**non·talk′a·tive** *adj.*
**non·tech′ni·cal** *adj.*
**non·tem′po·ral** *adj.*
**non′ter·ri·to′ri·al** *adj.*
**non·tex′tu·al** *adj.*
**non′the·at′ri·cal** *adj.*
**non′ther·a·peu′tic** *adj.*
**non·ther′mal** *adj.*
**non·think′er** *n.*
**non·tox′ic** *adj.*
**non′tra·di′tion·al** *adj.*
**non·trag′ic** *adj.*
**non·trans·fer′a·ble** *adj.*
**non·tran·si′tion·al** *adj.*
**non′trans·par′ent** *adj.*
**non·truth′** *n.*
**non′tu·ber′cu·lar** *adj.*
**non·typ′i·cal** *adj.*
**non′ty·po·graph′ic** *adj.*
**non′ty·po·graph′i·cal** *adj.*
**non′un·der·stand′a·ble** *adj.*
**non′un·der·stand′ing** *n.*
**non′u·ni·form′** *adj.*
**non′u·ni·ver′sal** *adj.* & *n.*
**non·ur′ban** *adj.*
**non′u·til′i·tar′i·an** *adj.*
**non′va·lid′i·ty** *n.*
**non·val′ue** *n.*
**non·vas′cu·lar** *adj.*
**non·ven′om·ous** *adj.*
**non·ver′ti·cal** *adj.*
**non·vet′er·an** *n.*
**non′vi·o·la′tion** *n.*
**non·vis′u·al** *adj.*
**non·vi′tal** *adj.*
**non·vo′cal** *adj.*
**non′vo·ca′tion·al** *adj.*
**non·vol′a·tile** *adj.*
**non·vol′un·tar′y** *adj.*
**non·white′** *n.* & *adj.*
**non·work′er** *n.*
**non·yield′ing** *adj.*

involving, or advocating cessation of the proliferation of nuclear weapons <a *nonproliferation* agreement>

**non·pros** (nŏn'prŏs') *vt.* **-prossed, -pros·sing, -pros·ses.** [Short for NON PROSEQUITUR.] *Law.* To enter a judgment of non prosequitur against (a plaintiff).

**non pro·se·qui·tur** (nŏn' prə-sĕk'wĭ-tər, nōn') *n.* [Lat., he does not prosecute.] *Law.* The judgment entered against a plaintiff who fails to appear in court to prosecute a suit.

**non·read·er** (nŏn-rē'dər) *n.* One who does not or cannot read.

**non·rep·re·sen·ta·tion·al** (nŏn-rĕp'rĭ-zĕn-tā'shə-nəl) *adj.* Nonobjective. **—non·rep're·sen·ta'tion·al·ism** *n.*

**non·re·pro·duc·tive** (nŏn'rē-prə-dŭk'tĭv) *adj.* Not reproducing or capable of reproducing.

**non·res·i·dent** (nŏn-rĕz'ĭ-dənt, -dĕnt') *adj.* Not living in a particular place <a *nonresident* student> **—non·res'i·dence, non·res'i·den·cy** *n.* **—non·res'i·dent** *n.*

**non·re·sis·tance** (nŏn'rĭ-zĭs'təns) *n.* **1.** The principle or practice of complete obedience to authority even if arbitrary or unjust. **2.** The principle or practice of refusing to resort to physical force or violence even in self-defense.

**non·re·sis·tant** (nŏn'rĭ-zĭs'tənt) *adj.* Not resistant, esp. unable to resist infection or illness : SUSCEPTIBLE. **—non're·sis'tance** *n.* **—non're·sis'tant** *n.*

**non·re·stric·tive** (nŏn'rĭ-strĭk'tĭv) *adj.* **1.** Not restrictive. **2.** Of or designating a word, clause, or phrase that is descriptive of but not essential to the basic meaning of the element it modifies.

**non·re·turn·a·ble** (nŏn'rĭ-tûr'nə-bəl) *adj.* **1.** Not capable of being returned. **2.** Not exchangeable for a deposit, as beverage containers.

**non·rig·id** (nŏn-rĭj'ĭd) *adj.* **1.** Not rigid. **2.** Designating a lighter-than-air aircraft that holds its shape by gas pressure.

**non·sched·uled** (nŏn-skĕj'o͞old) *adj.* Operating without a regular schedule of passenger or cargo flights.

**non·sec·tar·i·an** (nŏn'sĕk-târ'ē-ən) *adj.* Not associated with or limited to a particular religious denomination <*nonsectarian* hymns> **—non'sec·tar'i·an·ism** *n.*

**non·sense** (nŏn'sĕns', -səns) *n.* **1.** Foolish or absurd language or behavior. **2.** Extravagant foolishness. **3.** Something of little or no importance or usefulness : TRIFLE. **4.** Pretentious or insolent behavior.

**nonsense verse** *n.* Humorous or whimsical verse often containing nonce words.

**non·sen·si·cal** (nŏn-sĕn'sĭ-kəl) *adj.* Foolish. **—non·sen'si·cal'i·ty** (-kăl'ĭ-tē), **non·sen'si·cal·ness** *n.* **—non·sen'si·cal·ly** *adv.*

**non se·qui·tur** (nŏn sĕk'wĭ-tər, -to͝or') *n.* [Lat., it does not follow.] **1.** An inference or conclusion that does not follow from the premises or evidence. **2.** A statement that does not follow logically from what preceded it.

**non·sex·ist** (nŏn-sĕk'sĭst) *adj.* Not discriminating against individuals, esp. women, on the basis of gender.

**non·sig·nif·i·cant** (nŏn'sĭg-nĭf'ĭ-kənt) *adj.* Having, producing, or being a value obtained from a statistical test that lies within the limits for being of random occurrence. **—non'sig·nif'i·cant·ly** *adv.*

**non·sked** (nŏn'skĕd') *n.* [Short for NONSCHEDULED.] *Informal.* A nonscheduled airline or cargo plane.

**non·skid** (nŏn'skĭd') *adj.* Designed to prevent or inhibit skidding <*nonskid* flooring>

**non·stan·dard** (nŏn-stăn'dərd) *adj.* **1.** Varying from or not conforming to the standard. **2.** Of, relating to, or indicating a level of language usage usu. avoided by educated speakers and writers.

**non·stick** (nŏn'stĭk') *adj.* Facilitating removal of adhered food particles <a *nonstick* coating on a saucepan>

**non·stop** (nŏn'stŏp') *adj.* **1.** Making or having made no stops <a *nonstop* airline flight> **2.** Unceasing : unremitting <*nonstop* chatter> **—non'stop'** *adv.*

**non·such** (nŭn'sŭch') *n. var. of* NONESUCH.

**non·suit** (nŏn-so͞ot') *n.* [ME, failure of a plaintiff to prosecute < AN *nounsute* : *non-*, no (< Lat. *non*) + *suite*, suit.] *Law.* A judgment against a plaintiff for failure to prosecute his or her case or to introduce sufficient evidence. *—vt.* **-suit·ed, -suit·ing, -suits.** To dismiss the lawsuit of.

**non·sup·port** (nŏn'sə-pôrt', -pōrt') *n. Law.* Failure to provide for the maintenance of one's legal dependents.

**non·tar·get** (nŏn-tär'gĭt) *adj.* Of, pertaining to, or being an object not meant to be acted upon by an agent.

**non·ten·ured** (nŏn-tĕn'yərd, -yo͝ord') *adj.* Not having or bringing about tenure <a *nontenured* professor><a *nontenured* teaching position>

**non·triv·i·al** (nŏn-trĭv'ē-əl) *adj. Math.* Of, pertaining to, or being an expression in which at least one variable is not equal to zero.

**non trop·po** (nŏn trô'pō, nōn) *adv. & adj.* [Ital., not too much.] *Mus.* In moderation. —Used to modify a direction.

**non-U** (nŏn-yo͞o') *adj. Chiefly Brit.* Not belonging or appropriate to upper-class custom.

**non·un·ion** (nŏn-yo͞on'yən) *adj.* **1.** Not belonging to a labor union. **2.** Not acknowledging or dealing with a labor union or employing union members.

**non·u·ple** (nŏn'yə-pəl) *adj.* [OFr. < Lat. *nonus*, nine.] **1.** Consisting of nine members, parts, or elements : NINEFOLD. **2.** Multiplied by nine. *—n.* A number or total that is nine times as great as another.

**non·use** nŏn'yo͞os' *n.* **1.** The fact or state of not being used. **2.** Failure to utilize : NEGLECT.

**non·us·er** (nŏn-yo͞o'zər) *n.* **1.** One who refrains from the use of something, as of alcohol or narcotic drugs. **2.** One who fails to take advantage of something, as a service.

**non·ver·bal** (nŏn-vûr'bəl) *adj.* **1.** Being other than verbal <*nonverbal* expression> **2. a.** Involving little use of language <a *nonverbal* aptitude test> **b.** Measuring low on a scale of verbal ability.

**non·vi·a·ble** (nŏn-vī'ə-bəl) *adj.* **1.** Not capable of living or developing, as a fetus. **2.** Not workable or practicable.

**non·vi·o·lence** (nŏn-vī'ə-ləns) *n.* **1.** Lack of violence. **2.** The doctrine, policy, or practice of rejecting violence in favor of peaceful tactics as a means of gaining esp. political objectives. **—non·vi'o·lent** *adj.* **—non·vi'o·lent·ly** *adv.*

**non·vot·er** (nŏn-vō'tər) *n.* One who does not vote or who has no right to vote.

**noo·dle¹** (no͞od'l) *n.* [Poss. alteration of NODDLE.] **1.** *Slang.* The human head. **2.** A stupid person.

**noo·dle²** (no͞od'l) *n.* [G. *Nudel.*] A typically ribbonlike strip of dried dough, usu. made of flour, eggs, and water.

**noo·dle³** (no͞od'l) *vi.* **-dled, -dling, -dles.** [Orig. unknown.] To improvise instrumental music idly and haphazardly.

**nook** (no͝ok) *n.* [ME *nok*, prob. of Scand. orig.] **1.** A small corner, alcove, or recess, esp. one that is part of a larger room. **2.** A hidden or secluded spot.

**noon** (no͞on) *n.* [ME *non* < OE, ninth hour after sunrise < LLat. *nona (hora)* < Lat. *nonus*, ninth.] **1. a.** Twelve o'clock in the daytime : MIDDAY. **b.** The time or the point in the sun's path when it is on the local meridian. **2.** The highest point : ZENITH. **3.** *Archaic.* Midnight.

**noon·day** (no͞on'dā') *n.* Noon.

**no one** *also* **no-one** (nō'wŭn') *pron.* No person : NOBODY.

**noon·tide** (no͞on'tīd') *n.* [ME *nontide* < OE *nōntīd* : *nōn*, noon + *tīd*, time.] Noon : noontime.

**noon·time** (no͞on'tīm') *n.* Noon.

**noose** (no͞os) *n.* [ME *nose.*] **1.** A loop formed in a rope by means of a slipknot so that it binds tighter as the rope is pulled. **2.** A snare or trap. *—vt.* **noosed, noos·ing, noos·es. 1.** To capture or hold by or as if by a noose. **2.** To make a noose of or in.

**Noot·ka** (no͞ot'kə, no͝ot'-) *n.* **1. a.** A tribe of Indians living on Vancouver Island in British Columbia and Cape Flattery in northwestern Washington. **b.** A member of this tribe. **2.** The Wakashan language of the Nootka. **—Noot'ka** *adj.*

**Nootka cypress** *n.* [After *Nootka* Sound, Canada.] A tall evergreen tree, *Chamaecyparis nootkatensis* of the northwestern coast of North America.

**no·pal** (nō'pəl, nō-päl', -päl') *n.* [Sp. < Nahuatl *nopalli.*] **1.** A cactus of the genus *Nopalea*, esp. *N. coccinellifera*, found chiefly in Mexico. **2.** A species of prickly pear, *Opuntia lindheimeri*, with yellow or red flowers and purple fruit.

**no-par** (nō'pär') *adj.* Having no face value <a *no-par* stock>

**no-par-val·ue** (nō'pär-văl'yo͞o) *adj.* No-par.

**nope** (nōp) *adv.* [Alteration of NO¹.] NO¹ 1.

**nor¹** (nôr; nər *when unstressed*) *conj.* [ME : *ne*, no + *or*, or.] And not : or not : not either <We have neither the time *nor* the money.>

**†nor²** (nôr; nər *when unstressed*) *conj.* [ME.] *Regional.* Than.

**NOR** (nôr) *n.* [N(OT) + OR.] A machine logic circuit that produces an output inverse to that of an OR circuit.

**nor-** *pref.* [< NORMAL.] An unaltered parent compound <*norepinephrine*>

**nor·a·dren·a·lin** (nôr'ə-drĕn'ə-lĭn) *n.* Norepinephrine.

**Nor·dic** (nôr'dĭk) *adj.* [Fr. *nordique* < OFr. *nord*, north < OE *norð.*] **1.** Of, pertaining to, or belonging to the subdivision of the Caucasoid ethnic group that is most predominant in Scandinavia. **2.** Of or pertaining to a class of people who are typically tall, longheaded, blond, and blue-eyed. **3.** Of or pertaining to ski competition featuring cross-country racing and ski jumping. **—Nor'dic** *n.*

**Nord·mann fir** (nôrd'mən) *n.* [After A. von *Nordmann* (d. 1866).] A widely planted evergreen tree, *Abies nordmanniana*, bearing reddish-brown, erect cones.

**nor·ep·i·neph·rine** (nôr'ĕp-ə-nĕf'rĭn, -rēn') *n.* A hormone, $(OH)_2C_6H_3{\cdot}CHOH{\cdot}CH_2{\cdot}NH_2$, that is a vasoconstrictor formed naturally in the body's sympathetic nerve endings.

**Nor·folk Island pine** (nôr'fək) *n.* An evergreen tree, *Araucaria excelsa*, native to Norfolk Island in the South Pacific.

**Norfolk jacket** *n.* [After *Norfolk*, England.] A belted jacket with two box pleats in front and back.

**no·ri·a** (nôr'ē-ə, nōr'-) *n.* [Sp. < Ar. *nā'ūrah* < *na'ara*, to creak.] A water wheel with buckets attached to its rim that are used to raise water from a stream, esp. for transferal to an irrigation trough.

**nor·ite** (nôr'īt') *n.* [Norw. *norit* < *Norge*, Norway.] Gabbro. **—nor·it'ic** (nô-rĭt'ĭk) *adj.*

**norm** (nôrm) *n.* [Lat. *norma*, carpenter's square.] **1.** A standard, model, or pattern considered to be as typical for a specific group. **2.** *Math.* **a.** A mode. **b.** An average. **c.** The length of a vector.

**Nor·ma** (nôr′mə) *n.* [NLat. < Lat. *norma*, carpenter's square.] A constellation in the Southern Hemisphere.

**nor·mal** (nôr′məl) *adj.* [Lat. *normalis*, made according to the square < *norma*, carpenter's square.] **1.** Conforming, adhering to, or constituting a typical or usual standard, pattern, level, or type. **2.** *Biol.* **a.** Not affected, immunized, or altered by experimentation. **b.** Functioning or occurring in a natural way. **3.** *Chem.* **a.** Describing a solution having one gram equivalent weight of solute per liter of solution. **b.** Describing an aliphatic hydrocarbon having a straight and unbranched chain of carbon atoms. **4.** *Math.* Being at right angles : PERPENDICULAR. **5. a.** Relating to or characterized by average intelligence or development. **b.** Free from physical or emotional disorder. —*n.* **1.** Something normal : STANDARD. **2.** The expected or usual state, form, amount, or degree. **3. a.** Correspondence to a norm. **b.** An average. **4.** *Math.* A perpendicular, esp. a perpendicular to a line tangent to a plane curve or to a plane tangent to a space curve. **—nor′mal·ly** *adv.*

**nor·mal·cy** (nôr′məl-sē) *n.* Normality.

**normal distribution** *n.* A theoretical frequency distribution for a set of variable data, usu. represented by a bell-shaped curve symmetrical about the mean.

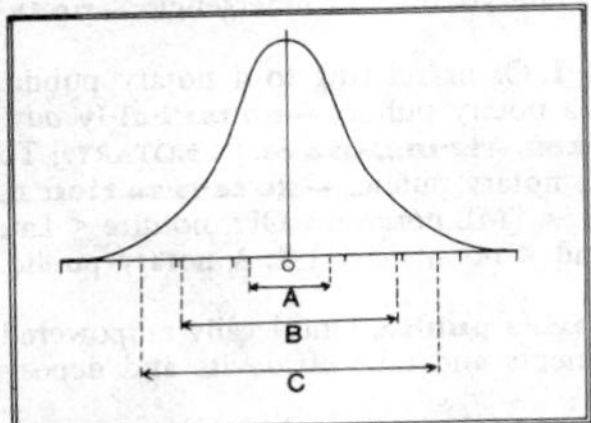


**normal distribution**
*A. 50% of area, B. 95% of area, C. 99% of area*

**nor·mal·i·ty** (nôr-măl′ĭ-tē) *n.* The quality or state of being normal.

**nor·mal·ize** (nôr′mə-līz′) *vt.* **-ized, -iz·ing, -iz·es. 1.** To cause to conform to a norm or standard. **2.** *Metallurgy.* To remove strains from and reduce coarse crystalline structures of by applying heat. **3.** To bring or return to a normal state <*normalize* diplomatic relations> **—nor′mal·i·za′tion** *n.* **—nor′mal·iz′er** *n.*

**normal pentane** *n.* A pentane.

**normal school** *n.* [Transl. of Fr. *école normale* (so called because the first school so named was intended as a model) < Lat. *normalis*, according to the square < *norma*, carpenter's square.] A school that trains teachers, mainly for the elementary grades.

**Nor·man** (nôr′mən) *n.* [ME < OFr. *Normant* (< ON *Norðmaðr* : *norðr*, north + *maðr*, man) and OE, var. of *Norðman* (*norð*, north + *man*, man).] **1.** A member of a Scandinavian people who conquered Normandy in the 10th cent. **2.** A member of a people of Norman and French blood who conquered England in 1066. **3.** A native or resident of Normandy. —*adj.* **1.** Of or pertaining to Normandy, the Normans, their culture, or their language. **2.** Of or designating a style of Romanesque architecture that was introduced from Normandy into England before 1066 and flourished until about 1200.

**Norman Conquest** *n.* The conquest of England by the Normans under William the Conqueror, beginning in 1066.

**Norman French** *n.* The dialect of Old French used in Normandy.

**nor·ma·tive** (nôr′mə-tĭv) *adj.* [Fr. *normatif* < *norme*, norm < Lat. *norma*, carpenter's square.] Of, pertaining to, or prescribing a norm or standard. **—nor′ma·tive·ly** *adv.* **—nor′ma·tive·ness** *n.*

**nor·mo·blast** (nôr′mə-blăst′) *n.* An immature red blood cell marked by abundant hemoglobin and a small nucleus.

**Norn** (nôrn) *n.* [ON.] *Norse Myth.* One of the three Fates.

**nor·nic·o·tine** (nôr-nĭk′ə-tēn′) *n.* A colorless liquid alkaloid, $C_9H_{12}N_2$, derived from tobacco and used as a plant insecticide.

**Norse** (nôrs) *adj.* [Du. *noorsch*, a Norwegian, alteration of *noordsch* < *noord*, north < MDu. *nort.*] **1.** Of or pertaining to ancient Scandinavia, its people, or their language. **2. a.** Of or pertaining to West Scandinavia or the languages of its inhabitants. **b.** Of or pertaining to Norway, its people, or their language. —*n.*, *pl.* **Norse. 1. a.** The people of Scandinavia. **b.** The people of West Scandinavia : the West Scandinavians, esp. the Norwegians. **c.** The ancient Norwegians. **2. a.** North Germanic. **b.** Any of the western Scandinavian languages, esp. Norwegian.

**Norse·man** (nôrs′mən) *n.* A member of one of the peoples of ancient Scandinavia.

**north** (nôrth) *n.* [ME < OE *norð*.] **1. a.** The direction along a meridian 90° counterclockwise from east : the direction to the left of sunrise. **b.** The cardinal point on the mariner's compass located at 0°. **2.** *often* **North.** The northern part of a country or region. —*adj.* **1.** To, toward, of, facing, or in the north. **2.** Coming from or originating in the north. —*adv.* In, from, or toward the north.

**north·bound** (nôrth′bound′) *adj.* Going toward the north.

**north by east** *n.* The direction or point on the mariner's compass halfway between due north and north-northeast that is 11°15′ east of due north. —*adv.* & *adj.* Toward or from north by east.

**north by west** *n.* The direction or point on the mariner's compass halfway between due north and north-northwest that is 11°15′ west of due north. —*adv.* & *adj.* Toward or from north by west.

**north·east** (nôrth-ēst′, nôr-ēst′) *n.* **1.** The direction or point on the mariner's compass halfway between north and east that is 45° east of due north. **2.** An area or region lying in the northeast. —*adj.* **1.** Situated toward, facing, or in the northeast. **2.** Coming from or originating in the northeast, as a wind. —*adv.* In, from, or toward the northeast. **—north·east′ern** *adj.*

**northeast by east** *n.* The direction or point on the mariner's compass halfway between northeast and east-northeast that is 56°15′ east of due north. —*adv.* & *adj.* Toward or from northeast by east.

**northeast by north** *n.* The direction or point on the mariner's compass halfway between northeast and north-northeast that is 33°45′ east of due north. —*adv.* & *adj.* Toward or from northeast by north.

**north·east·er** (nôrth-ē′stər, nôr-ē′-) *n.* **1.** A strong northeast wind. **2.** A storm having northeast winds.

**north·east·er·ly** (nôrth-ē′stər-lē, nôr-ē′-) *adj.* **1.** Located toward or in the northeast. **2.** From the northeast. **—north·east′er·ly** *adv.*

**north·east·ward** (nôrth-ēst′wərd, nôr-ēst′-) *adv.* Toward the northeast. —*adj.* Located toward or facing the northeast. —*n.* A direction or region to the northeast. **—north·east′ward·ly** *adj.* & *adv.* **—north·east′wards** (-wərdz) *adv.*

**north·er** (nôr′*th*ər) *n.* A sudden cold gale from the north.

**north·er·ly** (nôr′*th*ər-lē) *adj.* **1.** Located toward the north. **2.** From the north, as a wind. —*n.*, *pl.* **-lies.** A wind or storm from the north. **—north′er·ly** *adv.*

**north·ern** (nôr′*th*ərn) *adj.* [ME *northerne* < OE *norðerne.*] **1.** Located toward, in, or facing the north. **2.** Coming from the north, as a wind. **3.** Growing in the north. **4.** *often* **Northern.** Of, pertaining to, or typical of northern regions or the North. **5.** *Astron.* North of the celestial equator.

**Northern Cross** *n.* Cygnus.

**Northern Crown** *n.* Corona Borealis.

**north·ern·er** (nôr′*th*ər-nər) *n.* A native or inhabitant of the north, esp. of the northeastern United States.

**Northern Hemisphere** *n.* The half of the earth north of the equator.

**northern lights** *pl.n.* Aurora borealis.

**north·ern·most** (nôr′*th*ərn-mōst′) *adj.* Farthest north.

**Northern Spy** *n.* A large, yellowish-red, late-ripening apple.

**North Germanic** *n.* A subdivision of the Germanic languages that includes Norwegian, Icelandic, Swedish, Danish, and Faroese.

**north·ing** (nôr′thĭng, -*th*ĭng) *n.* **1.** The difference in latitude between two positions as a result of a movement to the north. **2.** Progress toward the north.

**north·land** (nôrth′lănd′, -lənd) *n.* *often* **Northland.** A region in the north, as of a territory or country. **—north′land′er** *n.*

**North·man** (nôrth′mən) *n.* A Norseman.

**north-north·east** (nôrth′nôrth-ēst′, nôr′nôr-ēst′) *n.* The direction or point on the mariner's compass halfway between due north and northeast that is 22°30′ east of due north. —*adv.* & *adj.* Toward, from, or in the north-northeast.

**north-north·west** (nôrth′nôrth-wĕst′, nôr′nôr-wĕst′) *n.* The direction or point on the mariner's compass halfway between due north and northwest that is 22°30′ west of due north. —*adv.* & *adj.* Toward, from, or in the north-northwest.

**North Pole** *n.* **1. a.** The northern end of the earth's axis of rotation. **b.** The celestial zenith of this terrestrial point, slightly more than 1° from Polaris. **2. north pole.** The north-seeking magnetic pole of a magnet.

**North Star** *n.* POLARIS 1.

**North·um·bri·an** (nôr-thŭm′brē-ən) *adj.* **1.** Of or pertaining to Northumbria or its Old English dialect. **2.** Of or pertaining to Northumberland. —*n.* **1.** A native or resident of Northumbria. **2.** A native or resident of Northumberland. **3.** The Old English dialect of Northumbria.

**north·ward** (nôrth′wərd) *adv.* & *adj.* Toward, to, or in the north. —*n.* A northern direction, point, or region. **—north′ward·ly** *adj.* & *adv.* **—north′wards** (-wərdz) *adv.*

**north·west** (nôrth-wĕst′, nôr-wĕst′) *n.* **1.** The direction or point on the mariner's compass halfway between north and west that is 45° west of due north. **2.** An area or region lying in the northwest. —*adj.* **1.** To, toward, of, facing, or in the northwest. **2.** Coming from the northwest. —*adv.* In, from, or toward the northwest. **—north·west′ern** *adj.*

**northwest by north** *n.* The direction or point on the mariner's compass halfway between northwest and north-northwest that is 33°45′ west of due north. —*adv.* & *adj.* Toward or from northwest by north.

**northwest by west** *n.* The direction or point on the mariner's compass halfway between northwest and west-northwest that is 56°15′ west of due north. —*adv. & adj.* Toward or from northwest by west.

**north·west·er** (nôrth-wĕs′tər) *n.* A storm or strong wind from the northwest.

**north·west·er·ly** (nôrth-wĕs′tər-lē, nôr-wĕs′-) *adj.* **1.** Toward or in the northwest. **2.** From the northwest. **—north·west′er·ly** *adv.*

**north·west·ward** (nôrth-wĕst′wərd) *adv.* Toward the northwest. —*adj.* Situated toward, facing, or in the northwest. —*n.* A direction or region toward the northwest. **—north·west′ward·ly** *adj. & adv.* **—north·west′wards** (-wərdz) *adv.*

**Nor·way maple** (nôr′wā′) *n.* A tall Eurasian tree, *Acer platanoides,* widely used in North America as a shade tree.

**Norway pine** *n.* [After *Norway,* Maine.] The red pine.

**Norway rat** *n.* The common brown rat, *Rattus norvegicus* found worldwide esp. in populated areas, and highly destructive.

**Norway spruce** *n.* A tall evergreen tree, *Picea abies* of northern regions, with long dark-green needles.

**Nor·we·gian** (nôr-wē′jən) *n.* [< Med. Lat. *Norwegia,* Norway < ON *Norvegr* : *norðr,* north + *vegr,* region.] **1.** A native or resident of Norway. **2.** The North Germanic language of the Norwegians. **—Nor·we′gian** *adj.*

**Norwegian elkhound** *n.* The elkhound.

**nose** (nōz) *n.* [ME < OE *nosu.*] **1. a.** The structure on the face or the forward part of the head of humans and other primates that contains the nostrils and organs of smell and forms the beginning of the respiratory tract. **b.** A similar feature or organ on the face, muzzle, snout, or front end of many other animals. **2. a.** The sense of smell <a hound with a keen *nose*> **b.** An aroma, as of wine : BOUQUET. **3.** Ability to detect, sense, or discover as if by smell <a *nose* for news> **4.** *Informal.* The nose regarded as a symbol of prying <kept my *nose* out of their business> **5.** Something resembling a nose in shape or position, as the forward end of an aircraft, rocket, or submarine. —*v.* **nosed, nos·ing, nos·es.** —*vt.* **1.** To find out by or as if by smell. **2.** To touch or examine with the nose : NUZZLE. **3.** To cause to move or advance cautiously <*nosed* the car out of the driveway> —*vi.* **1.** To smell or sniff. **2.** *Informal.* To pry inquisitively or in a meddlesome way. **3.** To push forward with caution <The plane *nosed* into its hangar.> **—follow (one's) nose. 1.** To move straight ahead. **2.** To be guided by instinct. **—look down (one's) nose at.** *Informal.* To regard or treat with arrogance, disapproval, or contempt. **—nose out.** To defeat by a very narrow margin. **—on the nose.** Exactly : precisely <guessed the cost *on the nose*>

**nose·band** (nōz′bănd′) *n.* The part of a bridle or halter that passes over an animal's nose.

**nose·bleed** (nōz′blēd′) *n.* A nasal hemorrhage.

**nose cone** *n.* The forwardmost and usu. separable section of a rocket or guided missile, shaped to offer minimum aerodynamic resistance and often bearing a protective cladding against heat.

**nose dive** *n.* **1.** A steep dive of an aircraft, nose toward the earth. **2.** A sudden swift plunge or drop.

**nose-dive** (nōz′dīv′) *vi.* **-dived** *or* **-dove** (-dōv′), **-div·ing, -dives.** To perform or undergo a nose dive.

**no-see-um** (nō-sē′əm) *n.* [Alteration of *no see them.*] The punkie.

**nose·gay** (nōz′gā′) *n.* [ME : *nose,* nose + *gay,* ornament < *gay,* joyous < OFr. *gai.*] A small bunch of flowers.

**nose job** *n. Slang.* Plastic surgery performed on the nose.

**nose·piece** (nōz′pēs′) *n.* **1.** A piece of armor that forms part of a helmet and protects the nose. **2.** The part of a pair of eyeglasses that fits across the nose. **3.** A noseband. **4.** The often rotatable part of a microscope to which one or more objective lenses are attached.

**nos·ey** (nō′zē) *adj. var. of* NOSY.

**nosh** (nŏsh) [Short for Yiddish *nosherai,* tidbits < OHG *hnascōn,* to nibble.] *Informal.* —*n.* A tidbit : snack. —*vi.* **noshed, nosh·ing, nosh·es.** To eat snacks between meals. **—nosh′er** *n.*

**no-show** (nō′shō′) *n. Slang.* **1. a.** A traveler who reserves a place, esp. on an airplane, but neither claims nor cancels the reservation before the time of departure. **b. no-shows.** Those who purchase tickets for an event but do not attend esp. out of disapproval or protest. **2.** A person who fails to keep an engagement or appointment.

**nos·ing** (nō′zĭng) *n.* **1. a.** The usu. rounded projecting edge of a stair tread. **b.** A shield covering this edge. **2.** A projecting edge of a molding.

**noso-** *pref.* [Gk. < *nosos,* a disease.] Disease <nosography>

**no·sog·ra·phy** (nō-sŏg′rə-fē, -zŏg′-) *n.* Systematization and description of diseases. **—no·sog′ra·pher** *n.* **—no′so·graph′ic** (nō′sə-grăf′ĭk), **no′so·graph′i·cal** *adj.*

**no·sol·o·gy** (nō-sŏl′ə-jē, -zŏl′-) *n.* **1.** The branch of medicine dealing with classification of diseases. **2.** A classification of diseases. **—no′so·log′i·cal** (-sə-lŏj′ĭ-kəl), **no′so·log′ic** *adj.* **—no′so·log′i·cal·ly** *adv.* **—no·sol′o·gist** *n.*

**nos·tal·gi·a** (nŏ-stăl′jə, nə-) *n.* [Gk. *nostos,* home + -ALGIA.] **1.** A bittersweet longing for persons, things, or situations of the past. **2.** Homesickness. **—nos·tal′gic** (-jĭk) *adj.* **—nos·tal′gi·cal·ly** *adv.*

**nos·toc** (nŏs′tŏk′) *n.* [Coined by Paracelsus (1493–1541).] A freshwater alga of the genus *Nostoc,* forming colonies of blue-green cells embedded in a jelly.

**nos·tril** (nŏs′trəl) *n.* [ME *nostrille* < OE *nosðyrl* : *nosu,* nose + *thyrl,* hole.] Either of the external openings of the nose.

**nos·trum** (nŏs′trəm) *n.* [Lat., our own, neuter of *noster,* ours.] **1.** A medicine, esp. a quack remedy, whose ingredients are usu. secret. **2.** A favorite but untested remedy for problems or evils.

**nos·y** *or* **nos·ey** (nō′zē) *adj.* **-i·er, -i·est.** [< NOSE.] *Informal.* Inquisitive : snoopy. **—nos′i·ly** *adv.* **—nos′i·ness** *n.*

**not** (nŏt) *adv.* [ME, alteration of *nought* < OE *nōwiht* : *nā,* no + *wiht,* thing.] In no way : to no degree. —Used to express negation, denial, refusal, or prohibition <We will *not* leave.><You may *not* watch TV.><sometimes happy, sometimes *not*>

**NOT** (nŏt) *n.* [< NOT.] A machine logic circuit that produces an output inverse to the input.

**no·ta be·ne** (nō′tə bē′nē, bĕn′ē) [Lat., note well.] —Used to direct attention to something esp. important.

**no·ta·bil·i·ty** (nō′tə-bĭl′ĭ-tē) *n., pl.* **-ties. 1.** The quality or state of being notable. **2.** A notable or prominent person.

**†no·ta·ble** (nō′tə-bəl) *adj.* [ME < OFr. < Lat. *notabilis* < *notare,* to note < *nota,* note.] **1. a.** Worthy of notice : STRIKING <*notable* charm><a *notable* achievement> **b.** Prominent or distinguished <Many *notable* diplomats attended.> **2.** (*also* nŏt′ə-bəl). *Archaic & Regional.* Diligent and efficient, esp. in household duties. —*n.* **1.** A person of distinction or eminent reputation. **2.** *often* **Notable.** One of a council of prominent persons before the French Revolution called into assembly to deliberate at times of emergency. **—no′ta·ble·ness** *n.* **—no′ta·bly** *adv.*

**no·tar·i·al** (nō-târ′ē-əl) *adj.* **1.** Of or relating to a notary public. **2.** Drawn up or executed by a notary public. **—no·tar′i·al·ly** *adv.*

**no·ta·rize** (nō′tə-rīz′) *vt.* **-rized, -riz·ing, -riz·es.** [< NOTARY.] To attest to or authenticate as a notary public. **—no′ta·ri·za′tion** *n.*

**no·ta·ry** (nō′tə-rē) *n., pl.* **-ries.** [ME *notarie* < OFr. *notaire* < Lat. *notarius* < *notarius,* shorthand < *nota,* mark.] **1.** A notary public. **2.** *Obs.* A stenographer.

**notary public** *n., pl.* **notaries public.** One legally empowered to witness and certify documents and take affidavits and depositions.

**no·tate** (nō′tāt′) *vt.* **-tat·ed, -tat·ing, -tates.** To put into notation <a ballet *notated* by choreographers>

**no·ta·tion** (nō-tā′shən) *n.* [Lat. *notatio* < *notare,* to note < *nota,* note.] **1. a.** A system of figures or symbols used in specialized fields to represent numbers, quantities, tones, or values <dance *notation*> **b.** The act or process of using such a system. **2.** A brief note : ANNOTATION. **—no·ta′tion·al** *adj.*

**notch** (nŏch) *n.* [*a notch,* alteration of *an otch* < OFr. *oche* < *ochier,* to notch < Lat. *absecare,* to cut off : *ab-,* off + *secare,* to cut.] **1. a.** A V-shaped cut. **b.** Such a cut used for keeping a record. **2.** A narrow pass between mountains. **3.** *Informal.* A level or degree <a *notch* more skillful than myself> —*vt.* **notched, notch·ing, notch·es. 1.** To cut a notch in. **2.** To record by or as if by making notches. **—notched** *adj.*

**notch·back** (nŏch′băk′) *n.* An automobile with a sloping roof and a distinctive rear deck.

**note** (nōt) *n.* [Lat. *nota,* mark.] **1.** A brief record, esp. one written down to aid the memory. **2.** A brief informal message or letter. **3.** A formal written diplomatic or official communication. **4.** A comment or explanation, as on a passage in a text. **5. a.** A piece of paper currency. **b.** A certificate issued by a government or a bank and occas. negotiable as money. **c.** A promissory note. **6.** *Mus.* **a.** A tone of definite pitch. **b.** A symbol for such a tone, indicating pitch by its position on the staff and duration by its shape. **c.** A key of a piano or similar instrument. **7.** A typical call or cry, as of a bird. **8.** A sign of a certain quality <a *note* of optimism> **9.** Importance : consequence <Nothing of *note* transpired at the meeting.> **10.** Notice : observation <take *note* of one's surroundings> **11.** *Archaic.* A song, melody, or tune. —*vt.* **not·ed, not·ing, notes. 1.** To observe carefully : NOTICE. **2.** To write down. **3.** To show : indicate. **4.** To make mention of : REMARK. **—compare notes.** To exchange ideas, views, or opinions. **—not′er** *n.*

**note·book** (nōt′bo͝ok′) *n.* A book of pages for notes.

**not·ed** (nō′tĭd) *adj.* Distinguished by reputation : FAMOUS <a *noted* sculptor> **—not′ed·ly** *adv.* **—not′ed·ness** *n.*

**note of hand** *n.* A promissory note.

**note·wor·thy** (nōt′wûr′thē) *adj.* Deserving notice or attention : REMARKABLE. **—note′wor′thi·ly** *adv.* **—note′wor′thi·ness** *n.*

**noth·ing** (nŭth′ĭng) *pron.* [ME < OE *nāthing* : *nān,* none + *thing,* thing.] **1.** No thing : not anything <knows *nothing* about golf> **2.** No significant or notable thing <There is *nothing* at the movies this week.> **3.** No part : no portion <*Nothing* is left of the cheesecake.> **4.** Insignificance : obscurity <rose from *nothing* to become President> **5.** Something or someone of no consequence, significance, or interest <The outcome is *nothing* to me.> **6.** Absence of anything perceptible : NONEXISTENCE <The music died away into *nothing.*> —*n.* **1.** Something that has no existence. **2.** Something that has no quantitative value : ZERO. **3. a.** Something trivial : TRIFLE

ă **pat** ā **pay** âr **care** ä **father** ĕ **pet** ē **be** hw **which** ĭ **pit** ī **tie** îr **pier** ŏ **pot** ō **toe** ô **paw, for** oi **noise** o͝o **took**

<The gift was a real *nothing.*> **b.** A trivial or inane word or remark. **c.** An inconsequential, insignificant, or uninteresting person : NONENTITY. *—adj. Slang.* Insignificant or worthless : TRIFLING <spent a *nothing* evening at home> *—adv.* In no way or degree : not at all <It's *nothing* short of miraculous.> **—nothing doing.** *Informal.* Certainly not.

☆ ***syns:*** NOTHING, NAUGHT, NIL, NIX, ZILCH, ZIP *pron. core meaning* : no thing : not anything <left *nothing* to the heirs><told the police *nothing*>

**noth·ing·ness** (nŭth′ĭng-nĭs) *n.* **1.** The quality or condition of being nothing : NONEXISTENCE. **2.** Empty space : VOID. **3.** Lack of consequence : INSIGNIFICANCE. **4.** Something unimportant or insignificant.

**no·tice** (nō′tĭs) *n.* [ME, knowledge < OFr. *notere* < Lat. *notitia* < *notus*, known, p.part. of *noscere*, to come to know.] **1.** Observation : attention <It has come to my *notice.*> **2.** Respectful attention or consideration. **3.** A written or printed announcement. **4. a.** A formal announcement of purpose, esp. of intention to withdraw from an agreement or leave a job. **b.** The condition of being notified of such purpose. **5.** A critical review, as of a play or book. **6.** An indication or warning of something. *—vt.* **-ticed, -tic·ing, -tic·es. 1.** To take note of : OBSERVE. **2.** To give attention to : MARK. **3.** To comment on : MENTION. **4.** To treat with courteous attention.

**no·tice·a·ble** (nō′tĭ-sə-bəl) *adj.* **1.** Attracting notice. **2.** Worthy of notice : SIGNIFICANT. **—no′tice·a·bly** *adv.*

☆ ***syns:*** NOTICEABLE, ARRESTING, ARRESTIVE, CONSPICUOUS, EYE-CATCHING, MARKED, OBSERVABLE, OUTSTANDING, POINTED, PROMINENT, REMARKABLE, SALIENT, SIGNAL, STRIKING *adj. core meaning* : readily attracting notice <a *noticeable* increase in temperature><a *noticeable* rise in the cost of living>

**no·ti·fi·ca·tion** (nō′tə-fĭ-kā′shən) *n.* **1.** The act or an instance of notifying. **2.** Something, as a letter, by which notice is given.

**no·ti·fy** (nō′tə-fī′) *vt.* **-fied, -fy·ing, -fies.** [ME *notifien* < OFr. *notifier* < Lat. *notificare*, to make known : *notus*, known (p.part. of *noscere*, to come to know) + *facere*, to make.] **1.** To give notice to : INFORM <*notified* the tenants of the rent increase> **2.** *Chiefly Brit.* To give notice of : make known. **—no′ti·fi′er** *n.*

**no·tion** (nō′shən) *n.* [Lat. *notio* < *noscere*, to come to know.] **1.** An opinion or belief. **2.** A mental image or representation : IDEA. **3.** A fanciful impulse : WHIM. **4. notions.** Small useful items, as needles, buttons, or thread.

**no·tion·al** (nō′shə-nəl) *adj.* **1.** Of, pertaining to, or constituting a notion. **2.** Speculative or theoretical. **3.** Existing in the mind : IMAGINARY. **4.** Conveying an idea of a thing or action : having full lexical meaning as distinguished from relational meaning <The word "did" is *notional* in "we did the work" and relational in "we did not agree."> **—no′tion·al·ly** *adv.*

**no·to·chord** (nō′tə-kôrd′) *n.* [Gk. *nōtos*, back + CHORD[3].] **1.** A flexible rodlike structure in some lower vertebrates that provides dorsal support : the primitive backbone. **2.** A structure similar to the notochord in embryos of higher vertebrates from which the spine develops. **—no′to·chord′al** *adj.*

**no·to·ri·e·ty** (nō′tə-rī′ĭ-tē) *n.* **a.** The quality or state of being notorious <The scandal gave them *notoriety.*> **2.** A notorious person.

**no·to·ri·ous** (nō-tôr′ē-əs, -tōr′-) *adj.* [Med. Lat. *notorius* < Lat. *notus*, known, p.part. of *noscere*, to come to know.] Known widely and usu. unfavorably <a *notorious* gambler> **—no·to′ri·ous·ly** *adv.* **—no·to′ri·ous·ness** *n.*

**no·tor·nis** (nō-tôr′nĭs) *n., pl.* **notornis.** [NLat. *Notornis*, genus name : Gk. *notos*, south + Gk. *ornis*, bird.] A rare flightless bird of the genus *Notornis*, found in New Zealand.

**notornis**

**no-trump** (nō′trŭmp′) *n.* **1.** A declaration to play a hand without a trump suit in bridge and other card games. **2.** A hand played without a trump suit. **—no′-trump′** *adj.*

**not·with·stand·ing** (nŏt′wĭth-stăn′dĭng, -wĭth-) *prep.* [ME *notwithstanding* : *not*, not + *withstanding*, pr.part. of *withstanden*, to resist. —see WITHSTAND.] In spite of. *—adv.* All the same : NEVERTHELESS. *—conj.* In spite of the fact that : ALTHOUGH.

**nou·gat** (nōō′gət) *n.* [Fr. < Prov. < Lat. *nux*, nut.] A confection made from a sugar or honey paste into which nuts are mixed.

**nought** (nôt) *n.* & *adj.* & *adv. var. of* NAUGHT.

**nou·me·non** (nōō′mə-nŏn′) *n., pl.* **-na** (-nə) [Gk., concept < *nouein*, to conceive < *nous*, mind.] *Philos.* **1.** An object of purely intellectual intuition as opposed to an object of sensuous perception. **2.** A thing-in-itself that is independent of the sensuous or intellectual perception of it. **—nou′men·al** *adj.*

**noun** (noun) *n.* [ME < AN < OFr. *non* < Lat. *nomen*, name.] A word that is used to name a person, place, thing, quality, or action and can function as the subject or object of a verb, the object of a preposition, or an appositive.

**nour·ish** (nûr′ĭsh, nŭr′-) *vt.* **-ished, -ish·ing, -ish·es.** [ME *norishen* < OFr. *norrir* < Lat. *nutrire.*] **1.** To provide with food or other nutriment necessary for life and growth : FEED. **2.** To foster the development of : PROMOTE <*nourish* one's creativity><*nourish* a hope> **3.** To keep alive : MAINTAIN. **—nour′ish·er** *n.* **—nour′ish·ment** (-mənt, nŭr′-) *n.*

**nous** (nōōs) *n.* [Gk.] *Philos.* Mind : reason, esp. the principle of divine reason.

**nou·veau riche** (nōō′vō rēsh′) *n., pl.* **nou·veaux riches** (nōō′vō rēsh′) [Fr., new rich.] One who has recently become rich, esp. one who flaunts one's wealth : PARVENU.

**nou·velle cuisine** (nōō-vĕl′) *n.* [Fr., new cuisine.] A contemporary school of French cooking that seeks to bring out the natural flavors of foods and advocates the use of light, low-calorie sauces and stocks.

**nouvelle vague** (väg′) *n.* [Fr.] NEW WAVE 1.

**no·va** (nō′və) *n., pl.* **-vae** (-vē) or **-vas.** [NLat. *(stella) nova*, new star < Lat. *novus*, new.] *Astron.* A variable star that suddenly increases greatly in brightness and then returns to its original appearance in a period of a few weeks to several months or years.

**no·vac·u·lite** (nō-văk′yə-līt′) *n.* [Lat. *novacula*, razor + -ITE.] A hard, dense, even-textured silica-bearing rock used in whetstones.

**no·vae** (nō′vē) *n. var. pl. of* NOVA.

**no·va·tion** (nō-vā′shən) *n.* [LLat. *novatio* < Lat. *novare*, to make new < *novus*, new.] *Law.* Substitution of a new obligation for an old obligation.

**nov·el**[1] (nŏv′əl) *n.* [Ital. *(storia) novella*, new story < *novello*, new < Lat. *novellus*, dim. of *novus.*] **1.** A fictional prose narrative of considerable length, typically having a plot that is unfolded by the actions, speech, and thoughts of the characters. **2.** The literary genre represented by novels.

**nov·el**[2] (nŏv′əl) *adj.* [ME < OFr. < Lat. *novellus*, dim. of *novus.*] Strikingly different, unusual, or new. **—nov′el·ly** *adv.*

**nov·el·ette** (nŏv′ə-lĕt′) *n.* A short novel.

**nov·el·ist** (nŏv′ə-lĭst) *n.* A writer of novels.

**nov·el·is·tic** (nŏv′ə-lĭs′tĭk) *adj.* Of, relating to, or typical of novels. **—nov′el·is′ti·cal·ly** *adv.*

**no·vel·la** (nō-vĕl′ə) *n., pl.* **-vel·las** or **-vel·le** (-vĕl′ē) [Ital. —see NOVEL[1].] **1.** A short prose tale marked by wit, terseness, or satire. **2.** A novelette.

**nov·el·ty** (nŏv′əl-tē) *n., pl.* **-ties.** [ME *novelte* < OFr. < *novel*, new. —see NOVEL[2].] **1.** The quality of being novel : NEWNESS. **2.** Something new and unusual : INNOVATION. **3. novelties.** Small mass-produced articles, as souvenirs or toys.

**No·vem·ber** (nō-vĕm′bər) *n.* [ME *Novembre* < OFr. < Lat. *November*, ninth month < *novem*, nine.] The 11th month of the Gregorian calendar. —See table at CALENDAR.

**no·ve·na** (nō-vē′nə) *n., pl.* **-nas** or **-nae** (-nē) [Med. Lat. < Lat. *novenus*, nine each < *novem*, nine.] *Rom. Cath. Ch.* Prayers and devotions for a special purpose, repeated for nine consecutive days.

**no·ver·cal** (nō-vûr′kəl) *adj.* [Lat. *novercalis* < *noverca*, stepmother.] Of, pertaining to, or typical of a stepmother.

**nov·ice** (nŏv′ĭs) *n.* [ME, probationary member of a religious order < OFr. < Med. Lat. *novicius* < Lat. *novus*, new.] **1.** One who is new to a field or activity : BEGINNER. **2.** One who has entered a religious order but has not yet taken final vows.

**no·vi·ti·ate** *also* **no·vi·ci·ate** (nō-vĭsh′ē-ĭt, -āt′) *n.* [Fr. *noviciat* < Med. Lat. *noviciatus* < *novicius*, novice < Lat. *novus*, new.] **1.** The period of being a novice. **2.** A place where novices live. **3.** NOVICE 2.

**No·vo·cain** (nō′və-kān′). A trademark for the anesthetic procaine hydrochloride.

**now** (nou) *adv.* [ME < OE *nū.*] **1.** At the present time <can't stop *now*> **2.** At once : IMMEDIATELY <Go *now.*> **3.** In the immediate past : RECENTLY <They arrived just *now.*> **4.** In the immediate future : SOON <We are leaving just *now.*> **5.** At this point in the series of events : THEN <It was *now* snowing very hard.> **6.** Nowadays. **7.** In these circumstances : as things are. **8.** —Used esp. to introduce a command, reproof, or request <*Now* hear this.> *—conj.* Seeing that : SINCE <*Now* that you're well, you can go back to work.> *—n.* The present time or moment. *—adj.* **1.** *Informal.* Of the present time : CURRENT <the *now* governor> **2.** *Slang.* In tune with the latest trends <the *now* look in fashion> **—now and again (or then).** Occasionally.

**NOW account** (nou) *n.* [N(EGOTIATED) O(RDER OF) W(ITHDRAWAL).] A savings account against which drafts can be written and which usu. bears interest.

**now·a·days** (nou′ə-dāz′) *adv.* [ME *nou a daies*, on this day, at the present time.] During the present time.

**no·way** (nō′wā′) *also* **no·ways** (-wāz′) *adv.* Nowise.
**no way** *interj. Informal.* —Used to indicate definite negation.
**no·ways** (nō′wāz′) *var. of* NOWAY.
**no·where** (nō′hwâr′, -wâr′) *adv.* **1.** Not anywhere. **2.** To no place or result <We got *nowhere* with our complaint.> —*n.* A remote or unknown place <lost in the middle of *nowhere*>
**no·wheres** (nō′hwârz′, -wârz′) *adv. Nonstandard.* Nowhere.
**no·whith·er** (nō′hwĭth′ər, -wĭth′-) *adv.* In no definite direction.
**no-win** (nō′wĭn′) *adj.* Incapable of affording victory or success <a *no-win* predicament>
**no·wise** (nō′wīz′) *adv.* [ME : *no*, no + *wise*, way < OE *wīse.*] In no way, manner, or degree : not at all.
**nox·ious** (nŏk′shəs) *adj.* [Lat. *noxius* < *noxa*, damage.] **1.** Injurious to physical health <*noxious* fumes> **2.** Harmful to the mind or morals <*noxious* propaganda><*noxious* gossip> **—nox′ious·ly** *adv.* **—nox′ious·ness** *n.*
**noz·zle** (nŏz′əl) *n.* [Dim. of NOSE.] **1.** A projecting part with an opening, as at the end of a hose, through which something is discharged. **2.** *Slang.* The nose.
**Np** *symbol for* NEPTUNIUM.
**nth** (ĕnth) *adj.* **1.** Pertaining to an indefinitely large ordinal number <four to the *nth* power> **2.** Highest : utmost <appalled to the *nth* degree>
**nth root** *n.* ROOT 9a.
**nu** (no͞o, nyo͞o) *n.* [Gk., of Semitic orig.; akin to Heb. *nūn*, nun.] The 13th letter of the Greek alphabet. —See table at ALPHABET.
**nu·ance** (no͞o-äns′, nyo͞o-, no͞o′äns′, nyo͞o′-) *n.* [Fr., shade < OFr. < *nuer*, to show different shades < *nue*, cloud < Lat. *nubes.*] **1.** A subtle or slight gradation or variation <*nuances* of mood and color> **2.** A subtle aspect or quality <detected *nuances* of humor in the writing> **3.** Awareness of, sensibility to, or the ability to express delicate shadings, as of feeling or meaning <"There was little opportunity for conceptual approaches, consecutive action, or a sense of *nuance*" —Henry Kissinger>
**nub** (nŭb) *n.* [Var. of *knub* < MLG *knubbe*, knot of a tree.] **1.** A protuberance or knob. **2.** A small lump. **3.** The essence : core <the *nub* of the matter> **—nub′by** *adj.*
**Nu·ba** (no͞o′bə, nyo͞o′-) *n., pl.* **Nuba. 1.** A Nubian. **2.** A member of any of several Negroid tribes of southern Sudan. **3.** The language of the Nuba.
**nub·bin** (nŭb′ĭn) *n.* [Dim. of NUB.] **1.** Something, as an ear of corn, that is imperfectly developed or stunted. **2.** A small chunk or piece. **3.** NUB 3.
**nub·ble** (nŭb′əl) *n.* [Dim. of NUB.] A small protuberance or lump. **—nub′bly** *adj.*
**Nu·bi·an** (no͞o′bē-ən, nyo͞o′-) *n.* **1.** A native or resident of Nubia. **2.** Any of the languages of Nubia. **—Nu′bi·an** *adj.*
**nu·bile** (no͞o′bĭl, -bīl′, nyo͞o′-) *adj.* [Fr. < Lat. *nubilis* < *nubere*, to take a husband.] Of a marriageable age or condition. —Used of young women. **—nu·bil′i·ty** (no͞o-bĭl′ĭ-tē, nyo͞o-) *n.*
**nu·cel·lus** (no͞o-sĕl′əs, nyo͞o-) *n., pl.* **-cel·li** (-sĕl′ī′) [NLat., alteration of Lat. *nucella*, dim. of *nux*, nut.] *Bot.* The center of the rudimentary seed of a plant, containing the plant's embryo sac. **—nu·cel′lar** *adj.*
**nu·cha** (no͞o′kə, nyo͞o′-) *n.* [ME < Med. Lat. < Ar. *nukhā′.*] The nape of the neck. **—nu′chal** *adj.*
**nucle-** *pref. var. of* NUCLEO-.
**nu·cle·ar** (no͞o′klē-ər, nyo͞o′-) *adj.* [< NUCLEUS.] **1.** *Biol.* Of, relating to, or forming a nucleus. **2.** *Physics.* Of or concerning atomic nuclei. **3.** Using or derived from the energy of atomic nuclei : ATOMIC <*nuclear*-powered submarines>> **4.** Of, using, or having atomic or hydrogen bombs.
**nuclear emulsion** *n. Physics.* A photographic emulsion used to detect and visually display the paths of charged subatomic particles, esp. of charged cosmic-ray particles.
**nuclear energy** *n.* The energy released by a nuclear reaction, esp. by fission, fusion, or radioactive decay.
**nuclear family** *n.* A self-contained family unit including a mother and father and their children.
**nuclear fission** *n. Physics.* FISSION 2.
**nuclear force** *n.* Strong interaction.
**nuclear fusion** *n. Physics.* FUSION 5.
**nuclear magnetic resonance** *n.* The magnetic resonance of an atomic nucleus.
**nuclear magneton** *n. Physics.* A unit of the magnetic moment of the nucleon.
**nuclear medicine** *n.* The branch of medicine that makes use of radioisotopes in diagnosis and therapy.
**nuclear physics** *n. (used with a sing. verb).* Scientific study of the forces, reactions, and internal structures of atomic nuclei.
**nuclear reaction** *n.* A reaction that alters the energy, composition, or structure of an atomic nucleus.
**nuclear reactor** *n.* Any of several devices in which a chain reaction is initiated and controlled with the consequent production of heat typically used for power generation and the neutrons and fission products used for experimental and medical purposes.
**nuclear resonance** *n.* The resonance absorption of a gamma ray by an atomic nucleus identical to the nucleus that emitted the ray.
**nu·cle·ase** (no͞o′klē-ās′, -āz′, nyo͞o′-) *n.* Any of several enzymes that hydrolize nucleic acids.
**nu·cle·ate** (no͞o′klē-ĭt, nyo͞o′-) *adj.* Having a nucleus or nuclei. —*v.* (-āt′) **-at·ed, -at·ing, -ates.** —*vt.* **1.** To bring together into a nucleus. **2.** To act as a nucleus for. —*vi.* To form a nucleus. **—nu′cle·a′tion** *n.* **—nu′cle·a′tor** *n.*
**nu·cle·i** (no͞o′klē-ī′, nyo͞o′-) *n. var. pl. of* NUCLEUS.
**nu·cle·ic acid** (no͞o-klē′ĭk, nyo͞o-) *n.* A member of either of two groups of complex compounds made up of purines, pyrimidines, carbohydrates, and phosphoric acid and found in all living cells.
**nucleo-** *or* **nucle-** *pref.* [< NUCLEUS.] **1.** Nucleus <*nucleo*plasm> **2.** Nucleic acid <*nucleo*protein>
**nu·cle·o·late** (no͞o′klē-ə-lāt′, nyo͞o′-) *also* **nu·cle·o·lat·ed** (-lā′tĭd) *adj.* [NUCLEOL(US) + -ATE.] Having a nucleolus or nucleoli.
**nu·cle·o·lus** (no͞o-klē′ə-ləs, nyo͞o-) *n., pl.* **-li** (-lī′) [Lat., dim. of *nucleus*, kernel < *nux*, nut.] *Biol.* **1.** A small, usu. round body in the nucleus of a cell, composed of protein and ribonucleic acid. **2.** A discrete cellular particle that resembles a nucleolus, other than a chromosome. **—nu·cle′o·lar** (-lər) *adj.*
**nu·cle·on** (no͞o′klē-ŏn′, nyo͞o′-) *n.* A proton or a neutron, esp. as part of an atomic nucleus. **—nu′cle·on′ic** *adj.*
**nu·cle·on·ics** (no͞o′klē-ŏn′ĭks, nyo͞o′-) *n.* [< NUCLEON.] *(sing. in number).* Nuclear energy technology.
**nucleon number** *n. Physics.* Mass number.
**nu·cle·o·phile** (no͞o′klē-ə-fīl′, nyo͞o′-) *n.* A substance that donates electrons.
**nu·cle·o·plasm** (no͞o′klē-ə-plăz′əm, nyo͞o′-) *n.* Protoplasm of a cell nucleus. **—nu′cle·o·plas′mic, nu′cle·o·plas·mat′ic** (-ō-plăz-măt′ĭk) *adj.*
**nu·cle·o·pro·tein** (no͞o′klē-ō-prō′tēn′, -prō′tē-ĭn, nyo͞o′-) *n.* Any of a group of substances composed of a protein and a nucleic acid and found in all living cells and viruses.
**nu·cle·o·side** (no͞o′klē-ə-sīd′, nyo͞o′-) *n.* A compound made of a sugar and a purine or pyrimidine base, esp. one obtained by hydrolysis of a nucleic acid, as adenosine.
**nu·cle·o·some** (no͞o′klē-ə-sōm′) *n.* Any of the basic globular subunits of chromatin composed of DNA and histone. **—nu′cle·o·som′al** (-sō′məl) *adj.*
**nu·cle·o·syn·the·sis** (no͞o′klē-ō-sĭn′thĭ-sĭs, nyo͞o′-) *n.* The process by which heavier chemical elements are synthesized from hydrogen nuclei in the interiors of stars. **—nu′cle·o·syn·thet′ic** (-sĭn-thĕt′ĭk) *adj.*
**nu·cle·o·tide** (no͞o′klē-ə-tīd′, nyo͞o′-) *n.* Any of various organic compounds composed of a nucleoside combined with phosphoric acid.
**nu·cle·us** (no͞o′klē-əs, nyo͞o′-) *n., pl.* **-cle·i** (-klē-ī′) *or* **-cle·us·es.** [Lat., kernel < *nux*, nut.] **1.** A central or essential part around which other parts are grouped or collected : CORE. **2.** Something regarded as a basis for future development and growth : KERNEL. **3.** *Biol.* A complex, usu. spherical protoplasmic body within a living cell that contains the cell's hereditary material and controls its metabolism, growth, and reproduction. **4.** *Bot.* **a.** The nucellus. **b.** The central kernel of a seed or nut. **c.** The central point of a starch granule. **5.** *Anat.* A group of nerve cells or localized mass of gray matter in the brain, where nerve fibers interconnect. **6.** *Physics.* The positively charged central region of an atom, made up of protons and neutrons and containing almost all of the mass of the atom. **7.** *Chem.* A group of atoms chemically bound in a structure resistant to alteration in chemical reactions. **8.** *Astron.* **a.** The central portion of the head of a comet. **b.** The central or brightest part of a nebula or of a galaxy. **9.** *Meteorol.* A particle on which water vapor molecules accumulate in free air to form a droplet or ice crystal.
**nu·clide** (no͞o′klīd′, nyo͞o′-) *n. Physics.* An atomic nucleus specified by its atomic number, atomic mass, and energy state. **—nu·clid′ic** (no͞o-klĭd′ĭk, nyo͞o-) *adj.*
**nude** (no͞od, nyo͞od) *adj.* **nud·er, nud·est.** [Lat. *nudus.*] **1.** Devoid of clothing : NAKED. **2.** Lacking any of various legal requisites, such as evidence. **3.** Of a neutral shade <*nude* stockings> —*n.* **1.** The nude human figure, esp. in artistic representation. **2.** The condition of being unclothed <swam in the *nude*> **—nude′ly** *adv.* **—nude′ness, nu′di·ty** (no͞o′dĭ-tē, nyo͞o′-) *n.*
**nudge**[1] (nŭj) *vt.* **nudged, nudg·ing, nudg·es.** [Prob. of Scand. orig.] **1.** To push against gently, esp. in order to attract attention or give a signal. **2.** To approach <*nudging* middle age> **3.** To urge to act : PROD <had to be *nudged* about the debt> —*n.* A gentle push.
**nudge**[2] *also* **nudzh** (no͝oj) [< Yiddish *nudyen*, to pester, bore < R. *nudnyi*, tedious, boring < *nudnost′*, tedium.] *Slang.* —*n.* One who persistently pesters, annoys, or complains. —*v.* **nudged, nudg·ing, nudg·es** *also* **nudzhed, nudzh·ing, nudzh·es.** —*vt.* To annoy or pester persistently. —*vi.* To carp or complain persistently.
**nudi-** *pref.* [< Lat. *nudus*, nude.] Naked : bare <*nudi*branch>
**nu·di·branch** (no͞o′də-brăngk′, nyo͞o′-) *n.* [< NLat. *Nudibranchia*, order name : NUDI- + Gk. *brankhia*, gills.] A mollusk of the order

Nudibranchia, as a sea slug. **—nu'di·bran'chi·ate** (-brăng'kē-ĭt), **nu'di·bran'chi·an** (-kē-ən) *adj.* & *n.*
**nud·ism** (no͞o'dĭz'əm, nyo͞o'-) *n.* The belief in or practice of going nude. **—nud'ist** *n.*
**nud·nik** *also* **nud·nick** (no͝od'nĭk) *n.* [Yiddish < Pol. *nudny,* tiresome < *nuda,* boredom.] *Slang.* A bothersome or boring person.
**nudzh** (no͝oj) *n.* & *v. Slang. var. of* NUDGE².
**nu·ga·to·ry** (no͞o'gə-tôr'ē, -tōr'ē, nyo͞o'-) *adj.* [Lat. *nugatorius* < *nugari,* to trifle < *nugae,* jokes.] **1.** Unimportant : trifling. **2.** Lacking force : INVALID.
**nug·get** (nŭg'ĭt) *n.* [Dial. E., dim. of *nug,* lump.] **1.** A small solid lump, esp. of gold. **2.** Something resembling a nugget, as in value <*nuggets* of insight>
**nui·sance** (no͞o'səns, nyo͞o'-) *n.* [ME *noisaunce* < OFr. *nuisance* < *nuire,* to harm < Lat. *nocēre.*] **1.** Something that is inconvenient or vexatious : BOTHER. **2.** *Law.* A use of property or course of conduct that interferes with the legal rights of others by causing damage, annoyance, or inconvenience.
**nuisance tax** *n.* A tax levied on separate purchases and collected directly from the purchaser.
**nuke** (no͞ok, nyo͞ok) [Shortening and alteration of NUCLEAR.] *Slang.* —*n.* **1.** A nuclear weapon. **2.** A nuclear-powered electric generating plant. —*vt.* **nuked, nuk·ing, nukes.** To attack with nuclear weapons <threatened to *nuke* the enemy>
**null** (nŭl) *adj.* [OFr. *nul* < Lat. *nullus* : *ne,* not + *ullus,* any.] **1.** Having no legal force : INVALID. **2.** Of no consequence, effect, or value : INSIGNIFICANT. **3.** Amounting to nothing : NONEXISTENT. **4.** *Math.* Of or relating to a set having no members or to zero magnitude. —*vt.* **nulled, null·ing, nulls.** To make null. —*n.* **1.** Nothing : zero. **2.** An instrumental reading of zero.
**nul·lah** (nŭl'ə) *n.* [Hindi *nālā,* rivulet, prob. of Dravidian orig.] A gully or ravine.
**null and void** *adj.* NULL 1.
**null character** *n.* A data control character that fills computer time by adding nonsignificant zeros to a data sequence.
**nul·li·fi·ca·tion** (nŭl'ə-fĭ-kā'shən) *n.* **1.** The act of nullifying or state of being nullified. **2.** A state's failure or refusal to recognize or enforce a U.S. law within its borders. **—nul'li·fi·ca'tion·ist** *n.*
**nul·li·fi·er** (nŭl'ə-fī'ər) *n.* **1.** One that nullifies. **2.** One who believes in nullification, esp. by a U.S. state.
**nul·li·fy** (nŭl'ə-fī') *vt.* **-fied, -fy·ing, -fies.** [LLat. *nullificare,* to despise : Lat. *nullus,* none + *facere,* to make.] **1.** To make null : INVALIDATE. **2.** To counteract the force or efficacy of.
**nul·lip·a·ra** (nə-lĭp'ər-ə) *n.* [Lat. *nullus,* none + -PARA.] A female who has not borne offspring. **—nul·lip'a·rous** *adj.*
**nul·li·ty** (nŭl'ĭ-tē) *n., pl.* **-ties. 1.** The quality or state of being null. **2.** Something that is null, esp. an act having no legal validity.
**numb** (nŭm) *adj.* **-er, -est.** [ME *nomen,* p. part. of *nimen,* to seize < OE *niman.*] **1.** Lacking the power to feel or move normally : BENUMBED <*numb* fingers and toes> **2.** Stunned or paralyzed, as from shock. —*vt.* & *vi.* **numbed, numb·ing, numbs.** To make or become numb. **—numb'ly** *adv.* **—numb'ness** *n.*
**num·ber** (nŭm'bər) *n.* [ME *nombre* < OFr. < Lat. *numerus.*] **1.** *Math.* **a.** A member of the set of positive integers, one of a series of symbols of unique meaning in a fixed order that can be derived by counting. **b.** A member of any of the further sets of mathematical objects, as negative integers and real numbers, that can be derived from the positive integers by mathematical induction. **2. numbers.** Arithmetic. **3. a.** A symbol used to represent a number. **b.** A numeral or series of numerals for reference or identification <a social security *number*> **4. a.** A total : sum <the *number* of inches in a foot> **b.** An indeterminate sum. ***usage:*** As a collective noun, *number* may take either a singular or plural verb. It takes a singular when preceded by *the,* as in *The number of students is small.* It takes a plural when preceded by *a,* as in *A number of the students are planning a demonstration.* **5.** Quantity of units or individuals <The gathering was small in *number.*> **6. numbers. a.** A large quantity or collection : MULTITUDE <*Numbers* of tourists visited the monument.> **b.** Numerical superiority. **7.** One item in a group or series considered esp. in numerical order. **8.** One of the separate offerings in a program of music. **9.** *Informal.* One singled out from a group for a particular characteristic. **10.** The indication, as by inflection, of the singularity or plurality of a linguistic form. **11. a. numbers.** Metrical periods or feet : VERSES. **b.** Measured rhythm in verse : METER. **12. numbers.** *Archaic.* Musical periods or measures. **13. numbers.** The numbers game. **14. Numbers** *(sing. in number).* —See table at BIBLE. —*v.* **-bered, -ber·ing, -bers.** —*vt.* **1.** To total in number or amount : add up to. **2.** To count or determine the number or amount of. **3.** To include in a group or category <We were *numbered* among the missing.> **4.** To mention one by one : ENUMERATE. **5.** To assign a number to. **6.** To limit or restrict in number <*numbered* the calories I was allowed> —*vi.* **1.** To count or call off numbers <*numbering* to five> **2.** To make up a group or number <The demonstrators *numbered* in the thousands.> **—do a number on.** *Slang.* **1.** To abuse or harm, esp. by trickery. **2.** To make fun of : RIDICULE. **—get (***or* **have) (someone's) number.** To determine or know someone's real character or motives. **—without (***or* **beyond) number.** Too many to be counted : COUNTLESS. **—num'ber·er** *n.*
**number cruncher** *n. Slang.* A computer able to perform complex and lengthy calculations. **—number crunching** *n.*
**num·ber·less** (nŭm'bər-lĭs) *adj.* Countless : innumerable.
**number line** *n.* A line that graphically expresses the real numbers as a series of points distributed about a point arbitrarily designated as zero and in which the magnitude of each number is represented by the distance of the corresponding point from zero.
**numbers game** *n.* A lottery in which bets are made on an unpredictable number, as a daily stock-exchange figure.
**numb·fish** (nŭm'fĭsh') *n.* The electric ray.
**numb·skull** (nŭm'skŭl') *n. var. of* NUMSKULL.
**nu·men** (no͞o'mən, nyo͞o'-) *n., pl.* **-mi·na** (-mə-nə) [Lat.] **1.** A presiding divinity or spirit of a place. **2.** A spirit held by animists to inhabit certain natural phenomena or objects. **3.** Creative energy.
**nu·mer·a·ble** (no͞o'mər-ə-bəl, nyo͞o'-) *adj.* [Lat. *numerabilis* < *numerare,* to count < *numerus,* number.] Capable of being counted.
**nu·mer·al** (no͞o'mər-əl, nyo͞o'-) *n.* [LLat. *numeralis* < Lat. *numerus,* number.] **1.** A symbol or mark that represents a number. **2. numerals. a.** The numbers, usu. the last two digits, indicating by year a graduating class in a school or college. **b.** Such numbers awarded for distinction, as in sports. —*adj.* Of, relating to, or expressing numbers. **—nu'mer·al·ly** *adv.*
**nu·mer·ar·y** (no͞o'mə-rĕr'ē, nyo͞o'-) *adj.* [Med. Lat. *numerarius* < Lat. *numerus,* number.] Of or relating to a number or numbers.
**nu·mer·ate** (no͞o'mə-rāt', nyo͞o'-) *vt.* **-at·ed, -at·ing, -ates.** [Lat. *numerare, numerat-,* to count < *numerus,* number.] To enumerate : count. **—nu'mer·a'tion** (-rā'shən, nyo͞o'-) *n.*
**nu·mer·a·tor** (no͞o'mə-rā'tər, nyo͞o'-) *n.* **1.** *Math.* **a.** The expression written above the line in a common fraction. **b.** An expression to be divided by another : DIVIDEND. **2.** One that numbers : ENUMERATOR.
**nu·mer·ic** (no͞o-mĕr'ĭk, nyo͞o-) *n.* A number or numeral.
**nu·mer·i·cal** (no͞o-mĕr'ĭ-kəl, nyo͞o-) *also* **nu·mer·ic** (-mĕr'ĭk) *adj.* [< Lat. *numerus,* number.] **1.** Of or relating to a number or series of numbers <arranged in *numerical* order> **2.** Denoting number or a number, as a symbol. **3.** Expressed in or counted by numbers <*numerical* force> **—nu·mer'i·cal·ly** *adv.*
**numerical analysis** *n.* The study of approximate solutions to mathematical problems, taking into account the extent of possible errors.
**numerical control** *n.* Control of a process or machine by a digital computer.
**numerical taxonomy** *n.* A branch of taxonomy that deals with the quantitative relationships between the elements classified.
**numerical value** *n. Math.* The absolute value of a number regardless of sign.
**nu·mer·ol·o·gy** (no͞o'mə-rŏl'ə-jē, nyo͞o'-) *n.* [Lat. *numerus,* number + -LOGY.] The study of the occult meanings of numbers and of their purported influence on human life. **—nu'mer·o·log'i·cal** (-mər-ə-lŏj'ĭ-kəl) *adj.* **—nu'mer·ol'o·gist** *n.*
**nu·mer·ous** (no͞o'mər-əs, nyo͞o'-) *adj.* [Lat. *numerosus* < *numerus,* number.] Consisting of many persons or items. ***usage:*** Since *numerous* is not a pronoun in standard English, expressions like *numerous of the firefighters* should be avoided. **—nu'mer·ous·ly** *adv.* **—nu'mer·ous·ness** *n.*
**nu·mi·na** (no͞o'mə-nə, nyo͞o'-) *n. pl. of* NUMEN.
**nu·mi·nous** (no͞o'mə-nəs, nyo͞o'-) *adj.* [< Lat. *numen,* numen.] **1.** Of or pertaining to a numen : SUPERNATURAL. **2.** Spiritually elevated : HOLY. **3.** Impossible to describe or understand : MYSTERIOUS.
**nu·mis·mat·ic** (no͞o'mĭz-măt'ĭk, -mĭs-, nyo͞o'-) *adj.* **1.** Of or relating to coins or currency. **2.** Of or relating to numismatics. **—nu'mis·mat'i·cal·ly** *adv.*
**nu·mis·mat·ics** (no͞o'mĭz-măt'ĭks, -mĭs-, nyo͞o'-) *n.* [< *numismatic,* pertaining to coins < Fr. *numismatique* < Lat. *numisma,* coin < Gk. *nomisma,* current coin < *nomizein,* to have in use < *nomos,* custom.] *(sing. in number).* The study or collection of currency, coins, and often medals. **—nu·mis'ma·tist** (no͞o-mĭz'mə-tĭst, -mĭs'-, nyo͞o-) *n.*
**nu·mis·ma·tol·o·gy** (no͞o-mĭz'mə-tŏl'ə-jē, -mĭs'-, nyo͞o-) *n.* Numismatics.
**num·mu·lar** (nŭm'yə-lər) *adj.* [Fr. *nummulaire* < Lat. *nummulus,* dim. of *nummus,* coin.] Coin-shaped.
**num·mu·lite** (nŭm'yə-līt') *n.* [NLat. *Nummulites,* genus name < Lat. *nummulus,* dim. of *nummus,* coin.] A tiny, chiefly marine, mostly extinct protozoan of the family Nummulitidae, with a coin-shaped shell. **—num'mu·lit'ic** (-lĭt'ĭk) *adj.*
**num·skull** *also* **numb·skull** (nŭm'skŭl') *n.* A stupid person.
**nun¹** (nŭn) *n.* [ME *nonne* < OE *nunne* and OFr. *nonne,* both < LLat. *nonna.*] A woman who belongs to a religious order devoted to religious service or meditation, usu. under vows of poverty, chastity, and obedience.
**nun²** (no͝on) *n.* [Heb. *nūn.*] The 14th letter of the Hebrew alphabet. —See table at ALPHABET.

**Nunc Di·mit·tis** (nŭngk′ dĭ-mĭt′ĭs, nōōngk′) *n.* The canticle of Simeon, beginning *"Nunc dimittis servum tuum"* ("Now lettest thou thy servant depart").

**nun·ci·a·ture** (nŭn′sē-ə-chōōr′, -chər, nōōn′-) *n.* [Ital. *nunciatura* < *nuncio,* nuncio < Lat. *nuntius,* messenger.] The office or term of office of a nuncio.

**nun·ci·o** (nŭn′sē-ō′, nōōn′-) *n., pl.* **-os.** [Ital. < Lat. *nuntius,* messenger.] A papal ambassador or representative.

**nun·cle** (nŭng′kəl) *n.* [Alteration of UNCLE.] *Archaic.* An uncle.

**nun·cu·pa·tive** (nŭn′kyə-pā′tĭv, nŭng′-, nŭn-kyōō′pə-tĭv) *adj.* [Med. Lat. *nuncupativus* < Lat. *nuncupare,* to name : *nomen,* name + *capere,* to take.] *Law.* Designating a will delivered orally to witnesses rather than written.

**nun·ner·y** (nŭn′ə-rē) *n., pl.* **-ies.** A convent.

**nup·tial** (nŭp′shəl, -chəl) *adj.* [Lat. *nuptialis* < *nuptiae,* wedding < *nubere,* to take a husband.] **1.** Of or relating to marriage or the wedding ceremony. **2.** Of, relating to, or occurring in the mating season. —*n. often* **nuptials.** A wedding ceremony. **—nup′tial·ly** *adv.*

**nurd** (nûrd) *n. var. of* NERD.

**nurse** (nûrs) *n.* [ME *norice,* nursemaid < OFr. *norrice* < LLat. *nutricia* < *nutrire,* to nourish.] **1.** One who is trained and licensed to care for the sick or disabled esp. under the supervision of a physician. **2. a.** A woman employed to take care of a child : NURSEMAID. **b.** A woman employed to suckle children other than her own : WET NURSE. **3.** One that serves as a nurturing or fostering means or influence. **4.** A worker ant or bee that cares for the young in an insect colony. —*v.* **nursed, nurs·ing, nurs·es.** —*vt.* **1.** To feed at the breast : SUCKLE. **2.** To serve as a nurse for. **3.** To try to cure or treat <*nurse* a cold> **4.** To take special care of : FOSTER <*nursed* the theater group through hard times> **5.** To bear in the mind : HARBOR <*nursing* resentment> **6.** To treat carefully, esp. in order to prevent pain : FAVOR <*nursed* a sore foot> **7.** To consume slowly, esp. in order to conserve. —*vi.* **1.** To take nourishment from the breast : SUCKLE. **2.** To serve as a nurse. **—nurs′er** *n.*

**nurse·maid** (nûrs′mād′) *n.* NURSE 2a.

**nurse practitioner** *n.* A registered nurse with preparation in a specialized education program enabling him or her to perform as a primary health care provider, as in obstetrics and gynecology.

**nurs·er·y** (nûr′sə-rē, nûrs′rē) *n., pl.* **-ies.** [ME *noricerie* < OFr. *norricerie* < *norrice,* nursemaid. —see NURSE.] **1.** A room or area set apart for the use of children. **2. a.** A nursery school. **b.** A place for the temporary care of children. **3.** A place where plants are grown for sale, transplanting, or experimentation. **4.** A place in which something is produced, fostered, or developed.

**nurs·er·y·maid** (nûr′sə-rē-mād′, nûrs′rē-) *n.* A nursemaid.

**nurs·er·y·man** (nûr′sə-rē-mən, nûrs′rē-) *n.* One who owns or works in a nursery for plants.

**nursery rhyme** *n.* A short rhymed poem or tale for children.

**nursery school** *n.* A school for children who are not old enough to attend kindergarten.

**nurse's aide** *n.* One who assists trained nurses, as by giving general patient care.

**nurs·ing** (nûr′sĭng) *n.* **1.** A nurse's occupation. **2.** A nurse's tasks.

**nursing home** *n.* A private hospital for the care of the aged or chronically ill.

**nurs·ling** (nûrs′lĭng) *n.* **1.** A nursing infant or young animal. **2.** A carefully nurtured person or thing.

**nur·tur·ance** (nûr′chər-əns) *n.* Provision of loving care and attention. **—nur′tur·ant** *adj.*

**nur·ture** (nûr′chər) *n.* [ME *norture* < OFr. < LLat. *nutritia,* a suckling < Lat. *nutrire,* to suckle.] **1.** Something that nourishes : SUSTENANCE. **2.** Upbringing : rearing. **3.** *Biol.* The sum of environmental influences and conditions acting on an organism. —*vt.* **-tured, -turing, -tures. 1.** To feed : nourish. **2.** To train : educate. **3.** To help grow or develop : CULTIVATE. **—nur′tur·er** *n.*

**nut** (nŭt) *n.* [ME *note* < OE *hnutu.*] **1. a.** A hard-shelled, solid-textured, one-celled fruit, as an acorn or a hazelnut, that does not split open. **b.** A seed borne in a fruit having a hard shell, as the peanut or almond. **c.** The kernel of such fruits or seeds. **2.** *Slang.* **a.** A crazy or eccentric person. **b.** An enthusiast : buff <a skiing *nut*> **3.** *Informal.* A difficult endeavor or problem. **4.** *Slang.* A person's head. **5. a.** A ridge of wood at the top of the fingerboard or neck of a stringed instrument over which the strings pass. **b.** A device at the lower end of a bow, as of a violin, used for adjusting the hairs. **6.** A small block of wood or metal with a central, threaded hole that is designed to fit around and secure a bolt or screw. —*vi.* **nutted, nut·ting, nuts.** To hunt for or gather nuts. **—nut′ter** *n.*

**nu·ta·tion** (nōō-tā′shən, nyōō-) *n.* [Lat. *nutatio* < *nutare,* to nod.] **1.** The act of nodding the head. **2.** *Astron.* A small periodic motion of the celestial pole of the earth with respect to the pole of the ecliptic. **3.** *Bot.* A slight circular or curving movement in the stem of a plant caused by irregular growth rates of different parts. **—nu·ta′tion·al** *adj.*

**nut·crack·er** (nŭt′krăk′ər) *n.* **1.** An implement used to crack nuts, typically having two hinged usu. metal levers between which the nut is squeezed. **2. a.** A bird, *Nucifraga caryocatactes* of northern Eurasia. **b.** A related bird, *N. columbianus* of western North America. **c.** The nuthatch.

**nut·gall** (nŭt′gôl′) *n.* A nutlike swelling produced on a tree, as an oak, by certain parasitic wasps.

**nut·hatch** (nŭt′hăch′) *n.* [ME *notehatch* : *note,* nut + *hache,* hatchet, of Germanic orig.] Any of several small birds of the family Sittidae, with long sharp bills, noted for their insectlike ability to maneuver on tree trunks and branches.

**nut house** *n. Slang.* A mental institution.

**nut·let** (nŭt′lĭt) *n.* **1.** A small nut. **2.** The stone or pit of certain fruits, as the peach or cherry.

**nut·meat** (nŭt′mēt′) *n.* The edible kernel of a nut.

**nut·meg** (nŭt′mĕg′) *n.* [ME *notemugge,* prob. ult. < OFr. *nois mugede* < VLat. **nuce muscata* : Lat. *nux,* nut + Lat. *muscus,* musk.] **1.** An evergreen tree, *Myristica fragrans,* native to the East Indies. **2.** The hard aromatic seed of the nutmeg, grated or ground for use as a spice. **3.** A grayish to moderate brown.

**nut pick** *also* **nut·pick** (nŭt′pĭk′) *n.* A small sharp-pointed tool used for digging the meat from nuts.

**nut pine** *n.* The piñon.

**nu·tri·a** (nōō′trē-ə, nyōō′-) *n.* [Sp., var. of *lutra,* otter < Lat.] **1.** The coypu. **2.** The light-brown fur of the coypu.

**nu·tri·ent** (nōō′trē-ənt, nyōō′-) *n.* [< Lat. *nutriens, nutrient-,* pr.part. of *nutrire,* to feed.] Something that nourishes, esp. a nourishing ingredient in a food. —*adj.* Nutritious.

**nu·tri·ment** (nōō′trə-mənt, nyōō′-) *n.* [Lat. *nutrimentum* < *nutrire,* to feed.] **1.** Something that nourishes : FOOD. **2.** Something that promotes growth or development. **—nu′tri·men′tal** (-mĕn′tl) *adj.*

**nu·tri·tion** (nōō-trĭsh′ən, nyōō-) *n.* [OFr. < LLat. *nutritio* < Lat. *nutrire,* to feed.] The process of nourishing or being nourished, esp. the process by which a living organism assimilates food and uses it for growth and tissue replacement. **—nu·tri′tion·al** *adj.* **—nu·tri′tion·al·ly** *adv.*

**nu·tri·tion·ist** (nōō-trĭsh′ə-nĭst, nyōō-) *n.* A specialist in the study of nutrition.

**nu·tri·tious** (nōō-trĭsh′əs, nyōō-) *adj.* [Lat. *nutritius* < *nutrix,* nurse.] Providing nourishment. **—nu·tri′tious·ly** *adv.* **—nu·tri′tious·ness** *n.*

☆ ***syns:*** NUTRITIOUS, NOURISHING, NUTRIENT, NUTRITIVE *adj. core meaning* : providing nourishment <always ate *nutritious* meals>

**nu·tri·tive** (nōō′trĭ-tĭv, nyōō′-) *adj.* [ME *nutritif* < OFr. < LLat. *nutritivus* < Lat. *nutrire,* to feed.] **1.** Nutritious. **2.** Of or relating to nutrition. **—nu′tri·tive·ly** *adv.*

**nuts** (nŭts) [< NUT.] *Slang.* —*adj.* **1.** Crazy : insane. **2.** Extremely enthusiastic <*nuts* about baseball> —*interj.* —Used to express disappointment, refusal, or contempt.

**nuts and bolts** *pl.n.* Basic working components or practical aspects <studied the *nuts and bolts* of the plan>

**nut·shell** (nŭt′shĕl′) *n.* The shell enclosing the meat of a nut. **—in a nutshell.** In a few words : CONCISELY.

**nut·ty** (nŭt′ē) *adj.* **-ti·er, -ti·est. 1.** Producing or containing nuts. **2.** Having a nutlike flavor. **3.** *Informal.* Crazy. **—nut′ti·ly** *adv.* **—nut′ti·ness** *n.*

**nux vom·i·ca** (nŭks vŏm′ĭ-kə) *n.* [Med. Lat., emetic nut : Lat. *nux,* nut + Lat. *vomere,* to vomit.] A tree native to southeastern Asia, *Strychnos nux-vomica,* bearing poisonous seeds that are the source of strychnine and brucine.

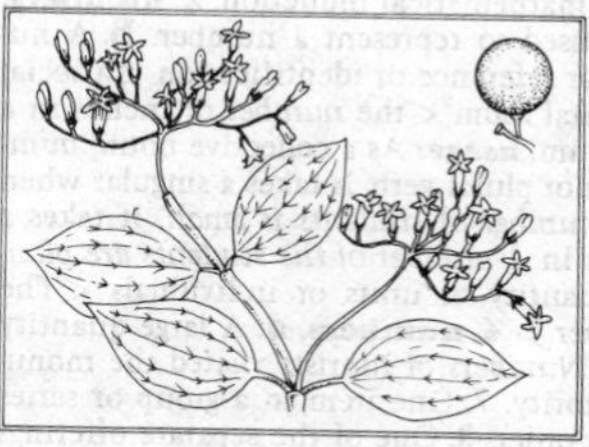
**nux vomica**

**nuz·zle** (nŭz′əl) *v.* **-zled, -zling, -zles.** [ME *noselen,* to bring the nose close to the ground < *nose,* nose.] —*vt.* **1.** To rub or push against gently with or as if with the nose or snout. **2.** To root or move with the snout. —*vi.* **1.** To make rubbing or pressing motions with or as if with the nose or snout. **2.** To nestle together. **—nuz′zler** *n.*

**Ny·an·ja** (nē-ăn′jə) *n.* A Bantu language of Malawi.

**nyc·ta·lo·pi·a** (nĭk′tə-lō′pē-ə) *n.* [LLat. < Gk. *nuktalōps,* night-blind : *nux,* night + *alaos,* blind + *ops,* eye.] Night blindness. **—nyc′ta·lo′pic** (-lō′pĭk, -lŏp′ĭk) *adj.*

**nyc·tit·ro·pism** (nĭk-tĭt′rə-pĭz′əm) *n.* [Gk. *nux, nukt-,* night + -TROPISM.] *Bot.* The tendency of the leaves of certain plants to change their position at nightfall. **—nyc′ti·tro′pic** (-tĭ-trō′pĭk, -trŏp′ĭk) *adj.*

**ny·lon** (nī'lŏn') *n.* [Coined by its inventors, E.I. duPont de Nemours & Co., Inc.] **1.** Any of a family of high-strength, resilient synthetic materials whose long-chain molecule contains the recurring amide group CONH. **2.** Cloth or yarn made from nylon. **3. nylons.** Stockings made of nylon.

**nymph** (nĭmf) *n.* [ME *nimphe* < Lat. *nympha* < Gk. *numphē.*] **1.** *Gk. & Rom. Myth.* One of numerous female spirits inhabiting and animistically representing features of nature, as woodlands and waters. **2.** A girl, esp. a beautiful one. **3.** One of an insect's young that undergoes incomplete metamorphosis. **—nymph'al** (nĭm'fəl) *adj.*

**nym·pha** (nĭm'fə) *n., pl.* **-phae** (-fē) [Lat., nymph < Gk. *numphē.*] **1.** NYMPH 3. **2. nymphae.** *Anat.* The labia minora.

**nym·pha·lid** (nĭm'fə-lĭd) *n.* [< NLat. *Nymphalidae,* family name < *Nymphalis,* genus name < Lat. *nympha,* nymph < Gk. *numphē.*] Any of various medium to large, often brilliantly colored butterflies of the family Nymphalidae, found worldwide.

**nym·phet** (nĭm-fĕt', nĭm'fĭt) *n.* **1.** A young nymph. **2.** A pubescent girl regarded as sexually desirable.

**nym·pho·lep·sy** (nĭm'fə-lĕp'sē) *n., pl.* **-sies.** [< NYMPHOLEPT.] **1.** A frenzy believed by ancient peoples to have been induced by nymphs. **2.** An emotional frenzy. **—nym'pho·lep'tic** (-lĕp'tĭk) *adj.*

**nym·pho·lept** (nĭm'fə-lĕpt') *n.* [Gk. *numpholēptos,* raptured : *numphē,* nymph + *lambanein,* to seize.] One who is in a state of nympholepsy.

**Ny·norsk** (nōō-nôrsk') *n.* [Norw., new Norwegian.] Landsmål.

**nys·tag·mus** (nĭ-stăg'məs) *n.* [Gk. *nustagmos,* drowsiness < *nustazein,* to be sleepy.] *Pathol.* An involuntary spasmodic motion of the eyeball. **—nys·tag'mic** *adj.*

# Oo

**o** *or* **O** (ō) *n., pl.* **o's** *or* **O's. 1.** The 15th letter of the English alphabet. **2.** A speech sound represented by the letter *o.* **3.** The 15th in a series. **4.** Something shaped like the letter O. **5.** A zero.

**O¹** (ō) *interj.* **1.** —Used before a name in formal address <Have mercy, O Lord.> **2.** —Used as an expression of surprise or strong emotion <O heavens!> ***usage:*** *O* as used in religious or poetic invocations is always capitalized and never set off by punctuation, as in *O mighty ocean.* The interjection *oh,* however, whether functioning as an independent element or part of a sentence, is capitalized only when it is the first word of a sentence and is usu. set off by punctuation: *Oh, I see.*

**O²** *symbol for* OXYGEN.

**-o-** [ME < OFr. < Lat. < Gk., thematic vowel of nouns and adjectives used in combination.] —Used as a connective to join word elements <acidophilic>

**oaf** *n.* [Obs. *aufe,* elf < ON *alfr.*] **1.** A dull, stupid person. **2.** A big clumsy person. **—oaf'ish** *adj.* **—oaf'ish·ly** *adv.* **—oaf'ish·ness** *n.*

**oak** (ōk) *n.* [ME *ok* < OE *āc.*] **1.** Any of various deciduous or evergreen trees or shrubs of the genus *Quercus,* bearing acorns as fruit. **2.** The wood of the oak. **3.** A tree or shrub, as the poison oak, that resembles the oak. **—oak'en** *adj.*

**oak apple** *n.* A gall that is produced on oak trees by the larva of a gall wasp.

**oak leaf cluster** *n.* A U.S. military decoration of bronze or silver oak leaves and acorns that is added to various medals in recognition of a subsequent decoration with the same medal.

**oa·kum** (ō'kəm) *n.* [ME *okum* < OE *ācumba* : *ā-,* off + *cemban,* to comb.] Loose hemp or jute fiber, occas. impregnated with tar or pitch, used principally to caulk seams in wooden ships and for packing pipe joints.

**oak wilt** *n.* A disease of oak trees caused by a fungus, *Chalara quercina,* and often resulting in wilting and dropping of leaves.

**oar** (ôr, ōr) *n.* [ME *or* < OE *ār.*] **1.** A long, thin, usu. wooden pole with a blade at one end, used to row and occas. to steer a boat. **2.** One using an oar : OARSMAN. *—v.* **oared, oar·ing, oars.** *—vt.* **1.** To propel with or as if with oars. **2.** To traverse with or as if with oars <*oared* the lake in half an hour> *—vi.* To move forward by or as if by rowing. **—oared** *adj.*

**oar·fish** (ôr'fĭsh', ōr'-) *n., pl.* **oarfish** or **-fish·es.** A marine fish, *Regalecus glesne,* with a slender body up to 30 feet or approx. 10 meters long, a dorsal fin extending the entire body length, and red-tipped rays above the head.

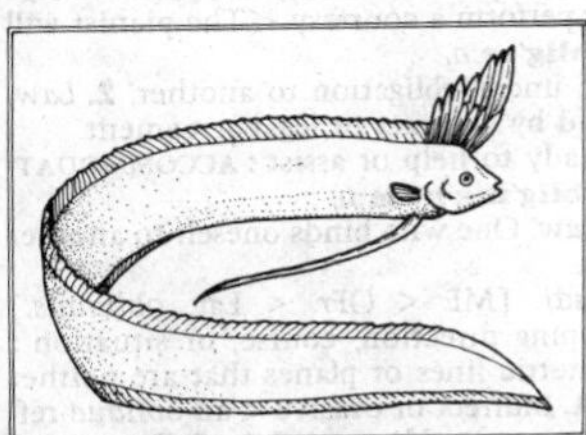

**oarfish**
*20–30 feet long*

**oar·lock** (ôr'lŏk', ōr'-) *n.* An often U-shaped device used as a fulcrum to hold an oar in place while rowing.

**oars·man** (ôrz'mən, ōrz'-) *n.* [*oar's,* possessive of OAR + MAN.] A rower.

**o·a·sis** (ō-ā'sĭs) *n., pl.* **-ses** (-sēz') [LLat. < Gk., of Egypt. orig.] **1.** A fertile or green spot, esp. one with water, in a desert. **2.** A shelter from surrounding unpleasantness <an *oasis* of calm in the busy airport>

**oast** (ōst) *n.* [ME *ost* < OE *āst.*] A kiln used to dry hops or malt or to dry and cure tobacco.

**oat** (ōt) *n.* [ME *ote* < OE *āte.*] **1. a.** A grass of the genus *Avena,* esp. *A. sativa,* widely cultivated for its edible seeds. **b.** *often* **oats** *(sing. or pl. in number).* The seeds of the oat, used as food and fodder. **2.** *Archaic.* A musical pipe made of an oat straw. **—oat·en** *adj.*

**oat·cake** (ōt'kāk') *n.* A flattened cake of baked oatmeal.

**oat grass** *n.* **1.** Any of various common meadow grasses of the genus *Arrhenatherum.* **2.** Any of several oatlike grasses.

**oath** (ōth) *n., pl.* **oaths** (ōthz, ōths) [ME *ooth* < OE *āð.*] **1. a.** A solemn declaration or promise, often calling upon God or a god as witness. **b.** The words or formula of an oath. **c.** Something declared or promised. **2.** An irreverent or blasphemous use of something held sacred, such as the name of God. **3.** An obscene utterance : SWEARWORD.

**oat·meal** (ōt'mēl') *n.* **1.** Meal of rolled or ground oats. **2.** A porridge made from rolled or ground oats.

**ob-** *pref.* [NLat. < Lat., toward, against < *ob,* toward.] Inverse : inversely <obcordate>

**O·ba·di·ah** (ō'bə-dī'ə) *n.* [Heb. *Ōbhadyāh.*] —See table at BIBLE.

**ob·bli·ga·to** (ŏb'lĭ-gä'tō) [Ital., p.part. of *obbligare,* to obligate < Lat. *obligare,* to oblige.] *Mus. —adj.* Not to be left out : COMPULSORY. —Used as direction. *—n., pl.* **-tos** *or* **-ti** (-tē). An obbligato musical accompaniment.

**ob·cor·date** (ŏb-kôr'dāt') *adj.* Heart-shaped, with the tapering end at the point of attachment <an *obcordate* leaf>

**ob·du·rate** (ŏb'dōō-rĭt, -dyōō-) *adj.* [ME *obdurat* < Lat. *obduratus,* p.part. of *obdurare,* to harden : *ob-* (intensive) + *durare,* to harden < *durus,* hard.] **1. a.** Persistent in wrongdoing. **b.** Hardened against sentiment : HARDHEARTED. **2.** Not yielding to persuasion : INTRACTABLE. **—ob'du·ra·cy** (-rə-sē, -dyōō-) *n.* **—ob'du·rate·ly** *adv.* **—ob'du·rate·ness** *n.*

**o·be·ah** (ō'bē-ə) *n.* [Efik *ubio,* something put in the ground to cause sickness or death.] Religious belief, prevalent in some parts of the West Indies and nearby tropical America, that is marked by witchcraft or sorcery.

**o·be·di·ence** (ō-bē'dē-əns) *n.* **1. a.** The quality or state of being obedient. **b.** An act or instance of obeying. **2.** A sphere of ecclesiastical authority.

**o·be·di·ent** (ō-bē'dē-ənt) *adj.* [ME < OFr. < Lat. *oboediens, oboedient-,* p.part. of *oboedire,* to obey.] Obeying or willing to obey a command or request. **—o·be'di·ent·ly** *adv.*

☆ ***syns:*** OBEDIENT, AMENABLE, BIDDABLE, COMPLIANT, COMPLYING, CONFORMABLE, SUBMISSIVE, TRACTABLE *adj. core meaning* : willing to carry out the wishes of others <an *obedient* child>

**o·bei·sance** (ō-bā'səns, ō-bē'-) *n.* [ME < OFr. *obeissance* < *obeissant,* p.part. of *obeir,* to obey.] **1.** A gesture or body movement, as a bow, expressing deference or respect. **2.** An attitude of deference or respect. **—o·bei'sant** *adj.*

**ob·e·li** (ŏb'ə-lī') *n. pl. of* OBELUS.

**o·be·lia** (ō-bēl′yə) *n.* [NLat. *Obelia*, genus name, prob. < Gk. *obelias*, a loaf baked on a spit < *obelos*, a spit.] Any of various colonial marine hydroids of the genus *Obelia*.

**ob·e·lisk** (ŏb′ə-lĭsk) *n.* [OFr. *obelisque* < Lat. *obeliscus* < Gk. *obeliskos*, dim. of *obelos*, a spit.] **1.** A four-sided usu. stone shaft that tapers to a pyramidal point. **2.** A dagger (†), used esp. as a reference mark. **—ob′e·lis′cal** (-lĭs′kəl) *adj.* **—ob′e·lis′koid′** (-koid′) *adj.*

**ob·e·lize** (ŏb′ə-līz′) *vt.* **-lized, -liz·ing, -liz·es.** [Gk. *obelizein* < *obelos*, a spit.] To annotate or mark with an obelus.

**ob·e·lus** (ŏb′ə-ləs) *n., pl.* **-li** (-lī′) [LLat. *obelus* < Gk. *obelos*, a spit.] **1.** A mark (— or ÷) used to indicate a spurious or doubtful passage in ancient manuscripts. **2.** OBELISK 2.

**O·ber·on** (ō′bə-rŏn′, -rən) *n.* [Fr. < OFr. *Auberon*, of Frankish orig.] The fairy king and husband of Titania in medieval folklore.

**o·bese** (ō-bēs′) *adj.* [Lat. *obesus*, grown fat from eating < p.part. of *obedere*, to eat away : *ob*, away + *edere*, to eat.] Extremely fat. **—o·be′si·ty** (ō-bē′sĭ-tē) *n.*

**o·bey** (ō-bā′) *v.* **o·beyed, o·bey·ing, o·beys.** [ME *obeien* < OFr. *obeir* < Lat. *oboediere*, to listen to : *ob*, to + *audire*, to hear.] *—vt.* **1.** To carry out or yield to the command, authority, or instruction of. **2.** To carry out or comply with (e.g., a command or regulation). *—vi.* To behave obediently. **—o·bey′er** *n.*

☆ ***syns:*** OBEY, COMPLY, FOLLOW, MIND *v. core meaning* **:** to act in conformity with a request, rule, or order. OBEY suggests an accepting of authority <*obeying* traffic regulations> COMPLY conveys an inclination to yield without protest <a singer who *complied* with the audience's request by singing an encore> FOLLOW suggests adhering to a prescribed course of action <*followed* the doctor's orders> MIND applies esp. to good behavior <*mind* one's parents>

**ob·fus·cate** (ŏb′fə-skāt′, ŏb-fŭs′kāt′) *vt.* **-cat·ed, -cat·ing, -cates.** [LLat. *obfuscare*, to darken : *ob* (intensive) + Lat. *fuscare*, to darken < *fuscus*, dark.] **1.** To make dark or obscure **:** CLOUD. **2.** To confuse <passions that *obfuscate* your reason> **—ob′fus·ca′tion** *n.* **—ob·fus′ca·to′ry** (ŏb-fŭs′kə-tôr′ē, -tōr′ē, əb-) *adj.*

**o·bi** (ō′bē) *n.* [J., belt.] A wide sash fastened behind with a bow, worn by women in Japan as a part of traditional dress.

**O·bie** (ō′bē) *n.* [< O.B., abbr. for OFF-BROADWAY.] An annual award for outstanding achievement in off-Broadway theater.

**o·bit** (ō′bĭt, ō-bĭt′) *n. Informal.* An obituary.

**o·bi·ter dic·tum** (ō′bĭ-tər dĭk′təm) *n., pl.* **obiter dic·ta** (dĭk′tə) [Lat., something said in passing.] **1.** *Law.* A judicial opinion that is not binding on the case in question. **2.** An incidental observation or remark.

**o·bit·u·ar·y** (ō-bĭch′o͞o-ĕr′ē) *n., pl.* **-ies.** [Med. Lat. *obituarius*, (report) of death < Lat. *obitus*, death < *obire*, to die : *ob*, down + *ire*, to go.] A published death notice, usu. with a brief biography of the deceased. **—o·bit′u·ar′y** *adj.*

**ob·ject**[1] (əb-jĕkt′) *v.* **-ject·ed, -ject·ing, -jects.** [ME *objecten* < Lat. *obicere*, to oppose : *ob*, toward + *jacere*, to throw.] *—vi.* **1.** To hold or present an opposing view **:** DISSENT. **2.** To feel adverse to or express disapproval of something <*object* to violence on television> *—vt.* To put forward in or as a reason for opposition <We *objected* that the rule was unfair.> **—ob·jec′tor** *n.*

**ob·ject**[2] (ŏb′jĭkt, -jĕkt′) *n.* [ME < Lat. *objectus*, p.part. of *obicere*, to throw before —see OBJECT[1].] **1.** Something perceptible esp. to the sense of touch or vision. **2.** *Philos.* Something intelligible or perceptible by the mind. **3.** A focus of attention, thought, feeling, or effort <an *object* of devotion> **4.** The purpose or goal of a specific action or endeavor <the *object* of the search> **5. a.** A noun or substantive that receives or is affected by the action of a verb within a sentence. **b.** A noun or substantive following and governed by a preposition.

**object glass** *n.* OBJECTIVE 4.

**ob·jec·ti·fy** (əb-jĕk′tə-fī′) *vt.* **-fied, -fy·ing, -fies.** [< OBJECT[2].] **1. a.** To present (something) as an object **:** EXTERNALIZE. **b.** To make objective. **2.** RATIONALIZE 3. **—ob·jec′ti·fi·ca′tion** *n.*

**ob·jec·tion** (əb-jĕk′shən) *n.* **1.** An act or instance of objecting. **2.** A statement offered in opposition. **3.** A reason, ground, or cause for expressing opposition.

**ob·jec·tion·a·ble** (əb-jĕk′shə-nə-bəl) *adj.* Provoking disapproval or opposition **:** OFFENSIVE <*objectionable* language> **—ob·jec′tion·a·bil′i·ty** *n.* **—ob·jec′tion·a·bly** *adv.*

**ob·jec·tive** (əb-jĕk′tĭv) *adj.* [Med. Lat. *objectivus* < Lat. *objectus*, object. —see OBJECT[2].] **1.** Of or pertaining to a material object as distinguished from a mental concept. **2.** Having actual existence or reality. **3. a.** Uninfluenced by emotion, surmise, or personal opinion. **b.** Based on observable phenomena <an *objective* forecast> **4.** *Med.* Indicating a symptom or abnormal condition perceived by someone other than the person afflicted. **5. a.** Denoting the case of a noun or pronoun serving as the object of the verb. **b.** Pertaining to a noun or pronoun used in such a case. *—n.* **1.** Something that actually exists as distinguished from something thought or felt to exist. **2.** Something worked toward or aspired to **:** GOAL. **3. a.** The objective case. **b.** A word in the objective case. **4. a.** The lens or lens system in a microscope or telescope that is closest to the object. **b.** A lens or lens system in a camera or projector that forms the image of the object. **—ob·jec′tive·ly** *adv.* **—ob·jec′tive·ness** *n.*

**objective complement** *n.* A noun, adjective, or pronoun serving as a complement to a verb and qualifying its direct object, as *chairwoman* in *They elected her chairwoman.*

**objective correlative** *n.* A situation that objectifies an emotion, often used as a literary device to elicit a particular emotional response in the reader.

**ob·jec·tiv·ism** (ŏb-jĕk′tĭ-vĭz′əm) *n.* **1.** *Philos.* One of several doctrines holding that all reality is objective and external to the mind and that knowledge is reliably based on observed phenomena. **2.** An emphasis on objective themes in literature and art. **—ob·jec′tiv·ist** *n.* **—ob·jec′tiv·is′tic** *adj.*

**ob·jec·tiv·i·ty** (ŏb′jĕk-tĭv′ĭ-tē) *n.* **1.** The quality or state of being objective. **2.** External reality.

**object language** *n.* A target language.

**object lens** *n.* OBJECTIVE 4.

**object lesson** *n.* **1.** A lesson that uses a material object as an aid to instruction. **2.** A concrete illustration of a moral or principle.

**ob·jet d'art** (ôb′zhĕ där′) *n., pl.* **ob·jets d'art** (ôb′zhĕ där′) [Fr., object of art.] An object, esp. a decorative one, that has artistic merit.

**ob·jet trou·vé** (ôb-zhā′ tro͞o-vā′) *n.* [Fr.] A found object.

**ob·jur·gate** (ŏb′jər-gāt′, ŏb-jûr′gāt′) *vt.* **-gat·ed, -gat·ing, -gates.** [Lat. *objurgare, objurgat-* : *ob*, against + *jurgare*, to scold (*jus*, law + *agere*, to carry on).] To scold or reprove sharply **:** BERATE. **—ob′jur·ga′tion** *n.* **—ob·jur′ga·to′ri·ly** (ŏb-jûr′gə-tôr′ə-lē, -tōr′-) *adv.* **—ob·jur′ga·to′ry** (-tôr′ē, -tōr′ē) *adj.*

**ob·lan·ce·o·late** (ŏb-lăn′sē-ə-lāt′) *adj.* Broad and rounded at the apex and tapering at the base <an *oblanceolate* leaf>

**o·blast** (ô′bləst, ŏ′bläst′) *n.* [R. *oblast′.*] A territorial administrative division within a republic in the U.S.S.R.

**ob·late**[1] (ŏb′lāt′, ŏ-blāt′) *adj.* [Prob. Med. Lat. *oblatus* : Lat. *ob* (intensive) + Lat. *latus*, p.part. of *ferre*, to carry.] **1.** Shaped like a spheroid. **2.** Compressed along or flattened at the poles <The earth is an *oblate* solid.> **—ob′late′ly** *adv.* **—ob′late′ness** *n.*

**ob·late**[2] (ŏb′lāt′) *n.* [Med. Lat. *oblatus* < Lat., p.part. of *offerre*, to offer : *ob*, toward + *ferre*, to carry.] **1.** A lay person dedicated to a religious, often monastic, life. **2. Oblate.** *Rom. Cath. Ch.* A member of a religious community for men or women.

**ob·la·tion** (ə-blā′shən, ō-blā′-) *n.* [ME *oblacioun* < OFr. *oblacion* < Lat. *oblatio* < *offerre*, to offer : *ob*, toward + *ferre*, to carry.] **1.** The act or ceremony of offering something in religious worship. **2. Oblation. a.** The act of offering the bread and wine of the Eucharist. **b.** Something offered, esp. the bread and wine of the Eucharist. **3.** A charitable offering or gift. **—ob·la′tion·al, ob′la·to′ry** (ŏb′lə-tôr′ē, -tōr′ē) *adj.*

**ob·li·gate** (ŏb′lĭ-gāt′) *vt.* **-gat·ed, -gat·ing, -gates.** [Lat. *obligare, obligat-.* —see OBLIGE.] **1. a.** To bind or compel by legal or moral constraint. **b.** OBLIGE 2. **2.** To commit (e.g., money) in fulfillment of an obligation. *—adj.* (-gĭt, -gāt′). **1.** *Biol.* Capable of surviving in only one environment. —Used of certain parasites. **2.** Absolutely required **:** ESSENTIAL. **—ob′li·ga·ble** (-gə-bəl) *adj.* **—ob′li·gate·ly** *adv.* **—ob′li·ga′tor** *n.*

**ob·li·ga·tion** (ŏb′lĭ-gā′shən) *n.* **1.** The act of binding oneself by a social, moral, or legal tie. **2. a.** A social, moral, or legal requirement, as a contract or promise, compelling one to a given course of action. **b.** A course of action imposed by law, society, or conscience by which one is bound or restricted. **3.** The constraining power of a law, promise, contract, or sense of duty. **4.** *Law.* **a.** A legal agreement stipulating a specified payment or action and often a specified penalty for failure to comply. **b.** The document expressing the terms of a legal obligation. **5. a.** Something owed as payment or in return for a special service or favor. **b.** The service or favor for which one is indebted to another. **6.** The state, fact, or feeling of being indebted to another for a special service or favor received.

**o·blig·a·to·ry** (ə-blĭg′ə-tôr′ē, -tōr′ē, ŏb′lĭ-gə-) *adj.* **1.** Legally or morally binding. **2.** Imposing or recording an obligation <a bill *obligatory*> **3.** Of the nature of an obligation **:** COMPULSORY <Attendance is *obligatory.*> **4.** *Biol.* OBLIGATE 1. **—o·blig′a·to′ri·ly** *adv.*

**o·blige** (ə-blīj′) *v.* **o·bliged, o·blig·ing, o·blig·es.** [ME *obligen* < OFr. *obligier* < Lat. *obligare* : *ob*, to + *ligare*, to bind.] *—vt.* **1.** To constrain by legal, social, moral, or physical means. **2.** To make indebted or grateful <were *obliged* to them for their kindness> **3.** To do a service or favor for <*obliged* me by doing the paperwork> *—vi.* To do a service or favor **:** perform a courtesy <The pianist will *oblige* with an encore.> **—o·blig′er** *n.*

**ob·li·gee** (ŏb′lə-jē′) *n.* **1.** One under obligation to another. **2.** *Law.* One to whom another is bound by contract or legal agreement.

**o·blig·ing** (ə-blī′jĭng) *adj.* Ready to help or assist **:** ACCOMMODATING. **—o·blig′ing·ly** *adv.* **—o·blig′ing·ness** *n.*

**ob·li·gor** (ŏb′lĭ-gôr′, -jôr′) *n. Law.* One who binds oneself to another by contract or legal agreement.

**o·blique** (ō-blēk′, ə-blēk′) *adj.* [ME < OFr. < Lat. *obliquus.*] **1. a.** Having a slanting or sloping direction, course, or situation **:** INCLINED. **b.** Designating geometric lines or planes that are neither perpendicular nor parallel. **2. a.** Indirect or evasive <an *oblique* reference> **b.** Devious or dishonest <an *oblique* reply> **3.** Descended by an indirect line : COLLATERAL. **4.** *Bot.* Having sides of unequal length or form <an *oblique* leaf> **5.** Designating a noun case that is neither the nominative or the vocative. —*n.* **1.** An oblique thing, as a line, direction, or muscle. **2.** *Naut.* The act of changing course by less than 90°. —*adv.* (ō-blīk′, ə-blīk). At an angle of 45° <Left *oblique*, march!> **—o·blique′ly** *adv.* **—o·blique′ness** *n.*

**oblique angle** *n.* An acute or obtuse angle.

**oblique triangle** *n.* A triangle having no right angle.

**o·bliq·ui·ty** (ō-blĭk′wĭ-tē, ə-blĭk′-) *n., pl.* **-ties.** [ME *obliquite* < Lat. *obliquitas* < *obliquus*, oblique.] **1.** The quality or state of being oblique. **2. a.** A deviation from a horizontal or vertical line, plane, position, or direction. **b.** The angle or extent of such a deviation. **3. a.** A mental deviation or aberration. **b.** Immoral behavior. **4. a.** Obscurity or indirectness in conduct or verbal expression. **b.** An obscure or devious statement. **—o·bliq′ui·tous** *adj.*

**o·blit·er·ate** (ə-blĭt′ə-rāt′, ō-blĭt′-) *vt.* **-at·ed, -at·ing, -ates.** [Lat. *oblitterare, oblitterat-*, to erase : *ob*, against + *littera*, letter.] **1.** To eliminate completely so as to leave no trace. **2.** To wipe out, wear away, or erase (e.g., an inscription). **—o·blit′er·a′tion** *n.* **—o·blit′er·a′tive** (-ə-rā′tĭv, -ər-ə-tĭv) *adj.* **—o·blit′er·a′tor** *n.*

**o·bliv·i·on** (ə-blĭv′ē-ən) *n.* [ME *oblivioun* < OFr. *oblivion* < Lat. *oblivio* < *oblivisci*, to forget.] **1.** The state or quality of being utterly forgotten. **2.** An act or instance of forgetting. **3.** Official disregard of offenses.

**o·bliv·i·ous** (ə-blĭv′ē-əs) *adj.* **1.** Lacking all memory : FORGETFUL. **2.** Lacking conscious awareness : UNMINDFUL. **—o·bliv′i·ous·ly** *adv.* **—o·bliv′i·ous·ness** *n.*

**ob·long** (ŏb′lông′, -lŏng′) *adj.* [ME < Lat. *oblongus* : *ob* (intensive) + *longus*, long.] **1.** Having a long dimension, esp. having one of two perpendicular dimensions, as width or length, greater than the other : ELONGATED. **2.** Having the shape of or like an ellipse or a rectangle. **3.** *Bot.* Having a rather elongated form with approx. parallel sides <an *oblong* leaf> **—ob′long′** *n.*

**ob·lo·quy** (ŏb′lə-kwē) *n., pl.* **-quies.** [ME *obloqui* < Med. Lat. *obloquium* < Lat. *obloqui*, to speak against : *ob*, against + *loqui*, to speak.] **1.** Abusively and defamatory language or utterance : CALUMNY. **2.** Damage to or loss of one's reputation.

**ob·nox·ious** (ŏb-nŏk′shəs, əb-) *adj.* [Lat. *obnoxiosus*, hurtful < *obnoxius*, punishable : *ob*, to + *noxa*, injury.] **1.** Highly offensive or disagreeable : REPUGNANT <an *obnoxious* boor> **2.** Likely to cause harm, injury, or evil. **3.** *Archaic.* Censurable. **—ob·nox′ious·ly** *adv.* **—ob·nox′ious·ness** *n.*

**o·boe** (ō′bō) *n.* [Ital. < Fr. *hautbois.* —see HAUTBOY.] **1.** A slender woodwind instrument with a conical bore and a double-reed mouthpiece, having a poignant penetrating sound. **2.** A reed stop in an organ with a tone similar to the oboe. **—o′bo·ist** *n.*

**ob·o·vate** (ŏb-ō′vāt′) *adj.* Egg-shaped, with the narrow end attached to the stalk <an *obovate* leaf>

**ob·o·void** (ŏb-ō′void′) *adj.* Egg-shaped, with the narrow end attached to the stem <an *obovoid* fruit>

**ob·scene** (ŏb-sēn′, əb-) *adj.* [Lat. *obscenus.*] **1.** Offensive to accepted standards of decency. **2.** Inciting lustful feelings : LEWD. **3.** Offensive or repulsive to the senses : LOATHSOME. **—ob·scene′ly** *adv.*

☆ **syns:** OBSCENE, BARNYARD, COARSE, CRUDE, DIRTY, FILTHY, FOUL, GROSS, INDECENT, LEWD, NASTY, PROFANE, RANK, RAUNCHY, RAW, SCATOLOGICAL, SCURRILOUS, SMUTTY *adj. core meaning* : offensive to accepted standards of decency <*obscene* photographs> <*obscene* gestures>

**ob·scen·i·ty** (ŏb-sĕn′ĭ-tē, əb-) *n., pl.* **-ties. 1.** The quality or state of being obscene. **2.** Indecency or offensiveness in expression, behavior, or appearance. **3.** Something obscene, as a word or act.

**ob·scur·ant** (ŏb-skyŏŏr′ənt, əb-) *n.* [Lat. *obscurans, obscurant-*, pr.part. of *obscurare*, to darken < *obscurus*, dark.] An opponent of intellectual development and political reform. —*adj.* **1.** Characteristic of an obscurant. **2.** Tending to make obscure.

**ob·scur·ant·ism** (ŏb-skyŏŏr′ən-tĭz′əm, əb-, ŏb′skyŏŏ-răn′-) *n.* **1.** The principles or practice of obscurants. **2.** A policy of withholding information from the public. **3. a.** An artistic or literary style marked by deliberate vagueness or obliqueness. **b.** An act or instance of this style. **—ob·scur′ant·ist** *n.*

**ob·scure** (ŏb-skyŏŏr′, əb-) *adj.* **-scur·er, -scur·est.** [ME < Lat. *obscurus.*] **1.** Deficient in light : DARK. **2. a.** Lacking clear delineation : INDISTINCT. **b.** Dimly sensed or perceived : FAINT. **c.** Having the mid-central unstressed sound represented by the schwa (ə). **3.** Remote from centers of human population <an *obscure* hamlet> **4.** Not readily apparent : INCONSPICUOUS <an *obscure* detail> **5.** Of humble or undistinguished station or reputation <an *obscure* civil servant> **6.** Not clearly expressed or easily understood : AMBIGUOUS. —*vt.* **-scured, -scur·ing, -scures. 1.** To make dim or indistinct <Haze *obscured* the valley.> **2.** To conceal in obscurity <the truth *obscured* with technicalities> **3.** To reduce (a vowel) to the mid-central unstressed sound represented by the schwa (ə). **—ob·scure′** *n.* **—ob·scure′ly** *adv.* **—ob·scure′ness** *n.*

**ob·scu·ri·ty** (ŏb-skyŏŏr′ĭ-tē, əb-) *n., pl.* **-ties.** [ME *obscurite* < OFr. < Lat. *obscuritas* < *obscurus*, dark.] **1.** Absence or deficiency of light : DARKNESS. **2. a.** The quality or state of being unknown <from *obscurity* to sudden renown> **b.** One that is obscure.

**ob·se·qui·ous** (ŏb-sē′kwē-əs, əb-) *adj.* [ME < Lat. *obsequiosus* < *obsequium*, compliance < *obsequi*, to comply : *ob*, to + *sequi*, to follow.] Fawning : servile <*obsequious* courtiers> **—ob·se′qui·ous·ly** *adv.* **—ob·se′qui·ous·ness** *n.*

**ob·se·quy** (ŏb′sĭ-kwē) *n., pl.* **-quies.** [ME *obsequi* < OFr. *obseque* < Med. Lat. *obsequiae* < Lat. *obsequium*, compliance. —see OBSEQUIOUS.] *often* **obsequies.** A funeral rite or ceremony.

**ob·serv·a·ble** (əb-zûr′və-bəl) *adj.* **1.** Capable of being observed : DISCERNIBLE. **2.** Deserving attention : NOTEWORTHY. —*n. Physics.* A physical property, as temperature or weight, that can be observed or measured directly. **—ob·serv′a·bly** *adv.*

**ob·ser·vance** (əb-zûr′vəns) *n.* **1.** The act or practice of complying with a law, custom, rule, or command. **2.** The act or custom of celebrating a ritual occasion, as a holiday. **3.** A customary rite or ceremony. **4.** The act of watching : OBSERVATION. **5.** *Rom. Cath. Ch.* The rule governing a religious order.

**ob·ser·vant** (əb-zûr′vənt) *adj.* [Fr. < Lat. *observans*, pr.part. of *observare*, to watch. —see OBSERVE.] **1.** Quick to apprehend or perceive : ALERT. **2.** Diligent in observing a law, custom, duty, or principle <*observant* of the dress code> **—ob·ser′vant·ly** *adv.*

**ob·ser·va·tion** (ŏb′zər-vā′shən) *n.* [Lat. *observatio* < *observare*, to watch. —see OBSERVE.] **1. a.** The act or faculty of observing. **b.** The fact of being observed. **2. a.** The act of noting and recording something, as a phenomenon, with instruments. **b.** The result or record of such notation <a seismological *observation*> **3.** A remark or comment. **4.** A judgment or inference based on observing. **—ob′ser·va′tion·al** *adj.* **—ob′ser·va′tion·al·ly** *adv.*

**ob·ser·va·to·ry** (əb-zûr′və-tôr′ē, -tōr′ē) *n., pl.* **-ries.** [Prob. Fr. *observatorie* < OFr. *observer*, to observe. —see OBSERVE.] **1.** A building designed and equipped for making observations esp. of astronomical or meteorological phenomena. **2.** A structure commanding an extensive view.

**ob·serve** (əb-zûrv′) *v.* **-served, -serv·ing, -serves.** [ME *observen* < OFr. *observer* < Lat. *observare* : *ob*, to + *servare*, to watch.] —*vt.* **1.** To notice : perceive. **2.** To watch attentively <*observe* a player's technique> **3.** To make a systematic or scientific observation of <*observe* the behavior of lions> **4.** To say casually : REMARK. **5.** To adbide by or adhere to <*observe* the speed limit> **6.** To keep or celebrate (e.g., a holiday). —*vi.* **1.** To take notice. **2.** To make a remark or comment. **3.** To watch or be present without taking active part. **—ob·serv′ing·ly** *adv.*

**ob·serv·er** (əb-zûr′vər) *n.* One that observes, as: **a.** A delegate sent to observe and report on a meeting or assembly without taking official part in the proceedings. **b.** A military aircraft crew member responsible for making observations. **c.** A soldier watching and reporting from an observation post.

**ob·sess** (əb-sĕs′, ŏb-) *vt.* **-sessed, -sess·ing, -sess·es.** [Lat. *obsidēre, obsess-*, to beset, possess : *ob*, on + *sedēre*, to sit.] To preoccupy the mind of excessively.

**ob·ses·sion** (əb-sĕsh′ən, ŏb-) *n.* **1.** Compulsive, often anxious preoccupation with a fixed idea or unwanted emotion. **2.** A compulsive, usu. irrational idea or emotion. **—ob·ses′sion·al** *adj.*

**ob·ses·sive** (əb-sĕs′ĭv, ŏb-) *adj.* **1.** Of, pertaining to, or characteristic of an obsession. **2.** Tending to cause an obsession. **3.** Excessive in nature or degree <an *obsessive* concern with appearance> **—ob·ses′sive** *n.* **—ob·ses′sive·ly** *adv.* **—ob·ses′sive·ness** *n.*

**ob·sid·i·an** (ŏb-sĭd′ē-ən) *n.* [Lat. *obsidianus*, alteration of *obsianus* (*lapis*), (stone) of Obsius, who reportedly discovered it.] A usu. black or banded acid-resistant volcanic glass, displaying curved lustrous surfaces when fractured.

**ob·so·lesce** (ŏb′sə-lĕs′) *vi.* **-lesced, -lesc·ing, -lesc·es.** To become obsolescent.

**ob·so·les·cent** (ŏb′sə-lĕs′ənt) *adj.* [Lat. *obsolescens, obsolescent-*, pr.part. of *obsolescere*, to wear out : *ob* (intensive) + *solēre*, to use.] Being in the process of becoming obsolete. **—ob′so·les′cence** *n.* **—ob′so·les′cent·ly** *adv.*

**ob·so·lete** (ŏb′sə-lēt′, ŏb′sə-lēt′) *adj.* [Lat. *obsoletus*, p.part. of *obsolescere*, to wear out. —see OBSOLESCENT.] **1.** No longer in use <an *obsolete* spelling> **2.** Outmoded in style, design, or construction <an *obsolete* engine> **3.** *Biol.* Increasingly vestigial or disappearing in each succeeding generation. —Used of plant or animal organs or characteristics. —*vt.* **-let·ed, -let·ing, -letes.** To cause to become obsolete. **—ob′so·lete′ly** *adv.* **—ob′so·lete′ness** *n.* **—ob′so·let′ism** *n.*

**ob·sta·cle** (ŏb′stə-kəl) *n.* [ME < OFr. < Lat. *obstaculum* < *obstare*, to hinder : *ob*, against + *stare*, to stand.] One that opposes, stands in the way of, or deters passage or progress.

**obstacle course** *n.* **1.** A military training course having obstacles, as walls and ditches, that must be surmounted. **2.** A situation full of obstacles that must be overcome.

**ob·stet·ric** (ŏb-stĕt′rĭk, əb-) *also* **ob·stet·ri·cal** (-rĭ-kəl) *adj.* [Lat. *obstetricus* < *obstetrix*, midwife < *obstare*, to stand before : *ob*, before + *stare*, to stand.] Of or relating to the medical profession of

ŏŏ **boot** ou **out** th **thin** *th* **this** ŭ **cut** ûr **urge** y **young**
yŏŏ **abuse** zh **vision** ə **about**, it**e**m, ed**i**ble, gall**o**p, circ**u**s

obstetrics or the care of women during and after pregnancy. **—ob·stet'ri·cal·ly** *adv.*

**ob·ste·tri·cian** (ŏb'stĭ-trĭsh'ən) *n.* A physician who specializes in obstetrics.

**ob·stet·rics** (ŏb-stĕt'rĭks, əb-) *n. (sing. or pl. in number).* The branch of medicine concerned with the care of women during pregnancy, childbirth, and the postpartum period.

**ob·sti·na·cy** (ŏb'stə-nə-sē) *n., pl.* **-cies. 1.** The state or quality of being obstinate. **2.** An act or instance of stubbornness.

**ob·sti·nate** (ŏb'stə-nĭt) *adj.* [ME < Lat. *obstinatus,* p.part. of *obstinare,* to persist.] **1.** Stubbornly adhering to an opinion, attitude, or course of action : tenaciously unwilling to yield. **2.** Difficult to manage or subdue : REFRACTORY. **3.** Difficult to cure or alleviate : PERSISTENT <an *obstinate* cough> <an *obstinate* cold> **—ob'sti·nate·ly** *adv.* **—ob'sti·nate·ness** *n.*

☆ **syns:** OBSTINATE, BULLHEADED, CLOSE-MINDED, DOGGED, HARDHEADED, HEADSTRONG, INCOMPLIANT, INTRACTABLE, INTRANSIGENT, PERTINACIOUS, PERVERSE, PIGHEADED, REFRACTORY, STIFF-NECKED, STUBBORN, TOUGH, WILLFUL *adj. core meaning* : tenaciously unwilling to yield <an *obstinate* person who never apologized> *ant:* pliable, pliant

**ob·strep·er·ous** (ŏb-strĕp'ər-əs, əb-) *adj.* [Lat. *obstreperus,* noisy < *obstrepere,* to clamor against : *ob,* against + *strepere,* to make a noise.] **1.** Noisily and aggressively defiant. **2.** Loud and unruly : BOISTEROUS. **—ob·strep'er·ous·ly** *adv.* **—ob·strep'er·ous·ness** *n.*

**ob·struct** (əb-strŭkt', ŏb-) *vt.* **-struct·ed, -struct·ing, -structs.** [Lat. *obstruere, obstruct-* : *ob,* against + *struere,* to pile up.] **1.** To clog or block (a passage) with obstacles. **2.** To impede, retard, or interfere with <*obstruct* legislation> **3.** To cut off from sight. **—ob·struct'er, ob·struc'tor** *n.* **—ob·struc'tive** *adj.* **—ob·struc'tive·ly** *adv.* **—ob·struc'tive·ness** *n.*

**ob·struc·tion** (əb-strŭk'shən, ŏb-) *n.* **1.** One that obstructs : OBSTACLE. **2.** An act or instance of obstructing. **3.** The act of impeding or an attempt to impede the conduct of esp. legislative business.

**ob·struc·tion·ist** (əb-strŭk'shə-nĭst, ŏb-) *n.* One who systematically hinders progress, esp. one who obstructs legislation, as by filibuster. **—ob·struc'tion·ism** *n.* **—ob·struc'tion·is'tic** *adj.*

**ob·tain** (əb-tān', ŏb-) *v.* **-tained, -tain·ing, -tains.** [ME *obteinen* < OFr. *obtenir* < Lat. *obtinēre* : *ob* (intensive) + *tenēre,* to hold.] *—vt.* To gain possession of, esp. by intention or endeavor : ACQUIRE. *—vi.* **1.** To be customary or widely accepted <outdated rules that no longer *obtain*> **2.** *Archaic.* To succeed. **—ob·tain'a·ble** *adj.* **—ob·tain'er** *n.*

**ob·tect** (ŏb-tĕkt') *also* **ob·tect·ed** (-tĕk'tĭd) *adj.* [Lat. *obtectus,* p.part. of *obtegere,* to cover over : *ob,* over + *tegere,* to cover.] Enclosed in or covered by a hardened secretion. —Used esp. of pupae having wings, legs, and antennae sealed against the body by such a covering.

**ob·test** (ŏb-tĕst') *vt.* **-test·ed, -test·ing, -tests.** [Lat. *obtestari* : *ob,* to + *testari,* to call as a witness < *testis,* witness.] To entreat : supplicate. **—ob'tes·ta'tion** *n.*

**ob·trude** (ŏb-trōōd', əb-) *v.* **-trud·ed, -trud·ing, -trudes.** [Lat. *obtrudere* : *ob-,* against + *trudere,* to thrust.] *—vt.* **1.** To force (oneself or one's ideas) upon others with undue insistence or without invitation. **2.** To thrust out. *—vi.* To force oneself upon others. **—ob·trud'er** *n.* **—ob·tru'sion** (-trōō'zhən) *n.*

**ob·tru·sive** (ŏb-trōō'sĭv, -zĭv, əb-) *adj.* [< Lat. *obtrudere, obtrus-,* to obtrude.] **1.** Protruding : projecting. **2.** Forward or assertive in a disruptive way : PUSHY <the *obtrusive* presence of hecklers> **3.** Undesirably conspicuous <an *obtrusive* stammer> **—ob·tru'sive·ly** *adv.* **—ob·tru'sive·ness** *n.*

**ob·tund** (ŏb-tŭnd') *vt.* **-tund·ed, -tund·ing, -tunds.** [ME *obtunden* < Lat. *obtundere* : *ob-,* against + *tundere,* to beat.] To lessen the force or intensity of : DEADEN. **—ob·tund'ent** *adj. & n.*

**ob·tu·rate** (ŏb'tə-rāt', -tyə-) *vt.* **-rat·ed, -rat·ing, -rates.** [Lat. *obturare, obturat-.*] To obstruct or close. **—ob'tu·ra'tion** *n.*

**ob·tu·ra·tor** (ŏb'tə-rā'tər, -tyə-) *n.* One that closes or obstructs, as: **a.** An organic structure, as the soft palate, that closes a body passage or opening. **b.** A prosthetic device that performs the same function.

**ob·tuse** (ŏb-tōōs', -tyōōs', əb-) *adj.* [Lat. *obtusus,* p.part. of *obtundere,* to blunt. —see OBTUND.] **1.** Not sharp, pointed, or acute in shape : BLUNT. **2.** *Bot.* Having a rounded or blunt tip <an *obtuse* leaf> **3.** Slow or dull in comprehension or discernment. **—ob·tuse'ly** *adv.* **—ob·tuse'ness** *n.*

**obtuse angle** *n.* An angle greater than 90° and less than 180°.

**ob·verse** (ŏb-vûrs', əb-, ŏb'vûrs') *adj.* [Lat. *obversus,* p.part. of *obvertere,* to turn toward. —see OBVERT.] **1.** Facing or turned toward the observer <the *obverse* side of a coin> **2.** *Bot.* Having a narrower base than top, as certain leaves : INVERSE. **3.** Serving as a counterpart or complement. *—n.* (ŏb'vûrs', ŏb-vûrs', əb-). **1.** The side of a coin or medallion bearing the principal stamp or design. **2.** The more evident of two possible alternatives, cases, or sides <the *obverse* of this dilemma> **3.** *Logic.* The counterpart of a proposition obtained by exchanging the affirmative for the negative quality of the whole proposition and then negating the predicate <The *obverse* of "every human is mortal" is "no human is immortal."> **—ob·verse'ly** *adv.*

**ob·vert** (ŏb-vûrt', əb-) *vt.* **-vert·ed, -vert·ing, -verts.** [Lat. *obvertere,* to turn toward : *ob,* toward + *vertere,* to turn.] **1.** To turn so as to bring another side or aspect to view. **2.** To alter the appearance of.

**ob·vi·ate** (ŏb'vē-āt') *vt.* **-at·ed, -at·ing, -ates.** [LLat. *obviare, obviat-,* to hinder : *ob-,* against + *via,* way.] To prevent or counteract by anticipating. **—ob'vi·a'tion** *n.* **—ob'vi·a'tor** *n.*

**ob·vi·ous** (ŏb'vē-əs) *adj.* [Lat. *obvius* : *ob,* against + *via,* way.] **1.** Easily understood or perceived : APPARENT. **2.** Lacking subtlety : TRANSPARENT <an *obvious* fraud> **3.** *Archaic.* Standing in the way or in front. **—ob'vi·ous·ly** *adv.* **—ob'vi·ous·ness** *n.*

▲ word history: *Obvious* is derived ultimately from the Latin phrase *ob viam,* literally "in the way." Metaphorical senses of the Latin phrase and the Latin adjective *obvius* derived from it were "at hand" and "exposed." The English word *obvious,* a borrowing from Latin, preserved these senses at first, but they are now obsolete. The current sense, "easily perceived," is a development of the English word.

**ob·vo·lute** (ŏb'və-lōōt', ŏb'və-lōōt') *adj.* [Lat. *obvolutus,* p.part. of *obvolvere,* to wrap around : *ob,* over + *volvere,* to wrap.] Folded together with overlapping edges. —Used of leaves and petals in a bud. **—ob'vo·lu'tion** *n.* **—ob'vo·lu'tive** *adj.*

**oc·a·ri·na** (ŏk'ə-rē'nə) *n.* [Ital., dim. of *oca,* goose < Lat. *avicula,* dim. of *avis,* bird.] A small wind instrument with a protruding mouthpiece, finger holes, and an oval shape.

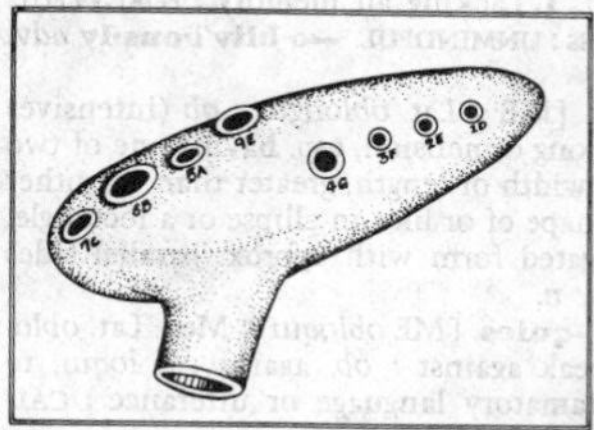

**ocarina**

**Oc·cam's razor** (ŏk'əmz) *n. var. of* OCKHAM'S RAZOR.

**oc·ca·sion** (ə-kā'zhən) *n.* [ME *occasioun* < OFr. *occasion* < Lat. *occasio* < *occidere,* to fall : *ob,* down + *cadere,* to fall.] **1. a.** An occurrence : incident. **b.** The time at which something occurs. **2.** A notable event. **3.** A favorable moment : OPPORTUNITY <had no *occasion* to meet> **4.** Something that brings on an action or event. **5.** Something that provides a cause or reason. **6.** A need created by a particular circumstance. **7. occasions.** *Archaic.* Personal requirements or necessities. **8. occasions.** Personal or business affairs. **9.** A large or important social gathering. *—vt.* **-sioned, -sion·ing, -sions.** To provide occasion for : CAUSE. **—on occasion.** From time to time.

**oc·ca·sion·al** (ə-kā'zhə-nəl) *adj.* **1. a.** Occurring from time to time : INFREQUENT. **b.** Occurring on a particular occasion. **2.** Created for a special occasion <an *occasional* poem> **3.** Designed for use as the occasion requires <*occasional* furniture> **4.** Acting as the cause of something. **5.** Acting in a specified capacity on an irregular or infrequent basis <an *occasional* golfer> **—oc·ca'sion·al·ly** *adv.*

**oc·ci·dent** (ŏk'sĭ-dənt, -dĕnt') *n.* [ME < OFr. *ocident* < Lat. *occidens* < pr.part. of *occidere,* to set (used of the sun).] **1.** The west. **2. Occident.** The countries of Europe and the Western Hemisphere.

**oc·ci·den·tal** *or* **Oc·ci·den·tal** (ŏk'sĭ-dĕnt'l). *—adj.* Of or relating to the countries of the Occident, their peoples, or their culture : WESTERN. *—n.* A native or inhabitant of a western country.

**Oc·ci·den·tal·ism** (ŏk'sĭ-dĕn'tl-ĭz'əm) *n.* A quality, custom, or trait characteristic of the Occident.

**oc·ci·den·tal·ize** *or* **Oc·ci·den·tal·ize** (ŏk'sĭ-dĕnt'l-īz') *vt.* **-ized, -iz·ing, -iz·es.** To make occidental in character, attitude, or habits. **—oc'ci·den'tal·i·za'tion** *n.*

**oc·cip·i·ta** (ŏk-sĭp'ĭ-tə) *n. var. pl. of* OCCIPUT.

**oc·cip·i·tal** (ŏk-sĭp'ĭ-tl) *adj.* [OFr. < Med. Lat. *occipitalis* < Lat. *occiput,* occiput.] Of or relating to the occiput or the occipital bone. *—n.* The occipital bone. **—oc·cip'i·tal·ly** *adv.*

**occipital bone** *n.* A curved, trapezoidal, compound bone forming the lower posterior part of the skull.

**occipital lobe** *n.* The posterior lobe of the cerebral hemisphere, shaped like a three-sided pyramid.

**oc·ci·put** (ŏk'sə-pŭt', -pət) *n., pl.* **oc·cip·i·ta** (ŏk-sĭp'ĭ-tə) *or* **oc·ci·puts.** [Lat. : *ob-,* against + *caput,* head.] The back of the skull, esp. the occipital area.

**oc·clude** (ə-klōōd') *v.* **-clud·ed, -clud·ing, -cludes.** [Lat. *occludere* : *ob* (intensive) + *claudere,* to close.] *—vt.* **1.** To cause to become closed : OBSTRUCT <*occlude* a blood vessel> **2.** To prevent the passage of <*occlude* radiation> **3.** *Chem.* To absorb or adsorb (a substance) in great quantity. **4.** *Meteorol.* To force (air) upward from the earth's surface, as when an advancing cold front undercuts a warm front. **5.** To bring together (the upper and lower teeth). —*vi.* To close so that the cusps come together. —Used of the upper and lower teeth. **—oc·clud′ent** *adj.*

**occluded front** *n.* The air front established when a cold front occludes a warm front.

**oc·clu·sal** (ə-klo͞o′zəl, -səl) *adj.* Of or pertaining to occlusion of the teeth.

**oc·clu·sion** (ə-klo͞o′zhən) *n.* **1. a.** The process of occluding. **b.** Something that occludes or obstructs. **2.** *Meteorol.* **a.** The process of occluding air masses. **b.** An occluded front. **3.** The fit of the teeth when brought together. **4. a.** The closing of the breath passage in the articulation of a stop. **b.** The blocking of the mouth passage in the articulation of a nasal consonant.

**oc·clu·sive** (ə-klo͞o′sĭv, -zĭv) *adj.* Occluding or tending to occlude. —*n.* **1.** Closure of the breath passage : STOP. **2.** A nasal consonant.

**oc·cult** (ə-kŭlt′, ō-kŭlt′, ŏk′ŭlt′) *adj.* [Lat. *occultus,* secret, p.part. of *occulere,* to conceal.] **1.** Of, relating to, or dealing with supernatural influences or phenomena. **2.** Beyond the realm of human comprehension. **3.** Available only to the initiate : SECRET <*occult* knowledge> **4.** Hidden from view : CONCEALED. —*n.* Occult practices or lore <dabbles in the *occult*> —*v.* (ə-kŭlt′, ō-kŭlt′) **-cult·ed, -cult·ing, -cults.** —*vt.* **1.** To conceal or cause to disappear from view. **2.** *Astron.* To conceal by occultation. —*vi.* To disappear from view esp. periodically, as a lighthouse beacon. **—oc·cult′ly** *adv.* **—oc·cult′ness** *n.*

**oc·cul·ta·tion** (ŏk′ŭl-tā′shən) *n.* [Lat. *occultatio* < *occultare,* freq. of *occulere,* to conceal.] **1.** *Astron.* **a.** Interposition of a celestial body between an observer and another celestial object, as when the moon passes between earth and sun in a solar eclipse. **b.** Progressive blocking of radiation, as light or radio waves from a celestial source during an occultation. **2.** The act of occulting or the state of being occulted.

**oc·cult·ism** (ə-kŭl′tĭz′əm, ō-kŭl′-, ŏk′ŭl-) *n.* **1.** Study of the occult. **2.** Belief in the supernatural. **—oc·cult′ist** *n.*

**oc·cu·pan·cy** (ŏk′yə-pən-sē) *n., pl.* **-cies. 1. a.** The act of taking or holding possession. **b.** The condition of being occupied. **2. a.** The period during which one rents, owns, or uses certain premises or land. **b.** The use to which something occupied is put <residential *occupancy*> **3.** The state of being an occupant or tenant. **4.** *Law.* The act of occupying previously unowned property with the intent of acquiring title to it.

**oc·cu·pant** (ŏk′yə-pənt) *n.* **1.** One that occupies a place or position, esp. a resident. **2.** *Law.* One who is the first to take possession of previously unowned land or premises.

**oc·cu·pa·tion** (ŏk′yə-pā′shən) *n.* [ME *occupacioun* < OFr. *ocupacion* < Lat. *occupatio* < *occupare,* to employ. —see OCCUPY.] **1. a.** An activity serving as one's regular employment : VOCATION. **b.** An activity engaged in esp. as a means of passing time. **2.** The act or process of occupying a place or the state of being occupied. **3. a.** Civil control of a nation or territory by a foreign military force. **b.** The military government exercising such control.

**oc·cu·pa·tion·al** (ŏk′yə-pā′shə-nəl) *adj.* Of, relating to, or caused by engagement in a particular occupation <an *occupational* hazard> **—oc′cu·pa′tion·al·ly** *adv.*

**occupational therapy** *n.* Therapy in which the principal element is some form of creative or productive activity. **—occupational therapist** *n.*

**oc·cu·py** (ŏk′yə-pī′) *vt.* **-pied, -py·ing, -pies.** [ME *occupien* < OFr. *ocuper* < Lat. *occupare,* to seize : *ob-* (intensive) + *capere,* to take.] **1.** To seize possession of and maintain control over by force. **2.** To fill up (space or time) <books that *occupied* two shelves> **3.** To reside in. **4.** To hold or fill (e.g., an office). **5. a.** To engage or busy (oneself) <*occupied* myself writing letters> **b.** To engross or keep busy <a puzzle that *occupies* the mind> **—oc′cu·pi′er** *n.*

**oc·cur** (ə-kûr′) *vi.* **-curred, -cur·ring, -curs.** [Lat. *occurrere* : *ob,* toward + *currere,* to run.] **1.** To take place : COME ABOUT. **2.** To be found to exist or appear <The error *occurs* twice on the same page.> **3.** To come to mind <The solution finally *occurred* to me.>

**oc·cur·rence** (ə-kûr′əns) *n.* **1.** An act or instance of occurring. **2.** An event : INCIDENT. **—oc·cur′rent** *adj.*

**o·cean** (ō′shən) *n.* [ME *occean* < OFr. < Lat. *oceanus* < Gk. *ōkeanos,* a great river encircling the earth.] **1. a.** The entire body of salt water that covers approx. 72% of the earth's surface. **b.** *often* **Ocean.** Any of the principal divisions of the ocean, including the Atlantic, Pacific, and Indian oceans, their southern extensions in Antarctica, and the Arctic Ocean. **2.** A great amount or expanse <*oceans* of food>

**o·cean·ar·i·um** (ō′shə-nâr′ē-əm) *n., pl.* **-i·ums** or **-i·a** (-ē-ə). A large aquarium for the study or exhibition of marine life.

**o·ce·an·ic** (ō′shē-ăn′ĭk) *adj.* **1.** Of or relating to an ocean. **2.** Produced by or living in an ocean, esp. in the open sea rather than in shallow coastal waters. **3.** Resembling an ocean in expanse : VAST.

**O·ce·a·nid** (ō-sē′ə-nĭd) *n., pl.* **O·ce·an·i·des** (ō′sē-ăn′ĭ-dēz′) [Gk. *ōkeanis, ōkeanid-* < *Ōkeanos,* Oceanus.] *Gk. Myth.* Any of the ocean nymphs held to be the daughters of Oceanus and Tethys.

**o·cean·og·ra·phy** (ō′shə-nŏg′rə-fē) *n.* Study and exploration of the oceans and their phenomena. **—o′cean·og′ra·pher** *n.* **—o′cean·o·graph′ic** (-nə-grăf′ĭk), **o′cean·o·graph′i·cal** *adj.* **—o′cean·o·graph′i·cal·ly** *adv.*

**o·cean·ol·o·gy** (ō′shə-nŏl′ə-jē) *n.* Oceanography. **—o′cean·o·log′ic** (-nə-lŏj′ĭk), **o′cean·o·log′i·cal** *adj.* **—o′cean·o·log′i·cal·ly** *adv.* **—o′cean·ol′o·gist** (-nŏl′ə-jĭst) *n.*

**ocean sunfish** *n.* A marine fish, *Mola mola,* found in warm seas and having a large globular body.

**O·ce·a·nus** (ō-sē′ə-nəs) *n.* [Gk. *Ōkeanos.*] *Gk. Myth.* A Titan, the god of the outer sea encircling the earth and the father of the Oceanides and the river gods.

**oc·el·lat·ed** (ŏs′ə-lā′tĭd, ō′sə-, ō-sĕl′ā′-) *also* **oc·el·late** (-lāt′) *adj.* [Lat. *ocellatus,* having little eyes < *ocellus,* dim. of *oculus,* eye.] **1.** Having an ocellus or ocelli. **2.** Resembling an ocellus. **3.** Having spotted markings. **—oc′el·la′tion** *n.*

**o·cel·lus** (ō-sĕl′əs) *n., pl.* **o·cel·li** (ō-sĕl′ī′) [Lat., dim. of *oculus,* eye.] **1.** A small simple eye, found in many invertebrates. **2.** A round marking that resembles an eye. **—o·cel′lar** (ō-sĕl′ər) *adj.*

**oc·e·lot** (ŏs′ə-lŏt′, ō′sə-) *n.* [Fr. < Nahuatl *ocelotl.*] A wild cat, *Felis pardalis* of the southwestern United States and Central and South America, having a black-spotted, tawny-gray coat.

**o·cher** or **o·chre** (ō′kər) *n.* [ME *oker* < OFr. *ocre* < Med. Lat. *ochra* < Gk. *ōkhra* < *ōkhros,* pale yellow.] **1.** An earthy mineral oxide of iron mingled with varying amounts of sand and clay, occurring in brown, yellow, or red and used as a pigment for color intensification. **2.** A moderate orange yellow, from moderate or deep orange to moderate or strong yellow. **—o′cher·ous** (ō′kər-əs), **o′cher·y** (ō′krē) *adj.*

**och·loc·ra·cy** (ŏk-lŏk′rə-sē) *n., pl.* **-cies.** [OFr. *ochlocratie* < Gk. *okhlokratia* : *okhlos,* mob + *kratos,* power.] Government by the masses : MOBOCRACY **—och′lo·crat′** (ŏk′lə-krăt′) *n.* **—och′lo·crat′ic, och′lo·crat′i·cal** *adj.* **—och′lo·crat′i·cal·ly** *adv.*

**och·lo·pho·bi·a** (ŏk′lə-fō′bē-ə) *n.* [Gk. *okhlos,* crowd + -PHOBIA.] Abnormal fear of crowds. **—och′lo·pho′bic** *adj.* & *n.*

**o·chre** (ō′kər) *n. var. of* OCHER.

**Ock·ham's razor** *also* **Oc·cam's razor** (ŏk′əmz) *n.* [After William of *Ockham* (1285?–1349).] A rule stating that entities should not be multiplied needlessly, which is interpreted to mean that the simplest of two or more competing theories is preferable or that an explanation for unknown phenomena should first be attempted in terms of what is already known.

**o'clock** (ə-klŏk′) *adv.* [Short for *of the clock.*] **1.** Of or according to the clock <five *o'clock*> **2.** According to an imaginary clock dial with the observer at the center and 12 o'clock regarded as straight ahead in horizontal position or directly overhead in vertical position <enemy aircraft at three *o'clock*>

**o·co·ti·llo** (ō′kə-tē′yō) *n., pl.* **-llos.** [Mex. Sp., dim. of *ocote,* a Mexican pine < Nahuatl *ocotl,* pitch pine.] A cactuslike tree, *Fouquieria splendens* of Mexico and the southwestern United States, bearing scarlet tubular flower clusters.

**oc·re·a** (ŏk′rē-ə) *n., pl.* **-re·ae** (-rē-ē′) [Lat., greave.] *Bot.* A stipular sheath enclosing the leafstalks of certain plants.

**oct-** or **octa-** *pref. vars. of* OCTO-.

**oc·ta·gon** (ŏk′tə-gŏn′) *n.* A polygon with eight sides and eight angles. **—oc·tag′o·nal** (ŏk-tăg′ə-nəl) *adj.* **—oc·tag′o·nal·ly** *adv.*

**oc·ta·he·dra** (ŏk′tə-hē′drə) *n. var. pl. of* OCTAHEDRON.

**oc·ta·he·dron** (ŏk′tə-hē′drən) *n., pl.* **-drons** or **-dra** (-drə). A polyhedron with eight plane surfaces. **—oc′ta·he′dral** *adj.* **—oc′ta·he′dral·ly** *adv.*

**oc·tal** (ŏk′təl) *adj.* Of, pertaining to, or being a number expressed in a numbering system of base eight.

**oc·tam·e·ter** (ŏk-tăm′ĭ-tər) *n.* A verse having eight measures or metrical feet to each line. **—oc·tam′e·ter** *adj.*

**oc·tan·dri·ous** (ŏk-tăn′drē-əs) *adj.* [OCT(O) + -ANDRY + -OUS.] *Bot.* Having eight stamens.

**oc·tane** (ŏk′tān′) *n.* **1.** Any of various isomeric paraffin hydrocarbons having the formula $C_8H_{18}$. **2.** A colorless inflammable hydrocarbon, $CH_3(CH_2)_6CH_3$, found in petroleum and used as a solvent. **3.** Octane number.

**octane number** *n.* A numerical measure of the antiknock properties of motor fuel, based on the percentage by volume of isooctane in a standard reference fuel.

**octane rating** *n.* Octane number.

**Oc·tans** (ŏk′tănz′) *n.* [Lat., half-quadrant < *octo,* eight.] A constellation in the Southern Hemisphere.

**oc·tant** (ŏk′tənt) *n.* [Lat. *octans, octant-,* half-quadrant < *octo,* eight.] **1.** One eighth of a circle: **a.** A 45° arc. **b.** The area enclosed by two radii at a 45° angle and the intersected arc. **2.** A navigational instrument based on the principle of the sextant but employing a 45° angle. **3.** *Astron.* The position of a celestial body when it is separated from another by a distance of 45°. **4.** One of eight parts into which three-dimensional space is divided by three usu. perpendicular coordinate planes. **—oc·tan′tal** (ŏk-tăn′təl) *adj.*

**oc·tave** (ŏk′tĭv, -tāv′) *n.* [OFr. < Lat. *octavus,* eighth < *octo,* eight.] **1.** *Mus.* **a.** The interval of eight diatonic degrees between two tones, the higher of which has twice as many vibrations per second as the lower. **b.** A tone eight full tones above or below another given tone. **c.** Two tones an octave apart that are sounded together. **d.** The consonance resulting when two such tones are sounded. **e.** A series of tones included within this interval or the keys of an instrument producing such a series. **f.** An organ stop producing tones an octave above those normally produced by the keys played. **2. a.** The eighth day after a feast day, counting the feast day as one. **b.** The period between a feast day and the eighth day following it. **3.** OCTET 3. **4. a.** A stanza of eight lines in poetry. **b.** OCTET 4. **5.** A rotating parry in fencing. **—oc·ta'val** (ŏk-tā'vəl, ŏk'tə-vəl) *adj.*

**oc·ta·vo** (ŏk-tā'vō, -tä') *n., pl.* **-vos.** [< Lat. *octavus,* eighth < *octo,* eight.] **1.** A page size equal to one eighth of a printer's sheet, orig. obtained by folding a sheet into eight leaves. **2.** A book made up of octavo pages.

**oc·ten·ni·al** (ŏk-tĕn'ē-əl) *adj.* [< LLat. *octennium,* period of eight years : Lat. *octo,* eight + Lat. *annus,* year.] **1.** Happening or recurring every eight years. **2.** Lasting eight years. **—oc·ten'ni·al** *n.* **—oc·ten'ni·al·ly** *adv.*

**oc·tet** (ŏk-tĕt') *n.* [Ital. *ottetto* < *otto,* eight < Lat. *octo.*] **1.** A musical composition written for eight voices or eight instruments. **2.** An ensemble of eight musicians. **3.** A group of eight. **4.** The first eight lines of an Italian sonnet.

**oc·til·lion** (ŏk-tĭl'yən) *n.* [OCT- + (M)ILLION.] **1.** The cardinal number equal to $10^{27}$. **2.** *Chiefly Brit.* The cardinal number equal to $10^{48}$.

**oc·til·lionth** (ŏk-tĭl'yənth) *n.* The ordinal number matching the number octillion in a series. **—oc·til'lionth** *adj.* & *adv.*

**octo-** *or* **octa-** *or* **oct-** *pref.* [Gk. *okta-, oktō-* (< *oktō*) and Lat. *octo-* (< *octo*).] Eight <*octane*>

**Oc·to·ber** (ŏk-tō'bər) *n.* [ME < Lat., eighth month < *octo,* eight.] **1.** The tenth month of the year in the Gregorian calendar. —See table at CALENDAR. **2.** *Chiefly Brit.* Ale brewed in Oct.

▲ word history: The Roman year originally began in March, and October was consequently the eighth month, as its name suggests: *October* is derived from Latin *octo,* "eight." The names of other months are also derived from the Latin names of numbers. *September,* the seventh month, is from *septem,* "seven"; *November,* the ninth month, is from *novem,* "nine"; *December,* the tenth month, is from *decem,* "ten." The months now known as July and August were originally named, respectively, *Quintilis,* "fifth month," and *Sextilis,* "sixth month."

**oc·to·dec·i·mo** (ŏk'tə-dĕs'ə-mō') *n., pl.* **-mos.** [< Lat. *octodecimus,* eighteenth < *octodecim,* eighteen : *octo,* eight + *decem,* ten.] **1.** A page size equal to one eighteenth of a printer's sheet, orig. obtained by folding a sheet into 18 leaves. **2.** A book made up of octodecimo pages.

**oc·to·ge·nar·i·an** (ŏk'tə-jə-nâr'ē-ən) *n.* [< Lat. *octogenarius,* containing eighty < *octogeni,* eighty each < *octoginta,* eighty : *octo,* eight + *-ginta,* times ten.] Being between 80 and 90 years of age. *—n.* A person between 80 and 90 years of age. **—oc'to·ge·nar'i·an** *adj.*

**oc·to·nar·y** (ŏk'tə-nĕr'ē) *adj.* [Lat. *octonarius,* containing eight < *octo,* eight.] **1.** Of or relating to the number eight. **2.** Having eight members or consisting of groups of eight. *—n., pl.* **-ies. 1.** OCTET 4. **2.** A group or set of eight.

**oc·to·pi** (ŏk'tə-pī') *n. var. pl. of* OCTOPUS.

**oc·to·ploid** (ŏk'tə-ploid') *adj.* Having eight haploid sets of chromosomes in a body cell. **—oc'to·ploid'** *n.*

**oc·to·pod** (ŏk'tə-pŏd') *n.* [NLat. *Octopoda,* order name < Gk. *oktopous,* octopus.] A mollusk of the order Octopoda, as an octopus, with eight arms or tentacles. **—oc'to·pod', oc'to·pod'ous** *adj.*

**oc·to·pus** (ŏk'tə-pəs) *n., pl.* **-pus·es** *or* **-pi** (-pī') [NLat. *Octopus,* genus name < Gk. *oktōpous,* eight-footed : *okto,* eight + *pous,* foot.] **1.** A carnivorous nocturnal marine mollusk of the genus *Octopus* or related genera, found worldwide, with a saclike body, a distinct head, and eight tentacles bearing double rows of suckers. **2.** Something, as a multinational corporation, that resembles an octopus in its many centrally controlled branches.

**oc·to·roon** (ŏk'tə-rōōn') *n.* [OCTO- + (QUAD)ROON.] One whose ancestry is one-eighth Negro.

**oc·to·syl·la·ble** (ŏk'tə-sĭl'ə-bəl) *n.* **1.** *also* **oc·to·syl·lab·ic** (ŏk'tō-sĭ-lăb'ĭk). **a.** A line of verse containing eight syllables. **b.** A poem written in octosyllables. **2.** A word of eight syllables. **—oc'to·syl·lab'ic** *adj.*

**oc·troi** (ŏk'troi', ôk-trwä') *n., pl.* **-trois** (ŏk'troiz'; ôk-trwä') [Fr. < OFr. < *octroyer,* to grant, perh. < Med. Lat. *auctorizare.* —see AUTHORIZE.] A tax, as in some European cities, levied on goods introduced from outside.

**oc·tu·ple** (ŏk'tə-pəl, -tōō'pəl, ŏk-tyōō'-) *adj.* **1.** Having eight parts, members, or copies. **2.** Multiplied by eight. *—vt. & vi.* **-pled, -pling, -ples.** To multiply or be multiplied by eight. **—oc'tu·ple** *n.*

**oc·u·lar** (ŏk'yə-lər) *adj.* [LLat. *ocularis* < Lat. *oculus,* eye.] **1. a.** Of or relating to the eye. **b.** Like the eye in form or function <an *ocular* design> **2.** Of or relating to the sense of sight. **3.** Seen by the eye : VISUAL <*ocular* evidence> *—n.* The eyepiece of an optical instrument.

**oc·u·list** (ŏk'yə-lĭst) *n.* [Fr. *oculiste* < Lat. *oculus,* eye.] **1.** An ophthalmologist. **2.** An optometrist.

**oc·u·lom·e·ter** (ŏk'yə-lŏm'ĭ-tər) *n.* [Lat. *oculus,* eye + -METER.] A device for measuring the direction, speed, and extent of eye movement.

**oc·u·lo·mo·tor** (ŏk'yə-lō-mō'tər) *adj.* [Lat. *oculus,* eye + MOTOR.] **1.** Relating to movement of the eyeball. **2.** Relating to the oculomotor nerve.

**oculomotor nerve** *n.* Either of the two cranial nerves that control the muscles of the eyeballs.

**Od** *or* **Odd** (ŏd) *interj.* [Alteration of GOD.] *Archaic.* —Used as a mild oath.

**OD** (ō'dē') [O(VER)D(OSE).] *n. Slang.* **1.** An overdose of a narcotic drug. **2.** One who has taken an overdose. **—OD** *v.* **(OD'd, OD'ing, OD's).**

**o·da·lisque** *also* **o·da·lisk** (ōd'ə-lĭsk') *n.* [Fr. < Turk. *ōdalik,* chambermaid : *ōdah,* room + *-lik,* suffix expressing function.] A woman slave or concubine in a harem.

**odd** (ŏd) *adj.* **-er, -est.** [ME *odde* < ON *oddi,* odd number.] **1.** Deviating from the customary or accepted : UNUSUAL. **2. a.** In excess of the indicated or approximate number, degree, or extent <20-*odd* students> ***usage:*** When used with round numbers to indicate an approximate figure, *odd* should be preceded by a hyphen to avoid confusion: *a career that spanned 40-odd years.* **b.** Being a remainder <a few *odd* tickets still unsold> **c.** Small or indefinite in amount <*odd* change> **3. a.** Being one of an incomplete pair or set <an *odd* glove> **b.** Remaining after others are paired or grouped <the *odd* guest at the dinner table> **4.** *Math.* Designating an integer not divisible by 2, as 1, 3, and 5. **5.** Not expected, regular, or frequent <worked *odd* hours> **6.** Remote : out-of-the-way. *—n.* **1.** Something odd. **2. a.** In U.S. play, a golf score one stroke higher than the score of one's opponent. **b.** In British play, a golfing handicap of one stroke given to a player as odds or an advantage of one stroke taken away from a player's score as odds. **—odd man out.** One who, by strangeness of behavior or belief, stands alone in a group. **—odd'ly** *adv.* **—odd'ness** *n.*

**Odd** *interj. var. of* OD.

**odd·ball** (ŏd'bôl') *n. Informal.* An eccentric person.

**Odd Fellow** *n.* A member of the Independent Order of Odd Fellows, a fraternal and benevolent secret society.

**odd·ish** (ŏd'ĭsh) *adj.* Somewhat odd.

**odd·i·ty** (ŏd'ĭ-tē) **1.** One that is odd. **2.** The state or quality of being odd : STRANGENESS.

**odd lot** *n.* A quantity different from a standard trading unit, esp. a block of fewer than 100 shares of stock.

**odd·ment** (ŏd'mənt) *n.* **1. a.** Something left over. **b. oddments.** Odds and ends. **2.** An oddity.

**odd-pin·nate** (ŏd'pĭn'āt') *adj.* Pinnate with a single, unpaired leaflet at the end of the leafstalk. **—odd'-pin'nate'ly** *adv.*

**odds** (ŏdz) *pl.n.* [Pl. of ODD.] **1.** An advantage, as score points, assigned to a weaker side in a contest to equalize the chances of all participants. **2.** A ratio expressing the probability of an event or outcome <The *odds* are three to one that our team will win.> **3.** A ratio between the amount of payment for a winning bet and the amount stated. **4.** The likelihood of something occurring or being so <The *odds* are that I will get the job.> **5.** An amount or degree by which one thing exceeds or falls short of another <lost the vote by substantial *odds*> **—at odds.** In conflict : in disagreement. **—by all odds.** Beyond any doubt : UNQUESTIONABLY <*by all odds* the greatest composer of our time>

**odds and ends** *pl.n.* Miscellaneous items or remnants.

**ode** (ōd) *n.* [Fr. < OFr. < LLat. *oda* < Gk. *aoidē,* song.] **1.** A classical poem intended to be sung by a chorus at a public festival or as part of a drama. **2.** A lyrical poem, often in praise of an object, person, or quality and usu. marked by exalted style. **—od'ic** (ō'dĭk) *adj.*

**-ode** *suff.* [Gk. *-odos* < *hodos,* way.] **1.** Way : path <electr*ode*> **2.** Electrode <dyn*ode*>

**o·de·um** (ō-dē'əm, ō'dē-) *n., pl.* **o·de·a** (ō-dē'ə, ō'dē-ə) [Lat. < Gk. *ōideion* < *aoidē,* song.] **1.** A small ancient Greek or Roman building used for public performances of music and poetry. **2.** A contemporary theater or music hall.

**O·din** (ō'dĭn) *n.* [ON *Ōdhinn.*] *Norse Myth.* The supreme deity and creator of the cosmos and man, the god of wisdom, war, art, culture, and the dead.

**o·di·ous** (ō'dē-əs) *adj.* [ME < OFr. *odios* < Lat. *odiosus* < *odium,* hatred < *odisse,* to hate.] Arousing hatred or repugnance : ABHORRENT. **—o'di·ous·ly** *adv.* **—o'di·ous·ness** *n.*

**o·di·um** (ō'dē-əm) *n.* [Lat., hatred < *odisse,* to hate.] **1.** The quality or state of being odious. **2.** Extreme disgust or contempt. **3.** Disgrace resulting from contemptible behavior.

**o·do·graph** (ō'də-grăf') *n.* [Gk. *hodos,* journey + -GRAPH.] **1.** A device for measuring speed and distance traveled on foot. **2.** An instrument for recording the distance and route traveled by a vehicle.

**o·dom·e·ter** (ō-dŏm'ĭ-tər) *n.* [Fr. *odomètre* : Gk. *hodos,* journey + Fr. *-mètre,* -meter.] An instrument indicating distance traveled by a vehicle. **—o·dom'e·try** *n.*

**-odon** *suff.* [NLat. < Gk. *odous,* tooth.] An animal having a specified kind of teeth <mast*odon*>

**o·do·nate** (ōd'n-āt', ō-dŏn'-) *n.* [NLat. *Odonata,* order name < Gk. *odous, odor,* tooth.] A predacious winged insect of the order Odonata, including the dragonflies and damselflies.

**odont-** *pref. var. of* ODONTO-.

**-odont** *suff.* [< Gk. *odous, odont-,* tooth.] Having teeth of a specified kind <pleur*odont*>

**-odontia** *suff.* [NLat. < Gk. *odous,* tooth.] The form of, condition of, or manner of treating the teeth <orth*odontia*>

**odonto-** or **odont-** *pref.* [Gk. < *odous,* tooth.] Tooth <*odonto*phore>

**o·don·to·blast** (ō-dŏn'tə-blăst') *n.* A tooth cell in the outer layer of dental pulp that produces dentine. **—o·don'to·blas'tic** *adj.*

**o·don·toid** (ō-dŏn'toid') *adj.* **1.** Resembling a tooth. **2.** Of or relating to the odontoid process.

**odontoid process** *n. Anat.* A small toothlike projection from the second vertebra of the neck around which the first vertebra rotates.

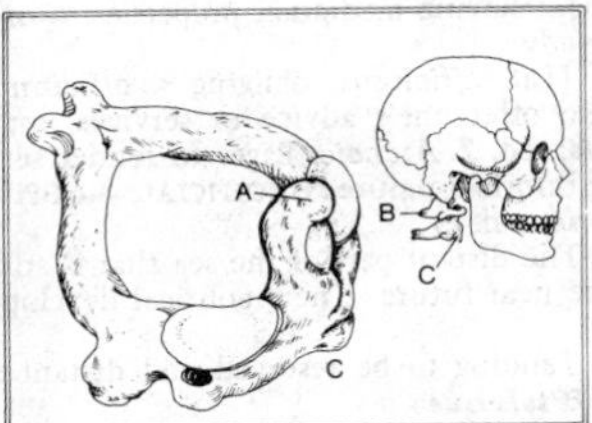


**odontoid process**
*A. odontoid process,*
*B. atlas, C. axis*

**o·don·tol·o·gy** (ō'dŏn-tŏl'ə-jē) *n.* Study of the anatomy, growth, and diseases of the teeth. **—o·don'to·log'i·cal** (-tə-lŏj'ĭ-kəl) *adj.* **—o·don'to·log'i·cal·ly** *adv.* **—o'don·tol'o·gist** *n.*

**o·don·to·phore** (ō-dŏn'tə-fôr', -fōr') *n.* A protrusile structure at the base of the mouth of most mollusks that supports the radula. **—o'don·toph'o·ral** (ō'dŏn-tŏf'ər-əl), **o'don·toph'o·rine'** (-ə-rīn', -rĭn), **o'don·toph'o·rous** (-ər-əs) *adj.*

**o·dor** (ō'dər) *n.* [ME *odour* < OFr. < Lat. *odor.*] **1.** The property or quality of a thing that stimulates or is perceived by the sense of smell <the foul *odor* of the chemical plant> **2.** A sensation, stimulation, or perception of the sense of smell. **3.** A pervasive quality : AIR <an *odor* of sanctity> **4.** Repute : esteem <a theory that is not currently in good *odor*> **—o'dored** *adj.*

**o·dor·if·er·ous** (ō'də-rĭf'ər-əs) *adj.* Having or giving off an odor. **—o'dor·if'er·ous·ly** *adv.* **—o'dor·if'er·ous·ness** *n.*

**o·dor·less** (ō'dər-lĭs) *adj.* Having no odor. **—o'dor·less·ly** *adv.* **—o'dor·less·ness** *n.*

**o·dor·ous** (ō'dər-əs) *adj.* Having a distinctive odor, as: **a.** Fragrant. **b.** Malodorous. **—o'dor·ous·ly** *adv.* **—o'dor·ous·ness** *n.*

**o·dour** (ō'dər) *n. Chiefly Brit. var. of* ODOR.

**O·dys·seus** (ō-dĭs'yōōs', ō-dĭs'ē-əs) *n.* [Gk. *Odusseus.*] *Gk. Myth.* The king of Ithaca, a leader of the Greeks in the Trojan War, who reached home after ten years of wandering.

**od·ys·sey** (ŏd'ĭ-sē) *n., pl.* **-seys.** [After the *Odyssey,* a Homeric epic recounting the wanderings of Odysseus after the fall of Troy < Fr. *Odyssée* < Lat. *Odyssea* < Gk. *Odusseia* < *Odusseus,* Odysseus.] **1.** An extended adventurous wandering. **2.** A spiritual or intellectual quest.

**oe·de·ma** (ĭ-dē'mə-) *n. var. of* EDEMA.

**oed·i·pal** (ĕd'ə-pəl, ē'də-) *adj. often* **Oedipal.** Of, pertaining to, or typical of the Oedipus complex. **—oed'i·pal·ly** *adv.*

**Oed·i·pus** (ĕd'ə-pəs, ē'də-) *n.* [Gk. *Oidipeus* : *oidan,* to swell + *pous,* foot.] *Gk. Myth.* A son of Laius and Jocasta, who was abandoned at birth and unintentionally killed his father and married his mother.

**Oedipus complex** *n.* Unconscious libidinal feelings in a child, esp. a male child, for the parent of the opposite sex, gen. appearing between the ages of three and five.

**oe·nol·o·gy** (ē-nŏl'ə-jē) *n.* [Gk. *oinos,* wine + -LOGY.] Study of wines and wine-making. **—oe'no·log'i·cal** (ē'nə-lŏj'ĭ-kəl) *adj.* **—oe·nol'o·gist** *n.*

**oe·no·mel** (ē'nə-mĕl') *n.* [Gk. *oinomeli* : *oinos,* wine + *meli,* honey.] A beverage of ancient Greece, made of wine and honey.

**o'er** (ôr, ōr) *prep. & adv.* Over.

**oer·sted** (ûr'stĕd') *n.* [After Hans Christian *Oersted* (1777–1851).] The centimeter-gram-second electromagnetic unit of magnetic intensity, equal to the magnetic intensity one centimeter from a unit magnetic pole.

**oe·soph·a·gus** (ĭ-sŏf'ə-gəs) *n. var. of* ESOPHAGUS.

**oes·tro·gen** (ĕs'trə-jən) *n. var. of* ESTROGEN.

**oes·trus** (ĕs'trəs) *n. var. of* ESTRUS.

**oeu·vre** (œ'vrə) *n., pl.* **oeu·vres** (œ'vrə) [Fr. < Lat. *opus,* work.] **1.** A work of art. **2.** The entire body of an artist's lifework.

**of** (ŭv, ŏv; *unstressed* əv) *prep.* [ME < OE.] **1.** Derived or coming from <tribes *of* the south> **2.** Caused by <the patient's slow death *of* cancer> **3.** At a distance from <a mile west *of* there> **4.** So as to be separated or relieved from <robbed *of* one's self-respect><cured *of* measles> **5.** From the total or group comprising <give *of* one's energies><two *of* my associates><most *of* the cases> **6.** Composed or made from <a coat *of* wool> **7.** Associated with or adhering to <a member *of* your religion> **8.** Belonging or connected to <the rungs *of* a ladder> **9. a.** Possessing : having <a person *of* courage> **b.** On the part of <very kind *of* you> **10.** Containing or carrying <a basket *of* peaches> **11.** Specified as <a depth *of* 50 meters><the Garden *of* Eden> **12.** Centering on : directed toward <a love *of* dogs> **13.** Produced by : issuing from <products *of* the vine> **14.** Characterized or identified by <a year *of* discord> **15. a.** With reference to : ABOUT <thought little *of* my proposals><speak *of* it tomorrow> **b.** In respect to <slow *of* speech> **16.** Set aside for <needed a day *of* rest> **17.** Before : until <ten minutes *of* one> **18.** During or on a specified time <*of* recent years> **19.** By <beloved *of* one's family> **20.** —Used to indicate an appositive <that idiot *of* a driver> **21.** *Archaic.* On <"A plague *of* all cowards, I say" —Shakespeare> ***usage:*** Grammarians have sometimes condemned the "double genitive" construction, as in *a friend of my sister's.* This usage is well supported by literary precedent, however, and should be regarded as entirely acceptable at all levels.

**off** (ôf, ŏf) *adv.* [ME *of* < OE.] **1.** From a place or position <drive *off*> **2. a.** At a certain distance in space or time <a kilometer *off*><a month *off*> **b.** From a given course or route : ASIDE <swerved *off* into a gulley> **c.** Into an unconscious state <nodded *off*> **3. a.** So as to be no longer on, attached, or connected <shaved *off* the beard> **b.** So as to be divided <marked *off* the field by yards> **4.** So as to be no longer continuing or operating <turn *off* the TV> **5.** So as to be completely removed, finished, or eliminated <write *off* a report><kill *off* the fleas> **6.** So as to be smaller, fewer, or less <Car sales dropped *off.*> **7.** So as to be away from work or duty <took a week *off*> **8.** Offstage. *—adj.* **1. a.** Distant or removed : FARTHER <the *off* side of the warehouse> **b.** Remote : slim <on the *off* chance that I'm not busy> **2.** Not on, attached, or connected <with my gloves *off*> **3.** Not operating or operational <The TV is *off.*> **4.** No longer taking place : CANCELED <The party is *off.*> **5.** Slack <Production is *off* this year.> **6. a.** Below standard <Your pitching is *off* today.> **b.** Inaccurate : incorrect <Your statistical results are *off.*> **c.** Somewhat crazy : ECCENTRIC <is a little bit *off*> **7.** Started on the way : GOING <I'm *off* to see the wizard.> **8. a.** Absent or away from work or duty <I'm *off* today.> **b.** Spent away from work or duty <My *off* day is Monday.> **9.** On the right side of a vehicle or draft team <The *off* horse is lame.> **10.** *Naut.* Farthest from the shore : SEAWARD. **11.** Designating or toward the side of the cricket field facing the batsman. **12.** Off-color. *—prep.* **1.** So as to be removed or distant from <The bird hopped *off* the twig.> **2.** Away or relieved from <*off* duty> **3. a.** By consuming <living *off* bark and weeds> **b.** With the means provided by <living *off* my investments> **c.** *Nonstandard.* From <got a loan *off* us> **4.** Extending or branching out from <an artery *off* the heart> **5.** Not up to the usual standard of <*off* your tennis game> **6.** So as to abstain from <went *off* drugs> **7.** To seaward of <a mile *off* Edisto Island> *—v.* **offed, off·ing, offs.** *—vi.* To go away : LEAVE. *—vt. Slang.* To murder. **—off and on.** Intermittently <slept *off and on* during the flight>

**of·fal** (ô'fəl, ŏf'əl) *n.* [ME : *of-,* off + *fal,* fall.] **1.** Waste parts or leavings, esp. of a butchered animal. **2.** Filth : refuse.

**off·beat** (ôf'bēt', ŏf'-) *n.* An unaccented beat in a musical measure. *—adj.* (ôf'bēt', ŏf'-) *Slang.* Not conforming to an ordinary pattern or type : UNCONVENTIONAL.

**off-Broad·way** (ôf'brôd'wā', ŏf'-) *adj.* **1.** Designating or relating to often experimental and low-cost theatrical activity presented in theaters outside the Broadway entertainment district of New York City. **2.** Located outside the Broadway entertainment district.

**off·cast** (ôf'kăst', ŏf'-) *adj.* Discarded : rejected.

**off-col·or** (ôf'kŭl'ər, ŏf'-) *adj.* **1.** Varying from the standard or desired color. **2.** In poor taste : INDECENT <an *off-color* remark> **3.** *Chiefly Brit.* Not in good spirits or health.

**of·fence** (ə-fĕns') *n. Chiefly Brit. var. of* OFFENSE.

**of·fend** (ə-fĕnd') *v.* **-fend·ed, -fend·ing, -fends.** [ME *offenden* < OFr. *ofendre* < Lat. *offendere.*] *—vt.* **1.** To arouse anger, resentment, or indignation in. **2.** To be disagreeable or displeasing to <crude language that *offended* our ears> **3. a.** To transgress : violate. **b.** *Obs.* To cause to sin. *—vi.* **1.** To cause displeasure or affront <Slovenly habits may *offend.*> **2. a.** To violate a moral or divine law <*offend* against decency> **b.** To violate a law or rule. **—of·fend'er** *n.*

**of·fense** (ə-fĕns') *n.* [ME < OFr. *ofense* < Lat. *offensa* < p.part. of *offendere,* to offend.] **1. a.** The act of offending or causing anger, displeasure, resentment, or affront. **b.** The state of being offended <took *offense* at the thoughtless remark> **2. a.** A breach of a social or moral code : SIN. **b.** A violation of law : CRIME. **3.** Something that outrages moral sensibilities <Torture is an *offense* to the human spirit.> **4.** (ô'fĕns'). The act of attacking or assaulting. **5.** (ô'fĕns').

ōō b**oo**t ou **ou**t th **th**in *th* **th**is ŭ c**u**t ûr **ur**ge y **y**oung
yōō ab**u**se zh vi**si**on ə **a**bout, it**e**m, ed**i**ble, gall**o**p, circ**u**s

**a.** An athletic team having possession of a ball or puck. **b.** The ability or potential of a team to score points. **c.** The means or tactics used in an attempt to score points.

**of·fen·sive** (ə-fĕn′sĭv) *adj.* **1.** Disagreeable to the senses <an *offensive* smell> **2.** Causing anger, resentment, displeasure, or affront <an *offensive* slur> **3. a.** Engaged in an attack. **b.** Of, relating to, or used for attack <*offensive* weapons> **c.** Of or relating to a team in possession of a ball or puck <the *offensive* backfield> —*n.* **1.** An attitude of attack <took the *offensive*> **2.** An attack : assault <opened *offensives* on two fronts> **—of·fen′sive·ly** *adv.* **—of·fen′sive·ness** *n.*

**of·fer** (ô′fər, ŏf′ər) *v.* **-fered, -fer·ing, -fers.** [ME *offren,* ult. < Lat. *offerre* : *ob,* to + *ferre,* to bring.] —*vt.* **1.** To present for acceptance or rejection : PROFFER. **2.** To put forward or submit for consideration : PROPOSE <*offer* a suggestion> **3.** To present for sale. **4.** To propose as : BID. **5.** To present as an act of worship or veneration <*offer* a blessing> **6.** To exhibit readiness or desire to do : VOLUNTEER <*offered* their assistance> **7.** To put up : DISPLAY <The opposing side *offered* a good fight.> **8.** To make available : EXTEND <*offer* credit> **9.** To furnish : provide <a resort that *offers* swimming and tennis> **10.** To present to an audience or to the public <a network *offering* all new programs> —*vi.* **1.** To present an offering in worship or devotion. **2.** To make an offer or proposal, esp. to make an offer of marriage. **3.** To present itself <waited years until a better job *offered*> —*n.* **1.** An act of offering <an *offer* of aid> **2.** Something offered, as a proposal, suggestion, bid, or invitation. **3.** *Law.* A proposal that if accepted constitutes a legally binding contract. **4.** The condition of being offered, esp. for sale <a new line of products on *offer*> **5. a.** An attempt : try. **b.** A show of intention. **—of′fer·er, of′fer·or** *n.*

☆ ***syns:*** OFFER, EXTEND, PRESENT, PROFFER, TENDER, VOLUNTEER *v. core meaning* : to put before another for acceptance or rejection <*offered* us another drink><*offered* help>

**of·fer·ing** (ô′fər-ĭng, ŏf′ər-) *n.* [ME *offring* < OE *offrung* < *offrian,* to offer a sacrifice or gift < Lat. *offerre,* to offer.] **1.** An act of making an offer. **2.** Something offered. **3.** A presentation made to a deity as an act of religious worship or sacrifice : OBLATION. **4.** A gift or contribution, esp. one made at a religious service.

**of·fer·to·ry** (ô′fər-tôr′ē, -tōr′ē, ŏf′ər-) *n., pl.* **-ries.** [Med. Lat. *offertorium* < Lat. *offerre,* to offer.] **1.** *often* **Offertory. a.** The part of the Eucharistic liturgy at which bread and wine are offered to God. **b.** A musical setting of the Offertory. **2.** A collection of offerings at a religious service.

**off·hand** (ôf′hănd′, ŏf′-) *adv.* Without forethought or preparation : EXTEMPORANEOUSLY. —*adj. also* **off·hand·ed** (-hăn′dĭd). In an offhand way. **—off′hand′ed·ly** *adv.* **—off′hand′ed·ness** *n.*

**off-hour** (ôf′our′, ŏf′-) *n.* A period of slack activity or demand.

**of·fice** (ô′fĭs, ŏf′ĭs) *n.* [ME < OFr., position of responsibility < Lat. *officium.*] **1. a.** A place in which business, professional, or clerical activities are conducted. **b.** The personnel working in an office. **2.** A function or duty assigned to or assumed by someone. **3.** A position of authority, duty, or trust, as in a government or institution <the *office* of mayor> **4. a.** A subdivision of a governmental department <the U.S. Patent *Office*> **b.** A major executive division of a government <the British Home *Office*> **5.** A public position <was elected to *office*> **6. offices.** *Chiefly Brit.* The areas of a house, as the kitchen and laundry, in which domestic chores are performed. **7.** *often* **offices.** A usu. beneficial act performed for another. **8.** An ecclesiastical ceremony or service usu. prescribed by liturgy, esp.: **a.** *Rom. Cath. Ch.* The canonical hours. **b.** A prayer service in the Anglican Church, as Morning or Evening Prayer. **c.** A ceremony for a special purpose, esp. a rite for the dead.

**office boy** *n.* A boy or man employed in a business office to do miscellaneous jobs.

**of·fice·hold·er** (ô′fĭs-hōl′dər, ŏf′ĭs-) *n.* A holder of public office.

**of·fi·cer** (ô′fĭ-sər, ŏf′ĭ-) *n.* [ME < OFr. *officier* < Med. Lat. *officarius* < Lat. *officium,* office.] **1.** One holding an office of authority or trust in an organization, as a government or corporation. **2.** One who holds a commission in the armed services. **3.** A licensed master, mate, chief engineer, or assistant engineer in the merchant marine. **4.** A member of a police force.

**officer of the day** *n.* A military officer who for a given day assumes responsibility for order, security, and the performance of the guard.

**officer of the deck** *n.* A naval officer assigned to represent the commanding officer of a vessel or installation for a specified period.

**of·fi·cial** (ə-fĭsh′əl) *adj.* [< ME, ecclesiastical officer < OFr. < LLat. *officialis* < Lat., of an office < *officium,* office.] **1.** Of or relating to an office or post of authority <*official* responsibilities> **2.** Authorized by a proper authority : AUTHORITATIVE <*official* recognition> **3.** Holding office or serving in a public capacity <an *official* delegate> **4.** Characteristic of or suitable to a position of authority : FORMAL <an *official* manner> **5.** Designating drugs authorized by or contained in the U.S. Pharmacopoeia or National Formulary. —*n.* **1.** One who holds an office or position, esp. one acting in a subordinate capacity for an institution such as a corporation or governmental agency. **2.** A sports referee or umpire. **—of·fi′cial·dom** *n.* **—of·fi′cial·ly** *adv.*

**of·fi·cial·ese** (ə-fĭsh′ə-lēz′, -lēs′) *n.* Language characteristic of official documents or statements, often regarded as obscure, wordy, or pretentiously formal in style.

**of·fi·cial·ism** (ə-fĭsh′ə-lĭz′əm) *n.* Rigid adherence to official regulations, procedures, and formalities.

**of·fi·ci·ant** (ə-fĭsh′ē-ənt) *n.* One who officiates at a religious service or ceremony : CELEBRANT.

**of·fi·ci·ar·y** (ə-fĭsh′ē-ĕr′ē) *n., pl.* **-ies. 1.** A body of officers or officials. **2.** An officer or official. —*adj.* **1.** Attached to or resulting from an office held. —Used of a title. **2.** Having a title resulting from the holding of an office. —Used of a dignitary.

**of·fi·ci·ate** (ə-fĭsh′ē-āt′) *vi.* **-at·ed, -at·ing, -ates.** [Med. Lat. *officiare, officiat-,* to conduct a religious service < Lat. *officium,* office.] **1.** To perform the functions and duties of an office. **2.** To minister at a religious service. **3.** To serve as a sports umpire or referee. **—of·fi′ci·a′tion** *n.* **—of·fi′ci·a′tor** *n.*

**of·fic·i·nal** (ə-fĭs′ə-nəl, ô′fĭ-sī′nəl, ŏf′ĭ-) *adj.* [Med. Lat. *officinalis,* kept in stock < Lat. *officina,* workshop < *opifex,* workman : *opus,* work + *facere,* to do.] **1.** Designating a drug available without prescription. **2.** Designating a plant having medicinal properties. **—of·fic′i·nal** *n.* **—of·fic′i·nal·ly** *adv.*

**of·fi·cious** (ə-fĭsh′əs) *adj.* [Lat. *officiosus,* obliging < *officium,* duty.] **1.** Excessively ready to offer one's advice or services, esp. where unnecessary and unwanted. **2.** *Archaic.* Eager to render services or help others. **3.** Of an informal nature : UNOFFICIAL. **—of·fi′cious·ly** *adv.* **—of·fi′cious·ness** *n.*

**off·ing** (ô′fĭng, ŏf′ĭng) *n.* **1.** The distant part of the sea that is still visible from the shore. **2.** The near future <new political developments in the *offing*>

**off·ish** (ô′fĭsh, ŏf′ĭsh) *adj.* Tending to be reserved and distant : ALOOF. **—off′ish·ly** *adv.* **—off′ish·ness** *n.*

**off-key** (ôf′-kē′, ŏf′-) *adj.* **1.** Pitched higher or lower than the correct notes of a melody. **2.** At odds with what is considered proper or normal : IRREGULAR. **—off′-key′** *adv.*

**off-lim·its** (ôf-lĭm′ĭts, ŏf-) *adj.* Prohibited to a designated group.

**off-line** (ôf′līn′, ŏf′-) *adj. Computer Sci.* Not under the control of a central computer, as in a manufacturing process.

**off-load** (ôf′lōd′, ŏf′-) *vt.* **-load·ed, -load·ing, -loads. 1.** To launch (a guided missile or rocket) with propellant tanks less than fully loaded in order to alter the center of gravity of the vehicle. **2.** To unload (a vehicle, esp. an aircraft).

**off of** *prep. Nonstandard.* Off.

**off-off-Broadway** (ôf′ôf-brôd′wā′, ŏf′ŏf-) *n.* A theatrical movement in New York City emphasizing experimental, avant-garde techniques and productions.

**off·print** (ôf′prĭnt′, ŏf′-) *n.* A reproduction of a printed article excerpted from a larger publication. **—off′print** *v.* **(-print·ed, -print·ing, -prints).**

**off-put·ting** (ôf′po͝ot′ĭng, ŏf′-) *adj.* Tending to repel or disconcert <a scornful manner that can be very *off-putting*>

**off-road vehicle** (ôf′rōd′, ŏf′-) *n.* A motor vehicle that is capable of being driven over unpaved surfaces.

**off·scour·ing** (ôf′skour′ĭng, ŏf′-) *n.* **1.** Something scoured off : REFUSE. **2.** A social misfit or outcast.

**off-screen** (ôf′-skrēn′, ŏf′-) *adj & adv.* **1.** Out of sight of the television or motion-picture viewer <a character who dies *off-screen*> **2.** In private.

**off-sea·son** (ôf′-sē′zən, ŏf′-) *n.* A period marked by the cessation or slackening of activity.

**off·set** (ôf′sĕt′, ŏf′-) *n.* **1.** Something that balances, compensates, or counteracts. **2.** Something deriving from but set off from something else. **3.** A ledge or recess in a wall formed by a reduction in thickness above : SETOFF. **4.** *Bot.* OFFSHOOT 3. **5.** *Geol.* A spur of a range of mountains or hills. **6.** A bend in a bar or pipe enabling it to pass around an obstruction. **7.** A crosscut or drift from a main level of a mine. **8.** A short distance measured perpendicularly from the main line, used in surveying to help in calculating the area of an irregular plot. **9.** OFFSHOOT 2. **10. a.** Offset printing. **b.** The unintentional transfer of fresh ink from one sheet or surface to another. —*v.* (ôf′sĕt′, ŏf′-, ôf-sĕt′, ŏf-) **-set, -set·ting, -sets.** —*vt.* **1.** To counterbalance, counteract, or compensate for <sufficient profits to *offset* earlier losses> **2. a.** To print by offset. **b.** To smear with an offset. **3.** To make or form an offset in (a wall, pipe, or bar). —*vi.* To develop as an offset. **—off′set′** *adj.*

**offset printing** *n.* Planographic printing by indirect image transfer, as: **a.** Printing from photomechanical plates. **b.** Printing from paper mats.

**off·shoot** (ôf′sho͞ot′, ŏf′-) *n.* **1.** Something that branches out or develops from a source. **2.** A branch, descendant, or member of a family or social group. **3.** A lateral shoot from the main stem of a plant.

**off·shore** (ôf′shôr′, -shōr′, ŏf′-) *adj.* **1.** Moving or directed away from the shore <an *offshore* current> **2. a.** Located or occurring at a distance from the shore <an *offshore* oil rig> **b.** Located or based outside the United States and not subject to U.S. tax laws <*offshore* corporations> —*adv.* **1.** Away from the shore <The wind was blowing *offshore.*> **2.** At a distance from the shore <a ship anchored *offshore*>

**off·side** *also* **off side** (ôf'sīd', ŏf'-) *adj.* Illegally forward or in advance of the ball or puck, as in football or hockey. **—off'side'** *adv.*

**off·spring** (ôf'sprĭng', ŏf'-) *n., pl.* **offspring.** [ME *ofspring* < OE : *of,* off + *springan,* to rise.] **1.** The progeny of a person, animal, or plant. **2.** A product : result.

**off-stage** (ôf'stāj', ŏf'-) *adj.* Located or occurring in the area of a stage invisible to the audience. —*adv.* **1.** Away from the area of a stage visible to the audience. **2. a.** Apart from public life <a performer known *off-stage* as a philanthropist> **b.** In private <The negotiations took place *off-stage.*>

**off-the-rack** (ôf'thə-răk', ŏf'-) *adj.* READY-MADE 1.

**off-the-rec·ord** (ôf'thə-rĕk'ərd, ŏf'-) *adj.* Not for publication or attribution <*off-the-record* remarks> **—off'-the-rec'ord** *adv.*

**off-the-shelf** (ôf'thə-shĕlf', ŏf'-) *adj.* Of or relating to merchandise carried in stock that is deliverable without alteration. **—off'-the-shelf'** *adv.*

**off-the-wall** (ôf'thə-wôl', ŏf'-) *adj. Informal.* Unconventional, unusual, or bizarre <*off-the-wall* behavior>

**off-track** (ôf'trăk', ŏf'-) *adj.* Of or relating to off-track betting.

**off-track betting** *n.* A system of gambling on horses in which bets are placed away from a racetrack.

**off-white** (ôf'hwīt', -wīt', ŏf'-) *n.* A grayish or yellowish white. **—off'-white'** *adj.*

**off year** *n.* **1.** A year in which no major political elections occur. **2.** A year of decreased production or activity <an *off year* for sales>

**oft** (ôft, ŏft) *adv.* [ME < OE.] Often.

**of·ten** (ô'fən, ŏf'ən) *adv.* [ME < *oft* < OE.] Many times : FREQUENTLY.

**of·ten·times** (ô'fən-tīmz', ŏf'ən-) *also* **oft·times** (ôft'tīmz'-, ŏf'-) *adv.* Repeatedly : frequently.

**O·ga·la·la** (ō'gə-lä'lə) *n. var. of* OGLALA.

**og·am** (ŏg'əm, ō'əm) *n. var. of* OGHAM.

**o·gee** (ō'jē') *n.* [Alteration of OGIVE.] **1.** A molding having an S-shaped profile. **2.** An arch formed by two curves meeting at a point.

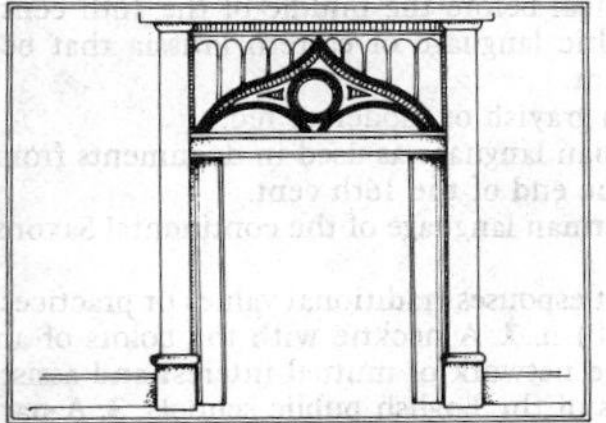

**ogee**
*Gothic ogee in the framework of a doorway*

**og·ham** *or* **og·am** (ŏg'əm, ō'əm) *n.* [Ir. Gael. < OIr. *ogom.*] **1.** An alphabet used for writing Irish from the 4th or 5th cent. A.D. to the early 7th cent. **2.** An inscription in the ogham alphabet.

**o·give** (ō'jīv') *n.* [< Fr., diagonal rib of a vault < OFr. *augive.*] **1.** *Statistics.* **a.** Graphic representation of a frequency distribution in which every ordinate represents the sum of frequencies in preceding intervals. **b.** A frequency distribution. **2. a.** A diagonal rib of a Gothic vault. **b.** A pointed arch. **—o·gi'val** *adj.*

**O·gla·la** (ō-glä'lə) *also* **O·ga·la·la** (ō'gə-lä'lə) *n., pl.* **Oglala** *or* **-las** *also* **Ogalala** *or* **-las. 1. a.** A tribe of Indians of the Teton Dakota group, inhabiting the area west of the Missouri River in South Dakota. **b.** A member of this tribe. **2.** The Siouan language of the Oglala.

**o·gle** (ō'gəl, ô'gəl) *v.* **o·gled, o·gling, o·gles.** [Perh. < LG *oegeln,* freq. of *oegen,* to eye < *oog,* eye.] —*vt.* **1.** To stare at. **2.** To stare at in an impertinent, provocative, or flirtatious way. —*vi.* To stare impertinently, provocatively, or flirtatiously. **—o·gle** *n.* **—o'gler** *n.*

**o·gre** (ō'gər) *n.* [Fr.] **1.** A man-eating giant or monster of folktale and legend. **2.** One who is particularly cruel, hideous, or brutish. **—o'gre·ish** (ō'gər-ĭsh, ō'grĭsh) *adj.*

**o·gress** (ō'grĭs) *n.* A woman who behaves like an ogre.

**oh** (ō) *interj.* **1.** —Used to express strong emotion, as surprise, fear, anger, or pain. **2.** —Used in direct address <*Oh,* sir! You forgot your coat.> **3.** —Used to indicate understanding or acknowledgment of a statement.

**O·hi·o buckeye** (ō-hī'ō) *n.* A tree, *Aesculus glabra* of the central United States, with compound leaves and yellowish-green flowers.

**ohm** (ōm) *n.* [After Georg S. *Ohm* (1787–1854).] A unit of electrical resistance equal to that of a conductor in which a current of one ampere is produced by a potential of one volt across its terminals. **—ohm'ic** (ō'mĭk) *adj.* **—ohm'i·cal·ly** *adv.*

**ohm·age** (ō'mĭj) *n.* Resistance expressed in ohms.

**ohm·me·ter** (ōm'mē'tər) *n.* An instrument for measuring the resistance of a conductor directly in ohms.

**o·ho** (ō-hō') *interj.* —Used esp. to express surprise or feigned astonishment.

**-oic** *suff.* [-O- + -IC.] Containing a carboxyl group or one of its derivatives <decano*ic* acid>

**-oid** *suff.* [Gk. *-oeidēs* < *eidos,* shape, form.] **1.** Having the appearance of : RESEMBLING <acanth*oid*> **2.** One that resembles something specified or has a specified quality <human*oid*>

**o·id·i·um** (ō-ĭd'ē-əm) *n., pl.* **-i·a** (-ē-ə) [NLat. *Oidium,* fungus genus < Gk. *oiōn,* egg.] A thin-walled spore in some filamentous fungi that is produced by fragmentation.

**oil** (oil) *n.* [ME < OFr. *oile* < Lat. *oleum,* olive oil < Gk. *elaion* < *elaia,* olive.] **1.** Any of numerous mineral, vegetable, and synthetic substances and vegetable and animal fats that are gen. slippery, combustible, viscous, liquid or liquefiable at room temperatures, soluble in various organic solvents such as ether but not in water, and used as food and in a wide variety of products, esp. fuels and lubricants. **2. a.** Petroleum. **b.** A petroleum derivative, as a motor oil. **3.** A substance having an oily consistency. **4.** An oil color. **5.** OIL PAINTING 1. **6.** *Informal.* Insincere flattery. —*v.* **oiled, oil·ing, oils.** —*vt.* To lubricate, coat, supply, or polish with oil. —*vi.* To take on fuel oil. **—oil (someone's) hand (***or* **palm).** *Informal.* **1.** To bribe. **2.** To give a tip to <*oiled* the cab driver's *hand*>

**oil beetle** *n.* An insect of the subfamily Meloinae that secretes an oily yellow substance when disturbed.

**oil·bird** (oil'bûrd') *n.* The guacharo.

**oil burner** *n.* **1.** A heating unit, boiler, or furnace that burns fuel oil. **2.** A device for injecting a fine spray of oil into an oil-burning heating unit prior to ignition.

**oil cake** *n.* The compact residue left after pressing the oil from certain seeds, as linseed and cottonseed, used as fertilizer or cattle feed.

**oil·can** (oil'kăn') *n.* A can for oil, esp. a can with a spout designed to dispense oil drop by drop, as for lubricating machinery.

**oil·cloth** (oil'klôth', -klŏth') *n.* A fabric treated with oil, clay, and pigments to make it waterproof.

**oil color** *n.* A color or paint consisting of pigment ground in oil.

**oiled** (oild) *adj.* **1.** Treated with oil. **2.** *Slang.* Drunk.

**oil·er** (oi'lər) *n.* **1.** One that oils something, as machinery and engines. **2.** An oil tanker. **3.** An oilcan. **4.** An oil-producing well. **5.** A ship fueled by oil. **6.** *Informal.* An oilskin garment.

**oil field** *n.* An area containing reserves of recoverable petroleum, esp. one that has been developed for production.

**oil gland** *n.* **1.** A gland that secretes oil. **2.** The uropygial gland.

**oil of turpentine** *n.* Refined turpentine.

**oil of vitriol** *n.* Sulfuric acid.

**oil paint** *n.* A paint in which the vehicle is a drying oil.

**oil painting** *n.* **1.** A painting done in oil colors. **2.** The art or practice of painting with oil colors.

**oil palm** *n.* **1.** A tall palm tree, *Elaeis guineensis* native to tropical Africa, bearing nutlike fruits that yield a commercially valuable oil. **2.** Any of several palms that yield oil.

**oil pan** *n.* The bottom of the crankcase in an internal-combustion engine, serving as an oil reservoir.

**oil·pa·per** (oil'pā'pər) *n.* Paper treated with oil to make it transparent and water-repellent.

**oil sand** *n.* **1.** A stratum or rock formation containing petroleum. **2.** A stratum of porous sandstone from which oil can be extracted through drilled wells.

**oil shale** *n.* A dark-brown or black shale containing hydrocarbons from which petroleum can be produced by distillation.

**oil·skin** (oil'skĭn') *n.* **1.** Cloth treated with oil to make it waterproof. **2.** A protective garment made of oilskin.

**oil slick** *n.* A thin film of oil on water.

**oil·stone** (oil'stōn') *n.* A smooth whetstone lubricated with oil.

**oil well** *n.* A hole drilled in the earth for extracting petroleum.

**oil·y** (oi'lē) *adj.* **-i·er, -i·est. 1.** Of, pertaining to, or resembling oil. **2.** Impregnated or coated with oil : GREASY. **3.** Excessively suave : UNCTUOUS <an *oily,* flattering tone of voice> **—oil'i·ly** *adv.* **—oil'i·ness** *n.*

**oink** (oingk) *n.* [Imit.] The characteristic grunting noise of a hog. **—oink** *v.* **(oinked, oink·ing, oinks).**

**oint·ment** (oint'mənt) *n.* [ME *oinement* < OFr. *oignement* < Lat. *unguentum* < *unguere,* to anoint.] A viscous substance used on the skin as an emollient, medicament, or cosmetic : SALVE.

**O·jib·wa** (ō-jĭb'wä', -wə) *also* **O·jib·way** (-wā') *n., pl.* **Ojibwa** *or* **-was** *also* **Ojibway** *or* **-ways.** [Ojibwa *Očipwee.*] **1. a.** A tribe of Indians inhabiting regions of the United States and Canada around Lake Superior. **b.** A member of this tribe. **2.** The Algonquian language of the Ojibwa.

**O.K.** *or* **OK** *or* **o·kay** (ō-kā') [Prob. abbreviation of *oll korrect,* slang respelling of *all correct.*] *Informal.* —*n., pl.* **O.K.'s** *or* **OK's** *or* **o·kays.** Approval : assent <got the doctor's *O.K.* to play sports again> —*vt.* **O.K.'d, O.K.'ing, O.K.'s** *or* **OK'd, OK'ing, OK's** *or* **o·kayed, o·kay·ing, o·kays.** To approve of or consent to : AUTHORIZE <The committee *O.K.'d* the expenditures.> —*interj.* —Used to express agreement or approval. **—O.K.** *adj. & adv.*

**o·ka·pi** (ō-kä'pē) *n., pl.* **okapi** *or* **-pis.** [Perh. of Mbuba orig.] A ruminant forest mammal, *Okapia johnstoni* of central Africa, related to the giraffe but smaller and having a short neck.

**o·kay** (ō-kā') *n.* & *v.* & *interj. var. of* O.K.

**O·kie** (ō'kē) *n.* [Shortening and alteration of *Oklahoma.*] *Slang.* A migrant farm worker, esp. one from Oklahoma in the 1930's.

**o·kra** (ō'krə) *n.* [Twi *nkruma.*] **1.** A tall tropical and semitropical plant, *Hibiscus esculentus,* having edible fruit. **2.** The mucilaginous green pods of the okra plant, used as a vegetable and in soups. **3.** GUMBO 2.

**Ok·to·ber·fest** (ŏk-tō'bər-fĕst') *n.* [G. : *Oktober,* October + *Fest,* festival.] A festival held in autumn that usu. features the consumption of beer.

**-ol**[1] *suff.* [< ALCOHOL.] An alcohol or a phenol <glycer*ol*>

**-ol**[2] *suff. var. of* -OLE.

**old** (ōld) *adj.* **-er, -est.** [ME < OE *eald.*] **1. a.** Having lived or existed for a relatively long time. **b.** Advanced in age. **2.** Made or originating in former times <an *old* painting> **3.** Of or relating to a long life <*old* age> **4.** Having or exhibiting the physical characteristics of age <*old* wrinkled hands> **5.** Having the wisdom or experience of age : MATURE <one who acts *old* for one's years> **6.** Having lived or existed for a specified time <18 years *old*> **7. a.** Belonging to a remote period in history : ANCIENT <*old* ruins> **b.** Belonging to or being of an earlier time : FORMER <my *old* high school> **8.** *often* **Old.** Being the earlier or earliest of two or more related objects, periods, versions, or stages <the *Old* Stone Age><*Old* Dutch> **9.** *Geol.* **a.** Having become slower in flow and less vigorous in action. —Used of rivers. **b.** Having become simpler in form and of lower relief. —Used of land forms. **10.** Exhibiting the effects of age or long use : WORN <an *old* battered suitcase> **11.** Known through long acquaintance <an *old* familiar face> **12.** Skilled or able through long experience : PRACTICED <an *old* expert at making speeches> **13.** *Informal.* —Used as an intensive <a grand *old* time><any *old* day> —*n.* **1.** Former times : YORE <knights of *old*> **2.** An individual of a specified age <a six-year-*old*> **—old'ness** *n.*

☆ ***syns:*** **1.** OLD, AGED, ELDERLY, SENIOR *adj. core meaning* : far along in life or time <*old* soldiers who had fought in World War I> ***ant:*** young, youthful **2.** OLD, AGED, ANCIENT, ANTIQUE, OLDEN, VENERABLE *adj. core meaning* : belonging to, existing in, or occurring in times long past <an *old* folk tale><an *old* burial ground surrounded by modern buildings> ***ant:*** new

**old boy** *n. Chiefly Brit.* A graduate of a boys' public school.

**old-boy** (ōld'boi') *adj.* Of, relating to, or made up of old boys <found a job through the *old-boy* network>

**Old Bulgarian** *n.* Old Church Slavonic.

**Old Catholic** *n.* A member of an independent religious sect formed by German Roman Catholics who refused to accept the doctrine of papal infallibility proclaimed by the Vatican Council in 1870.

**Old Church Slavonic** *n.* The literary language of the oldest Slavic manuscripts and the liturgical language of several Eastern churches.

**old country** *n.* The native country of an immigrant.

**Old Danish** *n.* The Danish language from the beginning of the 12th to the end of the 14th cent.

**Old Dutch** *n.* The Dutch language from the beginning of the 12th to the middle of the 13th cent.

**old·en** (ōl'dən) *adj.* [ME < *old,* old < OE *eald.*] Old : ancient <in *olden* days>

**Old English** *n.* **1.** The English language from the middle of the 5th to the beginning of the 12th cent. **2.** Black letter.

**Old English sheepdog** *n.* A sturdy dog orig. bred from Scottish and Russian ancestors, having a thick, shaggy, bluish-gray and white coat that hangs over the eyes.

**old fashioned** *n.* A cocktail made of whiskey, fruit, sugar, and bitters.

**old-fash·ioned** (ōld'făsh'ənd) *adj.* **1.** Of a method or style formerly in vogue : OUTDATED. **2.** Attached to or favoring customs, ideas, or methods of an earlier time.

**old fogy** *also* **old fogey** *n.* [Sc. < obs. *fogey,* invalid soldier.] One who is tiresomely old-fashioned or conservative. **—old'fo'gy·ish, old'fo'gey·ish** (ōld'fō'gē-ĭsh) *adj.*

**Old French** *n.* The French language from the 9th to the early 16th cent.

**old girl** *n. Chiefly Brit.* A graduate of a girls' school or college.

**old-girl** (ōld'gûrl') *adj.* Of, relating to, or made up of old girls <found a job through the *old-girl* network>

**Old Glory** *n.* The U.S. flag.

**old gold** *n.* A dark yellow, from light olive or olive brown to deep or strong yellow. **—old'-gold'** *adj.*

**old guard** *n. often* **Old Guard.** A conservative, often reactionary element of a particular society, class, or political group.

**old hand** *n.* One who has long experience : VETERAN.

**old hat** *adj. Informal.* **1.** Out-of-date : old-fashioned. **2.** Trite.

**Old High German** *n.* High German from the middle of the 9th to the end of the 11th cent.

**Old Icelandic** *n.* Icelandic from the middle of the 12th to the middle of the 16th cent.

**old·ie** (ōl'dē) *n.* Something old or outdated, esp. a song popular in an earlier time.

**Old Iranian** *n.* Any of the Iranian languages in use before the beginning of the Christian era.

**Old Irish** *n.* The Irish language from 725 to about 950.

**Old Italian** *n.* The Italian language as used until the middle of the 16th cent.

**old lady** *n. Slang.* **1.** One's mother. **2. a.** One's wife **b.** One's girlfriend.

**old-line** (ōld'līn') *adj.* **1.** Adhering to conservative or reactionary ideas. **2.** Long established : TRADITIONAL.

**old maid** *n.* **1.** *Informal.* A woman, esp. an older woman, who has remained unmarried : SPINSTER. **2.** *Informal.* A primly fastidious person. **3.** A card game in which the player left holding a designated card at the end is the loser or "old maid." **—old'-maid'ish** *adj.*

**old man** *n.* **1.** *Slang.* **a.** One's father. **b.** One's husband. **c.** One's boyfriend. **d.** A man in authority. **2.** The southernwood.

**old-man-and-wom·an** (ōld'măn'ənd-wŏŏm'ən) *n.* A plant, the houseleek.

**old-man cactus** (ōld'măn') *n.* A treelike cactus, *Cephalocereus senilis,* having branches tipped with long, white hair.

**old-man's-beard** (ōld'mănz-bîrd') *n.* A plant having parts that resemble a beard, as Spanish moss and virgin's bower.

**old master** *n.* **1.** An eminent European artist of the period from around 1500 to the early 1700's, esp. one of the great painters of this period. **2.** A work created by an old master.

**old moon** *n.* The last quarter of the moon.

**Old Nick** *n.* DEVIL 1.

**Old Norse** *n.* **1.** The North Germanic languages until the middle of the 14th cent. **2. a.** Old Icelandic. **b.** Old Norwegian.

**Old North French** *n.* The dialects of Old French spoken in northern France, esp. in Normandy and Picardy.

**Old Norwegian** *n.* The Norwegian language from the middle of the 12th to the end of the 14th cent.

**Old Persian** *n.* An Old Iranian language attested in cuneiform inscriptions dating from the 6th to the 5th cent. B.C.

**Old Portuguese** *n.* The Portuguese language until the middle of the 16th cent.

**Old Provençal** *n.* Provençal before the middle of the 16th cent.

**Old Prussian** *n.* The Baltic language of eastern Prussia that became extinct in the 18th cent.

**old rose** *n.* A dark pink to grayish or moderate red.

**Old Russian** *n.* The Russian language as used in documents from the middle of the 11th to the end of the 16th cent.

**Old Saxon** *n.* The Low German language of the continental Saxons until the 12th cent.

**old school** *n.* A group that espouses traditional values or practices.

**old-school tie** (ōld'skōōl') *n.* **1.** A necktie with the colors of an English public school. **2.** The network of mutual interest and assistance attributed to graduates of the English public schools. **3.** A narrow, exclusive attitude among members of a clique.

**Old Spanish** *n.* Spanish before the middle of the 16th cent.

**old squaw** *n.* A marine duck, *Clangula hyemalis* of Arctic and North Temperate regions, having black plumage with a white breast.

**old·ster** (ōld'stər) *n. Informal.* An old person.

**Old Stone Age** *n.* The Paleolithic Age.

**old style** *n.* **1.** A style of printing type originating in the 18th cent., having slanting serifs and slight contrast between light and heavy strokes. **2. Old Style.** The method of reckoning time according to the Julian calendar.

**Old Swedish** *n.* Swedish from the early 13th to the late 14th cent.

**Old Testament** *n.* [ME, transl. of LLat. *Vetus Testamentum,* transl. of Gk. *Palaia Diathēkē.*] **1.** The first of the two principal divisions of the Christian Bible, containing the Hebrew Scriptures. —See table at BIBLE. **2.** The covenant of God with Israel as distinguished in Christianity from the dispensation of Christ that constitutes the New Testament.

**old-time** (ōld'tīm') *adj.* Of, relating to, or characteristic of an earlier time.

**old-tim·er** (ōld'tī'mər) *n. Informal.* **1. a.** An old hand. **b.** An oldster. **2.** Something very old or antiquated.

**Old Welsh** *n.* The Welsh language before the 12th cent.

**old·wife** (ōld'wīf') *n.* **1.** Old squaw. **2.** A fish such as the menhaden or the alewife.

**old wives' tale** *n.* A piece of superstitious folklore.

**Old World** *n.* The Eastern Hemisphere, esp. Europe.

**old-world** (ōld'wûrld') *adj.* **1.** Antique and picturesque. **2.** *often* **Old-World.** Of, relating to, or native to the Old World.

**o·lé** (ō-lā') *interj.* [Sp. < Ar. *wa-llāh,* by God!, used as an expression of admiration.] —Used to express excited approval. **—o·lé'** *n.*

**ole-** *pref. var. of* OLEO-.

**-ole** *or* **-ol** *suff.* [< Lat. *oleum,* oil. —see OLEO-.] **1.** A usu. heterocyclic chemical compound containing a five-membered ring <pyr-

role> **2.** A chemical compound, esp. an ether, that does not contain hydroxyl <eucalyptol>

**o·le·a** (ō′lē-ə) *n. var. pl. of* OLEUM.

**o·le·ag·i·nous** (ō′lē-ăj′ə-nəs) *adj.* [Fr. *oléagineux* < Lat. *oleaginus,* of the olive tree < *olea,* olive tree < Gk. *elaia.*] **1.** Of or relating to oil. **2.** Unctuous : oily <*oleaginous* flattery> **—o′le·ag′i·nous·ly** *adv.* **—o′le·ag′i·nous·ness** *n.*

**o·le·an·der** (ō′lē-ăn′dər, ō′lē-ăn′dər) *n.* [Med. Lat.] A poisonous evergreen shrub of the genus *Nerium,* found in warm climates, esp. *N. oleander,* bearing fragrant white, purple, or rose flowers.

**o·le·an·do·my·cin** (ō′lē-ăn′də-mī′sĭn) *n.* [OLEAND(ER) + -MYCIN.] An antibiotic, $C_{35}H_{61}NO_{12}$, produced by *Streptomyces antibioticus,* effective primarily against gram-positive microorganisms.

**o·le·as·ter** (ō′lē-ăs′tər) *n.* [Lat. < *olea,* olive tree < Gk. *elaia.*] A small Eurasian tree, *Elaeagnus angustifolia,* with oblong silvery leaves, fragrant greenish flowers, and olivelike fruit.

**o·le·ate** (ō′lē-āt′) *n.* A salt or ester of oleic acid.

**o·lec·ra·non** (ō-lĕk′rə-nŏn′) *n.* [Gk. *ōlekranon* : *ōlenē,* elbow + *kranion,* head.] The upper end of the ulna projecting behind the elbow joint and forming the point of the elbow. **—o·lec′ra·nal, o′le·cra′ni·al** (ō′lĭ-krā′nē-əl), **o′le·cra′ni·an** (-nē-ən) *adj.*

**o·le·fin** (ō′lə-fĭn) *n.* [Fr. (*gaz*) *oléfiant,* oil-forming (gas) < Lat. *oleum,* oil < Gk. *elaion,* olive oil < *elaia,* olive.] Any of a class of unsaturated hydrocarbons, as ethylene, with the general formula $C_nH_{2n}$ and marked by relatively great chemical activity. **—o′le·fin′ic** (-fĭn′ĭk) *adj.*

**o·le·ic** (ō-lē′ĭk) *adj. Chem.* Of, relating to, or derived from oil.

**oleic acid** *n.* An oily liquid, $CH_3(CH_2)_7CH:CH(CH_2)_7COOH$, found in vegetable and animal oils.

**o·le·in** (ō′lē-ĭn) *also* **o·le·ine** (-ĭn, -ēn′) *n.* A yellow oily liquid, $(C_{17}H_{33}COO)_3C_3H_5$, occurring in most fats and oils and as the major constituent of olive oil.

**o·le·o** (ō′lē-ō′) *n.* Margarine.

**oleo-** *or* **ole-** *pref.* [Fr. *oléo-* < Lat. *oleo-* < *oleum,* oil < Gk. *elaion* < *elaia,* olive.] Oil <oleoresin>

**o·le·o·graph** (ō′lē-ə-grăf′) *n.* **1.** A chromolithograph printed so as to resemble an oil painting. **2.** The wavy pattern formed by a drop of oil on the surface of water. **—o′le·og′ra·pher** (-ŏg′rə-fər) *n.* **—o′le·o·graph′ic** *adj.* **—o′le·og′ra·phy** *n.*

**o·le·o·mar·ga·rine** (ō′lē-ō-mär′jə-rĭn, -rēn′) *n.* Margarine.

**o·le·o·res·in** (ō′lē-ō-rĕz′ĭn) *n.* **1.** A naturally occurring mixture of an oil and resin, as the exudate from pine trees. **2.** An oil-resin mixture extracted from plants, as capsicum. **—o′le·o·res′in·ous** *adj.*

**o·le·um** (ō′lē-əm) *n., pl.* **-le·a** (-lē-ə) *or* **-le·ums.** [Lat., olive oil. —see OIL.] A corrosive solution of sulfur trioxide in sulfuric acid.

**ol·fac·tion** (ŏl-făk′shən, ōl-) *n.* [< Lat. *olfacere,* to smell. —see OLFACTORY.] **1.** The sense of smell. **2.** An act of smelling.

**ol·fac·tom·e·ter** (ŏl′făk-tŏm′ĭ-tər, ōl′-) *n.* [OLFACT(ION) + -METER.] An instrument for measuring the keenness of the sense of smell. **—ol·fac′to·met′ric** (-tə-mĕt′rĭk) *adj.* **—ol′fac·tom′e·try** *n.*

**ol·fac·to·ry** (ŏl-făk′tə-rē, -trē, ōl-) *adj.* [< Lat. *olfacere,* to smell : *olēre,* to smell + *facere,* to do.] Of or relating to the sense of smell.

**olfactory lobe** *n.* A projection of the lower anterior portion of each cerebral hemisphere.

**olfactory nerve** *n.* Either of two bundles of nerve fibers, one on each side of the nasal cavity, that conduct chemical indications of smell.

**ol·fac·tron·ics** (ŏl′făk-trŏn′ĭks, ōl′-) *n.* [Blend of OLFACTION and ELECTRONICS.] (*sing. in number*). Study of the detection and identification of odors.

**olig-** *pref. var. of* OLIGO-.

**ol·i·garch** (ŏl′ĭ-gärk′, ō′lĭ-) *n.* [Gk. *oligarkhēs* : *oligos,* few + *arkhein,* to rule.] A ruling or influential member of an oligarchy.

**ol·i·gar·chy** (ŏl′ĭ-gär′kē, ō′lĭ-) *n., pl.* **-chies. 1. a.** Government by a few, esp. by a small group or class. **b.** The persons or families making up such a group or class. **2.** A state governed by oligarchy. **—ol′i·gar′chic, ol′i·gar′chi·cal** *adj.*

**oligo-** *or* **olig-** *pref.* [Gk. < *oligos,* few, little.] Few <oligosaccharide>

**Ol·i·go·cene** (ŏl′ĭ-gō-sēn′, ō′lĭ-) *adj.* Of or designating the geologic time and deposits of the epoch in the Tertiary period of the Cenozoic era that extended from the Eocene to the Miocene. —*n.* The Oligocene epoch.

**ol·i·go·chaete** (ŏl′ĭ-gō-kēt′, ō′lĭ-) *n.* [NLat. *Oligochaeta,* class name : OLIGO- + Gk. *khaitē,* long hair.] A worm of the class Oligochaeta, such as the earthworm. **—ol′i·go·chae′tous** (-kē′təs) *adj.*

**ol·i·go·clase** (ŏl′ĭ-gō-klās′, -klāz′, ō′lĭ-) *n.* Plagioclase.

**ol·i·go·cy·the·mi·a** *also* **ol·i·go·cy·thae·mi·a** (ŏl′ĭ-gō-sī-thē′mē-ə, ō′lĭ-) *n.* [OLIGO- + CYT(O)- + -(H)EMIA.] Deficiency of the cellular elements of the blood that causes a form of anemia.

**ol·i·go·gene** (ŏl′ĭ-gō-jēn′, ō′lĭ-) *n.* A gene determining major qualitative hereditary traits. **—ol′i·go·gen′ic** (-jĕn′ĭk) *adj.*

**o·lig·o·mer** (ə-lĭg′ə-mər) *n.* A polymer consisting of two, three, or four monomers. **—o·lig′o·mer′ic** *adj.* **—o·lig′o·mer′i·za′tion** *n.*

**ol·i·goph·a·gous** (ŏl′ĭ-gŏf′ə-gəs, ō′lĭ-) *adj.* Feeding on a limited variety of food substances. **—ol′i·goph′a·gy** (-jē) *n.*

**ol·i·go·phre·ni·a** (ŏl′ĭ-gō-frē′nē-ə, ō′lĭ-) *n.* Mental deficiency. **—ol′i·go·phren′ic** (-frĕn′ĭk) *adj.*

**ol·i·gop·o·ly** (ŏl′ĭ-gŏp′ə-lē, ō′lĭ-) *n., pl.* **-lies.** [OLIGO- + (MONO)POLY.] A market condition in which sellers are so few that the actions of any one of them can affect price and hence have a measurable impact on competitors. **—ol′i·gop′o·lis′tic** (-lĭs′tĭk) *adj.*

**ol·i·gop·so·ny** (ŏl′ĭ-gŏp′sə-nē, ō′lĭ-) *n., pl.* **-nies.** [OLIG(O)- + (MON)OPSONY.] A market condition in which purchasers are so few that the actions of any one of them can affect price and hence the costs that competitors must pay. **—ol′i·gop′so·nis′tic** (-nĭs′tĭk) *adj.*

**ol·i·go·sac·cha·ride** (ŏl′ĭ-gō-săk′ə-rīd′, ō′lĭ-) *n.* A sugar consisting of a small number of monosaccharide units.

**ol·i·go·tro·phic** (ŏl′ĭ-gō-trŏ′fĭk, -trŏf′ĭk, ō′lĭ-) *adj.* Lacking in plant nutrients and abundantly supplied with dissolved oxygen throughout, as a lake or pond.

**o·li·o** (ō′lē-ō′) *n., pl.* **-os.** [Alteration of Sp. *olla,* pot. —see OLLA.] **1.** OLLA PODRIDA 1. **2. a.** OLLA PODRIDA 2. **b.** A collection of various literary or artistic works or musical pieces. **3.** Vaudeville or musical entertainment presented between the acts of a burlesque or minstrel show.

**ol·i·va·ceous** (ŏl′ə-vā′shəs) *adj.* Olive-green.

**ol·i·var·y** (ŏl′ə-vĕr′ē) *adj.* [Lat. *olivarius,* of olives < *oliva,* olive < Gk. *elaia.*] **1.** Having the shape of an olive. **2.** *Anat.* Of, relating to, or designating one of the two oval bodies of nervous tissue located on either side of the medulla oblongata.

**ol·ive** (ŏl′ĭv) *n.* [ME < Lat. *oliva* < Gk. *elaia.*] **1.** An Old World semitropical evergreen tree, *Olea europaea,* having leathery leaves, yellow flowers, and an edible fruit. **2.** The small green oval fruit of the olive tree, an important food and a source of oil. **3.** A yellow green of low to medium lightness. **—ol′ive** *adj.*

**olive branch** *n.* **1.** A branch of an olive tree regarded as a symbol of peace. **2.** A peace offering.

**olive drab** *n.* **1.** A grayish olive to dark olive brown or olive gray. **2. a.** Cloth of an olive-drab color, often used in military uniforms. **b.** *also* **olive drabs.** A uniform of olive-drab cloth <wore *olive drabs* on maneuvers> **—ol′ive-drab′** *adj.*

**olive green** *n.* A green yellow hue of low to medium lightness. **—ol′ive-green′** *adj.*

**o·liv·e·nite** (ō-lĭv′ə-nīt′) *n.* [G. *Olivenite* : *olive,* olive + *-it,* -ite.] A basic arsenate of copper, $Cu_2(AsO_4)(OH)$, having an olive green, brown, or gray color and occurring in copper deposits.

**olive oil** *n.* Oil pressed from olives, widely used for cooking and as a salad oil, an ingredient of soaps, and an emollient.

**ol·i·vine** (ŏl′ə-vēn′) *n.* [G. *Olivin* < *olive,* olive, so called because of its color.] A mineral silicate of iron and magnesium, chiefly $Fe_2SiO_4$ and $Mg_2SiO_4$, found in metamorphic and igneous rocks and used as a structural material in refractories and in cements. **—ol′i·vin′ic** (-vĭn′ĭk), **ol′i·vi·nit′ic** (-və-nĭt′ĭk) *adj.*

**o·lla** (ŏl′ə, oi′ə) *n.* [Sp. < OSp. < Lat., var. of *aulla,* jar.] **1.** A wide-mouthed jar or pot made of earthenware. **2.** An olla podrida.

**olla po·dri·da** (pə-drē′də) *n.* [Sp.: *olla,* olla + *podrida,* rotten < Lat. *putridus* < *putrere,* to rot < *puter,* decaying.] **1.** A spicy stew of meat and vegetables. **2.** A mixture or medley : MISCELLANY.

**ol·o·gy** (ŏl′ə-jē) *n., pl.* **-gies.** [< -LOGY.] *Informal.* A branch of learning <"amphibology, parisology, and other *ologies*" —Evan Esar>

**O·lym·pi·ad** (ō-lĭm′pē-ăd′) *n.* [ME *olimpias* < Lat. *Olympics* < Gk. *Olumpias* < *Olumpios,* Olympian < *Olumpas,* Olympus, a mountain in Greece and fabled abode of the gods.] **1.** The four-year interval between the Olympic games, used by the ancient Greeks in reckoning dates. **2.** A celebration of the modern Olympic games.

**O·lym·pi·an** (ō-lĭm′pē-ən) *adj.* **1.** *Gk. Myth.* Of or relating to the principal gods of the ancient Greek pantheon, whose abode was Olympus. **2. a.** Majestic in bearing or manner. **b.** Superior to mundane affairs. **3.** Of or relating to the Olympic games. —*n.* **1.** *Gk. Myth.* One of the 12 major gods dwelling on Olympus. **2.** A contestant in the ancient or modern Olympic games.

**Olympian games** *pl.n.* OLYMPIC GAMES 1.

**O·lym·pic** (ō-lĭm′pĭk) *adj.* Of or relating to the Olympic games.

**Olympic games** *pl.n.* **1.** A Pan-Hellenic festival of athletic games and contests of dance and choral poetry, first held in 776 B.C. and celebrated at four-year intervals until A.D. 393 on the plain of Olympia in honor of the Olympian Zeus. **2.** A modern international revival of athletic contests patterned after the Olympic games and held every four years.

**O·lym·pics** (ō-lĭm′pĭks) *pl.n.* OLYMPIC GAMES 2.

**-oma** *suff.* [NLat. < Gk. *-ōma,* n. suffix.] Tumor <lipoma>

**O·ma·ha** (ō′mə-hô′, -hä′) *n., pl.* **Omaha** *or* **-has.** [Dhegia *umāhā.*] **1. a.** A tribe of Indians of northeastern Nebraska. **b.** A member of this tribe **2.** The Siouan language of the Omaha. **—O′ma·ha′** *adj.*

**o·ma·sum** (ō-mā′səm) *n., pl.* **-sa** (-sə) [Lat., bullock's tripe, prob. of Celtic orig.] The third stomach compartment of a ruminant animal, between the abomasum and the reticulum.

**om·bre** *also* **om·ber** (ŏm′bər) *n.* [Sp. *hombre,* man < Lat. *homo.*] A card game popular in Europe during the 17th and 18th cent, played by three players using a deck of 40 cards.

**om·buds·man** (ŏm′bŭdz′mən, -bədz-, -bo͝odz′-) *n.* [Norw. < ON *umboðsmaðr,* steward, manager: *um,* about + *boð,* command + *maðr,* man.] **1.** A government official, esp. in Scandinavian countries, charged with investigating citizens' complaints against the government. **2.** One who investigates complaints, as from consumers, and assists in achieving fair settlements. **—om′buds′man·ship′** *n.*

**-ome** *suff.* [NLat. *-oma* < Gk. *-ōma,* n. suffix.] Mass <biome>

**o·me·ga** (ō-mĕg′ə, ō-mē′gə, ō-mā′-) *n.* [Gk. *ō mega,* large o.] **1.** The 24th letter of the Greek alphabet. —See table at ALPHABET. **2.** The last : end. **3.** *Physics.* A subatomic particle in the baryon family having a mass 3,276 times that of the electron, a negative electric charge, and a mean lifetime of 1.5 × $10^{-10}$ second. **4.** An omega meson.

**omega meson** *n.* An unstable meson having a mass 1,532 times that of the electron.

**om·e·let** *also* **om·e·lette** (ŏm′ə-lĭt, ŏm′lĭt) *n.* [Fr. *omelette* < OFr. *amlette,* alteration of *alumette,* var. of *alumelle* < *lemelle,* knife blade < Lat. *lamella,* dim. of *lamina,* thin plate.] A dish made of beaten and cooked eggs that are folded often around a filling.

**o·men** (ō′mən) *n.* [Lat.] **1.** A phenomenon regarded as presaging good or evil. **2.** Portent : prognostication <signs of ill *omen*> *—vt.* **o·mened, o·men·ing, o·mens.** To be an omen of : PORTEND.

☆ ***syns:*** OMEN, AUGURY, FORETOKEN, PORTENT, PRESAGE, PROGNOSTIC, PROGNOSTICATION *n. core meaning* : a phenomenon that serves as a sign or warning of future good or evil <Their chance meeting was an *omen* of better times.>

**o·men·tum** (ō-mĕn′təm) *n., pl.* **-ta** (-tə) *or* **-tums.** [Lat.] *Anat.* One of two pairs of peritoneal folds: **a.** The greater omentum, consisting of a double fold of peritoneum, passing from the stomach to the transverse colon. **b.** The lesser omentum, doubled to join the lesser curve of the stomach and duodenum to the liver. **—o·men′tal** *adj.*

**o·mer** (ō′mər) *n.* [Heb. *'ōmer.*] An ancient Hebrew dry measure approx. equal to 3.7 quarts.

**om·i·cron** (ŏm′ĭ-krŏn′, ō′mĭ-) *n.* [Gk. *o mikron,* small o.] The 15th letter of the Greek alphabet. —See table at ALPHABET.

**om·i·nous** (ŏm′ə-nəs) *adj.* [Lat. *ominosus* < *omen,* omen.] **1.** Of or being an omen, esp. an evil one. **2.** Threatening : menacing <*ominous* dark sky> **—om′i·nous·ly** *adv.* **—om′i·nous·ness** *n.*

**o·mis·si·ble** (ō-mĭs′ə-bəl) *adj.* [< Lat. *omittere, omiss-,* to disregard. —see OMIT.] **1.** Capable of being omitted. **2.** Suitable for omission.

**o·mis·sion** (ō-mĭsh′ən) *n.* [ME *omissioun* < OFr. < LLat. *omissio* < Lat. *omittere,* to disregard.] **1. a.** An act or instance of omitting. **b.** The state of being omitted. **2.** Something left out or neglected.

**o·mis·sive** (ō-mĭs′ĭv) *adj.* Characterized by omission.

**o·mit** (ō-mĭt′) *vt.* **o·mit·ted, o·mit·ting, o·mits.** [ME *omitten* < Lat. *omittere* : *ob,* away + *mittere,* to send.] **1.** To leave out. **2. a.** To neglect : overlook. **b.** To leave undone : SKIP <*omitted* my homework>

**om·ma·tid·i·um** (ŏm′ə-tĭd′ē-əm) *n., pl.* **-i·a** (-ē-ə) [NLat., dim. of Gk. *omma,* eye.] One of the elements, resembling a single simplified eye, that make up the compound eye of arthropods. **—om′ma·tid′i·al** *adj.*

**om·mat·o·phore** (ō-măt′ə-fôr′, -fōr′) *n.* [Gk. *omma, ommat-* + -PHORE.] A movable eyestalk, as that of a snail. **—om′ma·toph′o·rous** (ŏm′ə-tŏf′ər-əs) *adj.*

**omni-** *pref.* [Lat. < *omnis,* all.] All <omnidirectional>

**om·ni·bus** (ŏm′nĭ-bŭs′) *n.* [Fr. < Lat., for all < *omnis,* all.] **1.** BUS 1. **2.** An anthology of the works of one author or of writings on a common theme. *—adj.* Including or providing for many items at the same time <an *omnibus* legislative bill>

**om·ni·di·rec·tion·al** (ŏm′nē-dĭ-rĕk′shə-nəl, -dī-) *adj.* Capable of receiving or transmitting signals in all directions.

**omnidirectional radio range** *n.* Omnirange.

**om·ni·far·i·ous** (ŏm′nĭ-fâr′ē-əs) *adj.* [LLat. *omnifarius* < *omnis,* all.] Of all kinds or varieties. **—om′ni·far′i·ous·ness** *n.*

**om·nip·o·tent** (ŏm-nĭp′ə-tənt) *adj.* [ME < OFr. < Lat. *omnipotens* : *omnis,* all + *potens,* pr.part. of *posse,* to be able.] Having unlimited or absolute power or authority : ALMIGHTY <an *omnipotent* dictator> *—n.* **Omnipotent.** GOD 1a, b. **—om·nip′o·tence, om·nip′o·ten·cy** *n.* **—om·nip′o·tent·ly** *adv.*

**om·ni·pres·ence** (ŏm′nĭ-prĕz′əns) *n.* [Med. Lat. *omnipraesentia* < *omnipraesens,* omnipresent : Lat. *omnis,* all + Lat. *praesens,* pr.part. of *praeese,* to be present. —see PRESENT¹.] The fact of being present or in existence everywhere at once. **—om′ni·pres′ent** *adj.*

**om·ni·range** (ŏm′nĭ-rānj′, -nē-) *n.* A radio network that broadcasts complete bearing information for aircraft.

**om·nis·cient** (ŏm-nĭsh′ənt) *adj.* [NLat. *omnisciens, omniscient-* < Med. Lat. *omniscientia* : Lat. *omnis,* all + Lat. *scientia,* knowledge < *scire,* to know.] Knowing everything. *—n.* **Omniscient.** GOD 1a, b. **—om·nis′cience, om·nis′cien·cy** *n.* **—om·nis′cient·ly** *adv.*

**om·ni·um-gath·er·um** (ŏm′nē-əm-găth′ər-əm) *n.* [Lat. *omnis,* all + GATHER.] A miscellaneous collection : HODGEPODGE.

**om·ni·vore** (ŏm′nə-vôr′, -vōr′) *n.* [< Lat. *omnivorus,* omnivorous.] An omnivorous animal.

**om·niv·o·rous** (ŏm-nĭv′ər-əs) *adj.* [Lat. *omnivorus* : *omnis,* all + *vorare,* to swallow up.] **1.** *Zool.* Eating both vegetable and animal substances. **2.** Eagerly absorbing everything available, as with the mind. **—om·niv′o·rous·ly** *adv.* **—om·niv′o·rous·ness** *n.*

**om·pha·los** (ŏm′fə-lŏs′, -ləs) *n., pl.* **-li** (-lī′) [Gk.] **1.** *Anat.* The navel. **2.** The central part : FOCUS.

**on** (ŏn, ôn) *prep.* [ME < OE.] **1.** —Used to indicate: **a.** Position above and in contact with <The vase is *on* the bureau.> **b.** Contact with a surface, regardless of position <a painting *on* the wall> **c.** Location at or along <a cottage *on* the river> **d.** Proximity <a city *on* the Polish border> **e.** Attachment to or suspension from <pearls *on* a string> **2.** —Used to indicate: **a.** Movement or direction toward a position <threw the books *on* the ground> **b.** Movement toward, against, or onto <jump *on* the platform><a march *on* Washington> **3.** —Used to indicate: **a.** Occurrence at a given time <*on* July first><*on* George Washington's birthday> **b.** The particular occasion or circumstance <*On* leaving the room, I heard them.> **c.** The exact moment or point of <*on* the hour> **4.** —Used to indicate: **a.** The object affected by actual, perceptible action <The spotlight fell *on* the soloist.> **b.** The agent or agency of a specified action <cut my hand *on* the broken glass><talked *on* the phone> **c.** The object affected by a figurative action <had pity *on* them> **d.** The object of an action directed, tending, or moving against it <an attack *on* the firebase> **5.** —Used to indicate a source or basis <live *on* bread and water> **6.** —Used to indicate: **a.** The state or process of <*on* leave><*on* fire> **b.** The purpose of <travel *on* company business> **c.** A means of conveyance <ride *on* a tram> **d.** Availability by means of <beer *on* tap><a physician *on* call> **e.** Association with <a pathologist *on* the hospital staff> **f.** Addition or repetition <error *on* error> **7. a.** Concerning : about <a book *on* plants> **b.** To the disadvantage of <We have some evidence *on* them.> **8.** In one's possession : WITH <not a cent *on* me> **9.** At the expense of <drinks *on* the house> ***usage:*** *On* and *upon* are often interchangeable when indicating location in space, as in *a bird sitting on* (or *upon*) *a branch.* However, when the relationship is not spatial, *upon* often cannot substitute for *on,* as in *I will arrive on* (not *upon*) *Tuesday.* *—adv.* **1.** In or into a position of being attached to or covering something <Put your shoes *on.*> **2.** In or into a position or condition of being supported by or in contact with something <Put the tea *on.*> **3.** In the direction of <They looked *on* while the spacecraft landed.> **4. a.** Toward or at a point lying ahead in space or time : FORWARD <The troops moved *on* to the next town.> **b.** At or to a more distant point in time or space <I'll do it later *on.*> **5.** In a continuous course <worked *on* quietly> **6. a.** In or into performance or operation <Turn *on* the TV.> **b.** In a state of activity <The show goes *on.*> **7.** In or at the present position <stay *on*><hang *on*> **8.** In a condition of being scheduled for or decided upon <There is a party *on* tomorrow.> *—adj.* **1.** In operation <The TV is *on.*> **2.** Engaged in a given function or activity, as a vocal or dramatic role. **3. a.** Planned : intended <has nothing *on* for this weekend> **b.** Taking place : HAPPENING <The parade is *on.*> **—and so on.** And like the preceding : and so forth. **—be on to.** *Informal.* To be aware of or have information about. **—on and off.** Intermittently. **—on and on.** Without stopping.

**-on¹** *suff.* [< ION.] **1. a.** Subatomic particle <baryon> **b.** Unit : quantum <photon> **2.** Basic hereditary unit <codon>

**-on²** *suff.* [NLat. < ARGON.] Inert gas <radon>

**-on³** *suff.* [Alteration of -ONE.] A chemical compound that is not a ketone or a compound containing oxygen in a carbonyl group <parathion>

**on·a·ger** (ŏn′ə-jər) *n.* [ME < Lat. < Gk. *onagros* : *onos,* ass + *agros,* field.] **1.** A wild ass, *Equus hemionus onager* of central Asia. **2.** A large catapult used in ancient and medieval combat.

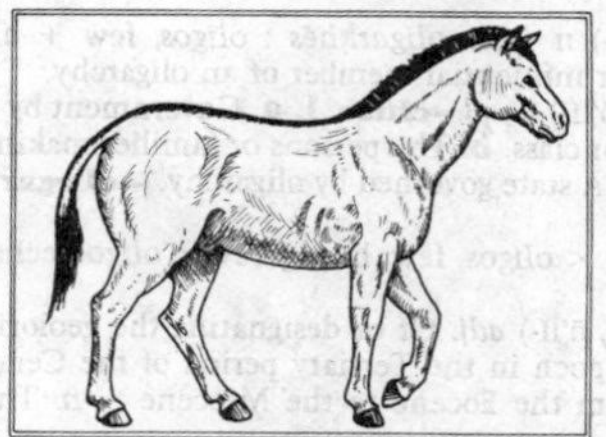

**onager**
*4–4½ feet high at shoulder*

**on-air** (ŏn′âr′) *adj.* **1.** Broadcast or broadcasting over the airwaves instead of being transmitted by cable. **2.** Appearing or working before the public, as on a radio or television broadcast, rather than behind the scenes <hired new *on-air* talent>

**on·board** (ŏn-bôrd′, -bōrd′, ôn-) *adj.* Carried aboard a vehicle <*onboard* flight recorders>

**once** (wŭns) *adv.* [ME *ones* < *on,* one < OE *ān.*] **1.** One time only <*once* a week> **2.** At one time in the past : FORMERLY <a *once* popular fashion> **3.** At any time : EVER <Time, *once* gone, will never return.> **4.** By one degree of relationship <my second cousin

*once* removed> —*n.* A single occurrence : one time <I will excuse you this *once.*> —*conj.* As soon as : WHEN <We can make plans *once* we arrive.> —*adj.* Having been at one time : FORMER <the *once* heavyweight champion>

**once-o·ver** (wŭns′ō′vər) *n. Informal.* A brief but comprehensive inspection or effort <gave the report a *once-over*>

**on·co·gen·e·sis** (ŏn′kō-jĕn′ĭ-sĭs, ŏng′-) *n.* [Gk. *onkos*, mass, tumor + -GENESIS.] Formation and development of tumors.

**on·co·gen·ic** (ŏn′kō-jĕn′ĭk, ŏng′-) *adj.* Tending to cause the formation of tumors. **—on′co·ge·nic′i·ty** (-jə-nĭs′ĭ-tē) *n.*

**on·col·o·gy** (ŏn-kŏl′ə-jē) *n.* [Gk. *onkos*, mass + -LOGY.] The medical study of tumors. **—on′co·log′i·cal** (-kə-lŏj′ĭ-kəl), **on′co·log′ic** *adj.* **—on·col′o·gist** *n.*

**on·com·ing** (ŏn′kŭm′ĭng, ôn′-) *adj.* Coming nearer : APPROACHING <*oncoming* traffic> —*n.* An advance : approach.

**one** (wŭn) *adj.* [ME *on* < OE *ān* : akin to G. *ein*, Lat. *unus*, Gk. *oinē* (ace on dice), Skt. *eka.*] **1.** Being a single unit, object, or entity <*one* mile> **2.** Characterized by unity : UNDIVIDED <We acted with *one* accord.> **3.** Occurring or existing as something indefinite, as in time or position <We will meet again *one* day.> **4. a.** Of the same kind or quality <two dogs of *one* breed> **b.** Forming a single entity of two or more components <three metals combined into *one* alloy> **5. a.** Being one in particular <*One* member was absent.> **b.** —Used as an intensifier of the quality specified <This was *one* happy day.> **6.** Being the only person of a specified or implied kind <the *one* friend I could trust> —*n.* **1. a.** A single person or thing : UNIT. **b.** The cardinal number, represented by the symbol 1, designating the first such unit in a series. **2.** A one-dollar bill. —*pron.* **1.** An indefinitely specified person or thing. **2.** An unspecified individual member of a group or class <*one* of the Victorians> **—at one.** In accord or unity. **—one and all.** Everyone. **—one by one.** Individually and in succession.

**-one** *suff.* [Alteration of -ENE.] **1.** A ketone or a related oxygen-containing compound <acet*one*> **2.** A chemical compound containing oxygen, esp. in a carbonyl group <lact*one*>

**one-base hit** (wŭn′bās′) *n. Baseball.* A hit that enables the batter to reach first base.

**one-di·men·sion·al** (wŭn′dĭ-mĕn′shə-nəl) *adj.* Lacking depth or vitality : SUPERFICIAL <a flat, *one-dimensional* performance>

**one-hand·ed** (wŭn′hăn′dĭd) *adj.* **1.** Having or using only one hand <a *one-handed* catch> **2.** Intended or designed for the use of only one hand <a *one-handed* piano exercise>

**one-horse** (wŭn′hôrs′) *adj.* **1.** Drawn by or using only one horse <a *one-horse* buggy> **2.** Small and unimportant <a *one-horse* town>

**O·nei·da** (ō-nī′də) *n., pl.* **Oneida** *or* **-das.** [Iroquois *oneyóte′.*] **1. a.** One of the five tribes belonging to the league of the Iroquois. **b.** A member of one of these tribes. **2.** The Iroquoian language of the Oneida.

**o·nei·ric** (ō-nī′rĭk) *adj.* [< Gk. *oneiros*, dream.] Of or relating to dreams.

**o·nei·ro·man·cy** (ō-nī′rə-măn′sē) *n.* [Gk. *oneiros*, dream + -MANCY.] Divination by dreams. **—o·nei′ro·man′cer** *n.*

**one-man** (wŭn′măn′) *adj.* **1.** Consisting of only one person <a *one-man* welcome committee> **2.** Featuring the work of one individual, as an artist or entertainer <a *one-man* exhibit> **3.** Designed for or restricted to one person <a *one-man* kayak>

**one·ness** (wŭn′nĭs) *n.* **1.** The state or quality of being one : SINGLENESS. **2.** Uniqueness : singularity. **3.** The condition of being undivided : WHOLENESS. **4.** Sameness of character <the insipid *oneness* of institutional food> **5.** Agreement : unison <*oneness* of mind>

**one-night stand** (wŭn′nīt′) *n.* **1. a.** A musical or dramatic performance in one place on one night only, usu. as part of a scheduled tour. **b.** The place at which such a performance is given. **2.** *Informal.* A sexual encounter limited to only one occasion.

**one-on-one** (wŭn′ŏn-wŭn′) *adj.* **1.** MAN-TO-MAN 2. **2.** Directly confronting or encountering another person.

**one-piece** (wŭn′pēs′) *adj.* Consisting of or fashioned in a single whole piece <a *one-piece* bathing suit>

**on·er·ous** (ŏn′ər-əs, ō′nər-) *adj.* [ME < OFr. *oneros* < Lat. *onerosus* < *onus*, burden.] **1.** Troublesome or oppressive : BURDENSOME. **2.** *Law.* Entailing obligations that exceed any advantage. **—on′er·ous·ly** *adv.* **—on′er·ous·ness** *n.*

**one·self** (wŭn-sĕlf′) *also* **one's self** (wŭn sĕlf′, wŭnz sĕlf′) *pron.* **1.** One's own self <have confidence in *oneself*> **2.** One's normal or healthy condition or state <come to *oneself*>

**one-shot** (wŭn′shŏt′) *adj.* **1.** Being effective after only one attempt <found a *one-shot* remedy to the problem> **2.** Being the only one and unlikely to be repeated <a *one-shot* opportunity>

**one-sid·ed** (wŭn′sī′dĭd) *adj.* **1.** Favoring one side or group : BIASED <a *one-sided* opinion> **2.** Larger or more developed on one side <a *one-sided* design> **3.** Existing or occurring on one side only. **—one′-sid′ed·ly** *adv.* **—one′-sid′ed·ness** *n.*

**one-step** (wŭn′stĕp′) *n.* A ballroom dance consisting of a series of unbroken rapid steps in 2/4 time. —*vi.* **-stepped, -step·ping, -steps.** To dance the one-step.

**one·time** (wŭn′tīm′) *adj.* Former <a *onetime* diving champion> ***usage:*** *Onetime* written as a single word means "former," whereas the hyphenated form *one-time* means "only once." Thus a *onetime champion* is a former champion, but a *one-time champion* has held the title only once.

**one-time** (wŭn′tīm′) *adj.* Having been such only once <a *one-time* medalist in 1980>

**one-to-one** (wŭn′tə-wŭn′) *adj.* **1. a.** Pairing each member of a class uniquely with a member of another class. **b.** *Math.* Pertaining to a correspondence that assigns to each member of one set a unique member of another set. **2.** Marked by proportional amounts on both sides.

**one-track** (wŭn′trăk′) *adj.* Obsessively intent on a single idea or goal <a *one-track* mind>

**one-up** (wŭn′ŭp′) *vt.* **-upped, -up·ping, -ups.** *Informal.* To practice one-upmanship on.

**one-up·man·ship** (wŭn-ŭp′mən-shĭp′) *n. Informal.* The practice and technique of keeping one step ahead of a competitor.

**one-way** (wŭn′wā′) *adj.* **1.** Moving or permitting movement in one direction only <*one-way* traffic> **2.** Of, relating to, or providing for travel in one direction only <a *one-way* trip>

**on·go·ing** (ŏn′gō-ĭng, ôn′-) *adj.* **1.** Currently taking place <an *ongoing* tournament> **2. a.** Progressing or developing. **b.** Persisting <an *ongoing* problem>

**on·ion** (ŭn′yən) *n.* [ME *oinyon* < OFr. *oignon* < Lat. *unio.*] **1.** A plant, *Allium cepa*, cultivated worldwide for its edible bulb. **2.** The rounded bulb of the onion plant, composed of tight, concentric layers and having a pungent odor and taste.

▲ word history: The Latin source of the English word *onion* is *unio*, which is derived from Latin *unus*, "one." *Unio* had several meanings in Latin, among them "unity" and "union." *Unio* was also used to designate a kind of large pearl, and in rustic speech it designated a kind of onion. The exact reason for calling an onion *unio* is unknown. *Unio* was chosen perhaps because of the onion's single bulb, in contrast to the multisegmented garlic and shallot, both related to the onion. *Unio* could also have been chosen because of some pearllike quality of the bulb, such as roundness.

**on·ion·skin** (ŭn′yən-skĭn′) *n.* A thin, strong, translucent paper.

**on-line** (ŏn′līn′, ôn′-) *adj.* **1.** *Computer Sci.* Under the control of a central computer, as in a manufacturing process. **2.** Being in progress : ONGOING <*on-line* development plans>

**on·look·er** (ŏn′lo͝ok′ər, ôn′-) *n.* One that looks on : SPECTATOR.

**on·ly** (ōn′lē) *adj.* [ME < OE *ānlīc* : *ān*, one + *-līc*, -ly.] **1.** Alone in class or kind : SOLE. **2.** Standing alone by reason of superiority or desirability <the *only* life for me> —*adv.* **1.** Without anything or anyone else : ALONE <*Only* two remained.> **2. a.** At the very least <I *only* ask that justice be done.> **b.** Merely <I was *only* joking.> **3.** Solely : exclusively : solely <answers *only* to the president> **4. a.** In the final analysis : ULTIMATELY <plans that can *only* end in disaster> **b.** With the final result : NEVERTHELESS <Prices fell last month *only* to rise again.> **5. a.** As recently as <bought it *only* yesterday> **b.** In the immediate past <*only* just left the office> —*conj.* **1.** Were it not that : EXCEPT. **2. a.** With the restriction that : BUT <You may visit, *only* don't stay long.> **b.** And yet : HOWEVER <The apartment is lovely, *only* I can't afford it.>

**on·o·mas·tic** (ŏn′ə-măs′tĭk) *adj.* [Gk. *onomastikos* < *onomazein*, to name < *onoma*, name.] Of or relating to a name or names.

**on·o·mas·tics** (ŏn′ə-măs′tĭks) *n. (sing. or pl. in number).* **1. a.** Study of the origin and form of words, esp. those used in specialized fields. **b.** Study of the origin and form of proper names. **2.** The system underlying the formation and use of words, esp. specialized vocabulary and proper names.

**on·o·mat·o·poe·ia** (ŏn′ə-măt′ə-pē′ə) *n.* [LLat. < Gk. *onomatopoiia* < *onomatopoiein*, to coin names: *onoma*, name + *poiein*, to make.] The formation or use of words, as *buzz* or *cuckoo*, having a sound that imitates what they denote. **—on′o·mat′o·poe′ic, on′o·mat′o·po·et′ic** (-pō-ĕt′ĭk) *adj.* **—on′o·mat′o·po·et′i·cal·ly** *adv.*

**On·on·da·ga** (ŏn′ən-dô′gə, -dä′gə) *n., pl.* **Onondaga** *or* **-gas.** [Iroquois *onǫ̃tà·ke′*, the name of their chief village.] **1. a.** A tribe of Indians once living in upper New York state and Ontario. **b.** A member of this tribe. **2.** The Iroquoian language of the Onondaga. **—On′on·da′gan** *adj.*

**on·rush** (ŏn′rŭsh′, ôn′-) *n.* **1.** A forward surge or flow. **2.** ASSAULT 1. **—on′rush′ing** *adj.*

**on·set** (ŏn′sĕt′, ôn′-) *n.* **1.** An assault : onslaught. **2.** A beginning : outset <the *onset* of night>

**on·shore** (ŏn′shôr′, -shōr′, ôn′-) *adj.* **1.** Moving or directed toward the shore <an *onshore* breeze> **2.** Based or operating on or along the shore <an *onshore* patrol> **3.** Domestic <*onshore* industry> **—on′shore′** *adv.*

**on·slaught** (ŏn′slôt′, ôn′-) *n.* [Obs. *anslaight* < MDu. *aenslag*, a striking at : *aan*, on + *slag*, a striking.] A violent attack.

**on-stage** (ŏn-stāj′, ôn-) *adj.* Located or occurring in the area of a stage visible to the audience. **—on·stage′** *adv.*

**ont-** *pref. var. of* ONTO-.

**-ont** *suff.* [< Gk. *ōn, ont-,* pr.part. of *einai,* to be.] Cell : organism <biont>

**on-the-job** (ŏn′thə-jŏb′, ôn′-) *adj.* Being something learned or experienced, often under supervision, while employed at a job <*on-the-job* instruction>

**on-the-scene** (ŏn′thə-sēn′, ôn′-) *adj.* Being present at the site of an event or action <an *on-the-scene* observer>

**on·to** (ŏn′tōō′, -tə, ôn′-) *prep.* **1.** On top of : UPON <The cat sprang *onto* the chair.> **2.** *Informal.* Aware of <I'm *onto* your schemes.> *—adj. Math.* Of, pertaining to, or being a mapping such that every element of the set referred to is the image of an element in another.

**onto-** or **ont-** *pref.* [LGk. < Gk. *ōn,* pr.part. of *einai,* to be.] **1.** Existence : being <*onto*logy> **2.** Organism <*onto*geny>

**on·tog·e·ny** (ŏn-tŏj′ə-nē) *n., pl.* **-nies.** The course of development of an individual organism. **—on′to·ge·net′ic** (ŏn′tō-jə-nĕt′ĭk) *adj.* **—on′to·ge·net′i·cal·ly** *adv.*

**on·tol·o·gy** (ŏn-tŏl′ə-jē) *n.* The philosophical study of the nature of being. **—on′to·log′i·cal** (ŏn′tə-lŏj′ĭ-kəl) *adj.* **—on′to·log′i·cal·ly** *adv.* **—on·tol′o·gist** *n.*

**o·nus** (ō′nəs) *n.* [Lat., burden.] **1.** Something burdensome, esp. an oppressive obligation or responsibility. **2. a.** A stigma. **b.** Blame.

**on·ward** (ŏn′wərd, ôn′-) *adj.* Moving or progressing forward in space or time. *—adv. also* **on·wards** (-wərdz). In an onward direction : FORWARD.

**-onym** *suff.* [Lat. *-onymum* < Gk. *-ōnumon* < *onuma,* name.] Word : name <acr*onym*>

**-onymy** *suff.* [Gk. *-ōnumia* < *-ōnumos,* having a specified kind of name < *onuma,* name.] A set of names : study of a kind of names <top*onymy*>

**on·yx** (ŏn′ĭks) *n.* [ME *oniche* < OFr. < Lat. *onyx* < Gk.] Chalcedony that occurs in colored bands and is used as a gemstone, esp. in intaglios and cameos.

**oo-** *pref.* [Gk. *ōio-* < *ōion,* egg.] Egg : ovum <*oo*phyte>

**o·o·cyst** (ō′ə-sĭst′) *n.* The encysted form of a sporozoan zygote.

**o·o·cyte** (ō′ə-sīt′) *n.* **1.** A cell, derived from an oogonium, that produces an ovum via meiosis. **2.** A female gamete in certain protozoa.

**oo·dles** (ōōd′lz) *pl.n.* [Orig. unknown.] *Informal.* A large amount : LOTS <*oodles* of time><*oodles* of money>

**o·o·ga·mete** (ō′ə-găm′ēt′, -gə-mēt′) *n.* A female gamete of sporozoans.

**o·og·a·mous** (ō-ŏg′ə-məs) *adj.* **1.** Marked by small, active male gametes and large, nonmotile female gametes. **2.** Relating to reproduction by oogamy.

**o·og·a·my** (ō-ŏg′ə-mē) *n., pl.* **-mies.** Fertilization of oogamous gametes.

**o·o·gen·e·sis** (ō′ə-jĕn′ə-sĭs) *n.* Enlargement and meiotic division of an oogonium to produce an ovum. **—o′o·ge·net′ic** (-jə-nĕt′ĭk) *adj.*

**o·o·go·ni·um** (ō′ə-gō′nē-əm) *n., pl.* **-ni·a** (-nē-ə) or **-ni·ums.** [OO- + NLat. *gonium,* cell < Gk. *gonos,* seed.] **1.** *Biol.* One of the cells that make up the bulk of ovarian tissue. **2.** *Bot.* A female reproductive structure in certain fungi, containing oospores.

**o·o·lite** (ō′ə-līt′) *also* **o·o·lith** (-lĭth′) *n.* **1.** A small, oval, calcareous grain occurring esp. in dolomites and limestones. **2.** Rock, usu. limestone, composed mainly of oolites. **—o′o·lit′ic** (ō′ə-lĭt′ĭk) *adj.*

**o·ol·o·gy** (ō-ŏl′ə-jē) *n.* The branch of ornithology concerned with the collection and study of birds' eggs. **—o′o·log′ic** (ō′ə-lŏj′ĭk), **o′o·log′i·cal** *adj.* **—o′o·log′i·cal·ly** *adv.* **—o·ol′o·gist** *n.*

**oo·long** (ōō′lông′, -lŏng′) *n.* [Chin. (Mandarin) *wu*[1] *long*[2] : *wu*[1], black + *long*[2], dragon.] A dark Chinese tea that is partly fermented before drying.

**oo·mi·ak** (ōō′mē-ăk′) *n. var. of* UMIAK.

**oomph** (ōōmf) *n.* [Orig. unknown.] *Slang.* **1.** Boundless enthusiasm or energy. **2.** Sex appeal.

**o·o·pho·rec·to·my** (ō′ə-fə-rĕk′tə-mē) *n., pl.* **-mies.** Surgical removal of one or both ovaries.

**o·o·pho·ri·tis** (ō′ə-fə-rī′tĭs) *n.* Ovarian inflammation.

**o·o·phyte** (ō′ə-fīt′) *n.* The stage in metagenetic plants marked by the development of sexual organs. **—o′o·phyt′ic** (-fĭt′ĭk) *adj.*

**oops** (ōōps, ŏŏps, wōōps, wŏŏps) *interj.* —Used to express mild surprise or dismay.

**o·o·sperm** (ō′ə-spûrm′) *n.* A fertilized ovum.

**o·o·sphere** (ō′ə-sfîr′) *n. Bot.* A nonmotile female gamete or egg, produced in an oogonium and ready for fertilization.

**o·o·spore** (ō′ə-spôr′, -spōr′) *n. Bot.* A thick-walled spore that develops from a fertilized oosphere or by parthenogenesis. **—o′o·spor′ic** (-spôr′ĭk, -spōr′-), **o·os′po·rous** (ō-ŏs′pər-əs, ō′ə-spôr′əs, -spōr′-) *adj.*

**o·o·the·ca** (ō′ə-thē′kə) *n., pl.* **-cae** (-sē). The egg case or capsule of certain insects and mollusks. **—o′o·the′cal** *adj.*

**o·o·tid** (ō′ə-tĭd′) *n.* [OO- + (SPERMA)TID.] One of the four sections into which a mature ovum divides.

**ooze**[1] (ōōz) *v.* **oozed, ooz·ing, ooz·es.** [ME *wosen* < *wose,* juice < OE *wōs.*] *—vi.* **1.** To flow or seep out slowly, as through small openings. **2.** To vanish or ebb slowly : DRAIN <felt my confidence *ooze* away> **3.** To move or advance slowly but steadily. **4.** To leak or exude moisture. *—vt.* **1.** To give off : EXUDE. **2.** To emit or manifest in abundance <*oozes* charm> *—n.* **1.** The act of oozing. **2.** Something that oozes. **3.** A solution of vegetable matter, as from oak bark, used in tanning.

☆ **syns:** OOZE, BLEED, EXUDE, LEACH, PERCOLATE, SEEP, TRANSUDE, WEEP *v. core meaning* : to flow or leak out slowly <sap *oozing* from a pine tree>

**ooze**[2] (ōōz) *n.* [ME *wose* < OE *wāse.*] **1.** Soft thin mud. **2.** Mudlike sediment covering the floor of oceans and lakes, composed mainly of remains of microscopic animals. **3.** Muddy ground : BOG.

**ooz·y**[1] (ōō′zē) *adj.* **-i·er, -i·est.** Slowly seeping. **—ooz′i·ness** *n.*

**ooz·y**[2] (ōō′zē) *adj.* **-i·er, -i·est.** Of, like, or composed of ooze <an *oozy* seabed> **—ooz′i·ly** *adv.* **—ooz′i·ness** *n.*

**o·pac·i·fi·er** (ō-păs′ə-fī′ər) *n.* A chemical agent added to a material to make it opaque.

**o·pac·i·ty** (ō-păs′ĭ-tē) *n., pl.* **-ties.** [Fr. *opacité* < Lat. *opacitas* < Lat. *opacus,* dark.] **1.** The quality or state of being opaque. **2.** Something opaque. **3.** Impenetrability : obscurity.

**o·pah** (ō′pə) *n.* [Of West African orig.] A large brightly colored marine fish, *Lampris regius,* with edible red flesh.

**opah**
*Approximately 5 feet long, average weight 160 pounds*

**o·pal** (ō′pəl) *n.* [Lat. *opalus* < Gk. *opallios* < Skt. *upalaḥ* < *upara-,* lower < *upa,* below.] A translucent, variously colored, occas. iridescent mineral of hydrated silicon dioxide, often used as a gem. **—o′pal·ine′** (ō′pə-līn′, -lēn′) *adj.*

**o·pal·esce** (ō′pə-lĕs′) *vi.* **-esced, -esc·ing, -esc·es.** [Back-formation < OPALESCENCE.] To shine or glisten with an iridescent play of color.

**o·pal·es·cence** (ō′pə-lĕs′əns) *n.* The quality or state of exhibiting a milky iridescence like that of an opal. **—o′pal·es′cent** *adj.*

**o·paque** (ō-pāk′) *adj.* [Partly < ME *opake,* and partly < OFr. *opaque,* both < Lat. *opacus,* dark.] **1. a.** Impervious to the passage of light. **b.** Not reflecting light : LUSTERLESS <an *opaque* paint> **2.** Impenetrable by a form of radiant energy other than visible light. **3. a.** Unperceptive : dense. **b.** So obscure as to be unintelligible <*opaque* language> *—n.* Something opaque, esp. an opaque pigment for darkening parts of a photographic print or negative. **—o·paque′ly** *adv.* **—o·paque′ness** *n.*

**op art** (ŏp) *n.* Optical art.

**op-ed page** (ŏp-ĕd′) *n.* [OP(POSITE) + ED(ITORIAL).] A newspaper section, usu. located opposite the editorial page, featuring articles and columns that present personal and often contrasting viewpoints.

**o·pen** (ō′pən) *adj.* [ME < OE.] **1. a.** Affording unobstructed entrance or exit. **b.** Affording unobstructed passage or view. **2. a.** Not covered, shut, or concealed <an *open* box><an *open* flame> **b.** Free from boundaries or limits <an *open* prairie> **3.** Not sealed, fastened, or locked <left the house wide *open*> **4. a.** Having interspersed spaces, gaps, or intervals : LOOSE <an *open* weave> **b.** Widely distributed <*open* population> **5. a.** Accessible to all : UNRESTRICTED <an *open* session><an *open* election> **b.** Lacking effective commercial regulation <an *open* town> **c.** Not legally controlled <*open* drug trafficking> **6.** Vulnerable : susceptible <*open* to criticism><*open* to doubt> **7. a.** Not taken or filled : AVAILABLE <The position is still *open.*> **b.** Available for use <the only path *open* to me> **8.** Ready to transact business <The coffee shop is *open.*> **9. a.** Unoccupied : unengaged <an *open* telephone line> **b.** Not yet resolved or decided <an *open* question> **10. a.** Marked by lack of pretense : CANDID <an *open* personality> **b.** Not reticent or restrained : BLUNT <*open* abuse> **c.** Free of prejudice <an *open* mind> **11. a.** Widely spaced or leaded. —Used of printed matter. **b.** Having constituent elements separated by space in printing or writing <"French window" is an *open* compound.> **12.** *Mus.* **a.** Not stopped by a finger. —Used of a string or hole of an instrument. **b.** Produced by an unstopped string or hole or without the use of valves, slides, or keys <an *open* chord on a guitar> **c.** Played without a mute. **13. a.** Articulated with the tongue in a low position <The vowel sound in the word "far" is *open.*> **b.** Ending in a vowel or diphthong <an *open* syllable> **14.** Having the lips parted <My mouth fell *open* with surprise.> **15.** Designating a method of punctuation in which commas and other pause marks are used sparingly. **16.** *Elect.* Having a gap which an electric current cannot cross. **17.** *Math.* **a.** Of, pertaining to, or being an interval that contains neither of its end points. **b.** Of, pertaining to, or being a set such that every point or at least one neighborhood of every point in the set is within the set. —*v.* **o·pened, o·pen·ing, o·pens.** —*vt.* **1.** To release from a fastened or closed position. **2.** To free from obstructions : CLEAR <*open* a path> **3.** To make or force an opening in <*open* an oyster> **4.** To form spaces or gaps between <soldiers *opening* ranks> **5. a.** To remove the cover or lid from : EXPOSE. **b.** To remove the wrapping from : UNDO. **6.** To unfold so the inner parts are displayed <*open* a magazine> **7. a.** To begin : initiate <*open* a session> **b.** To commence the operation of : ESTABLISH <*open* a restaurant> **c.** To begin (the action in a game of cards) by placing a first bet, by making the first bid, or by playing the first lead. **8.** To permit the use of. **9.** To make more understanding or responsive <*opened* their hearts to the victims> **10.** To reveal the secrets of : BARE. **11.** *Law.* To recall (an order or judgment) for a reexamination of its merits. —*vi.* **1.** To become open or unfastened. **2.** To draw apart : SEPARATE <A gap *opened* in the clouds.> **3.** To spread apart : UNFOLD. **4.** To become revealed <The valley *opened* before us.> **5.** To become sympathetic or receptive. **6. a.** To begin : commence <The meeting *opened* with a lively debate.> **b.** To begin business or operation. **c.** To give the first public performance <*opened* on Broadway> **d.** To make a bet, bid, or lead in starting a card game. **7.** To give access or view <The doors *open* onto a balcony.> **—open up. 1.** To spread out or become unobstructed <The view *opens up* when you get to the top.> **2.** To begin firing a gun. **3.** *Informal.* To speak or act freely and unrestrainedly. **4.** To make an opening in by cutting <The surgeon *opened up* the patient.> **5.** To make available <*open up* new markets> —*n.* **1. a.** An unobstructed area of land or water. **b.** The outdoors <camped in the *open*> **2.** An unconcealed or undisguised state <bring a problem out into the *open*> **3.** A contest or tournament in which both professional and amateur players may participate. **—o'pen·ly** *adv.* **—o'pen·ness** *n.*

☆ ***syns:*** **1.** OPEN, EXPOSED, UNCOVERED, UNPROTECTED *adj. core meaning* : having no protecting or concealing cover <an *open* campsite> **2.** OPEN, ACCESSIBLE, EMPLOYABLE, OPERATIVE, PRACTICABLE, USABLE *adj. core meaning* : available for use <only one alternative *open* to them> ***ant:*** closed

**open admission** *n.* Open enrollment.

**o·pen-air** (ō'pən-âr') *adj.* Outdoor <an *open-air* theater>

**o·pen-and-shut** (ō'pən-ən-shŭt') *adj.* Easily proved or decided <an *open-and-shut* case>

**open chain** *n. Chem.* A linear arrangement of atoms representing the basic structure of various carbon and silicon compounds.

**open-circuit** (ō'pən-sûr'kĭt) *adj.* Of, pertaining to, or being an open transmission circuit.

**open city** *n.* A city declared demilitarized during a war, thus, under international law, gaining immunity from attack.

**open classroom** *n.* A system of elementary education emphasizing informal, flexible, and individualized methods of instruction.

**open door** *n.* **1.** An unrestricted opportunity. **2.** Admission or access to all on equal terms. **3.** A policy by which a nation opens its trade to all other nations on equal and unrestrictive terms. **—o'pen-door'** (ō'pən-dôr', -dōr) *adj.*

**o·pen-end** (ō'pən-ĕnd') *adj.* **1.** Not explicitly limited as to amount or duration <an *open-end* contract> **2.** Permitting the borrowing of additional funds under existing terms <an *open-end* mortgage>

**o·pen-end·ed** (ō'pən-ĕn'dĭd) *adj.* **1.** Not restrained by definite limits, restrictions, or structure. **2.** Open or liable to change. **3.** Indefinite or inconclusive. **4.** Allowing for an expansive or unstructured response <an *open-ended* question>

**open-end investment company** *n.* A mutual fund.

**open enrollment** *n.* A policy permitting enrollment of a student in a college or university without regard to academic qualifications.

**o·pen·er** (ō'pə-nər) *n.* **1.** One that opens, esp. a device for opening cans or removing bottle caps. **2. a.** The player who starts the betting in poker. **b. openers.** A poker hand of sufficient value for the holder to open the betting. **3. a.** The first act in a theatrical variety show. **b.** The first game in a series. **4. openers.** A start : beginning <I'll have the shrimp cocktail for *openers.*>

**o·pen-eyed** (ō'pən-īd') *adj.* **1.** With the eyes wide open, as in surprise. **2.** Attentive and aware.

**o·pen-faced** (ō'pən-fāst') *adj.* **1.** Having an honest or unfeigned expression. **2.** Being a sandwich with the top side uncovered.

**o·pen·hand·ed** (ō'pən-hăn'dĭd) *adj.* Giving freely : GENEROUS. **—o'pen·hand'ed·ly** *adv.* **—o'pen·hand'ed·ness** *n.*

**o·pen-heart** (ō'pən-härt') *adj.* Of, relating to, or designating surgery in which the heart is open while its normal functions in the circulatory system are assumed by external apparatus.

**o·pen·heart·ed** (ō'pən-här'tĭd) *adj.* **1.** Frank and sincere. **2.** Compassionate. **—o'pen·heart'ed·ly** *adv.* **—o'pen·heart'ed·ness** *n.*

**o·pen-hearth** (ō'pən-härth') *adj.* **1.** Designating a reverberatory furnace used in producing steel. **2.** Of or relating to the high-quality steel produced in an open-hearth furnace.

**open house** *n.* An event in which hospitality is extended to all.

**o·pen·ing** (ō'pə-nĭng) *n.* **1.** An open space affording passage or view : APERTURE. **2.** A gap or breach. **3.** An initial stage or period. **4.** The first occasion <the *opening* of a new show> **5.** A specific series of beginning moves in certain games, esp. chess. **6.** A favorable opportunity or approach <an *opening* to further talks> **7.** An unfilled position or job : VACANCY. **8.** The act of becoming open or being made to open.

**open letter** *n.* A letter on a topic of general concern, addressed to an individual but intended for publication or distribution.

**open loop** *n. Computer Sci.* A computer control system lacking the capability to correct or alter itself.

**open marriage** *n.* A marriage in which the spouses agree to permit extramarital sexual activity.

**o·pen-mind·ed** (ō'pən-mīn'dĭd) *adj.* Receptive to new ideas or to reason. **—o'pen-mind'ed·ly** *adv.* **—o'pen-mind'ed·ness** *n.*

**o·pen-mouthed** (ō'pən-mouthd', -moutht') *adj.* **1.** With the mouth open. **2.** With the mouth involuntarily open, as in astonishment or awe. **3.** Vociferous : clamorous. **—o'pen-mouth'ed·ly** *adv.* **—o'pen-mouth'ed·ness** *n.*

**open season** *n.* **1.** A period when hunting is permitted for a specified game animal. **2.** *Informal.* A situation in which criticism of specified persons or things is unconstrained <*open season* on the new tax laws>

**open secret** *n.* Something ostensibly secret but in fact widely known.

**open sentence** *n.* A mathematical expression that contains at least one unknown quantity and that becomes true or false when a test value is substituted for the unknown.

**open ses·a·me** (sĕs'ə-mē) *n.* [From the magical formula used by Ali Baba in the *Arabian Nights* to open the door of the robbers' cave.] A dependable means of achieving a desired end.

**open shop** *n.* A business or industrial establishment in which workers are employed without regard to union membership.

**open stock** *n.* **1.** A merchandising system in which replacements for items sold in sets, as place settings, are carried at all times. **2.** Replacement items of merchandise as a whole.

**o·pen·work** (ō'pən-wûrk') *n.* Ornamental or structural work containing numerous openings, usu. in set patterns.

**op·er·a**[1] (ŏp'ər-ə, ŏp'rə) *n.* [Ital. < Lat., work produced < *opus,* work.] **1.** A form of theatrical presentation in which a dramatic performance is set to music. **2.** A work of opera. **3.** An opera house.

**o·pe·ra**[2] (ō'pər-ə, ŏp'ər-ə) *n. var. pl. of* OPUS.

**op·er·a·ble** (ŏp'ər-ə-bəl, ŏp'rə-) *adj.* **1.** Capable of being used or operated <an *operable* motor> **2.** Capable of being put into practice : PRACTICABLE <an *operable* scheme> **3.** Capable of being treated by surgical operation <an *operable* tumor> **—op'er·a·bil'i·ty** *n.* **—op'er·a·bly** *adv.*

**opera bouffe** (bo͞of) *n.* [Fr. < Ital. *opera buffa.*] Comic opera, often of a satirical nature.

**o·pe·ra buf·fa** (ŏp'ər-ə bo͞o'fə, ŏp'rə) *also* **o·pé·ra bouffe** (ô-pā-rä bo͞of') *n.* [Ital., comic opera.] A comic opera, esp. a farcical one of the 18th cent.

**o·pé·ra co·mique** (ŏp'ər-ə kŏ-mēk', ŏp'rə, ô-pā-rä kô-mēk') *n.* [Fr., comic opera.] Opera in which portions of the dialogue are spoken rather than sung.

**opera glasses** *pl.n.* Small, low-powered binoculars for use esp. at a theatrical performance.

**op·er·a·go·er** (ŏp'ər-ə-gō'ər, ŏp'rə-) *n.* One who attends operas, esp. often or regularly.

**opera hat** *n.* A collapsible top hat.

**opera house** *n.* A theater designed primarily for operas.

**op·er·and** (ŏp'ər-ənd) *n.* [< Lat. *operandum,* neuter gerund. of *operari,* to operate < *opus,* work.] A quantity on which a mathematical operation is performed.

**op·er·ant** (ŏp'ər-ənt) *adj.* [Lat. *operans, operant-,* pr.part. of *operari,* to work < *opus,* work.] **1.** Operating to produce effects : EFFECTIVE. **2.** *Psychol.* Of or relating to a behavior or response that operates on the environment to cause reinforcing and rewarding effects. —*n.* **1.** One that operates. **2.** *Psychol.* An element of operant behavior. **—op'er·ant·ly** *adv.*

**op·er·ate** (ŏp'ə-rāt') *v.* **-at·ed, -at·ing, -ates.** [Lat. *operari, operat-* < *opus,* work.] —*vi.* **1.** To run or function effectively : WORK. **2.** To have an effect or influence <Several factors *operated* to change their decision.> **3.** To perform surgery. **4.** To carry on a military or naval action or campaign. —*vt.* **1.** To control or direct the functioning of. **2.** To conduct the affairs of : MANAGE <*operate* a firm> **3.** To bring about or effect.

☆ ***syns:*** OPERATE, HANDLE, RUN, USE, WORK *v. core meaning* : to control or direct the functioning of <knew how to *operate* the machine efficiently>

**op·er·at·ic** (ŏp'ə-răt'ĭk) *adj.* [< OPERA.] **1.** Of, pertaining to, or typical of opera. **2.** Implausible or histrionic in a way regarded as characteristic of grand opera. **—op'er·at'i·cal·ly** *adv.*

**op·er·at·ics** (ŏp'ə-răt'ĭks) *n. (sing. or pl. in number).* Histrionics.

**operating system** *n.* Computer software designed to complement or support the hardware of a data processing system.

**op·er·a·tion** (ŏp′ə-rā′shən) *n.* [ME *operacioun* < OFr. *operacion* < Lat. *operatio* < *operari,* to work < *opus,* work.] **1.** An act, process, or way of operating. **2.** The condition of being operative or functioning <in full *operation*> **3.** A process or series of acts aimed at producing a desired result or effect <the *operation* of cleaning the house for the party> **4.** A method or process of productive activity. **5.** *Med.* A procedure for remedying an ailment, injury, or dysfunction in a living body, esp. one performed with instruments. **6.** *Math.* A procedure, as addition, substitution, or differentiation, performed in a specified sequence and in accordance with specific rules. **7.** *Computer Sci.* An action resulting from a single computer instruction. **8. a.** A military or naval action or campaign. **b. operations.** The office at an airport or air base where pilots file flight plans and where flying from the field is controlled. **c. operations.** The office or agency, as of a corporation, that carries out overall planning and operating functions.

**op·er·a·tion·al** (ŏp′ə-rā′shə-nəl) *adj.* **1.** Of or relating to an operation or a series of operations. **2.** Of, for, or engaged in military operations. **3. a.** Serviced and declared ready for use <an *operational* aircraft> **b.** Functioning properly. **—op′er·a′tion·al·ly** *adv.*

**operations research** *n.* Mathematical or scientific analysis of a governmental, military, or commercial operation in terms of its systematic performance and efficiency.

**op·er·a·tive** (ŏp′ər-ə-tĭv, ŏp′rə-, -ə-rā′tĭv) *adj.* **1.** Exerting influence or force. **2.** Functioning effectively : EFFICIENT. **3.** Being in effect or in operation <*operative* rules> **4.** Related to, concerned with, or engaged in mechanical or physical activity. **5.** Of, relating to, or resulting from a surgical operation. —*n.* **1.** A skilled worker, esp. in industry. **2. a.** A secret or trusted agent. **b.** A private detective. **—op′er·a·tive·ly** *adv.*

**op·er·a·tor** (ŏp′ə-rā′tər) *n.* **1.** One that operates a mechanical device <a crane *operator*> **2.** The owner or manager of a business. **3.** A dealer in stocks or commodities. **4.** A symbol, as a minus sign, that represents a mathematical operation. **5.** *Informal.* A shrewd and occas. unscrupulous person who is adept at manipulating other persons or rules to his or her advantage. **6.** A chromosomal sequence that is the region of an operon responsible for regulation of structural genes.

**o·per·cu·la** (ō-pûr′kyə-lə) *n. var. pl. of* OPERCULUM.

**o·per·cu·late** (ō-pûr′kyə-lĭt) *also* **o·per·cu·lat·ed** (-lā′tĭd) *adj.* Having an operculum.

**o·per·cu·lum** (ō-pûr′kyə-ləm) *n., pl.* **-la** (-lə) *or* **-lums.** [Lat., lid < *operire,* to cover.] **1.** *Biol.* A flap or lid that covers or closes an aperture, as the gill cover in some fishes or the horny plate in certain mollusks that seals the shell opening. **2.** *Anat.* A lid or flap, as the layer of tissue over an erupting tooth. **—o·per′cu·lar** *adj.* **—o·per′cu·lar·ly** *adv.*

**op·er·et·ta** (ŏp′ə-rĕt′ə) *n.* [Ital., dim. of *opera,* opera.] A theatrical production having many of the musical elements of opera but lighter in subject and style.

**op·er·on** (ŏp′ə-rŏn′) *n.* [< OPERATE.] A cluster of genes together with a distant gene that regulates the cluster's production of a set of different but functionally related enzymes.

**op·er·ose** (ŏp′ə-rōs′) *adj.* [Lat. *operosus* < *opus,* work.] **1.** Involving great labor : LABORIOUS. **2.** Diligent : industrious. **—op′er·ose′ly** *adv.* **—op′er·ose′ness** *n.*

**o·phid·i·an** (ō-fĭd′ē-ən) *n.* [< NLat. *Ophidia,* suborder name < Gk. *ophis,* snake.] A limbless reptile of the suborder Ophidia or Serpentes : SNAKE. **—o·phid′i·an** *adj.*

**oph·i·ol·o·gy** (ŏf′ē-ŏl′ə-jē, ō′fē-) *n.* [Gk. *ophis,* snake + -LOGY.] The branch of herpetology concerned with snakes. **—oph′i·o·log′i·cal** (-ə-lŏj′ĭ-kəl) *adj.* **—oph′i·ol′o·gist** *n.*

**o·phi·oph·a·gous** (ō′fē-ŏf′ə-gəs) *adj.* [Gk. *ophis,* snake + -PHAGOUS.] Feeding on snakes.

**oph·ite** (ŏf′īt′, ō′fīt′) *n.* [Lat. *ophites* < Gk. *ophitēs (lithos),* serpentlike (stone) < *ophis,* serpent.] **1.** A mottled-green rock composed of diabase. **2.** Any of various green rocks, as serpentine.

**o·phit·ic** (ŏ-fĭt′ĭk, ō-fĭt′-) *adj.* **1.** Of or relating to ophite. **2.** Having a texture of plagioclase crystals in a matrix of pyroxene crystals.

**Oph·i·u·chus** (ŏf′ē-yōō′kəs, ō′fē-) *n.* [Lat. < Gk. *ophioukhos* : *ophis,* serpent + *ekhein,* to hold.] A constellation in the equatorial region.

**ophthalm-** *pref. var. of* OPHTHALMO-.

**oph·thal·mia** (ŏf-thăl′mē-ə, ŏp-) *also* **oph·thal·mi·tis** (ŏf′thəl-mī′tĭs, -thăl-) *n.* [ME *obtalmia* < LLat. *ophthalmia* < Gk. < *ophthalmos,* eye.] Inflammation of the eye, esp. of the conjunctiva.

**oph·thal·mic** (ŏf-thăl′mĭk, ŏp-) *adj.* [Gk. *ophthalmikos* < *ophthalmos,* eye.] **1.** Of or relating to the eye : OCULAR. **2.** Having ophthalmia.

**oph·thal·mi·tis** (ŏf′thəl-mī′tĭs, -thăl-) *n. var. of* OPHTHALMIA.

**ophthalmo-** *or* **ophthalm-** *pref.* [Gk. *ophthalmos,* eye.] Eye : eyeball <*ophthalmo*scope>

**oph·thal·mol·o·gist** (ŏf′thăl-mŏl′ə-jĭst, ŏf′thəl-, ŏp′-) *n.* A physician specializing in the treatment of diseases of the eye.

**oph·thal·mol·o·gy** (ŏf′thăl-mŏl′ə-jē, ŏf′thəl-, ŏp′-) *n.* The medical specialty encompassing the anatomy, functions, pathology, and treatment of the eye. **—oph·thal′mo·log′ic** (-thăl′mə-lŏj′ĭk), **oph·thal′mo·log′i·cal** *adj.* **—oph·thal′mo·log′i·cal·ly** *adv.*

**oph·thal·mom·e·ter** (ŏf′thăl-mŏm′ĭ-tər, ŏf′thəl-, ŏp′-) *n.* An optical instrument for measuring astigmatism. **—oph·thal′mo·met′ric** (ŏf-thăl′mə-mĕt′rĭk, ŏp-), **oph·thal′mo·met′ri·cal** *adj.*

**oph·thal·mo·scope** (ŏf-thăl′mə-skōp′, ŏp-) *n.* An instrument consisting of a mirror with a central hole through which the eye is examined. **—oph·thal′mo·scop′ic** (-skŏp′ĭk), **oph·thal′mo·scop′i·cal** *adj.* **—oph′thal·mos′co·py** (ŏf′thăl-mŏs′kə-pē, ŏp′-) *n.*

**-opia** *suff.* [NLat. < Gk. *-ōpia* < *ōps,* eye.] A visual condition or defect of a specified kind <anisometr*opia*>

**o·pi·ate** (ō′pē-ĭt, -āt′) *n.* [Med. Lat. *opiatum* < *opiatus,* treated with opium < Lat. *opium,* opium.] **1.** A sedative narcotic containing opium or an opium derivative. **2.** Any sedative or narcotic drug. **3.** Something that numbs the senses or the mind. —*adj.* **1.** Consisting of or containing opium. **2.** Causing or producing sleep or sedation. **3.** Dulling the senses or mental processes : DEADENING. —*vt.* (ō′pē-āt′) **-at·ed, -at·ing, -ates. 1.** To subject to the action of an opiate. **2.** To deaden or dull as if with a narcotic drug.

**o·pine** (ō-pīn′) *vt.* **o·pined, o·pin·ing, o·pines.** [OFr. *opiner* < Lat. *opinari,* to suppose.] To hold or offer as an opinion : THINK.

**o·pin·ion** (ə-pĭn′yən) *n.* [ME *opinioun* < Lat. *opinio* < *opinari,* to suppose.] **1.** A belief or idea held with confidence but not substantiated by direct proof or knowledge. **2.** An evaluation or conclusion based on special knowledge or expertise <a medical *opinion*> **3.** A judgment or estimation of the value or worth of a person or thing <had a high *opinion* of the new director> **4.** The common, usual, or prevailing view or sentiment <public *opinion*> **5.** *Law.* A formal statement by a judge or jury of the legal reasons and principles for the conclusions of the court.

**o·pin·ion·at·ed** (ə-pĭn′yə-nā′tĭd) *adj.* Holding stubbornly and often unreasonably to one's personal opinions. **—o·pin′ion·at′ed·ly** *adv.* **—o·pin′ion·at′ed·ness** *n.*

**o·pin·ion·a·tive** (ə-pĭn′yə-nā′tĭv) *adj.* **1.** Relating to, based on, or of the nature of an opinion <*opinionative* reasoning> **2.** Opinionated. **—o·pin′ion·a′tive·ly** *adv.*

**o·pis·tho·branch** (ə-pĭs′thə-brăngk′) *n., pl.* **-branchs.** [NLat. *Opisthobranchia,* order name : Gk. *opisthen,* behind + Gk. *brankhion,* gill.] A marine gastropod of the order Opisthobranchia, marked by gills, a reduced or absent shell, and two pairs of tentacles.

**op·is·thog·na·thous** (ŏp′ĭs-thŏg′nə-thəs) *adj.* [Gk. *opisthen,* behind + -GNATHOUS.] Having receding jaws.

**o·pi·um** (ō′pē-əm) *n.* [ME < Lat. < Gk. *opion,* dim. of *opos,* juice.] **1.** A bitter, yellowish-brown, highly addictive drug derived from the opium poppy, containing alkaloids, as morphine, codeine, narcotine, and papaverine, with strong anesthetic properties. **2.** OPIATE 3.

**opium poppy** *n.* A plant, *Papaver somniferum* native to Asia Minor, having grayish-green leaves and variously colored flowers; the dried juice of its unripe pods is the original source of opium.

**o·pos·sum** (ə-pŏs′əm, pŏs′əm) *n., pl.* **opossum** *or* **-sums.** [Powhatan *aposoum.*] **1.** A nocturnal, arboreal marsupial of the family Didelphidae, esp. *Didelphis marsupialis* of the Western Hemisphere. **2.** A phalanger.

**op·po·nent** (ə-pō′nənt) *n.* [Lat. *opponens, opponent-,* pr.part. of *opponere,* oppose.] One that opposes another or others in a battle, contest, dispute, or debate. —*adj.* **1.** Acting against an antagonist or an opposing force <*opponent* armies> **2.** Located in front. **—op·po′nen·cy** *n.*

☆ ***syns:*** OPPONENT, ADVERSARY, ANTAGONIST, OPPOSER, OPPOSITION *n. core meaning* : one that opposes another in a battle, contest, controversy, or debate <had many *opponents* after the speech>

**op·por·tune** (ŏp′ər-tōōn′, -tyōōn′) *adj.* [ME < OFr. *oportune* < Lat. *opportunus* : *ob,* to + *portus,* harbor.] **1.** Favorable to or suited for a particular purpose <found an *opportune* moment to discuss my problem> **2.** Occurring at a fitting or advantageous time <an *opportune* encounter> **—op′por·tune′ly** *adv.* **—op′por·tune′ness** *n.*

**op·por·tun·ist** (ŏp′ər-tōō′nĭst, -tyōō′-) *n.* One who takes advantage of conditions or circumstances for self-serving purposes, usu. without concern for principles or consequences. **—op′por·tun′ism** *n.* **—op′por·tun·is′tic** *adj.*

**op·por·tu·ni·ty** (ŏp′ər-tōō′nĭ-tē, -tyōō′-) *n., pl.* **-ties.** [ME *opportunite* < OFr. < Lat. *opportunitas* < *opportunus,* opportune.] **1.** A favorable or promising combination of circumstances. **2.** A chance for advancement or improvement <a job *opportunity*>

☆ ***syns:*** OPPORTUNITY, BREAK, CHANCE, OCCASION, OPENING, SHOT *n. core meaning* : a favorable time or circumstance <a good *opportunity* to buy stock>

**op·pos·a·ble** (ə-pō′zə-bəl) *adj.* **1.** Capable of being opposed. **2.** Capable of being placed opposite or in opposition to something <an *opposable* thumb> **—op·pos′a·bil′i·ty** *n.*

**op·pose** (ə-pōz′) *v.* **-posed, -pos·ing, -pos·es.** [Fr. *opposer* < OFr. < Lat. *opponere* : *ob-,* against + *ponere,* to put.] —*vt.* **1.** To be in conflict or contention with <*oppose* the invading army> **2.** To be in disagreement with or resistant to <*oppose* all radical social changes> **3.** To place in opposition or be in opposition to. **4.** To move so as to be opposite something else. —*vi.* To act or be in opposition. **—op·pos'er** *n.*

**op·po·site** (ŏp'ə-zĭt) *adj.* [ME < OFr. *oposite* < Lat. *oppositus,* p.part. of *opponere,* oppose.] **1.** Located or placed directly across from something else or from each other <*opposite* ends of a street> **2.** Moving or tending away from each other <*opposite* directions> **3. a.** Directly contrary in nature or kind : ANTITHETICAL. **b.** Sharply and usu. antagonistically contrasting <*opposite* sides of an issue> **4.** *Bot.* Arranged in pairs on either side of a stem <*opposite* leaves> —*n.* **1.** One that is opposite or contrary to another. **2.** An opponent or antagonist. **3.** An antonym. —*adv.* In an opposite position or positions <standing *opposite* in the courtroom> —*prep.* **1.** Across from or facing <a hotel *opposite* the park> **2.** In a complementary dramatic role to <played *opposite* Gielgud> **—op'po·site·ly** *adv.* **—op'po·site·ness** *n.*

**opposite number** *n.* One holding a position in an organization that corresponds to that of a person in another organization.

**op·po·si·tion** (ŏp'ə-zĭsh'ən) *n.* [ME *opposicioun,* a placement opposite, a contradiction < Lat. *oppositio,* an opposing < *opponere,* to oppose.] **1.** The act or condition of opposing or of being in conflict. **2. a.** A location or position opposite to or facing another. **b.** Placement in such a location or position. **3.** An act or attitude of resistance or obstruction <strong *opposition* from union officials> **4.** *often* **Opposition.** A political party or organized group opposed to the party, group, or government in power. **5.** *Astron.* **a.** A geometric configuration in which the earth lies on a straight line between the sun and another planet. **b.** The position of the outer planet in this configuration. **6.** *Logic.* The relation between two propositions having an identical subject and predicate but differing in quantity, quality, or both. **7.** Contrast between two elements of a language, as phonemes, having a relationship such that the contrast is significant. **—op'po·si'tion·al** *adj.* **—op'po·si'tion·ist** *n.*

**op·press** (ə-prĕs') *vt.* **-pressed, -press·ing, -press·es.** [ME *oppressen,* to crush < OFr. *oppresser* < Med. Lat. *oppressare,* freq. of Lat. *opprimere* : *ob-,* against + *premere,* to press.] **1.** To persecute or subjugate by unjust use of force or authority. **2.** To weigh heavily upon, esp. so as to depress the mind or spirits. **3.** *Obs.* To crush or overwhelm. **—op·pres'sor** *n.*

**op·pres·sion** (ə-prĕsh'ən) *n.* **1.** An act of oppressing or the state of being oppressed. **2.** Something that oppresses. **3.** A feeling of being heavily weighed down.

**op·pres·sive** (ə-prĕs'ĭv) *adj.* [Med. Lat. *oppressivus* < Lat. *opprimere,* to press down. —see OPPRESS.] **1. a.** Difficult to bear : HARSH <*oppressive* regulations> **b.** Tyrannical <an *oppressive* dictatorship> **2.** Weighing heavily on the senses or spirit <*oppressive* humidity> **—op·pres'sive·ly** *adv.* **—op·pres'sive·ness** *n.*

**op·pro·bri·ous** (ə-prō'brē-əs) *adj.* [ME < LLat. *opprobriosus* < Lat. *opprobrium,* opprobrium.] **1.** Expressing disgrace or contemptuous reproach <*opprobrious* language> **2.** Disgraceful : ignoble <*opprobrious* conduct> **—op·pro'bri·ous·ly** *adv.*

**op·pro·bri·um** (ə-prō'brē-əm) *n.* [Lat. : *ob-,* against + *probrum,* reproach.] **1.** Disgrace or humiliation resulting from shameful conduct : IGNOMINY. **2.** Scornful contempt or reproach <a term of *opprobrium*> **3.** A cause of shame or disgrace.

**op·pugn** (ə-pyo͞on') *vt.* **-pugned, -pugn·ing, -pugns.** [ME *oppugnen* < Lat. *oppugnare,* to attack : *ob-,* against + *pugnare,* to fight.] To oppose, contradict, or call into question. **—op·pugn'er** *n.*

**op·sin** (ŏp'sĭn) *n.* [Gk. *opsis,* sight + -IN.] The protein constituent of rhodopsin.

**-opsis** *suff.* [NLat. < Gk. < *opsis,* sight, appearance.] Something resembling a specified thing <caryo*psis*>

**op·son·ic** (ŏp-sŏn'ĭk) *adj.* [OPSON(IN) + -IC.] Of, relating to, or having the effect of opsonin.

**op·son·i·fy** (ŏp-sŏn'ə-fī') *vt.* **-fied, -fy·ing, -fies.** [OPSON(IN) + -FY.] To make (bacteria) susceptible to phagocytosis : OPSONIZE. **—op·son'i·fi·ca'tion** *n.*

**op·so·nin** (ŏp'sə-nĭn) *n.* [< Lat. *opsonium,* relish < Gk. *opsōnion* < *opson.*] A substance occurring naturally in the blood that acts on bacteria to make it susceptible to phagocytosis.

**op·so·nize** (ŏp'sə-nīz') *vt.* **-nized, -niz·ing, -niz·es.** [< OPSONIN.] **1.** To form opsonins in. **2.** To opsonify. **—op'so·ni·za'tion** *n.*

**-opsy** *suff.* [NLat. *-opsia* < Gk. < *opsis,* sight.] Examination <bi*opsy*>

**opt** (ŏpt) *vi.* **opt·ed, opt·ing, opts.** [Fr. *opter* < Lat. *optare.*] To make a choice or decision <*opted* for the later date> **—opt out.** To choose not to engage in <*opted* out of a career in management>

**op·ta·tive** (ŏp'tə-tĭv) *adj.* [ME *optatif* < OFr. < LLat. *optativus* < Lat. *optare,* to wish.] **1.** Expressing a choice or wish. **2. a.** Denoting a mood of verbs in some languages, such as Greek, used to express a wish. **b.** Designating a statement using a verb in the subjunctive mood to indicate a wish or desire; e.g., *Had I the money, I would do it.* —*n.* **1.** The optative mood. **2.** A verb or expression in the optative mood. **—op'ta·tive·ly** *adv.*

**op·tic** (ŏp'tĭk) *adj.* [OFr. *optique* < Med. Lat. *opticus* < Gk. *optikos* < *optos,* visible.] **1.** Of or relating to the eye or to vision. **2.** Of or relating to the science of optics. —*n.* **1.** An eye. **2.** A component of an optical instrument.

**op·ti·cal** (ŏp'tĭ-kəl) *adj.* **1.** Of or relating to sight. **2.** Designed to assist vision <*optical* equipment> **3.** Of or relating to optics. **4.** Relating to or utilizing light <*optical* astronomy> **5.** Using light-sensitive devices <*optical* character recognition> **—op'ti·cal·ly** *adv.*

**optical activity** *n. Chem.* A property of a substance that enables it to rotate the plane of incident polarized light to the right or left.

**optical art** *n.* Abstract art employing geometric patterns or designs, esp. to create optical illusions.

**optical fiber** *n.* A flexible, optically transparent fiber, as of glass or plastic, able to transmit light by successive internal reflections.

**optical illusion** *n.* A deceptive or misleading visual image.

**optical maser** *n.* A laser, esp. one producing visible radiation.

**optic axis** *n.* A line or path through a crystal along which a ray of light can pass without undergoing double refraction.

**optic chiasm** *n.* Optic chiasma.

**optic chiasma** *n.* The partial decussation of the optic nerve fibers on the undersurface of the hypothalamus in the brain.

**optic disk** *n. Anat.* BLIND SPOT 1.

**op·ti·cian** (ŏp-tĭsh'ən) *n.* [Fr. *opticien* < Med. Lat. *optica,* optics.] **1.** One who makes lenses and eyeglasses. **2.** One who sells optical instruments and articles, as lenses and eyeglasses.

**optic nerve** *n.* Either of two sensory nerves connecting the retinas of the eyes with the brain.

**op·tics** (ŏp'tĭks) *n.* [Med. Lat. *optica* < Gk. *optikos,* of sight < *optos,* visible.] *(sing. in number). Physics.* Study of light and vision, chiefly of the generation, propagation, and detection of electromagnetic radiation having wavelengths greater than x-rays and shorter than microwaves.

**op·ti·ma** (ŏp'tə-mə) *n. var. pl. of* OPTIMUM.

**op·ti·mal** (ŏp'tə-məl) *adj.* Most desirable or favorable : OPTIMUM. **—op'ti·mal·ly** *adv.*

**op·ti·mism** (ŏp'tə-mĭz'əm) *n.* [Fr. *optimisme* < Lat. *optimus,* best.] **1.** A disposition to expect the best possible outcome or to emphasize the most positive aspects of a situation. **2.** *Philos.* **a.** The doctrine, asserted by Leibnitz, that our world is the best of all possible worlds. **b.** The belief that the universe is improving and that good ultimately triumphs over evil.

**op·ti·mist** (ŏp'tə-mĭst) *n.* **1.** One who habitually or in a particular case expects a favorable outcome. **2.** An adherent of philosophical optimism. **—op'ti·mis'tic** *adj.* **—op'ti·mis'ti·cal·ly** *adv.*

**op·ti·mize** (ŏp'tə-mīz') *vt.* **-mized, -miz·ing, -miz·es. 1.** To improve or develop as far as possible. **2.** To make the most effective use of <*optimize* our energy resources>

**op·ti·mum** (ŏp'tə-məm) *n., pl.* **-ma** (-mə) *or* **-mums.** [Lat. < *optimus,* best.] **1.** The best or most advantageous condition, degree, or amount. **2.** The most favorable conditions for growth and reproduction. —*adj.* Most advantageous or favorable : BEST.

**op·tion** (ŏp'shən) *n.* [Fr. < Lat. *optio.*] **1.** An act of choosing : CHOICE. **2.** Freedom to choose <gave us the *option* to stay home> **3. a.** The right to buy or sell something, as securities or property, within a specified time and at a specified price. **b.** The privilege of demanding fulfillment of a contract during a specified future time. **c.** A clause in an insurance policy permitting the policyholder to specify the manner in which payments are to be made or credited to him or her. **4.** Something chosen or available as a choice <Power steering is an *option* on all new models.> —*vt.* **-tioned, -tion·ing, -tions. 1. a.** To acquire an option on <*optioned* the adjoining property> **b.** To grant an option on. **2.** *Baseball.* To transfer (a player) to a minor-league club with the option of recalling him within a specified period of time.

**op·tion·al** (ŏp'shə-nəl) *adj.* Not compulsory, standard, or automatic. **—op'tion·al·ly** *adv.*

☆ ***syns:*** OPTIONAL, DISCRETIONARY, ELECTIVE, FACULTATIVE *adj. core meaning* : not compulsory or automatic <Power steering was *optional* equipment.> ***ant:*** compulsory, mandatory

**option play** *n. Football.* An offensive play in which a back has the choice of running with the ball or throwing a forward pass.

**op·tom·e·trist** (ŏp-tŏm'ĭ-trĭst) *n.* A specialist in optometry.

**op·tom·e·try** (ŏp-tŏm'ĭ-trē) *n.* [Gk. *optos,* visible + -METRY.] The profession of examining the eyes, measuring vision, and treating certain defects by means of corrective lenses or other methods that do not require license as a physician. **—op'to·met'ric** (ŏp'tə-mĕt'rĭk), **op'to·met'ri·cal** *adj.*

**op·u·lent** (ŏp'yə-lənt) *adj.* [Lat. *opulentus* < *ops,* wealth.] **1.** Having or marked by great wealth : AFFLUENT. **2. a.** Extravagant : lavish. **b.** Marked by vitality and abundance. **—op'u·lence, op'u·len·cy** *n.* **—op'u·lent·ly** *adv.*

**o·pun·ti·a** (ō-pŭn'shē-ə, -shə) *n.* [NLat. *Opuntio,* genus name < Lat. *(herba) Opuntia,* (herb) of Opus, an ancient city in Greece.] **1.** A cactus of the genus *Opuntia.* **2.** A prickly pear.

**o·pus** (ō'pəs) *n., pl.* **o·pe·ra** (ō'pər-ə, ŏp'ər-ə) *or* **o·pus·es.** [Lat.] A creative work, esp. a musical composition numbered to designate the order of a composer's works.

**o·pus·cule** (ō-pŭs′kyōōl) *n.* [Fr. < Lat. *opusculum,* dim. of *opus,* work.] A small and minor work, as of literature.

**o·quas·sa** (ō-kwăs′ə, ō-kwä′sə) *n., pl.* **oquassa** *or* **-sas.** [After Lake *Oquassa,* Maine.] A freshwater fish, *Salvelinus oquassa* of the Rangeley Lakes in Maine.

**or**[1] (ôr; *unstressed* ər) *conj.* [ME, contraction of *other,* perh. < OE *oððe.*] —Used to indicate: **1. a.** An alternative, usu. only before the last term of a series <cold *or* hot><this, that, *or* the other> **b.** The second of two alternatives, the first being preceded by *either* or *whether* <Your answer is either right *or* wrong.><I didn't know whether to laugh *or* cry.> **c.** *Archaic.* Either. **2.** A synonymous or equivalent expression <claustrophobia, *or* fear of enclosed places> **3.** Uncertainty or indefiniteness <three *or* four> ***usage:*** When a series connected by or is the subject of a verb, the verb is singular if all the elements of the series are singular (*Beer or wine is the only choice*) and plural if all the elements of the series are plural (*Any books or articles this author wrote were biased*). When the elements of a series are mixed in number, some have suggested that the verb be governed by the element to which it is nearer (*Candy or flowers are always welcome*), but other grammarians have argued that such constructions should be avoided by rewording (*Candy is always acceptable and so are flowers*).

**or**[2] (ôr) [ME < OE *ǣr* and ON *ār.*] *Archaic.* —*conj.* Before. —Followed by *ever* or *ere* <"I doubt he will be dead *or ere* I come" —Shakespeare> —*prep.* Before.

**or**[3] (ôr) *n.* [OFr. < Lat. *aurum.*] Gold, represented in heraldic engraving by a sprinkling of small dots over a white field.

**-or**[1] *suff.* [ME *-our* < OFr. *-eor, -eur,* partly < Lat. *-or,* and partly < Lat. *-ator.*] One that performs a specified action <accelerator>

**-or**[2] *suff.* [ME *-our* < OFr. *-eur* < Lat. *-or.*] State : quality : activity <behavior>

**o·ra** (ôr′ə, ōr′ə) *n. pl. of* OS[1].

**or·ach** *also* **or·ache** (ôr′ĭch, ŏr′-) *n.* [ME *arage* < OFr. *arrache* < Lat. *atriplex* < Gk. *atraphaxus.*] A plant of the genus *Atriplex,* esp. *A. hortensis,* with edible leaves resembling spinach.

**or·a·cle** (ôr′ə-kəl, ŏr′-) *n.* [ME < Lat. *oraculum* < *orare,* to speak.] **1. a.** A shrine consecrated to a prophetic deity. **b.** One who serves a deity, esp. as a medium for prophecies, at such a shrine. **c.** A divine prophecy, often in the form of an enigmatic statement or allegory. **2. a.** One regarded as a source of wise counsel or prophetic statements. **b.** An authoritative declaration from such an agency. **3.** A command or revelation from God. **4.** The sanctuary of the Temple in the Old Testament.

**o·rac·u·lar** (ô-răk′yə-lər, ō-răk′-) *adj.* [< Lat. *oraculum,* oracle < *orare,* to speak.] **1.** Of or relating to an oracle. **2.** Resembling or characteristic of an oracle, as: **a.** Solemnly prophetic. **b.** Brief and enigmatic : CRYPTIC <*oracular* advice> **—o·rac′u·lar·ly** *adv.*

**o·ral** (ôr′əl, ōr′-) *adj.* [< Lat. *ōs, ōr-,* mouth.] **1.** Spoken rather than written. **2.** Of or relating to the mouth <*oral* surgery> **3.** Used in or administered through the mouth <an *oral* vaccine> **4.** Consisting of or using speech <an *oral* agreement> **5.** Designating a speech sound emitted through the mouth only, with the nasal passages closed. **6.** *Psychol.* Of, relating to, or denoting the earliest stage of psychosexual development of the infant when gratification is derived chiefly from stimulation of the mouth and lips. —*n. often* **orals.** A school or college examination in which questions and answers are spoken rather than written. **—o′ral·ly** *adv.*

☆ ***syns:*** ORAL, SPOKEN, VERBAL *adj. core meaning* **:** expressed or transmitted in speech <an *oral* message>

**oral contraceptive** *n.* A hormone compound in pill form, to prevent ovulation taken esp. on a monthly schedule.

**oral history** *n.* **1.** Historical information obtained directly, as in tape-recorded interviews, from persons having firsthand knowledge. **2.** A written account based on oral history.

**or·ange** (ôr′ĭnj, ŏr′-) *n.* [ME < OFr. < Ar. *nāranj* < Pers. *nārang* < Skt. *nārangaḥ.*] **1. a.** A semitropical evergreen tree of the genus *Citrus,* having fragrant white flowers and round, yellowish-red fruit with a sectioned interior, esp. *C. sinensis,* the sweet orange, and *C. aurantium,* the Seville or sour orange. **b.** The pulpy edible fruit of the orange, having a sweet, acid juice. **2.** Any of various trees or plants resembling the orange, as the mock orange and the Osage orange. **3.** Any of a group of colors between red and yellow in hue, of medium lightness and moderate saturation. **—or′ange** *adj.*

**or·ange·ade** (ôr′ĭn-jād′, ŏr′-) *n.* A sweetened drink made of orange juice and plain or carbonated water.

**orange hawkweed** *n.* A plant, *Hieracium aurantiacum,* native to Europe, with hairy leaves and orange-red flower clusters.

**Or·ange·man** (ôr′ĭnj-mən, ŏr′-) *n.* [After William, Prince of *Orange,* later King William III of England (1650–1702).] **1.** A member of a Protestant secret society founded in Northern Ireland in 1795. **2.** A Protestant Irishman.

**orange milkweed** *n.* Butterfly weed.

**orange pekoe** *n.* **1.** A grade of black tea consisting of the terminal buds and surrounding leaves. **2.** A grade of black tea consisting of small leaves obtained by screening. **3.** A grade of black tea consisting of the first two full leaves of the shoot.

**or·ange·ry** (ôr′ĭnj-rē, ŏr′-) *n., pl.* **-ries.** [Fr. *orangerie* < *orange,* orange < OFr. —see ORANGE.] A sheltered or enclosed place, as a greenhouse, where orange trees are cultivated.

**orange stick** *n.* A stick orig. of orangewood, used in manicuring.

**or·ange·wood** (ôr′ĭnj-wŏŏd′, ŏr′-) *n.* The fine-grained wood of the orange tree, used in woodworking.

**o·rang·u·tan** (ō-răng′ə-tăn′, ə-răng′-) *also* **o·rang·ou·tan** (-tăng′) *n.* [Malay *ōrang hūtan* : *ōrang,* man + *hūtan,* forest.] An arboreal anthropoid ape, *Pongo pygmaeus* of Borneo and Sumatra, having a shaggy, reddish-brown coat, very long arms, and no tail.

**o·rate** (ô-rāt′, ō-rāt′, ôr′āt′, ōr′-) *vi.* **o·rat·ed, o·rat·ing, o·rates.** [Back-formation < *oration.*] To speak in a pompous, oratorical way.

**o·ra·tion** (ô-rā′shən, ō-rā′-) *n.* [Lat. *oratio, oration-* < *orare,* to speak.] **1.** A formal speech or address, esp. one given on a special occasion, as a civic holiday, academic celebration, or funeral. **2.** A pompous, rhetorical speech.

**or·a·tor** (ôr′ə-tər, ŏr′-) *n.* [ME *oratour* < Lat. *orator* < *orare,* to speak.] **1.** One who delivers an oration. **2.** One skilled in the art of public speaking. **—or′a·tor·ship′** *n.*

**or·a·tor·i·cal** (ôr′ə-tôr′ĭ-kəl, -tŏr′-) *adj.* Of or relating to an orator or to oratory. **—or′a·tor′i·cal·ly** *adv.*

☆ ***syns:*** ORATORICAL, DECLAMATORY, ELOCUTIONARY, RHETORICAL *adj. core meaning* **:** of or relating to the art of public speaking <Churchill's *oratorical* expertise is legendary.>

**or·a·to·ri·o** (ôr′ə-tôr′ē-ō′, -tōr′-, ŏr′-) *n., pl.* **-os.** [Ital. < *Oratorio,* the Oratory of St. Philip Neri at Rome, where famous musical services were held in the 16th cent. < LLat. *oratorium,* oratory, chapel.] A musical composition for voices and instruments, narrating a usu. sacred story without dramatic action or costumes.

**or·a·to·ry**[1] (ôr′ə-tôr′ē, -tōr′ē, ŏr′-) *n.* [Lat. *(ars) oratoria,* (art) of speaking < *oratorius,* oratorical < *orator,* speaker < *orare,* to speak.] **1.** The art of public speaking **:** RHETORIC. **2.** Rhetorical skill or style. **3.** Public speaking, esp. when flowery or emotive.

**or·a·to·ry**[2] (ôr′ə-tôr′ē, -tōr′ē, ŏr′-) *n., pl.* **-ries.** [ME *oratorie* < OFr. *oratori* < Lat. *(templum) oratorium,* (place) of prayer < *oratorius,* of prayer. —see ORATORY[1].] A place for prayer, as a small private chapel.

**orb** (ôrb) *n.* [OFr. *orbe* < Lat. *orbis.*] **1. a.** A sphere, esp. a celestial sphere. **b.** A range of endeavor, activity, or influence **:** PROVINCE. **2.** A celestial body. **3.** One of the concentric transparent spheres thought by ancient and medieval astronomers to revolve about the earth and support the heavenly bodies. **4.** A jeweled globe surmounted by a cross, as an emblem of royal power and justice. **5.** An eye. **6.** *Archaic.* A circular shape or object. —*v.* **orbed, orb·ing, orbs.** —*vt.* **1.** To shape into a sphere or circle. **2.** *Archaic.* To enclose **:** encircle. —*vi. Archaic.* To move in an orbit.

**or·bic·u·lar** (ôr-bĭk′yə-lər) *adj.* [ME *orbiculer* < LLat. *orbicularis* < Lat. *orbiculus,* dim. of *orbis,* orb.] **1.** Spherical or circular. **2.** *Bot.* Circular and flat <an *orbicular* leaf> **—or·bic′u·lar′i·ty** (-lăr′ĭ-tē) *n.* **—or·bic′u·lar·ly** *adv.*

**or·bic·u·late** (ôr-bĭk′yə-lĭt, -lāt′) *also* **or·bic·u·lat·ed** (-lā′tĭd) *adj.* [Lat. *orbiculatus* < *orbiculus,* dim. of *orbis,* orb.] Orbicular. **—or·bic′u·late·ly** *adv.*

**or·bit** (ôr′bĭt) *n.* [Lat. *orbita* < *orbitus,* circular < *orbis,* circle.] **1. a.** The path of a celestial body or manmade satellite as it revolves around another body. **b.** One complete revolution of such a body. **2.** The path of a body in a field of force surrounding another body, as the movement of an atomic electron in relation to a nucleus. **3. a.** A range of experience, activity, or knowledge <one's social *orbit*> **b.** A range of influence or control <countries in the Soviet *orbit*> **4.** Either of two bony cavities in the skull that hold the eye and its external structures. —*v.* **-bit·ed, -bit·ing, -bits.** —*vt.* **1.** To put into or propel in an orbit <*orbit* a new satellite> **2.** To revolve around (a center of attraction) <The earth *orbits* the sun.> —*vi.* To move in an orbit. **—or′bit·al** *adj.* **—or′bit·al·ly** *adv.*

**orbital velocity** *n.* The minimum velocity required to place a satellite in orbit about a celestial body.

**orc** (ôrk) *n.* [Fr. *orque* < Lat. *orca,* whale.] The killer whale.

**or·chard** (ôr′chərd) *n.* [ME < OE *ortgeard* : Lat. *hortus,* garden + OE *geard,* yard.] **1.** Land used for the cultivation of fruit or nut trees. **2.** The trees cultivated in an orchard.

**orchard grass** *n.* An Old World grass, *Dactylis glomerata,* widely planted as pasture.

**or·ches·tra** (ôr′kĭ-strə, ôr′kĕs′trə) *n.* [Lat., the semicircular space in front of the stage < Gk. *orkhēstra* < *orkheisthai,* to dance.] **1. a.** A large group of musicians performing together on various instruments, usu. organized into sections, as of strings, woodwinds, brass, and percussion instruments. **b.** The instruments played by an orchestra. **2.** The area in theaters and concert halls where the musicians sit, immediately in front of and below the stage. **3. a.** The section of seats in a theater nearest the stage. **b.** The entire main floor of a theater. **4.** A semicircular space in front of the stage in ancient Greek theaters on which the chorus danced. **—or·ches′tral** (ôr-kĕs′trəl) *adj.* **—or·ches′tral·ly** *adv.*

**or·ches·trate** (ôr′kĭ-strāt′) *vt.* **-trat·ed, -trat·ing, -trates. 1.** To compose or arrange (music) for orchestral performance. **2.** To organize, manage, or arrange so as to achieve a desired or effective combination <*orchestrate* a nationwide project>

**or·ches·tra·tion** (ôr′kĭ-strā′shən) *n.* **1.** An orchestrated musical composition. **2.** Arrangement of music for an orchestra.

**or·ches·tri·on** (ôr-kĕs′trē-ən) *also* **or·ches·tri·na** (ôr′kĭ-strē′nə) *n.* [ORCHESTR(A) + (MELOD)ION.] A mechanical musical instrument resembling a large barrel organ and producing sound suggestive of an orchestra.

**or·chid** (ôr′kĭd) *n.* [Lat. *orchis* < Gk. *orkhis*, orchid, testicle.] **1. a.** Any of numerous epiphytic or terrestrial plants of the family Orchidaceae, found worldwide but chiefly in the tropics and often having colorful, showy flowers of unusual shape. **b.** The flower of an orchid, esp. one cultivated for ornament or personal adornment. **2.** A pale to light purple, from grayish to purplish pink to strong reddish purple. **—or′chid** *adj.*

**or·chi·da·ceous** (ôr′kĭ-dā′shəs) *adj.* [< NLat. *Orchidaceae*, family name < Lat. *orchis*, orchid.] **1.** Of, relating to, or characteristic of the orchid family of plants. **2.** Suggesting ostentatious luxury : SHOWY.

**orchid tree** *n.* **1.** A small Asian tree, *Bauhinia variegata*, having showy purple or lavender flowers. **2.** An Indian tree, *Amherstia nobilis*, having compound leaves and a profusion of large, yellow-spotted, scarlet flowers.

**or·chil** (ôr′kĭl, -chĭl) *n.* [OFr. *orchel*.] **1.** Any of various lichens, chiefly of the genera *Roccella* and *Lecanora*, that yield a reddish dye. **2.** The dyestuff obtained from the orchil.

**or·chis** (ôr′kĭs) *n.* [NLat. *Orchis*, genus name < Lat., orchid.] An orchid of the genus *Orchis*, having magenta, white, or magenta-spotted flowers.

**Or·cus** (ôr′kəs) *n.* [Lat.] *Rom. Myth.* **1.** The world of the dead : HADES. **2.** PLUTO 1.

**or·dain** (ôr-dān′) *vt.* **-dained, -dain·ing, -dains.** [ME *ordeinen* < OFr. *ordener* < Lat. *ordinare*, to organize < *ordo*, order.] **1. a.** To invest with ministerial or priestly authority. **b.** To authorize as a rabbi. **2.** To order by virtue of established authority : DECREE. **3.** To prearrange unalterably : PREDESTINE <a fate *ordained* by the gods> **—or·dain′er** *n.* **—or·dain′ment** *n.*

**or·deal** (ôr-dēl′) *n.* [ME *ordal*, trial by ordeal < OE *ordāl*.] **1.** A difficult or painful experience, esp. one that severely tests endurance or character. **2.** A method of trial in which the accused was subjected to painful or hazardous tests, the result being regarded as a divine judgment of guilt or innocence.

**ordeal bark** *n.* [From its use in trials by ordeal.] The poisonous bark of an African tree, *Erythrophloeum guineense*.

**ordeal tree** *n.* UPAS 1.

**or·der** (ôr′dər) *n.* [ME *ordre* < OFr. < Lat. *ordo*.] **1.** A condition of logical or coherent arrangement among the individual elements of a group. **2. a.** A condition of standard or prescribed arrangement among component parts, such that proper functioning or appearance is achieved <a machine in working *order*> **b.** Systematic arrangement and design. **3. a.** The established organization or structure of society. **b.** The rule of law and custom or the observance of prescribed procedure <The streets returned to *order* after the looting.> **4.** A sequential arrangement in space or time <the *order* of events> **5.** An established sequence or procedure <the *order* of worship> **6.** An authoritative indication to be obeyed : COMMAND. **7. a.** A command issued by a superior military officer. **b. orders.** Formal written instructions to report for military duty at a specified time and place. **8. a.** A commission or instruction to buy, sell, or supply something <place an *order* for office equipment> **b.** That which is supplied, bought, or sold. **9.** *Informal.* An assigned task : UNDERTAKING <a tall *order* for a beginner> **10.** A meal or portion of food requested by a customer at a restaurant. **11.** *Law.* A direction or command delivered by a court and entered into the court record but not included in the final judgment. **12. a.** A grade or rank of the Christian ministry <the *order* of priesthood> **b. orders.** The office of an ordained minister or priest. **c. orders.** Holy orders. **13.** Any of the nine grades or choirs of angels. **14.** A group of persons living under a religious rule <*Order* of St. Francis> **15.** An organization of people united by some common fraternal bond or social aim. **16. a.** A group of persons on whom a government or sovereign has formally conferred honor for unusual service or merit <the *Order* of the British Empire> **b.** The special insignia such persons are entitled to wear. **17.** *often* **orders.** A social class. **18.** Degree of importance or quality : RANK <music of the highest *order*> **19. a.** Any of several styles of classical architecture characterized by the type of column employed. **b.** A style of architecture <a basilica of the Romanesque *order*> **20.** *Biol.* A taxonomic category of plants and animals ranking above the family and below the class. **21.** *Math.* **a.** An indicated number of successive differentiations to be performed. **b.** The number of elements in a finite group. **c.** The number of rows or columns in a matrix or determinant. **22.** A class defined by the attribute or attributes shared by all its members : KIND. —*v.* **-dered, -der·ing, -ders.** —*vt.* **1.** To give a command or instruction to. **2.** To issue a command or instruction for <The President *ordered* tighter security for the visit.> **3.** To request to be supplied with. **4.** To put into a systematic and methodical arrangement. **5.** To ordain <was *ordered* priest> —*vi.* To request that something be done or supplied <Check the specials before you *order.*> **—in order.** Called for : APPROPRIATE <A second opinion is *in order.*> **—in order that.** So that. **—in order to.** For the purpose of : so that. **—in short order.** With no delay : QUICKLY. **—on order.** Requested but not yet delivered. **—on the order of. 1.** Similar to : LIKE <a car *on the order of* the classic roadster> **2.** Approximately : about <a budget *on the order of* two million dollars> **—order up.** To summon (personnel) for active military duty <*ordered up* the reservists> **—to order.** According to the buyer's specifications. **—or′der·er** *n.*

**order arms** *n.* **1.** A position in the military manual of arms in which the rifle is held vertically next to the right leg, with the butt resting on the ground. **2.** A command to assume order arms.

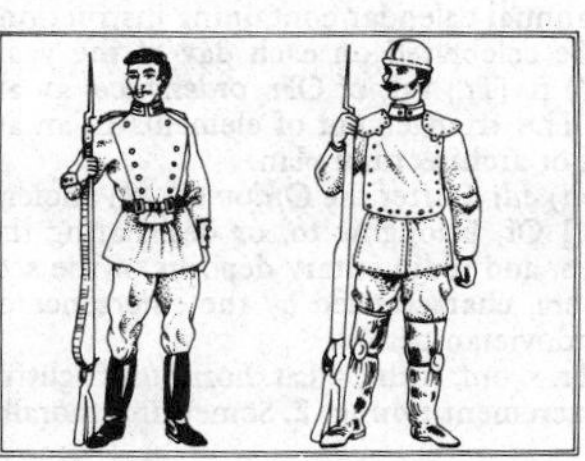
**order arms**

**or·der·ly** (ôr′dər-lē) *adj.* **1.** Having a methodical and systematic arrangement : NEAT <an *orderly* kitchen> **2.** Devoid of disruption or violence : PEACEFUL <an *orderly* protest demonstration> —*n., pl.* **-lies. 1.** An attendant who performs unskilled work in a hospital. **2.** A soldier assigned as messenger or attendant to a superior officer. —*adv.* Regularly : systematically. **—or′der·li·ness** *n.*

**order of business** *n.* Something, as a task, that must be addressed.

**order of magnitude** *n.* **1.** An estimate of size or magnitude expressed as a power of ten <The earth's mass is of the *order of magnitude* of $10^{22}$ tons; that of the sun is $10^{27}$ tons.> **2.** A range of values between a designated lower value and an upper value ten times as large <The masses of the earth and the sun differ by five *orders of magnitude.*>

**order of the day** *n.* **1.** The business or tasks scheduled to be taken up by a group for a particular day. **2.** The most important or characteristic aspect or concern <Getting the contracts signed is the *order of the day.*>

**or·di·nal** (ôr′dn-əl) *adj.* [LLat. *ordinalis* < Lat. *ordo*, order.] **1.** Being of a specified position in a numbered series <an *ordinal* rank of fifth> **2.** Relating to a biological order. —*n.* **1.** An ordinal number. **2. a.** A book of instructions for daily religious services. **b.** A book of forms for use in ordination.

**ordinal number** *n.* A number indicating position in a series or order, as first (1st), second (2nd), third (3rd), and so on.

**or·di·nance** (ôr′dn-əns) *n.* [ME *ordinaunce* < OFr. *ordenance* < Med. Lat. *ordinantia* < Lat. *ordinare*, to ordain < *ordo*, order.] **1.** An authoritative order or command. **2.** A practice or custom established by long usage. **3.** A religious rite, esp. Holy Communion. **4.** A statute or regulation, esp. one enacted by a municipal government.

**or·di·nar·i·ly** (ôr′dn-âr′ə-lē, ôr′dn-ĕr′-) *adv.* **1.** As a general rule. **2.** In the usual or regular way <*ordinarily* behaved> **3.** To the usual degree or extent <*ordinarily* high unemployment>

**or·di·nar·y** (ôr′dn-ĕr′ē) *adj.* [ME *ordinarie* < Lat. *ordinarius* < *ordo*, order.] **1.** Commonly encountered : USUAL. **2.** *Math.* Designating a differential equation with no more than two variables and derivatives of one with respect to the other. **3.** Of no exceptional kind, quality, or degree : AVERAGE <an *ordinary* day> **4.** Having immediate rather than delegated jurisdiction, as a judge. —*n., pl.* **-ies. 1.** The normal or usual condition or course of events <nothing out of the *ordinary* to be seen> **2.** *Law.* An official, as a judge, with immediate rather than delegated jurisdiction. **3.** The judge of a probate court in some U.S. states. **4.** *often* **Ordinary. a.** The part of the Mass remaining unchanged from day to day. **b.** A division of the Roman Breviary containing the unchangeable parts of the office other than the Psalms. **c.** A cleric, as the residential bishop of a diocese, with ordinary jurisdiction in the external forum over a specified territory. **5.** *Heraldry.* A simple or common charge, as the bend and the cross. **—or′di·nar′i·ness** *n.*

☆ ***syns:*** **1.** ORDINARY, AVERAGE, COMMON, COMMONPLACE, GARDEN, GARDEN-VARIETY, PLAIN, RUN-OF-THE-MILL, STOCK, UNEXCEPTIONAL *adj. core meaning* : of no special quality or type <an *ordinary* weed><an *ordinary* rock> ***ant:*** extraordinary **2.** ORDINARY, COMMON, NORMAL, TYPICAL, USUAL *adj. core meaning* : that is to be expected <*ordinary* problems of the city life> ***ant:*** unusual

**ordinary seaman** *n.* A seaman of lowest rank in the merchant marine.

**or·di·nate** (ôr′dn-ĭt, -āt′) *adj.* [Lat. *ordinatus,* p.part. of *ordinare,* to set in order < *ordo,* order.] Arranged in regular rows, as spots on an insect's wings. —*n. Math.* The plane Cartesian coordinate representing the distance from a specified point to the x-axis, measured parallel to the *y*-axis.

**or·di·na·tion** (ôr′dn-ā′shən) *n.* **1.** The act of ordaining or the condition of being ordained. **2. a.** The ceremony of admission to the ministry of a church. **b.** Admission to church ministry itself. **3.** A proper arrangement.

**or·di·nes** (ôr′də-nēz′) *n. var. pl. of* ORDO.

**ord·nance** (ôrd′nəns) *n.* [ME *ordinaunce.* —see ORDINANCE.] **1. a.** Military weapons as a whole, along with ammunition and the equipment to maintain them. **b.** The branch of a military force that designs, procures, maintains, and issues weapons. **2.** Heavy guns : ARTILLERY.

**or·do** (ôr′dō) *n., pl.* **-di·nes** (-də-nēz′) *or* **-dos.** [Med. Lat. < Lat., order.] *Rom. Cath. Ch.* An annual calendar containing instructions for the Mass and office to be celebrated on each day of the year.

**or·don·nance** (ôr′də-näns′) *n.* [Fr., var. of OFr. *ordenance,* an arranging. —see ORDINANCE.] The arrangement of elements in an artistic or literary composition or architectural plan.

**Or·do·vi·cian** (ôr′də-vĭsh′ən) *adj.* [After the *Ordovices,* an ancient Celtic tribe of North Wales.] Of, belonging to, or designating the geologic time, system of rocks, and sedimentary deposits of the second period of the Paleozoic era, characterized by the emergence of primitive fishes. —*n.* The Ordovician period.

**or·dure** (ôr′jər) *n.* [ME < OFr. < *ord,* filthy < Lat. *horridus,* frightful < *horrēre,* to shudder.] **1.** Excrement : dung. **2.** Something morally offensive : FILTH.

**ore** (ôr, ōr) *n.* [ME < OE *ōra.*] A mineral or aggregate of minerals from which a valuable constituent, esp. a metal, can be profitably extracted.

**ö·re** (œ′rə) *n., pl.* **öre.** [Dan. and Norw. *ore* and Swed. *öre,* all < Lat. *aureus,* gold coin < *aurum,* gold.] —See table at CURRENCY.

**o·re·ad** (ôr′ē-ăd′, ōr′-) *n.* [Gk. *Oreias, Oreiad-* < *oreios,* of a mountain < *oros,* mountain.] *Gk. Myth.* A mountain nymph.

**o·reg·a·no** (ə-rĕg′ə-nō′, ô-rĕg′-) *n.* [Sp. *orégano,* marjoram < Lat. *origanum* < Gk. *origanon,* prob. of North African orig.] An aromatic plant, *Origanum vulgare,* having pinkish flower spikes and leaves used as a seasoning.

**Or·e·gon fir** (ôr′ə-gən, -gŏn′, ŏr′-) *n.* The Douglas fir.

**Oregon grape** *n.* An evergreen shrub, *Mahonia aquifolium* of the Pacific coast of North America, with fragrant yellow flowers and small, bluish, edible berries.

**Oregon maple** *n.* A tree, *Acer macrophyllum* of northwestern North America, having large, lobed leaves.

**Oregon myrtle** *n.* The California laurel.

**Oregon pine** *n.* The wood of the Douglas fir.

**o·re·ide** (ôr′ē-īd′, ōr′-) *n. var. of* OROIDE.

**O·res·tes** (ô-rĕs′tēz) *n.* [Gk. *Orestēs* < *orestēs,* mountaineer < *oros,* mountain.] *Gk. Myth.* The son of Agamemnon and Clytemnestra, who, with his sister Electra, avenged the murder of his father by killing his mother and Aegisthus.

**or·fray** (ôr′frā′) *n. var. of* ORPHREY.

**or·gan** (ôr′gən) *n.* [ME < OFr. *organe* < LLat. *organum* < Lat., instrument < Gk. *organon.*] **1.** A musical instrument consisting of pipes of varying sizes that sound tones when supplied with air and a keyboard and pedals that direct the flow of air from a bellows to the pipes. **2.** Any of various instruments, as the electronic and barrel organs, that resemble the organ either in sound or mechanism. **3.** *Biol.* A differentiated part of an organism, adapted for a specific function. **4.** A group or branch performing specialized functions as part of a larger organization <The Security Council is an *organ* of the United Nations.> **5.** An instrument of communication, esp. a periodical publication issued by an organization, as a political party or business firm.

**organ-** *pref. var. of* ORGANO-.

**or·ga·na**[1] (ôr′gə-nə) *n. var. pl. of* ORGANON.

**or·ga·na**[2] (ôr′gə-nə) *n. var. pl. of* ORGANUM[1].

**or·gan·dy** *also* **or·gan·die** (ôr′gən-dē) *n., pl.* **-dies.** [Fr. *organdi.*] A stiff, transparent fabric of cotton or silk, used for trim, curtains, and light apparel.

**or·gan·elle** (ôr′gə-nĕl′) *n.* [NLat. *organella* < Lat., dim. of *organum,* implement < Gk. *organon.*] A specialized part of a cell that resembles and functions as an organ.

**organ grinder** *n.* A street musician who plays a hurdy-gurdy.

**or·gan·ic** (ôr-găn′ĭk) *adj.* [OFr. *organique* < Lat. *organicus,* of an implement < Gk. *organikos* < *organon,* implement.] **1.** Of, relating to, or affecting an organ of the body. **2.** Of, relating to, or derived from living organisms <*organic* waste> **3. a.** Using or grown with fertilizers consisting only of animal or plant matter, with no use of synthetic chemicals or pesticides <*organic* gardening> **b.** Free from chemical additives <*organic* foods> **c.** Simple, basic, and in harmony with nature. **4.** Having properties associated with living organisms. **5.** Likened to an organism in overall organization or development <The community functioned as an *organic* whole.> **6. a.** Constituting a basic part : INTEGRAL. **b.** *Law.* Designating or pertaining to the fundamental laws and principles of a government or organization. **7.** *Chem.* Of or designating carbon compounds. **—or·gan′i·cal·ly** *adv.*

**organic chemistry** *n.* The chemistry of carbon compounds.

**or·gan·i·cism** (ôr-găn′ĭ-sĭz′əm) *n.* **1.** The theory that all disease is associated with structural alterations of organs. **2.** The theory that the overall organization of an organism, rather than the functioning of individual organs, is the principal or exclusive determinant of every life process. **3.** The concept that society is analogous to a biological organism. **—or·gan′i·cist** *n.*

**or·gan·ism** (ôr′gə-nĭz′əm) *n.* **1.** A plant or animal <microscopic *organisms*> **2.** A system regarded as analogous to a living body <the social *organism*> **—or′gan·is′mal** (-nĭz′məl), **or′gan·is′mic** *adj.* **—or′gan·is′mi·cal·ly** *adv.*

**or·gan·ist** (ôr′gə-nĭst) *n.* One who plays the organ.

**or·gan·i·za·tion** (ôr′gə-nĭ-zā′shən) *n.* **1.** An act or instance of organizing or the process of being organized. **2.** The state or manner of being organized <a loose *organization* of political interest groups> **3.** Something organized or made into an ordered whole. **4.** Something comprising elements with varied functions that contribute to the whole and to collective functions : ORGANISM. **5.** A number of persons or groups having specific responsibilities and united for a particular purpose. **—or′gan·i·za′tion·al** *adj.* **—or′gan·i·za′tion·al·ly** *adv.*

**or·gan·ize** (ôr′gə-nīz′) *v.* **-ized, -iz·ing, -iz·es.** [ME *organisen* < Med. Lat. *organizare* < Lat. *organum,* instrument < Gk. *organon.*] —*vt.* **1.** To arrange or assemble into an orderly, structured, functional whole. **2. a.** To give a coherent form to : SYSTEMATIZE <*organize* one's time> **b.** To compose in a desired pattern or structure <"The painting is *organized* about a young reaper enjoying his noonday rest" —William Carlos Williams> **3.** To manage or arrange systematically for united or harmonious action <*organize* a protest> **4. a.** To establish as an organization <*organize* a political party> **b.** To persuade (employees) to form or join a labor union. **c.** To induce the employees of (a business or industry) to form or join a union <*organize* a steel mill> —*vi.* **1.** To develop into or assume an organic structure. **2. a.** To form a group, as a labor union. **b.** To join such a group. **—or′gan·iz′er** *n.*

**organo-** or **organ-** *pref.* [ME < Med. Lat. *organum,* organ of the body < Lat., implement. —see ORGAN.] **1.** Organ <*organo*therapy> **2.** Organic <*organo*mercurial>

**or·gan·o·chlo·rine** (ôr-găn′ə-klôr′ēn′, -ĭn, -klōr′-) *n.* Any of various hydrocarbon pesticides, as DDT, that contain chlorine.

**organ of Cor·ti** (kôr′tē) *n.* [After Alfonso *Corti* (1822–1888).] A specialized structure located on the inner surface of the basilar membrane of the cochlea that contains a series of sensory receptors responsive to sound vibrations.

**or·gan·o·gen·e·sis** (ôr′gə-nō-jĕn′ĭ-sĭs, ôr-găn′ə-) *n., pl.* **-ses** (-sēz′). Origin and development of biological organs. **—or′gan·o·ge·net′ic** (-jə-nĕt′ĭk) *adj.* **—or′gan·o·ge·net′i·cal·ly** *adv.*

**or·ga·nog·ra·phy** (ôr′gə-nŏg′rə-fē) *n.* Scientific description of the organs of plants and animals. **—or′gan·o·graph′ic** *adj.* **—or′ga·no·graph′i·cal·ly** *adv.*

**or·gan·o·lep·tic** (ôr′gə-nō-lĕp′tĭk, ôr-găn′ə-) *adj.* [Fr. *organoleptique* : *organe,* organ (< Lat. *organum,* implement < Gk. *organon*) + Gk. *lēptikos,* receptive < *lambanein,* to take.] Relating to or perceived by a sensory organ. **—or′gan·o·lep′ti·cal·ly** *adv.*

**or·gan·ol·o·gy** (ôr′gə-nŏl′ə-jē) *n.* Study of the structure and functions of plant and animal organs. **—or′gan·o·log′ic** (ôr′gə-nə-lŏj′ĭk, ôr-găn′ə-), **or′gan·o·log′i·cal** *adj.*

**or·gan·o·mer·cu·ri·al** (ôr-găn′ō-mər-kyŏŏr′ē-əl) *n.* An organic substance containing mercury.

**or·ga·non** (ôr′gə-nŏn′) *also* **or·ga·num** (-nəm) *n., pl.* **-na** (-nə) *or* **-nons.** [Gk., tool.] A set of methods or principles used in scientific or philosophical investigation.

**or·gan·o·ther·a·py** (ôr′gə-nō-thĕr′ə-pē, ôr-găn′ō-) *n.* Treatment of disease with animal organs or extracts such as thyroxin and insulin. **—or′gan·o·ther′a·peu′tic** (-thĕr′ə-pyōō′tĭk) *adj.*

**or·gan·ot·ro·pism** (ôr′gə-nŏt′rə-pĭz′əm) *also* **or·gan·ot·ro·py** (-pē) *n. Med.* Attraction of a chemical compound or a microorganism to specific bodily tissues or organs. **—or′gan·o·trop′ic** (ôr′gə-nō-trŏp′ĭk, ôr-găn′ō-) *adj.* **—or′gan·o·trop′i·cal·ly** *adv.*

**or·gan-pipe cactus** (ôr′gən-pīp′) *n.* A tall cactus, *Pachycereus marginatus,* found in Mexico and the southwestern United States.

**organ point** *n.* A pedal point.

**or·ga·num**[1] (ôr′gə-nəm) *n., pl.* **-na** (-nə) *or* **-nums.** [Med. Lat. < LLat., organ. —see ORGAN.] Vocal polyphonic music of the 9th to 13th cent, having two, three, or four voice parts.

**or·ga·num**[2] (ôr′gə-nəm) *n. var. of* ORGANON.

**or·gan·za** (ôr-găn′zə) *n.* [Orig. unknown.] A sheer, stiff, silk or synthetic fabric used for trimming, neckwear, or evening dresses.

**or·gan·zine** (ôr′gən-zēn′) *n.* [Fr. *organsin* < Ital. *organzino.*] A raw-silk thread, usu. used as a warp thread.

ă **pat** ā **pay** âr **care** ä **father** ĕ **pet** ē **be** hw **which** ĭ **pit** ī **tie** îr **pier** ŏ **pot** ō **toe** ô **paw, for** oi **noise** ŏŏ **took**

**or·gasm** (ôr′găz′əm) *n.* [Fr. *orgasme* < Gk. *orgasmos* < *organ,* to swell up, to be excited.] The climax of sexual excitement, marked normally by ejaculation of semen by the male and by the release of tumescence in erectile organs of both sexes. **—or·gas′mic** (ôr-găz′mĭk), **or·gas′tic** *adj.*

**or·geat** (ôr′zhä′) *n.* [Fr. < OFr. < OProv. *orjat* < *ordī,* barley < Lat. *hordeum.*] A sweet syrup of almond and orange used to flavor cocktails and food.

**or·gi·as·tic** (ôr′jē-ăs′tĭk) *adj.* [Gk. *orgiastikos* < *orgia,* secret rites.] Of, relating to, or characteristic of an orgy.

**or·gy** (ôr′jē) *n., pl.* **-gies.** [< *orgies,* secret rites < OFr. < Lat. *orgia* < Gk.] **1.** A secret rite in the cults of ancient Greek or Roman deities, typically involving frenzied dancing, singing, drinking, and sexual activity. **2.** A revel involving unrestrained indulgence, esp. sexual activity. **3.** Excessive indulgence in a specific activity <an *orgy* of crime>

**or·i·bi** (ôr′ə-bē, ŏr′-) *n., pl.* **oribi** *or* **or·i·bis.** [Afr. < Hottentot *arab* : *ara,* to provide with strips + *-b,* masc. noun-forming suffix.] A small brownish African antelope, *Ourebia ourebia.*

**o·ri·el** (ôr′ē-əl, ŏr′-) *n.* [ME < Med. Lat. *oriolum,* porch.] A projecting bay window, supported from below with a corbel or bracket.

**o·ri·ent** (ôr′ē-ənt, -ĕnt′, ŏr′-) *n.* [ME < OFr. < Lat. *oriens,* rising sun, east, pr.part. of *oriri,* to rise.] **1.** The east. **2. Orient. a.** The countries of Asia, esp. of eastern Asia. **b.** *Archaic.* The territories east of the Mediterranean. **3. a.** The luster characteristic of a pearl of high quality. **b.** A pearl having exceptional luster. *—adj.* **1.** *Archaic.* Eastern : oriental. **2.** Having a high luster <*orient* gemstones> **3.** *Archaic.* Rising in the sky : ASCENDING. *—v.* (ôr′ē-ĕnt′, ŏr′-) **-ent·ed, -ent·ing, -ents.** *—vt.* **1.** To place or locate in a specified relation to the points of the compass <*orient* the windows toward the south> **2. a.** To place or locate so as to face the east. **b.** To build (a church) with the nave laid out west to east and the altar at the eastern end. **3.** To align or position relative to a reference system. **4.** To determine the bearings of <*orient* oneself by using a map> **5.** To cause to become familiar with or adjusted to a situation or circumstance <*orient* a new employee to company policy> *—vi.* **1.** To turn toward the east. **2.** To become adjusted or aligned.

**o·ri·en·tal** (ôr′ē-ĕn′tl, ŏr′-) *adj.* **1.** Eastern. **2.** *often* **Oriental.** Relating to the countries or regions of the Orient or to their peoples, languages, or culture. **3. Oriental.** *Ecol.* Of or designating the zoographic region that includes tropical Asia and the adjacent islands of the Malay Archipelago. **4.** Lustrous and valuable <*an oriental* pearl> **5.** Designating precious varieties of corundum <an *oriental* ruby> *—n. often* **Oriental.** A native or inhabitant of the Orient. **—o′ri·en′tal·ly** *adv.*

**O·ri·en·tal·ism** *also* **o·ri·en·tal·ism** (ôr′ē-ĕn′tl-ĭz′əm, ŏr′-) *n.* **1.** A quality or custom peculiar to or characteristic of the Orient. **2.** Scholarly knowledge of eastern cultures, languages, and peoples. **—O′ri·en′tal·ist** *n.*

**O·ri·en·tal·ize** *also* **o·ri·en·tal·ize** (ôr′ē-ĕn′tl-īz′, ŏr′-) *v.* **-ized, -iz·ing, -iz·es.** *—vt.* To give an Oriental character or appearance to. *—vi.* To become Oriental.

**Oriental poppy** *n.* A plant, *Papaver orientale,* orig. of the Mediterranean region, widely cultivated for its brilliant scarlet and black flowers.

**Oriental rug** *n.* A rug made by hand in the Orient.

**or·i·en·tate** (ôr′ē-ĕn-tāt′, -ən-, ŏr′-) *vt. & vi.* **-tat·ed, -tat·ing, -tates.** To orient.

**o·ri·en·ta·tion** (ôr′ē-ĕn-tā′shən, -ən-, ŏr′-) *n.* **1.** An act of orienting or the state of being oriented. **2.** Location or position relative to the points of the compass. **3.** Construction of a church so that its longitudinal axis is from west to east and its main altar at the eastern end. **4.** The line or direction followed in the course of a trend, movement, or development. **5.** A general inclination <a Freudian *orientation*> **6.** Adjustment or adaptation to a new environment, situation, or belief. **7.** *Psychol.* Individual awareness of the objective world in its relation to the self. **8.** Introductory instruction concerning a new situation.

**o·ri·en·teer·ing** (ôr′ē-ĕn-tîr′ĭng, -ən-, ŏr′-) *n.* [Orig. a trademark.] A cross-country race in which competitors follow a course through unfamiliar territory using a map and compass.

**or·i·fice** (ôr′ə-fĭs, ŏr′-) *n.* [OFr. < LLat. *orificium* : Lat. *ōs,* mouth + Lat. *facere,* to make.] A mouth or vent : OPENING. **—or′i·fi′cial** *adj.*

**or·i·flamme** (ôr′ə-flăm′, ŏr′-) *n.* [ME *oriflamble* < OFr. *oriflambe* < Med. Lat. *auriflamma* : Lat. *aurum,* gold + Lat. *flamma,* flame.] **1.** The red flag of the Abbey of St. Denis, France, once a royal standard of France. **2.** An inspiring symbol or standard.

**o·ri·ga·mi** (ôr′ĭ-gä′mē) *n.* [J. : *ori,* a folding + *kami,* paper.] The Japanese art of folding paper into representational or decorative shapes.

**or·i·gin** (ôr′ə-jĭn, ŏr′-) *n.* [ME *origine,* ancestry < Lat. *origo* < *oriri,* to rise.] **1.** A point of origination : SOURCE. **2.** Ancestry : derivation <"We cannot escape our *origins,* however hard we try" —James Baldwin> **3.** A coming into being <the *origins* of World War I and World War II> **4.** *Anat.* The point of attachment of a muscle. **5.** *Math.* The point of intersection of coordinate axes, as in the Cartesian coordinate system.

☆ ***syns:*** ORIGIN, DERIVATION, FOUNTAIN, PROVENANCE, PROVENIENCE, ROOT, SOURCE, SPRING, WELL *n. core meaning* : a point of origination <the *origin* of Western civilization>

**o·rig·i·nal** (ə-rĭj′ə-nəl) *adj.* [ME < OFr. < Lat. *originalis* < *origo,* source < *oriri,* to rise.] **1.** Preceding all others in time : FIRST. **2. a.** Not derived from something else <an *original* script, not an adaptation> **b.** Showing a marked departure from previous practice : NEW <a truly *original* design> **3.** Productive of new things or new ideas : INVENTIVE. **4.** Being the source from which a copy, reproduction, or translation is made. *—n.* **1.** A first form from which varieties arise or imitations are made <Newer models are much lighter than the *original.*> **2.** An authentic work of art, as distinguished from an imitation or reproduction. **3.** One that is the model for a literary or artistic creation. **4.** One that is odd or singular : ECCENTRIC.

**o·rig·i·nal·i·ty** (ə-rĭj′ə-năl′ĭ-tē) *n., pl.* **-ties. 1.** The quality of being original. **2.** Capacity for independent thought or action. **3.** Something original.

**o·rig·i·nal·ly** (ə-rĭj′ə-nə-lē) *adv.* **1.** With reference to origin. **2.** At first. **3.** In a highly distinctive way <*originally* dressed>

**original sin** *n.* The condition of sin that marks all human beings as a result of Adam's first act of disobedience.

**o·rig·i·nate** (ə-rĭj′ə-nāt′) **-nat·ed, -nat·ing, -nates.** *—vt. & vi.* To bring or come into being. **—o·rig′i·na′tion** *n.* **—o·rig′i·na′tive** *adj.* **—o·rig′i·na′tive·ly** *adv.* **—o·rig′i·na′tor** *n.*

**o·ri·na·sal** (ôr′ə-nā′zəl, ŏr′-) *n.* [Lat. *ōs, ōr-,* mouth + NASAL.] A speech sound, as a French nasal vowel pronounced with both oral and nasal passages open. **—o′ri·na′sal** *adj.*

**o·ri·ole** (ôr′ē-ōl′, ŏr′-) *n.* [Fr. *oriol* < OFr. < Med. Lat. *oriolus* < Lat. *aureolus,* golden < *aureus* < *aurum,* gold.] **1.** Any of various Old World birds of the family Oriolidae, having bright yellow and black plumage in the males. **2.** Any of various New World birds of the family Icteridae, having black and orange or yellow plumage in the males.

**O·ri·on** (ō-rī′ən) *n.* [Gk. *Ōriōn.*] **1.** *Gk. Myth.* A giant hunter, pursuer of the Pleiades and lover of Eos, killed by Artemis. **2.** A constellation in the celestial equator.

**or·i·son** (ôr′ĭ-sən, -zən, ŏr′-) *n.* [ME *orisoun* < OFr. *oraison* < Lat. *oratio,* speech < *orare,* to speak.] A prayer.

**O·ri·ya** (ô-rē′yə) *n.* The Indic language of Orissa, a state in eastern India.

**Or·le·an·ist** (ôr′lē-ə-nĭst) *n.* A supporter of the Orléans branch of the French royal family, descended from a younger brother of Louis XIV.

**Or·lon** (ôr′lŏn′). A trademark for a synthetic acrylic fiber.

**or·lop** (ôr′lŏp′) *n.* [ME *overloper,* floor covering a ship's hold < MDu. *overloop* : *over,* over + *loopen,* to leap.] *Naut.* The lowest deck of a ship, esp. a warship.

**Or·mazd** *also* **Or·muzd** (ôr′məzd) *n.* [Pers. *Ormazd* < Avestan *Ahura-Mazda* : *ahura,* spirit + *mazdā,* wise.] The supreme deity of Zoroastrianism, the creator of the world, the source of light, and the embodiment of good.

**or·mer** (ôr′mər) *n.* [Dial. Fr. < Fr. *ormier,* short for *oreille-de-mer,* ear of the sea < Lat. *auris maris.*] *Chiefly Brit.* An abalone shell, esp. the shell of an edible species, *Haliotis tuberculata,* found chiefly in the Channel Islands.

**or·mo·lu** (ôr′mə-lōō′) *n.* [Fr. *or moulu,* ground gold.] **1.** A copper and tin or zinc alloy resembling gold in appearance and used to decorate furniture and jewelry. **2.** An imitation of gold.

**Or·muzd** (ôr′məzd) *n. var. of* ORMAZD.

**or·na·ment** (ôr′nə-mənt) *n.* [ME *ournement* < OFr. *ornement* < Lat. *ornamentum* < *ornare,* to embellish.] **1.** Something that decorates or adorns : EMBELLISHMENT. **2.** One regarded as a source of credit, honor, or pride <an *ornament* to one's school> **3.** *Mus.* A note or group of notes that embellishes a melody. *—vt.* (-mĕnt′) **-ment·ed, -ment·ing, -ments. 1.** To furnish with ornaments. **2.** To be an ornament to. **—or′na·ment′er** *n.*

**or·na·men·tal** (ôr′nə-mĕn′tl) *adj.* Of, relating to, or being an ornament. *—n.* Something ornamental, esp. a plant grown for its beauty. **—or′na·men′tal·ly** *adv.*

**or·na·men·ta·tion** (ôr′nə-mĕn-tā′shən) *n.* **1. a.** The act or process of ornamenting. **b.** The state of being ornamented. **2.** ORNAMENT 1.

**or·nate** (ôr-nāt′) *adj.* [ME < Lat. *ornatus,* p.part. of *ornare,* to embellish.] **1.** Elaborately and heavily ornamented. **2.** Showy or florid in style or manner : FLOWERY. **—or·nate′ly** *adv.* **—or·nate′ness** *n.*

☆ ***syns:*** ORNATE, BAROQUE, FLAMBOYANT, FLORID, ROCOCO *adj. core meaning* : elaborately and heavily ornamented <an *ornate* Bavarian palace> ***ant:*** austere.

**or·ner·y** (ôr′nə-rē) *adj.* **-i·er, -i·est.** [Alteration of ORDINARY.] Stubborn and ill-tempered.

**ornith-** *pref. var. of* ORNITHO-.

**or·nith·ic** (ôr-nĭth′ĭk) *adj.* [Gk. *ornithikos* < *ornis,* bird.] Of, pertaining to, or characteristic of birds.

**or·ni·thine** (ôr′nə-thēn′) *n.* [E. *ornith(uric acid),* an acid found in birds' urine + -INE².] An amino acid, $C_5H_{12}N_2O_{12}$, that functions in urea formation.

**ornitho-** or **ornith-** *pref.* [NLat. < Gk. < *ornis,* bird.] Bird <*ornithosis*><*ornithic*>

**or·ni·thol·o·gy** (ôr′nə-thŏl′ə-jē) *n.* The branch of zoology that is concerned with the study of birds. **—or′ni·tho·log′ic** (-thə-lŏj′ĭk), **or′ni·tho·log′i·cal** *adj.* **—or′ni·tho·log′i·cal·ly** *adv.* **—or′ni·thol′o·gist** *n.*

**or·ni·thop·ter** (ôr′nə-thŏp′tər) *n.* [ORNITHO- + (HELICO)PTER.] A hypothetical aircraft held aloft and propelled by wing movements.

**or·ni·tho·sis** (ôr′nə-thō′sĭs) *n.* A virus disease, resembling psittacosis, that infects domestic fowl and other birds and is communicable to humans. **—or′ni·thot′ic** (-thŏt′ĭk) *adj.*

**oro-** *pref.* [< Gk. *oros,* mountain.] Mountain <*orogeny*>

**o·rog·e·ny** (ô-rŏj′ə-nē) *also* **or·o·gen·e·sis** (ôr′ə-jĕn′ĭ-sĭs, ŏr′-) *n.* Formation of mountains, esp. by a folding of the earth's crust. **—or′o·gen′ic** (ôr′ə-jĕn′ĭk, ŏr′-) *adj.* **—or′o·gen′i·cal·ly** *adv.*

**o·rog·ra·phy** (ô-rŏg′rə-fē) *n.* Study of the physical geography of mountains and mountain ranges. **—or′o·graph′ic** (ôr′ə-grăf′ĭk, ŏr′-), **or′o·graph′i·cal** *adj.* **—or′o·graph′i·cal·ly** *adv.*

**o·ro·ide** (ôr′ō-īd′, ŏr′-) *also* **o·re·ide** (ôr′ē-īd′, ŏr′-) *n.* [Fr. *oréide* : *or,* gold (< Lat. *aurum*) + *-éide,* -oid.] An inexpensive alloy of copper, tin, and zinc, used in imitation gold jewelry.

**o·rol·o·gy** (ô-rŏl′ə-jē) *n.* The study of mountains. **—o′ro·log′i·cal** (ôr′ə-lŏj′ĭ-kəl, ŏr′-) *adj.* **—o′ro·log′i·cal·ly** *adv.* **—o·rol′o·gist** *n.*

**o·ro·tund** (ôr′ə-tŭnd′, ŏr′-) *adj.* [Lat. *ōre rotundo,* with a round mouth.] **1.** Full in sound : SONOROUS <an *orotund* voice> **2.** Pompous and bombastic <*orotund* language> **—o′ro·tun′di·ty** *n.*

**or·phan** (ôr′fən) *n.* [LLat. *orphanus* < Gk. *orphanos,* without parents.] A child whose parents are dead. *—adj.* **1.** Being an orphan. **2.** Intended for orphans <an *orphan* home> *—vt.* **-phaned, -phan·ing, -phans.** To deprive (a child) of one or both parents. **—or′phan·hood′** *n.*

**or·phan·age** (ôr′fə-nĭj) *n.* **1.** An institution for the care of orphans and abandoned children. **2.** The condition of being an orphan.

**Or·phe·us** (ôr′fē-əs, -fyo͞os′) *n.* [Gk.] *Gk. Myth.* A Thracian musician and poet whose music had the power to move inanimate objects. **—Or·phe′an** (ôr-fē′ən, ôr′fē-ən) *adj.*

**Or·phic** (ôr′fĭk) *adj.* [Gk. *Orphikos* < *Orpheus,* Orpheus.] **1.** Of or ascribed to Orpheus <*Orphic* mysteries> **2.** Of or characteristic of the philosophical and mystic principles set forth in the poems ascribed to Orpheus. **3.** Capable of casting a spell as Orpheus did by his music. **4.** *often* **orphic.** Mystic or occult. **—Or′phi·cal·ly** *adv.*

**Or·phism** (ôr′fĭz′əm) *n.* [Fr. *orphisme* < *Orphée,* Orpheus < Gk. *Orpheus.*] An ancient Greek mystery religion arising in the 6th cent. B.C. from a synthesis of pre-Hellenic beliefs with the Thracian cult of Zagreus. **—Or′phist** *n.*

**or·phrey** (ôr′frē) *also* **or·fray** (ôr′frā′) *n., pl.* **-phreys** *also* **-frays.** [ME *orfrey* < OFr. *orfrois* < Med. Lat. *aurifrigium* : Lat. *aurum,* gold + Lat. *Phrygius,* Phrygian.] **1.** A decorative embroidered band on the front of certain ecclesiastical vestments. **2.** Elaborate embroidery, esp. when made of gold.

**or·pi·ment** (ôr′pə-mənt) *n.* [ME < OFr. < Lat. *auripigmentum* : *aurum,* gold + *pigmentum,* pigment < *pingere,* to paint.] Arsenic trisulfide, $As_2S_3$, a lemon-yellow pigment used in tanning and in the manufacture of linoleum.

**or·pine** (ôr′pĭn) *n.* [ME *orpin* < OFr. < *orpiment,* orpiment.] A plant of the genus *Sedum,* esp. *S. telephium,* native to Eurasia, having reddish-purple flower clusters.

**Or·ping·ton** (ôr′pĭng-tən) *n.* [After *Orpington,* England.] A breed of large domestic fowls having a single comb and unfeathered legs.

**or·re·ry** (ôr′ə-rē, ŏr′-) *n., pl.* **-ries.** [After Charles Boyle (1676–1731), fourth Earl of *Orrery.*] A mechanical model of the solar system.

**or·ris** (ôr′ĭs, ŏr′-) *n.* [Var. of IRIS.] **1.** Any of several species of iris, esp. *Iris florentina,* having a fragrant rootstock. **2.** Orrisroot.

**or·ris·root** (ôr′ĭs-ro͞ot′, -ro͝ot′, ŏr′-) *n.* The fragrant rootstock of the orris, used in cosmetics and perfumes.

**ort** (ôrt) *n.* [ME *orte,* food left by animals < MDu. : *oor,* out + *eten,* to eat.] **1.** A scrap or leaving of food after a meal. **2.** A bit : scrap.

**orth-** *pref. var. of* ORTHO-.

**or·thi·con** (ôr′thĭ-kŏn′) *n.* [ORTH(O)- + ICON(OSCOPE).] A television camera pickup tube that uses a low-velocity electron beam to scan a photoactive mosaic.

**or·tho** (ôr′thō) *adj.* [< ORTHO-.] **1.** *Chem.* **a.** Designating the most fully hydrated form of an acid or of its salts. **b.** Of, relating to, or designating adjacent carbon positions in a benzene ring. **2.** *Physics.* Designating diatomic molecules in which the nuclei have the same spin directions. **3.** Orthochromatic.

**ortho-** or **orth-** *pref.* [ME < OFr. < Lat. < Gk. < *orthos,* straight, correct, right.] **1.** Straight : upright : vertical <*orthotropous*> **2.** Perpendicular <*orthorhombic*> **3.** Correct : correction <*orthopsychiatry*>

**or·tho·cen·ter** (ôr′thō-sĕn′tər) *n.* The point at which the three altitudes of a triangle intersect.

**or·tho·ce·phal·ic** (ôr′thō-sə-făl′ĭk) *also* **or·tho·ceph·a·lous** (-sĕf′ə-ləs) *adj.* Having a ratio of skull height to skull length between 0.70 and 0.75. **—or′tho·ceph′a·ly** (-sĕf′ə-lē) *n.*

**or·tho·chro·mat·ic** (ôr′thō-krō-măt′ĭk) *adj.* **1.** Of, having, or accurately reproducing natural or lifelike colors. **2.** Of or relating to a film, plate, or emulsion that renders all colors, except red, in tones of gray approximating the relative brilliance of these colors. **—or′tho·chro′ma·tism** (-krō′mə-tĭz′əm) *n.*

**or·tho·clase** (ôr′thə-klās′, -klāz′) *n.* [ORTHO- + Gk. *klasis,* a breaking < *klan,* to break.] A feldspar, essentially potassium aluminum silicate, $KAlSi_3O_8$, characterized by a monoclinic crystalline structure and occurring in granitic or igneous rock.

**or·tho·clas·tic** (ôr′thə-klăs′tĭk) *adj. Geol.* Having right-angled cleavage <an *orthoclastic* mineral>

**or·tho·don·tia** (ôr′thə-dŏn′shə) *n.* The branch of dentistry concerned with correcting abnormally aligned or positioned teeth. **—or′tho·don′tic** (-tĭk) *adj.* **—or′tho·don′tist** *n.*

**or·tho·don·tics** (ôr′thə-dŏn′tĭks) *n. (sing. in number).* Orthodontia.

**or·tho·dox** (ôr′thə-dŏks′) *adj.* [OFr. *orthodoxe* < LLat. *orthodoxus* < Gk. *orthodoxos* : *orthos,* correct + *doxa,* opinion < *dokein,* to think.] **1.** Adhering to the established and traditional faith, esp. in religion. **2.** Adhering to the Christian faith as set forth in the early Christian ecumenical creeds. **3.** Conforming to accepted standards or established practice. **4. Orthodox. a.** Of, relating to, or designating any of the churches of the Eastern Orthodox Church. **b.** Of, relating to, or denoting Orthodox Judaism. *—n.* **1.** One that is orthodox. **2. Orthodox.** A member of an Eastern Orthodox Church. **—or′tho·dox′ly** *adv.*

**Orthodox Church** *n.* The Eastern Orthodox Church.

**Orthodox Judaism** *n.* The branch of the Jewish faith that adheres to the Mosaic Law as interpreted in the Talmud and considers it binding in modern as well as ancient times.

**or·tho·dox·y** (ôr′thə-dŏk′sē) *n., pl.* **-ies. 1.** The quality or state of being orthodox. **2.** Orthodox practice, custom, or belief.

**or·tho·e·py** (ôr-thō′ə-pē, ôr′thō-ĕp′ē) *n.* [Gk. *orthoepeia,* correctness of diction : *orthos,* correct + *epos,* word.] **1.** The study of pronunciation. **2.** Standard pronunciation of a language. **—or′tho·ep′ic** (-ĕp′ĭk), **or′tho·ep′i·cal** (-ĕp′ĭ-kəl) *adj.* **—or·tho′e·pist** *n.*

**or·tho·gen·e·sis** (ôr′thō-jĕn′ĭ-sĭs) *n.* **1.** *Biol.* The theory that evolutionary change is predetermined by the constitution of germ plasm and unaffected by external factors. **2.** The theory that all cultures undergo progessive development in the same ordered stages. **—or′tho·ge·net′ic** (-jə-nĕt′ĭk) *adj.* **—or′tho·ge·net′i·cal·ly** *adv.*

**or·tho·gen·ic** (ôr′thō-jĕn′ĭk) *adj.* Relating to the treatment or correction of mental and emotional abnormalities in children.

**or·thog·na·thous** (ôr-thŏg′nə-thəs) *also* **or·thog·nath·ic** (ôr′thŏg-năth′ĭk) *adj.* Having the lower jaw aligned with the upper so that it does not protrude or recede. **—or·thog′na·thism, or·thog′na·thy** *n.*

**or·thog·o·nal** (ôr-thŏg′ə-nəl) *adj.* [Gk. *orthogōnios* : *orthos,* right + *gōnia,* angle.] *Math.* Relating to or composed of right angles. **—or·thog′o·nal·ly** *adv.*

**orthogonal projection** *n.* The two-dimensional graphic representation of an object formed by the perpendicular intersections of lines drawn from points on the object to a plane of projection.

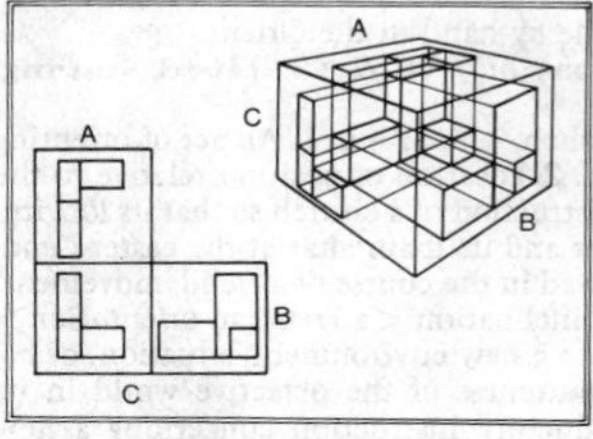


**orthogonal projection**
*A. top, B. side, C. front*

**or·tho·graph·ic** (ôr′thə-grăf′ĭk) *also* **or·tho·graph·i·cal** (-kəl) *adj.* **1.** Of or relating to orthography. **2.** Spelled correctly. **3.** *Math.* Having perpendicular lines. **—or′tho·graph′i·cal·ly** *adv.*

**orthographic projection** *n.* Orthogonal projection.

**or·thog·ra·phy** (ôr-thŏg′rə-fē) *n., pl.* **-phies. 1.** The art or technique of correct spelling according to established usage. **2.** The area of language study dealing with letters and their sequences in words. **3.** Symbolic representation of the sounds of language by letters or characters. **—or·thog′ra·pher, or·thog′ra·phist** *n.*

**or·tho·mo·lec·u·lar** (ôr′thō-mə-lĕk′yə-lər) *adj.* Of, relating to, or being a theory holding that mental illness is related to chemical deficiencies or imbalances and can be cured by restoring proper levels of chemical substances in the body.

**or·tho·pe·dics** *also* **or·tho·pae·dics** (ôr′thə-pē′dĭks) *n.* [< Fr. *orthopedique,* orthopedic : Gk. *orthos,* correct + Gk. *pais,* child.] *(sing. in number).* Surgical or manipulative treatment of disorders of the skeletal system and associated motor organs. **—or'tho·pe'dic** *adj.* **—or'tho·pe'di·cal·ly** *adv.* **—or'tho·pe'dist** *n.*

**or·tho·psy·chi·a·try** (ôr'thō-sī-kī'ə-trē, -sī-) *n. Psychiat.* Study and treatment of incipient and borderline mental disorders, esp. their development in children. **—or'tho·psy'chi·at'ric** (-sī'kē-ăt'rĭk), **or'tho·psy'chi·at'ri·cal** *adj.* **—or'tho·psy·chi'a·trist** *n.*

**or·thop·ter·an** (ôr-thŏp'tər-ən) *also* **or·thop·ter·on** (-tə-rŏn', -tər-ən) *n.* [NLat. *Orthoptera*, order name : Gk. *orthos*, straight + Gk. *pteron*, wing.] An insect of the order Orthoptera, including the grasshoppers, cockroaches, crickets, and locusts, marked by membranous folded hind wings covered by leathery narrow forewings. **—or·thop'ter·an, or·thop'ter·ous, or·thop'ter·al** *adj.*

**or·tho·rhom·bic** (ôr'thō-rŏm'bĭk) *adj.* Of or relating to a crystalline structure of three mutually perpendicular axes of different length.

**or·tho·scope** (ôr'thə-skōp') *n.* An instrument for examining the eye through a layer of water that compensates for the curvature of the cornea.

**or·tho·scop·ic** (ôr'thə-skŏp'ĭk) *adj.* **1.** Having normal vision. **2.** Relating to the use of the orthoscope. **3.** Giving an undistorted image.

**or·thos·ti·chous** (ôr-thŏs'tĭ-kəs) *adj.* [ORTHO- + Gk. *stikhos*, row.] *Biol.* Arranged in a vertical row <*orthostichous* flowers> **—or·thos'ti·chy** *n.*

**or·thot·ics** (ôr-thŏt'ĭks) *n.* [< Gk. *orthōsis*, a straightening < *orthoun*, to straighten < *orthos*, straight.] *(sing. in number).* A branch of medicine concerned with the design and use of mechanical devices to support or supplement weakened joints or limbs. **—or·thot'ic** *adj.* **—or·thot'ist** (ôr-thŏt'ĭst, ôr'thə-tĭst) *n.*

**or·tho·trop·ic** (ôr'thə-trŏp'ĭk) *adj.* Growing or tending to form in vertical direction <an *orthotropic* root> **—or'tho·trop'i·cal·ly** *adv.* **—or·thot'ro·pism** (ôr-thŏt'rə-pĭz'əm) *n.*

**or·thot·ro·pous** (ôr-thŏt'rə-pəs) *adj.* Growing straight, so the micropyle is at the side opposite the stalk <an *orthotropous* ovule>

**or·to·lan** (ôr'tə-lən) *n.* [Fr. < Prov., gardener < Lat. *hortulanus* < *hortus*, garden.] **1.** A small Old World bird, *Emberiza hortulana*, eaten as a delicacy. **2.** Any of several New World birds, as the sora and the bobolink.

**-ory** *suff.* [ME *-orie* < OFr. < Lat. *-orium*, n. suffix and *-orius*, adj. suffix.] **1.** Of, relating to, or marked by <advis*ory*> **2.** A place or thing used for or connected with <cremat*ory*>

**o·ryx** (ôr'ĭks, ōr'-, ŏr'-) *n., pl.* **o·ryx·es** *or* **oryx.** [Lat. < Gk. *orux.*] An antelope of the genus *Oryx*, of Africa and southwestern Asia, having long, straight or arching horns.

**os¹** (ŏs) *n., pl.* **o·ra** (ôr'ə, ōr'ə) [Lat. *ōs*, mouth.] A mouth or opening.

**os²** (ŏs) *n., pl.* **os·sa** (ŏs'ə) [Lat., bone.] A bone.

**os³** (ŏs) *n., pl.* **os·ar** (ō'sär') [Swed. *ås*, ridge < ON *áss.*] An esker.

**Os** *symbol for* OSMIUM.

**O·sage** (ō'sāj', ō-sāj') *n., pl.* **Osage** *or* **O·sag·es.** [Fr. < Osage *Wazházhe*, tribal name.] **1. a.** A tribe of Indians once inhabiting the region between the Missouri and Arkansas rivers. **b.** A member of this tribe. **2.** The Siouan language of the Osage. **—O'sage'** *adj.*

**Osage orange** *n.* An ornamental tree, *Maclura pomifera*, native to central North America, bearing inedible, orangelike fruit.

**os·ar** (ō'sär') *n. pl. of* OS³.

**Os·can** (ŏs'kən) *n.* **1.** One of an ancient people of Campania. **2.** The Italic language of the Oscans. **—Os'can** *adj.*

**Os·car** (ŏs'kər). A trademark for any of the golden statuettes awarded annually by the Academy of Motion Picture Arts and Sciences for achievement in motion pictures.

**os·cil·late** (ŏs'ə-lāt') *vi.* **-lat·ed, -lat·ing, -lates.** [Lat. *oscillare, oscillat-* < *oscillum*, swing.] **1.** To swing back and forth with a regular uninterrupted motion. **2.** To waver between two or more choices : VACILLATE. **3.** *Physics.* To vary between alternate extremes, usu. with a definable period. **—os'cil·la'tor** *n.* **—os'cil·la·to'ry** (-lə-tôr'ē, -tōr'ē) *adj.*

**os·cil·la·tion** (ŏs'ə-lā'shən) *n.* **1.** The act or state of oscillating. **2.** A single oscillatory cycle. **—os'cil·la'tion·al** *adj.*

**os·cil·lo·gram** (ŏ-sĭl'ə-grăm', ə-sĭl'-) *n.* [OSCILLO(GRAPH) + -GRAM.] **1.** The graph traced by an oscillograph. **2.** An instantaneous oscilloscope trace or photograph.

**os·cil·lo·graph** (ŏ-sĭl'ə-grăf') *n.* [OSCILL(ATION) + -GRAPH.] A device for recording oscillations in an electric current as a continuous graph of corresponding variation. **—os·cil'lo·graph'ic** *adj.* **—os·cil'lo·graph'i·cal·ly** *adv.* **—os'cil·log'ra·phy** (ŏs'ə-lŏg'rə-fē) *n.*

**os·cil·lo·scope** (ŏ-sĭl'ə-skōp', ə-sĭl'-) *n.* [OSCILL(ATION) + -SCOPE.] An instrument producing an instantaneous visual display on a fluorescent screen corresponding to oscillations in an electrical quantity. **—os·cil'lo·scop'ic** (-skŏp'ĭk) *adj.*

**os·cine** (ŏs'īn') *adj.* [NLat. *Oscines*, suborder name < Lat. *oscines*, pl. of *oscen*, bird used in augury : *ob-*, before + *canere*, to sing.] Of or relating to the Oscines, a large suborder of the passerine birds that includes most songbirds. **—os'cine'** *n.*

**os·ci·tan·cy** (ŏs'ĭ-tən-sē) *n., pl.* **-cies.** [< Lat. *oscitans*, pr.part. of *oscitare*, to yawn : *ōs*, mouth + *citare*, to move.] **1.** The act of yawning. **2.** The state of being inattentive or drowsy : DULLNESS. **—os'ci·tant** *adj.*

**Os·co-Um·bri·an** (ŏs'kō-ŭm'brē-ən) *n.* A subdivision of the Italic languages consisting of Oscan and Umbrian.

**os·cu·la** (ŏs'kyə-lə) *n. pl. of* OSCULUM.

**os·cu·lant** (ŏs'kyə-lənt) *adj.* [Lat. *osculans, osculant-*, pr.part. of *osculari*, to kiss. —see OSCULATE.] **1.** *Biol.* Intermediate in characteristics between two related or similar taxonomic groups. **2.** Closely adhering or joined : EMBRACING.

**os·cu·late** (ŏs'kyə-lāt') *v.* **-lat·ed, -lat·ing, -lates.** [Lat. *osculari, osculat-* < *osculum*, kiss, dim. of *ōs*, mouth.] *—vt.* To kiss. *—vi. Biol.* To have or share osculant characteristics.

**os·cu·la·tion** (ŏs'kyə-lā'shən) *n.* **1. a.** The act of kissing. **b.** A kiss. **2.** *Math.* A point where two branches of a curve have a common tangent and extend in both directions of the tangent. **—os'cu·la·to'ry** (ŏs'kyə-lə-tôr'ē, -tōr'ē) *adj.*

**os·cu·lum** (ŏs'kyə-ləm) *also* **os·cule** (ŏs'kyo͞ol') *n., pl.* **-cu·la** (-kyə-lə) *also* **-cules.** [Lat., dim. of *ōs*, mouth.] An opening in a sponge through which water is expelled. **—os'cu·lar** *adj.*

**-ose¹** *suff.* [ME < Lat. *-osus.*] Possessing : having the characteristics of : full of <cym*ose*>

**-ose²** *suff.* [< GLUCOSE.] **1.** Carbohydrate <fruct*ose*> **2.** Product of protein hydrolysis <prote*ose*>

**o·sier** (ō'zhər) *n.* [ME < OE *oser* and OFr. *osier*, both < Med. Lat. *osaria*, willow bed.] **1. a.** A willow having long, wandlike branches used in basketry and wickerwork, esp. *Salix viminalis* or *S. purpurea.* **b.** A branch or stem of an osier. **2.** A tree resembling the osier.

**O·si·ris** (ō-sī'rĭs) *n.* [Lat. < Gk., of Egypt. orig.] *Myth.* An ancient Egyptian god, brother and husband to Isis, whose annual death and rebirth symbolized the self-renewing fertility of nature.

**-osis** *suff.* [ME < Lat. < Gk. *-ōsis*, n. suffix.] **1.** Condition : process <osm*osis*> **2.** Diseased or abnormal condition <neur*osis*> **3.** Increase : formation <leukocyt*osis*>

**Os·man·li** (ŏz-măn'lē, ŏs-) *n., pl.* **-lis.** [Turk. : *Osman*, Osman I (1259–1326) + *-li*, adj. suffix.] **1.** An Ottoman Turk. **2.** The Turkish language. *—adj.* Ottoman.

**os·mat·ic** (ŏz-măt'ĭk) *also* **os·mic** (ŏz'mĭk) *adj.* [Gk. *osmē*, odor + -ATE¹ + -IC.] Having or marked by a sense of smell.

**os·mic¹** (ŏz'mĭk) *adj.* [OSM(IUM) + -IC.] Of, relating to, or containing osmium in a compound with a valence higher than that in a comparable osmous compound.

**os·mic²** (ŏz'mĭk) *adj. var. of* OSMATIC.

**os·mi·ous** (ŏz'mē-əs) *adj. var. of* OSMOUS.

**os·mi·rid·i·um** (ŏz'mə-rĭd'ē-əm) *n.* [OSM(IUM) + IRIDIUM.] A natural alloy of osmium and iridium with small inclusions of other metals such as platinum and rhodium, used in needles.

**os·mi·um** (ŏz'mē-əm) *n.* [NLat. < Gk. *osmē*, odor (from the odor of osmium tetroxide).] *Symbol* **Os** A bluish-gray, hard, metallic element, used in platinum alloys, as a catalyst, and in making pen points and phonograph needles; atomic number 76; atomic weight 190.2.

**os·mom·e·ter** (ŏz-mŏm'ĭ-tər, ŏs-) *n.* [OSMO(SIS) + -METER.] An instrument for measuring osmotic pressure. **—os'mo·met'ric** (ŏz'mə-mĕt'rĭk, ŏs'-) *adj.* **—os·mom'e·try** *n.*

**os·mo·reg·u·la·tion** (ŏz'mə-rĕg'yə-lā'shən, ŏs'-) *n.* [OSMO(SIS) + REGULATION.] Maintenance of optimal and constant osmotic pressure in the body of a living organism. **—os'mo·reg'u·la·to'ry** (-lə-tôr'ē, -tōr'ē) *adj.*

**os·mose** (ŏz'mōs', ŏs'-) *vi.* & *vt.* **-mosed, -mos·ing, -mos·es.** [Back-formation < OSMOSIS.] To diffuse or cause to diffuse by the process of osmosis.

**os·mo·sis** (ŏz-mō'sĭs, ŏs-) *n.* [< obs. *osmose*, ult. < Gk. *ōsmos*, thrust < *ōthein*, to push.] **1. a.** Diffusion of fluid through a semipermeable membrane, as the wall of a living cell. **b.** The tendency of a solution to diffuse in such a manner until equally concentrated on both sides of a membrane. **2.** A gradual, often unconscious process of absorption resembling fluid osmosis <learned German by *osmosis*> **—os·mot'ic** (-mŏt'ĭk) *adj.* **—os·mot'i·cal·ly** *adv.*

**os·mous** (ŏz'məs) *also* **os·mi·ous** (ŏz'mē-əs) *adj.* [OSM(IUM) + -OUS.] Of, relating to, or containing osmium in a compound having a valence lower than that in a comparable osmic compound.

**os·mun·da** (ŏz-mŭn'də) *also* **os·mund** (ŏz'mənd) *n.* [NLat. *Osmunda*, genus name < ME *osmunde*, a kind of fern < OFr. *osmonde.*] A fern of the genus *Osmunda*, esp. any of several species having fibrous roots used as a potting medium for cultivated plants.

**os·na·burg** (ŏz'nə-bûrg') *n.* [After *Osnaburg*, Osnabrück, West Germany.] A coarse cotton fabric, used as sacking and for draperies and upholstery.

**os·prey** (ŏs'prē, -prā') *n., pl.* **-preys.** [ME *osprai* < AN < Lat. *avis praedae*, bird of prey.] **1.** A large, fish-eating hawk, *Pandion haliaetus*, having plumage that is dark on the back and white below. **2.** A plume used as trimming on women's hats.

**os·sa** (ŏs'ə) *n. pl. of* OS².

**os·se·in** (ŏs'ē-ĭn) *n.* [OSSE(OUS) + -IN.] Bone residue left after acid dissolution, used in gelatin and glue.

**os·se·ous** (ŏs'ē-əs) *adj.* [Lat. *osseus* < *os*, bone.] Composed of, containing, or resembling bone : BONY. **—os'se·ous·ly** *adv.*

**Os·set** (ŏs′ĭt, ŏ-sĕt′) *also* **Os·sete** (ŏs′ēt′, ŏ-sĕt′) *n.* One of a people of Iranian ancestry living in Ossetia. **—Os·set′ic** *adj.*

**os·si·a** (ō-sē′ə) *conj.* [Ital. < *o sia,* or whether.] *Mus.* Or else. —Used as a direction to designate an alternative passage or section.

**Os·sian** (ŏsh′ən, ŏs′ē-ən) *n.* [Gael. *Oisin.*] A legendary Gaelic hero and bard of the 3rd cent. A.D.

**os·si·cle** (ŏs′ĭ-kəl) *n.* [Lat. *ossiculum,* dim. of *os,* bone.] *Anat.* A small bone, esp. one of the three bones of the inner ear. **—os·sic′u·lar** (ŏ-sĭk′yə-lər), **os·sic′u·late** (-lĭt) *adj.*

**os·si·fi·ca·tion** (ŏs′ə-fĭ-kā′shən) *n.* **1.** The natural process of bone formation. **2. a.** The abnormal hardening of soft tissue into a bonelike material : CALCIFICATION. **b.** A mass or deposit of such material. **3.** The state of being or process of becoming fixed in a rigidly conventional pattern <deplored the *ossification* of big bureaucracies> **—os·sif′i·ca·to′ry** (ŏ-sĭf′ĭ-kə-tôr′ē, -tōr′ē) *adj.*

**os·si·frage** (ŏs′ə-frĭj, -frāj′) *n.* [Lat. *ossifraga* < *ossifragus,* bone-breaking : *os,* bone + *frangere,* to break.] **1.** Osprey. **2.** Lammergeier.

**os·si·fy** (ŏs′ə-fī′) *v.* **-fied, -fy·ing, -fies.** [Lat. *os, oss-,* bone + -FY.] —*vi.* **1.** To become bony. **2.** To become set in a rigidly conventional pattern. —*vt.* **1.** To convert (e.g., cartilage) into bone. **2.** To mold into a rigidly conventional pattern. **—os·sif′ic** (ŏ-sĭf′ĭk) *adj.*

**os·so bu·co** (ō′sō bōō′kō) *n.* [Ital. *ossobuco,* marrowbone.] An Italian dish of veal shanks braised in white wine.

**os·su·ar·y** (ŏsh′ōō-ĕr′ē, ŏs′yōō-) *n., pl.* **-ies.** [LLat. *ossuarium,* neut. of Lat. *ossuarius,* of bones < *os,* bone.] A receptacle, as a vault or urn, for holding the bones of the dead.

**oste-** *pref. var. of* OSTEO-.

**os·te·al** (ŏs′tē-əl) *adj.* **1.** Osseous : bony. **2.** Of or pertaining to bone or to the skeleton.

**os·te·i·tis** (ŏs′tē-ī′tĭs) *n.* Inflammation of bone or bony tissue.

**os·ten·si·ble** (ŏ-stĕn′sə-bəl) *adj.* [Fr. < Med. Lat. *ostensibilis* < Lat. *ostendere,* to show : *ob-,* before + *tendere,* to stretch.] Represented or appearing as such : PROFESSED <Their *ostensible* friendliness was really an attempt to curry favor.> **—os·ten′si·bly** *adv.*

**os·ten·sive** (ŏ-stĕn′sĭv) *adj.* [LLat. *ostensivus* < Lat. *ostendere,* to show. —see OSTENSIBLE.] **1.** Ostensible. **2.** Obviously or manifestly demonstrative. **—os·ten′sive·ly** *adv.*

**os·ten·so·ri·um** (ŏs′tən-sôr′ē-əm, -sōr′ē-) *also* **os·ten·so·ry** (ŏ-stĕn′sə-rē) *n., pl.* **-so·ri·a** (-sôr′ē-ə, -sōr′-) *also* **-so·ries.** [Med. Lat. < Lat. *ostendere,* to show. —see OSTENSIBLE.] *Rom. Cath. Ch.* A monstrance.

**os·ten·ta·tion** (ŏs′tĕn-tā′shən, -tən-) *n.* [ME *ostentacioun* < Lat. *ostentatio* < *ostentare,* freq. of *ostendere,* to show. —see OSTENSIBLE.] **1.** Lavish or pretentious display meant to impress others. **2.** *Archaic.* An act of showing : EXHIBITION.

**os·ten·ta·tious** (ŏs′tĕn-tā′shəs, -tən-) *adj.* Marked by or given to ostentation : PRETENTIOUS. **—os′ten·ta′tious·ly** *adv.*

**osteo-** *or* **oste-** *pref.* [Gk. < *osteon,* bone.] Bone <*osteo*cranium>

**os·te·o·ar·thri·tis** (ŏs′tē-ō-är-thrī′tĭs) *n.* Degenerative joint disease. **—os′te·o·ar·thrit′ic** *adj.*

**os·te·o·blast** (ŏs′tē-ə-blăst′) *n.* A cell from which bone develops. **—os′te·o·blas′tic** *adj.*

**os·te·oc·la·sis** (ŏs′tē-ŏk′lə-sĭs) *n.* [OSTEO- + Gk. *klasis,* breakage < *klan,* to break.] **1.** Surgical fracture of a bone, performed to correct a deformity. **2.** Dissolution and resorption of bony tissue during the regeneration of bone.

**os·te·o·clast** (ŏs′tē-ə-klăst′) *n.* [OSTEO- + Med. Lat. *-clastes,* breaker < Med. Gk. *-klastēs* < Gk. *klan,* to break.] **1.** An instrument used in surgical osteoclasis. **2.** A large multinuclear cell that resorbs bony tissue in osteoclasis. **—os′te·o·clas′tic** *adj.*

**os·te·o·cra·ni·um** (ŏs′tē-ō-krā′nē-əm) *n.* The ossified embryonic cranium.

**os·te·o·cyte** (ŏs′tē-ə-sīt′) *n.* A bone cell.

**os·te·o·gen·ic sarcoma** (ŏs′tē-ə-jĕn′ĭk) *n.* Osteosarcoma.

**os·te·oid** (ŏs′tē-oid′) *adj.* Resembling bone. —*n.* The bone matrix, esp. before calcification.

**os·te·ol·o·gy** (ŏs′tē-ŏl′ə-jē) *n.* **1.** Anatomical study of bones. **2.** The bone structure of an animal. **—os′te·o·log′i·cal** (-ə-lŏj′ĭ-kəl) *adj.* **—os′te·ol′o·gist** *n.*

**os·te·o·ma** (ŏs′tē-ō′mə) *n., pl.* **-mas** *or* **-ma·ta** (-mə-tə). A benign bony tumor.

**os·te·o·ma·la·cia** (ŏs′tē-ō-mə-lā′shə, -shē-ə) *n.* [OSTEO- + Gk. *malakia,* softness < *malakos,* soft.] A deficiency disease resulting form a lack of vitamin D or of calcium and marked by a softening of the bones.

**os·te·o·ma·ta** (ŏs′tē-ō′mə-tə) *n. var. pl. of* OSTEOMA.

**os·te·o·my·e·li·tis** (ŏs′tē-ō-mī′ə-lī′tĭs) *n.* Infectious inflammation of the bone marrow.

**os·te·o·path** (ŏs′tē-ə-păth′) *also* **os·te·op·a·thist** (ŏs′tē-ŏp′ə-thĭst) *n.* One who practices osteopathy.

**os·te·op·a·thy** (ŏs′tē-ŏp′ə-thē) *n.* Medical therapy involving manipulative techniques for correcting abnormalities of the musculoskeletal system that are thought to cause disease and inhibit recovery. **—os′te·o·path′ic** (-ə-păth′ĭk) *adj.* **—os′te·o·path′i·cal·ly** *adv.*

**os·te·o·phyte** (ŏs′tē-ə-fīt′) *n.* A small abnormal bony outgrowth. **—os′te·o·phyt′ic** (-fĭt′ĭk) *adj.*

**os·te·o·plas·tic** (ŏs′tē-ə-plăs′tĭk) *adj.* **1.** *Med.* Of or relating to osteoplasty. **2.** *Physiol.* Relating or contributing to in bone formation.

**os·te·o·plas·ty** (ŏs′tē-ə-plăs′tē) *n., pl.* **-ties.** Surgical repair or replacement of bone.

**os·te·o·sar·co·ma** (ŏs′tē-ō-sär-kō′mə) *n.* A malignant sarcoma of the bone.

**os·te·ot·o·my** (ŏs′tē-ŏt′ə-mē) *n., pl.* **-mies.** Surgical division or sectioning of bone. **—os′te·ot′o·mist** *n.*

**os·ti·a** (ŏs′tē-ə) *n. pl. of* OSTIUM.

**Os·ti·ak** (ŏs′tē-ăk′) *n. var. of* OSTYAK.

**os·ti·ar·y** (ŏs′tē-ĕr′ē) *n., pl.* **-ies.** [Lat. *ostiarius,* doorkeeper < *ostium,* door < *ōs,* mouth.] **1.** *Rom. Cath. Ch.* One ordained in the lowest of the four minor orders. **2.** A doorkeeper at a church.

**os·ti·na·to** (ŏs′tĭ-nä′tō) *n., pl.* **-tos.** [Ital. < Lat. *obstinatus,* stubborn, p.part. of *obstinare,* to persist.] *Mus.* A brief, unvarying melody or figure repeated continually throughout a composition.

**os·ti·ole** (ŏs′tē-ōl′) *n.* [Lat. *ostiolum,* dim. of *ostium,* opening < *ōs,* mouth.] A small opening or pore. **—os′ti·o′lar** *adj.*

**os·ti·um** (ŏs′tē-əm) *n., pl.* **-ti·a** (-tē-ə) [Lat. < *ōs,* mouth.] A small opening : OSTIOLE.

**os·tler** (ŏs′lər) *n. var. of* HOSTLER.

**ost·mark** (ôst′märk′, ŏst′-) *n.* [G. : *Ost,* east (< OHG *ōstan*) + *Mark,* Deutsche mark.] The East German mark.

**os·to·my** (ŏs′tə-mē) *n., pl.* **-mies.** [< COLOSTOMY.] Surgical construction of an artificial excretory opening, as a colostomy.

**os·tra·cism** (ŏs′trə-sĭz′əm) *n.* [Fr. *ostracisme* < Gk. *ostrakismos* < *ostrakizein,* to ostracize. —see OSTRACIZE.] **1.** Banishment or exclusion from a group : REJECTION. **2.** The act of ostracizing or the state of being ostracized. **3.** The temporary banishment by popular vote of a citizen regarded as dangerous to the state in ancient Greece.

**os·tra·cize** (ŏs′trə-sīz′) *vt.* **-cized, -ciz·ing, -ciz·es.** [Gk. *ostrakizein* < *ostrakon,* potsherd, from the potsherds used as ballots in voting for ostracism.] **1.** To banish or exclude from a group, as in disgrace or disfavor : SHUN. **2.** To banish by ostracism, as in ancient Greece.

**os·tra·cod** (ŏs′trə-kŏd′) *n.* [NLat. *Ostracoda,* order name < Gk. *ostrakōdēs,* testaceous < *ostrakon,* shell.] Any of various minute, mainly freshwater crustaceans of the order Ostracoda, marked by a bivalve carapace completely enclosing the body.

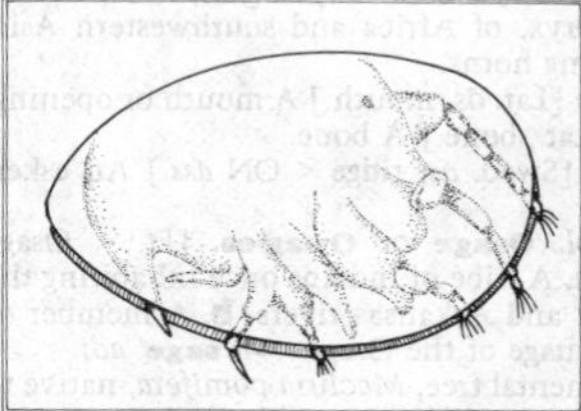

**ostracod**
*Less than one-quarter inch in diameter*

**os·trich** (ŏs′trĭch, ôs′-) *n., pl.* **-trich·es** *or* **ostrich.** [ME < OFr. *ostruce* and Med. Lat. *ostrica,* both < VLat. **avis struthio* : Lat. *avis,* bird + LLat. *struthio,* ostrich. —see STRUTHIOUS.] **1. a.** A large swift flightless African bird of the genus *Struthio,* having a long bare neck, powerful legs, two-toed feet, and large wing and tail feathers used for decoration. **b.** The rhea. **2.** One who tries to avoid threatening or unpleasant situations by refusing to face them.

**ostrich fern** *n.* A fern, *Matteuccia Struthiopteris* of northern temperate regions, having long plumelike fronds.

**Os·tro·goth** (ŏs′trə-gŏth′) *n.* [LLat. *Ostrogothi,* Ostrogoths : *ostro-,* eastward (of Germanic orig.) + *Gothi,* Goths, of Germanic orig.] One of a tribe of eastern Goths that conquered and ruled Italy from A.D. 493 to 555. **—Os′tro·goth′ic** *adj.*

**Os·ty·ak** *also* **Os·ti·ak** (ŏs′tē-ăk′) *n.* [R. < Ostyak *əsyakh,* dwellers on the Ob River < *Ās,* the Ob River.] **1.** One of a Finno-Ugric people inhabiting western Siberia. **2.** The Ugric language of the Ostyaks.

**Os·we·go tea** (ŏs-wē′gō) *n.* [After the *Oswego* River, New York.] An aromatic plant, *Monarda didyma* of North America, having fragrant scarlet flower clusters.

**ot-** *pref. var. of* OTO-.

**O·ta·hei·te orange** (ō′tə-hē′tē) *n.* [After *Otaheite,* var. of *Tahiti.*] An ornamental plant, *Citrus taitensis,* resembling a miniature orange tree, and having oval, insipid fruit.

**oth·er** (ŭth′ər) *adj.* [ME < OE *ōðer.*] **1. a.** Being or designating the remaining one of two or more <the *other* arm> **b.** Being or designating the remaining ones of several <*other* toys in the box> **2.** Different from that or those specified or implied <some *other* color for a change> **3.** Of a different quality or character <We have no way to get there *other* than walking.> **4.** Of a different time or era either past or future <dreaming of *other* days> **5.** Additional : extra <I have no *other* suit.> **6.** Opposite or contrary : REVERSE <the *other* direction> **7.** Alternate : second <every *other* year> **8.** Of the recent past <just the *other* day> —*n.* **1. a.** The remaining one of two or more <One was broken, but the *other* still worked.> **b. others.** The remaining ones of several <The first book was so good, I read all the *others.*> **2. a.** A different person or thing <one wave after the *other*> **b.** An additional person or thing <Are any *others* signed up?> —*pron.* **1.** A different or another person or thing <some problem or *other*> **2. others.** People aside from oneself <rely on *others* for help> —*adv.* In another way : OTHERWISE <an *other* than satisfactory job>

**oth·er-di·rect·ed** (ŭth′ər-dĭ-rĕk′tĭd, -dī-) *adj.* Directed or influenced mainly by external standards as opposed to one's own ideals or values. **—oth′er-di·rect′ed·ness** *n.*

**oth·er·wise** (ŭth′ər-wīz′) *adv.* [ME < OE *(on) ōðre wīsan,* (in) another manner.] **1.** In another way : DIFFERENTLY <believed *otherwise*> **2.** Under different circumstances <*Otherwise* I would have called.> **3.** In other respects <an *otherwise* perfect vacation> —*adj.* Other than supposed : DIFFERENT <Unfortunately, the facts were *otherwise.*>

**oth·er·world** (ŭth′ər-wûrld′) *n.* A world or existence beyond earthly reality.

**oth·er·world·ly** (ŭth′ər-wûrld′lē) *adj.* **1.** Of or characteristic of another esp. transcendent world : SUPERNATURAL. **2.** Concerned with spiritual, intellectual, or imaginative matters. **—oth′er·world′li·ness** *n.*

**o·tic** (ō′tĭk) *adj.* [Gk. *ōtikos* < *ous,* ear.] Of, relating to, or located near the ear : AURICULAR.

**-otic** *suff.* [Fr. *-otique* < Lat. *-oticus* < Gk. *-ōtikos,* adj. suffix.] **1.** Of, pertaining to, or characterized by a specified condition or process <anabi*otic*> **2.** Having a specified disease or abnormal condition <epizo*otic*> **3.** Characterized by an increase or formation of a specified kind <leukocyt*otic*>

**o·ti·ose** (ō′shē-ōs′, ō′tē-) *adj.* [Lat. *otiosus,* idle < *otium,* leisure.] **1.** Lazy : idle. **2.** Of no use. **3.** Futile : ineffective. **—o′ti·ose′ly** *adv.* **—o′ti·os′i·ty** (-ŏs′ĭ-tē) *n.*

**o·ti·tis** (ō-tī′tĭs) *n.* Inflammation of the ear. **—o·tit′ic** (ō-tĭt′ĭk) *adj.*

**oto-** or **ot-** *pref.* [NLat. < Gk. *ous,* ear.] Ear <*oto*logy>

**o·to·cyst** (ō′tə-sĭst′) *n.* **1.** The auditory capsule in a vertebrate embryo. **2.** A statocyst. **—o′to·cys′tic** *adj.*

**o·to·lar·yn·gol·o·gy** (ō′tō-lăr′ĭng-gŏl′ə-jē) *n.* The branch of medicine that combines treatment of the ear and throat. **—o′to·lar·yn′-go·log′i·cal** (-lə-rĭng′gə-lŏj′ĭ-kəl) *adj.* **—o′to·lar′yn·gol′o·gist** *n.*

**o·to·lith** (ō′tə-lĭth′) *n.* A minute calcareous particle found in the inner ear of certain vertebrates and in the statocysts of many invertebrates. **—o′to·lith′ic** *adj.*

**o·tol·o·gy** (ō-tŏl′ə-jē) *n.* The anatomy, physiology, and pathology of the ear. **—o′to·log′i·cal** (ō′tə-lŏj′ĭ-kəl) *adj.* **—o·tol′o·gist** *n.*

**o·to·rhi·no·lar·yn·gol·o·gy** (ō′tō-rī′nō-lăr′ĭng-gŏl′ə-jē) *n.* The branch of medicine that combines treatment of the ear, nose, and throat. **—o′to·rhi′no·la·ryn′go·log′i·cal** (-lə-rĭng′gə-lŏj′ĭ-kəl) *adj.* **—o′to·rhi′no·lar′yn·gol′o·gist** *n.*

**ot·ta·va** (ō-tä′və) *adv.* [Ital.] *Mus.* At an octave higher or lower than written. —Used as a direction.

**ottava ri·ma** (rē′mə) *n.* [Ital., eighth rhyme.] A stanza of verse made up of 8 lines of 11 syllables each in iambic pentameter and having a rhyme pattern *abababcc.*

**Ot·ta·wa** (ŏt′ə-wə, -wä′, -wô′) *n., pl.* **-was** or **-wa.** [After the *Ottawa* River.] **1. a.** A group of Indians inhabiting Michigan and southern Ontario. **b.** A member of this group. **2.** The dialect of Ojibwa spoken by the Ottawas.

**ot·ter** (ŏt′ər) *n., pl.* **otter** or **-ters.** [ME *oter* < OE *otor.*] **1.** Any of several aquatic, carnivorous mammals of the genus *Lutra* and related genera, having webbed feet and thick dark-brown fur. **2.** The fur of the otter, used in garments and as trimming.

**ot·to** (ŏt′ō) *n. var. of* ATTAR.

**ot·to·man** (ŏt′ə-mən) *n., pl.* **-mans.** [Fr. *ottomane,* fem. of *ottoman,* Ottoman.] **1. a.** An upholstered sofa without a back or arms. **b.** An upholstered or cushioned footstool. **2.** A heavy, corded fabric of silk or rayon, usu. used for coats and trimmings.

**Ot·to·man** (ŏt′ə-mən) *n., pl.* **-mans.** [Fr. < Med. Lat. *Ottomanus* < Ar. *Othmānī,* Turkish < Turk. *Osman,* Osman I (1259–1326).] A Turk. —*adj.* Of or relating to the Turks : TURKISH.

**oua·ba·in** (wä-bä′ĭn) *n.* [< Fr. *ouabaio* < Somali *wabayo.*] A white toxic glucoside, $C_{29}H_{44}O_{12}\cdot 8H_2O$, derived from the seeds of the African trees *Strophanthus gratus* and *Acokanthera ouabaio,* and used as a cardiac stimulant and by some African tribes as a poison.

**ou·bli·ette** (o͞o′blē-ĕt′) *n.* [Fr. < *oublier,* to forget < OFr. *oblider* < Lat. *oblivisci.*] A dungeon having a trap door in the ceiling as its only means of access or communication.

**ouch¹** (ouch) *interj.* —Used to express sudden pain or distress.

**ouch²** (ouch) *n.* [ME *an ouche,* an ouch, alteration of *a nouche* < AN *nouch,* brooch, of Germanic orig.] **1.** A setting for a gem. **2.** A buckle or a brooch set with jewels. **3.** *Obs.* A clasp : brooch.

**oud** (o͞od) *n.* [Ar. *'ūd.*] A lutelike musical instrument of northern Africa and southwest Asia.

**ought¹** (ôt) *aux.v.* [ME *oughten,* to be obliged to < *oughte,* owned < OE *āhte,* p.t. of *āgan,* to possess.] —Used to indicate: **1.** Duty or obligation <You *ought* to try harder than that.> **2.** Advisability or prudence <You *ought* to wear a coat.> **3.** Desirability <You *ought* to have been there; it was interesting.> **4.** Probability or likelihood <I *ought* to finish by next month.>

**ought²** (ôt) *pron. & adv. var. of* AUGHT¹.

**ought³** (ôt) *n. var. of* AUGHT².

**ought⁴** (ôt) *v. Obs. var. p.p. of* OWE.

**ou·gui·ya** (o͞o-gē′yə) *n.* [Native word in Mauritania.] —See table at CURRENCY.

**Oui·ja** (wē′jə, -jē). A trademark for a board with the alphabet and other symbols on it and a planchette that is thought, when touched with the fingers, to move in such a way as to spell out spiritualistic and telepathic messages on the board.

**ounce¹** (ouns) *n.* [ME *unce* < OFr. < Lat. *uncia* < *unus,* unit.] **1. a.** A unit of weight in the U.S. Customary System, an avoirdupois unit equal to 437.5 grains or 28.350 grams. **b.** A unit of apothecary weight, equal to 480 grains or 31.103 grams. **2. a.** A unit of volume or capacity in the U.S. Customary System, used in liquid measure, equal to 8 fluid drams or 29.573 milliliters. **b.** A unit of volume or capacity in the British Imperial System, used in dry and liquid measure, equal to 1.734 cubic inches or 28.412 milliliters. **3.** A tiny bit.

**ounce²** (ouns) *n.* [ME *once* < OFr., alteration of *lonce* < Lat. *lynx,* lynx < Gk. *lunx.*] The snow leopard.

▲ word history: The word *ounce²* is currently a name for the snow leopard, a large Asian mountain cat, but in earlier times it was used for other large cats, especially the lynx. *Ounce²* in fact is derived from the same source as the word *lynx,* which is Greek *lunx.* The Greek word was borrowed into Latin as *lynx;* English adopted the Latin form directly. The form *ounce* comes from Old French. The normal development of Latin *lynx* in Old French was *lonce.* The initial *l* in this word, however, was interpreted as the definite article *la,* whose vowel would be dropped before a word beginning with a vowel. The Old French form *once* arose from this mistake. English borrowed the French form *once,* which by a regular change in English became the modern form *ounce.*

**our** (our) *adj.* [ME *oure* < OE *ūre.*] Of or pertaining to us, ourself, or ourselves, esp. as possessors or possessor, agents or agent, or object or objects of an action <*our* house><*our* meeting by chance><*our* biggest success><*our* being delayed>

**Our Father** *n.* The Lord's Prayer.

**Our Lady** *n.* The Virgin Mary.

**ours** (ourz) *pron.* [ME *oures* < *oure,* our < OE *ūre.*] *(sing. or pl. in number).* That or those belonging to us <That boat is *ours.*><*Ours* were the only suggestions adopted.>

**our·self** (our-sĕlf′, är-) *pron.* Myself. —Used as a reflexive when *we* is used instead of *I* by a singular speaker or author, as in an editorial or royal proclamation.

**our·selves** (our-sĕlvz′, är-) *pl. pron.* **1.** Those identical with us. —Used: **a.** Reflexively as the direct or indirect object of a verb or the object of a preposition <We hurt *ourselves.*> **b.** For emphasis <We *ourselves* are excluded.> **c.** In an absolute construction <In difficulty *ourselves,* we were unable to help the others.> **2.** Our normal or healthy condition or state <We have not been *ourselves* since you left>.

**-ous** *suff.* [ME < OFr. *-ous, -eus, -eux* < Lat. *-osus* and *-us,* adj. suffixes.] **1.** Possessing : full of : characterized by <joy*ous*> **2.** Having a valence lower than that of a specified element in compounds or ions named with adjectives ending in *-ic* <ferr*ous*>

**ou·sel** (o͞o′zəl) *n. var. of* OUZEL.

**oust** (oust) *vt.* **oust·ed, oust·ing, ousts.** [Norman Fr. *ouster* < Lat. *obstare,* to hinder. —see OBSTACLE.] **1.** To force out or eject from a place or position : EXPEL <"the American Revolution, which *ousted* the English" —Virginia S. Eifert> **2.** To take the place of, esp. by force : SUPPLANT <One general merely *ousted* another.>

**oust·er** (ous′tər) *n.* [Norman Fr. < *ouster,* to oust.] **1.** The act of ousting or the state of being ousted. **2.** One that ousts. **3.** *Law.* The act of forcing one out of occupancy or possession of material property to which he or she is entitled : illegal or wrongful dispossession.

**out** (out) *adv.* [ME < OE *ūt.*] **1.** In a direction away from the inside <step *out* of the room> **2.** Away from the middle or center <The legs fold *out.*> **3. a.** Away from a home or business <go *out* for dinner> **b.** To an abnormal position <threw my knee *out*> **4.** Outside of a house or building : OUTDOORS <Put the dog *out.*> **5. a.** From within a container or source <The water leaked *out.*> **b.** From among others. **6. a.** To depletion or exhaustion <run *out* of gas> **b.** Into extinction or imperceptibility <The light went *out.*> **c.** To a finish or conclusion <serve my time *out*> **d.** To the fullest degree or extent <completely worn *out*> **e.** In or into competition <tried *out* for the soccer team> **7. a.** Into being or availability <The latest models are *out.*> **b.** Into public circulation <The new issue came *out* last week.> **8.** Into view <The sun came *out.*> **9.** Without inhibition : BOLDLY <Speak *out.*> **10.** Into possession of another or others : into distribution <hand *out* pamphlets> **11. a.** Into disuse or an unfashionable status <Wide ties have gone *out.*> **b.** Into a state of deprivation or loss <threw the corrupt officials *out*> **12.** *Baseball.* So as to be retired, or counted as an out <lined *out* to third base> **13.** —Used in two-way radio transmission to indicate that a message is terminated. —*adj.* **1.** Exterior : external. **2.** Located outside a building or shelter <The laundry was *out* all night.> **3.** Located away from home or business : ABSENT <The doctor is *out.*> **4.** Directed away from a place or center : OUTGOING. **5.** In evidence or view <The stars will soon be *out.*> **6.** Exhausted : depleted <The provisions were almost *out.*> **7.** Extinguished. **8.** Not functioning, operating, or flowing <The power is *out.*> **9.** Not to be considered : IMPOSSIBLE <Walking is *out*, because my feet are too sore.> **10.** No longer in office or power <After the election, the hard-liners were *out.*> **11.** No longer fashionable. **12.** *Informal.* Without an amount possessed previously <I'm *out* five dollars.> **13.** Bare or threadbare <a sweater *out* at the elbow> **14.** *Baseball.* Not allowed to continue to bat or run : RETIRED. **15.** Determined : intent <*out* for revenge> —*prep.* **1.** Forth from : THROUGH <fell *out* the window> **2.** Beyond or outside <*Out* this door is the garage.> —*n.* **1.** One that is out, esp. out of power. **2.** A means of escape or evasion : RESORT <Playing sick was my only *out.*> **3.** *Baseball.* **a.** A play in which a batter or base runner is retired. **b.** The player retired in such a play. **4.** A serve or return that falls out of bounds in a court game. **5.** A word or portion of a manuscript omitted from the printed copy. —*v.* **out·ed, out·ing, outs.** —*vi.* To be revealed or disclosed : COME OUT <Scandal will *out.*> —*vt.* **1.** To put (a person or thing) out. **2.** *Chiefly Brit.* To knock unconscious. **—on the outs.** *Informal.* Not on friendly terms : at odds.

**out-** *pref.* [< OUT.] In a way that surpasses, exceeds, or goes beyond <*outdistance*>

**out·age** (ou'tĭj) *n.* **1.** A quantity or portion that is lost during transportation, delivery, or storage. **2.** A temporary suspension of operation, esp. of electric power.

**out-and-out** (out'n-out') *adj.* Complete : absolute.

**out-and-outer** (out'ən-ou'tər) *n.* An extremist.

**out·back** (out'băk') *n.* The remote rural part of a country, esp. of Australia or New Zealand. **—out'back'** *adv.* **—out'back'er** *n.*

**out·bid** (out-bĭd') *vt.* **-bid, -bid·den** (-bĭd'n) *or* **-bid, -bid·ding, -bids.** To bid higher than.

**out·board** (out'bôrd', -bōrd') *adj.* **1.** *Naut.* **a.** Located outside the hull of a vessel. **b.** Being away from the center line of the hull of a ship. **2.** Situated toward or nearer the end of a wing of an aircraft. **—out'board'** *adv.*

**outboard motor** *n.* A detachable engine mounted on the stern of a boat.

**out·bound** (out'bound') *adj.* Outward bound : headed away.

**out·break** (out'brāk') *n.* **1.** A sudden rise of or increase <an *outbreak* of malaria> **2.** A sudden eruption : OUTBURST <an *outbreak* of hostilities>

☆ ***syns:*** OUTBREAK, EPIDEMIC, PLAGUE, RASH *n. core meaning* : a sudden increase, as in the occurrence of a disease <an *outbreak* of influenza>

**out·breed** (out'brēd') *vt.* **-bred** (-brĕd'), **-breed·ing, -breeds.** To subject to outbreeding.

**out·breed·ing** (out'brē'dĭng) *n.* **1.** The breeding of distantly related or unrelated stocks of animals. **2.** Marriage outside a particular group, as a tribe or family.

**out·build·ing** (out'bĭl'dĭng) *n.* A building, as a barn or shed, separate from but associated with a main building.

**out·burst** (out'bûrst') *n.* A sudden, violent display, as of emotion or activity : FLARE-UP.

**out·cast** (out'kăst') *n.* One that has been excluded from a group or society. **—out'cast'** *adj.*

**out·caste** (out'kăst') *n.* A Hindu who has been expelled from or has abandoned his or her caste.

**out·class** (out-klăs') *vt.* **-classed, -class·ing, -class·es.** To surpass decisively, so as to appear of a superior class.

**out·come** (out'kŭm') *n.* A final consequence : RESULT.

**out·crop** (out'krŏp') *n.* **1.** *Geol.* A stratum or formation, as of bedrock, that protrudes through the soil level. **2.** An outbreak. —*vi.* (out-krŏp') **-cropped, -crop·ping, -crops.** *Geol.* To protrude above the soil.

**out·cross** (out'krôs', -krŏs') *vt.* **-crossed, -cross·ing, -cross·es.** To breed (an animal) with an individual from a different strain of the same breed. —*n.* **1.** The process of outcrossing. **2.** An offspring produced by outcrossing.

**out·cry** (out'krī') *n.* **1.** A loud cry or clamor. **2.** A vociferous protest or objection <public *outcry* over the scandal> **3.** An auction.

**out·curve** (out'kûrv') *n. Baseball.* A pitched ball that curves away from the batter.

**out·date** (out-dāt') *vt.* **-dat·ed, -dat·ing, -dates.** To supersede or make obsolete.

**out·dat·ed** (out-dā'tĭd) *adj.* Old-fashioned : out-of-date.

**out·dis·tance** (out-dĭs'təns) *vt.* **-tanced, -tanc·ing, -tanc·es.** **1.** To outrun, esp. in a long-distance race. **2.** To surpass by a wide margin.

**out·do** (out-dōō') *vt.* **-did** (-dĭd'), **-done** (-dŭn'), **-do·ing, -does** (-dŭz'). To exceed in performance or achievement.

**out·door** (out'dôr', -dōr') *also* **out-of-door** (out'əv-dôr', -dōr') *adj.* Located in, done in, or suited to the open air <an *outdoor* theater>

**out·doors** (out-dôrz', -dōrz') *also* **out-of-doors** (out'əv-dôrz', -dōrz') —*adv.* In or into the open : outside of a house or shelter <sleeping *outdoors* under the stars> —*n.* The open air.

**out·doors·man** (out-dôrz'mən, -dōrz'-) *n.* One who spends considerable leisure time in outdoor activities, as hiking or fishing.

**out·er** (ou'tər) *adj.* **1.** Located on the outside : EXTERNAL. **2.** Farther from the center or middle. **3.** Relating to the body or its appearance rather than the mind or spirit.

**outer ear** *n.* The external ear.

**out·er·most** (ou'tər-mōst') *adj.* Most distant from the center or inside <Pluto is the *outermost* planet in our solar system.>

**outer space** *n.* Any region of space beyond limits determined with reference to the boundaries of a celestial body or system.

**out·face** (out-fās') *vt.* **-faced, -fac·ing, -fac·es.** **1.** To overcome with a bold or self-assured look : STARE DOWN. **2.** To resist or defy.

**out·fall** (out'fôl') *n.* A mouth or drainage outlet of an effluent, as a sewer or stream.

**out·field** (out'fēld') *n. Baseball* **1.** The playing area extending outward from the infield, divided into right, center, and left field. **2.** The players in the outfield. **—out'field'er** *n.*

**out·fit** (out'fĭt') *n.* **1.** A set of equipment or tools for a specialized purpose <a backpacker's *outfit*> **2.** A set of coordinated or specially designed clothing with accessories <a riding *outfit*> **3.** *Informal.* An association of persons, esp. a military unit or a business organization. **4.** The act of equipping. —*vt.* **-fit·ted, -fit·ting, -fits.** To provide with specialized equipment <a store that *outfits* scuba divers>

**out·flank** (out-flăngk') *vt.* **-flanked, -flank·ing, -flanks.** **1.** To maneuver around and behind the flank of (an opposing force). **2.** To gain a tactical advantage over.

**out·flow** (out'flō') *n.* **1.** The act of flowing out. **2.** Something that flows out. **3.** The amount flowing out : OUTGO.

**out·fox** (out-fŏks') *vt.* **-foxed, -fox·ing, -fox·es.** To outsmart.

**out·gas** (out'găs') *vt.* **-gassed, -gas·sing, -gas·ses.** To remove embedded gas from (a solid) by heating.

**out·go** (out-gō') *vt.* **-went** (-wĕnt'), **-gone** (-gôn', -gŏn'), **-go·ing, -goes** (-gōz'). To surpass : exceed. —*n.* (out'gō'), *pl.* **-goes. 1.** Something that goes out, esp. expenditures. **2.** The act of going out.

**out·go·ing** (out'gō'ĭng) *adj.* **1. a.** Departing : outbound <an *outgoing* ocean liner> **b.** Retiring from or relinquishing a position or office <the *outgoing* director> **2.** Sociable : friendly.

**out-group** (out'grōōp') *n.* A group of people excluded from or not belonging to an in-group.

**out·grow** (out-grō') *vt.* **-grew** (-grōō'), **-grown** (-grōn'), **-grow·ing, -grows. 1.** To grow too large for. **2.** To lose or discard in the course of maturation <*outgrew* my love of sweets> **3.** To surpass in growth.

**out·growth** (out'grōth') *n.* **1.** Something that grows out of something else : OFFSHOOT. **2.** A consequence or result.

**out·guess** (out-gĕs') *vt.* **-guessed, -guess·ing, -guess·es. 1.** To anticipate correctly the actions of. **2.** To gain the advantage over by cleverness or forethought : OUTWIT.

**out·haul** (out'hôl') *n. Naut.* A rope for extending a sail along a spar or boom.

**out·house** (out'hous') *n.* **1.** An outbuilding. **2.** An outdoor toilet housed in a small structure.

**out·ing** (ou'tĭng) *n.* **1.** An excursion or pleasure trip. **2.** A walk outdoors : AIRING.

**outing flannel** *n.* A soft, lightweight cotton fabric, usu. having a short nap on both sides.

**out·land** (out'lănd', -lənd) *n.* [ME < OE *ūtland* : *ūt*, out + *land*, land.] **1.** A foreign land. **2. outlands.** The remote or sparsely populated areas of a country. **—out'land'** *adj.* **—out'land'er** *n.*

**out·land·ish** (out-lăn'dĭsh) *adj.* [ME < OE *ūtlandisc* : *ūtland*, outland + *-isc*, -ish.] **1.** Strikingly foreign : UNFAMILIAR. **2.** Of or relating to a foreign country : EXOTIC. **3.** Geographically remote from the familiar world. **4.** Conspicuously unconventional : BIZARRE <an *outlandish* hairdo> **—out·land'ish·ly** *adv.* **—out·land'ish·ness** *n.*

**out·last** (out-lăst') *vt.* **-last·ed, -last·ing, -lasts.** To live, exist, or remain longer than.

☆ ***syns:*** OUTLAST, OUTLIVE, OUTWEAR, SURVIVE *v. core meaning* : to live, exist, or remain longer than <plastic that *outlasts* metal>

**out·law** (out'lô') *n.* [ME *outlawe* < OE *ūtlaga* < ON *ūtlagi* : *ūt*, out + *lǫg*, law.] **1.** A habitual or confirmed criminal. **2.** A fugitive from the law. **3.** A person excluded from normal legal rights and protection. **4.** A wild or vicious animal. —*vt.* **-lawed, -law·ing, -laws. 1.** To declare illegal <*outlawed* private possession of handguns> **2.** To ban. **3.** To deprive of the protection of the law.

**out·law·ry** (out'lô'rē) *n., pl.* **-ries.** [ME *outlauerie* < AN *utlagerie*

< OE ūtlaga, outlaw.] **1.** The act or process of outlawing. **2.** The state of being outlawed. **3.** Defiance of the law.

**out·lay** (out′lā′) *n.* **1.** The spending or disbursing of money. **2.** An amount spent : EXPENDITURE <large *outlays* for social programs> —*vt.* (out-lā′), **-laid** (-lād′), **-lay·ing, -lays.** To expend (money).

**out·let** (out′lĕt′, -lĭt) *n.* **1.** A passage for exit or escape : VENT. **2. a.** A means of release, as for energies, or emotions. **b.** A means of achieving self-expression <The piano was my only creative *outlet.*> **3. a.** A commercial market for goods or services. **b.** A store that sells the goods of a particular manufacturer or wholesaler. **4.** *Elect.* A receptacle, esp. one mounted in a wall, that is connected to a power supply and equipped with a socket for a plug.

**out·li·er** (out′lī′ər) *n.* **1.** One who lives relatively far from one's place of business. **2.** *Geol.* A stratum or rock mass separated from a main formation by erosion.

**out·line** (out′līn′) *n.* **1. a.** A line marking the contours or outer boundaries of a figure or object. **b.** The shape of an object or figure. **2.** A style of drawing in which objects are delineated in contours without shading. **3. a.** A general description covering the main points of a subject. **b.** An abstract. **c.** A summary of a written work or speech, usu. analyzed in headings and subheadings. **d.** A preliminary plan or draft. —*vt.* **-lined, -lin·ing, -lines. 1.** To draw an outline of. **2.** To display or accentuate the outline of. **3.** To give the main points of : SUMMARIZE.

☆ ***syns:*** OUTLINE, CONTOUR, DELINEATION, PROFILE, SILHOUETTE *n. core meaning* : a line marking and shaping the outer form of an object <saw the *outline* of a deer through the mist>

**out·live** (out-lĭv′) *vt.* **-lived, -liv·ing, -lives. 1.** To live or exist longer than : OUTLAST. **2.** To live through : SURVIVE <*outlive* the harsh winter>

**out·look** (out′lo͝ok′) *n.* **1.** The act of looking out. **2. a.** A place that offers a view of something. **b.** The view from an outlook. **3.** A point of view : ATTITUDE. **4.** Prospect <a gloomy economic *outlook*>

**out loud** *adv.* Aloud.

**out·ly·ing** (out′lī′ĭng) *adj.* Comparatively distant from or inaccessible to a center or middle.

**out·ma·neu·ver** (out′mə-no͞o′vər, -nyo͞o′-) *vt.* **-vered, -ver·ing, -vers. 1.** To triumph over by more artful maneuvering. **2.** To surpass in maneuverability <a car that *outmaneuvers* all other models>

**out·mod·ed** (out-mō′dĭd) *adj.* **1.** No longer in fashion. **2.** No longer practical or usable : OBSOLETE <*outmoded* ideas>

**out·most** (out′mōst′) *adj.* Outermost.

**out·num·ber** (out-nŭm′bər) *vt.* **-bered, -ber·ing, -bers.** To be more numerous than.

**out of** *prep.* **1. a.** From within to the outside of <ran *out of* the house> **b.** From a given condition <came *out of* the coma> **c.** From an origin, source, or cause <made *out of* plastic> **2. a.** In a position or situation beyond the range, boundaries, limits, or sphere of <*out of* sight> **b.** In a state or position away from the expected or usual <*out of* practice> **3.** Because of <did it *out of* jealousy> **4.** With headquarters in <works *out of* the central office> **5.** From among <nine *out of* ten votes>

**out-of-bounds** (out′əv-boundz′) *adv. & adj.* Beyond the designated or established boundaries.

**out-of-date** (out′əv-dāt′) *adj.* Old-fashioned : outmoded.

**out-of-door** (out′əv-dôr′, -dōr′) *adj. var. of* OUTDOOR.

**out-of-doors** (out′əv-dôrz′, -dōrz′) *adv. var. of* OUTDOORS.

**out-of-pock·et** (out′əv-pŏk′ĭt) *adj.* Designating incidental expenses usu. entailing the outlay of ready cash.

**out-of-stat·er** (out′əv-stā′tər) *n.* **1.** A visitor, as a tourist, from another state. **2.** A legal resident of one state who lives for a period of time in another state, as to attend school.

**out-of-the-way** (out′əv-*th*ə-wā′) *adj.* **1.** Remote : secluded. **2.** Out of the ordinary : UNUSUAL.

**out-of-town·er** (out′əv-tou′nər) *n.* A visitor from another town or city.

**out·pa·tient** (out′pā′shənt) *n.* A patient who receives treatment at a hospital or clinic without being admitted.

**out·per·form** (out′pər-fôrm′) *vt.* **-formed, -form·ing, -forms.** To surpass in performance.

**out·play** (out-plā′) *vt.* **-played, -play·ing, -plays.** To play better than (one's opponent) in a game.

**out·post** (out′pōst′) *n.* **1. a.** A detachment of troops stationed at a distance from a main unit of forces. **b.** The station occupied by such troops. **2.** An outlying settlement.

**out·pour** (out-pôr′, -pōr′) —*vi. & vt.* **-poured, -pour·ing, -pours.** To flow or pour out rapidly. —*n.* (out′pôr′, -pōr′). A rapid outflow : OUTPOURING. **—out·pour′er** *n.*

**out·pour·ing** (out′pôr′ĭng, -pōr′-) *n.* **1.** The act of pouring out. **2.** Something that pours out or is poured out : OUTFLOW.

**out·put** (out′po͝ot′) *n.* **1.** The act of producing : PRODUCTION. **2.** An amount produced, esp. during a given time. **3. a.** The power, energy, or work produced by a system. **b.** *Computer Sci.* The data produced by a computer from a specific input. —*vt.* **-put·ted** *or* **-put, -put·ting, -puts.** To produce or manufacture as output.

**out·rage** (out′rāj′) *n.* [ME, excess < OFr. < *outre,* beyond < Lat. *ultra.*] **1.** An extremely violent or vicious act. **2.** An act grossly offensive to morality, decency, or good taste. **3.** A severe insult or affront. **4.** Resentful anger aroused by an offensive or injurious act. —*vt.* **-raged, -rag·ing, -rag·es. 1.** To commit an outrage upon. **2.** RAPE[1] 1. **3.** To produce anger or resentment in <The attempted fraud *outraged* the entire community.>

**out·ra·geous** (out-rā′jəs) *adj.* **1. a.** Constituting an outrage : grossly offensive. **b.** Shamefully immoral : DISGRACEFUL. **2.** Showing no regard for accepted standards or norms <*outrageous* attire> **3.** Violent or unrestrained in behavior or temperament. **4.** Beyond all reason <spent an *outrageous* amount on clothes> **—out·ra′geous·ly** *adv.* **—out·ra′geous·ness** *n.*

☆ ***syns:*** **1.** OUTRAGEOUS, ATROCIOUS, CRYING, FLAGRANT, HEINOUS, MONSTROUS, SCANDALOUS, SHOCKING *adj. core meaning* : disgracefully and grossly offensive <an *outrageous* violation of the victim's rights> **2.** OUTRAGEOUS, PREPOSTEROUS, RIDICULOUS, UNREASONABLE *adj. core meaning* : beyond all reason <The repair bill was *outrageous.*>

**out·rank** (out-răngk′) *vt.* **-ranked, -rank·ing, -ranks.** To rank higher than.

**ou·tré** (o͞o-trā′) *adj.* [Fr., p.part. of *outrer,* to go beyond < *outre,* beyond < Lat. *ultra.*] Highly unconventional : ECCENTRIC.

**out·reach** (out-rēch′) *v.* **-reached, -reach·ing, -reach·es.** —*vt.* **1.** To reach or go beyond : SURPASS. **2.** To extend (something) outward. —*vi.* To reach out. —*n.* (out′rēch′). **1.** An act of reaching out. **2.** The extent of reach. **3.** An organized effort to extend services beyond usual limits, as to particular segments of a community.

**out·ride**[1] (out-rīd′) *vt.* **-rode** (-rōd′), **-rid·den** (-rĭd′n), **-rid·ing, -rides.** To surpass in riding or horsemanship : OUTSTRIP.

**out·ride**[2] (out′rīd′) *n.* [Coined by Gerard Manley Hopkins (1844–1889).] An unstressed syllable or cluster of syllables within a given metrical unit that is omitted from the scansion pattern in sprung rhythm.

**out·rid·er** (out′rī′dər) *n.* **1.** A mounted attendant who rides in front of or beside a vehicle. **2.** An escort : scout.

**out·rig·ger** (out′rĭg′ər) *n.* **1. a.** A narrow float attached to a seagoing canoe by projecting spars as a stabilizing support. **b.** A vessel fitted with an outrigger. **2.** A projecting frame extending laterally beyond the main structure of a vessel, vehicle, aircraft, or machine to stabilize the structure or to support an extending part.

**outrigger**

**out·right** (out′rīt′, -rīt′) *adv.* **1.** Without qualification or reservation : OPENLY. **2.** Utterly : entirely <accepted our conditions *outright*> **3.** Then and there : STRAIGHTWAY <slaughtered them *outright*> —*adj.* (out′rīt′). **1.** Being completely without reservation : UNQUALIFIED <an *outright* gift> **2. a.** Open : straightforward <an *outright* bribe> **b.** Complete : thoroughgoing <*outright* happiness> **3.** *Archaic.* Directed straight on : moving straight onward.

**out·run** (out-rŭn′) *vt.* **-ran** (-răn′), **-run, -run·ning, -runs. 1.** To run faster than. **2.** To escape from <*outrun* the law> **3.** To go beyond : EXCEED.

**out·sell** (out-sĕl′) *vt.* **-sold** (-sōld′), **-sell·ing, -sells. 1.** To exceed in sales <a book that *outsells* all others> **2.** To surpass in selling.

**out·set** (out′sĕt′) *n.* **1.** Start : beginning. **2.** An initial stage, as of an activity.

**out·shine** (out-shīn′) *v.* **-shone** (-shōn′), **-shin·ing, -shines.** —*vt.* **1.** To shine brighter than. **2.** To surpass in beauty or obvious excellence. —*vi.* To shine forth.

**out·shoot** (out-sho͞ot′) *v.* **-shot** (-shŏt′), **-shoot·ing, -shoots.** —*vt.* **1.** To shoot better than. **2.** To extend beyond : OVERHANG. —*vi.* To project or protrude. —*n.* (out′sho͞ot′). **1.** A projection or outgrowth. **2.** A flowing or gushing forth.

**out·shout** (out-shout′) *vt.* **-shout·ed, -shout·ing, -shouts.** To shout louder or longer than.

**out·side** (out-sīd′, out′sīd′) *n.* **1.** The part or parts that face out : EXTERIOR. **2. a.** The part or side of an object presented to the viewer : external aspect. **b.** An apparent or superficial aspect. **3.** The space beyond a boundary or confining structure. **4.** An utmost limit : MAXIMUM <I'll be gone two hours at the *outside.*> —*adj.* **1.** Acting, originating, or existing at a place beyond certain limits <*outside* interference> **2.** Of, restricted to, or situated on the outside of an enclosure or boundary : EXTERNAL <an *outside* antenna> **3.** Extreme : uttermost <an *outside* limit> **4.** Slim : slight <an *outside* chance> —*adv.* **1.** On or into the outside. **2.** Outdoors. —*prep.* **1.** On or to the outer side of. **2.** Beyond the limits of. **3.** With the exception of : EXCEPT <no information *outside* the published figures>

**outside of** *prep.* Outside.

**out·sid·er** (out-sī′dər) *n.* **1. a.** One who is excluded from a particular party, association, or set. **b.** One who is detached or isolated from the concerns or activities of his or her community. **2.** A contestant given little chance of winning : LONG SHOT.

**out·size** (out′sīz′) *n.* **1.** An unusual size, esp. a very large size. **2.** An outsize garment. —*adj. also* **out·sized** (out′sīzd′). Unusually large.

**out·skirts** (out′skûrts′) *pl.n.* The parts or regions on the edge of a central district : peripheral areas.

**out·smart** (out-smärt′) *vt.* **-smart·ed, -smart·ing, -smarts.** To gain the advantage over by cunning : OUTWIT.

**out·speak** (out-spēk′) *v.* **-spoke** (-spōk′), **-spok·en** (-spō′kən), **-speak·ing, -speaks.** —*vt.* **1.** To outperform in speech or debate : speak more persuasively than. **2.** To say candidly and frankly. —*vi.* To speak out.

**out·spend** (out-spĕnd′) *vt.* **-spent** (-spĕnt′), **-spend·ing, -spends. 1.** OVERSPEND 1. **2.** To outdo in spending <*outspend* the neighbors on decorations>

**out·spo·ken** (out-spō′kən) *adj.* **1.** Spoken without reserve : CANDID. **2.** Frank and unsparing in speech. **—out·spo′ken·ly** *adv.* **—out·spo′ken·ness** *n.*

**out·spread** (out-sprĕd′) *vi. & vt.* **-spread, -spread·ing, -spreads.** To spread out or cause to spread out. —*n.* (out′sprĕd′). The act of spreading out. —*adj.* Spread out : EXTENDED.

**out·stand** (out-stănd′) *vi.* **-stood** (-stŏŏd′), **-stand·ing, -stands. 1.** To stand out plainly : be outstanding. **2.** *Naut.* To set sail.

**out·stand·ing** (out-stăn′dĭng, out′stăn′-) *adj.* **1.** Standing out : projecting outward or upward. **2.** Conspicuous among others of its kind : PROMINENT. **3.** Pre-eminent among others of its kind : DISTINGUISHED. **4.** Still not resolved or settled <*outstanding* debts><a long *outstanding* quarrel>

**out·stare** (out-stâr′) *vt.* **-stared, -star·ing, -stares.** To stare out of countenance : FACE DOWN.

**out·sta·tion** (out′stā′shən) *n.* A remote station or post.

**out·stay** (out-stā′) *vt.* **-stayed, -stay·ing, -stays. 1.** To stay longer than : OVERSTAY. **2.** To show greater endurance than : OUTLAST.

**out·stretch** (out-strĕch′) *vt.* **-stretched, -stretch·ing, -stretch·es. 1.** To stretch out : EXTEND. **2.** To stretch beyond.

**out·strip** (out-strĭp′) *vt.* **-stripped, -strip·ping, -strips. 1.** To leave behind : OUTDISTANCE. **2.** To surpass : exceed.

**out·stroke** (out′strōk′) *n.* An outward stroke, esp. the stroke of an engine piston moving toward the crankshaft.

**out·take** (out′tāk′) *n.* **1. a.** A film take or scene, as of a motion picture, that is eliminated from the final version. **b.** A complete version, as of a musical recording, that is dropped in favor of another version. **2.** An opening for outward discharge : OUTLET.

**out·talk** (out-tôk′) *vt.* **-talked, -talk·ing, talks. 1.** To talk, louder, longer, or more ably than. **2.** To outwit by talking.

**out·think** (out-thĭngk′) *vt.* **-thought** (-thôt′), **-think·ing, -thinks. 1.** To outdo in thinking. **2.** To outwit by thinking.

**out·turn** (out′tûrn′) *n.* A total amount produced during a given period : OUTPUT.

**out·ward** (out′wərd) *adj.* [ME < OE *ūtweard* : *ūt*, out + *-weard*, -ward.] **1.** Of, located on, or moving toward the outside or exterior : OUTER. **2.** Relating to the physical self. **3.** Purely external : SUPERFICIAL <an *outward* calm that hid my fear> —*adv. also* **out·wards** (-wərdz). **1.** Toward the outside : away from a central point. **2.** On the outside : EXTERNALLY. **3.** Superficially : apparently. —*n.* **1.** The outside : exterior. **2.** Outward appearance. **3.** The external or material world. **—out′ward·ly** *adv.* **—out′ward·ness** *n.*

**out·wear** (out-wâr′) *vt.* **-wore** (-wôr′, -wōr′), **-worn** (-wôrn′, -wōrn′), **-wear·ing, -wears. 1.** To wear out : exhaust by using. **2.** To endure longer than : OUTLAST. **3.** To outlive or outgrow.

**out·weigh** (out-wā′) *vt.* **-weighed, -weigh·ing, -weighs. 1.** To weigh more than. **2.** To be more weighty or significant than.

**out·wit** (out-wĭt′) *vt.* **-wit·ted, -wit·ting, -wits. 1.** To get the better of by cleverness or cunning. **2.** *Archaic.* To surpass in intelligence.

☆ ***syns:*** OUTWIT, OUTMANEUVER, OUTSMART, OUTTHINK, OVERREACH *v. core meaning* : to get the better of by cleverness or cunning <Sherlock Holmes always *outwitted* his adversaries.>

**out·work¹** (out-wûrk′) *vt.* **-worked** *or* **-wrought** (-rôt′), **-work·ing, -works. 1.** To work better or faster than <*outworked* their colleagues> **2.** To work out : FINISH.

**out·work²** (out′wûrk′) *n.* A fortification constructed beyond a main defensive position or fortification.

**ou·zel** *also* **ou·sel** (ōō′zəl) *n.* [ME *osel* < OE *ōsle.*] **1.** Any of various European birds of the genus *Turdus.* **2.** The water ouzel.

**ou·zo** (ōō′zō) *n., pl.* **-zos.** [Mod. Gk.] A colorless Greek liqueur flavored with aniseed.

**ov-** *pref. var. of* OVI-.

**o·va** (ō′və) *n. pl. of* OVUM.

**o·val** (ō′vəl) *adj.* [Med. Lat. *ovalis* < Lat. *ovum*, egg.] **1.** Shaped like an egg. **2.** Shaped like an ellipse. —*n.* **1.** An oval figure or form. **2.** An oval track or field, as for horse racing or athletic events. **—o′val·ly** *adv.* **—o′val·ness** *n.*

---

**o′ver·a·bound′** *v.*
**o′ver·anx′ious** *adj.*
**o′ver·at·tached′** *adj.*
**o′ver·care′ful** *adj.*
**o′ver·chill′** *v.*
**o′ver·cook′** *v.*
**o′ver·cool′** *adj.*
**o′ver·cour′te·ous** *adj.*
**o′ver·crit′i·cal** *adj.*
**o′ver·crowd′** *v.*
**o′ver·dar′ing** *adj.*
**o′ver·dec′o·rate′** *v.*
**o′ver·de·mand′** *v.*
**o′ver·de·vot′ed** *adj.*
**o′ver·de·vo′tion** *n.*
**o′ver·dil′i·gence** *n.*
**o′ver·dil′i·gent** *adj.*
**o′ver·dis′ci·pline** *v.*
**o′ver·dra·mat′ic** *adj.*
**o′ver·drink′** *v.*
**o′ver·ea′ger** *adj.*
**o′ver·ear′nest** *adj.*
**o′ver·ear′nest·ly** *adv.*
**o′ver·eas′y** *adj.*
**o′ver·ed′u·cate′** *v.*
**o′ver·e·lab′o·rate** *adj.*
**o′ver·el′e·gant** *adj.*
**o′ver·em·bel′lish** *v.*
**o′ver·e·mo′tion·al** *adj.*
**o′ver·em·phat′ic** *adj.*
**o′ver·en·thu′si·as′tic** *adj.*
**o′ver·ex·cit′a·ble** *adj.*
**o′ver·ex·cite′** *v.*
**o′ver·ex·cite′ment** *n.*
**o′ver·ex·pand′** *v.*
**o′ver·ex·pan′sion** *n.*
**o′ver·ex·pect′ant** *adj.*
**o′ver·ex·u′ber·ant** *adj.*
**o′ver·faith′ful** *adj.*
**o′ver·fa·mil′iar** *adj.*
**o′ver·fan′ci·ful** *adj.*
**o′ver·fas·tid′i·ous** *adj.*
**o′ver·fat′ten** *v.*
**o′ver·fed′** *adj.*
**o′ver·feed′** *v.*
**o′ver·fem′i·nine** *adj.*
**o′ver·fierce′** *adj.*
**o′ver·fill′** *v.*
**o′ver·fond′ness** *n.*
**o′ver·frank′** *adj.*
**o′ver·free′ly** *adv.*
**o′ver·fre′quent** *adj.*
**o′ver·full′** *adj.*
**o′ver·full′ness** *n.*
**o′ver·fur′nish** *v.*
**o′ver·gen′er·al·ize′** *v.*
**o′ver·gen′er·ous** *adj.*
**o′ver·gift′ed** *adj.*
**o′ver·glad′** *adj.*
**o′ver·gra′cious** *adj.*
**o′ver·grate′ful** *adj.*
**o′ver·greas′y** *adj.*
**o′ver·hard′** *adj.*
**o′ver·hard′en** *v.*
**o′ver·harsh′** *adj.*
**o′ver·hast′y** *adj.*
**o′ver·heat′** *v.*
**o′ver·help′ful** *adj.*
**o′ver·hon′est** *adj.*
**o′ver·i·de′al·ism** *n.*
**o′ver·im·ag′i·na′tive** *adj.*
**o′ver·im·press′** *v.*
**o′ver·in·clined′** *adj.*
**o′ver·in·flate′** *v.*
**o′ver·in·fla′tion** *n.*
**o′ver·in′flu·en′tial** *adj.*
**o′ver·in·sis′tence** *n.*
**o′ver·in·sis′tent** *adj.*
**o′ver·in·sure′** *v.*
**o′ver·in′tel·lec′tu·al** *adj.*
**o′ver·in·tense′** *adj.*
**o′ver·in′ter·est** *n.*
**o′ver·in·vest′** *v.*
**o′ver·jeal′ous** *adj.*
**o′ver·keen′** *adj.*
**o′ver·kind′** *adj.*
**o′ver·large′** *adj.*
**o′ver·late′** *adj.*
**o′ver·lav′ish** *adj.*
**o′ver·lax′** *adj.*
**o′ver·lib′er·al** *adj.*
**o′ver·live′ly** *adj.*
**o′ver·loud′** *adj.*
**o′ver·loy′al** *adj.*
**o′ver·mag′ni·fy′** *v.*
**o′ver·man′y** *adj.*
**o′ver·ma·ture′** *adj.*
**o′ver·meek′** *adj.*
**o′ver·mer′ci·ful** *adj.*
**o′ver·might′y** *adj.*
**o′ver·mix′** *v.*
**o′ver·mod′est** *adj.*
**o′ver·moist′** *adj.*
**o′ver·mois′ten** *v.*
**o′ver·mort′gage** *v.*
**o′ver·near′** *adj.*
**o′ver·neat′** *adj.*
**o′ver·neg·lect′** *v.*
**o′ver·nerv′ous** *adj.*
**o′ver·nour′ish** *v.*
**o′ver·o·bese′** *adj.*
**o′ver·o·blige′** *v.*
**o′ver·ob·se′qui·ous** *adj.*
**o′ver·of·fi′cious** *adj.*
**o′ver·op′ti·mis′tic** *adj.*
**o′ver·or·nate′** *adj.*
**o′ver·par·tic′u·lar** *adj.*
**o′ver·pas′sion·ate** *adj.*
**o′ver·pas′sion·ate·ly** *adv.*
**o′ver·pa′tri·ot′ic** *adj.*
**o′ver·pay′** *v.*
**o′ver·pes′si·mis′tic** *adj.*

**oval window** *n.* The oval-shaped aperture in the middle ear to which the ossicles are connected.

**o·var·i·ec·to·my** (ō-vâr'ē-ĕk'tə-mē) *n., pl.* **-mies.** [OVAR(Y) + -ECTOMY.] Surgical excision of an ovary.

**o·var·i·ot·o·my** (ō-vâr'ē-ŏt'ə-mē) *n., pl.* **-mies.** [OVAR(Y) + -TOMY.] **1.** Ovariectomy. **2.** Surgical incision into an ovary.

**o·va·ri·tis** (ō'və-rī'tĭs) *n.* [OVAR(Y) + -ITIS.] Oophoritis.

**o·va·ry** (ō'və-rē) *n., pl.* **-ries.** [NLat. *ovarium* < Lat. *ovum*, egg.] **1.** The female reproductive gland, typically occurring in pairs, in which ova are produced. **2.** *Bot.* The part of a pistil containing the ovules. **—o·var'i·an** (ō-vâr'ē-ən), **o·var'i·al** *adj.*

**o·vate** (ō'vāt') *adj.* [Lat. *ovatus* < *ovum*, egg.] **1.** OVAL 1. **2.** *Bot.* Broad and rounded at the base and tapering toward the end <an *ovate* leaf> **—o'vate'ly** *adv.*

**o·va·tion** (ō-vā'shən) *n.* [Lat. *ovatio*, a Roman victory ceremony < *ovare*, to rejoice.] **1.** Enthusiastic and prolonged applause. **2.** A display of public homage or welcome. **3.** A victory celebration in ancient Rome on a somewhat smaller scale than a triumph.

**ov·en** (ŭv'ən) *n.* [ME < OE *ofen*.] An enclosed compartment supplied with heat and used for cooking food and for heating or drying objects placed within.

**ov·en·bird** (ŭv'ən-bûrd') *n.* [From its oven-shaped nest.] **1.** A thrushlike North American warbler, *Seiurus aurocapillus*, that builds a domed nest on the ground. **2.** Any of various South American birds of the family Furnariidae, often building domelike nests.

**ov·en·proof** (ŭv'ən-prōōf') *adj.* Resistant to heat produced in an oven.

**o·ver** (ō'vər) *prep.* [ME < OE *ofer*.] **1.** In or at a position above or higher than <a sign *over* the window> **2.** Above and across from one end or side to the other <a jump *over* the rails> **3.** On the other side of <a village *over* the border> **4.** Upon the surface of <a coat of varnish *over* the wood> **5.** All through <a tour *over* the country> **6.** So as to cover or close <put rocks *over* a cave entrance> **7.** Up to or higher than the level or height of <water *over* the dam> **8. a.** Through the period or duration of <records maintained *over* 20 years> **b.** Until or beyond the end of <stay *over* the holidays> **9.** More than in degree, quantity, or extent <*over* 50 miles> **10.** In preference to <respected *over* all other candidates> **11.** In a position to control or rule <preside *over* the meeting> **12.** Directed toward : UPON <influence *over* students> **13.** While occupied with or engaged in <a chat *over* dessert> **14.** With reference to : CONCERNING <an argument *over* technique> *—adv.* **1.** Above the top or surface. **2. a.** Across to another or opposite side. **b.** Across the edge or brim <The milk spilled *over.*> **3. a.** Across a distance in a specific direction or at a location <*over* in Asia> **b.** To another specified place or position <Move your chair *over* toward the window.> **4.** Throughout an entire area or region <travel all *over*> **5. a.** To a different opinion or allegiance <won us *over*> **b.** So as to have been transmitted to another : ACROSS <got my point *over*> **6.** To a different person, condition, or title <sign *over* land> **7.** So as to be totally enclosed or covered <The river froze *over.*> **8.** Completely through from beginning to end <thought the problem *over*> **9. a.** From an upright position <The tree fell *over.*> **b.** From an upward position to an inverted or reversed position <turn a magazine *over*> **10.** Another time : AGAIN <Count your cards *over.*> **11.** In repetition <100 times *over*> **12.** In addition or excess <three cents left *over*> **13.** Beyond or until a specified time <stay a week *over*> **14. a.** At an end. **b.** —Used in two-way radio transmissions to indicate that a message is complete and a reply is awaited. *—adj.* **1. a.** Upper : higher. **b.** External : outer. **2. a.** In excess : EXCESSIVE. **b.** Not yet used up. *—n.* A series of six balls bowled from one end of a cricket pitch. *—vt.* **o·vered, o·ver·ing, o·vers.** To jump over. **—over a barrel.** At the mercy of others. **—over (one's) head.** Beyond one's comprehension or abilities.

**over-** *pref.* [< OVER.] Excessive : excessively <*over*extend>

**o·ver·a·bun·dance** (ō'vər-ə-bŭn'dəns) *n.* Lavish abundance : EXCESS. **—o'ver·a·bun'dant** *adj.*

**o·ver·a·chieve** (ō'vər-ə-chēv') *vi.* **-chieved, -chiev·ing, -chieves.** To perform or accomplish beyond the usual or required level. **—o'ver·a·chiev'er** *n.* **—o'ver·a·chieve'ment** *n.*

**o·ver·act** (ō'vər-ăkt') *v.* **-act·ed, -act·ing, -acts.** *—vt.* To act (a part) with unnecessary exaggeration. *—vi.* To exaggerate a role.

**o·ver·ac·tive** (ō'vər-ăk'tĭv) *adj.* Active to an excessive or abnormal degree <an *overactive* child>

**o·ver·age¹** (ō'vər-ĭj) *n.* **1.** An actual amount, as of goods or money, that exceeds the listed or expected amount. **2.** A surplus or excess.

**o·ver·age²** (ō'vər-āj') *adj.* Over the required or proper age.

**o·ver·ag·gres·sive** (ō'vər-ə-grĕs'ĭv) *adj.* Excessively aggressive.

**o·ver·all** *also* **o·ver-all** (ō'vər-ôl') *adj.* **1.** From one end to the other. **2.** Including everything : COMPREHENSIVE. *—adv.* Considering everything : in general. *—n.* A loose-fitting protective outer garment : SMOCK.

**o·ver·alls** (ō'vər-ôlz') *pl.n.* Loose-fitting trousers with a bib front and shoulder straps, worn by themselves or over regular clothing as protection from dirt and wear.

**o·ver·arch** (ō'vər-ärch') *vt.* **-arched, -arch·ing, -arch·es.** **1.** To form an arch over. **2.** To extend over or throughout. **3.** To be decisive or central in : dominate and subordinate (all other elements) <"a . . . rhetoric of liberalism *overarching* local bread-and-butter issues" —Garry Wills>

**o·ver·arm** (ō'vər-ärm') *adj.* Executed with the arm raised above the shoulder <an *overarm* swimming stroke>

**o'ver·plain'** *adj.*
**o'ver·pol'ish** *v.*
**o'ver·pop'u·lar** *adj.*
**o'ver·pop'u·lous** *adj.*
**o'ver·pre·cise'** *adj.*
**o'ver·press'** *v.*
**o'ver·pro·cras'ti·na'tion** *n.*
**o'ver·pro·lif'ic** *adj.*
**o'ver·prom'i·nent** *adj.*
**o'ver·prompt'** *adj.*
**o'ver·proud'** *adj.*
**o'ver·pro·vide'** *v.*
**o'ver·pro·voke'** *v.*
**o'ver·pub'lic** *adj.*
**o'ver·pun'ish** *v.*
**o'ver·pun'ish·ment** *n.*
**o'ver·quan'ti·ty** *n.*
**o'ver·quick'** *adj.*
**o'ver·qui'et** *adj.*
**o'ver·ra'tion·al** *adj.*
**o'ver·read'y** *adj.*
**o'ver·re'al·is'tic** *adj.*
**o'ver·re·flec'tive** *adj.*
**o'ver·re·li'ant** *adj.*
**o'ver·re·lig'ious** *adj.*
**o'ver·re·served'** *adj.*
**o'ver·re·strain'** *v.*
**o'ver·rich'** *adj.*
**o'ver·rife'** *adj.*
**o'ver·right'eous** *adj.*
**o'ver·rig'id** *adj.*
**o'ver·rig'or·ous** *adj.*
**o'ver·rough'** *adj.*
**o'ver·rude'** *adj.*
**o'ver·sad'** *adj.*
**o'ver·sale'** *n.*
**o'ver·salt'** *v.*
**o'ver·salt'y** *adj.*
**o'ver·sat'u·rate'** *v.*
**o'ver·sat'u·ra'tion** *n.*
**o'ver·scent'ed** *adj.*
**o'ver·scru'pu·lous** *adj.*
**o'ver·scru'pu·lous·ness** *n.*
**o'ver·sea'son** *v.*
**o'ver·sea'soned** *adj.*
**o'ver·se·cure'** *adj.*
**o'ver·sen'ti·men'tal** *adj.*
**o'ver·se'ri·ous** *adj.*
**o'ver·se·vere'** *adj.*
**o'ver·sharp'** *adj.*
**o'ver·short'** *adj.*
**o'ver·short'en** *v.*
**o'ver·shrink'** *v.*
**o'ver·si'lent** *adj.*
**o'ver·skep'ti·cal** *adj.*
**o'ver·slow'** *adj.*
**o'ver·small'** *adj.*
**o'ver·smooth'** *adj.*
**o'ver·soak'** *v.*
**o'ver·soft'** *adj.*
**o'ver·sol'emn** *adj.*
**o'ver·so·lic'i·tous** *adj.*
**o'ver·so·phis'ti·cat'ed** *adj.*
**o'ver·so·phis'ti·ca'tion** *n.*
**o'ver·spar'ing** *adj.*
**o'ver·spar'ing·ly** *adv.*
**o'ver·spe'cial·i·za'tion** *n.*
**o'ver·spe'cial·ize'** *v.*
**o'ver·spec'u·late'** *v.*
**o'ver·spec'u·la'tion** *n.*
**o'ver·spec'u·la'tive** *adj.*
**o'ver·stim'u·late'** *v.*
**o'ver·stim'u·la'tion** *n.*
**o'ver·strain'** *v.*
**o'ver·stress'** *v.*
**o'ver·stretch'** *v.*
**o'ver·striv'ing** *adj.* & *n.*
**o'ver·sub'tle** *adj.*
**o'ver·sub'tle·ty** *n.*
**o'ver·suf·fi'cient** *adj.*
**o'ver·su'per·sti'tious** *adj.*
**o'ver·sure'** *adj.*
**o'ver·sus·pi'cious** *adj.*
**o'ver·sus·pi'cious·ly** *adv.*
**o'ver·sweet'** *adj.*
**o'ver·sys'tem·at'ic** *adj.*
**o'ver·talk'a·tive** *adj.*
**o'ver·talk'a·tive·ness** *n.*
**o'ver·teach'** *v.*
**o'ver·tech'ni·cal** *adj.*
**o'ver·te·na'cious** *adj.*
**o'ver·ten'der** *adj.*
**o'ver·ten'der·ness** *n.*
**o'ver·tense'** *adj.*
**o'ver·ten'sion** *n.*
**o'ver·thick'** *adj.*
**o'ver·thin'** *adj.*
**o'ver·thought'ful** *adj.*
**o'ver·thrift'y** *adj.*
**o'ver·tight'** *adj.*
**o'ver·tire'** *v.*
**o'ver·truth'ful** *adj.*
**o'ver·val'u·a·ble** *adj.*
**o'ver·va·ri'e·ty** *n.*
**o'ver·ve'he·ment** *adj.*
**o'ver·ven'ti·late'** *v.*
**o'ver·ven'tur·ous** *adj.*
**o'ver·vig'or·ous** *adj.*
**o'ver·vi'o·lent** *adj.*
**o'ver·warm'** *v.*
**o'ver·warmed'** *adj.*
**o'ver·weak'** *adj.*
**o'ver·wet'** *adj.*
**o'ver·will'ing** *adj.*
**o'ver·wise'** *adj.*
**o'ver·wor'ry** *v.*
**o'ver·young'** *adj.*
**o'ver·youth'ful** *adj.*
**o'ver·zeal'** *n.*
**o'ver·zeal'ous** *adj.*
**o'ver·zeal'ous·ly** *adv.*
**o'ver·zeal'ous·ness** *n.*

**o·ver·awe** (ō'vər-ô') *vt.* **-awed, -aw·ing, -awes.** To overcome by inspiring awe : subdue with awe.
**o·ver·bal·ance** (ō'vər-băl'əns) *v.* **-anced, -anc·ing, -anc·es.** —*vt.* **1.** To outweigh. **2.** To throw off balance. —*vi.* To lose one's balance. —*n.* (ō'vər-băl'əns). Something that overbalances.
**o·ver·bear** (ō'vər-bâr') *v.* **-bore** (-bôr'), **-borne** (bôrn'), **-bear·ing, -bears.** —*vt.* **1.** To crush or bear down upon with physical force. **2.** To prevail over, as if by superior weight or force : DOMINATE. **3.** To be more important than : OUTWEIGH. —*vi.* To bear too much fruit or offspring.
**o·ver·bear·ing** (ō'vər-bâr'ĭng) *adj.* **1.** Overwhelming in power or authority : PREDOMINANT. **2.** Arrogant. **—o'ver·bear'ing·ly** *adv.*
**o·ver·bid** (ō'vər-bĭd') *v.* **-bid, -bid·den** (-bĭd'n) *or* **-bid, -bid·ding, -bids.** —*vt.* To outbid (a competitor). —*vi.* To bid higher than the actual value of something. **—o'ver·bid'** *n.*
**o·ver·bite** (ō'vər-bīt') *n.* Dental malocclusion in which the front upper incisor and canine teeth project over the lower.
**o·ver·blow** (ō'vər-blō') *vt.* **-blew** (-blōō'), **-blown** (-blōn'), **-blow·ing, -blows.** To blow (a wind instrument) so as to produce an overtone.
**o·ver·blown** (ō'vər-blōn') *adj.* **1.** Blown over or down. **2.** Exaggerated in importance or consequence : INFLATED <an *overblown* account of their achievements> **3.** Past the stage of full bloom. **4.** Obese.
**o·ver·board** (ō'vər-bôrd', -bōrd') *adv.* Over the side of a boat or ship. **—go overboard.** *Informal.* To exhibit excessive enthusiasm.
**o·ver·book** (ō'vər-bŏŏk') *v.* **-booked, -book·ing, -books.** —*vt.* To book passengers for (e.g., an airline flight) beyond the seating capacity. —*vi.* To book passengers beyond the seating capacity.
**o·ver·build** (ō'vər-bĭld') *v.* **-built** (-bĭlt'), **-build·ing, -builds.** —*vt.* **1.** To build over or on top of. **2.** To construct more buildings in (an area) than necessary or desirable. **3.** To build with excessive size or elaboration. —*vi.* To construct more buildings than needed.
**o·ver·bur·den** (ō'vər-bûr'dn) *vt.* **-dened, -den·ing, -dens. 1.** To burden with too much weight. **2.** To burden with too much work, care, or responsibility. —*n.* (ō'vər-bûr'dn). **1.** *Geol.* **a.** Material overlying a valuable mineral deposit. **b.** Sedimentary rock covering older crystalline layers. **2.** A sterile stratum overlying a stratum bearing traces of the archaeological culture being studied.
**o·ver·buy** (ō'vər-bī') *v.* **-bought** (-bôt'), **-buy·ing, -buys.** —*vt.* **1.** To buy in excess of what is needed. **2.** To buy (stock) on margin in excess of one's ability to provide further security if prices drop. —*vi.* To buy goods beyond one's needs or means.
**o·ver·call** (ō'vər-kôl') *vt.* **-called, -call·ing, -calls. 1.** To overbid. **2.** To bid higher than (one's opponent in a bridge game) when one's partner has not bid. —*n.* (ō'vər-kôl'). **1.** An overbid. **2.** An instance of overcalling in bridge.
**o·ver·cap·i·tal·ize** (ō'vər-kăp'ĭ-tl-īz') *vt.* **-ized, -iz·ing, -iz·es. 1.** To provide (a business enterprise) with an excessive amount of capital. **2.** To estimate the value of (property) too highly. **3.** To place an unlawfully or unreasonably high value on the nominal capital of (a corporation). **—o'ver·cap'i·tal·i·za'tion** *n.*
**o·ver·cast** (ō'vər-kăst', ō'vər-kăst') *adj.* **1. a.** Covered or obscured, as with clouds or haze. **b.** Cloudy <an *overcast* day> **2.** Melancholy : gloomy. **3.** Sewn with long, overlying stitches in order to prevent raveling, as the edges of fabric. —*n.* (ō'vər-kăst'). **1.** A covering, as of haze or clouds. **2.** An arch supporting one passage over another in a mine. **3.** A fishing cast falling beyond the point intended. **4.** An overcast seam or stitch. —*vt.* (ō'vər-kăst', ō'vər-kăst') **-cast·ed, -cast·ing, -casts. 1.** To make cloudy or gloomy. **2.** To cast beyond (the intended point) with a fishing rod. **3.** To sew with an overcast stitch.
**o·ver·cau·tious** (ō'vər-kô'shəs) *adj.* Excessively cautious. **—o'ver·cau'tious·ly** *adv.* **—o'ver·cau'tious·ness** *n.*
**o·ver·charge** (ō'vər-chärj') *vt.* **-charged, -charg·ing, -charg·es. 1.** To charge (a person or account) an excessive price for something. **2.** To fill too full : OVERLOAD. **3.** To exaggerate. —*n.* (ō'vər-chärj'). **1.** An excessive price or charge. **2.** A load or burden too full or heavy.
**o·ver·cloud** (ō'vər-kloud') *v.* **-cloud·ed, -cloud·ing, -clouds.** —*vt.* **1.** To cover with clouds. **2.** To make dark and gloomy. —*vi.* To become cloudy.
**o·ver·coat** (ō'vər-kōt') *n.* A heavy outer coat.
**o·ver·come** (ō'vər-kŭm') *v.* **-came** (-kām'), **-come, -com·ing, -comes.** [ME *overcomen* < OE *ofercuman.*] —*vt.* **1.** To defeat in competition or conflict : CONQUER. **2.** To prevail over by mental or moral effort : SURMOUNT <*overcome* one's shyness> **3.** To overpower, as with emotion : affect deeply. —*vi.* To surmount opposition : WIN OUT.
**o·ver·com·pen·sate** (ō'vər-kŏm'pən-sāt') *v.* **-sat·ed, -sat·ing, -sates.** —*vi.* **1.** To make a greater effort than required to achieve compensation. **2.** *Psychol.* To engage in overcompensation. —*vt.* To compensate excessively. **—o'ver·com·pen'sa·to'ry** (-kəm-pĕn'sə-tôr'ē, -tōr'ē) *adj.*
**o·ver·com·pen·sa·tion** (ō'vər-kŏm'pən-sā'shən) *n.* *Psychol.* Exertion of effort in excess of that needed to compensate for a real or imagined defect.
**o·ver·con·fi·dent** (ō'vər-kŏn'fĭ-dənt) *adj.* Too confident. **—o'ver·con'fi·dence** *n.* **—o'ver·con'fi·dent·ly** *adv.*
**o·ver·crop** (ō'vər-krŏp') *vt.* **-cropped, -crop·ping, -crops.** To exhaust the fertility of (land) by overcultivation.
**o·ver·crowd** (ō'vər-kroud') *vt.* **-crowd·ed, -crowd·ing, -crowds.** To cause to be excessively crowded.
**o·ver·de·vel·op** (ō'vər-dĭ-vĕl'əp) *vt.* **-oped, -op·ing, -ops. 1.** To develop to excess. **2.** To process (a photographic plate or film) too long or in too concentrated a solution. **—o'ver·de·vel'op·ment** *n.*
**o·ver·do** (ō'vər-dōō') *v.* **-did** (-dĭd'), **-done** (-dŭn'), **-do·ing, -does** (-dŭz'). —*vt.* **1.** To do, use, or stress to excess : take to extremes. **2.** To wear out the strength of : OVERTAX. **3.** To cook too long. —*vi.* To do something to excess.
**o·ver·dom·i·nance** (ō'vər-dŏm'ə-nəns) *n.* The state of a heterozygote whose phenotype is more pronounced or better adapted than that of either homozygote. **—o'ver·dom'i·nant** *adj.*
**o·ver·dose** (ō'vər-dōs') *n.* An excessive dose, esp. of a narcotic. —*vi.* & *vt.* (ō'vər-dōs') **-dosed, -dos·ing, -dos·es.** To take or administer an overdose.
**o·ver·draft** (ō'vər-drăft', -dräft') *n.* **1. a.** The act of overdrawing a bank account. **b.** The amount overdrawn. **2.** *also* **o·ver·draught. a.** A stream of air blown over the ignited fuel in a furnace. **b.** A series of flues in a kiln by which air is drawn downward. **c.** The air so drawn.
**o·ver·draw** (ō'vər-drô') *vt.* **-drew** (-drōō'), **-drawn** (-drôn'), **-draw·ing, -draws. 1.** To draw against (a bank account) in excess of credit. **2.** To draw (a bow) back too far. **3.** To so exaggerate (e.g., a description) as to spoil the effect.
**over·drawn** (ō'vər-drôn') *adj.* Having an account that has been drawn against in excess of credit.
**o·ver·dress** (ō'vər-drĕs') *vi.* **-dressed, -dress·ing, -dress·es.** To dress more formally or elaborately than desirable. —*n.* A skirted garment, as a pinafore, worn over other clothing.
**o·ver·drive** (ō'vər-drīv') *n.* A gearing mechanism of an automotive engine that reduces the power output required to maintain driving speed in a specific range by increasing the ratio of drive shaft to engine speed. —*vt.* (ō'vər-drīv') **-drove** (-drōv'), **-driv·en** (-drĭv'ən), **-driv·ing, -drives. 1.** To drive (a vehicle) too long or too hard. **2.** To push (oneself) too far : OVERWORK.
**o·ver·dub** (ō'vər-dŭb') *vt.* **-dubbed, -dub·bing, -dubs.** To blend, as recorded sound, with previously recorded sound to produce a multiple effect. **—o'ver·dub'** *n.*
**o·ver·due** (ō'vər-dōō', -dyōō') *adj.* **1.** Being unpaid after becoming due. **2.** Expected or deserved but not yet come <an *overdue* raise>
**o·ver·eat** (ō'vər-ēt') *vi.* **-ate** (-āt'), **-eat·ing, -eats.** To eat too much, esp. habitually. **—o'ver·eat'er** *n.*
**o·ver·em·pha·size** (ō'vər-ĕm'fə-sīz') *v.* **-sized, -siz·ing, -siz·es.** —*vt.* To place too much emphasis on. —*vi.* To employ too much emphasis. **—o'ver·em'pha·sis** (-sĭs) *n.*
**o·ver·es·ti·mate** (ō'vər-ĕs'tə-māt') *vt.* **-mat·ed, -mat·ing, -mates. 1.** To make too high an estimate of the amount or degree of. **2.** To esteem too greatly. **—o'ver·es'ti·mate** (-mĭt) *n.* **—o'ver·es'ti·ma'tion** *n.*
**o·ver·ex·ert** (ō'vər-ĭg-zûrt') *vt.* **-ert·ed, -ert·ing, -erts.** To exert too much : OVERTAX. **—o'ver·ex·er'tion** *n.*
**o·ver·ex·pose** (ō'vər-ĭk-spōz') *vt.* **-posed, -pos·ing, -pos·es. 1.** To expose too long or too much. **2.** To expose (a photographic film or plate) too long or with too much light. **—o'ver·ex·po'sure** *n.*
**o·ver·ex·tend** (ō'vər-ĭk-stĕnd') *vt.* **-tend·ed, -tend·ing, -tends.** To expand or extend beyond a safe or prudent limit <*overextend* one's credit> **—o'ver·ex·ten'sion** *n.*
**o·ver·fa·tigue** (ō'vər-fə-tēg') *n.* Extreme fatigue often to a degree beyond an individual's power to recover.
**o·ver·fish** (ō'vər-fĭsh') *vt.* **-fished, -fish·ing, -fish·es.** To fish (a body of water) to such a degree as to upset the ecological balance or cause depletion of living creatures.
**o·ver·flow** (ō'vər-flō') *v.* **-flowed, -flow·ing, -flows.** —*vi.* **1.** To flow or run over the brim, top, or banks. **2.** To be filled beyond capacity, as a container or waterway. **3.** To have an overabundant supply : be unbounded <*overflowing* with sympathy> —*vt.* **1.** To flow over the top, brim, or banks of. **2.** To spread or pour over : FLOOD. **3.** To cause to fill beyond capacity. —*n.* (ō'vər-flō'). **1.** The act of overflowing. **2.** Something that flows over : EXCESS <find extra rooms for the *overflow* of conventioneers> **3.** An outlet through which excess liquid may escape.
**o·ver·gar·ment** (ō'vər-gär'mənt) *n.* An outer garment, as a coat.
**o·ver·glaze** (ō'vər-glāz') *n.* An outer coat of glaze on ceramic or pottery. —*vt.* (ō'vər-glāz', ō'vər-glāz') **-glazed, -glaz·ing, -glaz·es.** To apply an overglaze to.
**o·ver·graze** (ō'vər-grāz') *vt.* **-grazed, -graz·ing, -graz·es.** To graze (animals or land) to excess.
**o·ver·grow** (ō'vər-grō', ō'vər-grō') *v.* **-grew** (-grōō'), **-grown** (-grōn'), **-grow·ing, -grows.** —*vt.* **1.** To spread over with growth. **2.** To outgrow. —*vi.* To grow beyond normal size.
**o·ver·growth** (ō'vər-grōth') *n.* **1.** A growth over or on something. **2.** Excessively luxuriant or abundant growth.

**o·ver·hand** (ō′vər-hănd′) *also* **o·ver·hand·ed** (ō′vər-hăn′dĭd) *adj.* **1.** Thrown, struck, or performed with the hand above the level of the shoulder <an *overhand* pitch> **2.** Sewn with stitches drawing two edges together, with each stitch passing over the seam formed by the edges. —*adv.* In an overhand way. —*n.* **1.** An overhand throw, stroke, or delivery. **2.** An overhand stitch or seam. —*vt.* **-hand·ed, -hand·ing, -hands.** To sew with an overhand seam or stitches.

**overhand knot** *n.* A simple knot formed by making a loop in a piece of cord and passing the end through it.

**o·ver·hang** (ō′vər-hăng′) *v.* **-hung** (-hŭng′), **-hang·ing, -hangs.** —*vt.* **1.** To project or extend beyond. **2.** To menace or threaten : loom over. **3.** To decorate with hangings. —*vi.* To hang or project over something. —*n.* (ō′vər-hăng′). **1.** A projecting part, as of a building or rock formation. **2.** Extent of projection <an *overhang* of ten inches> **3.** The part of a ship's bow or stern that projects over the water.

**o·ver·haul** (ō′vər-hôl′, ō′vər-hôl′) *vt.* **-hauled, -haul·ing, -hauls.** **1. a.** To examine thoroughly for needed repairs. **b.** To dismantle in order to make repairs. **c.** *Naut.* To slacken (a line) or to release and separate the blocks of (a tackle). **2.** To make all needed repairs on : SERVICE. **3.** To catch up with : OVERTAKE. —*n.* (ō′vər-hôl′). **1.** An act of overhauling. **2.** A repair job.

**o·ver·head** (ō′vər-hĕd′) *adj.* **1.** Located above the level of the head <an *overhead* fan> **2.** Of or relating to overhead expenses. —*n.* (ō′vər-hĕd′). **1.** The regular operating expenses of a business, including the costs of rent, utilities, upkeep, and taxes, and excluding labor and materials. **2.** The ceiling of a ship's cabin. **3.** Something, such as a light fixture, that is located above head level. —*adv.* (ō′vər-hĕd′). Over or above the level of the head <branches hanging *overhead*>

**o·ver·hear** (ō′vər-hîr′) *vt.* **-heard** (-hûrd′), **-hear·ing, -hears.** To hear (something spoken or someone speaking) without the speaker's knowledge or intention. **—o′ver·hear′er** *n.*

**o·ver·heat** (ō′vər-hēt′) *v.* **-heat·ed, -heat·ing, -heats.** —*vt.* **1.** To heat too hot. **2.** To cause to become excited or overwrought <*overheated* by a family quarrel> **3.** To overstimulate (e.g., the economy). —*vi.* To become overheated.

**o·ver·in·dulge** (ō′vər-ĭn-dŭlj′) *v.* **-dulged, -dulg·ing, -dulg·es.** —*vt.* To indulge excessively. —*vi.* To indulge in something to excess. **—o′ver·in·dul′gence** *n.* **—o′ver·in·dul′gent** *adj.* **—o′ver·in·dul′gent·ly** *adv.*

**o·ver·joyed** (ō′vər-joid′) *adj.* Filled with joy : ELATED.

**o·ver·kill** (ō′vər-kĭl′) *n.* **1.** Nuclear destructive capacity beyond what is needed to destroy a target. **2.** Excessive killing. **3.** An action, response, or amount in excess of what is called for <judicial *overkill* in sentencing minors> —*vt.* (ō′vər-kĭl′) **-killed, -kill·ing, -kills.** To destroy (an enemy target) with greater nuclear force than is needed.

**o·ver·lad·en** (ō′vər-lād′n) *adj.* Overloaded : overburdened.

**o·ver·land** (ō′vər-lănd′, -lənd) *adj.* Passing over or by way of land <an *overland* supply route> —*adv.* By way of land.

**o·ver·lap** (ō′vər-lăp′) *v.* **-lapped, -lap·ping, -laps.** —*vt.* **1.** To lie or extend over and cover part of. **2.** To have an area or range in common with : coincide partly with. —*vi.* **1.** To lie over and partly cover something. **2.** To coincide partly <Their territories *overlap.*> —*n.* (ō′vər-lăp′). **1.** A part that overlaps or is overlapped. **2.** An instance of overlapping.

**o·ver·lay** (ō′vər-lā′) *vt.* **-laid, -lay·ing, -lays.** **1.** To lay or spread over or on. **2. a.** To cover the surface of with a decorative layer or design <*overlay* silver with gold> **b.** To embellish superficially <a simple tune *overlaid* with ornate harmonies> **3.** To put an overlay upon in printing. —*n.* (ō′vər-lā′). Something laid over or covering something else, as: **a.** A layer of decoration, as veneer or gold leaf, applied to a surface. **b.** A sheet of paper used on a press tympan to vary the pressure that produces light and dark tones. **c.** A transparent sheet containing graphic matter, as labels or colored areas, placed on illustrative matter to be incorporated into it.

**o·ver·leap** (ō′vər-lēp′) *vt.* **-leaped** *or* **-leapt** (-lĕpt′), **-leap·ing, -leaps.** **1.** To leap over or across. **2.** OVERREACH 3.

**o·ver·learn** (ō′vər-lûrn′) *vt.* **-learned** *or* **-learnt** (-lûrnt′), **-learn·ing, -learns.** To continue studying or working at (e.g., a skill) after becoming proficient.

**o·ver·lie** (ō′vər-lī′) *vt.* **-lay** (-lā′), **-lain** (-lān′), **-lay·ing, -lies.** **1.** To lie over or upon. **2.** To kill by lying upon. —Used esp. of animals.

**o·ver·load** (ō′vər-lōd′) *vt.* **-load·ed, -load·ing, -loads.** To load too heavily. —*n.* (ō′vər-lōd′). An excessive load.

**o·ver·long** (ō′vər-lông′, -lŏng′) *adj.* Being or lasting too long <an *overlong* performance> —*adv.* For too long.

**o·ver·look** (ō′vər-lo͝ok′) *vt.* **-looked, -look·ing, -looks.** **1.** To look over or at from a higher position. **2.** To be located or rise above, esp. so as to afford a view over <windows *overlooking* the river> **3.** To fail to notice or consider : MISS. **4.** To ignore deliberately or indulgently : DISREGARD. **5.** To look over : EXAMINE. **6.** To watch over : OVERSEE. **7.** To cast a spell over with an evil eye. —*n.* (ō′vər-lo͝ok′). **1.** An elevated place affording an extensive view. **2.** An act of overlooking something.

**o·ver·lord** (ō′vər-lôrd′) *n.* **1.** A lord having power or authority over other lords. **2.** One who is in a position of domination or supremacy over others. **—o′ver·lord′ship′** *n.*

**o·ver·ly** (ō′vər-lē) *adv.* To an excessive degree.

**o·ver·man** *n.* **1.** (ō′vər-mən, -măn′). A person having authority over others, esp. a foreman or overseer. **2.** (ō′vər-măn′). SUPERMAN 2. —*vt.* (ō′vər-măn′) **-manned, -man·ning, -mans.** To provide (e.g., a ship) with more personnel than necessary.

**o·ver·mas·ter** (ō′vər-măs′tər) *vt.* **-tered, -ter·ing, -ters.** To overcome : overpower.

**o·ver·match** (ō′vər-măch′) *vt.* **-matched, -match·ing, -match·es.** **1.** To be more than the match of : OUTSTRIP. **2.** To match or pair with a superior opponent. —*n.* (ō′vər-măch′). A contest in which one opponent is distinctly superior.

**o·ver·much** (ō′vər-mŭch′) *adj.* Too much : EXCESSIVE. —*adv.* In excess. —*n.* (ō′vər-mŭch′, ō′vər-mŭch′). An excessive amount.

**o·ver·night** (ō′vər-nīt′) *adj.* **1.** Lasting for, extending over, or remaining during a night <an *overnight* guest> **2.** For use over a single night or for a short journey <*overnight* provisions> **3.** Immediate : sudden <an *overnight* success> —*adv.* (ō′vər-nīt′). **1.** During or lasting the night. **2.** On the preceding night or evening. **3.** In or as if in a single night : SUDDENLY <The city grew *overnight.*>

**overnight bag** *n.* A small bag or suitcase for carrying items needed on a short journey.

**o·ver·nu·tri·tion** (ō′vər-no͞o-trĭsh′ən, -nyo͞o-) *n.* Excessive consumption of food or nutrition.

**o·ver·op·ti·mis·tic** (ō′vər-ŏp′tə-mĭs′tĭk) *adj.* Excessively optimistic. **—o′ver·op′ti·mism** *n.* **—o′ver·op′ti·mis′ti·cal·ly** *adv.*

**o·ver·pass** (ō′vər-păs′) *n.* A passage, as an elevated roadway or bridge that crosses above another thoroughfare. —*vt.* (ō′vər-păs′, -păs′), **-passed** *or* **-past, -pass·ing, -pass·es.** **1.** To pass over or across : TRAVERSE. **2.** To go beyond : SURPASS. **3.** To disregard.

**o·ver·pay** (ō′vər-pā′) *v.* **-paid** (-pād′), **-pay·ing, -pays.** —*vt.* **1.** To pay (someone) too much. **2.** To pay an amount in excess of (a sum due). —*vi.* To pay too much. **—o′ver·pay′ment** *n.*

**o·ver·per·suade** (ō′vər-pər-swād′) *vt.* **-suad·ed, -suad·ing, -suades.** To persuade (someone) to act contrary to inclination. **—o′ver·per·sua′sion** *n.*

**o·ver·play** (ō′vər-plā′) *vt.* **-played, -play·ing, -plays.** **1.** To play (a dramatic role) in an exaggerated manner : OVERACT. **2.** To overestimate the strength of (one's holdings or position) and thus contribute to one's own defeat <*overplay* one's hand> **3.** To hit (a golf ball) beyond the green.

**o·ver·plus** (ō′vər-plŭs′) *n.* An excessive amount : SURPLUS.

**o·ver·pop·u·la·tion** (ō′vər-pŏp′yə-lā′shən) *n.* Excessive population of an area to the point of social detriment or environmental deterioration. **—o′ver·pop′u·lat′ed** (-lā′tĭd) *adj.*

**o·ver·pow·er** (ō′vər-pou′ər) *vt.* **-ered, -er·ing, -ers.** **1.** To overcome by superior force : SUBDUE. **2.** To affect so strongly as to make ineffective or helpless : OVERWHELM. **3.** To supply with excessive mechanical power.

**o·ver·pow·er·ing** (ō′vər-pou′ər-ĭng) *adj.* So strong as to overpower : OVERWHELMING. **—o′ver·pow′er·ing·ly** *adv.*

**o·ver·praise** (ō′vər-prāz′) *vt.* **-praised, -prais·ing, -prais·es.** To praise too highly.

**o·ver·price** (ō′vər-prīs′) *vt.* **-priced, -pric·ing, -pric·es.** To put too high a price or value on.

**o·ver·print** (ō′vər-prĭnt′) *vt.* **-print·ed, -print·ing, -prints.** To print over something already printed, esp. to print over (a graphic image) with another color. —*n.* (ō′vər-prĭnt′). **1.** A mark or impression made by overprinting. **2. a.** A mark or message printed over a postage stamp to note a special occasion or a change in use. **b.** A stamp so marked.

**o·ver·prize** (ō′vər-prīz′) *vt.* **-prized, -priz·ing, -priz·es.** To prize too much.

**o·ver·pro·duce** (ō′vər-prə-do͞os′, -dyo͞os′) *vt.* **-duced, -duc·ing, -duc·es.** To produce too much of. **—o′ver·pro·duc′tion** (-dŭk′shən) *n.* **—o′ver·pro·duc′tive** *adj.*

**o·ver·proof** (ō′vər-pro͞of′) *adj.* Having an alcohol content higher than proof spirit, esp. containing more than 50% alcohol by volume.

**o·ver·pro·tect** (ō′vər-prə-tĕkt′) *vt.* **-tect·ed, -tect·ing, -tects.** To protect too much. **—o′ver·pro·tec′tive** *adj.*

**o·ver·qual·i·fied** (ō′vər-kwŏl′ə-fīd′) *adj.* Having qualifications beyond what is necessary or desired.

**o·ver·rate** (ō′vər-rāt′) *vt.* **-rat·ed, -rat·ing, -rates.** To rate too highly.

**o·ver·reach** (ō′vər-rēch′) *v.* **-reached, -reach·ing, -reach·es.** —*vt.* **1.** To reach or extend over or beyond. **2.** To miss or lose by attempting too much <*overreaching* a goal> **3.** To defeat (oneself) by doing or trying to gain too much. **4.** To get the better of. —*vi.* **1.** To reach or go too far. **2.** To outwit or get the better of others. **3.** To strike the hind foot against the forefoot or foreleg on the same side of the body. —Used of a horse. **—o′ver·reach′er** *n.*

**o·ver·re·act** (ō′vər-rē-ăkt′) *vi.* **-act·ed, -act·ing, -acts.** To react with undue force or emotion. **—o′ver·re·ac′tion** *n.*

**o·ver·re·fine·ment** (ō′vər-rĭ-fīn′mənt) *n.* Excessive or unnecessary refinement. **—o′ver·re·fined′** *adj.*

**o·ver·ride** (ō′vər-rīd′) *vt.* **-rode** (-rōd′), **-rid·den** (-rĭd′n), **-rid·ing, -rides. 1.** To ride across. **2.** To trample upon. **3.** To ride (a horse) too hard. **4.** To prevail or take precedence over <Tax reform *overrode* all other issues.> **5.** To declare null and void : SET ASIDE. —*n.* (ō′vər-rīd′). A sales commission to an executive over and above any commission to subordinate salespeople.

**o·ver·ripe** (ō′vər-rīp′) *adj.* **1.** Too ripe. **2.** Decadent : unsavory. **—o′ver·ripe′ness** *n.*

**o·ver·rule** (ō′vər-ro͞ol′) *vt.* **-ruled, -rul·ing, -rules. 1. a.** To disallow the action or arguments of (a person), esp. by virtue of higher authority. **b.** To rule or decide against <*overrule* a decision> **c.** To declare null and void : REVERSE. **2.** To dominate by strong influence.

**o·ver·run** (ō′vər-rŭn′) *v.* **-ran** (-răn′), **-run, -run·ning, -runs.** —*vt.* **1.** To defeat conclusively and seize the positions of <The firebase was *overrun* by the enemy.> **2.** To spread or swarm over destructively <Weevils *overran* the garden.> **3.** To spread quickly throughout <Political hacks *overran* the new government.> **4.** To overflow <The river *overran* its banks.> **5.** To run or extend beyond. **6. a.** To rearrange or move (set type or pictures) from one column, line, or page to another. **b.** To print (a job order) in a quantity larger than that ordered. —*vi.* **1.** To run over : OVERFLOW. **2.** To go beyond the normal or desired limit. —*n.* (ō′vər-rŭn′). **1.** An act of overrunning. **2.** The amount by which something overruns. **3. a.** The act of exceeding estimated costs for work covered by contract. **b.** The amount by which actual costs exceed estimates.

**o·ver·score** (ō′vər-skôr′, -skōr′) *vt.* **-scored, -scor·ing, -scores.** To delete or cross out by drawing a line over or through.

**o·ver·seas** (ō′vər-sēz′, ō′vər-sēz′) *also* **o·ver·sea** (ō′vər-sē′, ō′vər-sē′) *adv.* Beyond the sea : ABROAD. —*adj.* Of, relating to, originating in, or situated in areas across the sea <*overseas* bases>

**overseas cap** *n.* A garrison cap.

**o·ver·see** (ō′vər-sē′) *vt.* **-saw** (-sô′), **-seen, -see·ing, -sees. 1.** To watch over and direct : SUPERVISE. **2.** To look over : INSPECT.

**o·ver·se·er** (ō′vər-sē′ər) *n.* **1.** One who keeps watch over and directs the work of others, esp. laborers. **2.** A supervisor.

**o·ver·sell** (ō′vər-sĕl′) *vt.* **-sold** (-sōld′), **-sell·ing, -sells. 1.** To contract to sell more of (a stock or commodity) than can be delivered. **2.** To be too insistent or eager in attempting to sell something to (someone). **3.** To make exaggerated or unwarranted claims for.

**o·ver·sen·si·tive** (ō′vər-sĕn′sĭ-tĭv) *adj.* Unduly sensitive. **—o′ver·sen′si·tive·ness, o′ver·sen′si·tiv′i·ty** *n.*

**o·ver·set** (ō′vər-sĕt′) *v.* **-set, -set·ting, -sets.** —*vt.* **1.** To push or tip over : OVERTURN. **2.** To distress or disturb : UPSET. **3. a.** To set too much (printing type). **b.** To set too much type for (a given space). —*vi.* **1.** To tip or fall over : OVERTURN. **2.** To set too much printed matter for a given space. —*n.* (ō′vər-sĕt′). **1.** An upset. **2.** An excess of set printing type.

**o·ver·sew** (ō′vər-sō′, ō′vər-sō′) *vt.* **-sewed, -sewn** (-sōn′) *or* **-sewed, -sew·ing, -sews.** To sew with overhand stitches.

**o·ver·sexed** (ō′vər-sĕkst′) *adj.* Having an immoderate sexual drive.

**o·ver·shad·ow** (ō′vər-shăd′ō) *vt.* **-owed, -ow·ing, -ows. 1.** To cast a shadow over. **2.** To make insignificant by comparison.

**o·ver·shoe** (ō′vər-sho͞o′) *n.* An item of footgear, as a galosh or rubber, worn over shoes as protection from water, snow, or cold.

**o·ver·shoot** (ō′vər-sho͞ot′) *v.* **-shot** (-shŏt′), **-shoot·ing, -shoots.** —*vt.* **1.** To shoot or pass over or beyond. **2.** To miss by or as if by shooting or propelling something too far. **3.** To go or fly beyond esp. when landing <The plane *overshot* the runway.> **4.** To go beyond : EXCEED. —*vi.* To go or shoot too far.

**o·ver·shot** (ō′vər-shŏt′) *adj.* **1.** Having an upper part projecting beyond the lower <an *overshot* jaw> **2.** Designating a water wheel propelled by the weight of water falling from the top.

**o·ver·sight** (ō′vər-sīt′) *n.* **1.** An unintentional omission or mistake. **2.** Watchful care or management : SUPERVISION.

**o·ver·sim·ple** (ō′vər-sĭm′pəl) *adj.* Too simple <an *oversimple* theory of behavior>

**o·ver·sim·pli·fy** (ō′vər-sĭm′plə-fī′) *v.* **-fied, -fy·ing, -fies.** —*vt.* To distort by describing too simply <*oversimplify* a complex set of economic problems> —*vi.* To cause distortions by presenting something in too simple a form. **—o′ver·sim′pli·fi·ca′tion** *n.*

**o·ver·size** (ō′vər-sīz′) *also* **o·ver·sized** (-sīzd′) *adj.* Larger in size than usual or necessary. —*n.* (ō′vər-sīz′). **1.** An unusually large size. **2.** An article made in an unusually large size.

**o·ver·skirt** (ō′vər-skûrt′) *n.* An outer skirt, esp. a shorter one worn draped over another skirt.

**o·ver·sleep** (ō′vər-slēp′) *v.* **-slept** (-slĕpt′), **-sleep·ing, -sleeps.** —*vi.* To sleep beyond one's usual or intended time for waking. —*vt.* To sleep beyond the time for <*overslept* my morning class>

**o·ver·soul** (ō′vər-soul′) *n.* A spiritual essence or vital force in the universe embracing all souls and thus transcending individual consciousness.

**o·ver·spend** (ō′vər-spĕnd′) *v.* **-spent** (-spĕnt′), **-spend·ing, -spends.** —*vi.* To spend more than is necessary or prudent. —*vt.* **1.** To spend beyond the limits of <*overspend* one's income> **2.** To exhaust <*overspent* with toil>

**o·ver·spread** (ō′vər-sprĕd′) *vt.* **-spread, -spread·ing, -spreads.** To spread over or overhead. **—o′ver·spread′** *n.*

**o·ver·state** (ō′vər-stāt′) *vt.* **-stat·ed, -stat·ing, -states.** To state in exaggerated terms. **—o′ver·state′ment** *n.*

**o·ver·stay** (ō′vər-stā′) *vt.* **-stayed, -stay·ing, -stays.** To stay beyond the set limits or expected duration of.

**o·ver·step** (ō′vər-stĕp′) *vt.* **-stepped, -step·ping, -steps.** To go beyond (a limit) <*overstep* one's authority>

**o·ver·stock** (ō′vər-stŏk′) *vt.* **-stocked, -stock·ing, -stocks.** To stock with too much of (a commodity). —*n.* (ō′vər-stŏk′). An excessive supply.

**o·ver·stuff** (ō′vər-stŭf′) *vt.* **-stuffed, -stuff·ing, -stuffs. 1.** To stuff too much into. **2.** To cover entirely with cushioned upholstery.

**o·ver·sub·scribe** (ō′vər-səb-skrīb′) *vt.* **-scribed, -scrib·ing, -scribes.** To subscribe to (e.g., a performance series) in excess of available accommodation or supply. **—o′ver·sub·scrip′tion** (-skrĭp′shən) *n.*

**o·ver·sub·tle** (ō′vər-sŭt′l) *adj.* Subtle to the point of confusion or irrelevance.

**o·ver·sup·ply** (ō′vər-sə-plī′) *n., pl.* **-plies.** A supply in excess of what is required. —*vt.* (ō′vər-sə-plī′) **-plied, -ply·ing, -plies.** To supply in excess.

**o·vert** (ō-vûrt′, ō′vûrt′) *adj.* [ME < OFr., p.part. of *ovrir,* to open < Lat. *aperire.*] Clearly evident <an *overt* dislike> **—o·vert′ly** *adv.*

**o·ver·take** (ō′vər-tāk′) *vt.* **-took** (-to͝ok′), **-tak·en, -tak·ing, -takes. 1.** To catch up with. **2.** To pass after catching up with. **3.** To come upon unexpectedly : SURPRISE.

**o·ver·tax** (ō′vər-tăks′) *vt.* **-taxed, -tax·ing, -tax·es. 1.** To impose an excessive tax or taxes on. **2.** To subject to an excessive burden or strain. **—o′ver·tax·a′tion** *n.*

**o·ver-the-count·er** (ō′vər-*th*ə-koun′tər) *adj.* **1.** Not available on an officially recognized stock exchange but traded in direct negotiation between buyers and sellers <*over-the-counter* stocks> **2.** Available for sale to the general public without a prescription <*over-the-counter* drugs> **3.** *Informal.* Aboveboard : legitimate.

**o·ver·throw** (ō′vər-thrō′) *vt.* **-threw** (-thro͞o′), **-thrown** (-thrōn′), **-throw·ing, -throws. 1.** To throw over : OVERTURN. **2.** To bring about the destruction or downfall of, esp. by force. **3.** To throw something over and beyond (a target) <The catcher *overthrew* second base.> —*n.* (ō′vər-thrō′). **1.** An instance of overthrowing. **2.** Downfall : destruction. **3.** The throwing of a ball over and beyond an intended mark, esp. in baseball.

☆ ***syns:*** OVERTHROW, BRING DOWN, OVERTURN, SUBVERT, TOPPLE, TUMBLE *v. core meaning* : to bring about the downfall of <Guerrillas *overthrew* the ruling junta.>

**o·ver·time** (ō′vər-tīm′) *n.* **1.** Time beyond an established limit, such as: **a.** Working hours in addition to those of the regular schedule. **b.** A period of playing time added after the expiration of the set time limit of an athletic contest. **2.** Payment for additional work done outside of regular working hours. —*adv.* Beyond the established time limit, esp. that of the normal working day <The staff worked *overtime.*> —*vt.* (ō′vər-tīm′) **-timed, -tim·ing, -times.** To exceed the desired timing for <*overtime* a photographic exposure>

**o·ver·tone** (ō′vər-tōn′) *n.* **1.** *Mus.* HARMONIC 1. **2.** *often* **overtones.** An indication : hint <threatening *overtones* in the ambassador's speech>

**o·ver·top** (ō′vər-tŏp′) *vt.* **-topped, -top·ping, -tops. 1.** To extend over or beyond the top of : rise above. **2.** To be superior to <"Religion *overtopped* the common affairs of life" —Albert C. Baugh>

**o·ver·train** (ō′vər-trān′) *v.* **-trained, -train·ing, -trains.** —*vt.* To train too much. —*vi.* To practice something excessively.

**o·ver·trick** (ō′vər-trĭk′) *n.* A trick won in excess of contract or game, as in bridge.

**o·ver·trump** (ō′vər-trŭmp′, ō′vər-trŭmp′) *v.* **-trumped, -trump·ing, -trumps.** —*vi.* To play a trump higher than one previously played on a trick, as in bridge. —*vt.* To trump with a higher trump.

**o·ver·ture** (ō′vər-cho͝or′) *n.* [ME < OFr. < Lat. *apertura,* opening < *aperire,* to open.] **1.** *Mus.* **a.** An instrumental introduction to an extended musical work, as an opera or an oratorio. **b.** A similar orchestral work, esp. one composed as introductory music to a play or as a concert piece. **2.** An introductory part or section, as of a poem. **3.** Something, as an offer or proposal, indicating readiness to initiate a move or to open a relationship <"I wanted revenge for her snub of my flirting *overture*" —John Updike> —*vt.* **-tured, -tur·ing, -tures. 1.** To present as an overture or proposal. **2.** To present or offer an overture to. **3.** To introduce with an overture or prelude.

**o·ver·turn** (ō′vər-tûrn′) *v.* **-turned, -turn·ing, -turns.** —*vt.* **1.** To cause to turn over or capsize : UPSET. **2.** To defeat : overthrow. —*vi.* To turn over or capsize : become upset. —*n.* (ō′vər-tûrn′). **1.** The act or process of overturning, or the state of being overturned. **2.** TURNOVER 5.

**o·ver·use** (ō′vər-yo͞oz′) *vt.* **-used, -us·ing, -us·es.** To use to excess. —*n.* (ō′vər-yo͞os′). Excessive use.

**o·ver·val·ue** (ō′vər-văl′yo͞o) *vt.* **-ued, -u·ing, -ues.** To place an excessive or unwarranted value on. **—o′ver·val′u·a·tion** *n.*
**o·ver·view** (ō′vər-vyo͞o′) *n.* **1.** A broad, comprehensive view. **2.** A review or summary.
**o·ver·wear** (ō′vər-wâr′) *vt.* **-wore** (-wôr′, -wōr′), **-worn** (-wôrn′, -wōrn′), **-wear·ing, -wears.** To wear out.
**o·ver·wea·ry** (ō′vər-wîr′ē) *adj.* Weary to the point of exhaustion. —*vt.* **-ried, -ry·ing, -ries.** To tire out : EXHAUST.
**o·ver·ween·ing** (ō′vər-wē′nĭng) *adj.* **1.** Presumptuously arrogant : OVERBEARING. **2.** Immoderate : excessive <*overweening* ambition>
**o·ver·weigh** (ō′vər-wā′) *vt.* **-weighed, -weigh·ing, -weighs. 1.** To outweigh. **2.** OVERWEIGHT 1.
**o·ver·weight** (ō′vər-wāt′) *adj.* Weighing more than is normal, necessary, or allowed. —*n.* (ō′vər-wāt′). **1.** More weight than is normal, necessary, or allowed. **2.** Greater weight or importance : PREPONDERANCE. —*vt.* (ō′vər-wāt′) **-weight·ed, -weight·ing, -weights. 1.** To weigh down too heavily : OVERLOAD. **2.** To give too much emphasis, consideration, or importance to <*overweight* the evidence>
**o·ver·whelm** (ō′vər-hwĕlm′, -wĕlm′) *vt.* **-whelmed, -whelm·ing, -whelms. 1.** To flood over and submerge : ENGULF <waves *overwhelming* the ship deck> **2.** To overcome utterly, as by physical or emotional force : OVERPOWER <*overwhelmed* us with kindness> **3.** To turn over : UPSET.
**o·ver·whelm·ing** (ō′vər-hwĕl′mĭng, -wĕl′) *adj.* Overpowering in strength or effect. **—o′ver·whelm′ing·ly** *adv.*
**o·ver·wind** (ō′vər-wīnd′) *vt.* **-wound** (-wound′), **-wind·ing, -winds.** To wind too tightly.
**o·ver·win·ter** (ō′vər-wĭn′tər) *vi.* **-tered, -ter·ing, -ters.** To spend or survive the winter. —*adj.* Occurring during winter.
**o·ver·work** (ō′vər-wûrk′) *v.* **-worked, -work·ing, -works.** —*vt.* **1.** To force to work too hard or too long. **2.** To use or rework to excess <*overwork* an analogy> —*vi.* To work too long or too hard. —*n.* (ō′vər-wûrk′). Excess work.
**o·ver·write** (ō′vər-rīt′) *v.* **-wrote** (-rōt′), **-writ·ten** (rĭt′n), **-writ·ing, -writes.** —*vt.* **1.** To write (something) over other writing. **2.** To write or describe in an excessively florid or prolix style. —*vi.* To write at unnecessarily great length.
**o·ver·wrought** (ō′vər-rôt′) *adj.* **1.** Excessively nervous or excited : AGITATED. **2.** Extremely ornate or elaborate : OVERDONE <an *overwrought* description>
**ovi-** *or* **ovo-** *or* **ov-** *pref.* [Lat. < *ovum*, egg.] Egg : ovum <*oviferous*>
**o·vi·bos** (ō′və-bŏs′) *n.* [Lat. *ovis*, sheep + Lat. *bos*, ox.] Musk ox.
**o·vi·ci·dal** (ō′vĭ-sīd′l) *adj.* Capable of killing eggs. **—o′vi·cide′** *n.*
**o·vi·duct** (ō′və-dŭkt′) *n.* A tube through which ova travel from an ovary. **—o′vi·duc′tal** *adj.*

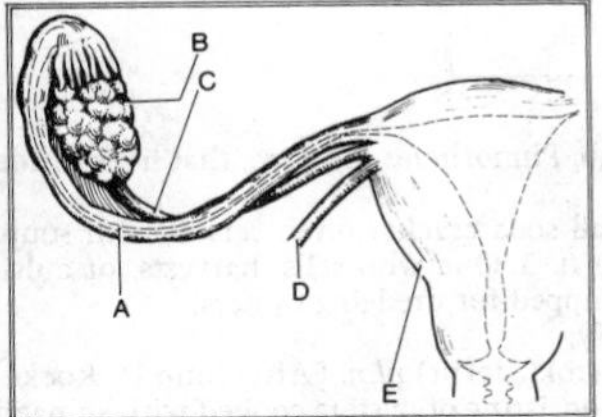


**oviduct**
*A. oviduct, B. ovary, C. ovarian ligament, D. uterine ligament, E. uterus*

**o·vif·er·ous** (ō-vĭf′ər-əs) *adj.* Bearing or producing ova.
**o·vi·form** (ō′və-fôrm′) *adj.* Egg-shaped.
**o·vine** (ō′vīn′) *adj.* [LLat. *ovinus* < Lat. *ovis*, sheep.] Of, resembling, or characteristic of sheep. —*n.* An ovine animal.
**o·vip·a·rous** (ō-vĭp′ər-əs) *adj.* Producing eggs that hatch outside the female's body <*oviparous* mammals> **—o′vi·par′i·ty** (ō′və-păr′ĭ-tē) *n.* **—o·vip′a·rous·ly** *adv.*
**o·vi·pos·it** (ō′və-pŏz′ĭt) *vi.* **-it·ed, -it·ing, -its.** To lay eggs, esp. with an ovipositor. **—o′vi·po·si′tion** (-pə-zĭsh′ən) *n.* **—o′vi·po·si′tion·al** *adj.*
**o·vi·pos·i·tor** (ō′və-pŏz′ə-tər) *n.* A tubular structure, occas. extending outside the abdomen, with which many insects lay eggs.
**o·vi·sac** (ō′və-săk′) *n.* An egg case or capsule, as an ootheca or a Graafian follicle.
**ovo-** *pref. var. of* OVI-.
**o·void** (ō′void′) *also* **o·voi·dal** (ō-void′l) *adj.* That is egg-shaped. **—o′void′** *n.*
**o·vo·lo** (ō′və-lō′) *n., pl.* **-li** (-lī′) [Ital., dim. of *ovo*, egg < Lat. *ovum*.] A rounded convex architectural molding, often a quarter section of a circle or ellipse.
**o·von·ic** (ō-vŏn′ĭk) *n.* [Ov(SHINSKY EFFECT) + (ELECTR)ONIC.] An electronic device that utilizes the Ovshinsky effect.
**o·vo·tes·tis** (ō′vō-tĕs′tĭs) *n., pl.* **-tes** (-tēz′). The hermaphroditic reproductive gland of certain gastropods.
**o·vo·vi·vip·a·rous** (ō′vō-vī-vĭp′ər-əs) *adj.* Producing eggs that hatch within the female's body, as do some reptiles and fishes. **—o′vo·vi·vip′a·rous·ly** *adv.* **—o′vo·vi′vi·par′i·ty** (-vī′və-păr′ĭ-tē), **o′vo·vi·vip′a·rous·ness** *n.*
**Ov·shin·sky effect** (ŏv-shĭn′skē, ōv-) *n.* [After Stanford R. *Ovshinsky* (b.1923).] The effect by which a specific glossy thin film switches from a nonconductor to a semi-conductor upon application of a minimum voltage.
**o·vu·late** (ŏ′vyə-lāt′, ōv′yə-) *vi.* **-lat·ed, -lat·ing, -lates.** [< OVULE.] **1.** To produce ova. **2.** To discharge ova. **—o′vu·la′tion** *n.*
**o·vule** (ō′vyo͞ol, ōv′yo͞ol) *n.* [Fr. < NLat. *ovulum*, dim. of Lat. *ovum*, egg.] **1.** *Zool.* An immature ovum. **2.** *Bot.* A minute rudimentary structure from which a plant seed develops after fertilization. **—o′vu·lar** (ō′vyə-lər, ŏv′yə-), **o′vu·lar′y** (-lĕr′ē) *adj.*
**o·vum** (ō′vəm) *n., pl.* **o·va** (ō′və) [Lat., egg.] The female reproductive cell of animals : EGG.
**owe** (ō) *v.* **owed, ow·ing, owes.** [ME *owen* < OE *āgan*, to possess.] —*vt.* **1.** To be indebted to the amount of <You *owe* me ten dollars.> **2.** To have a moral obligation to render or offer <We *owe* them our thanks.> **3.** To be in debt to. **4. a.** To be indebted or obliged for. **b.** To be indebted for the origin or existence of <I *owe* my life to quick thinking.> **5.** To bear (a certain feeling) toward a person <*owes* them a grudge> **6.** *Obs.* To own : have. —*vi.* To be in debt <I still *owe* for my car.>
**ow·ing** (ō′ĭng) *adj.* Still to be paid : DUE.
**owing to** *prep.* Because of : on account of.
**owl** (oul) *n.* [ME *owle* < OE *ūle*.] **1.** Any of various often nocturnal birds of prey of the order Strigiformes, characterized by large heads with short hooked beaks, eyes set in a frontal facial plane, and hooked and feathered talons. **2.** Any of a breed of domestic pigeons resembling owls.
**owl·et** (ou′lĭt) *n.* A young owl.
**owl·ish** (ou′lĭsh) *adj.* Resembling an owl. **—owl′ish·ly** *adv.* **—owl′ish·ness** *n.*
**owl's claws** (oulz) *n.* (*sing. or pl. in number*). A plant, *Helenium hoopesii* of western North America, with large yellow flower heads.
**owl's-clo·ver** (oulz′klō′vər) *n.* [From the resemblance of the flowers of some species to owl's faces.] A New World plant of the genus *Orthocarpus*, having red or purple flower spikes.
**own** (ōn) *adj.* [ME *owen* < OE *āgen*.] Of or belonging to oneself or itself <bake one's *own* bread> —*pron.* (*sing. or pl. in number*). The one or ones belonging to oneself. —Used after a possessive without a following noun as a pronoun equivalent in meaning to the adjective *own* <gave out uniforms so that each team member had his or her *own*> —*v.* **owned, own·ing, owns.** —*vt.* **1.** To have or possess <*owns* the firm> **2.** To admit or acknowledge <They *owned* their mistake.> —*vi.* To confess <*owned* to being annoyed> **—come into (one's) own.** To obtain rightful recognition and prosperity. **—hold (one's) own.** To maintain one's place in spite of adversity. **—of one's own.** Belonging to oneself. **—on one's own.** For or by oneself : INDEPENDENTLY. **—own up.** To confess fully and openly. **—own′er** *n.*
**own·er·ship** (ō′nər-shĭp′) *n.* **1.** The state or fact of being an owner. **2.** Legal right to the possession of a thing.
**ox** (ŏks) *n., pl.* **ox·en** (ŏk′sən) [ME < OE *oxa*.] **1.** An adult castrated bull of the genus *Bos*, used as a draft animal. **2.** A bovine mammal.
**ox-** *pref. var. of* OXO-.
**ox·a·cil·lin** (ŏk′sə-sĭl′ĭn) *n.* [OX(O) + A(ZOLE) + (PENI)CILLIN.] A semisynthetic penicillin effective against penicillinase-producing staphylococci.
**ox·a·late** (ŏk′sə-lāt′) *n.* [Fr. : *oxalique*, oxalic acid + *-ate*, -ate.] A salt or ester of oxalic acid. —*vt.* **-lat·ed, -lat·ing, -lates.** To treat (a specimen) with an oxalate or oxalic acid.
**ox·al·ic acid** (ŏk-săl′ĭk) *n.* [Fr. *oxalique* < Lat. *oxalis*, wood sorrel. —see OXALIS.] A poisonous, crystalline organic acid, $HOOCCOOH \cdot 2H_2O$, used in textile finishing and as a laundry bleach and a cleansing agent for metals.
**ox·a·lis** (ŏk′sə-lĭs, ŏk-săl′ĭs) *n.* [NLat. *Oxalis*, genus name < Lat. *oxalis*, wood sorrel < Gk. < *oxus*, sour.] A plant of the genus *Oxalis*, bearing pink, yellow, or white flowers.
**ox·blood red** (ŏks′blŭd′) *n.* A dark or deep red to medium reddish brown.
**ox·bow** (ŏks′bō′) *n.* **1.** A U-shaped, usu. wooden piece that fits around the neck of an ox, with its upper ends attached to the crossbar of the yoke. **2. a.** A U-shaped bend in a river. **b.** The land enclosed by an oxbow.
**oxen** (ŏk′sən) *n. pl. of* OX.
**ox·eye** (ŏk′sī′) *n.* [ME.] **1.** A Eurasian plant of the genus *Buphthalmum*, bearing daisylike flowers with yellow rays and dark centers. **2.** A North American plant of the genus *Heliopsis*, bearing flowers resembling those of the oxeye. **3.** A round or oval dormer window.
**oxeye daisy** *n.* DAISY 1.
**ox·ford** (ŏks′fərd) *n.* [After *Oxford*, England.] **1.** A sturdy low-cut shoe that laces over the instep. **2.** A cotton fabric of a tight basket weave, used mainly for shirts.

**oxford gray** *n.* A dark gray.
**Oxford movement** *n.* A movement within the Church of England, originating at Oxford University in 1833, that called for closer ties between the Anglican and Roman Catholic Churches.
**ox·heart** (ŏks′härt′) *n.* A variety of cultivated cherry bearing large sweet fruit.
**ox·i·dant** (ŏk′sĭ-dənt) *n.* A chemical reagent that oxidizes.
**ox·i·dase** (ŏk′sə-dās′) *n.* [OXID(ATION) + -ASE.] An animal or plant enzyme that acts as an oxidant. **—ox′i·da′sic** *adj.*
**ox·i·da·tion** (ŏk′sĭ-dā′shən) *n.* [Fr. < *oxider,* to oxidize < *oxide,* oxide. —see OXIDE.] **1.** Combination of a substance with oxygen. **2.** A reaction in which an element's valence is increased as a result of losing electrons. **—ox′i·da′tive** *adj.* **—ox′i·da′tive·ly** *adv.*
**ox·i·da·tion-re·duc·tion** (ŏk′sĭ-dā′shən-rĭ-dŭk′shən) *n.* A chemical reaction involving the transfer of electrons from one atom or molecule to another.
**ox·i·da·tive phos·pho·ryl·a·tion** (ok′sĭ-dā′tĭv fŏs′fə-rə-lā′-shən) *n.* A vital process of intracellular respiration occurring within the mitochondria of the cell, responsible for most adenosine triphosphate formation.
**ox·ide** (ŏk′sīd′) *n.* [Fr. < *oxygène,* oxygen. —see OXYGEN.] A binary compound of an element or radical with oxygen. **—ox·id′ic** (ŏk-sĭd′-ĭk) *adj.*
**ox·i·dize** (ŏk′sĭ-dīz′) *v.* **-dized, -diz·ing, -diz·es.** [< OXIDE.] *Chem.* —*vt.* **1.** To combine with oxygen or make into an oxide. **2.** To increase the positive charge or valence of (an element) by removing electrons. **3.** To coat with oxide. —*vi.* To become oxidized. **—ox′i·diz′a·ble** *adj.* **—ox′i·di·za′tion** *n.* **—ox′i·diz′er** (-dī′zər) *n.*
**ox·i·do·re·duc·tase** (ŏk′sĭ-dō-rĭ-dŭk′tās) *n.* [OXID(ATION) + REDUC(TION) + -ASE.] An enzyme that catalyzes an oxidation-reduction reaction.
**ox·lip** (ŏks′lĭp′) *n.* [OE *oxanslyppe* : *oxa,* ox + *slipa,* slimy substance.] A Eurasian primrose, *Primula elatior,* bearing yellow flower clusters.
**oxo-** *or* **ox-** *pref.* [Fr. < *oxygène,* oxygen.] Oxygen <oxophenarsine>
**Ox·o·ni·an** (ŏk-sō′nē-ən) *n.* [< Med. Lat. *Oxonia,* Oxford < OE *Oxnaford.*] **1.** A native or inhabitant of Oxford. **2.** A student or graduate of Oxford University. **—Ox·o′ni·an** *adj.*
**ox·peck·er** (ŏks′pĕk′ər) *n.* Either of two African birds, *Buphagus africanus* or *B. erythrorhyncus,* feeding chiefly on ticks found on the hides of cattle and herd animals.
**ox·tail** (ŏks′tāl′) *n.* The tail of an ox, esp. when used in soup.
**oxy-** *pref.* [< OXYGEN.] Oxygen, esp. additional oxygen <oxyacetylene>
**ox·y·a·cet·y·lene** (ŏk′sē-ə-sĕt′l-ĭn, -ēn′) *adj.* Of or containing a mixture of oxygen and acetylene, as commonly used in metal welding and cutting torches.
**ox·y·ac·id** (ŏk′sē-ăs′ĭd) *n.* An acid that contains oxygen.
**ox·y·ceph·a·ly** (ŏk′sē-sĕf′ə-lē) *n.* [Gk. *oxus,* sharp + -CEPHALY.] A congenital cephalic abnormality in which the skull assumes a conical shape. **—ox′y·ce·phal′ic** (-sə-făl′ĭk), **ox′y·ceph′a·lous** *adj.*
**ox·y·gen** (ŏk′sĭ-jən) *n.* [Fr. *oxygène* : Gk. *oxus,* sharp, acid + *-gène,* -gen.] *Symbol* **O** A colorless, tasteless, odorless gaseous element constituting 21% of the atmosphere by volume and essential to most combustion and combustive processes; atomic number 8; atomic weight 15.999. **—ox′y·gen′ic** (ŏk′sĭ-jĕn′ĭk) *adj.* **—ox′y·gen′i·cal·ly** *adv.* **—ox·yg′e·nous** (ŏk-sĭj′ə-nəs) *adj.*

▲ word history: The word *oxygen* was coined in French by the chemist Lavoisier in the 18th century, soon after the element was isolated. The French word *oxygène* was intended to mean "acid-producing," from Greek *oxus,* "sharp," used in the sense "acid," and the Greek suffix *-genēs,* "born," misinterpreted as "producing." Oxygen was considered at the time to be an essential component of an acid. Although this is not the case, the name *oxygen* has persisted for the element.

**ox·y·gen·ase** (ok′sĭ-jə-nās′) *n.* An oxidoreductase that catalyzes the transfer of free oxygen to its substrate.
**ox·y·gen·ate** (ŏk′sĭ-jə-nāt′) *also* **ox·y·gen·ize** (-nīz′) *vt.* **-at·ed, -at·ing, -ates** *also* **-ized, -iz·ing, -iz·es.** To treat, combine, or infuse with oxygen. **—ox′y·gen·a′tion** *n.*
**oxygen cycle** *n.* The cycle by which oxygen in the atmosphere is converted to carbon dioxide in animal and human respiration and regenerated by green plants in photosynthesis.
**oxygen mask** *n.* A masklike breathing device fitting over the mouth and nose and supplied with oxygen from a tank.
**oxygen tent** *n.* A transparent canopy covering the head and shoulders of a patient, used to regulate the supply of oxygen.
**ox·y·he·mo·glo·bin** (ŏk′sē-hē′mə-glō′bĭn, -hĕm′ə-) *n.* A bright-red chemical complex of hemoglobin and oxygen that transports oxygen from the lungs to the tissues via the blood.
**ox·y·mo·ron** (ŏk′sē-môr′ŏn′, -mōr′-) *n., pl.* **-mo·ra** (-môr′ə, -mōr′ə) [Gk. *oxumōron,* neuter of *oxumōros,* pointedly foolish : *oxus,* sharp + *mōros,* foolish.] A figure of speech in which antithetical incongruous terms are combined, as in *a deafening silence* or *a mournful optimist.*
**ox·y·sul·fide** (ŏk′sē-sŭl′fīd′) *n.* A compound consisting of oxygen and sulfur combined with a metal or positive radical in which part of the sulfur has been replaced by oxygen.
**ox·y·tet·ra·cy·cline** (ŏk′sē-tĕt′rə-sī′klĭn, -klēn′) *n.* An antibiotic, $C_{22}H_{24}N_2O_9 \cdot 2H_2O$, derived from the mold *Streptomyces rimosus* and used to treat bacterial infection in people and animals.
**ox·y·to·cic** (ŏk′sĭ-tō′sĭk) *n.* [Gk. *oxus,* sharp + *tokos,* childbirth < *tiktein,* to give birth.] A drug used to hasten the process of childbirth, esp. by inducing uterine contraction. **—ox′y·to′cic** *adj.*
**ox·y·to·cin** (ŏk′sĭ-tō′sĭn) *n.* An oxytocic pituitary hormone.
**ox·y·tone** (ŏk′sĭ-tōn′) *adj.* [Gk. *oxutonos* : *oxus,* sharp + *tonos,* tone.] **1.** Designating a Greek word that has an acute accent on its last syllable. **2.** Designating a word that has a heavy stress accent on its last syllable. **—ox′y·tone′** *n.*
**ox·y·u·ri·a·sis** (ŏk′sē-yŏŏ-rī′ə-sĭs) *n.* [NLat. *Oxyuris,* worm genus (Gk. *oxus,* sharp + Gk. *oura,* tail) + -IASIS.] Infestation with pinworms of the family Oxyuridae.
**oy·er and ter·mi·ner** (oi′ər; tûr′mə-nər) *n.* [ME < Norman Fr. *oyer et terminer,* to hear and determine.] *Law.* **1.** A trial or hearing. **2.** A high criminal court in some U.S. states. **3.** *Chiefly Brit.* **a.** A commission empowering a judge to hear and rule on a criminal case at the assizes. **b.** The court in which this hearing is held.
**o·yez** (ō′yĕs′, ō′yĕz′, ō′yā′) *also* **o·yes** (ō′yĕs′) *interj.* [ME < Norman Fr., hear ye, imper. pl. of *oyer,* to hear < OFr. *oïr* < Lat. *audire.*] —Used three times in succession to open the proceedings in a court of law. —*n., pl.* **o·yes·ses** (ō′yĕs′ĭz). The cry "oyez."
**oys·ter** (oi′stər) *n.* [ME *oistre* < OFr. < Lat. *ostrea* < Gk. *ostreon.*] **1. a.** An edible bivalve mollusk of the genus *Ostrea,* mainly of coastal marine waters, marked by an irregularly shaped shell. **b.** A similar or related bivalve mollusk, as the pearl oyster. **2.** A small pocket of flesh located along the pelvic bone of a fowl and regarded as a delicacy. **3. a.** A special delicacy. **b.** Something regarded as yielding special benefits. **4.** *Slang.* A close-mouthed person. —*vi.* **-tered, -ter·ing, -ters.** To gather, dredge for, or cultivate oysters.
**oyster bed** *n.* A place where oysters breed or are raised.
**oys·ter·catch·er** (oi′stər-kăch′ər) *n.* A shore bird of the genus *Haematopus,* with black and white plumage and a long orange-red bill.

**oystercatcher**
*Approximately 20 inches long*

**oyster crab** *n.* A small crab, *Pinnotheres ostreum,* that lives inside the shells of living oysters.
**oyster cracker** *n.* A small soda cracker often served with soup.
**oys·ter·man** (oi′stər-mən) *n.* **1.** One who sells, harvests, or cultivates oysters. **2.** A vessel equipped for dredging oysters.
**oyster plant** *n.* The salsify.
**oysters Rock·e·fel·ler** (rŏk′ə-fĕl′ər) *pl.n.* [After John D. *Rockefeller* (1839–1937).] A dish consisting of oysters cooked with spinach and a seasoned cream sauce.
**oyster white** *n.* A pale yellowish green to light gray.
**o·zo·ce·rite** (ō′zō-sîr′īt′) *also* **o·zo·ke·rite** (-kîr′-) *n.* [G. *Ozokerit* : Gk. *ozein,* to smell + Gk. *kēros,* wax + G. *-it,* ite.] A yellow-brown to green or black mineral hydrocarbon wax, used in making inks, lubricants, and electrical insulation.
**o·zone** (ō′zōn′) *n.* [G. *Ozon* < Gk. *ozōn,* pr.part. of *ozein,* to smell.] **1.** A blue gaseous allotrope of oxygen, $O_3$, derived from diatomic oxygen by electric discharge or exposure to ultraviolet radiation. **2.** *Informal.* Pure, fresh air. **—o·zo′nic** (ō-zō′nĭk, ō-zŏn′ĭk), **o′zon′·ous** (ō′zō′nəs) *adj.*
**ozone layer** *n.* Ozonosphere.
**o·zo·nide** (ō′zō-nīd′) *n.* Any of various often explosive chemicals formed by attachment of ozone to the double bond of an unsaturated compound and used in analytical chemistry to locate such bonds.
**o·zo·nize** (ō′zō-nīz′) *vt.* **-nized, -niz·ing, -niz·es. 1.** To treat or impregnate with ozone. **2.** To convert (oxygen) to ozone. **—o′zon·iz′er** *n.*
**o·zo·no·sphere** (ō-zō′nə-sfîr′) *n.* A region of the upper atmosphere, between 10 and 20 miles or 16 and 32 kilometers in altitude, containing a relatively high concentration of ozone that absorbs solar ultraviolet radiation in a wavelength range not screened by other atmospheric components. **—o·zo′no·spher′ic** (-sfîr′ĭk, -sfĕr′-), **o·zo′no·spher′i·cal** *adj.*

# Pp

**p** or **P** (pē) *n., pl.* **p's** or **P's. 1.** The 16th letter of the English alphabet. **2.** A speech sound represented by the letter *p.* **3.** The 16th in a series.

**P** *symbol for* PHOSPHORUS.

**pa** (pä) *n.* [Short for PAPA.] *Informal.* Papa : father.

**Pa** *symbol for* PROTACTINIUM.

**pa'an·ga** (päng'gə) *n.* [Tongan.] —See table at CURRENCY.

**PABA** (pä'bə) *n.* Para-aminobenzoic acid.

**Pab·lum** (păb'ləm). A trademark for a bland cereal for infants.

**pab·u·lum** (păb'yə-ləm) *n.* [Lat.] **1.** A substance that gives nourishment : FOOD. **2.** Insipid thought or writing.

**pac** *also* **pack** (păk) *n.* [Delaware *paku.*] **1.** A moccasin or soft shoe worn inside a boot. **2.** A high, laced, waterproof boot.

**pa·ca** (pä'kə, păk'ə) *n.* [Port. (Brazil) and Am. Sp., both < Tupi *páca.*] A nocturnal tropical American rodent of the genus *Cuniculus,* with a short tail and blackish or brown fur.

**paca**
*26 inches long*

**pace[1]** (pās) *n.* [ME < OFr. *pas* < Lat. *passus* < *pandere,* to stretch.] **1.** A step made in walking : STRIDE. **2.** The distance spanned by a step or stride, esp.: **a.** A unit of length equal to 30 inches. **b.** Thirty inches at quick marching time or 36 inches at double time. **c.** Five Roman feet or 58.1 English inches, measured from the point at which the heel of one foot is raised to the point at which it is set down again after an intervening step by the other foot. **d.** The modern version of the Roman pace, measuring five English feet. **3. a.** The rate of speed of walking or running. **b.** The rate of movement or progress. **4.** A manner of walking or running <a springy *pace*> **5.** A gait of a horse in which both feet on one side leave and return to the ground together. —*v.* **paced, pac·ing, pac·es.** —*vt.* **1.** To walk or stride back and forth across. **2.** To measure by counting the number of steps needed to cover a distance. **3.** To set or regulate the rate of speed for. **4.** To train (a horse) in a particular gait, esp. the pace. —*vi.* **1.** To walk with long, deliberate steps. **2.** To go at the pace. —Used of a horse or rider. **—put (one) through (one's) paces.** To test or demonstrate one's abilities or skills. **—set the pace. 1.** To go at a speed that other competitors attempt to equal or surpass. **2.** To behave or perform in a way that others try to emulate.

**pa·ce[2]** (pā'sē) *prep.* [Lat. < ablative of *pax,* peace.] With deference to : with all due respect to. **—pa'ce** *adv.*

**pace car** *n.* A usu. high-performance car that does not participate in a race but leads competing cars through the pace lap.

**pace lap** *n.* The initial lap of an automobile race in which the participants warm up their engines and get set for a fast start.

**pace·mak·er** (pās'mā'kər) *n.* **1.** One that sets the pace in a race. **2.** A leader in a field <the *pacemaker* in computer software> **3. a.** *Physiol.* A mass of specialized muscle fibers of the heart that regulate the heartbeat. **b.** *Med.* A usu. miniaturized, surgically implanted electronic device used to stimulate or stabilize heartbeat. **4.** *Biochem.* A substance that regulates a series of related reactions.

**pac·er** (pā'sər) *n.* **1.** A horse trained to pace. **2.** PACEMAKER 1.

**pace·set·ter** (pās'sĕt'ər) *n.* PACEMAKER 1, 2.

**pa·cha** (pä'shə, păsh'ə, pə-shä') *n. var. of* PASHA.

**pa·chi·si** (pə-chē'zē) *n.* [Hindi *pacīsī* < *pacīs,* twenty-five : Skt. *pañcan,* five + *viṃsatiḥ,* twenty.] **1.** An ancient game of India similar to backgammon but using cowry shells instead of dice. **2.** Parcheesi.

**pach·ou·li** (păch'ə-lē, pə-chōō'lē) *n. var. of* PATCHOULI.

**pach·y·derm** (păk'ĭ-dûrm') *n.* [Fr. *pachyderme* < Gk. *pakhudermos,* thick-skinned : *pakhus,* thick + *derma,* skin.] A large, hoofed, thick-skinned mammal, as the elephant, rhinoceros, or hippopotamus. **—pach'y·der'ma·tous** (-dûr'mə-təs), **pach'y·der'mous** *adj.*

**pach·y·san·dra** (păk'ĭ-săn'drə) *n.* [NLat. *Pachysandra,* genus name : Gk. *pakhus,* thick + NLat. *-andrus,* -androus.] A plant of the genus *Pachysandra,* esp. *P. terminalis,* with evergreen leaves and inconspicuous white flowers.

**pa·cif·ic** (pə-sĭf'ĭk) *also* **pa·cif·i·cal** (-ĭ-kəl) *adj.* [OFr. *pacifique* < Lat. *pacificus* : *pax,* peace + *facere,* to make.] **1.** Tending to diminish or end conflict. **2.** Peaceful : serene. **—pa·cif'i·cal·ly** *adv.*

**pac·i·fi·ca·tion** (păs'ə-fĭ-kā'shən) *n.* [OFr. < Lat. *pacificatio* < *pacificare,* to pacify. —see PACIFY.] **1.** The act of pacifying or the condition of being pacified. **2. a.** Reduction, as of a rebellious district, to peaceful submission <"Real *pacification* is hard to get in the Vietnamese countryside" —McGeorge Bundy> **b.** Policy or practical measures aiming to effect this submission. **3.** *often* **Pacification.** A peace treaty. **—pa·cif'i·ca'tor** (pə-sĭf'ĭ-kā'tər) *n.* **—pa·cif'i·ca·to'ry** (-kə-tôr'ē, -tōr'ē) *adj.*

**Pacific Standard Time** *n.* One of the four standard time zones of North America, operative from longitude 120° to 140°, with wide local variations in boundary.

**pac·i·fi·er** (păs'ə-fī'ər) *n.* **1.** One that pacifies. **2.** A plastic or rubber nipple for a baby to suck on.

**pac·i·fism** (păs'ə-fĭz'əm) *n.* [Fr. *pacifisme* < *pacifique,* pacific < OFr. —see PACIFIC.] **1.** The belief that disputes between nations should and can be settled peacefully. **2. a.** Opposition to war or violence as a means of resolving disputes. **b.** Such opposition demonstrated by refusal to participate in military action. **—pac'i·fist** *n.* **—pac'i·fis'tic** *adj.* **—pac'i·fis'ti·cal·ly** *adv.*

**pac·i·fy** (păs'ə-fī') *vt.* **-fied, -fy·ing, -fies.** [ME *pacifien* < Lat. *pacificare* : *pax,* peace + *facere,* to make.] **1.** To ease the anger or distress of. **2.** To end war or violence in. **—pac'i·fi'a·ble** *adj.*

**Pa·cin·i·an corpuscle** (pə-sĭn'ē-ən) *n.* [After Filippo *Pacini* (1812–1888).] An encapsulated sensory nerve ending in deep layers of the skin that functions as a receptor for heavy pressure and touch.

**pack[1]** (păk) *n.* [ME, of LG orig.] **1. a.** A collection of items tied up or wrapped : BUNDLE. **b.** A container designed to be carried on the back. **2.** An amount, as of food, that is processed and packaged at one time or in one season. **3.** A small package containing a standard number of identical or similar items <a *pack* of gum> **4. a.** A complete set of related items <a *pack* of playing cards> **b.** A large amount : HEAP. **5. a.** A group of animals, as dogs or wolves, that run and hunt together. **b.** A group or large number of people <a *pack* of delinquents> **c.** An organized troop having common interests <a Scout *pack*> **6.** A mass of large pieces of floating ice driven together. **7.** *Med.* **a.** The swathing of a patient in hot, cold, wet, or dry sheets or blankets. **b.** The sheets or blankets so used. **c.** A material, as gauze, therapeutically inserted into a body cavity or wound. **8.** ICE PACK 2. **9.** A cosmetic paste applied to the skin and allowed to dry before removal. —*v.* **packed, pack·ing, packs.** —*vt.* **1.** To fold, roll, or combine into a bundle : WRAP UP. **2. a.** To put into a container for transporting or storing <*pack* one's clothes> **b.** To fill up with items <*pack* one's bag> **3.** To process and put into containers in order to preserve, transport, or sell. **4. a.** To crowd together <*packed* us into the van> **b.** To fill up tight : CRAM <*packed* the elevator> **5.** *Med.* To wrap (a patient) in a pack. **6.** To wrap tightly for protection or to prevent leakage. **7.** To press together : compact firmly. **8.** *Informal.* To have available for action <*pack* a firearm> **9.** To send peremptorily <*packed* them off to jail> **10.** To rig (a voting panel) so as to bring about a certain result. —*vi.* **1.** To place one's belongings in boxes or luggage for transporting or storing. **2.** To be susceptible of compact storage <Paperbacks *pack* well.> **3.** To form lumps or masses : become compacted. **4.** To depart suddenly. **—pack it in.** To call it quits. **—send packing.** To dismiss (someone) abruptly. **—pack'a·bil'i·ty** *n.* **—pack'a·ble** *adj.*

**pack[2]** (păk) *n. var. of* PAC.

**pack·age** (păk'ĭj) *n.* **1.** A wrapped or boxed object : PARCEL. **2.** A container in which something is packed. **3. a.** A preassembled unit. **b.** A commodity, as food, uniformly processed and placed in containers. **4.** A proposition or offer consisting of several items each of which must be accepted <a vacation *package*> —*vt.* **-aged, -ag·ing, -ag·es.** To place in a package or make a package of.

**package store** *n.* A store that sells sealed bottles of alcoholic beverages for consumption away from its premises.
**pack·ag·ing** (păk′ə-jĭng) *n.* The act, process, industry, art, or style of packing.
**pack animal** *n.* An animal, as a mule, used to carry loads.
**packed** (păkt) *adj.* **1.** Crowded to capacity <a *packed* concert hall> **2.** Being compressed, as snow or mud. **3.** *Informal.* Filled with <action-*packed* adventure>
**pack·er** (păk′ər) *n.* One that packs, esp. one whose occupation is the processing and packing of wholesale goods, usu. meat products.
**pack·et** (păk′ĭt) *n.* [OFr. *pacquet,* of Germanic orig.] **1.** A small package or bundle. **2.** *Slang.* A sizable sum of money. **3.** A boat, usu. a coastal or river steamer, that plies a regular route carrying passengers, freight, and mail.
**pack ice** *n.* Floating ice driven together into a single mass.
**pack·ing** (păk′ĭng) *n.* **1.** The processing and packaging of goods, esp. food. **2.** A material used to prevent leakage or seepage, as around a pipe joint. **3.** Application of a medical pack.
**packing fraction** *n. Physics.* The quotient of the algebraic difference between the isotopic mass and the mass number of a nuclide, divided by its mass number, that is often interpreted as an indicator of stability.
**packing house** *n.* **1.** A firm that slaughters, processes, and packs livestock into meat and meat products. **2.** A firm that processes and packs other food products.
**pack·man** (păk′măn′, -mən) *n.* A peddler.
†**pack rat** *n.* **1.** A small North American rodent of the genus *Neotoma* that collects in its nests a wide variety of small objects. **2.** *Western U.S. Slang.* A petty thief. **3.** *Slang.* A collector or hoarder of miscellaneous objects.
**pack·sack** (păk′săk′) *n.* A canvas or leather traveling sack strapped to the shoulders.
**pack·sad·dle** (păk′săd′l) *n.* A saddle for a pack animal on which loads can be secured.
**pack·thread** (păk′thrĕd′) *n.* A strong two or three-ply twine for sewing or tying packages or bundles.
**pack train** *n.* A line of animals, as horses or mules, loaded with supplies for an expedition.
**pact** (păkt) *n.* [ME < OFr. < Lat. *pactum* < *pascisci,* to agree.] **1.** A formal agreement, as between nations : TREATY. **2.** BARGAIN 1a.
**pad¹** (păd) *n.* [Orig. unknown.] **1.** A thin, cushionlike mass of soft material used as filling or for protection against injury, as jarring or scraping. **2.** A flexible saddle without a frame. **3.** An ink-soaked cushion for inking a rubber stamp. **4.** A number of sheets of paper of the same size stacked one on top of the other and glued together at one end : TABLET. **5.** The broad, floating leaf of an aquatic plant such as the water lily. **6. a.** The cushionlike flesh on the underpart of the toes and feet of many animals. **b.** The foot of such an animal. **7.** The fleshy underside of the end of a finger or toe. **8.** A launch pad. **9.** *Slang.* One's lodgings. —*vt.* **pad·ded, pad·ding, pads. 1.** To line or stuff with soft material. **2. a.** To lengthen with extraneous material <*pad* a speech> **b.** To add to, esp. fraudulently <*pad* expenses> **—on the pad.** *Slang.* Taking bribes.
**pad²** (păd) *v.* **pad·ded, pad·ding, pads.** [Prob. < MDu. *paden,* to walk along a path < *pad,* path.] —*vi.* **1.** To go about on foot. **2.** To move or walk about quietly. —*vt.* To go along (a route) on foot. —*n.* **1.** A muffled sound like that of soft footsteps. **2.** A horse with a plodding gait. **—pad′der** *n.*
**pa·dauk** (pə-dôk′) *also* **pa·douk** (-do͝ok′) *n.* [Native word in Burma.] **1.** A tropical tree of the genus *Pterocarpus,* yielding reddish wood with a mottled or striped grain. **2.** The wood of the padauk tree, used for decorative cabinetwork.
**pad·ding** (păd′ĭng) *n.* **1.** The act of stuffing, filling, or lining something. **2.** A soft material used to make a pad. **3.** Matter added to a speech or written work to lengthen it.
**pad·dle¹** (păd′l) *n.* [ME *padell,* implement used for cleaning a plowshare.] **1.** A wooden implement with a blade at one end or sometimes at both ends, used without an oarlock to propel a canoe or small boat. **2.** An implement resembling a paddle, as: **a.** An iron tool for stirring molten ore in a furnace. **b.** A tool with a shovellike blade used to mix materials in glassmaking. **c.** A pallet with which to mix and shape clay. **d.** A narrow board used to beat clothes in hand laundering. **e.** A flattened board used to administer physical punishment. **f.** A light wooden racket used in playing table tennis. **3.** A board of a paddle wheel. **4.** An animal's flipper or flattened appendage. **5.** The act of paddling. —*v.* **-dled, -dling, -dles.** —*vi.* **1.** To propel a watercraft with a paddle. **2.** To row slowly and gently. **3.** To move through water by means of repeated short strokes of the limbs. —*vt.* **1.** To propel (a watercraft) with a paddle. **2.** To convey in a watercraft propelled by paddles. **3.** To beat with a paddle, esp. to punish by spanking with a paddle. **4.** To stir or shape with a paddle <*paddle* clay> **—pad′dler** *n.*
**pad·dle²** (păd′l) *vi.* **-dled, -dling, -dles.** [Orig. unknown.] **1.** To dabble about in shallow water. **2.** To waddle : toddle.
**pad·dle·ball** (păd′l-bôl′) *n.* **1.** A game for two to four players, using a wooden or plastic paddle and a ball similar to a tennis ball on a one-, three-, or four-walled court. **2.** The ball used in paddleball.
**pad·dle·board** (păd′l-bôrd′) *n.* A long narrow floatable board used in riding the surf or rescuing swimmers.
**paddle boat** *n.* A steamship propelled through the water by paddle wheels on each side or by one paddle wheel astern.
**pad·dle·fish** (păd′l-fĭsh′) *n., pl.* **paddlefish** or **-fish·es.** A large fish of the Mississippi River basin, *Polyodon spathula,* with a long paddle-shaped snout.
**paddle wheel** *n.* A steam-driven wheel with boards or paddles affixed around its circumference, used to propel a ship.
**pad·dling** (păd′lĭng, păd′l-ĭng) *n.* **1.** The act of moving a boat by a paddle. **2.** A spanking with a paddle.
**pad·dock** (păd′ək) *n.* [Alteration of ME *parrok* < OE *pearroc.*] **1.** A fenced area, usu. near a stable, used for grazing or exercising horses. **2.** An enclosure at a racetrack where the horses are assembled, saddled, and paraded before each race. **3.** An area at an automobile racetrack where cars are prepared before a race. **4.** *Austral.* Fenced-in land. —*vt.* **-docked, -dock·ing, -docks.** To confine in a paddock.
**pad·dy** (păd′ē) *n., pl.* **-dies.** [Malay *padi.*] **1.** Rice, esp. in the husk, whether gathered or growing. **2.** An irrigated or flooded field where rice is grown.
**paddy wagon** *n.* [< *Paddy,* offensive slang for an Irishman < the name *Patrick.*] *Slang.* A police van for suspects.
**Pa·di·shah** (pä′dĭ-shä′) *n.* [Pers. *pādshāh* : OPers. *pati,* master + *shāh,* king.] **1.** A title of the former monarch of Iran. **2.** A title of the former sultan of Turkey.
**pad·lock** (păd′lŏk′) *n.* [ME *padlok* : *pad-* (of unknown orig.) + *lok,* lock.] A detachable lock with a U-shaped bar hinged at one end, designed to be passed through the staple of a hasp or a link in a chain and then closed. —*vt.* **-locked, -lock·ing, -locks.** To lock up with or as if with a padlock.
**pa·douk** (pə-do͝ok′) *n. var. of* PADAUK.
**pa·dre** (pä′drā, -drē) *n.* [Sp., Ital., or Port., all < Lat. *pater,* father.] **1.** Father. —Used as a title of address for a priest in Italy, Spain, Portugal, and Latin America. **2.** *Informal.* A military chaplain. **3.** *Chiefly Brit.* A parson.
**pa·dro·ne** (pə-drō′nē, -nā) *n., pl.* **-nes** (-nēz, -nāz) or **-ni** (-nē) [Ital. < Lat. *patronus,* patron. —see PATRON.] **1.** A proprietor, esp. of an inn. **2.** One who exploitatively employs or finds work for Italian immigrants in America. **—pa·dro′nism** (pə-drō′nĭz′əm) *n.*
**pad·u·a·soy** (păj′o͞o-ə-soi′) *n.* [Alteration of Fr. *pou-de-soie.*] **1.** A rich, heavy, corded silk fabric. **2.** A hanging or garment made of paduasoy.
**pae·an** *also* **pe·an** (pē′ən) *n.* [Lat. *paean,* hymn of thanksgiving, often addresed to Apollo < Gk. *paian, paiōn* < *Paiōn,* a title of Apollo.] **1.** A song of joyful exultation or praise. **2.** A fervent expression of joy or praise. **3.** An ancient Greek hymn of thanksgiving to a god, esp. to Apollo.
**paedo-** or **paed-** *pref. vars. of* PEDO-².
**pae·do·gen·e·sis** (pē′dō-jĕn′ĭ-sĭs) *n.* Reproduction of young during the larval or preadult stage, occurring chiefly in insects. **—pae′do·ge·net′ic** (-jə-nĕt′ĭk) *adj.*
**pae·do·mor·phism** (pē′də-môr′fĭz′əm) *n.* Retention of juvenile characteristics in the adult. **—pae′do·mor′phic** (-fĭk) *adj.*
**pae·do·mor·pho·sis** (pē′də-môr′fə-sĭs) *n.* Phylogenetic change in which juvenile characteristics are retained by the adults.
**pa·el·la** (pä-ĕl′ə, pä-ā′lyä, -ā′yä) *n.* [Catalan < OFr. *paelle,* pan < Lat. *patella,* dim. of *patina,* pan.] A saffron-flavored Spanish dish made with rice, vegetables, meat, chicken, and seafood.
**pae·on** (pē′ən, -ŏn′) *n.* [Lat. < Gk. *paiōn.* —see PAEAN.] A metrical foot of poetry having one long syllable and three short syllables occurring in random order.
**pa·gan** (pā′gən) *n.* [ME < LLat. *paganus* < Lat., country-dweller < *pagus,* country.] **1.** One who is not a Christian, Moslem, or Jew : HEATHEN. **2.** One who has no religion. **3.** A non-Christian. **4.** A hedonist. —*adj.* **1.** Not Christian, Moslem, or Jewish. **2.** Not religious : HEATHEN. **—pa′gan·dom** *n.* **—pa′gan·ish** *adj.* **—pa′gan·ism** *n.*
**pa·gan·ize** (pā′gə-nīz′) *vt.* & *vi.* **-ized, -iz·ing, -iz·es.** To make or become pagan. **—pa′gan·i·za′tion** *n.*
**page¹** (pāj) *n.* [ME < OFr. < Ital. *paggio.*] **1.** A boy attending a knight as the initial stage of training for knighthood. **2.** A youth in ceremonial attendance or employment at court. **3. a.** A person employed to run errands, carry messages, or act as a guide, as in a hotel, theater, or club. **b.** A person similarly employed in a legislature. **4.** A boy who carries the bride's train at a wedding. —*vt.* **paged, pag·ing, pag·es. 1.** To summon or call (a person) by name. **2.** To attend as a page.
**page²** (pāj) *n.* [OFr. < Lat. *pagina.*] **1.** One side of a leaf, as of a book, letter, newspaper, or manuscript, esp. the entire leaf. **2.** The writing or printing on one side of a page. **3.** The type set for printing a page. **4.** A memorable or noteworthy event <a new *page* in my life> **5.** *Computer Sci.* **a.** A quantity of computer memory storage equal to between 512 and 4,096 bytes. **b.** A quantity of source program coding equal to between 8 and 64 lines. **6. pages.** A source or record of knowledge <in the *pages* of medicine> —*v.* **paged, pag·ing, pag·**

**es.** —*vt.* To number the pages of : PAGINATE. —*vi.* To turn pages.
**pag·eant** (păj′ənt) *n.* [ME *padgin,* scene of a play < Med. Lat. *pagina* < Lat., page.] **1.** An elaborate public dramatic presentation, usu. depicting some historical or legendary event. **2.** A spectacular procession or celebration. **3.** Colorful display. **4.** Showy display : POMP.
**pag·eant·ry** (păj′ən-trē) *n., pl.* **-ries. 1.** Pageants and their presentation. **2.** Grand display : pomp. **3.** Showy but empty display.
**page·boy** (pāj′boi′) *n.* [< PAGE[1].] A woman's usu. shoulder-length hair style in which the hair is turned under in a smooth roll.
**pag·er** (pā′jər) *n.* One that pages, esp. a beeper.
**pag·i·nal** (păj′ə-nəl) *adj.* [LLat. *paginalis* < Lat. *pagina,* page.] **1.** Of, pertaining to, or made up of pages. **2.** Page for page.
**pag·i·nate** (păj′ə-nāt′) *vt.* **-nat·ed, -nat·ing, -nates.** [< Lat. *pagina,* page.] To number the pages of. **—pag′i·na′tion** (-nā′shən) *n.*
**pag·ing** (pā′jĭng) *n. Computer Sci.* The transfer of pages of data from a computer's main memory to an auxiliary memory.
**pa·go·da** (pə-gō′də) *n.* [Port. *pagode,* poss. of Dravidian orig.] **1.** A religious building of the Far East, esp. a many-storied Buddhist tower erected as a memorial or shrine. **2.** A pagodalike structure, as a garden pavilion.
**pah** (pä) *interj.* —Used to express irritation or disgust.
**pah·la·vi** (pä′lə-vē′) *n., pl.* **-vis.** [Pers. *pahlawī,* after Mohammed Reza *Pahlavi* (1919–1980), Shah of Iran.] A gold coin once used in Iran.
**Pah·la·vi** (pä′lə-vē′) *also* **Peh·le·vi** (pā′lə-vē′) *n.* [Pers. *pahlawī* < *Pahlav,* Parthia.] An Iranian language used in Persia during the reign of the Sassanids.
**paid[1]** (pād) *v. p.t. & p.p. of* PAY[1].
**paid[2]** (pād) *v. var. p.t. & p.p. of* PAY[2].
**pail** (pāl) *n.* [ME *payle* < OE *pægel.*] **1.** A watertight cylindrical vessel, open at the top and fitted with a handle : BUCKET. **2.** The amount contained in a pail. **—pail′ful′** *adj.*
**pail·lasse** *also* **pal·liasse** (păl-yăs′, păl′yăs′) *n.* [Fr. < *paille,* straw < Lat. *palea.*] A thin mattress filled with sawdust or straw.
**pail·lette** (pä-yĕt′, pā-, pă-lĕt′) *n.* [Fr., dim. of *paille,* straw. —see PAILLASSE.] **1.** A small piece of metal or foil, used in enamel painting. **2.** A spangle used to decorate a dress or costume.
**pain** (pān) *n.* [ME < OFr. *peine* < Lat. *poena* < Gk. *poinē,* penalty.] **1.** An unpleasant sensation, occurring in varying degrees of severity as a consequence of injury, disease, or emotional disorder. **2.** Suffering or distress. **3. pains.** The pangs of childbirth. **4. pains.** Great care or effort <took *pains* to do it right> **5.** *Informal.* A nuisance. —*v.* **pained, pain·ing, pains.** —*vt.* To cause pain to : HURT. —*vi.* To be the cause of pain. **—on (**or **upon** or **under) pain of.** Subject to the penalty of a given punishment, as death. **—pain in the neck.** A source of annoyance.
**pain·ful** (pān′fəl) *adj.* **1.** Causing pain. **2.** Full of pain. **3.** Requiring care and labor : IRKSOME <a *painful* chore> **4.** *Archaic.* Diligent : careful. **—pain′ful·ly** *adv.* **—pain′ful·ness** *n.*

☆ ***syns:*** PAINFUL, ACHING, AFFLICTIVE, HURTFUL, SMARTING, SORE *adj. core meaning* : marked by, causing, or experiencing physical pain <a *painful* back injury> ***ant:*** painless

**pain·kill·er** (pān′kĭl′ər) *n.* An agent, as a drug, that relieves pain. **—pain′kill′ing** *adj.*
**pain·less** (pān′lĭs) *adj.* Free from pain or complication <a *painless* surgical procedure> **—pain′less·ly** *adv.* **—pain′less·ness** *n.*
**pain principle** *n. Psychoanal.* The unconscious desire for pain or destruction.
**pains·tak·ing** (pānz′tā′kĭng) *adj.* Diligent and attentive to detail : CAREFUL. —*n.* Diligence and attentiveness. **—pains′tak′ing·ly** *adv.*
**paint** (pānt) *n.* [< ME *painten,* to paint < OFr. *peindre* < Lat. *pingere.*] **1. a.** A liquid mixture, usu. of a solid pigment in a liquid vehicle, used as a protective or decorative coating. **b.** The thin dry film formed by such a mixture applied to a surface. **c.** The solid pigment before it is mixed with a vehicle. **2.** A cosmetic, esp. one that colors, as rouge. **3.** A pinto. —*v.* **paint·ed, paint·ing, paints.** —*vt.* **1.** To make with paints <*paint* a picture> **2. a.** To represent in a picture with paints. **b.** To depict vividly in words. **3.** To coat or decorate with paint <*paint* a room> **4.** To apply cosmetics to. **5.** To apply medicine to : SWAB <*paint* a cut> —*vi.* **1.** To practice the art of painting pictures. **2.** To cover something with paint. **3.** To apply make-up. **4.** To serve as a surface to be coated with paint <a surface that *paints* well> **—paint the town red.** *Slang.* To go on a spree.
**paint·brush** (pānt′brŭsh′) *n.* A brush for applying paint.
**paint·ed** (pān′tĭd) *adj.* **1.** Represented in paint. **2. a.** Covered or decorated with paint. **b.** Brightly colored : GAUDY. **3.** Excessively made up with cosmetics. **4.** Having no reality : ARTIFICIAL.
**painted bunting** *n.* A small bird of the southern United States, *Passerina ciris,* with brilliant multicolored plumage.
**painted cup** *n.* The Indian paintbrush.
**painted lady** *n.* A widely distributed butterfly, *Vanessa cardui,* with brown, black, and orange markings.

**paint·er[1]** (pān′tər) *n.* One who paints, either as an artist or a worker.
**pain·ter[2]** (pān′tər) *n.* [ME *paynter,* poss. < OFr. *pentoir,* strong rope < *pendre,* to hand < Lat. *pendēre.*] *Naut.* A rope attached to the bow of a boat, used for tying up.
**†pain·ter[3]** (pān′tər) *n.* [Alteration of PANTHER.] *Regional.* A mountain lion or lynx.
**paint·er·ly** (pān′tər-lē) *adj.* **1.** Of, relating to, or typical of a painter : ARTISTIC. **2. a.** Having qualities unique to the art of painting. **b.** Designating a style of painting characterized by openness of form, with shapes distinguished by variations of color rather than by outline or contour. **—paint′er·li·ness** *n.*
**painter's colic** *n.* [So called because the disease is often caused by exposure to lead-base paint.] Chronic intestinal pains and constipation resulting usu. from lead poisoning.
**paint·ing** (pān′tĭng) *n.* **1.** The art, process, or occupation of coating surfaces with paint. **2.** A picture or design in paint.
**pair** (pâr) *n., pl.* **pair** *also* **pairs.** [ME < OFr. *paire* < Lat. *paria* < *par,* equal.] **1.** Two corresponding persons or items, similar in form or function and matched or associated. **2.** One object consisting of two joined, similar parts that are dependent on each other <a *pair* of scissors> **3. a.** Two persons joined together in marriage or engagement. **b.** Two persons having something in common and considered together <a *pair* of golfers> **c.** Two mated animals. **d.** Two animals joined together in work. **4.** Two playing cards of the same denomination. **5.** Two members of a deliberative body with opposing opinions on a given issue who offset each other by agreeing to abstain from voting on the issue. **6.** *Chem.* An electron pair. —*v.* **paired, pair·ing, pairs.** —*vt.* **1.** To arrange in sets of two : COUPLE. **2.** To join in a pair : MATE. **3.** To provide a partner for. —*vi.* **1.** To form a pair or pairs. **2.** To join as mates in marriage.
**pair of compasses** *n.* COMPASS 1c.
**pair of virginals** *n.* A virginal.
**pair production** *n.* The simultaneous creation of a positron and electron from a high-energy gamma ray in a very strong electric field, esp. in that of an atomic nucleus.
**pai·sa** (pī-sä′) *n., pl.* **pai·se** (pī-sā′) *or* **pai·sas.** [Hindi *paisā.*] —See table at CURRENCY.
**pai·sa·no** (pī-zä′nō) *also* **pai·san** (pī-zän′) *n., pl.* **-sa·nos** *also* **-sans.** [Sp. < Fr. *paysan* < OFr. *païsant,* peasant. —see PEASANT.] **1.** Countryman : compatriot. **2.** *Slang.* Friend : pal.
**pais·ley** (pāz′lē) *adj.* [After *Paisley,* Scotland.] **1.** Made of a soft wool fabric with a woven or printed colorful, swirled pattern of abstract, curved shapes, derived from the palmette motif of Persian rugs. **2.** Marked with a paisley pattern. —*n., pl.* **-leys.** An article of clothing, as a shawl, made of paisley.
**Pai·ute** *also* **Pi·ute** (pī′yo͞ot′) *n., pl.* **Paiute** *or* **-utes** *also* **Piute** *or* **-utes. 1. a.** Either of two distinct Indian peoples, the Northern Paiute and the Southern Paiute, belonging to the Shoshonean subfamily of the Uto-Aztecan language family, who once lived in the southwestern United States. **b.** A member of either of these peoples. **2.** The Shoshonean language of the Paiute. **—Pai′ute′** *adj.*
**pa·ja·mas** (pə-jä′məz, -jăm′əz) *pl.n.* [Hindi *pāejāma,* loose-fitting trousers : Pers. *pāī,* leg + Pers. *jāmah,* garment.] **1.** A loose-fitting garment having of trousers and a jacket, worn for sleeping or lounging. **2.** Loose-fitting trousers worn by both sexes in the Orient.
**pak choi** (bäk′ choi′) *n.* [Chin. *bai[2] cai[4]* : *bai[2],* white + *cai[4],* vegetable.] A Chinese plant, *Brassica chinensis,* similar to the common cabbage and used as a vegetable.
**pal** (păl) [Romany *phal, phral* < Skt. *bhrātṛ,* brother.] *Informal.* —*n.* A friend : chum. —*vi.* **palled, pal·ling, pals.** To associate as pals <*palled* around together>
**pal·ace** (păl′ĭs) *n.* [ME < OFr. *palais* < Lat. *palatium* < *Palatium,* Palatine Hill, Rome, Italy (from its being the site where emperors built their homes).] **1.** The official residence of a royal person. **2.** *Chiefly Brit.* The official residence of a high-ranking dignitary, as a bishop or archbishop. **3. a.** A large or splendid residence. **b.** A large, often ornate building used for entertainment or exhibitions.
**pal·a·din** (păl′ə-dĭn) *n.* [Fr. < Ital. *paladino* < Lat. *palatinus,* palatine.] **1.** One of the 12 peers of Charlemagne's court. **2.** A heroic paragon of chivalry.
**palaeo-** *or* **palae-** *pref. vars. of* PALEO-.
**pa·laes·tra** (pə-lĕs′trə) *n. var. of* PALESTRA.
**pal·an·quin** *also* **pal·an·keen** (păl′ən-kēn′) *n.* [Port. *palanquin* < Javanese *pĕlangki* < Skt. *palyankaḷ, paryankaḷ,* bed : *pari,* around + *añcati,* he bends.] An eastern Asian covered litter, carried on poles on the shoulders of two or four men.
**pal·at·a·ble** (păl′ə-tə-bəl) *adj.* [< PALATE.] **1.** Sufficiently agreeable in flavor to be eaten. **2.** Acceptable to the mind or sensibilities <a *palatable* resolution of the dispute> **—pal′at·a·bil′i·ty, pal′at·a·ble·ness** *n.* **—pal′at·a·bly** *adv.*
**pal·a·tal** (păl′ə-təl) *adj.* **1.** Of or pertaining to the palate. **2. a.** Produced with the front of the tongue near or against the hard palate, as the (y) in English *youth.* **b.** Produced with the blade of the tongue near the hard palate, as the (ch) in English *chair.* **c.** Produced with the front of the tongue in a forward position. —Used of a vowel. —*n.* A palatal sound. **—pal′a·tal·ly** *adv.*

**pal·a·tal·ize** (păl′ə-tə-līz′) *vt.* **-ized, -iz·ing, -iz·es.** To pronounce as or alter to a palatal sound. **—pal′a·tal·i·za′tion** *n.*

**pal·ate** (păl′ĭt) *n.* [ME < Lat. *palatum.*] **1.** The roof of the mouth in vertebrates having a complete or partial separation of the mouth cavity and nasal passage, consisting of a bony front, the hard palate, backed by the fleshy soft palate. **2.** A part like a palate, as in a lipped flower. **3.** The sense of taste <a delicate *palate*>

**pa·la·tial** (pə-lā′shəl) *adj.* [< Lat. *palatium,* palace.] **1.** Of or suitable for a palace. **2.** Spacious and ornate like a palace. **—pa·la′tial·ly** *adv.* **—pa·la′tial·ness** *n.*

**pa·lat·i·nate** (pə-lăt′n-āt′, -ĭt) *n.* [Med. Lat. *palatinatus* < Lat. *palatinus,* a palatine < *palatinus,* palatine.] The office, powers, or territory of a palatine.

**pal·a·tine[1]** (păl′ə-tīn′) *n.* [Lat. *palatinus* < *palatinus,* of a palace < *palatium,* palace. —see PALACE.] **1. a.** A soldier of the palace guard of the Roman emperors formed in the time of Diocletian. **b.** A soldier of a major division of the Roman army formed in the time of Constantine. **2.** A title of various administrative officials of the late Roman and Byzantine empires. **3.** A feudal lord exercising sovereign power over his lands. —*adj.* **1.** Belonging to or suitable for a palace. **2.** Relating to or designating a palatine or palatinate.

**pal·a·tine[2]** (păl′ə-tēn′) *n.* [Fr., after Princess *Palatine,* Charlotte Elizabeth of Bavaria (1652–1722).] A woman's fur cape and hood.

**pal·a·tine[3]** (păl′ə-tīn′) *adj.* Of or relating to the palate. —*n.* Either of the two bones that make up the hard palate.

**pa·lav·er** (pə-lăv′ər, -lä′vər) *n.* [Port. *palavra,* speech < LLat. *parabola.* —see PARABLE.] **1. a.** Idle chatter. **b.** Beguiling talk. **2.** *Obs.* A parley between European explorers and representatives of local populations, esp. in Africa. —*v.* **-ered, -er·ing, -ers.** —*vt.* To flatter or cajole. —*vi.* To chatter idly.

**pale[1]** (pāl) *n.* [ME < OFr. *pal* < Lat. *palus.*] **1.** A stake or pointed stick : PICKET. **2.** A fence enclosing an area. **3.** The area enclosed by a fence or boundary. **4.** *Heraldry.* A wide vertical band in the center of an escutcheon. **5. the Pale.** The medieval dominions of the English in Ireland. —*vt.* **paled, pal·ing, pales.** To enclose with pales : fence in. **—beyond the pale.** Irrevocably unreasonable or unacceptable.

**pale[2]** (pāl) *adj.* **pal·er, pal·est.** [ME < OFr. < Lat. *pallidus* < *pallēre,* to be pale.] **1.** Whitish : pallid <a *pale* complexion> **2.** Of a low intensity of color : LIGHT <a *pale* shade> **3.** Of a color having high lightness and low saturation <*pale* blue> **4.** Of a low intensity of light : DIM <a *pale* moon> **5.** Feeble : weak <a *pale* substitute> —*v.* **paled, pal·ing, pales.** —*vt.* To cause to turn pale. —*vi.* **1.** To become pale : BLANCH. **2.** To decrease in relative importance. **—pale′ly** *adv.* **—pale′ness** *n.*

**pale-** *pref. var. of* PALEO-.

**pa·le·a** (pā′lē-ə) *n., pl.* **-le·ae** (-lē-ē′) [NLat. < Lat., chaff.] A small chafflike bract enclosing the flower of a grass spikelet.

**Pa·le·arc·tic** (pā′lē-ärk′tĭk, -är′tĭk) *adj.* Of or designating the zoogeographic region that includes Europe, the northwestern coast of Africa, and Asia north of the Himalayas.

**pale-dry** (pāl′drī′) *adj.* Light-colored and dry in flavor <*pale-dry* ginger ale>

**pa·le·eth·nol·o·gy** *also* **pa·le·ëth·nol·o·gy** (pā′lē-ĕth-nŏl′ə-jē) *n.* The ethnology of early human beings. **—pa′le·eth′no·log′ic** (-ĕth′nə-lŏj′ĭk), **pa′le·eth′no·log′i·cal** *adj.*

**paleo-** *or* **pale-** *or* **palaeo-** *or* **palae-** *pref.* [Gk. *palaio-* < *palaios,* ancient < *palai,* long ago.] **1.** Ancient : prehistoric <*paleo*botany> **2.** Early : primitive <*paleo*anthropology>

**pa·le·o·an·throp·ic** (pā′lē-ō-ăn-thrŏp′ĭk) *adj.* Of or relating to extinct members of the genus *Homo* that preceded *H. sapiens.*

**pa·le·o·an·thro·pol·o·gy** (pā′lē-ō-ăn′thrə-pŏl′ə-jē) *n.* The study of humanlike creatures more primitive than *Homo sapiens.* **—pa′le·o·an′thro·po·log′ic** (-pə-lŏj′ĭk), **pa′le·o·an′thro·po·log′i·cal** *adj.* **—pa′le·o·an′thro·pol′o·gist** *n.*

**pa·le·o·bot·a·ny** (pā′lē-ō-bŏt′n-ē) *n.* Study of plant fossils and ancient vegetation. **—pa′le·o·bo·tan′ic** (-bə-tăn′ĭk), **pa′le·o·bo·tan′i·cal** *adj.* **—pa′le·o·bo·tan′i·cal·ly** *adv.* **—pa′le·o·bot′a·nist** *n.*

**Pa·le·o·cene** (pā′lē-ə-sēn′) *adj.* Of, belonging to, or designating the geologic time, rock series, and sedimentary deposits of the first epoch of the Tertiary period, preceding the Eocene and marked by the appearance of placental mammals. —*n.* The Paleocene epoch.

**pa·le·o·e·col·o·gy** (pā′lē-ō-ĭ-kŏl′ə-jē) *n.* A branch of ecology that deals with the interaction between ancient organisms and their environment. **—pa′le·o·ec′o·log′i·cal** (-ĕk′ə-lŏj′ĭ-kəl, -ē′kə-), **pa′le·o·ec′o·log′ic** *adj.* **—pa′le·o·e·col′o·gist** *n.*

**pa·le·og·ra·phy** (pā′lē-ŏg′rə-fē) *n.* **1.** The study and scholarly interpretation of ancient written documents. **2.** The documents studied in paleography. **—pa′le·og′ra·pher** *n.* **—pa′le·o·graph′ic** (-ə-grăf′ĭk), **pa′le·o·graph′i·cal** *adj.*

**pa·le·o·lith** (pā′lē-ə-lĭth′) *n.* A Paleolithic stone implement.

**Pa·le·o·lith·ic** (pā′lē-ə-lĭth′ĭk) *adj.* Of, belonging to, or designating the cultural period beginning with the earliest chipped stone tools, about 750,000 years ago, until the beginning of the Mesolithic, about 15,000 years ago. —*n.* The Paleolithic Age.

**pa·le·on·tol·o·gy** (pā′lē-ŏn-tŏl′ə-jē) *n.* **1.** The study of fossils. **2.** Paleozoology. **—pa′le·on′to·log′ic** (-ŏn′tə-lŏj′ĭk), **pa′le·on′to·log′i·cal** *adj.* **—pa′le·on·tol′o·gist** *n.*

**Pa·le·o·zo·ic** (pā′lē-ə-zō′ĭk) *adj.* Of, belonging to, or designating the era of geologic time that includes the Cambrian, Ordovician, Silurian, Devonian, Mississippian, Pennsylvanian, and Permian periods and is marked by the appearance of marine invertebrates, primitive fishes, land plants, and primitive reptiles. —*n.* The Paleozoic era.

**pa·le·o·zo·ol·o·gy** (pā′lē-ō-zō-ŏl′ə-jē) *n.* The study of animal fossils and ancient animal life. **—pa′le·o·zo′o·log′i·cal** (-zō′ə-lŏj′ĭ-kəl) *adj.* **—pa′le·o·zo·ol′o·gist** *n.*

**pa·les·tra** *also* **pa·laes·tra** (pə-lĕs′trə) *n., pl.* **-trae** (-trē) *or* **-tras.** [Lat. *palaestra* < Gk. *palaistra* < *palaiein,* to wrestle.] A public place in ancient Greece for training and practice in athletics, esp. wrestling. **—pa·les′tral, pa·les′tri·an** *adj.*

**pal·ette** (păl′ĭt) *n.* [Fr. < OFr., small potter's shovel, dim. of *pale,* shovel < Lat. *pala.*] **1.** A small board, usu. with a hole for the thumb, upon which an artist mixes colors. **2. a.** The range of colors used in a given painting or by a given artist <a brilliant *palette*> **b.** The range of qualities inherent in other art forms, as music.

**palette knife** *n.* An artist's knife with a thin flexible blade, used for mixing, scraping, or applying paint.

**pal·frey** (pôl′frē) *n., pl.* **-freys.** [ME < OFr. *palefrei* < Med. Lat. *palefredus* < LLat. *paraveredus,* extra post horse : Gk. *para,* extra + Lat. *veredus,* post horse.] *Archaic.* A woman's saddle horse.

**Pa·li** (pä′lē) *n.* [Skt. *pāliḥ,* row, of Dravidian orig.] An ancient Indic language that is a scriptural and liturgical language of Hinayana Buddhism.

**pal·i·mo·ny** (păl′ə-mō′nē) *n.* [Blend of PAL and ALIMONY.] *Informal.* An allowance for support made under court order and given usu. by a man to his former mistress or live-in companion after they have separated.

**pal·imp·sest** (păl′ĭmp-sĕst′) *n.* [Lat. *palimpsestus* < Gk. *palimpsēstos,* scraped again : *palin,* again + *psēn,* to scrape.] A written document, usu. on vellum or parchment, that has been written upon several times, often with remnants of erased writing still visible.

**pal·in·drome** (păl′ĭn-drōm′) *n.* [Gk. *palindromos,* running back again : *palin,* again + *dromos,* a running.] A word, phrase, verse, or sentence that reads the same backward or forward; e.g., *A man, a plan, a canal, Panama!* **—pal′in·dro′mic** (-drō′mĭk, -drŏm′ĭk) *adj.*

**pal·ing** (pā′lĭng) *n.* **1.** One of a row of upright, pointed sticks forming a fence. **2.** Pales used in making fences. **3.** A fence made of pales or pickets.

**pal·in·gen·e·sis** (păl′ĭn-jĕn′ĭ-sĭs) *n., pl.* **-ses** (-sēz′) [Gk. *palin,* again + -GENESIS.] **1.** The doctrine of transmigration of souls : METEMPSYCHOSIS. **2.** *Biol.* The repetition by a single organism of various stages in the evolution of its species during embryonic development. **—pal′in·ge·net′ic** (-jə-nĕt′ĭk) *adj.* **—pal′in·ge·net′i·cal·ly** *adv.*

**pal·i·node** (păl′ə-nōd′) *n.* [LLat. *palinodia* < Gk. *palinōidia* : *palin,* again + *ōidē,* song.] **1.** A poem in which the poet retracts something said in a previous poem. **2.** A formal statement of retraction.

**pal·i·sade** (păl′ĭ-sād′) *n.* [Fr. *palissade* < OProv. *palissada* < *palissa,* stake < Lat. *palus.*] **1. a.** A fence of pales forming a defense barrier or fortification. **b.** One of the pales of such a fence. **2. palisades.** A line of steep, lofty cliffs, usu. along a river. —*vt.* **-sad·ed, -sad·ing, -sades.** To fortify or surround with a palisade.

**palisade cell** *n. Bot.* One of the cells of palisade parenchyma.

**palisade parenchyma** *n. Bot.* A leaf tissue made up of columnar cells that contain numerous chloroplasts.

**pal·ish** (pā′lĭsh) *adj.* Slightly pale.

**pall[1]** (pôl) *n.* [ME *palle* < OE *pæll,* purple robe < Lat. *pallium,* cloak.] **1.** A cover for a coffin, bier, or tomb, often of black, purple, or white velvet. **2.** A coffin, esp. one being borne to a grave or tomb. **3. a.** A covering that darkens or obscures <a *pall* of smog> **b.** A gloomy atmosphere or effect <The bad news cast a *pall* over the room.> **4. a.** A linen cloth or a square of cardboard faced with cloth used to cover the Eucharistic chalice. **b.** PALLIUM 2. —*vt.* **palled, pall·ing, palls.** To cover with or as if with a pall.

**pall[2]** (pôl) *v.* **palled, pall·ing, palls.** [ME *pallen,* short for *appallen,* to grow faint. —see APPALL.] —*vi.* **1.** To become boring or tedious <The party began to *pall.*> **2.** To have a dulling, wearisome, or unpleasant effect <Their carping started to *pall* on us.> **3.** To become cloyed or satiated. —*vt.* **1.** To satiate : cloy. **2.** To make vapid or wearisome.

**pal·la·di·a** (pə-lā′dē-ə) *n. var. pl. of* PALLADIUM[2].

**Pal·la·di·an[1]** (pə-lā′dē-ən) *adj.* [< Lat. *palladius* < Gk. *palladios* < *Pallas,* Pallas Athena.] **1.** Of, relating to, or typical of Athena. **2.** Of, relating to, or marked by wisdom or learning.

**Pal·la·di·an[2]** (pə-lā′dē-ən) *adj.* **1.** Of or designating the Renaissance architectural style of Andrea Palladio. **2.** Of or designating a mid-18th-cent. architectural style derived from that of Palladio, esp. in Britain.

**pal·la·dic** (pə-lā′dĭk, -lăd′ĭk) *adj.* Of or designating compounds containing trivalent or tetravalent palladium.

**pal·la·di·um[1]** (pə-lā′dē-əm) *n.* [< PALLAS.] *Symbol* **Pd** A soft, ductile, steel-white, tarnish-resistant, metallic element alloyed for use in electric contacts, jewelry, nonmagnetic watch parts, and surgical instruments; atomic number 46; atomic weight 106.4.

**pal·la·di·um²** (pə-lā'dē-əm) *n., pl.* **-di·a** (-dē-ə) *or* **-di·ums.** [Lat. *Palladium,* a statue of Pallas Athena believed to protect Troy < Gk. *Palladion* < *Pallas,* Pallas Athena.] **1.** A sacred object having the power to preserve or protect a city or state possessing it. **2.** A safeguard, esp. one viewed as a guarantee of the integrity of social institutions.

**pal·la·dous** (pə-lā'dəs, păl'ə-dəs) *adj.* Of, relating to, or containing palladium, esp. bivalent palladium.

**Pal·las** (păl'əs) *n.* [Peter S. *Pallas* (1741–1811), its discoverer.] The second-largest asteroid of the solar system, approx. 300 miles or 483 kilometers in diameter.

**Pallas Athena** *or* **Pallas Athene** *n. Gk. Myth.* Athena.

**pall·bear·er** (pôl'bâr'ər) *n.* A person who carries or attends a coffin at a funeral.

**pal·let¹** (păl'ĭt) *n.* [OFr. *palette,* small potter's shovel. —see PALETTE.] **1.** A machine part that converts reciprocating motion to rotary motion or vice versa, as a click or pawl for controlling the motion of a ratchet wheel in a watch escapement. **2.** The projection or lip of a pawl for engaging the teeth on a ratchet wheel. **3.** A wooden paddlelike potter's tool for mixing and shaping clay. **4.** A tool for printing or gilding letters on book bindings or for taking up and applying gold leaf. **5.** A portable platform for moving or storing cargo or freight. **6.** A painter's palette.

**pal·let²** (păl'ĭt) *n.* [ME *pailet* < AN *paillete,* bundle of straw < *paille,* straw < Lat. *palea,* chaff.] A thin straw-filled mattress or small hard bed.

**pal·lette** (pă-lĕt') *n.* [Alteration of PALETTE.] A plate on a suit of armor that protects the armpit.

**pal·li·a** (păl'ē-ə) *n. var. pl. of* PALLIUM.

**pal·li·al** (păl'ē-əl) *adj.* [< NLat. *pallium,* cerebral cortex < Lat., cloak.] **1.** Of or relating to the cerebral cortex. **2.** Of or relating to the mantle of a mollusk or brachiopod.

**pal·liasse** (păl-yăs', păl'yăs') *n. var. of* PAILLASSE.

**pal·li·ate** (păl'ē-āt') *vt.* **-at·ed, -at·ing, -ates.** [LLat. *palliare, palliat-,* to cloak < Lat. *pallium,* cloak.] **1.** To make (e.g., an offense) seem less serious : EXTENUATE. **2.** To make less intense or severe : MITIGATE <tried to *palliate* our misery> **—pal'li·a'tion** *n.* **—pal'li·a'tor** *n.*

**pal·li·a·tive** (păl'ē-ā'tĭv, -ē-ə-tĭv) *adj.* Serving or tending to palliate. **—pal'li·a·tive** *n.* **—pal'li·a'tive·ly** *adv.*

**pal·lid** (păl'ĭd) *adj.* [Lat. *pallidus* < *pallēre,* to be pale.] **1.** Abnormally pale or wan <the sick child's *pallid* face> **2.** Lacking intensity of hue or brightness. **3.** Lacking in spirit or vitality : DULL <*pallid* amusements> **—pal'lid·ly** *adv.* **—pal'lid·ness** *n.*

**pal·li·um** (păl'ē-əm) *n., pl.* **-li·ums** *or* **-li·a** (-ē-ə) [Lat.] **1.** A cloak worn by the Romans. **2.** A vestment worn by the pope and conferred by him on archbishops and sometimes on bishops. **3. a.** *Biol.* The cerebral cortex. **b.** The mantle of a mollusk or brachiopod.

**pall-mall** (pĕl'mĕl', păl'măl', pôl'môl') *n.* [Obs. Fr. *pallemaille* < Ital. *pallamaglio* : *palla,* ball (of Germanic orig.) + *maglio,* mallet (< Lat. *malleus*).] **1.** A 17th-cent. game in which a boxwood ball was struck with a mallet to propel it through an iron ring suspended at the end of an alley. **2.** The alley in which pall-mall was played.

**pal·lor** (păl'ər) *n.* [Lat. < *pallēre,* to be pale.] Extreme or abnormal paleness <a deathly *pallor*>

**pal·ly** (păl'ē) *adj.* **-li·er, -li·est.** *Informal.* Being pals : FRIENDLY.

**palm¹** (päm) *n.* [ME *paume* < OFr. < Lat. *palma,* palm of the hand, palm tree.] **1. a.** The inner surface of the hand from the wrist to the base of the fingers. **b.** The similar part of the forefoot of a quadruped. **2.** A unit of length equal to either the width or the length of the hand. **3.** The part of a mitten or glove that covers the palm of the hand. **4.** A metal shield worn by sailmakers over the palm of the hand and used to force a needle through heavy canvas. **5.** The blade of an oar or paddle. **6.** The flattened part of the antlers of an animal, as a deer or moose. —*vt.* **palmed, palm·ing, palms. 1.** To conceal (something) in the palm of the hand, as in a sleight-of-hand trick. **2.** To pick up furtively. **3.** *Basketball.* To commit a violation by allowing (the ball) to rest momentarily in the palm of the hand while dribbling. **—cross one's palm.** To pay, tip, or bribe. **—grease the palm of.** To bribe. **—have an itching palm.** To have a craving for money. **—palm off.** To dispose of or pass off by deception <*palmed off* the copy as the original painting>

**palm²** (päm) *n.* [ME *palme* < OE *palm* < Lat. *palma,* palm tree, palm of the hand.] **1.** Any of various chiefly tropical evergreen trees or shrubs of the family Palmaceae, with unbranched trunks bearing a crown of large pinnate or palmate leaves. **2.** A leaf or frond of a palm tree, carried as an emblem of victory, success, or joy. **3.** Victory : triumph. **4.** A small metallic representation of a palm leaf added to a military decoration awarded a second time. **—bear (***or* **carry off) the palm.** To be the winner in a contest. **—palm'like'** *adj.*

▲ word history: Even though *palm¹,* "palm of the hand," and *palm²,* "palm tree," are descended from the same Latin word, they are considered separate words in Modern English because they entered the language by different routes. The Latin word *palma* meant basically "palm of the hand"; it was also used to denote the palm tree because the leaves of the palm tree resemble an outspread hand. Old English borrowed *palma* to mean "palm tree"; the language already had a word, *folm* (which is cognate with *palma*), for "palm of the hand." In Middle English times the French descendent of *palma, paume,* was borrowed for "palm of the hand" and supplanted the native word. An *l* was added to English *paume* to reflect the spelling of the Latin word.

**pal·mar** (păl'mər, păl'-, pä'mər) *adj.* [NLat. *palmaris* < Lat. *palma,* palm.] Of, relating to, or corresponding to the palm of the hand or an animal's paw.

**pal·ma·ry** (păl'mə-rē, păl'-, pä'mə-) *adj.* [Lat. *palmarius,* deserving the palm < *palma,* palm tree.] Highly praiseworthy : OUTSTANDING.

**pal·mate** (păl'māt', păl'-, pä'māt') *also* **pal·mat·ed** (-mā'tĭd) *adj.* [Lat. *palmatus* < *palma,* palm.] **1.** Resembling a hand with the fingers extended, as coral or antlers. **2.** *Bot.* Having leaflets or lobes radiating from one point. **3.** *Zool.* Having webbed toes, as the feet of many water birds. **—pal'mate'ly** *adv.*

**pal·ma·tion** (păl-mā'shən, păl-, pä-mā'-) *n.* **1.** The state of being palmate. **2. a.** A palmate form or structure. **b.** A part or section of a palmate structure.

**palm·er** (pä'mər) *n.* [ME < AN < Med. Lat. *palmarius* < Lat. *palma,* palm.] A medieval European pilgrim who carried a palm branch as a token of having visited the Holy Land.

**palm·er·worm** (pä'mər-wûrm') *n.* A caterpillar that injures fruit trees by feeding on their leaves, esp. the small green caterpillar of a North American moth, *Dichomeris ligulella.*

**pal·mette** (păl-mĕt') *n.* [Fr., dim. of *palme,* palm < Lat. *palma.*] A stylized palm leaf used as a decorative element, notably in Persian rugs and in classical moldings, reliefs, frescoes, and vase paintings.

**pal·met·to** (păl-mĕt'ō) *n., pl.* **-tos** *or* **-toes.** [Sp. *palmito,* dim. of *palma,* palm < Lat.] A small, mostly tropical palm with fan-shaped leaves, esp. *Sabal palmetto* of the southeastern United States.

**palm·ist** (pä'mĭst) *also* **palm·is·ter** (-mĭ-stər) *n.* [Prob. back-formation < PALMISTRY.] One who practices palmistry.

**palm·is·try** (pä'mĭ-strē) *n.* [ME *pawmestrie* < *paume,* palm. —see PALM¹.] The art or practice of telling fortunes from the lines, marks, and patterns on the palms of the hands : CHIROMANCY.

**pal·mi·tate** (păl'mĭ-tāt', păl'-, pä'mĭ-) *n.* [PALMIT(IN) + -ATE².] A salt or ester of palmitic acid.

**pal·mit·ic acid** (păl-mĭt'ĭk, păl-, pä-mĭt'-) *n.* [< PALMITIN.] A common fatty acid, $CH_3(CH_2)_{14}COOH$, found in many natural oils and fats and used in making soaps.

**pal·mi·tin** (păl'mĭ-tĭn, păl'-, pä'mĭ-) *n.* [Fr. *palmitine,* perh. < *palmite,* pith of the palm tree < *palme,* palm < Lat. *palma.*] The glyceryl ester, $C_3H_5(OC_{16}H_{31}O)_3$, of palmitic acid, found in palm oil and animal fats and used in making soaps.

**palm oil** *n.* **1.** A yellowish fatty oil obtained esp. from the crushed nuts of the West African palm, *Elaeis guineensis,* and used to make soaps, chocolates, cosmetics, and candles. **2.** A reddish-yellow, butterlike fatty oil derived from the fermented pulp of this palm and used as a lubricant and in the manufacture of soaps and candles.

**palm sugar** *n.* Sugar made from the sap of various palm trees.

**Palm Sunday** *n.* The Sunday before Easter, commemorating Christ's entry into Jerusalem when palm branches were scattered before Him.

**palm·y** (pä'mē) *adj.* **-i·er, -i·est. 1.** Of or relating to palm trees. **2.** Covered with palm trees. **3.** Prosperous : flourishing.

**pal·my·ra** (păl-mī'rə) *n.* [Port. *palmeira* < *palma,* palm < Lat.] A tall tropical Asian palm, *Borassus flabellifera,* with large fanlike leaves.

**palmyra palm** *n.* The palmyra.

**pal·o·mi·no** (păl'ə-mē'nō) *n., pl.* **-nos.** [Am. Sp. < Sp., dove-colored < Lat. *palumbinus,* pertaining to ringdoves < *palumbes,* ringdove.] A horse with a golden or tan coat and a white or cream-colored mane and tail.

**pa·loo·ka** (pə-lōō'kə) *n.* [Orig. unknown.] *Slang.* An incompetent or easily defeated person, esp. a prizefighter.

**pa·lo·ver·de** (păl'ō-vûr'dē, -vûrd') *n.* [Mex. Sp. : Sp. *palo,* tree (< Lat. *pallus,* stake) + Sp. *verde,* green (< Lat. *virdis*).] **1.** A spiny, nearly leafless shrub, *Cercidium torreyanum* of southwestern North America, bearing small yellow flowers. **2.** Any of several shrubs similar to the paloverde.

**palp** (pălp) *n.* [Fr. *palpe* < Lat. *palpus,* a touching.] *Zool.* An elongated sensory organ, usu. near the mouth, in invertebrate organisms such as mollusks, crustaceans, and insects.

**pal·pa·ble** (păl'pə-bəl) *adj.* [ME < LLat. *palpabilis* < Lat. *palpare,* to touch.] **1.** Capable of being touched, felt, or handled : TANGIBLE. **2.** Easily perceived : OBVIOUS. **3.** *Med.* Perceptible by palpation <a *palpable* growth> **—pal'pa·bil'i·ty** *n.* **—pal'pa·bly** *adv.*

**pal·pate¹** (păl'pāt') *vt.* **-pat·ed, -pat·ing, -pates.** [Lat. *palpare, palpat-,* to touch.] To examine or explore by touching (an organ or bodily area) as a diagnostic aid. **—pal·pa'tion** *n.* **—pal'pa'tor** *n.*

**pal·pate²** (păl'pāt') *adj.* Having a palp or palps.

**pal·pe·bral** (păl'pə-brəl, păl-pē'brəl, -pĕb'rəl) *adj.* [LLat. *palpebralis* < Lat. *palpebra,* eyelid.] Of or pertaining to the eyelids.

**pal·pi** (păl′pī) *n. pl. of* PALPUS.
**pal·pi·tant** (păl′pĭ-tənt) *adj.* [Lat. *palpitans, palpitant-*, pr.part. of *palpitare*, to palpitate.] Palpitating : quivering.
**pal·pi·tate** (păl′pĭ-tāt′) *vi.* **-tat·ed, -tat·ing, -tates.** [Lat. *palpitare, palpitat-*, freq. of *palpare*, to touch.] **1.** To quiver : flutter. **2.** To beat more quickly than normal : THROB. —Used of the heart. **—pal′pi·tat′ing·ly** *adv.* **—pal′pi·ta′tion** (-tā′shən) *n.*
**pal·pus** (păl′pəs) *n., pl.* **-pi** (-pī′) [NLat. < Lat., a feeling.] *Zool.* A palp.
**pals·grave** (pôlz′grāv′) *n.* [MDu. *paltsgrave* : *palts*, palatine + *grave*, count.] PALATINE[1] 3.
**pal·sied** (pôl′zēd) *adj.* **1.** Afflicted with palsy. **2.** Shaking : trembling.
**pal·sy** (pôl′zē) *n., pl.* **-sies.** [ME < OFr. *paralisie* < Lat. *paralysis.* —see PARALYSIS.] **1.** Paralysis. **2.** A condition characterized by loss of power to feel or to control movement in any part of the body. **3. a.** A weakening or debilitating influence. **b.** An enfeebled condition or debilitated state believed to result from such an influence. **4.** A fit of strong emotion characterized by an inability to act <a *palsy* of terror> —*vt.* **-sied, -sy·ing, -sies. 1. a.** To paralyze. **b.** To deprive of strength. **2.** To make helpless, as with apprehension.
**pal·sy-wal·sy** (păl′zē-wăl′zē) *adj.* [Redup. of *palsy*, alteration of PALLY.] *Slang.* Pally.
**pal·ter** (pôl′tər) *vi.* **-tered, -ter·ing, -ters.** [Orig. unknown.] **1.** To behave or talk insincerely : EQUIVOCATE. **2.** To be capricious : TRIFLE. **3.** To quibble, esp. in bargaining. **—pal′ter·er** *n.*
**pal·try** (pôl′trē) *adj.* **-tri·er, -tri·est.** [< obs. *paltry*, trash.] **1.** Insignificant : trifling. **2.** Worthless : trashy. **3.** Contemptible. **4.** Stingy. **—pal′tri·ly** *adv.* **—pal′tri·ness** *n.*
**pa·lu·dal** (pə-lo͞od′l, păl′yə-dəl) *adj.* [< Lat. *palus, palud-*, marsh.] Of or relating to a swamp : MARSHY.
**pal·u·dism** (păl′yə-dĭz′əm) *n.* [< Lat. *palus, palud-*, marsh.] The infectious disease malaria.
**pal·y** (pā′lē) *adj.* Somewhat pale.
**pal·y·nol·o·gy** (păl′ə-nŏl′ə-jē) *n.* [Gk. *palunein*, to sprinkle (< *palē*, dust) + -LOGY.] The scientific study of spores and pollen. **—pal′y·no·log′i·cal** (-nə-lŏj′ĭ-kəl), **pal′y·no·log′ic** *adj.* **—pal′y·no·log′i·cal·ly** *adv.* **—pal′y·nol′o·gist** *n.*
**pam** (păm) *n.* [Prob. short for Gk. *pamphilos*, loved by all : *pan*, all + *philos*, beloved.] The jack of clubs and highest trump in certain variations of loo.
**pam·pa** (păm′pə) *n., pl.* **-pas** (-pəz, -pəs) [Am. Sp. < Quechua, flat field.] A nearly treeless grassland region of South America, mainly in central Argentina and Uruguay between the Andes and the Atlantic.
**pam·pas grass** (păm′pəs) *n.* A tall grass, *Cortaderia argentea* of southern South America, with silvery plumes.
**pam·pe·an** (păm′pē-ən, păm-pē′ən) *adj.* Of or relating to the pampas or the Indian people who live there. —*n.* **Pampean.** An Indian of the pampas.
**pam·per** (păm′pər) *vt.* **-pered, -per·ing, -pers.** [ME *pamperem*, of Du. orig.] **1.** To treat with excessive care and indulgence : CODDLE. **2.** *Archaic.* To indulge with rich food : GLUT. **—pam′per·er** *n.*
**pam·pe·ro** (păm-pâr′ō, păm-) *n., pl.* **-ros.** [Am. Sp. < *pampa*, pampas.] A strong cold southwest wind blowing across the pampas.
**pam·phlet** (păm′flĭt) *n.* [ME *pamflet* < Med. Lat. *Pamphilus*, a short amatory Latin poem of the 12th cent.] **1.** An unbound printed work, usu. with a paper cover. **2.** An unbound short essay or treatise, usu. on a current topic. **—pam′phlet·ar′y** *adj.*
**pam·phlet·eer** (păm′flĭ-tîr′) *n.* A writer of pamphlets or other short works taking a partisan stand on an issue. —*vi.* **-eered, -eer·ing, -eers.** To write and publish pamphlets.
**pan[1]** (păn) *n.* [ME < OE *panne*, poss. < Lat. *patina* < Gk. *patanē.*] **1.** A shallow, wide, open container used for holding liquids or cooking. **2.** A vessel similar to a pan, esp.: **a.** An open metal dish used to separate gold or other metal from gravel or waste by washing. **b.** Either of the receptacles on a balance or pair of scales. **c.** A vessel used for boiling and evaporating liquids. **3. a.** A basin or depression in the earth, often containing water or mud. **b.** A natural or artificial basin used to obtain salt by evaporating brine. **4.** A piece of drift ice broken off from a larger floe. **5.** A small cavity in the lock of a flintlock firearm that is used to hold powder. **6.** *Slang.* FACE 1. **7.** *Informal.* Severe criticism, esp. a negative review. —*v.* **panned, pan·ning, pans.** —*vt.* **1.** To wash (e.g., gravel) in a pan for precious metal. **2.** To cook (food) in a pan. **3.** *Informal.* To review or criticize harshly <*panned* the new film> —*vi.* **1.** To wash gravel, sand, or other sediments in a pan. **2.** To yield gold as a result of washing in a pan. **—pan out.** *Informal.* To be successful <"If I don't *pan out* as an actor I can still go back to school" —Saul Bellow>
**pan[2]** (păn) *n.* [Hindi *pān* < Skt. *parṇam*, feathery leaf.] **1.** The leaf of the betel tree. **2.** A preparation of pan with betel nuts and lime chewed in the Orient.
**pan[3]** (păn) *v.* **panned, pan·ning, pans.** [Short for PANORAMA.] —*vi.* To move a motion-picture or television camera to follow a moving object or create a panoramic effect. —*vt.* To move (a camera) in such a manner.
**Pan** (păn) *n.* [Lat. < Gk.] *Gk. Myth.* The god of woods, fields, and flocks, with a human torso and goat's legs, horns, and ears.
**pan-** *pref.* [Gk. < *pas*, all.] **1.** All <*panorama*> **2.** Involving all of or the union of a specified group <*Pan*-Hellenic> **3.** General : whole <*pan*leucopenia>
**pan·a·ce·a** (păn′ə-sē′ə) *n.* [Lat. < Gk. *panakeia* < *panakēs*, all-healing : *pan-*, all + *akos*, cure.] **1.** A remedy for all diseases, evils, or troubles : CURE-ALL. **2.** ELIXIR 2. **—pan′a·ce′an** *adj.*
☆ ***syns:*** PANACEA, CATHOLICON, CURE-ALL *n. core meaning* : something believed to cure all human disorders <sought a *panacea* for urban blight>
**pa·nache** (pə-năsh′, -näsh′) *n.* [Fr. < Ital. *pennachio* < LLat. *pinnaculum*, dim. of Lat. *pinna*, feather.] **1.** A bunch of feathers or a plume, esp. on a helmet. **2.** Dash : verve.
**pa·na·da** (pə-nä′də) *n.* [Sp. < *pan*, bread < Lat. *panis.*] A paste or gruel of bread crumbs, toast, or flour combined with liquid, used for making soups, binding forcemeats, or thickening sauces.
**pan·ag·glu·ti·nin** (păn′ə-glo͞ot′n-ĭn) *n.* An agglutinin capable of agglutinating red blood cells of every blood group.
**Pan·a·ma hat** (păn′ə-mä′) *n.* **1.** A natural-colored, hand-plaited hat made from leaves of the jipijapa plant of South and Central America. **2.** A machine-made imitation of a Panama hat.
**Panama Red** *n.* Very potent, reddish marijuana of Panamanian origin.
**Pan-A·mer·i·can** (păn′ə-mĕr′ĭ-kən) *adj.* Of or relating to North, South, and Central America as a whole.
**pan·a·tel·a** *also* **pan·e·tel·a** *or* **pan·e·tel·la** (păn′ə-tĕl′ə) *n.* [Sp., cigar, biscuit < Ital. *panatella*, biscuit < *panata*, panada < *pane*, bread < Lat. *panis.*] A long, slender cigar.
**pan-broil** (păn′broil′) *vt.* **-broiled, -broil·ing, -broils.** To cook (e.g., meat) over direct heat in an uncovered, usu. ungreased skillet.
**pan·cake** (păn′kāk′) *n.* A thin cake made of batter poured on a hot, greased skillet and cooked on both sides until brown. —*v.* **-caked, -cak·ing, -cakes.** —*vi.* To make a pancake landing in an aircraft. —*vt.* To cause (an airplane) to make a pancake landing.
**pancake landing** *n.* An emergency or irregular landing in which an aircraft drops flat to the ground from a low altitude.
**Pan-Cake Make-Up** (păn′kāk′). A trademark for a semisolid face-powder cosmetic pressed into a flat cake.
**pan·chax** (păn′chăks′) *n.* [NLat.] A small brightly colored Old World tropical fish of the genus *Aplocheilus* or related genera, often kept in home aquariums.
**Pan·chen La·ma** (pän′chən lä′mə) *n.* [Tibetan : *paṇchen*, great scholar (Skt. *paṇḍitaḥ*, scholar + Tibetan *chen-po*, great) + *bla-ma*, monk.] One of Tibet's two grand lamas, ranking next in line to the Dalai Lama and serving as his chief spiritual counselor.
**pan·chro·mat·ic** (păn′krō-măt′ĭk) *adj.* Sensitive to all colors <*panchromatic* film> **—pan·chro′ma·tism** (-krō′mə-tĭz′əm) *n.*
**pan·cra·ti·um** (păn-krā′shē-əm) *n.* [Lat. < Gk. *pankration* : *pan-*, all + *kratos*, strength.] An athletic contest in ancient Greece that involved boxing and wrestling.
**pan·cre·as** (păng′krē-əs, păn′-) *n.* [Gk. *pankreas* : *pan-*, all + *kreas*, flesh.] *Anat.* A long, soft, irregularly shaped gland lying behind the stomach that secretes digestive enzymes and produces insulin. **—pan′cre·at′ic** (păng′krē-ăt′ĭk, păn′-) *adj.*
**pancreat-** *pref. var. of* PANCREATO-.
**pan·cre·a·tec·to·my** (păng′krē-ə-tĕk′tə-mē, păn′-) *n., pl.* **-mies.** Surgical excision of the pancreas.
**pancreatic juice** *n.* A clear alkaline secretion of the pancreas containing enzymes that aid in the digestion of proteins, carbohydrates, and fats.
**pan·cre·a·tin** (păng′krē-ə-tĭn, păn′-, păn-krē′ə-tĭn) *n.* A mixture of enzymes extracted from cattle or hog pancreases and used as a digestive aid.
**pan·cre·a·ti·tis** (păng′krē-ə-tī′tĭs, păn′-) *n.* Inflammation of the pancreas.
**pancreato-** *or* **pancreat-** *pref.* [Gk. *pankreas, pankreat-*, pancreas.] Pancreas <*pancreatin*>
**pan·cre·o·zy·min** (păng′krē-ō-zī′mĭn, păn′-) *n.* [PANCRE(AS) + -O- + ZYM(O)- + -IN.] A hormone produced by the intestinal mucosa that stimulates the secretion of pancreatic juice.
**pan·da** (păn′də) *n.* [Fr., perh. < a native word in Nepal.] **1.** A large bearlike mammal, *Ailuropoda melanoleuca* of the mountains of China and Tibet, with distinctively marked, black-and-white woolly fur. **2.** A small raccoonlike mammal, *Ailurus fulgens* of northeastern Asia, with reddish fur and a long ringed tail.
**panda car** *n.* [From its orig. black-and-white markings.] *Chiefly Brit.* A police cruiser.
**pan·dae·mo·ni·um** (păn′də-mō′nē-əm) *n. var. of* PANDEMONIUM.
**pan·da·nus** (păn-dā′nəs, -dăn′əs) *n.* [NLat. *Pandanus*, genus name < Malay *pandan*, screw pine.] Any of various palmlike trees and shrubs of the genus *Pandanus* of southeastern Asia, with large prop roots and a crown of narrow leaves that yield a fiber used in woven articles, as mats. **—pan′da·na′ceous** (păn′də-nā′shəs) *adj.*
**Pan·da·rus** (păn′dər-əs) *also* **Pan·dar** (păn′dər) *n.* [Lat. < Gk. *Pandaros.*] **1.** The leader of the Lycians, slain by Diomedes in the *Iliad.* **2.** The procurer of Cressida for Troilus in medieval romance.

**Pandean pipes** (păn-dē′ən) *pl.n.* [< PAN.] A panpipe.

**pan·dect** (păn′dĕkt′) *n.* [LLat. *Pandectae,* the Pandects, a digest of Roman civil law < *pandecta,* encyclopedia < Gk. *pandektēs,* all-receiving : *pan-,* all + *dektēs,* receiver < *dekheisthai,* to receive.] **1.** A comprehensive digest or complete treatise. **2. pandects.** A complete legal code. **3. Pandects.** A digest of Roman civil law, compiled for the emperor Justinian in the 6th cent. A.D.

**pan·dem·ic** (păn-dĕm′ĭk) *adj.* [LLat. *pandemus* < Gk. *pandēmos,* of all the people : *pan-,* all + *dēmos,* people.] **1.** Widespread : general <*pandemic* discontent> **2.** *Med.* Epidemic over an esp. wide geographic area. —*n.* A pandemic disease.

**pan·de·mo·ni·um** *also* **pan·dae·mo·ni·um** (păn′də-mō′nē-əm) *n.* [After *Pandæmonium,* capital of Hell in *Paradise Lost,* an epic poem by John Milton (1608–1674).] **1.** A place marked by great noise and disorder <"The whole lobby was a perfect *pandemonium,* and the din was terrific" —Jerome K. Jerome> **2.** Wild uproar or noise <*pandemonium* in the grandstands> **—pan′de·mo′ni·ac** (-mō′nē-ăk′) *adj.*

**pan·der** (păn′dər) *also* **pan·der·er** (păn′dər-ər) *n.* [ME *Pandare,* Pandarus < OItal. *Pandaro* < Lat. *Pandarus* < Gk. *Pandaros.*] **1.** A go-between or liaison in sexual intrigues : PROCURER. **2.** One who caters to the base desires and tastes of others or exploits their weaknesses. —*v.* **-dered, -der·ing, -ders.** —*vt.* To act as a pander for. —*vi.* To act as a pander. **—pan′der·ism** *n.*

**Pan·do·ra's box** (păn-dôr′əz, -dōr′-) *n.* [Gk. *Pandōra* : *pan,* all + *-dōra,* giving (from the myth that the first woman, *Pandora,* came to man with a box which was not to be opened, but which she opened out of curiosity and thereby released all that is evil to mankind).] A source of many unexpected troubles.

**pan·dore** (păn′dôr′, -dōr′) *n.* [Gk. *pandoura.*] A bandore.

**pan·dow·dy** (păn-dou′dē) *n., pl.* **-dies.** [Orig. unknown.] Sliced fruit mixed with sugar and spices, topped with a thick crust, and baked in a deep dish.

**pan·du·rate** (păn-dŏŏr′ĭt, -dyŏŏr′-) *also* **pan·du·ri·form** (-ə-fôrm′) *adj.* [NLat. *panduratus* < LLat. *pandura,* three-string lute < Gk. *pandoura.*] Shaped like a violin <a *pandurate* leaf>

**pane** (pān) *n.* [ME, section < OFr. *pan,* piece of cloth < Lat. *pannus,* cloth.] **1. a.** A glass-filled section of a window or door. **b.** The glass used in such a division. **2.** A panel, as of a door or wall. **3.** A flat surface or facet of a many-sided object, as a bolt.

**pan·e·gyr·ic** (păn′ə-jĭr′ĭk, -jī′rĭk) *n.* [Fr. *panégyrique* < Lat. *panegyricus* < Gk. *panēgurikos* < *panēgurikos,* of a public assembly < *panēguris,* public assembly : *pan-,* all + *aguris,* assembly.] **1.** A formal eulogistic composition intended as a public compliment. **2.** Elaborate praise : ENCOMIUM. **—pan′e·gyr′i·cal** *adj.* **—pan′e·gyr′i·cal·ly** *adv.*

**pan·e·gy·rize** (păn′ə-jə-rīz′) *v.* **-rized, -riz·ing, -riz·es.** —*vt.* To eulogize. —*vi.* To compose, deliver, or indulge in panegyrics. **—pan′e·gyr′ist** (-jĭr′ĭst, -jī′rĭst) *n.*

**pan·el** (păn′əl) *n.* [ME, piece of cloth < OFr. < Lat. *pannus,* cloth.] **1.** A flat, usu. rectangular piece forming a part of a surface in which it is set and being raised, recessed, or framed. **2.** The space or section in a railing or fence between two posts. **3.** A vertical section of fabric : GORE. **4. a.** A thin wooden board used as a surface for oil painting. **b.** A painting on such a board. **5. a.** A board having switches to control parts of an electric device. **b.** An instrument panel. **6.** A section of a telephone switchboard. **7. a.** The total list of persons summoned for jury duty. **b.** Those persons selected from the list to compose a jury. **c.** A jury. **8. a.** A group of people gathered to plan, discuss, or decide something or to participate in a game show or other radio or television program. **b.** A discussion by such a group. —*vt.* **-eled, -el·ing, -els** *also* **-elled, -el·ling, -els. 1.** To furnish or cover with panels. **2.** To decorate with panels. **3.** To separate into panels. **4.** To select or impanel (a jury).

**panel discussion** *n.* A usu. formal discussion of a subject of public interest by a group of persons forming a panel.

**pan·el·ing** (păn′ə-lĭng) *n.* A section of panels or paneled wall.

**pan·el·ist** (păn′ə-lĭst) *n.* A member of a panel.

**panel truck** *n.* A small delivery truck with a fully enclosed body.

**pan·e·tel·a** *or* **pan·e·tel·la** (păn′ə-tĕl′ə) *n. vars. of* PANATELA.

**pan·et·to·ne** (păn′ə-tō′nē) *n., pl.* **-nes** *or* **-ni** (-nē) [Ital. < *panetto,* a small loaf, dim. of *pane,* bread < Lat. *panis.*] A festive Italian yeast cake made with candied fruit peels and raisins.

**pan fish** *n.* A fish small enough to be fried whole in a pan.

**pan-fry** (păn′frī′) *vt.* **-fried, -fry·ing, -fries.** To sauté.

**pan·ful** (păn′fŏŏl′) *n.* The amount that a pan will hold.

**pang** (păng) *n.* [Orig. unknown.] **1.** A sudden sharp spasm of pain. **2.** A sudden sharp feeling of emotional distress <a *pang* of guilt> —*vt.* **panged, pang·ing, pangs.** To cause anguish.

**pan·gen·e·sis** (păn-jĕn′ĭ-sĭs) *n.* The now discredited hypothesis that every somatic cell generates self-representative hereditary materials that enter the blood stream and eventually coalesce in reproductive cells, enabling inheritance of acquired characteristics. **—pan′ge·net′ic** (-jə-nĕt′ĭk) *adj.* **—pan′ge·net′i·cal·ly** *adv.*

**Pan·gloss·i·an** (păn-glŏs′ē-ən, -glôs′-, păng-) *adj.* [After *Pangloss,* an optimist in *Candide,* a satire by Voltaire (1694–1778).] Naively or blindly optimistic.

**pan·go·lin** (păng′gə-lĭn, păn′-) *n.* [Malay *pĕngguling* < *guling,* to roll.] A long-tailed, scale-covered mammal of the genus *Manis* of tropical Africa and Asia, with a long snout and a sticky tongue for catching and eating ants.

**pan·han·dle[1]** (păn′hăn′dl) *v.* **-dled, -dling, -dles.** [Back-formation < *panhandler,* beggar, from the resemblance of a beggar's outstretched arm to the handle of a pan.] *Informal.* —*vi.* To beg, esp. on the streets. —*vt.* **1.** To beg from. **2.** To obtain by panhandling. **—pan′han′dler** *n.*

**pan·han·dle[2]** (păn′hăn′dl) *n.* **1.** The handle of a pan. **2.** *often* **Panhandle.** A narrow strip of territory projecting from a larger, broader area <the Texas *Panhandle*>

**Pan-Hel·len·ic** *also* **Pan·hel·len·ic** (păn′hə-lĕn′ĭk) *adj.* **1.** Of or relating to Greek peoples or a movement to unify them. **2.** Of or relating to Greek-letter fraternities and sororities.

**pan·hu·man** (păn-hyōō′mən) *adj.* Of or relating to all humanity.

**pan·ic** (păn′ĭk) *n.* [< Fr. *panique,* terrified < Gk. *panikos* < *Pan,* Pan.] **1.** Sudden overwhelming terror, often affecting many people at once. **2.** Mass alarm about the economy, often resulting in a depression. **3.** *Slang.* One that is uproariously funny. —*adj.* **1.** Of, relating to, or resulting from sudden overpowering terror. **2.** *often* **Panic.** Of or pertaining to Pan. —*vt.* & *vi.* **-icked, -ick·ing, -ics.** To affect or be affected with panic. **—pan′ick·y** *adj.*

**panic button** *n.* **1.** A device for setting off an alarm in an emergency. **2.** *Slang.* A hasty irrational response.

**panic grass** *n.* [ME *panik* < Lat. *panicum.*] Any of numerous grasses of the genus *Panicum,* many of which are grown for fodder.

**pan·i·cle** (păn′ĭ-kəl) *n.* [Lat. *panicula,* dim. of *panus,* a swelling.] *Bot.* An irregularly branched flower cluster. **—pan′i·cled** *adj.*

**pan·ic-strick·en** (păn′ĭk-strĭk′ən) *also* **pan·ic-struck** (-strŭk′) *adj.* Overwhelmed with panic.

**pa·nic·u·late** (pə-nĭk′yə-lĭt, -lāt′) *also* **pa·nic·u·lat·ed** (-lā′tĭd) *adj.* [NLat. *paniculatus* < Lat. *panicula,* panicle. —see PANICLE.] *Bot.* Growing or arranged in a panicle. **—pa·nic′u·late·ly** *adv.*

**Pan·ja·bi** (pŭn-jä′bē, -jăb′ē) *n. var. of* PUNJABI.

**pan·jan·drum** (păn-jăn′drəm) *n.* [After the Grand *Panjandrum,* a character in a nonsense farrago written by Samuel Foote (1720–1777).] An important person <"Once an editor of ladies' magazines and now a *panjandrum* of the publishing business" —Nat Hentoff>

**pan·mic·tic** (păn-mĭk′tĭk) *adj.* [PAN- + Gk. *miktos,* mixed < *mignunai,* to mix.] Of or relating to panmixis.

**pan·mix·is** (păn-mĭk′sĭs) *n.* [PAN- + Gk. *mixis,* act of mingling < *mignunai,* to mix.] Random mating within a breeding population.

**panne** (păn) *n.* [Fr. < OFr., fur lining < Lat. *penna,* feather.] A special finish for velvet and satin that produces a high luster.

**pan·nier** (păn′yər, păn′ē-ər) *n.* [ME *panier* < OFr. < Lat. *panarium,* breadbasket < *panis,* bread.] **1.** A large wicker basket, esp.: **a.** One of a pair of baskets carried on either side of a pack animal or over a person's shoulders. **b.** A basket carried on a person's back. **2. a.** A framework of rigid material, as wire or bone, once used to expand a woman's skirt at the hips. **b.** A skirt or overskirt puffed out at the hips. **—pan′niered** *adj.*

**pannier**

**pan·ni·kin** (păn′ĭ-kĭn) *n.* [Dim. of PAN[1].] *Chiefly Brit.* A small saucepan or metal cup.

**pa·no·cha** (pə-nō′chə) *also* **pa·no·che** (-chē) *n.* [Mex. Sp., pressed cakes of brown sugar, dim. of Sp. *pan,* bread < Lat. *panis.*] **1.** A coarse grade of Mexican sugar. **2.** Penuche.

**pan·o·pho·bi·a** (păn′ə-fō′bē-ə) *n.* [PAN- + -PHOBIA.] Fear of everything.

**pan·o·ply** (păn′ə-plē) *n., pl.* **-plies.** [Gk. *panoplia* : *pan-,* all + *hoplon,* weapon.] **1.** The complete array of warrior's armor and weapons. **2.** A protective covering. **3.** An impressive display : POMP.

**pan·op·tic** (păn-ŏp′tĭk) *also* **pan·op·ti·cal** (-tĭ-kəl) *adj.* [< Gk. *panoptēs,* all-seeing : *pan-,* all + *optos,* visible.] Including everything visible in one view.

**pan·o·ram·a** (păn′ə-răm′ə, -rä′mə) *n.* [PAN- + Gk. *horama,* sight < *horan,* to see.] **1.** An unlimited view of all visible objects over a wide area. **2.** A comprehensive picture of a chain of events or a given subject <a *panorama* of Egyptian civilization> **3.** A picture or series of pictures representing a continuous scene, unrolled and passed before the spectator a section at a time. **—pan′o·ram′ic** (-răm′ĭk) *adj.* **—pan′o·ram′i·cal·ly** *adv.*

**pan·pipe** (păn′pīp′) *n.* [PAN + PIPE.] *often* **panpipes.** A primitive wind instrument consisting of a series of pipes or reeds of graduated length bound together, played by blowing across the top open ends.

**pan·sy** (păn′zē) *n., pl.* **-sies.** [OFr. *pensée* < *pensée,* thought < *penser,* to think. —see PENSIVE.] **1.** A garden plant hybridized from *Viola tricolor hortensis,* bearing rounded, variously colored velvety petals. **2.** A deep to strong violet.

**pant** (pănt) *v.* **pant·ed, pant·ing, pants.** [ME *panten* < OFr. *pantaisier* < VLat. **phantasiare* < Gk. *phantasioun,* to cause to imagine < *phantasia,* appearance. —see FANTASY.] *—vi.* **1.** To breathe rapidly in short gasps, as after exertion. **2.** To give off or emit in loud puffs. **3.** To pulsate rapidly : THROB. **4.** To yearn eagerly <*panting* for recognition> *—vt.* To utter breathlessly or hurriedly. *—n.* **1.** The act of panting. **2.** A short labored breath : GASP. **3.** A short loud puff, as of steam from an engine. **4.** A pulsation. **—pant′ing·ly** *adv.*

**pan·ta·lets** *also* **pan·ta·lettes** (păn′tə-lĕts′) *pl.n.* [< PANTALOON.] **1.** Long ruffled underpants extending below the skirt, worn by women in the mid-19th cent. **2.** Frills attached to the legs of underpants.

**pan·ta·loon** (păn′tə-lo͞on′) *n.* [OFr. *pantalon,* a kind of trouser, after *Pantalon,* Pantaloon < OItal. *Pantalone.*] **1. pantaloons. a.** Men's wide breeches extending from waist to ankle, worn in England during the reign of Charles II. **b.** Tight trousers extending from waist to ankle with straps passing under the instep, worn esp. in the 19th cent. **2. Pantaloon.** A character in the commedia dell'arte, portrayed as a foolish old man in slippers and tight trousers. **3.** The butt of a clown's jokes, a stock character in modern pantomime.

**pant·dress** (pănt′drĕs′) *n.* A dress with a skirt divided like culottes.

**pan·tech·ni·con** (păn-tĕk′nĭ-kŏn′, -kən) *n.* [After *Pantechnicon,* London, England, a 19th-cent. bazaar that became a furniture warehouse.] *Chiefly Brit.* **1.** A storage warehouse, esp. for furniture. **2.** A large truck, esp. a furniture van.

**pan·the·ism** (păn′thē-ĭz′əm) *n.* **1.** The doctrine identifying the Deity with the various forces and workings of nature. **2.** Belief in and worship of all gods. **—pan′the·ist** *n.* **—pan′the·is′tic, pan′the·is′ti·cal** *adj.* **—pan′the·is′ti·cal·ly** *adv.*

**pan·the·on** (păn′thē-ŏn′, -ən) *n.* [ME *panteon* < Lat. *Pantheon,* a Roman temple < Gk. *pantheion,* a temple sacred to all the gods : *pan-,* all + *theos,* god.] **1.** A temple dedicated to all gods. **2.** All the gods of a people. **3.** A public building commemorating and dedicated to the great persons of a nation. **4.** A group of famous persons.

**pan·ther** (păn′thər) *n.* [ME *pantere* < OFr. < Lat. *panthera* < Gk. *panthēr.*] **1.** The leopard, *Panthera pardus,* esp. in its unspotted black form. **2.** The mountain lion.

**pant·ies** (păn′tēz) *pl.n.* [< PANTS.] Short underpants for women or girls.

**pan·tile** (păn′tīl′) *n.* [PAN[1] + TILE.] An S-curved roofing tile, laid so that the down curve of one tile overlaps the up curve of the next one. **—pan′tiled′** *adj.*

**pan·tof·fle** *also* **pan·to·fle** (păn-tŏf′əl, -tō′fəl, -to͞o′fəl, păn′tə-fəl) *n.* [ME *pantufle* < OFr. *pantoufle.*] A slipper.

**pan·to·graph** (păn′tə-grăf′) *n.* [Fr. *pantographe* : Gk. *panto-,* all + Gk. *graphein,* to write.] **1.** An instrument for copying a plane figure to any desired scale, having styluses mounted on four jointed rods in the form of a parallelogram with extended sides. **2.** A linked framework, as an extensible telephone arm or a power-collecting trolley on an electric locomotive. **—pan′to·graph′ic** *adj.*

**pan·to·mime** (păn′tə-mīm′) *n.* [Lat. *pantomimus* < Gk. *pantomimos,* pantomimic actor : *pas,* all + *mimos,* mime.] **1. a.** A genre of theatrical performance in ancient Rome in which one actor played all the parts in dumb show, accompanied by music and singing. **b.** The actor in this genre. **c.** A specified revival or derivative of this genre. **2.** A British genre of musical plays for children, usu. based on fairy stories, having specified conventions derived from the commedia dell'arte. **3.** Acting that consists primarily of gesture. **4.** Dumb show used for expressive communication. *—vt & vi.* **-mimed, -mim·ing, -mimes.** To represent by pantomime or express oneself in pantomime. **—pan′to·mim′ic** (-mĭm′ĭk), **pan′to·mim′i·cal** *adj.* **—pan′to·mim′ist** (-mī′mĭst) *n.*

**pan·to·then·ate** (păn′tə-thĕn′āt′, păn-tŏth′ə-nāt′) *n.* [PANTOTHEN(IC ACID) + -ATE.] A salt or ester of pantothenic acid.

**pan·to·then·ic acid** (păn′tə-thĕn′ĭk) *n.* [< Gk. *pantothen,* from all sides < *pas,* all.] A component of the vitamin B complex, $C_9H_{17}NO_5$, common in liver but found in all living tissue.

**pan·toum** (păn-to͞om′) *n.* [Fr. < Malay *pantun.*] A verse form composed of quatrains in which the second and fourth lines are repeated as the first and third lines of the following quatrain and the final line of the poem repeats the opening line.

**pan·try** (păn′trē) *n., pl.* **-tries.** [ME *pantrie* < AN *panetrie,* bread-closet < OFr. *panetier,* pantry servant < *pan,* bread < Lat. *panis.*] **1.** A small room, usu. off a kitchen, esp. for food storage. **2.** A small room, as in a hospital, for preparing food on order.

**pants** (pănts) *pl.n.* [Short for *pantaloons,* plural of PANTALOON.] **1.** Trousers. **2.** Underpants. **—with one's pants down.** *Informal.* In an embarrassing position or situation.

**pant·suit** *also* **pants suit** (pănt′so͞ot′) *n.* A woman's suit having trousers instead of a skirt.

**pant·y·hose** (păn′tē-hōz′) *n., pl.* **pantyhose.** A one-piece garment consisting of stretchable stockings and underpants.

**pant·y raid** (păn′tē) *n.* A raid on a girls' dormitory usu. by college boys ostensibly to obtain undergarments as trophies.

**pant·y·waist** (păn′tē-wāst′) *n.* **1.** A child's undergarment having a shirt and pants buttoned together at the waist. **2.** *Slang.* A weak effeminate man. **—pant′y·waist′** *adj.*

**pan·zer** (păn′zər, pänt′sər) *adj.* [G. < *Panzer,* armor < MHG *panzier* < OFr. *pancier* < *pance,* body. —see PAUNCH.] **1.** Protected by armor : ARMORED. **2.** Using or equipped with armored or mechanized units. *—n.* An armored tank.

**†pap[1]** (păp) *n.* [ME.] **1.** *Chiefly Regional.* A teat or nipple. **2.** Something resembling a nipple.

**pap[2]** (păp) *n.* [ME.] **1.** Soft or semiliquid, easily digested food, as for infants. **2.** Something lacking real value or substance. **3.** *Slang.* Money and favors obtained as political patronage <"uncouth, self-seeking politicians primarily interested in patronage, privilege, and *pap*"—Fiorello LaGuardia>

**pa·pa** (pä′pə, pə-pä′) *also* **pop·pa** (pä′pə) *n.* [Fr.] Father.

**pa·pa·cy** (pā′pə-sē) *n., pl.* **-cies.** [ME *papacie* < Med. Lat. *papatia* < LLat. *papa,* pope. —see POPE.] **1.** The office and jurisdiction of a pope. **2.** A succession of popes. **3.** The period of time during which a pope is in office. **4. Papacy.** *Rom. Cath. Ch.* The system of church government headed by the pope.

**pa·pa·in** (pə-pā′ĭn, -pī′ĭn) *n.* [< PAPAYA.] A protein-digesting enzyme obtained from the unripe fruit of the papaya and used as a meat tenderizer and medically as a protein digestant.

**pa·pal** (pā′pəl) *adj.* [ME < OFr. < Med. Lat. *papalis* < LLat. *papa,* pope. —see POPE.] **1.** Of, relating to, or issued by the pope <a *papal* audience> **2.** Of or relating to the papacy. **3.** Of or relating to the Roman Catholic Church. **—pa′pal·ly** *adv.*

**Pa·pa·ni·co·laou's test** (pä′pə-nē′kə-louz′) *n.* A Pap test.

**pa·pa·raz·zo** (pä′pə-rät′sō) *n., pl.* **-zi** (-sē) [After Signor *Paparazzo,* a character in *La Dolce Vita,* a movie by Federico Fellini (b. 1920) < dial. Ital. *paparazzo,* a kind of buzzing insect.] A free-lance reporter or photographer who doggedly searches for sensational stories about or takes candid pictures of celebrities.

**pa·pav·er·ine** (pə-păv′ə-rēn′, -ər-ĭn) *n.* [Lat. *papaver,* poppy + -INE.] A nonaddictive opium derivative, $C_{20}H_{21}NO_4$, used medicinally as an antispasmodic.

**pa·paw** *also* **paw·paw** (pô′pô′) *n.* [Prob. < Sp. *papaya,* papaya. —see PAPAYA.] **1.** A central North American tree, *Asimina triloba,* yielding small fleshy edible fruit. **2.** The fruit of the papaw tree.

**pa·pa·ya** (pə-pä′yə) *n.* [Sp., of Cariban orig.] **1.** An evergreen tropical American tree, *Carica papaya,* yielding large yellow edible fruit. **2.** The fruit of the papaya tree.

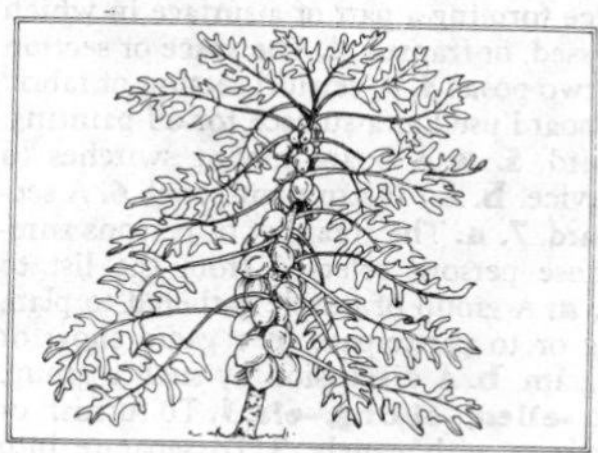

**papaya**

**pa·per** (pā′pər) *n.* [ME *papir* < OFr. *papier* < Lat. *papyrus,* papyrus paper < Gk. *papuros,* papyrus.] **1. a.** A thin material made of cellulose pulp, derived mainly from wood, rags, and certain grasses, processed into flexible leaves or rolls by deposit from an aqueous suspension, and used mainly for writing, printing, drawing, wrapping, and covering walls. **b.** A single sheet of this material. **2.** One or more sheets of paper bearing writing or printing, as: **a.** An official document. **b.** An essay, treatise, or scholarly dissertation. **c.** A written academic assignment or examination. **d.** A newspaper. **3. papers.** A collection of letters, diaries, and other writings, esp. those produced by one person. **4. papers.** Documents establishing the identity of the bearer. **5. papers.** Ship's papers. **6.** A negotiable note, as a bill, check, or letter of credit : COMMERCIAL PAPER. **7.** *Slang.* **a.** A free ticket or pass. **b.** The audience admitted with free passes. *—vt.* **-pered, -per·ing, -pers. 1.** To cover or wrap in paper. **2.** To supply with paper. **3.** To cover with wallpaper. **4.** *Slang.* To issue free passes for (e.g., a play). **—paper over. 1.** To put or keep out of sight : CONCEAL. **2.** To minimize the significance of (e.g, a dispute) by treating lightly or glossing over. *—adj.* **1.** Made of paper. **2.** Resembling paper in thinness or flimsiness. **3.** Existing only in theory : planned but not realized <*paper* gains> **—on paper. 1.** In print or writing.

**2.** In theory <a great idea *on paper*, but difficult to realize> **—pa′per·er** *n.* **—pa′per·y** *adj.*

**pa·per·back** (pā′pər-băk′) *n.* A book with a flexible paper binding. **—pa′per·backed′** *adj.*

**paper birch** *n.* A North American birch tree, *Betula papyrifera*, having paperlike white bark used esp. to make decorative baskets.

**pa·per·board** (pā′pər-bôrd′, -bōrd′) *n.* Cardboard.

**pa·per·bound** (pā′pər-bound′) *adj.* Bound in paper.

**paper boy** *n.* One who delivers or sells newspapers : NEWSBOY.

**paper cutter** *n.* A usu. square, often calibrated board with a long pivoting cutting knife attached to one side, used to trim paper to specific dimensions.

**pa·per·hang·er** (pā′pər-hăng′ər) *n.* **1.** A person whose occupation is covering or decorating walls with wallpaper. **2.** *Slang.* A person who passes bad checks. **—pa′per·hang′ing** *n.*

**pa·per·knife** (pā′pər-nīf′) *n.* A thin dull knife used esp. for opening sealed envelopes.

**paper money** *n.* Currency in the form of government notes and bank notes.

**paper mulberry** *n.* An Asian shade tree, *Broussonetia papyrifera*, with bark that can be processed into a paperlike fabric.

**paper nautilus** *n.* A marine mollusk, *Argonauta argo*, with a paper-thin spiral shell.

**pa·per-thin** (pā′pər-thĭn′) *adj.* Very thin <*paper-thin* partitions>

**paper tiger** *n.* One that appears powerful and dangerous but is actually impotent and weak.

**pa·per-train** (pā′pər-trān′) *vt.* **-trained, -train·ing, -trains.** To train (e.g., a dog) to urinate and defecate on paper indoors.

**paper wasp** *n.* A wasp, as a hornet, that builds paperlike nests.

**pa·per·weight** (pā′pər-wāt′) *n.* A small, heavy, often decorative object placed on top of loose papers to keep them in place.

**pa·per·work** *also* **paper work** (pā′pər-wûrk′) *n.* Work involving the routine handling of documents and forms.

**pap·e·terie** (păp′ĭ-trē, păp-trē′) *n.* [Fr. < *papier*, paper < OFr. —see PAPER.] A box used to hold stationery and writing materials.

**pa·pier-mâ·ché** (pā′pər-mə-shā′, pă-pyā′-) *n.* [Fr. : *papier*, paper + *mâché*, p.part of *mâcher*, to chew.] A material made from paper pulp or shreds of paper mixed with glue or paste that can be molded when wet and painted and varnished when dry. *—adj.* **1.** That is made of papier-mâché. **2.** Not real : ARTIFICIAL.

**pa·pil·la** (pə-pĭl′ə) *n., pl.* **-pil·lae** (-pĭl′ē) [Lat., nipple.] A small nipplelike projection, as a protuberance on the top of the tongue. **—pap′il·lar′y** (păp′ə-lĕr′ē, pə-pĭl′ə-rē) *adj.* **—pap′il·late′** (păp′ə-lāt′, pə-pĭl′ĭt) *adj.* **—pap′il·lose′** (păp′ə-lōs′, pə-pĭl′ōs′) *adj.*

**pap·il·lo·ma** (păp′ə-lō′mə) *n., pl.* **-ma·ta** (-mə-tə) *or* **-mas.** [PAPILL(A) + -OMA.] A small benign epithelial tumor in the breast, intestine, mucous membrane, or skin, appearing as an overgrowth of cells on a core of smooth connective tissue. **—pap′il·lom′a·tous** *adj.*

**pap·il·lote** (pä′pē-yōt′, păp′ē-) *n.* [Fr. < *papillon*, butterfly < Lat. *papilio*.] **1.** A frilled paper cover for decorating the bone end of a cooked chop or cutlet. **2.** Oiled parchment in which foods such as meat and fish are baked.

**pa·poose** (pă-po͞os′, pə-) *n.* [Algonquian *papoos*.] A North American Indian infant or young child.

**pa·po·va·vi·rus** (pə-pō′və-vī′rəs) *n.* [PA(PILLOMA) + PO(LYOMA) + VA(CUOLATION) + VIRUS.] Any of a group of animal viruses that are associated with various tumors, as warts or papillomas.

**pap·pus** (păp′əs) *n., pl.* **pap·pi** (păp′ī′) [Lat. < Gk. *pappos*.] A tuft of bristles or a similar structure topping the achene of plants such as dandelions and thistles. **—pap′pose** (-ōs), **pap′pous** (-əs) *adj.*

**pap·py**[1] (păp′ē) *adj.* **-pi·er, -pi·est.** Of or like pap : MUSHY.

**pap·py**[2] (păp′ē) *n., pl.* **-pies.** [Dim. of PAPA.] *Informal.* Father.

**pa·pri·ka** (pă-prē′kə, pə-, păp′rĭ-kə) *n.* [Hung. < Serbian < *papar*, pepper < Gk. *peperi*.] **1.** A mild, powdered seasoning made from sweet red peppers. **2.** A dark to deep or vivid reddish orange.

**Pap smear** (păp) *n.* A Pap test.

**Pap test** *n.* [After George *Papanicolaou* (1883–1962), its inventor.] *Med.* A test in which a smear of a bodily secretion, esp. from the cervix or vagina, is immediately examined for exfoliated cells to detect cancer in an early stage or to evaluate hormonal condition.

**Pap·u·an** (păp′yo͞o-ən) *n.* **1.** A native or inhabitant of Papua New Guinea. **2.** A member of a subgroup of an Oceanic Negroid people of Melanesia. **3.** Any of the indigenous languages of New Guinea, New Britain, and the Solomon Islands.

**pap·ule** (păp′yo͞ol) *also* **pap·u·la** (-yə-lə) *n., pl.* **-ules** *also* **-u·lae** (-yə-lē′) [Lat. *papula*.] A small, inflammatory, congested spot on the skin. **—pap′u·lar** (-yə-lər), **pap′u·lif′er·ous** (-yə-lĭf′ər-əs) *adj.*

**pa·py·ri** (pə-pī′rī′) *n. var. pl. of* PAPYRUS.

**pap·y·rol·o·gy** (păp′ə-rŏl′ə-jē) *n.* The study of papyrus manuscripts.

**pa·py·rus** (pə-pī′rəs) *n., pl.* **-rus·es** *or* **-ri** (-rī′) [ME < Lat. < Gk. *papuros*.] **1.** A tall aquatic sedge, *Cyperus papyrus* of southern Europe and northern Africa. **2. a.** A paper made from the pith or stems of the papyrus, used as a writing material in ancient times. **b.** A document written on papyrus.

**par** (pär) *n.* [< Lat., equal.] **1.** An accepted or normal standard <not feeling up to *par*> **2.** An equality of status, level, or value <on a *par* with the competition> **3. a.** The established face value of a monetary unit expressed in terms of a monetary unit of another country using the same metal standard. **b.** A condition of equality between the face value of a negotiable instrument, as a stock or bond, and its current market value. **4.** The number of golf strokes regarded necessary to complete a hole or course in expert play. *—vt.* **parred, par·ring, pars.** To score par on (a hole or course) in golf. *—adj.* **1.** Equal to the standard : NORMAL. **2.** Of or pertaining to monetary face value.

**par-** *pref. var. of* PARA-[1].

**par·a**[1] (păr′ə) *adj.* [< PARA-[1].] **1.** Of, relating to, or designating positions in a benzene ring separated by two carbon atoms. **2.** Designating a diatomic molecule in which the nuclei have opposite spin directions.

**pa·ra**[2] (pä-rä′, pä′rä) *n.* [Turk. < Pers. *pārah*.] —See table at CURRENCY.

**para-**[1] *or* **par-** *pref.* [ME < OFr. < Lat. < Gk. < *para*, beside.] **1.** Beside : near : alongside <*para*thyroid> **2.** Beyond <*para*normal> **3.** Incorrect : abnormal <*par*esthesia> **4.** Resembling : similar to <*para*typhoid> **5.** Subsidiary : assistant <*para*professional> **6.** Isomeric : polymeric <*par*aldehyde> **7.** Substitution of radicals at two opposite ends of the benzene ring <*para*dichlorobenzene>

**para-**[2] *pref.* [< PARACHUTE.] Parachute : parachutist <*para*troops>

**-para** *suff.* [Lat. < *parere*, to give birth.] A woman who has given birth to a specified number of children <multi*para*>

**par·a-a·mi·no·ben·zo·ic acid** (păr′ə-ə-mē′nō-bĕn-zō′ĭk, -ăm′ə-) *n.* A crystalline para form of aminobenzoic acid that is part of the vitamin B complex.

**par·a·bi·o·sis** (păr′ə-bī-ō′sĭs) *n.* Artificial or natural fusion of two organisms, as in the experimental joining of animals for research or the development of Siamese twins. **—par′a·bi·ot′ic** *adj.* **—par′a·bi·ot′i·cal·ly** *adv.*

**par·a·blast** (păr′ə-blăst′) *n.* The food yolk of a meroblastic egg. **—par′a·blas′tic** *adj.*

**par·a·ble** (păr′ə-bəl) *n.* [ME < OFr. *parabole* < LLat. *parabola* < Gk. *parabolē* < *paraballein*, to compare : *para*, beside + *ballein*, to throw.] A simple story illustrating a moral or religious lesson.

**pa·rab·o·la** (pə-răb′ə-lə) *n.* [NLat. < Gk. *parabolē* < *paraballein*, to compare. —see PARABLE.] A plane curve formed by: **a.** A conic section taken parallel to an element of the intersected cone. **b.** The locus of points equidistant from a fixed line and a fixed point not on the line.

**par·a·bol·ic** (păr′ə-bŏl′ĭk) *also* **par·a·bol·i·cal** (-ĭ-kəl) *adj.* **1.** Of or like a parable. **2.** Of or shaped like a parabola <a *parabolic* trajectory> **—par′a·bol′i·cal·ly** *adv.*

**pa·rab·o·loid** (pə-răb′ə-loid′) *n.* [PARABOL(A) + -OID.] A surface having parabolic sections parallel to a single coordinate axis, as a paraboloid of revolution. **—pa·rab′o·loi′dal** (-loid′l) *adj.*

**par·a·chute** (păr′ə-sho͞ot′) *n.* [Fr., blend of *parasol*, parasol + *chute*, fall.] **1.** A folded hemispherical canopy attached by a harness and designed to open to retard free fall from an aircraft. **2.** Any of various unpowered devices similar to a parachute that are used for retarding free-falling or free-speeding motion. **3.** A membranous, winglike extension between the limbs of flying squirrels and certain lizards : PATAGIUM. *—v.* **-chut·ed, -chut·ing, -chutes.** *—vt.* To drop (e.g., supplies) by means of a parachute. *—vi.* To descend by means of a parachute. **—par′a·chut′ic** *adj.* **—par′a·chut′ist** *n.*

**parachute spinnaker** *n. Naut.* An oversized spinnaker used on racing yachts.

**Par·a·clete** (păr′ə-klēt′) *n.* [ME *Paraclit* < OFr. *Paraclet* < LLat. *Paracletus* < Gk. *Paraklētos* < *parakalein*, to invoke : *para*, to the side of + *kalein*, to call.] The Holy Ghost.

**pa·rade** (pə-rād′) *n.* [Fr. < Ital. *parata* < Lat. *parare*, to prepare.] **1. a.** A public procession on a ceremonial or festive occasion. **b.** The occasion or action of making such a procession. **c.** The event itself or the persons involved. **d.** An assembly or congregation, as of promenaders. **2. a.** A ceremonial review of troops. **b.** The troops participating in a review. **c.** The place of assembly for a review of troops. **3.** A succession <a *parade* of incompetent workers> **4.** A movement acquiring momentum and strength from the increasing popularity of its approach to a political issue : BANDWAGON. **5.** An ostentatious display. **6.** A public square or promenade. *—v.* **-rad·ed, -rad·ing, -rades.** *—vt.* **1.** To assemble (troops) for a formal display or review. **2.** To march or walk through or around. **3.** To exhibit ostentatiously : FLAUNT <*paraded* their wealth> *—vi.* **1.** To assemble for a formal military review. **2.** To take part in a parade. **3.** To promenade in a public place. **—pa·rad′er** *n.*

**par·a·di·chlo·ro·ben·zene** (păr′ə-dī-klôr′ə-bĕn′zēn′, -bĕn-zēn′, -klōr′-) *n.* A white crystalline compound, $C_6H_4Cl_2$, used as a germicide and insecticide.

**par·a·digm** (păr′ə-dīm′, -dĭm′) *n.* [ME, example < LLat. *paradigma* < Gk. *paradeigma* < *paradeiknunai*, to exibit : *para*, alongside + *deikunai*, to show.] **1.** A list of all the inflectional forms of a word taken as an example of the conjugation or declension to which it belongs. **2.** An example. **—par′a·dig·mat′ic** (-dĭg-măt′ĭk) *adj.*

**par·a·dise** (păr'ə-dīs', -dīz') *n.* [ME *paradis* < OFr. < LLat. *paradisus* < Gk. *paradeisos*, garden < Avestan *pairi-daēza-* : *pairi*, around + *daēza-*, wall.] **1.** *often* **Paradise.** The Garden of Eden. **2. a.** HEAVEN 2. **b.** An intermediate resting place for righteous souls awaiting the Resurrection. **3.** A place of ideal beauty or loveliness <an island *paradise*> **4.** A state of delight. **—par'a·di·si'a·cal** (păr'ə-dĭ-sī'ə-kəl, -zī'ə-kəl), **par'a·di·sa'i·cal** (-dĭ-sā'ĭ-kəl, -zā'ĭ-kəl) *adj.* **—par'a·di·si'a·cal·ly, par'a·di·sa'i·cal·ly** *adv.*

**par·a·dox** (păr'ə-dŏks') *n., pl.* **-dox·es.** [Lat. *paradoxum* < Gk. *paradoxon* < *paradoxos*, conflicting with expectation : *para*, beyond + *doxa*, opinion < *dokein*, to think.] **1.** A seemingly contradictory statement that may nonetheless be true. **2.** One exhibiting contradictory or inexplicable aspects or qualities. **3.** An essentially self-contradictory assertion based on valid deduction from acceptable premises. **4.** A statement contrary to popular opinion. **—par'a·dox'i·cal** *adj.* **—par'a·dox'i·cal·ly** *adv.* **—par'a·dox'i·cal·ness** *n.*

**par·a·drop** (păr'ə-drŏp') *n.* The delivery of supplies by parachute. *—vt.* **-dropped, -drop·ping, -drops.** To deliver by parachute.

**par·aes·the·sia** (păr'ĭs-thē'zhə) *n. var. of* PARESTHESIA.

**par·af·fin** (păr'ə-fĭn) *n.* [G. : Lat. *parum*, too little + Lat. *affinis*, associated with (from its lack of affinity with other material).] **1.** *Chem.* A white or colorless, waxy, solid hydrocarbon mixture used to make candles, wax paper, lubricants, and sealing materials. **2.** *Chem.* A member of the paraffin series. **3.** *Chiefly Brit.* Kerosene. *—vt.* **-fined, -fin·ing, -fins.** To saturate, impregnate, or coat with paraffin. **—par'af·fin'ic** *adj.*

**paraffin series** *n. Chem.* A homologous group of saturated aliphatic hydrocarbons having the general formula $C_nH_{2n+2}$, the simplest and most abundant of which is methane.

**paraffin wax** *n.* PARAFFIN 1.

**par·a·for·mal·de·hyde** (păr'ə-fôr-măl'də-hīd') *n.* A white solid polymer of formaldehyde, $(HCHO)_n$, where *n* is at least 6, used as a fungicide, disinfectant, and fumigant.

**par·a·gen·e·sis** (păr'ə-jĕn'ĭ-sĭs) *also* **par·a·ge·ne·sia** (-jə-nē'zhə, -zhē-ə) *n.* The successive order in which a formation of associated minerals is generated. **—par'a·ge·net'ic** (-jə-nĕt'ĭk) *adj.*

**par·a·gon** (păr'ə-gŏn', -gən) *n.* [OFr. < OItal. *paragone* < *paragonare*, to test on a touchstone < Gk. *parakonan*, to sharpen : *para*, alongside + *akonē*, whetstone.] **1.** A pattern or model of excellence or perfection <a *paragon* of integrity> **2. a.** An unflawed diamond weighing at least 100 carats. **b.** A very large spherical pearl. **3.** A printing type size of 20 points. *—vt.* **-goned, -gon·ing, -gons.** To parallel or match.

**par·a·graph** (păr'ə-grăf') *n.* [OFr. *paragraphe* < Med. Lat. *paragraphus*, sign marking a new paragraph < Gk. *paragraphos*, line in a dialogue showing a change of speakers < *paragraphein*, to write beside : *para*, beside + *graphein*, to write.] **1.** A distinct division of a written work or composition that expresses a thought or point relevant to the whole but is complete in itself and may consist of one or more sentences. **2.** A mark (¶) used to indicate where a new paragraph should begin or to serve as a reference mark. **3.** A brief article, notice, or announcement, as in a newspaper. *—vt.* **-graphed, -graph·ing, -graphs.** To arrange in or divide into paragraphs. **—par'a·graph'ic, par'a·graph'i·cal** *adj.*

**Par·a·guay tea** (păr'ə-gwī', -gwā') *n.* MATÉ 2.

**par·a·jour·nal·ism** (păr'ə-jûr'nə-lĭz'əm) *n.* Highly subjective journalism. **—par'a·jour'nal·ist** *n.* **—par'a·jour'nal·is'tic** *adj.*

**par·a·keet** (păr'ə-kēt') *n.* [OFr. *paroquet.*] A small parrot, usu. with a long tapering tail.

**par·a·kite** (păr'ə-kīt') *n.* A specialized parachute that is towed by an automobile or motorboat and lifts a person in its harness up and through the air.

**par·al·de·hyde** (pă-răl'də-hīd') *n.* [PAR(A)-[1] + (ACET)ALDEHYDE.] A colorless aromatic liquid polymer, $C_6H_{12}O_3$, of acetaldehyde, used as a solvent and sedative.

**par·a·le·gal** (păr'ə-lē'gəl) *adj.* Of, pertaining to, or being a person with specialized training who assists a lawyer. **—par'a·le'gal** *n.*

**par·al·lax** (păr'ə-lăks') *n.* [Fr. *parallaxe* < Gk. *parallaxis* < *parallassein*, to change : *para*, among + *allassein*, to exchange < *allos*, other.] An apparent change in the direction of an object, caused by a change in observational position that provides a new line of sight. **—par'al·lac'tic** (-lăk'tĭk) *adj.*

**par·al·lel** (păr'ə-lĕl') *adj.* [Lat. *parallelus* < Gk. *parallēlos* : *para*, beside + *allēlōn*, of one another < *allos*, other.] **1.** Being an equal distance apart at every point. **2. a.** Designating two or more straight coplanar lines that do not intersect. **b.** Designating two or more planes, or a line and a plane, that do not intersect. **c.** Designating curves or surfaces everywhere equidistant. **3. a.** Having readily recognized similarities, analogous aspects, or comparable parts. **b.** Having the same direction or tendency. **4.** Containing or marked by corresponding syntactical constructions or forms. **5.** *Mus.* Moving consistently by the same intervals. —Used of two or more melodies. **6.** *Elect.* Designating a circuit or part of a circuit connected in parallel. *—adv.* In a parallel manner or relationship. *—n.* **1.** A line or surface equidistant from another. **2.** One of a set of parallel geometric figures, usu. lines. **3. a.** Something that is analogous to or closely resembles something else. **b.** A comparison indicating analogy or similarity. **4.** The condition of being parallel : near similarity or exact agreement in particulars. **5.** Any of the imaginary lines representing degrees of latitude that encircle the earth parallel to the plane of the equator. **6.** A sign (||) indicating material referred to in a note or reference. **7.** *Elect.* A configuration of two or more two-terminal components connected between two points in a circuit with one terminal of each connected to each of the two points <circuits in *parallel*> *—vt.* **-leled, -lel·ing, -lels. 1.** To make or place parallel to. **2.** To be or extend parallel to. **3.** To be similar or analogous to. **4.** To be or provide an equal or match for. **5.** To demonstrate to be analogous.

☆ ***syns:*** PARALLEL, COEXTENSIVE, COLLATERAL, CONCURRENT *adj. core meaning* : lying in the same plane and not intersecting <*parallel* railroad tracks>

**parallel bars** *pl.n.* **1.** Two horizontal poles set parallel to each other in adjustable upright supports and used in gymnastic exercises. **2.** A competitive gymnastics event in which parallel bars are used.

**par·al·lel·e·pi·ped** (păr'ə-lĕl'ə-pī'pĭd, -pĭp'ĭd) *n.* [Gk. *parallēlepipedon* : *parallēlos*, parallel + *epipedon*, plane surface < *epipedos*, level (*epi*, on + *pedon*, ground).] A solid having six faces, each a parallelogram.

**par·al·lel·ism** (păr'ə-lĕl-ĭz'əm) *n.* **1.** The state or position of parallel relationship. **2.** Likeness, correspondence, or similarity in aspect, course, or tendency. **3. a.** The use of corresponding syntactic forms. **b.** An instance of such use. **4.** *Philos.* The doctrine that to every mental change there corresponds a concomitant, but causally unconnected physical alteration.

**par·al·lel·o·gram** (păr'ə-lĕl'ə-grăm') *n.* [LLat. *parallelogrammum*, Gk. *parallēlogrammon* < *parallēlogrammos*, bound by parallel lines : *parallēlos*, parallel + *grammē*, line.] A four-sided plane figure with opposite sides parallel.

**pa·ral·o·gism** (pə-răl'ə-jĭz'əm) *n.* [Fr. *paralogisme* < LLat. *paralogismus* < Gk. *paralogismos* < *paralogos*, unreasonable : *para*, beyond + *logos*, reason.] *Logic.* Illogical or fallacious reasoning, esp. a faulty argument of whose fallacy the reasoner is unaware. **—pa·ral'o·gist** *n.* **—pa·ral'o·gis'tic** *adj.*

**par·a·lyse** (păr'ə-līz') *v. Chiefly Brit. var. of* PARALYZE.

**pa·ral·y·sis** (pə-răl'ĭ-sĭs) *n., pl.* **-ses** (-sēz') [Lat. < Gk. *paralusis* < *paraluein*, to disable : *para-*, in an injurious way + *luein*, to release.] **1.** Partial or complete loss of the ability to move or have sensation in a bodily part as a result of injury to or disease of its nerve supply. **2.** Stoppage or impairment of activity <a *paralysis* of industrial production> **—par'a·lyt'ic** (păr'ə-lĭt'ĭk) *adj.* & *n.*

**paralysis ag·i·tans** (ăj'ĭ-tănz') *n.* [NLat. : Lat. *paralysis*, palsy + Lat. *agitans*, pr.part. of *agitare*, to shake.] Parkinson's disease.

**par·a·lyze** (păr'ə-līz') *vt.* **-lyzed, -lyz·ing, -lyz·es.** [Fr. *paralyser* < *paralysie*, paralysis < Lat. *paralysis.*] **1.** To affect with paralysis. **2.** To make helpless or unable to move. **3.** To impair the functioning or progress of <The blizzard *paralyzed* traffic.> **—par'a·ly·za'tion** *n.* **—par'a·lyz'er** *n.*

☆ ***syns:*** PARALYZE, BENUMB, NUMB, STUN, STUPEFY *v. core meaning* : to make helpless, as by emotion <Sudden fear *paralyzed* them.>

**par·a·mag·net** (păr'ə-măg'nĭt) *n.* A paramagnetic substance.

**par·a·mag·net·ic** (păr'ə-măg-nĕt'ĭk) *adj.* Pertaining to or denoting a substance in which an induced magnetic field is in the same direction as and greater in strength than the magnetizing field, but much weaker than in ferromagnetic materials. **—par'a·mag·net'i·cal·ly** *adv.* **—par'a·mag'net·ism** (-măg'nĭ-tĭz'əm) *n.*

**par·a·mat·ta** *or* **par·ra·mat·ta** (păr'ə-măt'ə) *n.* [After *Parramatta*, Australia.] A fine light silk-and-wool or cotton-and-wool dress fabric.

**par·a·me·ci·um** (păr'ə-mē'shē-əm, -sē-əm) *n., pl.* **-ci·a** (-shē-ə, -sē-ə) *or* **-ci·ums.** [NLat. *Paramecium*, genus name < Gk. *paramēkes*, oblong : *para*, alongside + *mēkos*, length.] Any of various usu. oval-shaped ciliate protozoans of the genus *Paramecium*, with an oral groove for feeding.

**par·a·med·ic** (păr'ə-mĕd'ĭk) *n.* One trained to supply emergency medical treatment or assist medical professionals.

**par·a·med·i·cal** (păr'ə-mĕd'ĭ-kəl) *adj.* Of, designating, or relating to paramedics or their work.

**par·a·ment** (păr'ə-mənt) *n., pl.* **-ments** *or* **-men·ta** (-mĕn'tə) [ME < Med. Lat. *paramentum* < *parare*, to decorate < Lat., to prepare.] *often* **paraments** *or* **paramenta.** Ecclesiastical hangings or vestments.

**pa·ram·e·ter** (pə-răm'ĭ-tər) *n.* **1.** A variable or arbitrary constant appearing in a mathematical expression, each value of which restricts or determines the specific form of the expression. **2.** *Informal.* **a.** A fixed boundary or limit : CONSTANT. **b.** A typical element <Violence and crime are some of the *parameters* of urban life.> **—par'a·met'ric** (păr'ə-mĕt'rĭk), **par'a·met'ri·cal** *adj.* **—par'a·met'ri·cal·ly** *adv.*

**par·a·mil·i·tar·y** (păr'ə-mĭl'ĭ-tĕr'ē) *adj.* Of, pertaining to, or designating forces organized after a military pattern, esp. as a potential auxiliary military force.

**par·am·ne·sia** (păr'ăm-nē'zhə) *n.* A distortion of memory in which experience and fantasy are confused : DÉJA VU.

**pa·ra·mo** (pä'rə-mō', păr'ə-) *n., pl.* **-mos.** [Sp. *páramo*, open desolate land.] A high treeless plain of tropical South America.

**par·a·morph** (păr'ə-môrf') *n.* A mineral crystal formed or affected by paramorphism.

**par·a·mor·phine** (păr'ə-môr'fēn') *n.* Thebaine.

**par·a·mor·phism** (păr'ə-môr'fĭz'əm) *n.* Structural alteration of a mineral without change of chemical composition. **—par'a·mor'phic, par'a·mor'phous** *adj.*

**par·a·mount** (păr'ə-mount') *adj.* [AN *paramont* : OFr. *par*, by (< Lat. *per*) + OFr. *amont*, above.] **1.** Of foremost importance or concern. **2.** Supreme in rank, power, or authority. *—n.* One of highest power or authority. **—par'a·mount'cy** *n.* **—par'a·mount'ly** *adv.*

**par·a·mour** (păr'ə-mo͝or') *n.* [ME < *par amour*, by way of love < OFr. : *par*, by (< Lat. *per*) + *amour*, love (< Lat. *amor* < *amare*, to love).] A lover, esp. one in an adulterous relationship.

**par·am·y·lum** (pă-răm'ə-ləm) *n.* A starchlike reserve carbohydrate found in various protozoans and algae.

**pa·rang** (pä'răng') *n.* [Malay.] A short, heavy, straight-edged knife used in Malaysia and Indonesia as a tool and weapon.

**par·a·noi·a** (păr'ə-noi'ə) *n.* [NLat. < Gk., *madness* < *paranoos*, demented : *para*, beyond + *nous*, mind.] A nondegenerative, limited, usu. chronic psychosis marked by delusions of persecution or of grandeur, strenuously defended by the patient with apparent logic and reason.

**par·a·noi·ac** (păr'ə-noi'ăk', -noi'ĭk) *n.* One afflicted with paranoia. *—adj.* Of, relating to, or resembling paranoia.

**par·a·noid** (păr'ə-noid') *adj.* **1.** Relating to, typical of, or afflicted with paranoia. **2.** Showing behavior suggestive of paranoia. *—n.* One afflicted with paranoia.

**par·a·nor·mal** (păr'ə-nôr'məl) *adj.* Not within the range of normal experience or scientifically explainable phenomena. **—par'a·nor·ma'li·ty** *n.* **—par'a·nor'mal·ly** *adv.*

**par·an·thro·pus** (păr'ən-thrō'pəs, pă-răn'thrə-pəs) *n., pl.* **-pus·es.** [NLat. *Paranthropus*, genus name : PARA-[1] + Gk. *anthropos*, man.] An extinct anthropoid ape of the genus *Paranthropus*, known from remains found in Olduvai Gorge, Tanzania.

**par·a·pet** (păr'ə-pĭt, -pĕt') *n.* [Fr. < Ital. *parapetto* : *parare*, to shield + *petto*, chest < Lat. *pectus*.] **1.** A low protective railing or wall along the edge of a roof or balcony. **2.** An earthen or stone embankment protecting soldiers from enemy fire. **—par'a·pet'ed** (-pĕt'ĭd) *adj.*

**par·aph** (păr'əf, pə-răf') *n.* [OFr. *parraphe* < Med. Lat. *paragraphus*, paragraph sign. —see PARAGRAPH.] A flourish made after or below a signature, orig. to prevent forgery.

**par·a·pher·na·lia** (păr'ə-fər-nāl'yə, -fə-nāl'yə) *pl.n.* [Med. Lat., a married woman's property exclusive of her dowry < Gk. *parapherna* : *para*, beyond + *phernē*, dowry.] *(sing. or pl. in number).* **1.** Personal belongings. **2.** The articles used in a given activity : EQUIPMENT <cooking *paraphernalia*> **3.** A married woman's personal property exclusive of her dowry, according to common law.

**par·a·phrase** (păr'ə-frāz') *n.* [OFr. < Lat. *paraphrasis* < Gk. < *paraphrazein*, to paraphrase : *para*, alongside + *phrazein*, to show.] **1.** A restatement of a text or passage in another form or other words, often to clarify meaning. **2.** The making of paraphrases, often used as a teaching device. *—vt. & vi.* **-phrased, -phras·ing, -phras·es.** To express in or compose a paraphrase. **—par'a·phras'a·ble** *adj.* **—par'a·phras'er** *n.*

☆ **syns:** PARAPHRASE, REPHRASE, RESTATE, REWORD, TRANSLATE *v. core meaning* : to express the meaning of in other, esp. simpler words <*paraphrased* a passage from Shakespeare>

**par·a·phras·tic** (păr'ə-frăs'tĭk) *or* **par·a·phras·ti·cal** (-tĭ-kəl) *adj.* [Fr. *paraphrastique* < Gk. *paraphrastikos* < *paraphrazein*, to paraphrase.] **1.** Like a paraphrase. **2.** Explaining or translating more amply or clearly. **—par'a·phras'ti·cal·ly** *adv.*

**pa·raph·y·sis** (pə-răf'ĭ-sĭs) *n., pl.* **-ses** (-sēz') [PARA-[1] + Gk. *phusis*, nature.] One of the sterile filaments accompanying the spore-carrying or sexual organs of certain fungi or other cryptogamic plants.

**par·a·ple·gi·a** (păr'ə-plē'jē-ə, -jə) *n.* [NLat. < Gk. *paraplēgia*, hemiplegia < *paraplēssein*, to strike on one side : *para*, beside + *plēssein*, to strike.] Complete paralysis of the lower half of the body including both legs that is caused by injury to or disease of the spinal cord. **—par'a·ple'gic** (-plē'jĭk) *adj. & n.*

**par·a·po·di·um** (păr'ə-pō'dē-əm) *n., pl.* **-di·a** (dē-ə). One of the fleshy paired appendages of each body segment in some annelids that function in locomotion and breathing.

**par·a·pro·fes·sion·al** (păr'ə-prə-fĕsh'ə-nəl) *n.* A trained worker who is not a member of a given profession but who assists a professional.

**par·a·psy·chol·o·gy** (păr'ə-sī-kŏl'ə-jē) *n.* The study of phenomena, as telepathy, clairvoyance, and psychokinesis, that are not explainable by known natural laws. **—par'a·psy'cho·log'i·cal** *adj.* **—par'a·psy·chol'o·gist** *n.*

**par·a·quat** (păr'ə-kwŏt') *n.* [PARA-[1] + QUAT(ERNARY).] A yellow compound, $C_{12}H_{14}N_2Cl_2$, used as a herbicide.

**Pa·rá rubber** (pə-rä', păr'ə) *n.* [After *Pará*, a state in Brazil.] Rubber obtained from various tropical South American trees of the genus *Hevea*, esp. *H. brasiliensis.*

**par·a·sang** (păr'ə-săng') *n.* [Lat. *parasanga* < Gk. *parasangas*, of Iranian orig.] An ancient Persian unit of distance, usu. estimated at 3½ miles.

**par·a·se·le·ne** (păr'ə-sĭ-lē'nē) *n., pl.* **-nae** (-nē) [PARA-[1] + Gk. *selēnē*, moon.] A luminous spot on a lunar halo. **—par'a·se·le'nic** (-lē'nĭk, -lĕn'ĭk) *adj.*

**par·a·site** (păr'ə-sīt') *n.* [OFr., a person who continually eats at the expense of another < Lat. *parasitus* < Gk. *parasitos* : *para*, beside + *sitos*, grain.] **1.** *Biol.* An organism that grows, feeds, and is sheltered on or in a different organism while contributing nothing to the survival of its host. **2.** A person who habitually exploits or takes advantage of the generosity of others. **3.** A sycophant. **4.** A professional dinner guest, esp. in ancient Greece.

**par·a·sit·ic** (păr'ə-sĭt'ĭk) *also* **par·a·sit·i·cal** (-ĭ-kəl) *adj.* **1.** Of, relating to, or typical of a parasite. **2.** Caused by a parasite, as a disease. **—par'a·sit'i·cal·ly** *adv.*

**par·a·sit·i·cide** (păr'ə-sĭt'ĭ-sīd') *n.* An agent or preparation used to destroy parasites. *—adj.* Destructive to parasites. **—par'a·sit'i·ci'dal** (-sīd'l) *adj.*

**par·a·sit·ism** (păr'ə-sĭ-tĭz'əm, -sī-) *n.* **1.** The typical mode of existence or behavior of a parasite. **2.** A diseased condition caused by parasitic infestation.

**par·a·sit·ize** (păr'ə-sĭ-tīz', -sī-) *vt.* **-ized, -iz·ing, -iz·es.** To live on (a host) as a parasite.

**par·a·si·tol·o·gy** (păr'ə-sĭ-tŏl'ə-jē, -sī-) *n.* The scientific study of parasitism. **—par'a·si'to·log'ic** (-sĭ'tə-lŏj'ĭk), **par'a·si'to·log'i·cal** *adj.* **—par'a·si·tol'o·gist** *n.*

**par·a·sit·o·sis** (păr'ə-sĭ-tō'sĭs, -sī-) *n.* Parasitic infestation or a disease resulting from it.

**par·a·sol** (păr'ə-sôl', -sŏl') *n.* [Fr. < OItal. *parasole* : *parare*, to shield + *sole*, sun < Lat. *sol.*] A light, usu. small umbrella used esp. by women as a sunshade.

**par·a·sym·pa·thet·ic nervous system** (păr'ə-sĭm'pə-thĕt'ĭk) *n. Anat.* The part of the autonomic nervous system originating in the central and back parts of the brain and in the lower part of the spinal cord that, in general, inhibits or opposes the physiological effects of the sympathetic nervous system, as in tending to stimulate digestive secretions, slowing the heart, and dilating blood vessels.

**par·a·syn·the·sis** (păr'ə-sĭn'thĭ-sĭs) *n.* The formation of words by a combination of compounding and adding an affix, as in *downhearted*, formed from *down* plus *heart* plus *-ed*, not *down* plus *hearted*. **—par'a·syn·thet'ic** (-thĕt'ĭk) *adj.*

**par·a·tax·is** (păr'ə-tăk'sĭs) *n.* [Gk., a placing side by side < *paratassein*, to arrange side by side : *para*, beside + *tassein*, to arrange.] The coordination of grammatical elements such as phrases or clauses without the use of coordinating elements such as conjunctions, as *It was hot; the grass turned brown.* **—par'a·tac'tic** (-tăk'tĭk), **par'a·tac'ti·cal** *adj.* **—par'a·tac'ti·cal·ly** *adv.*

**par·a·thi·on** (păr'ə-thī'ŏn') *n.* [PARA-[1] + thio(phosphate) + -ON.] A highly poisonous yellowish liquid, $(C_2H_5O)_2P(S)OC_6H_4NO_2$, used as an agricultural insecticide.

**par·a·thy·roid·ec·to·my** (păr'ə-thī'roi-dĕk'tə-mē) *n., pl.* **-mies.** Surgical excision of the parathyroid glands.

**par·a·thy·roid gland** (păr'ə-thī'roid') *n.* Any of four small kidney-shaped glands that lie in pairs near the lateral lobes of the thyroid gland and secrete a hormone necessary for calcium and phosphorus metabolism.

**par·a·troop** (păr'ə-tro͞op') *adj.* Of or relating to paratroops.

**par·a·troop·er** (păr'ə-tro͞o'pər) *n.* A member of the paratroops.

**par·a·troops** (păr'ə-tro͞ops') *pl.n.* Infantry trained and equipped to parachute from an aircraft.

**par·a·ty·phoid fever** (păr'ə-tī'foid') *n.* An acute intestinal disease that resembles typhoid fever and is caused by any of three bacteria of the genus *Salmonella.*

**par·a·vane** (păr'ə-vān') *n.* A device equipped with sharp teeth and towed alongside a ship to cut the mooring cables of submerged mines.

**par·boil** (pär'boil') *vt.* **-boiled, -boil·ing, -boils.** [ME *parboilen* < *parboilen*, to boil thoroughly < OFr. *parboillir* < LLat. *perbullire* : Lat. *per*, thoroughly + *bullire*, to boil.] **1.** To cook partially by boiling briefly. **2.** To subject to intense, usu. uncomfortable heat.

**par·buck·le** (pär'bŭk'əl) *n.* [Orig. unknown.] **1.** A rope sling for rolling cylindrical objects up or down an inclined plane. **2.** A sling for lowering or raising an object vertically. *—vt.* **-led, -ling, -les.** To lower or raise with a parbuckle.

**Par·cae** (pär'sē) *pl.n.* [Lat.] *Rom. Myth.* The Three Fates.

**par·cel** (pär'səl) *n.* [ME, portion < OFr. *parcelle* < Lat. *particula*, dim. of *pars*, part.] **1.** Something wrapped up or packaged. **2.** A portion or plot of land, usu. a division of a larger area. **3.** A quantity of merchandise offered for sale. **4.** A group or company : PACK <a *par-*

*cel* of fools> —*vt.* **-celed, -cel·ing, -cels. 1.** To divide into parts and distribute. **2.** To make into a parcel : PACKAGE. **3.** *Naut.* To wind protective strips of canvas around (rope).

**parcel post** *n.* The branch of the postal service that handles and delivers parcels sent through the mail.

**par·ce·nar·y** (pär′sə-nĕr′ē) *n., pl.* **-ies.** [AN *parcenarie* < OFr. *parçonerie* < *parçonier,* partner. —see PARCENER.] COPARCENARY 1.

**par·ce·ner** (pär′sə-nər) *n.* [ME < AN < OFr. *parçonier* < *parçon,* partition < Lat. *partitio.* —see PARTITION.] *Law.* A coparcener.

**parch** (pärch) *v.* **parched, parch·ing, parch·es.** [ME *parchen.*] —*vt.* **1.** To make very dry, esp. by the action of heat <soil *parched* by the sun> **2.** To make thirsty. **3.** To dry or roast (e.g., corn) by exposing to heat. —*vi.* **1.** To become very dry. **2.** To become thirsty.

**Par·chee·si** (pär-chē′zē). A trademark for a board game based on the ancient game of pachisi.

**parch·ment** (pärch′mənt) *n.* [ME *perchement* < OFr. *parchemin,* alteration of Lat. *pergamina* < Gk. *pergamēnē* < *Pergamēnos,* of Pergamum, after *Pergamon* (Pergamum), an ancient Greek city in Asia Minor.] **1.** The skin of a sheep or goat prepared for writing or painting on. **2. a.** A written text or drawing on a sheet of parchment. **b.** An academic diploma. **3.** Paper made in imitation of parchment.

**pard** (pärd) *n.* [ME *parde* < OFr. < Lat. *pardus* < Gk. *pardos.*] A large cat, esp. a leopard.

**par·don** (pär′dn) *vt.* **-doned, -don·ing, -dons.** [ME *pardonen* < OFr. *pardoner* < LLat. *perdonare,* to give wholeheartedly : Lat. *per* (intensive) + Lat. *donare,* to give.] **1.** To release from punishment : FORGIVE. **2.** To pass over (an offense) without punishment. **3.** To make courteous allowance for : EXCUSE. —*n.* **1.** The act of forgiving. **2.** *Law.* **a.** The exemption of a convicted person from the penalties of an offense or crime by the power of the executor of the laws. **b.** The official document or warrant declaring such an exemption. **3.** *Rom. Cath. Ch.* INDULGENCE 5. **—par′don·a·ble** *adj.* **—par′don·a·ble·ness** *n.* **—par′don·a·bly** *adv.*

☆ ***syns:*** PARDON, EXCUSE, FORGIVE *v. core meaning* : to release from punishment for an offense <a criminal *pardoned* by the governor> PARDON also may be used to mean overlooking (an offense) without demanding punishment, placing blame, or subjecting the perpetrator to disfavor <Please *pardon* my rudeness.> Similarly, EXCUSE suggests overlooking an offense and making allowances for what led up to it <asked them to *excuse* us for what we said yesterday> FORGIVE implies giving up all resentment as well as all claim to retribution <*forgave* them for stealing the family silver>

**par·don·er** (pär′dn-ər) *n.* **1.** One who pardons. **2.** A medieval ecclesiastic authorized to raise money for religious works by granting papal indulgences to contributors.

**pare** (pâr) *vt.* **pared, par·ing, pares.** [ME *paren* < OFr. *parer,* to prepare, adorn < Lat. *parare.*] **1.** To remove the outer covering or skin of by peeling <*pare* apples> **2.** To remove by or as if by cutting, clipping, or shaving <*pare* one's nails> **3.** To whittle away. **—par′er** *n.*

**par·e·gor·ic** (păr′ə-gôr′ĭk, -gŏr′-) *n.* [LLat. *paregoricus,* soothing < Gk. *parēgorikos* < *parēgoros,* consoling.] Camphorated tincture of opium, used esp. to relieve diarrhea and intestinal pain.

**pa·ren·chy·ma** (pə-rĕng′kə-mə) *n.* [NLat. < Gk. *parenkhuma,* visceral flesh < *parenkhein,* to pour in beside : *para,* beside + *en,* in + *khein,* to pour.] **1.** *Anat.* The tissue characteristic of an organ. **2.** *Bot.* Tissue consisting of soft unspecialized thin-walled cells. **—pa·ren′chy·mal, par′en·chym′a·tous** (păr′ĕn-kĭm′ə-təs) *adj.* **—par′en·chym′a·tous·ly** *adv.*

**par·ent** (pâr′ənt) *n.* [ME < OFr. < Lat. *parens* < pr.part. of *parere,* to give birth.] **1.** A mother or father. **2.** A forefather : ancestor. **3.** An organism that produces or generates another. **4.** A guardian : protector. **5.** The cause or source : ORIGIN. —*v.* **-ent·ed, -ent·ing, -ents. 1.** To act or serve as a parent to. **2.** To cause to come into being : ORIGINATE. **—par′ent·hood′** *n.*

**par·ent·age** (pâr′ən-tĭj) *n.* **1.** Descent or derivation from parents or ancestors : LINEAGE. **2.** Derivation from a source. **3.** The state or relationship of being a parent.

**pa·ren·tal** (pə-rĕn′tl) *adj.* **1.** Of, pertaining to, or typical of a parent. **2.** *Genetics.* Designating the generation from which a genetic experiment begins. **—pa·ren′tal·ly** *adv.*

**par·en·ter·al** (pă-rĕn′tər-əl) *adj.* **1.** *Physiol.* Located outside the alimentary canal. **2.** *Med.* Taken into the body or administered in a manner other than through the digestive tract, as by intravenous or intramuscular injection.

**pa·ren·the·sis** (pə-rĕn′thə-sĭs) *n., pl.* **-ses** (-sēz′) [LLat. < Gk. < *parentithenai,* to insert : *para,* beside + *en,* in + *tithenai,* to put.] **1.** Either or both of the upright curved lines, ( ), used to mark off explanatory or qualifying remarks in writing or printing. **2.** *Math.* A parenthesis used as one of a pair to enclose a sum, product, or other expression considered or treated as a collective entity in a mathematical operation. **3. a.** An amplifying or qualifying phrase occurring within a sentence in such a way as to form an interpolation independent of the surrounding syntactical structure. **b.** A comment departing from the theme of discourse : DIGRESSION. **4.** An interruption of continuity.

**par·en·thet·i·cal** (păr′ən-thĕt′ĭ-kəl) *also* **par·en·thet·ic** (-ĭk) *adj.* **1.** Contained in or as if in parentheses : qualifying or explanatory <a *parenthetical* aside> **2.** Having or using parentheses. **—par′en·thet′i·cal·ly** *adv.*

**par·ent·ing** (pâr′ən-tĭng, păr′-) *n.* The rearing of children, esp. the care, love, and guidance given by parents.

**pa·re·sis** (pə-rē′sĭs, păr′ĭ-sĭs) *n.* [NLat. < Gk. < *parienai,* to let fall : *para,* beside + *hienai,* to throw.] *Pathol.* **1.** Partial paralysis. **2.** General paresis. **—pa·ret′ic** (pə-rĕt′ĭk) *n.* & *adj.* **—pa·ret′i·cal·ly** *adv.*

**par·es·the·sia** *also* **par·aes·the·sia** (păr′ĭs-thē′zhə) *n. Pathol.* Abnormal or impaired skin sensation, as burning, prickling, itching, or tingling. **—par′es·thet′ic** (-thĕt′ĭk) *adj.*

**pa·re·u** (pä′rā-o͞o′) *n.* [Tahitian.] A rectangular piece of cloth worn in Polynesia as a loincloth or wraparound skirt.

**pa·re·ve** (pä′rə-və) *also* **par·ve** (pär′və) *adj.* [Yiddish *parev*†.] Designating or relating to foods that are made without meat, milk, or derived products and therefore may be eaten with meat or dairy dishes, according to Jewish dietary laws.

**par ex·cel·lence** (pär ĕk-sə-läNS′) *adj.* [Fr. : *par,* by + *excellence,* excellence.] Pre-eminent <a musician *par excellence*>

**par·fait** (pär-fā′) *n.* [Fr. < *parfait,* perfect < Lat. *perfectus.* —see PERFECT.] **1.** A dessert of cream, eggs, sugar, and flavoring frozen together and served in a tall glass. **2.** A dessert of variously flavored layers of ice cream or ices served in a tall glass.

**parfait glass** *n.* A tall slender glass with a short stem, used in serving a parfait.

**par·fleche** (pär′flĕsh′) *n.* [Canadian Fr.] **1.** Rawhide soaked in lye and water to remove the hair and then dried on a stretcher. **2.** An article, as a shield or case, made of parfleche.

**par·get** (pär′jĭt) *n.* [ME < *pargetten,* to parget < OFr. *parjeter,* to throw onto a surface : *par,* onto (< Lat. *per*) + *jeter,* to throw (< Lat. *jacere*).] **1.** Plaster, roughcast, or a similar mixture used to coat walls or line chimneys. **2.** Ornamental plasterwork. **3.** A cement mixture for waterproofing outer walls. —*vt.* **-get·ed, -get·ing, -gets.** To cover or ornament with parget. **—par′get·ing** *n.*

**par·he·li·a** (pär-hē′lē-ə, -hēl′yə) *n. pl. of* PARHELION.

**par·he·lic circle** *also* **par·he·lic ring** (pär-hē′lĭk) *n.* A luminous halo visible at the height of the sun and parallel to the horizon, caused by the sun's rays reflecting off atmospheric ice crystals.

**par·he·li·on** (pär-hē′lē-ən, -hēl′yən) *n., pl.* **-he·li·a** (-hē′lē-ə, -hēl′yə) [Lat. *parelion* < Gk. *parēlion* : *para,* beside + *hēlios,* sun.] A bright spot sometimes appearing to either side of the sun, often on a luminous ring or halo. **—par·he′lic** (-hē′lĭk) *adj.*

**pa·ri·ah** (pə-rī′ə) *n.* [Tamil *paṛaiyan* < *paṛai,* drum.] **1.** A member of a low caste of domestic and agricultural workers in southern India and Burma. **2.** A social outcast.

**pa·ri·es** (pâr′ē-ēz′) *n., pl.* **pa·ri·e·tes** (pə-rī′ə-tēz′) [NLat. < Lat., wall.] *Biol. often* **parietes.** The wall of an organ.

**pa·ri·e·tal** (pə-rī′ĭ-təl) *adj.* [Fr. *pariétal* < LLat. *parietalis* < Lat. *paries,* wall.] **1.** *Biol.* Relating to or forming the wall of a hollow structure. **2.** *Anat.* Of or relating to either of the parietal bones. **3.** *Bot.* Attached to the ovary wall. —Used of the ovules or placenta in certain plants. **4.** Dwelling within or having authority within the environs or buildings of a college or university. —*n.* **1.** A parietal part, as a wall, bone, or plate. **2. parietals.** The rules governing the visiting privileges of members of the opposite sex in campus dormitories.

**parietal bone** *n. Anat.* Either of two large irregularly quadrilateral bones between the frontal and occipital bones that together form the sides and top of the skull.

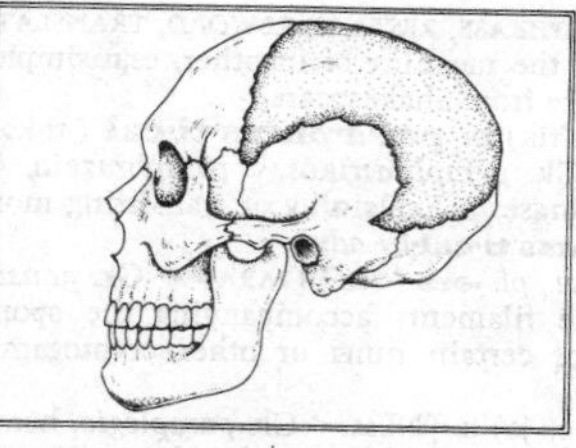

**parietal bone**

**parietal cell** *n.* One of the large peripheral cells of the gastric mucosa that secrete hydrochloric acid.

**parietal lobe** *n. Anat.* The division of each hemisphere of the brain that lies beneath each parietal bone.

**pa·ri·e·tes** (pə-rī′ə-tēz′) *n. pl. of* PARIES.

**par·i-mu·tu·el** (păr′ĭ-myo͞o′cho͞o-əl) *n.* [Fr. *pari mutuel* : *pari,* wager + *mutuel,* mutual.] **1.** A system of betting on races whereby the winners divide the total amount bet after deducting management expenses, in proportion to the sums they have wagered individually. **2.** A machine that records pari-mutuel bets and computes the payoffs.

**par·ing** (pâr′ĭng) *n.* **1.** Something, as a skin or peeling, that has been pared off. **2.** The act of cutting back or shaving off <a radical *paring* of expenses>

**paring knife** *n.* A small knife with a short blade and a firm handle used esp. in paring fruit or vegetables.

**pa·ri pas·su** (păr′ē päs′o͞o) *adv.* [Lat.] With equal pace, speed, or progress : side by side.

**Par·is** (păr′ĭs) *n.* [Lat. < Gk.] *Gk. Myth.* The prince of Troy whose abduction of Helen provoked the Trojan War.

**Paris daisy** *n.* MARGUERITE 1.

**Paris green** *n.* [After *Paris,* France.] A poisonous emerald-green powder, $(CuO)_3As_2O_3 \cdot Cu(C_2H_3O_2)_2$, used as a pigment, wood preservative, and insecticide.

**par·ish** (păr′ĭsh) *n.* [ME *parisshe* < OFr. *paroche* < LLat. *parochia* < LGk. *paroikia* < *paroikos,* Christian < Gk., neighbor : *para,* beside + *oikos,* house.] **1.** An administrative part of a diocese that has its own church, as in the Anglican, Roman Catholic, and some other churches. **2.** *Chiefly Brit.* A political division of a county for local civil government, usu. corresponding to the ecclesiastical parish. **3.** An administrative subdivision in Louisiana that corresponds to a county in other U.S. states. **4.** Parishioners.

**pa·rish·ion·er** (pə-rĭsh′ə-nər) *n.* [ME *parisshoner* < *paroschien,* parishioner < OFr. *paroissien* < *paroisse,* parish, var. of *paroche.* —see PARISH.] A member of a parish.

**par·i·ty¹** (păr′ĭ-tē) *n., pl.* **-ties.** [LLat. *paritas* < *par,* equal.] **1.** Equality, as in value, position, or amount. **2.** Functional equivalence, as in the development of strategic arms. **3.** The equivalent in value of a sum of money expressed in terms of a different currency at a fixed, official rate of exchange. **4.** Equality of prices of goods or securities in two different markets. **5.** A level for farm-product prices maintained by governmental support and intended to give farmers the same purchasing power they had during a chosen base period. **6.** *Math.* The comparative odd-even relationship between two integers. **7. a.** An intrinsic symmetry property of subatomic particles that is marked by the behavior of the wave function of such particles under reflection through the origin of spatial coordinates. **b.** A quantum number, either +1 (even) or −1 (odd), that mathematically describes this property. **8.** *Computer Sci.* **a.** The state of a binary-coded character under a system in which a character having an even number of digits is assigned the code 0 and one having an odd number is assigned the code 1. **b.** Parity bit.

**par·i·ty²** (păr′ĭ-tē) *n.* [Lat. *parere,* to give birth, bring forth + -ITY.] *Med.* **1.** The condition of having borne offspring. **2.** The number of children borne by one woman.

**parity bit** *n. Computer Sci.* A bit added to a group of bits that indicates parity and is used to check the accuracy of data.

**park** (pärk) *n.* [ME, royal land set aside for hunting < OFr. *parc.*] **1.** A tract of land reserved for public use, as: **a.** An expanse of enclosed grounds for recreational use within or adjoining a town. **b.** A landscaped city square. **c.** A tract of land kept in its natural state. **2.** A stadium or enclosed playing field. **3.** A country estate, esp. one that has extensive gardens, woods, pastures, and game preserves. **4. a.** An area where military vehicles and artillery are serviced and stored. **b.** The materiel kept in such an area. —*v.* **parked, park·ing, parks.** —*vt.* **1.** To put or leave (a vehicle) for a period of time in a certain location. **2.** *Informal.* To place, put, set, or leave, esp. temporarily <You can *park* your books at my house.> **3.** To assemble (artillery or other equipment) in order. —*vi.* To station a vehicle in a parking space. **—park′er** *n.*

**par·ka** (pär′kə) *n.* [Aleut. < R., pelt < Samoyed.] **1.** A hooded fur jacket worn as an outer garment by Eskimos. **2.** A warm usu. hooded jacket.

**Par·ker House roll** (pär′kər) *n.* [After the *Parker House,* a hotel in Boston, Massachusetts.] A yeast-leavened roll, shaped by folding a flat round of dough in half.

**parking lot** *n.* An area for parking motor vehicles.

**parking meter** *n.* A coin-operated device that registers the amount of time purchased for the parking of a motor vehicle.

**parking orbit** *n. Aerospace.* A temporary orbit for spacecraft.

**Par·kin·son·ism** (pär′kĭn-sə-nĭz′əm) *n.* [< PARKINSON'S DISEASE.] **1.** A chronic neurological condition characterized by muscular rigidity, tremor, and impaired motor control. **2.** Parkinson's disease.

**Par·kin·son's disease** (pär′kĭn-sənz) *n.* [After James *Parkinson* (1755–1824).] A progressive nervous disease of later life, characterized by muscular tremor, slowing of movement, partial facial paralysis, peculiarity of gait and posture, and weakness.

**Par·kin·son's Law** (pär′kĭn-sənz) *n.* [After C. Northcote *Parkinson* (b. 1909).] Any of several satirical observations propounded as economic laws, as "work expands to fill the time available for its completion."

**park·way** (pärk′wā′) *n.* A broad highway, often divided by landscaped median strips.

**par·lance** (pär′ləns) *n.* [OFr. < *parler,* to speak. —see PARLEY.] **1.** A particular manner of speaking : IDIOM <medical *parlance*> **2.** Conversation, esp. a parley or debate.

**par·lan·do** (pär-län′dō) *also* **par·lan·te** (-tā) *adj.* [Ital. < *parlare,* to speak < Med. Lat. *parabolare.* —see PARLEY.] *Mus.* Sung in a style suggestive of speech. —Used as a direction. **—par·lan′do** *adv.*

**par·lay** (pär′lā′, -lē) *vt.* **-layed, -lay·ing, -lays.** [Fr. *paroli* < dial. Ital., pl. of *parolo,* a set of dice < *paro,* pair < Lat. *par,* equal.] **1.** To bet (an original wager and its winnings) on a subsequent event. **2.** To maneuver (an asset) to great advantage. —*n.* A bet totaling the sum of an original wager plus its winnings or a series of bets made in such a way.

**parle** (pärl) *n. & v. Archaic. var. of* PARLEY.

**par·ley** (pär′lē) *n., pl.* **-leys.** [Fr. *parlée* < *parler,* to talk < OFr. < Med. Lat. *parabolare* < LLat. *parabola,* discourse. —see PARABLE.] A discussion or conference, esp. between opponents or enemies. —*vi.* **-leyed, -ley·ing, -leys.** To discuss, debate, or confer with another, esp. with an enemy.

**par·lia·ment** (pär′lə-mənt) *n.* [ME, a meeting about national concerns < OFr. *parlement* < *parler,* to talk. —see PARLEY.] **1.** A national representative body having supreme legislative powers within the state. **2. Parliament.** The legislative body of various countries, esp. that of the United Kingdom, consisting of the House of Lords and the House of Commons.

**par·lia·men·tar·i·an** (pär′lə-mĕn-târ′ē-ən) *n.* **1.** An expert in parliamentary procedures, rules, or debate. **2. Parliamentarian.** A supporter of the Long Parliament during the English Civil War and the Commonwealth : ROUNDHEAD. —*adj.* Of or relating to the Long Parliament or to the Roundheads.

**par·lia·men·ta·ry** (pär′lə-mĕn′tə-rē, -mĕn′trē) *adj.* **1.** Of, relating to, or like a parliament. **2.** Stemming from, passed, or decreed by a parliament. **3.** Following the rules and customs of a parliament. **4.** Having a parliament.

**parliamentary law** *n.* A body of rules governing procedure in legislative and deliberative assemblies.

**par·lor** (pär′lər) *n.* [ME *parlur* < AN < OFr. *parler,* to talk. —see PARLEY.] **1.** A room in a private home for entertaining visitors. **2.** A small lounge or sitting room offering limited privacy, as at an inn or tavern. **3.** A room equipped and furnished for a special function or business <a beauty *parlor*><an ice-cream *parlor*>

**parlor car** *n.* A railroad car for day travel having individual reserved seats.

**parlor game** *n.* A game that can be played indoors.

**parlor grand** *n.* A grand piano intermediate in length between a concert grand and a baby grand.

**par·lour** (pär′lər) *n. Chiefly Brit. var. of* PARLOR.

**par·lous** (pär′ləs) *adj.* [ME, alteration of *perilous,* perilous < *peril,* peril. —see PERIL.] **1.** Hazardous : dangerous. **2.** *Obs.* Dangerously cunning. **—par′lous·ly** *adv.*

**Par·ma violet** (pär′mə) *n.* [After *Parma,* Italy.] A variety of violet, *Viola odorata sempervirens,* cultivated for its sweet-smelling pale-lavender flowers.

**Par·me·san** (pär′mə-zän′, -zän′, -zən) *n.* [Orig. a trademark.] A hard, sharp, dry Italian cheese made from skim milk and usu. served grated as a garnish.

**par·mi·gia·na** (pär′mĭ-zhä′nə, -jä′-) *adj.* [Ital., fem. of *parmigiano,* of *Parma,* after *Parma,* Italy.] Made or covered with Parmesan cheese <chicken *parmigiana*>

**Par·nas·si·an¹** (pär-năs′ē-ən) *adj.* [Lat. *parnassius,* of Parnassus < *parnasios,* after *Parnasos* (Parnassus), a mountain in Greece sacred to Apollo and the Muses.] Of or relating to poetry.

**Par·nas·si·an²** (pär-năs′ē-ən) *n.* [Fr. *parnassien,* after *Le Parnasse contemporain,* The Contemporary Parnassian, the group's first anthology of poetry (1866).] A member of a school of late 19th-cent. French poets whose work is marked by detachment and emphasis on metrical form. **—Par·nas′si·an** *adj.*

**pa·ro·chi·al** (pə-rō′kē-əl) *adj.* [ME *parochiell* < OFr. *parochial* < LLat. *parochialis* < *parochia,* parish. —see PARISH.] **1.** Of, relating to, supported by, or located in a parish. **2.** Narrowly restricted : PROVINCIAL <a *parochial* outlook> **—pa·ro′chi·al·ism** *n.* **—pa·ro′chi·al·ist** *n.* **—pa·ro′chi·al·ly** *adv.*

**parochial school** *n.* A school supported by a church parish.

**par·o·dy** (păr′ə-dē) *n., pl.* **-dies.** [Lat. *parodia* < Gk. *parōidia* : *para-,* subsidiary to + *ōidē,* song.] **1.** A literary or artistic work that broadly mimics an author's style and holds it up to ridicule. **2.** A highly inferior imitation : TRAVESTY. —*vt.* **-died, -dy·ing, -dies.** To make a parody of. **—pa·rod′ic** (pə-rŏd′ĭk), **pa·rod′i·cal** *adj.* **—par′o·dist** *n.*

**pa·rol** (pə-rōl′, păr′əl) *n.* [OFr. *parole* < LLat. *parabola,* discourse. —see PARABLE.] *Law.* An oral utterance : word of mouth. —*adj. Law.* Given by parol.

**pa·role** (pə-rōl′) *n.* [Fr., promise < OFr. —see PAROL.] **1.** *Law.* **a.** The conditional release of a prisoner before his or her term has expired. **b.** The duration of such conditional release. **2.** A password used by a military officer of the day or an officer on guard. **3.** Word of honor : PROMISE. —*vt.* **-roled, -rol·ing, -roles.** To release (a prisoner) on parole.

**pa·rol·ee** (pə-rō′lē′) *n.* One who is released on parole.

**par·o·no·ma·sia** (păr′ə-nō-mā′zhə, -zhē-ə) *n.* [Lat. < Gk. < *paronomazein,* to call by a different name : *para,* besides + *onomazein,* to name < *onoma,* name.] Word play, esp. a pun. **—par′o·no·mas′tic** (-măs′tĭk), **par′o·no·ma′sial** *adj.* **—par′o·no·mas′ti·cal·ly** *adv.*

**par·o·nych·i·a** (păr′ə-nĭk′ē-ə) *n.* [Lat. < Gk. *parōnukhia* : *para-,* around + *onux,* nail.] *Med.* Inflammation of the tissue around a nail.

**par·o·nym** (păr′ə-nĭm′) *n.* [Gk. *parōnumon* < *parōnumos,* derivative. —see PARONYMOUS.] A paronymous word. **—par′o·nym′ic** *adj.*

**pa·ron·y·mous** (pə-rŏn′ə-məs) *adj.* [Gk. *parōnumos,* derivative : *para,* besides + *onuma,* name.] Allied by derivation from the same root; e.g., *dutiful* and *duteous.*

**pa·rot·id** (pə-rŏt′ĭd) *n.* A parotid gland. **—pa·rot′id** *adj.*

**pa·rot·i·dec·to·my** (pə-rŏt′ĭ-dĕk′tə-mē) *n., pl.* **-mies.** Surgical excision of the parotid gland.

**pa·rot·id gland** (pə-rŏt′ĭd) *n.* [< Gk. *parōtis, parōtid-,* tumor near the ear : *para,* beside + *ous,* ear.] Either of the largest of the paired salivary glands, situated below and in front of each ear.

**par·o·ti·tis** (păr′ə-tī′tĭs) *also* **pa·rot·i·di·tis** (pə-rŏt′ĭ-dī′tĭs) *n.* Inflammation of the parotid glands, as in mumps. **—par′o·tit′ic** (-tĭt′ĭk) *adj.*

**-parous** *suff.* [Lat. *-parous* < *parere,* to give birth.] Giving birth to : PRODUCING <multi*parous*>

**Par·ou·si·a** (pär′ōō-sē′ə, pə-rōō′zē-ə) *n.* [Gk., presence, arrival < *pareinai,* to be present.] The Second Coming.

**par·ox·ysm** (păr′ək-sĭz′əm) *n.* [Fr. *paroxysme* < Med. Lat. *paroxysmus* < Gk. *paroxusmos* < *paroxunein,* to stimulate : *para* (intensive) + *oxunein,* to goad < *oxus,* sharp.] **1.** A sudden outburst of action or emotion <a *paroxysm* of fear> **2.** *Pathol.* **a.** A crisis in or recurrent intensification of a disease. **b.** A spasm or fit : CONVULSION. **—par′ox·ys′mal** (-sĭz′məl) *adj.* **—par′ox·ys′mal·ly** *adv.*

**par·ox·y·tone** (pă-rŏk′sĭ-tōn′) *adj.* [Gk. *paroxutonos* : *para,* beside + *oxutonos,* oxytone.] Having an acute accent on the penultimate syllable. —Used of certain words in Greek and of certain Romance languages, as French and Portuguese. —*n.* A paroxytone word.

**par·quet** (pär-kā′) *n.* [Fr., parquetry < OFr., dim. of *parc,* enclosure.] **1. a.** The part of the main floor of a theater between the orchestra pit and the parquet circle. **b.** ORCHESTRA 3b. **2.** A parquetry floor. —*vt.* **-quet·ed** (-kād′), **-quet·ing** (-kā′ĭng), **-quets** (-kāz′). **1.** To furnish (a room) with a parquetry floor. **2.** To make (a floor) of parquetry.

**parquet circle** *n.* The section of a theater parquet underneath a rear balcony.

**par·quet·ry** (pär′kĭ-trē) *n., pl.* **-ries.** [Fr. *parqueterie* < *parquet,* parquetry. —see PARQUET.] Wood, often of contrasting colors, worked into an inlaid mosaic and used esp. for floors.

**parr** (pär) *n., pl.* **parr** or **parrs.** [Orig. unknown.] **1.** A salmon during the first two years of its life when it lives in fresh water. **2.** The young of various fishes.

**par·ral** (păr′əl) *n. var. of* PARREL.

**par·ra·mat·ta** (păr′ə-măt′ə) *n. var. of* PARAMATTA.

**par·rel** *also* **par·ral** (păr′əl) *n.* [ME *perell,* alteration of *parail,* equipment, short for *appareil,* apparel. —see APPAREL.] *Naut.* A sliding loop of rope or chain to which a running yard or gaff is fastened, allowing movement of the yard up and down the mast.

**par·ri·cide** (păr′ĭ-sīd′) *n.* [Lat. *parricida.*] **1. a.** One who murders his or her parent or other near relative. **b.** The act of committing such a murder. **2.** One who murders someone to whom he or she owes reverence. **—par′ri·cid′al** (-sīd′l) *adj.* **—par′ri·cid′al·ly** *adv.*

**par·rot** (păr′ət) *n.* [Prob. from *Perrot,* dim. of *Pierre,* Peter.] **1.** Any of numerous tropical and semitropical birds of the order Psittaciformes, having short, hooked bills, brightly colored plumage, and in some species the ability to imitate human speech or other sounds. **2.** A mindless imitator of another's words or actions. —*vt.* **-rot·ed, -rot·ing, -rots.** To repeat or imitate by rote. **—par′rot·er** *n.*

**parrot fever** *n.* Psittacosis.

**par·rot·fish** (păr′ət-fĭsh′) *n., pl.* **parrotfish** *or* **-fish·es.** Any of various brightly colored tropical marine fishes of the family Scaridae, with jaws resembling a parrot's beak.

**par·ry** (păr′ē) *v.* **-ried, -ry·ing, -ries.** [Prob. < Fr. *parez,* imper. of *parer,* to defend < Ital. *parare* < Lat., to prepare.] —*vt.* **1.** To ward off : DEFLECT (e.g., a fencing thrust). **2.** To avoid skillfully : EVADE <*parry* difficult questions> —*vi.* To deflect or ward off a blow. —*n., pl.* **-ries. 1.** The act of warding off or deflecting a blow, esp. in fencing. **2.** An evasion.

**parse** (pärs) *v.* **parsed, pars·ing, pars·es.** [< Lat. *pars orationis,* part of speech.] —*vt.* **1.** To break (a sentence) down into component parts of speech with an analysis of the form, function, and syntactical relationship of each part. **2.** To describe (a word) by stating the part of speech, form, and syntactical relationships within a sentence. —*vi.* To admit of being parsed. **—pars′er** *n.*

**par·sec** (pär′sĕk′) *n.* [PAR(ALLAX) + SEC(OND).] A unit of astronomical length based on the distance from earth at which stellar parallax is one second of arc and equal to 3.258 light years, $1.918 \times 10^{13}$ miles, or $3.086 \times 10^{13}$ kilometers.

**Par·see** *also* **Par·si** (pär′sē, pär-sē′) *n.* [Pers. *Pārsī* < *Pārs,* Persia.] **1.** A member of a Zoroastrian sect in India, descended from Persians. **2.** The Iranian dialect used in the religious literature of the Parsees. **—Par′see′ism** *n.*

**par·si·mo·ni·ous** (pär′sə-mō′nē-əs) *adj.* Characterized by parsimony. **—par′si·mo′ni·ous·ly** *adv.* **—par′si·mo′ni·ous·ness** *n.*

**par·si·mo·ny** (pär′sə-mō′nē) *n.* [ME *parcimony* < Lat. *parsimonia* < *parcere,* to spare.] **1.** Excessive thrift or frugality : STINGINESS. **2.** Economy or simplicity of assumptions in logical formulation.

**pars·ley** (pär′slē) *n.* [ME *persely* < OE *petersilie* < LLat. *petrosilium* < Lat. *petroselinum* < Gk. *petroselinon* : *petra,* rock + *celinon,* celery.] A cultivated herb, *Petroselinum crispum,* with much-divided, curled leaves that are used in cookery.

**pars·nip** (pär′snĭp′) *n.* [ME *pasnepe,* alteration of OFr. *pasnaie* < Lat. *pastinaca* < *pastinum,* a kind of two-pronged dibble.] **1.** A strong-scented plant, *Pastinaca sativa,* cultivated for its long, white, edible root. **2.** The root of the parsnip.

**par·son** (pär′sən) *n.* [ME, rector < OFr. *persone* < Med. Lat. *persona* < Lat., character.] **1.** An Anglican cleric with full authority over a parish under ecclesiastical law. **2.** A cleric in the Reformed tradition.

**par·son·age** (pär′sə-nĭj) *n.* The official residence of a parson.

**parson's nose** *n.* A pope's nose.

**Parsons table** (pär′sənz) *n.* [Prob. < the name *Parsons.*] A usu. rectangular table with straight legs that form its four corners.

**part** (pärt) *n.* [ME < OFr. and OE, both < Lat. *pars.*] **1.** A portion, division, or segment of a whole : PIECE. **2.** Any of several equal portions or fractions into which a whole may be divided. **3.** *Math.* An aliquot part. **4. a.** A division of an animal or plant, as an organ or root. **b. parts.** The external genitals. **5.** A component capable of being separated from a system <an engine *part*> **6.** A role. **7.** One's proper or expected share in responsibility or obligation : DUTY. **8.** *often* **parts.** Individual endowment or ability <a person of many *parts*> **9.** *often* **parts.** A region, land, or territory <unknown *parts*> **10.** The line where the hair on the head is parted. **11.** *Mus.* **a.** One of the melodic lines in concerted music or harmony. **b.** A written representation of such a melodic line. —*v.* **part·ed, part·ing, parts.** —*vt.* **1.** To divide or break into separate pieces. **2.** To break up or end (a relationship) by separating. **3.** To separate by or as if by coming between : put or keep apart. **4.** To comb (the hair) away from a dividing line on the scalp. **5.** *Archaic.* To divide into shares or portions. —*vi.* **1.** To divide or break : come apart. **2.** To go away from one another : SEPARATE <We *parted* amicably.> **3.** To separate into ways going in different directions, as a road. **4.** To depart : leave. **5.** To die. **—part with.** To give up : RELINQUISH. —*adv.* In part : PARTIALLY <*part* terrier, *part* spaniel> —*adj.* Not full or complete : PARTIAL <a *part* proprietor> **—for (one's) part.** So far as one is concerned. **—for the most part.** To the greater extent : GENERALLY. **—in good part.** Without taking offense <took our practical joke *in good part*> **—in part.** To some extent : PARTLY. **—on the part of.** Regarding the one specified. **—part and parcel.** A basic part or necessary function. **—take part.** To join in : PARTICIPATE. **—take (someone's) part.** To side with someone in a dispute : SUPPORT.

**par·take** (pär-tāk′) *v.* **-took** (-tŏŏk′), **-tak·en, -tak·ing, -takes.** [Back-formation < *partaker,* one who partakes : PART + TAKER.] —*vi.* **1.** To take part or have a share : PARTICIPATE <*partake* in the celebration> **2.** To take or be given part or portion <*partook* of food and drink> **3.** To have a specified quality or characteristic : show evidence <an occurrence that *partakes* of the mysterious> —*vt.* To take or have part of : SHARE. **—par·tak′er** *n.*

**part·ed** (pär′tĭd) *adj.* **1.** Separated or divided into parts : CLEFT. **2.** Kept apart : SEPARATED. **3.** *Bot.* Cleft almost to the base, so as to have distinct divisions or lobes. **4.** *Archaic.* Deceased.

**par·terre** (pär-târ′) *n.* [Fr. < OFr., ornamental garden < *par terre,* on the ground.] **1.** A parquet circle. **2.** A flower garden having beds and paths arranged in a pattern.

**par·the·no·car·py** (pär′thə-nō-kär′pē) *n.* [Gk. *parthenos,* virgin + *karpos,* fruit.] The production of fruit without fertilization. **—par′the·no·car′pic** *adj.* **—par′the·no·car′pi·cal·ly** *adv.*

**par·the·no·gen·e·sis** (pär′thə-nō-jĕn′ĭ-sĭs) *n.* [Gk. *parthenos,* virgin + GENESIS.] Reproduction of organisms without conjunction of gametes of opposite sexes. **—par′the·no·ge·net′ic** (-jə-nĕt′ĭk) *adj.* **—par′the·no·ge·net′i·cal·ly** *adv.*

**Par·thi·an** (pär′thē-ən) *adj.* [After *Parthia,* an ancient country in Western Asia (from the tactics of Parthian archers, who shot at the enemy while feigning flight).] Of or pertaining to a shot fired by one who is in actual or feigned retreat. **—Par′thi·an** *n.*

**par·tial** (pär′shəl) *adj.* [ME *parcial,* biased < OFr. *partial* < LLat. *partialis* < Lat. *pars,* part.] **1.** Relating to or affecting only part : INCOMPLETE. **2.** Favoring one person or side over another : BIASED. **3.** Having a particular liking for someone or something <*partial* to sports cars> **4.** *Math.* Of, designating, or relating to operations or sequences of operations, such as differentiation and integration, when applied to only one of several variables at a time. —*n.* **1.** *Mus.* HARMONIC 1. **2.** A partial derivative. **—par′tial·ness** *n.*

☆ ***syns:*** PARTIAL, FRACTIONAL, FRAGMENTARY, PART *adj. core meaning* : relating to or affecting only a part <*partial* success><a *partial* solution> ***ant:*** whole

**partial derivative** *n.* The derivative with respect to a single variable of a function of two or more variables, regarding other variables as constants.

**partial differential equation** *n.* A differential equation having at least one partial derivative.

**partial differentiation** *n.* Differentiation with respect to a single variable in a function of several variables, regarding other variables as constants.

**partial fraction** *n.* One of a set of fractions having an algebraic sum equal to a specified fraction.

**par·ti·al·i·ty** (pär′shē-ăl′ĭ-tē, pär-shăl′-) *n., pl.* **-ties. 1.** The state of being partial. **2.** Prejudice or bias. **3.** A special liking : PREDILECTION.

**par·tial·ly** (pär′shə-lē) *adv.* To a degree : not completely.

**partial pressure** *n.* The pressure that one component of a mixture of gases would exert if it were alone in a container.

**partial tone** *n. Mus.* HARMONIC 1.

**par·ti·ble** (pär′tə-bəl) *adj.* [LLat. *partibilis* < Lat. *partiri,* to divide < *pars,* part.] Capable of being parted, divided, or separated : DIVISIBLE.

**par·tic·i·pant** (pär-tĭs′ə-pənt) *n.* One that participates or takes part. *—adj.* Taking part : PARTICIPATING.

**par·tic·i·pate** (pär-tĭs′ə-pāt′) *v.* **-pat·ed, -pat·ing, -pates.** [Lat. *participare, participat-* < *particeps,* partaker : *pars,* part + *capere,* to take.] *—vi.* To join or share with others : take part <*participate* in a bike race> *—vt. Archaic.* To share in : partake of. **—par·tic′i·pance** *n.* **—par·tic′i·pa′tive** *adj.* **—par·tic′i·pa′tor** *n.*

**par·tic·i·pa·tion** (pär-tĭs′ə-pā′shən) *n.* **1.** The act or fact of participating <*participation* in a contest> **2.** A taking part or sharing <each student's *participation* in the class project>

☆ ***syns:*** PARTICIPATION, INVOLVEMENT, SHARING *n. core meaning* : the act or fact of participating <increased *participation* in the political process>

**par·tic·i·pa·to·ry** (pär-tĭs′ə-pə-tôr′ē, -tōr′ē) *adj.* Characterized by or involving participation, esp. affording the opportunity for individual participation, as in the democratic process.

**par·ti·cip·i·al** (pär′tĭ-sĭp′ē-əl) *adj.* [Lat. *participialis* < *participium,* participle.] Of, relating to, consisting of, or formed with a participle. **—par′ti·cip′i·al·ly** *adv.*

**par·ti·ci·ple** (pär′tĭ-sĭp′əl) *n.* [ME < OFr. < Lat. *participium.*] A nominal verb form used with an auxiliary verb to indicate certain tenses and also functioning independently as an adjective.

**par·ti·cle** (pär′tĭ-kəl) *n.* [ME < Lat. *particula,* dim. of *pars,* part.] **1.** A tiny piece or part : SPECK. **2.** A minuscule amount, trace, or degree <not a *particle* of truth> **3.** *Physics.* **a.** A body whose spatial extent and internal motion and structure, if any, are irrelevant in a specific problem. **b.** An elementary particle. **4. a.** One of a class of forms, as prepositions or conjunctions, consisting of a single word that has no inflection. **b.** A suffix or prefix, as *-ly* or *non-.* **5.** A small division or section of something written, as a clause of a document. **6.** *Rom. Cath. Ch.* **a.** A small piece of the consecrated Host. **b.** One of the smaller, individual Hosts.

## SUBATOMIC PARTICLES

| Family Name | Particle Name | Particle Symbol* | Antiparticle Symbol | Mass** | Particle's Electric Charge† | Average Lifetime‡ (sec) |
|---|---|---|---|---|---|---|
| | photon | $\gamma$ | $(\gamma)$ | 0 | 0 | stable |
| lepton | electron's neutrino | $\nu_e$ | $\bar{\nu}_e$ | 0 | 0 | stable |
| | muon's neutrino | $\nu_\mu$ | $\bar{\nu}_\mu$ | 0 | 0 | stable |
| | electron | $e^-$ | $e^+$ | 1 | −1 | stable |
| | muon | $\mu^-$ | $\mu^+$ | 207 | −1 | $2.2 \times 10^{-6}$ |
| meson | pion | $\pi^0$ | $\pi^0$ | 264 | 0 | $0.8 \times 10^{-16}$ |
| | | $\pi^+$ | $\pi^-$ | 273 | +1 | $2.6 \times 10^{-8}$ |
| | kaon | $K^+$ | $K^-$ | 966 | +1 | $1.2 \times 10^{-8}$ |
| | | $K^0$ | $\bar{K}^0$ | 975 | 0 | $0.9 \times 10^{-10}$ or $5.2 \times 10^{-8}$ |
| baryon | ***nucleon*** | | | | | |
| | proton | p | $\bar{p}$ | 1,836 | +1 | stable |
| | neutron | n | $\bar{n}$ | 1,839 | 0 | $1.0 \times 10^{+3}$ |
| | ***hyperon*** | | | | | |
| | lambda | $\Lambda$ | $\bar{\Lambda}$ | 2,183 | 0 | $2.6 \times 10^{-10}$ |
| | sigma | $\Sigma^+$ | $\bar{\Sigma}^-$ | 2,328 | +1 | $0.8 \times 10^{-10}$ |
| | | $\Sigma^0$ | $\bar{\Sigma}^0$ | 2,333 | 0 | $< 1.0 \times 10^{-14}$ |
| | | $\Sigma^-$ | $\bar{\Sigma}^+$ | 2,343 | −1 | $1.7 \times 10^{-10}$ |
| | xi | $\Xi^0$ | $\bar{\Xi}^0$ | 2,572 | 0 | $3.0 \times 10^{-10}$ |
| | | $\Xi^-$ | $\bar{\Xi}^-$ | 2,585 | −1 | $1.7 \times 10^{-10}$ |
| | omega | $\Omega^-$ | $\Omega^+$ | 3,276 | −1 | $1.5 \times 10^{-10}$ |

*Particle symbols are frequently written without superscripts indicating charge; antiparticles are commonly identified by a bar over the particle symbol, but variations using parentheses, or no distinguishing marks at all, are also widely used.

**Approximate masses expressed in terms of the electron's mass.

†Given in terms of the electron's charge, and for particles only; antiparticles have the opposite charge or are neutral if the particle is neutral.

‡Stable means the particle lasts an indefinitely long time and has no known decay mode; the neutral kaon is a composite of two particles, the average lifetime of each of which is given; when bound in a nucleus, the neutron is stable.

**par·ti-col·ored** (pär′tē-kŭl′ərd) *adj.* [Obs. *party,* parti-colored + COLORED.] Having parts or sections colored differently : PIED.

**par·tic·u·lar** (pər-tĭk′yə-lər) *adj.* [ME < *particuler* < OFr. < LLat. *particularis* < Lat. *particula,* dim. of *pars,* part.] **1.** Of, belonging to, or associated with a single person, group, thing, or category. **2.** Separate and distinct from others : SPECIFIC. **3.** Worthy of note : EXCEPTIONAL. **4.** Especially or excessively concerned with details or niceties : FASTIDIOUS. **5.** *Logic.* Encompassing some but not all of a class or group : RESTRICTED. —Used of a proposition. **6.** *Math.* Designating a solution of a differential equation, as distinguished from the general representation of the set of all solutions. *—n.* **1.** An individual item, fact, or detail <mistaken in every *particular*> **2.** *often* **particulars.** Items or details of news or information. **3.** *often* **particulars.** A separate case or individual instance. **4.** *Logic.* A particular proposition. **—in particular.** Particularly : especially <Eat fresh fruit *in particular.*>

▲ word history: The word *particular* is derived from Latin *particula,* "a little part." The derived Latin adjective, *particularis,* originally meant "partial; pertaining to one part." The word was frequently used in contrast to *universalis,* "universal," and in this way came to denote a single specific person or thing.

**par·tic·u·lar·ism** (pər-tĭk′yə-lə-rĭz′əm) *n.* **1.** Exclusive interest in or adherence to one's own group, party, sect, or nation. **2.** A policy of permitting each state in a nation or federation to act independently. **3.** The belief that an individual is elected to salvation and grace by God's free choice rather than by God's foreseeing the individual's response. **—par·tic′u·lar·ist** *n.* **—par·tic′u·lar·is′tic** *adj.*

**par·tic·u·lar·i·ty** (pər-tĭk′yə-lăr′ĭ-tē) *n., pl.* **-ties. 1.** The quality or state of being particular or distinct rather than general. **2.** Exactitude of detail, esp. in description. **3.** Attention to or concern with details : FASTIDIOUSNESS. **4.** A specific detail or point : PARTICULAR. **5.** An individual characteristic : PECULIARITY.

**par·tic·u·lar·ize** (pər-tĭk′yə-lə-rīz′) *v.* **-ized, -iz·ing, -iz·es.** *—vt.* **1.** To enumerate or state in detail : ITEMIZE. **2.** To treat or mention individually : SPECIFY. *—vi.* To give details or particulars. **—par·tic′u·lar·i·za′tion** *n.* **—par·tic′u·lar·iz′er** *n.*

**par·tic·u·lar·ly** (pər-tĭk′yə-lər-lē) *adv.* **1.** To a great degree : ESPECIALLY. **2.** With particular emphasis or reference : SPECIFICALLY. **3.** In a particular manner : INDIVIDUALLY. **4.** In detail.

**par·tic·u·late** (pər-tĭk′yə-lĭt, -lāt′) *adj.* Of, relating to, or made up of separate particles. *—n.* A particulate substance.

**part·ing** (pär′tĭng) *n.* **1.** The act or process of separating or dividing. **2.** A separation or division. **3.** A departure or leave-taking. *—adj.* Relating to, done, given, or said on separating or departing <a *parting* jest> **—parting of the ways.** A point of divergence.

**par·ti pris** (pär′tē prē′) *n.* [Fr. : *parti,* decision + *pris,* p.part. of *prendre,* to take.] An inclination for or against one that inhibits impartial judgment : BIAS.

**par·ti·san¹** (pär′tĭ-zən) *n.* [Fr. < OItal. *partigiano* < *parte,* part < Lat. *pars.*] **1.** A usu. militant advocate or supporter of a party, cause, faction, person, or idea <"I avow myself the *partisan* of truth alone" —William Harvey> **2.** A member of a detached, often unofficially organized body of fighters who attack an enemy within occupied territory : GUERRILLA. *—adj.* **1.** Of, relating to, or typical of a partisan or partisans. **2.** Devoted to or biased in support of a single party or cause. **—par′ti·san·ship′** *n.*

**par·ti·san²** *also* **par·ti·zan** (pär′tĭ-zən) *n.* [OFr. *partizane* < OItal. *partesana,* var. of *partigiano,* supporter. —see PARTISAN¹.] A weapon having a long shaft surmounted by a blade with broad, horizontally projecting cutting edges, used mainly in the 16th and 17th cent.

**par·ti·ta** (pär-tē′tə) *n.* [Ital. < *partire,* divide < Lat. —see PARTITE.] *Mus.* A set of related instrumental pieces, as a suite or series of variations.

**par·tite** (pär′tīt′) *adj.* [Lat. *partitus* < *partire,* to divide < *pars,* part.] Divided into parts.

**par·ti·tion** (pär-tĭsh′ən) *n.* [ME *particioun* < OFr. *partition* < Lat. *partitio* < *partire,* to divide < *pars,* part.] **1. a.** The act or process of dividing into parts. **b.** The state of being so divided. **2.** Something that separates, as a panel, screen, or partial wall dividing a larger area. **3.** A part or section into which something has been divided. **4.** *Math.* **a.** An expression of a positive integer as a sum of positive integers. **b.** The decomposition of a set into a family of mutually exclusive sets. **5.** *Logic.* Analysis of a class into its component parts. *—vt.* **-tioned, -tion·ing, -tions. 1.** To divide into parts, pieces, or sections. **2.** To divide or separate by a partition <*partition* off a dining area> **—par·ti′tion·ist** *n.* **—par·ti′tion·ment** *n.*

**par·ti·tive** (pär′tĭ-tĭv) *adj.* [Lat. *partitivus* < *partire,* to divide. —see PARTITE.] **1.** Serving to divide into parts. **2.** Indicating a part as distinct from a whole; e.g., in the sentence *We spent some of the money, some of the money* is a partitive construction. —*n.* **1.** A partitive word, as *some* or *many.* **2.** A partitive construction or case. **—par'ti·tive·ly** *adv.*

**par·ti·zan** (pär'tĭ-zən) *n. var. of* PARTISAN².

**part·let** (pärt'lĭt) *n.* [ME *patelet* < OFr. *patelete,* band of cloth, dim. of *patte,* paw.] A woman's garment worn esp. in the 16th cent., consisting of a covering for the neck and shoulders with a band or ruffle at the neck.

partlet

**part·ly** (pärt'lē) *adv.* In part : in some degree <*partly* right>

**part·ner** (pärt'nər) *n.* [ME *partener,* alteration of *parcener,* parcener. —see PARCENER.] A person associated with another or others in a common activity or interest, esp.: **a.** A member of a business partnership. **b.** A spouse. **c.** Either of two persons dancing together. **d.** One of a pair or a team in a game or sport, as bridge or tennis. —*vt.* **-nered, -ner·ing, -ners. 1.** To make a partner of. **2.** To bring together as partners. **3.** To be the partner of.

☆ ***syns:*** PARTNER, ALLY, ASSOCIATE, COLLEAGUE, CONFEDERATE *n. core meaning* : one who cooperates with another in a venture, occupation, or challenge <*partners* in business> PARTNER implies a relationship, frequently between two people, in which each person has equal status and a certain independence but also has unspoken or formal obligation to the other or others <law *partners*> A COLLEAGUE is a fellow member of a staff or organization <my editorial *colleagues*> An ALLY is one who, out of a common cause, has taken one's side and can be relied on <were *allies* in the argument><the western *Allies*> CONFEDERATE often has a negative connotation and suggests guilt by association <worked with a *confederate* in the robberies> An ASSOCIATE is anyone who works in the same place as another, usu. in direct contact <one of my *associates* in the office>

**part·ner·ship** (pärt'nər-shĭp') *n.* **1.** The state of being a partner. **2.** A contract entered into by two or more persons in which each agrees to furnish a part of the capital and labor for a business enterprise and by which each shares in some fixed proportion in profits and losses.

**part of speech** *n.* **1.** One of a group of traditional classifications of words according to their functions in context, including the noun, pronoun, verb, adjective, adverb, preposition, conjunction, and interjection, and sometimes the article. **2.** A word considered as a part of speech.

**par·ton** (pär'tŏn') *n.* A hypothetical elementary particle thought to be a constituent of hadrons.

**par·tridge** (pär'trĭj) *n., pl.* **partridge** or **-tridg·es.** [ME *partrich* < OFr. *perdriz* < Lat. *perdix* < Gk.] **1.** A plump-bodied Old World game bird, esp. one of the genera *Perdix* and *Alectoris.* **2.** A bird, as the ruffed grouse or the bobwhite, similar to the partridge.

**par·tridge·ber·ry** (pär'trĭj-bĕr'ē) *n., pl.* **-ries.** A woody, creeping, evergreen plant, *Mitchella repens* of eastern North America, bearing small white flowers and scarlet berries.

**partridge pea** *n.* A plant, *Cassia fasciculata* of eastern and central North America, with yellow flowers.

**part song** *n.* A song with two or more voice parts.

**part-time** (pärt'tīm') *adj.* For or during less than the usual time <*part-time* jobs> **—part'-time'** *adv.* **—part'-tim'er** *n.*

**par·tu·ri·ent** (pär-tyŏŏr'ē-ənt, -tŏŏr'-) *adj.* [Lat. *parturiens, parturient-,* pr.part. of *parturire,* to be in labor < *parere,* to bear.] **1.** Being in labor and about to bring forth young. **2.** Of or relating to giving birth. **3.** Being about to produce or come forth with something, as an idea or discovery. **—par·tu'ri·en·cy** *n.*

**par·tu·ri·fa·cient** (pär-tyŏŏr'ə-fā'shənt, pär-tŏŏr'-) *adj.* [PARTURI(TION) + FACIENT.] Facilitating or inducing labor during childbirth. —*n.* A drug facilitating or inducing childbirth.

**par·tu·ri·tion** (pär'tyŏŏ-rĭsh'ən, -tŏŏ-, pär'chə-) *n.* [LLat. *parturitio* < Lat. *parturire,* to be in labor. —see PARTURIENT.] The act of giving birth : CHILDBIRTH.

**part·way** (pärt'wā') *adv. Informal.* To a certain degree : in part.

**par·ty** (pär'tē) *n., pl.* **-ties.** [ME, a company of persons < OFr. *partie* < *partir,* to divide < Lat. *partire.* —see PARTITE.] **1. a.** A social gathering esp. for pleasure or amusement <a birthday *party*> **b.** A group of persons gathered together to participate in an activity <a camping *party*> **2.** A political group organized to promote and support its principles and candidates for public office. **3.** *Law.* A person or group involved in a legal proceeding. **4.** A participant or accessory <a *party* to the crime> **5.** *Informal.* A person. **6.** A selected group of soldiers <a scouting *party*> —*vi.* **-tied, -ty·ing, -ties.** *Informal.* **1.** To give or attend a party. **2.** To carouse.

**party line** *n.* **1.** A telephone circuit connecting two or more subscribers with a central exchange. **2.** The official policies and principles of a political party to which loyal members are expected to adhere. **—party liner** *n.*

**par·ty-poop·er** (pär'tē-pōō'pər) *n. Slang.* One who declines to participate enthusiastically in the recreation of a group : WET BLANKET.

**party wall** *n. Law.* A wall on the boundary line of adjoining properties that is shared by two owners or tenants.

**pa·rure** (pə-rŏŏr') *n.* [Fr. < OFr., adornment < *parer,* to adorn. —see PARE.] A set of matched jewelry.

**par value** *n.* The value imprinted on a stock certificate or bond that provides the basis for bond interest, preferred stock dividend, or share of equity capital : FACE VALUE.

**par·ve** (pär'və) *adj. var. of* PAREVE.

**par·ve·nu** (pär'və-nōō', -nyōō') *n.* [Fr. < p.part of *parvenir,* to arrive < Lat. *pervenire* : *per,* through + *venire,* to come.] One who has suddenly risen above one's social and economic class without the qualifications or background for the new status. **—par've·nu'** *adj.*

**par·vis** (pär'vĭs) *n.* [ME < OFr. < LLat. *paradisus,* garden, paradise. —see PARADISE.] **1.** An enclosed courtyard or area in front of a church or palace. **2.** A colonnade or portico in front of a church.

**par·vo·vi·rus** (pär'vō-vī'rəs) *n.* [Lat. *parvus,* small + VIRUS.] Any of a group of DNA-containing animal viruses.

**pas** (pä) *n., pl.* **pas** (pä) [Fr. < Lat. *passus,* step. —see PACE¹.] **1. a.** A ballet step or a series of such steps. **b.** A ballet dance. **2.** The right of precedence <would not yield *pas* to the other princes>

**pas·cal** (păs-kăl', pä-skäl') *n.* [After Blaise *Pascal* (1623–1662).] A unit of pressure equal to one newton per square meter.

**pas·cal celery** *also* **Pas·chal celery** (păs'kəl) *n.* [Orig. unknown.] Any of several types of commercially grown celery with unblanched, green stalks.

**pas·chal** (păs'kəl) *adj.* [ME *paskal* < OFr. *pascal* < LLat. *paschalis* < *pascha,* Passover, Easter < LGk. *paskha* < Heb. *pesaḥ,* Pesach.] Of or relating to Passover or to Easter.

**paschal lamb** *n.* **1.** A lamb eaten at the feast of the Passover. **2. Paschal Lamb.** AGNUS DEI 1.

**pas de deux** (pä də dœ') *n., pl.* **pas de deux.** [Fr. *pas,* step + *de,* of + *deux,* two.] A ballet figure or dance for two persons.

**pa·se** (pä'sā) *n.* [Sp. < *pasar,* to pass < VLat. ***passare.* —see PASS.] A movement of the cape by the matador to attract and direct the charge of the bull.

**pa·sha** *also* **pa·cha** (pä'shə, păsh'ə, pə-shä') *n.* [Turk. *paşa.*] An honorary title once placed after the name of a Turkish military or civil official.

**Pash·to** (pŭsh'tō) *also* **Push·tu** (-tōō) *n.* [Pers. *pashtu* < Pashto.] An Iranian language that is the principal vernacular language of Afghanistan and parts of western Pakistan.

**Pa·siph·a·ë** (pə-sĭf'ə-ē') *n.* [Lat. < Gk. *Pasiphaē.*] *Gk. Myth.* The wife of Minos and mother, by a white bull, of the Minotaur.

**pasque·flow·er** (păsk'flou'ər) *n.* [Blend of OFr. *pasque,* Easter, and E. *passeflower,* pasqueflower < OFr. *passefleur* : *passer,* to pass + *fleur,* flower.] A plant of the genus *Anemone,* with large blue, purple, or white flowers and plumed fruit.

**pas·qui·nade** (păs'kwə-nād') *n.* [Fr. < Ital. *pasquinata,* after *Pasquino,* the nickname given to a statue in Rome, Italy on which lampoons were posted.] **1.** A lampoon posted in a public place. **2.** Satire. —*vt.* **-nad·ed, -nad·ing, -nades.** To lampoon with a pasquinade. **—pas'qui·nad'er** *n.*

**pass** (păs) *v.* **passed, pass·ing, pass·es.** [ME *passen* < OFr. *passer* < VLat. **passare* < Lat. *passus,* step. —see PACE¹.] —*vi.* **1.** To move on or ahead : PROCEED. **2.** To extend : run <The stream *passes* through the meadow.> **3.** To gain passage despite obstacles <*pass* through hard times> **4.** To catch up with and move past another vehicle. **5.** To move past in time : ELAPSE <The hours *passed* slowly.> **6.** To be transferred from one to another : CIRCULATE <The gossip *passed* from desk to desk.> **7.** To be communicated or exchanged between persons <Angry words *passed.*> **8.** To be transferred or conveyed to another, as by will or deed. **9.** To undergo transition from one condition, form, quality, or characteristic to another <Twilight *passed* into evening.> **10.** To come to an end <My headache finally *passed.*> **11.** To cease to live : DIE. **12.** To happen : take place <What *passed* during the night?> **13.** To be allowed to occur without notice or challenge <I let their insolence *pass.*> **14.** To undergo an examination or trial with favorable results. **15.** To be approved or adopted <The resolution *passed.*> **16.** *Law.* **a.** To sit in judicial or legal investigation. **b.** To pronounce an opinion, judgment, or sentence. **c.** To sit in adjudication. **17.** To transfer a ball or puck to a teammate in a sport. **18.** To thrust or lunge in fencing. **19.** To let one's turn to play or bid go by in a game. —*vt.* **1.** To go by without stopping. **2. a.** To go by without paying attention to. **b.** To fail to pay (a dividend). **3.** To go beyond : EXCEED <The inheritance *passed* my wildest dreams.> **4.** To go across : go through <*passed* the border into Canada> **5. a.** To undergo (a trial or examination) with favorable results. **b.** To cause or allow to go through a trial, test, or examination successfully. **6. a.** To cause to move <*passed* our hands over the wood> **b.** To cause to move into a certain position <*pass* a ribbon around a package> **c.** To cause to move as part of a process <*pass* gravy through a strainer> **7.** To cause to go by <*pass* the infantry in review> **8.** To allow to go by or elapse : SPEND <*passed* some time in England> **9. a.** To cause to be transferred from one to another : CIRCULATE <*passed* the rumor quickly> **b.** To hand over to someone else <*pass* the salt> **c.** To circulate fraudulently <*pass* rubber checks> **d.** *Law.* To transfer title or ownership of. **10. a.** To throw (a ball) to a teammate in a sport. **b.** *Baseball.* To walk (a batter). **11.** To cross over : issue from <No information will *pass* my lips.> **12.** To discharge (bodily waste) : VOID. **13. a.** To approve : adopt <The committee *passed* the proposal.> **b.** To be sanctioned, ratified, or approved by <The bill *passed* the Senate.> **14.** To pronounce : utter <The judge *passed* sentence.> **15.** To go past without noticing. ***usage:*** *Passed* is the correct spelling for the past tense and past participle of the verb *pass,* as in *days that passed quickly* or the *eons that have passed since life began. Past* is the correct spelling for the corresponding adjective (*in past centuries*), adverb (*raced past*), and preposition (*past noon*). **—pass away. 1.** To go away in time : END. **2.** To die. **3.** To spend or while away (time). **—pass for.** To be accepted as being something one is not <*pass for* a teen-ager> **—pass off.** To offer, sell, or put into circulation (an imitation of something) as genuine. **—pass out. 1.** To lose consciousness. **2.** To die. **—pass over.** To leave out : DISREGARD. **—pass up.** *Informal.* To let go by : REJECT <*pass up* a chance for promotion> *—n.* **1.** The act of passing : PASSAGE. **2.** A way through or on which one can move or travel, esp. a narrow gap between mountain peaks. **3. a.** A permit, ticket, or authorization to come and go at will. **b.** A free ticket entitling one to transportation or admission. **c.** Written leave of absence from military duty. **4.** A sweep or run by an aircraft over an area or target. **5.** An often critical condition or situation : PREDICAMENT. **6.** A sexual overture or invitation. **7.** A motion of the hand or the waving of a wand for magic. **8. a.** A transfer of a ball or puck between teammates. **b.** A lunge or thrust in fencing. **c.** *Baseball.* A walk. **9.** A refusal to bid, draw, bet, or play in certain games. **10.** A winning throw of the dice in the game of craps. **11.** A pase. **—bring to pass.** To cause to happen. **—come to pass.** To happen. **—in passing.** By the way : PARENTHETICALLY. **—pass muster.** To measure up to a given standard. **—pass the buck.** To shift responsibility or blame to another. **—pass the hat.** To take up a collection. **—pass'er** *n.*

**pass·a·ble** (păs'ə-bəl) *adj.* **1.** Able to be passed, traversed, or crossed, as a road or stream. **2.** Acceptable for general circulation, as currency. **3.** Satisfactory but not outstanding. **—pass'a·ble·ness** *n.* **—pass'a·bly** *adv.*

**pas·sa·ca·glia** (pä'sə-käl'yə, păs'ə-käl'yə) *n.* [Ital. < Sp. *passacalle* : *pasar,* to pass + *calle,* street < Lat. *callis,* path.] *Mus.* A 17th- and 18th-cent. musical form with continuous variations on a ground bass in slow triple meter.

**pas·sage** (păs'ĭj) *n.* [ME < OFr. < *passer,* to pass.] **1.** The act or process of passing, esp.: **a.** A movement from one place to another : TRANSIT. **b.** The process of elapsing. **c.** The process of passing from one stage or condition to another : TRANSITION. **d.** *Obs.* Death. **e.** The enactment into law of a legislative measure. **2.** A journey, esp. by air or water. **3. a.** The right to travel on something, esp. a ship. **b.** The price paid for this right. **4.** The right, permission, or power to come and go freely. **5. a.** A path, channel, or duct through, over, or along which something may pass <the respiratory *passage*> **b.** A corridor. **6.** An occurrence between two persons, esp.: **a.** An exchange of words, arguments, or vows. **b.** An exchange of blows. **7.** A segment of a literary work <a famous Biblical *passage*> **8.** *Mus.* A segment of a composition. **9.** *Med.* Evacuation of the bowels.

**pas·sage·way** (păs'ĭj-wā') *n.* A corridor.

**Pas·sa·ma·quod·dy** (păs'ə-mə-kwŏd'ē) *n., pl.* **Passamaquoddy** or **-dies. 1. a.** A tribe of Indians once inhabiting Maine and New Brunswick, Canada. **b.** A member of this tribe. **2.** The Algonquian language of the Passamaquoddy.

**pas·sant** (păs'ənt) *adj.* [ME < OFr. < pr.part. of *passer,* to pass.] *Heraldry.* Designating a beast facing and walking toward the viewer's right with one front leg raised.

**pass·book** (păs'bo͝ok') *n.* **1.** A bankbook. **2.** A book in which a merchant records credit sales.

**pas·sé** (pă-sā') *adj.* [Fr. p.part. of *passer,* to pass < OFr.] **1.** No longer current or in fashion : OUT-OF-DATE. **2.** Past the prime : AGED.

**passed ball** *n. Baseball.* A fieldable pitch missed by the catcher that allows a runner to advance a base.

**pas·sel** (păs'əl) *n.* [Alteration of PARCEL.] *Informal.* A large number or quantity.

**passe·men·terie** (păs-mĕn'trē) *n.* [Fr. < *passement,* ornamental braid < *passer,* to pass < OFr.] Decorative trimming for a garment, as braid, lace, or metallic beads.

**pas·sen·ger** (păs'ən-jər) *n.* [ME *passager* < OFr. < *passager,* passing < *passage,* passage < *passer,* to pass.] **1.** A traveler in a conveyance, as a train, automobile, or bus, who does not participate in its operation. **2.** A wayfarer or traveler.

**passenger pigeon** *n.* An extinct migratory bird, *Ectopistes migratorius,* abundant in North America until the later 19th cent.

**passe par·tout** (păs pär-to͞o') *n.* [Fr. : *passe,* imper. of *passer,* to pass + *partout,* everywhere.] **1.** Something enabling one to pass or go everywhere, esp. a master key. **2. a.** A mounting for a picture in which colored tape forms the frame. **b.** The tape so used. **3.** A mat used in mounting a picture.

**pas·ser-by** *also* **pas·ser·by** (păs'ər-bī') *n., pl.* **pas·sers-by** *also* **pas·sers·by.** One who passes by, often by chance.

**pas·ser·ine** (păs'ə-rīn') *adj.* [Lat. *passerinus,* of sparrows < *passer,* sparrow.] Relating to or designating birds of the order Passeriformes, which includes perching birds and songbirds, as the jays, blackbirds, finches, warblers, and sparrows. *—n.* A bird of the order Passeriformes.

**pas seul** (pä sœl') *n.* [Fr.] A solo ballet dance or figure.

**pass-fail** (păs'fāl') *n.* A system of grading in which traditional grades are not used and the student either passes or fails.

**pas·si·ble** (păs'ə-bəl) *adj.* [ME < OFr. < Med. Lat. *passibilis* < LLat. < Lat. *pati,* to suffer.] Capable of suffering. **—pas'si·bil'i·ty** *n.*

**pas·sim** (păs'ĭm) *adv.* [Lat. < *passus,* scattered < *pandere,* to scatter.] Frequently : throughout. —Used in textual annotation to indicate that the word or passage occurs frequently in the work cited.

**pass·ing** (păs'ĭng) *adj.* **1.** Of brief duration : TRANSITORY <a *passing* caprice> **2.** Cursory : casual <a *passing* remark> **3.** Passable : satisfactory <C is a *passing* grade.> **4.** *Obs.* Very : great <" 'Tis a *passing* shame" —Shakespeare> *—adv.* Very : surpassingly <a bride who was *passing* fair> *—n.* **1.** The act of one that passes or the fact of having passed. **2.** A place where or a means by which one can pass. **3.** Death <mourned the *passing* of a friend>

**passing note** *n. Mus.* A note that is not part of a particular chord but is placed between two chords to provide a smooth transition from one to the other.

**pas·sion** (păsh'ən) *n.* [ME < OFr. < LLat. *passio,* suffering < Lat. *passus,* p.part. of *pati,* to suffer.] **1.** A powerful emotion or appetite, as love, joy, hatred, anger, or greed. **2. a.** Ardent, adoring love. **b.** Strong sexual desire : LUST. **c.** The object of such love or desire. **3. a.** Boundless enthusiasm <a *passion* for music> **b.** The object of such enthusiasm <Music is my only *passion.*> **4.** An emotional display, esp. of anger. **5. Passion. a.** The sufferings of Christ in the period following the Last Supper and including the Crucifixion. **b.** A Gospel narrative, musical setting, or serial pictorial representation of the Passion. **—pas'sion·less** *adj.*

☆ ***syns:*** PASSION, ARDOR, ENTHUSIASM, FERVOR, FIRE, ZEAL *n. core meaning* : powerful feeling for or about someone or something <loved me with great *passion*><delivered the sermon with *passion*> PASSION is a deep, overwhelming feeling or emotion; when directed toward a person, it usu. indicates love and sexual desire, although it can also refer to hostile emotions <loathed them with *passion*> Used lightly, it suggests avid interest <a *passion* for fast cars> ENTHUSIASM also reflects excitement and responsiveness to specific things <supported the team with *enthusiasm*> ARDOR can suggest great devotion to a cause but commonly means a warm, loving feeling directed toward persons <embraced with *ardor*> FERVOR and FIRE indicate a highly intense, sustained emotional condition frequently with potential loss of control implied <fought with *fervor*><the *fire* of a true revolutionary> ZEAL, which sometimes reflects strong, forceful devotion to a cause, expresses a driving motivation or attitude <the religious *zeal* of the Puritans>

**pas·sion·al** (păsh'ə-nəl) *adj.* Of or relating to passion. *—n.* A book of the sufferings of saints and martyrs.

**pas·sion·ate** (păsh'ə-nĭt) *adj.* **1.** Having or capable of intense feelings. **2.** Wrathful by temperament : CHOLERIC. **3.** Amorous : lustful. **4.** Showing or expressing strong emotion : ARDENT <a *passionate* denunciation of racism> **5.** Arising from or marked by passion <*passionate* fury> **—pas'sion·ate·ly** *adv.* **—pas'sion·ate·ness** *n.*

**pas·sion·flow·er** (păsh'ən-flou'ər) *n.* [So called from the resemblance of its parts to the instruments of the Passion.] A chiefly tropical American vine of the genus *Passiflora,* usu. bearing large variously colored flowers.

**passion fruit** *n.* The edible fruit of the passionflower.

**Passion play** *n.* A play representing the Passion of Christ.

**Passion Sunday** *n.* The second Sunday before Easter.

**Pas·sion·tide** (păsh'ən-tīd') *n.* The fortnight between Passion Sunday and Easter.

**Passion Week** *n.* **1.** The week between Passion Sunday and Palm Sunday. **2.** Holy Week.

**pas·sive** (păs'ĭv) *adj.* [ME < Lat. *passivus,* capable of suffering < *passus,* p.part. of *pati,* to suffer.] **1.** Receiving or subjected to an action without responding or initiating a corresponding action. **2.** Accepting without resistance or objection. **3.** Not participating, acting, or operating : INERT. **4.** Designating certain bonds or shares that do not bear financial interest. **5.** Denoting a verb form or voice used to indicate that the grammatical subject is the object of the action or the effect of the verb; e.g., in the sentence *I was appalled by their incompetence, was appalled* is in the passive voice. **6.** *Chem.* INERT 3. **7.** *Electron.* Exhibiting no gain or contributing no energy <a *passive* capacitor> —*n.* **1.** The passive voice. **2.** A verb or construction in the passive voice. **—pas'sive·ly** *adv.* **—pas'sive·ness, pas·siv'i·ty** (pă-sĭv'ĭ-tē) *n.*

☆ **syns:** PASSIVE, ACQUIESCENT, NONRESISTANT, RESIGNED, SUBMISSIVE, YIELDING *adj. core meaning* : submitting without objection or resistance <remained *passive* when arrested>

**passive immunity** *n.* Immunity acquired by an individual after the transfer of antibodies through injection or by natural means, as by placental transfer to a fetus. **—passive immunization** *n.*

**passive resistance** *n.* Resistance to authority or law by nonviolent methods, as peaceful demonstrations, fasting, or refusal to comply. **—passive resister** *n.*

**passive restraint** *n.* An automatic safety device, as an air bag, that acts to protect an automobile rider during a crash.

**pas·siv·ism** (păs'ĭv-ĭz'əm) *n.* Passive quality, character, or behavior. **—pas'siv·ist** *n.*

**pass·key** (păs'kē') *n.* A master key.

**Pass·o·ver** (păs'ō'vər) *n.* [Transl. of Heb. *pesaḥ*, Pesach.] A Jewish festival commemorating the escape of the Jews from Egypt, traditionally celebrated for eight days in the spring.

**pass·port** (păs'pôrt', -pōrt') *n.* [Fr. *passeport* : *passer*, to pass + *port*, port.] **1.** An official governmental document certifying the identity and citizenship of a person traveling abroad. **2.** A permit issued by a foreign country permitting one to travel or to transport goods through that country. **3.** An official document issued to a ship, esp. a neutral merchant ship in wartime, authorizing it to leave port or enter certain waters freely. **4.** Something enabling one to be admitted or accepted <talent as a *passport* to an acting career>

**pass·word** (păs'wûrd') *n.* A secret word or phrase spoken to gain admittance.

**past** (păst) *adj.* [ME < p.part. of *passen*, to pass. —see PASS.] **1.** No longer current : gone by <My childhood is *past.*> **2.** Having existed or occurred in an earlier time : BYGONE <*past* tragedies><*past* decades> **3. a.** Earlier than the present time : AGO <20 years *past*> **b.** Recently gone by or elapsed <in the *past* week> **4.** Having served formerly in a given official capacity <a *past* governor> **5.** Of, relating to, or denoting a verb tense or form used to express an action or condition prior to the time it is expressed. —*n.* **1.** The time before the present. **2. a.** Former background, career, experiences, and activities <a scientist with an illustrious *past*> **b.** A former, secret period of a person's life. **3. a.** The past tense. **b.** A verb form in the past tense. —*adv.* So as to pass by or go beyond <A dog ran *past.*> —*prep.* **1.** Beyond in time : later than <It is *past* noon.> **2.** Beyond in position <the house *past* the orchard> **3.** Beyond the power, scope, extent, or influence of <Your attitude is *past* comprehending.> **4.** Beyond the number or amount of <can't go *past* 40 miles per hour>

**pas·ta** (päs'tə) *n.* [Ital. < LLat.] **1.** Paste or dough made of flour and water, used dried, as in macaroni, or fresh, as in ravioli. **2.** A prepared dish of pasta.

**paste¹** (pāst) *n.* [ME < OFr. < LLat. *pasta.*] **1.** A smooth viscous adhesive, as flour and water or starch and water, used to fasten things together. **2.** A soft, smooth, thick mixture similar to paste. **3.** A smooth dough of water, flour, and butter or shortening, used in making pastry. **4.** A food that has been pounded or processed until it is reduced to a smooth, creamy mass <liver *paste*> **5.** A sweet, doughy candy or confection. **6.** Moistened clay used to make porcelain or pottery. **7. a.** A hard, brilliant glass used to make artificial gems. **b.** A gem made of this glass. —*vt.* **past·ed, past·ing, pastes. 1.** To cause to adhere by applying paste. **2.** To cover with something to which paste has been applied <*paste* a wall with posters> **—paste up.** To prepare a printing mechanical of.

**paste²** (pāst) [Alteration of BASTE³.] —*vt.* **past·ed, past·ing, pastes.** To punch. —*n.* A hard blow.

**paste·board** (pāst'bôrd', -bōrd') *n.* **1.** A thin firm board made of pressed paper pulp or sheets of paper pasted together : PAPERBOARD. **2. a.** A ticket. **b.** A playing card. **c.** A visiting card.

**pas·tel** (pă-stĕl') *n.* [Fr. < Ital. < LLat. *pastellus*, woad, dim. of *pasta*, paste.] **1. a.** A dried paste made of ground and mixed pigment, chalk, water, and gum, used to make crayons. **b.** A crayon of this material. **c.** A picture or sketch drawn with pastels. **2.** The art or process of drawing with pastels. **3.** A light, soft hue or tint. **4.** A brief or sketchy prose work. **—pas·tel'ist** *n.*

**pas·tern** (păs'tərn) *n.* [ME *pastron*, hobble < OFr. *pasturon* < *pasture*, pasture.] **1.** The part of a horse's foot between the fetlock and hoof. **2.** A comparable part of the leg of another quadruped, as a dog.

**paste-up** (pāst'ŭp') *n.* **1.** A composition of light flat objects pasted on a backing : COLLAGE. **2.** A mechanical.

**pas·teur·i·za·tion** (păs'chər-ĭ-zā'shən, păs'tər-) *n.* [After Louis *Pasteur* (1822–1895).] The act or process of destroying most disease-producing microorganisms and limiting fermentation in liquids, as milk or beer, by partial or complete sterilization. **—pas'teur·ize'** (păs'chə-rīz', păs'tə-) *v.* **(-ized, -iz·ing, -iz·es). —pas'teur·iz'er** *n.*

**Pas·teur treatment** (pă-stûr') *n.* [After Louis *Pasteur* (1822–1895).] A rabies treatment in which growth of antibodies is stimulated during incubation of the disease by increasingly strong inoculations of attenuated rabies virus.

**pas·tic·cio** (pă-stē'chō, -chē-ō, päs-) *n., pl.* **-ci** (-chē) [Ital. < Med. Lat. *pasticius*, pasty < Lat. *pasta*, paste.] A work, esp. a musical work, produced by borrowing fragments or motifs from various sources.

**pas·tiche** (pă-stēsh', pä-) *n.* [Fr. < Ital. *pasticcio*, pasticcio.] **1.** A dramatic, literary, or musical piece openly imitating the work of another artist, often with satirical intent. **2.** A hodgepodge.

**pas·tille** (pă-stēl') *also* **pas·til** (păs'tĭl) *n.* [Fr. < Lat. *pastillus.*] **1.** A small flavored or medicated tablet : TROCHE. **2.** An aromatic tablet burned to fumigate or deodorize the air. **3.** PASTEL 1.

**pas·time** (păs'tīm') *n.* An activity that pleasantly occupies one's spare time : DIVERSION.

**pas·ti·na** (pă-stē'nə) *n.* [Ital., dim. of *pasta*, pasta.] Tiny pieces of macaroni, usu. cooked in soups or used as baby food.

**past master** *n.* **1.** One formerly holding the position of master in an organization, as a lodge or club. **2.** An expert in an activity.

**pas·tor** (păs'tər) *n.* [ME *pastour* < AN < Lat. *pastor*, shepherd < *pascere*, to pasture.] **1.** A Christian minister having spiritual charge over a congregation or parish. **2.** A shepherd. **—pas'tor·ship'** *n.*

**pas·tor·al** (păs'tər-əl) *adj.* [ME < Lat. *pastoralis* < *pastor*, shepherd. —see PASTOR.] **1.** Of or relating to shepherds, herdsmen, and others directly involved in animal husbandry. **2. a.** Of or relating to the country or country life : RURAL. **b.** Having the qualities of idealized country life, as charming simplicity and a slow, carefree pace. **3.** Of or designating an artistic work that idealizes country life. **4.** Of or relating to a pastor or a pastor's duties. —*n.* **1.** A literary or artistic work that portrays idealized rural life. **2.** *Mus.* A pastorale. **—pas'tor·al·ism** *n.* **—pas'tor·al·ist** *n.* **—pas'tor·al·ly** *adv.*

**pas·to·rale** (păs'tə-räl', -răl', -rä'lē, päs'-) *n., pl.* **-ra·li** (-rä'lē) or **-rales.** [Ital. < *pastorale*, pastoral < Lat. *pastoralis* < *pastor*, shepherd. —see PASTOR.] *Mus.* **1.** An opera or other vocal composition based on a rural subject or theme. **2.** An instrumental composition having a tender melody in a moderately slow rhythm suggestive of idyllic rural life.

**pas·tor·ate** (păs'tər-ĭt) *n.* **1.** The office, rank, or jurisdiction of a pastor. **2.** A pastor's term of office with one congregation. **3.** Pastors as a group.

**†pas·to·ri·um** (pă-stôr'ē-əm, -stōr'-) *n. Chiefly Southern U.S.* The residence of a pastor : PARSONAGE.

**past participle** *n.* A verb form indicating past or completed action or time that is used as a verbal adjective in phrases such as *completed tasks* and *molded clay* and with auxiliaries to form the passive voice or perfect and pluperfect tenses in constructions such as *The tasks were completed* and *The sculptor had molded the clay.*

**past perfect** *n.* The pluperfect.

**pas·tra·mi** (pə-strä'mē) *n.* [Yiddish < Rum. *pastramă* < *păstra*, to preserve.] A highly seasoned smoked cut of beef, usu. from the breast or shoulder.

**pas·try** (pās'trē) *n., pl.* **-tries.** [< PASTE¹.] **1.** A usu. rich, baked paste of flour, water, and shortening, used for the crusts of foods such as pies and turnovers. **2.** Baked foods made with pastry.

**past tense** *n.* A verb tense used to express an action or condition that occurred in or during the past; e.g., in *While I was working, they watched the news,* the verbs *was working* and *watched* are in the past tense.

**pas·tur·age** (păs'chər-ĭj) *n.* **1.** The vegetation, as grass, eaten by grazing animals. **2. a.** Land covered with grass or vegetation suitable for grazing animals. **b.** The right to graze animals on such land. **3.** The business of grazing cattle.

**pas·ture** (păs'chər) *n.* [ME < OFr. < LLat. *pastura* < Lat. *pascere*, to pasture.] **1. a.** Vegetation, as grass, eaten as food by grazing animals. **b.** Ground on which such vegetation grows. **2.** The feeding or grazing of animals. —*v.* **-tured, -tur·ing, -tures.** —*vt.* **1.** To herd (animals) into a pasture to graze. **2.** To provide (animals) with pasturage. —Used of land. —*vi.* To graze in a pasture. **—pas'tur·a·ble** *adj.* **—pas'tur·er** *n.*

**pas·ture·land** (păs'chər-lănd') *n.* Land suitable for grazing.

**past·y¹** (pā'stē) *adj.* **-i·er, -i·est. 1.** Like paste in color or consistency. **2.** Having a pale and lifeless appearance. **—past'i·ness** *n.*

**pas·ty²** (păs'tē) *n., pl.* **-ties.** [ME < OFr. *pastee* < *paste*, paste.] A pie with a filling of seasoned meat or fish.

▲ word history: The noun *pasty²*, "a meat pie," was borrowed from Old French *pastee* in the 13th century. *Pastee* is derived from Old French *paste*, literally "paste," which originally referred to an edible dough of flour and liquid. Old French *pastee* is descended from the same word as Italian *pasta.* The modern French form of *pastee* is *pâté*, which was borrowed into English in the 18th century as a synonym for *pasty².* *Pâté* in English now denotes the meat alone, without its pastry crust. *Patty* is an anglicized variant of *paté.*

**PA system** (pē'ā') *n.* A public-address system.

**pat** (păt) *v.* **pat·ted, pat·ting, pats.** [< ME *patte,* a light blow.] —*vt.* **1. a.** To tap gently with the open hand or with something flat. **b.** To stroke lightly as a gesture of affection. **2.** To mold by tapping gently with the hands or a flat implement <*pat* dough into a pan> —*vi.* **1.** To run or walk with a tapping sound. **2.** To hit or beat against something gently or lightly. —*n.* **1.** A light stroke or tap. **2.** The sound made by a light stroke or tap or by light footsteps. **3.** A small mass of something, as butter, shaped into an individual portion. —*adj.* **1.** Opportune : fitting. **2.** Needing no change : exactly right. **3.** Contrived <was ready with a *pat* reply> —*adv. Informal.* **1.** Without changing position : STEADFASTLY <stood *pat* on my decision> **2.** Perfectly : precisely : aptly. **—down pat.** *Informal.* Known and understood completely. **—pat'ly** *adv.* **—pat'ness** *n.*

**pa·ta·ca** (pə-tä'kə) *n.* [Port.] —See table at CURRENCY.

**pa·ta·gi·um** (pə-tā'jē-əm) *n., pl.* **-gi·a** (-jē-ə) [NLat. < Lat., gold edging on a woman's tunic < Gk. **patageion* < *patogos,* clatter.] **1.** *Zool.* A thin membrane extending between the fore and hind limb to form a wing or winglike extension, as in flying squirrels and bats. **2.** An expandable membranous fold of skin between the wing and body of a bird.

**patch** (păch) *n.* [ME *pacche.*] **1.** A small piece of material affixed to another, larger piece to reinforce or conceal a weakened or worn area. **2. a.** A small piece of cloth used for patchwork. **b.** A small cloth badge affixed to a sleeve to indicate the military unit to which one belongs. **3.** A covering or dressing applied to protect a wound or sore. **4.** A small pad or shield of cloth worn over an injured eye. **5.** A beauty spot. **6. a.** A small piece of land. **b.** The produce grown on such a piece of land <a *patch* of cucumbers> **7.** A small part or section of a surface that differs from or contrasts with the whole <The snow made white *patches* on the pavement.> **8.** A small piece or portion. —*vt.* **patched, patch·ing, patch·es. 1.** To put a patch or patches on. **2. a.** To make (e.g., a quilt) by sewing scraps of material together. **b.** To make by piecing various elements together, esp. hastily <*patched* together a report> **3.** To mend, repair, or put together, esp. hastily or shoddily. **—patch up.** To settle <We *patched up* our differences.> **—patch'a·ble** *adj.* **—patch'er** *n.*

**patch·ou·li** *also* **patch·ou·ly** *or* **pach·ou·li** (păch'ə-lē, pə-chōō'lē) *n., pl.* **-lis** *also* **-lies** *or* **-lis.** [Tamil *paccilai.*] **1.** An Asiatic tree of the genus *Pogostemon,* esp. *P. patchouly* or *P. cablin,* bearing leaves that yield a fragrant oil used in perfumery. **2.** A perfume made from the oil of the patchouli.

**patch pocket** *n.* An unfitted flat outside pocket on a garment.

**patch test** *n.* A test for allergic reaction made by applying a suspected allergen to the skin in a small surgical pad.

**patch·work** (păch'wûrk') *n.* **1.** Needlework made of varicolored patches of material sewed together, as in a quilt. **2.** A miscellany : jumble.

**patch·y** (păch'ē) *adj.* **-i·er, -i·est. 1.** Made up of or marked by patches <*patchy* jeans> **2.** Uneven in quality or performance. **—patch'i·ly** *adv.* **—patch'i·ness** *n.*

**pate** (pāt) *n.* [ME.] **1.** The head, esp. the top of the head <a shiny *pate*> **2.** The brain : intellect. **—pat'ed** *adj*

**pâte** (pät) *n.* [Fr. < OFr. *paste,* paste.] Paste used in making pottery and porcelain.

**pâ·té** (pä-tā') *n.* [Fr. < OFr. *pastee,* pasty (pie) < *paste,* paste.] **1.** A meat paste, esp. pâté de foie gras. **2.** A small pastry filled with meat or fish.

**pâ·té de foie gras** (pä-tā' də fwä grä') *n.* [Fr. : *pâté,* pâté + *de,* of + *foie,* liver + *gras,* fat.] A paste made from goose liver, usu. with truffles.

**pa·tel·la** (pə-tĕl'ə) *n., pl.* **-lae** (-ē) [Lat. < dim. of *patina,* plate. —see PATEN.] **1. a.** A flat, triangular bone at the front of the knee joint. **b.** A dish-shaped anatomical formation. **2.** An ancient Roman pan or dish. **—pa·tel'lar, pa·tel'late** (-ĭt, -āt') *adj.*

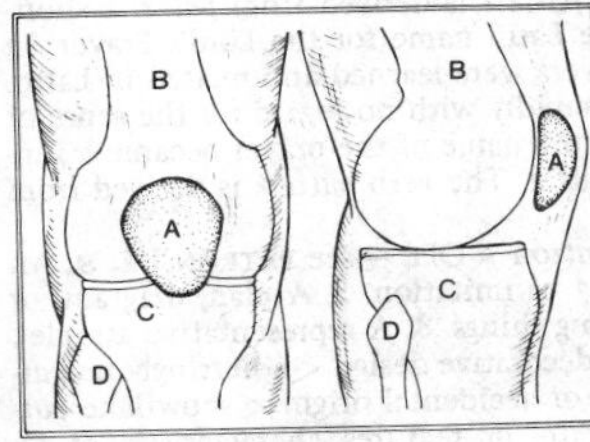


**patella**
*A. patella, B. femur, C. tibia, D. fibula*

**pa·tel·li·form** (pə-tĕl'ə-fôrm') *adj.* [PATELL(A) + -FORM.] Shaped like a pan, dish, or cup.

**pat·en** *also* **pat·in** (păt'n) *or* **pa·tine** (pă-tēn') *n.* [ME < OFr. *patene* < Lat. *patina,* plate < Gk. *patanē.*] **1.** A plate used to hold the Eucharistic bread. **2.** A plate. **3.** A thin metal disk.

**pa·ten·cy** (pāt'n-sē) *n.* The quality or state of being obvious.

**pat·ent** (păt'nt) *n.* [ME, unsealed < OFr. < Lat. *patens* < pr.part. of *patēre,* to be opened.] **1. a.** A grant made by a government to an inventor, assuring the sole right to make, use, and sell the invention for a certain period of time. **b.** Letters patent. **c.** Something protected by such a grant. **2. a.** A grant made by a government to an individual, conveying fee-simple title to public lands. **b.** The official document of such a grant. **c.** The land so granted. **3.** An exclusive title or right. —*adj.* **1.** Plain : obvious. **2.** Protected by a patent. **3.** Of, relating to, or dealing in patents. **4.** (pāt'nt) *Biol.* Spreading open : EXPANDED. **5.** Of high quality. —Used of flour. —*vt.* **-ent·ed, -ent·ing, -ents. 1.** To obtain a patent on. **2.** To grant a patent to. **—pat'·ent·a·bil'i·ty** *n.* **—pat'ent·a·ble** *adj.* **—pat'ent·ly** (pāt'nt-lē) *adv.*

**pat·ent·ee** (păt'n-tē') *n.* One who has been granted a patent.

**pat·ent leather** (păt'nt) *n.* [So called because it is made by a once-patented process.] **1.** Black leather finished to a hard, shiny surface. **2.** Any of several synthetic materials resembling patent leather.

**patent log** *n. Naut.* A screw log.

**patent medicine** *n.* A medical preparation, as a drug, that is protected by a patent and can be purchased without a prescription.

**patent office** *n.* A government bureau that studies claims for and grants patents.

**pat·en·tor** (păt'n-tər, păt'n-tôr') *n.* One that grants a patent.

**patent right** *n.* The right granted by a patent, esp. the right to have exclusive manufacture and sale of an invention.

**pa·ter** (pā'tər) *n.* [Lat.] *Chiefly Brit.* Father.

**pa·ter·fa·mil·i·as** (pā'tər-fə-mĭl'ē-əs, pä'tər-) *n., pl.* **pa·tres·fa·mil·i·as** (pā'trēz-fə-mĭl'ē-əs, pä'-) [Lat. : *pater,* father + *familia,* family.] **1.** The male head of a household. **2.** The father of a family.

**pa·ter·nal** (pə-tûr'nəl) *adj.* [Med. Lat. *paternalis* < Lat. *paternus* < *pater,* father.] **1.** Of, relating to, or typical of a father : FATHERLY. **2.** Inherited or received from a father. **3.** Of or relating to the father's side of a family. **—pa·ter'nal·ly** *adv.*

**pa·ter·nal·ism** (pə-tûr'nə-lĭz'əm) *n.* A policy or practice of treating or governing people in a paternal manner, esp. by taking care of their needs without giving them responsibility. **—pa·ter'nal·is'tic** *adj.* **—pa·ter'nal·is'ti·cal·ly** *adv.*

**pa·ter·ni·ty** (pə-tûr'nĭ-tē) *n.* [OFr. *paternite* < LLat. *paternitas* < Lat. *paternus,* paternal < *pater,* father.] **1.** The fact or condition of being a father. **2.** Paternal descent. **3.** Origin : authorship.

**paternity test** *n.* A test using blood group identification of a mother, child, and suspected father to establish probable paternity.

**pa·ter·nos·ter** (pā'tər-nŏs'tər, pä'tər-, păt'ər-) *n.* [ME < Med. Lat. < Lat. *pater noster,* our father.] **1.** *often* **Paternoster.** The Lord's Prayer. **2.** One of the large beads on a rosary on which the Lord's Prayer is said. **3.** A sequence of words spoken as a prayer or magic formula. **4.** A weighted fishing line with several jointed attachments for hooks connected by beadlike swivels.

**path** (păth, päth) *n., pl.* **paths** (păthz, päthz, păths, päths) [ME < OE *pæð.*] **1.** A track or way made by footsteps. **2.** A road or way made for a particular purpose <a hiking *path*> **3.** The route or course along which something moves <the *path* of a cyclone> **4.** A course of action or conduct <the *path* of virtuousness>

**path-** *pref. var. of* PATHO-.

**-path** *suff.* [Back-formation < -PATHY.] **1.** A practitioner of a given type of medical treatment <naturo*path*> **2.** One suffering from a given type kind of disorder <socio*path*>

**Pa·than** (pə-tän') *n.* [Hindi *Paṭhan* < Pashto *Pĕṣtana,* pl. of *Pĕṣtūn,* an Afghan < *pashtu,* Pashto.] An Afghan, esp. one of Indo-Iranian stock and Moslem religion.

**pa·thet·ic** (pə-thĕt'ĭk) *also* **pa·thet·i·cal** (-ĭ-kəl) *adj.* [Fr. *pathétique* < LLat. *patheticus* < Gk. *pathētikos,* sensitive < *pathētos,* liable to suffer < *pathos,* suffering.] **1.** Of, relating to, expressing, or arousing pity, sympathy, or tenderness. **2.** Inadequate : pitiful <a *pathetic* attempt at humor> **—pa·thet'i·cal·ly** *adv.*

**pathetic fallacy** *n.* The attribution of human emotions or characteristics to things; e.g., *angry seas, kind rain.*

**path·find·er** (păth'fīn'dər, päth'-) *n.* One who discovers a way into or through unexplored regions.

**path·less** (păth'lĭs) *adj.* Unmarked by paths or trails.

**patho-** *or* **path-** *pref.* [NLat. < Gk. < *pathos,* suffering.] Disease : suffering <*path*ogen>

**path·o·gen** (păth'ə-jən) *n.* An agent that causes disease, esp. a microorganism such as a bacterium or fungus.

**path·o·gen·e·sis** (păth'ə-jĕn'ĭ-sĭs) *also* **pa·thog·e·ny** (pă-thŏj'ə-nē) *n.* The development of a diseased or morbid condition.

**path·o·gen·ic** (păth'ə-jĕn'ĭk) *also* **path·o·ge·net·ic** (păth'ō-jə-nĕt'ĭk) *adj.* Capable of causing disease. **—path'o·gen'i·cal·ly** *adv.* **—path'o·ge·nic'i·ty** (-jə-nĭs'ĭ-tē) *n.*

**pa·thog·no·mon·ic** (pə-thŏg'nə-mŏn'ĭk, păth'əg-nō-) *adj.* [Gk. *pathognōmonikos* : *pathos,* suffering + *gnōmonikos,* able to judge < *gnōmon,* interpreter.] Typical of a particular disease.

**path·o·log·i·cal** (păth'ə-lŏj'ĭ-kəl) *also* **path·o·log·ic** (-ĭk) *adj.* **1.** Of or relating to pathology. **2.** Relating to or caused by disease. **3.** Disordered in behavior. **—path'o·log'i·cal·ly** *adv.*

**pa·thol·o·gy** (pă-thŏl'ə-jē) *n., pl.* **-gies. 1.** The branch of medicine concerned with the study of the nature of disease and its causes, processes, development, and consequences. **2.** Anatomic or functional manifestations of disease. **—pa·thol'o·gist** *n.*

**pa·thos** (pā'thŏs', -thôs') *n.* [Gk., suffering.] **1.** A quality arousing feelings of pity, sympathy, tenderness, or sorrow. **2.** A feeling of sympathy or pity. **3.** The transient, emotional, or subjective elements in a work of art.

**path·way** (păth'wā', päth'-) *n.* A path.

**-pathy** *suff.* [Lat. *-pathia* < Gk. *-patheia* < *pathos,* suffering.] **1.** Feeling : suffering : perception <tele*pathy*> **2. a.** Disease <neuro*pathy*> **b.** A system of treating disease <homeo*pathy*>

**pa·tience** (pā'shəns) *n.* **1.** The quality of being patient : capacity of calm endurance. **2.** *Chiefly Brit.* SOLITAIRE 2.

**pa·tient** (pā'shənt) *adj.* [ME *pacient* < OFr. < Lat. *patiens* < *pati,* to endure.] **1.** Capable of bearing affliction calmly. **2.** Understanding : tolerant. **3.** Persevering : constant <a *patient* student> **4.** Capable of bearing delay. —*n.* **1.** One under medical treatment. **2.** *Archaic.* One who suffers. **—pa'tient·ly** *adv.*

☆ ***syns:*** PATIENT, FORBEARING, LONG-SUFFERING, RESIGNED *adj. core meaning* : enduring or capable of enduring hardship or inconvenience without complaint <a *patient* parent> ***ant:*** impatient

**pat·in** (păt'n) *n. var. of* PATEN.

**pat·i·na¹** (păt'ə-nə) *n., pl.* **-nae** (-nē) [Med. Lat. < Lat., plate. —see PATEN.] PATEN 3.

**pat·i·na²** (păt'ə-nə, pə-tē'nə) *also* **pa·tine** (pă-tēn') *n.* [Ital. < Lat., plate. —see PATEN.] **1.** A thin layer of usu. brown or green corrosion that appears on copper or copper alloys, as bronze, as a result of natural or artificial oxidation. **2.** The sheen on an antique surface produced by use and age.

**pa·tine** (pă-tēn') *vt.* **-tined, -tin·ing, -tines.** [< PATINA².] To coat with a patina. —*n.* **1.** *var. of* PATEN. **2.** *var. of* PATINA².

**pat·i·o** (păt'ē-ō', pä'tē-ō') *n., pl.* **-os.** [Sp. < OSp. < Lat. *patēre,* to be open.] **1.** An inner roofless courtyard. **2.** A usu. paved space that adjoins a residence and is used for dining or recreation.

**pa·tis·se·rie** (pä-tēs-rē') *n.* [Fr. *pâtisserie* < OFr. *pâtissier,* pastry cook, ult. < LLat. *pasta,* dough.] A bakery specializing in French pastry.

**pat·ois** (păt'wä', pä-twä') *n., pl.* **pat·ois** (păt'wäz', pä-twä') [Fr. < OFr.] **1. a.** A regional dialect. **b.** Substandard or illiterate speech. **2.** The special jargon of a group : CANT.

**patr-** *pref. var. of* PATRI-.

**pa·tres·fa·mil·i·as** (pā'trēz-fə-mĭl'ē-əs, pä-) *n. pl. of* PATERFAMILIAS.

**patri-** *or* **patr-** *pref.* [Lat. < *pater,* father.] Father <*patri*lineal>

**pa·tri·arch** (pā'trē-ärk') *n.* [ME *patriarche* < OFr. < LLat. *patriarcha* < Gk. *patriarkhēs* : *patria,* family (< *patēr,* father) + *arkhos,* ruler (< *arkhein,* to rule).] **1.** The male leader of a family or tribe. **2. a.** Any of the Old Testament fathers of the human race. **b.** One of the founders of the 12 tribes of Israel. **3.** A former title for the bishops of Rome, Constantinople, Jerusalem, Antioch, and Alexandria. **4.** *Rom. Cath. Ch.* A bishop who holds the highest episcopal rank after the pope. **5.** The head of an Eastern Orthodox or Greek Orthodox church. **6.** *Mormon Ch.* A high dignitary of the priesthood empowered to invoke blessings. **7.** One considered to be the founder or original head of an enterprise, organization, or tradition. **8.** An aged and venerable man : ELDER. **9.** The older or most venerable specimen in a group <*patriarch* of the flock>

**pa·tri·ar·chal** (pā'trē-är'kəl) *also* **pa·tri·ar·chic** (-är'kĭk) *adj.* **1.** Relating to or typical of a patriarch : VENERABLE. **2.** Of or relating to a patriarchy. **3.** Ruled by a patriarch. **—pa'tri·ar'chal·ism** *n.* **—pa'tri·ar'chal·ly** *adv.*

**patriarchal cross** *n.* A Latin cross having two horizontal bars, of which the upper is the shorter.

**pa·tri·ar·chate** (pā'trē-är'kĭt, -kāt') *n.* **1.** The territory, rule, or rank of a patriarch. **2.** A patriarchy.

**pa·tri·ar·chy** (pā'trē-är'kē) *n., pl.* **-chies. 1.** A system of social organization in which descent and succession are traced through the male line. **2.** The rule of a family or tribe by men.

**pa·tri·cian** (pə-trĭsh'ən) *n.* [ME *patricion* < OFr. *patricien* < Lat. *patricius* < *patres,* senators < pl. of *pater,* father.] **1.** A member of one of the noble families of the ancient Roman Republic. **2.** An aristocrat. **3.** A refined person. **—pa·tri'cian** *adj.* **—pa·tri'cian·ly** *adv.*

**pa·tri·ci·ate** (pə-trĭsh'ē-ĭt, -āt') *n.* [Lat. *patriciatus* < *patricius,* patrician.] **1.** The rank of patrician. **2.** Patricians as a class.

**pat·ri·cide** (păt'rĭ-sīd') *n.* [LLat. *patricidium* : Lat. *pater,* father + Lat. *caedere,* to kill.] **1.** The act of murdering one's father. **2.** One who murders one's father. **—pat'ri·cid'al** (păt'rə-sīd'l) *adj.*

**pat·ri·cli·nous** *also* **pat·ro·cli·nous** (păt'rĭ-klī'nəs) *adj.* [PATRI- + Gk. *klinein,* to lean.] Derived from the male line.

**pat·ri·lin·e·al** (păt'rə-lĭn'ē-əl) *adj.* Relating to, based on, or tracing descent through the male line.

**pat·ri·lo·cal** (păt'rə-lō'kəl) *adj.* Relating to the home territory of a husband's family or tribe in primitive societies.

**pat·ri·mo·ny** (păt'rə-mō'nē) *n., pl.* **-nies.** [ME < OFr. *patrimoine* < Lat. *patrimonium* < *pater,* father.] **1.** An inheritance from a father or other ancestor. **2.** Heritage. **3.** An endowment or estate belonging to a church. **—pat'ri·mo'ni·al** *adj.* **—pat'ri·mo'ni·al·ly** *adv.*

**pa·tri·ot** (pā'trē-ət, -ŏt') *n.* [OFr. *patriote,* compatriot < LLat. *patriota* < Gk. *patriōtēs* < *patrios,* of one's fathers < *patris,* fatherland < *patēr,* father.] One who loves, supports, and defends one's country. **—pa'tri·ot'ic** *adj.* **—pa'tri·ot'i·cal·ly** *adv.*

**pa·tri·ot·ism** (pā'trē-ə-tĭz'əm) *n.* Love of and devotion to one's own country.

**Patriots' Day** *n.* Apr. 19, the anniversary of the Battles of Lexington and Concord in 1775, celebrated as a legal holiday in Maine and Massachusetts.

**pat·ro·cli·nous** (păt'rə-klī'nəs) *adj. var. of* PATRICLINOUS.

**Pa·tro·clus** (pə-trō'kləs) *n.* [Lat. < Gk. *Patroklos.*] *Gk. Myth.* A Greek warrior, the friend of Achilles.

**pa·trol** (pə-trōl') *n.* [Fr. *patrouille* < *patrouiller,* to patrol < *patouiller,* to paddle about in mud < *patte,* paw.] **1. a.** The action of moving about an area for purposes of security or observation. **b.** A person or group of persons who carry out such an action. **2. a.** A military unit sent on a reconnaissance mission. **b.** One or more vehicles, vessels, or aircraft assigned to guard or reconnoiter a given area. **3.** A group of eight Boy Scouts forming a division of a troop. —*v.* **-trolled, -trol·ling, -trols.** —*vt.* To engage in a patrol of. —*vi.* To carry out a patrol. **—pa·trol'ler** *n.*

**patrol car** *n.* A squad car.

**pa·trol·man** (pə-trōl'mən) *n.* A police officer or guard who patrols an assigned area.

**patrol torpedo boat** *n.* A PT boat.

**patrol wagon** *n.* A police truck used to convey prisoners.

**pa·tron** (pā'trən) *n.* [ME < OFr. < Med. Lat. *patronus* < Lat. < *pater,* father.] **1.** One that supports, protects, or champions : BENEFACTOR <a *patron* of the ballet> **2.** A regular customer. **—pa'tron·al** (pā'trə-nəl) *adj.*

**pa·tron·age** (pā'trə-nĭj, păt'rə-) *n.* **1.** Support or encouragement from a patron. **2.** A patronizing manner. **3.** The trade given to a commercial establishment by its customers. **4.** Customers or patrons as a group : CLIENTELE. **5. a.** The power of appointing people to governmental or political positions. **b.** The positions so distributed.

**pa·tron·ize** (pā'trə-nīz', păt'rə-) *vt.* **-ized, -iz·ing, -iz·es. 1.** To act as a patron to : SUPPORT. **2.** To visit regularly as a customer. **3.** To treat in an offensively condescending manner. **—pa'tron·iz'er** *n.* **—pa'tron·iz'ing·ly** *adv.*

☆ ***syns:*** PATRONIZE, SPONSOR, SUPPORT *v. core meaning* : to act as a patron to <*patronized* the arts>

**patron saint** *n.* The guardian saint of a person, class, place, or activity.

**pat·ro·nym·ic** (păt'rə-nĭm'ĭk) *n.* [LLat. *patronymicum* < *patronymicus,* of a patronymic < Gk. *patrōnumikos* < *patrōnumia* : *pater,* father + *onuma,* name.] A name received from a paternal ancestor, esp. one formed by an affix, as in *Robertson,* the son of Robert. **—pat'ro·nym'ic** *adj.* **—pat'ro·nym'i·cal·ly** *adv.*

**pa·troon** (pə-trōōn') *n.* [Du. < Fr. *patron,* patron < OFr.] A landholder in New York and New Jersey who was granted certain proprietary and manorial powers under Dutch colonial rule.

**pat·sy** (păt'sē) *n., pl.* **-sies.** [Orig. unknown.] *Slang.* One who is cheated, victimized, or made the butt of a joke.

**pat·ten** (păt'n) *n.* [ME *patin* < OFr. < *patte,* paw.] A wooden sandal or shoe.

**pat·ter¹** (păt'ər) *v.* **-tered, -ter·ing, -ters.** [Freq. of PAT.] —*vi.* **1.** To make a succession of quick, light, soft taps. **2.** To move with quick, light, softly audible steps. —*vt.* To cause to patter. —*n.* A succession of quick, light, tapping sounds.

**pat·ter²** (păt'ər) *n.* [ME *pater,* chatter < *paternoster,* paternoster (from the mechanical and rapid recitation of the prayer).] **1.** The jargon of a particular group : CANT. **2. a.** Glib, rapid-fire speech, as of an auctioneer, salesperson, or comedian. **b.** The words of an entertainer's humorous song or monologue. **3.** Meaningless talk : CHATTER. —*v.* **-tered, -ter·ing, -ters.** —*vi.* **1.** To chatter rapidly and glibly. **2.** To mumble prayers in a mechanical manner. —*vt.* To utter in a glib, rapid, or mechanical manner. **—pat'ter·er** *n.*

▲ word history: The word *patter²* is derived from *pater,* a shortened form of *paternoster,* the Latin name for the Lord's Prayer. In medieval times Christian prayers were learned and recited in Latin. They were also often recited rapidly with no regard for the sense of the words. From this practice the name of the prayer became a general word for meaningless chatter. The verb *patter²* is derived from the noun.

**pat·tern** (păt'ərn) *n.* [ME *patron* < OFr. —see PATRON.] **1. a.** An archetype. **b.** An ideal worthy of imitation. **2.** A plan, diagram, or model to be followed in making things. **3.** A representative sample : SPECIMEN. **4. a.** An artistic or decorative design <a herringbone *pattern*> **b.** A design of natural or accidental origin <snowflake *patterns*> **5.** A composite of traits or features characteristic of an individual <emotional *patterns*> **6.** Form and style in an artistic work or body of works. **7. a.** The configuration of identically aimed rifle shots on a target. **b.** Distribution and spread of shot from a shotgun. **8.** Enough material to make a complete garment. **9.** A standardized diagram transmitted to test television picture quality. **10.** The ordered flight path of an aircraft about to land. —*vt.*

**-terned, -tern·ing, -terns. 1.** To make, mold, or design by following a pattern. **2.** To cover or decorate with a design or pattern.

**pat·ty** (păt'ē) *n., pl.* **-ties.** [Fr. pâté < OFr. *paste,* paste. —see PASTE[1].] **1. a.** A small, oval, flattened cake of chopped or minced food. **b.** A similarly shaped candy. **2.** A patty shell.

**pat·ty·pan squash** (păt'ē-păn') *n.* The cymling.

**patty shell** *n.* A shell of baked puff pastry made to be filled with creamed meat, seafood, vegetables, or fruit.

**pat·u·lous** (păch'ə-ləs) *also* **pat·u·lent** (-lənt) *adj.* [Lat. *patulus* < *patēre,* to be open.] *Bot.* Spreading or expanded, as tree branches. **—pat'u·lous·ly** *adv.* **—pat'u·lous·ness** *n.*

**pau·ci·ty** (pô'sə-tē) *n.* [ME *paucite* < OFr. < Lat. *paucitas* < *paucus,* few.] **1.** Smallness of number : FEWNESS <a *paucity* of customers> **2.** Smallness of quantity : SCARCITY <a *paucity* of fuel oil>

**Paul Bunyan** (pôl'bŭn'yən) *n.* An extraordinarily huge lumberjack in American folklore.

**Pau·li exclusion principle** (pô'lē, pou'-) *n.* [After Wolfgang *Pauli* (1900–1958).] Exclusion principle.

**Paul·ist** (pô'lĭst) *n.* A priest belonging to the Roman Catholic Missionary Society of Saint Paul the Apostle.

**pau·low·ni·a** (pô-lō'nē-ə) *n.* [After Princess Anna *Paulovna* (1795–1865), queen of William II of the Netherlands.] A tree of the genus *Paulownia,* native to the Orient, with large heart-shaped leaves and purplish or white flower clusters.

**paunch** (pônch) *n.* [ME *paunche* < AN, var. of OFr. *pance* < Lat. *pantex.*] **1.** A potbelly. **2.** The rumen.

**paunch·y** (pôn'chē, pän'-) *adj.* **-i·er, -i·est.** Having a potbelly. **—paunch'i·ness** *n.*

**pau·per** (pô'pər) *n.* [< Lat., poor.] **1.** One who is extremely poor. **2.** One living on public charity. *—vt.* **-pered, -per·ing, -pers.** To make a pauper of. **—pau'per·i·za'tion** *n.* **—pau'per·ize'** (pô'pə-rīz') *v.* **(-ized, -iz·ing, -iz·es).**

☆ ***syns:*** PAUPER, BEGGAR, HAVE-NOT, INDIGENT *n. core meaning* : an impoverished person <the increasing number of *paupers* throughout the land>

**pau·per·ism** (pô'pə-rĭz'əm) *n.* **1.** The quality or state of being a pauper. **2.** Paupers as a group.

**pause** (pôz) *vi.* **paused, paus·ing, paus·es.** [ME, pause < Lat. *pausa* < Gk. *pausis* < *pauein,* to stop.] **1.** To suspend or cease an action for a time. **2.** To stop temporarily and remain : LINGER <*pausing* to chat with a neighbor> **3.** To hesitate <*paused* before answering> *—n.* **1.** A temporary stop. **2.** A suspended reaction or delay, as from uncertainty : HESITATION. **3.** A break, stop, or rest for a calculated purpose or effect. **4.** *Mus.* **a.** A sign indicating that a note or rest is to be held. **b.** A measured break or rest : CAESURA. **5.** A reason for hesitation <Your objection gave me *pause.*>

☆ ***syns:*** PAUSE, ABIDE, BIDE, LINGER, TARRY, WAIT *v. core meaning* : to stop temporarily and remain, as if reluctant to leave <*paused* to watch the brilliant sunset>

**pa·vane** *also* **pa·van** (pə-vän', -văn') *n.* [OFr. *pavane* < OSp. *pavana* < OItal.] **1.** A slow, stately court dance of the 16th cent. **2.** Music for the pavane.

**pave** (pāv) *vt.* **paved, pav·ing, paves.** [ME *paven* < OFr. *paver* < Lat. *pavire,* to stamp.] **1.** To cover with a hard, smooth surface that will bear travel. **2.** To cover uniformly, as if with pavement. **3.** To be or compose the pavement of. **—pave the way.** To make development or progress easier <breakthroughs that *paved the way* for disease control> **—pav'er** *n.*

**pa·vé** (pă-vā') *n.* [Fr. < p.part. of *paver,* to pave.] A setting of precious stones placed together so closely that no metal shows <diamonds in *pavé*> **—pa·vé** *adj.*

**pave·ment** (pāv'mənt) *n.* **1. a.** A hard, paved surface, esp. of a public area or thoroughfare. **b.** The material of which such a surface is made. **2.** *Chiefly Brit.* A sidewalk.

**pav·id** (păv'ĭd) *adj.* [Lat. *pavidus* < *pavēre,* to fear.] Timid : fearful.

**pa·vil·ion** (pə-vĭl'yən) *n.* [ME *pavilon* < OFr. *pavillon* < Lat. *papilio.*] **1.** An ornate tent. **2. a.** A light, sometimes ornamental roofed structure, used at parks or fairs for amusement or shelter. **b.** A usu. temporary structure erected at a fair or show for use by an exhibitor. **3.** A structure connected to a larger building : ANNEX. **4.** One of a group of related buildings forming a complex, as of a hospital. **5.** The surface of a brilliant-cut gem that slants outward from girdle to culet. *—vt.* **-ioned, -ion·ing, -ions.** To shelter in or as if in a pavilion.

**pav·ing** (pā'vĭng) *n.* **1.** The laying of pavement. **2.** A pavement. **3.** Material used for pavement.

**pav·ior** (pāv'yər) *n.* [ME *pavier* < *paven,* to pave.] **1.** One that paves. **2.** Material or tools used for paving.

**pav·is** *also* **pav·isse** (păv'ĭs) *n.* [ME < OFr. *pavais* < OItal. *pavese,* after *Pavia,* Italy.] A large medieval body shield.

**Pa·vo** (pā'vō) *n.* [Lat. *pavo,* peacock.] A constellation in the Southern Hemisphere.

**pav·o·nine** (păv'ə-nīn') *adj.* [Lat. *pavoninus* < *pavo,* peacock.] Of or like a peacock or a peacock's tail.

**paw** (pô) *n.* [ME *pawe* < OFr. *powe,* of Germanic orig.] **1.** The nailed or clawed foot of an animal. **2.** *Informal.* A human hand, esp. a large clumsy one. *—v.* **pawed, paw·ing, paws.** *—vt.* **1.** To strike with the paw or paws. **2.** To strike or scrape with a beating motion. **3.** To handle clumsily, rudely, or with too much familiarity. *—vi.* **1.** To scrape the ground with the forefeet. **2.** To make clumsy grasping motions with the hands. **—paw'er** *n.*

**pawl** (pôl) *n.* [Poss. < Du. *pal.*] A hinged or pivoted device adapted to fit into a notch of a ratchet wheel to impart forward motion or prevent backward motion.

**pawn[1]** (pôn) *n.* [ME *paun* < OFr. *pan.*] **1.** Something given as security for a loan : PLEDGE. **2.** The condition of being held as a pledge against the payment of a loan <diamonds at *pawn*> **3.** A person serving as security : HOSTAGE. **4.** The act of pawning. *—vt.* **pawned, pawn·ing, pawns. 1.** To give or deposit as security for the payment of a loan. **2.** To hazard : risk <*pawn* one's reputation> **—pawn'a·ble** *adj.* **—pawn'age** *n.* **—pawn'er** (pô'nər), **paw'nor** (-nôr') *n.*

**pawn[2]** (pôn) *n.* [ME < OFr. *paon* < Med. Lat. *pedo,* foot soldier < LLat., one who has wide feet < Lat. *pes,* foot.] **1.** A chess piece of lowest value, able to move forward one square at a time, or two squares for the first move, and capture on a one-space diagonal forward move. **2.** One used to further the purposes of another <a *pawn* in the power struggle>

▲ word history: The word *pawn*[2] denoting one of the pieces used in chess is a doublet of *peon,* since both are derived from Medieval Latin *pedo,* "foot soldier." *Pawn* comes from Old French *paon,* "foot soldier," "pawn in chess," which is a variant of *pion,* the ancestor of the modern French word *pion* with the same meanings. *Peon* is the Spanish descendent of *pedo.* Like French *pion,* Spanish *peon* means "foot soldier" and "pawn," but in American Spanish it came to denote a day laborer as well.

**pawn·bro·ker** (pôn'brō'kər) *n.* One who lends money at interest in exchange for valuable personal property left as security. **—pawn'bro'king** *n.*

**Paw·nee** (pô-nē') *n., pl.* **Pawnee** *or* **-nees. 1. a.** A confederation of four Plains Indian tribes living formerly in the region of Kansas and Nebraska and now on a reservation in Oklahoma. **b.** A member of this confederation. **2.** The Caddoan language of the Pawnee.

**pawn·shop** (pôn'shŏp') *n.* The shop of a pawnbroker.

**pawn ticket** *n.* A receipt for pawned goods.

**paw·paw** (pô'pô') *n. var. of* PAPAW.

**pay[1]** (pā) *v.* **paid, pay·ing, pays.** [ME *payen* < OFr. *paier* < Med. Lat. *pacare* < Lat., to pacify < *pax,* peace.] *—vt.* **1.** To give money to in return for goods or services rendered. **2.** To give (money) in exchange for goods or services. **3.** To give the indicated amount of <*pay* rent> **4.** To gain revenge for or upon : REQUITE <*pay* someone back for an injury> **5.** To yield as a return <dividends *paying* 13%> **6.** To bear the cost of <I *paid* their bill.> **7.** To afford an advantage to : PROFIT <It *paid* me to shop around.> **8.** To give or bestow (e.g., a compliment). **9.** To make (a visit or call). **10.** *p.t.* **payed.** *Naut.* To let out (a line or cable) by slackening. *—vi.* **1.** To make payment. **2.** To discharge a debt or obligation. **3.** To be profitable or worthwhile. **—pay off. 1. a.** To pay the full amount on (a debt). **b.** To get revenge for or on : REQUITE. **2.** To pay the wages due to (an employee) and discharge. **3.** *Informal.* To bribe. **4.** To allow (e.g., a rope) to run off a reel or spool. **5.** *Naut.* To turn or cause to turn (a vessel) to leeward. **—pay up.** To pay the full amount demanded. *—adj.* **1.** Of, relating to, giving, or receiving payments. **2.** Requiring payment to operate <a *pay* clothes dryer> **3.** Yielding valuable metal in mining. *—n.* **1.** The act of paying or state of being paid. **2.** Money given in return for work done : WAGES. **3. a.** Recompense or reward <Your smile was *pay* enough.> **b.** Retribution or punishment. **4.** Paid employment <the workers in our *pay*> **5.** One considered with regard to one's credit or willingness to pay. **—pay (one's) dues.** To earn a right or position through hard work, experience, or suffering. **—pay (one's) way.** To contribute one's own share : pay for oneself. **—pay the piper.** To bear the consequences. **—pay through the nose.** To pay excessively.

**pay[2]** (pā) *vt.* **payed** *or* **paid, pay·ing, pays.** [Obs. Fr. *peier* < Lat. *picare* < *pix,* pitch.] *Naut.* To coat or cover (e.g., seams of a ship) with waterproof materials, as tar or asphalt.

**pay·a·ble** (pā'ə-bəl) *adj.* **1.** Requiring payment on a certain date : DUE. **2.** Specifying payment to a particular person. **3.** Capable of producing profit : PROFITABLE. **—pay'a·bly** *adv.*

**pay cable** *n.* Pay-TV received over a cable.

**pay·check** (pā'chĕk') *n.* **1.** A check issued to an employee in payment of salary or wages. **2.** Salary or wages.

**pay·day** (pā'dā') *n.* The day on which wages are paid.

**pay dirt** *n.* **1.** Earth, ore, or gravel with a metal content rich enough to make mining profitable. **2.** *Slang.* A profitable or useful discovery.

**payed** (pād) *v.* **1.** *p.t.* & *p.p. of* PAY[1] 10. **2.** *var. p.t.* & *p.p. of* PAY[2].

**pay·ee** (pā-ē') *n.* The one to whom money is paid.

**pay·er** (pā'ər) *n.* **1.** One that pays. **2.** The one responsible for paying a bill or note.

**pay·load** (pā'lōd') *n.* **1.** The revenue-producing part of a cargo. **2.** The explosive charge in the warhead of a missile. **3. a.** The total weight of passengers and cargo that an aircraft carries or can carry. **b.** *Aerospace.* The total weight of the instruments, crew, and life-support systems that a spacecraft carries or can carry. **c.** The passengers, crew, instruments, or equipment carried by an aircraft, spacecraft, or rocket.

**pay·mas·ter** (pā′măs′tər) *n.* A person in charge of paying wages and salaries.

**pay·ment** (pā′mənt) *n.* **1.** The act of paying or state of being paid. **2.** The amount paid. **3.** One's due, reward, or punishment : REQUITAL.

**pay·nim** (pā′nĭm) *n.* [ME *painim* < OFr. *paienisme*, heathendom < LLat. *paganismus* < *paganus*, pagan. —see PAGAN.] *Archaic.* **1. a.** A non-Christian, esp. a Moslem. **b.** A pagan. **2.** The pagan world.

**pay·off** (pā′ôf′, -ŏf′) *n.* **1. a.** Full payment of a salary or wages. **b.** The time of such payment. **2.** *Informal.* **a.** Final reckoning or settlement. **b.** The climax of a narrative or sequence of events. **3.** Final retribution or revenge. **4.** *Informal.* A bribe. **5.** *Math.* The amount gained or lost by a player in game theory.

**pay·o·la** (pā-ō′lə) *n.* [PAY + *(Victr)ola,* a trademark for a phonograph.] *Slang.* **1.** Bribery, esp. the bribing of disc jockeys to promote records. **2.** A bribe, esp. one given to a disc jockey.

**pay·out** (pā′out′) *n.* A percentage of corporate earnings paid as dividends to shareholders.

**pay·roll** *also* **pay roll** (pā′rōl′) *n.* **1.** A list of employees receiving wages, with the amounts due to each. **2.** The total amount of money to be paid out to employees at a given time.

**pay station** *n.* A coin-operated public telephone.

**pay television** *n.* Pay-TV.

**pay toilet** *n.* A public toilet in a booth with a coin-operated door.

**pay-TV** (pā′tē-vē′) *n.* A system for receiving television broadcasts that requires a monthly subscription payment.

**Pb** [Lat. *plumbum*, lead.] *symbol for* LEAD.

**Pd** *symbol for* PALLADIUM.

**pe** (pā) *n.* [Heb. *peh.*] The 17th letter of the Hebrew alphabet. —See table at ALPHABET.

**pea** (pē) *n.* [Back-formation < ME *pease* (taken as pl.) < OE *pise* < LLat. *pisa*, ult. < Gk. *pison.*] **1.** A climbing annual vine, *Pisum sativum*, cultivated in all temperate zones and bearing compound leaves, small white flowers, and edible seeds in a green, elongated pod. **2.** One of the rounded green seeds of the pea, used as a vegetable. **3. peas.** The unopened pods of the pea plant. **4.** A plant of the genus *Lathyrus*, as the sweet pea or the beach pea.

▲ word history: The contrast between singular and plural for nouns is so important in English that there are very few nouns that have identical singular and plural forms. Those that do tend to be survivals rather than borrowings or new formations: *sheep* and *deer* are as old as English itself. Many words that originally had identical forms for the two numbers eventually developed new plurals ending with the regular plural suffix *-s*: *daughter* had an Old English plural *dohtor*, which was the same as the singular. Other words developed new singular forms; the word *pea* is an example. The Old English ancestor of *pea* was *pise*, ultimately from Greek *pison*. The regular Old English plural of *pise* was *pisan*. By the 15th century the singular and plural had fallen together in the common form *pese*, pronounced like *peas*, the modern plural of *pea*. Because of its pronunciation the *-s* in *pese* was interpreted as the plural noun suffix and a new singular without *-s* was developed to conform to the usual pattern of English nouns. The new singular is spelled *pea* in Modern English.

**pea bean** *n.* The navy bean.

**pea·bod·y bird** (pē′bŏd′ē, -bə-dē) *n.* [Prob. imit. of its song.] The white-throated sparrow.

**peace** (pēs) *n.* [ME *pees* < OFr. *pais* < Lat. *pax.*] **1.** The absence of war or other hostilities. **2.** An agreement or treaty to end hostilities. **3.** Freedom from quarrels and disagreement : harmonious relations. **4.** Public security and order <arrested for disturbing the *peace*> **5.** Inner contentment : SERENITY. *—interj.* —Used as a greeting or farewell and as a request for silence. **—at peace. 1.** In a state of tranquillity : SERENE. **2.** Free from strife <a nation *at peace*> **—hold (***or* **keep) one's peace.** To be silent. **—keep the peace.** To maintain or observe law and order.

**peace·a·ble** (pē′sə-bəl) *adj.* **1.** Inclined or disposed to peace. **2.** Undisturbed : peaceful. **—peace′a·ble·ness** *n.* **—peace′a·bly** *adv.*

**Peace Corps** *n.* A Federal government organization set up in 1961, that trains and sends American volunteers abroad to work with people of developing countries on projects for agricultural, technological, and educational improvement.

**peace·ful** (pēs′fəl) *adj.* **1.** Undisturbed by strife, turmoil, or disagreement : TRANQUIL. **2.** Opposed to strife : PEACEABLE. **3.** Of or typical of a state of peace. **—peace′ful·ly** *adv.* **—peace′ful·ness** *n.*

**peaceful coexistence** *n.* Existence together peacefully rather than in a state of hostility or war <*peaceful coexistence* between the superpowers>

**peace-keeping** *also* **peace·keep·ing** (pēs′kē′pĭng) *adj.* Of or pertaining to the preservation of peace, esp. the international supervision of a truce between hostile nations.

**peace·mak·er** (pēs′mā′kər) *n.* One who makes peace, esp. by settling the disputes of others. **—peace′mak′ing** *n. & adj.*

**peace offering** *n.* An offering made to an adversary in the interests of peace or reconciliation.

**peace officer** *n.* A law officer, as a sheriff, responsible for maintaining civil peace.

**peace pipe** *n.* The calumet.

**peace sign** *n.* A hand sign made with the palm forward and the middle and index fingers forming a V, used to express a desire for peace.

**peace·time** (pēs′tīm′) *n.* A time of peace.

**peach** (pēch) *n.* [ME *peche* < OFr., a peach < LLat. *persica* < Lat. *Persicus*, Persian.] **1. a.** A small tree, *Prunus persica*, native to China but widely cultivated throughout the temperate zones, with pink flowers and edible fruit. **b.** The soft, juicy, single-seeded fruit of the peach tree, with yellow flesh and downy, red-tinted, yellow skin. **2.** A light moderate to strong yellowish pink to light orange. **3.** *Slang.* A particularly admirable or pleasing person or thing.

**peach·y** (pē′chē) *adj.* **-i·er, -i·est. 1.** Like a peach, esp. in color or texture. **2.** *Slang.* Very pleasing : FINE. **—peach′i·ness** *n.*

**pea coat** *n.* A pea jacket.

**pea·cock** (pē′kŏk′) *n.* [ME *pecok* : OE *pēa*, peafowl (< Lat. *pavo*, peacock) + *cok*, cock < OE *coc.*] **1.** The male peafowl, distinguished by its crested head, brilliant blue or green feathers, and long tail feathers that have eyelike, iridescent spots and can be spread in a fanlike form. **2.** A vain person given to self-display. *—vi.* **-cocked, -cock·ing, -cocks.** To exhibit oneself vainly. **—pea′cock′ish, pea′cock′y** *adj.*

**peacock blue** *n.* A moderate to dark or strong greenish blue. **—pea′cock′-blue′** (pē′kŏk′blōō′) *adj.*

**pea·fowl** (pē′foul′) *n., pl.* **peafowl** *or* **-fowls.** [PEA(COCK) + FOWL.] Either of two large pheasants, *Pavo cristatus* of India and Ceylon or *P. muticus* of southeastern Asia.

**peag** *also* **peage** (pēg) *n.* [Narraganset *wamponpeag.*] Wampum.

**pea green** *n.* A moderate, strong, or brilliant yellow green to moderate yellowish green. **—pea′-green′** (pē′grēn′) *adj.*

**pea·hen** (pē′hĕn′) *n.* The female peafowl.

**pea jacket** *n.* [By folk etymology < Du. *pijjekker* : *pij*, a kind of coarse cloth + *jekker*, jacket.] A short, warm, double-breasted coat of heavy wool, worn esp. by sailors.

**peak[1]** (pēk) *n.* [Prob. alteration of PIKE[4].] **1.** A tapering, projecting point. **2. a.** The pointed summit of a mountain. **b.** The mountain itself. **3. a.** The point of a beard. **b.** A widow's peak. **4.** The point of greatest development, value, height, or intensity : CREST. **5.** *Physics.* The highest value attained by a varying quantity. **6.** *Naut.* **a.** The narrow portion of a ship's hull at the bow or stern. **b.** The upper after corner of a fore-and-aft sail. **c.** The outermost end of a gaff. *—v.* **peaked, peak·ing, peaks.** *—vt.* **1.** *Naut.* To raise (a gaff) above the horizontal. **2.** To bring to a maximum of development, growth, value, or intensity. *—vi.* **1.** To be formed into a peak or peaks <Whip the cream until it *peaks.*> **2.** To achieve a maximum of development, growth, value, or intensity. *—adj.* Approaching or being the maximum <*peak* traffic>

**peak[2]** (pēk) *vi.* **peaked, peak·ing, peaks.** [Orig. unknown.] To become pale, sickly, or emaciated.

**peaked[1]** (pēkt, pē′kĭd) *adj.* Ending in a peak : POINTED.

**peak·ed[2]** (pē′kĭd) *adj.* Having a sickly appearance.

**peal** (pēl) *n.* [ME *pele*, summons to church by bell, short for *apel*, appeal. —see APPEAL.] **1.** A ringing of a set of bells, esp. a change or set of changes rung on bells. **2.** A set of bells tuned to each other : CHIME. **3.** A loud burst of noise <*peals* of childish laughter> *—v.* **pealed, peal·ing, peals.** *—vi.* To sound in a peal : RING. *—vt.* To utter loudly and sonorously.

**pe·an** (pē′ən) *n. var. of* PAEAN.

**pea·nut** (pē′nŭt′) *n.* **1.** A tropical American vine, *Arachis hypogaea*, widely cultivated in semitropical regions, with yellow flowers on stalks that bend over so that the seed pods ripen underground. **2.** The edible, nutlike, oily seed of the peanut, used for food and as a source of oil. **3.** *Slang.* A small or insignificant person. **4. peanuts.** *Slang.* A very small amount of money. *—adj. Slang.* Having little or no importance : INSIGNIFICANT.

**peanut brittle** *n.* A hard candy containing peanuts.

**peanut butter** *n.* A paste made from roasted ground peanuts.

**peanut oil** *n.* The oil pressed from peanuts, used as a pharmaceutical vehicle, in soaps, and in cooking.

**pear** (pâr) *n.* [ME *pere* < OE *peru*, a pear < Lat. *pirum.*] **1.** A widely cultivated tree, *Pyrus communis*, bearing glossy leaves, white flowers, and edible fruit. **2.** The fruit of the pear, spherical at the base and tapering toward the top.

**pear haw** *n.* A shrub or small tree, *Crataegus uniflora* of southeastern North America, with white flowers and small, orange-red, pear-shaped fruit.

**pearl[1]** (pûrl) *n.* [ME *perle* < OFr. < VLat. *pernula*, dim. of Lat. *perna*, sea-mussel.] **1.** A smooth, lustrous, variously colored deposit, chiefly calcium carbonate, formed around a grain of sand or other foreign matter in the shells of certain mollusks and valued as a gem. **2.** Mother-of-pearl. **3.** One likened to a pearl in beauty or value. **4.** A printing type size measuring five points. **5.** A yellowish white. *—v.*

**pearled, pearl·ing, pearls.** —*vt.* **1.** To cover or decorate with or as if with pearls. **2.** To make into the shape or color of pearls. —*vi.* **1.** To dive or fish for pearls or pearl-bearing mollusks. **2.** To form beads resembling pearls. —*adj.* Having the shape or color of pearls.

**pearl²** (pûrl) *v.* & *n.* *var. of* PURL.

**pearl ash** *n.* Potassium carbonate.

**pearl danio** *n.* [NLat. *Danio,* genus name.] A slender freshwater tropical fish, *Brachydanio albolineatus,* bearing silvery scales and popular as an aquarium fish.

**pearl diver** *n.* A person who dives in search of pearl-bearing mollusks.

**pearl·er** (pûr′lər) *n.* **1.** A pearl diver. **2.** A boat whose crew is engaged in seeking or trading pearls.

**pearl·es·cent** (pûr-lĕs′ənt) *adj.* Having a pearly gloss or shine.

**pearl gray** *n.* A light gray, from yellowish to light bluish gray. **—pearl′-gray′** (pûrl′grā′) *adj.*

**Pearl Harbor** *n.* [After *Pearl Harbor,* Oahu, Hawaii, from the surprise attack there by the Japanese in 1941.] A swift surprise attack that usu. causes great destruction.

**pearl·ite** (pûr′līt′) *n.* [Fr. *perlite* < *perle,* pearl.] **1.** A mixture of ferrite and cementite forming distinct layers or bands in slowly cooled carbon steels. **2.** *var. of* PERLITE.

**pearl onion** *n.* A small often pickled onion used esp. as a garnish.

**pearl millet** *n.* A tropical grass, *Pennisetum glaucum,* with long dense flowering spikes and whitish seeds used as food.

**pearl oyster** *n.* A bivalve tropical marine mollusk of the genus *Pinctada* or related genera, esp. *P. margaritifera,* a major commercial source of pearls.

**pearl·y** (pûr′lē) *adj.* **-i·er, -i·est. 1.** Resembling pearls <*pearly* teeth> **2.** Covered or adorned with pearls or mother-of-pearl.

**pearly everlasting** *n.* A plant, *Anaphalis margaritaceae,* with woolly gray-green foliage and long-lasting whitish flowers.

**pearly nautilus** *n.* The chambered nautilus.

**pear·main** (pâr′mān′) *n.* [ME *parmain,* a kind of pear < OFr.] An old variety of red-skinned apple.

**peas·ant** (pĕz′ənt) *n.* [ME *paissaunt* < OFr. *païsant* < *païs,* country < LLat. *pagensis,* inhabitant of a district < Lat. *pagus,* district.] **1.** A member of the class comprising small farmers and tenants, sharecroppers, and laborers on the land where these constitute the primary labor force in agriculture. **2.** A rustic. **3.** An uncouth, crude, or ill-bred person : BOOR. **—peas′ant·ry** *n.*

**peas·cod** (pēz′kŏd′) *n.* *var. of* PEASECOD.

**pease** (pēz) *n., pl.* **pease** *or* **peas·en** (pē′zən). *Obs.* A pea.

**pease·cod** *also* **peas·cod** (pēz′kŏd′) *n.* [ME *pesecod* : *pese,* pea + *cod,* cod. —see COD².] *Obs.* The pod of the pea.

**peas·en** (pē′zən) *n.* *var. pl. of* PEASE.

**pea·shoot·er** (pē′shōō′tər) *n.* A toy having of a small tube through which pellets, as dried peas, are blown.

**pea soup** *n.* **1.** A thick soup or purée made of dried peas. **2.** *Slang.* Dense fog.

**peat** (pēt) *n.* [ME *pete* < Med. Lat. *peta.*] Partially carbonized vegetable matter, usu. mosses, found in bogs and used as fertilizer and fuel. **—peat′y** *adj.*

**peat bog** *n.* A bog or swamp where peat has accumulated.

**peat moss** *n.* **1.** A moss of the genus *Sphagnum,* growing in wet places. **2.** The partly carbonized remains of peat moss, used as a mulch and plant food.

**pea·vey** *also* **pea·vy** (pē′vē) *n., pl.* **-veys** *also* **-vies.** [After Joseph *Peavey* (fl. 1875), American inventor.] A wooden lever with a metal point and a hinged metal hook near the end, used by lumberjacks to handle logs.

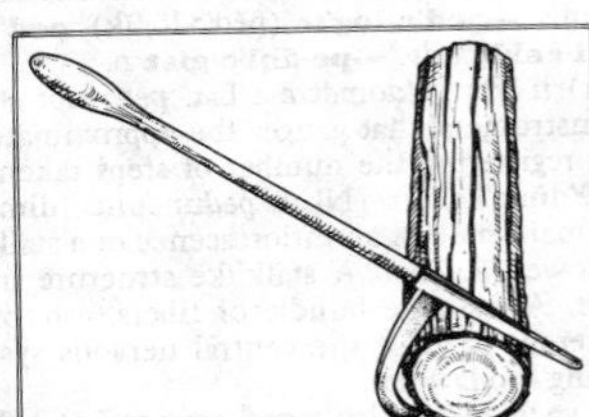

peavey

**peb·ble** (pĕb′əl) *n.* [ME *pobble* < OE *papolstān.*] **1.** A small stone worn smooth by erosion. **2. a.** Clear, colorless quartz : ROCK CRYSTAL. **b.** A lens made of such quartz. **3.** A crinkled surface, as on paper or leather. —*vt.* **-bled, -bling, -bles. 1.** To pave or pelt with pebbles. **2.** To impart an irregularly rough, grainy surface to (paper or leather). **—peb′bly** *adj.*

**pe·can** (pĭ-kän′, -kăn′) *n.* [Algonquian *paccan.*] **1.** A tree, *Carya illinoensis* of the southern United States, with deeply furrowed bark and edible nuts. **2.** The smooth, thin-shelled, oval nut of the pecan.

**pec·ca·ble** (pĕk′ə-bəl) *adj.* [OFr. < Lat. *peccare,* to sin.] Liable to sin. **—pec′ca·bil′i·ty** *n.*

**pec·ca·dil·lo** (pĕk′ə-dĭl′ō) *n., pl.* **-loes** *or* **-los.** [Sp. *pecadillo,* dim. of *pecado,* sin < Lat. *peccatum* < *peccare,* to sin.] A minor sin or fault.

**pec·cant** (pĕk′ənt) *adj.* [Lat. *peccans, peccant-,* pr.part. of *peccare,* to sin.] **1.** Sinful : guilty. **2.** Violating a rule or accepted practice : ERRING. **—pec′can·cy** *n.* **—pec′cant·ly** *adv.*

**pec·ca·ry** (pĕk′ə-rē) *n., pl.* **-ries.** [Sp. *pécari* < Cariban *pakira.*] Either of two piglike hoofed mammals, *Tayassu tajacu* or *T. pecari* of southern North America, Central America, and South America, with long, dark, dense bristles.

**pec·ca·vi** (pĕ-kä′wē, -vē, -kā′vī′) *n., pl.* **-vis.** [Lat., I have sinned < *peccare,* to sin.] A confession of sin.

**peck¹** (pĕk) *v.* **pecked, peck·ing, pecks.** [ME *pecken,* prob. < MLG *pekken.*] —*vt.* **1.** To strike with a beak or pointed instrument. **2.** To make (e.g., a hole) by striking repeatedly with the beak or a pointed instrument. **3.** To grasp and pick up with the beak <a hen *pecking* grain> **4.** *Informal.* To kiss briefly and casually. —*vi.* **1.** To make strokes with or as if with the beak. **2.** To eat in small, sparing bits : NIBBLE. **3.** To criticize incessantly : CARP. —*n.* **1. a.** A stroke or light blow with the beak. **b.** A mark or hole made by such a stroke. **2.** *Informal.* A quick light kiss.

**peck²** (pĕk) *n.* [ME < OFr. *pek.*] **1. a.** A unit of volume or capacity in the U.S. system, used in dry measure, equal to 8 quarts or 537.605 cubic inches. **b.** A unit of volume or capacity in the British Imperial System, used in dry and liquid measure, equal to 554.84 cubic inches. **2.** A container measuring or holding a peck. **3.** *Informal.* A large quantity : LOT <a *peck* of problems>

**peck·er** (pĕk′ər) *n.* **1.** One that pecks. **2.** *Chiefly Brit.* Courage.

**pecking order** *n.* **1.** A hierarchy within poultry flocks, according to which each member submits to pecking and domination by the stronger or more aggressive members and has the privilege of pecking and dominating the weaker ones. **2.** A human hierarchy.

**Peck's bad boy** (pĕks) *n.* [After *Peck's Bad Boy and His Pa,* a book by George Wilbur Peck (1840–1916).] One whose bad behavior annoys and embarrasses others.

**Peck·sniff·i·an** (pĕk-snĭf′ē-ən) *adj.* [After Seth *Pecksniff,* a character in *Martin Chuzzlewit,* a novel by Charles Dickens (1812–1870).] Hypocritically benevolent : SANCTIMONIOUS.

**pec·tase** (pĕk′tās′, -tāz′) *n.* [PECT(IN) + -ASE.] An enzyme found in certain fruits that catalyzes the conversion of pectins to pectic acids.

**pec·tate** (pĕk′tāt′) *n.* [PECT(IC ACID) + -ATE.] A salt or ester of pectic acid.

**pec·ten** (pĕk′tən) *n., pl.* **-ti·nes** (-tə-nēz′) [NLat. < Lat., comb.] *Zool.* **1.** A comblike body structure or organ, as the ridged part of the eyelid of birds and reptiles. **2.** A scallop of the genus *Pecten.*

**pec·tic acid** (pĕk′tĭk) *n.* [Fr. *pectique,* related to pectin. —see PECTIN.] Any of several colloidal substances, essentially complex organic acids, derived from pectin.

**pec·tin** (pĕk′tĭn) *n.* [Fr. *pectine* < *pectique,* related to pectin < Gk. *pēktikos,* coagulating < *pēktos,* coagulated, *pēgnunai,* to coagulate.] Any of a group of complex colloidal substances of high molecular weight found in ripe fruits, as apples, and used to jell various foods, drugs, and cosmetics. **—pec′tic, pec′tin·ous** *adj.*

**pec·tin·ase** (pĕk′tə-nās′, -nāz′) *n.* A plant enzyme that catalyzes the hydrolysis of pectin.

**pec·ti·nate** (pĕk′tə-nāt′) *also* **pec·ti·nat·ed** (-nā′tĭd) *adj.* [Lat. *pecten, pectin-,* comb + -ATE.] Having teeth or projections like a comb : COMBLIKE. **—pec′ti·na′tion** *n.*

**pec·tin·es·ter·ase** (pĕk′tə-nĕs′tə-rās′, -rāz′) *n.* An enzyme that catalyzes the hydrolysis of pectins into pectic acids.

**pec·to·ral** (pĕk′tər-əl) *adj.* [ME, something worn on the chest < OFr. < Lat. *pectorale,* breastplate < *pectoralis,* pectoral < *pectus,* breast.] **1.** *Anat.* Pertaining to the breast or chest. **2.** *Med.* Useful in diseases of the chest. **3.** Worn on the chest or breast, as a prelate's cross. —*n.* **1.** A chest organ or muscle. **2.** A pectoral fin. **3.** A medicine for chest diseases. **4.** An ornament worn on the chest.

**pectoral arch** *n.* The pectoral girdle.

**pectoral fin** *n.* Either of the anterior pair of fins attached to the pectoral girdle of fishes.

**pectoral girdle** *n.* A skeletal structure in vertebrates, attached to and supporting the forelimbs or fins.

**pectoral muscle** *n.* One of the four muscles of the upper anterior chest.

**pectoral sandpiper** *n.* A New World sandpiper, *Erolia melanotos,* with brownish streaks on the upper part of the breast.

**pec·u·late** (pĕk′yə-lāt′) *v.* **-lat·ed, -lat·ing, -lates.** [Lat. *peculari, peculat-* < *peculium,* private property < *pecu,* cattle.] —*vt.* To embezzle or take wrongfully for one's own use. —*vi.* To steal money or goods entrusted to one. **—pec′u·la′tion** *n.* **—pec′u·la′tor** *n.*

**pe·cu·liar** (pĭ-kyōōl′yər) *adj.* [ME *peculier,* of private property < Lat. *peculiaris* < *peculium,* private property < *pecu,* cattle.] **1.** Unusual or eccentric : ODD <*peculiar* habits> **2.** Distinct from others <a subject of *peculiar* delicacy> **3. a.** Exclusive : unique. **b.** Belonging primarily or distinctively to one person, group, or kind. —*n.* **1.** A property or privilege that is exclusively one's own. **2.** *Chiefly Brit.* A church or parish under the jurisdiction of a diocese different from that in which it lies. **—pe·cu'liar·ly** *adv.*

▲ word history: The sense "odd" for *peculiar* has developed since the word was borrowed into English, and it is possible to reconstruct the steps to its appearance. *Peculiar* is derived from Latin *peculiaris,* which originally meant "pertaining to private property" but developed the extended sense "belonging to oneself alone." The English word was used with the senses of Latin *peculiaris,* but *peculiar* came to mean "exclusive, unique" and then "unique" in the sense "singular, unusual, odd."

**peculiar galaxy** *n., pl.* **-ies.** Any of a group of galaxies that have highly unusual shapes.

**pe·cu·li·ar·i·ty** (pĭ-kyo͞o'lē-ăr'ĭ-tē, -kyo͞ol-yăr'-) *n., pl.* **-ties. 1.** The quality or state of being peculiar. **2.** A distinctive or notable feature or characteristic. **3.** An idiosyncrasy : eccentricity.

**pe·cu·ni·ar·y** (pĭ-kyo͞o'nē-ĕr'ē) *adj.* [Lat. *pecuniarius* < *pecunia,* money.] **1.** Consisting of or relating to money. **2.** Requiring the payment of money <a *pecuniary* violation>

**ped-[1]** *pref. var. of* PEDO-[1].

**ped-[2]** *pref. var. of* PEDO-[2].

**-ped** or **-pede** *suff.* [< Lat. *pes, ped-,* foot.] Foot <maxilli*ped*>

**ped·a·gog·ic** (pĕd'ə-gŏj'ĭk, -gō'jĭk) *also* **ped·a·gog·i·cal** (-gŏj'ĭ-kəl, -gō'jĭ-kəl) *adj.* **1.** Of, relating to, or typical of teaching. **2.** Marked by pedantic formality. **—ped'a·gog'i·cal·ly** *adv.*

**ped·a·gog·ics** (pĕd'ə-gŏj'ĭks, -gō'jĭks) *n. (sing. in number).* The art of teaching : PEDAGOGY.

**ped·a·gogue** (pĕd'ə-gŏg', -gôg') *n.* [ME *pedagoge* < OFr. < Lat. *paedagogus* < Gk. *paidagōgos* : *pais,* boy + *agōgos,* leader < *agein,* to lead.] **1.** One who teaches : EDUCATOR. **2.** One who instructs in a pedantic or dogmatic manner. **—ped'a·gogu'ish** *adj.*

**ped·a·go·gy** (pĕd'ə-gō'jē, -gŏj'ē) *n.* **1.** The art or profession of teaching. **2.** Preparatory instruction or training.

**ped·al** (pĕd'l) *n.* [Fr. *pedale* < Ital. < Lat. *pedalis,* of the foot < *pes,* foot.] **1.** A lever operated by the foot on various musical instruments, as the piano or harp. **2.** A pedal point. **3.** A lever worked by the foot in a mechanism, as a bicycle or sewing machine. *—adj.* **1.** Of or relating to a foot or footlike part. **2.** *Mus.* Of or relating to a pedal. *—v.* **-aled, -al·ing, -als** or **-alled, -al·ling, -als.** *—vi.* **1.** To use or operate a pedal or pedals. **2.** To ride a bicycle. *—vt.* To operate the pedals of.

**pe·dal·fer** (pĭ-dăl'fər) *n.* [PED(O)-[1] + AL(UM) + Lat. *ferrum,* iron.] Soil rich in alumina and iron and deficient in carbonates, usu. found in humid, high-temperature regions with forest cover.

**pedal keyboard** *n.* A keyboard of pedals in a musical instrument such as a pipe organ.

**pedal piano** *n.* A piano with a pedal keyboard.

**pedal point** *n. Mus.* A note, usu. in the bass and on the tonic or the dominant, sustained through harmonic changes in the other parts.

**pedal pushers** *pl.n.* Women's and girls' calf-length slacks.

**ped·ant** (pĕd'nt) *n.* [OFr. < OItal. *pedante,* poss. < Lat. *paedagogans,* pr.part. of *paedagogare,* to instruct < *paedagogus,* pedagogue.] **1.** One unduly concerned with book learning and formal rules without an understanding or experience of practical affairs. **2.** One who displays one's learning ostentatiously. **3.** *Obs.* A schoolmaster.

**pe·dan·tic** (pə-dăn'tĭk) *adj.* Marked by a narrow, often ostentatious concern for book learning and formal rules : DONNISH. **—pe·dan'ti·cal·ly** *adv.*

☆ ***syns:*** PEDANTIC, ACADEMIC, BOOKISH, DONNISH, FORMALISTIC, INKHORN, SCHOLASTIC *adj. core meaning* : marked by a narrow, often ostentatious concern for book learning and formal rules <a *pedantic* attention to details>

**ped·ant·ry** (pĕd'n-trē) *n., pl.* **-ries. 1.** Pedantic attention to rules or detail. **2. a.** The habit of mind or manner characteristic of a pedant. **b.** An instance of pedantic behavior.

**ped·ate** (pĕd'āt') *adj.* [Lat. *pedatus,* p.part. of *pedare,* to furnish with feet < *pes,* foot.] **1.** *Zool.* Having feet. **2.** Resembling or functioning as a foot. **3.** *Bot.* Having radiating lobes or divisions, with the lateral lobes cleft or divided.

**ped·dle** (pĕd'l) *v.* **-dled, -dling, -dles.** [Back-formation < PEDDLER.] *—vt.* **1.** To travel about selling (goods). **2.** To seek to disseminate : give out <*peddling* propaganda> **3.** To engage in the illicit sale of (drugs). *—vi.* **1.** To travel about selling goods. **2.** To occupy oneself with trifles.

☆ ***syns:*** PEDDLE, HAWK, HUCKSTER, VEND *v. core meaning* : to travel about selling (goods) <*peddled* trinkets at country fairs>

**ped·dler** (pĕd'lər) *n.* [ME *pedlere,* prob. alteration of *peddere* < *pedde,* covered basket.] One who peddles goods for a living : HAWKER.

**-pede** *suff. var. of* -PED.

**pe·des** (pĕd'ās') *n. pl. of* PES.

**ped·es·tal** (pĕd'ĭ-stəl) *n.* [OFr. *piedestal* < OItal. *piedestallo* < *pie di stallo,* foot of a stall.] **1.** An architectural support or base, as for a column or statue. **2.** A foundation : SUPPORT. **3.** A position of high regard <put my college professor on a *pedestal*> *—vt.* **-taled, -tal·ing, -tals** or **-talled, -tal·ling, -tals.** To provide with or place on a pedestal.

**pe·des·tri·an** (pə-dĕs'trē-ən) *n.* [Lat. *pedester,* going on foot < *pedes,* a pedestrian < *pes,* foot.] A person traveling on foot. *—adj.* **1.** Of, pertaining to, or made for pedestrians. **2.** Going or performed on foot <*pedestrian* errands> **3.** Lacking distinction : ORDINARY <*pedestrian* lyrics> **—pe·des'tri·an·ism** *n.*

**pe·des·tri·an·ize** (pĭ-dĕs'trē-ə-nīz') *vt.* **-ized, -iz·ing, -iz·es.** To convert (a street or area) into a pedestrian walkway or mall. **—pe·des'tri·an·i·za'tion** *n.*

**pedi-** *pref.* [< Lat. *pes, ped-,* foot.] Foot <*pedi*form>

**pe·di·a·tri·cian** (pē'dē-ə-trĭsh'ən) *also* **pe·di·at·rist** (pē'dē-ăt'rĭst) *n.* A physician specializing in pediatrics.

**pe·di·at·rics** (pē'dē-ăt'rĭks) *n. (sing. in number).* The branch of medicine dealing with the care of infants and children and the treatment of their diseases. **—pe'di·at'ric** *adj.*

**ped·i·cab** (pĕd'ĭ-kăb') *n.* A small three-wheeled, hooded passenger vehicle that is pedaled.

**ped·i·cel** (pĕd'ĭ-səl, -sĕl') *also* **ped·i·cle** (-kəl) *n.* [NLat. *pedicellus* < Lat. *pediculus,* dim. of *pes,* foot.] **1.** A small organ, part, or stalk, esp. one serving as a support. **2.** *Bot.* **a.** Any of several small stalks bearing a single flower in an inflorescence. **b.** A support for a fern sporangium or moss capsule. **—ped'i·cel'lar** (pĕd'ĭ-sĕl'ər) *adj.*

**ped·i·cel·late** (pĕd'ĭ-sĕl'ĭt, -āt') *adj.* Having or supported by a pedicel <*pedicellate* plants>

**ped·i·cle** (pĕd'ĭ-kəl) *n. var. of* PEDICEL.

**pe·dic·u·lar** (pə-dĭk'yə-lər) *adj.* [Lat. *pedicularis* < *pediculus,* dim. of *pedis,* louse.] Of, relating to, or caused by lice.

**pe·dic·u·late** (pə-dĭk'yə-lĭt, -lāt') *adj.* [< NLat. *Pediculati,* order name < Lat. *pediculus,* dim. of *pes,* foot.] Of or relating to marine fishes of the order Pediculati or Lophiiformes, which includes the anglerfishes. *—n.* A pediculate fish.

**pe·dic·u·lo·sis** (pə-dĭk'yə-lō'sĭs) *n.* [Lat. *pediculus,* dim of *pedis,* louse + -OSIS.] Infestation with lice. **—pe·dic'u·lous** (-ləs) *adj.*

**ped·i·cure** (pĕd'ĭ-kyo͝or') *n.* [Fr. *pédicure* : Lat. *pes,* foot + Lat. *curare,* to take care of < *cura,* care.] **1. a.** Podiatry. **b.** A podiatrist. **2. a.** Cosmetic care of the feet and toenails. **b.** A cosmetic treatment of the feet and toenails. *—vt.* **-cured, -cur·ing, -cures.** To give a pedicure to. **—ped'i·cur'ist** *n.*

**ped·i·form** (pĕd'ə-fôrm') *adj.* Foot-shaped.

**ped·i·gree** (pĕd'ĭ-grē') *n.* [ME *pedegru* < OFr. *pie de grue,* crane's foot (from the resemblance of a crane's foot to the lines of succession on a genealogical chart).] **1.** A line of ancestors : LINEAGE. **2.** A list of ancestors : FAMILY TREE. **3.** A list of the ancestors of a purebred animal. **—ped'i·greed'** *adj.*

**ped·i·ment** (pĕd'ə-mənt) *n.* [Obs. *perement,* prob. alteration of PYRAMID.] **1.** A wide, low-pitched gable surmounting the facade of a Grecian-style building. **2.** An element similar to or derivative of a pediment, used widely in architecture and decoration. **—ped'i·men'tal** (-mĕn'tl) *adj.* **—ped'i·ment'ed** (-mĕn'tĭd) *adj.*

**ped·i·palp** (pĕd'ə-pălp') *n. Zool.* One of the second pair of appendages of an arachnid that are modified for various sensory functions.

**ped·lar** (pĕd'lər) *n. Chiefly Brit. var. of* PEDDLER.

**pedo-[1]** or **ped-** *pref.* [< Gk. *pedon,* soil, earth.] Soil <*pedo*cal>

**pedo-[2]** or **ped-** or **paedo-** or **paed-** *pref.* [Gk. *paido-* < *pais,* child, boy.] Child <*pedo*dontia>

**ped·o·cal** (pĕd'ə-kăl') *n.* [PEDO-[1] + Lat. *calx,* lime < Gk. *khalix,* pebble.] A lime-rich soil of cool, semiarid, and arid regions. **—ped'o·cal'ic** *adj.*

**pe·do·don·tia** (pē'də-dŏn'shə) *n.* The dentistry of children's teeth. **—pe'do·don'tist** (-dŏn'tĭst) *n.*

**ped·o·gen·e·sis** (pĕd'ə-jĕn'ĭ-sĭs) *n.* The process of soil formation.

**pe·dol·o·gy[1]** (pē-dŏl'ə-jē) *n.* The study of the development and behavior of children. **—pe'do·log'ic** (-də-lŏj'ĭk), **pe'do·log'i·cal** *adj.* **—pe'do·log'i·cal·ly** *adv.* **—pe·dol'o·gist** *n.*

**pe·dol·o·gy[2]** (pĭ-dŏl'ə-jē, pĕ-) *n.* The scientific study of the origins, characteristics, and uses of soils. **—ped'o·log'ic** (pĕd'ə-lŏj'ĭk), **ped'o·log'i·cal** *adj.* **—ped'o·log'i·cal·ly** *adv.* **—pe·dol'o·gist** *n.*

**pe·dom·e·ter** (pĭ-dŏm'ĭ-tər) *n.* [Fr. *pédomètre* : Lat. *pes,* foot + Gk. *metron,* measure.] An instrument that gauges the approximate distance traveled on foot by registering the number of steps taken.

**pe·dun·cle** (pĭ-dŭng'kəl, pē'dŭng'kəl) *n.* [NLat. *pedunculus,* dim. of Lat. *pes,* foot.] **1.** *Bot.* The main stalk of an inflorescence or a stalk or stem bearing a solitary flower. **2.** *Zool.* A stalklike structure in invertebrate animals. **3.** *Anat.* A stalklike bundle of fibers, esp. of nerve fibers, connecting different parts of the central nervous system. **—pe·dun'cu·lar** (pĭ-dŭng'kyə-lər) *adj.*

**pe·dun·cu·late** (pĭ-dŭng'kyə-lĭt, -lāt') *also* **pe·dun·cu·lat·ed** (-lā'tĭd) *adj.* Having or supported on a peduncle.

**pee** (pē) *n.* The letter *p.*

**peek** (pēk) *vi.* **peeked, peek·ing, peeks.** [ME *piken.*] **1.** To glance quickly. **2.** To look or peer furtively, as through a hole or crevice or from a place of concealment. **3.** To become visible gradually, as if emerging from hiding <wild mushrooms *peeking* through the leaves> *—n.* A brief or furtive look.

**peek·a·boo** (pēk'ə-bo͞o') *n.* [Orig. unknown.] A game for amusing a child, in which one repeatedly covers and exposes one's face, exclaiming "Peekaboo!" *—adj.* Made of a sheer or transparent fabric.

**peel[1]** (pēl) *n.* [< ME *pelen,* to peel < OFr. *peler* < Lat. *pilare,* to deprive of hair < *pilus,* hair.] The skin or rind of certain fruits, as the orange or banana. —*v.* **peeled, peel·ing, peels.** —*vt.* **1.** To strip or cut away the skin, rind, or bark from : PARE. **2.** To pull off : strip away <*peel* the tape from a package> —*vi.* **1.** To lose or shed a covering, as skin or bark. **2.** To come off in thin pieces or strips, as bark, skin, or paint. **3.** *Slang.* To remove one's clothes : UNDRESS. **—peel off. 1.** To leave flight formation in order to land or make a dive. **2.** To take leave or depart, esp. hurriedly.

**peel[2]** (pēl) *n.* [ME < OFr. *pele* < Lat. *pala.*] **1.** A long-handled shovel-like tool used by bakers to move bread or pastries into and out of an oven. **2.** A T-shaped pole used by printers for hanging freshly printed sheets of paper to dry.

**peel[3]** (pēl) *n.* [ME *pel,* small castle < AN < Lat. *palus,* stake.] A fortified house or tower constructed in the borderland of Scotland and England in the 16th cent.

**peel·er[1]** (pē′lər) *n.* One that peels, esp. a kitchen implement for peeling the skin or rind from a vegetable or fruit.

**peel·er[2]** (pē′lər) *n.* [After Sir Robert *Peel* (1788–1850).] *Chiefly Brit.* A police officer.

**peen** (pēn) *n.* [Prob. of Scand. orig.] The often wedge- or ball-shaped end of a hammerhead opposite the flat striking surface, used for chipping, indenting, and metalworking. —*vt.* **peened, peen·ing, peens.** To hammer, bend, or shape with a peen.

**peep[1]** (pēp) *vi.* **peeped, peep·ing, peeps.** [ME *pepen.*] **1.** To utter short, soft, high-pitched sounds, like those of a bird : CHEEP. **2.** To speak in a hesitant, thin, high-pitched voice. —*n.* **1.** A weak, shrill sound or utterance, like that of a young bird. **2.** A slight sound or utterance, esp. of objection or complaint. **3.** Any of various small North American sandpipers.

**peep[2]** (pēp) *v.* **peeped, peep·ing, peeps.** [ME *pepen,* prob. alteration of *piken,* to peek.] —*vi.* **1.** To steal a quick, furtive glance. **2.** To peer from behind something or through a small aperture. **3.** To become visible gradually, as if emerging from hiding <The moon *peeped* through the clouds.> —*vt.* To cause to emerge or become partly visible. —*n.* **1.** A quick or furtive look : GLANCE. **2.** A first glimpse or appearance, as of dawn.

**peep·er[1]** (pē′pər) *n.* A creature that peeps, esp. a tree frog.

**peep·er[2]** (pē′pər) *n.* **1.** One who looks furtively. **2.** *Slang.* An eye.

**peep·hole** (pēp′hōl′) *n.* A small hole or crevice through which one may peep.

**peeping Tom** *n.* [After the legendary *Peeping Tom* of Coventry who was the only person to see the naked Lady Godiva.] A person who gets pleasure esp. of a sexual nature from secretly watching others : VOYEUR.

**peep·show** *also* **peep show** (pēp′shō′) *n.* An exhibition of pictures or objects viewed through a small aperture or magnifying glass.

**peep sight** *n.* A rear sight of a firearm having of an adjustable eyepiece with a small opening through which the front sight and the target are aligned.

**pee·pul** *also* **pi·pal** (pē′pəl) *n.* [Hindi *pīpal* < Skt. *pippalam.*] A fig tree, *Ficus religiosa* of India, regarded as sacred by Buddhists.

**peer[1]** (pîr) *vi.* **peered, peer·ing, peers.** [Perh. alteration of AP-PEAR.] **1.** To look intently, searchingly, or with difficulty. **2.** To be partially visible : SHOW <A faint sun *peered* through the fog.>

**peer[2]** (pîr) *n.* [ME < OFr. *per* < *per,* equal < Lat. *par.*] **1.** A person who has equal standing with another, as in status, class, or age. **2.** *Archaic.* A companion : fellow. **3. a.** A nobleman. **b.** A member of the British peerage, as a duke, marquis, earl, viscount, or baron.

☆ ***syns:*** PEER, COEQUAL, COLLEAGUE, COMPEER, EQUAL, EQUIVALENT, FELLOW *n. core meaning* : one very similar to another in rank or position <judged by a jury of one's *peers*>

**peer·age** (pîr′ĭj) *n.* **1.** The rank or title of a peer. **2.** Peers as a group. **3.** A book listing peers and their families.

**peer·ess** (pîr′ĭs) *n.* **1.** The wife or widow of a peer. **2.** A woman who holds a peerage by descent or appointment.

**peer·less** (pîr′lĭs) *adj.* Having no peer : UNMATCHED <*peerless* beauty> **—peer′less·ly** *adv.* **—peer′less·ness** *n.*

**peet·weet** (pēt′wēt′) *n.* [Imit. of its song.] The spotted sandpiper.

**peeve** (pēv) *vt.* **peeved, peev·ing, peeves.** [Back-formation < PEEVISH.] To annoy or make resentful : VEX. —*n.* **1.** A grievance : vexation. **2.** A resentful mood.

**pee·vish** (pē′vĭsh) *adj.* [ME *pevish,* spiteful.] **1.** Querulous : discontented. **2.** Ill-tempered. **3.** Fractious : contrary. **—pee′vish·ly** *adv.* **—pee′vish·ness** *n.*

**pee·wee** (pē′wē) *n.* [Prob. redup. of WEE.] **1.** *Informal.* One that is small, esp. a small child. **2.** *var. of* PEWEE. **—pee′wee** *adj.*

**peg** (pĕg) *n.* [ME *pegge.*] **1. a.** A small cylindrical or tapered pin, as of wood, used to plug a hole or to fasten things. **b.** A similar pin forming a projection that may be used as a support or boundary marker. **2.** One of the pins of a stringed musical instrument that are turned to tighten or slacken the strings so as to regulate their pitch. **3.** An implement fitted with a pointed prong or claw for tearing or catching. **4.** A degree or notch, esp. in estimation. **5.** *Chiefly Brit.* A shot of liquor. **6.** *Baseball.* A low and fast throw made to put a base-runner out. **7.** *Informal.* A wooden leg. —*v.* **pegged, peg·ging, pegs.** —*vt.* **1.** To plug or fasten with a peg. **2.** To designate or mark by means of pegs. **3.** To fix (a price) at a certain level or within a certain range. **4.** *Informal.* To categorize : classify <*pegged* them as liberals> **5.** *Informal.* To throw. —*vi.* To work steadily : PERSIST <*pegging* away at the assignment> **—take (someone) down a peg.** To reduce the pride of : HUMBLE.

**Peg·a·sus** (pĕg′ə-səs) *n.* [Gk. *Pēgasos.*] **1.** *Gk. Myth.* A winged horse that with a stroke of his hoof caused the fountain Hippocrene to spring forth from Mount Helicon. **2.** A constellation in the Northern Hemisphere.

**peg leg** *n. Informal.* An artificial leg.

**peg·ma·tite** (pĕg′mə-tīt′) *n.* [Fr. < Gk. *pēgma,* framework < *pēgnunai,* to fasten.] A coarse-grained igneous rock, largely granite, sometimes rich in rare elements such as uranium, tungsten, and tantalum. **—peg′ma·tit′ic** (-tĭt′ĭk) *adj.*

**Peh·le·vi** (pā′lə-vē′) *n. var. of* PAHLAVI.

**pei·gnoir** (pān-wär′, pĕn-) *n.* [Fr. < OFr. *peigner,* to comb the hair < Lat. *pectinare* < *pecten,* comb.] A woman's loose-fitting dressing gown or negligee.

**pej·o·ra·tion** (pĕj′ə-rā′shən, pē′jə-) *n.* [Med. Lat. *pejoratio* < LLat. *pejorare,* to make worse < Lat. *pejor,* worse.] **1.** The process or condition of worsening or degenerating. **2.** The process by which the semantic status of a word changes for the worse over a period of time.

**pe·jor·a·tive** (pĭ-jôr′ə-tĭv, -jŏr′-, pĕj′ə-rā′tĭv, pē′jə-) *adj.* Tending to make or become worse : DISPARAGING. —*n.* A pejorative word. **—pe·jor′a·tive·ly** *adv.*

**pe·kin** (pē′kĭn′) *n.* [Fr. *pékin,* after *Pékin* (Beijing), China.] **1.** A striped or figured silk fabric. **2. Pekin.** A large white duck orig. bred in the Orient, widely raised in the United States for food.

**Pe·king·ese** (pē′kĭng-ēz′, -ēs′) *also* **Pe·kin·ese** (pē′kə-nēz′, -nēs′) *n., pl.* **Pekingese** *also* **Pekinese. 1.** A native or resident of Peking (now Beijing), China. **2.** The Chinese dialect of Peking. **3.** (pē′-kə-nēz′, -nēs′). A long-haired toy dog of a breed developed in China, with a flat nose and short bowed forelegs. **—Pe′king·ese′** *adj.*

**Pe·king man** (pē′kĭng) *n.* [After *Peking* (Beijing), China.] An extinct hominid primate of the genus *Sinanthropus,* known from fossil remains of the Pleistocene epoch.

**pe·koe** (pē′kō) *n.* [Chin. (Amoy) *peh ho* : *peh,* white + *ho,* down, fine feathers.] A black tea made from the leaves around the buds.

**pel·age** (pĕl′ĭj) *n.* [Fr. < OFr. *poil,* hair < Lat. *pilus.*] The coat of a mammal, consisting of hair, fur, wool, or other soft covering.

**Pe·la·gi·an·ism** (pə-lā′jē-ə-nĭz′əm) *n.* The theological doctrine of Pelagius, a British or Irish monk, and condemned as heresy by the Roman Catholic Church in A.D. 416; included in its tenets were denial of original sin and affirmation of an individual's ability to be righteous by the exercise of free will. **—Pe·la′gi·an** *adj.* & *n.*

**pe·lag·ic** (pə-lăj′ĭk) *adj.* [Lat. *pelagicus* < Gk. *pelagikos* < *pelagos,* sea.] Of, relating to, or living in open seas or oceans rather than waters adjacent to land or inland waters.

**pel·ar·gon·ic acid** (pĕl′är-gŏn′ĭk, -gō′nĭk) *n.* [PELARGON(IUM) + -IC.] A colorless or yellow oil, $CH_3(CH_2)_7COOH$, used as a gasoline additive and in making lacquers, plastics, and pharmaceuticals.

**pel·ar·go·ni·um** (pĕl′är-gō′nē-əm) *n.* [NLat. *Pelargonium,* genus name < Gk. *pelargos,* stork.] A plant or shrub of the genus *Pelargonium,* which includes the geraniums.

**Pe·las·gi·an** (pə-lăz′jē-ən) *n.* [< Gk. *Pelasgoi.*] One of a people living in the region of the Aegean Sea before the coming of the Greeks. **—Pe·las′gi·an, Pe·las′gic** *adj.*

**pe·lec·y·pod** (pə-lĕs′ə-pŏd′) *n.* [< NLat. *Pelecypoda,* class name : Gk. *pelekus,* ax + Gk. *pous,* foot.] A lamellibranch.

**pel·er·ine** (pĕl′ə-rēn′, pĕl′ər-ĭn) *n.* [Fr. *pèlerine* < fem. of *pèlerin,* pilgrim < LLat. *pelegrinus.* —see PILGRIM.] A woman's usu. short cape with points in front.

**pelerine**

**Pe·le·us** (pē′lē-əs, pĕl′yōōs′) *n.* [Lat. < Gk. *Pēleus.*] *Gk. Myth.* A son of Aeacus and father of Achilles.

**pelf** (pĕlf) *n.* [ME < OFr. *pelfre.*] Riches or wealth, esp. when fraudulently obtained.

**pel·i·can** (pĕl′ĭ-kən) *n.* [ME < OE *pellican* < LLat. *pelicanus* < Gk. *pelekan, pelekinos* < *pelekus,* ax.] A large web-footed bird of the genus *Pelecanus* of tropical and warm regions, with a large pouch under the lower bill for catching and holding fish.

**pe·lisse** (pə-lēsʹ) *n.* [Fr. < Med. Lat. *pellicia* < Lat. *pellicius,* made of skin < *pellis,* skin.] **1.** A long cloak or outer robe, usu. of fur or with a fur lining. **2.** A woman's loose light cloak, often with arm openings.

**pe·lite** (pēʹlītʹ) *n.* [Gk. *pēlos,* clay + -ITE.] Rock made up of fine fragments, as of clay, quartz particles, or rock flour. **—pe·litʹic** (pĭ-lĭtʹĭk) *adj.*

**pel·lag·ra** (pə-lăgʹrə, -lāʹgrə, -läʹgrə) *n.* [Ital. : *pelle,* skin (< Lat. *pellis*) + Gk. *agra,* seizure.] A chronic disease caused by niacin deficiency and marked by skin eruptions, digestive and nervous disturbances, and eventual mental deterioration. **—pel·lagʹrous** *adj.*

**pel·lag·rin** (pə-lăgʹrĭn, -lāʹgrĭn, -läʹgrĭn) *n.* [< PELLAGRA.] A person afflicted with pellagra.

**pel·let** (pĕlʹĭt) *n.* [ME *pelet* < OFr. *pelote* < VLat. **pilotta,* dim. of Lat. *pila,* ball.] **1.** A small solid or densely packed ball or mass, as of bread, wax, or medicine. **2.** A bullet or piece of small shot. **3.** A stone ball used as a catapult missile or as a primitive cannonball. —*vt.* **-let·ed, -let·ing, -lets. 1.** To make or form into pellets. **2.** To strike with pellets.

**pel·li·cle** (pĕlʹĭ-kəl) *n.* [OFr. < Med. Lat. *pellicula* < Lat., dim. of *pellis,* skin.] A thin skin or film, as an organic membrane or a liquid film. **—pel·licʹu·lar** (pə-lĭkʹyə-lər) *adj.*

**pel·li·to·ry¹** (pĕlʹĭ-tôrʹē, -tōrʹē) *n., pl.* **-ries.** [ME *peletre* < OFr. *piretre* < Lat. *pyrethrum.*] A small Mediterranean plant, *Anacyclus pyrethrum,* containing a volatile oil once used for the relief of toothache and facial neuralgia.

**pel·li·to·ry²** (pĕlʹĭ-tôrʹē, -tōrʹē) *n., pl.* **-ries.** [ME *peritorie* < OFr. *paritaire* < LLat. *parietaria* < *parietarius,* belonging to walls < Lat. *paries,* wall.] A plant of the genus *Parietaria,* bearing long, narrow leaves with hairy tufts at the base.

**pell-mell** *also* **pell·mell** (pĕlʹmĕlʹ) *adv.* [Fr. *pêle-mêle* < OFr. *pesle mesle,* prob. redup. of *mesle,* imper. of *mesler,* to mix. —see MEDDLE.] **1.** In a jumbled, confused manner : HELTER-SKELTER. **2.** In frantic, disorderly haste : HEADLONG. **—pell'-mell'** *adj.* & *n.*

**pel·lu·cid** (pə-lo͞oʹsĭd) *adj.* [Lat. *pellucidus* < *pellucēre,* to shine through : *per,* through + *lucēre,* to shine.] **1.** Admitting the maximum passage of light : TRANSPARENT. **2.** Clear in style or meaning, as a prose work. **—pel·lu·cidʹi·ty** (-sĭdʹĭ-tē), **pel·luʹcid·ness** *n.* **—pel·luʹcid·ly** *adv.*

**Pe·lops** (pēʹlŏpsʹ) *n.* [Lat. < Gk. : *pelios,* dark + *ops,* face.] *Gk. Myth.* The son of Tantalus and father of Atreus.

**pe·lo·ri·a** (pə-lôrʹē-ə, -lōrʹ-) *n.* [NLat. < Gk. *pelōros,* monstrous < *pelōr,* monster.] Unusual regularity in the form of a flower that is normally irregular. **—pe·lorʹic** (pə-lôrʹĭk, -lŏrʹ-) *adj.*

**pe·lo·rus** (pə-lôrʹəs, -lōrʹ-) *n., pl.* **-rus·es.** [Orig. unknown.] A fixed compass card on which a ship's bearings are taken.

**pe·lo·ta** (pə-lōʹtə) *n.* [Sp. < OFr. *pelote,* pellet.] **1.** Jai alai. **2.** The ball used in jai alai.

**pelt¹** (pĕlt) *n.* [ME.] **1.** The skin of an animal with the hair or fur still on it. **2.** A stripped animal skin ready for tanning.

**pelt²** (pĕlt) *v.* **pelt·ed, pelt·ing, pelts.** [ME *pelten.*] —*vt.* **1.** To strike or assail repeatedly with or as if with blows or missiles : BOMBARD <*pelted* each other with pebbles> **2.** To cast, hurl, or throw (missiles). **3.** To strike repeatedly <Heavy rain *pelted* the cars.> —*vi.* **1.** To beat or strike heavily and repeatedly. **2.** To move at a rapid pace. —*n.* **1.** A sharp blow. **2.** A rapid pace : SPEED <riding at full *pelt*> **—peltʹer** *n.*

**pel·tate** (pĕlʹtātʹ) *adj.* [NLat. *peltatus* < Lat. *pelta,* small shield < Gk. *peltē.*] Having the leaf stalk attached near the center of the surface rather than at or near the margin. **—pelʹtateʹly** *adv.*

**pelt·ing** (pĕlʹtĭng) *adj.* [Orig. unknown.] *Archaic.* Paltry : petty.

**pel·try** (pĕlʹtrē) *n.* [ME < AN *pelterie* < OFr. *peletier,* furrier, prob. < *pel,* skin < Lat. *pellis.*] Undressed pelts.

**pel·vic** (pĕlʹvĭk) *adj.* Of, in, near, or relating to the pelvis.

**pelvic arch** *n.* Pelvic girdle.

**pelvic fin** *n.* Either of a pair of lateral hind fins of fishes, attached to the pelvic girdle.

**pelvic girdle** *n.* The skeletal structure of bone or cartilage by which the hind limbs or analogous parts are supported and joined to the vertebral column.

**pel·vis** (pĕlʹvĭs) *n., pl.* **-vis·es** *or* **-ves** (-vēz) [NLat. < Lat., basin.] *Anat.* **1.** A basin-shaped skeletal structure, consisting of the innominate bones on the sides, the pubis in front, and the sacrum and coccyx behind, that rests on the lower limbs and supports the spinal column. **2.** The hollow funnel in the outlet of the kidney into which urine is discharged before entering the ureter.

**pem·mi·can** *also* **pem·i·can** (pĕmʹĭ-kən) *n.* [Cree *pimikân* < *pimii,* fat.] **1.** A food prepared by North American Indians from lean, dried strips of meat pounded into paste, mixed with fat and berries, and pressed into small cakes. **2.** A food like pemmican used esp. as emergency rations.

**pem·phi·gus** (pĕmʹfĭ-gəs, pĕm-fīʹgəs) *n.* [NLat. < Gk. *pemphix,* pustule.] An acute or chronic skin disease marked by groups of itching or burning blisters.

**pen¹** (pĕn) *n.* [ME *penne* < OFr. < Lat. *penna,* feather.] **1.** An instrument for writing or drawing with ink or similar fluid, esp.: **a.** A quill. **b.** A pen point. **c.** A penholder and its pen point. **d.** A ballpoint pen. **e.** A fountain pen. **2.** An instrument for writing that is a means of expression. **3.** A writer or author <a *pen* for hire> **4.** A style of writing <a humorous *pen*> **5. pens.** Pinions. **6.** *Zool.* The hornlike internal shell of a squid. —*vt.* **penned, pen·ning, pens.** To compose or write with a pen. **—penʹner** *n.*

▲ word history: The English word *pen¹,* "a writing implement," is derived from the Latin word *penna,* which originally meant only "feather." The ancient Romans applied ink to paper and parchment with a reed pen, or *calamus.* In time, the reed pen was replaced by sharpened and split feathers, and in later Latin *penna* acquired the meaning "feather used to write with; pen." *Penna* descended to Old French as *penne,* which was borrowed into English with both senses, "feather" and "pen," the former of which is now obsolete.

**pen²** (pĕn) *n.* [ME < OE *penn.*] **1. a.** A fenced enclosure for animals. **b.** The animals kept in a pen. **c.** A small enclosure, as a bullpen or playpen. **2.** A repair dock for submarines. —*vt.* **penned** *or* **pent** (pĕnt), **pen·ning, pens.** To confine in or as if in a pen.

**pen³** (pĕn) *n.* [Orig. unknown.] A female swan.

**pen⁴** (pĕn) *n.* [Short for PENITENTIARY.] *Slang.* PENITENTIARY 1.

**pe·nal** (pēʹnəl) *adj.* [ME < OFr. < Lat. *poenalis* < *poena,* penalty.] **1.** Of, pertaining to, or prescribing punishment, as for breaking the law. **2.** Subject to punishment <a *penal* felony> **3.** Serving as or being a means or place of punishment. **—peʹnal·ly** *adv.*

**penal code** *n.* The body of laws pertaining to crimes and offenses and the penalties for their commission.

**pe·nal·ize** (pēʹnə-līzʹ, pĕnʹə-) *vt.* **-ized, -iz·ing, -iz·es. 1.** To subject to a penalty, esp. for infringement of a law or official regulation. **2.** To impose a handicap on. **—peʹnal·i·zaʹtion** *n.*

**pen·al·ty** (pĕnʹəl-tē) *n., pl.* **-ties.** [Med. Lat. *poenalitas* < Lat. *poenalis,* penal.] **1.** A punishment established by law or authority for a crime or offense. **2.** A forfeit, esp. a sum of money, required for an offense. **3.** The disadvantage or painful consequences resulting from an action or condition <ate too much and had to pay the *penalty*> **4.** A punishment, handicap, or loss of advantage imposed on a competitor or team for infraction of a rule. **5.** *often* **penalties.** Points scored in contract bridge by the opponents when the declarer fails to make his or her bid.

**penalty box** *n.* An area to the side of an ice-hockey rink in which penalized players remain for the duration of their penalties.

**pen·ance** (pĕnʹəns) *n.* [ME < OFr. < Lat. *paenitentia,* penitence < *paenitens,* penitent.] **1.** A voluntary act of devotion or self-mortification to show sorrow for a sin or misdeed. **2.** A sacrament in some Christian churches that includes contrition, confession to a priest, acceptance of punishment, and absolution. —*vt.* **-anced, -anc·ing, -ances.** To impose penance on.

**Pe·na·tes** (pə-nāʹtēz, -näʹ-) *pl.n.* [Lat.] The Roman gods of the household, whose cult was closely connected and often identified with that of the Lares.

**pence** (pĕns) *n. Chiefly Brit. var. pl. of* PENNY 1.

**pen·cel** *also* **pen·sil** (pĕnʹsəl) *n.* [ME < OFr. *penoncel,* dim. of *penon,* pennon.] A narrow flag, streamer, or pennon, esp. one carried at the top of a spear or lance.

**pen·chant** (pĕnʹchənt) *n.* [Fr. < pr.part. of *pencher,* to incline < Lat. *pendēre,* to hang.] A definite liking : INCLINATION.

**pen·cil** (pĕnʹsəl) *n.* [ME *pencel,* artist's brush < OFr. *pincel* < Lat. *penicillus* < *peniculus,* brush, dim. of *penis,* tail.] **1.** A narrow, gen. cylindrical implement for writing, drawing, or marking, having a thin rod of graphite, crayon, or similar substance encased in wood or held in a plastic or metal mechanical device. **2. a.** A narrow medicated or cosmetic stick <styptic *pencil*><eyebrow *pencil*> **b.** Something shaped like a pencil. **3. a.** *Archaic.* An artist's brush, esp. a fine one. **b.** An artist's particular style or technique. **c.** Descriptive skill. **4.** A cluster of rays, esp. light rays, radiating from or converging on a single point. **5.** *Math.* A one-parameter family of three-dimensional or plane figures, such as all straight lines in a plane that pass through a fixed point. —*vt.* **-ciled, -cil·ing, -cils** *or* **-cilled, -cil·ling, -cils.** To write, produce, mark, or color with or as if with a pencil. **—penʹcil·er** *n.*

**pencil pusher** *n. Informal.* One whose job involves paperwork.

**pen·dant** *also* **pen·dent** (pĕnʹdənt) *n.* [ME *pendaunt* < OFr. *pendant* < pr.part. of *pendre,* to hang < Lat. *pendēre.*] **1.** Something suspended from something else, esp. an ornament or piece of jewelry attached to a necklace or bracelet. **2.** A hanging lamp or chandelier. **3.** A sculptured ornament suspended from a vaulted Gothic ceiling or roof. **4. a.** One of a matched pair. **b.** One that supplements or complements another : COMPLEMENT.

**pen·dent** *also* **pen·dant** (pĕnʹdənt) *adj.* [ME *pendant* < OFr. —see PENDANT.] **1.** Hanging down : SUSPENDED. **2.** Projecting : overhanging. **3.** Awaiting settlement : PENDING. **—penʹdent·ly** *adv.*

**pen·den·tive** (pĕn-dĕnʹtĭv) *n.* [Fr. *pendentif* < Lat. *pendens,* hanging < *pendēre,* to hang.] An overhanging, triangular section of vaulting between the rim of a dome and each adjacent pair of the arches that support it.

**pend·ing** (pĕnʹdĭng) *adj.* [Fr. *pendant,* pendant, pending (< OFr.) + -ING¹.] **1.** Awaiting action, confirmation, or decision. **2.** Impending : imminent. —*prep.* **1.** While in the process of : DURING. **2.** While awaiting : UNTIL <*pending* further developments>

**pen·drag·on** (pĕn-drăg'ən) *n.* [ME < Welsh : *pen,* chief + *dragon,* leader.] The title of the supreme war leader of the post-Roman Celts of southern Britain. **—pen·drag'on·ship'** *n.*

**pen·du·lar** (pĕn'jə-lər, pĕn'dyə-) *adj.* Of or resembling the motion of a pendulum.

**pen·du·lous** (pĕn'jə-ləs, pĕn'dyə-) *adj.* [Lat. *pendulus* < *pendēre,* to hang.] **1.** Suspended so as to swing or sway. **2.** Undecided : hesitant. **—pen'du·lous·ly** *adv.* **—pen'du·lous·ness** *n.*

**pen·du·lum** (pĕn'jə-ləm, pĕn'dyə-, pĕn'də-) *n.* [NLat. < Lat. *pendulus,* hanging < *pendēre,* to hang.] **1.** An object suspended from a fixed support so that it swings freely back and forth under the influence of gravity, commonly used to regulate various devices, esp. clocks. **2.** Something that swings back and forth from one course, opinion, or condition to another <the *pendulum* of fashion>

**Pe·nel·o·pe** (pə-nĕl'ə-pē) *n.* [Lat. < Gk. *Pēnelopeia.*] *Gk. Myth.* Odysseus' wife and Telemachus' mother, renowned for her faithfulness.

**pe·ne·plain** *also* **pe·ne·plane** (pē'nə-plān') *n.* [Lat. *paene, pene,* almost + PLAIN.] *Geol.* A nearly flat land surface representing an advanced stage of erosion.

**pe·nes** (pē'nēz) *n. var. pl. of* PENIS.

**pen·e·tra·ble** (pĕn'ĭ-trə-bəl) *adj.* [Lat. *penetrabilis* < *penetrare,* to penetrate.] Capable of being penetrated. **—pen'e·tra·bil'i·ty** *n.* **—pen'e·tra·bly** *adv.*

**pen·e·tra·li·a** (pĕn'ĭ-trā'lē-ə) *pl.n.* [Lat. < *penetralis,* inner < *penetrare,* to penetrate.] The innermost or most secret parts.

**pen·e·tram·e·ter** (pĕn'ĭ-trăm'ĭ-tər) *n. var. of* PENETROMETER.

**pen·e·trance** (pĕn'ĭ-trəns) *n.* [< PENETRANT.] The degree or frequency with which a gene manifests its effect.

**pen·e·trant** (pĕn'ĭ-trənt) *adj.* [Lat. *penetrans, penetrant-,* pr.part. of *penetrare,* to enter.] Penetrating : piercing. —*n.* Something that penetrates or is capable of penetrating.

**pen·e·trate** (pĕn'ĭ-trāt') *v.* **-trat·ed, -trat·ing, -trates.** [Lat. *penetrare, penetrat-.*] —*vt.* **1.** To enter or force a way into : PIERCE. **2. a.** To enter into and permeate. **b.** To cause to be permeated or diffused : STEEP. **3.** To grasp the inner meaning of : UNDERSTAND. **4.** To see through <My eyes couldn't *penetrate* the gloom.> **5.** To affect deeply, as by piercing the consciousness or emotions. —*vi.* **1.** To pierce or enter into something. **2.** To gain admittance or access. **3.** To gain insight. **—pen'e·tra'tive** *adj.*

☆ ***syns:*** PENETRATE, BREAK (through), PERFORATE, PIERCE, PUNCTURE *v. core meaning* : to pass into or through by overcoming resistance <The arrow *penetrated* the target.>

**pen·e·trat·ing** (pĕn'ĭ-trā'tĭng) *adj.* **1.** Able or seeming to penetrate <a *penetrating* scream> **2.** Keenly perceptive or understanding <a *penetrating* intellect> **—pen'e·trat'ing·ly** *adv.*

**pen·e·tra·tion** (pĕn'ĭ-trā'shən) *n.* **1.** The act or process of penetrating, esp.: **a.** The act of entering a country so as to establish influence <*penetration* of undercover agents> **b.** An attack that penetrates enemy territory or a military front. **2.** The ability or power to penetrate. **3.** The depth reached by a projectile after hitting its target. **4.** The capacity or action of understanding : INSIGHT.

**pen·e·trom·e·ter** (pĕn'ĭ-trŏm'ĭ-tər) *also* **pen·e·tram·e·ter** (-trăm'ĭ-tər) *n.* **1.** A device for measuring the penetrating power of x-rays. **2.** A device for measuring the penetrability of semisolids.

**pen·guin** (pĕn'gwĭn, pĕng'gwĭn) *n.* [Poss. < Welsh *pen gwyn,* white head.] **1.** A flightless marine bird of the family Spheniscidae of cool regions of the Southern Hemisphere, with flipperlike wings, webbed feet, and scalelike barbless feathers. **2.** *Obs.* The great auk.

**pen·hold·er** (pĕn'hōl'dər) *n.* A holder for a pen point.

**-penia** *suff.* [NLat. < Gk. *penia,* poverty, lack.] Lack : deficiency <leuko*penia*>

**pen·i·cil·la·mine** (pĕn'ĭ-sĭl'ə-mēn') *n.* [PENICILL(IN) + AMINE.] A degradation product of penicillin, $C_5H_{11}NO_2S$, used in the medical treatment of rheumatoid arthritis.

**pen·i·cil·late** (pĕn'ĭ-sĭl'ĭt, -āt') *adj.* [Lat. *penicillus,* brush + -ATE.] Having or resembling a tuft or brush of fine hairs, as those on caterpillars and certain grasses. **—pen'i·cil'late·ly** *adv.* **—pen'i·cil·la'tion** (-sə-lā'shən) *n.*

**pen·i·cil·lin** (pĕn'ĭ-sĭl'ĭn) *n.* [PENICILL(IUM) + -IN.] An isomeric antibiotic compound obtained from penicillium molds, esp. *Penicillium notatum* and *P. chrysogenum,* or produced biosynthetically and used to prevent or treat a wide variety of infections and diseases.

**pen·i·cil·lin·ase** (pĕn'ĭ-sĭl'ĭ-nās') *n.* A bacterial enzyme that inactivates penicillin by hydrolysis.

**pen·i·cil·li·um** (pĕn'ĭ-sĭl'ē-əm) *n., pl.* **-cil·li·ums** *or* **-cil·li·a** (-sĭl'ē-ə) [NLat. *Penicillium,* genus name < Lat. *penicillus,* brush. —see PENCIL.] A blue-green mold of the genus *Penicillium,* producing tufts of fine filaments that grow on decaying fruits and ripening cheese and used in making penicillin and cheese.

**pen·in·su·la** (pə-nĭn'syə-lə, -sə-lə) *n.* [Lat. : *paene,* almost + *insula,* island.] **1.** A long projection of land into water, connected with the mainland by an isthmus. **2.** A long projection of land into water with or without a well-defined isthmus. **—pen·in'su·lar** *adj.*

**pe·nis** (pē'nĭs) *n., pl.* **-nis·es** *or* **-nes** (-nēz) [Lat.] *Anat.* **1.** The male organ of copulation in higher vertebrates and usu. of urinary excretion in mammals. **2.** Any of various copulatory organs in males of lower animals.

**pen·i·tent** (pĕn'ĭ-tənt) *adj.* [ME < OFr. < Lat. *paenitens* < *paenitēre,* to repent.] Feeling or exhibiting remorse for one's sins or misdeeds. —*n.* **1.** One who is penitent. **2.** A person performing penance under the direction of a confessor. **—pen'i·tence** *n.* **—pen'i·tent·ly** *adv.*

**pen·i·ten·tial** (pĕn'ĭ-tĕn'shəl) *adj.* **1.** Of, relating to, or displaying penitence. **2.** Relating to or like penance. —*n.* **1.** A book or set of church rules dealing with the sacrament of penance. **2.** A penitent. **—pen'i·ten'tial·ly** *adv.*

**pen·i·ten·tia·ry** (pĕn'ĭ-tĕn'shə-rē) *n., pl.* **-ries.** [ME *penitenciary,* penance officer < Med. Lat. *penitentiarius* < Lat. *paenitentia,* penitence < *paenitens,* penitent.] **1.** A prison for those convicted of major crimes. **2.** *Rom. Cath. Ch.* **a.** A tribunal of the Roman Curia, presided over by a cardinal designated in this office as the Grand Penitentiary and having jurisdiction in matters relating to penance, dispensations, and papal absolutions. **b.** A canon whose special function is the administration of the sacrament of penance in cathedrals or churches having a chapter of canons. —*adj.* **1.** Of or for the purpose of penance : PENITENTIAL. **2.** Relating to or used for punishment or reform of criminals or wrongdoers. **3.** Resulting in or punishable by imprisonment in a penitentiary <a *penitentiary* felony>

**pen·knife** (pĕn'nīf') *n.* A small pocketknife.

**pen·light** (pĕn'līt') *n.* A small flashlight approx. the size and shape of a fountain pen.

**pen·man** (pĕn'mən) *n.* **1.** A copyist : scribe. **2.** An expert in penmanship. **3.** An author : writer.

**pen·man·ship** (pĕn'mən-shĭp') *n.* The art, skill, style, or manner of handwriting.

**pen·na** (pĕn'ə) *n., pl.* **pen·nae** (pĕn'ē) [Lat., feather.] One of the larger feathers forming the visible plumage of a bird. **—pen·na'ceous** (pĕ-nā'shəs) *adj.*

**pen name** *also* **pen·name** (pĕn'nām') *n.* A writer's pseudonym.

**pen·nant** (pĕn'ənt) *n.* [Blend of PENDANT and PENNON.] **1.** *Naut.* A long, often triangular, tapering flag used on ships for signaling or identification. **2.** A flag or emblem similar to a ship's pennant. **3. a.** A flag that serves as the emblem of the championship in a professional baseball league. **b.** The yearly championship in such a league.

**pen·nate** (pĕn'āt') *also* **pen·nat·ed** (pĕn'ā'tĭd) *adj.* [Lat. *penatus* < *penna,* wing.] Feathered or winged : PINNATE.

**pen·ni** (pĕn'ē) *n., pl.* **pen·nis** *or* **pen·ni·a** (pĕn'ē-ə) [Finn.] —See table at CURRENCY.

**pen·ni·less** (pĕn'ē-lĭs, pĕn'ə-lĭs) *adj.* **1.** Lacking money. **2.** Extremely poor. **—pen'ni·less·ly** *adv.* **—pen'ni·less·ness** *n.*

**pen·non** (pĕn'ən) *n.* [ME < OFr. *penon,* aug. of *penne,* wing < Lat. *penna.*] **1.** A long narrow streamer or banner borne on a lance. **2.** A pennant, banner, or flag. **3.** A pinion : wing. **—pen'noned** *adj.*

**pen·non·cel** *also* **pen·on·cel** *or* **pen·non·celle** (pĕn'ən-sĕl') *n.* [ME *penoncelle* < OFr. *penoncel,* dim. of *penon,* pennon.] A small pennon, flag, or streamer borne on a lance.

**Penn·syl·va·nia Dutch** (pĕn'səl-vān'yə, -vā'nē-ə) *n.* **1.** The descendants of German and Swiss immigrants who settled in Pennsylvania in the 17th and 18th cent. **2.** The dialect of High German spoken by the Pennsylvania Dutch.

**Pennsylvania German** *n.* PENNSYLVANIA DUTCH 2.

**Penn·syl·va·nian** (pĕn'səl-vān'yən, -vā'nē-ən) *adj.* [After *Pennsylvania.*] Of, belonging to, or designating the geologic time, system of rocks, and sedimentary deposits of the sixth period of the Paleozoic era, marked by the development of coal-bearing rock formations. —*n.* The Pennsylvanian period.

**pen·ny** (pĕn'ē) *n., pl.* **-nies.** [ME, an English coin < OE *penig.*] **1.** *pl.* **-nies** *also chiefly Brit.* **pence** (pĕns).—See table at CURRENCY. **2.** Any of various coins of small denomination. **3.** A sum of money. **—pretty penny.** *Informal.* A large sum of money.

**penny ante** *n.* **1.** A poker game in which the highest bet is limited to a very small sum. **2.** *Informal.* A small-scale business transaction. **—pen'ny-an'te** *adj.*

**penny arcade** *n.* A public place, as in an amusement park, where coin-operated devices are played for entertainment.

**pen·ny·cress** (pĕn'ē-krĕs') *n.* A plant of the genus *Thlaspi,* indigenous to Europe, bearing small, winged seed pods, esp. *T. arvense,* which grows as a weed throughout North America.

**pen·ny-pinch** (pĕn'ē-pĭnch') *vi.* **-pinched, -pinch·ing, -pinch·es.** *Informal.* To be stingy.

**penny pincher** *n. Informal.* A parsimonious person. **—pen'ny-pinch'ing** *adj.* & *n.*

**pen·ny·roy·al** (pĕn'ē-roi'əl) *n.* [Prob. by folk etymology < AN *puliol real* : OFr. *poliol,* thyme (< Lat. *pulegium*) + AN *real,* royal (< OFr. *roial* < Lat. *regalis* < *rex,* king).] **1.** A Eurasian plant, *Mentha pulegium,* with hairy leaves and small lilac-blue flowers, that yields a useful aromatic oil. **2.** An aromatic plant, *Hedeoma pulegioides* of eastern North America, similar to the pennyroyal.

**pen·ny·weight** (pĕn′ē-wāt′) *n.* A unit of troy weight equal to 24 grains, 1/20 of a troy ounce, or approx. 1.555 grams.

**pen·ny-wise** (pĕn′ē-wīz′) *adj.* Careful only in dealing with small sums of money or small matters.

**pen·ny·wort** (pĕn′ē-wûrt′, -wôrt′) *n.* Any of several plants with rounded leaves suggestive of pennies, as: **a.** A North American plant, *Obolaria virginica*, with fleshy leaves and small white or purplish flowers. **b.** A Eurasian plant, *Cotyledon umbilicus*, with thick, rounded leaves and yellowish-green flowers.

**pen·ny·worth** (pĕn′ē-wûrth′) *n.* **1.** The amount that a penny will buy. **2.** A small amount : MODICUM. **3.** A bargain.

**Pe·nob·scot** (pə-nŏb′skət, -skŏt′) *n., pl.* **Penobscot** or **-scots.** **1. a.** A tribe of Indians who were part of the Algonquin federation and once inhabited central Maine. **b.** A member of this tribe. **2.** The Algonquian language of the Penobscot. **—Pe·nob′scot** *adj.*

**pe·nol·o·gy** *also* **poe·nol·o·gy** (pē-nŏl′ə-jē) *n.* [Gk. *poinē*, penalty + -LOGY.] The theory and practice of prison management and criminal rehabilitation. **—pe′no·log′i·cal** (pē′nə-lŏj′ĭ-kəl) *adj.* **—pe′no·log′i·cal·ly** *adv.* **—pe·nol′o·gist** *n.*

**pen·on·cel** (pĕn′ən-sĕl′) *n. var. of* PENNONCEL.

**pen pal** *n.* A friend with whom one is acquainted only by correspondence.

**pen point** *n.* A tapering metal device with a split point that fits into a holder and is used for writing.

**pen·sil** (pĕn′səl) *n. var. of* PENCEL.

**pen·sile** (pĕn′sīl′) *adj.* [Lat. *pensilis* < *pendēre*, to hang.] **1.** Hanging down loosely <a *pensile* bird nest> **2.** Building a hanging nest. **—pen′sile′ness, pen·sil′i·ty** (pĕn-sĭl′ĭ-tē) *n.*

**pen·sion**[1] (pĕn′shən) *n.* [ME, payment < OFr. < Lat. *pensio* < *pendere*, to pay.] A sum of money paid regularly as a retirement benefit or by way of patronage. —*vt.* **-sioned, -sion·ing, -sions. 1.** To grant a pension to. **2.** To retire or dismiss with a pension.

**pen·sion**[2] (päN-syôN′) *n.* [Fr. < OFr., payment. —see PENSION[1].] A small hotel or boarding house in Europe.

**pen·sion·ar·y** (pĕn′shə-nĕr′ē) *adj.* **1.** Constituting a pension. **2.** Venal : mercenary. —*n., pl.* **-ies. 1.** A pensioner. **2.** A hireling.

**pen·sion·er** (pĕn′shə-nər) *n.* **1.** One who receives a pension. **2.** One who is dependent on the bounty of another. **3.** *Obs.* A pensionary.

**pen·sive** (pĕn′sĭv) *adj.* [ME *pensif* < OFr. < *penser*, to think < Lat. *pensare*, freq. of *pendere*, to weigh.] **1.** Deep in thought. **2.** Suggesting or expressing deep, often melancholy thought. **—pen′sive·ly** *adv.* **—pen′sive·ness** *n.*

☆ **syns:** PENSIVE, MEDITATIVE, MUSING, WISTFUL *adj. core meaning* : suggestive of or expressing deep, often melancholy thought <a faraway, *pensive look*>

**pen·ste·mon** (pĕn-stē′mən, pĕn′stə-mən) *n.* [NLat. *Penstemon*, genus name : Gk. *pente*, five + Gk. *stēmōn*, thread.] A plant of the genus *Penstemon*, which includes the beard-tongues.

**pen·stock** (pĕn′stŏk′) *n.* **1.** A gate or sluice for controlling a flow of water. **2.** A pipe or conduit for carrying water to a water wheel or turbine.

**pent** (pĕnt) *adj.* [P.part. of PEN[2].] Closely penned or confined.

**penta-** or **pent-** *pref.* [Gk. < *pente*, five.] Five <*penta*merous>

**pen·ta·chlo·ro·phe·nol** (pĕn′tə-klôr′ə-fē′nōl′, -nôl′, -klōr′-) *n.* A compound, $C_6Cl_5OH$, used in solution as a wood preservative, fungicide, and disinfectant.

**pen·ta·cle** (pĕn′tə-kəl) *n.* [Med. Lat. **pentaculum* : Gk. *penta-*, five + Lat. *-culum*, dim. suffix.] A five-pointed star formed by five straight lines connecting the vertices of a pentagon and enclosing another pentagon in the completed figure.

**pen·tad** (pĕn′tăd′) *n.* [Gk. *pentas*, *pentad-*, the number five < *pente*, five.] A group of five.

**pen·ta·dac·tyl** (pĕn′tə-dăk′təl) *also* **pen·ta·dac·ty·late** (-dak′tə-lĭt, -lāt′) *adj.* [Lat. *pentadactylus* < Gk. *pentadaktulos* : *penta-*, five + *daktulos*, finger.] Having five digits or toes on each hand or foot. **—pen′ta·dac′tyl·ism** *n.*

**pen·ta·gon** (pĕn′tə-gŏn′) *n.* [Lat. *pentagonum* < Gk. *pentagōnon* : *penta-*, five + *gōnia*, angle.] **1.** A polygon having five sides and five interior angles. **2. Pentagon.** The U.S. military establishment. **—pen·tag′o·nal** (pĕn-tăg′ə-nəl) *adj.* **—pen·tag′o·nal·ly** *adv.*

**Pen·ta·gon·ese** (pĕn′tə-gŏn-ēz′, -ēs′) *n.* Military jargon.

**pen·ta·gram** (pĕn′tə-grăm′) *n.* A pentacle.

**pen·ta·he·dron** (pĕn′tə-hē′drən) *n., pl.* **-drons** or **-dra** (-drə). A solid having five plane faces. **—pen′ta·he′dral** *adj.*

**pen·tam·er·ous** (pĕn-tăm′ər-əs) *adj.* **1.** Having five similar parts. **2.** *Bot.* Having flower parts, as petals, sepals, and stamens, in sets of five. **—pen·tam′er·ism** *n.*

**pen·tam·e·ter** (pĕn-tăm′ĭ-tər) *n.* [Lat. < Gk. *pentametros* : *penta-*, five + *metron*, measure.] **1.** A line of verse having five metrical feet. **2.** English verse composed in iambic pentameter.

**pen·tane** (pĕn′tān′) *n.* Any of three isomeric hydrocarbons, $C_5H_{12}$, of the methane series: **a.** Normal pentane, a colorless flammable liquid used in the making of artificial ice and as an anesthetic and a general solvent. **b.** Isopentane, a colorless flammable liquid used as a solvent and in making polystyrene foam. **c.** Neopentane, a colorless gas used in making synthetic rubber.

**pen·tan·gu·lar** (pĕn-tăng′gyə-lər) *adj.* Having five angles.

**pen·ta·ploid** (pĕn′tə-ploid′) *adj.* Having five haploid sets of chromosomes in each nucleus. —*n.* An organism having pentaploid chromosomes.

**pen·ta·quine** (pĕn′tə-kwēn′, -kwĭn) *also* **pen·ta·quin** (-kwĭn) *n.* [PENTA- + QUIN(OLIN)E.] A drug used with quinine in the treatment of malaria.

**pen·tar·chy** (pĕn′tär′kē) *n., pl.* **-chies.** [Gk. *pentarkhia* : *penta-*, five + *arkhein*, to rule.] **1.** Government by five rulers. **2.** A body of five rulers governing jointly. **3.** An association or federation of five governments, each ruled by a different leader. **—pen·tar′chi·cal** (pĕn-tär′kĭ-kəl) *adj.*

**pen·ta·stich** (pĕn′tə-stĭk′) *n.* [< LGk. *pentastikhos*, of five lines : *penta-*, five + *stikhos*, line.] A stanza or poem having five lines.

**Pen·ta·teuch** (pĕn′tə-to͝ok′, -tyo͝ok′) *n.* [LLat. *Pentateuchus* < Gk. *Pentateukos* : *penta-*, five + *teukhos*, scroll.] The first five books of the Old Testament. —See table at BIBLE. **—Pen′ta·teuch′al** *adj.*

**pen·tath·lon** (pĕn-tăth′lən, -lŏn′) *n.* [Gk. : *pente*, five + *athlon*, contest.] An athletic contest consisting of five events for each participant, usu. running, horseback riding, swimming, fencing, and pistol shooting. **—pen·tath′lete′** (pĕn-tăth′lēt′) *n.*

**pen·ta·ton·ic scale** (pĕn′tə-tŏn′ĭk) *n.* Any of various five-tone musical scales, esp. one consisting of the first, second, third, fifth, and sixth tones of a diatonic scale.

**pen·ta·va·lent** (pĕn′tə-vā′lənt) *adj.* Having a valence of 5.

**pen·taz·o·cine** (pĕn-tăz′ə-sēn′) *n.* [Orig. unknown.] A synthetic drug, $C_{19}H_{27}NO$, used to relieve pain.

**Pen·te·cost** (pĕn′tĭ-kôst′, -kŏst′) *n.* [ME < OE *Pentecosten* < LLat. *Pentecoste* < Gk. *pentēkostē (hēmera)*, fiftieth (day) < *pentēkostos*, fiftieth < *pentēkonta*, fifty : *pente*, five + *-konta*, times ten.] **1.** A festival of the Christian Church occurring on the seventh Sunday after Easter, to commemorate the descent of the Holy Ghost upon the disciples. **2.** Shavuot.

**Pen·te·cos·tal** (pĕn′tĭ-kŏs′təl, -kôs′-) *adj.* **1.** Of, relating to, or occurring at Pentecost. **2.** Of, relating to, or designating any of various Christian religious congregations that seek to be filled with the Holy Ghost, in emulation of the disciples at Pentecost. **—Pen′te·cos′tal** *n.* **—Pen′te·cos′tal·ism** *n.* **—Pen′te·cos′tal·ist** *n.*

**pent·house** (pĕnt′hous′) *n.* [ME *pentis*, a shed attached to a wall of a building < OFr. *appentis* < Med. Lat. *appenticium*, appendage < Lat. *appendix*. —see APPENDIX.] **1. a.** An apartment or dwelling on the roof of a building. **b.** A residence, often with a terrace, comprising the top floor of an apartment house. **c.** A structure housing machinery on the roof of a building. **2.** A shed or sloping roof attached to the side of a wall or building.

**pent·land·ite** (pĕnt′lən-dīt′) *n.* [Fr., after Joseph B. *Pentland* (1797–1873).] A light-brown nickel iron sulfide, the principal ore of nickel.

**pen·to·bar·bi·tal sodium** (pĕn′tə-bär′bĭ-tôl′) *n.* A white crystalline or powdery barbiturate, $C_{11}H_{17}N_2O_3Na$, used as a sedative.

**pen·tode** (pĕn′tōd′) *n.* A vacuum tube with five electrodes that can be used as a logic element in a computer.

**pen·to·san** (pĕn′tə-săn′) *n.* Any of a group of complex carbohydrates found with cellulose in many woody plants and yielding pentoses on hydrolysis.

**pen·tose** (pĕn′tōs′, -tōz′) *n.* A sugar having five carbon atoms per molecule.

**pen·to·side** (pĕn′tə-sīd′) *n.* A glycoside that produces a pentose on hydrolysis.

**Pen·to·thal** (pĕn′tə-thôl′). A trademark for a thiopental sodium.

**pent·ox·ide** (pĕnt-ŏk′sīd′) *n.* An oxide having five atoms of oxygen in the molecule.

**pent-up** (pĕnt′ŭp′) *adj.* Not given expression <*pent-up* anger>

**pen·tyl** (pĕn′təl) *n.* AMYL 1.

**pen·tyl·ene·tet·ra·zol** (pĕn′tə-lēn′tĕt′rə-zôl′, -zōl′) *n.* [PENT(A)- + (METH)YLENE + TETR(A)- + AZ(O)- + -OL.] A drug, $C_6H_{10}N_4$, used as a stimulant of the central nervous system.

**pe·nu·che** *also* **pe·nu·chi** (pə-no͞o′chē) *n.* [Var. of PANOCHA.] A fudgelike confection of brown sugar, butter, milk or cream, and nuts.

**pe·nuch·le** or **pe·nuck·le** (pē′nŭk′əl) *n. vars. of* PINOCHLE.

**pe·nult** (pē′nŭlt′, pĭ-nŭlt′) *also* **pe·nul·ti·ma** (pĭ-nŭl′tə-mə) *n.* [Lat. *paenultima* < *paenultimus*, next to last : *paene*, almost + *ultimus*, last.] **1.** The next to last syllable in a word. **2.** The next to last item in a series.

**pe·nul·ti·mate** (pĭ-nŭl′tə-mĭt) *adj.* [Lat. *paenultimus*. —see PENULT.] **1.** Next to last. **2.** Of or relating to the penult of a word. —*n.* The next to the last.

**pe·num·bra** (pĭ-nŭm′brə) *n., pl.* **-brae** (-brē) or **-bras.** [Lat. *paene*, almost + Lat. *umbra*, shadow.] **1.** A partial shadow, as in an eclipse, between regions of total shadow and total illumination. **2.** A partially darkened fringe around a sunspot. **3.** An outlying, surrounding region : PERIPHERY. **—pe·num′bral, pe·num′brous** *adj.*

**pe·nu·ri·ous** (pə-no͝or′ē-əs, -nyo͝or′-) *adj.* [Med. Lat. *penuriosus* < Lat. *penuria*, want.] **1.** Stingy : miserly. **2.** Yielding little : BARREN. **3.** Destitute. **—pe·nu′ri·ous·ly** *adv.* **—pe·nu′ri·ous·ness** *n.*

**pen·u·ry** (pĕn′yə-rē) *n.* [ME < Lat. *penuria*, want.] **1.** Extreme want or poverty : DESTITUTION. **2.** Extreme dearth or insufficiency.

**Pe·nu·ti·an** (pə-no͞o′tē-ən, -shən) *n.* A stock of North American Indian languages spoken in Pacific coastal areas from California into British Columbia.

**pe·on** (pē′ŏn′, pē′ən) *n.* [Am. Sp. *peón* < Med. Lat. *pedo*, foot soldier. —see PAWN².] **1. a.** An unskilled laborer or farm worker of Latin America or the southwestern United States. **b.** Such a worker bound in servitude to a landlord creditor. **2.** (pyo͞on). A native Indian or Ceylonese messenger, servant, or foot soldier. **3.** A menial worker.

**pe·on·age** (pē′ə-nĭj) *also* **pe·on·ism** (-nĭz′əm) *n.* **1.** The condition of being a peon. **2.** A system by which debtors are bound in servitude to their creditors until their debts are paid.

**pe·o·ny** (pē′ə-nē) *n., pl.* **-nies.** [ME *pione* < OE *peonie* < Lat. *peonia* < Gk. *paiōnia*, after *Paiōn*, Apollo's title as physician.] A garden plant of the genus *Paeonia*, with large pink, red, white, or cream-colored flowers.

**peo·ple** (pē′pəl) *n., pl.* **people.** [ME < OFr. *pueple* < Lat. *populus.*] **1.** A body of persons living in the same country under one government : NATIONALITY. **2.** *pl.* **peo·ples.** A body of persons sharing a common religion, culture, language, or inherited condition of life. **3.** Persons with regard to their residence, class, profession, or group. **4. the people.** The mass of ordinary persons : POPULACE. **5.** The citizens of a political unit, as a nation or state : ELECTORATE. **6.** Persons loyal or subordinate to a ruler, superior, or employer. **7.** Family, relatives, or ancestors. **8.** Persons regarded as members of a community <the *people's* court> **9.** Human beings regarded as distinct from lower animals or inanimate things. **10.** A class or kind of beings distinct from human beings <feline and canine *people*> ***usage:*** *People* and *persons* are sometimes distinguished in usage. *People* refers to a large group of individuals, collectively and indefinitely, as in *People will not accept a candidate whose private life has been tainted by scandal. Persons* refers to a specific and relatively small number, as in *Six persons were absent.* In modern usage, however, *people* is also acceptable in such contexts, as in *We counted 18 people.* —*vt.* **-pled, -pling, -ples.** To furnish with a population : POPULATE. **—peo′pler** *n.*

**people mover** *n.* A transport system, as a moving sidewalk or an automated monorail, for conveying people short distances.

**People's Party** *n.* The Populist Party.

**People's Republic** *n.* A political unit controlled and organized by a national Communist party.

**pep** (pĕp) [Short for PEPPER.] *Informal.* —*n.* Energetic high spirits : VIM. —*vt.* **pepped, pep·ping, peps.** To infuse with energy or liveliness : INVIGORATE <The brisk walk *pepped* us up.> **—pep′pi·ness** *n.* **—pep′py** *adj.*

**pep·los** (pĕp′ləs, -lŏs′) *also* **pep·lus** (-ləs) *n., pl.* **-los·es** *also* **-lus·es.** [Gk.] A loose outer robe worn by women in ancient Greece.

**peplos**

**pep·lum** (pĕp′ləm) *n., pl.* **-lums.** [Lat., robe of state < Gk. *peplos*, peplos.] **1.** A short overskirt or ruffle attached at the waistline of a jacket, blouse, or dress. **2.** A peplos.

**pep·lus** (pĕp′ləs) *n. var. of* PEPLOS.

**pe·po** (pē′pō) *n., pl.* **-pos.** [Lat., a kind of melon < Gk. *pepōn.*] The fruit of any of various related plants, as the cucumber, squash, pumpkin, and melon, with a hard rind, fleshy pulp, and many seeds.

**pep·per** (pĕp′ər) *n.* [ME *peper* < OE *pipor* < Lat. *piper* < Gk. *peperi*, < Skt. *pippalī*, berry.] **1. a.** A woody vine, *Piper nigrum* of the East Indies, yielding small, berrylike fruit. **b.** The dried, blackish fruit of this vine, used as a pungent condiment. **2.** Any of several other plants of the genus *Piper*, as the cubeb, betel, and kava. **3. a.** Any of several varieties of a woody plant, *Capsicum frutescens* or *C. annuum*, of tropical origin. **b.** The podlike fruit of any of these plants, varying in size, shape, and degree of pungency, with the milder types including the bell pepper and pimiento and the more pungent types including the cherry pepper. **4.** Any of various condiments made from the more pungent varieties of *C. frutescens*, as chili and cayenne pepper. —*vt.* **-pered, -per·ing, -pers. 1.** To season or sprinkle with pepper. **2.** To sprinkle liberally <dark hair *peppered* with silver> **3.** To pelt or shower with small missiles. **4.** To make lively and vivid with wit or invective, as a speech.

**pep·per-and-salt** (pĕp′ər-ən-sôlt′) *adj.* Having a close mixture of black and white <*pepper-and-salt* hair>

**pep·per·box** (pĕp′ər-bŏks′) *n.* A container with small holes in the top for sprinkling ground pepper.

**pep·per·bush** (pĕp′ər-bo͝osh′) *n.* Sweet pepperbush.

**pep·per·corn** (pĕp′ər-kôrn′) *n.* **1.** A dried berry of the pepper vine, *Piper nigrum.* **2.** Something small or insignificant.

**pep·per·grass** (pĕp′ər-grăs′) *n.* A plant of the genus *Lepidium*, esp. *L. virginicum*, bearing small white flowers and pungent seeds.

**pep·per·idge** (pĕp′ər-ĭj) *n.* [Orig. unknown.] The sour gum.

**pepper mill** *n.* A utensil for grinding peppercorns.

**pep·per·mint** (pĕp′ər-mĭnt′) *n.* **1.** A plant, *Mentha piperita*, bearing small purple or white flowers and downy leaves that yield a pungent oil. **2.** The oil from the peppermint, used as flavoring. **3.** A lozenge or candy with peppermint flavoring.

**pep·per·o·ni** (pĕp′ə-rō′nē) *n.* [Ital. *peperoni*, pl. of *peperone*, chili, aug. of *pepe*, pepper < Lat. *piper.* —see PEPPER.] A highly spiced pork and beef sausage.

**pepper pot** *n.* **1.** A soup made with vegetables and meat, esp. tripe, seasoned with pepper and often containing dumplings. **2.** A thick West Indian stew of meat or fish, vegetables, and condiments. **3.** A pepperbox.

**pepper tree** *n.* Any of several South American trees of the genus *Schinus*, esp. *S. molle*, with compound leaves and yellowish-white flowers.

**pep·per·wort** (pĕp′ər-wûrt′, -wôrt′) *n.* **1.** An aquatic or marsh plant of the genus *Marsilea*, with floating leaves rising from long runners. **2.** Peppergrass.

**pep·per·y** (pĕp′ə-rē) *adj.* **1.** Of, like, or containing pepper : PUNGENT. **2.** Quick-tempered : testy. **3.** Sharp and stinging in content or style : FIERY <a *peppery* denunciation> **—pep′per·i·ness** *n.*

**pep pill** *n. Slang.* A tablet or capsule with an ingredient that stimulates the central nervous system, esp. one of the amphetamines.

**pep·sin** *also* **pep·sine** (pĕp′sĭn) *n.* [G. < Gk. *pepsis*, digestion < *peptein*, to digest.] **1.** A digestive enzyme found in gastric juice that catalyzes the breakdown of protein to peptides. **2.** A substance containing pepsin, obtained from the stomachs of hogs and used as a digestive aid.

**pep·sin·o·gen** (pĕp-sĭn′ə-jən) *n.* An inert substance found in the cells of the gastric mucosa that is converted to pepsin during digestion by the action of hydrochloric acid.

**pep talk** *n.* A usu. short, emotional speech of exhortation delivered by a leader, as to a team or staff.

**pep·tic** (pĕp′tĭk) *adj.* [Lat. *pepticus* < Gk. *peptikos* < *peptein*, to digest.] **1. a.** Of, relating to, or assisting digestion. **b.** Induced by or associated with the action of digestive secretions <treated for a *peptic* ulcer> **2.** Of, relating to, or involving pepsin. **3.** Capable of digesting. —*n.* A digestive agent.

**pep·tid** (pĕp′tĭd) *n. var. of* PEPTIDE.

**pep·ti·dase** (pĕp′tĭ-dās′, -dāz′) *n.* An enzyme that hydrolyzes peptides, releasing amino acids.

**pep·tide** (pĕp′tīd′) *also* **pep·tid** (pĕp′tĭd) *n.* [PEPT(ONE) + -IDE.] A natural or synthetic compound containing two or more amino acids linked by the carboxyl group of one amino acid and the amino group of another. **—pep·tid′ic** (-tĭd′ĭk) *adj.* **—pep·tid′i·cal·ly** *adv.*

**peptide bond** *n.* The chemical bond between the organic acid groups and amine groups of neighboring amino acids, constituting the primary linkage of all protein structures.

**pep·ti·do·gly·can** (pĕp′tĭ-dō-glī′kən, -kăn′) *n.* [PEPTIDE + *glycan*, a polysaccharide.] A polymer found in the cell walls of procaryotes and consisting of polysaccharide and peptide chains in a strong molecular network.

**pep·tize** (pĕp′tīz′) *vt.* **-tized, -tiz·ing, -tiz·es.** [Gk. *peptein*, to digest + -IZE.] *Chem.* **1.** To increase the dispersion of (a colloidal solution). **2.** To liquefy (a colloidal gel) to form a sol. **—pep′ti·za′tion** *n.* **—pep′tiz′er** *n.*

**pep·tone** (pĕp′tōn′) *n.* [G. *Pepton* < Gk. *peptos*, digested.] Any of various protein compounds obtained by enzyme or acid hydrolysis of natural protein and used as nutrients and culture media. **—pep·ton′ic** (-tŏn′ĭk) *adj.*

**pep·to·nize** (pĕp′tə-nīz′) *vt.* **-nized, -niz·ing, -niz·es. 1.** To convert (protein) into a peptone. **2.** To dissolve (food) by means of a proteolytic enzyme. **3.** To combine with peptone. **—pep′to·ni·za′tion** *n.*

**Pe·quot** (pē′kwŏt′) *n., pl.* **Pequot** *or* **-quots.** [Prob. alteration of Narraganset *paquatanog*, destroyers.] **1. a.** A tribe of Indians formerly living in southern New England. **b.** A member of this tribe. **2.** The Algonquian language of the Pequot. **—Pe′quot** *adj.*

**per** (pûr) *prep.* [Lat.] **1.** By means of : THROUGH. **2.** To, for, or by each : for every <$12 dollars *per* hour> **3.** According to : BY <*per* our agreement> ***usage:*** *Per* is used with reference to units of measurement, as in *per hour* or *per mile.* Its more general use, as in *per the terms of the contract*, is best left to business correspondence or legal documents.

**per-** *pref.* [< Lat. *per*, through.] **1.** Thoroughly : completely : intensely <*perfervid*> **2.** Containing an element in its highest oxidation state <*per*chloric acid> **3.** Containing a large or the largest possible proportion of an element <*per*oxide> **4.** Containing the peroxy group <*per*acid>

**per·ac·id** (pûr′ăs′ĭd) *n.* **1.** An acid containing the peroxy group. **2.** An inorganic acid, as perchloric acid, containing the largest proportion of oxygen in a series of related acids.

**per·ad·ven·ture** (pûr′əd-vĕn′chər, pĕr′-) *adv.* [ME *per aventure* < OFr., by chance.] *Archaic.* Perhaps : perchance. —*n.* Chance or uncertainty : DOUBT.

**per·am·bu·late** (pə-răm′byə-lāt′) *v.* **-lat·ed, -lat·ing, -lates.** [Lat. *perambulare, perambulat-* : *per-*, through + *ambulare*, to walk.] —*vt.* To walk through, esp. so as to inspect. —*vi.* To walk about : STROLL. **—per·am′bu·la′tion** *n.* **—per·am′bu·la·to′ry** (-lə-tôr′ē, -tōr′ē) *adj.*

**per·am·bu·la·tor** (pə-răm′byə-lā′tər) *n. Chiefly Brit.* A baby carriage : PRAM.

**per an·num** (pər ăn′əm) *adv.* [Lat.] By the year : ANNUALLY.

**per·bo·rate** (pər-bôr′āt′, -bōr′-) *n.* A salt containing the radical $BO_3$, formed from a borate and hydrogen peroxide.

**per·cale** (pər-kāl′) *n.* [Fr. < Pers. *pargālah.*] An opaque cotton fabric used to make sheets, pillowcases, and clothing.

**per·ca·line** (pûr′kə-lēn′) *n.* [Fr., dim. of *percale*, percale.] A glazed fine cotton fabric used for dress goods, shirting, and linings.

**per cap·i·ta** (pər kăp′ĭ-tə) *adv. & adj.* [Med. Lat., by heads.] **1.** Per unit of population <lowest income *per capita* of any nation> **2.** Equally to each individual.

**per·ceive** (pər-sēv′) *vt.* **-ceived, -ceiv·ing, -ceives.** [ME *perceiven* < OFr. *perceivre* < Lat. *percipere* : *per-* (intensive) + *capere*, to seize.] **1.** To become aware of directly through the senses, esp. to see or hear. **2.** To take notice of : OBSERVE. **3.** To achieve understanding of. **—per·ceiv′a·ble** *adj.* **—per·ceiv′a·bly** *adv.*

**per cent** *also* **per·cent** (pər-sĕnt′) [Lat. *per centum*, by the hundred.] —*adv.* Out of each hundred : per hundred. —*n., pl.* **per cent** *also* **percent. 1.** One part in a hundred. **2.** A percentage <only a small *per cent* attended> —*adj.* Paying interest at a specified percentage. ***usage:*** *Per cent* is gen. used with a specific figure. The number of its verb is governed by the number of the following noun, whether expressed or understood, as in *Ten per cent of the crop was lost* or *Six per cent of the workers were laid off.*

**per·cent·age** (pər-sĕn′tĭj) *n.* **1.** A fraction or ratio with 100 fixed and understood as the denominator, formed by multiplying a decimal equivalent of a fraction by 100; e.g., 0.98 equals a percentage of 98. **2.** A proportion or share in relation to a whole : PART. **3.** An amount, as an allowance, duty, or commission, that varies in proportion to a larger sum, as total sales. **4.** *Informal.* Advantage : gain <There is no *percentage* in cheating on exams.>

**per·cen·tile** (pər-sĕn′tīl′) *n.* A number that corresponds to one of 100 equal divisions of the range of a statistic in a given sample and characterizes a contained value of the statistic as not exceeded by a specified percentage of all the values in the sample; e.g., a score higher than 97% of those attained on an examination is said to be in the 97th percentile.

**per cen·tum** (pər sĕn′təm) *n.* [Lat.] Per cent.

**per·cept** (pûr′sĕpt′) *n.* [Back-formation < PERCEPTION.] **1.** The object of perception. **2.** An impression in the mind of something perceived by the senses, viewed as the basic component in the formation of concepts.

**per·cep·ti·ble** (pər-sĕp′tə-bəl) *adj.* Capable of being perceived : DISCERNIBLE. **—per·cep′ti·bil′i·ty** *n.* **—per·cep′ti·bly** *adv.*

☆ ***syns:*** PERCEPTIBLE, APPRECIABLE, DISCERNIBLE, NOTICEABLE, PALPABLE *adj. core meaning* : capable of being noticed or detected <a *perceptible* improvement in the patient> ***ant:*** imperceptible

**per·cep·tion** (pər-sĕp′shən) *n.* [Lat. *perceptio* < *percipere*, to perceive.] **1.** The act, process, or faculty of perceiving. **2.** The effect or product of perceiving. **3. a.** Insight, intuition, or knowledge gained by perceiving. **b.** Capacity for such insight. **—per·cep′tion·al** *adj.*

**per·cep·tive** (pər-sĕp′tĭv) *adj.* **1.** Of or relating to perception. **2. a.** Capable of perceiving. **b.** Marked by understanding and discernment : SENSITIVE. **—per·cep′tive·ly** *adv.* **—per′cep·tiv′i·ty** (pûr′sĕp-tĭv′ĭ-tē) *n.*

**per·cep·tu·al** (pər-sĕp′cho͞o-əl) *adj.* Of, based on, or involving perception. **—per·cep′tu·al·ly** *adv.*

**perch**[1] (pûrch) *n.* [ME *perche* < OFr. < Lat. *pertica.*] **1.** A branch or rod serving as a roost for a bird. **2.** A place for sitting or resting. **3.** A pole used in acrobatics. **4.** *Chiefly Brit.* **a.** ROD 12a. **b.** One square rod of land. **5.** A unit of cubic measure used in stonework, usu. 16.5 feet (one rod) by 1 foot by 1.5 feet, or .70 cubic meter. **6.** A frame on which cloth is laid for inspection of quality. —*v.* **perched, perch·ing, perch·es.** —*vi.* **1.** To alight or rest on a perch : ROOST. **2.** To stand, sit, or rest on an elevated place. —*vt.* **1.** To place on or as if on a perch. **2.** To lay (cloth) on a perch for examination.

**perch**[2] (pûrch) *n., pl.* **perch** or **perch·es.** [ME *perche* < OFr. < Lat. *perca* < Gk. *perkē.*] **1.** A freshwater fish of the genus *Perca*, esp. either of two edible species, *P. flavescens* of North America and *P. fluviatilis* of Europe. **2.** A fish, as the pike perch, related or similar to the perch.

**per·chance** (pər-chăns′) *adv.* [ME < AN *par chance* : *par*, by + *chance*, chance.] Possibly : perhaps.

**per·chlo·rate** (pər-klôr′āt′, -klōr′-) *n.* A salt or ester of perchloric acid.

**per·chlo·ric acid** (pər-klôr′ĭk, -klōr′-) *n.* A clear, colorless liquid, $HClO_4$, explosively unstable under some conditions, that is a powerful oxidant used as a catalyst and in explosives.

**per·chlo·ride** (pər-klôr′īd′, -klōr′-) *also* **per·chlo·rid** (-klôr′ĭd, -klōr′-) *n.* A chloride having more chlorine than other chlorides of the same element.

**per·chlor·o·eth·yl·ene** (pər-klôr′ō-ĕth′ə-lēn′, -klōr′-) *n.* A colorless, nonflammable organic solvent, $Cl_2C{:}CCl_2$, used in dry-cleaning solutions and to dissolve a variety of waxes, tars, rubbers, and gums.

**per·cip·i·ent** (pər-sĭp′ē-ənt) *adj.* [Lat. *percipiens, percipient-*, pr.part. of *percipere*, to perceive.] Having the power of perceiving, esp. perceiving readily and keenly. —*n.* One that perceives. **—per·cip′i·ence, per·cip′i·en·cy** *n.*

**per·coid** (pûr′koid′) *also* **per·coi·de·an** (pər-koi′dē-ən) *adj.* [NLat. *Percoidea*, suborder name < Lat. *perca*, perch. —see PERCH[2].] Of or relating to the Percoidea, a large suborder of fishes that includes the perches, sunfishes, groupers, and grunts. —*n.* A fish of the percoid group.

**per·co·late** (pûr′kə-lāt′) *v.* **-lat·ed, -lat·ing, -lates.** [Lat. *percolare, percolat-* : *per*, through + *colare*, to filter < *colum*, sieve.] —*vt.* **1.** To cause (liquid, powder, or small particles) to pass through a porous substance or small holes : FILTER. **2.** To pass or ooze through : PERMEATE. **3.** To make (coffee) in a percolator. —*vi.* **1.** To seep or drain through a porous substance or filter. **2.** *Informal.* To become active or lively. —*n.* (pûr′kə-lĭt, -lāt′). A liquid that has been percolated. **—per′co·la′tion** *n.*

**per·co·la·tor** (pûr′kə-lā′tər) *n.* A coffeepot in which boiling water is forced repeatedly up through a center tube to filter back down through a basket of ground coffee.

**per con·tra** (pər kŏn′trə) *adv.* [Lat.] **1. a.** On the contrary. **b.** By way of contrast. **2.** As an offset.

**per·cur·rent** (pər-kûr′ənt) *adj.* [Lat. *percurrens, percurrent-*, pr.part. of *percurrere*, to run through : *per*, through + *currere*, to run.] Designating a midrib of a leaf that extends from base to apex.

**per·cuss** (pər-kŭs′) *vt.* **-cussed, -cuss·ing, -cuss·es.** [Lat. *percutere, percuss-*, to strike hard : *per-* (intensive) + *quatere*, to strike.] To strike or tap firmly, esp. to practice medical percussion on.

**per·cus·sion** (pər-kŭsh′ən) *n.* [Lat. *percussio* < *percutere*, to percuss.] **1.** The striking together of two bodies, esp. when noise is produced. **2.** The sound, vibration, or shock caused by the percussion of two bodies. **3.** Detonation of a percussion cap in a firearm. **4.** A method of medical diagnosis in which various areas of the body, esp. the chest, back, and abdomen, are tapped to determine by resonance the condition of internal organs. **5.** Musical percussion instruments as a whole.

**percussion cap** *n.* A thin metal cap containing a detonator, as gunpowder, that explodes on being struck.

**percussion instrument** *n.* A musical instrument in which sound is produced by striking, as a drum, xylophone, or piano, or by shaking, as a tambourine or maraca.

**per·cus·sion·ist** (pər-kŭsh′ə-nĭst) *n.* One who plays percussion instruments.

**per·cus·sive** (pər-kŭs′ĭv) *adj.* Of, relating to, or marked by percussion. **—per·cus′sive·ly** *adv.* **—per·cus′sive·ness** *n.*

**per·cu·ta·ne·ous** (pûr′kyo͞o-tā′nē-əs) *adj. Med.* Effected, passed, or performed through or by means of the skin. **—per′cu·ta′ne·ous·ly** *adv.*

**per di·em** (pər dē′əm, dī′əm) *adv.* [Lat.] By the day : per day. —*n., pl.* **per diems.** An allowance for daily expenses. —*adj.* **1.** Reckoned on a daily basis. **2.** Paid by the day.

**per·di·tion** (pər-dĭsh′ən) *n.* [ME *perdicion* < LLat. *perditio* < Lat. *perdere*, to lose : *per*, away + *dare*, to give.] **1. a.** The loss of the soul : eternal damnation. **b.** Hell. **2.** *Archaic.* Utter ruin.

**per·du** *or* **per·due** (pər-do͞o′, -dyo͞o′) *n.* [Fr. *sentinelle perdue* : *sentinelle*, sentinel + *perdre*, to lose.] *Obs.* A soldier sent on a dangerous mission.

**per·du·ra·ble** (pər-do͝or′ə-bəl, -dyo͝or′-) *adj.* [ME < OFr. < LLat. *perdurabilis* < Lat. *perdurare*, to endure : *per-*, throughout + *durare*, to last.] Extremely durable : PERMANENT. **—per·du′ra·bil′i·ty** *n.* **—per·du′ra·bly** *adv.*

**per·e·gri·nate** (pĕr′ĭ-grə-nāt′) *v.* **-nat·ed, -nat·ing, -nates.** [Lat. *peregrinari, peregrinat-* < *peregrinus*, foreigner. —see PEREGRINE.] —*vi.* To travel or journey from place to place. —*vt.* To travel through or over. **—per′e·gri·na′tion** *n.* **—per′e·gri·na′tor** *n.*

**per·e·grine** (pĕr′ə-grĭn, -grēn′) *adj.* [Med. Lat. *peregrinus* < Lat., foreigner < *pereger*, being abroad : *per*, through + *ager*, land.] **1.** Alien : foreign. **2.** Migratory. —*n.* The peregrine falcon.

**peregrine falcon** *n.* [ME, transl. of Med. Lat. *falco peregrinus* (so called because the falcons were caught in passage rather than taken from the nest, as were eyas falcons).] A widely distributed bird of prey, *Falco peregrinus*, with gray and white plumage, once much used in falconry.

**pe·rei·ra bark** (pə-râr′ə) *n.* [After Jonathan *Pereira* (1804–1853).] The bark of a South American tree, *Geissospermum vellosii,* the source of a substance used at one time in the treatment of malaria.

**per·emp·to·ry** (pə-rĕmp′tə-rē) *adj.* [LLat. *peremptorius* < *perimere,* to take away : *per-* (intensive) + *emere,* to obtain.] **1.** Terminating all debate or action <a *peremptory* edict> **2.** Not allowing refusal or contradiction : IMPERATIVE <*peremptory* orders> **3.** Having the nature of or expressing a command : URGENT <a *peremptory* tone of voice> **4.** Offensively self-assured : DICTATORIAL. **—per·emp′to·ri·ly** *adv.* **—per·emp′to·ri·ness** *n.*

**per·en·ni·al** (pə-rĕn′ē-əl) *adj.* [Lat. *perennis* : *per-,* throughout + *annus,* year.] **1.** Lasting or active through the year or through many years. **2. a.** Lasting indefinitely : PERPETUAL <*perennial* tranquillity> **b.** Appearing again and again : RECURRENT. **3.** *Bot.* Having a life span of more than two years. —*n. Bot.* A perennial plant. **—per·en′ni·al·ly** *adv.*

**per·fect** (pûr′fĭkt) *adj.* [ME *perfit* < OFr. *parfit* < Lat. *perfectus* < p.part. of *perficere,* to finish : *per-* (intensive) + *facere,* to do.] **1.** Lacking nothing essential to the whole : complete of its nature or kind. **2.** Being in a state of undiminished or highest excellence : FLAWLESS <a *perfect* example> **3.** Completely adept or talented in a certain field or area <a *perfect* musician> **4.** Completely reproducing or corresponding to a type or original : EXACT <a *perfect* copy> **5.** Thorough : complete <a *perfect* ninny> **6.** Undiluted : pure <*perfect* blue> **7.** Excellent and delightful in all respects <a *perfect* wedding> **8.** *Bot.* Having both stamens and pistils in the same flower : MONOCLINOUS. **9.** Of, relating to, or constituting a verb form expressing action completed prior to a fixed point of reference in time. **10.** *Mus.* **a.** Designating the three basic intervals of the octave, fourth, and fifth. **b.** Designating a cadence in which the final chord has its root in both bass and soprano. ***usage:*** Traditionalists consider *perfect* to be an absolute term and therefore reject its use with modifiers of degree such as *more* or *less.* Nonetheless such usage is entirely acceptable, esp. when *perfect* is used in the sense of "excellent in all respects," as in *A more perfect example could not be found.* —*n.* **1.** The perfect tense. **2.** A verb or verb form in the perfect tense. —*vt.* (pər-fĕkt′) **-fect·ed, -fect·ing, -fects.** To bring to perfection or completion. **—per·fect′er** *n.* **—per′fect·ness** *n.*

☆ ***syns:*** PERFECT, CONSUMMATE, FAULTLESS, FLAWLESS, IMPECCABLE, INDEFECTIBLE *adj. core meaning* : supremely excellent <a *perfect* diamond><a *perfect* performance> ***ant:*** imperfect

**per·fec·ta** (pər-fĕk′tə) *n.* [Short for Am. Sp. *quiniela perfecta,* perfect quinella.] An exacta.

**per·fect·i·ble** (pər-fĕk′tə-bəl) *adj.* Capable of becoming or being made perfect. **—per·fect′i·bil′i·ty** *n.*

**per·fec·tion** (pər-fĕk′shən) *n.* **1.** The quality, state, or condition of being perfect. **2.** The act or process of perfecting. **3.** One considered perfect. **4.** An instance of excellence.

**per·fec·tion·ism** (pər-fĕk′shə-nĭz′əm) *n.* **1.** A predilection for setting extremely high standards and being displeased with anything less. **2.** A belief that moral or spiritual perfection can be achieved by people in this life. **—per·fec′tion·ist** *n.*

**per·fec·tive** (pər-fĕk′tĭv) *adj.* **1.** Tending toward perfection. **2.** Of or designating a verb in the perfective aspect. —*n.* **1.** The perfective aspect. **2.** A verb in the perfective aspect. **—per·fec′tive·ly** *adv.* **—per·fec′tive·ness, per′fec·tiv′i·ty** (pûr′fĕk-tĭv′ĭ-tē) *n.*

**perfective aspect** *n.* An aspect of verbs that expresses a completed action as distinct from a continuing or not necessarily completed action.

**per·fect·ly** (pûr′fĭkt-lē) *adv.* **1.** In a perfect way or to a perfect degree. **2.** Wholly : completely <*perfectly* satisfied>

**perfect number** *n.* A positive integer that is equal to the sum of its integral factors, including 1 but excluding itself.

**per·fec·to** (pər-fĕk′tō) *n., pl.* **-tos.** [< Sp., perfect < Lat. *perfectus.*] A cigar of standard length, thick in the center and tapering at each end.

**perfect participle** *n.* The past participle.

**perfect pitch** *n.* ABSOLUTE PITCH 2.

**perfect rhyme** *n.* **1.** The commonest English rhyme, having identity in sound for the last accented vowel and any final consonants or syllables but with variation in the preceding consonant; e.g., *great, late; rider, beside her; dutiful, unbeautiful.* **2.** A rhyme of two words pronounced identically but differing in meaning; e.g., *sight, site.*

**perfect square** *n.* An integer that is the square of an integer.

**perfect year** *n.* A year having 355 days or a leap year having 385 days in the Hebrew calendar.

**per·fer·vid** (pər-fûr′vĭd) *adj.* Extremely or extravagantly eager : IMPASSIONED. **—per·fer′vid·ly** *adv.* **—per·fer′vid·ness** *n.*

**per·fi·dy** (pûr′fĭ-dē) *n., pl.* **-dies.** [Lat. *perfidia* < *perfidus,* treacherous : *per,* through + *fides,* faith.] Deliberate breach of faith. **—per·fid′i·ous** (pər-fĭd′ē-əs) *adj.* **—per·fid′i·ous·ly** *adv.*

**per·fo·li·ate** (pər-fō′lē-ĭt) *adj.* [NLat. *perfoliatus* : Lat. *per,* through + Lat. *foliatus,* bearing leaves < *folium,* leaf.] Designating a leaf that completely clasps the stem and is apparently pierced by it. **—per·fo′li·a′tion** *n.*

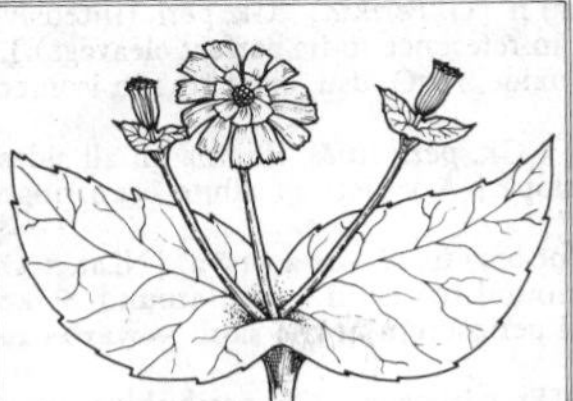

**perfoliate**
*Perfoliate leaves*

**per·fo·rate** (pûr′fə-rāt′) *vt.* **-rat·ed, -rat·ing, -rates.** [Lat. *perforare, perforat-* : *per* (intensive) + *forare,* to bore.] **1.** To pierce, punch, or bore a hole in : PENETRATE. **2.** To pierce or stamp with rows of holes, as those between postage stamps, to allow easy separation. —*adj.* (pûr′fər-ĭt, -fə-rāt′). Perforated. **—per′fo·ra·ble** (-fər-ə-bəl) *adj.* **—per′fo·ra′tion** (-rā′shən) *n.* **—per′fo·ra′tive, per′fo·ra·to′ry** (-fər-ə-tôr′ē, -tōr′ē) *adj.* **—per′fo·ra′tor** *n.*

**per·force** (pər-fôrs′, -fōrs′) *adv.* [ME *par force* < OFr. : *par,* by + *force,* force.] By necessity <You must *perforce* pay off the debt.>

**per·form** (pər-fôrm′) *v.* **-formed, -form·ing, -forms.** [ME *performen* < AN *parformer* < OFr. *parfornir* : *par-* (intensive < Lat. *per-*) + *fornir,* to furnish, of Germanic orig.] —*vt.* **1.** To begin and carry through to completion <*perform* a wedding ceremony> **2.** To take action in accordance with the requirements of : FULFILL <*perform* one's civil duties> **3. a.** To enact (a feat or role) before an audience. **b.** To give a public presentation of (e.g., a play). —*vi.* **1.** To carry on : FUNCTION <doesn't *perform* well on tests> **2.** To fulfill an obligation or requirement. **3.** To portray a role or demonstrate a skill before an audience. **4.** To present an entertainment, as a dramatic work, before an audience. **—per·form′a·ble** *adj.* **—per·form′er** *n.*

**per·form·ance** (pər-fôr′məns) *n.* **1.** The act of performing or state of being performed. **2.** Style of performing a work or role before an audience. **3.** Manner of functioning <an engine's top *performance*> **4.** A presentation, esp. a theatrical one, before an audience. **5.** Something performed : ACCOMPLISHMENT.

**per·fume** (pûr′fyo͞om′, pər-fyo͞om′) *n.* [OFr. *parfum* < OItal. *parfumo* < *parfumare,* to fill with smoke : *par-* (intensive < Lat. *per-*) + *fumare,* to smoke < Lat. < *fumus,* smoke.] **1.** A volatile liquid, distilled from flowers or prepared synthetically, that emits and diffuses a fragrant odor. **2.** A pleasing scent. —*vt.* (pər-fyo͞om′) **-fumed, -fum·ing, -fumes.** To impart a pleasant odor to. **—per·fum′er** *n.*

**per·fum·er·y** (pər-fyo͞o′mə-rē) *n., pl.* **-ies. 1.** Perfumes in general. **2.** An establishment that specializes in making or selling perfume. **3.** The art of making perfume.

**per·func·to·ry** (pər-fŭngk′tə-rē) *adj.* [LLat. *perfunctorius* < Lat. *perfungi,* to get through with : *per-* (intensive) + *fungi,* to perform.] **1.** Done or acting routinely and impersonally. **2.** Apathetic : indifferent. **—per·func′to·ri·ly** *adv.* **—per·func′to·ri·ness** *n.*

☆ ***syns:*** PERFUNCTORY, AUTOMATIC, MECHANICAL *adj. core meaning* : performed or performing automatically and impersonally <signaled approval with a *perfunctory* wave of the hand>

**per·fuse** (pər-fyo͞oz′) *vt.* **-fused, -fus·ing, -fus·es.** [Lat. *perfundere, perfus-,* to pour over : *per-* (intensive) + *fundere,* to pour.] **1.** To coat, suffuse, or permeate with liquid, color, or light. **2.** To pour or diffuse (a liquid) over or through something. **—per·fu′sive** (pər-fyo͞o′sĭv, -zĭv) *adj.*

**per·fu·sion** (pər-fyo͞o′zhən) *n.* Injection of fluid into an artery in order to reach tissues.

**per·go·la** (pûr′gə-lə) *n.* [Ital. < Lat. *pergula.*] An arbor or passageway with a roof of trelliswork on which climbing plants are grown.

**per·haps** (pər-hăps′) *adv.* Possibly : maybe.

**peri-** *pref.* [Lat. < Gk. < *peri,* around, near.] **1.** Around : about : enclosing <*perimysium*> **2.** Near <*perinatal*>

**per·i·anth** (pĕr′ē-ănth′) *n.* [NLat. *perianthus* : Gk. *peri,* around + Gk. *anthos,* flower.] The outer envelope of a flower, consisting of the calyx and corolla or of one of these if the other is absent.

**per·i·apt** (pĕr′ē-ăpt′) *n.* [OFr. *periapte* < Gk. *periapton* : *peri,* around + *haptos,* fastened < *haptein,* to fasten.] An amulet or charm worn as protection against evil and disease.

**per·i·car·di·tis** (pĕr′ĭ-kär-dī′tĭs) *n.* Inflammation of the pericardium.

**per·i·car·di·um** (pĕr′ĭ-kär′dē-əm) *n., pl.* **-di·a** (-dē-ə) [NLat. < Gk. *perikardion* < *perikardios,* around the heart : *peri,* around + *kardia,* heart.] The membranous sac enclosing the heart. **—per′i·car′di·al, per′i·car′di·ac′** *adj.*

**per·i·carp** (pĕr′ĭ-kärp′) *n.* [NLat. *pericarpium* < Gk. *perikarpion,* pod : *peri,* around + *karpos,* fruit.] The wall of a ripened fruit or ovary. **—per′i·car′pi·al** *adj.*

**per·i·chon·dri·um** (pĕr′ĭ-kŏn′drē-əm) *n., pl.* **-dri·a** (-drē-ə) [NLat. : PERI- + Gk. *khondros,* cartilage.] *Anat.* The fibrous membrane covering the surface of cartilage except at joint endings. **—per′i·chon′dri·al** *adj.*

**per·i·clase** (pĕr′ĭ-klās′, -klāz′) *n.* [G. *Periklas* : Gk. *peri-* (intensive) + *klasis,* breaking (so called in reference to its perfect cleavage).] A mineral form of magnesium oxide, MgO, usu. occurring in isomeric grains or crystals.

**per·i·cline** (pĕr′ĭ-klīn′) *n.* [< Gk. *periklinēs,* sloping on all sides : *peri,* around + *klinein,* to slope.] A variety of albite occurring as elongated white crystals.

**per·i·cra·ni·um** (pĕr′ĭ-krā′nē-əm) *n., pl.* **-ni·a** (-nē-ə) [NLat. < Gk. *perikranion* < *perikranios,* around the skull : *peri,* around + *kranion,* cranium.] The external periosteum of the skull. **—per′i·cra′ni·al** *adj.*

**per·i·cy·cle** (pĕr′ĭ-sī′kəl) *n.* [Fr. *péricycle* < Gk. *perikuklos,* spherical : *peri,* around + *kuklos,* circle.] The growing layer of parenchyma cells and fibers between the endodermis and the conducting tissue in plant roots and stems. **—per′i·cy′clic** (-sī′klĭk, -sĭk′lĭk) *adj.*

**per·i·derm** (pĕr′ĭ-dûrm′) *n.* An outer layer of tissue of plant roots and stems, including the bark and the layer of growing tissue beneath the bark. **—per′i·der′mal** (-dûr′məl), **per′i·der′mic** *adj.*

**pe·rid·i·um** (pə-rĭd′ē-əm) *n., pl.* **-i·a** (-ē-ə) [NLat. < Gk. *pēridion,* dim. of *pēra,* leather pouch.] The covering of the spore-bearing organ in many fungi. **—pe·rid′i·al** *adj.*

**per·i·dot** (pĕr′ĭ-dŏt′, -dō′) *n.* [Fr. *péridot* < OFr. *peritot.*] A green olivine used as a gem. **—per′i·dot′ic** (-dŏt′ĭk, -dō′tĭk) *adj.*

**per·i·do·tite** (pĕr′ĭ-dō-tīt′, pə-rĭd′ə-tīt′) *n.* Any of a group of igneous rocks having a granitelike texture and composed mainly of olivine and various pyroxenes and amphiboles.

**per·i·gee** (pĕr′ə-jē) *n.* [OFr. < Med. Lat. *perigeum* < LGk. *perigeion* : Gk. *peri,* near + Gk. *gē,* earth.] The point nearest the earth in the orbit of the moon or a satellite. **—per′i·ge′al, per′i·ge′an** *adj.*

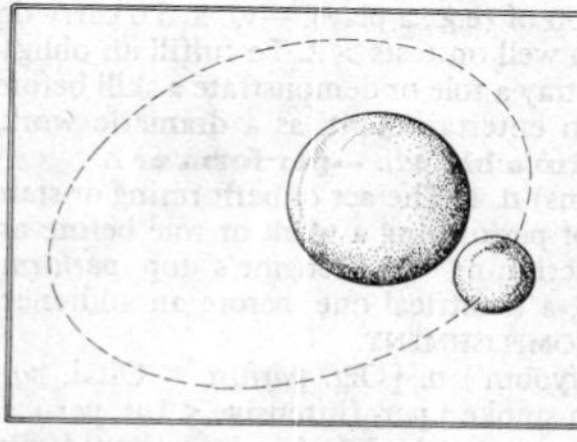

**perigee**

**pe·rig·y·nous** (pə-rĭj′ə-nəs) *adj. Bot.* **1.** Having sepals, petals, and stamens around the edge of a cuplike receptacle containing the ovary. **2.** Designating perigynous flower parts <*perigynous* petals and stamens> **—pe·rig′y·ny** (pə-rĭj′ə-nē) *n.*

**per·i·he·li·on** (pĕr′ə-hē′lē-ən, -hēl′yən) *n., pl.* **-he·li·a** (-hē′lē-ə, -hēl′yə) [PERI- + Gk. *hēlios,* sun.] The point nearest the sun in the orbit of a celestial body, as a planet. **—per′i·he′li·al** *adj.*

**per·i·kar·y·on** (pĕr′ĭ-kăr′ē-ŏn′, -ən) *n., pl.* **-kar·y·a** (kăr′ē-ə). The cell body of a neuron containing the nucleus. **—per′i·kar′y·al** *adj.*

**per·il** (pĕr′əl) *n.* [ME < OFr. < Lat. *periculum.*] **1.** Exposure to the risk of harm or loss : DANGER. **2.** Something that endangers : HAZARD. *—vt.* **-iled, -il·ing, -ils** *also* **-illed, -il·ling, -ils.** To expose to danger or the chance of injury. **—per′il·ous** *adj.* **—per′il·ous·ly** *adv.*

**per·i·lymph** (pĕr′ə-lĭmf′) *n.* The fluid in the space between the membranous and bony labyrinths of the internal ear.

**pe·rim·e·ter** (pə-rĭm′ĭ-tər) *n.* [Fr. *périmètre* < Lat. *perimetros* < Gk. : *peri,* around + *metron,* measure.] **1. a.** *Math.* A closed curve bounding a plane area. **b.** The length of such a boundary. **2.** A fortified strip or boundary usu. protecting a military position. **3.** The outer limits of an area. **—per′i·met′ric** (pĕr′ə-mĕt′rĭk), **per′i·met′ri·cal** *adj.* **—per′i·met′ri·cal·ly** *adv.*

**per·i·morph** (pĕr′ə-môrf′) *n.* A mineral enclosing another mineral. **—per′i·mor′phic, per′i·mor′phous** *adj.* **—per′i·mor′phism** *n.*

**per·i·my·si·um** (pĕr′ə-mĭzh′ē-əm, -mĭz′ē-əm) *n., pl.* **-my·si·a** (-mĭzh′ē-ə, -mĭz′ē-ə) [NLat. : PERI- + Gk. *mus,* muscle.] A sheath of connective tissue enveloping bundles of muscle fibers.

**per·i·na·tal** (pĕr′ə-nāt′l) *adj.* Occurring near the time of birth <*perinatal* mortality> **—per′i·na′tal·ly** *adv.*

**per·i·neph·ri·um** (pĕr′ə-nĕf′rē-əm) *n., pl.* **-ri·a** (-rē-ə) [NLat. < Gk. *perinephros,* fat around the kidneys : *peri,* around + *nephros,* kidney.] The connective and fatty tissue surrounding the kidney. **—per′i·neph′ral, per′i·neph′ri·al, per′i·neph′ric** *adj.*

**per·i·ne·um** (pĕr′ə-nē′əm) *n., pl.* **-ne·a** (-nē′ə) [NLat. < LLat. *perinaion* < Gk. : *peri,* around + *inan,* to excrete.] **1.** The portion of the body in the pelvis occupied by urogenital passages and the rectum, bounded in front by the pubic arch, in the back by the coccyx, and laterally by part of the hipbone. **2.** The region between the scrotum and the anus in males and between the posterior vulva junction and the anus in females. **—per′i·ne′al** *adj.*

**per·i·neu·ri·um** (pĕr′ə-nŏŏr′ē-əm, -nyŏŏr′-) *n., pl.* **-neu·ri·a** (-nŏŏr′ē-ə, -nyŏŏr′-). A sheath of connective tissue enclosing a primary bundle of nerve fibers. **—per′i·neu′ri·al** *adj.*

**pe·ri·od** (pĭr′ē-əd) *n.* [ME *paryode* < OFr. *periode* < Med. Lat. *periodus* < Lat., cycle < Gk. *periodos,* circuit : *peri,* around + *hodos,* way.] **1.** An interval of time marked by the occurrence of certain conditions or events <a *period* of six weeks> **2.** An interval of time marked by the prevalence of a specified culture, ideology, or technology <the *period* of the industrial revolution> **3.** A unit of geologic time longer than an epoch and shorter than an era. **4.** A distinct evolutionary or developmental phase : STAGE <Shakespeare's early comic *period*> **5.** An arbitrary temporal unit, esp.: **a.** A division of an academic day <history class during first *period*> **b.** A division of the playing time of a game. **6.** *Physics & Astron.* The time interval between two successive occurrences of a recurrent event : CYCLE. **7.** An instance or occurrence of menstruation. **8.** A point or portion of time at which something is ended. **9.** The full pause at the end of a spoken sentence. **10.** A punctuation mark ( . ) indicating a full stop, placed at the end of declarative sentences and other statements thought to be complete and after many abbreviations. **11.** A sentence of several carefully balanced clauses in formal writing. **12.** A metrical unit of Greek verse having two or more cola. **13.** *Mus.* A group of two or more phrases in a composition, made up of 8 or 16 measures and ending with a cadence. **14.** *Math.* **a.** The least interval in the range of the independent variable of a periodic function of a real variable in which all possible values of the dependent variable are assumed. **b.** A group of digits separated by commas in a written number. **c.** The number of digits that repeat in a repeating decimal; e.g., $1/7 = 0.142857142857\ldots$ has a six-digit period. *—adj.* Of, belonging to, or representing a certain historical age or time <a *period* drama><*period* dress>

**pe·ri·od·ic** (pĭr′ē-ŏd′ĭk) *adj.* [Lat. *periodicus* < Gk. *periodikos* < *periodos,* circuit. —see PERIOD.] **1.** Having periods or repeated cycles. **2.** Occurring or appearing at regular intervals. **3.** Taking place now and then : INTERMITTENT <*periodic* mood swings> **—pe′ri·od′i·cal·ly** *adv.* **—pe′ri·o·dic′i·ty** (-ə-dĭs′ĭ-tē) *n.*

**per·i·od·ic acid** (pûr′ĭ-ŏd′ĭk) *n.* A white, crystalline inorganic acid, $H_5IO_4{\cdot}2H_2O$, used as an oxidizer.

**pe·ri·od·i·cal** (pĭr′ē-ŏd′ĭ-kəl) *adj.* **1.** Periodic. **2. a.** Published at regular intervals of more than one day. **b.** Of or relating to a publication issued at such intervals. *—n.* A periodical publication.

**periodical cicada** *n.* Seventeen-year locust.

**periodic law** *n. Chem.* The principle that the properties of the elements recur periodically with increasing atomic number.

**periodic table** *n. Chem.* A tabular arrangement of the elements according to their atomic number.

**per·i·o·don·tal** (pĕr′ē-ō-dŏn′tl) *adj.* Of or designating tissue and structures surrounding and supporting the teeth. **—per′i·o·don′tal·ly** *adv.*

**per·i·o·don·tics** (pĕr′ē-ō-dŏn′tĭks) *n. (sing. in number).* The dental specialty of periodontal disease. **—per′i·o·don′tic, per′i·o·don′ti·cal** *adj.* **—per′i·o·don′tist** *n.*

**period piece** *n.* An artistic, literary, or other work that evokes a given historical period.

**per·i·o·nych·i·um** (pĕr′ē-ō-nĭk′ē-əm) *n., pl.* **-i·a** (-ē-ə) [NLat. : PERI- + Gk. *onux,* nail.] The border tissue surrounding the nail.

**per·i·os·te·um** (pĕr′ē-ŏs′tē-əm) *n., pl.* **-te·a** (-tē-ə) [NLat. < LLat. *periosteon* < Gk. < *periosteos,* around the bone : *peri,* around + *osteon,* bone.] A fibrous membrane covering all bones, except at points of articulation. **—per′i·os′te·al, per′i·os′te·ous** *adj.*

**per·i·os·ti·tis** (pĕr′ē-ŏs-tī′tĭs) *n.* Inflammation of the periosteum. **—per′i·os·tit′ic** (-tĭt′ĭk) *adj.*

**per·i·o·tic** (pĕr′ē-ō′tĭk) *adj.* **1.** Located around the ear. **2.** Of or designating the bones immediately around the inner ear.

**per·i·pa·tet·ic** (pĕr′ə-pə-tĕt′ĭk) *adj.* [OFr. *peripatetique* < Lat. *peripateticus* < Gk. *peripatētikos* < *peripatein,* to walk about : *peri,* around + *patein,* to walk.] **1.** Of or relating to walking. **2. Peripatetic.** Of or relating to Aristotle's philosophy or methods of teaching. *—n.* **1.** A person who walks from place to place : ITINERANT. **2. Peripatetic.** A follower of the philosophy of Aristotle.

**per·i·pe·te·ia** (pĕr′ə-pə-tē′ə, -tī′ə) *n.* [Gk. < *peripiptein,* to change suddenly : *peri,* around + *piptein,* to fall.] An unexpected or abrupt change in a situation or course of events esp. in a literary work.

**pe·rip·e·ty** (pə-rĭp′ĭ-tē) *n.* Peripeteia.

**pe·riph·er·al** (pə-rĭf′ər-əl) *adj.* **1.** Relating to, situated on, or comprising the periphery. **2.** Auxiliary. **3.** Of, relating to, or being the outer area of the visual field. *—n.* An auxiliary device that works in conjunction with a computer. **—pe·riph′er·al·ly** *adv.*

**peripheral nervous system** *n.* The part of the nervous system comprising the cranial nerves, the spinal nerves, and the sympathetic nervous system.

**pe·riph·er·y** (pə-rĭf′ə-rē) *n., pl.* **-ies.** [ME *peripherie* < LLat. *peripheria* < Gk. *periphereia* < *peripherēs,* carrying around : *peri,* around + *pherein,* to carry.] **1. a.** The outermost region or part within a precise boundary. **b.** The area or region immediately beyond a precise boundary. **c.** A zone constituting an imprecise boundary. **2.** *Math.* **a.** PERIMETER 1a. **b.** The surface of a solid. **3.** *Anat.* A region in which nerves end.

**pe·riph·ra·sis** (pə-rĭf′rə-sĭs) *n., pl.* **-ses** (-sēz′) [Lat. < Gk. < *peri-*

*phrazein,* to express periphrastically : *peri,* around + *phrazein,* to say.] **1.** The use of circumlocution. **2.** A circumlocution.

**per·i·phras·tic** (pĕr′ə-frăs′tĭk) *adj.* **1.** Having the nature of or marked by periphrasis. **2.** Constructed by using an auxiliary word rather than an inflected form; e.g., the phrases *the word of my teacher* and *my teacher did say* are periphrastic, while *my teacher's word* and *my teacher said* are inflected. **—per′i·phras′ti·cal·ly** *adv.*

**pe·riph·y·ton** (pə-rĭf′ĭ-tŏn′) *n.* [NLat. < Gk. *periphutos,* planted all over : *peri,* around + *phuein,* to grow.] Sessile organisms that live attached to surfaces projecting from the bottom in a freshwater aquatic environment.

**pe·rip·ter·al** (pə-rĭp′tər-əl) *adj.* [Lat. *peripteros* < Gk. : *peri,* around + *pteron,* wing.] Erected with a row of columns on all sides. *—n.* A structure with rows of columns on all sides.

**pe·rique** (pə-rēk′) *n.* [Louisiana Fr.] A black, strongly flavored aromatic tobacco grown in Louisiana and used in various blends.

**per·i·sarc** (pĕr′ĭ-särk′) *n.* [PERI- + Gk. *sarx,* flesh.] A horny external covering that encloses the polyp colonies of certain hydrozoans. **—per′i·sar′cal, per′i·sar′cous** *adj.*

**per·i·scope** (pĕr′ĭ-skōp′) *n.* A tubular optical instrument that contains reflecting elements, as mirrors and prisms, to permit observation from a position displaced from a direct line of sight. **—per′i·scop′ic** (-skŏp′ĭk), **per′i·scop′i·cal** *adj.*

**†per·ish** (pĕr′ĭsh) *vi.* **-ished, -ish·ing, -ish·es.** [ME *perisshen* < OFr. *perir, periss-,* to perish < Lat. *perire* : *per-,* away + *ire,* to go.] **1.** To die, esp. in a violent or untimely manner. **2.** To pass gradually from existence. **3.** *Regional.* To deteriorate or spoil.

**per·ish·a·ble** (pĕr′ĭ-shə-bəl) *adj.* Liable to perish, decay, or spoil : easily damaged or ruined. *—n. often* **perishables.** Something, esp. foodstuff, apt to spoil or decay. **—per′ish·a·bil′i·ty, per′ish·a·ble·ness** *n.* **—per′ish·a·bly** *adv.*

**pe·ris·so·dac·tyl** (pə-rĭs′ō-dăk′təl) [Gk. *perissodaktulos* : *perissos,* uneven (< *peri,* beyond) + *daktulos,* finger.] *Zool. —adj.* **1.** Having an odd number of toes. **2.** Of or relating to certain hoofed mammals, as rhinoceroses and horses, of the order Perissodactyla, that have an odd number of toes. *—n.* A hoofed mammal of the order Perissodactyla. **—pe·ris′so·dac′ty·lous** (-dăk′tə-ləs) *adj.*

**per·i·stal·sis** (pĕr′ĭ-stôl′sĭs, -stăl′-) *n., pl.* **-ses** (-sēz) [NLat. < Gk. *peristaltikos,* peristaltic < *peristellein,* to wrap around : *peri,* around + *stellein,* to place.] Wavelike muscular contractions that push contained matter along tubular organs, as in the alimentary canal. **—per′i·stal′tic** (-stôl′tĭk, -stăl′-) *adj.* **—per′i·stal′ti·cal·ly** *adv.*

**per·i·stome** (pĕr′ĭ-stōm′) *n.* [PERI- + Gk. *stoma,* mouth.] **1.** *Bot.* A circular row of toothlike appendages surrounding the mouth of a moss capsule. **2.** *Zool.* The area around the mouth in certain invertebrates. **—per′i·sto′mal** (-stō′məl), **per′i·sto′mi·al** (-stō′mē-əl) *adj.*

**per·i·style** (pĕr′ĭ-stīl′) *n.* [Fr. *péristyle* < Lat. *peristylum* < Gk. *peristulon* < *peristulos,* surrounded by columns : *peri,* around + *stulos,* pillar.] **1.** A series of columns surrounding a structure, as a temple, or enclosing a court. **2.** A court enclosed by a colonnade. **—per′i·sty′lar** (-stī′lər) *adj.*

**per·i·the·ci·um** (pĕr′ə-thē′shē-əm, -sē-əm) *n., pl.* **-ci·a** (-shē-ə, -sē-ə) [NLat. : PERI- + Gk. *thēkion,* dim. of *thēkē,* case.] A small fruiting body in certain fungi, containing ascospores.

**per·i·to·ne·um** *also* **per·i·to·nae·um** (pĕr′ĭ-tə-nē′əm) *n., pl.* **-ne·a** *also* **-nae·a** (-nē′ə) [LLat. < Gk. *peritonaion* < *peritonaios,* stretched across < *peritonos,* stretched around : *peri,* around + *teinein,* to stretch.] The membrane lining the walls of the abdominal cavity and enclosing the viscera. **—per′i·to·ne′al** *adj.* **—per′i·to·ne′al·ly** *adv.*

**per·i·to·ni·tis** (pĕr′ĭ-tə-nī′tĭs) *n.* Inflammation of the peritoneum.

**per·i·trich** (pĕr′ĭ-trĭk′) *n., pl.* **pe·rit·ri·cha** (pə-rĭt′rĭ-kə) [NLat. *Peritrichida,* order name : PERI- + Gk. *thrix,* hair.] A bell-shaped or tubular microorganism of the order Peritrichida, with a wide oral opening surrounded by cilia. **—pe·rit′ri·chous** (pə-rĭt′rĭ-kəs) *adj.*

**per·i·wig** (pĕr′ĭ-wĭg′) *n.* [Alteration of OFr. *perruque.* —see PERUKE.] A wig or peruke. **—per′i·wigged′** (-wĭgd′) *adj.*

▲ word history: For more than a century (from about 1660 to 1780) decorative heads of false hair were almost universally worn by fashionable men and women in Europe. In English such headdresses were called *perukes* or *periwigs.* Both words are derived from Italian *perruca,* which originally meant "bushy head of hair" and later "wig." *Perruca* was borrowed into French as *perruque,* which developed into two forms in English: *peruke* and *periwig,* which are synonymous. *Periwig* was shortened to *wig,* which is the form now in common use.

**per·i·win·kle[1]** (pĕr′ĭ-wĭng′kəl) *n.* [ME *periwinkle, prob. alteration of OE *pīnewincle* : Lat. *pīna,* mussel (< Gk. *pinē*) + OE *-wincel,* snail shell.] **1.** A small edible marine snail esp. of the genus *Littorina,* with a thick cone-shaped whorled shell. **2.** The shell of any of the periwinkles.

**per·i·win·kle[2]** (pĕr′ĭ-wĭng′kəl) *n.* [ME *pervenke* < OFr. *pervenche* < Lat. *pervinca.*] A trailing evergreen plant of the genus *Vinca,* esp. *V. minor,* bearing dark-green glossy leaves and blue flowers.

**per·jure** (pûr′jər) *vt.* **-jured, -jur·ing, -jures.** [ME *perjuren* < OFr. *perjurer* < Lat. *perjurare* : *per,* through + *jurare,* to swear.] To render (oneself) guilty of perjury by willfully testifying falsely under oath. **—per′jur·er** *n.*

**per·ju·ry** (pûr′jə-rē) *n., pl.* **-ries.** [ME *perjurie* < AN < Lat. *perjurium* < *perjurare,* to perjure.] The deliberate, willful giving of incomplete, misleading, or false testimony under oath. **—per·ju′ri·ous** (pər-jŏŏr′ē-əs) *adj.* **—per·ju′ri·ous·ly** *adv.*

**perk[1]** (pûrk) *v.* **perked, perk·ing, perks.** [ME *perken,* to be lively < ONFr. *perquer,* to perch < *perque,* perch < Lat. *pertica.*] *—vi.* **1.** To stick up or jut out, as an animal's ears. **2.** To carry oneself in a lively and jaunty manner. *—vt.* To cause to stick up or jut out <The cat *perked* its ears at the sound.> **—perk up. 1.** To regain or cause to regain one's good spirits or animation. **2.** To improve the appearance of : spruce up. *—adj.* Perky.

**perk[2]** (pûrk) *vi.* To percolate.

**perk[3]** (pûrk) *n. often* **perks.** A perquisite <"The high-flying corporate *perks*—office refrigerators stocked with Perrier water, first-class trips for employees" *—Newsweek*>

**perk·y** (pûr′kē) *adj.* **-i·er, -i·est.** Cheerful and brisk : LIVELY. **—perk′i·ly** *adv.* **—perk′i·ness** *n.*

**per·lite** *also* **pearl·ite** (pûr′līt′) *n.* [Fr. < *perle,* pearl < OFr. —see PEARL[1].] A natural volcanic glass similar to obsidian but having distinctive concentric cracks and a relatively high water content that in a fluffy heat-expanded form is used as a lightweight aggregate in plaster and concrete and in thermal and acoustic insulation. **—per·lit′ic** (pər-lĭt′ĭk) *adj.*

**perm** (pûrm) *n.* A permanent wave. *—vt. & vi.* **permed, perm·ing, perms.** To treat (hair) with a permanent wave.

**per·ma·frost** (pûr′mə-frôst′, -frŏst′) *n.* [PERMA(NENT) + FROST.] Permanently frozen subsoil continuous in underlying polar regions and occurring locally in perennially frigid areas.

**perm·al·loy** (pûr′mə-loi′, pûrm-ăl′oi′) *n.* [PERM(EABLE) + ALLOY.] Any of several alloys of nickel and iron having high magnetic permeability.

**per·ma·nent** (pûr′mə-nənt) *adj.* [ME < OFr. < Lat. *permanens,* pr.part. of *permanēre,* to endure : *per,* throughout + *manēre,* to remain.] **1.** Lasting or meant to last indefinitely : ENDURING. **2.** Not expected to change in status, condition, or place <a *permanent* residence> *—n.* **1.** A permanent wave. **2.** A long-lasting hair setting. **—per′ma·nence** (-nəns), **per′ma·nen·cy** *n.* **—per′ma·nent·ly** *adv.* **—per′ma·nent·ness** *n.*

**permanent magnet** *n.* A material that retains induced magnetic properties after it is removed from a magnetic field.

**permanent press** *n.* Durable press. **—per′ma·nent-press′** *adj.*

**permanent tooth** *n.* One of the 32 teeth of the second set of teeth in mammals that grow as the milk teeth are shed.

**permanent wave** *n.* **1.** Artificial waves in the hair produced by applying chemicals to it while wet, winding it on curlers, and drying with heat. **2.** The process used in making permanent waves. **3.** A preparation used in making permanent waves.

**per·man·ga·nate** (pər-măn′gə-nāt′) *n.* [PERMANGAN(IC ACID) + -ATE[2].] Any of the salts of permanganic acid, all of which are strong oxidizing agents.

**per·man·gan·ic acid** (pûr′măn-găn′ĭk) *n.* An unstable inorganic acid, $HMnO_4$, existing primarily as a strongly oxidizing, purple aqueous solution.

**per·me·a·bil·i·ty** (pûr′mē-ə-bĭl′ĭ-tē) *n.* **1.** The condition or property of being permeable. **2.** *Physics.* Magnetic permeability. **3.** The rate of diffusion of a pressurized gas through a porous material.

**per·me·a·ble** (pûr′mē-ə-bəl) *adj.* [LLat. *permeabilis* < Lat. *permeare,* to penetrate. —see PERMEATE.] Capable of being permeated : PENETRABLE. **—per′me·a·bly** *adv.*

**per·me·ance** (pûr′mē-əns) *n.* [< Lat. *permeans,* pr.part. of *permeare,* to penetrate. —see PERMEATE.] A measure of the ability of a magnetic circuit to conduct magnetic flux.

**per·me·ase** (pûr′mē-ās′) *n.* [PERME(ATE) + -ASE.] Any of a group of enzymes that regulate the transport of other substances across a cell membrane.

**per·me·ate** (pûr′mē-āt′) *v.* **-at·ed, -at·ing, -ates.** [Lat. *permeare, permeat-,* to penetrate : *per,* through + *meare,* to pass.] *—vt.* **1.** To flow or spread throughout : PERVADE. **2.** To pass through the openings or interstices of (e.g., a membrane). *—vi.* To spread : diffuse. **—per′me·ant** (-ənt), **per′me·a′tive** *adj.* **—per′me·a′tion** *n.*

**Per·mi·an** (pûr′mē-ən, pĕr′-) *adj.* [After *Perm,* an oblast of the USSR.] Of, belonging to, or designating the geologic time, system of rocks, and sedimentary deposits of the seventh and last period of the Paleozoic era. *—n.* The Permian period.

**per·mis·si·ble** (pər-mĭs′ə-bəl) *adj.* That can be permitted. **—per·mis′si·bil′i·ty, per·mis′si·ble·ness** *n.* **—per·mis′si·bly** *adv.*

**per·mis·sion** (pər-mĭsh′ən) *n.* [ME < OFr. < Lat. *permissio* < *permittere,* to permit.] **1.** The act of permitting. **2.** Consent, esp. formal consent.

**per·mis·sive** (pər-mĭs′ĭv) *adj.* **1.** Granting permission. **2.** Not forbidden : PERMITTED. **3.** Tolerant : lenient. **4.** Permitting discretion : OPTIONAL. **—per·mis′sive·ly** *adv.* **—per·mis′sive·ness** *n.*

**per·mit** (pər-mĭt′) *v.* **-mit·ted, -mit·ting, -mits.** [Lat. *permittere* : *per,* through + *mittere,* to let go.] *—vt.* **1.** To consent to : ALLOW.

2. To give permission to or for : AUTHORIZE. 3. To afford opportunity to. —*vi.* To afford opportunity : ALLOW <if the situation *permits*> —*n.* (pûr′mĭt, pər-mĭt′). **1.** Permission, esp. in written form. **2.** A document or certificate giving permission to do something. **—per·mit′ter** *n.*

☆ ***syns:*** **1.** PERMIT, ALLOW, HAVE, LET, TOLERATE *v. core meaning* : to neither forbid nor prevent <*permitted* the children to misbehave> ***ant:*** forbid, prohibit **2.** PERMIT, ALLOW, AUTHORIZE, LET, SANCTION *v. core meaning* : to give consent to <*permitted* five days of sick leave>

**per·mit·tiv·i·ty** (pûr′mĭ-tĭv′ĭ-tē) *n., pl.* **-ties.** *Physics.* The ratio of electric flux density produced by an electric field in a medium to that produced in a vacuum by the same field.

**per·mu·ta·tion** (pûr′myoo-tā′shən) *n.* **1.** A complete change : TRANSFORMATION. **2.** The act of altering a given set of objects in a group. **3.** *Math.* An ordered arrangement of some or all of the elements of a set. **—per′mu·ta′tion·al** *adj.*

**per·mute** (pər-myoot′) *vt.* **-mut·ed, -mut·ing, -mutes.** [ME *permuten* < OFr. *permuter* < Lat. *permutare* : *per-* (intensive) + *mutare,* to change.] **1.** To change the order of. **2.** *Math.* To subject to permutation. **—per·mut′a·ble** *adj.*

**per·ni·cious** (pər-nĭsh′əs) *adj.* [Lat. *perniciosus* < *pernicies,* destruction : *per-* (intensive) + *nex,* violent death.] **1. a.** Tending to cause serious injury or death : DEADLY <a *pernicious* disease> **b.** Causing great harm : RUINOUS <*pernicious* rumors> **2.** *Archaic.* Evil : wicked. **—per·ni′cious·ly** *adv.* **—per·ni′cious·ness** *n.*

**pernicious anemia** *n.* A severe anemia associated with failure to absorb vitamin $B_{12}$ and marked by the presence of abnormally large red blood cells, gastrointestinal disturbances, and lesions of the spinal cord.

**per·nick·e·ty** (pər-nĭk′ĭ-tē) *adj. var. of* PERSNICKETY.

**per·o·ne·al** (pĕr′ə-nē′əl) *adj.* [NLat. *peroneus* < *perone,* fibula < Gk. *peronē.*] Of or relating to the fibula or to the outer portion of the leg.

**per·o·ral** (pər-ôr′əl, -ōr′) *adj.* [Lat. *per,* through + Lat. *ōs, ōr-,* mouth.] Administered by way of the mouth. **—per·o′ral·ly** *adv.*

**per·o·rate** (pĕr′ə-rāt′) *vi.* **-rat·ed, -rat·ing, -rates.** [Lat. *perorare, perorat-,* to harangue at length : *per-* (intensive) + *orare,* to speak.] **1.** To conclude a speech with a formal recapitulation. **2.** To speak at great length, often in a grandiloquent manner. **—per′o·ra′tion** *n.* **—per′o·ra′tion·al** *adj.*

**per·ox·i·dase** (pə-rŏk′sĭ-dās′, -dāz′) *n.* An enzyme found in some animal cells and most plant cells that catalyzes peroxide oxidation reactions.

**per·ox·ide** (pə-rŏk′sīd′) *also* **per·ox·id** (-sĭd) *n.* **1.** Hydrogen peroxide. **2.** A compound containing oxygen that yields hydrogen peroxide with an acid, as sodium peroxide, $Na_2O_2$. —*vt.* **-id·ed, -id·ing, -ides. 1.** To treat with peroxide. **2.** To bleach (hair) with hydrogen peroxide. **—per·ox′ide′** *adj.*

**per·ox·i·some** (pə-rŏk′sĭ-sōm′) *n.* [PEROXI(DE) + -SOME[3].] A cell organelle having enzymes that catalyze the production and breakdown of hydrogen peroxide. **—per·ox′i·som′al** (-sō′məl) *adj.*

**per·ox·y** (pə-rŏk′sē) *adj.* [PER- + OXY-.] Containing the bivalent group $O_2$.

**per·pend** (pər-pĕnd′) *v.* **-pend·ed, -pend·ing, -pends.** [Lat. *perpendere,* to consider carefully : *per-* (intensive) + *pendere,* to consider.] —*vt.* To wonder about : PONDER. —*vi.* To be meditative : REFLECT.

**per·pen·dic·u·lar** (pûr′pən-dĭk′yə-lər) *adj.* [ME *perpendiculer* < OFr. < Lat. *perpendicularius* < *perpendiculum,* plumb line : *per-* (intensive) + *pendĕre,* to hang.] **1.** *Math.* Intersecting at or forming right angles. **2.** At right angles to the horizontal : VERTICAL. **3.** *often* **Perpendicular.** Designating a style of English Gothic architecture of the 14th and 15th cent., marked by emphasis of the vertical element. —*n.* **1.** A line or plane perpendicular to a line or plane. **2.** A vertical position. **3.** A device, as a plumb line, used in marking the vertical from a point. **4.** A vertical or nearly vertical line or plane. **—per′pen·dic′u·lar′i·ty** (-lăr′ĭ-tē) *n.* **—per′pen·dic′u·lar·ly** *adv.*

**per·pe·trate** (pûr′pĭ-trāt′) *vt.* **-trat·ed, -trat·ing, -trates.** [Lat. *perpetrare, perpetrat-,* to accomplish : *per-* (intensive) + *patrare,* to bring about.] To be guilty of : COMMIT <*perpetrate* a crime> <*perpetrate* a hoax> **—per′pe·tra′tion** *n.* **—per′pe·tra′tor** *n.*

**per·pet·u·al** (pər-pĕch′oo-əl) *adj.* [ME *perpetuel* < OFr. < Lat. *perpetualis* < *perpetuus,* continuous < *perpes,* uninterrupted : *per-* (intensive) + *petere,* to go toward.] **1.** Lasting for eternity : never ending. **2.** Lasting for an indefinitely long duration. **3.** Instituted to be valid for an unlimited duration, as a peace treaty. **4.** Continuing without interruption or surcease <*perpetual* complaining> **5.** Flowering throughout the growing season. **—per·pet′u·al·ly** *adv.*

**perpetual calendar** *n.* A chart or mechanical device that indicates the day of the week corresponding to any given date over a period of many years.

**perpetual motion** *n.* The hypothetical continuous operation of an isolated mechanical device or other closed system without a sustaining energy source.

**per·pet·u·ate** (pər-pĕch′oo-āt′) *vt.* **-at·ed, -at·ing, -ates.** [Lat. *perpetuare, perpetuat-* < *perpetuus,* continuous. —see PERPETUAL.] **1.** To make perpetual. **2.** To prolong the existence or remembrance of <*perpetuate* a childhood fear> **—per·pet′u·ance, per·pet′u·a′tion** *n.* **—per·pet′u·a′tor** *n.*

**per·pe·tu·i·ty** (pûr′pĭ-too′ĭ-tē, -tyoo′-) *n., pl.* **-ties. 1.** The quality or state of being perpetual. **2.** Time without end : ETERNITY. **3.** *Law.* **a.** The condition of an estate that is limited so as to be inalienable either perpetually or longer than the period determined by law. **b.** An estate so limited. **4.** An annuity payable indefinitely.

**per·plex** (pər-plĕks′) *vt.* **-plexed, -plex·ing, -plex·es.** [< obs. *perplex,* perplexed < Lat. *perplexus* : *per-* (intensive) + *plectere,* to entwine.] **1.** To bewilder or confuse : PUZZLE. **2.** To make complex.

**per·plexed** (pər-plĕkst′) *adj.* **1.** Bewildered : PUZZLED. **2.** Complicated : involved. **—per·plex′ed·ly** (pər-plĕk′sĭd-lē) *adv.*

**per·plex·i·ty** (pər-plĕk′sĭ-tē) *n., pl.* **-ties. 1.** The state of being perplexed or puzzled. **2.** The condition of being complicated or intricate <"the *perplexity* of life in twentieth-century America" —Daniel J. Boorstin> **3.** Something that perplexes.

**per·qui·site** (pûr′kwĭ-zĭt) *n.* [ME, property acquired otherwise than by inheritance < Med. Lat. *perquisitum,* acquisition < p.part. of Lat. *perquirere,* to search diligently for : *per-* (intensive) + *quaerere,* to seek.] **1.** A profit or payment received in addition to a regular wage or salary, esp. a benefit expected as one's due. **2.** A gratuity : tip. **3.** Something claimed as an exclusive right <"Politics was the *perquisite* of the upper class" —Richard B. Sewall>

**per·ry** (pĕr′ē) *n., pl.* **-ries.** [ME *perrye* < OFr. *pere* < VLat. **piratum* < Lat. *pirum,* pear.] A fermented ciderlike beverage made from pears.

**per se** (pər sā′, sē′) *adv.* [Lat.] In or by itself : INTRINSICALLY.

**per·se·cute** (pûr′sĭ-kyoot′) *vt.* **-cut·ed, -cut·ing, -cutes.** [OFr. *persecuter* < LLat. *persequi* < Lat., to pursue : *per-* (intensive) + *sequi,* to follow.] **1.** To harass or oppress with ill-treatment. **2.** To bother persistently : ANNOY. **—per′se·cu′tion** (-kyoo′shən) *n.* **—per′se·cu′tion·al, per′se·cu′tive, per′se·cu·to′ry** (-kyoo-tôr′ē, -tōr′ē, -kyoo′tə-rē) *adj.* **—per′se·cu′tor** *n.*

**Per·se·id** (pûr′sē-ĭd) *n., pl.* **-ids** *or* **Per·se·i·des** (pər-sē′ĭ-dēz′) [< Lat. *Perseus,* the constellation Perseus.] A meteor shower that appears to originate near Perseus during the second week of August.

**Per·seph·o·ne** (pər-sĕf′ə-nē) *n.* [Lat. < Gk. *Persephonē.*] *Gk. Myth.* The daughter of Demeter and queen of the underworld as the wife of Hades.

**Per·se·us** (pûr′sē-əs, -soos′) *n.* [Lat. < Gk.] **1.** *Gk. Myth.* The son of Zeus and Danae who slew Medusa and rescued Andromeda. **2.** A constellation in the Northern Hemisphere.

**Perseus**

**per·se·ver·ance** (pûr′sə-vîr′əns) *n.* **1.** Steadfast adherence to a course of action, belief, or purpose. **2.** The Calvinistic doctrine that those who have been chosen by God will continue in a state of grace to the end and will finally be saved.

**per·sev·er·a·tion** (pər-sĕv′ə-rā′shən) *n. Psychol.* Continued or repetitive activity or actions, esp.: **a.** The uncontrollable repetition of a gesture, word, phrase, or expression. **b.** The spontaneous recurrence of a thought, image, phrase, or tune in the mind.

**per·se·vere** (pûr′sə-vîr′) *vi.* **-vered, -ver·ing, -veres.** [ME *perseveren* < OFr. *perseverer* < Lat. *perseverare* < *perseverus,* very serious : *per-* (intensive) + *severus,* severe.] To persist in or remain constant to a purpose, idea, or task in the face of discouragement or opposition. **—per′se·ver′ing·ly** *adv.*

**Per·sian** (pûr′zhən) *adj.* Of or relating to Persia or Iran, its people, language, or culture. —*n.* **1.** A native or resident of ancient Persia or modern Iran. **2. a.** Any of the Iranian languages of the Persians in use during various historical periods. **b.** The modern Iranian language of Iran and western Afghanistan.

**Persian cat** *n.* A stocky domestic cat with long silky fur.

**Persian lamb** *n.* **1.** The lamb of the karakul sheep of Asia. **2.** The glossy, tightly curled fur obtained from the Persian lamb, usu. when it is three or four days old.

**Persian melon** *n.* A melon, *Cucumis melo inodorus,* with an unridged, light-colored rind and orange flesh.

**per·si·flage** (pûr′sə-fläzh′) *n.* [Fr. < *persifler,* to banter : *per-* (intensive < Lat.) + *siffler,* to whistle (< Lat. *sibilare).*] **1.** Light bantering style in writing or speaking. **2.** Good-natured raillery.

**per·sim·mon** (pər-sĭm′ən) *n.* [Of Algonquian orig.] **1.** A chiefly tropical tree of the genus *Diospyros,* with hard wood and orange-red fruit that is edible only when completely ripe. **2.** The fruit of a persimmon tree.

**per·sist** (pər-sĭst′, -zĭst′) *vi.* **-sist·ed, -sist·ing, -sists.** [Lat. *persistere* : *per-* (intensive) + *sistere,* to stand.] **1.** To be obstinately insistent, repetitious, or tenacious. **2.** To hold steadfastly and firmly to a purpose, state, or undertaking, despite obstacles, warnings, or setbacks. **3.** To continue in existence : LAST. **—per·sis′ter** *n.*

**per·sist·ence** (pər-sĭs′təns, -zĭs′-) *also* **per·sist·en·cy** (-tən-sē) *n.* **1.** The act of persisting. **2.** The quality of being persistent : PERSEVERANCE. **3.** The continuance of an effect after the cause is removed <*persistence* of vision>

**per·sist·ent** (pər-sĭs′tənt, -zĭs′-) *adj.* **1.** Refusing to give up or let go : TENACIOUS. **2.** Insistently repetitive or continuous <a *persistent* dizziness> **3.** Enduring. **4.** *Bot.* Lasting past maturity without falling off, as certain leaves or flowers. **5.** *Zool.* Retained permanently rather than disappearing in an early stage of development <*persistent* gills> **—per·sist′ent·ly** *adv.*

**per·snick·e·ty** (pər-snĭk′ĭ-tē) *also* **per·nick·e·ty** (-nĭk′ĭ-tē) *adj.* [Orig. unknown.] *Informal.* **1.** Fastidious. **2.** Requiring strict attention to detail. **—per·snick′e·ti·ness** *n.*

**per·son** (pûr′sən) *n.* [ME < OFr. *persone* < Lat. *persona,* prob. < Etruscan *phersu,* mask.] **1.** A living human being. **2.** The composite of characteristics that make up an individual personality. **3.** An individual of specified character <a *person* of eminence> **4.** The living body of a human being <a gun carried on one's *person*> **5.** Guise : character <"Well, in her *person,* I say I will not have you" —Shakespeare> **6.** Physique and general appearance. **7.** *Law.* A human being or organization with legal rights and duties. **8.** The separate individualities of the Father, Son, and Holy Spirit, as distinguished from the essence of the Godhead that unites them. **9. a.** Any of three groups of pronoun forms with corresponding verb inflections that distinguish between the speaker, the individual addressed, and the individual or thing spoken of. **b.** Any of the different forms or inflections expressing these distinctions. ***usage:*** *Person* has come into increasing use in creating compounds that may refer to either a man or a woman, as *chairperson* or *spokesperson.*

**per·so·na** (pər-sō′nə, -nä′) *n.* [Lat. —see PERSON.] **1.** *pl.* **-nae** (-nē). A character in a literary or dramatic work. **2.** *pl.* **-nas.** *Psychol.* The role that a person assumes in order to display his or her conscious intentions.

**per·son·a·ble** (pûr′sə-nə-bəl) *adj.* Pleasing in appearance or personality : ATTRACTIVE. **—per′son·a·ble·ness** *n.*

**per·son·age** (pûr′sə-nĭj) *n.* [ME < OFr. < *persone,* person. —see PERSON.] **1.** PERSON 1. **2.** A person of distinction or note.

**per·so·na gra·ta** (pər-sō′nə grä′tə, grăt′ə) *adj.* [Lat., an acceptable person.] Fully acceptable or welcome, esp. to a foreign government <ambassadors who were *persona grata*>

**per·son·al** (pûr′sə-nəl) *adj.* **1.** Of or relating to a particular person : PRIVATE <*personal* concerns> **2. a.** Done, made, or performed in person <a *personal* visit> **b.** Done to or for or directed toward a particular person <a *personal* request> **3.** Concerning a particular individual's intimate affairs, interests, or activities. **4. a.** Aimed pointedly at the most intimate aspects of a person, esp. in a critical or hostile manner <resented the highly *personal* remark> **b.** Tending to make remarks, or be unduly questioning, about another's affairs <becomes *personal* in a dispute> **5.** Of or pertaining to the body or physical being <*personal* hygiene> **6.** Pertaining to or having the nature of a person or self-conscious being <a *personal* Deity> **7.** *Law.* Pertaining to a person's movable property <*personal* belongings> **8.** Indicating grammatical person. *—n.* A personal item or notice in a newspaper paragraph or classified ad.

**personal effects** *pl.n.* Privately owned items, as keys, an identification card, or a wallet or watch, that are regularly carried or worn on one's person.

**personal equation** *n.* The characteristics of a person as they tend to cause variation in observation, judgment, and reasoning.

**per·son·al·ism** (pûr′sə-nə-lĭz′əm) *n.* **1.** The quality of being characterized by purely personal modes of expression or behavior : IDIOSYNCRASY. **2.** *Philos.* Any of various trends of subjective idealism regarding personality as the key to the interpretation of reality. **—per′son·al·ist** *n.* & *adj.* **—per′son·al·is′tic** *adj.*

**per·son·al·i·ty** (pûr′sə-năl′ĭ-tē) *n., pl.* **-ties.** [ME *personalite* < LLat. *personalitas* < *personalis,* personal < Lat. *persona,* person. —see PERSON.] **1.** The quality or state of being a person. **2. a.** The totality of qualities and traits, as of character or behavior, that are peculiar to an individual. **b.** A person as the embodiment of distinctive traits of mind and behavior <a powerful military *personality*> **3.** The pattern of collective character, behavioral, temperamental, emotional, and mental traits of an individual. **4.** Distinctive qualities of an individual, esp. those distinguishing personal characteristics that make one socially appealing. **5.** *Informal.* A person of prominence or renown <*personalities* of stage and screen> ***usage:*** Although *personality* in the sense of "celebrity" occurs often in speech and journalism, its use in more formal styles is unacceptable to many. **6.** *often* **personalities.** An offensively personal remark. **7.** The characteristics of a place or situation that give it a distinctive character <artwork that lends an office *personality*>

**per·son·al·ize** (pûr′sə-nə-līz′) *vt.* **-ized, -iz·ing, -iz·es.** **1.** To take (a remark or characterization) personally. **2.** To personify. **3.** To have printed, engraved, or monogrammed with one's name or initials <*personalized* stationery> **—per′son·al·i·za′tion** *n.*

**per·son·al·ly** (pûr′sə-nə-lē) *adv.* **1.** In person <I congratulated them *personally.*> **2.** As far as oneself is concerned <*Personally,* I like the new styles.> **3.** As a person <I dislike them *personally* but respect their work.> **4.** In a personal manner <Try not to take the layoff *personally.*>

**personal pronoun** *n.* A pronoun denoting the person speaking, the person spoken to, or the person or thing spoken about.

**personal property** *n. Law.* Temporary or movable property as distinguished from real property.

**per·son·al·ty** (pûr′sə-nəl-tē) *n., pl.* **-ties.** [AN *personalte* < LLat. *personalitas,* personality. —see PERSONALITY.] *Law.* Personal property.

**per·so·na non gra·ta** (pər-sō′nə nŏn grä′tə, grăt′ə) *adj.* [Lat., an unacceptable person.] Unacceptable or unwelcome, esp. to a foreign government <ambassadors who were *persona non grata*>

**per·son·ate¹** (pûr′sə-nāt′) *vt.* **-at·ed, -at·ing, -ates.** [< PERSON.] **1.** To portray the part or play the role of : IMPERSONATE. **2.** To endow with personal qualities : PERSONIFY. **3.** *Law.* To assume the identity of with intent to deceive. **—per′son·a′tion** *n.* **—per′son·a′tive** *adj.* **—per′son·a′tor** *n.*

**per·son·ate²** (pûr′sə-nĭt) *adj.* [Lat. *personatus,* masked < *persona,* mask. —see PERSON.] *Bot.* Two-lipped, with the base closed by a prominent palate <a *personate* corolla>

**per·son·hood** (pûr′sən-hŏŏd′) *n.* The condition of being a person, esp. those qualities that confer distinct individuality.

**per·son·i·fi·ca·tion** (pər-sŏn′ə-fĭ-kā′shən) *n.* **1.** The act of personifying. **2.** A person or thing typifying a certain quality or idea that is outstanding : EXEMPLIFICATION. **3.** A rhetorical figure of speech in which inanimate objects or abstractions are endowed with human qualities or are represented as possessing human form, as in *Winter lay dying* or *Mystery stalked the deserted palace.* **4.** The artistic representation of an abstract quality or idea as a person.

**per·son·i·fy** (pər-sŏn′ə-fī′) *vt.* **-fied, -fy·ing, -fies.** [Fr. *personnifier* < *personne,* person < OFr. *persone.* —see PERSON.] **1.** To think of or represent (an inanimate object or abstraction) as having personality or the qualities, thoughts, or movements of a living being. **2.** To represent (an object or abstraction) by a human figure. **3.** To represent (an abstract quality or idea) <This figure *personifies* goodness.> **4.** To be the embodiment or perfect example of <"Stalin now *personified* bolshevism in the eyes of the world" —A.J.P. Taylor> **—per·son′i·fi′er** *n.*

**per·son·nel** (pûr′sən-ĕl′) *n.* [Fr. < *personnel,* personal < OFr. *personal* < LLat. *personalis.* —see PERSONALITY.] **1.** The body of persons employed by or active in an organization, business, or service. ***usage:*** *Personnel* is a collective noun and therefore is never used with a specific number. It is acceptable, however, to use another qualifying word, as in *A number of armed forces personnel* (not *six armed forces personnel*) *gave testimony at the court-martial.* **2.** An administrative division of an organization concerned with the body of persons employed by or active in it.

**per·spec·tive** (pər-spĕk′tĭv) *n.* [OItal. *perspectiva* < LLat., of a view < Lat. *perspicere,* to inspect : *per-* (intensive) + *specere,* to look.] **1.** The technique of representing three-dimensional objects and depth relationships on a two-dimensional surface. **2.** A vista or view. **3.** The appearance of objects in depth as perceived by normal binocular vision. **4.** The relationship of aspects of a subject to each other and to a whole <put American history in *perspective*> **5.** Subjective evaluation of relative significance : POINT OF VIEW. *—adj.* Of, seen, or represented in perspective. **—per·spec′tive·ly** *adv.*

**per·spi·ca·cious** (pûr′spĭ-kā′shəs) *adj.* [< Lat. *perspicax, perspicac-* < *perspicere,* to look through. —see PERSPECTIVE.] Able to perceive or understand keenly <a *perspicacious* observer of foreign events> **—per′spi·ca′cious·ly** *adv.* **—per′spi·ca′cious·ness** *n.*

**per·spi·cac·i·ty** (pûr′spĭ-kăs′ĭ-tē) *n.* Acuteness of perception, discernment, or understanding.

**per·spi·cu·i·ty** (pûr′spĭ-kyōō′ĭ-tē) *n.* **1.** The quality of being perspicuous. **2.** Perspicacity.

**per·spic·u·ous** (pər-spĭk′yōō-əs) *adj.* [Lat. *perspicuus* < *perspicere,* to see through. —see PERSPECTIVE.] Lucidly presented or expressed : CLEAR. **—per·spic′u·ous·ly** *adv.* **—per·spic′u·ous·ness** *n.*

**per·spi·ra·tion** (pûr′spə-rā′shən) *n.* **1.** The saline moisture excreted through the pores of the skin by the sweat glands : SWEAT. **2.** The act or process of perspiring. **—per·spir′a·to′ry** (pər-spīr′ə-tôr′ē, -tōr′ē, pûr′spər-ə-) *adj.*

**per·spire** (pər-spīr′) *v.* **-spired, -spir·ing, -spires.** [Fr. *perspirer* < OFr. < Lat. *perspirare,* to breathe through : *per,* through + *spirare,* to breathe.] *—vi.* To excrete perspiration through the pores of the skin. *—vt.* To expel through external pores : EXUDE.

**per·suade** (pər-swād′) *vt.* **-suad·ed, -suad·ing, -suades.** [Lat. *persuadēre* : *per-* (intensive) + *suadēre,* to urge.] **1. a.** To cause (someone) to do something by means of entreaty, argument, or reasoning. **b.** To win over (someone) to a course of action by reasoning or inducement. **2.** To make (someone) believe something : CONVINCE <"to make children fit to live in a society by *persuading* them to learn and accept its codes" —Alan W. Watts> **—per·suad′a·ble** *adj.* **—per·suad′er** *n.*

**per·sua·si·ble** (pər-swā′zə-bəl, -sə-bəl) *adj.* Capable of being persuaded. **—per·sua′si·bil′i·ty, per·sua′si·ble·ness** *n.*

**per·sua·sion** (pər-swā′zhən) *n.* [Lat. *persuasio* < *persuadēre,* to persuade. —see PERSUADE.] **1.** The act of persuading or state of being persuaded <"The *persuasion* of a democracy to big changes is at best a slow process" —Harold J. Laski> **2.** Power or ability to persuade. **3.** A strong belief or conviction. **4. a.** A body of religious beliefs : RELIGION. **b.** A group devoted to such a religion. **5.** A faction : party.

**per·sua·sive** (pər-swā′sĭv, -zĭv) *adj.* Tending or having the power to persuade <a *persuasive* speaker> **—per·sua′sive·ly** *adv.* **—per·sua′sive·ness** *n.*

**pert** (pûrt) *adj.* **-er, -est.** [ME < OFr. *apert* < Lat. *apertus,* open, p.part. of *aperire,* to open.] **1.** Impudent : impertinent. **2.** High-spirited : lively. **3.** Jaunty <a *pert* ponytail> **—pert′ly** *adv.* **—pert′ness** *n.*

**per·tain** (pər-tān′) *vi.* **-tained, -tain·ing, -tains.** [ME *pertenen* < OFr. *partenir* < Lat. *pertinēre* : *per-* (intensive) + *tenēre,* to hold.] **1.** To have relation or reference <doubts *pertaining* to their qualifications> **2.** To belong as an adjunct or accessory <the suffering that *pertains* to poverty> **3.** To be fitting or appropriate.

**per·ti·na·cious** (pûr′tn-ā′shəs) *adj.* [< Lat. *pertinax, pertinac-* : *per-* (intensive) + *tenax,* tenacious < *tenēre,* to hold.] **1.** Holding firmly or tenaciously to a purpose, belief, or opinion. **2.** Stubbornly or perversely persistent. **—per′ti·na′cious·ly** *adv.* **—per′ti·na′cious·ness, per′ti·nac′i·ty** (-ăs′ĭ-tē) *n.*

**per·ti·nent** (pûr′tn-ənt) *adj.* [ME < OFr. < Lat. *pertinens,* pr.part. of *pertinēre,* to pertain. —see PERTAIN.] Of, pertaining to, or connected with a specific matter : RELEVANT. **—per′ti·nence, per′ti·nen·cy** *n.* **—per′ti·nent·ly** *adv.*

**per·turb** (pər-tûrb′) *vt.* **-turbed, -turbing, -turbs.** [ME *perturben* < OFr. *perturber* < Lat. *perturbare* : *per-* (intensive) + *turbare,* to throw into disorder.] **1.** To disturb greatly : DISCOMPOSE. **2.** To throw into great confusion : DISRUPT. **3.** *Physics.* To cause perturbation, as of an electronic or celestial orbit. **—per·turb′a·ble** *adj.*

**per·tur·ba·tion** (pûr′tər-bā′shən) *n.* **1.** The act of perturbing or state of being perturbed. **2.** Variation in a designated orbit, as of an electron or planet, resulting from the influence of one or more external bodies. **—per′tur·ba′tion·al** *adj.*

**per·tus·sis** (pər-tŭs′ĭs) *n.* [NLat. : Lat. *per-* (intensive) + Lat. *tussis,* cough.] Whooping cough. **—per·tus′sal** *adj.*

**Pe·ru balsam** (pə-ro͞o′) *n.* Balsam of Peru.

**pe·ruke** (pə-ro͞ok′) *n.* [Fr. *perruque* < OFr., wig < OItal. *perruca.*] A wig, esp. one worn by men in the 17th and 18th cent.

**pe·ruse** (pə-ro͞oz′) *vt.* **-rused, -rus·ing, -rus·es.** [ME *perusen,* to use up : Lat. *per-* (intensive) + ME *usen,* to use.] To read or examine, esp. very carefully. **—pe·rus′a·ble** *adj.* **—pe·rus′al** *n.*

**Pe·ru·vi·an bark** (pə-ro͞o′vē-ən) *n.* CINCHONA 2.

**per·vade** (pər-vād′) *vt.* **-vad·ed, -vad·ing, -vades.** [Lat. *pervadere* : *per,* through + *vadere,* to go.] To be spread throughout : PERMEATE. **—per·vad′er** *n.* **—per·va′sion** (-vā′zhən) *n.*

**per·va·sive** (pər-vā′sĭv, -zĭv) *adj.* [< Lat. *pervadere, pervas-,* to pervade.] **1.** Having the quality of pervasiveness. **2.** Tending to pervade. **—per·va′sive·ly** *adv.* **—per·va′sive·ness** *n.*

**per·verse** (pər-vûrs′, pûr′vûrs′) *adj.* [ME *pervers* < OFr. < Lat. *pervertere,* to pervert. —see PERVERT.] **1.** Deviating from what is right or good : PERVERTED. **2.** Obstinately persisting in an error or fault : wrongly self-willed. **3. a.** Disposed to contradict and oppose. **b.** Marked by or arising from opposition and contradiction. **4.** Peevish : cranky. **—per·verse′ly** *adv.* **—per·verse′ness** *n.*

**per·ver·sion** (pər-vûr′zhən, -shən) *n.* **1.** The act of perverting or state of being perverted. **2.** A sexual practice or act considered deviant. **—per·ver′sive** (-sĭv, -zĭv) *adj.*

**per·ver·si·ty** (pər-vûr′sĭ-tē) *n., pl.* **-ties. 1.** The quality or state of being perverse. **2.** An act or instance of perversity.

**per·vert** (pər-vûrt′) *vt.* **-vert·ed, -vert·ing, -verts.** [ME *perverten* < OFr. *pervertir* < Lat. *pervertere* : *per-* (intensive) + *vertere,* to turn.] **1.** To cause to turn from what is considered morally right : CORRUPT. **2.** To reduce to a lower condition : DEBASE. **3.** To use incorrectly : MISUSE. **4.** To interpret incorrectly : MISCONSTRUE <chose to *pervert* the meaning of the ritual> —*n.* (pûr′vûrt′). One who practices sexual perversion. **—per·vert′er** *n.* **—per·vert′i·ble** *adj.*

**per·vert·ed** (pər-vûr′tĭd) *adj.* **1.** Deviating greatly from what is considered right and correct. **2.** Of, pertaining to, or practicing sexual perversion. **3.** Misinterpreted : misconstrued <a *perverted* reading of a liturgical passage> **—per·vert′ed·ly** *adv.* **—per·vert′ed·ness** *n.*

**per·vi·ous** (pûr′vē-əs) *adj.* [Lat. *pervius* : *per,* through + *via,* way.] **1.** Open to passage or entrance : PERMEABLE. **2.** Accessible to arguments, ideas, or change. **—per′vi·ous·ly** *adv.* **—per′vi·ous·ness** *n.*

**pes** (pās) *n., pl.* **pe·des** (pĕd′ās′) [Lat.] A foot or footlike part, esp. the foot of a quadruped vertebrate.

**Pe·sach** (pā′säKH′) *n.* [Heb. *pesaḥ* < *pāsaḥ,* he passed over.] Passover.

**pe·sade** (pə-säd′, -zäd′) *n.* [Fr., var. of obs. *posade* < OFr. < OItal. *posata* < *posare,* to pause < LLat. *pausare* < Lat. *pausa,* pause. —see PAUSE.] The act or position of a horse when rearing on its hind legs with its forelegs in the air.

**pe·se·ta** (pə-sā′tə) *n.* [Sp., dim. of *peso,* peso.] —See table at CURRENCY.

**pe·se·wa** (pā-sā′wä) *n., pl.* **pesewa** or **-was.** [Native word in Ghana.] —See table at CURRENCY.

**pes·ky** (pĕs′kē) *adj.* **-ki·er, -ki·est.** [Prob. alteration of PEST.] *Informal.* Annoying : bothersome. **—pes′ki·ly** *adv.* **—pes′ki·ness** *n.*

**pe·so** (pā′sō) *n., pl.* **-sos.** [Sp. < Lat. *pensum,* something weighed < *pendere,* to weigh.] —See table at CURRENCY.

**pes·sa·ry** (pĕs′ə-rē) *n., pl.* **-ries.** [ME *pessarie* < Med. Lat. *pessarium* < LLat. *pessum* < Gk. *pessos.*] **1.** A contraceptive or supportive device placed and worn in the vagina. **2.** A medicated vaginal suppository.

**pes·si·mism** (pĕs′ə-mĭz′əm) *n.* [Fr. *pessimisme* < Lat. *pessimus,* worst.] **1.** A tendency to take the least hopeful view of a situation. **2.** The doctrine or belief that this is the worst of all possible worlds and that all things ultimately tend toward evil. **3.** The doctrine or belief that the evil in the world outweighs the good. **—pes′si·mist** *n.* **—pes′si·mis′tic** *adj.* **—pes′si·mis′ti·cal·ly** *adv.*

**pest** (pĕst) *n.* [OFr. *peste,* pestilence < Lat. *pestis.*] **1.** A nuisance. **2.** An injurious animal or plant, esp. one harmful to humans. **3.** A pestilence.

**pes·ter** (pĕs′tər) *vt.* **-tered, -ter·ing, -ters.** [Prob. < OFr. *empestrer,* to hobble < VLat. **impastoriare* : Lat. *in-,* in + LLat. *pastura,* pasture < Lat. *pascere,* to pasture.] To beset with petty annoyances : BOTHER. **—pes′ter·er** *n.*

**pest house** *n.* A hospital for patients suffering from infectious disease, esp. from plague.

**pes·ti·cide** (pĕs′tĭ-sīd′) *n.* A chemical used to kill pests, esp. insects and rodents.

**pes·tif·er·ous** (pĕs-tĭf′ər-əs) *adj.* [ME < Lat. *pestiferus* : *pestis,* pestilence + *ferre,* to carry.] **1.** Producing or breeding infectious disease. **2.** Contaminated by or infected with an epidemic disease. **3.** Morally evil or deadly : PERNICIOUS. **4.** *Informal.* Troublesome : annoying. **—pes·tif′er·ous·ly** *adv.* **—pes·tif′er·ous·ness** *n.*

**pes·ti·lence** (pĕs′tə-ləns) *n.* **1. a.** A usu. fatal epidemic disease, esp. bubonic plague. **b.** An epidemic of such a disease. **2.** An evil, pernicious agent or influence.

**pes·ti·lent** (pĕs′tə-lənt) *also* **pes·ti·len·tial** (pĕs′tə-lĕn′shəl) *adj.* [ME < Lat. *pestilens* < *pestis,* pestilence.] **1.** Tending to cause death : DEADLY. **2.** Likely to cause an epidemic disease. **3.** Contaminated or infected with a contagious disease. **4.** Morally, socially, or politically harmful : PERNICIOUS. **5.** Causing disfavor or annoyance.

**pes·tle** (pĕs′əl, pĕs′təl) *n.* [ME *pestel* < OFr. < Lat. *pistillum.*] **1.** A club-shaped hand tool for mashing or grinding substances in a mortar. **2.** A large bar moved vertically to stamp or pound, as in a press or mill. —*v.* **-tled, -tling, -tles.** —*vt.* To pound, grind, or mash with a pestle. —*vi.* To use a pestle.

**pet**[1] (pĕt) *n.* [Orig. unknown.] **1.** An animal kept for pleasure or companionship. **2.** An object of the affections. **3.** A person esp. loved or indulged : FAVORITE <The youngest was my father's *pet.*> —*adj.* **1.** Kept as a pet <a *pet* gerbil> **2.** Particularly indulged or cherished : FAVORITE <a *pet* pupil><a *pet* peeve> —*v.* **pet·ted, pet·ting, pets.** —*vt.* To stroke or caress gently : PAT. —*vi. Informal.* To make love by fondling and caressing. **—pet′ter** *n.*

**pet**[2] (pĕt) *n.* [Orig. unknown.] A fit of bad temper : PIQUE. —*vi.* **pet·ted, pet·ting, pets.** To be peevish and sulky.

**pet·al** (pĕt′l) *n.* [NLat. *petalum* < Gk. *petalon,* leaf.] A separate, often brightly colored segment of a corolla. **—pet′aled, pet′alled** *adj.*

**-petal** *suff.* [< NLat. *-petus* < Lat. *petere,* to seek.] Moving toward <basi*petal*>

**pet·al·if·er·ous** (pĕt′l-ĭf′ər-əs) *adj.* Bearing petals.

**pet·al·ine** (pĕt′l-ĭn, -īn′) *adj.* Of or resembling a petal.

**pet·al·oid** (pĕt′l-oid′) *adj.* Like a petal.

**pet·al·ous** (pĕt′l-əs) *adj.* **1.** Having petals. **2.** Having a given kind or a particular number of petals.

**pe·tard** (pĭ-tärd′) *n.* [Fr. *pétard* < OFr. *peter,* to break wind < *pet,* a breaking of wind < Lat. *peditum* < *pedere,* to break wind.] **1.** A small bell-shaped bomb used to breach a gate or wall. **2.** A firecracker.

**pet·cock** *also* **pet cock** (pĕt′kŏk′) *n.* [Perh. PET(TY) + COCK[1].] A small valve or faucet used to drain or reduce pressure from pipes, radiators, and boilers.

**pe·te·chi·a** (pə-tē′kē-ə) *n., pl.* **-chi·ae** (-kē-ī′) [NLat. < Ital. *petecchia,* skin spot.] A small spot on a body surface, as the skin or mucous membrane, caused by a minute hemorrhage and often seen in typhus. **—pe·te′chi·al** *adj.* **—pe·te′chi·ate** (-ĭt) *adj.*

**pe·ter** (pē′tər) *vi.* **-tered, -ter·ing, -ters.** [Orig. unknown.] **1.** To come to an end slowly : DWINDLE. **2.** To become exhausted.

**Pe·ter** (pē′tər) *n.* —See table at BIBLE.
**Peter Principle** *n.* [After Laurence Johnston *Peter* (b. 1919), its formulator.] The idea that an employee within an organization will advance to his or her highest level of incompetence and remain there.
**Peter's pence** *also* **Peter pence** *n.* [After St. *Peter,* from the tradition that he founded the papacy.] **1.** A tax of one penny per household paid in medieval England to the Papal See. **2.** An annual voluntary contribution made by Roman Catholics toward the expenses of the Holy See.
**pet·i·o·lar** (pĕt′ē-ō′lər) *adj.* Of, relating to, or growing on a petiole.
**pet·i·o·late** (pĕt′ē-ə-lāt′, pĕt′ē-ō′lĭt) *adj.* Having a petiole.
**pet·i·ole** (pĕt′ē-ōl′) *n.* [NLat. *petiolus* < LLat., fruit stalk, dim. of Lat. *pes,* foot.] **1.** *Bot.* The stalk by which a leaf is attached to a stem : LEAFSTALK. **2.** *Zool.* The slender stalklike connection between the thorax and abdomen in certain insects.
**pet·i·o·lule** (pĕt′ē-ō-lo͞ol′, pĕt′ē-ŏl′yo͞ol) *n.* [NLat. *petiolulus,* dim. of *petiolus,* petiole. —see PETIOLE.] The stalk of a leaflet in a compound leaf.
**pet·it** *also* **pet·ty** (pĕt′ē) *adj.* [ME, insignificant < OFr., small.] *Law.* Lesser : minor.
**pe·tit bour·geois** *also* **pet·ty bour·geois** (pĕt′ē bo͝or-zhwä′) *n.* [Fr. *petit-bourgeois* : *petit,* small + *bourgeois,* bourgeois.] **1.** A member of the petty bourgeoisie. **2.** Petite bourgeoisie.
**pe·tite** (pə-tēt′) *adj.* [Fr., fem. of *petit* < OFr.] Small, slender, and trim. —Used of a girl or woman. —*n.* A clothing size for short slim women. **—pe·tite′ness** *n.*
**pe·tite bour·geoi·sie** (pə-tēt′ bo͝or′zhwä-zē′) *n.* [Fr. *petite-bourgeoise* : *petite,* small + *bourgeoisie,* bourgeoisie.] The petty bourgeoisie.
**pe·tite mar·mite** (pə-tēt′ mär-mēt′) *n.* [Fr. : *petite,* little + *marmite,* kettle.] A rich consommé made and served in a small covered earthenware casserole.
**pe·tit four** (pĕt′ē fôr′, fōr′) *n., pl.* **pe·tits fours** *or* **pe·tit fours** (pĕt′ē fôrz′, fōrz′) [Fr. : *petit,* little + *four,* oven.] A small rich frosted and decorated tea cake.
**pe·ti·tion** (pə-tĭsh′ən) *n.* [ME *peticion* < OFr. *petition* < Lat. *petitio* < *petere,* to request.] **1.** A solemn entreaty or request to a superior authority : SUPPLICATION. **2.** A formal written document requesting a right or a benefit from a person or group in authority. **3.** *Law.* **a.** A formal written application requesting a court for a specific judicial action, as an appeal. **b.** The judicial action that is asked for in any such request. —*v.* **-tioned, -tion·ing, -tions.** —*vt.* **1.** To address a petition to. **2.** To request formally by petition. —*vi.* To make an entreaty. **—pe·ti′tion·ar′y** (pə-tĭsh′ə-nĕr′ē) *adj.* **—pe·ti′tion·er** *n.*
**pe·ti·ti·o prin·ci·pi·i** (pə-tĭsh′ē-ō′ prĭn-sĭp′ē-ē′) *n.* [Med. Lat. : Lat. *petitio,* request + Lat. *principii,* of the beginning.] *Logic.* The fallacy of assuming in the premise of an argument that which one wishes to prove in the conclusion : begging the question.
**pet·it juror** *also* **pet·ty juror** (pĕt′ē) *n.* A member of a petit jury.
**pet·it jury** *also* **pet·ty jury** (pĕt′ē) *n.* A jury of 12 persons that sits at civil and criminal trials.
**pet·it larceny** *also* **pet·ty larceny** (pĕt′ē) *n.* The theft of objects whose value is below a certain arbitrary standard.
**pe·tit mal** (pĕt′ē mäl′, măl′) *n.* [Fr. : *petit,* small + *mal,* illness.] A form of epilepsy marked by frequent but transient lapses of consciousness and only rare, mild seizures.
**pet·it point** (pĕt′ē point′) *n.* [Fr.] **1.** A small stitch used in needlepoint. **2.** Needlepoint done with a small stitch.
**pet·nap** (pĕt′năp′) *vt.* **-naped, -nap·ing, -naps** *or* **-napped, -nap·ping, -naps.** [PET + (KID)NAP.] To steal (a pet), usu. for profit. **—pet′nap′er, pet′nap′per** *n.*
**petr-** *pref. var. of* PETRO-.
**Pe·trar·chan sonnet** (pĭ-trär′kən) *n.* [After *Petrarch* (1304–1374).] A sonnet form of Italian origin having an octave with the rhyme pattern *abbaabba* and a sestet of various rhyme patterns such as *cdcdcd* or *cdecde.*
**pet·rel** (pĕt′rəl) *n.* [Orig. unknown.] A sea bird of the order Procellariiformes, esp. the storm petrel.

**petrel**
*The black petrel, up to 9 inches long*

**petri-** *pref. var. of* PETRO-.
**Pe·tri dish** (pē′trē) *n.* [After Julius R. *Petri* (1852–1921).] A shallow dish with a loose-fitting cover, used esp. to culture microorganisms for research.
**pet·ri·fac·tion** (pĕt′rə-făk′shən) *also* **pet·ri·fi·ca·tion** (-fĭ-kā′shən) *n.* **1.** The process of petrifying, esp. the conversion of organic matter into stone or a stony substance. **2.** The quality or state of being petrified, as by terror.
**pet·ri·fy** (pĕt′rə-fī′) *v.* **-fied, -fy·ing, -fies.** [OFr. *petrifier* : Lat. *petra,* rock (< Gk.) + Lat. *facere,* to make.] —*vt.* **1.** To convert (organic matters, as wood) into a stony replica by structural impregnation with dissolved minerals. **2.** To cause to become stonelike or stiff : DEADEN. **3.** To stun or paralyze with great fear : TERRIFY. —*vi.* To become stony, esp. by mineral replacement of organic matter.
**Pe·trine** (pē′trīn′) *adj.* [LLat. *Petrus,* Peter + -INE[1].] Of or pertaining to Saint Peter.
**petro-** *or* **petri-** *or* **petr-** *pref.* [< Gk. *petros,* stone and *petra,* rock.] **1.** Rock : stone <*petro*glyph> **2.** Petroleum <*petro*chemistry>
**pet·ro·chem·i·cal** (pĕt′rō-kĕm′ĭ-kəl) *n.* A chemical derived from petroleum or natural gas. **—pet′ro·chem′i·cal** *adj.*
**pet·ro·chem·is·try** (pĕt′rō-kĕm′ĭ-strē) *n.* The chemistry of petroleum and its derivatives.
**pet·ro·dol·lar** (pĕt′rō-dŏl′ər) *n.* A unit of hard currency, as a dollar, held by oil-exporting countries as a result of the sharp increases in oil prices that generated a transfer of purchasing power to oil-exporting nations and a balance-of-payments deficit for oil-importing nations.
**pet·ro·gen·e·sis** (pĕt′rō-jĕn′ĭ-sĭs) *n.* A branch of petrology that deals with the origin of rocks. **—pet′ro·ge·net′ic** (-jə-nĕt′ĭk) *adj.*
**pet·ro·glyph** (pĕt′rə-glĭf′) *n.* An ancient line drawing or carving on rock. **—pet′ro·glyph′ic** *adj.*
**pe·trog·ra·phy** (pə-trŏg′rə-fē) *n.* Description and classification of rocks. **—pe·trog′ra·pher** *n.* **—pet′ro·graph′ic** (pĕt′rə-grăf′ĭk), **pet′ro·graph′i·cal** *adj.* **—pet′ro·graph′i·cal·ly** *adv.*
**pet·rol** (pĕt′rəl) *n.* [Fr. *(essence de) petrole,* (essence of) petroleum < OFr. *petrole,* petroleum < Med. Lat. *petroleum.*] *Chiefly Brit.* Gasoline.
**pet·ro·la·tum** (pĕt′rə-lā′təm, -lä′təm) *n.* [NLat. < Med. Lat. *petroleum,* petroleum.] A colorless-to-amber gelatinous semisolid, obtained from petroleum, made up of various methane and olefin hydrocarbons, and used in medicinal ointments and lubricants.
**pe·tro·le·um** (pə-trō′lē-əm) *n.* [Med. Lat. : Lat. *petra,* rock (< Gk.) + Lat. *oleum,* oil.] A natural, yellow-to-black, thick, flammable liquid hydrocarbon mixture found principally beneath the earth's surface and processed for fractions including natural gas, gasoline, naphtha, kerosene, fuel and lubricating oils, paraffin wax, asphalt, and varied derivative products.
**petroleum jelly** *n.* Petrolatum.
**pe·trol·ic** (pə-trŏl′ĭk) *adj.* [PETROL(EUM) + -IC.] Of, relating to, or derived from petroleum.
**pe·trol·o·gy** (pə-trŏl′ə-jē) *n.* The study of the origin, composition, structure, and alteration of rocks. **—pet′ro·log′ic** (pĕt′rə-lŏj′ĭk), **pet′ro·log′i·cal** *adj.* **—pet′ro·log′i·cal·ly** *adv.* **—pe·trol′o·gist** *n.*
**pet·ro·pol·i·tics** (pĕt′rō-pŏl′ĭ-tĭks) *n. (sing. or pl. in number).* The strategic practice of controlling petroleum sales so as to achieve international political and economic ends and goals.
**pe·tro·sal** (pə-trō′səl) *also* **pet·rous** (pĕt′rəs) *adj.* [< Lat. *petrosus,* rocky. —see PETROUS.] *Anat.* Relating to or located near the portion of the temporal bone that surrounds the inner ear.
**pet·rous** (pĕt′rəs) *adj.* [Lat. *petrosus,* rocky < *petra,* rock < Gk.] **1.** Of, relating to, or resembling rock : HARD. **2.** *var. of* PETROSAL.
**pet·ti·coat** (pĕt′ē-kōt′) *n.* [ME *petycote* : *pety,* small + *cote,* coat.] **1.** An underskirt, esp. a woman's slip. **2.** Something, as a decorative hanging, that resembles a woman's slip. **3.** *Slang.* A woman or girl. —*adj.* **1.** Female : feminine. **2.** Of or by women, as governmental rule. **—pet′ti·coat′ed** *adj.*
**petticoat narcissus** *n.* A small daffodil, *Narcissus bulbocodium,* native to the Mediterranean region and having white or yellow flowers.
**pet·ti·fog** (pĕt′ē-fŏg′, -fôg′) *vi.* **-fogged, -fog·ging, -fogs.** [Back-formation < PETTIFOGGER.] To behave like a pettifogger.
**pet·ti·fog·ger** (pĕt′ē-fŏg′ər, -fô′gər) *n.* [Orig. unknown.] **1.** A petty, quibbling, unscrupulous lawyer. **2.** One who quibbles over trivia.
**pet·ting** (pĕt′ĭng) *n. Informal.* Caressing and kissing.
**pet·ting zoo** (pĕt′ĭng) *n.* A collection of usu. young farm animals and sometimes docile wild animals for children to feed and pet.
**pet·tish** (pĕt′ĭsh) *adj.* [Prob. < PET[2].] Ill-tempered : peevish. **—pet′tish·ly** *adv.* **—pet′tish·ness** *n.*
**pet·ty** (pĕt′ē) *adj.* **-ti·er, -ti·est.** [ME *pety,* alteration of *petit.* —see PETIT.] **1.** Small, trivial, or insignificant <*petty* complaints> **2.** Of contemptibly narrow mind or views. **3.** *Informal.* Mean : spiteful. **4.** Of subordinate or inferior rank. **5.** *Law. var. of* PETIT. **—pet′ti·ly** *adv.* **—pet′ti·ness** *n.*
**petty bourgeois** (bo͝or-zhwä) *n. var. of* PETIT BOURGEOIS.
**petty bour·geoi·sie** (bo͝or′zhwä-zē′) *n.* The class of small business people and tradesmen.

**pet·ty cash** *n.* A small fund of money for incidental expenses, as in an office.
**petty juror** *n. var. of* PETIT JUROR.
**petty jury** *n. var. of* PETIT JURY.
**petty larceny** *n. var. of* PETIT LARCENY.
**petty officer** *n.* A noncommissioned naval officer.
**pet·u·lant** (pĕch′ə-lənt) *adj.* [OFr. < Lat. *petulans* < *petulare, dim. of *petere,* to assail.] **1.** Unreasonably ill-tempered or irritable : PEEVISH. **2.** Contemptuous in speech or behavior. **—pet′u·lance, pet′u·lan·cy** *n.* **—pet′u·lant·ly** *adv.*
**pe·tu·nia** (pĭ-to͞on′yə, -tyo͞on′-) *n.* [NLat. *Petunia,* genus name < obs. Fr. *petun,* tobacco < Tupi *petyn.*] **1.** A widely cultivated plant of the genus *Petunia,* native to South America, with funnel-shaped flowers in colors from white to purple. **2.** A moderate to dark purple.
**pe·tun·tze** *or* **pe·tun·tse** (pə-to͞on′tsē) *n.* [Chin. (Mandarin) *bai¹ dun¹zi⁵* : *bai¹,* white + *dun¹zi⁵,* block of stone.] A feldspar sometimes mixed with kaolin in Chinese porcelain.
**pew** (pyo͞o) *n.* [ME *pewe* < OFr. *puie,* raised seat < Lat. *podia,* pl. of *podium,* balcony < Gk. *podion,* base, dim. of *pous,* foot.] **1.** A bench for seating in a church. **2.** A compartment that provides seating for a number of people in a church.
**pe·wee** *also* **pee·wee** (pē′wē) *n.* [Imit. of its cry.] A small olive-brown North American woodland bird of the genus *Contopus.*
**pe·wit** (pē′wĭt′, pyo͞o′ĭt) *n.* [Imit. of its cry.] The lapwing.
**pew·ter** (pyo͞o′tər) *n.* [ME *pewtre* < OFr. *peutre.*] **1.** Any of numerous silver-gray alloys of tin with various amounts of antimony, copper, and lead, once used widely for fine kitchen utensils and tableware. **2.** Pewter articles as a whole. *—adj.* Made of pewter.
**pew·ter·er** (pyo͞o′tər-ər) *n.* One who makes pewter objects, as candlesticks or tableware.
**pe·yo·te** (pā-ō′tē) *also* **pe·yo·tl** (pā-ōt′l) *n.* [Mex. Sp. < Nahuatl *peyotl.*] **1.** MESCAL 1. **2.** A hallucinatory drug derived from the tubercles of mescal.
**pfen·nig** (fĕn′ĭg) *n., pl.* **pfen·nigs** *or* **pfen·ni·ge** (fĕn′ĭ-gə) [G. < OHG *pfenning.*] —See table at CURRENCY.
**PG** (pē′jē′) *adj.* [Short for P(ARENTAL) G(UIDANCE).] Indicating a motion-picture rating that allows admission of persons of all ages but suggests parental guidance in the case of children.
**pH** (pē′āch′) *n.* [P(OTENTIAL OF) H(YDROGEN).] *Chem.* A measure of the acidity or alkalinity of a solution, numerically equal to 7 for neutral solutions, increasing with increasing alkalinity and decreasing with increasing acidity.
**Phae·dra** (fē′drə, fē′-) *n.* [Lat. < Gk. *Phaidra* < fem. of *phaidros,* shining.] *Gk. Myth.* The wife of Theseus who fell in love with her stepson, Hippolytus.
**Pha·ë·thon** (fā′ə-thŏn′) *n.* [Lat. < Gk. *Phaethōn* < *phaethōn,* shining.] *Gk. Myth.* A son of the sun god Helios who was killed when he drove his father's chariot across the sky.
**pha·e·ton** (fā′ĭ-tən) *n.* [After PHAËTHON.] **1.** A light, open, four-wheeled carriage, usu. drawn by a pair of horses. **2.** A touring car.
**phage** (fāj) *n.* A bacteriophage.
**-phage** *suff.* [Gk. *-phagos* < *phagein,* to eat.] One that eats <macrophage>
**-phagia** *suff.* [Gk. < *phagein,* to eat.] The eating of a specified substance or eating in a specified manner <dys*phagia*>
**phago-** *pref.* [Gk. < *phagein,* to eat.] Eating : consuming <*phago*cyte>
**phag·o·cyte** (făg′ə-sīt′) *n.* A cell, as a leukocyte, that engulfs and digests foreign bodies, as cells and microorganisms, in the bloodstream and tissues. **—phag′o·cyt′ic** (-sĭt′ĭk) *adj.*
**phag·o·cy·tin** (făg′ə-sī′tĭn) *n.* A bactericidal substance liberated by disintegration of phagocytes.
**phag·o·cy·tize** (făg′ə-sĭ-tīz′, -sī′-) *vt.* **-tized, -tiz·ing, -tiz·es.** To ingest by phagocytosis.
**phag·o·cy·to·sis** (făg′ə-sī-tō′sĭs) *n.* [PHAGOCYT(E) + -OSIS.] Envelopment and digestion of foreign bodies, as bacteria, by phagocytes. **—phag′o·cy·tot′ic** (-tŏt′ĭk) *adj.*
**phag·o·some** (făg′ə-sōm) *n.* An intracellular membrane-bound vesicle containing material ingested by the process of phagocytosis.
**-phagous** *suff.* [Lat. *-phagus* < Gk. *-phagos* < *phagein,* to eat.] Feeding on : EATING <ichthyo*phagous*>
**-phagy** *suff.* -PHAGIA.
**pha·lange** (fā′lānj′, fə-lănj′) *n.* [Fr. < Gk. *phalanx.*] PHALANX 3.
**pha·lan·ge·al** (fə-lăn′jē-əl, fā-) *also* **pha·lan·gal** (fə-lăng′gəl, fā-) *or* **pha·lan·ge·an** (fə-lăn′jē-ən, fā-) *adj. Anat.* Of or relating to a phalanx.
**pha·lan·ger** (fə-lăn′jər) *n.* [Prob. < Fr. < Gk. *phalanx,* toe bone.] A small arboreal marsupial of the family Phalangeridae of Australia and neighboring islands, with dense fur and a long tail.
**pha·lanx** (fā′lăngks′) *n., pl.* **pha·lanx·es** *or* **pha·lan·ges** (fə-lăn′jēz, fā-) [Lat. < Gk.] **1.** A close formation of spearmen carrying overlapping shields, developed by Philip II of Macedonia and used by Alexander the Great. **2.** A close-knit or compact body of people, esp. one unified by a common goal <"formed a solid *phalanx* in defence of the Constitution and Protestant religion" —G.M. Trevelyan> **3.** *pl.* **phalanges.** *Anat.* A bone of a finger or toe.
**phal·a·rope** (făl′ə-rōp′) *n.* [Fr. < NLat. *phalaropus* : Gk. *phalaris,* coot + Gk. *pous,* foot.] A wading bird of the family Phalaropodidae, having lobed toes to facilitate swimming.
**phal·lic** (făl′ĭk) *adj.* [Gk. *phallikos* < *phallos,* phallus.] **1.** Of, relating to, or resembling a phallus. **2.** Of or relating to the cult of the phallus as an embodiment of generative power. **—phal′li·cal·ly** *adj.*
**phal·lus** (făl′əs) *n., pl.* **phal·li** (făl′ī′) *or* **phal·lus·es.** [LLat. < Gk. *phallos.*] **1.** *Anat.* **a.** The penis. **b.** The embryonic tissue that develops into either the penis or clitoris. **2.** A representation of the penis and testes as an embodiment of generative power. **3.** *Psychoanal.* The immature penis regarded as the libidinal object of infantile sexuality.
**-phane** *or* **-phan** *suff.* [Gk. *-phanēs,* appearing < *phainesthai,* to appear.] A substance resembling something specified <trypto*phan*>
**phan·er·o·gam** (făn′ər-ə-găm′, fə-nâr′ə-) *n.* [NLat. *phanerogramus* : Gk. *phaneros,* visible + Gk. *gamos,* marriage.] A plant that bears flowers and true seeds. **—phan′er·o·gam′ic** (făn′ər-ə-găm′ĭk, fə-nâr′ə-), **phan′er·og′a·mous** (făn′ə-rŏg′ə-məs) *adj.*
**phan·tasm** (făn′tăz′əm) *n.* [ME *fantasme* < OFr. < Lat. *phantasma* < Gk. < *phantazein,* to make visible < *phainein,* to show.] **1.** PHANTOM 1. **2.** An illusory mental image. **3.** Objective reality as perceived and distorted by the five senses according to Platonic philosophy. **—phan·tas′mal** (-tăz′məl), **phan·tas′mic** (-tăz′mĭk) *adj.*
**phan·tas·ma** (făn-tăz′mə) *n., pl.* **-ma·ta** (-mə-tə). A phantasm.
**phan·tas·ma·go·ri·a** (făn-tăz′mə-gôr′ē-ə, -gōr′-) *also* **phan·tas·ma·go·ry** (-gôr′ē, -gōr′ē) *n.* [Poss. < Gk. *phantasma,* phantasm + *agora,* assembly.] **1.** A rapid, bewildering sequence of fantastic images, as seen in fever or dreams. **2.** Fantastic imagery in art. **—phan·tas′ma·gor′ic** (-gôr′ĭk, -gŏr′-), **phan·tas′ma·gor′i·cal** *adj.*
**phan·tas·ma·ta** (făn-tăz′mə-tə) *n. pl. of* PHANTASMA.
**phan·ta·sy** (făn′tə-sē) *n. var. of* FANTASY.
**phan·tom** (făn′təm) *n.* [ME *fantom* < OFr. *fantosme* < Lat. *phantasma.* —see PHANTASM.] **1.** Something apparently seen, heard, or sensed, but having no physical reality : APPARITION. **2.** An image, as in a daydream or fantasy, having substance only in the mind. *—adj.* Illusory : ghostlike.
**Phar·aoh** *also* **phar·aoh** (fâr′ō, fā′rō) *n.* [LLat. *Pharao* < Gk. *Pharaō* < Heb. *Par'ōh,* of Egyptian orig.] **1.** A king of ancient Egypt. **2.** A tyrant. **—Phar′a·on′ic** (fâr′ā-ŏn′ĭk) *adj.*
**pharaoh ant** *n.* A small reddish ant, *Monomorium pharaonis,* commonly infesting human dwellings.
**phar·i·sa·ic** (făr′ĭ-sā′ĭk) *also* **phar·i·sa·i·cal** (-sā′ĭ-kəl) *adj.* **1. Pharisaic** *also* **Pharisaical.** Of, pertaining to, or typical of the ancient Pharisees. **2.** Hypocritically self-righteous and reproachful. **—phar′i·sa′i·cal·ly** *adv.*
**phar·i·sa·ism** (făr′ĭ-sā-ĭz′əm) *also* **phar·i·see·ism** (-sē-ĭz′əm) *n.* **1. Pharisaism** *also* **Phariseeism.** The doctrines and practices of the Pharisees. **2.** Hypocritical observance of the letter of religious or moral law without regard for the spirit : SANCTIMONIOUSNESS.
**phar·i·see** (făr′ĭ-sē) *n.* [ME *pharise* < OE *farise* < LLat. *pharisaeus* < Gk. *pharisaios* < Aram. *perīshayyā.*] **1. Pharisee.** A member of an ancient Jewish sect emphasizing strict observance of the Mosaic law in both its oral and written form. **2.** A sanctimonious person.
**phar·i·see·ism** (făr′ĭ-sē-ĭz′əm) *n. var. of* PHARISAISM.
**phar·ma·ceu·ti·cal** (fär′mə-so͞o′tĭ-kəl) *also* **phar·ma·ceu·tic** (-tĭk) [< LLat. *pharmaceuticus* < Gk. *pharmakeutikos* < *pharmakeuein,* to administer drugs < *pharmakon,* drug.] *—adj.* Of or relating to pharmacy or pharmacists. *—n.* A drug or medication. **—phar′ma·ceu′ti·cal·ly** *adv.*
**phar·ma·ceu·tics** (fär′mə-so͞o′tĭks) *n.* (*sing. in number*). The science of preparing and dispensing drugs.
**phar·ma·cist** (fär′mə-sĭst) *n.* A person trained in pharmacy : DRUGGIST.
**pharmaco-** *pref.* [< Gk. *pharmakon,* drug, poison.] Drug : medicine <*pharmaco*gnosy>
**phar·ma·co·dy·nam·ics** (fär′mə-kō′dī-năm′ĭks) *n.* (*sing. in number*). Study of drug action on living organisms. **—phar′ma·co′dy·nam′ic** *adj.* **—phar′ma·co′dy·nam′i·cal·ly** *adv.*
**phar·ma·co·ge·net·ics** (fär′mə-kō-jə-nĕt′ĭks) *n.* (*sing. in number*). Study of hereditary influences to drug response. **—phar′ma·co·ge·net′ic** *adj.*
**phar·ma·cog·no·sy** (fär′mə-kŏg′nə-sē) *n.* [PHARMACO- + Gk. *gnōsis,* knowledge.] The branch of pharmacology concerned with crude natural drugs. **—phar′ma·cog′no·sist** *n.* **—phar′ma·cog·nos′tic** (-kŏg-nŏs′tĭk) *adj.*
**phar·ma·co·ki·net·ics** (fär′mə-kō-kĭ-nĕt′ĭks) *n.* (*sing. in number*). Study of the physical absorption, metabolism, and excretion of drugs. **—phar′ma·co·ki·net′ic** *adj.*
**phar·ma·col·o·gy** (fär′mə-kŏl′ə-jē) *n.* The scientific study of drugs, including their composition, uses, and physiological effects. **—phar′ma·co·log′ic** (-kə-lŏj′ĭk), **phar′ma·co·log′i·cal** *adj.* **—phar′ma·co·log′i·cal·ly** *adv.* **—phar′ma·col′o·gist** *n.*
**phar·ma·co·poe·ia** (fär′mə-kə-pē′ə) *n., pl.* **-ias.** [Gk. *pharmakopoiia,* preparation of drugs < *pharmakopoios,* preparing drugs :

*pharmakon*, drug + *poiein*, to make.] **1.** A book containing an official or standard list of drugs along with recommended procedures for their preparation and use. **2.** A collection or stock of drugs. **—phar′ma·co·poe′ial** (-pē′əl) *adj.* **—phar′ma·co·poe′ist** (-pē′ĭst) *n.*
**phar·ma·cy** (fär′mə-sē) *n.*, *pl.* **-cies.** [ME *farmacie* < OFr. < LLat. *pharmacia* < Gk. *pharmakeia* < *pharmakon*, drug.] **1.** The art of preparing and dispensing drugs. **2.** A place where drugs are sold : DRUGSTORE.
**pha·ros** (fâr′ŏs′) *n.* [Lat. < Gk., after *Pharos*, an island in the bay of Alexandria, Egypt, and the site of a famous ancient lighthouse.] A lighthouse.
**pharyng-** *pref. var. of* PHARYNGO-.
**pha·ryn·ge·al** (fə-rĭn′jē-əl, -jəl, făr′ĭn-jē′əl) *also* **pha·ryn·gal** (fə-rĭng′gəl) [< NLat. *pharyngeus* < *pharynx*, pharynx.] *—adj.* Of, relating to, or produced by the pharynx <*pharyngeal* speech sounds> *—n.* A speech sound produced in the pharynx.
**pha·ryn·ges** (fə-rĭn′jēz) *n. var. pl. of* PHARYNX.
**phar·yn·gi·tis** (făr′ĭn-jī′tĭs) *n.* Inflammation of the pharynx.
**pharyngo-** *or* **pharyng-** *pref.* [NLat. < Gk. *pharungo-* < *pharunx*, throat.] Pharynx <*pharyngo*scope>
**phar·yn·gol·o·gy** (făr′ĭn-gŏl′ə-jē) *n.* Medical study of the physiology and pathology of the pharynx.
**pha·ryn·go·scope** (fə-rĭng′gə-skōp′) *n.* An instrument for examining the pharynx. **—phar′yn·gos′co·py** (făr′ĭn-gŏs′kə-pē) *n.*
**phar·yn·got·o·my** (făr′ĭn-gŏt′ə-mē) *n.*, *pl.* **-mies.** Surgical incision of the pharynx.
**phar·ynx** (făr′ĭngks) *n.*, *pl.* **pha·ryn·ges** (fə-rĭn′jēz) *or* **phar·ynx·es.** [NLat. *pharynx, pharyng-* < Gk. *pharunx*, throat.] The section of the digestive tract from the oral cavity to the larynx, where it becomes continuous with the esophagus.
**phase** (fāz) *n.* [NLat. *phasis*, phase of the moon < Gk. < *phainein*, to show.] **1.** A distinct stage of development <"The American occupation of Japan fell into three successive *phases*" —Edwin Reischauer> **2.** A temporary attitude, manner, or behavior pattern <My interest in comic books was a passing *phase.*> **3.** An aspect : facet <every *phase* of the project> **4.** *Astron.* Any of the cyclically recurring apparent forms of the moon or a planet. **5.** *Physics.* **a.** A particular stage in a periodic phenomenon or process. **b.** The elapsed fraction of a complete cycle as measured from a specified reference point and often expressed as an angle. **6.** *Chem.* A discrete homogeneous part of a material system that is mechanically separable from the rest, as steam from water. **7.** *Biol.* A typical form or appearance that occurs in a cycle or distinguishes certain individuals within a group. *—vt.* **phased, phas·ing, phas·es.** To plan or carry out systematically by phases. **—in phase.** In a correlated or synchronized way. **—out of phase.** In an unsynchronized or uncorrelated way. **—phase in.** To introduce by stages. **—phase out.** To eliminate by stages. **—pha′sic** (fā′zĭk) *adj.*
**phase contrast microscope** *also* **phase microscope** *n.* A microscope that represents differences in the phase of light transmitted or reflected by a specimen as variations in contrast in the image.
**phase modulation** *n.* Variation of the phase of a carrier wave by an amount proportional to the amplitude of a modulating signal.
**phase·out** (fāz′out′) *n.* Systematic discontinuation of production or operations through phases <a *phaseout* of government regulation of airline fares>
**phase rule** *n.* A rule stating that the number of degrees of freedom in a material system at equilibrium is equal to the number of components minus the number of phases plus the constant 2; e.g., the system of water vapor, water, and ice has zero degrees of freedom, since three phases of one component coexist.
**-phasia** *suff.* [NLat. < Gk., speech < *phasis*, utterance < *phanai*, to say, speak.] A speech disorder of a specified kind <*aphasia*>
**pheas·ant** (fĕz′ənt) *n.*, *pl.* **pheas·ants** *or* **pheasant.** [ME *fesant* < OFr. *fesan* < Lat. *phasianus* < Gk. *phasianos (ornis)*, (bird) of the Phasis River < *Phasis*, the Phasis River in the Caucasus.] **1.** Any of various chickenlike birds of the family Phasianidae, indigenous to the Old World, with long tails and, in the males of many species, highly colorful plumage. **2.** A ruffed grouse.
**phel·lem** (fĕl′əm, -ĕm′) *n.* [< Gk. *phellos*, cork.] CORK 4.
**phel·lo·derm** (fĕl′ə-dûrm′) *n.* [Gk. *phellos*, cork + -DERM.] The soft green cortex tissue that forms on the inner side of the phellogen of some trees. **—phel′lo·der′mal** *adj.*
**phel·lo·gen** (fĕl′ə-jən) *n.* [Gk. *phellos*, cork + -GEN.] Tissue in woody plants from which cork and phelloderm develop. **—phel′lo·ge·net′ic** (-jə-nĕt′ĭk), **phel′lo·gen′ic** *adj.*
**phen-** *pref. var. of* PHENO-.
**phe·na·caine** (fē′nə-kān′, fĕn′ə-) *n.* [Alteration of PHENO- + -CAINE.] A white crystalline compound, $C_{18}H_{22}N_2O_2$, used in its hydrochloride form as a local anesthetic.
**phe·nac·e·tin** (fə-năs′ĭ-tĭn) *also* **phe·nac·e·tine** (-tēn′) *n.* Acetophenetidin.
**phen·a·cite** (fĕn′ə-sīt′) *n.* [< Gk. *phenax, phenak-*, imposter.] A natural beryllium silicate, $Be_2SiO_4$, occurring as vitreous crystals sometimes used as gems.
**phe·nan·threne** (fə-năn′thrēn′) *n.* [PHEN(O)- + ANTHR(AC)ENE.] A colorless crystalline compound, $C_{14}H_{10}$, derived from coal-tar oils and used in drugs, dyes, and explosives.
**phen·ar·sa·zine chloride** (fə-när′sə-zēn′) *n.* [PHEN(O)- + ARS(ENIC) + AZINE.] A highly poisonous yellow crystalline compound, $C_{12}H_9AsClN$, used as a poison gas and occas. with tear gas.
**phen·a·zine** (fĕn′ə-zēn′) *also* **phen·a·zin** (-zĭn) *n.* A yellow crystalline compound, $C_6H_4N_2C_6H_4$, used to manufacture azine dyes.
**phen·cy·cli·dine** (fĕn-sī′klĭ-dēn′, -dĭn, -sĭk′lĭ-) *n.* [PHEN(O)- + CYCL(O)- + -ID(E) + -INE.] A drug, $C_{17}H_{25}N$, used in veterinary medicine as an anesthetic and illegally as a hallucinogen.
**phe·nix** (fē′nĭks) *n. var. of* PHOENIX.
**pheno-** *or* **phen-** *pref.* [< Gk. *phainein*, to show.] **1.** Showing : displaying <*pheno*cryst> **2. a.** Related to or derived from benzene <*phen*ol> **b.** Containing phenyl <*pheno*thiazine>
**Phe·no·bar·bi·tal** (fē′nō-bär′bĭ-tôl′). A trademark for a crystalline compound, $C_{12}H_{12}N_2O_3$, used in medicine as a sedative and hypnotic.
**phe·no·cop·y** (fē′nə-kŏp′ē) *n.*, *pl.* **-ies.** [PHENO(TYPE) + COPY.] An environmentally induced phenotypic variation bearing close resemblance to a genotype other than its own.
**phe·no·cryst** (fē′nə-krĭst′) *n.* [PHENO- + CRYST(AL).] A relatively large isolated crystal embedded in porphyry. **—phe′no·crys′tic** *adj.*
**phe·nol** (fē′nōl′, -nōl′) *n.* **1.** A caustic, poisonous, white crystalline compound, $C_6H_5OH$, derived from benzene and used in resins, disinfectants, plastics, and pharmaceuticals. **2.** Any of a class of aromatic organic compounds having at least one hydroxyl group attached directly to the benzene ring.
**phe·no·lic** (fĭ-nō′lĭk, -nŏl′ĭk) *adj.* Of, relating to, containing, or derived from phenol.
**phenolic resin** *n.* Any of various synthetic thermosetting resins, obtained by the reaction of phenols with simple aldehydes and used as coatings and adhesives and in making molded products.
**phe·nol·o·gy** (fĭ-nŏl′ə-jē) *n.* [PHENO(MENON) + -LOGY.] Study of periodic biological phenomena, as breeding, flowering, and migration, esp. as related to climate. **—phe′no·log′i·cal** (fē′nə-lŏj′ĭ-kəl) *adj.* **—phe·nol′o·gist** *n.*
**phe·nol·phthal·ein** (fē′nōl-thăl′ēn′, -thăl′ē-ĭn, -thā′lēn′, -thā′-lē-ĭn) *n.* A pale-yellow crystalline powder, $(C_6H_4OH)_2C_2O_2C_6H_4$, used as an acid-base indicator, in making dyes, and as a cathartic.
**phe·nom·e·nal** (fĭ-nŏm′ə-nəl) *adj.* **1.** Of, relating to, or being a phenomenon or phenomena. **2.** Extraordinary or outstanding : EXCEPTIONAL <a *phenomenal* vocabulary for a two-year-old> **3.** *Philos.* Known or derived through the senses rather than through the mind. **—phe·nom′e·nal·ly** *adv.*
**phe·nom·e·nal·ism** (fĭ-nŏm′ə-nə-lĭz′əm) *n. Philos.* The theory, proposed by Hume and his successors, that concepts and precepts constitute the sole object of knowledge, the objects of perception themselves, their origin outside the mind, or the nature of the mind itself remaining beyond inquiry. **—phe·nom′e·nal·ist** *n.* **—phe·nom′e·nal·is′tic** *adj.* **—phe·nom′e·nal·is′ti·cal·ly** *adv.*
**phe·nom·e·nol·o·gy** (fĭ-nŏm′ə-nŏl′ə-jē) *n.* **1.** The study of human awareness in which considerations of objective reality and purely subjective response are temporarily left out of account. **2.** *Philos.* A movement based on phenomenology, originated about 1905 by the German philosopher Edmund Husserl. **—phe·nom′e·no·log′i·cal** (-nə-lŏj′ĭ-kəl) *adj.* **—phe·nom′e·no·log′i·cal·ly** *adv.* **—phe·nom′e·nol′o·gist** *n.*
**phe·nom·e·non** (fĭ-nŏm′ə-nŏn′, -nən) *n.*, *pl.* **-na** (-nə) *or* **-nons.** [LLat. *phaenomenon* < Gk. *phainomenon* < *phainomenos*, pr.part. of *phainesthai*, to appear < *phainein*, to show.] **1.** *pl.* **phenomena.** An occurrence or fact directly perceptible by the senses. **2.** *pl.* **phenomenons. a.** An unusual, unaccountable, or remarkable fact or occurrence. **b.** One outstanding for a quality or achievement : PARAGON <a runner who was a *phenomenon* of endurance> **3.** *pl.* **phenomena.** *Philos.* That which appears real to the senses, regardless of whether its underlying existence is proved or its nature understood. **4.** *pl.* **phenomena.** *Physics.* An observable event.
**phe·no·thi·a·zine** (fē′nō-thī′ə-zēn′) *n.* A greenish organic compound, $C_{12}H_9NS$, used in dyes, anthelmintics, and insecticides.
**phe·no·type** (fē′nə-tīp′) *n.* [G. *Phänotypus* : Gk. *phainein*, to show + Gk. *typos*, type.] **1.** The genetically and environmentally determined physical appearance of an organism, esp. as considered with respect to all possible genetically influenced expressions of one specific character. **2.** An individual or group of organisms displaying a particular phenotype. **—phe′no·typ′ic** (-tĭp′ĭk), **phe′no·typ′i·cal** *adj.* **—phe′no·typ′i·cal·ly** *adv.*
**phe·nox·ide** (fĭ-nŏk′sīd′) *n.* A salt of carbolic acid.
**phen·yl** (fĕn′əl, fē′nəl) *n.* An organic radical, $C_6H_5$, derived from benzene by the removal of one hydrogen atom. **—phe·nyl′ic** (fĭ-nĭl′ĭk) *adj.*
**phen·yl·al·a·nine** (fĕn′əl-ăl′ə-nēn′, fē′nəl-) *n.* An amino acid, $C_6H_5CH_2CH(NH_2)COOH$, occuring naturally in many proteins and extracted for use as a dietary supplement.
**phen·yl·bu·ta·zone** (fĕn′əl-byōō′tə-zōn′) *n.* A white compound,

$C_{19}H_{20}N_2O_2$, used as an anti-inflammatory drug and an analgesic in the treatment of arthritis.

**phen·yl·ene** (fĕn′ə-lēn′, fē′nə-) *n.* The organic radical $C_6H_4$, derived from benzene by removal of two hydrogen atoms.

**phen·yl·ke·to·nu·ri·a** (fĕn′əl-kēt′n-o͝or′ē-ə, -yo͝or′-, fē′nəl-) *n.* A hereditary disorder of phenylalanine metabolism marked by mental retardation and brain damage due to accumulation of toxic metabolic products.

**pher·o·mone** (fĕr′ə-mōn′) *n.* [Gk. *pherein,* to carry + (HOR)MONE.] A chemical substance secreted by an animal that influences specific patterns of behavior by other members of the same species.

**phew** (fyo͞o) *interj.* —Used to express relief, fatigue, discomfort, surprise, or disgust.

**phi** (fī) *n.* [Med. Gk. < Gk. *phei.*] The 21st letter of the Greek alphabet. —See table at ALPHABET.

**phi·al** (fī′əl) *n.* [ME *fiol* < OFr. *fiole* < Lat. *phiala,* shallow vessel < Gk. *phialē.*] A vial.

**Phi Be·ta Kap·pa** (fī′ bā′tə kăp′ə) *n.* [< the initials of the society's Gk. motto *philosophia biou kubernētēs,* philosophy the guide of life.] **1.** An honorary society of college students and graduates whose members are chosen on the basis of academic achievement. **2.** A member of Phi Beta Kappa.

**phil-** *pref. var. of* PHILO-.

**-phil** *suff. var. of* -PHILE.

**Phil·a·del·phi·a lawyer** (fĭl′ə-dĕl′fē-ə) *n.* [After *Philadelphia,* Pennsylvania.] A shrewd and unscrupulous lawyer who manipulates legal technicalities.

**Philadelphia pepper pot** *n.* PEPPER POT 1.

**phi·lan·der** (fĭ-lăn′dər) *vi.* **-dered, -der·ing, -ders.** [< obs. *philander,* lover < Gk. *philandros,* loving men : *philos,* loving + *anēr,* man.] To engage in casual love affairs. **—phi·lan′der·er** *n.*

**phi·lan·thro·py** (fĭ-lăn′thrə-pē) *n., pl.* **-pies.** [LLat. *philanthropia* < Gk. *philanthrōpia* < *philanthrōpos,* loving mankind : *philos,* loving + *anthrōpos,* mankind.] **1.** Works or endeavor, as charitable aid or endowments, intended to increase the well-being of humanity. **2.** Love of humanity in general. **3.** An institution established to promote human welfare. **—phil′an·throp′ic** (fĭl′ən-thrŏp′ĭk), **phil′an·throp′i·cal** *adj.* **—phi·lan′thro·pist** *n.*

**phi·lat·e·ly** (fĭ-lăt′l-ē) *n.* [Fr. *philatélie* : Gk. *philos,* loving + Gk. *ateleia,* exemption from payment (*a-,* without + *telos,* charge).] Postage stamp collecting. **—phil′a·tel′ic** (fĭl′ə-tĕl′ĭk), **phil′a·tel′i·cal** *adj.* **—phil′a·tel′i·cal·ly** *adv.* **—phi·lat′e·list** *n.*

**-phile** *or* **-phil** *suff.* [Partly < Fr. *-phile,* and partly < NLat. *-philus* (< Lat.), both < Gk. *-philos* < *philos,* beloved, dear.] **1.** One that loves or has a strong affinity or preference for <audio*phile*> **2.** Having a strong affinity or preference for : LOVING <Franco*phile*>

**Phi·le·mon** (fĭ-lē′mən, fī-) *n.* —See table at BIBLE.

**phil·har·mon·ic** (fĭl′här-mŏn′ĭk, fĭl′ər-) *adj.* [Fr. *philharmonique* < Ital. *filarmonico* : Gk. *philos,* loving + *armonico,* harmonic < Lat. *harmonicus.* —see HARMONIC.] **1.** Dedicated to or appreciative of music. **2.** Relating to a symphony orchestra. —*n. also* **Philharmonic.** A symphony orchestra or its supporting organization.

**phil·hel·lene** (fĭl-hĕl′ēn′) *also* **phil·hel·len·ist** (fĭl-hĕl′ə-nĭst) *n.* [Gk. *philellēn* : *philos,* loving + *Hellēn,* a Greek.] An admirer of Greece or the Greeks. **—phil′hel·len′ic** (fĭl′hĕ-lĕn′ĭk) *adj.* **—phil·hel′len·ism** *n.*

**-philia** *suff.* [NLat. < Gk. *philia,* friendship < *philos,* loving.] **1.** Tendency toward <hemo*philia*> **2.** Abnormal attraction to <necro*philia*>

**-philiac** *suff.* [NLat. *-philia,* -philia + Gk. *-akos,* adj. suffix.] **1.** One that has a tendency toward <hemo*philiac*> **2.** One that has an abnormal attraction to <copro*philiac*>

**-philic** *suff.* -PHILOUS.

**Phi·lip·pi·ans** (fĭ-lĭp′ē-ənz) *n.* (*sing. in number*). —See table at BIBLE.

**Phi·lip·pic** (fĭ-lĭp′ĭk) *n.* **1.** Any of the orations of Demosthenes against Philip of Macedonia in the 4th cent. B.C. **2.** Any of the orations of Cicero against Antony in 44 B.C. **3. philippic.** An angry or bitter verbal denunciation : DIATRIBE.

**Phil·ip·pine mahogany** (fĭl′ə-pēn′) *n.* A Philippine hardwood tree of the genus *Shorea* or related genera, with wood similar to the mahogany.

**Phil·is·tine** (fĭl′ĭ-stēn′, fĭ-lĭs′tĭn, -tēn′) *n.* [ME < LLat. *Philistini,* Philistines < LGk. *Philistinoi* < Heb. *pelishtīm* < *pelesheth,* Philistia.] **1.** One of the people of ancient Philistia. **2. a.** A smug, ignorant person regarded as antagonistic or indifferent to artistic and cultural values. **b.** One who lacks knowledge in a given area. —*adj. often* **philistine.** Boorish : ignorant.

**Phil·is·tin·ism** *also* **phi·lis·tin·ism** (fĭl′ĭ-stē-nĭz′əm, fĭ-lĭs′tə-nĭz′əm, -tē-nĭz′əm) *n.* Smug conventionalism.

**phil·lu·men·ist** (fə-lo͞o′mə-nĭst) *n.* [PHIL(O)- + Lat. *lumen,* light + -IST.] A collector of matchbooks or matchboxes.

**philo-** *or* **phil-** *pref.* [NLat. < Gk. < *philos,* beloved, dear.] Having a strong affinity or preference for : LOVING <*philo*progenitive>

**phil·o·den·dron** (fĭl′ə-dĕn′drən) *n., pl.* **-drons** *or* **-dra** (-drə) [NLat. *Philodendron,* genus name < Gk., neuter of *philodendros,* loving trees : *philos,* loving + *dendron,* tree.] A climbing tropical American vine of the genus *Philodendron,* widely cultivated as a house plant.

▲ word history: The name *philodendron* literally means "fond of trees." It was given to a genus of tropical climbing plants because in their native habitat they twine around trees.

**phi·lol·o·gy** (fĭ-lŏl′ə-jē) *n.* [Fr. *philologie* < Lat. *philologia,* love of learning < Gk. < *philologos,* loving learning : *philos,* loving + *logos,* reason, speech.] **1.** Historical linguistics. **2.** Literary or classical scholarship. **—phi·lol′o·ger, phi·lol′o·gist** *n.* **—phil′o·log′ic** (fĭl′ə-lŏj′ĭk), **phil′o·log′i·cal** *adj.* **—phil′o·log′i·cal·ly** *adv.*

**phil·o·mel** (fĭl′ə-mĕl′) *n.* [After PHILOMELA.] A nightingale.

**Phil·o·me·la** (fĭl′ə-mē′lə) *n.* [Lat. < Gk. *Philomelē.*] *Gk. Myth.* An Athenian princess who, after being raped and having her tongue cut out by Tereus, was turned into a nightingale.

**phil·o·pro·gen·i·tive** (fĭl′ō-prō-jĕn′ĭ-tĭv) *adj.* **1.** Producing many offspring : PROLIFIC. **2.** Loving one's own offspring or children as a whole. **3.** Of or pertaining to love of children. **—phil′o·pro·gen′i·tive·ly** *adv.* **—phil′o·pro·gen′i·tive·ness** *n.*

**phi·lo·sophe** (fē′lə-zôf′) *n.* [Fr. < OFr., philosopher.] One of the prominent philosophical, social, and political thinkers of the 18th-cent. French Enlightenment.

**phi·los·o·pher** (fĭ-lŏs′ə-fər) *n.* [ME < OFr. *philosophe* < Lat. *philosophus* < Gk. *philosophos,* loving wisdom, philosopher : *philos,* loving + *sophia,* wisdom.] **1.** A student or scholar of philosophy. **2.** An author or adherent of a particular philosophy. **3.** One who is calm and rational under any circumstances.

**philosophers' stone** *also* **philosopher's stone** *n.* An imagined substance regarded as having the power to transmute base metal into gold.

**phil·o·soph·i·cal** (fĭl′ə-sŏf′ĭ-kəl) *also* **phil·o·soph·ic** (-ĭk) *adj.* **1.** Of, relating to, or based on a philosophy. **2.** Typical of a philosopher, as in enlightenment and wisdom. **—phil′o·soph′i·cal·ly** *adv.*

**phi·los·o·phize** (fĭ-lŏs′ə-fīz′) *v.* **-phized, -phiz·ing, -phiz·es.** —*vi.* **1.** To speculate in a philosophical way. **2.** To set forth or express a moralistic, often superficial philosophy. —*vt.* To consider (a matter) from a philosophical standpoint. **—phi·los′o·phiz′er** *n.*

**phi·los·o·phy** (fĭ-lŏs′ə-fē) *n., pl.* **-phies.** [ME *philosophie* < OFr. < Lat. *philosophia* < Gk. < *philosophos,* loving wisdom, philosopher.] **1. a.** Love and pursuit of wisdom by intellectual investigation and moral self-discipline. **b.** Inquiry into laws and causes underlying reality. **c.** A system of philosophical investigation or exposition. **2.** Inquiry into the nature of things based on logical reasoning rather than empirical methods. **3.** Analysis and critique of fundamental beliefs as they come to be conceptualized and formulated. **4.** The synthesis of all learning. **5.** Investigation of natural phenomena and its systematization in theory and experiment, as in astrology, astronomy, or alchemy <hermetic *philosophy*><natural *philosophy*> **6.** All learning except practical arts and technical precepts. **7.** All the disciplines presented in university curriculums of science and the liberal arts, except law, medicine, and theology <Doctor of *Philosophy*> **8.** The science that encompasses aesthetics, ethics, metaphysics, logic, and epistemology. **9.** A system of fundamental or motivating principles : basis of action or belief <a *philosophy* of government> **10.** A general viewpoint : THEORY <My *philosophy* of travel is to go in style.> **11.** The overall values by which one lives <a *philosophy* of life> **12.** The calmness, equanimity, and detachment regarded as befitting a philosopher.

**-philous** *suff.* [< Gk. *philos,* beloved, dear.] Having a strong affinity or preference for : LOVING <anemo*philous*>

**phil·ter** *also* **phil·tre** (fĭl′tər) [Fr. *philtre* < OFr. < Lat. *philtrum* < Gk. *philtron* < *philein,* to love.] —*n.* **1.** A love potion. **2.** A magic charm or potion. —*vt.* **-tered, -ter·ing, -ters** *also* **-tred, -tring, -tres.** To charm or seduce with or as if with a philter.

**phleb-** *pref. var. of* PHLEBO-.

**phle·bi·tis** (flĭ-bī′tĭs) *n.* Inflammation of a vein or veins. **—phle·bit′ic** (-bĭt′ĭk) *adj.*

**phlebo-** *or* **phleb-** *pref.* [Gk. < *phleps,* blood vessel.] Vein <*phlebo*logy>

**phle·bog·ra·phy** (flĭ-bŏg′rə-fē) *n.* Roentgenography of a vein following injection of a radiopaque substance. **—phle′bo·gram′** (flē′bə-grăm′) *n.* **—phle′bo·graph′ic** *adj.*

**phle·bol·o·gy** (flĭ-băl′ə-jē) *n.* Medical study of the anatomy, functioning, and diseases of veins.

**phle·bot·o·mize** (flĭ-bŏt′ə-mīz′) *vt.* **-mized, -miz·ing, -miz·es.** To perform a phlebotomy on.

**phle·bot·o·my** (flĭ-bŏt′ə-mē) *n., pl.* **-mies.** [ME *flebotomie* < OFr. < LLat. *phlebotomia* < Gk. *phlebotomia* < *phlebotomos,* opening a vein : *phleps,* vein + *-tomos,* cutting < *temnein,* to cut.] The medical practice of opening a vein to draw blood. **—phleb′o·tom′ic** (flĕb′ə-tŏm′ĭk), **phleb′o·tom′i·cal** *adj.* **—phle·bot′o·mist** *n.*

**Phleg·e·thon** (flĕg′ə-thŏn′) *n.* [Lat. < Gk. *Phlegethōn* < *phlegethein,* to blaze < *phlegein,* to burn.] *Gk. Myth.* A river of fire, one of the six rivers of Hades.

**phlegm** (flĕm) *n.* [ME *fleume* < OFr. < LLat. *phlegma,* clammy humor of the body < Gk. *phlegma,* clammy humor of the body, heat < *phlegein,* to burn.] **1.** *Physiol.* Thick, heavy mucus secreted by the respiratory mucosa. **2.** One of the four humors of ancient and medieval physiology. **3.** Sluggish or torpid temperament. **4.** Calm self-possession : EQUANIMITY. **—phlegm'y** (flĕm'ē) *adj.*

**phleg·mat·ic** (flĕg-măt'ĭk) *also* **phleg·mat·i·cal** (-ĭ-kəl) *adj.* [ME *fleumatike* < OFr. *flaumatique* < LLat. *phlegmaticus,* full of phlegm < Gk. *phlegmatikos* < *phlegma,* clammy humor of the body, heat < *phlegein,* to burn.] **1.** Of or relating to phlegm : PHLEGMY. **2.** Having or indicating a calm, sluggish temperament.

**phlo·em** (flō'ĕm') *n.* [G. < Gk. *phloios,* bark.] The food-conducting tissue of vascular plants, made up of sieve tubes and other cellular material.

**phlo·gis·tic** (flō-jĭs'tĭk) *adj.* **1.** Of or pertaining to phlogiston. **2.** *Med.* Of or relating to fever or inflammation.

**phlo·gis·ton** (flō-jĭs'tŏn', -tən) *n.* [NLat. < Gk., neuter of *phlogistos,* inflammable < *phlogizein,* to set on fire < *phlox,* flame.] A hypothetical substance formerly regarded as a volatile constituent of all combustible material and released in combustion as flame.

**phlog·o·pite** (flŏg'ə-pīt') *n.* [G. *Phlogopit* < Gk. *phlogōpos,* fiery-looking : *phlox,* flame + *ōps,* face.] A yellow to dark-brown mica, $KMg_3AlSi_3O_{10}(OH)_2$, used in insulation.

**phlox** (flŏks) *n., pl.* **phlox** *or* **phlox·es.** [NLat. *Phlox,* genus name < Lat. *phlox,* a kind of flower < Gk. wallflower, flame.] A New World plant of the genus *Phlox,* with lance-shaped leaves and white, red, or purple flower clusters.

**phlyc·te·na** *also* **phlyc·tae·na** (flĭk-tē'nə) *n., pl.* **-nae** (-nē) [Gk. *phluktaina* < *phluzein,* to boil over.] A small blister : VESICLE.

**-phobe** *suff.* [< Gk. *-phobos,* fearing < *phobos,* fear.] One that fears or is averse to a specified thing <ailuro*phobe*>

**pho·bi·a** (fō'bē-ə) *n.* [NLat. < LLat. *-phobia,* -phobia.] **1.** A persistent, illogical, or abnormal fear of a specific thing or situation. **2.** A strong dislike or aversion. **—pho'bic** (fō'bĭk) *adj.*

**-phobia** *suff.* [NLat. < LLat. < Gk. < *phobos,* fear.] An intense, illogical, or abnormal fear of a specified thing <xeno*phobia*>

**-phobic** *suff.* [Fr. *-phobique* < LLat. *-phobicus* < Gk. *-phobikos* < *-phobia,* -phobia.] **1.** Having a fear of or an aversion for <xeno*phobic*> **2.** Lacking an affinity for <lyo*phobic*>

**Pho·bos** (fō'bŏs', fŏb'ŏs') *n.* [Gk. *phobos,* fear.] The inner and larger of the two satellites of the planet Mars.

**-phobous** *suff.* -PHOBIC.

**phoe·be** (fē'bē) *n.* [Poss. alteration of PEWIT.] A small dull-colored North American bird of the genus *Sayornis.*

**Phoe·be** (fē'bē) *n.* [Lat. < Gk. *Phoibē* < *phoibos,* shining.] **1.** *Gk. Myth.* The goddess Artemis. **2.** The moon. **3.** *Astron.* The ninth satellite of Saturn.

**Phoe·bus** (fē'bəs) *n.* [Lat. < Gk. *Phoibos* < *phoibos,* shining.] **1.** *Gk. Myth.* The god of the sun. **2.** The sun.

**Phoe·ni·cian** (fĭ-nĭsh'ən, -nē'shən) *n.* [ME *Phenecien* < OFr. < Lat. *Phoenicius* < Gk. *phoinix,* Phoenician, phoenix.] **1.** A native, resident, or subject of ancient Phoenicia. **2.** The Semitic language of ancient Phoenicia. **—Phoe·ni'cian** *adj.*

**phoe·nix** *also* **phe·nix** (fē'nĭks) *n.* [ME *fenix* < OFr. < Lat. *phoenix* < Gk. *phoinix.*] **1.** A mythological Egyptian bird that after 500 years consumed itself by fire and rose renewed from its ashes. **2.** One of unsurpassed beauty or excellence : PARAGON. **3. Phoenix.** A constellation in the Southern Hemisphere. **—phoe'nix·like'** *adj.*

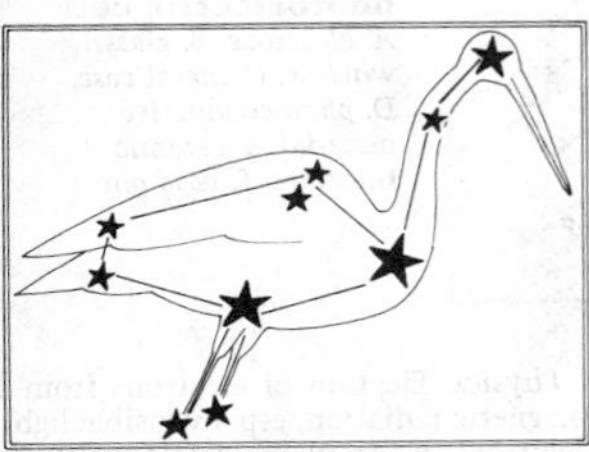

**Phoenix**

**phon-** *pref. var. of* PHONO-.

**pho·nate** (fō'nāt') *v.* **-nat·ed, -nat·ing, -nates.** *—vi.* To utter speech sounds. *—vt.* To utter (a sound). **—pho·na'tion** *n.*

**phone¹** (fōn) *n.* [Gk. *phōnē,* sound.] An individual speech sound.

**phone²** (fōn) [Short for TELEPHONE.] *Informal. —n.* **1.** A telephone. **2.** An earphone. *—v.* **phoned, phon·ing, phones.** *—vi.* To telephone. *—vt.* **1.** To telephone (someone). **2.** To impart (e.g., news or information) by telephone <*phoned* my story to the city editor>

**-phone** *suff.* [< Gk. *phōnē,* sound, voice.] **1.** Sound <allo*phone*> **2.** Device that receives or emits sound <geo*phone*>

**pho·ne·mat·ic** (fō'nĭ-măt'ĭk) *adj.* Phonemic.

**pho·neme** (fō'nēm') *n.* [Fr. *phonème* < Gk. *phōnēma,* utterance < *phōnein,* to speak.] One of the set of the smallest units of speech, as the *m* of *mat* and the *b* of *bat* in English, that distinguish one utterance or word from another in a given language.

**pho·ne·mic** (fə-nē'mĭk, fō-) *adj.* **1.** Of or pertaining to phonemes. **2.** Of or pertaining to phonemics. **3.** Serving to make a unit of speech distinctive. **—pho·ne'mi·cal·ly** *adv.*

**pho·ne·mics** (fə-nē'mĭks, fō-) *n. (sing. in number).* The study and description of the phonemes of a language. **—pho·ne'mi·cist** (-mĭ-sĭst) *n.*

**pho·net·ic** (fə-nĕt'ĭk) *adj.* [Gk. *phōnētikos* < *phōein,* to speak.] **1.** Of or relating to phonetics. **2.** Relating to a system for representing speech sounds in which each symbol denotes only one sound <*phonetic* spelling> **—pho·net'i·cal·ly** *adv.*

**phonetic alphabet** *n.* **1.** A standardized set of symbols used in phonetic transcription. **2.** Any of various systems of code words for identifying letters of the alphabet in voice communication.

**pho·ne·ti·cian** (fō'nĭ-tĭsh'ən) *also* **pho·net·i·cist** (fə-nĕt'ĭ-sĭst) *n.* An expert in phonetics.

**pho·net·ics** (fə-nĕt'ĭks) *n. (sing. in number).* **1.** The branch of linguistics concerned with the study of speech sounds and their production, description, combination, and representation by written symbols. **2.** The system of sounds of a particular language.

**pho·ney** (fō'nē) *adj. & n. var. of* PHONY.

**-phonia** *suff.* [Gk. *-phōnia,* sound < *phōnē.*] Speech disorder of a specified kind <dys*phonia*>

**phon·ic** (fŏn'ĭk) *adj.* Of, relating to, or having the nature of sound, esp. speech sounds. **—phon'i·cal·ly** *adv.*

**phon·ics** (fŏn'ĭks) *n. (sing. in number).* **1.** The scientific study of sound : ACOUSTICS. **2.** Use of elementary phonetics in the teaching of reading.

**phono-** *or* **phon-** *pref.* [< Gk. *phōnē,* sound, voice.] Sound : voice : speech <*phono*logy>

**pho·no·car·di·o·gram** (fō'nə-kär'dē-ə-grăm') *n.* A graphic record of heart sounds made by a phonocardiograph.

**pho·no·car·di·o·graph** (fō'nə-kär'dē-ə-grăf') *n.* An instrument for recording heart sounds, used in medical diagnosis. **—pho'no·car'di·o·graph'ic** *adj.* **—pho'no·car'di·og'ra·phy** (-ŏg'rə-fē) *n.*

**pho·no·gram** (fō'nə-grăm') *n.* A symbol or character, as in a phonetic alphabet, representing a word or phoneme in speech. **—pho'no·gram'ic, pho'no·gram'mic** *adj.* **—pho'no·gram'i·cal·ly, pho'no·gram'mi·cal·ly** *adv.*

**pho·no·graph** (fō'nə-grăf') *n.* An instrument that uses a vibrating needle to reproduce recorded sound from the grooves of a disc or cylinder. **—pho'no·graph'ic** *adj.* **—pho'no·graph'i·cal·ly** *adv.*

**pho·nog·ra·phy** (fə-nŏg'rə-fē, fō-) *n.* **1.** Representation of speech or speech sounds by means of phonograms : phonetic transcription. **2.** A system of shorthand based on phonetic transcription. **—pho·nog'ra·pher, pho·nog'ra·phist** *n.*

**pho·no·lite** (fō'nə-līt') *n.* [Fr. < G. *Phonolith* : Gk. *phōnē,* sound + Gk. *lithos,* stone.] *Mineral.* Volcanic rock made up chiefly of nepheline and orthoclase. **—pho'no·lit'ic** (fō'nə-lĭt'ĭk) *adj.*

**pho·nol·o·gy** (fə-nŏl'ə-jē, fō-) *n.* Study of speech sounds, including phonetics and phonemics. **—pho·no·log'ic** (fō'nə-lŏj'ĭk), **pho'no·log'i·cal** *adj.* **—pho'no·log'i·cal·ly** *adv.* **—pho·nol'o·gist** *n.*

**pho·non** (fō'nŏn') *n. Physics.* The quantum of acoustic or vibrational energy, regarded as a discrete particle and used esp. in mathematical models to calculate vibrational and thermal properties of solids.

**pho·no·re·cep·tion** (fō'nō-rĭ-sĕp'shən) *n.* Perception of or response to sound waves. **—pho'no·re·cep'tor** (-tər) *n.*

**pho·no·scope** (fō'nə-skōp') *n.* A device producing a visible display of the mechanical properties of a sounding body, esp. of musical instruments.

**pho·no·type** (fō'nə-tīp') *n.* **1.** A phonetic symbol used in printing. **2.** Text printed in phonetic symbols. **—pho'no·typ'ic** (fō'nə-tĭp'ĭk), **pho'no·typ'i·cal** *adj.* **—pho'no·typ'i·cal·ly** *adv.*

**pho·no·typ·y** (fō'nə-tī'pē) *n.* Transcription of speech sounds by means of phonetic symbols. **—pho'no·typ'ist** *n.*

**pho·ny** *also* **pho·ney** (fō'nē) [Orig. unknown.] *Informal. —adj.* **-ni·er, -ni·est.** Not real or genuine : FALSE <a *phony* promise> *—n., pl.* **-nies** *also* **-neys.** **1.** Something not genuine : FAKE. **2.** An insincere or hypocritical person. **—pho'ni·ly** *adv.* **—pho'ni·ness** *n.*

**-phony** *suff.* [Gk. *-phōnia* < *phōnē,* sound.] Sound <tele*phony*>

**phoo·ey** (fo͞o'ē) *interj.* —Used to express disgust or contempt.

**pho·rate** (fôr'āt', fōr'-) *n.* [(PHOS)PHOR(OUS) + (THION)ATE.] A toxic liquid, $C_7H_{17}O_2PS_3$, used as an insecticide.

**-phore** *suff.* [< Gk. *-phoros,* bearing < *pherein,* to bear.] Bearer : carrier <chromato*phore*>

**-phoresis** *suff.* [Gk. *phorēsis,* act of bearing < *phorein,* to carry, freq. of *pherein,* to bear.] Transmission <electro*phoresis*>

**-phorous** *suff.* [Gk. *-phoros* < *pherein,* to bear.] Bearing or producing <gono*phorous*>

**phos-** *pref.* [< Gk. *phōs,* light.] Light <*phos*gene>

**phos·gene** (fŏs'jēn', fŏz'-) *n.* A colorless volatile liquid or gas, $COCl_2$, used as a poison gas and in making glass, resins, dyes, and plastics.

**phosph-** *pref. var. of* PHOSPHO-.

**phos·pha·tase** (fŏs′fə-tās′, -tāz′) *n.* [PHOSPHAT(E) + -ASE.] Any of numerous enzymes that catalyze the hydrolysis of esters to phosphoric acid and are distinguished by activity in carbohydrate and nucleotide metabolism and in bone formation.

**phos·phate** (fŏs′fāt′) *n.* [Fr. < *acide phosphorique*, phosphoric acid.] **1.** *Chem.* A salt or ester of phosphoric acid containing chiefly pentavalent phosphorus and oxygen. **2.** A fertilizer including phosphorus compounds. **3.** A beverage of carbonated water, flavoring, and a small amount of phosphoric acid. **—phos·phat′ic** (fŏs-făt′ĭk) *adj.*

**phosphate rock** *n.* A sedimentary rock consisting chiefly of apatite, used as fertilizer and as a source of phosphorous compounds.

**phos·pha·tide** (fŏs′fə-tīd′) *n.* [PHOSPHAT(E) + -IDE.] Any of a group of lipid compounds, as cephalin and lecithin, consisting primarily of glycerol and phosphoric acid and occurring abundantly in plant and animal tissues with stored fats.

**phos·pha·tize** (fŏs′fə-tīz′) *vt.* **-tized, -tiz·ing, -tiz·es. 1.** To convert into a phosphate or phosphates. **2.** To treat with phosphate or phosphoric acid. **—phos′pha·ti·za′tion** *n.*

**phos·pha·tu·ri·a** (fŏs′fə-tŏŏr′ē-ə, -tyŏŏr′-) *n.* [PHOSPHAT(E) + -URIA.] Excessive phosphate discharge in urine. **—phos′pha·tu′ric** *adj.*

**phos·phene** (fŏs′fēn′) *n.* [PHOS- + Gk. *phainein*, to show.] A visual sensation of light experienced when the eyeball is pressed.

**phos·phide** (fŏs′fīd′) *also* **phos·phid** (-fĭd) *n.* A compound of phosphorus and a more electropositive element.

**phos·phine** (fŏs′fēn′) *also* **phos·phin** (-fĭn) *n.* **1.** A colorless, poisonous, spontaneously flammable gas, $PH_3$, having a garliclike odor and used as a doping agent for solid-state components. **2.** A synthetic yellow dye.

**phos·phite** (fŏs′fīt′) *n.* A salt of phosphorous acid.

**phospho-** *or* **phosph-** *pref.* [Fr. < *phosphore*, phosphorus < NLat. *phosphorus.*] **1.** Phosphorus <*phosphine*> **2.** Phosphate <*phospholipid*>

**phos·pho·cre·a·tine** (fŏs′fō-krē′ə-tēn′) *also* **phos·pho·cre·a·tin** (-tĭn) *n.* An organic compound, $C_4H_{10}N_3O_5$, capable of providing physiologic energy, as in muscular contraction.

**phos·pho·lip·id** (fŏs′fō-lĭp′ĭd) *n.* A phosphatide.

**phos·pho·ni·um** (fŏs-fō′nē-əm) *n.* [PHOSPH(O)- + (AMM)ONIUM.] A univalent radical, $PH_4$, derived from phosphine.

**phos·pho·pro·tein** (fŏs′fō-prō′tēn′, -prō′tē-ĭn) *n.* Any of a group of proteins, as casein, having chemically bound phosphoric acid.

**phos·phor** (fŏs′fər, -fôr′) *n.* [Fr. *phosphore* < NLat. *phosphorus*, phosphorus. —see PHOSPHORUS.] **1.** A substance capable of emitting light when stimulated by incident radiation. **2.** Something having phosphorescent or fluorescent qualities.

**phosphor bronze** *n.* A strong, durable, corrosion-resistant bronze containing up to 0.5% phosphorus and used in electric switches, chains, and springs.

**phos·pho·resce** (fŏs′fə-rĕs′) *vi.* **-resced, -resc·ing, -resc·es.** [Prob. back-formation < PHOSPHORESCENCE.] To persist in emitting light, unaccompanied by sensible heat or combustion, after exposure to and removal of a source of radiation.

**phos·pho·res·cence** (fŏs′fə-rĕs′əns) *n.* [< PHOSPHOR.] **1.** Persistent emission of light following exposure to and removal of incident radiation. **2.** Organically generated emission of light : BIOLUMINESCENCE. **—phos′pho·res′cent** *adj.* **—phos′pho·res′cent·ly** *adv.*

**phos·phor·ic** (fŏs-fôr′ĭk, -fŏr′-) *adj.* Of, relating to, or including phosphorus, esp. in a valence state higher than that of a comparable phosphorous compound.

**phosphoric acid** *n.* A clear colorless liquid, $H_3PO_4$, used in soaps and detergents, pharmaceuticals, food flavoring, and animal feed.

**phos·pho·rism** (fŏs′fə-rĭz′əm) *n.* [PHOSPHOR(US) + -ISM.] Chronic phosphorus poisoning from inhalation or ingestion.

**phos·pho·rite** (fŏs′fə-rīt′) *n.* [PHOSPHOR(US) + -ITE.] **1.** A fibrous variety of apatite. **2.** A concretionary mass of rock composed principally of calcium phosphate.

**phos·pho·rous** (fŏs′fər-əs, fŏs-fôr′əs, -fŏr′-) *adj.* Of, relating to, or containing phosphorus, esp. with valence 3.

**phosphorous acid** *n.* A white or yellowish hygroscopic crystalline solid, $H_3PO_3$, used to make phosphite salts and as a reducing agent.

**phos·pho·rus** (fŏs′fər-əs) *n.* [NLat. < Gk. *phōsphoros*, bringing light : *phōs*, light + *-phoros*, bearing < *pherein*, to bear.] **1.** *Symbol* **P** A highly reactive, poisonous nonmetallic element used in fireworks, safety matches, incendiary shells, fertilizers, steel, and glass; atomic number 15; atomic weight 30.9738. **2.** A phosphorescent substance.

**phos·pho·ryl·ase** (fŏs′fər-ə-lās′, -lāz′) *n.* [PHOSPHOR(US) + -YL + -ASE.] An enzyme that catalyzes the production of phosphates from glycogen.

**phos·pho·ryl·ate** (fŏs′fər-ə-lāt′) *vt.* **-at·ed, -at·ing, -ates.** [PHOSPHOR(US) + -YL + -ATE.] To convert (an organic substance) into an organic phosphate. **—phos′pho·ry·la′tion** *n.*

**phot** (fōt) *n.* [Gk. *phōs*, *phōt-*, light.] *Physics.* A unit of illumination equal to one lumen per square centimeter.

**phot-** *pref. var. of* PHOTO-.

**pho·tic** (fō′tĭk) *adj.* **1.** Of or relating to light. **2.** *Biol.* Relating to the emission of visible light by organisms. **3.** Designating the upper zone or layer of a body of water, which is penetrable by sunlight in sufficient quantity to allow photosynthesis.

**pho·to** (fō′tō) *Informal.* —*n., pl.* **-tos.** A photograph. —*vt. & vi.* **-toed, -to·ing, -tos.** To photograph.

**photo-** *or* **phot-** *pref.* [< Gk. *phōs*, *phōt-*, light.] **1.** Light : radiant energy <*photo*synthesis> **2.** Photographic <*photo*montage> **3.** Photoelectric <*photo*emission>

**pho·to·ac·tive** (fō′tō-ăk′tĭv) *adj.* **1.** Capable of responding to light photoelectrically. **2.** Capable of responding to light by chemical reaction. **—pho′to·ac·tiv′i·ty** *n.*

**pho·to·au·to·tro·phic** (fō′tō-ô′tə-trō′fĭk, -trŏf′ĭk) *adj. Biol.* Capable of synthesizing food from inorganic materials by using light as a source of energy. **—pho′to·au′to·troph′** *n.*

**pho·to·bi·ot·ic** (fō′tō-bī-ŏt′ĭk) *adj. Biol.* Needing light for the continued life and growth.

**pho·to·cell** (fō′tō-sĕl′) *n.* A photoelectric cell.

**pho·to·chem·is·try** (fō′tō-kĕm′ĭ-strē) *n.* The chemistry of the interactions of radiant energy and chemical systems. **—pho′to·chem′i·cal** *adj.* **—pho′to·chem′i·cal·ly** *adv.*

**pho·to·co·ag·u·la·tion** (fō′tō-kō-ăg′yə-lā′shən) *n.* Surgical coagulation of tissue using intense light energy, as a laser beam.

**pho·to·com·pose** (fō′tō-kəm-pōz′) *vt.* **-posed, -pos·ing, -pos·es.** To prepare (manuscript) for printing by photocomposition. **—pho′to·com·pos′er** *n.*

**pho·to·com·po·si·tion** (fō′tō-kŏm′pə-zĭsh′ən) *n.* Preparation of written material for printing by the projection of images of type characters onto photosensitive film or paper, from which printing plates are made.

**pho·to·con·duc·tiv·i·ty** (fō′tō-kŏn′dŭk-tĭv′ĭ-tē) *n.* Electrical conductivity that is affected by illumination. **—pho′to·con·duc′tion** *n.* **—pho′to·con·duc′tive** *adj.*

**pho·to·cop·i·er** (fō′tə-kŏp′ē-ər) *n.* A machine for making photographic reproductions of printed or graphic material.

**pho·to·cop·y** (fō′tə-kŏp′ē) *vt.* **-cop·ied, -cop·y·ing, -cop·ies.** To reproduce (printed matter) by a photographic process such as xerography. —*n., pl.* **-cop·ies.** A photographic reproduction.

**pho·to·cur·rent** (fō′tō-kûr′ənt) *n. Physics.* An electric current produced by illumination of a photoelectric material.

**pho·to·de·com·po·si·tion** (fō′tō-dē-kŏm′pə-zĭsh′ən) *n.* Photolysis.

**pho·to·dis·in·te·gra·tion** (fō′tō-dĭs-ĭn′tĭ-grā′shən) *n.* Nuclear disintegration or transformation resulting from absorption of high-energy radiation, as of gamma rays.

**pho·to·dra·ma** (fō′tə-drä′mə, -drăm′ə) *n.* A photoplay.

**pho·to·du·pli·cate** (fō′tō-dōō′plĭ-kāt′, -dyōō′-) *vt.* **-cat·ed, -cat·ing, -cates.** To photocopy. **—pho′to·du′pli·cate** (-kĭt) *n.* **—pho′to·du′pli·ca′tion** *n.*

**pho·to·dy·nam·ic** (fō′tō-dī-năm′ĭk) *adj.* Producing or intensifying a toxic reaction to light in organisms.

**pho·to·e·lec·tric** (fō′tō-ĭ-lĕk′trĭk) *also* **pho·to·e·lec·tri·cal** (-trĭ-kəl) *adj.* Of or relating to electric effects, esp. increased electrical conduction, caused by illumination. **—pho′to·e·lec′tri·cal·ly** *adv.*

**photoelectric cell** *n.* An electronic device having an electrical output that varies in response to light.

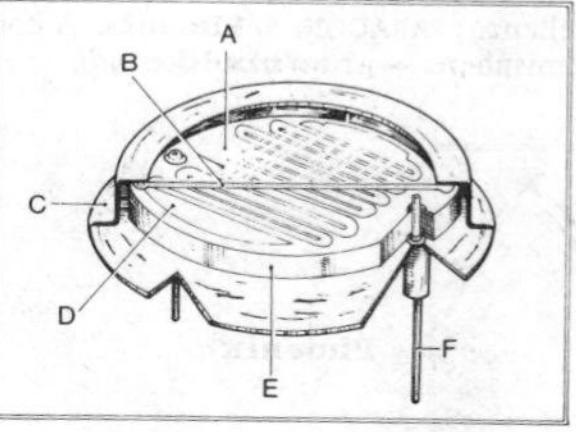


**photoelectric cell**
*A. electrode, B. glass window, C. metal case, D. photoconductive material, E. ceramic substrate, F. base pin*

**photoelectric effect** *n. Physics.* Ejection of electrons from a substance by incident electromagnetic radiation, esp. by visible light.

**pho·to·e·lec·tron** (fō′tō-ĭ-lĕk′trŏn′) *n.* An electron released from a substance by the photoelectric effect.

**pho·to·e·mis·sion** (fō′tō-ĭ-mĭsh′ən) *n.* Emission of photoelectrons, esp. from metallic surfaces.

**pho·to·en·grave** (fō′tō-ĕn-grāv′) *vt.* **-graved, -grav·ing, -graves.** To reproduce (graphic matter) by photoengraving. **—pho′to·en·grav′er** *n.*

**pho·to·en·grav·ing** (fō′tō-ĕn-grā′vĭng) *n.* **1.** The process of reproducing graphic material by photomechanical transference of the image onto a plate in etched relief for printing. **2.** A plate prepared by photoengraving. **3.** A print made by photoengraving.

**pho·to-es·say** (fō′tō-ĕs′ā′) *n.* A journalistic feature treating a subject mainly through photographs. **—pho′to-es′say·ist** *n.*

**photo finish** *n.* **1.** A finish in a race in which the contestants are so close together that the winner must be determined from a photograph taken at the finish line. **2.** *Informal.* An extremely close competition.

**pho·to·flash** (fō'tō-flăsh') *n.* A flash bulb.

**pho·to·flood** (fō'tō-flŭd') *n.* A reusable electric floodlight giving bright continuous light, used for photographic illumination.

**pho·to·fluor·o·gram** (fō'tə-flŏŏr'ə-grăm') *n.* A photograph made by photofluorography.

**pho·to·fluor·og·ra·phy** (fō'tō-flŏŏ-rŏg'rə-fē) *n. Med.* Photography of fluoroscopic images. **—pho'to·fluor'o·graph'ic** (-flŏŏr'ə-grăf'ĭk) *adj.*

**pho·to·gel·a·tin process** (fō'tō-jĕl'ə-tĭn) *n.* Collotype.

**pho·to·gene** (fō'tə-jēn') *n. Physiol.* An afterimage.

**pho·to·gen·ic** (fō'tə-jĕn'ĭk) *adj.* **1.** Pleasing as a subject for photography <a *photogenic* face> **2.** *Biol.* Producing or emitting light : PHOSPHORESCENT. **3.** Caused or produced by light. **—pho'to·gen'i·cal·ly** *adv.*

**pho·to·gram** (fō'tə-grăm') *n.* **1.** An indistinct image obtained by placing an object in contact with film or photosensitive paper and exposing it to light. **2.** A photograph.

**pho·to·gram·me·try** (fō'tə-grăm'ĭ-trē) *n.* **1.** The art or process of using esp. aerial photography to produce accurate maps or charts. **2.** The science of making precise measurements by the use of photography. **—pho'to·gram·met'ric** (-grə-mĕt'rĭk) *adj.* **—pho'to·gram'me·trist** *n.*

**pho·to·graph** (fō'tə-grăf') *n.* An image, esp. a positive print, recorded by a camera and reproduced on a photosensitive surface. —*v.* **-graphed, -graph·ing, -graphs.** —*vt.* To take a photograph of. —*vi.* **1.** To practice photography. **2.** To be the subject for photographs <I never *photograph* well.> **—pho·tog'ra·pher** *n.*

**pho·to·graph·ic** (fō'tə-grăf'ĭk) *also* **pho·to·graph·i·cal** (-ĭ-kəl) *adj.* **1.** Of, relating to, or consisting of photography or a photograph. **2.** Used in photography <a *photographic* plate> **3.** Like a photograph, esp. in accuracy of representation <a portrait painted in *photographic* detail> **4.** Capable of forming exact and lasting impressions <a *photographic* memory> **—pho'to·graph'i·cal·ly** *adv.*

**pho·tog·ra·phy** (fə-tŏg'rə-fē) *n.* **1.** The process or technique of rendering optical images on photosensitive surfaces. **2.** The art, practice, or occupation of taking and printing photographs. **3.** A body of photographs.

**pho·to·gra·vure** (fō'tə-grə-vyŏŏr') *n.* The art or process of printing from an intaglio plate, etched according to a photographic image.

**pho·to·he·li·o·graph** (fō'tō-hē'lē-ə-grăf') *n.* A telescope designed or modified for solar photography.

**pho·to·jour·nal·ism** (fō'tō-jûr'nə-lĭz'əm) *n.* Journalism making primary use of photographs in presenting news or features. **—pho'to·jour'nal·ist** *n.* **—pho'to·jour'nal·is'tic** *adj.*

**pho·to·ki·ne·sis** (fō'tō-kĭ-nē'sĭs, -kī-) *n. Biol.* Movement in response to light. **—pho'to·ki·net'ic** (-nĕt'ĭk) *adj.*

**pho·to·lith·o·graph** (fō'tō-lĭth'ə-grăf') *vt.* **-graphed, -graph·ing, -graphs.** To reproduce (an image) by means of photolithography. **—pho'to·lith'o·graph'** *n.* **—pho'to·li·thog'ra·pher** (fō'tō-lĭ-thŏg'rə-fər) *n.*

**pho·to·li·thog·ra·phy** (fō'tō-lĭ-thŏg'rə-fē) *n.* A planographic printing process using plates made according to a photographic image. **—pho'to·lith'o·graph'ic** (-lĭth'ə-grăf'ĭk) *adj.*

**pho·tol·y·sis** (fō-tŏl'ĭ-sĭs) *n.* Chemical decomposition induced by radiant energy, esp. light. **—pho'to·lyt'ic** (fō'tə-lĭt'ĭk) *adj.* **—pho'to·lyt'i·cal·ly** *adv.*

**pho·to·map** (fō'tə-măp') *n.* A map produced by superimposing orienting data, as a grid, on an aerial photograph. —*v.* **-mapped, -map·ping, -maps.** —*vt.* To make a photomap of. —*vi.* To make a photomap.

**pho·to·me·chan·i·cal** (fō'tō-mĭ-kăn'ĭ-kəl) *adj.* Of, relating to, or designating any of various methods by which plates are prepared for printing by means of photography. **—pho'to·me·chan'i·cal·ly** *adv.*

**pho·tom·e·ter** (fō-tŏm'ĭ-tər) *n.* An instrument for measuring a property of light, esp. luminous intensity or flux.

**pho·tom·e·try** (fō-tŏm'ĭ-trē) *n. Physics.* Measurement of the properties of light, esp. of luminous intensity. **—pho'to·met'ric** (fō'tə-mĕt'rĭk), **pho'to·met'ri·cal** *adj.* **—pho·tom'e·trist** *n.*

**pho·to·mi·cro·graph** (fō'tō-mī'krə-grăf') *n.* A photograph made using a photomicroscope. —*vt.* **-graphed, -graph·ing, -graphs.** To photograph (a magnified image) by means of a photomicroscope. **—pho'to·mi·crog'ra·pher** (-mī-krŏg'rə-fər) *n.* **—pho'to·mi'cro·graph'ic** *adj.* **—pho'to·mi·crog'ra·phy** *n.*

**pho·to·mi·cro·scope** (fō'tō-mī'krə-skōp') *n.* An instrument for making photomicrographs. **—pho'to·mi'cro·scop'ic** (-skŏp'ĭk) *adj.*

**pho·to·mon·tage** (fō'tō-mŏn-täzh', -môn-) *n.* **1.** The art or technique of making a picture by assembling photographic images, often in combination with other types of graphic material. **2.** A composite picture made by photomontage.

**pho·ton** (fō'tŏn') *n.* The quantum of electromagnetic energy, gen. regarded as a discrete particle having zero mass, no electric charge, and an indefinitely long lifetime. **—pho·ton'ic** *adj.*

**pho·to·nu·cle·ar** (fō'tō-nōō'klē-ər, -nyōō'-) *adj. Physics.* Designating a nuclear reaction induced by photons.

**pho·to-off·set** (fō'tō-ôf'sĕt', -ŏf'-) *n.* Offset printing.

**pho·to·pe·ri·od** (fō'tō-pîr'ē-əd) *n.* The varying length of exposure of an organism to daylight as a proportion of the total day, considered esp. as to the effect on biological processes. **—pho'to·pe'ri·od'ic** (-ŏd'ĭk), **pho'to·pe'ri·od'i·cal** *adj.* **—pho'to·pe'ri·od·ism** *n.*

**pho·toph·i·lous** (fō-tŏf'ə-ləs) *also* **pho·to·phil·ic** (fō'tə-fĭl'ĭk) *adj. Biol.* Growing or functioning optimally when exposed to full or prolonged light. **—pho·toph'i·ly** *n.*

**pho·to·pho·bi·a** (fō'tə-fō'bē-ə) *n.* Abnormal intolerance of light. **—pho'to·pho'bic** (-fō'bĭk) *adj.*

**pho·to·phos·phor·y·la·tion** (fō'tō-fŏs'fôr-ə-lā'shən, -fər-) *n.* Phosphorylation induced by radiant energy in photosynthesis.

**pho·to·pi·a** (fō-tō'pē-ə) *n.* Adaptation of the eyes to light : daylight vision. **—pho·to'pic** (-tō'pĭk, -tŏp'ĭk) *adj.*

**pho·to·play** (fō'tə-plā') *n.* A drama filmed or adapted for filming as a motion picture.

**pho·to·re·al·ism** (fō'tō-rē'ə-lĭz'əm) *n.* A painting style resembling photography in close attention to detail. **—pho'to·re'al·ist** *n.*

**pho·to·re·cep·tion** (fō'tō-rĭ-sĕp'shən) *n.* The detection or perception of visible light : VISION. **—pho'to·re·cep'tive** *adj.*

**pho·to·re·cep·tor** (fō'tō-rĭ-sĕp'tər) *n.* A photoreceptive nerve.

**pho·to·re·con·nais·sance** (fō'tō-rĭ-kŏn'ə-səns, -zəns) *n.* Photographic aerial reconnaissance, esp. of military targets.

**pho·to·res·pi·ra·tion** (fō'tō-rĕs'pə-rā'shən) *n.* Oxidation of carbohydrates in plants accompanied by the release of carbon dioxide during photosynthesis.

**pho·to·sen·si·tive** (fō'tō-sĕn'sĭ-tĭv) *adj.* Sensitive to light <a *photosensitive* emulsion><a *photosensitive* organism> **—pho'to·sen'si·tiv'i·ty** *n.*

**pho·to·sen·si·ti·za·tion** (fō'tō-sĕn'sĭ-tĭ-zā'shən) *n.* **1.** The act or process of photosensitizing something. **2.** *Med.* Hypersensitivity of the skin to sunlight or ultraviolet radiation, caused by ingestion of endocrine products, small amounts of heavy metals, or fluorescent dyes and resulting in skin eruptions.

**pho·to·sen·si·tize** (fō'tō-sĕn'sĭ-tīz') *vt.* **-tized, -tiz·ing, -tiz·es.** To make (a substance or organism) sensitive to light.

**pho·to·set** (fō'tō-sĕt') *vt.* **-set, -set·ting, -sets.** To photocompose. **—pho'to·set'ter** *n.*

**pho·to·sphere** (fō'tə-sfîr') *n.* The surface of a star, esp. of the sun. **—pho'to·spher'ic** (-sfîr'ĭk, -sfĕr'ĭk) *adj.*

**Pho·to·stat** (fō'tə-stăt'). A trademark for a device used to make quick, direct-reading negative or positive copies.

**pho·to·syn·the·sis** (fō'tō-sĭn'thĭ-sĭs) *n.* The process by which chlorophyll-containing cells in green plants convert incident light to chemical energy and synthesize organic compounds from inorganic compounds, esp. carbohydrates from carbon dioxide and water, accompanied by the simultaneous release of oxygen. **—pho'to·syn·thet'ic** (-sĭn-thĕt'ĭk) *adj.* **—pho'to·syn·thet'i·cal·ly** *adv.*

**pho·to·syn·the·size** (fō'tō-sĭn'thĭ-sīz') *v.* **-sized, -siz·ing, -siz·es.** —*vt.* To synthesize (organic compounds) by the process of photosynthesis. —*vi.* To perform the process of photosynthesis.

**pho·to·tax·is** (fō'tō-tăk'sĭs) *also* **pho·to·tax·y** (fō'tō-tăk'sē) *n. Biol.* Movement of an organism in response to a source of light. **—pho'to·tac'tic** (-tăk'tĭk) *adj.*

**pho·to·tel·e·graph** (fō'tō-tĕl'ə-grăf') *vt.* **-graphed, -graph·ing, -graphs.** To transmit (images or printed matter) by facsimile. **—pho'to·tel'e·graph'ic, pho'to·tel'e·graph'i·cal** *adj.* **—pho'to·tel'e·graph'i·cal·ly** *adv.* **—pho'to·te·leg'ra·phy** (-tə-lĕg'rə-fē) *n.*

**pho·to·ther·a·peu·tics** (fō'tō-thĕr'ə-pyōō'tĭks) *n.* (*sing. or pl. in number*). Phototherapy.

**pho·to·ther·a·py** (fō'tō-thĕr'ə-pē) *n.* Treatment of disease, esp. certain skin conditions, with light, including ultraviolet and infrared radiation.

**pho·tot·o·nus** (fō-tŏt'ə-nəs) *n. Biol.* Photosensitivity. **—pho'to·ton'ic** (fō'tə-tŏn'ĭk) *adj.*

**pho·to·tran·sis·tor** (fō'tō-trăn-zĭs'tər, -sĭs'-) *n.* A transistor having highly photosensitive electrical characteristics.

**pho·to·troph** (fō'tə-trŏf') *n.* [PHOTO- + Gk. *trophē*, food.] An organism that obtains metabolic energy through photosynthesis.

**pho·tot·ro·pism** (fō-tŏt'rə-pĭz'əm) *also* **pho·tot·ro·py** (-pē) *n.* Growth or movement in response to a source of light. **—pho'to·tro'pic** (fō'tə-trō'pĭk, -trŏp'ĭk) *adj.* **—pho'to·tro'pi·cal·ly** *adv.*

**pho·to·tube** (fō'tō-tōōb', -tyōōb') *n.* An electron tube equipped with a photosensitive cathode.

**pho·to·type·set·ter** (fō'tō-tīp'sĕt'ər) *n.* A machine used in photocomposition.

**pho·to·type·set·ting** (fō'tō-tīp'sĕt'ĭng) *n.* Photocomposition.

**pho·to·ty·pog·ra·phy** (fō'tō-tī-pŏg'rə-fē) *n.* Photomechanical printing that resembles metal typography. **—pho'to·ty·pog'ra·pher** *n.* **—pho'to·ty'po·graph'ic** (-tī'pə-grăf'ĭk), **pho'to·ty'po·graph'i·cal** *adj.* **—pho'to·ty'po·graph'i·cal·ly** *adv.*

**pho·to·vol·ta·ic** (fō'tō-vŏl-tā'ĭk, -vōl-) *adj.* Capable of producing a voltage when exposed to radiant energy, esp. light.

**pho·to·zin·co·graph** (fō'tō-zĭng'kə-grăf') *vt.* **-graphed, -graph·ing, -graphs.** To make (prints) by photozincography. **—pho'to·zin'co·graph'** *n.*

**pho·to·zin·cog·ra·phy** (fō'tō-zĭng-kŏg'rə-fē) *n.* A photoengraving process using sensitized zinc plates.

**phrase** (frāz) *n.* [Lat. *phrasis,* style of speech < Gk. < *phrazein,* to explain.] **1.** A sequence of words regarded as a meaningful unit. **2.** A concise or familiar expression : CATCH PHRASE. **3.** A word or group of words read or spoken as a unit and separated by pauses or other junctures. **4.** Two or more words in sequence comprising a syntactic unit or group of syntactic units, less completely predicated than a sentence <a noun *phrase*> **5.** A series of dance movements forming a unit in a choreographic pattern. **6.** *Mus.* A brief, expressive passage, usu. having four or eight measures. —*v.* **phrased, phras·ing, phras·es.** —*vt.* **1.** To express orally or in writing <*phrased* my answer carefully> **2.** To pace or mark off (e.g., a spoken passage) by pauses. **3.** *Mus.* **a.** To divide (a composition) into phrases. **b.** To combine (notes) in a phrase. —*vi.* To make or play phrases. **—phras'al** *adj.* **—phras'al·ly** *adv.*

**phrase book** *n.* A book of idiomatic foreign language expressions and their translations.

**phrase·maker** (frāz'mā'kər) *n.* **1.** One who coins apt or memorable phrases. **2.** One who makes seemingly profound but often inane and meaningless phrases.

**phra·se·o·gram** (frā'zē-ə-grăm') *n.* A symbol, as in a shorthand system, used to represent esp. a common or repeated phrase.

**phra·se·o·graph** (frā'zē-ə-grăf') *n.* A phrase represented by a phraseogram. **—phra'se·o·graph'ic** *adj.*

**phra·se·ol·o·gy** (frā'zē-ŏl'ə-jē) *n., pl.* **-gies. 1.** The manner in which words and phrases are used in speech or writing : STYLE. **2.** A set of expressions used by a particular group or person <medical *phraseology*> **—phra'se·o·log'i·cal** (-ə-lŏj'ĭ-kəl) *adj.* **—phra'se·ol'o·gist** *n.*

**phra·try** (frā'trē) *n., pl.* **-tries.** [Gk. *phratria* < *phratēr,* member of a phratry.] **1.** A kinship group constituting an intermediate division in the primitive structure of the Hellenic tribe or phyle, made up of several patrilinear clans, and surviving in classical times as a territorial subdivision in the political and military organization of the Athenian state. **2.** An exogamous tribal subdivision, consisting of two or more related clans. **—phra'tric** *adj.*

**phre·at·ic** (frē-ăt'ĭk) *adj.* [< Gk. *phrear,* well.] Of or relating to ground water.

**phren-** *pref. var. of* PHRENO-.

**phre·net·ic** (frə-nĕt'ĭk) *also* **phre·net·i·cal** (-ĭ-kəl) *adj. vars. of* FRENETIC.

**-phrenia** *suff.* [< Gk. *phrēn,* mind.] Mental disorder <schizo*phrenia*>

**phren·ic** (frĕn'ĭk, frē'nĭk) *adj.* [PHREN(O) + -IC.] **1.** Of or relating to the mind. **2.** *Anat.* Of or pertaining to the diaphragm.

**phre·ni·tis** (frĭ-nī'tĭs) *n. Pathol.* **1.** Inflammation of the diaphragm. **2.** Delirium : frenzy. **—phre·nit'ic** (-nĭt'ĭk) *adj.*

**phreno-** *or* **phren-** *pref.* [< Gk. *phrēn,* diaphragm, mind.] **1.** Mind <*phreno*logy> **2.** Diaphragm <*phren*itis>

**phre·nol·o·gy** (frĭ-nŏl'ə-jē) *n.* Study of the conformation of the skull based on a belief that it is indicative of mental ability and character. **—phren'o·log'ic** (frĕn'ə-lŏj'ĭk, frē'nə-), **phren'o·log'i·cal** *adj.* **—phre·nol'o·gist** *n.*

**phren·sy** (frĕn'zē) *n. & v. var. of* FRENZY.

**Phryg·i·an** (frĭj'ē-ən) *adj.* Of or relating to Phrygia or its people, language, or culture. —*n.* **1.** A native or resident of Phrygia. **2.** The Indo-European language of the Phrygians.

**phthal·ein** *also* **phthal·eine** (thăl'ēn', thăl'ē-ĭn, thā'lēn', thā'lē-ĭn) *n.* [PHTHAL(IC) + -EIN.] Any of a group of chemical compounds formed by a combination of phthalic anhydride with a phenol, used to make synthetic dyes.

**phthal·ic** (thăl'ĭk, thā'lĭk) *adj.* [Short for *naphthalic* : NAPHTH(A) + AL(COHOL) + -IC.] **1.** Of, relating to, or derived from naphthalene. **2.** Of or relating to phthalic acid.

**phthalic acid** *n.* A colorless, crystalline organic acid, $C_6H_4(COOH)_2$, derived from naphthalene and used in the synthesis of dyes, perfumes, and medicines.

**phthalic anhydride** *n.* A white crystalline compound, $C_6H_4(CO)_2O$, prepared by oxidizing naphthalene and used in the manufacture of phthaleins, plasticizers, resins, and insecticides.

**phthal·in** (thăl'ĭn, thā'lĭn) *n.* A colorless compound derived from the reduction of phthaleins.

**phthal·o·cy·a·nine** (thăl'ō-sī'ə-nēn', thā'lō-) *n.* [PHTHAL(IC) + CYANINE.] Any of several stable, light-fast, green or blue organic pigments derived from the basic compound $(C_8H_4C_2N)_4N_4$ and used in enamels, linoleum, printing inks, and plastics.

**phthi·ri·a·sis** (thĭ-rī'ə-sĭs, thī-) *n.* [Lat. < Gk. *phtheiriasis* : *phtheir,* louse + *-iasis,* -iasis.] Infestation with lice.

**phthis·ic** (tĭz'ĭk) *n.* [ME *ptisike* < OFr. *tisique* < Lat. *phthisicus,* consumptive < Gk. *phthisikos* < *phthisis,* phthisis.] **1.** *var. of* PHTHISIS. **2.** *Archaic.* Asthma. **—phthis'ic, phthis'i·cal** *adj.*

**phthi·sis** (thī'sĭs) *also* **phthis·ic** (tĭz'ĭk) *n.* [Lat. < Gk. < *phthinein,* to waste away.] **1.** Tuberculosis of the lungs. **2.** Emaciation and atrophy of the body or a bodily part.

**phyco-** *pref.* [< Gk. *phukos,* seaweed.] Seaweed <*phyco*logy>

**phy·co·bi·lin** (fī'kō-bī'lĭn) *n.* [PHYCO- + Lat. *bilis,* bile + -IN.] Any of a group of water-soluble pigments found in certain algae.

**phy·co·cy·a·nin** (fī'kō-sī'ə-nĭn) *n.* A blue phycobilin occurring in the cells of blue-green algae.

**phy·co·er·y·thrin** (fī'kō-ĕr'ĭ-thrĭn) *n.* A red phycobilin occurring in the cells of red algae.

**phy·col·o·gy** (fī-kŏl'ə-jē) *n.* The branch of botany dealing with the study of seaweeds and algae. **—phy'co·log'i·cal** (fī'kə-lŏj'ĭ-kəl) *adj.* **—phy·col'o·gist** *n.*

**phy·co·my·cete** (fī'kō-mī'sēt', -mī-sēt') *n.* [NLat. *Phycomycetes,* class name : PHYCO- + -MYCETE.] Any of numerous fungi that resemble algae, including certain mildews and molds. **—phy'co·my·ce'·tous** *adj.*

**phy·la** (fī'lə) *n. pl. of* PHYLUM.

**phy·lac·ter·y** (fī-lăk'tə-rē) *n., pl.* **-ies.** [ME *filakterie* < LLat. *phylacterium* < Gk. *phulaktērion,* phylactery, safeguard < *phulaktēr,* guard < *phulassein,* to guard.] **1.** Either of two small leather boxes containing strips inscribed with quotations from the Hebrew Scriptures, one of which is strapped to the forehead and the other to the left arm by Jewish men during weekday morning worship. **2. a.** An amulet. **b.** A reminder.

**phy·lax·is** (fī-lăk'sĭs) *n.* [Gk. *phulaxis,* act of guarding < *phulassein,* to guard.] Suppression of infection by the body. **—phy·lac'tic** *adj.*

**phy·le** (fī'lē) *n., pl.* **-lae** (-lē) [Gk. *phulē,* tribe.] A kinship organization constituting the largest political subdivision of an ancient Greek city-state. **—phy'lic** *adj.*

**phy·let·ic** (fī-lĕt'ĭk) *adj.* [< NLat. *phylesis,* course of evolutionary development < Gk. *phulon,* race.] Of or relating to phylogeny or phylogenetic development. **—phy·let'i·cal·ly** *adv.*

**phyll-** *pref. var. of* PHYLLO-.

**-phyll** *suff.* [< Gk. *phullon,* leaf.] Leaf <sporo*phyll*>

**phyl·lite** (fĭl'īt') *n.* A gray, green, or red metamorphic rock, similar to slate but often having a wavy surface and a micaceous luster.

**phyllo-** *or* **phyll-** *pref.* [< Gk. *phullon,* leaf.] Leaf <*phyll*oid>

**phyl·lo·clade** (fĭl'ə-klād') *also* **phyl·lo·clad** (-klăd') *n.* [NLat. *phyllocladium* : PHYLLO- + Gk. *klados,* branch.] A phyllome derived from a flattened branch or stem, as in certain cacti.

**phyl·lode** (fĭl'ōd') *also* **phyl·lo·di·um** (fĭ-lō'dē-əm) *n., pl.* **-lodes** *also* **-lo·di·a** (-lō'dē-ə) [< Gk. *phullōdes,* like leaves : *phullon,* leaf + *eidos,* shape.] A phyllome derived from a flattened leafstalk. **—phyl·lo'di·al** *adj.*

**phyl·loid** (fĭl'oid') *adj.* Resembling a leaf : LEAFLIKE.

**phyl·lome** (fĭl'ōm') *n.* A leaf or a plant structure that performs the functions of a leaf. **—phyl·lo'mic** (fĭ-lō'mĭk, -lŏm'ĭk) *adj.*

**phyl·loph·a·gous** (fĭ-lŏf'ə-gəs) *adj.* Feeding on leaves.

**phyl·lo·pod** (fĭl'ə-pŏd') *n.* [NLat. *Phyllopoda,* order name : PHYLLO- + -POD.] A crustacean of the order Phyllopoda, having swimming and respiratory appendages suggestive of leaves. —*adj. also* **phyl·lop·o·dous** (fĭ-lŏp'ə-dəs). **1.** Having leaflike appendages. **2.** Of or pertaining to the phyllopods. **—phyl·lop'o·dan** (fĭ-lŏp'ə-dən) *adj. & n.*

**phyl·lo·tax·y** (fĭl'ə-tăk'sē) *also* **phyl·lo·tax·is** (fĭl'ə-tăk'sĭs) *n.* [NLat. *phyllotaxis* : PHYLLO- + -TAXIS.] **1.** The arrangement of leaves on a stem. **2.** The principles governing leaf arrangement. **—phyl'lo·tac'tic** (-tăk'tĭk), **phyl'lo·tac'ti·cal** *adj.*

**-phyllous** *suff.* [NLat. *-phyllus* < Gk. *phullon,* leaf.] Having a specified kind or number of leaves <gamo*phyllous*>

**phyl·lox·e·ra** (fĭl'ŏk-sîr'ə, fĭ-lŏk'sər-ə) *n., pl.* **-rae** (-rē) [NLat. *Phylloxera,* genus name : PHYLLO- + Gk. *xēros,* dry.] An aphidlike insect of the genus *Phylloxera,* esp. *P. vitifoliae,* a widely distributed species that attacks grapevines. **—phyl'lox·e'ran** *adj. & n.*

**phy·log·e·ny** (fī-lŏj'ə-nē) *n., pl.* **-nies.** [Gk. *phulon,* race, class + -GENY.] **1.** Evolutionary development of a plant or animal species. **2.** Historical or cultural evolution, as of a language or society. **—phy'lo·ge·net'ic** (fī'lō-jə-nĕt'ĭk), **phy'lo·gen'ic** *adj.* **—phy'lo·ge·net'i·cal·ly** *adv.*

**phy·lum** (fī'ləm) *n., pl.* **-la** (-lə) [NLat. < Gk. *phulon,* class.] **1.** *Biol.* A major taxonomic division of the animal kingdom or, less commonly, the plant kingdom, next above a class in size. **2.** A large division of genetically related families of languages or linguistic stocks.

**physi-** *pref. var. of* PHYSIO-.

**phys·i·at·rics** (fĭz'ē-ăt'rĭks) *n.* (*sing. in number*). The branch of medicine specializing in physical medicine or physical therapy. **—phys'i·at'rist** (-rĭst) *n.*

**phys·i·at·ry** (fĭz'ē-ăt'rē) *n. Med.* Physiatrics.

**phys·ic** (fĭz'ĭk) *n.* [ME *phisik* < OFr. *fisique,* medical science, natural science < Lat. *physica* < Gk. *phusikē,* fem. of *phusikos,* of nature < *phusis,* nature.] **1.** A medicine or drug. **2.** A cathartic. **3.** *Archaic.* The profession of medicine. —*vt.* **-icked, -ick·ing, -ics. 1.** *Archaic.* To treat with or as if with medicine. **2.** To act upon as a cathartic. **3.** To heal or cure.

**phys·i·cal** (fĭz′ĭ-kəl) *adj.* [ME *phisycal,* medical < Med. Lat. *physicalis* < Lat. *physica,* physics. —see PHYSICS.] **1.** Of or relating to the body : CORPOREAL <*physical* fitness> **2.** Of or relating to material things <*physical* environment> **3.** Of or relating to matter and energy or the sciences dealing with them, esp. physics. —*n.* A physical examination. **—phys′i·cal·ly** *adv.*

☆ ***syns:*** **1.** PHYSICAL, CONCRETE, CORPOREAL, MATERIAL, OBJECTIVE, PHENOMENAL, SENSIBLE, SUBSTANTIAL, TANGIBLE *adj. core meaning* : composed of or relating to things that occupy space and can be perceived by the senses <a *physical* barrier> ***ant:*** spiritual **2.** PHYSICAL, ANIMAL, CARNAL, FLESHLY, SENSUAL *adj. core meaning* : relating to the desires and appetites of the body <a purely *physical* attraction>

**physical anthropology** *n.* The science of human evolutionary biology, racial variation, and classification. **—physical anthropologist** *n.*

**physical chemistry** *n.* The branch of chemistry concerned with analysis of the properties and behavior of chemical systems primarily by physical theory and technique, as the thermodynamic analysis of macroscopic chemical phenomena.

**physical education** *n.* Education in the development and care of the human body, stressing athletics and including hygiene.

**physical examination** *n.* A medical examination to detect illness or dysfunction and esp. to determine physical fitness for a specified service or activity.

**physical geography** *n.* Study of the structure and phenomena of the earth's surface, esp. in its current aspects, including land formation, currents, climate, and distribution of flora and fauna.

**phys·i·cal·ism** (fĭz′ĭ-kə-lĭz′əm) *n. Philos.* The doctrine that all phenomena can be described in spatiotemporal terms and consequently that any descriptive scientific statement can in principle be reduced to an empirically verifiable physical statement. **—phys′i·cal·ist** *n.* **—phys′i·cal·is′tic** *adj.*

**physical medicine** *n.* Diagnosis and treatment of disease by essentially physical means, including exercise, manipulation, massage, and the application of heat, cold, radiation, electricity, and water.

**physical science** *n.* Any of the sciences, as chemistry, physics, astronomy, and geology, that investigate the nature and properties of energy and nonliving matter.

**physical therapy** *n.* Treatment of injury and disease by mechanical means, as heat, light, exercise, and massage.

**phy·si·cian** (fĭ-zĭsh′ən) *n.* [ME *fisicien* < OFr. < *fisique,* medical science. —see PHYSIC.] **1.** One licensed to practice medicine : medical doctor. **2.** One who heals, restores, or ministers to. **—phy·si′cian·ly** (-lē) *adj.*

**phys·i·cist** (fĭz′ĭ-sĭst) *n.* A scientist specializing in physics.

**phys·i·co·chem·i·cal** (fĭz′ĭ-kō-kĕm′ĭ-kəl) *adj.* **1.** Relating to properties that are both physical and chemical. **2.** Relating to physical chemistry.

**phys·ics** (fĭz′ĭks) *n.* [Lat. *physica* < Gk. *phusika,* neuter pl. of *phusikos,* of nature < *phusis,* nature.] *(sing. in number).* **1.** The science of matter and energy and of interactions between the two, grouped in traditional fields such as optics, acoustics, mechanics, electromagnetism, and thermodynamics, as well as in modern extensions including atomic and nuclear physics, particle physics, solid-state physics, plasma physics, and cryogenics. **2.** Physical properties, interactions, laws, or processes <the *physics* of jet propulsion> **3.** *Archaic.* The study of the natural or material world and phenomena : NATURAL PHILOSOPHY.

**physio-** *or* **physi-** *pref.* [Gk. *phusio-* < *phusis,* nature, origin, growth < *phuein,* to bring forth.] **1.** Nature : natural <*physiography*> **2.** Physical <*physiatry*>

**phys·i·og·no·my** (fĭz′ē-ŏg′nə-mē, -ŏn′ə-mē) *n., pl.* **-mies.** [ME *phisonomie* < OFr. < LLat. *physiognomia* < Gk. *phusiognomonia* : *phusis,* appearance + *gnōmōn,* interpreter < *gignōskein,* to know.] **1. a.** The art of appraising character or personality from facial features. **b.** Divination based on facial features. **2. a.** Facial features, esp. when regarded as revealing character. **b.** Outward appearance and character of an abstract or inanimate entity <the *physiognomy* of the moon> **—phys′i·og·nom′ic** (fĭz′ē-ŏg-nŏm′ĭk, fĭz′ē-ə-nŏm′ĭk), **phys′i·og·nom′i·cal** *adj.* **—phys′i·og·nom′i·cal·ly** *adv.* **—phys′i·og′no·mist** *n.*

**phys·i·og·ra·phy** (fĭz′ē-ŏg′rə-fē) *n.* Physical geography. **—phys′i·og′ra·pher** *n.* **—phys′i·o·graph′ic** (fĭz′ē-ə-grăf′ĭk), **phys′i·o·graph′i·cal** *adj.* **—phys′i·o·graph′i·cal·ly** *adv.*

**phys·i·o·log·i·cal** (fĭz′ē-ə-lŏj′ĭ-kəl) *also* **phys·i·o·log·ic** (-ĭk) *adj.* **1.** Of or relating to physiology. **2.** Characteristic of or in accord with the normal and healthy functioning of a living organism. **—phys′i·o·log′i·cal·ly** *adv.*

**physiological saline** *n.* A sterile salt isosmotic to body fluids.

**phys·i·ol·o·gy** (fĭz′ē-ŏl′ə-jē) *n.* [Lat. *physiologia* < Gk. *phusiologia* : *phusis,* nature + *logos,* account.] **1.** The biological science of essential and typical life processes, functions, and activities. **2.** An organism's vital processes. **—phys′i·ol′o·gist** *n.*

**phys·i·o·ther·a·py** (fĭz′ē-ō-thĕr′ə-pē) *n.* Physical therapy. **—phys′i·o·ther′a·peu′tic** (-thĕr′ə-pyōō′tĭk) *adj.*

**phy·sique** (fĭ-zēk′) *n.* [Fr. < *physique,* physical < Lat. *physicus,* of nature < Gk. *phusikos* < *phusis,* nature.] The human body considered with reference to its proportions, muscular development, and appearance <the *physique* of a runner> **—phy·siqued′** *adj.*

**phy·so·stig·mine** (fī′sō-stĭg′mēn′) *also* **phy·so·stig·min** (-mĭn) *n.* [< NLat. *Physostigma,* genus name of the Calabar bean : Gk. *phusa,* bellows + Gk. *stigma,* tattoo.] A colorless or pink poisonous crystalline compound, $C_{15}H_{21}N_3O_2$, obtained from the Calabar bean and used in medicinal drugs.

**phy·sos·to·mous** (fī-sŏs′tə-məs) *adj.* [Gk. *phusa,* bladder + -STOM(E) + -OUS.] Having a connecting tube between the air bladder and the alimentary canal, as in certain fishes.

**phyt-** *pref. var. of* PHYTO-.

**-phyte** *suff.* [Gk. *phuton,* plant < *phuein,* to grow.] **1.** A plant with a specified character or habitat <halo*phyte*> **2.** A pathological growth <osteo*phyte*>

**phyto-** *or* **phyt-** *pref.* [NLat. < Gk. *phuto-* < *phuton,* plant < *phuein,* to grow.] Plant <*phyto*genesis>

**phy·to·chem·is·try** (fī′tō-kĕm′ĭ-strē) *n.* The chemistry of plants. **—phy′to·chem′i·cal** (-ĭ-kəl) *adj.* **—phy′to·chem′i·cal·ly** *adv.* **—phy′to·chem′ist** *n.*

**phy·to·gen·e·sis** (fī′tō-jĕn′ĭ-sĭs) *also* **phy·tog·e·ny** (fī-tŏj′ə-nē) *n.* Plant origin and evolutionary development. **—phy′to·ge·net′ic** (-jə-nĕt′ĭk), **phy′to·ge·net′i·cal** *adj.* **—phy′to·ge·net′i·cal·ly** *adv.*

**phy·to·gen·ic** (fī′tō-jĕn′ĭk) *also* **phy·tog·e·nous** (fī-tŏj′ə-nəs) *adj.* Having a plant origin, as coal.

**phy·tog·e·ny** (fī-tŏj′ə-nē) *n. var. of* PHYTOGENESIS.

**phy·to·ge·og·ra·phy** (fī′tō-jē-ŏg′rə-fē) *n.* Scientific study of the distribution of plants. **—phy′to·ge·og′ra·pher** *n.* **—phy′to·ge′o·graph′i·cal** (-jē′ə-grăf′ĭ-kəl), **phy′to·ge′o·graph′ic** *adj.*

**phy·tog·ra·phy** (fī-tŏg′rə-fē) *n.* The science of plant description.

**phy·to·hor·mone** (fī′tō-hôr′mōn′) *n.* A hormone produced by a plant, esp. one that regulates plant growth.

**phy·tol** (fī′tôl′) *n.* A liquid alcohol, $C_{20}H_{40}O$, used in synthesizing vitamins E and K.

**phy·to·lite** (fī′tə-līt′) *also* **phy·to·lith** (-lĭth′) *n.* A fossil plant.

**phy·tol·o·gy** (fī-tŏl′ə-jē) *n.* The study of plants : BOTANY. **—phy′to·log′ic** (fī′tə-lŏj′ĭk), **phy′to·log′i·cal** *adj.*

**phy·ton** (fī′tŏn′) *n.* [NLat. < Gk. *phuton,* plant < *phuein,* to grow.] A unit of plant structure, esp. the smallest part of a plant that when severed is able to regenerate the entire organism. **—phy·ton′ic** *adj.*

**phy·to·path·o·gen** (fī′tō-păth′ə-jən) *n.* An organism, as a parasite, that is an agent of plant disease. **—phy′to·path′o·gen′ic** (-jĕn′ĭk) *adj.*

**phy·to·pa·thol·o·gy** (fī′tō-pə-thŏl′ə-jē) *n.* The science of plant diseases. **—phy′to·path′o·log′ic** (-păth′ə-lŏj′ĭk), **phy′to·path′o·log′i·cal** *adj.* **—phy′to·pa·thol′o·gist** *n.*

**phy·toph·a·gous** (fī-tŏf′ə-gəs) *adj.* Feeding on plants, including shrubs and trees. —Used esp. of certain insects.

**phy·to·plank·ton** (fī′tō-plăngk′tən) *n.* Microscopic floating aquatic plants. **—phy′to·plank·ton′ic** (-plăngk-tŏn′ĭk) *adj.*

**phy·to·so·ci·ol·o·gy** (fī′tō-sō′sē-ŏl′ə-jē, -shē-) *n.* The branch of ecology concerned with the characteristics, relationships, and distribution of associated plants. **—phy′to·so′ci·o·log′i·cal** (-sō′sē-ə-lŏj′ĭ-kəl, -shē-) *adj.* **—phy′to·so′ci·o·log′i·cal·ly** *adv.* **—phy′to·so′ci·ol′o·gist** *n.*

**phy·to·tox·ic** (fī′tō-tŏk′sĭk) *adj.* Poisonous to plants. **—phy′to·tox·ic′i·ty** (-tŏk-sĭs′ĭ-tē) *n.*

**pi¹** (pī) *n., pl.* **pis.** [Med. Gk. < Gk. *pei,* of Phoenician orig. : akin to Heb. *pē.*] **1.** The 16th letter of the Greek alphabet. —See table at ALPHABET. **2.** *Symbol* $\pi$ *Math.* A transcendental number, approx. 3.14159, representing the ratio of the circumference to the diameter of a circle and appearing as a constant in a wide range of mathematical problems.

**pi²** *also* **pie** (pī) [Orig. unknown.] —*n., pl.* **pis** *also* **pies.** A random or jumbled assortment of printing type. —*v.* **pied, pi·ing, pies** *also* **pied, pie·ing, pies.** —*vt.* To jumble or mix up (type). —*vi.* To become jumbled.

**pia** (pī′ə, pē′ə) *n. Anat.* The pia mater. **—pi′al** *adj.*

**pi·ac·u·lar** (pī-ăk′yə-lər) *adj.* [Lat. *piacularis* < *piaculum,* propitiatory sacrifice < *piare,* to appease < *pius,* dutiful.] **1.** Serving as expiation or atonement for a sacrilege <*piacular* sacrifice> **2.** Requiring expiation : BLAMEWORTHY.

**piaffe** (pyăf) *vi.* **piaffed, piaf·fing, piaffes.** [Fr. *piaffer.*] To perform the piaffer.

**piaf·fer** (pyăf′ər) *n.* [Fr. < *piaffer,* to piaffe.] A dressage movement in which a horse trots in place with high action of the legs.

**pia ma·ter** (mā′tər, mä′tər) *n.* [ME < Med. Lat., tender mother.] *Anat.* The fine vascular membrane enveloping the brain and spinal cord under the arachnoid membrane and the dura mater.

**pi·an·ism** (pē-ăn′ĭz′əm, pē′ə-nĭz′əm) *n.* The art or technique of playing the piano.

**pi·a·nis·si·mo** (pē′ə-nĭs′ə-mō′) [Ital., superl. of *piano,* soft < Lat.

*planus*, level.] *Mus.* —*adv. & adj.* Very softly. —Used as a direction. —*n.*, *pl.* **-mos.** A passage marked or performed pianissimo.

**pi·an·ist** (pē-ăn′ĭst, pē′ə-nĭst) *n.* One who plays the piano.

**pi·a·nis·tic** (pē′ə-nĭs′tĭk) *adj.* **1.** Of or relating to the piano. **2.** Well-suited to the piano <a *pianistic* melody> **—pi′a·nis′ti·cal·ly** *adv.*

**pi·a·nis·tics** (pē′ə-nĭs′tĭks) *n.* (*sing. or pl. in number*). Pianism.

**pi·an·o**[1] (pē-ăn′ō) *n.*, *pl.* **-os.** [Ital., short for *pianoforte.* —see PIANOFORTE.] A keyboard musical instrument, as the grand piano or the upright piano, with hammers that strike wire strings, producing sounds that may be softened or sustained by means of pedals.

**pi·a·no**[2] (pē-ä′nō) [Ital. < Lat. *planus*, flat.] *Mus.* —*adv. & adj.* Softly : quietly. —Used as a direction. —*n.*, *pl.* **-nos.** A passage marked or performed piano.

**piano bar** *n.* A cocktail lounge featuring piano entertainment.

**pi·an·o·for·te** (pē-ăn′ō-fôr′tā, -fôr′tē, -fôrt′) *n.* [Ital. < *piano e forte*, soft and loud.] A piano.

**piano hinge** *n.* A long narrow hinge with a pin running the entire length of its joint.

**pi·as·sa·va** (pē′ə-sä′və) *also* **pi·as·sa·ba** (-sä′bə) *n.* [Port. *piassaba* < Tupi *piaçaba.*] **1.** Either of two South American palm trees, *Attalea funifera* or *Leopoldina piassaba*, from which a strong, coarse fiber is obtained. **2.** The fiber of the piassava, used in ropes, brooms, and brushes.

**pi·as·ter** *also* **pi·as·tre** (pē-ăs′tər, -ä′stər) *n.* [Fr. *piastre* < Ital. *piastra*, thin metal plate < Lat. *emplastrum*, medical dressing. —see PLASTER.] **1.** —See table at CURRENCY. **2.** A piece of eight.

**pi·az·za** (pē-ăz′ə, -ä′zə) *n.*, *pl.* **-zas.** [Ital. < Lat. *platea*, street < Gk. *plateia (hodos)*, broad (way).] **1.** (*also* -ät′sə) *pl.* **pi·az·ze** (-ät′sā). A public square in an Italian town. **2.** A roofed and arcaded passageway : COLONNADE. **3.** A verandah.

**pi·broch** (pē′brŏKH′) *n.* [Sc. Gael. *piobaireachd*, pipe music.] A series of variations on a traditional dirge or martial theme for the Highland bagpipes.

**pic** (pĭk) *n.*, *pl.* **pics** *or* **pix.** [Short for PICTURE.] *Slang.* **1.** A photograph. **2.** A movie.

**pi·ca**[1] (pī′kə) *n.* [Prob. < Med. Lat., list of church services.] **1. a.** A unit of type size, equal to 12 points or approx. 1/6 inch. **b.** An equivalent unit of measurement used in composing printed material. **2.** A type size for typewriters, providing ten characters to the inch.

**pi·ca**[2] (pī′kə) *n.* [NLat. < Lat. *pica*, magpie.] A craving for unnatural food, as earth or ashes, occurring esp. in pregnancy or hysteria.

**pic·a·dor** (pĭk′ə-dôr′) *n.*, *pl.* **pic·a·dors** *or* **pic·a·do·res** (pĭk′ə-dôr′ās) [Sp. < *picar*, to prick.] A horseman in a bullfight who lances the bull's neck muscles so that it will tend to hold its head low.

**pi·ca·ra** (pē′kä-rä′) *n.*, *pl.* **-ras** (-räz′, -räs′). An adventuress.

**pic·a·resque** (pĭk′ə-rĕsk′, pē′kə-) *adj.* [Fr. < Sp. *picaresco* < *pícaro*, picaro.] **1.** Of or involving clever rogues or adventurers. **2.** Of, belonging to, or typical of a literary genre in which the rogue-hero's escapades are often depicted in a context of sharp social satire. —*n.* One that is picaresque.

**pi·ca·ro** (pē′kä-rō) *n.*, *pl.* **-ros** (-rōz′, -rōs′) [Sp. *pícaro* < *picar*, to prick < VLat. *piccare < Lat. *picus*, woodpecker.] A clever adventurer.

**pic·a·roon** (pĭk′ə-ro͞on′) *n.* [Sp. *picarón*, aug. of *pícaro*, picaro.] **1. a.** A pirate. **b.** A picaro. **2.** A pirate ship. —*vi.* **-rooned, -roon·ing, -roons.** To act as a pirate.

**pic·a·yune** (pĭk′ə-yo͞on′) *adj.* [Fr. *picaillon*, small coin < Prov. *picaioun.*] **1.** Of little importance or value : PALTRY. **2.** Mean : petty. —*n.* **1.** A Spanish-American half-real piece once used in parts of the southern United States. **2.** A five-cent piece. **3.** Something of insignificant value : TRIFLE. **—pic′a·yun′ish** *adj.*

**pic·ca·lil·li** (pĭk′ə-lĭl′ē) *n.*, *pl.* **-lis.** [Prob. alteration of PICKLE.] A spicy relish made of chopped and pickled vegetables.

**pic·co·lo**[1] (pĭk′ə-lō′) *n.*, *pl.* **-los.** [Ital., short for *piccolo flauto*, small flute.] A small flute pitched an octave above a standard flute. **—pic′co·lo′ist** *n.*

**pic·co·lo**[2] (pĭk′ə-lō′) *adj.* [Ital., small.] Designating a musical instrument notably smaller than the standard size <a *piccolo* trumpet>

**pice** (pīs) *n.*, *pl.* **pice.** [Hindi *paisā.*] —See table at CURRENCY.

**pi·ce·ous** (pī′sē-əs) *adj.* [Lat. *piceus* < *pix*, pitch.] **1.** Of or pertaining to pitch. **2.** Having a glossy black color.

**pich·i·ci·e·go** (pĭch′ĭ-sē-ā′gō) *also* **pich·i·ci·a·go** (-ä′gō, -ā′gō) *n.*, *pl.* **-gos.** [Poss. < Allentiac.] **1.** A small armadillo, *Chlamyphorus truncatus* of Argentina, with pale-pink armor and silky white hair. **2.** A South American armadillo, *Burmeisteria retusa*, resembling the pichiciego but with yellow-brown armor.

**pick**[1] (pĭk) *v.* **picked, pick·ing, picks.** [ME *piken*, to prick, prob. < OFr. *piquer* < VLat. *piccare < Lat. *picus*, woodpecker.] —*vt.* **1.** To select from a group <*pick* the winner> **2. a.** To select or cull. **b.** To gather in : HARVEST <*pick* raspberries> **c.** To gather the harvest from <needed two days to *pick* the whole orchard> **3. a.** To remove the outer covering of : PLUCK <*pick* a turkey clean of feathers> **b.** To tear off piece by piece <*pick* petals from a flower> **4.** To remove foreign matter from (the teeth). **5.** To poke and pull at with the fingers. **6.** To break up or separate with a sharp, pointed instrument. **7.** To dig or make (a hole) with a sharp instrument. **8.** To seek and discover (a flaw) <*picked* holes in their presentation> **9.** To take up (food) with the beak : PECK. **10.** To steal the contents of <*pick* a pocket> **11.** To open (a lock) without using the key. **12.** To make (one's way) carefully <*picked* their way over the rocky ground> **13.** To provoke <*pick* a fight> **14. a.** To pluck (the strings of a musical instrument). **b.** To play (a tune) by plucking strings. —*vi.* **1.** To decide or choose <couldn't *pick* between them> **2.** To work with a pick. **3.** To find fault or make petty criticisms : CARP <always *picking* about something> **4.** To be gathered or harvested <The corn *picked* quickly this year.> **—pick apart. 1.** To separate into pieces by picking. **2.** To refute or find flaws in by close examination <The lawyer *picked apart* the plaintiff's testimony.> **—pick at. 1.** To pluck or pull at with the fingers. **2.** To eat sparingly or without appetite. **3.** *Informal.* To nag. **—pick off. 1.** To shoot after singling out. **2.** *Baseball.* To put (a base runner) out with a quick throw, as from the catcher or pitcher. **3.** To intercept, as a football pass. **—pick on.** To tease or bully. **—pick out. 1.** To select or choose. **2.** To discern from the surroundings : DISTINGUISH <The airplane was too far away to pick out.> **3.** To play (music) uncertainly by ear <*pick out* a melody on the piano> **—pick over.** To examine or sort out item by item. **—pick up. 1. a.** To take up (something) by hand. **b.** To gather or collect <*picked up* the scattered toys> **c.** To tidy up <*picked up* the apartment> **2.** To take on (e.g., freight or passengers). **3.** *Informal.* **a.** To acquire casually or by accident <*picked up* some burrs on my socks> **b.** To acquire (knowledge) by study or experience <*picked up* Italian during my travels> **c.** To claim <*picked up* their clothes from the cleaners> **d.** To buy <*picked up* some bread after work> **e.** To accept (a bill) in order to pay it <Let me *pick up* the tab.> **f.** To come down with (a disease) <*picked up* a bad cough> **g.** To gain <*picked up* a first down on that play> **4.** *Slang.* To take into legal custody <The police *picked up* two suspects.> **5.** *Slang.* To make casual acquaintance with, usu. in anticipation of sexual relations. **6.** To come upon and follow <*pick up* a scent> **7.** To continue after a break <Let's *pick up* the discussion after lunch.> **8.** *Informal.* To improve in activity or condition <Business is finally *picking up* again.> **9.** *Informal.* To pack one's belongings <just *picked up* and left> —*n.* **1. a.** The act of picking, esp. with a pointed instrument. **b.** A long-toothed comb used in grooming Afros and some perms. **2.** The act of choosing or selecting : CHOICE <Take your *pick.*> **3.** Something selected as the best or choicest part <the *pick* of the crop> **4.** The amount or quantity of a crop picked by hand. **—pick and choose.** To select with great care. **—pick′er** *n.*

**pick**[2] (pĭk) *n.* [ME *pik.*] **1.** A tool for breaking hard surfaces, having a curved bar sharpened at both ends and fitted to a long handle. **2.** Something used for picking, as a toothpick, an ice pick, or a picklock. **3.** *Mus.* A plectrum.

**pick**[3] (pĭk) *n.* [< ME *pykken*, to stick in the ground.] **1.** A weft thread in weaving. **2.** A passage or throw of the shuttle in a loom. —*vt.* **picked, pick·ing, picks. 1.** To throw (a shuttle) across the loom. **2.** *Archaic.* To cast : pitch.

**pick·a·back** (pĭk′ə-băk′) *adv.* & *n.* & *v. var. of* PIGGYBACK.

**pick·ax** *or* **pick·axe** (pĭk′ăks′) *n.* [Alteration of ME *pikois* < OFr. *picois* < *pic*, pick, prob. < Lat. *picus*, woodpecker.] A pick, esp. with one end of the head pointed and the other end with a chisel edge for cutting through roots. —*v.* **-axed, -ax·ing, -ax·es.** —*vi.* To use a pickax. —*vt.* To use a pickax on.

**picked**[1] (pĭkt) *adj.* **1.** Chosen by careful selection. **2.** Cleaned by picking out damaged or undesirable parts. **3.** Harvested : gathered. **4.** Worked upon with a pick. **5.** Ornamented with a hand-worked line of short running stitches along the edges.

**†picked**[2] (pĭkt) *adj.* [< PICK[2].] *Regional.* Pointed <a *picked* cap>

**pick·er·el** (pĭk′ər-əl, pĭk′rəl) *n.*, *pl.* **pickerel** *or* **-els.** [ME *pikerel*, dim. of *pike*, pike.] **1.** A North American freshwater food and game fish of the genus *Esox*, esp. *E. niger* or *E. vermiculatus.* **2.** A fish, as the walleye, related to or resembling the pickerel. **3.** *Chiefly Brit.* A small or young pike.

**pick·er·el·weed** (pĭk′ər-əl-wēd′, pĭk′rəl-) *n.* A North American plant, *Pontederia cordata*, found in freshwater shallows and having heart-shaped leaves and violet-blue flower spikes.

**pick·et** (pĭk′ĭt) *n.* [Fr. *piquet* < OFr. < *piquer*, to prick. —see PICK[1].] **1.** A pointed stake driven into the ground, as to secure a tent, support a fence, tether animals, mark points in surveying, or serve as a defense. **2.** A detachment of one or more soldiers advanced or held in readiness to warn of an enemy's approach. **3. a.** A person stationed outside a place of employment, usu. during a strike, to express protest or grievance and discourage entry by customers or nonstriking employees. **b.** A protester or demonstrator, esp. one carrying a sign. —*v.* **-et·ed, -et·ing, -ets.** —*vt.* **1.** To secure, enclose, tether, mark out, or fortify with pickets. **2. a.** To post as a picket. **b.** To guard with a picket. **3.** To station (a picket or pickets) during a strike or demonstration. —*vi.* To serve as a picket. **—pick′et·er** *n.*

**picket fence** *n.* A fence of pointed, upright pickets.

**picket line** *n.* A line or procession of people picketing a company, business establishment, institution, etc.

**pick·ing** (pĭk′ĭng) *n.* **1.** The act of one that picks. **2. pickings.** Something that is or may be picked. **3.** *often* **pickings. a.** Leftovers. **b.** A share of spoils.

**pick·le** (pĭk′əl) *n.* [ME *pekille,* prob. < MDu. *pekel.*] **1.** An item of food, as a cucumber, preserved and flavored in a solution of brine or vinegar. **2.** A solution of brine or vinegar, often spiced, for preserving and flavoring food. **3.** An acid or other chemical solution used as a bath to remove oxides and scale from the surface of metals before plating or finishing. **4.** *Informal.* A troublesome, difficult, or embarrassing situation. —*vt.* **-led, -ling, -les. 1.** To preserve or flavor in brine or vinegar. **2.** To treat (metal) in a chemical bath.

**pick·led** (pĭk′əld) *adj.* **1.** Preserved in or treated with pickle <*pickled* eggs> **2.** *Slang.* Drunk.

**pick·lock** (pĭk′lŏk′) *n.* **1.** One who picks locks, esp. a thief. **2.** An instrument for picking a lock.

**pick-me-up** (pĭk′mē-ŭp′) *n. Informal.* An often alcoholic drink taken as a stimulant or to relieve a hangover.

**pick·pock·et** (pĭk′pŏk′ĭt) *n.* One who steals from pockets or purses.

**pick·proof** (pĭk′pro͞of′) *adj.* Designed to prevent picking <a *pickproof* lock>

**pick·up** (pĭk′ŭp′) *n.* **1. a.** The action or process of picking something up <the *pickup* and delivery of packages> **b.** The act of striking or fielding a ball after it has touched the ground <a good *pickup* by the shortstop> **c.** Capacity for acceleration <a sports car with good *pickup*> **d.** *Informal.* An improvement in activity or condition <a *pickup* in new orders> **e.** *Slang.* An arrest. **2.** One that is picked up, esp.: **a.** Freight or passengers. **b.** A balance brought forward in accounting. **c.** Previous journalistic copy to which later copy is added. **d.** *Mus.* An unstressed note or notes leading into a passage or melody. **e.** *Informal.* A hitchhiker. **f.** *Slang.* A stranger with whom casual acquaintance is made, usu. in anticipation of sexual relations. **3.** One that picks up, esp.: **a.** A pickup truck. **b.** The rotary rake on machinery, as a harvester, that picks up windrowed hay or straw. **4.** *Electron.* **a.** A device for converting the oscillations of a phonograph needle into electrical impulses. **b.** The tone arm of a record player. **5. a.** Reception of sound or light waves for conversion into electrical impulses. **b.** The apparatus used for such reception. **c.** A telecast originating outside a studio. **d.** The apparatus for transmitting a broadcast from an outside location to the broadcasting station.

**pickup truck** *n.* A light truck with an open body and low sides.

**pick·y** (pĭk′ē) *adj.* **-i·er, -i·est.** *Informal.* Excessively fastidious.

**pic·lo·ram** (pĭk′lə-răm′, pī′klə-) *n.* [PIC(OLINE) + (CH)LOR(O)- + AM(INE).] A colorless compound, $C_6H_3Cl_3N_2O_2$, used as a herbicide.

**pic·nic** (pĭk′nĭk) *n.* [Fr. *pique-nique,* prob. redup. of *piquer,* to pick.] **1.** An open-air meal, esp. one eaten on an excursion. **2.** *Slang.* An easy task or pleasant experience <The exam was no *picnic.*> **3.** A shoulder of pork from which most of the butt has been removed. —*vi.* **-nicked, -nick·ing, -nics.** To go on or participate in a picnic. **—pic′nick·er** *n.* **—pic′nick·y** *adj.*

▲ word history: The word *picnic* is most probably derived from French *pique-nique,* whose origin is not known with absolute certainty. The word seems to have originated in the 18th century when it referred to a meal for which each guest contributed a dish to be served. At some point it became common to hold such parties outdoors, and by the 19th century *picnic* referred to any open-air meal.

**pico-** *pref.* [Sp. *pico,* small quantity < *picar,* to prick. —see PICARO.] **1.** One trillionth ($10^{-12}$) <*pico*second> **2.** Very small <*pico*rnavirus>

**pi·co·gram** (pē′kə-grăm′, pī′-) *n.* One trillionth ($10^{-12}$) of a gram.

**pic·o·line** (pĭk′ə-lēn′, pī′kə-) *n.* [Lat. *pix,* pitch + -OL + -INE.] Any of three isomeric liquid methylpyridine bases, $C_6H_7N$, obtained from coal tar, bone oil, and horse urine and used as an industrial solvent.

**pi·cor·na·vi·rus** (pē-kôr′nə-vī′rəs, pī-) *n.* [PICO- + RNA + VIRUS.] Any of a group of very small RNA-containing viruses that includes the Coxsackie virus.

**pi·co·sec·ond** (pē′kə-sĕk′ənd, pī′-) *n.* One trillionth ($10^{-12}$) of a second.

**pi·cot** (pē′kō, pē-kō′) *n.* [Fr. < OFr. < *pic,* point < *piquer,* to prick. —see PICK[1].] A small embroidered loop forming an ornamental edging on some lace and ribbon. —*vt.* **-coted** (-kōd), **-cot·ing** (-kō-ĭng), **-cots** (-kōz). To trim with edging.

**pic·o·tee** (pĭk′ə-tē′) *n.* [Fr. *picoté,* marked with points < *picoter,* to mark with points < *picot,* point, picot.] A carnation having pale petals edged with a darker color.

**pic·quet** (pĭ-kā′) *n. var. of* PIQUET.

**picr-** *pref. var. of* PICRO-.

**pic·rate** (pĭk′rāt′) *n.* A salt or ester of picric acid.

**pic·ric acid** (pĭk′rĭk). A poisonous, explosive yellow crystalline solid, $C_6H_2(NO_2)_3OH$, used in explosives, antiseptics, and dyes.

**picro-** *or* **picr-** *pref.* [Gk. *pikro-* < *pikros,* bitter.] **1.** Bitter <*picro*toxin> **2.** Picric acid <*picr*ate>

**pic·ro·tox·in** (pĭk′rə-tŏk′sĭn) *n.* A bitter powder, $C_{30}H_{34}O_{13}$, used as a stimulant and antidote for barbiturate poisoning.

**Pict** (pĭkt) *n.* [ME < LLat. *Picti,* Picts.] One of the ancient inhabitants of North Britain, who were assimilated by the invading Scots between the 6th and 9th cent. A.D.

**pic·to·gram** (pĭk′tə-grăm′) *n.* [PICTO(GRAPH) + -GRAM.] A pictorial representation of numerical data or relationships.

**pic·to·graph** (pĭk′tə-grăf′) *n.* [Lat. *pictus,* p.part. of *pingere,* to paint + -GRAPH.] **1.** A stylized picture representing a word or idea : HIEROGLYPH. **2.** A record in hieroglyphics. **—pic′to·graph′ic** *adj.* **—pic′to·graph′i·cal·ly** *adv.* **—pic·tog′ra·phy** (pĭk-tŏg′rə-fē) *n.*

**Pic·tor** (pĭk′tər) *n.* [Lat. *pictor,* painter < *pingere,* to paint.] A constellation in the Southern Hemisphere.

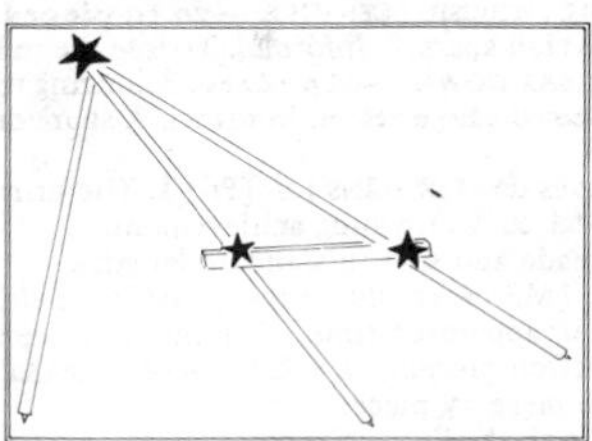

Pictor

**pic·to·ri·al** (pĭk-tôr′ē-əl, -tōr′-) *adj.* [< LLat. *pictorius* < Lat. *pictor,* painter < *pingere,* to paint.] **1.** Relating to, marked by, or consisting of pictures. **2.** Graphic : depictive <*pictorial* writing> **3.** Illustrated by pictures. **—pic·to′ri·al** *n.* **—pic·to′ri·al·ly** *adv.* **—pic·to′ri·al·ness** *n.*

**pic·ture** (pĭk′chər) *n.* [ME < Lat. *pictura* < *pingere,* to paint.] **1.** A visual representation or design rendered on a surface, as by drawing, painting, photography, or printmaking. **2.** A visible image, esp. one on a flat surface <saw my *picture* reflected in the glass> **3.** A vivid or realistic verbal description <a novel that gives a *picture* of another era> **4.** One that bears a striking resemblance to another : LIKENESS. **5.** One that embodies or typifies an emotion, mood, or state of being <You're the *picture* of health.> **6.** The chief circumstances of an event or time : SITUATION <gave me the whole *picture*> **7.** A motion picture. **8.** A tableau vivant. —*vt.* **-tured, -tur·ing, -tures. 1.** To make a visual representation or picture of. **2.** To form a mental image of : VISUALIZE <I can't quite *picture* it.> **3.** To describe vividly in words.

**Pic·ture·phone** (pĭk′chər-fōn′). A trademark for a device that combines telephone and television communications.

**picture puzzle** *n.* A jigsaw puzzle.

**pic·tur·esque** (pĭk′chə-rĕsk′) *adj.* [Fr. *pittoresque* < Ital. *pittoresco* < *pittore,* painter < Lat. *pictor* < *pingere,* to paint.] **1.** Of, suggesting, or suitable for a picture <*picturesque* mountain scenery> **2.** Interesting or quaintly attractive : CHARMING <a *picturesque* country inn> **3.** Strikingly expressive : VIVID <*picturesque* language> **—pic′tur·esque′ly** *adv.* **—pic′tur·esque′ness** *n.*

**picture window** *n.* A large, usu. single-paned window offering a broad outside view.

**pic·ul** (pĭk′əl) *n.* [Malay *pikul.*] Any of various Oriental units of weight, esp. a Chinese unit equal to 133.33 pounds or 214.53 kilograms.

**pid·dle** (pĭd′l) *v.* **-dled, -dling, -dles.** [Orig. unknown.] —*vt.* To use triflingly : SQUANDER <*piddled* away my fortune> —*vi.* To spend time aimlessly : DIDDLE.

**pid·dling** (pĭd′lĭng) *adj.* Beneath consideration : TRIFLING.

**pid·dock** (pĭd′ək) *n.* [Orig. unknown.] A marine bivalve mollusk of the family Pholadidae, capable of boring into wood, rock, and clay.

**pidg·in** (pĭj′ən) *n.* [< PIDGIN ENGLISH.] A simplified medium of speech, usu. a mixture of several languages, having a rudimentary vocabulary and grammar and used for communication between groups speaking different languages. **—pidg′in·i·za′tion** *n.* **—pidg′in·ize′** *v.* **(-ized, -iz·ing, -iz·es).**

**Pidgin English** *also* **pidgin English** *n.* [Alteration of *business English.*] A pidgin based on English, used as a trade language in parts of Eastern Asia and Melanesia.

**pie[1]** (pī) *n.* [ME.] **1.** A baked dish consisting of a pastry shell filled with fruit, cheese, meat, or other ingredients and usu. covered with a pastry crust. **2.** A layer cake with custard, cream, or jelly filling. **—pie in the sky.** An empty promise.

**pie[2]** (pī) *n.* [ME < OFr. < Lat. *pica.*] MAGPIE 1.

**pie[3]** (pī) *n.* [Hindi *pā′ī* < Skt. *pādika-,* quarter < *pādah,* foot, leg.] A monetary unit once used in India and Pakistan.

**pie[4]** (pī) *n.* [Med. Lat. *pica.*] An almanac of services used in the English church before the Reformation.

**pie[5]** (pī) *n.* & *v. var. of* PI[2].

**pie·bald** (pī′bôld′) *adj.* [PIE[2] + BALD.] Having variegated markings, esp. spotted in black and white <a *piebald* horse> —*n.* A piebald animal, esp. a horse.

**piece** (pēs) *n.* [ME *pece* < OFr. < Med. Lat. *pecia*, prob. of Gaulish orig.] **1.** A thing regarded as a unit or element of a larger quantity or class : PORTION <a *piece* of pie><a *piece* of a puzzle> **2.** A part separated or broken off from a whole : FRAGMENT <a *piece* of glass> **3.** An object that is one member of a group or class <a *piece* of furniture> **4.** An artistic, literary, or musical work or composition. **5.** An instance : specimen <a *piece* of advice> **6.** One's fully expressed opinion <speak one's *piece*> **7.** A coin <a 50¢ *piece*> **8.** One of the counters or men used in playing board games. **9.** Any of the chess figures other than a pawn. **10.** A firearm. **11.** A short distance <"There was farm country down the road on the right a *piece*" —James Agee> —*vt.* **pieced, piec·ing, piec·es. 1.** To mend or restore by adding pieces to. **2.** To join the pieces of : ASSEMBLE <*pieced* together the model airplane> **—a piece of (one's) mind.** *Informal.* Frank and unsparing criticism : CENSURE. **—go to pieces. 1.** To shatter into small pieces : fall apart. **2.** *Informal.* To lose mental and emotional self-control : BREAK DOWN. **—of a piece.** Belonging to the same kind or class. **—piece of the action.** *Informal.* A share of an activity or of profits.

**pièce de ré·sis·tance** (pyĕs də rā-zē-stäNs′) *n.* [Fr.] **1.** The principal or featured dish of a meal. **2.** A crowning achievement.

**piece goods** *pl.n.* Fabrics made and sold in standard lengths.

**piece·meal** (pēs′mēl′) *adv.* [ME *pecemele* : *pece*, piece + *-mele*, by a fixed measure < OE *mǣl*, appointed time.] **1.** Piece by piece : GRADUALLY <built my collection *piecemeal*> **2.** In pieces : APART. —*adj.* Accomplished or made piece by piece.

**piece of eight** *n.* An old Spanish silver coin.

**piece·work** (pēs′wûrk′) *n.* Work paid for according to the number of items produced. **—piece′work′er** *n.*

**pie chart** *n.* A circular chart having radii dividing the circle into areas proportional to the relative size of the quantities represented.

**pied**[1] (pīd) *adj.* [ME < *pie*, magpie. —see PIE[2].] Spotted or patched with color : SPLOTCHED.

**pied**[2] (pīd) *v. var. p.t. & p.p. of* PI[2].

**pied-à-terre** (pyā-dä-târ′) *n., pl.* **pieds-à-terre** (pyā-dä-târ′) [Fr. : *pied*, foot + *à*, to + *terre*, ground.] A second or temporary lodging.

**pied·mont** (pēd′mŏnt′) *n.* [After *Piedmont*, a region in Italy.] A region or area situated at the foot of a mountain range. **—pied′mont′** *adj.*

**pied piper** *n.* [After *The Pied Piper of Hamelin*, title and hero of a poem by Robert Browning (1812–1889).] **1.** One who entices others with delusive promises. **2.** An appealing but irresponsible leader.

**pie·plant** (pī′plănt′) *n.* RHUBARB 1.

**pier** (pîr) *n.* [ME *per* < OE < Med. Lat. *pera.*] **1. a.** A platform extending from a shore over water and supported by piles or pillars, used to secure, shelter, and provide access to vessels. **b.** Such a structure used mainly for public recreation. **2.** A supporting structure at the junction of connecting spans of a bridge. **3.** Any of various vertical supporting structures, esp.: **a.** A pillar, rectangular in cross section, supporting a roof or arch. **b.** A section of wall between windows or doors. **c.** A reinforcing structure projecting from a wall : BUTTRESS.

**pierce** (pîrs) *v.* **pierced, pierc·ing, pierc·es.** [ME *percen* < OFr. *percer* < VLat. **pertusiare* < Lat. *pertundere*, to bore through : *per-*, through + *tundere*, to beat.] —*vt.* **1.** To puncture or pass through with or as if with a sharp instrument : STAB. **2.** To make a hole or opening in : PERFORATE. **3.** To make a way through <a trail that *pierced* the mountains> **4.** To sound sharply through : PENETRATE <A shout *pierced* the fog.> **5.** To succeed in understanding or discerning <*pierced* the complexities of the problem> **6.** To affect penetratingly : move deeply <was *pierced* by guilt> —*vi.* To penetrate into or through something. **—pierc′er** *n.* **—pierc′ing** *adj.* **—pierc′ing·ly** *adv.*

**Pi·e·ri·an Spring** (pī-îr′ē-ən) *n.* [< Lat. *Pierius*, sacred to the Muses < Gk. *Pieria*, a region of Macedonia.] **1.** *Gk. Myth.* A spring in Macedonia, sacred to the Muses. **2.** A source of inspiration.

**Pier·rot** (pē′ə-rō′, pyĕ-rō′) *n.* [Fr., dim. of the name *Pierre*, Peter < Lat. *Petrus.*] A comic character in traditional French pantomime, dressed in a floppy white costume.

**pie·tà** *also* **Pie·tà** (pyā-tä′) *n.* [Ital., pity < Lat. *pietas.* —see PIETY.] A representation of the Virgin Mary holding and mourning over the dead body of Jesus.

**pi·e·tism** (pī′ĭ-tĭz′əm) *n.* [G. *Pietismus* < Lat. *pietas*, piety.] **1.** Piety. **2.** Exaggerated or affected piety. **3. Pietism.** A reform movement in the German Lutheran Church during the 17th and 18th cent., which sought to renew the devotional ideal in the Protestant religion. **—pi′e·tist** *n.* **—pi′e·tis′tic, pi′e·tis′ti·cal** *adj.* **—pi′e·tis′ti·cal·ly** *adv.*

**pi·e·ty** (pī′ĭ-tē) *n., pl.* **-ties.** [Fr. *pieté* < Lat. *pietas*, dutiful conduct < *pius*, dutiful.] **1.** Religious devotion and reverence to God. **2.** Devotion and fidelity to parents and family. **3.** A pious thought or act. **4.** The quality or state of being pious.

**piezo-** *pref.* [< Gk. *piezein*, to squeeze.] Pressure <*piezometer*>

**pi·e·zo·e·lec·tric·i·ty** (pī-ē′zō-ə-lĕk-trĭs′ĭ-tē, pē-ā′zō-) *n. Physics.* Generation of electricity or of electric polarity in dielectric crystals subjected to mechanical stress, and conversely, generation of stress in such crystals subjected to an applied voltage. **—pi·e′zo·e·lec′tric, pi·e′zo·e·lec′tri·cal** *adj.* **—pi·e′zo·e·lec′tri·cal·ly** *adv.*

**pi·e·zom·e·ter** (pī′ĭ-zŏm′ĭ-tər, pē′ĭ-) *n.* An instrument for measuring pressure, esp. high pressure. **—pi·e′zo·met′ric** (pī-ē′zə-mĕt′rĭk, pē-ā′zə-), **pi′e′zo·met′ri·cal** *adj.* **—pi·e′zom′e·try** *n.*

**pif·fle** (pĭf′əl) *vi.* **-fled, -fling, -fles.** [Orig. unknown.] To talk or act in a foolish or futile way. —*n.* Inane ideas or talk : NONSENSE.

**pig** (pĭg) *n.* [ME *pigge.*] **1.** A stout-bodied mammal of the family Suidae, with short legs, cloven hoofs, bristly hair, and a cartilaginous snout used for rooting, esp. the domesticated hog, *Sus scrofa*, when young or relatively small. **2.** The edible parts of a pig. **3.** *Informal.* One considered to be greedy, dirty, or messy. **4.** GUINEA PIG 1. **5. a.** An oblong block of metal, chiefly iron or lead, poured from a smelting furnace. **b.** A mold in which such metal is cast. **c.** Pig iron. —*vi.* **pigged, pig·ging, pigs.** To give birth to pigs : FARROW. **—pig in a poke.** Something offered in a manner that conceals its true value or nature. **—pig it.** To live or eat in a piglike manner. **—pig out.** *Slang.* To eat to surfeit.

**pig bed** *n.* A bed of sand in which pigs of iron are cast.

**pig·boat** (pĭg′bōt′) *n. Slang.* A submarine.

**pi·geon** (pĭj′ən) *n.* [ME < OFr. *pijon* < Lat. *pipio*, young chirping bird < *pipire*, to chirp.] **1.** A widely distributed dove of the family Columbidae, having a prominent chest, short legs, and a small head, esp. the rock dove, *Columba livia*, or any of its domesticated varieties. **2.** *Slang.* One easily swindled : DUPE.

**pigeon breast** *n.* Chicken breast. **—pi′geon-breast′ed** *adj.*

**pigeon hawk** *n.* A small falcon, *Falco columbarius.*

**pi·geon·hole** (pĭj′ən-hōl′) *n.* **1.** A hole for nesting in a pigeon loft. **2.** A small compartment or recess, as in a desk, for holding papers : CUBBYHOLE. **3.** A specific, often misleading stereotyped category. —*vt.* **-holed, -hol·ing, -holes. 1.** To place or file in a pigeonhole. **2.** To classify mentally : CATEGORIZE. **3.** To set aside and ignore : SHELVE.

**pigeon pea** *n.* **1.** A tropical shrub, *Cajanus indicus*, bearing orange-yellow flowers. **2.** The edible brown seed of the pigeon pea.

**pi·geon-toed** (pĭj′ən-tōd′) *adj.* Having the toes turned inward.

**pi·geon·wing** (pĭj′ən-wĭng′) *n.* A dance step performed by jumping and clapping the feet together.

**pig·fish** (pĭg′fĭsh′) *n., pl.* **pigfish** *or* **-fish·es.** A marine food fish, *Orthopristis chrysopterus* of U.S. coastal Atlantic waters.

**pig·ger·y** (pĭg′ə-rē) *n., pl.* **-ies.** A place where pigs are kept.

**pig·gin** (pĭg′ĭn) *n.* [Orig. unknown.] A small wooden bucket with one stave projecting above the rim for use as a handle.

**pig·gish** (pĭg′ĭsh) *adj.* **1.** Like a pig. **2.** Stubborn : pigheaded. **—pig′gish·ly** *adv.* **—pig′gish·ness** *n.*

**pig·gy** (pĭg′ē) *n., pl.* **-gies.** A little pig.

**pig·gy·back** (pĭg′ē-băk′) *also* **pick·a·back** (pĭk′ə-băk′) [Alteration of PICKABACK.] —*adv.* **1.** On the back or shoulders <carry a child *piggyback*> **2.** By a method of transportation in which truck trailers are carried on trains or cars on specially designed trucks. —*n.* The act of transporting piggyback. —*v.* **-backed, -back·ing, -backs.** —*vt.* To cause to be aligned with something, as an issue, that is larger or more important. —*vi.* To function as if carried on the back of another. **—pig′gy·back′** *adj.*

**piggy bank** *n.* A coin bank shaped like a pig.

**pig·head·ed** (pĭg′hĕd′ĭd) *adj.* Stubborn. **—pig′head′ed·ly** *adv.* **—pig′head′ed·ness** *n.*

**pig iron** *n.* Crude iron cast in blocks.

**pig Latin** *n.* A jargon systematically formed by transposing the initial consonant to the end of the word so as to form an additional syllable, as *igpay atinlay* for *pig Latin.*

**pig lead** *n.* Crude lead cast in blocks.

**pig·let** (pĭg′lĭt) *n.* A young pig.

**pig·ment** (pĭg′mənt) *n.* [Lat. *pigmentum* < *pingere*, to paint.] **1.** A substance used as coloring. **2.** Dry coloring matter, usu. an insoluble powder to be mixed with a base, as oil or water, to make paint and similar products. **3.** *Biol.* A substance, as hemoglobin or chlorophyll, that imparts a characteristic color to animal or plant tissue. —*vt.* **-ment·ed, -ment·ing, -ments.** To color with pigment. **—pig′men·tar′y** (pĭg′mən-tĕr′ē) *adj.*

**pig·men·ta·tion** (pĭg′mən-tā′shən) *n. Biol.* **1.** Coloration of tissues by pigment. **2.** Deposition of pigment by cells.

**Pig·my** (pĭg′mē) *n. & adj. var. of* PYGMY.

**pig·nut** (pĭg′nŭt′) *n.* **1.** A hickory tree, *Carya glabra* or *C. ovalis* of the eastern United States, bearing edible nuts. **2.** The nut of either of the pignut trees, having a rather bitter kernel. **3.** The earthnut.

**pig·pen** (pĭg′pĕn′) *n.* **1.** A pen for pigs. **2.** A dirty place.

**pig·skin** (pĭg′skĭn′) *n.* **1.** The skin of a pig. **2.** Leather made from pigskin. **3.** *Informal.* **a.** A football. **b.** A saddle.

**pigs·ney** (pĭgz′nē) *n.* [ME *piggesnye* : *pigge*, pig + *nye*, alteration of *eye*, eye.] *Obs.* **1.** A darling. **2.** An eye.

**pig·sty** (pĭg′stī′) *n., pl.* **-sties.** A pigpen.

**pig·tail** (pĭg′tāl′) *n.* **1.** A plait of tightly braided hair. **2.** A twisted strand of tobacco. **—pig′tailed′** *adj.*

**pig·weed** (pĭg′wēd′) *n.* **1.** A common wild plant, *Chenopodium album*, having leaves with a mealy texture and small green flowers.

2. A coarse weed, *Amaranthus retroflexus,* with hairy leaves and green flower spikes.

**pi·ka** (pē′kə) *n.* [Tungus *piika.*] A small, tailless, harelike mammal of the genus *Ochotona,* found in mountainous areas of North America and Eurasia.

**pike[1]** (pīk) *n.* [OFr. *pique* < *piquer,* to prick. —see PICK[1].] A long spear once used by infantry. —*vt.* **piked, pik·ing, pikes.** To pierce with a pike. **—piked** *adj.*

**pike[2]** (pīk) *n., pl.* **pike** *or* **pikes.** [ME.] **1.** A large freshwater food and game fish, *Esox lucius* of the Northern Hemisphere, having a long snout and a narrow body. **2.** A fish similar to the pike.

**pike[3]** (pīk) *n.* [Short for TURNPIKE.] **1.** A turnpike. **2. a.** A tollgate on a turnpike. **b.** The toll charged. —*vi.* **piked, pik·ing, pikes.** To move quickly.

**pike[4]** (pīk) *n.* [ME, poss. of Scand. orig.] *Chiefly Brit.* A hill with a pointed summit.

**pike[5]** (pīk) *n.* [ME < OE *pīc.*] A sharp point or spike, as the tip of a spear.

**pike perch** *n.* A fish, as the walleye, related to the perch and resembling the pike.

**pik·er** (pī′kər) *n.* [Orig. unknown.] *Slang.* A stingy, small-minded person, esp. an overcautious gambler.

**pike·staff** (pīk′stăf′) *n.* **1.** The shaft of a pike. **2.** A walking stick tipped with a metal spike.

**pi·laf** *or* **pi·laff** (pĭ-läf′, pē-) *also* **pi·lau** (pĭ-lô′, pē-) *n.* [Turk. *pilāw.*] A dish made with rice steamed in a seasoned broth and mixed with meat, vegetables, or fish.

**pi·lar** (pī′lər) *adj.* [NLat. *pilaris* < Lat. *pilus,* hair.] Of, relating to, or covered with hair.

**pi·las·ter** (pĭ-lăs′tər) *n.* [OFr. *pilastre* < Ital. *pilastro* < Med. Lat. *pilastrum* < Lat. *pila,* pillar.] A supporting column or pillar with a capital and base, set partially into a wall as an ornamental motif.

**pi·lau** (pĭ-lô′, pē-) *n. var. of* PILAF.

**pil·chard** (pĭl′chərd) *n.* [Orig. unknown.] A sardinelike fish related to the herring, esp. a commercially important edible species, *Sardina pilchardus* of European waters.

**pile[1]** (pīl) *n.* [ME < OFr. < Lat. *pila,* pillar.] **1.** A mass of material or quantity of objects lying or thrown together in a heap. **2.** *Informal.* A great deal or amount <a *pile* of work> **3.** *Slang.* A large sum of money : FORTUNE. **4.** A funeral pyre. **5.** A large building or complex of buildings. **6.** *Physics.* A nuclear reactor. **7.** *Elect.* Voltaic pile. —*v.* **piled, pil·ing, piles.** —*vt.* **1.** To place or stack in a pile. **2.** To load with a pile <*piled* my plate with food> —*vi.* **1.** To form a heap or pile. **2.** To move in a disorderly group or mass <*pile* out of the classroom> **—pile up. 1.** To accumulate. **2.** To undergo a serious vehicular collision.

**pile[2]** (pīl) *n.* [ME < OE *pīl* < Lat. *pilum,* spear.] **1.** A long, sturdy post or column of timber, steel, or concrete, driven into the earth as a foundation or support for a structure. **2.** *Heraldry.* A wedge-shaped charge pointing downward. **3.** A Roman javelin. —*vt.* **piled, pil·ing, piles. 1.** To drive piles into. **2.** To support with piles.

**pile[3]** (pīl) *n.* [ME < Lat. *pilus,* hair.] **1.** The soft, resilient surface of certain fabrics, as plush, velvet, and carpeting, consisting of closely spaced, cut or uncut loops of yarn. **2.** Soft, fine fur, hair, or wool. **—piled** *adj.*

**pi·le·a** (pī′lē-ə) *n. pl. of* PILEUM.

**pi·le·at·ed** (pī′lē-ā′tĭd) *also* **pi·le·ate** (-ĭt) *adj.* [< Lat. *pileatus,* wearing a pileus < *pileus,* felt cap.] **1.** *Bot.* Having a pileus. **2.** Having a crest covering the pileum.

**pileated woodpecker** *n.* A large North American woodpecker, *Dryocopus pileatus,* with black and white plumage and a bright red crest.

**pile driver** *n.* **1.** A machine that drives piles by raising a weight between guideposts and dropping it on the head of the pile. **2.** An operator of a pile driver.

**piles** (pīlz) *pl.n.* [< Lat. *pila,* ball.] HEMORRHOIDS 2.

**pi·le·um** (pī′lē-əm) *n., pl.* **-le·a** (-lē-ə) [NLat. < Lat. *pileus,* felt cap.] The top of a bird's head from the base of the bill to the nape.

**pile·up** (pīl′ŭp′) *n.* A serious traffic collision usu. involving several motor vehicles.

**pi·le·us** (pī′lē-əs) *n., pl.* **-le·i** (-lē-ī′) [NLat. < Lat., cap.] **1.** *Bot.* The umbrellalike cap of a stalked, fleshy fungus, as a mushroom. **2.** A round brimless skullcap worn by ancient Romans.

**pile·wort** (pīl′wûrt′, -wôrt′) *n.* A plant, as the lesser celandine or fireweed, reputed to be effective in treating piles.

**pil·fer** (pĭl′fər) *v.* **-fered, -fer·ing, -fers.** [OFr. *pelfrer,* to rob < *pelfre,* booty.] —*vt.* To steal, esp. repeatedly and in small amounts : FILCH. —*vi.* To steal. **—pil′fer·age** (-ĭj) *n.* **—pil′fer·er** *n.*

**pil·grim** (pĭl′grəm) *n.* [ME < OFr. *peligrin* < LLat. *pelegrinus,* alteration of Lat. *peregrinus,* foreigner.] **1.** A person who journeys to a shrine or sacred place as an act of religious devotion. **2.** One who undertakes a quest for a goal believed to be sacred. **3.** A traveler : wanderer. **4. Pilgrim.** One of the English Puritans who founded the colony of Plymouth in New England in 1620.

**pil·grim·age** (pĭl′grə-mĭj) *n.* **1.** A journey to a holy place or shrine. **2.** An extended journey or search, esp. one of exalted purpose. —*vi.* **-aged, -ag·ing, -ag·es.** To make or go on a pilgrimage.

**pi·li** (pī′lī′) *n. pl. of* PILUS.

**pil·ing** (pī′lĭng) *n.* **1.** The act of driving piles. **2.** Piles collectively. **3.** A structure consisting of piles.

**Pil·i·pi·no** (pĭl′ə-pē′nō) *n.* [Tagalog < *pilipino,* Filipino < Sp.] A language based on Tagalog that is the official language of the Republic of the Philippines.

**pill[1]** (pĭl) *n.* [Lat. *pilula,* dim. of *pila,* ball.] **1.** A small, often coated tablet or pellet of medicine, taken by swallowing whole or chewing. **2. the pill.** *Informal.* An oral contraceptive. **3.** *Slang.* Something, as a baseball, that resembles a pill. **4.** Something unpleasant or distasteful but inevitable <Losing the election was a bitter *pill.*> **5.** *Slang.* An ill-natured, disagreeable person. —*v.* **pilled, pil·ling, pills.** —*vt.* **1.** To dose with pills. **2.** To make into pills. **3.** *Slang.* To blackball. —*vi.* To form small balls resembling pills <a cardigan that *pills*>

**pill[2]** (pĭl) *v.* **pilled, pil·ling, pills.** [ME *pillen* < OFr. *piller.*] —*vt.* *Archaic.* To subject to extortion. —*vi.* *Chiefly Brit.* To come off, as in scales or flakes.

**pil·lage** (pĭl′ĭj) *v.* **-laged, -lag·ing, -lag·es.** [ME < OFr. < *piller,* to plunder < *peille,* rag < Lat. *pilleus, pileus,* felt cap.] —*vt.* **1.** To rob of goods or property by force, esp. in time of war : PLUNDER. **2.** To take as spoils. —*vi.* To take spoils by force. —*n.* **1.** The act of pillaging. **2.** Something pillaged : SPOILS. **—pil′lag·er** *n.*

**pil·lar** (pĭl′ər) *n.* [ME *piller* < OFr. *pilier* < Med. Lat. *pilare* < Lat. *pila.*] **1.** A slender freestanding vertical support : COLUMN. **2.** A decorative pillar or shaft. **3.** A person of central importance or responsibility <a *pillar* of the community> —*vt.* **-lared, -lar·ing, -lars.** To support or decorate with pillars. **—from pillar to post.** From one place, situation, or resource to another.

**pill·box** (pĭl′bŏks′) *n.* **1.** A small box for pills. **2.** A woman's small round flat-topped hat. **3.** A fortified concrete gun emplacement with a low flat roof.

**pill bug** *n.* A small terrestrial crustacean of the genus *Armadillidium* or related genera, with a segmented body capable of being curled into a ball.

**pil·lion** (pĭl′yən) *n.* [Prob. < Sc. Gael. *pillean,* dim. of *peall,* covering < Lat. *pellis,* skin.] **1.** A cushion for an extra rider behind the saddle on a horse or motorcycle. **2.** A bicycle or motorcycle saddle.

**pil·lo·ry** (pĭl′ə-rē) *n., pl.* **-ries.** [ME < OFr. *pilori.*] A wooden framework with holes for the head and hands, in which offenders were once locked to be exposed to public ridicule as punishment. —*vt.* **-ried, -ry·ing, -ries. 1.** To punish by putting in a pillory. **2.** To expose to scorn and abuse.

**pil·low** (pĭl′ō) *n.* [ME *pilwe* < OE *pyle* < Lat. *pulvinus.*] **1.** A cloth case filled with a soft stuffing, as down, feathers, or foam rubber, and used to cushion the head esp. during sleep. **2.** A decorative cushion. **3.** The pad on which bobbin lace is made. —*v.* **-lowed, -low·ing, -lows.** —*vt.* **1.** To rest (one's head) on or as if on a pillow. **2.** To act as a pillow for <Leaves *pillowed* my head.> —*vi.* To rest on or as if on a pillow. **—pil′low·y** *adj.*

**pillow block** *n.* A block that encloses and supports a journal or shaft : BEARING.

**pil·low·case** (pĭl′ō-kās′) *n.* A removable covering for a pillow.

**pillow lace** *n.* Bobbin lace.

**pil·low·slip** (pĭl′ō-slĭp′) *n.* A pillowcase.

**pi·lo·car·pine** (pī′lō-kär′pēn′) *n.* [< NLat. *Pilocarpus,* jaborandi genus : Gk. *pilos,* wool + Gk. *karpos,* fruit.] A poisonous, colorless or yellow compound, $C_{11}H_{16}N_2O_2$, derived from the leaves of the jaborandi tree and used to induce sweating.

**pi·lose** (pī′lōs′) *adj.* [Lat. *pilosus* < *pilus,* hair.] Covered with fine downy hair. **—pi·los′i·ty** (-lŏs′ĭ-tē) *n.*

**pi·lot** (pī′lət) *n.* [OFr. *pilote* < OItal. *pilota,* alteration of *pedota,* prob. < Med. Gk. **pedōtēs* < Gk. *pedon,* rudder.] **1.** One who operates or is licensed to operate an aircraft in flight. **2. a.** One who is licensed or employed to conduct a ship into and out of port or through dangerous waters. **b.** A helmsman. **3.** One who guides or directs others. **4.** The part of a tool or machine that guides the whole. **5.** PILOT LIGHT 1. **6.** A television program produced as a prototype of a series being considered for adoption by a network. —*vt.* **-lot·ed, -lot·ing, -lots. 1.** To serve as the pilot of. **2.** To steer or direct the course of. —*adj.* **1.** Serving as a tentative model for future experiment or development. **2.** Serving or leading as guide <a *pilot* parachute> **—pi′lot·less** *adj.*

☆ ***syns:*** PILOT, EXPERIMENTAL, TEST, TRIAL *adj. core meaning* : serving as a tentative model for future experiment or development <a *pilot* project for urban renewal>

**pi·lot·age** (pī′lə-tĭj) *n.* **1.** *Naut.* **a.** The business or technique of piloting. **b.** The fee paid to a pilot. **2.** Aerial navigation by visual identification of landmarks.

**pilot balloon** *n.* A small balloon used to determine wind direction and velocity.

**pilot bread** *n.* Hardtack.

**pilot burner** *n.* **1.** A small service burner, as in a boiler system, kept lighted to ignite a primary burner. **2.** PILOT LIGHT 1.

**pilot cell** *n.* A storage battery cell tested to determine the condition of the entire battery.

**pilot engine** *n.* A locomotive sent ahead of a train to check the track for safety and clearance.
**pilot fish** *n.* A marine fish, *Naucrates ductor,* that often accompanies larger fishes, esp. sharks.
**pi·lot·house** (pī′lət-hous′) *n.* An enclosed area on the deck or bridge of a vessel from which the vessel is controlled.
**pi·lot·ing** (pī′lə-tĭng) *n.* **1.** The service or occupation of a pilot. **2.** *Naut.* Coastal navigation by reference to landmarks, as buoys and soundings.
**pilot lamp** *n.* A small electric lamp wired to light in response to specified conditions in an electric circuit.
**pilot light** *n.* **1.** A small jet of gas, as in a stove, kept lighted in order to ignite a gas burner or oven. **2.** A pilot lamp.
**pilot whale** *n.* A small whale of the genus *Globicephala.*
**Pilt·down man** (pĭlt′doun′) *n.* [After *Piltdown* Common, East Sussex, England.] A supposedly early human genus and species, *Eoanthropus dawsoni,* postulated from bones allegedly found in an early Pleistocene gravel bed between 1909 and 1915 and proved in 1953 to have been a hoax.
**pil·ule** (pĭl′yo͞ol) *n.* [OFr. < Lat. *pilula,* dim. of *pila,* ball.] A small pill. **—pil′u·lar** (pĭl′yə-lər) *adj.*
**pi·lus** (pī′ləs) *n., pl.* **-li** (-lī′) [Lat.] A hair or a hairlike structure.
**Pi·ma** (pē′mə) *n., pl.* **Pima** *or* **-mas. 1. a.** An Indian tribe living in southern Arizona and northern Mexico. **b.** A member of this tribe. **2.** The Uto-Aztecan language of the Pima. **—Pi′man** *adj.*
**pi·ma cotton** (pē′mə) *n.* [After *Pima* County, Arizona, where it was developed.] A very strong, medium-staple cotton developed from selected Egyptian cottons in the southwestern United States.
**pi·men·to** (pĭ-mĕn′tō) *n., pl.* **-tos.** [Sp. *pimienta,* pepper < Med. Lat. *pigmentum* < Lat., pigment < *pingere,* to paint.] **1.** Allspice. **2.** The pimiento.
**pi mes·on** (pī mĕz′ŏn′, mē′zŏn′, mĕs′ŏn′, mē′sŏn′) *n.* A pion.
**pi·mien·to** (pĭ-mĕn′tō, -myĕn′tō) *n., pl.* **-tos.** [Sp. < *pimienta,* pepper. —see PIMENTO.] A garden pepper, *Capsicum anuum,* or its mild, ripe, red fruit, used in cooking and as stuffing for green olives.
**pimp** (pĭmp) *n.* [Orig. unknown.] A pander : procurer. *—vi.* **pimped, pimp·ing, pimps.** To serve as a pimp.
**pim·per·nel** (pĭm′pər-nĕl′, -nəl) *n.* [ME *pympernel* < OFr. *pimpernelle* < LLat. *pimpinella* < Lat. *piper,* pepper.] A plant of the genus *Anagallis,* esp. the scarlet pimpernel, *A. arvensis,* whose red, purple, or white flowers close in bad weather.
**pim·ple** (pĭm′pəl) *n.* [ME *pinple.*] A small swelling of the skin, occas. containing pus : PUSTULE. **—pim′pled, pim′ply** *adj.*
**pin** (pĭn) *n.* [ME < OE *pinn,* prob. < Lat. *pinna,* feather.] **1.** A short, straight, stiff piece of wire with a sharp point and a blunt head, used esp. for fastening. **2.** Something, as a safety pin or a hairpin, that resembles a pin in shape or use. **3.** An ornament fastened to the clothing by a clasp. **4.** Something of little or no value <didn't care a *pin* about the insult> **5.** A slender cylindrical wooden or metal piece for holding or fastening parts together or serving as a support for suspending one thing from another, as: **a.** A thin rod for securing the ends of fractured bones. **b.** A peg for fixing the crown to the root of a tooth. **c.** A cotter pin. **6.** *Naut.* **a.** A belaying pin. **b.** A thole pin. **7.** *Mus.* One of the pegs securing the strings and regulating their tension on a stringed instrument. **8.** The portion of a key stem entering a lock. **9.** A rolling pin. **10.** One of the wooden clubs at which the ball is aimed in bowling. **11.** The pole bearing a pennant to mark a hole in golf. **12. pins.** *Informal.* The legs. *—vt.* **pinned, pin·ning, pins. 1.** To secure or fasten with or as if with a pin or pins. **2. a.** To transfix. **b.** To hang or attach as if by pinning <*pinned* our hopes on winning the lottery> **3. a.** To win a fall from in wrestling. **b.** To hold fast : IMMOBILIZE <was *pinned* under the debris> **4.** To give (someone) a fraternity pin in token of attachment. **—pin down.** To fix or establish clearly <couldn't *pin down* the source of the rumor> **—pin on.** To attribute (a crime or wrongdoing) <The robbery was *pinned on* the wrong person.> *—adj.* Having a grain suggestive of the heads of pins.
**pi·ña cloth** (pēn′yə) *n.* [Sp. *piña,* pineapple < Lat. *pinea,* pine cone < *pinus,* pine.] A soft, sheer fabric made from the fibers of pineapple leaves.
**pi·ña co·la·da** (pēn′yə kō-lä′də, kə-, pĭn′-) *n.* [Sp., strained pineapple.] A mixed drink of rum, coconut cream, and unsweetened pineapple juice.
**pin·a·fore** (pĭn′ə-fôr′, -fōr′) *n.* [PIN + AFORE.] A sleeveless, apronlike garment, worn esp. by small girls as a dress or overdress.
**pi·nas·ter** (pī-năs′tər) *n.* [Lat., wild pine < *pinus,* pine.] A Mediterranean pine tree, *Pinus pinaster,* with a pyramidal shape.
**pi·ña·ta** (pēn-yä′tə) *n.* A decorated container filled with candy and gifts and suspended from the ceiling, to be broken with a stick as part of traditional festivities in certain Latin American countries.
**pin·ball** (pĭn′bôl′) *n.* A game played on a device in which the player manipulates a ball over a slanted surface having obstacles and targets.
**pince-nez** (păns′nā′, pĭns′-) *n., pl.* **pince-nez** (-nāz′, -nā′) [Fr. : *pincer,* to pinch (< OFr. *pincier*) + *nez,* nose < Lat. *nasus.*] Eyeglasses that are worn clipped to the bridge of the nose.
**pin·cer** (pĭn′sər) *n.* Something that resembles one of the grasping parts of pincers.
**pin·cers** (pĭn′sərz) *also* **pinch·ers** (pĭn′chərz) *pl.n.* [ME *pynsour,* pincer < OFr. *pincier,* to pinch.] *(sing. or pl. in number).* **1.** A grasping tool consisting of a pair of jaws and handles pivoted together to work in opposition. **2.** The articulated prehensile claws of certain arthropods, as the lobster. **3.** A military maneuver in which the enemy is attacked from two flanks and the front.
**pinch** (pĭnch) *v.* **pinched, pinch·ing, pinch·es.** [ME *pinchen* < OFr. *pincier.*] *—vt.* **1.** To squeeze between the thumb and a finger, the jaws of a tool, or other edges. **2.** To squeeze or bind in such a way as to cause pain or discomfort <These boots *pinch* my toes.> **3.** To nip, wither, or shrivel <a face *pinched* by the cold> **4.** To subject to straitened circumstances : DEPRIVE. **5.** *Slang.* STEAL 1. **6.** *Slang.* To arrest. **7.** To move by means of a pinch bar. **8.** *Naut.* To head (a vessel) very close into the wind. *—vi.* **1.** To press, squeeze, or bind painfully <My collar *pinches.*> **2.** To be miserly. **3.** To drag an oar at the end of a stroke. *—n.* **1.** The act of pinching. **2.** An amount that can be held between thumb and forefinger <a *pinch* of baking soda> **3.** A painful, difficult, or straitened circumstance <felt the *pinch* of unemployment> **4.** An emergency situation <You can rely on us in a *pinch.*> **5.** *Informal.* A robbery or theft. **6.** *Slang.* An arrest. **—pinch pennies.** To be frugal or miserly.
**pinch bar** *n.* A crowbar with a pointed projection at one end.
**pinch·beck** (pĭnch′bĕk′) *n.* [After its inventor, Christopher *Pinchbeck* (1670?–1732).] **1.** An alloy of zinc and copper used as imitation gold. **2.** A cheap imitation. *—adj.* **1.** Made of pinchbeck. **2.** Imitation : spurious.
**pinch·cock** (pĭnch′kŏk′) *n.* A clamp used to close or regulate a flexible tube, esp. in laboratory apparatus.
**pinch effect** *n. Physics.* The radial constriction of a plasma, caused by the interaction of its internal electric currents and its self-generated magnetic field.
**pinch·ers** (pĭn′chərz) *n. var. of* PINCERS.
**pinch-hit** (pĭnch′hĭt′) *vi.* **-hit, -hit·ting, -hits. 1.** *Baseball.* To bat in place of a player scheduled to bat, esp. when a hit is badly needed. **2.** *Informal.* To substitute for another in an emergency or on short notice. **—pinch hitter** *n.*
**pin curl** *n.* A coiled strand of usu. damp hair secured with a bobby pin or clip and combed into a wave or curl when dry.
**pin·cush·ion** (pĭn′ko͝osh′ən) *n.* A small, firm cushion in which pins are stuck when not in use.
**Pin·dar·ic** (pĭn-dăr′ĭk) *adj.* **1.** Relating to or typical of the poetic style of Pindar. **2.** Of or typical of a Pindaric ode. *—n.* A Pindaric ode.
**Pindaric ode** *n.* **1.** An ode in the form developed by Pindar, having a series of triads formed by the strophe, antistrophe, and epode. **2.** An adaptation of the ode developed by Pindar, with irregular stanzas and rhyme schemes, esp. as practiced by English poets of the 17th and 18th cent.
**pine¹** (pīn) *n.* [ME < OE *pīn* < Lat. *pinus.*] **1.** An evergreen tree of the genus *Pinus,* bearing cones and needle-shaped leaf clusters and valued for shade and ornament and for its wood and resinous sap, which yields pine tar and turpentine. **2. a.** A coniferous tree, esp. of the family Pinaceae, as the cedar, spruce, or fir. **b.** The wood of a coniferous tree.
**pine²** (pīn) *v.* **pined, pin·ing, pines.** [ME *pinen* < OE *pīnian,* prob. < Lat. *poena,* punishment < Gk. *poinē.*] *—vi.* **1.** To suffer strong longing or yearning <*pine* for one's lost love> **2.** To wither or waste away from longing or grief. *—vt. Archaic.* To grieve or mourn for. *—n. Archaic.* Intense longing or grief.
**pin·e·al** (pĭn′ē-əl, pī′nē-) *adj.* [Fr. *pinéal* < OFr. *pineal* < Lat. *pinea,* pine cone < *pineus,* of pine < *pinus,* pine.] **1.** Shaped like a pine cone. **2.** Relating to the pineal body.
**pineal body** *n.* A small rudimentary glandular body of uncertain function, located in the brain at the roof of the third ventricle.
**pine·ap·ple** (pīn′ăp′əl) *n.* [ME, pine cone : *pine,* pine + *apple,* apple.] **1. a.** A tropical American plant, *Ananas comosus,* with large swordlike leaves and a large, fleshy, edible fruit consisting of the flowers fused into a compound whole with a terminal tuft of leaves. **b.** The fruit of this plant. **2.** *Slang.* A small hand grenade.
**pineapple weed** *n.* A North American plant, *Matricaria matricarioides,* with greenish-yellow rayless flower heads and an odor of pineapple when crushed.
**pine·drops** (pīn′drŏps′) *pl.n. (sing. or pl. in number).* A purplish-brown, leafless, parasitic plant, *Pterospora andromedea,* bearing white or reddish flowers.
**pine finch** *n.* The pine siskin.
**pine mouse** *n.* A vole of the genus *Pitymys,* esp. *P. pinetorum,* a tiny forest animal of eastern North America.
**pi·nene** (pī′nēn′) *n.* Either of two isomeric terpene liquids, $C_{10}H_{16}$, that are the main constituents of oil or spirits of turpentine.
**pine needle** *n.* The needle-shaped leaf of a pine tree.
**pine nut** *n.* The edible seed of certain pines, as the piñon.
**pin·er·y** (pī′nə-rē) *n., pl.* **-ies. 1.** A hothouse or plantation for growing pineapples. **2.** A forest of pine trees.

**pine·sap** (pĭn′săp′) *n.* A fleshy white or reddish plant, *Monotropa hypopithys,* growing as a saprophyte or parasite on tree roots.
**pine siskin** *n.* A North American finch, *Spinus pinus,* with streaked brownish plumage.
**pine tar** *n.* A viscous or semisolid brown to black substance produced by the destructive distillation of pine wood and used as an antiseptic and in roofing compositions, paints, varnishes, and expectorants.
**pi·ne·tum** (pī-nē′təm) *n., pl.* **-ta** (-tə) [Lat., pine grove < *pinus,* pine.] An area planted with pine trees or related conifers, esp. for botanical study.
**pine vole** *n.* The pine mouse.
**pine warbler** *n.* A small yellow-breasted songbird, *Dendroica pinus* of eastern North America.
**pine·y** (pī′nē) *adj. var. of* PINY.
**pin·feath·er** (pĭn′fĕth′ər) *n.* A growing feather still enclosed in its horny sheath, esp. one just emerging through the skin.
**pin·fish** (pĭn′fĭsh′) *n., pl.* **pinfish** *or* **-fish·es.** A small, spiny-finned fish, *Lagodon rhomboides* of southeastern coastal waters of the United States.
**pin·fold** (pĭn′fōld′) *n.* [ME *pynfold* < OE *pundfald* : *pund-,* enclosure + *fald,* fold.] A pound for stray animals. —*vt.* **-fold·ed, -fold·ing, -folds.** To confine in or as if in a pinfold.
**ping** (pĭng) *n.* [Imit.] A brief high-pitched sound, as of a bullet striking metal. —*vi.* **pinged, ping·ing, pings.** To produce a ping.
**Ping-Pong** (pĭng′pông′, -pŏng′). A trademark for table tennis.
**pin·head** (pĭn′hĕd′) *n.* **1.** The head of a pin. **2.** Something small, trifling, or insignificant. **3.** *Slang.* A stupid person. **—pin′head′ed** *adj.* **—pin′head′ed·ness** *n.*
**pin·hole** (pĭn′hōl′) *n.* A tiny puncture made by or as if by a pin.
**pin·ion[1]** (pĭn′yən) *n.* [ME *pynyon* < OFr. *pignon* < VLat. **pinnio* < Lat. *penna,* feather.] **1.** A bird's wing. **2.** The outer rear edge of a bird's wing, containing the primary feathers. **3.** A bird's primary feather. —*vt.* **-ioned, -ion·ing, -ions. 1. a.** To remove or bind the wing feathers of (a bird) to prevent flight. **b.** To cut or bind (the wings of a bird). **2.** To immobilize or restrain by binding the arms. **3.** To fix in one place : CONFINE.
**pin·ion[2]** (pĭn′yən) *n.* [Fr. *pignon* < OFr. < *peigne,* comb < Lat. *pecten* < *pectere,* to comb.] A small cogwheel that engages or is engaged by a larger cogwheel or a rack.
**pin·ite** (pĭn′īt′, pē′nīt′) *n.* [G. *Pinit,* after *Pini,* a mine in Saxony where it was found.] A hydrous, usu. amorphous mineral silicate of aluminum and potassium.
**pink[1]** (pĭngk) *n.* [Orig. unknown.] **1. a.** A plant of the genus *Dianthus,* often cultivated for its sweet-smelling flowers. **b.** Any of various plants such as the wild pink and the moss pink. **c.** A flower of any of these plants. **2.** The highest degree of perfection or excellence <feeling in the *pink*> **3.** Any of a group of colors reddish in hue, of medium to high lightness and low to moderate saturation. **4. pinks.** Light-brown trousers once worn as part of the winter semidress uniform by U.S. Army officers. **5.** *Slang.* One regarded as influenced by or sympathetic with Communist doctrine. —*adj.* **-er, -est. 1.** Having a pink color. **2.** Designating the scarlet coat worn by fox hunters. **3.** *Slang.* Influenced by or sympathetic with Communist doctrine. **—pink′ness** *n.*
**pink[2]** (pĭngk) *vt.* **pinked, pink·ing, pinks.** [ME *pinken,* prob. of LG orig.] **1. a.** To pierce with a pointed weapon : PRICK. **b.** To hurt the feelings of with criticism or scorn. **2.** To decorate with a perforated pattern. **3.** To cut with pinking shears.
**pink[3]** (pĭngk) *also* **pink·ie** (pĭng′kē) *or* **pink·y** *n., pl.* **pinks** *also* **pink·ies.** [ME < MDu. *pinke.*] *Naut.* A sailing vessel with a narrow stern.
**pink·eye** *also* **pink eye** (pĭngk′ī′) *n.* [Partial trans. of obs. Du. *pinck oogen,* small eyes.] Acute contagious conjunctivitis, marked by inflamed eyelids and eyeballs.
**pink·ie[1]** *also* **pink·y** (pĭng′kē) *n., pl.* **-ies.** [Prob. < Du. *pinkje,* dim. of *pink,* little finger.] *Informal.* The fifth or little finger.
**pink·ie[2]** (pĭng′kē) *n. var. of* PINK[3].
**pinking shears** *pl.n.* Sewing scissors with serrated or notched blades, used to finish edges of cloth with a scalloped or zigzag pattern for decoration or to prevent raveling.
**pink·ish** (pĭng′kĭsh) *adj.* Somewhat pink. **—pink′ish·ness** *n.*
**pink lady** *n.* A cocktail of gin, brandy, lemon or lime juice, egg white, and grenadine, shaken with cracked ice and strained.
**pink·o** (pĭng′kō) *n., pl.* **-os.** *Slang.* PINK[1] 5.
**pink·root** (pĭngk′rōōt′, -rŏŏt′) *n.* A plant, *Spigelia marilandica* of eastern North America, with red and yellow flowers and whose root was once used as an anthelmintic.
**pink root** *n.* A disease of onions and related plants caused by a fungus, *Pyrenochaeta terrestris,* and resulting in stunted plants with shriveled pink roots.
**†Pink·ster** *also* **Pinx·ter** (pĭngk′stər) *n.* [Du. < MDu. *pinxter* < Goth. *paintekuste* < Gk. *pentēkostē,* fiftieth. —see PENTECOST.] *Regional.* Whitsunday or Whitsuntide.

**pinkster flower** *also* **pinxter flower** *n.* A North American shrub, *Rhododendron nudiflorum,* bearing fragrant pink flowers that bloom before the leaves appear.
**pink·y[1]** (pĭng′kē) *n. var. of* PINKIE[1].
**pink·y[2]** (pĭng′kē) *n. var. of* PINK[3].
**pin money** *n.* Money for incidental expenses, as in a household.
**pin·na** (pĭn′ə) *n., pl.* **pin·nae** (pĭn′ē) *or* **pin·nas.** [NLat. < Lat., feather.] **1.** *Bot.* One of the leaflets of a pinnate leaf. **2.** *Zool.* A feather, wing, fin, or similar appendage. **3.** *Anat.* The external part of the ear : AURICLE. **—pin′nal** *adj.*
**pin·nace** (pĭn′ĭs) *n.* [OFr. *pinace,* prob. < OSp. *pinaza* < *pino,* pine < Lat. *pinus.*] *Naut.* **1.** A small sailing boat previously used as a tender for merchant and war vessels. **2.** A small ship or ship's boat.
**pin·na·cle** (pĭn′ə-kəl) *n.* [ME < OFr. *pinacle* < LLat. *pinnaculum* < Lat. *pinna,* pinnacle, feather.] **1.** A small spire on a roof or buttress. **2.** A tall pointed formation, as a mountain peak. **3.** The highest point : ACME <the *pinnacle* of success> —*vt.* **-cled, -cling, -cles. 1.** To furnish with a pinnacle. **2.** To place on or as if on a pinnacle.

pinnacle

**pin·nae** (pĭn′ē) *n. var. pl. of* PINNA.
**pin·nate** (pĭn′āt′) *also* **pin·nat·ed** (pĭn′ā′tĭd) *adj.* [Lat. *pinnatus,* feathered < *pinna,* feather.] **1.** Resembling a feather : PENNATE. **2.** *Bot.* Having leaflets, lobes, or divisions in a featherlike arrangement on each side of a common axis, as many compound leaves do. **—pin′nate′ly** *adv.*
**pinnati-** *pref.* [< Lat. *pinnatus,* feathered < *pinna,* feather.] Resembling a feather <*pinnati*fid>
**pin·nat·i·fid** (pĭ-năt′ə-fĭd) *adj.* Having pinnately cleft divisions or lobes. —Used of certain leaves. **—pin·nat′i·fid·ly** *adv.*
**pin·nat·i·sect** (pĭ-năt′ĭ-sĕkt′) *adj.* Divided nearly to the midrib. —Used of certain leaves.
**pin·ni·ped** (pĭn′ə-pĕd′) *adj.* [NLat. *Pinnipedia,* order name : Lat. *pinna,* feather + Lat. *pes,* foot.] Of or belonging to the Pinnipedia, an order of aquatic mammals including the seals, walruses, and similar animals bearing finlike flippers as organs of locomotion. —*n.* A mammal belonging to the Pinnipedia.
**pin·nule** (pĭn′yōōl) *also* **pin·nu·la** (pĭn′yə-lə) *n., pl.* **pin·nules** *also* **pin·nu·lae** (pĭn′yə-lē′) [NLat. *pinnula* < Lat., little feather, dim. of *pinna,* feather.] **1.** *Bot.* One of the leaflets of a pinnately compound leaf. **2.** *Zool.* A featherlike organ or part, as a small fin, or one of the appendages of a crinoid. **—pin′nu·lar** (pĭn′yə-lər) *adj.*
**pin oak** *n.* A tree, *Quercus palustris* of eastern North America, with horizontal or drooping branches and sharply lobed leaves.
**pi·noch·le** *or* **pi·noc·le** (pē′nŭk′əl, -nŏk′əl) *also* **pe·nuch·le** *or* **pe·nuck·le** (pē′nŭk′əl) *n.* [Orig. unknown.] **1.** A game of cards for 2 to 4 persons, played with a special deck of 48 cards, with points being scored by taking tricks and forming certain combinations. **2.** The combination of the queen of spades and jack of diamonds in the game of pinochle.
**pin·o·cy·to·sis** (pĭn′ə-sī-tō′sĭs, -sī-, pī′nə-) *n.* [Gk. *pinein,* to drink + CYT(O)- + -OSIS.] The introduction of fluids into a cell by invagination of the cell membrane, followed by the growth of vesicles within the cells. **—pin′o·cy·tot′ic** (-tŏt′ĭk) *adj.* **—pin′o·cy·tot′i·cal·ly** *adv.*
**pi·no·le** (pĭ-nō′lē) *n.* [Mex. Sp. < Nahuatl *pinolli.*] Meal of ground corn or wheat and mesquite beans.
**pi·ñon** *also* **pin·yon** (pĭn′yōn′, -yən) *n., pl.* **pi·ñons** *or* **pi·ño·nes** (pĭn-yō′nēz) *also* **pin·yons.** [Sp., pine cone < *piña* < Lat. *pinea,* fem. of *pineus,* of pine < *pinus,* pine.] A pine tree bearing nutlike edible seeds, esp. *Pinus cembroides edulis* of western North America.
**piñon jay** *also* **pinyon jay** *n.* A small, uncrested, dull-blue jay, *Gymnorhinus cyanocephala* of western North America.
**pin·point** (pĭn′point′) *n.* **1.** Something extremely small : PARTICLE <a *pinpoint* of sunlight> **2. a.** A point on a map designating a strictly defined military target. **b.** A precisely identified and limited target. —*vt.* **-point·ed, -point·ing, -points. 1.** To pierce with or as if with a pin <The lights on the bridge *pinpointed* the gloom.> **2.** To locate and identify precisely <*pinpoint* the source of an infection> **3.** To take precise aim at (e.g., a target). —*adj.* **1.** Marked by meticulous accuracy <*pinpoint* precision> **2.** Minuscule <*pinpoint* insects>
**pin·prick** (pĭn′prĭk′) *n.* **1.** A slight puncture made by a pin. **2.** An insignificant wound. **3.** A minor annoyance. —*v.* **-pricked, -prick·**

**ing, -pricks.** —*vt.* To puncture with a pin. —*vi.* To make a slight puncture with a pin.

**pins and needles** *pl.n.* Tingling felt in a part of the body, as a hand or foot, that has been numbed from lack of circulation. **—on pins and needles.** In a state of anxiety or tense anticipation.

**pin·scher** (pĭn′shər) *n.* A Doberman pinscher.

**pin·set·ter** (pĭn′sĕt′ər) *n.* One employed to set up pins in a bowling alley.

**pin·stripe** (pĭn′strīp′) *n.* **1.** A thin stripe on a fabric. **2.** A kind of fabric with thin stripes, often used for suits. **—pin′-striped′** *adj.*

**pint** (pīnt) *n.* [ME *pinte,* a unit of volume < OFr.] **1. a.** A unit of volume or capacity in the U.S. Customary System, used in liquid measure, equal to 16 fluid ounces or .473 liter. **b.** A unit of volume or capacity in the U.S. Customary System, used in dry measure, equal to ½ quart or .551 liter. **c.** A unit of volume or capacity in the British Imperial System, used in dry and liquid measure, equal to .568 liter. **2. a.** A container with a pint capacity. **b.** The amount of a substance that a pint container can hold.

**pin·ta** (pĭn′tə, pēn′tä) *n.* [Am. Sp. < Sp., painted mark < Lat. *pingere,* to paint.] A contagious skin disease prevalent in tropical America, caused by spirochete microorganisms and marked by extreme thickening and spotty discoloration of the skin.

**pin·tail** (pĭn′tāl′) *n., pl.* **pintail** or **-tails.** A duck, *Anas acuta* of the Northern Hemisphere, with gray, brown, and white plumage and a sharply pointed tail.

**pin·ta·no** (pĭn-tä′nō) *n., pl.* **pintano** or **-nos.** [Am. Sp.] A dark-banded fish, *Abudefduf marginatus* of southern Atlantic waters.

**pin·tle** (pĭn′tl) *n.* [ME *pintel* < OE, penis.] An upright pin or bolt used as a pivot, esp.: **a.** *Naut.* The pin on which a rudder turns. **b.** The pin on a gun carriage.

**pin·to** (pĭn′tō) *n., pl.* **-tos** or **-toes.** [Sp., piebald < VLat. *pinctus < Lat. *pictus,* p.part. of *pingere,* to paint.] A horse with irregular markings or spots. —*adj.* Irregularly marked : PIEBALD.

**pinto bean** *n.* A form of the common string bean that has mottled seeds and is grown primarily in the southwestern United States.

**pint·size** (pīnt′sīz′) *also* **pint·sized** (-sīzd′) *adj. Informal.* Of a small size : DIMINUTIVE.

**pin·up** (pĭn′ŭp′) *n.* **1.** A picture to be pinned up on a wall, esp. a photograph of a sexually attractive woman or movie star. **2.** A woman regarded as a suitable model for a pinup. —*adj.* **1.** Relating to or suitable for a pinup. **2.** Designed for wall attachment, as a lamp.

**pin·wale** (pĭn′wāl′) *n.* A corduroy made with narrow wales or ribs. —*adj.* Of, pertaining to, or designating pinwale.

**pin·weed** (pĭn′wēd′) *n.* A plant of the genus *Lechea,* with narrow leaves and numerous small flowers.

**pin·wheel** (pĭn′hwēl′, -wēl′) *n.* **1.** A toy consisting of a stick to which revolving vanes of colored paper or plastic are fastened. **2.** A firework that forms a rotating wheel of colored flames. **3.** A wheel with a circle of pins at right angles to its face, used as a tripping device.

**pin·work** (pĭn′wûrk′) *n.* The fine stitches raised in needlepoint lace from the surface of a motif.

**pin·worm** (pĭn′wûrm′) *n.* A small nematode worm, *Enterobius vermicularis,* that infests the human intestines and rectum.

**pin·wrench** (pĭn′rĕnch′) *n.* A wrench with a projection designed to fit a hole in the object to be turned.

**Pinx·ter** (pĭngk′stər) *n. var. of* PINKSTER.

**pinxter flower** *n. var. of* PINKSTER FLOWER.

**pin·y** *also* **pine·y** (pī′nē) *adj.* **-i·er, -i·est.** Resembling, typical of, composed of, or covered with pines.

**Pin·yin** or **pin·yin** (pĭn′yĭn′, -yĭn) *n.* [Chin. (Mandarin) *pin[1] yin[1],* spelling of the sound : *pin[1],* to combine + *yin[1],* sound.] A system for transliterating Chinese ideograms into the roman alphabet.

**pin·yon** (pĭn′yŏn′, -yən) *n. var. of* PIÑON.

**pinyon jay** *n. var. of* PIÑON JAY.

**pi·o·let** (pē′ə-lā′) *n.* [Fr. < dial. Fr. *piola,* ax < OProv. *apcha,* of Germanic orig.] An ice ax used in mountain climbing.

**pi·on** (pī′ŏn′) *n.* [Contraction of PI MESON.] Either of two subatomic particles in the meson family: pi zero, having a mass 264 times that of the electron, zero electric charge, and a mean lifetime of 0.9 × $10^{-16}$ second; and pi plus, having a mass 273 times that of the electron, a positive electric charge, and a mean lifetime of 2.6 × $10^{-8}$ second.

**pi·o·neer** (pī′ə-nîr′) *n.* [OFr. *pionier,* foot soldier < *pion* < Med. Lat. *pedo* < Lat. *pes,* foot.] **1.** One who ventures into unexplored or unclaimed territory to settle. **2.** An innovator <a *pioneer* in electronics> **3.** A military engineer employed in the construction and fortification of positions and the maintenance of communication lines. **4.** *Ecol.* An animal or plant species that establishes itself in a previously barren environment. —*adj.* **1.** Like a pioneer : INNOVATIVE <a *pioneer* physicist> **2.** Of or pertaining to early settlers or their time. —*v.* **-neered, -neer·ing, -neers.** —*vt.* **1.** To explore, open up, or settle (a region). **2.** To innovate or participate in the development of <*pioneered* the radio industry> —*vi.* To act as a pioneer.

▲ word history: The word *pioneer* is derived from French *pion,* "foot soldier." *Pioneer* originally denoted a soldier whose task was to prepare the way for the main body of troops marching to a new area. From this use the word was applied to anyone who ventures into an unknown region.

**pi·os·i·ty** (pī-ŏs′ĭ-tē) *n., pl.* **-ties.** [< PIOUS.] Ostentatious or exaggerated piousness.

**pi·ous** (pī′əs) *adj.* [Lat. *pius,* dutiful.] **1.** Having or displaying reverence and earnest compliance in the observance of religion : DEVOUT. **2. a.** Characterized by conspicuous devoutness. **b.** Marked by false devoutness : HYPOCRITICAL. **3.** Devotional <*pious* readings of Scripture> **4.** Professing or displaying a strict, traditional sense of virtue and morality : HIGH-MINDED. **5.** Worthy : commendable <*pious* donations> **—pi′ous·ly** *adv.* **—pi′ous·ness** *n.*

**pip[1]** (pĭp) *n.* [Short for PIPPIN.] **1.** The small seed of a fruit, as an orange or apple. **2.** *Informal.* Something remarkable of its kind.

**pip[2]** (pĭp) *vt.* **pipped, pip·ping, pips.** [Poss. < PIP[3].] *Chiefly Brit.* **1.** To strike with a gunshot. **2.** To blackball.

**pip[3]** (pĭp) *n.* [Orig. unknown.] **1. a.** A dot indicating a unit of numerical value on dice or dominoes. **b.** A speck or spot. **2.** A rootstock of a flowering plant, esp. the lily of the valley. **3.** *Informal.* A shoulder insignia of certain British army officers. **4.** A radar signal.

**pip[4]** (pĭp) *v.* **pipped, pip·ping, pips.** [Orig. unknown.] —*vt.* To break through (an eggshell) in hatching. —*vi.* To chirp or peep, as a chick or young bird does. —*n.* A short, high-pitched radio signal.

**pip[5]** (pĭp) *n.* [ME *pippe* < MDu. < Lat. *pituita.*] **1. a.** A disease of birds, marked by a thick mucous discharge that forms a crust in the mouth and throat. **b.** The crust symptomatic of pip. **2.** *Slang.* A minor or imaginary human ailment.

**pip·age** *also* **pipe·age** (pī′pĭj) *n.* **1. a.** Transmission of liquids through pipes. **b.** The charge for such transmission. **2.** PIPING 1.

**pi·pal** (pē′pəl) *n. var. of* PEEPUL.

**pipe** (pīp) *n.* [ME < OE *pīpe* < VLat. *pipa < Lat. *pipare,* to chirp.] **1. a.** A hollow cylinder or tube for conveying a fluid or gas. **b.** A section or piece of such a tube. **2. a.** An instrument for smoking, having a tube of wood or clay with a mouthpiece at one end and a small bowl at the other. **b.** A pipeful. **3. a.** *Biol.* A tubular organ or part. **b. pipes.** *Informal.* The human respiratory system. **4. a.** A wine cask with a capacity of 126 gallons. **b.** This volume as a unit of liquid measure. **5.** *Mus.* **a.** A tubular wind instrument, as a flute. **b.** Any of the tubes in an organ. **c. pipes.** A small wind instrument, having tubes of different lengths bound together. **d. pipes.** A bagpipe. **6.** The sound of the voice, esp. as used in singing or acting. **7.** A birdcall. **8.** *Naut.* A whistle used for signaling crewmen. **9. a.** A vertical, cylindrical vein of ore. **b.** One of the vertical veins of eruptive origin in which diamonds are found in South Africa. **10.** *Geol.* An eruptive passageway opening into the crater of a volcano. **11.** *Metallurgy.* A cone-shaped cavity in a steel ingot, formed during cooling by escaping gases. **12.** *Slang.* An easy task, esp. an easy academic course. —*v.* **piped, pip·ing, pipes.** —*vt.* **1.** To convey (liquid or gas) by means of pipes. **2.** To connect or provide with pipes. **3. a.** To play (a tune) on a pipe or pipes. **b.** To lead (e.g., children) by playing on pipes. **4.** *Naut.* **a.** To call (the crew) by sounding a boatswain's pipe. **b.** To receive aboard or mark the departure of by playing a boatswain's pipe <*piped* the officer aboard> **5.** To utter in a shrill, reedy tone. **6.** To furnish (a garment) with piping. —*vi.* **1.** To play on a pipe. **2.** To speak shrilly. **3.** To chirp or whistle, as a bird does. **4.** *Naut.* To signal the crew with a boatswain's pipe. **5.** *Metallurgy.* To develop conical cavities. **—pipe down.** *Slang.* To stop talking : be quiet. **—pipe up.** To speak up, esp. in a small shrill voice.

**pipe·age** (pī′pĭj) *n. var. of* PIPAGE.

**pipe clay** *n.* A fine white clay used in making tobacco pipes and pottery, in calico printing, and in whitening leather.

**pipe cleaner** *n.* A pliant, tufted rod used to clean the stem of a tobacco pipe.

**pipe dream** *n.* [From the fantasies induced by smoking a pipe of opium.] A fantastic, impractical notion or hope.

**pipe·fish** (pīp′fĭsh′) *n., pl.* **pipefish** or **-fish·es.** Any of various slim, elongated marine or freshwater fishes of the family Syngnathidae, having an external covering of bony plates and a tubelike snout.

**pipe fitter** *n.* One who installs and repairs piping systems.

**pipe·fit·ting** (pīp′fĭt′ĭng) *n.* **1. a.** The act or work of joining pipes together. **b.** A branch of the plumbing trade dealing esp. with the installation and repair of piping systems. **2.** A section of pipe used to join two or more pipes together.

**pipe·ful** (pīp′fo͝ol′) *n.* An amount sufficient to fill a pipe.

**pipe·line** (pīp′līn′) *n.* **1.** A conduit of pipe for conveying water or petroleum products. **2.** A channel by which confidential or secret information is transmitted. **3.** A line of communication or route of supply. —*vt.* **-lined, -lin·ing, -lines.** **1.** To convey by means of a pipeline. **2.** To lay a pipeline through.

**pipe organ** *n.* ORGAN 1.

**pip·er** (pī′pər) *n.* **1.** One who plays on a pipe. **2.** One who lays or installs piping. **3.** One who applies piping to fabric or a garment.

**pi·per·a·zine** (pī-pĕr′ə-zēn′, pĭ-) *n.* [PIPER(INE) + AZ(O)- + -INE[2].] A colorless, crystalline compound, $C_4H_{10}N_2$, used to inhibit corrosion, in insecticides, and as a vermifuge.

**pi·per·i·dine** (pī-pĕr'ĭ-dēn', pĭ-) *n.* [PIPER(INE) + -IDE + -INE².] A strongly basic, colorless liquid, $C_5H_{10}NH$, used to make rubber and as a curing agent in epoxy resins.

**pip·er·ine** (pĭp'ə-rēn') *n.* [Lat. *piper*, pepper + -INE².] A crystalline solid, $C_{17}H_{19}NO_3$, extracted from black pepper and used as an insecticide and as flavoring.

**pi·per·o·nal** (pĭ-pĕr'ə-năl', pī-) *n.* [PIPER(INE) + -ON(E) + -AL³.] A white powder, $C_8H_6O_3$, having a floral odor, used as flavoring and in perfume.

**pipe·stone** (pīp'stōn') *n.* A heat-hardened, compacted red clay used by American Indians for making tobacco pipes.

**pi·pette** *also* **pi·pet** (pī-pĕt') *n.* [Fr., dim. of *pipe*, pipe < VLat. **pipa*. —see PIPE.] A usu. calibrated glass tube of varying length, open at both ends and used to transfer small volumes of liquid.

**pipe vine** *n.* A woody vine, *Aristolochia durior* of the eastern United States, bearing greenish, brown-mottled flowers shaped like a curved pipe.

**pipe wrench** *n.* A wrench with two serrated jaws, one adjustable, for gripping and turning pipe.

**pip·ing** (pī'pĭng) *n.* **1.** An amount or system of pipes. **2.** *Mus.* **a.** The act of playing on a pipe. **b.** The music made by a pipe. **3.** A shrill high-pitched sound. **4.** A rounded strip of cloth trimming for garments or decorative fabrics. **5.** A rounded ribbon of icing on a pastry. —*adj.* **1.** Playing on a pipe. **2.** Making a high-pitched sound with little resonance, as a pipe does. —*adv.* Extremely <*piping* hot>

**pip·it** (pĭp'ĭt) *n.* [Imit. of its song.] A songbird of the genus *Anthus*, with brownish upper parts and a streaked breast.

**pip·kin** (pĭp'kĭn) *n.* [Poss. PIPE + -KIN.] **1.** A small earthenware or metal cooking pot. **2.** A piggin.

**pip·pin** (pĭp'ĭn) *n.* [ME *pipin* < OFr. *pepin*.] **1.** Any of several varieties of apple. **2.** PIP¹ 1. **3.** *Slang.* One that is greatly admired.

**pip·sis·se·wa** (pĭp-sĭs'ə-wô') *n.* [Cree *pipisisikweu*.] A North American evergreen plant of the genus *Chimaphila*, esp. *C. umbellata*, bearing pinkish or white flowers.

**pip-squeak** (pĭp'skwēk') *n.* A person regarded as small and insignificant.

**pi·quant** (pē'kənt, -känt', pē-känt') *adj.* [OFr. < pr.part. *piquer*, to prick. —see PICK¹.] **1.** Pleasantly pungent in taste or odor : SPICY. **2.** Pleasantly or appealingly provocative : CHARMING. **3.** *Archaic.* Causing injured pride or feelings : STINGING. **—pi'quan·cy, pi'quant·ness** *n.* **—pi'quant·ly** *adv.*

**pique** (pēk) *n.* [OFr., animosity < *piquer*, to prick. —see PICK¹.] Vexation or resentment arising from wounded pride or vanity. —*vt.* **piqued, piqu·ing, piques. 1.** To cause to feel vexation or resentment. **2.** To arouse : provoke <The epitaph *piqued* our curiosity.> **3.** To pride (oneself) <*piqued* ourselves on our stylishness>

**pi·qué** (pĭ-kā', pē-) *n.* [Fr. < *piquer*, to quilt < OFr., to backstitch, prick. —see PICK¹.] A tightly woven fabric with various patterns of wales, produced esp. by a double warp.

**pi·quet** *also* **pic·quet** (pĭ-kā') *n.* [Fr.] A card game for two people, played with a deck from which all cards below the seven, aces being high, are omitted.

**pi·ra·cy** (pī'rə-sē) *n.*, *pl.* **-cies.** [Med. Lat. *piratia* < LGk. *peirateia* < Gk. *peiratēs*, pirate.] The act or practice of pirating.

**pi·ra·gua** (pĭ-rä'gwə) *n.* [Sp. < Carib.] **1.** DUGOUT 1. **2.** A flat-bottomed, two-masted sailing boat.

**pi·ra·nha** *also* **pi·ra·ña** (pĭ-rän'yə, -răn'yə) *n.* [Port. < Tupi : *pirá*, fish + *sainha*, tooth.] A tropical American freshwater fish of the genus *Serrasalmus*, voraciously carnivorous and often predatory on living animals.

**pi·ra·ru·cu** (pĭ-rär'ə-ko͞o') *n.* [Port. *pirarucú* < Tupi *pirá-rucú* : *pirá*, fish + *urucú*, red.] The arapaima.

**pi·rate** (pī'rĭt) *n.* [ME < Lat. *pirata* < Gk. *peiratēs* < *peiran*, to attempt.] **1. a.** One who robs at sea or plunders the land from the sea without commission from a sovereign nation. **b.** A ship used for this purpose. **2.** One who makes use of or reproduces another's work, esp. literary work, illicitly or without permission. —*v.* **-rat·ed, -rat·ing, -rates.** —*vt.* **1.** To subject to attack and robbery <*pirate* a ship at sea> **2.** To seize (goods) by piracy. **3.** To make use of or reproduce (another's work) illicitly. —*vi.* To practice piracy. **—pi·rat'ic** (pī-răt'ĭk), **pi·rat'i·cal** *adj.* **—pi·rat'i·cal·ly** *adv.*

**pirate perch** *n.* A small North American freshwater fish, *Aphredoderus sayanus*, having its anal opening in the throat.

**pi·rog** (pĭ-rōg') *n.*, *pl.* **-ro·ghi** (-rō'gē) *or* **-ro·gi.** [R., prob. < *pir*, feast.] A large Russian pastry made of dough stuffed with various combinations of ground meat, eggs, or cabbage.

**pi·rogue** (pĭ-rōg') *n.* [Fr. < Sp. *piragua* < Carib.] DUGOUT 1.

**pi·rosh·ki** *or* **pi·roj·ki** (pĭ-rôzh'kē) *pl.n. vars. of* PIROZHKI.

**pir·ou·ette** (pĭr'o͞o-ĕt') *n.* [Fr. < OFr. *pirouet*, spinning top.] A full turn of the body on the tip of the toe or on the ball of the foot in ballet. —*vi.* **-et·ted, -et·ting, -ettes.** To perform a pirouette.

**pi·rozh·ki** *also* **pi·rosh·ki** *or* **pi·roj·ki** (pĭ-rôzh'kē) *pl.n.* [Yiddish *pirozshke* < R. *pirozhki*, pl. of *pirozhok*, small pocket of pastry, dim. of *pirog*, pirog.] Small Russian pastries usu. filled with ground meat or cabbage.

**pis al·ler** (pē ză-lā') *n.* [Fr., to go worse.] The final recourse or expedient.

**pis·ca·ry** (pĭs'kə-rē) *n.*, *pl.* **-ries.** [ME < Med. Lat. *piscaria* < Lat., neuter pl. of *piscarius*, of fish < *piscis*, fish.] **1.** The right to fish in waters owned by another. **2.** A fishery.

**pis·ca·to·ri·al** (pĭs'kə-tôr'ē-əl, -tōr'-) *also* **pis·ca·to·ry** (pĭs'kə-tôr'ē, -tōr'ē) *adj.* [< Lat. *piscatorius* < *piscator*, fisherman < *piscari*, to fish < *piscis*, fish.] **1.** Of or relating to fish, fishing, or fishermen. **2.** Involved in fishing. **—pis'ca·to'ri·al·ly** *adv.*

**Pi·sces** (pī'sēz) *n.* [ME < Lat. < pl. of *piscis*, fish.] **1.** A constellation in the Northern Hemisphere. **2. a.** The 12th sign of the zodiac. **b.** One born under this sign. **—Pi'sce·an** (pī'sē-ən) *adj.*

**pisci-** *pref.* [< Lat. *piscis*, fish.] Fish <*piscivorous*>

**pi·sci·cul·ture** (pī'sĭ-kŭl'chər, pĭs'ĭ-) *n.* The breeding, hatching, and rearing of fish under controlled conditions. **—pi'sci·cul'tur·al** *adj.* **—pi'sci·cul'tur·ist** *n.*

**pi·sci·form** (pī'sĭ-fôrm', pĭs'ĭ-) *adj.* Shaped like a fish.

**pi·sci·na** (pĭ-sē'nə, -sī'nə, -shē'nə) *n.*, *pl.* **-nae** (-nē) [Med. Lat. < Lat., fish-pond < *piscis*, fish.] A stone basin with a drain for carrying away sacramental water in a church. **—pis'ci·nal** (pĭs'ə-nəl) *adj.*

**pi·scine** (pī'sēn', pĭs'īn') *adj.* [Med. Lat. *piscinus* < Lat. *piscis*, fish.] Of, relating to, or typical of fish.

**Pi·scis Aus·tri·nus** (pī'sĭs ôs-trī'nəs) *n.* [NLat., the Southern Fish.] A constellation in the Southern Hemisphere.

**pi·sciv·o·rous** (pĭ-sĭv'ər-əs, pī-) *adj.* Habitually feeding on fish.

**pish** (pĭsh) *interj.* —Used to express disdain.

**pi·shogue** *also* **pi·shoge** (pĭ-shōg') *n.* [Ir. Gael. *pīseog* < MIr. *pisōc*.] *Ir.* **1.** Witchcraft : black magic. **2.** An evil spell : INCANTATION.

**pi·si·form** (pī'sə-fôrm') *adj.* [Lat. *pisum*, pea + -FORM.] Resembling a pea in size or shape. —*n. Anat.* A small bone at the junction of the ulna and the wrist.

**pis·mire** (pĭs'mīr', pĭz'-) *n.* [ME *pissemyre* : *pisse*, urine + *mire*, ant, of Scand. orig.] An ant.

**pis·mo clam** (pĭz'mō) *n.* [After *Pismo* Beach, California.] A North American edible marine clam, *Tivela stultorum* of the southern Pacific coast.

**pi·so·lite** (pī'sə-līt') *n.* [Gk. *pisos*, pea + -LITE.] A small round accretionary limestone mass. **—pi'so·lit'ic** (-lĭt'ĭk) *adj.*

**pis·soir** (pē-swär') *n.* [Fr. < *pisser*, to urinate < OFr. *pissier*.] A public urinal located on streets in some European countries.

**pis·ta·chi·o** (pĭ-stăsh'ē-ō', -stä'shē-ō') *n.*, *pl.* **-os.** [Ital. *pistacchio* < Lat. *pistacium*, pistachio nut < Gk. *pistakion* < *pistakē*, pistachio tree < Pers. *pistah*.] **1. a.** A tree, *Pistacia vera* of the Mediterranean region and western Asia, bearing small hard-shelled nuts. **b.** The nut of this tree, with an edible, oily, green kernel. **2.** The flavor of pistachio nuts.

**pistachio green** *n.* A moderate to light yellowish or yellow green.

**pis·ta·reen** (pĭs'tə-rēn') *n.* [Prob. alteration of Sp. *peseta*, peseta. —see PESETA.] A small silver coin used in America and the West Indies during the 18th cent.

**pis·til** (pĭs'təl) *n.* [Fr. < Lat. *pistillum*, pestle.] The seed-bearing organ of a flower, including the stigma, style, and ovary.

**pis·til·late** (pĭs'tə-lāt', -lĭt) *adj.* **1.** Having a pistil or pistils. **2.** Bearing pistils but no stamens, as certain flowers.

**pis·tol** (pĭs'təl) *n.* [Fr. *pistole* < G. *Pistole* < Czech *pištala*, pipe.] A firearm designed to be held and fired with one hand. —*vt.* **-toled, -tol·ing, -tols.** To shoot with a pistol.

**pis·tole** (pĭ-stōl') *n.* [Fr.] An old gold coin, used in some European countries until the late 19th cent.

**pis·to·leer** (pĭs'tə-lîr') *n.* One armed with a pistol.

**pistol grip** *n.* **1. a.** The grip of a pistol, shaped to fit the hand. **b.** A similar grip sometimes used on a rifle or other firearm. **2.** A grip used on certain tools, as a saw, shaped to fit the hand.

**pis·tol-whip** (pĭs'təl-hwĭp', -wĭp') *vt.* **-whipped, -whip·ping, -whips. 1.** To beat with a pistol barrel. **2.** To beat while threatening with a gun. **3.** To attack violently.

**pis·ton** (pĭs'tən) *n.* [Fr. < OFr. < OItal. *pistone*, large pestle < *pistare*, to pound < Lat.] **1.** A solid cylinder or disk that fits snugly into a larger cylinder and moves back and forth under fluid pressure, as in a reciprocating engine, or displaces or compresses fluids, as in pumps and compressors. **2.** *Mus.* A valve mechanism in brass instruments for altering pitch.

**piston ring** *n.* An adjustable metal ring fitting around a piston and closing the gap between the piston and cylinder wall.

**piston rod** *n.* A connecting rod that transmits power to or is powered by a piston.

**pit¹** (pĭt) *n.* [ME < OE *pytt*.] **1.** A relatively deep hole in the ground, either natural, as a sinkhole or pothole, or manmade, as a mine shaft. **2.** A trap consisting of a concealed hole in the ground : PITFALL. **3.** An unexpected trouble or hidden danger. **4. a.** Hell. **b.** An extremely dirty, unattractive place. **c. the pits.** *Slang.* The worst imaginable <"New York politics are *the pits*" —*Washington Star*> **5.** An enclosed space, often one dug in the ground, in which animals, as dogs or gamecocks, are placed for fighting. **6. a.** A natural depression in the surface of a body, organ, or part. **b.** A small indentation in the skin left by injury or disease : POCKMARK. **7. a.** The section of a theater in which the musicians sit, directly in front of the stage. **b.** The area behind the stalls of a theater. **8.** The section of an exchange where trading in a specific commodity is carried on. **9.** A refueling area at an auto racecourse. **10.** *Football.* The middle areas of the defensive and offensive lines. **11.** *Bot.* A thin-walled spot or depression in the wall of some plant cells. —*v.* **pit·ted, pit·ting, pits.** —*vt.* **1.** To make cavities, depressions, or scars in. **2.** To set into contest or opposition <*pitted* the champions against each other> —*vi.* **1.** To become marked with small pits. **2.** To stop at one's refueling area during an auto race.

**pit²** (pĭt) *n.* [Du. < MDu.] The single, central kernel of certain fruits, as a cherry or peach : STONE. —*vt.* **pit·ted, pit·ting, pits.** To extract pits from (fruit).

**pi·ta¹** (pē'tə) *n.* [Mod. Gk., pie, cake.] Round, flat bread that is hollow inside and forms a pocket when cut in half.

**pi·ta²** (pē'tə) *n.* [Sp. < Quechua, to complicate.] **1.** A plant of the genus *Agave* that yields a strong fiber. **2.** A species of pineapple, *Ananas magdalenae,* whose leaves yield a fine whitish fiber. **3.** The fiber of the pita plant, used to make cordage and paper.

**pit·a·pat** (pĭt'ə-păt') *vi.* **-pat·ted, -pat·ting, -pats.** [Imit.] **1.** To move with quick, tapping steps. **2.** To make a repeated tapping sound. —*n.* A series of quick steps, taps, or beats. **—pit'a·pat'** *adv.*

**pitch¹** (pĭch) *n.* [ME *pich* < OE *pic* < Lat. *pix.*] **1.** A thick, dark, sticky substance obtained from the distillation residue of coal tar, wood tar, or petroleum and used for waterproofing, caulking, roofing, and paving. **2.** A natural bitumen, as mineral pitch or asphalt. **3.** A resin derived from the sap of various coniferous trees, as the pines. —*vt.* **pitched, pitch·ing, pitch·es.** To cover or smear with or as if with pitch.

**pitch²** (pĭch) *v.* **pitched, pitch·ing, pitch·es.** [ME *pichen.*] —*vt.* **1.** To throw, usu. in a specific direction : TOSS. **2.** *Baseball.* **a.** To throw (the ball) from the mound to the batter. **b.** To play (a game or part of one) in the position of pitcher. **3.** To put up or in position : ESTABLISH <*pitch* camp> **4.** To set firmly : IMPLANT <*pitch* fence poles> **5. a.** To fix the level of <*pitched* our hopes high> **b.** To set the character and course of <*pitched* the talk to teen-agers> **6. a.** To set at a specified downward slant, as the angle of a roof. **b.** *Mus.* To set in a particular key. **7.** To lead (a card), thus establishing the trump suit. **8.** To sell or present in a high-pressure fashion. **9.** To throw away : discard. —*vi.* **1.** To throw or toss something, as a ball, horseshoe, or bale. **2.** *Baseball.* To play in the position of pitcher. **3.** To take a plunge or fall, esp. forward <*pitched* over the cliff> **4. a.** To lurch or stumble. **b.** To buck, as a horse. **5. a.** *Naut.* To dip bow and stern alternately. **b.** To revolve about a lateral axis so that the nose lifts or descends in relation to the tail. —Used of an airplane. **c.** To revolve about a lateral axis that is both perpendicular to the longitudinal axis and horizontal to the earth. —Used of a space vehicle. **6.** To slope downward. **7.** To set up living quarters : ENCAMP. **8.** To make a casual, usu. hurried choice or decision <They *pitched* on that solution at the last minute.> **—pitch in.** *Informal.* **1.** To set to work vigorously. **2.** To join forces with others : COOPERATE. **—pitch into.** *Informal.* To attack verbally or physically : ASSAULT. —*n.* **1.** An act or instance of pitching. **2.** *Baseball.* **a.** A throw of the ball by the pitcher for action by the batter. **b.** A ball so thrown. **3.** The rectangular area between the wickets in cricket, 20 meters by 3 meters or 22 yards by 10 feet. **4.** *Naut.* The alternate dip and rise of a ship's bow and stern. **5. a.** A steep downward slant. **b.** The degree of such a slant. **6. a.** The angle of a roof. **b.** The highest point of a structure <the *pitch* of a bridge> **7.** A level or degree of development or intensity <work proceeded at a hectic *pitch*> **8.** The subjective quality of a complex sound, dependent on frequency, loudness, and intensity and often measured as the frequency of a pure tone of specified intensity judged equivalent to the complex sound by a normal ear. **9.** *Mus.* **a.** The relative position of a tone in a scale, as determined by its frequency. **b.** A standard that establishes a frequency for each tone, used in the tuning of instruments. **10. a.** The distance traveled by a machine screw in a single revolution. **b.** The distance between two corresponding points on adjacent screw threads or gear teeth. **11.** The distance a propeller would travel in an ideal medium during one complete revolution, measured parallel to the shaft of the propeller. **12. a.** *Slang.* An often hard-sell talk, as by a salesperson. **b.** An advertisement. **c.** The stand of a vender or hawker. **13.** The game of seven-up.

**pitch accent** *n.* Tonic accent.

**pitch-black** (pĭch'blăk') *adj.* Extremely black or dark.

**pitch·blende** (pĭch'blĕnd') *n.* [G. *Pechblend* : *Pech,* pitch (< Lat. *pix*) + *Blende,* blende. —see BLENDE.] The essential ore of uranium, a brownish-black mineral of uraninite and uranium trioxide with small amounts of water and uranium decay products.

**pitch-dark** (pĭch'därk') *adj.* Totally dark <a *pitch-dark* night>

**pitched battle** *n.* **1.** A fierce battle fought by opponents in close contact. **2.** An intense dispute.

**pitch·er¹** (pĭch'ər) *n.* **1.** One that pitches. **2.** *Baseball.* The player who throws the ball from the mound for action by the batter. **3.** A golf iron with a sharply inclined head.

**pitch·er²** (pĭch'ər) *n.* [ME *picher* < OFr. *pichier* < Med. Lat. *bicarium,* drinking cup < Gk. *bikos,* jar.] **1.** A vessel for liquids, with a handle and a lip or spout for pouring. **2.** *Bot.* A pitcherlike part, as the leaf of a pitcher plant.

**pitcher plant** *n.* An insectivorous plant of the genera *Sarracenia, Nepenthes,* or *Darlingtonia,* bearing leaves modified to form pitcherlike organs that attract and trap insects.

**pitch·fork** (pĭch'fôrk') *n.* [ME *pikforke* : *pik,* pick + *forke,* fork.] A large fork with sharp, widely spaced prongs for pitching hay and breaking ground. —*vt.* **-forked, -fork·ing, -forks.** To lift or toss with a pitchfork.

**pitch·man** (pĭch'mən) *n.* A vender or peddler of small wares, esp. one with a colorful or high-powered sales talk.

**pitch·out** (pĭch'out') *n.* **1.** *Baseball.* A pitch deliberately thrown high and away from the batter to make it easier for the catcher to throw out a base runner attempting to steal. **2.** *Football.* A lateral pass from the back receiving the snap from the center to another back behind the line of scrimmage.

**pitch pine** *n.* A pine tree yielding pitch or turpentine, as *Pinus rigida* or *P. echinata* of eastern North America.

**pitch pipe** *n. Mus.* A small pipe that gives the standard pitch for a piece of music or for tuning an instrument.

**pitch·stone** (pĭch'stōn') *n.* [Transl. of G. *Pechstein.*] Any of various volcanic glasses distinguished by their pitchlike luster and relatively high water content.

**pitch·y** (pĭch'ē) *adj.* **-i·er, -i·est. 1.** Full of or smeared with pitch. **2.** Resembling pitch. **3.** Pitch-black. **—pitch'i·ness** *n.*

**pit·e·ous** (pĭt'ē-əs) *adj.* [ME *piteus* < OFr. < *pite,* pity. —see PITY.] **1.** PITIFUL 1. **2.** Having pity : COMPASSIONATE. **—pit'e·ous·ly** *adv.* **—pit'e·ous·ness** *n.*

**pit·fall** (pĭt'fôl') *n.* **1.** A danger or difficulty not easily anticipated or avoided. **2.** A trap made by digging a hole in the ground and concealing its opening.

☆ ***syns:*** PITFALL, BOOBY TRAP, TRAP *n. core meaning* : a source of danger or difficulty not easily foreseen and avoided <the *pitfalls* of stock market speculation>

**pith** (pĭth) *n.* [ME < OE *piða.*] **1.** *Bot.* The soft, spongelike substance in the center of stems and branches of most vascular plants. **2. a.** The most central and material part : HEART. **b.** A basic trait that defines and establishes character. **3.** Force : vigor. **4.** Importance or significance. —*vt.* **pithed, pith·ing, piths. 1.** To remove the pith from (a plant stem). **2.** To sever or destroy the spinal cord of (a laboratory animal), usu. by means of a needle inserted into the vertebral canal. **3.** To kill (cattle) by cutting the spinal cord.

**pith·e·can·thro·pus** (pĭth'ĭ-kăn'thrə-pəs, -kăn-thrō'pəs) *n.* [NLat. *Pithecanthropus,* genus name : Gk. *pithēkos,* ape + Gk. *anthrōpos,* man.] A member of a genus formerly designated *Pithecanthropus,* based on bone fragments found in Java and thought to indicate the existence of a primate between humans and apes but now reclassified in the extinct species *Homo erectus.*

**pithecanthropus**
*The Java man*

**pith helmet** *n.* A light sun hat made from dried pith : TOPI.

**pith·y** (pĭth'ē) *adj.* **-i·er, -i·est. 1.** Consisting of or like pith. **2.** Pointedly meaningful : COGENT. **—pith'i·ly** *adv.* **—pith'i·ness** *n.*

**pit·i·a·ble** (pĭt'ē-ə-bəl) *adj.* **1.** Arousing or deserving pity or compassion : LAMENTABLE. **2.** Arousing disdainful pity : CONTEMPTIBLE. **—pit'i·a·ble·ness** *n.* **—pit'i·a·bly** *adv.*

**pit·i·ful** (pĭt'ĭ-fəl) *adj.* **1.** Arousing pity : PATHETIC. **2.** So inferior or trivial as to be contemptible : PALTRY. **3.** *Archaic.* Filled with pity or compassion. **—pit'i·ful·ly** *adv.* **—pit'i·ful·ness** *n.*

**pit·i·less** (pĭt'ĭ-lĭs) *adj.* **1.** Having no pity : MERCILESS. **2.** Totally uncompromising. **—pit'i·less·ly** *adv.* **—pit'i·less·ness** *n.*

**pit·man** (pĭt'mən) *n.* **1.** *pl.* **pit·men.** A worker employed near or inside a pit, as a coal mine. **2.** *pl.* **pit·mans.** A connecting rod.

**pi·ton** (pē'tŏn') *n.* [Fr. < OFr., nail.] A metal spike fitted at one end with an eye or ring through which to pass a rope and driven into rock or ice for support in mountain climbing.

**Pi·tot-stat·ic tube** (pē'tō-stăt'ĭk, pē-tō'-) *n.* A device having a Pitot tube and static tube combined so that total and static pressure in a fluid stream can be measured simultaneously, used in aircraft to determine relative wind speed.

ă **pat** ā **pay** âr **care** ä **father** ĕ **pet** ē **be** hw **which** ĭ **pit**
ī **tie** îr **pier** ŏ **pot** ō **toe** ô **paw, for** oi **noise** ŏŏ **took**

**Pi·tot tube** (pē'tō, pē-tō') *n.* [After Henri *Pitot* (1695–1771), its inventor.] **1.** A device used to measure the total pressure of a fluid stream that is essentially a tube attached to a manometer at one end and pointed upstream at the other. **2.** A Pitot-static tube.

**pit·saw** *also* **pit saw** (pĭt'sô') *n.* A large saw for cutting logs, hand-operated by two people, one of whom stands on the log and the other in a pit underneath.

**pit stop** *n.* **1.** A stop at a pit during an auto race, usu. for fuel or a change of tires. **2. a.** *Informal.* A stop during a journey for rest, food, or fuel. **b.** A place where such a stop is made.

**pit·tance** (pĭt'ns) *n.* [ME *pitaunce* < OFr. *pitance*, donation to a monastery < Med. Lat. *pietantia* < Lat. *pietas*, piety < *pius*, dutiful.] **1.** A meager monetary allowance. **2.** A very small salary or remuneration. **3.** A small portion or amount.

**pit·ted** (pĭt'ĭd) *adj.* **1.** Characterized by pits. **2.** Having the pit removed <*pitted* olives>

**pit·ter-pat·ter** (pĭt'ər-păt'ər) *n.* [Imit.] A rapid series of light, tapping sounds. **—pit'ter-pat'ter** *v.* **(-tered, -ter·ing, -ters).**

**pi·tu·i·cyte** (pĭ-tōō'ĭ-sīt', -tyōō'-) *n.* [PITUI(TARY) + -CYTE.] A modified neuroglia cell of the posterior lobe of the pituitary gland.

**pi·tu·i·tar·y** (pĭ-tōō'ĭ-tĕr'ē, -tyōō'-) *n.*, *pl.* **-ies.** [< Lat. *pituitarius*, of phlegm < *pituita*, phlegm.] **1.** *Anat.* The pituitary gland. **2.** *Med.* An extract from the anterior or posterior lobes of the pituitary gland, prepared for therapeutic use. *—adj.* **1.** Of the pituitary gland. **2.** Of or designating a physique marked by obesity, enlarged bones, and soft parts of arms, legs, and face, thought to be caused by excessive secretion from the pituitary gland. **3.** Secreting phlegm or mucus.

**pituitary gland** *n.* A small, oval endocrine gland attached to the base of the vertebrate brain, whose secretions control the other endocrine glands and influence growth, metabolism, and maturation.

**pi·tu·i·tous** (pĭ-tōō'ĭ-təs, -tyōō'-) *adj.* [Lat. *pituitosus* < *pituita*, phlegm.] Containing, discharging, or resembling mucus.

**pit viper** *n.* A venomous snake of the family Crotalidae, as a rattlesnake or copperhead, with a small pit on each side of the head.

**pit·y** (pĭt'ē) *n.*, *pl.* **-ies.** [ME *pite* < OFr. < Lat. *pietas* < *pius*, dutiful.] **1.** Compassion for suffering or one who suffers. **2.** A regrettable or disagreeable fact or necessity <It's a *pity* that you lost.> *—v.* **-ied, -y·ing, -ies.** *—vt.* To feel pity for. *—vi.* To feel pity. **—pit'y·ing·ly** *adv.*

☆ ***syns:*** PITY, COMMISERATION, COMPASSION, SYMPATHY *n. core meaning* : sympathetic, sad concern for suffering or one who suffers <felt *pity* for the refugees>

**pit·y·ri·a·sis** (pĭt'ĭ-rī'ə-sĭs) *n.* [NLat. < Gk. *pituriasis* < *pituron*, grain husk.] Any of various skin diseases of humans and animals, marked by epidermal shedding of flaky scales.

**più** (pyōō) *adv.* [Ital. < Lat. *plus.*] *Mus.* More. —Used to qualify an adverb or adjective in directions.

**Pi·ute** (pī'yōōt') *n. var. of* PAIUTE.

**piv·ot** (pĭv'ət) *n.* [Fr. < OFr.] **1.** A short rod or shaft about which a related part rotates or swings. **2. a.** An essential component that determines direction or effect. **b.** A key position or player, as in basketball. **3.** The act of turning on or as if on a pivot. *—v.* **-ot·ed, -ot·ing, -ots.** *—vt.* **1.** To mount on, attach by, or furnish with a pivot or pivots. **2.** To cause to turn on a pivot, esp. to place under the control of a determining factor. *—vi.* To turn on or as if on a pivot. **—piv'ot·a·ble** *adj.* **—piv'ot·al** *adj.* **—piv'ot·al·ly** *adv.*

**pix**[1] (pĭks) *n. var. pl. of* PIC.

**pix**[2] (pĭks) *n. var. of* PYX.

**pix·ie** (pĭk'sē) *n. var. of* PIXY.

**pix·i·lat·ed** (pĭk'sə-lā'tĭd) *adj.* [< PIXY.] **1.** Behaving as if somewhat mentally unbalanced. **2.** Whimsical. **3.** *Slang.* Drunk. **—pix'i·la'tion** (-lā'shən) *n.*

**pix·y** or **pix·ie** (pĭk'sē) *n.*, *pl.* **-ies.** [Orig. unknown.] A fairylike or elfin creature. *—adj.* Playfully mischievous. **—pix'y·ish, pix'ie·ish** *adj.*

**piz·za** (pēt'sə) *n.* [Ital., prob. < VLat.* *picea* < Lat., fem. of *piceus*, of pitch < *pix*, pitch.] A baked dish of Italian origin consisting of bread dough covered with spiced tomato sauce and usu. cheese and often with a variety of toppings.

**piz·zazz** (pĭ-zăz') *n.* [Orig. unknown.] *Slang.* Zest : flair.

**piz·ze·ri·a** (pēt'sə-rē'ə) *n.* A place where pizzas are made and sold.

**piz·zi·ca·to** (pĭt'sĭ-kä'tō) [Ital., p.part. of *pizzicare*, to pluck < *pizzare*, to prick.] *Mus.* *—adj.* Played by plucking rather than bowing strings. *—n.*, *pl.* **-tos.** A pizzicato note or passage. **—piz'zi·ca'to** *adv.*

**PL/1** (pē'ĕl-wŭn') *n.* [P(ROGRAMMING) L(ANGUAGE) 1.] *Computer Sci.* A programming language designed for scientific and commercial uses at varying levels of complexity.

**plac·a·ble** (plăk'ə-bəl, plā'kə-) *adj.* [ME, agreeable < OFr. < Lat. *placabilis* < *placare*, to calm.] Easily calmed or pacified : TRACTABLE. **—plac'a·bil'i·ty** *n.* **—plac'a·bly** *adv.*

**plac·ard** (plăk'ärd', -ərd) *n.* [ME *placquart*, official document < OFr. < *plaquier*, to plate < MDu. *placken*, to patch.] **1.** A printed or written announcement for display in a public place : POSTER. **2.** A nameplate, as on the door of a house. *—vt.* **-ard·ed, -ard·ing, -ards.** **1.** To announce (e.g., a message) on a placard. **2.** To post placards on or in. **3.** To display as or as if a placard. **—plac'ard'er** *n.*

**pla·cate** (plā'kāt', plăk'āt') *vt.* **-cat·ed, -cat·ing, -cates.** [Lat. *placare, placat-.*] To appease the anger of, esp. by yielding concessions : PACIFY. **—pla'cat'er** *n.* **—pla·ca'tion** (plā-kā'shən) *n.* **—pla'ca·to'ry** (-tôr'ē, -tōr'ē), **pla'ca'tive** *adj.*

**place** (plās) *n.* [ME < OFr., open space < Lat. *platea*, broad street < Gk. *plateia (hodos)*, broad (street).] **1.** An area with definite or indefinite boundaries. **2.** An area occupied by or set aside for a specific person or purpose. **3.** A definite location, esp.: **a.** A dwelling, as a house or an apartment. **b.** A business establishment or office. **c.** A particular city or town. **4. Place.** A public square, thoroughfare, or short street in a town. **5. a.** A space for one person to sit or stand, as a passenger or spectator. **b.** A table setting for one person. **6.** A position regarded as held by someone or something else : STEAD <took me in your *place*> **7.** A particular point up to which one has read in a book <a marker to keep one's *place*> **8.** A position figuratively occupied by a thing, group, or activity in a larger complex : FUNCTION. **9.** Proper or customary order or location <a *place* for everything> **10.** A social station entailing a certain mode of behavior. **11.** High rank or office. **12.** A relative position in a series : STANDING <first *place*> **13.** *Math.* The position of a number in relation to other numbers in a series. *—v.* **placed, plac·ing, plac·es.** *—vt.* **1. a.** To put in a particular position : SET. **b.** To offer for consideration <*placed* the question before the committee> **2.** To put in a relation or order <*Place* the numbers in descending order.> **3.** To find living quarters for (someone). **4.** To arrange for the publication or production of (a literary or dramatic work). **5.** To appoint to a post <was *placed* in a position of political importance> **6.** To rank (someone or something) in an order or sequence. **7.** To date or identify in a particular context <We *placed* the mask as pre-Columbian.> **8.** To identify by association : REMEMBER <couldn't *place* your face at first> **9.** To give an order for <*place* a wager> **10.** To apply for : request formally, as a commercial order. **11.** To invest (money). **12.** To adjust (one's voice) for the best possible effects. *—vi.* **1.** To arrive among the first three finishers in a race. **2.** To finish second in a race. **—in place of.** Instead of. **—place'a·ble** *adj.*

**pla·ce·bo** (plə-sē'bō) *n.*, *pl.* **-bos** *or* **-boes.** [ME < Lat., I shall please, the first word of the first antiphon of the service.] **1.** (plä-chā'bō). *Rom. Cath. Ch.* The service or office of vespers for the dead. **2. a.** *Med.* A substance containing no medication and given usu. to indulge a patient. **b.** An inactive substance used as a control in an experiment. **3.** Something lacking intrinsic remedial value done or given to humor or pacify another.

**place kick** *n.* *Football.* A kick, as for a field goal, for which the ball is held or propped up in a fixed position. **—place'-kick'** *v.* **(-kicked, -kick·ing, -kicks).** **—place'-kick'er** *n.*

**place mat** *n.* A protective and decorative mat for a single setting of dishes and silverware.

**place·ment** (plās'mənt) *n.* **1.** The act of placing or state of being placed. **2.** The act or business of finding jobs, lodgings, or other positions for applicants. **3.** *Football.* **a.** The positioning of the ball for a place kick. **b.** A place kick.

**pla·cen·ta** (plə-sĕn'tə) *n.*, *pl.* **-tas** *or* **-tae** (-tē) [NLat. < Lat., flat cake < Gk. *plakous* < *plax*, flat surface.] **1. a.** *Anat.* A vascular membranous organ that develops in female mammals during pregnancy, lining the uterine wall and partially enveloping the fetus, to which it is attached by the umbilical cord; following birth, the placenta is expelled. **b.** An organ in various other animals, including certain reptiles and sharks, with similar functions. **2.** *Bot.* **a.** The part of the ovary to which the ovules are attached. **b.** The tissue that bears the spore cases in nonflowering plants. **—pla·cen'tal** *adj.*

**plac·en·ta·tion** (plăs'ən-tā'shən) *n.* **1.** *Zool.* **a.** Formation of a placenta. **b.** The type or structure of a placenta. **2.** *Bot.* The way in which the placenta is arranged in or attached to the ovary.

**plac·er** (plăs'ər) *n.* [Sp. < *plaza*, place < Lat. *platea*, broad street. —see PLACE.] **1.** A glacial or alluvial deposit of gravel or sand containing eroded particles of valuable minerals. **2.** A place where a placer deposit is washed to remove its mineral content.

**placer mining** *n.* The obtaining of minerals from placers by washing or dredging. **—placer miner** *n.*

**place setting** *n.* A table setting for one person.

**plac·id** (plăs'ĭd) *adj.* [Lat. *placidus* < *placēre*, to please.] **1.** Having an undisturbed surface or aspect : CALM. **2.** Complacent : self-satisfied. **—pla·cid'i·ty** (plə-sĭd'ĭ-tē), **plac'id·ness** *n.* **—plac'id·ly** *adv.*

**plack·et** (plăk'ĭt) *n.* [Orig. unknown.] **1.** A slit in a dress, blouse, or skirt to make the garment easy to put on or take off. **2.** A pocket, esp. in a woman's skirt.

**plac·oid** (plăk'oid) *adj.* [Gk. *plax, plak-*, flat surface + -OID.] Platelike, as the hard toothlike scales of sharks, skates, and rays.

**pla·gal** (plā'gəl) *adj.* [Med. Lat. *plagalis* < Gk. *plagios*, oblique < *plagos*, side.] *Mus.* **1.** Designating a medieval mode having a range from the fourth below to the fifth above its final tone. **2.** Designating a cadence with the subdominant chord immediately preceding the tonic chord.

**pla·gia·rism** (plā'jə-rĭz'əm) *n.* [< PLAGIARY.] **1.** The act of plagiarizing. **2.** Material that has been plagiarized. **—pla'gia·rist** *n.* **—pla'gia·ris'tic** *adj.*

**pla·gia·rize** (plā′jə-rīz′) *v.* **-rized, -riz·ing, -riz·es.** [< PLAGIARY.] —*vt.* **1.** To steal and use (the ideas or writings of another) as one's own. **2.** To take passages or ideas from and use them as one's own. —*vi.* To practice plagiarism. **—pla′gia·riz′er** *n.*

☆ ***syns:*** PLAGIARIZE, CRIB, PIRATE *v. core meaning* : to steal and use (the work of another) as one's own <*plagiarized* whole sections of another writer's novel>

**pla·gia·ry** (plā′jə-rē) *n., pl.* **-ries.** [Lat. *plagiarius,* plunderer < *plagium,* kidnapping < *plaga,* net.] **1.** Plagiarism. **2.** *Archaic.* One who plagiarizes.

**plagio-** *pref.* [< Gk. *plagios,* oblique < *plagos,* side.] Slanting : inclining <*plagio*tropism>

**pla·gi·o·clase** (plā′jē-ə-klās′, -klāz′, plăj′ē-) *n.* [G. *Plagioklas* : Gk. *plagios,* oblique (< *plagos,* side) + Gk. *klasis,* breaking < *klan,* to break.] One of a common rock-forming series of triclinic feldspars, made up of mixtures of sodium and calcium aluminum silicates.

**pla·gi·ot·ro·pism** (plā′jē-ŏt′rə-pĭz′əm) *n. Biol.* The tendency to grow at an oblique or horizontal angle. —Used chiefly of roots, stems, or branches. **—pla′gi·o·tro′pic** (-ə-trō′pĭk, -trŏp′ĭk) *adj.* **—pla′gi·o·tro′pi·cal·ly** *adv.*

**plague** (plāg) *n.* [ME *plage* < OFr., wound < Lat. *plaga* < Gk. *plēgē.*] **1.** A great affliction or calamity, orig. one of divine retribution. **2.** A sudden influx, as of injurious or destructive insects. **3. a.** A nuisance <"the *plague* of social jabbering" —George Santayana> **b.** A sudden, undesirable epidemic <a *plague* of child-abuse cases> **4.** A highly infectious, usu. fatal epidemic disease, esp. the bubonic plague. —*vt.* **plagued, plagu·ing, plagues. 1.** To harass or annoy. **2.** To afflict with or as if with plague : DISTRESS <"Runaway inflation further *plagued* the wage- or salary-earner" —Edwin Reischauer> **—plagu′er** *n.* **—plague′some** *adj.*

**pla·guy** *also* **pla·guey** (plā′gē) *adj. Informal.* Irritating : annoying. **—pla′guy, pla′gui·ly** *adv.*

**plaice** (plās) *n., pl.* **plaice** *or* **plaic·es.** [ME < OFr. *plaïs* < LLat. *platessa,* ult. < Gk. *platus,* broad.] **1.** An edible marine flatfish, *Pleuronectes platessa* of western European waters. **2.** A flatfish, as *Hippoglossoides platessoides* of North American Atlantic waters, related to the plaice.

**plaid** (plăd) *n.* [Sc. Gael. *plaide.*] **1.** A rectangular woolen scarf of a checked or tartan pattern worn over one shoulder by Scottish Highlanders. **2.** Cloth with a checked or tartan pattern. **3.** A checked or tartan pattern. **—plaid′ed** *adj.*

**plain** (plān) *adj.* **-er, -est.** [ME < OFr. < Lat. *planus,* flat.] **1.** Free from obstructions to sight <in *plain* view> **2.** *Archaic.* Having no discernible elevation or depression : LEVEL. **3.** Clearly evident : easily understood <made my meaning *plain*> **4.** Uncomplicated : simple <*plain* sewing> **5.** Open and without pretense. **6.** Not mixed with other substances : PURE <*plain* quinine water> **7.** Common in rank or station : ORDINARY. **8.** Unpretentious. **9.** Not rich or elaborate <*plain* meat and potatoes> **10.** With little decoration or ornamentation <a *plain* tailored suit> **11.** Not dyed, twilled, or patterned. **12.** Not beautiful or handsome. **13.** Utter : sheer <*plain* foolishness> —*n.* **1.** An extensive, level, treeless land region. **2.** Something devoid of extraneous matter or ornamentation. —*adv.* In a clear or intelligible way. **—plain′ly** *adv.* **—plain′ness** *n.*

☆ ***syns:*** **1.** PLAIN, SIMPLE, UNADORNED *adj. core meaning* : with little decoration <a *plain* gray skirt> ***ant:*** ornate **2.** PLAIN, FORTHRIGHT, STRAIGHTFORWARD *adj. core meaning* : done openly and without pretense <*plain* dealing> **3.** PLAIN, HOMELY *adj. core meaning* : not physically attractive <a *plain* individual>

**plain·chant** (plān′chănt′) *n.* [Fr., transl. of Med. Lat. *cantus planus,* plain song.] PLAINSONG 2.

**plain·clothes man** (plān′klōz′) *also* **plain·clothes·man** (plān′klōz′mən) *n.* A police officer or detective who wears civilian clothes on duty.

**plain-laid** (plān′lād′) *adj.* Designating a rope made of three strands laid together with a right-hand twist.

**Plain People** *pl.n.* Members of the Mennonite, Amish, or Dunker sects, noted for their custom of wearing plain dress.

**plain sailing** *n.* Easy progress over a straightforward course.

**Plains Indian** *n.* A member of any of the tribes of Indians that once inhabited the Great Plains of the United States and Canada.

**plains·man** (plānz′mən) *n.* An inhabitant or settler of U.S. prairie regions.

**plain·song** (plān′sông′, -sŏng′) *n.* [Transl. of Med. Lat. *cantus planus.*] **1.** Gregorian chant. **2.** Any of various bodies of medieval liturgical music without strict meter and sung without accompaniment.

**plain·spo·ken** (plān′spō′kən) *adj.* Straightforward : candid <a *plainspoken* commentator> **—plain′spo′ken·ness** *n.*

**plaint** (plānt) *n.* [ME < OFr. *plainte* < Lat. *planctus,* lament < p.part. of *plangere,* to beat one's breast.] **1.** A complaint. **2.** An utterance of sorrow or grief : LAMENTATION. **3.** *Law.* A statement of grievance submitted to a court as a request for redress.

**plain·tiff** (plān′tĭf) *n.* [ME *playntif* < OFr. *plaintif* < *plaintif,* plaintive.] *Law.* The party that institutes a suit in a court.

**plain·tive** (plān′tĭv) *adj.* [ME *playntif* < OFr. *plaintif* < *plainte,* plaint.] Expressing sorrow : MOURNFUL. **—plain′tive·ly** *adv.*

**plain weave** *n.* A weave in which the filling threads and the warp threads interlace alternately, forming a checkerboard pattern.

**plait** (plāt, plăt) *n.* [ME, fold < OFr. *pleit* < Lat. *plicare,* to fold.] **1.** A braid, esp. of hair. **2.** A pleat. —*vt.* **plait·ed, plait·ing, plaits. 1.** To braid <*plaited* the child's long hair> **2.** To pleat. **3.** To make by braiding or pleating. **—plait′er** *n.*

**plan** (plăn) *n.* [Fr. < *planter,* to plant < Lat. *plantare* < *planta,* sole of the foot.] **1.** A detailed scheme, program, or method worked out beforehand for the accomplishment of an object <a career *plan*> **2.** A proposed or tentative project or purpose : INTENTION <made *plans* for the weekend> **3.** An outline or sketch, as of a story. **4. a.** A drawing or diagram made to scale showing structure or arrangement <an architectural *plan*> **b.** Such a plan showing how to build or assemble something. **5.** One of several imaginary planes perpendicular to the line of vision between the viewer and the object being depicted. —*v.* **planned, plan·ning, plans.** —*vt.* **1.** To formulate a scheme or program for the accomplishment or attainment of <*plan* an outreach program> **2.** To have as a specific goal or purpose : INTEND <They *plan* to visit Disneyland.> **3.** To draw or make a graphic representation of. —*vi.* To make plans. **—plan′ner** *n.*

**plan-** *pref. var. of* PLANO-.

**pla·nar** (plā′nər, -när′) *adj.* [LLat. *planaris,* flat < Lat. *planus.*] **1.** Of, pertaining to, or located in a plane. **2.** Flat. **3.** Having a two-dimensional characteristic. **—pla·nar′i·ty** (plā-năr′ĭ-tē) *n.*

**pla·nar·i·a** (plə-nâr′ē-ə) *n.* A planarian.

**pla·nar·i·an** (plə-nâr′ē-ən) *n.* [< NLat. *Planaria,* genus name < LLat. *planarius,* on level ground < *planus,* flat.] A flatworm of the order Tricladida, with a broad, ciliated body and a three-branched digestive cavity.

**pla·na·tion** (plā-nā′shən) *n.* [< PLANE[1].] Lateral mechanical erosion, as of a valley, by a running stream.

**planch·et** (plăn′chĭt) *n.* [Dim. of dial. *planch,* board < ME *plaunche* < OFr. *planche* < Lat. *planca.*] **1.** A flat disk of metal ready for stamping as a coin. **2.** A small disk of metal on which a radioactive substance is deposited for measurement of its activity.

**plan·chette** (plăn-shĕt′) *n.* [Fr., dim. of OFr. *planche,* board. —see PLANCHET.] A small triangular board with a pointer supported by two casters and a vertical pencil reputed to spell out messages from the spirit world when the operator's fingers are placed lightly upon it.

**Planck's constant** (plängks) *n.* [After Max K.E.L. *Planck* (1858–1947).] *Physics.* The constant of proportionality relating the quantum of energy that can be possessed by radiation to the frequency of that radiation; its value is approx. $6.625 \times 10^{-27}$ erg-second.

**plane[1]** (plān) *n.* [Lat. *planum,* flat surface < *planus,* flat.] **1.** *Math.* A surface containing all the straight lines connecting any two points on it. **2.** A level or flat surface. **3.** A stage or level of existence, development, or achievement <lectures on a spiritual *plane*> **4.** An airplane or hydroplane. **5.** A supporting surface of an airplane, as an airfoil or wing. —*adj.* **1.** *Math.* Designating a figure lying in a plane. **2.** Flat. **—plane′ness** *n.*

**plane[2]** (plān) *n.* [ME < OFr. < LLat. *plana* < *planare,* to plane < *planus,* flat.] **1.** A carpenter's tool with an adjustable blade for leveling and smoothing wood. **2.** A trowel-shaped tool for smoothing the surface of clay, sand, or plaster in a mold. —*v.* **planed, plan·ing, planes.** —*vt.* **1.** To smooth with or as if with a plane. **2.** To remove with a plane. —*vi.* **1.** To undergo planing. **2.** To act as a plane.

**plane[3]** (plān) *vi.* **planed, plan·ing, planes.** [Fr. *planer,* to glide < *plan,* level surface < Lat. *planum* < *planus,* flat.] **1.** To rise partly out of the water, as a hydroplane does at high speeds. **2.** To soar or glide. **3.** To travel by airplane.

**plane[4]** (plān) *n.* [ME < OFr. < Lat. *platanus* < Gk. *platanos* < *platus,* broad.] The plane tree.

**plane angle** *n.* An angle formed by two straight lines.

**plane geometry** *n.* The geometry of planar figures.

**plane·load** (plān′lōd′) *n.* The load an airplane can carry.

**plan·er** (plā′nər) *n.* **1.** One that planes. **2.** A machine tool for leveling and smoothing the surfaces of wood or metal. **3.** A smooth block of wood used to level a form of type.

**pla·ner tree** (plā′nər) *n.* [After J.J. *Planer* (1743–1789).] A small swamp tree, *Planera aquatica* of the southern United States, bearing small, rough, nutlike fruit.

**plane·side** (plān′sīd′) *n.* The area next to an airplane.

**plan·et** (plăn′ĭt) *n.* [ME < OFr. *planete* < LLat. *planeta* < Gk. *planēs,* wanderer < *planasthai,* to wander.] **1.** A nonluminous celestial body illuminated by light from a star, as the sun, around which it revolves. **2.** One of the seven celestial bodies visible to the naked eye and thought to revolve in the heavens about a fixed Earth and among fixed stars in ancient astronomy. **3.** One of the seven revolving celestial bodies that in conjunction with the stars are supposed to influence human personalities and concerns.

**plane table** *n.* A portable surveying instrument consisting primarily of a drawing board and a ruler mounted on a tripod and used to sight and map topographical details.

**plan·e·tar·i·um** (plăn′ĭ-târ′ē-əm) *n., pl.* **-i·ums** *or* **-i·a** (-ē-ə). **1.** An apparatus or model representing the solar system. **2.** A device

ă **pat** ā **pay** âr **care** ä **father** ĕ **pet** ē **be** hw **which** ĭ **pit** ī **tie** îr **pier** ŏ **pot** ō **toe** ô **paw,** for oi **noise** ŏŏ **took**

for projecting images of celestial bodies in their courses onto the inner surface of a hemispherical dome. **3.** A building or room containing a planetarium.

**plan·e·tar·y** (plăn′ĭ-tĕr′ē) *adj.* **1.** Of, relating to, or like the physical or orbital characteristics of a planet or the planets. **2. a.** Worldwide: global. **b.** Terrestrial: mundane. **3.** Erratic: wandering. **4.** Denoting or relating to a gear train consisting of a central gear with an internal ring gear and one or more pinions. **5.** Of enormous size or scope.

**planetary nebula** *n.* Any of several objects in the Galaxy having a hot, blue-white, central star surrounded by an envelope of expanding gas.

**plan·e·tes·i·mal** (plăn′ĭ-tĕs′ə-məl, -tĕz′-) *n.* [PLANET + (INFINIT)ESIMAL.] One of innumerable small bodies held to have orbited the sun during the formation of the planets. **—plan′e·tes′i·mal** *adj.*

**planetesimal hypothesis** *n.* The hypothesis that the planets and satellites of the solar system were formed by gravitational aggregation of planetesimals.

**plan·e·toid** (plăn′ĭ-toid′) *n.* ASTEROID 1. **—plan′e·toi′dal** *adj.*

**plane tree** *n.* A tree of the genus *Platanus,* with ball-shaped clusters of fruit and outer bark that usu. flakes off in patches.

**planet wheel** *n.* One of the small gear wheels in an epicyclic train.

**plan·gent** (plăn′jənt) *adj.* [Lat. *plangens, plangent-,* pr.part. of *plangere,* to strike.] **1. a.** Striking with a deep, reverberating sound, as waves against the shore. **b.** Loud and sonorous, as the sound of bells: RESONANT. **2.** Expressing sadness: PLAINTIVE. **—plan′gen·cy** *n.* **—plan′gent·ly** *adv.*

**plani-** *pref. var. of* PLANO-.

**pla·nim·e·ter** (plə-nĭm′ĭ-tər, plā-) *n.* [Fr. *planimètre* : Lat. *planum,* level surface (< *planus,* flat) + *mètre,* meter < Gk. *metron,* measure.] An instrument that measures the area of a plane figure as a mechanically coupled pointer traverses the figure's perimeter. **—pla′ni·met′ric** (plā′nə-mĕt′rĭk), **pla′ni·met′ri·cal** *adj.* **—pla′ni·met′ri·cal·ly** *adv.* **—pla·nim′e·try** *n.*

**plan·ish** (plăn′ĭsh) *vt.* **-ished, -ish·ing, -ish·es.** [OFr. *planir, planiss-,* to make smooth < *plan,* level < Lat. *planus.*] To smooth, flatten, toughen, or polish (metal) by hammering or rolling. **—plan′ish·er** *n.*

**pla·ni·sphere** (plā′nĭ-sfîr′) *n.* [ME *planisperie* < Med. Lat. *planisphaerium* : Lat. *planus,* flat + Lat. *sphaera,* sphere < Gk. *sphaira.*] **1.** A representation of a sphere or part of a sphere on a plane surface. **2.** *Astron.* A polar projection of the celestial sphere on a chart equipped with an adjustable overlay to show the stars visible at a given time and place. **—pla′ni·spher′ic** (-sfîr′ĭk, -sfĕr′-), **pla′ni·spher′i·cal** *adj.*

**plank** (plăngk) *n.* [ME < ONFr. *planke* < Lat. *planca.*] **1. a.** A piece of lumber cut thicker than a board. **b.** PLANKING 2. **2.** A support: foundation. **3.** One of the articles of a political platform. —*vt.* **planked, plank·ing, planks. 1.** To furnish or cover with planks. **2.** To bake or broil and serve (fish or meat) on a plank. **3.** To put or set down emphatically or with force. **4.** *Informal.* To pay immediately <*planked* down $100 for the ticket>

**plank·ing** (plăng′kĭng) *n.* **1.** The act of laying planks. **2.** Planks as a whole. **3.** Something made of planks.

**plank-sheer** (plăngk′shîr′) *n.* [Alteration of obs. *plancher,* planking < ME *plauncher* < OFr. *planchier* < *planche,* plank < Lat. *planca.*] A horizontal timber forming the outer edge of the upper deck of a wooden ship.

**plank·ter** (plăngk′tər) *n.* [< Gk. *planktēr,* wanderer < *planktos,* wandering. —see PLANKTON.] One of the minute organisms that collectively make up plankton.

**plank·ton** (plăngk′tən) *n.* [G. < Gk., neuter of *planktos,* wandering < *plazein,* to drive astray.] Gen. microscopic plant and animal organisms that float or drift in great numbers in fresh or salt water. **—plank·ton′ic** (-tŏn′ĭk) *adj.*

**Planned Parenthood** (plănd). A service mark for an organization that provides family planning services.

**plano-** *or* **plani-** *or* **plan-** *pref.* [< Lat. *planus,* flat.] Flat <*plano*convex>

**pla·no·con·cave** (plā′nō-kŏn-kāv′, -kŏn′kāv′) *adj.* Flat or plane on one side and concave on the other.

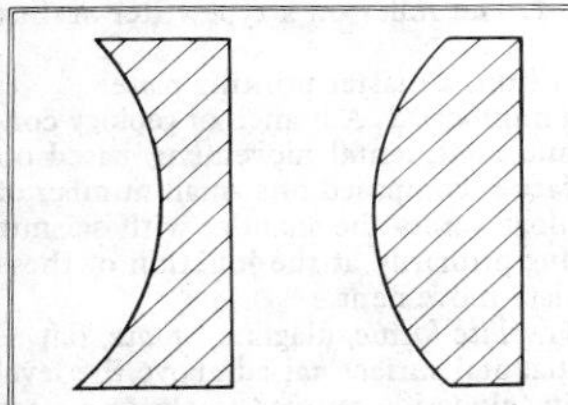

(Left) **planoconcave** and (right) **planoconvex**

**pla·no·con·vex** (plā′nō-kŏn-vĕks′, -kŏn′vĕks′) *adj.* Flat or plane on one side and convex on the other.

**pla·nog·ra·phy** (plə-nŏg′rə-fē, plā-) *n.* A process for printing from a smooth surface, as offset or lithography. **—pla′no·graph′ic** (plā′nə-grăf′ĭk) *adj.* **—pla′no·graph′i·cal·ly** *adv.*

**pla·nom·e·ter** (plə-nŏm′ĭ-tər, plā-) *n.* A flat metal plate for gauging the accuracy of a plane surface in precision metalworking. **—pla·nom′e·try** *n.*

**plant** (plănt) *n.* [ME < OE < Lat. *planta,* shoot.] **1.** An organism of the vegetable kingdom having cellulose cell walls, growing by synthesis of inorganic substances, and lacking the power of locomotion. **2.** A plant, as an herb, that lacks a permanent woody stem. **3.** An industrial or manufacturing establishment: FACTORY. **4.** The buildings, equipment, and fixtures of an institution. **5.** An apparently trivial line or passage in a play or story that becomes important later. **6. a.** One placed in an audience to encourage applause or contribute to the action of the play. **b.** *Slang.* One placed or stationed in a given location in order to observe, spy, or inform. **7.** A misleading piece of evidence placed so as to be discovered. **8.** *Slang.* A scheming trick: SWINDLE. —*vt.* **plant·ed, plant·ing, plants. 1.** To place or set (e.g., seeds) in the ground to grow. **2. a.** To furnish or supply (a plot of land) with seeds or plants. **b.** To stock (a body of water) with fish or spawn. **c.** To introduce (an animal) into an area. **3.** To fix or set firmly in position <*planted* my feet and refused to budge> **4.** To establish or set up: FOUND <*plant* a settlement> **5.** To implant (e.g., an idea) in the mind. **6.** *Slang.* **a.** To place or station (a person) to observe, spy, or inform <*planted* a government agent near the hideout> **b.** To place (e.g., false evidence) for the purpose of deception. **7.** *Informal.* To hide by burying. **8.** *Slang.* To deliver (a blow or punch) forcibly. **—plant′a·ble** *adj.* **—plant′like′** *adj.*

**plan·tain[1]** (plăn′tən) *n.* [ME < OFr. < Lat. *plantago* < *planta,* sole of the foot.] A plant of the genus *Plantago,* esp. *P. major,* a weed with broad leaves and a spike of small greenish flowers.

**plan·tain[2]** (plăn′tən) *n.* [Sp. *plántano* < Med. Lat. *plantanus,* plane tree < Lat. *platanus.* —see PLANE[4].] **1.** A large tropical plant, *Musa paradisiaca,* resembling the banana and bearing similar fruit. **2.** The fruit of the plantain.

**plantain lily** *n.* A widely cultivated Asian plant of the genus *Hosta,* bearing white, blue, or lilac flowers.

**plan·tar** (plăn′tər, -tär′) *adj.* [Lat. *plantaris* < *planta,* sole of the foot.] Of, relating to, or occurring on the sole of the foot.

**plan·ta·tion** (plăn-tā′shən) *n.* **1.** An area under cultivation. **2.** A group of cultivated plants or trees. **3.** A large estate or farm on which crops are raised, often by resident workers. **4.** A newly established settlement or colony.

**plant·er** (plăn′tər) *n.* **1. a.** One who plants. **b.** A machine or tool for planting or sowing seeds. **2.** The manager or owner of a plantation. **3.** An early colonist or settler. **4.** A decorative container for house plants.

**planter's punch** (plăn′tərz) *n.* A drink of rum with lemon or lime juice, sugar syrup, water or soda, bitters, and grenadine.

**plant hormone** *n.* Phytohormone.

**plan·ti·grade** (plăn′tĭ-grād′) *adj.* [Fr. : Lat. *planta,* sole of the foot + Lat. *-gradus,* going < *gradi,* to step.] Walking with the entire lower surface of the foot on the ground, as humans and bears do. —*n.* A plantigrade animal.

**plant louse** *n.* An aphid or similar small insect.

**plan·u·la** (plăn′yə-lə) *n., pl.* **-lae** (-lē′) [NLat. < Lat., little plane < *planus,* flat.] The free-swimming, ciliated larva of a coelenterate. **—plan′u·lar** *adj.*

**plaque** (plăk) *n.* [Fr. < OFr., metal plate < MDu. *placke* < *placken,* to patch.] **1.** A flat plate, slab, or disk engraved or ornamented for mounting, as on a monument or wall. **2.** A small brooch or pin worn as an ornament or badge of membership. **3.** *Pathol.* **a.** A small, disk-shaped formation or growth: PATCH. **b.** A thin film of mucus and microorganisms on a tooth surface.

**plash** (plăsh) *n.* [Prob. imit.] **1.** A light splash. **2.** The sound of a plash. —*v.* **plashed, plash·ing, plash·es.** —*vt.* To spatter (liquid) about: SPLASH. —*vi.* To splash lightly.

**-plasia** *suff.* [NLat. < Gk. *plasis,* molding < *plassein,* to mold.] Development: growth <achondro*plasia*>

**plasm** (plăz′əm) *n.* **1.** GERM PLASM 3. **2.** *var. of* PLASMA.

**plasm-** *pref. var. of* PLASMO-.

**-plasm** *suff.* [< PLASMA.] Material forming tissue or cells <cyto*plasm*>

**plas·ma** (plăz′mə) *also* **plasm** (plăz′əm) *n.* [NLat. < LLat., image < Gk. < *plassein,* to mold.] **1. a.** *Physiol.* The clear yellowish fluid portion of blood, lymph, or intramuscular fluid in which cells are suspended. **b.** *Med.* Cell-free sterilized blood plasma used in transfusions. **2.** Cytoplasm or protoplasm. **3.** The fluid portion of milk from which the curd has been separated by coagulation: WHEY. **4.** *Physics.* An electrically neutral, highly ionized gas made up of ions, electrons, and neutral particles. **—plas·mat′ic** (plăz-măt′ĭk), **plas′mic** *adj.*

**plasma cell** *n.* A large oval cell that contains deeply staining chromatin material and aids in production of immunoglobulins.

**plas·ma·gel** (plăz′mə-jĕl′) *n.* A jellylike state of cytoplasm, occurring in the periphery of the amoeba.

**plas·ma·gene** (plăz′mə-jēn′) *n. Genetics.* A self-reproducing hereditary structure held to exist in cytoplasm and function in a manner analogous to, but independent of, chromosomal genes. **—plas′ma·gen′ic** (-jĕ′nĭk, -jĕn′ĭk) *adj.*

**plasma membrane** *n. Biol.* The semipermeable membrane enclosing the cytoplasm of a cell.

**plas·ma·pher·e·sis** (plăz′mə-fĕr′ĭ-sĭs) *n.* [PLASMA + Gk. *aphairesis,* removal. —see APHAERESIS.] A process in which blood is withdrawn from a donor, the plasma and erythrocytes are separated from the blood, and the erythrocytes are returned to the circulatory system of the donor.

**plas·ma·sol** (plăz′mə-sôl′, -sŏl′, -sōl′) *n.* [PLASMA + SOL⁴.] *Biol.* A state of cytoplasm that is more liquid than plasmagel.

**plas·mid** (plăz′mĭd) *n.* [PLASM(A) + -ID.] A genetic element occurring outside of the nucleus that is present in the cytoplasm of some bacterial cells.

**plas·min** (plăz′mĭn) *n.* A proteolytic enzyme in plasma that dissolves the clotting factors, esp. fibrin, in blood.

**plas·min·o·gen** (plăz-mĭn′ə-jən) *n.* The precursor to plasmin that is found in blood plasma and body fluids.

**plasmo-** *or* **plasm-** *pref.* [< PLASMA.] Plasma <*plasmin*>

**plas·mo·des·ma** (plăz′mə-dĕz′mə) *also* **plas·mo·desm** (plăz′mə-dĕz′əm) *n., pl.* **-ma·ta** (-mə-tə) *or* **-mas.** [PLASMO- + Gk. *desma,* bond < *dein,* to bind.] *Biol.* A strand of living cytoplasm connecting two cells that are otherwise functionally separate.

**plas·mo·di·um** (plăz-mō′dē-əm) *n., pl.* **-di·a** (-dē-ə) [NLat. *Plasmodium,* genus name: PLASM(O)- + *-odium,* resembling < Gk. *-oeidēs,* -oid.] **1.** A protozoan of the genus *Plasmodium,* which includes the parasites that cause malaria. **2.** A naked multinucleate mass of protoplasm, as that characteristic of the vegetative phase of the slime molds.

**plas·mol·y·sis** (plăz-mŏl′ĭ-sĭs) *n.* Contraction or shrinkage of cellular protoplasm, esp. a plant cell, caused by water loss via osmosis. **—plas′mo·lyt′ic** (-mə-lĭt′ĭk) *adj.* **—plas′mo·lyt′i·cal·ly** *adv.*

**plas·mo·lyze** (plăz′mə-līz′) *vt.* & *vi.* **-lyzed, -lyz·ing, -lyz·es.** [< PLASMOLYSIS.] To subject to or undergo plasmolysis.

**plas·mo·some** (plăz′mə-sōm′) *n.* A nucleolus.

**-plast** *suff.* [< Gk. *plastos,* molded < *plassein,* to mold.] An organized unit of living matter **:** CELL <chloro*plast*>

**plas·ter** (plăs′tər) *n.* [ME < OE < Lat. *emplastrum,* medical dressing < Gk. *emplastron* < *emplassein,* to plaster on : *en-,* in + *plassein,* to mold.] **1.** A mixture of lime, sand, and water, occas. containing fiber, that hardens to a smooth solid and is used for coating walls and ceilings. **2.** Plaster of Paris. **3.** A pastelike mixture applied to a part of the body for healing or cosmetic purposes. **4.** Mustard plaster. *—vt.* **-tered, -ter·ing, -ters. 1.** To cover, coat, or repair with plaster. **2.** To cover by or as if by pasting, esp. to cover conspicuously or excessively. **3.** To apply a plaster to. **4.** To cause to adhere to another surface. **5.** To smooth down (e.g., hair) by applying a sticky substance. **6.** *Informal.* **a.** To inflict heavy injury or damage on. **b.** To defeat decisively. **—plas′ter·er** *n.* **—plas′ter·y** *adj.*

**plas·ter·board** (plăs′tər-bôrd′, -bōrd′) *n.* A thin rigid board or sheet of layers of fiberboard or paper, usu. bonded to a plaster core and used to cover walls and ceilings.

**plaster cast** *n.* **1.** A sculptured plaster of Paris cast, mold, or object. **2.** CAST 11.

**plas·tered** (plăs′tərd) *adj. Slang.* Drunk.

**plas·ter·ing** (plăs′tər-ĭng) *n.* **1.** The act of applying or working with plaster. **2.** A coating or layer of plaster. **3.** A resounding defeat.

**plaster of Paris** *n.* [After *Paris,* France, where it was originally made.] Any of a group of gypsum cements, chiefly hemihydrated calcium sulfate, $CaSO_4 \cdot 1/2H_2O$, a white powder that forms a paste when mixed with water and hardens into a solid, used in making casts, molds, and sculpture.

**plas·tic** (plăs′tĭk) *adj.* [Lat. *plasticus* < Gk. *plastikos* < *plassein,* to mold.] **1.** Capable of being formed or shaped **:** PLIABLE. **2.** Relating to or dealing with shaping or modeling. **3.** Giving form or shape to a substance. **4.** Easily influenced **:** IMPRESSIONABLE. **5.** Having sculptural qualities. **6.** Made of plastic. **7.** *Physics.* Capable of undergoing continuous deformation without rupture or relaxation. **8.** *Biol.* Capable of building tissue **:** FORMATIVE. **9.** Characterized by artificiality, pretension, or lack of originality <the *plastic* world of Madison Avenue hype> *—n.* Any of various complex organic compounds produced by polymerization, capable of being molded, extruded, cast into various shapes and films, or drawn into filaments used as textile fibers. **—plas′ti·cal·ly** *adv.* **—plas·tic′i·ty** (plăs-tĭs′ĭ-tē) *n.*

**-plastic** *suff.* [Gk. *plastikos,* fit for molding. —see PLASTIC.] Forming **:** growing **:** developing <cyto*plastic*>

**plastic art** *n.* **1.** Three-dimensional art, as sculpture. **2.** A visual art form, as a film or painting, as distinguished from a written art form, as literature or music.

**plastic explosive** *n.* A pliable explosive material that adheres to various surfaces and is usu. detonated from a distance by a fuse or radio signal.

**plas·ti·cize** (plăs′tĭ-sīz′) *vt.* & *vi.* **-cized, -ciz·ing, -ciz·es.** To make or become plastic. **—plas′ti·ci·za′tion** *n.*

**plas·ti·ciz·er** (plăs′tĭ-sī′zər) *n.* Any of various substances added to plastics or other materials to keep them pliable or soft.

**plastic surgery** *n.* Surgery to remodel, repair, or restore lost, injured, or defective body parts. **—plastic surgeon** *n.*

**plas·tid** (plăs′tĭd) *n.* [G. < Gk. *plastos,* molded < *plassein,* to mold.] Any of several specialized cytoplasmic structures found in plant cells and some plantlike organisms and having various physiological functions. **—plas·tid′i·al** (plăs-tĭd′ē-əl) *adj.*

**plas·tique** (plăs-tēk′) *n.* [Fr. < Lat. *plasticus,* plastic.] A plastic explosive.

**plas·to·mer** (plăs′tə-mər) *n.* [Gk. *plastos,* molded + (POLY)MER.] A hard, tough polymer, as acrylate resin.

**plas·tron** (plăs′trən) *n.* [OFr. < OItal. *piastrone,* aug. of *piastra,* thin metal plate. —see PIASTER.] **1.** A breastplate worn under a coat of mail. **2.** A protective breastplate worn by fencers. **3.** A trimming on the front of a bodice. **4.** The front of a formal dress shirt. **5.** *Zool.* The ventral surface of the shell of a tortoise or turtle. **—plas′tral** (-trəl) *adj.*

**-plasty** *suff.* [Gk. *-plastia* < *plastos,* molded < *plassein,* to mold.] Plastic surgery <dermato*plasty*>

**-plasy** *suff.* -PLASIA.

**plat** (plăt) *vt.* **plat·ted, plat·ting, plats.** [ME *platen,* alteration of *plaiten,* to fold < *plait,* fold. —see PLAIT.] To plait or braid. *—n.* A braid.

**plate** (plāt) *n.* [ME < OFr. < fem. of *plat,* flat < VLat. **plattus* < Gk. *platus.*] **1.** A flat, smooth, relatively thin, rigid body of uniform thickness. **2. a.** A sheet of rolled, hammered, or cast metal. **b.** A very thin plated coat or layer of metal. **3. a.** A flat piece of metal forming a machine part. **b.** A flat piece of metal on which something is engraved. **4. a.** A thin piece of metal used for armor. **b.** Armor made of plate. **5. a.** A sheet of material, as metal, plastic, rubber, or paperboard, converted into a printing surface, as an electrotype or stereotype. **b.** A print of engraved material, as a woodcut or lithograph, esp. when reproduced in a book. **c.** A full-page book illustration, often in color and printed on paper different from that used on the text pages. **6.** A light-sensitive sheet of metal or glass on which a photographic image can be recorded. **7.** A thin metallic or plastic support fitted to the gums to secure artificial teeth. **8.** A horizontal member capping the exterior wall studs, upon which the roof rafters rest in wood-frame construction. **9.** *Baseball.* Home plate. **10. a.** A shallow dish on which food is served or from which it is eaten. **b.** PLATEFUL 1. **c.** A main course served on a plate. **11.** Service and food for one person at a meal <brunch at $12 per *plate*> **12.** Household articles, as hollowware, covered with a precious metal, as gold or silver. **13.** A dish passed among a congregation for the collection of offerings. **14. a.** An article of silver or gold, as a cup or bowl, offered as a prize. **b.** A contest, esp. a horse race, offering such a prize. **15.** A thin cut of beef from the brisket. **16.** *Anat.* & *Zool.* **a.** A thin flat layer or scale. **b.** A platelike organ or part. **17.** *Electron.* **a.** An electrode, as in a storage battery or capacitor. **b.** The anode in an electron tube. **18.** *Geol.* A large rigid section of the earth's lithosphere that floats upon the earth's mantle according to the theory of plate tectonics. *—vt.* **plat·ed, plat·ing, plates. 1.** To cover or coat with a thin layer of metal. **2.** To armor. **3.** To make an electrotype or stereotype from. **4.** To give a glossy finish to (paper) by pressing between metal sheets or rollers. **—plate′like′** *adj.*

**pla·teau** (plă-tō′) *n., pl.* **-teaus** *or* **-teaux** (-tōz′) [Fr. < OFr., platter < *plat,* flat. —see PLATE.] **1.** An elevated and fairly level expanse of land **:** TABLELAND. **2. a.** A level or stage of growth or development. **b.** A relatively stable or quiescent period or state.

**plat·ed** (plā′tĭd) *adj.* **1.** Coated with a thin layer of metal <silver-*plated*> **2.** Covered or furnished with plates or sheets of metal, as armor. **3.** Knitted with two kinds of yarn, one on the face and one on the back.

**plate·ful** (plāt′fo͝ol′) *n., pl.* **-fuls. 1.** The amount that a plate will hold. **2.** A generous portion of food.

**plate glass** *n.* A strong rolled and polished glass containing few impurities and used for mirrors and large windows.

**plate·let** (plāt′lĭt) *n.* A minute protoplasmic disk, smaller than a red blood cell, found in the blood of vertebrates and held to promote coagulation.

**plat·en** (plăt′n) *n.* [ME *plateyne,* paten < OFr. *platine,* metal plate < *plate,* plate. —see PLATE.] **1.** One of the two flat members of the printing unit of a printing press that serves to position the paper and hold it against the inked type. **2.** The roller on a typewriter against which the keys strike.

**plate proof** *n.* A proof taken from a master printing plate.

**plate tectonics** *n. (sing. in number).* **1.** A branch of geology concerned with seismic activity and continental movement, based on the theory that the earth's surface is composed of a small number of large, semirigid sections that float across the mantle, with seismic activity and volcanism occurring primarily at the junction of these sections. **2.** The dynamics of plate movement.

**plat·form** (plăt′fôrm′) *n.* [OFr. *plate-forme,* diagram : *plate,* flat + *forme,* form.] **1.** A floor or horizontal surface raised above the level of the adjacent area, as a landing alongside railroad tracks or a stage for public speaking. **2.** A vestibule at the end of a railway car. **3.** A formal declaration of principles, as by a political party or candidate. **4. a.** A layer of leather between the inner and outer soles of a shoe. **b.** A shoe having a platform.

**platform balance** *n.* An equal-arm balance having two flat platforms above the beam and frequently using a sliding rider instead of weights.

**platform car** *n.* A railroad car having no roof or sides.

**platform scale** *n.* An industrial weighing instrument having a platform coupled to an automatic system of levers and adjustable weights and used to weigh large or heavy objects.

**platin-** *pref. var. of* PLATINO-.

**pla·ti·na** (plə-tē′nə) *n.* [Sp., dim. of *plata,* silver, flat < VLat. **plattus* < Gk. *platus,* flat.] Platinum, esp. as found naturally.

**plat·ing** (plā′tĭng) *n.* **1.** A thin coating or layer of metal, as silver or gold. **2.** A covering or layer of metal sheets or plates.

**platini-** *pref. var. of* PLATINO-.

**pla·tin·ic** (plə-tĭn′ĭk) *adj. Chem.* Of, relating to, or containing platinum, esp. with valence 4.

**plat·i·nize** (plăt′n-īz′) *vt.* **-nized, -niz·ing, -niz·es.** To electroplate with platinum.

**platino-** *or* **platini-** *or* **platin-** *pref.* [< PLATINUM.] Platinum <*platino*type>

**plat·i·no·cy·a·nide** (plăt′n-ō-sī′ə-nīd′) *n.* A complex salt of platinous cyanide and another cyanide.

**plat·i·noid** (plăt′n-oid′) *adj.* Like platinum. —*n.* **1.** An alloy of copper, nickel, tungsten, and zinc, formerly used in electric coils. **2.** A metal resembling platinum chemically, esp. osmium, iridium, or palladium.

**plat·i·nous** (plăt′n-əs) *adj.* Of, relating to, or containing platinum, esp. with valence 2.

**plat·i·num** (plăt′n-əm) *n.* [NLat. < Sp. *platina,* platinum. —see PLATINA.] **1.** *Symbol* **Pt** A silver-white, corrosive-resistant metallic element used in electrical components, electroplating, jewelry, dentistry, and as a catalyst; atomic number 78; atomic weight 195.09. **2.** A medium to light gray.

**platinum black** *n.* A fine black powder of metallic platinum, used as a gas absorbent and a catalyst.

**platinum blond** *n.* **1.** A very light silver-blond hair color, esp. when artificially produced. **2.** One having platinum blond hair.

**plat·i·tude** (plăt′ĭ-to͞od′, -tyo͞od′) *n.* [Fr. < *plat,* flat < OFr. —see PLATE.] **1.** A trite or banal expression. **2.** Lack of originality. **—plat′i·tu′di·nal, plat′i·tu′di·nous** *adj.* **—plat′i·tu′di·nous·ly** *adv.*

**plat·i·tu·di·nar·i·an** (plăt′ĭ-to͞od′n-âr′ē-ən, -tyo͞od′-) *n.* [PLATITUDIN(OUS) + -ARIAN.] One who habitually uses platitudes.

**plat·i·tu·di·nize** (plăt′ĭ-to͞od′n-īz′, -tyo͞od′-) *vi.* **-nized, -niz·ing, -niz·es.** To use platitudes.

**Pla·ton·ic** (plə-tŏn′ĭk, plā-) *adj.* [After *Plato* (427?–347 B.C.), Greek philosopher.] **1.** *also* **Pla·ton·i·cal** (-ĭ-kəl). Of, relating to, or typical of Plato or his philosophy. **2.** *often* **platonic.** Transcending physical desire and tending toward the purely spiritual or ideal <*platonic* love> **3.** *often* **platonic.** Theoretical or speculative. **—Pla·ton′i·cal·ly** *adv.*

**Pla·to·nism** (plāt′n-ĭz′əm) *n.* The philosophy of Plato, esp. insofar as it asserts ideal forms as an absolute and eternal reality of which the phenomena of the world are an imperfect and transitory reflection. **—Pla′to·nist** *n.* **—Pla′to·nis′tic** *adj.*

**pla·toon** (plə-to͞on′) *n.* [Fr. *peloton* < OFr. *pelote,* little ball. —see PELLET.] **1.** A subdivision of a military company usu. consisting of squads or sections. **2.** A body of persons working together. **3.** A group of players within a team, esp. a football team, used for offense or defense. —*v.* **-tooned, -toon·ing, -toons.** —*vt.* To alternate (a player) with another player in the same position. —*vi.* **1.** To use different players at the same position. **2.** To be platooned with another player.

**platoon sergeant** *n.* The senior noncommissioned officer in an army platoon or comparable unit.

**Platt·deutsch** (plät′doich′) *n.* [G. < Du. *platduits,* Low German : *plat,* low, flat (< MDu. < OFr.) + *Duitsch,* German < MDu. *duutsch.*] The Low German vernacular of northern Germany.

**plat·ter** (plăt′ər) *n.* [ME *plater* < AN < OFr. *plate,* plate.] **1.** A large, shallow dish, used esp. for serving food. **2.** A meal or course served on a platter. **3.** *Slang.* A phonograph record. **—on a platter.** Effortlessly : easily <could have had the nomination *on a platter*>

**plat·y**[1] (plā′tē) *adj.* **-i·er, -i·est.** Designating soil or minerals occurring in flaky layers.

**plat·y**[2] (plăt′ē) *n., pl.* **-ys** *or* **-ies.** [< NLat. *Platypoecilus,* genus name of platys : PLATY- + Gk. *poikilos,* many-colored.] A small freshwater fish of the genus *Xiphophorus* of southern North America, esp. *X. maculatus,* a colorful aquarium fish.

**platy-** *pref.* [< Gk. *platus,* broad, flat.] Flat <*platy*helminth>

**plat·y·fish** (plăt′ē-fĭsh′) *n.* PLATY[2].

**plat·y·hel·minth** (plăt′ĭ-hĕl′mĭnth) *n.* [PLATY- + Gk. *helmis,* parasitic worm.] A parasitic or nonparasitic worm of the phylum Platyhelminthes, as a tapeworm or a planarian, with a flattened body. **—plat′y·hel·min′thic** *adj.*

**plat·y·pus** (plăt′ĭ-pəs) *n., pl.* **-pus·es.** [NLat. < Gk. *platupous,* flat-footed : *platus,* flat + *pous,* foot.] A semiaquatic, web-footed egg-laying mammal, *Ornithorhynchus anatinus* of Australia and Tasmania, with a broad flat tail and a snout resembling a duck's bill.

**plat·yr·rhine** (plăt′ĭ-rīn′) *also* **plat·yr·rhin·i·an** (plăt′ĭ-rĭn′ē-ən) *adj.* [NLat. *Platyrrhina,* group name < Gk. *platurrhis,* broad-nosed : *platus,* wide + *rhis,* nose.] **1.** Having a broad flat nose. **2.** *Zool.* Of or designating the New World monkeys, many of which have widely separated nostrils. —*n.* A platyrrhine individual. **—plat′yr·rhi′ny** (-rī′nē) *n.*

**plau·dit** (plô′dĭt) *n.* [< Lat. *plaudite,* pl. imper. of *plaudere,* to applaud.] **1.** A round of applause. **2.** *often* **plaudits.** Enthusiastic expression of approval or praise.

**plau·si·ble** (plô′zə-bəl) *adj.* [Lat. *plausibilis,* deserving applause < *plaudere,* to applaud.] **1.** Seemingly or apparently valid, likely, or acceptable <a *plausible* motive> **2.** Giving a deceptive impression of truth, acceptability, or reliability : SPECIOUS. **—plau′si·bil′i·ty, plau′si·ble·ness** *n.* **—plau′si·bly** *adv.*

**plau·sive** (plô′zĭv, -sĭv) *adj.* [< Lat. *plaudere, plaus-,* to applaud.] **1.** Showing or expressing approval or praise. **2.** *Obs.* Plausible.

**play** (plā) *v.* **played, play·ing, plays.** [ME *playen* < OE *plegan.*] —*vi.* **1.** To occupy oneself in recreation, amusement, or sport. **2. a.** To take part in a game. **b.** To participate in a betting game : GAMBLE. **3.** To act in sport or jest. **4.** To deal or behave carelessly or indifferently : TRIFLE. **5.** To make love in a playful manner. **6.** To act or behave in a specified way <*play* hard to get> **7.** To act or perform, esp. in a dramatic production. **8.** To perform on a musical instrument. **9.** To emit sound or be sounded in performance <The orchestra is *playing.*> **10.** To be performed, as in a theater or on television <What's *playing* tonight?> **11.** To move or seem to move quickly, lightly, or irregularly <lights *playing* on the celebrities> **12.** To function or operate uninterruptedly, esp. to discharge a steady stream, as a fountain. **13.** To move or operate freely within a bounded space, as machine parts do. —*vt.* **1. a.** To perform or act (a role or part) in a dramatic production. **b.** To assume the role of : act as <*play* the fool> **2.** To put on or perform (a theatrical work) on or as if on the stage. **3.** To put on or produce a theatrical performance in <*play* the Palace> **4.** To pretend to be <The children *played* cops and robbers.> **5.** To participate in (a game or sport). **6.** To compete against in a game or sport. **7. a.** To occupy or work at (a position) in a game <*plays* shortstop> **b.** To employ (a player) in a game or position. **c.** To use or move (a card, piece, or ball) in a game or sport. **8.** To perform or put into effect, esp. as a jest or deception <*play* a prank> **9.** To use or manipulate, esp. for one's own interests. **10. a.** To wager : bet. **b.** To make a wager on. **11. a.** To perform on (a musical instrument). **b.** To perform (music) on an instrument. **12.** To cause (e.g., a record or phonograph) to emit recorded sounds. **13.** To discharge, set off, or cause to operate in or as if in a continuous stream <*play* a garden hose on a lawn> **14.** To cause to move quickly, lightly, or irregularly <*play* searchlights over the harbor> **15.** To exhaust (a hooked fish) by allowing it to pull on the line. **—play along.** *Informal.* To agree to cooperate with or participate in an activity or plan. **—play at. 1.** To participate in : engage in. **2.** To do or take part in half-heartedly. **—play back.** To perform a playback of (a newly recorded tape or disc). **—play down.** To minimize the importance of <*play down* one's faults> **—play off. 1. a.** To establish the winner of (a tie) by playing in an additional game or series of games. **b.** To participate in a play-off. **2.** To set (one individual or party) in opposition to another, so as to advance one's own interests. **—play on (or upon).** To take advantage of (another's attitudes or feelings) for one's own interests. **—play out. 1.** To play or do until completed : FINISH. **2.** To use up : EXHAUST. **—play up.** *Informal.* To publicize or emphasize <*play up* one's virtues> —*n.* **1. a.** A literary work written for performance on the stage : DRAMA. **b.** The performance of such a work. **2.** Activity engaged in for recreation or enjoyment, esp. the natural activities of children. **3.** Absence of harmful intent : JESTING <done in *play*> **4. a.** The act of carrying on or engaging in a game or sport. **b.** The manner of playing a game or sport. **5.** A manner or method of dealing with people generally <fair *play*> **6.** A move or action in a game <It's my *play.*> **7.** Participation in betting : GAMBLING. **8.** Action, motion, or use <the *play* of ocean waves> **9.** Quick, often irregular movement or action, esp. of light or color. **10.** Movement or space for movement, as of mechanical parts. **—in play.** In a position to be played, as in a game or sport. **—make a play for.** *Informal.* To make an attempt to attract or obtain by using skill, artifice, or wiles. **—out of play.** Not in play. **—play ball.** *Informal.* To cooperate. **—play both ends against the middle.** To set opposing parties or interests against one another so as to advance one's own goals. **—play by ear.** To deal with extemporaneously. **—play catch up.** *Informal.* To act rapidly in order to move ahead from a lagging position. **—play down to.** To simplify one's manner or the meaning of, esp. to win support or favor. **—play games.** *Informal.* To conceal the truth from by deception. **—play hardball.** To use rough, unscrupulous tactics in order to achieve a goal <"This is a judicious way of accusing the Court of *playing* political *hardball* on the sly" —Edwin M. Yoder, Jr.>

o͞o **boot**   ou **out**   th **thin**   *th* **this**   ŭ **cut**   ûr **urge**   y **young**
yo͞o **abuse**   zh **vision**   ə **about,** it**e**m, edib**i**le, gall**o**p, circ**u**s

**—play into the hands of.** To act or behave so as to give an advantage to (an opponent). **—play (one's) cards.** *Informal.* To act with all the means at one's disposal. **—play possum.** To pretend to be sleeping or dead. **—play the field.** To date more than one person. **—play the game.** To behave according to conventional standards or custom. **—play up to.** *Informal.* To curry favor with. **—play'a·bil'i·ty** *n.* **—play'a·ble** *adj.*

**pla·ya** (plī'ə) *n.* [Sp. < Med. Lat. *plagia,* prob. < Gk., sides.] **1.** The beach or bank of a river. **2.** A nearly level area at the bottom of a desert basin, sometimes temporarily covered with water.

**play-act** (plā'ăkt') *vi.* **-act·ed, -act·ing, -acts. 1.** To play a role in a dramatic performance. **2.** To make believe. **3.** To behave in a histrionic or artificial way.

**play·back** (plā'băk') *n.* **1.** The act or process of replaying a newly made record or tape. **2.** A method of or apparatus for reproducing sound recordings.

**play·bill** (plā'bĭl') *n.* A poster announcing a theatrical performance.

**Play·bill** (plā'bĭl'). A trademark for a theater program.

**play·boy** (plā'boi') *n.* A man devoted to the pursuit of pleasure.

**play-by-play** (plā'bī-plā') *adj.* Consisting of a detailed running commentary or account, as of the action of a sports event.

**play·date** (plā'dāt') *n.* The scheduled showing of a production, as a movie, esp. on pay-TV.

**play·er** (plā'ər) *n.* **1.** One who participates in a game or sport. **2.** One who performs in theatrical roles. **3.** One who plays a musical instrument. **4.** The mechanism actuating a player piano. **5.** A phonograph. **6.** A gambler. **7.** A trifler.

**player piano** *n.* A mechanically operated piano whose keys are actuated by a perforated paper roll.

**play·ful** (plā'fəl) *adj.* **1.** Full of good spirits and fun : FROLICSOME <a *playful* puppy> **2.** Jesting : humorous. **—play'ful·ly** *adv.* **—play'ful·ness** *n.*

**play·girl** (plā'gûrl') *n.* A woman devoted to the pursuit of pleasure.

**play·go·er** (plā'gō'ər) *n.* One who attends the theater, esp. frequently.

**play·ground** (plā'ground') *n.* **1.** An outdoor area set aside for recreation and play, esp. one having equipment such as seesaws and swings. **2.** A field or area of unrestricted activity.

**play·house** (plā'hous') *n.* **1.** A theater. **2.** A small house for children to play in. **3.** A doll house.

**playing card** *n.* A card, usu. one of a deck of 52, marked with its rank and suit and used in playing various games.

**playing field** *n.* A field for games such as cricket and soccer.

**play·let** (plā'lĭt) *n.* A short play.

**play·mate** (plā'māt') *n.* A companion in play or recreation.

**play-off** (plā'ôf', -ŏf') *n.* **1.** A final game or series of games played to break a tie. **2.** A series of games played to determine a championship.

**play·pen** (plā'pĕn') *n.* A portable enclosure in which a baby or young child can be left to play.

**play·room** (plā'ro͞om', -ro͝om') *n.* A room designed or set aside for playing or recreation.

**play·thing** (plā'thĭng') *n.* **1.** Something to play with : TOY. **2.** One regarded or treated as a toy.

**play·wear** (plā'wâr') *n.* Dress suitable for recreational activities.

**play·wright** (plā'rīt') *n.* A writer of plays : DRAMATIST.

**pla·za** (plä'zə, plăz'ə) *n.* [Sp. < Lat. *platea,* broad street. —see PLACE.] **1.** A public square or similar open area in a city or town. **2.** A broad paved area for automobiles, esp.: **a.** The widened roadway forming the approach to tollbooths on a highway. **b.** A parking or servicing area next to a highway. **3.** A shopping center.

**plea** (plē) *n.* [ME *plai,* lawsuit < OFr. *plaid* < Med. Lat. *placitum* < Lat., decision < *placēre,* to please.] **1.** An entreaty or appeal <a *plea* for mercy> **2.** A pretext or justification : EXCUSE. **3.** *Law.* **a.** An allegation offered in pleading a case. **b.** A defendant's establishment of an allegation of fact in answer to the declaration made by the plaintiff in common law. **c.** The answer of the accused to a charge or indictment in criminal law. **d.** A special answer in equity law depending on or demonstrating one or more reasons why a suit should be delayed, dismissed, or barred. **e.** An action or suit.

**plea-bar·gain** (plē'bär'gən) *vi.* **-gained, -gain·ing, -gains.** To make an agreement to plead guilty or enter a plea of guilty to a lesser charge so as to avoid trial for a more serious offense. **—plea'-bar'-gain·ing** *n.*

**pleach** (plēch, plāch) *vt.* **pleached, pleach·ing, pleach·es.** [ME *plechen* < ONFr. *plechier* < Lat. *plectere.*] To interlace or plait (e.g., twigs or branches), esp. in making an arbor.

**plead** (plēd) *v.* **plead·ed** or **pled** (plĕd), **plead·ing, pleads.** [ME *pleden* < OFr. *plaidier* < Med. Lat. *placitare* < *placitum,* lawsuit. —see PLEA.] —*vi.* **1.** To appeal earnestly : BEG. **2.** To offer persuasive reasons or argue for or against something. **3.** To furnish or provide an argument or appeal <Our adversities *plead* for us.> **4.** *Law.* **a.** To put forward a plea of a specific nature in a court of law <*plead* not guilty> **b.** To file an answer or pleading on behalf of a defendant or as part of the prosecution in a law action. **c.** To address a court as a lawyer or advocate. —*vt.* **1.** To assert or urge as defense, vindication, or excuse <*plead* insanity> **2.** To present as an answer to a charge, indictment, or declaration made against one. **3.** To argue or present (a case) in a court or similar tribunal or to an authorized person.

***usage:*** Technically, one is said to *plead guilty* or *plead not guilty,* but not to *plead innocent.* In nonlegal contexts, however, *plead innocent* is well established at all levels. **—plead'a·ble** *adj.* **—plead'er** *n.* **—plead'ing·ly** *adv.*

**plead·ing** (plē'dĭng) *n.* **1. a.** The act of entreating or making a plea. **b.** An entreaty or plea thus made. **2.** *Law.* **a.** The act or procedure of one who acts as an advocate in a law court. **b.** The act or technique of drawing up or presenting pleas in legal cases. **c.** A formal, gen. written statement propounding the cause of action or the defense of a legal case. **d. pleadings.** The consecutive statements, allegations, and counter allegations made in turn by plaintiff and defendant, or prosecutor and accused, until a single issue is reached upon which the trial may be held.

**pleas·ance** (plĕz'əns) *n.* [ME < OFr. *plaisance* < *plaisant,* pleasant.] **1.** A secluded garden or landscaped area, as on an estate. **2.** Pleasure or a source of pleasure.

**pleas·ant** (plĕz'ənt) *adj.* [ME *plesant* < OFr. *plaisant,* pr.part. of *plaisir,* to please < Lat. *placēre.*] **1.** Providing pleasure. **2.** Pleasing, as in manner or appearance. **3.** Agreeably fair <a *pleasant* climate> **4.** Lively : merry. **—pleas'ant·ly** *adv.* **—pleas'ant·ness** *n.*

☆ ***syns:*** PLEASANT, ENJOYABLE, PLEASING, PLEASURABLE *adj. core meaning* : providing enjoyment <a *pleasant* way to spend an evening> ***ant:*** unpleasant

**pleas·ant·ry** (plĕz'ən-trē) *n., pl.* **-ries.** [Fr. *plaisanterie* < OFr. < *plaisant,* pleasant.] **1.** A humorous, jesting, or entertaining remark or action. **2.** Pleasingly humorous style or manner in conversation or social situations.

**please** (plēz) *v.* **pleased, pleas·ing, pleas·es.** [ME *plesen* < OFr. *plaisir* < Lat. *placēre.*] —*vt.* **1.** To give enjoyment, pleasure, or satisfaction to : DELIGHT. **2.** To be the will or desire of <May it *please* your Honor.> **3.** To be willing to : be so kind as to <*Please* call a taxi.> —*vi.* **1.** To give pleasure or satisfaction <Our business aims to *please.*> **2.** To have the will or desire : WISH <I do as I *please.*> **—if you please. 1.** If it is your will, desire, or pleasure. **2.** If you can believe or imagine it. —Used as an ironical expression of surprise or indignation.

☆ ***syns:*** PLEASE, SATISFY, SUIT *v. core meaning* : to be satisfactory to <an arrangement that *pleased* everyone>

**pleas·ing** (plē'zĭng) *adj.* Giving pleasure or enjoyment : AGREEABLE. **—pleas'ing·ly** *adv.* **—pleas'ing·ness** *n.*

**pleas·ur·a·ble** (plĕzh'ər-ə-bəl) *adj.* Pleasing : gratifying. **—pleas'-ur·a·bil'i·ty, pleas'ur·a·ble·ness** *n.* **—pleas'ur·a·bly** *adv.*

**pleas·ure** (plĕzh'ər) *n.* [ME *plesure* < OFr. *plaisir* < *plaisir,* to please < Lat. *placēre.*] **1.** An enjoyable sensation or emotion : DELIGHT. **2.** A source of enjoyment, gratification, or delight. **3.** Amusement, diversion, or worldly enjoyment. **4.** Sensual indulgence or gratification. **5.** One's preference, wish, or choice <What is your *pleasure?*> —*v.* **-ured, -ur·ing, -ures.** —*vt.* To give pleasure or enjoyment to : GRATIFY. —*vi.* **1.** To take pleasure : DELIGHT <*pleasures* in the warmth of the sun> **2.** *Informal.* To go in search of pleasure or enjoyment. **—pleas'ure·less** *adj.*

**pleasure principle** *n.* [Transl. of G. *Lustprinzip.*] The tendency to reduce pain and seek immediate gratification of instinctual needs.

**pleat** (plēt) *n.* [ME *plete,* var. of *plait,* pleat, fold. —see PLAIT.] A fold in material made by doubling the material upon itself and then pressing or stitching into place : PLAIT. —*vt.* **pleat·ed, pleat·ing, pleats.** To arrange or press in pleats : PLAIT.

**pleat·er** (plē'tər) *n.* A sewing-machine attachment that pleats.

**pleb** (plĕb) *n.* [Short for PLEBEIAN.] **1.** A plebeian : commoner. **2.** A plebe.

**plebe** (plēb) *n.* [Obs. *plebe,* common people < Fr. *plèbe* < Lat. *plebs.*] A freshman at the U.S. Military or Naval Academy.

**ple·be·ian** (plĭ-bē'ən) *adj.* [Lat. *plebeius* < *plebs,* common people.] **1.** Of or relating to the Roman plebs. **2.** Of, belonging to, or typical of commoners. **3.** Crude in style or manner : COARSE. —*n.* **1.** One of the Roman plebs. **2.** A member of the lower classes. **3.** One who is coarse or vulgar. **—ple·be'ian·ism** *n.* **—ple·be'ian·ly** *adv.*

**ple·bes** (plē'bēz) *n. pl. of* PLEBS.

**pleb·i·scite** (plĕb'ĭ-sīt', -sĭt) *n.* [Fr. *plébiscite* < Lat. *plebiscitum* : *plebs,* common people + *scitum,* decree < *sciscere,* to decree.] **1.** A direct vote in which the entire electorate can accept or refuse the measure, program, or government of the person or party initiating the consultation. **2.** A consultation whereby a population exercises the right of national self-determination. **—ple·bis'ci·tar'y** (plə-bĭs'ĭ-tĕr'ē, plĕb'ĭ-sī'tə-rē) *adj.*

**plebs** (plĕbz) *n., pl.* **ple·bes** (plē'bēz) [Lat.] **1.** The common people of ancient Rome. **2.** The common people : POPULACE.

**ple·cop·ter·an** (plĭ-kŏp'tər-ən) *n.* [< NLat. *Plecoptera,* order name : Gk. *plekein,* to braid + Gk. *pteron,* wing.] The stonefly. **—ple·cop'ter·an** *adj.*

**plec·tog·nath** (plĕk'tŏg-năth') *n.* [NLat. *Plectognathi,* order name : Gk. *plektos,* twisted + Gk. *gnathos,* jaw.] A tropical marine fish of the order Tetraodontiformes or Plectognathi, which includes the triggerfishes, puffers, and trunkfishes. **—plec'tog·nath'** *adj.*

**plec·trum** (plĕk′trəm) *n., pl.* **-trums** *or* **-tra** (-trə) [Lat. < Gk. *plēktron* < *plēssein*, to strike.] **1.** A small, thin, flexible piece of material, as metal, plastic, or bone, for plucking the strings of certain musical instruments, as the lute or guitar. **2.** QUILL 4.

**pled** (plĕd) *v. var. p.t. & p.p. of* PLEAD.

**pledge** (plĕj) *n.* [ME < OFr. *plege* < LLat. *plevium*, a security, of Germanic orig.] **1.** A formal promise to do or not to do something. **2. a.** Something given or held as security to guarantee payment of a debt or fulfillment of an obligation. **b.** The condition of something thus given or held <put jewelry in *pledge*> **3.** *Law.* **a.** A delivery of goods or personal property as security for a debt or obligation. **b.** The contract by which such delivery is made. **4.** A sign or token <gave a ring as a *pledge* of faithfulness> **5.** One who has been accepted for membership in a fraternity or similar organization but has not yet been initiated. **6.** The act of drinking to someone : TOAST. *—v.* **pledged, pledg·ing, pledg·es.** *—vt.* **1.** To offer or guarantee by a solemn promise. **2.** To bind or secure by or as if by a pledge. **3.** To deposit as security : PAWN. **4.** To drink a toast to. **5. a.** To promise to join (e.g., a fraternity). **b.** To accept as a prospective member. *—vi.* **1.** To make a solemn promise. **2.** To drink a toast to someone.

**pledg·ee** (plĕj-ē′) *n.* **1.** One to whom something is pledged. **2.** One with whom something is deposited as a pledge.

**pledge·or** (plĕj′ər, plĕj-ôr′) *n. var. of* PLEDGOR.

**pledg·er** (plĕj′ər) *n.* One who makes or gives a pledge.

**pledg·or** *also* **pledge·or** (plĕj′ər, plĕj-ôr′) *n. Law.* One who deposits his property as a pledge.

**-plegia** *suff.* [NLat. < Gk. *plēgē*, stroke < *plēssein*, to strike.] Paralysis <mono*plegia*>

**Ple·iad** (plē′əd,-ăd′) *n., pl.* **Ple·ia·des** (plē′ə-dēz′) [Back-formation < PLEIADES.] **1.** One of the Pleiades. **2.** *also* **ple·iad.** A group of seven illustrious persons.

**Ple·ia·des** (plē′ə-dēz′) *pl. n.* [Lat. < Gk.] **1.** *Gk. Myth.* The seven daughters of Atlas who were changed into stars. **2.** An open star cluster in the constellation Taurus, having several hundred stars of which six are visible to the naked eye.

**pleio-** *pref. var. of* PLEO-.

**plei·o·tax·y** (plī′ə-tăk′sē) *n.* [Gk. *pleiōn*, more + -TAXY.] An increase in the number of whorls in an inflorescence.

**plei·ot·ro·pism** (plī-ŏt′rə-pĭz′əm) *also* **plei·ot·ro·py** (-pē) *n.* [Gk. *pleiōn*, more + -TROPISM.] Control or determination of more than one characteristic or function by a single gene. **—plei′o·tro′pic** (plī′ə-trō′pĭk, -trŏp′ĭk) *adj.* **—plei′o·tro′pi·cal·ly** *adv.*

**Pleis·to·cene** (plī′stə-sēn′) *adj.* [Gk. *pleistos*, most + -CENE.] Of, belonging to, or designating the geologic time, rock series, and sedimentary deposits of the earlier of the two epochs of the Quaternary period, marked by the alternate appearance and recession of northern glaciation and the appearance of the progenitors of humans. *—n.* The Pleistocene epoch or system of deposits.

**ple·na** (plē′nə, plĕn′ə) *n. var. pl. of* PLENUM.

**ple·na·ry** (plē′nə-rē, plĕn′ə-) *adj.* [LLat. *plenarius* < Lat. *plenus*, full.] **1.** Complete in all aspects or essentials : FULL <*plenary* powers> **2.** Fully attended by all qualified members <a *plenary* council session> **—ple′na·ri·ly** *adv.* **—ple′na·ri·ness** *n.*

**plenary indulgence** *n. Rom. Cath. Ch.* An indulgence that remits the full temporal punishment incurred by a sinner.

**plench** (plĕnch) *n.* [PL(IERS) + (WR)ENCH.] A hand tool used for gripping, pulling, and turning objects under zero gravity, operated by squeezing the handle.

**plen·i·po·ten·ti·ar·y** (plĕn′ē-pə-tĕn′shē-ĕr′ē, -shə-rē) *adj.* [Med. Lat. *plenipotentarius* < LLat. *plenipotens*, invested with full power : Lat. *plenus*, full + Lat. *potens*, powerful.] Conferring or invested with full powers. *—n., pl.* **-ies.** A diplomatic agent, as an ambassador, fully authorized to represent a government.

**plen·i·tude** (plĕn′ĭ-to͞od′, -tyo͞od′) *n.* [ME < OFr. < Lat. *plenitudo* < *plenus*, full.] **1.** Abundance. **2.** The condition of being full, ample, or complete. **—plen′i·tu′di·nous** (-to͞od′n-əs, -tyo͞od′-) *adj.*

**plen·te·ous** (plĕn′tē-əs) *adj.* [ME *plentivous* < OFr. *plentiveus* < *plentif* < *plente*, plenty.] **1.** Abundant : ample. **2.** Producing or yielding in abundance. **—plen′te·ous·ly** *adv.* **—plen′te·ous·ness** *n.*

**plen·ti·ful** (plĕn′tĭ-fəl) *adj.* **1.** Existing in great quantity or ample supply. **2.** Producing or providing an abundance. **—plen′ti·ful·ly** *adv.* **—plen′ti·ful·ness** *n.*

**plen·ty** (plĕn′tē) *n.* [ME *plentie* < OFr. *plente* < Lat. *plenitas* < *plenus*, full.] **1.** A full or ample amount or supply <*plenty* of room> **2.** A large quantity or amount : ABUNDANCE <materials in *plenty*> **3.** A condition of general abundance or prosperity. *—adj.* Plentiful : abundant. *—adv. Informal.* Very : quite <*plenty* hot for June>

**ple·num** (plē′nəm, plĕn′əm) *n., pl.* **ple·nums** *or* **ple·na** (plē′nə, plĕn′ə) [Lat. < neuter of *plenus*, full.] **1.** An enclosure in which air or other gas is at a pressure greater than that outside the enclosure. **2.** An assembly or meeting attended by all members. **3.** Fullness.

**pleo-** *or* **pleio-** *or* **plio-** *pref.* [< Gk. *pleiōn*, *pleōn*, more.] More <*pleo*pod>

**ple·och·ro·ism** (plē-ŏk′rō-ĭz′əm) *n.* [PLEO- + Gk. *khrōs*, color + -ISM.] The property possessed by some crystals of exhibiting different colors, esp. three different colors, when viewed along different axes. **—ple′o·chro′ic** (plē′ə-krō′ĭk) *adj.*

**ple·o·mor·phism** (plē′ə-môr′fĭz′əm) *n.* **1.** *Chem.* POLYMORPHISM 2. **2.** *Biol.* The occurrence of two or more structural forms during a life cycle, esp. of certain plants. **—ple′o·mor′phic** *adj.*

**ple·o·nasm** (plē′ə-năz′əm) *n.* [LLat. *pleonasmus* < Gk. *pleonasmos* < *pleonazein*, to be excessive < *pleōn*, more.] **1. a.** The use of more words than are needed to express an idea : REDUNDANCY. **b.** An instance of redundancy. **2.** A superfluous word or phrase. **—ple′o·nas′tic** *adj.* **—ple′o·nas′ti·cal·ly** *adv.*

**ple·o·pod** (plē′ə-pŏd′) *n.* A swimmeret.

**ple·ro·cer·coid** (plîr′ō-sûr′koid′) *n.* [< Gk. *plērēs*, full + *kerkos*, tail.] The infective larva of some tapeworms, having a solid elongated body.

**ple·si·o·sau·rus** (plē′sē-ə-sôr′əs, plē′zē-) *also* **ple·si·o·saur** (plē′sē-ə-sôr′, plē′zē-) *n., pl.* **-sau·ri** (-sôr′ī′) *also* **-saurs.** [Gk. *plēsios*, near + -SAURUS.] A large marine reptile of the extinct suborder Plesiosauria, common in Europe and North America during the Mesozoic era.

**ples·sim·e·ter** (plĕ-sĭm′ĭ-tər) *n. var. of* PLEXIMETER.

**ples·sor** (plĕs′ər) *n. var. of* PLEXOR.

**pleth·o·ra** (plĕth′ər-ə) *n.* [Med. Lat. < Gk. *plēthōra* < *plēthein*, to be full.] **1.** Excess : overabundance <a *plethora* of duties> **2.** An excess of blood in the circulatory system or in one organ or area.

**ple·thor·ic** (plĕ-thôr′ĭk, -thŏr′-, plĕth′ər-ĭk) *adj.* **1. a.** Excessive in quantity : OVERABUNDANT. **b.** Excessive in style : TURGID. **2.** Marked by an excess of blood <a *plethoric* bodily condition> **—ple·thor′i·cal·ly** *adv.*

**ple·thys·mo·graph** (plĕ-thĭz′mə-grăf′, plə-) *n.* [Gk. *plēthusmos*, increase (< *plēthunein*, to increase < *plēthus*, quantity < *plēthein*, to be full) + -GRAPH.] An instrument for measuring variations in the size of an organ or body part on the basis of the amount of blood passing through or present in the part. **—ple·thys′mo·graph′ic** *adj.* **—ple·thys′mo·graph′i·cal·ly** *adv.* **—pleth′ys·mog′ra·phy** (plĕth′ĭz-mŏg′rə-fē) *n.*

**pleur-** *pref. var. of* PLEURO-.

**pleu·ra[1]** (plo͝or′ə) *n., pl.* **pleu·rae** (plo͝or′ē) [Med. Lat. < Gk., side.] Either of two membranous sacs, each of which lines one side of the thoracic cavity and envelops the adjacent lung, reducing the friction of respiratory movements to a minimum. **—pleu′ral** *adj.*

**pleu·ra[2]** (plo͝or′ə) *n. pl. of* PLEURON.

**pleu·ri·sy** (plo͝or′ĭ-sē) *n.* [ME *pluresy* < OFr. *pleuresie* < Med. Lat. *pleuresis* < LLat. *pleuritis* < Gk. < *pleura*, side.] *Pathol.* Inflammation of the pleura, often marked when acute by exudation into the pleural cavity and production of adhesions that may be permanent or, if infected, result in empyema. **—pleu·rit′ic** (plo͞o-rĭt′ĭk) *adj.*

**pleurisy root** *n.* Butterfly weed.

**pleuro-** *or* **pleur-** *pref.* [NLat. < Gk. *pleura*, side, rib.] **1.** Side : lateral <*pleuro*dont> **2.** Pleura : pleural <*pleuro*tomy>

**pleu·ro·dont** (plo͝or′ə-dŏnt′) *adj.* Having the teeth attached by their sides to the inner side of the jaw, as in some lizards. *—n.* A lizard with pleurodont teeth.

**pleu·ron** (plo͝or′ŏn′) *n., pl.* **pleu·ra** (plo͝or′ə) [NLat. < Gk., side.] An external lateral part of the body segments of arthropods.

**pleu·ro·pneu·mo·nia** (plo͝or′ō-no͞o-mōn′yə, -nyo͞o-) *n.* Pneumonia aggravated by pleurisy.

**pleu·rot·o·my** (plo͞o-rŏt′ə-mē) *n., pl.* **-mies.** Surgical incision of the pleura.

**pleus·ton** (plo͞o′stən, -stŏn′) *n.* [Gk. *pleusis*, sailing + (PLANK)TON.] Plants that float on the surface of bodies of fresh water. **—pleus·ton′ic** (plo͞o-stŏn′ĭk) *adj.*

**plex·i·form** (plĕk′sə-fôrm′) *adj.* [PLEX(US) + -FORM.] Similar to or formed like a plexus : structurally complicated.

**Plex·i·glas** (plĕk′sĭ-glăs′). A trademark for a light, transparent, weather-resistant thermoplastic.

**plex·im·e·ter** (plĕk-sĭm′ĭ-tər) *also* **ples·sim·e·ter** (plĕ-sĭm′-) *n.* [PLEX(OR) + -METER.] A small, thin plate held against the body and struck with a plexor. **—plex′i·met′ric** (plĕk′sə-mĕt′rĭk) *adj.* **—plex·im′e·try** *n.*

**plex·or** (plĕk′sər) *also* **ples·sor** (plĕs′ər) *n.* [< Gk. *plexis*, stroke < *plēssein*, to strike.] A small rubber-headed hammer used in diagnosis by percussion.

**plex·us** (plĕk′səs) *n., pl.* **plexus** *or* **-us·es.** [NLat. < Lat., braid < *plectere*, to plait.] **1.** A structure in the form of a network, esp. of nerves, blood vessels, or lymphatics. **2.** An interlacing of parts : NETWORK.

**pli·a·ble** (plī′ə-bəl) *adj.* **1.** Easily bent or shaped : MALLEABLE. **2. a.** Adaptable to change : RECEPTIVE. **b.** Easily influenced, persuaded, or swayed : FLEXIBLE. **—pli′a·bil′i·ty, pli′a·ble·ness** *n.* **—pli′a·bly** *adv.*

**pli·ant** (plī′ənt) *adj.* [ME *plyante* < OFr. *pliant*, pr.part. of *plier*, to fold < Lat. *plicare.*] **1.** Easily bent or flexed : SUPPLE. **2.** Easily modified to fit conditions : ADAPTABLE. **3.** Yielding readily to domination or influence : COMPLIANT. **—pli′an·cy** *n.* **—pli′ant·ly** *adv.* **—pli′ant·ness** *n.*

**pli·ca** (plī′kə) *n., pl.* **pli·cae** (plī′sē) [Med. Lat., fold < Lat. *plicare*, to fold.] *Zool.* A fold or ridge, as of skin, membrane, or shell. **—pli′cal** (plī′kəl) *adj.*

**pli·cate** (plī′kāt′) *also* **pli·cat·ed** (-kā′tĭd) *adj.* [Lat. *plicatus,* p.part. of *plicare,* to fold.] Arranged in fanlike folds : PLEATED. **—pli′cate′ly** *adv.* **—pli′cate′ness** *n.*

**pli·ca·tion** (plī-kā′shən) *also* **plic·a·ture** (plĭk′ə-cho͝or′) *n.* **1. a.** The act or process of folding. **b.** The state of being folded. **2.** A fold.

**pli·er** (plī′ər) *n.* One who plies a trade.

**pli·ers** (plī′ərz) *pl.n.* A tool having a pair of pivoted jaws, used for bending, cutting, or holding.

**plight¹** (plīt) *n.* [ME *plit,* fold, condition < Norman Fr. < Lat. *plicitum,* p.part. of *plicare,* to fold.] A difficult, embarrassing, or adverse situation : PREDICAMENT.

**plight²** (plīt) *vt.* **plight·ed, plight·ing, plights.** [ME *plighten* < OE *plihtan,* to endanger < *pliht,* danger.] **1.** To promise or bind by a solemn pledge, esp. to betroth. **2.** To give or pledge (e.g., one's word or oath). *—n.* A solemn pledge, as of fidelity : ENGAGEMENT. **—plight one's troth. 1.** To become engaged to marry. **2.** To give one's solemn oath. **—plight′er** *n.*

**plim·soll** (plĭm′səl, -sôl′) *also* **plim·sol** *or* **plim·sole** (plĭm′sōl′) *n.* [Prob. from the resemblance of its mudguard to a PLIMSOLL MARK.] *Chiefly Brit.* SNEAKER 2.

**Plimsoll mark** *n.* [After Samuel *Plimsoll* (1824–1898).] One of a set of lines on the hull of a merchant ship that indicate the depth to which it may be legally loaded under specified conditions.

**plink** (plĭngk) *v.* **plinked, plink·ing, plinks.** [Imit.] *—vt.* **1.** To cause to make a metallic clinking sound. **2.** To shoot at in a random, casual manner. *—vi.* **1.** To make a clinking sound. **2.** To shoot at randomly selected targets.

**plinth** (plĭnth) *n.* [Fr. *plinthe* < Lat. *plinthus* < Gk. *plinthos,* tile.] **1.** A slab or block on which a pedestal, column, or statue is placed. **2.** The base block at the junction of the horizontal baseboard and vertical trim around an opening. **3.** A continuous course of stones supporting a wall. **4.** A square base, as of a vase.

**plio-** *pref. var. of* PLEO-.

**Pli·o·cene** (plī′ə-sēn′) *adj.* [Gk. *pleiōn,* more + -CENE.] Of, belonging to, or designating the geologic time, rock series, and sedimentary deposits of the last of the five epochs of the Tertiary period, marked by the appearance of distinctly modern plants and animals. *—n.* The Pliocene epoch or system of deposits.

**Pli·o·film** (plī′ə-fĭlm′). A trademark for a pliant, transparent rubber compound used to make waterproof items.

**pli·o·tron** (plī′ə-trŏn′) *n.* [Orig. a trademark.] A vacuum tube running with a high-temperature cathode that has one or more grids.

**plis·sé** *also* **plis·se** (plĭ-sā′) *n.* [Fr. < p.part. of *plisser,* to pleat < OFr. < *pli,* fold < *plier,* to fold < Lat. *plicare.*] **1.** A puckered texture of cloth produced by treating fabric with a caustic soda. **2.** Fabric having a plissé texture.

**plod** (plŏd) *v.* **plod·ded, plod·ding, plods.** [Imit.] *—vi.* **1.** To move or walk heavily or laboriously : TRUDGE. **2.** To act or work steadily or monotonously : DRUDGE. *—vt.* To trudge heavily and slowly along or over. *—n.* **1.** The act of moving or walking heavily and slowly. **2.** The sound made by a heavy step. **—plod′der** *n.* **—plod′ding·ly** *adv.*

**-ploid** *suff.* [Back-formation < DIPLOID and HAPLOID.] Having a number of chromosomes bearing a specified relationship to the basic number of chromosomes of a group <hetero*ploid*>

**ploi·dy** (ploi′dē) *n.* [Back-formation < DIPLOIDY and HAPLOIDY.] A multiple of a set of chromosomes.

**plonk** (plŏngk, plŭngk) *v.* & *n.* & *adv. var. of* PLUNK.

**plop** (plŏp) *v.* **plopped, plop·ping, plops.** [Imit.] *—vi.* **1.** To fall with a sound like that of a rather heavy object falling into water. **2.** To drop or sink heavily <*plop* onto a couch> *—vt.* To drop or move so as to make a plopping sound. *—n.* A plopping sound or movement. **—plop** *adv.*

**plo·sion** (plō′zhən) *n.* [< EXPLOSION.] **1.** The articulation of a plosive sound. **2.** The sudden release of occluded air occurring in the articulation of certain stop consonants.

**plo·sive** (plō′sĭv, -zĭv) *adj.* [Fr.] Pronounced with a complete closure and abrupt opening of the oral passage, as the (p) in *top* or (d) in *adorn.* *—n.* A plosive speech sound.

**plot** (plŏt) *n.* [ME < OE.] **1. a.** A small piece of ground, gen. used for a particular purpose <a burial *plot*> **b.** A measured area of land : LOT. **2.** A ground plan, as for a building : DIAGRAM. **3.** The outline or plan of action of a narrative or drama. **4.** A secret plan to accomplish an illegal or hostile purpose. *—v.* **plot·ted, plot·ting, plots.** *—vt.* **1.** To represent graphically, as on a chart <*plot* an aircraft's course> **2.** To prearrange secretly or deviously <*plot* a crime> **3.** To conceive and arrange the action and incidents of <*plot* a story> **4.** *Math.* **a.** To locate (points or other figures) on a graph by means of coordinates. **b.** To draw (a curve) connecting points on a graph. *—vi.* **1.** To devise secretly : CONSPIRE. **2.** To be located by means of coordinates, as on a chart or with data. **—plot′less** *adj.* **—plot′less·ness** *n.* **—plot′ter** *n.*

☆ ***syns:*** **1.** PLOT, INTRIGUE, STORY *n. core meaning* : the series of events and relationships that form the basis of a composition <a novel with a complex *plot*> **2.** PLOT, COLLUSION, CONSPIRACY, INTRIGUE, MACHINATION, SCHEME *n. core meaning* : a secret plan to achieve an evil or illegal goal <a *plot* to steal government secrets>

**plot line** *n.* A story line.

**plottage** (plŏt′ĭj) *n.* The area of land in a plot.

**plotting board** *n. Computer Sci.* An output device that plots the curves of functions of variables.

**plotting table** *n.* A plotting board.

**plough** (plou) *n.* & *v. Chiefly Brit. var. of* PLOW.

**plov·er** (plŭv′ər, plō′vər) *n., pl.* **plover** *or* **-ers.** [ME < OFr. *plovier* < VLat. **plovarius* < *pluvia,* rain. —see PLUVIAL.] **1.** Any of various widely distributed wading birds of the family Charadriidae, with rounded bodies, short tails, and short bills. **2.** A bird related to or resembling the plover.

**plover**
*The American golden plover, up to 11 inches long*

**plow** (plou) *n.* [ME < OE *plōg,* plowland.] **1. a.** A farm implement having a heavy blade at the end of a beam, usu. hitched to a draft team or motor vehicle and used for breaking up soil and cutting furrows in preparation for sowing. **b.** A plowlike implement, as a snowplow. **2. Plow.** The Big Dipper. *—v.* **plowed, plow·ing, plows.** *—vt.* **1.** To break and turn over (earth) with a plow. **2. a.** To form (e.g., a furrow) with a plow. **b.** To make or form with driving force <*plowed* our way through the mob> **3.** To make furrows or indentations in. **4.** To cut through (water) <a ship *plowing* the ocean> *—vi.* **1.** To break and turn up earth with a plow. **2.** To be capable of being plowed <earth that *plows* easily> **3.** To move or progress with driving force <*plowed* through the spectators> **4.** To proceed with difficulty : PLOD <*plowing* through paperwork> **—plow back.** To reinvest (e.g., profits) in one's business. **—plow into.** *Informal.* **1.** To strike with force. **2.** To undertake (e.g., a task) eagerly and vigorously. **—plow under. 1.** To overwhelm. **2.** To cause to vanish. **—plow′a·ble** *adj.* **—plow′er** *n.*

**plow·boy** (plou′boi′) *n.* **1.** A boy who leads or directs a team of animals in plowing. **2.** A country boy.

**plow·head** (plou′hĕd′) *n.* The metal shackle at the leading end of the beam of a plow, used to attach the plow to a tractor or draft animal.

**plow·land** (plou′lănd′) *n.* **1.** A unit of land area in medieval England roughly equal to the area capable of being plowed by a team of eight oxen in a single year. **2.** Land under cultivation or suitable for cultivation.

**plow·man** (plou′mən) *n.* **1.** One who plows. **2. a.** A farmer. **b.** A rustic.

**plow·share** (plou′shâr′) *n.* The cutting blade of a plow.

**plow steel** *n.* A high-strength steel having a carbon content of 0.5–0.95% and used chiefly to make wire rope.

**ploy** (ploi) *n.* [Orig. unknown.] An artifice or stratagem to obtain an advantage over one's opponent.

**pluck** (plŭk) *v.* **plucked, pluck·ing, plucks.** [ME *plukken* < OE *pluccian,* prob. < VLat. **piluccare* < Lat. *pilus,* hair.] *—vt.* **1.** To detach by grasping and pulling abruptly with the fingers : PICK <*pluck* a rose> **2.** To pull out the feathers or hair of <*pluck* one's eyebrows> **3.** To give an abrupt pull to : tug at. **4.** *Mus.* To sound (the strings of an instrument) by pulling and releasing with the fingers or a plectrum. **5.** *Slang.* To swindle or rob. *—vi.* To give an abrupt pull : TUG. *—n.* **1.** The act of plucking : TUG. **2.** Resourceful courage and daring in the face of hardship or adversities : SPIRIT. **3.** The heart, liver, windpipe, and lungs of a slaughtered animal. **—pluck′er** *n.*

**pluck·y** (plŭk′ē) *adj.* **-i·er, -i·est.** Having or displaying courage or spirited resourcefulness in trying circumstances : BRAVE. **—pluck′i·ly** *adv.* **—pluck′i·ness** *n.*

**plug** (plŭg) *n.* [Du. < MDu. *plugge.*] **1.** An object, as a cork or wad of cloth, used to close a gap or hole. **2.** *Elect.* **a.** A fitting, commonly having two metal prongs for insertion into a fixed socket, used to connect an appliance to a power supply. **b.** A spark plug. **3.** A fireplug. **4. a.** A flat cake of pressed or twisted tobacco. **b.** A portion of chewing tobacco. **5.** *Geol.* A mass of igneous rock filling the opening or vent of a volcano. **6.** *Informal.* A favorable public mention of a product, business, or performance, esp. when spoken over radio or television. **7.** *Slang.* Something defective, inferior, or useless, esp. an old, worn-out horse. **8.** *Slang.* A gunshot or bullet. **9.** *Slang.* A plug hat. —*v.* **plugged, plug·ging, plugs.** —*vt.* **1.** To fill (a hole) tightly with or as if with a plug or stopper. **2.** To use as a plug <*plugged* a stopper in the sink> **3.** To connect (an electrical appliance) to a socket <*plugged* in the television> **4.** *Slang.* **a.** To hit with a bullet : SHOOT. **b.** To hit with the fist : PUNCH. **5.** *Informal.* **a.** To make favorable public mention of (e.g., a product). **b.** To publicize (e.g., a song) by performing or playing repeatedly. —*vi.* **1.** *Informal.* To work persistently and doggedly at an activity. **2.** *Informal.* To work for a particular person or cause <*plug* for a candidate> **3.** *Slang.* To fire bullets. **—plug in. 1.** To function by being connected to an electrical outlet. **2.** *Slang.* To be attuned or responsive to. **—plug'ger** *n.*

**plug board** *n.* **1.** A control panel or wiring panel. **2.** A removable panel in a computing device that may be rewired at will to sort data by a prescribed pattern.

**plug-com·pat·i·ble** (plŭg'kəm-păt'ə-bəl) *adj. Computer Sci.* Capable of being connected peripherally without modification.

**plug hat** *n. Slang.* A man's high silk hat.

**plug-in** (plŭg'ĭn') *adj.* Designed to be plugged in to an electric circuit <a *plug-in* frying pan>

**plug-ug·ly** (plŭg'ŭg'lē) *n., pl.* **-lies.** [Orig. unknown.] *Slang.* A gangster, esp. one hired to intimidate.

**plum¹** (plŭm) *n.* [ME < OE *plūme* < Lat. *prunum* < Gk. *proumnon.*] **1. a.** A shrub or small tree of the genus *Prunus,* bearing smooth-skinned, fleshy, edible fruit with a single hard-shelled seed. **b.** The fruit of a plum tree. **2. a.** Any of several trees bearing plumlike fruit. **b.** The fruit of such a tree. **3.** A raisin when added to a pudding or cake. **4.** A sugarplum. **5.** A dark purple to deep reddish purple. **6.** Something esp. desirable, as a good position. **—plum'like'** *adj.*

**plum²** (plŭm) *adj. & adv. Informal. var. of* PLUMB.

**plum·age** (plōō'mĭj) *n.* [ME < OFr. < *plume,* plume < Lat. *pluma.*] **1.** The feathers of a bird. **2.** Feathers used as decoration. **3.** Elaborate dress : FINERY. **—plum'aged** *adj.*

**plu·mate** (plōō'māt') *adj.* [Lat. *plumatus,* feathered < *pluma,* feather.] Resembling a plume or feather.

**plumb** (plŭm) *n.* [ME < OFr. *plomme* < Lat. *plumbum,* lead.] **1.** A weight suspended from the end of a line, used to determine water depth. **2.** A plumb used to test or establish vertical alignment. —*adj.* **1.** Exactly vertical. **2.** *also* **plum.** *Informal.* Utter : complete <a *plumb* idiot> —*adv.* **1.** In a vertical or perpendicular line. **2.** *also* **plum.** *Informal.* Completely : utterly <*plumb* crazy> —*v.* **plumbed, plumb·ing, plumbs.** —*vt.* **1.** To test the angle or alignment of with a plumb. **2.** To straighten or make perpendicular, as a wall. **3.** To determine the depth of : SOUND. **4.** To examine or investigate closely <"Shallow ideas are *plumbed* and discarded" —Gilbert Highet> **5.** To seal with lead. —*vi.* To work as a plumber. **—out of (or off) plumb.** Not vertical. **—plumb'a·ble** *adj.*

**plum·ba·go** (plŭm-bā'gō) *n., pl.* **-gos.** [Lat., lead ore < *plumbum,* lead.] **1.** Graphite. **2.** A plant of the genus *Plumbago,* as the leadwort.

**plumb bob** *n.* A usu. conical piece of metal attached to the end of a plumb line.

**plumb·er** (plŭm'ər) *n.* [ME *plummer* < OFr. *plommier* < LLat. *plumbarius,* lead worker < Lat. *plumbum,* lead.] **1.** One whose occupation is installing and repairing pipes and plumbing. **2.** *Slang.* One who investigates and tries to stop leaks of sensitive information <"The *plumbers* pushed the departments to investigate with interviews and polygraph tests" —Richard M. Nixon>

**plumber's helper** *n.* PLUNGER 3.

**plumber's snake** *n.* SNAKE 4.

**plumb·er·y** (plŭm'ə-rē) *n., pl.* **-ies. 1.** A plumber's workshop or place of business. **2.** A plumber's work : PLUMBING.

**plum·bif·er·ous** (plŭm-bĭf'ər-əs) *adj.* [Lat. *plumbum,* lead + -FEROUS.] Containing lead.

**plumb·ing** (plŭm'ĭng) *n.* **1.** The pipes, fixtures, and other apparatus of a water, gas, or sewage system. **2.** The trade or work of a plumber. **3.** The act of using a plumb line.

**plum·bism** (plŭm'bĭz'əm) *n.* [< Lat. *plumbum,* lead.] Chronic lead poisoning.

**plumb line** *n.* **1.** A line from which a weight is suspended to determine verticality or depth. **2.** A line regarded as directed exactly toward the earth's center of gravity.

**plumb rule** *n.* A narrow strip of wood with a plumb line and bob attached, used to test or establish verticality.

**plum duff** *n.* A flour pudding with raisins or currants, boiled in a cloth bag.

**plume** (plōōm) *n.* [ME < OFr. < Lat. *pluma.*] **1.** A feather, esp. a large and ornamental one. **2.** A large feather or cluster of feathers worn as an ornament or symbol of rank, as on a helmet. **3.** A token of achievement or honor. **4.** A featherlike structure, form, or object <smoke rising in *plumes*> **5.** *Geol.* A column of molten rock hypothesized to rise from the earth's lower mantle and held to be the driving force in plate tectonics. —*vt.* **plumed, plum·ing, plumes. 1.** To cover, decorate, or provide with or as if with plumes. **2.** To smooth (feathers) in preening. **3.** To pride or congratulate (oneself).

**plume·let** (plōōm'lĭt) *n.* A small plume.

**plum·met** (plŭm'ĭt) *n.* [ME *plomet* < OFr. *plombet,* ball of lead, dim. of *plomme,* lead < Lat. *plumbum.*] **1.** A plumb bob. **2.** Something that weighs down or oppresses. —*vi.* **-met·ed, -met·ing, -mets.** To drop straight down : PLUNGE <an eagle *plummeting* to earth>

**plum·my** (plŭm'ē) *adj.* **-mi·er, -mi·est. 1.** Filled with plums. **2.** Very desirable : CHOICE <a *plummy* sinecure> **3.** Rich and mellow, often to the point of sounding affected <a *plummy* contralto>

**plu·mose** (plōō'mōs') *adj.* [Lat. *plumosus* < *pluma,* feather.] **1.** Having plumes or feathers : FEATHERED. **2.** Resembling a plume : FEATHERY. **—plu'mose'ly** *adv.* **—plu·mos'i·ty** (-mŏs'ĭ-tē) *n.*

**plump¹** (plŭmp) *adj.* **-er, -est.** [ME, dull, prob. < MLG *plomp.*] **1.** Well-rounded and full in form : CHUBBY. **2.** Abundant : ample <a *plump* bonus> —*vt. & vi.* **plumped, plump·ing, plumps.** To make or become plump. **—plump'ly** *adv.* **—plump'ness** *n.*

**plump²** (plŭmp) *v.* **plumped, plump·ing, plumps.** [MLG *plumpen.*] —*vi.* **1.** To drop abruptly or heavily : PLOP <*plumped* onto the grass> **2.** To come or go abruptly or hurriedly. **3.** To give full support or approval <*plumped* for the professor's appointment as chairperson> —*vt.* To throw down or drop heavily or abruptly <*plump* stones into the well> —*n.* **1.** A heavy or abrupt fall or collision. **2.** The sound of a heavy fall or collision. —*adj.* Direct : blunt. —*adv.* **1.** With a heavy or abrupt impact. **2.** Straight down. **3.** Without qualification : BLUNTLY. **—plump'ly** *adv.* **—plump'ness** *n.*

**plump·ish** (plŭm'pĭsh) *adj.* Somewhat plump.

**plum pudding** *n.* A rich boiled or steamed pudding made with flour, suet, raisins, currants, citron, and spices.

**plum tomato** *n.* A form of the cherry tomato with oval or somewhat oblong fruit.

**plu·mule** (plōōm'yōōl) *n.* [Lat. *plumula,* dim. of *pluma,* feather.] **1.** A down feather. **2.** *Bot.* The rudimentary bud of a plant embryo. **—plu'mu·lose'** *adj.*

**plum·y** (plōō'mē) *adj.* **1.** Consisting of or covered with feathers. **2.** Resembling a plume or feather.

**plun·der** (plŭn'dər) *v.* **-dered, -der·ing, -ders.** [G. *plündern* < MHG *plundern* < *plunder,* household goods.] —*vt.* **1.** To rob of goods by force, esp. in time of war : PILLAGE. **2.** To seize wrongfully or forcibly : STEAL. —*vi.* To take booty : ROB. —*n.* **1.** Property or goods stolen by fraud or force : BOOTY. **2.** The act or practice of plundering. **—plun'der·a·ble** *adj.* **—plun'der·er** *n.* **—plun'der·ous** *adj.*

☆ ***syns:*** PLUNDER, BOOTY, LOOT, PILLAGE, PRIZE, SPOILS *n. core meaning* : goods or property seized unlawfully, esp. by a victor in wartime <a pirate's *plunder*>

**plun·der·age** (plŭn'dər-ĭj) *n.* **1.** Robbery. **2.** *Law.* **a.** The embezzling of goods on board a ship. **b.** The goods so acquired.

**plunge** (plŭnj) *v.* **plunged, plung·ing, plung·es.** [ME *plungen* < OFr. *plonger* < VLat. **plumbicare* < Lat. *plumbum,* lead.] —*vt.* **1.** To thrust or throw forcefully into a place or substance. **2.** To cast abruptly or violently into a given state or situation. —*vi.* **1.** To throw oneself into a place or substance. **2.** To throw oneself earnestly or wholeheartedly into a given state or activity. **3.** To enter violently or speedily. **4.** To descend steeply, as a road or cliff. **5.** To move forward and downward violently. **6.** *Informal.* To gamble or speculate extravagantly. —*n.* **1.** An act or instance of plunging. **2. a.** A place or area for plunging or diving, as a swimming pool. **b.** A swim : dip.

**plung·er** (plŭn'jər) *n.* **1.** One that plunges. **2.** A part that operates with a repeated plunging or thrusting movement, as a piston. **3.** A device having a rubber suction cup attached to the end of a stick, used to unstop clogged pipes and drains.

**plunk** (plŭngk) *also* **plonk** (plŏngk, plŭngk) [Imit.] *Informal.* —*v.* **plunked, plunk·ing, plunks** *also* **plonked, plonk·ing, plonks.** —*vt.* **1.** To strum or pluck (the strings of a musical instrument). **2.** To throw or place heavily or abruptly <*plunked* the glass down> —*vi.* **1.** To emit a hollow, twanging sound. **2.** To drop or fall abruptly or heavily : PLUMP. —*n.* **1.** A brief, hollow, twanging sound. **2.** A heavy stroke or blow. —*adv.* **1.** With a brief, hollow thud. **2.** Exactly : precisely <*plunk* in the middle of the street> **—plunk'er** *n.*

**plu·per·fect** (plōō-pûr'fĭkt) *adj.* [Alteration of LLat. *plus quam perfectum,* more than perfect.] Of or designating a verb tense used to express action completed prior to a specified or implied past time. —*n.* **1.** The pluperfect tense, formed in English with the past participle of a verb and one or more auxiliaries, as *had been done* in the sentence *The housework had been done an hour before the guests arrived.* **2.** A verb or form in the pluperfect tense.

**plu·ral** (plŏŏr'əl) *adj.* [ME < OFr. *plurel* < Lat. *pluralis* < *plus,* more.] **1.** Pertaining to or having more than one member, set, or kind. **2.** Of or relating to a grammatical form that designates more than one of the things specified. —*n.* **1.** The plural number or form. **2.** A word or term in the plural form. **—plu'ral·ly** *adv.*

**plu·ral·ism** (plŏŏr'ə-lĭz'əm) *n.* **1.** The state of being plural. **2.** A condition of society in which numerous distinct ethnic, religious, or cultural groups coexist within one nation. **3.** The holding by one person of more than one office or position, esp. two or more ecclesiastical benefices, at the same time. **4.** *Philos.* **a.** The doctrine that reality is made up of many ultimate substances. **b.** The belief that no single explanatory system or view of reality can account for all the phenomena of life.

**plu·ral·ist** (plo͝or′ə-lĭst) *n.* **1.** One who holds more than one office, esp. two or more ecclesiastical benefices, at the same time. **2.** One who adheres to philosophical pluralism. **—plu′ral·is′tic** *adj.*

**plu·ral·i·ty** (plo͞o-răl′ĭ-tē) *n., pl.* **-ties. 1.** The state or fact of being plural. **2.** A large number or quantity : MULTITUDE. **3. a.** PLURALISM 3. **b.** The offices or benefices held by a pluralist. **4. a.** The number of votes cast for the winning alternative in a contest of more than two alternatives if this number is not more than one half of the total votes cast. **b.** The number by which the vote of a winning candidate exceeds that of the closest opponent. **5.** The greater or larger part.

**plu·ral·ize** (plo͝or′ə-līz′) *v.* **-ized, -iz·ing, -iz·es.** *—vt.* **1.** To make plural. **2.** To express in the plural. *—vi.* **1.** To become plural. **2.** To hold more than one position or ecclesiastical benefice at one time. **—plu′ral·i·za′tion** *n.*

**plus** (plŭs) *prep.* [Lat., more.] **1.** Added to. **2.** Increased by : along with <income *plus* benefits> ***usage:*** As a preposition, *plus* does not have the conjunctive force of *and.* Therefore when *plus* is used after a singular subject, the verb remains in the singular, as in *Your knowledge plus your experience makes you a formidable opponent.* *—adj.* **1. a.** Involving or pertaining to addition. **b.** Positive, as on a scale. **2.** Added or extra <a *plus* advantage> **3.** *Informal.* Increased to a further degree <enthusiasm *plus*> **4.** Slightly more than <a grade of B *plus*> **5.** POSITIVE 12. *—n., pl.* **plus·es** *or* **plus·ses. 1.** The plus sign (+). **2.** A positive quantity. **3.** A favorable factor <Knowledge of computers is a *plus.*> *—conj. Informal.* And.

**plus fours** *pl.n.* [From the fact that they are four inches longer than ordinary knickerbockers.] Loose knickers bagging below the knees, worn esp. for sports.

**plush** (plŭsh) *n.* [OFr. *pluche* < *peluchier,* to pluck, prob. < VLat. **piluccare* < Lat. *pilus,* hair.] A silk, cotton, or synthetic fabric having a deep, thick pile. *—adj.* **1.** Made of or covered with plush. **2.** *Informal.* Luxurious. **—plush′ly** *adv.* **—plush′ness** *n.*

**plush·y** (plŭsh′ē) *adj.* **-i·er, -i·est. 1.** Resembling plush in texture. **2.** *Informal.* Ostentatiously luxurious <a *plushy* penthouse apartment> **—plush′i·ly** *adv.* **—plush′i·ness** *n.*

**plus sign** *n.* The symbol (+), as in 2 + 2 = 4, used to indicate addition or a positive quantity.

**Plu·to** (plo͞o′tō) *n.* [Lat. < Gk. *ploutōn* < *ploutos,* wealth.] **1.** *Rom. Myth.* The god of the dead and ruler of the underworld. **2.** The ninth and farthest planet from the sun, having a sidereal period of revolution about the sun of 248.4 years, 2.8 billion miles or 4.5 billion kilometers distant at perihelion and 4.6 billion miles or 7.4 billion kilometers at aphelion, and a diameter approx. half that of Earth.

**plu·toc·ra·cy** (plo͞o-tŏk′rə-sē) *n., pl.* **-cies.** [Gk. *ploutokratia* : *ploutos,* wealth + *kratos,* strength.] **1.** Government by the wealthy. **2.** A wealthy class that controls a government. **3.** A government or state in which the wealthy rule.

**plu·to·crat** (plo͞o′tə-krăt′) *n.* [< PLUTOCRACY.] **1.** A member of a governing wealthy class. **2.** One having political control or influence because of wealth. **—plu′to·crat′ic** *adj.* **—plu′to·crat′i·cal·ly** *adv.*

**plu·ton** (plo͞o′tŏn′) *n.* [Prob. back-formation < PLUTONIC.] Igneous rock formed beneath the surface of the earth by consolidation of magma.

**Plu·to·ni·an** (plo͞o-tō′nē-ən) *also* **Plu·ton·ic** (-tŏn′ĭk) *adj.* **1.** Of or relating to the god Pluto or the underworld. **2.** Of or pertaining to the planet Pluto.

**plu·ton·ic** (plo͞o-tŏn′ĭk) *adj.* [< PLUTO.] Of deep igneous or magmatic origin, as rock or water.

**plu·to·ni·um** (plo͞o-tō′nē-əm) *n.* [Lat. *Pluto, Pluton-,* Pluto + -IUM.] *Symbol* **Pu** A naturally radioactive, silvery metallic element, used as a reactor fuel and in nuclear weapons; atomic number 94; longest-lived isotope Pu 244.

**plu·vi·al** (plo͞o′vē-əl) *also* **plu·vi·an** (-ən) *adj.* [Lat. *pluvialis* < *pluvia,* rain < *pluvius,* rainy < *pluvere,* to rain.] **1.** Of or pertaining to rain : RAINY. **2.** *Geol.* Caused by the action of rain.

**plu·vi·om·e·ter** (plo͞o′vē-ŏm′ĭ-tər) *n.* [Fr. *pluviomètre* : Lat. *pluvia,* rain < *pluvius,* rainy < *pluvere,* to rain + Gk. *metron,* measure.] A rain gauge. **—plu′vi·o·met′ric** (plo͞o′vē-ə-mĕt′rĭk), **plu′vi·o·met′ri·cal** *adj.* **—plu′vi·o·met′ri·cal·ly** *adv.* **—plu′vi·om′e·try** *n.*

**plu·vi·ous** (plo͞o′vē-əs) *also* **plu·vi·ose** (-ōs′) *adj.* [ME *pluvyous* < Lat. *pluviosus* < *pluvia,* rain. —see PLUVIAL.] Marked by heavy rainfall : RAINY. **—plu′vi·os′i·ty** (-ŏs′ĭ-tē) *n.*

**ply¹** (plī) *vt.* **plied, ply·ing, plies.** [ME *plien* < OFr. *plier* < Lat. *plicare,* to fold.] **1.** To join together, as by twisting or molding. **2.** To double over (e.g., cloth). *—n., pl.* **plies. 1.** A layer, as of cloth or paperboard. **2.** One of the sheets of wood glued together to form plywood. **3.** One of the strands twisted together to make yarn, rope, or thread. **4.** An inclination : bias.

**ply²** (plī) *v.* **plied, ply·ing, plies.** [ME *plien* < *applien,* to apply. —see APPLY.] *—vt.* **1.** To use diligently as a tool or weapon : WIELD <*ply* an ax> **2.** To engage in (e.g., a trade) : PRACTICE. **3.** To cross or sail over regularly <Clipper ships *plied* the Atlantic.> **4.** To continue offering to or supplying <*plied* us with cookies> **5.** To assail vigorously. *—vi.* **1.** To traverse a route or course regularly. **2.** To perform or work diligently or regularly <*plied* at the baker's trade> **3.** *Naut.* To work against the wind by a zigzag course : TACK.

**Plym·outh Rock** (plĭm′əth) *n.* [After *Plymouth Rock,* legendary landing place of the Pilgrims in 1620.] An American breed of fowl raised for both meat and eggs.

**Plymouth Rock**
*27 inches high*

**ply·wood** (plī′wo͝od′) *n.* [PLY¹ + WOOD.] A structural material consisting of layers of wood glued tightly together, usu. with the grains of adjoining layers at right angles to each other.

**Pm** *symbol for* PROMETHIUM.

**pneum-** *pref. var. of* PNEUMO-.

**pneu·ma** (no͞o′mə, nyo͞o′-) *n.* [Gk.] The soul or vital spirit.

**pneumat-** *pref. var. of* PNEUMATO-.

**pneu·mat·ic** (no͞o-măt′ĭk, nyo͞o-) *also* **pneu·mat·i·cal** (-ĭ-kəl) *adj.* [Fr. *pneumatique* < Lat. *pneumaticus* < Gk. *pneumatikos* < *pneuma,* wind.] **1.** Of or relating to air or other gases. **2.** Of or relating to pneumatics. **3.** Run by or using compressed air, as a drill. **4.** Filled with air, esp. compressed air, as a tire. **5.** *Zool.* Having air cavities, as the bones of certain birds. **6.** Having or relating to an ample, shapely bust. **7.** Of or relating to the pneuma : SPIRITUAL. **—pneu·mat′i·cal·ly** *adv.* **—pneu′ma·tic′i·ty** (no͞o′mə-tĭs′ĭ-tē, nyo͞o′-) *n.*

**pneu·mat·ics** (no͞o-măt′ĭks, nyo͞o-) *n.* *(sing. in number).* The study of the mechanical properties of air and other gases.

**pneumato-** *or* **pneumat-** *pref.* [< Gk. *pneuma, pneumat-,* wind, breath.] **1.** Air : gas <*pneumato*lysis> **2.** Breath : respiration <*pneumato*meter>

**pneu·mat·o·graph** (no͞o-măt′ə-grăf′, nyo͞o-) *n. var. of* PNEUMOGRAPH.

**pneu·ma·tol·o·gy** (no͞o′mə-tŏl′ə-jē, nyo͞o′-) *n.* **1.** The doctrine or study of spiritual beings and phenomena, esp. the belief in spirits intervening between human beings and God. **2.** The Christian doctrine of the Holy Ghost. **—pneu′ma·to·log′ic** (-tə-lŏj′ĭk), **pneu′ma·to·log′i·cal** *adj.* **—pneu′ma·tol′o·gist** *n.*

**pneu·ma·tol·y·sis** (no͞o′mə-tŏl′ĭ-sĭs, nyo͞o′-) *n.* A process of mineral formation or rock alteration caused by the action of gases emitted from solidifying magma. **—pneu′ma·to·lyt′ic** (-tə-lĭt′ĭk) *adj.*

**pneu·ma·tom·e·ter** (no͞o′mə-tŏm′ĭ-tər, nyo͞o′-) *n.* A device for measuring the pressure of inhalation or expiration in the lungs. **—pneu′ma·tom′e·try** *n.*

**pneu·mat·o·phore** (no͞o-măt′ə-fôr′, -fōr′, nyo͞o-) *n.* **1.** *Zool.* A gas-filled sac serving as a float in some colonial organisms, as the Portuguese man-of-war. **2.** *Bot.* A specialized respiratory root structure in certain aquatic plants. **—pneu·mat′o·phor′ic** *adj.*

**pneu·mec·to·my** (no͞o-mĕk′tə-mē, nyo͞o-) *n., pl.* **-mies.** Surgical excision of a lung or part of a lung.

**pneumo-** *or* **pneum-** *pref.* [< Gk. *pneuma,* wind, breath.] **1.** Air : gas <*pneumo*thorax> **2.** Lung : pulmonary <*pneumo*coniosis> **3.** Respiration <*pneumo*graph> **4.** Pneumonia <*pneumo*coccus>

**pneu·mo·ba·cil·lus** (no͞o′mō-bə-sĭl′əs, nyo͞o′-) *n., pl.* **-cil·li** (-sĭl′ī′). A bacterium, *Klebsiella pneumoniae,* associated with respiratory infections, esp. pneumonia.

**pneu·mo·coc·cus** (no͞o′mə-kŏk′əs, nyo͞o′-) *n., pl.* **-coc·ci** (-kŏk′sī′, -kŏk′ī′). A bacterium, *Diplococcus pneumoniae,* that causes pneumonia. **—pneu′mo·coc′cal** (-kŏk′əl) *adj.*

**pneu·mo·co·ni·o·sis** (no͞o′mō-kō′nē-ō′sĭs, nyo͞o′-) *n.* [PNEUMO- + Gk. *konis,* dust + -OSIS.] A lung disease caused by prolonged inhalation of metallic or mineral dusts.

**pneu·mo·gas·tric** (no͞o′mō-găs′trĭk, nyo͞o′-) *adj.* **1.** Of or involving the lungs and the stomach. **2.** Pertaining to the vagus nerve.

**pneumogastric nerve** *n. Anat.* The vagus.

**pneu·mo·graph** (no͞o′mə-grăf′, nyo͞o′-) *also* **pneu·mat·o·graph** (no͞o-măt′ə-grăf′, nyo͞o-) *n.* A device for recording chest movements during respiration. **—pneu′mo·graph′ic** *adj.*

**pneu·mo·nec·to·my** (no͞o′mə-nĕk′tə-mē, nyo͞o′-) *n., pl.* **-mies.** [Gk. *pneumōn,* lung + -ECTOMY.] Surgical removal of lung tissue.

**pneu·mo·nia** (no͞o-mōn′yə, nyo͞o-) *n.* [NLat. < Gk., alteration of *pleumonia* < *pleumōn,* lung.] An acute or chronic disease caused by viruses, bacteria, or physical and chemical agents and characterized by inflammation of the lungs.

**pneu·mon·ic** (no͞o-mŏn′ĭk, nyo͞o-) *adj.* [NLat. *pneumonicus* < Gk. *pneumonikos,* of the lungs < *pneumōn,* lung.] **1.** Relating to, affected by, or similar to pneumonia. **2.** Of, affecting, or relating to the lungs : PULMONARY.

**pneu·mo·tho·rax** (no͞o'mō-thôr'ăks', -thōr'-, nyo͞o'-) *n.* Accumulation of air or gas in the pleural cavity, occurring as a result of injury or disease and sometimes induced to collapse the lung in the treatment of tuberculosis and other lung diseases.

**Po** *symbol for* POLONIUM.

**poach**[1] (pōch) *vt.* **poached, poach·ing, poach·es.** [ME *pochen* < OFr. *pochier,* to put in a bag < *poche,* bag, of Germanic orig.] To cook in a boiling or simmering liquid.

**poach**[2] (pōch) *v.* **poached, poach·ing, poach·es.** [OFr. *pocher,* to trample, of Germanic orig.] *—vi.* **1.** To trespass on another's property in order to take fish or game. **2.** To take fish or game in a prohibited area. **3.** To become broken up or muddy from being trampled, as land. **4.** To sink into soft earth when walking. *—vt.* **1.** To trespass on (another's property) for fishing or hunting. **2.** To take (fish or game) illegally. **3.** To poach (land) by trampling.

**poach·er**[1] (pō'chər) *n.* A vessel or dish for poaching food, as eggs or fish.

**poach·er**[2] (pō'chər) *n.* **1.** One who hunts or fishes illegally on the property of another. **2.** Any of various marine fishes of the family Agonidae, chiefly of northern Pacific waters, with an external covering of bony plates.

**po·chard** (pō'chərd) *n.* [Orig. unknown.] A duck of the genera *Aythya* or *Netta,* esp. the European species *A. ferina,* with a reddish head and gray and black plumage.

**pock** (pŏk) *n.* [ME *pokke* < OE *pocc.*] **1.** A pustule caused by an eruptive disease, as smallpox or chicken pox. **2.** A mark or scar left in the skin by a pock : POCKMARK. *—vt.* **pocked, pock·ing, pocks.** To mark with pocks : PIT.

**pock·et** (pŏk'ĭt) *n.* [ME *poket* < ONFr. *pokete,* dim. of *poke,* bag, of Germanic orig.] **1.** A small, flat, often decorative pouch sewed into or onto a garment. **2.** A small sack or bag. **3.** A receptacle, cavity, or opening. **4.** Financial means or supply. **5. a.** A small cavity in the earth containing ore. **b.** A small body or accumulation of ore. **6.** One of the pouchlike receptacles at the corners and sides of a billiard or pool table. **7.** A racing position in which a contestant has no room to pass a group of contestants immediately in front or to the side. **8.** A small, isolated, or protected area or group. **9.** An air pocket. **10.** A bin for storing materials, as ore or grain. *—adj.* **1.** Suitable for or capable of being carried in one's pocket <a *pocket* dictionary> **2.** Tiny : miniature. *—vt.* **-et·ed, -et·ing, -ets. 1.** To place in or as if in one's pocket. **2.** To take for oneself, esp. dishonestly. **3.** To accept or tolerate (e.g., an insult). **4.** To conceal or suppress (e.g., one's pride). **5.** To prevent (a bill) from becoming law by delaying its signing until the legislature adjourns. **6.** To hem in (a competitor) in a race. **7.** To hit (a ball) into a pocket of a pool or billiard table. **—in one's pocket.** In one's power or possession. **—in pocket. 1.** Having sufficient funds. **2.** Having made a profit. **—out of pocket. 1.** Low on funds. **2.** Having experienced a loss. **—pock'et·a·ble** *adj.* **—pock'et·er** *n.*

**pocket billiards** *pl.n.* POOL[2] 6.

**pock·et·book** (pŏk'ĭt-bo͝ok') *n.* **1.** A pocket-sized case or folder used to hold money and papers : BILLFOLD. **2.** A bag for carrying money, papers, and various small articles : PURSE. **3.** Financial resources. **4.** A usu. paperbound pocket-sized book.

**pocket borough** *n.* A borough in England, prior to the Parliamentary reform of 1832, whose representation was controlled by a single person or family.

**pocket bread** *n.* PITA[1].

**pocket calculator** *n.* A small calculator that can be carried in a pocket.

**pocket edition** *n.* POCKETBOOK 4.

**pock·et·ful** (pŏk'ĭt-fo͝ol') *n., pl.* **pock·et·fuls** or **pock·ets·ful.** The amount that a pocket will hold.

**pocket gopher** *n.* GOPHER 1.

**pock·et·knife** (pŏk'ĭt-nīf') *n.* A small knife with a blade or blades folding into the handle.

**pocket money** *n.* Money for small or incidental expenses.

**pocket mouse** *n.* A small, burrowing North American rodent of the genus *Perognathus,* with fur-lined external cheek pouches.

**pock·et-sized** (pŏk'ĭt-sīzd') or **pock·et·size** (-sīz') *adj.* Small enough to be carried in a pocket <a *pocket-sized* notebook>

**pocket veto** *n.* The indirect veto of a bill presented to a chief executive within ten days of a legislature's adjournment and held unsigned by that official until the actual date of adjournment. **—pocket veto** *v.* **(-toed, -to·ing, -toes).**

**pock·mark** (pŏk'märk') *n.* A pitlike scar left on the skin by an eruptive disease, as smallpox or chicken pox. *—vt.* To cover with pockmarks : PIT. **—pock'marked'** *adj.*

**pock·y** (pŏk'ē) *adj.* **-i·er, -i·est. 1.** Relating to, resembling, or having pocks. **2.** Pertaining to, having, or resembling syphilis.

**po·co** (pō'kō) *adv.* [Ital. < Lat. *paucus.*] *Mus.* A little : SOMEWHAT. —Used as a direction.

o͞o **boot** ou **out** th **thin** *th* **this** ŭ **cut** ûr **urge** y **young**
yo͞o ab**use** zh vi**sion** ə **a**bout, it**e**m, ed**i**ble, gall**o**p, circ**u**s

**po·co a po·co** (pō'kō ä pō'kō) *adv.* [Ital.] *Mus.* Little by little : GRADUALLY. —Used as a direction.

**po·co·cu·ran·te** (pō'kō-ko͞o-răn'tē, -rän'tē) *adj.* [Ital. : *poco,* little (< Lat. *paucus*) + *curante,* pr.part. of *curare,* to care for (< Lat. < *cura,* care).] Apathetic : indifferent. *—n.* One who does not care. **—po'co·cu·ran'tism** *n.*

**†po·co·sin** (pə-kō'sĭn) *n.* [Delaware *pâkwesen.*] *Chiefly Southeastern U.S.* A swamp in an upland coastal region.

**pod**[1] (pŏd) *n.* [Prob. alteration of obs. *cod* < ME < OE *codd.*] **1.** *Bot.* **a.** A dehiscent seed vessel or fruit of a leguminous plant, as the pea. **b.** A fruit containing several seeds that usu. dries and splits open. **2.** A podlike protective covering. **3.** A streamlined external housing that encloses aircraft engines, machine guns, or fuel. **4.** *Aerospace.* A detachable compartment in a spacecraft for personnel or instrumentation. *—v.* **pod·ded, pod·ding, pods.** *—vi.* **1.** To bear or produce pods. **2.** To swell or expand like a pod. *—vt.* To remove (seeds) from a pod.

**pod**[2] (pŏd) *n.* [Orig. unknown.] A school of seals or whales.

**pod**[3] (pŏd) *n.* [Orig. unknown.] **1.** The lengthwise groove in a boring tool, as an auger. **2.** The socket for holding the bit in a boring tool.

**-pod** or **-pode** *suff.* [NLat. *-podius, -poda* < Gk. *pous,* foot.] Foot : footlike part <pleo*pod*>

**po·dag·ra** (pə-dăg'rə) *n.* [ME < Lat. < Gk. : *pous,* foot + *agra,* trap, seizure.] Gout, esp. of the big toe. **—po·dag'ral, po·dag'ric** *adj.*

**-pode** *suff. var. of* -POD.

**po·des·ta** (pō-dĕs'tə, pō'dĕ-stä') *n.* [Ital. *podestà* < OItal. *podestate* < Lat. *potestas,* power.] The chief magistrate of a medieval Italian republic.

**po·di·a** (pō'dē-ə) *n. var. pl. of* PODIUM.

**po·di·a·try** (pə-dī'ə-trē) *n.* The study and treatment of foot ailments. **—po'di·at'ric** (pō'dē-ăt'rĭk) *adj.* **—po·di'a·trist** *n.*

**pod·ite** (pŏd'īt') *n.* [Gk. *pous, pod-,* foot + -ITE[1].] A limb segment of an arthropod appendage. **—po·dit'ic** (pə-dĭt'ĭk) *adj.*

**po·di·um** (pō'dē-əm) *n., pl.* **-di·a** (-dē-ə) *or* **-di·ums.** [Lat. < Gk. *podion,* base, dim. of *pous,* foot.] **1.** An elevated platform, as for a lecturer or orchestra conductor. **2.** A low wall serving as an architectural foundation. **3.** A wall enclosing the arena of an ancient amphitheater. **4.** *Biol.* A structure functioning as or resembling a foot.

**pod·o·phyl·lin** (pŏd'ə-fĭl'ĭn) *n.* [< NLat. *Podophyllum,* genus name : Gk. *pous,* foot + Gk. *phullon,* leaf.] A bitter-tasting resin extracted from the root of the May apple and used as a cathartic.

**-podous** *suff.* [-POD + -OUS.] Having a specified number or kind of feet or footlike parts <gastro*podous*>

**Po·dunk** (pō'dŭngk') *n.* [After *Podunk,* name of two New England towns.] *Slang.* A small, provincial, insignificant town.

**pod·zol** (pŏd'zôl') *n.* [R. : *pod,* ground + *zola,* ashes.] A leached soil formed mainly in cool, humid climates. **—pod·zol'ic** (pŏd-zŏl'ĭk, -zô'lĭk) *adj.*

**pod·zol·i·za·tion** (pŏd'zô-lĭ-zā'shən) *n.* **1.** The process by which soils become acidic through depletion of bases. **2.** The development of a podzol.

**po·em** (pō'əm, -ĭm) *n.* [OFr. *poème* < Lat. *poema* < Gk. *poiēma* < *poiein,* to create.] **1.** A composition designed to convey a vivid and imaginative sense of experience, esp. by the use of condensed language chosen for its sound and suggestive power as well as for its meaning and by the use of such literary techniques as structured meter, natural cadences, rhyme, or metaphor. **2.** A composition in verse rather than in prose. **3.** A literary composition written with an intensity or beauty of language more characteristic of poetry than of prose. **4.** A creation, object, or experience that embodies the lyrical beauty or structural perfection characteristic of traditional poetry.

**po·e·nol·o·gy** (pē-nŏl'ə-jē) *n. var. of* PENOLOGY.

**po·e·sy** (pō'ĭ-zē, -sē) *n., pl.* **-sies.** [ME *poesie* < OFr. < Lat. *poesis* < Gk. *poiēsis* < *poiein,* to create.] **1.** Poetry. **2.** The art or practice of composing poems. **3.** The inspiration involved in composing poetry.

**po·et** (pō'ĭt) *n.* [ME < OFr. *poete* < Lat. *poeta* < Gk. *poiētēs* < *poiein,* to create.] **1.** A writer of poems. **2.** One who is esp. gifted in the perception and expression of the beautiful or lyrical.

**po·et·as·ter** (pō'ĭt-ăs'tər) *n.* [NLat. : Lat. *poeta,* poet + Lat. *-aster,* partially resembling.] An inferior poet.

**po·et·ess** (pō'ĭ-tĭs) *n.* A woman who writes poems.

**po·et·ic** (pō-ĕt'ĭk) *adj.* [OFr. *poetique* < Lat. *poeticus* < Gk. *poiētikos,* inventive < *poiein,* to make.] **1.** Of or relating to poetry. **2.** Having a quality or style characteristic of poetry. **3.** Suitable as a subject for poetry <a *poetic* romance> **4.** Of, relating to, or befitting a poet <*poetic* imagination> **5.** Characterized by romantic imagery.

**po·et·i·cal** (pō-ĕt'ĭ-kəl) *adj.* **1.** Poetic. **2.** Fancifully depicted or embellished : IDEALIZED. **—po·et'i·cal·ly** *adv.* **—po·et'i·cal·ness** *n.*

**po·et·i·cism** (pō-ĕt'ĭ-sĭz'əm) *n.* An outdated or trite poetic term or expression.

**poetic justice** *n.* An outcome whereby one receives one's just deserts in a manner peculiarly or ironically appropriate.

**poetic license** *n.* The liberty taken, esp. by an artist or writer, in deviating from conventional fact or form to achieve a desired effect.

**po·et·ics** (pō-ĕt'ĭks) *n. (sing. in number).* **1.** Literary criticism dealing with the nature, forms, and laws of poetry. **2.** A study of or treatise on poetry or aesthetics. **3.** Poetic utterances or feelings.

**po·et·ize** (pō′ĭ-tīz′) *v.* **-ized, -iz·ing, -iz·es.** —*vi.* To write or express oneself in poetry. —*vt.* To give poetic expression to. **—po′et·iz′er** *n.*

**poet laureate** *n., pl.* **poets laureate** *or* **poet laureates. 1.** A poet appointed by the British sovereign to a lifetime position as chief poet of the kingdom. **2.** A poet acclaimed as the best or most representative of a group or locality. **3.** A poet honored for excellence.

**po·et·ry** (pō′ĭ-trē) *n.* [ME *poetrie* < OFr. < Med. Lat. *poetria* < Lat. *poeta*, poet. —see POET.] **1.** The art or work of a poet. **2. a.** Poems regarded as forming a division of literature. **b.** The poetic works of a given author, group, nation, or kind. **3.** A piece of literature written in meter : VERSE. **4.** Prose resembling a poem in some respect, as in form or sound. **5.** The essence of or characteristic quality possessed by a poem. **6.** The lyrical quality or structural perfection of an object, act, or experience <the *poetry* of a pas de deux>

**po·go·ni·a** (pə-gō′nē-ə, -gōn′yə) *n.* [NLat. *Pogonia*, genus name < Gk. *pōgōn*, beard.] A small terrestrial orchid of the genus *Pogonia*, growing in the North Temperate Zone and bearing pink or whitish flowers.

**pog·o·nip** (pŏg′ə-nĭp′) *n.* [Paiute.] A dense fog of suspended ice particles, occurring esp. in mountain valleys of the western United States.

**po·go stick** (pō′gō) *n.* [< *Pogo*, a former trademark.] A strong stick with footrests and a heavy spring set into the bottom end, propelled along the ground by hopping.

**po·grom** (pə-grŏm′, pō′grəm) *n.* [R. : *po-*, like < *po*, next to + *grom*, thunder.] An organized and often officially encouraged persecution or massacre of a minority group, esp. one conducted against the Jews. —*vt.* To massacre in a pogrom.

**po·gy** (pō′gē) *n., pl.* **pogy** *or* **-gies.** [Of Algonquian orig.] The menhaden.

**poi** (poi) *n.* [Hawaiian.] A pastelike Hawaiian food made from cooked fermented taro root.

**-poiesis** *suff.* [< Gk. *poiēsis*, creation < *poiein*, to make.] Production : creation : formation <hemato*poiesis*>

**-poietic** *suff.* [< Gk. *poiētikos*, creative < *poiētēs*, maker < *poiein*, to make.] Productive : formative <galacto*poietic*>

**poign·ant** (poin′yənt) *adj.* [ME *poinaunt* < OFr. *poignant*, pr.part. of *poindre*, to sting < Lat. *pungere*.] **1. a.** Physically painful. **b.** Keenly distressing to the mind <*poignant* worry> **c.** Appealing to the emotions : TOUCHING <a *poignant* love story> **2.** Piercing : incisive <a *poignant* critique> **3. a.** Neat, skillful, and to the point <*poignant* textual illustrations> **b.** Astute and pertinent : RELEVANT <a *poignant* proposal> **4.** Agreeably stimulating or intense <*poignant* pleasure> **5. a.** *Archaic.* Sharp or sour to the taste : PIQUANT. **b.** Sharp or pungent to the smell <a *poignant* aftershave> **—poign′ance, poign′an·cy** *n.* **—poign′ant·ly** *adv.*

**poi·kil·o·therm** (poi-kĭl′ə-thûrm′) *n.* [Gk. *poikilos*, variegated + -THERM.] A poikilothermous organism, as a reptile or fish.

**poi·kil·o·ther·mous** (poi′kĭl-ə-thûr′məs) *also* **poi·kil·o·ther·mal** (-məl) *adj.* Having a body temperature that varies with the external environment : COLD-BLOODED. **—poi′kil·o·ther′mism** *n.*

**poi·lu** (pwä-lü′) *n.* [Fr. < *poilu*, hairy < OFr. < *poil*, hair < Lat. *pilus*, hair.] *Slang.* A French front-line soldier in World War I.

**poin·ci·an·a** (poin′sē-ăn′ə, -ä′nə) *n.* [NLat. *Poinciana*, genus name, after M. de *Poinci*, 17th-cent. governor of French Antilles.] **1.** A tropical tree of the genus *Poinciana*, bearing large red or orange flowers. **2.** The royal poinciana.

**poin·set·ti·a** (poin-sĕt′ē-ə, -sĕt′ə) *n.* [NLat., after Joel R. *Poinsett* (1799–1851), its discoverer.] A tropical American shrub, *Euphorbia pulcherrima*, having usu. scarlet bracts under small yellow flowers.

**point** (point) *n.* [ME < OFr. < VLat. **puncta* < Lat. *punctus* < *pungere*, to prick.] **1.** A sharp or tapered end. **2.** Something that has a sharp or tapered end. **3.** A tapering extension of land projecting into water : CAPE. **4.** A mark formed by or as if by the sharp end of something. **5.** A mark or dot used in printing or writing. **6.** A mark used in punctuation, esp. a period. **7.** A decimal point. **8.** The vowel point. **9.** One of the protruding marks used in certain methods of writing and printing for the blind. **10.** *Math.* A dimensionless geometric object having no property but location. **11.** A position, place, or locality : SPOT <connections to New York and *points* east> **12.** A specified degree, condition, or limit, as in a scale or course. **13. a.** Any of the 32 equal divisions marked at the circumference of a mariner's compass card that indicate direction. **b.** The distance or interval of 11°15′ between any two adjacent markings. **14.** A distinct condition or degree <at the *point* of death> **15.** A specific moment in time <We've accomplished nothing up to this *point.*> **16.** A crucial situation. **17.** An important, essential, or primary factor. **18.** A purpose, goal, advantage, or reason <no *point* in arguing> **19.** The major idea or essential part of a narrative or concept. **20.** A significant, outstanding, or effective idea, argument, or suggestion. **21.** A separate or individual item or element : DETAIL. **22.** An important or distinctive quality or characteristic, esp. a standard characteristic used to judge an animal. **23.** A single unit, as in counting, rating, or measuring. **24. a.** A unit of academic credit usu. equal to one hour of class work per week during one semester. **b.** A numerical unit equal to a letter grade in grading academic achievement. **25.** A unit of counting or scoring in a game or sport. **26.** The stiff and attentive stance taken by a hunting dog. **27.** *Elect.* **a.** An electrical contact, esp. one in the distributor of an automobile engine. **b.** *Chiefly Brit.* An outlet or socket. **28. a.** A unit equal to one dollar and used to quote or state the current prices of commodities or stocks. **b.** A unit equal to one percentage point, used in reference to ownership. **c. points.** A percentage of the face value of a mortgage or loan charged as a placement or service fee. **29.** *Mus.* A phrase, as a fugue subject, in contrapuntal music. **30.** A unit of type size equal to 0.01384 inch or approx. 1/72 of an inch. **31.** A jeweler's unit of mass equal to 2 milligrams or 0.01 carat. **32. a.** Needlepoint. **b.** Bobbin lace. **33. a.** A movable rail, tapered at the end, as that used in a railroad switch. **b.** The tip or vertex of the angle created by the intersection of rails in a frog or switch. **34.** A ribbon or cord with a metal tag at the end, used to fasten clothing in the 16th and 17th cent. —*v.* **point·ed, point·ing, points.** —*vt.* **1.** To direct or aim <*point* a rifle> **2.** To bring to notice <*pointed* out our mistake> **3.** To indicate the position or direction of. **4.** To provide with a point : SHARPEN. **5.** To separate with a decimal point. **6.** To mark with a point or period : PUNCTUATE. **7.** To mark (a consonant) with a vowel point. **8.** To give emphasis to (e.g., a remark) : STRESS. **9.** To indicate the presence and position of (game) by standing immobile and pointing the muzzle, as a hunting dog does. **10.** To fill and finish the joints of (brickwork) with mortar or cement. —*vi.* **1.** To direct attention or indicate position with or as if with the finger. **2.** To turn the mind or thought in a particular direction. **3.** To be turned or faced in a given direction : AIM. **4.** To show the location of animals hunted as game by standing still and facing in that direction, as a hunting dog does. **5.** *Naut.* To sail close to the wind. **—beside the point.** Irrelevant to the matter at hand. **—stretch (*or* strain) a point. 1.** To make an exception. **2.** To exaggerate. **—to the point.** Relevant to the matter at hand.

**point·blank** (point′blăngk′) *adj.* [Orig. unknown.] **1.** Aimed straight at the mark or target, esp. aimed straight without allowing for the drop in a projectile's course. **2. a.** So close to a target that a weapon may be aimed directly at it. **b.** Close enough so that missing the target is unlikely or impossible. **3.** Straightforward : blunt <a *pointblank* rejection> —*adv.* **1.** With a straight aim : DIRECTLY. **2.** Without hesitation, deliberation, or equivocation <denied it *pointblank*>

**point defect** *n.* A departure from symmetry in the alignment of atoms in a crystal that affects only one or two lattice sites.

**point-de·vice** (point′dĭ-vīs′) *adj.* [ME *at point devis*, prob. < OFr. **a point devis*, to the arranged point.] *Archaic.* Scrupulously correct or neat : PRECISE. **—point′-de·vice′** *adv.*

**point·ed** (poin′tĭd) *adj.* **1.** Having an end coming to a point. **2.** Sharp : incisive <a *pointed* observation> **3.** Obviously directed at or making reference to a particular person or thing <a *pointed* remark> **4.** Clearly evident or conspicuous : MARKED <a *pointed* lack of enthusiasm> **5.** Marked by the use of a pointed crown, as in Gothic architecture. **—point′ed·ly** *adv.* **—point′ed·ness** *n.*

**point·er** (poin′tər) *n.* **1.** One that sharpens, directs, indicates, or points. **2.** A scale indicator on a measuring instrument, as a watch or balance. **3.** A long, tapered stick for pointing to objects, as on a chart or blackboard. **4.** One of a breed of hunting dogs having a short-haired coat that is usu. white with black or brownish spots. **5.** A piece of advice : SUGGESTION. **6.** *Computer Sci.* A computer word that directs the user to the address of a core storage location.

**point estimate** *n.* The single value assigned to a parameter in point estimation.

**point estimation** *n.* A method of making a mathematical estimate by assigning a single value to a parameter.

**poin·til·lism** (pwăn′tē-ĭz′əm, point′l-ĭz′əm) *n.* [Fr. *pointillisme* < *pointiller*, to paint small dots < *pointille*, small dot < Ital. *puntiglio*, dim. of *punto*, point < Lat. *punctum* < *pungere*, to prick.] A postimpressionist school of painting associated with Seurat and his followers in late 19th-cent. France and marked by the application of paint in small dots and brush strokes that blend together when seen from a distance. **—poin′til·list** *n.* **—poin′til·lis′tic** *adj.*

**point lace** *n.* Needlepoint.

**point·less** (point′lĭs) *adj.* **1.** Irrelevant : meaningless. **2.** Ineffectual : futile. **—point′less·ly** *adv.* **—point′less·ness** *n.*

**point man** *n.* **1.** A soldier assigned some distance ahead of a patrol as a lookout. **2.** One who speaks for or represents another, esp. in a confrontation with opponents.

**point of accumulation** *n.* A limit point.

**point of honor** *n.* A matter affecting one's honor.

**point of no return** *n.* **1.** The point in the flight of an aircraft beyond which the aircraft must proceed, there being insufficient fuel for return to the starting point. **2.** The point in a course of action beyond which reversal is not possible.

**point of order** *n.* A question as to whether what is being discussed is in order or allowed by the rules.

**point-of-sale** (point′əv-sāl′) *adj.* Of, relating to, or being the physical location where an item or service is purchased.

**point of view** *n.* **1.** The position from which something is considered or observed : STANDPOINT. **2.** One's way of viewing things.

**point system** *n.* **1.** A system of measurement by the point in printing. **2.** A system of printing or writing for the blind that uses an alphabet of raised symbols or dots that correspond to letters, as Braille. **3.** A system of evaluating a student's academic achievement by using numerical units or points equivalent to letter grades.

**point·y** (poin'tē) *adj.* **-i·er, -i·est.** Having a pointed end.

**poise**[1] (poiz) *v.* **poised, pois·ing, pois·es.** [ME *poisen*, to weigh < OFr. *poiser* < VLat. **pesare* < Lat. *pensare*.] —*vt.* To hold or carry in equilibrium : BALANCE. —*vi.* To be balanced or held in suspension : HOVER <*poised* on the verge of fame> —*n.* **1.** The state or condition of being balanced or held in equilibrium : STABILITY. **2.** Freedom from constraint, embarrassment, or awkwardness. **3.** The bearing of the head or body : MIEN. **4.** A state of hovering or being suspended.

**poise**[2] (poiz) *n.* [Fr., after Jean Louis Marie *Poiseuille* (1799–1869).] A centimeter-gram-second unit of dynamic viscosity equal to one dyne-second per square centimeter.

**poi·son** (poi'zən) *n.* [ME < OFr. < Lat. *potio*, drink < *potare*, to drink.] **1.** A substance that causes injury, illness, or death, esp. by chemical means. **2.** Something destructive or fatal. **3.** *Chem.* A substance that retards or inhibits a chemical reaction. —*vt.* **-soned, -son·ing, -sons. 1.** To injure or kill with poison. **2.** To put poison on or into <*poison* food> **3. a.** To pollute, as by noxious fumes. **b.** To have a harmful influence on : CORRUPT <Sibling rivalry *poisoned* our relationship.> **4.** *Chem.* To retard or inhibit (a chemical reaction). —*adj.* Poisonous. **—poi'son·er** *n.*

**poison elder** *n.* Poison sumac.

**poison gas** *n.* A crippling or lethal vapor used in warfare.

**poison hemlock** *n.* A poisonous Eurasian plant, *Conium maculatum*, naturalized in North America, with compound leaves and umbels of small white flowers.

**poison ivy** *n.* A North American shrub or vine, *Rhus radicans*, bearing trifoliate leaflets, small green flowers, and whitish berries and causing a rash on contact.

**poison oak** *n.* **1.** A shrub, *Rhus toxicodendron* of the southeastern United States or *R. diversiloba* of western North America, related to poison ivy and causing a rash on contact. **2.** Poison ivy.

**poi·son·ous** (poi'zə-nəs) *adj.* **1.** Capable of injuring or killing by or as if by poison. **2.** Containing a poison. **3.** Marked by ill will <a *poisonous* look> **—poi'son·ous·ly** *adv.* **—poi'son·ous·ness** *n.*

☆ ***syns:*** POISONOUS, MEPHITIC, MEPHITICAL, POISON, TOXIC, TOXICANT, VENOMOUS, VIRULENT *adj. core meaning* : capable of injuring or killing by poison <*poisonous* waste dumps> <a *poisonous* snake> ***ant:*** nonpoisonous

**poison sumac** *n.* A swamp shrub, *Rhus vernix* of the southeastern United States, bearing compound leaves and greenish-white berries and causing an itching rash on contact with the skin.

**Pois·son distribution** (pwä-sôN') *n.* [After Siméon Denis *Poisson* (1781–1840).] A probability distribution used to describe the occurrence of unlikely events in a large number of independent repeated trials.

**poke**[1] (pōk) *v.* **poked, pok·ing, pokes.** [ME *poken*, prob. < MLG.] —*vt.* **1.** To push or jab at, as with a finger or arm : PROD. **2.** To make (e.g., a hole or pathway) by or as if by poking, thrusting, or prodding. **3.** To cause to stick out <The monkey *poked* its head through the bars.> **4.** To stir (a fire) by prodding the wood or coal with a poker or stick. **5.** To strike : punch. —*vi.* **1.** To make thrusts or jabs, as with a stick or poker. **2.** To pry or meddle : INTRUDE <Quit *poking* into my affairs.> **3.** To look or search curiously in a desultory manner <just *poking* around> **4.** To live or proceed in a slow or lazy manner : PUTTER. **5.** To thrust forward : APPEAR <The seal's head *poked* from under the water.> —*n.* **1.** A push, thrust, or jab. **2.** A punch or blow with the fist. **3.** One who moves slowly or aimlessly. **—poke fun at.** To ridicule mischievously : TEASE.

☆ ***syns:*** POKE, PUSH, SHOVE, THRUST *v. core meaning* : to cause to stick out <*poked* their heads out the window>

**poke**[2] (pōk) *n.* [< POKE[1].] **1.** A large bonnet having a projecting brim at the front. **2.** The brim of a poke.

**†poke**[3] (pōk) *n.* [ME < ONFr. *poke*, of Germanic orig.] *Regional.* A bag or sack.

**poke**[4] (pōk) *n.* [Algonquian (Virginia) *pakon*, any plant used for dyeing < *pak*, blood.] Pokeweed.

**poke·ber·ry** (pōk'bĕr'ē) *n., pl.* **-ries. 1.** The blackish-red berry of the pokeweed. **2.** Pokeweed.

**pok·er**[1] (pō'kər) *n.* One that pokes, esp. a metal rod used to stir a fire.

**pok·er**[2] (pō'kər) *n.* [Orig. unknown.] A card game played by two or more players who bet on the value of their hands.

**poker face** *n.* An expressionless face, as that of an expert poker player. **—pok'er·faced'** (pō'kər-fāst') *adj.*

**poke·root** (pōk'ro͞ot', -ro͝ot') *n.* Pokeweed.

**poke·weed** (pōk'wēd') *n.* [POKE[4] + WEED.] A tall North American plant, *Phytolacca americana*, with blackish-red berries, small white flowers, and a poisonous root.

**po·key** (pō'kē) *n., pl.* **-keys.** [Orig. unknown.] *Slang.* Jail.

**pok·y** *also* **poke·y** (pō'kē) *adj.* **pok·i·er, pok·i·est.** [< POKE[1].] *Informal.* **1.** Dilatory : slow. **2.** Frumpish : shabby. **3.** Small and cramped <a *poky* little kitchen> **—pok'i·ly** *adv.* **—pok'i·ness** *n.*

**pol** (pôl) *n.* A politician.

**Poland China** *n.* A large black-and-white pig orig. bred in North America.

**Poland China**
*3½ feet high at shoulder*

**po·lar** (pō'lər) *adj.* **1. a.** Of, relating to, or designating a pole. **b.** Measured from or referred to a pole. **2.** Relating to, connected with, or situated near the North Pole or South Pole. **3. a.** Passing over a planet's north and south poles. **b.** Traveling in a polar orbit. **4.** Occupying or marked by opposite extremes <*polar* opposites> **5.** Serving as a guide, as a polestar or a pole of the earth. **6.** Pivotal or central <the *polar* issue>

**polar angle** *n.* The angle formed by the polar axis and the radius vector in a polar coordinate system.

**polar axis** *n.* The fixed reference axis from which the polar angle is measured in a polar coordinate system.

**polar bear** *n.* A large white-furred bear, *Thalarctos maritimus* of Arctic regions.

**polar body** *n.* A minute cell produced and ultimately discarded in the development of an oocyte, containing little or no cytoplasm but having one of the nuclei derived from the first or second meiotic division.

**polar cap** *n.* **1. a.** A high-latitude icecap. **b.** The polar regions of ice. **2.** *Astron.* A differentiated polar region of a planet.

**polar circle** *n.* Either the Arctic Circle or Antarctic Circle.

**polar coordinate** *n.* Either of two coordinates, the polar angle or radius vector, that together specify the position of a point in a plane.

**po·lar·im·e·ter** (pō'lə-rĭm'ĭ-tər) *n.* An instrument for measuring the rotation of the plane of polarization of polarized light or the degree of polarization of light passing through an optical structure or sample. **—po'lar·i·met'ric** (-lər-ə-mĕt'rĭk) *adj.* **—po'lar·im'e·try** *n.*

**Po·lar·is** (pə-lăr'ĭs, -lär'ĭs) *n.* [NLat. (*Stella*) *Polaris*, polar (star).] **1.** A star of the second magnitude, at the end of the handle of the Little Dipper and almost at the north celestial pole. **2.** A U.S. Navy intermediate range surface-to-surface ballistic missile.

**po·lar·i·scope** (pō-lăr'ĭ-skōp') *n.* An instrument for detecting or exhibiting the properties of polarized light or for studying the interactions of polarized light with optically transparent media.

**po·lar·i·ty** (pō-lăr'ĭ-tē) *n., pl.* **-ties. 1.** Intrinsic polar separation, alignment, or orientation, esp. of a physical property. **2.** The possession or manifestation of two opposing attributes, tendencies, or principles. **3.** An indicated polar extreme <an electric terminal with negative *polarity*>

**po·lar·i·za·tion** (pō'lər-ĭ-zā'shən) *n.* **1.** Production or the condition of polarity, as: **a.** Uniform and nonrandom elliptical, circular, or linear variation of a wave characteristic, esp. of vibrational orientation, in light or other radiation. **b.** *Physics & Chem.* Partial or complete polar separation of positive and negative electric charge in a nuclear, atomic, molecular, or chemical system. **2.** A concentration, as of groups, forces, or interests, about two contrasting or conflicting positions.

**po·lar·ize** (pō'lə-rīz') *v.* **-ized, -iz·ing, -iz·es.** —*vt.* **1.** To impart polarity to or induce polarization in. **2.** To cause to concentrate about two contrasting or conflicting positions. —*vi.* To become polarized. **—po'lar·iz'a·ble** *adj.* **—po'lar·iz'er** *n.*

**polarizing microscope** *n.* A microscope in which the object viewed is illuminated by polarized light.

**polar nucleus** *n.* Either of two nuclei located centrally in a seed plant embryo sac that fuse to form the endosperm nucleus.

**po·lar·og·ra·phy** (pō'lə-rŏg'rə-fē) *n.* [POLAR(IZATION) + -GRAPHY.] An electrochemical method of quantitative or qualitative analysis based on the relationship between an increasing current passing through the solution being analyzed and the increasing voltage used to produce the current. **—po·lar'o·graph'ic** (-lăr'ə-grăf'ĭk) *adj.* **—po·lar'o·graph'i·cal·ly** *adv.*

**Po·lar·oid** (pō'lə-roid'). A trademark for a specially treated, transparent plastic capable of polarizing light passing through it, used in glare-reducing optical devices.

**Polar Regions** *pl.n.* The land and water areas surrounding the North and South Poles.

**polar star** *n.* POLARIS 1.

**pol·der** (pōl′dər) *n.* [Du. < MDu.] An area of low-lying land, esp. in the Netherlands, that has been reclaimed from a body of water and is protected by dikes.

**pole**[1] (pōl) *n.* [ME < Lat. *polus* < Gk. *polos.*] **1.** Either axial extremity of an axis through a sphere. **2.** Either of the regions contiguous to the extremities of the earth's rotational axis, the North Pole or the South Pole. **3.** *Physics.* A magnetic pole. **4.** *Elect.* Either of two oppositely charged terminals, as in an electric cell or battery. **5.** *Astron.* A celestial pole. **6.** *Biol.* A physiologically or structurally distinct region at either axial extremity of a nucleus, cell, or organism. **7.** Either of two antithetical ideas, propensities, forces, or positions. **8.** A fixed point of reference. **9.** *Math.* The origin in a polar coordinate system : the polar angle vertex.

**pole**[2] (pōl) *n.* [ME < OE *pāl* < Lat. *palus,* stake.] **1.** A long, relatively slender, and gen. rounded piece of material, as wood or metal. **2.** The long, tapering, wooden shaft extending up from the front axle of a vehicle to the collars of the animals drawing it : TONGUE. **3. a.** ROD 12a. **b.** A unit of area equal to a square rod. **4.** *Naut.* A small or light spar. —*v.* **poled, pol·ing, poles.** —*vt.* **1.** To propel with a pole. **2.** To support (plants) with a pole. **3.** To strike, poke, or stir with a pole. —*vi.* **1.** To propel a boat or raft with a pole. **2.** To use ski poles to gain speed.

**Pole** (pōl) *n.* A native or resident of Poland.

**pole·ax** *or* **pole·axe** (pōl′ăks′) [ME *pollax* : *poll,* head (< MLG *polle*) + *ax,* ax < OE *æxa.*] —*n.* **1.** A medieval battle-ax having an ax, or an ax, hammer, and pick combination, with a long shaft. **2.** An ax having a hammer face opposite the blade, used to slaughter cattle. —*vt.* **-axed, -ax·ing, -ax·es.** To strike or fell with or as if with a poleax.

**pole bean** *n.* Any of various cultivated climbing beans trained to grow on poles or supports.

**pole·cat** (pōl′kăt′) *n.* [ME *polcat.*] **1.** A carnivorous mammal, *Mustela putorius* of Eurasia and northern Africa, with dark-brown or black fur. **2.** SKUNK 1.

**pole horse** *n.* A horse harnessed to the pole of a vehicle.

**po·leis** (pō′lās′) *n. pl. of* POLIS.

**pole jump** *n.* A pole vault.

**pole lamp** *n.* A usu. spring-loaded pole extending from ceiling to floor and having attached lamp fixtures.

**po·lem·ic** (pə-lĕm′ĭk) *n.* [Med. Lat. *polemicus,* controversialist < Gk. *polemikos,* hostile < *polemos,* war.] **1.** A controversial argument, esp. one that attacks or attempts to refute an accepted opinion or doctrine. **2. polemics.** *(sing. in number).* **a.** The art or practice of argumentation or controversy. **b.** The practice of theological controversy to refute doctrinal errors. **3.** One inclined to or engaged in controversy, argument, or refutation. —*adj. also* **po·lem·i·cal** (-ĭ-kəl). **1.** Of or relating to a controversy, argument, or refutation. **2.** Tending to argue : ARGUMENTATIVE. **—po·lem′i·cal·ly** *adv.*

**po·lem·i·cist** (pə-lĕm′ĭ-sĭst) *also* **po·lem·ist** (pə-lĕm′ĭst, pŏl′ə-mĭst) *n.* One skilled or involved in polemics.

**pol·er** (pō′lər) *n.* **1.** One that propels, supports, conveys, or strikes with a pole. **2.** A pole horse.

**pole·star** (pōl′stär′) *n.* **1.** POLARIS 1. **2.** A guiding principle.

**pole vault** *n.* **1.** A field event in which the contestant vaults over a high crossbar with the aid of a long pole. **2.** A vault made with the aid of a long pole.

**pole-vault** (pōl′vôlt′) *vi.* **-vault·ed, -vault·ing, -vaults.** To perform or complete a pole vault. **—pole′-vault′er** *n.*

**po·lice** (pə-lēs′) *n., pl.* **police.** [OFr., government < LLat. *politia* < Gk. *politeia* < *polis,* city.] **1.** The regulation and control of the affairs of a community, esp. with respect to the maintenance of order, law, health, morals, safety, and other matters affecting general welfare. **2. a.** The governmental department charged with the regulation and control of the affairs of a community, now chiefly the department established to maintain order, enforce the law, and prevent and detect crime. **b.** The official civil force or body of individuals established and maintained for this purpose : POLICE FORCE. **c.** *(pl. in number).* The members of such a force : POLICE OFFICERS. **3. a.** A group of individuals resembling the police force of a community in organization or function <security *police*> **b.** The members of such a group. **4. a.** The cleaning of a military base or other military area. **b.** The soldiers assigned to a particular maintenance duty. —*vt.* **-liced, -lic·ing, -lic·es. 1.** To regulate, control, or keep in order with or as if with police. **2.** To make (a military area) clean and neat.

**police action** *n.* A localized military action undertaken without a formal declaration of war.

**police court** *n.* An inferior court having the power to prosecute minor criminal offenses and detain for trial persons charged with more serious offenses.

**police dog** *n.* A dog, as a German shepherd, trained to aid the police.

**police force** *n.* A body of individuals trained in methods of law enforcement and crime prevention and detection, and authorized to maintain peace, order, and safety within the community.

**po·lice·man** (pə-lēs′mən) *n.* A man who is a member of a police force.

**police officer** *n.* A policeman or policewoman.

**police power** *n.* The inherent authority of a government to impose restrictions on private rights for the sake of public welfare, order, and security.

**police reporter** *n.* A newspaper reporter assigned to obtain and cover news in a local police department.

**police state** *n.* A political unit in which the government exercises rigid and repressive controls over the social, economic, and political life, esp. by means of a secret police force.

**police station** *n.* The headquarters of a unit of a police force where those under arrest are first charged.

**po·lice·wom·an** (pə-lēs′wo͝om′ən) *n.* A woman who is a member of a police force.

**pol·i·clin·ic** (pŏl′ē-klĭn′ĭk) *n.* [G. *Poliklinik* : Gk. *polis,* city + Gk. *klinikos,* of a bed < *klinein,* to lie down.] The department of a hospital that treats outpatients.

**pol·i·cy**[1] (pŏl′ĭ-sē) *n., pl.* **-cies.** [ME *policie* < OFr., government < Lat. *politia.* —see POLICE.] **1.** A plan or course of action, as of a government, political party, or business, designed to influence and determine decisions and actions <foreign *policy*><company *policy*> **2. a.** A course of action, guiding principle, or procedure considered to be expedient, prudent, or advantageous <a *policy* of fair play> **b.** Prudence, shrewdness, or sagacity in practical matters.

**pol·i·cy**[2] (pŏl′ĭ-sē) *n., pl.* **-cies.** [OFr. *police,* certificate < OItal. *polizza* < Med. Lat. *apodixa,* receipt < Med. Gk. *apodeixis* < Gk., proof < *apodeiknunai,* to demonstrate. —see APODICTIC.] **1.** A written contract or certificate of insurance. **2.** The numbers game.

**pol·i·cy·hold·er** (pŏl′ĭ-sē-hōl′dər) *n.* One that holds an insurance contract or policy.

**pol·i·cy·mak·ing** *or* **pol·i·cy-mak·ing** (pŏl′ĭ-sē-mā′kĭng) *n.* The top-level formation and implementation of policy, esp. official government policy. **—pol′i·cy·mak′er** *n.*

**po·li·o** (pō′lē-ō′) *n.* Poliomyelitis.

**po·li·o·my·e·li·tis** (pō′lē-ō-mī′ə-lī′tĭs) *n.* [Gk. *polios,* gray + MYELITIS.] An infectious viral disease occurring chiefly in children and in its acute forms attacking the central nervous system and producing paralysis, muscular atrophy, and often deformity. **—po′li·o·my′e·lit′ic** (-lĭt′ĭk) *adj.*

**po·li·o·vi·rus** (pō′lē-ō-vī′rəs) *n.* A virus separable into three serotypes that is the causative agent of poliomyelitis.

**po·lis** (pō′lĭs) *n., pl.* **-leis** (-lās′) [Gk.] A city-state of ancient Greece.

**pol·ish** (pŏl′ĭsh) *v.* **-ished, -ish·ing, -ish·es.** [ME *polisshen* < OFr. *polir, poliss-* < Lat. *polire.*] —*vt.* **1.** To make smooth and shiny by rubbing or chemical action. **2.** To free from coarseness : REFINE. **3.** To complete or perfect by removing flaws from. —*vi.* **1.** To become smooth or shiny by or as if by rubbing. **2.** To become refined or perfected. **—polish off.** *Informal.* To dispose of or finish quickly and easily. —*n.* **1.** Smoothness or shininess of surface or finish. **2.** A substance containing abrasive particles or chemical agents applied to smooth or shine a surface. **3.** The act or process of polishing. **4.** Refinement : elegance. **—pol′ish·er** *n.*

**Po·lish** (pō′lĭsh) *adj.* Of or relating to Poland, its inhabitants, or their language or culture. —*n.* The Slavic language of the Poles.

**pol·ished** (pŏl′ĭsht) *adj.* **1. a.** Made smooth and shiny. **b.** Naturally smooth and shiny. **2.** Cultured : refined. **3.** Having no errors or imperfections : FLAWLESS.

**pol·it·bu·ro** (pŏl′ĭt-byo͝or′ō, pə-lĭt′-) *n.* [R., contraction of *politisheskoe buro,* political bureau.] The chief political and executive committee of a Communist party.

**po·lite** (pə-līt′) *adj.* **-lit·er, -lit·est.** [ME *polyt,* polished < Lat. *polire,* to polish.] **1.** Marked by consideration for others, good manners, or tact. **2.** Refined : civilized. **—po·lite′ly** *adv.* **—po·lite′ness** *n.*

**pol·i·tesse** (pŏl′ĭ-tĕs′, pô′lĭ-) *n.* [Fr. < OFr., cleanliness < OItal. *politezza* < *pulire,* to polish < Lat. *polire.*] Courteous formality.

**pol·i·tic** (pŏl′ĭ-tĭk) *adj.* [ME *polytyk* < OFr. *politique* < Lat. *politicus,* political < Gk. *politikos* < *politēs,* citizen < *polis,* city.] **1.** Artful : shrewd <a *politic* diplomat> **2.** Using, exhibiting, or proceeding from policy : JUDICIOUS. **3.** Crafty : cunning. **—pol′i·tic·ly** *adv.*

**po·lit·i·cal** (pə-lĭt′ĭ-kəl) *adj.* **1.** Of, relating to, or dealing with the study, structure, or affairs of government, politics, or the state. **2.** Having a definite or organized policy or structure of government. **3.** Typical of or resembling politics, political parties, or politicians. **—po·lit′i·cal·ly** *adv.*

**political economy** *n.* The science of economics.

**political science** *n.* The study of the principles, processes, and structure of government and political institutions.

**pol·i·ti·cian** (pŏl′ĭ-tĭsh′ən) *n.* **1. a.** One actively involved in politics, esp. party politics. **b.** One who holds or seeks a political office. **2.** One who seeks personal or partisan gain, often by crafty or dishonest means. **3.** One who is skilled or experienced in the science or administration of government.

**po·lit·i·cize** (pə-lĭt′ĭ-sīz′) *v.* **-cized, -ciz·ing, -ciz·es.** —*vi.* To engage in or discuss politics. —*vt.* To make political. **—po·lit′i·ci·za′tion** *n.*

**pol·i·tick** (pŏl′ĭ-tĭk) *vi.* **-ticked, -tick·ing, -ticks.** To engage in or talk politics.

**po·lit·i·co** (pə-lĭt′ĭ-kō′) *n., pl.* **-cos.** [< Ital., political < Lat. *politicus.* —see POLITIC.] A politician.

**pol·i·tics** (pŏl′ĭ-tĭks) *n.* **1.** *(sing. in number).* The art or science of government : POLITICAL SCIENCE. **2.** *(sing. in number).* The activities or affairs of a government, politician, or political party. **3.** *(sing. in number).* **a.** Conduct of or participation in political affairs, often professionally. **b.** The business, activities, or profession of one so involved. **4.** *(sing. in number).* The methods or tactics involved in managing a government or state. **5.** *(pl. in number).* Intrigue or maneuvering within a group <company *politics*> **6.** *(pl. in number).* One's general position or attitude on political subjects <My *politics* are liberal.>

**pol·i·ty** (pŏl′ĭ-tē) *n., pl.* **-ties.** [OFr. *politie* < Lat. *politia,* government. —see POLICE.] **1.** The form of government of a nation, state, church, or organization. **2.** An organized society, as a nation, having a specific form of government.

**pol·ka** (pōl′kə, pō′kə) *n.* [Czech < Pol. < *Polka,* Polish woman, fem. of *Polak,* Pole.] **1.** A lively round dance of central European origin, performed by couples in duple meter. **2.** Music for the polka. —*vi.* **-kaed, -ka·ing, -kas.** To dance the polka.

**pol·ka dot** (pō′kə) *n.* **1.** One of a number of regularly spaced dots or round spots forming a pattern on cloth. **2.** A pattern or fabric that has polka dots.

**poll** (pōl) *n.* [ME *pol,* head < MLG *polle.*] **1.** The casting and registering of votes in an election. **2.** The number of votes cast or recorded. **3.** *often* **polls.** The place where votes are cast and registered. **4.** A survey of the public or a sample of the public to record opinion or acquire information. **5.** The head, esp. the top or back of the head. **6.** The broad or blunt end of a tool, as an ax or hammer. —*v.* **polled, poll·ing, polls.** —*vt.* **1.** To receive (a given number of votes) in an election. **2.** To receive or record the votes of (e.g., a jury). **3.** To cast (a vote or ballot) at the polls or in an election. **4.** To question in a survey : CANVASS. **5.** To clip or trim (e.g., hair, horns, or wool). **6.** To trim or clip the hair, wool, branches, or horns of : SHEAR. —*vi.* To vote at the polls or in an election. **—poll′er** *n.*

**pol·lack** (pŏl′ək) *n. var. of* POLLOCK.

**pol·lard** (pŏl′ərd) *n.* [< POLL.] **1.** A tree whose top branches have been cut back to the trunk so that it may produce a dense growth of new shoots. **2.** An animal, as an ox, goat, or sheep, that no longer has its horns. —*vt.* **-lard·ed, -lard·ing, -lards.** To change or convert into a pollard.

**polled** (pōld) *adj.* Having no horns : HORNLESS.

**pol·len** (pŏl′ən) *n.* [NLat. < Lat., flour.] The fine powderlike material produced by the anthers of flowering plants and functioning as the male element in fertilization.

**pol·len·ate** (pŏl′ə-nāt′) *v. var. of* POLLINATE.

**pollen count** *n.* The average number of pollen grains, usu. of ragweed, in a cubic yard or other standard volume of air over a 24-hour period at a given time and place, used to estimate the possible severity of hay-fever attacks.

**pol·len·if·er·ous** (pŏl′ə-nĭf′ər-əs) *adj. var. of* POLLINIFEROUS.

**pol·len·o·sis** (pŏl′ə-nō′sĭs) *n. var. of* POLLINOSIS.

**pollen tube** *n. Bot.* The slender tube emitted by a grain of pollen that penetrates an ovule and fertilizes it.

**pol·lex** (pŏl′ĕks′) *n., pl.* **pol·li·ces** (pŏl′ĭ-sēz′) [Lat.] The innermost forelimb digit : THUMB.

**pol·li·cal** (pŏl′ĭ-kəl) *adj.* Of or relating to the thumb.

**pol·li·ces** (pŏl′ĭ-sēz′) *n. pl. of* POLLEX.

**pollin-** *pref. var. of* POLLINI-.

**pol·li·nate** *also* **pol·len·ate** (pŏl′ə-nāt′) *vt.* **-nat·ed, -nat·ing, -nates** *also* **-at·ed, -at·ing, -ates.** [< NLat. *pollen, pollin-,* pollen < Lat., flour.] *Bot.* To transfer or convey pollen from an anther to a stigma of (a plant or flower) in the process of fertilization. **—pol′li·na′tion** *n.* **—pol′li·na′tor** *n.*

**pollini-** *or* **pollin-** *pref.* [< NLat. *pollen, pollin-,* pollen.] Pollen <*polliniferous*>

**pol·lin·i·a** (pŏ-lĭn′ē-ə) *n. pl. of* POLLINIUM.

**pol·lin·ic** (pŏ-lĭn′ĭk) *adj.* Of or relating to pollen.

**pol·li·nif·er·ous** *also* **pol·len·if·er·ous** (pŏl′ə-nĭf′ər-əs) *adj.* **1.** Yielding or producing pollen. **2.** Adapted for carrying pollen.

**pol·lin·i·um** (pŏ-lĭn′ē-əm) *n., pl.* **-i·a** (-ē-ə). An adhesive mass of agglutinated pollen grains, found in the flowers of most orchids and milkweeds.

**pol·li·nize** (pŏl′ə-nīz′) *vt.* **-nized, -niz·ing, -niz·es.** To pollinate. **—pol′li·ni·za′tion** *n.* **—pol′li·niz′er** *n.*

**pol·li·no·sis** *also* **pol·len·o·sis** (pŏl′ə-nō′sĭs) *n.* Allergic reaction to pollen, as in hay fever or asthma.

**pol·li·wog** *also* **pol·ly·wog** (pŏl′ē-wŏg′, -wôg′) *n.* [ME *polwygle* : *pol,* head (< MLG *polle*) + *wiglen,* to wiggle. —see WIGGLE.] A tadpole.

**pol·lock** *also* **pol·lack** (pŏl′ək) *n., pl.* **pollock** *or* **-locks** *also* **pollack** *or* **-lacks.** [Sc. *podlok.*] A marine food fish, *Pollachius virens* of northern Atlantic waters.

**poll·ster** (pōl′stər) *n.* One who takes public-opinion surveys.

**poll tax** *n.* A tax levied on persons rather than on property, often as a requirement for voting.

**pol·lut·ant** (pə-lo͞ot′nt) *n.* Something that pollutes, esp. a waste material that contaminates air, soil, or water.

**pol·lute** (pə-lo͞ot′) *vt.* **-lut·ed, -lut·ing, -lutes.** [ME *polluten* < Lat. *polluere.*] **1.** To make morally impure : CORRUPT. **2.** To make unfit for or harmful to living things, esp. by the addition of waste matter. **—pol·lut′er** *n.* **—pol·lut′ive** *adj.*

**pol·lu·tion** (pə-lo͞o′shən) *n.* **1. a.** The act or process of polluting. **b.** The state of being polluted. **2.** Contamination of air, soil, or water by the discharge of harmful substances.

**Pol·lux** (pŏl′əks) *n.* [Lat. < Gk. *Polludeukēs.*] **1.** *Gk. Myth.* One of the Dioscuri. **2.** A bright star in the constellation Gemini.

**Pol·ly·an·na** (pŏl′ē-ăn′ə) *n.* [After *Pollyanna,* the heroine of *Pollyanna,* novel by Eleanor Porter (1868–1920).] A blindly or foolishly optimistic person.

**pol·ly·wog** (pŏl′ē-wŏg′, -wôg′) *n. var. of* POLLIWOG.

**po·lo** (pō′lō) *n.* [Of Tibeto-Burman orig.] **1.** A game played on horseback by two teams of three or four players equipped with long-handled mallets for driving a small wooden ball through the opponents' goal. **2.** Water polo. **—po′lo·ist** *n.*

**polo coat** *n.* A tailored, usu. loose-fitting overcoat made from camel's hair or a similar material.

**pol·o·naise** (pŏl′ə-nāz′, pō′lə-) *n.* [Fr. < fem. of *polonais,* Polish < Med. Lat. *Polonia,* Poland.] **1.** A stately, marchlike Polish dance in triple time. **2.** Music for or in the style of the polonaise. **3.** A dress of the 18th cent., with a fitted bodice and draped cutaway skirt worn over an elaborate underskirt.

**po·lo·ni·um** (pə-lō′nē-əm) *n.* [< Med. Lat. *Polonia,* Poland.] *Symbol* **Po** A naturally radioactive metallic element, occurring in minute quantities as a product of radium disintegration and produced by bombarding bismuth or lead with neutrons; atomic number 84; longest-lived isotope Po 210.

**polo shirt** *n.* A knit pullover sport shirt.

**pol·ter·geist** (pōl′tər-gīst′) *n.* [G. : *poltern,* to make noises (< MHG *boldern*) + *Geist,* ghost < OHG.] A usu. mischievous ghost that manifests itself by noises and rappings.

**pol·troon** (pŏl-tro͞on′) *n.* [OFr. *poltron* < OItal. *poltrone,* aug. of *poltro,* colt < Lat. *pullus,* young animal.] *Archaic.* A base coward. **—pol·troon′er·y** *n.*

**poly-** *pref.* [< Gk. *polus,* much, many.] **1.** Many : much <*poly*atomic> **2.** Excessive : abnormal <*poly*dipsia> **3.** Polymer : polymeric <*poly*ethylene>

**pol·y·a·cryl·a·mide** (pŏl′ē-ə-krĭl′ə-mīd′) *n.* [POLY- + ACRYL(IC ACID) + AMIDE.] A white polyamide ($-CH_2CHCONH_2-$) of acrylic acid.

**polyacrylamide gel** *n.* A hydrated polyacrylamide of stiff consistency used as a medium for electrophoresis.

**pol·y·al·co·hol** (pŏl′ē-ăl′kə-hôl′) *n.* An alcohol having more than one hydroxy group.

**pol·y·am·ide** (pŏl′ē-ăm′īd′) *n.* A polymer having repeated amide groups, as in various kinds of nylon.

**pol·y·an·dric** (pŏl′ē-ăn′drĭk) *adj.* Of or pertaining to polyandry.

**pol·y·an·drous** (pŏl′ē-ăn′drəs) *adj.* **1.** Relating to or practicing polyandry. **2.** *Bot.* Having an indefinite number of stamens.

**pol·y·an·dry** (pŏl′ē-ăn′drē) *n.* **1.** The state or practice of having more than one husband at a time. **2.** *Bot.* The condition of being polyandrous.

**pol·y·an·thus** (pŏl′ē-ăn′thəs) *n., pl.* **-thus·es.** [NLat. < Gk. *poluanthos,* having many flowers : *polus,* many + *anthos,* flower.] Any of a group of hybrid garden primroses with variously colored flower clusters.

**polyanthus narcissus** *n.* A bulbous Eurasian plant, *Narcissus tazetta,* with fragrant white or yellow flower clusters.

**pol·y·a·tom·ic** (pŏl′ē-ə-tŏm′ĭk) *adj.* Having three or more atoms as constituents. —Used esp. of molecules.

**pol·y·ba·sic** (pŏl′ē-bā′sĭk) *adj.* Polyprotic.

**pol·y·ba·site** (pŏl′ē-bā′sīt′) *n.* [G. *Polybasit* : Gk. *polus,* many + Gk. *basis,* base.] A black mineral with a metallic luster, essentially $(Ag, Cu)_{16}Sb_2S_{11}$, often found in veins of silver.

**pol·y·car·bon·ate** (pŏl′ĭ-kär′bə-nāt′) *n.* Any of a family of thermoplastics marked by a high softening temperature and high impact strength.

**pol·y·car·pel·lar·y** (pŏl′ē-kär′pə-lĕr′ē) *adj.* [POLY- + CARPEL + -ARY.] Having or composed of many carpels.

**pol·y·car·pous** (pŏl′ē-kär′pəs) *also* **po·ly·car·pic** (-pĭk) *adj.* Having fruit with two or more carpels. **—pol′y·car′py** *n.*

**pol·y·chaete** *also* **pol·y·chete** (pŏl′ĭ-kēt′) *n.* [NLat. *Polychaeta,* class name < Gk. *polukhaitēs,* with much hair : *polus,* much + *khaitē,* long hair.] A marine worm of the class Polychaeta, with paired, flattened, bristle-tipped organs of locomotion. **—pol′y·chaete′, pol′y·chae′tous** *adj.*

**pol·y·chlo·rin·at·ed biphenyl** (pŏl'ĭ-klôr'ə-nā'tĭd, -klōr'-) *n.* Any of a family of industrial compounds produced by chlorination of biphenyl, noted chiefly as an environmental pollutant that accumulates in animal tissue with resultant pathogenic and teratogenic effects.

**pol·y·chro·mat·ic** (pŏl'ē-krō-măt'ĭk) *also* **pol·y·chro·mic** (-krō'mĭk) *or* **pol·y·chro·mous** (-krō'məs) *adj.* **1.** Having many colors or manifesting changes of color. **2.** Designating radiation of more than one wavelength.

**pol·y·chro·mat·o·phil·i·a** (pŏl'ē-krō-măt'ə-fĭl'ē-ə) *also* **pol·y·chro·mo·phil·i·a** (-krō'mə-) *n.* Susceptibility to staining with more than one type of dye, as seen in diseased red blood cells. **—pol'y·chro·mat'o·phil'ic** *adj.*

**pol·y·chrome** (pŏl'ē-krōm') *adj.* **1.** POLYCHROMATIC 1. **2.** Made or decorated in many colors. —*n.* An object, as a piece of pottery, having or decorated in many colors.

**pol·y·chro·mic** (pŏl'ē-krō'mĭk) *or* **pol·y·chro·mous** (-krō'məs) *n. var. of* POLYCHROMATIC.

**pol·y·chro·my** (pŏl'ē-krō'mē) *n.* The art of employing many colors in decoration, esp. in sculpture and architecture.

**pol·y·clin·ic** (pŏl'ē-klĭn'ĭk) *n.* A clinic or hospital that treats many types of injuries and diseases.

**pol·y·con·ic projection** (pŏl'ē-kŏn'ĭk) *n.* A conic map projection having distances between meridians along every parallel equal to those distances on a globe, the central geographic meridian is a straight line and the others are curved, while the parallels are arcs of circles.

**pol·y·cot·y·le·don** (pŏl'ē-kŏt'l-ēd'n) *also* **pol·y·cot** (pŏl'ē-kŏt') *n.* A plant having several cotyledons. **—pol'y·cot'y·le'don·ous** *adj.*

**pol·y·cy·clic** (pŏl'ĭ-sī'klĭk, -sĭk'lĭk) *adj. Chem.* Having two or more atomic rings in a molecule.

**pol·y·cy·the·mi·a** (pŏl'ē-sī-thē'mē-ə) *n. Pathol.* A condition characterized by an abnormally large number of red cells in the blood.

**pol·y·dac·tyl** (pŏl'ē-dăk'təl) *also* **pol·y·dac·ty·lous** (-tə-ləs) —*adj.* Having more than the normal number of fingers or toes. —*n.* A polydactyl individual. **—pol'y·dac'tyl·ism, pol'y·dac'ty·ly** *n.*

**pol·y·dem·ic** (pŏl'ē-dĕm'ĭk) *adj.* [POLY- + (EN)DEMIC.] Inhabiting or occurring in two or more regions.

**pol·y·dip·si·a** (pŏl'ē-dĭp'sē-ə) *n.* [NLat. : POLY- + Gk. *dipsa,* thirst.] Excessive or abnormal thirst. **—pol'y·dip'sic** *adj.*

**pol·y·e·lec·tro·lyte** (pŏl'ē-ĭ-lĕk'trə-līt') *n.* An electrolyte having a high molecular weight, as a polysaccharide or protein.

**pol·y·em·bry·o·ny** (pŏl'ē-ĕm'brē-ə-nē, -ĕm-brī'-) *n.* Development of more than one embryo from a single egg or ovule. **—pol'y·em'·bry·on'ic** (-brē-ŏn'ĭk) *adj.*

**pol·y·ene** (pŏl'ē-ēn') *n.* An organic compound having many double bonds.

**pol·y·es·ter** (pŏl'ē-ĕs'tər) *n.* A synthetic resin, produced mainly by reaction of dibasic acids with dihydric alcohols and used in waterproof fibers, plastics, boat hulls, swimming pools, and adhesives. **—pol'y·es'ter·i·fi·ca'tion** (-ə-fĭ-kā'shən) *n.*

**pol·y·e·ther** (pŏl'ē-ē'thər) *n.* A polymer having a carbon-oxygen bond in the repeating unit, esp. when derived from an epoxide or an aldehyde.

**pol·y·eth·yl·ene** (pŏl'ē-ĕth'ə-lēn') *n. Chem.* A polymerized ethylene resin, used esp. in films and sheets for packaging or molded for kitchenware, tubing, and containers.

**polyethylene glycol** *n.* Any of a number of high molecular weight liquids that are colorless and soluble in water and in many organic solvents, used chiefly in detergents and as emulsifying agents and plasticizers.

**po·lyg·a·la** (pə-lĭg'ə-lə) *n.* [NLat. *Polygala,* genus name < Gk. *polugalon,* milkwort : *polus,* much + *gala,* milk.] The milkwort.

**po·lyg·a·mist** (pə-lĭg'ə-mĭst) *n.* One who practices polygamy.

**po·lyg·a·mous** (pə-lĭg'ə-məs) *adj.* **1.** Of, pertaining to, engaged in, or marked by polygamy. **2.** *Bot.* **a.** Having both unisexual and hermaphroditic flowers on the same plant. **b.** Having either unisexual or hermaphroditic flowers on different plants of the same species. **—po·lyg'a·mous·ly** *adv.*

**po·lyg·a·my** (pə-lĭg'ə-mē) *n.* [Fr. *polygamie* < LLat. *polygamia* < Gk. *polugamia* : *polus,* many + *gamos,* marriage.] The state or practice of having more than one spouse at a time.

**pol·y·gene** (pŏl'ē-jēn') *n.* One of a set of cooperating genes, each producing a small quantitative effect.

**pol·y·gen·e·sis** (pŏl'ē-jĕn'ĭ-sĭs) *n.* Derivation of a species or type from more than one ancestor. **—pol'y·gen'e·sist** *n.* **—pol'y·ge·net'ic** (-jə-nĕt'ĭk), **pol'y·gen'ic** (-jĕn'ĭk), **po·lyg'e·nous** (pə-lĭj'ə-nəs) *adj.*

**pol·y·glot** (pŏl'ē-glŏt') *adj.* [Fr. *polyglotte* < Gk. *poluglōttos* : *polus,* many + *glōtta,* tongue.] Speaking, writing, written in, or composed of several languages. —*n.* **1.** One with a reading, writing, or speaking knowledge of several languages. **2.** A book, esp. a Bible, containing several versions of the same text in different languages. **3.** A mixture or confusion of languages. **—pol'y·glot'ism, pol'y·glot'tism** *n.*

**pol·y·gon** (pŏl'ē-gŏn') *n.* [LLat. *polygonum* < Gk. *polugōnon* < *polugōnos,* having many angles : *polus,* many + *gōnia,* angle.] A closed plane figure bounded by three or more line segments. **—po·lyg'o·nal** (pə-lĭg'ə-nəl) *adj.* **—po·lyg'o·nal·ly** *adv.*

**po·lyg·o·num** (pə-lĭg'ə-nəm) *n.* [NLat. *Polygonum,* genus name < Gk. *polugonon,* knotgrass : *polus,* many + *gonu,* knee.] A plant of the widely distributed genus *Polygonum,* with knotlike stem joints.

**pol·y·graph** (pŏl'ē-grăf') *n.* An instrument that simultaneously records variations in such physiological processes as heartbeat, blood pressure, and respiration, often used as a lie detector. **—pol'y·graph'ic** *adj.*

**po·lyg·y·ny** (pə-lĭj'ə-nē) *n.* [POLY- + Gk. *gunē,* woman.] The state or practice of having more than one wife at a time <tribal groups that practiced *polygyny*> **—po·lyg'y·nous** *adj.*

**pol·y·he·dra** (pŏl'ē-hē'drə) *n. var. pl. of* POLYHEDRON.

**pol·y·he·dral angle** (pŏl'ē-hē'drəl) *n.* The configuration of three or more planes having intersections that form a common vertex.

**pol·y·he·dron** (pŏl'ē-hē'drən) *n., pl.* **-drons** *or* **-dra** (-drə). A solid bounded by polygons. **—pol'y·he'dral** *adj.*

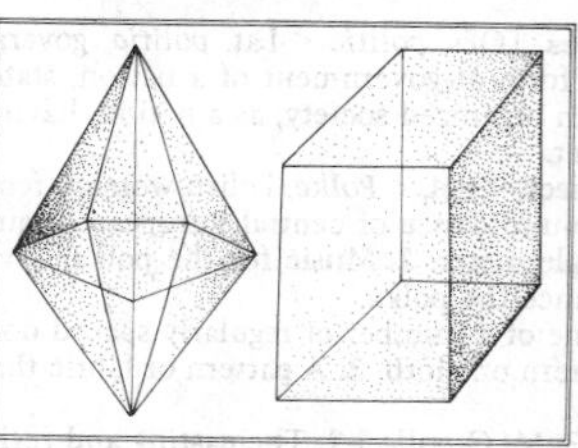

**polyhedron**
*Two types of polyhedrons*

**pol·y·his·tor** (pŏl'ē-hĭs'tər) *n.* [< Gk. *poluistōr,* very learned : *polus,* much + *histōr,* learned.] A polymath. **—pol'y·his·tor'ic** (-hĭ-stôr'ĭk, -stŏr'-) *adj.*

**pol·y·hy·dric** (pŏl'ē-hī'drĭk) *adj. Chem.* Having at least two hydroxyl groups.

**Pol·y·hym·ni·a** (pŏl'ē-hĭm'nē-ə) *also* **Po·lym·ni·a** (pə-lĭm'nē-ə) *n.* [Lat. < Gk. *Polumnia* < *polumnos,* abounding in songs : *polus,* many + *humnos,* hymn.] *Gk. Myth.* The Muse of sacred song, rhetoric, and mime.

**pol·y·im·ide** (pŏl'ē-ĭm'īd') *n.* A synthetic polymeric resin of a class resistant to high temperatures, wear, and corrosion, used mainly as a film or coating on a substrate substance.

**pol·y·mas·ti·gote** (pŏl'ē-măs'tĭ-gōt') *adj.* [Alteration of E. *polymastigate* : POLY- + Gk. *mastix,* whip + -ATE¹.] Having a tuftlike arrangement of flagella.

**pol·y·math** (pŏl'ē-măth') *n.* [Gk. *polumathēs* : *polus,* much + *manthanein,* to learn.] One of great or varied learning. **—pol'y·math', pol'y·math'ic** *adj.* **—po·lym'a·thy** (pə-lĭm'ə-thē, pŏl'ə-măth'ē) *n.*

**pol·y·mer** (pŏl'ə-mər) *n.* [Back-formation < POLYMERIC.] Any of numerous natural and synthetic compounds of usu. high molecular weight composed of up to millions of repeated linked units, each a relatively light and simple molecule.

**pol·y·mer·ase** (pŏl'ə-mə-rās', -rāz') *n.* Any of various enzymes that aid in the linkage of nucleotides in the formation of DNA or RNA with an existing strand of DNA or RNA acting as a template.

**pol·y·mer·ic** (pŏl'ə-mĕr'ĭk) *adj.* [Gk. *polumerēs,* having many parts : *polus,* many + *meros,* part.] Of, relating to, or composed of a polymer. **—pol'y·mer'i·cal·ly** *adv.* **—po·lym'er·ism** (pə-lĭm'ə-rĭz'əm, pŏl'ə-mə-) *n.*

**po·lym·er·i·za·tion** (pə-lĭm'ər-ĭ-zā'shən, pŏl'ə-mər-) *n.* **1.** The uniting of two or more monomers to form a polymer. **2.** A chemical process that brings about polymerization.

**pol·y·mer·ize** (pŏl'ə-mə-rīz', pə-lĭm'ə-) *vi.* & *vt.* **-ized, -iz·ing, -iz·es.** To undergo or subject to polymerization.

**po·lym·er·ous** (pə-lĭm'ər-əs) *adj.* Made up of numerous parts.

**Po·lym·ni·a** (pə-lĭm'nē-ə) *n. var. of* POLYHYMNIA.

**pol·y·morph** (pŏl'ē-môrf') *n.* **1.** *Biol.* An organism marked by polymorphism. **2.** *Chem.* A specific crystalline form of a compound that is capable of crystallizing in different forms.

**pol·y·mor·phism** (pŏl'ē-môr'fĭz'əm) *n.* **1.** *Biol.* The occurrence of different forms, stages, or color types in individual organisms or in organisms of the same species. **2.** *Chem.* Crystallization of a compound in at least two distinct forms. **—pol'y·mor'phic, pol'y·mor'phous** *adj.*

**pol·y·mor·pho·nu·cle·ar** (pŏl'ĭ-môr'fə-no͞o'klē-ər, -nyo͞o'-) *adj.* Having a lobed nucleus. —Used of leukocytes.

**pol·y·myx·in** (pŏl'ē-mĭk'sĭn) *n.* [< NLat. *polymyxa,* specific epithet of *Bacillus polymyxa* : POLY- + Gk. *muxa,* slime.] Any of various chiefly toxic antibiotics derived from the soil bacterium *Bacillus polymyxa* and active against Gram-negative bacteria.

**Pol·y·ne·sian** (pŏl'ə-nē'zhən, -shən) *adj.* Of or relating to Polynesia or its inhabitants, culture, or languages. —*n.* **1.** A native or resident of Polynesia. **2.** A subfamily of the Austronesian language family spoken in Polynesia.

**Pol·y·ni·ces** (pŏl'ə-nī'sēz) *n.* [Lat. < Gk. *Poluneikēs.*] *Gk. Myth.* A son of Oedipus for whom an expedition against Thebes was raised.

**pol·y·no·mi·al** (pŏl'ē-nō'mē-əl) *adj.* [POLY- + (BI)NOMIAL.] Of, relating to, or composed of more than two names or terms. —*n.* **1.** *Biol.* A taxonomic designation having more than two terms. **2.** *Math.* **a.** An algebraic function of two or more summed terms, each term consisting of a constant multiplier and one or more variables raised, in general, to integral powers, e.g., the general form of a polynomial of degree *n* in a single real variable x is $a_0x^n + a_1x^{n-1} + \cdots + a_{n-1}x + a_n$ where $a_0, a_1, \cdots, a_n$ are real numbers with $a_0 \neq 0$ and *n* is a positive integer. **b.** A mathematical expression of two or more terms.

**pol·y·nu·cle·o·tide** (pŏl'ĭ-nōō'klē-ə-tīd', -nyōō'-) *n.* A polymeric chain of nucleotides.

**po·lyn·ya** (pŏl'ən-yä') *n.* [R. *polyn'ya* < *polyĭ*, open.] A large area of open water surrounded by sea ice.

**pol·yp** (pŏl'ĭp) *n.* [Fr. *polype*, octopus < Lat. *polypus* < Gk. *polupous* : *polus*, many + *pous*, foot.] **1.** *Zool.* A coelenterate with a cylindrical body and an oral opening usu. surrounded by tentacles, as a coral or hydra. **2.** *Pathol.* A growth protruding from the mucous lining of an organ, as the nose. **—pol'yp·oid'** *adj.*

**pol·y·par·y** (pŏl'ə-pĕr'ē) *also* **pol·y·par·i·um** (pŏl'ə-pâr'ē-əm) *n., pl.* **-ies** *also* **-i·a** (-ē-ə). *Zool.* The common framework and base of a polyp colony, esp. of coral.

**pol·y·pep·tide** (pŏl'ē-pĕp'tīd') *n.* A peptide having between 10 and 100 amino acids. **—pol'y·pep·tid'ic** (-tĭd'ĭk) *adj.*

**pol·y·pet·al·ous** (pŏl'ē-pĕt'l-əs) *adj. Bot.* Having separate petals.

**pol·y·pha·gi·a** (pŏl'ē-fā'jē-ə, -jə) *n.* An excessive or pathological desire to eat. **—pol'y·pha'gi·an** *adj.*

**po·lyph·a·gous** (pə-lĭf'ə-gəs) *adj.* Feeding on or utilizing a wide variety of foods. **—po·lyph'a·gy** *n.*

**Pol·y·phe·mus** (pŏl'ə-fē'məs) *n.* [Lat. < Gk. *poluphēmos.*] *Gk. Myth.* The Cyclops who confined Odysseus and his companions in a cave until Odysseus blinded him and escaped.

**pol·y·phe·mus moth** (pŏl'ə-fē'məs) *n.* [After POLYPHEMUS.] A large North American moth, *Antheraea polyphemus*, with an eyelike spot on each hind wing.

**pol·y·phone** (pŏl'ē-fōn') *n.* A written character or combination of characters having two or more phonetic values, as the letter *a.*

**pol·y·phon·ic** (pŏl'ē-fŏn'ĭk) *adj.* Of, relating to, or typical of polyphony. **—pol'y·phon'i·cal·ly** *adv.*

**po·lyph·o·ny** (pə-lĭf'ə-nē) *n., pl.* **-nies.** Music with two or more independent melodic parts sounded together. **—po·lyph'o·nous** *adj.* **—po·lyph'o·nous·ly** *adv.*

**pol·y·phy·let·ic** (pŏl'ē-fī-lĕt'ĭk) *adj.* Relating to or characterized by development from more than one ancestral type. **—pol'y·phy·let'i·cal·ly** *adv.*

**pol·y·pi** (pŏl'ə-pī') *n. var. pl. of* POLYPUS.

**pol·y·ploid** (pŏl'ē-ploid') *adj.* Having more than twice the normal haploid chromosome number. —*n.* An organism with more than two sets of chromosomes. **—pol'y·ploi'dy** *n.*

**pol·yp·ne·a** (pŏl'ĭp-nē'ə) *n.* [NLat. : POLY- + Gk. *pnoia*, breathing.] Very rapid breathing : PANTING. **—pol'yp·ne'ic** (-nē'ĭk) *adj.*

**pol·y·pod** (pŏl'ē-pŏd') *also* **po·lyp·o·dous** (pə-lĭp'ə-dəs) *adj.* Having numerous feet.

**pol·y·po·dy** (pŏl'ē-pō'dē) *n., pl.* **-dies.** [ME *polypodie* < Lat. *polypodium* < Gk. *polupodion* < dim. of *polupous*, octopus. —see POLYP.] A fern of the widely distributed genus *Polypodium*, bearing simple or compound fronds and creeping rootstocks.

**pol·y·pore** (pŏl'ē-pôr', -pōr') *n.* A pore fungus.

**pol·y·pro·pyl·ene** (pŏl'ĭ-prō'pə-lēn') *n.* Any of various thermoplastic resins that are polymers of propylene.

**pol·y·pro·tic** (pŏl'ē-prō'tĭk) *adj.* [POLY- + PROT(ON) + -IC.] Designating an acid with two or more replaceable hydrogen atoms in each molecule.

**pol·yp·tych** (pŏl'ĭp-tĭk') *n.* [< Gk. *poluptukhos*, having many folds : *polus*, many + *ptukhē*, fold.] A decorated altarpiece or panel having three or more hinged sections that can be folded together.

**pol·y·pus** (pŏl'ə-pəs) *n., pl.* **-pi** (-pī') or **-pus·es.** [Lat., polypus, octopus. —see POLYP.] POLYP 2.

**pol·y·ri·bo·some** (pŏl'ĭ-rī'bə-sōm') *n.* A cluster of ribosomes connected by a molecule of messenger RNA.

**pol·y·sac·cha·ride** (pŏl'ē-săk'ə-rīd') *also* **pol·y·sac·cha·rid** (-rĭd) *or* **pol·y·sac·cha·rose** (-rōs', -rōz') *n.* A group of nine or more monosaccharides joined by glycosidic bonds, as cellulose and starch.

**pol·y·sep·al·ous** (pŏl'ē-sĕp'ə-ləs) *adj. Bot.* Having distinctly separated sepals.

**pol·y·some** (pŏl'ĭ-sōm') *n.* Polyribosome.

**pol·y·so·mic** (pŏl'ē-sō'mĭk) *adj.* [POLY- + (CHROMO)SOM(E) + -IC.] Having an excess number of one or more chromosomes.

**pol·y·sor·bate** (pŏl'ĭ-sôr'bāt') *n.* Any of a class of emulsifiers used in some pharmaceuticals and in food preparation.

**pol·y·sper·my** (pŏl'ē-spûr'mē) *n.* The entry of several sperms into an ovum during fertilization. **—pol'y·sper'mic** *adj.*

**po·lys·ti·chous** (pə-lĭs'tĭ-kəs) *adj.* [POLY- + LLat. *-stichus* < Gk. *-stikhos* < *stikhos*, row.] Arranged in two or more series or rows.

**pol·y·sty·rene** (pŏl'ē-stī'rēn') *n.* A hard, rigid, dimensionally stable, easily molded, clear thermoplastic polymer used esp. as a thermal or electrical insulating material.

**pol·y·sul·fide** (pŏl'ē-sŭl'fīd') *n.* A sulfide compound having at least two sulfur atoms per molecule.

**pol·y·syl·lab·ic** (pŏl'ē-sĭ-lăb'ĭk) *adj.* **1.** Having more than three syllables. **2.** Marked by words having more than three syllables. **—pol'y·syl·lab'i·cal·ly** *adv.*

**pol·y·syl·la·ble** (pŏl'ē-sĭl'ə-bəl) *n.* A word made up of more than three syllables.

**pol·y·syn·de·ton** (pŏl'ē-sĭn'dĭ-tŏn') *n.* [LGk. *polusundeton*, neuter of *polusundetos*, using many connectives : *polus*, many + *sundetos*, bound together. —see SYNDETIC.] Repetition of connectives or conjunctions in close succession for rhetorical effect, as in the phrase *neither rain nor snow nor sleet nor hail.*

**pol·y·syn·thet·ic** (pŏl'ē-sĭn-thĕt'ĭk) *adj.* Designating a language, as Eskimo, in which many of the elements of a phrase or sentence are combined into one utterance and do not exist separately.

**pol·y·tech·nic** (pŏl'ē-tĕk'nĭk) *adj.* [Fr. *polytechnique* < Gk. *polutekhnos*, skilled in many arts : *polus*, many + *tekhnē*, skill.] Relating to or dealing with many arts or sciences. —*n.* A school specializing in the teaching of industrial arts and applied sciences.

**pol·y·tene** (pŏl'ĭ-tēn') *adj.* [POLY- + Lat. *taenia*, band < Gk. *tainia*, ribbon.] Having or relating to large, multistranded chromosomes that have corresponding chromomeres in contact.

**pol·y·tet·ra·fluor·o·eth·yl·ene** (pŏl'ē-tĕt'rə-flŏŏr'ō-ĕth'ə-lēn') *n.* A waxy, opaque-white, thermoplastic resin, $(C_2F_4)_n$, thermally stable, resistant to acids, alkalis, and oxidizing agents, and having a low coefficient of friction, used as a low-friction coating esp. for cooking vessels and for chemical-resistant gaskets, seals, and hoses.

**pol·y·the·ism** (pŏl'ē-thē-ĭz'əm) *n.* [Fr. *polythéisme* < Gk. *polutheos*, polytheistic : *polus*, many + *theos*, god.] Belief in or worship of more than one god. **—pol'y·the'ist** *n.* **—pol'y·the·is'tic** *adj.*

**pol·y·thene** (pŏl'ə-thēn') *n. Chiefly Brit. var. of* POLYETHYLENE.

**po·lyt·o·cous** (pə-lĭt'ə-kəs) *adj.* [Gk. *polutokos*, bearing many offspring : *polus*, many + *tokos*, offspring.] *Biol.* Producing many offspring or ova at a single time.

**pol·y·to·nal·i·ty** (pŏl'ē-tō-năl'ĭ-tē) *n. Mus.* Simultaneity of two or more tonalities. **—pol'y·to'nal** *adj.* **—pol'y·to'nal·ly** *adv.*

**pol·y·tro·phic** (pŏl'ē-trō'fĭk, -trŏf'ĭk) *adj.* **1.** *Biol.* Subsisting on various types of organic material. **2.** *Pathol.* Marked by or pertaining to excessive nutrition.

**pol·y·typ·ic** (pŏl'ē-tĭp'ĭk) *also* **pol·y·typ·i·cal** (-ĭ-kəl) *adj.* Existing in, having, or involving many different types or forms.

**pol·y·un·sat·u·rat·ed** (pŏl'ē-ŭn-săch'ə-rā'tĭd) *adj.* Relating to long-chain carbon compounds, esp. fats, having many unsaturated bonds.

**pol·y·ure·thane** (pŏl'ē-yŏŏr'ə-thān') *n.* Any of various thermoplastic or thermosetting resins, widely varying in flexibility and used in tough chemical-resistant coatings, adhesives, foams, and electrical insulation.

**pol·y·u·ri·a** (pŏl'ē-yŏŏr'ē-ə) *n.* Excessive excretion of urine, as in diabetes. **—pol'y·u'ric** *adj.*

**pol·y·va·lent** (pŏl'ē-vā'lənt) *adj.* **1.** *Microbiol.* Having, sensitive to, or interacting with more than one kind of antigen, antibody, toxin, or microorganism. **2.** *Chem.* **a.** Having more than one valence. **b.** Having a valence of 3 or higher. **—pol'y·va'lence** *n.*

**pol·y·vi·nyl** (pŏl'ē-vī'nəl) *adj.* Designating any of a group of polymerized thermoplastic vinyls, as polyvinyl chloride.

**polyvinyl chloride** *n.* A common thermoplastic resin, used in a wide variety of products, as rainwear, garden hoses, phonograph records, and floor tiles.

**pol·y·zo·an** (pŏl'ē-zō'ən) *n.* [< NLat. *Polyzoa*, phylum name : POLY- + Gk. *zōia*, pl. of *zōion*, animal.] A bryozoan. **—pol'y·zo'an** *adj.*

**pol·y·zo·ar·i·um** (pŏl'ē-zō-âr'ē-əm) *also* **pol·y·zo·a·ry** (-zō'ə-rē) *n., pl.* **-ar·i·a** (-âr'ē-ə) *also* **-a·ries.** [NLat. < *Polyzoa*, phylum name. —see POLYZOAN.] A bryozoan colony or its supporting skeletal structure.

**pol·y·zo·ic** (pŏl'ē-zō'ĭk) *adj. Biol.* **1.** Forming or composed of a colony of zooids. **2.** Having numerous sporozoites.

**pom·ace** (pŭm'ĭs, pŏm'-) *n.* [Med. Lat. *pomacium*, cider < Lat. *pomum*, fruit.] **1.** The pulpy refuse remaining after the juice has been pressed from fruit, as apples or pears. **2.** Pulpy material remaining after the extraction of oil from fish, seeds, or nuts.

**pomace fly** *n.* The fruit fly.

**po·ma·ceous** (pō-mā'shəs) *adj.* [< Lat. *pomum*, fruit.] **1.** Of, relating to, or typical of apples. **2.** Of, relating to, or bearing pomes.

**po·made** (pō-mād', -mäd') *n.* [Fr. *pommade* < Ital. *pomata* < *pomo*, apple < Lat. *pomum*, fruit.] A perfumed ointment for the hair. —*vt.* **-mad·ed, -mad·ing, -mades.** To apply pomade to.

**po·man·der** (pō′măn′dər, pō-măn′-) *n.* [ME, alteration of OFr. *pome d'embre,* apple of amber.] **1.** A mixture of aromatic substances, once worn enclosed in a bag or box as a protection against odor and infection. **2.** A container for pomander.

**pome** (pōm) *n.* [ME < OFr., apple < Lat. *pomum,* fruit.] A fleshy fruit having seeds but no stone, as an apple, pear, or quince.

**pome·gran·ate** (pŏm′grăn′ĭt, pŏm′ĭ-, pŭm′-, pŭm′ĭ-) *n.* [ME *pomegranard* < OFr. *pome grenate* : *pome,* apple + *grenate,* having many seeds.] **1.** A semitropical Asian shrub or small tree, *Punica granatum,* widely cultivated for its edible fruit. **2.** The fruit of the pomegranate tree, having a tough, reddish rind and containing many seeds enclosed in a juicy red pulp with a mildly acid flavor.

**pom·e·lo** (pŏm′ə-lō′) *n., pl.* **-los.** [Alteration of POMPELMOUS.] The grapefruit.

**Pom·er·a·ni·an** (pŏm′ə-rā′nē-ən, -rān′yən) *adj.* Of or pertaining to Pomerania or its people. —*n.* **1.** A native of Pomerania. **2.** One of a breed of small dogs with long, silky hair.

**Pomeranian**
*5–7 inches high at shoulder*

**po·mi·cul·ture** (pō′mĭ-kŭl′chər) *n.* [Lat. *pomum,* fruit + CULTURE.] Cultivation of fruit.

**po·mif·er·ous** (pō-mĭf′ər-əs) *adj.* [Lat. *pomifer,* fruit-bearing : *pomum,* fruit + *ferre,* to bear.] Bearing pomes.

**pom·mel** (pŭm′əl, pŏm′-) *n.* [ME *pomel* < OFr. < VLat. **pomellum* < Lat. *pomum,* fruit.] **1.** A knob on the hilt of a weapon, as a sword. **2.** The upper front part of a saddle : SADDLEBOW. —*vt.* **-meled, -mel·ing, -mels** *also* **-melled, -mel·ling, -mels.** To pummel.

**po·mol·o·gy** (pō-mŏl′ə-jē) *n.* [Lat. *pomum,* fruit + -LOGY.] The scientific study and cultivation of fruit. **—po′mo·log′i·cal** (pō′mə-lŏj′ĭ-kəl) *adj.* **—po′mo·log′i·cal·ly** *adv.* **—po·mol′o·gist** *n.*

**pomp** (pŏmp) *n.* [ME < OFr. *pompe* < Lat. *pompa* < Gk. *pompē,* procession < *pempein,* to send.] **1.** Dignified or magnificent display : SPLENDOR. **2.** Vain or ostentatious display.

**pom·pa·dour** (pŏm′pə-dôr′, -dōr′) *n.* [After Jeanne Antoinette Poisson (1721–1764), Marquise de *Pompadour,* its inventor.] **1. a.** A woman's hair style formed by sweeping the hair straight up from the forehead. **b.** A man's hair style with the hair brushed up from the forehead. **2.** Hair styled in a pompadour.

**pom·pa·no** (pŏm′pə-nō′, pŭm′-) *n., pl.* **pompano** *or* **-nos.** [Sp. *pámpano salpa,* a kind of fish.] **1.** A marine food fish of the genus *Trachinotus,* esp. *T. carolinus* of tropical and temperate Atlantic waters. **2.** A small butterfish, *Palometa simillima* of American Pacific coastal waters, valued as a food fish.

**Pom·pe·ian red** (pŏm-pā′ən) *n.* A grayish to moderate red.

**pom·pel·mous** (pŏm′pəl-mōōs′) *n.* [Du. *pompelmoes.*] SHADDOCK 2.

**pom·pom** (pŏm′pŏm′) *also* **pom·pon** (-pŏn′) *n.* [Fr. *pompon.*] **1.** A tuft or ball of material, as cotton, wool, or feathers, worn as decoration. **2.** A small, buttonlike flower of some chrysanthemums and dahlias.

**pom-pom** (pŏm′pŏm′) *n.* [Imit.] An automatic, rapid-fire antiaircraft cannon usu. mounted on ships.

**pom·pous** (pŏm′pəs) *adj.* [ME < OFr. *pompeux* < LLat. *pomposus* < Lat. *pompa,* pomp. —see POMP.] **1.** Marked by an exaggerated show of dignity or self-importance : PRETENTIOUS. **2.** Bombastic or self-important. **3.** Marked by stately display : CEREMONIOUS. **—pom·pos′i·ty** (-pŏs′ĭ-tē), **pom′pous·ness** *n.* **—pom′pous·ly** *adv.*

**pon·cho** (pŏn′chō) *n., pl.* **-chos.** [Am. Sp. < Araucanian *pontho,* woolen fabric.] **1.** A blanketlike cloak having a hole in the center for the head. **2.** A poncholike garment used as a raincoat.

**pond** (pŏnd) *n.* [ME *ponde* < OE *pund-,* enclosure.] A still body of water smaller than a lake.

**pon·der** (pŏn′dər) *v.* **-dered, -der·ing, -ders.** [ME *ponderen* < OFr. *ponderer* < Lat. *ponderare* < *pondus,* weight.] —*vt.* **1.** To weigh or appraise carefully. **2.** To think about : CONSIDER. —*vi.* To meditate : reflect. **—pon′der·a·bil′i·ty** *n.* **—pon′der·a·ble** (-ə-bəl) *adj.* **—pon′der·er** *n.*

**pon·der·o·sa pine** (pŏn′də-rō′sə) *n.* [NLat. *Pinus ponderosa,* ponderous pine.] A tall timber tree, *Pinus ponderosa* of western North America, with long dark-green needles.

**pon·der·ous** (pŏn′dər-əs) *adj.* [ME < OFr. *pondereux* < Lat. *ponderosus* < *pondus,* weight.] **1.** Having great weight : MASSIVE. **2.** Unwieldy or awkward from weight. **3.** Lacking fluency : LABORED. **—pon′der·os′i·ty** (-ŏs′ĭ-tē), **pon′der·ous·ness** *n.* **—pon′der·ous·ly** *adv.*

☆ ***syns:*** PONDEROUS, ELEPHANTINE, HEAVY-HANDED, LABORED *adj. core meaning* : lacking fluency or gracefulness <a *ponderous* history of education in America>

**pond lily** *n.* The water lily.

**pond scum** *n.* Any of various freshwater algae that form a usu. greenish scum on the surface of stagnant water.

**pond·weed** (pŏnd′wēd′) *n.* A floating or submerged aquatic plant of the genus *Potamogeton.*

**pone** (pōn) *n.* [Of Algonquian orig.] Corn pone.

**pon·gee** (pŏn-jē′, pŏn′jē) *n.* [Chin. (Mandarin) *ben$^3$zhi$^1$* : *ben$^3$,* one's own + *zhi$^1$,* to weave, spin.] A thin, soft cloth of Chinese or Indian silk with a knotty weave.

**pon·iard** (pŏn′yərd) *n.* [Fr. *poignard* < *poing,* fist < OFr. < Lat. *pugnus.*] A dagger with a usu. slender blade. —*vt.* **-iard·ed, -iard·ing, -iards.** To stab or kill with a poniard.

**pons** (pŏnz) *n., pl.* **pon·tes** (pŏn′tēz) [NLat. < Lat., bridge.] *Anat.* **1.** A slender tissue joining two parts of an organ. **2.** The pons varolii.

**pons as·i·no·rum** (ăs′ə-nôr′əm, -nōr′əm) *n.* [Lat., asses' bridge.] A difficult or critical test of ability for one who is inexperienced or lacking in knowledge.

**pons va·ro·li·i** (və-rō′lē-ī′) *n.* [NLat., bridge of Varoli, after Constanzo *Varoli* (1542–1575).] A band of nerve fibers in the brain connecting the medulla oblongata and the mesencephalon below the cerebellum.

**pon·tes** (pŏn′tēz) *n. pl. of* PONS.

**pon·ti·fex** (pŏn′tə-fĕks′) *n., pl.* **pon·tif·i·ces** (pŏn-tĭf′ĭ-sēz′) [Lat.] A member of the ancient Roman Pontifical College, the highest college of priests, headed by the Pontifex Maximus.

**pon·tiff** (pŏn′tĭf) *n.* [Fr. < Lat. *pontifex,* pontifex.] **1. a.** The pope. **b.** A bishop. **2.** A pontifex.

**pon·tif·i·cal** (pŏn-tĭf′ĭ-kəl) *adj.* [Lat. *pontificalis,* of a pontifex < *pontifex,* pontifex.] **1.** Relating to, typical of, or befitting a pope or bishop. **2.** Having the dignity, pomp, or authority of a pontiff. **3.** Pompously authoritative or dogmatic. —*n.* **1. pontificals.** The vestments and insignia of a pontiff. **2.** A book of rites and ceremonies for a bishop. **—pon·tif′i·cal·ly** *adv.*

**Pontifical Mass** *n.* A celebration of the Eucharist performed by a bishop in all Roman Catholic churches, many Anglican churches, and some Lutheran churches.

**pon·tif·i·cate** (pŏn-tĭf′ĭ-kĭt, -kāt′) *n.* [Lat. *pontificatus* < *pontifex,* pontifex.] The office or term of office of a pontiff. —*vi.* (pŏn-tĭf′ĭ-kāt′) **-cat·ed, -cat·ing, -cates. 1.** To administer the office of a pontiff. **2.** To speak or behave with pompous authority or dogmatism <*pontificated* about their great responsibilities>

**pon·tif·i·ces** (pŏn-tĭf′ĭ-sēz′) *n. pl. of* PONTIFEX.

**pon·til** (pŏn′tĭl) *n.* [Fr., poss. < Ital. *puntello,* dim. of *punto,* point < Lat. *punctum* < *pungere,* to prick.] A punty.

**pon·tine** (pŏn′tīn′) *adj.* [< Lat. *pons, pont-,* bridge.] **1.** Of or relating to bridges. **2.** *Anat.* Of or relating to the pons varolii.

**Pont l'É·vêque** (pôN′ lā-vĕk′) *n.* [After *Pont L'Évêque,* France.] A mild, soft-centered French cheese made of whole milk.

**pon·to·nier** (pŏn′tə-nîr′) *n.* [Fr. *pontonnier* < *ponton,* pontoon.] One in charge of pontoons or engaged in the construction of pontoon bridges.

**pon·toon** (pŏn-tōōn′) *n.* [Fr. *ponton* < OFr. < Lat. *ponto,* boat bridge < *pons,* bridge.] **1. a.** A flat-bottomed boat or portable float used to support a pontoon bridge. **b.** A floating structure serving as a dock. **2.** A float on a seaplane.

**pontoon bridge** *n.* A temporary floating bridge using pontoons for support.

**po·ny** (pō′nē) *n., pl.* **-nies.** [Prob. < obs. Fr. *poulenet,* dim. of *poulain,* colt < Lat. *pullus,* foal.] **1.** Any of several types or breeds of horses that remain small when full grown. **2.** *Informal.* A racehorse. **3.** Something small for its kind. **4.** *Informal.* A translation of a foreign language text, esp. one used secretly by students. **5.** *Chiefly Brit.* The sum of 25 pounds. —*v.* **-nied, -ny·ing, -nies.** *Slang.* —*vt.* To prepare (lessons) with a pony. —*vi.* To use a pony in preparing lessons. **—pony up.** *Slang.* To pay money owed or due.

**pony express** *n.* A system of mail transportation by relays of ponies, esp. the system in operation from St. Joseph, Missouri, to Sacramento, California (1860–61).

**po·ny·tail** (pō′nē-tāl′) *n.* A hair style in which the hair is drawn back and fastened with a band or clasp so that it hangs down like a pony's tail.

**pooch** (pōōch) *n.* [Orig. unknown.] *Slang.* DOG 1.

**pood** (pōōd) *n.* [R. *pud* < ON *pund,* pound < OE < Lat. *pondo.*] A Russian weight equivalent to approx. 36 pounds avoirdupois.

**poo·dle** (pōōd′l) *n.* [G. *Pudel,* short for *Pudelhund* : LG *pudeln,* to splash + *hund,* dog.] Any of a breed of small to large-sized dogs with thick curly hair, orig. developed in Europe as hunting dogs.

**poof** (pōōf, pŏŏf) *interj.* —Used to express contempt or indicate abrupt appearance or disappearance.

**pooh** (pōō) *interj.* —Used to express disdain or contempt.

ă **pat** ā **pay** âr **care** ä **father** ĕ **pet** ē **be** hw **which** ĭ **pit** ī **tie** îr **pier** ŏ **pot** ō **toe** ô **paw, for** oi **noise** ŏŏ **took**

**Pooh-Bah** (po͞o'bä') *n.* [After *Pooh-Bah,* Lord-High-Everything-Else, character in *The Mikado* by Sir William Schwenck Gilbert (1836–1911) and Sir Arthur Seymour Sullivan (1842–1900).] **1.** A pompous official, esp. one who holds many offices but fulfills none of them. **2.** One who holds high office.

**pooh-pooh** (po͞o'po͞o') *vt.* **-poohed, -pooh·ing, -poohs.** [Redup. of POOH.] *Informal.* To express disdain or contempt for.

**pool¹** (po͞ol) *n.* [ME *pool* < OE *pōl.*] **1. a.** A small body of still water. **b.** Something resembling a pool <a *pool* of moonlight> **2.** A puddle of standing liquid, as blood. **3.** A deep place in a river or stream. **4.** A swimming pool.

**pool²** (po͞ol) *n.* [Fr. *poule,* stakes < Lat. *pullus,* young of an animal.] **1. a.** A game of chance in which the contestants put money into a common fund that is later paid to the winner. **b.** The fund containing the bets made in a game of chance or on a horse race. **2.** A grouping of resources for the common advantage of the participants. **3. a.** A fund established by a group of stockholders for speculating in or manipulating prices of securities. **b.** The persons or parties participating in such a combination. **4.** An agreement between competing business concerns to establish controls over production, market, and prices for common profit. **5.** A fencing match in which each member of a team fences successively with each member of an opposing team. **6.** Any of several games played on a six-pocket billiard table usu. with 15 object balls and a cue ball. —*v.* **pooled, pool·ing, pools.** —*vt.* To put into a common fund for use by all <*pool* our experience> —*vi.* To join or form a pool.

**pool·room** (po͞ol'ro͞om', -ro͝om') *n.* A commercial establishment or room for the playing of pool or billiards.

**pool table** *n.* A six-pocket billiard table on which pool is played.

**poon** (po͞on) *n.* [Singhalese *pūna.*] An Asian tree of the genus *Calophyllum,* yielding light hard wood used for masts and spars.

**poop¹** (po͞op) *n.* [OFr. *poupe* < Lat. *puppis.*] *Naut.* **1.** The stern superstructure of a ship. **2.** The poop deck.

**poop²** (po͞op) *vt.* **pooped, poop·ing, poops.** [Orig. unknown.] *Slang.* To tire out : EXHAUST. **—poop out. 1.** To quit (e.g., a race or contest) because of exhaustion. **2.** To decide not to participate, esp. at the last moment.

**poop³** (po͞op) *n.* [Orig. unknown.] *Slang.* Inside information.

**poop deck** *n. Naut.* The aftermost deck of a ship.

**poor** (po͝or) *adj.* **-er, -est.** [ME *poure* < OFr. *povre* < Lat. *pauper.*] **1. a.** Having little or no money and few or no possessions. **b.** *Law.* Dependent on charity or public funds : DESTITUTE. **c.** Lacking in financial or other resources <a region *poor* in minerals> **2. a.** Lacking in mental or moral quality. **b.** Inadequate : inferior <a *poor* substitute> **3. a.** Lacking fertility <*poor* land> **b.** Undernourished : lean. **4. a.** Lacking in value : TRIVIAL. **b.** Lacking in quantity <Attendance has been *poor.*> **5. a.** Humble <in my *poor* estimation> **b.** Pitiable <the *poor* kid> ***usage:*** *Poor* is an adjective, not an adverb; therefore such sentences as *We did poor* and *They never played poorer* are correctly expressed as *We did poorly* and *They never played more poorly.* **—poor'ness** *n.*

☆ ***syns:*** POOR, BROKE, BUSTED, DESTITUTE, IMPOVERISHED, INDIGENT, NEEDY, PENNILESS, PENURIOUS, STRAPPED *adj. core meaning* : having little or no money <too *poor* to eat well> ***ant:*** rich

**poor box** *n.* A box, esp. in a church, for collecting alms.

**poor boy** *n.* HERO 5.

**poor farm** *n.* A farm that houses, supports, and employs paupers at public expense.

**poor·house** (po͝or'hous') *n.* An establishment maintained at public expense as housing for the needy or poverty-stricken.

**poo·ri** *also* **pu·ri** (po͝or'ē) *n.* [Hindi *puri* < Skt. *pūraḥ,* cake.] A light, flat, usu. deep-fried wheat cake of Pakistan and northern India.

**poor·ish** (po͝or'ĭsh) *adj.* Somewhat poor.

**poor law** *n.* A law or system of laws providing for public relief and support of the poor.

**†poor·ly** (po͝or'lē) *adv.* In a poor manner <did *poorly* on the exam> —*adj. Regional.* In poor health : ILL.

**poor·mouth** (po͝or'mouth', -mouth') *v.* **-mouthed, -mouth·ing, -mouths.** —*vt.* To speak ill of : BADMOUTH. —*vi.* To claim poverty as an excuse for one's position or actions. —*n.* An exaggerated declaration of poverty.

**pop¹** (pŏp) *v.* **popped, pop·ping, pops.** [ME *poppen.*] —*vi.* **1.** To make a sudden sharp, explosive sound. **2.** To burst open with a sudden sharp, explosive sound. **3.** To move or appear quickly or unexpectedly <*popped* into sight> **4.** To open wide (e.g., one's eyes) suddenly. **5.** *Baseball.* To hit a short high fly ball, esp. one that can be caught by an infielder. **6.** To shoot a firearm, as a pistol. —*vt.* **1.** To cause to make a sharp explosive sound. **2.** To cause to explode with a sharp sound. **3.** To put or thrust suddenly or unexpectedly <*popped* the baby into bed> **4.** To fire (e.g., a pistol). **5.** To fire at : SHOOT. **6.** To strike or hit. **7.** *Baseball.* To hit (a ball) high in the air but not far. **8.** *Slang.* **a.** To use (drugs), esp. orally <*popping* uppers> **b.** To drink (e.g., beer). **—pop off.** *Informal.* **1. a.** To leave hurriedly or abruptly. **b.** To die suddenly and unexpectedly. **2.** To speak in a burst of vehement anger. —*n.* **1.** A sudden sharp, explosive sound. **2.** A shot with a firearm. **3.** A nonalcoholic, flavored, carbonated beverage : SODA POP. **4.** An alcoholic drink. **5.** *Baseball.* A pop fly. —*adv.* **1.** With a popping sound. **2.** Suddenly or unexpectedly. **—pop the question.** *Informal.* To propose marriage.

**pop²** (pŏp) *n.* [Short for *poppa,* var. of PAPA.] FATHER 1.

**pop³** (pŏp) *adj.* [Short for POPULAR.] *Informal.* **1.** Of, relating to, or specializing in popular music <a *pop* hit> **2.** Of or pertaining to culture produced or influenced by the mass media <*pop* psychology> **3.** Suggestive of pop art.

**pop art** *n.* A form of art that adapts techniques of commercial art, as comic strips, to depict objects of everyday life. **—pop artist** *n.*

**pop·corn** (pŏp'kôrn') *n.* [Contraction of *popped corn.*] **1.** A variety of corn, *Zea mays everta,* with hard kernels that burst to form white, irregularly shaped puffs when heated. **2.** The popped, edible kernels of popcorn.

**popcorn flower** *n.* A plant of the genus *Plagiobothrys* of northwestern North America, with small white flower clusters.

**pope** (pōp) *n.* [ME < OE *papa* < LLat., title of bishops < LGk. *pappas,* father.] **1.** *often* **Pope.** The bishop of Rome and head of the Roman Catholic Church, acting by apostolic succession from Saint Peter as vicar of Christ on earth. **2.** A priest of the Eastern Orthodox Church. **3.** One regarded as having unquestioned authority.

**pope·dom** (pōp'dəm) *n.* The office, jurisdiction, or tenure of a pope : PAPACY.

**pope's nose** *n. Informal.* The rump of a cooked fowl.

**pop·eyed** (pŏp'īd') *adj.* **1.** Having bulging eyes. **2.** Amazed.

**pop fly** *n. Baseball.* A short high fly ball.

**pop·gun** (pŏp'gŭn') *n.* A toy gun that usu. shoots corks and makes a popping noise.

**pop·in·jay** (pŏp'ĭn-jā') *n.* [ME *papenjay,* parrot < OFr. *papegai* < Ar. *babaghā.*] A vain, garrulous person.

**pop·lar** (pŏp'lər) *n.* [ME *poplere* < OFr. *poplier* < Lat. *populus.*] **1. a.** A fast-growing deciduous tree of the genus *Populus.* **b.** The wood of the poplar. **2.** The tulip tree.

**pop·lin** (pŏp'lĭn) *n.* [Fr. *papeline* < Ital. *papalina,* fem. of *papalino,* papal < Med. Lat. *papalis* < LLat. *papa,* pope. —see POPE.] A ribbed fabric of silk, rayon, wool, or cotton, used in making clothing and upholstery.

**pop·lit·e·al** (pŏp-lĭt'ē-əl, pŏp'lĭ-tē'əl) *adj.* [< NLat. *popliteus* < Lat. *poples,* ham of the knee.] Of or relating to the part of the leg behind the knee joint.

**pop·o·ver** (pŏp'ō'vər) *n.* A light, puffy, hollow muffin made with eggs, milk, and flour.

**pop·pa** (pä'pə) *n. var. of* PAPA.

**pop·per** (pŏp'ər) *n.* **1.** One that pops. **2.** A pan or container for making popcorn. **3.** *Slang.* An ampule of amyl nitrate used illicitly as a stimulant.

**pop·pet** (pŏp'ĭt) *n.* [ME *popet,* small child, ult. < Lat. *pupa,* doll.] **1.** A poppet valve. **2.** *Naut.* **a.** A wooden strip on a boat's gunwale that forms or supports the oarlocks. **b.** One of the beams of a launching cradle supporting a ship's hull. **3.** *Chiefly Brit.* A darling : dear.

**poppet valve** *n.* An intake or exhaust valve operated by springs and cams that plugs and unplugs its opening by axial motion.

**pop·ple¹** (pŏp'əl) *vi.* **-pled, -pling, -ples.** [ME *poplen.*] To move in a tossing, bubbling, or rippling manner, as choppy water. —*n.* **1.** Choppy water. **2.** The sound made by boiling liquid.

**pop·ple²** (pŏp'əl) *n.* [ME *popul* < OE < Lat. *populus.*] *Informal.* A poplar.

**pop·py** (pŏp'ē) *n., pl.* **-pies.** [ME *popi* < OE *popig* < Lat. *papaver.*] **1.** A plant of the genus *Papaver* of temperate regions, with usu. large red, orange, or white flowers and a milky white juice. **2.** Any of several plants related to the poppy, as the California poppy. **3.** The narcotic extracted from the opium poppy. **4.** A vivid red to reddish orange.

**pop·py·cock** (pŏp'ē-kŏk') *n.* [Du. dial. *pappekak* : *pap,* pap (< MDu. *pappe*) + *kak,* dung < *kakken,* to defecate < Lat. *cacare.*] Nonsense.

**Pop·si·cle** (pŏp'sĭ-kəl, -sĭk'əl). A trademark for colored, flavored ice molded into a rectangle with two flat sticks for handles.

**pop-top** (pŏp'tŏp') *adj.* Designating or made with a tab that can be pulled up or off to make an opening. **—pop'-top'** *n.*

**pop·u·lace** (pŏp'yə-lĭs) *n.* [Fr. < Ital. *popolaccio,* rabble < *popolo,* the people < Lat. *populus.*] **1.** The common people : MASSES. **2.** A population.

**pop·u·lar** (pŏp'yə-lər) *adj.* [Lat. *popularis,* of the people < *populus,* the people.] **1.** Widely liked or appreciated : FAVORITE <a *popular* song><a *popular* vacation spot> **2.** Liked and sought after for company by others. **3.** Of, representing, or carried on by the people at large <*popular* democracy> **4.** Suitable for or reflecting the taste and intelligence of the people at large <*popular* magazines> **5.** Accepted by or prevalent among the people in general <a *popular* misconception> **6.** Suited to or within the means of ordinary people <meals at *popular* prices> **7.** Originating among the people <*popular* fables> **—pop'u·lar'i·ty** (-lăr'ĭ-tē) *n.* **—pop'u·lar·ly** *adv.*

**popular front** *n.* A political coalition of a kind formed in European countries during the 1930's as an alliance of democratic and revolutionary parties in the struggle against reaction and fascism.

**pop·u·lar·ize** (pŏp′yə-lə-rīz′) *vt.* **-ized, -iz·ing, -iz·es.** To make popular, esp. to make readily intelligible to the layperson <*popularize* physics> **—pop′u·lar·i·za′tion** *n.* **—pop′u·lar·iz′er** *n.*

**pop·u·late** (pŏp′yə-lāt′) *vt.* **-lat·ed, -lat·ing, -lates.** [Med. Lat. *populare, populat-* < Lat. *populus,* the people.] **1.** To supply with inhabitants, as by colonization : PEOPLE. **2.** To inhabit or become inhabitants of.

**pop·u·la·tion** (pŏp′yə-lā′shən) *n.* [LLat. *populatio* < Lat. *populus,* the people.] **1. a.** All of the people inhabiting a specified area. **b.** The total number of such people. **2.** The total number of inhabitants of a particular group, class, or race in a given area. **3.** The act or process of furnishing with inhabitants. **4.** *Ecol.* All the organisms that make up a specific group or occur in a specified habitat. **5.** The set of individuals, items, or data from which a statistical sample is taken.

**population explosion** *n.* Geometric expansion of a biological population, esp. unrestrained growth in human population resulting from a decrease in infant mortality and an increase in longevity.

**Pop·u·lism** (pŏp′yə-lĭz′əm) *n.* **1.** The philosophy of the Populist Party. **2. populism.** A political philosophy directed to the needs of the common people and advocating a more equitable distribution of wealth and power.

**Pop·u·list** (pŏp′yə-lĭst) *n.* **1.** A member or supporter of the Populist Party. **2. populist.** An advocate of populism. *—adj.* **1.** Of or relating to the Populist Party. **2. populist.** Of, relating to, or typical of populism or its advocates.

**Populist Party** *n.* An American political party that sought to represent the interests of farmers and laborers in the 1890's, advocating increased currency issue, free coinage of gold and silver, public ownership of railroads, and a graduated federal income tax.

**pop·u·lous** (pŏp′yə-ləs) *adj.* [ME *populus* < Lat. *populosus* < *populus,* the people.] **1.** Having many people or inhabitants. **2. a.** Numerous. **b.** Crowded. **—pop′u·lous·ly** *adv.* **—pop′u·lous·ness** *n.*

**pop-up** (pŏp′ŭp′) *n. Baseball.* A pop fly.

**pop wine** *n.* A sweet, often fruit-flavored, inexpensive wine.

**por·bea·gle** (pôr′bē′gəl) *n.* [Cornish *porghbugel.*] A small shark, *Lamna nasus* of temperate Atlantic waters, with a pointed nose.

**porbeagle**
*Up to 10 feet long*

**por·ce·lain** (pôr′sə-lĭn, pōr′-) *n.* [OFr. *porcelaine,* cowry shell, porcelain < OItal. *porcellana* < *porcello,* little pig, vulva < Lat. *porcellus,* dim. of *porcus,* pig, vulva.] **1.** A hard, white, translucent ceramic made by firing a pure clay and then glazing with variously colored fusible materials : CHINA. **2.** An object made of porcelain. **—por′ce·la′ne·ous** (-lā′nē-əs) *adj.*

**porcelain enamel** *n.* A silicate glass fired on metal.

**porch** (pôrch, pōrch) *n.* [ME *porche* < OFr. < Lat. *porticus,* portico < *porta,* gate.] **1.** A covered platform, usu. with a separate roof, at an entrance to a house. **2.** An open or enclosed gallery or room attached to the outside of a structure. **3.** *Obs.* A portico or covered walk.

**por·cine** (pôr′sīn′) *adj.* [Lat. *porcinus* < *porcus,* pig.] Of or similar to swine or a pig.

**por·cu·pine** (pôr′kyə-pīn′) *n.* [ME *porkepin* < OFr. *porc espin,* spiny pig.] Any of various rodents, including members of the Old World genus *Hystrix,* the New World genus *Erethizon,* and related genera, that are covered with long, sharp quills or spines.

**porcupine fish** *n.* A spiny tropical marine fish, *Diodon hystrix,* capable of inflating itself when attacked.

**pore¹** (pôr, pōr) *vi.* **pored, por·ing, pores.** [ME *pouren.*] **1.** To gaze steadily or earnestly. **2.** To read or study carefully and attentively <*poring* over the map> **3.** To meditate deeply : PONDER.

**pore²** (pôr, pōr) *n.* [ME < OFr. < Lat. *porus* < Gk. *poros.*] **1.** A minute orifice, as one in the skin of an animal, serving as an outlet for perspiration, or in a plant stem or leaf, serving as a means of absorption and transpiration. **2.** A minute surface opening or passageway, as in a rock.

**pore fungus** *n.* A fungus having a crustlike fruiting body with a porous or pitted surface.

**por·gy** (pôr′gē) *n., pl.* **porgy** or **-gies.** [Sp. *pargo* < Lat. *pagrus,* a kind of fish < Gk. *pagros.*] **1.** A deep-bodied marine fish of the family Sparidae. **2.** A fish related to or resembling the porgy.

**po·rif·er·an** (pə-rĭf′ər-ən) *n.* [NLat. *Porifera,* phylum name : Lat. *porus,* pore (< Gk. *poros*) + Lat. *ferre,* to bear.] SPONGE 1a. **—po·rif′er·al, po·rif′er·an** *adj.*

**po·rif·er·ous** (pə-rĭf′ər-əs) *adj.* [Lat. *porus,* pore (< Gk. *poros*) + -FEROUS.] **1.** Having pores. **2.** *Zool.* Of or pertaining to the phylum Porifera.

**pork** (pôrk, pōrk) *n.* [ME < OFr. *porc,* pig < Lat. *porcus.*] **1.** The flesh of swine used as food. **2.** *Slang.* Government funds, appointments, or other favors acquired by a representative for his or her constituency as political patronage.

**pork barrel** *n. Slang.* A government project or appropriation aiding a specific legislator's constituency.

**pork·er** (pôr′kər, pōr′-) *n.* A fattened young pig.

**pork·pie** *also* **pork pie** (pôrk′pī′, pōrk′-) *n.* **1.** A thick-crusted pie filled with chopped pork. **2.** A man's hat with a low, flat crown and a snap brim.

**por·ky** (pôr′kē) *n., pl.* **-kies.** [Shortening and alteration of PORCUPINE.] *Informal.* A porcupine. *—adj.* **-ki·er, -ki·est.** Fat : pudgy.

**por·no** (pôr′nō) *also* **porn** (pôrn) *n. Slang.* **1.** Pornography. **2.** Pornographic material, as a picture, movie, or book.

**por·nog·ra·phy** (pôr-nŏg′rə-fē) *n.* [< Gk. *pornographos,* writing about prostitutes : *pornē,* prostitute + *graphein,* to write.] The presentation of sexually explicit behavior, as in a photograph, intended to arouse sexual excitement. **—por·nog′ra·pher** *n.* **—por′no·graph′ic** (pôr′nə-grăf′ĭk) *adj.*

**po·ro·mer·ic** (pôr′ə-mĕr′ĭk, pōr′-) *n.* [Gk. *poros,* pore + (POLY)MERIC.] Any of several tough, porous substitutes for leather.

**po·ros·i·ty** (pə-rŏs′ĭ-tē, pô-) *n., pl.* **-ties.** [Med. Lat. *porositas* < *porosus,* porous.] **1.** The quality, state, or degree of being porous. **2.** A porous structure or part.

**po·rous** (pôr′əs, pōr′-) *adj.* [ME < Med. Lat. *porosus* < Lat. *porus,* pore < Gk. *poros.*] **1.** Having or full of pores. **2.** Allowing the passage of gas or liquid through pores or interstices. **—po′rous·ly** *adv.* **—po′rous·ness** *n.*

**por·phyr·i·a** (pôr-fîr′ē-ə) *n.* [NLat. < PORPHYRIN.] A hereditary pathological disorder of porphyrin metabolism marked by photosensitivity and the excretion of porphyrins in the urine.

**por·phy·rin** (pôr′fə-rĭn) *n.* [< Gk. *porphura,* purple.] Any of various nitrogen-containing organic compounds occurring universally in protoplasm and providing the foundation structure for hemoglobin, chlorophyll, and certain enzymes.

**por·phy·rit·ic** (pôr′fə-rĭt′ĭk) *also* **por·phy·rit·i·cal** (-ĭ-kəl) *adj.* **1.** Of or containing porphyry. **2.** Having relatively large isolated crystals in a mass of fine texture.

**por·phy·roid** (pôr′fə-roid′) *n.* Metamorphic rock having porphyritic texture.

**por·phy·ry** (pôr′fə-rē) *n., pl.* **-ries.** [ME *porfurie* < Med. Lat. *porphyrium* < Lat. *porphyrites* < Gk. *porphuritēs* < *porphura,* purple.] Rock having relatively large conspicuous crystals, esp. feldspar, in a fine-grained igneous matrix.

**por·poise** (pôr′pəs) *n., pl.* **porpoise** or **-pois·es.** [ME *porpoys* < OFr. *porpois* < Med. Lat. *porcopiscis* : Lat. *porcus,* pig + Lat. *piscis,* fish.] **1.** A gregarious marine mammal of the genus *Phocaena* or related genera, with a blunt snout and a triangular dorsal fin. **2.** An aquatic mammal, as the dolphin, related to the porpoise.

**por·ridge** (pôr′ĭj, pŏr′-) *n.* [Alteration of POTTAGE.] Cereal, as oatmeal, boiled until thick and usu. eaten with milk.

**por·rin·ger** (pôr′ĭn-jər, pŏr′-) *n.* [Alteration of ME *potinger* < OFr. *potager* < *potage,* soup < *pot,* pot.] A shallow bowl or cup with a handle.

**port¹** (pôrt, pōrt) *n.* [ME < OE < Lat. *portus.*] **1. a.** A town or city having a harbor for ships taking on or discharging cargo. **b.** A place on a waterway that provides a harbor for a nearby city. **c.** The harbor or waterfront district of a city. **2.** A place of shelter or anchorage : HAVEN. **3.** A port of entry. **4.** *Computer Sci.* An entrance to or exit for a data network.

**port²** (pôrt, pōrt) *n.* [Orig. unknown.] The left-hand side of a ship or aircraft facing forward. *—adj.* Of, relating to, or situated on the port. *—vt.* **port·ed, port·ing, ports.** To turn or shift (the helm of a vessel) to the left.

**port³** (pôrt, pōrt) *n.* [ME < OFr. *porte,* gate < Lat. *porta.*] **1.** *Naut.* **a.** A porthole. **b.** A covering for a porthole. **2.** An opening, as in a cylinder or valve face, for the passage of fluid or steam. **3.** A hole in an armored vehicle or fortified structure through which weapons may be fired. **4.** *Scot.* A gateway or portal, as to a town.

**port⁴** (pôrt, pōrt) *n.* [After *Oporto,* Portugal.] A rich sweet fortified red wine.

**port⁵** (pôrt, pōrt) *vt.* **port·ed, port·ing, ports.** [OFr. *porter,* to carry < Lat. *portare.*] To carry (a weapon) diagonally across the body, with the muzzle or blade near the left shoulder. *—n.* **1.** The position of a ported weapon. **2.** Carriage or bearing : POSTURE.

**port·a·ble** (pôr′tə-bəl, pōr′-) *adj.* [ME < OFr. < LLat. *portabilis* < Lat. *portare,* to carry.] **1.** Capable of being carried. **2.** Easily carried or moved. **3.** *Archaic.* Endurable : bearable. *—n.* Something portable, as a light typewriter or television. **—port′a·bil′i·ty, port′a·ble·ness** *n.* **—port′a·bly** *adv.*

**port·age** (pôr′tĭj, pōr′-, pôr-täzh′) *n.* [ME < OFr. < *porter*, to carry < Lat. *portare*.] **1.** The carrying of boats and supplies overland between two waterways or around an obstacle, as a waterfall. **2.** A track or route used for portage. —*v.* **-aged, -ag·ing, -ag·es.** —*vt.* To transport by portage : PACK <"They had illegally *portaged* back to Canada a small fortune in beaver skins" —Irving Stone> —*vi.* To make a portage.

**por·tal** (pôr′tl, pōr′-) *n.* [ME < OFr. < Med. Lat. *portale*, city gate < *portalis*, of a gate < Lat. *porta*, gate.] **1.** A large and imposing doorway, entrance, or gate. **2.** An entrance or means of entrance <a *portal* of enlightenment> **3.** The portal vein. —*adj.* Of or relating to the portal vein.

**por·tal-to-por·tal** (pôr′tl-tə-pôr′tl, pōr′tl-tə-pōr′tl) *adj.* Of or based on the total time spent by a worker on the employer's property regardless of actual work time <*portal-to-portal* pay>

**portal vein** *n. Anat.* A vein that conducts blood from the digestive organs, spleen, pancreas, and gallbladder to the liver.

**por·ta·men·to** (pôr′tə-mĕn′tō, pōr′-) *n., pl.* **-ti** (-tē) [Ital. < *portare*, to carry < Lat.] *Mus.* A smooth constant glide in passing from one tone to another, esp. with the voice or with a bowed stringed instrument.

**por·ta·pak** *or* **por·ta·pack** (pôr′-tə-păk′, pōr′-) *n.* [Blend of PORTABLE and PACK.] A small videotape recorder and camera combined in a portable unit.

**por·ta·tive** (pôr′tə-tĭv, pōr′-) *adj.* [ME *portatif* < OFr. < Lat. *portare*, to carry.] **1.** Portable. **2.** Capable of carrying.

**port·cul·lis** (pôrt-kŭl′ĭs, pōrt-) *n.* [ME *portculis* < OFr. *porte coleïce*, sliding door.] A sliding wood or iron grille suspended in the gateway of a fortified place in such a way that it can be quickly lowered in case of attack.

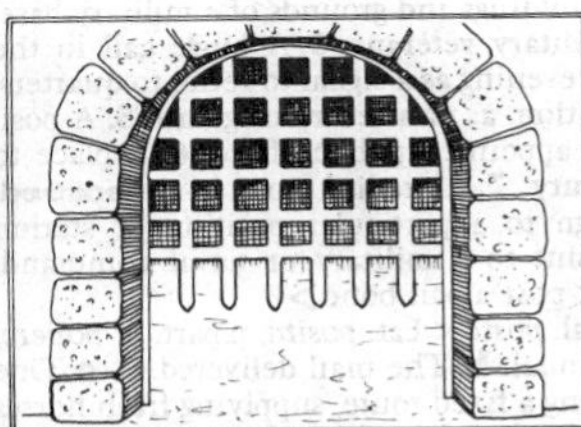

**portcullis**

**Port du Sa·lut** (pôr′ dü să-lo͞o′) *n. var. of* PORT SALUT.

**Porte** (pôrt, pōrt) *n.* [Fr., short for *la Sublime Porte*, the High Gate.] The government of the Ottoman Empire.

**porte-co·chère** *or* **porte-co·chere** (pôrt′kō-shâr′, pōrt′-) *n.* [Fr. *porte cochère*, coach door.] **1.** A carriage entrance leading into the courtyard of a town house. **2.** A porch roof projecting over a driveway at the entrance to a building.

**por·tend** (pôr-tĕnd′, pōr-) *vt.* **-tend·ed, -tend·ing, -tends.** [ME *portenden* < Lat. *portendere*.] **1.** To serve as a sign or warning of : PRESAGE. **2.** To suggest or indicate <rumors that *portend* unrest>

**por·tent** (pôr′tĕnt′, pōr′-) *n.* [Lat. *portentum* < *portendere*, to portend.] **1.** An indication of something important, calamitous, or evil about to occur : OMEN. **2.** Prophetic or threatening significance. **3.** Something marvelous : PRODIGY.

**por·ten·tous** (pôr-tĕn′təs, pōr-) *adj.* **1.** Like or constituting a portent : FOREBODING. **2.** Arousing wonder and awe. **3.** Characterized by pompousness. **—por·ten′tous·ly** *adv.* **—por·ten′tous·ness** *n.*

**por·ter¹** (pôr′tər, pōr′-) *n.* [ME *portour* < OFr. *porteur* < LLat. *portator* < Lat. *portare*, to carry.] **1.** One employed to carry travelers' luggage. **2.** One employed to wait on passengers in a railway parlor or sleeping car. **3.** One employed to do routine cleaning, as in an office or institution.

**por·ter²** (pôr′tər, pōr′-) *n.* [ME < OFr. *portier* < LLat. *portarius* < Lat. *porta*, gate.] *Chiefly Brit.* A gatekeeper : doorman.

**por·ter³** (pôr′tər, pōr′-) *n.* [Short for *porter's beer*.] A dark beer resembling light stout, made from browned or charred malt.

**por·ter·age** (pôr′tər-ĭj, pōr′-) *n.* **1.** The carrying of parcels or goods as done by porters. **2.** The charge for such service.

**por·ter·ess** (pôr′tər-ĭs, pōr′-) *n. var. of* PORTRESS.

**por·ter·house** (pôr′tər-hous′, pōr′-) *n.* **1.** A 19th-cent. American alehouse or chophouse. **2.** A cut of beef from the thick end of the short loin, having a T-bone and a sizable piece of tenderloin.

**porterhouse steak** *n.* PORTERHOUSE 2.

**port·fo·li·o** (pôrt-fō′lē-ō′, pōrt-) *n., pl.* **-os.** [Ital. *portafoglio* : *portare*, to carry (< Lat.) + *foglio*, sheet < Lat. *folium*, leaf.] **1. a.** A portable case for holding papers, drawings, or photographs. **b.** The materials included in such a case, esp. when representative of one's work <a designer's *portfolio*> **2.** The office or post of a cabinet member or minister of state. **3.** A list of investments, securities, and commercial paper owned, as by a bank or individual investor.

**port·hole** (pôrt′hōl′, pōrt′-) *n.* **1.** A small, usu. circular window in a ship's side. **2.** An embrasure.

**por·ti·co** (pôr′tĭ-kō′, pōr′-) *n., pl.* **-coes** or **-cos.** [Ital. < Lat. *porticus* < *porta*, gate.] A walkway or porch with a roof supported by columns, often at the entrance of a building. **—por′ti·coed′** *adj.*

**por·tière** *or* **por·tiere** (pôr-tyâr′, pōr-) *n.* [Fr. < OFr., fem. of *portier*, porter < LLat. *portarius* < Lat. *porta*, gate.] A heavy curtain hung across a doorway.

**por·tion** (pôr′shən, pōr′-) *n.* [ME < OFr. < Lat. *portio*.] **1.** A part of a whole. **2.** A part separated from a whole. **3.** A part allotted to a person or group, as: **a.** A helping of food. **b.** The part of an estate received by an heir. **c.** A woman's dowry. **4.** One's destiny or fate. —*vt.* **-tioned, -tion·ing, -tions. 1.** To divide into parts or shares for distribution. **2.** To provide with a share, inheritance, or dowry. **—por′tion·a·ble** *adj.* **—por′tion·er** *n.* **—por′tion·less** *adj.*

**Port·land cement** (pôrt′lənd, pōrt′-) *n.* [After *Portland*, England, from its resemblance to limestone quarried there.] A hydraulic cement made by heating a mixture of limestone and clay, containing oxides of calcium, aluminum, iron, and silicon, in a kiln and pulverizing the resultant clinker.

**port·ly** (pôrt′lē, pōrt′-) *adj.* **-li·er, -li·est.** [< PORT⁵.] **1.** Corpulent : stout. **2.** *Archaic.* Stately : imposing. **—port′li·ness** *n.*

**port·man·teau** (pôrt-măn′tō, pōrt-, pôrt′măn-tō′, pōrt′-) *n., pl.* **-teaus** or **-teaux** (-tōz) [Fr. *portemanteau* < OFr. : *porter*, to carry (< Lat. *portare*) + *manteau*, cloak < Lat. *mantellum*.] *Chiefly Brit.* A large leather suitcase that opens into two hinged compartments.

**portmanteau word** *n.* A word formed by merging the sounds and meanings of two different words, e.g., *chortle*, from *chuckle* and *snort*.

**port of call** *n.* A port where ships dock in the course of voyages to load or unload cargo, obtain supplies, or undergo repairs.

**port of entry** *n.* A place where travelers or goods may enter or leave a country under official supervision.

**por·trait** (pôr′trĭt, -trāt′, pōr′-) *n.* [Fr. < OFr. < *portraire*, to portray.] **1.** A likeness of a person, as a painting or photograph, esp. one showing the face. **2.** A verbal picture or description, esp. of a person.

**por·trait·ist** (pôr′trə-tĭst, pōr-) *n.* One who makes portraits, esp. a painter or photographer.

**por·trai·ture** (pôr′trĭ-cho͝or′, pōr′-) *n.* **1.** The art or practice of making portraits. **2.** A portrait. **3.** A group of portraits.

**por·tray** (pôr-trā′, pōr-) *vt.* **-trayed, -tray·ing, -trays.** [ME *portraien* < OFr. *portraire* < Lat. *protrahere*, to reveal : *pro-*, forth + *trahere*, to draw.] **1.** To make a picture of. **2.** To depict or describe in words. **3.** To represent dramatically, as on the stage. **—por·tray′a·ble** *adj.* **—por·tray′al** (-trā′əl, pōr-) *n.* **—por·tray′er** *n.*

**por·tress** (pôr′trĭs, pōr′-) *also* **por·ter·ess** (pôr′tər-ĭs, pōr′-) *n.* **1.** A woman porter or doorkeeper, esp. in a convent. **2.** A charwoman.

**Port Sa·lut** (pôr′ să-lo͞o) *also* **Port du Sa·lut** (pôr′ dü să-lo͞o′) *n.* [After *Port du Salut*, Trappist abbey in France.] A semihard fermented cheese made orig. by Trappist monks in France.

**Por·tu·guese** (pôr′chə-gēz′, -gēs′, pōr′-) *adj.* Of or relating to Portugal, its people, or their language. —*n., pl.* **Portuguese. 1. a.** A native or resident of Portugal. **b.** One of Portuguese descent. **2.** The Romance language of Portugal and Brazil.

**Portuguese man-of-war** *n.* A complex colonial organism of the genus *Physalia* of warm seas, with a bluish bladderlike float from which are suspended numerous long stinging tentacles capable of inflicting severe injury.

**por·tu·lac·a** (pôr′chə-lăk′ə, pōr′-) *n.* [NLat. *Portulaca*, genus name < Lat. *portulaca*, purslane < *portula*, dim. of *porta*, gate.] A plant of the genus *Portulaca*, bearing fleshy stems and leaves, esp. *P. grandiflora*, cultivated for its variously colored flowers that open only in sunlight.

**pose¹** (pōz) *v.* **posed, pos·ing, pos·es.** [ME *posen* < OFr. *poser* < LLat. *pausare*, to rest < Lat. *pausa*, pause. —see PAUSE.] —*vi.* **1.** To assume or hold a position or posture, as in sitting for a portrait. **2.** To affect a particular mental attitude. **3.** To pretend to be other than what one is. —*vt.* **1.** To place (e.g., a model) in a specific position. **2.** To advance or put forward <*pose* a problem> —*n.* **1.** A bodily posture or position, esp. one assumed for an artist or photographer. **2.** An affected physical or mental attitude.

**pose²** (pōz) *vt.* **posed, pos·ing, pos·es.** [ME *apposen*, alteration of *opposen* < OFr. *opposer*, to oppose. —see OPPOSE.] To puzzle or confuse with a difficult question or problem.

**Po·sei·don** (pō-sīd′n) *n.* [Lat. < Gk. *Poseidōn*.] *Gk. Myth.* The god of the waters, earthquakes, and horses.

**pos·er¹** (pō′zər) *n.* One who poses.

**pos·er²** (pō′zər) *n.* A baffling question or problem.

**po·seur** (pō-zœr′) *n.* [Fr. < OFr. *poser*, to pose. —see POSE¹.] One who assumes an attitude, character, or manner to impress others.

**posh** (pŏsh) *adj.* [Orig. unknown.] Fashionable and expensive.

**pos·i·grade** (pŏz′ĭ-grād′) *adj.* [POSI(TIVE) + (RETRO)GRADE.] Of, pertaining to, or being an auxiliary rocket on a spacecraft that provides additional thrust in the direction of the spacecraft's motion.

**pos·it** (pŏz′ĭt) *vt.* **-it·ed, -it·ing, -its.** [< Lat. *positus,* p.part. of *ponere,* to place.] **1.** To place in position. **2.** To present as a fact or assumption : POSTULATE.

**po·si·tion** (pə-zĭsh′ən) *n.* [OFr. < Lat. *positio* < *ponere,* to place.] **1.** A place or location. **2.** The right or appropriate place <The contestants were in *position.*> **3. a.** The way in which one is placed <in an inconspicuous *position*> **b.** The arrangement of bodily parts : POSTURE <a prone *position*> **4.** An advantageous place or location <race cars jockeying for *position*> **5.** A situation as it relates to the surrounding circumstances <not in a *position* to quibble> **6.** An attitude or point of view on a certain question. **7.** Social status. **8.** A post of employment : JOB. **9.** The area for which a particular player is responsible in a sport. **10. a.** The act or process of positing. **b.** The principle or proposition posited. —*vt.* **-tioned, -tion·ing, -tions.** To place in proper position. **—po·si′tion·al** *adj.* **—po·si′tion·er** *n.*

**position paper** *n.* **1.** A detailed policy report that usu. explains, justifies, or recommends a course of action. **2.** An aide-mémoire.

**pos·i·tive** (pŏz′ĭ-tĭv) *adj.* [ME < OFr. *positif* < Lat. *positivus,* formally laid down < *ponere,* to place.] **1.** Marked by or exhibiting certainty, acceptance, or affirmation <a *positive* reply> **2.** Measured or moving in a direction of increase, progress, or forward motion. **3.** Openly or explicitly laid down or expressed <a *positive* claim> **4.** Admitting of no doubt : IRREFUTABLE. **5. a.** Determined or settled in opinion or assertion : CONFIDENT <a *positive* attitude> **b.** Overconfident : dogmatic. **6.** Formally or arbitrarily determined : PRESCRIBED. **7.** Concerned with practical rather than theoretical matters. **8.** Composed of or marked by the presence of distinctive qualities or attributes : REAL. **9.** *Philos.* Of or relating to positivism. **10.** *Informal.* Complete : utter <a *positive* angel> **11.** *Math.* Relating to or designating: **a.** A quantity greater than zero. **b.** The sign (+). **c.** A quantity, number, angle, or direction opposite to another designated as negative. **12.** *Physics.* Relating to or designating electric charge of a sign opposite to that of an electron. **13.** *Med.* Indicating the presence of a particular disease, condition, or organism <a *positive* TB test> **14.** *Biol.* Indicating or marked by response or motion toward the source of a stimulus. **15.** Having the areas of light and dark in their original and normal relationship, as in a photographic print made from a negative. **16.** Of, relating to, or denoting the simple uncompared degree of an adjective or adverb. **17.** Driven by or generating power directly through intermediate machine parts having little or no play. —*n.* **1.** Something positive. **2.** *Philos.* Something perceptible to the senses. **3.** *Math.* A quantity greater than zero. **4.** *Physics.* A positive electric charge. **5.** A photographic image in which the lights and darks appear as they do naturally. **6. a.** The uncompared degree of an adjective or adverb. **b.** A word in this degree. **—pos′i·tive·ly** *adv.* **—pos′i·tive·ness** *n.*

**positive prescription** *n. Law.* PRESCRIPTION 4a.

**pos·i·tiv·ism** (pŏz′ĭ-tĭ-vĭz′əm) *n.* **1. a.** A philosophical doctrine contending that sense perceptions are the only admissible basis of human knowledge and precise thought. **b.** The application of this doctrine in logic, epistemology, and ethics. **2.** The system of Auguste Comte designed to supersede theology and metaphysics and depending on a hierarchy of the sciences, beginning with mathematics and culminating in sociology. **3.** The quality or state of being positive. **—pos′i·tiv·ist** *n.* **—pos′i·tiv·is′tic** *adj.*

**pos·i·tron** (pŏz′ĭ-trŏn′) *n.* [POSI(TIVE) + (ELEC)TRON.] The antiparticle of the electron.

**pos·i·tro·ni·um** (pŏz′ĭ-trō′nē-əm) *n.* [NLat. < POSITRON.] A short-lived association of an electron and a positron bound together in a configuration resembling the hydrogen atom.

**pos·se** (pŏs′ē) *n.* [Short for Med. Lat. *posse comitatus,* power of the county.] **1.** A group of persons deputized by a sheriff to aid in law enforcement. **2.** A search party.

**pos·sess** (pə-zĕs′) *vt.* **-sessed, -sess·ing, -sess·es.** [ME *possessen* < OFr. *possesser* < Lat. *possidere* : *potis,* capable + *sedere,* to sit.] **1.** To have as property : OWN. **2.** To have as a quality, characteristic, or attribute <*possessed* much courage> **3.** To acquire mastery of or have knowledge of <*possess* secret information> **4.** To gain or exert influence over : DOMINATE <Rage *possessed* me.> **5.** To control or maintain in a given condition <*possessed* my equanimity despite the turmoil> **6.** To cause to have, hold, or master something, as property or knowledge. **7.** To cause to be influenced or controlled, as by an idea or emotion. **8.** *Obs.* To gain or seize. **—pos·ses′sor** *n.*

**pos·sessed** (pə-zĕst′) *adj.* **1.** Having as a possession. **2.** Controlled by or as if by a supernatural force : OBSESSED. **3.** Self-possessed.

**pos·ses·sion** (pə-zĕsh′ən) *n.* **1.** The act or fact of possessing. **2.** The state of being possessed. **3.** Something possessed or owned. **4. possessions.** Wealth or property. **5.** *Law.* Actual holding or occupancy with or without rightful ownership. **6.** A territory subject to foreign control. **7.** Self-control. **8.** The state of being dominated by or as if by evil spirits or an obsession.

**pos·ses·sive** (pə-zĕs′ĭv) *adj.* **1.** Of or relating to possession or ownership. **2.** Having or displaying a desire to control or dominate <a *possessive* parent> **3.** Of, relating to, or designating a noun or pronoun case that indicates possession. —*n.* **1.** The possessive case. **2.** A possessive grammatical form or construction. **—pos·ses′sive·ly** *adv.* **—pos·ses′sive·ness** *n.*

**possessive adjective** *n.* A pronominal adjective that expresses possession.

**possessive pronoun** *n.* A pronoun denoting possession and capable of substituting for a noun phrase.

**pos·ses·so·ry** (pə-zĕs′ə-rē) *adj.* **1.** Of, relating to, or having possession. **2.** *Law.* Depending on or arising from possession.

**pos·set** (pŏs′ĭt) *n.* [ME *poshet.*] A spiced drink of hot sweetened milk curdled with ale or wine.

**pos·si·bil·i·ty** (pŏs′ə-bĭl′ĭ-tē) *n., pl.* **-ties. 1.** The state or fact of being possible. **2.** Something possible. **3. possibilities.** Potentially favorable results <This old house has great *possibilities.*>

**pos·si·ble** (pŏs′ə-bəl) *adj.* [ME < OFr. < Lat. *possibilis* < *posse,* to be able.] **1.** Capable of existing, happening, or being true without contradicting proven facts, laws, or circumstances. **2.** Capable of taking place or being done without offense to nature, character, or custom. **3.** Capable of favorable development : POTENTIAL <a *possible* building site> **4.** Of uncertain likelihood. **—pos′si·bly** *adv.*

**pos·sum** (pŏs′əm) *n. var. of* OPOSSUM.

**possum haw** *n.* **1.** A holly, *Ilex decidua* of the southeastern United States, with bright-red fruit. **2.** A shrub, *Viburnum nudum* of the eastern United States, with white flowers and bluish-black fruit.

**post[1]** (pōst) *n.* [ME < OE < Lat. *postis.*] **1.** A stake of material, as wood, set upright into the ground to serve as a marker or support. **2.** A goal post. **3.** The starting gate at a racetrack. —*vt.* **post·ed, post·ing, posts. 1. a.** To put up (an announcement) in a place of public view. **b.** To cover (e.g., a wall) with posters. **2.** To announce by or as if by posters <*post* marital banns> **3.** To put up signs on (property) warning against trespassing. **4.** To denounce publicly <*post* a person as an embezzler> **5.** To publish (a name) on a list.

**post[2]** (pōst) *n.* [Fr. *poste* < OItal. *posto* < VLat. **postum* < Lat. *positum,* p.part. of *ponere,* to place.] **1. a.** A military base where troops are stationed. **b.** The buildings and grounds of a military base. **2.** A local organization of military veterans. **3.** A bugle call in the British Army, sounded in the evening as a signal to retire to quarters. **4.** An assigned position or station, as of a sentry or guard. **5.** A position of employment, esp. an appointed public office. **6.** A place to which one is assigned for duty. **7.** A trading post. —*vt.* **post·ed, post·ing, posts. 1.** To assign to a particular position or station <*post* a guard> **2.** To appoint to a military or naval command. **3.** To put forward : PRESENT <*post* a bail bond>

**post[3]** (pōst) *n.* [Fr. *poste* < Ital. *posta* < Lat. *posita,* p.part. of *ponere,* to place.] **1. a.** A delivery of mail. **b.** The mail delivered. **2. a.** One of a series of relay stations along a fixed route, supplying fresh horses and riders for the delivery of mail on horseback. **b.** A rider on such a route : COURIER. **3.** *Chiefly Brit.* **a.** A governmental system for transporting and delivering the mail. **b.** A post office. —*v.* **post·ed, post·ing, posts.** —*vi.* **1.** To travel in stages or relays. **2.** To travel quickly. **3.** To bob up and down in the saddle in rhythm with a horse's trotting gait. —*vt.* **1.** To send by mail in a system of relays on horseback. **2.** To mail (e.g., a letter). **3.** To inform of the latest news <kept me *posted*> **4. a.** To transfer (an item) to a ledger in bookkeeping. **b.** To make the necessary entries in (a ledger). **5.** To place on a list or in a record. **6.** *Computer Sci.* To enter a unit of information on a record or into a section of computer storage. —*adv.* **1.** By post horse. **2.** By mail. **3.** With great speed : RAPIDLY.

☆ ***syns:*** POST, ENTER, INSERT, RECORD, REGISTER *v. core meaning* : to place on a list or in a record <*posted* the names of the major contributors>

▲ word history: The word *post*³, meaning "mail," is ultimately derived from Latin *ponere,* "to place, put in position." This meaning of the word *post* was a result of the method of delivering mail. In the 16th century horsemen were stationed at designated places along certain roads to ride in relays with royal dispatches and other papers. These couriers were called "posts." As the system of mail delivery expanded during the next two centuries, *post* was applied to a delivery of mail and then to the organization responsible for the entire system of delivering mail.

**post-** *pref.* [Lat. < *post,* behind, after.] **1.** After : later <*post*millenial> **2.** Behind : posterior to <*post*axial>

**post·age** (pō′stĭj) *n.* The charge for mailing an item.

**postage meter** *n.* A machine used to print the correct amount of postage on each piece of mail.

**postage stamp** *n.* A small engraved, usu. adhesive label issued by a government and sold in various denominations to be affixed to items of mail as proof of the payment of postage.

**post·al** (pō′stəl) *adj.* Of or relating to the post office or mail service. —*n.* A postal card. **—post′al·ly** *adv.*

**postal card** *n.* A card printed with a postage stamp, issued and sold by a government, for sending messages at low rates.

**postal order** *n. Chiefly Brit.* A money order.

**postal service** *n.* POST OFFICE 1.

**post·ax·i·al** (pōst-ăk′sē-əl) *adj.* Situated behind an axis of the body, esp. posterior to the ulna of the arm or the fibula of the leg. **—post·ax′i·al·ly** *adv.*

**post·bel·lum** (pōst-bĕl′əm) *adj.* [Lat. *post,* after + Lat. *bellum,* war.] Happening after a war, esp. the American Civil War.
**post·box** *also* **post box** (pōst′bŏks′) *n.* A mailbox.
**post·card** *also* **post card** (pōst′kärd′) *n.* **1.** An unofficial card, usu. bearing a picture on one side, with space for an address, postage stamp, and short message. **2.** A postal card.
**post·ca·va** (pōst-kā′və) *n. Anat.* The inferior vena cava. **—post·ca′val** *adj.*
**post chaise** *n.* A closed, four-wheeled, horse-drawn carriage, once used to transport mail and passengers.
**post·clas·si·cal** (pōst-klăs′ĭ-kəl) *adj.* Of, relating to, or being a time following a classical period, as in art, literature, or culture.
**post·co·lo·ni·al** (pōst′kə-lō′nē-əl) *adj.* Of, pertaining to, or being the time following the establishment of independence in a colony.
**post·date** (pōst-dāt′) *vt.* **-dat·ed, -dat·ing, -dates. 1.** To put a date on (e.g., a check) that is later than the actual date. **2.** To follow in time : occur later than.
**post·di·lu·vi·an** (pōst′dĭ-lo͞o′vē-ən) *also* **post·di·lu·vi·al** (-əl) *adj.* [POST + Lat. *diluvium,* flood. —see DILUVIAL.] Existing or happening after the Biblical Flood. *—n.* One living after the Biblical Flood.
**post·doc·tor·al** (pōst-dŏk′tər-əl) *also* **post·doc·tor·ate** (-ĭt) *adj.* Of, relating to, or engaged in academic study beyond the level of a doctor's degree.
**post·er[1]** (pō′stər) *n.* **1.** A large printed and often illustrated placard, bill, or announcement posted to advertise or publicize something. **2.** One who posts bills or notices.
**post·er[2]** (pō′stər) *n. Archaic.* One who traveled post.
**poster color** *n.* TEMPERA 1.
**poste res·tante** (pōst′ rĕ-stänt′) *n.* [Fr. : *poste,* mail + *restante,* pr.part. of *rester,* to remain.] A notation written on a letter indicating that the letter should be held at the post office until claimed by the addressee.
**pos·te·ri·or** (pŏ-stîr′ē-ər, pō-) *adj.* [Lat., comp. of *posterus,* coming after < *post,* after.] **1.** Situated behind a part or toward the rear of a structure. **2.** Relating to the caudal end of the body in an animal or the dorsal side in humans. **3.** *Bot.* Adjacent to or nearest the main axis or stem. **4.** Coming after in order : FOLLOWING. **5.** Following in time : SUBSEQUENT. *—n. often* **posteriors.** The buttocks. **—pos·te′ri·or·ly** *adv.*
**pos·te·ri·or·i·ty** (pŏ-stîr′ē-ôr′ĭ-tē, -ŏr′-, pō-) *n.* The state of being posterior in location or time.
**pos·ter·i·ty** (pŏ-stĕr′ĭ-tē) *n.* [ME *posterite* < OFr. < Lat. *posteritas* < *posterus,* coming after < *post,* after.] **1.** Future generations. **2.** One's descendants as a group.
**pos·tern** (pō′stərn, pŏs′tərn) *n.* [ME *posterne* < OFr., alteration of *posterle* < LLat. *posterula,* dim. of Lat. *posterus,* behind < *post,* after.] A small rear gate, esp. one in a castle or fort. *—adj.* Located in back or at the side.
**poster paint** *n.* TEMPERA 1.
**Post Exchange.** A trademark for a store on a military base that sells goods to military personnel and their dependents.
**post·ex·il·i·an** (pōst′ĕg-zĭl′ē-ən, -zĭl′yən, -ĕk-sĭl′ē-ən, -sĭl′yən) *also* **post·ex·il·ic** (-ĕg-zĭl′ĭk, -ĕk-sĭl′ĭk) *adj.* Of or relating to the period of Jewish history following the Babylonian captivity (after 586 B.C.).
**post·fix** (pōst-fĭks′) *vt.* **-fixed, -fix·ing, -fix·es.** To suffix. *—n.* (pōst′fĭks′). A suffix. **—post·fix′al, post·fix′i·al** *adj.*
**post-free** (pōst′frē′) *adj. Chiefly Brit.* Postpaid.
**post·gan·gli·on·ic** (pōst′găng-glē-ŏn′ĭk) *adj.* Located posterior or distal to a ganglion.
**post·gla·cial** (pōst-glā′shəl) *adj.* Relating to or happening during the time following a glacial period.
**post·grad·u·ate** (pōst-grăj′o͞o-ĭt, -āt′) *adj.* Of, relating to, or pursuing advanced study after graduation from high school or college. *—n.* One engaged in postgraduate study.
**post·haste** (pōst′hāst′) *adv.* [From the phrase *post, haste,* a direction on letters.] With great speed : RAPIDLY. *—n. Archaic.* Great speed.
**post·hole** (pōst′hōl′) *n.* A hole dug for a fence post.
**post·hu·mous** (pŏs′chə-məs) *adj.* [Lat. *posthumus,* alteration of *postumus,* superl. of *posterus,* coming after < *post,* after.] **1.** Taking place or continuing after one's death <*posthumous* acclaim> **2.** Published after the author's death. **3.** Born after the death of the father. **—post′hu·mous·ly** *adv.* **—post′hu·mous·ness** *n.*
**post·hyp·not·ic** (pōst′hĭp-nŏt′ĭk) *adj.* Of, pertaining to, or taking effect during the waking state after a hypnotic trance <*posthypnotic* suggestion>
**pos·tiche** (pô-stēsh′, pŏ-) *n.* [Fr., out of place < Ital. *posticcio,* counterfeit < *posto,* added < Lat. *positus,* p.part. of *ponere,* to place.] **1.** Something false : SHAM. **2.** A small hair piece : TOUPEE.
**pos·til·ion** *also* **pos·til·lion** (pō-stĭl′yən, pŏ-) *n.* [Fr. *postillion* < Ital. *postiglione* < *posta,* mail coach. —see POST[3].] One who rides the left-hand lead horse to guide the horses drawing a coach.

**post·im·pres·sion·ism** (pōst′ĭm-prĕsh′ə-nĭz′əm) *n.* A 19th-cent. French school of painting that rejected the objective naturalism of impressionism in favor of more formal or subjective styles. **—post′im·pres′sion·ist** *n.* **—post′im·pres′sion·is′tic** *adj.*
**post·ir·ra·di·a·tion** (pōst′ĭ-rā′dē-ā′shən) *adj.* Taking place after irradiation.
**post·lude** (pōst′lo͞od′) *n.* [POST- + (PRE)LUDE.] **1. a.** An organ voluntary performed at the end of a church service. **b.** A concluding piece of music. **2.** A final chapter or phase.
**post·man** (pōst′mən) *n.* A letter carrier.
**post·mark** (pōst′märk′) *n.* An official mark printed over the stamp on a piece of mail, esp. one that cancels the stamp and records the date and place of mailing. *—vt.* **-marked, -mark·ing, -marks.** To stamp with a postmark.
**post·mas·ter** (pōst′măs′tər) *n.* A government official in charge of the operations of a local post office. **—post′mas′ter·ship′** *n.*
**postmaster general** *n., pl.* **postmasters general.** The executive head of a national postal service.
**post·me·rid·i·an** (pōst′mə-rĭd′ē-ən) *adj.* [Lat. *postmeridianus* : *post,* after + *meridianus,* of midday. —see MERIDIAN.] Of, relating to, or occurring in the afternoon.
**post me·rid·i·em** (pōst′ mə-rĭd′ē-əm) *adj.* [Lat., after midday.] After noon. —Used chiefly in the abbreviated form to specify the hour <7:30 P.M.>
**post·mil·le·nar·i·an** (pōst′mĭl-ə-nâr′ē-ən) *adj.* Of or relating to postmillennialism. *—n.* One who believes in postmillennialism.
**post·mil·le·nar·i·an·ism** (pōst′mĭl-ə-nâr′ē-ə-nĭz′əm) *n.* Postmillennialism.
**post·mil·len·ni·al** (pōst′mə-lĕn′ē-əl) *also* **post·mil·len·ni·an** (-ən) *adj.* Existing or occurring after the millennium.
**post·mil·len·ni·al·ism** (pōst′mə-lĕn′ē-ə-lĭz′əm) *n.* The doctrine that the Second Coming of Christ will follow the millennium. **—post′mil·len′ni·al·ist** *n.*
**post·mis·tress** (pōst′mĭs′trĭs) *n.* A woman government official in charge of the operations of a local post office.
**post·mor·tem** (pōst-môr′təm) *adj.* [Lat. *post mortem,* after death.] **1.** Happening, done, or obtained after death. **2.** Of or relating to a postmortem examination. *—n.* **1.** An autopsy. **2.** *Informal.* An analysis or review of a completed event.
**postmortem examination** *n.* An autopsy.
**post·na·sal** (pōst-nā′zəl) *adj.* Posterior to the nose.
**postnasal drip** *n.* Chronic secretion of mucus from the posterior nasal cavities, resulting in congestion and soreness of the throat.
**post·na·tal** (pōst-nāt′l) *adj.* Of or taking place during the period immediately after birth. **—post·na′tal·ly** *adv.*
**post·nup·tial** (pōst-nŭp′shəl, -chəl) *adj.* Occurring after marriage. **—post·nup′tial·ly** *adv.*
**post-o·bit** (pōst-ō′bĭt) *adj.* [Lat. *post obitum,* after death.] Occurring or taking effect after one's death. *—n.* A bond given by a borrower promising to repay a debt after the death of a specified person from whose estate he or she expects to inherit.
**post-obit bond** *n.* A post-obit.
**post office** *n.* **1.** The public department responsible for the transportation and delivery of the mails. **2.** A local office where mail is received, sorted, and readied for delivery and where postal materials, as stamps and money orders, are sold.
**post·op·er·a·tive** (pōst-ŏp′ər-ə-tĭv, -ŏp′rə-tĭv, -ŏp′ə-rā′tĭv) *adj.* Occurring or performed after surgery. **—post·op′er·a·tive·ly** *adv.*
**post·or·bi·tal** (pōst-ôr′bĭt-l) *adj.* Located behind the eye socket.
**post·paid** (pōst′pād′) *adj.* Having the postage paid in advance.
**post·par·tum** (pōst-pär′təm) *adj.* [Lat. *post partum,* after birth.] Of or taking place in the period shortly after childbirth.
**post·pone** (pōst-pōn′, pōs-pōn′) *vt.* **-poned, -pon·ing, -pones.** [Lat. *postponere* : *post,* after + *ponere,* to put.] **1.** To put off until a future time : DELAY. **2.** To place after in importance : SUBORDINATE. **—post·pon′a·ble** *adj.* **—post·pone′ment** *n.* **—post·pon′er** *n.*
**post·po·si·tion** (pōst′pə-zĭsh′ən) *n.* [Fr. < OFr. *postposer,* to put after < Lat. *postponere.* —see POSTPONE.] **1.** Placement of a word or suffixed element after the word to which it is grammatically related. **2.** A word or suffixed element placed postpositionally, as a preposition placed after its object. **—post′po·si′tion·al** *adj.* **—post′po·si′tion·al·ly** *adv.*
**post·pos·i·tive** (pōst-pŏz′ĭ-tĭv) *adj.* [LLat. *postpositivus* < Lat. *postponere,* to put after. —see POSTPONE.] Placed after or suffixed to another word. *—n.* An appended or suffixed word or word element. **—post·pos′i·tive·ly** *adv.*
**post·pran·di·al** *also* **post-pran·di·al** (pōst-prăn′dē-əl) *adj.* Following a meal, esp. dinner <a *postprandial* cigar>
**post·script** (pōst′skrĭpt′, pōs′skrĭpt′) *n.* [Lat. *postscriptum,* p.part. of *postscribere,* to write after : *post,* after + *scribere,* to write.] **1.** A message added at the end of a letter after the writer's signature. **2.** Information added to a manuscript, as to an article or book.
**post·test** (pōst′tĕst′) *n.* An evaluative test, often in conjunction with a pretest, given after a unit of instruction or course of study.
**post time** *n.* The time set just before the official start of a race after which point no further betting is allowed.
**post·trau·mat·ic** (pōst′trou-măt′ĭk, -trô-) *adj.* Following injury or resulting from it <*posttraumatic* paralysis>

**pos·tu·lant** (pŏs′chə-lənt) *n.* [Fr. < OFr. < Lat. *postulans*, pr.part. of Lat. *postulare*, to request.] **1.** One submitting a request or application : PETITIONER. **2.** A candidate for admission into a religious order. **—pos′tu·lan·cy, pos′tu·lant·ship′** *n.*

**pos·tu·late** (pŏs′chə-lāt′) *vt.* **-lat·ed, -lat·ing, -lates.** [Lat. *postulare, postulat-*, to request.] **1.** To make claim for : DEMAND. **2.** To assume or claim as real or true, esp. as a basis of an argument. **3.** To assume as an axiom or premise : take for granted. *—n.* (pŏs′cho͞o-lĭt, -lāt′). **1.** Something assumed without proof as being self-evident or gen. accepted, esp. when used as a basis for an argument. **2.** A basic principle : fundamental element. **3.** *Math.* An axiom. **4.** A prerequisite : requirement. **—pos′tu·la′tion** *n.*

**pos·tu·la·tor** (pŏs′chə-lā′tər) *n.* **1.** One who postulates. **2.** *Rom. Cath. Ch.* A church official who presents a plea for canonization or beatification.

**pos·ture** (pŏs′chər) *n.* [Fr. < Ital. *postura* < Lat. *positura*, position < *ponere*, to place.] **1.** A position or attitude of the body or of bodily parts <a standing *posture*> **2.** A characteristic manner of bearing one's body : CARRIAGE <good *posture*> **3.** A position assumed by an artist's model. **4.** Present condition or tendency <put the nation in a *posture* of rebellion> **5.** A frame of mind : OUTLOOK. *—v.* **-tured, -tur·ing, -tures.** *—vi.* To assume an exaggerated or unnatural pose or mental attitude. *—vt.* To put in a posture : POSE. **—pos′tur·al** *adj.* **—pos′tur·er, pos′tur·ist** *n.*

**post·vo·cal·ic** (pōst′vō-kăl′ĭk) *adj.* Designating a consonant or consonantal sound directly following a vowel.

**post·war** (pōst′wôr′) *adj.* Taking place after a war.

**po·sy** (pō′zē) *n., pl.* **-sies.** [Alteration of POESY.] **1.** A flower or nosegay. **2.** *Archaic.* A brief verse or sentimental phrase, esp. one inscribed on a trinket.

**pot** (pŏt) *n.* [ME < OE *pott.*] **1. a.** A round, rather deep household vessel used esp. for holding liquids or cooking. **b.** Such a vessel and its contents <a *pot* of chili> **c.** The amount such a vessel will hold : POTFUL. **2. a.** A large drinking cup : TANKARD. **b.** A drink of liquor contained in such a cup. **3.** An artistic or decorative ceramic vessel. **4.** A flowerpot. **5.** Something resembling a round cooking vessel in appearance or function, as a chimney pot or chamber pot. **6.** A wicker or wire basket used as a trap for fish, crustaceans, or eels. **7. a.** The total amount staked by all the players in one hand of a card game. **b.** The area on a card table where stakes are placed. **8.** *Informal.* A common fund to which members of a group contribute. **9.** *Computer Sci.* A section of computer storage set aside for storing accumulated data. **10.** *Informal.* A pot shot. **11.** *Informal.* A potbelly. **12.** *Slang.* Marijuana. *—v.* **pot·ted, pot·ting, pots.** *—vt.* **1.** To place or plant in a pot. **2.** To preserve (food) in a pot. **3.** To cook in a pot. **4.** To shoot (game) for food rather than for sport. **5.** *Informal.* To shoot with a pot shot. **6.** *Informal.* To win or capture : BAG. *—vi. Informal.* To take a pot shot. **—go to pot.** *Informal.* To deteriorate.

**po·ta·ble** (pō′tə-bəl) *adj.* [Fr. < LLat. *potabilis* < Lat. *potare*, to drink.] Fit to drink. *—n.* **potables.** Drinkables. **—po′ta·bil′i·ty, po′ta·ble·ness** *n.*

**pot·ash** (pŏt′ăsh′) *n.* [Sing. of obs. *pot ashes.*] **1.** Potassium carbonate. **2.** Potassium hydroxide. **3.** Any of several compounds having potassium, esp. soluble compounds, as potassium oxide, potassium chloride, and various potassium sulfates, used mainly in fertilizers.

**potash feldspar** *n.* Orthoclase.

**potash muriate** *n.* Potassium chloride.

**po·tas·si·um** (pə-tăs′ē-əm) *n.* [< POTASH.] *Symbol* **K** A soft, silver-white, light, highly reactive metallic element found in or converted to a wide variety of salts used esp. in fertilizers and soaps; atomic number 19; atomic weight 39.102. **—po·tas′sic** *adj.*

**po·tas·si·um-ar·gon** (pə-tăs′ē-əm-är′gŏn′) *adj.* Of, pertaining to, or designating a geologic dating method relying on the percentage of potassium in a specimen that has radioactively decayed to argon.

**potassium bitartrate** *n.* A white crystalline solid or powder, $KHC_4H_4O_6$, used esp. in baking powder and laxatives.

**potassium bromide** *n.* A white crystalline solid or powder, KBr, used as a sedative and in spectroscopy and photographic emulsion.

**potassium carbonate** *n.* A transparent, white, deliquescent, granular powder, $K_2CO_3$, used in making soaps, glass, ceramics, and pigments.

**potassium chlorate** *n.* A poisonous crystalline compound, $KClO_3$, used in explosives, matches, and fireworks and as an oxidizing agent, bleach, and disinfectant.

**potassium chloride** *n.* A colorless crystalline solid or powder, KCl, used in fertilizers and a wide variety of potassium compounds.

**potassium cyanide** *n.* An extremely poisonous white compound, KCN, used as a fumigant and insecticide and in electroplating, photography, and the extraction of gold and silver from ores.

**potassium dichromate** *n.* A bright yellowish-red crystalline compound, $K_2Cr_2O_7$, used as an oxidizing agent and in pyrotechnics, explosives, and safety matches.

**potassium hydroxide** *n.* A caustic deliquescent solid, KOH, used as a bleach and in making liquid detergents and soaps, oxalic acid, matches, and many potassium compounds.

**potassium iodide** *n.* A white crystalline compound, KI, used in photography and as an analytical reagent.

**potassium muriate** *n.* Potassium chloride.

**potassium nitrate** *n.* A transparent white crystalline compound, $KNO_3$, used in pickling meat and in the manufacture of pyrotechnics, explosives, matches, rocket propellants, and fertilizers.

**potassium permanganate** *n.* A dark purple crystalline compound, $KMnO_4$, used in deodorizers and dyes and as an oxidizing agent and disinfectant.

**potassium sulfate** *n.* A colorless or white crystalline compound, $K_2SO_4$, used as a reagent in analytical chemistry and in medicine, glassmaking, and fertilizers.

**po·ta·tion** (pō-tā′shən) *n.* [ME *potacion* < OFr. < Lat. *potatio* < *potare*, to drink.] **1.** The act of drinking. **2.** A drink, esp. of an alcoholic beverage.

**po·ta·to** (pə-tā′tō) *n., pl.* **-toes.** [Sp. *patata* < Taino *batata.*] **1.** A plant, *Solanum tuberosum*, native to South America and widely cultivated for its starchy, edible tubers. **2.** A tuber of the potato plant.

**potato beetle** *n.* A small yellow-and-black striped beetle, *Leptinotarsa decemlineata*, that is a major agricultural pest.

**potato blight** *n.* Any of various highly destructive fungal diseases of the potato.

**potato bug** *n.* The potato beetle.

**potato chip** *n.* A thin slice of potato deep-fried until crisp and usu. salted.

**po·ta·to·ry** (pō′tə-tôr′ē, -tōr′ē) *adj.* [Lat. *potatorius* < *potare*, to drink.] Of, relating to, or given to drinking.

**pot-au-feu** (pô-tō-fœ′) *n.* [Fr. : *pot*, pot + *au*, on the + *feu*, fire.] A French dish of boiled meats and vegetables.

**Pot·a·wat·o·mi** (pŏt′ə-wŏt′ə-mē) *n., pl.* **Potawatomi** *or* **-mis.** **1. a.** A tribe of Indians inhabiting Michigan. **b.** A member of this tribe. **2.** The Algonquian language of the Potawatomi.

**pot·bel·lied stove** (pŏt′bĕl′ēd) *n.* A potbelly stove.

**pot·bel·ly** (pŏt′bĕl′ē) *n., pl.* **-lies.** **1.** A protruding abdominal region. **2.** A potbelly stove. **—pot′bel′lied** *adj.*

**potbelly stove** *n.* A short rounded stove in which wood or coal is burned.

**pot·boil·er** (pŏt′boi′lər) *n.* A sensational, usu. inferior literary or artistic work produced quickly for profit.

**pot cheese** *n.* Cottage cheese.

**po·teen** (pō-tēn′) *n.* [Ir. Gael. *poitīn* < *pota*, pot.] Irish whiskey distilled unlawfully.

**po·tence** (pōt′ns) *n.* Potency.

**po·ten·cy** (pōt′n-sē) *n., pl.* **-cies.** **1.** The quality or state of being potent. **2.** Inherent capacity for growth and development : POTENTIALITY.

**po·tent** (pōt′nt) *adj.* [ME < Lat. *potens*, pr.part. of *posse*, to be able.] **1.** Possessing inner or physical strength : POWERFUL. **2.** Having a strong influence or effect : COGENT <a *potent* defense> **3.** Having great authority or control <"The police were *potent* only so long as they were feared" —Thomas Burke> **4.** Capable of causing strong physiological or chemical effects, as medicines or alcoholic beverages. **5.** Able to perform sexually. —Used of a male. **—po′tent·ly** *adv.* **—po′tent·ness** *n.*

**po·ten·tate** (pōt′n-tāt′) *n.* [ME *potentat* < OFr. < LLat. *potentatus* < Lat., power < *potens*, pr.part. of *posse*, to be able.] **1.** One who has the power and position to rule over others : MONARCH. **2.** One who dominates or leads a group or endeavor <*potentates* of industry>

**po·ten·tial** (pə-tĕn′shəl) *adj.* [ME *potencial* < OFr. < LLat. *potentialis*, powerful < Lat. *potentia*, power < *potens*, pr.part. of *posse*, to be able.] **1.** Capable of being but not yet in existence. **2.** Denoting possibility, capability, or power. —Used to designate a verb form with auxiliaries such as *may* or *can*; e.g., *It may snow.* *—n.* **1.** The inherent ability or capacity for growth, development, or coming into being. **2.** Something having the capacity for development or growth. **3.** A potential verb form. **4.** *Physics.* The work required to bring a unit electric charge, magnetic pole, or mass from an infinitely distant position to a designated point in a static electric, magnetic, or gravitational field, respectively. **5.** *Elect.* The potential energy of a unit charge at any point in an electric circuit measured with respect to a given reference point in the circuit or to ground : VOLTAGE. **—po·ten′tial·ly** *adv.*

☆ ***syns:*** POTENTIAL, EVENTUAL, LATENT, POSSIBLE *adj. core meaning* : capable of being but not yet in existence <a *potential* buyer><a *potential* threat> ***ant:*** actual, real

**potential energy** *n.* The energy of a particle or system of particles derived from position rather than motion, with respect to a specified datum in a field of force.

**po·ten·ti·al·i·ty** (pə-tĕn′shē-ăl′ĭ-tē) *n., pl.* **-ties.** **1.** Inherent capacity for growth, development, or coming into existence. **2.** Something having potentiality.

**po·ten·til·la** (pōt′n-tĭl′ə) *n.* [Med. Lat., garden valerian < Lat. *potens*, pr.part. of *posse*, to be able.] A plant or shrub of the genus *Potentilla* of the North Temperate Zone.

**po·ten·ti·om·e·ter** (pə-tĕn'shē-ŏm'ĭ-tər) *n.* [POTENTI(AL) + -METER.] **1.** An instrument for measuring an unknown voltage or potential difference by comparison to a standard voltage. **2.** A three-terminal resistor with an adjustable center connection, used for volume control in radio and television receivers. **—po·ten'ti·o·met'ric** (-shē-ō-mĕt'rĭk) *adj.*
**pot·ful** (pŏt'fo͝ol') *n.* **1.** The amount a pot will hold. **2.** *Informal.* A large amount, as of money.
**pot·head** (pŏt'hĕd') *n. Slang.* A habitual user of marijuana.
**poth·er** (pŏth'ər) *n.* [Orig. unknown.] **1.** A disturbance : commotion. **2.** A state of nervous activity : FUSS. **3.** A cloud of dust or smoke that smothers or chokes. *—v.* **-ered, -er·ing, -ers.** *—vt.* To make confused or troubled. *—vi.* To be concerned with trifles : FUSS.
**pot·herb** (pŏt'ûrb', -hûrb') *n.* A plant whose leaves, stems, or flowers are cooked and eaten or used as seasoning.
**pot·hold·er** (pŏt'hōl'dər) *n.* A small fabric pad used to handle hot cooking utensils.
**†pot·hole** (pŏt'hōl') *n.* **1.** A deep hole or pit, esp. one in a road surface. **2.** A deep, round hole worn in rock by loose stones whirling in strong rapids or waterfalls. **3.** *Western U.S.* A place filled with mud or quicksand that is a hazard to cattle.
**pot·hook** (pŏt'ho͝ok') *n.* **1.** A hooked or bent piece of iron for hanging a pot or kettle over a fire. **2.** A curved iron rod with a hooked end used for lifting hot pots, irons, or stove lids. **3.** A curved, S-shaped mark made in writing. **4.** *often* **pothooks. a.** Illegible handwriting or aimless doodling. **b.** *Informal.* Stenographic writing.
**pot·house** (pŏt'hous') *n. Chiefly Brit.* A tavern.
**pot·hunt·er** (pŏt'hŭn'tər) *n.* **1.** One who hunts game for food, ignoring the rules of sport. **2.** One who participates in contests simply to win prizes. **3.** A nonprofessional archaeologist. **—pot'hunt'ing** *n.*
**po·tiche** (pô-tēsh') *n.* [Fr. < *pot,* pot < OFr.] A jar or vase with a removable cover and a round or polygonal body tapering at the neck.
**po·tion** (pō'shən) *n.* [ME *pocion* < OFr. < Lat. *potio* < *potare,* to drink.] A liquid dose, esp. one of medicinal, magic, or poisonous content.
**pot·latch** (pŏt'lăch') *n.* [Chinook < Nootka *patshatl,* gift.] A ceremonial feast among Indian tribes living on the northwest Pacific coast in which the host distributes gifts requiring reciprocation.
**pot·luck** (pŏt'lŭk') *n.* Whatever food happens to be available for a meal, esp. when offered to a guest.
**pot marigold** *n.* A plant, *Calendula officinalis,* often grown for its bright orange or yellow flowers, the dried florets of which were once used for seasoning.
**pot marjoram** *n.* MARJORAM 2.
**pot·pie** (pŏt'pī') *n.* **1.** A mixture of meat or poultry and vegetables covered with a pastry crust and baked in a deep dish. **2.** A meat or poultry stew with dumplings.
**pot·pour·ri** (pō'po͞o-rē') *n., pl.* **-ris.** [Fr. *pot pourri,* transl. of Sp. *olla podrida.* —see OLLA PODRIDA.] **1.** A combination of various incongruous elements. **2.** A miscellaneous collection or anthology. **3.** A mixture of dried flower petals and spices kept in a jar and used esp. to scent the air.
**pot roast** *n.* A cut of beef browned and then cooked until tender, often with vegetables, in a covered pot.
**pot·sherd** (pŏt'shûrd') *also* **pot·shard** (-shärd') *n.* [ME *potschoord* : *pot,* pot + *schord,* var. of *shard,* shard < OE *sceard.*] A piece of broken pottery, esp. one having archaeological significance.
**pot shot** *n.* [So called because such a shot is fired by a hunter whose main purpose is to get food for his pot.] **1.** A shot fired aimlessly or at a target within easy range. **2.** A criticism made without careful thought and aimed at a handy target for attack <political opponents taking *pot shots* at the governor>
**pot·stone** (pŏt'stōn') *n.* An impure variety of steatite once used to make cooking vessels.
**pot·tage** (pŏt'ĭj) *n.* [ME *potage* < OFr. < *pot,* pot.] **1.** A thick soup or stew of vegetables and sometimes meat. **2.** *Archaic.* Porridge.
**pot·ted** (pŏt'ĭd) *adj.* **1. a.** Placed in a pot. **b.** Grown in a pot, as a plant. **2.** Preserved in a pot, can, or jar, as meat. **3.** *Slang.* Stoned.
**pot·ter¹** (pŏt'ər) *n.* [ME *pottere* < OE < *pott,* pot.] A maker of earthenware pots, dishes, or other vessels.
**pot·ter²** (pŏt'ər) *v. Chiefly Brit. var. of* PUTTER.
**potter's clay** *n.* A clay low in iron content that is suitable for modeling or making pottery.
**potter's earth** *n.* Potter's clay.
**potter's field** *n.* [From the potter's field mentioned in the Gospel according to St. Matthew.] A burial ground for indigent or unknown persons.
**potter's wheel** *n.* A device having a revolving, often treadle-operated horizontal disk upon which clay is shaped by hand.
**potter wasp** *n.* A wasp of the genus *Eumenes* that builds pot-shaped nests of clay.
**pot·ter·y** (pŏt'ə-rē) *n., pl.* **-ies.** [OFr. *poterie* < *potier,* potter, prob. < *pot,* pot.] **1.** Ware, as vases, pots, bowls, or plates, shaped from moist clay and hardened by heat. **2.** The craft or occupation of a potter. **3.** A potter's workshop.
**pot·tle** (pŏt'l) *n.* [ME *potel* < OFr. < *pot,* pot.] **1. a.** A pot or drinking vessel with a two-quart capacity. **b.** The liquid contained in a pottle. **2.** A former liquid measure equivalent to about two quarts.
**pot·to** (pŏt'ō) *n., pl.* **-tos.** [Of Niger-Congo orig.] A small African primate of the genera *Perodicticus* or *Arctocebus,* with hands and feet adapted for grasping and woolly fur.

**potto**
*Up to 18 inches long*

**Pott's disease** (pŏts) *n.* [After Percival *Pott* (1714–1788).] Partial destruction of the bones of the vertebrae, usu. caused by a tuberculous infection and often producing curvature of the spine.
**pot·ty¹** (pŏt'ē) *adj.* **-ti·er, -ti·est.** [Poss. < POT.] *Chiefly Brit.* **1.** Of little importance : TRIVIAL. **2.** Slightly intoxicated. **3.** Silly or crazy.
**pot·ty²** (pŏt'ē) *n., pl.* **-ties.** A small pot for use as a toilet by a young child.
**pot·ty-chair** (pŏt'ē-châr') *n.* A small chair with an opening in the seat and a receptacle beneath, used for toilet-training young children.
**pouch** (pouch) *n.* [ME *pouche* < OFr., of Germanic orig.] **1.** A small bag closed with a drawstring and used esp. for carrying loose pipe tobacco in one's pocket. **2.** A small or medium-sized bag of flexible material used for holding or carrying various items, esp. one used to transport mail or diplomatic dispatches. **3.** *Archaic.* A purse for small coins. **4.** A leather bag for carrying powder or small-arms ammunition. **5.** Something shaped like a small bag <*pouches* under the eyes> **6.** *Zool.* A saclike structure, as the cheek pockets of the gopher or the external abdominal pocket in which marsupials carry their young. **7.** *Scot.* A pocket. *—v.* **pouched, pouch·ing, pouch·es.** *—vt.* **1.** To place in or as if in a pouch : POCKET <*pouched* all the profits> **2.** To cause to resemble a pouch. **3.** To swallow. —Used of certain birds or fishes. *—vi.* To take on the form of a pouch or pouchlike cavity. **—pouch'y** *adj.*
**pouf** (po͞of) *n.* [Fr.] **1.** A woman's hair style popular in the 18th cent., having high rolled puffs. **2.** A part of a garment, as a dress, gathered into a puff. **3.** A rounded ottoman.
**pouil·ly-fuis·sé** (po͞o-yē'fwē-sā') *n.* [After Solutré-*Pouilly* and *Fuissé,* villages in France.] A dry white Burgundy.
**pou·lard** *also* **pou·larde** (po͞o-lärd') *n.* [Fr. *poularde* < OFr. *pollarde* < *polle,* hen < Lat. *pullus,* chicken.] A young hen that has been spayed for fattening.
**poult** (pōlt) *n.* [ME *pult,* short for *polet* < OFr. *poulet,* dim. of *poule, polle,* hen. —see POULARD.] A young fowl, esp. a turkey.
**poul·ter·er** (pōl'tər-ər) *n.* [Alteration of ME *pulter* < OFr. *pouletier* < *poulet,* young fowl. —see POULT.] *Chiefly Brit.* A poultry dealer.
**poul·ter's measure** (pōl'tərz) *n.* [< obs. *poulter,* poulterer, from the practice of giving a few extra eggs in the dozen.] A metrical pattern orig. composed of a couplet whose first line had 12 syllables and the second 14.
**poul·tice** (pōl'tĭs) *n.* [Med. Lat. *pultes,* thick paste < Lat., pl. of *puls,* pap, poss. < Gk. *poltos,* porridge.] A usu. heated, moist, soft mass of an adhesive substance, as meal or clay, spread on cloth and applied to warm, moisten, or stimulate a sore or inflamed part of the body. *—vt.* **-ticed, -tic·ing, -tic·es.** To apply a poultice to.
**poul·try** (pōl'trē) *n.* [ME *pultrie* < OFr. *pouletrie* < *pouletier,* poulterer. —see POULTERER.] Domestic fowl, as chickens, turkeys, ducks, or geese, raised for eggs or flesh.
**pounce¹** (pouns) *v.* **pounced, pounc·ing, pounc·es.** [< ME, talon of a bird of prey, perh. var. of *ponson,* a pointed tool. —see PUNCHEON.] *—vi.* **1.** To swoop or spring with intent to seize someone or something. **2.** To attack or grab suddenly and unexpectedly <*pounce* on a chance to win> *—vt.* To seize with or as if with talons. *—n.* **1.** The act of pouncing. **2.** The talon or claw of a bird of prey. **—pounc'er** *n.*
**pounce²** (pouns) *n.* [Fr. *ponce* < Lat. *pumex,* pumice.] **1.** A fine powder once used to smooth and finish writing paper and soak up ink. **2.** A fine powder, as pulverized charcoal, dusted over a stencil to transfer a design to an underlying surface. *—vt.* **pounced, pounc·ing, pounc·es. 1.** To sprinkle, smooth, or treat with pounce. **2.** To transfer (a stenciled design) with pounce. **—pounc'er** *n.*
**pounce³** (pouns) *vt.* **pounced, pounc·ing, pounc·es.** [ME *poun-*

*sen,* prob. < OFr. *poiçonner,* to prick < *poinon,* a pointed tool. —see PUNCHEON.] To ornament (e.g., metal) by perforating from the back with a pointed implement.

**pounce box** *n.* A small box with a perforated top, once used to sprinkle pounce or sand on writing paper to dry the ink.

**poun·cet box** (poun'sĭt) *n.* [Prob. alteration of *pounce box.*] A small perfume box with a perforated top.

**pound**[1] (pound) *n., pl.* **pound** *or* **pounds.** [ME < OE *pund* < Lat. *pondo,* a unit of weight.] **1. a.** A unit of weight equal to 16 ounces, 7,000 grains, or 453.592 grams. **b.** A unit of apothecary weight equal to 5,760 grains or 373.242 grams. **2.** Any of several units of weight differing in various times and countries. **3.** A British unit of force equal to the weight of a standard one-pound mass where the local acceleration of gravity is 32.174 feet per second per second. **4.** —See table at CURRENCY. **5.** A monetary unit of Scotland before the Union, usu. worth a small fraction of the pound sterling.

▲ word history: The word *pound*[1], "a unit of currency," is a specialized use of the word *pound* meaning "a unit of weight." The word *pound*[1] is derived from Latin *pondo,* "a measure of weight." The original Latin word for "a pound weight" was *libra; pondo,* which is actually the ablative case of *pondus,* "a weight," developed into a separate noun from the phrase *libra pondo,* "a pound by weight." *Pondo* descended into Old English as *pund,* which had the same meanings as modern *pound*[1]. The monetary unit was originally the value of a pound weight of silver. The symbol £ still used for the British pound is a stylized L that stands for the Latin word *libra.*

**pound**[2] (pound) *v.* **pound·ed, pound·ing, pounds.** [ME *pounen* < OE *pūnian.*] —*vt.* **1.** To strike or hammer with a heavy blow or blows. **2.** To drive (something) in or out with repeated blows. **3.** To beat to a pulp or powder : PULVERIZE. **4.** To instill by emphatic and persistent repetition <*pound* the lessons into their heads> —*vi.* **1.** To strike vigorous, repeated blows. **2.** To move along noisily and heavily <*pounded* down the corridor> **3.** To pulsate rapidly and heavily <My heart was *pounding.*> **4.** To work or move laboriously <a freighter *pounding* through stormy seas> —*n.* **1.** A heavy blow. **2.** The sound of a heavy blow : THUMP. **3.** The act of pounding. **—pound'er** *n.*

**pound**[3] (pound) *n.* [ME.] **1.** A public enclosure for the confinement of stray dogs or livestock. **2.** A place in which impounded property is held until redeemed. **3.** An enclosure in which animals or fish are trapped or kept. **4.** A place of confinement for lawbreakers. —*vt.* **pound·ed, pound·ing, pounds.** To confine in or as if in a pound.

**pound·age**[1] (poun'dĭj) *n.* **1.** A tax or commission based on value per pound sterling. **2.** A rate or charge based on weight in pounds. **3.** Weight calculated in pounds.

**pound·age**[2] (poun'dĭj) *n.* **1.** Confinement of animals in a pound. **2.** A fee charged for the redemption of impounded animals or other property.

**pound·al** (poun'dl) *n.* [POUND + (QUINT)AL.] A unit of force in the foot-pound-second system of measurement, equal to the force required to accelerate a standard one-pound mass one foot per second per second.

**pound cake** *n.* A rich cake made with eggs, flour, butter, and sugar.

**pound-fool·ish** (pound'fo͞o'lĭsh) *adj.* Unwise in dealing with large sums of money or important matters.

**pound of flesh** *n.* [From Antonio's debt to Shylock in *The Merchant of Venice* by William Shakespeare (1564–1616).] A debt harshly insisted on.

**pound scots** *n.* POUND[1] 5.

**pound sterling** *n.* POUND[1] 4.

**pour** (pôr, pōr) *v.* **poured, pour·ing, pours.** [ME *pouren.*] —*vt.* **1.** To cause (a liquid or granular solid) to stream or flow. **2.** To send forth, produce, express, or utter voluminously, as if in a stream or flood <*poured* out my sorrows> —*vi.* **1.** To stream or flow profusely or continuously. **2.** To rain heavily or hard. **3.** To go forth or stream in large numbers or quantity <The children *poured* into the gym.> —*n.* A pouring forth, esp. a downpour of rain. **—pour'er** *n.*

**pour·boire** (po͝or-bwär') *n.* [Fr. < *pour boire,* for drinking.] TIP[4] 1.

**pour·par·ler** (po͝or'pär-lā') *n.* [Fr. < *pour parler,* for speaking.] Conversation or discussion preliminary to negotiation.

**pour point** *n.* The lowest temperature at which a liquid, as oil, will pour when cooled under given conditions.

**pousse-ca·fé** (po͞os'kă-fā') *n.* [Fr. : *pousse,* a push + *café,* coffee.] **1.** A drink made of several liqueurs of different densities poured to form differently colored layers. **2.** A brandy or liqueur served after dinner with coffee.

**pous·sette** (po͞o-sĕt') *n.* [Fr. < *pousse,* a push < *pousser,* to push < OFr. < Lat. *pulsare,* freq. of *pellere,* to impel.] A country-dance figure in which a couple or couples join hands and swing around the floor.

**pout**[1] (pout) *v.* **pout·ed, pout·ing, pouts.** [ME *pouten.*] —*vi.* **1.** To protrude the lips in an expression of sullen displeasure. **2.** To show displeasure or disappointment : SULK. **3.** To protrude or project. —*vt.* **1.** To protrude (the lips). **2.** To express or utter with a pout. —*n.* **1.** A protrusion of the lips, esp. as an expression of sullen discontent. **2.** *often* **pouts.** Petulant sulkiness.

**pout**[2] (pout) *n., pl.* **pout** *or* **pouts.** [ME **poute* < OE *-pūte,* as in *aelepūte,* eelpout.] A marine or freshwater fish, esp. an eelpout.

**pout·er** (pou'tər) *n.* One of a breed of pigeons capable of distending the crop until the breast becomes puffed out.

**pov·er·ty** (pŏv'ər-tē) *n.* [ME *poverte* < OFr. < Lat. *paupertas* < *pauper,* poor.] **1.** Lack of the means of providing material needs or comforts. **2.** Deficiency in amount : SCANTINESS <*poverty* of insight> **3.** Unproductiveness or infertility, as of soil. **4.** The renunciation made by a member of a religious order of the right to own property.

**poverty grass** *n.* Any of several North American grasses that grow in poor or sandy soil.

**poverty level** *n.* A minimum income level below which one is classified as lacking adequate subsistence and living in poverty.

**pov·er·ty-strick·en** (pŏv'ər-tē-strĭk'ən) *adj.* Very poor.

**pow** (pou) *n.* A loud sound, as of an explosion or a blow with the fist.

**POW** (pē'ō'dŭb'əl-yo͞o, -yo͞o) *n.* A prisoner of war.

**pow·der** (pou'dər) *n.* [ME *poudre* < OFr. < Lat. *pulvis.*] **1.** A substance composed of finely dispersed solid particles. **2.** Any of various preparations in the form of powder, as certain medicines and cosmetics. **3.** Gunpowder or a similar explosive mixture. **4.** Light, dry snow. —*v.* **-dered, -der·ing, -ders.** —*vt.* **1.** To reduce to powder : PULVERIZE. **2.** To cover or dust with or as if with powder. **3.** *Slang.* To defeat easily or decisively. —*vi.* **1.** To become pulverized. **2.** To use powder as a cosmetic. **—take a powder.** *Slang.* To depart quickly. **—pow'der·er** *n.*

**powder blue** *n.* [From the color of powdered smalt.] A moderate to pale blue or purplish blue.

**powder flask** *n.* A small flask for carrying gunpowder.

**powder horn** *n.* A container made of an animal's horn capped at the open end and used to carry gunpowder.

**powder keg** *n.* **1.** A small cask for holding explosives, esp. gunpowder. **2.** Something potentially explosive.

**powder metallurgy** *n.* The technology of powdered metals, esp. the production and utilization of metallic powders for massive materials and shaped objects.

**powder puff** *n.* A soft pad for applying powder to the skin.

**pow·der-puff** (pou'dər-pŭf') *adj.* Of, pertaining to, or being a usu. competitive activity in which only women take part <*powder-puff* basketball>

**powder room** *n.* A lavatory for women.

**pow·der·y** (pou'də-rē) *adj.* **1.** Consisting of or similar to powder. **2.** Covered or dusted with or as if with powder. **3.** Easily pulverized : FRIABLE.

**powdery mildew** *n.* **1.** Any of various fungi of the family Erysiphaceae or genus *Oidium* that produce powdery conidia on the host. **2.** A plant disease caused by a powdery mildew.

**†pow·er** (pou'ər) *n.* [ME *pouer* < OFr. *poeir* < *poeir,* to be able < Lat. **potēre* < *potis,* able.] **1.** The ability or capacity to act or perform effectively. **2.** *often* **powers.** A specific capacity, faculty, or aptitude <strong decisive *powers*> **3.** Strength or force exerted or capable of being exerted : MIGHT. **4.** The ability or official capacity to exercise control over others. **5.** A person, group, or nation having great influence or control over others. **6.** The might of a nation, political organization, or similar group. **7.** Effectiveness : forcefulness <a film of extraordinary *power*> **8.** *Regional.* A large number or amount. **9.** *Physics.* The rate at which work is performed, mathematically expressed as the first derivative of work with respect to time and commonly measured in units such as the watt and horsepower. **10.** *Elect.* **a.** The product of applied potential difference and current in a direct-current circuit. **b.** The product of the effective values of the voltage and current with the cosine of the phase angle between current and voltage in an alternating-current circuit. **11.** *Math.* **a.** EXPONENT 3. **b.** The number of elements in a finite set. **12.** *Statistics.* The probability of rejecting the null hypothesis where it is false. **13.** A measure of the magnification of an optical instrument, as a telescope or microscope. **14. powers.** The sixth group of angels in the hierarchical order of nine. **15.** *Archaic.* An armed force. —*vt.* **-ered, -er·ing, -ers.** To supply with power, esp. mechanical power.

☆ **syns:** POWER, AUTHORITY, CLOUT, CONTROL *n. core meaning* : the right or ability to dominate or rule others <the *power* of an emperor> POWER is the most general, being applicable to strong influence based on rank, position, character, or another advantage <an executive with the *power* to hire and fire at will> AUTHORITY suggests legitimate and recognized power <the mayor's *authority* to dismiss dishonest commissioners> CONTROL stresses the right to regulate or direct as well as dominate <guards having total *control* over the prisoners> CLOUT stresses the existence of strong influence over others <a politician with a lot of *clout* on Capitol Hill>

**pow·er·boat** (pou'ər-bōt') *n.* A motorboat.

**power brake** *n.* A motor vehicle brake that is assisted by a power mechanism operated by the engine that amplifies pressure applied to the brake pedal.

**power broker** *n.* One, esp. a politician, who exerts strong influence by virtue of the individuals and votes he or she controls.

**power dive** *n.* A downward plunge of an aircraft accelerated by both gravity and engine power.

**pow·er-dive** (pou'ər-dīv') *vi.* & *vt.* **-dived** or **-dove** (-dōv'), **-div·ing, -dives.** To make or cause to make a power dive.

**power drill** *n.* **1.** A portable electric drill. **2.** A large drilling machine with a vertical, motorized drill set in a table stand.

**†pow·er·ful** (pou'ər-fəl) *adj.* **1.** Having or capable of exerting power. **2.** Effective or potent <a *powerful* antibiotic> **3.** *Regional.* Great <The storm did a *powerful* lot of harm.> *—adv. Regional.* Very <It was *powerful* humid.> **—pow'er·ful·ly** *adv.* **—pow'er·ful·ness** *n.*

**pow·er·house** (pou'ər-hous') *n.* **1.** A station for the generating of electricity. **2.** One who has great energy or force.

**pow·er·less** (pou'ər-lĭs) *adj.* **1. a.** Unable to manage for oneself. **b.** Lacking assistance or protection. **2.** Lacking legal or other authority. **—pow'er·less·ly** *adv.* **—pow'er·less·ness** *n.*

☆ ***syns:*** **1.** POWERLESS, HELPLESS, IMPOTENT, INCAPABLE *adj. core meaning* : unable to manage for oneself <a dying patient *powerless* to speak> **2.** POWERLESS, DEFENSELESS, HELPLESS *adj. core meaning* : lacking help or protection <*powerless* hostages at the mercy of their captors>

**power mower** *n.* A lawn mower powered by a gasoline or electric motor.

**power of appointment** *n. Law.* Authority granted to one person by another to succeed to property upon the death of the latter.

**power of attorney** *n. Law.* A legal instrument authorizing one to act as another's attorney or agent.

**power pack** *n.* A portable device that converts supply current to direct or alternating current as required by specific equipment.

**power plant** *n.* **1.** All the equipment, including structural members, that makes up a unit power source <the *power plant* of a spacecraft> **2.** A complex of structures, machinery, and associated equipment for generating power, esp. electric power.

**power play** *n.* **1. a.** An offensive maneuver in a team game, esp. in football, in which a massive concentration of players is applied in a certain area. **b.** A situation in ice hockey in which one team has a temporary numerical advantage because the other team has one or more players in the penalty box. **2.** A strategic action or maneuver, as in politics or business, based on the use or threat of power or superior strength as a means of coercion.

**power politics** *n.* [Transl. of G. *Machtpolitik.*] *(sing. or pl. in number).* International diplomacy in which nations use or threaten to use military or economic power to further their own interests.

**power series** *n.* A sum of successively higher integral powers of a variable or combination of variables, each multiplied by a constant coefficient.

**power shovel** *n.* A large, usu. mobile machine having a boom, a dipper stick, and a bucket for excavating.

**power station** *n.* POWER PLANT 2.

**power steering** *n.* A power-assisted device on a car that facilitates the turning of the steering wheel by the driver.

**power structure** *n.* ESTABLISHMENT 5.

**power take-off** *n.* A mechanism attached to a motor vehicle engine that supplies power to nonvehicular devices, as a backhoe or pump.

**power train** *n.* An assembly of gears and associated parts by which power is transmitted from an engine to a driving axle.

**Pow·ha·tan** (pou'ə-tăn') *n., pl.* **Powhatan** or **-tans. 1. a.** One of various tribes of Indians formerly living in eastern Virginia. **b.** A member of one of these tribes. **2.** The Algonquian language of the Powhatan.

**pow·wow** (pou'wou') *n.* [Of Algonquian orig.] **1.** A North American Indian medicine man. **2.** A North American Indian ceremony in which incantations and dancing are used to invoke divine aid, as for success in hunting or battle. **3.** A conference or meeting with or of North American Indians. **4.** *Informal.* A gathering or conference. *—vi.* **-wowed, -wow·ing, -wows.** To hold a powwow.

**pox** (pŏks) *n.* [Alteration of *pocks,* pl. of POCK.] **1.** A disease marked by purulent skin eruptions, as smallpox or chicken pox. **2.** Syphilis. **3.** *Archaic.* Misfortune and calamity.

**pox·vi·rus** (pŏks'vī'rəs) *n.* Any of a group of DNA-containing animal viruses, including the causative agents of smallpox and vaccinia.

**poz·zuo·la·na** (pŏt'swə-lä'nə) *also* **poz·zo·la·na** (pŏt'sə-) *n.* [Ital. *pozzolana* < *Pozzuoli,* Italy.] **1.** A siliceous volcanic ash used to produce hydraulic cement. **2.** An artificially produced substance resembling pozzuolana ash. **—poz'zuo·la'nic** *adj.*

**PPLO** (pē'pē-ĕl-ō') *n., pl.* **PPLO.** [P(LEURO)-P(NEUMONIA)L(IKE) + O(RGANISM).] Mycoplasma.

**Pr** *symbol for* PRASEODYMIUM.

**praam** *also* **pram** (präm) *n.* [Du. < MDu. *praem* < Czech.] **1.** A flat-bottomed boat used esp. in the Baltic as a barge. **2.** *Chiefly Brit.* A small dinghy with a flat, snub-nosed bow.

**praam**
*A British praam*

**prac·ti·ca·ble** (prăk'tĭ-kə-bəl) *adj.* [Fr. < *pratiquer,* to practice. —see PRACTICE.] **1.** Capable of being done : FEASIBLE. **2.** Capable of being used for a specified purpose <a *practicable* source of power> ***usage:*** *Practicable* refers to something that can be put into effect. *Practical* refers to something that is also sensible and worthwhile. Thus, it might be *practicable* to transport children to school by balloon, but it would not be *practical.* **—prac'ti·ca·bil'i·ty** *n.* **—prac'ti·ca·bly** *adv.*

**prac·ti·cal** (prăk'tĭ-kəl) *adj.* [< LLat. *practicus* < Gk. *praktikos* < *prattein,* to act.] **1.** Of, pertaining to, governed by, or gained through practice or action rather than theory, speculation, or ideals. **2.** Manifested in or involving practice. **3.** Actually engaged in some work or occupation. **4.** Capable of being used or put into effect : USEFUL <*practical* knowledge of auto repair> **5.** Designed to serve a useful purpose <*practical* shoes> **6.** Concerned with the production or operation of something useful <Metalworking is a *practical* art.> **7.** Having or displaying good judgment : SENSIBLE. **8.** Being actually so in almost every respect : VIRTUAL <a *practical* catastrophe> **—prac'ti·cal'i·ty** (-kăl'ĭ-tē), **prac'ti·cal·ness** *n.*

☆ ***syns:*** PRACTICAL, FUNCTIONAL, HANDY, SERVICEABLE, USEFUL, UTILITARIAN *adj. core meaning* : serving or capable of serving a useful purpose <a *practical* kitchen device—not a worthless gadget> ***ant:*** impractical

**practical joke** *n.* A mischievous trick or prank played on a person esp. to cause embarrassment.

**prac·ti·cal·ly** (prăk'tĭk-lē) *adv.* **1.** In a way that is practical. **2.** In every important respect : VIRTUALLY. **3.** Almost <*Practically* everyone contributed.>

**practical nurse** *n.* A licensed practical nurse.

**prac·tice** (prăk'tĭs) *v.* **-ticed, -tic·ing, -tic·es.** [ME *practisen* < OFr. *practiser* < Med. Lat. *practicare* < LLat. *practicus,* practical.] *—vt.* **1.** To do or perform customarily or habitually <*practice* moderation in eating> **2.** To perform or exercise repeatedly in order to acquire or perfect a skill <*practice* musical scales> **3.** To give lessons or repeated instructions to : DRILL. **4.** To work at, esp. as a profession <*practice* medicine> **5.** To carry out in action : OBSERVE <*practice* one's faith> **6.** *Obs.* To plot (something evil). *—vi.* **1.** To do or perform something repeatedly or habitually. **2.** To do something repeatedly in order to acquire or perfect a skill. **3.** To work at a profession. **4.** *Obs.* To intrigue or plot. *—n.* **1.** A habitual or customary action or manner of doing something <made a *practice* of being thrifty> **2. a.** Repeated performance of an activity in order to acquire or perfect a skill. **b.** *Archaic.* The skill so acquired or perfected. **c.** The condition of being skilled through repeated exercise <keep in *practice*> **3.** The act or process of doing something : PERFORMANCE. **4.** The exercise of an occupation or profession <the *practice* of dentistry> **5.** The business of a professional person. **6.** *often* **practices.** A habitual act or action <standard business *practices*> **7.** The methods of procedure used in a court of law. **8.** *Archaic.* **a.** The act of tricking. **b.** A trick. **—prac'tic·er** *n.*

☆ ***syns:*** **1.** PRACTICE, DRILL, EXERCISE, REHEARSAL, STUDY, TRAINING *n. core meaning* : repetition of an action so as to develop or maintain one's skill <took years of *practice* to ski well> **2.** PRACTICE, PURSUIT *n. core meaning* : the exercise of a profession or occupation <began the *practice* of medicine>

**prac·ticed** (prăk'tĭst) *adj.* **1.** Skilled : proficient. **2.** Acquired or perfected by practice.

**practice teaching** *n.* Classroom teaching done by a college student under the supervision of an experienced teacher as an internship in teaching methodology prior to certification as a professional.

**prac·tic·ing** (prăk'tĭ-sĭng) *adj.* Actively engaged in a particular occupation or way of life <a *practicing* psychiatrist>

**prac·ti·cum** (prăk'tĭ-kəm) *n.* [G. *Praktikum* < LLat. *practicum,* neuter of *practicus,* practical.] Supervised practical application of a previously studied theory <an advanced *practicum* for teaching arithmetic to learning-disabled pupils>

**prac·tise** (prăk'tĭs) *v.* & *n. Chiefly Brit. var. of* PRACTICE.

**prac·ti·tion·er** (prăk-tĭsh'ə-nər) *n.* [OFr. *practicien* < *pratique,* practice < LLat. *practicus,* practical.] **1.** One who practices an occupation, profession, or technique. **2.** One engaged in the public ministry of spiritual healing in Christian Science.

**prae·di·al** *also* **pre·di·al** (prē'dē-əl) *adj.* [Med. Lat. *praedialis,* of an estate < Lat. *praedium,* estate < *praes,* surety.] **1.** Relating to land or its products. **2.** Attached to or arising from land or landed property <*praedial* tenants>

**prae·fect** (prē′fĕkt′) *n. var. of* PREFECT.

**prae·lect** (prĭ-lĕkt′) *v. var. of* PRELECT.

**prae·mu·ni·re** (prē′myo͞o-nī′rē) *n.* [ME *premunire facias* < Med. Lat. *praemunire facias,* that you cause to warn, words used in the writ.] **1.** The offense in English history of appealing to or obeying a foreign court or authority, thus challenging the supremacy of the Crown. **2. a.** The writ charging praemunire. **b.** The penalty for praemunire.

**prae·no·men** (prē-nō′mən) *n., pl.* **-nom·i·na** (-nŏm′ə-nə, -nō′mə-nə) *or* **-no·mens.** [Lat. : *prae-,* before + *nomen,* name.] A first or given name. **—prae·nom′i·nal** (-nŏm′ə-nəl) *adj.*

**prae·tor** (prē′tər) *n.* [Lat. < *praeire,* to go before : *prae-,* before + *ire,* to go.] A high elected magistrate of the ancient Roman Republic, ranking below a consul and serving as a judge. **—prae′tor·ship′** *n.*

**prae·to·ri·an** (prē-tôr′ē-ən, -tōr′-) *adj.* **1.** Of or relating to a praetor. **2. Praetorian.** Of, being, or belonging to the elite bodyguard of the Roman emperors. —*n.* **1.** A praetor. **2. Praetorian.** A member of the bodyguard of the Roman emperors.

**prag·mat·ic** (prăg-măt′ĭk) *adj.* [Lat. *pragmaticus,* skilled in business < Gk. *pragmatikos* < *pragma,* deed < *prattein,* to do.] **1.** Concerned with causes and effects or needs and results rather than ideas or theories : PRACTICAL. **2.** Of or relating to pragmatism. —*n.* **1.** A pragmatic sanction. **2.** A busybody : meddler. **—prag·mat′i·cal** *adj.* **—prag·mat′i·cal·ly** *adv.*

**prag·mat·ics** (prăg-măt′ĭks) *n.* (*sing. or pl. in number*). The branch of semiotics concerned with the relations between signs or expressions and their users.

**pragmatic sanction** *n.* An edict issued by a sovereign that becomes part of the fundamental law of the land.

**prag·ma·tism** (prăg′mə-tĭz′əm) *n.* **1.** *Philos.* The theory, developed by Charles S. Peirce and William James, that the meaning of a proposition or course of action lies in its observable consequences and that the sum of these consequences constitutes its meaning. **2.** A practical way of solving problems. **—prag′ma·tist** *n.* **—prag′ma·tis′tic** *adj.*

**prai·rie** (prâr′ē) *n.* [Fr. < OFr. *praerie* < Lat. *pratum,* meadow.] An extensive tract of flat or rolling grassland, esp. the plain of central North America.

**prairie breaker** *n.* A plow that cuts a wide furrow and turns the earth completely over.

**prairie chicken** *n.* Either of two birds, *Tympanuchus cupido* or *T. pallidicinctus* of western North America, with deep-chested bodies and mottled brownish plumage.

**prairie dog** *n.* A burrowing rodent of the genus *Cynomys* of west-central North America, having yellowish fur and a barklike call and living in large communities.

**†prairie oyster** *n.* **1.** *Slang.* A raw egg immersed in liquid and swallowed whole, esp. to relieve a hangover. **2.** *Chiefly Regional.* The cooked testis of a calf used as food.

**prairie schooner** *n.* A canvas-covered wagon used by pioneers crossing the North American prairies.

**praise** (prāz) *n.* [ME *preisen* < OFr. *presier* < LLat. *pretiare,* to prize < Lat. *pretium,* price.] **1.** An expression of approval or admiration : COMMENDATION. **2.** The extolling of a deity, ruler, or hero. **3.** *Archaic.* A reason for praise : MERIT. —*vt.* **praised, prais·ing, prais·es. 1.** To express warm approval of or admiration for : COMMEND. **2.** To exalt or extol : WORSHIP. **—prais′er** *n.*

☆ ***syns:*** PRAISE, ACCLAIM, APPLAUD, COMMEND, LAUD *v. core meaning* : to express warm approval of <*praised* the food> To PRAISE is to express one's esteem or admiration <*praised* their good sense and learning> ACCLAIM and APPLAUD are often but not always used literally to indicate actual applause or cheering <The audience *acclaimed* the artist's performance.> <critics *applauding* a new novel> COMMEND implies speaking well of and is usu. more formal and official <*commended* the commission for its thorough report> LAUD is also a formal term meaning to give the highest praise to <prisoners of war *lauded* for their bravery>

**praise·wor·thy** (prāz′wûr′thē) *adj.* Meriting praise : ADMIRABLE. **—praise′wor′thi·ly** *adv.* **—praise′wor′thi·ness** *n.*

**Pra·krit** (prä′krĭt) *n.* [Skt. *prākrtam* < *prākrta-,* natural, vulgar, vernacular.] Any of the ancient or modern vernacular Indic languages as opposed to the literary language Sanskrit. **—Pra·krit′ic** *adj.*

**pra·line** (prä′lēn′, prā′-, prô′-) *n.* [Fr., after César de Choiseul, Count du Plessis-*Praslin* (1598–1675).] A crisp confection made of nut kernels, esp. pecans, stirred in boiling sugar syrup until brown.

**prall·tril·ler** (präl′trĭl′ər) *n.* [G. : *prallen,* to rebound + *Triller,* trill < Ital. *trillo* < *trillare,* to trill.] *Mus.* A mordent using the auxiliary note above the principal note.

**pram**[1] (prăm) *n. Chiefly Brit.* A perambulator.

**pram**[2] (präm) *n. var. of* PRAAM.

**prance** (prăns) *v.* **pranced, pranc·ing, pranc·es.** [ME *prauncen.*] —*vi.* **1. a.** To spring forward on the hind legs. —Used of a horse. **b.** To move with a succession of such springs or bounds. **2.** To ride a prancing horse. **3.** To walk or move about in a lively manner : STRUT. —*vt.* To cause (a horse) to prance. —*n.* An act of prancing : CAPER. **—pranc′er** *n.* **—pranc′ing·ly** *adv.*

**pran·di·al** (prăn′dē-əl) *adj.* [< Lat. *prandium,* late breakfast.] Of or pertaining to a meal. **—pran′di·al·ly** *adv.*

**prank**[1] (prăngk) *n.* [Orig. unknown.] A mischievous act.

☆ ***syns:*** PRANK, ANTIC, CAPER, FROLIC, JOKE, LARK, MONKEYSHINE, SHENANIGAN, TOMFOOLERY, TRICK *n. core meaning* : a mischievous act <Halloween *pranks*>

**prank**[2] (prăngk) *v.* **pranked, prank·ing, pranks.** [Of Germanic orig.] —*vt.* To dress or decorate gaudily or ostentatiously. —*vi.* To make an ostentatious display.

**prank·ish** (prăng′kĭsh) *adj.* Marked by mischievous, impish behavior. **—prank′ish·ly** *adv.* **—prank′ish·ness** *n.*

**prank·ster** (prăngk′stər) *n.* One who plays pranks or tricks.

**pra·se·o·dym·i·um** (prā′zē-ō-dĭm′ē-əm, prā′sē-) *n.* [Gk. *prasios,* leek-green (< *prason,* leek) + (DI)DYMIUM.] *Symbol* **Pr** A soft, silvery, malleable, ductile rare-earth element, used to color glass yellow and in metallic alloys; atomic number 59; atomic weight 140.907.

**prat** (prăt) *n.* [Orig. unknown.] *Slang.* The buttocks.

**prate** (prāt) *v.* **prat·ed, prat·ing, prates.** [ME *praten.*] —*vi.* To talk volubly and usu. inconsequentially : CHATTER. —*vt.* To utter idly or to little purpose. —*n.* Empty, foolish, or trivial talk. **—prat′er** *n.* **—prat′ing·ly** *adv.*

**prat·fall** (prăt′fôl′) *n.* A fall on the buttocks.

**prat·in·cole** (prăt′n-kōl′, prāt′-, prăt′ĭng-, prā′tĭng-) *n.* [NLat. *pratincola* : Lat. *pratum,* meadow + Lat. *incola,* inhabitant.] An Old World bird of the genus *Glareola,* with brown and black plumage, long, pointed wings, and a forked tail.

**pratincole**
*9–10 inches long*

**pra·tique** (pră-tēk′) *n.* [Fr., ult. < Lat. *practicus,* practical. —see PRACTICAL.] *Naut.* Clearance granted to a ship to proceed into port after compliance with health or quarantine regulations.

**prat·tle** (prăt′l) *v.* **-tled, -tling, -tles.** [Freq. of PRATE.] —*vi.* To talk idly or meaninglessly : BABBLE. —*vt.* To utter in a silly or childish way. —*n.* Childish or meaningless sounds : BABBLE. **—prat′tler** *n.*

**prawn** (prôn) *n.* [ME *prayne.*] An edible crustacean of the genus *Palaemonetes* and related genera, closely related to and resembling the shrimps. —*vi.* **prawned, prawn·ing, prawns.** To fish for prawns. **—prawn′er** *n.*

**prax·es** (prăk′sēz′) *n. pl. of* PRAXIS.

**prax·i·ol·o·gy** *or* **prax·e·ol·o·gy** (prăk′sē-ŏl′ə-jē) *n.* [PRAXI(S) + -LOGY.] The study of human conduct.

**prax·is** (prăk′sĭs) *n., pl.* **-es** (-sēz′) [Med. Lat. < Gk., action < *prattein,* to do.] **1.** Practical application or exercise of a branch of learning. **2.** Habitual or established practice : CUSTOM.

**pray** (prā) *v.* **prayed, pray·ing, prays.** [ME *preyen* < OFr. *preier* < Lat. *precari* < *prex,* prayer.] —*vi.* **1.** To utter or address a petition to God or another deity. **2.** To make a fervent request : PLEAD. —*vt.* **1.** To say a prayer or prayers to. **2.** To beseech : implore <*Pray* return safely.> **3.** To make a devout or earnest request for <I *pray* your understanding.> **4.** To move or bring by prayer or entreaty.

**pray·er**[1] (prā′ər) *n.* One who prays.

**prayer**[2] (prâr) *n.* [ME *preiere* < OFr. < Med. Lat. *precaria* < Lat. *precarius,* obtained by entreaty < *precari,* to entreat < *prex,* prayer.] **1. a.** A reverent petition made to God or another deity. **b.** The act of making such a petition. **2.** An act of communion with God, as a confession, praise, or thanksgiving. **3.** A specially worded form used to address God. **4. prayers.** A religious service in which praying predominates. **5. a.** An earnest request. **b.** The thing requested. **6.** The slightest chance or hope <hadn't a *prayer* of surviving> **7.** *Law.* **a.** The request of a complainant, as stated in a bill in equity, that the court grant the aid or relief solicited. **b.** The section of the bill that contains this request.

**prayer beads** *pl.n.* A string of beads for keeping count of the prayers one is saying : ROSARY.

**prayer book** *n.* A book of prayers and other forms of worship.

**prayer·ful** (prâr′fəl) *adj.* Inclined to pray : DEVOUT. **—prayer′ful·ly** *adv.* **—prayer′ful·ness** *n.*

**prayer meeting** *n.* An evangelical service, esp. one held on a weekday evening, in which the laity participate by singing, praying, or testifying.

**prayer shawl** *n.* A tallith.

**prayer wheel** *n.* A cylinder inscribed with or containing written prayers and revolved on an axis, used esp. by Tibetan Buddhists.

**praying mantis** *n.* A green or brownish predatory insect, *Mantis religiosa,* that while at rest folds its front legs as if in prayer.

**pre-** *pref.* [ME < OFr. < Lat. *prae-* < *prae,* before, in front.] **1. a.** Earlier : before : prior to <*pre*historic> **b.** Preparatory : preliminary <*pre*medical> **c.** In advance <*pre*pay> **2.** Anterior : in front of <*pre*axial>

**preach** (prēch) *v.* **preached, preach·ing, preach·es.** [ME *prechen* < OFr. *prechier* < LLat. *praedicare* < Lat., to announce : *prae-,* before + *dicare,* to proclaim.] —*vt.* **1.** To expound upon, esp. to urge acceptance of or obedience to <*preach* love> **2.** To deliver or put forth (e.g., religious instruction). —*vi.* **1.** To deliver a sermon. **2.** To give moral or religious instruction, esp. at length and tiresomely.

**preach·er** (prē′chər) *n.* **1.** A Protestant minister. **2.** One who preaches.

**preach·i·fy** (prē′chə-fī′) *vi.* **-fied, -fy·ing, -fies.** *Informal.* To preach tediously and didactically. **—preach′i·fi·ca′tion** *n.*

**preach·ment** (prēch′mənt) *n.* **1.** The act of preaching. **2.** A tiresome or unwelcome moral lecture.

**preach·y** (prē′chē) *adj.* **-i·er, -i·est.** Given to preaching.

**pre·ad·o·les·cence** (prē′ăd-l-ĕs′əns) *n.* The period between childhood and adolescence, usu. designated as between the ages of ten and twelve. **—pre′ad·o·les′cent** *n.* & *adj.*

**pre·ag·ri·cul·tur·al** (prē′ăg-rĭ-kŭl′chər-əl) *adj.* Existing or happening before the advent of agriculture.

**pre·am·ble** (prē′ăm′bəl, prē-ăm′-) *n.* [ME < OFr. *preambule* < Med. Lat. *praeambulum* < LLat. *praeambulus,* walking in front : *prae-,* in front + *ambulare,* to walk.] **1.** A preliminary statement, esp. the introduction to a formal document that explains its purpose. **2.** An introductory fact or occurrence : PRELIMINARY. **—pre·am′bu·lar′y** (-byə-lĕr′ē) *adj.*

**pre·am·pli·fi·er** (prē-ăm′plə-fī′ər) *n.* An electronic circuit or device that detects and sufficiently amplifies weak signals, esp. from a radio receiver, for subsequent amplification stages.

**pre·ar·range** (prē′ə-rānj′) *vt.* **-ranged, -rang·ing, -rang·es.** To arrange in advance. **—pre′ar·range′ment** *n.*

**pre·as·signed** (prē′ə-sīnd′) *adj.* Assigned beforehand.

**pre·a·tom·ic** (prē′ə-tŏm′ĭk) *adj.* Of or relating to the period before the use of atomic energy.

**pre·ax·i·al** (prē-ăk′sē-əl) *adj.* Anatomically positioned in front of a body axis. **—pre·ax′i·al·ly** *adv.*

**preb·end** (prĕb′ənd) *n.* [ME *prebende* < OFr. < Med. Lat. *praebenda* < Lat., state allowance, neuter pl. gerund. of *praebere,* to grant : *prae-,* forward + *habēre,* to hold.] **1.** A stipend drawn from a special endowment granted by a cathedral or church to a member of the clergy. **2.** The property or tithe providing the endowment for a stipend. **3.** PREBENDARY 1. **—pre·ben′dal** (prĭ-bĕn′dl, prĕb′ən-dəl) *adj.*

**preb·en·dar·y** (prĕb′ən-dĕr′ē) *n., pl.* **-ies. 1.** A member of the clergy who receives a prebend. **2.** A member of the Anglican clergy holding the honorary title of prebend without a stipend.

**Pre·cam·bri·an** (prē-kăm′brē-ən) *adj.* Of, belonging to, or designating the oldest and largest division of geologic time, preceding the Cambrian, often subdivided into the Archeozoic and Proterozoic eras, and marked by the appearance of primitive forms of life. —*n.* The Precambrian era.

**pre·can·cel** (prē-kăn′səl) *vt.* **-celed, -cel·ing, -cels** or **-celled, -cel·ling, -cels.** To cancel (a postage stamp) before mailing. —*n.* A precanceled stamp or envelope. **—pre·can′cel·la′tion** *n.*

**pre·can·cer·ous** (prē-kăn′sər-əs) *adj.* Likely or tending to become cancerous.

**pre·car·i·ous** (prĭ-kâr′ē-əs) *adj.* [Lat. *precarius,* obtained by entreaty < *precari,* to entreat < *prex,* prayer.] **1.** Dangerously lacking in stability or security. **2.** Subject to chance. **3.** Based on unproved or uncertain premises. **4.** *Archaic.* Dependent on the will or favor of another. **—pre·car′i·ous·ly** *adv.* **—pre·car′i·ous·ness** *n.*

**prec·a·to·ry** (prĕk′ə-tôr′ē, -tōr′ē) *also* **prec·a·tive** (-tĭv) *adj.* [LLat. *precatorius* < *precari,* to entreat < *prex,* prayer.] Pertaining to or expressing supplication.

**pre·cau·tion** (prĭ-kô′shən) *n.* [Fr. *précaution* < LLat. *praecautio* < Lat. *praecavēre,* to guard against : *prae-,* before + *cavēre,* to beware.] **1.** An action taken in advance to protect against possible danger or failure : SAFEGUARD. **2.** Caution practiced in advance : CIRCUMSPECTION. **—pre·cau′tion·ar′y, pre·cau′tion·al** *adj.*

**pre·cau·tious** (prĭ-kô′shəs) *adj.* Exercising precaution. **—pre·cau′tious·ly** *adv.* **—pre·cau′tious·ness** *n.*

**pre·ca·va** (prē-kā′və, -kä′-) *n., pl.* **-vae** (-vē) [PRE- + (VENA) CAVA.] The superior vena cava. **—pre·ca′val** *adj.*

**pre·cede** (prĭ-sēd′) *v.* **-ced·ed, -ced·ing, -cedes.** [ME *preceden* < OFr. *preceder* < Lat. *praecedere* : *prae-,* before + *cedere,* to go.] —*vt.* **1.** To come before in time. **2.** To come before in rank or order : OUTRANK. **3.** To be in a position in front of. **4.** To introduce : preface <*precede* a sermon with announcements> —*vi.* To come, exist, or occur before.

☆ ***syns:*** PRECEDE, ANTECEDE, ANTEDATE, FORERUN, PREDATE *v. core meaning* : to come, exist, or occur prior to in time <The steam engine *preceded* the diesel engine.> ***ant:*** follow, succeed

**prec·e·dence** (prĕs′ĭ-dəns, prĭ-sēd′ns) *also* **prec·e·den·cy** (prĕs′ĭ-dən-sē, prĭ-sēd′n-sē) *n.* **1.** The act, state, or right of preceding : PRIORITY. **2.** An order of rank observed on formal occasions.

**prec·e·dent** (prĕs′ĭ-dənt) *n.* [ME < OFr. < Lat. *praecedens,* pr.part. of *praecedere,* to go before. —see PRECEDE.] **1. a.** An act or instance used as an example in dealing with subsequent similar cases. **b.** *Law.* A judicial decision used as a standard in subsequent similar cases. **2.** Custom or convention. —*adj.* **pre·ced·ent** (prĭ-sēd′nt, prĕs′ĭ-dənt). Prior : preceding.

**prec·e·den·tial** (prĕs′ĭ-dĕn′shəl) *adj.* **1.** Of or relating to a precedent. **2.** Having precedence.

**pre·ced·ing** (prĭ-sē′dĭng) *adj.* Existing or coming before in time, place, rank, or sequence : PREVIOUS.

**pre·cen·sor** (prē-sĕn′sər) *vt.* **-sored, -sor·ing, -sors.** To censor (e.g., a publication) prior to public release.

**pre·cen·tor** (prĭ-sĕn′tər) *n.* [LLat. *praecentor* < Lat. *praecinere,* to sing before : *prae-,* before + *canere,* to sing.] One who directs a church choir. **—pre′cen·to′ri·al** (prē′sĕn-tôr′ē-əl, -tōr′-) *adj.*

**pre·cept** (prē′sĕpt′) *n.* [ME < Lat. *praeceptum* < *praecipere,* to advise, teach : *prae-,* before + *capere,* to take.] **1.** A principle or rule imposing a standard of action or conduct. **2.** *Law.* A writ.

**pre·cep·tive** (prĭ-sĕp′tĭv) *adj.* **1.** Of or expressing a precept. **2.** Giving precepts : DIDACTIC. **—pre·cep′tive·ly** *adv.*

**pre·cep·tor** (prĭ-sĕp′tər, prē′sĕp′tər) *n.* [ME *preceptur* < Lat. *praeceptor* < *praecipere,* to teach. —see PRECEPT.] An instructor. **—pre′cep·to′ri·al** (prē′sĕp-tôr′ē-əl, -tōr′-) *adj.* **—pre′cep·to′ri·al·ly** *adv.*

**pre·cess** (prē-sĕs′, prē′sĕs′) *vi.* **-cessed, -cess·ing, -cess·es.** [Back-formation < PRECESSION.] To move in or be subjected to precession.

**pre·ces·sion** (prē-sĕsh′ən) *n.* [LLat. *praecessio* < Lat. *praecedere,* to go before. —see PRECEDE.] **1.** The act or state of preceding : PRECEDENCE. **2.** *Physics.* A complex motion executed by a rotating body subjected to a torque tending to change its axis of rotation, marked for constant speed of rotation and constant magnitude of the applied torque by a conical locus of the axis. **3.** *Astron.* Precession of the equinoxes. **—pre·ces′sion·al** *adj.*

**precession of the equinoxes** *n. Astron.* A slow westward shift of the equinoctial points along the plane of the ecliptic at a rate of 50.27 seconds of arc per year, caused by precession of the earth's axis of rotation.

**pre-Chris·tian** (prē-krĭs′chən) *adj.* Of, relating to, or being the time prior to the beginning of Christianity.

**pre·cinct** (prē′sĭngkt′) *n.* [ME *precincte,* an enclosed space < Med. Lat. *praecinctum* < Lat. *praecingere,* to encircle : *prae-,* before + *cingere,* to gird.] **1. a.** A district or subdivision of a city patrolled by a unit of its police. **b.** The police station in such a district. **2.** An election district of a town or city. **3.** *often* **precincts. a.** An enclosure or place marked off by specific limits. **b.** A boundary. **4. precincts.** Environs. **5. precincts.** A province of thought or action.

**pre·ci·os·i·ty** (prĕsh′ē-ŏs′ĭ-tē, prĕs′-) *n., pl.* **-ties.** [ME *preciousite* < OFr. *preciosite* < Lat. *pretiositas* < *pretiosus,* precious < *pretium,* price.] Extreme overrefinement or meticulousness, as in language.

**pre·cious** (prĕsh′əs) *adj.* [ME < OFr. *precios* < Lat. *pretiosus* < *pretium,* price.] **1.** Highly valuable or very costly. **2.** Highly cherished. **3.** Dear : beloved. **4.** Affectedly dainty or overrefined. **5.** *Informal.* Arrant : thoroughgoing. —*adv.* —Used as an intensifier <had *precious* little right to request favors> **—pre′cious·ly** *adv.* **—pre′cious·ness** *n.*

**precious stone** *n.* A mineral, as a diamond, emerald, ruby, or sapphire, valued for its rarity or appearance.

**prec·i·pice** (prĕs′ə-pĭs) *n.* [OFr. < Lat. *praecipitium* < *praeceps,* headlong. —see PRECIPITATE.] **1.** A very steep or overhanging mass of rock, as a crag or the face of a cliff. **2.** The brink of a dangerous situation <nations on the *precipice* of nuclear confrontation>

**pre·cip·i·ta·ble** (prĭ-sĭp′ĭ-tə-bəl) *adj.* [< PRECIPITATE.] Capable of being precipitated.

**pre·cip·i·tant** (prĭ-sĭp′ĭ-tənt) *adj.* [Lat. *praecipitans, praecipitant-,* pr.part. of *praecipitare,* to throw headlong. —see PRECIPITATE.] **1.** Rushing headlong. **2.** Impulsive : rash. **3.** Abrupt and unexpected : SUDDEN. —*n.* An agent causing precipitation. **—pre·cip′i·tance** (-təns), **pre·cip′i·tan·cy** (-tən-sē) *n.* **—pre·cip′i·tant·ly** *adv.*

**pre·cip·i·tate** (prĭ-sĭp′ĭ-tāt′) *v.* **-tat·ed, -tat·ing, -tates.** [Lat. *praecipitare, praecipitat-,* to throw headlong < *praeceps,* headlong : *prae-,* in front + *caput,* head.] —*vt.* **1.** To hurl downward, from or as if from a great height. **2.** To cause to occur before expected or needed : ACCELERATE. **3.** To cause (water vapor) to condense and fall as rain or snow. **4.** *Chem.* To cause (a solid substance) to be separated from a solution. —*vi.* **1.** To condense and fall as snow or rain. **2.** *Chem.* To be separated from a solution as a precipitate. **3.** To fall headlong. —*adj.* (-tĭt). **1.** Moving heedlessly and rapidly. **2.** Acting with excessive haste or impulse. **3.** Happening abruptly or without warning. ***usage:*** *Precipitate* applies to rash and overly hasty acts. *Precipitous* is used to refer to steepness either of a physical nature, as in *a precipitous slope,* or in a figurative sense, as in *a precipitous rise in interest rates.* —*n.* (-tāt′, -tĭt). *Chem.* A solid or solid phase separated from a solution. **—pre·cip'i·tate·ly** (-tĭt-lē) *adv.* **—pre·cip'i·tate·ness** *n.* **—pre·cip'i·ta'tive** *adj.* **—pre·cip'i·ta'tor** *n.*

**pre·cip·i·ta·tion** (prĭ-sĭp'ĭ-tā'shən) *n.* **1.** A headlong rush or fall. **2.** Abrupt, impulsive haste. **3. a.** Water droplets or ice particles, as rain or snow, condensed from atmospheric water vapor and massive enough to fall to the earth's surface. **b.** The amount of precipitation falling in a given area within a given time. **4.** *Chem.* Production of a precipitate.

**pre·cip·i·tin** (prĭ-sĭp'ĭ-tĭn) *n.* [PRECIPIT(ATE) + -IN.] An antibody that reacts with an antigen to produce a precipitate.

**pre·cip·i·tin·o·gen** (prĭ-sĭp'ĭ-tĭn'ə-jən) *n.* An antigen that causes the formation of a specific precipitin.

**pre·cip·i·tous** (prĭ-sĭp'ĭ-təs) *adj.* [Fr. *précipiteux* < OFr. < Lat. *praecipitium,* precipice. —see PRECIPICE.] **1.** Resembling a precipice: very steep. **2.** Having several precipices <*a precipitous* canyon> **3.** PRECIPITATE 2. **—pre·cip'i·tous·ly** *adv.* **—pre·cip'i·tous·ness** *n.*

**pré·cis** (prā'sē, prā-sē') *n., pl.* **pré·cis** (prā'sēz, prā-sēz') [Fr. < *précis,* condensed < OFr. *precis.* —see PRECISE.] A concise summary of the essential facts of a text, as a book or article: ABSTRACT. *—vt.* **-cised, -cis·ing, -cis·es.** To make a précis of.

**pre·cise** (prĭ-sīs') *adj.* [OFr. *precis,* condensed < Lat. *praecisus,* p.part. of *praecidere,* to shorten : *prae-,* in front + *caedere,* to cut.] **1.** Clearly stated or depicted: DEFINITE <*a precise* account> **2.** Capable of, caused by, or designating an action, performance, or process carried out or successively repeated within close specified limits <a *precise* weight> **3.** Exactly corresponding to what is indicated <the *precise* time of arrival> **4.** Strictly distinguished from others. **5.** Distinct and correct in sound or statement <*precise* articulation> **6.** Conforming strictly to rule or proper form <*precise* manners> **—pre·cise'ly** *adv.* **—pre·cise'ness** *n.*

☆ ***syns:*** PRECISE, EXACT, IDENTICAL, VERY *adj. core meaning:* strictly distinguished from others <at that *precise* moment>

**pre·ci·sian** (prĭ-sĭzh'ən) *n.* [< PRECISE.] **1.** A strict and precise adherent to established rules, forms, or standards. **2.** One who is very exacting about the forms of religious observance or moral behavior, esp. an English Puritan of the 16th or 17th cent. **—pre·ci'sian·ism** *n.* **—pre·ci'sian·ist** *n.*

**pre·ci·sion** (prĭ-sĭzh'ən) *n.* [Fr. *précision* < Lat. *praecisio,* a cutting off < *praecidere,* to cut off. —see PRECISE.] The quality or state of being precise. *—adj.* **1.** Used or intended for precise measurement <*precision* instruments> **2.** Made for the least variation from a set standard <*precision* parts> **—pre·ci'sion·ism** *n.*

**pre·ci·sion·ist** (prĭ-sĭzh'ə-nĭst) *n.* One who values precision.

**pre·clin·i·cal** (prē-klĭn'ĭ-kəl) *adj.* Occurring before the diagnosis of disease is possible.

**pre·clude** (prĭ-klōōd') *vt.* **-clud·ed, -clud·ing, -cludes.** [Lat. *praecludere* : *prae-,* in front + *claudere,* to close.] To make impossible or impracticable by prior action: PREVENT. **—pre·clu'sion** (-klōō'zhən) *n.* **—pre·clu'sive** (-klōō'sĭv, -zĭv) *adj.* **—pre·clu'sive·ly** *adv.*

**pre·co·cial** (prĭ-kō'shəl) *adj.* [< NLat. *praecoces,* precocial birds < Lat. *praecox,* premature. —see PRECOCIOUS.] Of or typifying birds that are covered with down and that are able to move about when first hatched.

**pre·co·cious** (prĭ-kō'shəs) *adj.* [Lat. *praecox,* premature < *praecoquere,* to boil before : *prae-,* before + *coquere,* to cook.] **1.** Exhibiting or marked by unusually early development or maturity, esp. in mental aptitude. **2.** *Bot.* Blossoming before the leaves sprout. **—pre·co'cious·ly** *adv.* **—pre·co'cious·ness, pre·coc'i·ty** (-kŏs'ĭ-tē) *n.*

**pre·cog·ni·tion** (prē'kŏg-nĭsh'ən) *n.* [LLat. *praecognitio* < Lat. *praecognoscere,* to foresee : *prae-,* before + *cognoscere,* to know. —see COGNITION.] Clairvoyant knowledge of something prior to its happening. **—pre·cog'ni·tive** *adj.*

**pre·co·lo·ni·al** (prē'kə-lō'nē-əl) *adj.* Of, relating to, or being the time before colonization.

**pre-Co·lum·bi·an** (prē'kə-lŭm'bē-ən) *adj.* Of, pertaining to, or originating in the Americas before the voyages of Columbus.

**pre·con·ceive** (prē'kən-sēv') *vt.* **-ceived, -ceiv·ing, -ceives.** To form an opinion or conception of in advance.

**pre·con·cep·tion** (prē'kən-sĕp'shən) *n.* **1.** An opinion or conception arrived at prior to actual knowledge. **2.** A prejudice.

**pre·con·cert** (prē'kən-sûrt') *vt.* **-cert·ed, -cert·ing, -certs.** To agree on or arrange in advance.

**pre·con·di·tion** (prē'kən-dĭsh'ən) *n.* A condition that must exist or be established before something can happen or be considered: PREREQUISITE. *—vt.* **-tioned, -tion·ing, -tions.** To condition, train, or accustom in advance.

**pre·con·scious** (prē-kŏn'shəs) *adj. Psychoanal.* That is capable of being recalled although not present in the conscious mind. **—pre·con'scious·ly** *adv.*

**pre·cook** (prē-kŏŏk') *vt.* **-cooked, -cook·ing, -cooks.** To cook in advance or cook partially before final cooking.

**pre·crit·i·cal** (prē-krĭt'ĭ-kəl) *adj.* Being or occurring prior to the occurrence of a critical condition.

**pre·cur·sive** (prĭ-kûr'sĭv) *adj.* Precursory.

**pre·cur·sor** (prĭ-kûr'sər, prē'kûr'sər) *n.* [Lat. *praecursor* < *praecurrere,* to run before : *prae-,* before + *currere,* to run.] **1.** One that precedes and indicates or announces another to come: FORERUNNER. **2.** One that precedes another: PREDECESSOR.

**pre·cur·so·ry** (prĭ-kûr'sə-rē) *adj.* **1.** Preceding as a precursor: PRELIMINARY. **2.** Suggesting or designating something to follow.

**pre·da·cious** *or* **pre·da·ceous** (prĭ-dā'shəs) *adj.* [< Lat. *praedari,* to plunder < *praeda,* booty.] PREDATORY 2. **—pre·da'cious·ness, pre·dac'i·ty** (-dăs'ĭ-tē) *n.*

**pre·date** (prē-dāt') *vt.* **-dat·ed, -dat·ing, -dates. 1.** To mark or designate with a date earlier than the actual one. **2.** To precede in time: ANTEDATE.

**pre·da·tion** (prĭ-dā'shən) *n.* [Lat. *praedatio* < *praedari,* to plunder < *praeda,* booty.] **1.** The act or practice of marauding or plundering. **2.** Capture of prey as a means of sustaining life.

**pred·a·tor** (prĕd'ə-tər, -tôr') *n.* [Lat. *praedator,* pillager < *praedari,* to plunder < *praeda,* booty.] **1.** An animal that lives by preying on others. **2.** One who abuses or plunders other people for his or her own gain.

**pred·a·to·ry** (prĕd'ə-tôr'ē, -tōr'ē) *adj.* [Lat. *praedatoris* < *praedari,* to plunder < *praeda,* booty.] **1.** Of, pertaining to, or characterized by plundering, pillaging, or marauding. **2.** Preying on other animals: PREDACIOUS. **3.** Marked by a tendency to victimize or destroy others for one's own benefit. **—pred'a·to'ri·ly** *adv.* **—pred'a·to'ri·ness** *n.*

**pre·dawn** (prē'dôn') *n.* The time just before dawn.

**pre·de·cease** (prē'dĭ-sēs') *vt.* **-ceased, -ceas·ing, -ceas·es.** To die before (another person).

**pred·e·ces·sor** (prĕd'ĭ-sĕs'ər, prē'dĭ-) *n.* [ME *predecessour* < OFr. *predecesseur* < LLat. *praedecessor* : Lat. *prae-,* before + Lat. *decessor,* someone who leaves < *decedere,* to depart (*de-,* away + *cedere,* to go).] **1.** One who comes before another in time, esp. in an office or position. **2.** Something that has been succeeded by another. **3.** An ancestor: forefather.

**pre·des·ti·nar·i·an** (prē-dĕs'tə-nâr'ē-ən) *adj.* **1.** Of or relating to predestination. **2.** Believing in or based on the doctrine of predestination. *—n.* A believer in the doctrine of predestination. **—pre·des'ti·nar'i·an·ism** *n.*

**pre·des·ti·nate** (prē-dĕs'tə-nāt') *vt.* **-nat·ed, -nat·ing, -nates.** [ME *predestinaten* < Lat. *praedestinare.* —see PREDESTINE.] **1.** To determine or destine in advance: FOREORDAIN. **2.** To predestine. *—adj.* (prē-dĕs'tə-nĭt, -nāt'). Foreordained: predestined.

**pre·des·ti·na·tion** (prē-dĕs'tə-nā'shən) *n.* **1.** The act of predestining or the state of being predestined. **2. a.** The act whereby God is believed to have foreordained all things. **b.** Relegation of all souls either to salvation or to damnation by this act. **c.** The doctrine that God has foreordained all things, esp. the salvation of individual souls. **3.** Destiny: fate.

**pre·des·tine** (prē-dĕs'tĭn) *vt.* **-tined, -tin·ing, -tines.** [ME *predestinen* < OFr. *predestiner* < Lat. *praedestinare* : *prae-,* before + *destinare,* to determine.] **1.** To fix upon, decide, or decree in advance: FOREORDAIN. **2.** To foreordain by divine decree or will.

**pre·de·ter·mine** (prē'dĭ-tûr'mĭn) *vt.* **-mined, -min·ing, -mines.** [LLat. *praedeterminare* : Lat. *prae-,* before + Lat. *determinare,* to limit. —see DETERMINE.] **1.** To determine, decide, or establish ahead of time <Weather *predetermines* crop production.> **2.** To influence or sway toward an opinion or action: PREDISPOSE. **—pre'de·ter'mi·na'tion** *n.* **—pre'de·ter'mi·na'tive** (-mə-nā'tĭv, -nə-tĭv) *adj.* **—pre'de·ter'min·er** *n.*

**pre·di·al** (prē'dē-əl) *adj. var. of* PRAEDIAL.

**pred·i·ca·ble** (prĕd'ĭ-kə-bəl) *adj.* [Med. Lat. *praedicabilis* < LLat. *praedicare,* to proclaim. —see PREACH.] Able to be stated or predicated. *—n.* **1.** Something that can be predicated: ATTRIBUTE. **2.** *Logic.* One of five general attributes of a class—genus, species, property, difference, and accident—designating the peculiar relation between a predicate and its subject regardless of the quantity or quality of a proposition. **—pred'i·ca·bil'i·ty, pred'i·ca·ble·ness** *n.*

**pre·dic·a·ment** (prĭ-dĭk'ə-mənt) *n.* [ME, something predicated < LLat. *praedicamentum* < *praedicare,* to proclaim. —see PREACH.] **1.** An embarrassing or troublesome situation. **2.** *Logic.* CATEGORY 2. **—pre·dic'a·men'tal** (-mĕn'tl) *adj.* **—pre·dic'a·men'tal·ly** *adv.*

**pred·i·cate** (prĕd'ĭ-kāt') *v.* **-cat·ed, -cat·ing, -cates.** [Lat. *praedicare, praedicat-,* to proclaim. —see PREACH.] *—vt.* **1.** To base or establish (e.g., a statement) <I *predicate* my opinion on these points.> **2.** To state or affirm as an attribute or quality <*predicate* the restlessness of youth> **3.** To carry the connotation of: IMPLY. **4.** *Logic.* To make (a term or expression) the predicate of a proposition. **5.** To proclaim: assert. *—vi.* To make a statement or assertion. *—n.* (prĕd'ĭ-kĭt). **1.** The part of a sentence or clause that states something about the subject, composed of a verb and often including objects, modifiers, or complements of the verb. **2.** *Logic.* Whatever is stated about the subject of a proposition. *—adj.* (prĕd'ĭ-kĭt). **1.** Of or belonging to the predicate of a sentence or clause. **2.** Stated: predicated. **—pred'i·ca'tive** *adj.* **—pred'i·ca'tive·ly** *adv.*

**predicate nominative** *n.* A noun or pronoun following a linking verb and referring to the same person or thing as the subject of the verb.

**pred·i·ca·tion** (prĕd′ĭ-kā′shən) *n.* **1.** The act or procedure of predicating, esp. a logical assertion or affirmation. **2.** Something predicated. **—pred′i·ca′tion·al** *adj.*

**pred·i·ca·to·ry** (prĕd′ĭ-kə-tôr′ē, -tōr′ē) *adj.* [LLat. *praedicatorius* < Lat. *praedicare,* to proclaim. —see PREACH.] Of, relating to, or typical of preaching or a preacher.

**pre·dict** (prĭ-dĭkt′) *v.* **-dict·ed, -dict·ing, -dicts.** [Lat. *praedicere, praedict-* : *prae-,* before + *dicere,* to say.] —*vt.* To state, tell about, or make known beforehand, esp. on the basis of special knowledge. —*vi.* To foretell what will happen : PROPHESY. **—pre·dict′a·bil′i·ty** *n.* **—pre·dict′a·ble** *adj.* **—pre·dict′a·bly** *adv.* **—pre·dic′tor** *n.*

☆ ***syns:*** PREDICT, FORECAST, FORETELL, PORTEND, PROGNOSTICATE *v. core meaning* : to tell about or make known (future events) in advance, esp. by means of special knowledge or inference <*predicted* the rise in gasoline prices>

**pre·dic·tion** (prĭ-dĭk′shən) *n.* **1.** An act of predicting. **2.** Something predicted : PROPHECY. **—pre·dic′tive** *adj.* **—pre·dic′tive·ly** *adv.* **—pre·dic′tive·ness** *n.*

**pre·di·gest** (prē′dĭ-jĕst′, -dī-jĕst′) *vt.* **-gest·ed, -gest·ing, -gests.** To subject to partial digestion. **—pre′di·ges′tion** *n.*

**pred·i·lec·tion** (prĕd′l-ĕk′shən, prēd′-) *n.* [Fr. *prédilection* < Med. Lat. *praediligere,* to prefer : Lat. *prae-,* before + Lat. *diligere,* to love. —see DILIGENT.] Partiality or disposition in favor of something.

**pre·dis·pose** (prē′dĭ-spōz′) *vt.* **-posed, -pos·ing, -pos·es. 1.** To make (someone) inclined to favor something in advance <was *predisposed* to like the new house> **2.** To make susceptible or liable <conditions that *predispose* migrant farm workers to health problems> **3.** *Archaic.* To settle in advance. **—pre′dis·pos′al** *n.* **—pre′dis·po·si′tion** (prē′dĭs-pə-zĭsh′ən) *n.*

**pred·ni·sone** (prĕd′nĭ-sōn′,-zōn′) *n.* [E. *pregnane,* a hydrocarbon + D(I)- + -(E)N(E) + (CORT)ISONE.] An analog of cortisone, $C_{21}H_{26}O_5$, used as an anti-inflammatory agent for treating arthritis.

**pre·doc·tor·al** (prē-dŏk′tər-əl) *adj.* Of, relating to, or engaged in advanced academic study for a doctorate.

**pre·dom·i·nant** (prĭ-dŏm′ə-nənt) *adj.* [OFr. < Med. Lat. *praedominans,* pr.part. of *praedominari,* to predominate.] **1.** Having greatest ascendancy, importance, influence, authority, or force : PREPONDERANT. **2.** Most common or conspicuous : PREVALENT <the *predominant* geometric pattern in a design> **—pre·dom′i·nance** (-nəns), **pre·dom′i·nan·cy** (-nən-sē) *n.* **—pre·dom′i·nant·ly** *adv.*

**pre·dom·i·nate** (prĭ-dŏm′ə-nāt′) *v.* **-nat·ed, -nat·ing, -nates.** [Med. Lat. *praedominari, praedominat-* : Lat. *prae-,* before + Lat. *dominari,* to rule < *dominus,* master.] —*vi.* **1.** To be of greater power, importance, or quantity : PREPONDERATE. **2.** To have authority, power, or controlling influence. —*vt.* To prevail over. **—pre·dom′i·nate·ly** (-nĭt-lē) *adv.* **—pre·dom′i·nat′ing·ly** *adv.* **—pre·dom′i·na′tion** *n.* **—pre·dom′i·na′tor** *n.*

**pree·mie** *also* **pre·mie** (prē′mē) *n.* [Shortening and alteration of PREMATURE.] *Informal.* An infant born prematurely.

**pre-em·i·nent** or **pre·em·i·nent** (prē-ĕm′ə-nənt) *adj.* [LLat. *praeeminens,* pr.part of Lat. *praeeminēre,* to excel : *prae-,* before + *eminēre,* to stand out.] Superior to or notable above all others : OUTSTANDING. **—pre-em′i·nence** *n.* **—pre-em′i·nent·ly** *adv.*

**pre·em·ploy·ment** (prē′ĕm-ploi′mənt) *adj.* Of or occurring at a time before employment <*preemployment* examinations>

**pre-empt** or **pre·empt** (prē-ĕmpt′) *v.* **-empt·ed, -empt·ing, -empts.** [Back-formation < PRE-EMPTION.] —*vt.* **1.** To acquire ownership of by prior right or opportunity, esp. to settle on (public land) so as to gain the right to buy before others. **2.** To appropriate, seize, or act for oneself before others. **3.** To be presented in place of : DISPLACE <Olympic games coverage *pre-empted* regular network programs.> **4.** To take precedence over <The problem of declining sales *pre-empted* discussion of other matters.> **5.** To gain a pre-eminent place in <a candidate who *pre-empted* all others in the race> —*vi.* To make a pre-emptive bid in bridge. **—pre-emp′tor** (-ĕmp′tôr′) *n.* **—pre·emp′to·ry** (-ĕmp′tə-rē) *adj.*

**pre-emp·tion** or **pre·emp·tion** (prē-ĕmp′shən) *n.* [< Med. Lat. *praeemere,* to buy before : Lat. *prae-,* before + Lat. *emere,* to buy.] **1. a.** The right to purchase something, esp. government-owned land, before others. **b.** A purchase made when such a right is granted. **2.** Acquisition or appropriation of something beforehand.

**pre-emp·tive** or **pre·emp·tive** (prē-ĕmp′tĭv) *adj.* **1.** Of, relating to, or typical of pre-emption. **2.** Having or granted by the right of pre-emption. **3.** Designating or typical of a bid in bridge that is unnecessarily high and is meant to keep the opposing players from bidding. **4.** Composing or pertaining to a military strike made so as to secure an advantage in the face of an impending enemy strike <*pre-emptive* bombing> **—pre-emp′tive·ly** *adv.*

**preen** (prēn) *v.* **preened, preen·ing, preens.** [ME *preinen,* poss. var. of *prouynen,* to prune. —see PRUNE².] —*vt.* **1.** To smooth or clean (feathers) with the bill. —Used of birds. **2.** To groom (oneself) with elaborate care or vanity : PRIMP. **3.** To take pride or satisfaction in (oneself) : GLOAT. —*vi.* To dress up : PRIMP. **—preen′er** *n.*

**pre-en·gi·neered** (prē′ĕn-jə-nîrd′) *adj.* Prefabricated.

**pre-es·tab·lish** *or* **pre·es·tab·lish** (prē′ĭ-stăb′lĭsh) *vt.* **-lished, -lish·ing, -lish·es.** To establish beforehand.

**pre-ex·il·i·an** (prē′ĕg-zĭl′ē-ən, -zĭl′yən, -ĕk-sĭl′ē-ən, -sĭl′yən) *also* **pre-ex·il·ic** (-ĕg-zĭl′ĭk, -ĕk-sĭl′ĭk) *adj.* Relating to the history of the Jewish people before their exile in Babylonia at the end of the 6th cent. B.C.

**pre-ex·ist** *or* **pre·ex·ist** (prē′ĭg-zĭst′) *v.* **-ist·ed, -ist·ing, -ists.** —*vi.* To exist before. —*vt.* To exist before (something) <dinosaurs that *pre-existed* mammals> **—pre′-ex·is′tence** *n.* **—pre′-ex·is′tent** *adj.*

**pre·fab** (prē′făb′) *n.* A prefabricated structure or part.

**pre·fab·ri·cate** (prē-făb′rĭ-kāt′) *vt.* **-cat·ed, -cat·ing, -cates. 1.** To construct or manufacture beforehand. **2.** To construct in standard, easily shipped and assembled sections. **—pre·fab′ri·ca′tion** *n.* **—pre·fab′ri·ca′tor** *n.*

**pref·ace** (prĕf′ĭs) *n.* [ME < OFr. < Med. Lat. *prefatia* < Lat. *praefatio,* something said before < *praefari,* to say before : *prae-,* before + *fari,* to say.] **1. a.** An introductory statement or essay, usu. by the author, explaining the scope, intention, or background of a book : FOREWORD. **b.** The introductory part of a speech. **2.** An introductory approach : PRELIMINARY. **3.** *often* **Preface.** *Rom. Cath. Ch.* A thanksgiving prayer ending with the Sanctus and introducing the canon of the Mass. —*vt.* **-aced, -ac·ing, -ac·es. 1.** To introduce by or supply with an introductory essay or statement. **2.** To serve as an introduction to. **—pref′ac·er** *n.*

**pref·a·to·ry** (prĕf′ə-tôr′ē, -tōr′ē) *also* **pref·a·to·ri·al** (prĕf′ə-tôr′ē-əl, -tōr′-) *adj.* [< Lat. *praefatio,* preface.] Being or functioning as an introductory statement or essay. **—pref′a·to′ri·ly** *adv.*

**pre·fect** *also* **prae·fect** (prē′fĕkt′) *n.* [ME < OFr. < Lat. *praefectus* < p.part. of *praeficere,* to place at the head of : *prae-,* before + *facere,* to make.] **1.** A high civil or military official, as a magistrate or administrator of ancient Rome. **2.** A high administrative police official in some European countries. **3.** The dean in a Jesuit school. **4.** A student officer, esp. in a private school. **—pre·fec′tur·al** (prĭ-fĕk′chər-əl) *adj.* **—pre′fec′ture** (prē′fĕk′chər) *n.*

**prefect apostolic** *n., pl.* **prefects apostolic.** *Rom. Cath. Ch.* A priest with broad jurisdiction in a missionary territory.

**pre·fer** (prĭ-fûr′) *vt.* **-ferred, -fer·ring, -fers.** [ME *preferren* < OFr. *preferer* < Lat. *praeferre* : *prae-,* before + *ferre,* to bear.] **1.** To choose as more desirable : like better <*prefers* sailing to fishing> **2.** *Law.* To give priority or precedence to (a creditor). **3.** *Law.* To file, prosecute, or offer for consideration or resolution, as before a magistrate or court <*prefer* charges> **4.** *Archaic.* To recommend for advancement or appointment.

**pref·er·a·ble** (prĕf′ər-ə-bəl, prĕf′rə-) *adj.* More desirable or worthy : PREFERRED <the *preferable* alternative> **—pref′er·a·bil′i·ty, pref′er·a·ble·ness** *n.* **—pref′er·a·bly** *adv.*

**pref·er·ence** (prĕf′ər-əns, prĕf′rəns) *n.* [Fr. *préférence* < Med. Lat. *praeferentia* < Lat. *praeferre,* to prefer.] **1. a.** An act of preferring or the state of being preferred. **b.** Exercise of choice. **c.** One preferred. **2.** *Law.* **a.** The paying of one or more creditors by an insolvent debtor before or to the exclusion of other creditors. **b.** The right to be so paid. **3.** The granting of precedence or advantage to one over others.

**pref·er·en·tial** (prĕf′ə-rĕn′shəl) *adj.* **1.** Of, having, providing, or securing advantage or preference <*preferential* consideration> **2.** Demonstrating or originating from partiality or preference <*preferential* import duties> **—pref′er·en′tial·ism** *n.* **—pref′er·en′tial·ist** *n.* **—pref′er·en′tial·ly** *adv.*

**preferential shop** *n.* A union shop whose management gives precedence to union members in hiring, promoting, or laying off.

**preferential voting** *n.* A system of voting in which the voter ranks his or her choices according to preference.

**pre·fer·ment** (prĭ-fûr′mənt) *n.* **1.** The act of advancing to a higher position or office : PROMOTION. **2.** A position, appointment, or rank giving advancement.

**preferred stock** *n.* The stock of a corporation with priority or preference over the common stock in the distribution of dividends and assets.

**pre·fig·u·ra·tion** (prē-fĭg′yə-rā′shən) *n.* **1.** The act of representing, suggesting, or imagining in advance : FORESHADOWING. **2.** Something that prefigures. **—pre·fig′ur·a·tive** (-fĭg′yər-ə-tĭv) *adj.* **—pre·fig′ur·a·tive·ly** *adv.* **—pre·fig′ur·a·tive·ness** *n.*

**pre·fig·ure** (prē-fĭg′yər) *vt.* **-ured, -ur·ing, -ures.** [ME *prefiguren* < LLat. *praefigurare* : Lat. *prae-,* before + Lat. *figurare,* to shape < *figura,* shape.] **1.** To suggest, indicate, or represent by an antecedent form or model : PRESAGE <The work of Defoe *prefigured* the development of the novel.> **2.** To imagine or picture to oneself in advance. **—pre·fig′ure·ment** *n.*

**pre·fix** (prē-fĭks′, prē′fĭks′ ) *vt.* **-fixed, -fix·ing, -fix·es.** [OFr. *prefixer* : *pre-,* before (< Lat. *prae-*) + *fixer,* to place < *fixe,* fastened < Lat. *fixus,* p.part. of *figere,* to fasten.] **1.** To put or fix before. **2.** (prē-fĭks′). To settle or arrange in advance. —*n.* (prē′fĭks′). **1.** An affix, such as *mis-* in *mistrust,* put before a word to produce a derivative word or an inflected form. **2.** A title placed before one's name. **—pre′fix′al** *adj.* **—pre′fix′al·ly** *adv.*

**pre·flight** (prē′flīt′) *adj.* Preparing for or occurring before flight <*preflight* instrument check-outs>

**pre·form** (prē′fôrm′) *vt.* **-formed, -form·ing, -forms.** To shape or form in advance.

**pre·for·ma·tion** (prē′fôr-mā′shən) *n.* **1.** An act of shaping or forming in advance. **2.** A now invalidated biological theory that all parts of a future organism exist completely formed in the germ cell and develop only by increasing in size. **—pre′for·ma′tion·ism** *n.*

**pre·fron·tal** (prē-frŭn′tl) *adj.* Of, relating to, or situated in the forward part of the frontal lobe of the brain.

**prefrontal lobotomy** *n.* An operation in which the white fibers joining the prefrontal and frontal lobes of the brain to the thalamus are severed.

**pre·gan·gli·on·ic** (prē-găng′glē-ŏn′ĭk) *adj.* Situated in a position that is proximal or anterior to a ganglion.

**preg·na·ble** (prĕg′nə-bəl) *adj.* [ME *prenable* < OFr. < *prendre,* to capture < Lat. *prehendere.*] Vulnerable to seizure, as a fort. **—preg′na·bil′i·ty** *n.*

**preg·nan·cy** (prĕg′nən-sē) *n., pl.* **-cies. 1. a.** The state of being pregnant. **b.** An instance of being pregnant. **c.** The period during which a developing fetus is carried within the uterus. **2.** The quality of being pregnant, as with significance.

**preg·nant**[1] (prĕg′nənt) *adj.* [ME < Lat. *praegnans.*] **1.** Carrying a developing fetus within the uterus. **2.** Inventive : creative. **3.** Fraught with implication or significance <a *pregnant* pause> **4. a.** Profuse : abounding. **b.** Overflowing : replete. **5.** Producing results : FRUITFUL <a *pregnant* undertaking> **—preg′nant·ly** *adv.*

**preg·nant**[2] (prĕg′nənt) *adj.* [ME *preignant* < OFr., pr.part. of *preindre,* to press < Lat. *premere.*] *Archaic.* Convincing : cogent. — Used of an argument or proof.

**preg·nen·o·lone** (prĕg-nĕn′ə-lōn′) *n.* [E. *pregnene,* a hydrocarbon + -OL[1] + -ONE.] A steroid ketone, $C_{21}H_{32}O_2$, resulting from the oxidation of cholesterol and other steroids.

**pre·heat** (prē-hēt′) *vt.* **-heat·ed, -heat·ing, -heats.** To heat beforehand. **—pre·heat′er** *n.*

**pre·hen·sile** (prē-hĕn′səl, -sīl′) *adj.* [Fr. *préhensile* < Lat. *prehendere,* to grasp.] Adapted for seizing or holding, esp. by wrapping around an object <a *prehensile* tail> **—pre′hen·sil′i·ty** (-sĭl′ĭ-tē) *n.*

**prehensile**
*Two examples of prehensile tails:* (left) *opossum and* (right) *spider monkey*

**pre·hen·sion** (prē-hĕn′shən) *n.* [Lat. *prehensio* < *prehendere,* to seize.] **1.** The act of grasping or seizing. **2. a.** Perception by the senses. **b.** Understanding.

**pre·his·tor·ic** (prē′hĭ-stôr′ĭk, -stŏr′-) *also* **pre·his·tor·i·cal** (-ĭ-kəl) *adj.* Of, relating to, or belonging to the era before recorded history. **—pre′his·tor′i·cal·ly** *adv.*

**pre·his·to·ry** (prē-hĭs′tə-rē) *n.* The history of humankind in the period before recorded history. **—pre′his·tor′i·an** (-hĭ-stôr′ē-ən, -stŏr′-) *n.*

**pre·ig·ni·tion** (prē′ĭg-nĭsh′ən) *n.* Ignition of fuel in an internal-combustion engine before the spark passes through the fuel, caused by a hot spot in the cylinder or by too great a compression ratio for the fuel.

**pre·in·duc·tion** (prē′ĭn-dŭk′shən) *adj.* Happening before induction into military service.

**pre·in·dus·tri·al** (prē′ĭn-dŭs′trē-əl) *adj.* Of, relating to, or occurring before industrialization.

**pre·judge** (prē-jŭj′) *vt.* **-judged, -judg·ing, -judg·es.** To judge beforehand without sufficient evidence. **—pre·judg′er** *n.* **—pre·judg′ment, pre·judge′ment** *n.*

**prej·u·dice** (prĕj′ə-dĭs) *n.* [ME < OFr. < Lat. *praejudicium* : *prae-,* before + *judicium,* judgment < *judex,* judge.] **1. a.** An adverse opinion or judgment formed beforehand or without full knowledge or complete examination of the facts. **b.** A preconceived idea or preference : BIAS. **2.** The act or state of holding unreasonable preconceived judgments or convictions. **3.** Irrational hatred or suspicion of a specific group, race, or religion. **4.** Detriment to one resulting from the preconceived and unfavorable conviction of another or others. *—vt.* **-diced, -dic·ing, -dic·es. 1.** To cause (someone) to judge prematurely and irrationally : BIAS. **2.** To affect injuriously or detrimentally by a judgment or act.

☆ ***syns:*** PREJUDICE, BIGOTRY, INTOLERANCE *n. core meaning* : irrational suspicion or hatred of a particular group, race, or religion <fought racial *prejudice*>

**prej·u·di·cial** (prĕj′ə-dĭsh′əl) *adj.* Of the nature of or causing prejudice. **—prej′u·di′cial·ly** *adv.* **—prej′u·di′cial·ness** *n.*

**prej·u·di·cious** (prĕj′ə-dĭsh′əs) *adj.* Detrimental : prejudicial. **—prej′u·di′cious·ly** *adv.*

**prel·a·cy** (prĕl′ə-sē) *n., pl.* **-cies. 1. a.** The office or station of a prelate. **b.** Prelates as a group. **2.** Church government administrated by prelates.

**pre·lap·sar·i·an** (prē′lăp-sâr′ē-ən) *adj.* [PRE- + Lat. *lapsus,* fall + -ARIAN.] Of or relating to the period before the fall of man.

**prel·ate** (prĕl′ĭt) *n.* [ME *prelat* < OFr. < Med. Lat. *praelatus* < Lat., p.part. of *praeferre,* to carry before : *prae-,* before + *ferre,* to carry.] A high-ranking ecclesiastic, as a bishop or an abbot. **—pre·lat′ic** (prĭ-lăt′ĭk) *adj.*

**prelate nul·li·us** (no͞o-lē′əs) *n.* [PRELATE + NLat. *nullius (dioecesis),* of no (diocese).] *Rom. Cath. Ch.* A prelate, gen. a titular bishop, who has jurisdiction over a territory not in a diocese but subject directly to the Holy See.

**pre·launch** (prē′lônch′, -lŏnch′) *adj.* Preparatory or preliminary to the launch of a missile or spacecraft <*prelaunch* flight tests>

**pre·lect** *also* **prae·lect** (prĭ-lĕkt′) *vi.* **-lect·ed, -lect·ing, -lects.** [Lat. *praelegere, praelect-* : *prae-,* in front of + *legere,* to read.] To discourse or lecture in public. **—pre·lec′tion** *n.* **—pre·lec′tor** *n.*

**pre·li·ba·tion** (prē′lī-bā′shən) *n.* [Lat. *praelibatio* < *praelibare,* to taste beforehand : *prae-,* before + *libare,* to taste.] A foretaste.

**pre·lim** (prē′lĭm′, prĭ-lĭm′) *n. Informal.* PRELIMINARY 2.

**pre·lim·i·nar·y** (prĭ-lĭm′ə-nĕr′ē) *adj.* [Fr. *préliminaire* < Med. Lat. *praeliminaris* : Lat. *prae-,* before + Lat. *limen,* threshold.] Before or preparing for the main matter, action, or business : PREFATORY. *—n., pl.* **-ies. 1.** Something preparatory or antecedent, as a statement or act. **2. a.** An academic test or examination preparatory to one that is longer, more complex, or more important. **b.** An event that precedes the main event of a program, esp. in boxing or wrestling. **3. preliminaries.** Matter, as the title page or preface, that precedes the text of a book. **—pre·lim′i·nar′i·ly** (-nâr′ə-lē) *adv.*

**pre·lit·er·ate** (prē-lĭt′ər-ĭt) *adj.* Of or relating to a culture lacking a written language. **—pre·lit′er·ate** *n.*

**prel·ude** (prĕl′yo͞od′, prā′lo͞od′, prē′-) *n.* [OFr. < Med. Lat. *praeludium* < Lat. *praeludere,* to play beforehand : *prae-,* before + *ludere,* to play < *ludus,* game.] **1.** An introductory performance, event, or act coming before a more important one. **2.** *Mus.* A piece or movement functioning as an introduction to a composition, esp.: **a.** An independent piece of moderate length that comes before a fugue. **b.** The opening section of a suite. **c.** The overture to an opera or oratorio, or a similar piece played before one of the acts of an opera. **d.** A piece played before a church service. **e.** A relatively short composition in a free style, usu. for piano or orchestra. *—v.* **-ud·ed, -ud·ing, -udes.** *—vt.* **1.** To function as a prelude to. **2.** To introduce with or as if with a prelude. *—vi.* To serve as a prelude or introduction. **—prel′ud′er** *n.* **—pre·lu′di·al** (prĭ-lo͞o′dē-əl) *adj.*

**pre·ma·lig·nant** (prē′mə-lĭg′nənt) *adj.* Precancerous.

**pre·mar·i·tal** (prē-măr′ĭ-tl) *adj.* Occurring or existing before marriage <*premarital* agreements> **—pre·mar′i·tal·ly** *adv.*

**pre·ma·ture** (prē′mə-tyo͝or′, -to͝or′, -cho͝or′) *adj.* [Lat. *praematurus* : *prae-,* before + *maturus,* ripe.] **1.** Occurring, growing, or existing before the customary, correct, or assigned time : uncommonly or unexpectedly early <a *premature* departure> **2.** Born after a gestation period of less than the normal time <a *premature* baby> **—pre′ma·ture′ly** *adv.* **—pre′ma·ture′ness, pre′ma·tu′ri·ty** *n.*

**pre·max·il·la** (prē′măk-sĭl′ə) *n., pl.* **-il·lae** (-sĭl′ē). Either of two bones in front of and between the maxillary bones in the upper jaw of vertebrates. **—pre·max′il·lar′y** (-măk′sə-lĕr′ē) *adj.*

**pre·med** (prē′mĕd′) *Informal.* *—adj.* Premedical. *—n.* A premedical student.

**pre·med·i·cal** (prē-mĕd′ĭ-kəl) *adj.* Preparing for or relating to the studies that prepare one for the study of medicine.

**pre·med·i·tate** (prē-mĕd′ĭ-tāt′) *v.* **-tat·ed, -tat·ing, -tates.** [Lat. *praemeditari, praemeditat-* : *prae-,* before + *meditari,* to consider.] *—vt.* To plan, arrange, or plot (a deed or events) in advance. *—vi.* To deliberate or meditate beforehand. **—pre·med′i·ta′tive** *adj.* **—pre·med′i·ta′tor** *n.*

**pre·med·i·tat·ed** (prē-mĕd′ĭ-tā′tĭd) *adj.* Marked by deliberate purpose, prior consideration, and planning. **—pre·med′i·tat′ed·ly** *adv.*

**pre·med·i·ta·tion** (prē-mĕd′ĭ-tā′shən) *n.* **1.** The act of speculating, arranging, or plotting in advance. **2.** Contemplation and plotting of a crime in advance, showing intent to commit the crime.

**pre·men·stru·al** (prē-mĕn′stro͞o-əl) *adj.* Of, relating to, or occurring just before menstruation. **—pre·men′stru·al·ly** *adv.*

**pre·mie** (prē′mē) *n. var. of* PREEMIE.

**pre·mi·er** (prē′mē-ər, prĕm′ē-, prĭ-mîr′) *adj.* [ME *primier* < OFr. *premier* < Lat. *primarius* < *primus,* first.] **1.** First in status or importance : CHIEF. **2.** First to happen or exist : EARLIEST. *—n.* (prĭ-mîr′). **1.** A prime minister. **2.** The chief executive of a Canadian province. **—pre·mier′ship′** *n.*

**pre·mière** (prĭ-mîr′, -myâr′) *n.* [Fr. < fem. of *premier,* first < OFr. —see PREMIER.] **1.** The first public performance, as of a movie or play. **2.** The leading lady of a theatrical company. *—v.* **-mièred,**

**-mièr·ing, -mières.** —*vt.* To present the first public performance of. —*vi.* To have the first public performance. —*adj.* Paramount.

**pre·mil·le·nar·i·an** (prē-mĭl'ə-nâr'ē-ən) *adj.* Of or relating to premillennialism. —*n.* A believer in premillennialism. **—pre·mil'le·nar'i·an·ism** *n.*

**pre·mil·len·ni·al** (prē'mĭ-lĕn'ē-əl) *adj.* Of or occurring prior to the millennium. **—pre'mil·len'ni·al·ly** *adv.*

**pre·mil·len·ni·al·ism** (prē'mĭ-lĕn'ē-ə-lĭz'əm) *n.* The belief that Christ's Second Coming will occur immediately before the millennium. **—pre'mil·len'ni·al·ist** *n.*

**prem·ise** (prĕm'ĭs) *n.* [ME *premisse* < OFr. < Med. Lat. *praemissa* < Lat., p.part. of *praemittere,* to set in front : *prae-,* before + *mittere,* to send.] **1. a.** A proposition on which an argument is based or from which a conclusion is drawn. **b.** *Logic.* One of the first two propositions of a syllogism, from which the conclusion is drawn. **2. premises.** *Law.* The preliminary or explanatory facts or statements of a document, as in an equity bill or deed. **3. premises. a.** Land and the buildings on it. **b.** A building or section of a building. —*v.* **-ised, -is·ing, -is·es.** —*vt.* **1.** To state in advance as an explanation or introduction. **2.** To state or assume as a proposition in an argument. —*vi.* To make a premise.

**pre·mi·um** (prē'mē-əm) *n.* [Lat. *praemium,* profit : *prae-,* before + *emere,* to take.] **1.** A prize given for a particular act. **2.** Something offered free or at a lowered price as an inducement to buy. **3.** A sum of money or bonus paid on top of a regular price, salary, or other amount. **4.** The amount paid, often in addition to the interest, to secure a loan. **5.** The amount paid or payable, often in installments, for an insurance policy. **6.** Valuation of something above its par or nominal value, as money or securities. **7.** Payment for training in a profession or trade. **8.** An unusual or high value <put a *premium* on integrity and diligence>

**pre·mix** (prē'mĭks') *n.* A packaged mixture, as powdered ingredients for a cake, that is prepared in advance and intended to be blended later with other ingredients, as eggs or water.

**pre·mo·lar** (prē-mō'lər) *n.* One of eight bicuspid teeth located in pairs on both sides of the upper and lower jaws behind the canines and in front of the molars. **—pre·mo'lar** *adj.*

**pre·mo·ni·tion** (prē'mə-nĭsh'ən, prĕm'ə-) *n.* [OFr. *premonicion* < LLat. *praemonitio* < Lat. *praemonēre,* to forewarn : *prae-,* before + *monēre,* to warn.] **1.** A forewarning. **2.** A presentiment of the future : FOREBODING. **—pre·mon'i·to'ri·ly** (-mŏn'ĭ-tôr'ə-lē, -tōr'-) *adv.* **—pre·mon'i·to'ry** *adj.*

**pre·morse** (prĭ-môrs') *adj.* [Lat. *praemorsus,* p.part. of *praemordēre,* to bite off in front : *prae-,* in front + *mordēre,* to bite.] *Biol.* Abruptly truncated, as though bitten or broken off.

**pre·mu·ni·tion** (prē'myoo-nĭsh'ən) *n.* [Lat. *praemunitio,* fortification beforehand : *prae-,* before + *munire,* to fortify.] Relative immunity to major infection as a result of inducing an active low-grade infection. **—pre·mune'** (prē-myoon') *adj.*

**pre·name** (prē'nām') *n.* A forename.

**pre·na·tal** (prē-nāt'l) *adj.* Existing or occurring before birth. **—pre·na'tal·ly** *adv.*

**pre·oc·cu·pan·cy** (prē-ŏk'yə-pən-sē) *n.* **1.** The act or right of taking possession or occupying before others. **2.** The state of being preoccupied or engrossed.

**pre·oc·cu·pa·tion** (prē-ŏk'yə-pā'shən) *n.* **1.** The state of being preoccupied : ABSORPTION. **2.** Something that engrosses the mind <Winning the tournament was their sole *preoccupation.*> **3.** Occupation or possession in advance.

**pre·oc·cu·pied** (prē-ŏk'yə-pīd') *adj.* **1. a.** Absorbed in thought : ENGROSSED. **b.** Excessively concerned with something : DISTRACTED. **2.** Formerly or already occupied. **3.** Already used and therefore unavailable for additional use. —Used of taxonomic names.

**pre·oc·cu·py** (prē-ŏk'yə-pī') *vt.* **-pied, -py·ing, -pies. 1.** To occupy the mind or attention of completely : ENGROSS. **2.** To take possession of in advance or before another.

**pre·op·er·a·tive** (prē-ŏp'ər-ə-tĭv, -ŏp'rə-, -ŏp'ə-rā'-) *adj.* Occurring or performed prior to surgery. **—pre·op'er·a·tive·ly** *adv.*

**pre·or·bit·al** (prē-ôr'bĭ-tl) *adj.* Occurring before establishment of an orbit, esp. the orbital flight of a spacecraft.

**pre·or·dain** (prē'ôr-dān') *vt.* **-dained, -dain·ing, -dains.** To appoint, decree, or ordain in advance : FOREORDAIN. **—pre'or·dain'·ment** *n.* **—pre·or'di·na'tion** (-ôr'dn-ā'shən) *n.*

**prep** (prĕp) *adj. Informal.* Preparatory <a *prep* program> —*n.* **1.** *Informal.* A preparatory school. **2.** *Chiefly Brit.* Preparation of lessons : HOMEWORK. —*v.* **prepped, prep·ping, preps.** *Informal.* —*vi.* **1.** To attend a preparatory school. **2.** To study or train in preparation for something <was *prepping* for law boards> —*vt.* To prepare (someone) for medical examination or surgery.

**pre·pack·age** (prē-păk'ĭj) *vt.* **-aged, -ag·ing, -ag·es.** To wrap or package (products) before selling.

**prep·a·ra·tion** (prĕp'ə-rā'shən) *n.* **1.** The act or process of preparing. **2.** The state of being prepared : READINESS. **3.** *often* **preparations.** Preliminary measures that serve to make ready for something <*preparations* for the banquet> **4.** A substance, as a medicine, prepared for a specific purpose. **5.** *Mus.* **a.** Anticipation of a dissonant tone by means of its introduction as a consonant tone in the preceding chord. **b.** The tone so used.

**pre·par·a·tive** (prĭ-păr'ə-tĭv, -pâr'-) *adj.* Serving or tending to prepare. **—pre·par'a·tive** *n.* **—pre·par'a·tive·ly** *adv.*

**pre·par·a·tor** (prĭ-păr'ə-tər, -pâr'-) *n.* One who prepares specimens for display or for scientific investigation.

**pre·par·a·to·ry** (prĭ-păr'ə-tôr'ē, -tōr'ē, -pâr'-) *adj.* **1.** Serving to make ready or prepare : INTRODUCTORY. **2.** Occupied in or relating to preparation, esp. for admission to college. **—pre·par'a·to'ri·ly** *adv.*

**preparatory school** *n.* A usu. private secondary school, preparing students for college or, in Great Britain, for public school.

**pre·pare** (prĭ-pâr') *v.* **-pared, -par·ing, -pares.** [ME *preparen* < OFr. *preparer* < Lat. *praeparare* : *prae-,* before + *parare,* to make ready.] —*vt.* **1.** To make ready in advance for a particular purpose, event, or occasion. **2.** To put together or make by combining various elements or ingredients : MANUFACTURE. **3.** To fit out : EQUIP <The hikers were *prepared* for two weeks on the trail.> **4.** *Mus.* To lead up to and soften (a dissonance or its impact) by means of preparation. —*vi.* To get ready. **—pre·par'ed·ly** (-pâr'ĭd-lē) *adv.* **—pre·par'er** *n.*

**pre·par·ed·ness** (prĭ-pâr'ĭd-nĭs) *n.* The state of being prepared <military *preparedness* for combat>

**pre·pay** (prē-pā') *vt.* **-paid, -pay·ing, -pays.** To pay or pay for in advance. **—pre·pay'ment** *n.*

**pre·pense** (prĭ-pĕns') *adj.* [Alteration of obs. *purpensed* < ME, p.part. of *purpensen,* to premeditate < OFr. *pourpenser* : *pour,* before (< Lat. *pro-*) + *penser,* to think < Lat. *pensare,* freq. of *pendere,* to weigh.] Contemplated or arranged beforehand : PREMEDITATED <malice *prepense*> **—pre·pense'ly** *adv.*

**pre·plan** (prē-plăn') *vi.* & *vt.* **-planned, -plan·ning, -plans.** To make plans or plan beforehand.

**pre·pon·der·ance** (prĭ-pŏn'dər-əns) *also* **pre·pon·der·an·cy** (-ən-sē) *n.* Superiority in weight, quantity, power, or importance.

**pre·pon·der·ant** (prĭ-pŏn'dər-ənt) *adj.* Having superior power, force, or importance : PREDOMINANT. **—pre·pon'der·ant·ly** *adv.*

**pre·pon·der·ate** (prĭ-pŏn'də-rāt') *vi.* **-at·ed, -at·ing, -ates.** [Lat. *praeponderare, praeponderat-* : *prae-,* in front of + *ponderare,* to weigh < *pondus,* weight.] **1.** To be greater than something else in weight. **2.** To be greater in power, force, quantity, or importance : PREDOMINATE. **3.** *Archaic.* To be weighed down, as one end of a balance. —*adj.* (-dər-ĭt). Preponderant. **—pre·pon'der·ate·ly, pre·pon'der·at'ing·ly** *adv.*

**prep·o·si·tion** (prĕp'ə-zĭsh'ən) *n.* [ME *preposicioun* < Lat. *praepositio* < *praeponere,* to put in front : *prae-,* in front + *ponere,* to put.] **1.** A word in some languages that shows the relation between a substantive and a verb, an adjective, or another substantive, as English *at, by, in, to, from,* and *with.* **2.** A word or construction having a function similar to a preposition, as *in regard to* or *concerning.* ***usage:*** There is nothing inherently incorrect about ending a sentence with a preposition, although such placement may be awkward (*the spoon to stir the soup with*) or it may provide a weak ending (*a place we spent a lot of time in*). But often the final position is the only natural one for the preposition, as in *We have much to be thankful for.* Rephrasing the example immediately preceding would result in an awkward or stilted sentence; in such cases the natural order should be kept. **—prep'o·si'tion·al** (-zĭsh'ə-nəl) *adj.* **—prep'o·si'·tion·al·ly** *adv.*

**prepositional phrase** *n.* A phrase made up of a preposition and the noun it governs and having adjectival or adverbial value.

**pre·pos·i·tive** (prĭ-pŏz'ĭ-tĭv) *adj.* [LLat. *praepositivus* < Lat. *praeponere,* to put in front. —see PREPOSITION.] Put before, as a word or particle : PREFIXED. —*n.* A word or particle put before another word. **—pre·pos'i·tive·ly** *adv.*

**pre·pos·sess** (prē'pə-zĕs') *vt.* **-sessed, -sess·ing, -sess·es. 1.** To preoccupy the mind of to the exclusion of other feelings or thoughts. **2.** To influence in advance against or in favor of : PREJUDICE. **3.** To impress favorably beforehand.

**pre·pos·sess·ing** (prē'pə-zĕs'ĭng) *adj.* **1.** Impressing favorably : PLEASING. **2.** *Archaic.* Causing prejudice. **—pre'pos·sess'ing·ly** *adv.* **—pre'pos·sess'ing·ness** *n.*

**pre·pos·ses·sion** (prē'pə-zĕsh'ən) *n.* **1.** A preconception or prejudice. **2.** The state of being preoccupied, as with thoughts.

**pre·pos·ter·ous** (prĭ-pŏs'tər-əs) *adj.* [Lat. *praeposterus* : *prae-,* before + *posterus,* coming after < *post,* after.] That is contrary to nature, reason, or common sense : ABSURD. **—pre·pos'ter·ous·ly** *adv.* **—pre·pos'ter·ous·ness** *n.*

**pre·po·tent** (prē-pōt'nt) *also* **pre·po·ten·tial** (prē'pə-tĕn'shəl) *adj.* [ME < Lat. *praepotens,* pr.part. of *praeposse,* to be more powerful : *prae-,* before + *posse,* to be powerful.] Very great in power, influence, or force : PREDOMINANT. **—pre·po'ten·cy** (prē-pōt'n-sē) *n.* **—pre·po'tent·ly** *adv.*

**prep·pie** or **prep·py** (prĕp'ē) *n., pl.* **-pies.** [Shortening and alteration of PREPARATORY SCHOOL.] *Informal.* **1.** A student in a preparatory school. **2.** A student or young adult whose dress and manner are traditional and conservative. **—prep'pie, prep'py** *adj.*

**pre·proc·ess** (prē-prŏs'ĕs', -prō'sĕs') *vt.* **-essed, -ess·ing, -ess·es.** To perform preliminary processing on.

**pre·pro·duc·tion** (prē'prə-dŭk'shən) *adj.* **1.** Occurring or existing before production <*preproduction* arrangements> **2.** Of, relating to, or being a prototype <*preproduction* models of next year's cars>

**pre·pro·fes·sion·al** (prē'prə-fĕsh'ə-nəl) *adj.* Preparatory to the practice of a profession or to its specialized field of study.

**pre·pro·gram** (prē-prō'grăm', -grəm) *vt.* **-grammed, -gram·ming, -grams** *or* **-gramed, -gram·ing, -grams.** To program beforehand.

**prep school** *n. Informal.* A preparatory school.

**pre·pu·ber·ty** (prē-pyōō'bər-tē) *n.* The period just before puberty.

**pre·pu·bes·cence** (prē'pyōō-bĕs'əns) *n.* Prepuberty.

**pre·pu·bes·cent** (prē'pyōō-bĕs'ənt) *adj.* Of or pertaining to prepuberty <*prepubescent* children>

**pre·pub·li·ca·tion** (prē-pŭb'lĭ-kā'shən) *adj.* Of or relating to a time just before the publication date of a book.

**pre·puce** (prē'pyōōs') *n.* [ME < OFr. < Lat. *praeputium.*] **1.** The loose fold of skin that covers the glans of the penis. **2.** A loose fold of skin covering the glans of the clitoris. **—pre·pu'tial** (-pyōō'shəl) *adj.*

**pre·punch** (prē-pŭnch') *vt.* **-punched, -punch·ing, -punch·es.** To punch data cards or tape prior to an anticipated use.

**pre-Raph·a·el·ite** (prē-răf'ē-ə-līt', -rā'fē-) *n.* A painter or writer belonging to or influenced by the pre-Raphaelite Brotherhood, a society founded in 1848 to revive the spirit and style of Italian painting before Raphael. *—adj.* Of, relating to, or typical of the pre-Raphaelites. **—pre-Raph'a·el·it'ism** *n.*

**pre·re·cord** (prē'rĭ-kôrd') *vt.* **-cord·ed, -cord·ing, -cords.** To record (e.g., a TV program) at an earlier time for later broadcasting.

**pre·reg·is·tra·tion** (prē-rĕj'ĭ-strā'shən) *n.* Early registration, as for returning college students, that occurs prior to general registration.

**pre·req·ui·site** (prē-rĕk'wĭ-zĭt) *adj.* Required as a prior condition. **—pre·req'ui·site** *n.*

**pre·rog·a·tive** (prĭ-rŏg'ə-tĭv) *n.* [ME < OFr. < Lat. *praerogativa* < *praerogativus,* asked first : *prae-,* before + *rogare,* to ask.] **1.** An exclusive, esp. hereditary or official right or privilege. **2.** Any exclusive right or privilege. **3.** A natural gift or advantage making one superior. **4.** Superiority : pre-eminence. *—adj.* Of, stemming from, or exercising a prerogative. **—pre·rog'a·tived** *adj.*

**pres·age** (prĕs'ĭj) *n.* [ME < Lat. *praesagium* < *praesagire,* to perceive beforehand : *prae-,* before + *sagire,* to perceive acutely.] **1.** An indication of a future happening : OMEN. **2.** A feeling of what is going to take place : PRESENTIMENT. **3.** Prophetic significance or meaning. **4.** A prediction. *—v.* **pre·sage** (prĭ-sāj', prĕs'ĭj) **-saged, -sag·ing, -sag·es.** *—vt.* **1.** To indicate or warn of in advance : PORTEND. **2.** To have a presentiment of. **3.** To foretell or predict. *—vi.* To make or utter a prediction. **—pre·sage'ful** (prĭ-sāj'fəl) *adj.* **—pre·sag'er** (prĭ-sā'jər) *n.*

**pres·by·o·pi·a** (prĕz'bē-ō'pē-ə, prĕs'-) *n.* [Gk. *presbus,* old man + -OPIA.] Inability of the eye to focus sharply on close objects, caused by hardening of the crystalline lens with advancing age. **—pres'by·op'ic** (-ŏp'ĭk, -ō'pĭk) *adj.*

**pres·by·ter** (prĕz'bĭ-tər, prĕs'-) *n.* [LLat. < Gk. *presbuteros* < comp. of *presbus,* old man.] **1.** An elder of the congregation in the early Christian Church. **2.** A priest in various hierarchical churches. **3.** A teaching or a ruling elder in the Presbyterian Church.

**pres·byt·er·ate** (prĕz-bĭt'ər-ĭt, -ə-rāt', prĕs-) *n.* **1.** The office of a presbyter. **2.** A body or order of presbyters.

**pres·by·te·ri·al** (prĕz'bĭ-tîr'ē-əl, prĕs'-) *adj.* Of or relating to a presbyter or the presbytery. **—pres'by·te'ri·al·ly** *adv.*

**pres·by·te·ri·an** (prĕz'bĭ-tîr'ē-ən, prĕs'-) *adj.* **1.** Of or relating to ecclesiastical government by presbyters. **2. Presbyterian.** Of or pertaining to a Presbyterian Church. *—n.* **Presbyterian.** A member or adherent of a Presbyterian Church. **—pres'by·te'ri·an·ism** *n.*

**Presbyterian Church** *n.* Any of various Protestant churches governed by presbyters and traditionally Calvinist in doctrine.

**pres·by·ter·y** (prĕz'bĭ-tĕr'ē, prĕs'-) *n., pl.* **-ies.** [LLat. *presbyterium,* council of elders < Gk. *presbuterion* < *presbuteros,* elder. —see PRESBYTER.] **1. a.** A court consisting of Presbyterian Church ministers and representative elders of a specific locality. **b.** The district represented by this court. **2.** Presbyters as a group. **3.** Church government by presbyters. **4.** The section of a church reserved for the clergy. **5.** *Rom. Cath. Ch.* A priest's residence.

**pre·school** (prē'skōōl') *adj.* Of, relating to, or designed for a child of nursery-school age. *—n.* A nursery school. **—pre'school'er** *n.*

**pre·sci·ence** (prē'shē-əns, -shəns, prĕsh'ē-əns, prĕsh'əns) *n.* Knowledge of events or actions before they happen : FORESIGHT.

**pre·sci·ent** (prē'shē-ənt, -shənt, prĕsh'ē-ənt, prĕsh'ənt) *adj.* [Lat. *praesciens, praescient-,* pr.part. of *praescire,* to know beforehand : *prae-,* before + *scire,* to know.] **1.** Of or relating to prescience. **2.** Having prescience. **—pre'sci·ent·ly** *adv.*

**pre·sci·en·tif·ic** (prē-sī'ən-tĭf'ĭk) *adj.* Of, relating to, or taking place at a time before the beginning of modern science and the application of its methods.

**pre·scind** (prĭ-sĭnd') *v.* **-scind·ed, -scind·ing, -scinds.** [Lat. *praescindere,* to cut off in front : *prae-,* in front + *scindere,* to cut off.] *—vt.* To consider individually. *—vi.* To withdraw one's attention.

**pre·screen** (prē-skrēn') *vt.* **-screened, -screen·ing, -screens.** **1.** To view (a motion picture) prior to release for public showing. **2.** To screen in advance <*prescreen* candidates for the position>

**pre·scribe** (prĭ-skrīb') *v.* **-scribed, -scrib·ing, -scribes.** [Lat. *praescribere* : *prae-,* before + *scribere,* to write.] *—vt.* **1.** To set down as a rule or guide : ENJOIN. **2.** *Med.* To recommend or order the use of (a drug or treatment). *—vi.* **1.** To establish rules, laws, or directions. **2.** *Med.* To recommend or order a drug, remedy, or treatment. **3.** *Law.* **a.** To assert a title or right to something on the grounds of prescription. **b.** To become invalidated or unenforceable by prescription. **—pre·scrib'er** *n.*

**pre·script** (prē'skrĭpt') *n.* [Lat. *praescriptum* < *praescribere,* to order. —see PRESCRIBE.] Something prescribed, esp. a rule of conduct. *—adj.* (prē'skrĭpt', prĭ-skrĭpt'). Established as a rule : PRESCRIBED.

**pre·scrip·ti·ble** (prĭ-skrĭp'tə-bəl) *adj.* **1.** Capable of being prescribed. **2.** Requiring or derived from prescription. **—pre·scrip'ti·bil'i·ty** *n.*

**pre·scrip·tion** (prĭ-skrĭp'shən) *n.* [Lat. *praescriptio,* precept < *praescribere,* to order. —see PRESCRIBE.] **1.** An act of prescribing. **2.** *Med.* **a.** A physician's written instruction for preparation and administration of a medication. **b.** A prescribed medication. **c.** An ophthalmologist's or optometrist's written instruction for grinding of corrective lenses. **3.** A rule. **4.** *Law.* **a.** The process of acquiring title to property by reason of uninterrupted possession of specified duration. **b.** The time limit beyond which an action, debt, or crime is no longer enforceable or valid.

**prescription drug** *n.* A controlled drug available only by the order of a physician's prescription.

**pre·scrip·tive** (prĭ-skrĭp'tĭv) *adj.* **1.** Sanctioned by custom or usage of long duration. **2.** Making or giving injunctions, directions, laws, or rules. **3.** *Law.* Acquired by or based on uninterrupted possession. **—pre·scrip'tive·ly** *adv.* **—pre·scrip'tive·ness** *n.*

**pre·sell** (prē-sĕl') *vt.* **-sold** (-sōld'), **-sell·ing, -sells.** To promote (a product not yet on the market).

**pres·ence** (prĕz'əns) *n.* **1.** The state or fact of being present. **2.** Immediate proximity in time or space. **3.** The area immediately surrounding a great personage, esp. a sovereign. **4.** One who is present. **5. a.** Manner of carrying oneself. **b.** Self-assurance and confidence. **6.** A supernatural influence felt to be close at hand.

**presence of mind** *n.* The ability to think and act efficiently, esp. in an emergency.

**pres·ent¹** (prĕz'ənt) *n.* [ME < OFr. < Lat. *praesens,* pr.part. of *praeesse,* to be present : *prae-,* in front + *esse,* to be.] **1.** A moment or period in time perceptible as intermediate between past and future : NOW. **2. a.** The present tense. **b.** A verb form in the present tense. **3. presents.** The document or instrument in question <be it known by these *presents*> *—adj.* **1.** Being, relating to, or happening at a moment or period in time regarded as the present. **2.** Being at hand. **3.** *Obs.* Alert to circumstances : ATTENTIVE. **4.** *Archaic.* Readily available : IMMEDIATE. **5.** Denoting a verb tense or form that expresses current time. **—at present.** Right now. **—for the present.** For the time being. **—pres'ent·ness** *n.*

**pre·sent²** (prĭ-zĕnt') *vt.* **-sent·ed, -sent·ing, -sents.** [ME *presenten* < OFr. *presentare,* to show < *praesens,* pr.part. of *praeesse,* to be in front of. —see PRESENT¹.] **1. a.** To introduce, esp. with formal ceremony. **b.** To introduce (a young woman) to society with conventional ceremony. **2.** To bring before the public <*present* a new play> **3. a.** To make a gift or award of. **b.** To make a gift to. **4.** To offer to view : DISPLAY <*present* one's diplomatic credentials> **5.** To offer for consideration. **6.** To salute with or aim (a weapon). **7.** To recommend (a cleric) for a benefice. **8.** *Law.* **a.** To offer to a legislature or court for consideration. **b.** To bring a charge or indictment against. *—n.* **pres·ent** (prĕz'ənt). A gift. **—pre·sent'er** *n.*

**pre·sent·a·ble** (prĭ-zĕn'tə-bəl) *adj.* **1.** Capable of being given, displayed, or offered. **2.** Suitable for introduction to others. **—pre·sent'a·bil'i·ty, pre·sent'a·ble·ness** *n.* **—pre·sent'a·bly** *adv.*

**pres·en·ta·tion** (prĕz'ən-tā'shən, prē'zən-) *n.* **1.** An act of presenting or the state of being presented. **2.** A performance, as of a drama. **3.** Something presented, as a military decoration. **4. a.** A formal introduction. **b.** A social debut. **5.** The act or right of naming a member of the clergy to a benefice. **6.** *Med.* The position of the fetus in the uterus at birth with respect to the mouth of the uterus <a breech *presentation*> **—pres'en·ta'tion·al** *adj.*

**pre·sent·a·tive** (prĭ-zĕn'tə-tĭv) *adj.* **1.** Having the capacity or function of bringing an idea or image to mind. **2. a.** Perceived or capable of being perceived directly. **b.** Having the ability to so perceive. **3.** Capable of naming or of being named to a benefice. **—pre·sent'a·tive·ness** *n.*

**pres·ent-day** (prĕz'ənt-dā') *adj.* Current <*present-day* issues>

**pres·ent·ee** (prĕz'ən-tē', prĭ-zĕn'-) *n.* **1.** One who is presented. **2.** One to whom something is given.

**pre·sen·tient** (prē-sĕn'shənt, -shē-ənt, -zĕn'-, prĭ-) *adj.* [Lat. *praesentiens, praesentient-,* pr.part. of *praesentire,* to feel beforehand. —see PRESENTIMENT.] Having a presentiment.

**pre·sen·ti·ment** (prĭ-zĕn'tə-mənt) *n.* [Obs. Fr. < OFr. *presentir,* to feel beforehand < Lat. *praesentire* : *prae-,* before + *sentire,* to feel.]

A sense of something about to occur : PREMONITION. **—pre·sen′ti·men′tal** (-mĕn′tl) *adj.*

**pres·ent·ly** (prĕz′ənt-lē) *adv.* **1.** In a short time : SOON <will leave *presently*> **2.** At this time or period : NOW <is *presently* traveling abroad> ***usage:*** Because *presently* may be used to mean not only "soon" but also "at the present time," care should be taken to see that the intended sense is clear from the context. **3.** *Archaic.* At once : IMMEDIATELY.

**pre·sent·ment** (prĭ-zĕnt′mənt) *n.* **1.** An act of presenting. **2.** Something presented. **3.** *Law.* **a.** The act of submitting or presenting a formal statement of a legal matter to a court or authorized person. **b.** The report concerning an offense written by a grand jury and based on the jury's own knowledge and observation. **4.** An act of presenting a note or bill for payment.

**present participle** *n.* A participle expressing present action, in English formed by the infinitive plus *-ing* and used to express present action in relation to the time indicated by the finite verb in its clause, to form progressive tenses with modal auxiliaries, and to function as a verbal adjective.

**present perfect** *n.* **1.** The verb tense expressing action completed at the present time, formed in English by combining the present tense of *have* with a past participle, as in *We have given.* **2.** A verb in the present perfect tense.

**present tense** *n.* The verb tense expressing action in the present time, as in *It stings.*

**pres·er·va·tion·ist** (prĕz′ər-vā′shə-nĭst) *n.* An advocate of preservation, as of a building with historical value.

**pre·ser·va·tive** (prĭ-zûr′və-tĭv) *adj.* Tending to preserve or capable of preserving. —*n.* A preserving agent, esp. a chemical used in foods to inhibit spoilage.

**pre·serve** (prĭ-zûrv′) *v.* **-served, -serv·ing, -serves.** [ME *preserven* < OFr. *preserver* < LLat. *praeservare* : Lat. *prae-*, before + Lat. *servare*, to guard.] —*vt.* **1.** To keep safe, as from injury or peril : PROTECT. **2.** To maintain unchanged. **3.** To keep or maintain intact <ought to *preserve* the public safety> **4.** To prepare (food) for future use, as by canning or salting. **5.** To prevent (organic bodies) from decaying or spoiling. **6.** To keep or protect (game or fish) for one's private hunting or fishing. —*vi.* **1.** To treat fruit or other foods so as to prevent decay. **2.** To maintain a private area stocked with fish or game. —*n.* **1.** A preservative. **2.** *often* **preserves.** Fruit cooked with sugar to guard against decay. **3.** An area set aside for the protection of wildlife or natural resources. **4.** Something held to be restricted to the use of certain persons <Ancient Greek is the *preserve* of scholars.> **—pre·serv′a·bil′i·ty** *n.* **—pre·serv′a·ble** *adj.* **—pres′er·va′tion** (prĕz′ər-vā′shən) *n.* **—pre·serv′er** *n.*

**pre·set** (prē-sĕt′) *vt.* **-set, -set·ting, -sets.** To set beforehand <*preset* an oven>

**pre·shrunk** *also* **pre-shrunk** (prē′shrŭngk′) *adj.* Shrunk during manufacture to reduce the chance of later shrinkage.

**pre·side** (prĭ-zīd′) *vi.* **-sid·ed, -sid·ing, -sides.** [Fr. *presider* < Lat. *praesidēre* : *prae-*, in front of + *sedēre*, to sit.] **1.** To act as chairperson or president, as at a meeting. **2.** To possess or use authority or control. **3.** *Mus.* To be the featured instrumentalist. **—pre·sid′er** *n.*

**pres·i·den·cy** (prĕz′ĭ-dən-sē, -dĕn′-) *n., pl.* **-cies. 1.** The office, function, or term of a president. **2.** *often* **Presidency.** The office of president of a republic, esp. of the United States. **3.** *Mormon Ch.* **a.** A governing body on a local level consisting of three men. **b.** *often* **Presidency.** The church's chief administrative body.

**pres·i·dent** (prĕz′ĭ-dənt, -dĕnt′) *n.* [ME < OFr. < Lat. *praesidens*, pr.part. of *praesidēre*, to preside. —see PRESIDE.] **1.** One appointed or elected to preside over an organized body of people, as an assembly or meeting. **2.** *often* **President.** The chief executive of a republic, esp. of the United States. **3.** The chief executive officer, as of a branch of government, a corporation, a board of trustees, or a university. **—pres′i·dent·ship′** *n.*

**pres·i·dent-e·lect** (prĕz′ĭ-dənt-ĭ-lĕkt′) *n.* One who has been elected president but has not yet started the term of office.

**pres·i·den·tial** (prĕz′ĭ-dĕn′shəl) *adj.* **1.** Of or pertaining to a president or presidency. **2.** Providing for a president elected independently of the legislature. **—pres′i·den′tial·ly** *adv.*

**president pro tem** (prō tĕm′) *n. Informal.* A president pro tempore.

**president pro tem·po·re** (prō tĕm′pə-rē) *n.* The senator who presides over the U.S. Senate in the absence of the Vice President.

**pre·sid·i·a** (prĭ-sĭd′ē-ə) *n. var. pl. of* PRESIDIUM.

**pre·sid·i·al** (prĭ-sĭd′ē-əl) *also* **pre·sid·i·ar·y** (-ĕr′ē) *adj.* Of, relating to, or having a presidio.

**pre·sid·i·o** (prĭ-sē′dē-ō′, -sĭd′ē-ō′) *n., pl.* **-os.** [Sp. < Lat. *praesidium* < *praesidēre*, to guard. —see PRESIDE.] A garrison, esp. a fortress of the kind established in the U.S. Southwest by the Spanish.

**pre·sid·i·um** (prĭ-sĭd′ē-əm) *n., pl.* **-i·a** (-ē-ə) or **-i·ums.** [R. *prezidium* < Lat. *praesidium*, garrison < *praesidēre*, to guard. —see PRESIDE.] **1.** A permanent executive committee in a Communist country holding power to act for a larger governing body. **2. Presidium.** A committee of the Supreme Soviet headed by the premier and constituting the highest policy-making body of the Soviet Union.

**pre·sig·ni·fy** (prē-sĭg′nə-fī′) *vt.* **-fied, -fy·ing, -fies.** To signify beforehand : PREFIGURE.

**pre·soak** (prē-sōk′) *vt.* **-soaked, -soak·ing, -soaks.** To soak (e.g., clothes) before washing. —*n.* (prē′sōk′). An automatic washer cycle for presoaking clothes.

**pre·sort** (prē-sôrt′) *vt.* **-sort·ed, -sort·ing, -sorts.** To sort (mail) according to Zip Codes before delivering to a post office.

**press**[1] (prĕs) *v.* **pressed, press·ing, press·es.** [ME *pressen* < OFr. *presser* < Lat. *pressare*, freq. of *premere*, to press.] —*vt.* **1.** To exert steady force or weight against. **2. a.** To squeeze the juice or other contents from <*press* grapes> **b.** To extract (e.g., juice) by squeezing or compressing. **3. a.** To compact or reshape by exerting steady force. **b.** To iron (e.g., clothing). **4.** To clasp or embrace closely. **5.** To seek to influence, as by unyielding arguments or entreaties. **6.** To urge on. **7.** To place in constraining or trying circumstances : HARASS. **8.** To lay stress on : EMPHASIZE. **9.** To advance or carry on energetically, as an attack. **10.** To put forward importunately or insistently. —*vi.* **1.** To exert force or pressure. **2.** To weigh heavily, as on the mind. **3.** To advance eagerly. **4.** To be urgent. **5.** To iron clothes or other material. **6.** To assemble closely and in large numbers : CROWD. **7.** To use urgent persuasion or entreaty. —*n.* **1.** A machine or device that exerts pressure. **2.** A printing press. **3.** A place where matter is printed. **4.** The method, art, or business of printing. **5. a.** Printed matter as a whole, esp. newspapers and periodicals. **b.** The individuals, as editors and reporters, involved with such publications. **c.** The matter dealt with in such publications, as news and criticism. **6.** An act of gathering in large numbers or of pushing forward. **7.** A large gathering : THRONG. **8. a.** An act of applying pressure. **b.** The state of being pressed. **9.** Urgency or haste <the *press* of business> **10.** The set of proper creases in a garment or fabric, formed by ironing. **11.** An upright closet or case used for storing clothing, books, or other articles. **—(hard) pressed for.** To be lacking in <was *pressed for* time> **—press (one's) luck.** To push for something in spite of opposing odds. **—press the flesh.** *Informal.* To gladhand <a political candidate eagerly *pressing the flesh*>

**press**[2] (prĕs) *vt.* **pressed, press·ing, press·es.** [Alteration of obs. *prest*, to hire for military service < ME, enlistment money < OFr. < *prester*, to lend < Med. Lat. *prestare* < Lat., to furnish.] **1.** To force into service in the army or navy : IMPRESS. **2.** To use in a way different from the intended or usual. —*n.* **1.** Conscription or impressment into service, esp. into the navy. **2.** An official warrant for impressment into military service.

**press agency** *n.* A news agency.

**press agent** *n.* One who arranges advertising and publicity, as for a performer or a business. **—press′-a′gent·ry** (-ā′jən-trē) *n.*

**press association** *n.* A news agency.

**press·board** (prĕs′bôrd′, -bōrd′) *n.* **1.** A heavy glazed paper or pasteboard used esp. for covering the platen or cylinder of a printing press. **2.** A small ironing board.

**press box** *n.* A section for reporters, as in a stadium.

**press conference** *n.* An interview held for reporters by a political figure, a government spokesperson, or a celebrity.

**press·er** (prĕs′ər) *n.* **1.** One who presses clothes. **2.** A device that applies pressure to a product in manufacturing or canning.

**press gang** *also* **press·gang** (prĕs′găng′) *n.* A company of men under an officer detailed to press men into military or naval service.

**press·ing** (prĕs′ĭng) *adj.* **1.** Demanding immediate attention : URGENT <a *pressing* requirement> **2.** Importunate : insistent <a *pressing* invitation> **—press′ing·ly** *adv.*

**press·man** (prĕs′mən, -măn′) *n.* **1.** A printing press operator. **2.** *Chiefly Brit.* A newspaperman.

**press·mark** (prĕs′märk′) *n.* **1.** A notation in or on a book showing where it should be placed in a library. **2.** A notation or figure in the margin of a printed sheet showing the press upon which it was printed.

**press of sail** *n. Naut.* The greatest amount of sail that a ship can carry safely.

**pres·sor** (prĕs′ôr′, -ər) *adj.* [< Lat. *pressus*, p.part. of *premere*, to press.] Inducing an increase in blood pressure.

**press release** *n.* An announcement of an event, performance, or other news or publicity item given to the press.

**press·room** (prĕs′ro͞om′, -ro͝om′) *n.* The room in a printing or newspaper publishing establishment where the presses are located.

**press·run** (prĕs′rŭn′) *n.* The specific number of copies printed during a continuous operation of a printing press.

**press secretary** *n.* One who manages the public affairs and press conferences of a public figure.

**pres·sure** (prĕsh′ər) *n.* [Lat. *pressura* < *premere*, to press.] **1.** An act of pressing or the state of being pressed. **2.** Application of continuous force by one body on another that it touches : COMPRESSION. **3.** *Physics.* Force applied over a surface, measured as force per unit of area. **4.** A constraining influence on the will or mind, as a moral force. **5.** Urgent claim or demand <under the *pressure* of official duties> **6.** A burdensome, distressing, or weighty state. **7.** *Archaic.* A mark made by application of force or weight : IMPRESSION. —*vt.* **-sured, -sur·ing, -sures.** To force, as by overpowering influence.

**pressure cabin** *n.* A pressurized section of an aircraft.

**pressure cooker** *n.* **1.** An airtight metal pot that uses steam under pressure at high temperature to cook food quickly. **2.** A position of difficulty, stress, or anxiety : HOT SEAT.

**pressure gauge** *n.* **1.** A device for measuring fluid pressure. **2.** A device for measuring the pressure of explosions.

**pressure group** *n.* A group that exerts pressure to promote or safeguard its interests.

**pressure point** *n.* Any of a number of areas on the body where an artery runs close to a bone so that applying pressure to the artery presses it against the bone, thus controlling bleeding.

**pressure suit** *n.* A garment worn in high-altitude aircraft or in spacecraft to compensate for low-pressure conditions.

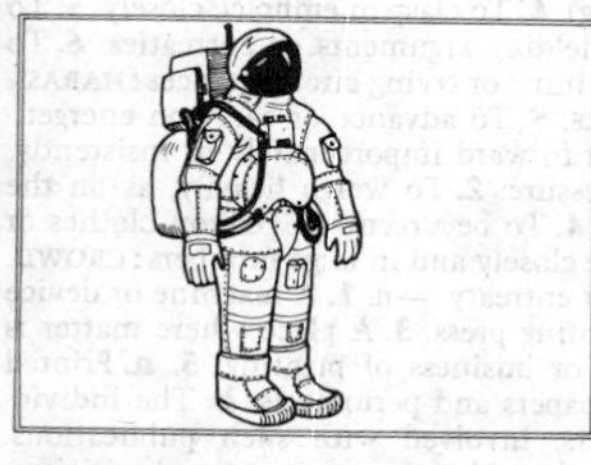

**pressure suit**

**pres·sur·ize** (prĕsh′ə-rīz′) *vt.* **-ized, -iz·ing, -iz·es. 1.** To maintain normal air pressure in (an enclosure, as an aircraft or submarine). **2.** To put (gas or liquid) under a greater than normal pressure. **3.** To design to resist pressure. **—pres′sur·i·za′tion** *n.* **—pres′sur·iz′er** *n.*

**press·work** (prĕs′wûrk′) *n.* **1.** Management or operation of a printing press. **2.** The matter printed by a printing press.

**Pres·ter John** (prĕs′tər jŏn′) *n.* [ME *prestre,* priest < OFr. < LLat. *presbyter.* —see PRESBYTER.] A legendary medieval Christian priest and king believed to have ruled a Christian kingdom in the Far East or in Ethiopia.

**pres·ti·dig·i·ta·tion** (prĕs′tĭ-dĭj′ĭ-tā′shən) *n.* [Fr. < *prestidigitateur,* conjurer : *preste,* nimble (< Ital. *presto* < LLat. *praestus* < Lat. *praesto,* at hand) + Lat. *digitus,* finger.] Manual dexterity in the execution of tricks : SLEIGHT OF HAND. **—pres′ti·dig′i·ta′tor** *n.*

**pres·tige** (prĕ-stēzh′, -stēj′) *n.* [Fr., illusion < Lat. *prestigiae,* tricks < *praestringere,* to dazzle : *prae-,* before + *stringere,* to bind.] **1.** Prominence or influential status gained through success, renown, or wealth. **2.** The power to command admiration in a group. **—pres·tige′ful** *adj.*

**pres·ti·gious** (prĕ-stē′jəs, -stĭj′əs) *adj.* Having prestige : ESTEEMED. **—pres·ti′gious·ly** *adv.* **—pres·ti′gious·ness** *n.*

**pres·tis·si·mo** (prĕ-stĭs′ə-mō′) *adv.* & *adj.* [Ital., superl. of *presto,* presto.] *Mus.* At as fast a tempo as possible. —Used as a direction. **—pres·tis′si·mo′** *n.*

**pres·to** (prĕs′tō) *adv.* & *adj.* [Ital. < LLat. *praestus,* quick < Lat. *praesto,* at hand.] **1.** *Mus.* In rapid tempo. —Used as a direction. **2.** Suddenly. **—pres′to** *n.*

**pre·sume** (prĭ-zo͞om′) *v.* **-sumed, -sum·ing, -sumes.** [ME *presumen* < OFr. *presumer* < LLat. *praesumere* < Lat., to anticipate : *prae-,* before + *sumere,* to take.] *—vt.* **1.** To assume to be true without proof to the contrary. **2.** To appear to prove. **3.** To engage oneself in without authority or permission : DARE <*presumed* to invite themselves to lunch> *—vi.* **1.** To act overconfidently : take liberties. **2.** To take unjustified advantage of something <Don't *presume* on their kindness.> **—pre·sum′a·ble** (-zo͞o′mə-bəl) *adj.* **—pre·sum′a·bly** *adv.* **—pre·sum′ed·ly** (-zo͞o′mĭd-lē) *adv.* **—pre·sum′er** *n.*

**pre·sum·ing** (prĭ-zo͞o′mĭng) *adj.* Having or displaying arrogant self-confidence : PRESUMPTUOUS. **—pre·sum′ing·ly** *adv.*

**pre·sump·tion** (prĭ-zŭmp′shən) *n.* [ME *presumpcion* < OFr. < LLat. *praesumptio* < *praesumere,* to presume.] **1.** Boldly arrogant or offensive behavior or language. **2. a.** An act of accepting as true. **b.** Acceptance or belief based on reasonable evidence. **3.** A condition or basis for accepting or presuming. **4.** *Law.* An inference as to the truth of an allegation or proposition, based on probable reasoning in the absence of or prior to actual proof or disproof.

**pre·sump·tive** (prĭ-zŭmp′tĭv) *adj.* **1.** Providing a reasonable basis for acceptance or belief. **2.** Based on probability or presumption <an heir *presumptive*> **—pre·sump′tive·ly** *adv.*

**pre·sump·tu·ous** (prĭ-zŭmp′cho͞o-əs) *adj.* [ME < OFr. *presumptueux* < LLat. *presumptuosus* < *praesumptio,* presumption.] Being overly forward or confident : ARROGANT. **—pre·sump′tu·ous·ly** *adv.* **—pre·sump′tu·ous·ness** *n.*

**pre·sup·pose** (prē′sə-pōz′) *vt.* **-posed, -pos·ing, -pos·es. 1.** To assume or suppose in advance. **2.** To require or involve necessarily as an antecedent condition. **—pre·sup′po·si′tion** (-sŭp′ə-zĭsh′ən) *n.*

**pre·tax** (prē′tăks′) *adj.* Existing before tax deductions <*pretax* profits>

**pre·teen** (prē′tēn′) *adj.* **1.** Relating to or intended for preadolescent children <*preteen* footwear> **2.** Being a preadolescent child. **—pre′teen′** *n.*

**pre·tence** (prē′tĕns′, prĭ-tĕns′) *n. Chiefly Brit. var. of* PRETENSE.

**pre·tend** (prĭ-tĕnd′) *v.* **-tend·ed, -tend·ing, -tends.** [ME *pretenden* < Lat. *praetendere* : *prae-,* in front + *tendere,* to extend.] *—vt.* **1.** To affect : feign. **2.** To allege or claim insincerely or falsely : PROFESS. **3.** To represent fictitiously in play : make believe. **4.** To take upon oneself : VENTURE <I cannot *pretend* to approve of what you did.> *—vi.* **1.** To feign an action or character, as in play. **2.** To put forward a claim.

**pre·tend·ed** (prĭ-tĕn′dĭd) *adj.* **1.** Reputed or asserted : ALLEGED. **2.** False : feigned. **—pre·tend′ed·ly** *adv.*

**pre·tend·er** (prĭ-tĕn′dər) *n.* **1.** One who simulates, pretends, or alleges falsely : HYPOCRITE. **2. a.** One who sets forth a claim. **b.** A claimant to a throne.

**pre·tense** (prē′tĕns′, prĭ-tĕns′) *n.* [ME < AN, ult. < Lat. *praetendere,* to pretend.] **1.** A false appearance or action designed to deceive. **2.** A false or studied show : AFFECTATION. **3.** A false reason or excuse : PRETEXT. **4.** Something imagined or pretended : MAKE-BELIEVE. **5.** Mere show without reality : SHAM. **6.** A right asserted with or without foundation : CLAIM. **7.** Ostentation : pretentiousness.

**pre·ten·sion** (prĭ-tĕn′shən) *n.* **1.** A specious allegation : PRETEXT. **2.** A claim to something, as a right or privilege. **3.** Advancement of a claim. **4.** PRETENSE 7.

**pre·ten·tious** (prĭ-tĕn′shəs) *adj.* **1.** Demanding or claiming a position of distinction or merit, esp. when unjustified. **2.** Making an extravagant outward show. **—pre·ten′tious·ly** *adv.* **—pre·ten′tious·ness** *n.*

**pret·er·it** or **pret·er·ite** (prĕt′ər-ĭt) *adj.* [ME < OFr. < Lat. *praeteritus,* p.part. of *praeterire,* to go by : *praeter,* beyond (comp. of *prae,* before) + *ire,* to go.] Denoting the verb tense expressing or describing a past or completed action or state. *—n.* **1.** The verb form expressing or describing a past or completed action or condition : PAST TENSE. **2.** A verb in the preterit form.

**pret·er·i·tion** (prĕt′ə-rĭsh′ən) *n.* [LLat. *praeteritio* < Lat. *praeterire,* to go by. —see PRETERIT.] **1.** An act of passing by, ignoring, or omitting. **2.** *Law.* Neglect of a testator to mention a legal heir in his or her will. **3.** The Calvinist doctrine that God neglected to designate those who would be damned, determining only the elect.

**pre·ter·mit** (prē′tər-mĭt′) *vt.* **-mit·ted, -mit·ting, -mits.** [Lat. *praetermittere* : *praeter,* beyond (comp. of *prae,* before) + *mittere,* to let go.] **1.** To ignore deliberately or permit to pass unnoticed or unmentioned. **2.** To fail to do or include : OMIT. **3.** To interrupt or terminate. **—pre′ter·mis′sion** (-mĭsh′ən) *n.* **—pre′ter·mit′ter** *n.*

**pre·ter·nat·u·ral** (prē′tər-năch′ər-əl, -năch′rəl) *adj.* [Med. Lat. *praeternaturalis* : Lat. *praeter,* beyond (comp. of *prae,* before) + Lat. *natura,* nature.] **1.** Beyond the normal course of nature : ABNORMAL. **2.** Transcending the natural or material order : SUPERNATURAL. **—pre′ter·nat′u·ral·ism** *n.* **—pre′ter·nat′u·ral·ly** *adv.* **—pre′ter·nat′u·ral·ness** *n.*

**pre·test** (prē′tĕst′) *n.* **1. a.** A test given to determine whether a class is adequately prepared for a new course. **b.** The condition of a sample before experimental modification. **2.** Advance testing of something, as a questionnaire, product, or idea. *—vt.* & *vi.* (prē-tĕst′) **-test·ed, -test·ing, -tests.** To subject to or conduct a pretest.

**pre·text** (prē′tĕkst′) *n.* [Lat. *praetextum* < p.part. of *praetexere,* to disguise : *prae-,* before + *texere,* to weave.] Ostensible purpose : EXCUSE. *—vt.* **-text·ed, -text·ing, -texts.** To allege as an excuse.

**pre·treat** (prē-trēt′) *vt.* **-treat·ed, -treat·ing, -treats.** To treat in advance. **—pre·treat′ment** *n.*

**pre·tri·al** (prē-trī′əl, -trīl′) *adj.* Existing or taking place before a trial <a *pretrial* conference>

**pret·ti·fy** (prĭt′ĭ-fī′) *vt.* **-fied, -fy·ing, -fies.** To make pretty. **—pret′ti·fi·ca′tion** *n.* **—pret′ti·fi′er** *n.*

**†pret·ty** (prĭt′ē) *adj.* **-ti·er, -ti·est.** [ME *prety,* clever, fine < OE *prættig,* cunning < *prætt,* trick.] **1.** Pleasing or attractive esp. in a graceful or delicate way. **2.** Excellent : good. **3.** *Archaic.* Elegant : fine. **4.** Effeminate. **5.** *Informal.* Considerable in size or extent <a *pretty* business deal> *—adv.* **1.** To a fair degree : MODERATELY <a *pretty* fair skater> **2.** *Regional.* Pleasingly : prettily. *—n., pl.* **-ties.** One that is pretty. *—vt.* **-tied, -ty·ing, -ties.** *Informal.* To make pretty <*pretty* up the apartment> **—sitting pretty.** *Informal.* In favorable circumstances. **—pret′ti·ly** *adv.* **—pret′ti·ness** *n.*

**pre·tu·ber·cu·lous** (prē′to͞o-bûr′kyə-ləs, -tyo͞o-) *adj.* Relating to lesions of tuberculosis occurring before the actual development of the disease.

**pret·zel** (prĕt′səl) *n.* [G. < OHG *brezitella* < Lat. *bracchium,* arm < Gk. *brakhiōn.*] A glazed biscuit salted on the outside and usu. baked in the shape of a loose knot or stick.

**pre·vail** (prĭ-vāl′) *vi.* **-vailed, -vail·ing, -vails.** [ME *prevayllen* < Lat. *praevalēre,* to be stronger : *prae-,* beyond + *valēre,* to be strong.] **1.** To be greater in influence or strength : TRIUMPH <*prevailed* over many adverse conditions> **2.** To be or become effective : WIN OUT. **3.** To be most frequent or common : PREDOMINATE. **4.** To be in current force, use, or effect. **—prevail on** (*or* **upon**). To persuade successfully <*prevailed* on us to go to the lecture> **—pre·vail′er** *n.*

**pre·vail·ing** (prĭ-vā'lĭng) *adj.* **1.** Most frequent or common : PREDOMINANT. **2.** Generally current : WIDESPREAD. **—pre·vail'ing·ly** *adv.* **—pre·vail'ing·ness** *n.*

☆ ***syns:*** PREVAILING, CURRENT, PREVALENT, REGNANT, RIFE, WIDESPREAD *adj. core meaning* : most commonly existing or encountered at a given time <the *prevailing* views in astrophysics>

**prev·a·lent** (prĕv'ə-lənt) *adj.* [Lat. *praevalens, praevalent-,* pr.part. of *praevalere,* to be stronger. —see PREVAIL.] Widely existing, accepted, or practiced. **—prev'a·lence** *n.* **—prev'a·lent·ly** *adv.*

**pre·var·i·cate** (prĭ-văr'ĭ-kāt') *vi.* **-cat·ed, -cat·ing, -cates.** [Lat. *praevaricari, praevaricat-* : *prae-,* before + *varicare,* to straddle < *varicus,* straddling < *varus,* bent.] To stray from or evade the truth : EQUIVOCATE. **—pre·var'i·ca'tion** *n.* **—pre·var'i·ca'tor** *n.*

**pre·ven·ience** (prĭ-vēn'yəns) *n.* **1.** The act or state of being antecedent or prevenient. **2.** Attention to another's needs.

**pre·ven·ient** (prĭ-vēn'yənt) *adj.* [Lat. *praeveniens, praevenient-,* pr.part. of *praevenire,* to precede : *prae-,* before + *venire,* to come.] **1.** Antecedent. **2.** Anticipatory. **—pre·ven'ient·ly** *adv.*

**pre·vent** (prĭ-vĕnt') *v.* **-vent·ed, -vent·ing, -vents.** [ME *preventen,* to anticipate < Lat. *praevenire* : *prae-,* before + *venire,* to come.] *—vt.* **1.** To keep from happening : AVERT. **2.** To keep (someone) from doing something : IMPEDE <*prevented* me from playing> **3.** To anticipate or counter in advance. **4.** *Archaic.* To come before : PRECEDE. *—vi.* To present an obstacle <There will be a game if nothing *prevents.*> **—pre·vent'a·bil'i·ty, pre·vent'i·bil'i·ty** *n.* **—pre·vent'a·ble, pre·vent'i·ble** *adj.* **—pre·vent'er** *n.* **—pre·ven'tion** (-vĕn'shən) *n.*

**pre·ven·ta·tive** (prĭ-vĕn'tə-tĭv) *adj.* & *n. var. of* PREVENTIVE.

**pre·ven·tive** (prĭ-vĕn'tĭv) *also* **pre·ven·ta·tive** (-tə-tĭv) *—adj.* **1.** Intended or used to hinder or prevent. **2.** *Med.* Thwarting or warding off illness or disease : PROPHYLACTIC. *—n.* **1.** Something that prevents : OBSTACLE. **2.** *Med.* Something for warding off illness. **—pre·ven'tive·ly** *adv.* **—pre·ven'tive·ness** *n.*

**preventive detention** *n.* Pretrial imprisonment without the right to bail of one accused of a felony and judged dangerous to society.

**pre·view** *also* **pre·vue** (prē'vyōō') *—n.* **1.** An advance showing, as of a film, to an invited audience before public presentation. **2.** An advance viewing or exhibition, esp. the presentation of several scenes advertising a forthcoming film. **3.** An introductory or limited experience. *—vt.* **-viewed, -view·ing, -views** *also* **-vued, -vu·ing, -vues.** To exhibit or view in advance.

**pre·vi·ous** (prē'vē-əs) *adj.* [Lat. *praevius,* going before : *prae-,* before + *via,* way.] **1.** Existing or happening prior to something else in time or order : ANTECEDENT. **2.** *Informal.* Hasty : premature. **—pre'vi·ous·ly** *adv.* **—pre'vi·ous·ness** *n.*

**previous question** *n.* The motion in parliamentary procedure to take an immediate vote on the main question under consideration or on any other questions so designated.

**previous to** *prep.* Prior to : BEFORE.

**pre·vise** (prĭ-vīz') *vt.* **-vised, -vis·ing, -vis·es.** [Lat. *praevidēre, praevis-* : *prae-,* before + *vidēre,* to see.] **1.** To foresee. **2.** To notify ahead of time. **—pre·vi'sion** (-vĭzh'ən) *n.* **—pre·vi'sion·al, pre·vi'sion·ar'y** (-ə-nĕr'ē) *adj.* **—pre·vi'sor** *n.*

**pre·vo·cal·ic** (prē'vō-kăl'ĭk) *adj.* Coming before a vowel.

**pre·vo·ca·tion·al** (prē'vō-kā'shə-nəl) *adj.* Of or relating to instruction given in preparation for vocational school.

**pre·vue** (prē'vyōō') *n.* & *v. var. of* PREVIEW.

**pre·war** (prē'wôr') *adj.* Existing or taking place before a war.

**prex·y** (prĕk'sē) *n., pl.* **-ies.** [Shortening and alteration of PRESIDENT.] *Slang.* A president, esp. of a college or university.

**prey** (prā) *n.* [ME *preie* < OFr. < Lat. *praeda.*] **1.** A creature hunted or caught for food : QUARRY. **2.** A victim. **3.** An act of preying. *—vi.* **preyed, prey·ing, preys. 1.** To hunt, catch, or eat as prey <Foxes *prey* on chickens.> **2.** To victimize or make a profit at someone else's expense. **3.** To plunder : pillage. **4.** To exert a baneful or injurious effect <Guilt *preyed* upon my mind.> **—prey'er** *n.*

**Pri·am** (prī'əm) *n.* [Lat. *Priamus* < Gk. *Priamos.*] *Gk. Myth.* King of Troy, the father of Paris and Hector, who was killed when the Greeks captured his city.

**pri·a·pic** (prī-ā'pĭk, -ăp'ĭk) *also* **pri·a·pe·an** (prī'ə-pē'ən) *adj.* [< PRIAPUS.] Phallic.

**pri·a·pism** (prī'ə-pĭz'əm) *n.* [Fr. *priapisme* < LLat. *priapismus* < Gk. *priapismos* < *priapizein,* to have an erection < *Priapos,* Priapus.] Persistent, usu. painful erection of the penis, esp. due to disease.

**pri·a·pus** (prī-ā'pəs) *n.* [Lat. < Gk. *Priapos.*] **1. Priapus.** *Gk.* & *Rom. Myth.* The god of procreation, guardian of vineyards and gardens, and personification of the erect phallus. **2.** An image of Priapus, often used in ancient gardens as a scarecrow. **3.** A phallic representation.

**price** (prīs) *n.* [ME *pris* < OFr. < Lat. *pretium.*] **1.** The sum of money or goods asked or given for something. **2.** The cost at which something is obtained. **3.** The cost of bribing someone. **4.** A reward offered for the killing or capture of a person. **5.** Value : worth. *—vt.* **priced, pric·ing, pric·es. 1.** To set a price for <socks *priced* at $50> **2.** To find out the price of <spent the day *pricing* cars> **—price out of the market.** To charge so much for goods or services that people no longer buy or use them.

**price cutting** *n.* Reduction of retail prices to a level so low as to eliminate competition.

**price-earn·ings ratio** (prīs'ûr'nĭngz) *n.* The ratio of a common stock's market price to its earnings per share, used as an indicator of a corporation's profitability.

**price index** *n.* A number relating prices of a group of commodities to their prices during an arbitrarily chosen base period.

**price·less** (prīs'lĭs) *adj.* **1.** Of inestimable worth : INVALUABLE. **2.** Highly amusing, absurd, or odd.

**price tag** *n.* **1.** A label on a piece of merchandise showing its price. **2.** Cost : price.

**price war** *n.* A period of sharp competition among businesses in which each competitor tries to cut retail prices below those of the others.

**pric·ey** (prī'sē) *adj.* **-i·er, -i·est.** *Informal.* Expensive.

**prick** (prĭk) *n.* [ME < OE *prica,* puncture.] **1. a.** An act of piercing or pricking. **b.** The sensation of being pierced or pricked. **2.** A painful or stinging feeling of remorse or sorrow. **3.** A small mark or puncture made by a pointed object. **4.** A hare's track or footprint. **5.** A pointed object, as an ice pick or thorn. *—v.* **pricked, prick·ing, pricks.** *—vt.* **1.** To puncture lightly. **2.** To sting with an emotional or mental pang. **3.** To impel : incite <"My duty *pricks* me on" —Shakespeare> **4.** To mark on a surface by means of small punctures <*prick* a design on leather> **5.** *Naut.* To measure with dividers on a chart. **6.** To pierce the quick of (a horse's hoof) while shoeing. **7.** To transplant (seedlings) before a final planting. *—vi.* **1.** To pierce or puncture something. **2.** To feel a stinging or pricking sensation. **3.** To ride at a gallop. **4.** To point upward. **—prick up (one's) ears.** To begin to listen attentively.

**prick·er** (prĭk'ər) *n.* **1.** One that pricks. **2.** A pricking tool. **3.** A prickle or thorn.

**prick·et** (prĭk'ĭt) *n.* [ME *priket* < *prik,* prick < OE *prica,* puncture.] **1. a.** A small point or spike for holding a candle upright. **b.** A candlestick with such a spike. **2.** A buck in its second year, prior to the branching of its horns.

**prick·le** (prĭk'əl) *n.* [ME *prikel* < OE *pricel.*] **1.** A small sharp point, spine, or thorn. **2.** A tingling or pricking sensation. *—v.* **-led, -ling, -les.** *—vt.* **1.** To prick with or as if with a thorn. **2.** To cause a tingling or pricking sensation in. *—vi.* **1.** To feel a tingling or pricking sensation. **2.** To stand up like prickles.

**prick·ly** (prĭk'lē) *adj.* **-li·er, -li·est. 1.** Having prickles. **2.** Tingling : smarting. **3.** Irritable. **—prick'li·ness** *n.*

**prickly ash** *n.* An aromatic shrub or small tree, *Zanthoxylum americanum* of eastern North America, with prickly stems and feathery leaves.

**prickly heat** *n.* Miliaria.

**prickly pear** *n.* **1.** Any of various cacti of the genus *Opuntia,* with bristly flattened or cylindrical joints, showy, usu. yellow flowers, and ovoid, occas. edible fruit. **2.** The fruit of a prickly pear.

**prickly poppy** *n.* Any of various plants of the genus *Argemone* chiefly of tropical America, with large yellow or white flowers and prickly leaves, stems, and pods.

**prick·y** (prĭk'ē) *adj.* **-i·er, -i·est.** Prickly.

**pride** (prīd) *n.* [ME < OE *prȳde* < *prūd,* proud < OFr.] **1.** A sense of one's own proper value or dignity : SELF-RESPECT. **2.** Pleasure or satisfaction taken in one's work, achievements, or possessions. **3. a.** A source of pride <These soldiers were their country's *pride.*> **b.** The most successful or thriving condition : PRIME <the *pride* of youth> **4. a.** An overly high opinion of oneself : CONCEIT. **b.** Consideration or personification of this condition as the first of the seven cardinal sins. **5.** Mettle or spirit in horses. **6.** A group of lions. *—v.* **prid·ed, prid·ing, prides.** *—vt.* To esteem (oneself) for <I *pride* myself on this cabinetwork.> *—vi.* To indulge in self-esteem. **—pride'ful** *adj.* **—pride'ful·ly** *adv.* **—pride'ful·ness** *n.*

**prie-dieu** (prē-dyœ') *n., pl.* **-dieus** or **-dieux** (-dyœz') [Fr. *prie Dieu,* pray God.] A low desk with space for a book above and with a foot piece below for kneeling in prayer.

**pri·er** *also* **pry·er** (prī'ər) *n.* One who pries.

**priest** (prēst) *n.* [ME *preost* < OE *prēost* < VLat. *prester < LLat. *presbyter.* —see PRESBYTER.] **1.** A member of the second grade of clergy ranking below a bishop but over a deacon and having authority to pronounce absolution and administer all sacraments save that of ordination in the Roman Catholic, Eastern Orthodox, Anglican, Armenian, and separated Catholic hierarchies. **2.** A non-Catholic cleric. **3.** One whose role is considered comparable to that of a priest <computers and their *priests*> *—vt.* **priest·ed, priest·ing, priests.** To ordain or admit to the priesthood. **—priest'hood'** *n.*

▲ word history: *Priest* and *presbyter* are both descended from the Greek word *presbuteros,* literally "elder," the comparative form of *presbus,* "old man." The Greek word was used in early Christian writing to denote a particular order of Christian minister. *Presbuteros* was borrowed into Latin and from there it was borrowed into the Germanic languages, appearing in Old English as *prēost,* modern *priest.* Because Old English *prēost* and its descendants were used to refer not only to Christian ministers but to Jewish and pagan priests as well, *presbuteros* was reborrowed as *presbyter* during the 16th-century Reformation to apply only to officials of the Christian church.

**priest·ess** (prē′stĭs) *n.* A woman priest.

**priest·ly** (prēst′lē) *adj.* **-li·er, -li·est.** Of, relating to, or suitable for a priest. **—priest′li·ness** *n.*

**prig** (prĭg) *n.* [Orig. unknown.] **1.** One regarded as overprecise, affectedly arrogant, smug, or narrow-minded. **2.** *Archaic.* A coxcomb. **3.** *Chiefly Brit.* A petty thief or pickpocket. *—vt.* **prigged, prig·ging, prigs.** *Chiefly Brit.* To steal : pilfer. **—prig′ger·y** *n.* **—prig′gish** *adj.* **—prig′gish·ly** *adv.* **—prig′gish·ness** *n.*

**prim**[1] (prĭm) *adj.* **prim·mer, prim·mest.** [Orig. unknown.] **1.** Affectedly precise or proper. **2.** Decorous : formal. **3.** Neat : trim. *—v.* **primmed, prim·ming, prims.** *—vt.* To fix (the face or mouth) in a prim expression. *—vi.* To assume a prim expression. **—prim′ly** *adv.* **—prim′ness** *n.*

**prim**[2] (prĭm) *n.* [Orig. unknown.] PRIVET 1.

**pri·ma ballerina** (prē′mə) *n., pl.* **prima ballerinas.** [Ital., first ballerina.] The leading woman dancer in a ballet company.

**pri·ma·cy** (prī′mə-sē) *n., pl.* **-cies. 1.** The state of being first or foremost. **2.** The office or province of an ecclesiastical primate.

**pri·ma donna** (prē′mə, prĭm′ə) *n., pl.* **prima donnas.** [Ital., first lady.] **1.** A leading woman singer, as in an opera company. **2.** A conceited and temperamental person.

**pri·ma fa·cie** (prī′mə fā′shē, fā′shə) *adv.* & *adj.* [Lat.] Before closer inspection : at first sight.

**prima-facie evidence** *n. Law.* Evidence that would, if uncontested, establish a fact or raise a presumption of a fact.

**pri·mal** (prī′məl) *adj.* [Med. Lat. *primalis* < Lat. *primus,* first.] **1.** First in time. **2.** Of first importance. **—pri·mal′i·ty** (-măl′ĭ-tē) *n.*

**pri·mar·i·ly** (prī-mâr′ə-lē, -mĕr′-) *adv.* **1.** At first : ORIGINALLY. **2.** Principally : chiefly.

**pri·mar·y** (prī′mĕr′ē, -mə-rē) *adj.* [Lat. *primarius,* chief < *primus,* first.] **1.** Happening first in time or sequence : ORIGINAL <a *primary* source> **2.** Primitive : unsophisticated <the *primary* instinct of motherhood> **3.** Being or standing first in a list, series, or sequence. **4.** First or best in degree, quality, or importance. **5.** *Geol.* Of, relating to, or indicating the earliest periods of geologic development up to and including the Paleozoic era : PRECAMBRIAN. **6.** Being a basic or fundamental part of an organized whole <a *primary* part> **7.** Immediate : direct <a *primary* result> **8.** Of or relating to the basic colors from which all other colors can be derived. **9. a.** Having a linguistic element such as a word root as a basis that cannot be further analyzed or broken down. —Used of the derivation of a word or word element. **b.** Referring to present or future time. —Used as a collective designation for various present and future verb tenses in Latin, Greek, and Sanskrit. **10.** *Elect.* Of, relating to, or designating an inducting current, circuit, or coil. **11.** Of, relating to, or indicating the main flight feathers projecting along the outer edge of a bird's wing. **12.** *Chem.* **a.** Relating to the replacement of one of several atoms or radicals in a compound by another atom or radical. **b.** Having a carbon atom attached solely to one other carbon atom in a molecule. **13.** *Biochem.* Of, relating to, or being the sequence of amino acids in a protein. *—n., pl.* **-ies. 1. a.** One that is first in time, order, or sequence. **b.** One that is first in degree, quality, or importance. **c.** A fundamental. **2. a.** A meeting of the registered voters of a political party to nominate candidates and to choose delegates to their party convention. **b.** A preliminary election in which the registered voters of a political party nominate candidates for office. **3.** A primary color. **4.** One of the main flight feathers projecting along the outer edge of a bird's wing. **5.** *Elect.* An inducting current, circuit, or coil. **6.** *Astron.* A celestial body, esp. a star, to which the orbit of a satellite, or secondary, is referred. **7.** A cosmic ray.

☆ ***syns:*** PRIMARY, CAPITAL, CARDINAL, CHIEF, DOMINANT, FIRST, FOREMOST, MAIN, PARAMOUNT, PRE-EMINENT, PREMIER, PRIME, PRINCIPAL, TOP *adj. core meaning* : most important, influential, or significant <a *primary* leader of the liberal opposition><*primary* responsibilities to the family>

**primary accent** *n.* **1.** The strongest degree of stress placed on a syllable in the pronunciation of a word. **2.** A mark (′) used to indicate the strongest degree of stress in the pronunciation of a word.

**primary atypical pneumonia** *n.* A mild pneumonia, prob. caused by a virus.

**primary cell** *n.* A cell in which an irreversible chemical reaction generates electricity.

**primary coil** *n.* An electrically conducting coil, as in a transformer, that transmits an inducting current.

**primary color** *n.* A color belonging to any three groups each of which is regarded as generating all colors: **a.** Additive, physiological, or light primaries—red, green, and blue; lights of red, green, and blue wavelengths may be mixed to produce all colors. **b.** Subtractive or colorant primaries—magenta, yellow, and cyan; substances that reflect light of one of these wavelengths and absorb other wavelengths may be mixed to produce all colors. **c.** Psychological primaries—red, yellow, green, and blue, plus the achromatic pair black and white; all colors may be subjectively conceived as mixtures of these.

**primary election** *n.* PRIMARY 2b.

**primary school** *n.* A school usu. including the first three or four grades of elementary school and occas. kindergarten.

**pri·mate** (prī′mĭt, -māt′) *n.* [ME *primat* < OFr. < Med. Lat. *primas* < Lat., principal < *primus,* first.] **1.** A bishop of highest rank in a province or country. **2.** (prī′māt′). Any of the order Primates, including the monkeys, apes, and humans. **—pri′mate·ship′** *n.* **—pri·ma′tial** (-mā′shəl) *adj.*

**pri·ma·ve·ra** (prē′mə-vĕr′ə) *n.* [Am. Sp. < Sp., spring < Lat. *prima vera* : *primus,* first + *ver,* spring.] **1.** A tree, *Cybistax donnellsmithii* of Central America, with yellow flowers and close-grained, light-colored wood. **2.** The wood of the primavera.

**prime** (prīm) *adj.* [ME, first in occurrence < OFr. < Lat. *primus.*] **1.** First in excellence, quality, or value. **2.** First in degree or rank : CHIEF. **3.** First in time, order, or sequence. **4.** Of the highest U.S. Government grade of meat. **5.** *Math.* Indicating a prime number. *—n.* **1.** The earliest hours of the day : DAWN. **2.** The first season of the year : SPRING. **3.** The years of one's physical perfection and intellectual vigor. **4.** The period or phase of ideal or peak condition <corn crop in its *prime*> **5.** The first position of thrust and parry in fencing. **6.** A mark (′) placed above and to the right of a letter in order to distinguish it from the same letter already in use or to designate a related quantity or thing, as feet, minutes of angle, or minutes of time. **7. a.** The second of the seven canonical hours. **b.** The time set aside for this prayer, usu. approx. 6:00 A.M. **8.** *Math.* A prime number. *—v.* **primed, prim·ing, primes.** *—vt.* **1.** To make ready : PREPARE. **2.** To prepare (a gun or mine) for firing by inserting a charge of gunpowder or a primer. **3.** To prepare for operation, as by pouring water into a pump or gasoline into a carburetor. **4.** To prepare (a surface) for painting by covering with size, primer, or an undercoat. **5.** To prepare with information : COACH. *—vi.* To prepare someone or something for future action or operation. **—prime the pump.** *Informal.* To stimulate the growth or operation of something. **—prime′ly** *adv.* **—prime′ness** *n.*

**prime meridian** *n.* The zero meridian (0°), used as a reference line from which longitude east and west is measured, and passing through Greenwich, England.

**prime minister** *n.* **1.** A chief minister appointed by a ruler. **2.** The head of the cabinet and often also the chief executive of a parliamentary democracy. **—prime ministership, prime ministry** *n.*

**prime mover** *n.* **1.** The initial force that moves or engages a machine, as electricity, wind, or gravity. **2.** Something considered as the initial source of energy directed toward a goal <Ambition was the *prime mover* of their careers.> **3.** A machine or mechanism that converts natural energy into work. **4.** A heavy-duty truck or tractor. **5.** The self-moved being in Aristotelian philosophy that causes all motion.

**prime number** *n.* A number that has itself and unity as its only factors.

**prim·er**[1] (prĭm′ər) *n.* [ME < Norman Fr. < Med. Lat. *primarius,* first < Lat. < *primus.*] **1.** An elementary textbook. **2.** A book covering the basic elements of a subject.

**prim·er**[2] (prī′mər) *n.* **1.** One that primes or causes to be primed. **2.** A cap or tube holding a small amount of explosive for detonating the main explosive charge of a firearm or mine. **3.** An undercoat of paint or size applied to prepare a surface for further work.

**prime rate** *n.* The lowest rate of interest on bank loans at a given time and place, offered to preferred borrowers.

**pri·me·ro** (prĭ-mâr′ō) *n.* [Alteration of Sp. *primera,* fem. of *primero,* first < Lat. *primarius,* principal < *primus,* first.] A gambling card game popular in Elizabethan England.

**prime time** *n.* The hours, usu. during the evening, when most television viewers are available.

**pri·me·val** (prī-mē′vəl) *adj.* [< Lat. *primaevus,* young : *primus,* first + *aevum,* age.] Belonging to the first or earliest age or ages : ORIGINAL. **—pri·me′val·ly** *adv.*

**pri·mi** (prē′mē) *n. pl. of* PRIMO.

**prim·ing** (prī′mĭng) *n.* **1.** The explosive for igniting a charge. **2.** A preliminary surface coating, as paint.

**pri·mip·a·ra** (prī-mĭp′ər-ə) *n., pl.* **-a·ras** *or* **-a·rae** (-ə-rē′) [Lat. : *primus,* first + *parere,* to give birth.] *Med.* **1.** A woman undergoing a first pregnancy. **2.** A woman who has given birth to only one child. **—pri′mi·par′i·ty** (-mĭ-păr′ĭ-tē) *n.* **—pri·mip′a·rous** *adj.*

**prim·i·tive** (prĭm′ĭ-tĭv) *adj.* [ME *primitif* < OFr. < Lat. *primitivus* < *primus,* first.] **1. a.** Of or relating to an earliest or original stage or state. **b.** Archetypal. **2.** Marked by simplicity or crudity : UNSOPHISTICATED <*primitive* implements> **3.** Of or relating to early stages in the development of human culture <*primitive* dwellings> **4. a.** Serving as the source for derived or inflected forms <"Pick" is the *primitive* word from which "picket" is derived.> **b.** Being a protolanguage <*primitive* Germanic> **5.** *Math.* A form in algebra or geometry from which another form is derived. **6. a.** Having a painting style of an unsophisticated or early culture. **b.** Self-taught. **c.** Of or relating to late medieval European painters. **7.** *Geol.* Of or relating to rocks formed by the first solidification of the earth's crust. **8.** *Biol.* Happening in or typical of an early stage of evolution or development. —*n.* **1.** A member of a primitive society or culture. **2.** One at a low or early developmental stage. **3. a.** One belonging to an early stage in the development of an artistic trend. **b.** An artist having or affecting a primitive style. **c.** A self-taught artist. **4.** A word or word element from which another word or inflected form of the word is derived. **5.** *Computer Sci.* A basic or fundamental unit of machine instruction or translation. **—prim'i·tive·ly** *adv.* **—prim'i·tive·ness, prim'i·tiv'i·ty** *n.*

**prim·i·tiv·ism** (prĭm'ĭ-tĭ-vĭz'əm) *n.* **1.** The quality or state of being primitive. **2.** A belief in primitive ideas or customs. **3.** The style of primitive painters. **4.** A belief that civilization is evil or that the earliest period of human history was the best. **—prim'i·tiv·ist** *n.* **—prim'i·tiv·is'tic** *adj.*

**pri·mo** (prē'mō) *n., pl.* **-mi** (-mē) [Ital. < Lat. *primus,* first.] *Mus.* The chief part in a duet or ensemble composition. **—pri'mo** *adj.*

**pri·mo·gen·i·tor** (prī'mō-jĕn'ĭ-tər) *n.* [Med. Lat. : Lat. *primus,* first + Lat. *genitor,* begetter < *gignere,* to beget.] An earliest ancestor or forefather.

**pri·mo·gen·i·ture** (prī'mō-jĕn'ĭ-chŏor') *n.* [Med. Lat. *primogenitura* : Lat. *primus,* first + Lat. *genitura,* birth < *gignere,* to beget.] **1.** The state of being the first-born or eldest child of the same parents. **2.** *Law.* The right of the eldest child, esp. the eldest son, to inherit the estate of one or both parents. **—pri'mo·gen'i·tar'y** (-jĕn'ĭ-tĕr'ē), **pri'mo·gen'i·tal** *adj.*

**pri·mor·di·al** (prī-môr'dē-əl) *adj.* [ME < LLat. *primordialis* < Lat. *primordium,* origin : *primus,* first + *ordiri,* to begin.] **1.** Being or occurring first in sequence of time : ORIGINAL. **2.** Fundamental or primary <fill a *primordial* role> **3.** *Biol.* Belonging to or typical of the earliest stage of development of an organism or part. —*n.* A basic principle. **—pri·mor'di·al·ly** *adv.*

**pri·mor·di·um** (prī-môr'dē-əm) *n., pl.* **-di·a** (-dē-ə) [Lat. —see PRIMORDIAL.] An organ or part in its most rudimentary form or stage.

**primp** (prĭmp) *v.* **primped, primp·ing, primps.** [Orig. unknown.] —*vt.* To neaten (one's appearance) with much attention to detail. —*vi.* To preen.

**prim·rose** (prĭm'rōz') *n.* [ME *primerose* < OFr. < Med. Lat. *prima rosa,* first rose.] **1.** A plant of the genus *Primula,* bearing tubular, variously colored, five-lobed flowers. **2.** The evening primrose.

**primrose path** *n.* A lifestyle of worldly ease or pleasure.

**pri·mum mo·bi·le** (prī'məm mō'bə-lē', prē'məm mō'bĭ-lā') *n.* [Med. Lat., first mover, transl. of Ar. *almuḥarrik alawwal.*] **1.** The tenth and outermost concentric sphere of the universe believed in medieval astronomy to revolve around the earth from east to west in 24 hours and believed to cause the other 9 spheres to revolve with it. **2.** PRIME MOVER 2, 5.

**pri·mus** (prī'məs) *n., pl.* **-mus·es.** [Med. Lat. < Lat., first.] The first in rank of the bishops of Scotland.

**pri·mus in·ter pa·res** (prī'məs ĭn'tər pâr'ēz, prē'mŏŏs ĭn'tər pä'rās') *n.* [Lat.] The first among equals.

**prince** (prĭns) *n.* [ME < OFr. < Lat. *princeps* < *primus,* first.] **1.** A hereditary ruler : KING. **2.** The ruler of a principality. **3.** A male member of a royal family other than the monarch. **4.** A nobleman. **5.** An outstanding man in a group or class <a merchant *prince*> **6.** *Rom. Cath. Ch.* A high-ranking cleric, esp. a cardinal. **—prince'ship'** *n.*

**Prince Al·bert** (ăl'bərt) *n.* [After *Prince Albert* Edward (1841–1910), later Edward VII.] A man's long double-breasted frock coat.

**Prince Albert**

**prince charming** *n.* [After *Prince Charming,* hero of *Cinderella,* a fairy tale.] A man who satisfies a woman's romantic dreams.

**prince consort** *n.* The husband of a sovereign queen.

**prince·dom** (prĭns'dəm) *n.* **1.** PRINCIPALITY 1. **2.** The rank or status of a prince.

**prince·ling** (prĭns'lĭng) *also* **prince·let** (prĭns'lĭt) *n.* A minor prince.

**prince·ly** (prĭns'lē) *adj.* **-li·er, -li·est. 1.** Of or relating to a prince : ROYAL. **2.** Suitable for a prince <a *princely* sum> **—prince'li·ness** *n.* **—prince'ly** *adv.*

**Prince of Wales** *n.* **1.** A title conferred by the sovereign on the male heir to the British throne. **2.** The male heir to the British throne, upon whom the title Prince of Wales has been conferred.

**prince regent** *n.* A prince who rules during the minority, absence, or incapacity of a sovereign.

**prince's-feath·er** (prĭn'sĭz-fĕth'ər) *n.* **1.** A tall plant, *Polygonum orientale,* with hairy stems and long pink or rose flower spikes. **2.** A plant, *Amaranthus hybridus hypochondriacus,* with reddish foliage and dense brownish-red flower clusters.

**prince's-pine** (prĭn'sĭz-pīn') *n.* The pipsissewa.

**prin·cess** (prĭn'sĭs, -sĕs', prĭn-sĕs') *n.* [ME *princesse* < OFr. < *prince,* prince.] **1.** *Archaic.* A hereditary woman ruler : QUEEN. **2.** The woman ruler of a principality. **3.** A woman member of a royal family other than the monarch. **4.** A noblewoman. **5.** The wife of a prince. **6.** A woman regarded as having the status or attributes of a princess. —*adj. also* **prin·cesse** (prĭn-sĕs'). Designed to hang in smooth, close-fitting, unbroken lines from shoulder to flared hem <a *princess* dress>

**Princess Royal** *n.* The eldest daughter of a British sovereign. —Used as a title when conferred by the sovereign.

**prin·ci·pal** (prĭn'sə-pəl) *adj.* [ME < OFr. < Lat. *principalis* < *princeps,* ruler < *primus,* first.] First, highest, or foremost in importance, rank, worth, or degree : CHIEF. —*n.* **1.** One who holds a post of presiding rank, esp. the head of an elementary school or high school. **2.** A primary participant. **3.** One having a main or starring role. **4. a.** The capital or main body of an estate or financial holding. **b.** A sum of money owed as a debt, on which interest is figured. **5.** *Law.* **a.** One who empowers another to act as his or her representative. **b.** The person with prime responsibility for an obligation as opposed to one who acts as surety or as an endorser. **c.** One who commits or is an accomplice to a crime. **6.** The main supporting truss or rafter that forms a roof. **—prin'ci·pal·ly** *adv.* **—prin'ci·pal·ship'** *n.*

**principal diagonal** *n.* The diagonal in a square matrix that goes from the upper left corner to the lower right corner.

**principal focus** *n.* A focal point.

**prin·ci·pal·i·ty** (prĭn'sə-păl'ĭ-tē) *n., pl.* **-ties. 1.** A territory ruled by a prince or from which the title of a prince is derived. **2.** The position, authority, or jurisdiction of a prince : SOVEREIGNTY. **3. principalities.** An angel of the third lowest rank in the nine orders of angels.

**principal parts** *pl.n.* The primary forms of a verb from which all other forms may be derived in traditional grammars of inflected languages, including in English the present infinitive (*play, eat*), the past tense (*played, ate*), the past participle (*played, eaten*), and the present participle (*playing, eating*).

**prin·cip·i·um** (prĭn-sĭp'ē-əm) *n., pl.* **-i·a** (-ē-ə) [Lat. —see PRINCIPLE.] A basic principle.

**prin·ci·ple** (prĭn'sə-pəl) *n.* [ME < OFr. *principe* < Lat. *principium* < *princeps,* first < *primus.*] **1.** A basic truth, law, or assumption <the *principles* of capitalism> **2. a.** A rule or standard, esp. of good behavior <a person of *principle*> **b.** Moral or ethical standards or judgments as a whole <actions founded on *principle* rather than desires> **3.** A fixed or predetermined policy or mode of action <acting on the *principle* of the golden rule> **4.** A basic or essential quality or element determining intrinsic nature or characteristic behavior <the *principle* of self-preservation natural to humankind> **5.** A rule or law concerning the operation of natural phenomena or mechanical processes <explained the *principle* of the internal-combustion engine> **6.** A basic source. **7. Principle.** *Christian Science.* GOD 1c. **—in principle.** With regard to basics <a proposal that seems satisfactory *in principle*>

**prin·ci·pled** (prĭn'sə-pəld) *adj.* Motivated by or based on ethical or moral principles.

**prink** (prĭngk) *v.* **prinked, prink·ing, prinks.** [Prob. alteration of PRANK².] —*vt.* To adorn (oneself) in a conspicuous manner. —*vi.* To primp. **—prink'er** *n.*

**print** (prĭnt) *n.* [ME *preinte* < OFr. < p.part. of *preindre,* to press < Lat. *premere.*] **1.** A mark or impression made in or on a surface by pressure <the *print* of boots in the snow> **2. a.** A device or implement, as a stamp, die, or seal, for pressing markings on or into a surface. **b.** Something formed or marked by such a device <a *print* of sealing wax> **3. a.** Lettering or other impressions made in ink from type by a printing press or other means. **b.** Matter so produced. **c.** The state or form of matter so produced. **4.** A design or picture transferred from a medium, as an engraved plate, wood block or lithographic stone. **5.** A photographic image transferred to paper or a similar surface, usu. from a negative. **6. a.** A fabric or garment with a dyed pattern that has been pressed onto it, usu. by engraved rollers. **b.** The pattern itself. —*v.* **print·ed, print·ing, prints.** —*vt.* **1.** To press (e.g., a mark or design) onto or into a surface. **2.** To make an impression on or in (a surface) with a device, as a stamp, seal, or die. **3.** To press (a stamp or similar device) onto or into a surface to leave a marking. **4.** To produce by means of pressed type on a paper surface, with or as if with a printing press. **5.** To make available in printed form : PUBLISH. **6.** To write (something) in characters similar to those usu. used in print. **7.** To impress firmly in the mind or memory. **8.** To produce (a positive photograph) by passing light through a negative onto sensitized paper. —*vi.* **1.** To work as a printer. **2.** To write characters similar to those usu. used in print. **3.** To produce or receive an impression, marking, or image. **—in print. 1.** In printed or published form. **2.** Still available from a publisher. **—out of print.** No longer available from a publisher. **—print out.** To print as a computer function : produce printout.

**print·a·ble** (prĭn′tə-bəl) *adj.* **1.** Capable of being printed or of producing a print. **2.** Fit for publication. **—print′a·bil′i·ty** *n.*

**print bar** *n.* A mechanism in a printing device that carries the template of the final form of the alphanumeric characters to be printed.

**printed circuit** *n.* An electric circuit in which the conducting connections are formed by depositing a conducting metal, as copper, in predetermined patterns on an insulating substrate, while other materials, esp. semiconductors, are deposited to form various electronic components.

**printed matter** *n.* Printed material, as a book or magazine, that is not regarded as first-class mail and qualifies for a special postal rate.

**print·er** (prĭn′tər) *n.* **1.** One that prints, esp. one whose occupation is printing. **2.** The part of a computer that produces printed matter.

**printer's devil** *n.* An apprentice printer.

**print·er·y** (prĭn′tə-rē) *n., pl.* **-ies. 1.** A place that does typographic printing. **2.** A factory that prints fabrics.

**print·ing** (prĭn′tĭng) *n.* **1.** The process, art, or business of producing printed material by means of inked type and a printing press or by similar means. **2. a.** The act of one that prints. **b.** Printed matter. **3.** All the copies of a book or other publication printed at one time. **4.** Written characters not joined to each other and similar to those in print.

**printing ink** *n.* Ink made esp. for use in printing.

**printing office** *n.* An establishment where material is printed, esp. an officially authorized one.

**printing press** *n.* A machine that transfers lettering or images by contact with various forms of inked surface onto paper or similar material.

**print·mak·ing** (prĭnt′mā′kĭng) *n.* The artistic design and production of prints, as woodcuts or silkscreens. **—print′mak′er** *n.*

**print·out** (prĭnt′out′) *n.* The printed output of a computer.

**print wheel** *n.* A disk-shaped mechanism in a printing device that carries the template of the characters to be printed around its rim and prints one character at a time, rotating after each character to the correct position for the next.

**pri·or**[1] (prī′ər) *adj.* [Lat.] **1.** Preceding in time or order <a *prior* engagement> **2.** Preceding in importance or value <a *prior* requirement> **—prior to.** Before. **—pri′or·ly** *adv.*

**pri·or**[2] (prī′ər) *n.* [ME *priour* < OE and OFr. *prior,* both < Med. Lat. < Lat., superior.] **1.** A monastic officer in charge of a priory or ranking next below an abbot. **2.** A ruling magistrate in medieval Florence. **—pri′or·ate** (-ĭt), **pri′or·ship′** *n.*

**pri·or·ess** (prī′ər-ĭs) *n.* [ME *prioresse* < OFr., fem. of *prior,* a prior.] A nun at the head of a priory or ranking next below an abbess.

**pri·or·i·tize** (prī-ôr′ĭ-tīz′, -ŏr′-) *vt.* **-tized, -tiz·ing, -tiz·es.** [PRIOR-IT(Y) + -IZE.] To arrange or deal with in order of importance <*prioritized* our needs> ***usage:*** Many condemn the use of *prioritize* on the grounds that it is jargon, but in recent times it has become firmly established in English at all levels of speech and writing.

**pri·or·i·ty** (prī-ôr′ĭ-tē, -ŏr′-) *n., pl.* **-ties.** [ME *priorite* < OFr. < Med. Lat. *prioritas* < Lat. *prior,* first.] **1.** Precedence, esp. established by order of urgency or importance. **2. a.** A recognized right to precedence. **b.** An authoritative rating setting such precedence. **3.** Something that merits prior attention.

**pri·o·ry** (prī′ə-rē) *n., pl.* **-ies.** [ME *priorie* < Norman Fr. < Med. Lat. *prioria,* a prior < Lat., superior.] A monastery or convent under the governance of a prior or prioress.

**prise** (prīz) *v.* & *n. var. of* PRIZE[3].

**pri·sere** (prī′sîr′) *n.* [PRI(MARY) + SERE[2].] The succession of vegetation that occurs in an area not previously occupied by a community as it passes from barren earth or water to a climax community.

**prism** (prĭz′əm) *n.* [LLat. < Gk. < *prizein,* to saw.] **1.** A polyhedron with congruent, parallel polygons as bases and parallelograms as sides. **2.** A homogeneous transparent solid, usu. with triangular bases and rectangular sides, for producing or analyzing a continuous spectrum. **3.** A cut-glass object, as a pendant of a chandelier. **4.** A crystalline solid with at least three similar faces paralleling a single axis.

**pris·mat·ic** (prĭz-măt′ĭk) *also* **pris·mat·i·cal** (-ĭ-kəl) *adj.* **1.** Of, relating to, or like a prism. **2.** Refracting light, as a prism. **3.** Iridescent : multicolored. **—pris·mat′i·cal·ly** *adv.*

**pris·ma·toid** (prĭz′mə-toid′) *n.* [Gk. *prisma, prismat-,* prism + -OID.] A polyhedron with all vertices lying in one of two parallel planes. **—pris′ma·toi′dal** *adj.*

**pris·moid** (prĭz′moid′) *n.* A prismatoid with polygons having the same number of sides as bases, and faces that are parallelograms or trapezoids. **—pris·moi′dal** *adj.*

**pris·on** (prĭz′ən) *n.* [ME < OFr. < Lat. *prensio,* a seizing, short for *prehensio* < *prehendere,* to seize.] **1. a.** A place for confining people awaiting trial or for punishment after trial and conviction : JAIL. **b.** An institution for confining people convicted of major crimes : PENITENTIARY. **2.** Confinement or forcible restraint. **3.** Imprisonment. *—vt.* **-oned, -on·ing, -ons.** To imprison.

**prison camp** *n.* **1.** A camp for prisoners of war. **2.** A minimum security facility where trustworthy prisoners are interned.

**pris·on·er** (prĭz′ə-nər, prĭz′nər) *n.* **1.** One kept in custody, captivity, or a condition of forcible restraint, esp. while on trial or serving a prison sentence. **2.** One deprived of freedom of action or expression <a *prisoner* of destiny>

**prisoner of war** *n.* One captured by or surrendering to enemy forces during wartime.

**prisoner's base** *n.* A children's game in which two competing teams try to capture opposing players by tagging them and bringing them to a base.

**prison fever** *n.* [So called because it formerly prevailed in prisons.] Typhus.

**pris·sy** (prĭs′ē) *adj.* **-si·er, -si·est.** [Blend of PRIM and SISSY.] Finicky, fussy, and prudish. **—pris′si·ly** *adv.* **—pris′si·ness** *n.*

**pris·tine** (prĭs′tēn′, prĭ-stēn′) *adj.* [Lat. *pristinus.*] **1.** Of, relating to, or characteristic of the earliest condition or time : PRIMITIVE. **2.** Remaining in a pure state : UNCORRUPTED. **—pris·tine′ly** *adv.*

**prith·ee** (prĭth′ē, prĭth′ē) *interj.* [Alteration of *(I) pray thee.*] *Archaic.* —Used to express a request or wish.

**pri·va·cy** (prī′və-sē) *n., pl.* **-cies. 1.** Seclusion or isolation from the view of, or from contact with, others. **2.** Secrecy : concealment.

**pri·vate** (prī′vĭt) *adj.* [ME *privat* < Lat. *privatus,* not in public life, p.part. of *privare,* to release < *privus,* individual.] **1.** Secluded from the sight, presence, or intrusion of others <a *private* office> **2.** Of or limited to one person : PERSONAL <*private* thoughts> **3.** Not available for public use, control, or participation <a *private* dining room> **4.** Belonging to a specific person or persons <*private* industry> **5.** Not in an official or public position <a *private* citizen> **6.** Not public : INTIMATE <a *private* affair> *—n.* **1. a.** An enlisted person ranking below private first class in the U.S. Army or Marine Corps. **b.** One having a similar rank in other military organizations. **2. privates.** The genitals. **—in private.** Confidentially or secretly. **—pri′vate·ly** *adv.* **—pri′vate·ness** *n.*

☆ ***syns:*** PRIVATE, PERSONAL, PRIVY *adj. core meaning* : belonging or confined to a particular person or group as opposed to the public or the government <*private* property> ***ant:*** public

▲ word history: The noun *private* denoting a soldier of the lowest rank developed from the adjective *private* in the sense "not holding an official or public position." Just as a private citizen is one who holds no public office, so a private soldier is one who has neither special responsibilities nor the rank that goes with them.

**private detective** *n.* A privately employed detective.

**private enterprise** *n.* **1.** Privately owned business activities unregulated by state ownership or control. **2.** A privately owned business enterprise, esp. one operating under a free-enterprise system or laissez-faire capitalism.

**pri·va·teer** (prī′və-tîr′) *n.* **1.** A ship privately owned and crewed but authorized by a government during wartime to attack and capture enemy vessels. **2.** The commander or a crew member of a privateer. *—vi.* **-teered, -teer·ing, -teers.** To sail as a privateer.

**private eye** *n. Informal.* A private detective.

**private first class** *n.* An enlisted person ranking below corporal and above private in the U.S. Army or Marine Corps.

**private investigator** *n.* A private detective.

**private law** *n.* The branch of law which deals with or affects the rights of, and the relations between, private individuals.

**private member** *n. Chiefly Brit.* A member of Parliament not holding office in the government or in his or her party.

**private parts** *pl.n.* The genitals.

**private school** *n.* A secondary or elementary school run and supported by private individuals or a corporation instead of by a government or public agency.

**pri·va·tion** (prī-vā′shən) *n.* [ME *privacion* < OFr. *privation* < Lat. *privatio* < *privare,* to deprive < *privus,* without.] **1. a.** Lack of the basic necessities or comforts of life. **b.** The condition produced by such lack. **2.** An act or result of deprivation or loss.

**pri·vat·ism** (prī′və-tĭz′əm) *n.* The social position of being noncommittal to or not involved with anything except one's own immediate interests and lifestyle. **—pri′va·tis′tic** *adj.*

**priv·a·tive** (prĭv′ə-tĭv) *adj.* [Lat. *privativus* < *privare,* to deprive < *privus,* without.] **1.** Causing deprivation, lack, or loss. **2.** Altering the meaning of a term from positive to negative. *—n.* A privative prefix or suffix, as *a-, non-, un-,* or *-less.* **—priv′a·tive·ly** *adv.*

**priv·et** (prĭv′ĭt) *n.* [Orig. unknown.] **1.** A shrub, *Ligustrum vulgare* or *L. ovalifolium,* with pointed leaves and white flower clusters, widely used for hedges. **2.** A plant similar or related to the privet.

**priv·i·lege** (prĭv′ə-lĭj) *n.* [ME < OFr. < Lat. *privilegium,* a law affecting one person : *privus,* single + *lex,* law.] **1. a.** A special advantage, immunity, permission, right, or benefit granted to or enjoyed by an individual, class, or caste. **b.** Such a right or advantage held because of one's status or rank, and exercised to the exclusion or detriment of others. **2.** The principle of granting and holding privileges <a ruling class dependent on *privilege*> **3.** An option to buy

or sell a stock, including put, call, spread, and straddle. —*vt.* **-leged, -leg·ing, -leg·es. 1.** To grant a privilege to. **2.** To free or exempt.
**priv·i·leged** (prĭv′ə-lĭjd) *adj.* Having privileges or enjoying a privilege <*privileged* club members>
**privileged communication** *n. Law.* **1.** A confidential communication that one cannot be forced to disclose. **2.** A communication not subject to charges of slander or libel.
**priv·i·ty** (prĭv′ĭ-tē) *n., pl.* **-ties.** [ME *privete,* secret < OFr. < Med. Lat. *privitas* < Lat. *privus,* private.] **1.** Knowledge of something private or secret shared between individuals, esp. with the implication of consent or approval. **2.** *Law.* **a.** A relation between parties held to be sufficiently close and direct to uphold a legal claim on behalf of or against another party with whom this relation exists. **b.** A successive or mutual interest in or relationship to the same property.
**priv·y** (prĭv′ē) *adj.* [ME *prive* < OFr. < Lat. *privatus,* private < *privus.*] **1.** Made a party to private or secret information <was *privy* to military plans> **2.** Belonging or proper to a person, as the British sovereign, in a private rather than official capacity. **3.** *Archaic.* Concealed: secret. —*n., pl.* **-ies. 1. a.** A latrine. **b.** An outhouse. **2.** *Law.* One of the parties having an interest in the same matter. **—priv′i·ly** (prĭv′ə-lē) *adv.*
**Privy Council** *n.* **1.** A group of dignitaries and officials chosen by the British sovereign as an advisory council to the Crown and usu. functioning through its committees. **2. privy council.** An advisory council to an executive. **—Privy Councillor** *n.*
**prix fixe** (prē′ fēks′) *n., pl.* **prix fixes** (prē′ fēks′) [Fr., fixed price.] **1.** TABLE D'HÔTE 2. **2.** The price at which a table d'hôte meal is offered.
**prize[1]** (prīz) *n.* [ME *pris.* —see PRICE.] **1.** Something offered or won as an award for superiority or excellence in competition. **2.** Something offered for winning in a game of chance. **3.** Something worth striving for or aspiring to. —*adj.* **1.** Offered or given as a prize <a *prize* statuette> **2.** Given or likely to win a prize <a *prize* dog> **3.** Deserving a prize : FIRST-CLASS. —*vt.* **prized, priz·ing, priz·es. 1.** To value highly. **2.** To estimate the worth of : EVALUATE.
▲ word history: The noun *prize*[1] emerged in the 16th century as a variant of *price. Price* was a borrowing of Old French *pris,* which had a range of senses related to the idea of "value," such as "excellence," "money paid for something," and "esteem." Many of these senses have become obsolete; only the sense "sum of money, cost," is common in modern use. In Middle English times *price* meant "pre-eminent position," especially that of the victor in a contest. By the 16th century the spelling *prize* had arisen for this sense, which was extended to mean the symbol of such a victory. In the 17th century the spelling *price* for the new senses was completely supplanted by *prize,* which is now a separate word.
**prize[2]** (prīz) *n.* [ME *prise* < OFr. < VLat. **presa* < Lat. *prehendere,* to seize.] **1.** Something seized by force or taken as booty, esp. an enemy ship and cargo captured at sea during wartime. **2.** An act of seizing : CAPTURE.
**†prize[3]** *also* **prise** (prīz) *vt.* **prized, priz·ing, priz·es** *also* **prised, pris·ing, pris·es.** [< obs. *prize,* instrument for prizing < ME *prise* < OFr., grasp. —see PRIZE[2].] To move or force with or as if with a lever : PRY. —*n.* **1.** Leverage. **2.** *Regional.* Something used as a lever.
**prize·fight** (prīz′fīt′) *n.* A match between professional boxers for money. **—prize′fight′er** *n.* **—prize′fight′ing** *n.*
**prize ring** *n.* **1.** The platform enclosed by ropes in which boxers fight. **2.** Professional boxing.
**prize winner** *n.* A winner of a prize.
**prize-win·ning** (prīz′wĭn′ĭng) *adj.* Having won a prize or being judged deserving of such <a *prize-winning* performance>
**pro[1]** (prō) *n., pl.* **pros.** [ME < Lat., for.] **1.** An argument in support of something. **2.** One who favors a proposal or takes the affirmative side in debate. —*adv.* In favor of. —*adj.* Supporting : favoring.
**pro[2]** (prō) [Short for PROFESSIONAL.] *Informal.* —*n., pl.* **pros. 1.** A professional, esp. in sports. **2.** An expert in a given field. —*adj.* Professional <*pro* tennis>
**pro-[1]** *pref.* [ME < Lat. *pro,* for.] **1.** Acting in the place of : substituting for <*pro*gesterone> **2.** Supporting : favoring <*pro*revolutionary>
**pro-[2]** *pref.* [< Gk. *pro,* before, in front of.] **1. a.** Earlier : prior to <*pro*thrombin> **b.** Rudimentary <*pro*nucleus> **2.** Anterior : in front of <*pro*cambium>
**pro·a** (prō′ə) *n.* [< Malay *pĕrāhū,* prob. Marathi *pąḍāv.*] A swift Malayan sailboat having a triangular sail and a single outrigger.
**pro-a·bor·tion** (prō′ə-bôr′shən) *adj.* Favoring or supporting legalized clinical abortions. **—pro-a·bor′tion·ist** *n.*
**prob·a·bi·lism** (prŏb′ə-bə-lĭz′əm) *n.* **1.** *Philos.* The doctrine that probability is adequate basis for belief and action, since certainty in knowledge cannot be attained. **2.** *Rom. Cath. Ch.* The doctrine that when there is uncertainty as to the moral rectitude of an action, the opinion that supports liberty may be followed, if it is substantially probable, even though the contrary may be equally or even more probable. **—prob′a·bi·list** *n.* **—prob′a·bi·lis′tic** *adj.*

**prob·a·bil·i·ty** (prŏb′ə-bĭl′ĭ-tē) *n., pl.* **-ties. 1.** The quality or state of being probable : LIKELIHOOD. **2.** A probable situation or event. **3.** *Math.* A number expressing the likelihood of occurrence of a specific event, such as the ratio of the number of experimental results that would produce the event to the total number of results considered possible. **—in all probability.** Most probably.
**probability density** *n.* **1.** A function of a continuous random variable, the integral of which, over a given interval, gives the probability that the value of the variable will fall within the interval. **2.** Calculated value of a probability density.
**probability distribution** *n.* **1.** Probability density. **2.** A function of a discrete random variable yielding the probability that the variable will have a given value.
**probability theory** *n.* A branch of mathematics that studies the likelihood of occurrence of random events for the purpose of predicting the behavior of defined systems.
**prob·a·ble** (prŏb′ə-bəl) *adj.* [ME, provable < OFr. < Lat. *probabilis* < *probare,* to prove.] **1.** Apt to occur or to be true. **2.** Relatively likely but not certain : PLAUSIBLE. **3.** Of or relating to moral opinions and actions for the lawfulness of which intrinsic reasons or extrinsic authority may be adduced : PROVABLE.
**probable cause** *n. Law.* Reasonable grounds for belief that an accused person is guilty as charged.
**prob·a·bly** (prŏb′ə-blē) *adv.* Most likely : PRESUMABLY.
**pro·bang** (prō′băng′) *n.* [Alteration of obs. *provang.*] A long slender flexible rod with a tuft or sponge at the end, for removing foreign bodies from the larynx or esophagus.
**pro·bate** (prō′bāt′) *n.* [ME *probat* < Lat. *probatum,* something proved < *probare,* to prove < *probus,* good.] **1. a.** Legal establishment of the validity of a will. **b.** The process of legally establishing the validity of a will. **2.** The right to validate wills. —*vt.* **-bat·ed, -bat·ing, -bates.** To establish the validity of (a will).
**probate court** *n.* A court with jurisdiction over probating wills and administering estates.
**pro·ba·tion** (prō-bā′shən) *n.* [ME *probacion* < OFr. < Lat. *probatio,* trial < *probare,* to test < *probus,* good.] **1.** A period for testing one's fitness for membership in a working or social group. **2.** *Law.* The action of suspending the sentence of one convicted of a minor offense and granting the offender provisional freedom on the promise of good behavior. **3.** A period when a student is allowed to redeem failing grades or bad conduct. **4.** The status of a probationer. **—pro·ba′tion·al, pro·ba′tion·ar′y** *adj.* **—pro·ba′tion·al·ly** *adv.*
**pro·ba·tion·er** (prō-bā′shə-nər) *n.* One on probation.
**probation officer** *n.* A court officer who investigates, reports on, and supervises convicted offenders on probation.
**pro·ba·tive** (prō′bə-tĭv) *also* **pro·ba·to·ry** (-tôr′ē, -tōr′ē) *adj.* [ME *probatiffe* < Lat. *probativus* < *probare,* to prove < *probus,* good.] **1.** Serving to test, try, or prove. **2.** Supplying evidence or proof.
**probe** (prōb) *n.* [< Med. Lat. *proba,* examination < Lat. *probare,* to test < *probus,* good.] **1.** An object or device for investigating an unknown configuration, condition, or region <a space *probe*> **2.** A slender flexible instrument for exploring a wound or body cavity. **3.** The act of exploring or searching with the aid of a probe. **4.** An investigation, esp. an investigation of illegal practices <a congressional *probe*> —*v.* **probed, prob·ing, probes.** —*vt.* **1.** To explore with a probe. **2.** To examine or investigate : delve into. —*vi.* To carry on an exploratory investigation : SEARCH. **—prob′er** *n.*
**pro·ben·e·cid** (prō-bĕn′ĭ-sĭd) *n.* [PRO(PYL) + BEN(ZOIC) + (A)CID.] A derivative of benzoic acid, $C_{13}H_{19}NO_4S$, used medically as a drug treating gout.
**pro·bi·ty** (prō′bĭ-tē) *n.* [OFr. *probite* < Lat. *probitas* < *probus,* honest.] Complete integrity : UPRIGHTNESS.
**prob·lem** (prŏb′ləm) *n.* [ME *probleme* < OFr. < Lat. *problema* < Gk. *problēma* < *proballein,* to throw out : *pro-,* forward + *ballein,* to throw.] **1.** A question or situation that presents doubt, perplexity, or difficulty. **2.** One who is difficult to deal with. **3.** A question offered for consideration, discussion, or solution.
**prob·lem·at·i·cal** (prŏb′lə-măt′ĭ-kəl) *also* **prob·lem·at·ic** (-ĭk) *adj.* **1.** Posing a problem. **2.** Open to doubt : DEBATABLE. **—prob′lem·at′i·cal·ly** *adv.*
**prob·lem-o·ri·ent·ed language** (prŏb′ləm-ôr′ē-ĕn′tĭd, -ōr′-) *n.* *Computer Sci.* A programming language intended for use in solving a given set of problems.
**pro·bos·cid·i·an** (prō′bŏs-ĭd′ē-ən) *also* **pro·bos·ci·de·an** (prō-bŏs′ĭ-dē′ən) *adj.* [< NLat. *Proboscidea,* order name < Lat. *proboscis,* proboscis.] Of or belonging to the Proboscidea, an order of mammals typified by a trunk or proboscis and including the elephant. —*n.* An animal of the order Proboscidea.
**pro·bos·cis** (prō-bŏs′ĭs) *n., pl.* **-cis·es** or **-ci·des** (-ĭ-dēz′) [Lat. < Gk. *proboskis* : *pro-,* in front + *boskein,* to feed.] **1.** A long flexible snout or trunk, as of an elephant. **2.** The slender tubular feeding and sucking structure of some insects and worms. **3.** A human nose, esp. a prominent one.
**pro-bus·ing** (prō-bŭs′ĭng) *adj.* Favoring or supporting the busing of children to schools outside their neighborhoods as a way of achieving racial integration.
**pro·caine hydrochloride** (prō′kān′) *n.* [PRO-[1] + (CO)CAIN.] A white crystalline powder, $C_{13}H_{20}O_2N_2 \cdot HCl$, used as a local anesthetic.

**pro·cam·bi·um** (prō-kăm′bē-əm) *n. Bot.* A layer of undifferentiated plant cells from which the vascular tissue is formed. **—pro·cam′bi·al** *adj.*

**pro·carp** (prō′kärp′) *n. Bot.* A specialized female sex organ in certain algae.

**pro·car·y·ote** *also* **pro·kar·y·ote** (prō-kăr′ē-ōt′) *n.* [PRO-[2] + Gk. *karuōtos,* having nuts.] A cellular organism, as a bacterium or blue-green alga, the nucleus of which has no limiting membrane. **—pro·car′y·ot′ic** (-ŏt′ĭk) *adj.*

**pro·ce·dur·al** (prə-sē′jər-əl) *adj.* Of or concerning procedure, esp. of a court of law or parliamentary body. **—pro·ce′dur·al·ly** *adv.*

**pro·ce·dure** (prə-sē′jər) *n.* [Fr. *procédure* < OFr. < *proceder,* to proceed.] **1.** A way of performing or effecting something. **2.** A course of action. **3.** A set of established forms or methods for carrying on the affairs of a business, legislative body, or court of law.

**pro·ceed** (prō-sēd′, prə-) *vi.* **-ceed·ed, -ceed·ing, -ceeds.** [ME *proceden* < OFr. *proceder* < Lat. *procedere* : *pro-,* forward + *cedere,* to go.] **1.** To advance, esp. after an interruption : CONTINUE. **2.** To undertake and carry on an act or process. **3.** To move on in a methodical way. **4.** To issue forth : ORIGINATE. **5.** To institute and conduct legal action. **—pro·ceed′er** *n.*

**pro·ceed·ing** (prō-sē′dĭng, prə-) *n.* **1.** A course of action. **2.** Continuation of an action. **3. proceedings.** A succession of events taking place at a specific place or occasion. **4. proceedings.** An official record of business conducted by an organization : MINUTES. **5.** *Law.* **a. proceedings.** Litigation. **b.** The act of instituting or conducting litigation.

**pro·ceeds** (prō′sēdz′) *pl.n.* The money obtained from a commercial or fund-raising venture : YIELD.

**pro·ce·phal·ic** (prō′sə-făl′ĭk) *adj. Anat.* Situated on or close to the front of the human head.

**proc·ess**[1] (prŏs′ĕs′, prō′sĕs′) *n., pl.* **proc·ess·es** (prŏs′ĕs′ĭz, prō′sĕs′-, prŏs′ĭ-sēz′, prō′sĭ-) [ME *proces* < OFr. < Lat. *processus,* advance < p.part. of *procedere,* to advance. —see PROCEED.] **1.** A system of operations in producing something. **2.** A series of actions, changes, or functions that achieve an end or result. **3.** Passage of time. **4.** Ongoing movement : PROGRESSION. **5.** *Law.* **a.** A summons or writ ordering a defendant to appear in court. **b.** The total quantity of writs or summonses issued in a particular proceeding. **c.** The whole course of a judicial proceeding. **6.** *Biol.* A part extending or projecting from an organ or organism : APPENDAGE. **7.** A photomechanical or photoengraving method. —*vt.* **-essed, -ess·ing, -ess·es. 1.** To put through the steps of a prescribed procedure. **2.** To prepare, treat, or convert by subjecting to a special process. **3.** *Law.* To serve with a summons or writ. **4.** To institute legal proceedings against : PROSECUTE. **5.** *Computer Sci.* To perform operations on data. —*adj.* **1.** Prepared or converted by a special treatment <*process* cheese> **2.** Made by or used in photomechanical or photoengraving methods <a *process* print>

**pro·cess**[2] (prə-sĕs′) *vi.* **-cessed, -cess·ing, -cess·es.** [Back-formation < PROCESSION.] To move in or as if in a procession.

**pro·ces·sion** (prə-sĕsh′ən) *n.* [ME < OFr. < Lat. *processio,* advance < *procedere,* to advance. —see PROCEED.] **1.** An act of proceeding, moving along, or issuing forth. **2. a.** A group moving along in an orderly and formal way, usu. in a long line. **b.** The movement of such a group. **3.** A continuous and orderly course <the *procession* of the months> —*vi.* **-sioned, -sion·ing, -sions.** To form or go in a procession.

**pro·ces·sion·al** (prə-sĕsh′ə-nəl) *adj.* Of, relating to, or fitting for a procession. —*n.* **1.** A book containing the ritual observed during a religious procession. **2.** A hymn sung at the beginning of a church service. **3.** Music meant to be played or sung during a procession. **—pro·ces′sion·al·ly** *adv.*

**proc·es·sor** (prŏs′ĕs′ər) *n.* **1.** One that processes. **2.** A computer. **3.** *Computer Sci.* **a.** A central processing unit. **b.** A program that translates another program into a form acceptable to the computer being used.

**process printing** *n.* Printing from multiple, usu. four, halftone images, each inked with a different color such that the composite impression will reproduce the colors of the original.

**pro·cès-ver·bal** (prō-sā′vĕr-bäl′) *n., pl.* **-ver·baux** (-vĕr-bō′) [Fr. : *procès,* proceedings + *verbal,* oral.] **1.** An official record of diplomatic negotiations. **2.** A detailed official record of legal proceedings.

**pro-choice** (prō-chois′) *adj.* Advocating a person's right to choose an abortion.

**pro·claim** (prō-klām′, prə-) *vt.* **-claimed, -claim·ing, -claims.** [ME *proclaymen* < OFr. *proclamer* < Lat. *proclamare* : *pro-,* forward + *clamare,* to cry out.] **1.** To announce officially and publicly : DECLARE. **2.** To indicate unmistakably. **3.** To praise : extol. **—pro·claim′er** *n.*

**proc·la·ma·tion** (prŏk′lə-mā′shən) *n.* **1.** An act of proclaiming. **2.** Something proclaimed, esp. an official public announcement.

**pro·clit·ic** (prō-klĭt′ĭk) *adj.* [NLat. *procliticus* < Gk. *proklinein,* to lean forward : *pro-,* forward + *klinein,* to lean.] Forming an accentual unit with the following word and thus having no independent accent. **—pro·clit′ic** *n.*

**pro·cliv·i·ty** (prō-klĭv′ĭ-tē) *n., pl.* **-ties.** [Lat. *proclivitas* < *proclivus,* inclined : *pro-,* forward + *clivus,* slope.] A natural propensity : PREDISPOSITION.

**Proc·ne** (prŏk′nē) *n.* [Lat. < Gk. *Proknē.*] *Gk. Myth.* The sister of Philomela and betrayed wife of Tereus who was changed into a swallow before Tereus could kill her.

**pro·con·sul** (prō-kŏn′səl) *n.* [ME < Lat. < *pro consule,* for the consul.] **1.** A Roman provincial governor of consular rank. **2.** A high administrator in a modern colony, dependency, or other occupied area, and having wide-ranging powers. **—pro·con′su·lar** (-sə-lər) *adj.* **—pro·con′su·late** (-sə-lĭt) *n.* **—pro·con′sul·ship′** *n.*

**pro·cras·ti·nate** (prō-krăs′tə-nāt′, prə-) *v.* **-nat·ed, -nat·ing, -nates.** [Lat. *procrastinare, procrastinat-* : *pro-,* forward + *cras,* tomorrow.] —*vi.* To put off doing something until a later date. —*vt.* To delay. **—pro·cras′ti·na′tion** *n.* **—pro·cras′ti·na′tor** *n.*

**pro·cre·ate** (prō′krē-āt′) *v.* **-at·ed, -at·ing, -ates.** [Lat. *procreare, procreat-* : *pro-,* forward + *creare,* to create.] —*vt.* **1.** To beget (offspring). **2.** To produce or create : ORIGINATE. —*vi.* To beget offspring. **—pro′cre·ant** (-ənt) *adj.* **—pro′cre·a′tion** *n.* **—pro′cre·a′tor** *n.*

**pro·cre·a·tive** (prō′krē-ā′tĭv) *adj.* **1.** Capable of reproducing : GENERATIVE. **2.** Of or directed to procreation <*procreative* impluse>

**pro·crus·te·an** *also* **Pro·crus·te·an** (prō-krŭs′tē-ən) *adj.* [After *Procrustes,* a mythical Greek giant who stretched or shortened captives to make them fit his beds < *prokrouein,* to stretch out : *pro-,* forth + *krouein,* to beat.] **1.** Producing or intended to produce conformity by arbitrary or ruthless methods. **2.** Having or exhibiting ruthless disregard for individual differences or special circumstances <*procrustean* legislative enactments>

**procrustean bed** *also* **Procrustean bed** *n.* An arbitrary standard to which precise conformity is forced.

**pro·cryp·tic** (prō-krĭp′tĭk) *adj.* [Prob. PRO(TECT) + CRYPTIC.] *Biol.* Having a pattern or coloration adapted for natural camouflage.

**proc·ti·tis** (prŏk-tī′tĭs) *n.* [Gk. *prōktos,* anus + -ITIS.] Inflammation of the anus or rectum.

**proc·tol·o·gy** (prŏk-tŏl′ə-jē) *n.* [Gk. *prōktos,* anus + -LOGY.] Physiology and pathology of the rectal area. **—proc′to·log′ic** (-tə-lŏj′ĭk), **proc′to·log′i·cal** *adj.* **—proc′to·log′i·cal·ly** *adv.* **—proc·tol′o·gist** *n.*

**proc·tor** (prŏk′tər) *n.* [ME, university officer < *procuratour,* procurator.] A dormitory and examination supervisor in a school. —*vt.* **-tored, -tor·ing, -tors.** To supervise (an examination). **—proc·to′ri·al** (-tôr′ē-əl, -tōr′-) *adj.* **—proc′tor·ship′** *n.*

**proc·to·scope** (prŏk′tə-skōp′) *n.* [Gk. *prōktos,* anus + -SCOPE.] An instrument for dilating and examining the rectum. **—proc′to·scop′ic** (-skŏp′ĭk) *adj.* **—proc·tos′co·py** (-tŏs′kə-pē) *n.*

**pro·cum·bent** (prō-kŭm′bənt) *adj.* [Lat. *procumbens, procumbent-,* pr.part. of *procumbere,* to bend down : *pro-,* forward + *cumbere,* to lie down.] **1.** *Bot.* Trailing on the ground <a *procumbent* vine> **2.** Lying face down : PRONE.

**proc·u·ra·tor** (prŏk′yə-rā′tər) *n.* [ME *procuratour* < OFr. < Lat. *procurator* < *procurare,* to take care of. —see PROCURE.] **1.** An agent with power of attorney. **2.** A Roman official acting as a financial agent of the emperor or as an administrator of a minor province. **—proc′u·ra·to′ri·al** (-yər-ə-tôr′ē-əl, -tōr′-) *adj.*

**pro·cure** (prō-kyoŏr′, prə-) *v.* **-cured, -cur·ing, -cures.** [ME *procuren* < OFr. *procurer* < Lat. *procurare,* to manage : *pro-,* for + *curare,* to care for.] —*vt.* **1.** To obtain : acquire. **2.** To bring about : EFFECT <*procure* a settlement of a strike> **3.** To obtain for another (a person) for sexual intercourse. —*vi.* To work as a procurer. **—pro·cur′a·ble** *adj.* **—pro·cur′ance, pro·cure′ment** *n.*

**pro·cur·er** (prō-kyoŏr′ər, prə-) *n.* **1.** One who procures <a *procurer* of new executive talent> **2.** PANDER 1.

**pro·cur·ess** (prō-kyoŏr′ĭs, prə-) *n.* A woman who procures.

**Pro·cy·on** (prō′sē-ŏn′) *n.* [Lat. < Gk. *Prokuon* : *pro-,* before + *kuon,* dog.] A double star in the constellation Canis Minor.

**prod** (prŏd) *vt.* **prod·ded, prod·ding, prods.** [Orig. unknown.] **1.** To poke or jab, as with a pointed instrument. **2.** To rouse to action : GOAD. —*n.* **1.** Something for prodding : GOAD. **2.** An incitement : stimulus. **—prod′der** *n.*

**prod·i·gal** (prŏd′ĭ-gəl) *adj.* [Lat. *prodigus* < *prodigere,* to squander : *prod-,* before (var. of *pro-*) + *agere,* to drive.] **1.** Recklessly wasteful : EXTRAVAGANT. **2. a.** Profuse in giving. **b.** Exceedingly abundant. **3.** Profuse : lavish <*prodigal* praise> —*n.* One inclined to luxury or extravagance. **—prod′i·gal·ly** *adv.*

**prod·i·gal·i·ty** (prŏd′ĭ-găl′ĭ-tē) *n., pl.* **-ties. 1.** Extravagant wastefulness. **2.** Profuse generosity. **3.** Extreme abundance.

**pro·di·gious** (prə-dĭj′əs) *adj.* [Lat. *prodigiousus* < *prodigium,* omen.] **1.** Awesomely great in size, force, or extent : ENORMOUS. **2.** Extraordinary : marvelous. **3.** Ominous : portentous. **—pro·di′gious·ly** *adv.* **—pro·di′gious·ness** *n.*

**prod·i·gy** (prŏd′ə-jē) *n., pl.* **-gies.** [Lat. *prodigium,* omen.] **1.** One with remarkable talents or powers <a child *prodigy*> **2.** An act or event so remarkable or rare as to inspire wonder. **3.** An omen.

**pro·drome** (prō′drōm′) *n., pl.* **-dromes** or **-dro·ma·ta** (-drō′mə-tə) [Fr. < Gk. *prodromos,* precursor : *pro-,* forward + *dromos,* running.] A symptom of the onset of a disease. **—pro·dro′mal** (-drō′məl), **pro·drom′ic** (-drŏm′ĭk) *adj.*

**pro·duce** (prə-do͞os′, -dyo͞os′, prō-) *v.* **-duced, -duc·ing, -duc·es.** [Lat. *producere* : *pro-*, forward + *ducere*, to lead.] —*vt.* **1.** To bring forth : YIELD. **2.** To create by physical or mental effort. **3.** To manufacture. **4.** To give rise to. **5.** To bring forward : EXHIBIT. **6.** To sponsor and present to the public <*produce* a musical revue> **7.** To extend (an area or volume) or lengthen (a line). —*vi.* To make or yield the customary product or products. —*n.* (prŏd′o͞os, prō′do͞os). A product, esp. farm products as a whole. **—pro·duc′i·ble** *adj.*

**pro·duc·er** (prə-do͞o′sər, -dyo͞o′-, prō-) *n.* **1.** One that produces, esp. one that manufactures or grows goods or services to sell. **2.** One who finances and supervises the production of public entertainment, as a play. **3.** A furnace for manufacturing producer gas. **4.** *Ecol.* An autotrophic organism in an ecosystem.

**producer gas** *n.* A gas used as fuel, generated by passing air with steam over burning coke or coal to yield a combustible mixture of nitrogen, carbon monoxide, and hydrogen.

**producer goods** *or* **producers′ goods** *pl.n.* Goods, as raw materials or tools, for making consumer goods.

**prod·uct** (prŏd′əkt) *n.* [Lat. *productum* < p.part. of *producere*, to bring forth. —see PRODUCE.] **1.** Something produced by human or mechanical effort or by a natural process. **2.** A direct result : CONSEQUENCE. **3.** *Chem.* A substance produced by a chemical change. **4.** *Math.* **a.** The result arrived at by performing multiplication. **b.** A scalar product. **c.** A vector product.

**pro·duc·tion** (prə-dŭk′shən, prō-) *n.* **1.** An act or process of producing. **2.** Creation of value or wealth by producing goods and services. **3.** PRODUCT 1. **4.** The total number of products : OUTPUT. **5.** A public performance, as of a play. **—pro·duc′tion·al** *adj.*

**pro·duc·tive** (prə-dŭk′tĭv, prō-) *adj.* **1.** Producing or able to produce. **2.** Producing abundantly : FERTILE. **3.** Yielding useful or favorable results : CONSTRUCTIVE <*productive* marketing ideas> **4.** Of or involved in the creation of goods and services to produce wealth or value. **5.** Resulting in <investigations *productive* of discoveries> **—pro·duc′tive·ly** *adv.* **—pro′duc·tiv′i·ty** (prō′dŭk-tĭv′ĭ-tē, prŏd′-ək-), **pro·duc′tive·ness** *n.*

**pro·em** (prō′ĕm′) *n.* [ME *proheme* < OFr. < Lat. *prooemium* < Gk. *prooimion*, prelude : *pro-*, before + *oimē*, song.] A short introduction : PREFACE. **—pro·e′mi·al** (prō-ē′mē-əl, -ĕm′ē-) *adj.*

**pro·en·zyme** (prō-ĕn′zīm′) *n.* Zymogen.

**pro·es·trus** (prō-ĕs′trəs) *n.* The period of preparation for pregnancy immediately preceding estrus in female mammals.

**prof** (prŏf) *n. Informal.* A professor.

**pro·fane** (prō-fān′, prə-) *adj.* [ME *prophane* < OFr. < Lat. *profanus* : *pro-*, before + *fanum*, temple.] **1.** Showing irreverence or contempt toward God or sacred things : BLASPHEMOUS. **2.** Nonreligious in subject matter, form, or use. **3.** Not initiated into the mysteries of ritual. **4.** Coarse : vulgar. —*vt.* **-faned, -fan·ing, -fanes. 1.** To treat irreverently. **2.** To put to an improper, unworthy, or degrading use : ABUSE. **—prof′a·na′tion** (prŏf′ə-nā′shən) *n.* **—pro·fan′a·to′ry** (prō-făn′ə-tôr′ē, -tōr′ē, prə-) *adj.* **—pro·fane′ly** *adv.* **—pro·fane′ness** *n.* **—pro·fan′er** *n.*

☆ ***syns:*** PROFANE, LAY, SECULAR, TEMPORAL, WORLDLY *adj. core meaning* : not religious in subject matter, form, or use <sacred and *profane* music> ***ant:*** sacred

**pro·fan·i·ty** (prō-făn′ĭ-tē, prə-) *n., pl.* **-ties. 1.** The quality or state of being profane. **2. a.** Abusive, vulgar, or irreverent language. **b.** Use of abusive, vulgar, or irreverent language.

**pro·fess** (prə-fĕs′, prō-) *v.* **-fessed, -fess·ing, -fess·es.** [Lat. *profiteri, profess-* : *pro-*, forth + *fateri*, to acknowledge.] —*vt.* **1.** To affirm openly : DECLARE <*professed* my ignorance> **2.** To make a pretense of <"... *professed* to despise everything that had happened since 1850" —Louis Auchincloss> **3.** To claim skill in or knowledge of <*profess* pharmacy> **4.** To affirm belief in <*profess* Hinduism> **5.** To receive into a religious order. —*vi.* **1.** To make an open affirmation. **2.** To take the vows of a religious order. **—pro·fess′ed·ly** (-fĕs′ĭd-lē) *adv.*

**pro·fes·sion** (prə-fĕsh′ən) *n.* [ME, vow made on entering a religious order < OFr. < Lat. *professio*, declaration < *profiteri*, to declare. —see PROFESS.] **1.** An occupation or vocation requiring training in the liberal arts or the sciences and advanced study in a specialized field. **2.** The body of qualified persons of a specific occupation or field. **3.** An act or instance of professing : DECLARATION. **4.** Acknowledgment of religious faith.

**pro·fes·sion·al** (prə-fĕsh′ə-nəl) *adj.* **1.** Of, pertaining to, engaged in, or appropriate for a profession. **2.** Engaged in a specific activity as a means of livelihood : CAREER <a *professional* musician> **3.** Performed by persons receiving pay <*professional* soccer> **4. a.** Possessing great skill or experience in a field or activity. **b.** *Informal.* Behaving in such a way as to appear professional <a *professional* grouch> —*n.* **1.** One following a profession. **2.** One who makes a living as an athlete. **3.** One with assured competence in a field. **—pro·fes′sion·al·ly** *adv.*

**pro·fes·sion·al·ism** (prə-fĕsh′ə-nə-lĭz′əm) *n.* **1.** Professional status, methods, character, or standards. **2.** Use of professional players in organized athletics.

**pro·fes·sion·al·ize** (prə-fĕsh′ə-nə-līz′) *vt.* **-ized, -iz·ing, -iz·es.** To make professional. **—pro·fes′sion·al·i·za′tion** *n.*

**pro·fes·sor** (prə-fĕs′ər) *n.* [ME *professour* < Lat. *professor* < *profiteri*, to profess.] **1. a.** A teacher of the highest rank in an institution of higher learning. **b.** A teacher : instructor. **2.** One who professes. **—pro′fes·so′ri·al** (prō′fĭ-sôr′ē-əl, -sōr′-, prŏf′ĭ-) *adj.* **—pro′fes·so′ri·al·ly** *adv.* **—pro·fes′sor·ship′** *n.*

**pro·fes·so·ri·ate** *or* **pro·fes·so·ri·at** (prō′fĭ-sôr′ē-ət, -ăt′, -sōr′-, prŏf′ĭ-) *n.* **1.** The office or rank of a professor. **2.** University and college professors as a group.

**prof·fer** (prŏf′ər) *vt.* **-fered, -fer·ing, -fers.** [ME *profren* < OFr. *poroffrir* : *por-*, forth (< Lat. *pro-*) + *offrir*, to offer < Lat. *offerre*. —see OFFER.] To offer : tender. **—prof′fer** *n.* **—prof′fer·er** *n.*

**pro·fi·cient** (prə-fĭsh′ənt) *adj.* [Lat. *proficiens, proficient-*, pr.part. of *proficere*, to advance. —see PROFIT.] Performing in a given art, skill, or branch of learning with correctness and facility. —*n.* An expert. **—pro·fi′cien·cy** (prə-fĭsh′ən-sē) *n.* **—pro·fi′cient·ly** *adv.*

**pro·file** (prō′fīl′) *n.* [Obs. Ital. *profilo* < *profilare*, to draw in outline : *pro-*, forward (< Lat.) + *filare*, to draw a line < LLat., to spin < Lat. *filum*, thread.] **1. a.** A side view of an object or structure, esp. of a human head. **b.** A representation of an object or structure seen from the side. **2.** An outline of an object. **3.** A biographical essay presenting the subject's most remarkable characteristics and accomplishments. **4.** A graph or table representing numerically the extent to which a person or thing shows various tested features <a corporation *profile*> —*vt.* **-filed, -fil·ing, -files. 1.** To draw or shape a profile of. **2.** To write a profile of.

**prof·it** (prŏf′ĭt) *n.* [ME < OFr. < Lat. *profectus* < p.part. of *proficere*, to gain : *pro-*, forward + *facere*, to make.] **1.** An advantageous gain or return : BENEFIT. **2.** The return received on a business undertaking after meeting all operating expenses. **3.** *often* **profits. a.** The return received on an investment after paying all charges. **b.** The rate of increase in the net worth of a business enterprise during a given accounting period. **c.** Income received from investments or property. **d.** The amount received for a commodity or service above the original cost. —*v.* **-it·ed, -it·ing, -its.** —*vi.* **1.** To make a gain or profit. **2.** To be advantageous : BENEFIT. —*vt.* To be beneficial to. **—prof′it·a·bil′i·ty** *n.* **—prof′it·less** *adj.*

**prof·it·a·ble** (prŏf′ĭ-tə-bəl) *adj.* Affording profit. **—prof′it·a·ble·ness** *n.* **—prof′it·a·bly** *adv.*

☆ ***syns:*** PROFITABLE, ADVANTAGEOUS, FAT, LUCRATIVE, MONEYMAKING, REMUNERATIVE *adj. core meaning* : affording profit <a *profitable* venture> ***ant:*** profitless, unprofitable

**profit and loss** *n.* An account showing net profit and loss during a given period.

**prof·i·teer** (prŏf′ĭ-tîr′) *n.* One who makes excessive profits on commodities in short supply. **—prof·i·teer′** *v.* **(-teered, -teer·ing, -teers).**

**profit sharing** *n.* A system by which employees share in the profits of a business.

**profit system** *n.* Free enterprise.

**prof·li·gate** (prŏf′lĭ-gĭt, -gāt′) *adj.* [Lat. *profligatus*, p.part. of *profligare*, to ruin : *pro-*, forward + *fligere*, to strike.] **1.** Given over to dissipation : DISSOLUTE. **2.** Recklessly extravagant. —*n.* A wastrel. **—prof′li·ga·cy** (-gə-sē) *n.* **—prof′li·gate·ly** *adv.*

**pro for·ma** (prō fôr′mə) *adj.* [Lat., according to form.] **1.** Made or done in a mechanical and unenthusiastic manner. **2.** Provided in advance so as to specify form or to describe items.

**pro·found** (prə-found′, prō-) *adj.* **-er, -est.** [ME *profounde* < OFr. *profund* < Lat. *profundus* : *pro-*, before + *fundus*, bottom.] **1.** Situated at, extending to, or coming from a great depth : DEEP. **2.** Coming as if from the depths of one's being <*profound* hatred> **3.** Thoroughgoing : far-reaching. **4.** Penetrating beyond the obvious or superficial. **5.** Unqualified : absolute <*profound* stillness on the lake> **—pro·found′ly** *adv.* **—pro·found′ness** *n.*

**pro·fun·di·ty** (prə-fŭn′dĭ-tē, prō-) *n., pl.* **-ties.** [ME *profundite* < OFr. < LLat. *profunditas* < Lat. *profundus*, deep. —see PROFOUND.] **1.** Great depth. **2.** Depth of intellect, feeling, or meaning. **3.** Something profound or abstruse.

**pro·fuse** (prə-fyo͞os′, prō-) *adj.* [ME < Lat. *profusus*, p.part. of *profundere*, to pour forth : *pro-*, forward + *fundere*, to pour.] **1.** Plentiful : copious <*profuse* vegetation> **2.** Giving or given freely and abundantly : EXTRAVAGANT <*profuse* in their admiration> **—pro·fuse′ly** *adv.* **—pro·fuse′ness** *n.*

**pro·fu·sion** (prə-fyo͞o′zhən, prō-) *n.* **1.** The state of being profuse : ABUNDANCE. **2.** Lavish or unrestrained expense : EXTRAVAGANCE. **3.** A profuse outpouring or display.

**pro·gen·i·tor** (prō-jĕn′ĭ-tər) *n.* [ME *progenitour* < OFr. *progeniteur* < Lat. *progenitor* < *progignere*, to beget : *pro-*, forward + *gignere*, to beget.] **1.** A direct ancestor. **2.** An originator of a line of descent.

**prog·e·ny** (prŏj′ə-nē) *n., pl.* **-nies.** [ME *progenie* < OFr. < Lat. *progenies* < *progignere*, to beget. —see PROGENITOR.] **1.** Children or descendants : OFFSPRING. **2.** A result of creative effort : PRODUCT.

**pro·ges·ta·tion·al** (prō′jĕs-tā′shə-nəl) *adj.* **1.** Preceding gestation. **2.** Preceding ovulation.

o͞o **boot** ou **out** th **th**in *th* **th**is ŭ **cu**t ûr **ur**ge y **y**oung
yo͞o **abu**se zh vi**s**ion ə **a**bout, it**e**m, ed**i**ble, gall**o**p, circ**u**s

**pro·ges·ter·one** (prō-jĕs'tə-rōn') *n.* [PRO-[1] + GES(TATION) + STER(OL) + -ONE.] A female hormone, $C_{21}H_{30}O_2$, secreted by the corpus luteum of the ovary before implantation of the fertilized ovum.

**pro·ges·to·gen** (prō-jĕs'tə-jən) *n.* [PROGEST(ATIONAL) + (ESTR)OGEN.] Any of several natural or synthetic progestational hormones.

**pro·glot·tid** (prō-glŏt'ĭd) *also* **pro·glot·tis** (-glŏt'ĭs) *n., pl.* **-glot·tids** *also* **-glot·ti·des** (-glŏt'ĭ-dēz') [Gk. *proglōttis, proglōttid-,* tip of the tongue (from its shape) : *pro-,* before + *glotta,* tongue.] A segment of a tapeworm, holding male and female reproductive organs. **—pro·glot'tic, pro'glot·ti·de'an** (-glŏt-ĭ-dē'ən, -glŏ-tĭd'ē-ən) *adj.*

**prog·na·thous** (prŏg'nə-thəs, prŏg-nā'-) *also* **prog·nath·ic** (prŏg-năth'ĭk, -nā'thĭk) *adj.* Having conspicuously projecting jaws. **—prog'na·thism** (-nə-thĭz'əm) *n.*

**prog·no·sis** (prŏg-nō'sĭs) *n., pl.* **-ses** (-sēz') [LLat. < Gk. *prognōsis* < *progignōskein,* to foreknow : *pro-,* before + *gignōskein,* to know.] **1. a.** A prediction of probable development and outcome of a disease. **b.** Likelihood of recovery from a disease. **2.** A prediction.

**prog·nos·tic** (prŏg-nŏs'tĭk) *adj.* [Med. Lat. *prognosticus* < Gk. *prognōstikos* < *progignōskein,* to foreknow. —see PROGNOSIS.] **1.** Of, pertaining to, or functioning as a prognosis. **2.** Predicting : foretelling. —*n.* **1.** A sign or omen of a future event. **2.** A symptom signifying the future development of a disease.

**prog·nos·ti·cate** (prŏg-nŏs'tĭ-kāt') *vt.* **-cat·ed, -cat·ing, -cates.** [Med. Lat. *prognosticare, prognosticat-* < Lat. *prognosticum,* sign of the future < Gk. *prognōstikon* < *prognōstikos,* foreknowing. —see PROGNOSTIC.] **1.** To predict, using current signs as a guide. **2.** To foreshadow : portend. **—prog·nos'ti·ca'tion** *n.* **—prog·nos'ti·ca'tive** *adj.* **—prog·nos'ti·ca'tor** *n.*

**pro·gram** (prō'grăm', -grəm) *n.* [Fr. *programme* < LLat. *programma,* public notice < Gk. < *prographein,* to set forth as a public notice : *pro-,* forth + *graphein,* to write.] **1. a.** A list of the order of events and other relevant facts for a public presentation. **b.** The presentation. **2.** A scheduled television or radio show. **3.** An organized list of procedures : SCHEDULE. **4.** *Computer Sci.* **a.** A procedure for solving a problem, including data collection and processing and presentation of results. **b.** This procedure coded for a computer. **5.** An instruction sequence in programmed instruction. —*v.* **-grammed, -gram·ming, -grams** *also* **-gramed, -gram·ing, -grams.** —*vt.* **1.** To include or schedule in a program. **2.** To train or regulate (e.g., the mind or the senses) to perform in a given way <continually trying to *program* their reactions> **3.** *Computer Sci.* To provide (a computer) with a set of instructions for solving a problem. —*vi.* **1.** To design or schedule programs. **2. a.** To prepare an instructional sequence in programmed instruction. **b.** To instruct by such a sequence. **—pro·gram'ma·bil'i·ty** *n.* **—pro'gram'ma·ble** *adj.* **—pro'gram·mat'ic** (prō'grə-măt'ĭk) *adj.* **—pro·gram·mat'i·cal·ly** *adv.*

**program director** *n.* A television or radio station director responsible for choosing, planning, and scheduling programs.

**pro·gramme** (prō'grăm', -grəm) *n.* & *v. Chiefly Brit. var. of* PROGRAM.

**pro·grammed** *or* **pro·gramed** (prō'grămd', -grəmd) *adj.* Of, relating to, or taking place by programmed instruction <*programmed* learning><a *programmed* language course>

**programmed instruction** *n.* A teaching method in which the information to be learned is presented in discrete units, with a correct response to each unit required from the learner before going on to the next unit.

**pro·gram·mer** *or* **pro·gram·er** (prō'grăm'ər) *n.* **1.** *Computer Sci.* One who prepares a computer program. **2.** One who prepares an instructional program.

**pro·gram·ming** *or* **pro·gram·ing** (prō'grăm'ĭng, -grə-mĭng) *n.* The designing, scheduling, or planning of a program.

**program music** *n.* Music meant to suggest images, incidents, or scenes in a story.

**prog·ress** (prŏg'rĕs', -rəs, prō'grĕs') *n.* [ME *progresse* < Lat. *progressus,* p.part. of *progredi,* to advance : *pro-,* forward + *gradi,* to step.] **1.** Movement toward a goal. **2.** Development : unfolding. **3.** Steady improvement, as of a society or civilization. **4.** A state journey made by a sovereign through his or her realm. —*vi.* **pro·gress** (prə-grĕs') **-gressed, -gress·ing, -gress·es. 1.** To proceed. **2.** To advance toward a more desirable form. **—in progress.** Going on or happening now.

**pro·gres·sion** (prə-grĕsh'ən) *n.* **1.** Progress. **2.** Advance. **3.** A sequence, as of events. **4.** *Math.* A series of numbers or quantities each derived from the one preceding by a consistent operation. **5.** *Mus.* **a.** A succession of tones or chords. **b.** A series of repetitions of a phrase, each in a new position on the scale. **—pro·gres'sion·al** *adj.*

**pro·gres·sive** (prə-grĕs'ĭv) *adj.* **1.** Moving forward : ADVANCING. **2.** Proceeding in steps <*progressive* deterioration in trade relations. **3.** Promoting or favoring political or social reform : LIBERAL. **4. Progressive.** Of or belonging to a Progressive Party. **5.** Of, pertaining to, or influenced by a theory of education marked by stressing the individual needs and capacities of each child and informality of curriculum. **6.** Of or denoting a tax system in which the rate of taxation increases as the taxable amount increases. **7.** *Pathol.* Continuously spreading or increasing in severity. **8.** Indicating a verb form that expresses an action or condition in progress. —*n.* **1.** One who favors or strives for reform in politics, education, or other fields. **2. Progressive.** A member of a Progressive Party. **3.** A progressive verb form. **—pro·gres'sive·ly** *adv.* **—pro·gres'sive·ness** *n.*

**Pro·gres·sive-Con·ser·va·tive Party** (prə-grĕs'ĭv-kən-sûr'və-tĭv) *n.* A leading Canadian political party advocating close ties with Britain and economic nationalism.

**Progressive Party** *n.* **1.** A chiefly agrarian American political party organized under the leadership of Theodore Roosevelt in 1912. **2.** A U.S. political party organized in 1924 and led by Robert M. La Follette. **3.** A U.S. political party formed in 1948, orig. led by Henry A. Wallace.

**pro·gres·siv·ism** (prə-grĕs'ĭ-vĭz'əm) *n.* The doctrines and practice of political or educational progressives. **—pro·gres'siv·ist** *n.* **—pro·gres'siv·is'tic** *adj.*

**pro·hib·it** (prō-hĭb'ĭt) *vt.* **-it·ed, -it·ing, -its.** [ME *prohibiten* < Lat. *prohibēre,* to prevent : *pro-,* in front + *habēre,* to hold.] **1.** To forbid by authority. **2.** To prevent : debar.

**pro·hi·bi·tion** (prō'ə-bĭsh'ən) *n.* **1.** An act of prohibiting. **2.** A law, order, or decree that forbids something. **3. a.** The forbidding by law of the manufacture, transportation, sale, and possession of alcoholic beverages. **b. Prohibition.** The period (1920–33) during which a law forbidding the manufacture and sale of alcoholic beverages was in force in the United States.

**pro·hi·bi·tion·ist** (prō'ə-bĭsh'ə-nĭst) *n.* **1.** One in favor of outlawing the manufacture and sale of alcoholic beverages. **2.** *often* **Prohibitionist.** A member of the Prohibition Party.

**Prohibition Party** *n.* A U.S. political party organized in 1869 that advocated prohibition.

**pro·hib·i·tive** (prō-hĭb'ĭ-tĭv) *also* **pro·hib·i·to·ry** (-tôr'ē, -tōr'ē) *adj.* **1.** Prohibiting : forbidding <*prohibitive* statutes> **2.** Preventing or hindering purchase or use <*prohibitive* costs> **—pro·hib'i·tive·ly** *adv.* **—pro·hib'i·tive·ness** *n.*

**proj·ect** (prŏj'ĕkt', -ĭkt) *n.* [ME *proiecte* < Lat. *projectus,* p.part. of *proicere,* to throw out : *pro-,* forth + *jacere,* to throw.] **1.** A plan or proposal : SCHEME. **2.** An undertaking requiring joint effort. **3.** A research undertaking. —*v.* **pro·ject** (prə-jĕkt'), **-ject·ed, -ject·ing, -jects.** —*vt.* **1.** To thrust outward or forward. **2.** To throw forward : HURL. **3.** To transport in one's imagination. **4.** To externalize and attribute (e.g., an emotion) to someone or something else. **5.** To direct (one's voice) so as to be heard easily at a distance. **6.** To form a plan or intention for. **7.** To cause (an image) to appear upon a surface. **8.** *Math.* To produce a projection. —*vi.* **1.** To extend forward or out : PROTRUDE. **2.** To direct one's voice so as to be heard easily at a distance. **—pro·ject'a·ble** *adj.*

**pro·jec·tile** (prə-jĕk'təl, -tīl') *n.* [NLat. *projectilis* < Lat. *proicere,* to throw out. —see PROJECT.] **1.** A fired, thrown, or otherwise projected object, as a bullet, lacking any capacity for self-propulsion. **2.** A self-propelling missile, as a rocket. —*adj.* **1.** Capable of being impelled or hurled forward. **2.** Driving forward : IMPELLING. **3.** *Zool.* Capable of being thrust outward : PROTRUSILE.

**pro·jec·tion** (prə-jĕk'shən) *n.* **1.** An act of projecting. **2.** Something that thrusts outward : PROTUBERANCE. **3.** A plan for a future course of action. **4. a.** The process of projecting a filmed image onto a screen or other viewing surface. **b.** The image so projected. **5.** The image of a geometric figure made by a coordinate mapping. **6.** A system of intersecting lines, as the grid of a map, on which part or all of the globe or the celestial sphere may be shown as a plane surface. **7.** *Psychol.* Naive or unconscious attribution of one's own feelings, attitudes, or desires to others.

**projection booth** *n.* A booth, as in a theater, from which a film projector is operated.

**pro·jec·tion·ist** (prə-jĕk'shə-nĭst) *n.* **1.** An operator of a film projector. **2.** A mapmaker.

**projection room** *n.* **1.** A room with facilities for the private viewing of a film. **2.** A projection booth.

**pro·jec·tive** (prə-jĕk'tĭv) *adj.* **1.** Relating to or made by projection. **2.** Extending outward : PROJECTING. **3.** Designating a property of a geometric figure that does not vary when the figure undergoes projection. **—pro·jec'tive·ly** *adv.*

**projective geometry** *n.* Study of geometric properties that are invariant under projection.

**projective test** *n.* A psychological test in which a subject's responses to relatively unstructured standard stimuli, as a series of cartoons, abstract patterns, or incomplete sentences, are analyzed for determinants of personality or sometimes cognition.

**pro·jec·tor** (prə-jĕk'tər) *n.* **1.** A machine for projecting an image onto a screen. **2.** A device for projecting a beam of light. **3.** One who devises plans or projects.

**pro·jec·tu·al** (prə-jĕk'cho͞o-əl) *n.* [PROJECT + (VIS)UAL.] A piece of instructional material designed for projection onto a screen by a projector.

**pro·kar·y·ote** (prō-kăr'ē-ōt') *n. var. of* PROCARYOTE.

**pro·lac·tin** (prō-lăk'tĭn) *n.* A pituitary hormone that stimulates the secretion of milk.

**pro·la·mine** *also* **pro·la·min** (prō′lə-mĭn, -mēn′) *n.* [PROL(INE) + AM(MONIA) + -INE.] Any of a class of simple proteins in wheat, rye, and other grains.

**pro·lan** (prō′lăn′) *n.* [G. < Lat. *proles,* offspring.] The gonadotropic hormone in pregnant womens' urine, used to indicate pregnancy.

**pro·lapse** (prō-lăps′) [LLat. *prolapsus,* a falling < Lat., p.part. of *prolabi,* to fall down : *pro-,* forward + *labi,* to fall.] *Med.* —*vi.* **-lapsed, -laps·ing, -laps·es.** To fall or slip out of place. —*n.* (prō′-lăps′, prō-lăps′) *also* **pro·lap·sus** (prō-lăp′səs). The falling down or slipping out of place of an organ or part, as the uterus.

**pro·late** (prō′lāt′) *adj.* [Lat. *prolatus,* p.part. of *proferre,* to stretch out : *pro-,* forth + *ferre,* to carry.] Designating the shape of a solid, esp. of a spheroid, with a polar axis longer than the equatorial diameter. **—pro′late′ly** *adv.* **—pro′late′ness** *n.*

**prole** (prōl) *n.* [Short for PROLETARIAN.] A proletarian.

**pro·leg** (prō′lĕg′) *n.* One of the stubby limbs on the abdominal segments of caterpillars and some other insect larvae.

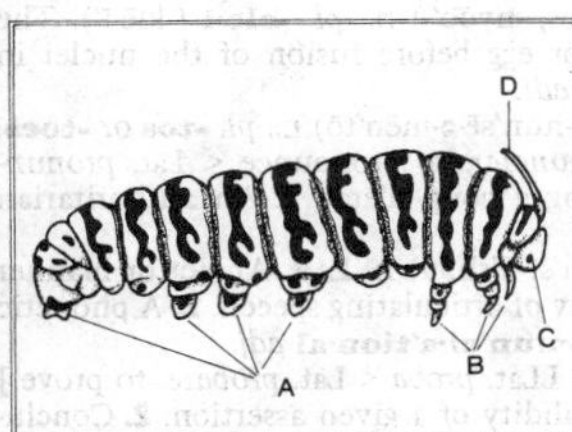


**proleg**
*Of a caterpillar: A. prolegs, B. thoracic legs, C. ocellus, D. everted scent gland*

**pro·le·gom·e·non** (prō′lĭ-gŏm′ə-nŏn′, -nən) *n., pl.* **-na** (-nə) [Gk. < passive neuter pr.part. of *prolegein,* to say beforehand : *pro-,* before + *legein,* to say.] A critical introduction. **—pro′le·gom′e·nous** *adj.*

**pro·lep·sis** (prō-lĕp′sĭs) *n., pl.* **-ses** (-sēz′) [LLat. < Gk. *prolēpsis* < *prolambanein,* to anticipate : *pro-,* before + *lambanein,* to take.] **1.** The anticipation and answering of an objection or argument in advance of its being put forward by one's opponent. **2.** Use of a descriptive word in anticipation of the act or circumstances that would make it applicable. **—pro·lep′tic** (-lĕp′tĭk), **pro·lep′ti·cal** *adj.*

**pro·le·tar·i·an** (prō′lĭ-târ′ē-ən) *adj.* [< Lat. *proletarius* < a Roman citizen of the lowest class < *proles,* offspring (from the fact that propertyless citizens were deemed by a Roman constitution to be of service to the state only by having children).] Of, relating to, or typical of the proletariat <*proletarian* states such as the Soviet Union> **—pro′le·tar′i·an** *n.* **—pro′le·tar′i·an·ism** *n.*

**pro·le·tar·i·at** (prō′lĭ-târ′ē-ĭt) *n.* [Fr. *prolétariat* < Lat. *proletarius,* a Roman citizen of the lowest class. —see PROLETARIAN.] **1.** The class of industrial wage earners who, lacking either capital or production means, must earn a living by selling their labor. **2.** The poorest class of working people.

**pro-life** (prō-līf′) *adj.* Opposed to legalized abortions <*pro-life* activists> **—pro-lif′er** *n.*

**pro·lif·er·ate** (prə-lĭf′ə-rāt′) *v.* **-at·ed, -at·ing, -ates.** [Back-formation < E. *proliferation,* the act of proliferating < Fr. *prolifération* < *prolifère,* procreative < Med. Lat. *prolifer.* —see PROLIFEROUS.] —*vi.* **1.** To reproduce or produce new growth or parts rapidly and repeatedly <*proliferating* cancer cells> **2.** To increase or spread rapidly. —*vt.* To cause to grow or increase rapidly. **—pro·lif′er·a′tion** *n.* **—pro·lif′er·a′tive** *adj.*

**pro·lif·er·ous** (prə-lĭf′ər-əs) *adj.* [Med. Lat. *prolifer* < Lat. *proles,* offspring.] **1.** *Zool.* Reproducing freely by means of buds and side branches, as coral. **2.** *Bot.* Freely producing buds or offshoots, occas. from unusual places. **—pro·lif′er·ous·ly** *adv.*

**pro·lif·ic** (prə-lĭf′ĭk) *adj.* [Fr. *prolifique* < Med. Lat. *prolificus* < Lat. *proles,* offspring.] **1.** Producing offspring or fruit in abundance : FERTILE. **2.** Producing abundant works or results. **—pro·lif′i·ca·cy** (-ĭ-kə-sē), **pro·lif′ic·ness** *n.* **—pro·lif′i·cal·ly** *adv.*

**pro·line** (prō′lēn′) *n.* [G. *Prolin* < *Pyrol,* pyrole.] An amino acid, $C_5H_9O_2N$, occurring in many proteins.

**pro·lix** (prō-lĭks′, prō′lĭks′) *adj.* [ME < OFr. *prolixe* < Lat. *prolixus,* extended.] **1.** Wordy and tedious. **2.** Tending to speak or write at great length. **—pro·lix′i·ty** (-lĭk′sĭ-tē) *n.* **—pro·lix′ly** *adv.*

**pro·loc·u·tor** (prō-lŏk′yə-tər) *n.* [ME < Lat., advocate < *proloqui,* to speak out : *pro-,* forth + *loqui,* to speak.] A presiding officer or chairperson, esp. of the lower house of a convocation in the Anglican Church.

**pro·logue** *also* **pro·log** (prō′lôg′, -lŏg′) *n.* [ME *prolog* < OFr. *prologue* < Lat. *prologus* < Gk. *prologos* : *pro-,* before + *legein,* to speak.] **1.** The lines introducing a discourse or play. **2.** An introductory act or event.

**pro·long** (prə-lông′, -lŏng′) *vt.* **-longed, -long·ing, -longs.** [ME *prolongen* < OFr. *prolonguer* < LLat. *prolongare* : Lat. *pro-,* out + Lat. *longus,* long.] **1.** To lengthen in duration : PROTRACT <*prolonged* the discussion> **2.** To lengthen in scope or extent. **—pro·long′er** *n.*

**pro·lon·gate** (prə-lông′gāt′, -lŏng′-, prō-) *vt.* **-gat·ed, -gat·ing, -gates.** To prolong. **—pro′lon·ga′tion** (prō′lông-gā′shən, -lŏng-) *n.*

**pro·lu·sion** (prō-lo͞o′zhən) *n.* [Lat. *prolusio* < *proludere,* to practice beforehand : *pro-,* before + *ludere,* to play.] **1.** A preliminary exercise. **2.** An essay written as a preface to a more detailed work. **—pro·lu′so·ry** (-sə-rē, -zə-) *adj.*

**prom** (prŏm) *n.* [Short for PROMENADE.] A formal dance for a high-school or college class.

**pro·me·carb** (prō′mĭ-kärb′) *n.* [PRO(PYL) + ME(THYL) + CARB(AMATE).] A colorless crystalline compound used as an insecticide on potatoes and fruit.

**prom·e·nade** (prŏm′ə-nād′, -näd′) *n.* [Fr. < *(se) promener,* to take a walk < LLat. *prominare,* to drive forward : *pro-,* forward + *minare,* to drive < *minari,* to threaten < *minae,* threats.] **1. a.** A leisurely walk, esp. one taken as a social activity in a public place. **b.** A public place for a promenade. **2. a.** A formal ball. **b.** A formal march by the guests at the start of a ball. **3.** A march performed between the figures of a square dance. —*v.* **-nad·ed, -nad·ing, -nades.** —*vi.* **1.** To go on a leisurely walk. **2.** To perform a promenade in square dancing. —*vt.* **1.** To take a promenade along or through. **2.** To take or display on or as if on a promenade. **—prom′e·nad′er** *n.*

**promenade deck** *n.* The upper deck or a section of the upper deck on a passenger ship where the passengers can promenade.

**Pro·me·the·an** (prə-mē′thē-ən) *adj.* **1.** Relating to or resembling Prometheus. **2.** Boldly creative : ORIGINAL. —*n.* One who is Promethean in actions or manner.

**Pro·me·the·us** (prə-mē′thē-əs, -thyo͞os′) *n.* [Lat. < Gk. *Promētheus.*] *Gk. Myth.* A Titan who stole fire from Olympus and gave it to human beings.

**pro·me·thi·um** (prə-mē′thē-əm) *n.* [< PROMETHEUS.] *Symbol* **Pm** A radioactive rare-earth element; atomic number 61; longest-lived isotope Pm 145.

**prom·i·nence** (prŏm′ə-nəns) *also* **prom·i·nen·cy** (-nən-sē) *n.* **1.** The quality or state of being prominent. **2.** Something prominent. **3.** *Astron.* A tonguelike cloud of flaming gas rising from the sun's surface, seen as part of the corona during a total eclipse.

**prom·i·nent** (prŏm′ə-nənt) *adj.* [Lat. *prominens, prominent-,* pr.part. of *prominēre,* to jut out.] **1.** Projecting outward from a line or surface : PROTUBERANT. **2.** Immediately noticeable : CONSPICUOUS. **3.** Widely known : EMINENT. **—prom′i·nent·ly** *adv.*

**prom·is·cu·i·ty** (prŏm′ĭ-skyo͞o′ĭ-tē, prō′mĭ-) *n., pl.* **-ties.** **1.** The state or character of being promiscuous. **2.** Promiscuous sexual intercourse. **3.** An indiscriminate mixture.

**pro·mis·cu·ous** (prə-mĭs′kyo͞o-əs) *adj.* [Lat. *promiscuus* : *pro-* (intensive) + *miscēre,* to mix.] **1.** Composed of varied and unrelated individuals or parts : CONFUSED. **2.** Without standards of selection : INDISCRIMINATE. **3.** Indiscriminate in sexual relations. **4.** Random : casual. **—pro·mis′cu·ous·ly** *adv.* **—pro·mis′cu·ous·ness** *n.*

**prom·ise** (prŏm′ĭs) *n.* [ME *promys* < Lat. *promissum,* neuter p.part. of *promittere,* to promise : *pro-,* forth + *mittere,* to send.] **1.** An assurance that one will or will not do something. **2.** Something promised. **3.** A sign of future excellence or success. —*v.* **-ised, -is·ing, -is·es.** —*vt.* **1.** To pledge or offer assurance <They *promise* to go.> **2.** To make a promise of. **3.** To provide a basis for expecting. —*vi.* **1.** To make a promise. **2.** To provide a basis for expectation <a future that *promises* well> **—prom′is·er** *n.*

☆ **syns:** PROMISE, COVENANT, GUARANTEE, PLEDGE, VOW *n.* *core meaning* : a declaration that one will or will not do a certain thing <made a *promise* to repay the loan>

**Promised Land** *n.* **1.** The land of Canaan that was promised to Abraham and his descendants. **2. promised land.** A place of anticipated happiness.

**prom·is·ee** (prŏm′ĭ-sē′) *n.* *Law.* One to whom a promise is made.

**prom·is·ing** (prŏm′ĭ-sĭng) *adj.* Apt to develop in a desirable manner. **—prom′is·ing·ly** *adv.*

**prom·i·sor** (prŏm′ĭ-sôr′) *n.* *Law.* A maker of a promise.

**prom·is·so·ry** (prŏm′ĭ-sôr′ē, -sōr′ē) *adj.* [Med. Lat. *promissorius* < Lat. *promissor,* one who promises < *promittere,* to promise.] **1.** Containing, concerning, or having the nature of a promise. **2.** Indicating how the provisions of an insurance contract will be carried out after it is signed.

**promissory note** *n.* A written promise to pay or repay a specified amount of money at a stated time or on demand.

**pro·mo** (prō′mō) *n., pl.* **-mos.** [Short for PROMOTION.] *Informal.* A promotional presentation, as a television spot, radio announcement, or personal appearance.

**prom·on·to·ry** (prŏm′ən-tôr′ē, -tōr′ē) *n., pl.* **-ries.** [Med. Lat. *promontorium,* alteration of Lat. *promuntorium.*] **1.** A high ridge of land or rock jutting out into a sea or other expanse of water. **2.** *Anat.* A projecting bodily part.

**pro·mot·a·ble** (prə-mō′tə-bəl) *adj.* **1.** Having the qualities needed for advancing to a higher position or rank. **2.** Suitable for consumer marketing and promotion <a *promotable* article> **—pro·mot′a·bil′i·ty** *n.*

**pro·mote** (prə-mōt′) *vt.* **-mot·ed, -mot·ing, -motes.** [ME *promoten* < Lat. *promovēre,* to advance : *pro-,* forward + *movēre,* to move.] **1. a.** To raise to a more important or responsible rank or job. **b.** To move (a student) to the next higher grade. **2.** To contribute to the progress or growth of : FURTHER. **3.** To urge the adoption of : ADVOCATE. **4.** To attempt to sell or popularize by advertising or by getting financial help <*promote* a new magazine><*promote* a Broadway show>

**pro·mot·er** (prə-mō′tər) *n.* **1.** An active supporter : ADVOCATE. **2.** A financial and publicity organizer, as of a boxing match.

**pro·mo·tion** (prə-mō′shən) *n.* **1.** An act of promoting. **2.** Advancement in responsibility or rank. **3.** Encouragement : furtherance. **4.** Advertising, publicity, and public relations. **—pro·mo′tion·al** *adj.* **—pro·mo′tion·al·ly** *adv.*

**pro·mo·tive** (prə-mō′tĭv) *adj.* Tending or serving to promote. **—pro·mo′tive·ness** *n.*

**prompt** (prŏmpt) *adj.* **-er, -est.** [ME < OFr. < Lat. *promptus,* p.part. of *promere,* to bring forth : *pro-,* forth + *emere,* to take.] **1.** On time : PUNCTUAL. **2.** Done without delay. *—vt.* **prompt·ed, prompt·ing, prompts. 1.** To press into action : INCITE. **2.** To give rise to : INSPIRE. **3.** To help with a reminder : REMIND. **4.** To give a cue to, as in a play. *—n.* **1. a.** The act of prompting or giving a cue. **b.** The information offered : CUE. **2. a.** A prompt note. **b.** The time limit stipulated in a prompt note. **—promp′ti·tude′** (prŏmp′tĭ-tōōd′, -tyōōd′), **prompt′ness** *n.* **—prompt′ly** *adv.*

**prompt·book** (prŏmpt′bŏŏk′) *n.* An annotated script used by a theater prompter.

**prompt·er** (prŏmp′tər) *n.* **1.** One who prompts. **2.** One who gives cues to actors.

**prompt·ing** (prŏmp′tĭng) *n. Computer Sci.* A language function alerting a user to the need for more input so as to continue processing data.

**prompt note** *n.* A notice sent to the purchaser of goods as a reminder of the amount owed the seller and the date payment is due.

**prom·ul·gate** (prŏm′əl-gāt′, prō-mŭl′gāt′) *vt.* **-gat·ed, -gat·ing, -gates.** [Lat. *promulgare, promulgat-.*] **1.** To make known (e.g., a decree) by public declaration. **2.** To put (a law) into effect by formal public announcement. **—prom′ul·ga′tion** (prŏm′əl-gā′shən, prō′məl-) *n.* **—prom′ul·ga′tor** *n.*

**pro·na·tal·ism** (prō-nāt′l-ĭz′əm) *n.* A policy or attitude that encourages having children, as by exalting or rewarding parenthood. **—pro·na′tal·ist** *n.* **—pro·na′tal·is′tic** *adj.*

**pro·nate** (prō′nāt′) *vt.* **-nat·ed, -nat·ing, -nates.** [LLat. *pronare, pronat-,* to bend forward < *pronus,* turned forward.] To turn (the palm or inner surface of the hand or forelimb) downward or backward. **—pro·na′tion** *n.*

**pro·na·tor** (prō′nā′tər) *n., pl.* **pro·na·to·res** (prō′nə-tôr′ēz, -tōr′-). The forearm or forelimb muscle that effects pronation.

**prone** (prōn) *adj.* [ME < Lat. *pronus,* inclined forward.] **1.** Lying with the front or face downward : PROSTRATE. **2.** Tending <*prone* to self-pity> *—adv.* In a prone manner <stretched *prone* on the floor> **—prone′ly** *adv.* **—prone′ness** *n.*

**pro·neph·ros** (prō-nĕf′rəs, -rŏs′) *n.* [PRO-² + Gk. *nephros,* kidney.] A primitive kidney that disappears early in the embryonic development of higher vertebrates. **—pro·neph′ric** (-rĭk) *adj.*

**prong** (prông, prŏng) *n.* [ME *pronge,* forked instrument.] **1.** A sharply pointed part of a tool or utensil, as a tine of a fork. **2.** A sharply pointed projection. *—vt.* **pronged, prong·ing, prongs.** To pierce with or as if with a prong.

**prong·horn** (prông′hôrn′, prŏng′-) *n., pl.* **pronghorn** or **-horns.** A small deer, *Antilocapra americana* of western North American plains, similar to an antelope and with small forked horns.

**pronghorn**
*3 feet high at shoulder*

**pro·nom·i·nal** (prō-nŏm′ə-nəl) *adj.* [LLat. *pronominalis* < Lat. *pronomen,* pronoun : *pro-,* in place of + *nomen,* name.] **1.** Of, relating to, or functioning as a pronoun. **2.** Like a pronoun, as by designating a person, place, or thing, while functioning mainly as another part of speech; e.g., *our* in *our choice* is a pronominal adjective. **—pro·nom′i·nal·ly** *adv.*

**pro·noun** (prō′noun′) *n.* One of a class of words that function as substitutes for nouns or noun phrases and denote persons or things asked for, previously designated, or understood from the context.

**pro·nounce** (prə-nouns′) *v.* **-nounced, -nounc·ing, -nounc·es.** [ME *pronouncen* < OFr. *prononcier* < Lat. *pronuntiare,* to declare : *pro-,* forth + *nuntiare,* to announce < *nuntius,* message.] *—vt.* **1.** To articulate (a word or speech sound). **2.** To transcribe (a word) in phonetic symbols. **3.** To state officially and formally : DECLARE. **4.** To declare to be in a specified condition. *—vi.* **1.** To declare one's opinion or make a pronouncement. **2.** To articulate words. **—pro·nounce′a·ble** *adj.* **—pro·nounc′er** *n.*

☆ ***syns:*** PRONOUNCE, ARTICULATE, ENUNCIATE, SAY, UTTER, VOCALIZE *v. core meaning :* to produce or make (speech sounds) <*pronounced* the vowels with a southern drawl>

**pro·nounced** (prə-nounst′) *adj.* **1.** Voiced : spoken. **2.** Strongly marked : DISTINCT <a *pronounced* improvement in the economy> **—pro·nounc′ed·ly** (-noun′sĭd-lē) *adv.* **—pro·nounc′ed·ness** *n.*

**pro·nounce·ment** (prə-nouns′mənt) *n.* **1.** A formal declaration. **2.** An authoritative statement.

**pro·nounc·ing** (prə-noun′sĭng) *adj.* Relating to, designed for, or showing pronunciation <a *pronouncing* dictionary>

**pron·to** (prŏn′tō) *adv.* [Sp. < Lat. *promptus.* —see PROMPT.] *Informal.* Without delay : QUICKLY.

**pro·nu·cle·us** (prō-nōō′klē-əs, -nyōō′-) *n., pl.* **-cle·i** (-klē-ī′). The haploid nucleus of a sperm or egg before fusion of the nuclei in fertilization. **—pro·nu′cle·ar** *adj.*

**pro·nun·ci·a·men·to** (prō-nŭn′sē-ə-mĕn′tō) *n., pl.* **-tos** or **-toes.** [Sp. *pronunciamiento* < *pronunciar,* to pronounce < Lat. *pronuntiare.*] **1.** An edict announcing a coup d'état. **2.** An authoritarian pronouncement : EDICT.

**pro·nun·ci·a·tion** (prə-nŭn′sē-ā′shən) *n.* **1. a.** An act or manner of articulating speech. **b.** A way of articulating speech. **2.** A phonetic transcription of a word. **—pro·nun′ci·a′tion·al** *adj.*

**proof** (prōōf) *n.* [ME *prove* < LLat. *proba* < Lat. *probare,* to prove.] **1.** Evidence establishing the validity of a given assertion. **2.** Conclusive demonstration. **3.** Proving of something by experiment, test, or trial <Durability was a *proof* of the tire's quality.> **4.** *Archaic.* Proven impenetrability. **5.** *Law.* The whole body of evidence that determines the judgment or verdict in a case. **6.** Substantiation of a proposition by application of specified rules, as of induction or deduction, to assumptions, axioms, and sequentially derived conclusions. **7.** Strength of a liquor in relation to proof spirit. **8. a.** A trial sheet of printed material that is checked against the original manuscript and on which corrections are made. **b.** A trial impression of a plate, stone, or block taken at any of various stages in engraving. **c.** A trial photographic print. *—adj.* **1.** Fully or successfully resistant : IMPERVIOUS <*proof* against earthquake damage> **2.** Of standard alcoholic strength. **3.** Used in proving or making corrections. *—v.* **proofed, proof·ing, proofs.** *—vt.* **1.** To run off (a printed or engraved proof). **2.** To proofread (copy). **3.** To work (dough) into suitable lightness. **4.** To make resistant. *—vi.* To proofread.

**-proof** *suff.* [< PROOF.] Impervious to : able to withstand <bullet-*proof*>

**proof·like** (prōōf′līk′) *adj.* Having the appearance of a coin, esp. its mint luster, that is not to be circulated.

**proof·read** (prōōf′rēd′) *v.* **-read** (-rĕd′), **-read·ing, -reads.** *—vt.* To read (copy or a printer's proof) against the original manuscript for corrections. *—vi.* To correct a printer's proof while reading against the original manuscript. **—proof′read′er** *n.*

**proof sheet** *n.* PROOF 8a.

**proof spirit** *n.* An alcohol-water mixture or a beverage with a standard amount of alcohol, the U.S. standard being 100 proof, or 50%, of ethyl alcohol by volume at 60°F.

**prop¹** (prŏp) *n.* [ME *proppe,* prob. of MDu. orig.] **1.** A device for showing something up. **2.** A support or stay. *—vt.* **propped, propping, props.** To keep from falling : SUPPORT.

**prop²** (prŏp) *n.* PROPERTY 4.

**prop³** (prŏp) *n. Informal.* A propeller.

**prop-** *pref.* [< PROPIONIC ACID.] Related to or derived from propionic acid <*propane*>

**pro·pae·deu·tic** (prō′pĭ-dōō′tĭk, -dyōō′-) *adj.* [< Gk. *propaideuein,* to teach beforehand : *pro-,* before + *paideuein,* to teach < *pais,* child.] Providing introductory instruction. **—pro′pae·deu′tic** *n.*

**prop·a·gan·da** (prŏp′ə-găn′də) *n.* [Ital., short for NLat. *Sacra Congregatio de Propaganda Fide,* Sacred Congregation for Propagating the Faith.] **1.** Methodical propagation of a particular doctrine or of allegations reflecting its views and interests <Communist *propaganda*> **2.** Material spread abroad by the advocates of a doctrine. **3. Propaganda.** The Congregation of the Roman Curia with authority in the matter of preaching the gospel and of establishing the Church in non-Christian countries and of administering Church missions in territories without any properly organized hierarchy. **—prop′a·gan′dism** *n.* **—prop′a·gan′dist** *n.* **—prop′a·gan·dis′tic** *adj.* **—prop′a·gan·dis′ti·cal·ly** *adv.*

**prop·a·gan·dize** (prŏp′ə-găn′dīz′) *v.* **-dized, -diz·ing, -diz·es.** *—vt.* **1.** To spread (a doctrine or opinion) by propaganda. **2.** To subject (a person or group of persons) to propaganda. *—vi.* To spread propaganda.

**prop·a·gate** (prŏp′ə-gāt′) *v.* **-gat·ed, -gat·ing, -gates.** [Lat. *propa-*

*gare, propagat-* < *propages,* offspring.] —*vt.* **1.** To cause (animals or plants) to multiply or breed. **2.** To breed (offspring). **3.** To hand down (characteristics) from one generation to the next. **4.** To make known : PUBLICIZE <*propagate* a canard> **5.** *Physics.* To cause (e.g., a wave) to move through a medium : TRANSMIT. —*vi.* **1.** *Physics.* To move through a medium. **2.** To breed or multiply. **—prop'a·ga·ble** (-gə-bəl) *adj.* **—prop'a·ga'tive** *adj.* **—prop'a·ga'tor** *n.*

**prop·a·ga·tion** (prŏp'ə-gā'shən) *n.* **1.** Increase or spread, as by natural reproduction. **2.** Dissemination, as of a belief. **—prop'a·ga'tion·al** *adj.*

**pro·pane** (prō'pān') *n.* A colorless gas, $C_3H_8$, occurring in natural gas and petroleum and much used as a fuel.

**pro·par·ox·y·tone** (prō'păr-ŏk'sĭ-tōn') *adj.* [Gk. *proparoxutonos* : *pro-,* before + *paroxutonos,* paroxytone. —see PAROXYTONE.] Bearing an acute accent on the antepenult in Classical Greek. —*n.* A proparoxytone word. **—pro'par·ox'y·ton'ic** (-tŏn'ĭk) *adj.*

**pro·pel** (prə-pĕl') *vt.* **-pelled, -pel·ling, -pels.** [ME *propellen* < Lat. *propellere* : *pro-,* forward + *pellere,* to drive.] **1.** To cause to move or sustain in motion. **2.** To urge on or persuade to do : MOTIVATE <was *propelled* by the family into law>

**pro·pel·lant** *also* **pro·pel·lent** (prə-pĕl'ənt) *n.* **1.** Something, as an explosive charge or a rocket fuel, that propels or supplies thrust. **2.** A gas, as a fluorocarbon, that serves as a vehicle for discharging the contents of an aerosol container. —*adj.* Serving to propel.

**pro·pel·ler** *also* **pro·pel·lor** (prə-pĕl'ər) *n.* A simple machine for propelling an aircraft or a boat, esp. one with radiating blades mounted on a revolving power-driven shaft.

**pro·pend** (prō-pĕnd') *vi.* **-pend·ed, -pend·ing, -pends.** [Lat. *propendēre* : *pro-,* forward + *pendēre,* to hang.] To have a propensity toward.

**pro·pene** (prō'pēn') *n.* Propylene.

**pro·pen·si·ty** (prə-pĕn'sĭ-tē) *n., pl.* **-ties.** [< obs. *propense,* inclined < Lat. *propensus,* p.part. of *propendēre,* to be inclined. —see PROPEND.] An inherent inclination : TENDENCY.

**prop·er** (prŏp'ər) *adj.* [ME *propre* < OFr. < Lat. *proprius,* one's own.] **1.** Appropriate : suitable <acted at the *proper* time> **2.** Out-and-out : thorough. **3.** Worthy of the name <a *proper* gentleman> **4.** Meeting a required standard of validity or competence. **5. a.** Within the precise limitation of the term <the village *proper*> **b.** Rigorously correct : EXACT. **6.** Typically belonging to the being or thing under consideration <an optical effect *proper* to fluids> **7. a.** Seemly : decorous. **b.** Showing excessive propriety or gentility. **8.** *Math.* Designating a subset of a given set when the latter has at least one element not in the subset. —*adv.* Thoroughly : completely. —*n. also* **Proper. 1.** The parts of the Mass that vary according to the particular day or feast. **2.** An ecclesiastical office to be said on an appointed day or feast. **—prop'er·ly** *adv.* **—prop'er·ness** *n.*

**proper adjective** *n.* An adjective formed from a proper noun.

**pro·per·din** (prō-pûr'dn) *n.* [Perh. PRO-[1] + Lat. *perdere,* to destroy + -IN.] A natural protein in human blood serum that helps to immunize against infectious diseases.

**proper fraction** *n.* **1.** A numerical fraction in which the numerator is less than the denominator. **2.** A polynomial fraction in which the numerator is of lower degree than the denominator.

**proper noun** *n.* A noun designating by name a being or thing lacking a limiting modifier, and gen. capitalized.

**prop·er·tied** (prŏp'ər-tēd) *adj.* Owning land or securities as a chief source of revenue.

**prop·er·ty** (prŏp'ər-tē) *n., pl.* **-ties.** [ME *proprete* < OFr. *propriete* < Lat. *proprietas,* ownership < *proprius,* one's own.] **1.** Ownership. **2. a.** A possession. **b.** Possessions as a whole. **3.** Something tangible or intangible to which its owner holds legal title. **4.** Something other than costumes and scenery used as part of a dramatic production. **5. a.** A characteristic trait. **b.** A special capability or power : VIRTUE. **c.** A quality that defines or describes an object or substance. **d.** A characteristic attribute common to all members of a class. **e.** *Logic.* A predicable that is common and peculiar to the whole of a species and is necessarily predicated of its essence without being part of that essence. **—prop'er·ty·less** *adj.*

**property damage insurance** *n.* Liability insurance for claims brought against one who causes damage to another's property, as by a car accident.

**property tax** *n.* A tax levied against a property owner.

**pro·phage** (prō'fāj') *n.* A noninfectious association between a bacterial virus and a bacterium in which the viral chromosomes link with the bacterial chromosomes but do not cause disruption of the bacterial cell or promote replication of the virus itself.

**pro·phase** (prō'fāz') *n.* The first stage in cell division by mitosis, during which chromosomes form from the chromatin of the nucleus. **—pro·pha'sic** (-fā'zĭk) *adj.*

**proph·e·cy** (prŏf'ĭ-sē) *n., pl.* **-cies.** [ME *prophecie* < OFr. < Lat. *prophetia* < Gk. *prophētēs,* prophet.] **1.** A prediction. **2. a.** The inspired utterance of a prophet, held to be a declaration of divine will. **b.** Such a revelation revealed orally or in writing.

## PROOFREADERS' MARKS

| Instruction | Mark in Margin | Mark in Type |
|---|---|---|
| Delete | ℯ | the ~~good~~ word |
| Insert indicated material | good | the ‸ word |
| Let it stand | stet | the ~~good~~ word |
| Make capital | cap | the word |
| Make lower case | lc | The Word |
| Set in small capitals | SC | See word. |
| Set in italic type | ital | The word is word. |
| Set in roman type | rom | the word |
| Set in boldface type | bf | the entry word |
| Set in lightface type | lf | the entry word |
| Transpose | tr | the word good |
| Close up space | ⁐ | the wo rd |
| Delete and close up space | ℯ | the woord |
| Spell out | sp | 2 words |
| Insert: space | # | theword |
| period | ⊙ | This is the word‸ |
| comma | ‸ | words‸ words, words |
| hyphen | =/ | word‸for‸word test |
| colon | ⊙ | The following words‸ |
| semicolon | ;‸ | Scan the words‸ skim the words. |
| apostrophe | ’ | John‸s words |
| quotation marks | “/”/ | the word word |
| parentheses | (/)/ | The word‸word‸is in parentheses. |
| brackets | [/]/ | He read from the Word‸the Bible.‸ |
| en dash | 1/N | 1964‸1972 |
| em dash | 1/M / 1/M / | The dictionary‸how often it is needed‸belongs in every home. |
| superior type | 2 | 2 = 4 |
| inferior type | 2 | HO |
| asterisk | * | word |
| dagger | † | a word‸ |
| double dagger | ‡ | words and words‸ |
| section symbol | § | ‸Book Reviews |
| virgule | / | either‸or |
| Start paragraph | ¶ | "Where is it?" ‸"It's on the shelf." |
| Run in | run in | The entry word is printed in boldface. The pronunciation follows. |
| Turn right side up | 9 | the word |
| Move left | ⊏ | ⊏ the word |
| Move right | ⊐ | the word |
| Move up | ⊓ | the word |
| Move down | ⊔ | the word |
| Align | ‖ | the word / the word / the word |
| Straighten line | ═ | the word |
| Wrong font | wf | the word |
| Broken type | × | the word |

**proph·e·sy** (prŏf′ĭ-sī′) *v.* **-sied, -sy·ing, -sies.** [ME *prophecien* < OFr. *prophecier* < *prophecie,* prophecy.] —*vt.* **1.** To reveal by divine inspiration. **2.** To predict. **3.** To prefigure. —*vi.* **1.** To reveal the will or message of God. **2.** To predict the future. **3.** To speak as a prophet. **—proph′e·si′er** *n.*

**proph·et** (prŏf′ĭt) *n.* [ME < OFr. < Lat. *propheta* < Gk. *prophētēs* : *pro-,* before + *phanai,* to speak.] **1.** One who speaks by divine inspiration or as the interpreter through whom the will of a god is expressed. **2.** A soothsayer : predictor. **3.** The chief spokesperson of a cause or movement.

**proph·et·ess** (prŏf′ĭ-tĭs) *n.* **1.** A woman who speaks by divine inspiration or as the interpreter through whom the will of a god is expressed. **2.** The chief spokesperson of a cause or movement.

**pro·phet·ic** (prə-fĕt′ĭk) *also* **pro·phet·i·cal** (-ĭ-kəl) *adj.* **1.** Of or associated with a prophet or prophecy. **2.** Having the nature of prophecy. **—pro·phet′i·cal·ly** *adv.* **—pro·phet′i·cal·ness** *n.*

**pro·phy·lac·tic** (prō′fə-lăk′tĭk, prŏf′ə-) *adj.* [Gk. *prophulaktikos* < *prophulassein,* to take precautions against : *pro-,* before + *phulassein,* to protect < *phulax,* guard.] Serving to defend against or prevent something, esp. disease : PROTECTIVE. —*n.* A prophylactic device or agent. **—pro′phy·lac′ti·cal·ly** *adv.*

**pro·phy·lax·is** (prō′fə-lăk′sĭs, prŏf′ə-) *n., pl.* **-lax·es** (-lăk′sēz′) [NLat. < Gk. *prophulaktikos,* prophylactic.] Protective treatment for or prevention of disease.

**pro·pin·qui·ty** (prə-pĭng′kwĭ-tē) *n.* [ME *propinquite* < Lat. *propinquitas* < *propinquus,* near.] **1.** Proximity : nearness. **2.** Kinship. **3.** Similarity in nature.

**pro·pi·on·al·de·hyde** (prō′pē-ŏn-ăl′də-hīd′) *n.* [PROPION(IC ACID) + ALDEHYDE.] A flammable liquid, $C_3H_6O$, used in making rubber chemicals and plastics.

**pro·pi·o·nate** (prō′pē-ə-nāt′) *n.* [PROPION(IC ACID) + -ATE.] A salt or ester of propionic acid.

**pro·pi·on·ic acid** (prō′pē-ŏn′ĭk) *n.* [Fr. *propionique* : Gk. *pro-,* first + Gk. *piōn,* fat (so called because it is first in order among the fatty acids).] A fatty acid, $CH_3CH_2CO_2H$, made synthetically and much used in a salt form as a mold inhibitor in bread.

**pro·pi·ti·ate** (prō-pĭsh′ē-āt′) *vt.* **-at·ed, -at·ing, -ates.** [Lat. *propitiare, propitiat-* < *propitius,* favorable.] To appease. **—pro·pi′ti·a·ble** (-pĭsh′ē-ə-bəl, -pĭsh′ə-bəl) *adj.* **—pro·pi′ti·at′ing·ly** *adv.* **—pro·pi′ti·a′tion** *n.* **—pro·pi′ti·a′tive** *adj.* **—pro·pi′ti·a′tor** *n.*

**pro·pi·ti·a·to·ry** (prō-pĭsh′ē-ə-tôr′ē, -tōr′ē, -pĭsh′ə-) *adj.* Of or offered in propitiation : CONCILIATORY. **—pro·pi′ti·a·to′ri·ly** *adv.*

**pro·pi·tious** (prə-pĭsh′əs) *adj.* [ME *propicius* < OFr. < Lat. *propitius.*] **1.** Presenting favorable circumstances : AUSPICIOUS. **2.** Gracious : kindly. **—pro·pi′tious·ly** *adv.* **—pro·pi′tious·ness** *n.*

**prop-jet** (prŏp′jĕt′) *n.* A turboprop.

**pro·plas·tid** (prō-plăs′tĭd) *n.* The precursor of a cell plastid.

**prop·o·lis** (prŏp′ə-lĭs) *n.* [Lat. < Gk., suburb, bee glue : *pro-,* before + *polis,* city.] A resinous substance gathered from a variety of plants by bees, and used with beeswax in constructing their hives.

**pro·po·nent** (prə-pō′nənt) *n.* [Lat. *proponens, proponent-,* pr.part. of *proponere,* to set forth. —see PROPOSE.] An advocate of something.

**pro·por·tion** (prə-pôr′shən, -pōr′-) *n.* [ME *proporcioun* < OFr. *proportion* < Lat. *proportio* < *pro portione,* for its share.] **1.** A part considered in relation to the whole. **2.** A relationship between things or parts of things with respect to relative magnitude, quantity, or degree. **3.** A relationship between quantities such that if one varies then another varies in a manner dependent on the first : RATIO. **4.** Harmonious relation. **5.** *often* **proportions.** Size : dimensions. **6.** *Math.* A relation of equality between two ratios; e.g., four quantities, *a, b, c, d,* are said to be in proportion if $a/b = c/d$. —*vt.* **-tioned, -tion·ing, -tions. 1.** To adjust for attainment of proper relations between parts. **2.** To form with proportion. **—pro·por′tion·a·ble** *adj.* **—pro·por′tion·a·bly** *adv.* **—pro·por′tion·er** *n.* **—pro·por′tion·ment** *n.*

**pro·por·tion·al** (prə-pôr′shə-nəl, -pōr′-) *adj.* **1.** Being in proportion. **2.** Properly related in size or other measurable characteristics. **3.** *Math.* Having a constant ratio. —*n.* One of the quantities in a mathematical proportion. **—pro·por′tion·al′i·ty** (-năl′ĭ-tē) *n.* **—pro·por′tion·al·ly** *adv.*

**proportional representation** *n.* Representation of all parties in a legislature in proportion to their popular vote.

**pro·por·tion·ate** (prə-pôr′shə-nĭt, -pōr′-) *adj.* PROPORTIONAL 1. —*vt.* (-nāt′) **-at·ed, -at·ing, -ates.** To make proportionate. **—pro·por′tion·ate·ly** *adv.* **—pro·por′tion·ate·ness** *n.*

**pro·pos·al** (prə-pō′zəl) *n.* **1.** An act of proposing. **2.** A plan or scheme proposed. **3.** An offer of marriage.

**pro·pose** (prə-pōz′) *v.* **-posed, -pos·ing, -pos·es.** [ME *proposen* < OFr. *proposer* < Lat. *proponere* : *pro-,* forth + *ponere,* to put.] —*vt.* **1.** To put forward for consideration, discussion, or adoption. **2.** To nominate or present (a person) for a position, office, or membership. **3.** To offer (a toast to be drunk). **4.** To purpose : intend. —*vi.* To form or make a proposal, esp. of marriage. **—pro·pos′er** *n.*

☆ ***syns:*** PROPOSE, OFFER, POSE, PROPOUND, PUT FORTH, SUBMIT, SUGGEST *v. core meaning* : to advance, as an idea, for consideration <*proposed* that they take a vacation>

**prop·o·si·tion** (prŏp′ə-zĭsh′ən) *n.* [ME *proposicioun* < OFr. *proposition* < Lat. *propositio* < *proponere,* to set forth. —see PROPOSE.] **1.** PROPOSAL 2. **2.** *Informal.* A matter requiring special handling <a controversial *proposition*> **3.** *Informal.* An immoral or dubious proposal. **4.** A subject for analysis or discussion. **5.** *Logic.* **a.** A statement in which the subject is affirmed or denied by the predicate. **b.** Something expressed in a statement, as opposed to the way it is expressed. **c.** A statement containing only logical constants and having a fixed truth-value. —*vt.* **-tioned, -tion·ing, -tions.** *Informal.* To propose a private bargain to, esp. to make a sexual overture to. **—prop′o·si′tion·al** *adj.* **—prop′o·si′tion·al·ly** *adv.*

**propositional function** *n. Logic.* An expression having the form of a proposition but containing undefined symbols for the substantive elements, and becoming a proposition when appropriate values are assigned to the symbols.

**pro·pos·i·tus** (prō-pŏz′ĭ-təs) *n., pl.* **-ti** (-tī′) [Lat., p.part. of *proponere,* to set forth. —see PROPOSE.] **1.** *Law.* One from whom a line of descent is traced. **2.** The person most directly affected by an action.

**pro·pound** (prə-pound′) *vt.* **-pound·ed, -pound·ing, -pounds.** [ME *proponen* < Lat. *proponere,* to set forth. —see PROPOSE.] To put forward for consideration : SET FORTH. **—pro·pound′er** *n.*

**pro·prae·tor** (prō-prē′tər) *n.* [Lat. : *pro-,* for + *praetor,* praetor. —see PRAETOR.] A Roman official appointed, usu. directly after being praetor, to be a chief provincial administrator. **—pro′prae·to′ri·al** (prō′prĭ-tôr′ē-əl, -tōr′-), **pro′prae·to′ri·an** *adj.*

**pro·pran·o·lol** (prō-prăn′ə-lôl′, -lōl′) *n.* [Alteration of PROPANE + -OL + -OL.] A drug, $C_{16}H_{21}NO_2$, for treating angina pectoris and cardiac arrhythmia.

**pro·pri·e·tar·y** (prə-prī′ĭ-tĕr′ē) *adj.* [LLat. *proprietarius* < Lat. *proprietas,* property < *proprius,* one's own.] **1.** Of or relating to a proprietor or to proprietors as a group. **2.** Exclusively owned : PRIVATE. **3.** Appropriate to an owner <a *proprietary* demeanor> **4.** Owned by a private individual or corporation under a trademark or patent <a *proprietary* drug> —*n., pl.* **-ies. 1.** A proprietor. **2.** A group of proprietors. **3.** Ownership. **4.** The governor of a proprietary colony. **5.** A proprietary medicine. **—pro·pri′e·tar′i·ly** *adv.*

**proprietary colony** *n.* A North American colony, as Maryland or Pennsylvania, organized in the 17th cent. in territory granted by the Crown to one or more Lords Proprietary who had full governing rights.

**pro·pri·e·tor** (prə-prī′ĭ-tər) *n.* **1.** One with legal title to something : OWNER. **2.** The owner or owner-manager, as of a business. **—pro·pri′e·to′ri·al** (-tôr′ē-əl, -tōr′-) *adj.* **—pro·pri′e·to′ri·al·ly** *adv.* **—pro·pri′e·tor·ship′** *n.*

**pro·pri·e·tress** (prə-prī′ĭ-trĭs) *n.* **1.** A woman with legal title to something : OWNER. **2.** The owner or owner-manager, as of a business.

**pro·pri·e·ty** (prə-prī′ĭ-tē) *n., pl.* **-ties.** [ME *propriete,* particular character < OFr. —see PROPERTY.] **1.** The quality of being proper : APPROPRIATENESS. **2.** Conformity to generally current usages and customs. **3. proprieties.** The usages and customs of polite society.

**pro·pri·o·cep·tion** (prō′prē-ō-sĕp′shən) *n.* [Lat. *proprius,* one's own + (RE)CEPTION.] Reception of stimuli originating within an organism.

**pro·pri·o·cep·tor** (prō′prē-ō-sĕp′tər) *n.* [Lat. *proprius,* one's own + (RE)CEPTOR.] A sensory receptor, chiefly in muscles, tendons, and joints, that responds to stimuli originating within the organism. **—pro′pri·o·cep′tive** *adj.*

**prop root** *n.* A root growing from above ground into the soil and providing support for the plant stem, as in corn.

**prop·to·sis** (prŏp-tō′sĭs) *n., pl.* **-ses** (-sēz′) [LLat. < Gk. *proptōsis,* prolapse : *pro-,* forward + *piptein,* to fall.] Forward displacement of an organ, esp. the eyeball.

**pro·pul·sion** (prə-pŭl′shən) *n.* [Med. Lat. *propulsio* < Lat. *propellere,* to drive forward. —see PROPEL.] **1.** The process of driving or propelling. **2.** A driving or propelling force. **—pro·pul′sive, pro·pul′so·ry** *adj.*

**pro·pyl** (prō′pĭl) *n.* A univalent organic radical with composition $C_3H_7$, obtained from propane. **—pro·pyl′ic** *adj.*

**prop·y·la** (prŏp′ə-lə, prō′pə-) *n. pl. of* PROPYLON.

**prop·y·lae·um** (prŏp′ə-lē′əm, prō′pə-) *n., pl.* **-lae·a** (-lē′ə) [Lat. < Gk. *propulaion* : *pro-,* before + *pulē,* gate.] An entranceway into a temple or group of buildings.

**propyl alcohol** *n.* A clear colorless liquid, $CH_3CH_2CH_2OH$, extensively used as a solvent.

**pro·pyl·ene** (prō′pə-lēn′) *n.* A flammable gas, $CH_3CH{:}CH_2$, derived from petroleum hydrocarbon cracking and used in organic synthesis.

**propylene glycol** *n.* A colorless viscous hygroscopic liquid, $CH_3CHOHCH_2OH$, used in antifreeze solutions, in hydraulic fluids, and as a solvent.

**prop·y·lon** (prŏp′ə-lŏn′, prō′pə-) *n., pl.* **-la** (-lə) [Gk. *propulon* : *pro-,* before + *pulē,* gate.] A propylaeum.

**pro ra·ta** (prō rā′tə, răt′ə, rä′tə) *adv.* [Lat. *pro rata (parte),* according to the calculated (share).] In proportion according to a precisely determined element, as share or liability.

**pro·rate** (prō-rāt′, prō′rāt′) *v.* **-rat·ed, -rat·ing, -rates.** [< PRO RATA.] —*vt.* To divide, distribute, or assess proportionately. —*vi.* To settle matters on the basis of proportional distribution. **—pro·rat′a·ble** *adj.* **—pro·ra′tion** *n.*

**pro·rogue** (prō-rōg′) *vt.* **-rogued, -rogu·ing, -rogues.** [ME *prorogen* < OFr. *proroger,* to postpone < Lat. *prorogare* : *pro-,* forward + *rogare,* to ask.] **1.** To terminate a session of (e.g., a parliament). **2.** To postpone. **—pro′ro·ga′tion** (prō′rō-gā′shən) *n.*

**pros-** *pref.* [Gk. < *pros,* near, at.] **1.** Near: toward <*pros*enchyma> **2.** In front of <*pros*encephalon>

**pro·sa·ic** (prō-zā′ĭk) *adj.* [LLat. *prosaicus* < Lat. *prosa,* prose. —see PROSE.] **1. a.** Of or like prose. **b.** Matter-of-fact : straightforward. **2.** Lacking imagination and spirit : ORDINARY. **—pro·sa′i·cal·ly** *adv.* **—pro·sa′ic·ness** *n.*

**pro·sa·ism** (prō′zā-ĭz′əm) *n.* **1.** A prosaic quality or style. **2.** A prosaic expression or term.

**pro·sce·ni·um** (prō-sē′nē-əm) *n.* [Lat. < Gk. *proskēnion* : *pro-,* before + *skēnē,* buildings at the back of the stage.] **1.** *pl.* **-ni·ums.** The area of a modern theater between the orchestra and the curtain. **2.** *pl.* **-ni·a** (-nē-ə). The stage of an ancient theater between the background and the orchestra.

**pro·sciut·to** (prō-shōō′tō) *n.* [Ital.] An aged, dry-cured, usu. thin-sliced Italian ham requiring no cooking.

**pro·scribe** (prō-skrīb′) *vt.* **-scribed, -scrib·ing, -scribes.** [Lat. *proscribere,* to put up someone's name as outlawed : *pro-,* in front + *scribere,* to write.] **1.** To denounce : condemn. **2.** To prohibit. **3.** To publish the name of (a person) as outlawed. **—pro·scrib′er** *n.*

**pro·scrip·tion** (prō-skrĭp′shən) *n.* **1.** An act of proscribing : PROHIBITION. **2.** The state of being proscribed : OUTLAWRY. **—pro·scrip′tive** *adj.* **—pro·scrip′tive·ly** *adv.*

**prose** (prōz) *n.* [ME < OFr. < Lat. *prosa (oratio),* straightforward (discourse) < *proversus,* p.part. of *provertere,* to turn forward : *pro-,* forward + *vertere,* to turn.] **1.** Ordinary speech or writing. **2.** Commonplace expression or quality. **3.** *Rom. Cath. Ch.* A hymn of irregular meter sung after the gradual. —*v.* **prosed, pros·ing, pros·es.** —*vt.* To put into prose. —*vi.* **1.** To write prose. **2.** To speak or write in a dull, monotonous style.

**pros·e·cute** (prŏs′ĭ-kyōōt′) *v.* **-cut·ed, -cut·ing, -cutes.** [ME *prosecuten* < Lat. *prosequi* : *pro-,* forward + *sequi,* to follow.] —*vt.* **1.** To pursue or persist in so as to finish. **2.** To carry on : PRACTICE. **3. a.** To initiate legal or criminal court action against. **b.** To seek to enforce or obtain by legal action. —*vi.* **1.** To initiate and conduct legal proceedings. **2.** To act as prosecutor. **—pros′e·cut′a·ble** *adj.*

**prosecuting attorney** *n.* An attorney authorized to prosecute cases on behalf of a government and the people.

**pros·e·cu·tion** (prŏs′ĭ-kyōō′shən) *n.* **1.** An act of prosecuting. **2.** Institution and conduct of a legal proceeding. **3.** A prosecuting attorney.

**pros·e·cu·tor** (prŏs′ĭ-kyōō′tər) *n.* **1.** One who prosecutes. **2.** One who initiates and carries out a legal action, esp. criminal proceedings. **3.** A prosecuting attorney.

**pros·e·cu·to·ri·al** (prŏs′ĭ-kyōō-tôr′ē-əl, -tōr′-) *adj.* Of, relating to, or concerned with prosecution.

**pros·e·lyte** (prŏs′ə-līt′) *n.* [ME *proselite* < LLat. *proselytus* < Gk. *proselutos.*] A new convert to a doctrine or religion. —*vt.* & *vi.* **-lyt·ed, -lyt·ing, -lytes.** To proselytize. **—pros′e·lyt′er** *n.*

**pros·e·ly·tism** (prŏs′ə-lĭ-tĭz′əm, -lī-) *n.* **1.** The practice of proselytizing. **2.** The state of being a proselyte.

**pros·e·ly·tize** (prŏs′ə-lĭ-tīz′) *v.* **-tized, -tiz·ing, -tiz·es.** —*vi.* To make proselytes. —*vt.* To convert from one faith or belief to another. **—pros′e·ly·ti·za′tion** *n.* **—pros′e·ly·tiz′er** *n.*

**pros·en·ceph·a·lon** (prŏs′ĕn-sĕf′ə-lŏn′) *n.* The forebrain. **—pros′en·ce·phal′ic** (-sə-făl′ĭk) *adj.*

**pros·en·chy·ma** (prŏ-sĕng′kĭ-mə) *n.* [PROS- + (PAR)ENCHYMA.] Tissue composed of elongated, unspecialized cells, found in most flowering plants. **—pros′en·chym′a·tous** (-kĭm′ə-təs) *adj.*

**prose poem** *n.* A prose work that has some features and qualities of poetry.

**Pro·ser·pi·na** (prō-sûr′pə-nə) *also* **Pros·er·pi·ne** (prŏs′ər-pīn′, prō-sûr′pə-nē) *n.* [Lat.] *Rom. Myth.* The wife of Pluto and daughter of Ceres and the goddess of the underworld.

**pro·sim·i·an** (prō-sĭm′ē-ən) *adj.* [< NLat. *Prosimii,* suborder name : Lat. *pro-,* before + Lat. *simia,* ape. —see SIMIAN.] Of or belonging to the Prosimii, a suborder of primates that includes the lemurs, lorises, and tarsiers. —*n.* A primate of the suborder Prosimii.

**pro·sit** (prōst, prō′zĭt) *interj.* [G. < Lat., may it benefit.] —Used as a drinking toast to someone's health.

**pros·o·dist** (prŏs′ə-dĭst) *n.* A specialist in prosody.

**pros·o·dy** (prŏs′ə-dē) *n.* [ME *prosodye* < Lat. *prosodia,* accent < Gk. *prosōidia,* singing in accord : *pros-,* corresponding to + *ōidē,* song.] **1.** Study of the metrical structure of verse. **2.** A system of versification. **—pro·sod′ic** (prə-sŏd′ĭk) *adj.* **—pro·sod′i·cal·ly** *adv.*

**pro·so·ma** (prō-sō′mə) *n.* [PRO-[2] + Gk. *sōma,* body.] *Zool.* The anterior portion of an invertebrate body when there is no evidence of primitive segmentation. **—pro·so′mal** *adj.*

**pro·so·po·pe·ia** *also* **pro·so·po·poe·ia** (prə-sō′pə-pē′ə) *n.* [Lat. *prosopopoeia* < Gk. *prosōpopoiia* : *pros,* toward + *ōpon,* face + *poiein,* to make.] **1.** Impersonation of an imaginary or absent speaker. **2.** PERSONIFICATION 3. **—pro·so′po·pe′ial** *adj.*

**pros·pect** (prŏs′pĕkt′) *n.* [ME *prospecte* < Lat. *prospectus,* distant view < p.part. of *prospicere,* to look out : *pro-,* forward + *specere,* to look.] **1.** Something expected : POSSIBILITY. **2. prospects.** Chances for success. **3. a.** A potential customer or purchaser. **b.** A potentially successful candidate. **4.** The direction in which an object, as a building, faces : OUTLOOK. **5.** Something presented to the eye : SCENE. **6.** The act of surveying or examining. **7. a.** The site or probable site of a mineral deposit. **b.** An actual or probable mineral deposit. **c.** The mineral yield obtained by working an ore. —*v.* **-pect·ed, -pect·ing, -pects.** —*vt.* To search for or explore (a region) for gold or other mineral deposits. —*vi.* To explore for mineral deposits.

**pro·spec·tive** (prə-spĕk′tĭv) *adj.* **1.** Apt to occur. **2.** Apt to become or be <a *prospective* patient> **—pro·spec′tive·ly** *adv.*

**pros·pec·tor** (prŏs′pĕk′tər) *n.* An explorer for natural deposits, as gold or oil, in an area.

**pro·spec·tus** (prə-spĕk′təs) *n.* [Lat., distant view. —see PROSPECT.] A formal description of a proposed commercial, literary, or financial venture distributed to prospective participants.

**pros·per** (prŏs′pər) *v.* **-pered, -per·ing, -pers.** [ME *prosperen* < OFr. *prosperer* < Lat. *prosperare,* to render fortunate < *prosperus,* fortunate.] —*vi.* To fare well. —*vt. Archaic.* To cause to thrive.

☆ ***syns:*** PROSPER, BOOM, FLOURISH, THRIVE *v. core meaning* : to fare well <an industry that *prospered*>

**pros·per·ous** (prŏs′pər-əs) *adj.* **1.** Having success : FLOURISHING. **2.** Enjoying financial security. **3.** Propitious : favorable. **—pros·per′i·ty** (prŏ-spĕr′ĭ-tē) *n.* **—pros′per·ous·ly** *adv.* **—pros′per·ous·ness** *n.*

☆ ***syns:*** PROSPEROUS, COMFORTABLE, EASY, WELL-FIXED, WELL-HEELED, WELL-OFF, WELL-TO-DO *adj. core meaning* : enjoying steady good fortune or financial security <a *prosperous* family>

**pros·ta·glan·din** (prŏs′tə-glăn′dĭn) *n.* [PROSTA(TE) + GLAND + -IN.] Any of a group of physiologically active hormonelike substances obtained from fatty acids, occurring in various human body tissues, and perhaps affecting blood pressure, metabolism, and smooth muscle activity.

**pros·tate** (prŏs′tāt′) *n.* [NLat. *prostata* < Gk. *prostatēs* < *proistanai,* to set before : *pro-,* in front + *histanai,* to cause to stand.] A gland in male mammals consisting of muscular and glandular tissue around the urethra at the bladder. **—pro·stat′ic** (prŏ-stăt′ĭk) *adj.*

**pros·ta·tec·to·my** (prŏs′tə-tĕk′tə-mē) *n., pl.* **-mies.** Surgical removal of all or part of the prostate.

**pros·ta·ti·tis** (prŏs′tə-tī′tĭs) *n.* Inflammation of the prostate.

**pros·the·sis** (prŏs-thē′sĭs) *n., pl.* **-ses** (-sēz′) [Gk., addition < *prostithenai,* to add : *pros-,* in addition + *tithenai,* to put.] **1.** Artificial replacement of a limb, tooth, or other part of the body. **2.** An artificial device used in prosthetic replacement.

**pros·thet·ic** (prŏs-thĕt′ĭk) *adj.* Of or relating to prosthetics or a prosthesis. **—pros·thet′i·cal·ly** *adv.*

**prosthetic group** *n.* A link other than an amino acid in a protein chain.

**pros·thet·ics** (prŏs-thĕt′ĭks) *n. (sing. in number).* Prosthetic surgery. **—pros′the·tist** (prŏs′thĭ-tĭst) *n.*

**pros·tho·don·tics** (prŏs′thə-dŏn′tĭks) *n.* [PROSTH(ESIS) + Gk. *odons, odont-,* tooth.] *(sing. in number).* Prosthetic dentistry, esp. replacement of missing teeth by bridges and dentures. **—pros′tho·don′tist** *n.*

**pros·ti·tute** (prŏs′tĭ-tōōt′, -tyōōt′) *n.* [Lat. *prostituta* < p.part. of *prostituere,* to prostitute : *pro-,* forward + *statuere,* to place.] **1.** One who solicits and takes payment for sexual intercourse. **2.** One who sells his or her abilities or name to an undeserving cause. —*vt.* **-tut·ed, -tut·ing, -tutes. 1.** To offer (oneself or another) for sexual hire. **2.** To sell (oneself or one's talents) to an undeserving cause. **—pros′ti·tu′tion** (-tōō′shən, -tyōō′-) *n.* **—pros′ti·tu′tor** *n.*

**pro·sto·mi·um** (prō-stō′mē-əm) *n., pl.* **-mi·a** (-mē-ə) [NLat. : PRO- + Gk. *stoma,* mouth.] The portion of the head in annelids and mollusks located anterior to the mouth.

**pros·trate** (prŏs′trāt′) *vt.* **-trat·ed, -trat·ing, -trates.** [Lat. *prosternere, prostrat-* : *pro-,* forward + *sternere,* to cast down.] **1.** To make (oneself) bow or kneel down in adoration or humility. **2.** To throw down flat. **3.** To lay low : OVERCOME <Agonizing pain *prostrated* the soldier.> —*adj.* **1.** Lying face down, as in adoration or submission. **2.** Lying down full-length. **3.** Physically or emotionally exhausted : INCAPACITATED. **4.** *Bot.* Growing flat along the ground. **—pros′tra′tor** *n.*

**pros·tra·tion** (prŏ-strā′shən) *n.* **1.** The act of prostrating oneself or the state of being prostrate. **2.** Complete exhaustion.

**pro·style** (prō′stīl′) *adj.* [Lat. *prostylos* < Gk. *prostulos* : *pro-,* in front + *stulos,* pillar.] Having a row of columns across the front only, as in some Greek temples.

**pros·y** (prō′zē) *adj.* **-i·er, -i·est.** [< PROSE.] **1.** Matter-of-fact : dry. **2.** Dull : commonplace. **—pros′i·ly** *adv.* **—pros′i·ness** *n.*

**prot-** *pref. var. of* PROTO-.

**pro·tac·tin·i·um** (prō′tăk-tĭn′ē-əm) *n. Symbol* **Pa** A rare radioactive element chemically similar to uranium; atomic number 91; longest-lived isotope Pa 231.

**pro·tag·o·nist** (prō-tăg′ə-nĭst) *n.* [Gk. *prōtagōnistēs* : *prōtos,* first + *agōnistēs,* actor < *agōnizesthai,* to contend < *agōn,* contest.] **1.** The leading character in Greek drama or other literary form. **2. a.** A leading or principal figure. **b.** *Informal.* The leader of a cause.

**pro·ta·mine** (prō′tə-mēn′, -mĭn) *also* **pro·ta·min** (-mĭn) *n.* Any of the group of the simplest proteins that are highly basic, soluble in water, not coagulated by heat, and yield only amino acids, mainly arginine, upon hydrolysis.

**pro·ta·no·pi·a** (prō′tə-nō′pē-ə) *n.* Colorblindness in which red and bluish-green stimuli are confused with neutral stimuli and with each other.

**prot·a·sis** (prŏt′ə-sĭs) *n., pl.* **-ses** (-sēz′) [LLat., proposition < Gk. < *proteinein,* to propose : *pro-,* forward + *teinein,* to stretch.] **1.** A subordinate clause, esp. in a conditional sentence. **2.** The introductory part of a classical drama. **—pro·tat′ic** (prō-tăt′ĭk, prō-) *adj.*

**prote-** *pref. var. of* PROTEO-.

**pro·te·an** (prō′tē-ən, prō-tē′-) *adj.* [< PROTEUS.] Easily assuming different shapes or forms : VARIABLE.

▲ word history: The word *protean* comes from *Proteus,* the name of a Greek sea god who had the power to change his shape. Proteus had the gift of prophecy, but those who wanted to consult him first had to bind him securely. Proteus would then change into various shapes, such as a wild boar, a tiger, a rush of water, and a raging fire. A questioner who could keep Proteus restrained until he returned to his original shape would receive an answer.

**pro·te·ase** (prō′tē-ās′, -āz′) *n.* An enzyme that catalyzes the hydrolytic breakdown of proteins.

**pro·tect** (prə-tĕkt′) *vt.* **-tect·ed, -tect·ing, -tects.** [Lat. *protegere, protect-* : *pro-,* in front + *tegere,* to cover.] **1.** To keep from harm, attack, or injury : GUARD. **2.** To aid (domestic industry) with tariffs on imports. **3.** To assure payment of (e.g., drafts or notes) by reserving funds. **—pro·tect′ing·ly** *adv.*

**pro·tec·tant** (prə-tĕk′tənt) *n.* One that protects <used a *protectant* before painting the fresh wood>

**pro·tect·er** (prə-tĕk′tər) *n. var. of* PROTECTOR.

**pro·tec·tion** (prə-tĕk′shən) *n.* **1.** An act of protecting or the state of being protected. **2.** One that protects. **3.** A pass ensuring safe-conduct to travelers. **4.** A tariff system protecting domestic industries from foreign competition. **5.** Money that racketeers extort in exchange for a promise of freedom from molestation. **—pro·tec′tion·al** *adj.*

**pro·tec·tion·ism** (prə-tĕk′shə-nĭz′əm) *n.* The economic theory and system of protection. **—pro·tec′tion·ist** *n.*

**pro·tec·tive** (prə-tĕk′tĭv) *adj.* Adapted or meant to give protection. **—pro·tec′tive** *n.* **—pro·tec′tive·ly** *adv.* **—pro·tec′tive·ness** *n.*

**pro·tec·tor** *also* **pro·tect·er** (prə-tĕk′tər) *n.* **1.** One who protects : GUARDIAN. **2.** A device that protects : GUARD. **3. Protector. a.** One who rules a kingdom during the monarch's minority. **b.** The head of the Commonwealth of England, Scotland, and Ireland from 1653 to 1659. **—pro·tec′tor·al** *adj.* **—pro·tec′tor·ship′** *n.*

**pro·tec·tor·ate** (prə-tĕk′tər-ĭt) *n.* **1. a.** A relationship of protection and partial control assumed by a superior power over a dependent country or region. **b.** The protected country or region. **2. Protectorate. a.** The government, office, or term of a protector. **b.** The government of England under Oliver Cromwell (1653–58) and his son Richard (1658–59).

**pro·tec·to·ry** (prə-tĕk′tə-rē) *n., pl.* **-ries.** An institution for the welfare of destitute children.

**pro·té·gé** (prō′tə-zhā′, prō′tə-zhā′) *n.* [Fr. < p.part. of *protéger,* to protect < Lat. *protegere.* —see PROTECT.] A man or boy whose welfare, training, or career is advanced by an influential person.

**pro·té·gée** (prō′tə-zhā′, prō′tə-zhā′) *n.* [Fr., fem. of *protégé,* protégé.] A woman or girl whose welfare, training, or career is advanced by an influential person.

**pro·te·i** (prō′tē-ī′) *n. pl. of* PROTEUS.

**pro·te·id** (prō′tē-ĭd) *n.* A protein.

**pro·tein** (prō′tēn′, -tē-ĭn) *n.* [Fr. *protéine* < LGk. *proteios,* primary < Gk. *prōtos,* first.] Any of a group of complex nitrogenous organic compounds of high molecular weight that have amino acids as their basic structural units and that are found in all living matter and are required for the growth and repair of animal tissue. **—pro′tein·a′ceous** (prōt′n-ā′shəs, prō′tē-nā′-), **pro′te·in′ic** (prō′tē-ĭn′ĭk) *adj.*

**pro·tein·ase** (prōt′n-ās′, -āz′, prō′tē-nās′, -nāz′) *n.* A protease that hydrolyzes proteins into polypeptides.

**pro·tein·ate** (prōt′n-āt′, prō′tē-nāt′) *n.* A protein compound.

**pro·tein·oid** (prōt′n-oid′, prō′tē-noid′) *n.* Any of several polypeptides prepared by polymerization of mixtures of amino acids, gen. through electrical stimulation, that resemble organized naturally produced proteins and may represent an early form of protein evolution.

**pro·tein·u·ri·a** (prōt′n-o͝or′ē-ə, -yo͝or′-, prō′tē-no͝or′-, -nyo͝or′-) *n.* Presence of protein in the urine, often due to kidney disease.

**pro tem** (prō tĕm′) *adv.* Pro tempore.

**pro tem·po·re** (prō tĕm′pə-rē) *adv.* [Lat.] For the time being.

**proteo-** *or* **prote-** *pref.* [< PROTEIN.] Protein <*proteolysis*>

**pro·te·o·clas·tic** (prō′tē-ō-klăs′tĭk) *adj.* [PROTEO- + Gk. *klastos,* broken < *klan,* to break.] Of, relating to, or inducing proteolysis.

**pro·te·ol·y·sis** (prō′tē-ŏl′ĭ-sĭs) *n.* Breakdown of proteins into simpler soluble substances. **—pro′te·o·lyt′ic** (-ə-lĭt′ĭk) *adj.*

**pro·te·ose** (prō′tē-ōs′, -ōz′) *n.* Any of several water-soluble proteins manufactured during digestion.

**Prot·er·o·zo·ic** (prŏt′ər-ə-zō′ĭk, prō′tər-) *adj.* [Gk. *proteros,* earlier + -ZOIC.] Of, belonging to, or designating the geologic time and deposits of the Precambrian era between the Archeozoic era and the Cambrian period of the Paleozoic era. *—n.* The Proterozoic era.

**pro·test** (prə-tĕst′, prō-, prō′tĕst′) *v.* **-test·ed, -test·ing, -tests.** [ME *protesten* < OFr. *protester* < Lat. *protestari* : *pro-,* forth + *testari,* to testify < *testis,* witness.] *—vt.* **1.** To object to, esp. in a formal statement. **2.** To affirm or promise with earnest solemnity. **3.** *Law.* To declare (a bill) dishonored or refused. **4.** *Archaic.* To proclaim or make known. *—vi.* **1.** To express strong objection. **2.** To make an earnest avowal or affirmation. *—n.* (prō′tĕst′). **1.** A formal declaration of objection or disapproval issued by a concerned party. **2. a.** A display or gesture of disapproval. **b.** Reluctance to accept or do something <went to the hospital under *protest*> **3.** *Law.* **a.** A formal statement drawn up by a notary for a creditor declaring that the debtor has refused to accept or honor a bill. **b.** A formal declaration made by a taxpayer stating that the tax demanded is excessive or illegal and retaining the right to contest it. **—pro·test′er** *n.* **—pro·test′ing·ly** *adv.*

**Prot·es·tant** (prŏt′ĭ-stənt) *n.* [< Lat. *protestans, protestant-,* pr. part. of *protestari,* to protest.] **1.** A Christian belonging to a sect descending from those that seceded from the Church of Rome during the Reformation. **2.** One of those who adhered to the doctrine of Luther and in 1529 protested against the decree of the Diet of Spires commanding submission to the authority of Rome. **3. protestant** (*also* prə-tĕs′tənt). One who makes a declaration or avowal. **—Prot′es·tant** *adj.*

**Protestant Episcopal Church** *n.* A U.S. church body orig. associated with the Church of England, but since 1789 organized as a separate entity.

**Prot·es·tant·ism** (prŏt′ĭ-stən-tĭz′əm) *n.* **1.** Adherence to a Protestant church. **2.** The religion and religious tendencies fostered by the Protestant movement. **3.** Protestants as a group.

**prot·es·ta·tion** (prŏt′ĭ-stā′shən, prō′tĭ-) *n.* **1.** An emphatic declaration. **2.** A strong or formal expression of dissent.

**pro·te·us** (prō′tē-əs) *n., pl.* **-te·i** (-tē-ī′) [NLat. *Proteus,* genus name < Lat., Proteus.] Any of various Gram-negative, rod-shaped bacteria of the genus *Proteus* that include some species associated with human enteritis.

**Pro·te·us** (prō′tē-əs, -tyo͞os′) *n.* [Lat. < Gk. *Prōteus.*] *Gk. Myth.* A sea god who could change his shape at will.

**pro·tha·la·mi·on** (prō′thə-lā′mē-ən, -ŏn′) *n., pl.* **-mi·a** (-mē-ə) [PRO-² + Gk. *epithalamion,* epithalamium. —see EPITHALAMIUM.] A song in celebration of a wedding : EPITHALAMIUM.

**pro·thal·lus** (prō-thăl′əs) *also* **pro·thal·li·um** (prō-thăl′ē-əm) *n., pl.* **-thal·li** (-thăl′ī′) *also* **-thal·li·a** (-thăl′ē-ə) [PRO-² + Gk. *thallos,* shoot.] A small flat mass of tissue produced by a germinating spore of ferns and some mosses and related plants that has sexual organs and eventually becomes a mature plant. **—pro·thal′li·al** *adj.*

**proth·e·sis** (prŏth′ĭ-sĭs) *n., pl.* **-ses** (-sēz′) [Gk., prefixing < *protithenai,* to put before : *pro-,* before + *tithenai,* to put.] **1.** Addition of a phoneme at the beginning of a word to ease pronunciation or to form a new word. **2.** Preparation of the Eucharistic elements for consecration in the Eastern Orthodox Church. **—pro·thet′ic** (prō-thĕt′ĭk) *adj.* **—pro·thet′i·cal·ly** *adv.*

**pro·thon·o·tar·y** (prō-thŏn′ə-tĕr′ē, prō′thə-nō′tə-rē) *also* **pro·ton·o·tar·y** (prō-tŏn′ə-tĕr′ē, prō′tə-nō′tə-rē) *n., pl.* **-ies.** [ME < LLat. *protonotarius* : Gk. *prōtos,* first + Lat. *notarius,* secretary < *nota,* mark.] **1.** The principal clerk in certain courts of law. **2.** *Rom. Cath. Ch.* One of a college of 12 ecclesiastics charged with the registry of important pontifical proceedings. **3.** A chief scribe. **—pro·thon′o·tar′i·al** (prō-thŏn′ə-târ′ē-əl, prō′thə-nō-târ′-) *adj.*

**prothonotary warbler** *n.* [Probably from the bright-yellow robes worn by ecclesiastics at important meetings.] A small bird, *Protonotaria citrea* of southeastern North America, with a deep-yellow head and breast and grayish wings.

**prothonotary warbler**
*5½ inches long*

**pro·tho·rax** (prō-thôr′ăks′, -thōr′-) *n., pl.* **-tho·rax·es** or **-tho·ra·ces** (-thôr′ə-sēz′, -thōr′-). The anterior division of an insect's thorax, bearing the first pair of legs. **—pro′tho·rac′ic** (prō′thə-răs′ĭk) *adj.*

**pro·throm·bin** (prō-thrŏm′bĭn) *n.* A plasma protein converted into thrombin during blood coagulation.

**pro·tist** (prō′tĭst) *n.* [NLat. *Protista,* kingdom name < Gk. *prōtista,* neuter pl. of *prōtistos,* the very first < *prōtos,* first.] Any of the unicellular organisms of the kingdom Protista, which includes protozoans, bacteria, some algae, and other forms not easily classified as either plants or animals. **—pro·tis′tan** (-tĭs′tən) *adj.* & *n.* **—pro′tis·tol′o·gy** (prō′tĭ-stŏl′ə-jē) *n.*

**pro·ti·um** (prō′tē-əm, prō′shē-) *n.* The most abundant isotope of hydrogen, $H^1$, with atomic mass 1.

**proto-** or **prot-** *pref.* [< Gk. *prōtos,* first.] **1.** Earliest: first in time <*proto*lithic><*proto*history> **2.** First formed: primitive <*proto*human> **3. Proto-.** Being a form of a language that is the ancestor of a language or group of related languages <*Proto*-Germanic><*Proto*-Algonquian> **4.** Having the least amount of a specified element or radical <*proto*porphyrin>

**Pro·to-Al·gon·qui·an** (prō′tō-ăl-gŏng′kwē-ən, -kē-ən) *n.* The earliest reconstructed ancestor of the Algonquian languages.

**pro·to·col** (prō′tə-kôl′, -kōl′, -kŏl′) *n.* [OFr. *prothocole,* formula for drawing up state documents < Med. Lat. *protocollum* < LGk. *prōtokollon,* table of contents : Gk. *prōtos,* first + LGk. *kollema,* sheets of a papyrus glued together < *kollan,* to glue together < *kolla,* glue.] **1.** The forms of ceremony and etiquette used by diplomats and heads of state. **2.** The first copy of a treaty or other state document before its ratification. **3.** A preliminary record or draft of a transaction. **4.** The plan for a scientific experiment or treatment. —*vi.* **-coled, -col·ing, -cols** or **-colled, -col·ling, -cols.** To form or issue protocols. **—pro′to·col′ar** (-kŏl′ər), **pro′to·col′a·ry** (-kŏl′ə-rē) *adj.*

**pro·to·derm** (prō′tə-dûrm′) *n. Bot.* Dermatogen. **—pro′to·derm′al** *adj.*

**Pro·to-Ger·man·ic** (prō′tō-jûr-măn′ĭk) *n.* The reconstructed prehistoric ancestor of the Germanic languages.

**pro·to·his·to·ry** (prō′tō-hĭs′tə-rē, -hĭs′trē) *n.* Study of a culture just before its earliest recorded history. **—pro′to·his·tor′i·an** (-hĭ-stôr′ē-ən, -stōr′-) *n.* **—pro′to·his·tor′ic** (-hĭ-stôr′ĭk, -stōr′-) *adj.*

**pro·to·hu·man** (prō′tō-hyo͞o′mən) *adj.* Of or relating to several species of prehistoric primates similar to modern human beings but more primitive in development. —*n.* A protohuman primate.

**Pro·to-In·do-Eur·o·pe·an** (prō′tō-ĭn′dō-yo͝or′ə-pē′ən) *n.* The earliest reconstructed stage of Indo-European.

**pro·to·lan·guage** (prō′tō-lăng′gwĭj) *n.* A language that is the recorded or hypothetical ancestor of another language or group of languages.

**pro·to·lith·ic** (prō′tə-lĭth′ĭk) *adj.* Of, relating to, or typical of the very beginning of the Stone Age: EOLITHIC.

**pro·to·mar·tyr** (prō′tō-mär′tər) *n.* The first martyr in a cause.

**pro·to·mor·phic** (prō′tə-môr′fĭk) *adj.* Primitive in form or structure.

**pro·ton** (prō′tŏn′) *n.* [Gk. *prōton,* neuter of *prōtos,* first.] A stable, positively charged subatomic particle in the baryon family with a mass 1,836 times that of the electron. **—pro·ton′ic** *adj.*

**pro·to·ne·ma** (prō′tə-nē′mə) *n., pl.* **-ne·ma·ta** (-nē′mə-tə, -nĕm′ə-tə) [PROTO- + Gk. *nēma,* thread.] *Bot.* A green threadlike structure that arises on germination of a moss spore and that eventually grows into a mature plant. **—pro′to·ne′mal** (-nē′məl), **pro′to·ne′ma·tal** (-nē′mə-təl, -nĕm′ə-təl) *adj.*

**pro·ton·o·tar·y** (prō-tŏn′ə-tĕr′ē, prō′tə-nō′tə-rē) *n. var. of* PROTHONOTARY.

**proton synchrotron** *n. Physics.* A ring-shaped synchrotron that uses a frequency-modulated accelerating voltage for accelerating protons to energies of several billion electron volts.

**pro·to·path·ic** (prō′tə-păth′ĭk) *adj.* [< LGk. *prōtpathēs,* affected first < Gk. *prōtopathein,* to feel first : *prōto-,* first + *paskhein,* to feel.] Of or designating the cutaneous sensory reception of gross pressure, pain, heat, or cold. **—pro·top′a·thy** (prō-tŏp′ə-thē) *n.*

**pro·to·plasm** (prō′tə-plăz′əm) *n.* A complex jellylike colloidal substance conceived of as constituting the living matter of plant and animal cells and performing the basic life functions. **—pro′to·plas′mic** (-plăz′mĭk), **pro′to·plas′mal, pro′to·plas·mat′ic** (-plăz-măt′ĭk) *adj.*

**pro·to·plast** (prō′tə-plăst′) *n.* [OFr. *protoplaste* < LLat. *protoplastus* < Gk. *prōtoplastos* : *prōtos,* first + *plassein,* to form.] **1.** Something made or formed first: PROTOTYPE. **2.** *Biol.* Living cellular material. **—pro′to·plas′tic** *adj.*

**pro·to·por·phy·rin** (prō′tō-pôr′fə-rĭn) *n.* A metal-free porphyrin, $C_{34}H_{34}N_4O_4$, derived from the hemin of blood.

**pro·to·stele** (prō′tə-stēl′, prō′tə-stē′lē) *n.* A stele lacking pith and having a solid core of xylem. **—pro′to·ste′lic** (-stē′lĭk) *adj.*

**pro·to·tro·phic** (prō′tə-trō′fĭk, -trŏf′ĭk) *adj.* Securing nourishment from assimilation of inorganic materials <*prototrophic* bacteria> **—pro′to·troph′** *n.*

**pro·to·type** (prō′tə-tīp′) *n.* [Fr. < Gk. *prōtotupon,* archetype : *prōtos,* first + *tupos,* model.] **1.** An original type, form, or instance that is a model on which later stages are based or judged <the V-1 as a *prototype* of modern rockets> **2.** An early typical example. **3.** *Biol.* A primitive or ancestral form or species. **—pro′to·typ′al** (-tī′pəl), **pro′to·typ′ic** (-tĭp′ĭk), **pro′to·typ′i·cal** *adj.*

**pro·to·xy·lem** (prō′tə-zī′ləm) *n. Bot.* The first formed xylem that differentiates from the procambium.

**pro·to·zo·an** (prō′tə-zō′ən) also **pro·to·zo·on** (-ŏn′) *n., pl.* **-zo·ans** also **-zo·a** (-zō′ə) [< NLat. *Protozoa,* subkingdom name : PROTO- + Gk. *zōia,* pl. of *zōion,* animal.] Any of the single-celled, usu. microscopic organisms of the phylum or subkingdom Protozoa, including the most primitive forms of animal life. **—pro′to·zo′an, pro′to·zo′ic** (-zō′ĭk) *adj.*

**pro·to·zo·ol·o·gy** (prō′tə-zō-ŏl′ə-jē) *n.* Study of protozoans. **—pro′to·zo′o·log′i·cal** (-zō′ə-lŏj′ĭ-kəl) *adj.* **—pro′to·zo·ol′o·gist** *n.*

**pro·to·zo·on** (prō′tə-zō′ŏn′) *n. var. of* PROTOZOAN.

**pro·tract** (prō-trăkt′, prə-) *vt.* **-tract·ed, -tract·ing, -tracts.** [Lat. *protrahere, protract-* : *pro-,* forth + *trahere,* to drag.] **1.** To prolong in time. **2.** To draw to scale by means of a scale and protractor: PLOT. **3.** *Anat.* To extend or protrude. **—pro·tract′ed·ly** (-trăk′tĭd-lē) *adv.* **—pro·tract′ed·ness** *n.* **—pro·trac′tive** *adj.*

**pro·trac·tile** (prō-trăk′təl, -tīl′, prə-) also **pro·tract·i·ble** (-tə-bəl) *adj.* Capable of being protracted: EXTENSIBLE. **—pro′trac·til′i·ty** (prō′trăk-tĭl′ĭ-tē) *n.*

**pro·trac·tion** (prō-trăk′shən, prə-) *n.* **1.** An act of protracting or the state of being protracted. **2.** Irregular lengthening of a normally short syllable.

**pro·trac·tor** (prō-trăk′tər, prə-) *n.* **1.** A semicircular instrument for measuring and constructing angles. **2.** An adjustable pattern used by tailors. **3.** *Anat.* An extensor. **4.** One that protracts or delays.

**pro·trude** (prō-tro͞od′) *v.* **-trud·ed, -trud·ing, -trudes.** [Lat. *protrudere* : *pro-,* forward + *trudere,* to thrust.] —*vt.* To push or thrust outward. —*vi.* To jut out: PROJECT. **—pro·trud′ent** (-tro͞od′nt) *adj.*

**pro·tru·sile** (prō-tro͞o′səl, -sīl′) also **pro·tru·si·ble** (prō-tro͞o′sə-bəl) *adj.* [PROTRUS(ION) + -ILE.] Capable of being thrust outward, as the tongue. **—pro′tru·sil′i·ty** (prō′tro͞o-sĭl′ĭ-tē) *n.*

**pro·tru·sion** (prō-tro͞o′zhən) *n.* [< Lat. *protrusus,* p.part. of *protrudere,* to protrude.] **1.** An act of protruding or the state of being protruded. **2.** Something that protrudes.

**pro·tru·sive** (prō-tro͞o′sĭv, -zĭv) *adj.* **1.** Tending to protrude. **2.** Unduly or unpleasantly conspicuous: OBTRUSIVE. **—pro·tru′sive·ly** *adv.* **—pro·tru′sive·ness** *n.*

**pro·tu·ber·ance** (prō-to͞o′bər-əns, -tyo͞o′-) *n.* **1.** Something, as a bulge or knob, that protrudes. **2.** The state of being protuberant.

**pro·tu·ber·an·cy** (prō-to͞o′bər-ən-sē, -tyo͞o′-) *n., pl.* **-cies.** Protuberance.

**pro·tu·ber·ant** (prō-to͞o′bər-ənt, -tyo͞o′-) *adj.* [LLat. *protuberans, protuberant-,* pr.part. of *protuberare,* to bulge out. —see PROTUBERATE.] Swelling outward: BULGING. **—pro·tu′ber·ant·ly** *adv.*

**pro·tu·ber·ate** (prō-to͞o′bə-rāt′, -tyo͞o′-) *vi.* **-at·ed, -at·ing, -ates.** [LLat. *protuberare* : Lat. *pro-,* forth + Lat. *tuber,* a swelling.] To swell or bulge out. **—pro·tu′ber·a′tion** *n.*

**proud** (proud) *adj.* **-er, -est.** [ME < OE *prūd* < OFr. *prud, preu,* brave, virtuous < LLat. *prode,* advantageous < Lat. *prodesse,* to be good : *prod-,* for (var. of *pro-*) + *esse,* to be.] **1.** Feeling pleasurable satisfaction over an attribute or act by which one's status or sense of self is measured <*proud* of one's children> **2.** Causing pride: GRATIFYING <a *proud* day> **3.** Characterized by self-respect. **4.** Haughty. **5.** Of great dignity: HONORED <a *proud* name> **6.** Majestic: magnificent. **7.** Spirited. **—proud′ly** *adv.* **—proud′ness** *n.*

☆ ***syns:*** PROUD, PRIDEFUL, SELF-ESTEEMING, SELF-RESPECTING *adj. core meaning*: properly valuing oneself, one's honor, or one's dignity <too *proud* to accept charity>

**proud flesh** *n.* [So called because of its swelling up.] *Pathol.* The swollen flesh around a healing wound due to granulation tissue.

**prove** (pro͞ov) *v.* **proved, proved** or **prov·en** (pro͞o′vən), **prov·ing, proves.** [ME *proven* < OFr. *prover* < Lat. *probare,* to test < *probus,* good.] —*vt.* **1.** To establish the truth or soundness of by presentation of argument or evidence. **2.** *Law.* To establish the genuineness of (a will). **3.** To ascertain the quality of by testing: TRY OUT. **4.** *Math.* **a.** To validate (a hypothesis or proposition) by a proof. **b.** To verify (the result of a calculation). **5.** To make a sample impression of (type). **6.** *Archaic.* To experience: undergo. —*vi.* To be shown to be: TURN OUT <a plot that *proved* unconvincing on the stage> ***usage:*** *Proved* is the preferred form of the past participle of the verb *prove,* as in *We have proved our point.* The form *proven* is a Scots variant made familiar through legal use (*The charges were not proven*). However, *proven* is more widely used as an adjective in the position immediately before the noun it modifies, as in *a proven talent.* **—prov′a·bil′i·ty, prov′a·ble·ness** *n.* **—prov′a·ble** *adj.* **—prov′a·bly** *adv.* **—prov′er** *n.*

**prov·en** (pro͞o′vən) *adj.* [P.part. of PROVE.] Proved <*proven* linguistic ability> **—prov′en·ly** *adv.*

**prov·e·nance** (prŏv′ə-nəns, -năns′) *n.* [Fr. < *provenir,* to originate < Lat. *provenire* : *pro-,* forth + *venire,* to come.] Place of origin: DERIVATION.

**Pro·ven·çal** (prō'vən-säl', prŏv'ən-) *n.* The Romance language of Provence. **—Pro'ven·çal'** *adj.*

**prov·en·der** (prŏv'ən-dər) *n.* [ME *provendre* < OFr. < Med. Lat. *probenda,* var. of *praebenda.* —see PREBEND.] **1.** Dry food, as hay, used as feed for livestock. **2.** Food or provisions.

**pro·ve·nience** (prə-vēn'yəns, -vē'nē-əns) *n.* [< Lat. *proveniens,* pr.part. of *provenire,* to originate. —see PROVENANCE.] An origin.

**pro·ven·tric·u·lus** (prō'vĕn-trĭk'yə-ləs) *n., pl.* **-li** (-lī') [PRO-² + Lat. *ventriculus,* stomach, dim. of *venter,* belly.] **1.** A division of the stomach anterior to the gizzard in birds. **2.** A digestive division in insects and some worms similar to the proventriculus. **—pro'ven·tric'u·lar** (-lər) *adj.*

**prov·erb** (prŏv'ûrb') *n.* [ME *proverbe* < OFr. < Lat. *proverbium* : *pro-,* forth + *verbum,* word.] **1.** A short, pithy, and much-used saying that expresses a well-known truth or fact. **2.** One accepted as a characteristic example. **3. Proverbs** *(sing. in number).* —See table at BIBLE.

**pro·ver·bi·al** (prə-vûr'bē-əl) *adj.* **1.** Of the nature of a proverb. **2.** Expressed in a proverb. **3.** Widely referred to <their *proverbial* finesse> **—pro·ver'bi·al·ly** *adv.*

**pro·vide** (prə-vīd') *v.* **-vid·ed, -vid·ing, -vides.** [ME *providen* < Lat. *providēre,* to prepare for : *pro-,* forward + *vidēre,* to see.] —*vt.* **1.** To furnish : supply. **2.** To make ready : PREPARE. **3.** To make available : AFFORD. **4.** To set down as a stipulation. —*vi.* **1.** To take measures in preparation. **2.** To supply means of subsistence. **3.** To make a stipulation or condition <The statute *provides* for stiff penalties for offenders.> **—pro·vid'er** *n.*

**pro·vid·ed** (prə-vī'dĭd) *conj.* On the condition : IF <will arrive Thursday *provided* the children are well>

**prov·i·dence** (prŏv'ĭ-dəns, -dĕns') *n.* **1.** Care or preparation beforehand : FORESIGHT. **2.** Prudent management : ECONOMY. **3.** Divine direction <"Some sought the key to history in the working of . . . . *providence*" —William Ebenstein> **4. Providence.** GOD 1a, b.

**prov·i·dent** (prŏv'ĭ-dənt, -dĕnt') *adj.* [ME < Lat. *providens,* pr.part.of *providēre,* to provide for. —see PROVIDE.] **1.** Providing for future events or needs. **2.** Economical : frugal. **—prov'i·dent·ly** *adv.*

**prov·i·den·tial** (prŏv'ĭ-dĕn'shəl) *adj.* **1.** Of or caused by divine providence. **2.** Occurring as if through divine intervention : OPPORTUNE. **—prov'i·den'tial·ly** *adv.*

**pro·vid·ing** (prə-vī'dĭng) *conj.* On the condition : PROVIDED.

**prov·ince** (prŏv'ĭns) *n.* [ME *provynce* < OFr. *province* < Lat. *provincia.*] **1.** A territory governed as a political or administrative unit of a country or empire. **2. a.** A division of territory under the jurisdiction of an archbishop. **b.** A territorial unit of a religious order. **3. provinces.** Regions of a country located away from the capital or population center. **4.** An extensive area of knowledge, activity, or interest <the *province* of modern European history> **5.** The range of one's proper duties and functions : JURISDICTION. **6.** *Ecol.* A subdivision of a region. **7.** A country or region conquered by the Romans and administered by them as a self-contained unit.

**pro·vin·cial** (prə-vĭn'shəl) *adj.* **1.** Of or relating to a province. **2.** Of or typical of people from the provinces : UNSOPHISTICATED. **3.** Limited in perspective. —*n.* **1.** A native or inhabitant of the provinces. **2.** One with provincial ideas or habits. **3.** *Rom. Cath. Ch.* The superior of a province of a religious order. **—pro·vin'cial·ism, pro·vin'ci·al'i·ty** (-shē-ăl'ĭ-tē) *n.* **—pro·vin'cial·ly** *adv.*

**pro·vin·cial·ize** (prə-vĭn'shə-līz') *vt.* **-ized, -iz·ing, -iz·es.** To make provincial. **—pro·vin'cial·i·za'tion** *n.*

**proving ground** *n.* A place for testing new devices, weapons, or theories.

**pro·vi·sion** (prə-vĭzh'ən) *n.* [ME < OFr., forethought < Lat. *provisio* < *providēre,* to foresee. —see PROVIDE.] **1.** An act of supplying or equipping. **2.** Something provided. **3.** A preparatory action or measure. **4. provisions.** A stock of supplies that are necessary, esp. food. **5.** A stipulation or qualification, esp. a clause in a document or agreement. —*vt.* **-sioned, -sion·ing, -sions.** To supply with provisions. **—pro·vi'sion·er** *n.*

**pro·vi·sion·al** (prə-vĭzh'ə-nəl) *also* **pro·vi·sion·ar·y** (-vĭzh'ə-nĕr'ē) *adj.* Provided temporarily, pending permanent arrangements <a *provisional* capital> **—pro·vi'sion·al·ly** *adv.*

**pro·vi·so** (prə-vī'zō) *n., pl.* **-sos** *or* **-soes.** [ME < Med. Lat. *proviso quod,* provided that.] A clause containing a qualification, condition, or restriction in a document.

**pro·vi·so·ry** (prə-vī'zə-rē) *adj.* Depending on a proviso : CONDITIONAL. **—pro·vi'so·ri·ly** *adv.*

**pro·vi·ta·min** (prō-vī'tə-mĭn) *n.* A substance converted to a vitamin within the body, as carotene into vitamin A.

**Pro·vo** (prō'vō) *n.* [Shortening and alteration of *provisional* (*wing*), name of the faction.] A member of the extremist faction of the Irish Republican Army.

**pro·vo·ca·teur** (prō-vŏk'ə-tûr') *n.* [Fr.] An agent provocateur.

**prov·o·ca·tion** (prŏv'ə-kā'shən) *n.* [ME *provocacioun* < OFr. *provocation* < Lat. *provocatio* < *provocare,* to challenge. —see PROVOKE.] **1.** An act of provoking. **2.** Something that provokes.

**pro·voc·a·tive** (prə-vŏk'ə-tĭv) *adj.* Likely to provoke. —*n.* PROVOCATION 2. **—pro·voc'a·tive·ly** *adv.* **—pro·voc'a·tive·ness** *n.*

**pro·voke** (prə-vōk') *vt.* **-voked, -vok·ing, -vokes.** [ME *provoken* < OFr. *provoker* < Lat. *provocare,* to challenge : *pro-,* forth + *vocare,* to call.] **1.** To cause anger, resentment, or deep feeling in. **2.** To cause to take action. **3.** To stir to action. **—pro·vok'ing·ly** *adv.*

☆ ***syns:*** PROVOKE, AROUSE, EXCITE, GALVANIZE, GOAD, IMPEL, INCITE, INFLAME, INSPIRE, INSTIGATE, KINDLE, MOTIVATE, MOVE, ROUSE, SPUR, STIMULATE *v. core meaning* : to stir to action or feeling <*provoked* a fight><carelessness that *provoked* anger>

**pro·vok·ing** (prə-vō'kĭng) *adj.* Vexatious. **—pro·vok'ing·ly** *adv.*

**pro·vo·lo·ne** (prō'və-lō'nē) *n.* A smooth, hard, often smoked cheese orig. from Italy.

**pro·vost** (prō'vōst', -vəst, prŏv'əst) *n.* [ME < OE *profost* and OFr. *provost,* both < Med. Lat. *propositus* < Lat. *praepositus,* superintendent < p.part. of *praeponere,* to place over : *prae-,* before + *ponere,* to put.] **1.** The chief magistrate of certain Scottish cities. **2.** The chief officer of some colleges. **3.** The highest official in certain collegiate churches or cathedrals. **4.** The keeper of a prison.

**pro·vost court** (prō'vō) *n.* A military court for the trial of minor offenses committed in occupied hostile territories.

**pro·vost guard** (prō'vō) *n.* A detail of troops doing police duty under a provost marshal.

**pro·vost marshal** (prō'vō) *n.* **1.** The head of a military police unit. **2.** A naval officer responsible for prisoners facing court-martial.

**prow** (prou) *n.* [OFr. *proue* < Lat. *prora* < Gk. *prōira.*] **1.** The forward part of a ship's hull : BOW. **2.** A projecting part similar in configuration to the prow of a ship, as the forward end of a ski.

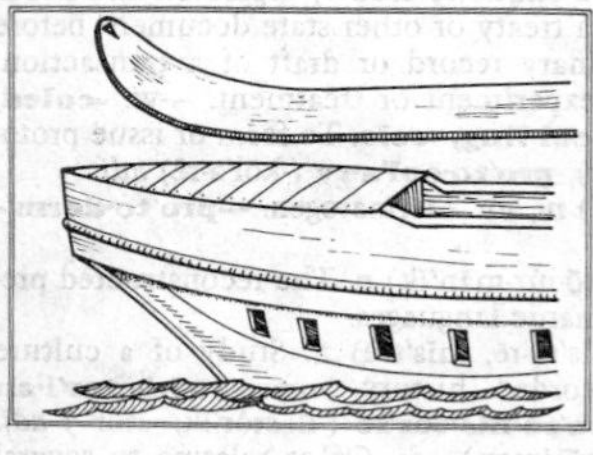

**prow**
*Two types of prows: (above) of a ski and (below) of a boat*

**prow·ess** (prou'ĭs) *n.* [ME *prowesse* < OFr. *proesse* < *prou,* var. of *prud,* brave. —see PROUD.] **1.** Superior skill or ability. **2.** Superior strength, courage, or daring, esp. in battle.

**prowl** (proul) *v.* **prowled, prowl·ing, prowls.** [ME *prollen.*] —*vt.* To roam through stealthily, as in searching for prey. —*vi.* To rove furtively or with predatory purpose. **—prowl** *n.* **—prowl'er** *n.*

**prowl car** *n.* A squad car.

**prox·i·mal** (prŏk'sə-məl) *adj.* [< Lat. *proximus,* superl. of *propior,* near.] **1.** Nearest : proximate. **2.** *Biol.* Near the central part of the body or a point of attachment or origin <the *proximal* end of a bone> **—prox'i·mal·ly** *adv.*

**prox·i·mate** (prŏk'sə-mĭt) *adj.* [Lat. *proximatus,* p.part. of *proximare,* to come near < *proximus,* superl. of *propior,* near.] **1.** Closely related in space, time, or order. **2.** Approximate. **—prox'i·mate·ly** *adv.* **—prox'i·mate·ness** *n.*

**prox·im·i·ty** (prŏk-sĭm'ĭ-tē) *n.* [OFr. *proximite* < Lat. *proximitas* < *proximus,* superl. of *propior,* near.] The state, quality, or fact of being near or next.

**proximity fuze** *n.* An electronic device for detonating a projectile as it nears a target, used in antiaircraft shells.

**prox·i·mo** (prŏk'sə-mō') *adv.* [Lat. *proximo mense,* in the next month.] *Archaic.* Of or in the following month.

**prox·y** (prŏk'sē) *n., pl.* **-ies.** [ME *proxcy* < Norman Fr. *procuracie* < Med. Lat. *procuratia* < Lat. *procurare,* to take care of. —see PROCURE.] **1.** An agent authorized to act for another. **2.** Authority to act for another. **3.** Written authorization to act in place of another.

**prude** (prōōd) *n.* [Fr., short for OFr. *preudefemme,* virtuous woman : *preu,* virtuous, proud + *de,* of (< Lat.) + *femme,* woman (< Lat. *femina*).] One who is overly anxious about being or seeming to be proper, modest, or righteous.

☆ ***syns:*** PRUDE, BLUENOSE, GOODY-GOODY, MRS. GRUNDY, OLD MAID, PRIG, PURITAN, VICTORIAN *n. core meaning* : a person who is too much concerned with being proper, modest, or righteous <*prudes* who opposed dancing and card playing>

**pru·dence** (prōōd'ns) *n.* **1.** The state, quality, or fact of being prudent. **2.** Careful management : ECONOMY.

**pru·dent** (prōōd'nt) *adj.* [ME < OFr. < Lat. *prudens,* short for *providens.* —see PROVIDENT.] **1.** Using good judgment or common sense in handling practical matters. **2.** Careful with respect to one's own interests : PROVIDENT. **3.** Careful about one's conduct : CIRCUMSPECT. **—pru'dent·ly** *adv.*

**pru·den·tial** (prōō-dĕn'shəl) *adj.* **1.** Arising from or marked by prudence. **2.** Using prudence, good judgment, or common sense. **—pru·den'tial·ly** *adv.*

**prud·er·y** (pro͞o′də-rē) *n., pl.* **-ies.** [Fr. *pruderie* < *prude*, prude. —see PRUDE.] **1.** The quality or state of being prudish. **2.** An instance of prudish talk or behavior.

**prud·ish** (pro͞o′dĭsh) *adj.* Marked by or displaying the characteristics of a prude : PRIGGISH. **—prud′ish·ly** *adv.* **—prud′ish·ness** *n.*

**pru·i·nose** (pro͞o′ə-nōs′) *adj.* [Lat. *pruinosus*, frosty < *pruina*, hoarfrost.] *Bot.* Having a powdery white covering or bloom.

**prune[1]** (pro͞on) *n.* [ME < OFr. < Lat. *prunum*, plum.] **1. a.** The partially dried fruit of the common plum, *Prunus domestica.* **b.** A plum that can be dried without spoiling. **2.** *Slang.* An ill-tempered person.

▲ word history: *Prune*[1] and *plum*[1] are doublets, having come by separate routes from Latin *prunum*, "a plum." *Prunum* was borrowed into the Germanic languages at a very early date, before the Angles and Saxons settled in Britain. The Old English form was *plūme*, which became *plum* in Modern English. In Old French Latin *prunum* became *prune.* English borrowed *prune* from French with the specialized meaning "dried plum."

**prune[2]** (pro͞on) *v.* **pruned, prun·ing, prunes.** [ME *prouynen* < OFr. *proignier* < VLat. **prorotundiare* : Lat. *pro-*, in front + Lat. *rotundus*, round.] —*vt.* **1.** To cut off or remove living or dead parts or branches of (e.g., a plant) to improve shape or stimulate growth. **2.** To remove or cut out as unnecessary. **3.** To reduce <*prune* the corporation's payroll> —*vi.* To remove the superfluous or undesirable. **—prun′er** *n.*

**pru·nel·la** (pro͞o-nĕl′ə) *also* **pru·nel·lo** (-nĕl′ō) *n., pl.* **-las** *also* **-los.** [Fr. *prunelle*, sloe, dim. of *prune*, prune.] A strong heavy fabric of worsted twill, used mainly for shoe uppers, clerical robes, and academic gowns.

**pru·nelle** (pro͞o-nĕl′) *n.* [Fr., dim. of *prune*, prune.] A brownish sloe-flavored French liqueur.

**pru·nel·lo** (pro͞o-nĕl′ō) *n. var. of* PRUNELLA.

**pruning hook** *n.* A long pole with a curved saw blade and usu. a clipping device on one end, used esp. for pruning small trees.

**pru·ri·ent** (pro͝or′ē-ənt) *adj.* [Lat. *pruriens*, *prurient-*, pr.part. of *prurire*, to feel desire, itch.] **1.** Obsessively interested in sexual matters. **2. a.** Marked by an obsessive interest in sex. **b.** Arousing or appealing to an obsessive interest in sex <*prurient* novels> **—pru′ri·ence, pru′ri·en·cy** *n.* **—pru′ri·ent·ly** *adv.*

**pru·ri·go** (pro͞o-rī′gō) *n.* [Lat., an itching < *prurire*, to itch.] A chronic inflammatory skin disease marked by eruption and severe itching. **—pru·rig′i·nous** (-rĭj′ə-nəs) *adj.*

**pru·ri·tus** (pro͞o-rī′təs) *n.* [Lat. < *prurire*, to itch.] Severe itching, usu. of undamaged skin. **—pru·rit′ic** (-rĭt′ĭk) *adj.*

**Prus·sian** (prŭsh′ən) *adj.* **1.** Of or relating to Prussia, its people, or their language and culture. **2.** Resembling the Junkers and the military class of Prussia. —*n.* **1.** One of the western Balts formerly inhabiting the region between the Vistula and Neman. **2.** A Baltic inhabitant of Prussia. **3.** A German inhabitant of Prussia.

**Prussian blue** *n.* [After *Prussia*, where the dye was discovered.] **1.** An insoluble dark-blue pigment and dye, ferric ferrocyanide or one of its modifications. **2.** Iron blue. **3.** A moderate to strong blue or deep greenish blue.

**prus·si·ate** (prŭs′ē-āt′) *n.* [Fr. < *(acide) prussique*, prussic acid.] **1.** A ferrocyanide or ferricyanide. **2.** A salt of hydrocyanic acid : CYANIDE.

**prus·sic acid** (prŭs′ĭk) *n.* [So called because it is obtained from Prussian blue.] Hydrocyanic acid.

**pry[1]** (prī) *vi.* **pried, pry·ing, pries.** [ME *prien.*] To inquire or look closely or curiously, often furtively : SNOOP. —*n., pl.* **pries. 1.** An act of prying. **2.** An overly curious person : SNOOP. **—pry′ing·ly** *adv.*

**pry[2]** (prī) *vt.* **pried, pry·ing, pries.** [Alteration of PRIZE[3].] **1.** To raise, move, or force open with a pry or lever. **2.** To secure with effort or trouble <*pried* the secret out of them> —*n., pl.* **pries. 1.** An implement, as a crowbar, for applying leverage. **2.** Leverage.

**pry·er** (prī′ər) *n. var. of* PRIER.

**psalm** (säm) *n.* [ME < OE *psealm* < LLat. *psalmus* < Gk. *psalmos* < *psallein*, to play the harp.] **1.** A sacred song : HYMN. **2. Psalms** *(sing. in number).* —See table at BIBLE. —*vt.* **psalmed, psalm·ing, psalms.** To sing of or celebrate in psalms. **—psalm′ist** (sä′mĭst) *n.*

**psalm·o·dy** (sä′mə-dē, săl′mə-) *n., pl.* **-dies.** [ME *psalmodie* < LLat. *psalmodia* < Gk. *psalmōdia*, singing to the harp : *psalmos*, psalm (< *psallein*, to play the harp) + *ōidē*, song.] **1.** The act or practice of singing psalms. **2.** Composition or arrangement of psalms for singing. **3.** A collection of psalms. **—psalm′o·dist** *n.*

**Psal·ter** *also* **psal·ter** (sôl′tər) *n.* [ME < OE *psaltere* and OFr. *psautier*, both < LLat. *psalterium* < Gk. *psaltērion*, harp. —see PSALTERY.] A book containing the Book of Psalms or a version of, musical setting for, or selection from it.

**psal·te·ri·um** (sôl-tîr′ē-əm) *n., pl.* **-ri·a** (-ē-ə) [LLat., psalter, so called because when slit open its folds fall apart like the leaves of a book.] The omasum. **—psal·te′ri·al** *adj.*

**psal·ter·y** (sôl′tə-rē) *also* **psal·try** (sôl′trē) *n., pl.* **-ter·ies** *also* **-tries.** [ME *psalterie* < OFr. < Lat. *psalterium* < Gk. *psaltērion* < *psallein*, to play the harp.] An ancient stringed musical instrument played by plucking the strings with a plectrum or the fingers.

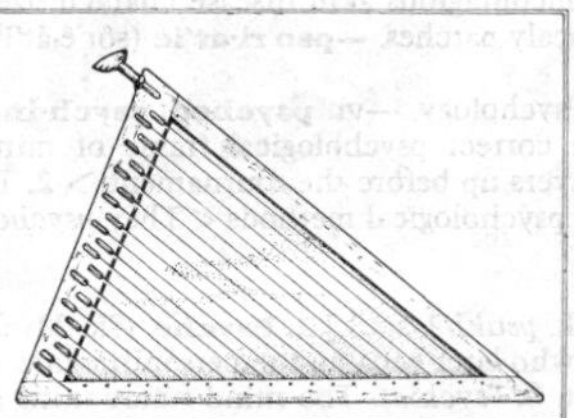

psaltery

**p's and q's** (pēz′ ən kyo͞oz′) *pl.n.* **1.** Socially proper conduct : MANNERS. **2.** Something, as manners or behavior, one should be aware of <Fear of punishment taught them to watch their *p's and q's.*>

**pse·phol·o·gy** (sē-fŏl′ə-jē) *n.* [Gk. *psēphos*, pebble, ballot (from the ancient Greeks' use of pebbles for voting) + -LOGY.] Study of political elections. **—pse′pho·log′i·cal** (sē′fə-lŏj′ĭ-kəl) *adj.* **—pse·phol′o·gist** *n.*

**pseud-** *pref. var. of* PSEUDO-.

**pseud·ax·is** (so͞o-dăk′sĭs) *n.* A sympodium.

**pseud·e·pig·ra·pha** (so͞o′dĭ-pĭg′rə-fə) *pl.n.* [Gk. : *pseudēs*, false (< *pseudein*, to lie) + *epigraphein*, to ascribe (*epi-*, upon + *graphein*, to write).] **1.** Spurious writings, esp. writings erroneously credited to Biblical characters or times. **2.** Jewish religious texts written between 200 B.C. and A.D. 200 and erroneously ascribed to various prophets and kings of Hebrew Scriptures. **—pseud′e·pig′ra·phal, pseud′ep·i·graph′ic** (so͞o′dĕp′ĭ-grăf′ĭk), **pseud′ep·i·graph′i·cal, pseud′e·pig′ra·phous** *adj.*

**pseudo-** *or* **pseud-** *pref.* [ME < LLat. < Gk. *pseudēs*, false < *pseudein*, to lie.] **1.** False : deceptive : sham <*pseudo*science> **2.** Apparently similar <*pseudo*carp>

**pseu·do·carp** (so͞o′də-kärp′) *n.* An accessory fruit. **—pseu′do·car′pous** *adj.*

**pseu·do·coel** (so͞o′də-sēl′) *also* **pseu·do·coe·lom** (so͞o′də-sē′ləm) *n.* A body cavity not formed by gastrulation and lacking a mesodermal lining.

**pseu·do·coe·lo·mate** (so͞o′dō-sē′lə-māt′) *adj.* Having a pseudocoel.

**pseu·do·cy·e·sis** (so͞o′dō-sī-ē′sĭs) *n.* A psychosomatic condition in which physical symptoms of pregnancy, as weight gain and amenorrhea, are manifested without conception.

**pseu·do-e·vent** (so͞o′dō-ĭ-vĕnt′) *n. Informal.* An event staged to attract attention <the *pseudo-events* of a political campaign>

**pseu·do·mo·nad** (so͞o′də-mō′năd′) *n.* [NLat. *Pseudomonas*, *Pseudomonad-*, genus name : PSEUDO- + Gk. *monas*, unit < *monos*, one.] Any of the Gram-negative, rod-shaped bacteria of the genus *Pseudomonas*, including some plant and animal pathogens.

**pseu·do·morph** (so͞o′də-môrf′) *n.* **1.** A false, deceptive, or irregular form. **2.** *Mineral.* A mineral with the crystalline form of another mineral rather than that usually typical of its composition. **—pseu′do·mor′phic, pseu′do·mor′phous** *adj.* **—pseu′do·mor′phism** *n.*

**pseu·do·nym** (so͞od′n-ĭm′) *n.* [Fr. *pseudonyme* < Gk. *pseudonumon* : *pseudēs*, false + *onoma*, name.] A fictitious name taken on by an author : PEN NAME. **—pseu·don′y·mous** (so͞o-dŏn′ə-məs) *adj.* **—pseu·don′y·mous·ly** *adv.* **—pseu·don′y·mous·ness** *n.*

**pseu·do·po·di·um** (so͞o′də-pō′dē-əm) *also* **pseu·do·pod** (so͞o′də-pŏd′) *n., pl.* **-po·di·a** (-pō′dē-ə) *also* **-pods.** [NLat. : PSEUDO- + Gk. *podion*, dim. of *pous*, foot.] A temporary protrusion of the cytoplasm of a cell, used as a means of locomotion and of surrounding and ingesting food in organisms such as the amoeba.

**pseu·do·preg·nan·cy** (so͞o′dō-prĕg′nən-sē) *n.* **1.** A condition resembling pregnancy that occurs in some mammals after infertile copulation. **2.** Pseudocyesis.

**pseu·do·ran·dom** (so͞o′dō-răn′dəm) *adj.* Of, pertaining to, or being random numbers generated by a definite, nonrandom computational process.

**pseu·do·sci·ence** (so͞o′dō-sī′əns) *n.* A theory, methodology, or activity that appears to be or is advanced as scientific. **—pseu′do·sci′en·tif′ic** (-ən-tĭf′ĭk) *adj.* **—pseu′do·sci′en·tist** *n.*

**pshaw** (shô) *interj.* —Used to indicate impatience, irritation, disapproval, or disbelief.

**psi** (sī, psī) *n.* [LGk. < Gk. *psei.*] The 23rd letter of the Greek alphabet. —See table at ALPHABET.

**psil·o·cy·bin** (sĭl′ə-sī′bĭn, sī′lə-) *n.* [NLat. *Psilocybe*, genus name + -IN.] A compound, $C_{12}H_{17}N_2O_4P$, derived from the mushroom *Psilocybe mexicana*, that is a strong hallucinogen.

**psi·lom·e·lane** (sī-lŏm′ə-lān′) *n.* [Gk. *psilos*, bare + Gk. *melas*, *melan-*, black.] A hard black hydrated oxide ore of manganese.

**psit·ta·cine** (sĭt′ə-sīn′) *adj.* [Lat. *psittacinus* < *psittacus*, parrot < Gk. *psittakos.*] Of, relating to, or typical of parrots.

**psit·ta·co·sis** (sĭt′ə-kō′sĭs) *n.* [Lat. *psittacus,* parrot (< Gk. *psittakos*) + -OSIS.] A virus disease of parrots and related birds communicable to human beings, in whom it produces high fever and pneumonialike complications. **—psit′ta·cot′ic** (-kŏt′ĭk, -kō′tĭk) *adj.*

**pso·ri·a·sis** (sə-rī′ə-sĭs) *n.* [Gk. *psōriasis* < *psōrian,* to have the itch < *psōra,* itch.] A chronic noncontagious skin disease characterized by inflammation and white scaly patches. **—pso′ri·at′ic** (sôr′ē-ăt′ĭk, sōr′-) *adj.*

**psych** (sīk) *Informal.* —*n.* Psychology. —*vt.* **psyched, psych·ing, psychs. 1.** To put into the correct psychological frame of mind <The coach *psyched* the players up before the tournament.> **2.** To weaken the confidence of by psychological methods <They *psyched* out all their opponents.>

**psych-** *pref. var. of* PSYCHO-.

**psy·che** (sī′kē) *n.* [Lat. < Gk. *psukhē,* soul.] **1. Psyche.** *Gk. Myth.* A maiden loved by Eros and who later became the personification of the soul. **2.** The soul or spirit. **3.** *Psychiat.* The mind functioning as the center of thought, feeling, and behavior and consciously or unconsciously adjusting and relating the body to its social and physical environment.

**psy·che·de·li·a** (sī′kĭ-dē′lē-ə, -dēl′yə) *n.* The subculture of people associated with psychedelic drugs.

**psy·che·del·ic** (sī′kĭ-dĕl′ĭk) *adj.* [PSYCHE + Gk. *dēlos,* visible.] Of, relating to, or producing hallucinations, distorted perceptions, and, occas., states similar to psychosis. **—psy′che·del′i·cal·ly** *adv.*

**psy·chi·a·try** (sĭ-kī′ə-trē, sī-) *n.* The branch of medicine concerned with the study, diagnosis, treatment, and prevention of mental illness. **—psy′chi·at′ric** (sī′kē-ăt′rĭk), **psy′chi·at′ri·cal** *adj.* **—psy′chi·at′ri·cal·ly** *adv.* **—psy·chi′a·trist** *n.*

**psy·chic** (sī′kĭk) *also* **psy·chi·cal** (-kĭ-kəl) *adj.* [Gk. *psukhikos* < *psukhē,* soul.] **1.** Of or relating to the human psyche. **2. a.** Of or relating to extraordinary, esp. extrasensory and nonphysical, mental processes, as extrasensory perception and mental telepathy. **b.** Originating in, brought on by, or reacting to such processes. —*n.* **psychic. 1.** An individual apparently responding to psychic forces. **2.** MEDIUM 5. **—psy′chi·cal·ly** *adv.*

**psy·cho** (sī′kō) *Slang.* —*n., pl.* **-chos.** A psychopath. —*adj.* Insane.

**psycho-** *or* **psych-** *pref.* [< Gk. *psukhē,* spirit, life.] **1. a.** Mind: mental <*psycho*genic> **b.** Mental activities or processes <*psycho*motor> **2.** Psychology: psychological <*psycho*history>

**psy·cho·ac·tive** (sī′kō-ăk′tĭv) *adj.* Influencing the mind or mental processes <*psychoactive* drugs>

**psy·cho·a·nal·y·sis** (sī′kō-ə-năl′ĭ-sĭs) *n.* **1.** The analytic technique created by Sigmund Freud that uses free association, dream interpretation, and analysis of resistance and transference to explore mental processes. **2.** The theory of human psychology founded by Freud on the concepts of the unconscious, resistance, repression, sexuality, and the Oedipus complex. **3.** A psychiatric therapy incorporating psychoanalysis. **—psy′cho·an′a·lyst** (-ăn′ə-lĭst) *n.* **—psy′cho·an′a·lyt′ic** (-ăn′ə-lĭt′ĭk), **psy′cho·an′a·lyt′i·cal** *adj.* **—psy′cho·an′a·lyt′i·cal·ly** *adv.*

**psy·cho·an·a·lyze** (sī′kō-ăn′ə-līz′) *vt.* **-lyzed, -lyz·ing, -lyz·es.** To analyze and treat by psychoanalysis.

**psy·cho·bab·ble** (sī′kō-băb′əl) *n.* Psychological jargon, esp. that of psychotherapy. **—psy′cho·bab′bler** *n.*

**psy·cho·bi·og·ra·phy** (sī′kō-bī-ŏg′rə-fē, -bē-) *n.* **1.** A biography that analyzes the psychological make-up or characteristics of an individual. **2.** Character analysis. **—psy′cho·bi·og′ra·pher** *n.* **—psy′cho·bi′o·graph′i·cal** (-bī′ə-grăf′ĭ-kəl) *adj.*

**psy·cho·bi·ol·o·gy** (sī′kō-bī-ŏl′ə-jē) *n.* Study of the interactions between mental and biological processes. **—psy′cho·bi′o·log′i·cal** (-bī′ə-lŏj′ĭ-kəl) *adj.* **—psy′cho·bi·ol′o·gist** *n.*

**psy·cho·chem·i·cal** (sī′kō-kĕm′ĭ-kəl) *n.* A psychoactive substance.

**psy·cho·dra·ma** (sī′kə-drä′mə, -drăm′ə) *n.* **1.** A psychotherapeutic and analytic technique in which individuals are given roles to be extemporaneously played within a dramatic context set up by a therapist. **2.** A dramatization that uses psychodrama. **—psy′cho·dra·mat′ic** (-drə-măt′ĭk) *adj.*

**psy·cho·dy·nam·ics** (sī′kō-dī-năm′ĭks, -dĭ-) *n. (sing. in number).* **1.** Interaction of various mental or emotional processes, esp. when considered as constituents of a system of interrelated forces. **2.** Behavioral analysis in terms of motives or drives. **—psy′cho·dy·nam′ic** *adj.* **—psy′cho·dy·nam′i·cal·ly** *adv.*

**psy·cho·gen·e·sis** (sī′kə-jĕn′ĭ-sĭs) *n.* **1.** Generation and development of psychological processes, personality, or behavior. **2.** The psychological source of a psychic process or event. **—psy′cho·ge·net′ic** (-jə-nĕt′ĭk) *adj.* **—psy′cho·ge·net′i·cal·ly** *adv.*

**psy·cho·gen·ic** (sī′kə-jĕn′ĭk) *adj.* Originating in the mind or in mental activities and conditions. **—psy′cho·gen′i·cal·ly** *adv.*

**psy·chog·no·sis** (sī′kŏg-nō′sĭs, -kəg-) *n. Psychiat.* Diagnosis of psychic disorders. **—psy′chog·nos′tic** (-nŏs′tĭk) *adj.*

**psy·cho·graph** (sī′kə-grăf′) *n.* **1.** A psychological profile of an individual or group. **2.** A psychobiography. **—psy′cho·graph′ic** *adj.*

**psy·cho·his·to·ry** (sī′kō-hĭs′tə-rē) *n., pl.* **-ries.** A psychoanalytic interpretation of a historical person or event <The book *A Prince of Our Disorder* is a *psychohistory* of T.E. Lawrence.> **—psy′cho·his·tor′i·an** (-hĭ-stôr′ē-ən, -stōr′-) *n.* **—psy′cho·his·tor′i·cal** (-stôr′ĭ-kəl, -stōr′-) *adj.*

**psy·cho·ki·ne·sis** (sī′kō-kĭ-nē′sĭs, -kī-) *n.* **1.** Production of motion, esp. in inanimate and remote objects, by means of psychic powers. **2.** *Psychiat.* Uninhibited maniacal motor response. **—psy′cho·ki·net′ic** (-kĭ-nĕt′ĭk, -kī-) *adj.* **—psy′cho·ki·net′i·cal·ly** *adv.*

**psy·cho·lin·guis·tics** (sī′kō-lĭng-gwĭs′tĭks) *n. (sing. in number).* Study of the interaction between psychological factors and linguistic behavior. **—psy′cho·lin′guist** *n.* **—psy′cho·lin·guis′tic** *adj.*

**psy·cho·log·i·cal** (sī′kə-lŏj′ĭ-kəl) *adj.* **1.** Of or relating to psychology. **2.** Of, relating to, or derived from the mind or emotions. **3.** Able to influence the mind or emotions. **—psy′cho·log′i·cal·ly** *adv.*

**psychological moment** *n.* The time when a person is most apt to react in the desired manner.

**psy·chol·o·gize** (sī-kŏl′ə-jīz′) *v.* **-gized, -giz·ing, -giz·es.** —*vt.* To explain (behavior) psychologically. —*vi.* **1.** To investigate psychologically. **2.** To reason or speculate psychologically.

**psy·chol·o·gy** (sī-kŏl′ə-jē) *n., pl.* **-gies. 1.** The science of mental processes and behavior. **2.** Emotional and behavioral characteristics of an individual, group, or activity <the *psychology* of war> **3.** Subtle tactical action or argument <used good *psychology* in getting the child to obey> **—psy·chol′o·gist** *n.*

**psy·cho·met·rics** (sī′kə-mĕt′rĭks) *n. (sing. in number).* **1.** Measurement of psychological variables, as intelligence, aptitude, and emotional disturbance. **2.** Mathematical, esp. statistical design of psychological tests and measures. **—psy′cho·met′ric, psy′cho·met′ri·cal** *adj.* **—psy′cho·met′ri·cal·ly** *adv.* **—psy·chom′e·tri′cian** (sī-kŏm′ĭ-trĭsh′ən), **psy·chom′e·trist** (sī-kŏm′ĭ-trĭst) *n.*

**psy·cho·mo·tor** (sī′kə-mō′tər) *adj.* Of or relating to muscular activity associated with mental processes.

**psy·cho·neu·ro·sis** (sī′kō-nŏŏ-rō′sĭs, -nyŏŏ-) *n.* Neurosis. **—psy′cho·neu·rot′ic** (-rŏt′ĭk) *adj. & n.*

**psy·cho·path** (sī′kə-păth′) *n.* [< PSYCHOPATHY.] A person having a personality disorder, esp. one manifested in aggressively antisocial behavior. **—psy′cho·path′ic** *adj.* **—psy′cho·path′i·cal·ly** *adv.*

**psy·cho·pa·thol·o·gy** (sī′kō-pə-thŏl′ə-jē, -pă-) *n.* Study of pathological mental conditions. **—psy′cho·path′o·log′i·cal** (-păth′ə-lŏj′ĭ-kəl), **psy′cho·path′o·log′ic** *adj.* **—psy′cho·pa·thol′o·gist** *n.*

**psy·chop·a·thy** (sī-kŏp′ə-thē) *n.* Mental disorder, esp. that of unknown etiology.

**psy·cho·phys·ics** (sī′kō-fĭz′ĭks) *n. (sing. in number).* Psychological study of the relationships between physical stimuli and sensory response. **—psy′cho·phys′i·cal** *adj.* **—psy′cho·phys′i·cal·ly** *adv.* **—psy′cho·phys′i·cist** (-fĭz′ĭ-sĭst) *n.*

**psy·cho·phys·i·ol·o·gy** (sī′kō-fĭz′ē-ŏl′ə-jē) *n.* Study of the correlations between behavior and physiology. **—psy′cho·phys′i·o·log′i·cal** (-fĭz′ē-ə-lŏj′ĭ-kəl), **psy′cho·phys′i·o·log′ic** *adj.* **—psy′cho·phys′i·ol′o·gist** *n.*

**psy·cho·sis** (sī-kō′sĭs) *n., pl.* **-ses** (-sēz′). Severe mental disorder, with or without organic damage, marked by degeneration of normal intellectual and social functioning and by complete or partial withdrawal from reality.

**psy·cho·so·cial** (sī′kō-sō′shəl) *adj.* Involving aspects of both social and psychological behavior <a child's *psychosocial* adjustment> **—psy′cho·so′cial·ly** *adv.*

**psy·cho·so·mat·ic** (sī′kə-sō-măt′ĭk) *adj.* **1.** Of or relating to phenomena that are both physiological and psychological. **2.** Of or relating to a partially or entirely psychogenic disease or physiological disorder. —*n.* One who experiences bodily symptoms because of mental conflict. **—psy′cho·so·mat′i·cal·ly** *adv.*

**psy·cho·sur·ger·y** (sī′kō-sûr′jə-rē) *n.* Brain surgery for treating mental disorders. **—psy′cho·sur′geon** *n.* **—psy′cho·sur′gi·cal** *adj.*

**psy·cho·tech·nics** (sī′kō-tĕk′nĭks) *n. (sing. in number).* Practical or technological use of psychology, as in analysis of social or industrial problems. **—psy′cho·tech′ni·cal** *adj.* **—psy′cho·tech·ni′cian** (-tĕk-nĭsh′ən) *n.*

**psy·cho·ther·a·peu·tics** (sī′kō-thĕr′ə-pyŏŏ′tĭks) *n. (sing. or pl. in number).* Psychotherapy.

**psy·cho·ther·a·py** (sī′kō-thĕr′ə-pē) *n.* Psychological treatment of mental, emotional, and nervous disorders. **—psy′cho·ther′a·peu′tic** (-pyŏŏ′tĭk) *adj.* **—psy′cho·ther′a·peu′ti·cal·ly** *adv.* **—psy′cho·ther′a·pist** *n.*

**psy·chot·ic** (sī-kŏt′ĭk) *n.* [< PSYCHOSIS.] One afflicted with a psychosis. **—psy·chot′ic** *adj.* **—psy·chot′i·cal·ly** *adv.*

**psy·chot·o·mi·met·ic** (sī-kŏt′ō-mə-mĕt′ĭk, -mī-) *adj.* [PSYCHOT(IC) + MIMETIC.] Pertaining to or inducing psychotic symptoms, as psychotic alteration of mind or behavior. —*n.* A psychotomimetic agent. **—psy·chot′o·mi·met′i·cal·ly** *adv.*

**psy·cho·tro·pic** (sī′kə-trō′pĭk, -trŏp′ĭk) *adj.* Having a mind-altering effect <a *psychotropic* drug>

**psych-out** (sīk′out′) *n. Informal.* An act or instance of undermining someone's confidence by psychological means.

**psychro-** *pref.* [< Gk. *psukhros,* cold.] Cold <*psychro*meter>

**psy·chrom·e·ter** (sī-krŏm′ĭ-tər) *n.* A hygrometer that uses the difference in readings between two thermometers, one having a wet bulb ventilated to cause evaporation and the other having a dry bulb, as a measure of atmospheric moisture.

**psy·chro·phil·ic** (sī′krō-fĭl′ĭk) *adj. Biol.* Thriving at relatively low temperatures, usu. at or below 15°C. —Used of certain bacteria.

**psyl·la** (sĭl′ə) *also* **psyl·lid** (sĭl′ĭd) *n.* [NLat. *Psylla*, genus name < Gk. *psulla*, flea.] A plant louse of the family Chermidae or Psyllidae, esp. *Psylla pyricola*, a pest that infests pear trees.

**Pt** *symbol for* PLATINUM.

**ptar·mi·gan** (tär′mĭ-gən) *n., pl.* **ptarmigan** *or* **-gans.** [Alteration of Sc. Gael. *tarmachan.*] A bird of the genus *Lagopus* of the arctic and subarctic regions of the Northern Hemisphere, with feathered feet and plumage that is brownish in summer and white in winter.

**PT boat** (pē-tē′) *n.* [P(ATROL) + T(ORPEDO) BOAT.] A fast maneuverable lightly armed vessel used to torpedo enemy shipping.

**-pter** *suff.* [< Gk. *pteron*, wing.] Wing: winglike part or appendage <ornitho*pter*>

**pter·i·dol·o·gy** (tĕr′ĭ-dŏl′ə-jē) *n.* [Gk. *pteris, pterid-*, fern (< *pteron*, feather) + -LOGY.] Study of ferns. **—pter′i·do·log′i·cal** (-də-lŏj′ĭ-kəl) *adj.* **—pter′i·dol′o·gist** *n.*

**pte·rid·o·phyte** (tə-rĭd′ə-fīt′, tĕr′ĭ-dō-) *n.* [NLat. *Pteridophyta*, division name : Gk. *pteris*, fern (< *pteron*, feather) + Gk. *phuton*, plant.] A plant of the division Pteridophyta, including the ferns, club mosses, and horsetails. **—pte·rid′o·phyt′ic** (tə-rĭd′ə-fĭt′ĭk, tĕr′ĭ-dō-), **pter′i·doph′y·tous** (tĕr′ĭ-dŏf′ĭ-təs) *adj.*

**pter·o·cer·coid** (tĕr′ə-sûr′koid′) *n.* [Gk. *pteron*, wing + *kerkos*, tail + -OID.] The infective larva of some tapeworms, marked by its solid elongated body.

**pter·o·dac·tyl** (tĕr′ə-dăk′təl) *n.* [NLat. *Pterodactylus*, reptile genus : Gk. *pteron*, wing + Gk. *daktulos*, finger.] Any of various extinct flying reptiles of the family Pterodactylidae. **—pter′o·dac′ty·loid′** *adj.* **—pter′o·dac′ty·lous** *adj.*

**pter·o·pod** (tĕr′ə-pŏd′) *n.* [NLat. *Pteropoda*, order name < Gk. *pteropous*, wing-footed : *pteron*, wing + *pous*, foot.] Any of various small marine gastropod mollusks of the order Pteropoda that swim with winglike expanded lobes of the foot. **—pter′o·pod′** *adj.* **—pte·rop′o·dan** (tə-rŏp′ə-dən) *adj.* & *n.*

**pter·o·saur** (tĕr′ə-sôr′) *n.* [NLat. *Pterosauria*, order name : Gk. *pteron*, wing + Gk. *sauros*, lizard.] Any of various extinct flying reptiles of the order Pterosauria, including the pterodactyls, of the Jurassic and Cretaceous periods, marked by wings made up of a flap of skin supported by the very long fourth digit on each front leg.

**pter·o·yl·glu·tam·ic acid** (tĕr′ō-ĭl-glo͞o-tăm′ĭk) *n.* [E. *pteroic acid*, an amino acid + -YL + GLUTAMIC ACID.] Folic acid.

**pter·y·goid** (tĕr′ĭ-goid′) *adj.* [< Gk. *pterugoeidēs*, winglike : *pterux*, wing (< *pteron*) + *-eidēs*, -oid.] *Anat.* Of or designating either of two processes in the skull attached like wings to the body of the sphenoid bone. —*n.* Either of the pterygoid anatomical processes.

**ptis·an** (tĭz′ən, tĭ-zăn′) *n.* [ME *tisan* < OFr. *tisane* < Lat. *ptisana* < Gk. *ptisanē* < *ptissein*, to crush.] A slightly medicinal infusion, as barley water.

**Ptol·e·ma·ic** (tŏl′ə-mā′ĭk) *adj.* **1.** Of or relating to the astronomer Ptolemy. **2.** Of or relating to the Ptolemies or to Egypt during their rule (323–30 B.C.).

**Ptolemaic system** *n.* Ptolemy's astronomical system, having the earth at the center of the universe, with the moon, planets, and stars revolving about it. **—Ptol′e·ma′ist** *n.*

**pto·maine** *also* **pto·main** (tō′mān′, tō-mān′) *n.* [Ital. *ptomaina* < Gk. *ptōma*, corpse < *piptein*, to fall.] Any of various basic nitrogenous materials, some poisonous, produced by putrefaction and decomposition of protein.

**ptomaine poisoning** *n.* Food poisoning caused by bacteria or bacterial toxins.

**pto·sis** (tō′sĭs) *n., pl.* **-ses** (-sēz′) [Gk. *ptōsis*, fall < *piptein*, to fall.] Abnormal and permanent lowering of an organ, esp. drooping of the upper eyelid due to muscle failure. **—pto′tic** *adj.*

**pty·a·lin** (tī′ə-lĭn) *n.* [Gk. *ptualon*, saliva (< *ptuein*, to spit) + -IN.] A salivary enzyme in humans and some lower animals that hydrolyzes starch into maltose and various dextrins.

**pty·a·lism** (tī′ə-lĭz′əm) *n.* [< Gk. *ptualon*, saliva < Gk. *ptuein*, to spit.] Excessive saliva flow.

**Pu** *symbol for* PLUTONIUM.

**pub** (pŭb) *n.* [Short for PUBLIC HOUSE.] A tavern: inn.

**pub-crawl** (pŭb′krôl′) *vi.* **-crawled, -crawl·ing, -crawls.** *Slang.* To barhop.

**pu·ber·ty** (pyo͞o′bər-tē) *n.* [ME *puberte* < Lat. *pubertas* < *puber*, adult < *pubes*, pubic hair.] **1.** The stage of maturation in which an individual becomes physiologically capable of sexual reproduction. **2.** Approach to maturity <"Mankind will not reach *puberty* for another hundred thousand years" —René Dubos> **—pu′ber·tal, pu′ber·al** (pyo͞o′bər-əl) *adj.*

**pu·ber·u·lent** (pyo͞o-bĕr′yə-lənt, -bĕr′ə-) *also* **pu·ber·u·lous** (-bĕr′yə-ləs, -bĕr′ə-) *adj.* [< Lat. *puber*, downy < *pubes*, pubic hair.] Covered with minute hairs or very fine down.

**pu·bes** (pyo͞o′bēz) *n., pl.* **pubes.** [Lat.] **1.** The pubic region. **2.** The pubic hair. **3.** *pl. of* PUBIS.

**pu·bes·cence** (pyo͞o-bĕs′əns) *n.* **1.** A covering of soft down or short hairs, as on certain plants and insects. **2.** The quality or state of being pubescent.

**pu·bes·cent** (pyo͞o-bĕs′ənt) *adj.* [Fr. < Lat. *pubescens*, pr.part of *pubescere*, to reach puberty < *puber*, adult < *pubes*, pubic hair.] **1.** Covered with short hairs or soft down. **2.** Reaching or having reached puberty.

**pu·bic** (pyo͞o′bĭk) *adj.* [< PUBES.] Of or in the region of the lower part of the abdomen, the pubis, or the pubes.

**pu·bis** (pyo͞o′bĭs) *n., pl.* **-bes** (-bēz) [NLat. *(os) pubis*, (bone) of the groin < *pubes*, groin.] The forward portion of either of the hipbones, at the juncture forming the front arch of the pelvis.

**pub·lic** (pŭb′lĭk) *adj.* [ME *publyk* < OFr. *public* < Lat. *publicus* < *populus*, people.] **1.** Of, affecting, or concerning the community or the people <the *public* good> **2.** Maintained for or used by the people or community <a *public* stadium> **3.** Participated in or attended by the people or community <a *public* concert> **4.** Connected with or acting on behalf of the people, community, or government <*public* officials> **5.** Open to the judgment or knowledge of all <a *public* statement> —*n.* **1.** The community or the people as a group. **2.** A group of people sharing a mutual interest. **3.** Admirers or followers, esp. of a celebrity. **—pub′lic·ness** *n.*

☆ ***syns:*** PUBLIC, DEMOCRATIC, GENERAL, POPULAR *adj. core meaning:* of, representing, or carried on by the people <*public* elections> ***ant:*** private

**public access** *n.* Availability of television or radio broadcast facilities, as provided by law, for use by the public for presentation of programs, as those of community interest.

**pub·lic-ad·dress system** (pŭb′lĭk-ə-drĕs′) *n.* An electronic amplification apparatus with loudspeakers for broadcasting in public areas.

**pub·li·can** (pŭb′lĭ-kən) *n.* [ME, tax collector < OFr. *publicain* < Lat. *publicanus* < *publicum*, public revenue < *publicus*, public < *populus*, people.] **1.** *Chiefly Brit.* The keeper of a public house: INNKEEPER. **2.** A collector of public taxes or tolls in the ancient Roman Empire. **3.** A collector of taxes or tribute.

**public assistance** *n.* RELIEF 3.

**pub·li·ca·tion** (pŭb′lĭ-kā′shən) *n.* [ME *publicacioun* < OFr. < LLat. *publicatio* < *publicare*, to make public < Lat. *publicus*, public < *populus*, people.] **1.** The act or process of publishing printed matter. **2.** An issue of printed material offered for distribution or sale. **3.** Communication of information to the public.

**public defender** *n.* A usu. publicly appointed attorney or staff of attorneys having responsibility for the legal defense of those unable to afford or obtain legal assistance.

**public domain** *n.* **1.** Land owned and controlled by the state or federal government. **2.** The status of products, publications, and processes unprotected by copyright or patent.

**public health** *n.* The art and science of protecting and improving community health by means of preventive medicine, health education, communicable disease control, and the application of the social and sanitary sciences.

**public house** *n. Chiefly Brit.* A place, as a tavern, licensed to sell alcoholic beverages.

**pub·li·cist** (pŭb′lĭ-sĭst) *n.* A press or publicity agent.

**pub·lic·i·ty** (pŭ-blĭs′ĭ-tē) *n.* [Fr. *publicité* < *public*, public < OFr. —see PUBLIC.] **1. a.** Information concerning a person, group, event, or product that is disseminated through various communications media to attract public attention. **b.** Public notice, interest, or notoriety achieved by the spreading of such information. **c.** The act, process, or occupation of disseminating information to gain public interest. **2.** The state of being public.

**pub·li·cize** (pŭb′lĭ-sīz′) *vt.* **-cized, -ciz·ing, -ciz·es.** To give publicity to <*publicize* a new movie>

**public law** *n.* **1.** The branch of law dealing with the state or government and its relationships with individuals or other governments. **2.** A law affecting the public.

**pub·lic·ly** (pŭb′lĭk-lē) *adv.* **1.** In a public manner: OPENLY. **2.** By or with consent of the public.

**public prosecutor** *n.* A government official who prosecutes criminal actions on behalf of the state or community.

**public relations** *pl.n.* **1. a.** The methods and activities employed in persuading the public to understand and regard favorably a person, business, or institution. **b.** The degree of understanding and favorable regard achieved. **2.** A staff hired to promote a favorable relationship with the public. **3.** The art or science of promoting favorable relations with the public.

**public sale** *n.* AUCTION 1.

**public school** *n.* **1.** A U.S. elementary or secondary school supported by public funds and providing free education for children of a community or district. **2.** *Chiefly Brit.* A private boarding school for pupils between the ages of 13 and 18.

**public servant** *n.* One elected or appointed to a government position <senators, judges, and other *public servants*>

**public service** *n.* **1.** Employment in a governmental system, esp. in the civil services. **2.** A service performed for public benefit.

**pub·lic-serv·ice corporation** (pŭb′lĭk-sûr′vĭs) *n.* A corporation providing utilities for the public.

**public speaking** *n.* The art or process of making speeches before an audience. **—public speaker** *n.*

**pub·lic-spir·it·ed** (pŭb′lĭk-spĭr′ĭ-tĭd) *adj.* Motivated by or exhibiting devotion to the public welfare. **—pub′lic-spir′it·ed·ness** *n.*

**public television** *n.* Noncommercial television providing esp. educational programs for the public.

**public utility** *n.* **1.** A private business organization, subject to governmental regulation, that provides an essential commodity or service, as water, electricity, or communication, to the public. **2.** *often* **public utilities.** Stock shares issued by a public utility.

**public works** *pl.n.* Construction projects, as highways or dams, financed by public funds and constructed by a government for the general public.

**pub·lish** (pŭb′lĭsh) *v.* **-lished, -lish·ing, -lish·es.** [ME *publishen* < OFr. *publier* < LLat. *publicare,* to make public. —see PUBLICATION.] —*vt.* **1.** To issue and prepare (printed material) for public distribution or sale. **2.** To bring to public notice : ANNOUNCE. —*vi.* **1.** To issue a publication. **2.** To be the author of a published work or works. **—pub′lish·a·ble** *adj.* **—pub′lish·er** (pŭb′lĭ-shər) *n.*

**puc·coon** (pə-ko͞on′) *n.* [Algonquian *pocoon.*] **1. a.** A North American plant of the genus *Lithospermum,* yielding a red or yellow dye, esp. *L. canescens,* having orange flowers. **b.** A plant yielding a reddish dye, as the bloodroot. **2.** The dye from a puccoon.

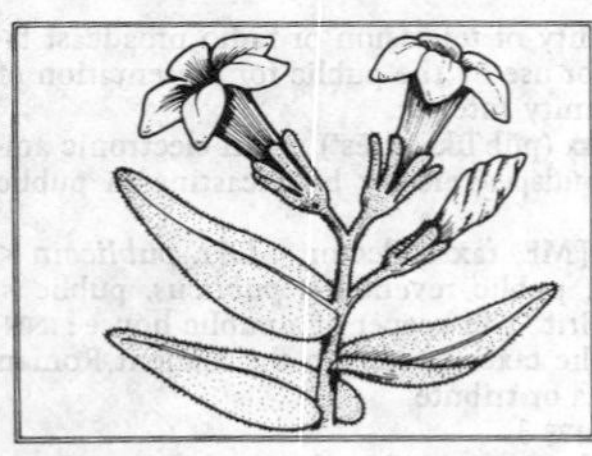

puccoon

**puce** (pyo͞os) *n.* [Fr. *(couleur) puce,* flea (color) < Lat. *pulex,* flea.] A deep red to dark grayish purple. **—puce** *adj.*

**puck** (pŭk) *n.* [Prob. < dial. *puck,* to strike, var. of POKE¹.] A hard rubber disk used in ice hockey.

**Puck** (pŭk) *n.* [ME *pouke* < OE *pūca.*] A mischievous sprite in English folklore.

**puck·a** (pŭk′ə) *n. var. of* PUKKA.

**puck·er** (pŭk′ər) *v.* **-ered, -er·ing, -ers.** [Perh. < POCKET.] —*vt.* To gather into small folds or wrinkles. —*vi.* To become contracted and wrinkled. **—puck′er** *n.*

**puck·ish** (pŭk′ĭsh) *adj.* Playful : impish. **—puck′ish·ly** *adv.* **—puck′ish·ness** *n.*

**pud·ding** (po͝od′ĭng) *n.* [ME < OFr. *boudin* < Lat. *botulus,* sausage.] **1. a.** A sweet dessert, usu. containing flour or a cereal product, that has been steamed, boiled, or baked. **b.** A mixture with a soft puddinglike consistency. **2.** A sausagelike preparation made with minced meat or other ingredients stuffed into a bag or skin and boiled.

**pudding stone** *n.* CONGLOMERATE 2.

**pud·dle** (pŭd′l) *n.* [ME *podel,* dim. of OE *pudd,* ditch.] **1.** A small pool of liquid, esp. of muddy water. **2.** A tempered paste of wet clay and sand used as waterproofing. —*v.* **-dled, -dling, -dles.** —*vt.* **1.** To make muddy. **2.** To work (clay or sand) into a thick, watertight paste. **3.** To process (impure metal) by puddling. —*vi.* To splash or dabble in or as if in a puddle. **—pud′dly** *adj.*

**pud·dler** (pŭd′lər) *n.* One who puddles iron or clay.

**pud·dling** (pŭd′lĭng) *n.* **1.** Purification of impure metal, esp. pig iron, by agitation of a molten bath of the metal in an oxidizing atmosphere. **2.** Compaction of wet material, as clay, to make a watertight paste.

**pu·den·cy** (pyo͞od′n-sē) *n.* [LLat. *pudentia* < Lat. *pudēre,* to be ashamed.] Modesty.

**pu·den·dum** (pyo͞o-dĕn′dəm) *n., pl.* **-da** (-də) [Lat., neuter gerund. of *pudēre,* to be ashamed.] **1.** A woman's external genital organs. **2. pudenda.** External genitals of either sex. **—pu·den′dal** *adj.*

**pudg·y** (pŭj′ē) *adj.* **-i·er, -i·est.** [Orig. unknown.] Short and fat : CHUBBY. **—pudg′i·ness** *n.*

**pueb·lo** (pwĕb′lō) *n., pl.* **-los.** [Sp., pueblo, people < Lat. *populus,* people.] **1.** A community dwelling, occas. several stories high, built of adobe or stone by Indian tribes of the southwestern United States. **2. Pueblo** *pl.* **Pueblo** or **-los.** A member of an Indian tribe, as the Hopi or Zuñi, inhabiting pueblos. **3.** An Indian village of the southwestern United States.

**pu·er·ile** (pyo͞o′ər-əl, pyo͝or′əl, -īl′) *adj.* [Fr. *puéril* < Lat. *puerilis* < *puer,* child.] **1.** Belonging to childhood : JUVENILE. **2.** Childish : immature. **—pu′er·ile·ly** *adv.* **—pu′er·ile·ness, pu′er·il′i·ty** (-ĭl′ĭ-tē) *n.*

**pu·er·per·al** (pyo͞o-ûr′pər-əl) *adj.* [< Lat. *puerperus,* bearing children : *puer,* child + *parēre,* to bear.] Associated with, resulting from, or occurring after childbirth.

**puerperal fever** *n.* Infection of the endometrium and of the blood stream after childbirth.

**pu·er·pe·ri·um** (pyo͞o′ər-pîr′ē-əm) *n., pl.* **-ri·a** (-ē-ə) [Lat., childbirth < *puerperus,* bearing children. —see PUERPERAL.] **1.** The state of a woman while bearing a child or just thereafter. **2.** The approximate six-week period from childbirth to return of normal uterine size.

**puff** (pŭf) *n.* [ME *puffen* < OE *pyffan.*] **1. a.** A short forceful exhalation of breath. **b.** A short sudden gust of wind. **c.** A sudden brief emission of air, smoke, or vapor. **d.** A short sibilant sound made by a puff. **2.** An amount of smoke, vapor, or similar material released in a puff. **3.** An act of drawing in and exhaling the breath, as in smoking tobacco. **4.** A swelling or rounded protuberance. **5.** A light inflated pastry, often filled with cream or custard. **6.** A soft light pad for applying cosmetic powder. **7.** A soft roll of hair forming part of a coiffure. **8.** A gathered and protruding section of fabric. **9.** A light padded covering, esp. for a bed. **10.** A commendable or flattering recommendation. —*v.* **puffed, puff·ing, puffs.** —*vi.* **1.** To blow in puffs. **2.** To come forth in puffs. **3.** To breathe rapidly and forcefully. **4.** To emit puffs. **5.** To take puffs on a cigarette, pipe, or cigar. **6.** To swell or seem to swell, as with air or pride. —*vt.* **1.** To emit in puffs. **2.** To impel with puffs. **3.** To smoke (e.g., a cigar). **4.** To inflate or distend. **5.** To fill with conceit or pride. **6.** To publicize with exaggerated praise : HYPE. **—puff′i·ly** *adv.* **—puff′i·ness** *n.* **—puff′y** *adj.*

**puff adder** *n.* [So called because it inflates its body when excited.] **1.** A venomous African viper, *Bitis arietans,* with crescent-shaped yellowish markings. **2.** The hognose snake.

**puff·ball** (pŭf′bôl′) *n.* **1.** Any of various fungi of the genus *Lycoperdon* and related genera, having a ball-shaped fruiting body that when broken open releases the enclosed spores in puffs of dust. **2.** *Informal.* The rounded head of a dandelion that has gone to seed.

**puffed-up** (pŭft′ŭp′) *adj.* Marked by exaggerated self-importance : POMPOUS.

**puff·er** (pŭf′ər) *n.* **1.** One that puffs. **2.** Any of various marine fishes of the family Tetraodontidae, capable of swelling up.

**puff·er·y** (pŭf′ə-rē) *n.* Favorable and often overly flattering publicity, esp. for promotional purposes.

**puf·fin** (pŭf′ĭn) *n.* [ME *poffoun.*] Any of several sea birds of the genera *Fratercula* and *Lunda* of northern regions, with black and white plumage and a vertically flattened, brightly colored bill.

**puff pastry** *n.* Dough rolled and folded in layers that expands in baking to form light, flaky pastry.

**pug¹** (pŭg) *n.* [Orig. unknown.] **1.** A small dog orig. bred in China, having a snub nose, a wrinkled face, a square body, short smooth hair, and a curled tail. **2.** A pug nose.

**pug²** (pŭg) *n.* [Orig. unknown.] **1.** Clay ground and kneaded with water into a plastic consistency for forming bricks or pottery. **2.** A machine for grinding and mixing clay. —*vt.* **pugged, pug·ging, pugs. 1.** To work or knead (clay) with water. **2.** To fill in with clay or mortar. **3.** To make soundproof by packing or covering with clay, sawdust, mortar, or felt.

**pug³** (pŭg) *n.* [Hindi *pag,* prob. < Skt. *padakam,* step, pace < *padam,* foot.] A footprint, esp. of an animal.

**pug⁴** (pŭg) *n.* [Short for PUGILIST.] *Slang.* A boxer.

**pug·gree** (pŭg′rē) *also* **pug·ga·ree** or **pug·a·ree** (pŭg′ə-rē) *n.* [Hindi *pagrī.*] A light scarf wrapped around the crown of a hat or sun helmet.

**pu·gi·lism** (pyo͞o′jə-lĭz′əm) *n.* [< Lat. *pugil,* boxer.] The skill or practice of fighting with the fists : BOXING. **—pu′gi·list** *n.* **—pu′gi·lis′tic** *adj.*

**pu·gil stick** (pyo͞o′jəl) *n.* [Lat. *pugil,* boxer + STICK.] A pole with padded ends used in the armed forces to simulate bayonet fighting.

**pug·mark** (pŭg′märk′) *n.* PUG³.

**pug·na·cious** (pŭg-nā′shəs) *adj.* [Lat. *pugnax, pugnac-* < *pugnare,* to fight < *pugnus,* fist.] Having a quarrelsome, combative disposition. **—pug·na′cious·ly** *adv.* **—pug·na′cious·ness, pug·nac′i·ty** (-năs′ĭ-tē) *n.*

**pug nose** *n.* [Prob. < PUG¹.] A short nose somewhat flattened and upturned. **—pug′-nosed′** *adj.*

**puis·ne** (pyo͞o′nē) [Norman Fr. *puisne* < OFr. : *puis,* afterward (< Lat. *post*) + *ne,* born < Lat. *nasci,* to be born.] *Chiefly Brit.* Lower in rank : JUNIOR. —*n.* One of lesser rank, esp. an associate judge.

**puis·sance** (pwĭs′əns, pyo͞o′ĭ-səns, pyo͞o-ĭs′əns) *n.* Strength : might. **—puis′sant** *adj.* **—puis′sant·ly** *adv.*

**puke** (pyo͞ok) [Prob. imit.] *Slang.* —*vi. & vt.* **puked, puk·ing, pukes.** To vomit or vomit up. —*n.* **1.** Vomit. **2.** An act of vomiting.

**puk·ka** *also* **puck·a** (pŭk′ə) *adj.* [Hindi *pakkā,* cooked, ripe, firm < Skt. *pakva-,* p.part. of *pacati,* he cooked.] **1.** Authentic : genuine. **2.** Superior : first-rate.

**pul** (po͞ol) *n., pl.* **puls** or **pu·li** (po͞o′lē) [Pers. *pūl.*] —See table at CURRENCY.

**pu·la** (po͞o'lə) *n., pl.* **pula** [Native word in Botswana.] —See table at CURRENCY.

**pul·chri·tude** (pŭl'krĭ-to͞od', -tyo͞od') *n.* [ME *pulcritude* < Lat. *pulchritūdō* < *pulcher,* beautiful.] Physical beauty. **—pul'chri·tu'di·nous** *adj.*

**pule** (pyo͞ol) *vi.* **puled, pul·ing, pules.** [Perh. < Fr. *piauler.*] To whimper : whine. **—pul'er** *n.*

**pu·li¹** (po͝ol'ē, pyo͞o'lē) *n., pl.* **pu·lis** *or* **pu·lik** (po͝ol'ēk, pyo͞o'lēk). [Hung.] A long-haired sheep dog of a Hungarian breed.

**pu·li²** (po͞o'lē) *n. var. pl. of* PUL.

**pu·li·cide** (pyo͞o'lĭ-sīd') *n.* [Lat. *pulex, pulic-,* flea + -CIDE.] A flea-killing agent.

**pu·lik** (po͝ol'ēk, pyo͞o'lēk) *n. var. pl. of* PULI¹.

**Pu·lit·zer Prize** (po͝ol'ĭt-sər, pyo͞o'lĭt-) *n.* Any of several awards established by Joseph Pulitzer and conferred annually for accomplishment in various fields of American journalism, music, and literature.

**pull** (po͝ol) *v.* **pulled, pull·ing, pulls.** [ME *pullen* < OE *pullian.*] *—vt.* **1.** To apply force to so as to cause or tend to cause motion toward the source of the force. **2.** To remove from a fixed position : EXTRACT <*pull* nails> **3.** To tug at : JERK. **4.** To rend or tear : RIP. **5.** To stretch (e.g., taffy) repeatedly. **6.** To strain (e.g., a muscle) injuriously. **7.** *Informal.* To draw : attract <an entertainer who *pulls* big crowds> **8.** *Slang.* To draw out (a knife or gun) in readiness for use. **9.** *Informal.* To use less than full force in delivering (a punch). **10.** *Baseball.* To hit (a ball) in the direction one is facing when the swing is carried through. **11. a.** To operate (an oar) in rowing. **b.** To transport or propel by rowing. **c.** To be rowed by <a boat that *pulls* six oars> **12.** To rein in (a horse) to keep from winning a race. **13.** To produce (a print or impression) from type. *—vi.* **1.** To exert force in pulling something. **2.** To move <The train *pulled* out of the station.> **3.** To drink or inhale deeply. **4.** To row a boat. **—pull away. 1.** To move away or backward : WITHDRAW. **2.** To move ahead <The runner *pulled away* and took the lead.> **—pull down. 1.** To destroy : demolish <*pull down* an old barn> **2.** To reduce to a lower level. **3.** To depress, as in spirits or health. **4.** *Informal.* To draw (money) as wages. **—pull for.** To work, hope, or cheer for the success of. **—pull in. 1.** To arrive at a destination <They *pulled in* at noon.> **2.** To rein in : RESTRAIN. **3.** *Informal.* To arrest <was *pulled in* for burglary> **—pull off.** *Informal.* To perform in spite of difficulties : BRING OFF. **—pull out.** To withdraw from a commitment or situation. **—pull over.** To bring a vehicle to a stop at a curb or at the side of a road. **—pull through.** To come or bring successfully through illness or trouble. **—pull up. 1.** To bring or come to a halt. **2.** To move to a place or position ahead, as in a race. *—n.* **1.** The act or process of pulling. **2.** Force exerted in pulling or necessary to overcome resistance in pulling. **3.** A sustained effort <a long *pull* up the hill> **4.** Something, as a knob on a drawer, used for pulling. **5.** A deep inhalation or draft, as on a cigarette. **6.** *Slang.* A means of gaining special influence : ADVANTAGE <has *pull* with the owner> **7.** *Informal.* Ability to draw or attract : APPEAL <box office *pull*> **—pull a fast one.** *Informal.* To play a trick : CHEAT. **—pull (one's) punch (*or* punches).** To refrain from using all the force and resources at one's disposal. **—pull (oneself) together.** To regain one's composure. **—pull (one's) weight.** To do one's own share, as of work. **—pull (someone's) leg.** To play a joke on : TEASE. **—pull strings (*or* wires).** To exert secret control or influence so as to gain an end. **—pull the rug (out) from under.** To remove all assistance and support from, usu. suddenly. **—pull the wool over (someone's) eyes.** To hoodwink : deceive. **—pull together.** To make a cooperative effort. **—pull up stakes.** To move out : LEAVE <*pulled up stakes* and went West> **—pull'er** *n.*

**pull·back** (po͝ol'băk') *n.* **1.** The act or process of pulling something back, esp. an orderly troop withdrawal. **2.** A device for holding or drawing back <curtain *pullbacks*>

**pull date** *n.* A date stamped on a packaged food product beyond which it should not be sold.

**pul·let** (po͝ol'ĭt) *n.* [ME *pulet* < OFr. *poulet,* dim. of *poul,* cock and *poule,* hen, both < Lat. *pullus,* chicken.] A young hen, esp. of the common domestic fowl, usu. less than one year old.

**pul·ley** (po͝ol'ē) *n., pl.* **-leys.** [ME *poley* < OFr. *polie,* ult. < Gk. *polos,* axis.] **1.** A simple device for changing the direction and point of application of a pulling force, esp. for lifting weight, consisting of a wheel with a grooved rim in which a pulled rope or chain is run. **2.** A wheel turned by or driving a belt.

**Pull·man** (po͝ol'mən) *n.* [After George M. *Pullman* (1831–1897).] **1.** A railroad parlor car or sleeping car. **2.** A large suitcase.

**pull-on** (po͝ol'ŏn', -ôn') *n.* A garment, as a sweater or pants, designed to be pulled on.

**pul·lo·rum disease** (pə-lôr'əm, -lōr'-) *n.* [NLat. *pullorum,* specific epithet of *Salmonella pullorum* < Lat., genitive pl. of *pullus,* chicken.] A severe contagious diarrhea of young poultry, caused by the bacterium *Salmonella pullorum.*

**pull·out** (po͝ol'out') *n.* **1.** A withdrawal, esp. of troops. **2.** The change from a dive to level flight in aviation. **3.** Something designed to be pulled out.

**pull·o·ver** (po͝ol'ō'vər) *n.* A garment, as a sweater, that must be put on by being drawn over the head.

**pul·lu·late** (pŭl'yə-lāt') *vi.* **-lat·ed, -lat·ing, -lates.** [Lat. *pullulare, pullulat-* < *pullulus,* dim. of *pullus,* chicken.] **1.** To put forth sprouts : GERMINATE. **2.** To breed abundantly or rapidly. **3.** To swarm : teem. **—pul'lu·la'tion** *n.* **—pul'lu·la'tive** *adj.*

**pull-up** (po͝ol'ŭp') *n.* An exercise for strengthening the arms, performed by hanging by the hands from an overhead bar and pulling the body upward until the chin is even with or above the bar.

**pul·mo·nar·y** (po͝ol'mə-nĕr'ē, pŭl'-) *adj.* [Lat. *pulmonarius* < *pulmo,* lung.] **1.** Of or relating to the lungs. **2.** PULMONATE 1.

**pulmonary artery** *n.* An artery in which blood travels directly from the heart to the lungs.

**pulmonary vein** *n.* One of four veins in which blood travels directly from the lungs to the heart.

**pul·mo·nate** (po͝ol'mə-nāt', pŭl'-) *adj.* [NLat. *pulmonatus* < Lat. *pulmo,* lung.] **1.** Having lungs or lunglike organs. **2.** Pertaining to the Pulmonata, an order of gastropods including snails and slugs, in which the mantle cavity is modified to function as a lung. *—n.* A member of the Pulmonata.

**pul·mon·ic** (po͝ol-mŏn'ĭk, pŭl-) *adj.* PULMONARY 1.

**pulp** (pŭlp) *n.* [Lat. *pulpa,* flesh.] **1.** A soft, moist, shapeless mass of matter. **2.** The soft moist part of fruit. **3.** A mass of pressed vegetable matter <orange *pulp*> **4.** The soft pith forming the contents of the stem of a plant. **5.** A mixture of cellulose material, as wood, paper, and rags, ground up and moistened to make paper. **6.** The soft inner structure of a tooth, consisting of nerve and blood vessels. **7.** A mixture of powdered ore and water. **8.** A magazine or book having rough-textured paper made from wood pulp and often containing lurid subject matter. *—v.* **pulped, pulp·ing, pulps.** *—vt.* **1.** To reduce to pulp. **2.** To remove the pulp from. *—vi.* To become reduced to pulp. **—pulp'i·ness** *n.* **—pulp'ous** (pŭl'pəs), **pulp'y** *adj.*

**pul·pit** (po͝ol'pĭt, pŭl'-) *n.* [ME < Lat. *pulpitum,* platform.] **1.** An elevated platform, lectern, or stand used in preaching or conducting a religious service. **2.** A raised platform, as one used by harpooners in a whaling boat. **3. a.** The clergy. **b.** The ministry of preaching.

**pulp·wood** (pŭlp'wo͝od') *n.* Soft wood, as spruce, aspen, or pine, used in making paper.

**pul·que** (po͝ol'kā', -kē, po͞ol'-) *n.* [Mex. Sp. < Nahuatl *poliuhqui,* decomposed.] A fermented milky beverage made in Mexico from various species of agave.

**pul·sar** (pŭl'sär') *n.* [PULSE + (STELL)AR.] *Astron.* Any of several very short-period variable galactic radio sources gen. believed to be rotating neutron stars.

**pul·sate** (pŭl'sāt') *vi.* **-sat·ed, -sat·ing, -sates.** [Lat. *pulsare, pulsat-,* freq. of *pellere,* to beat.] **1.** To expand and contract rhythmically : THROB. **2.** To quiver.

**pul·sa·tile** (pŭl'sə-təl, -tīl') *adj.* [Med. Lat. *pulsatilis* < Lat. *pulsare,* freq. of *pellere,* to beat.] Vibrating : pulsating.

**pul·sa·tion** (pŭl-sā'shən) *n.* **1.** An act of pulsating. **2.** A single beat, throb, or vibration.

**pul·sa·tor** (pŭl'sā'tər, pŭl-sā'-) *n.* A pulsating device.

**pul·sa·to·ry** (pŭl'sə-tôr'ē, -tōr'ē) *adj.* Having rhythmical vibration or movement.

**pulse¹** (pŭls) *n.* [ME *pous* < OFr. < Lat. *pulsus* < *pellere,* to beat.] **1.** *Physiol.* Rhythmical throbbing of arteries generated by regular contractions of the heart. **2.** A regular or rhythmical beating. **3.** A single beat or throb. **4.** *Physics & Electron.* A transient amplification or intensification of a characteristic of a system, esp. of a wave characteristic, followed by return to equilibrium or steady state <a signal *pulse*><beam *pulse*> **5.** Perceptible emotions or opinions of a group of people. *—vi.* **pulsed, puls·ing, puls·es.** To pulsate. **—pulse'less** *adj.*

**pulse²** (pŭls) *n.* [ME *pols* < OFr. < Lat. *puls,* pottage of meal and pulse.] **1.** The edible seeds of certain pod-bearing plants, as peas and beans. **2.** A plant yielding pulse.

**pulse·jet** (pŭls'jĕt') *n.* A jet engine in which air intake and combustion occur intermittently, producing rapid periodic bursts of thrusts.

**pulse modulation** *n.* Modulation by coded variation of the amplitude or other characteristic of wave pulses.

**pul·sim·e·ter** (pŭl-sĭm'ĭ-tər) *also* **pul·som·e·ter** (-sŏm'-) *n. Med.* An instrument for measuring the frequency or strength of the pulse.

**pul·som·e·ter** (pŭl-sŏm'ĭ-tər) *n.* **1.** A pump for raising water by the pulsed condensation of steam. **2.** *var. of* PULSIMETER.

**pul·ver·a·ble** (pŭl'vər-ə-bəl) *adj.* Capable of being pulverized.

**pul·ver·ize** (pŭl'və-rīz') *v.* **-ized, -iz·ing, -iz·es.** [LLat. *pulverizare* < Lat. *pulvis,* dust.] *—vt.* **1.** To crush, pound, or grind to a powder or dust. **2.** To demolish <*pulverized* the old building> *—vi.* To be ground or reduced to powder or dust. **—pul'ver·iz'a·ble** *adj.* **—pul'ver·i·za'tion** *n.* **—pul'ver·iz'er** *n.*

**pul·ver·u·lent** (pŭl-vĕr'yə-lənt, -vĕr'ə-) *adj.* [Lat. *pulverulentus,* dusty < *pulvis,* dust.] **1.** Made of, covered with, or crumbling to fine powder or dust. **2.** Dusty : crumbly.

**pul·vil·lus** (pŭl-vĭl′əs) *n., pl.* **-vil·li** (-vĭl′ī′) [Lat., dim. of *pulvinus,* cushion.] One of the soft, cushionlike pads between the claws of an insect's foot.

**pul·vi·nate** (pŭl′və-nāt′) *also* **pul·vi·nat·ed** (-nā′tĭd) *adj.* [Lat. *pulvinatus* < *pulvinus,* cushion.] **1.** Shaped like a cushion. **2.** *Bot.* Having a swelling at the base. —Used of a leafstalk.

**pul·vi·nus** (pŭl-vī′nəs, -vē′-) *n., pl.* **-ni** (nī′) [Lat., cushion.] *Bot.* A swelling of the stem at the base of a leafstalk.

**pu·ma** (pyo͞o′mə, po͞o′-) *n.* [Sp. < Quechua *poma.*] The mountain lion.

**pum·ice** (pŭm′ĭs) *n.* [ME *pomys* < OFr. *pomis* < Lat. *pumex.*] A porous lightweight volcanic rock used in solid form as an abrasive and in powdered form as a polish and abrasive. —*vt.* **-iced, -ic·ing, -ic·es.** To clean, polish, or smooth with pumice. **—pu·mi′ceous** (pyo͞o-mĭsh′əs, pə-) *adj.* **—pum′ic·er** *n.*

**pum·mel** (pŭm′əl) *vt.* **-meled, -mel·ing, -mels** *also* **-melled, -mel·ling, -mels.** To beat : pommel. **—pum′mel** *n.*

**pump¹** (pŭmp) *n.* [ME *pumpe.*] **1.** A device or machine for transferring a gas or liquid from a source or container through tubes or pipes to another container or receiver. **2. a.** *Biochem.* A biochemical mechanism for transporting ions, atoms, or molecules against a concentration gradient by an expenditure of energy. **b.** The process of such transport. **3.** *Physics.* Electromagnetic radiation used to raise atoms or molecules to a higher energy level. **4.** HEART 1a. —*v.* **pumped, pump·ing, pumps.** —*vt.* **1.** To raise or cause to flow by means of a pump. **2.** To inflate with gas by means of a pump <*pump* up an air mattress> **3.** To remove the water from. **4.** To cause to operate with the up-and-down motion of a pump handle. **5.** To eject, propel, or insert with or as if with a pump. **6.** *Physics.* To supply (a laser) with sufficient energy to achieve population inversion. **7.** *Physics.* To raise atoms or molecules to a higher energy level by exposing them to electromagnetic radiation at a resonant frequency. **8.** *Biochem.* To transport ions, atoms, or molecules against a concentration gradient by the expenditure of chemically stored energy. **9.** To question persistently or closely <The police *pumped* the suspect.> —*vi.* **1.** To operate a pump. **2.** To raise or move gas or liquid with a pump. **3.** To move up and down in the manner of a pump handle. **—pump′er** *n.*

**pump²** (pŭmp) *n.* [Orig. unknown.] A low-cut shoe without fastenings.

**pumped storage** *n.* A system of generating electricity using hydroelectric power in which electricity is generated during the hours of peak consumption by using water that has been pumped into an elevated reservoir during the hours of low consumption.

**pum·per·nick·el** (pŭm′pər-nĭk′əl) *n.* [G.] A dark sourdough bread made from whole coarsely ground rye.

**pump·kin** (pŭmp′kĭn, pŭm′-, pŭng′-) *n.* [Alteration of obs. *pumpion* < OFr. *pompon* < Lat. *pepo* < Gk. *pepōn,* large melon < *pepōn,* ripe < *peptein,* to ripen.] **1. a.** A coarse trailing vine, *Cucurbita pepo,* widely cultivated for its fruit. **b.** The large pulpy round fruit of the pumpkin vine, having a thick, orange-yellow rind and numerous seeds. **c.** A vine, *C. maxima* or *C. moschata,* bearing large pumpkin-like squashes. **2.** A moderate to strong orange.

**pump·kin·seed** (pŭmp′kĭn-sēd′, pŭm′-, pŭng′-) *n.* **1.** The seed of the pumpkin. **2.** A North American sunfish, *Lepomis gibbosus,* having brightly colored markings.

**pun** (pŭn) *n.* [Orig. unknown.] A play on words, occas. on different senses of the same word and occas. on the similar sense or sound of different words. —*vi.* **punned, pun·ning, puns.** To make a pun.

**punch¹** (pŭnch) *n.* [Short for PUNCHEON¹.] **1.** A tool for circular or shaped piercing. **2.** A tool for forcing a pin, bolt, or rivet in or out of a hole. **3.** A tool for stamping a design on a surface. **4.** A countersink. —*v.* **punched, punch·ing, punch·es.** —*vt.* To use a punch on : mark or perforate. —*vi.* To use a punch.

**†punch²** (pŭnch) *vt.* **punched, punch·ing, punch·es.** [ME *punchen.*] **1.** To hit with a sharp blow of the fist. **2.** To prod or poke with a stick. **3.** *Western U.S.* To herd (cattle). **—punch in.** To check in formally at a job before a day's work. **—punch out. 1.** To check out formally at a job before leaving after a day's work. **2.** *Slang.* To eject from a military aircraft. —*n.* **1.** A blow with the fist. **2.** Energy or drive. **—punch′er** *n.* **—punch′less** *adj.*

**punch³** (pŭnch) *n.* [Perh. < Hindi *pānch* < Skt. *pañca-,* five (from its orig. having been prepared from five ingredients).] A sweetened beverage of fruit juices, often spiced, usu. with a wine or liquor base.

**Punch** (pŭnch) *n.* [Short for PUNCHINELLO.] The contentious hook-nosed husband of Judy in the comic puppet show *Punch and Judy.* **—pleased as Punch.** Highly pleased.

**punch·board** (pŭnch′bôrd′, -bōrd′) *n.* A small, usu. rectangular board, used as a game of chance, that contains many holes each filled with a folded slip of paper that when punched out indicates a designated prize, win, or loss.

**punch bowl** *n.* A large bowl for serving a beverage, as punch.

**punch card** *also* **punched card** *n.* A computer card punched with holes or notches to represent letters and numbers or with a pattern of holes to represent related data.

**punch-drunk** (pŭnch′drŭngk′) *adj.* **1.** Acting confused, bewildered, or dazed. **2.** Showing signs of brain damage caused by repeated blows to the head in boxing.

**pun·cheon¹** (pŭn′chən) *n.* [ME *ponson,* a sharp tool < OFr. *poinçon* < Lat. *pungere,* to prick.] **1.** A short wooden upright used in structural framing. **2.** A piece of broad roughly dressed heavy timber, with one face finished flat. **3.** A punching, stamping, or perforating tool, esp. one used by a goldsmith.

**pun·cheon²** (pŭn′chən) *n.* [OFr. *poinçon.*] **1.** A cask with a capacity of approx. 318 liters, or 84 U.S. gallons. **2.** The amount of liquid contained in a puncheon.

**punch·er** (pŭn′chər) *n.* **1.** One that punches. **2.** A cowboy.

**Pun·chi·nel·lo** (pŭn′chə-nĕl′ō) *n., pl.* **-los** *or* **-loes.** [Orig. unknown.] **1.** The short fat clown or buffoon in an Italian puppet show. **2.** One thought to resemble a short fat puppet.

**Punchinello**

**punching bag** *n.* A stuffed or inflated bag that is usu. hung up so that it can be punched with the fists for exercise.

**punch line** *n.* The climax of a joke or humorous story.

**punch-out** (pŭnch′out′) *n.* A section of material, as cardboard, scored or perforated so that it may easily be pushed out.

**punch press** *n.* A power press that can be fitted with various dies, as for metalworking.

**punch tape** *n.* Paper tape in which holes representing data to be processed by computer are punched.

**punch-up** (pŭnch′ŭp′) *n.* *Chiefly Brit.* A fist fight.

**punch·y** (pŭn′chē) *adj.* **-i·er, -i·est.** Punch-drunk. **—punch′i·ly** *adv.* **—punch′i·ness** *n.*

**punc·tate** (pŭngk′tāt′) *also* **punc·tat·ed** (-tā′tĭd) *adj.* [NLat. *punctatus* < Lat. *punctum,* prick mark < *pungere,* to prick.] Having tiny spots, points, or depressions <a *punctate* leaf> **—punc·ta′tion** *n.*

**punc·til·i·o** (pŭngk-tĭl′ē-ō′) *n., pl.* **-os.** [Ital. *punctiglio,* dim. of *punto,* point < Lat. *punctum* < *pungere,* to prick.] **1.** A fine point of etiquette. **2.** Meticulous observance of formalities.

**punc·til·i·ous** (pŭngk-tĭl′ē-əs) *adj.* **1.** Attentive to the finer points of etiquette and formal conduct. **2.** Exact : scrupulous. **—punc·til′i·ous·ly** *adv.* **—punc·til′i·ous·ness** *n.*

**punc·tu·al** (pŭngk′cho͞o-əl) *adj.* [Med. Lat. *punctualis* < Lat. *punctum,* point < *pungere,* to prick.] **1.** Acting or arriving exactly at the designated time : PROMPT. **2.** Paid or accomplished at or by the appointed time. **3.** Precise : exact. **4.** Confined to or having the nature of a point in space. **—punc′tu·al′i·ty** (-ăl′ĭ-tē), **punc′tu·al·ness** *n.* **—punc′tu·al·ly** *adv.*

☆ ***syns:*** PUNCTUAL, PROMPT, TIMELY *adj. core meaning* : occurring, acting, or performed exactly at the time appointed <a *punctual* arrival> ***ant:*** unpunctual

**punc·tu·ate** (pŭngk′cho͞o-āt′) *v.* **-at·ed, -at·ing, -ates.** [Med. Lat. *punctuare, punctuat-* < Lat. *punctum,* point < *pungere,* to prick.] —*vt.* **1.** To provide (a text) with punctuation marks. **2.** To interrupt periodically. **3.** To emphasize or stress. —*vi.* To use punctuation. **—punc′tu·a′tive** *adj.* **—punc′tu·a′tor** *n.*

**punc·tu·a·tion** (pŭngk′cho͞o-ā′shən) *n.* **1. a.** Use of standard signs and marks in writing and printing to separate words into sentences, clauses, and phrases in order to clarify meaning. **b.** The marks so used. **2.** An act or instance of punctuating.

**punctuation mark** *n.* A sign or mark, as the comma (,), used to punctuate a text.

**punc·ture** (pŭngk′chər) *v.* **-tured, -tur·ing, -tures.** [< Lat. *punctura,* a pricking < *pungere,* to prick.] —*vt.* **1.** To pierce with a pointed object. **2.** To make (a hole) by piercing. **3.** To cause to collapse by piercing. **4.** To depreciate or deflate <remarks that *punctured* their enthusiasm> —*vi.* To be pierced or punctured. —*n.* **1.** An act or instance of puncturing. **2.** A depression or hole made by a sharp object, esp. a hole in a pneumatic tire. **—punc′tur·a·ble** *adj.*

**puncture weed** *n.* A prostrate weed, *Tribulus terrestris* native to Europe, bearing fruit with stout divergent spines.

**pun·dit** (pŭn′dĭt) *n.* [Hindi *paṇḍit* < Skt. *paṇḍitaḥ,* a learned man, of Dravidian orig.] **1.** A Brahmanic scholar. **2.** A learned person. **3.** An authority or critic. **—pun′dit·ry** *n.*

**†pung** (pŭng) *n.* [Short for *tompong,* of Algonquian orig.] *New England.* A low one-horse box sleigh.

**pun·gent** (pŭn′jənt) *adj.* [Lat. *pungens, pungent-*, pr. part. of *pungere*, to sting.] **1.** Affecting the organs of smell or taste with a sharp, acrid sensation. **2.** Incisive : caustic <*pungent* criticism> **3.** Pointed <a *pungent* leaf> **—pun′gen·cy** *n.* **—pun′gent·ly** *adv.*

☆ ***syns:*** PUNGENT, PIQUANT, SHARP, SPICY, ZESTY *adj. core meaning* : affecting the organs of taste or smell with a strong, acrid sensation <a *pungent* aroma><a *pungent* sauce> ***ant:*** bland

**Pu·nic** (pyōō′nĭk) *adj.* [Lat. *Punicus* < *Poenus*, a Carthaginian < Gk. *Phoinix*, Phoenician.] **1.** Of or relating to ancient Carthage or its people. **2.** Treacherous : untrustworthy. —*n.* The dialect of Phoenician spoken in ancient Carthage.

**pun·ish** (pŭn′ĭsh) *v.* **-ished, -ish·ing, -ish·es.** [ME *punissen* < OFr. *punir* < Lat. *poenire* < *poena*, punishment < Gk. *poinē*.] —*vt.* **1.** To subject (someone) to a penalty for a crime, fault, or bad behavior. **2.** To inflict a penalty on a wrongdoer for (an offense). **3.** To handle harshly : INJURE. **4.** *Informal.* To deplete (a stock or supply) heavily. —*vi.* To give punishment. **—pun′ish·a·bil′i·ty** *n.* **—pun′ish·a·ble** *adj.* **—pun′ish·er** *n.*

☆ ***syns:*** PUNISH, CORRECT, DISCIPLINE *v. core meaning* : to subject (someone) to a penalty for a wrong <was *punished* for stealing automobiles>

**pun·ish·ment** (pŭn′ĭsh-mənt) *n.* **1.** An act of punishing or the state of being punished. **2.** A penalty imposed for wrongdoing. **3.** *Informal.* Rough handling.

**pu·ni·tive** (pyōō′nĭ-tĭv) *adj.* [Fr. *punitif* < Med. Lat. *punitivus* < Lat. *punire*, to punish, var. of *poenire*. —see PUNISH.] Inflicting or aiming to inflict punishment <*punitive* measures> **—pu′ni·tive·ly** *adv.* **—pu′ni·tive·ness** *n.*

**punitive damages** *pl.n. Law.* Damages awarded by a court to a plaintiff as additional punishment to a defendant for a serious wrong.

**pu·ni·to·ry** (pyōō′nĭ-tôr′ē, -tōr′ē) *adj.* [Lat. *punitus*, punished (< p.part. of *punire*) + -ORY.] Punitive.

**Pun·ja·bi** *also* **Pan·ja·bi** (pŭn-jä′bē, -jăb′ē) *n.* **1.** A native of the Punjab. **2.** An Indic language spoken in the Punjab.

**punk**[1] (pŭngk) *n.* [Orig. unknown.] **1.** Dry, decayed wood, used as tinder. **2.** A substance that smolders when ignited, used for lighting fireworks. **3.** Chinese incense.

**punk**[2] (pŭngk) *n.* [Orig. unknown.] **1.** *Slang.* **a.** An inexperienced or immature person. **b.** A young tough. **2.** *Slang.* **a.** Punk rock. **b.** A punk rocker. **3.** *Archaic.* A whore. —*adj. Slang.* **1.** Of inferior quality : VALUELESS. **2.** Weak in spirits or health. **3.** Of or pertaining to a style of dress worn by punk rockers and typified by bizarre make-up and outlandish, shocking clothing.

**pun·ka** *or* **pun·kah** (pŭng′kə) *n.* [Hindi *pankhá* < Skt. *pakṣakaḥ*, fan < *pakṣaḥ*, wing.] A fan used esp. in India, made of a palm frond or strip of cloth hung from the ceiling and moved by a servant.

**punk·ie** *also* **punk·y** *or* **punk·ey** (pŭng′kē) *n., pl.* **-ies** *also* **-eys.** [< PUNK[1].] A tiny winged biting insect of the genus *Culicoides* and related genera.

†**pun·kin** (pŭng′kĭn) *n. Regional.* A pumpkin.

**punk rock** *n.* Hard-driving rock music marked by extremely bitter treatment of alienation and social discontent.

**punk rocker** *n. Slang.* **1.** A performer of punk rock music. **2.** A wearer of punk fashions.

**punk·y** (pŭng′kē) *n. var. of* PUNKIE.

**pun·ster** (pŭn′stər) *n.* A maker of puns.

**punt**[1] (pŭnt) *n.* [MLG *punte*, ferryboat < Lat *ponto*, pontoon < *pons*, bridge.] An open flat-bottomed boat with squared ends, moved by a long pole and used in shallow waters. —*v.* **punt·ed, punt·ing, punts.** —*vt.* **1.** To propel (a boat) with a pole. **2.** To carry in a punt. —*vi.* To go in a punt. **—punt′er** *n.*

**punt**[2] (pŭnt) [Orig. unknown.] *Football.* —*n.* A kick in which the ball is dropped from the hands and kicked before it touches the ground. —*v.* **punt·ed, punt·ing, punts.** —*vt.* To propel (a football) by means of a punt. —*vi.* To execute a punt. **—punt′er** *n.*

**punt**[3] (pŭnt) *vi.* **punt·ed, punt·ing, punts.** [Fr. *ponter* < *ponte*, point < Sp. *punto* < Lat. *punctum* < *pungere*, to prick.] **1.** To lay a bet against the bank, as in roulette. **2.** *Chiefly Brit.* To gamble. **—punt′er** *n.*

**pun·ty** (pŭn′tē) *n., pl.* **-ties.** [Var. of PONTIL.] An iron rod on which molten glass is handled.

**pu·ny** (pyōō′nē) *adj.* **-ni·er, -ni·est.** [OFr. *puisne*. —see PUISNE.] Inferior in strength, size, or importance. **—pu′ni·ly** *adv.* **—pu′ni·ness** *n.*

**pup** (pŭp) *n.* [Back-formation < PUPPY.] **1.** A young dog : PUPPY. **2.** The young of certain animals, as the seal. **3.** PUPPY 2. —*vi.* **pupped, pup·ping, pups.** To give birth to pups.

**pu·pa** (pyōō′pə) *n., pl.* **-pae** (-pē) *or* **-pas.** [Lat., girl.] The inactive stage in the metamorphosis of many insects, following the larval stage and preceding the adult form. **—pu′pal** *adj.*

**pu·pate** (pyōō′pāt′) *vi.* **-pat·ed, -pat·ing, -pates.** To become a pupa. **—pu·pa′tion** *n.*

**pu·pil**[1] (pyōō′pəl) *n.* [ME *pupille*, orphan < OFr. < Lat. *pupillus*, dim. of *pupus*, boy.] **1.** A student under the supervision of a teacher. **2.** *Law.* A minor under the supervision of a guardian.

**pu·pil**[2] (pyōō′pəl) *n.* [ME *pupilla* < Lat.] The apparently black circular aperture in the center of the iris of the eye. **—pu′pi·lar** *adj.*

▲ word history: The words *pupil*[1] and *pupil*[2] are ultimately derived from the same Latin source, *pupus*, meaning "a child." *Pupil*[1], "a student," is derived from *pupillus*, a diminutive formation from *pupus*. *Pupil*[2], "pupil of the eye," is derived from *pupilla*, the feminine form of the masculine *pupillus*. The pupil of the eye is so called because of the tiny image that can be seen reflected in it.

**pu·pil·age** *also* **pu·pil·lage** (pyōō′pə-lĭj) *n.* The state or period of being a pupil.

**pu·pil·lar·y**[1] (pyōō′pə-lĕr′ē) *adj.* Of or relating to a ward or a student.

**pu·pil·lar·y**[2] (pyōō′pə-lĕr′ē) *adj.* Of or affecting the pupil of the eye.

**pu·pip·a·rous** (pyōō-pĭp′ər-əs) *adj.* Producing well-developed young that are ready to pupate, as certain parasitic flies.

**pup·pet** (pŭp′ĭt) *n.* [ME *popet*, doll, ult. < Lat. *pupa*.] **1.** A small figure of a person or animal, having jointed parts animated from above by strings or wires : MARIONETTE. **2.** A figure having a cloth body and hollow head, designed to be fitted over and manipulated by the hand. **3.** A toy representing a human figure : DOLL. **4.** One whose behavior is controlled by the will of others.

**pup·pet·eer** (pŭp′ĭ-tîr′) *n.* One who operates and entertains with puppets or marionettes.

**pup·pet·ry** (pŭp′ĭ-trē) *n., pl.* **-ries.** **1.** The art of making puppets and presenting puppet shows. **2.** The actions of puppets.

**Pup·pis** (pŭp′ĭs) *n.* [Lat. < *puppis*, stern.] A constellation in the Southern Hemisphere.

**pup·py** (pŭp′ē) *n., pl.* **-pies.** [ME *popi* < OFr. *popee*, doll < Lat. *pupa*.] **1.** A young dog : PUP. **2.** A conceited or insolent youth.

**pup·py·ish** (pŭp′ē-ĭsh) *adj.* Like or characteristic of a puppy.

**puppy love** *n.* Adolescent love or infatuation.

**pup tent** *n.* A shelter tent.

**pur·blind** (pûr′blīnd′) *adj.* [ME *pur blind*, totally blind.] **1.** Having such poor vision as to be partly or nearly blind. **2.** Slow in comprehension. **—pur′blind′ly** *adv.* **—pur′blind′ness** *n.*

**pur·chas·a·ble** (pûr′chĭ-sə-bəl) *adj.* **1.** Capable of being bought. **2.** Capable of being bribed : VENAL. **—pur′chas·a·bil′i·ty** *n.*

**pur·chase** (pûr′chĭs) *vt.* **-chased, -chas·ing, -chas·es.** [ME *pourchasen* < OFr. *pourchacier*, to pursue : *pour-*, for (< Lat. *pro-*) + *chacier*, to chase. —see CHASE[1].] **1.** To obtain in exchange for money or its equivalent : BUY. **2.** To acquire by effort : EARN. **3.** *Law.* To acquire (property) legally by means other than inheritance. **4.** To move or hold with a mechanical device, as a lever or wrench. —*n.* **1.** Something bought. **2. a.** The act of buying. **b.** Acquisition through the payment of money or its equivalent. **3.** *Law.* Acquisition of property other than by inheritance. **4.** A grip applied manually or mechanically to move something or prevent it from slipping. **5.** A device, as a tackle or lever, that is used to obtain mechanical advantage. **6.** A position, as of a lever or one's feet, affording means to move or secure a weight. **7.** A means of increasing influence, power, or advantage. **—pur′chas·er** *n.*

**purchasing power** *n.* **1.** The ability to purchase, gen. measured by income. **2.** The value of a specified monetary unit in terms of the goods or services that can be purchased with it.

**pur·dah** (pûr′də) *n.* [Hindi *parda*, veil < Pers. *pardah*.] **1.** A curtain used to screen women in India from men or strangers. **2.** The Hindu practice of secluding women.

**pure** (pyŏŏr) *adj.* **pur·er, pur·est.** [ME *pur* < OFr. < Lat. *purus*, clean.] **1.** Having a homogeneous or uniform composition : not mixed <*pure* nitrogen> **2.** Free from impurities or adulterants : FULL-STRENGTH. **3.** Free from contamination : CLEAN. **4.** Free from foreign elements. **5.** Containing nothing inappropriate or unnecessary. **6.** Complete : total <*pure* ignorance of our plight> **7.** Without faults : PERFECT. **8.** Virgin : chaste. **9.** Of unmixed blood or ancestry. **10.** *Genetics.* Breeding true to parental type : HOMOZYGOUS. **11.** *Mus.* Free from discordant qualities. **12.** Articulated with a single unchanging speech sound : MONOPHTHONGAL. **13.** Theoretical <*pure* science> **14.** *Philos.* Free from empirical elements. **—pure′ly** *adv.* **—pure′ness** *n.*

☆ ***syns:*** PURE, PERFECT, PLAIN, SHEER, UNADULTERATED, UNMITIGATED *adj. core meaning* : free from extra, unneeded elements <*pure* gold><*pure* nitrogen> ***ant:*** impure

**pure·blood** (pyŏŏr′blŭd′) *also* **pure·blood·ed** (-blŭd′ĭd) *adj.* Of pure breeding stock : PUREBRED. **—pure′blood′** *n.*

**pure·bred** (pyŏŏr′brĕd′) *adj.* Of a strain established through breeding many generations of unmixed stock. —*n.* (pyŏŏr′brĕd′). A purebred animal.

**pure democracy** *n.* A democracy in which the power to govern lies directly in the hands of the people rather than being exercised through their representatives.

**pu·rée** (pyŏŏ-rā′, pyŏŏr′ā) *vt.* **-réed, -rée·ing, -rées.** [Fr. < OFr. *purer*, to strain < Lat. *purare*, to purify < *purus*, clean.] To rub (food) through a strainer. —*n.* Food prepared by puréeing.

**pur·fle** (pûr′fəl) *vt.* **-fled, -fling, -fles.** [ME *purfilen* < OFr. *porfiler* < VLat. *profilare : Lat. *pro-*, forth + Lat. *filum*, thread.] To finish or decorate the border or edge of. —*n. also* **purfling** (-flĭng). An ornamental edging or border.

**pur·ga·tion** (pûr-gā′shən) *n.* An act of purging or purifying.

**pur·ga·tive** (pûr′gə-tĭv) *adj.* Tending to cleanse or purge, esp. tending to cause evacuation of the bowels. —*n.* A purgative agent : CATHARTIC.

**pur·ga·to·ri·al** (pûr′gə-tôr′ē-əl, -tōr′-) *adj.* **1.** Serving to purify of sin : EXPIATORY. **2.** Of, relating to, or like purgatory.

**pur·ga·to·ry** (pûr′gə-tôr′ē, -tōr′ē) *n., pl.* **-ries.** [ME *purgatorie* < Med. Lat. *purgatorium* < Lat. *purgare*, to cleanse < *purus*, clean.] **1.** *Rom. Cath. Ch.* A state in which the souls of those who have died in grace must expiate their sins before attaining heaven. **2.** A place or condition of suffering, expiation, or remorse. —*adj.* Purgative.

**purge** (pûrj) *v.* **purged, purg·ing, purg·es.** [ME *purgen* < OFr. *purger* < Lat. *purgare*, to cleanse < *purus*, clean.] —*vt.* **1. a.** To free from impurities : PURIFY. **b.** To remove (impurities and other elements) by or as if by cleansing. **2.** To rid of guilt, sin, or defilement. **3.** *Law.* To clear (a person) of a charge or imputation. **4.** To rid (a group, as a nation or political party) of persons believed to be undesirable. **5.** *Med.* **a.** To cause evacuation of (the bowels). **b.** To induce evacuation of the bowels in (a patient). —*vi.* **1.** To become clean or pure. **2.** To undergo or cause an emptying of the bowels. —*n.* **1.** An act or process of purging. **2.** Something that purges, esp. a medicinal purgative. **—purg′er** *n.*

**pu·ri** (po͝or′ē) *n. var. of* POORI.

**pu·ri·fi·ca·tor** (pyo͝or′ə-fĭ-kā′tər) *n.* **1.** A cloth for cleaning the chalice after the celebration of the Eucharist. **2.** A purifying agent.

**pu·ri·fy** (pyo͝or′ə-fī′) *v.* **-fied, -fy·ing, -fies.** [ME *purifien* < OFr. *purifier* < Lat. *purificare* : *purus*, clean + *facere*, to make.] —*vt.* **1.** To rid of impurities : CLEANSE. **2.** To rid of foreign or unwanted elements. **3.** To free from guilt, sin, or defilement. —*vi.* To become pure or clean. **—pu′ri·fi·ca′tion** *n.* **—pu·rif′i·ca·to′ry** (pyo͞o-rĭf′ĭ-kə-tôr′ē, -tōr′ē) *adj.* **—pu′ri·fi′er** *n.*

**Pu·rim** (po͝or′ĭm, po͞o-rēm′) *n.* [Heb. *pūrīm*, pl. of *pūr*, lot (from the lots Haman cast to decide the day of the massacre), perh. < Akkadian *pūru*, stone.] A Jewish holiday in the month of Adar, celebrating the deliverance of the Jews from massacre by Haman.

**pu·rine** (pyo͝or′ēn′) *n.* [G. *Purin*, a blend of Lat. *purus*, clean and NLat. *uricus*, *uricum*, uric (acid).] **1.** A colorless crystalline compound, $C_5H_4N_4$, used in organic synthesis and metabolism studies. **2.** Any of a group of naturally occurring organic compounds derived from or having molecular structures related to purine, including uric acid, adenine, guanine, and caffeine.

**pur·ism** (pyo͝or′ĭz′əm) *n.* **1.** Strict observance of or insistence on traditional correctness, esp. of language. **2.** An example of purism.

**pur·ist** (pyo͝or′ĭst) *n.* An advocate of strict correctness, esp. in language use. **—pu·ris′tic** (pyo͞o-rĭs′tĭk) *adj.* **—pu·ris′ti·cal·ly** *adv.*

**Pu·ri·tan** (pyo͝or′ĭ-tn) *n.* [< LLat. *puritas*, purity < Lat. *purus*, pure.] **1.** A member of a group of English Protestants who in the 16th and 17th cent. advocated simplification of the creeds and ceremonies of the Church of England and strict religious discipline. **2. puritan.** One who lives in accordance with Protestant precepts, esp. one who views luxury or pleasure as sinful. —*adj.* **1.** Of or relating to the Puritans or Puritanism. **2. puritan.** Characteristic of a puritan : PURITANICAL.

**pu·ri·tan·i·cal** (pyo͝or′ĭ-tăn′ĭ-kəl) *adj.* **1.** Austerely moral and stern in religious observance. **2. Puritanical.** Of, relating to, or characteristic of the Puritans. **—pu′ri·tan′i·cal·ly** *adv.* **—pu′ri·tan′i·cal·ness** *n.*

**Pu·ri·tan·ism** (pyo͝or′ĭ-tn-ĭz′əm) *n.* **1.** The doctrines and practices of the Puritans. **2. puritanism.** Scrupulous moral sternness, esp. aversion to social pleasures and indulgences.

**pu·ri·ty** (pyo͝or′ĭ-tē) *n.* **1.** The quality or state of being pure. **2.** A quantitative assessment of homogeneity or uniformity. **3.** Freedom from guilt or sin : INNOCENCE. **4.** Absence in speech or writing of elements considered inappropriate to good style, as slang or foreign words. **5.** The proportion of a single-frequency spectral component in a mixture of achromatic and spectral colors.

**Pur·kin·je cell** (pûr-kĭn′jē) *n.* [After Johannes E. *Purkinje* (1787–1869).] Any of numerous nerve cells of the cerebellar cortex distinguished by a large rounded cell body.

**Purkinje fiber** *n.* Any of the large modified cardiac muscle fibers that compose the cardiac impulse-conducting system of the heart.

**purl¹** (pûrl) *vi.* **purled, purl·ing, purls.** [Prob. of Scand. orig.] To flow or ripple with a murmuring sound. —*n.* The sound made by rippling water.

**purl²** *also* **pearl** (pûrl) *v.* **purled, purl·ing, purls** *also* **pearled, pearl·ing, pearls.** [Orig. unknown.] —*vt.* **1.** To knit with a purl stitch. **2.** To edge or finish something with lace or embroidery. —*vi.* **1.** To do knitting with a purl stitch. **2.** To edge or finish with embroidery or lace. —*n.* **1.** The inversion of a knit stitch : PURL STITCH. **2.** An ornamental edging of embroidery or lace. **3.** Gold or silver wire used in embroidery.

**pur·lieu** (pûrl′yo͞o, pûr′lo͞o) *n.* [ME *perlew*, alteration of Norman Fr. *puralée*, perambulation < OFr. *poraler*, to traverse : *por-*, forth (< Lat. *pro-*) + *aler*, to go.] **1.** An outlying or neighboring area. **2. purlieus.** Outskirts. **3.** A place that one frequents : HAUNT.

**pur·lin** *also* **pur·line** (pûr′lĭn) *n.* [ME *purlyn.*] One of several horizontal timbers supporting the rafters of a roof.

**pur·loin** (pər-loin′, pûr′loin′) *vt. & vi.* **-loined, -loin·ing, -loins.** [ME *purloynen*, to remove < Norman Fr. *purloigner* : *pur-*, away (< Lat. *pro-*) + *loign*, far < Lat. *longē* < *longus*, long.] To steal or to commit theft. **—pur·loin′er** *n.*

**purl stitch** *n.* An inverted knitting stitch, often alternated with the plain stitch to produce a ribbed effect.

**pur·ple** (pûr′pəl) *n.* [ME *purpel* < OE *purple* < *purpura*, purple cloth < Lat., purple < Gk. *porphura*, a shellfish yielding purple dye.] **1.** Any of a group of colors with a hue between that of violet and red. **2.** Cloth of a color between violet and red, once worn as a symbol of royalty or high office. **3.** Imperial power : high station. **4.** The office or rank of a cardinal. **5.** The office or rank of a bishop. —*adj.* **1.** Of the color purple. **2.** Regal or imperial : ROYAL. **3.** Ornate and elaborate. —*vt. & vi.* **-pled, -pling, -ples.** To make or become purple.

**pur·ple·heart** (pûr′pəl-härt′) *n.* **1.** Any of several tropical American trees of the genus *Peltogyne*, valued for their decorative wood. **2.** The purple heartwood of the purpleheart.

**Purple Heart** *n.* A U.S. military decoration awarded to service personnel wounded in action.

**purple loosestrife** *n.* A marsh plant, *Lythrum salicaria*, with long purple flower spikes.

**purple martin** *n.* A North American bird, *Progne subis*, related to the swallows, having a glossy blue-black back and, in the male, a dark breast.

**purple salt** *n.* Potassium permanganate.

**pur·plish** (pûr′plĭsh) *adj.* Somewhat purple.

**pur·port** (pər-pôrt′, -pōrt′) *vt.* **-port·ed, -port·ing, -ports.** [ME *purporten* < OFr. *purporter*, to contain < Med. Lat. *proportare*, to extend : Lat. *pro-*, forth + Lat. *portare*, to carry.] **1.** To have or present the appearance, often false, of being or intending : PROFESS <a reporter who *purports* to be objective> **2.** To have the intention of doing : PURPOSE. —*n.* (pûr′pôrt′, -pōrt′). **1.** Meaning presented, intended, or implied : IMPORT. **2.** Purpose or intention.

**pur·port·ed** (pər-pôr′tĭd, -pōr′-) *adj.* Assumed to be such : SUPPOSED. **—pur·port′ed·ly** *adv.*

**pur·pose** (pûr′pəs) *n.* [ME *purpos* < OFr. < *purposer*, to intend < Lat. *proponere*, to put forward. —see PROPOSE.] **1.** The object toward which one strives or for which something exists : GOAL. **2.** A desired or intended result or effect. **3.** Determination : resolution. **4.** The matter at hand. —*vt.* **-posed, -pos·ing, -pos·es.** To resolve or intend to accomplish or perform. **—on purpose.** Willfully : deliberately. **—to good purpose.** With favorable results. **—to little (or no) purpose.** With few or no results. **—pur′pose·ly** (-lē) *adv.*

**pur·pose·ful** (pûr′pəs-fəl) *adj.* **1.** Having a purpose : INTENTIONAL. **2.** Having or showing purpose : DETERMINED. **—pur′pose·ful·ly** *adv.* **—pur′pose·ful·ness** *n.*

**pur·pose·less** (pûr′pəs-lĭs) *adj.* Without any purpose : POINTLESS. **—pur′pose·less·ly** *adv.* **—pur′pose·less·ness** *n.*

**pur·po·sive** (pûr′pə-sĭv) *adj.* **1.** Having or serving a purpose. **2.** Purposeful as opposed to random or aimless <*purposive* actions> **—pur′po·sive·ly** *adv.* **—pur′po·sive·ness** *n.*

**pur·pu·ra** (pûr′pə-rə, -pyə-) *n.* [NLat. < Lat., purple.] A condition marked by purplish discolorations of the skin and mucous membranes caused by hemorrhages into the affected tissues. **—pur·pu′ric** (-pyo͝or′ĭk) *adj.*

**purr** (pûr) *n.* [Imit.] **1.** The softly vibrant sound made by a cat to express contentment or pleasure. **2.** A sound similar to that made by a contented cat. —*v.* **purred, purr·ing, purrs.** —*vi.* To make or utter a soft, vibrant sound <The motor *purred* along.> —*vt.* To express by a soft vibrant sound.

**purse** (pûrs) *n.* [ME *purs* < OE < LLat. *bursa*, bag < Gk., leather.] **1.** A small pouch or bag for carrying money. **2.** A pocketbook or handbag. **3.** Something that resembles a pouch or bag. **4.** Available wealth or resources : MONEY. **5.** A sum of money collected as a present or offered as a prize. —*vt.* **pursed, purs·ing, purs·es.** To contract (the lips or brow) into folds or wrinkles. **—purse′like′** *adj.*

**purs·er** (pûr′sər) *n.* [ME < *purs*, purse.] The officer in charge of money matters on board a ship.

**purse seine** (sān) *n.* A fishing seine pursed or drawn into the shape of a bag to enclose the catch.

**purse strings** *pl.n.* Financial support or resources.

**purs·lane** (pûrs′lĭn, -lān′) *n.* [ME < OFr. *porcelaine*, cowrie shell. —see PORCELAIN.] A trailing weed, *Portulaca oleracea*, with small yellow flowers, reddish stems, and fleshy leaves occas. used in salads.

**pur·su·ance** (pər-so͞o′əns) *n.* The carrying out or putting into effect of something : PROSECUTION.

**pur·su·ant to** (pər-so͞o′ənt) *prep.* [ME *poursuiant* < OFr., pr.part. of *poursuivre*, to pursue < Lat. *prosequi*. —see PROSECUTE.] In accordance with : as a follow-up to.

**pur·sue** (pər-so͞o′) *v.* **-sued, -su·ing, -sues.** [ME *pursuen* < Norman Fr. *pursuer* < Lat. *prosequi.* —see PROSECUTE.] —*vt.* **1.** To follow in an effort to capture or overtake. **2.** To strive to obtain or accomplish. **3.** To proceed along the course of : FOLLOW. **4.** To carry further : ADVANCE. **5.** To be engaged in (e.g., a hobby). **6.** To persecute : harass. —*vi.* **1.** To follow in an effort to capture or overtake : CHASE. **2.** To carry on : CONTINUE. **—pur·su′a·ble** *adj.* **—pur·su′er** *n.*

☆ ***syns:*** PURSUE, CHASE, RUN AFTER *v. core meaning* : to follow (another) with the intent of overtaking and capturing <The police *pursued* the kidnapper across the state.>

**pur·suit** (pər-so͞ot′) *n.* [ME *pursuite* < OFr. *poursuite* < *poursuivre,* to pursue < Lat. *prosequi.* —see PROSECUTE.] **1.** An act or instance of pursuing or chasing. **2.** The act of striving <the *pursuit* of happiness> **3.** An activity, as a hobby or vocation, regularly engaged in.

**pursuit plane** *n.* A high-speed fighter aircraft designed and equipped to pursue and attack enemy aircraft.

**pur·sui·vant** (pûr′swĭ-vənt) *n.* [ME *pursevant,* attendant < OFr. *poursuivant,* follower < pr.part. of *poursuivre,* to follow < Lat. *prosequi.* —see PROSECUTE.] **1.** An officer ranking below a herald in the British Colleges of Heralds. **2.** A follower : attendant.

**pur·te·nance** (pûr′tn-əns) *n.* [ME *portenaunce* < OFr. *partenance,* accessory < *partenir,* to pertain. —see PERTAIN.] An animal's viscera or inner organs, esp. the heart, liver, and lungs.

**pu·ru·lence** (pyo͝or′ə-ləns, pyo͝or′yə-) *n.* **1.** The state of secreting or containing pus. **2.** Pus.

**pu·ru·lent** (pyo͝or′ə-lənt, pyo͝or′yə-) *adj.* [Lat. *purulentus* < *pus,* pus.] Containing or secreting pus. **—pu′ru·lent·ly** *adv.*

**pur·vey** (pər-vā′, pûr′vā′) *vt.* **-veyed, -vey·ing, -veys.** [ME *purveien* < OFr. *porveioir* < Lat. *providēre.* —see PROVIDE.] **1.** To supply (e.g., food) : FURNISH. **2.** To circulate or advertise. **—pur·vey′ance** *n.*

**pur·vey·or** (pər-vā′ər) *n.* **1.** One who furnishes provisions, esp. food. **2.** A distributor : dispenser <a *purveyor* of rumors>

**pur·view** (pûr′vyo͞o′) *n.* [ME *purveu,* proviso < Norman Fr. *purveu,* it is provided (from the use of this word to introduce a proviso) < OFr. *porveu* < *porveioir,* to provide. —see PURVEY.] **1.** Extent or range of power, function, or competence : SCOPE. **2.** Range of vision, understanding, or experience : OUTLOOK. **3.** *Law.* The scope, body, or limit of a statute.

**pus** (pŭs) *n.* [Lat.] A viscous yellowish-white fluid formed in infected tissue, consisting chiefly of leucocytes, cellular debris, and liquefied tissue elements.

**Pu·sey·ism** (pyo͞o′zē-ĭz′əm, pyo͞o′sē-) *n.* [After Edward B. *Pusey* (1800–1882).] Tractarianism. **—Pu′sey·ite′** (-īt′) *n.*

**push** (po͝osh) *v.* **pushed, push·ing, push·es.** [ME *pusshen* < OFr. *pouser* < Lat. *pulsare,* freq. of *pellere,* to beat.] —*vt.* **1.** To exert force against (an object) in order to move. **2.** To move (an object) by exerting force : SHOVE. **3.** To force (one's way) <*pushed* their way through the mob> **4.** To urge forward. **5.** To urge insistently to do something : PRESSURE. **6.** To bear hard on : PRESS. **7.** To extend or enlarge. **8.** *Slang.* **a.** To promote or sell (a product). **b.** To sell (a narcotic) illegally. —*vi.* **1.** To exert outward force against something. **2.** To press forward or advance despite difficulty or opposition. **3.** To expend vigorous effort. **—push around.** *Informal.* To treat or threaten to treat harmfully : INTIMIDATE. **—push off.** *Informal.* To set out : DEPART. **—push on.** To continue : proceed. —*n.* **1.** The act of pushing : THRUST. **2.** A vigorous or persistent effort toward an end : DRIVE. **3.** A stimulus to action : PROVOCATION. **4.** *Informal.* Persevering energy : ENTERPRISE. **—push (one's) luck.** To take ever increasing risks or chances.

**push·ball** (po͝osh′bôl′) *n.* **1.** A game in which two opposing teams attempt to push a heavy ball, six feet in diameter, across a goal. **2.** The ball used in the game of pushball.

**push broom** *n.* A broom with a wide brush perpendicular to the end of a long handle designed to be pushed in sweeping.

**push button** *n.* A small button that activates an electric circuit.

**push-but·ton** (po͝osh′bŭt′n) *adj.* Equipped with or operated by a push button.

**push·cart** (po͝osh′kärt′) *n.* A light cart pushed by hand.

**push·down** (po͝osh′doun′) *n. Computer Sci.* A section of stored data from which the most recently stored material must be the first to be utilized.

**push·er** (po͝osh′ər) *n.* **1.** One that pushes. **2.** *Slang.* One who sells drugs illegally.

**push·ful** (po͝osh′fo͝ol′) *adj.* Pushing. **—push′ful·ness** *n.*

**push·ing** (po͝osh′ĭng) *adj.* **1.** Ambitious : enterprising. **2.** Aggressively forward. **—push′ing·ly** *adv.*

**push·o·ver** (po͝osh′ō′vər) *n.* **1.** An easily accomplished activity. **2.** One easily defeated or taken advantage of.

**push·pin** (po͝osh′pĭn′) *n.* **1.** A tacklike pin with a large head that is easily inserted into a wall or board. **2.** A game played by children with pins.

**push·rod** *also* **push rod** (po͝osh′rŏd′) *n.* A rod moved by a cam to operate the valves in an internal-combustion engine.

**Push·tu** (pŭsh′to͞o) *n. var. of* PASHTO.

**push·up** (po͝osh′ŭp′) *n.* **1.** An exercise for strengthening arm muscles, performed by lying with the face and palms to the floor and by pushing the body up and down with the arms. **2.** *Computer Sci.* A section of stored data from which the earliest stored material must be the first to be utilized.

**push·y** (po͝osh′ē) *adj.* **-i·er, -i·est.** *Informal.* Objectionably forward or aggressive. **—push′i·ly** *adv.* **—push′i·ness** *n.*

**pu·sil·la·nim·i·ty** (pyo͞o′sə-lə-nĭm′ĭ-tē) *n.* The quality or state of being pusillanimous : COWARDICE.

**pu·sil·lan·i·mous** (pyo͞o′sə-lăn′ə-məs) *adj.* [LLat. *pusillanimis* : Lat. *pusillus,* weak (< *pusus,* boy) + *animus,* spirit.] Lacking courage : COWARDLY. **—pu′sil·lan′i·mous·ly** *adv.*

**puss**[1] (po͝os) *n.* [Prob. of Germanic orig.] *Informal.* **1.** A cat. **2.** A girl or young woman.

**puss**[2] (po͝os) *n.* [Ir. Gael. *bus,* mouth < OIr., lip.] *Slang.* **1.** The human mouth. **2.** The human face.

**puss·ley** (pŭs′lē) *n.* [Alteration of PURSLANE.] Purslane.

**puss·y**[1] (po͝os′ē) *n., pl.* **-ies.** *Informal.* **1.** A cat. **2.** A fuzzy catkin, esp. of the pussy willow.

**pus·sy**[2] (pŭs′ē) *adj.* **-si·er, -si·est.** Like or containing pus.

**puss·y·cat** (po͝os′ē-kăt′) *n.* **1.** A cat. **2.** *Informal.* One who is easygoing, amiable, or mild-mannered.

**puss·y·foot** (po͝os′ē-fo͝ot′) *vi.* **-foot·ed, -foot·ing, -foots.** **1.** To move cautiously or stealthily. **2.** *Slang.* To act or proceed timidly to avoid committing oneself. **—puss′y·foot′er** *n.*

**puss·y·toes** (po͝os′ē-tōz′) *pl.n.* [From the cluster's resemblance to a cat's paw.] *(sing. or pl. in number).* Any of several low-growing plants of the genus *Antennaria,* having leaves with whitish down and small white flower clusters.

**pussy willow** *n.* **1.** A North American shrub or small tree, *Salix discolor,* having silky catkins. **2.** Any of several willows similar to the pussy willow.

**pus·tu·lant** (pŭs′chə-lənt, pŭs′tyə-) *adj.* Causing pustules to form. —*n.* A pustule-producing agent.

**pus·tu·lar** (pŭs′chə-lər, pŭs′tyə-) *adj.* Of, pertaining to, or having pustules.

**pus·tu·late** (pŭs′chə-lāt′, pŭs′tyə-) *vi.* & *vt.* **-lat·ed, -lat·ing, -lates.** To form or cause to form pustules. —*adj.* Covered with pustules or pustulelike blisters. **—pus′tu·la′tion** *n.*

**pus·tule** (pŭs′cho͞ol, pŭs′tyo͞ol) *n.* [ME < OFr. < Lat. *pustula,* blister.] **1.** A slight, inflamed elevation of the skin filled with pus. **2.** A small swelling similar to a blister or pimple.

**put** (po͝ot) *v.* **put, put·ting, puts.** [ME *putten.*] —*vt.* **1.** To place in a designated location : SET. **2.** To cause to be in a specified condition <*put* one's desk in order> **3.** To cause to undergo something : SUBJECT. **4.** To assign : attribute <*put* a new meaning on the event> **5.** To estimate <*put* the value at $100> **6.** To levy or impose <*put* a tariff on imported goods> **7.** To wager (a stake) : BET. **8.** To hurl with an overhand pushing motion. **9.** To bring up for consideration or judgment <*put* a question to the teacher> **10.** To express : state <*putting* it mildly> **11.** To render in a specified language or literary form <*put* the passage into German> **12.** To adapt <words *put* to music> **13.** To urge or force to some action <*put* a burglar to flight> **14.** To apply <*put* their energies to it> —*vi.* **1.** To begin to move, esp. in a hurry. **2.** *Naut.* To proceed <The boat *put* into the harbor.> **—put about.** *Naut.* To go or cause to go from one tack to another. **—put across. 1.** To state so as to be clearly or readily understood or accepted. **2.** To attain or carry through by stratagem. **—put away. 1.** To renounce : discard <*put away* all distracting thoughts> **2.** *Informal.* To consume (food or drink) readily and quickly. **3. a.** *Informal.* To confine to a mental institution. **b.** *Informal.* To kill. **c.** To bury. **—put by.** To save for later use <*put by* a supply of canned goods> **—put down. 1. a.** To write down. **b.** To enter in a list. **2. a.** To bring to an end : REPRESS <*put down* a riot> **b.** To make ineffective <*put down* wild stories> **3.** *Slang.* **a.** To criticize <*put* us *down* for asking too many questions> **b.** To disparage : belittle <*put down* our attempts to win> **c.** To humiliate <*put* the student *down* with a sarcastic comment> **4. a.** To assign to a category <*put* everyone *down* as liars> **b.** To attribute <*put* it *down* to inexperience> **5.** To consume (food or drink) readily. **—put forth. 1.** To grow <plants *putting forth* leaves> **2.** To bring to bear : EXERT. **3.** To offer for consideration. **—put forward.** To propose for consideration <*put forward* a new plan> **—put in. 1.** To make a formal offer of <*put in* a bid> **2.** To interpose. **3.** To spend (time) at a given location or at a job <*put in* a full day's work> **4.** To plant <*put in* peas and carrots> **5.** *Naut.* To enter a port or harbor. **6.** To enter a request, offer, or application <*put in* for a two-week vacation> **—put off. 1. a.** To postpone : delay <*put off* seeing the dentist> **b.** To persuade to wait <managed to *put off* the tax collector> **2.** To take off : discard. **3.** To repel or repulse, as from bad manners. **4.** To pass (money) or sell (merchandise) fraudulently. **—put on. 1.** To clothe oneself with : DON. **2.** To apply : activate <*put* the gas *on*> **3.** To assume affectedly <*put on* a foreign accent> **4.** *Slang.* To mislead or tease (another) <Don't *put* me *on.*> **5.** To add <*put on* 15 pounds> **6.** To produce : perform <*put on* a magic show> **—put out. 1.** To extinguish. **2.** *Naut.* To leave, as a port or harbor : DEPART. **3.** To expel <*put out* a heckler> **4.** To publish <*put out* a new edition> **5.** To inconvenience <Their

early arrival *put* us *out*.> **6.** To irritate. **7.** *Baseball.* To retire (a runner). **—put over. 1.** To postpone : delay. **2.** To put across, esp. to deceive <*put* one *over* on me> **—put through. 1.** To bring to a successful end <*put* a number of projects *through*> **2.** To cause to undergo <*put* us *through* a lot of suffering> **3. a.** To make a telephone connection for. **b.** To obtain a connection for (a telephone call). **—put to.** *Naut.* To head for shore. **—put together. 1.** To construct : build. **2.** To add : combine. **—put up. 1.** To erect : build. **2.** To preserve : can. **3.** To nominate. **4.** To provide (funds) in advance. **5.** To provide lodgings for <*put* the tourists *up* for the night> **6.** To incite to an action <wondering who *put* them *up* to stealing> **7.** To start (game animals) from cover <*put up* a pheasant> **8.** To offer for sale. **9. a.** To make a display or the appearance of <*put up* a good front> **b.** To engage in : CARRY ON <*put up* a good argument> **—put upon.** To impose on : OVERBURDEN. *—n.* **1.** An act of putting the shot. **2.** An option to sell a stipulated amount of stock or securities within a designated time and at a fixed price. *—adj. Informal.* Fixed : stationary. **—put down roots.** To establish a permanent residence. **—put in mind.** To remind. **—put (one's) finger on.** To identify. **—put (one's) house in order.** To organize one's affairs. **—put on the dog.** *Slang.* To give oneself airs. **—put the arm (**or **bite) on.** *Slang.* To ask (someone) for money. **—put the finger on.** *Slang.* To inform on. **—put the make on.** *Slang.* To make sexual advances to. **—put the screws to (**or **on).** To pressure (another) in an extreme manner. **—put to bed.** To make final preparations for the printing of (e.g., a newspaper). **—put to it.** To give extreme difficulty to. **—put two and two together.** To draw the proper conclusion from given evidence or indications. **—put up with.** To endure without complaint <had to *put up with* constant nagging>

**pu·ta·men** (pyo͞o-tā′mən) *n., pl.* **-tam·i·na** (-tăm′ə-nə) [Lat., husk < *putare*, to prune.] A hard shell-like covering, as that enclosing the kernel of a peach. **—pu·tam′i·nous** (-tăm′ə-nəs) *adj.*

**pu·ta·tive** (pyo͞o′tə-tĭv) *adj.* [ME < OFr. *putatif* < LLat. *putativus* < Lat. *putare*, to consider.] **1.** Regarded as such : SUPPOSED. **2.** Assumed to exist or to have once existed. **—pu′ta·tive·ly** *adv.*

**put-down** (po͝ot′doun′) *n. Slang.* A rejection or dismissal, esp. in the form of a critical or humiliating remark.

**put·log** (po͝ot′lôg′, -lŏg′, pŭt′-) *n.* [Alteration of obs. *pullock*, perh. < PUT.] One of the short pieces of lumber that support a scaffolding floor.

**put·off** (po͝ot′ôf′, -ŏf′) *n.* A pretext for inaction : EXCUSE.

**put-on** (po͝ot′ŏn′, -ôn′) *adj.* Assumed : feigned. *—n. Slang.* **1.** The act of misleading or teasing someone, esp. for amusement. **2.** Something, as a prank or book, intended as a hoax or joke : SPOOF. **3.** A deceptive outward appearance.

**Pu·tong·hua** (po͞o′tông′hwä′, -wä′, -tông′-) *n.* [Chin. (Mandarin) *pu*[3] *tong*[1] *hua*[4] : *pu*[3], common + *tong*[1], through + *hua*[4], words.] MANDARIN 3.

**put·out** (po͝ot′out′) *n. Baseball.* A play in which a batter or base runner is retired.

**put-put** (pŭt′pŭt′) *n.* [Imit. of a running engine.] *Slang.* **1.** A small gasoline engine. **2.** A vehicle or boat operated by a put-put.

**pu·tre·fac·tion** (pyo͞o′trə-făk′shən) *n.* [ME *putrefaccioun* < LLat. *putrefactio* < Lat. *putrefacere*, to make rotten. —see PUTREFY.] **1.** Partial decomposition of organic matter by microorganisms, producing foul-smelling matter. **2.** Putrefied matter. **3.** The state of having been putrefied. **—pu′tre·fac′tive** *adj.*

**pu·tre·fy** (pyo͞o′trə-fī′) *v.* **-fied, -fy·ing, -fies.** [ME *putrefien* < OFr. *putrefier* < Lat. *putrefacere* : *puter*, rotten + *facere*, to make.] *—vt.* **1.** To cause to decay. **2.** To make gangrenous. *—vi.* **1.** To decompose. **2.** To become gangrenous.

**pu·tres·cent** (pyo͞o-trĕs′ənt) *adj.* [Lat. *putrescens*, *putrescent-*, pr.part. of *putrescere*, to rot, inceptive of *putrēre*, to be rotten < *puter*, rotten.] **1.** Becoming putrid : PUTREFYING. **2.** Of or relating to putrefaction. **—pu·tres′cence** (-trĕs′əns) *n.*

**pu·tres·ci·ble** (pyo͞o-trĕs′ə-bəl) *adj.* [Fr. < Lat. *putrescere*, to rot. —see PUTRESCENT.] Subject to putrefaction.

**pu·trid** (pyo͞o′trĭd) *adj.* [Lat. *putridus* < *putrere*, to be rotten < *puter*, rotten.] **1.** Decomposed and foul-smelling : ROTTEN. **2.** Proceeding from, relating to, or displaying putrefaction. **3.** Morally rotten : CORRUPT. **4.** Extremely objectionable : VILE. **—pu·trid′i·ty** (-trĭd′ĭ-tē), **pu′trid·ness** *n.* **—pu′trid·ly** *adv.*

**putsch** *also* **Putsch** (po͝och) *n.* [G.] A sudden attempt by a group to overthrow a government. **—putsch′ist** *n.*

**putt** (pŭt) *n.* [Var. of *put*.] A light golf stroke made on the putting green in an effort to place the ball into the hole. *—v.* **putt·ed, putt·ing, putts.** *—vt.* To hit (a golf ball) with a light stroke on the green. *—vi.* To putt a golf ball.

**put·tee** (pŭ-tē′, pŭt′ē) *also* **put·ty** (pŭt′ē) *n., pl.* **-tees** *also* **-ties.** [Hindi *paṭṭī* < Skt. *paṭṭikā*, fem. of *paṭṭakaḥ*, bandage, ribbon < *paṭṭaḥ*, strip of cloth.] **1.** A strip of cloth wound spirally around the leg from ankle to knee. **2.** A gaiter covering the lower leg.

**put·ter**[1] (pŭt′ər) *n.* **1.** A short stiff-shafted golf club used for putting. **2.** A golfer who is putting.

**put·ter**[2] (pŭt′ər) *v.* **-tered, -ter·ing, -ters.** [Var. of dial. *potler*, perh. < *pote*, to push.] *—vi.* To occupy oneself in an idle or aimless manner. *—vt.* To waste (time) in idling <*puttering* around in the garage> **—put′ter·er** *n.*

**putting green** *n.* **1.** The smooth, closely mowed area at the end of a fairway on a golf course in which the hole is placed. **2.** An area for putting practicing.

**put·ty**[1] (pŭt′ē) *n., pl.* **-ties.** [Fr. *potée* < OFr., a potful < *pot*, pot.] **1. a.** A doughlike cement made by mixing whiting and linseed oil, used to fill holes in woodwork and secure panes of glass. **b.** A substance with a similar consistency or function. **2.** A fine lime cement used as a finishing coat on plaster. **3.** A light brownish gray to light grayish brown. *—vt.* **-tied, -ty·ing, -ties.** To cover, fill, or secure with putty.

**put·ty**[2] (pŭt′ē) *n. var. of* PUTTEE.

**put·ty·root** (pŭt′ē-ro͞ot′, -ro͝ot′) *n.* [From the use of the sticky substance in its corm as a cement.] A North American orchid, *Aplectrum hyemale*, bearing a single leaf and yellowish-brown or purplish flowers.

**put-up** (po͝ot′ŭp′) *adj. Informal.* Secretly prearranged.

**puz·zle** (pŭz′əl) *v.* **-zled, -zling, -zles.** [Orig. unknown.] *—vt.* **1.** To cause doubt and indecision in : PERPLEX. **2.** To solve or clarify (something confusing) by reasoning or study <*puzzled* out the import of that comment> *—vi.* **1.** To be perplexed. **2.** To ponder over a problem in an effort to solve or comprehend it. *—n.* **1.** Something that puzzles. **2.** A game, toy, or testing device that tests ingenuity. **3.** Perplexity. **—puz′zle·ment** (-mənt) *n.* **—puz′zler** *n.*

**py-** *pref. var. of* PYO-.

**py·a** (pē-ä′) *n.* [Burmese.] —See table at CURRENCY.

**pyc·nid·i·um** (pĭk-nĭd′ē-əm) *n., pl.* **-i·a** (-ē-ə) [NLat. : Gk. *puknos*, thick + Gk. *-idion*, dim. suffix.] A rounded or flask-shaped asexual fruiting body containing spores that occurs in certain fungi. **—pyc·nid′i·al** *adj.*

**pyc·nom·e·ter** (pĭk-nŏm′ĭ-tər) *n.* [Gk. *puknos*, dense + METER.] A standard vessel for measuring the specific gravity or density of materials.

**py·e·li·tis** (pī′ə-lī′tĭs) *n.* [Gk. *puelos*, basin + -ITIS.] Pyelonephritis.

**py·e·lo·ne·phri·tis** (pī′ə-lō-nĭ-frī′tĭs) *n.* [Gk. *puelos*, basin + NEPHRITIS.] Inflammation of both the kidney and its pelvis. **—py′e·lo·ne·phrit′ic** (-frĭt′ĭk) *adj.*

**py·e·mi·a** (pī-ē′mē-ə) *n.* Pus in the blood. **—py·e′mic** *adj.*

**py·gid·i·um** (pī-jĭd′ē-əm) *n., pl.* **-i·a** (-ē-ə) [Gk. *pugidion*, dim. of *pugē*, buttocks.] The posterior body region of certain arthropods. **—py·gid′i·al** *adj.*

**pyg·mae·an** *or* **pyg·me·an** (pĭg-mē′ən, pĭg′mē-) *adj.* [< Lat. *pygmaeus*.] Pygmy.

**Pyg·ma·lion** (pĭg-māl′yən, -mā′lē-ən) *n.* [Lat. < Gk. *Pugmaliōn*.] *Gk. Myth.* A king of Cyprus who carved and then fell in love with a statue of a woman, which Aphrodite brought to life as Galatea.

**pyg·moid** (pĭg′moid′) *adj.* Like or characteristic of a Pygmy.

**Pyg·my** *also* **Pig·my** (pĭg′mē) *n., pl.* **-mies.** [ME *pigmie* < Lat. *pygmaeus*, dwarfish < Gk. *pugmaios* < *pugmē*, the length from the elbow to the knuckles.] **1.** *Gk. Myth.* A member of a race of dwarfs. **2.** A member of any of several African and Asian peoples with a hereditary stature of from four to five feet. **3. pygmy.** A small insignificant individual. *—adj.* **1.** Of or relating to the Pygmies. **2. pygmy. a.** Atypically or unusually small. **b.** Insignificant : trivial.

**py·ja·mas** (pə-jä′məz, -jăm′əz) *pl.n. Chiefly Brit. var. of* PAJAMAS.

**pyk·nic** (pĭk′nĭk) *adj.* [< Gk. *puknos*, compact.] Having a short, stocky, and powerful physique : ENDOMORPHIC. **—pyk′nic** *n.*

**py·lon** (pī′lŏn′) *n.* [Gk. *pulōn*, gateway < *pulē*, gate.] **1.** A monumental gateway in the form of a pair of truncated pyramids serving as the entrance to an ancient Egyptian temple. **2.** A large structure or structures marking an approach or entrance. **3.** A tower marking a turning point in a race among aircraft. **4.** A steel tower supporting high-tension wires. **5.** A temporary artificial leg.

**py·lo·rus** (pī-lôr′əs, -lōr′-, pĭ-) *n., pl.* **-lo·ri** (-lôr′ī′, -lōr′ī′) [LLat. *pylorus* < Gk. *pulōros* : *pulē*, gate + *ouros*, watcher < *horan*, to see.] The passage linking the stomach and duodenum. **—py·lo′ric** (-ĭk) *adj.*

**pyo-** *or* **py-** *pref.* [< Gk. *puon*, pus.] Pus <pyoderma>

**py·o·der·ma** (pī′ə-dûr′mə) *n.* A pus-causing skin disease. **—py′o·der′mic** *adj.*

**py·o·gen·e·sis** (pī′ə-jĕn′ĭ-sĭs) *n.* Pyosis. **—py′o·gen′ic** *adj.*

**py·or·rhe·a** *or* **py·or·rhoe·a** (pī′ə-rē′ə) *n.* **1.** A discharge of pus. **2.** Inflammation of the gum and tooth sockets leading to loosening of the teeth. **—py′or·rhe′al** *adj.*

**py·o·sis** (pī-ō′sĭs) *n.* Formation of pus.

**pyr-** *pref. var. of* PYRO-.

**py·ra·can·tha** (pī′rə-kăn′thə) *n.* [Gk. *purakantha*, a shrub : *pur*, fire + *akantha*, thorn.] A shrub of the genus *Pyracantha*, the fire thorn.

**py·ral·id** (pī-răl′ĭd) *also* **py·ral·i·did** (pī-răl′ĭ-dĭd) *n.* [NLat. *Pyralididae*, family name < Gk. *puralis*, an insect said to live in fire < *pur*, fire.] One of various small or medium-sized moths of the large, widely distributed family Pyralididae. *—adj.* Of or belonging to the Pyralididae.

**pyr·a·mid** (pĭr′ə-mĭd) *n.* [Lat. *pyramis, pyramid-* < Gk. *puramis.*] **1.** A polyhedron with a polygonal base and triangular faces meeting in a common vertex. **2.** Something shaped like a pyramid. **3.** A massive monument found esp. in Egypt, with a rectangular base and four triangular faces culminating in a single apex, and serving as a temple or tomb. **4.** The transactions involved in pyramiding stock. —*v.* **-mid·ed, -mid·ing, -mids.** —*vt.* **1.** To build or place in the shape of a pyramid. **2.** To build (e.g., an argument or thesis) progressively from a basic general premise. **3.** To speculate in (stock) by making a series of buying and selling transactions in which paper profits are used as margin for buying more stock. —*vi.* **1.** To take the shape of a pyramid. **2.** To increase rapidly and on a widening base. **3.** To pyramid stocks. **—py·ram′i·dal** (pĭ-răm′ĭ-dəl), **pyr′a·mid′ic, pyr′a·mid′i·cal** *adj.* **—py·ram′i·dal·ly** *adv.*

**Pyr·a·mus** (pĭr′ə-məs) *n.* [Lat.] *Rom. Myth.* The legendary Babylonian youth who killed himself when he mistakenly thought his lover Thisbe was dead.

**py·rar·gy·rite** (pī-rär′jə-rīt′, pī-) *n.* [G. *Pyrargyrit* : Gk. *pur,* fire + Gk. *arguros,* silver + *-it,* -ite.] A deep red to black silver ore with composition $Ag_3SbS_3$.

**pyre** (pīr) *n.* [Lat. *pyra* < Gk. *pura* < *pur,* fire.] **1.** A heap of combustibles for burning a corpse as a funeral rite. **2.** A pile of combustibles.

**py·rene** (pī′rēn′, pī-rēn′) *n.* [NLat. *pyrena* < Gk. *purēn.*] The stone of certain fruits.

**py·re·noid** (pī-rē′noid′, pī′rə-) *n.* [NLat. *pyrena,* fruit stone + -OID.] One of the protein granules of certain algae and similar plants in which starch is formed.

**py·re·thrin** (pī-rē′thrĭn, -rĕth′rĭn) *n.* [PYRETHR(UM) + -IN.] Either of two viscous liquid esters, $C_{21}H_{28}O_3$ or $C_{22}H_{28}O_5$, extracted from pyrethrum flowers and used as insecticides.

**py·re·thrum** (pī-rē′thrəm, -rĕth′rəm) *n.* [Lat., pellitory < Gk. *purethron,* feverfew, ult. < *pur,* fire.] **1.** An Old World plant of the genus *Chrysanthemum* and related genera, as *C. coccineum,* cultivated for its showy flowers. **2.** The dried flowers of *C. cinerariaefolium* or *C. coccineum,* used as an insecticide.

**py·ret·ic** (pī-rĕt′ĭk) *adj.* [Gk. *puretikos* < *puretos,* fever < *pur,* fire.] Affected or marked by fever : FEVERISH.

**Py·rex** (pī′rĕks′). A trademark for any of various types of heat-resistant and chemical-resistant glass.

**py·rex·i·a** (pī-rĕk′sē-ə) *n.* [NLat. < Gk. *purexis* < *puressein,* to have a fever < *puretos,* fever < *pur,* fire.] Fever. **—py·rex′i·al, py·rex′ic** *adj.*

**pyr·he·li·om·e·ter** (pīr′hē-lē-ŏm′ĭ-tər, pĭr′-) *n.* A device for measuring all or restricted components of incident solar radiation. **—pyr′he·li·o·met′ric** (-ə-mĕt′rĭk) *adj.*

**py·ric** (pī′rĭk, pĭr′ĭk) *adj.* [Fr. *pyrique* < Gk. *pur,* fire.] Of, relating to, or resulting from burning.

**pyr·i·dine** (pĭr′ĭ-dēn′) *n.* A flammable, colorless or yellowish liquid base, $C_5H_5N$, used to synthesize vitamins and drugs, as a solvent, and as a denaturant for alcohol. **—py·rid′ic** (pĭ-rĭd′ĭk) *adj.*

**pyr·i·dox·a·mine** (pĭr′ĭ-dŏk′sə-mēn′) *n.* [PYRIDOX(INE) + -AMINE.] An amine, $C_8H_{12}N_2O_2$, of the vitamin $B_6$ group.

**pyr·i·dox·ine** (pĭr′ĭ-dŏk′sēn′, -sĭn) *also* **pyr·i·dox·in** (-dŏk′sĭn) *n.* [PYRID(INE) + OX- + -INE.] A pyridine derivative, $C_8H_{11}O_3N$, occurring in plant and animal tissues and active in various metabolic processes.

**pyr·i·form** (pĭr′ə-fôrm′) *adj.* [Med. Lat. *pyrum,* pear (alteration of Lat. *pirum*) + -FORM.] Shaped like a pear.

**py·rim·i·dine** (pī-rĭm′ĭ-dēn′, pĭ-) *n.* [Alteration of PYRIDINE.] **1.** A liquid and crystalline organic base, $C_4H_4N_2$. **2.** One of several basic compounds, as uracil, having a molecular structure similar to pyrimidine and found in living matter as a nucleotide component.

**py·rite** (pī′rīt′) *n.* [< Lat. *pyrites,* flint. —see PYRITES.] A yellow to brown, widely occurring mineral sulfide, $FeS_2$, used as an iron ore and to produce sulfur dioxide for sulfuric acid. **—py·rit′ic** (-rĭt′ĭk), **py·rit′i·cal** *adj.*

**py·ri·tes** (pī-rī′tēz, pī′rīts′) *n., pl.* **pyrites.** [Lat., flint < Gk. *puritēs (lithos),* fire (stone) < *pur,* fire.] Any of various natural metallic sulfides, esp. of iron.

**pyro-** or **pyr-** *pref.* [< Gk. *pur,* fire.] **1.** Fire : heat <*pyro*technic> **2.** Resulting from or as if from the action of fire or heat <*pyro*graphy> **3.** Fever <*pyro*gen> **4.** Derived from an acid by the loss of a water molecule <*pyro*sulfuric acid>

**py·ro·cat·e·chol** (pī′rō-kăt′ĭ-kôl′, -kōl′) *n.* [PYRO- + CATECH(U) + -OL.] A colorless crystalline organic compound, $C_6H_4(OH)_2$, used as an antiseptic and photographic developer.

**py·ro·cel·lu·lose** (pī′rō-sĕl′yə-lōs′, -lōz′) *n.* A cellulose nitrate used as a component of smokeless powder.

**py·ro·chem·i·cal** (pī′rō-kĕm′ĭ-kəl) *adj.* Referring to chemical activity at elevated temperatures. **—py′ro·chem′i·cal·ly** *adv.*

**py·ro·clas·tic** (pī′rō-klăs′tĭk) *adj.* Formed by rock fragmentation resulting from volcanic ejection.

**py·ro·e·lec·tric** (pī′rō-ĭ-lĕk′trĭk) *adj.* Showing or relating to pyroelectricity. —*n.* A pyroelectric material.

**py·ro·e·lec·tric·i·ty** (pī′rō-ĭ-lĕk-trĭs′ĭ-tē) *n.* Generation of electric charge on a crystal by change of temperature.

**py·ro·gal·lic acid** (pī′rō-găl′ĭk, -gô′lĭk) *n.* Pyrogallol.

**py·ro·gal·lol** (pī′rō-găl′ôl′, -ōl′, -gô′lôl′, -lōl′) *n.* [PYRO- + GALL(IC) + -OL[1].] A white lustrous crystalline compound, $C_6H_3(OH)_3$, used as a photographic developer and in the treatment of skin diseases. **—py′ro·gal′lic** (-găl′ĭk, -gô′lĭk) *adj.*

**py·ro·gen** (pī′rə-jən) *n.* A fever-producing substance.

**py·ro·gen·ic** (pī′rō-jĕn′ĭk) *also* **py·rog·e·nous** (pī-rŏj′ə-nəs) *adj.* **1.** Producing or produced by fever. **2.** Caused by or generating heat. **3.** IGNEOUS 2. **—py′ro·ge·nic′i·ty** (-rō-jə-nĭs′ĭ-tē) *n.*

**py·rog·ra·phy** (pī-rŏg′rə-fē) *n.* **1.** The art or process of producing designs on material, as wood or leather, by using heated tools or a fine flame. **2.** A design made by pyrography. **—py′ro·graph′** (pī′rə-grăf′) *n.* **—py·rog′ra·pher** *n.* **—py′ro·graph′ic** *adj.*

**py·ro·lig·ne·ous** (pī′rō-lĭg′nē-əs) *adj.* Made by destructive distillation of wood.

**pyroligneous acid** *n.* A reddish-brown wood distillate containing acetic acid, methyl alcohol, acetone, and a tarry residue.

**py·ro·lu·site** (pī′rō-lo͞o′sīt′) *n.* [G. *Pyrolusit* : Gk. *pur,* fire + Gk. *lousis,* a washing (< *louein,* to wash) + *-it,* -ite.] A soft, black to dark-gray ore of manganese, consisting mainly of manganese dioxide.

**py·rol·y·sis** (pī-rŏl′ĭ-sĭs) *n.* Chemical change due to heat. **—py′ro·lyt′ic** (-rə-lĭt′ĭk) *adj.* **—py′ro·lyt′i·cal·ly** *adv.*

**py·ro·lyze** (pī′rə-līz′) *vt.* **-lyzed, -lyz·ing, -lyz·es.** To subject to pyrolysis.

**py·ro·man·cy** (pī′rə-măn′sē) *n.* [ME *piromance* < OFr. *pyromancie* < LLat. *pyromantia* < Gk. *puromanteia* : *pur-,* fire + *manteia,* divination. —see -MANCY.] Divination by fire or flames. **—py′ro·man′tic** (-măn′tĭk) *adj.*

**py·ro·ma·ni·a** (pī′rō-mā′nē-ə, -măn′yə) *n.* An irresistible urge to start fires. **—py′ro·ma′ni·ac′** (-mā′nē-ăk′) *adj.* & *n.* **—py′ro·ma·ni′a·cal** (-mə-nī′ə-kəl) *adj.*

**py·ro·met·al·lur·gy** (pī′rō-mĕt′l-ûr′jē) *n.* Metallurgy that depends on the action of heat, as smelting. **—py′ro·met′al·lur′gi·cal** (-mĕt′l-ûr′jĭ-kəl) *adj.*

**py·rom·e·ter** (pī-rŏm′ĭ-tər) *n.* An electrical thermometer for measuring high temperatures. **—py′ro·met′ric** (-rə-mĕt′rĭk), **py′ro·met′ri·cal** *adj.* **—py′ro·met′ri·cal·ly** *adv.* **—py·rom′e·try** *n.*

**py·ro·mor·phite** (pī′rə-môr′fīt′) *n.* [G. *Pyromorphit* : Gk. *pur,* fire + Gk. *morphē,* form + *-it,* -ite.] A lead ore with composition $Pb_5(PO_4,AsO_4)_3Cl$, occurring in green, brown, or yellow crystals.

**py·rope** (pī′rōp′) *n.* [ME *pirope* < OFr. < Lat. *pyropus,* gold-bronze < Gk. *purōpos* : *pur,* fire + *ōps,* eye.] A deep-red garnet, $Mg_3Al_2Si_3O_{12}$, used as a gem.

**py·ro·phor·ic** (pī′rə-fôr′ĭk, -fŏr′-) *adj.* [< Gk. *purophoros,* fire-bearing : *pur,* fire + *pherein,* to carry.] **1.** Igniting spontaneously in air. **2.** Producing sparks by friction.

**py·ro·phos·phate** (pī′rə-fŏs′fāt′) *n.* A salt or ester of pyrophosphoric acid. **—py′ro·phos·phat′ic** (-făt′ĭk) *adj.*

**py·ro·phos·phor·ic acid** (pī′rō-fŏs-fôr′ĭk, -fŏr′-) *n.* A syrupy viscous liquid, $H_4P_2O_7$, used as a catalyst and in making of organic chemicals.

**py·ro·phyl·lite** (pī′rō-fĭl′īt′, pī-rŏf′ə-līt′) *n.* A silvery white or pale-green mineral aluminum silicate, $Al_2Si_4O_{10}(OH)_2$, occurring naturally in soft, compact masses.

**py·ro·sis** (pī-rō′sĭs) *n.* [Gk. *purōsis,* a burning < *puroun,* to burn < *pur,* fire.] Heartburn.

**py·ro·stat** (pī′rə-stăt′) *n.* **1.** An automatic sensing device that activates an alarm or extinguisher in case of fire. **2.** A high-temperature thermostat.

**py·ro·sul·fate** (pī′rō-sŭl′fāt′) *n.* [PYROSULF(URIC ACID) + -ATE[2].] A salt of pyrosulfuric acid.

**py·ro·sul·fu·ric acid** (pī′rō-sŭl-fyo͝or′ĭk) *n.* A heavy, oily, colorless to dark-brown liquid, $H_2S_2O_7$, made by adding sulfur trioxide to concentrated sulfuric acid and used in petroleum refining and explosives.

**py·ro·tech·nic** (pī′rə-tĕk′nĭk) *also* **py·ro·tech·ni·cal** (-nĭ-kəl) *adj.* **1.** Of or relating to fireworks. **2.** Suggestive of fireworks : BRILLIANT. **—py′ro·tech′ni·cal·ly** *adv.*

**py·ro·tech·nics** (pī′rə-tĕk′nĭks) *n.* **1.** *(sing. in number).* The art of manufacturing or setting off fireworks. **2.** *(sing. or pl. in number).* A fireworks display. **3.** *(sing. or pl. in number).* A brilliant display, as of rhetoric or wit, or of virtuosity in the performing arts. **—py′ro·tech′nist** *n.*

**py·ro·tech·ny** (pī′rə-tĕk′nē) *n.* PYROTECHNICS 1.

**py·rox·ene** (pī-rŏk′sēn′) *n.* [Fr. *pyroxène* : Gk. *pur,* fire + Gk. *xenos,* stranger.] Any of a group of crystalline mineral silicates common in igneous and metamorphic rocks and containing two metallic oxides, as of magnesium, calcium, iron, or sodium. **—py′rox·en′ic** (pī′rŏk-sē′nĭk, -sĕn′ĭk) *adj.*

**py·rox·e·nite** (pī-rŏk′sə-nīt′) *n.* An igneous rock consisting mainly of pyroxenes. **—py·rox′e·nit′ic** (-nĭt′ĭk) *adj.*

**py·rox·y·lin** (pī-rŏk′sə-lĭn) *also* **py·rox·y·line** (-lēn′, -lĭn) *n.* A highly flammable nitrocellulose used in the manufacture of collodion, plastics, and lacquers.

**pyr·rhic** (pĭr′ĭk) *n.* [Gk. *purrhikhios* < *purrhikhē,* a war dance.] A Greek metrical foot composed of two short syllables. **—pyr′rhic** *adj.*

**Pyr·rhic victory** (pĭr′ĭk) *n.* [From the victory of *Pyrrhus* (319–272 B.C.), king of Epirus, over the Romans at Asculum in 279 B.C.] A victory won at a staggering cost.

**pyr·rho·tite** (pĭr′ə-tīt′) *also* **pyr·rho·tine** (-tīn′) *n.* [G. *Pyrrohotin* < Gk. *purrhotēs,* redness < *purrhos,* fiery < *pur,* fire.] A naturally occurring brownish-bronze iron sulfide, FeS, characterized by weak magnetic properties and used as an iron ore and in making sulfuric acid.

**pyr·rhu·lox·i·a** (pĭr′ə-lŏk′sē-ə, pĭr′yə-) *n.* [NLat. *Pyrrhuloxia,* genus name : *Pyrrhula,* finch genus (< Gk. *purrhoulas,* red bird < *purrhos,* red < *pur,* fire) + *Loxia,* crossbill genus < Gk. *loxos,* oblique.] A crested gray and red bird, *Pyrrhuloxia sinuata* of the southwestern United States and Mexico, having a short thick bill.

**pyr·role** (pîr′ōl′) *n.* [Gk. *purrhos,* red (< *pur,* fire) + -OLE.] A yellowish or brown liquid, $C_4H_5N$, having a characteristic odor similar to chloroform and used to manufacture a wide variety of drugs. **—pyr·ro′lic** (pĭ-rō′lĭk) *adj.*

**py·ru·vic acid** (pī-rōō′vĭk, pĭ-) *n.* [PYR(O)- + Lat. *uva,* grape.] A colorless liquid, $CH_3COCOOH$, formed as a fundamental intermediate in protein and carbohydrate metabolism.

**Py·thag·o·re·an·ism** (pĭ-thăg′ə-rē′ə-nĭz′əm) *n.* Pythagoras' syncretistic philosophy, mainly distinguished by its description of reality in terms of arithmetical relationships. **—Py·thag′o·re′an** (-rē′ən) *n. & adj.*

**Pythagorean theorem** *n.* The theorem that the sum of the squares of the lengths of the sides of a right triangle is equal to the square of the length of the hypotenuse.

**Pyth·i·an** (pĭth′ē-ən) *adj.* [Lat. *Pythius* < Gk. *Puthios* < *puthō,* ancient name of Delphi, Greece.] **1.** Of or relating to Delphi, the temple of Apollo at Delphi, or its oracle. **2.** Of or relating to the Pythian games. **—Pyth′ic** *adj.*

**Pythian games** *pl.n.* A pan-Hellenic athletic tournament held every four years at Delphi in ancient Greece in honor of Apollo.

**Pyth·i·as** (pĭth′ē-əs) *n.* [Lat., alteration of Gk. *Phintias.*] *Rom. Myth.* A Greek who barely escaped execution while being held as a voluntary hostage in place of his condemned friend Damon.

**Py·thon** (pī′thŏn′, -thən) *n.* [Gk. *Puthōn.*] **1.** *Gk. Myth.* A huge serpent monster that lived in the caves of Parnassus and was killed by Apollo. **2. python.** Any of various nonvenomous Old World snakes of the family Pythonidae that coil around and suffocate their prey. **3. python. a.** A soothsaying demon or spirit. **b.** One possessed by such a spirit.

**py·tho·ness** (pī′thə-nĭs, pĭth′ə-) *n.* [ME *phitonesse* < OFr. *phitonise* < LLat. *pythonissa* < Gk. *Puthōn,* Python.] **1.** *Gk. Myth.* The priestess of Apollo at Delphi. **2.** A prophetess.

**py·thon·ic** (pī-thŏn′ĭk) *adj.* **1.** Of, relating to, or like a python. **2.** Of or resembling an oracle : PROPHETIC. **3.** Extraordinarily huge and powerful.

**py·u·ri·a** (pī-yōōr′ē-ə) *n.* Abnormal presence of pus in the urine.

**pyx** *also* **pix** (pĭks) *n.* [ME *pyxe* < Lat. *pyxis,* box < Gk. *puxis.*] **1. a.** A container in which supplies of wafers for the Eucharist are kept. **b.** A container in which the Eucharist is carried to the sick. **2.** A chest in a mint in which specimen coins are deposited to await assay.

**pyx·i·des** (pĭk′sĭ-dēz′) *n. pl. of* PYXIS.

**pyx·id·i·um** (pĭk-sĭd′ē-əm) *n., pl.* **-i·a** (-ē-ə) [NLat. < Gk. *puxidion,* dim. of *puxis,* box.] A pyxis.

**pyx·ie** (pĭk′sē) *n.* [NLat. *Pyxidanthera,* genus name : PYXIS + Med. Lat. *anthera,* pollen. —see ANTHER.] A creeping evergreen shrub, *Pyxidanthera barbulata* native to eastern U.S. pine barrens, having small white or pinkish flowers.

**pyx·is** (pĭk′sĭs) *n., pl.* **pyx·i·des** (-sĭ-dēz′) [Gk. *puxis,* box.] *Bot.* A seed capsule having a circular lid that falls off to release the seeds.

**Pyx·is** (pĭk′sĭs) *n.* [NLat. *Pyxis (nautica),* (mariner's) compass < Gk. *puxis,* box.] A constellation in the Southern Hemisphere.

# Qq

**q** *or* **Q** (kyōō) *n., pl.* **q's** *or* **Q's. 1.** The 17th letter of the English alphabet. **2.** A speech sound represented by the letter *q.* **3.** The 17th in a series.

**Q fever** *n.* [Q(UERY) + FEVER.] An infectious disease caused by the rickettsia *Coxiella burnetii,* marked by fever and muscle pains.

**qin·tar** (kĭn-tär′) *n. var. of* QUINTAR.

**qoph** (kôf) *n.* [Heb. *qōph.*] The 19th letter of the Hebrew alphabet. —See table at ALPHABET.

**q.t.** (kyōō′tē′) *n.* [Short for QUIET.] *Slang.* Quiet <on the *q.t.*>

**qua** (kwā, kwä) *prep.* [Lat., ablative fem. sing. of *qui,* who.] In the character or capacity of : AS <The general *qua* chief of staff signed the order.>

**Quaa·lude** (kwā′lōōd′). A trademark for methaqualone.

**quack¹** (kwăk) *n.* [Imit.] The sound uttered by a duck. *—vi.* **quacked, quack·ing, quacks.** To utter a quack.

**quack²** (kwăk) *n.* [Short for QUACKSALVER.] **1.** One who pretends to have medical expertise. **2.** A charlatan. *—vi.* **quacked, quack·ing, quacks.** To act or do business as a quack. **—quack′ery** *n.*

**quack grass** *n.* [Var. of QUITCH GRASS.] Couch grass.

**quack·sal·ver** (kwăk′săl′vər) *n.* [Obs. Du.] *Archaic.* QUACK².

**quad¹** (kwŏd) *n.* QUADRANGLE 2.

**quad²** (kwŏd) *n.* A quadrat.

**quad³** (kwŏd) *n.* A quadruplet.

**quadr-** *pref. var. of* QUADRI-.

**quad·ran·gle** (kwŏd′răng′gəl) *n.* [ME < OFr. < LLat. *quadrangulum,* neuter of *quadrangulus,* four-cornered, var. of Lat. *quadriangulus* : *quadri-,* four + *angulus,* angle.] **1.** *Math.* A plane figure composed of four points, no three of which are collinear, connected by straight lines. **2. a.** A rectangular area, as a courtyard, enclosed by buildings. **b.** The buildings surrounding this area. **3.** The land area shown on one atlas sheet charted by the U.S. Geological Survey.

**quad·rant** (kwŏd′rənt) *n.* [Lat. *quadrans, quadrant-,* quarter.] **1.** *Math.* **a.** A circular arc subtending a central angle of 90° : one fourth of the circumference of a circle. **b.** The plane area bounded by two perpendicular radii and the arc they subtend. **c.** Any of the four areas into which a plane is divided by the reference axes in a Cartesian coordinate system, designated first, second, third, and fourth, counting clockwise from the area in which both coordinates are positive. **2.** A mechanical device, as a machine part, shaped like a quarter circle. **3.** An early instrument for determining altitudes, composed of a 90° graduated arc with a movable radius for measuring angles.

**quad·ra·phon·ic** (kwŏd′rə-fŏn′ĭk) *adj.* Of or for an extension of stereophonic sound reproduction having two additional channels for greater distibution and fidelity of sound. **—qua·draph′o·ny** (kwŏ-drăf′ə-nē) *n.*

**quad·ra·son·ic** (kwŏd′rə-sŏn′ĭk) *adj.* Quadraphonic.

**quad·rat** (kwŏd′rət, -răt′) *n.* [Var. of QUADRATE.] A piece of printing type metal lower than the raised typeface, used for filling spaces and blank lines.

**quad·rate** (kwŏd′rāt′, -rĭt) *n.* [ME < Lat. *quadratus,* p.part. of *quadrare,* to make square < *quadrus,* square.] **1. a.** A square or cube. **b.** An approx. square or cubic area, volume, or object. **2.** *Zool.* A bone or cartilaginous structure of the skull that joins the upper and lower jaws in fish, birds, reptiles, and amphibians. *—adj.* **1.** Having four sides and four angles. **2.** *Zool.* Designating the quadrate bone or cartilage. *—vi.* **-rat·ed, -rat·ing, -rates.** *Archaic.* To correspond : agree.

**quad·rat·ic** (kwŏ-drăt′ĭk) *adj.* [< QUADRATE.] Of, relating to, or possessing mathematical quantities of the second degree or less. **—quad·rat′ic** *n.* **—quad·rat′i·cal·ly** *adv.*

**quadratic equation** *n.* An equation of the second degree, having the general form $ax^2 + bx + c = 0$, where *a, b,* and *c* are constants.

**quadratic formula** *n.* The formula $x = [-b \pm \sqrt{(b^2 - 4ac)}]/2a$, used for computing the roots of a quadratic equation.

**quad·rat·ics** (kwŏ-drăt′ĭks) *n. (sing. in number).* The algebra of quadratic equations.

**quad·ra·ture** (kwŏd′rə-chŏŏr′) *n.* **1.** The process of making something square. **2.** *Math.* The process of constructing a square equal in area to a given surface. **3.** A configuration in which the angular separation of two celestial bodies, as measured from a third, is 90°

**quad·ren·ni·al** (kwŏ-drĕn′ē-əl) *adj.* **1.** Occurring once in four years. **2.** Lasting for four years. **—quad·ren'ni·al** *n.* **—quad·ren'ni·al·ly** *adv.*

**quad·ren·ni·um** (kwŏ-drĕn'ē-əm) *n., pl.* **-ni·ums** *or* **-ni·a** (-ē-ə) [Lat. *quadriennium* : *quadri-*, four + *annus*, year.] A period of four years.

**quadri-** *or* **quadru-** *or* **quadr-** *pref.* [ME < Lat.] **1.** Four <*quadri*lateral> **2.** Square <*quadr*ate>

**quad·ric** (kwŏd'rĭk) *adj.* Of or relating to geometric surfaces defined by quadratic equations.

**quad·ri·cen·ten·ni·al** (kwŏd'rĭ-sĕn-tĕn'ē-əl) *n.* A 400th anniversary. **—quad'ri·cen·ten'ni·al** *adj.*

**quad·ri·ceps** (kwŏd'rĭ-sĕps') *n.* [QUADRI- + (BI)CEPS.] The large four-part extensor muscle at the front of the thigh. **—quad'ri·cip'i·tal** (-sĭp'ĭ-təl) *adj.*

**quad·ri·fid** (kwăd'rə-fĭd') *adj. Bot.* Divided into four parts <a *quadrifid* leaf>

**quad·ri·ga** (kwŏd'rī-gə) *n., pl.* **-gae** (-gē) [Lat., sing. of *quadrigae*, team of four horses, short for *quadrijugae*, fem. pl. of *quadrijugus*, of a team of four : *quadri-*, four + *jungere*, to yoke.] A two-wheeled chariot drawn by four horses harnessed abreast.

**quad·ri·lat·er·al** (kwŏd'rə-lăt'ər-əl) *n. Math.* A four-sided polygon. **—quad'ri·lat'er·al** *adj.*

**qua·drille¹** (kwŏ-drĭl', kwə-, kə-) *n.* [Fr. < *quadrille*, one of four divisions of an army < Sp. *cuadrilla*, dim. of *cuadra*, square < Lat. *quadra*.] **1.** A square dance composed of five figures and performed by four couples. **2.** Music for the quadrille in 6/8 and 2/4 time.

**qua·drille²** (kwŏ-drĭl', kwə-, kə-) *n.* [Fr., perh. < Sp. *cuartillo* < *cuarto*, fourth < Lat. *quartus*.] An 18th-cent. card game played by four people, using a deck of 40 cards.

**quad·ril·lion** (kwŏ-drĭl'yən) *n.* [QUADR(I)- + (M)ILLION.] **1.** The cardinal number equal to $10^{15}$. **2.** *Chiefly Brit.* SEPTILLION 1. **—quad·ril'lion** *adj.*

**quad·ril·lionth** (kwŏ-drĭl'yənth) *n.* The ordinal number matching quadrillion in a series. **—quad·ril'lionth** *adj.* & *adv.*

**quad·ri·par·tite** (kwŏd'rə-pär'tīt') *adj.* **1.** Composed of or divided into four parts. **2.** Involving four participants.

**quad·ri·phon·ic** (kwŏd'rə-fŏn'ĭk) *adj.* Quadraphonic. **—quad'ri·phon'y** *n.*

**quad·ri·ple·gi·a** (kwŏd'rə-plē'jē-ə, -jə) *n.* Total paralysis from the neck down. **—quad'ri·ple'gic** (-jĭk) *adj.* & *n.*

**quad·ri·va·lent** (kwŏd'rə-vā'lənt) *adj. Chem.* **1.** Having four valences. **2.** Having a valence of four : TETRAVALENT. **—quad'ri·va'lence, quad'ri·va'len·cy** *n.*

**quad·riv·i·um** (kwŏ-drĭv'ē-əm) *n., pl.* **-i·a** (-ē-ə) [LLat. < Lat., place where four roads meet : *quadri-*, four + *via*, road.] The higher division of the seven liberal arts in the Middle Ages, comprising arithmetic, geometry, astronomy, and music.

**quad·roon** (kwŏ-dro͞on') *n.* [Sp. *cuarteron* < *cuarto*, quarter < Lat. *quartus*.] One having a quarter Negro ancestry.

**quadru-** *pref. var. of* QUADRI-

**quad·ru·ma·nous** (kwŏ-dro͞o'mə-nəs) *also* **quad·ru·ma·nal** (-nəl) *adj.* [QUADRU- + Lat. *manus*, hand.] *Zool.* Having four feet with opposable first digits, as primates other than humans.

**quad·rum·vi·rate** (kwŏd'rəm-və-rĭt) *n.* [QUADR(I)- + (TRI)UMVIRATE.] A group of four persons sharing office or authority, esp. a government of four persons. **—quad'rum·vir** *n.*

**quad·ru·ped** (kwŏd'rə-pĕd') *n.* A four-footed animal. **—quad·ru'pe·dal** (kwŏ-dro͞o'pə-dəl, kwŏd'rə-pĕd'l) *adj.*

**quad·ru·ple** (kwŏ-dro͞o'pəl, -drŭp'əl, kwŏd'ro͞o-pəl) *adj.* [Fr. < Lat. *quadruplus*.] **1.** Consisting of four parts or members. **2.** Multiplied by four : FOURFOLD. **3.** *Mus.* Having four beats to the measure. —*v.* **-pled, -pling, -ples.** —*vt.* To multiply by four : increase fourfold. —*vi.* To become quadrupled. **—quad·ru'ple** *n.*

**quad·ru·plet** (kwŏ-drŭp'lĭt, -dro͞o'plĭt, kwŏd'rə-plĭt) *n.* **1.** A group of four related by common properties or behavior. **2.** One of four born at a single birth.

**quad·ru·pli·cate** (kwŏ-dro͞o'plĭ-kĭt) *adj.* [Lat. *quadruplicatus*, p.part. of *quadruplicare*, to multiply by four < *quadruplex*, fourfold : *quadru-*, four + *-plex*, -fold.] **1. a.** QUADRUPLE 2. **b.** Reproduced or copied four times. **2.** Fourth in a group of four identical things. —*v.* (-kāt') **-cat·ed, -cat·ing, -cates.** —*vt.* To multiply or copy four times. —*vi.* To become quadruplicated. **—quad·ru'pli·cate** *n.* **—quad·ru'pli·cate·ly** (-kĭt-lē) *adv.* **—quad·ru'pli·ca'tion** *n.*

**quaes·tor** (kwĕs'tər, kwē'stər) *n.* [ME *questor* < Lat. *quaestor* < *quaerere*, to inquire.] A government or military official in ancient Rome responsible for administration and finance. **—quaes·to'ri·al** (kwĕ-stôr'ē-əl, -stōr'-, kwē-) *adj.* **—quaes'tor·ship'** *n.*

**quaff** (kwŏf, kwăf, kwôf) *v.* **quaffed, quaff·ing, quaffs.** [Orig. unknown.] —*vt.* & *vi.* To drink heartily. —*n.* A hearty draft. **—quaff'er** *n.*

**quag** (kwăg, kwŏg) *n.* [Orig. unknown.] A quagmire.

**quag·ga** (kwăg'ə, kwŏg'ə) *n.* [Afr. < Xhosa *i-qwara*, perh. < Hottentot *qúagga*.] A zebralike mammal, *Equus quagga* of southern Africa, extinct since the late 19th cent.

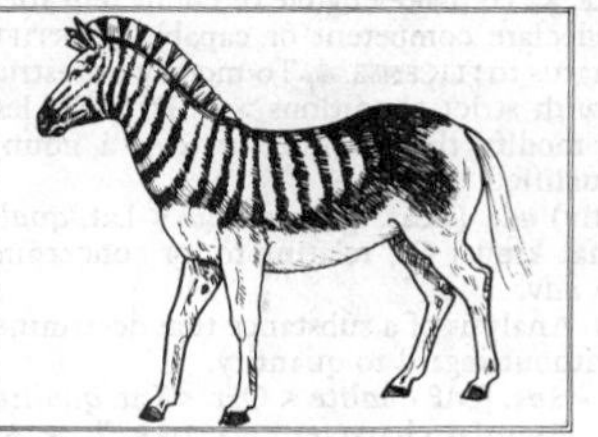

quagga

**quag·gy** (kwăg'ē, kwŏg'ē) *adj.* **-gi·er, -gi·est.** **1.** Having a soft, spongy surface : MARSHY. **2.** Lacking firmness : FLABBY.

**quag·mire** (kwăg'mīr', kwŏg'-) *n.* **1.** Land with a soft, yielding surface. **2.** A difficult or irksome situation : PREDICAMENT <the *quagmire* of bureaucratic regulations>

**qua·hog** *also* **qua·haug** (kwô'hôg', -hŏg', kwō'-, kō'-) *n.* [Narraganset *poquaûhock*.] A hard-shelled edible clam, *Venus mercenaria*, of the Atlantic coast of North America.

**quai** (kā, kē) *n.* A quay.

**quaich** *also* **quaigh** (kwāKH) *n.* [Sc. Gael. *cuach*.] A two-handled Scottish drinking cup.

**quail¹** (kwāl) *n., pl.* **quail** *or* **quails.** [ME *quaille* < OFr. < Med. Lat. *quaccula*.] **1.** A small chickenlike Old World bird of the genus *Coturnix*, esp. *C. coturnix*, with mottled brown plumage and a short tail. **2.** A New World bird, as the bobwhite, resembling or related to the quail.

**quail²** (kwāl) *vi.* **quailed, quail·ing, quails.** [ME *quailen*, to give way.] To shrink back in fear : COWER.

**quaint** (kwānt) *adj.* **-er, -est.** [ME *cointe*, strange < OFr. < Lat. *cognitus*, p.part of *cognoscere*, to learn. —see COGNITION.] **1.** Charmingly old-fashioned <*quaint* Irish villages> **2.** Unusual or unfamiliar : STRANGE <foreigners with *quaint* speech> **—quaint'ly** *adv.* **—quaint'ness** *n.*

☆ ***syns:*** QUAINT, FUNNY, ODD, ODDBALL *adj. core meaning* : charmingly curious, esp. in an old-fashioned or unusual way <an old house with *quaint* meandering stairways>

▲ word history: The adjective *quaint* is ultimately derived from Latin *cognitus*, "known," the past participle of *cognoscere*, "to know, learn." English did not borrow the Latin word directly but rather its Old French descendent *cointe* or *queinte*, which had developed senses greatly removed from that of the Latin word. Old French *queinte* meant basically "wise," "skilled," and "clever." From this use the word developed the senses "cleverly or ingeniously made or done" and "strange, curious." The usual modern sense, "charmingly old-fashioned," is probably not a direct outgrowth of the now obsolete senses borrowed from French but stems from an artificial revival of *quaint* by writers of the Romantic period. Many of these writers were interested in medieval customs, traditions, and literature, and they occasionally used obsolete and charmingly old-fashioned words—not always correctly—when writing on medieval topics.

**quake** (kwāk) *vi.* **quaked, quak·ing, quakes.** [ME *quaken* < OE *cwacian*.] **1.** To tremble or shake with instability or shock. **2.** To shiver or shudder, as with cold or fright. —*n.* **1.** An instance of quaking. **2.** An earthquake. **—quak'y** *adj.*

**Quak·er** (kwā'kər) *n.* [< QUAKE.] *Informal.* FRIEND 5. **—Quak'er·ism** *n.* **—Quak'er·ly** *adv.* & *adj.*

**Quaker gun** *n.* [From the *Quakers'* opposition to war.] A dummy gun or artillery piece, often made of wood.

**Quak·er-la·dies** (kwā'kər-lā'dēz) *pl.n.* Bluets.

**qua·le** (kwä'lē) *n., pl.* **-li·a** (-lē-ə) [Lat., neuter of *qualis*, of what kind.] A property, as coldness or softness, considered independently from things having the property.

**qual·i·fi·ca·tion** (kwŏl'ə-fĭ-kā'shən) *n.* **1.** An act of qualifying or the state of being qualified. **2.** A quality or ability that makes one eligible or suitable <has good *qualifications* to be a manager> **3.** A condition or circumstance that must be satisfied. **4.** A restriction or modification.

☆ ***syns:*** QUALIFICATION, ELIGIBILITY, ELIGIBLENESS, FITNESS, SUITABLENESS *n. core meaning* : the quality or state of being eligible <a person of unquestionable *qualification*>

**qual·i·fied** (kwŏl'ə-fīd') *adj.* **1. a.** Able to perform as required : COMPETENT. **b.** Having met the requirements for a specific position or task. **2.** Not total or wholehearted : LIMITED. **—qual'i·fied'ly** (-fīd'lē, -fī'ĭd-lē) *adv.*

☆ ***syns:*** QUALIFIED, LIMITED, MODIFIED, RESERVED, RESTRICTED *adj. core meaning* : not total, unlimited, or wholehearted <a *qualified* plan for expansion> ***ant:*** absolute, unqualified

**qual·i·fi·er** (kwŏl'ə-fī'ər) *n.* **1.** One that qualifies. **2.** A word or phrase that qualifies or modifies the meaning of another word or phrase.

**qual·i·fy** (kwŏl′ə-fī′) *v.* **-fied, -fy·ing, -fies.** [OFr. *qualifier* < Med. Lat. *qualificare,* to attribute a quality to : Lat. *qualis,* of such a kind + Lat. *facere,* to make.] —*vt.* **1.** To describe by specifying the characteristics of : CHARACTERIZE. **2.** To make eligible or competent for a task, or position. **3. a.** To declare competent or capable : CERTIFY. **b.** To give official or legal status to : LICENSE. **4.** To modify or restrict <*qualified* their approval with strict conditions> **5.** To make less intense : MODERATE. **6.** To modify the meaning of (e.g., a noun). —*vi.* To be or to become qualified.

**qual·i·ta·tive** (kwŏl′ĭ-tā′tĭv) *adj.* [LLat. *qualitativus* < Lat. *qualitas,* quality < *qualis,* of what kind.] Of, relating to, or concerning quality. **—qual′i·ta′tive·ly** *adv.*

**qualitative analysis** *n.* Analysis of a substance that determines its chemical constituents without regard to quantity.

**qual·i·ty** (kwŏl′ĭ-tē) *n., pl.* **-ties.** [ME *qualite* < OFr. < Lat. *qualitas* < *qualis,* of what kind.] **1.** Essential character : NATURE. **2. a.** An inherent or distinguishing attribute : PROPERTY. **b.** A character trait <has many admirable *qualities*> **3. a.** Superiority of kind <can't afford to buy *quality*> **b.** Degree or grade of excellence <service of poor *quality*> **4. a.** High social standing. **b.** The upper class. **5.** *Mus.* Timbre, as determined by overtones. **6.** The character of a vowel sound determined by the size and shape of the oral cavity and the amount of resonance with which the sound is produced. **7.** *Logic.* The positive or negative character of a proposition. —*adj.* Superior of its kind <*quality* workmanship>

**quality control** *n.* A system for maintaining proper standards in manufactured goods, esp. by regular inspection of the product.

**qualm** (kwäm, kwôm) *n.* [Orig. unknown.] **1.** A sudden feeling of faintness or nausea. **2.** A sensation of doubt or uneasiness : MISGIVING. **3.** A pang of conscience.

☆ ***syns:*** QUALM, COMPUNCTION, MISGIVING, RESERVATION, SCRUPLE *n. core meaning* : a feeling of uncertainty about the fitness or correctness of an action <had no *qualms* about asking for a raise>

**qualm·ish** (kwä′mĭsh, kwôm′-ĭsh) *adj.* **1.** Feeling qualms. **2.** Of or producing qualms. **—qualm′ish·ly** *adv.*

**quam·ash** (kwŏm′ăsh′) *n. var. of* CAMAS.

**quan·da·ry** (kwŏn′də-rē, -drē) *n., pl.* **-ries.** [Orig. unknown.] A perplexing situation or state : DILEMMA.

**quan·ta** (kwŏn′tə) *n. pl. of* QUANTUM.

**quan·ta·some** (kwŏn′tə-sōm′) *n.* One of numerous granules found on the inner lamellar surface of a chloroplast.

**quan·ti·fy** (kwŏn′tə-fī′) *vt.* **-fied, -fy·ing, -fies.** [Med. Lat. *quantificare* : Lat. *quantus,* how great + Lat. *facere,* to make.] **1.** To measure or express the quantity of. **2.** *Logic.* To limit the variables of (a proposition) by means of an operator such as *all* or *some.* **—quan′ti·fi′a·ble** *adj.* **—quan′ti·fi·ca′tion** (-fĭ-kā′shən) *n.* **—quan′ti·fi′er** *n.*

**quan·ti·tate** (kwŏn′tĭ-tāt′) *vt.* **-tat·ed, -tat·ing, -tates.** QUANTIFY 1. **—quan′ti·ta′tion** *n.*

**quan·ti·ta·tive** (kwŏn′tĭ-tā′tĭv) *adj.* [Med. Lat. *quantitativus* < Lat. *quantitas,* quantity < *quantus,* how great.] **1. a.** Expressed or capable of expression as a quantity. **b.** Of, relating to, or susceptible of measurement. **c.** Of or relating to number or measurement. **2.** Relating to syllables esp. in classical verse that are based on duration of sound rather than stress. **—quan′ti·ta′tive·ly** *adv.* **—quan′ti·ta′tive·ness** *n.*

**quantitative analysis** *n.* Analysis of a substance that determines the amounts or proportions of its chemical constituents.

**quantitative gene** *n. Genetics.* A polygene.

**quan·ti·ty** (kwŏn′tĭ-tē) *n., pl.* **-ties.** [ME *quantite* < OFr. < Lat. *quantitas* < *quantus,* how great.] **1. a.** A specified or undetermined number or amount. **b.** A large number or amount <bought office supplies in *quantity*> **c.** An exact amount or number. **2.** The property or aspect of a thing that can be measured, counted, or compared. **3.** *Math.* The object of a mathematical operation. **4.** The length of a vowel or consonant sound expressed in terms of its temporal duration. **5.** *Logic.* The exact character of a proposition in reference to its universality, singularity, or particularity.

☆ ***syns:*** QUANTITY, AMOUNT, BODY, BUDGET, BULK, CORPUS, QUANTUM *n. core meaning* : a measurable whole <a large *quantity* of coal>

**quan·tize** (kwŏn′tīz′) *vt.* **-tized, -tiz·ing, -tiz·es.** *Physics.* **1.** To limit the possible values of (e.g., a quantity) to a discrete set of values by quantum mechanical rules. **2.** To replace the dynamic variables of a system by the corresponding quantum mechanical operators in order to calculate the behavior of the system. **—quan′ti·za′tion** *n.*

**quan·tum** (kwŏn′təm) *n., pl.* **-ta** (-tə) [Lat., neuter of *quantus,* how great.] **1.** A quantity or amount. **2.** A specified portion. **3.** Something that can be measured or counted. **4.** *Physics.* **a.** An indivisible unit of energy, equal for radiation of frequency *v* to the product h*v*, where h is Planck's constant. **b.** The particle mediating a specific type of fundamental interaction.

**quantum chromodynamics** *n. (sing. in number).* Chromodynamics.

**quantum electrodynamics** *n. Physics. (sing. in number).* The quantum mechanical theory of the properties and interactions of charged elementary particles, esp. of the electron, with the electromagnetic field.

**quantum fluid** *n. Physics.* Quantum liquid.

**quantum jump** *n.* **1.** *Physics.* Transition of an atomic or molecular system from one discrete energy level to another with concomitant emission or absorption of radiation having energy equal to the difference between the two levels. **2.** A sudden change or increase, as in knowledge or information.

**quantum liquid** *n. Physics.* A fluid exhibiting thermal, kinetic, or conductive behavior attributable to the quantum statistics obeyed by the particles of the fluid, esp. a superfluid.

**quantum mechanics** *n. (sing. or pl. in number). Physics.* Quantum theory, esp. the quantum theory of the structure and behavior of atoms and molecules.

**quantum number** *n. Physics.* Any of a set of real numbers that individually characterize the properties and collectively specify the state of a particle or of an atomic system.

**quantum state** *n. Physics.* Any of the possible states of a system described by quantum theory.

**quantum theory** *n. Physics.* A mathematical theory of dynamic systems in which dynamic variables are represented by abstract mathematical operators having properties that specify the behavior of the system.

**quar·an·tine** (kwôr′ən-tēn′, kwŏr′-) *n.* [Ital. *quarantine* < *quaranta,* forty < Lat. *quadraginta.*] **1. a.** A period of enforced isolation at a port of entry that is imposed on a vehicle, a person, or material suspected of carrying a contagious disease. **b.** A place for such isolation. **2.** Enforced isolation or restriction of the spread of movement imposed to prevent contagious disease. **3.** A state of enforced isolation or detention. **4.** A period of 40 days. —*vt.* **-tined, -tin·ing, -tines. 1.** To detain in or as if in quarantine. **2.** To isolate politically or economically.

**quark** (kwôrk) *n.* [Poss. < *Three quarks for Mister Marks!,* a line in *Finnegans Wake* by James Joyce (1882–1941).] *Physics.* Any of a group of hypothetical subatomic particles having electric charges of magnitude one-third or two-thirds that of the electron, proposed as the fundamental units of matter.

**quar·rel¹** (kwôr′əl, kwŏr′-) *n.* [ME *querele,* complaint < OFr. < Lat. *querela* < *queri,* to complain.] **1.** An angry disagreement or dispute : ARGUMENT. **2.** A cause for complaint or dispute <My *quarrel* is with the grading system.> —*vi.* **-reled, -rel·ing, -rels** *or* **-relled, -rel·ling, -rels. 1.** To take part in a quarrel : argue angrily. **2.** To differ : disagree <Let's not *quarrel* over details.> **3.** To find fault : COMPLAIN. **—quar′rel·er, quar′rel·ler** *n.*

**quar·rel²** (kwôr′əl, kwŏr′-) *n.* [ME *quarel* < OFr. < VLat. **quadrellus,* dim. of Lat. *quadrus,* square.] **1.** A bolt for a crossbow. **2.** A tool, as a stonemason's chisel, having a squared head. **3.** A small pane or tile in the shape of a diamond or square.

**quar·rel·some** (kwôr′əl-səm, kwŏr′-) *adj.* Inclined to quarrel.

**quar·ry¹** (kwôr′ē, kwŏr′ē) *n., pl.* **-ries.** [ME *querre,* entrails of a deer given to hounds as a reward < OFr. *cuiree* < LLat. *corata,* viscera < Lat. *cor,* heart.] **1.** An animal hunted for food or sport : PREY. **2.** The object of a hunt or pursuit.

**quar·ry²** (kwôr′ē, kwŏr′ē) *n., pl.* **-ries.** [ME *quarey* < OFr. *quarriere* < Lat. *quadrus,* square.] **1.** An open excavation or pit from which stone is obtained. **2.** A rich or productive source. —*vt.* **-ried, -ry·ing, -ries. 1.** To extract (stone) from a quarry, as by cutting or blasting. **2.** To use (land) as a quarry. **—quar′ri·er** *n.*

**quar·ry³** (kwôr′ē, kwŏr′ē) *n., pl.* **-ries.** [Var. of QUARREL².] QUARREL² 3.

**quart** (kwôrt) *n.* [ME < OFr. *quarte* < Lat. *quartus,* fourth.] **1. a.** A unit of liquid volume or capacity in the U.S. Customary System equal to 2 pints or .946 liter. **b.** A unit of dry volume or capacity in the U.S. Customary System equal to 2 pints or 1.101 liters. **c.** A unit of volume or capacity in the British Imperial System equal to 1.201 U.S. liquid quarts, 1.032 U.S. dry quarts, or 69.354 cubic inches. **2.** A container with a capacity of one quart.

**quar·tan** (kwôrt′n) *adj.* [ME *quartain* < OFr. *quartaine* < Lat. *quartana* < *quartanus,* of the fourth < *quartus,* fourth.] Recurring at four-day intervals, or approx. every 72 hours. —Used of a fever. **—quar′tan** *n.*

**quar·ter** (kwôr′tər) *n.* [ME < OFr. *quartier* < Lat. *quartarius* < *quartus,* fourth.] **1.** One of four equal parts. **2.** A coin valued at one fourth of the U.S. or Canadian dollar. **3.** One fourth of an hour. ***usage:*** When referring to the time, *quarter* may occur with or without *a* in the following phrases: (*a*) *quarter of* (or *to* or *before* ) *ten;* (*a*) *quarter past* (or *after*) *six.* **4. a.** One fourth of a year : three months <Earnings rose in the final *quarter.*> **b.** An academic term lasting for approx. three months. **5.** *Astron.* One fourth of the period of the moon's revolution around the earth. **6.** One of the four equal periods of playing time of a game. **7.** One fourth of a yard : nine inches. **8.** One fourth of a mile : two furlongs. **9.** One fourth of a pound : four ounces. **10.** One fourth of a ton : 500 pounds. —Used as a measure of grain. **11.** *Chiefly Brit.* A measure of grain equal to approx. eight bushels. **12. a.** One fourth of a hundredweight : 25 pounds. **b.** One fourth of a British hundredweight : 28 pounds. **13. a.** Any of the four divisions of the compass as defined by the cardinal points. **b.** *Naut.* The general direction on either side of a ship located 45° off the stern. **c.** Any of the four major divisions of the horizon corresponding to the four quarters of the compass. **d.** An area of the earth regarded as included in a specified quarter. **14.** *Naut.* **a.** The upper part of the after side of a ship, usu. between the aftermost mast and the stern. **b.** The portion of a yard between the yardarm and the slings. **15.** *Heraldry.* Any of four equal sections of a shield. **16.** One leg of an animal carcass, usu. including the adjoining parts. **17.** Either side of a horse's hoof. **18.** The back part of a shoe between the vamp and the heel. **19. quarters. a.** A place of lodging or residence. **b.** Housing for military personnel : BARRACKS. **c.** Working or living space <an office with cramped *quarters*> **20.** *often* **quarters.** An assigned station, as for officers and crew on a warship. **21.** A distinctive section of a city or town <the immigrant *quarter*> **22.** *often* **quarters.** An unspecified person or group <received praise from all *quarters*> **23.** Mercy or clemency, esp. toward an enemy. —*v.* **-tered, -ter·ing, -ters.** —*vt.* **1. a.** To cut or divide into four equal or equivalent parts. **b.** To quartersaw. **2.** To divide or separate into segments <*quarter* an orange> **3.** To dismember (a carcass) into four parts. **4.** *Heraldry.* To divide (a shield) into symmetrical equal areas. **5. a.** To mark or place (e.g., holes) a fourth of a circle apart. **b.** To position (machine parts) at right angles to one another. **6.** To provide (e.g., troops) with housing or shelter. **7.** To cover or search (terrain) by ranging from side to side while slowly advancing. —*vi.* **1.** To take up or be assigned lodgings. **2.** To cover an area of ground by ranging over it from side to side. —*adj.* **1.** Being one of four equal or equivalent parts. **2.** Being one fourth of a standard or usual value. **—at close quarters.** At close range.

**quar·ter·age** (kwôr′tər-ĭj) *n.* A monetary payment, as a wage or allowance, that is made or received quarterly.

**quar·ter·back** (kwôr′tər-băk′) *n. Football.* The offensive backfield player who directs and usu. initiates the plays. —*v.* **-backed, -back·ing, -backs.** —*vt.* **1.** To direct the offense of. **2.** To lead or direct the operations of. —*vi.* To play the quarterback position.

**quarter day** *n.* Any of the four days of the year considered to be the beginning of a new season or quarter when most quarterly payments are due.

**quar·ter-deck** (kwôr′tər-dĕk′) *n.* The stern portion of the upper deck of a sailing ship, usu. reserved for officers.

**quar·ter·fi·nal** (kwôr′tər-fī′nəl) *adj.* Designating one of four competitions in a tournament, whose winners go on to play in semifinal competitions. —*n.* **1. quarterfinals.** A quarterfinal round. **2.** A quarterfinal match. **—quar′ter·fi′nal·ist** *n.*

**quarter horse** *n.* [So called because it was formerly trained for races up to a quarter mile.] A strong saddle horse orig. bred in the western United States.

**quar·ter-hour** *also* **quarter hour** (kwôr′tər-our′) *n.* **1.** Fifteen minutes. **2.** The point on a clock's face marking either 15 minutes after or 15 minutes before an hour.

**quar·ter·ly** (kwôr′tər-lē) *adj.* **1.** Of or pertaining to a quarter or quarters. **2.** Being one of four parts. **3.** Occurring at three-month intervals. —*n., pl.* **-lies. 1.** A publication issued regularly four times a year. **2.** An examination taken at three-month intervals. —*adv.* In or by quarters.

**quar·ter·mas·ter** (kwôr′tər-măs′tər) *n.* **1.** A military officer in charge of administering provisions and supplies. **2.** A petty officer responsible for the navigation of a ship.

**quar·tern** (kwôr′tərn) *n.* [ME *quartron* < OFr. < *quartier,* quarter. —see QUARTER.] One fourth of something.

**quarter note** *n. Mus.* A note having one fourth the time value of a whole note.

**quar·ter-phase** (kwôr′tər-fāz′) *adj.* Two-phase.

**quar·ter·saw** (kwôr′tər-sô′) *vt.* **-sawed, -sawed** or **-sawn** (-sôn′) **-saw·ing, -saws.** To saw (a log) into quarters lengthwise along its axis.

**quarter section** *n.* A quarter of a square mile of land or 160 acres.

**quarter sessions** *pl.n. Law.* **1.** A British local court of limited jurisdiction that sits quarterly. **2.** A U.S. local court having criminal jurisdiction and occas. administrative functions, that sits quarterly.

**quar·ter·staff** (kwôr′tər-stăf′) *n., pl.* **-staves** (-stāvz′). A long, sturdy staff, once used as a weapon.

**quar·ter·tone** (kwôr′tər-tōn′) *n. Mus.* Half a semitone.

**quar·tet** *also* **quar·tette** (kwôr-tĕt′) *n.* [Ital. *quartetto,* dim. of *quarto,* fourth < Lat. *quartus.*] **1.** A musical composition for four voices or instruments. **2.** An ensemble of four musicians. **3.** A set of four persons or things.

**quar·tic** (kwôr′tĭk) *adj.* [< Lat. *quartus,* fourth.] *Math.* Of or designating the fourth degree. **—quar′tic** *n.*

**quar·tile** (kwôr′tīl′, -tĭl) *n.* [< Lat. *quartus,* fourth.] *Statistics.* The value of the boundary at the 25th, 50th, or 75th percentiles of a frequency distribution divided into four parts, each containing a quarter of the population.

**quar·to** (kwôr′tō) *n., pl.* **-tos.** [< Lat. *quartus,* fourth.] **1.** A page size equal to one fourth of a whole sheet, orig. obtained by folding a sheet into quarters. **2.** A book made up of quarto pages.

**quartz** (kwôrts) *n.* [G. *Quarz* < MHG, of Slav. orig.] A hard, crystalline, vitreous mineral silicon dioxide, $SiO_2$, occurring abundantly as a component of granite and sandstone or as various pure crystals such as agate, flint, chert, opal, and chalcedony. **—quartz′ose′** (kwôrt′sōs′) *adj.*

**quartz glass** *n.* A pure silica glass, highly transparent to ultraviolet radiations.

**quartz·if·er·ous** (kwôrt-sĭf′ər-əs) *adj.* Containing quartz.

**quartz·ite** (kwôrt′sīt′) *n.* A metamorphic rock formed by the recrystallization of quartz sandstone.

**quartz lamp** *n.* An incandescent lamp enclosed by a quartz envelope containing mercury vapor that emits ultraviolet radiation when heated by a filament.

**qua·sar** (kwā′zär′, -sär′, -zər, -sər) *n.* [QUAS(I) + (STELL)AR.] A quasi-stellar object.

**quash** (kwŏsh) *vt.* **quashed, quash·ing, quash·es.** [ME *quassen* < OFr. *casser* < LLat. *cassare* < Lat. *cassus,* void.] **1.** To set aside or annul, esp. by judicial action. **2.** To crush or forcibly suppress <*quash* a revolt>

**qua·si** (kwā′zī′, -sī′, kwä′zē, -sē) *adj.* [Lat., as if.] Having a likeness to something : RESEMBLING <a *quasi* government>

**quasi-** *pref.* [< Lat. *quasi,* as if.] To some degree : in some manner <*quasi*-scientific>

**qua·si-stel·lar object** (kwā′zī-stĕl′ər, -sī′-, kwä′zē-, -sē-) *n.* A member of any of several classes of starlike objects having exceptionally large red shifts that are often emitters of radio frequency as well as visible radiation and have apparently immense speeds, energies, and distances from earth.

**quasi-stellar radio source** *n.* A quasi-stellar object.

**quas·sia** (kwŏsh′ə) *n.* [NLat., after Graman *Quassi,* an 18th-cent. Surinamian.] **1.** A tropical American tree, *Quassia amara,* bearing bright scarlet flowers. **2.** A bitter substance derived from the wood and bark of the quassia and related trees, used in medicine and as an insecticide.

**qua·ter·nar·y** (kwŏt′ər-nĕr′ē, kwə-tûr′nə-rē) *adj.* [Lat. *quaternarius* < *quaterni,* by four < *quater,* four times.] **1.** Having four parts : being in fours. **2.** *Chem.* **a.** Designating a compound having four alkyl groups bonded to a nitrogen or phosphorus atom. **b.** Designating a compound composed of four different atoms or radicals. **3. Quaternary.** Of, belonging to, or designating the geologic time, system of rocks, and sedimentary deposits of the second period of the Cenozoic era, extending from the close of the Tertiary through the present, divided into the Pleistocene and Holocene epochs and marked by the appearance of humans. —*n., pl.* **-nar·ies. 1.** The number four. **2.** That member of a group which is fourth in order or degree. **3. Quaternary.** The Quaternary period or system of deposits.

**quaternary ammonium compound** *n.* Any of a group of compounds in which organic radicals replace the hydrogen atoms of the ammonium radical, used as solvents, antiseptics, and emulsifying agents.

**qua·ter·ni·on** (kwə-tûr′nē-ən) *n.* [ME *quaternioun* < LLat. *quaternio* < Lat. *quaterni,* by four < *quater,* four times.] **1.** A set of four persons or things. **2.** *Math.* An element of a system of four dimensional vectors obeying laws similar to those of complex numbers.

**quat·rain** (kwŏt′rān′, kwŏ-trān′) *n.* [Fr. < OFr. < *quatre,* four < Lat. *quattuor.*] A stanza or poem consisting of four lines.

**quat·re·foil** (kăt′ər-foil′, kăt′rə-) *n.* [ME *quaterfoile* : *quater-,* four (< OFr. *quatre* < Lat. *quattuor*) + *foil,* leaf < OFr. < Lat. *folium.*] **1.** A stylized representation of a four-petaled flower or a four-lobed leaf, esp. in heraldry. **2.** An ornament or tracery with four foils or lobes.

**quatrefoil**

**quat·tro·cen·to** (kwŏt′rō-chĕn′tō) *n.* [Ital., short for *millequattrocento,* one thousand four hundred.] The 15th-cent. period of Italian art, architecture, and literature.

**qua·ver** (kwā′vər) *v.* **-vered, -ver·ing, -vers.** [ME *quaveren,* freq. of *quaren,* to tremble.] —*vi.* **1.** To tremble, as from emotion or weakness : QUIVER. **2.** To speak in a tremulous voice or utter a tremulous sound. **3.** To produce a trill on a musical instrument or with the voice. —*vt.* To sing or utter with a trill. —*n.* **1.** A quivering sound. **2.** A trill. **3.** *Mus. Chiefly Brit.* An eighth note. **—qua'ver·ing·ly** *adv.* **—qua'ver·y** *adj.*

**quay** (kē, kā) *n.* [ME *keye* < OFr. *quai,* of Celt. orig.] A wharf or paved embankment for loading and unloading ships.

**quay·age** (kē'ĭj) *n.* **1.** A charge for the use of a quay. **2.** Room or space on quay. **3.** Quays as a whole.

**quean** (kwēn) *n.* [ME *quen* < OE *cwene,* woman.] **1.** A disreputable woman, esp. a prostitute. **2.** *Chiefly Scot.* A young woman.

**quea·sy** *also* **quea·zy** (kwē'zē) *adj.* **-si·er, -si·est** *also* **-zi·er, -zi·est.** [ME *coisy.*] **1.** Nauseated. **2.** Easily nauseated. **3.** Causing nausea : SICKENING <a *queasy* flight> **4. a.** Causing uneasiness. **b.** Nervous : troubled. **5. a.** Easily upset. **b.** Squeamish : ill at ease. **—quea'si·ly** *adv.* **—quea'si·ness** *n.*

**Qué·be·cois** (kā'bĕ-kwä') *n., pl.* **Québecois.** [Fr. < *Québec,* Quebec.] A native or resident of Quebec, esp. a French-speaking one.

**que·bra·cho** (kā-brä'chō) *n., pl.* **-chos.** [Am. Sp., var. of *quiebrahacha* : *quebrar,* to break (< Lat. *crepare,* to crack) + *hacha,* ax < Fr. *hache* < OFr., of Germanic orig.] **1.** A South American tree with very hard wood, esp. *Aspidosperma quebracho-blanco,* whose bark is used in medicine, and *Schinopsis lorentzii,* whose wood is a source of tannin. **2.** The wood or bark of a quebracho tree.

**Quech·ua** (kĕch'wə, -wä') *n., pl.* **Quechua** *or* **-uas.** [Sp. < Quechua *kkechúwa,* robber.] **1. a.** A tribe of South American Indians orig. forming the ruling class of the Incan Empire. **b.** A member of this tribe. **2. a.** The language of the Quechua spoken also by other Indian peoples of Peru, Ecuador, Bolivia, Chile, and Argentina. **b.** A language family consisting of the Quechua language. **—Quech'uan** *adj.*

**queen** (kwēn) *n.* [ME *quene* < OE *cwēn.*] **1.** The wife or widow of a king. **2.** A woman monarch. **3. a.** A woman or girl regarded as pre-eminent in a given domain <a *queen* of rock 'n' roll> **b.** A woman or girl who is the winner of a competition, esp. of a beauty contest. **c.** Something personified as a woman and regarded as preeminent in a given domain <The orchid is *queen* of flowers.> **4.** The most powerful chess piece, able to move across unoccupied squares in a straight line in any direction. **5.** A playing card bearing the figure of a queen, next below the king and above the jack in each suit. **6.** The fertile, fully developed female in a colony of social insects, as bees or termites. —*v.* **queened, queen·ing, queens.** —*vt.* **1.** To make (a woman) a queen. **2.** To raise (a pawn) to queen in chess. —*vi.* **1.** To reign as queen. **2.** To play the queen : DOMINEER <*queens* it over the first-year students> **—queen'like** *adj.*

▲ word history: The word *queen* is a native English word; its Old English form was *cwēn.* It is related to Greek *gunē,* "woman," which is the source of the English prefix *gyno–* and the suffixes *–gynous* and *–gyny.* The spelling *qu–* for *queen* was introduced by French scribes after the Normans conquered England in 1066. All Old English words beginning with *cw–* were respelled with *qu–*; others besides *queen* are *quick, quell,* and *qualm.*

**Queen Anne** *n.* The style in English architecture and furniture that was widespread during the reign of Queen Anne (1702–14).

**Queen Anne's lace** *n.* A widely distributed plant, *Daucus carota* orig. of Eurasia, with finely divided leaves and small white flowers in flat clusters.

**queen consort** *n., pl.* **queens consort.** A reigning king's wife.

**queen-cup** (kwēn'kŭp') *n.* A plant, *Clintonia uniflora* of northwestern North America, bearing a solitary white flower and a blue berry.

**queen·ly** (kwēn'lē) *adj.* **-li·er, -li·est. 1.** Of or like a queen. **2.** Relating to or suitable for a queen. **—queen'li·ness** *n.*

**queen mother** *n.* A dowager queen who is the mother of a reigning monarch.

**queen of the prairie** *n.* A plant, *Filipendula rubra* of the central United States, having compound leaves and small pink flower clusters.

**queen olive** *n.* An olive bearing large fruit, used for eating rather than as a source of oil.

**queen post** *n.* Either of two vertical members between the tie beam and straining beam that support the rafters in a pitched roof.

**queen regnant** *n., pl.* **queens regnant.** A queen reigning in her own right.

**Queen's Bench** *n.* A division of the British superior courts system that hears criminal and civil cases. —Used when the monarch is a woman.

**Queen's Counsel** *n.* A barrister serving as counsel to the British crown. —Used when the monarch is a woman.

**queen·ship** (kwēn'shĭp') *n.* **1.** The rank or condition of being a queen. **2.** A regal quality, as of a queen.

**queen-size** (kwēn'sīz') *adj.* **1.** Of, relating to, or being a bed with dimensions of approx. 60 inches by 80 inches, or 152.4 centimeters by 203.2 centimeters <*queen-size* mattress> **2.** Extra large <*queen-size* pantyhose>

**queen truss** *n.* A building truss utilizing queen posts.

**queer** (kwîr) *adj.* **-er, -est.** [Orig. unknown.] **1.** Deviating from the normal or expected : STRANGE. **2.** Odd or unconventional : ECCENTRIC. **3.** Questionable in nature or character : SUSPICIOUS <something *queer* about the deal> **4.** *Slang.* Fake : counterfeit. **5.** Feeling slightly ill : QUEASY. —*vt.* **queered, queer·ing, queers.** *Slang.* **1.** To thwart or ruin <*queered* the deal at the last minute> **2.** To put into a bad position. **—queer'ish** *adj.* **—queer'ly** *adv.* **—queer'ness** *n.*

**quell** (kwĕl) *vt.* **quelled, quell·ing, quells.** [ME *quellen,* to kill < OE *cwellan.*] **1.** To put down forcibly : SUPPRESS. **2.** To pacify : quiet.

**quench** (kwĕnch) *vt.* **quenched, quench·ing, quench·es.** [ME *quenchen* < OE *ācwencan.*] **1.** To put out (e.g., a fire) : EXTINGUISH. **2.** To suppress : dampen <couldn't *quench* our spirits> **3.** To put an end to : DESTROY. **4.** To satisfy : slake. **5.** To cool (hot metal) by immersing in water or other liquid. **—quench'a·ble** *adj.* **—quench'er** *n.* **—quench'less** *adj.*

**que·nelle** (kə-nĕl') *n.* [Fr.] A forcemeat dumpling poached in stock or water.

**quer·ce·tin** (kwûr'sĭ-tĭn) *n.* [< Lat. *quercus,* oak.] A yellow, powdered crystalline compound, $C_{15}H_{10}O_7$, synthesized or occurring as a glycoside in the cortex of numerous plants and used as a drug to treat abnormal capillary fragility.

**quer·ci·tron** (kwûr'sĭ-trən, -trŏn', kwər-sĭt'rən) *n.* [Lat. *quercus,* oak + CITRON.] **1.** The black oak. **2. a.** The bright-orange inner bark of the black oak. **b.** A yellow dye derived from quercitron.

**quern** (kwûrn) *n.* [ME *querne* < OE *cweorn.*] A primitive hand-turned grain mill.

**quer·u·lous** (kwĕr'ə-ləs, kwĕr'yə-) *adj.* [Lat. *querulus* < *queri,* to complain.] **1.** Given to complaining : PEEVISH. **2.** Expressing a complaint : WHINING <a *querulous* tone> **—quer'u·lous·ly** *adv.* **—quer'u·lous·ness** *n.*

**que·ry** (kwîr'ē) *n., pl.* **-ries.** [Alteration of obs. *quaere* < Lat., imper. of *quaerere,* to ask.] **1. a.** A request for information : QUESTION. **b.** An inquiry. **2.** A doubt in the mind. **3.** A notation, usu. a question mark, calling attention to an item to question its accuracy or validity. —*vt.* **-ried, -ry·ing, -ries. 1.** To express doubt or uncertainty about : QUESTION. **2.** To put a question to (a person). **3.** To mark (an item) with a query. **—que'ri·er, que'rist** *n.*

☆ ***syns:*** QUERY, INTERROGATION, INTERROGATORY, QUESTION *n. core meaning* : a request for data <editors answering language *queries*>

**quest** (kwĕst) *n.* [ME *queste* < OFr. < Lat. *questa,* fem. p.part. of *quaerere,* to seek.] **1.** An act or instance of seeking or inquiring : SEARCH. **2.** An adventurous pursuit esp. of a spiritual goal undertaken by a knight in medieval romance. **3.** *Archaic.* A jury of inquest. —*v.* **quest·ed, quest·ing, quests.** —*vi.* **1.** To go on a quest. **2.** To search for game. —*vt.* To search for : SEEK. **—quest'er** *n.*

**ques·tion** (kwĕs'chən) *n.* [ME < OFr. < Lat. *quaestio* < *quaerere,* to ask.] **1. a.** An expression of inquiry that requires or invites an informative reply. **b.** An interrogative sentence or phrase. **2.** A subject or point open to debate : ISSUE. **3.** An unresolved matter : PROBLEM <a *question* of morals> **4.** An item on an examination or questionnaire. **5. a.** A proposition brought up for consideration, as by an assembly. **b.** The act of bringing such a proposal to vote. **6.** Uncertainty : doubt <some *question* as to who won the election> **7.** Possibility : chance <no *question* of turning back now> —*v.* **-tioned, -tion·ing, -tions.** —*vt.* **1.** To put a question to. **2.** To interrogate, as a witness or suspect. **3.** To express doubt about : DISPUTE. **4.** To analyze : examine. —*vi.* To ask questions. **—ques'tion·er** *n.* **—ques'tion·ing·ly** *adv.*

**ques·tion·a·ble** (kwĕs'chə-nə-bəl) *adj.* **1.** Open to question or doubt : PROBLEMATICAL. **2.** Of dubious morality or respectability. **—ques'tion·a·ble·ness, ques'tion·a·bil'i·ty** *n.* **—ques'tion·a·bly** *adv.*

**question mark** *n.* A punctuation mark (?) placed at the end of a sentence or phrase to indicate a direct question.

**ques·tion·naire** (kwĕs'chə-nâr') *n.* [Fr. < *questionner,* to ask < *question,* question < OFr.] A form containing a set of related questions, esp. one designed to gather statistical data from a sample population, as for a survey.

**quet·zal** (kĕt-säl') *n., pl.* **-zals** *or* **-za·les** (-sä'lās) [Am. Sp. < Nahuatl *quetzalli,* large brilliant tail feather.] **1.** A Central American bird, *Pharomacrus mocino,* having bronze-green and red plumage and, in the male, long flowing tail feathers. **2.** —See table at CURRENCY.

**quetzal**
*24 inches long*

**Quet·zal·co·a·tl** (kĕt-säl'kō-ät'l) *n.* [Nahuatl.] A god of the Toltecs and Aztecs, represented as a plumed serpent.
**queue** (kyōō) *n.* [Fr. < OFr. *cue*, tail < Lat. *cauda.*] **1.** A line of people or vehicles awaiting individual turns. **2.** A braid of hair worn hanging down the back of the neck : PIGTAIL. **3.** *Computer Sci.* A sequence of stored computer data or programs awaiting processing. —*vi.* **queued, queu·ing, queues.** To form or wait in a line <*queue* up at the post office>
**quib·ble** (kwĭb'əl) *vi.* **-bled, -bling, -bles.** [Prob. < obs. *quib*, perh. < Lat. *quibus*, pl. of *qui*, who, from its frequent use in legal documents.] To raise trivial objections or distinctions, esp. so as to avoid admitting the truth or importance of something. —*n.* **1.** A petty distinction or irrelevant objection. **2.** *Archaic.* A pun. **—quib'bler** *n.*
**quiche** (kēsh) *n.* [Fr. < dial. G. *küche*, dim. of *kuche*, cake < OHG *kuocho.*] A dish consisting of unsweetened custard baked in a pastry shell often with other ingredients, as vegetables or seafood.
▲ word history: A quiche may seem to be a quintessentially French dish, but the word *quiche* is actually a Gallicized German word. Quiche was originally a specialty of Lorraine, a region in northeastern France bordering on Germany. The region was claimed by both countries at various times and both French and German are spoken there. The word *quiche* is a borrowing of Alsatian German *küche*, a diminutive of *kuche*, "cake." The form *quiche* is a French spelling of the German word.
**Qui·ché** (kē-chā') *n.* **1.** An Indian people of Guatemala. **2.** The Mayan language of the Quiché.
**quiche Lor·raine** (lə-rān', lô-) *n.* [Fr., after *Lorraine*, a region of northeastern France.] A quiche made with bacon and cheese.
**quick** (kwĭk) *adj.* **-er, -est.** [ME *quicke*, swift, alive < OE *cwicu*, alive.] **1.** Moving or performing with speed and agility : FAST. **2.** Thinking or understanding rapidly and easily : BRIGHT <a *quick* learner> **3. a.** Perceiving or responding with speed and sensitivity : KEEN. **b.** Reacting immediately and sharply <a *quick* temper> **4. a.** Occurring or achieved in a relatively short time <a *quick* trip> **b.** Done or occurring immediately : PROMPT <*quick* service> **5.** Tending to react hastily or impulsively <*quick* to forgive> **6.** *Archaic.* Alive. **7.** *Archaic.* Pregnant. —*n.* **1.** Raw or sensitive exposed flesh, as under the fingernails. **2.** The most intimate and sensitive aspect of the emotions <pierced me to the *quick*> **3.** The living <the *quick* and the dead> **4.** The vital core of a thing : ESSENCE <the *quick* of the matter> —*adv.* Fast : promptly. ***usage:*** Both *quick* and *quickly* can be used as adverbs. In speech, *quick* is more frequent, as in *Come quick!* In writing, *quickly* is preferred, as in *They returned quickly when they heard the news.* **—quick'ly** *adv.* **—quick'ness** *n.*
**quick-and-dirt·y** (kwĭk'ən-dûr'tē) *adj.* Shoddily made or done : cheap <a *quick-and-dirty* construction project>
**quick assets** *pl.n.* Liquid assets, including cash on hand and assets readily convertible to cash.
**quick bread** *n.* A bread made with a leavening agent, as baking powder, that does not require a leavening period before baking.
**quick·en** (kwĭk'ən) *v.* **-ened, -en·ing, -ens.** —*vt.* **1.** To make more rapid : ACCELERATE. **2.** To make or bring alive : VITALIZE. **3.** To stimulate : stir <The good news *quickened* our interest.> **4.** To make steeper. —*vi.* **1.** To become more rapid. **2.** To come or return to life. **3.** To reach the stage of pregnancy when the fetus can be felt to move. **—quick'en·er** *n.*
**quick-freeze** (kwĭk'frēz') *vt.* **-froze** (-frōz'), **-froz·en** (-frō'zən), **-freez·ing, -freez·es.** To freeze (food) by a process sufficiently rapid to retain desirable properties, as flavor and nutritional value.
**quick grass** *n.* [Var. of QUITCHGRASS.] Couch grass.
**quick·ie** (kwĭk'ē) *n. Informal.* Something done or made hastily.
**quick·lime** (kwĭk'līm') *n.* [ME *quykke lyme*, transl. of Lat. *calx viva.*] Calcium oxide.
**quick·sand** (kwĭk'sănd') *n.* A bed of loose sand mixed with water forming a soft, shifting mass that does not support heavy objects.
**quick·set** (kwĭk'sĕt') *n. Chiefly Brit.* **1.** Cuttings or slips of a plant, as hawthorn, capable of rooting when set in the ground. **2.** A hedge grown from quickset.
**quick·sil·ver** (kwĭk'sĭl'vər) *n.* [ME < OE *cwicseolfor*, transl. of Lat. *argentum vivum.*] MERCURY 2. —*adj.* Unpredictable : mercurial.
▲ word history: The name *quicksilver* for the element mercury is a translation of Latin *argentum vivum*, literally "living silver." Mercury was so called because it is a silvery colored metal that is liquid at ordinary temperatures. In *quicksilver* the word *quick* preserves its original but now archaic sense "living, alive."
**quick·step** (kwĭk'stĕp') *n. Mus.* A march for accompanying military quick time.
**quick-tem·pered** (kwĭk'tĕm'pərd) *adj.* Easily aroused to anger.
**quick time** *n.* A military marching pace of 120 steps per minute.
**quick-wit·ted** (kwĭk'wĭt'ĭd) *adj.* Mentally alert and sharp : CLEVER. **—quick'-wit'ted·ly** *adv.* **—quick'-wit'ted·ness** *n.*
**quid¹** (kwĭd) *n.* [ME *quide*, cud < OE *cwidu.*] A cut of something to be chewed, esp. a plug of tobacco.
**quid²** (kwĭd) *n., pl.* **quid** *or* **quids.** [Orig. unknown.] *Chiefly Brit.* A pound sterling.
**quid·di·ty** (kwĭd'ĭ-tē) *n., pl.* **-ties.** [Med. Lat. *quidditas* < Lat. *quid*, what.] **1.** The inherent nature of a thing : ESSENCE. **2.** An unreasonably fine distinction : QUIBBLE.
**quid·nunc** (kwĭd'nŭngk') *n.* [Lat. *quid nunc?* what now?] A prying or meddlesome person : BUSYBODY.
**quid pro quo** (kwĭd' prō kwō') *n.* [Lat., something for something.] An equal exchange or substitution.
**qui·es·cent** (kwī-ĕs'ənt, kwē-) *adj.* [Lat. *quiescens, quiescent-*, pr. part. of *quiescere*, to rest < *quies*, rest.] Inactive or still : DORMANT. **—qui·es'cence** *n.* **—qui·es'cent·ly** *adv.*
**qui·et** (kwī'ĭt) *adj.* **-et·er, -et·est.** [ME < OFr. *quiete* < Lat. *quietus*, p.part. of *quiescere*, to rest < *quies*, rest.] **1.** Making little or no sound : SILENT. **2.** Free of noise : HUSHED. **3.** Calm and unmoving : STILL <a *quiet* woodland pool> **4.** Free of agitation and turmoil : UNTROUBLED <a *quiet* life> **5.** Restful : soothing. **6.** Marked by tranquillity : PEACEFUL. **7.** Not showy or obtrusive : RESTRAINED. —*n.* The condition or quality of being quiet : TRANQUILLITY. —*v.* **-et·ed, -et·ing, -ets.** —*vt.* **1.** To cause to become quiet. **2.** *Law.* To make (a title) secure by freeing from all questions or claims. —*vi.* To become quiet <The heckler finally *quieted* down.> **—qui'et·ly** *adv.* **—qui'et·ness** *n.*
☆ ***syns:*** QUIET, INOBTRUSIVE, RESTRAINED, SUBDUED, UNOBTRUSIVE *adj. core meaning* : not showy or obtrusive <decor that is subtle and *quiet*> ***ant:*** gaudy, loud
**qui·et·ism** (kwī'ĭ-tĭz'əm) *n.* **1.** Christian mysticism calling for passive contemplation and the beatific annihilation of individual will. **2.** Quietness and passivity. **—qui'et·ist** *n.* **—qui'et·is'tic** *adj.*
**qui·e·tude** (kwī'ĭ-tōōd', -tyōōd') *n.* [LLat. *quietudo* < Lat. *quietus*, p.part. of *quiescere*, to rest < *quies*, rest.] Quiet tranquillity.
**qui·e·tus** (kwī-ē'təs) *n.* [Short for Med. Lat. *quietus est*, he is discharged (of an obligation).] **1.** Something that suppresses, terminates, or allays. **2.** Release from life : DEATH. **3.** A final discharge, as of a debt or obligation.
**quiff** (kwĭf) *n.* [Orig. unknown.] *Chiefly Brit.* A tuft of hair, esp. a forelock.
**quill** (kwĭl) *n.* [ME *quil*, of Germanic orig.] **1.** The hollow stemlike main shaft of a feather. **2.** A large wing or tail feather. **3.** A writing pen made from a quill. **4.** A plectrum for a stringed musical instrument such as a harpsichord. **5.** A toothpick made from the stem of a feather. **6.** One of the sharp hollow spines of a hedgehog or porcupine. **7.** A musical pipe having a hollow stem. **8.** A spool or bobbin for holding or winding yarn. **9.** A mechanical device consisting of a hollow cylinder that rotates on a solid shaft. —*vt.* **quilled, quill·ing, quills. 1.** To wind (yarn or thread) onto a quill. **2.** To imprint (fabric) with textured ridges.
**quill·back** (kwĭl'băk') *n., pl.* **-backs** *or* **quillback.** A North American freshwater fish, *Carpiodes cyprinus*, having a dorsal fin with one ray conspicuously extended.
**quill·wort** (kwĭl'wûrt', -wôrt') *n.* An aquatic plant of the genus *Isoetes*, having short, fleshy stems and grasslike leaves.
**quilt** (kwĭlt) *n.* [ME *quilte* < OFr. *cuilte* < Lat. *culcita*, mattress.] **1.** A bed covering consisting of two layers of fabric with a layer of batting or feathers between and stitched firmly together, usu. in a decorative pattern. **2.** A padded cover similar to a quilt. —*v.* **quilt·ed, quilt·ing, quilts.** —*vt.* **1.** To make into a quilt by stitching together (layers of fabric). **2.** To construct like a quilt <*quilt* a vest> **3.** To pad and stitch in decorative designs. —*vi.* **1.** To make a quilt. **2.** To do quilted work.
**quilt·ing** (kwĭl'tĭng) *n.* **1.** The act or process of doing quilted work. **2. a.** Material used to make quilts. **b.** Quilted material.
**quin-** *pref. var. of* QUINO-.
**quin·a·crine hydrochloride** (kwĭn'ə-krēn') *n.* [QUIN- + ACR(ID)INE.] A bright yellow, bitter, crystalline compound, used chiefly to treat malaria.
**quin·a·liz·a·rin** (kwĭn'ə-lĭz'ə-rĭn) *n.* A reddish crystalline compound, $C_{14}H_8O_6$.
**qui·nate** (kwī'nāt') *adj.* [< Lat. *quini*, five each.] Arranged in groups of five <*quinate* leaflets>
**quince** (kwĭns) *n.* [ME *quynce*, pl. of *quyn*, quince < OFr. *coin* < Lat. *cotoneum*, var of *cydoneum*, after *Cydonia*, an ancient town in Crete.] **1.** A tree, *Cydonia oblonga*, orig. of western Asia, bearing white flowers and edible fruit. **2.** The yellow applelike fruit of the quince.
**quin·cun·cial** *also* **quin·cunx·ial** (kwĭn-kŭn'shəl) *adj.* Of, relating to, or forming a quincunx. **—quin·cun'cial·ly** *adv.*
**quin·cunx** (kwĭn'kŭngks') *n.* [Lat., five twelfths : *quinque*, five + *uncia*, twelfth < *unus*, unit.] An arrangement of five things with one at each corner of a square or rectangle and one at the center.
**quin·cunx·ial** (kwĭn-kŭn'shəl) *adj. var. of* QUINCUNCIAL.
**quin·de·cen·ni·al** (kwĭn'dĭ-sĕn'ē-əl) *adj.* [Lat. *quindecim*, fifteen (*quinque*, five + *decem*, ten) + *annus*, year.] **1.** Occurring once every 15 years. **2.** Lasting 15 years. —*n.* A 15th anniversary.
**qui·nel·la** (kwĭ-nĕl'ə, kē-) *or* **qui·nie·la** (kēn-yĕl'ə) *n.* [Am. Sp., a lottery-like game.] A system of betting in which the bettor wins by correctly picking the first two finishers of a race regardless of their order.

**quin·i·dine** (kwĭn′ĭ-dēn′) *n.* A colorless crystalline alkaloid, $C_{20}H_{24}N_2O_2$, resembling quinine and used in treating certain heart disorders and malaria.
**qui·nie·la** (kēn-yĕl′ə) *n. var. of* QUINELLA.
**qui·nine** (kwī′nīn′) *n.* **1.** A bitter, colorless, amorphous powder or crystalline alkaloid, $C_{20}H_{24}N_2O_2{\cdot}3H_2O$, obtained from cinchona and used as an antimalarial drug. **2.** A compound or salt of quinine, used medicinally or as a flavoring.
**quinine water** *n.* A carbonated beverage flavored with quinine.
**quin·nat salmon** (kwĭn′ăt′) *n.* [Salish *t′kwinnat.*] The Chinook salmon.
**quino-** *or* **quin-** *pref.* [Sp. *quina,* cinchona bark, short for *quinaquina* < Quechua.] **1.** Cinchona : cinchona bark <*quinoidine*> **2.** Quinone <*quinoid*>
**quin·oid** (kwĭn′oid′) *n.* A substance resembling quinone in structure or physical properties.
**qui·noi·dine** (kwĭ-noi′dēn′, -dĭn) *n.* A brownish-black alkaloidal residue left after extraction of crystalline alkaloids from cinchona bark, used as a quinine substitute.
**quin·o·line** (kwĭn′ə-lēn′, -lĭn) *n.* An aromatic organic base, $C_9H_7N$, having a pungent tarlike odor, synthesized or derived from coal tar, and used as a food preservative and in making antiseptics and dyes.
**qui·none** (kwĭ-nōn′, kwĭn′ōn′) *n.* Any of a class of aromatic compounds occurring widely in plants, esp. the yellow crystalline form, $C_6H_4O_2$, used in tanning, dye-making and photography.
**quin·o·noid** (kwĭn′ə-noid′, kwĭ-nō′noid′) *adj.* Of, containing, or resembling quinone, in structure or properties.
**quin·qua·ge·nar·i·an** (kwĭng′kwə-jə-nâr′ē-ən) *n.* [< Lat. *quinquagenarius,* containing fifty < *quinquageni,* fifty each < *quinquaginta,* fifty : *quinque,* five + *-ginta,* times ten.] A person 50 years old or between the ages of 50 and 60. **—quin′qua·ge·nar′i·an** *adj.*
**Quin·qua·ges·i·ma** (kwĭng′kwə-jĕs′ə-mə) *n.* [Med. Lat. < Lat., fiftieth < *quinquaginta,* fifty. —see QUINQUAGENARIAN.] Shrove Sunday, about 50 days before Easter.
**quinque-** *pref.* [< Lat. *quinque,* five.] Five <*quinque*valent>
**quin·que·fo·li·ate** (kwĭng′kwə-fō′lē-ĭt, -āt′) *adj.* Having five leaves, leaflets, or leaflike parts.
**quin·quen·ni·al** (kwĭn-kwĕn′ē-əl, kwĭng-) *adj.* **1.** Occurring once every five years. **2.** Lasting for five years. —*n.* **1.** A fifth anniversary. **2.** A quinquennium. **—quin·quen′ni·al·ly** *adv.*
**quin·quen·ni·um** (kwĭn-kwĕn′ē-əm, kwĭng-) *n., pl.* **-quen·ni·ums** *or* **-quen·ni·a** (-kwĕn′ē-ə) [Lat. : *quinque,* five + *annus,* year.] A period of five years.
**quin·que·va·lent** (kwĭng′kwə-vā′lənt) *adj.* Pentavalent. **—quin′que·va′lence** *n.*
**quin·sy** (kwĭn′zē) *n.* [ME *quinesye* < Med. Lat. *quinancia* < Gk. *kunankhē* : *kuōn,* dog + *ankhein,* to strangle.] Acute tonsillitis, often accompanied by fever and formation of abscess.
**quint¹** (kwĭnt) *n.* [Fr. < Lat. *quintus,* fifth.] A sequence of five cards of the same suit in one hand in piquet.
**quint²** (kwĭnt) *n.* A quintuplet.
**quin·tain** (kwĭnt′n) *n.* [ME *quintaine* < OFr., prob. < Lat. *quintana,* a street in a Roman camp < *quintus,* fifth.] A target, esp. a revolving device mounted on a post, used in tilting.
**quin·tal** (kwĭnt′l) *n.* [ME, a unit of weight < OFr. < Med. Lat. *quintale* < Ar. *qinṭār* < LGk. *kentēnarion* < LLat. *centenarium* < Lat. *centum,* hundred.] **1.** A metric unit of mass equal to 100 kilograms. **2.** HUNDREDWEIGHT 2.
**quin·tar** (kēn-tär′) *also* **qin·tar** (kĭn-) *n.* [Albanian *qintar.*] —See table at CURRENCY.
**quin·tes·sence** (kwĭn-tĕs′əns) *n.* [ME < OFr. *quinte essence,* fifth essence < Med. Lat. *quinta essentia,* transl. of Gk. *pemptē ousia.*] **1.** Pure, undiluted essence. **2.** The purest or most characteristic instance <the *quintessence* of mercy> **3.** The fifth and highest essence, after the four elements of earth, air, fire, and water, thought to be the substance of the heavenly bodies and latent in all things in ancient and medieval philosophy. **—quin′tes·sen′tial** (kwĭn′tĭ-sĕn′shəl) *adj.* **—quin′tes·sen′tial·ly** *adv.*
**quin·tet** *also* **quin·tette** (kwĭn-tĕt′) *n.* [Ital. *quintetto* < *quinto,* fifth < Lat. *quintus.*] **1.** A musical composition for five voices or instruments. **2.** An ensemble of five musicians. **3.** A group of five persons or things.
**quin·tile** (kwĭn′tīl′, kwĭnt′l) *n.* [< Lat. *quintus,* fifth.] **1.** The astrological aspect of planets separated in the sky by 72° or one fifth of the zodiac. **2.** *Statistics.* **a.** One of five usu. equal portions of a frequency distribution. **b.** Any of the four values dividing a sample into quintiles.
**quin·til·lion** (kwĭn-tĭl′yən) *n.* [Lat. *quintus,* fifth + (M)ILLION.] **1.** The cardinal number equal to $10^{18}$. **2.** *Chiefly Brit.* The cardinal number equal to $10^{30}$. **—quin·til′lion** *adj.*
**quin·til·lionth** (kwĭn-tĭl′yənth) *n.* The ordinal number matching quintillion in a series. **—quin·til′lionth** *adj.*
**quin·tu·ple** (kwĭn-tōō′pəl, -tyōō′-, -tŭp′əl, kwĭn′tə-pəl) *adj.* [OFr. < Lat. *quintus,* fifth.] **1.** Consisting of five parts or units. **2.** Multiplied by five : FIVEFOLD. —*v.* **-pled, -pling, -ples.** —*vt.* To multiply by five. —*vi.* To be multiplied fivefold. **—quin·tu′ple** *n.*
**quin·tu·plet** (kwĭn-tŭp′lĭt, -tōō′plĭt, -tyōō′plĭt, kwĭn′tə-plĭt) *n.* [< QUINTUPLE.] **1.** A group of five things related by common properties or behavior. **2.** One of five born at a single birth.
**quin·tu·pli·cate** (kwĭn-tōō′plĭ-kĭt, -tyōō′-) *adj.* [Prob. < QUINTUPLE.] **1.** Multiplied or copied five times : QUINTUPLE. **2.** Fifth in a group of five identical items. —*vt.* (-kāt′) **-cat·ed, -cat·ing, -cates.** To multiply or copy five times. **—quin·tu′pli·cate** *n.*
**quip** (kwĭp) *n.* [Alteration of obs. *quippy,* perh. < Lat. *quippe,* indeed < *quid,* what.] **1.** A witty, offhand remark. **2.** A cleverly sarcastic remark : GIBE. **3.** QUIBBLE 1. **4.** Something curious or odd. —*vi.* **quipped, quip·ping, quips.** To make a quip.
**quip·ster** (kwĭp′stər) *n.* One who makes quips.
**qui·pu** (kē′pōō) *n.* [Sp. *quipo* < Quechua *quipu.*] An Incan device consisting of variously knotted and colored cords attached to a base rope, used for calculating and recording numbers.
**quire¹** (kwīr) *n.* [ME *quayer,* four doubled sheets of paper < OFr. *quaer* < Lat. *quaterni,* set of four < *quater,* four times.] A uniform set of 24 or sometimes 25 sheets of paper : one twentieth of a ream.
**quire²** (kwīr) *n.* & *v. Archaic. var. of* CHOIR.
**quirk** (kwûrk) *n.* [Orig. unknown.] **1.** A sudden sharp bend or crook. **2.** A peculiarity : idiosyncrasy. **3.** An unexpected or unaccountable act or event : VAGARY. **4.** An equivocation : quibble. **5.** A lengthwise groove on an architectural molding between the convex upper part and the soffit. **—quirk′i·ly** *adv.* **—quirk′i·ness** *n.* **—quirk′y** *adj.*
**quirt** (kwûrt) *n.* [Perh. < Sp. *cuerda,* whip < Lat. *chorda,* cord < Gk. *khordē.*] A short-handled riding whip with a lash of braided rawhide.

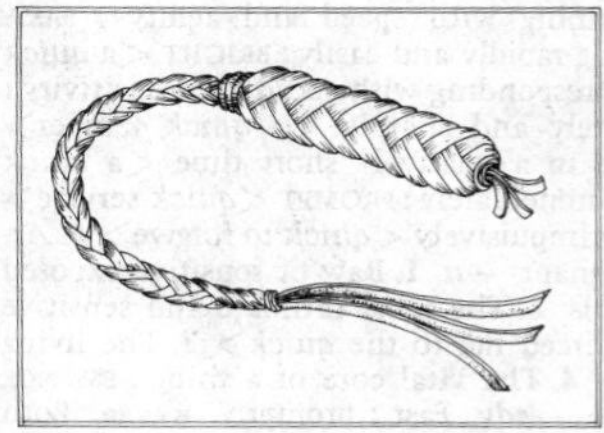
**quirt**

**quis·ling** (kwĭz′lĭng) *n.* [After Vidkun *Quisling* (1887–1945).] A traitor, esp. one serving as the puppet of an occupying enemy force.
**quit** (kwĭt) *v.* **quit** *or* **quit·ted** (kwĭt′ĭd), **quit·ting, quits.** [ME *quiten,* to release < OFr. *quiter* < Med. Lat. *quietare* < Lat. *quietus,* at rest.] —*vt.* **1.** To depart from : LEAVE. **2.** To leave the company of. **3.** To give up or withdraw from : RELINQUISH <*quit* school> **4.** To abandon or put aside : DISCONTINUE <*quit* smoking> **5.** To leave off : cease. **6.** To rid oneself of by paying <*quit* a debt> **7.** To release from a burden or responsibility. **8.** To conduct (oneself) in a specified way <*quit* ourselves like professionals> —*vi.* **1.** To cease to perform : STOP. **2.** To give up, as in defeat. **3.** To leave a job. —*adj.* Absolved of a duty or obligation : FREE.
☆ ***syns:*** QUIT, LEAVE, RESIGN, TERMINATE *v. core meaning* : to relinquish one's engagement in or occupation with <*quit* drinking><*quit* their jobs>
**quitch grass** (kwĭch) *n.* [Ult. < OE *cwice.*] Couch grass.
**quit·claim** (kwĭt′klām′) [AN *quiteclame* < *quiteclamer,* to release < OFr. : *quite,* free (< Lat. *quietus,* freed of) + *clamer,* to proclaim < Lat. *clamare.*] *Law.* —*n.* The transfer of a title, right, or claim to another. —*vt.* **-claimed, -claim·ing, -claims.** To renounce all claim to (a possession or right).
**quite** (kwīt) *adv.* [< ME, rid of < OFr. < Lat. *quietus,* freed of.] **1.** To the fullest extent : completely <not *quite* empty> **2.** Actually : really <*quite* a good dinner> **3.** To a degree : RATHER <*quite* late in arriving>
**quit·rent** (kwĭt′rĕnt′) *n.* [ME *quiterent* : *quite,* free + *rent,* rent.] A rent paid by a feudal tenant in lieu of customary labor or services.
**quits** (kwĭts) *adj.* [ME, prob. < Med. Lat. *quittus,* alteration of Lat. *quietus,* freed of.] Being on equal terms, as by payment or requital.
**quit·tance** (kwĭt′ns) *n.* [ME *quitance* < OFr. < *quiter,* to free. —see QUIT.] **1. a.** Release from a debt, obligation, or penalty. **b.** A document certifying such a release. **2.** Something given or done as recompense : REPAYMENT.
**quit·ter** (kwĭt′ər) *n.* One who gives up easily.
**quit·tor** (kwĭt′ər) *n.* [ME *quiture,* perh. < OFr., decoction < Lat. *coctura* < *coquere,* to cook.] An inflammation of the hoof cartilage of horses and other solid-hoofed animals, marked by degeneration of tissue, formation of a slough, and fistulous sores.
**quiv·er¹** (kwĭv′ər) *vi.* **-ered, -er·ing, -ers.** [ME *quiveren,* perh. < *quiver,* nimble.] To shake with a rapid slight motion : TREMBLE. —*n.* The act or motion of quivering.

**quiv·er²** (kwĭv'ər) *n.* [ME < AN *quiveir*, var. of OFr. *cuivre*, of Germanic orig.] **1.** A portable case for arrows. **2.** The arrows held in a quiver.
**qui vive** (kē vēv') *n.* [Fr., (long) live who? (a sentry's challenge to determine a person's political sympathies).] A sentry's challenge.
**quix·ot·ic** (kwĭk-sŏt'ĭk) *also* **quix·ot·i·cal** (-ĭ-kəl) *adj.* [After *Don Quixote*, hero and title of a romance by Miguel de Cervantes (1547–1616).] Idealistic in a romantic and impractical way. **—quix·ot'i·cal·ly** *adv.* **—quix'o·tism** (kwĭk'sə-tĭz'əm) *n.*
**quiz** (kwĭz) *vt.* **quizzed, quiz·zing, quiz·zes.** [Orig. unknown.] **1.** To question closely or repeatedly : INTERROGATE. **2.** To test the knowledge of by posing questions. **3.** *Chiefly Brit.* To poke fun at : MOCK. *—n., pl.* **quiz·zes. 1.** A questioning or inquiry. **2.** A short written or oral test. **3.** A practical joke. **—quiz'zer** *n.*
**quiz·mas·ter** (kwĭz'măs'tər) *n.* A master of ceremonies in a quiz show who puts questions to the contestants.
**quiz show** *n.* A radio or television show in which contestants compete by answering questions.
**quiz·zi·cal** (kwĭz'ĭ-kəl) *adj.* **1.** Suggesting puzzlement : QUESTIONING. **2.** Teasing : mocking. **3.** Eccentric : odd. **—quiz'zi·cal'i·ty** (-kăl'ĭ-tē) *n.* **—quiz'zi·cal·ly** *adv.*
**quod** (kwŏd) *n.* [Orig. unknown.] *Chiefly Brit.* Prison.
**quod·li·bet** (kwŏd'lə-bĕt') *n.* [Lat., anything at all : *quod*, what + *libet*, it pleases < *libere*, to please.] **1. a.** A theological or philosophical issue presented for formal argument or disputation. **b.** The disputation itself. **2.** A usu. humorous musical medley.
**quoin** (koin, kwoin) *n.* [Var. of COIN.] **1. a.** An exterior structural angle, as of a masonry wall. **b.** A stone serving to form a quoin : CORNERSTONE. **2.** A keystone. **3.** A wedge-shaped block used to lock printing type in a chase. **4.** A wedge used to elevate a gun. *—vt.* **quoined, quoin·ing, quoins.** To provide, secure, or elevate with a quoin.

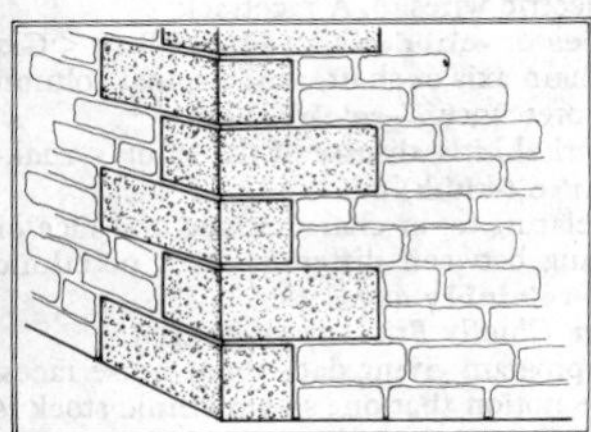
quoin

**quoit** (kwoit, koit) *n.* [ME *coite.*] **1. quoits** *(sing. in number).* A game in which iron or rope rings are pitched at a stake with the object of encircling it. **2.** A ring used in quoits.
**quon·dam** (kwŏn'dəm, -dăm') *adj.* [Lat. < *quom*, when.] That once was : FORMER <"the *quondam* drunkard, now perfectly sober" —Bret Harte>
**Quon·set** (kwŏn'sĭt). A trademark for a prefabricated portable hut having a semicircular roof of corrugated metal that curves down to form walls.
**quo·rum** (kwôr'əm, kwōr'-) *n.* [ME, quorum of justices of the peace < Lat., of whom, from the wording of a commission naming certain persons as members of a body (as the bench).] **1.** The minimum number of officers and members of a constituted body who must be present for the valid transaction of business. **2.** An exclusive group.
**quo·ta** (kwō'tə) *n.* [Med. Lat. < Lat., fem. of *quotus*, of what number < *quot*, how many.] **1. a.** A proportional share, as of goods, assigned to a group or to each member of a group : ALLOTMENT. **b.** A production assignment. **2.** The highest number or proportion, esp. of people, permitted admission, as to a nation, group, or institution.
**quot·a·ble** (kwō'tə-bəl) *adj.* Suitable for or worthy of quoting. **—quot'a·bil'i·ty** *n.*
**quo·ta·tion** (kwō-tā'shən) *n.* **1.** The act of quoting. **2.** A passage quoted. **3. a.** The quoting of current bids and prices for goods and securities. **b.** The prices or bids cited. **—quo·ta'tion·al** *adj.* **—quo·ta'tion·al·ly** *adv.*
**quotation mark** *n.* Either of a pair of punctuation marks (" ") or (' ') used to mark the beginning and end of a passage attributed to another and repeated word for word.
**quote** (kwōt) *v.* **quot·ed, quot·ing, quotes.** [ME *coten*, to mark a book with numbers or marginal references < Med. Lat. *quotare* < Lat. *quotus*, of what number < *quot*, how many.] *—vt.* **1.** To repeat or copy the words of (another), usu. with acknowledgment of the source. **2.** To cite or refer to for authority or illustration. **3.** To state (a price) for securities, goods, or services. *—vi.* To give a quotation, as from a book. *—n. Informal.* **1.** A quotation. ***usage:*** Many consider *quote* unacceptable as a substitute for *quotation*, esp. in formal contexts. **2.** A quotation mark. **—quot'er** *n.*
**quoth** (kwōth) [ME < OE *cwæð*, he said < *cweðan*, to say.] *vt. Archaic* Uttered : said. —Used only in the first and third persons, with the subject following <"*Quoth* the raven 'Nevermore!' " —Poe>

▲ word history: The archaic verb *quoth*, which persists in English in large part because of Poe's encounter with the raven, is unrelated in origin to the modern verb *quote* in spite of the similarity of sound. *Quoth* means "said" and is derived from Old English *cwæð*, the past singular form of *cweðan*, "to say, speak." The present tense of *cweðan* has been completely lost in Modern English except in the compound *bequeath*. The verb *quote* is derived ultimately from Latin *quot*, "how many." In medieval times sections of books were marked with numbers, as chapters are today, for easy reference. The Latin verb *quotare*, derived from *quot*, meant "to mark with numbers," referring especially to the marking of books. English borrowed *quotare* as *quote* with its original meaning, but *quote* developed more general senses than *quotare*. The English verb was used to mean "to give a number as a reference" and then "to cite, refer to." Finally *quote* came to mean "to repeat a passage word for word" rather than "to refer to a passage obliquely, by numbers or other shorthand."

**quo·tha** (kwō'thə) *interj.* [Contraction of *quoth he.*] *Archaic.* —Used to express sarcasm or surprise, after quoting the word or statement of another.
**quo·tid·i·an** (kwō-tĭd'ē-ən) *adj.* [ME *cotidian* < OFr. *cotidien* < Lat. *quotidianus* < *quotidie*, each day : *quot*, as many as + *dies*, day.] **1.** Recurring daily. **2.** Ordinary : everyday.
**quo·tient** (kwō'shənt) *n.* [ME *quocient* < Lat. *quotiens*, how many times < *quot*, how many.] The quantity resulting from division of one quantity by another.
**qu·rush** (kōō'rəsh) *n., pl.* **qurush** *or* **-es.** [Ar. *qurūš.*] —See table at CURRENCY.

# Rr

**r** *or* **R** (är) *n., pl.* **r's** *or* **R's. 1.** The 18th letter of the English alphabet. **2.** A speech sound represented by the letter *r*. **3.** The 18th in a series.
**R** (är) *adj.* [Short for RESTRICTED.] Indicating a motion-picture rating of such nature that no one under the age of 17 can be admitted unless accompanied by a parent or guardian.
**Ra¹** (rä) *also* **Re** (rā) *n.* [Of Egypt. orig.] *Myth.* The ancient Egyptian sun god, the supreme deity depicted as a man with the head of a hawk crowned with a solar disk and uraeus.
**Ra²** *symbol for* RADIUM.
**ra·ba·to** (rə-bä'tō) *n. var. of* REBATO.
**rab·bet** (răb'ĭt) *also* **re·bate** (rē'bāt', răb'ĭt) [ME *rabet* < OFr. *rabat*, act of beating down < *rabattre*, to beat down again. —see REBATE¹.] *—n.* **1.** A groove or cut along the edge of a board that receives or interlocks with another piece to form a joint. **2.** A joint made with a rabbet. *—v.* **-bet·ed, -bet·ing, -bets** *also* **-bat·ed, -bat·ing, -bates.** *—vt.* **1.** To cut a rabbet in. **2.** To join by a rabbet. *—vi.* To be joined by a rabbet.
**rab·bi** (răb'ī) *also* **rab·bin** (răb'ĭn) *n., pl.* **-bis** *also* **-bins.** [LLat. < Gk. *rhabbi* < Heb. *rabbi* : *rabh*, master + *-ī*, my.] **1.** An ordained spiritual leader of a Jewish congregation. **2.** One once authorized to interpret Jewish law.
**rab·bin·ate** (răb'ĭn-āt', -ĭt) *n.* **1.** The office or function of a rabbi. **2.** Rabbis as a group.
**rab·bin·i·cal** (rə-bĭn'ĭ-kəl) *also* **rab·bin·ic** (-ĭk) *adj.* [< Fr. *rab-*

*bin,* rabbi.] Of, relating to, or characteristic of rabbis or their teachings, learning, writings, or language. **—rab·bin'i·cal·ly** *adv.*

**Rab·bin·ic Hebrew** (rə-bĭn'ĭk) *n.* The Hebrew language as used in the learned writings of esp. medieval rabbis.

**rab·bin·ism** (răb'ĭn-ĭz'əm) *n.* Rabbinical teachings and traditions.

**rab·bin·ist** (răb'ĭn-ĭst) *n.* A strict observer of the Talmud and of rabbinical traditions. **—rab'bin·is'tic, rab'bin·it'ic** (-ĭt'ĭk) *adj.*

**rab·bit** (răb'ĭt) *n., pl.* **rabbit** *or* **-bits.** [ME *rabet.*] **1.** A furry, long-eared, burrowing mammal of the family Leporidae, including the cottontail, or the commonly domesticated Old World species *Oryctolagus cuniculus.* **2.** A hare. **3.** The fur of a rabbit or hare. **4.** Welsh rabbit. *—vi.* **-bit·ed, -bit·ing, -bits.** To hunt rabbits or hares. **—rab'bit·er** *n.*

**rabbit ears** *pl.n. Informal.* An indoor television antenna composed of two usu. adjustable rods connected to a base and swiveling apart at a V-shaped angle.

**rabbit fever** *n.* Tularemia.

**rab·bit-foot clover** (răb'ĭt-fo͝ot') *n.* An Old World clover, *Trifolium arvense,* bearing pinkish-gray furlike flowers similar to rabbits' paws.

**rabbit punch** *n.* A chopping blow to the back of the neck.

**rab·ble¹** (răb'əl) *n.* [ME, pack of animals.] **1.** A tumultuous mob. **2.** The lower classes.

**rab·ble²** (răb'əl) *n.* [Fr. *râble,* fire shovel < OFr. *roable* < Med. Lat. *rotabulum* < Lat. *rutabulum* < *ruere,* to rake up.] *Metallurgy.* **1.** An iron bar with one end bent like a rake, used to skim and stir molten iron in puddling. **2.** A tool or mechanically operated device similar to a rabble used in refining or roasting furnaces. *—vt.* **-bled, -bling, -bles.** To stir or skim (molten iron) with a rabble.

**rab·bler** (răb'lər) *n.* RABBLE².

**rab·ble-rous·er** (răb'əl-rou'zər) *n.* A demagogue.

**Rab·e·lai·si·an** (răb'ə-lā'zē-ən, -zhən) *adj.* **1.** Of or pertaining to François Rabelais or to his works. **2.** Marked by broad caricature and coarse, ribald humor.

**Ra·bi** (ru̇'bē) *also* **Ra·bi·a** (rə-bē'ə) *n.* [Ar. *rabī',* spring.] Either the third or the fourth month of the Moslem year. —See table at CALENDAR.

**rab·id** (răb'ĭd) *adj.* [Lat. *rabidus* < *rabere,* to rave.] **1.** Of or afflicted with rabies. **2.** Fanatical : overzealous <a *rabid* sports fan> **3.** Raging : violent <*rabid* hostility> **—ra·bid'i·ty** (rə-bĭd'ə-tē, rā-), **rab'id·ness** *n.* **—rab'id·ly** *adv.*

**ra·bies** (rā'bēz) *n.* [NLat. < Lat. *rabies,* rage < *rabere,* to rave.] An acute, infectious, often fatal viral disease of most mammals that attacks the central nervous system and is transmitted by the bite of an infected animal. **—ra'bi·et'ic** (-ĕt'ĭk) *adj.*

**rac·coon** *also* **ra·coon** (ră-ko͞on') *n., pl.* **-coons** *or* **raccoon** *also* **racoon.** [Algonquian (Virginia) *arathkone.*] **1.** A carnivorous North American mammal, *Procyon lotor,* with black masklike facial markings, grayish-brown fur, and a bushy black-ringed tail. **2.** The fur of the raccoon. **3.** An animal resembling or related to the raccoon.

**race¹** (rās) *n.* [Fr., generation < OItal. *razza.*] **1.** A local geographic or global human population distinguished as a more or less distinct group by genetically transmitted physical characteristics. **2.** Humanity as a whole. **3.** A group of people united or classified together on the basis of common history, nationality, or geographic distribution <the Spanish *race*> **4.** A genealogical line : LINEAGE. **5.** *Biol.* **a.** An animal or plant population that differs from others of the same species in the frequency of hereditary traits : SUBSPECIES. **b.** A breed or strain, as of domestic animals. **6.** A characteristic quality, as the flavor of a wine.

**race²** (rās) *n.* [ME *ras* < ON *rās.*] **1. a.** A contest of speed, as in running, driving, or riding. **b. races.** A scheduled series of such contests held on a regular course. **2.** A rivalry or competition for supremacy <the mayoral *race*> **3.** Rapid or steady onward movement <the *race* of time> **4. a.** A swift or strong current of water. **b.** The channel of such a current. **c.** An artificial channel built to transport water and utilize its energy. **5.** A groove or track in which a machine part slides or rolls. **6.** A slipstream. *—v.* **raced, rac·ing, rac·es.** *—vi.* **1.** To compete in a contest of speed. **2.** To move rapidly or at top speed. **3.** To run too rapidly because of decreased resistance <a motor that was *racing*> *—vt.* **1.** To compete against in a race. **2.** To place or enter in a race <I'll *race* my boat against yours.> **3.** To cause (e.g., an engine with the gears disengaged) to run swiftly or too swiftly.

**race³** (rās) *n.* [OFr. *rais,* root < Lat. *radix.*] A root, esp. of ginger.

**race·course** (rās'kôrs') *n.* A racetrack.

**race·horse** (rās'hôrs') *n.* A horse bred and trained to race.

**ra·ceme** (rā-sēm', rə-) *n.* [Lat. *racemus,* a bunch of grapes.] *Bot.* An inflorescence in which stalked flowers are arranged singly along a central stem, as in the lily of the valley.

**ra·ce·mic** (rā-sē'mĭk, -sĕm'ĭk, rə-) *adj.* Of or relating to a chemical compound containing equal quantities of dextrorotatory and levorotatory isomers so that it does not rotate the plane of incident polarized light. **—rac'e·mism'** (răs'ə-mĭz'əm, rā-sē'-) *n.*

**racemic acid** *n.* An optically inactive form of tartaric acid, $C_4H_6O_6 \cdot H_2O$, that can be separated into dextrorotatory and levorotatory components and is occas. found in grape juice during winemaking.

**ra·ce·mi·form** (rā-sē'mə-fôrm') *adj. Bot.* Racemelike in form.

**rac·e·mi·za·tion** (răs'ə-mĭ-zā'shən) *n.* Conversion of an optically active substance to a racemic form.

**rac·e·mose** (răs'ə-mōs') *adj.* [Lat. *racemosus,* full of clusters < *racemus,* bunch of grapes.] **1.** *Bot.* Resembling or growing in a raceme. **2.** *Anat.* Having a structure of clustered parts <*racemose* glands> **—rac'e·mose'ly** *adv.*

**rac·er** (rā'sər) *n.* **1.** One that engages in races or is capable of great speed. **2.** A swift, nonvenomous North American snake of the genus *Coluber.*

**race riot** *n.* A riot caused by racial hatred or unrest.

**race·run·ner** (rās'rŭn'ər) *n.* A swift, highly active New World lizard of the genus *Cnemidophorus.*

**racerunner**
*9 inches long including tail*

**race·track** (rās'trăk') *n.* An often oval course designed for racing.

**race·way** (rās'wā') *n.* **1.** RACE² 4c. **2.** A usu. rectangular conduit in a building for safeguarding electric wires. **3.** A racetrack.

**ra·chis** (rā'kĭs) *n., pl.* **-chis·es** *or* **-chi·des** (-kə-dēz') [NLat. < Gk. *rhakhis,* backbone.] *Biol.* A main axis or shaft, as the spinal column or the central stem of an inflorescence. **—ra'chi·al** *adj.*

**ra·chi·tis** (rə-kī'tĭs) *n.* [Gk. *rhakhitis,* disease of the spine < *rhakhis,* spine.] Rickets. **—ra·chit'ic** (-kĭt'ĭk) *adj.*

**ra·cial** (rā'shəl) *adj.* **1.** Of, relating to, or characteristic of a race or ethnic group. **2.** Of or existing between different races or ethnic groups <*racial* harmony> **—ra'cial·ly** *adv.*

**ra·cial·ism** (rā'shə-lĭz'əm) *n. Chiefly Brit. var. of* RACISM.

**racing form** *n.* A printed program giving data about horse races.

**ra·cism** (rā'sĭz'əm) *n.* **1.** The notion that one's own ethnic stock is superior. **2.** Prejudice or discrimination based on racism. **—rac'ist** *n.*

**rack¹** (răk) *n.* [ME *rakke,* prob. < MDu., framework.] **1.** A framework or stand intended to hold or display certain articles, esp.: **a.** A triangular frame for arranging billiard balls at the start of a game. **b.** A receptacle for livestock feed. **c.** A frame for holding bombs in an aircraft. **d.** An upright framework for holding cases of printing type or galley proof. **2.** A toothed bar designed to mesh with another toothed machine part, as a gearwheel or pinion. **3.** An instrument of torture for stretching and gradually dislocating the victim's body. **4. a.** Intense anguish. **b.** A cause of intense anguish. **5.** A set of antlers. *—vt.* **racked, rack·ing, racks. 1.** To place (e.g., billiard balls) in a rack. **2.** To torture by means of the rack. **3.** To torment <Pain *racked* my body.> **4.** To strain with great effort <*racked* their brains over the puzzle> **—on the rack.** Under great stress or strain. **—rack up.** *Slang.* To accumulate or score <*rack up* points> **—rack'er** *n.*

**rack²** (răk) *n.* [Orig. unknown.] Either of two gaits of horses, the pace or the single-foot. *—vi.* **racked, rack·ing, racks.** To go or move in a rack.

**rack³** (răk) *n.* [ME *rak,* prob. of Scand. orig.] A thin or broken layer of wind-driven clouds. *—vi.* **racked, rack·ing, racks.** To be driven by the wind, as clouds.

**rack⁴** (răk) *n.* [Var. of WRACK¹.] Destruction <*rack* and ruin> ***usage:*** In modern usage, *rack* is an acceptable variant of *wrack,* meaning "severe damage." Thus, it is correct to write either *rack and ruin* or *wrack and ruin.*

**rack⁵** (răk) *vt.* **racked, rack·ing, racks.** [ME *rakken* < OProv. *arracar* < *raca,* stems and husks of grapes.] To drain (cider or wine) from the dregs.

**rack⁶** (răk) *n.* [Prob. < RACK¹.] **1.** A wholesale rib cut of lamb between the shoulder and the loin. **2.** A crown roast of lamb.

**rack and pinion** *n.* A device for the interconversion of linear and rotary motion, consisting of a pinion and a mated rack.

**rack·et¹** *also* **rac·quet** (răk'ĭt) *n.* [OFr. *raquette* < Ar. *râhet,* palm of the hand.] **1.** A piece of sports equipment consisting of a round or oval frame with a network of tightly laced strings and a handle, used to strike a ball or shuttlecock. **2.** A wooden paddle, as one used in table tennis. **3. rackets** *(sing. in number).* A game similar to tennis, played in a four-walled court.

**rack·et²** (răk′ĭt) *n.* [Orig. unknown.] **1.** An uproar : din. **2.** *often* **rackets.** A dishonest business or practice, esp. one drawing illegal profits from organized fraud or extortion. **3.** *Slang.* A business or job. **4.** An easy and profitable means of livelihood. —*vi.* **-et·ed, -et·ing, -ets.** To lead an active social life.
**rack·et·eer** (răk′ĭ-tîr′) *n.* One engaged in an illegal business. —*vi.* **-eered, -eer·ing, -eers.** To engage in a racket.
**rack·et·y** (răk′ĭ-tē) *adj.* Noisy : clamorous.
**rack railway** *n.* A cog railway.
**rack-rent** (răk′rĕnt′) *n.* [< RACK¹.] Exorbitant rent. —*vt.* **-rent·ed, -rent·ing, -rents.** To exact exorbitant rent for or from.
**ra·clette** (rä-klĕt′, ră-) *n.* [Fr. < *racler,* to scrape.] **1.** A Swiss dish made with cheese melted and served on boiled bread or potatoes. **2.** A firm cheese used in making raclette.
**rac·on·teur** (răk′ŏn-tûr′) *n.* [Fr. < OFr. < *raconter,* to tell : *re-,* again (< Lat.) + *cunter,* to tell, to reckon. —see ACCOUNT.] An accomplished and witty storyteller.
**ra·coon** (ră-ko͞on′) *n. var. of* RACCOON.
**rac·quet** (răk′ĭt) *n. var. of* RACKET¹.
**rac·quet·ball** (răk′ĭt-bôl′) *n.* A court game identical to handball but utilizing a short racquet and a larger, softer ball.
**rac·y** (rā′sē) *adj.* **-i·er, -i·est.** [< RACE¹.] **1.** Having a distinctive quality or taste. **2. a.** Piquant or pungent. **b.** Bordering on impropriety or indelicacy : RISQUÉ. **c.** Lively : vigorous. **—rac′i·ly** *adv.* **—rac′i·ness** *n.*
☆ ***syns:*** RACY, OFF-COLOR, RISQUÉ, SPICY, SUGGESTIVE *adj. core meaning* : bordering on indelicacy or impropriety <*racy* jokes>
**rad** (răd) *n.* [Short for RADIATION.] *Physics.* A unit of energy absorbed from ionizing radiation, equal to 100 ergs per gram of irradiated material.
**ra·dar** (rā′där) *n.* [RA(DIO) D(ETECTING) A(ND) R(ANGING).] **1.** A method of detecting distant objects or phenomena and determining information such as their velocity or position by analysis of very high frequency radio waves reflected from their surfaces. **2.** The equipment used in radar detection.
**radar astronomy** *n.* Astronomy using reflected radio waves to investigate characteristics of celestial bodies.
**radar beacon** *n.* A fixed device that sends or receives, amplifies, alters, and returns a radar signal, permitting a distant receiver to determine its bearing and occas. its range.
**ra·dar·scope** (rā′där-skōp′) *n.* [RADAR + (OSCILLO)SCOPE.] The oscilloscope viewing screen of a radar receiver.
**radar telescope** *n.* A large radar antenna used in radar astronomy.
**rad·dle** (răd′l) *vt.* **-dled, -dling, -dles.** [< dial. *raddle,* stick interwoven with others in a fence < OFr. *reddalle,* poss. < MHG *reidel.*] To twist together or interweave.
**rad·dled** (răd′ld) *adj.* [Orig. unknown.] **1.** Broken-down and worn-out. **2.** Confused or befuddled.
**radi-** *pref. var. of* RADIO-.
**ra·di·al** (rā′dē-əl) *adj.* [Med. Lat. < Lat. *radius,* ray.] **1. a.** Of, relating to, or arranged like rays or radii. **b.** Radiating from or converging to a common center. **c.** Having or marked by radial parts or a radial arrangement. **2.** Moving or directed along a radius. **3.** *Anat.* Of, relating to, or near the radius or forearm. **4.** Developing symmetrically around a central point. —*n.* **1.** A radial part, as a ray or spoke. **2.** A radial tire. **—ra′di·al·ly** *adv.*
**radial engine** *n.* An internal-combustion engine, as once used in propeller-driven aircraft, with radially arrayed cylinders.
**radial symmetry** *n.* Symmetric arrangement of elements, esp. of radiating parts, around a central point or axis.
**radial tire** *also* **radial ply tire** *n.* A pneumatic tire in which the ply cords extending to beads are laid at approx. right angles to the center line of the tread.
**ra·di·an** (rā′dē-ən) *n.* [RADI(US) + -AN.] A unit of angular measure equal to the angle subtended at the center of a circle by an arc of length equal to the radius of the circle, equal to $^{360}/_{2\pi}$°, or approx. 57°17′44.6′.
**ra·di·ance** (rā′dē-əns) *also* **ra·di·an·cy** (-ən-sē) *n.* **1.** Quality or state of being radiant. **2.** *Physics.* Radiant energy emitted per unit time in a specified direction by a unit area of a radiating surface.
**ra·di·ant** (rā′dē-ənt) *adj.* [Lat. *radians, radiant-,* pr.part. of *radiare,* to radiate.] **1.** Emitting heat or light. **2.** Being or emitted as radiation <*radiant* heat> **3.** Filled with or expressing elated emotion, as love or happiness. **4.** Glowing : brilliant <a *radiant* gem> —*n.* **1.** An object or point from which heat or light rays are emitted. **2.** *Astron.* The apparent celestial origin of a meteoric shower. **—ra′di·ant·ly** *adv.*
**radiant energy** *n. Physics.* Energy transferred by radiation, esp. by an electromagnetic wave.
**radiant flux** *n.* Rate of flow of radiant energy.
**ra·di·ate** (rā′dē-āt′) *v.* **-at·ed, -at·ing, -ates.** [Lat. *radiare, radiat-,* to emit beams < *radius,* ray.] —*vi.* **1.** To emit radiation. **2.** To issue or emerge in rays. **3.** To diverge or converge radially, as the spokes of a wheel. —*vt.* **1.** To emit (e.g., light). **2.** To spread or disseminate as if from a center <*radiate* the good news> **3.** To irradiate or illuminate (an object). **4.** To manifest in a glowing way <*radiated* joy> —*adj.* (rā′dē-ĭt). **1.** *Bot.* Having rays or raylike parts <*radiate* flowers> **2.** *Zool.* Marked by radial symmetry. **3.** Surrounded with rays, as a head represented on a coin. **—ra′di·a′tive** *adj.*
**ra·di·a·tion** (rā′dē-ā′shən) *n.* **1.** An act or process of radiating. **2.** *Physics.* **a.** Emission and propagation of waves or particles. **b.** The propagating waves or particles, as light, sound, radiant heat, or particles, emitted by radioactivity. **3.** *Anat.* Radial arrangement of parts, as of a group of nerve fibers connecting different areas of the brain. **4.** *Biol.* Adaptive radiation.
**radiation sickness** *n.* Illness induced by exposure to ionizing radiation, ranging in severity from nausea, diarrhea, and headache to sterility, loss of teeth and hair, reduction in red and white blood cell count, extensive hemorrhaging, and death.
**ra·di·a·tor** (rā′dē-ā′tər) *n.* **1.** A heating device composed of a framework of connected pipes for the circulation of steam or hot water. **2.** A cooling device, as in motor vehicular engines, through which water or other fluids circulate as a coolant. **3.** *Physics.* A body that emits radiation. **4.** A transmitting antenna.
**rad·i·cal** (răd′ĭ-kəl) *adj.* [ME, of a root < LLat. *radicalis,* having roots < Lat. *radix,* root.] **1.** Arising from or going to a root or source : BASIC. **2.** Drastic or sweeping : EXTREME <a *radical* shift in policy> **3.** Favoring or resulting in extreme or revolutionary changes, as in political organization <a *radical* faction> **4.** Of or designating a word root. **5.** *Bot.* Of, relating to, or growing from the root. —*n.* **1.** An advocate of political and social revolution. **2.** *Math.* The root of a quantity as indicated by the radical sign. **3.** *Chem.* An atom or group of atoms with at least one unpaired electron. **4.** ROOT¹ 8. **—rad′i·cal·ly** *adv.* **—rad′i·cal·ness** *n.*
**radical expression** *n. Math.* An expression or form in which radical signs appear.
**rad·i·cal·ism** (răd′ĭ-kə-lĭz′əm) *n.* The doctrines or practices esp. of political radicals.
**rad·i·cal·ize** (răd′ĭ-kə-līz′) *vt.* **-ized, -iz·ing, -iz·es.** To make radical or more radical. **—rad′i·cal·i·za′tion** *n.*
**radical sign** *n. Math.* **1.** The sign $\sqrt{}$ placed before a quantity, indicating extraction of the root designated by a raised integral index. **2.** The radical sign together with a horizontal bar extending from its top to the end of the expression from which a root is to be extracted.
**rad·i·cand** (răd′ĭ-kănd′) *n.* [Lat. *radicandum,* neuter gerund. of *radicare,* to take root < *radix,* root.] *Math.* The quantity under a radical sign <3 is the *radicand* of $\sqrt{3}$>
**rad·i·ces** (răd′ĭ-sēz′, rā′də-) *n. var. pl. of* RADIX.
**rad·i·cle** (răd′ĭ-kəl) *n.* [Lat. *radicula,* dim. of *radix,* root.] **1.** *Bot.* The part of the plant embryo that develops into the primary root. **2.** *Anat.* A small structure resembling a root, as a fibril of a nerve.
**ra·di·i** (rā′dē-ī′) *n. var. pl. of* RADIUS.
**ra·di·o** (rā′dē-ō) *n., pl.* **-os.** [Short for RADIOTELEGRAPHY.] **1.** Use of electromagnetic waves in the radio frequency range to transmit or receive electric signals without wires connecting the points of transmission and reception. **2.** Communication of audible signals encoded in electromagnetic waves transmitted and received by radio. **3.** Broadcast of programmed material, as information or entertainment, by radio transmission. **4. a.** Equipment used to transmit radio signals : TRANSMITTER. **b.** Equipment used to receive radio signals : RECEIVER. **c.** A complex of radio equipment combining both transmitter and receiver. **5. a.** A station for radio transmitting. **b.** A radio broadcast organization or network. **c.** The radio broadcast industry. **6.** A message sent by radio. —*v.* **-oed, -o·ing, -os.** —*vt.* To transmit or communicate by radio. —*vi.* To transmit a message by radio.
**radio-** *or* **radi-** *pref.* [< RADIATION.] **1.** Radiation : radiant energy <*radio*meter> **2.** Radioactive <*radio*chemistry> **3.** Radio <*radio*telephone>
**ra·di·o·ac·tive** (rā′dē-ō-ăk′tĭv) *adj.* Of, exhibiting, or emitting radioactivity. **—ra′di·o·ac′tive·ly** *adv.*
**radioactive decay** *n.* Progressive decrease in the number of radioactive atoms in a substance as a result of spontaneous nuclear disintegration or transformation.
**radioactive series** *n.* A group of isotopes related by a process of radioactive decay in which the heavier members of the group are transformed into successively lighter ones, the lightest being stable.
**ra·di·o·ac·tiv·i·ty** (rā′dē-ō-ăk′tĭv′ĭ-tē) *n.* **1.** Spontaneous emission of radiation, either directly from unstable atomic nuclei or as a consequence of a nuclear reaction. **2.** Radiation emitted by radioactivity, including alpha particles, nucleons, electrons, and gamma rays.
**radio astronomy** *n.* Study of celestial bodies and phenomena by observation and analysis of their associated radio-frequency emissions.
**radio beacon** *n.* A fixed radio transmitter that broadcasts distinctive signals as a navigational aid.
**radio beam** *n.* A focused beam of radio signals transmitted by a radio beacon to guide ships or aircraft.

**ra·di·o·bi·ol·o·gy** (rā'dē-ō-bī-ŏl'ə-jē) *n.* **1.** Study of the effects of radiation on living organisms. **2.** Use of radioactive tracers to study biological processes. **—ra'di·o·bi'o·log'i·cal** (-ə-lŏj'ĭ-kəl) *adj.* **—ra'di·o·bi·ol'o·gist** *n.*
**ra·di·o·broad·cast** (rā'dē-ō-brôd'kăst') *v.* **-cast** *or* **-cast·ed, -cast·ing, -casts.** —*vt.* To broadcast (e.g., a program) by radio. —*vi.* To broadcast by radio. **—ra'di·o·broad'cast'er** *n.*
**ra·di·o·car·bon** (rā'dē-ō-kär'bən) *n.* Radioactive carbon, esp. carbon 14.
**radiocarbon dating** *n.* Carbon dating.
**ra·di·o·cast** (rā'dē-ō-kăst') *vt.* & *vi.* **-cast** *or* **-cast·ed, -cast·ing, -casts.** [RADIO + (BROAD)CAST.] To radiobroadcast. **—ra'di·o·cast'er** *n.*
**ra·di·o·chem·is·try** (rā'dē-ō-kĕm'ĭ-strē) *n.* The chemistry of radioactive materials. **—ra'di·o·chem'i·cal** *adj.*
**radio compass** *n.* A navigational aid composed of an automatic radio receiver that analyzes incoming radio signals to determine the direction of transmission.
**ra·di·o·el·e·ment** (rā'dē-ō-ĕl'ə-mənt) *n.* A naturally occurring or artificially produced radioactive element.
**radio frequency** *n.* **1.** The wave band frequency used by or assigned to a specific radio station. **2.** A frequency in the range within which radio waves may be transmitted, from about 10 kilocycles/second to about 300,000 megacycles/second.

## TABLE OF RADIO FREQUENCIES

| Classification | Abbreviation | Range |
|---|---|---|
| very low frequency | vlf *or* VLF | 10–30 kilocycles/second |
| low frequency | lf *or* LF | 30–300 kilocycles/second |
| medium frequency | mf *or* MF | 300–3,000 kilocycles/second |
| high frequency | hf *or* HF | 3,000–30,000 kilocycles/second |
| very high frequency | vhf *or* VHF | 30–300 megacycles/second |
| ultrahigh frequency | uhf *or* UHF | 300–3,000 megacycles/second |
| superhigh frequency | shf *or* SHF | 3,000–30,000 megacycles/second |
| extremely high frequency | ehf *or* EHF | 30,000–300,000 megacycles/second |

**radio galaxy** *n.* A galaxy emitting large amounts of radio energy.
**ra·di·o·ge·nic** (rā'dē-ō-jĕn'ĭk) *adj.* Caused by radioactivity.
**ra·di·o·gram** (rā'dē-ō-grăm') *n.* **1.** A message transmitted by wireless telegraphy. **2.** A radiograph.
**ra·di·o·graph** (rā'dē-ō-grăf') *n.* An image produced on a radiosensitive surface, as a photographic film, by radiation other than visible light, esp. by x-rays passed through an object. —*vt.* **-graphed, -graph·ing, -graphs.** To make a radiograph of. **—ra'di·og'ra·pher** (-ŏg'rə-fər) *n.* **—ra'di·o·graph'ic** *adj.* **—ra'di·o·graph'i·cal·ly** *adv.* **—ra'di·og'ra·phy** *n.*
**ra·di·o·im·mu·no·as·say** (rā'dē-ō-ĭm'yə-nō-ăs'ā, -ĭm-yōō'-) *n.* Immunoassay of a radioactively labeled substance, as an enzyme or hormone.
**ra·di·o·i·so·tope** (rā'dē-ō-ī'sə-tōp') *n.* A naturally occurring or artificially produced radioactive isotope of an element.
**ra·di·o·lar·i·an** (rā'dē-ō-lâr'ē-ən) *n.* [< NLat. *Radiolaria*, order name < LLat. *radiolus*, dim. of *radius*, ray.] A marine protozoan of the order Radiolaria, having a siliceous skeleton and spicules.
**ra·di·o·lo·ca·tion** (rā'dē-ō-lō-kā'shən) *n.* Detection and positional location of distant objects by radar.
**ra·di·ol·o·gy** (rā'dē-ŏl'ə-jē) *n.* **1.** Use of ionizing radiation for medical diagnosis, esp. the use of x-rays in medical radiography or fluoroscopy. **2.** Use of radiation for scientific examination of material structures : RADIOSCOPY. **—ra'di·o·log'i·cal** (-ə-lŏj'ĭ-kəl) *adj.* **—ra'di·o·log'i·cal·ly** *adv.* **—ra'di·ol'o·gist** *n.*
**ra·di·ol·y·sis** (rā'dē-ŏl'ĭ-sĭs) *n.* Chemical dissociation of molecules due to radiation. **—ra'di·o·lyt'ic** (-ə-lĭt'ĭk) *adj.*
**ra·di·o·man** (rā'dē-ō-măn') *n.* A radio operator or technician.
**ra·di·om·e·ter** (rā'dē-ŏm'ĭ-tər) *n.* A device detecting and measuring radiation, composed of a set of lightweight vanes blackened on one side and suspended in a partial vacuum so as to revolve about a central axis when exposed to varying amounts of incident radiation. **—ra'di·o·met'ric** (-ō-mĕt'rĭk) *adj.* **—ra'di·om'e·try** *n.*

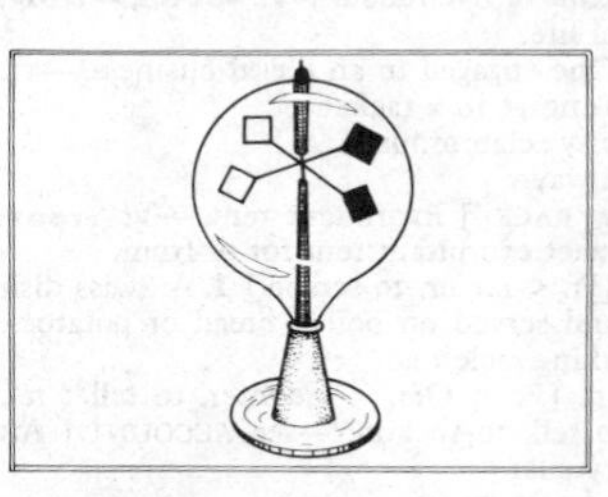
radiometer

**ra·di·o·nu·clide** (rā'dē-ō-nōō'klīd', -nyōō'-) *n.* A radioactive nuclide.
**ra·di·o·paque** (rā'dē-ō-pāk') *adj.* Impervious to radiation, as x-rays.
**ra·di·o·phone** (rā'dē-ō-fōn') *n.* A radiotelephone. **—ra'di·o·phon'ic** (-fŏn'ĭk) *adj.*
**ra·di·o·pho·to·graph** (rā'dē-ō-fō'tə-grăf') *also* **ra·di·o·pho·to** (-fō'tō) *n.* A photograph transmitted by radio waves, with each image point being reproduced by a received electric impulse. **—ra'di·o·pho·tog'ra·phy** (-fə-tŏg'rə-fē) *n.*
**ra·di·os·co·py** (rā'dē-ŏs'kə-pē) *n.* Examination of the internal structure of optically opaque objects by use of penetrating radiation, esp. by x-rays : RADIOLOGY. **—ra'di·o·scop'ic** (-ō-skŏp'ĭk), **ra'di·o·scop'i·cal** *adj.*
**ra·di·o·sen·si·tive** (rā'dē-ō-sĕn'sĭ-tĭv) *adj.* Sensitive to radiation. —Used esp. of living structures.
**ra·di·o·sonde** (rā'dē-ō-sŏnd') *n.* [RADIO- + Fr. *sonde*, sounding line < OFr., prob. of Germanic orig.] An instrument for gathering and transmitting meteorological data from the upper atmosphere, carried aloft usu. by balloon.
**radio spectrum** *n.* The complete range of electromagnetic communications frequencies, including those used for radio, radar, and television.
**ra·di·o·tel·e·graph** (rā'dē-ō-tĕl'ĭ-grăf') *n.* Transmission of messages by radiotelegraphy. **—ra'di·o·tel'e·graph'ic** *adj.*
**ra·di·o·te·leg·ra·phy** (rā'dē-ō-tə-lĕg'rə-fē) *n.* Wireless telegraphy.
**ra·di·o·tel·e·phone** (rā'dē-ō-tĕl'ə-fōn') *n.* A telephone enabling two-way audible communication by means of radio. **—ra'di·o·tel'e·phon'ic** (-ə-fŏn'ĭk) *adj.* **—ra'di·o·te·leph'o·ny** (-tə-lĕf'ə-nē) *n.*
**radio telescope** *n.* A directional radio-antenna system for intercepting and analyzing radio waves of extraterrestrial origin.
**ra·di·o·ther·a·py** (rā'dē-ō-thĕr'ə-pē) *n.* Treatment of disease with radiation, esp. by selective irradiation with x-rays or other ionizing radiation and by ingestion of radioisotopes.
**ra·di·o·tho·ri·um** (rā'dē-ō-thôr'ē-əm, -thōr'-) *n.* A radioactive isotope of thorium with mass number 228.
**ra·di·o·trac·er** (rā'dē-ō-trā'sər) *n.* A radioactive tracer.
**radio wave** *n.* A radio-frequency electromagnetic wave.
**rad·ish** (răd'ĭsh) *n.* [ME *radiche* < OE *rædic* < Lat. *radix*, root.] **1.** A plant of the genus *Raphanus*, esp. *R. sativus*, having a thickened edible root. **2.** The pungent root of the radish, eaten usu. raw as an appetizer or salad vegetable.
**ra·di·um** (rā'dē-əm) *n.* [< Lat. *radius*, ray.] *Symbol* **Ra** A rare brilliant-white, luminescent, highly radioactive metallic element used in radiotherapy, as a neutron source, and as a constituent of luminescent paints; atomic number 88; longest-lived isotope Ra 226.
**radium therapy** *n.* Use of radium in radiotherapy, esp. in treating cancer.
**ra·di·us** (rā'dē-əs) *n., pl.* **-di·i** (-dē-ī') *or* **-di·us·es.** [Lat., ray.] **1. a.** A line segment joining the center of a circle with any point on its circumference. **b.** A line segment joining the center of a sphere with any point on its surface. **c.** A line segment joining the center of a regular polygon to any of its vertices. **d.** The length of any such line segment. **2.** A measure of circular area or extent <every town within a *radius* of 50 miles> **3.** A measure of range or scope, as of experience or influence <within the *radius* of known facts> **4.** A radial structure or part, as a wheel spoke or a mechanically pivoted arm. **5.** *Anat.* **a.** A long, prismatic, slightly curved bone, the shorter and thicker of the two forearm bones, located on the lateral side of the ulna. **b.** A similar bone in many vertebrates.
**radius vector** *n.* **1.** *Math.* **a.** A line segment that joins a variable point to the origin of polar or spherical coordinates. **b.** The length of such a line segment. **2.** *Astron.* A line segment joining the center of a satellite to the focus of its orbit.
**ra·dix** (rā'dĭks) *n., pl.* **rad·i·ces** (răd'ə-sēz', rā'də-) *or* **ra·dix·es.** [Lat.] **1.** *Biol.* A root or point of origin. **2.** *Math.* The base of a system

of numbers, as are 2 of the binary system and 10 of the decimal system.
**ra·dome** (rā′dōm) *n.* [RA(DAR) + DOME.] A domelike protective housing for a radar antenna used esp. in certain aircraft.
**ra·don** (rā′dŏn) *n.* [RAD(IUM) + -ON[2].] *Symbol* **Rn** A colorless, radioactive, inert gaseous element formed by disintegration of radium and used in radiotherapy; atomic number 86; atomic weight 222.
**rad·u·la** (răj′ōō-lə) *n., pl.* **-lae** (-lē′) [NLat. < Lat., scraper < *radere*, to scrape.] *Zool.* A flexible tonguelike organ in mollusks, having rows of horny teeth used to shred and ingest food. **—rad′u·lar** *adj.*
**raf·fi·a** *also* **raph·i·a** (răf′ē-ə) *n.* [Malagasy *rafia.*] **1.** An African palm tree, *Raphia ruffia*, with large leaves yielding a useful fiber. **2.** The fiber of the leaves of the raffia.
**raf·fi·nose** (răf′ə-nōs′) *n.* [Fr. < *raffiner*, to refine : *re-*, again + *affiner*, to refine (*a-*, to + *fin*, fine).] A white crystalline sugar, $C_{18}H_{32}O_{16}·5H_2O$, obtained from sugar beets and cottonseed meal.
**raff·ish** (răf′ĭsh) *adj.* [Prob. < dial. *raff*, rubbish < ME *raf.*] **1.** Vulgar : flashy. **2.** Rakish. **—raff′ish·ly** *adv.* **—raff′ish·ness** *n.*
**raf·fle[1]** (răf′əl) *n.* [ME *rafle*, a game using dice < OFr. *raffle*, act of seizing.] A lottery in which a number of persons buy chances on a prize. *—v.* **-fled, -fling, -fles.** *—vt.* To offer as a prize in a raffle <*raffle* off a fur coat> *—vi.* To participate in a raffle. **—raf′fler** *n.*
**raf·fle[2]** (răf′əl) *n.* [Prob. < Fr. *rafle*, act of seizing < OFr.] Rubbish : jumble.
**raf·fle·sia** (ră-flē′zhə) *n.* [NLat., genus name, after Sir Stamford *Raffles* (1781–1826).] A parasitic leafless plant of the genus *Rafflesia*, found in tropical Asia, bearing large, often malodorous flowers.
**raft[1]** (răft) *n.* [ME < ON *raptr*, beam.] **1.** A level floating structure or watercraft, typically made of logs, planks, or barrels, used for transport or as a platform for swimmers. **2.** A life raft. *—v.* **raft·ed, raft·ing, rafts.** *—vt.* **1.** To convey on a raft. **2.** To make into a raft <*rafted* balsa logs together> *—vi.* To go or travel by raft.
**raft[2]** (răft) *n.* [Alteration of dial. *raff*, rubbish < ME *raf.*] *Informal.* A great number, amount, or collection <*a raft* of troubles>
**raft·er** (răf′tər) *n.* [ME < OE *ræfter.*] One of the sloping, parallel beams that support a pitched roof.
**rag[1]** (răg) *n.* [ME *ragge* < OE **ragg* < ON *rögg.*] **1.** A scrap of cloth. **2.** Cloth converted to pulp for papermaking <high-quality *rag* paper> **3.** A remnant : scrap. **4.** *Slang.* **a.** A newspaper. **b.** A periodical specializing in sensationalism or gossip. **5. rags. a.** Tattered or threadbare clothing. **b.** *Informal.* Glad rags. **6.** The stringy central portion and membranous walls of citrus fruits.
**rag[2]** (răg) *vt.* **ragged, rag·ging, rags.** [Orig. unknown.] **1.** *Slang.* To taunt : tease. **2.** *Slang.* To scold. **3.** *Chiefly Brit.* To play a joke on. *—n. Chiefly Brit.* A practical joke : PRANK.
**rag[3]** (răg) *n.* [Orig. unknown.] **1.** A roofing slate with one rough surface. **2.** *Chiefly Brit.* A coarsely textured rock.
**rag[4]** (răg) *vt.* **ragged, rag·ging, rags.** [Short for RAGTIME.] To play or compose (a piece of music) in ragtime. *—n.* A piece of music written in ragtime.
**ra·ga** (rä′gə) *n.* [Skt. *rāgaḥ*, color, musical mode.] A traditional form in Hindu music, consisting of a theme that expresses some aspect of religious feeling and sets forth a tonal system on which variations are improvised within a prescribed framework of typical progressions, melodic formulas, and rhythmic patterns.
**rag·a·muf·fin** (răg′ə-mŭf′ĭn) *n.* [After *Ragamoffyn*, demon in *Piers Plowman*, a 14th-cent. allegorical poem.] A dirty or unkempt child.
**rag·bag** (răg′băg′) *n.* **1.** A bag for storing rags. **2.** A jumbled assortment : HODGE PODGE.
**rage** (rāj) *n.* [ME < OFr. < LLat. *rabia* < Lat. *rabies* < *rabere*, to be mad.] **1. a.** Violent anger. **b.** A fit of anger. **2.** Furious intensity, as of a storm or disease. **3.** Burning desire or passion. **4.** A craze : fad <the latest *rage*> *—vi.* **raged, rag·ing, rag·es. 1.** To speak or behave furiously. **2.** To move or act with great violence or intensity <The hurricane *raged* for three days.> **3.** To prevail or spread unchecked <The fighting *raged* throughout the city.>
**rag·ged** (răg′ĭd) *adj.* [ME < *ragge*, rag.] **1.** Torn, frayed, or tattered. **2. a.** Dressed in tattered or threadbare clothes. **b.** Unkempt or shaggy. **3.** Mismatched or jumbled <a *ragged* assortment> **4.** Having a rough surface or edges. **5.** Uneven : sloppy <a *ragged* performance> **6.** Harsh : rasping <a *ragged* cry> **—rag′ged·ly** *adv.* **—rag′ged·ness** *n.*
**ragged robin** *n.* A plant, *Lychnis flos-cuculi*, orig. of Eurasia, bearing reddish or white flowers with deeply lobed petals.
**ra·gi** (răg′ē) *n.* [Hindi *rāgī* < Skt.] A grass, *Eleusine coracana* of Africa and Asia, cultivated for its edible grain.
**rag·lan** (răg′lən) *n.* [After Fitzroy James Henry Somerset (1788–1855), 1st Baron *Raglan.*] A loose garment with slanted shoulder seams and sleeves extending in one piece to the neckline.
**rag·man** (răg′măn′) *n.* A dealer in rags and refuse.
**ra·gout** (ră-gōō′) *n.* [Fr. *ragoût* < *ragoûter*, to revive the taste : *re-*, again (< Lat.) + *à*, to (< Lat. *ad*) + *goût*, taste (< Lat. *gustus*).] A thick spicy stew of vegetables and meat.
**rag picker** *n.* A ragman.
**rag·tag** (răg′tăg′) *adj.* **1.** Tattered or unkempt : RAGGED. **2.** Falling apart : RAMSHACKLE.
**ragtag and bobtail** *n.* Rabble : riffraff.
**rag·time** (răg′tīm′) *n.* [Prob. < *ragged time*, syncopated rhythm.] *Mus.* A jazz style marked by elaborately syncopated rhythm in the melody and a steadily accented accompaniment.
**rag·weed** (răg′wēd′) *n.* [From the ragged shape of its leaves.] **1.** A weed of the genus *Ambrosia*, esp. *A. artemisiifolia* or *A. trifida*, whose airborne pollen is one of the principal causes of hay fever. **2.** *Chiefly Brit.* The ragwort.
**rag·wort** (răg′wûrt′, -wôrt′) *n.* [From the ragged shape of its leaves.] A plant of the genus *Senecio*, bearing yellow flowers, esp. *S. aureus*, the golden ragwort of eastern North America, and *S. jacobaea* of Europe.
**rah** (rä) *interj.* [Short for HURRAH.] —Used to express enthusiastic approval or pleasure.
**raid** (rād) *n.* [Sc., raid on horseback < ME *rade* < OE *rād*, ride, raid.] **1.** A surprise attack, as one made by a commando force. **2.** Sudden and forcible entry, as of an illegal establishment, by law officers. **3.** A bold predatory operation mounted against a competition, esp. in business. **4.** An attempt by speculators to drive stock prices down by mutual selling. *—v.* **raid·ed, raid·ing, raids.** *—vt.* To make a raid on. *—vi.* To conduct or participate in a raid. **—raid′er** *n.*
☆ ***syns:*** RAID, HARASS, HARRY, MARAUD *v. core meaning* : to make a surprise attack on <The soldiers *raided* the villages at night.>
**rail[1]** (rāl) *n.* [ME *raile* < OFr. *reille* < Lat. *regula*, rod, ruler < *regere*, to rule.] **1.** A horizontal bar supported by vertical posts, as in a fence. **2.** A railing or balustrade. **3.** A steel bar usu. used in pairs as a track for railroad cars and similar vehicles. **4.** The railroad as a means of transportation <shipments made by *rail*> **5. rails.** The stocks and bonds issued by railroads. **6.** A horizontal piece of wood in a door or in paneling. *—vt.* **railed, rail·ing, rails.** To enclose or supply with a rail or rails.
**rail[2]** (rāl) *n.* [ME *raile* < OFr. *raale.*] A small brownish marsh bird of the family Rallidae, having long toes and stubby wings adapted for only short flights.
**rail[3]** (rāl) *vi.* **railed, rail·ing, rails.** [ME *railen* < OFr. *railler* < VLat. **ragulare* < LLat. *ragere*, to bray.] To condemn or attack in bitter or abusive language <*railed* against big oil> **—rail′er** *n.*
**rail fence** *n.* A fence of split logs secured to posts.
**rail·head** (rāl′hĕd′) *n.* **1.** The farthest point on a railroad to which rails have been laid. **2.** The section of a railroad where military supplies are unloaded.
**rail·ing** (rā′lĭng) *n.* **1. a.** A banister, balustrade, or fence made of rails. **b.** The upper longitudinal part of a balustrade. **2.** Rails as a whole. **3.** Material for making rails.
**rail·ler·y** (rā′lə-rē) *n., pl.* **-ies.** [Fr. *raillerie* < OFr. *railler*, to rail.] Good-natured ridicule or teasing : BANTER.
**rail·road** (rāl′rōd′) *n.* **1.** A road composed of parallel steel rails supported by ties and providing a track for locomotive-drawn trains and other rolling stock. **2.** The complete system of railroad track, along with the land, stations, rolling stock, and other property used in rail transportation. *—v.* **-road·ed, -road·ing, -roads.** *—vt.* **1.** To transport by railroad. **2.** To supply (an area) with railroads. **3.** *Informal.* **a.** To push through quickly so as to forestall discussion <*railroad* a bill through the Senate> **b.** To cause (someone) to be condemned or imprisoned without due process of law. *—vi.* To work for a railroad company.
**railroad flat** *n.* A narrow apartment in which the rooms are connected in a line, with doors at the front and rear.
**rail·road·ing** (rāl′rō′dĭng) *n.* The railroad business.
**rail-split·ter** (rāl′splĭt′ər) *n.* One that splits logs for fences.
**rail·way** (rāl′wā′) *n.* **1.** A railroad, esp. one operated over a limited area, as a street railway. **2.** A track providing a runway for wheeled equipment.
**rai·ment** (rā′mənt) *n.* [ME *rayment*, short for *arrayment* < OFr. *araiement*, array < *arrayer*, to array.] Clothing : garments.
**rain** (rān) *n.* [ME < OE *rēn.*] **1. a.** Water condensed from atmospheric vapor, falling to earth in drops. **b.** A fall of such water : RAINSTORM. **c.** The descent of such water. **d.** Rainy weather. **2.** An abundant outpouring : SHOWER <a *rain* of abuse> **3. rains.** The rainy season or seasonal rainfalls, as in certain tropical areas. *—v.* **rained, rain·ing, rains.** *—vi.* **1.** To fall as rain from the clouds. **2.** To fall like rain <Confetti *rained* all around.> **3.** To release rain. *—vt.* **1.** To send or pour down. **2.** To give or spread profusely. **—rain out.** To force the postponement of (an outdoor event) because of rain.
**rain·bow** (rān′bō′) *n.* [ME < OE *rēnboga* : *rēn*, rain + *boga*, bow.] **1. a.** An arc of spectral colors appearing in the sky opposite the sun as a result of the refractive dispersion of sunlight in raindrops or mist. **b.** A similar arc, as in a waterfall mist or graded display of colors. **2.** An illusory hope.
**rainbow cactus** *n.* A tall spiny cylindrical cactus, *Echinocereus*

*rigidissimus* of the southwestern United States and Mexico, bearing showy pink flowers.

**rainbow trout** *n.* A North American food and game fish, *Salmo gairdneri,* having a reddish longitudinal band and black spots.

**rain check** *n.* **1.** A ticket stub for an outdoor event entitling the holder to admission at a future date if the scheduled event is canceled due to bad weather. **2.** A postponement of the acceptance or fulfillment of an offer, esp. a guarantee that a sale item temporarily out of stock can be purchased later at the sale price.

**rain·coat** (rān'kōt') *n.* A waterproof or water-resistant coat.

**rain date** *n.* A second date scheduled for an outdoor event in case the first date is canceled or rained out.

**rain·drop** (rān'drŏp') *n.* A drop of rain.

**rain·fall** (rān'fôl') *n.* **1.** A fall of rain. **2.** PRECIPITATION 3b.

**rain forest** *n.* A dense, usu. tropical evergreen forest with an annual rainfall of at least 100 inches.

**rain gauge** *also* **rain gage** *n.* A device for measuring rainfall.

**rain·mak·er** (rān'mā'kər) *n.* One who professes an ability to produce or bring on rain.

**rain·mak·ing** (rān'mā'kĭng) *n.* **1.** The rituals and ceremony practiced by a rainmaker. **2.** *Informal.* Cloud seeding.

**rain·spout** (rān'spout') *n.* A spout draining a roof gutter.

**rain·squall** (rān'skwôl') *n.* A squall accompanied by rain.

**rain·storm** (rān'stôrm) *n.* A storm accompanied by rain.

**rain-wash** (rān'wŏsh') *Geol.* —*n.* **1.** Rock debris carried downhill by rain. **2.** Disintegrated rock and similar matter washed down into the soil by rain. —*vt.* **-washed, -wash·ing, -wash·es.** To wash (material) down a slope by rain.

**rain·wat·er** (rān'wô'tər, -wŏt'ər) *n.* Water precipitated as rain and having low dissolved mineral content.

**rain·wear** (rān'wâr') *n.* Waterproof clothing.

**rain·y** (rā'nē) *adj.* **-i·er, -i·est.** Marked by, full of, or bringing rain. **—rain'i·ness** *n.*

**rainy day** *n.* A time of trouble or need.

**raise** (rāz) *v.* **raised, rais·ing, rais·es.** [ME *raisen* < ON *reisa.*] —*vt.* **1.** To move or cause to move upward or to a higher position : LIFT. **2.** To set or place upright : STAND. **3.** To build or erect. **4. a.** To cause to appear, arise, or exist <The sting *raised* a bump.> **b.** To awaken from or as if from death. **5.** To increase in quantity, size, or worth <*raised* their prices> **6.** To increase in intensity, strength, degree, or pitch <*raise* one's voice> **7.** To improve in rank or dignity : PROMOTE. **8. a.** To grow or breed <*raise* Shetland ponies> **b.** To bring up : REAR <*raise* a family> **9.** To put forward for consideration <*raise* an issue> **10.** *Law.* To begin or set (a lawsuit) in operation. **11.** To utter or express (e.g., a cry). **12. a.** To bring about : PROVOKE <*raise* a smile> **b.** To arouse or stir up <*raise* a fuss over nothing> **13.** To gather together : COLLECT <*raise* funds> **14.** To cause (dough) to rise. **15.** To end (a siege) by withdrawing troops or forcing the enemy troops to withdraw. **16.** To remove or withdraw (an order). **17. a.** To increase (a poker bet). **b.** To bet more than (a preceding bettor in poker). **c.** To increase the bid of (one's bridge partner). **18.** *Naut.* To bring into sight by approaching nearer <*raise* land> **19.** To alter and increase illegally the written value of (e.g., a check). **20.** To cough up (phlegm). **21.** *Scot.* To make angry : ENRAGE. **22.** To make contact with esp. by radio. —*vi.* To increase the stakes in poker or gambling. —*n.* **1.** An act of raising or increasing. **2.** An increase in wages. **—raise cain (or hell).** **1.** To behave in a rowdy or disruptive way. **2.** To reprimand someone severely and loudly. **—raise eyebrows.** To cause surprise, amazement, or consternation.

▲ word history: The verb *raise* is a doublet of *rear*[2]. The former is a borrowing of Old Norse *reisa* and the latter is a native English word, but both are descended from the same Germanic ancestor, *raizjan.* Under certain conditions Germanic *z* became *r* in the West Germanic languages, of which English is one. The original consonant (spelled *s*) remained in the other Germanic languages, including Old Norse. Both *raise* and *rear*[2] are related to *rise.* No words related to these three verbs have been found outside of Germanic.

**raised** (rāzd) *adj.* **1.** Represented in relief, as a surface design : EMBOSSED. **2.** Made light and high by leavening such as yeast.

**rai·sin** (rā'zən) *n.* [ME < OFr., grape < Lat. *racemus,* bunch of grapes.] Any of several sweet, usu. seedless grapes, dried either in the sun or artificially.

**rai·son d'ê·tre** (rĕ-zôɴ' dĕ'tr) *n.* [Fr.] Reason or purpose for existence.

**raj** (räj) *n.* [Hindi *rāj* < Skt. *rājan,* king < *rājati,* he rules.] Sovereignty : dominion.

**ra·ja** (rä'jə) *n. var. of* RAJAH.

**Raj·ab** (rŭj'əb) *n.* [Ar.] The seventh month of the Moslem year. —See table at CALENDAR.

**ra·jah** *or* **ra·ja** (rä'jə) *n.* [Hindi *rājā* < Skt. *rājan,* king < *rājati,* he rules.] A chief, prince, or ruler in India or the East Indies.

**Raj·put** *also* **Raj·poot** (räj'po͝ot) *n.* One of a Hindu people claiming descent from the warlike and powerful rulers of northern India from the 8th to the 13th cent.

**rake**[1] (rāk) *n.* [ME < OE *raca.*] **1.** A long-handled implement with rigid or flexible teeth at its head, used esp. to gather grass and leaves or to loosen or smooth earth. **2.** An implement, as a clam rake, that resembles a rake. —*v.* **raked, rak·ing, rakes.** —*vt.* **1.** To gather or manipulate with or as if with a rake. **2.** To cultivate or till (e.g., earth) with a rake or similar implement. **3.** To gain in abundance <*rake* in the cash> **4.** To examine or search thoroughly <*rake* over the evidence> **5.** To scratch : scrap. **6.** To aim heavy gunfire along the length of. —*vi.* **1.** To use a rake. **2.** To conduct a search <*raked* through the debris for valuables> **—rake up.** To revive or bring to light : UNCOVER <*rake up* old scandals> **—rak'er** *n.*

**rake**[2] (rāk) *n.* [Short for RAKEHELL.] A libertine : roué.

**rake**[3] (rāk) *v.* **raked, rak·ing, rakes.** [Orig. unknown.] —*vi.* To slant or incline from the vertical, as a ship's mast. —*vt.* To cause to lean or slant. —*n.* **1.** Inclination from the vertical or from the horizontal. **2.** The angle between the cutting edge of a tool and a plane perpendicular to the working surface to which the tool is applied.

**rak·ee** (răk'ē, rä'kē, rä'kə) *n. var. of* RAKI.

**rake·hell** (rāk'hĕl') *n.* [Poss. by folk etymology < obs. *rackle,* headstrong.] RAKE[2]. **—rake'hell'y** *adj.*

**rake-off** (rāk'ôf', -ŏf') *n.* [From the rake used by a croupier in a gambling house.] *Slang.* A share or percentage of the profits of an enterprise, esp. one given or accepted as a bribe.

**rak·i** *also* **rak·ee** (răk'ē, rä'kē, rä'kə) *n.* [Turk. *rāqī.*] An anise-flavored grape or plum brandy of Turkey and the Balkan Peninsula.

**rak·ish**[1] (rā'kĭsh) *adj.* [Prob. < RAKE[3] (from the raking masts of pirate ships).] **1.** *Naut.* Trim and streamlined. **2.** Jaunty or showy. **—rak'ish·ly** *adv.* **—rak'ish·ness** *n.*

**rak·ish**[2] (rā'kĭsh) *adj.* Like a rake : DISSOLUTE. **—rak'ish·ly** *adv.* **—rak'ish·ness** *n.*

**rale** *also* **râle** (räl) *n.* [Fr. *râle* < *râler,* to make a rattling sound in the throat.] An abnormal or pathological respiratory sound.

**ral·len·tan·do** (räl'ən-tän'dō, räl'lĕn-tän'dō) [Ital., pr.part. of *rallentare,* to slow down : *re-,* (intensive < Lat.) + *allentare,* to slow down < LLat. (Lat. *ad-,* to + Lat. *lentus,* slow).] —*adv. & adj. Mus.* With gradual slackening in tempo. —Used as a direction. **—ral'len·tan'do** *n.*

**ral·ly**[1] (răl'ē) *v.* **-lied, -ly·ing, -lies.** [Fr. *rallier* < OFr. *ralier* : *re-,* again + *alier,* to unite, ally. —see ALLY.] —*vt.* **1.** To call together for a common purpose : ASSEMBLE. **2.** To reassemble and restore to order <*rally* retreating troops> **3.** To rouse or revive (e.g., one's strength) from decline or inactivity. —*vi.* **1.** To gather for a common purpose. **2.** To unite in an effort for a common cause. **3.** To recover abruptly from a setback or disadvantage <The racehorse *rallied* from behind to win by a length.> **4.** To show sudden improvement in health or spirits. **5.** To exchange several strokes in net games. —*n., pl.* **-lies.** **1.** An assembly, esp. one intended to inspire enthusiasm for a cause <a sports *rally*> **2. a.** A reassembling, as of dispersed troops. **b.** The signal ordering this. **3.** A sharp improvement in health, vigor, or spirits. **4.** A notable rise in stock market prices and active trading after a decline. **5.** An abrupt recovery from a setback or disadvantage. **6.** An exchange of several strokes in net games before one side scores a point. **7.** A competition in which cars are driven over public roads and under normal traffic regulations but with specified rules as to speed, time, and route. **—ral'li·er** *n.*

**ral·ly**[2] (răl'ē) *v.* **-lied, -ly·ing, -lies.** [Fr. *railler* < OFr., to rail. —see RAIL[3].] —*vt.* To tease good-humoredly : BANTER. —*vi.* To banter or jest. **—ral'li·er** *n.*

**ralph** (rălf) *vi.* **ralphed, ralph·ing, ralphs.** [Imit.] *Slang.* To vomit.

**ram** (răm) *n.* [ME < OE *ramm.*] **1.** A male sheep. **2. Ram.** Aries. **3.** A device used to batter, drive, or crush by forceful impact, esp.: **a.** A battering-ram. **b.** The weight that drops in a pile driver. **c.** The plunger or piston of a force pump or hydraulic press. **4. a.** A projection on the prow of a warship, used to batter or pierce an enemy vessel. **b.** A ship having such a projection. **5.** A hydraulic ram. —*vt.* **rammed, ram·ming, rams.** **1.** To drive or strike against with a heavy impact : BUTT. **2.** To force or knock into place. **3.** To stuff : cram. **4.** To force passage or acceptance of <*rammed* the budget through committee> **—ram'mer** *n.*

**Ra·ma** (rä'mə) *n.* [Skt. *Ramaḥ* < *rāma-,* dark, beautiful.] A deified hero worshiped as an incarnation of Vishnu in Hinduism.

**Ram·a·dan** (răm'ə-dän') *n.* [Ar. *Ramaḍān* < *ramaḍ,* dryness.] **1.** The ninth month of the Moslem year. —See table at CALENDAR. **2.** The fast from sunrise to sunset that is observed during Ramadan.

**Ra·man effect** (rä'mən) *n.* [After Sir Chandrasekhara Venkata *Raman* (1888–1970), its discoverer.] *Physics.* Alteration in frequency and random alteration in phase of light scattered in a material medium.

**ra·mate** (rā'māt') *adj.* [< Lat. *ramus,* branch.] Having branches : BRANCHED.

**ram·ble** (răm'bəl) *vi.* **-bled, -bling, -bles.** [Prob. < ME *romblen* < *romen,* to roam.] **1.** To walk or wander aimlessly. **2.** To follow an irregularly winding course of motion or growth. **3.** To speak or write at length and with many digressions. **—ram'ble** *n.*

**ram·bler** (răm'blər) *n.* **1.** One that rambles. **2.** A climbing rose bearing numerous red, pink, or white flowers.

**ram·bling** (răm′blĭng) *adj.* **1.** Wandering : roaming. **2.** Large and irregular in shape or extent : SPRAWLING <an old *rambling* farmhouse> **3.** Lengthy and drawn out <a *rambling* story> **—ram′bling·ly** *adv.*

**Ram·bouil·let** (răm′bo͝o-lā, răm′bo͞o-yā′) *n.* [After *Rambouillet*, town in northern France.] A merino sheep orig. bred in France, raised for wool and meat.

**Rambouillet**
*3 feet high at shoulder*

**ram·bunc·tious** (răm-bŭngk′shəs) *adj.* [Prob. alteration of E. *rumbustious*, alteration of *robustious* < Lat. *robustus*, strong < *robur*, oak, strength.] Boisterous : unruly.

**ram·bu·tan** (răm-bo͞ot′ən) *n.* [Malay < *rambut*, hair.] **1.** A tree, *Nephelium lappaceum* of southeastern Asia, bearing edible fruit. **2.** The oval, red, soft-spined fruit of the rambutan.

**ram·e·kin** *also* **ram·e·quin** (răm′ĭ-kĭn) *n.* [Fr. *ramequin* < LG *ramken*, dim. of *ram*, cream < MLG *rōme.*] **1.** A cheese preparation made with eggs and bread crumbs or unsweetened puff pastry, baked and served in individual dishes. **2.** A small individual dish used for both baking and serving.

**ra·mi** (rā′mī′) *n. pl. of* RAMUS.

**ram·ie** (răm′ē) *n.* [Malay *rami.*] **1.** A nettlelike Asian plant, *Boehmeria nivea*, having fibrous stem. **2.** The flaxlike fiber or bast of the ramie, used in making fabrics and cordage.

**ram·i·fi·ca·tion** (răm′ə-fĭ-kā′shən) *n.* **1.** The act or process of branching out or dividing into branches. **2.** A branch or other subordinate part extending from a main body. **3.** An arrangement of branches or branching parts. **4.** A consequence or development stemming from and often complicating something, as a plan or action <the *ramifications* of a change in policy>

**ram·i·form** (răm′ə-fôrm′) *adj.* [Lat. *ramus*, branch + -FORM.] Branchlike or branching.

**ram·i·fy** (răm′ə-fī′) *v.* **-fied, -fy·ing, -fies.** [OFr. *ramifier* : Lat. *ramus*, branch + Lat. *facere*, to make.] *—vt.* To divide or form into branches or branchlike extensions. *—vi.* To branch out.

**ram·jet** (răm′jĕt′) *n.* A jet engine that propels aircraft by igniting fuel with air taken and compressed by the engine in a fashion that produces greater exhaust than intake velocity.

**ra·mose** (rā′mōs′, rə-mōs′) *adj.* [Lat. *ramosus* < *ramus*, branch.] Having many branches.

**ra·mous** (rā′məs) *adj.* [Lat. *ramosus*, ramose.] **1.** Of or resembling branches. **2.** Branching : ramose.

**ramp¹** (rămp) *n.* [Fr. *rampe* < *ramper*, to slope < OFr., to ramp.] **1.** An inclined passage or roadway connecting different levels, as of a building or road. **2.** A concave bend of a handrail where a sharp change in level or direction occurs, as at a stair landing. **3.** A mobile staircase for boarding and leaving an aircraft.

**ramp²** (rămp) *vi.* **ramped, ramp·ing, ramps.** [ME *rampen* < OFr. *ramper*, to rear up, of Germanic orig.] **1.** To stand menacingly with the arms or forelegs raised. **2.** To act or move furiously or violently. *—n.* The act of ramping.

**ram·page** (răm′pāj′) *n.* [Sc., poss. < RAMP².] A course of violent, frenzied action or behavior. *—vi.* (răm-pāj′). **-paged, -pag·ing, -pag·es.** To move about wildly or violently. **—ram·pa′geous** (-pā′jəs) *adj.* **—ram·pa′geous·ly** *adv.* **—ram·pa′geous·ness** *n.* **—ram·pag′er** *n.*

**ram·pant** (răm′pənt) *adj.* [ME *rampaunt* < OFr. *rampant*, pr.part. of *ramper*, to ramp.] **1.** Growing or extending unchecked : PROFUSE. **2.** Marked by uncontrolled violence, extravagance, or lack of restraint <*rampant* crime> **3. a.** Rearing or ramping on the hind legs. **b.** *Heraldry.* Rearing on the left hind leg with the forelegs elevated, the right above the left, and usu. with the head in profile. **4.** Springing from a structural support or abutment higher at one side than at the other <a *rampant* arch> **—ram′pan·cy** *n.* **—ram′pant·ly** *adv.*

**ram·part** (răm′pärt′, -pərt) *n.* [OFr. < *ramparer*, to fortify : *re-* (intensive < Lat.) + *emparer*, to take possession of < OProv. *amparar* (Lat. *ante-*, before + Lat. *parare*, to prepare).] **1.** A fortification composed of an elevation or embankment, often provided with a parapet. **2.** A protection or defense. *—vt.* **-part·ed, -part·ing, -parts.** To defend with a rampart.

**ram·pike** (răm′pīk′) *n.* [Orig. unknown.] A standing dead tree or tree stump, esp. one killed by fire.

**ram·pi·on** (răm′pē-ən) *n.* [Prob. < OFr. *raiponce* < OItal. *raponzo*, prob. < *rapa*, turnip < Lat. *rapum.*] **1.** A Eurasian plant, *Campanula rapunculus*, with bluish flower clusters and an edible root used in salads. **2.** A plant of the genus *Phyteuma*, similar to the rampion.

**ram·rod** (răm′rŏd′) *n.* **1.** A metal rod for plunging the charge into a muzzle-loading firearm. **2.** A rod for cleaning the barrel of a firearm. *—adj.* Characterized by rigidity, stiffness, or severity. *—vt.* **-rod·ded, -rod·ding, -rods.** *Informal.* To force through to completion.

**ram·shack·le** (răm′shăk′əl) *adj.* [Back-formation < E. *ramshackled*, ramshackle < *ransackled*, p.part. of obs. *ransackle*, to ransack, freq. of ME *ransaken*, to ransack. —see RANSACK.] Apt to fall apart due to shoddy construction or upkeep : RICKETY.

**ram's horn** *n.* A shofar.

**ram·son** (răm′zən, -sən) *n.* [ME *ramsyn* < OE *hramsan*, ramsons, pl. of *hransa.*] *often* **ramsons.** A broad-leaved Eurasian garlic, *Allium ursinum*, with a bulbous root used in relishes and salads.

**ram·til** (răm′tĭl) *n.* [Hindi *rāmtil* : Skt. *rāma-*, dark + Skt. *tilaḥ*, sesame.] An African plant, *Guizotia abyssinica*, with oil-rich seeds.

**ram·u·lose** (răm′yə-lōs′) *adj.* [Lat. *ramulosus* < *ramulus*, dim. of *ramus*, branch.] Having numerous small branches.

**ra·mus** (rā′məs) *n., pl.* **-mi** (-mī′) [NLat. < Lat., ramus.] *Biol.* A branchlike part of a structure.

**ran** (răn) *v. p.t. of* RUN.

**Ran** (răn) *n.* [ON *Rān.*] *Norse Myth.* The goddess of the sea.

**ranch** (rănch) *n.* [Mex. Sp. *rancho*, small ranch < Sp., hut < OSp. *rancher*, to be billeted < OFr. *ranger*, to put in place < *renc*, place, line.] **1.** An extensive farm, esp. in the American West, devoted to raising livestock, as beef cattle, sheep, or horses. **2.** A large farm on which a crop or a kind of animal is raised <a mink *ranch*> *—vi.* **ranched, ranch·ing, ranch·es.** To manage or work on a ranch.

**ranch·er** (răn′chər) *n.* **1.** An owner or manager of a ranch. **2.** RANCH HOUSE 2.

**†ran·che·ri·a** (răn′chə-rē′ə) *n.* [Mex. Sp. < *ranchero*, ranchero.] *Southwestern U.S.* **1. a.** A Mexican herdsman's hut. **b.** A village of rancherias. **2.** An Indian village.

**†ran·che·ro** (răn-châr′ō) *n., pl.* **-ros.** [Mex. Sp. < *rancho*, small ranch. —see RANCH.] *Southwestern U.S.* A rancher.

**ranch house** *n.* **1.** A dwelling on a ranch. **2.** A rectangular one-story house with a low-pitched roof.

**ranch·man** (rănch′mən) *n.* RANCHER 1.

**ranch mink** *n.* A mink bred in captivity from Alaskan and Labrador strains for special pelt colors and qualities.

**†ran·cho** (răn′chō) *n., pl.* **-chos.** [Mex. Sp., small ranch. —see RANCH.] *Southwestern U.S.* **1.** A hut or group of huts for housing ranch workers. **2.** A ranch.

**ran·cid** (răn′sĭd) *adj.* [Lat. *rancidus* < *rancēre*, to stink.] **1.** Having the rank odor or taste of spoiled fats or oils : PUTRID. **2.** Offensive : foul <*rancid* humor> **—ran·cid′i·ty, ran′cid·ness** *n.*

**ran·cor** (răng′kər) *n.* [ME *rancour* < OFr. < Lat. *rancor*, rancid smell < *rancēre*, to stink.] Bitter, long-lasting resentment : ILL WILL. **—ran′cor·ous** *adj.* **—ran′cor·ous·ly** *adv.* **—ran′cor·ous·ness** *n.*

**ran·cour** (răng′kər) *n. Chiefly Brit. var. of* RANCOR.

**rand** (rănd, rănd) *n.* [Afr. < Du., edge.] —See table at CURRENCY.

**ran·dan** (răn′dăn′) *n.* [Orig. unknown.] **1.** A boat rowed by three persons. **2.** The method of rowing a randan, in which the persons fore and aft use one oar each and the person amidships uses two.

**ran·dom** (răn′dəm) *adj.* [ME *randoun* < OFr. *randon* < *randir*, to run, of Germanic orig.] **1.** Having no specific pattern, purpose, organization, or structure : HAPHAZARD. **2.** *Statistics.* **a.** Of or designating a phenomenon that does not yield the same results every time it occurs under identical circumstances. **b.** Of or designating an event having a relative frequency of occurrence that approaches a stable limit as the number of observations of the event increases to infinity. **c.** Of or designating a sample drawn from a population so that each member of the population has an equal chance to be drawn. **d.** Of or relating to a member of such a sample <a *random* number> **—at random.** Without a definite system or plan : INDISCRIMINATELY. **—ran′dom·ly** *adv.*

☆ ***syns:*** RANDOM, HAPHAZARD, INDISCRIMINATE, PURPOSELESS *adj. core meaning* : having no particular pattern, purpose, organization, or structure <shocked at the *random* violence>

**ran·dom-ac·cess** (răn′dəm-ăk′sĕs′) *adj. Computer Sci.* Allowing access to stored data without regard to data sequence.

**ran·dom·ize** (răn′də-mīz′) *vt.* **-ized, -iz·ing, -iz·es.** To make random, esp. for scientific experimentation. **—ran′dom·i·za′tion** *n.*

**random variable** *n. Statistics.* A variable whose numerical values are determined by the results of a chance experiment.

**random walk** *n. Math.* A series of sequential movements in which the direction and size of each move is randomly determined.

**ran·dy** (răn′dē) *adj.* **-di·er, -di·est.** [Poss. < obs. *rand*, alteration of RANT.] **1.** Lecherous : lascivious. **2.** *Chiefly Scot.* Ill-mannered.

**ra·nee** (rä′nē) *n. var. of* RANI.

**rang** (răng) *v. p.t. of* RING².

**range** (rānj) *n.* [ME, series < OFr. *renge* < *rengier*, to put in a row < *renc*, line, of Germanic orig.] **1. a.** Extent of experience, knowledge, perception, or ability. **b.** The area or sphere in which an activity takes place. **c.** The full extent covered <within the *range* of probability> **2. a.** An extent or amount of variation <a seasonal *range* of temperatures> **b.** *Mus.* The gamut of tones within the capacity of a voice or instrument. **3. a.** The maximum or effective distance that can be traversed, as by bullets, sound, or a radio signal. **b.** The distance to a target. **4.** The maximum distance that a vehicle, as a ship or aircraft, can travel before exhausting its fuel supply. **5.** A place for shooting at targets. **6.** *Aerospace.* An area for testing rockets and missiles. **7.** An extensive tract of open land on which livestock wander and graze. **8.** The geographic region in which a given plant or animal normally lives or grows. **9.** An act of roaming or wandering over a large area. **10.** *Math.* The totality of points in a set established by a mapping. **11.** *Statistics.* A measure of dispersion equal to the difference or interval between the smallest and largest of a set of quantities. **12.** A rank, class, or order. **13.** A chain or extended series, esp. one of mountains. **14.** One of a series of double-faced bookcases in a library stack room. **15.** A single series or row of townships, each six miles square, extending parallel to and numbered east and west from a survey base meridian line. **16.** A cooking stove with an oven and a number of separate burners or heating units. —*v.* **ranged, rang·ing, rang·es.** —*vt.* **1.** To arrange or dispose in a particular order, esp. in lines or rows. **2.** To assign to a particular category : CLASSIFY. **3.** To align (e.g., a gun) with a target. **4. a.** To determine the distance of (a target). **b.** To be capable of reaching (a maximum distance). **5.** To roam or travel over or through (a region), as in exploration. **6.** To turn (livestock) on a range to graze. **7.** *Naut.* To uncoil (an anchor cable) on deck so the anchor may descend easily. —*vi.* **1.** To vary within specified limits <prices *ranging* from 20 to 30 dollars> **2.** To extend in a particular direction <a trail *ranging* westward> **3.** To extend in the same direction. **4.** To travel over or through a given area, as in exploration. **5.** To wander or roam. **6.** To live or grow within a particular region <a species that *ranges* throughout the northwest>

☆ ***syns:*** RANGE, EXTENT, ORBIT, REACH, REALM, SCOPE, SPHERE, SWEEP *n. core meaning* : an area within which something or someone exists, acts, or has influence or power <the *range* of a missile><the *range* of human intellect>

**range finder** *n.* An optical, acoustical, or electronic instrument used to ascertain the distance of an object or target.

**range·land** (rānj'lănd', -lənd) *n.* A tract of land suitable for range.

**rang·er** (rān'jər) *n.* **1.** A rover : wanderer. **2.** An armed, usu. mounted law officer patrolling a specific region. **3. Ranger.** One of a unit of U.S. soldiers trained esp. for making raids. **4. a.** A forest ranger. **b.** *Chiefly Brit.* The keeper of a royal forest or park. **5.** One of a cattle herd that grazes on a range.

**rang·y** (rān'jē) *adj.* **-i·er, -i·est. 1.** Inclined to roam. **2.** Long and lean in physique. **3.** Offering ample range : ROOMY.

**ra·ni** *also* **ra·nee** (rä'nē) *n., pl.* **-nis.** [Hindi *rānī* < Skt. *rājñī*, fem. of *rājān*, rajah. —see RAJAH.] **1.** The wife of a rajah. **2.** A reigning Hindu queen or princess.

**rank¹** (răngk) *n.* [OFr. *renc*, of Germanic orig.] **1. a.** A relative position in society. **b.** An official grade or position <the *rank* of colonel> **c.** A relative position or degree of value in a hierarchy. **d.** High or eminent status or position <persons of *rank*> **2.** A row, line, range, or series. **3. a.** A line of soldiers or military vehicles standing side by side in close order. **b. ranks.** Enlisted armed forces <a career in the *ranks*> **c. ranks.** Enlisted military personnel. **4. ranks.** A body of people classed together : NUMBERS <joined the *ranks* of the unemployed> **5.** Any of the horizontal lines of squares on a chessboard. —*v.* **ranked, rank·ing, ranks.** —*vt.* **1.** To place in a row or rows. **2.** To give a particular order or position to : CLASSIFY. **3.** To take precedence over : OUTRANK. —*vi.* **1.** To hold a given position or rank <*rank* second in popularity> **2.** To form or stand in ranks. **—pull rank.** *Slang.* To use one's superior rank to gain an advantage.

**rank²** (răngk) *adj.* **-er, -est.** [ME < OE *ranc*, strong, full-grown.] **1.** Growing profusely or unmanageably <*rank* weeds> **2.** Yielding an abundant, often excessive crop <*rank* soil> **3.** Strong and offensive in odor or taste. **4.** Disgusting : indecent. **5.** Absolute : utter <a *rank* amateur> **—rank'ly** *adv.* **—rank'ness** *n.*

**rank and file** *n.* **1.** The troops of an army. **2.** The ordinary members of a group or organization, excluding the leaders and officers.

**rank·er** (răng'kər) *n. Chiefly Brit.* **1.** An enlisted soldier. **2.** A commissioned officer who has been promoted from enlisted status.

**Ran·kine scale** (răng'kĭn) *n.* [After William J.M. *Rankine* (1820–1872).] The scale of absolute temperature using Fahrenheit degrees, in which the freezing point of water is 491.69° and the boiling point, 671.69°.

**rank·ing** (răng'kĭng) *adj.* Of the highest rank : PRE-EMINENT.

**ran·kle** (răng'kəl) *v.* **-kled, -kling, -kles.** [ME *ranclen* < OFr. *rancler*, alteration of *draoncler* < *draoncle*, festering sore < LLat. *dracunculus*, small serpent, dim. of Lat. *draco*, serpent.] —*vi.* **1.** To be a source of persistent irritation or resentment. **2.** To become sore or inflamed : FESTER. —*vt.* To irritate : embitter.

**ran·sack** (răn'săk') *vt.* **-sacked, -sack·ing, -sacks.** [ME *ransaken* < ON *rannsaka* : *rann*, house + *-saka*, to search.] **1.** To search or rummage thoroughly. **2.** To search or scour for plunder : PILLAGE. **—ran'sack'er** *n.*

**ran·som** (răn'səm) *n.* [ME *ransoun* < OFr. *rançon* < Lat. *redemptio*, a buying back < *redimere*, to redeem. —see REDEEM.] **1. a.** Release of a person or property in return for payment of a demanded price. **b.** The price demanded or the payment made. **2.** Redemption from sin and its consequences. —*vt.* **-somed, -som·ing, -soms. 1. a.** To obtain the release of by paying a certain price. **b.** To release after receiving such a payment. **2.** To deliver from sin and its consequences. **—ran'som·er** *n.*

**rant** (rănt) *v.* **rant·ed, rant·ing, rants.** [Prob. < Du. *ranten.*] —*vi.* To speak or declaim violently or vehemently : RAVE. —*vt.* To exclaim with vehemence or extravagance. —*n.* **1.** Violent, loud, or raving speech. **2.** *Chiefly Brit. Regional.* Wild, uproarious merriment. **—rant'er** *n.*

☆ ***syns:*** RANT, HARANGUE, MOUTH, RAVE *v. core meaning* : to speak in a loud, violent, or vehement manner <*ranted* on and on about high taxes>

**ran·u·la** (răn'yə-lə) *n.* [NLat. < Lat., swelling on the tongue, dim. of *rana*, frog.] A cyst on the underside of tongue caused by obstruction of a salivary gland duct.

**ra·nun·cu·lus** (rə-nŭng'kyə-ləs) *n., pl.* **-lus·es** *or* **-li** (-lī') [NLat., *Ranunculus*, genus name < Lat., a kind of medicinal plant, dim. of *rana*, frog.] A plant of the genus *Ranunculus*, which includes the buttercups.

**rap¹** (răp) *v.* **rapped, rap·ping, raps.** [ME *rappen.*] —*vt.* **1.** To hit sharply and swiftly : STRIKE <*rapped* the floor with a cane> **2.** To utter sharply <*rap* out an angry reply> **3.** To blame or criticize. —*vi.* To strike a quick light blow <*rapped* on the glass> —*n.* **1.** A quick light knock or blow. **2.** A knocking or tapping sound. **3.** *Slang.* **a.** Reprimand or censure. **b.** A prison sentence. **4.** *Slang.* A negative quality or characteristic associated with a person or object. **—beat the rap.** *Slang.* To escape punishment or be acquitted of a charge. **—take the rap.** *Slang.* To accept punishment for a crime, esp. in order to protect another.

**rap²** (răp) *vt.* **rapt** *or* **rapped, rap·ping, raps.** [Back-formation < RAPT.] **1.** *p.p.* **rapt.** To overcome with rapture. **2.** To snatch.

**rap³** (răp) *n.* [< obs. *rap*, 18th-cent. Irish counterfeit halfpenny.] *Informal.* The least bit <I don't care a *rap*.>

**rap⁴** (răp) [Poss. < RAPPORT.] *Slang.* —*vi.* **rap·ped, rap·ping, raps.** To discuss freely and at length. —*n.* A conversation or discussion.

**ra·pa·cious** (rə-pā'shəs) *adj.* [< Lat. *rapax*, *rapac-* < *rapere*, to seize.] **1.** Taking by force : PLUNDERING. **2.** Ravenous : greedy. **3.** RAPTORIAL 1. **—ra·pa'cious·ly** *adv.* **—ra·pa'cious·ness, ra·pac'i·ty** (rə-păs'ĭ-tē) *n.*

**rape¹** (rāp) *n.* [ME < *rapen*, to rape < OFr. *raper* < Lat. *rapere*, to seize.] **1.** The crime of forcing another person to submit to sexual intercourse. **2.** The act of seizing and carrying off by force : ABDUCTION. **3.** Abusive or improper treatment : VIOLATION <a *rape* of justice> —*vt.* **raped, rap·ing, rapes. 1.** To force (another person) to submit to sexual intercourse. **2.** To seize and carry off by force. **3.** To pillage or plunder. **—rap'er, rap'ist** *n.*

**rape²** (rāp) *n.* [ME < Lat. *rapa*, turnip.] A Eurasian plant, *Brassica napus*, cultivated as fodder and for its oil-rich seeds.

**rape³** (rāp) *n.* [Fr. *râpe*, grape stalk < OFr. < *rasper*, to scrape, of Germanic orig.] The pulpy refuse remaining after pressing the juice from grapes in wine-making.

**rape oil** *n.* The edible oil extracted from rapeseed, also used as a lubricant and in the manufacture of various products.

**rape·seed** (rāp'sēd') *n.* The seed of the rape plant.

**Raph·a·el** (răf'ē-əl, rā'fē-) *n.* [Heb. *Rĕphā'ēl.*] One of the archangels, mentioned in the Apocrypha.

**ra·phe** *also* **rha·phe** (rā'fē) *n., pl.* **-phae** (-fē') [NLat. < Gk. *rhaphē*, seam < *rhaptein*, to sew.] *Biol.* A seamlike line or ridge between two similar parts, as in the scrotum, the valves of a diatom, or certain seed coats.

**raph·i·a** (răf'ē-ə) *n. var. of* RAFFIA.

**ra·phide** (rā'fīd) *also* **ra·phis** (-fĭs) *n., pl.* **raph·i·des** (răf'ĭ-dēz') [Back-formation < *rapides*, pl. < NLat. < Gk. *rhaphides*, pl. of *rhaphis*, needle < *rhaptein*, to sew.] *Bot.* One of a bundle of needle-shaped crystals, consisting mainly of calcium oxalate, found in many plant cells.

**rap·id** (răp'ĭd) *adj.* **-er, -est.** [Lat. *rapidus* < *rapere*, to seize.] Moving, acting, or occurring with great speed : SWIFT. —*n. often* **rapids.** A very fast-moving section of a river, caused by a steep descent in the riverbed. **—ra·pid'i·ty** (rə-pĭd'ĭ-tē), **rap'id·ness** *n.* **—rap'id·ly** *adv.*

**rapid eye movement** *n.* The rapid, periodic, jerky movement of the eyes during certain stages of the sleep cycle when dreaming occurs.

**rap·id-fire** (răp'ĭd-fīr') *adj.* **1.** Firing or designed to fire shots in rapid succession. **2.** Marked by rapid, continuous occurrence <a string of *rapid-fire* insults>

**rapid transit** *n.* An urban passenger transportation system using underground or elevated trains or a combination of both.

**ra·pi·er** (rā'pē-ər, răp'yər) *n.* [Fr. *rapière* < OFr. *(espee) rapiere*, rapier (sword).] **1.** A long, slender, two-edged 16th- and 17th-cent.

ă **pat** ā **pay** âr **care** ä **father** ĕ **pet** ē **be** hw **which** ĭ **pit**
ī **tie** îr **pier** ŏ **pot** ō **toe** ô **paw, for** oi **noise** ŏŏ **took**

sword with a cuplike hilt. **2.** A sharp-pointed, light sword lacking a cutting edge and used only for thrusting.

**rap·ine** (răp′ĭn) *n.* [ME *rapyne* < Lat. *rapina* < *rapere,* to seize.] Forcible seizure of another's property : PLUNDER.

**rap·pa·ree** (răp′ə-rē′) *n.* [Ir. Gael. *rapaire.*] **1.** A 17th-cent. Irish freebooting soldier. **2.** A robber or bandit.

**rap·pee** (ră-pē′) *n.* [Fr. *(tabac) râpé,* grated (tobacco) < *râper,* to grate < OFr. *rasper,* to scrape, of Germanic orig.] A pungent snuff made from a dark strong tobacco.

**rap·pel** (ră-pĕl′) *n.* [Fr. < OFr., recall < *rapeler,* to recall : *re-,* back + *appeler,* to summon.] The act or method of descending a sheer face, as a cliff, by means of a double rope passed under one thigh and over the opposite shoulder. —*vi.* **-pelled, -pel·ling, -pels.** To descend from a steep height by rappel.

rappel

**rap·pen** (rä′pən) *n., pl.* **rappen.** [G. < MHG *rappe,* raven.] —See table at CURRENCY.

**rap·port** (ră-pôr′, -pōr′, rə-) *n.* [Fr. < *rapporter,* to bring back < OFr. *raporter* : *re-,* back + *aporter,* to bring < Lat. *apportare* (*ad-,* to + *portare,* to carry).] Relationship, esp. one of mutual trust or emotional affinity.

**rap·proche·ment** (rä′prôsh-mäN′) *n.* [Fr. < *rapprocher,* to bring together : *re-* (intensive < Lat.) + *approcher,* to approach (Lat. *ad-,* to + Lat. *prope,* near).] **1.** Reestablishment of cordial relations, as between two governments or factions. **2.** The state of reconciliation or of the resumption of cordial relations.

**rap·scal·lion** (răp-skăl′yən) *n.* [Alteration of E. *rascallion* < RASCAL.] A scamp : rascal.

**rap session** *n. Slang.* An open informal discussion.

**rap sheet** *n. Informal.* A police arrest record.

**rapt** (răpt) *adj.* [P.part. of RAP[2].] **1.** Deeply moved or delighted : ENRAPTURED <gazed with *rapt* wonder> **2.** Deeply absorbed : ENGROSSED <*rapt* attention>

**rap·tor** (răp′tər) *n.* [Lat., one who seizes < *rapere,* to seize.] A bird of prey, as a hawk or owl.

**rap·to·ri·al** (răp-tôr′ē-əl, -tōr′-) *adj.* **1.** Subsisting on live prey : PREDATORY. **2.** Adapted for the seizing of prey. **3.** Of, pertaining to, or characteristic of birds of prey.

**rap·ture** (răp′chər) *n.* [Med. Lat. *raptura,* ecstasy < Lat. *rapere,* to seize.] **1.** The state of being transported by a lofty emotion : ECSTASY. **2.** An expression of ecstatic feeling <spoke in *raptures* about the performance> —*vt.* **-tured, -tur·ing, -tures.** To enrapture.

**rap·tur·ous** (răp′chər-əs) *adj.* Filled with rapture : ECSTATIC. **—rap′tur·ous·ly** *adv.* **—rap′tur·ous·ness** *n.*

**ra·ra a·vis** (râr′ə ā′vĭs) *n., pl.* **rara a·vis·es** *or* **ra·rae a·ves** (râr′ē ā′vēz) [Lat., rare bird.] One that is rare or unique.

**rare[1]** (râr) *adj.* **rar·er, rar·est.** [ME < Lat. *rarus.*] **1.** Occurring infrequently : UNCOMMON. **2.** Highly valued owing to scarceness or uncommonness : SPECIAL <a *rare* manuscript> **3.** Thin in density : RAREFIED. —Used of gases. **—rare′ness** *n.*

**rare[2]** (râr) *adj.* **rar·er, rar·est.** [ME *rere,* lightly boiled < OE *hrēr.*] Cooked a short time to retain juice and redness <*rare* steak> **—rare′ness** *n.*

▲ word history: The adjective *rare*[2] meaning "lightly cooked" referred only to eggs until modern times. Although its application to other foods, especially meat, occurs in some English dialects, this use of *rare* was considered an Americanism until very recently, when it was adopted into standard British usage.

**rare·bit** (râr′bĭt) *n.* [Prob. alteration of (WELSH) RABBIT.] A cheese dish, Welsh rabbit.

**rare earth** *n.* **1.** Any of various oxides of the rare-earth elements. **2.** A rare-earth element.

**rare-earth element** (râr′ûrth′) *n.* [So called because they were orig. thought to be rare.] Any of the abundant metallic elements of atomic number 57 through 71.

**rar·ee show** (râr′ē) *n.* [Alteration of *rare show.*] **1.** A peepshow. **2.** A show or spectacle, esp. one given in the street.

**rar·e·fac·tion** (râr′ə-făk′shən) *also* **rar·e·fi·ca·tion** (-fĭ-kā′shən) *n.* The act or process of rarefying or the state of being rarefied. **—rar′e·fac′tive** *adj.*

**rar·e·fied** (râr′ə-fīd) *adj.* **1.** Appealing to or intended for a small and select group : ESOTERIC. **2.** Elevated in character or style : LOFTY.

**rar·e·fy** (râr′ə-fī′) *v.* **-fied, -fy·ing, -fies.** [ME *rarefien* < OFr. *rarefier* < Lat. *rarefacere* : *rarus,* rare + *facere,* to make.] —*vt.* **1.** To make thin, light, or less dense. **2.** To refine or purify. —*vi.* To become thin, purer, or less dense. **—rar′e·fi′a·ble** *adj.*

**rare·ly** (râr′lē) *adv.* **1.** Not often : INFREQUENTLY. **2.** In an unusual degree : EXCEPTIONALLY. **3.** With uncommon excellence.

**rare·ripe** (râr′rīp′) *n.* [Dial. *rare,* early + RIPE.] A fruit or vegetable that ripens early. **—rare′ripe′** *adj.*

**rar·ing** (râr′ĭng) *adj.* [< pr.part. of dial. *rare,* var. of REAR[2].] *Informal.* Full of enthusiasm : EAGER <*raring* to go>

**rar·i·ty** (râr′ĭ-tē) *n., pl.* **-ties.** **1.** Something rare. **2.** The quality or state of being rare.

**ras·bo·ra** (răz-bôr′ə, -bōr′ə) *n.* [NLat. *Rasbora,* genus name < a native word in the East Indies.] A tropical fish of the genus *Rasbora,* esp. any of several brightly colored species kept in home aquariums.

**ras·cal** (răs′kəl) *n.* [ME, man of low station < OFr. *rascaille,* prob. < OFr. *rasque,* mud.] **1.** One who is unscrupulous and dishonest : SCOUNDREL. **2.** One that is playfully mischievous. **—ras·cal′i·ty** (-kăl′ĭ-tē) *n.* **—ras′cal·ly** *adj. & adv.*

**rase** (rāz) *vt.* **rased, ras·ing, ras·es.** **1.** To erase. **2.** *var. of* RAZE.

**rash[1]** (răsh) *adj.* **-er, -est.** [ME *rasche,* active.] **1.** Marked by ill-considered boldness or haste : IMPETUOUS. **2.** *Obs.* Quick in producing an effect. **—rash′ly** *adv.* **—rash′ness** *n.*

**rash[2]** (răsh) *n.* [Poss. < obs. Fr. *rache* < OFr. *rasche* < *raschier,* to scratch < VLat. **rasciare* < Lat. *radere.*] **1.** A skin eruption. **2.** An outbreak of certain activities or incidents within a brief period <a *rash* of house fires>

**rash·er** (răsh′ər) *n.* [Orig. unknown.] **1.** A thin slice of fried or broiled bacon. **2.** A dish or serving of rashers.

**ra·so·ri·al** (rə-zôr′ē-əl, -zōr′-, -sôr′-, -sōr′-) *adj.* [< LLat. *rasor,* scraper < Lat. *radere,* to scrape.] Characteristically scratching the ground for food. —Used of chickens and similar birds.

**rasp** (răsp) *v.* **rasped, rasp·ing, rasps.** [ME *raspen* < OFr. *rasper,* of Germanic orig.] —*vt.* **1.** To scrape or grind down with a coarse, sharp-textured file. **2.** To utter in a grating voice. **3.** To grate on (nerves or feelings). —*vi.* **1.** To scrape harshly : GRATE. **2.** To make a harsh grating sound. —*n.* **1.** A coarse file with sharp, raised, pointed projections. **2.** The act of filing with a rasp. **3.** A grating, abrasive sound. **—rasp′er** *n.* **—rasp′ing·ly** *adv.*

**rasp·ber·ry** (răz′bĕr′ē) *n.* [Obs. *raspis,* raspberry + BERRY.] **1.** A shrubby, usu. thorny plant of the genus *Rubus,* as *R. strigosus* of eastern North America and *R. idaeus* of Europe, bearing edible, usu. reddish berries. **2.** The fruit of the raspberry, composed of a mass of small fleshy drupelets. **3.** A moderate to dark or deep purplish red. **4.** *Slang.* A derisive or contemptuous sound made by vibrating the extended tongue and the lips while exhaling.

**rasp·y** (răs′pē) *adj.* **-i·er, -i·est.** Rough : grating.

**Ras·ta·fa·ri·an·ism** (rä′stə-făr′ē-ə-nĭz′əm, răs′tə-făr′-) *n.* [After *Ras Tafari,* former name of Haile Selassie (1891–1975).] A black Jamaican religious cult whose members worship Haile Selassie. **—Ras′ta** (rä′stə, răs′tə), **Ras′ta·fa′ri·an** *n.*

**rat** (răt) *n.* [ME < OE *ræt.*] **1.** A long-tailed rodent similar to the mouse but larger, esp. one of the genus *Rattus.* **2.** An animal similar to a rat. **3.** *Slang.* One who is despicable and sneaky, esp. one who betrays or informs on one's associates. **4.** *Informal.* A pad of hair or other material worn as part of a woman's coiffure to puff out the hair. —*vi.* **rat·ted, rat·ting, rats.** **1.** To hunt for or catch rats, esp. with dogs. **2.** *Slang.* To desert or betray one's comrades by giving information <*rat* to the principal on one's friends>

**rat·a·ble** (rā′tə-bəl) *adj.* **1.** Capable of being rated, estimated, or appraised. **2.** Proportional. **3.** *Chiefly Brit.* Liable to assessment : TAXABLE. **—rat′a·bil′i·ty, rat′a·ble·ness** *n.* **—rat′a·bly** *adv.*

**rat·a·fi·a** (răt′ə-fē′ə) *also* **rat·a·fee** (-ə-fē′) *n.* [Fr.] **1.** A cordial flavored with fruit kernels or almonds. **2.** A ratafia-flavored biscuit.

**rat·a·plan** (răt′ə-plăn′) *n.* [Fr.] A tattoo, as of a drum, horses' hoofs, or rapid gunfire.

**rat-a-tat-tat** (răt′ə-tăt′tăt′) *n.* [Imit.] A series of short sharp sounds, as those made by knocking on a door.

**rat-bite fever** (răt′bīt′) *n.* Either of two infectious diseases contractible from the bite of a rat: **a.** That caused by *Streptobacillus moniliformis* and marked by skin rash, joint pains, headache, and vomiting. **b.** That caused by *Spirillum minus,* with ulceration at the site of the bite, a purplish rash, and recurrent fever.

**rat cheese** *n.* Cheddar.

**ratch·et** (răch′ĭt) *n.* [Fr. *rochet* < OFr. *rocquet,* head of a lance, of Germanic orig.] **1.** A mechanism consisting of a pawl that engages the sloping teeth of a bar or wheel, permitting motion in one direction only. **2.** The pawl, bar, or wheel of a ratchet.

**rate[1]** (rāt) *n.* [ME < OFr. < Med. Lat. *rata,* proportion, short for *pro rata parte,* according to an estimated part.] **1.** A quantity measured in terms of another measured quantity <a *rate* of pressure of 20 pounds per square inch> **2.** A measure of a part with respect to a whole : PROPORTION <the birth *rate*> **3.** The cost per unit of a service or commodity <shipping *rates*> **4.** A charge or payment calculated by means of a particular ratio or formula <interest *rates*> **5.** Level of quality. **6. rates.** *Chiefly Brit.* A local property tax. —*v.* **rat·ed, rat·ing, rates.** —*vt.* **1.** To calculate the value of : APPRAISE. **2.** To assign to a particular rank or class : GRADE. **3.** To judge or regard <*rated* the book quite high> **4.** To value for purposes of taxation. **5.** To set a rate for (goods to be shipped). **6.** To specify the performance limits of (e.g., a machine). **7.** *Informal.* To deserve or merit <*rate* a second look> —*vi.* **1.** To be ranked in a particular class. **2.** *Informal.* To have status, influence, or importance. **—at any rate. 1.** Whatever the case may be. **2.** At least.

**rate**[2] (rāt) *v.* **rat·ed, rat·ing, rates.** [ME *raten.*] —*vt.* To berate. —*vi.* To express angry reproach.

**ra·tel** (rāt'l, rät'l) *n.* [Afr. < MDu., rattle.] The honey badger.

**rate of exchange** *n.* The ratio at which a unit of one currency may be exchanged for a unit of another currency.

**rate·pay·er** (rāt'pā'ər) *n. Chiefly Brit.* One who pays rates.

**rat·er** (rā'tər) *n.* **1.** One that rates, esp. one that establishes or applies a rating. **2.** One having an indicated rating <a second-*rater*>

**rat·fink** (răt'fĭngk') *n. Slang.* RAT 3.

**rat·fish** (răt'fĭsh') *n., pl.* **ratfish** or **-fish·es.** A long, narrow-tailed fish, *Hydrolagus colliei,* of coastal Pacific waters.

**rathe** (rāth, răth) *adj.* [ME, quick < OE *hraðe.*] *Archaic.* Appearing or ripening early in the year.

**rath·er** (răth'ər, rä'thər) *adv.* [ME < OE *hraðor,* comp. of *hrað,* quickly.] **1.** More readily : PREFERABLY <We'd *rather* stay home.> **2.** With more reason, wisdom, logic, or other justification. **3.** More accurately : EXACTLY <I read the report, or *rather* I skimmed it.> **4.** To a certain extent : SOMEWHAT <*rather* cool> **5.** On the contrary. **6.** (răth'ûr', rä'-). *Chiefly Brit.* Most certainly. —Used as an emphatic affirmative reply.

**raths·kel·ler** (rät'skĕl'ər, răt'-, răth'-) *n.* [Obs. G., restaurant in the city hall basement : *Rat,* council + *Keller,* cellar < Lat. *cellarium.*] A restaurant in the style of a German cellar tavern that features the serving of beer.

**rat·i·fy** (răt'ə-fī') *vt.* **-fied, -fy·ing, -fies.** [ME *ratifien* < OFr. *ratifier* < Med. Lat. *ratificare* : Lat. *ratus,* fixed (< *reri,* to reckon) + Lat. *facere,* to make.] To give formal sanction or approval to and thereby validate. **—rat'i·fi·ca'tion** (-fĭ-kā'shən) *n.* **—rat'i·fi'er** *n.*

**rat·i·né** (răt'ə-nā') *n.* [Fr., p.part. of *ratiner,* to adorn.] A loosely woven nubby fabric.

**rat·ing**[1] (rā'tĭng) *n.* **1.** A position assigned on a scale : STANDING. **2.** A classification according to specialty or proficiency, as of a member of the armed forces. **3.** An evaluation of financial status <a credit *rating*> **4.** A specified performance limit, as of range, capacity, or operational capability <efficiency *ratings*> **5.** Popularity of a television or radio program as estimated by an audience poll. **6.** *Chiefly Brit.* An enlisted person in the navy.

**rat·ing**[2] (rā'tĭng) *n.* A scolding.

**ra·tio** (rā'shō, rā'shē-ō') *n., pl.* **-tios.** [Lat., calculation < *reri,* to reckon.] **1.** Relation in number or degree between two similar things. **2.** The relative value of gold and silver in a bimetallic currency system. **3.** *Math.* The relative size of two quantities expressed as the quotient of one divided by the other <The *ratio* of 7 to 4 is written 7:4 or 7/4.>

**ra·ti·oc·i·nate** (răsh'ē-ŏs'ə-nāt') *vi.* **-nat·ed, -nat·ing, -nates.** [Lat. *ratiocinari, ratiocinat-* < *ratio,* calculation < *reri,* to reckon.] To reason logically and methodically. **—ra'ti·oc'i·na'tion** *n.* **—ra'ti·oc'i·na'tive** *adj.* **—ra'ti·oc'i·na'tor** *n.*

**ra·tion** (răsh'ən, rā'shən) *n.* [Fr. < Lat. *ratio,* calculation < *reri,* to reckon.] **1.** A fixed portion, esp. an amount of food allotted to soldiers or to civilians in times of scarcity. **2. rations.** Food issued or available to members of a group. —*vt.* **-tioned, -tion·ing, -tions. 1.** To supply with rations. **2.** To distribute as rations <*rationed* out bread and milk> **3.** To restrict to limited amounts <had to *ration* gasoline>

**ra·tion·al** (răsh'ən-əl) *adj.* [Lat. *rationalis* < *ratio,* reason < *reri,* to reckon.] **1.** Having or exercising the ability to reason. **2.** Of sound mind : SANE. **3.** Consistent with or based on reason : LOGICAL <*rational* answers> **4.** *Math.* Designating an algebraic expression no variable of which appears in an irreducible radical or with a fractional exponent. **—ra'tion·al·ly** *adv.* **—ra'tion·al·ness** *n.*

**ra·tion·ale** (răsh'ə-năl') *n.* [Lat. *rationale,* neuter of *rationalis,* rational.] **1.** The fundamental reasons for something : BASIS. **2.** An exposition of guiding principles or justifications.

**rational function** *n.* A function that is a quotient of two polynomials.

**ra·tion·al·ism** (răsh'ə-nə-lĭz'əm) *n.* The view that reason or intellectual conception provides the only valid basis for action or belief and that reason is the prime source of knowledge and of spiritual truth. **—ra'tion·al·ist** *n.* **—ra'tion·al·is'tic** *adj.* **—ra'tion·al·is'ti·cal·ly** *adv.*

**ra·tion·al·i·ty** (răsh'ə-năl'ĭ-tē) *n., pl.* **-ties. 1.** The quality or condition of being rational. **2.** A rational practice or belief.

**ra·tion·al·ize** (răsh'ə-nə-līz') *v.* **-ized, -iz·ing, -iz·es.** —*vt.* **1.** To make rational. **2.** To interpret from a rational standpoint. **3.** To devise self-satisfying but incorrect reasons for (one's behavior). **4.** *Math.* To remove radicals without changing the value of (an expression) or roots of (an equation). **5.** *Chiefly Brit.* To bring modern, efficient methods to (e.g., an industry). —*vi.* **1.** To think in a rational or rationalistic way. **2.** To devise self-satisfying but incorrect reasons for one's behavior. **—ra'tion·al·i·za'tion** *n.* **—ra'tion·al·iz'er** *n.*

**rational number** *n.* A number capable of being expressed as an integer or quotient of integers.

**rat·ite** (răt'īt') *adj.* [< Lat. *ratitus,* marked with the figure of a raft < *ratis,* raft.] Designating a flightless bird, as the ostrich or emu, having a flat breastbone without the keellike prominence characteristic of most flying birds. **—rat'ite** *n.*

**rat·line** *also* **rat·lin** (răt'lĭn) *n.* [Orig. unknown.] *Naut.* **1.** Any of the small ropes fastened horizontally to the shrouds and forming a ladder for going aloft. **2.** The rope used for ratline.

**ra·toon** (ră-tōōn') *n.* [Sp. *retoño,* sprout < *retoñar* : *re-,* again (< Lat.) + *otoñar,* to grow in autumn < Lat. *autumnus,* autumn.] A basal shoot or sucker on certain plants, as the banana, pineapple, or sugar cane. —*v.* **-tooned, -toon·ing, -toons.** —*vi.* To sprout or grow as a ratoon. —*vt.* To propagate (a crop) from ratoons.

**rat race** *n. Slang.* A tiring, harrassing, usu. competitive activity.

**rats·bane** (răts'bān') *n.* Rat poison, esp. arsenic trioxide.

**rat snake** *n.* A nonvenomous snake of the genus *Elaphe.*

**rat-tail** (răt'tāl') *also* **rat-tailed** (-tāld') *adj.* Shaped like a rat's tail <a *rat-tail* file> —*n.* GRENADIER 2.

**rat-tail cactus** *n.* A tropical American cactus, *Aporocactus flagelliformis,* with thin, hanging or creeping stems and red flowers.

**rat·tan** (ră-tăn', rə-) *n.* [Malay *rotan.*] **1.** An Asian climbing palm of the genera *Calamus, Daemonorops,* or *Plectomia,* with long tough pliant stems. **2.** The stems of the rattan, used to make wickerwork. **3.** A cane or switch made from rattan.

**rat·teen** (ră-tēn') *n.* [Fr. *ratine.*] *Archaic.* A thick twilled woolen cloth.

**rat·ter** (răt'ər) *n.* **1.** A catcher or killer of rats. **2.** *Slang.* A deserter, betrayer, or traitor.

**rat·tle**[1] (răt'l) *v.* **-tled, -tling, -tles.** [ME *rattelen* < MLG *rattelen.*] —*vi.* **1. a.** To make a quick succession of short, sharp sounds. **b.** To move with such sounds <a truck *rattling* over the bumpy road> **2.** To talk uninterruptedly and at length, usu. on a trivial subject <*rattled* on about their trip> —*vt.* **1.** To cause to rattle. **2.** To utter or perform rapidly or effortlessly <*rattle* off a list of state capitals> **3.** *Informal.* To unnerve : fluster <The booing began to *rattle* me.> —*n.* **1.** Short percussive sounds produced in rapid succession. **2.** A device, as a baby's toy, that produces a rattle when shaken. **3.** A rattling sound in the throat caused by obstructed breathing, esp. near the time of death. **4.** The series of horny structures at the end of a rattlesnake's tail. **5.** Loud or rapid talk.

**rat·tle**[2] (răt'l) *vt.* **-tled, -tling, -tles.** [Back-formation < *rattling,* var. of RATLINE.] *Naut.* To secure ratlines to (shrouds).

**rat·tle·box** (răt'l-bŏks') *n.* A plant or shrub of the genus *Crotalaria,* having inflated pods within which the seeds rattle.

**rat·tle·brained** (răt'l-brānd') *adj.* Giddy and talkative : THOUGHTLESS. **—rat'tle·brain'** *n.*

**rat·tler** (răt'lər) *n.* **1.** One that rattles. **2.** A rattlesnake. **3.** *Informal.* A freight train.

**rat·tle·snake** (răt'l-snāk') *n.* A venomous New World snake of the genera *Crotalus* or *Sistrurus,* having at the end of the tail a series of loosely attached horny segments that can be vibrated to produce a rattling sound.

**rattlesnake master** *n.* A plant supposedly effective against snakebites, as *Eryngium yuccifolium* of the southeastern United States, having narrow spiny leaves and pale blue or white flowers.

**rattlesnake plantain** *n.* [From the resemblance of its leaves to a rattlesnake's skin.] A small orchid of the genus *Goodyera,* having striped or mottled leaves and whitish flower spikes.

**rattlesnake root** *n.* [From the belief that the root cured a rattlesnake's bite.] A plant of the genus *Prenanthes,* with thick bitter-tasting roots.

**rattlesnake weed** *n.* A plant, *Hieracium venosum,* with red-veined or purple-veined leaves and yellow flowers.

**rat·tle·trap** (răt'l-trăp') *n.* A dilapidated, worn-out vehicle.

**rat·tling** (răt'lĭng) *Informal.* —*adj.* **1.** Animated : brisk <a *rattling* pace> **2.** Very good. —*adv.* Especially : very <a *rattling* good time>

**rat·tly** (răt'lē) *adj.* Rattling or apt to rattle : CLATTERING.

**rat·trap** (răt'trăp') *n.* **1.** A trap for catching rats. **2.** A run-down or unsanitary dwelling.

**rat trap cheese** *n.* Cheddar.

**rat·ty** (răt'ē) *adj.* **-ti·er, -ti·est. 1.** Of or characteristic of rats. **2.** Overrun by rats. **3.** *Slang.* Dilapidated : shabby. **4.** Irritable.

**rau·cous** (rô'kəs) *adj.* [Lat. *raucus.*] **1.** Loud and hoarse <*raucous* cries> **2.** Noisy : boisterous <a *raucous* crowd> **—rau'cous·ly** *adv.* **—rau'ci·ty** (rô'sə-tē), **rau'cous·ness** *n.*

**raun·chy** (rôn'chē, rän'-) *adj.* **-chi·er, -chi·est.** [Orig. unknown.] *Slang.* **1.** Filthy : grubby. **2. a.** Obscene : vulgar <*raunchy* jokes> **—raun'chi·ly** *adv.* **—raun'chi·ness** *n.*

**rau·wol·fi·a** (rou-wo͝ol'fē-ə, rô-) *n.* [NLat. *Rauwolfia,* genus name, after Leonhard *Rauwolf* (d. 1596).] A tropical tree or shrub of the genus *Rauwolfia,* esp. *R. serpentina* of southeastern Asia.

**rav·age** (răv′ĭj) *v.* **-aged, -ag·ing, -ag·es.** [Fr. *ravager* < *ravir,* to ravish.] —*vt.* **1.** To bring heavy destruction on : DEVASTATE <An earthquake *ravaged* the city.> **2.** To sack : pillage. —*vi.* To wreak destruction. —*n.* **1.** The act or practice of ravaging. **2.** Grievous damage : DETERIORATION <the *ravages* of disease> **—rav′ag·er** *n.*

**rave** (rāv) *v.* **raved, rav·ing, raves.** [ME *raven* < ONFr. *raver.*] —*vi.* **1.** To speak wildly, incoherently, or irrationally <*raving* like a maniac> **2.** To rage : roar <The wind *raved* outside the house.> **3.** To speak with wild enthusiasm <We *raved* about the concert.> —*vt.* To utter frenziedly. —*n.* **1.** The act of raving. **2.** *Informal.* A highly enthusiastic review or opinion <The book got *raves.*>

**rav·el** (răv′əl) *v.* **-eled, -el·ing, -els** *also* **-elled, -el·ling, -els.** [Du. *rafelen* < *rafel,* loose thread.] —*vt.* **1.** To separate the strands or fibers of (e.g., rope) : UNRAVEL. **2.** To clarify by separating the aspects of. **3.** To tangle or complicate. —*vi.* **1.** To become unraveled or undone. **2.** To become tangled or confused. —*n.* **1.** A raveling. **2.** A broken or discarded thread. **3.** A tangle. **—rav′el·er** *n.*

**rav·el·ing** *also* **rav·el·ling** (răv′ə-lĭng) *n.* A thread or fiber that has become separated from a woven material.

**rav·el·ment** (răv′əl-mənt) *n.* Complexity or confusion.

**ra·ven[1]** (rā′vən) *n.* [ME < OE *hræfn.*] A large bird, *Corvus corax,* with black plumage and a croaking cry. —*adj.* Glossy black.

**rav·en[2]** (răv′ən) *v.* **-ened, -en·ing, -ens.** [OFr. *raviner,* to take by force < VLat. **rapinare* < Lat. *rapina,* rapine < *rapere,* to seize.] —*vt.* **1.** To consume greedily : DEVOUR. **2.** To seek or seize as prey or plunder. —*vi.* **1.** To seek or seize prey or plunder. **2.** To eat ravenously. —*n. var. of* RAVIN. **—rav′en·er** *n.*

**rav·en·ous** (răv′ə-nəs) *adj.* [ME < OFr. *ravineux* < *raviner,* to take by force. —see RAVEN[2].] **1.** Extremely or insatiably hungry. **2.** Rapacious. **3.** Greedy for gratification <*ravenous* for fame> **—rav′en·ous·ly** *adv.* **—rav′en·ous·ness** *n.*

☆ ***syns:*** RAVENOUS, FAMISHED, HUNGRY, STARVING *adj. core meaning* : desiring or craving food <was *ravenous* after the long trip>

**ra·vi·gote** (rä-vē-gôt′) *n.* [Fr. < *ravigoter,* to add new vigor : *re-,* again (< Lat.) + *a-,* to (< Lat. *ad-*) + *vigeur,* vigor < Lat. *vigor.*] A vinegar sauce flavored with minced onion, capers, and herbs.

**rav·in** *also* **rav·en** (răv′ən) *n.* [ME *ravine* < OFr., rapine < Lat. *rapina* < *rapere,* to seize.] **1.** Voracity : rapaciousness. **2.** Something taken as prey. **3.** The act or practice of preying.

**ra·vine** (rə-vēn′) *n.* [Fr. < OFr., violent rush < Lat. *rapina,* rapine < *rapere,* to seize.] A steep narrow gorge or cleft in the earth's surface, esp. one worn by a creek or stream.

**rav·ing** (rā′vĭng) *adj.* **1.** Talking or behaving deliriously : WILD <a *raving* lunatic> **2.** *Informal.* Exciting admiration <a *raving* success> —*n.* Delirious irrational speech. **—rav′ing·ly** *adv.*

**ra·vi·o·li** (răv′ē-ō′lē, rä′vē-) *pl.n.* [Ital., pl. of dial. *raviolo,* dim. of *rava,* turnip < Lat. *rapa.*] **1.** Fresh pasta shaped into small cases and stuffed with chopped meat or cheese. **2.** A dish made with ravioli.

**rav·ish** (răv′ĭsh) *vt.* **-ished, -ish·ing, -ish·es.** [ME *ravisshen* < OFr. *ravir, raviss-* < VLat. **rapire* < Lat. *rapere,* to seize.] **1.** To seize and carry off by force. **2.** To rape : violate. **3.** To overwhelm with emotion : ENRAPTURE. **—rav′ish·er** *n.*

**rav·ish·ing** (răv′ĭ-shĭng) *adj.* Alluring. **—rav′ish·ing·ly** *adv.*

**rav·ish·ment** (răv′ĭsh-mənt) *n.* **1.** RAPE[1] 1, 2. **2.** Rapture : bewitchment.

**raw** (rô) *adj.* **-er, -est.** [ME < OE *hrēaw.*] **1.** Uncooked. **2. a.** Being in a natural condition : not refined or processed <*raw* silk> **b.** Not finished, coated, or treated <*raw* wood> **3.** Untrained and inexperienced <*raw* recruits> **4.** Recently applied : FRESH <*raw* plaster> **5.** Exposed and irritated <a *raw* wound> **6.** Inflamed : sore <a *raw* throat> **7.** Unpleasantly damp and chilly <a *raw* wind> **8.** Cruel and unfair <a *raw* punishment> **9.** Stark : crude <a *raw* depiction of violence> **—in the raw.** In a crude or unrefined state <nature *in the raw*> **—raw′ly** *adv.* **—raw′ness** *n.*

**raw·boned** (rô′bōnd′) *adj.* Having a lean, gaunt frame with prominent bones.

**raw data** *pl.n.* Unprocessed or unanalyzed information.

**raw deal** *n. Slang.* Unfair or objectionable treatment.

**raw·hide** (rô′hīd′) *n.* **1.** The untanned hide of an animal, esp. untanned cowhide. **2.** A whip or thong made of rawhide. —*vt.* **-hid·ed, -hid·ing, -hides.** To beat with a rawhide whip.

**ra·win·sonde** (rā′wĭn-sŏnd′) *n.* [RA(DAR) + WIN(D) + (RADIO)SONDE.] A radiosonde for observing upper-air wind velocity and direction that is tracked by a radio direction finder or radar.

**raw material** *n.* **1.** Unfinished materials, esp. unprocessed natural products, used in manufacture or assembly. **2.** Unprocessed data.

**raw sienna** *n.* **1.** A brownish-yellow pigment. **2.** A brownish orange to light brown.

**raw silk** *n.* **1.** Untreated silk as reeled from the cocoon. **2.** Fabric woven from raw silk.

**ray[1]** (rā) *n.* [ME < OFr. *rai* < Lat. *radius.*] **1. a.** A narrow beam or apparent thin line of radiation, esp. one of visible light. **b.** A graphic or other representation of such a line. **2.** A small amount : TRACE <a *ray* of happiness> **3.** A straight line extending from a point. **4.** A structure having the form of a straight line extending from a given point. **5.** *Bot.* A ray flower. **6.** *Zool.* One of the bony spines supporting the membrane of a fish's fin. —*vt.* **rayed, ray·ing, rays. 1.** To send out as rays : EMIT. **2.** To supply with rays or radiating lines. **3.** To cast rays on : IRRADIATE.

**ray[2]** (rā) *n.* [ME *raye* < OFr. *raie* < Lat. *raia.*] Any of various marine fishes of the order Rajiformes or Batoidei, marked by cartilaginous skeletons, horizontally flattened bodies, and narrow tails.

**ray flower** *n.* Any of the flat, strap-shaped marginal flowers around the head of certain flowers, as the daisy.

**Ray·leigh scattering** (rā′lē) *n.* [After John William Strutt (1842–1919), 3rd Baron *Rayleigh.*] Scattering of light waves by particles with dimensions much smaller than their wavelengths, resulting in angular separation of colors and responsible for the reddish color of sunset and the blue of the sky.

**ray·less** (rā′lĭs) *adj.* **1.** Lacking rays <a *rayless* flower> **2.** Dark and gloomy <a *rayless* cavern>

**ray·on** (rā′ŏn) *n.* [< RAY[1].] **1.** A synthetic textile fiber produced by forcing a cellulose solution through fine spinnerets and solidifying the resulting filaments. **2.** A fabric woven or knit from rayon.

**raze** *also* **rase** (rāz) *vt.* **razed, raz·ing, raz·es** *also* **rased, ras·ing, ras·es.** [ME *rasen,* to scrape < OFr. *raser* < VLat. **rasare* < Lat. *radere.*] **1.** To tear down or demolish <*raze* an old house> **2.** To scrape or shave off. **3.** *Archaic.* To erase.

**ra·zor** (rā′zər) *n.* [ME *rasor* < OFr. < *raser,* to scrape. —see RAZE.] A sharp-edged cutting implement used esp. for shaving.

**ra·zor·back** (rā′zər-băk′) *n.* **1.** A semiwild hog of the southeastern United States, having a narrow body with a ridged back. **2.** The rorqual. **3.** A sharp ridged hill.

**razorback**
*Approximately 3 feet high at shoulder*

**ra·zor·bill** (rā′zər-bĭl′) *n.* The razor-billed auk.

**ra·zor-billed auk** (rā′zər-bĭld′) *n.* A northern Atlantic sea bird, *Alca torda* with black-and-white plumage and a flattened, white-ringed bill.

**razor clam** *n.* Any of various clams of the family Solenidae, having long narrow shells.

**razz** (răz) *n.* [Shortening and alteration of RASPBERRY.] *Slang.* RASPBERRY 4. —*vt.* **razzed, razz·ing, razz·es.** *Slang.* To taunt, heckle, or tease.

**raz·zle-daz·zle** (răz′əl-dăz′əl) *n.* [Redup. of DAZZLE.] *Slang.* A flashy action or display intended to bewilder, deceive, or overwhelm.

**razz·ma·tazz** (răz′mə-tăz′) *n.* [Prob. alteration of RAZZLE-DAZZLE.] *Slang.* **1.** Razzle-dazzle. **2.** Ambiguous or evasive language : DOUBLE TALK. **3.** Ebullient energy : VIM.

**Rb** *symbol for* RUBIDIUM.

**re[1]** (rā) *n. Mus.* [ME < Med. Lat. —see GAMUT.] A solmization syllable representing the second tone of the diatonic scale.

**re[2]** (rē) *prep.* [Lat., ablative of *res,* thing.] In reference : CONCERNING.

**Re[1]** *symbol for* RHENIUM.

**Re[2]** (rā) *n. var. of* RA.

**re-** *pref.* [ME < OFr. < Lat.] **1.** Again : anew <*re*build> **2.** Backwards : back <*re*act> **3.** —Used as an intensive <*re*fine>

**-'re.** Are <They'*re* not at home.>

**reach** (rēch) *v.* **reached, reach·ing, reach·es.** [ME *rechen* < OE *ræcan.*] —*vt.* **1.** To stretch out or put forth (a bodily part) : EXTEND <*reach* out a hand> **2.** To touch or grasp by stretching out or extending <couldn't *reach* the top branch> **3.** To arrive at : ATTAIN <*reach* a decision> **4. a.** To succeed in communicating with <You can *reach* me by telephone.> **b.** To succeed in having an effect on <Parents can't always *reach* their children.> **5. a.** To extend as far as <The road *reached* the coast.> **b.** To carry as far as <The news *reached* every town.> **6.** To aggregate or amount to <The population *reached* two million.> **7.** *Informal.* To give or hand over to someone <*Reach* me the butter.> **8.** To score (a hit), as with a weapon. —*vi.* **1.** To extend or thrust out something. **2.** To try to touch or grasp something <*reach* for a pen> **3. a.** To have coextension in time or space. **b.** To be extensive in effect or influence. **4.** *Naut.* To sail with the wind abeam. —*n.* **1.** The act or power of stretching or thrusting out. **2.** The distance or extent something can reach. **3. a.** The range of one's understanding : COMPREHENSION <an explanation within my *reach*> **b.** The range or scope of influence or effect. **4.** An unbroken expanse <a *reach* of open water> **5.** A shaft connecting the rear axle of a vehicle with the front.

6. *Naut.* The tack of a sailing vessel with the wind abeam. **7.** The stretch of water visible between bends in a river or channel. **—reach'er** *n.*

**re·act** (rē-ăkt') *vi.* **-act·ed, -act·ing, -acts. 1.** To act in response to a stimulus <*react* to the encouraging news> **2.** To act in opposition to a force or condition <*reacted* against the influence of television> **3.** *Chem.* To undergo chemical change.

**re·ac·tance** (rē-ăk'təns) *n.* Opposition to alternating electric current flow caused by inductance and capacitance in a circuit.

**re·ac·tant** (rē-ăk'tənt) *n.* A substance involved in a chemical reaction, esp. a directly reacting substance present at the initiation of the reaction.

**re·ac·tion** (rē-ăk'shən) *n.* **1. a.** Response to a stimulus. **b.** The state resulting from such a response. **2.** A contrary or opposing action. **3. a.** A tendency to revert to an earlier state. **b.** Opposition to progress or liberalism. **4.** A chemical transformation in which a substance decomposes, combines with other substances, or interchanges constituents with other substances. **5.** *Physics.* A nuclear reaction.

**re·ac·tion·ar·y** (rē-ăk'shə-nĕr'ē) *adj.* Marked by reaction, esp. opposing progress or liberalism. **—re·ac'tion·ar'y** *n.*

**reaction engine** *n.* An engine that develops thrust by the focused expulsion of matter, esp. ignited fuel gases.

**reaction time** *n.* The time interval between application of a stimulus and detection of a response.

**re·ac·ti·vate** (rē-ăk'tə-vāt') *vt.* **-vat·ed, -vat·ing, -vates. 1.** To make active again. **2.** To restore the effectiveness or functioning of. **—re·ac'ti·va'tion** *n.*

**re·ac·tive** (rē-ăk'tĭv) *adj.* **1.** Tending to react in response to a stimulus. **2.** Marked by reaction. **3.** *Chem. & Physics.* Tending to participate in reactions.

**re·ac·tor** (rē-ăk'tər) *n.* **1.** One that reacts. **2.** *Elect.* A circuit element, such as a coil, used to introduce reactance. **3.** *Physics.* A nuclear reactor.

**read** (rēd) *v.* **read** (rĕd), **read·ing, reads.** [ME *reden* < OE *rǣdan*, to advise.] *—vt.* **1.** To examine and grasp the meaning of (written or printed characters, words, or sentences). **2.** To utter or express aloud (written or printed material) <*read* an essay to the class> **3.** To interpret the meaning or nature of through close examination or observation <*read* the tracks in the snow> **4.** To determine the intent, mood, or thoughts of <*read* someone's mind> **5. a.** To attribute (a particular meaning) to something read, heard, or considered <*read* a different motive into their actions> **b.** To interpret (something written, said, or done) as having a particular meaning or significance <*read* the statement as a peace offering> **6.** To predict or foretell (the future). **7.** To receive or comprehend (e.g., a radio message). **8.** To study or make a study of <*read* philosophy> **9.** To learn or get knowledge of from something written or printed <*read* that the election was a dead heat> **10.** To proofread. **11.** To have or use as a preferred reading in a particular passage <For "flavour" *read* "flavor."> **12.** To indicate, register, or show <The speedometer *reads* 60.> *—vi.* **1.** To read printed or written characters, as of words or music. **2.** To speak aloud the words that one is reading <*read* to the students every week> **3.** To learn by reading <*read* about dinosaurs> **4.** To study. **5.** To have a particular wording <How does the title *read?*> **6.** To contain a specific meaning <As the law *reads*, I'm innocent.> **7.** To have a specified character or quality for the reader <a novel that *reads* smoothly> **—read out. 1.** To read aloud. **2.** To expel by proclamation from a political, social, or other group <*read* the dissidents *out* of the party> *—n.* Something read <That story is a good *read.*> *—adj* (rĕd). Informed by reading **:** LEARNED. **—read between the lines.** To perceive or detect a meaning or implication that is obscure or unexpressed. **—read the riot act. 1.** To reprimand scathingly. **2.** To object strongly. **3.** To order a mob to disperse.

▲ word history: Anyone who can read these words probably takes that ability as a matter of course, as a necessary skill for successfully coping with modern life. Yet this attitude is only a recent one, as the history of the word *read* shows. The basic meaning of Old English *rǣdan*, the ancestor of *read*, was "to advise, counsel." The sense still exists as an archaism, although in this use the word is spelled *rede*. Behind this meaning is the notion of achieving control over the

**re'a·ban'don** *v.*
**re'ab·sorb'** *v.*
**re'ab·sorp'tion** *n.*
**re'ac·cept'** *v.*
**re'ac·com'mo·date'** *v.*
**re'ac·com'pa·ny** *v.*
**re'ac·cuse'** *v.*
**re'ac·quire'** *v.*
**re'a·dapt'** *v.*
**re'ad·dress'** *v.*
**re'ad·journ'** *v.*
**re'ad·journ'ment** *n.*
**re'ad·mis'sion** *n.*
**re'ad·mit'** *v.*
**re'ad·mit'tance** *n.*
**re'a·dopt'** *v.*
**re'a·dorn'** *v.*
**re'ad·vance'** *v.*
**re·an'i·mate'** *v.*
**re·an'i·ma'tion** *n.*
**re'an·nex'** *v.*
**re'a·noint'** *v.*
**re'ap·pear'** *v.*
**re'ap·pear'ance** *n.*
**re'ap·ply'** *v.*
**re'ap·point'** *v.*
**re'ap·point'ment** *n.*
**re·ar'gue** *v.*
**re·ar'gu·ment** *n.*
**re'as·cend'** *v.*
**re'as·cent'** *n.*
**re'as·sem'ble** *v.*
**re'as·sem'bly** *n.*
**re'as·sert'** *v.*
**re'as·ser'tion** *n.*
**re'as·sign'** *v.*
**re'as·sim'i·late'** *v.*
**re'as·sim'i·la'tion** *n.*
**re'as·so'ci·ate'** *v.*
**re'as·sume'** *v.*
**re'as·sump'tion** *n.*
**re'at·tach'** *v.*
**re'at·tack'** *v.*
**re'at·tain'** *v.*
**re'at·tempt'** *v.*
**re'a·vow'** *v.*
**re'a·wake'** *v.*
**re'a·wak'en** *v.*
**re·boil'** *v.*
**re·bur'y** *v.*
**re·car'ry** *v.*
**re·cel'e·brate'** *v.*
**re·chal'lenge** *v.*
**re·char'ter** *v.*
**re·check'** *v.*
**re·choose'** *v.*
**re·chris'ten** *v.*
**re·cir'cle** *v.*
**re·cir'cu·late'** *v.*
**re·clasp'** *v.*
**re·clean'** *v.*
**re·clothe'** *v.*
**re·coin'** *v.*
**re·coin'age** *n.*
**re·col'o·nize'** *v.*
**re·col'or** *v.*
**re'com·bine'** *v.*
**re'com·mence'** *v.*
**re'com·mis'sion** *n.*
**re'con·dense'** *v.*
**re'con·duct'** *v.*
**re·con'quer** *v.*
**re·con'quest** *n.*
**re·con'se·crate'** *v.*
**re'con·sol'i·date'** *v.*
**re·con'sti·tute** *v.*
**re·con'sti·tu'tion** *n.*
**re'con·vene'** *v.*
**re·cop'y** *v.*
**re'cor·o·na'tion** *n.*
**re·cross'** *v.*
**re·crown'** *v.*
**re'crys·tal·li·za'tion** *n.*
**re·crys'tal·lize'** *v.*
**re·cul'ti·vate'** *v.*
**re'ded·i·ca'tion** *n.*
**re'de·feat'** *v. & n.*
**re'de·fine'** *v.*
**re'de·mand'** *v.*
**re'dem·on'strate'** *v.*
**re'de·ny'** *v.*
**re'de·pos'it** *v. & n.*
**re'de·scend'** *v.*
**re'de·scent'** *n.*
**re'de·scribe'** *v.*
**re'de·ter'mine** *v.*
**re'di·gest'** *v.*
**re'dis·cov'er** *v.*
**re'dis·cov'er·y** *n.*
**re'dis·solve'** *v.*
**re'dis·till'** *v.*
**re'di·vide'** *v.*
**re'di·vi'sion** *n.*
**re·draft'** *n.*
**re·draw'** *v.*
**re·drive'** *v.*
**re·dry'** *v.*
**re·dye'** *v.*
**re-ed'it** *v.*
**re-el'e·vate'** *v.*
**re'-em·bark'** *v.*
**re'-em·bod'y** *v.*
**re'em·brace'** *v.*
**re'-e·merge'** *v.*
**re'-e·mer'gence** *n.*
**re-em'i·grate'** *v.*
**re'-en·cour'age** *v.*
**re'-en·cour'age·ment** *n.*
**re'-en·dow** *v.*
**re'-en·gage'** *v.*
**re'-en·gage'ment** *n.*
**re'-en·grave'** *v.*
**re'-en·list'** *v.*
**re'-en·list'ment** *n.*
**re'-en·slave'** *v.*
**re'-e·rect'** *v.*
**re'-es·tab'lish** *v.*
**re'-es·tab'lish·ment** *n.*
**re'-e·val'u·ate'** *v.*
**re'-e·val'u·a'tion** *n.*
**re'-ex·change'** *v.*
**re'-ex·hib'it** *v.*
**re'-ex·pel'** *v.*
**re'-ex·pe'ri·ence** *v.*
**re'-ex·port'** *v.*
**re·face'** *v.*
**re·fash'ion** *v.*
**re·fas'ten** *v.*
**re·fer'ti·lize'** *v.*
**re·fire'** *v.*
**re·flow'** *v.*
**re·flow'er** *v.*
**re·fold'** *v.*
**re·forge'** *v.*
**re·for'mu·late'** *v.*
**re'for·ti·fi·ca'tion** *n.*
**re·for'ti·fy'** *v.*
**re·frame'** *v.*
**re·freeze'** *v.*
**re·fur'nish** *v.*
**re·gath'er** *v.*
**re·gear'** *v.*
**re·ger'mi·nate'** *v.*
**re'ger·mi·na'tion** *n.*
**re·gild'** *v.*
**re·glaze'** *v.*

uncertain or the unknown, and other senses of *rǣdan* contain this idea. The Old English verb also meant "to interpret," especially to interpret dreams and signs; the interpretation of marks representing words is just an extension of this sense. The notion of "interpretation" is still very much alive in *riddle,* a noun related to *read.*

**read·a·ble** (rē′də-bəl) *adj.* **1.** Capable of being read easily : LEGIBLE. **2.** Pleasurable or interesting to read. **—read′a·bil′i·ty, read′a·ble·ness** *n.* **—read′a·bly** *adv.*

**read·er** (rē′dər) *n.* **1.** One who reads. **2.** A professional reciter of literary works. **3.** A minor cleric or layperson who recites lessons or prayers in church services. **4.** One employed by a publisher to read and evaluate manuscripts. **5.** A corrector of printers' proofs. **6.** A teaching assistant who reads and grades examination papers. **7.** *Chiefly Brit.* A university lecturer. **8. a.** A textbook of reading exercises. **b.** An anthology, esp. a literary anthology.

**read·er·ship** (rē′dər-shĭp′) *n.* **1.** The readers of a publication. **2.** *Chiefly Brit.* The office of a reader.

**read·i·ly** (rĕd′ə-lē) *adv.* **1.** Promptly. **2.** Willingly. **3.** Easily.

**read·ing** (rē′dĭng) *n.* **1.** The act or practice of reading. **2.** Written or printed matter. **3.** An oral rendition or recital of a written work <a poetry *reading*> **4.** An official or public recitation of written material <the *reading* of a verdict> **5.** An interpretation from a specified point of view <a personal *reading* of events> **6.** The specific form or variation of a passage or text. **7.** The information indicated by a gauge or graduated instrument.

**re·ad·just** (rē′ə-jŭst′) *vt.* **-just·ed, -just·ing, -justs.** To adjust or arrange again. **—re′ad·just′er** *n.* **—re′ad·just′ment** *n.*

**read-only memory** (rēd′ōn′lē) *n. Computer Sci.* A small memory that allows fast retrieval of permanently stored data.

**read·out** *also* **read-out** (rēd′out′) *n.* Presentation of computer data, usu. in digital form, from calculations or storage.

**read·y** (rĕd′ē) *adj.* **-i·er, -i·est.** [ME *redy* < OE *rǣde.*] **1.** Prepared or available for service or action. **2.** Mentally disposed : WILLING <*ready* to compromise> **3.** Liable or about to do something <ripe fruit *ready* to fall> **4.** Prompt in apprehending or reacting <a *ready* intelligence> **5.** Available <*ready* cash> *—vt.* **read·ied, read·y·ing, read·ies.** To cause to be ready. **—read′i·ness** *n.*

**read·y-made** (rĕd′ē-mād′) *adj.* **1.** Already made, prepared, or available <*ready-made* pie crust><*ready-made* suits> **2.** Conceived in advance <a *ready-made* excuse>

**re·af·firm** (rē′ə-fûrm′) *vt.* **-firmed, -firm·ing, -firms.** To affirm or assert again. **—re′af·fir·ma′tion** (rē′ăf-ər-mā′shən) *n.*

**re·a·gent** (rē-ā′jənt) *n.* A substance used in a chemical reaction to detect, examine, measure, or produce other substances.

**re·a·gin** (rē-ā′jĭn) *n.* [REAG(ENT) + -IN.] **1.** An antibody found in the blood of individuals having a genetic predisposition to allergies such as asthma and hay fever. **2.** A substance present in the blood of individuals who have a positive serological test for syphilis. **—re′a·gin′ic** (rē′ə-jĭn′ĭk) *adj.* **—re′a·gin′i·cal·ly** *adv.*

**re·al**[1] (rē′əl, rēl) *adj.* [ME, of property or things < AN < Med. Lat. *realis* < LLat., real < Lat. *res,* thing.] **1.** Not imaginary, fictional, or pretended : ACTUAL. **2.** Genuine or authentic <*real* gold> **3.** Essential : basic <The *real* culprit is greed.> **4.** Being no less than what is stated : TRUE <a *real* threat> **5.** Not to be taken lightly : SERIOUS <in *real* trouble> **6.** *Philos.* Existing actually and objectively. **7.** Of, relating to, or designating an image formed by light rays that converge in space. **8.** *Math.* Of, relating to, or designating the nonimaginary part of a complex quantity. **9.** *Law.* Of or relating to fixed or stationary property, as buildings or land. *—adv. Informal.* Very <*real* upset> **—real′ness** *n.*

☆ ***syns:*** REAL, SUBSTANTIAL, SUBSTANTIVE, TANGIBLE *adj. core meaning* : having reality <*real* evidence><*real,* not imagined pain>

**re·al**[2] (rā-äl′) *n., pl.* **-als** *or* **-a·les** (-ä′lās) [Sp. < *real,* royal < Lat. *regalis* < *rex,* king.] An obsolete Spanish silver monetary unit.

**re·al**[3] (rā-äl′) *n., pl.* **reals** *or* **reis** (rās) [Port. < *real,* royal < Lat. *regalis* < *rex,* king.] Either of two monetary units once in circulation in Portugal and Brazil.

**real estate** *n.* Land owned as property, along with natural resources and permanent buildings on it.

**re·glue′** *v.*
**re·grade′** *v.*
**re·graft′** *v.*
**re·grant′** *v.*
**re·han′dle** *v.*
**re·heat′** *v.*
**re·heel′** *v.*
**re′hire′** *v.*
**re′ig·nite′** *v.*
**re′im·plant′** *v.*
**re′im·pose′** *v.*
**re′im·po·si′tion** *n.*
**re′im·preg′nate** *v.*
**re′im·press′** *v.*
**re′im·print′** *v.*
**re′im·pris′on** *v.*
**re′im·pris′on·ment** *n.*
**re′in·au′gu·rate′** *v.*
**re′in·cite′** *v.*
**re′in·cor′po·rate′** *v.*
**re′in·cur′** *v.*
**re′in·duce′** *v.*
**re′in·fect′** *v.*
**re′in·flame′** *v.*
**re′in·form′** *v.*
**re′in·fuse′** *v.*
**re′in·hab′it** *v.*
**re′in·oc′u·late′** *v.*
**re′in·oc′u·la′tion** *n.*
**re′in·scribe′** *v.*
**re′in·sert′** *v.*
**re′in·ser′tion** *n.*
**re′in·spect′** *v.*
**re′in·spec′tion** *n.*
**re′in·spire′** *v.*
**re′in·stall′** *v.*
**re′in·stal·la′tion** *n.*
**re′in·struct′** *v.*
**re′in·ter′** *v.*
**re′in·ter′ro·gate′** *v.*
**re′in·tro·duce′** *v.*
**re′in·tro·duc′tion** *n.*
**re′in·vent′** *v.*
**re′in·ves′ti·gate′** *v.*
**re′in·ves·ti·ga′tion** *n.*
**re′in·vest′ment** *n.*
**re′in·vite′** *v.*
**re′in·volve′** *v.*
**re·judge′** *v.*
**re·kin′dle** *v.*
**re·la′bel** *v.*
**re·lace′** *v.*
**re·launch′** *v.*
**re·learn′** *v.*
**re·light′** *v.*
**re·liq′ui·date′** *v.*
**re′liq·ui·da′tion** *n.*
**re·load′** *n.* & *v.*
**re·loan′** *n.* & *v.*
**re′man·u·fac′ture** *v.*
**re·mar′ry** *v.*
**re·meas′ure** *v.*
**re·melt′** *v.*
**re·merge′** *v.*
**re·mi′grate** *v.*
**re′mi·gra′tion** *n.*
**re·mix′** *v.*
**re′mod·i·fi·ca′tion** *n.*
**re·mod′i·fy′** *v.*
**re·mold′** *v.*
**re·name′** *v.*
**re·nav′i·gate′** *v.*
**re·no′ti·fy′** *v.*
**re′ob·tain′** *v.*
**re′ob·tain′a·ble** *adj.*
**re′oc·cu·pa′tion** *n.*
**re·oc′cu·py′** *v.*
**re′oc·cur′** *v.*
**re′oc·cur′rence** *n.*
**re′op·pose′** *v.*
**re′or·dain′** *v.*
**re′or·di·na′tion** *n.*
**re·o′ri·ent′** *v.*
**re·pac′i·fy′** *v.*
**re·pack′** *v.*
**re·paint′** *v.* & *n.*
**re·pa′per** *v.*
**re·pave′** *v.*
**re·pe′nal·ize′** *v.*
**re·pledge′** *v.*
**re·plunge′** *v.*
**re·pol′ish** *v.*
**re·pop′u·late′** *v.*
**re′pop·u·la′tion** *n.*
**re·pour′** *v.*
**re′pro·claim′** *v.*
**re·pur′chase** *v.*
**re·pu′ri·fy′** *v.*
**re′pur·sue′** *v.*
**re·ra′di·ate′** *v.*
**re·read′** *v.*
**re′re·cord′** *v.*
**re·rise′** *v.*
**re·roll′** *v.*
**re·route′** *v.*
**re·sad′dle** *v.*
**re·sail′** *v.*
**re·sa·lute′** *v.*
**re·seal′** *v.*
**re·seat′** *v.*
**re·seed′** *v.*
**re·seek′** *v.*
**re·seg′re·gate′** *v.*
**re·seize′** *v.*
**re·sei′zure** *n.*
**re·sell′** *v.*
**re·set′tle** *v.*
**re·set′tle·ment** *n.*
**re·sharp′en** *v.*
**re·ship′** *v.*
**re·ship′ment** *n.*
**re·sift′** *v.*
**re·sol′der** *v.*
**re′so·lid′i·fy′** *v.*
**re·sow′** *v.*
**re·spread′** *v.*
**re·stack′** *v.*
**re·stip′u·late′** *v.*
**re′stip·u·la′tion** *n.*
**re·strength′en** *v.*
**re·string′** *v.*
**re·strive′** *v.*
**re·stud′y** *n.* & *v.*
**re′sub·ject′** *v.*
**re′sub·jec′tion** *n.*
**re·sum′mon** *v.*
**re·sum′mons** *n.*
**re′sup·ply′** *v.*
**re·teach′** *v.*
**re·test′** *v.*
**re·tie′** *v.*
**re′trans·late′** *v.*
**re′tra·verse′** *v.*
**re·type′** *v.*
**re·u′til·ize′** *v.*
**re·ut′ter** *v.*
**re·val′ue** *v.*
**re·var′nish** *v.*
**re′ver·i·fi·ca′tion** *n.*
**re·ver′i·fy′** *v.*
**re·vin′di·cate** *v.*
**re′vin·di·ca′tion** *n.*
**re·voice′** *v.*
**re·warm′** *v.*
**re·wash′** *v.*
**re·weigh′** *v.*

**re·al·gar** (rē-ăl′gär′, -gər) *n.* [ME < Med. Lat. < Catalan < Ar. *rajh alghār*, powder of the mine.] A soft orange-red arsenic ore, $As_2S_2$, used in tanning, pyrotechnics, and as a pigment.

**re·a·lign** (rē′ə-līn′) *vt.* **-ligned, -lign·ing, -ligns. 1.** To put back into proper order or alignment. **2.** To make new groupings of or working arrangements between. **—re′a·lign′ment** *n.*

**re·al·ism** (rē′ə-lĭz′əm) *n.* **1.** An inclination toward factual truth and pragmatism. **2.** Artistic or literary representation intended as an accurate and unidealized portrayal of real life or of the objective world. **3.** *Philos.* **a.** The doctrine that universal principles are more real than objects as sensed. **b.** The doctrine that names somehow denote the essences of things or categories of things. **c.** The doctrine that the objects of perception exist independently of the perceiver.

**re·al·ist** (rē′ə-lĭst) *n.* **1.** One inclined to factual truth and pragmatism. **2.** A practitioner of philosophic or artistic realism.

**re·al·is·tic** (rē′ə-lĭs′tĭc) *adj.* **1.** Tending to or expressing an awareness of things as they really are <a *realistic* goal for a beginner> **2.** Accurately representing what is depicted or described <a *realistic* portrayal of pioneer life> **—re′al·is′ti·cal·ly** *adv.*

**re·al·i·ty** (rē-ăl′ĭ-tē) *n., pl.* **-ties. 1.** The quality or state of being actual or true. **2.** A person, entity, or event that is real <a promise that soon became a *reality*> **3.** The totality of all things possessing existence or essence. **4.** The domain of actual or practical experience <One must learn to face *reality.*> **5.** *Philos.* The sum of all that is real, absolute, and unchangeable.

**reality principle** *n. Psychoanal.* Awareness of and adjustment to environmental demands in a way that assures ultimate satisfaction of instinctual needs.

**re·al·ize** (rē′ə-līz′) *v.* **-ized, -iz·ing, -iz·es.** [Fr. *réaliser* < OFr. *realiser* < *real*, real < LLat. *realis* < Lat. *res*, thing.] *—vt.* **1.** To comprehend fully or correctly. **2.** To bring about or make real : FULFILL <*realized* one's potential> **3.** To make realistic. **4.** To obtain or achieve <*realize* a profit> **5.** To bring in (a sum) as profit by sale. *—vi.* To exchange holdings or goods for money. **—re′al·iz′a·ble** *adj.* **—re′al·i·za′tion** (-lĭ-zā′shən) *n.* **—re′al·iz′er** *n.*

**re·al·ly** (rē′ə-lē′, rē′lē) *adv.* **1.** In actual truth or fact <The horseshoe crab isn't *really* a crab at all.> **2.** Truly <a *really* enjoyable evening> **3.** Indeed <*Really*, you shouldn't have done it.>

**realm** (rĕlm) *n.* [ME *realme* < OFr. < Lat. *regimen*, government < *regere*, to rule.] **1.** A kingdom. **2.** A field, sphere, or domain <the *realm* of government>

**real number** *n.* A rational or irrational number.

**re·al·po·li·tik** (rā-äl′pō′lĭ-tēk′) *n.* [G. : *real*, practical + *Politik*, politics.] A diplomatic policy based on the aggressive pursuit of national interests without regard for ethical or philosophical considerations. **—re·al′po′li·tik′er** *n.*

**real time** *n. Computer Sci.* **1.** The actual time in which a physical process under computer study or control occurs. **2.** The time required for a computer to solve a problem, measured from the time data are fed in to the time a solution is received.

**Re·al·tor** (rē′əl-tər, -tôr′). A collective mark for a real-estate agent affiliated with the National Association of Realtors.

**re·al·ty** (rē′əl-tē) *n., pl.* **-ties.** Real estate.

**ream¹** (rēm) *n.* [ME *reme* < OFr. *remme* < Ar. *rizmah*, bundle.] **1.** A quantity of paper, formerly 480 sheets, now 500 sheets or, in a printer's ream, 516 sheets. **2.** *often* **reams.** A great number or amount.

**ream²** (rēm) *vt.* **reamed, ream·ing, reams.** [Orig. unknown.] **1.** To enlarge or shape (a hole or bore) with or as if with a reamer. **2.** To remove (material) by reaming. **3.** To squeeze the juice out of (fruit) with a reamer.

**ream·er** (rē′mər) *n.* **1.** A tool for enlarging or shaping holes. **2.** A utensil with a conical, ridged center, used for extracting citrus-fruit juice. **3.** One that reams.

**reap** (rēp) *v.* **reaped, reap·ing, reaps.** [ME *repen* < OE *rīpan.*] *—vt.* **1.** To cut (grain or pulse) for harvest with a scythe, sickle, or reaper. **2.** To harvest (a crop). **3.** To harvest a crop from <*reap* a meadow> **4.** To obtain as a result of effort <*reaped* benefits from my education> *—vi.* **1.** To cut or harvest grain or pulse. **2.** To obtain a return or reward.

**reap·er** (rē′pər) *n.* **1.** One that reaps. **2.** A machine for harvesting crops, as grain.

**re·ap·por·tion** (rē′ə-pôr′shən) *vt.* **-tioned, -tion·ing, -tions.** To apportion or distribute anew.

**re·ap·por·tion·ment** (rē′ə-pôr′shən-mənt) *n.* **1.** An act of reapportioning or the state of being reapportioned. **2.** Redistribution of representation in a legislative body, esp. periodic reallotment of U.S. Congressional seats to reflect demographic changes as required by the Constitution.

**re·ap·prais·al** (rē′ə-prā′zəl) *n.* A new appraisal.

**rear¹** (rîr) *n.* [Prob. short for REARGUARD or REARWARD².] **1.** A hind part. **2.** The point or area farthest from the front <the *rear* of a building> **3.** The part of a military deployment usu. farthest from the fighting front. **4.** *Informal.* The buttocks. *—adj.* Of, at, or located in the rear <a *rear* entrance>

**rear²** (rîr) *v.* **reared, rear·ing, rears.** [ME *reren*, to raise < OE *ræran.*] *—vt.* **1.** To care for (a child) during the early stages of life : BRING UP. **2.** To lift upright : RAISE. **3.** To build : erect. **4.** To tend (growing plants or animals). *—vi.* **1.** To rise on the hind legs, as a horse. **2.** To rise high in the air : TOWER. **—rear′er** *n.*

**rear admiral** *n.* A naval flag officer ranking below a vice admiral and above a captain.

**rear guard** *n.* [ME *reregarde* < OFr. : *rere*, backward (< Lat. *retro*) + *garde*, guard < *garder*, to defend, of Germanic orig.] A detachment of troops assigned to protect the rear of a military force.

**rear-guard** (rîr′gärd′) *adj.* Of or relating to resistance, either military or social <fought *rear-guard* actions against social change>

**re·arm** (rē-ärm′) *v.* **-armed, -arm·ing, -arms.** *—vt.* **1.** To arm again. **2.** To equip with better weapons. *—vi.* To arm oneself again. **—re·ar′ma·ment** (rē-är′mə-mənt) *n.*

**rear·most** (rîr′mōst′) *adj.* Farthest in the rear : LAST.

**re·ar·range** (rē′ə-rānj′) *vt.* **-ranged, -rang·ing, -rang·es.** To change the arrangement of. **—re′ar·range′ment** *n.*

**rear·view mirror** (rîr′vyōō′) *n.* A mirror, as one attached to a motor vehicle, providing a view of what is behind it.

**rear·ward¹** (rîr′wərd) *adj.* & *adv.* At or toward the rear. *—n.* A position or place at the rear. **—rear′wards** *adv.*

**rear·ward²** (rîr′wôrd′) *n.* [ME *rerewarde* < AN : *rere*, behind (< Lat. *retro*) + *warde*, guard, of Germanic orig.] The rear guard of an armed force.

**rea·son** (rē′zən) *n.* [ME < OFr. *raison* < Lat. *ratio* < *reri*, to think.] **1.** The basis or motive for an action, decision, or belief. **2.** A declaration explaining or justifying an action, decision, or belief <stated my *reason* for quitting> **3.** An underlying fact or motive that provides logical sense for a premise or occurrence <There is *reason* to hope for a peaceful solution.> **4.** The capacity for rational thought, inference, or discrimination. **5.** Sound or sensible judgment : REASONABILITY. **6.** Normal mental state : SANITY. **7.** *Logic.* A premise, usu. the minor premise, of an argument. *—v.* **-soned, -son·ing, -sons.** *—vi.* **1.** To use the faculty of reason : think logically. **2.** To talk or argue logically and persuasively. **3.** *Archaic.* To engage in discussion or conversation. *—vt.* **1.** To determine or conclude by logical thinking <*reasoned* out a plan> **2.** To persuade or dissuade (someone) with reasons. **3.** To debate : discuss. **—by reason of.** Because of. **—within reason.** Within the bounds of good sense or practicality. **—with reason.** With good cause. **—rea′son·er** *n.*

**rea·son·a·ble** (rē′zə-nə-bəl) *adj.* **1.** Capable of reasoning : RATIONAL. **2.** Governed by or in accordance with reason or sound thinking. **3.** Within the bounds of common sense <allowed a *reasonable* time for the trip> **4.** Not extreme or excessive : FAIR <*reasonable* fuel rates> **—rea′son·a·bil′i·ty, rea′son·a·ble·ness** *n.* **—rea′son·a·bly** *adv.*

**rea·son·ing** (rē′zə-nĭng) *n.* **1.** Use of reason, esp. to form conclusions, judgments, or inferences. **2.** Arguments or evidence used in reasoning.

**re·as·sure** (rē′ə-shŏŏr′) *vt.* **-sured, -sur·ing, -sures. 1.** To restore confidence to. **2.** To assure again. **3.** To reinsure. **—re′as·sur′ance** *n.* **—re′as·sur′ing·ly** *adv.*

**re·a·ta** (rē-ä′tə) *n. var. of* RIATA.

**Ré·au·mur** *or* **Re·au·mur** (rā′ō-myŏŏr′) *adj.* [After René Antoine de *Réaumur* (1683–1757).] Indicated on a thermometer scale that registers the freezing point of water as 0° and the boiling point as 80°.

**reave¹** (rēv) *v.* **reaved** *or* **reft** (rĕft), **reav·ing, reaves.** [ME *reven* < OE *rēafian.*] *Archaic.* *—vt.* **1.** To seize and carry off forcibly. **2.** To deprive of : BEREAVE. *—vi.* To rob, plunder, or pillage.

**reave²** (rēv) *vt.* **reaved** *or* **reft** (rĕft), **reav·ing, reaves.** [ME *reven*, poss. alteration of ON *rifa*, to rive.] *Archaic.* To break or tear apart.

**Reb¹** *also* **reb** (rĕb) *n.* [Short for REBEL.] *Informal.* A Confederate soldier in the American Civil War.

**Reb²** (rĕb) *n.* [Yiddish < Heb. *rabbi*, my master. —see RABBI.] A Jewish title of respect, approx. equivalent to "Mr." or "Sir," but used with the given name rather than the surname.

**re·bar·ba·tive** (rē-bär′bə-tĭv) *adj.* [Fr. *rébarbatif* < OFr. *rebarber*, to be repellent : *re-*, against (< Lat.) + *barbe*, beard < Lat. *barba.*] Tending to irritate : REPELLENT.

**re·bate¹** (rē′bāt′) *n.* [ME *rebaten* < OFr. *rebattre*, to reduce, to beat down again : *re-*, again (< Lat.) + *abattre*, to beat down (*a-*, to + *battre*, to batter).] A deduction from an amount charged or a return of part of a price paid. *—vt.* (rē′bāt′, rĭ-bāt′) **-bat·ed, -bat·ing, -bates. 1.** To deduct or return (an amount) from a payment or bill. **2.** To dull or blunt (e.g., a weapon). **3.** To diminish. **—re′bat·er** *n.*

**re·bate²** (rē′bāt′, răb′ĭt) *n.* & *v. var. of* RABBET.

**re·ba·to** (rĭ-bä′tō) *also* **ra·ba·to** (rə-) *n., pl.* **-tos.** [OFr. *rabat* < *rabattre*, to turn down again, reduce. —see REBATE¹.] A stiff flaring collar, as of lace, worn by both sexes early in the 17th cent.

**re·bec** *also* **re·beck** (rē′bĕk′) *n.* [OFr., alteration of *rebebe* <

OProv. *rebab* < Ar. *rebāb.*] A pear-shaped, two- or three-stringed medieval musical instrument, played with a bow.

**re·bel** (rĭ-bĕl′) *vi.* **-belled, -bel·ling, -bels.** [ME *rebellen* < OFr. *rebeller* < Lat. *rebellare* : *re-*, against + *bellare,* to make war < *bellum,* war.] **1.** To refuse allegiance to and oppose by force an established government or ruling authority. **2.** To defy or resist an authority or convention. **3.** To feel or express strong unwillingness or repugnance <*rebelled* at the thought of the dentist's chair> —*n.* **reb·el** (rĕb′əl). One who rebels or is rebellious.

**re·bel·lion** (rĭ-bĕl′yən) *n.* [ME < OFr. < Lat. *rebellio* < *rebellare,* to rebel.] **1.** An uprising or organized opposition intended to change or overthrow an existing government or ruling authority. **2.** An act or show of defiance toward an authority or established convention.

☆ ***syns:*** REBELLION, INSURGENCE, INSURRECTION, MUTINY, REVOLT, UPRISING *n. core meaning* : organized opposition intended to change or overthrow existing authority <an armed *rebellion* against the right-wing junta>

**re·bel·lious** (rĭ-bĕl′yəs) *adj.* **1.** Participating in or favoring a rebellion. **2.** Of or characteristic of a rebel. **3.** Resisting direction or control : UNRULY. **—re·bel′lious·ly** *adv.* **—re·bel′lious·ness** *n.*

**re·bind** (rē-bīnd′) *vt.* **-bound** (-bound′), **-bind·ing, -binds.** To bind again, esp. to put a new binding on (a book). —*n.* (rē′bīnd′). A book that has been rebound.

**re·birth** (rē-bûrth′, rē′bûrth′) *n.* **1.** A second or new birth : REINCARNATION. **2.** A revival : renaissance.

**re·born** (rē-bôrn′) *adj.* Emotionally or spiritually renewed.

**re·bound¹** (rē′bound′, rĭ-) *v.* **-bound·ed, -bound·ing, -bounds.** [ME *rebounden* < OFr. *rebondir* : *re-*, back (< Lat.) + *bondir,* to resound. —see BOUND¹.] —*vi.* **1.** To bounce or spring back after colliding with something. **2.** To recover, as from adversity or a slump <Stocks *rebounded* after earlier losses.> **3.** To resound : re-echo. **4.** *Basketball.* To retrieve the ball as it bounces off the backboard or rim after an unsuccessful shot. —*vt.* To cause to rebound. —*n.* (rē′bound′, rĭ-bound′). **1.** A springing or bounding back : RECOIL. **2. a.** A rebounding ball or hockey puck. **b.** *Basketball.* The act or an instance of retrieving a rebounding ball. **3.** A quick recovery from reversal or disappointment <They were married on the *rebound.*> **—re·bound′er** *n.*

**re·bound²** (rē-bound′) *v. p.t. & p.p. of* REBIND.

**re·bo·zo** (rĭ-bō′sō) *n., pl.* **-zos.** [Sp. < *rebosar,* to muffle with a shawl.] A long shawl worn over the head and shoulders mainly by Mexican women.

**re·broad·cast** (rē-brôd′kăst′) *vt.* **-cast** *or* **-cast·ed, -cast·ing, -casts. 1.** To repeat the broadcast of (a recorded program). **2.** To receive and relay (a live broadcast). **—re·broad′cast′** *n.*

**re·buff** (rĭ-bŭf′) *n.* [OFr. *rebuffer* < OItal. *ribuffare* < *ribuffo,* reprimand : *ri-*, back (< Lat. *re-*) + *buffo,* gust.] **1.** A blunt or abrupt repulse or refusal. **2.** A check or abrupt setback to action or progress. —*vt.* **-buffed, -buff·ing, -buffs. 1.** To refuse abruptly or bluntly : SNUB. **2.** To drive back : REPEL.

**re·build** (rē-bĭld′) *vt.* **-built** (-bĭlt′), **-build·ing, -builds. 1.** To build again. **2.** To make extensive structural repairs on. **3.** To remodel or make extensive changes in <*rebuilding* society>

**re·buke** (rĭ-byo͞ok′) *vt.* **-buked, -buk·ing, -bukes.** [ME *rebuken* < ONFr. *rebuker.*] **1.** To criticize or reprove sharply : REPRIMAND. **2.** To check or repress. **—re·buke′** *n.*

**re·bus** (rē′bəs) *n., pl.* **-bus·es.** [Lat., ablative pl. of *res,* thing.] A puzzle consisting of words or syllables represented as pictures.

**re·but** (rĭ-bŭt′) *v.* **-but·ted, -but·ting, -buts.** [ME *rebuten* < OFr. *reboter* : *re-*, back (< Lat.) + *boter,* to butt, of Germanic orig.] —*vt.* **1.** To refute, esp. by offering opposing arguments, as in a legal case. **2.** To repel. —*vi.* To present opposing evidence or arguments.

**re·but·tal** (rĭ-bŭt′l) *n.* **1.** An act of rebutting. **2.** A statement made in rebutting.

**re·but·ter** (rĭ-bŭt′ər) *n.* **1.** One who rebuts or refutes. **2.** *Law.* The defendant's answer to the plaintiff's surrejoinder.

**re·cal·ci·trant** (rĭ-kăl′sĭ-trənt) *adj.* [Lat. *recalcitrans, recalcitrant-*, pr.part. of *recalcitrare,* to be disobedient : *re-*, back + *calcitrare,* to kick < *calx,* heel.] Stubbornly resistant to authority, domination, or guidance : REFRACTORY. —*n.* A recalcitrant person. **—re·cal′ci·trance, re·cal′ci·tran·cy** *n.*

**re·ca·les·cence** (rē′kə-lĕs′əns) *n.* [< Lat. *recalescere,* to grow warm again : *re-*, again + *calescere,* to become warm < *calēre,* to be warm.] Sudden increase of heat in a cooling metal caused by an exothermic structural change. **—re′ca·les′cent** *adj.*

**re·call** (rĭ-kôl′) *vt.* **-called, -cal·ling, -calls. 1.** To ask or order to return <*recall* laid-off employees> **2.** To summon back to awareness of or concern with the subject or situation at hand. **3.** To recollect : remember. **4.** To cancel, take back, or revoke. **5.** To bring back : RESTORE. —*n.* (*also* rē′kôl′). **1.** An act of recalling, esp. an official order to return. **2.** A signal, as a bugle call, used to summon military personnel back to their posts. **3.** The ability to remember information or experiences <gifted with total *recall*> **4.** An act of revoking. **5.** The procedure by which a public official may be removed from office by popular vote or the right to use this procedure. **6.** A request by the maker of a product specified as defective for its return to the dealer for repairs or adjustments. **—re·call′a·ble** *adj.*

**re·cant** (rĭ-kănt′) *v.* **-cant·ed, -cant·ing, -cants.** [Lat. *recantare* : *re-*, back + *cantare,* to sing < *canere.*] —*vt.* To make a disavowal or formal retraction of (a statement or belief to which one has previously committed oneself). —*vi.* To make a formal disavowal or retraction of a previously held belief <was forced by the new evidence to *recant*> **—re′can·ta′tion** *n.* **—re·cant′er** *n.*

**re·cap¹** (rē-kăp′) *vt.* **-capped, -cap·ping, -caps. 1.** To replace a cap or caplike covering on <*recap* a bottle> **2.** To restore (a worn automotive tire) to usable condition by bonding new rubber onto the old casing : RETREAD. —*n.* (rē′kăp′). A recapped tire.

**re·cap²** (rē′kăp′) *vt.* **-capped, -cap·ping, -caps.** To recapitulate, as a news story. **—re′cap′** *n.*

**re·cap·i·tal·ize** (rē-kăp′ĭ-tl-īz′) *vt.* **-ized, -iz·ing, -iz·es.** To change the capital structure of (a corporation). **—re·cap′i·tal·i·za′tion** *n.*

**re·ca·pit·u·late** (rē′kə-pĭch′ə-lāt′) *v.* **-lat·ed, -lat·ing, -lates.** [LLat. *recapitulare, recapitulat-* : *re-*, again + *capitulare,* to put under headings < Lat. *capitulum,* heading, dim. of *caput,* head.] —*vt.* **1.** To repeat in concise form : SUMMARIZE. **2.** *Biol.* To appear to repeat (the evolutionary stages of the species) during embryonic development of the individual organism. —*vi.* To summarize <*recapitulated* the account> **—re·ca·pit′u·la′tive, re·ca·pit′u·la·to′ry** (-lə-tôr′ē, -tōr′ē) *adj.*

**re·ca·pit·u·la·tion** (rē′kə-pĭch′ə-lā′shən) *n.* **1.** The act or process of recapitulating. **2.** A concise review : SUMMARY. **3.** *Biol.* Apparent repetition of some of the evolutionary stages of the species during embryonic development of the individual organism. **4.** *Mus.* Restatement of the exposition that makes up the third section of the typical sonata form.

**re·cap·ture** (rē-kăp′chər) *n.* **1. a.** An act of recovering or retaking. **b.** The condition of being recovered or retaken. **2.** *Law.* The act or an instance of retaking booty or goods. **3.** Something recaptured. **4.** The lawful taking by a government of a set amount of the profits of a public-service corporation above a stipulated rate of return. —*vt.* **-tured, -tur·ing, -tures. 1.** To capture again. **2.** To recall <hoping to *recapture* old memories> **3.** To get possession of by the government procedure of recapture.

**re·cast** (rē-kăst′) *vt.* **-cast, -cast·ing, -casts. 1.** To mold again <*recast* a statue> **2.** To set down or present (e.g., ideas) in a new or different arrangement. **3.** To change the cast of (a theatrical production). —*n.* (rē′kăst′). **1.** The act or process of recasting. **2.** Something made by recasting.

**re·cede** (rĭ-sēd′) *vi.* **-ced·ed, -ced·ing, -cedes.** [Lat. *recedere* : *re-*, back + *cedere,* to go away.] **1.** To move back or away from a limit, point, or mark. **2.** To slope backward. **3.** To become or seem to become farther away. **4.** To withdraw or retreat.

☆ ***syns:*** RECEDE, EBB, RETRACT, RETREAT *v. core meaning* : to move back or away from a point, limit, or mark <tidal waters *receding*> ***ant:*** advance, proceed

**re·cede** (rē-sēd′) *vt.* **-ced·ed, -ced·ing, -cedes.** To yield or grant to one previously in possession : cede back.

**†re·ceipt** (rĭ-sēt′) *n.* [ME *receite* < ONFr. < Med. Lat. *recepta* < Lat. *recipere,* to receive.] **1. a.** An act of receiving something. **b.** The fact of being received. **2.** *often* **receipts.** The amount or quantity received <box office *receipts*> **3.** A written acknowledgment that a specified article, sum of money, or delivery of merchandise has been received. **4.** *Regional.* A recipe. —*v.* **-ceipt·ed, -ceipt·ing, -ceipts.** —*vt.* **1.** To mark (a bill) as having been paid. **2.** To give or write a receipt for (money paid or goods delivered). —*vi.* To give a receipt.

**re·ceiv·a·ble** (rĭ-sē′və-bəl) *adj.* **1.** Appropriate for being accepted, esp. as payment. **2.** Waiting for or requiring payment : DUE <accounts *receivable*> —*n.* **receivables.** Business assets representing the total amounts due from others.

**re·ceive** (rĭ-sēv′) *v.* **-ceived, -ceiv·ing, -ceives.** [ME *receiven* < ONFr. *receivre* < Lat. *recipere* : *re-*, again + *capere,* to take.] —*vt.* **1.** To acquire or take (something given, offered, or transmitted) : GET. **2.** To acquire knowledge of or information about <eager to *receive* more facts> **3.** To have bestowed on oneself, as a title. **4.** To meet with : EXPERIENCE <*receive* kind treatment> **5.** To have (something) inflicted or imposed on oneself <*receive* a punishment> **6.** To bear the weight or force of : SUPPORT. **7.** To take or intercept the impact of, as a blow. **8.** To take in or hold. **9.** To admit <*receive* all ticket holders> **10.** To greet or welcome <*receive* New Year's callers> **11.** To acquire or perceive mentally : UNDERSTAND <*receive* a feeling of hostility> **12.** To regard in a specified way <a proposal well *received*><a new symphony that was poorly *received*> **13.** To listen to and formally and authoritatively acknowledge <*receive* a pledge of loyalty> —*vi.* **1.** To acquire or get something. **2.** To welcome or admit guests or visitors : HOST. **3.** To partake of the Eucharist. **4.** *Electron.* To convert incoming electromagnetic waves into visible or audible signals. **5.** *Football.* To catch or take possession of a kicked ball.

**Received Pronunciation** *n.* The pronunciation of British English that reflects the cultural and social influence of southern English speech and was at one time typical of the English spoken at the public schools and Oxford and Cambridge universities.

**Received Standard English** *n.* British English marked esp. by Received Pronunciation.

**re·ceiv·er** (rĭ-sē′vər) *n.* **1.** One that receives something. **2.** An official appointed to receive and account for money due. **3.** *Law.* One appointed by a court administrator to take into custody the property or funds of others, pending litigation. **4.** One who knowingly buys or receives stolen goods. **5.** A receptacle meant for a particular purpose. **6.** *Electron.* An apparatus, as a part of a radio, television set, or telephone, that receives incoming electromagnetic signals and converts them to perceptible forms. **7. a.** *Football.* A member of the offensive team eligible to catch a forward pass. **b.** *Baseball.* The catcher.

**re·ceiv·er·ship** (rĭ-sē′vər-shĭp′) *n. Law.* **1.** The office or functions of a receiver. **2.** The state of being held by a receiver.

**receiving blanket** *n.* A lightweight blanket for wrapping a baby.

**receiving line** *n.* A group of people who stand in line to greet guests as they arrive at a formal affair.

**re·cen·sion** (rĭ-sĕn′shən) *n.* [Lat. *recensio*, a reviewing < *recensēre*, to review : *re-*, again + *censēre*, to estimate.] **1.** A critical revision of a text incorporating the most plausible elements from varying sources. **2.** A text revised through recension.

**re·cent** (rē′sənt) *adj.* [Lat. *recens.*] **1.** Of, belonging to, or happening at a time just before the present. **2.** Modern : new. **3. Recent.** *Geol.* Of, belonging to, or designating the Holocene epoch. **—re′cen·cy, re′cent·ness** *n.* **—re′cent·ly** *adv.*

**re·cept** (rē′sĕpt′) *n.* [RE- + (PER)CEPT.] A mental image formed from what is common to a succession of perceptions.

**re·cep·ta·cle** (rĭ-sĕp′tə-kəl) *n.* [Lat. *receptaculum* < *receptare*, to receive again, freq. of *recipere*, to receive.] **1.** A container. **2.** *Bot.* The part of a flower stalk that bears and holds up the floral organs. **3.** *Elect.* A fitting connected to a power supply and equipped to take a plug.

**re·cep·tion** (rĭ-sĕp′shən) *n.* [Lat. *receptio* < *recipere*, to receive.] **1.** The act or process of receiving or of being received. **2.** A welcome, greeting, or acceptance <a warm *reception*> **3.** A social function <a state *reception*> **4.** Mental approval or acceptance <*reception* of a new book> **5.** *Electron.* **a.** The action of receiving electromagnetic signals. **b.** The quality or condition of received signals.

**re·cep·tion·ist** (rĭ-sĕp′shə-nĭst) *n.* An office employee whose chief task is to receive callers and answer the telephone.

**re·cep·tive** (rĭ-sĕp′tĭv) *adj.* **1.** Capable of or qualified for receiving. **2.** Ready or willing to receive favorably <*receptive* to our plan> **—re·cep′tive·ly** *adv.* **—re′cep·tiv′i·ty, re·cep′tive·ness** *n.*

**re·cep·tor** (rĭ-sĕp′tər) *n.* A nerve ending specialized to sense or receive stimuli.

**re·cer·ti·fi·ca·tion** (rē′sûr-tə-fĭ-kā′shən) *n.* Renewal of certification, esp. by a licensing board. **—re·cer′ti·fy** *v.* **(-fied, -fy·ing, -fies).**

**re·cess** (rē′sĕs′, rĭ-sĕs′) *n.* [Lat. *recessus* < *recedere*, to recede. —see RECEDE.] **1. a.** A temporary halt in the usual activities of an engagement, occupation, or pursuit. **b.** The period of such a halt. **2.** *often* **recesses.** A remote, secret, or isolated spot. **3. a.** An indentation or small hollow. **b.** An alcove. —*v.* **-cessed, -cess·ing, -cess·es.** —*vt.* **1.** To put in a recess. **2.** To make a recess in. **3.** To halt for a recess. —*vi.* To take a recess.

**re·ces·sion** (rĭ-sĕsh′ən) *n.* [Lat. *recessio* < *recessus*, recess.] **1.** The act of withdrawing or returning. **2.** The procession of clergy and choir members leaving church at the conclusion of a series. **3.** A temporary slowdown in economic activity.

**re-ces·sion** (rē-sĕsh′ən) *n.* The act of giving something back to a former owner.

**re·ces·sion·al** (rĭ-sĕsh′ə-nəl) *adj.* Of or relating to recession. —*n.* **1.** A hymn sung during the exit of the clergy and choir after a service. **2.** A recession from a church.

**re·ces·sive** (rĭ-sĕs′ĭv) *adj.* **1.** Tending to recede. **2.** *Genetics.* Of, relating to, or designating an allele that does not produce a phenotypic effect when heterozygous with a dominant allele. —*n. Genetics.* **1.** A recessive allele or trait. **2.** An organism having a recessive trait. **—re·ces′sive·ly** *adv.*

**re·charge** (rē-chärj′) *vt.* **-charged, -charg·ing, -charg·es.** To charge again, esp. to re-energize a storage battery. **—re′charge′** *n.* **—re·charge′a·ble** *adj.* **—re·charg′er** *n.*

**ré·chauf·fé** (rā′shō-fā′) *n.* [Fr. < *réchauffer*, to reheat : *re-*, again (< Lat.) + *chauffer*, to warm < OFr. —see CHAFE.] **1.** Warmed leftover food. **2.** Old material reprocessed.

**re·cher·ché** (rə-shĕr′shā′) *adj.* [Fr. < *rechercher*, to research < OFr. *rercercher.* —see RESEARCH.] **1.** Uncommonly rare and exquisite : CHOICE. **2.** Having contrived elegance and refinement : AFFECTED. **3.** Refined and elegant to an extreme : PRECIOUS. **4.** Overdone and ostentatious : PRETENTIOUS.

**re·cid·i·vism** (rĭ-sĭd′ə-vĭz′əm) *n.* [< Fr. *récidiviste*, recidivist < *récidiver*, to relapse < Med. Lat. *recidivare* < Lat. *recidivus*, returning < *recidere*, to fall back : *re-*, back + *cadere*, to fall.] A tendency to slip back into a previous behavior pattern, esp. a tendency to return to criminal habits and activities. **—re·cid′i·vist** *n.* **—re·cid′i·vis′tic, re·cid′i·vous** *adj.*

**rec·i·pe** (rĕs′ə-pē′) *n.* [Lat., imper. of *recipere*, to take, receive. —see RECEIVE.] **1.** Directions and a list of ingredients for preparing or making something, esp. food. **2.** A medical prescription. **3.** A formula for a desired end <a *recipe* for winning>

**re·cip·i·ence** (rĭ-sĭp′ē-əns) *also* **re·cip·i·en·cy** (-ən-sē) *n.* The capacity to receive : RECEPTIVITY.

**re·cip·i·ent** (rĭ-sĭp′ē-ənt) *adj.* [Lat. *recipiens, recipient-*, pr.part. of *recipere*, to receive.] Operating as a receiver : RECEPTIVE. —*n.* One that receives or is receptive.

**re·cip·ro·cal** (rĭ-sĭp′rə-kəl) *adj.* [Lat. *reciprocus*, alternating.] **1.** Concerning each of two or more persons or things. **2.** Interchanged, given, or owed mutually <*reciprocal* resources> **3.** Performed, experienced, or felt by both sides <*reciprocal* regard> **4.** Interchangeable : complementary. **5.** Expressing mutual action or relationship. —Used of some verbs and compound pronouns. **6.** *Math.* Of or relating to a quantity divided into 1. —*n.* **1.** Something reciprocal to something else. **2.** *Math.* The quotient of a specific quantity divided into 1; e.g., the reciprocal of 7 being 1/7 and the reciprocal of 2/3 being 3/2. **—re·cip′ro·cal′i·ty** (-kăl′ĭ-tē), **re·cip′ro·cal·ness** *n.* **—re·cip′ro·cal·ly** *adv.*

**reciprocal pronoun** *n.* A pronoun or pronominal phrase expressing mutual action or relationship.

**re·cip·ro·cate** (rĭ-sĭp′rə-kāt′) *v.* **-cat·ed, -cat·ing, -cates.** [Lat. *reciprocare, reciprocat-*, to move back and forth < *reciprocus*, alternating.] —*vt.* **1.** To give or take mutually : INTERCHANGE. **2.** To show or feel in return or response. —*vi.* **1.** To move back and forth alternately. **2.** To give and take something mutually. **3.** To make a return for something given or done. **4.** To be equivalent or complementary. **—re·cip′ro·ca′tive** *adj.* **—re·cip′ro·ca′tor** *n.*

**reciprocating engine** *n.* An engine with a crankshaft turned by linearly reciprocating pistons.

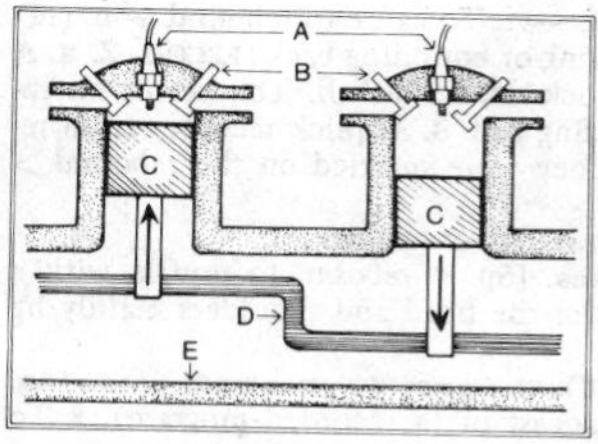


**reciprocating engine**
*A. spark plug, B. valve, C. piston, D. crank, E. crankcase*

**re·cip·ro·ca·tion** (rĭ-sĭp′rə-kā′shən) *n.* **1.** An alternating back-and-forth movement. **2.** The act or fact of reciprocating : INTERCHANGE.

**rec·i·proc·i·ty** (rĕs′ə-prŏs′ĭ-tē) *n., pl.* **-ties. 1.** A reciprocal condition or relationship. **2.** A mutual or cooperative interchange of favors or privileges, esp. the exchange of rights or privileges of trade between nations.

**re·ci·sion** (rĭ-sĭzh′ən) *n.* [OFr. < Lat. *recidere*, to cut away : *re-*, back + *caedere*, to cut.] An act of rescinding : ANNULMENT.

**re·cit·al** (rĭ-sīt′l) *n.* **1.** An act of reading or reciting in a public performance. **2.** A report or account filled with details : NARRATION. **3.** A performance of music or dance, esp. by a soloist.

**rec·i·ta·tion** (rĕs′ĭ-tā′shən) *n.* **1.** An act of reciting memorized materials in a public performance. **2. a.** Oral delivery of prepared lessons by a pupil. **b.** The class period within which this recitation takes place. **3.** The material included in a recitation.

**rec·i·ta·tive**[1] (rĕs′ĭ-tā′tĭv, rĭ-sī′tə-tĭv) *adj.* Of, relating to, or having the nature of a recital or recitation.

**rec·i·ta·tive**[2] (rĕs′ĭ-tə-tēv′) *n.* [Ital. *recitativo* < Lat. *recitare*, to recite.] *Mus.* **1.** A style used in opera and oratorio in which the text is declaimed in the rhythm of natural speech with slight melodic variation. **2.** A passage performed in recitative.

**re·ci·ta·ti·vo** (rĕs′ĭ-tə-tē′vō, rä′chē-tä-) *n., pl.* **-vi** (-vē) *or* **-vos** (-vōz) [Ital.] *Mus.* RECITATIVE[2].

**re·cite** (rĭ-sīt′) *v.* **-cit·ed, -cit·ing, -cites.** [ME *reciten* < OFr. *reciter* < Lat. *recitare*, to read out : *re-* (intensive) + *citare*, to quote < *ciēre*, to call.] —*vt.* **1.** To repeat or say aloud (something rehearsed or memorized), esp. to do so before an audience. **2.** To tell in detail. **3.** To list : enumerate. —*vi.* **1.** To give a recitation. **2.** To repeat lessons prepared or memorized. **—re·cit′er** *n.*

**reck** (rĕk) *vt.* & *vi.* **recked, reck·ing, recks.** [ME *recken* < OE *reccan.*] To pay attention to or to have caution.

**reck·less** (rĕk′lĭs) *adj.* [ME *reckeles* < OE *rēcelēas.*] **1. a.** Heedless or careless. **b.** Headstrong : rash. **2.** Having no regard for consequences : WILD <a *reckless* driver> **—reck′less·ly** *adv.* **—reck′less·ness** *n.*

**reck·on** (rĕk′ən) *v.* **-oned, -on·ing, -ons.** [ME *reknen* < OE *gerecenian*, to explain.] —*vt.* **1.** To count : compute <*reckon* costs> **2.** To regard as being : consider as. **3.** *Informal.* To think or assume

<I *reckon* I'll be there.> —*vi.* **1.** To make a calculation : FIGURE. **2.** To rely <*reckon* on the kindness of others> **—reckon with.** To come to terms or settle accounts with. **—reck'on·er** *n.*

**reck·on·ing** (rĕk'ə-nĭng) *n.* **1.** An act of counting. **2.** An itemized bill or statement of an amount to be paid. **3.** A settlement of accounts <a final *reckoning*> **4. a.** The act or process of ascertaining the position of a ship or aircraft by computation. **b.** The position so computed.

**re·claim** (rĭ-klām') *vt.* **-claimed, -claim·ing, -claims.** [ME *reclamen,* to call back < OFr. *reclamer,* to entreat < Lat. *reclamare* : *re-,* against + *clamare,* call out.] **1.** To prepare for cultivation or habitation, as by filling, irrigating, or fertilizing <*reclaim* wetlands> **2.** To obtain (usable substances) from trash or waste products. **3.** To turn from error, evil, or barbarism : REFORM. **4.** To tame (a falcon). **—re·claim'a·ble** *adj.* **—re·claim'ant, re·claim'er** *n.*

**re-claim** (rē-klām') *vt.* **-claimed, -claim·ing, -claims.** To demand the restoration or return of.

**rec·la·ma·tion** (rĕk'lə-mā'shən) *n.* [OFr. < Lat. *reclamatio,* cry of opposition < *reclamare,* to exclaim against. —see RECLAIM.] **1.** The act or process of reclaiming. **2.** A restoration, as to fertility, usefulness, or morality.

**ré·clame** (rā-kläm') *n.* [Fr., advertising < *réclamer,* to claim < OFr. *reclamer,* to exclaim against. —see RECLAIM.] **1.** Public acclaim. **2.** A liking or talent for publicity.

**rec·li·nate** (rĕk'lə-nāt') *adj.* [Lat. *reclinatus,* p.part. of *reclinare,* to recline.] *Bot.* Bent or turned downward toward the base.

**re·cline** (rĭ-klīn') *v.* **-clined, -clin·ing, -clines.** [ME *reclinen* < OFr. *recliner* < Lat. *reclinare* : *re-,* back + *clinare,* to bend.] —*vt.* To cause to assume a prone or leaning position. —*vi.* To lie back or down. **—rec'li·na'tion** (rĕk'lə-nā'shən) *n.* **—re·clin'er** *n.*

**re·cluse** (rĕk'lo͞os', rĭ-klo͞os') *n.* [ME < OFr. *reclus* < Lat. *reclusus,* p.part. of *recludere,* to close up : *re-* (intensive) + *claudere,* to close.] One who retreats from the world to live in solitude. —*adj.* Withdrawn from the world : SOLITARY.

**re·clu·sion** (rĭ-klo͞o'zhən) *n.* **1.** The state of being a recluse. **2.** The state of being in solitary confinement.

**re·clu·sive** (rĭ-klo͞o'sĭv, -zĭv) *adj.* **1.** Seeking or preferring seclusion or isolation. **2.** Furnishing seclusion <a *reclusive* cabin>

**rec·og·ni·tion** (rĕk'əg-nĭsh'ən) *n.* [Lat. *recognitio* < *recognoscere,* to recognize.] **1.** An act of recognizing or the state of being recognized. **2.** Awareness that something perceived has been perceived before. **3.** An acknowledgment, as of a claim <*recognition* of the new government's legitimacy> **4.** Attention or favorable notice <*recognition* for our contribution> **—re·cog'ni·to'ry** (rĭ-kŏg'nə-tôr'ē, -tōr'ē), **re·cog'ni·tive** *adj.*

**re·cog·ni·zance** (rĭ-kŏg'nĭ-zəns, -kŏn'ĭ-) *n.* [ME < OFr. *reconoissance* < *reconoistre,* to recognize.] **1.** *Law.* **a.** An obligation of record entered into before a court or magistrate with the condition to perform a particular act, as to make a court appearance. **b.** A sum of money pledged to guarantee the performance of such an act. **2.** A recognition. **3.** *Archaic.* A pledge : token. **—re·cog'ni·zant** *adj.*

**rec·og·nize** (rĕk'əg-nīz') *vt.* **-nized, -niz·ing, -niz·es.** [OFr. *reconoistre, reconoiss-* < Lat. *recognoscere* : *re-,* again + *cognoscere,* to know.] **1.** To know to be something that has been perceived before <*recognize* a face> **2.** To identify or know from previous experience or knowledge <*recognize* jealousy> **3.** To perceive or acknowledge the validity or reality of <*recognize* a request> **4.** To acknowledge as a speaker. **5.** To acknowledge or accept the national status of as a new government. **6.** To acknowledge, sanction, or be grateful for <*recognize* services rendered> **7.** To admit the acquaintance of, as by salutation <*recognize* our new neighbor> **8.** *Law.* To enter into a recognizance. **—rec'og·niz'a·ble** *adj.* **—rec'og·niz'a·bly** *adv.* **—rec'og·niz'er** *n.*

**re·coil** (rĭ-koil') *vi.* **-coiled, -coil·ing, -coils.** [ME *recoilen* < OFr. *reculer* : *re-,* back (< Lat.) + *cul,* buttocks < Lat. *culus.*] **1.** To kick or spring back, as upon firing. **2.** To shrink back in fear or loathing. **3.** To fall back : RETURN. —*n.* (*also* rē'koil'). **1.** The backward action or kick of a firearm upon firing. **2.** The act or state of recoiling : REACTION. **—re·coil'er** *n.*

**rec·ol·lect** (rĕk'ə-lĕkt') *vt.* & *vi.* **-lect·ed, -lect·ing, -lects.** [Med. Lat. *recolligere, recollect-* < Lat., to gather up : *re-,* again + *colligere,* to collect. —see COLLECT.] To recall to one's mind or to have a memory. **—rec'ol·lec'tion** *n.* **—rec'ol·lec'tive** *adj.* **—rec'ol·lec'tive·ly** *adv.*

**re-col·lect** (rē'kə-lĕkt') *vt.* **-lect·ed, -lect·ing, -lects.** **1.** To collect again. **2.** To calm or control (oneself). **—re'-col·lec'tion** *n.*

**re·com·bi·nant** (rē-kŏm'bə-nənt) *n.* An organism in which genetic recombination has occurred. **—re·com'bi·nant** *adj.*

**recombinant DNA** *n.* DNA prepared through laboratory manipulation in which genes from one species of an organism are transplanted or spliced to another organism.

**re·com·bi·na·tion** (rē'kŏm-bə-nā'shən) *n.* Formation in offspring of genetic combinations not present in parents.

**rec·om·mend** (rĕk'ə-mĕnd') *vt.* **-mend·ed, -mend·ing, -mends.** [ME *recommenden* < Med. Lat. *recommendare* : Lat. *re-* (intensive) + Lat. *commendare,* to entrust (*com-,* together + *mandare,* to order).] **1.** To praise or commend to another as being desirable or worthy : ENDORSE <*recommended* you for the position> **2.** To make attractive or acceptable <Diligence *recommends* any person.> **3.** To commit to the charge of another : ENTRUST. **4.** To counsel or advise (that something be done). **—rec'om·mend'a·ble** *adj.* **—rec'om·mend'er** *n.*

**rec·om·men·da·tion** (rĕk'ə-mĕn-dā'shən) *n.* **1.** The act of recommending. **2.** A positive statement regarding a person's character or qualifications. **—rec'om·men'da·to'ry** (-də-tôr'ē, -tōr'ē) *adj.*

**re·com·mit** (rē'kə-mĭt') *vt.* **-mit·ted, -mit·ting, -mits.** **1.** To commit again. **2.** To refer (e.g., proposed legislation) to a committee again. **—re'com·mit'ment, re'com·mit'tal** (-mĭt'l) *n.*

**rec·om·pense** (rĕk'əm-pĕns') *vt.* **-pensed, -pens·ing, -pens·es.** [ME *recompensen* < OFr. *recompenser* < LLat. *recompensare* : Lat. *re-,* back + Lat. *compensare,* to compensate. —see COMPENSATE.] **1.** To award compensation to. **2.** To award compensation for. —*n.* **1.** Amends made for something, as damage or loss. **2.** Payment in return for something given or done, as services.

**re·com·pose** (rē'kəm-pōz') *vt.* **-posed, -pos·ing, -pos·es.** **1.** To compose again : REARRANGE. **2.** To restore to composure : CALM. **—re'com·po·si'tion** (rē'kŏm-pə-zĭsh'ən) *n.*

**re·con** (rē'kŏn') *n.* [REC(OMBINATION) + -ON[1].] The smallest genetic unit capable of recombination.

**rec·on·cil·a·ble** (rĕk'ən-sī'lə-bəl, rĕk'ən-sī'-) *adj.* Capable of or qualified for reconciliation. **—rec'on·cil'a·bil'i·ty, rec'on·cil'a·ble·ness** *n.* **—rec'on·cil'a·bly** *adv.*

**rec·on·cile** (rĕk'ən-sīl') *vt.* **-ciled, -cil·ing, -ciles.** [ME *reconcilen* < OFr. *reconcilier* < Lat. *reconciliare* : *re-,* back + *conciliare,* to conciliate < *concilium,* meeting.] **1.** To re-establish friendship between. **2.** To settle or resolve, as a dispute. **3.** To bring (oneself) to accept. **4.** To make consistent or compatible <*reconcile* their opposing views> **—rec'on·cile'ment, rec'on·cil'i·a'tion** (-sĭl'ē-ā'shən) *n.* **—rec'on·cil'er** *n.* **—rec'on·cil'i·a·to'ry** (-sĭl'ē-ə-tôr'ē, -tōr'ē) *adj.*

☆ ***syns:*** RECONCILE, CONCILIATE, REUNITE *v. core meaning* : to re-establish friendship between <two siblings who were *reconciled* after their long separation>

**rec·on·dite** (rĕk'ən-dīt', rĭ-kŏn'dīt') *adj.* [Lat. *reconditus,* p.part. of *recondere,* to put away : *re-,* back + *condere,* to hide (*com-,* together + *dare,* to give).] **1.** Not easily understood : ABSTRUSE. **2.** Concerned with or treating the abstruse or obscure <*recondite* research> **3.** Concealed : hidden. **—rec'on·dite'ly** *adv.* **—rec'on·dite'ness** *n.*

**re·con·di·tion** (rē'kən-dĭsh'ən) *vt.* **-tioned, -tion·ing, -tions.** To restore by repairing, renovating, or rebuilding.

**re·con·firm** (rē'kən-fûrm') *vt.* **-firmed, -firm·ing, -firms.** To confirm again, esp. to support or establish more firmly. **—re'con·fir·ma'tion** (rē'kŏn-fər-mā'shən) *n.*

**re·con·nais·sance** *also* **re·con·nois·sance** (rĭ-kŏn'ə-səns, -zəns) *n.* [Fr. < OFr. *reconoissance* < *reconnoistre,* to recognize < Lat. *recognoscere.* —see RECOGNIZE.] Inspection or exploration of an area, esp. one made to obtain military intelligence.

**re·con·noi·ter** (rē'kə-noi'tər, rĕk'ə-) *v.* **-tered, -ter·ing, -ters.** [Obs. Fr. *reconnoître* < OFr. *reconnoistre,* to recognize. —see RECONNAISSANCE.] —*vt.* To make a preliminary inspection of. —*vi.* To make a reconnaissance. **—re'con·noi'ter·er** *n.*

**re·con·sid·er** (rē'kən-sĭd'ər) *v.* **-ered, -er·ing, -ers.** —*vt.* **1.** To consider again, esp. with intent to modify an earlier decision. **2.** To take up for reconsideration, as a matter already acted on by a legislature. —*vi.* To consider again. **—re'con·sid'er·a'tion** *n.*

☆ ***syns:*** RECONSIDER, RE-EXAMINE, RETHINK, REVIEW, THINK (over) *v. core meaning* : to consider again, esp. with the possibility of change <*reconsidered* changing careers>

**re·con·struct** (rē'kən-strŭkt') *vt.* **-struct·ed, -struct·ing, -structs.** To construct again.

**re·con·struc·tion** (rē'kən-strŭk'shən) *n.* **1.** The act or result of reconstructing. **2. Reconstruction.** The period (1865–77) during which the Federal government controlled the states of the former Southern Confederacy before they were readmitted to the Union. **—re'con·struc'tive** *adj.*

**re·con·vert** (rē'kən-vûrt') *v.* **-vert·ed, -vert·ing, -verts.** —*vt.* To cause to undergo conversion to a former state. —*vi.* To undergo conversion to a former state. **—re'con·ver'sion** (-vûr'zhən, -shən) *n.*

**re·con·vey** (rē'kən-vā') *vt.* **-veyed, -vey·ing, -veys.** To convey back to a former owner or place. **—re'con·vey'ance** *n.*

**re·cord** (rĭ-kôrd') *v.* **-cord·ed, -cord·ing, -cords.** [ME *recorden* < OFr. *recorder* < Lat. *recordari,* to remember : *re-,* again + *cor,* heart.] —*vt.* **1.** To set down for preservation in writing or other permanent form. **2.** To register : indicate <The registrar *records* all students' grades.> **3.** To register (sound) in permanent form by electrical or mechanical means for reproduction. —*vi.* To record something. —*n.* **rec·ord** (rĕk'ərd). **1. a.** An account, as of information, set down esp. in writing as a way of preserving knowledge. **b.** Something on which such an account is made. **c.** Something that records <a *record* of land transactions> **2.** Information or data on a specific subject collected and preserved <the fastest mile on *record*> **3.** Known history of performance or achievement <your employment *record*> **4.** The best performance known, as in a sport <the world *record* in the shot put> **5.** *Law.* **a.** An account officially written and kept as evidence or testimony. **b.** An account of judicial or legislative proceedings written and kept as evidence. **c.** The documents or volumes holding such evidence. **6. a.** A disk designed for playing on a phonograph. **b.** Something, as magnetic tape, on which sound or visual images have been recorded. **—off the record.** Not for publication <an interview that was *off the record*> **—on record.** Known to have stated or taken a certain position <went *on record* as opposing the new legislation>

**re·cord·er** (rĭ-kôr′dər) *n.* **1.** One that records, as a tape recorder. **2.** A judge with criminal jurisdiction in a city. **3.** A flute with eight finger holes and a whistlelike mouthpiece.

**re·cord·ing** (rĭ-kôr′dĭng) *n.* **1.** Something on which sound or visual images have been recorded. **2.** A recorded sound or picture.

**re·count** (rĭ-kount′) *vt.* **-count·ed, -count·ing, -counts.** [ME *recounten* < OFr. *reconter* : *re-*, again + *conter*, relate. —see COUNT[1].] **1.** To narrate the facts or details of. **2.** To enumerate. **—re·count′al** *n.*

**re-count** (rē-kount′) *vt.* **-count·ed, -count·ing, -counts.** To count again. —*n.* (*also* rē′kount′). An additional count, esp. a second count of election votes.

**re·coup** (rĭ-ko͞op′) *v.* **-couped, -coup·ing, -coups.** [ME *recoupen* < OFr. *recouper*, to cut back : *re-*, back (< Lat.) + *couper*, to cut < *coup*, blow. —see COUP.] —*vt.* **1.** To receive an equivalent for : make up for <*recoup* our losses> **2.** To return as an equivalent for : REIMBURSE. **3.** *Law.* To deduct or withhold (part of something due) for an equitable reason. —*vi.* To regain an earlier favorable position. —*n.* The act of recouping. **—re·coup′a·ble** *adj.* **—re·coup′ment** *n.*

**re·course** (rē′kôrs′, -kōrs′, rĭ-kôrs′, -kōrs′) *n.* [ME *recours* < OFr. < Lat. *recursus*, a running back < *recurrere*, to run back : *re-*, back + *currere*, to run.] **1.** An act or instance of turning to one for help or security <have *recourse* to the police> **2.** A source of help or security <My only *recourse* was the the court.> **3.** *Law.* The right to demand payment from the endorser of a commercial paper when the first party liable fails to pay.

**re·cov·er** (rĭ-kŭv′ər) *v.* **-ered, -er·ing, -ers.** [ME *recoveren* < OFr. *recoverer* < Lat. *recuperare.* —see RECUPERATE.] —*vt.* **1.** To get back : REGAIN. **2.** To restore (oneself) to a normal state. **3.** To make up for. —*vi.* **1.** To regain a normal or usual condition, as of health. **2.** To win a favorable judgment in a lawsuit. **—re·cov′er·a·ble** *adj.* **—re·cov′er·er** *n.*

☆ ***syns:*** **1.** RECOVER, RECOUP, REGAIN, RETRIEVE *v. core meaning* : to get back <hoped to *recover* the stolen car> **2.** RECOVER, MEND, RALLY, RECUPERATE *v. core meaning* : to regain one's health <*recovered* from the illness>

**re-cov·er** (rē-kŭv′ər) *vt.* **-ered, -er·ing, -ers.** To cover anew.

**re·cov·er·y** (rĭ-kŭv′ə-rē) *n.*, *pl.* **-ies. 1.** An act, instance, process, or period of recovering. **2.** A return to a normal condition. **3.** Something gained or restored in recovering. **4.** The act of obtaining usable substances from unusable sources, as waste material.

**recovery room** *n.* A hospital room fitted out for the care and observation of postoperative patients.

**rec·re·ant** (rĕk′rē-ənt) *adj.* [ME < OFr., pr.part. of *recroire*, to remember < Med. Lat. *recredere* : Lat. *re-*, back + Lat. *credere*, to entrust.] **1.** Disloyal or unfaithful to a belief, duty, or cause. **2.** Lacking in courage : COWARDLY. —*n.* **1.** A disloyal person. **2.** A coward. **—rec′re·ance, rec′re·an·cy** *n.* **—rec′re·ant·ly** *adv.*

**rec·re·ate** (rĕk′rē-āt′) *v.* **-at·ed, -at·ing, -ates.** [Lat. *recreare*, *recreat-* : *re-*, anew + *creare*, to create.] —*vt.* To refresh mentally or physically. —*vi.* To take recreation. **—rec′re·a′tive** *adj.*

**re-cre·ate** (rē′krē-āt′) *vt.* **-at·ed, -at·ing, -ates.** To create anew.

**rec·re·a·tion** (rĕk′rē-ā′shən) *n.* Refreshment of one's mind or body after work through an amusing or stimulating activity : PLAY. **—rec′re·a′tion·al** *adj.*

☆ ***syns:*** RECREATION, DIVERSION, ENTERTAINMENT *n. core meaning* : activity that refreshes the mind or body after work <jogging and other forms of *recreation*> RECREATION implies something that restores one's strength, spirits, or vitality <played tennis for *recreation*> DIVERSION suggests something to take one's attention off customary affairs <went shopping for *diversion*> ENTERTAINMENT shares these meanings but esp. suggests a performance or show designed to amuse or divert <sought *entertainment* in night clubs>

**recreational vehicle** *n.* A motor vehicle, as a camper or a mobile home, used for traveling and recreational activities.

**rec·re·ment** (rĕk′rə-mənt) *n.* [Lat. *recrementum* : *re-*, back + *cernere*, to separate.] Waste matter : DROSS <*recrement* of an ore> **—rec′re·men′tal** (-mĕn′tl) *adj.*

**re·crim·i·nate** (rĭ-krĭm′ə-nāt′) *v.* **-nat·ed, -nat·ing, -nates.** [Med. Lat. *recriminare*, *recriminat-* : Lat. *re-*, against + *criminare*, to accuse < *crimen*, accusation.] —*vt.* To accuse in return. —*vi.* To oppose one accusation with another. **—re·crim′i·na′tive, re·crim′i·na·to′ry** (-nə-tôr′ē, -tōr′ē) *adj.* **—re·crim′i·na′tor** *n.*

**re·crim·i·na·tion** (rĭ-krĭm′ə-nā′shən) *n.* **1.** An act of recriminating. **2.** A countercharge.

**re·cru·desce** (rē′kro͞o-dĕs′) *vi.* **-desced, -desc·ing, -desc·es.** [Lat. *recrudescere*, to grow raw again : *re-*, again + *crudescere*, to get worse < *crudus*, raw.] To break out again after a dormant or inactive period. **—re′cru·des′cence** *n.* **—re′cru·des′cent** *adj.*

**re·cruit** (rĭ-kro͞ot′) *v.* **-cruit·ed, -cruit·ing, -cruits.** [Fr. *recruter* < obs. *recrute*, recruit < OFr. *recroistre*, to grow again < Lat. *recrescere* : *re-*, again + *crescere*, to grow.] —*vt.* **1.** To engage (persons) for duty in the armed forces. **2.** To strengthen or raise (an armed force) by enlistment. **3.** To provide with new members or employees. **4.** To enroll or seek to enroll <companies *recruiting* engineering graduates> **5.** To replenish. —*vi.* **1.** To raise a military force. **2.** To obtain replacements for or new supplies of something lost, wasted, or needed. —*n.* **1.** A new member of a military force, esp. one of the lowest rank. **2.** A new member of an organization or body. **—re·cruit′er** *n.* **—re·cruit′ment** *n.*

**rec·ta** (rĕk′tə) *n. var. pl. of* RECTUM.

**rec·tal** (rĕk′təl) *adj.* Of, relating to, or close to the rectum.

**rec·tan·gle** (rĕk′tăng′gəl) *n.* [Med. Lat. *rectangulum* : Lat. *rectus*, right + Lat. *angulus*, angle.] A parallelogram with a right angle.

**rec·tan·gu·lar** (rĕk-tăng′gyə-lər) *adj.* **1.** Shaped like a rectangle. **2.** Having right angles. **3.** Designating a geometric coordinate system with mutually perpendicular axes. **—rec·tan′gu·lar′i·ty** (-lăr′ĭ-tē) *n.* **—rec·tan′gu·lar·ly** *adv.*

**rectangular coordinate** *n. Math.* A coordinate in a rectangular Cartesian coordinate system.

**rec·ti** (rĕk′tī′) *n. pl. of* RECTUS.

**rec·ti·fi·er** (rĕk′tə-fī′ər) *n.* **1.** One that rectifies. **2.** *Elect.* A device, as a diode, that converts alternating current to direct current. **3.** A worker who blends or dilutes whiskey or other alcoholic beverages.

**rec·ti·fy** (rĕk′tə-fī′) *vt.* **-fied, -fy·ing, -fies.** [ME *rectifien* < OFr. *rectifier* < Med. Lat. *rectificare* : *rectus*, right + *facere*, to make.] **1.** To set right : CORRECT. **2.** To correct by calculation or adjustment. **3.** *Chem.* To refine or purify, esp. by distillation. **4.** *Elect.* To convert (alternating current) into direct current. **5.** To adjust (the proof of alcoholic beverages) by adding water or other liquids. **—rec′ti·fi′a·ble** *adj.* **—rec′ti·fi·ca′tion** (-fə-kā′shən) *n.*

**rec·ti·lin·e·ar** (rĕk′tə-lĭn′ē-ər) *adj.* [LLat. *rectilineus* : Lat. *rectus*, right + Lat. *linea*, line.] Moving in, made up of, bounded by, or marked by a straight line or lines. **—rec′ti·lin′e·ar·ly** *adv.*

**rec·ti·tude** (rĕk′tĭ-to͞od′, -tyo͞od′) *n.* [ME < OFr. < LLat. *rectitudo* < Lat. *rectus*, straight.] **1.** Moral integrity. **2.** The quality or state of being correct in intellectual judgment. **3.** Straightness.

**rec·to** (rĕk′tō) *n.*, *pl.* **-tos.** [Lat. *recto* (*folio*), (the page) being right.] The right-hand page of a book or the front side of a leaf.

**rec·tor** (rĕk′tər) *n.* [Lat., director < *regere*, to rule.] **1.** A member of the clergy in charge of a parish in the Protestant Episcopal Church. **2.** A member of the Anglican clergy who has charge of a parish and owns the tithes from it. **3.** A Roman Catholic priest appointed to be managerial as well as spiritual head of a church or other institution, such as a seminary or university. **4.** The principal of certain schools, colleges, and universities. **—rec′tor·ate** (-ĭt) *n.* **—rec·to′ri·al** (rĕk-tôr′ē-əl, -tōr′-) *adj.*

**rec·to·ry** (rĕk′tə-rē) *n.*, *pl.* **-ries. 1.** The house in which a parish priest or minister lives. **2. a.** An Anglican rector's dwelling. **b.** An Anglican rector's office and benefice.

**rec·trix** (rĕk′trĭks) *n.*, *pl.* **rec·tri·ces** (rĕk′trə-sēz′, rĕk-trī′sēz) [NLat. < Lat., fem. of *rector*, director < *regere*, to rule.] One of the stiff main feathers of a bird's tail.

**rec·tum** (rĕk′təm) *n.*, *pl.* **-tums** *or* **-ta** (-tə) [NLat. *rectum* (*intestinum*), straight (intestine).] The portion of the large intestine from the sigmoid flexure to the anal canal.

**rec·tus** (rĕk′təs) *n.*, *pl.* **-ti** (-tī′) [NLat. < Lat., straight.] Any of various straight muscles, as of the abdomen, eye, neck, and thigh.

**re·cum·bent** (rĭ-kŭm′bənt) *adj.* [Lat. *recumbens*, pr.part. of *recumbere*, to lie down : *re-*, back + *cumbere*, to lie.] **1.** Lying down : RECLINING <a *recumbent* statue> **2.** Resting : idle. **3.** *Biol.* Resting upon the surface from which it arises <a *recumbent* organ> **—re·cum′bence, re·cum′ben·cy** *n.* **—re·cum′bent·ly** *adv.*

**re·cu·per·ate** (rĭ-ko͞o′pə-rāt′, -kyo͞o′-) *v.* **-at·ed, -at·ing, -ates.** [Lat. *recuperare*, *recuperat-* : *re-*, back + *capere*, to take.] —*vi.* **1.** To return to strength or health : RECOVER. **2.** To recover from financial loss. —*vt.* **1.** To restore to strength or health. **2.** To regain. **—re·cu′per·a′tion** *n.* **—re·cu′per·a′tive** (-pə-rā′tĭv, -pər-ə-tĭv, -kyo͞o′-), **re·cu′per·a·to′ry** (-pər-ə-tôr′ē, -tōr′ē) *adj.*

**re·cur** (rĭ-kûr′) *vi.* **-curred, -cur·ring, -curs.** [Lat. *recurrere* : *re-*, back + *currere*, to run.] **1.** To occur or come up again or repeatedly. **2.** To return to one's attention or memory. **3.** To return in thought or speech. **4.** To have recourse. **—re·cur′rence** *n.*

**re·cur·rent** (rĭ-kûr′ənt, -kŭr′-) *adj.* **1.** Occurring or appearing again or repeatedly. **2.** *Anat.* Running in a reverse direction. —Used of arteries and nerves. **—re·cur′rent·ly** *adv.*

**recurrent fever** *n.* Relapsing fever.

**recurring decimal** *n. Math.* A repeating decimal.

**re·cur·sion** (rĭ-kûr′zhən) *n.* [LLat. *recursio*, a running back < Lat. *recurrere*, to run back, recur.] **1.** A mathematical expression, as a polynomial, each term of which is determined by application of a formula to preceding terms. **2.** A formula that generates the successive terms of a recursion. **—re·cur'sive** *adj.*

**re·cur·vate** (rĭ-kûr'văt', -vĭt) *adj.* [Lat. *recurvatus,* p.part. of *recurvare,* to recurve.] Curved or bent backward.

**re·curve** (rē-kûrv') *vt.* & *vi.* **-curved, -curv·ing, -curves.** [Lat. *recurvare* : Lat. *re-,* back + *curvare,* to curve.] To curve backward or downward or to become curved backward or downward. **—re'cur·va'tion** (rē'kûr-vā'shən) *n.*

**rec·u·sant** (rĕk'yə-zənt, rĭ-kyōō'-) *n.* [Lat. *recusans,* pr.part. *recusare,* to refuse : *re-,* against + *causa,* cause.] **1.** A Roman Catholic who refused to attend the services of the Church of England between the reigns of Henry VIII and George II. **2.** A nonconformist : dissenter. **—rec'u·san·cy** *n.* **—rec'u·sant** *adj.*

**re·cy·cle** (rē-sī'kəl) *vt.* **-cled, -cling, -cles. 1.** To put or pass through a cycle again, as for additional processing or treatment. **2.** To begin a separate cycle in. **3. a.** To recover useful materials from (garbage or waste). **b.** To extract and reuse (useful substances recovered from waste). **4. a.** To use again, esp. to reprocess so as to use again <*recycle* glass bottles> **b.** To recondition and adapt to a new use or function <*recycling* old factories into condominiums>

**red** (rĕd) *n.* [ME < OE *rēad.*] **1.** Any of a group of colors that may vary in lightness and saturation and whose hue is similar to that of blood; the hue of the long-wave end of the spectrum; one of the additive or light primaries; one of the psychological primary hues evoked in the normal observer by the long-wave end of the spectrum. **2.** A pigment or dye with a red hue. **3.** Something that has a red hue <was awarded the *red* for second place> **4. a.** *often* **Red.** A Communist. **b.** A revolutionary activist. *—adj.* **red·der, red·dest. 1.** With a color similar to that of blood. **2.** Reddish in color or with portions that are reddish in color <a *red* horse><a *red* leaf> **3.** With a coppery skin tone. **4.** With a ruddy or flushed complexion <*red* with rage> **—in the red.** Operating at a financial loss or being in debt. **—see red.** To be or become enraged. **—red'ly** *adv.* **—red'ness** *n.*

**re·dact** (rĭ-dăkt') *vt.* **-dact·ed, -dact·ing, -dacts.** [Lat. *redigere, redact-,* to drive back : *re-, red-,* back + *agere,* to drive.] **1.** To write in correct form or formulate (e.g., a proclamation). **2.** To make ready (a document) for publication : EDIT. **—re·dac'tion** *n.* **—re·dac'tor** (-dăk'tər, -tôr') *n.*

**red algae** *pl.n.* Any of the algae of the division Rhodophyta that are red or reddish in color.

**red·bait** (rĕd'bāt') *vt.* **-bait·ed, -bait·ing, -baits.** To attack or denounce (a person or group) as Communist. **—red'bait'er** *n.* **—red'bait'ing** *n.*

**red·bird** (rĕd'bûrd') *n.* A bird, as the cardinal or the scarlet tanager, with red plumage.

**red blood cell** *n.* An erythrocyte.

**red-blood·ed** (rĕd'blŭd'ĭd) *adj.* Strong and virile.

**red·breast** (rĕd'brĕst') *n.* **1.** A bird with a red breast, as the robin. **2.** A freshwater sunfish, *Lepomis auritus* of eastern North America.

**red·brick** (rĕd'brĭk') *adj.* [So-called because many of the buildings of such universities were built of red bricks.] Of, relating to, or being the British universities other than Oxford and Cambridge.

**red·bud** (rĕd'bŭd') *n.* Any of several shrubs or small trees of the genus *Cercis,* bearing pinkish blossoms that appear before the leaves.

**red·cap** (rĕd'kăp') *n.* A porter, usu. in a railroad station.

**red carpet** *n.* A carpet spread out for important visitors. **—roll out the red carpet.** To welcome with great hospitality or ceremony.

**red cedar** *n.* **1.** An evergreen tree, *Juniperus virginiana* of eastern North America. **2.** A tall evergreen tree, *Thuja plicata* of western North America. **3.** The reddish, aromatic, durable wood of the red cedar or similar trees.

**red cent** *n. Informal.* A penny <not a *red cent* in my wallet>

**red clover** *n.* A Eurasian plant, *Trifolium pratense,* widely naturalized in North America and often planted as a forage or cover crop, bearing leaflets in groups of three and globular heads of fragrant rose-purple flowers.

**red·coat** (rĕd'kōt') *n.* A British soldier during the American Revolution and the War of 1812.

**Red Cross** *n.* The emblem of the International Red Cross, a Geneva cross or a red Greek cross on a white background.

**†redd** (rĕd) *vt.* **redd·ed** *or* **redd, redd·ing, redds.** [ME *redden,* to clear, prob. alteration of *ridden,* to rid< ON *ryðja.*] *Regional.* To put in order : TIDY <*redded* up the front room>

**red deer** *n.* A common European deer, *Cervus elaphus,* with a reddish-brown coat and many-branched antlers.

**red·den** (rĕd'n) *vt.* & *vi.* **-dened, -den·ing, -dens.** To make or become red <eyes *reddened* by lack of sleep>

**red·dish** (rĕd'ĭsh) *adj.* Somewhat red. **—red'dish·ness** *n.*

**red-dog** (rĕd'dôg', -dŏg') *v.* **-dogged, -dog·ging, -dogs.** *Football.* *—vi.* To charge across the line of scrimmage in an effort to overpower the opposing quarterback before he can throw a forward pass. —Used of linebackers and occas. of defensive backs. *—vt.* To charge (the passer). **—red'-dog'** *n.*

**red drum** *n.* A food fish, *Sciaenops ocellata* of southern Atlantic coastal waters.

**†rede** (rēd) *vt.* **red·ed, red·ing, redes.** [ME *reden* < OE *rǣdan.*] *Regional.* **1.** To give advice to : COUNSEL. **2.** To explain : interpret. *—n.* **1.** *Regional.* Advice or counsel. **2.** *Archaic.* A narration.

**re·dec·o·rate** (rē-dĕk'ə-rāt') *v.* **-rat·ed, -rat·ing, -rates.** *—vt.* To change the appearance or décor of : REFURBISH. *—vi.* To change a decorative scheme. **—re·dec'o·ra'tion** *n.* **—re·dec'o·ra'tor** *n.*

**re·deem** (rĭ-dēm') *vt.* **-deemed, -deem·ing, -deems.** [ME *redemen* < Lat. *redimere* : *re-, red-,* back + *emere,* to buy.] **1.** To regain possession of by payment of a specified amount. **2.** To pay off, as a promissory note. **3.** To turn in (e.g., coupons or trading stamps) and receive something in exchange. **4.** To fulfill (e.g., a pledge). **5.** To convert into cash <*redeem* bond coupons> **6.** To rescue or ransom : LIBERATE. **7.** To save from sin and its consequences. **8.** To make up for. **—re·deem'a·ble** *adj.*

**re·deem·er** (rĭ-dē'mər) *n.* **1.** One who redeems. **2. Redeemer.** Christ.

**re·de·liv·er** (rē'dĭ-lĭv'ər) *vt.* **-ered, -er·ing, -ers. 1.** To deliver again. **2.** To deliver in return : GIVE BACK.

**re·demp·tion** (rĭ-dĕmp'shən) *n.* [ME *redempcioun* < OFr. *redemption* < Lat. *redemptio* < *redimere,* to redeem.] **1.** An act of redeeming or the state of being redeemed. **2.** Recovery of something pawned or mortgaged. **3.** Payment of an obligation, as a government's payment of the value of its bonds. **4.** Deliverance upon payment of ransom : RESCUE. **5.** Salvation from sin through Christ's sacrifice. **—re·demp'tion·al, re·demp'tive, re·demp'to·ry** (-tə-rē) *adj.*

**re·demp·tion·er** (rĭ-dĕmp'shə-nər) *n.* A colonial emigrant from Europe to America who paid for the voyage by serving as an indentured servant for a specified period.

**Re·demp·tor·ist** (rĭ-dĕmp'tər-ĭst) *n.* A member of the Congregation of the Most Holy Redeemer, a Roman Catholic order founded by St. Alphonsus de Liguori in 1732.

**re·de·ploy** (rē'dĭ-ploi') *vt.* **-ployed, -ploy·ing, -ploys.** To move (military forces) from one zone to another. **—re'de·ploy'ment** *n.*

**re·de·sign** (rē'dĭ-zīn') *vt.* **-signed, -sign·ing, -signs.** To alter the appearance or function of. **—re'de·sign'** *n.*

**re·de·vel·op** (rē'dĭ-vĕl'əp) *v.* **-oped, -op·ing, -ops.** *—vt.* **1.** To develop (something) again. **2.** To tone or intensify (a developed photographic negative) by a second developing process. **3.** To restore to a better condition <*redeveloping* the inner city> *—vi.* To develop again. **—re'de·vel'op·er** *n.* **—re'de·vel'op·ment** *n.*

**red·eye** (rĕd'ī') *n.* **1.** *Informal.* A danger signal on a railroad. **2.** *Slang.* Inferior whiskey. **3.** A red-eyed fish, as the rock bass. **4.** *Slang.* A late-night or overnight flight <caught the *redeye* from Los Angeles to New York>

**redeye gravy** *n.* Gravy made from the juices of a cooked ham.

**red fir** *n.* **1.** An evergreen tree, *Abies magnifica* of California and Oregon, having reddish wood valued as timber. **2.** The wood of the red fir or similar trees.

**red fire** *n.* A combustible compound, as salts of lithium or strontium, that burns bright red and is used in fireworks and flares.

**red·fish** (rĕd'fĭsh') *n., pl.* **redfish** *or* **-fish·es.** A reddish fish.

**red fox** *n.* Any of several foxes of the genus *Vulpes,* with reddish fur, esp. *V. fulva* of North America.

**Red Guard** *n.* A member of a radical political group with Maoist tendencies.

**red gum**[1] *n.* An Australian tree of the genus *Eucalyptus,* esp. *E. rostrata,* naturalized in California.

**red gum**[2] *n.* Strophulus.

**red-hand·ed** (rĕd'hăn'dĭd) *adv.* & *adj.* In the act of doing something wrong <caught *red-handed* with the stolen car> **—red'hand'ed·ly** *adv.*

**red·head** (rĕd'hĕd') *n.* **1.** A person with red hair. **2.** A North American duck, *Aythya americana,* the male of which has black and gray plumage and a reddish head.

**redhead**
*18–23 inches long*

**red heat** *n.* **1.** The temperature of a red-hot substance. **2.** The physical condition of a red-hot substance.

**red herring** *n.* [From its use to distract hunting dogs from the trail.] **1.** A smoked herring with a reddish color. **2.** Something that detracts attention from the matter or issue at hand.

**red-hot** (rĕd′hŏt′) *adj.* **1.** Glowing hot. **2.** Heated, as with excitement, anger, or enthusiasm <a *red-hot* confrontation with the press> **3.** Very recent : NEW <*red-hot* news bulletins> —*n.* **1.** A hot dog. **2.** A small usu. round red candy with a strong cinnamon flavor.
**re·di·a** (rē′dē-ə) *n., pl.* **-di·ae** (-dē-ē′) [NLat., after Francesco *Redi*, 17th-cent. Italian naturalist.] A larva of certain trematodes that is produced within the sporocyst and that can generate additional rediae or cercariae.
**Red Indian** *n.* A North American Indian.
**red·in·gote** (rĕd′ĭng-gōt′) *n.* [Fr. < E. *riding coat.*] **1.** A man's long double-breasted topcoat with full skirt. **2.** A woman's full-length unlined coat or dress open down the front to display a dress or underdress.
**red ink** *n.* [From the use of red ink to record debits and losses in financial records.] **1.** A business financial loss. **2.** The state of showing a deficit in business operations.
**re·di·rect** (rē′dĭ-rĕkt′, -dī-) *vt.* **-rect·ed, -rect·ing, -rects.** To change the direction or course of. **—re′di·rec′tion** *n.*
**re·dis·count** (rē-dĭs′kount′) *vt.* **-count·ed, -count·ing, -counts.** To discount again. —*n.* **1.** The act of rediscounting. **2.** *often* **rediscounts.** Commercial paper discounted a second time.
**re·dis·trib·ute** (rē′dĭ-strĭb′yŏŏt) *vt.* **-ut·ed, -ut·ing, -utes.** To distribute again in a different way : REALLOCATE. **—re·dis′tri·bu′tion** *n.* **—re′dis·trib′u·tive** *adj.*
**re·dis·trict** (rē-dĭs′trĭkt) *vt.* **-trict·ed, -trict·ing, -tricts.** To divide again into districts, esp. to draw up new boundaries for administrative or election districts.
**red lead** *n.* A bright-red powder, $Pb_3O_4$, used in paints, glass, pottery, and pipe-joint packing.
**red-let·ter** (rĕd′lĕt′ər) *adj.* [From the practice of marking in red the holy days in church calendars.] Memorably happy <a *red-letter* occasion>
**red light** *n.* **1.** A red traffic or danger signal indicating stop. **2.** A sign of caution : DETERRENT.
**red-light district** (rĕd′līt′) *n.* An area with many brothels.
**red·line** (rĕd′līn′) *v.* **-lined, -lin·ing, -lines.** —*vi.* To refuse home mortgages or insurance to areas or neighborhoods judged to be poor financial risks. —*vt.* **1.** To discriminate against by refusing to grant loans, mortgages, or insurance to. **2.** To remove from operational status due to the presence of mechanical defects <*redlined* three fighter aircraft> **—red′lin′ing** *n.*
**red maple** *n.* A medium-sized American maple, *Acer rubrum*, with reddish twigs and buds.
**red mulberry** *n.* A tree, *Morus rubra* of eastern and central North America, with irregularly lobed leaves and edible, blackberrylike fruit.
**red mullet** *n.* A fish of the family Mullidae : GOATFISH.
**re·do** (rē-dŏŏ′) *vt.* **-did** (-dĭd′), **-done** (-dŭn′), **-do·ing, -does** (-dŭz′). **1.** To do over again. **2.** To redecorate.
**red ocher** *n.* **1.** A natural red mixture of clay and iron oxide. **2.** A refined form of red ocher used as pigment.
**red·o·lent** (rĕd′ə-lənt) *adj.* [ME < OFr. < Lat. *redolens*, pr. part. of *redolēre*, to smell : *re-, red-* (intensive) + *olēre*, to smell.] **1.** Having or emitting fragrance : AROMATIC. **2.** Suggestive : reminiscent <a corporate takeover *redolent* of Machiavellian maneuvers> **—red′o·lence, red′o·len·cy** *n.* **—red′o·lent·ly** *adv.*
**red osier** *n.* A North American shrub, *Cornus stolonifera*, often growing in dense clumps and with red branches, white flowers, and bluish-white, berrylike fruit.
**re·dou·ble** (rē-dŭb′əl) *v.* **-bled, -bling, -bles.** —*vt.* **1.** To double. **2.** To repeat. **3.** To double the doubling bid of (an opponent) in bridge. —*vi.* **1.** To become twice as great. **2.** To double a double in bridge.
**re·doubt** (rĭ-dout′) *n.* [OFr. *redoute* < OItal. *ridotto* < Med. Lat. *reductus*, concealed place < Lat. *reducere*, to withdraw : *re-*, back + *ducere*, to lead.] **1.** A small, often temporary defensive fortification. **2.** A reinforcing breastwork or earthwork within a permanent rampart. **3.** A protected place of defense or refuge.
**re·doubt·a·ble** (rĭ-dou′tə-bəl) *adj.* [ME < OFr. *redoutable* < *redouter*, to dread : *re-* (intensive < Lat.) + *douter*, to doubt < Lat. *dubitare.*] **1.** Arousing awe or fear : FORMIDABLE. **2.** Worthy of respect or honor. **—re·doubt′a·bly** *adv.*
**re·dound** (rĭ-dound′) *vi.* **-dound·ed, -dound·ing, -dounds.** [ME *redounden*, to abound < OFr. *redonder* < Lat. *redundare*, to overflow. —see REDUNDANT.] **1.** To have an effect <actions that *redound* to one's discredit> **2.** To return : recoil. **3.** To contribute : accrue.
**red·out** (rĕd′out′) *n.* A reddening of the visual field produced by blood that is forced into the head when an individual is subjected to negative gravitational force.
**re·dox** (rē′dŏks′) *n.* [RED(UCTION) + OX(IDATION).] *Chem.* Oxidation-reduction.
**red-pen·cil** (rĕd′pĕn′səl) *vt.* **-ciled, -cil·ing, -cils** *also* **-cilled, -cil·ling, -cils.** To censor, cut, revise, or correct with or as if with a red pencil.
**red pepper** *n.* **1.** The pungent podlike fruit of any of several varieties of the pepper plant, *Capsicum frutescens.* **2.** Cayenne pepper.
**red pine** *n.* An evergreen timber tree, *Pinus resinosa* of northeastern North America.
**red·poll** (rĕd′pōl′) *n.* Any of several finches of the genus *Acanthis*, bearing brownish plumage and a red crown.
**Red Poll** *n.* Any of a reddish, hornless cattle bred in England and raised for dairy and meat products.
**red puccoon** *n.* The bloodroot.
**re·dress** (rĭ-drĕs′) *vt.* **-dressed, -dress·ing, -dress·es.** [ME *redressen* < OFr. *redresser* : *re-*, back (< Lat.) + *dresser*, to arrange. —see DRESS.] **1.** To remedy or rectify. **2.** To make amends to. **3.** To make amends for. **4.** To adjust (e.g., a balance). —*n.* (*also* rē′drĕs). **1.** Satisfaction or amends for wrong done. **2.** Correction or reformation. **—re·dress′er, re·dres′sor** *n.*
▲ word history: The meanings of *dress* and *redress* have diverged so far in Modern English that it is difficult to see that they were once related in the same way as *arrange* and *rearrange*: the prefix *re-* means "again" in each case. Both *dress* and *redress* were medieval borrowings from Old French, in which the verb *dresser* meant merely "to arrange." In Modern English *dress* has become narrowed to meaning primarily "to clothe," although it still has the sense "to prepare" or "to arrange" in reference to certain specific items. *Redress* originally meant "to arrange or set in order again," but in English it now means "to set right" in the sense "to rectify" and "to make amends or restitution for an injury."
**red·root** (rĕd′rŏŏt′) *n.* **1.** A bog plant, *Lachnanthes tinctoria* of eastern North America, with red roots and woolly yellow flowers. **2.** PIGWEED 2.
**red salmon** *n.* The sockeye salmon.
**red·shank** (rĕd′shăngk′) *n.* An Old World wading bird, *Tringa totanus*, with long red legs.
**red shift** *n.* **1.** An apparent increase in the wavelength of radiation emitted by a receding celestial body as a result of the Doppler effect. **2.** An increase in wavelength resulting from loss of energy by radiation moving against a gravitational field.
**red·shirt** (rĕd′shûrt′) *n.* [From the red jersey worn by such athletes to distinguish them from the regular players.] A college athlete who does not take part in varsity competition for one year so as to lengthen his or her four-year period of eligibility. **—red′shirt′** *v.* **(-shirt·ed, -shirt·ing, -shirts).**
**red snapper** *n.* Any of several marine food fishes of the genus *Lutjanus* of tropical and semitropical waters, with reddish bodies.
**red spider** *n.* Any of various small red mites of the family Tetranychidae that feed on vegetation.
**red squill** *n.* **1.** SEA ONION 1. **2.** A powder prepared from the bulbs of the red squill and used as a rat poison.
**red squirrel** *n.* A North American squirrel, *Tamiasciurus hudsonicus*, with reddish or tawny fur.
**red·start** (rĕd′stärt′) *n.* [RED + obs. *start*, tail < ME *stert* < OE *steort.*] **1.** A small North American bird, *Setophaga ruticilla*, the male of which has black plumage with orange patches on the wings and tail. **2.** A European bird, *Phoenicurus phoenicurus*, with grayish plumage and a rust-red breast and tail.
**red tape** *n.* [From its former use in tying British official documents.] Official forms and bureaucratic procedures.
**red tide** *n.* Ocean waters colored by the increase of one-celled, red, plantlike animals in numbers great enough to kill fish.
**red·top** (rĕd′tŏp′) *n.* A widely cultivated grass, *Agrostis alba*, native to Europe, with reddish flower clusters.
**re·duce** (rĭ-dŏŏs′, -dyŏŏs′) *v.* **-duced, -duc·ing, -duc·es.** [ME *reducen*, to bring back < Lat. *reducere* : *re-*, back + *ducere*, to lead.] —*vt.* **1.** To lessen in extent, amount, number, degree, or price. **2.** To gain control of : CONQUER. **3.** To put in order or arrange systematically. **4.** To separate into orderly components by analysis. **5.** To bring to a certain state or condition <War conditions *reduced* the populace to starvation and disease.> **6.** To make into powder, as by pounding or grinding : PULVERIZE. **7.** To thin (paint) with a solvent. **8.** *Chem.* **a.** To decrease the valence of (an atom) by adding electrons. **b.** To deoxidize. **c.** To add hydrogen to. **d.** To change to a metallic state by removing nonmetallic constituents : SMELT. **9.** *Math.* To change the form of (an expression) without changing the value. **10.** *Med.* To restore (a fractured or displaced body part) to a normal condition. —*vi.* **1.** To become diminished. **2.** To lose weight, as by dieting. **—re·duc′er** *n.* **—re·duc′i·bil′i·ty** *n.* **—re·duc′i·ble** *adj.* **—re·duc′i·bly** *adv.*
**reducing agent** *n.* An agent that chemically reduces substances.
**re·duc·tant** (rĭ-dŭk′tənt) *n.* A reducing agent.
**re·duc·tase** (rĭ-dŭk′tās′, -tāz′) *n.* [REDUCT(ION) + -ASE.] An enzyme that catalyzes biochemical reduction reactions.
**re·duc·ti·o ad ab·sur·dum** (rĭ-dŭk′tē-ō ăd əb-sûr′dəm) *n.* [Lat., reduction to absurdity.] Disproof of a proposition by demonstrating the absurdity of its inevitable conclusion.
**re·duc·tion** (rĭ-dŭk′shən) *n.* [ME *reduccion*, restoration < OFr. *reduction* < Lat. *reductio* < *reducere*, to bring back : *re-*, back + *ducere*, to lead.] **1.** An act or process of reducing. **2.** The result of reducing. **3.** The amount by which something is lessened or diminished. **4.** *Biol.* The first meiotic division, in which the chromosome

number is reduced. **5.** *Chem.* A decrease in positive valence or an increase in negative valence by the gaining of electrons. **6.** *Math.* **a.** The canceling of common factors in the numerator and denominator of a fraction. **b.** Conversion of a fraction to its decimal equivalent. **—re·duc'tion·al, re·duc'tive** *adj.*

**reduction division** *n. Biol.* **1.** REDUCTION 4. **2.** MEIOSIS 1.

**re·duc·tiv·ism** (rĭ-dŭk'tə-vĭz'əm) *n.* Minimal art.

**re·dun·dan·cy** (rĭ-dŭn'dən-sē) *n., pl.* **-cies. 1.** The state of being redundant. **2.** Excess or superfluity. **3.** Unnecessary repetition : WORDINESS. **4. a.** Duplication or repetition of elements in electronic or mechanical equipment to provide alternative functional channels in case of failure. **b.** Repetition of parts or all of a message to circumvent transmission errors.

**re·dun·dant** (rĭ-dŭn'dənt) *adj.* [Lat. *redundans,* pr.part. of *redundare,* to overflow : *re-, red-* (intensive) + *undare,* to surge < *unda,* wave.] **1.** Exceeding what is required or natural : SUPERFLUOUS. **2.** Needlessly repetitive : VERBOSE. **—re·dun'dant·ly** *adv.*

**re·du·pli·cate** (rĭ-dōō'plə-kāt', -dyōō'-) *v.* **-cat·ed, -cat·ing, -cates.** [LLat. *reduplicare, reduplicat-* : *re-,* again + *duplicare,* to duplicate.] *—vt.* **1.** To repeat over and again : REDOUBLE. **2. a.** To double (the initial syllable or all of a root word) to make an inflectional or derivational form. **b.** To form (a new word) by doubling all or part of a word. *—vi.* To be doubled. *—adj.* (rĭ-dōō'plə-kĭt, -dyōō'-). Doubled.

**re·du·pli·ca·tion** (rĭ-dōō'plĭ-kā'shən, -dyōō'-) *n.* **1.** An act of reduplicating or the state of being reduplicated. **2.** A result of reduplicating. **3. a.** A word formed by or having a reduplicated element. **b.** The additional element in a reduplicated word form. **—re·du'pli·ca'tive** *adj.* **—re·du'pli·ca'tive·ly** *adv.*

**re·du·vi·id** (rĭ-dōō'vē-ĭd, -dyōō'-) *n.* [< NLat. *Reduviidae,* family name < *Reduvius,* type genus < Lat. *reduvia,* hangnail, fragment.] Any of various hemipterous insects of the family Reduviidae that have curved beaks adapted for sucking blood.

**red·wing** (rĕd'wĭng') *n.* **1.** A North American blackbird, *Agelaius phoeniceus,* the male of which has scarlet patches on the wings. **2.** A European thrush, *Turdus iliacus,* with reddish feathers beneath the wings.

**red-winged black·bird** (rĕd'wĭngd' blăk'bûrd') *n.* REDWING 1.

**red·wood** (rĕd'wŏŏd') *n.* **1.** A very tall evergreen tree, *Sequoia sempervirens* of coastal and northern California. **2.** The soft reddish wood of the redwood tree. **3.** A reddish wood from which a red dye is derived.

**re-ech·o** (rĭ-ĕk'ō) *v.* **-oed, -o·ing, -oes.** *—vi.* To reverberate again. *—vt.* To echo back : REPEAT.

**reed** (rēd) *n.* [ME *rede* < OE *hrēod.*] **1. a.** A tall grass with hollow, jointed stalks, esp. one of the genera *Phragmites* or *Arundo.* **b.** The stalk of one of these plants. **2.** A primitive wind instrument made of a hollow reed stalk. **3.** *Mus.* **a.** A flexible strip of cane or metal set into the mouthpiece of certain musical instruments to create tone by vibrating in response to a stream of air. **b.** An instrument, as an oboe or clarinet, equipped with a reed. **4.** A narrow movable frame fitted with reed or metal strips that separate the warp threads in weaving. **5.** REEDING 1a.

**reed·bird** (rēd'bûrd') *n.* The bobolink.

**reed·buck** (rēd'bŭk') *n.* [Transl. of Afr. *rietbok.*] An African antelope of the genus *Redunca.*

**reed·ing** (rē'dĭng) *n.* **1. a.** A convex decorative molding with parallel strips similar to thin reeds. **b.** Decoration by use of reedings. **2.** Corrugations on the edge of a coin.

**reed·ling** (rēd'lĭng) *n.* A small Eurasian marsh bird, *Panurus biarmicus,* the male of which has mustachelike black markings.

**reed mace** *n.* The cattail.

**reed organ** *n.* A keyboard instrument in which free-beating reeds emit tones when currents of air act upon them.

**reed pipe** *n.* An organ pipe with a reed that vibrates and emits a tone when air is forced through it.

**reed stop** *n.* A stop on an organ made up of reed pipes with various tonal characteristics.

**re·ed·u·cate** (rē-ĕj'ə-kāt') *vt.* **-cat·ed, -cat·ing, -cates. 1.** To instruct again. **2.** To retrain (a person) to function effectively : REHABILITATE. **—re·ed'u·ca'tion** *n.*

**reed·y** (rē'dē) *adj.* **-i·er, -i·est. 1.** Full of reeds. **2.** Made of reeds. **3.** Similar to a reed. **4.** Producing a tone like that of a reed instrument. **—reed'i·ness** *n.*

**reef¹** (rēf) *n.* [MDu. *rif,* poss. < ON, ridge.] **1.** *Geol.* A strip or ridge of rocks, sand, or coral that rises to or close to the surface of a body of water. **2.** A vein of ore. **—reef'y** *adj.*

**reef²** (rēf) [ME *riff* < ON *rif,* ridge.] *Naut. —n.* A portion of a sail rolled and tied down to reduce the area exposed to the wind. *—vt.* **reefed, reef·ing, reefs. 1.** To reduce the size of (a sail) by tucking in a part and tying it to or rolling it around a yard. **2.** To shorten (a topmast or bowsprit) by taking part of it in.

**reef·er¹** (rē'fər) *n.* **1.** A person who reefs, as a midshipman. **2.** A short, heavy, close-fitting, double-breasted jacket.

**reef·er²** (rē'fər) *n.* [Prob. < REEF².] *Slang.* Marijuana, esp. a marijuana cigarette.

**reef knot** *n.* A square knot.

**reek** (rēk) *v.* **reeked, reek·ing, reeks.** [ME *reken* < OE *rēocan.*] *—vi.* **1.** To smoke, steam, or fume. **2.** To be permeated by something disagreeable <an arrangement *reeking* of cronyism> **3.** To give off or become permeated with a strong, unpleasant odor. *—vt.* **1.** To emit or exude (e.g., smoke). **2.** To process or treat by subjecting to the action of smoke. *—n.* **1.** A strong, unpleasant odor : STENCH. **2.** Vapor : steam. **—reek'er** *n.* **—reek'y** *adj.*

**reel¹** (rēl) *n.* [ME < OE *hrēol.*] **1.** A device, as a cylinder, spool, or frame, that spins on an axis and is used for winding rope, tape, or other flexible materials. **2.** A cylindrical device fastened to a fishing rod to let out or wind up the line. **3.** The quantity of wire, film, or other material wound on one reel. **4.** A fast dance of Scottish origin. **5.** The Virginia reel. **6.** The music for a reel. *—vt.* **reeled, reel·ing, reels. 1.** To wind on a reel. **2.** To recover by winding on a reel <*reeling* in a lot of fish> **—reel off.** To recite effortlessly and usu. at length <*reeled off* dates and statistics> **—reel'a·ble** *adj.*

**reel²** (rēl) *v.* **reeled, reel·ing, reels.** [ME *relen,* prob. < *reel,* spool < OE *hrēol.*] *—vi.* **1.** To be thrown off balance or fall back. **2.** To stagger, lurch, or sway, as from inebriation. **3.** To go round and round in a spinning motion. **4.** To feel dizzy. *—vt.* To cause to reel. *—n.* A staggering, swaying, or whirling movement. **—reel'er** *n.*

**re-e·lect** *or* **re·e·lect** (rē'ĭ-lĕkt') *vt.* **-lect·ed, -lect·ing, -lects.** To elect again. **—re'-e·lec'tion** *n.*

**re-en·act** *or* **re·en·act** (rē'ĕn-ăkt', -ə-năkt') *vt.* **-act·ed, -act·ing, -acts. 1.** To enact again <*re-enact* an ordinance> **2.** To perform again <*re-enact* the first scene> **3.** To create a second time : RECREATE <*re-enact* the crime> **—re'en·act'ment** *n.*

**re-en·force** (rē'ĭn-fôrs', -fōrs') *v. var. of* REINFORCE.

**re-en·ter** *or* **re·en·ter** (rē-ĕn'tər) *v.* **-tered, -ter·ing, -ters.** *—vi.* To come in or enter again. *—vt.* To record again on a list or ledger. **—re-en'trance** *n.*

**re-en·trant** *or* **re·en·trant** (rē-ĕn'trənt) *adj.* Pointing inward. *—n.* A re-entrant angle or part.

**re-entrant angle** *n.* An interior angle of a polygon that is greater than 180°.

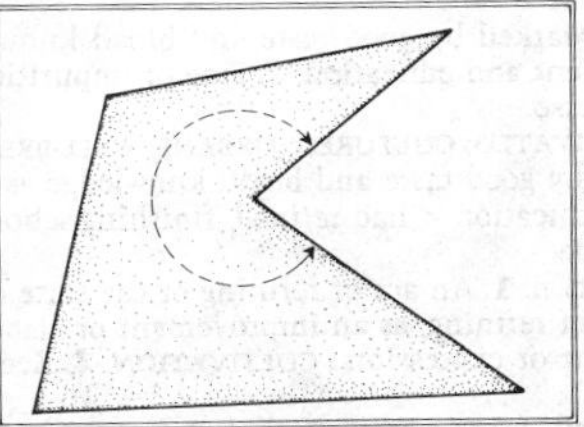

**re-entrant angle**

**re-en·try** *or* **re·en·try** (rē-ĕn'trē) *n., pl.* **-tries. 1.** An act of re-entering. **2.** *Law.* Recovery of ownership under a right reserved in a previous property transaction. **3. a.** An act of regaining the lead by taking a trick in bridge and whist. **b.** The card that will take a trick and thus regain the lead. **4.** *Aerospace.* Return of a missile or spacecraft to the earth's atmosphere.

**reeve¹** (rēv) *n.* [ME < OE *gerēfa.*] **1.** A high local administrative officer appointed by the Anglo-Saxon kings. **2.** A bailiff or steward of a manor in the later medieval period. **3.** A minor officer of a parish or another local authority. **4.** The elected president of a town council in some parts of Canada.

**reeve²** (rēv) *vt.* **reeved** *or* **rove** (rōv), **reev·ing, reeves.** [Orig. unknown.] *Naut.* **1.** To pass (a rope or rod) through a hole, ring, pulley, or block. **2.** To fasten by passing through or around.

**reeve³** (rēv) *n.* [Prob. alteration of RUFF¹.] The female of the ruff, a sandpiper.

**re-ex·am·ine** *or* **re·ex·am·ine** (rē'ĭg-zăm'ĭn) *vt.* **-ined, -in·ing, -ines. 1.** To examine again or anew : REVIEW. **2.** *Law.* To question (a witness) again after cross-examination. **—re'-ex·am'i·na'tion** *n.*

**re·fect** (rĭ-fĕkt') *vt.* **-fect·ed, -fect·ing, -fects.** [Lat. *reficere, refect-,* to refresh : *re-,* back + *facere,* to make.] *Archaic.* To refresh with food and drink.

**re·fec·tion** (rĭ-fĕk'shən) *n.* [ME *refeccioun* < OFr. *refection* < Lat. *refectio* < *reficere,* to refresh. —see REFECT.] **1.** Refreshment with food and drink. **2.** A light meal or repast.

**re·fec·to·ry** (rĭ-fĕk'tə-rē) *n., pl.* **-ries.** [LLat. *refectorium* < Lat. *reficere,* to refresh. —see REFECT.] A room where meals are served, esp. in a monastery or college.

**refectory table** *n.* A long narrow table with heavy legs.

**re·fer** (rĭ-fûr') *v.* **-ferred, -fer·ring, -fers.** [ME *referen* < OFr. *referer* < Lat. *referre* : *re-,* back + *ferre,* to carry.] *—vt.* **1.** To direct to a source for help or information <*referred* me to a good tax lawyer> **2.** To assign or attribute to. **3.** To assign to or regard as belonging within a particular kind or class. **4.** To submit (a matter in dispute) to an authority for arbitration, decision, or examination. **5.** To direct the attention of. —*vi.* **1.** To pertain : concern <comments *referring* to the author's last book> **2.** To allude or make reference <*referred* to Virginia as the Old Dominion> **3.** To turn to, as for information or authority. **—ref·er·a·ble** (rĕf′ər-ə-bəl, rĭ-fûr′-) *adj.* **—re·fer·ral** (rĭ-fûr′əl) *n.* **—re·fer·rer** *n.*

**ref·e·ree** (rĕf′ə-rē′) *n.* **1.** One to whom something is referred, esp. for settlement or decision. **2.** An official supervising the play in a sport : UMPIRE. **3.** *Law.* One appointed by a court to examine and report on a case. —*v.* **-reed, -ree·ing, -rees.** —*vt.* To judge as referee. —*vi.* To act as referee.

**ref·er·ence** (rĕf′ər-əns, rĕf′rəns) *n.* **1.** An act of referring. **2. a.** One that is referred to. **b.** Significance in a designated context. **c.** Meaning or denotation. **3.** The state of being related or referred <with *reference* to><in *reference* to> **4.** An allusion to an event or situation <without making *references* to their earlier misfortune> **5. a.** A note in a publication guiding the reader to another passage or source. **b.** The passage or source so indicated. **c.** A mark or footnote for directing a reader elsewhere for more information. **6.** *Law.* **a.** Submission of a case to a referee. **b.** Legal actions conducted before or by a referee. **7. a.** One in a position to recommend another or to vouch for his or her fitness, as for a job. **b.** A written statement about someone's qualifications, character, and dependability. **—ref·er·enc·er** *n.* **—ref·er·en·tial** (-ə-rĕn′shəl) *adj.*

**ref·er·en·dum** (rĕf′ə-rĕn′dəm) *n., pl.* **-dums** *or* **-da** (-də) [Lat., neuter gerund. of *referre,* to refer.] **1. a.** Submission of a proposed public measure or actual statute to a direct popular vote. **b.** Such a vote. **2.** A note from a diplomat to his or her government requesting instructions.

**ref·er·ent** (rĕf′ər-ənt, rĭ-fûr′ənt) *n.* **1.** Something that refers, esp. a linguistic item in its capacity of referring to a meaning. **2.** Something referred to.

**re·fill** (rē-fĭl′) *vt.* **-filled, -fill·ing, -fills.** To fill again. —*n.* (rē′fĭl′). **1.** A product packaged to replace the used contents of a container. **2.** A second or subsequent filling.

**re·fine** (rĭ-fīn′) *v.* **-fined, -fin·ing, -fines.** —*vt.* **1.** To reduce to a pure state : PURIFY. **2.** To remove by purifying. **3.** To free from coarse characteristics. —*vi.* **1.** To become free of impurities. **2.** To acquire polish or elegance. **3.** To use subtlety and precise distinctions in thought or speech. **—re·fin·er** *n.*

**re·fined** (rĭ-fīnd′) *adj.* **1.** Marked by good taste and broad knowledge as a result of development and education. **2.** Free of impurities : PURIFIED. **3.** Extremely precise.

☆ ***syns:*** REFINED, CULTIVATED, CULTURED, URBANE, WELL-BRED *adj. core meaning* : marked by good taste and broad knowledge as a result of development and education <had *refined,* finishing-school manners> ***ant:*** coarse, crude

**re·fine·ment** (rĭ-fīn′mənt) *n.* **1.** An act of refining or the state of being refined. **2.** The result of refining, as an improvement or elaboration. **3.** Fineness of thought or expression : CULTIVATION. **4.** Keen or precise phrasing.

**re·fin·er·y** (rĭ-fī′nə-rē) *n., pl.* **-ies.** An industrial plant for purifying a crude substance, as petroleum.

**re·fin·ish** (rē-fĭn′ĭsh) *vt.* **-ished, -ish·ing, -ish·es.** To apply a new finish to (furniture). **—re·fin·ish·er** *n.*

**re·fit** (rē-fĭt′) *v.* **-fit·ted, -fit·ting, -fits.** —*vt.* To prepare and equip for additional use. —*vi.* To be made fit again. —*n.* (rē′fĭt′, rē-fĭt′). **1.** Repair of damage or wear. **2.** A secondary or subsequent preparation of supplies and equipment.

**re·flect** (rĭ-flĕkt′) *v.* **-flect·ed, -flect·ing, -flects.** [ME *reflecten* < OFr. *reflecter* < Lat. *reflectere,* to bend back : *re-,* back + *flectere,* to bend.] —*vt.* **1.** To throw or bend back (e.g., light) from a surface. **2.** To form an image of (an object) : MIRROR. **3.** To manifest as a result of one's actions <This work *reflects* carelessness.> **4.** *Archaic.* To bend back. —*vi.* **1.** To be bent back. **2.** To give back a likeness. **3.** To think or consider seriously. **4.** To bring blame or reproach.

☆ ***syns:*** REFLECT, IMAGE, MIRROR *v. core meaning* : to send back or form an image of <a pool that *reflects* nearby buildings>

**re·flec·tance** (rĭ-flĕk′təns) *n.* The ratio of the total radiant flux, as of light, reflected by a surface to the total incident on the surface.

**reflecting telescope** *n.* An optical telescope in which the principal image-forming element is a parabolic or spherical mirror.

**re·flec·tion** (rĭ-flĕk′shən) *n.* **1.** An act or instance of reflecting or the state of being reflected. **2.** Something reflected, as light, radiant heat, sound, or an image. **3. a.** Careful consideration : MEDITATION. **b.** The results of such consideration. **4.** An imputation of censure or discredit. **—re·flec·tion·al** *adj.*

**re·flec·tive** (rĭ-flĕk′tĭv) *adj.* **1.** Of, relating to, produced by, or resulting from reflection. **2.** Meditative. **—re·flec·tive·ly** *adv.*

**re·flec·tiv·i·ty** (rē′flĕk-tĭv′ĭ-tē) *n., pl.* **-ties.** **1.** The quality of being reflective. **2.** Ability to reflect. **3.** *Physics.* The ratio of the intensity of the total radiation, as of light, reflected from a surface to the total incident on the surface.

**re·flec·tom·e·ter** (rē′flĕk-tŏm′ĭ-tər) *n.* An instrument for measuring the reflectance of a surface.

**re·flec·tor** (rĭ-flĕk′tər) *n.* **1.** Something that reflects. **2.** A surface that reflects radiation. **3.** A reflecting telescope.

**re·flex** (rē′flĕks′) *adj.* [Lat. *reflexus,* p.part. of *reflectere,* to bend back. —see REFLECT.] **1.** Turned, thrown, or bent backward. **2.** *Physiol.* Designating an involuntary action or response, as a sneeze, blink, or hiccup. —*n.* (rē′flĕks′). **1.** Reflection or an image caused by reflection. **2.** *Physiol.* Involuntary response to a stimulus. **3.** *Psychol.* An unlearned or instinctive response to a stimulus. **4.** A linguistic form or feature that reflects or represents an earlier, often reconstructed form or feature having undergone phonetic or other change. —*vt.* (rĭ-flĕks′) **-flexed, -flex·ing, -flex·es.** **1.** To bend, turn back, or reflect. **2.** To cause to undergo a reflex process.

**reflex angle** *n.* An angle greater than 180° and less than 360°.

**reflex arc** *n. Physiol.* The neural path of a simple reflex.

**reflex camera** *n.* A camera fitted with a mirror to reflect the exact focused image recordable onto a coupled viewing screen.

**re·flex·ion** (rĭ-flĕk′shən) *n. Chiefly Brit. var. of* REFLECTION.

**re·flex·ive** (rĭ-flĕk′sĭv) *adj.* **1. a.** Designating a verb with an identical subject and direct object, as *dressed* in the sentence *I dressed myself.* **b.** Designating the pronoun used as direct object of a reflexive verb, as *myself* in *I dressed myself.* **2.** Of or relating to a reflex. —*n.* A reflexive verb or pronoun. **—re·flex′ive·ly** *adv.* **—re·flex′ive·ness, re′flex·iv′i·ty** (rē′flĕk-sĭv′ĭ-tē) *n.*

**ref·lu·ent** (rĕf′lo͞o-ənt) *adj.* [Lat. *refluens,* pr.part. of *refluere,* to flow back : *re-,* back + *fluere,* to flow.] Flowing back : EBBING. **—ref′lu·ence** *n.*

**re·flux** (rē′flŭks′) *n.* [ME < Med. Lat. *refluxus* : Lat. *re-,* back + *fluxus,* flow < *fluere,* to flow.] A flowing back : EBB.

**re·for·est** (rē-fôr′ĭst, -fŏr′ĭst) *vt.* **-est·ed, -est·ing, -ests.** To replant (an area) with forest trees. **—re′for·es·ta′tion** *n.*

**re·form** (rĭ-fôrm′) *v.* **-formed, -form·ing, -forms.** [ME *reformen* < OFr. *reformer* < Lat. *reformare* : *re-,* again + *formare,* to form < *forma,* form.] —*vt.* **1.** To improve by altering, correcting errors, or removing defects. **2.** To eliminate abuse or malpractice in <*reform* the penal system> **3.** To cause (a person) to give up irresponsible or immoral practices. —*vi.* To give up irresponsible or immoral practices. —*n.* **1.** A change for the better. **2.** A movement that tries to improve a sociopolitical situation without revolutionary change. **3.** Moral improvement. **—re·for′ma·tive** *adj.* **—re·form′er** *n.*

**re-form** (rē-fôrm′) *vt. & vi.* **-formed, -form·ing, -forms.** To form again or to become formed again.

**ref·or·ma·tion** (rĕf′ər-mā′shən) *n.* **1.** An act of reforming or the state of being reformed. **2. Reformation.** The 16th-cent. movement leading to separation of the Protestant churches from the Roman Catholic Church. **—ref′or·ma′tion·al** *adj.*

**re·for·ma·to·ry** (rĭ-fôr′mə-tôr′ē, -tōr′ē) *n., pl.* **-ries.** A penal institution for the discipline, reformation, and training of young offenders. —*adj.* Serving or intending to reform.

**re·formed** (rĭ-fôrmd′) *adj.* **1.** Improved by removal of faults. **2.** Improved in conduct or character. **3. Reformed.** Of, relating to, or denoting the Protestant churches that follow the teachings of John Calvin and Ulrich Zwingli. **4. Reformed.** Of, relating to, or denoting Reform Judaism.

**re·form·ism** (rĭ-fôr′mĭz′əm) *n.* A doctrine or movement of reform. **—re·form′ist** *n.*

**Reform Judaism** *n.* A branch of Judaism dating from the 19th cent. that aims for reconciliation of historical Judaism with modern life and does not demand strict adherence to traditional law.

**reform school** *n.* A reformatory.

**re·fract** (rĭ-frăkt′) *vt.* **-fract·ed, -fract·ing, -fracts.** [Lat. *refringere, refract-,* to break up : *re-,* back + *frangere,* to break.] To deflect (e.g., light) from a straight path by refraction.

**refracting telescope** *n.* A telescope in which lenses alone produce the final image.

**re·frac·tion** (rĭ-frăk′shən) *n.* **1.** *Physics.* Deflection of a propagating wave, as of light or sound, at the boundary between two mediums with different refractive indices or in passage through a medium of nonuniform density. **2.** *Astron.* Apparent positional elevation of celestial objects resulting from deflection of light entering the earth's atmosphere. **—re·frac′tion·al, re·frac′tive** *adj.* **—re·frac′tive·ly** *adv.* **—re·frac′tive·ness, re′frac·tiv′i·ty** (rē′frăk-tĭv′ĭ-tē) *n.*

**refractive index** *n. Physics.* Index of refraction.

**re·frac·tom·e·ter** (rē′frăk-tŏm′ĭ-tər) *n.* An instrument for measuring indices of refraction.

**re·frac·tor** (rĭ-frăk′tər) *n.* **1.** One that refracts. **2.** A refracting telescope.

**re·frac·to·ry** (rĭ-frăk′tə-rē) *adj.* [Obs. *refractary* < Lat. *refractarius* < *refringere,* to break up. —see REFRACT.] **1.** Obstinate : unmanageable <a *refractory* child> **2.** Difficult to melt or work : resistant to heat <a *refractory* metal> **3.** Not responsive to treatment <a *refractory* illness> —*n., pl.* **-ries.** **1.** One that is refractory. **2.** Material resistant to high temperatures. **—re·frac′to·ri·ly** *adv.* **—re·frac′to·ri·ness** *n.*

**re·frain**[1] (rĭ-frān′) *v.* **-frained, -frain·ing, -frains.** [ME *refrenen* < OFr. *refrener* < Lat. *refrenare,* to restrain : *re-,* back + *frenum,* bridle.] —*vi.* To hold oneself back : FORBEAR. —*vt. Archaic.* To restrain or hold back : CURB. **—re·frain'er** *n.* **—re·frain'ment** *n.*

☆ ***syns:*** REFRAIN, ABSTAIN, FORBEAR, KEEP, WITHHOLD *v. core meaning* : to hold oneself back <*refrain* from smoking>

**re·frain²** (rĭ-frān') *n.* [ME *refreyn* < OFr. *refrain* < *refraindre,* to resound < Lat. *refringere,* to break up. —see REFRACT.] **1. a.** A phrase or verse repeated at intervals throughout a song or poem, esp. at the end of each stanza. **b.** Music for the refrain of a poem. **2.** A repeated theme.

**re·fran·gi·ble** (rĭ-frăn'jə-bəl) *adj.* Capable of being refracted. **—re·fran'gi·bil'i·ty, re·fran'gi·ble·ness** *n.*

**re·fresh** (rĭ-frĕsh') *v.* **-freshed, -fresh·ing, -fresh·es.** [ME *refresshen* < OFr. *refreschir* : *re-,* anew (< Lat.) + *fres,* fresh, of Germanic orig.] —*vt.* **1.** To revive (one) with or as if with rest, food, or drink. **2.** To make cool, clean, or damp : FRESHEN. **3.** To renew by stimulation <*refresh* my memory> —*vi.* **1.** To take refreshment. **2.** To become revived : REINVIGORATE.

**re·fresh·en** (rĭ-frĕsh'ən) *vt.* & *vi.* **-fresh·ened, -fresh·en·ing, -fresh·ens.** To refresh.

**re·fresh·er** (rĭ-frĕsh'ər) *n.* One that refreshes. —*adj.* Serving to reacquaint one with previously studied material <a *refresher* course>

**re·fresh·ing** (rĭ-frĕsh'ĭng) *adj.* **1.** Serving to refresh. **2.** Pleasantly fresh and unusual. **—re·fresh'ing·ly** *adv.*

**re·fresh·ment** (rĭ-frĕsh'mənt) *n.* **1.** An act of refreshing or the state of being refreshed. **2.** Something, as food or drink, that refreshes. **3. refreshments. a.** An assortment of light foods. **b.** A light meal or snack.

**re·frig·er·ant** (rĭ-frĭj'ər-ənt) *adj.* **1.** Cooling or freezing. **2.** *Med.* Reducing fever. —*n.* **1.** A substance, as air, ammonia, water, or carbon dioxide, for producing refrigeration, either as the working substance of a refrigerator or by direct absorption of heat. **2.** *Med.* A fever-reducing agent.

**re·frig·er·ate** (rĭ-frĭj'ə-rāt') *vt.* **-at·ed, -at·ing, -ates.** [Lat. *refrigerare, refrigerat-* : *re-,* anew + *frigerare,* to make cool < *frigus,* cool.] **1.** To cool or chill (a substance). **2.** To preserve (food) by chilling. **—re·frig'er·a'tion** *n.* **—re·frig'er·a'tive** *adj.* & *n.* **—re·frig'er·a·to'ry** (-ər-ə-tôr'ē, -tōr'ē) *adj.*

**re·frig·er·a·tor** (rĭ-frĭj'ə-rā'tər) *n.* A cabinet or room for storing substances, as food, at a low temperature.

**re·frin·gence** (rĭ-frĭn'jəns) *n.* Refractive power.

**re·frin·gent** (rĭ-frĭn'jənt) *adj.* [Lat. *refringens,* pr.part. of *refringere,* to break up. —see REFRACT.] Of, relating to, or producing refraction : REFRACTIVE.

**reft¹** (rĕft) *v. var. p.t.* & *p.p. of* REAVE¹.

**reft²** (rĕft) *v. var. p.t.* & *p.p. of* REAVE².

**re·fu·el** (rē-fyōō'əl) *vt.* & *vi.* **-eled, -el·ing, -els** *also* **-elled, -el·ling, -els.** To supply again with fuel or to take on a fresh fuel supply.

**ref·uge** (rĕf'yōōj) *n.* [ME < OFr. < Lat. *refugium* < *refugere,* to run away : *re-,* back + *fugere,* to flee.] **1. a.** The state of being protected, as from danger or hardship. **b.** Protection so provided. **2.** A place that provides protection or shelter : HAVEN. **3.** Something to which one may turn for help, relief, or escape <Silence was my only *refuge.*> —*v.* **-uged, -ug·ing, -ug·es.** *Archaic.* —*vt.* To give refuge to. —*vi.* To take refuge.

☆ ***syns:*** REFUGE, ASYLUM, SANCTUARY, SHELTER *n. core meaning* : the state of being protected, as from danger <sought *refuge* in the embassy>

**ref·u·gee** (rĕf'yōō-jē') *n.* [Fr. *réfugié* < p.part. of *réfugier,* to take refuge < OFr. *refuge,* refuge.] One who flees, usu. to another country for refuge, esp. from invasion, oppression, or persecution.

**re·ful·gent** (rĭ-fōōl'jənt, -fŭl'-) *adj.* [Lat. *refulgens* < pr.part. of *refulgere,* to flash back : *re-,* back + *fulgere,* to flash.] Shining radiantly : RESPLENDENT <admired the *refulgent* sunset> **—re·ful'gence, re·ful'gen·cy** *n.* **—re·ful'gent·ly** *adv.*

**re·fund** (rĭ-fŭnd', rē'fŭnd') *v.* **-fund·ed, -fund·ing, -funds.** [ME *refunden,* to pour back < OFr. *refunder* < Lat. *refundere* : *re-,* back + *fundere,* to pour.] —*vt.* **1.** To give back : RETURN. **2.** To repay (money). —*vi.* To make repayment. —*n.* (rē'fŭnd'). **1.** A repayment of funds. **2.** An amount repaid. **—re·fund'a·ble** *adj.* **—re·fund'er** *n.* **—re·fund'ment** *n.*

**re-fund** (rē-fŭnd') *vt.* **-fund·ed, -fund·ing, -funds. 1.** To fund anew. **2.** To pay back (a debt) with new borrowing, esp. to replace (a bond issue) with a new bond issue.

**re·fur·bish** (rē-fûr'bĭsh) *vt.* **-bished, -bish·ing, -bish·es.** To make clean, bright, or fresh again. **—re·fur'bish·ment** *n.*

**re·fus·al** (rĭ-fyōō'zəl) *n.* **1.** An act of refusing. **2.** The opportunity or right to accept or reject.

**re·fuse¹** (rĭ-fyōōz') *v.* **-fused, -fus·ing, -fus·es.** [ME *refusen* < OFr. *refuser* < VLat. **refusare* < Lat. *refundere,* to pour back. —see REFUND.] —*vt.* **1.** To decline to do, accept, give, or allow. **2.** To decline to jump (an obstacle). —Used of a horse. —*vi.* To decline to do, accept, allow, or give something. **—re·fus'er** *n.*

**ref·use²** (rĕf'yōōs) *n.* [ME < OFr. *refus,* rejection < *refuser,* to refuse.] Something rejected or discarded as worthless or useless : TRASH.

**re·fuse·nik** (rĭ-fyōōz'nĭk) *n., pl.* **-niks.** *n.* A Soviet citizen who has been refused the right to emigrate from the U.S.S.R.

**ref·u·ta·tion** (rĕf'yōō-tā'shən) *also* **re·fut·al** (rĭ-fyōōt'l) *n.* **1.** An act of refuting. **2.** Something that refutes.

**re·fute** (rĭ-fyōōt') *vt.* **-fut·ed, -fut·ing, -futes.** [Lat. *refutare* : *re-,* back + *-futare,* to beat.] **1.** To prove to be false or mistaken : DISPROVE <*refute* previous statements> **2.** To deny the accuracy or truth of. **—re·fut'a·bil'i·ty** (rĭ-fyōō'tə-bĭl'ĭ-tē, rĕf'yə-tə-) *n.* **—re·fut'a·ble** *adj.* **—re·fut'a·bly** *adv.* **—re·fut'er** *n.*

**re·gain** (rē-gān') *vt.* **-gained, -gain·ing, -gains. 1.** To recover ownership of. **2.** To reach again. **—re·gain'er** *n.*

**re·gal** (rē'gəl) *adj.* [ME < OFr. < Lat. *regalis* < *rex,* king.] **1.** Of or relating to a monarch : ROYAL. **2.** Belonging to or appropriate for a monarch <a *regal* coronation procession> **3.** Of great magnificence : SPLENDID. **—re·gal'i·ty** (rĭ-găl'ĭ-tē) *n.* **—re'gal·ly** *adv.*

**re·gale** (rĭ-gāl') *v.* **-galed, -gal·ing, -gales.** [Fr. *régaler* < OFr. *regaler* : *re-* (intensive < Lat.) + *gale,* pleasure.] —*vt.* **1.** To delight or entertain : give pleasure to. **2.** To entertain lavishly with food and drink : provide a feast for. —*vi.* To feast oneself. —*n.* **1.** A great feast. **2.** Choice food or drink. **3.** Refreshment. **—re·gale'ment** *n.*

**re·ga·lia** (rĭ-gāl'yə, -gā'lē-ə) *pl.n.* [Med. Lat. < Lat., neuter pl. of *regalis,* regal.] *(sing. or pl. in number).* **1.** The emblems and symbols of royalty, as a crown and scepter. **2.** Rights and privileges of royalty. **3.** The distinctive symbols of a rank, office, order, or society. **4.** Splendid raiment.

**re·gard** (rĭ-gärd') *v.* **-gard·ed, -gard·ing, -gards.** [ME *regarden* < OFr. *regarder* : *re-,* back + *guarder,* to guard, of Germanic orig.] —*vt.* **1.** To observe closely. **2.** To look on or consider in a specific way <I *regard* them as fools.> **3.** To have great esteem for. **4.** *Archaic.* To relate, concern, or refer to. **5.** To consider or take into account. **6.** *Obs.* To take care of. —*vi.* **1.** To look : gaze. **2.** To give heed. —*n.* **1.** A look : gaze. **2.** Careful thought or attention : HEED. **3.** Respect, affection, or esteem <earned the *regard* of all> **4. regards.** Greetings that show respect or affection <send our *regards*> **5.** A specific point or respect <a good administrator in this *regard*> **6.** Basis for action : MOTIVE. **7.** *Obs.* Appearance or aspect. **—as regards.** Concerning. **—in (**or **with) regard to.** With respect to : CONCERNING. ***usage:*** *Regard* is customarily used in the singular in the phrase *in* (or *with*) *regard to.*

**re·gar·dant** (rĭ-gär'dnt) *adj.* [ME < OFr., pr.part. of *regarder,* to regard.] *Heraldry.* With the face turned backward in profile.

**re·gard·ful** (rĭ-gärd'fəl) *adj.* **1.** Attentive : heedful. **2.** Deferential : respectful. **—re·gard'ful·ly** *adv.* **—re·gard'ful·ness** *n.*

**re·gard·ing** (rĭ-gär'dĭng) *prep.* In reference to.

**re·gard·less** (rĭ-gärd'lĭs) *adj.* Unmindful. —*adv.* In spite of everything : ANYWAY. **—re·gard'less·ly** *adv.* **—re·gard'less·ness** *n.*

**re·gard·less of** *prep.* In spite of.

**re·gat·ta** (rĭ-gä'tə, -găt'ə) *n.* [Ital.] A boat race or an organized series of boat races.

**re·ge·late** (rē'jə-lāt', rē'jə-lāt') *vi.* **-lat·ed, -lat·ing, -lates.** [RE- + Lat. *gelare, gelat-,* to freeze.] To undergo regelation.

**re·ge·la·tion** (rē'jə-lā'shən) *n.* **1.** Fusion of two blocks of ice by pressure. **2.** Successive melting under pressure and freezing when pressure is relaxed at the interface of two blocks of ice.

**re·gen·cy** (rē'jən-sē) *n., pl.* **-cies. 1.** A person or group chosen to head the government in case of a monarch's minority, absence, incompetence, or sickness. **2.** The period during which a regent governs. **3.** The office, area of jurisdiction, or government of a regent or regents. —*adj.* **—Re·gen·cy.** Of, pertaining to, or typical of the style, esp. in furniture, common during the regency (1811–20) of George, Prince of Wales.

**re·gen·er·ate** (rĭ-jĕn'ə-rāt') *v.* **-at·ed, -at·ing, -ates.** [Lat. *regenerare, regenerat-,* to reproduce : *re-,* again + *generare,* to beget < *genus,* birth.] —*vt.* **1.** To reform spiritually or morally. **2.** To form, construct, or create anew. **3.** *Biol.* To replace (a lost or damaged organ or part) by formation of new tissue. —*vi.* **1.** To become formed or constructed again. **2.** To undergo spiritual conversion or rebirth. **3.** To effect regeneration. —*n.* (rĭ-jĕn'ər-ĭt). One spiritually reborn or reformed. —*adj.* (rĭ-jĕn'ər-ĭt). **1.** Spiritually or morally revitalized. **2.** Restored : renewed. **—re·gen'er·a·ble** (-rə-bəl) *adj.* **—re·gen'er·a·cy** (-jĕn'ər-ə-sē) *n.* **—re·gen'er·ate·ly** *adv.* **—re·gen'er·a'tor** *n.*

**re·gen·er·a·tion** (rĭ-jĕn'ə-rā'shən) *n.* **1.** An act or process of regenerating or state of being regenerated. **2.** Spiritual or moral revival or rebirth. **3.** *Biol.* Regrowth of lost or destroyed parts or organs.

**re·gen·er·a·tive** (rĭ-jĕn'ə-rā'tĭv, -ər-ə-tĭv) *adj.* **1.** Of, relating to, or characterized by regeneration. **2.** Apt to regenerate. **—re·gen'er·a'tive·ly** *adv.*

**re·gent** (rē'jənt) *n.* [ME < OFr., ruling < Lat. *regens,* pr.part. of *regere,* to rule.] **1.** One who rules during the minority, absence, or disability of a monarch. **2.** One acting as a ruler or governor. **3.** One serving on a governing board of an institution, as a university. **—re'gent·al** (-jən'tl) *adj.*

**reg·gae** (rĕg'ā') *n.* [Orig. unknown.] Popular music of Jamaican origin combining elements of calypso, soul, and rock 'n' roll and marked by a simple syncopated rhythm.

**re·gi·cide** (rĕj′ĭ-sīd′) *n.* [Lat. *rex, regi-,* king + -CIDE.] **1.** Murder of a king. **2.** One who murders a king. **—reg′i·ci′dal** (-sīd′l) *adj.*

**re·gime** *also* **ré·gime** (rā-zhēm′, rĭ-) *n.* [Fr. *régime* < Lat. *regimen* < *regere,* to rule.] **1. a.** A management system. **b.** A government in power : ADMINISTRATION <the Thatcher *regime*> **2.** A social system or pattern. **3.** REGIMEN 2.

**reg·i·men** (rĕj′ə-mən, -mĕn′) *n.* [ME < Lat., regime.] **1.** Governmental rule or control. **2.** A system, as of therapy or diet.

**reg·i·ment** (rĕj′ə-mənt) *n.* [ME < OFr. < LLat. *regimentum,* rule < Lat. *regere,* to rule.] A military unit of ground or airborne troops composed of at least two battalions <an artillery *regiment*><a paratroop *regiment*> *—vt.* (rĕj′ə-mĕnt′) **-ment·ed, -ment·ing, -ments. 1.** To organize into a regiment. **2.** To appoint to a regiment. **3.** To put into systematic order. **4.** To impose uniformity and rigid order upon. **—reg′i·men′tal** (-mĕn′tl) *adj.* **—reg′i·men′tal·ly** *adv.* **—reg′i·men·ta′tion** *n.*

**reg·i·men·tals** (rĕj′ə-mĕnt′lz) *pl.n.* **1.** The uniform and insignia typical of a specific regiment. **2.** Military dress.

**re·gion** (rē′jən) *n.* [ME *regioun* < OFr. *region* < Lat. *regio* < *regere,* to direct.] **1.** A large, usu. continuous portion of a surface or space : AREA. **2.** A large, indefinite area of the earth's surface. **3.** A particular district or territory. **4.** A field of interest or activity : SPHERE. **5.** A part of the earth marked by distinctive animal or plant life. **6.** An area of the body with natural or arbitrarily assigned boundaries <the *region* of the lower back>

**re·gion·al** (rē′jə-nəl) *adj.* **1.** Of, relating to, or characteristic of a large geographic region. **2.** Of, relating to, or typical of a specific region or district : LOCALIZED. **3.** Of, belonging to, or typical of a form of a language that is distributed in identifiable geographic areas and has identifiable phonetic, structural, and other differences from the standard form of the language : DIALECTAL. *—n.* **1.** Something that serves a region, as a branch of an organization. **2.** One that is regional. **—re′gion·al·ly** *adv.*

**ré·gis·seur** (rā′zhē-sûr′) *n., pl.* **-seurs** (-sûr′) [Fr. < *régir,* to direct < Lat. *regere* < *rex,* king.] The director of a ballet.

**reg·is·ter** (rĕj′ĭ-stər) *n.* [ME *registre* < OFr. < Med. Lat. *registrum* < LLat. *regesta* < Lat., neuter pl. of *regestus,* p.part. of *regerere,* to record: *re-,* back + *gerere,* to carry.] **1. a.** A formal or official recording of items, names, or actions. **b.** A book for such entries. **2.** An entry in a register. **3.** A device that automatically registers an amount or number. **4.** An adjustable grill-like device through which heated or cooled air is released into a room. **5.** Correct alignment or positioning. **6.** Registration : registry. **7.** *Mus.* **a.** The range of an instrument or voice. **b.** A part of such a range that has similar quality. **c.** A group of matched organ pipes : STOP. *—v.* **-tered, -ter·ing, -ters.** *—vt.* **1. a.** To enter in an official register. **b.** To enroll formally or officially. **2.** To indicate, as on an instrument or scale. **3.** To show (emotion). **4.** To cause (mail) to be officially recorded by payment of a fee. *—vi.* **1.** To place or cause placement of one's name in a register. **2.** To have one's name officially placed on a list of eligible voters. **3.** To make an impression. **4.** To be properly aligned. **—reg′is·ter·er** *n.* **—reg′is·tra·ble** (-ĭ-strə-bəl) *adj.*

**reg·is·tered** (rĕj′ĭ-stərd) *adj.* **1.** Having the owner's name listed in a register <*registered* bonds> **2.** Having a pedigree recorded in a breed association studbook <a *registered* cocker spaniel> **3.** Officially qualified <a *registered* pharmacist>

**registered mail** *n.* Mail that is recorded by the post office when sent and at each point on its route so as to assure safe delivery.

**registered nurse** *n.* A graduate trained nurse who has passed a state registration examination.

**reg·is·trant** (rĕj′ĭ-strənt) *n.* One who registers or is registered.

**reg·is·trar** (rĕj′ĭ-strär′, rĕj′ĭ-strär′) *n.* **1.** An officer in a college or university who keeps records on the enrollment and academic standing of students. **2.** An officer of a corporation charged with maintaining records of ownership of its securities. **3.** A hospital admitting officer.

**reg·is·tra·tion** (rĕj′ĭ-strā′shən) *n.* **1.** An act of registering. **2.** The number of persons registered : ENROLLMENT. **3.** An entry in a register. **4.** A document certifying an act of registering. **5.** *Mus.* **a.** A combination of organ stops chosen to be used in playing a piece. **b.** The technique of choosing and adjusting organ stops.

**reg·is·try** (rĕj′ĭ-strē) *n., pl.* **-tries. 1.** Registration. **2.** A ship's registered nationality : FLAG. **3.** A place for registering. **4.** A book for official records.

**re·gius professor** (rē′jəs, -jē-əs) *n.* [< Lat. *regius,* royal < *rex,* king.] One holding a professorship set up by royal subsidy at certain British universities.

**reg·let** (rĕg′lĭt) *n.* [OFr. < *regle,* ruler < Lat. *regula* < *regere,* to rule.] **1.** A narrow, flat molding. **2.** A flat piece of wood for separating lines of printing type.

**reg·nal** (rĕg′nəl) *adj.* [Med. Lat. *regnalis,* royal < Lat. *regnum,* reign < *rex,* king.] Specifying a particular year of a monarch's reign reckoned from the date of accession.

**reg·nant** (rĕg′nənt) *adj.* [Lat. *regnans,* pr.part. of *regnare,* to reign < *regnum,* reign < *rex,* king.] **1.** Reigning : ruling. **2.** Predominant. **3.** Widespread : prevalent.

**reg·o·lith** (rĕg′ə-lĭth′) *n.* [Gk. *rhēgos,* blanket + -LITH.] The layer of loose rock material resting on bedrock, making up the surface of most land.

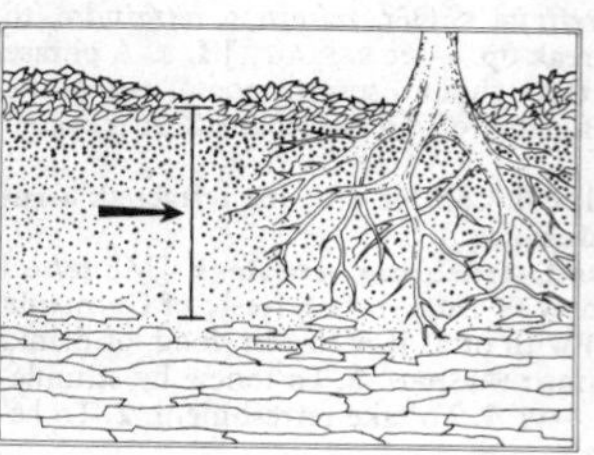

**regolith**

**re·gorge** (rē-gôrj′) *vt.* **-gorged, -gorg·ing, -gorg·es.** [Fr. *regorger* < OFr. : *re-,* back + *gorger,* to gorge < *gorge,* throat.] To disgorge.

**re·gress** (rĭ-grĕs′) *v.* **-gressed, -gress·ing, -gress·es.** [Lat. *regredi, regress-* : *re-,* back + *gradi,* to go.] *—vi.* **1.** To revert or return to a former and usu. worse condition. **2.** To have a tendency to approach or return to a statistical mean. *—vt.* To induce a state of regression in. *—n.* (rē′grĕs′). **1.** Return or withdrawal. **2.** An act of reasoning backward from an effect to a cause. **—re·gres′sor** *n.*

**re·gres·sion** (rĭ-grĕsh′ən) *n.* **1.** Reversion : retrogression. **2.** Relapse to a less perfect or developed state. **3.** *Psychoanal.* Reversion to a more primitive or less mature behavior pattern. **4.** *Statistics.* The tendency for the expected value of one of two jointly correlated random variables to approach more closely the mean value of its set than the other. **5.** *Astron.* Retrogradation.

**re·gres·sive** (rĭ-grĕs′ĭv) *adj.* **1.** Likely to return or revert. **2.** Marked by regression or a tendency to regress. **3.** Having the rate decrease as the taxable amount increases. **—re·gres′sive·ly** *adv.* **—re·gres′sive·ness** *n.*

**re·gret** (rĭ-grĕt′) *v.* **-gret·ted, -gret·ting, -grets.** [ME *regretten* < OFr. *regretter,* to lament.] *—vt.* **1.** To feel sorry, disappointed, or distressed about. **2.** To feel sorrow or grief over : MOURN. *—vi.* To feel regret. *—n.* **1.** A sense of loss and longing for someone gone. **2.** Distress over a desire unfulfilled or an action performed or not performed. **3.** An expression of disappointment or grief. **4. regrets.** A polite declining of an invitation. **—re·gret′ta·ble** *adj.* **—re·gret′ta·bly** *adv.* **—re·gret′ter** *n.*

**re·gret·ful** (rĭ-grĕt′fəl) *adj.* Full of regret or sorrow. **—re·gret′ful·ly** *adv.* **—re·gret′ful·ness** *n.*

**re·group** (rē-gro͞op′) *v.* **-grouped, -group·ing, -groups.** *—vt.* To arrange in a new grouping. *—vi.* **1.** To reassemble or change a military formation, as after a battle. **2.** To reorganize for a fresh attempt, as after a defeat.

**reg·u·lar** (rĕg′yə-lər) *adj.* [ME, under religious rule < OFr. *reguler* < Lat. *regularis,* of a bar, according to rule < *regula,* bar, ruler < *regere,* to rule.] **1.** Customary, usual, or normal. **2.** Orderly or symmetric. **3.** Conforming to set procedure, principle, or discipline. **4.** Methodical : well-ordered. **5.** Happening at fixed intervals : PERIODIC. **6.** Not varying : CONSTANT. **7.** Formally correct : PROPER. **8.** Having the necessary qualifications for an occupation. **9.** Thorough : out-and-out <a *regular* scoundrel> **10.** *Informal.* Good : nice <a *regular* person> **11.** *Bot.* Bearing symmetrically arranged parts of similar size and shape <*regular* flowers> **12.** Belonging to a standard mode of conjugation or inflection. **13.** Belonging to a religious order and bound by its rules <the *regular* clergy> **14. a.** Having equal sides and equal angles. —Used of polygons. **b.** Having faces that are congruent regular polygons and congruent polyhedral angles. —Used of polyhedrons. **15.** Belonging to or constituting a nation's permanent army. *—n.* **1.** A member of the clergy or of a religious order. **2.** A soldier belonging to a regular army. **3.** A loyal, dependable person <one of the party *regulars*> **4.** A clothing size for persons of average height. **5.** *Informal.* A habitual customer. **—reg′u·lar′i·ty** (-lăr′ĭ-tē) *n.* **—reg′u·lar·ly** *adv.*

**regular army** *n.* A nation's permanent standing army.

**reg·u·lar·ize** (rĕg′yə-lə-rīz′) *vt.* **-ized, -iz·ing, -iz·es.** To make regular or cause to conform. **—reg′u·lar·i·za′tion** *n.* **—reg′u·lar·iz′er** *n.*

**regular year** *n.* In the Jewish calendar: **a.** An ordinary year of 354 days. **b.** A leap year of 384 days.

**reg·u·late** (rĕg′yə-lāt′) *vt.* **-lat·ed, -lat·ing, -lates.** [Lat. *regulare, regulat-* < Lat. *regula,* a rule.] **1.** To control or direct in agreement with a rule. **2.** To adjust in conformity to a requirement or specification. **3.** To adjust (a mechanism) for accurate and correct operation. **—reg′u·la′tive, reg′u·la·to′ry** (-lə-tôr′ē, -tōr′ē) *adj.*

**reg·u·la·tion** (rĕg′yə-lā′shən) *n.* **1.** An act of regulating. **2.** A principle, rule, or law designed for controlling or governing behavior. **3.** A governmental order with the force of law.

**reg·u·la·tor** (rĕg'yə-lā'tər) *n.* One that regulates, as: **a.** The mechanism in a watch that governs its speed. **b.** An accurate clock used as a standard for timing other clocks. **c.** A device to maintain uniform speed in a machine : GOVERNOR. **d.** A device for controlling the flow of gases, liquids, or electric current.
**regulator gene** *n.* A gene that represses the activity of another gene in an operon.
**Reg·u·lus** (rĕg'yə-ləs) *n., pl.* **-li** (-lī') *or* **-lus·es.** [Lat., ruler of a small country, dim. of *rex,* king.] **1.** A bright double star in the constellation Leo. **2. regulus.** *Metallurgy.* **a.** The relatively pure metallic part of a charge that settles on the bottom of a furnace or crucible. **b.** A relatively impure product of various ores in smelting. **—reg'u·line** (rĕg'yə-lĭn, -līn') *adj.*
**re·gur·gi·tate** (rē-gûr'jĭ-tāt') *v.* **-tat·ed, -tat·ing, -tates.** [Med. Lat. *regurgitare, regurgitat-* : *re-,* back + LLat. *gurgitare,* to engulf < Lat. *gurges,* whirlpool.] *—vi.* To rush or surge back : REGORGE. *—vt.* To cause to pour back, esp. to cast up (partially digested food) : VOMIT. **—re·gur'gi·ta'tion** *n.* **—re·gur'gi·ta·tive** *adj.*
**re·hab** (rē'hăb') *n.* Something, esp. a building, that has undergone structural rehabilitation. *—vt. & vi.* **-habbed, -hab·bing, -habs.** To rehabilitate structurally or to undergo structural rehabilitation.
**re·ha·bil·i·tate** (rē'hə-bĭl'ĭ-tāt') *vt.* **-tat·ed, -tat·ing, -tates.** [Med. Lat. *rehabilitare, rehabilitat-* : Lat. *re-,* again + LLat. *habilitare,* to habilitate. —see HABILITATE.] **1.** To restore (e.g., a handicapped person) to customary activity through education and therapy. **2.** To reinstate the good name of. **3.** To restore the former rank, privileges, or rights of. **4.** To restore to a former state. **—re'ha·bil'i·ta'tion** *n.* **—re'ha·bil'i·ta'tive** *adj.*
**re·hash** (rē-hăsh') *vt.* **-hashed, -hash·ing, -hash·es.** To go over, repeat, or rework (old material). *—n.* (rē'hăsh'). **1.** The process or act of rehashing. **2.** The result of rehashing.
**re·hear** (rē-hîr') *vt.* **-heard** (-hûrd'), **-hear·ing, -hears. 1.** To hear again. **2.** *Law.* To give a second or new judicial hearing to.
**re·hear·ing** (rē-hîr'ĭng) *n. Law.* A second or new hearing of a case, plea, or suit in the same court.
**re·hears·al** (rĭ-hûr'səl) *n.* **1.** The act or process of practicing for a forthcoming performance. **2.** A verbal repetition or recital.
**re·hearse** (rĭ-hûrs') *v.* **-hearsed, -hears·ing, -hears·es.** [ME *rehercen,* to repeat < OFr. *rehercer* : *re-,* again (< Lat.) + *hercer,* to harrow < *herce,* harrow < Lat. *hirpex.*] *—vt.* **1.** To practice (a song, play, or dance), esp. in preparation for a performance. **2.** To perfect or cause to perfect (an action) by repetition. **3.** To retell or recite. *—vi.* To rehearse a song, play, or dance. **—re·hears'er** *n.*
**re·house** (rē-houz') *vt.* **-housed, -hous·ing, -hous·es.** To put or re-establish in a new, usu. improved dwelling.
**reichs·mark** (rīks'märk') *n., pl.* **reichsmark** *or* **-marks.** [G. : *Reichs,* genitive of *Reich,* realm + *Mark,* unit of currency < MHG *marke.*] A monetary unit of Germany from 1925–48.
**re·i·fy** (rē'ə-fī', rā'-) *vt.* **-fied, -fy·ing, -fies.** [Lat. *res,* thing + -FY.] To regard or treat (an abstraction) as if it had concrete or material existence. **—re'i·fi·ca'tion** (-fĭ-kā'shən) *n.* **—re'i·fi'er** *n.*
**reign** (rān) *n.* [ME *reigne* < OFr. < Lat. *regnum* < *rex,* king.] **1.** Exercise of sovereign power, as by a monarch. **2.** The term during which a sovereign rules. **3.** Widespread influence <the *reign* of reason> *—vi.* **reigned, reign·ing, reigns. 1.** To exercise sovereign power. **2.** To hold the title of sovereign, but with restricted power. **3.** To have the greatest authority or importance : PREVAIL.
**reign of terror** *n.* [After the *Reign of Terror,* a period during the French revolution when thousands of persons were executed.] A time of violence committed esp. by those in power that results in widespread terror.
**re·im·burse** (rē'ĭm-bûrs') *vt.* **-bursed, -burs·ing, -burs·es.** [RE- + obs. *imburse,* to pay < Med. Lat. *imbursare,* to appropriate (Lat. *in-,* in + Lat. *bursa,* bag < Gk., skin).] **1.** To repay <*reimbursed* their travel expenses> **2.** To pay back or compensate (a person) for money spent or for losses or damages incurred. **—re'im·burs'a·ble** *adj.* **—re'im·burse'ment** *n.*
**re·im·port** (rē'ĭm-pôrt', -pōrt', rē-ĭm'pôrt', -pōrt') *vt.* **-port·ed, -port·ing, -ports.** To bring back into a country (goods made from raw materials orig. exported from that country). *—n.* (rē-ĭm'pôrt', -pōrt'). **1.** An act of reimporting. **2.** Reimported goods. **—re'im·por·ta'tion** *n.*
**re·im·pres·sion** (rē'ĭm-prĕsh'ən) *n.* REPRINT 1a.
**rein** (rān) *n.* [ME *reine* < OFr. *resne* < VLat. **retina* < Lat. *retinēre,* to retain.] **1.** *often* **reins.** A long narrow leather strap fastened to the bit of a bridle and used by a rider or driver for controlling a horse or other animal. **2.** A means of restraint, check, or guidance. *—v.* **reined, rein·ing, reins.** *—vt.* **1.** To check or hold back. **2.** To guide or control. **3.** To equip with reins. *—vi.* **1.** To control a horse with reins. **2.** To control oneself as if with reins. **—give (free) rein to.** To free from controls. **—tight rein.** Close control.
**re·in·car·nate** (rē'ĭn-kär'nāt') *vt.* **-nat·ed, -nat·ing, -nates.** To incarnate again. **—re'in·car·na'tion** *n.* **—re'in·car·na'tion·ist** *n.*

**rein·deer** (rān'dîr') *n., pl.* **reindeer** *or* **-deers.** [ME *reyndere* < ON *hreindýri* : *hreinn,* reindeer + *dyr,* deer.] A large deer, *Rangifer tarandus* of Arctic regions of the Old World and Greenland, with branched antlers in both sexes.
**reindeer moss** *n.* An erect, grayish, branching lichen, *Cladonia rangiferina* of Arctic regions, the chief source of food for reindeer.
**re·in·fec·tion** (rē'ĭn-fĕk'shən) *n.* A second infection that occurs after recovery from an earlier infection by the same causative agent.
**re·in·force** *also* **re·en·force** (rē'ĭn-fôrs', -fōrs') *vt.* **-forced, -forc·ing, -forc·es.** [RE- + *inforce,* var. of ENFORCE.] **1.** To give more force or effectiveness to : STRENGTHEN. **2.** To strengthen militarily with supplementary manpower or materiel. **3.** To strengthen, as by adding extra support or padding. **4.** To increase in number. **5.** *Psychol.* **a.** To reward (e.g., an experimental subject) with reinforcement following a desired response. **b.** To stimulate (a response) with the use of a reinforcer. **—re'in·force'a·ble** *adj.*
**reinforced concrete** *n.* Poured concrete containing steel bars or metal netting to enhance its tensile strength.
**re·in·force·ment** (rē'ĭn-fôrs'mənt, -fōrs'-) *n.* **1.** The act or process of reinforcing or the condition of being reinforced. **2.** Something that reinforces. **3.** *often* **reinforcements.** Additional troops, vessels, or materiel sent to aid a military operation. **4.** *Psychol.* **a.** The occurrence or experimental introduction of an unconditioned stimulus along with a conditioned stimulus. **b.** The strengthening of a conditioned response by this method. **c.** The strengthening of an instrumental or operant conditioned response leading to satisfaction. **d.** An event, circumstance, or condition that increases the chance that a particular response will recur in a situation like that in which the reinforcing condition orig. occurred.
**re·in·forc·er** (rē'ĭn-fôr'sər, -fōr'-) *n. Psychol.* A stimulus, as a reward, that in operant conditioning strengthens a desired response.
**reins** (rānz) *pl.n.* [ME < OFr. < Lat. *renes.*] **1.** The kidneys, loins, or lower back region. **2.** The seat of affections and passions.
**re·in·state** (rē'ĭn-stāt') *vt.* **-stat·ed, -stat·ing, -states. 1.** To bring back into use or existence. **2.** To restore to a former condition or position. **—re'in·state'ment** *n.*
**re·in·sure** (rē-ĭn-shoor') *vt.* **-sured, -sur·ing, -sures. 1.** To insure again. **2.** To insure by contracting to transfer in whole or in part a risk or contingent liability already covered under an existing contract. **—re'in·sur'ance** *n.* **—re'in·sur'er** *n.*
**re·in·te·grate** (rē-ĭn'tĭ-grāt') *vt.* **-grat·ed, -grat·ing, -grates.** To restore to a state of integration or unity. **—re'in·te·gra'tion** *n.* **—re·in'te·gra'tive** *adj.*
**re·in·ter·pret** (rē'ĭn-tûr'prĭt) *vt.* **-pret·ed, -pret·ing, -prets.** To interpret again. **—re'in·ter'pre·ta'tion** (-tûr'prĭ-tā'shən) *n.*
**re·in·vest** (rē'ĭn-vĕst') *vt.* **-vest·ed, -vest·ing, -vests.** To invest (capital or earnings) again. **—re'in·vest'ment** *n.*
**re·in·vig·o·rate** (rē'ĭn-vĭg'ə-rāt') *vt.* **-rat·ed, -rat·ing, -rates.** To make vigorous again <government grants that *reinvigorated* our national arts> **—re'in·vig'o·ra'tion** *n.* **—re'in·vig'o·ra'tor** *n.*
**reis** (rās) *n. var. pl. of* REAL[3].
**re·is·sue** (rē-ĭsh'oo) *v.* **-sued, -su·ing, -sues.** *—vt.* To issue again. *—vi.* To come forth again. *—n.* **1.** A second or subsequent issue, as of a book. **2.** A reprinting of postage stamps from unchanged plates.
**re·it·er·ate** (rē-ĭt'ə-rāt') *vt.* **-at·ed, -at·ing, -ates.** [Lat. *reiterare, reiterat-* : *re-,* again + *iterare,* to iterate < *iterum,* again.] To state again. **—re·it'er·a'tion** *n.* **—re·it'er·a'tive** *adj.* **—re·it'er·a'tive·ly** *adv.* **—re·it'er·a'tive·ness** *n.* **—re·it'er·a'tor** *n.*
**re·ject** (rĭ-jĕkt') *vt.* **-ject·ed, -ject·ing, -jects.** [ME *rejecten* < Lat. *reicere,* to throw back : *re-,* back + *jacere,* to throw.] **1.** To refuse to accept, recognize, or make use of : REPUDIATE. **2.** To refuse to consider or grant : DENY. **3.** To refuse affection or recognition to (a person). **4.** To throw away : DISCARD. **5.** To spit out or vomit. *—n.* (rē'jĕkt'). One rejected. **—re·ject'er, re·jec'tor** *n.* **—re·jec'tion** (-jĕk'shən) *n.* **—re·jec'tive** *adj.*
☆ ***syns:*** REJECT, DECLINE, DISMISS, REFUSE, SPURN, TURN DOWN *v. core meaning* : to be unwilling to accept, consider, or receive <*rejected* all their suggestions> ***ant:*** accept
**rejection slip** *n.* A printed form or note enclosed with an author's manuscript that has been rejected for publication.
**rejective art** *n.* Minimal art. **—re·jec'tiv·ist** (rĭ-jĕk'tĭ-vĭst') *adj.*
**re·joice** (rĭ-jois') *v.* **-joiced, -joic·ing, -joic·es.** [ME *rejoicen* < OFr. *rejoir, rejoiss-* : *re-* (intensive) + *joir,* to be joyful < Lat. *gaudēre.*] *—vi.* To feel or be joyful. *—vt.* To fill with joy : GLADDEN. **—rejoice in.** To have or possess <*rejoice in* a large and loving family> **—re·joic'er** *n.* **—re·joic'ing·ly** *adv.*
☆ ***syns:*** REJOICE, DELIGHT, EXULT *v. core meaning* : to feel or take joy or pleasure <*rejoiced* at the sight of their lost kitten>
**re·join** (rĭ-join') *v.* **-joined, -join·ing, -joins.** [ME *rejoinen* < OFr. *rejoindre* : *re-,* back (< Lat.) + *joindre,* to join < Lat. *jungere.*] *—vt.* To say as a reply. *—vi.* **1.** To answer : respond. **2.** *Law.* To answer a plaintiff's replication.
**re-join** (rē-join') *v.* **-joined, -join·ing, -joins.** *—vt.* **1.** To come together again in company with. **2.** To join or put together again : REUNITE. *—vi.* To be or become joined again.
**re·join·der** (rĭ-join'dər) *n.* [ME *rejoyner* < OFr. *rejoindre,* to answer, rejoin.] **1.** An answer, esp. to a reply. **2.** *Law.* A second pleading by a defendant, in answer to a plaintiff's replication.

**re·ju·ve·nate** (rĭ-jo͞o′və-nāt′) *vt.* **-nat·ed, -nat·ing, -nates.** [RE- + Lat. *juvenis,* a youth.] **1.** To restore the youthful appearance or vigor of. **2.** *Geol.* **a.** To stimulate (a stream) to renewed erosive activity, as by uplift. **b.** To develop youthful topographical features in (a previously leveled area). **—re·ju′ve·na′tion** *n.* **—re·ju′ve·na′tor** (-tər) *n.*

**re·ju·ve·nes·cence** (rĭ-jo͞o′və-nĕs′əns) *n.* Renewal of youthful appearance or character. **—re·ju′ve·nes′cent** *adj.*

**re·laid** (rē-lād′, rē′lād′) *v. p.t. & p.p. of* RE-LAY.

**re·lapse** (rĭ-lăps′) *vi.* **-lapsed, -laps·ing, -laps·es.** [Lat. *relabi, relaps-* : *re-,* back + *labi,* to slide.] **1.** To fall back or revert to an earlier state. **2.** To regress after partial recovery from illness. **3.** To slip back into bad ways : BACKSLIDE. —*n.* (rē′lăps, rĭ-lăps′). An act, instance, or result of relapsing. **—re·laps′er** *n.*

**relapsing fever** *n.* An infectious disease marked by chills and fever, and caused by spirochetes transmitted by lice and ticks.

**re·late** (rĭ-lāt′) *v.* **-lat·ed, -lat·ing, -lates.** [Lat. *referre, relat-* : *re-,* back + *ferre,* to bear.] —*vt.* **1.** To tell or narrate. **2.** To bring into logical or natural association. —*vi.* **1.** To have connection, relation, or reference. **2.** To interact with others in a significant or coherent way <always *related* easily to my peers> **3.** To respond, esp. in a favorable way. **—re·lat′a·ble** *adj.* **—re·lat′er** *n.*

**re·lat·ed** (rĭ-lā′tĭd) *adj.* **1.** Connected : associated. **2.** Connected by kinship, marriage, or common origin. **3.** Having a close harmonic connection. **—re·lat′ed·ly** *adv.* **—re·lat′ed·ness** *n.*

**re·la·tion** (rĭ-lā′shən) *n.* **1.** A logical or natural association between two or more things : CONNECTION. **2.** Connection of people by blood or marriage : KINSHIP. **3.** One connected to another by blood or marriage : RELATIVE. **4.** The mode which connects one with another <the *relation* of parent to child> **5. relations.** The connections, dealings, or associations bringing together persons, groups, or nations in personal, business, or diplomatic affairs. **6.** Reference : regard. **7. a.** An act of telling or narrating. **b.** A narrative : account. **8.** *Law.* Assumption that an act or proceeding has occurred in advance of its completion or official enactment. **9. relations.** Sexual intercourse.

**re·la·tion·al** (rĭ-lā′shə-nəl) *adj.* **1.** Of or arising from kinship. **2.** Designating or constituting relations. **3.** Expressing a syntactic relation. **—re·la′tion·al·ly** *adv.*

**re·la·tion·ship** (rĭ-lā′shən-shĭp′) *n.* **1.** The state or fact of being related. **2.** Connection by blood or marriage : KINSHIP. **3.** A particular state of affairs among people related to or dealing with one another <have a close *relationship* with their cousins>

**rel·a·tive** (rĕl′ə-tĭv) *adj.* [ME *relatif* < LLat. *relativus* < Lat. *referre,* to relate.] **1.** Having relevance. **2.** Regarded in comparison to or relationship with something else. **3.** Dependent on or interconnected with something else for intelligibility or significance : not absolute. **4.** Referring to or qualifying a grammatical antecedent. **5.** *Mus.* Bearing the same key signature. —Used of major and minor scales and keys. —*n.* **1.** One related to another by kinship. **2.** One that is relative. **3.** A relative term. **—rel′a·tive·ly** *adv.* **—rel′a·tive·ness** *n.*

**relative biological effectiveness** *n. Physics.* A measure of the capacity of a specific ionizing radiation to produce a specific biological effect, usu. expressed as the dose of radium gamma rays or 200-kilovolt x-rays relative to the dose of the ionization in question required to produce the effect.

**relative clause** *n.* A dependent clause introduced by a relative pronoun.

**relative humidity** *n.* The ratio of the amount of water vapor in the air at a specific temperature to the maximum capacity of the air at that temperature.

**relative permittivity** *n.* Permittivity.

**relative pitch** *n.* **1.** The pitch of a tone as determined by its location in a scale. **2.** Ability to produce or recognize a tone by mentally establishing a relationship between its pitch and that of a recently heard tone.

**relative pronoun** *n.* A pronoun introducing a relative clause and referring to an antecedent.

**relative to** *prep.* With regard to.

**rel·a·tiv·ism** (rĕl′ə-tĭ-vĭz′əm) *n. Philos.* The theory that truth is an ethical relative to the group or individual that holds it.

**rel·a·tiv·ist** (rĕl′ə-tĭ-vĭst) *n.* **1.** A proponent of relativism. **2.** A physicist concentrating on the theories of relativity.

**rel·a·tiv·is·tic** (rĕl′ə-tĭ-vĭs′tĭk) *adj.* **1.** Of or relating to relativism. **2.** *Physics.* **a.** Of, relating to, or caused by speeds that are large with respect to the speed of light <*relativistic* increase in mass> **b.** Of or relating to phenomena that can be explained by special or general relativity <*relativistic* mechanics>

**rel·a·tiv·i·ty** (rĕl′ə-tĭv′ĭ-tē) *n.* **1.** The quality or state of being relative. **2.** *Philos.* Existence dependent entirely on relation to a thinking mind. **3.** A state of dependence in which the existence or significance of one entity is entirely dependent on that of another. **4.** *Physics.* **a.** Special relativity. **b.** General relativity.

**re·la·tor** (rĭ-lā′tər) *n.* One who relates or narrates.

**re·lax** (rĭ-lăks′) *v.* **-laxed, -lax·ing, -lax·es.** [ME *relaxen* < Lat. *relaxare* : *re-,* back + *laxare,* to loosen < *laxus,* loose.] —*vt.* **1.** To make lax or loose <*relax* my hold on the line> **2.** To make less strict or severe. **3.** To reduce in intensity : SLACKEN. **4.** To relieve from effort or strain. —*vi.* **1.** To take one's ease : REST. **2.** To become lax or loose. **3.** To become less strict or severe. **4.** To become less formal, aloof, or tense. **—re·lax′a·ble** *adj.* **—re·lax′er** *n.*

**re·lax·ant** (rĭ-lăk′sənt) *n.* Something, as a drug or therapeutic treatment, that relaxes or relieves nervous and muscular tension. —*adj.* Tending to relax or to relieve tension.

**re·lax·a·tion** (rē′lăk-sā′shən) *n.* **1.** An act of relaxing or the state of being relaxed. **2.** Mental or physical refreshment : RECREATION <swim for *relaxation*> **3.** A loosening or slackening. **4.** Reduction in strictness or severity. **5.** *Physiol.* The lengthening of inactive muscle or muscle fibers. **6.** *Physics.* Return or adjustment of a system to equilibrium following displacement or abrupt change. **7.** *Math.* A numerical method in which the errors following an initial approximation are reduced by succeeding approximations until all errors are within specified limits. **—re·lax′a·tive** *adj. & n.*

**relaxation time** *n. Physics.* The time needed for an exponential variable to decrease to $1/e$ (0.368) of its initial value.

**re·laxed** (rĭ-lăkst′) *adj.* **1.** Not strict or rigorous. **2.** Not rigid or tense. **3.** Informal and easy in manner.

**re·lax·in** (rĭ-lăk′sən) *n.* A female hormone secreted by the corpus luteum that helps soften the cervix and relax the pelvic ligaments in childbirth.

**re·lay** (rē′lā, rĭ-lā′) *n.* [ME *relai* < OFr. < *relaier,* to relay : *re-,* back + *laier,* to leave, alteration of *laissier* < Lat. *laxare,* to loosen < *laxus,* loose.] **1.** A fresh team, as of horses or dogs, to relieve weary animals in a hunt, task, or journey. **2.** A labor crew that relieves another crew at work : SHIFT. **3.** An act of passing something along from one person, group, or station to another. **4. a.** A relay race. **b.** A division of a relay race. **5.** *Elect.* An automatic electromagnetic or electromechanical device that responds to a small current or voltage change by activating switches or other devices in an electric circuit. —*vt.* **-layed, -lay·ing, -lays. 1.** To pass or send along by or as if by relay <*relay* the good news> **2.** To equip with fresh relays. **3.** *Elect.* To control or retransmit by way of a relay.

**re-lay** (rē-lā′) *vt.* **-laid** (-lād′), **-lay·ing, -lays.** To lay again.

**relay race** *n.* A race between two or more teams, in which each team member runs only a specific stretch of the race, and then is relieved by another member of his or her team.

**re·leas·a·ble** (rĭ-lē′sə-bəl) *adj.* **1.** Capable of being released. **2.** Designed to release <*releasable* ski bindings> **—re·leas·a·bil′i·ty** *n.* **—re·leas′a·bly** *adv.*

**re·lease** (rĭ-lēs′) *vt.* **-leased, -leas·ing, -leas·es.** [ME *relesen* < OFr. *relaissier* < Lat. *relaxare,* to relax.] **1.** To set free from confinement, restraint, or bondage : LIBERATE. **2.** To unfasten, free, or let go of. **3.** To relieve from obligation or debt. **4.** To permit performance, sale, publication, or circulation of. **5.** To give up (e.g., a right). —*n.* **1.** A deliverance : liberation. **2.** An authoritative discharge, as from an obligation or from prison. **3.** An unfastening or letting go of something caught or held fast. **4.** A device or catch to lock or release a mechanism. **5. a.** A freeing of something for general publication, use, or circulation. **b.** Something thus released <a news *release*> **6.** *Law.* **a.** Abandonment of a right, title, or claim to another. **b.** The document granting such a relinquishment. **—re·leas′er** *n.*

**re-lease** (rē-lēs′) *vt.* **-leased, -leas·ing, -leas·es.** To lease again.

**released time** *n.* A part of a regular school day in some U.S. public schools when children are excused from class to receive outside religious instruction.

**rel·e·gate** (rĕl′ə-gāt′) *vt.* **-gat·ed, -gat·ing, -gates.** [Lat. *relegare, relegat-,* to send away : *re-,* back + *legare,* to send.] **1.** To send or consign, esp. to an obscure place, position, or condition. **2.** To assign to a specified class. **3.** To refer or assign (e.g., a task) for performance or decision. **4.** To banish : exile. **—rel′e·ga′tion** *n.*

**re·lent** (rĭ-lĕnt′) *v.* **-lent·ed, -lent·ing, -lents.** [ME *relenten* < Med. Lat. **relentare* : *re-* (intensive < Lat.) + *lentare,* to soften < Lat., to bend < *lentus,* flexible.] —*vi.* To become softened or gentler in attitude or temper. —*vt.* To cause to slacken or abate.

**re·lent·less** (rĭ-lĕnt′lĭs) *adj.* **1.** Unyielding : pitiless <*relentless* punishment> **2.** Steady and persistent : UNREMITTING <*relentless* cold> **—re·lent′less·ly** *adv.* **—re·lent′less·ness** *n.*

**rel·e·vance** (rĕl′ə-vəns) *also* **rel·e·van·cy** (-vən-sē) *n.* **1. a.** Pertinence to the matter at hand. **b.** Social applicability <a college curriculum with no *relevance*> **2.** The capability of an information retrieval system to select and retrieve data suitable for a user's needs.

**rel·e·vant** (rĕl′ə-vənt) *adj.* [Med. Lat. *relevans, relevant-* < Lat., pr.part. of *relevare,* to relieve.] Related to the matter at hand. **—rel′e·vant·ly** *adv.*

☆ ***syns:*** RELEVANT, APROPOS, GERMANE, MATERIAL, PERTINENT *adj. core meaning* : related to the matter at hand <*relevant* questions><*relevant* issues> ***ant:*** extraneous, irrelevant

**re·li·a·ble** (rĭ-lī′ə-bəl) *adj.* That can be relied upon : DEPENDABLE. **—re·li′a·bil′i·ty, re·li′a·ble·ness** *n.* **—re·li′a·bly** *adv.*

**re·li·ance** (rĭ-lī′əns) *n.* **1.** The act of relying. **2.** Confidence : trust. **3.** One that can be depended on.

**re·li·ant** (rĭ-lī′ənt) *adj.* Having or demonstrating reliance : TRUSTING. **—re·li′ant·ly** *adv.*

**rel·ic** (rĕl′ĭk) *n.* [ME *relik* < OFr. *relique* < LLat. *reliquiae,* sacred relics < Lat., remains < *relinquere,* to leave behind. —see RELINQUISH.] **1. a.** Something that has survived decay or deterioration. **b.** A belief or custom remaining as a trace of an earlier culture or an obsolete practice. **2.** Something held dear for its age or associations with a person, place, or event : KEEPSAKE. **3.** An object of religious veneration, esp. an article supposed to be associated with a saint or martyr. **4. relics.** A corpse : remains.

**rel·ict** (rĕl′ĭkt, rĭ-lĭkt′) *n.* [Lat. *relictus,* p.part. of *relinquere,* to leave behind.—see RELINQUISH.] **1.** *Ecol.* An organism or species of an earlier era surviving in an environment that has changed considerably. **2.** A widow. —*adj. Geol.* Relating to something that has survived, as structures or minerals after destructive processes.

**re·lic·tion** (rĭ-lĭk′shən) *n. Geol.* Gradual recession of water, leaving permanently dry land.

**re·lief** (rĭ-lēf′) *n.* [ME < OFr. < *relever,* to relieve.] **1.** Ease from or lessening of discomfort or pain. **2.** Something that reduces pain, discomfort, fear, or anxiety. **3.** Aid, as money or food, given to the needy, aged, or to the inhabitants of a disaster area. **4. a.** A release from a job, post, or duty, as of a sentinel. **b.** One who assumes the duties of another. **5. a.** The projection of figures or forms from a flat background, or such a projection that is apparent only, as in painting. **b.** A work of art done in this way. **6.** Variations in elevation of a portion of the earth's surface. **7.** Distinction or prominence caused by contrast <"the light brought the white church . . . into *relief* from the flat ledges" —Willa Cather> **8.** A payment made by the heir of a deceased tenant to a lord for the privilege of succeeding to the tenant's estate in feudal law. **—on relief.** Receiving government funds because of poverty or need.

☆ ***syns:*** RELIEF, DOLE, HANDOUT, WELFARE *n. core meaning* : assistance, esp. money, food, and other necessities, given to the needy or dispossessed <received *relief* following the flood>

**relief**
(Top) *Low relief and* (bottom) *high relief*

**relief map** *n.* A map showing land configuration, as with contour lines, shading, or colors.

**relief pitcher** *n. Baseball.* A pitcher who is ready to replace another pitcher during a game.

**re·lieve** (rĭ-lēv′) *vt.* **-lieved, -liev·ing, -lieves.** [ME *releven* < OFr. *relever* < Lat. *relevare* : *re-* (intensive) + *levare,* to raise < *levis,* light.] **1.** To lessen or alleviate <a drug that *relieved* all of the symptoms> **2.** To free from pain or trouble <*relieved* our suffering> **3.** To furnish assistance to. **4.** To release (a person) from an obligation, restriction, or burden, as by judicial action or legislation. **5.** To free from a designated duty by providing or acting as a substitute. **6.** To make less unpleasant or tiresome. **7.** To make clear or more effective through contrast : SET OFF <A black sash *relieves* a white gown.> **8.** To excrete the bodily wastes of (oneself). **9. a.** To take away from. **b.** To rob <*relieved* me of my watch and money> **—re·liev·a·ble** *adj.* **—re·liev′er** *n.*

**re·lie·vo** (rĭ-lē′vō) *n., pl.* **-vos.** [Ital. < *rilievare,* to raise < Lat. *relevare.* —see RELIEVE.] RELIEF 5.

**re·li·gion** (rĭ-lĭj′ən) *n.* [ME *religioun* < OFr. *religion* < Lat. *religio.*] **1. a.** Belief in and reverence for a supernatural power accepted as the creator and governor of the universe. **b.** A specific unified system of this expression <the Buddhist *religion*> **2.** The spiritual or emotional attitude of one who recognizes the existence of a superhuman power or powers. **3.** An objective pursued with fervor or conscientious devotion <make a *religion* of a hobby>

☆ ***syns:*** RELIGION, CREED, DENOMINATION, FAITH, PERSUASION, SECT *n. core meaning* : a system of religious belief <many *religions* throughout the world>

**re·lig·ion·ism** (rĭ-lĭj′ən-ĭz′əm) *n.* Overdone or affected religious zeal. **—re·lig′ion·ist** *n.*

**re·li·gi·ose** (rĭ-lĭj′ē-ōs′) *adj.* [< RELIGIOUS.] Overly religious, esp. in a conspicuous or sentimental manner.

**re·lig·i·os·i·ty** (rĭ-lĭj′ē-ŏs′ĭ-tē) *n., pl.* **-ties. 1.** The state of being religious. **2.** Excessive or affected piety.

**re·lig·ious** (rĭ-lĭj′əs) *adj.* [ME < OFr. < Lat. *religiosus* < *religio,* religion.] **1.** Of, relating to, or teaching religion. **2.** Adhering to or manifesting religion : PIOUS. **3.** Extremely faithful : CONSCIENTIOUS <*religious* devotion to duty> —*n., pl.* **religious.** One, as a sister or friar, belonging to a religious order. **—re·li′gious·ly** *adv.* **—re·li′gious·ness** *n.*

**re·line** (rē-līn′) *vt.* **-lined, -lin·ing, -lines. 1.** To make new lines on. **2.** To put a new lining in.

**re·lin·quish** (rĭ-lĭng′kwĭsh) *vt.* **-quished, -quish·ing, -quish·es.** [ME *relinquisshen* < OFr. *relinquir, relinquiss-* < Lat. *relinquere,* to leave behind : *re-,* behind + *linquere,* to leave.] **1.** To retire from : ABANDON. **2.** To put aside or desist from (something practiced, professed, or intended). **3.** To surrender : renounce. **4.** To cease holding physically : RELEASE. **—re·lin′quish·er** *n.* **—re·lin′quish·ment** *n.*

**rel·i·quar·y** (rĕl′ə-kwĕr′ē) *n., pl.* **-ies.** [Fr. *reliquaire* < Med. Lat. *reliquiarium* < LLat. *reliquiae,* sacred relics. —see RELIC.] A receptacle, as a coffer or shrine, for keeping or displaying sacred relics.

**rel·ique** (rĕl′ĭk) *n. Archaic. var. of* RELIC.

**re·liq·ui·ae** (rĭ-lĭk′wĭ-ē′) *pl.n.* [Lat., remains. —see RELIC.] Remains, esp. of fossil organisms.

**rel·ish** (rĕl′ĭsh) *n.* [ME *reles,* taste < OFr., something remaining < *relaissier,* to leave behind < Lat. *relaxare,* to relax.] **1.** An appetite for something : LIKING. **2. a.** Pleasure : zest. **b.** Something that adds pleasure or zest. **3.** A savory or spicy condiment served with food. **4.** The flavor of a food, esp. when appetizing. **5.** A trace or suggestion of an important quality. —*v.* **-ished, -ish·ing, -ish·es.** —*vt.* **1.** To take pleasure in : ENJOY. **2.** To like the flavor of. **3.** To give spice or flavor to. —*vi.* To have a distinctive or pleasing taste. **—rel′ish·a·ble** *adj.*

**re·live** (rē-lĭv′) *v.* **-lived, -liv·ing, -lives.** —*vt.* To undergo again, esp. via the imagination <*relived* the accident> —*vi.* To live again.

**re·lo·cate** (rē-lō′kāt′) *v.* **-cat·ed, -cat·ing, -cates.** —*vt.* To establish in a new place. —*vi.* To become established in a new residence or place of business <*relocated* in Ohio> **—re′lo·ca′tion** *n.*

**re·lu·cent** (rĭ-lo͞o′sənt) *adj.* [Lat. *relucens,* pr.part. of *relucēre,* to shine back : *re-,* back + *lucēre,* to shine.] Reflecting light : SHINING.

**re·luct** (rĭ-lŭkt′) *vi.* **-luct·ed, -luct·ing, -lucts.** [Lat. *reluctari* : *re-,* against + *luctari,* to struggle.] To show reluctance or repugnance <readers who *reluct* at explicit violence in novels>

**re·luc·tance** (rĭ-lŭk′təns) *also* **re·luc·tan·cy** (-tən-sē) *n.* **1.** The state of being reluctant : DISINCLINATION. **2.** *Physics.* A magnetic quantity analogous to electric resistance and equal in a closed magnetic circuit to the ratio of circuit length to the product of cross-sectional area and permeability.

**re·luc·tant** (rĭ-lŭk′tənt) *adj.* [Lat. *reluctans, reluctant-,* pr.part. of *reluctari,* to reluct.] **1.** Unwilling : averse <*reluctant* to go> **2.** Marked by unwillingness <*reluctant* to help me> **3.** Offering resistance : OPPOSING. **—re·luc′tant·ly** *adv.*

**rel·uc·tiv·i·ty** (rĕl′ək-tĭv′ĭ-tē) *n., pl.* **-ties.** [RELUCT(ANCE) + (CONDUCT)IVITY.] *Physics.* A measure of the resistance of a material to the establishment of a magnetic field within it, equal to the reciprocal of magnetic permeability.

**re·lume** (rĭ-lo͞om′) *vt.* **-lumed, -lum·ing, -lumes.** [RE- + (IL)LUME.] To illuminate again.

**re·ly** (rĭ-lī′) *vi.* **-lied, -ly·ing, -lies.** [ME *relien,* to rally < OFr. *relier* < Lat. *religare,* to bind fast : *re-* (intensive) + *ligare,* to bind.] **1.** To depend <*rely* on one's doctor for help with health problems> **2.** To trust confidently <*rely* on the child to do as told> **—re·li′er** *n.*

**rem** (rĕm) *n.* [R(OENTGEN) E(QUIVALENT IN) M(AN).] *Physics.* The amount of ionizing radiation needed to produce the same biological effect as one roentgen of high-penetration x-rays.

**REM** (rĕm) *n.* [R(APID) E(YE) M(OVEMENT).] Rapid, periodic, jerky movement of the eyes during particular phases of the sleep cycle when dreaming occurs.

**re·main** (rĭ-mān′) *vi.* **-mained, -main·ing, -mains.** [ME *remaynen* < OFr. *remaindre* < Lat. *remanere* : *re-,* back + *manere,* to stay.] **1.** To continue without change of condition, quality, or place. **2.** To stay or be left over after the removal, departure, loss, or destruction of others. **3.** To be left as still to be dealt with <Repairs *remain* to be completed.> **4.** To persist or endure.

**re·main·der** (rĭ-mān′dər) *n.* [ME < AN < OFr. *remaindre,* to remain.] **1.** Something left over after other parts have been taken away. **2.** *Math.* **a.** The dividend less the product of the divisor and quotient in division. **b.** The difference in subtraction. **3.** *Law.* An estate effective and enjoyable only after the settlement of another estate. **4.** A book remaining with a publisher after sales have dropped, sold at a reduced price. —*vt.* **-dered, -der·ing, -ders.** To sell (books) as remainders.

**re·mains** (rĭ-mānz′) *pl.n.* **1.** All that is left after other parts have been taken away, used up, or destroyed. **2.** A corpse. **3.** A deceased author's unpublished writings. **4.** Ancient ruins or fossils.

**re·make** (rē-māk′) *vt.* **-made** (-mād′), **-mak·ing, -makes.** To make anew : RECONSTRUCT. —*n.* **1.** An instance of making anew. **2.** Something made again <a *remake* of an old film>

**re·man** (rē-măn′) *vt.* **-manned, -man·ning, -mans. 1.** To supply with new personnel, as for work. **2.** To renew the courage of.

**re·mand** (rĭ-mănd′) *vt.* **-mand·ed, -mand·ing, -mands.** [ME *remaunden* < OFr. *remander* < LLat. *remandere,* to send back : *re-,* back + *mandare,* to order.] **1.** To send or order back. **2.** *Law.* **a.** To send back (one in custody) to prison, to another court, or to another agency for further action. **b.** To send back (a case) to a lower court with orders about further action. —*n.* **1.** An act of remanding or the state of being remanded. **2.** One remanded. **—re·mand′ment** *n.*

**rem·a·nence** (rĕm′ə-nəns) *n.* [< ME *remanent,* remaining < Lat. *remanens,* pr.part. of *remanēre,* to remain.] *Physics.* Magnetic induction remaining in a material after removal of the magnetizing force. **—rem′a·nent** *adj.*

**re·mark** (rĭ-märk′) *v.* **-marked, -mark·ing, -marks.** [Fr. *remarquer* < OFr. : *re-* (intensive < Lat.) + *marquer,* to mark, ult. of Germanic orig.] —*vt.* **1.** To say or write casually and briefly as a comment. **2.** To take notice of : OBSERVE. —*vi.* To make a comment or observation. —*n.* **1.** An act of noticing and mentioning <a picture deserving of *remark*> **2.** A casual or brief expression of opinion : COMMENT. **—re·mark′er** *n.*

**re·mark·a·ble** (rĭ-mär′kə-bəl) *adj.* **1.** Worthy of notice. **2.** Extraordinary. **—re·mark′a·ble·ness** *n.* **—re·mark′a·bly** *adv.*

**re·marque** (rĭ-märk′) *n.* [Fr., remark < *remarquer,* to remark.] **1.** A mark put in the margin of an engraving plate to show its stage of development in advance of completion. **2.** A print or proof from a plate bearing such a mark.

**re·match** (rē-măch′, rē′măch′) *n.* A second contest between the same opponents.

**re·me·di·a·ble** (rĭ-mē′dē-ə-bəl) *adj.* Capable of being remedied. **—re·me′di·a·ble·ness** *n.* **—re·me′di·a·bly** *adv.*

**re·me·di·al** (rĭ-mē′dē-əl) *adj.* **1.** Providing a remedy. **2.** Meant to correct, esp. poor study or reading habits. **—re·me′di·al·ly** *adv.*

**re·me·di·ate** (rĭ-mē′dē-āt′) *vt.* **-at·ed, -at·ing, -ates.** To provide remedial aid for (e.g., a learning disability). **—re·me′di·a·tion** *n.*

**rem·e·dy** (rĕm′ə-dē) *n., pl.* **-dies.** [ME < AN *remedie* < Lat. *remedium* : *re-,* against + *mederi,* to heal.] **1.** Something, as medicine or therapy, that relieves pain, cures disease, or corrects a disorder. **2.** Something that corrects an evil, fault, or error <tried to find a *remedy* for child abuse> **3.** *Law.* A legal way of preventing or correcting a wrong or enforcing a right. **4.** A mint's allowance for deviation from the standard weight or quality of coins. —*vt.* **-died, -dy·ing, -dies. 1.** To relieve or cure (a disease or disorder). **2.** To set right or rectify (an error).

**re·mem·ber** (rĭ-mĕm′bər) *v.* **-bered, -ber·ing, -bers.** [ME *remembren* < OFr. *remembrer* < LLat. *rememorari,* to remember again : Lat. *re-,* again + LLat. *memorari,* to be mindful of < Lat. *memor,* mindful.] —*vt.* **1.** To recall to the mind **2.** To retain in the mind <*remember* all the lines of the leading role> **3.** To keep (someone) in mind as deserving of affection or recognition. **4.** To reward with a gift or tip. **5.** To give greetings from. **6.** *Archaic.* To remind. —*vi.* To have or use the faculty of memory. **—re·mem′ber·a·bil′i·ty** *n.* **—re·mem′ber·a·ble** *adj.* **—re·mem′ber·er** *n.*

☆ ***syns:*** REMEMBER, RECALL, RECOLLECT *v. core meaning* **:** to bring to mind <*remembered* their happy childhood> ***ant:*** forget

**re·mem·brance** (rĭ-mĕm′brəns) *n.* [ME < OFr. < *remembrer,* to remember.] **1.** An act of remembering or the state of being remembered. **2.** Something serving to celebrate or honor the memory of a person or event : MEMORIAL. **3.** The length of time spanned by one's memory. **4.** Something remembered : REMINISCENCE. **5.** A memento : souvenir. **6.** A greeting.

**re·mem·branc·er** (rĭ-mĕm′brən-sər) *n.* One that causes another to remember : REMINDER.

**re·mex** (rē′mĕks′) *n., pl.* **rem·i·ges** (rĕm′ə-jēz′) [NLat. < Lat., rower : *remus,* oar + *agere,* to drive.] A quill or flight feather of a bird's wing. **—re·mig′i·al** (rĭ-mĭj′ē-əl) *adj.*

**re·mil·i·ta·rize** (rē-mĭl′ĭ-tə-rīz′) *vt.* **-rized, -riz·ing, -riz·es.** To militarize again, as with forces. **—re·mil′i·ta·ri·za′tion** *n.*

**re·mind** (rĭ-mīnd′) *vt.* **-mind·ed, -mind·ing, -minds.** To cause to remember. **—re·mind′er** *n.*

**rem·i·nisce** (rĕm′ə-nĭs′) *vi.* **-nisced, -nisc·ing, -nisc·es.** [Back-formation < REMINISCENT.] To recollect and tell of bygone experiences or events.

**rem·i·nis·cence** (rĕm′ə-nĭs′əns) *n.* **1.** The act or process of recalling the past. **2.** A thing remembered : MEMORY. **3.** *often* **reminiscences.** An account of past experiences. **4.** An occurrence that brings to mind a similar, previous occurrence.

**rem·i·nis·cent** (rĕm′ə-nĭs′ənt) *adj.* [Lat. *reminiscens, reminiscent-,* pr.part. of *reminisci,* to recollect.] **1.** Having the quality of or containing reminiscence. **2.** Tending to recall or talk of the past. **—rem′i·nis′cent·ly** *adv.*

**re·mint** (rē-mĭnt′) *vt.* **-mint·ed, -mint·ing, -mints.** To make into new coin by melting down and reprocessing.

**re·mise** (rĭ-mīz′) *vt.* **-mised, -mis·ing, -mis·es.** [ME *remisen* < OFr. *remis,* p.part. of *remettre,* to remit < Lat. *remittere.* —see REMIT.] *Law.* To relinquish a claim to : surrender by deed.

**re·miss** (rĭ-mĭs′) *adj.* [ME < Lat. *remissus,* slack, p.part. of *remittere,* to remit.] **1.** Lax in attending to duty : NEGLIGENT. **2.** Showing carelessness or slackness. **—re·miss′ly** *adv.* **—re·miss′ness** *n.*

**re·mis·si·ble** (rĭ-mĭs′ə-bəl) *adj.* Capable of being remitted <*remissible* sins> **—re·mis′si·bil′i·ty** *n.* **—re·mis′si·bly** *adv.*

**re·mis·sion** (rĭ-mĭsh′ən) *n.* **1. a.** An act of remitting. **b.** The state or period in which something, as disease symptoms, is remitted. **2.** Release, as from an obligation or debt. **3.** A lessening of intensity : ABATEMENT.

**re·mit** (rĭ-mĭt′) *v.* **-mit·ted, -mit·ting, -mits.** [ME *remitten* < Lat. *remittere* : *re-,* back + *mittere,* to send.] —*vt.* **1.** To send (money) <*remit* payment herewith> **2. a.** To refrain from exacting (e.g., a penalty). **b.** To forgive <*remit* a sin> **3.** To restore to an original state. **4.** *Law.* To refer (a case) back to a lower court for further consideration. **5.** To allow (e.g., attention or diligence) to slacken. **6.** To defer : postpone. —*vi.* **1.** To send money. **2.** To diminish, as in intensity : ABATE. **—re·mit′ta·ble** *adj.* **—re·mit′ter** *n.*

**re·mit·tal** (rĭ-mĭt′l) *n.* Remission.

**re·mit·tance** (rĭ-mĭt′ns) *n.* **1.** Money or credit remitted. **2.** An act of sending money or credit.

**re·mit·tent** (rĭ-mĭt′nt) *adj.* Marked by temporary abatements in severity. —Used esp. of diseases. **—re·mit′tence, re·mit′ten·cy** *n.* **—re·mit′tent·ly** *adv.*

**rem·nant** (rĕm′nənt) *n.* [ME *remanent.* —see REMANENCE.] **1.** Something left over : REMAINDER. **2.** A piece of fabric left over after the rest of the bolt has been sold. **3.** A surviving trace. **4.** *often* **remnants.** A small remaining group of people.

**re·mod·el** (rē-mŏd′l) *vt.* **-eled, -el·ing, -els** *also* **-elled, -el·ling, -els.** To remake with a new structure : RECONSTRUCT.

**re·mon·e·tize** (rē-mŏn′ə-tīz′, rē-mŭn′-) *vt.* **-tized, -tiz·ing, -tiz·es.** To restore (e.g., silver) to use as legal tender. **—re·mon′e·ti·za′tion** *n.*

**re·mon·strance** (rĭ-mŏn′strəns) *n.* **1.** An act of remonstrating. **2.** An expression of protest, opposition, or reproof.

**re·mon·strant** (rĭ-mŏn′strənt) *adj.* Marked by remonstrance : EXPOSTULATORY. —*n.* **1.** One who remonstrates. **2. Remonstrant. a.** One of the Dutch Arminians who, in 1610, made a formal statement about the grounds of their dissent from strict Calvinism. **b.** A member of the Protestant denomination that these dissenters founded. **—re·mon′strant·ly** *adv.*

**re·mon·strate** (rĭ-mŏn′strāt′) *v.* **-strat·ed, -strat·ing, -strates.** [Med. Lat. *remonstrare, remonstrat-,* to demonstrate : Lat. *re-* (intensive) + Lat. *monstrare,* to show < *monstrum,* portent.] —*vt.* To say in protest, objection, or reproof. —*vi.* To protest, object, or express a reprove. **—re′mon·stra′tion** (rē′mŏn-strā′shən, rĕm′ən-) *n.* **—re·mon′stra·tive** (rĭ-mŏn′strə-tĭv) *adj.* **—re·mon′stra·tive·ly** *adv.* **—re·mon′stra′tor** *n.*

**re·mon·tant** (rĭ-mŏn′tənt) *adj.* [Fr., pr.part. of *remonter,* to rise again < OFr. —see REMOUNT.] Blooming more than once in a season. —*n.* A remontant rose.

**rem·o·ra** (rĕm′ər-ə) *n.* [Lat., delay < *remorari,* to delay : *re-,* back + *morari,* to delay < *mora,* delay.] Any of several marine fishes of the family Echeneidae, having on the head a sucking disk with which they fasten themselves to sharks, whales, sea turtles, or ship hulls.

**remora**
*Approximately 2 feet long*

**re·morse** (rĭ-môrs′) *n.* [ME < OFr. *remors* < Med. Lat. *remorsus* < Lat. *remordēre,* to torment : *re-* (intensive) + *mordēre,* to bite.] **1.** Moral anguish and bitter regret arising from repentance for past misdeeds. **2.** *Obs.* Compassion. **—re·morse′ful** *adj.* **—re·morse′ful·ly** *adv.* **—re·morse′ful·ness** *n.*

**re·morse·less** (rĭ-môrs′lĭs) *adj.* Lacking remorse. **—re·morse′less·ly** *adv.* **—re·morse′less·ness** *n.*

**re·mote** (rĭ-mōt′) *adj.* **-mot·er, -mot·est.** [Lat. *remotus,* p.part. of *removere,* to remove.] **1.** Situated relatively far away in space. **2.** Distant in time <the *remote* days of yore> **3.** Barely discernible : SLIGHT <a *remote* chance> **4.** Being distantly related by blood or marriage <a *remote* family connection> **5.** Distant in manner : ALOOF. **—re·mote′ly** *adv.* **—re·mote′ness** *n.*

**remote control** *n.* The direction of a remote activity, process, or machine, as by radioed instructions or coded signals.

**re·mo·tion** (rĭ-mō′shən) *n.* [ME *remocion* < Lat. *remotio* < *removēre,* to remove.] **1.** An act of removing. **2.** *Obs.* Departure.

**ré·mou·lade** (rā′mo͞o-läd′) *n.* [Fr.] A piquant cold sauce for cold poultry, meat, and shellfish, made of mayonnaise mixed with chopped pickles, capers, anchovies, and herbs.

**re·mount** (rē-mount′) *vt.* **-mount·ed, -mount·ing, -mounts.** [ME *remounten* < OFr. *remonter,* to remount, to rise again : *re-,* again (< Lat.) + *monter,* to mount. —see MOUNT[1].] **1.** To mount again. **2.** To supply with fresh horses. —*n.* (rē′mount′, rē-mount′). A fresh horse.

**re·mov·a·ble** (rĭ-mo͞o′və-bəl) *adj.* Able to be removed. **—re·mov′a·bil′i·ty, re·mov′a·ble·ness** *n.* **—re·mov′a·bly** *adv.*

**re·mov·al** (rĭ-mo͞o′vəl) *n.* **1.** An act of removing or the fact of having been removed. **2.** Relocation, as of a residence or business. **3.** Dismissal, as from office.

**re·move** (rĭ-mo͞ov′) *v.* **-moved, -mov·ing, -moves.** [ME *removen* < OFr. *remouvoir* < Lat. *removēre,* to move back : *re-,* back + *movēre,* to move.] —*vt.* **1.** To move from a position occupied <*remove* the plates> **2.** To convey from one place to another. **3.** To take from one's person : DOFF <*remove* one's coat> **4.** To take away <*remove* blemishes> **5.** To do away with : ELIMINATE. **6.** To dismiss from office. —*vi.* **1.** To change one's place of residence or business : MOVE. **2.** To depart. —*n.* **1.** REMOVAL 1. **2.** The distance or degree of space, time, or status that separates persons or things <were at a safe *remove* from the explosion> **—re·mov′er** *n.*

**re·moved** (rĭ-mo͞ovd′) *adj.* **1.** REMOTE 1, 2. **2.** Separated in relationship by a particular degree of descent <My first cousin's child is my first cousin once *removed.*> **—re·mov′ed·ly** (-mo͞o′vĭd-lē) *adv.* **—re·mov′ed·ness** (-mo͞o′vĭd-nĭs) *n.*

**re·mu·da** (rĭ-mo͞o′də) *n.* [Mex. Sp., change of horses < Sp., exchange < *remudar,* to exchange : *re-,* in return (< Lat.) + *mudar,* to change < Lat. *mutare.*] A herd of horses from which ranch hands choose their mounts.

**re·mu·ner·ate** (rĭ-myo͞o′nə-rāt′) *vt.* **-at·ed, -at·ing, -ates.** [Lat. *remunerare, remunerat-* : *re-,* in response + *munerare,* to give < *munus,* gift.] **1.** To pay to (one) for goods provided, services rendered, or losses incurred <*remunerated* us for our travel expenses> **2.** To make payment for <*remunerate* their work> **—re·mu′ner·a·bil′i·ty** (-nər-ə-bĭl′ĭ-tē) *n.* **—re·mu′ner·a·ble** *adj.* **—re·mu′ner·a′tion** (-rā′shən) *n.* **—re·mu′ner·a′tor** *n.*

**re·mu·ner·a·tive** (rĭ-myo͞o′nə-rā′tĭv) *adj.* **1.** Apt to be well remunerated : PROFITABLE. **2.** Serving to remunerate. **—re·mu′ner·a′tive·ly** *adv.* **—re·mu′ner·a′tive·ness** *n.*

**Re·mus** (rē′məs) *n.* [Lat.] *Rom. Myth.* Romulus' twin brother.

**ren·ais·sance** (rĕn′ĭ-säns′, -zäns′, rĭ-nā′səns) *n.* [Fr. < OFr. < *renaistre,* to be born again < Lat. *renasci* : *re-,* again + *nasci,* to be born.] **1.** A revival : rebirth. **2. Renaissance. a.** The humanistic revival of classical art, literature, and learning originating in 14th-cent. Italy and later spreading through Europe. **b.** The period of this revival, approx. from the 14th through the 16th cent. **3.** *often* **Renaissance.** A period of revived artistic or intellectual achievement or enthusiasm <the Celtic *Renaissance*> —*adj.* **Renaissance. 1.** Of, relating to, or typical of the Renaissance or its artistic and intellectual works and styles. **2.** Of or indicating the style of architecture and decoration common during the Renaissance.

**Renaissance man** *n.* A man having varied interests and expertise in several areas.

**re·nal** (rē′nəl) *adj.* [Fr. *rénal* < LLat. *renalis* < Lat. *renes,* kidneys.] Of, relating to, or in the region of the kidneys.

**re·nas·cence** (rĭ-năs′əns, -nā′səns) *n.* **1.** A new life or birth : REBIRTH. **2. Renascence.** Renaissance.

**re·nas·cent** (rĭ-năs′ənt, -nā′sənt) *adj.* [Lat. *renascens, renascent-,* pr.part. of *renasci,* to be born again : *re-,* again + *nasci,* to be born.] **1.** Coming into being again. **2.** Showing fresh growth or vigor.

**ren·coun·ter** (rĕn-koun′tər) *n.* [OFr. *rencontre* < *rencontrer,* to meet : *re-,* again (< Lat.) + *encontrer,* to meet.] An unplanned meeting. —*vt.* & *vi.* **-tered, -ter·ing, -ters.** To meet unexpectedly or to have an unexpected meeting.

**rend** (rĕnd) *v.* **rent** (rĕnt) or **rend·ed, rend·ing, rends.** [ME *renden* < OE *rendan.*] —*vt.* **1.** To tear apart or into pieces violently : SPLIT. **2.** To remove forcibly : WREST. **3.** To penetrate and disturb as if by tearing <howls *rending* the night stillness> **4.** To distress (e.g., the heart) painfully. —*vi.* To come apart : BURST.

**ren·der** (rĕn′dər) *vt.* **-dered, -der·ing, -ders.** [ME *rendren,* to give in return < OFr. *rendre,* to give back < VLat. **rendere* < Lat. *reddere* : *re-, red-,* back + *dare,* to give.] **1.** To submit for consideration or approval <*render* a bill> **2.** To give or make available <*render* aid> **3.** To give what is due or correct <*rendered* respect> **4.** To give in return or retribution <*render* regrets for being absent> **5.** To surrender or relinquish : YIELD. **6.** To represent in a verbal or artistic form : DEPICT <"Joyce has attempted . . . to *render* . . . what our participation in life is like" —Edmund Wilson> **7.** To perform an interpretation of (e.g., a musical piece). **8.** To express in another language or form : TRANSLATE. **9.** To pronounce formally <The court *rendered* its ruling.> **10.** To cause to become : MAKE. **11.** To reduce, convert, or melt down (fat) by heating. **12.** To coat (e.g., brick) with cement or plaster. —*n.* A payment in kind, services, or cash from a tenant to a feudal lord. **—ren′der·a·ble** *adj.* **—ren′der·er** *n.*

**ren·dez·vous** (rän′dā-vo͞o′, -də-) *n., pl.* **-vous** (-vo͞oz′) [Fr. < OFr. < the phrase *rendez vous,* present yourselves.] **1.** A meeting place arranged in advance. **2.** A meeting arranged in advance. **3.** A popular gathering place. **4.** The process of maneuvering two spacecraft together. —*vt.* & *vi.* **-voused** (-vo͞od′), **-vous·ing** (-vo͞o′ĭng), **-vous** (-vo͞oz′). To bring or meet together at a specified time and place.

**ren·di·tion** (rĕn-dĭsh′ən) *n.* [Obs. Fr. < OFr. *rendre,* to give back. —see RENDER.] An act of rendering, as: **a.** An interpretation of a dramatic piece or musical score. **b.** A musical or dramatic performance. **c.** An often interpretive translation. **d.** A surrender.

**ren·e·gade** (rĕn′ĭ-gād′) *n.* [Sp. *renegado* < Med. Lat. *renegatus* < p.part. of *renegare,* to deny : Lat. *re-* (intensive) + *negare,* to deny.] **1.** One who rejects one's religion, cause, allegiance, or group for another : TRAITOR. **2.** An outlaw. —*vi.* **-gad·ed, -gad·ing, -gades.** To become a renegade.

**re·nege** (rĭ-nĭg′, -nĕg′, -nēg′) *v.* **-neged, -neg·ing, -neges.** [Med. Lat. *renegare,* to deny. —see RENEGADE.] —*vi.* **1.** To fail to adhere to a promise or commitment. **2.** To fail to follow suit in a card game when able and obliged by the rules to do so. —*vt.* To renounce : disown. **—re·nege′** *n.* **—re·neg′er** *n.*

**re·ne·go·ti·ate** (rē′nĭ-gō′shē-āt′) *vt.* **-at·ed, -at·ing, -ates.** To negotiate anew, esp. to modify the terms of (a contract) so as to limit or get back the contractor's excess profits. **—re′ne·go′ti·a·ble** *adj.* **—re′ne·go·ti·a′tion** *n.*

**re·new** (rĭ-no͞o′, -nyo͞o′) *v.* **-newed, -new·ing, -news.** [ME *renewen* : *re-,* again + *new,* new.] —*vt.* **1.** To make new or as if new again : RESTORE. **2.** To take up again : RESUME. **3.** To repeat so as to reaffirm. **4.** To regain (vigor) : REVIVE. **5.** To arrange for the extension of <*renew* a lease> **6.** To replenish. **7.** To bring into being again : RE-ESTABLISH. —*vi.* **1.** To become new again. **2.** To start over. **—re·new′a·bil′i·ty** (-ə-bĭl′ĭ-tē) *n.* **—re·new′a·ble** *adj.* **—re·new′a·bly** *adv.* **—re·new′al** *n.* **—re·new′er** *n.*

**re·new·ed·ly** (rĭ-no͞o′ĭd-lē, -nyo͞o′-) *adv.* Over again : ANEW.

**ren·i·form** (rĕn′ə-fôrm′, rē′nə-) *adj.* [Lat. *renes,* kidneys + -FORM.] Shaped like a kidney.

**ren·in** (rĕn′ĭn) *n.* [Lat. *renes,* kidneys + -IN.] A protein-digesting enzyme released by the kidney that raises blood pressure.

**ren·i·tent** (rĕn′ĭ-tənt, rĭ-nīt′nt) *adj.* [Lat. *renitens, renitent-,* pr.part. of *reniti,* to resist : *re-,* against + *niti,* to press forward.] **1.** Resistant to physical pressure. **2.** Unwilling to yield or be influenced : RECALCITRANT. **—ren′i·tence, ren′i·ten·cy** *n.*

**ren·min·bi** (rĕn′mĭn-bē′) *n.* [Chin. (Mandarin) *ren*[2] *min*[2] *bi*[4] : *ren*[2] *min*[2], the people (*ren*[2], human being + *min*[2], people) + *bi*[4], money.] —See table at CURRENCY.

**ren·nase** (rĕn′ās) *n.* Rennin.

**ren·net** (rĕn′ĭt) *n.* [ME.] **1.** The inner lining of the fourth stomach of young ruminants, as calves. **2.** A dried extract made from the stomach lining of a ruminant, for curdling milk. **3.** Rennin.

**ren·nin** (rĕn′ĭn) *n.* [RENN(ET) + -IN.] A milk-coagulating enzyme produced from rennet and used in making cheeses and junkets.

**ren·nin·o·gen** (rĕ-nĭn′ə-jən) *n.* The zymogenic precursor of rennin.

**re·nom·i·nate** (rē-nŏm′ə-nāt′) *vt.* **-nat·ed, -nat·ing, -nates.** To nominate again. **—re′nom·i·na′tion** *n.*

**re·nounce** (rĭ-nouns′) *v.* **-nounced, -nounc·ing, -nounc·es.** [ME *renouncen* < OFr. *renoncer* < Lat. *renuntiare,* to report : *re-,* back + *nuntiare,* to announce < *nuntius,* messenger.] —*vt.* **1.** To give up, esp. by making a formal announcement. **2.** To reject : disown. —*vi.* To revoke in cards. —*n.* A revoke in cards. **—re·nounce′ment** *n.* **—re·nounc′er** *n.*

**ren·o·vate** (rĕn′ə-vāt′) *vt.* **-vat·ed, -vat·ing, -vates.** [Lat. *renovare, renovat-* : *re-,* again + *novare,* to make new < *novus,* new.] **1.** To restore to a previous condition, as by remodeling. **2.** To give new vigor to : REVIVE. **—ren′o·va′tion** *n.* **—ren′o·va′tor** *n.*

**re·nown** (rĭ-noun′) *n.* [ME *renowne* < OFr. *renon* < *renomer,* to make famous : *re-,* again (< Lat.) + *nomer,* to name < Lat. *nominare* < *nomen,* name.] **1.** The quality of being widely honored and acclaimed : FAME. **2.** *Obs.* Report : rumor.

**re·nowned** (rĭ-nound′) *adj.* Famous.

**rent**[1] (rĕnt) *n.* [ME < OFr. *rente* < VLat. **rendita,* fem. p.part. of **rendere.* —see RENDER.] **1.** Payment, usu. of an amount set by contract, made by a tenant at designated intervals in return for the right to occupy or use another's property. **2. a.** The return received from cultivated or improved land after deducting all production costs. **b.** The revenue from a piece of land exceeding the amount of revenue from the poorest or least favorably situated land, under equal market conditions. —*v.* **rent·ed, rent·ing, rents.** —*vt.* **1.** To obtain occupancy or use of (another's property) in return for periodic payments. **2.** To grant temporary occupancy or use of (one's own property) in return for periodic payments. —*vi.* To be for rent. **—for rent.** Available for use or service in return for payment. **—rent·a·bil′i·ty** *n.* **—rent′a·ble** *adj.*

**rent**[2] (rĕnt) *n.* [< p.part. of REND.] **1.** An opening made by rending : RIP. **2.** A breach of relations between persons.

**rent·al** (rĕn′tl) *n.* **1.** An amount paid out or received as rent. **2.** A list of tenants and schedule of rents. **3.** Property available for rent. **4.** An act of renting. **5.** An agency renting something. **—rent′al** *adj.*

**rent control** *n.* Official, esp. governmental, control and regulation of the amount charged for rented housing.

**rent·er** (rĕn′tər) *n.* One who pays rent to use another's property : TENANT.

**rent strike** *n.* A tenants' agreement to refuse to pay rent, as in protest of poor services.

**re·num·ber** (rē-nŭm′bər) *vt.* **-bered, -ber·ing, -bers.** To number again or in a different order.

**re·nun·ci·a·tion** (rĭ-nŭn′sē-ā′shən) *n.* [ME < Lat. *renuntiatio* < *renuntiare,* to renounce.] **1.** The act or practice of renouncing <*renunciation* of all pleasure> **2.** A declaration in which something is renounced. **—re·nun′ci·a′tive, re·nun′ci·a·to′ry** (-ə-tôr′ē) *adj.*

**re·o·pen** (rē-ō′pən) *v.* **-pened, -pen·ing, -pens.** —*vt.* To open or take up again. —*vi.* To start over : RESUME.

**re·or·der** (rē-ôr′dər) *v.* **-dered, -der·ing, -ders.** —*vt.* **1.** To order again. **2.** To put in order again. **3.** To rearrange. —*vi.* To order the same goods again. —*n.* An order of goods like an earlier one from the same supplier.

**re·or·gan·i·za·tion** (rē-ôr′gə-nĭ-zā′shən) *n.* **1.** The act or process of organizing again or differently. **2.** A complete change in the structure of a business corporation. **—re·or′gan·i·za′tion·al** *adj.*

**re·or·gan·ize** (rē-ôr′gə-nīz′) *v.* **-ized, -iz·ing, -iz·es.** —*vt.* To organize again. —*vi.* To undergo or make changes in organization. **—re·or′gan·iz′er** *n.*

**re·o·vi·rus** (rē′ō-vī′rəs) *n.* [R(ESPIRATORY) + E(NTERIC) + O(RPHAN) + VIRUS.] Any of a group of RNA-containing animal viruses that occas. occur in the respiratory and digestive tracts of healthy individuals.

**rep**[1] *also* **repp** (rĕp) *n.* [Fr. *reps.*] A ribbed or corded fabric of various materials, as cotton, wool, or silk.

**rep**[2] (rĕp) *n. Informal.* A representative.

**rep**[3] (rĕp) *n.* [R(OENTGEN) + E(QUIVALENT) + P(HYSICAL).] *Physics.* A unit of absorbed radiation dose, equal to the absorbed dose in water having been exposed to one roentgen.

**rep**[4] (rĕp) *n. Informal.* A repertory theater.

**re·pack·age** (rē-păk′ĭj) *vt.* **-aged, -ag·ing, -ag·es.** To package again, esp. to put in a better package. **—re·pack′ag·er** *n.*

**re·pair**[1] (rĭ-pâr′) *v.* **-paired, -pair·ing, -pairs.** [ME *repairen* < OFr. *reparer* < Lat. *reparare* : *re-,* back + *parare,* to put in order.] —*vt.* **1.** To restore to sound condition after damage or injury : FIX. **2.** To set right : REMEDY <*repair* a mistake> **3.** To renew or refresh. **4.** To compensate for (e.g., a loss or wrong). —*vi.* To make repairs. —*n.* **1.** The work, act, or process of repairing. **2.** General condition after use or repairing <an automobile in excellent *repair*> **3.** An instance of repairing. **—re·pair′er** *n.*

**re·pair**[2] (rĭ-pâr′) *vi.* **-paired, -pair·ing, -pairs.** [ME *reparen,* to return < OFr. *repairer* < LLat. *repatriare,* to return to one's country. —see REPATRIATE.] To betake oneself : GO <*repair* to the drawing room> —*n.* **1.** An act of repairing. **2.** A place to which one goes often or habitually : HAUNT. **—re·pair′a·ble** *adj.*

**re·pair·man** (rĭ-pâr′măn′, -mən) *n.* One whose occupation is making repairs.

**re·pand** (rĭ-pănd′) *adj.* [Lat. *repandus,* bent backward : *re-,* back + *pandus,* p.part. of *pandere,* to spread.] *Bot.* Having a rather wavy margin <a *repand* leaf>

**rep·a·ra·ble** (rĕp′ər-ə-bəl) *adj.* Capable of being repaired. **—rep′a·ra·bil′i·ty** *n.* **—rep′a·ra·bly** *adv.*

**rep·a·ra·tion** (rĕp′ə-rā′shən) *n.* [ME < OFr. < LLat. *reparatio* < Lat. *reparare,* to repair. —see REPAIR[1].] **1.** The act or process of repairing or the state of being repaired. **2.** The act or process of making amends : EXPIATION. **3.** Something done or paid as amends : COMPENSATION. **4. reparations.** Compensation that a defeated nation is obliged to pay as indemnity for damage or injury during a war.

**re·par·a·tive** (rĭ-păr′ə-tĭv) *also* **re·par·a·to·ry** (-tôr′ē, -tōr′ē) *adj.* **1.** Tending to repair. **2.** Of, pertaining to, or having the nature of reparations.

**rep·ar·tee** (rĕp′ər-tē′, -tā′, -är-) *n.* [Fr. *repartie* < *repartir,* to retort < OFr., to retort, to depart again : *re-,* again (< Lat.) + *partir,* to depart < Lat. *partire,* to divide < *pars,* part.] **1. a.** A quick, witty reply or retort. **b.** Witty, lively, conversation full of such replies. **2.** Skill and cleverness in conversational repartee.

**re·par·ti·tion** (rē′pär-tĭsh′ən) *n.* **1.** Apportionment : distribution. **2.** A partitioning again or in another way. —*vt.* **-tioned, -tion·ing, -tions.** To partition again : REDIVIDE.

**re·pass** (rē-păs′) *v.* **-passed, -pass·ing, -pass·es.** —*vt.* **1.** To pass (something) again. **2.** To cause to pass again in the opposite direction. —*vi.* To pass again. **—re·pas′sage** (-ĭj) *n.*

**re·past** (rĭ-păst′) *n.* [ME < OFr. < *repaistre,* to feed < LLat. *repascere* : *re-* (intensive) + *pascere,* to feed.] **1.** A meal. **2.** The food eaten or offered at a meal. —*v.* **-past·ed, -past·ing, -pasts.** —*vi.* To take food. —*vt. Obs.* To give food to.

**re·pa·tri·ate** (rē-pā′trē-āt′) *vt.* **-at·ed, -at·ing, -ates.** [LLat. *repatriare,* to return to one's country : *re-,* back + *patria,* native country.] To restore or return to the country of birth or citizenship <*repatriate* the exiles> —*n.* (-ĭt, -āt′). One who has been repatriated. **—re·pa′tri·a′tion** *n.*

**re·pay** (rĭ-pā′) *v.* **-paid** (-pād′), **-pay·ing, -pays.** —*vt.* **1.** To pay back <*repay* a loan> **2.** To give back, either in return or in requital <*repay* ill will with ill will> **3.** To make compensation or a return for <a candidate who *repays* staff loyalty with gratitude> **4.** To make or do in return <*repay* a visit> —*vi.* To make repayment or requital. **—re·pay′a·ble** *adj.* **—re·pay′ment** *n.*

**re·peal** (rĭ-pēl′) *vt.* **-pealed, -peal·ing, -peals.** [ME *repelen* < AN *repeler* < OFr. *rapeler* : *re-,* back (< Lat.) + *apeler,* to appeal. —see APPEAL.] **1.** To revoke or rescind, esp. by an official or formal act. **2.** *Obs.* To summon back or recall, esp. from exile. **—re·peal′** *n.* **—re·peal′a·ble** *adj.* **—re·peal′er** *n.*

**re·peat** (rĭ-pēt′) *v.* **-peat·ed, -peat·ing, -peats.** [ME *repeten* < OFr. *repeter* < Lat. *repetere,* to seek again : *re-,* again + *petere,* to seek.] —*vt.* **1.** To state or utter again. **2.** To utter in duplication of another's utterance. **3.** To recite from memory. **4.** To tell to another. **5.** To do, experience, or produce again. **6.** To express (oneself) in the same way or words. —*vi.* **1.** To do or say something again. **2.** To commit the offense of voting more than once in a single election. —*n.* **1.** An act of repeating. **2.** Something repeated <a *repeat* of a successful concert> **3.** *Mus.* **a.** A section or passage repeated. **b.** A sign usu. composed of two vertical dots, designating a passage to be repeated. **—re·peat′a·bil′i·ty** *n.* **—re·peat′a·ble** *adj.*

**re·peat·ed** (rĭ-pē′tĭd) *adj.* Said, done, or occurring again and again. **—re·peat′ed·ly** *adv.*

**re·peat·er** (rĭ-pē′tər) *n.* **1.** One that repeats. **2.** A clock or watch with a pressure-activated mechanism that strikes the hour. **3.** A repeating firearm. **4.** A student who repeats a course, usu. one that has been failed. **5.** One who fraudulently votes more than once in a single election. **6.** One convicted of repeated wrongdoing.

**repeating decimal** *n. Math.* A decimal in which after a certain digit there is indefinite repetition of a pattern of one or more digits.

**repeating firearm** *n.* A firearm capable of firing several times without being reloaded.

**re·pel** (rĭ-pĕl′) *v.* **-pelled, -pel·ling, -pels.** [ME *repellen* < Lat. *repellere* : *re-,* back + *pellere,* to drive.] —*vt.* **1.** To drive back : WARD OFF <*repel* mosquitoes> **2.** To offer resistance to : fight against <*repel* an attack> **3.** To refuse to accept : REJECT <*repel* an invitation> **4.** To turn away from : SPURN. **5.** To cause distaste or aversion in <Their behavior *repels* everyone.> **6.** To be unable to absorb or mix with. **7.** To present an opposing force to <Electric charges of like signs *repel* one another.> —*vi.* **1.** To offer a resistant force to something. **2.** To cause distaste or aversion. ***usage:*** Both *repel* and *repulse* mean "to drive back or away." Only *repel,* however, has the sense "to cause distaste or aversion in," as in *They were repelled by the filth and noise.* **—re·pel′ler** *n.*

☆ ***syns:*** **1.** REPEL, COMBAT, FEND (off), REPULSE, WARD OFF *v. core meaning* : to turn or drive away <*repelled* the advancing troops> **2.** REPEL, DISGUST, NAUSEATE, REVOLT, SICKEN *v. core meaning* : to cause aversion in <behavior that *repelled* all their friends> ***ant:*** attract

**re·pel·lent** (rĭ-pĕl′ənt) *adj.* **1. a.** Serving or tending to repel. **b.** Capable of repelling. **2.** Inspiring distaste or aversion : REPULSIVE. **3.** Resistant or impervious to a substance <a water-*repellent* coat> —*n.* Something that repels, esp.: **a.** A substance for repelling insects. **b.** A substance or treatment for making a fabric or surface impervious or resistant to a potential source of damage. **—re·pel′lence, re·pel′len·cy** *n.* **—re·pel′lent·ly** *adv.*

**re·pent**[1] (rĭ-pĕnt′) *v.* **-pent·ed, -pent·ing, -pents.** [ME *repenten* < OFr. *repentir* : *re-,* in response (< Lat.) + *pentir,* to be sorry < Lat. *paenitēre.*] —*vi.* **1. a.** To feel remorse, contrition, or self-reproach. **b.** To feel such regret for previous behavior as to change one's mind about it. **2.** To make a change for the better because of remorse or contrition for one's sins. —*vt.* **1.** To feel regret or self-reproach for. **2.** To cause to feel remorse or regret. **—re·pent′er** *n.*

**re·pent**[2] (rē′pənt) *adj.* [Lat. *repens, repent-,* pr.part. of *repere,* to creep.] *Biol.* Creeping along the ground : PROSTRATE.

**re·pen·tance** (rĭ-pĕn′təns) *n.* **1.** Remorse or contrition for past conduct or sin. **2.** The act or process of repenting.

**re·pen·tant** (rĭ-pĕn′tənt) *adj.* Marked by or showing repentance : PENITENT. **—re·pen′tant·ly** *adv.*

**re·per·cus·sion** (rē′pər-kŭsh′ən, rĕp′ər-) *n.* [Lat. *repercussio* < *repercutere,* to cause to rebound : *re-,* back + *percutere,* to strike.] **1.** An indirect effect, influence, or result created by an action or event. **2.** A recoil, rebounding, or reciprocal motion after impact. **3.** A reflection, esp. of sound. **—re′per·cus′sive** *adj.*

**rep·er·toire** (rĕp′ər-twär′) *n.* [Fr. < LLat. *repertorium.* —see REPERTORY.] **1.** The stock of songs, plays, operas, readings, or other pieces that a player or company is prepared to perform. **2.** The range or number of skills, aptitudes, or special accomplishments, as of a person or group.

**rep·er·to·ry** (rĕp′ər-tôr′ē, -tōr′ē) *n., pl.* **-ries.** [LLat. *repertorium* < Lat. *repertus,* p.part. of *reperire,* to find out : *re-,* again + *parire,* to produce.] **1.** A repertoire. **2.** A theater in which a resident company presents plays from a specified repertoire, usu. in alternation. **3.** A place, as a storehouse, where a stock of things is kept : REPOSITORY. **—rep′er·to′ri·al** *adj.*

**rep·e·tend** (rĕp′ĭ-tĕnd′, rĕp′ĭ-tĕnd′) *n.* [Lat. *repetendum,* neuter gerund. of *repetere,* to repeat.] **1.** A repeated word, sound, or phrase : REFRAIN. **2.** *Math.* The digit or group of digits that repeats infinitely in a repeating decimal.

**rep·e·ti·tion** (rĕp′ĭ-tĭsh′ən) *n.* [Lat. *repetitio* < *repetere,* to repeat.] **1.** The act or process or an instance of repeating or being repeated. **2.** A recitation or recital, esp. of prepared or memorized material. **—rep′e·ti′tion·al** *adj.*

**rep·e·ti·tious** (rĕp′ĭ-tĭsh′əs) *adj.* Marked by or filled with repetition, esp. unnecessary or tiresome repetition. **—rep′e·ti′tious·ly** *adv.* **—rep′e·ti′tious·ness** *n.*

**re·pet·i·tive** (rĭ-pĕt′ĭ-tĭv) *adj.* Repetitious. **—re·pet′i·tive·ly** *adv.* **—re·pet′i·tive·ness** *n.*

**re·phrase** (rē-frāz′) *vt.* **-phrased, -phras·ing, -phras·es.** To phrase again or in a different way.

**re·pine** (rĭ-pīn′) *vi.* **-pined, -pin·ing, -pines. 1.** To be discontented or in low spirits : COMPLAIN. **2.** To yearn after something. **—re·pin′er** *n.*

**re·place** (rĭ-plās′) *vt.* **-placed, -plac·ing, -plac·es. 1.** To put back in a previous place or position. **2.** To take or fill the place of. **3.** To be or provide a replacement for. **4.** To return or pay back : REFUND. **—re·place′a·ble** *adj.* **—re·plac′er** *n.*

**re·place·ment** (rĭ-plās′mənt) *n.* **1.** The act or process of replacing or of being replaced : SUBSTITUTION. **2.** One that replaces, esp. one assigned to a vacant military position. **—re·place′ment** *adj.*

**re·plant** (rē-plănt′) *vt.* **-plant·ed, -plant·ing, -plants. 1.** To plant (something) again or in a new place. **2.** To supply with new plants <*replant* a flower bed> *—n.* (rē′plănt′). Something replanted.

**re·play** (rē-plā′) *vt.* **-played, -play·ing, -plays.** To play over again. *—n.* (rē′plā′). **1.** The act or process of replaying something, as by a videotape. **2.** Something replayed.

**re·plead·er** (rĭ-plē′dər) *n. Law.* **1.** A court order obliging parties to plead their case again because of a prior miscarried or erroneous pleading. **2.** The right of pleading again.

**re·plen·ish** (rĭ-plĕn′ĭsh) *v.* **-ished, -ish·ing, -ish·es.** [ME *replenisshen* < OFr. *replenir, repleniss-* : *re-,* again (< Lat.) + *plenir,* to fill < *plein,* full < Lat. *plenus.*] *—vt.* **1.** To fill or make complete again : add a new supply to <*replenish* the cooking supplies> **2.** To inspire or nourish <mountain views that *replenish* jaded spirits> *—vi.* To become full again. **—re·plen′ish·er** *n.* **—re·plen′ish·ment** *n.*

**re·plete** (rĭ-plēt′) *adj.* [ME *replet* < OFr. < Lat. *repletus,* p.part. of *replēre,* to refill : *re-,* again + *plēre,* to fill.] **1.** Abundantly provided for : ABOUNDING. **2.** Filled to satiation : GORGED. **—re·plete′ness** *n.*

**re·ple·tion** (rĭ-plē′shən) *n.* **1.** The state of being fully supplied or completely filled. **2.** Excessive fullness.

**re·plev·i·a·ble** (rĭ-plĕv′ē-ə-bəl) *adj. Law.* Capable of being recovered by replevin.

**re·plev·in** (rĭ-plĕv′ĭn) [ME < AN *replevine* < *replevir,* to give as a security < OFr. : *re-,* back + *plevir,* to pledge.] *Law. —n.* **1.** An action to recover personal property said or claimed to be unlawfully taken. **2.** The writ or procedure of replevin. *—vt.* **-ined, -in·ing, -ins.** To replevy.

**re·plev·y** (rĭ-plĕv′ē) [AN *replevir* < OFr., to give as a security. —see REPLEVIN.] *Law. —vt.* **-ied, -y·ing, -ies.** To regain possession of by a writ of replevin. *—n., pl.* **-ies.** Replevin.

**rep·li·ca** (rĕp′lĭ-kə) *n.* [Ital. < *replicare,* to repeat < Lat., to fold back. —see REPLICATE.] **1.** A copy or reproduction of a work of art, esp. one done by the original artist. **2.** A copy or reproduction.

**rep·li·cate** (rĕp′lĭ-kāt′) *v.* **-cat·ed, -cat·ing, -cates.** [Lat. *replicare, replicat-,* to fold back : *re-,* back + *plicare,* to fold.] *—vt.* **1.** To duplicate, copy, or repeat. **2.** To fold over or bend back upon itself. *—vi.* To become replicated. *—adj.* (-kĭt). Folded over or bent back upon itself <a *replicate* leaf> **—rep′li·ca′tive** *adj.*

**rep·li·ca·tion** (rĕp′lĭ-kā′shən) *n.* **1.** A fold or a folding back. **2.** A reply to an answer : REJOINDER. **3.** *Law.* The plaintiff's response to the defendant's answer or plea. **4.** An echo : reverberation. **5.** A copy or reproduction. **6.** An act or process of reproducing or duplicating.

**re·ply** (rĭ-plī′) *v.* **-plied, -ply·ing, -plies.** [ME *replien* < OFr. *replier* < Lat. *replicare,* to fold back. —see REPLICATE.] *—vi.* **1.** To give an answer in writing or speech. **2.** To respond by an action or gesture <*replied* with a shrug of the shoulders> **3. a.** To echo. **b.** To return gunfire or an attack <The howitzers *replied.*> **4.** *Law.* To answer a defendant's plea. *—vt.* To say or give as an answer <I *replied* that I was not interested in the program.> *—n., pl.* **-plies. 1.** An answer in speech or writing. **2.** A response by action or gesture. **3.** *Law.* A plaintiff's speech or argument in answer to a defendant's. **—re·pli′er** *n.*

**re·po** (rē′pō′) *n.* [Shortening and alteration of REPURCHASE AGREEMENT.] *Informal.* A repurchase agreement.

**re·po·lar·i·za·tion** (rē-pō′lər-ĭ-zā′shən) *n.* Restoration of a polarized state in a muscle fiber or membrane after contraction.

**re·port** (rĭ-pôrt′, -pōrt′) *n.* [ME *reporten* < OFr. *reporter* < Lat. *reportare* : *re-,* back + *portare,* to carry.] **1.** A usu. detailed account. **2.** A formal account of the proceedings or transactions of a group. **3.** An account of a judicial decision or court case. **4.** Common talk : RUMOR. **5.** Reputation : repute <a person of excellent *report*> **6.** An explosive noise <the firecracker's *report*> *—v.* **-port·ed, -port·ing, -ports.** *—vt.* **1.** To make or present an account of, often officially, formally, or periodically. **2.** To relate or tell about : PRESENT <*report* on the results of one's research> **3.** To write or supply an account or summation of for publication or broadcast. **4.** To submit or relate the results of considerations regarding <The executive committee *reported* the new personnel policy.> **5.** To carry back and repeat to another. **6.** To complain about or denounce <dissidents who were *reported* to the secret police> *—vi.* **1.** To make a report. **2.** To work as a reporter for a newspaper or similar publication. **3.** To present oneself <*report* for assignment> **—on report.** Liable to disciplinary action. **—report out.** To return after deliberation to a legislative body for action <The committee *reported* the new appropriations bill *out.*> **—re·port′a·ble** *adj.*

**re·port·age** (rĕp′ər-täzh′, rĭ-pôr′tĭj, -pōr′-) *n.* **1.** The reporting of news or information of general interest. **2.** Something reported.

**report card** *n.* A report of a student's progress presented at regular intervals to a parent or guardian.

**re·port·ed·ly** (rĭ-pôr′tĭd-lē, -pōr′-) *adv.* By report : SUPPOSEDLY.

**re·port·er** (rĭ-pôr′tər, -pōr′-) *n.* **1.** One who reports. **2.** A writer of news stories. **3.** One with authority to write and issue official accounts of judicial or legislative proceedings. **—rep′or·to′ri·al** (rĕp′ər-tôr′ē-əl, -tōr′-, rē′pər-) *adj.* **—rep′or·to′ri·al·ly** *adv.*

**re·pose¹** (rĭ-pōz′) *n.* [< ME *reposen* < OFr. *reposer* < LLat. *repausare* : Lat. *re-* (intensive) + Lat. *pausare,* to rest < *pausa,* rest. —see PAUSE.] **1.** An act of resting or the state of being at rest. **2.** Poise : composure. **3.** Tranquillity. *—v.* **-posed, -pos·ing, -pos·es.** *—vt.* To lay (oneself) down. *—vi.* **1.** To lie at rest. **2.** To lie supported by something. **—re·pos′al** *n.* **—re·pos′er** *n.*

**re·pose²** (rĭ-pōz′) *vt.* **-posed, -pos·ing, -pos·es.** [ME *reposen* < Lat. *reponere* : *re-,* back + *ponere,* to place.] To place (e.g., trust) in <Management had *reposed* its hopes in new products.>

**re·pose·ful** (rĭ-pōz′fəl) *adj.* Characterized by or contributive to repose. **—re·pose′ful·ly** *adv.* **—re·pose′ful·ness** *n.*

**re·pos·it** (rĭ-pŏz′ĭt) *vt.* **-it·ed, -it·ing, -its.** [Lat. *reponere, reposit-* : *re-,* back + *ponere,* to place.] To put away : STORE. **—re′po·si′tion** (rē′pə-zĭsh′ən, rĕp′ə-) *n.*

**re·pos·i·to·ry** (rĭ-pŏz′ĭ-tôr′ē, -tōr′ē) *n., pl.* **-ries. 1.** A place where things may be stored for safekeeping. **2.** A warehouse. **3.** A museum. **4.** A burial vault : TOMB. **5.** A recipient of a confidence.

**re·pos·sess** (rē′pə-zĕs′) *vt.* **-sessed, -sess·ing, -sess·es. 1. a.** To regain ownership of. **b.** To reclaim ownership of for failure to pay installments due. **2.** To return ownership to. **—re′pos·ses′sion** *n.*

**re·pous·sé** (rə-pōō-sā′) *adj.* [Fr., p.part. of *repousser,* to push back < OFr. : *re-,* back (< Lat.) + *pousser,* to push < Lat. *pulsare,* to beat, freq. of *pellere,* to push.] **1.** Shaped or decorated with patterns in relief made by hammering and pressing on the reverse side. —Used esp. of metal. **2.** Raised in relief. *—n.* **1.** A repoussé design. **2.** The technique of hammering and pressing repoussé designs.

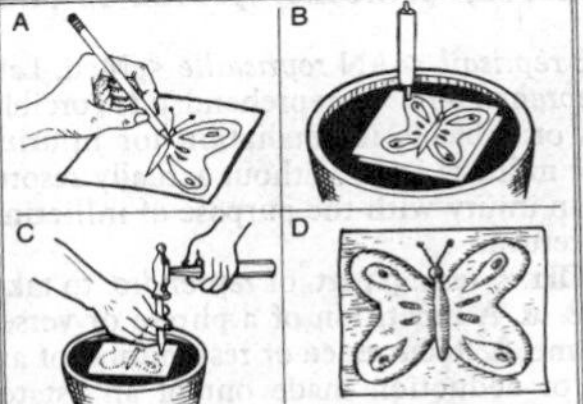


**repoussé**
*A. A design is drawn on a piece of paper. B. It is then traced onto a sheet of warm, somewhat malleable metal. C. Using various punches, the artist presses the design into the metal. D. The result is repoussé.*

**repp** (rĕp) *n. var. of* REP¹.

**rep·re·hend** (rĕp′rĭ-hĕnd′) *vt.* **-hend·ed, -hend·ing, -hends.** [ME *reprehenden* < Lat. *reprehendere* : *re-* (intensive) + *prehendere,* to seize.] To censure : reprove <The colonel was severely *reprehended* by the general.> **—rep′re·hen′sion** (-hĕn′shən) *n.*

**rep·re·hen·si·ble** (rĕp′rĭ-hĕn′sə-bəl) *adj.* [LLat. *reprehensibilis* < Lat. *reprehendere,* to reprehend.] Worthy of censure or rebuke. **—rep′re·hen′si·bil′i·ty, rep′re·hen′si·ble·ness** *n.* **—rep′re·hen′si·bly** *adv.*

**rep·re·sent** (rĕp′rĭ-zĕnt′) *vt.* **-sent·ed, -sent·ing, -sents.** [ME *representen* < Lat. *repraesentare,* to show : *re-,* again + *praesentare,* to present.] **1.** To stand for : SYMBOLIZE <The maple leaf *represents* Canada.> **2.** To depict in art : PORTRAY. **3.** To present clearly to the mind. **4.** To draw attention to by way of protest or remonstrance. **5.** To describe as an embodiment of a specified quality. **6. a.** To function as the official and authorized delegate or agent for. **b.** To act as a spokesperson for. **7.** To be an example of <a marsupial *represented* by six species> **8.** To be the equivalent of. **9. a.** To stage (e.g., a play) : PRODUCE. **b.** To act the part or role of. **—rep′re·sent′a·bil′i·ty** *n.* **—rep′re·sent′a·ble** *adj.* **—rep′re·sent′er** *n.*

☆ ***syns:*** REPRESENT, EMBODY, EPITOMIZE, EXEMPLIFY, PERSONIFY, SYMBOLIZE, TYPIFY *v. core meaning* : to serve as the image of <a President who *represented* the ideals of democracy>

**rep·re·sen·ta·tion** (rĕp′rĭ-zĕn-tā′shən, -zən-) *n.* **1.** An act of representing or the state of being represented. **2.** Something that represents. **3. a.** An account or statement, as of facts, allegations, or arguments. **b.** An expostulation : protest. **4.** A production or presentation, as of a play. **5.** Service as an official delegate, agent, or spokesperson. **6.** The right or privilege of being represented by delegates having a voice in a legislative body. **7.** *Law.* A statement of fact made by one party so as to induce another party to enter into a contract.

**rep·re·sen·ta·tion·al** (rĕp′rĭ-zĕn-tā′shə-nəl, -zən-) *adj.* Of or relating to representation, esp. to realistic graphic representation. **—rep′re·sen·ta′tion·al·ism** *n.*

**rep·re·sen·ta·tive** (rĕp′rĭ-zĕn′tə-tĭv) *n.* **1.** One that exemplifies or typifies others of the same class. **2.** A delegate or agent for another. **3. a.** A member of a usu. legislative body selected by popular vote. **b.** A member of the House of Representatives or of a U.S. state legislature. —*adj.* **1.** Representing or capable of representing, depicting, or portraying. **2.** Authorized to act as an official delegate or agent. **3.** Of, relating to, or typical of government by representation. **4.** Typical of others of the same class. **—rep′re·sen′ta·tive·ly** *adv.* **—rep′re·sen′ta·tive·ness** *n.*

☆ ***syns:*** REPRESENTATIVE, DELEGATE, DEPUTY *n. core meaning:* one who serves as an agent for another <*representatives* at a political convention>

**re·press** (rĭ-prĕs′) *v.* **-pressed, -press·ing, -press·es.** [ME *repressen* < Lat. *reprimere* : *re-*, back + *premere*, to press.] —*vt.* **1.** To hold back : RESTRAIN <*repress* a yawn> **2.** To suppress : quell <*repress* student uprisings> **3.** *Psychoanal.* To exclude (e.g., memories) from the conscious mind. —*vi.* To take repressive action. **—re·press′i·bil′i·ty** *n.* **—re·press′i·ble** *adj.*

**re·pres·sion** (rĭ-prĕsh′ən) *n.* **1.** An act of repressing or the state of being repressed <political *repression*> **2.** *Psychoanal.* Unconscious exclusion of painful impulses, desires, or fears from the conscious mind. **—re·pres′sion·ist** *adj.*

**re·pres·sive** (rĭ-prĕs′ĭv) *adj.* Causing or likely to cause repression. **—re·pres′sive·ly** *adv.* **—re·pres′sive·ness** *n.*

**re·pres·sor** (rĭ-prĕs′ər) *n.* **1.** One that represses. **2.** *Biol.* A chemical compound that prevents the synthesis of a protein by interfering with the action of DNA.

**re·prieve** (rĭ-prēv′) *vt.* **-prieved, -priev·ing, -prieves.** [ME *repryen* < OFr. *reprendre*, to take back < Lat. *reprehendere*, to hold back. —see REPREHEND.] **1.** To delay the punishment of. **2.** To give temporary relief to. —*n.* **1. a.** Delaying of punishment. **b.** A warrant for such a delay. **2.** Temporary relief, as from danger or pain. **—re·priev′a·ble** *adj.*

**rep·ri·mand** (rĕp′rə-mănd′) *vt.* **-mand·ed, -mand·ing, -mands.** [OFr. *reprimander* < *reprimende*, a reprimand < Lat. *reprimenda*, neuter pl. gerund. of *reprimere*, to repress.] To censure or rebuke harshly or formally. **—rep′ri·mand′** *n.*

**re·print** (rē′prĭnt′) *n.* **1.** Something printed again, esp.: **a.** A new printing exactly like an original. **b.** A separately printed excerpt. **2.** A facsimile of a stamp printed after the original issue of the stamp has stopped. —*vt.* (rē-prĭnt′) **-print·ed, -print·ing, -prints.** To print again. **—re·print′er** *n.*

**re·pri·sal** (rĭ-prī′zəl) *n.* [ME *reprisail* < AN *reprisaille* < Med. Lat. *reprisalia*, reprisals < Lat. *reprehendere*, to reprehend.] **1.** Forcible seizure of an enemy's goods or subjects in retaliation for injuries inflicted. **2.** Use of political or military force without actually resorting to war. **3.** Retaliation for an injury with the purpose of inflicting comparable or like injury in return.

**re·prise** (rĭ-prēz′) *n.* [ME < OFr. < fem. p.part. of *reprendre*, to take back. —see REPRIEVE.] **1.** *Mus.* **a.** A repetition of a phrase or verse. **b.** A return to an original theme. **2.** Recurrence or resumption of an action. **3.** An annual charge or deduction made out of an estate.

**re·pro** (rē′prō) *n., pl.* **-pros.** A reproduction proof.

**re·proach** (rĭ-prōch′) *vt.* **-proached, -proach·ing, -proach·es.** [ME *reprochen* < OFr. *reprochier* < VLat. **repropiare* : Lat. *re-*, back + Lat. *prope*, near.] **1.** To blame for something : REBUKE. **2.** To bring shame on : DISGRACE. —*n.* **1.** Blame : rebuke. **2.** A cause of blame or rebuke. **3.** Disgrace : shame. **—re·proach′a·ble** *adj.* **—re·proach′a·ble·ness** *n.* **—re·proach′a·bly** *adv.* **—re·proach′er** *n.*

**re·proach·ful** (rĭ-prōch′fəl) *adj.* Expressing reproach. **—re·proach′ful·ly** *adv.* **—re·proach′ful·ness** *n.*

**rep·ro·bate** (rĕp′rə-bāt′) *n.* [LLat. *reprobatus*, p.part. of *reprobare*, to reprove.] **1.** One who is morally unprincipled. **2.** One who is predestined to damnation. —*adj.* **1.** Morally unprincipled : PROFLIGATE. **2.** Rejected by God and with no hope of salvation. —*vt.* **-bat·ed, -bat·ing, -bates. 1.** To disapprove of : CONDEMN. **2.** To abandon to eternal damnation. **—rep′ro·ba′tion** *n.* **—rep′ro·ba′tive** *adj.*

**re·proc·ess** (rē-prŏs′ĕs′, -prō′sĕs′) *vt.* **-essed, -ess·ing, -ess·es.** To cause to undergo additional or special processing before reuse.

**re·pro·duce** (rē′prə-do͞os′, -dyo͞os′) *v.* **-duced, -duc·ing, -duc·es.** —*vt.* **1.** To produce a counterpart, image, or copy of. **2.** *Biol.* To generate (offspring) by sexual or asexual means. **3.** To produce again : RE-CREATE. **4.** To bring (e.g., a memory) to mind again. —*vi.* **1.** To generate offspring. **2.** To undergo copying. **—re′pro·duc′er** *n.* **—re′pro·duc′i·bil′i·ty** *n.* **—re′pro·duc′i·ble** *adj.*

**re·pro·duc·tion** (rē′prə-dŭk′shən) *n.* **1.** The act of reproducing or process of being reproduced. **2.** Something reproduced. **3.** *Biol.* The sexual or asexual process by which organisms generate others of the same kind.

**reproduction proof** *n.* A proof of typeset material made for reproduction through a photographic process.

**re·pro·duc·tive** (rē′prə-dŭk′tĭv) *adj.* **1.** Of or relating to reproduction. **2.** Likely to reproduce. **—re′pro·duc′tive·ly** *adv.* **—re′pro·duc′tive·ness** *n.*

**re·prog·ra·phy** (rĭ-prŏg′rə-fē) *n.* [REPRO(DUCTION) + -GRAPHY.] Exact reproduction, as by offset printing or photocopying, of graphic material. **—re·prog′ra·pher** *n.* **—re′pro·graph′ic** (rē′prə-grăf′ĭk, rĕp′rə-) *adj.* **—re′pro·graph′ics** *n.*

**re·proof** (rĭ-pro͞of′) *n.* An act or expression of reproving.

**re·prove** (rĭ-pro͞ov′) *vt.* **-proved, -prov·ing, -proves.** [ME *reproven* < OFr. *reprover* < LLat. *reprobare*, to disapprove : Lat. *re-* (reversal) + Lat. *probare*, to approve.] **1.** To rebuke for a misdeed or fault : SCOLD. **2.** To find fault with. **—re·prov′a·ble** *adj.* **—re·prov′er** *n.* **—re·prov′ing·ly** *adv.*

**rep·tile** (rĕp′tĭl, -tīl′) *n.* [ME *reptil* < OFr. *reptile* < Lat. *reptilis*, creeping < *repere*, to creep.] **1.** A cold-blooded, usu. egg-laying vertebrate of the class Reptilia, as a snake, lizard, crocodile, turtle, or dinosaur, having an outer covering of scales or horny plates and breathing with lungs. **2.** One who is despicable or treacherous.

**rep·til·i·an** (rĕp-tĭl′ē-ən, -tĭl′yən) *adj.* **1.** Of or relating to reptiles. **2.** Similar to or typical of a reptile. **—rep·til′i·an** *n.*

**re·pub·lic** (rĭ-pŭb′lĭk) *n.* [OFr. *republique* < Lat. *respublica* : *res*, thing + *publica*, fem. of *publicus*, of the people.] **1. a.** A political order whose head of state is not a monarch and in modern times is usu. a president. **b.** A nation with such a political order. **2. a.** A political order in which the supreme power is held by a body of citizens who are entitled to vote for officers and representatives responsible to them. **b.** A nation with such a political order <the Federal *Republic* of Germany> **3.** A specific republican government of a nation. **4.** A group of people working as equals in the same sphere or field <the *republic* of letters>

▲ word history: The word *republic* is from Latin *respublica*, "public affairs, the state," which is a compound of *res*, "thing, affairs, business," and *publica*, "public, common." The Romans' idea of the state is summed up by the word *respublica*, for they regarded government as being the concern of all the people (*publica* is derived from *populus*, "the people"). Although the Roman state was never truly democratic like some of the Greek city-states, the Romans strongly believed that is was everyone's duty to serve the common good. After lasting almost 500 years the Roman republic was finally replaced by imperial rule, but the republican idea survived to be influential in modern European and American history.

**re·pub·li·can** (rĭ-pŭb′lĭ-kən) *adj.* **1.** Of, relating to, or typical of a republic. **2.** Favoring a republican form of government. **3. Republican.** Of, relating to, typical of, or belonging to the U.S. Republican Party. —*n.* **1.** One who favors a republican form of government. **2. Republican.** A member of the U.S. Republican Party. **—re·pub′li·can·ism** *n.*

**re·pub·li·can·ize** (rĭ-pŭb′lĭ-kə-nīz′) *vt.* **-ized, -iz·ing, -iz·es.** To make republican. **—re·pub′li·can·i·za′tion** *n.*

**Republican Party** *n.* **1.** One of the two chief U.S. political parties organized in 1854 to oppose slavery. **2.** The Democratic-Republican Party, a U.S. political party organized by Thomas Jefferson in 1792.

**re·pub·lish** (rē-pŭb′lĭsh) *vt.* **-lished, -lish·ing, -lish·es. 1.** To publish again. **2.** *Law.* To revive (a canceled will). **—re·pub′li·ca′tion** (rē-pŭb′lĭ-kā′shən) *n.* **—re·pub′lish·er** *n.*

**re·pu·di·ate** (rĭ-pyo͞o′dē-āt′) *vt.* **-at·ed, -at·ing, -ates.** [Lat. *repudiare*, *repudiat-* < *repudium*, divorce.] **1.** To reject the validity of. **2.** To refuse to recognize or pay. **3.** To reject as untrue. **4.** To disown (e.g., a child). **—re·pu′di·a′tive** *adj.* **—re·pu′di·a′tor** *n.*

☆ ***syns:*** REPUDIATE, DENY, DISAVOW, DISCLAIM, DISOWN *v. core meaning:* to refuse to recognize or acknowledge the truth or validity of <*repudiated* the previous testimony>

**re·pu·di·a·tion** (rĭ-pyo͞o′dē-ā′shən) *n.* **1.** An act of repudiating or the state of being repudiated. **2.** Refusal, esp. by public authorities, to acknowledge a debt or contract. **—re·pu′di·a′tion·ist** *n.*

**re·pugn** (rĭ-pyo͞on′) *v.* **-pugned, -pugn·ing, -pugns.** [ME *repugnen* < OFr. *repugner* < Lat. *repugnare*, to fight against : *re-*, against + *pugnare*, to fight.] —*vt.* To resist or oppose. —*vi. Archaic.* To be opposed : CONFLICT.

**re·pug·nance** (rĭ-pŭg′nəns) *also* **re·pug·nan·cy** (-nən-sē) *n.* **1.** Extreme dislike or aversion. **2.** *Logic.* The relationship of contradictory terms : INCONSISTENCY.

**re·pug·nant** (rĭ-pŭg′nənt) *adj.* [ME, resisting < OFr. < Lat. *repugnans*, pr.part. of *repugnare*, to fight against. —see REPUGN.] **1.** Eliciting extreme dislike or aversion. **2.** *Logic.* Contradictory : inconsistent. **—re·pug′nant·ly** *adv.*

**re·pulse** (rĭ-pŭls′) *vt.* **-pulsed, -puls·ing, -puls·es.** [Lat. *repellere*, *repuls-* : *re-*, back + *pellere*, to drive.] **1.** To drive back : REPEL. **2.** To rebuff or reject rudely or coldly. —*n.* **1.** An act of repulsing or the state of being repulsed. **2.** Rejection : refusal. **—re·puls′er** *n.* **—re·pul′sion** (-pŭl′shən) *n.*

**re·pul·sive** (rĭ-pŭl′sĭv) *adj.* **1.** So loathsome as to elicit extreme aversion. **2.** Tending to repel or drive off. **—re·pul'sive·ly** *adv.* **—re·pul'sive·ness** *n.*

**re·pur·chase agreement** (rē-pûr'chĭs) *n.* A contract giving the seller of property the right or obligation to buy back the property under specified terms.

**rep·u·ta·ble** (rĕp'yə-tə-bəl) *adj.* **1.** Having a good reputation <a *reputable* used car dealer> **2.** In correct usage. —Used of words. **—rep'u·ta·bil'i·ty** *n.* **—rep'u·ta·bly** *adv.*

**rep·u·ta·tion** (rĕp'yə-tā'shən) *n.* [ME *reputacion* < Lat. *reputatio*, a reckoning < *reputare*, to think over. —see REPUTE.] **1.** The general estimation held of one by the public. **2.** The state or situation of being held in high repute. **3.** A specific characteristic or trait attributed to one <a *reputation* for generosity>

**re·pute** (rĭ-pyo͞ot') *vt.* **-put·ed, -put·ing, -putes.** [ME *reputen* < OFr. *reputer* < Lat. *reputare*, to think over : *re-*, again + *putare*, to think.] **1.** To attribute a specific fact or characteristic to. **2.** To regard or suppose. —*n.* **1.** Reputation. **2.** A good reputation.

**re·put·ed** (rĭ-pyo͞o'tĭd) *adj.* Assumed to be such <*reputed* gangsters> **—re·put'ed·ly** *adv.*

☆ ***syns:*** REPUTED, ALLEGED, CONJECTURAL, PUTATIVE, SUPPOSED *adj. core meaning* : assumed to be such <had lunch with *reputed* crime figures>

**re·quest** (rĭ-kwĕst') *vt.* **-quest·ed, -quest·ing, -quests.** [OFr. *requester* < *requeste*, a request < VLat. **requaestia* < Lat. *requirere*, to ask for. —see REQUIRE.] **1.** To ask for. **2.** To ask (one) to do something. —*n.* **1.** The act of asking. **2.** Something requested. **3.** The fact or condition of being requested. **—by request.** In response to an expressed desire. **—re·quest'er** *n.*

**req·ui·em** (rĕk'wē-əm, rē'kwē-) *n.* [ME < Lat., accusative of *requies*, rest, the first word of the mass for the dead.] **1. Requiem.** *Rom. Cath. Ch.* **a.** A mass for a deceased person. **b.** A musical composition for such a mass. **2.** A hymn, composition, or service for commemorating the dead.

**req·ui·es·cat** (rĕk'wē-ĕs'kăt', -kät') *n.* [Lat., may he rest.] A prayer for the repose of the souls of the dead.

**re·quire** (rĭ-kwīr') *vt.* **-quired, -quir·ing, -quires.** [ME *requiren* < OFr. *requere* < VLat. **requaerere* < Lat. *requirere* : *re-*, again + *quaerere*, to seek.] **1.** To have as a requisite : NEED. **2.** To call for as appropriate : DEMAND <a problem *requiring* immediate remedy> **3.** To impose an obligation on : COMPEL. **4.** To command : order. **—re·quir'a·ble** *adj.* **—re·quire'ment** *n.* **—re·quir'er** *n.*

**req·ui·site** (rĕk'wĭ-zĭt) *adj.* [ME < Lat. *requisitus*, p.part. of *requirere*, to require.] Essential : required. **—req'ui·site** *n.* **—req'ui·site·ly** *adv.* **—req'ui·site·ness** *n.*

**req·ui·si·tion** (rĕk'wĭ-zĭsh'ən) *n.* **1.** A formal written request for something needed. **2.** A necessity. **3.** The state of being needed. **4.** A formal request of one government to another demanding return of a fugitive or criminal. —*vt.* **-tioned, -tion·ing, -tions. 1.** To demand, as for military needs. **2.** To make demands of.

**re·quit·al** (rĭ-kwīt'l) *n.* **1.** An act of requiting. **2.** Return, as for an injury or a friendly act.

**re·quite** (rĭ-kwīt') *vt.* **-quit·ed, -quit·ing, -quites.** [RE- + obs. *quite*, to pay, var. of QUIT.] **1.** To make repayment or return for <*requite* another's love> **2.** To avenge. **—re·quit'a·ble** *adj.* **—re·quit'er** *n.*

**re·ra·di·a·tion** (rē-rā'dē-ā'shən) *n. Physics.* Radiation emission caused by radiation absorption.

**rer·e·dos** (rĕr'ĭ-dŏs', rîr'-, rîr'dŏs') *n.* [ME < OFr. *areredos* : *arere*, behind (Lat. *ad*, to + *retro*, backward) + *dos*, back < Lat. *dorsum*.] **1.** A retable. **2.** The back of an open hearth of a fireplace.

**re·re·lease** (rē'rĭ-lēs') *vt.* **-leased, -leas·ing, -leas·es.** To release (e.g., a film) again. **—re're·lease'** *n.*

**re·run** (rē'rŭn') *n.* A repetition of a recorded motion-picture or television performance. —*vt.* (rē-rŭn') **-ran** (-răn'), **-run, -run·ning, -runs.** To present a rerun of.

**res ad·ju·di·ca·ta** (rēz' ə-jo͞o'dĭ-kä'tə) *n. var. of* RES JUDICATA.

**re·sale** (rē'sāl', rē-sāl') *n.* An act of selling again. **—re·sal'a·ble** *adj.*

**re·scind** (rĭ-sĭnd') *vt.* **-scind·ed, -scind·ing, -scinds.** [Lat. *rescindere* : *re-* (intensive) + *scindere*, to cut.] To make void : ANNUL. **—re·scind'a·ble** *adj.* **—re·scind'er** *n.* **—re·scind'ment** *n.*

**re·scis·sion** (rĭ-sĭzh'ən) *n.* [LLat. *rescissio* < Lat. *rescindere*, to rescind.] An act of rescinding. **—re·scis'so·ry** (-sĭz'ə-rē, -sĭs-) *adj.*

**re·scis·so·ry** (rĭ-sĭz'ə-rē, -sĭs'-) *adj.* Of or relating to rescission.

**re·script** (rē'skrĭpt') *n.* [Lat. *rescriptum* < neuter p.part. of *rescribere*, to write back : *re-*, back + *scribere*, to write.] **1.** *Rom. Cath. Ch.* A response from the pope or another ecclesiastical superior to a question concerning doctrine or discipline. **2.** A formal decree : EDICT. **3.** An act of rewriting. **4.** A reply from a Roman emperor to a magistrate's question about a point of law.

**res·cue** (rĕs'kyo͞o) *vt.* **-cued, -cu·ing, -cues.** [ME *rescuen* < OFr. *rescourre* : *re-*, back (< Lat.) + *escourre*, to shake < Lat. *excutere* (*ex-*, out + *quatere*, to shake).] **1.** To save, as from imprisonment or danger. **2.** *Law.* To deliver from legal custody by force. —*n.* **1.** An act of rescuing : DELIVERANCE. **2.** *Law.* Removal from legal custody by force. **—res'cu·a·ble** *adj.* **—res'cu·er** *n.*

**rescue grass** *n.* [Prob. alteration of FESCUE GRASS.] A tall grass native to tropical America, *Bromus catharticus*, cultivated in warm regions for hay.

**rescue mission** *n.* An esp. urban religious mission that helps and tries to convert destitute individuals.

**re·search** (rĭ-sûrch', rē'sûrch') *n.* [OFr. *recerche* < *recercher*, to research : *re-*, again (< Lat. *re-*) + *cerchier*, to search. —see SEARCH.] **1.** Scientific or scholarly investigation. **2.** Close careful study. —*v.* **-searched, -search·ing, -search·es.** —*vi.* To engage in or do research. —*vt.* To study thoroughly. **—re·search'a·ble** *adj.* **—re·search'er, re·search'ist** *n.*

**ré·seau** or **re·seau** (rā-zō', rĭ-) *n., pl.* **-seaus** or **-seaux** (-zōz') [Fr. < OFr. *reseuil*, dim. of *raiz*, net < Lat. *rete*.] **1.** A net or mesh foundation for lace. **2.** *Astron.* A reference grid of fine lines making uniform squares on a photographic plate or print, for aiding in measurement. **3.** A mosaic screen of fine lines of three colors, used in color photography.

**re·sect** (rĭ-sĕkt') *vt.* **-sect·ed, -sect·ing, -sects.** [Lat. *resecare*, *resect-*, to cut off : *re-*, back + *secare*, to cut.] To perform a resection on. **—re·sect'a·bil'i·ty** *n.* **—re·sect'a·ble** *adj.*

**re·sec·tion** (rĭ-sĕk'shən) *n.* Surgical excision of part of an organ or structure.

**re·se·da** (rĭ-sē'də, -sĕd'ə) *n.* [NLat. *Reseda*, genus name < Lat. *reseda*, a kind of plant.] **1.** A plant of the genus *Reseda*, including the mignonette. **2.** A grayish or dark green to yellow green or light olive. **—re·se'da** *adj.*

**re·seg·re·ga·tion** (rē-sĕg'rĭ-gā'shən) *n.* A renewal of segregation, as in a school system, after a period of desegregation.

**re·sem·blance** (rĭ-zĕm'bləns) *n.* **1.** Similarity in nature, form, or appearance. **2.** Something similar to another : LIKENESS.

**re·sem·ble** (rĭ-zĕm'bəl) *vt.* **-bled, -bling, -bles.** [ME *resemblen* < OFr. *resembler* : *re-* (intensive < Lat.) + *sembler*, to be like < Lat. *simulare*, to imitate < *similis*, like.] To have a similarity to <a child who *resembles* its parents> **—re·sem'bler** *n.*

**re·sent** (rĭ-zĕnt') *vt.* **-sent·ed, -sent·ing, -sents.** [Obs. Fr. *resentir*, to feel strongly : *re-* (intensive) + *sentir*, to feel < Lat. *sentire*.] To feel indignantly aggrieved at.

**re·sent·ful** (rĭ-zĕnt'fəl) *adj.* Full of deep hostility due to a real or imagined offense. **—re·sent'ful·ly** *adv.* **—re·sent'ful·ness** *n.*

☆ ***syns:*** RESENTFUL, ACRIMONIOUS, BITTER, EMBITTERED *adj. core meaning* : feeling or expressing deep hostility due to a real or imagined offense <*resentful* for having been excluded from the group>

**re·sent·ment** (rĭ-zĕnt'mənt) *n.* Indignation or ill will felt as a result of a real or imagined offense.

**re·ser·pine** (rĭ-sûr'pēn', -pĭn, rĕs'ər-pĭn, -pēn', rĕz'-) *n.* [G. *Reserpin* < NLat. *Rauwolfia serpentina*, species of snakeroot : RAUWOLFIA + LLat. *serpentina*, fem. of *serpentinus*, serpentine < Lat. *serpere*, to creep.] A white to yellowish powder, $C_{33}H_{40}N_2O_9$, isolated from the roots of certain species of rauwolfia and used as a tranquilizer and sedative.

**res·er·va·tion** (rĕz'ər-vā'shən) *n.* **1.** An act of reserving. **2.** Something reserved. **3.** A limiting qualification, condition, or exception <has *reservations* about the scheme> **4.** Land set apart by the federal government for a special purpose, esp. one for the use of an American Indian people or tribe. **5. a.** An arrangement for securing accommodations ahead of time, as in a hotel. **b.** Accommodations so secured. **c.** The record or promise of such an arrangement.

☆ ***syns:*** RESERVATION, PRESERVE, RESERVE *n. core meaning* : public land kept for a special purpose <a wildlife *reservation*>

**re·serve** (rĭ-zûrv') *vt.* **-served, -serv·ing, -serves.** [ME *reserven* < OFr. *reserver* < Lat. *reservare*, to keep back : *re-*, back + *servare*, to keep.] **1.** To save for future use or a special purpose. **2.** To set apart for a specific person or use. **3.** To keep or secure for oneself : RETAIN <I *reserve* the right to object.> —*n.* **1.** Something saved for future use or a special purpose. **2.** An act of reserving. **3.** The keeping of one's feelings, thoughts, or affairs to oneself. **4.** Self-restraint in expression : RETICENCE. **5.** Skeptical caution. **6.** An amount of capital held back from investment by a bank or company in order to meet possible or probable demands. **7.** RESERVATION 4 <a grazing *reserve*> **8.** *often* **reserves. a.** A fighting force kept uncommitted until strategic need arises. **b.** The part of a country's armed forces not on active duty but subject to emergency call-up. **—re·serv'a·ble** *adj.* **—re·serv'er** *n.*

**reserve bank** *n.* A central bank holding other banks' reserves.

**re·served** (rĭ-zûrvd') *adj.* **1.** Kept back or set aside. **2.** Self-restrained in behavior and expression. **—re·serv'ed·ly** (-zûr'vĭd-lē) *adv.* **—re·serv'ed·ness** *n.*

**re·serv·ist** (rĭ-zûr'vĭst) *n.* A member of a military reserve.

**res·er·voir** (rĕz'ər-vwär', -vwôr', -vôr') *n.* [Fr. *réservoir* < *réserver*, to reserve.] **1.** A body of water collected and stored for future use in a natural or artificial lake. **2.** A receptacle for storing a fluid. **3.** *Anat.* A cisterna. **4.** A large supply : RESERVE <a *reservoir* of good will>

**re·set** (rē-sĕt') *vt.* **-set, -set·ting, -sets. 1.** To set again <*reset* a broken leg> **2.** To change the reading of <*reset* an odometer> —*n.*

(rē′sĕt′). **1.** An act of resetting. **2.** Something reset. **—re·set′ta·ble** *adj.* **—re·set′ter** *n.*

**res ges·tae** (rās′ gĕs′tī′, rēz′ jĕs′tē) *pl.n.* [Lat., things done.] **1.** Things done : DEEDS. **2.** *Law.* The facts of a particular case that are admissible evidence.

**resh** (rĕsh) *n.* [Heb. *rēsh* < *rōsh,* head.] The 20th letter of the Hebrew alphabet. —See table at ALPHABET.

**re·shape** (rē-shāp′) *vt.* **-shaped, -shap·ing, -shapes.** To shape, form, or organize again. **—re·shap′er** *n.*

**re·shuf·fle** (rē-shŭf′əl) *vt.* **-fled, -fling, -fles. 1.** To shuffle again <*reshuffling* the deck> **2.** To organize or arrange anew <The president *reshuffled* members of the study group.> **—re·shuf′fle** *n.*

**re·sid** (rĭ-zĭd′) *n. Informal.* Residual oil.

**re·side** (rĭ-zīd′) *vi.* **-sid·ed, -sid·ing, -sides.** [ME *residen* < OFr. *resider* < Lat. *residēre,* to sit back : *re-,* back + *sedēre,* to sit.] **1.** To live in a place for a permanent or extended time. **2.** To be inherently present. **3.** To be vested, as a power or right. **—re·sid′er** *n.*

**res·i·dence** (rĕz′ĭ-dəns, -dĕns′) *n.* **1.** The place in which one lives. **2.** The act or a period of residing somewhere. **3.** RESIDENCY 1. **4.** A corporation's official home or location. **—in residence.** Committed to live and work in a specific place, often for a particular length of time <musician *in residence* on campus>

**res·i·den·cy** (rĕz′ĭ-dən-sē, -dĕn′-) *n., pl.* **-cies. 1.** The period during which a physician receives specialized clinical training. **2. a.** The house of a colonial resident. **b.** The sphere of authority of a colonial resident. **3.** RESIDENCE 1, 2.

**res·i·dent** (rĕz′ĭ-dənt, -dĕnt′) *n.* **1.** One whose home is in a particular location. **2.** A colonial official acting as adviser to the ruler of a protected state, often with quasi-gubernatorial powers. **3.** A nonmigratory animal, as a bird. **4.** A physician serving a period of residency. *—adj.* **1.** Living in a particular location : RESIDING. **2.** Living somewhere because of duty or work. **3.** Inherently present. **4.** Nonmigratory, as an animal.

**res·i·den·tial** (rĕz′ĭ-dĕn′shəl) *adj.* **1.** Of, pertaining to, or having residence. **2.** Of or appropriate for residences <*residential* sections of the city> **—res′i·den′tial·ly** *adv.*

**res·i·den·ti·ar·y** (rĕz′ĭ-dĕn′shē-ĕr′ē, -shə-rē) *adj.* **1.** Having a residence, esp. an official one. **2.** Concerning or requiring official residence. *—n., pl.* **-ies. 1.** A resident. **2.** A member of the clergy required to have an official residence.

**re·sid·u·a** (rĭ-zĭj′o͞o-ə) *n. pl. of* RESIDUUM.

**re·sid·u·al** (rĭ-zĭj′o͞o-əl) *adj.* **1.** Of, relating to, or typical of a residue. **2.** Left over as a residue. *—n.* **1.** The amount remaining at the end of a process : REMAINDER. **2.** *often* **residuals.** A payment made to a performer for each repeat showing of a recorded television show. **—re·sid′u·al·ly** *adv.*

**residual oil** *n.* The low-grade oil products that are left following petroleum distillation.

**re·sid·u·ar·y** (rĭ-zĭj′o͞o-ĕr′ē) *adj.* **1.** Of, relating to, or making a residue. **2.** *Law.* Entitled to the residue of an estate.

**res·i·due** (rĕz′ĭ-do͞o′, -dyo͞o′) *n.* [ME < OFr. *residu* < Lat. *residuum,* neuter of *residuus,* remaining < *residēre,* to sit back. —see RESIDE.] **1.** The remainder of something after removal of a part. **2.** Matter left after completion of an abstractive chemical or physical process, such as evaporation, combustion, distillation, or filtration. **3.** *Law.* The remainder of a testator's estate following satisfaction of all claims, debts, and bequests.

**re·sid·u·um** (rĭ-zĭj′o͞o-əm) *n., pl.* **-u·a** (-o͞o-ə) [Lat., residue.] **1.** RESIDUE 1. **2.** *Law.* RESIDUE 3.

**re·sign** (rĭ-zīn′) *v.* **-signed, -sign·ing, -signs.** [ME *resigner* < OFr. *resigner* < Lat. *resignare,* to unseal : *re-* (reversal) + *signare,* to seal < *signum,* mark.] *—vt.* **1.** To submit (oneself) passively : accept as unavoidable. **2.** To give up (a position), esp. by formal notification : QUIT. **3.** To relinquish (a right, privilege, or claim). *—vi.* To give up one's job or office, esp. by formal notification : QUIT. **—re·sign′er** *n.*

**re-sign** (rē-sīn′) *vt.* **-signed, -sign·ing, -signs.** To sign again.

**res·ig·na·tion** (rĕz′ĭg-nā′shən) *n.* **1. a.** An act or instance of resigning. **b.** An oral or written statement that one is resigning a position or office. **2.** Passive acceptance.

**re·signed** (rĭ-zīnd′) *adj.* Feeling or characterized by resignation : ACQUIESCENT. **—re·sign′ed·ly** (-zī′nĭd-lē) *adv.* **—re·sign′ed·ness** *n.*

**re·sile** (rĭ-zīl′) *vi.* **-siled, -sil·ing, -siles.** [Lat. *resilire,* to leap back : *re-,* back + *salire,* to leap.] To spring back, esp. to resume a former position or shape after having been stretched or pressed.

**re·sil·ience** (rĭ-zĭl′yəns) *also* **re·sil·ien·cy** (-yən-sē) *n.* **1.** Ability to recover rapidly from illness, change, or misfortune : BUOYANCY. **2.** The property of a material that enables it to regain its original shape or position after having been bent, stretched, or compressed : ELASTICITY. **—re·sil′ient** *adj.* **—re·sil′ient·ly** *adv.*

**res·in** (rĕz′ĭn) *n.* [ME < OFr. *resine* < Lat. *resina.*] **1.** Any of numerous clear to translucent yellow or brown solid or semisolid viscous substances of plant origin, as copal, rosin, and amber, used chiefly in lacquers, varnishes, inks, adhesives, synthetic plastics, and pharmaceuticals. **2.** Any of numerous physically similar polymerized synthetics or chemically modified natural resins including thermoplastic materials, as polyvinyl, polystyrene, and polyethylene, and thermosetting materials, as polyesters, epoxies, and silicones, that are used with fillers, stabilizers, pigments, and other components to form plastics. *—vt.* **-ined, -in·ing, -ins.** To treat or rub with resin. **—res′in·ous** (rĕz′ə-nəs) *adj.*

**res·in·ate** (rĕz′ə-nāt′) *vt.* **-at·ed, -at·ing, -ates.** To impregnate, permeate, or flavor with resin.

**resin canal** *n.* A tubular intercellular space with resin-secreting cells, often found in gymnosperms.

**res·in·if·er·ous** (rĕz′ə-nĭf′ər-əs) *adj.* Yielding resin.

**res·in·oid** (rĕz′ə-noid′) *adj.* Typical of, relating to, or containing resin. *—n.* A resinoid synthetic, esp. a thermosetting resin.

**re·sist** (rĭ-zĭst′) *v.* **-sist·ed, -sist·ing, -sists.** [ME *resisten* < Lat. *resistere* : *re-,* against + *sistere,* to place.] *—vt.* **1. a.** To oppose actively. **b.** To oppose with force. **2.** To remain firm in opposing the action or effect of : WITHSTAND. **3.** To refrain from giving in to or enjoying <*resisted* all affection> *—vi.* To offer resistance. *—n.* A substance that can coat and protect a surface, as from corrosion. **—re·sist′er** *n.*

☆ **syns:** RESIST, FIGHT, WITHSTAND *v. core meaning* : to oppose actively and with force <Rebels *resisted* the government troops.>

**re·sis·tance** (rĭ-zĭs′təns) *n.* **1. a.** An act of resisting. **b.** Capacity to resist. **2.** A force that tends to oppose or retard motion. **3.** *Elect.* Opposition to electric current characteristic of a medium, substance, or circuit element. **4.** An underground organization struggling for national liberation in a country under totalitarian control. **5.** *Psychoanal.* A process in which the ego opposes the conscious recall of unpleasant experiences. **—re·sis′tant** *adj.*

**re·sist·i·ble** (rĭ-zĭs′tə-bəl) *adj.* Capable of being resisted. **—re·sist′i·bil′i·ty** *n.* **—re·sist′i·bly** *adv.*

**re·sis·tive** (rĭ-zĭs′tĭv) *adj.* Capable of, tending toward, or marked by resistance <a cat *resistive* to travel> **—re·sis′tive·ly** *adv.* **—re·sis′tive·ness** *n.*

**re·sis·tiv·i·ty** (rē′zĭs-tĭv′ĭ-tē) *n.* **1.** Capacity for or tendency toward resistance. **2.** *Elect.* The resistance per unit length of a substance with uniform cross section.

**re·sist·less** (rĭ-zĭst′lĭs) *adj.* **1.** Incapable of being resisted : IRRESISTIBLE. **2.** Unable to resist : UNRESISTING. **—re·sist′less·ly** *adv.* **—re·sist′less·ness** *n.*

**re·sis·tor** (rĭ-zĭs′tər) *n.* An electric circuit element used for providing resistance.

**res ju·di·ca·ta** (rēz′ jo͞o′dĭ-kä′tə) *also* **res ad·ju·di·ca·ta** (rēz′ ə-jo͞o′-) *n.* [Lat., thing decided.] *Law.* An adjudicated precedent.

**re·sole** (rē-sōl′) *vt.* **-soled, -sol·ing, -soles.** To put a new sole on (footwear).

**re·sol·u·ble** (rĭ-zŏl′yə-bəl) *adj.* [LLat. *resolubilis* < Lat. *resolvere,* to resolve.] Capable of being resolved. **—re·sol′u·bil′i·ty** *n.*

**res·o·lute** (rĕz′ə-lo͞ot′) *adj.* [Lat. *resolutus,* p.part. of *resolvere,* to resolve.] Marked by firmness or determination : UNWAVERING. **—res′o·lute′ly** *adv.* **—res′o·lute′ness** *n.*

**res·o·lu·tion** (rĕz′ə-lo͞o′shən) *n.* **1.** Firm determination. **2.** An act of resolving to do something. **3.** A course of action determined or decided on. **4.** A formal statement of a decision or expression of opinion put before or adopted by an assembly such as the U.S. Congress. **5.** Separation or reduction of something into its constituent parts <the prismatic *resolution* of sunlight into its spectral colors> **6.** *Med.* Abatement or termination of an abnormal condition, as a fever or inflammation. **7.** A decision of a court of law. **8.** An explanation, as of a puzzle or problem : SOLUTION. **9.** *Mus.* **a.** The progression of a dissonant tone or chord to a consonant tone or chord. **b.** The tone or chord to which such a progression is made.

**re·solve** (rĭ-zŏlv′) *v.* **-solved, -solv·ing, -solves.** [ME *resolven,* to dissolve < Lat. *resolvere,* to untie : *re-,* back + *solvere,* to untie.] *—vt.* **1.** To make a firm decision about. **2.** To cause (one) to arrive at a decision. **3.** To decide or express by formal vote. **4.** To separate (something) into constituent parts. **5.** To change or convert <My resentment *resolved* itself into resignation.> **6.** To reach a solution to : SOLVE. **7.** To dispel or remove (doubts) : explain away. **8.** To bring to a usu. successful conclusion <*resolve* a dilemma> **9.** *Med.* To cause reduction of (e.g., an inflammation). **10.** *Mus.* To cause (a tone or chord) to progress from dissonance to consonance. **11.** *Chem.* To separate (a racemic compound or mixture) into its optically active constituents. **12.** To make visible and distinguish parts of (an image). **13.** *Math.* To separate (e.g., a vector) into coordinate components. **14.** To melt or dissolve (something). *—vi.* **1.** To arrive at a decision or make a determination <*resolve* on a new proposal> **2.** To become separated or reduced to components. **3.** *Mus.* To undergo resolution. *—n.* **1.** Steadfastness of purpose : RESOLUTION. **2.** A fixed intention or decision. **3.** A formal resolution made by a deliberative body. **—re·solv′a·bil′i·ty, re·solv′a·ble·ness** *n.* **—re·solv′a·ble** *adj.* **—re·solv′er** *n.*

**re·sol·vent** (rĭ-zŏl′vənt) *adj.* Causing or capable of causing separation into constituents : SOLVENT. *—n.* A resolvent substance, esp. an anti-inflammatory medicine.

**res·o·nance** (rĕz′ə-nəns) *n.* **1.** The quality or state of being resonant. **2.** *Physics.* **a.** Enhancement of the response of an electric or mechanical system to a periodic driving force when the driving frequency is equal to the natural undamped frequency of the system. **b.** The condition of a system of subatomic particles in which the probability of a particular reaction, as for nuclear capture of a neutron, is a maximum. **c.** The event corresponding to such a maximum, esp. the formation of a particle state with but a few possible modes of decay and marked by a lifetime considerably longer than neighboring states. **3.** Intensification and prolongation of sound, esp. of a musical tone, caused by sympathetic vibration. **4.** *Med.* The sound caused by diagnostic percussion of the normal chest. **5.** *Chem.* The phenomenon of interrelated alternative bond structures in certain molecules, caused by redistribution of valence electrons without change in the relative positions of bound atoms and resulting in highly stable compounds. **6.** Intensification of vocal tones during articulation.

**res·o·nant** (rĕz'ə-nənt) *adj.* [Lat. *resonans, resonant-,* pr.part. of *resonare,* to resound.] **1.** Of, relating to, or showing resonance. **2. a.** Producing resonance. **b.** Having or producing a full, deep, or rich sound <a *resonant* baritone> **3.** Resulting from or as if from resonance <*resonant* amplification> **—res'o·nant·ly** *adv.*

☆ ***syns:*** RESONANT, MELLOW, RESOUNDING, SONOROUS, VIBRANT *adj. core meaning* **:** having or producing a full, deep, or rich sound <a *resonant* baritone>

**resonant circuit** *n.* An electric circuit with inductance and capacitance selected to produce a specified value of the natural frequency of the circuit.

**res·o·nate** (rĕz'ə-nāt') *v.* **-nat·ed, -nat·ing, -nates.** [Lat. *resonare, resonat-.* —see RESOUND.] —*vi.* **1.** To exhibit or produce resonance or resonant effects. **2.** To resound. —*vt.* To subject to resonating. **—res'o·na'tion** *n.*

**res·o·na·tor** (rĕz'ə-nā'tər) *n.* **1.** A resonating system. **2. a.** A hollow chamber or cavity with dimensions selected to allow internal resonant oscillation of electromagnetic or acoustical waves of specific frequencies. **b.** *Electron.* A microwave-generating tube or device containing such resonant chambers or cavities. **c.** *Electron.* A resonant circuit.

**re·sorb** (rē-sôrb', -zôrb') *v.* **-sorbed, -sorb·ing, -sorbs.** [Lat. *resorbere,* to suck back : *re-,* back + *sorbere,* to suck up.] —*vt.* **1.** To absorb again. **2.** *Biol.* To dissolve and assimilate (e.g., bone tissue). —*vi.* To undergo resorption. **—re·sorp'tion** (-sôrp'shən, -zôrp'-) *n.*

**res·or·cin·ol** (rə-zôr'sə-nôl', -nōl') *also* **res·or·cin** (rə-zôr'sĭn) *n.* [RES(IN) + ORC(HIL) + -IN + -OL.] A white crystalline compound, $C_6H_4(OH)_2$, for treating certain skin diseases and in dyes, resin adhesives, and pharmaceuticals.

**re·sort** (rĭ-zôrt') *vi.* **-sort·ed, -sort·ing, -sorts.** [ME *resorten,* to return < OFr. *resortir,* to go out again : *re-,* again + *sortir,* to go out.] **1.** To have recourse <The government *resorted* to rationing.> **2.** To go commonly or often **:** REPAIR. —*n.* **1.** A place visited by people for relaxation or recreation <a ski *resort*> **2.** A customary or frequent going or gathering <a well-known spot of *resort*> **3.** Recourse. **4.** One turned to for aid or relief.

**re·sound** (rĭ-zound') *v.* **-sound·ed, -sound·ing, -sounds.** [ME *resounen* < OFr. *resoner* < Lat. *resonare* : *re-,* again + *sonare,* to sound.] —*vi.* **1.** To be filled with sound **:** REVERBERATE. **2.** To make a loud, long, or reverberating sound. **3.** To sound loudly **:** RING. **4.** To become famous or celebrated. —*vt.* **1.** To send back (sound). **2.** To utter or emit loudly. **3.** To proclaim widely **:** CELEBRATE. **—re·sound'ing** *adj.* **—re·sound'ing·ly** *adv.*

**re·source** (rē'sôrs', -sōrs', -zôrs', -zōrs', rĭ-sôrs', -sōrs', -zôrs', -zōrs') *n.* [Fr. *ressource* < OFr. *ressourse* < *resourdre,* to rise again < Lat. *resurgere* : *re-,* again + *surgere,* to rise.] **1.** Something that can be looked to for support or aid. **2.** An accessible supply that can be withdrawn from when necessary. **3.** An ability to handle a situation in an effective manner. **4.** *often* **resources.** Means that can be used profitably. **5.** *often* **resources.** Available capital **:** ASSETS. **6.** A way of spending one's leisure time.

**re·source·ful** (rĭ-sôrs'fəl, -sōrs'-, -zôrs'-, -zōrs'-) *adj.* Able to use the means at one's disposal to meet situations effectively. **—re·source'ful·ly** *adv.* **—re·source'ful·ness** *n.*

☆ ***syns:*** RESOURCEFUL, FERTILE, INGENIOUS, INVENTIVE *adj. core meaning* **:** able to use the means at one's disposal to meet situations effectively <an intelligent, *resourceful* commander>

**re·spect** (rĭ-spĕkt') *vt.* **-spect·ed, -spect·ing, -spects.** [Lat. *respicere, respect-* : *re-,* back + *specere,* to look.] **1.** To feel or show deferential regard for **:** ESTEEM. **2.** To avoid violation of or interference with. **3.** To relate or refer to **:** CONCERN. —*n.* **1.** Deferential regard **:** ESTEEM. **2.** The state of being regarded with deference or esteem. **3.** Willingness to show consideration or appreciation. **4. respects.** Polite expressions of consideration or deference <paid our *respects*> **5.** A specific aspect, feature, or detail. **6.** Relation **:** reference. **—re·spect'er** *n.*

**re·spect·a·ble** (rĭ-spĕk'tə-bəl) *adj.* **1.** Deserving respect or esteem **:** WORTHY. **2.** Of or good or fitting for correct behavior or conventional conduct. **3.** Of reasonably good quality <*respectable* painting> **4.** Considerable in amount, number, or size <a *respectable* sum of money> **5.** Acceptable in appearance **:** PRESENTABLE <a *respectable* pair of shoes> **—re·spect'a·bil'i·ty** (-bĭl'ĭ-tē), **re·spect'a·ble·ness** *n.* **—re·spect'a·bly** *adv.*

**re·spect·ful** (rĭ-spĕkt'fəl) *adj.* Displaying or characterized by appropriate respect. **—re·spect'ful·ly** *adv.* **—re·spect'ful·ness** *n.*

**re·spect·ing** (rĭ-spĕk'tĭng) *prep.* With respect to.

**re·spec·tive** (rĭ-spĕk'tĭv) *adj.* Relating to two or more persons or things considered individually **:** PARTICULAR <our *respective* contributions> **—re·spec'tive·ness** *n.*

**re·spec·tive·ly** (rĭ-spĕk'tĭv-lē) *adv.* Singly in the order indicated or mentioned <speaking to each camper *respectively*>

**re·spell** (rē-spĕl') *vt.* **-spelled** *or* **-spelt** (-spĕlt'), **-spell·ing, -spells.** To spell again or in a new way, esp. by using a phonetic alphabet.

**res·pi·ra·ble** (rĕs'pər-ə-bəl, rĭ-spīr'-) *adj.* **1.** Suitable for breathing. **2.** Capable of being inhaled. **—res'pi·ra·bil'i·ty** *n.*

**res·pi·ra·tion** (rĕs'pə-rā'shən) *n.* **1.** The act or process of inhaling and exhaling **:** BREATHING. **2.** The metabolic process by which an organism assimilates oxygen and releases carbon dioxide and other products of oxidation. **—res'pi·ra'tion·al** *adj.*

**res·pi·ra·tor** (rĕs'pə-rā'tər) *n.* **1.** An apparatus used in administering artificial respiration. **2.** A screenlike device worn over the mouth or nose or both to protect the respiratory tract.

**res·pi·ra·to·ry** (rĕs'pər-ə-tôr'ē, -tōr'ē, rĭ-spīr'ə-) *adj.* Of, relating to, affecting, or used in respiration.

**respiratory pigment** *n.* A conjugated colored protein, as hemoglobin, found in living organisms and functioning in oxygen transfer in cellular respiration.

**respiratory system** *n.* The integrated system of organs involved in the intake and exchange of oxygen and carbon dioxide between an organism and the atmosphere.

**re·spire** (rĭ-spīr') *v.* **-spired, -spir·ing, -spires.** [ME *respiren,* to breathe again < Lat. *respirare* : *re-,* again + *spirare,* to breathe.] —*vi.* **1.** To inhale and exhale **:** BREATHE. **2.** To undergo the metabolic process of respiration. **3.** To breathe easily again, as after exertion. —*vt.* To inhale and exhale (air) **:** BREATHE.

**res·pite** (rĕs'pĭt) *n.* [ME < OFr. *respit* < Lat. *respectus,* refuge < *respicere,* to respect.] **1.** A usu. short time of rest or relief. **2.** *Law.* Temporary suspension of a death sentence **:** REPRIEVE. —*vt.* **-pit·ed, -pit·ing, -pites. 1.** To delay **:** postpone. **2.** To grant a reprieve from (a punishment or sentence).

**re·splen·dent** (rĭ-splĕn'dənt) *adj.* [ME < Lat. *resplendens,* pr.part. of *resplendēre,* to shine brightly : *re-* (intensive) + *splendēre,* to shine.] Full of splendor. **—re·splen'dence, re·splen'den·cy** *n.* **—re·splen'dent·ly** *adv.*

**re·spond** (rĭ-spŏnd') *v.* **-spond·ed, -spond·ing, -sponds.** [Lat. *respondēre* : *re-,* in return + *spondēre,* to promise.] —*vi.* **1.** To reply **:** answer. **2.** To act in return or in answer. **3.** To react positively or favorably. —*vt.* To say in reply **:** ANSWER. —*n.* A pilaster supporting an arch in a building. **—re·spond'er** *n.*

**re·spon·dent** (rĭ-spŏn'dənt) *adj.* **1.** Giving or given as an answer **:** RESPONSIVE. **2.** *Law.* Being a defendant. —*n.* **1.** One who responds. **2.** *Law.* A defendant, esp. in a divorce or equity case. **—re·spon'dence, re·spon'den·cy** *n.*

**re·sponse** (rĭ-spŏns') *n.* [ME *respons* < OFr. < Lat. *responsum* < *respondēre,* to respond.] **1.** An act of responding. **2.** A reply **:** answer. **3.** A reaction, as that of an organism or mechanism, to a particular stimulus. **4. a.** Words sung or spoken by a congregation or choir in answer to the officiating cleric. **b.** An anthem sung after a reading **:** RESPONSORY.

**re·spon·si·bil·i·ty** (rĭ-spŏn'sə-bĭl'ĭ-tē) *n., pl.* **-ties. 1.** The quality, state, or fact of being responsible. **2.** Something for which one is accountable **:** DUTY.

**re·spon·si·ble** (rĭ-spŏn'sə-bəl) *adj.* [Obs. Fr., corresponding to < Lat. *respondēre,* to respond.] **1.** Being legally or ethically accountable for the welfare or care of another. **2.** Involving personal accountability or ability to act free from guidance or higher authority. **3.** Being a source or cause. **4.** Capable of making moral or rational decisions on one's own, thereby being answerable for one's behavior. **5.** Capable of being trusted or depended on **:** RELIABLE. **6.** Based on or marked by good judgment. **7.** Having the means to pay debts or meet obligations. **8.** Required to render account **:** ANSWERABLE <The cabinet is *responsible* to the President.> **—re·spon'si·ble·ness** *n.* **—re·spon'si·bly** *adv.*

**re·spon·sive** (rĭ-spŏn'sĭv) *adj.* **1.** Answering or replying **:** RESPONDING. **2.** Promptly reacting to suggestions, influences, appeals, or efforts. **3.** Having or using responses <*responsive* anthems> **—re·spon'sive·ly** *adv.* **—re·spon'sive·ness** *n.*

**re·spon·so·ry** (rĭ-spŏn'sə-rē) *n., pl.* **-ries.** [ME < LLat. *responsoria* < Lat. *respondēre,* to respond.] A responsive reading or anthem in a religious service.

**res pub·li·ca** (rās pōō'blē-kä') *n.* [Lat. —see REPUBLIC.] **1.** A state, republic, or commonwealth. **2.** Public good or welfare.

**res·sen·ti·ment** (rə-säN'tē-mäN) *n.* [G. < Fr., resentment.] General resentment and often hostility harbored by one person or group against another person or group, esp. chronically and with no means of direct expression.

**rest¹** (rĕst) *n.* [ME < OE.] **1.** The act or state of ceasing from work, activity, or motion : QUIET. **2.** Peace, ease, or refreshment as a consequence of sleep or the cessation of an activity. **3.** SLEEP 1. **4.** DEATH 2. **5.** Relief or freedom from disturbance or disquiet. **6.** Mental or emotional serenity. **7.** Cessation or lack of motion. **8.** *Mus.* **a.** An interval of silence equivalent to one of the possible time values within a measure. **b.** The mark or symbol designating such a pause and its length. **9.** A short pause in a line of poetry : CAESURA. **10.** A supporting device. —*v.* **rest·ed, rest·ing, rests.** —*vi.* **1.** To stop motion, work, or activity. **2.** To lie down, esp. to sleep. **3.** To be at peace or ease. **4.** To be, become, or remain still, quiet, or inactive for a time. **5.** To be supported or based. **6.** To be imposed or placed, as a responsibility <The task *rests* with them.> **7.** To depend or rely. **8.** To be in a designated location. **9.** To be fixed or directed on something. **10.** To remain : linger. **11.** *Law.* To cease voluntarily the presentation of evidence in a case. —*vt.* **1.** To give rest or repose to. **2.** To place, lay, or lean for ease, support, or repose. **3.** To base or ground. **4.** To fix or direct (e.g., the gaze). **5.** To bring to rest : HALT. **6.** *Law.* To cease voluntarily the introduction of evidence in (a case). **—at rest. 1.** In a state or condition of repose, esp.: **a.** ASLEEP 1. **b.** DEAD 1. **2.** Motionless : inactive. **3.** Free from anxiety or distress : SERENE. **—lay to rest. 1.** To bury (the dead). **2.** To put an end to (e.g., a false assertion) <*lay* a rumor *to rest*> **—rest′er** *n.*

**rest²** (rĕst) *n.* [ME < OFr. *reste* < *rester,* to remain < Lat. *restare,* to stay behind : *re-,* back + *stare,* to stay.] **1.** The part left over following removal of something : REMAINDER. **2.** Those remaining <The *rest* are coming later.> —*vi.* **rest·ed, rest·ing, rests. 1.** To be or continue to be : REMAIN <*rest* easy> **2.** To remain or be left over.

**rest³** (rĕst) *n.* [ME < *arest* < OFr. < *arester,* to arrest. —see ARREST.] A support for a lance on the side of the breastplate of medieval armor.

**re·start** (rē-stärt′) *v.* **-start·ed, -start·ing, -starts.** —*vt.* To start again <*restart* a motor> —*vi.* To begin operation again. **—re′start′** *n.* **—re·start′a·ble** *adj.*

**re·state** (rē-stāt′) *vt.* **-stat·ed, -stat·ing, -states.** To state again or anew. **—re·state′ment** *n.*

**res·tau·rant** (rĕs′tər-ənt, -tə-ränt′) *n.* [Fr. < pr.part. of *restaurer,* to restore < OFr. *restorer.* —see RESTORE.] A place for serving meals to the public.

**res·tau·ra·teur** (rĕs′tər-ə-tûr′) *also* **res·tau·ran·teur** (-tə-rän-tûr′) *n.* [Fr. < *restaurer,* to restore.] The owner or manager of a restaurant.

**rest energy** *n.* The energy equivalent of the rest mass of a body, equal to the rest mass multiplied by the speed of light squared.

**rest·ful** (rĕst′fəl) *adj.* **1.** Providing, characterized by, or suggesting rest : TRANQUIL. **2.** Being at rest : QUIET. **—rest′ful·ly** *adv.* **—rest′ful·ness** *n.*

**rest·har·row** (rĕst′hăr′ō) *n.* [Obs. *rest,* to check, short for ARREST + HARROW.] Any of several Old World plants of the genus *Ononis,* with tough woody stems and roots and purplish or pink flowers.

**rest home** *n.* An establishment where elderly or sick people live and are cared for.

**rest·ing** (rĕs′tĭng) *adj.* **1. a.** Being in a state of rest or inactivity. **b.** DEAD 1. **2.** *Bot.* Dormant. —Used esp. of spores that germinate after a prolonged period.

**resting cell** *n.* A cell not actively undergoing division.

**res·ti·tute** (rĕs′tĭ-to͞ot′, -tyo͞ot′) *v.* **-tut·ed, -tut·ing, -tutes.** [Lat. *restituere, restitut-* : *re-,* back + *statuere,* to set up < *stare,* to stand.] —*vt.* **1.** To bring back to a previous condition : RESTORE. **2.** To refund. —*vi.* To undergo restitution.

**res·ti·tu·tion** (rĕs′tĭ-to͞o′shən, -tyo͞o′-) *n.* **1.** An act of restoring to the proper owner something taken away, lost, or surrendered. **2.** An act of repaying or compensating for loss, damage, or injury : INDEMNIFICATION. **3.** A return to or restoration of a former state or position.

**res·tive** (rĕs′tĭv) *adj.* [ME *restyf,* stationary < OFr. *restif* < VLat. **restivus* < Lat. *restare,* to keep back : *re-,* back + *stare,* to stand.] **1.** Impatient or nervous under restriction, delay, or pressure : UNEASY. **2.** Difficult to control : REFRACTORY. **3.** Refusing to move, as a horse. **—res′tive·ly** *adv.* **—res′tive·ness** *n.*

**rest·less** (rĕst′lĭs) *adj.* **1.** Without quiet, repose, or rest <*restless* sleep> **2.** Unable to rest, relax, or be still <a *restless* puppy> **3.** Constantly in motion <the *restless* ocean waves> **—rest′less·ly** *adv.* **—rest′less·ness** *n.*

**rest mass** *n.* The physical mass of a body as observed in a reference system with respect to which the body is at rest.

**re·stock** (rē-stŏk′) *vt.* **-stocked, -stock·ing, -stocks.** To furnish new stock for.

**res·to·ra·tion** (rĕs′tə-rā′shən) *n.* **1.** An act of restoring <*Restoration* of the icon was difficult.> **2.** The state of being restored. **3.** A renovated structure, as a building. **4. Restoration. a.** Charles II's return to the British throne in 1660. **b.** The period between Charles II's return and the Revolution of 1688.

**re·stor·a·tive** (rĭ-stôr′ə-tĭv, -stōr′-) *adj.* **1.** Of or relating to restoration. **2.** Tending or having the power to restore. **—re·stor′a·tive** *n.* **—re·stor′a·tive·ly** *adv.* **—re·stor′a·tive·ness** *n.*

**re·store** (rĭ-stôr′, -stōr′) *vt.* **-stored, -stor·ing, -stores.** [ME *restoren* < OFr. *restorer* < Lat. *restaurare.*] **1.** To bring back into existence or use. **2.** To bring back to an original state. **3.** To put (someone) back in a former position <*restore* the monarch to the throne> **4.** To give make restitution of : give back <*restore* the stolen goods to the rightful owner> **—re·stor′er** *n.*

☆ ***syns:*** **1.** RESTORE, REINSTATE, RENEW, REVIVE *v. core meaning* : to bring back into existence or use <*restore* law and order> **2.** RESTORE, REBUILD, RECONDITION, RECONSTRUCT, REHABILITATE *v. core meaning* : to bring back to a previous condition <*restored* a Victorian mansion> **3.** RESTORE, REINSTATE, RETURN *v. core meaning* : to give (something) back to its owner or possessor <an exiled monarch hoping to be *restored* to the throne>

**re·strain** (rĭ-strān′) *vt.* **-strained, -strain·ing, -strains.** [ME *restreyer* < OFr. *restraindre* < Lat. *restringere,* to bind back : *re-,* back + *stringere,* to bind.] **1.** To control : check. **2.** To take away the freedom or liberty of. **3.** To restrict or limit. **—re·strain′a·ble** *adj.* **—re·strain′ed·ly** (-strā′nĭd-lē) *adv.* **—re·strain′er** *n.*

**re·straint** (rĭ-strānt′) *n.* [ME *restreinte* < OFr. *restrainte* < p.part. of *restraindre,* to restrain.] **1.** An act of restraining or the state of being restrained. **2.** Loss or curtailment of freedom. **3.** An inhibiting influence : LIMITATION. **4.** An instrument or means of restraining. **5.** Control or suppression of feelings : CONSTRAINT.

**re·strict** (rĭ-strĭkt′) *vt.* **-strict·ed, -strict·ing, -stricts.** [Lat. *restringere, restrict-,* to restrain.] To hold within limits : CONFINE. **—re·stric′tion** *n.*

**re·strict·ed** (rĭ-strĭk′tĭd) *adj.* **1.** Confined : limited. **2.** Excluding or unavailable to members of particular groups <*restricted* housing> **—re·strict′ed·ly** *adv.*

**re·stric·tive** (rĭ-strĭk′tĭv) *adj.* **1.** Serving or tending to restrict. **2.** Denoting a subordinate clause, phrase, or term regarded as limiting the use of the word or word group that it modifies, thus being essential to the meaning of the sentence <a *restrictive* clause> **—re·stric′tive·ly** *adv.* **—re·stric′tive·ness** *n.*

**rest room** *n.* A public lavatory.

**re·struc·ture** (rē-strŭk′chər) *v.* **-tured, -tur·ing, -tures.** —*vt.* To alter the make-up of. —*vi.* To alter the structure of something.

**re·sult** (rĭ-zŭlt′) *vi.* **-sult·ed, -sult·ing, -sults.** [ME *resulten* < Med. Lat. *resultare* < Lat., to leap back : *re-,* back + *saltare,* to dance, freq. of *salire,* to leap.] **1.** To happen or exist as a result of a cause. **2.** To end in a given way. —*n.* The consequence or outcome of an action. **—re·sult′ful** *adj.* **—re·sult′ful·ness** *n.* **—re·sult′less** *adj.*

**re·sul·tant** (rĭ-zŭl′tənt) *adj.* Issuing as a result. —*n.* **1.** A result : consequence. **2.** *Math.* A vectorial sum. **—re·sul′tant·ly** *adv.*

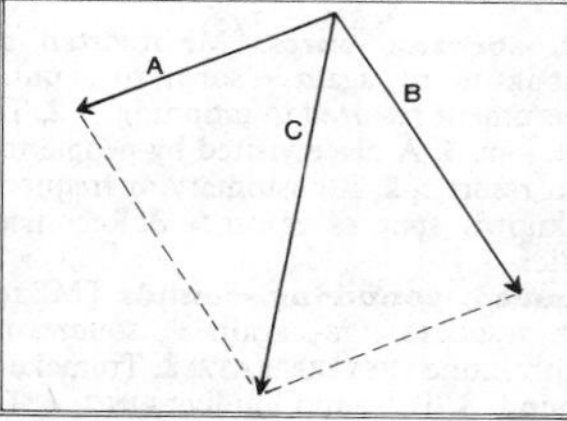


**resultant**
*C, the resultant of vectors A and B*

**re·sume** (rĭ-zo͞om′) *v.* **-sumed, -sum·ing, -sumes.** [ME *resumer* < OFr. *resumer* < Lat. *resumere* : *re-,* again + *sumere,* to take up.] —*vt.* **1.** To begin or take up again following interruption. **2.** To assume or take again. **3.** To take on or take back again. —*vi.* To start again or continue after interruption. **—re·sum′a·ble** *adj.* **—re·sum′er** *n.*

**rés·u·mé** (rĕz′o͞o-mā′, rĕz′o͞o-mā′) *n.* [Fr. < p.part. of *résumer,* to summarize < OFr. *resumer,* to resume.] A summary, esp. a record of one's personal history, educational background, and employment experience.

**re·sump·tion** (rĭ-zŭmp′shən) *n.* [ME < OFr. < LLat. *resumptio* < Lat. *resumere,* to resume.] An act of resuming <*resumption* of Mideast hostilities>

**re·su·pi·nate** (rĭ-so͞o′pə-nāt′, -nĭt) *adj.* [Lat. *resupinatus,* bent back, p.part. of *resupinare,* to bend back : *re-,* back + *supinus,* supine.] *Biol.* Inverted or turned upside-down. **—re·su′pi·na′tion** *n.*

**re·su·pine** (rĕs′ə-pīn′) *adj.* [Lat. *resupinus* < *resupinare,* to bend back. —see RESUPINATE.] Lying on the back : SUPINE.

**re·sur·face** (rē-sûr′fəs) *v.* **-faced, -fac·ing, -fac·es.** —*vt.* To cover with a new surface. —*vi.* To come to the surface again : REAPPEAR.

**re·surge** (rĭ-sûrj′) *vi.* **-surged, -surg·ing, -surg·es.** [Lat. *resurgere* : *re-,* again + *surgere,* to rise.] **1.** To rise again. **2.** To surge back again.

**re·sur·gent** (rĭ-sûr′jənt) *adj.* **1.** Rising or tending to rise again. **2.** Sweeping or surging back again. **—re·sur′gence** *n.*

**res·ur·rect** (rĕz′ə-rĕkt′) *v.* **-rect·ed, -rect·ing, -rects.** [Back-formation < RESURRECTION.] —*vt.* **1.** To raise from the dead. **2.** To bring back into practice, notice, or use. —*vi.* To return to life.

**res·ur·rec·tion** (rĕz′ə-rĕk′shən) *n.* [ME < OFr. < LLat. *resurrectio* < Lat. *resurgere,* to rise again. —see RESURGE.] **1.** An act of rising from the dead or returning to life. **2.** The state of those who have returned to life. **3.** The act of bringing back to practice, notice, or use : REVIVAL. **4. Resurrection. a.** The rising again of Christ on the third day after the Crucifixion. **b.** The rising again of the dead at the Last Judgment. **—res′ur·rec′tion·al** *adj.*

**resurrection fern** *n.* An American fern, *Polypodium polypoides* of warm regions, with fronds that curl up and apparently die in prolonged dry weather but expand under moist conditions.

**res·ur·rec·tion·ist** (rĕz′ə-rĕk′shə-nĭst) *n.* **1.** One who steals bodies from graves for the purpose of selling them for dissection : BODY SNATCHER. **2.** One who resurrects.

**resurrection plant** *n.* A plant, esp. the rose of Jericho, that appears dead during dry periods but expands and resumes growing under moist conditions.

**re·sur·vey** (rē′sər-vā′, rē-sûr′vā′) *vt.* **-veyed, -vey·ing, -veys.** To survey anew. *—n.* (rē-sûr′vā′). A new survey.

**re·sus·ci·tate** (rĭ-sŭs′ĭ-tāt′) *v.* **-tat·ed, -tat·ing, -tates.** [Lat. *resuscitare, resuscitat-* : *re-,* again + *suscitare,* to stir up (*sub-,* up from below + *citare,* to move violently, freq. of *ciēre,* to put in motion).] *—vt.* To restore consciousness, vigor, or life to : REVIVE <*resuscitate* a drowning victim> *—vi.* To return to consciousness, vigor, or life. **—re·sus′ci·ta·ble** (-tə-bəl) *adj.* **—re·sus′ci·ta′tion** *n.* **—re·sus′ci·ta′tive** *adj.* **—re·sus′ci·ta′tor** *n.*

**ret** (rĕt) *v.* **ret·ted, ret·ting, rets.** [ME *reten.*] *—vt.* To moisten or soak (e.g., flax) to soften and separate the fibers by partial rotting. *—vi.* To become retted.

**re·ta·ble** (rē′tā′bəl, rĕt′ə-) *n.* [Fr. < Sp. *retablo* : Lat. *retro-,* back + *tabulum,* table.] A structure forming the back of an altar, esp.: **a.** An overhanging shelf for ornaments and lights. **b.** A frame around painted panels.

**re·tail** (rē′tāl′) *n.* [ME < OFr. *retaille,* piece cut off < *retailler,* to cut up : *re-* (intensive) + *tailler,* to cut. —see TAILOR.] The sale of goods in small quantities to consumers. *—adj.* Of, relating to, or engaged in selling goods at retail. *—v.* **-tailed, -tail·ing, -tails.** *—vt.* **1.** To sell in small quantities directly to consumers. **2.** (*also* rĭ-tāl′). To tell and retell. *—vi.* To sell at retail. **—re′tail′** *adv.* **—re′tail′er** *n.*

**re·tain** (rĭ-tān′) *vt.* **-tained, -tain·ing, -tains.** [ME *reteinen* < OFr. *retenir* < Lat. *retinēre* : *re-,* back + *tenēre,* to hold.] **1.** To keep or hold in one's possession. **2.** To keep or hold in a specific location, condition, or position. **3.** To keep in mind : REMEMBER. **4.** To hire (e.g., an attorney) by paying a fee. **5.** To keep in one's service or pay. **—re·tain′a·ble** *adj.* **—re·tain′ment** *n.*

**retained object** *n.* An object in a passive construction that is identical to the object in the corresponding active construction, as *book* in *I was given the book by you.*

**re·tain·er¹** (rĭ-tā′nər) *n.* **1.** One that retains. **2. a.** One who served in a noble household, as in the feudal period, but who held a higher rank than a servant : ATTENDANT. **b.** A domestic servant. **c.** An employee. **3.** A device, frame, or groove that guides or restrains something else.

**re·tain·er²** (rĭ-tā′nər) *n.* **1.** An act of retaining a professional adviser, as an attorney. **2.** The fee paid to engage the services of a professional adviser.

**re·take** (rē-tāk′) *vt.* **-took** (-to͝ok′), **-tak·en** (-tā′kən), **-tak·ing, -takes. 1.** To take again. **2.** To photograph again. *—n.* (rē′tāk′). A rephotographed scene.

**re·tal·i·ate** (rĭ-tăl′ē-āt′) *v.* **-at·ed, -at·ing, -ates.** [Lat. *retaliare, retaliat-* : *re-,* back + *talio,* punishment in kind.] *—vi.* To return like for like, esp. to return evil for evil. *—vt.* To pay back (an injury) in kind. **—re·tal′i·a′tion** *n.* **—re·tal′i·a′tive, re·tal′i·a·to′ry** (-ə-tôr′ē, -tōr′ē) *adj.*

**re·tard** (rĭ-tärd′) *v.* **-tard·ed, -tard·ing, -tards.** [ME *retarden* < OFr. *retarder* < Lat. *retardare* : *re-,* back + *tardare,* to delay < *tardus,* slow.] *—vt.* To slow the progress of : DELAY. *—vi.* To become delayed. *—n.* **1.** Delay. **2.** *Mus.* A slackening of tempo. **—re·tar′dant** *adj.* & *n.* **—re·tard′er** *n.*

**re·tar·date** (rĭ-tär′dāt′, -dĭt) *n.* A mentally retarded person.

**re·tar·da·tion** (rē′tär-dā′shən) *n.* **1.** The act or process of retarding. **2.** The state of being retarded. **3.** Something that retards : DELAY. **4.** *Mus.* RETARD 2. **5.** *Psychol.* Mental deficiency.

**re·tard·ed** (rĭ-tär′dĭd) *adj.* Slow or backward in mental or emotional development or in academic achievement.

**retch** (rĕch) *v.* **retched, retch·ing, retch·es.** [Alteration of obs. *reach,* ult. < OE *hrǣcan.*] *—vi.* To try to vomit : HEAVE. *—vt.* To vomit. **—retch** *n.*

**re·te** (rē′tē) *n., pl.* **re·ti·a** (rē′tē-ə, rē′shə) [NLat. < Lat., net.] An anatomical mesh or network, as of veins, arteries, or nerves.

**re·tell** (rē-tĕl′) *vt.* **-told** (-tōld′), **-tell·ing, -tells. 1.** To relate again <*retell* an old story> **2.** To count again.

**re·tell·ing** (rē-tĕl′ĭng) *n.* A new version or adaptation of a story <a *retelling* of a Norse myth>

o͞o **boot** ou **out** th **thin** *th* **this** ŭ **cut** û**r urge** y **young**
yo͞o ab**use** zh vi**si**on ə **a**bout, it**e**m, ed**i**ble, gall**o**p, circ**u**s

**re·tene** (rē′tēn′, rĕt′ēn′) *n.* [Gk. *rhētinē,* resin.] A crystalline compound, $C_{18}H_{18}$, obtained from pine tar, fossil resins, and tar oils.

**re·ten·tion** (rĭ-tĕn′shən) *n.* [ME *retencion* < OFr. < Lat. *retentio* < *retinēre,* to retain : *re-,* back + *tenēre,* to hold.] **1.** The act of retaining or the state of being retained. **2.** Ability to retain. **3.** Capacity to remember : MEMORY. **4.** Something retained. **5.** *Pathol.* Involuntary withholding of normally excreted wastes or secretions.

**re·ten·tive** (rĭ-tĕn′tĭv) *adj.* Having the ability or capacity to retain <a *retentive* mind> **—re·ten′tive·ly** *adv.* **—re·ten′tive·ness** *n.* **—re′ten·tiv′i·ty** (rē′tĕn-tĭv′ĭ-tē) *n.*

**re·think** (rē-thĭngk′) *vt.* & *vi.* **-thought** (-thôt′), **-think·ing, -thinks.** To reconsider or involve oneself in reconsideration. **—re′think′** *n.* **—re·think′er** *n.*

**re·ti·a** (rē′tē-ə, rē′shə) *n. pl. of* RETE.

**re·ti·ar·y** (rē′shē-ĕr′ē) *adj.* [< Lat. *rete,* net.] Of, similar to, or constituting a net or web.

**ret·i·cence** (rĕt′ĭ-səns) *n.* **1.** The quality or state of being reticent : RESERVE. **2.** An instance of being reticent.

**ret·i·cent** (rĕt′ĭ-sənt) *adj.* [Lat. *reticens, reticent-,* to keep silent : *re-* (intensive) + *tacēre,* to be silent.] **1.** Hesitant or reluctant to talk freely : RESERVED <was *reticent* about telling them the truth> **2.** Restrained or reserved in style. **—ret′i·cent·ly** *adv.*

**ret·i·cle** (rĕt′ĭ-kəl) *n.* [Lat. *reticulum,* dim. of *rete,* net.] A grid or pattern for establishing scale or position in the eyepiece of an optical instrument.

**re·tic·u·la** (rĭ-tĭk′yə-lə) *n. pl. of* RETICULUM.

**re·tic·u·lar** (rĭ-tĭk′yə-lər) *adj.* [NLat. *reticularis* < Lat. *reticulum,* dim. of *rete,* net.] **1.** Netlike. **2.** Intricate : complex.

**re·tic·u·late** (rĭ-tĭk′yə-lĭt, -lāt′) *adj.* [Lat. *reticulatus* < *reticulum,* dim of *rete,* net.] Resembling or forming a network <*reticulate* veins in leaves> *—v.* (-lāt′) **-lat·ed, -lat·ing, -lates.** *—vt.* **1.** To make a net or network of. **2.** To mark with lines having similarity to a network. *—vi.* To form a net or network. **—re·tic′u·late·ly** *adv.* **—re·tic′u·la′tion** *n.*

**ret·i·cule** (rĕt′ĭ-kyo͞ol′) *n.* [Fr. *réticule* < Lat. *reticulum,* dim. of *rete,* net.] **1.** A woman's drawstring handbag or purse. **2.** A reticle.

**re·tic·u·lo·cyte** (rĭ-tĭk′yə-lō-sīt′) *n.* [RETICUL(UM) + -CYTE.] An immature erythrocyte that contains a network of basophilic filaments. **—re·tic′u·lo·cyt′ic** (-sĭt′ĭk) *adj.*

**re·tic·u·lo·en·do·the·li·al system** (rĭ-tĭk′yə-lō-ĕn′də-thē′lē-əl) *n.* [< RETICUL(UM) + ENDOTHELIAL.] The widely diffused bodily system comprising all phagocytic cells except the leukocytes.

**re·tic·u·lum** (rĭ-tĭk′yə-ləm) *n., pl.* **-la** (-lə) [Lat., net, dim. of *rete,* net.] **1.** A netlike structure or formation : NETWORK. **2.** *Zool.* The second compartment of a ruminant stomach, lined with a membrane having honeycombed ridges. **3. Reticulum.** A constellation in the Southern Hemisphere.

**re·ti·form** (rē′tə-fôrm′, rĕt′ə-) *adj.* [Lat. *rete,* net + -FORM.] Arranged like a net : RETICULATE.

**retin-** *pref. var. of* RETINO-.

**ret·i·na** (rĕt′n-ə) *n., pl.* **ret·i·nas** *or* **ret·i·nae** (rĕt′n-ē′) [ME *rethina* < Med. Lat. *retina,* prob. < Lat. *rete,* net.] A delicate multilayer light-sensitive membrane lining the inner eyeball and connected by the optic nerve to the brain. **—ret′i·nal** *adj.*

**ret·i·nac·u·lum** (rĕt′n-ăk′yə-ləm) *n., pl.* **-la.** [NLat. < Lat., band, tether < *retinēre,* to restrain. —see RETAIN.] *Biol.* A band or bandlike structure that keeps an organ or part in place. **—ret′i·nac′u·lar** *adj.*

**ret·i·nae** (rĕt′n-ē′) *n. var. pl. of* RETINA.

**ret·i·nene** (rĕt′n-ēn′) *n.* A crystalline retinal pigment, $C_{19}H_{27}CHO$, a component of rhodopsin.

**ret·i·ni·tis** (rĕt′n-ī′tĭs) *n.* Inflammation of the retina.

**retino-** *or* **retin-** *pref.* [< RETINA.] Retina <*retinoscopy*>

**ret·i·nol** (rĕt′n-ôl′, -ōl′) *n.* Vitamin A.

**ret·i·nop·a·thy** (rĕt′n-ŏp′ə-thē) *n.* A pathological retinal disorder. **—ret′i·no·path′ic** (-ō-păth′ĭk) *adj.*

**ret·i·no·scope** (rĕt′n-ə-skōp′) *n., pl.* **-pies.** An optical instrument for examining refraction of light in the eye.

**ret·i·nos·co·py** (rĕt′n-ŏs′kə-pē) *n., pl.* **-pies.** Medical examination and analysis of the refractive properties of the eye. **—ret′i·no·scop′ic** (-ə-skŏp′ĭk) *adj.*

**ret·i·nue** (rĕt′n-o͞o′, rĕt′n-yo͞o′) *n.* [ME *retenue* < OFr. < fem. p.part. of *retenir,* to retain.] A group of attendants or followers of a high-ranking person.

☆ **syns:** RETINUE, ENTOURAGE, FOLLOWING, SUITE, TRAIN *n.* *core meaning* : a group of attendants or followers of a high-ranking person <special rooms for the monarch's *retinue*>

**re·tire** (rĭ-tīr′) *v.* **-tired, -tir·ing, -tires.** [OFr. *retirer* : *re-,* back + *tirer,* to draw.] *—vi.* **1.** To withdraw, as for rest, seclusion, or shelter. **2.** To go to bed. **3.** To give up business or public life and live on one's income, savings, or pension. **4.** To fall back : RETREAT <Napoleon's army *retired* after taking Moscow.> *—vt.* **1.** To remove from active service <*retire* a faithful servant> **2.** To lead (e.g., troops) away from action : WITHDRAW. **3.** To remove from circulation <*retire* old railroad bonds> **4.** *Baseball.* To put out (a batter).

**re·tired** (rĭ-tīrd′) *adj.* **1.** Withdrawn : secluded. **2.** Withdrawn from business or public life. **3.** Received by a person in retirement <*retired* benefits> **—re·tired·ly** *adv.* **—re·tired′ness** *n.*

**re·tir·ee** (rĭ-tīr′ē′) *n.* One who has retired from an occupation.

**re·tire·ment** (rĭ-tīr′mənt) *n.* **1.** An act of retiring. **2.** The state of being retired from one's occupation **3.** Privacy or seclusion. **4.** A place of seclusion : RETREAT.

**re·tir·ing** (rĭ-tīr′ĭng) *adj.* Shy and reserved : MODEST. **—re·tir′ing·ly** *adv.* **—re·tir′ing·ness** *n.*

**re·tool** (rē-tōōl′) *vt.* **-tooled, -tool·ing, -tools. 1.** To fit out with new tools <*retooled* the factory for peacetime production> **2.** To revise and reorganize.

**re·tort¹** (rĭ-tôrt′) *v.* **-tort·ed, -tort·ing, -torts.** [Lat. *retorquere, retort-*, to bend back : *re-*, back + *torquere*, to bend.] *—vt.* **1.** To return in kind : REPAY. **2. a.** To reply promptly and frankly. **b.** To present a counterargument to. *—vi.* **1.** To make a reply, esp. one that is prompt and forthright. **2.** To present a counterargument. *—n.* **1.** A prompt, frank reply, esp. one that turns the first speaker's words to his or her own disadvantage. **2.** An act of retorting. **—re·tort′er** *n.*

☆ ***syns:*** RETORT, COMEBACK, COUNTER, REPARTEE, RIPOSTE *n. core meaning* : a spirited, incisive reply <a clever *retort*>

**re·tort²** (rĭ-tôrt′, rē′tôrt′) *n.* [OFr. *retorte* < Med. Lat. *retorta*, fem. of Lat. *retortus*, p.part. of *retorquere*, to bend back. —see RETORT¹.] A closed laboratory vessel with an outlet tube, used for distillation, sublimation, or decomposition by heat.

**re·touch** (rē-tŭch′) *v.* **-touched, -touch·ing, -touch·es.** *—vt.* **1.** To add new details or touches to for correction or improvement. **2.** To improve or change (a photographic negative or print) by adding details or taking out defects. *—vi.* To give retouches. *—n.* (rē′tŭch′, rē-tŭch′). The act or process of retouching. **—re·touch′er** *n.*

**re·trace** (rē-trās′) *vt.* **-traced, -trac·ing, -trac·es. 1.** To trace again. **2.** To go back over, as one's steps. **—re·trace′a·ble** *adj.*

**re·tract** (rĭ-trăkt′) *v.* **-tract·ed, -tract·ing, -tracts.** [ME *retracten* < OFr. *retracter* < Lat. *retractare*, to handle again, freq. of *retrahere*, to draw back : *re-*, back + *trahere*, to draw.] *—vt.* **1.** To take back : DISAVOW <glad to *retract* my remarks> **2.** To draw back or in <a plane unable to *retract* its landing gear> **3. a.** To utter (a sound) with the tongue drawn back. **b.** To draw back (the tongue). *—vi.* **1.** To disavow something. **2.** To draw back. **—re·tract′a·ble, re·tract′i·ble** *adj.* **—re′trac·ta′tion** (rē′trăk-tā′shən) *n.*

**re·trac·tile** (rĭ-trăk′tĭl, -tīl′) *adj.* Capable of being drawn back or in <a cat's *retractile* claws> **—re′trac·til′i·ty** (rē′trăk-tĭl′ĭ-tē) *n.*

**re·trac·tion** (rĭ-trăk′shən) *n.* **1.** An act of recanting or something retracted <a *retraction* of a previous statement> **2.** The power of drawing back or of being drawn back.

**re·trac·tive** (rĭ-trăk′tĭv) *adj.* Tending to retract.

**re·trac·tor** (rĭ-trăk′tər) *n.* **1.** One that retracts. **2.** *Anat.* A muscle, as a flexor, that retracts an organ or part. **3.** *Med.* An instrument that holds back the edges of a wound.

**re·train** (rē-trān′) *vt. & vi.* **-trained, -train·ing, -trains.** To train or undergo training again. **—re·train′a·ble** *adj.*

**re·tral** (rē′trəl, rĕt′rəl) *adj.* [< Lat. *retro*, back.] **1.** Situated at, close to, or toward the back. **2.** Backward : reverse. **—re′tral·ly** *adv.*

**re·tread** (rē-trĕd′) *vt.* **-tread·ed, -tread·ing, -treads.** To fit (a worn automotive tire) with a new tread. *—n.* (rē′trĕd′). **1.** A retreaded tire. **2.** *Informal.* One retrained for work.

**re·tread** (rē-trĕd′) *vt.* **-trod** (-trŏd′), **-trod·den** (-trŏd′n), **-tread·ing, -treads.** To tread again.

**re·treat** (rĭ-trēt′) *n.* [ME *retret* < OFr. *retrait* < p.part. of *retraire*, to draw back < *retrahere*. —see RETRACT.] **1.** An act of going backward or of withdrawing. **2.** A quiet, private, or secure place : REFUGE. **3. a.** A time of seclusion, retirement, or solitude. **b.** A time of group withdrawal for prayer, meditation, and study. **4. a.** Withdrawal of a military force from a dangerous situation or from an enemy attack. **b.** The signal for such withdrawal. **c.** A bugle call signaling the lowering of the flag at sunset, as on a military post. *—v.* **-treat·ed, -treat·ing, -treats.** *—vi.* **1.** To fall back or withdraw because of danger or an enemy attack. **2.** To slope backward. *—vt.* To move (a chess piece) back. **—re·treat′er** *n.*

**re·trench** (rĭ-trĕnch′) *v.* **-trenched, -trench·ing, -trench·es.** [Obs. Fr. *retrencher* < OFr. *retrenchier* : *re-* (intensive) + *trenchier*, to cut. —see TRENCH.] *—vt.* **1.** To cut down : REDUCE. **2.** To remove, delete, or omit. *—vi.* To economize. **—re·trench′ment** *n.*

**re·tri·al** (rē-trī′əl, -trīl′, rē′trī′əl, -trīl′) *n.* Another trial, as of a legal case <returned the case to a lower court for *retrial*>

**ret·ri·bu·tion** (rĕt′rə-byōō′shən) *n.* [ME *retribucion* < OFr. *retribution* < LLat. *retributio* < Lat. *retribuere*, to pay back : *re-*, back + *tribuere*, to grant.] **1.** Something given or demanded in repayment, esp. punishment. **2.** Punishment or reward meted out in a future life based on performance in this one.

**re·trib·u·tive** (rĭ-trĭb′yə-tĭv) *adj.* Of, concerning, or marked by retribution. **—re·trib′u·tive·ly** *adv.*

**re·trib·u·to·ry** (rĭ-trĭb′yə-tôr′ē, -tōr′ē) *adj.* Retributive.

**re·triev·al** (rĭ-trē′vəl) *n.* **1.** The act or process of retrieving. **2.** Possibility of being retrieved <lost possessions beyond *retrieval*>

**re·trieve** (rĭ-trēv′) *v.* **-trieved, -triev·ing, -trieves.** [ME *retreven* < OFr. *retrover* : *re-*, again + *trover*, to find.] *—vt.* **1.** To get back : REGAIN. **2.** To restore : revive. **3.** To put right : RECTIFY. **4.** To recall to mind : REMEMBER. **5.** To find and bring back : FETCH. *—vi.* To find and bring back game. **—re·trieve′** *n.* **—re·triev′a·bil′i·ty** *n.* **—re·triev′a·ble** *adj.* **—re·triev′a·bly** *adv.*

**re·triev·er** (rĭ-trē′vər) *n.* **1.** One that retrieves. **2.** A dog trained to retrieve game.

**retro-** *pref.* [< Lat. *retro*, backward, behind.] **1.** Backward : back <*retro*rocket> **2.** Situated behind <*retro*lental> **3.** Pertaining to the past : retroactive <*retro*fit>

**ret·ro·ac·tion** (rĕt′rō-ăk′shən) *n.* **1.** A retroactive action. **2.** An opposing or reciprocal action : REACTION.

**ret·ro·ac·tive** (rĕt′rō-ăk′tĭv) *adj.* [Fr. *rétroactif* < Lat. *retroagere*, to drive back : *retro*, back + *agere*, to drive.] Influencing or applying to a period prior to enactment <*retroactive* pay increases> **—ret′ro·ac′tive·ly** *adv.* **—ret′ro·ac·tiv′i·ty** (-ĭ-tē) *n.*

**ret·ro·cede** (rĕt′rō-sēd′) *v.* **-ced·ed, -ced·ing, -cedes.** [Lat. *retrocedere* : *retro*, back + *cedere*, to go.] *—vi.* To go back : RECEDE. *—vt.* To give back : RETURN. **—ret′ro·ces′sion** (-sĕsh′ən) *n.*

**ret·ro·en·gine** (rĕt′rō-ĕn′jĭn) *n.* A spacecraft's rocket engine producing thrust in a direction opposite to that of the spacecraft's motion, used to decelerate.

**ret·ro·fire** (rĕt′rō-fīr′) *v.* **-fired, -fir·ing, -fires.** [RETRO(ENGINE) + FIRE.] *—vi.* To be ignited. *—vt.* To cause to ignite. —Used of a retroengine.

**ret·ro·fit** (rĕt′rō-fĭt′) *vt.* **-fit·ted, -fit·ting, -fits.** To furnish or provide with new equipment or parts unavailable at the time of original manufacture or construction.

**ret·ro·flex** (rĕt′rə-flĕks′) *also* **ret·ro·flexed** (-flĕkst′) *adj.* [NLat. *retroflexus* < LLat. *retroflectere*, to bend back : *retro*, back + *flectere*, to bend.] **1.** Bent, curved, or turned backward. **2.** Pronounced with the tip of the tongue turned back against the roof of the mouth. **—ret′ro·flex′ion, ret′ro·flec′tion** *n.*

**ret·ro·grade** (rĕt′rə-grād′) *adj.* [ME < Lat. *retrogradus* : *retro-*, back + *gradus*, step.] **1.** Moving or tending backward. **2.** Inverted or reversed. **3.** Reverting to an earlier or inferior condition. **4.** *Astron.* Having a direction of motion opposite to that of the earth on its axis or of the planets around the sun. **5.** *Archaic.* Opposed : contrary. *—vi.* **-grad·ed, -grad·ing, -grades. 1.** To move or appear to move backward <a glacier *retrograding*> **2.** To degenerate : decline. **—ret′ro·gra·da′tion** (-rō-grā-dā′shən) *n.* **—ret′ro·grade′ly** *adv.*

**ret·ro·gress** (rĕt′rə-grĕs′, rĕt′rə-grĕs′) *vi.* **-gressed, -gress·ing, -gress·es.** [Lat. *retrogradi, retrogress-* : *retro*, back + *gradi*, to go.] **1.** To return to an earlier, inferior, or simpler condition. **2.** To go or move backward. **—ret′ro·gres′sive** *adj.* **—re′tro·gres′sive·ly** *adv.*

**ret·ro·gres·sion** (rĕt′rə-grĕsh′ən) *n.* **1.** The act or process of deteriorating or declining. **2.** *Biol.* A return to a simpler or more primitive stage or state.

**ret·ro·len·tal** (rĕt′rō-lĕn′tl) *adj.* [RETRO- + NLat. *lens*, lens + -AL.] Behind a lens. —Used of the eye.

**ret·ro·oc·u·lar** (rĕt′rō-ŏk′yə-lər) *adj.* Situated behind the eye.

**ret·ro·pha·ryn·ge·al** (rĕt′rō-fə-rĭn′jē-əl, -jəl, -făr′ĭn-jē′əl) *adj.* Behind the pharynx.

**ret·ro·rock·et** (rĕt′rō-rŏk′ĭt) *n.* A retroengine.

**re·trorse** (rĭ-trôrs′, rē′trôrs′) *adj.* [Lat. *retrorsus*, contraction of *retroversus* : *retro*, back + *versus*, p.part. of *vertere*, to turn.] Directed or turned backward or downward. **—re·trorse′ly** *adv.*

**ret·ro·spect** (rĕt′rə-spĕkt′) *n.* [< Lat. *retrospectus*, p.part. of *retrospicere*, to look back at : *retro*, back + *specere*, to look at.] A review, survey, or contemplation of things in the past. *—v.* **-spect·ed, -spect·ing, -spects.** *—vi.* **1.** To contemplate the past. **2.** To refer back. *—vt.* To look back on or contemplate (things past). **—in retrospect.** Looking backward or considering the past. **—ret′ro·spec′tion** *n.*

**ret·ro·spec·tive** (rĕt′rə-spĕk′tĭv) *adj.* **1.** Looking back on, contemplating, or directed to the past. **2.** Looking or directed backward. **3.** Applying to or influencing the past : RETROACTIVE. **4.** Of or relating to an exhibition of the work of an artist or school over a period of years. *—n.* A retrospective art exhibition. **—ret′ro·spec′tive·ly** *adv.*

**re·trous·sé** (rə-trōō-sā′, rĕt′rōō-sā′) *adj.* [Fr., p.part. of *retrousser*, to turn back < OFr. : *re-*, back + *trousser*, to tie in a bundle.] Turned up at the end. —Used of the nose.

**ret·ro·ver·sion** (rĕt′rō-vûr′zhən, -shən) *n.* [< Lat. *retroversus*, retrorse.] **1.** A turning or tilting backward. **2.** The state of being turned or tilted back.

**re·try** (rē-trī′) *vt.* **-tried, -try·ing, -tries.** To try again <*retry* a felony case>

**ret·si·na** (rĕt′sĭ-nə, rĕt-sē′nə) *n.* [Mod. Gk., prob. < Ital. *retsina*, resin.] A resinated Greek wine.

**re·turn** (rĭ-tûrn′) *v.* **-turned, -turn·ing, -turns.** [ME *retornen* < OFr. *retourner* : *re-*, back (< Lat.) + *tourner*, to turn. —see TURN.] *—vi.* **1.** To go or come back, as to a former condition or place. **2.** To revert in speech, thought, or practice. **3.** To revert to a former owner. **4.** To answer : respond. *—vt.* **1.** To send, put, or carry back <*return* containers to the store> **2. a.** To give or send back in reciprocation <I *returned* their compliments.> **b.** To reflect <*return* an echo> **3.** To produce or yield (profit or interest) as a payment for labor, investment, or expenditure. **4.** To reflect or send back (light or sound). **5.** To submit (e.g., a report) to a judge or other official. **6.** To render or deliver (a verdict). **7.** To re-elect, as to a governorship. **8.** To respond to (a partner's lead) by leading the same suit in certain card games. **9.** To turn away from or place at an angle to the previous line of direction. —*n.* **1.** The act or state of going, coming, bringing, or sending back. **2. a.** Something brought or sent back. **b.** Something that goes or comes back. **3.** A recurrence, as of a periodic occasion or event. **4.** Something exchanged for that received : REPAYMENT. **5.** A reply : response. **6. a.** Profit made on an exchange of goods. **b.** *often* **returns.** A yield or profit, as from investments or labor. **c.** Output or yield per unit instead of cost per unit, as in the manufacturing of a specific product. **7. a.** A report, list, or set of statistics, esp. one that is official or formal. **b.** *often* **returns.** A report on the vote in an election. **c.** *Chiefly Brit.* An election. **8.** A lead in certain card games that responds to the lead of one's partner. **9.** In tennis and certain other sports: **a.** The act of sending the ball back to one's opponent. **b.** The ball so returned. **10.** *Football.* **a.** The act of running back the ball after a kickoff, punt, interception, or fumble. **b.** The yardage gained in a return. **11. a.** Extension of an architectural part, as a molding or projection, at a usu. 90° angle to the main part. **b.** A part of a building set at an angle to the façade. **12.** A round-trip ticket. **13.** *Law.* **a.** The bringing or sending back of a writ, subpoena, or other document, gen. with a short written report on it, by a sheriff or other officer, to the court of original jurisdiction. **b.** A certified report by an official, as an assessor, election officer, or collector. **14.** A statement on the required official form indicating taxable income, allowed deductions, exemptions, and the computed tax due. **—in return.** In repayment. **—re·turn'a·ble** *asj.* **—re·turn'er** *n.*

☆ ***syns:*** RETURN, RECUR, REVERT *v. core meaning* : to come back to a former condition <a disease that *returned*>

**re·turn·ee** (rĭ-tûr'nē') *n.* One that has returned, as from a voyage.

**re·tuse** (rĭ-to͞os', -tyo͞os') *adj.* [Lat. *retusus,* p.part. of *retundere,* to beat back : *re-,* back + *tundere,* to beat.] Having a rounded or blunt apex with a shallow notch. —Used chiefly of leaves.

**Reu·ben**[1] (ro͞o'bən) *n.* [Heb. *Re'ū-bēn.*] **1.** Jacob's eldest son, the ancestor of one of the tribes of Israel in the Old Testament. **2.** The tribe of Israel descended from Reuben.

**Reu·ben**[2] (ro͞o'bən) *n.* [< the name *Reuben.*] A hot sandwich of corned beef, Swiss cheese, and sauerkraut usu. served on rye bread.

**re·u·ni·fy** (rē-yo͞o'nə-fī') *vt.* **-fied, -fy·ing, -fies.** To restore unity to. **—re·u'ni·fi·ca'tion** (-fĭ-kā'shən) *n.*

**re·un·ion** (rē-yo͞on'yən) *n.* **1.** The act of reuniting or the state of being reunited. **2.** A gathering of the members of a group who have been previously separated.

**re·u·nite** (rē'yo͞o-nīt') *vt.* & *vi.* **-nit·ed, -nit·ing, -nites.** [Med. Lat. *reunire, reunit-* : Lat. *re-,* again + LLat. *unire,* to unite < Lat. *unus,* one.] To bring or come together again.

**re-up** (rē-ŭp') *vi. Informal.* **-upped, -up·ping, -ups.** To enlist again <*re-upped* in the army>

**re·use** (rē-yo͞oz') *vt.* **-used, -us·ing, -us·es.** To use again. **—re·us'a·ble** *adj.* **—re·use'** (-yo͞os') *n.*

**rev** (rĕv) *Informal.* —*n.* A revolution, as of a motor. —*v.* **revved, rev·ving, revs.** —*vt.* **1.** To increase the speed of (a motor). **2.** To rouse or excite as preparation for action <*revving* ourselves up for the tennis finals> —*vi.* **1.** To operate at an increased speed. **2.** To be roused or excited about something.

**re·val·i·date** (rē-văl'ĭ-dāt') *vt.* **-dat·ed, -dat·ing, -dates.** To declare valid again. **—re·val'i·da'tion** *n.*

**re·val·u·ate** (rē-văl'yo͞o-āt') *vt.* **-at·ed, -at·ing, -ates.** To make a new valuation of. **—re·val'u·a'tion** *n.*

**re·vamp** (rē-vămp') *vt.* **-vamped, -vamp·ing, -vamps.** **1.** To patch up or restore : RENOVATE. **2.** To revise or reconstruct (e.g., a manuscript). **3.** To vamp (footwear) anew.

▲ word history: The word *revamp* is formed from the English prefix *re-,* "again," and the verb *vamp,* "to refurbish." The verb *vamp* is derived from the noun *vamp,* which denotes a part of a shoe. The simple verb *vamp* originally meant "to provide a shoe with a new vamp" and was used figuratively to denote any kind of restoration and refurbishing. Although the prefix *re-* in *revamp* means "again", it functions primarily as an intensifier.

**re·vanch·ism** (rĭ-vănch'ĭz'əm) *n.* [< Fr. *revanche,* revenge < *revancher,* to revenge < OFr. *revencher.* —see REVENGE.] A foreign policy motivated by a desire to win back territory previously lost to an enemy. **—re·vanch'ist** *n.*

**re·veal** (rĭ-vēl') *vt.* **-vealed, -veal·ing, -veals.** [ME *revelen* < OFr. *reveler* < Lat. *revelare* : *re-* (reversal) + *velare,* to cover < *velum,* veil.] **1.** To make known : DIVULGE. **2.** To bring to view : SHOW. **—re·veal'a·ble** *adj.* **—re·veal'er** *n.* **—re·veal'ment** *n.*

**re·veil·le** (rĕv'ə-lē) *n.* [Alteration of Fr. *réveillez,* imper. pl. of *réveiller,* to wake < OFr. *reveiller* : *re-,* again (< Lat.) + *eveiller,* to awake < Lat. *evigilare* (*ex-,* out + *vigilare,* to stay awake < *vigil,* awake).] **1.** The blowing of a bugle early in the morning to awaken and summon persons in a camp or garrison. **2.** The first military formation of the day.

**rev·el** (rĕv'əl) *vi.* **-eled, -el·ing, -els** *also* **-elled, -el·ling, -els.** [ME *revelen* < OFr. *reveler,* to make noise, to rebel < Lat. *rebellare,* to rebel.] **1.** To take great pleasure <I *reveled* in my hobbies.> **2.** To make merry. —*n. often* **revels.** Merrymaking. **—rev'el·er** *n.*

**rev·e·la·tion** (rĕv'ə-lā'shən) *n.* [ME < OFr. < LLat. *revelatio* < Lat. *revelare,* to reveal.] **1.** Something revealed. **2.** An act of revealing, esp. a dramatic disclosure of something not formerly known or realized. **3.** A manifestation of divine will or truth. **4. Revelation.** —See table at BIBLE.

☆ ***syns:*** REVELATION, DISCLOSURE, EXPOSE, EXPOSURE *n. core meaning* : something disclosed, esp. something not previously known or realized <a *revelation* about the dangers of toxic waste>

**rev·e·la·to·ry** (rĕv'ə-lə-tôr'ē, -tōr'ē, rĭ-vĕl'ə-) *adj.* Of, relating to, or containing a revelation.

**rev·el·ry** (rĕv'əl-rē) *n., pl.* **-ries.** Merrymaking <New Year's *revelry*> **—rev'el·rous** (-rəs) *adj.*

**rev·e·nant** (rĕv'ə-nənt) *n.* [Fr. < pr.part. of *revenir,* to return < Lat. *revenire* : *re-,* back + *venire,* to come.] **1.** One that comes back following an absence. **2.** One who returns after death.

**re·venge** (rĭ-vĕnj') *vt.* **-venged, -veng·ing, -veng·es.** [ME *revenger* < OFr. *revenger,* var. of *revencher* < LLat. *revindicare* : *re-,* in response + *vindicare,* to avenge < *vindex,* avenger.] **1.** To inflict punishment in return for (injury or insult) : AVENGE. **2.** To seek or take vengeance for (oneself or another). —*n.* **1.** Vengeance : retaliation. **2.** An act of taking vengeance. **3.** Desire for revenge. **—re·veng'er** *n.*

**re·venge·ful** (rĭ-vĕnj'fəl) *adj.* Full of or inclined to revenge. **—re·venge'ful·ly** *adv.* **—re·venge'ful·ness** *n.*

**rev·e·nue** (rĕv'ə-no͞o, -nyo͞o) *n.* [ME < OFr. < fem. p.part. of *revenir,* to return < Lat. *revenire.* —see REVENANT.] **1.** Government income of from all sources appropriated for paying public expenses. **2.** Yield from investment or property : INCOME. **3.** A single source of income. **4.** A governmental department established for collecting funds.

**rev·e·nu·er** (rĕv'ə-no͞o'ər, -nyo͞o'-) *n.* **1.** A revenue agent of the U.S. Treasury Department, esp. one responsible for policing illegal alcohol distilleries and bootlegging. **2.** A lightly armed motorboat used in collecting revenue.

**re·ver·ber·ate** (rĭ-vûr'bə-rāt') *v.* **-at·ed, -at·ing, -ates.** [Lat. *reverberare, reverberat-,* to cause to rebound : *re-,* back + *verbarare,* to beat < *verber,* whip.] —*vi.* **1.** To re-echo : resound. **2.** To be repeatedly reflected. **3.** To rebound or recoil : REDOUND. —*vt.* **1.** To re-echo (a sound). **2.** To reflect (heat or light) repeatedly. **—re·ver'ber·ant·ly** (-bər-ənt-lē) *adv.* **—re·ver'ber·a'tion** *n.* **—re·ver'ber·a·to'ry** (-bə-rə-tôr'ē, -tōr'ē), **re·ver'ber·ant, re·ver'ber·a'tive** *adj.*

**re·vere**[1] (rĭ-vîr') *vt.* **-vered, -ver·ing, -veres.** [Lat. *revereri* : *re-* (intensive) + *vereri,* to respect.] To regard with awe, great respect, or devotion.

**re·vere**[2] (rĭ-vîr', -vâr') *n. var. of* REVERS.

**rev·er·ence** (rĕv'ər-əns) *n.* **1.** Profound awe and respect and often love : VENERATION. **2.** An act of showing respect, esp. an obeisance. **3.** The state of being revered. **4. Reverence.** —Used as a title of respect for a cleric. —*vt.* **-enced, -enc·ing, -enc·es.** To venerate. **—rev'er·enc·er** *n.*

**rev·er·end** (rĕv'ər-ənd) *adj.* [ME < OFr. < Lat. *reverendus,* gerund. of *revereri,* to revere.] **1.** Worthy of reverence. **2.** Relating to or characteristic of the clergy : CLERICAL. **3.** Reverend. —Used as a title for a cleric. —*n. Informal.* A cleric.

**rev·er·ent** (rĕv'ər-ənt) *adj.* [ME < Lat. *reverens,* pr.part. of *revereri,* to revere.] REVERENTIAL 1. **—rev'er·ent·ly** *adv.*

**rev·er·en·tial** (rĕv'ə-rĕn'shəl) *adj.* **1.** Expressing reverence. **2.** Inspiring reverence. **—rev'er·en'tial·ly** *adv.*

**rev·er·ie** (rĕv'ər-ē) *n.* [ME < OFr. < *rever,* to dream.] **1.** Abstracted musing : DAYDREAMING. **2.** A daydream <caught up in a *reverie* of years long past>

**re·vers** *also* **re·vere** (rĭ-vîr', -vâr') *n., pl.* **revers** *also* **-veres** (-vîrz', -vârz') [Fr. < OFr., reverse.] A part of a garment folded back to reveal the reverse side, as a lapel.

revers

**re·ver·sal** (rĭ-vûr'səl) *n.* **1.** An act or instance of reversing or the state of being reversed. **2.** *Law.* An act or instance of changing or setting aside a lower court's decision by a higher court.

**re·verse** (rĭ-vûrs') *adj.* [ME *revers* < OFr. < Lat. *reversus,* turned back, p.part. of *revertere,* to revert.] **1.** Turned backward in position, direction, or order. **2.** Moving or acting in a manner contrary to the custom. **3.** Producing backward movement <a *reverse* gear> —*n.* **1.** The opposite or contrary of something. **2.** A back or rear part. **3.** A change to an opposite position, condition, or direction. **4.** A change in fortune from better to worse : SETBACK <financial *reverses*> **5. a.** A mechanism for reversing movement, as a gear in a motor vehicle. **b.** The reverse position or operating condition of such a mechanism. —*v.* **-versed, -vers·ing, -vers·es.** —*vt.* **1.** To turn to the opposite direction or tendency. **2.** To turn inside out or upside down. **3.** To switch the locations of : TRANSPOSE. **4.** *Law.* To revoke or annul (a decision or decree). —*vi.* **1.** To turn or move in the opposite direction. **2.** To reverse the action of an engine. **—in reverse.** In an opposite direction. **—re·verse′ly** *adv.* **—re·vers′er** *n.*

☆ ***syns:*** REVERSE, INVERT, TRANSPOSE *v. core meaning* : to change to the opposite position, course, or direction <*reversed* the order of the pictures on the wall>

**reverse discrimination** *n.* Discrimination against members of a dominant group, esp. discrimination against whites or men in employment.

**re·vers·i·ble** (rĭ-vûr′sə-bəl) *adj.* **1.** Capable of being reversed. **2.** *Chem. & Physics.* Capable of successively assuming or producing either of two states <a *reversible* cell><a *reversible* reaction> —*n.* A reversible garment. **—re·vers′i·bil′i·ty, re·vers′i·ble·ness** *n.* **—re·vers′i·bly** *adv.*

**re·ver·sion** (rĭ-vûr′zhən) *n.* **1.** A return to a prior condition, belief, or interest. **2.** A turning away or in the opposite direction : REVERSAL. **3.** *Genetics.* Atavism. **4.** *Law.* **a.** Return of an estate to the grantor or to his or her estate after expiration of the grant. **b.** The estate thus returned. **c.** The right to succeed to an estate.

**re·ver·sion·ar·y** (rĭ-vûr′zhə-nĕr′ē) *adj. Law.* Of or connected with the reversion of an estate.

**re·ver·sion·er** (rĭ-vûr′zhə-nər) *n. Law.* One entitled to receive an estate in reversion.

**re·vert** (rĭ-vûrt′) *vi.* **-vert·ed, -vert·ing, -verts.** [ME *reverten* < OFr. *revertir* < Lat. *revertere* : *re-*, back + *vertere*, to turn.] **1.** To return to a former condition, practice, subject, or belief. **2.** *Law.* To return to the previous owner or the previous owner's heirs. —Used of money or property. **3.** To return to an ancestral type. **—re·vert′er** *n.* **—re·vert′i·ble** *adj.* **—re·ver′tive** *adj.*

**re·vest** (rē-vĕst′) *vt.* **-vest·ed, -vest·ing, -vests.** [ME *revesten*, to clothe < OFr. *revestir* < LLat. *revestire*, to clothe again : Lat. *re-*, again + *vestire*, to clothe < *vestis*, garment.] **1.** To invest (one) again with power or ownership : REINSTATE. **2.** To vest (e.g., power) once again in a person or agency.

**re·vet** (rĭ-vĕt′) *v.* **-vet·ted, -vet·ting, -vets.** [Fr. *revêtir* < OFr. *revestir*, to clothe again. —see REVEST.] —*vt.* To retain (a wall of earth) with a layer of other material, as stone. —*vi.* To construct a revetment.

**re·vet·ment** (rĭ-vĕt′mənt) *n.* **1.** A facing, as of masonry, for supporting an embankment. **2.** A barricade against explosives.

**re·view** (rĭ-vyo͞o′) *v.* **-viewed, -view·ing, -views.** [OFr. *revoir* : *re-*, again (< Lat.) + *voir*, to see < Lat. *vidēre*.] —*vt.* **1.** To study or examine again. **2.** To consider retrospectively. **3.** To examine with an eye to correction or criticism <*review* a manuscript for style> **4.** To write or give a critical report on (a new work or performance). **5.** *Law.* To examine (an action or determination), esp. in a higher court, for the purpose of correcting possible errors. **6.** To subject to a formal inspection, esp. a military inspection. —*vi.* **1.** To peruse material. **2.** To act as a reviewer, as for a newspaper. —*n.* **1.** Reexamination : reconsideration. **2.** A retrospective view or survey. **3.** Restudy of subject matter. **4.** An inspection or examination with the intention of evaluating. **5. a.** A report or essay giving a critical estimate of a work or performance. **b.** A periodical devoted to such reports. **6.** A formal military inspection. **7.** *Law.* An examination of an action or determination, esp. by a higher court, so as to correct possible errors. **8.** A revue. **—re·view′a·ble** *adj.* **—re·view′er** *n.*

☆ ***syns:*** REVIEW, COMMENT, COMMENTARY, CRITICISM, CRITIQUE, NOTICE *n. core meaning* : evaluative and critical discourse <The new book received rave *reviews*.>

**re·vile** (rĭ-vīl′) *v.* **-viled, -vil·ing, -viles.** [ME *revilen* < OFr. *reviler* : *re-* (intensive < Lat.) + *vil*, vile < Lat. *vilis*.] —*vt.* To denounce abusively. —*vi.* To use abusive language. **—re·vile′ment** *n.* **—re·vil′er** *n.* **—re·vil′ing·ly** *adv.*

**re·vise** (rĭ-vīz′) *vt.* **-vised, -vis·ing, -vis·es.** [Lat. *revisere*, to visit again : *re-*, again < *visere*, freq. of *vidēre*, to see.] **1.** To prepare a newly edited version of (a text). **2.** To change : modify <*revise* a previous opinion> —*n.* (rē′vīz′, rĭ-vīz′). A printing proof made from an earlier proof on which alterations have been made. **—re·vis′a·ble** *adj.* **—re·vis′al** (-vī′zəl) *n.* **—re·vis′er, re·vi′sor** *n.*

☆ ***syns:*** REVISE, AMEND, EMEND, EMENDATE, REVAMP *v. core meaning* : to prepare a new version of <an almanac that is *revised* each year>

**Revised Standard Version** *n.* A modern American version of the English Bible in the King James tradition.

**Revised Version** *n.* An English and American revision of the King James Version of the Bible.

**re·vi·sion** (rĭ-vĭzh′ən) *n.* **1.** The act or procedure of revising. **2.** A revised version. **—re·vi′sion·ar′y** *adj.*

**re·vi·sion·ism** (rĭ-vĭzh′ə-nĭz′əm) *n.* **1.** Advocacy of the revision of an accepted, usu. long-standing view, theory, or doctrine, esp. a revision of historical events and movements. **2.** A recurrent tendency in the Communist movement to revise Marxist theory in such a way as to provide reasons for a retreat from the revolutionary to the reformist position. **—re·vi′sion·ist** *n.*

**re·vis·it** (rē-vĭz′ĭt) *vt.* **-it·ed, -it·ing, -its.** To visit again. **—re·vis′it** *n.* **—re′vis·i·ta′tion** *n.*

**re·vi·so·ry** (rĭ-vī′zə-rē) *adj.* Of, relating to, effecting, or having the power of revision.

**re·vi·tal·ize** (rē-vīt′l-īz′) *vt.* **-ized, -iz·ing, -iz·es.** To give new life or vigor to <Their work will *revitalize* inner-city neighborhoods.> **—re·vi′tal·i·za′tion** *n.*

**re·viv·al** (rĭ-vī′vəl) *n.* **1.** An act of reviving or the state of being revived. **2.** Restoration to use, acceptance, activity, or vigor after obscurity or quiescence. **3.** A new presentation, as of an old play, film, opera, or ballet. **4.** Renewed interest in religion. **5.** A meeting or series of meetings in order to reawaken religious faith, often marked by impassioned evangelism and public professions of faith.

**re·viv·al·ism** (rĭ-vī′və-lĭz′əm) *n.* **1.** The spirit or activities typical of religious revivals. **2.** A desire to revive.

**re·viv·al·ist** (rĭ-vī′və-lĭst) *n.* **1.** A promoter or leader of religious revivals. **2.** One who revives practices or ideas of an earlier time. **—re·viv′al·is′tic** *adj.*

**re·vive** (rĭ-vīv′) *v.* **-vived, -viv·ing, -vives.** [ME *reviven* < OFr. *revivre* < LLat. *revivere*, to live again : Lat. *re-*, again + Lat. *vivere*, to live.] —*vt.* **1.** To resuscitate. **2.** To impart new health, vigor, or spirit to. **3.** To restore to use, currency, activity, or notice. **4.** To restore the validity or effectiveness of. **5.** To renew in the mind : RECALL. **6.** To present (e.g., an old play) again. —*vi.* **1.** To return to life or consciousness. **2.** To regain health, vigor, or good spirits. **3.** To restore to use, notice, or currency. **4.** To restore to effectiveness, validity, or operative condition. **—re·viv′er** *n.*

**re·viv·i·fy** (rē-vĭv′ə-fī′) *vt.* **-fied, -fy·ing, -fies.** [OFr. *revivifier* < LLat. *revivificare* : Lat. *re-*, again + LLat. *vivificare*, to vivify. —see VIVIFY.] REVIVE 2. **—re·viv′i·fi·ca′tion** *n.*

**rev·o·ca·tion** (rĕv′ə-kā′shən) *n.* An act or instance of revoking or the state of having been revoked. **—rev′o·ca·to′ry** (rĕv′ə-kə-tôr′ē, -tōr′ē) *adj.*

**re·voke** (rĭ-vōk′) *v.* **-voked, -vok·ing, -vokes.** [ME *revoken* < OFr. *revoquer* < Lat. *revocare*, to call back : *re-*, back + *vocare*, to call.] —*vt.* To nullify by withdrawing, recalling, or reversing <*revoke* a permit> —*vi.* To fail to follow suit in a card game when one is required and able to do so. —*n.* A failure to follow suit in a card game. **—rev′o·ca·ble** (rĕv′ə-kə-bəl), **re·vok′a·ble** (rĭ-vō′-) *adj.* **—re·vok′er** *n.*

**re·volt** (rĭ-vōlt′) *v.* **-volt·ed, -volt·ing, -volts.** [OFr. *revolter* < OItal. *rivoltare*, to overthrow < VLat. **revolvitare* < Lat. *revolvere*, to turn over. —see REVOLVE.] —*vi.* **1.** To attempt to overthrow the authority of the state : REBEL. **2.** To oppose or refuse to accept something <*revolt* against new taxes> —*vt.* To fill with disgust : REPEL. —*n.* **1.** An uprising, esp. against established authority : REBELLION. **2.** An act of rejection or protest. **3.** The state of a person or persons in rebellion. **—re·volt′er** *n.*

**re·volt·ing** (rĭ-vōl′tĭng) *adj.* Disgustingly offensive : LOATHSOME. **—re·volt′ing·ly** *adv.*

**rev·o·lute** (rĕv′ə-lo͞ot′) *adj.* [Lat. *revolutus*, p.part. of *revolvere*, to roll back. —see REVOLVE.] *Bot.* Rolled back on the undersurface from the tip or margins, as some leaves before they are expanded.

**rev·o·lu·tion** (rĕv′ə-lo͞o′shən) *n.* [ME *revolucioun* < OFr. *revolution* < LLat. *revolutio* < Lat. *revolvere*, to turn over. —see REVOLVE.] **1. a.** Orbital motion about a point <a planet's *revolution* around the sun> **b.** A turning or rotational motion about an axis. **c.** A single complete cycle of such axial or orbital motion. **2.** A sudden or radical change in a situation <the *revolution* in medicine> **3. a.** An abrupt political overthrow or seizure of power brought about from within a given system. **b.** Activities directed toward effecting basic changes in the socioeconomic structure, as of a cultural or minority segment of the population. **—rev′o·lu′tion·ist** (-lo͞o′shə-nĭst) *n.*

**rev·o·lu·tion·ar·y** (rĕv′ə-lo͞o′shə-nĕr′ē) *adj.* **1.** Of, relating to, or effecting a social or political revolution. **2.** Marked by or resulting in extreme change <a *revolutionary* invention> —*n., pl.* **-ies.** A militant in the quest for revolution.

**Revolutionary War** *n.* The American Revolution.

**rev·o·lu·tion·ize** (rĕv′ə-lo͞o′shə-nīz′) *vt.* **-ized, -iz·ing, -iz·es.** **1.** To effect a radical change in. **2.** To cause (a country) to undergo a social or political revolution. **3.** To imbue with revolutionary principles <agitators *revolutionizing* the workers>

**re·volve** (rĭ-vŏlv′) *v.* **-volved, -volv·ing, -volves.** [ME *revolven* < Lat. *revolvere*, to turn over, to roll back : *re-*, back + *volvere*, to roll.] —*vi.* **1.** To orbit a central point. **2.** To turn on an axis : ROTATE. **3.** To recur at periodic intervals or in cycles. **4.** To be held in the mind and considered in turn. —*vt.* **1.** To cause to revolve. **2.** To reflect on : PONDER. **—re·volv'a·ble** *adj.*

**re·volv·er** (rĭ-vŏl'vər) *n.* **1.** A handgun having a revolving cylinder with several cartridge chambers. **2.** One that revolves.

**re·vue** (rĭ-vyōō') *n.* [Fr. < OFr., p.part. of *revoir,* to review.] An often satirical musical show consisting of skits, songs, and dances.

**re·vul·sion** (rĭ-vŭl'shən) *n.* [Lat. *revulsio* < *revellere,* to tear back : *re-,* back + *vellere,* to tear.] **1.** A sudden and strong change or reaction in feeling, esp. a feeling of extreme loathing or disgust. **2.** Withdrawal from or aversion to something. **—re·vul'sive** *adj.*

**re·ward** (rĭ-wôrd') *n.* [ME *rewarden* < ONFr. *rewarder* : *re-* (intensive < Lat.) + *warder,* to watch over, of Germanic orig.] **1.** Something, as money, given or offered esp. for a special service, as the return of a lost article or the capture of a criminal. **2.** A satisfying result. —*vt.* **-ward·ed, -ward·ing, -wards. 1.** To give a reward to or for. **2.** To gratify or satisfy : RECOMPENSE. **—re·ward'er** *n.*

**re·wind** (rē-wīnd') *vt.* **-wound** (-wound'), **-wind·ing, -winds.** To wind again. —*n.* (rē'wīnd', rē-wīnd'). An act or process of rewinding. **—re·wind'er** *n.*

**re·wire** (rē-wīr') *vt.* **-wired, -wir·ing, -wires.** To provide (e.g., a house or room) with new wiring.

**re·word** (rē-wûrd') *vt.* **-word·ed, -word·ing, -words. 1.** To state or express again in different words. **2.** To state or express again in the same words : REPEAT.

**re·work** (rē-wûrk') *vt.* **-worked, -work·ing, -works. 1.** To work over anew : REVISE. **2.** To subject to a new or repeated process. —*n.* (rē'wûrk'). Something reworked <a *rework* of an old play>

**re·write** (rē-rīt') *vt.* **-wrote** (-rōt'), **-writ·ten** (-rĭt'n), **-writ·ing, -writes. 1.** To write again, esp. in an improved form. **2.** To write (an account given by a reporter) in a form appropriate for publishing. —*n.* (rē'rīt'). Rewritten material. **—re·writ'er** *n.*

**Reye's syndrome** (rāz) *n.* [After R.D.K. *Reye* (d. 1977).] Acute encephalopathy marked by fever, vomiting, fatty infiltration of the liver, disorientation, and coma that usu. follows a viral infection, as influenza, and occurs mainly in children.

**Rey·nard** (rā'nərd, -närd', rĕn'ərd) *n.* [ME *Reynard* < OFr. *Renart,* the fox in *Roman de Renart,* the Romance of Reynard.] FOX 1.

**re·zone** (rē-zōn') *vt.* **-zoned, -zon·ing, -zones.** To alter the zoning of, as for industrial or business purposes.

**R factor** *n.* [R(ESISTANCE) FACTOR.] A genetic factor of bacteria that transmits resistance to antibiotics from one bacterium to another by conjugation.

**Rh** *symbol for* RHODIUM.

**rhab·do·man·cy** (răb'də-măn'sē) *n.* [LGk. *rhabdomanteia* : Gk. *rhabdos,* rod + Gk. *-manteia,* -mancy.] Divination by means of a wand or a rod, esp. for locating underground water or ores. **—rhab'do·man'cer** *n.*

**rhab·do·my·o·ma** (răb'dō-mī-ō'mə) *n., pl.* **-mas** *or* **-ma·ta** (-mə-tə). [Gk. *rhabdos,* rod + MYOMA.] *Pathol.* A tumor in the striated muscles.

**rhab·do·vi·rus** (răb'də-vī'rəs) *n.* [Gk. *rhabdos,* rod + VIRUS.] Any of a group of RNA-containing plant and animal viruses that include the rabies virus.

**Rhad·a·man·thine** (răd'ə-măn'thĭn, -thīn') *adj.* [< Gk. *Rhadamanthos,* a judge in the underworld.] Extraordinarily just.

**Rhae·to-Ro·man·ic** (rē'tō-rō-măn'ĭk) *also* **Rhae·to-Ro·mance** (-rō-măns') *n.* [Lat. *Rhaetus,* of Rhaetia, a Roman province + ROMANIC.] A Romance language of southern Switzerland, northern Italy, and the Tyrol. **—Rhae'to-Ro·man'ic, Rhae'to-Ro·mance'** *adj.*

**rha·phe** (rā'fē) *n. var. of* RAPHE.

**rhap·sod·ic** (răp-sŏd'ĭk) *also* **rhap·sod·i·cal** (-ĭ-kəl) *adj.* **1.** Of, resembling, or characteristic of a rhapsody. **2.** Excessively impassioned or enthusiastic. **—rhap·sod'i·cal·ly** *adv.*

**rhap·so·dist** (răp'sə-dĭst) *n.* **1.** An ancient Greek epic singer. **2.** A user of excessively enthusiastic or impassioned language.

**rhap·so·dize** (răp'sə-dīz') *v.* **-dized, -diz·ing, -diz·es.** —*vi.* To express oneself in an extravagantly enthusiastic way. —*vt.* To recite (something) in the manner of a rhapsody.

**rhap·so·dy** (răp'sə-dē) *n., pl.* **-dies.** [Lat. *rhapsodia,* epic poem < Gk. *rhapsōidia* < *rhapsōidos,* singer of epic poems : *rhaptein,* to sew together + *ōidē,* song.] **1.** Exalted or immoderately enthusiastic expression of feeling in speech or writing. **2.** A literary work written in an exalted or impassioned style. **3.** *Mus.* A composition of irregular form and an often improvisational nature. **4.** An ancient Greek epic poem or a part of one suitable for uninterrupted recitation.

**rhat·a·ny** (răt'n-ē) *n., pl.* **-nies.** [Am. Sp. *ratania* < Quechua *ratanya.*] **1.** A South American shrub, *Krameria triandra* or *K. argentea,* having thick fleshy roots. **2.** The dried root of the rhatany, once used as an astringent.

**rhe·a** (rē'ə) *n.* [NLat. *Rhea,* genus name < Lat., the mother of Romulus and Remus.] Any of several flightless South American birds of the genus *Rhea,* resembling the ostrich but somewhat smaller and having three toes instead of two.

**rhea**
*4½ feet long*

**Rhe·a** (rē'ə) *n.* [Gk. *Rhea.*] *Gk. Myth.* The wife of Cronus and mother of Zeus.

**Rhen·ish** (rĕn'ĭsh) *adj.* [< Lat. *Rhenus,* the Rhine.] Of or relating to the river Rhine or the lands bordering on it. —*n.* RHINE WINE 1.

**rhe·ni·um** (rē'nē-əm) *n.* [< Lat. *Rhenus,* the Rhine.] *Symbol* **Re** A rare, dense, silvery-white metallic element with a high melting point, used for electrical contacts and with tungsten for high-temperature thermocouples; atomic number 75; atomic weight 186.2.

**rheo-** *pref.* [< Gk. *rheos,* stream < *rhein,* to flow.] Current : flow <*rheo*taxis>

**rhe·ol·o·gy** (rē-ŏl'ə-jē) *n.* Study of the deformation and flow of matter. **—rhe'o·log'i·cal** (rē'ə-lŏj'ĭ-kəl) *adj.* **—rhe·ol'o·gist** *n.*

**rhe·om·e·ter** (rē-ŏm'ĭ-tər) *n.* An instrument for measuring the flow of viscous liquids, as of blood.

**rhe·o·stat** (rē'ə-stăt') *n.* A continuously variable electrical resistor for regulating current. **—rhe'o·stat'ic** *adj.*

**rhe·o·tax·is** (rē'ə-tăk'sĭs) *n.* Movement of an organism in response to the flow of a current. **—rhe'o·tac'tic** (-tăk'tĭk) *adj.*

**rhe·sus monkey** (rē'səs) *n.* [< NLat. *Rhesus* < Gk. *Rhēsos,* a mythical king of Thrace.] A brownish monkey, *Macaca mulatta* of India, used in biological experiments.

**rhe·tor** (rē'tôr', -tər) *n.* [ME *rether* < Med. Lat. < Lat. *rhetor* < Gk. *rhētōr.*] *Obs.* **1.** A teacher of rhetoric. **2.** An orator.

**rhet·o·ric** (rĕt'ər-ĭk) *n.* [ME *rethorik* < OFr. *rethorique* < Lat. *rhetorica* < Gk. *rhētorikē (tekhnē),* rhetorical (art) < *rhētorikos,* rhetorical < *rhētōr,* rhetor.] **1.** Study of the elements, as structure or style, used in writing and speaking. **2.** The art of effective expression and the persuasive use of language. **3.** Insincere or pretentious language <campaign *rhetoric*> **4.** Verbal communication : DISCOURSE.

**rhe·tor·i·cal** (rĭ-tôr'ĭ-kəl, -tŏr'-) *adj.* **1.** Concerned primarily with style or effect. **2.** Like rhetoric : ORATORICAL. **—rhe·tor'i·cal·ly** *adv.*

**rhetorical question** *n.* A question to which no answer is expected.

**rhet·o·ri·cian** (rĕt'ə-rĭsh'ən) *n.* **1.** An expert in or a teacher of rhetoric. **2.** An eloquent speaker or writer. **3.** One given to verbal extravagance.

**rheum** (rōōm) *n.* [ME *reume* < OFr. < Lat. *rheuma* < Gk. < *rhein,* to flow.] A thin or watery mucous discharge from the eyes or nose. **—rheum'y** *adj.*

**rheu·mat·ic** (rōō-măt'ĭk) *adj.* [ME *rhewmatyk,* of rheum < Lat. *rheumaticus,* a person suffering from rheum < Gk. *rheumatikos,* subject to rheum < *rheuma,* stream < *rhein,* to flow.] Of, relating to, or afflicted with rheumatism. —*n.* **1.** One afflicted with rheumatism. **2. rheumatics.** *Informal.* Pains due to rheumatism.

**rheumatic fever** *n.* A severe infectious disease occurring chiefly in children, marked by fever and painful inflammation of the joints and often resulting in permanent damage to the heart valves.

**rheu·ma·tism** (rōō'mə-tĭz'əm) *n.* [Lat. *rheumatismus,* rheum < Gk. *rheumatismos* < *rheumatizesthai,* to suffer from rheum < *rheuma,* stream < *rhein,* to flow.] **1.** Any of several pathological conditions of the muscles, tendons, joints, bones, or nerves, marked by discomfort and disability. **2.** Rheumatoid arthritis.

**rheu·ma·toid** (rōō'mə-toid') *also* **rheu·ma·toi·dal** (rōō'mə-toid'l) *adj.* **1.** Of or resembling rheumatism. **2.** Afflicted with rheumatism. **—rheu'ma·toi'dal·ly** *adv.*

**rheumatoid arthritis** *n.* A chronic disease characterized by stiffness and inflammation of the joints, loss of mobility, weakness, and deformity.

**rheumatoid factor** *n.* An immunoglobulin present in the blood serum of many people afflicted with rheumatoid arthritis, used as a means of diagnosing the disease.

**Rh factor** (är'ăch') *n.* [< RH(ESUS MONKEY), from its being first detected in the blood of this animal.] Any of several substances on the surface of red blood cells that induce antigenic reactions with Rh- negative blood cells.

**rhin-** *pref. var. of* RHINO-.

**rhi·nal** (rī'nəl) *adj.* Of or relating to the nose : NASAL.

**rhi·nen·ceph·a·lon** (rī'nĕn-sĕf'ə-lŏn', -lən) *n., pl.* **-la** (-lə). The olfactory region of the brain, located in the cerebrum. **—rhi'nen·ce·phal'ic** (-sə-făl'ĭk) *adj.*

**rhine·stone** (rīn'stōn') *n.* [After the *Rhine,* a river in Europe.] A colorless artificial gem of paste or glass, often with facets that sparkle in imitation of diamond.

**Rhine wine** (rīn) *n.* **1.** Any of several dry white wines produced in the Rhine valley. **2.** A light dry wine similar to Rhine wine.

**rhi·ni·tis** (rī-nī′tĭs) *n.* Inflammation of the nasal mucous membranes.
**rhi·no¹** (rī′nō) *n., pl.* **-nos.** *Informal.* A rhinoceros.
**rhi·no²** (rī′nō) *n.* [Orig. unknown.] *Chiefly Brit.* Money.
**rhino-** or **rhin-** *pref.* [< Gk. *rhis, rhin-*, nose.] Nose : nasal <*rhinitis*> <*rhinology*>
**rhi·noc·er·os** (rī-nŏs′ər-əs) *n., pl.* **rhinoceros** or **-os·es.** [ME *rinoceros* < Lat. *rhinoceros* < Gk. *rhinokerōs* : *rhis*, nose + *keras*, horn.] Any of several large thick-skinned herbivorous mammals of the family Rhinocerotidae of Africa and Asia, having one or two upright horns on the snout.
**rhi·nol·o·gy** (rī-nŏl′ə-jē) *n.* The anatomy, physiology, and pathology of the nose. **—rhi·nol′o·gist** *n.*
**rhi·no·phar·yn·gi·tis** (rī′nō-făr′ĭn-jī′tĭs) *n.* Inflammation of the nasal and pharyngeal mucous membrane.
**rhi·no·plas·ty** (rī′nō-plăs′tē, -nə-) *n.* Plastic surgery of the nose. **—rhi′no·plas′tic** *adj.*
**rhi·nos·co·py** (rī-nŏs′kə-pē) *n.* Examination of the nasal passages.
**rhi·no·vi·rus** (rī′nō-vī′rəs) *n.* Any of a group of picornaviruses that are causative agents of disorders of the respiratory tract, as the common cold.
**rhiz-** *pref. var. of* RHIZO-.
**rhi·zan·thous** (rī-zăn′thəs) *adj.* Bearing flowers directly from the root.
**rhizo-** or **rhiz-** *pref.* [< Gk. *rhiza*, root.] Root <*rhizogenic*>
**rhi·zo·bi·um** (rī-zō′bē-əm) *n., pl.* **-bi·a** (-bē-ə) [NLat. *Rhizobium*, genus name : RHIZO- + Gk. *bios*, life.] Any of various nitrogen-fixing bacteria of the genus *Rhizobium* that form nodules on the roots of leguminous plants, as clover and beans.
**rhi·zo·ceph·a·lan** (rī′zō-sĕf′ə-lən) *n.* [< NLat. *Rhizocephala*, order name : RHIZO- + Gk. *kephalē*, head.] Any of various small aquatic crustaceans of the order Rhizocephala that are parasitic on other crustaceans. **—rhi′zo·ceph′a·lous** *adj.*
**rhi·zo·gen·ic** (rī′zō-jĕn′ĭk) also **rhi·zo·ge·net·ic** (-jə-nĕt′ĭk) *adj. Bot.* Generating roots <*rhizogenic* plant tissue>
**rhi·zoid** (rī′zoid′) *n.* **1.** A slender rootlike filament by which mosses, liverworts, and ferns attach themselves to the substratum and absorb nourishment. **2.** A rootlike extension of the thallus of a fungus. **—rhi′zoid′, rhi·zoi′dal** (-zoid′l) *adj.*
**rhi·zome** (rī′zōm′) *n.* [NLat. *rhizoma* < Gk. *rhizōma*, mass of roots < *rhizoun*, to cause to take root < *rhiza*, root.] A rootlike, usu. horizontal stem growing under or along the ground that sends out roots from its lower surface and leaves or shoots from its upper surface. **—rhi·zom′a·tous** (-zŏm′ə-təs, -zō′mə-təs) *adj.*
**rhi·zo·morph** (rī′zō-môrf′) *n.* A rootlike part, as hyphae in fungi.
**rhi·zo·mor·phous** (rī′zō-môr′fəs) *adj. Bot.* Formed like a root.
**rhi·zoph·a·gous** (rī-zŏf′ə-gəs) *adj.* Feeding on roots.
**rhi·zo·pod** (rī′zō-pŏd′) *n.* [< NLat. *Rhizopoda*, class name : RHIZO- + Gk. *pous*, foot.] A protozoan of the class or subclass Rhizopoda, as an amoeba, moving and ingesting food by means of pseudopodia. **—rhi·zop′o·dan** (-zŏp′ə-dən) *adj.* & *n.* **—rhi·zop′o·dous** *adj.*
**rhi·zo·pus** (rī′zō-pəs) *n.* [NLat. *Rhizopus*, genus name : RHIZO- + Gk. *pous*, foot.] An often destructive fungus of the genus *Rhizopus*, as *R. nigricans*, the common bread mold.
**rhi·zo·sphere** (rī′zə-sfîr′) *n.* The soil zone of increased microbial growth and activity that surrounds the roots of a plant.
**rhi·zot·o·my** (rī-zŏt′ə-mē) *n., pl.* **-mies.** Surgical severance of spinal nerve roots to relieve pain or hypertension.
**Rh-neg·a·tive** (är′āch-nĕg′ə-tĭv) *adj.* Lacking an Rh factor.
**rho** (rō) *n.* [Gk. *rhō*, of Phoenician orig.; akin to Heb. *rēsh*.] The 17th letter of the Greek alphabet. —See table at ALPHABET.
**rhod-** *pref. var. of* RHODO-.
**rho·da·mine** (rō′də-mēn′) *n.* A synthetic red to pink dye.
**Rhode Island Red** (rōd) *n.* Any of an American breed of domestic fowls with dark reddish-brown feathers.
**Rho·de·sian man** (rō-dē′zhən) *n.* A fossil man, *Homo rhodesiensis* or *Cyphanthropus rhodesiensis*, found in south-central Africa, having a large low skull with massive brow ridges and skeletal bones similar to modern man.
**Rhodesian ridgeback** *n.* A large dog orig. bred in Africa, having short, yellowish-tan hair forming a ridge along the back.
**Rhodes scholar** (rōdz) *n.* A British Commonwealth or U.S. student who holds a scholarship established by the will of Cecil J. Rhodes that permits attendance at Oxford University for two or three years. **—Rhodes scholarship** *n.*
**rho·di·um** (rō′dē-əm) *n.* [Gk. *rhodon*, rose + -IUM.] *Symbol* **Rh** A hard, durable, silvery-white metallic element that is used to form high-temperature alloys with platinum and is plated on other metals to produce a durable corrosion-resistant coating; atomic number 45; atomic weight 102.905.
**rhodo-** or **rhod-** *pref.* [< Gk. *rhodon*, rose.] Rose : rosy : red <*rhodolite*>
**rho·do·chro·site** (rō′də-krō′sīt′) *n.* [G. *Rhodochrosit* : Gk. *rhodon*, rose + Gk. *khrōs*, color + G. *-it*, -ite.] A naturally occurring impure light-pink to rose-red manganese carbonate, $MnCO_3$, with a pearly or vitreous luster, used as a manganese ore.
**rho·do·den·dron** (rō′də-dĕn′drən) *n.* [NLat. *Rhododendron*, genus name < Lat., oleander < Gk. : *rhodon*, rose + *dendron*, tree.] Any of various evergreen shrubs of the genus *Rhododendron* of the North Temperate Zone, having variously colored flower clusters.
**rho·do·lite** (rō′də-līt′) *n.* A rose-red or pink garnet, a silicate mineral used as a gem.
**rho·do·mon·tade** (rŏd′ə-mŏn-tād′, -täd, rō′də) *n. var. of* RODOMONTADE.
**rho·do·nite** (rō′də-nīt′) *n.* [G. *Rhodonit* < Gk. *rhodon*, rose.] A pink to rose-red mineral, essentially a glassy crystalline manganese silicate, $MnSiO_3$, used as an ornamental stone.
**rho·do·plast** (rō′də-plăst′) *n.* A reddish chromatophore found in red algae.
**rho·dop·sin** (rō-dŏp′sĭn) *n.* The pigment sensitive to red light in the retinal rods of the eyes, consisting of retinene and opsin.
**rho·do·ra** (rō-dôr′ə, -dōr′ə) *n.* [NLat. *Rhodora*, genus name < Lat., a kind of plant.] A shrub, *Rhododendron canadense* of eastern North America, with rose-purple flowers that bloom before the leaves appear.
**rhomb-** *pref. var. of* RHOMBO-.
**rhom·ben·ceph·a·lon** (rŏm′bĕn-sĕf′ə-lŏn′, -lən) *n.* The section of the embryonic brain from which the metencephalon, myelencephalon, and subsequently the cerebellum, pons, and medulla oblongata develop.
**rhom·bi** (rŏm′bī′) *n. var. of* RHOMBUS.
**rhom·bic** (rŏm′bĭk) *adj.* **1.** Shaped like a rhombus <a *rhombic* geometrical figure> **2.** Orthorhombic.
**rhombo-** or **rhomb-** *pref.* [< RHOMBUS.] Rhombus <*rhombohedron*><*rhomboid*>
**rhom·bo·he·dron** (rŏm′bō-hē′drən) *n., pl.* **-drons** or **-dra** (-drə). A prism with six faces, each a rhombus. **—rhom′bo·he′dral** *adj.*
**rhom·boid** (rŏm′boid′) *n.* [LLat. *rhomboides* < Gk. *rhomboeidēs*, resembling a rhombus : *rhombos*, rhombus + *-eidēs*, -oid.] A parallelogram with unequal adjacent sides. —*adj.* Shaped like a rhomboid. **—rhom·boi′dal** (-boid′l) *adj.*
**rhom·bus** (rŏm′bəs) *n., pl.* **-bus·es** or **-bi** (-bī′) [Lat. < Gk. *rhombos*.] An equilateral parallelogram.
**rhon·chus** (rŏng′kəs) *n., pl.* **-chi** (-kī′) [LLat., a snoring < Gk. *rhonkos* < *rhenkein*, to snore.] A coarse rattling sound somewhat like snoring, usu. caused by secretion in the bronchial tube. **—rhon′chal, rhon′chi·al** (-kē-əl) *adj.*
**Rh-pos·i·tive** (är′āch-pŏz′ĭ-tĭv) *adj.* Containing an Rh factor.
**rhu·barb** (ro͞o′bärb′) *n.* [ME *rubarbe* < OFr. < Med. Lat. *rubarbum*, prob. < LLat. *rha barbarum*, barbarian rhubarb.] **1.** A plant of the genus *Rheum*, marked by large long-stalked leaves, esp. *R. rhaponticum*, the common garden rhubarb, having long green or reddish acid leafstalks that are edible when cooked and sweetened. **2.** The dried, bitter-tasting rhizome and roots of *R. palmatum* or *R. officinale* of central Asia, used as a laxative. **3.** *Slang.* A heated debate; quarrel, or fight.
**rhumb** (rŭm, rŭmb) *n.* [Sp. *rumbo*.] **1.** A rhumb line. **2.** One of the points of the mariner's compass.
**rhum·ba** (rŭm′bə, ro͝om′-, ro͞om′-) *n. var. of* RUMBA.
**rhumb line** *n.* The path of a ship that maintains a fixed compass direction, shown on a map as a line crossing all meridians at the same angle.
**rhyme** also **rime** (rīm) [ME *rime* < OFr. < Med. Lat. *rithmus*, rhythm < Lat. *rhythmos* < Gk. *rhuthmos*.] —*n.* **1.** Correspondence of terminal sounds of words or of lines of verse. **2. a.** A poem or verse with a regular correspondence of sounds, esp. at the ends of lines. **b.** Poetry or verse of this kind. **3.** A word that corresponds with another in terminal sound, as *forlorn* and *outworn*. —*v.* **rhymed, rhym·ing, rhymes** also **rimed, rim·ing, rimes.** —*vi.* **1.** To form a rhyme. **2.** To compose rhymes or verse. **3.** To use rhymes in composing verse. —*vt.* **1.** To put into rhyme or compose with rhymes. **2.** To use (a word or words) as a rhyme.
**rhym·er** also **rim·er** (rī′mər) *n.* A composer of rhyming verses, esp. a rhymester.
**rhyme scheme** *n.* Arrangement of rhymes in a stanza or poem.
**rhyme·ster** also **rime·ster** (rīm′stər) *n.* **1.** A composer of light verse. **2.** An inferior poet.
**rhyming slang** *n.* Slang in which a word is replaced by a word or phrase that rhymes with it or by part of a rhyming phrase; e.g., *jimmygrant* for *immigrant*.
**rhyn·cho·ce·pha·lian** (rĭng′kō-sə-fāl′yən) *adj.* [< NLat. *Rhynchocephalia*, order name : Gk. *rhunkhos*, beak + *kephalē*, head.] Of or belonging to the Rhynchocephalia, an order of mostly extinct lizardlike reptiles. —*n.* A rhynchocephalian reptile.
**rhy·o·lite** (rī′ə-līt′) *n.* [G. *Rhyolit* < Gk. *rhuax*, stream < *rhein*, to flow.] A glassy volcanic rock, similar in composition to granite and usu. exhibiting flow lines.
**rhythm** (rĭth′əm) *n.* [OFr. *rhythme* < Lat. *rhythmus* < Gk. *rhuthmos*.] **1.** Movement or variation marked by regular recurrence or alternation of different quantities or conditions. **2.** Patterned, recurring alternations of contrasting elements of sound or speech. **3.** *Mus.* **a.** A regular pattern formed by a series of notes of differing stress and

duration. **b.** A specific kind of such a pattern. **c.** A group of instruments in a band supplying the rhythm. **4. a.** The metrical flow of sound with a regulated pattern of long and short or accented and unaccented syllables in poetry. **b.** A specific kind of such a metrical flow <dactylic *rhythm*> **5.** A harmonious or regular pattern created by lines, colors, and forms in the visual arts, as painting or sculpture.

**rhythm and blues** *n.* Music developed by black Americans that combines blues and jazz, marked by a strong simple rhythm.

**rhyth·mi·cal** (rĭth′mĭ-kəl) *also* **rhyth·mic** (-mĭk) *adj.* Of, relating to, or having rhythm. **—rhyth′mi·cal·ly** *adv.*

**rhyth·mics** (rĭth′mĭks) *n.* *(sing. in number).* The study of rhythm.

**rhyth·mist** (rĭth′mĭst) *n.* **1.** One who is expert in or has a keen sense of rhythm. **2.** One who studies or produces rhythm.

**rhythm method** *n.* A birth-control method dependent on continence during the period of female ovulation.

**ri·al**[1] (rē-ôl′, -äl′) *n.* [Pers. < Ar. *riyāl.*] —See table at CURRENCY.

**ri·al**[2] (rē-ôl′, -äl′) *n. var. of* RIYAL.

**ri·al·to** (rē-äl′tō, rā-äl′). [After *Rialto,* an island in Venice on which a market was situated.] **1.** A theatrical district. **2.** MARKETPLACE 1.

**ri·a·ta** *also* **re·a·ta** (rē-ä′tə) *n.* [Sp., rope.] A lariat : lasso.

**rib** (rĭb) *n.* [ME < OE.] **1.** *Anat.* **a.** One of a series of long, curved bones, occurring in 12 pairs in human beings and extending from the spine to or toward the sternum. **b.** A similar bone in most vertebrates. **2.** A part or piece serving to shape or support <the *ribs* of an umbrella> **3.** A cut of meat enclosing one or more ribs. **4.** One of numerous curved members attached to a boat or ship's keel and extending upward and outward to form the framework of the hull. **5.** One of many formed transverse pieces along the length of an aircraft wing used to establish shape. **6. a.** An arch or a projecting arched member of a vault. **b.** One of the curved pieces of an arch. **7.** A raised ridge or wale in cloth or knitted material. **8.** *Bot.* One of the main veins of a leaf or other organ. **9.** *Slang.* A joke. *—vt.* **ribbed, rib·bing, ribs. 1.** To support, shape, or provide with a rib or ribs. **2.** To make with ridges or raised markings. **3.** *Slang.* To poke fun at <*ribbed* me about my Halloween costume>

**rib·ald** (rĭb′əld) *adj.* [ME *ribaud,* ribald person < OFr. *ribauld* < *riber,* to be wanton, of Germanic orig.] Relating to or indulging in coarse, licentious humor. *—n.* A ribald person.

**rib·ald·ry** (rĭb′əl-drē) *n., pl.* **-ries.** Ribald language or joking.

**rib·and** (rĭb′ənd) *n.* [ME *riban* < OFr. —see RIBBON.] A ribbon, esp. one used as a decoration.

**rib·band** (rĭb′ănd, -ənd, -ən) *n.* A length of flexible metal or wood for holding the ribs of a ship in place while the exterior planking or plating is being applied.

**rib·bon** (rĭb′ən) *n.* [ME *riban* < OFr., of Germanic orig.] **1.** A narrow strip or band of fine fabric, finished at the edges and used for trimming or tying. **2.** Something resembling a ribbon, as a measuring tape. **3. ribbons.** Ragged or tattered strips <a shirt torn to *ribbons*> **4.** A strip, usu. of inked cloth, for making the impression, as in a typewriter. **5. a.** A band of colored cloth signifying membership in an order or the award of a prize. **b.** A strip of colored cloth worn on the left breast of a uniform to indicate the award of a medal or decoration. **6. ribbons.** *Informal.* Reins for driving horses. *—vt.* **-boned, -bon·ing, -bons. 1.** To tie or decorate with ribbons. **2.** To tear into shreds.

**rib·bon·fish** (rĭb′ən-fĭsh′) *n., pl.* **ribbonfish** *or* **-fish·es.** Any of several marine fishes, chiefly of the genus *Trachipterus,* with long narrow compressed bodies.

**ribbon snake** *n.* A nonvenomous North American snake, *Thamnophis sauritus,* having yellow or reddish stripes.

**ribbon worm** *n.* A nemertean.

**rib cage** *n.* The enclosing structure formed by the ribs and the bones to which they are attached.

**rib eye** *n.* A cut of meat taken from the outside of the rib.

**rib·grass** (rĭb′grăs′) *n.* A weedy plant, *Plantago lanceolata,* with ribbed lancelike leaves and a small dense whitish flower spike.

**ri·bo·fla·vin** (rī′bō-flā′vĭn) *n.* [RIBO(SE) + FLAVIN.] A crystalline orange-yellow pigment, $C_{17}H_{20}O_6N_4$, the principal growth-promoting factor in the vitamin B complex, found in leafy vegetables, milk, egg yolks, and fresh meat, and produced synthetically.

**ri·bo·nu·cle·ase** (rī′bō-no͞o′klē-ās, -nyo͞o′-, -āz) *n.* Any of various enzymes that decompose ribonucleic acids.

**ri·bo·nu·cle·ic acid** (rī′bō-no͞o-klē′ĭk, -nyo͞o-) *n.* [RIBO(SE) + NUCLEIC ACID.] A universal polymeric constituent of all living cells, composed of a single-strand chain of alternating phosphate and ribose units with the bases adenine, guanine, cytosine, and uracil bonded to the ribose, the structure and base sequence of which are determinants of protein synthesis.

**ri·bo·nu·cle·o·pro·tein** (rī′bō-no͞o′klē-ō-prō′tēn, -tē-ĭn, -nyo͞o′-) *n.* [RIBONUCLE(IC ACID) + PROTEIN.] A nucleoprotein containing RNA.

**ri·bo·nu·cle·o·side** (rī′bō-no͞o′klē-ə-sīd′, -nyo͞o′-) *n.* [RIBO(SE) + NUCLEOSIDE.] A nucleoside containing ribose as its sugar component.

**ri·bose** (rī′bōs′) *n.* [< G. *Ribonsäure,* a tetrahydroxyl acid from which ribose is obtained.] A pentose sugar, $C_5H_{10}O_5$, occurring as a component of nucleic acids.

**ribosomal RNA** *n.* The RNA that is a permanent structural element of a ribosome.

**ri·bo·some** (rī′bə-sōm) *n.* [RIBO(SE) + -SOME[3].] A spherical cytoplasmic RNA-containing particle active in the synthesis of protein. **—ri·bo·so′mal** (-sō′məl) *adj.*

**rib roast** *n.* A cut of beef containing the sizable piece located along the outside of the rib.

**rib·wort** (rĭb′wûrt′, -wôrt′) *n.* Ribgrass.

**rice** (rīs) *n.* [ME *ryce* < OFr. *ris* < OItal. *riso* < Lat. *oryza* < Gk. *oruza.*] **1.** A cereal grass, *Oryza sativa,* cultivated extensively in warm climates and used as a staple food throughout the world. **2.** The starchy edible seed of rice. *—vt.* **riced, ric·ing, ric·es.** To sieve (food) to the consistency of rice.

**†rice·bird** (rīs′bûrd′) *n.* **1.** *Southern U.S.* The bobolink. **2.** A bird that frequents rice fields.

**rice paper** *n.* A thin paper made chiefly from the pith of the rice-paper tree.

**rice-pa·per plant** (rīs′pā′pər) *n.* Rice-paper tree.

**rice-paper tree** *n.* A shrub or small tree, *Tetrapanax papyriferum* of eastern Asia, grown as a source of fiber for rice paper.

**ric·er** (rī′sər) *n.* A kitchen utensil used for ricing soft foods by extrusion through small holes.

**rice weevil** *n.* A small destructive insect, *Sitophilus oryzae,* that infests stored grain and cereal products.

**rich** (rĭch) *adj.* **-er, -est.** [ME *riche* < OE *rīce.*] **1.** Having extensive material wealth. **2.** Having great value or worth. **3.** Magnificent : luxurious. **4. a.** Abundantly supplied. **b.** Abounding, esp. in natural resources. **5.** Extremely productive. **6.** Containing a large amount of choice ingredients, as butter, sugar, or eggs <a *rich* pastry> **7. a.** Pleasantly full and mellow <a *rich* voice> **b.** Warm and strong in color. **8.** Containing a large proportion of fuel to air. **9.** *Informal.* Very funny. **—rich′ly** *adv.* **—rich′ness** *n.*

☆ ***syns:*** RICH, AFFLUENT, LOADED, MONEYED, WEALTHY *adj. core meaning* : having a large amount of money, land, or other material possessions <a *rich* politician> ***ant:*** poor

▲ word history: The close connection between wealth and power can be seen in the semantics of the word *rich.* The Old English ancestor of *rich, rīce,* meant basically "powerful," although it also meant "wealthy" and "of high rank." In Old English times it was likely that a person would possess either all three attributes or none. In Middle English the sense "wealthy" came to predominate because Old French *rich,* which meant "wealthy," was also in use as an English word. Both the French word and the English word have a common Germanic ancestor, which is a borrowing through Celtic of Latin *rex,* "king." There are many English words related to *rich,* among them *realm, reign, regime,* and *rajah.*

**Rich·ard Roe** (rĭch′ərd) *n.* A name used in legal proceedings to refer to a fictitious or unidentified person.

**rich·en** (rĭch′ən) *vt.* **-ened, -en·ing, -ens.** To make rich.

**rich·es** (rĭch′ĭz) *pl.n.* [ME < *richesse,* wealth < OFr. < *riche,* rich, of Germanic orig.] **1.** Abundant wealth. **2.** Precious possessions.

**Rich·ter scale** (rĭk′tər) *n.* [After Charles F. *Richter* (b. 1900).] A logarithmic scale ranging from 1 to 10, for expressing the magnitude or total energy of an earthquake.

**ri·cin** (rī′sĭn, rĭs′ĭn) *n.* [Lat. *ricinus,* castor-oil plant.] A poisonous protein extracted from the castor bean and used as a biochemical reagent.

**ri·cin·o·le·ic acid** (rĭs′ĭn-ō-lē′ĭk) *n.* [Lat. *ricinus,* castor-oil plant + OLEIC.] An unsaturated fatty acid, $C_{18}H_{34}O_3$, prepared from castor oil and used in making soaps and in textile finishing.

**rick** (rĭk) *n.* [ME *reke* < OE *hrēac.*] A stack, as of straw or hay, esp. when covered or thatched for protection from the weather. *—vt.* **ricked, rick·ing, ricks.** To pile in ricks.

**rick·ets** (rĭk′ĭts) *n.* [Orig. unknown.] *(sing. in number).* A deficiency disease resulting from a lack of vitamin D and from insufficient exposure to sunlight, marked by defective bone growth and occurring chiefly in children.

**rick·ett·si·a** (rĭ-kĕt′sē-ə) *n., pl.* **-si·ae** (-sē-ē′) [NLat. *Rickettsia,* genus name, after Howard T. *Ricketts* (1871–1910).] Any of various microorganisms of the genus *Rickettsia,* carried as parasites by many ticks, lice, and fleas, that cause diseases such as typhus and Rocky Mountain spotted fever in human beings. **—rick·ett′si·al** *adj.*

**rick·ett·si·o·sis** (rĭ-kĕt′sē-ō′sĭs) *n., pl.* **-ses** (-sēz′). Infection with or disease caused by rickettsiae.

**rick·et·y** (rĭk′ĭt-ē) *adj.* **-i·er, -i·est.** [< RICKETS.] **1.** Likely to break or fall apart : SHAKY. **2.** Feeble with age : INFIRM. **3.** Of, having, or resembling rickets. **—rick′et·i·ness** *n.*

**rick·ey** (rĭk′ē) *n., pl.* **-eys.** [Prob. < the name *Rickey.*] A drink of soda water, lime juice, sugar, and usu. gin.

**rick·rack** (rĭk′răk′) *n.* [Redup. of RACK[1].] A narrow flat zigzag braid used as a trimming.

**rick·sha** *or* **rick·shaw** (rĭk′shô) *n.* [Short for JINRIKSHA.] A small two-wheeled Oriental carriage drawn by one or two persons.

**ric·o·chet** (rĭk′ə-shā′, -shĕt′) *vi.* **-cheted** (-shād′), **-chet·ing** (-shā′-ĭng), **-chets** *also* **-chet·ted** (-shĕt′ĭd), **-chet·ting** (-shĕt′ĭng), **-chets.** [Fr.] To rebound from a surface. **—ric′o·chet′** *n.*
**ri·cot·ta** (rē-kôt′tä, rĭ-kŏt′ə) *n.* [Ital. < Lat. *recocta*, fem. p.part. of *recoquere*, to cook again : *re-*, again + *coquere*, to cook.] **1.** An Italian soft cheese resembling cottage cheese. **2.** A U.S. cheese similar to ricotta.
**ric·tus** (rĭk′təs) *n.* [Lat. < p.part. of *ringi*, to open the mouth wide.] **1. a.** The expanse of an open mouth, a bird's beak, or a similar structure. **b.** A gaping grimace <"his lips drawn back in a silent *rictus* of ecstatic agony" —Anthony Grey> **2.** A cleft, split, or gap. **—ric′tal** *adj.*
**rid** (rĭd) *vt.* **rid** *or* **rid·ded, rid·ding, rids.** [ME < *ridden* < ON *ryðja.*] To free from <*rid* myself of indebtedness> **—rid′der** *n.*
**rid·dance** (rĭd′ns) *n.* **1.** An act of ridding. **2.** A deliverance from or removal of something.
**rid·den** (rĭd′n) *adj.* [< p.part. of RIDE.] Dominated : obsessed <poverty-*ridden*><guilt-*ridden*>
**rid·dle¹** (rĭd′l) *vt.* **-dled, -dling, -dles.** [ME *riddlen*, to sift < *riddil*, sieve < OE *hriddel.*] **1.** To pierce with numerous holes : PERFORATE <*riddled* the plane with bullets> **2.** To put through a coarse sieve. **3.** To spread throughout <a proposal *riddled* with faults> *—n.* A coarse sieve, as for gravel. **—rid′dler** *n.*
**rid·dle²** (rĭd′l) *n.* [ME *redeles* < OE *rǣdelse.*] **1.** A question or statement requiring thought to answer or comprehend : CONUNDRUM. **2.** One that is puzzling : ENIGMA. *—v.* **rid·dled, -dling, -dles.** *—vt.* To solve or explain. *—vi.* **1.** To propound or solve riddles. **2.** To speak in riddles. **—rid′dler** *n.*
**ride** (rīd) *v.* **rode** (rōd), **rid·den** (rĭd′n), **rid·ing, rides.** [ME *riden* < OE *rīdan.*] *—vi.* **1.** To be carried, as in a vehicle or on horseback. **2. a.** To travel over a surface <a car that *rides* well.> **b.** To proceed or move. **3.** To float or move on or as if on water <*rode* into office on a wave of patriotism> **4.** To lie at anchor. **5.** To seem to float. **6.** To be sustained or supported. **7.** To depend <The election *rode* on popular support.> **8.** To continue without interference <hoped to let the matter *ride*> **9.** To work or move from the proper place, esp. on the body <a shirt that *rides* up> *—vt.* **1.** To sit on and drive. **2.** To be supported or carried on <a surfer *riding* the waves> **3.** To travel along, over, or through <*ride* the turnpike> **4.** To rest on by overlapping : OVERLIE. **5.** To take part in or do by riding <*rode* their final race> **6.** To dominate or control. **7.** To cause to ride, esp. to cause to be carried <*rode* them out of town> **8.** To keep (a vessel) at anchor. **9.** *Informal.* To ridicule or tease. **10.** To dominate the mind or thoughts of <*ridden* with doubt> **11.** To keep partially engaged by slightly depressing a pedal with the foot <*riding* the brake> **—ride out.** To survive : outlast <*rode out* the hurricane> *—n.* **1.** An act of riding. **2.** A path made for riding on horseback, esp. through woodlands. **3.** A device, as at an amusement park, that one rides for excitement or pleasure. **4.** A means of transport <Their *ride* never came.> **—ride for a fall.** To court danger. **—ride herd on.** To keep watch or control over. **—ride high.** To experience success. **—ride roughshod over.** To take a course of action regardless of the opinions, feeling, or welfare of others. **—ride shotgun. 1.** To guard a person or thing while in transit. **2.** To ride in the passenger seat of a car or truck. **—take for a ride.** *Slang.* **1.** To transport to a place and kill (someone). **2.** To deceive : cheat.
**rid·er** (rī′dər) *n.* **1.** One that rides. **2.** A person who rides horses. **3.** A clause, usu. having little relevance to the main issue, added to a legislative bill. **4.** An amendment or addition to a document or record. **5.** Something resting on or supported by something else.
**rid·er·ship** (rī′dər-shĭp′) *n.* The number of passengers who ride a public transit system.
**ridge** (rĭj) *n.* [ME *rigge* < OE *hrycg.*] **1.** A long narrow upper section or crest <the *ridge* of a snowdrift> **2.** A long narrow land elevation. **3.** A long, narrow, or crested part of the body. **4.** The horizontal line formed by the juncture of two sloping planes, esp. the line formed by the surfaces of a roof. **5.** A narrow raised strip, as in cloth or on plowed ground. *—v.* **ridged, ridg·ing, ridg·es.** *—vt.* To form into, mark with, or provide with ridges. *—vi.* To form ridges.
**ridge·back** (rĭj′băk′) *n.* A Rhodesian ridgeback.
**ridge·ling** *also* **ridg·ling** (rĭj′lĭng) *n.* [Orig. unknown.] A male animal with one or two undescended testicles.
**ridge·pole** (rĭj′pōl′) *n.* **1.** A horizontal beam at the ridge of a roof to which the rafters are attached. **2.** The horizontal pole at the top of a tent.

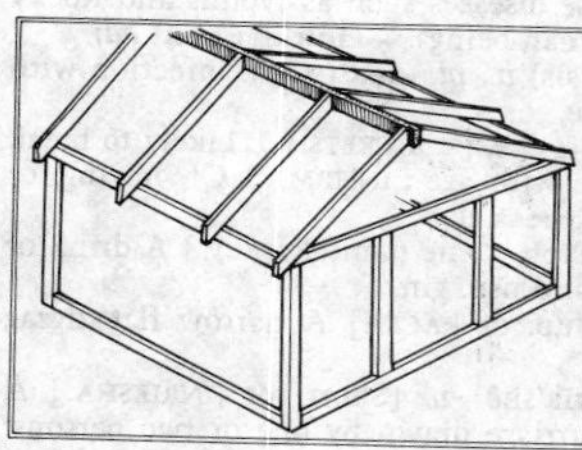

**ridgepole**

**ridg·y** (rĭj′ē) *adj.* **-i·er, -i·est.** Having or forming ridges.
**rid·i·cule** (rĭd′ə-kyōōl′) *n.* [Fr. < Lat. *ridiculum*, joke < *ridiculus*, laughable, ridiculous.] Words or actions intended to evoke sardonic laughter at or feelings toward one. *—vt.* **-culed, -cul·ing, -cules.** To make fun of. **—rid′i·cul′er** *n.*
☆ ***syns:*** RIDICULE, DERIDE, GIBE, MOCK, TAUNT, TWIT *v. core meaning* : to make fun of <*ridiculed* us for losing the game>
**ri·dic·u·lous** (rĭ-dĭk′yə-ləs) *adj.* [Lat. *ridiculus*, laughable < *ridēre*, to laugh.] Deserving or inspiring ridicule : LUDICROUS. **—ri·dic′u·lous·ly** *adv.* **—ri·dic′u·lous·ness** *n.*
**rid·ing¹** (rī′dĭng) *n.* **1.** An act of riding. **2.** Horseback riding.
**rid·ing²** (rī′dĭng) *n.* [ME *rithing* < OE **ðriðing* < ON *ðriðjungr*, third part < *ðriði*, third.] **1.** One of three former administrative divisions of Yorkshire, England. **2.** A Canadian administrative or electoral division.
**riding habit** *n.* The outfit worn by a horseback rider.
**rid·ley** (rĭd′lē) *n., pl.* **-leys.** [Prob. < the name *Ridley.*] A marine turtle, *Lepidochelys kempi* of the Gulf of Mexico and Atlantic coastal waters.
**ri·el** (rē-ĕl′) *n.* [Orig. unknown.] —See table at CURRENCY.
**Rie·mann·i·an geometry** (rē-män′ē-ən) *n.* [After Bernhard *Riemann* (1826–1866).] A non-Euclidean geometry based on the postulate that there are no parallel lines.
**Ries·ling** (rēs′lĭng) *n.* [G.] A dry white wine similar to Rhine wine.
**ri·fam·pi·cin** (rĭ-făm′pĭ-sĭn) *n.* [Blend of E. *rifamycin*, an antibiotic, and AMPICILLIN.] An antibiotic that has both antibacterial and antiviral action.
**rife** (rīf) *adj.* **rif·er, rif·est.** [ME *rif* < OE *ryfe.*] **1.** Common or frequent in occurrence : WIDESPREAD. **2.** Abounding : full <a government *rife* with corruption>
**riff** (rĭf) *n.* [Orig. ] *Mus.* A short rhythmic phrase constantly repeated.
**Riff** (rĭf) *also* **Rif·fi·an** (rĭf′ē-ən) *n.* A Berber tribesman of the Rif country in northern Morocco.
**rif·fle** (rĭf′əl) *n.* [Orig. unknown.] **1. a.** A rocky shoal or sandbar lying just below the surface of a waterway. **b.** A stretch of choppy water caused by such a shoal or sandbar : RAPID. **2. a.** The sectional stone or wood bottom lining of a sluice, arranged to trap mineral particles, as of gold. **b.** A block or groove in such a lining. **3.** An act of shuffling playing cards. *—v.* **-fled, -fling, -fles.** *—vt.* **1.** To shuffle (playing cards) by holding part of a deck in each hand and raising up the edges before releasing them to fall alternately in one stack. **2.** To thumb through (e.g., the pages of a book). *—vi.* **1.** To shuffle playing cards. **2.** To become choppy, as water.
**rif·fler** (rĭf′lər) *n.* [OFr. *riffloir* < *riffler*, to scratch.] A file with curved ends suitable for scraping.
**riff·raff** (rĭf′răf′) *n.* [ME *rif and raf*, one and all < OFr. *rif et raf.*] **1.** Disreputable people. **2.** Trash : rubbish.
**ri·fle¹** (rī′fəl) *n.* [< *rifle*, to cut spiral grooves < OFr., to scratch.] **1. a.** A firearm with a rifled bore designed to be fired from the shoulder. **b.** An artillery piece or naval gun with such spiral grooves. **2. rifles.** Troops armed with rifles. *—vt.* **-fled, -fling, -fles.** To cut spiral grooves within (e.g., a gun barrel).
**ri·fle²** (rī′fəl) *vt.* **-fled, -fling, -fles.** [ME *riflen*, to plunder < OFr. *rifler.*] **1.** To search with intent to steal. **2.** To ransack : pillage. **3.** To rob. **—ri′fler** *n.*
**ri·fle·bird** (rī′fəl-bûrd′) *n.* Any of several birds of paradise of the genera *Craspedophora* and *Ptiloris* of Australia and New Guinea.
**ri·fle·man** (rī′fəl-mən) *n.* **1.** A soldier armed with a rifle. **2.** One who shoots a rifle skillfully.
**ri·fle·ry** (rī′fəl-rē) *n.* **1.** The art and practice of marksmanship. **2.** Rifle fire.
**ri·fle·scope** (rī′fəl-skōp′) *n.* A telescopic sight for a rifle.
**ri·fling** (rī′flĭng) *n.* **1.** The process or operation of cutting spiral grooves in a rifle barrel. **2.** Grooves cut in a rifle barrel.
**rift¹** (rĭft) *n.* [ME, of Scand. orig.] **1. a.** *Geol.* FAULT 3. **b.** A narrow rock fissure. **2.** A break in friendly relations. *—v.* **rift·ed, rift·ing, rifts.** *—vi.* To split open : BREAK. *—vt.* To cause to break or split open.
**rift²** (rĭft) *n.* [Prob. alteration of dial. *riff*, reef.] **1.** A shallow area in a waterway. **2.** The backwash of a wave that has broken on a beach.
**rift valley** *n.* A long narrow depression in the earth's surface formed when the land sinks between two fairly parallel faults.
**†rig** (rĭg) *vt.* **rigged, rig·ging, rigs.** [ME *riggen*, prob. of Scand orig.] **1.** To fit out with harness or equipment. **2.** *Naut.* **a.** To equip (a ship) with shrouds, sails, and yards. **b.** To fit (e.g., sails or shrouds) to masts and yards. **3.** *Informal.* To dress or clothe <*rigged* out in their formal clothes> **4.** To make or construct hastily or in a makeshift manner <*rig* up a shelter for the night> **5.** To manipulate dishonestly for personal gain <*rig* the election> *—n.* **1.** *Naut.* The arrangement of masts, sails, and spars on a sailing vessel. **2.** Special equipment or gear. **3.** A vehicle with one or more horses harnessed to it. **4.** The special apparatus used for drilling oil wells. **5.** *Western U.S.* A saddle. **6.** *Informal.* A dress or costume. **7.** Fishing tackle.

ă **pat** ā **pay** âr **care** ä **father** ĕ **pet** ē **be** hw **which** ĭ **pit**
ī **tie** îr **pier** ŏ **pot** ō **toe** ô **paw, for** oi **noise** o͝o **took**

**rig·a·doon** (rĭg′ə-do͞on′) *n.* [Fr. *rigaudon.*] **1.** A lively jumping quickstep for one couple. **2.** Music for the rigadoon, usu. in rapid duple meter.

**rig·a·ma·role** (rĭg′ə-mə-rōl′) *n. var. of* RIGMAROLE.

**rig·a·to·ni** (rĭg′ə-tō′nē) *n.* [Ital. < *regato,* p.part. of *regare,* to draw a line < *riga,* line, of Germanic orig.] Slightly curved, ribbed macaroni tubes cut into short lengths.

**Ri·gel** (rī′jəl) *n.* [Ar. *rijl,* foot.] A bright double star in the constellation Orion.

**rig·ger** (rĭg′ər) *n.* **1.** One who rigs. **2.** *Naut.* A ship with a specific kind of rigging.

**rig·ging** (rĭg′ĭng) *n.* **1.** *Naut.* The system of chains, ropes, and tackle for supporting and controlling the sails, masts, and yards of a sailing vessel. **2.** The supporting material for construction work.

**right** (rīt) *adj.* **-er, -est.** [ME < OE *riht.*] **1.** Conforming with or conformable to law, justice, or morality. **2.** Being in accord with fact, reason, or truth : CORRECT <the *right* response> **3.** Fitting, proper, or appropriate. **4.** Most beneficial, desirable, or convenient <the *right* time to ask> **5.** Being in a satisfactory state or condition <made things *right*> **6.** Being in good mental or physical health or order. **7.** Intended to be worn facing outward or toward an observer <was sure to wear the sweater *right* side out> **8.** *Archaic.* Not spurious : GENUINE. **9. a.** Of, relating to, or toward that side of the human body away from the heart. **b.** Of or located on the side opposite the left. **c.** Toward this side <made a *right* turn> **10.** Of or tending toward conservative or reactionary political views or policies. **11.** *Math.* **a.** Formed by or in reference to a plane or line perpendicular to another plane or line. **b.** Having the axis perpendicular to the base <a *right* cone> **12.** Straight : uncurved <a *right* edge> —*n.* **1.** That which is just, legal, morally good, or appropriate. **2. a.** The right-hand side or direction. **b.** Something on or toward the right-hand side. **3.** A political group, as a faction or party, whose policies are conservative or reactionary. **4. a.** The right hand. **b.** A blow with the right hand <a *right* to the nose> **5.** Something due to one by law, custom, or nature. **6.** A just or legal title or claim. **7. a.** A stockholder's privilege of buying additional stock in a corporation at a special price, usu. at par or at a price below the current market value. **b.** The negotiable certificate on which this privilege is indicated. **c.** *often* **rights.** A privilege of subscribing for a given stock or bond. —*adv.* **1.** In a straight line : DIRECTLY <passed *right* by the window> **2.** Properly : well <The shirt doesn't fit *right.*> **3.** Precisely : just <I dropped it *right* over there.> **4.** Immediately <*right* after lunch> **5.** Thoroughly : quite <The cold air went *right* through me.> **6.** According to morality, law, or justice. **7.** Accurately : correctly. **8.** On or toward the right side or direction <turned left and then turned *right*> **9.** To a high degree : VERY <a *right* comfortable apartment> **10.** —Used as an intensive <kept *right* on lying> **11.** Very <the *Right* Honorable Prime Minister> —*v.* **right·ed, right·ing, rights.** —*vt.* **1.** To put in or restore to an upright or proper position <They *righted* their canoe.> **2.** To put in order or set right : CORRECT <finally *righted* many unsafe working conditions> **3.** To make amends for : REDRESS <*right* an injustice> —*vi.* To regain an upright or proper position. **—by rights.** Justly : properly. **—right′er** *n.* **—right′ness** *n.*

▲ word history: The political sense of *right,* meaning "conservative" or "reactionary," goes back to the French Revolution. In the French National Assembly of 1789 the nobles sat on the president's right and the commoners sat on the left. The nobility as a group tended to be politically more conservative than the commoners. In later assemblies and parliaments seating continued to be assigned on the basis of political views as established in the first assembly.

**right angle** *n.* An angle formed by the perpendicular intersection of two straight lines, equal to 90°.

**right-an·gled** (rīt′ăng′gəld) *adj.* Forming or containing one or more right angles <a *right-angled* bend>

**right ascension** *n.* The angular distance of a celestial body or point on the celestial sphere, measured eastward from the vernal equinox along the celestial equator to the hour circle of the body or point and expressed in degrees or in hours.

**right away** *adv.* Without delay : IMMEDIATELY.

**right circular cone** *n.* CONE 1b.

**right·eous** (rī′chəs) *adj.* [ME *ryghtuous* < OE *rihtwīs* : *rihgt,* right + *wīs,* wise, manner.] **1.** Meeting the standards of what is right and just : morally right. **2.** *Slang.* Genuine : true. —*n.* Righteous individuals as a group. **—right′eous·ly** *adv.* **—right′eous·ness** *n.*

**right face** *n.* A military command to turn 90° to the right.

**right field** *n. Baseball.* The part of the outfield to the right as viewed from home plate.

**right fielder** *n. Baseball.* An outfielder who defends right field.

**right·ful** (rīt′fəl) *adj.* **1.** Proper or right : JUST. **2.** Having a just or proper claim <returned the wallet to its *rightful* owner> **3.** Held or owned by just or proper claim <This land is my *rightful* property.> **—right′ful·ly** *adv.* **—right′ful·ness** *n.*

**right-hand** (rīt′hănd′) *adj.* **1.** Located on the right side. **2.** Directed toward the right side <a *right-hand* movement> **3.** Of, for, or done by the right hand. **4.** Helpful : reliable <my *right-hand* associate> **—right′hand′er** *n.*

**right-hand·ed** (rīt′hăn′dĭd) *adj.* **1.** Using the right hand more easily or skillfully than the left. **2.** Done with the right hand. **3.** Made to be used by the right hand. **4.** Turning or spiraling from left to right : CLOCKWISE. **—right′-hand′ed, right′-hand′ed·ly** *adv.* **—right′-hand′ed·ness** *n.*

**right·ism** *also* **Right·ism** (rī′tĭz′əm) *n.* Conservative or reactionary political ideas or activities. **—right′ist** *n.*

**right·ly** (rīt′lē) *adv.* **1.** In a proper manner : CORRECTLY <behave *rightly*> **2.** With honesty : JUSTLY. **3.** *Informal.* Really <wasn't *rightly* sure>

**right-mind·ed** (rīt′mīn′dĭd) *adj.* Having views and ideas based on what is right. **—right′-mind′ed·ness** *n.*

**right of asylum** *n. Law.* The right of receiving protection within a foreign embassy or other place recognized by law, custom, or treaty.

**right off** *adv.* Straightaway : immediately. **—right off the bat.** Right away.

**right of search** *n. Law.* The right of a warring nation to stop a neutral vessel on the high seas and search it for contraband.

**right of way** *also* **right-of-way** (rīt′əv-wā′) *n., pl.* **rights-of-way** *or* **right-of-ways. 1.** *Law.* **a.** The right to pass over property owned by another party. **b.** The path or thoroughfare on which such passage is made. **2.** The strip of land over which facilities as highways, railroads, or power lines are built. **3.** The legal or customary right of a person, vessel, or vehicle to pass in front of another.

**right on** *interj. Slang.* —Used as an exclamation of encouragement, support, or approval.

**right-on** (rīt′ŏn′, -ôn′) *adj. Slang.* **1.** Modern and sophisticated : TRENDY. **2.** Absolutely correct : perfectly true.

**right-to-life** (rīt′tə-līf′) *adj.* Of, pertaining to, or advocating laws that prohibit abortion on demand. **—right′-to-lif′er** *n.*

**right-to-work law** (rīt′tə-wûrk′) *n.* A state law that prohibits union shops.

**right triangle** *n.* A triangle containing an angle of 90°.

**right whale** *n.* Any of several whales of the family Balaenidae, marked by a large head, absence of a dorsal fin, and whalebone plates in the mouth.

**right wing** *n.* A division of a larger political group holding relatively conservative views. **—right winger** *n.*

**rig·id** (rĭj′ĭd) *adj.* [OFr. *rigide* < Lat. *rigidus* < *rigēre,* to be stiff.] **1.** Not bending : INFLEXIBLE. **2.** Not moving : STATIONARY. **3.** Difficult <a *rigid* test> **4.** Scrupulously strict <a *rigid* religious order> **—ri·gid′i·ty** (rĭ-jĭd′ĭ-tē) *n.* **—rig′id·ly** *adv.* **—rig′id·ness** *n.*

☆ ***syns:*** RIGID, INELASTIC, INFLEXIBLE, STIFF, UNBENDING, UNYIELDING *adj. core meaning* : not changing shape or bending <*rigid* iron bars> ***ant:*** elastic, flexible

**rig·ma·role** (rĭg′mə-rōl) *also* **rig·a·ma·role** (-ə-mə-rōl) *n.* [Alteration of obs. *ragman roll,* catalogue < ME *rageman rolle,* scroll used in Ragman, a game of chance.] **1.** Confused or incoherent talk : NONSENSE. **2.** A complicated set of petty procedures.

**rig·or** (rĭg′ər) *n.* [ME *rigour* < OFr. < Lat. *rigor* < *rigēre,* to be stiff.] **1.** Strictness or sterness, as in action, temperament, or judgment. **2.** Hardship. **3.** A cruel or severe act. **4.** *Med.* Shivering or trembling, as caused by a chill. **5.** *Physiol.* Rigidity in living tissues or organs that prevents response to stimuli. **6.** *Obs.* Stiffness : rigidity.

**rig·or·ism** (rĭg′ə-rĭz′əm) *n.* Severity or strictness in conduct, judgment, or practice. **—rig′or·ist** *n.* **—rig′or·is′tic** *adj.*

**rigor mor·tis** (môr′tĭs) *n.* [Lat., stiffness of death.] Muscular stiffening following death.

**rig·or·ous** (rĭg′ər-əs) *adj.* **1.** Marked by or acting with rigor. **2.** Full of rigors : HARSH. **3.** Extremely accurate : PRECISE. **—rig′or·ous·ly** *adv.* **—rig′or·ous·ness** *n.*

**rig·our** (rĭg′ər) *n. Chiefly Brit. var. of* RIGOR.

**Rig-Ve·da** (rĭg-vā′də, -vē′də) *n.* [Skt. *rgvedaḥ* : *rc,* verse, sacred text + *vedah,* veda.] The most ancient collection of Hindu sacred verses.

**Riks·mal** (rēks′mōl) *n.* [Norw. : *rik,* kingdom + *mål,* speech.] An official literary form of Norwegian based on written Danish.

**rile** (rīl) *vt.* **riled, ril·ing, riles.** [Alteration of ROIL.] **1.** To irritate : vex. **2.** To stir up (liquid) : ROIL.

**ril·ey** (rī′lē) *adj.* **1.** Riled : upset. **2.** Roiled : turbid.

**rill** *also* **rille** (rĭl) *n.* [LG *rille.*] **1.** A small brook : RIVULET. **2.** A long, narrow, straight depression on the moon's surface.

**rill·et** (rĭl′ĭt) *n.* A small rill.

**rim** (rĭm) *n.* [ME *rym* < OE *rima.*] **1.** The edge, border, or margin of an object. **2.** The circular outer part of a wheel, furthest from the axle. **3.** A circular metal structure around which a wheel tire is fitted. —*vt.* **rimmed, rim·ming, rims. 1.** To furnish with a rim. **2.** To roll around the rim of (e.g., a basket) without falling in. —Used of a ball in certain sports.

**rime**[1] (rīm) *n.* [ME *rim* < OE *hrīma.*] A frost or granular ice coating, as on grass and trees : HOARFROST. —*vt.* **rimed, rim·ing, rimes.** To cover with or as if with rime. **—rim′y** *adj.*

**rime**[2] (rīm) *n.* & *v. var. of* RHYME.

**rim·er** (rī′mər) *n. var. of* RHYMER.

**rime riche** (rēm rēsh′) *n., pl.* **rimes riches** (rēm rēsh′) [Fr. : *rime,* rhyme + *riche,* rich.] Perfect rhyme.

**rime·ster** (rīm'stər) *n. var. of* RHYMESTER.

**ri·mose** (rī'mōs', rī-mōs') *adj.* [Lat. *rimosus* < *rima,* fissure.] Full of cracks or crevices. **—ri'mose·ly** *adv.* **—ri·mos'i·ty** (-mŏs'ĭ-tē) *n.*

**rim·ple** (rĭm'pəl) *n.* [ME *rymple* < OE *hrympel.*] A fold : wrinkle. *—v.* **-pled, -pling, -ples.** *—vt.* To rumple : wrinkle. *—vi.* To form wrinkles or creases.

**rind** (rīnd) *n.* [ME < OE.] A tough outer covering, as bark, the skin of a fruit, or the coating on cheese or bacon.

**rin·der·pest** (rĭn'dər-pĕst') *n.* [G. : *Rinder,* cattle + *Pest,* plague < Lat. *pestis.*] An acute contagious virus disease, chiefly of cattle, marked by ulceration of the intestinal tract.

**rin·for·zan·do** (rēn'fôr-tsän'dō) *adj.* [Ital., pr.part. of *rinforzare,* to reinforce : *ri-,* again (< Lat. *re-*) + *inforzare,* to enforce (< OFr. *enforcier*). —see ENFORCE.] *Mus.* With a sudden increase of emphasis. —Used as a direction.

**ring[1]** (rĭng) *n.* [ME < OE *hring.*] **1.** A circular object, form, or arrangement with a vacant circular center. **2.** A small circular band, gen. made of precious metal and often set with jewels, worn on a finger. **3.** A circular band used for holding, carrying, or containing something <a wooden napkin *ring*> **4.** A circular movement or course, as in dancing. **5.** An enclosed, usu. circular area in which exhibitions, sports, or contests take place. **6. a.** A rectangular arena set off by stakes and ropes in which boxing or wrestling is held. **b.** The sport of boxing. **7. a.** An enclosed area in which bets are placed at a racetrack. **b.** Bookmakers as a group. **8.** An exclusive group of people acting privately or illegally to advance their own gain <a cocaine *ring*> **9.** A political contest : RACE. **10.** *Bot.* An annual ring. **11.** *Math.* The planar area between two concentric circles : ANNULUS. **12.** *Math.* An algebraic system consisting of a set with two binary operations in the set such that the set together with one operation, usu. denoted *addition,* is a commutative group, together with the second, usu. denoted *multiplication,* is a semigroup, and multiplication is distributive over addition. **13.** Any of the turns comprising a spiral or helix. **14.** *Chem.* A group of atoms chemically bound in a manner graphically representable as a circular form. *—v.* **ringed, ring·ing, rings.** *—vt.* **1.** To surround with a ring : ENCIRCLE. **2.** To form into a ring or rings. **3.** To supply or ornament with a ring or rings. **4.** To remove a circular strip of bark around the circumference of (a tree trunk or branch) : GIRDLE. **5.** To put a ring in the nose of (an animal). **6.** To hem in (animals) by riding in a circle around them. **7.** To toss a ring over (a peg), as in horseshoes. *—vi.* **1.** To form a ring or rings. **2.** To move, fly, or run in a spiral or circular course.

**ring[2]** (rĭng) *v.* **rang** (răng), **rung** (rŭng), **ring·ing, rings.** [ME *ringen* < OE *hringan.*] *—vi.* **1.** To give forth a clear resonant sound. **2.** To cause something to ring. **3.** To sound a bell in order to summon someone. **4.** To have a character suggestive of a particular quality <a tale that *rings* true> **5.** To be filled with sound : RESOUND <The room *rang* with merrymaking.> **6.** To hear a persistent humming or buzzing <ears *ringing* from the explosion> **7.** To be filled with rumor or talk <a town that *rang* with false rumors> *—vt.* **1.** To cause (e.g., a bell) to ring. **2.** To produce (a sound) by or as if by ringing. **3.** To announce, signal, or proclaim by or as if by ringing <The clock *rang* the hour.> **4.** To call (someone) on the telephone. **5.** To test (e.g., a coin) for quality by the sound it produces when struck against something. **—ring up.** To record, esp. by means of a cash register <*rang up* all cash sales> *—n.* **1.** The sound created by a sonorous vibrating object, as a struck bell. **2.** A loud sound, esp. one continued or repeated. **3.** A telephone call. **4.** A suggestion of a particular quality <a plan with a phoney *ring* to it> **5.** A set of bells. **6.** An act or instance of sounding a bell. **—ring a bell.** *Informal.* To arouse an often indistinct memory.

**ring-billed gull** (rĭng'bĭld') *n.* A North American gull, *Larus delawarensis,* with a black ring around its bill.

**ring·bolt** (rĭng'bōlt') *n.* A bolt having a ring fitted through an eye at its head.

**ring·bone** (rĭng'bōn') *n.* A bony growth on the fetlock, pastern, or coffin bone of a horse's foot, usu. causing lameness.

**ring·dove** (rĭng'dŭv') *n.* **1.** An Old World pigeon, *Streptopelia risoria,* having black markings forming a half circle on the neck. **2.** The wood pigeon.

**ringed** (rĭngd) *adj.* **1.** Wearing or marked with a ring or rings. **2.** Encircled by rings.

**rin·gent** (rĭn'jənt) *adj.* [Lat. *ringens, ringent-,* pr.part. of *ringi,* to open the mouth wide.] *Biol.* Having gaping liplike parts, as the corolla of some flowers or the shells of certain bivalves.

**ring·er[1]** (rĭng'ər) *n.* A horseshoe or quoit thrown so that it encircles the peg.

**ring·er[2]** (rĭng'ər) *n.* **1.** One that sounds a bell or chime. **2.** *Slang.* A contestant entered dishonestly into a competition. **3.** *Slang.* One bearing a striking resemblance to another.

**Ring·er's solution** (rĭng'ərz) *also* **Ring·er solution** (rĭng'ər) *n.* [After Sydney *Ringer* (1835–1910).] An aqueous solution of the chlorides of sodium, potassium, and calcium that is isotonic to animal tissue and is used topically as a physiological saline.

**ring finger** *n.* The third finger of the left hand.

**ring·hals** (rĭng'häls') *n.* [Afr. : *ring,* ring (< MDu. *rinc*) + *hals,* neck < MDu.] An African snake, *Haemachates haemachatus,* that spits venom at its victims.

**ring·lead·er** (rĭng'lē'dər) *n.* A leader of others, esp. in improper or illegal activities.

**ring·let** (rĭng'lĭt) *n.* **1.** A long spirally curled lock of hair. **2.** A small circle or ring. **—ring'let·ed** *adj.*

**ring·mas·ter** (rĭng'măs'tər) *n.* One in charge of the performances in a circus ring.

**Ring Nebula** *n.* A planetary nebula in the constellation Lyra.

**ring-necked pheasant** (rĭng'nĕkt') *n.* A widely distributed bird, *Phasianus colchicus* native to the Old World, of which the male has brightly colored plumage, a long pointed tail, and a white ring around the neck.

**ring·side** (rĭng'sīd') *n.* **1.** The area just outside an arena or ring, as at a prizefight. **2.** A place that provides a close view of an event.

**ring·tail** (rĭng'tāl') *n.* A ring-tailed animal, as the cacomistle.

**ring-tailed** (rĭng'tāld') *adj.* **1.** Having a tail with ringlike markings. **2.** Having a tail that curls to form a ring.

**ring·worm** (rĭng'wûrm') *n.* Any of a number of contagious skin diseases caused by several related fungi, characterized by ring-shaped, scaly, itching patches on the skin.

**rink** (rĭngk) *n.* [ME *renk,* racecourse < OFr. *renc,* line, of Germanic orig.] **1.** An area surfaced with smooth ice for hockey, skating, or curling. **2.** A smooth floor suited for roller-skating. **3.** A building that houses a surface prepared for skating. **4.** A section of a bowling green large enough for holding a match. **5.** A team of players in quoits, bowling, or curling.

**rin·ky-dink** (rĭng'kē-dĭngk') *adj.* [Orig. unkown.] *Slang.* **1.** Old-fashioned : outmoded. **2.** Not important : INSIGNIFICANT. **—rin'ky-dink'** *n.*

**rinse** (rĭns) *vt.* **rinsed, rins·ing, rins·es.** [ME *ryncen* < OFr. *rincer* < VLat. **recentiare* < Lat. *recens,* fresh.] **1.** To wash lightly with water. **2.** To remove (e.g., soap) by washing lightly in water. *—n.* **1.** An act of rinsing. **2.** Liquid used in rinsing. **3.** A cosmetic solution used in tinting or conditioning the hair. **—rins'a·ble, rins'i·ble** *adj.* **—rins'er** *n.*

**ri·ot** (rī'ət) *n.* [ME < OFr., dispute < *rioter,* to quarrel, poss. < *ruire,* to roar < Lat. *rugire.*] **1.** A raucous or violent disturbance created by a large number of people. **2.** *Law.* A violent disturbance of the public peace by three or more persons assembled for a common private purpose. **3.** An uncontrolled outbreak, as of laughter. **4.** A profusion, as of colors. **5. a.** Unrestrained revelry : MERRYMAKING. **b.** Debauchery. **6.** *Slang.* One that is irresistibly funny. *—v.* **-ot·ed, -ot·ing, -ots.** *—vi.* **1.** To take part in a riot. **2.** To live riotously or engage in unrestrained revelry. *—vt.* To waste (money or time) in wild or wanton living. **—run riot. 1.** To move or act with wild abandon. **2.** To grow luxuriantly. **—ri'ot·er** *n.*

**riot act** *n.* **1.** A forceful warning or reproach <read me the *riot act* for being late> **2. Riot Act.** An English law, enacted in 1715, providing that if 12 or more persons unlawfully assemble and disturb the public peace, they must disperse upon proclamation or be considered guilty of a felony.

**ri·ot·ous** (rī'ət-əs) *adj.* **1.** Of, relating to, or resembling a riot. **2.** Taking part in or inciting to riot or uproar. **3.** Unrestrained : disorderly. **4.** Dissolute : wanton. **5.** Profuse or luxuriant <a *riotous* growth of kudzu> **—ri'ot·ous·ly** *adv.* **—ri'ot·ous·ness** *n.*

**rip[1]** (rĭp) *v.* **ripped, rip·ping, rips.** [ME *rippen.*] *—vt.* **1.** To tear or cut apart harshly or energetically : SLASH. **2.** To remove by cutting or tearing roughly. **3.** To split or saw (wood) along the grain. **4.** *Informal.* To display, produce, or utter suddenly <*ripped* out a bitter epithet> *—vi.* **1.** To become split apart or torn. **2.** *Informal.* To move rapidly or violently. **—rip into.** To attack intensely : CENSURE <*ripped into* the candidate's record> **—rip off.** *Slang.* **1.** To steal from : ROB <*ripped off* a drugstore> **2.** To steal <*ripped off* a portable radio> **3.** To exploit, swindle or defraud <were *ripped off* by a door-to-door salesperson> *—n.* **1.** A split or torn place, esp. along a seam. **2.** The act of ripping. **3.** A ripsaw. **—rip'per** *n.*

**rip[2]** (rĭp) *n.* [Prob. < RIP[1].] **1.** A stretch of broken water in a river, estuary, or tidal channel. **2.** A rip current.

**rip[3]** (rĭp) *n.* [Poss. shortening and alteration of REPROBATE.] **1.** A dissolute person. **2.** An old or worthless horse.

**ri·par·i·an** (rĭ-pâr'ē-ən) *adj.* [Lat. *riparius* < *ripa,* bank.] Of, on, or relating to the bank of a natural course of water.

**riparian right** *n. Law.* The right, as to fishing or to the use of a riverbed, of one who owns riparian land.

**rip·cord** (rĭp'kôrd') *n.* **1.** A cord pulled to release the pack of a parachute. **2.** A cord pulled to release gas from a balloon.

**rip current** *n.* A current of water disturbed by an opposing current, esp. in tidal waters or by passage over an irregular bottom.

**ripe** (rīp) *adj.* **rip·er, rip·est.** [ME < OE *rīpe.*] **1.** Fully developed : MATURE. **2.** Resembling matured fruit, as in fullness. **3.** Sufficiently advanced in preparation or aging to be used <*ripe* cheddar> **4.** Thoroughly matured, as by study or experience : SEASONED <*ripe* understanding> **5.** Advanced in years <a *ripe* old age> **6.** Fully prepared to do or undergo something : READY. **7.** Sufficiently advanced :

OPPORTUNE <The time was *ripe* for revolution.> **—ripe'ly** *adv.* **—ripe'ness** *n.*

**rip·en** (rī'pən) *vt.* & *vi.* **-ened, -en·ing, -ens.** To make or become ripe : MATURE. **—rip'en·er** *n.*

**rip-off** (rĭp'ôf', -ŏf') *n. Slang.* **1.** A theft. **2.** A thief. **3.** Exploitation. **4.** Something, as a story or film, that is clearly imitative of or based on something else.

**ri·poste** (rĭ-pōst') *n.* [Fr. < Ital. *risposta,* answer, fem. p.part. of *rispondere,* to answer < Lat. *respondēre* : *re-,* in return + *spondēre,* to promise.] **1.** A quick thrust given after parrying an opponent's lunge in fencing. **2.** A retaliatory maneuver, action, or retort. *—vi.* **-post·ed, -post·ing, -postes. 1.** To make a return thrust. **2.** To retort quickly.

**rip·ping** (rĭp'ĭng) *adj.* [Prob. < RIP[1].] *Informal.* Excellent : wonderful <had a *ripping* time at the fair>

**rip·ple**[1] (rĭp'əl) *v.* **-pled, -pling, -ples.** [Poss. freq. of RIP[1].] *—vi.* **1. a.** To form or display little undulations or waves on a surface. **b.** To flow with such undulations or waves on a surface. **2.** To rise and fall gently in volume or tone. *—vt.* To cause to form small waves or undulations. *—n.* **1.** A small wave. **2.** A wavelike motion. **3.** A sound like rippling water <a soft *ripple* of laughter> **—rip'pler** *n.* **—rip'pling·ly** *adv.*

**rip·ple**[2] (rĭp'əl) *n.* [ME *ripelen,* to remove seeds.] A comblike toothed instrument for removing seeds from fibers, as flax. *—vt.* **-pled, -pling, -ples.** To remove seeds from with a ripple.

**ripple effect** *n.* A gradually spreading effect or influence <"warned of a far-reaching *ripple effect* inflating the economy" — *Newsweek*>

**rip·ply** (rĭp'lē) *adj.* **-pli·er, -pli·est.** Marked by or sounding in ripples.

**rip·rap** (rĭp'răp') *n.* [Redup. of RAP[1].] **1.** A loose assemblage of broken stones erected in water or on soft ground as a foundation. **2.** The broken stones used for a riprap. *—vt.* **-rapped, -rap·ping, -raps. 1.** To construct a riprap in or on. **2.** To strengthen with a riprap.

**rip-roar·ing** (rĭp'rôr'ĭng, -rōr'ĭng) *also* **rip-roar·i·ous** (rĭp'rôr'ē-əs, -rōr'ē-əs) *adj.* [RIP[1] + (UP)ROAR(IOUS) + -ING.] Noisy, lively, and exciting.

**rip·saw** (rĭp'sô') *n.* A coarse-toothed saw for cutting wood along the grain.

**rip·snort·er** (rĭp'snôr'tər) *n. Slang.* One extraordinary for intensity, strength, or excellence. **—rip'snort'ing** *adj.*

**rip tide** *n.* A rip current.

**Rip·u·ar·i·an** (rĭp'yo͞o-âr'ē-ən) *adj.* [Med. Lat. *Ripuarius.*] Of or designating a group of 4th-cent. Franks who lived along the Rhine near Cologne. **—Rip'u·ar'i·an** *n.*

**rise** (rīz) *v.* **rose** (rōz), **ris·en** (rĭz'ən), **ris·ing, ris·es.** [ME *risen* < OE *rīsan.*] *—vi.* **1.** To assume a standing position after lying, sitting, or kneeling. **2.** To get out of bed. ***usage:*** Either *rise* or *arise* is correct in this sense. **3.** To move from a lower to a higher position : ASCEND <Warm air *rises.*> **4.** To increase in volume, size, or level <The pond *rises* every spring.> **5.** To increase in amount, number, or value. **6.** To increase in intensity, force, or speed <The wind started to *rise.*> **7.** To increase in volume or pitch. **8.** To appear above the horizon <The sun *rises* earlier in the summer.> **9.** To extend upward <The lone tree *rose* above the hill.> **10.** To slope or slant upward <Mount Everest *rises* to 29,028 feet.> **11.** To come into existence <A hurricane *rose* in the south.> **12.** To be erected <new apartments *rising* in the city> **13.** To appear at the surface of the water or the earth : EMERGE. **14.** To puff up or become larger : SWELL <dough *rising*> **15.** To become erect and stiff. **16.** To attain a higher status <a recruit *rising* through the ranks> **17.** To become apparent to the mind or senses <Doubt *rose* to haunt me.> **18.** To exert oneself to meet a challenge or demand. **19.** To return to life. **20.** To rebel. **21.** To close a session of an official assembly : ADJOURN. *—vt.* **1.** To cause to rise. **2.** To cause (a distant object at sea) to become visible above the horizon by advancing closer. *—n.* **1.** An act of rising : ASCENT. **2.** Degree of elevation or ascent. **3.** The appearance of a celestial body, as the sun, above the horizon. **4.** An increase in height, as of water level. **5.** A gently sloped hill. **6.** An origin, source, or beginning <followed the river to its *rise*> **7.** Occasion or opportunity. **8.** Emergence of a fish seeking food or bait at the water's surface. **9.** An increase in worth, price, number, or degree. **10.** Increase in intensity, volume, or pitch. **11.** Elevation in social rank, fortune, or importance. **12.** The height of a flight of stairs or of a single riser. **13.** *Chiefly Brit.* An increase in wages : RAISE. **14.** *Informal.* An angry reaction <got an unexpected *rise* out of them>

☆ ***syns:*** RISE, ASCEND, CLIMB, MOUNT, SOAR *v. core meaning* : to move upward <hot air that *rises*> RISE is applied to a great range of events, chiefly involving steady or customary upward movement <the sun *rising* over the eastern horizon><prices that *rise* and fall> ASCEND means rising step by step, literally or figuratively <*ascend* a staircase><*ascend* through the ranks> CLIMB suggests steady progress against gravity or other resistance <a rocket *climbing* rapidly><*climbed* to the top of their profession> MOUNT often implies reaching a level or limit <a death toll that *mounted*><*mounting* to the top of the hill> SOAR suggests the effortless attainment of great height <eagles *soaring* in the sky>; often it refers to what rises rapidly and suddenly, esp. above what is normal <the *soaring* cost of living> ***ant:*** fall

**ris·er** (rī'zər) *n.* **1.** A person who rises, esp. from sleep <an early *riser*> **2.** The vertical part of a step in a stairway.

**ris·i·bil·i·ty** (rĭz'ə-bĭl'ĭ-tē) *n., pl.* **-ties. 1.** Tendency or ability to laugh. **2. risibilities.** A sense of the ludicrous. **3.** Laughter.

**ris·i·ble** (rĭz'ə-bəl) *adj.* [LLat. *risibilis* < Lat. *ridēre,* to laugh.] **1.** Capable of laughing or inclined to laugh. **2.** Relating to laughter. **3.** Apt to excite laughter : LUDICROUS. **—ris'i·bly** *adv.*

**ris·ing** (rī'zĭng) *adj.* **1.** Ascending, sloping upward, or advancing. **2.** Coming to maturity : EMERGING. *—n.* **1.** The act of one that rises. **2.** An insurrection : uprising. **3.** A prominence : projection. **4.** The leaven or yeast used to make dough rise in baking.

**risk** (rĭsk) *n.* [Fr. *risque* < Ital. *risco.*] **1.** Possibility of suffering harm or loss : DANGER. **2.** A factor, course, or element involving uncertain danger : HAZARD. **3. a.** The danger or probability of loss to an insurer. **b.** The amount that an insurance company stands to lose. **c.** One considered with respect to the possibility of loss to an insurer <a good *risk*> *—vt.* **risked, risk·ing, risks. 1.** To expose to a chance of loss or damage : ENDANGER. **2.** To incur the risk of <Their comments *risked* a bitter reply.> **—risk'er** *n.*

**risk·y** (rĭs'kē) *adj.* **-i·er, -i·est.** Accompanied by or involving risk or hazard : DANGEROUS. **—risk'i·ness** *n.*

**Ri·sor·gi·men·to** (rē-sôr'jē-mĕn'tō) *n.* [Ital., resurrection < *risorgere,* to rise again < Lat. *resurgere* : *re-,* again + *surgere,* to rise (*sub-,* up from under + *regere,* to direct).] The period of or the movement for Italian liberation and political unification, beginning about 1750 and lasting until 1870.

**ri·sot·to** (rē-sôt'ō, rĭ-sŏt'ō) *n.* [Ital.] Rice cooked in broth with grated cheese and seasonings.

**ris·qué** (rĭs-kā') *adj.* [Fr. < *risquer,* to risk < *risque,* risk.] Suggestive of or bordering on impropriety or indelicacy : RACY.

**ris·sole** (rĭ'sŏl, rē-sôl') *n.* [Fr. < OFr. *ruissole* < VLat. **russeola* (pasta), reddish paste < fem. of LLat. *russeolus,* reddish < Lat. *russus,* red.] A small pastry-enclosed croquette with a minced meat or fish filling, usu. fried in deep fat.

**ris·so·lé** (rē-sô-lā') *adj.* [Fr. < *rissoler,* to brown < *rissole,* rissole.] Browned by frying.

**ri·tar·dan·do** (rē'tär-dän'dō) *adj.* & *adv.* [Ital., pr.part. of *ritardare,* to slow down < Lat. *retardare,* to retard. —see RETARD.] *Mus.* Rallentando. —Used as a direction.

**rite** (rīt) *n.* [ME < Lat. *ritus.* ] **1.** The customary or prescribed form for conducting esp. a religious ceremony <the *rite* of marriage> **2.** A ceremonial act or series of acts <ancient fertility *rites*> **3. Rite.** A branch of the Christian church distinguished by its own liturgy <Eastern *Rite*>

**rite of passage** *n.* A significant event in a persons's life that indicates a transition from one stage to another, as from adolescence to adulthood, and that may be marked by a ceremony or ritual.

**ri·tor·nel·lo** (rē'tôr-nĕl'lō) *n., pl.* **-li** (-lē) or **-los.** [Ital., dim. of *ritorno,* return < *ritornare,* to return.] *Mus.* **1.** An instrumental interlude recurring after each stanza in a vocal work. **2.** A passage for full orchestra in a baroque concerto grosso. **3.** An instrumental interlude in early 17th-cent. opera. **4.** The refrain of a rondo.

**rit·ter** (rĭt'ər) *n., pl.* **ritter.** [G. < MHG *riter* < MDu. *ridder.*] A knight.

**rit·u·al** (rĭch'o͞o-əl) *n.* [Lat. *ritualis,* of rites < *ritus,* rite.] **1.** The prescribed form or order of conducting a religious or solemn ceremony. **2.** A body of ceremonies or rites, as those used in a church or fraternal organization. **3.** A book of rites or ceremonial forms. **4. rituals. a.** A ceremonial act or a series of such acts. **b.** Performance of such acts. **5.** A detailed method of procedure faithfully or consistently followed <the *ritual* of proofreading> **—rit'u·al·is'tic** (-ə-lĭs'tĭk) *adj.* **—rit'u·al·is'ti·cal·ly** *adv.* **—rit'u·al·ly** *adv.*

**rit·u·al·ism** (rĭch'o͞o-ə-lĭz'əm) *n.* **1.** The practice or observance of religious ritual. **2.** Insistence on or adherence to ritual.

**rit·u·al·ist** (rĭch'o͞o-ə-lĭst) *n.* **1.** An authority on or student of ritual. **2.** A practitioner or advocate of ritual.

**rit·u·al·ize** (rĭch'o͞o-ə-līz') *v.* **-ized, -iz·ing, -iz·es.** *—vi.* To engage in ritualism. *—vt.* **1.** To make a ritual of. **2.** To force a ritual on. **—rit'u·al·i·za'tion** *n.*

**ritz·y** (rĭt'sē) *adj.* **-i·er, -i·est.** [After the *Ritz* hotels.] *Slang.* Elegant : swank.

▲ word history: The word *ritzy* comes from the name of various celebrated hotels called *Ritz,* such as the *Hotel Ritz* in Paris and the *Ritz Hotel* in London, which have a reputation for opulent elegance. They were founded by César Ritz, a Swiss hotelier who lived from 1850 to 1918.

**riv·age** (rĭv'ĭj) *n.* [ME < OFr. < *rive,* bank < Lat. *ripa.*] *Archaic.* A coast, shore, or bank.

**ri·val** (rī'vəl) *n.* [OFr. < Lat. *rivalis,* a rival, one using the same stream as another < *rivus,* stream.] **1.** One who tries to surpass or equal another or who pursues the same object as another : COMPETITOR. **2.** One that equals or almost equals another in a specific respect. **3.** *Obs.* A companion or associate in a particular duty. *—v.*

**-valed, -val·ing, -vals** *also* **-valled, -val·ling, -vals.** —*vt.* **1.** To attempt to equal or surpass. **2.** To be the equal of : MATCH. —*vi.* To be a competitor : COMPETE. **—ri'val·ry** *n.*

**rive** (rīv) *v.* **rived, rived** *or* **riv·en** (rĭv'ən), **riv·ing, rives.** [ME *riven* < ON *rīfa.*] —*vt.* **1.** To tear apart or render. **2.** To break into pieces, as by a blow : CLEAVE. **3.** To break or distress (e.g., the spirit). —*vi.* To be or become split.

**riv·er** (rĭv'ər) *n.* [ME < OFr. *rivere* < VLat. **riparia* < fem. of Lat. *riparius,* of a bank < *ripa,* bank.] **1.** A large natural stream of water emptying into a large body of water, as an ocean or lake, and usu. fed along its course by converging tributaries. **2.** A stream or abundant flow resembling a river <a *river* of mud> **—up the river.** *Slang.* In or to prison.

▲ word history: The Latin source of English *river, ripa,* meant "riverbank" and did not denote the water flowing between the banks. The shift in meaning took place in Old French, which was an intermediate stage between the Latin and English words: Old French *rivere* meant both "river" and "riverbank." The Italian descendent of *ripa, riviera,* means "shore" and is used as the name of the Mediterranean coast of France and Italy.

**riv·er·bank** (rĭv'ər-băngk') *n.* The bank of a river.

**river basin** *n.* The land area drained by a river and its tributaries.

**riv·er·bed** (rĭv'ər-bĕd') *n.* The area between the banks of a river covered or once covered by water.

**riv·er·boat** (rĭv'ər-bōt') *n.* A boat suitable for use on a river.

**riv·er·head** (rĭv'ər-hĕd') *n.* The source of a river.

**river horse** *n.* HIPPOPOTAMUS 1.

**riv·er·ine** (rĭv'ə-rīn', -rēn') *adj.* **1.** Relating to or resembling a river. **2.** Located on or inhabiting the banks of a river : RIPARIAN.

**riv·er·side** (rĭv'ər-sīd') *n.* The bank or side of a river.

**riv·er·ward** (rĭv'ər-wərd) *also* **riv·er·wards** (-wərdz) *adv.* In a direction towards a river.

**riv·er·weed** (rĭv'ər-wēd') *n.* A North American plant, *Podostemum ceratophyllum,* resembling seaweed and growing on rocks in rapidly flowing streams.

**riv·et** (rĭv'ĭt) *n.* [ME *ryvette* < OFr. *river,* to attach.] A metal bolt or pin having a head on one end, used to fasten metal plates or other objects together by inserting the shank through a hole in each piece and hammering down the plain end so as to form a new head. —*vt.* **-et·ed, -et·ing, -ets. 1.** To fasten with or as if with a rivet. **2.** To hammer the headless end of so as to form a head and fasten something. **3.** To secure firmly : FIX. **4.** To engross or hold (e.g., the attention). **—riv'et·er** *n.*

**ri·vière** (rē-vyâr') *n.* [Fr., river < OFr. *rivere.* —see RIVER.] A necklace of precious stones, as diamonds, gen. in one strand.

**riv·u·let** (rĭv'yə-lĭt) *n.* [Ital. *rivoletto,* dim. of *rivolo,* small stream < Lat. *rivulus,* dim. of *rivus,* stream.] A small stream.

**ri·yal** *also* **ri·al** (rē-ôl', -äl') *n.* [Ar. *riyāl.*] —See table at CURRENCY.

**ri·yal-o·man·i** (rē-ôl'ō-mä'nē, rē-äl'-) *n.* [Ar. *riyāl 'umānīy,* riyal of Oman.] —See table at CURRENCY.

**RK galaxy** (är'kā') *n.* A ring galaxy composed of a ring structure with a large bright knot of incandescent material on the ring itself.

**Rn** *symbol for* RADON.

**RNA** (är'ĕn-ā') *n.* Ribonucleic acid.

**RN·ase** (är'ĕn-ās') *also* **RNA·ase** (är'ĕn-ā'ās', -āz') *n.* Ribonuclease.

**roach¹** (rōch) *n., pl.* **roach** *or* **roach·es.** [ME *roche* < OFr.] **1.** A northern European freshwater fish, *Rutilus rutilus.* **2.** A fish similar or related to the roach, as one of the North American sunfishes.

**roach²** (rōch) *n.* The cockroach.

**roach³** (rōch) *n.* [Orig. unknown.] **1.** A roll of hair brushed up from the forehead or temple. **2.** *Naut.* A cut on the edge of a sail to prevent chafing. **3.** *Slang.* The butt of a marijuana cigarette. —*vt.* **roached, roach·ing, roach·es. 1.** To brush (hair) in a roach. **2.** To shave (the mane of a horse) to a short bristle.

**roach clip** *n.* A device for holding the butt of a marijuana cigarette.

**road** (rōd) *n.* [ME *rood,* act of riding < OE *rād.*] **1. a.** An open, gen. public way for the passage of persons, vehicles, and animals. **b.** The surface of a road : ROADBED. **2.** A course or path. **3.** A railroad. **4.** *often* **roads.** A roadstead. **—on the road. 1.** On tour <a rock band *on the road*> **2.** Traveling <sales representatives *on the road*> **3.** Wandering, as a vagabond.

**road agent** *n.* A robber of stagecoaches.

**road·bed** (rōd'bĕd') *n.* **1. a.** The foundation on which the ties, rails, and ballast of a railroad are laid. **b.** A layer of ballast directly under the ties. **2.** The surface and foundation of a road.

**road·block** (rōd'blŏk') *n.* **1. a.** A barricade or obstruction across a road set up by the military to prevent the passage of enemy troops. **b.** A road barricade or checkpoint set up by the police. **2.** An obstruction in a road, as rocks or a fallen tree. **3.** A situation or condition preventing further progress toward a goal.

**road hog** *n.* A driver whose vehicle crowds another's traffic lane.

**road·house** (rōd'hous') *n.* An inn, nightclub, or restaurant located on a road outside a city or town.

**road metal** *n.* Crushed or broken stone or cinders used in the construction and repair of roads and roadbeds.

**road·run·ner** (rōd'rŭn'ər) *n.* A swift-running crested bird, *Geococcyx californianus* of southwestern North America, having streaked brownish plumage and a long tail.

**road show** *n.* **1.** A show presented by a touring troupe of theatrical performers. **2.** A new motion picture shown at selected theaters usu. for higher ticket prices.

**road·side** (rōd'sīd') *n.* The side of a road.

**road·stead** (rōd'stĕd') *n.* *Naut.* A sheltered offshore anchorage area for ships.

**road·ster** (rōd'stər) *n.* **1.** An open car having a single seat in the front for two or three people and a rumble seat or luggage compartment in the back. **2.** A horse for riding on a road.

**road test** *n.* **1.** A test of a motor vehicle's performance capability under actual road conditions. **2.** A test of driving proficiency on the road required of a candidate for a driver's license.

**road·way** (rōd'wā') *n.* A road, esp. the part over which vehicles travel.

**road·work** (rōd'wûrk') *n.* **1.** Road construction or repairs. **2.** Outdoor long-distance running as a form of physical exercise or conditioning.

**roam** (rōm) *v.* **roamed, roam·ing, roams.** [ME *romen.*] —*vi.* To move or travel aimlessly : WANDER. —*vt.* To wander over or through <*roamed* the streets and alleys> **—roam** *n.* **—roam'er** *n.*

**roan** (rōn) *adj.* [OFr. < OSp. *roano.*] Having a bay, chestnut, or sorrel coat thickly sprinkled with white or gray <a *roan* colt> —*n.* **1.** The coloring of a roan horse. **2.** A roan animal. **3.** A soft flexible sheepskin leather, often treated to resemble morocco and used in bookbinding.

**roar** (rôr) *v.* **roared, roar·ing, roars.** [ME *roren* < OE *rārian.*] —*vi.* **1.** To utter a deep loud prolonged sound, esp. in rage or excitement. **2.** To laugh loudly. **3.** To produce or make a loud noise <motors *roaring*> **4.** To be disorderly or rowdy. **5.** To breathe with a rasping sound. —Used of a horse. —*vt.* **1.** To utter with a deep loud prolonged sound. **2.** To cause to roar <stopped at the traffic light and impatiently *roared* the engine> —*n.* **1.** A deep loud prolonged sound or cry, as of a person in distress or rage. **2.** The deep loud cry of a wild animal. **3.** A loud prolonged noise <the *roar* of breaking> **4.** A loud burst of laughter. **—roar'er** *n.*

**roar·ing** (rôr'ĭng) *adj.* Briskly successful : THRIVING <a *roaring* new business> —*adv.* Extremely : very <got *roaring* drunk>

**roast** (rōst) *v.* **roast·ed, roast·ing, roasts.** [ME *rosten* < OFr. *rostir,* of Germanic orig.] —*vt.* **1.** To cook with dry heat, as in an oven or near hot coals. **2.** To brown, dry, or parch by exposing to heat. **3.** To expose to intense or excessive heat. **4.** *Metallurgy.* To heat (ore) in a furnace in order to purify, dehydrate, or oxidize. **5.** *Informal.* To criticize harshly : RIDICULE. —*vi.* **1.** To cook food, as meat, in an oven. **2.** To undergo roasting. —*n.* **1.** Something roasted. **2.** A cut of meat suitable or prepared for roasting. **3.** The act or process of roasting or the state of being roasted. **4.** Severe ridicule or criticism.

**roast·er** (rō'stər) *n.* **1.** One that roasts. **2.** A pan or apparatus for roasting. **3.** Something, esp. a young chicken, fit for roasting.

**rob** (rŏb) *v.* **robbed, rob·bing, robs.** [ME *robben* < OFr. *rober,* of Germanic orig.] —*vt.* **1.** To take property from (a person or persons) unlawfully by intimidation or force : commit robbery on. **2.** To take valuable or desired articles illegally from <*robbed* a gas station> **3. a.** To deprive (a person) unjustly of something desired, due, or anticipated <*robbed* me of my good name> **b.** To deprive of something injuriously <a disease that *robs* a tree of its nutrients> **4.** To steal. —*vi.* To commit or engage in robbery. **—rob'ber** *n.*

☆ **syns:** ROB, HEIST, HIT, HOLD UP, KNOCK OFF, RIP OFF, STICK UP *v. core meaning* : to take property or possessions from (another) unlawfully and forcibly <*robbed* a liquor store><*robbing* pedestrians>

**ro·ba·lo** (rō-bä'lō) *n., pl.* **-los** *or* **robalo.** [Sp. *róbalo.*] A chiefly tropical marine food fish, as the snook, belonging to the family Centropomidae.

**robber baron** *n.* **1.** An American industrial or financial magnate of the latter 19th cent. who became wealthy by unethical means, such as questionable stock-market operations or exploitation of labor or political connections. **2.** A feudal lord who robbed travelers passing through his domain.

**robber fly** *n.* Any of various predatory flies of the family Asilidae, having long bristly legs.

**rob·ber·y** (rŏb'ə-rē) *n., pl.* **-ies.** An act or instance of illegally taking another's property by the use of intimidation or violent force.

**robe** (rōb) *n.* [ME < OFr., of Germanic orig.] **1.** A long, loose, flowing outer garment, esp.: **a.** An official garment worn on formal occasions to indicate rank or office, as by a judge or high church official. **b.** An academic gown. **c.** A dressing gown or bathrobe. **2. robes.** Clothes in general : APPAREL. **3.** A blanket or covering. —*vt.* & *vi.* **robed, rob·ing, robes.** To dress in or don a robe.

**rob·in** (rŏb'ĭn) *n.* [From the name *Robin.*] **1.** A North American songbird, *Turdus migratorius,* having a rust-red breast and gray and black upper plumage. **2.** A small Old World bird, *Erithacus rubecula,* having an orange breast and a brown back. **3.** A bird resembling a robin.

**Robin Good·fel·low** (go͝od′fĕl′ō) *n.* Puck.

**Robin Hood** (ho͝od) *n.* A legendary 12th-cent. English outlaw famous for chivalry, courage, and the practice of robbing the wealthy to aid the poor.

**rob·in's-egg blue** (rŏb′ĭnz-ĕg′) *n.* A pale to green to light bluish green to greenish or grayish blue.

**Rob·in·son Cru·soe** (rŏb′ĭn-sən kro͞o′sō) *n.* A shipwrecked English sailor who lived for years on a small tropical island, the hero of Daniel Defoe's novel *Robinson Crusoe* (1719).

**Robin's plantain** *n.* A plant, *Erigeron pulchellus* of eastern North America, having many-rayed purplish flowers.

**ro·ble** (rō′blā) *n.* [Sp., oak < Lat. *robur.*] **1.** An oak, *Quercus lobata* of California, having leathery leaves and slender, pointed acorns. **2.** Any of various trees that are similar or related to the roble.

**ro·bot** (rō′bət, -bŏt′) *n.* [Czech < *robota,* work.] **1.** A mechanical apparatus that resembles a human being and is capable of performing human tasks or behaving in a human manner. **2.** A person who works mechanically without original thought. **3.** A machine or device that works automatically or by remote control. **—ro′bot·is′tic** (-bə-tĭs′tĭk) *adj.*

▲ word history: The word *robot* was invented by the Czech author Karel Čapek in his play *R.U.R.*, which stands for "Rossum's Universal Robots." The play was written in 1920 and was translated into English a few years later. The word *robot* quickly gained currency in English. Čapek derived the word from Czech *robota,* "forced labor, drudgery."

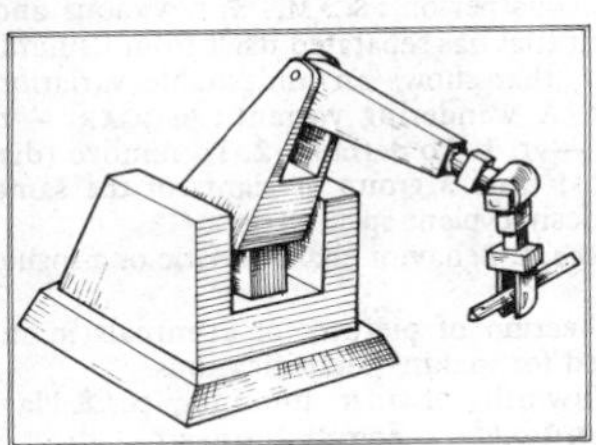

**robot**

**robot bomb** *n.* A small, explosive-carrying, jet-propelled gyroscopically guided winged missile.

**ro·bot·ics** (rō-bŏt′ĭks) *n.* (*sing. in number*). Study and application of the technology of robots.

**robot pilot** *n.* An automatic pilot.

**rob roy** (rŏb roi′) *n.* [< *Rob Roy,* nickname of Robert Macgregor (1671–1734).] A mixed drink made with Scotch whisky, sweet vermouth, and bitters.

**ro·bust** (rō-bŭst′, rō′bŭst′) *adj.* [Lat. *robustus* < *robur,* strength, oak.] **1.** Full of health and strength : VIGOROUS. **2.** Powerfully built : STURDY. **3.** Requiring or suited to physical strength or endurance <*robust* manual labor> **4.** Boisterous : rough. **5.** Marked by richness and fullness : FULL-BODIED <a *robust* wine> **—ro·bust′ly** *adv.* **—ro·bust′ness** *n.*

**roc** (rŏk) *n.* [Ar. *rukkh.*] An enormous, powerful, legendary bird of prey.

**roc·am·bole** (rŏk′əm-bōl′) *n.* [Fr. < G. *Rockenbolle* : *Rocken,* distaff (< OHG *rocko*) + *Bolle,* bulb < OHG *bolla,* ball.] **1.** A European plant, *Allium scorodoprasum,* with a garliclike bulb. **2.** The bulb of the rocambole, used as a seasoning.

**Ro·chelle powder** (rō-shĕl′) *n.* Seidlitz powder.

**Rochelle salt** *n.* [After La *Rochelle,* France.] A colorless efflorescent crystalline compound, $KNaC_4H_4O_6{\cdot}4H_2O$, used in making mirrors, in electronics, and as a laxative.

**roche mou·ton·née** (rôsh′ mo͞ot′n-ā′) *n., pl.* **roches mou·ton·nées** (rôsh′ mo͞ot′n-ā′, -āz′) [Fr., fleecy rock.] A bedrock outcrop worn smooth by glacial abrasion.

**roch·et** (rŏch′ĭt) *n.* [ME < OFr., of Germanic orig.] A ceremonial linen vestment worn by church dignitaries, as bishops.

**rock¹** (rŏk) *n.* [ME < ONFr. *roque.*] **1.** A relatively hard naturally formed mass of mineral or petrified matter : STONE. **2. a.** A relatively small fragment or piece of rock. **b.** A relatively large body of such material, as a cliff or peak. **3.** *Geol.* A naturally formed mineral mass or aggregate that constitutes a significant part of the earth's crust. **4.** One that is stable, firm, or dependable. **5. rocks.** *Slang.* Money. **6.** *Slang.* A large gem, esp. a diamond. **7. a.** A varicolored stick candy. **b.** Rock candy. **—on the rocks. 1.** In a state of destruction or ruin <The business was *on the rocks.*> **2.** Without money : BANKRUPT. **3.** Served over ice cubes without water or mix <bourbon *on the rocks*>

**rock²** (rŏk) *v.* **rocked, rock·ing, rocks.** [ME *rokken* < OE *roccian.*] *—vi.* **1.** To move back and forth or from side to side, esp. gently or rhythmically. **2.** To sway severely, as from a shock or blow : SHAKE. **3.** To be washed and panned in a cradle or in a rocker. —Used of ores. *—vt.* **1.** To sway back and forth or from side to side, esp. so as to soothe or lull to sleep. **2.** To cause to sway or shake violently. **3.** To disturb the emotions or mind of : UPSET <The heinous crime *rocked* the town.> **4.** To wash or pan (ore) in a cradle or rocker. **5.** To roughen (a metal plate) with various rockers and roulettes in mezzotint engraving. *—n.* **1.** An act of rocking. **2.** A rocking motion. **3.** Rock 'n' roll. **—rock the boat.** *Slang.* To disturb the balance of a situation. **—rock′ing·ly** *adv.*

**rock·a·bil·ly** (rŏk′ə-bĭl′ē) *n.* [ROCK ('N' ROLL) + (HILL)BILLY.] Popular music combining rock 'n' roll and country and western styles.

**rock·a·by** *also* **rock·a·bye** *or* **rock-a-bye** (rŏk′ə-bī′) [ROCK + (LULL)ABY.] *interj.* —Used to lull an infant to sleep.

**rock-and-roll** (rŏk′ən-rōl′) *n. var. of* ROCK 'N' ROLL.

**rock and rye** *n.* A rye whiskey bottled commercially with rock candy and fruit.

**rock·a·way** (rŏk′ə-wā′) *n.* [After *Rockaway,* a town in New Jersey where it was first made.] A four-wheeled carriage with two seats and a standing top.

**rock bass** *n.* **1.** A freshwater food fish, *Ambloplites rupestris* of eastern and central North America. **2.** A fish similar or related to the rock bass.

**rock bottom** *n.* The absolute bottom : NADIR. **—rock′-bot′tom** (rŏk′bŏt′əm) *adj.*

**rock-bound** (rŏk′bound′) *adj.* Hemmed in by or bordered with rocks.

**rock candy** *n.* Sugar in the form of large hard clear crystals.

**Rock Cornish hen** *n.* A small fowl of a breed developed by crossing white Plymouth Rock and Cornish strains, used esp. as a roasting chicken.

**rock crystal** *n.* Transparent quartz, esp. when colorless.

**rock dove** *n.* A bird, *Columba livia* native to Europe but widely distributed elsewhere, having variously colored plumage with iridescent neck markings.

**rock·er** (rŏk′ər) *n.* **1.** A person who rocks something, as a cradle. **2.** A rocking chair. **3.** A rocking horse. **4.** One of the two curved pieces on which a cradle, rocking chair, or a similar device rocks. **5.** A cradle for washing or panning ores. **6.** A small steel plate with a toothed, curved edge used to roughen a copper plate for a mezzotint. **7.** An ice skate with a curved blade. **8.** A curved stripe at the bottom part of a chevron worn by some noncommissioned officers. **9. a.** A rock song, singer, or musician. **b.** A follower of rock music. **—off (one's) rocker.** *Slang.* Not in one's right mind : CRAZY.

**rocker arm** *n.* A pivoted lever, as in a car engine, used to transfer cam or pushrod motion to a valve stem.

**rocker cam** *n.* A cam on a rockshaft.

**rock·er·y** (rŏk′ə-rē) *n., pl.* **-ies.** A rock garden.

**rock·et¹** (rŏk′ĭt) *n.* [Ital. *rocchetta,* dim. of *rocca,* distaff, of Germanic orig.] **1. a.** A device propelled by ejection of matter, esp. by the high-velocity ejection of the gaseous combustion products produced by internal ignition of solid or liquid fuels. **b.** An engine that propels in this manner. **2. a.** A rocket-powered weapon equipped with a warhead. **b.** An obsolete incendiary weapon with a rounded hollow warhead filled with explosives fired from a ship. **3.** A firework for aerial display. *—v.* **-et·ed, -et·ing, -ets.** *—vi.* **1.** To move rapidly, as a rocket does. **2.** To fly swiftly straight up, as a frightened game bird. **3.** To rise swiftly or unexpectedly : SKYROCKET. *—vt.* **1.** To assault with rockets. **2.** To carry by means of a rocket.

**rock·et²** (rŏk′ĭt) *n.* [OFr. *roquette* < Ital. *ruchetta,* dim. of *ruca* < Lat. *eruca,* a kind of cabbage.] **1.** A plant, *Eruca sativa* native to Eurasia, having yellowish-white flowers and leaves occas. used in salads. **2.** A plant, esp. one of the genus *Hesperis,* related to the rocket.

**rock·et·eer** (rŏk′ĭ-tîr′) *n.* A designer, student, launch crew member, or pilot of a propulsion rocket.

**rocket engine** *n.* A propulsion engine that functions by means of rockets, esp. one on a spacecraft or aircraft.

**rocket motor** *n.* A rocket engine, esp. one using solid propellants.

**rock·et·ry** (rŏk′ĭ-trē) *n.* The science and technology of rocket design, construction, and flight.

**rock·et·sonde** (rŏk′ĭt-sŏnd′) *n.* [ROCKET + Fr. *sonde,* sounding line < OFr., prob. of Germanic orig.] A rocket used for upper atmospheric observation.

**rock·fish** (rŏk′fĭsh′) *n., pl.* **rockfish** *or* **-fish·es. 1.** Any of various fishes living among rocks. **2.** Any of various fishes, chiefly of the genus *Sebastodes,* of Pacific waters. **3.** The striped bass.

**rock flour** *n.* Pulverized rock produced by glacial abrasion.

**rock garden** *n.* A rocky area in which plants particularly adapted to such terrain are cultivated.

**rock hound** *n.* **1.** A specialist in geology. **2.** A rock and mineral collector.

**rocking chair** *n.* A chair mounted on rockers or springs.

**rocking horse** *n.* A toy horse large enough for a child to ride, mounted on rockers or springs.

**rock·ling** (rŏk′lĭng) *n., pl.* **rockling** or **-lings.** A small marine fish of the family Gadidae, of North Atlantic coastal waters.

**rock lobster** *n.* The spiny lobster.

**rock maple** *n.* **1.** The sugar maple. **2.** The wood of the sugar maple.

**rock 'n' roll** (rŏk′ ən rōl′) *also* **rock-and-roll** (rŏk′ən-rōl′) *n.* Popular music combining elements of rhythm and blues with country and western music and having a heavily accented beat.

**rock oil** *n. Chiefly Brit.* Petroleum.

**rock·oon** (rŏ-ko͞on′) *n.* [ROCK(ET) + (BALL)OON.] A device used for high-altitude research sounding, composed of a small solid-propellant rocket carried aloft by a balloon.

**rock-ribbed** (rŏk′rĭbd′) *adj.* **1.** Having rocks or rock outcroppings. **2.** Not yielding : STERN.

**rock·rose** (rŏk′rōz′) *n.* Any of various plants or shrubs of the genus *Helianthemum* and related genera, having roselike white, yellow, or reddish flowers.

**rock salt** *n.* Common salt, essentially sodium chloride, occurring in large solid masses.

**rock·shaft** (rŏk′shăft′) *n.* A shaft that oscillates or rocks on its bearings but does not revolve.

**rock·weed** (rŏk′wēd′) *n.* A coarse brownish seaweed of the genera *Fucus* or *Ascophyllum* that grows on coastal rocks.

**rock·work** (rŏk′wûrk′) *n.* **1.** A natural mass of rocks. **2.** Stonework that imitates the irregular surface of natural rocks.

**rock·y**[1] (rŏk′ē) *adj.* **-i·er, -i·est. 1.** Made of, containing, or abounding in rock or rocks. **2.** Resembling rock : HARD. **3.** Characterized by impediments or difficulties <the *rocky* road to social and economic equality> **—rock′i·ness** *n.*

**rock·y**[2] (rŏk′ē) *adj.* **-i·er, -i·est. 1.** Inclined or prone to sway unsteadily : SHAKY. **2.** Weak, dizzy, or nauseated. **—rock′i·ness** *n.*

**Rocky Mountain goat** *n.* The mountain goat.

**Rocky Mountain sheep** *n.* The bighorn.

**Rocky Mountain spotted fever** *n.* An acute infectious disease caused by a microorganism, *Rickettsia rickettsii,* transmitted by ticks, marked by high fever, muscular pains, and skin eruptions, and endemic throughout North America.

**ro·co·co** (rə-kō′kō, rō′kə-kō′) *adj.* [Fr., alteration of *rocaille,* rockwork < *roc,* rock, var. of *roche* < OFr.] **1.** Of or relating to an artistic style originating in 18th-cent. France and marked by fanciful asymmetric ornamentation. **2.** Very elaborate : ORNATE. **3.** *Mus.* Of or relating to an 18th-cent. style immediately following the baroque in Europe. **—ro·co′co** *n.*

**rod** (rŏd) *n.* [ME *rodd* < OE.] **1.** A thin straight piece or bar of material, as of wood or metal. **2.** A shoot or stem cut from or growing as part of a woody plant. **3. a.** A stick or bundle of sticks used for whipping. **b.** Punishment : correction. **4.** A fishing rod. **5.** A scepter or staff symbolizing authority or power. **6.** Power or dominion, esp. when tyrannical. **7.** A metal bar in a machine. **8.** A measuring stick. **9.** A leveling rod. **10.** A lightning rod. **11.** A divining rod. **12. a.** A linear measure equal to 5.5 yards, 16.5 feet, or 5.03 meters. **b.** A unit of measure equal to 30.25 square yards or 22.41 square meters. **13.** *Anat.* Any of various rod-shaped cells in the retina that respond to dim light. **14.** *Microbiol.* An elongated microorganism. **15.** *Slang.* A handgun. **16.** A drawbar beneath a freight car.

**rode** (rōd) *v. p.t. of* RIDE.

**ro·dent** (rōd′nt) *n.* [< NLat. *Rodentia,* order name < Lat. *rodens,* pr.part. of *rodere,* to gnaw.] A mammal of the order Rodentia, as a mouse, squirrel, rat, or beaver, characterized by large incisors adapted for gnawing or nibbling. *—adj.* **1.** Gnawing. **2.** Of or relating to rodents.

**ro·den·ti·cide** (rō-dĕn′tĭ-sīd′) *n.* An agent for killing rodents.

**ro·de·o** (rō′dē-ō′, rō-dā′ō) *n., pl.* **-os.** [Sp. < *rodear,* to surround < Lat. *rotare,* to rotate < *rota,* wheel.] **1.** A cattle roundup. **2.** An enclosure for cattle that have been rounded up. **3.** A competition featuring such skills as bronco riding and steer wrestling.

**rod·o·mon·tade** *also* **rhod·o·mon·tade** (rŏd′ə-mŏn-tād′, -täd′, rō′də-) *n.* [OFr. < OItal. *rodomontada* < *Rodomonte,* a character in *Orlando Innamorata,* by Matteo Boiardo (1434–1494).] Pretentious bragging : BLUSTER. *—adj.* Pretentiously boastful : BRAGGING. *—vi.* **-tad·ed, -tad·ing, -tades.** To boast or brag : BLUSTER.

**roe**[1] (rō) *n.* [ME *row.*] **1.** A fish's egg-laden ovary. **2.** The egg mass of certain crustaceans, such as the lobster. **3.** Soft roe.

**roe**[2] (rō) *n.* [ME *ro* < OE *rā.*] The roe deer.

**roe·buck** (rō′bŭk′) *n.* A male roe deer.

**roe deer** *n.* A rather small, delicately formed Eurasian deer, *Capreolus capreolus,* having a brownish coat and short branched antlers in the male.

**roent·gen** *also* **rönt·gen** (rĕnt′gən, -jən, rŭnt′-) *n.* [After Wilhelm Konrad *Roentgen* (1845–1923).] *Physics.* An obsolete unit of radiation dosage equal to the quantity of ionizing radiation that will produce one electrostatic unit of electricity in one cubic centimeter of dry air at 0°C and standard atmospheric pressure. **—roent′gen** *adj.*

**roent·gen·ize** (rĕnt′gə-nīz′, -jə-, rŭnt′-) *vt.* **-ized, -iz·ing, -iz·es.** To subject to x-rays.

**roentgeno-** *pref.* [< ROENTGEN.] X-ray <*roentgenography*>

**roent·gen·o·gram** (rĕnt′gə-nə-grăm′, -jə-, rŭnt′-) *also* **roent·gen·o·graph** (-grăf′) *n.* A photograph made with x-rays.

**roent·gen·o·graph** (rĕnt′gə-nə-grăf′, -jə-, rŭnt′-) *n.* A roentgenogram.

**roent·gen·og·ra·phy** (rĕnt′gə-nŏg′rə-fē, -jə-, rŭnt′-) *n.* X-ray photography. **—roent′gen·o·graph′ic** (-nə-grăf′ĭk) *adj.* **—roent′gen·o·graph′ic·al·ly** *adv.*

**roent·gen·ol·o·gy** (rĕnt′gə-nŏl′ə-jē, -jə-, rŭnt′-) *n.* Radiology with x-rays. **—roent′gen·o·log′ic** (-ə-lŏj′ĭk), **roent′gen·o·log′i·cal** *adj.* **—roent′gen·o·log′i·cal·ly** *adv.* **—roent′gen·ol′o·gist** *n.*

**roent·gen·o·scope** (rĕnt′gə-nə-skōp′, -jə-, rŭnt′-) *n.* A fluoroscope. **—roent′gen·o·scop′ic** (-skŏp′ĭk) *adj.* **—roent′gen·os′co·py** (-nŏs′kə-pē) *n.*

**roent·gen·o·ther·a·py** (rĕnt′gə-nə-thĕr′ə-pē, -jə-, rŭnt′-) *n.* Therapeutic use of x-rays in treating disease.

**Roentgen ray** *n.* X-RAY 1b.

**ro·ga·tion** (rō-gā′shən) *n.* [ME *rogacioun* < Lat. *rogatio* < *rogare,* to ask.] **1.** *often* **rogations.** Solemn prayer or supplication, esp. as chanted during the rites of Rogation Day. **2. a.** The proposal of a law by consul or tribune to the ancient Roman people for acceptance or rejection. **b.** A law proposed in this way.

**Rogation Day** *n.* One of the three days of prayer preceding Ascension Day.

**ro·ga·to·ry** (rŏ′gə-tôr′ē, -tōr′ē) *adj.* [Fr. *rogatoire* < Med. Lat. *rogatorius* < Lat. *rogare,* to ask.] Requesting information : QUESTIONING.

**rog·er** (rŏj′ər) *interj.* [< the name *Roger,* code word for the letter *r.*] —Used in radio communications to indicate that a message has been received.

**rogue** (rōg) *n.* [Orig. unknown.] **1.** An unprincipled person : SCOUNDREL. **2.** A playfully mischievous person : SCAMP. **3.** A vicious and solitary animal, as an elephant that has separated itself from its herd. **4.** An organism, esp. a plant, that shows an undesirable variation from a standard. **5.** *Archaic.* A wandering vagrant : BEGGAR. *—v.* **rogued, rogu·ing, rogues.** *—vt.* **1.** To defraud. **2.** To remove (diseased or abnormal specimens) from a group of plants of the same variety. *—vi.* To remove undesired plant specimens.

**rogu·er·y** (rō′gə-rē) *n., pl.* **-ies. 1.** Behavior characteristic of a rogue. **2.** A mischievous act.

**rogues' gallery** *n.* A collection of pictures of criminals maintained in police files and used for making identifications.

**rogu·ish** (rō′gĭsh) *adj.* **1.** Unworthy of trust : unprincipled. **2.** Playfully mischievous. **—rogu′ish·ly** *adv.* **—rogu′ish·ness** *n.*

**roil** (roil) *v.* **roiled, roil·ing, roils.** [Orig. unknown.] *—vt.* **1.** To make (a liquid) cloudy by stirring up sediment. **2.** To displease or disturb : VEX. *—vi.* To be in a state of agitation or turbulence.

**roil·y** (roi′lē) *adj.* **-i·er, -i·est. 1.** Cloudy : muddy. **2.** Agitated.

**rois·ter** (roi′stər) *vi.* **-tered, -ter·ing, -ters.** [< obs. *roister,* roisterer, prob. < OFr. *rustre,* ruffian, alteration of *ruste* < Lat. *rusticus,* rustic < *rus,* country.] **1.** To engage in boisterous revelry. **2.** To behave in a swaggering way. **—rois′ter·er** *n.* **—rois′ter·ous** *adj.*

**ro·la·mite** (rō′lə-mīt′) *n.* [ROL(L) + *-amite,* of unknown orig.] A device consisting of two or more hard cylindrical rollers fitted between two parallel constraints with a flexible nonstretching band looped around them in which the rollers move against each other within the constraints with very little friction to perform various functions.

**Ro·land** (rō′lənd, rô-läN′) *n.* A legendary defender of the Christians and nephew of Charlemagne, killed in the battle against the Saracens at Roncesvalles in A.D. 778.

**role** *also* **rôle** (rōl) *n.* [Fr. *rôle* < OFr. *rolle,* roll of parchment < Lat. *rotula,* little wheel, dim. of *rota,* wheel.] **1.** A part played by an actor or actress in a dramatic performance. **2.** The characteristic and expected social behavior of an individual. **3.** A position or function.

**role model** *n.* An individual who serves as a model in a particular behavioral role for another individual to emulate.

**role-play** (rōl′plā′) *v.* **-played, -play·ing, -plays.** *—vt.* To play the part of : ACT OUT. *—vi.* To play a role.

**Rolf·ing** (rôl′fĭng, rŏl′-). A trademark for a technique of deep, often painful muscular manipulation and massage designed to relieve both bodily and emotional tension.

**roll** (rōl) *v.* **rolled, roll·ing, rolls.** [ME *rollen* < OFr. *roler* < VLat. **rotulare* < Lat. *rotula,* little wheel, dim. of *rota,* wheel.] *—vi.* **1.** To move forward along a surface by revolving on an axis or by repeatedly turning over. **2.** To travel or be moved on rollers or wheels. **3.** To travel around : WANDER <*roll* from city to city> **4. a.** To travel or be carried in a vehicle. **b.** To be carried on a stream. **5.** To gain momentum <The new project began to *roll.*> **6.** To go by : ELAPSE <The months *rolled* by.> **7.** To recur periodically, in or as if in cycles <Christmas has *rolled* around again.> **8.** To move in a periodic revolution, as a planet in its orbit. **9.** To turn over and over <The puppy *rolled* in the grass.> **10.** To shift the eyes, usu. quickly and continually. **11.** To turn around or revolve on or as if on an axis. **12.** To advance with a rising and falling motion : UNDULATE <waves *rolling* toward shore> **13.** To extend or appear to extend in gentle undulations <grassland *rolling* to the sea> **14.** To rock from side to side, as a ship. **15.** To walk with an unsteady, swaying motion. **16.** To take the shape of a cylinder or ball. **17.** To become flattened by or as if by pressure applied by a roller. **18.** To make a deep, extended, echoing sound <thunder *rolling* in the distance> **19.** To make a sustained trilling sound, as certain birds. **20.** To beat a drum in a continuous series of short blows. **21.** To pour or flow in a continual stream : FLOW <shoppers *rolling* in> **22.** To enjoy abundant amounts <*rolling* in cash> *—vt.* **1.** To cause to move forward along a surface by revolving on an axis or by repeatedly turning over. **2.** To move or push along on rollers or wheels. **3.** To impel or send onward in a continuous swelling motion. **4.** To give a swaying, rocking motion to <heavy seas *rolling* huge ships> **5.** To turn around or partly turn around : ROTATE. **6.** To cause to begin operating or moving <*roll* the presses> **7.** To lay out or extend <*rolled* out an extension cord> **8.** To utter or pronounce with a trill. **9.** To utter or emit in full, swelling tones. **10.** To beat (a drum) with a continuous series of short blows. **11.** To wrap (an object) round and round onto itself or around something else <*roll* up a poster> **12. a.** To enfold or envelop in a covering <*roll* cheese in a tortilla> **b.** To make by shaping into a ball or cylinder. **13.** To compress, spread, or flatten by applying pressure with a roller <*roll* cookie dough> **14.** To apply ink to (printing type) with a roller or rollers. **15.** To throw (dice), as in craps. **16.** *Slang.* To rob (a drunken, sleeping, or helplessly unaware person). **—roll back. 1.** To reduce (prices or wages) to a previous lower level. **2.** To cause to turn back : RETREAT. **—roll out.** To get out of bed. **—roll up. 1.** *Informal.* To arrive in a vehicle. **2.** To accumulate : amass <*rolled up* quite a fortune> *—n.* **1.** The act or an instance of rolling. **2.** Something rolled up in the form of a cylinder <a *roll* of masking tape> **3.** A quantity, as of cloth or wallpaper, rolled into a cylinder and often regarded as a unit of measure. **4.** A piece of parchment or paper that may be or is rolled up : SCROLL. **5.** A catalogue : register. **6.** A list of names of persons belonging to a given group. **7.** A cylindrical or rounded mass. **8. a.** A small rounded portion of bread. **b.** A portion of food shaped like a tube with a filling. **9.** A swaying, rolling, or rocking motion. **10.** A gentle swell or undulation of a surface <the *roll* of the prairie> **11.** A deep rumble or reverberation. **12.** A rapid succession of short sounds <a drum *roll*> **13.** A trill. **14.** A rhythmical resonant flow of words. **15.** A roller, esp. a cylinder on which to roll material up or with which to flatten material. **16.** A maneuver in which an aircraft makes a single complete rotation about its longitudinal axis without changing direction or losing altitude. **17.** *Slang.* Money, esp. a wad of paper money. **—roll the bones.** To cast dice, esp. in craps.

**roll·a·way** (rōl'ə-wā') *adj.* Set on rollers for easy movement and storage.

**roll·back** (rōl'băk') *n.* A reduction of wages or prices to a previous lower level by governmental action or direction.

**roll bar** *n.* A sturdy metal bar built into or next to the inside roof of a motor vehicle to prevent or reduce injury in case of a rollover.

**roll call** *n.* **1.** The reading aloud of a list of names of people, as in a classroom or military post, to ascertain absentees. **2.** The time fixed for a roll call.

**roll·er** (rō'lər) *n.* **1.** One that rolls. **2.** A cylindrical device, esp.: **a.** A small spokeless wheel, as that of a roller skate or caster. **b.** An elongated cylinder on which something such as a window shade or towel is wound. **c.** A heavy cylinder for leveling or crushing operations. **d.** A hard rubber cylinder for inking printing type before the paper is impressed. **e.** A cylinder, as of wire mesh or foam rubber, around which a strand of hair is wound to produce a soft curl or wave. **3.** A long rolled bandage. **4.** A heavy swelling wave breaking on a coast. **5.** Any of various African, Far Eastern, and Australian birds of the family Coraciidae, having bright blue wings, stocky bodies, and hooked bills and noted for their aggressiveness.

**roller bearing** *n.* A bearing utilizing rollers to reduce friction between machine parts.

**roller coaster** *n.* A steep sharply banked elevated railway with small open passenger cars, operated as an amusement-park ride.

**roller skate** *n.* A shoe with four small wheels attached to it for skating on a hard, usu. flat surface, as a sidewalk or a floor.

**roll·er-skate** (rō'lər-skāt') *vi.* **-skat·ed, -skat·ing, -skates.** To skate on roller skates. **—roller skater** *n.*

**Rolle's theorem** (rôlz, rōlz) *n.* [After Michel *Rolle* (d. 1719).] A theorem in mathematics stating that if a curve is continuous, has two x-intercepts, and has a tangent at every point between the intercepts, at least one of these tangents is parallel to the x-axis.

**roll film** *n.* Photographic film rolled on a spool.

**rol·lick** (rŏl'ĭk) *vi.* **-licked, -lick·ing, -licks.** [Orig. unknown.] To behave or move in a playful, carefree manner : ROMP. *—n.* A frolicsome escapade : LARK. **—rol'lick·some, rol'lick·y** *adj.*

**rol·lick·ing** (rŏl'ĭ-kĭng) *adj.* Carefree and high-spirited. **—rol'lick·ing·ly** *adv.*

**rolling mill** *n.* **1.** A factory in which metal is rolled into sheets, bars, or other shapes. **2.** A machine for rolling metal.

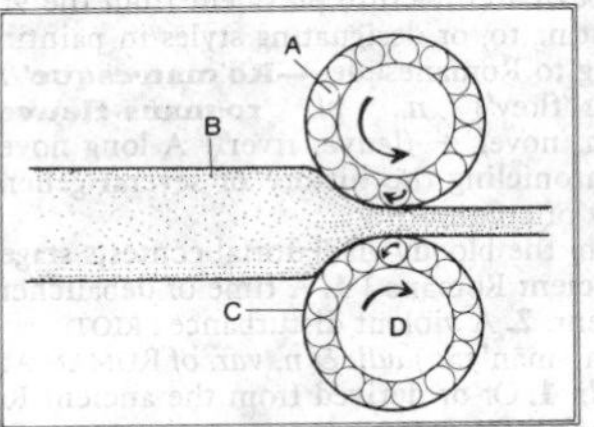


**rolling mill**
*A. work roll, B. hot strip, C. cage, D. backup roll*

**rolling pin** *n.* A smooth usu. wood cylinder with a handle at each end, used for rolling out dough.

**rolling stock** *n.* A railroad's wheeled vehicles.

**roll·mops** (rōl'mŏps') *n., pl.* **rollmops.** [G. : *rollen,* to roll + *Mops,* pug dog.] A marinated fillet of herring wrapped around a gherkin or onion and served as an hors d'oeuvre.

**roll·o·ver** (rōl'ō'vər) *n.* **1.** The act or process of rolling over. **2.** An accident in which a motor vehicle overturns. **3. a.** Refinance of a maturing note. **b.** Reinvestment of funds in such a way as to avoid payment of taxes.

**roll-top desk** (rōl'tŏp') *n.* A desk equipped with a flexible sliding top of parallel slats.

**roll·way** (rōl'wā') *n.* A surface along which cylinders or objects on rollers may be moved.

**ro·ly-po·ly** (rō'lē-pō'lē) *adj.* [Redup. of *roly* < ROLL.] Short and plump : PUDGY. *—n., pl.* **-lies. 1.** A roly-poly creature. **2.** *Chiefly Brit.* A pudding made by rolling up jam or fruit in pastry dough and cooking it.

**Ro·ma·ic** (rō-mā'ĭk) *n.* [Mod. Gk. *Rhōmaikos* < Gk., Roman, pertaining to the eastern Roman Empire < *Rhōmē,* Rome, Constantinople < Lat. *Roma,* Rome.] Modern vernacular Greek. **—Ro·ma'ic** *adj.*

**ro·maine** (rō-mān') *n.* [Fr. < fem. of *Romain,* Roman < OFr. < Lat. *Romanus* < *Roma,* Rome.] A variety of lettuce, *Lactuca sativa longifolia,* having long crisp leaves forming a slender head.

**ro·man** (rō-mäN') *n.* [Fr. < OFr. *romans,* romance.] A metrical medieval French narrative derived from the ancient epic poems.

**Ro·man** (rō'mən) *adj.* [ME < OE *Rōmān* < Lat. *Romanus* < *Roma,* Rome.] **1.** Of, relating to, derived from, or characteristic of Rome and its people, esp. ancient Rome. **2.** Of, relating to, composed in, or characteristic of the Latin language. **3.** Of or relating to the Roman Catholic Church. **4.** Of or designating a style of architecture developed by the ancient Romans and marked by great round arches and barrel vaults, concrete masonry construction, and classical orders as decorative features. **5. roman.** Of, set, or printed in type characterized by upright letters. *—n.* **1.** A resident, native, or citizen of Rome, esp. ancient Rome. **2.** The Italian language as spoken in Rome. **3.** One belonging to the Roman Catholic Church. **4. roman.** Roman type or letters. **5. Romans** *(sing. in number).* —See table at BIBLE.

**ro·man à clef** (rō-mäN' ä klā') *n., pl.* **ro·mans à clef** (rō-mäN' ä klā') [Fr. : *roman,* novel + *à,* with + *clef,* key.] A novel in which actual persons or places are fictionally depicted.

**Roman alphabet** *n.* The Latin alphabet.

**Roman calendar** *n.* The lunar calendar used by the ancient Romans until the introduction of the Julian calendar in 46 B.C.

**Roman candle** *n.* A firework consisting of a tube from which colored balls of fire are ejected.

**Roman Catholic** *adj.* Of, designating, or relating to the Roman Catholic Church. **—Roman Catholic** *n.*

**Roman Catholic Church** *n.* The Christian church that is characterized by a hierarchic structure of priests and bishops in which doctrinal and disciplinary authority are dependent upon apostolic succession, with the pope as head of the episcopal college.

**Roman Catholicism** *n.* The practices, doctrines, and organization of the Roman Catholic Church.

**ro·mance** (rō-măns', rō'măns') *n.* [ME < OFr. *romans,* romance, work written in French < Lat. *Romanicus,* Roman < *Romanus* < *Roma,* Rome.] **1. a.** A long medieval narrative in prose or verse telling of the adventures of chivalric heroes. **b.** A long fictitious tale of heroes and mysterious or extraordinary events. **c.** The class of literature of such tales. **d.** A quality suggestive of the adventure and idealized exploits found in such tales. **2.** A story, novel, or film dealing with a love affair. **3.** The style or class of fictional works about idealized love. **4. a.** A love affair. **b.** Romantic involvement : LOVE. **c.** A strong, usu. ephemeral interest or enthusiasm. **5.** Inclination toward the romantic or adventurous. **6.** A fictitiously embellished explanation or account. **7.** A short lyrical song or instrumental piece. **8. Romance.** The Romance languages. *—adj.* **Romance.** Of, relating to, or constituting the languages that developed from Latin. *—v.* (rō-măns') **-manced, -manc·ing, -manc·es.** *—vi.* **1.** To invent, tell, or write romances. **2.** To think or act in a romantic way. *—vt. Informal.*

1. To make romantic love to : WOO. 2. To have a love affair with. **—ro·manc'er** *n.*

**Ro·man·esque** (rō'mə-nĕsk') *adj.* 1. Of, relating to, or designating a transitional style of European architecture prevalent from the 9th to the 12th cent. 2. Of, relating to, or designating styles in painting and sculpture corresponding to Romanesque. **—Ro'man·esque'** *n.*

**ro·man-fleuve** (rō-mäN'flœv') *n., pl.* **ro·mans-fleuves** (rō-mäN'flœv') [Fr. : *roman,* novel + *fleuve,* river.] A long novel, often in many volumes, chronicling the history of several generations of a group, as a family or community.

**Roman holiday** *n.* [From the bloody gladiatorial contests staged as entertainment for the ancient Romans.] 1. A time of debauchery or of often sadistic enjoyment. 2. A violent disturbance : RIOT.

**Ro·ma·ni·an** (rō-mā'nē-ən, -mān'yən) *adj. & n. var. of* RUMANIAN.

**Ro·man·ic** (rō-măn'ĭk) *adj.* 1. Of or derived from the ancient Romans. 2. Romance. **—Ro·man'ic** *n.*

**Ro·man·ism** (rō'mə-nĭz'əm) *n.* Roman Catholicism.

**Ro·man·ist** (rō'mə-nĭst) *n.* 1. One who professes Roman Catholicism. 2. A student or specialist on Roman culture, law, and institutions.

**Ro·man·ize** (rō'mə-nīz') *vt.* **-ized, -iz·ing, -iz·es.** 1. To convert (someone) to Roman Catholicism. 2. To make Roman in character, allegiance, or style. 3. To write or transliterate in the Latin alphabet. **—Ro'man·i·za'tion** *n.*

**Roman law** *n.* The system of laws of ancient Rome, on which the legal systems of many countries are based.

**Roman nose** *n.* A nose with a high, prominent bridge.

**Roman numeral** *n.* A numeral formed with the characters I, V, X, L, C, D, and M in the ancient Roman system of numeration.

## ROMAN NUMERALS

| | | | |
|---|---|---|---|
| I | 1 | XXI | 21 |
| II | 2 | XXIX | 29 |
| III | 3 | XXX | 30 |
| IV | 4 | XL | 40 |
| V | 5 | XLVIII | 48 |
| VI | 6 | IL | 49 |
| VII | 7 | L | 50 |
| VIII | 8 | LX | 60 |
| IX | 9 | XC | 90 |
| X | 10 | XCVIII | 98 |
| XI | 11 | IC | 99 |
| XII | 12 | C | 100 |
| XIII | 13 | CI | 101 |
| XIV | 14 | CC | 200 |
| XV | 15 | D | 500 |
| XVI | 16 | DC | 600 |
| XVII | 17 | CM | 900 |
| XVIII | 18 | M | 1,000 |
| XIX | 19 | MDCLXVI | 1666 |
| XX | 20 | MCMLXX | 1970 |

**Ro·ma·no** (rə-mä'nō, rō-) *n.* [Ital., Roman < Lat. *Romanus.*] A dry hard Italian cheese similar to but sharper than Parmesan.

**Ro·mansch** *also* **Ro·mansh** (rō-mänsh', -mănsh') *n.* [Romansch *Romantsch* < Lat. *Romanicus,* Roman. —see ROMANCE.] The Rhaeto-Romanic dialects of eastern Switzerland and neighboring parts of Italy.

**ro·man·tic** (rō-măn'tĭk) *adj.* [Fr. *romantique* < OFr. *romans,* romance.] 1. Of, relating to, or characteristic of romance. 2. Given to feelings or thoughts of romance. 3. Conducive to romance. 4. Imaginative but impractical : VISIONARY. 5. Not based on fact : IMAGINARY. 6. Of or characteristic of romanticism in the arts. —*n.* A romantic person. **—ro·man'ti·cal·ly** *adv.*

**ro·man·ti·cism** (rō-măn'tĭ-sĭz'əm) *n.* 1. *also* **Romanticism.** An intellectual and artistic movement that originated in the late 18th cent. and emphasized strong emotion, imagination, freedom from classical correctness in art forms, and rebellion against social conventions. 2. The attitudes and spirits characteristic of romantic thought. **—ro·man'ti·cist** *n.*

**ro·man·ti·cize** (rō-măn'tĭ-sīz') *v.* **-cized, -ciz·ing, -ciz·es.** —*vt.* To interpret romantically. —*vi.* To think in a romantic way. **—ro·man'ti·ci·za'tion** *n.*

**Rom·a·ny** (rŏm'ə-nē, rō'mə-) *n., pl.* **Romany** *or* **-nies.** [Romany *romani,* pl. of *romano,* gypsy < *rom,* man < Skt. *ḍomaḥ,* man of a low caste.] 1. A Gypsy. 2. The Indic language of the Gypsies. **—Rom'a·ny** *adj.*

**ro·maunt** (rō-mônt', -mŏnt') *n.* [ME < OFr. *romant,* romance < Lat. *Romanicus,* Roman. —see ROMANCE.] *Archaic.* A verse romance.

**Ro·me·o** (rō'mē-ō') *n., pl.* **-os.** [After *Romeo,* the hero of *Romeo and Juliet* by William Shakespeare (1564–1616).] A male lover.

**Rom·ish** (rō'mĭsh) *adj.* Of or relating to the Roman Catholic Church. **—Rom'ish·ly** *adv.* **—Rom'ish·ness** *n.*

**romp** (rŏmp) *vi.* **romped, romp·ing, romps.** [Alteration of RAMP².] 1. To frolic or play boisterously. 2. *Slang.* To win easily. —*n.* 1. Spirited, merry play : FROLIC. 2. One that sports and frolics. 3. *Slang.* An easy win.

**romp·er** (rŏm'pər) *n.* 1. One that romps. 2. **rompers.** A loose-fitting playsuit with short bloomers worn esp. by small children.

**Rom·u·lus** (rŏm'yə-ləs) *n.* [Lat.] *Rom. Myth.* The son of Mars, legendary founder of Rome, and brother of Remus..

**ron·deau** (rŏn'dō, rŏn-dō') *n., pl.* **-deaux** (-dōz, -dōz') [OFr., alteration of *rondel.* —see RONDEL.] 1. A lyrical poem of French origin having 13 or occas. 10 lines with two rhymes throughout and with the opening phrase repeated twice as a refrain. 2. *Mus.* A monophonic trouvère song.

**ron·del** (rŏn'dəl, rŏn-dĕl') *n.* [ME < OFr., dim. of *ronde,* circle, round. —see ROUND.] A rondeau usu. having 14 lines.

**ron·de·let** (rŏn'dl-ĕt', -dl-ā') *n.* [OFr., dim. of *rondel,* rondel.] A short rondeau with five or seven lines and one refrain in one stanza.

**ron·do** (rŏn'dō, rŏn-dō') *n., pl.* **-dos.** [Ital. *rondò* < OFr. *rondeau,* rondeau.] A musical composition with a refrain that occurs at least three times in its original key between contrasting couplets.

**ron·dure** (rŏn'jər, -dyŏŏr') *n.* [OFr. *rondeur,* roundness < *ronde,* round. —see ROUND.] Something circular or gracefully rounded.

**ron·nel** (rŏn'əl) *n.* [< *Ronnel,* a non-U.S. trademark.] 1. A solid, light-brown compound, $C_8H_8Cl_3O_3PS$, used as an insecticide esp. against flies and cockroaches. 2. **Ronnel.** A trademark for ronnel outside the United States.

**rönt·gen** (rĕnt'gən, -jən, rŭnt'-) *n. var. of* ROENTGEN.

**rood** (rōōd) *n.* [ME < OE *rōd.*] 1. **a.** A crucifix symbolizing the cross on which Christ was crucified. **b.** A large crucifix or the representation of one over the altar screen of a medieval church. 2. *Chiefly Brit.* A measure of length that varies from 5½ to 8 yards. 3. A measure of land equal to ¼ acre or 40 square rods.

**rood screen** *n.* An ornamented altar screen, usu. surmounted by a crucifix, separating the choir of a church from the nave.

**roof** (rōōf, rŏŏf) *n.* [ME < OE *hrōf.*] 1. The exterior surface and its supporting structures on the top of a building. 2. The top covering of something <the *roof* of a vehicle> 3. **a.** A vaulted inner structure <the *roof* of a cave> **b.** The highest point : SUMMIT. 4. A house or home. —*vt.* **roofed, roof·ing, roofs.** To furnish or cover with or as if with a roof. **—raise the roof.** *Slang.* 1. To be boisterously noisy. 2. To complain vehemently.

**roof·er** (rōō'fər, rŏŏf'ər) *n.* One who lays or repairs roofs.

**roof garden** *n.* 1. A garden on the roof of an urban building. 2. A restaurant at the top or on the roof of a building that often features music and dancing.

**roof·ing** (rōō'fĭng, rŏŏf'ĭng) *n.* Materials used in building a roof.

**roof·less** (rōōf'lĭs, rŏŏf'-) *adj.* 1. Lacking a roof. 2. Having no home or shelter : HOMELESS.

**roof·top** (rōōf'tŏp', rŏŏf'-) *n.* The surface of a roof.

**roof·tree** (rōōf'trē', rŏŏf'-) *n.* 1. A long horizontal beam extending along the ridge of a roof : RIDGEPOLE. 2. A roof.

**rook¹** (rŏŏk) *n.* [ME *rok* < OE *hrōc.*] A crowlike Old World bird, *Corvus frugilegus,* that nests in colonies near the tops of trees. —*vt.* **rooked, rook·ing, rooks.** *Slang.* To swindle : cheat.

**rook²** (rŏŏk) *n.* [ME *rok* < OFr. *roc* < Ar. *rukh* < Pers.] A chess piece that may move in a straight line over any number of empty squares in a rank or file.

**rook·er·y** (rŏŏk'ə-rē) *n., pl.* **-ies.** 1. **a.** A place where rooks nest and breed. **b.** The breeding ground of certain other birds and animals, such as seals. 2. *Informal.* A crowded, run-down tenement.

**rook·ie** (rŏŏk'ē) *n.* [Alteration of RECRUIT.] *Slang.* 1. An untrained recruit. 2. A novice player in sports. 3. An inexperienced person.

**room** (rōōm, rŏŏm) *n.* [ME *roum* < OE *rūm.*] 1. A space that is or may be occupied <not enough *room* for all the furniture> 2. **a.** An area separated by walls or partitions from other similar parts of the structure or building in which it is located. **b.** The people present in such an area <The whole *room* looked surprised.> 3. **rooms.** Living quarters. 4. Suitable opportunity. —*vi.* **roomed, room·ing, rooms.** To occupy a room : LODGE.

☆ ***syns:*** ROOM, LATITUDE, LEEWAY, MARGIN, PLAN, SCOPE *n. core meaning* : suitable opportunity to accept or allow something <no *room* for error>

**room and board** *n.* Meals and lodging either provided or earned.

**room·er** (rōō'mər, rŏŏm'ər) *n.* A lodger.

**room·ette** (rōō-mĕt', rŏŏm-ĕt') *n.* 1. A small private compartment in a railroad sleeping car. 2. A small room, as in a dormitory.

**room·ful** (rōōm'fŏŏl', rŏŏm'-) *n., pl.* **-fuls.** The amount or number that a room will hold.

**rooming house** *n.* A house where lodgers may rent rooms.

**room·mate** (rōōm'māt', rŏŏm'-) *n.* A person with whom one shares a room or apartment.

ă **pat** ā **pay** âr **care** ä **father** ĕ **pet** ē **be** hw **which** ĭ **pit** ī **tie** îr **pier** ŏ **pot** ō **toe** ô **paw, for** oi **noise** ŏŏ **took**

**room·y** (ro͞o'mē, ro͝om'ē) *adj.* **-i·er, -i·est.** Having plenty of room : SPACIOUS. **—room'i·ly** *adv.* **—room'i·ness** *n.*

**roor·back** (ro͝or'băk') *n.* [After Baron von *Roorback,* imaginary author of an imaginary book, *Roorback's Tour Through the Western and Southern States,* from which a passage was purportedly quoted in an attempt to disparage Presidential candidate James K. Polk in 1844.] A false or slanderous story used for political advantage.

**roost** (ro͞ost) *n.* [ME *rooste* < OE *hrōst.*] **1.** A perch on which domestic fowl or other birds rest or sleep. **2.** A place with perches for fowl or other birds. **3.** A place for sleep or rest. *—vi.* **roost·ed, roost·ing, roosts.** To rest or sleep on or as if on a perch or roost. **—rule the roost.** To be in charge : DOMINATE.

**roost·er** (ro͞o'stər) *n.* **1. a.** The adult male of the common domestic fowl. **b.** The adult male of other birds : COCK. **2.** A belligerent and cocky person.

**root**[1] (ro͞ot, ro͝ot) *n.* [ME *rot* < OE *rōt* < ON.] **1. a.** The usu. underground part of a plant that serves as support, draws food and water from the surrounding soil, and stores food. **b.** A similar underground plant part such as a rhizome, corm, or tuber. **c.** One of many small hairlike growths that serve to attach and support plants such as the ivy and other vines. **2.** The embedded part of an organ or structure such as a tooth, hair, or nerve. **3.** A support : base. **4.** An essential part or element : CORE <the *root* of the matter> **5.** A primary source : ORIGIN. **6.** An ancestor or antecedent. **7.** *often* **roots.** The condition of being settled and of belonging to a particular place or society <put down *roots* in a new city> **8.** An element constituting the basis from which a word is derived by phonetic change or by the addition of other elements, such as inflectional endings or affixes. **9.** *Math.* **a.** A number that when multiplied by itself an indicated number of times forms a product equal to a specified number <A fourth *root* of 4 is $\sqrt{2}$.> **b.** A number that reduces a polynomial equation in one variable to an identity when it is substituted for the variable. **c.** A root *a* of the polynomial equation $f(x) = 0$ in which $(x-a)$ occurs at least twice as a factor of $f(x)$. **10.** *Mus.* **a.** The note from which a chord is built. **b.** The first or lowest note of a triad or chord. *—v.* **root·ed, root·ing, roots.** *—vi.* **1.** To grow a root or roots. **2.** To become firmly settled, established, or entrenched. *—vt.* **1.** To cause to put out roots and grow. **2.** To implant by or as if by the roots. **3.** To furnish a primary source or origin to. **4.** To remove by or as if by the roots <*rooting* out malcontents from the company> **—root'er** *n.*

**root**[2] (ro͞ot, ro͝ot) *v.* **root·ed, root·ing, roots.** [ME *wroten* < OE *wrōtan.*] *—vt.* To dig with or as if with the snout or nose. *—vi.* **1.** To dig in the earth with or as if with the snout or nose. **2.** To rummage for something <*rooting* around in the attic> **—root'er** *n.*

**root**[3] (ro͞ot, ro͝ot) *vi.* **root·ed, root·ing, roots.** [Poss. alteration of ROUT[3].] **1.** To give encouragement to a team or contestant : CHEER. **2.** To lend support to a person or cause. **—root'er** *n.*

**root·age** (ro͞o'tĭj, ro͝ot-ĭj) *n.* **1.** A system or growth of roots. **2.** Establishment by or as if by roots.

**root beer** *n.* A carbonated soft drink made from extracts of the roots of certain plants.

**root canal** *n.* A pulp-filled cavity in a root of a tooth.

**root cap** *n. Bot.* A thimble-shaped mass of cells that covers and protects the tip of a growing root.

**root cellar** *n.* A cellar, usu. covered with earth for storing root crops and other vegetables.

**root hair** *n. Bot.* A thin hairlike outgrowth of a plant root that absorbs water and minerals from the soil.

**root knot** *n.* A disease of plants caused by a nematode and marked by protuberant enlargements on the roots.

**root·less** (ro͞ot'lĭs, ro͝ot'-) *adj.* **1.** Lacking roots. **2.** Not belonging to a particular place or society <*rootless* vagrants> **—root'less·ness** *n.*

**root·let** (ro͞ot'lĭt, ro͝ot'-) *n.* A small root or division of a root.

**root mean square** *n.* The square root of the arithmetic mean of the squares of a set of numbers.

**root·stalk** (ro͞ot'stôk', ro͝ot'-) *n.* A rhizome.

**root·stock** (ro͞ot'stŏk', ro͝ot'-) *n.* **1.** A rhizome. **2.** A source : origin.

**root·y** (ro͞o'tē, ro͝ot'ē) *adj.* **-i·er, -i·est. 1.** Full of roots. **2.** Consisting of a root or roots. **3.** Like roots. **—root'i·ness** *n.*

**rope** (rōp) *n.* [ME < OE *rāp.*] **1.** A heavy flexible cord of twisted fiber, as hemp. **2.** A cord with a noose at one end for hanging a person. **3.** Death by hanging. **4.** A lasso : lariat. **5. ropes.** Several cords strung between poles to enclose a boxing or wrestling ring. **6.** A string of items attached in one line by twisting or braiding. **7.** A sticky glutinous formation of stringy matter in a liquid. **8. ropes.** *Informal.* Specialized procedures or techniques <just learning the *ropes*> *—v.* **roped, rop·ing, ropes.** *—vt.* **1.** To tie or fasten with or as if with rope. **2.** To enclose with a rope <*rope* off the accident scene> **3.** To catch with a lasso or rope. **4.** *Informal.* To deceive : trick <*roped* us into buying worthless land> *—vi.* To become ropy and sticky. **—rop'er** *n.*

**rope tow** *n.* A ski tow consisting of an endless rope.

**rope·walk** (rōp'wôk') *n.* **1.** A long, usu. covered path or alley where ropes are made. **2.** A long narrow building containing a ropewalk.

**rop·y** (rō'pē) *adj.* **-i·er, -i·est. 1.** Resembling a rope or ropes. **2.** Forming sticky glutinous strings or threads, as some liquids. **—rop'i·ly** *adv.* **—rop'i·ness** *n.*

**roque** (rōk) *n.* [Alteration of CROQUET.] Croquet played on a hard court.

**Roque·fort** (rōk'fərt). A trademark for a cheese that is made from ewes' milk and ripened in caves.

**ro·que·laure** (rō'kə-lôr', -lōr', rŏk'ə-) *n.* [After Antoine Gaston Jean-Baptiste (1656–1738), Duc de *Roquelaure.*] A man's knee-length cloak popular during the 18th and early 19th cent.

**roquelaure**

**ror·qual** (rôr'kwəl) *n.* [Fr. < Norw. *røyrhval* < ON *reytharhvalr* : *reythr,* rorqual + *hvalr,* whale.] Any of several whalebone-bearing whales of the genus *Balaenoptera,* having longitudinal grooves on the throat and a small, pointed dorsal fin.

**Ror·schach test** (rôr'shäk', -shäKH') *n.* [After Hermann *Rorschach* (1884–1922).] A psychological test of personality in which a subject's interpretations of standardized inkblot designs are analyzed as a measure of emotional and intellectual functioning and integration.

**ro·sa·ceous** (rō-zā'shəs) *adj.* [< NLat. *Rosaceae,* family name < Lat. *rosaceus,* made of roses < *rosa,* rose.] **1.** *Bot.* Of or belonging to the Rosaceae, the plant family that includes roses. **2.** Like a rose.

**ros·an·i·line** *also* **ros·an·i·lin** (rō-zăn'ə-lĭn) *n.* [ROS(E) + ANILINE.] A brownish-red crystalline organic compound, $C_{20}H_{21}N_3O$, derived from aniline and used to make dyes.

**ro·sa·ry** (rō'zə-rē) *n., pl.* **-ries.** [Med. Lat. *rosarium* < Lat., rose garden < neuter of *rosarius,* of roses < *rosa,* rose.] **1.** *Rom. Cath. Ch.* **a.** A form of devotion to the Virgin Mary, consisting of three sets of five decades each of the "Hail Mary," each decade preceded by an "Our Father" and ending with a "Glory Be to the Father." **b.** A string of beads on which these prayers are counted. **2.** Beads similar to a rosary used by other religious groups.

**rosary pea** *n.* A woody vine, *Abrus precatorius* of tropical Asia, having scarlet, black-spotted, poisonous seeds used as beads.

**rose**[1] (rōz) *n.* [ME < OE < Lat. *rosa.*] **1. a.** Any of numerous shrubs or vines of the genus *Rosa,* usu. having prickly stems, compound leaves, and variously colored, often fragrant flowers. **b.** The flower of any of these plants, occurring in a wide variety of colors such as pink, red, yellow, and white. **2.** A plant related to or resembling the rose. **3.** A dark pink to purplish pink to moderate red or purplish red, covering a variable range of medium lightness and moderate saturation. **4.** A rosy color of the cheeks. **5.** An ornament shaped like a rose : ROSETTE. **6.** A perfume obtained from or having the odor of roses. **7.** A perforated nozzle for spraying water from a hose or sprinkling can. **8. a.** A gem cut marked by a flat base and a faceted hemispheric upper surface. **b.** A gem, esp. a diamond, with a rose cut. **9.** A rose window. **10.** A compass card. *—adj.* Of the color rose.

**rose**[2] (rōz) *v. p.t. of* RISE.

**ro·sé** (rō-zā') *n.* [Fr., pink < OFr. *rose,* rose < Lat. *rosa.*] A light pink wine made of red grapes whose skins have been removed during fermentation.

**rose acacia** *n.* A shrub, *Robinia hispida* of the southern United States, with bristly branches and rose-colored flower clusters.

**ro·se·ate** (rō'zē-ĭt, -āt') *adj.* [< Lat. *roseus,* rosy < *rosa,* rose.] **1.** Rose-colored. **2.** Highly optimistic : CHEERFUL. **—ro'se·ate·ly** *adv.*

**rose·bay** (rōz'bā') *n.* **1.** A shrub of the genus *Rhododendron,* esp. *R. maximum* of northeastern North America, having large glossy leaves and white or pink flower clusters. **2.** The oleander. **3.** *Chiefly Brit.* The willow herb.

**rose beetle** *n.* The rose chafer.

**rose-breast·ed grosbeak** (rōz'brĕs'tĭd) *n.* A North American bird, *Pheucticus ludovicianus,* the male of which is black and white with a rose patch on the breast.

**rose·bud** (rōz'bŭd') *n.* A bud of a rose.

**rose·bush** (rōz'bo͝osh') *n.* A shrub that bears roses.

**rose campion** *n.* A European plant, *Lychnis coronaria,* naturalized in northeastern North America, that is covered with woolly white down and has rose-red flowers.

**rose chafer** *n.* A long-legged gray beetle, *Macrodactylus subspinosus,* that damages garden plants, esp. roses and grapes.

**rose-col·ored** (rōz'kŭl'ərd) *adj.* **1.** Of the color rose. **2.** Overly optimistic.

o͞o **boot** ou **out** th **thin** *th* **this** ŭ **cut** ûr **urge** y **young**
yo͞o **abuse** zh **vision** ə **about,** it**e**m, edib**i**le, gall**o**p, circ**u**s

**rose·fish** (rōz′fĭsh′) *n., pl.* **rosefish** *or* **-fish·es.** A bright-red marine food fish, *Sebastes marinus* of North Atlantic waters.

**rose geranium** *n.* A woody plant, *Pelargonium graveolens*, having rose-pink flowers and fragrant leaves used for flavoring and in perfumery.

**ro·selle** (rō-zĕl′) *n.* [Orig. unknown.] A tropical Old World plant, *Hibiscus sabdariffa*, with yellow flowers, whose floral bracts have a pleasantly acid flavor when immature and are used in making jelly and beverages.

**rose mallow** *n.* A tall plant, *Hibiscus moscheutos*, growing in brackish marshes of eastern North America and having leaves covered with whitish down and pink or white flowers.

**rose·mar·y** (rōz′mĕr′ē) *n., pl.* **-ies.** [Alteration of ME *rosmarine* < Lat. *ros marinus*, sea dew.] An aromatic evergreen shrub, *Rosmarinus officinalis* of southern Europe, having light blue flowers and grayish-green leaves used in cooking and making perfumes.

**rose moss** *n.* **1.** A moss of the genus *Rhodobryum*, esp. *R. roseum*, having conspicuous terminal leaf rosettes. **2.** Portulaca.

**rose of Jer·i·cho** (jĕr′ĭ-kō′) *n.* A fernlike desert plant, *Anastatica hierochuntica*, that forms a tight ball when dry and unfolds and blooms under moist conditions.

**rose of Shar·on** (shăr′ən, shâr′-) *n.* **1.** A tall shrub, *Hibiscus syriacus*, having large reddish, white, or purple flowers. **2.** A shrubby plant, *Hypericum calycinum* native to Eurasia, having evergreen leaves and yellow flowers.

**ro·se·o·la** (rō-zē′ə-lə, rō′zē-ō′lə) *n.* [NLat. < Lat. *roseus*, rosy < *rosa*, rose.] A rose-colored skin rash. **—ro·se′o·lar** *adj.*

**rose pink** *n.* A light purplish pink to moderate or strong pink. **—rose′-pink′** (rōz′pĭngk′) *adj.*

**rose quartz** *n.* A pinkish quartz used as a gemstone.

**rose·root** (rōz′ro͞ot′, -ro͝ot′) *n.* A plant, *Sedum roseum* of the Northern Hemisphere, having fleshy leaves and greenish-yellow or purple flowers.

**Ro·set·ta stone** (rō-zĕt′ə) *n.* A basalt tablet bearing an inscription in Greek, Egyptian hieroglyphics, and Demotic that was discovered in 1799 near the town of Rosetta, Egypt, and provided the key to the decipherment of hieroglyphics.

**ro·sette** (rō-zĕt′) *n.* [Fr. < OFr. < dim. of *rose*, rose < Lat. *rosa*.] **1.** An ornament or badge made of silk or ribbon that is pleated or gathered to resemble a rose and is worn as a clothing decoration or as a part of a medal. **2.** A roselike formation or marking, as one of the clusters of spots on a leopard's fur. **3.** A carved, painted, or sculptured architectural ornament having a circular arrangement of parts resembling the petals of a rose. **4.** *Bot.* A circular cluster of plant parts, as leaves.

**rose water** *n.* A fragrant solution made by distilling or steeping rose petals in water, used in cosmetics and in cookery.

**rose window** *n.* A circular, usu. stained glass window with radiating, roselike tracery.

**rose·wood** (rōz′wo͝od′) *n.* **1.** A tropical or semitropical tree, chiefly of the genus *Dalbergia*, having hard reddish or dark wood with a strongly marked grain. **2.** The wood of a rosewood tree.

**Rosh Ha·sha·nah** *also* **Rosh Ha·sha·na** *or* **Rosh Ha·sho·na** *or* **Rosh Ha·sho·nah** (rŏsh′ hə-shä′nə, -shō′-, rōsh′-) *n.* [Heb. *rōsh hasshānāh*, beginning of the year.] The Jewish New Year, celebrated by Orthodox and Conservative Jews on the first and second of Tishri and by Reform Jews on the first of Tishri only.

**Ro·si·cru·cian** (rō′zĭ-kro͞o′shən, rŏz′ĭ-) *n.* [< Med. Lat. *Rosae Crucis*, transl. of G. *Rosenkreutz*, surname of the reputed founder of the society in the 15th cent.] **1.** A member of an international fraternity of religious mystics devoted to application of esoteric religious doctrine in modern life. **2.** A member of any of several secret religious organizations similar to the Rosicrucian fraternity, active in the 17th and 18th cent. **—Ro′si·cru′cian** *adj.* **—Ro′si·cru′cian·ism** *n.*

**ros·in** (rŏz′ĭn) *n.* [ME, var. of *resin*, resin.] A translucent yellowish to dark brown resin derived from the sap of various pine trees and used to increase sliding friction on the bows of certain stringed instruments, in manufactured products such as varnishes, inks, and linoleum, and in soldering compounds. *—vt.* **-ined, -in·ing, -ins.** To coat or rub with rosin. **—ros′in·y** *adj.*

**rosin oil** *n.* A white to brown viscous liquid obtained by fractional distillation of rosin and used in electrical insulation, lubricants, and printing inks.

**ros·in·ol** (rŏz′ĭ-nôl′, -nōl′) *n.* Rosin oil.

**ros·in·weed** (rŏz′ĭn-wēd′) *n.* A North American plant of the genus *Silphium* and related genera, esp. the compass plant, the cup plant, and the gum plant, having a resinous juice.

**ros·tel·la** (rŏ-stĕl′ə) *n. pl. of* ROSTELLUM.

**ros·tel·late** (rŏs′tə-lāt′, rŏ-stĕl′ĭt) *adj.* [NLat. *rostellatus* < *rostellum*, rostellum.] Having a rostellum.

**ros·tel·lum** (rŏ-stĕl′əm) *n., pl.* **ros·tel·la** (rŏ-stĕl′ə) [NLat. < Lat., dim. of *rostrum*, beak.] *Biol.* A small beaklike part, as a projection on the stigma of an orchid, a tubular mouth part on some insects, or the hooked projection on the head of a tapeworm. **—ros·tel′lar** *adj.*

**ros·ter** (rŏs′tər, rô′stər) *n.* [Du. *rooster*, roster, gridiron < *rooster*, to roast.] **1.** A list of names. **2.** A list of the names of military personnel enrolled for active duty.

**ros·tra** (rŏs′trə, rô′strə) *n. var. pl. of* ROSTRUM.

**ros·trate** (rŏs′trāt′, -trĭt, rô′strāt′, -strĭt) *adj.* [Lat. *rostratus* < *rostrum*, beak.] *Biol.* Having a beaklike part.

**ros·trum** (rŏs′trəm, rô′strəm) *n., pl.* **ros·trums** *or* **ros·tra** (rŏs′trə, rô′strə) [Lat., beak.] **1.** A raised platform for public speaking. **2.** In ancient Rome: **a.** A ship's curved beaklike prow. **b.** The speakers' platform in the Forum, decorated with the prows of captured enemy ships. **3.** *Biol.* A snoutlike or beaklike projection. **—ros′tral** *adj.*

**ros·y** (rō′zē) *adj.* **-i·er, -i·est. 1.** Having the pink or red color of a rose. **2.** Consisting of, decorated with, or suggestive of roses. **3.** Flushed with a healthy glow. **4.** Marked by optimism : CHEERFUL. **—ros′i·ly** *adv.* **—ros′i·ness** *n.*

**rot** (rŏt) *v.* **rot·ted, rot·ting, rots.** [ME *roten* < OE *rotian*.] *—vi.* **1.** To undergo decomposition, esp. organic decomposition : DECAY. **2.** To become useless or damaged because of decay. **3.** To decay morally. *—vt.* To cause to decay or decompose. *—n.* **1. a.** The process of rotting. **b.** The state of being rotten. **2. a.** Foot rot. **b.** Liver fluke. **3.** Any of several plant diseases marked by the breakdown of tissue and caused by various fungi, bacteria, or microorganisms. **4.** *Archaic.* A disease causing the decay of flesh. **5.** Nonsense. *—interj.* —Used to express contempt or impatience.

**ro·ta** (rō′tə) *n.* [Lat., wheel.] **1.** *Chiefly Brit.* A roster or roll call of names. **2.** *Chiefly Brit.* A round or rotation of duties. **3. Rota.** *Rom. Cath. Ch.* A tribunal of prelates that serves as an ecclesiastical court.

**Ro·tar·i·an** (rō-târ′ē-ən) *n.* A member of the Rotary Club, a major national and international service club.

**ro·ta·ry** (rō′tə-rē) *adj.* [Med. Lat. *rotarius* < Lat. *rota*, wheel.] Of, relating to, causing, or marked by rotation, esp. axial rotation. *—n., pl.* **-ries. 1.** A part or device that rotates around an axis. **2.** A traffic circle.

**rotary engine** *n.* **1.** An engine, as a turbine, in which power is supplied directly to rotary, parts, as vanes. **2.** A radial engine whose cylinders revolve around a stationary crankshaft.

**rotary harrow** *n.* A harrow consisting of a series of freely turning spike-rimmed wheels.

**rotary plow** *n.* A plow having a series of hoes arranged on a revolving power-driven shaft.

**rotary press** *n.* A printing press having a cylinder to which curved plates are attached so that when they revolve they will print onto a continuous roll of paper.

**rotary tiller** *n.* A rotary plow.

**ro·tate** (rō′tāt′) *v.* **-tat·ed, -tat·ing, -tates.** [Lat. *rotare, rotat-* < *rota*, wheel.] *—vi.* **1.** To turn on an axis. **2.** To proceed in sequence : ALTERNATE. *—vt.* **1.** To cause rotation. **2.** To grow or plant (crops) in a fixed order of succession. *—adj.* Having radiating parts : WHEEL-SHAPED. **—ro′tat′a·ble** *adj.*

**ro·ta·tion** (rō-tā′shən) *n.* **1. a.** Motion in which the path of every point in the moving object is a circle or circular arc centered on a specified axis, esp. on an internal axis <the axial *rotation* of a planet> **b.** A single complete cycle of such motion : REVOLUTION. **2.** *Math.* A coordinate transformation consisting of an angular displacement or successive angular displacements of coordinate axes with the origin remaining fixed. **3.** Uniform sequential variation. **—ro·ta′tion·al** *adj.*

**ro·ta·tive** (rō′tā′tĭv) *adj.* **1.** Of, relating to, causing, or marked by rotation. **2.** Marked by or occurring in succession or alternation. **—ro′ta′tive·ly** *adv.*

**ro·ta·tor** (rō′tā′tər) *n.* One that rotates.

**ro·ta·to·ry** (rō′tə-tôr′ē, -tōr′ē) *adj.* **1.** Of, relating to, causing, or marked by rotation. **2.** Proceeding or occurring in alternation or succession.

**rote**[1] (rōt) *n.* [ME.] **1.** Memorization using routine or repetition without full comprehension <students learning by *rote*> **2.** Thoughtless, mechanical repetition or routine.

**rote**[2] (rōt) *n.* [Prob. of Scand. orig.] The sound of surf breaking on the shore.

**rote**[3] (rōt) *n.* [ME < OFr., of Germanic orig.] A medieval stringed instrument.

**ro·te·none** (rōt′ə-nōn′) *n.* [J. *rōten*, derris + -ONE.] A white crystalline compound, $C_{23}H_{22}O_6$, extracted from the roots of derris and cubé and used as an insecticide.

**rot·gut** (rŏt′gŭt′) *n. Slang.* Inferior liquor.

**ro·ti·fer** (rō′tə-fər) *n.* [< NLat. *Rotifera*, phylum name < Lat. *rota*, wheel.] Any of various minute multicellular aquatic organisms of the phylum Rotifera, having at the anterior end a wheellike ring of cilia. **—ro·tif′er·al** (-tĭf′ər-əl), **ro·tif′er·ous** *adj.*

**ro·ti·form** (rō′tə-fôrm′) *adj.* [Lat. *rota*, wheel + -FORM.] Shaped like a wheel.

**ro·tis·se·rie** (rō-tĭs′ə-rē) *n.* [Fr. *rôtisserie* < OFr. *rostisserie* < *rostir*, to roast, of Germanic orig.] **1.** A cooking device equipped with a rotating spit on which food, as meat, is roasted. **2.** A shop or restaurant where meats are roasted to order.

**rot·l** (rŏt′l) *n.* [Ar. *raṭl*.] A unit of weight used in countries bordering on the Mediterranean and in nearby areas, varying in different regions from about one to five pounds.

**ro·to·gra·vure** (rō'tə-grə-vyo͝or') *n.* [Lat. *rota,* wheel + GRAVURE.] **1.** An intaglio printing process in which letters and pictures are transferred from an etched copper cylinder to a web of material, as paper or plastic, in a rotary press. **2.** Printed material, as a newspaper section, produced by rotogravure.

**ro·tor** (rō'tər) *n.* [Contraction of ROTATOR.] **1.** A rotating part of an electrical or mechanical device. **2.** An assembly of rotating horizontal airfoils, as that of a helicopter.

**rotor ship** *n.* A ship propelled by one or more tall cylindrical rotors operated by wind power.

**ro·to·till** (rō'tə-tĭl') *vt.* **-tilled, -till·ing, -tills.** [Back-formation < ROTOTILLER.] To turn over with a rotary plow.

**Ro·to·till·er** (rō'tə-tĭl'ər). A trademark for a rotary cultivator.

**rot·ten** (rŏt'n) *adj.* **-er, -est.** [ME *roten* < ON *rotinn.*] **1.** Putrefied or decayed : DECOMPOSED. **2.** Having a foul odor resulting from or suggestive of decay : PUTRID. **3.** Made unsound or weak by rot. **4.** Morally corrupt : DESPICABLE. **5.** Very bad : WRETCHED <a *rotten* climate> **—rot'ten·ly** *adv.* **—rot'ten·ness** *n.*

**rotten borough** *n.* An election district having only a few voters but the same voting power as other more populous districts.

**rot·ten·stone** (rŏt'n-stōn') *n.* A friable variety of tripoli, the product of decomposed siliceous limestone, used for polishing.

**rott·wei·ler** (rŏt'wī'lər, rôt'vī'-) *n.* [G. < *Rottweil,* city in southwestern Germany where it was orig. bred.] A German breed of dog having a stocky body, short black fur, and tan face markings.

**ro·tund** (rō-tŭnd') *adj.* [Lat. *rotundus* < *rota,* wheel.] **1.** Characterized by roundness : ROUNDED. **2.** Sonorous. **3.** Quite plump : CHUBBY. **—ro·tund'ly** *adv.* **—ro·tund'ness** *n.*

**ro·tun·da** (rō-tŭn'də) *n.* [Ital. < Lat., fem. of *rotundus,* rotund.] **1.** A circular hall or building, esp. one with a dome. **2.** A large room with a high ceiling, as a hotel lobby.

**ro·tun·di·ty** (rō-tŭn'dĭ-tē) *n., pl.* **-ties. 1.** The state of being round or plump. **2.** A rotund object or protrusion.

**ro·tu·rier** (rō-to͝or'ē-ā', -tyo͝or'-) *n.* [Fr. < OFr. < *roture,* newly cultivated land < Lat. *ruptura,* action of breaking < *rumpere,* to break.] A commoner.

**rou·ble** *or* **ru·ble** (ro͞o'bəl) *n.* [R. *rubl'.*] —See table at CURRENCY.

**rou·é** (ro͞o-ā') *n.* [Fr. < p.part. of *rouer,* to break on a wheel < Lat. *rotare,* to rotate < *rota,* wheel.] A lecherous man : RAKE.

**rouge** (ro͞ozh) *n.* [Fr. < OFr., of red color < Lat. *rubeus.*] **1.** A red or pink cosmetic for coloring the cheeks or lips. **2.** A reddish powder, chiefly ferric oxide, used to polish metals or glass. *—v.* **rouged, roug·ing, roug·es.** *—vt.* **1.** To put rouge on. **2.** To color with rouge. *—vi.* To use rouge as a cosmetic.

**rouge et noir** (ro͞ozh' ā nwär') *n.* [Fr., red and black.] A gambling card game played at a table marked with two red and two black diamond-shaped spots on which bets are placed.

**rough** (rŭf) *adj.* **-er, -est.** [ME < OE *rūh.*] **1.** Having an uneven bumpy surface. **2.** Coarse or shaggy to the touch <a *rough* woolen shirt> **3.** Marked by violent motion : TURBULENT <*rough* seas> **4.** Severely inclement : STORMY. **5.** Boisterous, unruly, or rowdy. **6.** Not gentle or careful : VIOLENT <*rough* treatment> **7.** Brutal : savage. **8.** Rude : uncouth. **9.** *Informal.* Trying or unpleasant to endure or do : DIFFICULT. **10.** Harsh to the ear <a *rough,* scraping sound> **11.** Harsh or sharp to the taste <a *rough,* immature wine> **12.** Lacking finesse or polish. **13.** Being in a natural state <*rough* gems> **14.** Not completed, perfected, or fully detailed <a *rough* sketch> **15.** Requiring physical strength rather than intelligence <*rough* labor> *—n.* **1.** Rugged overgrown ground. **2.** The part of a golf course left unmowed and uncultivated. **3.** A rough, irregular, or difficult part or state. **4.** Something hastily worked out or unfinished. **5.** A crude ill-mannered person : BOOR. *—vt.* **roughed, rough·ing, roughs. 1.** To make rough. **2.** To treat roughly or violently : MANHANDLE <*roughed* up the prisoner> **3.** To indicate or prepare in a rough or unfinished form <*rough* out a building plan> *—adv.* In a rough way. **—in the rough.** In a crude or unfinished state. **—rough it.** To do without the usual conveniences and comforts<*roughed it* for three days in the woods> **—rough'er** *n.* **—rough'ly** *adv.* **—rough'ness** *n.*

**rough·age** (rŭf'ĭj) *n.* The relatively coarse, indigestible parts of certain foods and fodder that contain cellulose and stimulate peristalsis.

**rough-and-read·y** (rŭf'ən-rĕd'ē) *adj.* Rough and crude, yet effective.

**rough-and-tum·ble** (rŭf'ən-tŭm'bəl) *adj.* Marked by roughness and disregard for rules or order <a *rough-and-tumble* fight>

**rough breathing** *n.* **1.** An aspirate sound in Greek like that of the letter *h* in English. **2.** The mark (ʽ) placed over initial sounds in Greek to indicate a preceding aspirate.

**rough·cast** (rŭf'kăst') *n.* **1.** A coarse plaster used for outside wall surfaces. **2.** A rough preliminary model. *—vt.* **-cast, -cast·ing, -casts. 1.** To plaster (e.g., a wall) with roughcast. **2.** To shape or work into a rough or preliminary form. **—rough'cast'er** *n.*

**rough·dry** (rŭf'drī') *vt.* **-dried, -dry·ing, -dries.** To dry (laundry) without smoothing out or ironing.

**rough·en** (rŭf'ən) *vt.* & *vi.* **-ened, -en·ing, -ens.** To make or become rough.

**rough·hew** (rŭf'hyo͞o') *vt.* **-hewed** *or* **-hewn** (-hyo͞on'), **-hew·ing, -hews. 1.** To hew (e.g., timber) roughly, without finishing. **2.** To make in rough form.

**rough·house** (rŭf'hous') *n.* Rowdy, boisterous play or behavior. *—v.* **-housed, -hous·ing, -hous·es.** *—vi.* To engage in roughhouse. *—vt.* To treat or handle roughly, usu. in fun.

**rough-leg·ged hawk** (rŭf'lĕg'ĭd) *n.* A hawk, *Buteo lagopus,* with dark plumage and whitish feathers covering the legs.

**rough·neck** (rŭf'nĕk') *n.* **1.** An uncouth person. **2.** A rowdy.

**rough·rid·er** (rŭf'rī'dər) *n.* **1.** A skilled rider of little-trained horses, esp. one who breaks horses for riding. **2. Roughrider.** A member of the 1st U.S. Volunteer Cavalry regiment under Theodore Roosevelt in the Spanish-American War.

**rough·shod** (rŭf'shŏd') *adj.* **1.** Shod with horseshoes having projecting nails or points to prevent slipping. **2.** Marked by brutal force.

**rou·lade** (ro͞o-läd') *n.* [Fr. < *rouler,* to roll < OFr. *roler.* —see ROLL.] **1.** *Mus.* **a.** An embellishment consisting of a rapid run of several notes sung to one syllable. **b.** A drum roll. **2.** A slice of meat rolled around a filling and cooked.

**rou·leau** (ro͞o-lō') *n., pl.* **-leaux** *or* **-leaus** (-lōz') [Fr. < OFr. *rolel,* dim. of *role,* roll < Lat. *rotula,* little wheel, dim. of *rota,* wheel.] **1.** A small roll, esp. of coins wrapped in paper. **2.** A roll or fold of ribbon for piping.

**rou·lette** (ro͞o-lĕt') *n.* [Fr. < OFr. < *rouelle,* dim. of *roue,* wheel < Lat. *rota.*] **1.** A gambling game in which the players bet on which slot of a rotating disk a small ball will come to rest in. **2.** A small toothed disk of tempered steel attached to a handle and used to make rows of dots, slits, or perforations, as in engraving or on a sheet of postage stamps. **3.** Short consecutive incisions made between individual stamps in a sheet for easy separation. *—vt.* **-lett·ed, -lett·ing, -lettes.** To divide or mark with a roulette.

**Rou·ma·ni·an** (ro͞o-mā'nē-ən, -mān'yən) *adj.* & *n. var. of* RUMANIAN.

**round¹** (round) *adj.* [ME < OFr. *ronde* < Lat. *rotundus* < *rota,* wheel.] **1.** Ball-shaped : spherical. **2.** Circular or circular in cross section. **3.** Having a curved surface or edge : not flat or angular. **4.** Pronounced with the lips in a rounded shape. **5.** Whole or complete : FULL. **6. a.** Expressed or designated as a whole number or integer : not fractional. **b.** Not exact : APPROXIMATE. **7.** Considerable : large <a *round* amount> **8.** Brought to a satisfying completion : FINISHED. **9.** Full in tone : SONOROUS. **10.** Brisk : lively <a *round* jogging pace> **11.** Blunt : outspoken <a *round* reprimand> **12.** Done with full force : UNRESTRAINED <a *round* spanking> *—n.* **1.** The state of being round. **2. a.** Something, as a circle, globe, disk, or ring, that is round. **b.** A rounded or curved form or part. **3.** A rung or crossbar, as on a ladder. **4.** A cut of beef from the part of the thigh between the shank and the rump. **5.** A gathering of people : GROUP. **6.** Movement around a circle or about an axis. **7.** A round dance. **8.** A complete course, succession, or series <a *round* of meetings> **9.** *often* **rounds.** A course of customary or prescribed duties, actions or places. **10.** A complete extent or range. **11.** One drink for each person in a group. **12.** A single outburst of cheering or applause. **13. a.** A single shot or volley. **b.** Ammunition for a single shot. **14.** A rounded slice of bread. **15.** A specified number of arrows shot from a specified distance to a target in archery. **16.** An interval of play in games and sports that occupies a specified time, comprises a certain number of plays, or allows each player a turn. **17.** *Mus.* A composition for two or more voices in which each voice enters at a different time with the same melody. *—v.* **round·ed, round·ing, rounds.** *—vt.* **1.** To make round. **2.** To pronounce with rounded lips : LABIALIZE. **3.** To make plump : FILL OUT. **4.** To bring to completion : FINISH. **5.** To express as a round number. **6.** To make a complete circuit of. **7.** To make a turn about or to the other side of <*rounded* a curve in the highway> **8.** To surround : encompass. **9.** To move or cause to proceed in a circular course. *—vi.* **1.** To become round. **2.** To take a circular course. **3.** To turn about, as on an axis : REVERSE. **4.** To become filled out, curved, or plump. **5.** To come to completion or perfection. **—round up. 1.** To seek out and bring together : GATHER. **2.** To herd (cattle) together from various places. *—adv.* Around. *—prep.* **1.** Around. **2.** From the beginning to the end of : THROUGHOUT <plants that grow *round* the year> **—in the round. 1.** With the stage in the center of the audience. **2.** Fully shaped so as to stand free of a background <a statue *in the round*> **—round'ness** *n.*

**round²** (round) *vt.* **round·ed, round·ing, rounds.** [ME *rounden* < OE *rūnian.*] *Archaic.* To whisper.

**round·a·bout** (round'ə-bout') *adj.* Circuitous : indirect. *—n.* **1.** A close-fitting short jacket. **2.** *Chiefly Brit.* A merry-go-round. **3.** *Chiefly Brit.* A traffic circle.

**round clam** *n.* The quahog.

**round dance** *n.* **1.** A folk dance performed with the dancers arranged in a circle. **2.** A ballroom dance in which couples proceed in a circular direction around the room.

**round·ed** (roun'dĭd) *adj.* **1.** Shaped in a circle or sphere. **2.** Pronounced with the lips shaped ovally : LABIALIZED. **3.** Balanced : complete. **—round'ed·ness** *n.*

**roun·del** (roun'dəl) *n.* [ME < OFr. *rondel*, dim. of *ronde*, circle, round. —see ROUND.] **1.** A curved form, esp. a semicircular panel, recess, or window. **2. a.** A rondel. **b.** A rondeau. **c.** An English variation of the rondeau consisting of three triplets with a refrain after the first and third.

**roun·de·lay** (roun'də-lā') *n.* [OFr. *rondelet*, dim. of *rondel*, rondel.] A song or poem with a frequently or regularly recurring refrain.

**round·er** (roun'dər) *n.* **1.** One that rounds, esp. a tool for rounding edges and corners. **2.** One who makes rounds, as a guard. **3. Rounder.** *Chiefly Brit.* A Methodist preacher who travels a circuit among congregations. **4.** *Informal.* A dishonest or dissolute person. **5. rounders** *(sing. in number).* An English game similar to baseball.

**round hand** *n.* A style of handwriting in which the letters are rounded and full rather than angular.

**Round·head** (round'hĕd') *n.* [From the close-cropped hair of the Puritans.] A member or supporter of the Parliamentary or Puritan party during the English Civil War.

**round·house** (round'hous') *n.* **1.** A circular building for housing and servicing locomotives. **2.** A cabin on the after part of the quarterdeck of a ship. **3.** A meld of four kings and four queens in pinochle. **4.** *Slang.* A blow or swing delivered with a sweeping sidearm movement.

**round·ish** (roun'dĭsh) *adj.* Somewhat round. **—round'ish·ness** *n.*

**round·let** (round'lĭt) *n.* [ME < OFr. *rondelet*, dim. of *rondel*, rondel.] **1.** A little circle. **2.** A small circular object.

**round·ly** (round'lē) *adv.* **1.** In the form of a sphere or circle. **2.** With full force or vigor : THOROUGHLY <*roundly* praised>

**round robin** *n.* **1.** A tournament in which each contestant is matched in turn against every other contestant. **2.** A petition or protest on which the signatures are arranged in the form of a circle in order to conceal the order of signing. **3.** A letter sent among members of a group, often with comments added by each person in turn.

**round-shoul·dered** (round'shōl'dərd) *adj.* Having the shoulders and upper back rounded.

**rounds·man** (roundz'mən) *n.* **1.** A police officer in charge of several patrolmen. **2.** A person who makes rounds, as a delivery person.

**round steak** *n.* A lean, oval cut of beef from between the rump and shank.

**Round Table** *n.* **1. a.** The circular table of King Arthur and his knights. **b.** The knights of King Arthur as a group. **2. round table.** A discussion or conference with several participants.

**round-the-clock** (round'thə-klŏk') *adj.* Lasting or continuing throughout the 24 hours of the day : CONTINUOUS.

**round trip** *n.* A trip from one place to another and back, usu. over the same route.

**round·up** (round'ŭp') *n.* **1. a.** The herding together of cattle for inspection, branding, or shipping. **b.** The cattle that are herded together. **c.** The persons and horses employed in such herding. **2.** A gathering up, as of persons under suspicion by the police. **3.** A summary <the late news *roundup*>

**round·worm** (round'wûrm') *n.* A nematode.

**roup** (ro͞op) *n.* [Orig. unknown.] An infectious disease of poultry and pigeons characterized by inflammation and discharge from the mouth and eyes.

**rouse** (rouz) *v.* **roused, rous·ing, rous·es.** [ME *rowsen*, to shake feathers.] —*vt.* **1.** To arouse from sleep, unconsciousness, or inactivity. **2.** To stir up, as to action or anger : EXCITE. —*vi.* **1.** To awaken. **2.** To become active. —*n.* An act or instance of rousing. **—rous'er** *n.*

**rous·ing** (rou'zĭng) *adj.* **1.** Inducing excitement : STIRRING <a *rousing* lecture> **2.** Full of vigor : LIVELY. **3.** *Informal.* Extraordinary : notable. **—rous'ing·ly** *adv.*

**Rous sarcoma** (rous) *n.* [After Francis P. *Rous* (1879–1970).] A malignant sarcoma that can be produced in chickens by inoculation with the specific viral causative agent.

**roust** (roust) *vt.* **roust·ed, roust·ing, rousts.** [Alteration of ROUSE.] To rout, esp. out of bed.

**roust·a·bout** (rous'tə-bout') *n.* **1.** A deck or wharf laborer, esp. on the Mississippi River. **2.** A laborer in a circus. **3.** A temporary or unskilled laborer, esp. one employed in an oil field.

**rout**[1] (rout) *n.* [ME *route* < OFr. < Lat. *rumpere*, to break.] **1.** A haphazard retreat or wild flight following defeat. **2.** An overwhelming defeat. **3. a.** A disorderly crowd of persons : MOB. **b.** People of the lowest class : RABBLE. **4.** A public disturbance : RIOT. **5.** *Archaic.* A company of people or animals, esp. of knights or wolves. **6.** *Archaic.* A wild party. —*vt.* **rout·ed, rout·ing, routs. 1.** To put to wild flight or retreat. **2.** To defeat decisively.

**rout**[2] (rout) *v.* **rout·ed, rout·ing, routs.** [Var. of ROOT[2].] —*vi.* **1.** To dig with the snout : ROOT. **2.** To poke around : RUMMAGE. —*vt.* **1.** To dig up with the snout. **2.** To expose to view : UNCOVER. **3.** To hollow, scoop, or gouge out. **4.** To drive or force out : EJECT.

**rout**[3] (rout, ro͞ot) *vi.* **rout·ed, rout·ing, routs.** [ME *routen* < ON *rauta*, to roar.] *Chiefly Brit.* To make a loud noise or clamor.

**route** (ro͞ot, rout) *n.* [ME < OFr. < Lat. *ruptus*, p.part. of *rumpere*, to break.] **1. a.** A road, course, or way for travel from one place to another. **b.** A highway. **c.** A means of reaching a goal. **2.** A customary line of travel. **3.** A specified course or territory assigned to a sales representative or a delivery person. —*vt.* **rout·ed, rout·ing, routes. 1.** To send along : FORWARD. **2.** To schedule the order of (a sequence of procedures).

**rout·er**[1] (ro͞o'tər, rou'tər) *n.* One that routes.

**rout·er**[2] (rou'tər) *n.* **1.** One that routs. **2.** A machine with a usu. high-speed, vertical cutting head for milling wood or metal.

**rou·tine** (ro͞o-tēn') *n.* [Fr. < OFr. < *route*, route.] **1.** A prescribed and detailed course of action to be followed regularly : standard procedure. **2.** A set of customary and often mechanically performed procedures or activities. **3.** A set piece of entertainment, esp. in a nightclub or theater. **4.** *Slang.* A particular kind of behavior or activity <went into your crying *routine*> —*adj.* **1.** In accordance with established procedure <a *routine* check of licenses> **2.** Habitual : regular. **3. a.** Not special : ORDINARY <a *routine* dinner> **b.** Lacking interest or originality <a *routine* production of the play> **—rou·tine'ly** *adv.* **—rou·tin'ism** *n.* **—rou·tin'ist** *n.*

☆ ***syns:*** ROUTINE, CUT-AND-DRIED, FORMULAIC, STANDARD, STOCK *adj. core meaning* : lacking interest or originality <a *routine* spy novel>

**rou·tin·ize** (ro͞o-tē'nīz', ro͞ot'n-īz') *vt.* **-ized, -iz·ing, -iz·es. 1.** To establish a routine for. **2.** To reduce to a routine. **—rou·tin'i·za'·tion** *n.*

**roux** (ro͞o) *n.* [Fr. *(beurre) roux*, browned (butter) < *roux*, reddish brown < OFr. *rous* < Lat. *russus*, red.] A mixture of flour and fat, as butter, cooked together and used as a thickening.

**rove**[1] (rōv) *v.* **roved, rov·ing, roves.** [ME *roven*, to shoot arrows at a mark.] —*vi.* To wander about aimlessly, esp. over a wide area : ROAM. —*vt.* To wander or roam around, over, or through. **—rove** *n.*

**rove**[2] (rōv) *vt.* **roved, rov·ing, roves.** [Orig. unknown.] **1.** To card (wool). **2.** To put (fibers) through an eye or opening. **3.** To stretch and twist (fibers) before spinning. —*n.* A slightly twisted and extended fiber or sliver.

**rove**[3] (rōv) *v. var. p.t. & p.p. of* REEVE[2].

**rove beetle** *n.* [Poss. < ROVE[1].] Any of numerous beetles of the family Staphylinidae, often found in decaying matter and having slender bodies and short wing covers.

**rov·er**[1] (rō'vər) *n.* **1.** A wanderer. **2.** A mark in archery selected by chance.

**ro·ver**[2] (rō'vər) *n.* [ME < MDu., robber < *roven*, to rob.] **1.** A pirate. **2.** A pirate vessel.

**row**[1] (rō) *n.* [ME < OE *rāw*.] **1.** A series of objects placed next to each other, usu. in a straight line. **2.** A succession without a break or gap in time <won the election for three years in a *row*> **3.** A continuous line of buildings along a street. —*vt.* **rowed, row·ing, rows.** To place in a row.

**row**[2] (rō) *v.* **rowed, row·ing, rows.** [ME *rowen* < OE *rōwan*.] —*vi.* To propel a boat with or as if with oars. —*vt.* **1.** To propel (a boat) with or as if with oars. **2.** To carry in or on a boat propelled by oars. **3.** To propel or convey in a way resembling rowing. **4.** To employ (a specified number of oars or oarsmen). **5.** To pull (an oar) as part of a racing crew. **6.** To race against by rowing. —*n.* **1. a.** An act or instance of rowing. **b.** A shift at the oars of a boat. **2.** A trip or excursion in a rowboat. **—row'er** *n.*

**row**[3] (rou) *n.* [Orig. unknown.] A noisy disturbance or quarrel : BRAWL. —*vi.* **rowed, row·ing, rows.** To take part in a row.

**row·an** (rou'ən) *n.* [Of Scand. orig.] A small deciduous tree, *Sorbus aucuparia* native to Europe, having white flower clusters and orange-red berries.

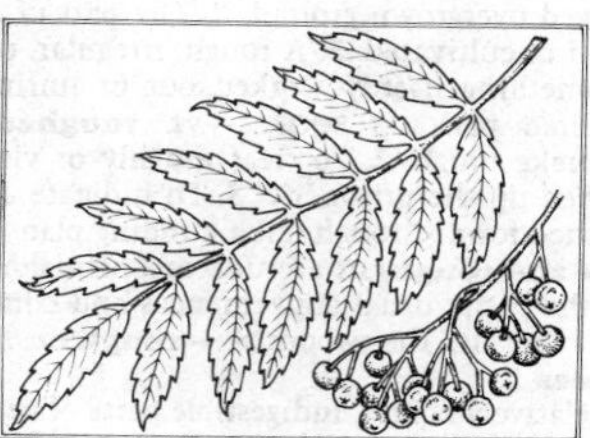

**rowan**
*Detail of leaves and berries*

**row·boat** (rō'bōt') *n.* A small boat propelled by oars.

**row·dy** (rou'dē) *n., pl.* **-dies.** [Prob. < ROW[3].] A disorderly person. —*adj.* **-di·er, -di·est.** Disorderly. **—row'di·ly** *adv.* **—row'di·ness** *n.* **—row'dy·ism** *n.*

**row·el** (rou'əl) *n.* [ME *rowelle* < OFr. *rouelle*, dim. of *roue*, wheel < Lat. *rota*.] A sharp-toothed wheel inserted into the end of the shank of a spur. —*vt.* **-eled, -el·ing, -els** *or* **-elled, -el·ling, -els.** To spur or urge with or as if with a rowel.

**row·en** (rou'ən) *n.* [ME *rewain* < OFr. *regain* : *re-*, again + *gaaignier*, to till.] A second crop, as of hay, in one season.

**row house** *n.* One of a series of identical houses situated side by side and joined by common walls.
**row·lock** (rō'lŏk') *n. Chiefly Brit.* An oarlock.
**roy·al** (roi'əl) *adj.* [ME < OFr. *roial* < Lat. *regalis* < *rex,* king.] **1.** Of or relating to a monarch. **2.** Of the rank of a monarch. **3.** Of, relating to, or in the service of a kingdom. **4.** Issued or performed by a sovereign <a *royal* edict> **5.** Founded, authorized, or chartered by a sovereign. **6.** Befitting royalty : STATELY. **7.** Superior in quality or size. —*n.* **1.** *Naut.* A sail set on the royalmast. **2.** A paper size, 20 by 25 inches for printing, 19 by 24 inches for writing. **3.** *Informal.* A member of a monarch's family. **—roy'al·ly** *adv.*
**royal blue** *n.* A deep to strong blue. **—roy'al-blue'** *adj.*
**royal fern** *n.* A deep-rooted fern, *Osmunda regalis,* of worldwide distribution, with tall upright fronds.
**royal flush** *n.* A poker hand consisting of the five highest cards of one suit.
**roy·al·ism** (roi'ə-lĭz'əm) *n.* Monarchism.
**roy·al·ist** (roi'ə-lĭst) *n.* **1.** A supporter of government by a monarch. **2. Royalist. a.** CAVALIER 3. **b.** An American loyal to British rule during the American Revolution : TORY.
**royal jelly** *n.* A nutritious substance secreted in the pharyngeal glands of worker bees, serving as food for the young larvae and as the only food for those that develop into queen bees.
**roy·al·mast** *also* **royal mast** (roi'əl-măst') *n. Naut.* The small mast just above the topgallant mast.

**royalmast**

**royal palm** *n.* Any of several palm trees of the genus *Roystonea,* having tall naked trunks surmounted by large tufts of pinnate leaves.
**royal poinciana** *n.* A tropical and semitropical tree, *Delonix regia* native to Madagascar, with large scarlet and yellow flower clusters and long pods.
**royal purple** *n.* A moderate or strong violet to deep purple or dark reddish purple. **—roy'al-pur'ple** *adj.*
**roy·al·ty** (roi'əl-tē) *n., pl.* **-ties. 1. a.** A person of royal rank or lineage. **b.** Monarchs and their families as a group. **2.** The lineage or rank of a monarch. **3.** The status, power, or authority of a monarch. **4.** Royal quality or bearing. **5.** A kingdom or possession ruled by a monarch. **6.** A right or prerogative of the crown, as that of receiving a percentage of the proceeds from mines in the royal domain. **7. a.** The granting of a right by a sovereign to a corporation or individual to exploit specified natural resources. **b.** The payment for such a right. **8. a.** A share paid to an author or composer out of the proceeds resulting from the sale or performance of his or her work. **b.** A share in the proceeds paid to an inventor or proprietor for the right to use his or her invention or services. **c.** A share of the profit or product reserved by the grantor, esp. of an oil or mining lease.
**RPG** (är'pē-jē') *n.* [R(EPORT) P(ROGRAM) G(ENERATOR).] *Computer Sci.* A programming language designed for business reporting that generates specific programs from the user's specifications.
**-rrhagia** *suff.* [NLat. < Gk. < *rhēgnunai,* to burst forth.] Abnormal or excessive flow or discharge <meno*rrhagia*>
**-rrhea** or **-rrhoea** *suff.* [ME *-ria* < LLat. *-rrhoea* < Gk. *-rrhoia* < *rhoia,* a flowing < *rhein,* to flow.] Flow : discharge <sebo*rrhea*>
**Ru** *symbol for* RUTHENIUM.
**rub** (rŭb) *v.* **rubbed, rub·bing, rubs.** [ME *rubben.*] —*vt.* **1. a.** To subject to the action of something that moves back and forth with friction and pressure. **b.** To cause to move along a surface with friction and pressure. **2.** To annoy : irritate. —*vi.* **1. a.** To move along a surface with friction and pressure. **b.** To chafe with friction. **c.** To cause annoyance or irritation. **2.** *Informal.* To continue in a given situation, usu. with difficulty. **3.** To admit rubbing <a chalkboard that *rubs* clean easily> **4.** To be transferred by proximity or contact <Their luck finally *rubbed* off on me.> **—rub in.** To carp on (an unpleasant subject). **—rub out. 1.** To obliterate by or as if by rubbing. **2.** *Slang.* To kill : murder. **—rub up. 1.** To refresh one's knowledge of. **2.** To improve or increase the keenness of (a mental faculty). —*n.* **1.** An act of rubbing. **2.** Application of friction and pressure <a body *rub*> **3.** Unevenness on a surface. **4.** An injurious or annoying act or remark. **5.** Difficulty <"Aye, there's the *rub*" —Shakespeare> **—rub elbows (**or **shoulders) with.** To mix or socialize closely with. **—rub the wrong way.** To annoy : vex.
**ru·basse** (rōō-băs', rōō'băs') *n.* [Fr. *rubace* < *rubis,* ruby.] A quartz colored ruby red by its iron-oxide content.
**ru·ba·to** (rōō-bä'tō) [Ital. *(tempo) rubato,* stolen (time) < *rubare,* to rob, of Germanic orig.] *n., pl.* **-tos.** *Mus.* Rhythmic flexibility within a phrase or measure. **—ru·ba'to** *adj.*
**rub·ber[1]** (rŭb'ər) *n.* **1.** A light-cream to dark-amber, amorphous, elastic, solid polymer of isoprene, $(C_5H_8)_n$, gen. prepared by coagulation and drying of the milky sap or latex of various tropical plants, esp. the rubber tree, and subsequently vulcanized, pigmented, and otherwise modified for finishing numerous products including electric insulation, elastic bands and belts, tires, and containers. **2.** Any of numerous synthetic elastic materials of varying chemical composition with properties similar to those of natural rubber. **3.** A low overshoe made of rubber. **4.** *Baseball.* The oblong piece of hard rubber on which the pitcher must stand when delivering the ball. **5. a.** One who rubs. **b.** One who gives a massage. **6.** Something made of rubber, as: **a.** An eraser. **b.** A tire. **c.** A set of tires on a vehicle.

▲ word history: The noun *rubber*[1], denoting a substance made from the sap of the rubber tree, is derived from the verb *rub.* One of the first discovered uses of this substance was rubbing out pencil marks. Although many other uses for rubber have since been found, its original name has stuck.

**rub·ber[2]** (rŭb'ər) *n.* [Orig. unknown.] **1.** A series of games of which two out of three or three out of five must be won to terminate the play. **2.** An odd game played to break a tie.
**rubber band** *n.* An elastic loop of natural or synthetic rubber for holding objects together.
**rub·ber-base paint** (rŭb'ər-bās') *n.* Latex paint.
**rubber check** *n.* A check returned by a bank because of insufficient funds in the account on which it is drawn.
**rub·ber·ize** (rŭb'ə-rīz') *vt.* **-ized, -iz·ing, -iz·es.** To treat, coat, or impregnate with rubber.
**rub·ber·neck** (rŭb'ər-nĕk') *Slang.* —*n.* A gawking sightseer or tourist. —*vi.* **-necked, -neck·ing, -necks.** To look about or survey with unsophisticated amazement or curiosity.
**rubber plant** *n.* **1.** Any of several tropical plants yielding sap that can be coagulated to form crude rubber. **2.** A small tree, *Ficus elastica,* having large, glossy, leathery leaves and popular as a house plant.
**rubber stamp** *n.* **1.** A piece of rubber affixed to a handle and bearing raised characters for making ink impressions, as of names, dates, or instructions. **2.** One that gives mechanical approval or endorsement of a policy without assessing its merit. **3.** Routine approval or endorsement.
**rub·ber-stamp** (rŭb'ər-stămp') *vt.* **-stamped, -stamp·ing, -stamps. 1.** To mark with the imprint of a rubber stamp. **2.** To endorse, vote for, or approve mechanically or routinely.
**rubber tree** *n.* A tree, *Hevea brasiliensis* native to tropical America but widely cultivated throughout the tropics, yielding latex, a major source of commercial rubber.
**rub·ber·y** (rŭb'ə-rē) *adj.* Of or like rubber : ELASTIC.
**rub·bing** (rŭb'ĭng) *n.* **1.** An act of polishing, drying, or cleaning. **2.** A representation of a raised or indented surface made by placing paper over the surface and rubbing the paper gently with a marking agent, as charcoal.
**rub·bish** (rŭb'ĭsh) *n.* [ME *robishe* < AN *robbous.*] **1.** Worthless material : TRASH. **2.** Foolish talk : NONSENSE.
**rub·bish·y** (rŭb'ĭ-shē) *adj.* **1.** Littered with rubbish. **2.** Of no value : WORTHLESS.
**rub·ble** (rŭb'əl) *n.* [ME *rubel.*] **1.** Fragments of rock or masonry crumbled by natural or human forces. **2. a.** Irregular fragments or pieces of rock used in masonry. **b.** The masonry made with such rocks. **—rub'bly** *adj.*
**rub·ble·work** (rŭb'əl-wôrk') *n.* RUBBLE 2b.
**rub·down** (rŭb'doun') *n.* A vigorous body massage.
**rube** (rōōb) *n.* [Prob. < *Rube,* nickname for *Reuben.*] *Slang.* An unsophisticated country fellow : BUMPKIN.
**ru·be·fa·cient** (rōō'bə-fā'shənt) *adj.* [Lat. *rubefaciens, rubefacient-,* pr.part. of *rubefacere,* to make red : *rubeus,* red + *facere,* to make.] Producing redness, as of the skin. —*n.* A skin irritant that causes redness. **—ru'be·fac'tion** (-făk'shən) *n.*
**Rube Gold·berg** (rōōb' gōld'bûrg') *adj.* [After Reuben (*Rube*) L. *Goldberg* (1883–1970), inventor of such contrivances.] Of, relating to, or being a contrivance that accomplishes by complicated means what apparently could have been accomplished simply.
**ru·bel·la** (rōō-bĕl'ə) *n.* [NLat., fem. of Lat. *rubellus,* red < *ruber.*] German measles.
**ru·bel·lite** (rōō'bə-līt', rōō-bĕl'īt') *n.* [Lat. *rubellus,* red (< *ruber*) + -ITE.] A red tourmaline used as a gemstone.
**ru·be·o·la** (rōō-bē'ə-lə, rōō'bē-ō'lə) *n.* [NLat., neuter pl. dim. of Lat. *rubeus,* red.] **1.** Measles. **2.** German measles. **—ru·be'o·lar** *adj.*
**ru·bes·cent** (rōō-bĕs'ənt) *adj.* [Lat. *rubescens, rubescent-,* pr.part. of *rubescere,* to grow red < *rubēre,* to be red.] Reddening. **—ru·bes'cence** *n.*
**Ru·bi·con** (rōō'bĭ-kŏn') *n.* [Lat. *Rubico, Rubicon-,* a river in Italy

the crossing of which by Julius Caesar and his army in 49 B.C. began a civil war.] A limit that when passed allows no return.

**ru·bi·cund** (ro͞o′bĭ-kənd) *adj.* [Lat. *rubicundus* < *rubēre,* to be red.] Inclined to a healthy rosiness : RUDDY. **—ru′bi·cun′di·ty** (-kŭn′dĭ-tē) *n.*

**ru·bid·i·um** (ro͞o-bĭd′ē-əm) *n.* [NLat. < Lat. *rubidus,* red < *rubēre,* to be red.] *Symbol* **Rb** A soft, silvery-white, highly reactive alkali element used in photocells and in the manufacture of vacuum tubes; atomic number 37; atomic weight 85.47.

**ru·big·i·nous** (ro͞o-bĭj′ə-nəs) *also* **ru·big·i·nose** (-nōs′) *adj.* [Lat. *rubiginosus* < *rubigo,* rust.] Reddish-brown.

**ru·bi·ous** (ro͞o′bē-əs) *adj.* Of the color of a ruby : RED.

**ru·ble** (ro͞o′bəl) *n. var. of* ROUBLE.

**ru·bric** (ro͞o′brĭk) *n.* [ME *rubrike* < OFr. *rubriche* < Lat. *rubrica,* rubric, red chalk < *ruber,* red.] **1.** A part of a manuscript or book, as a heading, title, or initial letter, that appears in decorative red lettering or is otherwise distinguished from the rest of the text. **2.** A title or heading of a statute or chapter in a code of law. **3. a.** A class or category. **b.** A title : name. **4.** A direction in a liturgical book, as a missal or hymnal. **5.** An authoritative direction or rule. **6.** A short commentary or explanation covering a broad subject. **7.** Red ocher. *—adj.* **1.** Red or reddish. **2.** Written in red. **—ru′bri·cal** *adj.*

**ru·bri·cate** (ro͞o′brĭ-kāt′) *vt.* **-cat·ed, -cat·ing, -cates.** [Lat. *rubricare, rubricat-,* to color red < *rubrica,* rubric.] **1.** To write, arrange, or print as a rubric. **2.** To provide with rubrics. **3.** To establish rules for. **—ru′bri·ca′tion** *n.* **—ru′bri·ca′tor** *n.*

**ru·bri·cian** (ro͞o-brĭsh′ən) *n.* One learned in the rubrics of ecclesiastical ritual.

**ru·by** (ro͞o′bē) *n., pl.* **-bies.** [ME < OFr. *rubi, rubis* < Med. Lat. *rubinus (lapis),* red (stone) < Lat. *rubeus,* red.] **1.** A deep-red translucent corundum highly valued as a precious stone. **2.** Something, as a watch bearing, made from a ruby. **3.** A dark or deep red to deep purplish red. *—adj.* Of the color ruby.

**ruby laser** *n.* A laser utilizing a ruby crystal to produce an intense beam of coherent red light, used in light-transmission communication and for localized heating.

**ru·by-throat·ed hummingbird** (ro͞o′bē-thrō′tĭd) *n.* A small bird, *Archilochus colubris* of eastern North America, having metallic-green upper plumage and in the male a brilliant red throat.

**ruche** (ro͞osh) *n.* [Fr. < OFr. < Med. Lat. *rusca,* bark of a tree, of Celt. orig.] A ruffle or pleat of fine fabric, as lace, used for trimming women's garments.

**ruch·ing** (ro͞o′shĭng) *n.* **1.** A ruche. **2.** Fabric for ruches.

**ruck**[1] (rŭk) *n.* [ME *ruke.*] **1.** A large number mixed together : JUMBLE. **2.** The multitude of common people.

**ruck**[2] (rŭk) *v.* **rucked, ruck·ing, rucks.** [Ult. < ON *hrukka,* wrinkle.] *—vt.* **1.** To make a fold in : CREASE. **2.** To ruffle or disturb : VEX. *—vi.* **1.** To become creased. **2.** To become irritated. *—n.* A crease or pucker, as in cloth.

**ruck·sack** (rŭk′săk′, ro͝ok′-) *n.* [G. : *Ruck,* back (< OHG *hrukki*) + *Sack,* sack < OHG *sac* < Lat. *saccus.*] A knapsack.

**ruck·us** (rŭk′əs) *n.* [Blend of RUCTION and RUMPUS.] *Informal.* A noisy disturbance.

**ruc·tion** (rŭk′shən) *n.* [Poss. alteration of INSURRECTION.] *Informal.* **1.** A riotous disturbance : UPROAR. **2.** A noisy quarrel.

**rudd** (rŭd) *n.* [Prob. < E. *rud,* red. —see RUDDLE.] A European freshwater fish, *Scardinius erythrophthalmus,* with a brownish body and red fins.

**rud·der** (rŭd′ər) *n.* [ME *rodyr* < OE *rōðer.*] **1. a.** A vertically hinged plate of material, as wood or metal, mounted at the stern of a vessel for directing its course. **b.** A similar structure at the tail of an aircraft, used for effecting horizontal changes in course. **2.** Something that controls direction : GUIDE.

**rud·der·post** (rŭd′ər-pōst′) *n. Naut.* A rudderstock.

**rud·der·stock** (rŭd′ər-stŏk′) *n. Naut.* The vertical shaft of a rudder that allows it to pivot when the tiller or steering gear is operated.

**rud·dle** (rŭd′l) *n.* [Dim. of E. *rud,* red < ME *rudde* < OE *rudu.*] Red ocherous iron ore, used in dyeing and marking. *—vt.* **-dled, -dling, -dles.** To dye or mark with or as if with red ocher <*ruddle* sheep>

**rud·dock** (rŭd′ək) *n.* [ME *ruddok* < OE *rudduc.*] *Chiefly Brit. Regional.* The Old World robin.

**rud·dy** (rŭd′ē) *adj.* **-di·er, -di·est.** [ME *rudie* < OE *rudig* < *rudu,* red.] **1.** Having a healthy reddish color. **2.** Reddish : rosy. **3.** *Chiefly Brit.* —Used as an intensive <a *ruddy* fool> **—rud′di·ly** *adv.* **—rud′di·ness** *n.*

**ruddy duck** *n.* A North American duck, *Oxyura jamaicensis,* with stiff pointed tail feathers and in the male brownish-red upper plumage and a black and white head.

**rude** (ro͞od) *adj.* **rud·er, rud·est.** [ME < OFr. < Lat. *rudis,* unformed.] **1.** Primitive : uncivilized <a *rude* and hostile land> **2.** Lowly : humble <a *rude* shepherd's hut> **3. a.** Lacking the graces of civilized life : UNCOUTH. **b.** Without education or knowledge : SIMPLE. **c.** Lacking good manners : DISCOURTEOUS. **4.** Formed without skill or precision : CRUDE <branches forming a *rude* shelter> **5.** Lively : robust. **6.** Sudden and jarring <a *rude* surprise> **—rude′ly** *adv.* **—rude′ness** *n.*

☆ ***syns:*** RUDE, DISCOURTEOUS, ILL-MANNERED, IMPOLITE, UNMANNERLY *adj. core meaning* : lacking good manners <a *rude* sales clerk><*rude* children> ***ant:*** courteous, polite

**ru·der·al** (ro͞o′dər-əl) [NLat. *ruderalis* < Lat. *rudus,* rubbish.] *Bot.* *—adj.* Growing in rubbish, poor land, or waste. *—n.* A ruderal plant.

**rudes·by** (ro͞odz′bē) *n., pl.* **-bies.** [RUDE + *-sby,* as in such names as *Grimsby.*] *Archaic.* An ill-bred or ill-tempered person.

**ru·di·ment** (ro͞o′də-mənt) *n.* [OFr. < Lat. *rudimentum* < *rudis,* rough, unformed.] **1.** *often* **rudiments.** A fundamental principle, element, or skill, as of a field of learning. **2.** *often* **rudiments.** Something in an incipient or undeveloped form <the *rudiments* of social behavior in young children> **3.** *Biol.* An imperfectly or incompletely developed organ or part. **—ru′di·men′tal** (-mĕn′tl) *adj.*

**ru·di·men·ta·ry** (ro͞o′də-mĕn′tə-rē, -mĕn′trē) *adj.* **1.** Of or relating to basic facts or principles that must be learned first. **2.** Being in the earliest stages of development : INCIPIENT. **3.** *Biol.* Imperfectly or incompletely developed : VESTIGIAL <*rudimentary* organs> **—ru′di·men·tar′i·ly** (-târ′ə-lē) *adv.* **—ru′di·men′ta·ri·ness** *n.*

**rue**[1] (ro͞o) *v.* **rued, ru·ing, rues.** [ME *ruen* < OE *hrēowan.*] *—vt.* To feel remorse, regret, or sorrow for. *—vi.* To be penitent. *—n. Archaic.* Sorrow : regret. **—ru′er** *n.*

**rue**[2] (ro͞o) *n.* [ME < OFr. < Lat. *ruta* < Gk. *rhutē.*] An aromatic Eurasian plant of the genus *Ruta,* esp. *R. graveolens,* with evergreen leaves that yield a volatile oil once used medicinally.

**rue anemone** *n.* A small North American woodland plant, *Anemonella thalictroides,* with pinkish or white flowers.

**rue·ful** (ro͞o′fəl) *adj.* **1.** Eliciting sympathy or compassion. **2.** Feeling or expressing regret. **—rue′ful·ly** *adv.* **—rue′ful·ness** *n.*

**ru·fes·cent** (ro͞o-fĕs′ənt) *adj.* [Lat. *rufescens, rufescent-,* pr.part. of *rufescere,* to become red < *rufus,* red.] Tinged with red. **—ru·fes′cence** *n.*

**ruff**[1] (rŭf) *n.* [Perh. short for RUFFLE[1].] **1.** A stiffly starched frilled or pleated circular collar of fabric, as lace, worn by 16th- and 17th-cent. men and women. **2.** A distinctive collarlike projection around the neck, as of feathers on a bird or of fur on a mammal. **3.** The male of a Eurasian sandpiper, *Philomachus pugnax,* with collarlike erectile feathers around the neck during the breeding season. **—ruffed** *adj.*

**ruff**[2] (rŭf) *n.* [OFr. *roffle.*] **1.** The playing of a trump card when one cannot follow suit. **2.** An old game resembling whist. *—vt. & vi.* **ruffed, ruff·ing, ruffs.** To trump or play a trump.

**ruff**[3] (rŭf) *n.* [ME *ruf,* poss. < *ruch,* rough < OE *rūh.*] A European freshwater fish, *Acerina cernua,* related to the perch.

**ruffed grouse** *n.* A chickenlike North American game bird, *Bonasa umbellus,* with mottled brownish plumage.

**ruf·fi·an** (rŭf′ē-ən, rŭf′yən) *n.* [OFr.] **1.** A rowdy fellow : TOUGH. **2.** A gangster or thug. **—ruf′fi·an·ism** *n.* **—ruf′fi·an·ly** *adj.*

**ruf·fle**[1] (rŭf′əl) *n.* [< ME *ruffelen,* to roughen.] **1.** A decorative strip of frilled or closely pleated fabric. **2.** A bird's ruff. **3. a.** A ruckus : fray **b.** Annoyance : vexation. **4.** An irregularity in or slight disturbance of a surface. *—v.* **-fled, -fling, -fles.** *—vt.* **1.** To disturb the smoothness or regularity of : RIPPLE. **2.** To pleat or gather (fabric) into a ruffle. **3.** To erect (the feathers). **4.** To discompose : fluster. **5.** To flip through (the pages of a book). **6.** To shuffle (cards). *—vi.* **1.** To become rough or irregular. **2.** To flutter. **3.** To become flustered.

**ruf·fle**[2] (rŭf′əl) *n.* [Freq. of obs. *ruff,* a drum roll.] A low continuous beating of a drum that is not as loud as a roll. *—vt.* **-fled, -fling, -fles.** To beat a ruffle on (a drum).

**ruf·fle**[3] (rŭf′əl) *vi.* **-fled, -fling, -fles.** [ME *ruffelen.*] To behave arrogantly : SWAGGER. **—ruf′fler** *n.*

**ru·fous** (ro͞o′fəs) *adj.* [Lat. *rufus,* red.] Being of a color that is strong yellowish pink to moderate orange.

**rug** (rŭg) *n.* [Of Scand. orig.] **1.** A piece of usu. heavy fabric for covering a portion of a floor. **2.** An animal skin used as a floor covering. **3.** *Chiefly Brit.* A piece of thick warm fabric or fur used as a coverlet or lap robe.

**ru·ga** (ro͞o′gə) *n., pl.* **-gae** (-gē′, -gī′) [Lat.] *often* **rugae.** *Biol.* A fold or crease, as in the stomach lining. **—ru′gate′** (-gāt′) *adj.*

**Rug·by** (rŭg′bē) *n.* [After *Rugby* School, England.] A form of football in which players on two competing teams may kick, dribble, or run with the ball and forward passing, substitution of players, and time-outs are not permitted.

**rug·ged** (rŭg′ĭd) *adj.* [ME, shaggy, of Scand. orig.] **1.** Having a rough irregular surface <the *rugged* landscape of the Falkland Islands> **2.** Having strong features marked with wrinkles or furrows. **3.** Tempestuous : stormy. **4.** Demanding great effort, ability, or perseverance. **5.** Lacking culture or polish. **6.** Vigorously healthy : HARDY. **—rug′ged·ly** *adv.* **—rug′ged·ness** *n.*

**rug·ger** (rŭg′ər) *n. Chiefly Brit.* Rugby.

**ru·gose** (ro͞o′gōs′) *also* **ru·gous** (-gəs) *adj.* [Lat. *rugosus* < *ruga,* wrinkle.] **1.** Having creases or wrinkles. **2.** *Bot.* Having a rough and ridged surface <*rugose* leaves> **—ru′gose′ly** *adv.* **—ru·gos′i·ty** (-gŏs′ĭ-tē) *n.*

**ru·in** (ro͞o′ĭn) *n.* [ME *ruine* < OFr. < Lat. *ruina,* ruin, a falling down < *ruere,* to rush.] **1.** Total destruction or disintegration. **2.** The cause of total destruction. **3.** A condition of total collapse <a neighbor-

hood falling into *ruin*> **4.** *often* **ruins.** The remains of something destroyed, disintegrated, or decayed. **5. a.** Total loss of one's health, position, fortune, or honor. **b.** Cause of such loss <Scandal was their *ruin.*> **6.** One that is ruined. —*v.* **-ined, -in·ing, -ins.** —*vt.* **1.** To demolish or destroy. **2.** To injure irreparably. **3.** To reduce to poverty or bankruptcy. **4.** To deprive of chastity. —*vi.* To fall into ruin. **—ru′in·a·ble** *adj.* **—ru′in·a′tion** (rōō′ə-nā′shən) *n.* **—ru′in·er** *n.*

**ru·in·ate** (rōō′ə-nāt′) *adj.* [Med. Lat. *ruinatus,* p.part. of *ruinare,* to ruin < Lat. *ruina,* ruin.] Ruined.

**ru·in·ous** (rōō′ə-nəs) *adj.* **1.** Causing or apt to cause ruin : DESTRUCTIVE. **2.** Falling to ruin. **—ru′in·ous·ly** *adv.* **—ru′in·ous·ness** *n.*

**rule** (rōōl) *n.* [ME *reule* < OFr. < Lat. *regula* < *regere,* to rule.] **1. a.** Governing power or its use or possession : AUTHORITY. **b.** The duration of such power. **2.** An authoritative direction for conduct, esp. one of the regulations governing procedure in a legislative body or a regulation observed by the players in a game, sport, or competition. **3.** A usual or customary course of action or behavior <Patience is the *rule* in that business.> **4.** A statement that describes what is true in most or all cases. **5.** A standard procedure or method for solving a class of mathematical problems. **6.** *Law.* **a.** A court order limited in application to a specific case. **b.** A subordinate regulation governing a specific matter. **7.** RULER 2. **8.** A thin metal strip of various widths and designs, used to print borders or lines, as between columns. —*v.* **ruled, rul·ing, rules.** —*vt.* **1.** To exercise control over : GOVERN. **2.** To dominate by powerful influence. **3.** To keep within proper limits : RESTRAIN. **4.** To determine judicially : DECREE. **5. a.** To mark with straight parallel lines <*rule* a sheet of paper> **b.** To mark (a straight line), as with a ruler. —*vi.* **1.** To exercise authority. **2.** To formulate and issue a decision or decree. **3.** To maintain a specified rate or level. **—as a rule.** In general <*As a rule,* we skip breakfast.> **—rule of thumb.** A useful principle having wide application but not intended to be strictly accurate. **—rule out.** To exclude from consideration <The rain *ruled out* the picnic.> **—rul′a·ble** *adj.*

**ruled surface** *n.* A surface, as a cone or a cylinder, generated by the motion of a straight line.

**rul·er** (rōō′lər) *n.* **1.** One that rules or governs, as a sovereign. **2.** A straight-edged strip, as of wood, metal, or plastic for drawing straight lines and measuring lengths.

**rul·ing** (rōō′lĭng) *adj.* **1.** Exercising authority or control. **2.** Predominant. —*n.* **1.** An act of governing or controlling. **2.** An authoritative or official decision <a judge's *ruling*>

**rum¹** (rŭm) *n.* [Prob. short for obs. *rumbullion.*] **1.** An alcoholic liquor distilled from fermented molasses or sugar cane. **2.** Intoxicating beverages.

**rum²** (rŭm) *adj.* **rum·mer, rum·mest.** [Orig. unknown.] *Chiefly Brit.* Odd : queer.

**ru·ma·ki** (rə-mä′kē) *n.* [Perh. of J. orig.] An appetizer of Japanese origin consisting of a marinated piece of chicken liver and a water chestnut that are wrapped in a slice of bacon and broiled or grilled.

**Ru·ma·ni·an** (rōō-mā′nē-ən, -mān′yən) *also* **Ro·ma·ni·an** (rō-) *or* **Rou·ma·ni·an** (rōō-) —*adj.* Of or relating to Rumania, its people, or their language. —*n.* **1.** A native or inhabitant of Rumania. **2.** The Romance language of the Rumanians.

**rum·ba** *also* **rhum·ba** (rŭm′bə, rōōm′-, rōōm′-) *n.* [Am. Sp. < *rumbo,* revelry < Sp., pomp.] **1.** A complex rhythmical dance originating in Cuba. **2.** A modern ballroom adaptation of the rumba.

**rum·ble** (rŭm′bəl) *v.* **-bled, -bling, -bles.** [ME *romblen,* prob. < MDu. *rommeler.*] —*vi.* **1.** To make a deep, prolonged rolling sound. **2.** To move or proceed with a rumbling sound. **3.** *Slang.* To take part in a gang fight. —*vt.* **1.** To utter with a rumbling sound. **2.** To mix or polish (metal parts) in a tumbling box. —*n.* **1.** A deep, prolonged rolling sound. **2.** A tumbling box. **3.** *Slang.* **a.** Widespread expression of unrest or discontent. **b.** A gang fight. **4.** A luggage compartment or servant's seat in the rear of a carriage. **—rum′bler** *n.* **—rum′bling·ly** *adv.* **—rum′bly** *adj.*

**rumble seat** *n.* An uncovered passenger seat that opens out from the rear of a car.

**ru·men** (rōō′mən) *n., pl.* **-mi·na** (-mə-nə) *or* **-mens.** [Lat., throat.] The first division of the stomach of a ruminant animal, in which food is partly digested before being regurgitated for further chewing.

**ru·mi·nant** (rōō′mə-nənt) *n.* [< NLat. *Ruminantia,* suborder name < Lat. *ruminare,* to ruminate.] A hoofed, even-toed, usu. horned mammal of the suborder Ruminantia, as a cow, sheep, goat, deer, or giraffe, having a stomach divided into four compartments, and chewing a cud consisting of regurgitated, partially digested food. —*adj.* **1.** That chews a cud. **2.** Of or belonging to the Ruminantia. **3.** Meditative : reflective.

**ru·mi·nate** (rōō′mə-nāt′) *v.* **-nat·ed, -nat·ing, -nates.** [Lat. *ruminare, ruminat-* < *rumen,* throat.] —*vi.* **1.** To chew a cud. **2.** To meditate at length : MUSE. —*vt.* To reflect or meditate on. **—ru′mi·na′tion** *n.* **—ru′mi·na′tive** *adj.* **—ru′mi·na′tive·ly** *adv.* **—ru′mi·na′tor** *n.*

**rum·mage** (rŭm′ĭj) *v.* **-maged, -mag·ing, -mag·es.** [Obs. *rummage,* act of packing cargo < OFr. *arrumage* < *arumer,* to stow : *a-,* to (< Lat. *ad-*) + *run,* ship's hold, of Germanic orig.] —*vt.* **1.** To search thoroughly by turning over, handling, or disarranging the contents of. **2.** To discover by searching thoroughly. —*vi.* To make a vigorous, usu. hasty search. —*n.* **1.** A thorough search among a number of things. **2.** A jumble of miscellaneous articles. **—rum′mag·er** *n.*

**rummage sale** *n.* **1.** A sale of assorted secondhand objects contributed by donors to raise money, as for a charity. **2.** A sale, esp. of unclaimed or excess goods, as at a warehouse or wharf.

**rum·mer** (rŭm′ər) *n.* [G. *Römer* < Du. *roemer* < *roemen,* to praise.] A large drinking cup or glass.

**rum·my¹** (rŭm′ē) *n.* [Orig. unknown.] A card game, played in many variations, the object of which is to obtain sets of three or more cards of the same rank or suit.

**rum·my²** (rŭm′ē) *n., pl.* **-mies.** *Slang.* A drunkard.

**rum·my³** (rŭm′ē) *adj.* **-mi·er, -mi·est.** *Chiefly Brit.* RUM².

**ru·mor** (rōō′mər) *n.* [ME < OFr. < Lat.] Unsubstantiated information of uncertain origin usu. spread by talk : HEARSAY. —*vt.* **-mored, -mor·ing, -mors.** To spread or tell by rumor.

**ru·mor·mon·ger** (rōō′mər-mŭng′gər, -mŏng′-) *n.* One who spreads rumors.

**ru·mour** (rōō′mər) *n.* & *v. Chiefly Brit. var. of* RUMOR.

**rump** (rŭmp) *n.* [ME *rumpe,* of Scand. orig.] **1.** An animal's fleshy hindquarters. **2.** A cut of veal or beef from the rump. **3.** The human buttocks. **4.** The part of a bird's back nearest the tail. **5.** The last or inferior part. **6.** A legislature having only a small part of its original membership and so being unrepresentative or lacking authority.

**rum·ple** (rŭm′pəl) *v.* **-pled, -pling, -ples.** [Du. *rompelen* < MDu.] —*vt.* To wrinkle or form into folds or creases. —*vi.* To become wrinkled or creased. **—rum′ple** *n.* **—rum′ply** *adj.*

**rum·pus** (rŭm′pəs) *n.* [Orig. unknown.] A noisy ruckus.

**rumpus room** *n.* A room for recreation and parties.

**rum·run·ner** (rŭm′rŭn′ər) *n.* **1.** One who smuggles liquor across a border. **2.** A boat for smuggling liquor across a border.

**run** (rŭn) *v.* **ran** (răn), **run, run·ning, runs.** [ME *runnen* < OE *rinnan.*] —*vi.* **1. a.** To move swiftly on foot so that both feet leave the ground for an instant during each stride. **b.** To move at a swift gallop. —Used of a horse. **2.** To retreat rapidly : FLEE <snatched the purse and *ran*> **3. a.** To move about at will <cattle *running* loose> **b.** To keep company <*ran* with a fast crowd> **c.** To go about from place to place : ROAM <always *running* around hatless> **4.** To migrate, esp. to move up a river in order to spawn <The shad are *running.*> **5. a.** To go quickly : HURRY <*run* for shelter> **b.** To resort to someone when in trouble or distress <always *running* to their lawyer> **c.** To make a short quick trip or visit <*ran* in for a cup of flour> **6. a.** To participate in a race or contest <*ran* in the Boston Marathon> **b.** To compete in a race for elected office <*ran* for city council> **c.** To finish a race or contest in a specified position <*ran* third> **7.** To move freely, as by rolling or sliding <The truck *ran* downhill.> **8.** To be in mechanical operation <an engine *running*> **9.** To go back and forth esp. on a regular basis <The ferry *runs* every two hours.> **10.** *Naut.* To sail or steer before the wind or on an indicated course <*run* before a storm> **11. a.** To flow in a steady stream. **b.** To emit pus or serous fluid. **12.** To melt and flow <The hot flame made the solder *run.*> **13.** To spread or dissolve, as dyes in fabric <Madras garments *run.*> **14.** To extend, stretch, or reach in a given direction or to a specific point <The freeway *runs* to the next city.> **15.** To extend, spread, or climb, due to growth <ivy *running* up the chimney> **16.** To spread rapidly <plague that *ran* through Europe> **17. a.** To be valid in a given area <The speed limit *runs* only to the city line.> **b.** To be present as a valid accompaniment <Fishing rights *run* with ownership of the land.> **18.** To unravel along a line <The stocking *ran.*> **19.** To continue in effect or operation <a contract with two years to *run*> **20.** To pass <Weeks *ran* into months.> **21.** To tend to persist or recur <Meanness *runs* in that family.> **22. a.** To accumulate : accrue <The interest *runs* from the first of the month.> **b.** To become payable. **23.** To take a specific form or order <reasoning that *runs* thus> **24.** To tend or incline in a given direction <Their tastes *run* to the expensive.> **25.** To occupy or exist in a certain range <Coat sizes *run* from small to large.> **26.** To be presented or performed for a continuous time period <The play *ran* for three months.> **27.** To pass into a specified condition <a family that *ran* into debt> —*vt.* **1. a.** To travel over on foot with speed <*ran* the entire distance in 15 minutes> **b.** To cause (an animal) to move quickly or rapidly. **2.** To allow to move without restraint. **3.** To do or accomplish by or as if by running <*run* errands in town> **4.** To hunt or chase <wild dogs *running* deer> **5.** To bring to a given state by or as if by running <a job that *ran* me ragged> **6.** To cause to move quickly. **7. a.** To cause to compete in or as if in a race. **b.** To present or nominate for elective office. **8.** To cause to move or progress freely. **9.** To cause to function : OPERATE. **10.** To transport <*ran* me into town> **11.** *Naut.* To cause to move on a course <We *ran* our boat into the marina.> **12. a.** To smuggle <*running* guns> **b.** To evade and pass through <*ran* the red light> **13.** To pass over or through <*run* the rapids> **14.** To cause to flow <*run* water> **15.** To stream with <The fountains *ran* pink champagne.> **16. a.** To melt, fuse, or smelt (metal). **b.** To cast or mold (molten metal). **17.** To cause to extend or pass <*run* a rope between poles> **18.** To mark or trace on a surface <*run* a pencil line between two points> **19.** To sew with a continuous line of stitches <*run* a new seam> **20.** To cause to unravel along a line. **21. a.** To cause to collide <*ran* the car into a brick wall> **b.** To cause to penetrate <*ran* a nail into my foot> **22.** To continue to present or perform. **23.** To publish in a periodical <*run* an ad> **24.** To subject or be subjected to <*run* risks> **25. a.** To score (balls or points) consecutively in billiard games <*run* 15 balls> **b.** To clear (the table) in pool by consecutive scores. **26.** To conduct : perform <*run* a lab experiment> **27.** To control, manage, or direct <*run* a business> **—run across.** To find by chance. **—run after. 1.** To pursue : chase. **2.** To seek the company or attention of. **—run against. 1.** To meet unexpectedly. **2.** To work against : OPPOSE. **—run along.** To go away : LEAVE. **—run away. 1. a.** To flee. **b.** To leave one's home, esp. to elope. **2.** To stampede. **—run away with. 1. a.** To make off with hurriedly. **b.** To steal. **2.** To be better than others in (e.g., a performance). **—run down. 1.** To stop because of lack of force or power. **2.** To become tired. **3. a.** To collide with and knock down. **b.** To collide with and cause to sink. **4.** To chase and capture. **5.** To trace the source of <*ran down* all leads> **6.** To disparage. **7.** To review <*run down* a list of options> **8.** *Baseball.* To put a runner out after trapping him or her between two bases. **—run in. 1.** To insert or include as an extra item. **2.** To print a solid body of text without a break. **3.** *Slang.* To take into legal custody. **—run into. 1.** To meet by chance <*ran into* an old classmate> **2.** To encounter <*ran into* trouble on the highway> **3.** To collide with <*ran into* another car> **4.** To amount to <Their net worth *ran into* the billions.> **—run off. 1.** To print, duplicate, or copy <*ran off* 300 copies> **2.** To run away : ELOPE. **3.** To spill over : OVERFLOW. **4.** To decide a competition or contest by a run-off. **5.** To drive off : EJECT <*run* trespassers off> **—run on. 1. a.** To keep going : CONTINUE. **b.** To talk volubly, persistently, and usu. inconsequentially. **2.** To continue a printing text without a formal break. **—run out. 1.** To become used up or exhausted. **2.** RUN OFF 5. **3.** To become void, esp. through the passage of time or an omission. **—run out of.** To exhaust a supply of <*ran out of* oil> **—run out on.** To abandon. **—run over. 1.** To collide with, knock down, and often pass over. **2.** To read or review quickly. **3.** To flow over. **4.** To exceed a limit. **—run through. 1.** To pierce <was *run through* by a sword> **2.** To use up quickly. **3.** To rehearse quickly. **4.** To go over the salient facts of <The crew *ran through* all preflight procedures.> **—run up.** To make or become greater or larger <*run up* a bill> *—n.* **1. a.** A pace faster than a walk. **b.** A gait faster than a canter. **2.** An act of running. **3. a.** A distance covered by or as if by running. **b.** The time required to cover such a distance. **4.** A running race <placed in the mile *run*> **5.** A quick trip or visit. **6.** *Baseball.* A point scored by advancing around the bases and reaching home plate safely. **7.** *Football.* An attempt to carry the ball past or through the opposing line, usu. for a specified distance <a 20-yard *run*> **8. a.** Migration of fish, esp. for spawning. **b.** A group or school of fish ascending a river for spawning. **9.** Unrestricted freedom or use <had the *run* of the house> **10.** A stretch or period of riding, as in a race or to hounds. **11.** A track or slope along or down which something can travel <a ski *run*> **12.** The distance a golf ball rolls after hitting the ground. **13. a.** A regular or scheduled route. **b.** A news reporter's territory. **14. a.** A continuous period of operation, esp. of a machine or factory. **b.** The production achieved during such a period <a press *run* of 5,000 copies> **15. a.** A movement or flow. **b.** The duration of such a flow. **c.** The amount of such a flow. **16.** A pipe or channel through which something flows. **17.** A small fast-flowing brook. **18.** A fall or slide, as of sand or mud. **19.** A continuous length or extent of material <a ten-foot *run* of plastic pipe> **20.** A vein or seam, as of ore or rock. **21.** Direction, configuration, or lie <the *run* of the grain in leather> **22. a.** A trail or way made or frequented by animals. **b.** An outdoor enclosure for domestic animals or poultry. **23. a.** A length of torn or unraveled stitches in a knitted fabric. **b.** A blemish caused by excessive paint flow. **24.** An unbroken series or sequence <a *run* of cold winters><a three-year *run* of performances> **25.** *Mus.* A rapid sequence of notes : ROULADE. **26.** A series of unexpected and urgent demands, as by depositors or customers <a massive *run* on the banks> **27. a.** A continuous set or sequence, as of playing cards in one suit. **b.** A successful sequence of shots or points. **28.** A sustained condition <a *run* of bad luck> **29.** A trend or tendency <the *run* of events> **30.** The average type, group, or category : MAJORITY <The broad *run* of voters want a liberal President.> **31. the runs.** *Slang.* Diarrhea. **—a run for (one's) money.** Strong competition. **—in the long run.** In the final analysis. **—in the short run.** In the immediate future. **—on the run. 1. a.** In rapid retreat. **b.** In hiding <bank robbers *on the run*> **2.** Hurrying busily from place to place. **—run rings around.** To be markedly superior to. **—run short. 1.** To become insufficient in supply <Fuel oil *ran short.*> **2.** To use up <*ran short* of staples>

**run·a·bout** (rŭn'ə-bout') *n.* **1. a.** A small, open automobile or carriage. **b.** A small motorboat. **c.** A light aircraft. **2.** A vagabond : wanderer.

**run·a·gate** (rŭn'ə-gāt') *n.* [Alteration of RENEGADE.] *Archaic.* **1.** A deserter or renegade. **2.** A vagabond.

**run·a·round** (rŭn'ə-round') *n.* **1.** Deception, usu. in the form of evasive excuses. **2.** Type set in a column narrower than the body of a printed text, as on either side of a picture.

**run·a·way** (rŭn'ə-wā') *n.* **1.** One who has run away, as a fugitive. **2.** Something that has escaped from proper confinement or control. **3.** *Informal.* An easy victory. *—adj.* **1.** Escaping or having escaped from control or captivity. **2.** Out of control or confinement <a *runaway* truck> **3.** Easily won. **4.** Of or relating to a rapid price rise.

**run·back** (rŭn'băk') *n. Football.* **1.** The act of returning a kickoff, punt, or intercepted pass. **2.** The distance covered in a runback.

**run·ci·ble spoon** (rŭn'sə-bəl) *n.* [Coined by Edward Lear (1812–1888).] A three-pronged fork, as a pickle fork, curved like a spoon and having a cutting edge.

**runcible spoon**

**run·ci·nate** (rŭn'sə-nāt') *adj.* [Lat. *runcinatus,* p.part. of *runcinare,* to plane < *runcina,* carpenter's plane.] *Bot.* Having saw-toothed divisions directed backward <*runcinate* leaves>

**run-down** (rŭn'doun') *n.* **1.** A point-by-point summary. **2.** *Baseball.* A play in which a runner is put out when trapped between bases. *—adj.* (rŭn'doun'). **1.** Being in poor physical condition. **2.** Unwound and not running.

**rune**[1] (rōōn) *n.* [Dan. < ON *rūn.*] **1.** One of the letters of an alphabet used by ancient Germanic peoples, esp. by the Scandinavians and Anglo-Saxons. **2.** A magic charm. **—run'ic** *adj.*

▲ word history: The word *rune*[1] meaning "a character in the runic alphabet" is probably a modern borrowing of Danish *rune* with the same meaning. The Old Norse ancestor of *rune* is *rūn,* and although it too meant "runic character," it also meant "mystery" and "secret lore." The sense "letter" belongs to this group because in early Germanic times the ability to read and write was a mysterious skill possessed by very few, who used runes to write magic spells and charms and to convey secret messages. The Old English cognate of Old Norse *rūn* was also *rūn,* but it did not survive the 16th century.

**rune**[2] (rōōn) *n.* [Finn. *runo,* of Germanic orig.] A Finnish poem or canto.

**rung**[1] (rŭng) *n.* [ME < OE *hrung.*] **1.** A rod or bar forming a step of a ladder. **2.** A crosspiece between the legs of a chair. **3.** The spoke in a wheel. **4.** *Naut.* One of the spokes or handles on a ship's steering wheel.

**rung**[2] (rŭng) *v. p.p.* of RING[2].

**run-in** (rŭn'ĭn') *n.* **1.** A quarrel : argument. **2.** Matter added to a printed text. *—adj.* (rŭn'ĭn). Added or inserted in a printed text.

**run·nel** (rŭn'əl) *n.* [ME *rynel* < OE < *rinnan,* to run.] **1.** A rivulet : brook. **2.** A narrow channel or course, as for water.

**run·ner** (rŭn'ər) *n.* **1.** A contestant in a race. **2. a.** *Baseball.* One who runs the bases. **b.** *Football.* A ball carrier. **3.** A fugitive. **4.** One who runs errands or delivers messages. **5.** An agent, as for a bank or brokerage house. **6.** One who solicits business, as for a hotel or store. **7.** A smuggler. **8.** A vessel engaged in smuggling. **9.** An operator or manager. **10.** A device in or on which a mechanism slides or moves, as: **a.** The blade of a skate. **b.** The supports on which a drawer slides. **11.** A long narrow carpet. **12.** A long narrow tablecloth. **13.** A roller towel. **14.** *Metallurgy.* A channel along which molten metal is poured into a mold : GATE. **15.** *Bot.* **a.** A slender creeping stem that puts forth roots from nodes spaced at intervals along its length. **b.** A plant, as the strawberry, having such a stem. **c.** A twining vine, as the scarlet runner. **16.** A marine fish of the family Carangidae, as the blue runner, *Caranx crysos* of temperate waters of the American Atlantic coast.

**run·ner-up** (rŭn'ər-ŭp') *n.* One that takes second place.

**run·ning** (rŭn'ĭng) *n.* **1.** The act of one that runs. **2.** Ability or power to run. **3.** The sport or exercise of one who runs. *—adv.* Consecutively <three years *running*> **—in the running. 1.** Entered as a contender in a competition. **2.** Having the possibility of winning or placing well in a competition. **—out of the running. 1.** Not entered as a contender in a competition. **2.** Having no possibility of winning a contest.

**running board** *n.* A narrow footboard extending under and beside the doors of some motor vehicles.

**running hand** *n.* Handwriting done rapidly without lifting the pen from the paper.

**running head** *n.* A title printed at the top of every page or every other page, as in a book.

**running knot** *n.* A slipknot.

**running light** *n.* A light on a vehicle, as a ship, kept lighted between dusk and dawn to indicate position and size.

**running mate** *n.* **1.** The candidate or nominee for the lesser of two closely associated political offices. **2.** A companion. **3.** A horse used to set the pace for another horse in a race.

**running start** *n.* A flying start.

**running stitch** *n.* One of a series of small even stitches.

**running title** *n.* A running head.

**run·ny** (rŭn′ē) *adj.* **-ni·er, -ni·est.** Inclined to flow or run.

**run-off** (rŭn′ôf′, -ŏf′) *n.* **1. a.** Overflow of a fluid from a container. **b.** Rainfall not absorbed by soil. **2.** Eliminated waste products from manufacturing processes. **3.** An extra competition held to break a tie.

**run-of-the-mill** (rŭn′əv-*thə*-mĭl′) *adj.* Not outstanding.

**run-on** (rŭn′ŏn′, -ôn′) *n.* Matter added or appended to text without a formal break. **—run′-on′** *adj.*

**runt** (rŭnt) *n.* [Orig. unknown.] **1.** An undersized animal, esp. the smallest animal of a litter. **2.** A person of small stature. **—runt′i·ness** *n.* **—runt′y** *adj.*

**run-through** (rŭn′thrōō′) *n.* A complete but rapid review or rehearsal, as of a theatrical work.

**run-up** (rŭn′ŭp′) *n.* An often sudden and rapid increase.

**run·way** (rŭn′wā′) *n.* **1.** A path, channel, or track over which something runs. **2.** The bed of a watercourse. **3.** A chute for sliding logs. **4.** A narrow track in a bowling lane on which balls are returned after they are bowled. **5.** A smooth ramp for wheeled vehicles. **6.** A narrow walkway extending from a stage into an auditorium. **7.** A strip of usu. paved level ground on which aircraft take off and land.

**ru·pee** (rōō-pē′, rōō′pē) *n.* [Hindi *rupaīyā* < Skt. *rupyam*, silver < *rūpam*, shape.] —See table at CURRENCY.

**ru·pi·ah** (rōō-pē′ə) *n., pl.* **rupiah** or **-ahs.** [Hindi *rupaīyā.* —see RUPEE.] —See table at CURRENCY.

**ru·pic·o·lous** (rōō-pĭk′ə-ləs) *adj.* [Lat. *rupes*, rock + -COLOUS.] Thriving among or inhabiting rocks.

**rup·ture** (rŭp′chər) *n.* [ME *ruptur* < OFr. *rupture* < Lat. *ruptura* < *rumpere*, to break.] **1. a.** The process of breaking open or bursting. **b.** The state of being broken open or burst. **2.** A break in friendly relations. **3.** *Pathol.* **a.** A hernia, esp. of the groin or intestines. **b.** A tear in bodily tissue. —*v.* **-tured, -tur·ing, -tures.** —*vt.* To break open. —*vi.* To suffer or undergo a rupture. **—rup′tur·a·ble** *adj.*

**ru·ral** (rōōr′əl) *adj.* [ME < OFr. < Lat. *ruralis* < *rus*, country.] **1.** Of or relating to the country : RUSTIC. **2.** Of or relating to people who live in the country. **3.** Of or relating to farming : AGRICULTURAL. **—ru′ral·ly** *adv.*

**rural free delivery** *n.* Free government delivery of mail in rural areas.

**ru·ral·ism** (rōōr′ə-lĭz′əm) *n.* Rurality.

**ru·ral·ist** (rōōr′ə-lĭst) *n.* **1.** One who lives in a rural area. **2.** An advocate of rural life.

**ru·ral·i·ty** (rōō-răl′ĭ-tē) *n., pl.* **-ties. 1.** The quality or state of being rural. **2.** A rural characteristic or trait.

**ru·ral·ize** (rōōr′ə-līz′) *vt.* & *vi.* **-ized, -iz·ing, -iz·es.** To make or become rural. **—ru′ral·i·za′tion** *n.*

**rural route** *n.* A rural mail route.

**ruse** (rōōs, rōōz) *n.* [ME, detour < OFr.< *ruser*, to drive back. —see RUSH.] An action or device intended to mislead or confuse.

**rush**[1] (rŭsh) *v.* **rushed, rush·ing, rush·es.** [ME *rushen* < OFr. *ruser*, to drive back < Lat. *recusare*, to reject : *re-*, back + *causari*, to give as a reason < *causa*, cause.] —*vi.* **1.** To move or act rapidly : HURRY. **2.** To make a sudden or swift attack or charge. **3.** To flow or surge swiftly, often with noise <water *rushing* over the falls> **4.** *Football.* To move the ball by running. —*vt.* **1.** To cause to move or act with vigorous haste or violence. **2.** To perform with great haste. **3.** To attack suddenly and swiftly. **4.** To transport or carry hastily <*rushed* them to school> **5.** To entertain or pay great attention to. —*n.* **1.** A sudden forward motion. **2.** An anxious and impatient movement to get to or from a place. **3.** General haste or busyness. **4.** A sudden attack : ONSLAUGHT. **5.** A rapid, often noisy flow or passage. **6.** *Football.* An attempt to move the ball by running. **7.** *often* **rushes.** The first, unedited print of a motion-picture scene. **8.** A great flurry of activity or press of business. **9.** The first sudden, often euphoric sensation experienced immediately after use of a narcotic. **—rush′er** *n.*

☆ ***syns:*** RUSH, DART, DASH, HASTEN, HURRY, HUSTLE, RACE, RUN, SCOOT, SCURRY, SPEED, TEAR, WHIZ, ZIP, ZOOM *v. core meaning* : to move swiftly <*rushed* to the hospital>

**rush**[2] (rŭsh) *n.* [ME < OE *rysc.*] **1. a.** Any of various grasslike marsh plants of the family Juncaceae, with pliant hollow or pithy stems. **b.** Any of various similar, usu. aquatic plants. **2.** The stem of a rush, used to make baskets, mats, and chair seats.

**rush hour** *n.* A period of heavy traffic. **—rush′-hour′** *adj.*

**rush·y** (rŭsh′ē) *adj.* **-i·er, -i·est. 1.** Resembling or characteristic of rushes. **2.** Abounding in rushes.

**rusk** (rŭsk) *n.* [Sp. *rosca*, coil.] **1.** A light, soft-textured sweetened biscuit. **2.** Sweet raised bread dried and browned in an oven.

**rus·set** (rŭs′ĭt) *n.* [ME < OFr. *rousset* < *rous*, red < Lat. *russus.*] **1.** A moderate to strong brown. **2.** A coarse reddish-brown to brown homespun cloth. **3.** A winter apple with a rough reddish-brown skin. —*adj.* Moderate brown to strong brown.

**Rus·sian** (rŭsh′ən) *n.* [Med. Lat. *Russi* < OR *Rus′.*] **1.** A native or inhabitant of Russia. **2.** A person of Russian descent. **3.** The Slavic language of the Russians that is the official language of the U.S.S.R. **—Rus′sian** *adj.*

**Russian dressing** *n.* Salad dressing, as mayonnaise, with chili sauce, chopped pickles, and pimientos.

**Rus·sian·ize** (rŭsh′ə-nīz′) *vt.* **-ized, -iz·ing, -iz·es.** To make Russian. **—Rus′sian·i·za′tion** *n.*

**Russian olive** *n.* The oleaster.

**Russian Orthodox Church** *n.* An independent branch of the Eastern Orthodox Church headed by the Patriarch of Moscow, or one of its autonomous branches outside Russia.

**Russian roulette** *n.* **1.** A stunt in which a person spins the cylinder of a revolver loaded with one bullet, aims the muzzle at his head, and pulls the trigger. **2.** An act of reckless bravado.

**Russian thistle** *n.* A red-stemmed prickly plant, *Salsola kali tenuifolia* native to Asia, that is a troublesome weed in western North America.

**Russian wolfhound** *n.* The borzoi.

**Russo-** *pref.* [< RUSSIA.] Russia : Russian <*Russo*phobe>

**Rus·so·phil·i·a** (rŭs′ə-fĭl′ē-ə) *n.* Interest in or enthusiasm for Russia and its people, culture, government, or language. **—Rus′so·phile′** (-fīl′, -fĭl) *n.*

**Rus·so·pho·bi·a** (rŭs′ə-fō′bē-ə) *n.* Dislike or fear of Russia or its policies. **—Rus′so·phobe′** (-fōb′) *n.*

**rust** (rŭst) *n.* [ME < OE *rūst.*] **1.** Any of various scaly or powdery reddish-brown or reddish-yellow hydrated ferric oxides formed on iron and iron-containing materials by low-temperature oxidation in the presence of water. **2.** Any of various metallic coatings, esp. oxides, formed by corrosion. **3.** A stain or coating resembling iron rust. **4.** Deterioration, as of ability, resulting from disuse or neglect. **5. a.** Any of various parasitic fungi of the order Uredinales that are harmful to a wide variety of plants. **b.** A plant disease caused by such fungi, characterized by reddish or brownish spots on leaves, stems, and other parts. **6.** A strong brown. —*v.* **rust·ed, rust·ing, rusts.** —*vi.* **1.** To become corroded. **2.** To deteriorate through neglect or inactivity. **3.** To become the color of rust. **4.** To develop a disease caused by a rust fungus. —*vt.* **1.** To corrode or subject (a metal) to rust formation. **2.** To spoil or impair, as by inactivity. **3.** To color something strong brown. **—rust** *adj.* **—rust′a·ble** *adj.*

**rus·tic** (rŭs′tĭk) *adj.* [ME *rustyk* < OFr. *rustique* < Lat. *rusticus* < *rus*, country.] **1.** Of, relating to, or typical of country life. **2.** Not sophisticated : SIMPLE. **3.** Made of rough tree branches. —*n.* **1.** A rural person. **2.** A coarse, crude, or simple person. **—rus′ti·cal·ly** *adv.*

**rus·ti·cate** (rŭs′tĭ-kāt′) *v.* **-cat·ed, -cat·ing, -cates.** [Lat. *rusticari*, *rusticat-*, to live in the country < *rus*, country.] —*vi.* To go to or live in the country. —*vt.* **1.** To send to the country. **2.** *Chiefly Brit.* To suspend (a student) from a university. **3.** To construct (masonry) with conspicuous, often beveled points. **—rus′ti·ca′tion** *n.* **—rus′ti·ca′tor** *n.*

**rus·tic·i·ty** (rŭ-stĭs′ĭ-tē) *n., pl.* **-ties. 1.** The state of being rustic. **2.** A rustic mannerism or trait.

**rus·tle** (rŭs′əl) *v.* **-tled, -tling, -tles.** [ME *rustlen.*] —*vi.* **1.** To move with soft fluttering or crackling sounds. **2.** To move or act vigorously or with speed. **3.** To forage food. **4.** To steal cattle. —*vt.* **1.** To cause to rustle. **2.** To obtain by rustling. **3.** To steal (cattle). **—rus′tler** *n.* **—rus′tling·ly** *adv.*

**rust·less** (rŭst′lĭs) *adj.* Free from rust.

**rust·proof** (rŭst′prōōf′) *adj.* Incapable of rusting.

**rust·y** (rŭs′tē) *adj.* **-i·er, -i·est. 1.** Covered with rust : CORRODED. **2.** Consisting of or produced by rust. **3.** Having a yellowish-red or brownish-red color. **4.** Working or functioning stiffly or incorrectly because of or as if because of rust. **5.** Weakened or impaired by disuse, neglect, or lack of practice. **—rust′i·ly** *adv.* **—rust′i·ness** *n.*

**rut**[1] (rŭt) *n.* [Poss. < OFr. *route*, way. —see ROUTE.] **1.** A sunken track or groove made by the passage of vehicles. **2.** A fixed and usu. boring routine. —*vt.* **rut·ted, rut·ting, ruts.** To furrow.

**rut**[2] (rŭt) *n.* [ME *rutte* < OFr. *rut*, bellowing < LLat. < Lat. *rugire*, to roar.] **1.** A cyclically recurring condition of sexual excitement and reproductive activity in male mammals, as deer. **2.** A periodic condition of mammalian sexual activity : ESTRUS. —*vi.* **rut·ted, rut·ting, ruts.** To be in rut.

**ru·ta·ba·ga** (rōō′tə-bā′gə, rōōt′ə-) *n.* [Dial. Swed. *rotabagge* : *rot*, root (< ON *rōt*) + *bagge*, bag (< ON *baggi*).] **1.** A plant, *Brassica napobrassica* native to Eurasia, having a thick bulbous root used as food and livestock feed. **2.** The edible root of the rutabaga.

**ruth** (ro͞oth) *n.* [ME < *rewen,* to rue < OE *hrēowan.*] **1.** Compassion or sympathy for another. **2.** Regret or misery about one's own misdeeds or flaws.

**Ruth** (ro͞oth) *n.* [Heb. *Rūth,* poss. < *rĕ'ūth,* companion.] —See table at BIBLE.

**Ru·the·ni·an** (ro͞o-thē'nē-ən, -thēn'yən) *n.* **1.** A member of a group of Ukrainians living in Ruthenia. **2.** The Ukrainian dialect of the Ruthenians. **—Ru·the'ni·an** *adj.*

**ru·then·ic** (ro͞o-thĕn'ĭk, -thē'nĭk) *adj.* Relating to or containing ruthenium with a high valence.

**ru·the·ni·ous** (ro͞o-thē'nē-əs) *adj.* [RUTHENI(UM) + -OUS.] Relating to or containing ruthenium with a low valence.

**ru·the·ni·um** (ro͞o-thē'nē-əm) *n.* [< Med. Lat. *Ruthenia,* Russia.] *Symbol* **Ru** A hard, white, acid-resistant metallic element used to harden platinum and palladium and in nonmagnetic wear-resistant alloys; atomic number 44; atomic weight 101.07.

**ruth·er·ford** (rŭth'ər-fərd) *n.* [After Ernest *Rutherford* (1871–1937), 1st Baron Rutherford.] A unit of radioactivity equal to the quantity of radioactive material that undergoes one million disintegrations per second.

**Rutherford scattering** (rŭth'ər-fərd) *n.* The scattering undergone by a stream of heavy charged particles fired at a sample of a heavy metal, due to exposure to coulombic forces in the atomic nuclei of the sample.

**ruth·ful** (ro͞oth'fəl) *adj.* **1.** Full of sorrow : RUEFUL. **2.** Causing sorrow or pity. **—ruth'ful·ly** *adv.* **—ruth'ful·ness** *n.*

**ruth·less** (ro͞oth'lĭs) *adj.* Having no compassion : MERCILESS. **—ruth'less·ly** *adv.* **—ruth'less·ness** *n.*

**ru·ti·lant** (ro͞ot'l-ənt) *adj.* [ME *rutilaunt* < Lat. *rutilans,* pr.part. of *rutilare,* to make red < *rutilus,* red.] Bright red.

**ru·tile** (ro͞o'tēl', -tīl') *n.* [G. *Rutil* < Lat. *rutilus,* red.] The lustrous red, reddish-brown, or black natural mineral form of titanium dioxide, $TiO_2$, used as a gemstone, as a source of titanium, and in paints and fillers.

**rut·tish** (rŭt'ĭsh) *adj.* Lustful : libidinous. **—rut'tish·ly** *adv.* **—rut'tish·ness** *n.*

**rut·ty** (rŭt'ē) *adj.* **-ti·er, -ti·est.** Full of ruts. **—rut'ti·ness** *n.*

**R-val·ue** (är'văl'yo͞o) *n.* [R(ESISTANCE) VALUE.] A measure of the capacity of a material, as insulation, to impede heat flow with increasing values indicating a greater capacity.

**Rx** (är'ĕks') *n.* [Alteration of ℞, symbol used in prescriptions, abbr. of Lat. *recipe,* imper. of *recipere,* to take. —see RECEIVE.] **1.** A prescription for medicine or medical appliances. **2.** A cure, remedy, or solution for a problem or a disorder.

▲ word history: There is no *x* in *Rx.* The spelling *Rx* is an attempt to represent in ordinary letters the symbol ℞, which is merely a capital *R* with a slash through the tail. ℞ is a symbol for *recipe,* which originally meant "a medicinal prescription" and only later came to mean "a formula used in cooking." *Recipe* is the imperative form of Latin *recipere,* "to take, receive." In Medieval Latin *recipe* often occurred as the first word of medicinal prescriptions directing one to take a certain quantity of a preparation.

**-ry** *suff. var. of* -ERY.

**ry·a** (rē'ə) *n.* [After *Rya,* a village in Sweden.] **1.** A handwoven Scandinavian rug with a thick pile and usu. very colorful abstract designs. **2.** The weaving pattern characteristic of rya rugs.

**rye**[1] (rī) *n.* [ME < OE *ryge.*] **1.** A widely cultivated cereal grass, *Secale cereale,* whose seeds are valued as grain. **2.** The grain of the rye, used in making flour and whiskey and for livestock feed. **3.** Whiskey made from rye.

**rye**[2] (rī) *n.* [Romany *rai* < Skt. *rājan,* king. —see RAJAH.] A male Gypsy.

**rye bread** *n.* Bread made partially or entirely from rye flour.

**rye grass** *n.* Any of several pasture or meadow grasses of the genus *Lolium* native to Eurasia.

**rynd** (rĭnd, rīnd) *n.* [ME.] An iron bar supporting an upper millstone.

**ry·ot** (rī'ət) *n.* [Hindi *ra'īyat* < Ar. *ra'īyah,* herd.] A peasant or tenant farmer in India.

# Ss

**s** or **S** (ĕs) *n., pl.* **s's** or **S's. 1.** The 19th letter of the English alphabet. **2.** A speech sound represented by the letter *s.* **3.** The 19th in a series. **4. S** A grade indicating that a student's work is satisfactory. **5.** Something shaped like the letter S.

**S** *symbol for* SULFUR.

**-s**[1] or **-es** *suff.* [ME < OE *-as,* nominative and accusative pl. suffix.] —Used to form plural nouns <book*s*>

**-s**[2] or **-es** *suff.* [ME < OE *-es, -as.*] —Used to form the 3rd person singular present of all regular and most irregular verbs <cough*s*><walk*s*><hit*s*>

**-s**[3] *suff.* [ME *-es,* genitive sing. suffix < OE.] —Used to form adverbs <caught unaware*s* in the storm>

**-'s**[1] *suff.* [ME *-es,* genitive sing. suffix < OE.] —Used to form the possessive case of singular nouns, plural nouns that do not end in *s,* certain pronouns, and phrases that function as nouns or pronouns <state'*s*><women'*s*><one'*s*><the family next door'*s* car>

**-'s**[2] **1.** Is <It'*s* a beautiful day.> **2.** Has <It'*s* rained for days.> **3.** Does <What'*s* your friend expect?> **4.** Us <Let'*s* eat.>

**Saa·nen** (sä'nən, zä'-) *n.* [After *Saanen,* Switzerland.] A dairy goat orig. bred in Switzerland, having a short-haired white coat and lacking horns.

**sab·a·dil·la** (săb'ə-dĭl'ə, -dē'ə) *n.* [Sp. *cebadilla,* dim. of *cebada,* barley < *cebo,* feed < Lat. *cibus,* food.] **1.** A tropical American plant, *Schoenocaulon officinale,* yielding poisonous seeds used in insecticides. **2.** The dry, ripe seeds of the sabadilla.

**sab·bat** (săb'ət) *n.* [Fr., sabbat, Sabbath < OFr., Sabbath.] The witches' Sabbath.

**Sab·ba·tar·i·an** (săb'ə-târ'ē-ən) *n.* [LLat. *sabbatarius* < Lat. *sabbatum,* Sabbath.] **1.** One who observes Saturday as the Sabbath, as in Judaism. **2.** One who believes in strict observance of the Sabbath. *—adj.* Pertaining to the Sabbath or to Sabbatarians. **—Sab'ba·tar'i·an·ism** *n.*

**Sab·bath** (săb'əth) *n.* [ME *sabath* < OFr. *sabbat* and OE *sabat,* both < Lat. *sabbatum* < Gk. *sabbaton* < Heb. *shabbāth* < *shābhath,* he rested.] **1.** Saturday, the seventh day of the week, observed as a day of rest and worship by Jews and some Christian sects. **2.** Sunday, the first day of the week, observed as a day of rest and worship by most Christians.

**sab·bat·i·cal** (sə-băt'ĭ-kəl) *also* **sab·bat·ic** (-ĭk) *adj.* [LLat. *sabbaticus* < Gk. *sabbatikos* < *sabbaton,* Sabbath.] **1. Sabbatical** *also* **Sabbatic.** Of or relating to the Sabbath as a day of rest. **2.** Relating to a sabbatical year. *—n.* A sabbatical year.

**sabbatical year** *n.* **1.** *often* **Sabbatical year.** A year during which land remained fallow, observed every seven years by the ancient Jews. **2.** A leave of absence, often with pay, usu. granted every seventh year for travel, research, or rest.

**Sa·bel·li·an** (sə-bĕl'ē-ən) *n.* [< Lat. *Sabellus,* Sabine.] **1.** A group of extinct Italic languages that includes Sabine. **2.** A speaker of one of these languages. **—Sa·bel'li·an** *adj.*

**sa·ber** (sā'bər) *n.* [Fr. *sabre* < obs. G. *sabel.*] **1.** A heavy cavalry sword having a slightly curved one-edged blade. **2.** A light dueling or fencing sword having an arched guard covering the hand and a tapered flexible blade with a cutting edge on one side and on the tip. *—vt.* **-bered, -ber·ing, -bers.** To strike, injure, or kill with a saber.

**saber rattling** *n.* Ostentatious display of military power.

**sa·ber-toothed tiger** (sā'bər-to͞otht') *n.* An extinct cat of the Oligocene to the Pleistocene epoch, esp. one of the larger members of the genus *Smilodon,* marked by long upper canine teeth.

**sa·bin** (sā'bĭn) *n.* [After Wallace Clement Ware *Sabine* (1868–1919).] A unit of acoustic absorption equivalent to the absorption by one square foot of a surface that absorbs all incident sound.

**Sa·bine** (sā'bīn') *n.* [ME *Sabyn* < Lat. *Sabinus.*] **1.** A member of an ancient tribe of central Italy, conquered and assimilated by the Romans in 290 B.C. **2.** The Italic language of the Sabines. *—adj.* Of or relating to the Sabine people or their language.

**Sa·bin vaccine** (sā'bĭn) *n.* [After Albert B. *Sabin* (b. 1906), its developer.] An oral vaccine consisting of a live attenuated virus, taken to immunize against poliomyelitis.

**sa·ble** (sā'bəl) *n.* [ME < OFr., of Slavic orig.] **1. a.** A carnivorous mammal, *Martes zibellina* of northern Europe and Asia, with soft, dark fur. **b.** The fur of the sable. **c.** The similar fur of other species of martens. **2. a.** The color black, esp. in heraldry. **b. sables.** Black

mourning garments. **3.** A grayish yellowish brown. **4.** A sablefish. —*adj.* **1.** Of the color black. **2.** Dark : somber.
**sable antelope** *n.* A large African antelope, *Hippotragus niger,* with a usu. dark coat and backward-curving horns.
**sa·ble·fish** (sā′bəl-fĭsh′) *n., pl.* **sablefish** or **-fish·es.** A dark-colored marine food fish, *Anoplopoma fimbria* of North American Pacific waters.
**sa·bot** (să-bō′, săb′ō) *n.* [Fr. < OFr.] **1.** A heavy wooden shoe worn in several European countries. **2.** (*also* săb′ət). A sandal or shoe with a band across the instep.

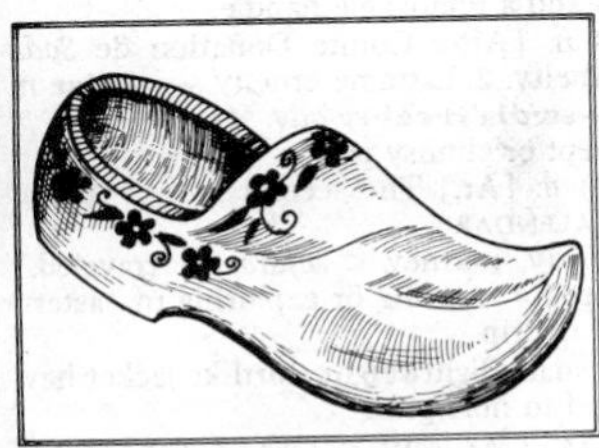
**sabot**

**sab·o·tage** (săb′ə-täzh′) *n.* [Fr. < *saboter,* to sabotage < *sabot,* sabot.] **1.** Destruction of property or obstruction of normal operations, as by enemy agents in a war. **2.** Treacherous action to defeat or hinder a cause or endeavor <industrial *sabotage*> —*vt.* **-taged, -tag·ing, -tag·es.** To commit sabotage against.
☆ ***syns:*** SABOTAGE, SUBVERT, UNDERMINE *v. core meaning* : to damage, destroy, or defeat by sabotage <*sabotaged* the project>
**sab·o·teur** (săb′ə-tûr′) *n.* [Fr. < *saboter,* to sabotage.] One who commits sabotage.
**sa·bra** (sä′brə) *n.* [Heb. *ṣabēr,* sabra, prickly pear.] A native-born Israeli.
**sa·bre** (sā′bər) *n.* & *v. Chiefly Brit. var. of* SABER.
**sab·u·lous** (săb′yə-ləs) *also* **sab·u·lose** (-lōs′) *adj.* [Lat. *sabulosus* < *sabulum,* coarse sand.] Gritty : sandy. **—sab′u·los′i·ty** (-lŏs′ĭ-tē) *n.*
**sac** (săk) *n.* [Fr., bag < Lat. *saccus.* —see SACK¹.] A pouch or pouchlike structure in a plant or animal, sometimes filled with fluid.
**Sac** (săk, sôk) *n. var. of* SAUK.
**sac·a·ton** (săk′ə-tōn′) *n.* [Am. Sp. *zacatón* < *zacate,* coarse grass < Nahuatl, grass.] A grass, *Sporobolus wrightii* of the southwestern United States, used for pasture and hay in saline areas.
**sac·cade** (să-käd′, sə-) *n.* [Fr., twitch < OFr. < *saquer,* to pull.] A rapid intermittent eye movement, as one that occurs when the eye fixes on one point after another in the visual field. **—sac·cad′ic** *adj.*
**sac·cate** (săk′āt′) *adj.* [< Lat. *saccus,* bag. —see SACK¹.] Shaped like or having a sac or pouch.
**sacchar-** *pref. var. of* SACCHARO-.
**sac·cha·rase** (săk′ə-rās′, -rāz′) *n.* Invertase.
**sac·cha·rate** (săk′ə-rāt′) *n.* [SACCHAR(IC ACID) + -ATE².] A salt or ester of saccharic acid.
**sac·char·ic acid** (sə-kăr′ĭk) *n.* A white crystalline acid, $COOH(CHOH)_4COOH$, formed by the oxidation of glucose, sucrose, or starch.
**sac·cha·ride** (săk′ə-rīd′) *n.* Any of a series of compounds of carbon, hydrogen, and oxygen in which the atoms of the latter two elements are in the ratio of 2:1, esp. those containing the group $C_6H_{10}O_5$.
**sac·char·i·fy** (sə-kăr′ə-fī′, să-) *vt.* **-fied, -fy·ing, -fies.** To convert (e.g., starch) into sugar. **—sac·char′i·fi·ca′tion** *n.*
**sac·cha·rim·e·ter** (săk′ə-rĭm′ĭ-tər) *n.* **1.** A polarimeter that indicates sugar concentration in a solution. **2.** An instrument that determines the sugar content of a fermenting sample from carbon dioxide measurements. **—sac′cha·rim′e·try** *n.*
**sac·cha·rin** (săk′ər-ĭn) *n.* A white crystalline powder, $C_7H_5NO_3S$, having a taste approx. 500 times sweeter than cane sugar and used as a calorie-free sweetener.
**sac·cha·rine** (săk′ər-ĭn, -ə-rēn′, -ə-rīn′) *adj.* **1.** Of or typical of sugar or saccharin : SWEET. **2.** Cloyingly sweet <a *saccharine* smirk> **—sac′cha·rine·ly** *adv.* **—sac′cha·rin′i·ty** (-ə-rĭn′ĭ-tē) *n.*
**saccharo-** or **sacchar-** *pref.* [< Lat. *saccharum,* sugar < Gk. *sakkharon* < Pali *sakkharā* < Skt. *śarkarā.*] Sugar <*saccharide*>
**sac·cha·roid** (săk′ə-roid′) or **sac·cha·roi·dal** (săk′ə-roid′l) *adj.* Designating rocks and minerals that have a granular structure similar to that of loaf sugar.
**sac·cha·rom·e·ter** (săk′ə-rŏm′ĭ-tər) *n.* A hydrometer that determines the amount of sugar in a solution from density measurements.
**sac·cha·ro·my·cete** (săk′ə-rō-mī′sēt′) *n.* Any of various yeast fungi, esp. of the genus *Saccharomyces,* many of which ferment sugar. **—sac′cha·ro·my·ce′tic, sac′cha·ro·my·ce′tous** *adj.*
**sac·cha·rose** (săk′ə-rōs′) *n.* Sucrose.
**sac·cu·late** (săk′yə-lāt′) or **sac·cu·lat·ed** (-lā′tĭd) *also* **sac·cu·lar** (-lər) *adj.* Divided into or formed of a series of saclike dilatations or pouches.
**sac·cule** (săk′yo͞ol) *also* **sac·cu·lus** (-yə-ləs) *n., pl.* **-cules** *also* **-cu·li** (-yə-lī′) [Lat., dim. of *saccus,* bag. —see SACK¹.] **1.** A small sac or pouch. **2.** The smaller of two membranous sacs in the vestibule of the labyrinth of the ear.
**sac·er·do·tal** (săs′ər-dōt′l, săk′-) *adj.* [ME < OFr. < Lat. *sacerdotalis* < *sacerdos,* priest < *sacer,* sacred.] **1.** Of or relating to priests or the priesthood : PRIESTLY. **2.** Of or relating to sacerdotalism. **—sac′er·do′tal·ly** *adv.*
**sac·er·do·tal·ism** (săs′ər-dōt′l-ĭz′əm, săk′-) *n.* The belief that priests act as mediators between God and human beings.
**sa·chem** (sā′chəm) *n.* [Narraganset *sâchim,* chief.] **1.** The chief of a North American Indian tribe or confederation, esp. an Algonquian chief. **2.** A high official of the Tammany Society.
**sa·cher torte** (sä′kər tôrt′, zä′KHər tôr′tə) *n.* [G. *Sachertorte* : *Sacher,* surname of a family of 19th- and 20th-cent. hoteliers + *Torte,* torte.] A rich chocolate cake usu. filled with apricot jam, topped with chocolate icing, and served with whipped cream.
**sa·chet** (să-shā′) *n.* [Fr. < OFr., dim. of *sac,* bag < Lat. *saccus.* —see SACK¹.] A small packet of perfumed powder used to scent clothes, as in a chest of drawers, a trunk, or a closet.
**sack¹** (săk) *n.* [ME < OE *sæcc* < Lat. *saccus* < Gk. *sakkos,* of Semitic orig.] **1. a.** A large bag of strong, coarse material for holding objects in bulk. **b.** A similar container of paper or plastic. **c.** The amount a sack will hold. **2.** *also* **sacque.** A loose-fitting garment for women and children. **3.** *Slang.* Dismissal from employment <got the *sack* for stealing> **4.** *Slang.* A bed, mattress, or sleeping bag. **5.** *Baseball.* A base. —*vt.* **sacked, sack·ing, sacks. 1.** To place in a sack. **2.** *Slang.* To discharge from employment. **—sack out.** *Slang.* To sleep.
**sack²** (săk) *vt.* **sacked, sack·ing, sacks.** [< OFr. *(mettre a) sac,* (to put in) a sack.] To rob of goods or valuable articles, esp. after capture. —*n.* **1.** The pillaging of a captured town. **2.** Plunder : loot.
**sack³** (săk) *n.* [< OFr. *(vin) sec,* dry (wine) < Lat. *siccus,* dry.] A light, dry, strong wine from Spain and the Canary Islands, imported to England in the 16th and 17th cent.
**sack·but** (săk′bŭt′) *n.* [OFr. *saqueboute* : *saquer,* to pull + *bouter,* to push, of Germanic orig.] **1.** A medieval musical instrument resembling the trombone. **2.** A triangular stringed instrument.
**sack·cloth** (săk′klôth′, -klŏth′) *n.* **1.** Sacking. **2. a.** A rough cloth of camel's hair, goat hair, hemp, cotton, or flax. **b.** Garments made of sackcloth, worn as a symbol of mourning or penitence.
**sack·ing** (săk′ĭng) *n.* A coarse, stout woven cloth, as burlap or gunny, used for making sacks.
**sacque** (săk) *n. var. of* SACK¹ 2.
**sa·cra** (sā′krə, săk′rə) *n. pl. of* SACRUM.
**sa·cral¹** (sā′krəl) *adj.* Of, near, or relating to the sacrum.
**sa·cral²** (sā′krəl) *adj.* [< Lat. *sacer, sacr-,* sacred.] Of or relating to sacred rites or observances.
**sac·ra·ment** (săk′rə-mənt) *n.* [ME < OFr. *sacrement* < LLat. *sacramentum* < Lat., oath < *sacrare,* to consecrate < *sacer,* sacred.] **1.** A formal Christian rite, as baptism or matrimony, esp. one thought to have been instituted by Jesus as a means of grace. **2.** *often* **Sacrament. a.** The Eucharist. **b.** The consecrated elements of the Eucharist, esp. the host.
**sac·ra·men·tal** (săk′rə-mĕn′tl) *adj.* **1.** Of, relating to, or used in a sacrament. **2.** Consecrated or bound by or as if by a sacrament <a *sacramental* duty> **3.** Having the force or efficacy of a sacrament. —*n.* A rite, action, or sacred object used in worship by some Christian churches. **—sac′ra·men′tal·ly** *adv.*
**sac·ra·men·tal·ism** (săk′rə-mĕn′tl-ĭz′əm) *n.* **1.** The doctrine that observance of the sacraments is necessary for salvation and that such participation can confer grace. **2.** Emphasis on the efficacy of a sacramental. **—sac′ra·men′tal·ist** *n.*
**Sac·ra·men·tar·i·an** (săk′rə-mĕn-târ′ē-ən) *n.* One who regards the sacraments, esp. the Eucharist, merely as visible symbols and not as a corporeal manifestation of Christ. —*adj.* **1.** Of or relating to Sacramentarians. **2.** Of or relating to sacramentalism or sacramentalists. **—Sac′ra·men·tar′i·an·ism** *n.*
**sa·crar·i·um** (sə-krâr′ē-əm, sā-, să-) *n., pl.* **-i·a** (-ē-ə) [Med. Lat. < Lat., shrine < *sacer,* sacred.] **1. a.** The sacristy of a church. **b.** SANCTUARY 1. **2.** A piscina.
**sa·cred** (sā′krĭd) *adj.* [ME, p.part. of *sacren,* to consecrate < OFr. *sacrer* < Lat. *sacrare* < *sacer,* sacred.] **1.** Dedicated to or reserved for the worship of a deity. **2.** Worthy of religious veneration <the *sacred* teachings of Jesus Christ> **3.** Made or declared holy <*sacred* bread and wine> **4.** Dedicated to or set apart for a single use, purpose, or person <an office *sacred* to the President> **5.** Worthy of respect : VENERABLE. **6.** Of or relating to religious objects, rites, or practices. **—sa′cred·ly** *adv.* **—sa′cred·ness** *n.*
**Sacred College** *n.* The College of Cardinals.
**sacred cow** *n.* [From the veneration of the cow by the Hindus.] One immune from criticism or attack.
**sac·ri·fice** (săk′rə-fīs′) *n.* [ME < OFr. < Lat. *sacrificium* : *sacer,* sacred + *facere,* to make.] **1. a.** The act of offering something to a deity in propitiation or homage, esp. the ritual slaughter of an animal or person. **b.** A victim offered in sacrifice. **2. a.** Forfeiture of some-

thing highly valued for the sake of one thought to have a greater value or claim. **b.** Something so forfeited. **3. a.** Relinquishment of something at less than its presumed value. **b.** Something so relinquished. **c.** A loss so sustained. **4.** *Baseball.* A sacrifice hit. —*v.* **-ficed, -fic·ing, -fic·es.** —*vt.* **1.** To offer as a sacrifice to a deity. **2.** To forfeit (one thing) for another thing thought to be of greater value. **3.** To sell or give away at a loss. —*vi.* **1.** To make or offer a sacrifice. **2.** *Baseball.* To make a sacrifice hit. **—sac′ri·fic′er** *n.* **—sac′ri·fi′cial** (-fĭsh′əl) *adj.* **—sac′ri·fi′cial·ly** *adv.*

**sacrifice fly** *n. Baseball.* A fly ball enabling a runner to score after it is caught by a fielder.

**sacrifice hit** *n. Baseball.* A bunt allowing a runner to advance a base while the batter is retired.

**sac·ri·lege** (săk′rə-lĭj) *n.* [ME < OFr. < Lat. *sacrilegium* < *sacrilegus,* one who steals sacred things : *sacer,* sacred + *legere,* to gather.] Desecration, profanation, or theft of something sacred. **—sac′ri·le′gist** (săk′rə-lē′jĭst) *n.*

**sac·ri·le·gious** (săk′rə-lē′jəs, -lĭj′əs) *adj.* **1.** Disrespectful toward something sacred : PROFANE. **2.** Guilty of sacrilege. **—sac′ri·le′gious·ly** *adv.* **—sac′ri·le′gious·ness** *n.*

☆ ***syns:*** SACRILEGIOUS, BLASPHEMOUS, PROFANE *adj. core meaning* : showing irreverence and contempt for something sacred <*sacrilegious* mockery of the Christian rites>

**sac·ris·tan** (săk′rĭ-stən) *n.* [ME < Med. Lat. *sacristanus* < *sacrista* < Lat. *sacer,* sacred.] **1.** One in charge of a sacristy. **2.** A sexton.

**sac·ris·ty** (săk′rĭ-stē) *n., pl.* **-ties.** [Fr. *sacristie* < Med. Lat. *sacristia* < *sacrista,* sacristan.] A room in a church for the storage of sacred vessels and vestments : VESTRY.

**sac·ro·il·i·ac** (săk′rō-ĭl′ē-ăk′, sā′krō-) *adj.* [SACR(UM) + ILI(UM) + -AC.] Of, relating to, or affecting the sacrum and ilium and their articulation or associated ligaments. —*n.* The sacroiliac region or cartilage.

**sac·ro·lum·bar** (săk′rō-lŭm′bər, -bär′, sā′krō-) *adj.* [SACR(UM) + LUMBAR.] Of, relating to, or affecting the sacrum and the lumbar region.

**sac·ro·sanct** (săk′rō-săngkt′) *adj.* [Lat. *sacrosanctus,* consecrated with religious ceremonies : *sacrum,* religious rite (< *sacer,* sacred) + *sanctus,* p.part. of *sancire,* to consecrate.] Sacred and inviolable. **—sac′ro·sanc′ti·ty** (-săngk′tĭ-tē) *n.*

**sa·crum** (sā′krəm, săk′rəm) *n., pl.* **sa·cra** (sā′krə, săk′rə) [NLat. < LLat. *(os) sacrum,* transl. of Gk. *hieron (osteon),* sacred (bone).] A triangular bone consisting of five fused vertebrae and forming the posterior section of the pelvis.

**sad** (săd) *adj.* **sad·der, sad·dest.** [ME, serious < OE *sæd,* sated.] **1.** Affected or marked by unhappiness or sorrow. **2.** Expressing sorrow or unhappiness. **3.** Causing sorrow or gloom. **4.** Deplorable : sorry <a *sad* situation to find myself in> **5.** Dark-hued : somber. **—sad′ly** *adv.* **—sad′ness** *n.*

**sad·den** (săd′n) *vt. & vi.* **-dened, -den·ing, -dens.** To make or become sad or unhappy.

**sad·dhu** (sä′do͞o) *n. var. of* SADHU.

**sad·dle** (săd′l) *n.* [ME *sadel* < OE *sadol.*] **1. a.** A leather seat for a rider, secured on an animal's back by a girth. **b.** A similar piece of equipment used to attach a pack to an animal. **c.** The padded part of a driving harness fitting over a horse's back. **d.** The seat of a bicycle, motorcycle, or similar vehicle. **e.** Something shaped like a saddle. **2. a.** A cut of meat including part of the backbone and both loins. **b.** The lower part of a male fowl's back. **3. a.** A saddle-shaped depression in the ridge of a hill. **b.** A ridge between two peaks. —*v.* **-dled, -dling, -dles.** —*vt.* **1.** To put a saddle on. **2.** To burden : encumber <*saddled* with too much work> —*vi.* **1.** To saddle a horse. **2.** To get into a saddle. **—in the saddle.** In control.

**sad·dle·bag** (săd′l-băg′) *n.* **1.** One of a pair of pouches that hang across the back of a horse behind the saddle. **2.** A pouch hanging from a saddle or over the rear wheel of a bicycle or motorcycle.

**saddle blanket** *n.* A blanket placed between a saddle and a horse's back to prevent chafing.

**sad·dle·bow** (săd′l-bō′) *n.* The arched upper front part of a saddle.

**sad·dle·cloth** (săd′l-klôth′, -klŏth′) *n.* A thick cloth placed under the saddle of a racehorse and bearing its number.

**saddle horse** *n.* A horse bred or trained for riding.

**sad·dler** (săd′lər) *n.* One that makes, repairs, or sells equipment for horses.

**saddle roof** *n.* A roof with a ridge and two gables.

**sad·dler·y** (săd′lə-rē) *n., pl.* **-ies.** **1.** Equipment, as saddles and harnesses, for horses : TACK. **2.** A shop selling saddlery. **3.** The business or craft of a saddler.

**saddle shoe** *n.* A flat shoe, usu. white, with a band of leather in a contrasting color across the instep.

**saddle soap** *n.* A preparation of mild soap and neat's-foot oil, used for cleaning and softening leather.

**saddle sore** *n.* **1.** A sore on a horse's back caused by an improperly fitted saddle. **2.** A sore on a rider caused by saddle chafing.

**saddle stitch** *n.* **1.** A simple overcasting stitch, usu. of thread contrasting in color with the fabric, used primarily to ornament clothing. **2.** A stitch used to sew together the leaves of a book, either with thread or wire.

**sad·dle·tree** (săd′l-trē′) *n.* The frame of a saddle.

**Sad·du·cee** (săj′ə-sē′, săd′yə-) *n.* [ME *Saducee* < OE *Sadducēas,* Sadducees < LLat. *Sadducaei* < LGk. *Saddoukaioi* < Heb. *Ṣĕddūqīm.*] A member of a Jewish sect of the 2nd cent. B.C. through the 1st cent. A.D. that retained the older interpretation of the written Mosaic law against the oral tradition and denied the resurrection of the dead. **—Sad′du·ce′an** (-sē′ən) *adj.* **—Sad′du·cee′ism** *n.*

**sa·de** *or* **sa·dhe** (sä′də, -dē, tsä′-) *n.* [Heb. *ṣadhe.*] The 18th letter of the Hebrew alphabet. —See table at ALPHABET.

**sa·dhu** *also* **sad·dhu** (sä′do͞o) *n.* [Skt. *sādhu-,* right, holy.] An ascetic Hindu holy man.

**sad·i·ron** (săd′ī′ərn) *n.* [SAD, heavy (obs.) + IRON.] A heavy flatiron with points at both ends and a removable handle.

**sa·dism** (sā′dĭz′əm, săd′ĭz′-) *n.* [After Comte Donation de *Sade* (1740–1814).] **1.** Delight in cruelty. **2.** Extreme cruelty. **—sa′dist** *n.* **—sa·dis′tic** (sə-dĭs′tĭk) *adj.* **—sa·dis′ti·cal·ly** *adv.*

**sad sack** *n. Informal.* An inept or clumsy person.

**Sa·far** *also* **Sa·phar** (sə-fär′) *n.* [Ar.] The second month of the Moslem year. —See table at CALENDAR.

**sa·fa·ri** (sə-fä′rē) *n.* [Ar. *safarīy,* journey < *safara,* he traveled.] **1.** An overland expedition, esp. for hunting or exploring in eastern Africa. **2.** *Informal.* A journey or trip.

**safari jacket** *n.* A belted, usually lightweight shirtlike jacket having pleated pockets that expand to hold gear.

**safe** (sāf) *adj.* **saf·er, saf·est.** [ME *sauf* < OFr. < Lat. *salvus,* healthy.] **1.** Secure from harm, danger, or evil. **2.** Free from injury or danger : UNHURT. **3.** Free from risk <a *safe* bet> **4.** Giving protection <a *safe* place> **5.** *Baseball.* Having reached a base without being put out. —*n.* **1.** A metal container usu. having a lock, used to store and protect valuables. **2.** A cooled compartment used to protect perishable foods. **—safe′ly** *adv.* **—safe′ness** *n.*

**safe-con·duct** (sāf′kŏn′dŭkt) *n.* **1.** An official document or escort assuring unmolested passage, as through enemy territory. **2.** The protection afforded by a safe-conduct.

**safe·crack·er** (sāf′krăk′ər) *n.* One who breaks into safes in order to steal. **—safe′crack′ing** *n.*

**safe-de·pos·it box** (sāf′dĭ-pŏz′ĭt) *n.* A fireproof metal box, usu. in a bank vault, for the safe storage of valuables.

**safe·guard** (sāf′gärd′) *n.* **1. a.** One that serves as a guard or protection. **b.** A mechanical device designed to prevent accidents. **c.** A safe-conduct. **2. a.** A protective stipulation, as in a contract. **b.** A precautionary measure. —*vt.* **-guard·ed, -guard·ing, -guards.** To protect : guard.

**safe house** *n. Informal.* A house or building used by intelligence agents or secret police as a place of refuge or as a place to avoid surveillance or interference.

**safe·keep·ing** (sāf′kē′pĭng) *n.* The act of keeping safe or state of being kept safe.

**safe·light** (sāf′līt′) *n.* A lamp with one or more color filters that permit moderate darkroom illumination without exposure of photosensitive film or paper.

**safe·ty** (sāf′tē) *n., pl.* **-ties.** **1.** The state of being safe. **2.** A device designed to prevent accidents, as a lock on a firearm that prevents accidental firing. **3.** *Football.* **a.** A play in which a member of the offensive team downs the ball, willingly or unwillingly, behind his own goal line, resulting in two points for the defensive team. **b.** One of two defensive football backs.

☆ ***syns:*** SAFETY, SECURITY *n. core meaning* : freedom from danger or harm <sought *safety* in the guarded fort> ***ant:*** danger

**safety circuit** *n.* An electronic circuit that prevents malfunction by either sounding an alert or activating a trip circuit on a protective device.

**safety glass** *n.* A shatterproof material consisting of two sheets of glass with an intermediate layer of transparent plastic.

**safety island** *n.* A marked-off area within a roadway from which traffic is banned, esp. for pedestrian safety.

**safety lamp** *n.* A miner's lamp with a protective wire gauze surrounding the flame to prevent ignition of flammable gases.

**safety match** *n.* A match that can be ignited only by being struck against a chemically prepared friction surface.

**safety net** *n.* **1.** A large net for catching one that falls or jumps, as from a circus trapeze. **2.** A guarantee, esp. of financial security.

**safety pin** *n.* **1.** A pin in the form of a clasp, with a sheath that covers and holds the point. **2.** A pin that prevents the premature or accidental detonation of an explosive device, as a bomb or grenade.

**safety razor** *n.* A razor in which the blade is fitted into a holder with guards to prevent cutting of the skin.

**safety valve** *n.* **1.** A valve in a pressure container, as in a steam boiler, that automatically opens when pressure reaches a dangerous level. **2.** An outlet for the release of repressed emotion or energy.

**saf·flow·er** (săf′lou′ər) *n.* [Du. *saffloer* < OFr. *saffleur* < OItal. *saffiore,* saffron < Ar. *aṣfar,* a yellow plant.] **1.** A native Asian plant, *Carthamus tinctorius,* with orange flowers yielding a dyestuff and seeds that are the source of an oil used in cooking, cosmetics, paints, and medicine. **2.** The dried flowers of the safflower.

**saf·fron** (săf′rən) *n.* [ME *saffran* < OFr. *safran* < Med. Lat. *safranum* < Ar. *za'farān.*] **1. a.** An Old World plant, *Crocus sativus,* with purple or white flowers having orange stigmas. **b.** The dried stigmas of the saffron, used to color foods and as a cooking spice and dyestuff. **2.** A moderate or strong orange-yellow to moderate orange.

**saf·ra·nine** (săf′rə-nēn′, -nĭn) *also* **saf·ra·nin** (-nĭn) *n.* [< Fr. *safran,* saffron < OFr.] A dye based on phenazine, used in the textile industry and as a biological stain.

**saf·role** (săf′rōl′) *n.* [< Fr. *safran,* saffron.] A colorless or pale-yellow oily liquid, $C_{10}H_{10}O_2$, derived esp. from oil of sassafras and used in the manufacture of perfume and soap.

**sag** (săg) *v.* **sagged, sag·ging, sags.** [ME *saggen,* prob. of Scand. orig.] —*vi.* **1.** To droop, sink, or settle from pressure or weight. **2.** To lose strength, firmness, or resilience. **3.** To decline, as in value or price. **4.** *Naut.* To drift to leeward. —*vt.* To cause to sag. —*n.* **1.** The act of sagging. **2.** An instance or amount of sagging. **3.** A sagging area : DEPRESSION. **4.** A temporary decline in monetary value. **5.** *Naut.* A drift to leeward.

**sa·ga** (sä′gə) *n.* [ON.] **1. a.** A 12th- and 13th-cent. prose narrative recounting historical and legendary events and exploits in Iceland or Norway. **b.** A modern prose narrative resembling a saga. **2.** A long, detailed account <the continuing *saga* of a soap opera>

**sa·ga·cious** (sə-gā′shəs) *adj.* [< Lat. *sagax, sagac-,* quick-witted.] Having or showing sound judgment and keen perception : WISE. **—sa·ga′cious·ly** *adv.* **—sa·ga′cious·ness** *n.*

**sa·gac·i·ty** (sə-găs′ĭ-tē) *n.* The quality of being sagacious : WISDOM.

**sag·a·more** (săg′ə-môr′, -mōr′) *n.* [Abnaki *sàkama.*] A subordinate Algonquian Indian chief.

**saga novel** *n.* A roman-fleuve.

**sage¹** (sāj) *n.* [ME < OFr. < VLat. **sapius* < Lat. *sapere,* to be wise.] One who is venerated for experience, judgment, and wisdom. —*adj.* **sag·er, sag·est. 1.** Possessing or displaying wisdom and calm judgment. **2.** Proceeding from or characterized by wisdom and calm judgment <*sage* counseling> **3.** *Archaic.* Serious : solemn. **—sage′ly** *adv.* **—sage′ness** *n.*

**sage²** (sāj) *n.* [ME *sauge* < OFr. < Lat. *salvia* < *salvus,* healthy.] **1.** A plant or shrub of the genus *Salvia,* esp. *S. officinalis,* with aromatic grayish-green leaves used in cooking. **2.** The leaves of the sage.

**sage·brush** (sāj′brŭsh′) *n.* An aromatic plant of the genus *Artemisia,* esp. *A. tridentata* of arid regions of western North America, with silver-green leaves and large clusters of small white flowers.

**sage grouse** *n.* A chickenlike bird, *Centrocercus urophasianus* of western North America, with long pointed tail feathers that can be spread fanwise.

**sag·ger** *also* **sag·gar** (săg′ər) *n.* [Perh. alteration of SAFEGUARD.] **1.** A protective casing of fire clay in which fine or delicate ceramic articles are fired. **2.** Clay used to make ceramic casings.

**Sa·git·ta** (sə-jĭt′ə) *n.* [Lat. < *sagitta,* arrow.] A constellation in the Northern Hemisphere.

**sag·it·tal** (săj′ĭ-tl) *adj.* [< Lat. *sagitta,* arrow.] **1.** *Anat.* Of or pertaining to the suture joining the two parietal bones of the skull. **2.** *Zool.* Of or pertaining to the sagittal plane. **—sag′it·tal·ly** *adv.*

**sagittal plane** *n.* The longitudinal vertical plane dividing the body of a bilaterally symmetric animal into right and left halves.

**Sag·it·tar·i·us** (săj′ĭ-târ′ē-əs) *n.* [ME < Lat. < *sagittarius,* archer < *sagitta,* arrow.] **1.** A constellation in the Southern Hemisphere. **2. a.** The ninth sign of the zodiac. **b.** One born under this sign.

**sag·it·tate** (săj′ĭ-tāt′) *adj.* [< Lat. *sagitta,* arrow.] *Bot.* Having the shape of an arrowhead <a *sagittate* leaf>

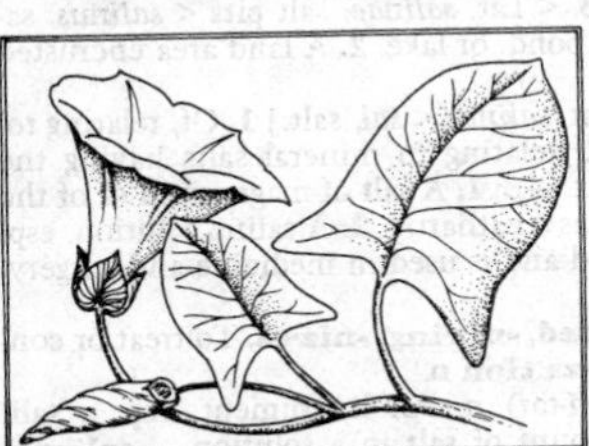

**sagittate**
*Sagittate leaves*

**sa·go** (sā′gō) *n., pl.* **-gos.** [Malay *sagu.*] A powdery starch obtained from the trunks of the sago palm and used as a food thickener and textile stiffener.

**sago palm** *n.* **1.** A palm tree of the genera *Metroxylon, Arenga,* or *Caryota* of tropical Asia. **2.** A palmlike cycad, *Cycas revoluta* of southeastern Asia.

**sa·gua·ro** (sə-gwär′ō, -wär′ō) *also* **sa·hua·ro** (sə-wär′ō) *n., pl.* **-ros.** [Mex. Sp.] **1.** An extremely large cactus, *Carnegiea gigantea* of the southwest United States and northern Mexico, with upward-curving branches, white flowers, and edible red fruit. **2.** The fruit of the saguaro.

**Sa·hap·tin** (sä-hăp′tĭn) *n., pl.* **Sahaptin** *or* **-tins. 1.** A member of an Indian people of Idaho, Washington, and Oregon. **2.** The language of the Sahaptin.

**sa·hib** (sä′ĭb) *n.* [Hindi *ṣāḥib,* master < Ar.] —Used as a title of respect equivalent to *master* or *sir* for Europeans in colonial India.

**sa·hua·ro** (sə-wär′ō) *n. var. of* SAGUARO.

**said** (sĕd) *adj.* [P.part of SAY.] Named or mentioned before : AFOREMENTIONED. ***usage:*** As an adjective, *said* is seldom appropriate to any but legal or business contexts, in which it is equivalent in meaning to *aforesaid,* as in *the said tenant* (previously named in a lease). In general usage, *said* is unnecessary and *the tenant* will suffice.

**sai·ga** (sī′gə) *n.* [R. *saiga,* of Turkic orig.] A small antelope, *Saiga tatarica* or *S. mongolia* of the plains of northern Eurasia, with a stubby proboscislike snout.

**sail** (sāl) *n.* [ME < OE *segl.*] **1. a.** A length of triangular or rectangular fabric attached to a ship to catch the wind and propel it. **b.** The sails of a ship or boat. **2.** *pl.* **sail** *or* **sails.** A sailing vessel. **3.** A trip or voyage in a sailing craft. **4.** Something that resembles a sail, as the blade of a windmill. —*v.* **sailed, sail·ing, sails.** —*vi.* **1.** To move across the surface of water, esp. by sailing vessel. **2.** To travel by water in a vessel. **3.** To start out on a voyage or trip. **4.** To operate a sailing craft, esp. for sport. **5.** To move quickly or effortlessly. —*vt.* **1.** To navigate or manage (a vessel). **2.** To voyage on or across <*sail* the Atlantic> **—sail into.** To attack or criticize vigorously.

**Sail·board** (sāl′bôrd′, -bōrd′). A trademark for a small light sailboat with a flat hull.

**sail·boat** (sāl′bōt′) *n.* A small boat propelled partly or wholly by sail.

**sail·cloth** (sāl′klôth′, -klŏth′) *n.* A strong fabric, as cotton canvas, used to make sails or tents.

**sail·er** (sā′lər) *n.* A usu. small, light ship or boat, esp. one having specified sailing qualities.

**sail·fish** (sāl′fĭsh′) *n., pl.* **sailfish** *or* **-fish·es.** A large marine fish of the genus *Istiophorus,* with the upper jaw prolonged into a spearlike bone and a large saillike dorsal fin.

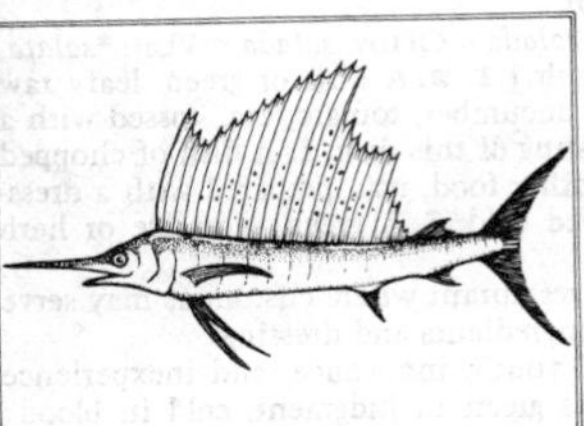

**sailfish**
*Up to 11 feet long*

**sail·ing** (sā′lĭng) *n.* **1.** Skill needed to operate and navigate a vessel : NAVIGATION. **2.** The sport of operating or riding in a sailboat. **3.** Departure or time of departure from a port.

**sail·or** (sā′lər) *n.* **1.** One who serves in a navy or works on a ship, esp. an ordinary seaman. **2.** One who travels by water. **3.** A straw hat with a flat top and flat brim.

**sail·or's-choice** (sā′lərz-chois′) *n., pl.* **sailor's-choice.** A fish of the North American Atlantic coast, as the pinfish or *Haemulon parrai* of more southerly waters.

**sail·plane** (sāl′plān′) *n.* A light glider used esp. for soaring. **—sail′plane′** *v.* **(-planed, -plan·ing, -planes). —sail′plan′er** *n.*

**sain·foin** (săn′foin′, săn′-) *n.* [Fr. < OFr. < Med. Lat. *sanum faenum* : Lat. *sanus,* healthy + Lat. *faenum,* hay.] A Eurasian plant, *Onobrychis viciaefolia,* with compound leaves and pink or white flowers, often used as fodder.

**saint** (sānt) *n.* [ME < OFr. < Lat. *sanctus,* holy, p.part. of *sancire,* to consecrate.] **1. a.** One officially recognized, esp. by canonization, as being entitled to public veneration and capable of interceding for people on earth. **b.** One who has died and gone to heaven. **c. Saint.** A member of certain religious groups, esp. a Latter-Day Saint. **2.** A highly virtuous person. —*vt.* **saint·ed, saint·ing, saints.** To recognize or venerate as a saint : CANONIZE. **—saint′dom** (sānt′dəm) *n.*

**Saint Ag·nes' Eve** (ăg′nĭs, -nĭ-sĭz) *n.* [After *St. Agnes* (d. A.D. 304).] The night of Jan. 20th, when, according to legend, a woman dreams of her future husband.

**Saint An·tho·ny's cross** (ăn′thə-nēz) *n.* A tau cross.

**Saint Ber·nard** (bər-närd′) *n.* A large strong dog orig. developed in Switzerland, with a heavy brown and white coat, orig. used by monks of the hospice of Saint Bernard in the Swiss Alps to help patrol the snow-covered region.

**saint·ed** (sān′tĭd) *adj.* **1.** Canonized. **2.** Of saintly character : HOLY.

**Saint El·mo's fire** (ĕl′mōz) *n.* [After *St. Elmo* (d. A.D. 303), patron saint of sailors.] An electric discharge seen emanating from a pointed object, as the mast of a ship or wing of an aircraft, during an electrical storm.

**saint·hood** (sānt′ho͝od′) *n.* **1.** The rank, character, or state of being a saint. **2.** Saints as a group.

**Saint Lo·uis encephalitis** (lōō′ĭs) *n.* [After *St. Louis,* Missouri, where an epidemic of the disease occurred in 1933.] A viral encephalitis transmitted by a North American culex mosquito.

**saint·ly** (sānt′lē) *adj.* **-li·er, -li·est.** Like, relating to, or befitting a saint. **—saint′li·ness** *n.*

**Saint Nich·o·las** (nĭk′ə-ləs) *or* **Saint Nick** (nĭk) *n.* [After *St. Nicholas* of Myra (d. A.D. 352?).] Santa Claus.

**Saint Pat·rick's Day** (păt′rĭks) *n.* Mar. 17, observed in honor of Saint Patrick, the legendary patron saint of Ireland.

**saint's day** *n.* A day in a liturgical calendar that honors a saint.

**Saint Val·en·tine's Day** (văl′ən-tīnz′) *n.* [After *St. Valentine* (d. A.D. 270?).] Feb. 14, on which valentines are traditionally exchanged.

**saith** (sĕth, sā′ĭth) *v. Archaic. 3rd person sing. present tense of* SAY.

**Sai·va** (sī′və, shī′-) *n.* [Skt. *śaiva-,* belonging to Shiva < *śivaḥ,* Shiva.] A member of the Hindu cult of the god Shiva. **—Sai′vism** *n.*

**sake**[1] (sāk) *n.* [ME, lawsuit, guilt < OE *sacu.*] **1.** Purpose : motive <a protest made for the *sake* of argument> **2.** Advantage : good <for the *sake* of the commonwealth> **3.** Personal benefit or interest : WELFARE <for one's own *sake*>

**sa·ke**[2] *also* **sa·ki** (sä′kē, -kĕ) *n.* [J., alcoholic drink.] A Japanese liquor made from fermented rice.

**sa·ker** (sā′kər) *n.* [ME *sagre* < OFr. *sacre* < Ar. *ṣaqr.*] A brown Eurasian falcon, *Falco cherrug,* often trained for falconry.

**sa·ki** (sä′kē, -kĕ) *n. var. of* SAKE[2].

**sal** (săl) *n.* [Lat.] Salt.

**sa·laam** (sə-läm′) *n.* [Ar. *salām,* peace.] **1.** A ceremonious act of deference or obeisance, esp. a low bow made while placing the right palm on the forehead. **2.** A respectful ceremonial greeting performed esp. in the East. *—vt. & vi.* **-laamed, -laam·ing, -laams.** To greet with or make a salaam.

**sal·a·ble** *also* **sale·a·ble** (sā′lə-bəl) *adj.* Offered or fit for sale : MARKETABLE. **—sal′a·bil′i·ty, sal′a·ble·ness** *n.* **—sal′a·bly** *adv.*

**sa·la·cious** (sə-lā′shəs) *adj.* [< Lat. *salax, salac-,* lustful, fond of leaping < *salire,* to leap.] **1.** Sexually appealing or stimulating : LASCIVIOUS. **2.** Lustful : bawdy. **—sa·la′cious·ly** *adv.* **—sa·la′cious·ness, sa·lac′i·ty** (sə-lăs′ĭ-tē) *n.*

**sal·ad** (săl′əd) *n.* [ME < OFr. *salade* < OProv. *salada* < VLat. ***salata,*** p.part. of ***salare*** < Lat. *sal,* salt.] **1. a.** A dish of green, leafy raw vegetables, often with radish, cucumber, tomato, etc., tossed with a dressing. **b.** The course consisting of this dish. **2.** A dish of chopped fruit, meat, seafood, eggs, or other food, usu. prepared with a dressing, as mayonnaise, and served cold. **3.** A green vegetable or herb used in salad, esp. lettuce.

**salad bar** *n.* A counter in a restaurant where customers may serve themselves a variety of salad ingredients and dressings.

**salad days** *pl.n.* A time of youth, innocence, and inexperience <"my *salad days* when I was green in judgment, cold in blood" —Shakespeare>

**salad dressing** *n.* A sauce, as of mayonnaise or oil and vinegar, served on salad.

**salad oil** *n.* An edible vegetable oil, as corn oil, that may be used in salad dressings.

**sal·a·man·der** (săl′ə-măn′dər) *n.* [ME *salamandre* < OFr. < Lat. *salamandra* < Gk.] **1.** A small lizardlike amphibian of the order Caudata, with porous, scaleless skin and four legs that are often weak or rudimentary. **2.** A mythical creature, gen. resembling a lizard, believed to be capable of living in or withstanding fire. **3.** An object, as a poker, used in fire or able to withstand heat. **4.** *Metallurgy.* A mass of solidified largely metallic material left in a blast-furnace hearth. **5.** A portable stove that is used to heat or dry buildings under construction. **—sal′a·man′drine** (-drĭn) *adj.*

**sa·la·mi** (sə-lä′mē) *n.* [Ital., pl. of *salame,* salami < *salare,* to salt < VLat. ***salare*** < Lat. *sal,* salt.] A highly spiced, salted sausage, made of pork or beef or a combination of meats.

**sal ammoniac** *n.* [ME *sal armoniak* < Lat. *sal ammoniacum,* salt of Ammon. —see AMMONIA.] Ammonium chloride.

**sal·a·ry** (săl′ə-rē, săl′rē) *n., pl.* **-ries.** [ME *salarie* < Lat. *salarium,* money given to Roman soldiers to buy salt < *sal,* salt.] Fixed compensation for services, paid on a regular basis. **—sal′a·ried** *adj.*

**sale** (sāl) *n.* [ME < OE *sala* < ON.] **1.** The act of selling : exchange of goods or services for an amount of money or its equivalent. **2.** An instance of selling property. **3.** An opportunity for selling or being sold : DEMAND. **4.** Availability for purchase <a shop where shoes are for *sale*> **5.** A selling of property to the highest bidder : AUCTION. **6.** A special disposal of goods at lowered prices <The sandals are on *sale* next week only.> **7. sales. a.** Activities involved in selling goods or services. **b.** Gross receipts.

**sale·a·ble** (sā′lə-bəl) *adj. var. of* SALABLE.

**sal·ep** (săl′əp, sə-lĕp′) *n.* [Fr. or Sp., both < Turk. *sālep* < Ar. *saḥleb,* a kind of orchid.] A starchy meal ground from the dried roots of various Old World orchids of the genera *Orchis* and *Eulophia,* used for food and formerly as medicine.

**sal·er·a·tus** (săl′ə-rā′təs) *n.* [NLat. *sal aeratus,* aerated salt.] Sodium or potassium bicarbonate used as a leavening agent.

**sales check** *n.* A slip of paper given by a store to serve as a record or receipt of something purchased.

**sales·clerk** (sālz′klûrk′) *n.* A person employed to sell merchandise in a store.

**sales·girl** (sālz′gûrl′) *n.* A saleswoman.

**Sa·le·sian** (sə-lē′zhən, sā-) *n.* A member of the Society of St. Francis de Sales, a Roman Catholic congregation founded in Turin in 1845 and dedicated primarily to education and missionary work. *—adj.* Of or relating to the Salesians.

**sales·la·dy** (sālz′lā′dē) *n.* A saleswoman.

**sales·man** (sālz′mən) *n.* A man employed to sell merchandise in a store or in a designated territory. **—sales′man·ship′** *n.*

**sales·peo·ple** (sālz′pē′pəl) *pl.n.* Persons engaged in selling merchandise or services.

**sales·per·son** (sālz′pûr′sən) *n.* A salesman or saleswoman.

**sales tax** *n.* A tax levied as a percentage of the retail price of merchandise and collected by the retailer.

**sales·wom·an** (sālz′wo͝om′ən) *n.* A woman employed to sell merchandise in a store or in a designated territory.

**sali-** *pref.* [< Lat. *sal,* salt.] Salt <*salinize*>

**Sa·li·an** (sā′lē-ən, sāl′yən) *adj.* [< LLat. *Salii,* the Salian Franks.] Of or relating to a tribe of Franks who settled in the Rhine region of the Netherlands in the 4th cent. A.D. *—n.* A Salian Frank.

**sal·ic** (săl′ĭk) *adj.* [S(ILICA) + AL(UMINA) + -IC.] Of or relating to igneous rocks, as quartz and the feldspars, that contain large amounts of silica and alumina.

**Sa·lic** (sā′lĭk, săl′ĭk) *also* **Sa·lique** (sā′lĭk, săl′ĭk, sə-lēk′, sā-) *adj.* [OFr. *salique* < Med. Lat. *Salicus* < LLat. *Salii,* the Salian Franks.] **1.** Designating or relating to the Salian Franks. **2.** Of or relating to the Salic law or the legal code of the Salian Franks.

**sal·i·cin** (săl′ĭ-sĭn) *n.* [Fr. *salicine* < Lat. *salix,* willow.] A bitter glucoside, $C_{13}H_{18}O_7$, obtained chiefly from the bark of willow and poplar trees and formerly used as an analgesic.

**Salic law** *n.* **1.** The legal code of the Salian Franks. **2.** A law, thought to derive from the code of laws of the Salian Franks, barring a woman from succeeding to a throne.

**sa·lic·y·late** (sə-lĭs′ə-lāt′, -lĭt, săl′ə-sĭl′ĭt) *n.* [SALICYL(IC ACID) + -ATE[2].] A salt or ester of salicylic acid.

**sal·i·cyl·ic acid** (săl′ĭ-sĭl′ĭk) *n.* [< Fr. *salicyle,* the radical of salicylic acid < *salicine,* salicin.] A white crystalline acid, $C_7H_6O_3$, used to make aspirin, as a preservative and flavoring agent, and in the external treatment of skin conditions, as eczema.

**sal·i·cyl·ism** (săl′ĭ-sə-lĭz′əm) *n.* [SALICYL(IC ACID) + -ISM.] A toxic syndrome caused by excessive doses of salicylic acid or salicylates.

**sa·li·ence** (sā′lē-əns, sāl′yəns) *also* **sa·li·en·cy** (sā′lē-ən-sē, sāl′yən-) *n.* **1.** The quality or state of being salient. **2.** A pronounced feature or part : HIGHLIGHT.

**sa·li·ent** (sā′lē-ənt, sāl′yənt) *adj.* [Lat. *saliens, salient-,* pr.part. of *salire,* to leap.] **1.** Projecting or protruding beyond a line or surface. **2.** Strikingly conspicuous : PROMINENT. **3.** Springing : jumping <a *salient* amphibian> *—n.* **1.** An area of a military defense, as a battle line, that projects closest to the enemy. **2.** A projecting angle or part. **—sa′li·ent·ly** *adv.* **—sa′li·ent·ness** *n.*

**sa·li·en·tian** (sā′lē-ĕn′shən) *n.* [< NLat. *Salientia,* order name < Lat. *saliens,* pr.part. of *salire,* to leap.] An amphibian of the order Salientia including the frogs and toads. *—adj.* Of or belonging to the Salientia.

**sa·lif·er·ous** (sə-lĭf′ər-əs) *adj.* Containing or producing salt.

**sa·lim·e·ter** (sə-lĭm′ĭ-tər) *n.* A specially graduated hydrometer that directly indicates the percentage of a salt in a solution. **—sal′i·met′ric** (săl′ə-mĕt′rĭk) *adj.* **—sa·lim′e·try** *n.*

**sa·li·na** (sə-lī′nə, -lē′-) *n.* [Sp. < Lat. *salinae,* salt pits < *salinus,* saline.] **1.** A salt marsh, spring, pond, or lake. **2.** A land area encrusted with salt.

**sa·line** (sā′lēn′, -līn′) *adj.* [Lat. *salinus* < *sal,* salt.] **1.** Of, relating to, or containing salt : SALTY. **2.** Relating to mineral salts having the characteristics of common salt. *—n.* **1.** A salt of magnesium or of the alkali metals, used medically as a cathartic. **2.** A saline solution, esp. one that is isotonic with blood and is used in medicine and surgery. **—sa·lin′i·ty** (sə-lĭn′ĭ-tē) *n.*

**sal·i·nize** (săl′ə-nīz′) *vt.* **-nized, -niz·ing, -niz·es.** To treat or contaminate with salt. **—sal′i·ni·za′tion** *n.*

**sal·i·nom·e·ter** (săl′ə-nŏm′ĭ-tər) *n.* An instrument, esp. a salimeter, that measures the amount of salt in a solution. **—sal′i·no·met′ric** (-nə-mĕt′rĭk) *adj.* **—sal′i·nom′e·try** *n.*

**Sa·lique** (sā′lĭk, săl′ĭk, sə-lēk′, sā-) *adj. var. of* SALIC.

**Salis·bur·y steak** (sôlz′bĕr′ē, -brē, sălz′-) *n.* [After J. H. *Salisbury,* 19th-cent. English nutritionist.] A patty of ground beef mixed with various seasonings and broiled or fried.

**Sa·lish** (sā′lĭsh) *also* **Sa·lish·an** (-lĭ-shən) *n.* **1.** A family of Indian languages of the northwestern United States and British Columbia. **2.** A member of a tribe speaking a Salish language. **—Sa′lish·an** *adj.*

**sa·li·va** (sə-lī′və) *n.* [Lat.] The watery, tasteless liquid mixture of salivary and oral mucous gland secretions that lubricates chewed food, moistens the oral walls, and contains the enzyme ptyalin, which functions in the predigestion of starches.

**sal·i·var·y** (săl′ə-vĕr′ē) *adj.* **1.** Of, relating to, or producing saliva. **2.** Of or relating to a salivary gland.
**salivary gland** *n.* A gland secreting saliva, esp. one of three pairs of large glands, the parotid, submandibular, and sublingual, whose secretions enter the mouth and mingle in saliva.
**sal·i·vate** (săl′ə-vāt′) *v.* **-vat·ed, -vat·ing, -vates.** —*vi.* To secrete or produce saliva. —*vt.* To produce excessive salivation in. **—sal′i·va′tion** (-vā′shən) *n.*
**Salk vaccine** (sôlk, sôk) *n.* [After Jonas *Salk* (b. 1914).] A killed-virus vaccine used to immunize against poliomyelitis.
**sal·let** (săl′ĭt) *n.* [ME < OFr. *salade.*] A light medieval helmet with a flaring brim in the back, occas. fitted with a visor.
**sal·low¹** (săl′ō) *adj.* **-er, -est.** [ME *salowe* < OE *salo.*] Sickly yellow in hue or complexion. —*vt.* **-lowed, -low·ing, -lows.** To make sallow. **—sal′low·ly** *adv.* **—sal′low·ness** *n.*
**sal·low²** (săl′ō) *n.* [ME *salwe* < OE *sealh.*] A European willow, esp. *Salix caprea,* whose wood is a source of charcoal.
**sal·ly** (săl′ē) *vi.* **-lied, -ly·ing, -lies.** [< OFr. *saillie,* a sally < *saillir,* to rush forward < Lat. *salire,* to leap.] **1.** To leap forth or rush out suddenly. **2.** To issue suddenly from a defensive or besieged position to attack an enemy. **3.** To set out on a journey or excursion <*sallied* forth on a shopping trip> —*n., pl.* **-lies. 1.** A sudden rush forward: LEAP. **2.** An assault from a defensive position: SORTIE. **3.** A sudden emergence into action or expression: OUTBURST. **4.** A witticism: quip. **5.** An excursion: jaunt.
**sal·ly lunn** (săl′ē lŭn′) *n.* [After *Sally Lunn,* 18th-cent. English baker.] A somewhat sweet bread leavened with yeast.
**sally port** *n.* A gate in a fortification designed for sorties.
**sal·ma·gun·di** (săl′mə-gŭn′dē) *n.* [Fr. *salmigondis.*] **1.** A salad of chopped meat, anchovies, eggs, and onions or other vegetables, often arranged in rows on lettuce and served with a dressing of vinegar and oil. **2.** An assortment: potpourri.
**sal·mi** (săl′mē) *n.* [Fr. *salmis,* short for *salmigondis,* salmagundi.] A highly spiced dish of roasted game birds minced and stewed in wine.
**salm·on** (săm′ən, sä′mən) *n., pl.* **salmon** *or* **-ons.** [ME < OFr. *saumon* < Lat. *salmo.*] **1.** A large food and game fish of the genera *Salmo* or *Oncorhynchus* of northern waters, swimming from salt to fresh water to spawn and having delicate pinkish flesh. **2.** A moderate, light, or strong yellowish pink to a moderate reddish orange or light orange.
**salm·on·ber·ry** (săm′ən-bĕr′ē, sä′mən-) *n.* **1.** A prickly shrub, *Rubus spectabilis* of western North America, with compound leaves and fragrant reddish flowers. **2.** The edible fruit of the salmonberry.
**sal·mo·nel·la** (săl′mə-nĕl′ə) *n., pl.* **-nel·lae** (-nĕl′ē) *or* **-nel·las** *or* **-nel·la.** [NLat. *Salmonella,* genus name, after Daniel E. *Salmon* (1850–1914).] Any of various rod-shaped bacteria of the genus *Salmonella,* many of which are pathogenic.
**sal·mo·nel·lo·sis** (săl′mə-nĕ-lō′sĭs) *n., pl.* **-ses** (-sēz′). Infection with salmonellae.
**salm·o·nid** (săm′ə-nĭd, săl′mə-) *adj.* Salmonoid. **—salm′o·nid** *n.*
**salm·o·noid** (săm′ə-noid′, săl′mə-) *adj.* **1.** Like or characteristic of a salmon. **2.** Of or belonging to the family Salmonidae, which includes the salmon, trout, and whitefishes. —*n.* A salmonoid fish.
**sal·ol** (săl′ôl′, -ōl′) *n.* [Orig. a trademark.] A white crystalline powder, $C_{13}H_{10}O_3$, derived from salicylic acid and used in making plastics and suntan oils and medicinally as an analgesic and antipyretic.
**sa·lom·e·ter** (să-lŏm′ĭ-tər, sə-) *n.* A salimeter.
**sa·lon** (sə-lŏn′, săl′ŏn′, sä-lôɴ′) *n.* [Fr. < Ital. *salone,* aug. of *sala,* hall, of Germanic orig.] **1.** A large room, as a drawing room, used for entertaining guests. **2.** A periodic gathering of socially or intellectually prominent persons. **3.** A hall or gallery for exhibiting works of art. **4.** A commercial establishment offering a product or service related to fashion <a *beauty* salon>
**sa·loon** (sə-lo͞on′) *n.* [Fr. *salon,* salon.] **1.** A place where alcoholic drinks are sold and drunk: TAVERN. **2.** A large room or hall for receptions, public entertainment, or exhibitions. **3. a.** The officers' dining and social room on a cargo ship. **b.** A large public lounge on a passenger ship. **4.** *Chiefly Brit.* A sedan automobile.
**sa·loon·keep·er** (sə-lo͞on′kē′pər) *n.* An owner or operator of a saloon for drinking.
**sa·loop** (sə-lo͞op′) *n.* [Alteration of SALEP.] A hot drink, formerly used medicinally, made from salep, sassafras, or similar aromatic herbs.
**salp** (sălp) *also* **sal·pa** (săl′pə) *n.* [NLat. *Salpa,* genus name < Lat., a kind of stockfish < Gk. *salpē.*] A free-swimming chordate of the genus *Salpa* of warm seas, with a translucent, somewhat flattened, keg-shaped body. **—sal′pi·form′** (săl′pə-fôrm′) *adj.*
**salping-** *pref.* [< Gk. *salpinx, salping-,* military trumpet.] Salpinx <*salpingitis*>
**sal·pin·gec·to·my** (săl′pĭn-jĕk′tə-mē) *n., pl.* **-mies.** Surgical removal of the Fallopian tube.
**sal·pin·ges** (săl-pĭn′jēz) *n. pl. of* SALPINX.
**sal·pin·gi·tis** (săl′pĭn-jī′tĭs) *n.* Inflammation of the Fallopian or Eustachian tube.
**sal·pinx** (săl′pĭngks) *n., pl.* **sal·pin·ges** (săl-pĭn′jēz) [Gk. *salpinx,* trumpet.] **1.** The Fallopian tube. **2.** The Eustachian tube. **—sal·pin′gi·an** (-pĭn′jē-ən, -jən) *adj.*
**sal·sa** (säl′sə) *n.* [Am. Sp. < Sp., sauce < Lat. *salsa.* —see SAUCE.] A popular Latin American dance music, marked by elements of jazz, blues, and rock.
**sal·si·fy** (săl′sə-fē, -fī′) *n.* [Fr. *salsifis* < obs. Ital. *salsifica.*] **1.** A native European plant, *Tragopogon porrifolius,* with grasslike leaves, purple flowers, and an edible taproot. **2.** The root of the salsify, eaten raw or cooked as a vegetable.
**sal soda** *n.* A hydrated sodium carbonate used as a general cleanser.
**salt** (sôlt) *n.* [ME < OE *sealt.*] **1.** A colorless or white crystalline solid, mainly sodium chloride, widely used as a food seasoning and preservative. **2.** A chemical compound formed by replacing all or part of the hydrogen ions of an acid with one or more cations of a base. **3. salts.** Any of various mineral salts used as a laxative or cathartic. **4. salts.** Smelling salts. **5. salts.** Epsom salts. **6.** An element giving flavor or zest. **7.** Sharp, lively wit. **8.** *Informal.* A sailor, esp. when old or experienced. **9.** A saltcellar. —*vt.* **salt·ed, salt·ing, salts. 1.** To add, treat, season, or sprinkle with salt. **2.** To cure or preserve by treating with salt or a salt solution. **3.** To provide salt for (e.g., deer or cattle). **4.** To add zest or liveliness to <*salt* a presentation with witty anecdotes> **5.** To give an appearance of value to by fraudulent means, esp. to place valuable minerals in (a mine) so as to deceive. **—salt away.** To put aside: SAVE. **—salt out.** To separate (a dissolved substance) by adding salt to the solution. **—worth (one's) salt.** Capable and efficient.
▲ word history: The importance of common salt in human life is perhaps less noticeable now than in the past, but the language records the role played by this simple chemical substance. The word *salt* is a native English word; over the centuries it has formed many compounds, most of which have obvious meanings. *Salt* is related to Latin *sal,* "salt," which has many derivatives in English whose origins are not immediately apparent. The word *salary* is perhaps the most notable word derived from *sal.* The Latin source of *salary* is *salarium,* which originally meant "money given to soldiers to buy salt" but later meant "a stipend, wages." Salt of course was put on food, both as a seasoning and as a preservative. *Salad* and *sauce* originally denoted salted accompaniments to a meal. *Salami* and *sausage* denote preparations of meat preserved with salt.
**salt-and-pep·per** (sôlt′ən-pĕp′ər) *adj.* Pepper-and-salt.
**sal·ta·rel·lo** (săl′tə-rĕl′ō, sôl′-) *n., pl.* **-rel·los** *or* **-rel·li** (-rĕl′ē) [Ital. < *saltare,* to leap < Lat. —see SALTATION.] A lively Italian dance with a skipping step at the start of each measure.
**sal·ta·tion** (săl-tā′shən, sôl-) *n.* [Lat. *saltatio* < *saltatus,* p.part. of *saltare,* to leap < freq. of *salire,* to jump.] **1.** The act of leaping, jumping, or dancing. **2.** Discontinuous movement or development, as if by leaps. **3.** *Biol.* A mutation or discontinuous variation.
**sal·ta·to·ri·al** (săl′tə-tôr′ē-əl, -tōr′-, sôl′-) *adj.* **1.** Of or pertaining to leaping or dancing. **2.** Adapted for or marked by leaping.
**sal·ta·to·ry** (săl′tə-tôr′ē, -tōr′ē, sôl′-) *adj.* **1.** Of, relating to, or adapted for leaping or dancing. **2.** Proceeding by leaps rather than by smooth, gradual transitions.
**salt·box** (sôlt′bŏks′) *n.* A frame house with two stories in front and one in back, topped by a roof with a long rear slope.
**salt·bush** (sôlt′bo͝osh′) *n.* A salt-tolerant plant of the genus *Atriplex,* esp. *A. hortensis.*
**salt cake** *n.* Impure sodium sulfate used in the manufacture of paper pulp, soaps and detergents, glass, ceramic glazes, and dyes.
**salt·cel·lar** (sôlt′sĕl′ər) *n.* [Alteration of ME *salt saler* : *salt,* salt + *saler,* saltcellar < OFr. *saliere* < Lat. *salarius,* of salt < *sal,* salt.] A small dish for holding salt at the table.
▲ word history: A saltcellar is not a basement storehouse for sodium chloride; the word in fact has nothing to do with the word *cellar* at all. The spelling *-cellar* is an alteration of the obsolete form *saler,* which meant "saltcellar." *Saler* is borrowed from Old French *saliere,* which is ultimately derived from Latin *sal,* "salt."
**salt·er** (sôl′tər) *n.* **1.** One that manufactures or sells salt. **2.** One that treats meat, fish, or other foods with salt.
**salt·ern** (sôl′tərn) *n.* [OE *sealtærn* : *sealt,* salt + *ærn,* house.] A building or place where salt is manufactured: SALTWORKS.
**salt grass** *n.* Any of various grasses, as those of the genus *Distichlis,* that grow in alkaline regions.
**salt hay** *n.* **1.** The wiry, tough stems of several species of salt-marsh rushes, esp. *Juncus gerardi,* used as a garden mulch and packing material. **2.** Hay prepared from salt grass.
**sal·tine** (sôl-tēn′) *n.* A thin, crisp cracker sprinkled with coarse salt.
**sal·tire** (sôl′tîr′, săl′-) *n.* [ME *sawtire* < OFr. *saultoir,* stile < *saulter,* to jump < Lat. *saltare.* —see SALTATION.] *Heraldry.* An X-shaped ordinary formed by the crossing of a bend and a bend sinister.
**salt·ish** (sôl′tĭsh) *adj.* Somewhat salty.
**salt lick** *n.* **1.** A natural deposit of exposed salt that animals lick. **2.** A block of salt or an artificial medicated saline preparation set out for cattle, sheep, or deer to lick.
**salt marsh** *n.* Low coastal grassland often covered by the tide.
**salt·pe·ter** (sôlt′pē′tər) *n.* [ME *salpetre* < OFr. < Med. Lat. *salpetra* : Lat. *sal,* salt + *petra,* rock < Gk.] **1.** Potassium nitrate. **2.** Sodium nitrate. **3.** Niter.

**salt rheum** *n.* Eczema.
**salt·shak·er** (sôlt′shā′kər) *n.* A container with a perforated top for sprinkling table salt.
**salt·wa·ter** (sôlt′wô′tər, -wŏt′ər) *adj.* Relating to, consisting of, or inhabiting salt water.
**salt·works** (sôlt′wûrks′) *pl.n. (used with a sing. or pl. verb).* A place where salt is manufactured commercially.
**salt·wort** (sôlt′wûrt′, -wôrt′) *n.* An Old World plant of the genus *Salsola*, esp. *S. kali*, having stiff prickly leaves and growing on sandy seashores.
**salt·y** (sôl′tē) *adj.* **-i·er, -i·est. 1.** Of, containing, or seasoned with salt. **2.** Suggestive of the sea or sailing life. **3. a.** Witty : piquant. **b.** EARTHY 3. **—salt′i·ly** *adv.* **—salt′i·ness** *n.*
**sa·lu·bri·ous** (sə-lōō′brē-əs) *adj.* [< Lat. *salubris* < *salus*, health.] Conducive or favorable to health or well-being. **—sa·lu′bri·ous·ly** *adv.* **—sa·lu′bri·ous·ness, sa·lu′bri·ty** (-brĭ-tē) *n.*
**sa·lu·ki** (sə-lōō′kē) *n.* [Ar. *salūqīy*, of Saluq, an ancient Arabian city.] A tall, slender dog orig. bred in Arabia and Egypt, with a smooth, silky, variously colored coat.
**sal·u·tar·y** (săl′yə-tĕr′ē) *adj.* [OFr. *salutaire* < Lat. *salutaris* < *salus*, health.] **1.** Effecting or designed to effect an improvement : REMEDIAL <*salutary* advice> **2.** Promoting health : WHOLESOME <a *salutary* diet> **—sal′u·tar′i·ly** (-târ′ə-lē) *adv.* **—sal′u·tar′i·ness** *n.*
**sal·u·ta·tion** (săl′yə-tā′shən) *n.* **1.** An expression of greeting or good will. **2.** A gesture of greeting, as a bow or kiss. **3.** A word or phrase of greeting, as *Dear Sir* in a letter.
**sa·lu·ta·to·ri·an** (sə-lōō′tə-tôr′ē-ən, -tōr′-) *n.* The student who delivers the salutatory at graduation exercises, usu. the one who ranks second in the class.
**sa·lu·ta·to·ry** (sə-lōō′tə-tôr′ē, -tōr′ē) *n., pl.* **-ries.** An opening or welcoming address. *—adj.* Of or expressing a salutation.
**sa·lute** (sə-lōōt′) *v.* **-lut·ed, -lut·ing, -lutes.** [ME *saluten* < Lat. *salutare* < *salus*, health.] *—vt.* **1.** To greet or address with an expression of welcome, good will, or respect. **2.** To recognize (a military superior) with a prescribed gesture, as by raising the hand to the forehead or cap. **3.** To honor formally and ceremoniously. **4.** To come forth as if to greet. *—vi.* To make a gesture of greeting or respect. *—n.* **1.** An act or gesture of welcome, honor, or courteous recognition. **2.** A formal military display of honor or greeting, as the firing of cannon. **—sa·lut′er** *n.*
**sal·va·ble** (săl′və-bəl) *adj.* [< LLat. *salvare*, to save < Lat. *salvus*, safe.] Capable of being saved or salvaged.
**Sal·va·do·ri·an** (săl′və-dôr′ē-ən, -dōr′-) *also* **Sal·va·do·ran** (-dôr′ən, -dōr′-) *n.* A native or resident of El Salvador. **—Sal′va·do′ri·an** *adj.*
**sal·vage** (săl′vĭj) *n.* [< Fr., act of saving < OFr. *salver*, to save < LLat. *salvare* < Lat. *salvus*, safe.] **1. a.** The rescue of a ship, its crew, or its cargo, as from fire or shipwreck. **b.** The ship, crew, or cargo so rescued. **c.** Compensation given to those who voluntarily aid in such a rescue. **2. a.** The act of saving imperiled property from loss. **b.** The property so saved. *—vt.* **-vaged, -vag·ing, -vag·es. 1.** To save from loss or destruction. **2.** To save (damaged or discarded material) for further use. **—sal′vage·a·ble** *adj.* **—sal′vag·er** *n.*
**sal·var·san** (săl′vər-săn′) *n.* [G.; orig. a trademark.] Arsphenamine.
**sal·va·tion** (săl-vā′shən) *n.* [ME < OFr. < LLat. *salvatio* < *salvare*, to save < Lat. *salvus*, safe.] **1. a.** Preservation or deliverance from danger, evil, or difficulty. **b.** A source, means, or cause of such salvation. **2. a.** Deliverance from the power or penalty of sin : REDEMPTION. **b.** *Christian Science.* The realization and demonstration of Life, Truth, and Love as supreme over all, carrying with it the destruction of the illusions of sin, sickness, and death. **—sal·va′tion·al** *adj.*
**Salvation Army** *n.* An international evangelical and charitable organization founded in England in 1865 by William Booth.
**sal·va·tion·ist** (săl-vā′shə-nĭst) *n.* **1.** *often* **Salvationist.** A member of the Salvation Army. **2.** An evangelist.
**salve**[1] (săv, säv) *n.* [ME < OE *sealf.*] **1.** An analgesic or medicinal ointment, as for burns or sores. **2.** A soothing or healing agent : BALM. **3.** Flattery or commendation. *—vt.* **salved, salv·ing, salves.** To heal or soothe with or as if with salve.
**salve**[2] (sălv) *vt.* **salved, salv·ing, salves.** [Back-formation < SALVAGE.] To salvage. **—salv′or** *n.*
**sal·ver** (săl′vər) *n.* [Alteration of Fr. *salve* < Sp. *salva*, tasting of food to detect poison < *salvar*, to save, taste food to detect poison < LLat. *salvare*, to save < Lat. *salvus*, safe.] A tray for serving food or drinks.
**sal·vi·a** (săl′vē-ə) *n.* [NLat. *Salvia*, genus name < Lat. *salvia*, sage. —see SAGE[2].] A plant or shrub of the genus *Salvia*, cultivated for its blue or bright red flowers.
**sal·vo** (săl′vō) *n., pl.* **-vos** *or* **-voes.** [Ital. *salva*, salute < Lat. *salve*, hail, imper. of *salvēre*, to be in good health < *salvus*, safe.] **1. a.** A simultaneous discharge of firearms. **b.** The release of a rack of bombs from an aircraft. **c.** The projectiles or bombs thus released. **2.** A sudden outburst of cheers or applause. **3.** A salute : tribute.
**sal vo·la·ti·le** (vō-lăt′ə-lē) *n.* [NLat., volatile salt.] A solution of ammonium carbonate in alcohol or ammonia water.
**sam·a·ra** (săm′ər-ə, sə-mâr′ə, -mär′ə) *n.* [Lat., elm seed.] A winged, indehiscent, usu. one-seeded fruit, as of the ash or maple.
**Sa·mar·i·tan** (sə-măr′ĭ-tn, -mâr′-) *n.* [ME < LLat. *Samaritanus* < Gk. *Samaritēs* < *Samareia*, Samaria.] **1.** A native or resident of Samaria. **2.** A Good Samaritan. *—adj.* Of or relating to Samaria or to Samaritans.
**sa·mar·i·um** (sə-mâr′ē-əm, -măr′-) *n.* [SAMAR(SKITE) + -IUM.] *Symbol* **Sm** A silvery or pale-gray metallic rare-earth element used in laser materials, in infrared absorbing glass, and as a neutron absorber; atomic number 62; atomic weight.
**sa·mar·skite** (sə-mär′skīt′, săm′ər-) *n.* [Fr., after Col. von *Samarski*, 19th-cent. Russian mine official.] A velvet-black mineral oxide with red-brown streaks, a source of several rare-earth metals.
**sam·ba** (săm′bə, säm′-) *n.* [Port.] **1.** An African dance modified in Brazil as a ballroom dance. **2.** Music in 4/4 time for dancing the samba. *—vi.* **-baed, -ba·ing, -bas.** To dance the samba.
**sam·bar** *also* **sam·bur** (săm′bər, säm′-) *n.* [Hindi *sāmbar* < Skt. *śambaraḥ.*] A large deer, *Cervus unicolor* of southeastern Asia, with a reddish-brown coat.
**Sam Browne belt** (săm′ broun′) *n.* [After Sir *Samuel* James *Browne* (1824–1901).] A belt with a shoulder strap running diagonally across the chest, worn as part of a military or police uniform.
**sam·bur** (săm′bər, säm′-) *n. var. of* SAMBAR.
**same** (sām) *adj.* [ME < ON *samr.*] **1.** Being the very one : IDENTICAL. **2.** Alike in kind, quality, quantity, or degree. **3.** Conforming in every detail <using the *same* procedures as before> **4.** Being the one previously mentioned or indicated : AFORESAID. *—adv.* In the same way. *—pron.* **1.** One identical with another. **2.** The one previously mentioned or described. ***usage:*** Only in legal contexts is *same* or *the same* used as a substitute for *it, they,* or *them.* Therefore, avoid sentences like *The charge is five dollars; please remit same.* ☆ ***syns:*** **1.** SAME, IDENTICAL, SELFSAME, VERY *adj. core meaning* : being one and not another or others <the *same* seat I had yesterday> ***ant:*** different **2.** SAME, EQUAL, EQUIVALENT, IDENTICAL *adj. core meaning* : agreeing exactly in value, quantity, or effect <the *same* rules expressed in different words> ***ant:*** different
**sa·mekh** (sä′mĕk′) *n.* [Heb. *sāmekh.*] The 15th letter of the Hebrew alphabet. —See table at ALPHABET.
**same·ness** (sām′nĭs) *n.* **1.** The quality or state of being the same. **2.** Lack of variety or change : MONOTONY.
**sam·i·sen** (săm′ĭ-sĕn′) *n.* [J. : *sami*, three + *sen*, string.] A Japanese musical instrument resembling a banjo, with a long neck and three strings played with a plectrum.
**sam·ite** (săm′īt′, sā′mīt′) *n.* [ME *samit* < OFr. < Med. Lat. *examitum* < Med. Gk. *hexamiton* < Gk. *hexamitos*, of six threads : *hexa-*, six + *mitos*, warp thread.] A heavy silk fabric, frequently interwoven with gold or silver, worn in the Middle Ages.
**sa·miz·dat** (sä′mēz-dät′) *n.* [R. : *sam*, self + *izdatel′stvo*, publisher < *izdat′*, to publish.] **1. a.** Secret publication and distribution of government-banned literature in the U.S.S.R. **b.** The literature produced by this system. **2.** An underground press.
**sam·let** (săm′lĭt) *n.* [Blend of SALMON + -LET.] A young salmon.
**Sa·mo·an** (sə-mō′ən) *adj.* Of or relating to Samoa, its Polynesian people, or their language. *—n.* **1.** A native or resident of Samoa. **2.** The Polynesian language of Samoa.
**sam·o·var** (săm′ə-vär′) *n.* [R. : *samo-*, self + *varit′*, to boil.] A metal urn with a spigot, used esp. in Russia to boil water for tea.
**Sam·o·yed** *also* **Sam·o·yede** (săm′ə-yĕd′, -oi-ĕd′) *n.* [R. *samoed* < Lapp *Sāme-Ǟednâma*, of Lapland.] **1.** A member of a Ural-Altaic people inhabiting the tundra lands of the northeastern European Soviet Union and northwestern Siberia. **2.** A branch of the Uralic language family that comprises the languages of the Samoyeds. **3.** A dog orig. bred in northern Eurasia, with a thick, long white coat. **—Sam′o·yed′ic** (-yĕd′ĭk) *adj.*
**samp** (sămp) *n.* [Narraganset *nasdump*, corn mush.] A coarse hominy or a porridge made from it.
**sam·pan** (săm′păn′) *n.* [Chin. *san*[1] *ban*[3] : *san*[1], small + *ban*[3], board.] A flat-bottomed Oriental skiff propelled usu. by two oars.
**sam·phire** (săm′fīr′) *n.* [Alteration of OFr. *(herbe de) Saint Pierre*, (herb of) St. Peter.] **1.** The glasswort. **2.** An Old World plant of coastal areas, esp. *Crithmium maritimum*, with fleshy divided leaves and small white or yellowish flowers.
**sam·ple** (săm′pəl) *n.* [ME < OFr. *essample.* —see EXAMPLE.] **1. a.** A part representative of a whole. **b.** An entity representative of a class : SPECIMEN. **2.** *Statistics.* A set of elements drawn from and analyzed to estimate the characteristics of a population. *—vt.* **-pled, -pling, -ples.** To take a sample of, esp. to test or examine by a sample.
**sam·pler** (săm′plər) *n.* **1.** One employed to take and appraise samples, as of a food product. **2.** A mechanical device for obtaining and analyzing samples. **3.** A piece of cloth embroidered with designs or mottoes in various stitches.
**sam·pling** (săm′plĭng) *n.* **1.** SAMPLE 2. **2.** The act or process of selecting a sample.
**sampling circuit** *n.* A circuit that yields an output suitable for use as an error or negative signal in a controller program that uses sampling action.

**sampling distribution** *n.* The distribution of a statistic, such as occurs when a number of sample means are calculated for a given population.
**sampling gate** *n.* A circuit that produces an output only when first activated by a preliminary pulse.
**sam·sa·ra** (səm-sä'rə) *n.* [Skt. *samsāraḥ* : *sam-*, together + *sarati*, it flows.] The repeated cycles of birth, suffering, death, and rebirth in Hinduism and Buddhism.
**Sam·u·el** (săm'yo͞o-əl) [LLat. < Gk. *Samouel* < Heb. *Shĕmū'ēl.*] —See table at BIBLE.
**sam·u·rai** (săm'ə-rī', -yə-) *n., pl.* **samurai** *or* **-rais.** [J., warrior.] **1.** The feudal military aristocracy of Japan. **2.** A professional warrior belonging to the samurai.
**san·a·to·ri·um** (săn'ə-tôr'ē-əm, -tōr'-) *also* **san·a·tar·i·um** (-târ'ē-əm) *n., pl.* **-to·ri·ums** *or* **-to·ri·a** (-tôr'ē-ə, -tōr'-) *also* **-tari·ums** *or* **-tari·a** (-târ'ē-ə) [NLat., neuter of LLat. *sanatorius*, curative < Lat. *sanatus*, p.part. of *sanare*, to heal < *sanus*, healthy.] **1.** An institution for the treatment of chronic diseases or for medically supervised recuperation. **2.** SANITARIUM 1.
**san·be·ni·to** (săn'bə-nē'tō) *n., pl.* **-tos.** [Sp. *sambenito* < *San Benito*, St. Benedict.] A sackcloth garment resembling a scapular, worn at an auto-da-fé of the Spanish Inquisition by condemned heretics, being yellow with red crosses for the penitent and black with painted flames and devils for the impenitent.
**sanc·ta** (săngk'tə) *n. var. pl. of* SANCTUM.
**sanc·ti·fy** (săngk'tə-fī') *vt.* **-fied, -fy·ing, -fies.** [ME *sanctifien* < OFr. *sanctifier* < LLat. *sanctificare* : Lat. *sanctus*, p.part. of *sancire*, to consecrate + Lat. *facere*, to make.] **1.** To set apart for sacred use : CONSECRATE. **2.** To make free from sin : PURIFY. **3.** To give religious sanction to, as with an oath. **4.** To give social or moral sanction to. **5.** To make productive of holiness or blessing. **—sanc'ti·fi·ca'tion** *n.* **—sanc'ti·fi'er** *n.*
**sanc·ti·mo·ni·ous** (săngk'tə-mō'nē-əs) *adj.* Feigning piety or righteousness. **—sanc'ti·mo'ni·ous·ly** *adv.* **—sanc'ti·mo'ni·ous·ness** *n.*
**sanc·ti·mo·ny** (săngk'tə-mō'nē) *n.* [OFr. *sanctimonie* < Lat. *sanctimonia*, sacredness < *sanctus*, holy. —see SANCTIFY.] Hypocritical piety or righteousness.
**sanc·tion** (săngk'shən) *n.* [OFr. < Lat. *sanctio*, an ordaining < *sanctus*, holy. —see SANCTIFY.] **1.** Authoritative approval or permission making a course of action valid. **2.** Support or encouragement, as from public opinion or established custom. **3.** A consideration, influence, or principle dictating an ethical choice. **4.** A law or decree. **5.** The penalty for noncompliance specified in a law or decree. **6.** A penalty, specified or in the form of moral pressure, that acts to ensure compliance or conformity. **7.** A coercive measure adopted usu. by several nations acting together against a nation violating international law. —*vt.* **-tioned, -tion·ing, -tions. 1.** To make legitimate : AUTHORIZE. **2.** To encourage by an indication of approval.
**sanc·ti·ty** (săngk'tĭ-tē) *n., pl.* **-ties.** [ME *saunctite* < OFr. *saintcite* < Lat. *sanctitas* < *sanctus*, sacred. —see SANCTIFY.] **1.** Holiness : saintliness. **2.** The quality or condition of being considered sacred : INVIOLABILITY. **3.** Something thought to be sacred.
**sanc·tu·ar·y** (săngk'cho͞o-ĕr'ē) *n., pl.* **-ies.** [ME *sanctuarie* < OFr. *saintuarie* < LLat. *sanctuarium* < Lat. *sanctus*, sacred. —see SANCTIFY.] **1. a.** A sacred place, as a church, temple, or mosque. **b.** The holiest of a sacred place. **2. a.** A place giving refuge, asylum, or immunity from arrest. **b.** The immunity provided by a sanctuary. **3.** A reserved area in which animals or birds are protected from hunting or molestation.
**sanc·tum** (săngk'təm) *n., pl.* **-tums** *or* **-ta** (-tə) [Lat., neuter of *sanctus*, sacred. —see SANCTIFY.] **1.** A sacred or holy place. **2.** A private place, as an office, where one is free from intrusion.
**sanctum sanc·to·rum** (săngk-tôr'əm, -tōr'-) *n.* [LLat. (transl. of Gk. *to hagion tōn hagiōn*, transl. of Heb. *qōdesh ha-qqodashīm*).] **1.** Holy of holies. **2.** A private place.
**Sanc·tus** (săngk'təs) *n.* [ME < Med. Lat. < Lat. *sanctus*, holy (the first word of the hymn). —see SANCTIFY.] A hymn of praise following the Preface in many liturgies.
**sand** (sănd) *n.* [ME < OE.] **1. a.** Loose, granular, gritty particles of worn or disintegrated rock, finer than gravel and coarser than dust. **b.** *often* **sands.** A land area covered with sand, as a beach or desert. **c.** The sand in an hourglass. **2. sands.** Moments of allotted time or duration. **3.** *Slang.* Courage : determination. **4.** A light grayish brown to yellowish gray. —*vt.* **sand·ed, sand·ing, sands. 1.** To sprinkle or cover with or as if with sand. **2.** To scrape or polish with sand or sandpaper. **3.** To mix with sand. **4.** To fill up (a harbor) with sand.
**san·dal¹** (săn'dl) *n.* [ME *sandalie* < Lat. *sandalium* < Gk. *sandalion*, dim. of *sandalon*, sandal.] **1.** A shoe with a sole fastened to the foot by thongs or straps. **2.** A light slipper or low-cut shoe fastened to the foot by an ankle strap. **3.** A low-cut rubber overshoe covering little more than the sole of the shoe. **4.** A strap or band for fastening a low shoe or slipper on the foot.
**san·dal²** (săn'dl) *n.* [ME < OFr. < Med. Lat. *sandalum, santalum* < Gk. *santalon, sandanon.*] Sandalwood.
**san·dal·wood** (săn'dl-wo͝od') *n.* **1. a.** An Asian tree of the genus *Santalum*, esp. *S. album*, with aromatic yellowish heartwood that is used in cabinetmaking and wood carving and yields an oil used in perfumery. **b.** The wood of the sandalwood or of similar trees. **2.** A light to moderate or grayish brown.
**san·da·rac** (săn'də-răk') *n.* [Lat., red pigment < Gk. *sandarakē*, realgar.] **1.** A tree, *Tetraclinis articulata* or *Callitris quadrivalvis* of northern Africa, with wood that yields a brittle, translucent resin used in varnishes. **2.** The resin of the sandarac.
**sand·bag** (sănd'băg') *n.* A bag filled with sand, used as ballast, to form protective or defensive barriers, or as a weapon. —*vt.* **-bagged, -bag·ging, -bags. 1.** To put sandbags in or around. **2. a.** To hit with or as if with a sandbag. **b.** *Informal.* To coerce.
**sand·bank** (sănd'băngk') *n.* A large mass of sand, as on a hillside or in a river.
**sand·bar** (sănd'bär') *n.* An offshore shoal of sand built up by the action of waves or currents.
**sand·blast** (sănd'blăst') *n.* **1. a.** A high-velocity blast of air or steam carrying sand to etch glass or clean stone or metal surfaces. **b.** A machine used to apply such a blast. **2.** A strong wind carrying sand along. —*vt.* **-blast·ed, -blast·ing, -blasts.** To apply a sandblast to (e.g., a building). **—sand'blast'er** *n.*
**sand-blind** (sănd'blīnd') *adj.* [ME < OE **sāmblind* : *sam-*, half + *blind*, blind.] Partially blind.
**sand·box** (sănd'bŏks') *n.* A low box filled with sand for children to play in.
**sandbox tree** *n.* [So called because the capsules were formerly used to hold sand for drying ink.] A tropical American tree, *Hura crepitans*, with a spiny trunk and woody seed capsules that split explosively when ripe.
**sand·bur** (sănd'bûr') *n.* **1.** A grass of the genus *Cenchrus*, esp. *C. tribuloides* of the eastern United States and tropical America, with spiny burlike fruiting clusters. **2.** A plant, *Solanum rostratum* of the western United States and Mexico, with prickly fruit.
**sand-cast** (sănd'kăst') *vt.* **-cast, -cast·ing, -casts.** To make (a casting) by pouring molten metal into a sand mold.
**sand crack** *n.* A fissure in the side of a horse's hoof, often causing lameness.
**sand dab** *n.* A small food fish of the genus *Citharichthys* of Pacific waters, related to and resembling the flounder.
**sand dollar** *n.* A thin circular echinoderm of the order Exocycloida or Clypeasteroidea, esp. *Echinarachnius parma* of sandy ocean bottoms of the northern Atlantic and Pacific.
**sand eel** *n.* The sand lance.
**sand·er** (săn'dər) *n.* **1. a.** One that spreads sand. **b.** One that sands surfaces, as of wood. **2.** A sanding machine.
**sand·er·ling** (săn'dər-lĭng) *n.* [Perh. < SAND + -LING¹.] A small shore bird, *Crocethia alba*, with gray and white plumage.
**sand flea** *n.* **1.** A beach flea. **2.** The chigoe.
**sand fly** *n.* Any of various small biting flies of the genus *Phlebotomus*, of tropical areas, some of which transmit diseases.
**sand·fly fever** (sănd'flī') *n.* A mild virus disease transmitted by the bite of a sand fly, *Phlebotomus papatasii*, marked by fever, malaise, eye pain, and headache.
**sand grouse** *n.* A pigeonlike bird of the genus *Pterocles* and related genera of Old World arid and semiarid regions.
**san·dhi** (săn'dē, sän'-) *n.* [Skt. *saṃdhiḥ*, union : *sam-*, together + *dadhāti*, he places.] Modification of the sound of a morpheme in certain phonetic contexts, e.g., the difference between the pronunciation of *the* in *the house* and in *the other house.*
**sand·hog** (sănd'hôg', -hŏg') *n.* One who works inside a caisson, as in the construction of underwater tunnels.
**sand hopper** *n.* A beach flea.
**sanding machine** *n.* A machine with a powered abrasive-covered disk or belt, used for smoothing, polishing, or refinishing.
**sand lance** *n.* A small marine fish of the genus *Ammodytes*, having a slender body with a forked tail fin and often burrowing in tideland sands.
**sand lily** *n.* A low-growing plant, *Leucocrinum montanum* of the western United States, with grasslike leaves and fragrant white star-shaped flowers.
**sand·lot** (sănd'lŏt') *n.* A vacant lot used esp. by children for unorganized sports.
**sand·man** (sănd'măn') *n.* A folklore character who puts children to sleep by sprinkling sand in their eyes.
**sand painting** *n.* **1.** A ceremonial design of the Navaho Indians made by trickling fine colored sand onto a base of neutral sand. **2.** The art of making designs with colored sand.
**sand·pa·per** (sănd'pā'pər) *n.* Heavy paper coated on one side with sand or other abrasive material, used for smoothing or polishing. —*vt.* **-pered, -per·ing, -pers.** To rub with sandpaper.
**sand·pi·per** (sănd'pī'pər) *n.* A small wading bird of the family Scolopacidae, usu. having a long, straight bill.
**sand·stone** (sănd'stōn') *n.* Variously colored sedimentary rock composed mainly of sandlike quartz grains cemented by lime, silica, or other materials.

**sand·storm** (sănd′stôrm′) *n.* A strong wind carrying clouds of sand.

**sand table** *n.* **1.** A table with raised edges, used for holding sand with which children play. **2.** A table on which a relief model of terrain is built out of sand for the study of military maneuvers.

**sand trap** *n.* A sand-filled hazard on a golf course.

**sand verbena** *n.* Any of several plants of the genus *Abronia* of western North America, with fragrant, usu. pink flowers.

**sand·wich** (sănd′wĭch, săn′-) *n.* [After John Montagu (1718–1792), 4th Earl of *Sandwich.*] **1.** Two or more slices of bread with a filling, as meat, cheese, jam, or various mixtures, placed between them. **2.** Something resembling a sandwich. —*vt.* **-wiched, -wich·ing, -wich·es. 1.** To make (something) into or as if into a sandwich. **2.** To insert (one thing) tightly between two other things of differing character or quality. **3.** To make room or time for <*sandwiched* the meeting into my schedule>

**sandwich board** *n.* Two large boards bearing placards, hinged at the top by straps for hanging over the shoulders with one board in front and the other behind, used mainly for advertising.

**sand·worm** (sănd′wûrm′) *n.* A segmented worm, esp. of the genera *Nereis* and *Arenicola,* gen. inhabiting coastal mud or sand and often used as fishing bait.

**sand·wort** (sănd′wûrt′, -wôrt′) *n.* A low-growing plant of the genus *Arenaria,* with small, usu. white flowers.

**sand·y** (săn′dē) *adj.* **-i·er, -i·est. 1.** Covered with, containing, or consisting of sand. **2.** Like sand : GRITTY. **3.** Of the color of sand. **—sand′i·ness** *n.*

**sand yacht** *n.* A three-wheeled beach vehicle with a sail or sails.

**sane** (sān) *adj.* **san·er, san·est.** [Lat. *sanus,* healthy.] **1.** Of sound mind. **2.** Having or showing sound judgment : REASONABLE. **—sane′ly** *adv.* **—sane′ness** *n.*

**San·for·ized** (săn′fə-rīzd′). A trademark for fabric preshrunk by a patented mechanical process so as to minimize later shrinkage.

**sang** (săng) *v. p.t. of* SING.

**san·ga·ree** (săng′gə-rē′) *n.* [Sp. *sangría,* act of bleeding < *sangre,* blood < Lat. *sanguis.*] A sweet, chilled beverage of wine or other alcoholic liquor and grated nutmeg.

**sang-froid** (säN-frwä′) *n.* [Fr. : *sang,* blood + *froid,* cold.] Composure : equanimity.

**san·gri·a** (săng-grē′ə, săn-) *n.* [Sp. *sangría,* act of bleeding < *sangre,* blood < Lat. *sanguis.*] A cold drink of red or white wine mixed with sugar, fruit juice, soda water, and sometimes brandy.

**san·gui·nar·i·a** (săng′gwə-nâr′ē-ə, -nĕr′-) *n.* [NLat. *Sanguinaria,* genus name < Lat., a plant that stanches blood, fem. of *sanguinarius,* sanguinary.] Bloodroot.

**san·gui·nar·y** (săng′gwə-nĕr′ē) *adj.* [Lat. *sanguinarius* < *sanguis,* blood.] **1.** Accompanied by carnage. **2.** Bloodthirsty. **3.** Consisting of blood. **—san′gui·nar′i·ly** (-nâr′ə-lē) *adv.*

**san·guine** (săng′gwĭn) *adj.* [ME *sanguin* < OFr. < Lat. *sanguineus* < *sanguis,* blood.] **1. a.** Of the color of blood : RED. **b.** Ruddy <a *sanguine* complexion> **2. a.** *Archaic.* Having blood as the dominant humor in terms of medieval physiology. **b.** Having the temperament and ruddy complexion formerly thought to be characteristic of a person dominated by this humor : PASSIONATE. **3.** Cheerfully confident : OPTIMISTIC <a *sanguine* outlook on life> **—san′guine·ly** *adv.* **—san′guine·ness, san·guin′i·ty** *n.*

**san·guin·e·ous** (săng-gwĭn′ē-əs) *adj.* [Lat. *sanguineus* < *sanguis,* blood.] **1.** Relating to or involving blood or bloodshed. **2.** Blood-red in color.

**san·guin·o·lent** (săng-gwĭn′ə-lənt) *adj.* [Lat. *sanguinolentus,* full of blood < *sanguis,* blood.] Mixed or colored with blood.

**San·hed·rin** (săn-hĕd′rĭn, -hē′drĭn, săn-) *n.* [Heb. *sanhedhrīn* < Gk. *sunedrion,* council < *sunedros,* sitting in council : *sun-,* together + *hedra,* seat.] The highest judicial and ecclesiastical tribunal of the ancient Jewish nation.

**san·i·cle** (săn′ĭ-kəl) *n.* [ME < OFr. < Med. Lat. *sanicula,* prob. < Lat. *sanus,* healthy.] A plant of the genus *Sanicula,* having small greenish flower clusters and reputed medicinal value as an astringent.

**sa·ni·es** (sā′nē-ēz′) *n., pl.* **sanies.** [Lat.] A thin, fetid, greenish discharge of serum and pus from a wound, ulcer, or fistula. **—sa′ni·ous** (-əs) *adj.*

**san·i·tar·i·a** (săn′ĭ-târ′ē-ə) *n. var. pl. of* SANITARIUM.

**san·i·tar·i·an** (săn′ĭ-târ′ē-ən) *n.* An expert in public health or sanitation.

**san·i·tar·i·um** (săn′ĭ-târ′ē-əm) *n., pl.* **-i·ums** *or* **-i·a** (-ē-ə). **1.** A health resort. **2.** SANATORIUM 1.

**san·i·tar·y** (săn′ĭ-tĕr′ē) *adj.* [Fr. *sanitaire* < Lat. *sanitas,* health < *sanus,* healthy.] **1.** Of or pertaining to health. **2.** Free from elements, as filth or bacteria, that endanger health : HYGIENIC. **—san′i·tar′i·ly** (-târ′ə-lē) *adv.*

**sanitary engineer** *n.* A civil engineer who specializes in the maintenance of conditions conducive to the preservation of public health, esp. in an urban area. **—sanitary engineering** *n.*

**sanitary landfill** *n.* Landfill.

**sanitary napkin** *n.* A disposable pad of absorbent material worn to absorb menstrual flow.

**san·i·ta·tion** (săn′ĭ-tā′shən) *n.* **1.** Formulation and application of measures designed to protect public health. **2.** Sewage disposal.

**san·i·tize** (săn′ĭ-tīz′) *vt.* **-tized, -tiz·ing, -tiz·es. 1.** To make sanitary. **2.** To make more acceptable by removing unpleasant or offensive features from <*sanitized* the film's language for a PG rating>

**san·i·ty** (săn′ĭ-tē) *n.* [ME *sanite* < OFr. < Lat. *sanitas,* health < *sanus,* healthy.] **1.** The state of being sane. **2.** Sound judgment or reason.

**San Jo·se scale** (săn′ hō-zā′) *n.* [After *San Jose,* California.] A scale insect, *Aspidiotus perniciosus,* that damages fruit trees and fruit-bearing plants.

**sank** (săngk) *v. p.t. of* SINK.

**San·khya** (săng′kyə) *n.* [Skt. *sāṁkhya-,* based on enumeration < *saṁkhyā,* enumeration : *sam-,* with + *khyāti,* he says.] A system of Hindu philosophy based on a dualism involving the ultimate principles of soul and potential matter.

**san·nup** (săn′əp) *n.* [Of Algonquian orig.] A married American Indian man.

**sans** (sănz) *prep.* [ME < OFr. < Lat. *sine.*] Without.

**sans-cu·lotte** (sănz′kyōō-lŏt′) *n.* [Fr. : *sans,* without + *culotte,* breeches.] **1.** An extreme radical republican during the French Revolution. **2.** A radical extremist. **—sans′-cu·lot′tic** (-lŏt′ĭk) *adj.* **—sans′-cu·lot′tism** *n.*

**san·sei** (săn′sā′, săn-sā′) *n., pl.* **sansei** *or* **-seis.** [J. : *san,* three + *sei,* generation.] The U.S.-born grandchild of Japanese immigrants to America.

**san·se·vie·ri·a** (săn′sə-vîr′ē-ə) *n.* [NLat. *Sanseveria,* genus name, after Raimondo di Sangro (1710–1771), Prince of *San Severo,* Italy.] A tropical Old World plant of the genus *Sansevieria,* having thick lance-shaped leaves and often grown as a house plant.

**San·skrit** (săn′skrĭt′) *n.* [Skt. *saṁskṛta-,* elegant : *sam-,* together + *karoti,* he makes.] An ancient Indic language that is the language of Hinduism and the Vedas and the classical literary language of India. **—San′skrit′ist** *n.*

**San·skrit·ic** (săn-skrĭt′ĭk) *n.* Indic. **—San·skrit′ic** *adj.*

**sans ser·if** (săn sĕr′ĭf) *n.* A letter or typeface without serifs.

**San·ta Claus** (săn′tə klôz′) *n.* [Alteration of dial. Du. *Sinterklaas* < MDu. *Sinterclaes,* St. Nicholas.] The personification of the spirit of Christmas, usu. represented as a jolly, fat, white-bearded old man dressed in a red suit.

**san·ta·lol** (săn′tə-lôl′, -lōl′) *n.* [NLat. *Santalum,* sandalwood genus + -OL[2].] A colorless liquid, $C_{15}H_{24}O$, obtained from sandalwood and used in perfumes.

**san·ton·i·ca** (săn-tŏn′ĭ-kə) *n.* [NLat. < Lat. *(herba) santonica,* fem. of *santonicus,* of the Santoni, a people of Aquitania.] **1.** An Old World wormwood, *Artemisia maritima,* with flowers that yield santonin. **2.** The dried unopened flowers of the santonica.

**san·to·nin** (săn′tə-nĭn) *n.* [SANTON(ICA) + -IN.] A colorless crystalline compound, $C_{15}H_{18}O_3$, obtained from species of wormwood, esp. santonica, and used as an anthelmintic.

**sap**[1] (săp) *n.* [ME < OE *sæp.*] **1. a.** The watery fluid that circulates through a plant, carrying nutrients to the tissues. **b.** A plant juice or fluid. **2.** An essential bodily fluid. **3.** Energy : vitality. **4.** *Slang.* A gullible person : DUPE. **5.** BLACKJACK[1] 1. —*vt.* **sapped, sap·ping, saps.** To hit or knock out with a sap.

**sap**[2] (săp) *n.* [OFr. *sappe* or OItal. *zappa.*] A covered trench or tunnel dug to a point within an enemy position. —*v.* **sapped, sap·ping, saps.** —*vt.* **1.** To undermine the foundations of (a fortification). **2.** To weaken gradually : DEVITALIZE. —*vi.* To dig a sap.

**sap·a·jou** (săp′ə-jōō) *n.* [Fr. < Tupi.] A capuchin monkey.

**Sa·phar** (sə-făr′) *n. var. of* SAFAR.

**sap·head** (săp′hĕd′) *n. Slang.* A fool. **—sap′head′ed** *adj.*

**sa·phe·na** (sə-fē′nə) *n., pl.* **-nae** (-nē′) [ME < Med. Lat. < Ar. *ṣāfin.*] Either of two large veins of the leg. **—sa·phe′nous** *adj.*

**sap·id** (săp′ĭd) *adj.* [Lat. *sapidus* < *sapere,* to taste.] **1. a.** Having flavor. **b.** Pleasantly flavorful : SAVORY. **2.** Pleasing to the mind : INTERESTING. **—sa·pid′i·ty** (să-pĭd′ĭ-tē, sə-) *n.*

**sa·pi·ent** (sā′pē-ənt) *adj.* [ME < OFr. < Lat. *sapiens,* pr.part. of *sapere,* to taste, be wise.] Extremely wise and discerning. **—sa′pi·ence** *n.* **—sa′pi·ent·ly** *adv.*

**sap·less** (săp′lĭs) *adj.* **1.** Lacking sap : DRY. **2.** Lacking spirit or vitality. **—sap′less·ness** *n.*

**sap·ling** (săp′lĭng) *n.* **1.** A young tree. **2.** YOUTH 3c.

**sap·o·dil·la** (săp′ə-dĭl′ə, -dē′yə) *n.* [Sp. *zapotillo,* dim. of *zapote,* sapodilla fruit < Nahuatl *tzapotl.*] **1.** An evergreen tree, *Achras zapota* of tropical America, with latex that yields chicle. **2.** The edible rough-skinned fruit of the sapodilla.

**sap·o·na·ceous** (săp′ə-nā′shəs) *adj.* [Lat. *sapo, sapon-,* soap + -ACEOUS.] Like soap. **—sap′o·na′ceous·ness** *n.*

**sap·o·nat·ed** (săp′ə-nā′tĭd) *adj.* [< Lat. *sapo, sapon-,* soap.] Combined or treated with a soap.

**sa·pon·i·fi·ca·tion** (sə-pŏn′ə-fĭ-kā′shən) *n.* Hydrolysis of an ester by an alkali, producing a free alcohol and an acid salt, esp. alkaline hydrolysis of fats to make soap.

**sa·pon·i·fy** (sə-pŏn′ə-fī′) *v.* **-fied, -fy·ing, -fies.** [Fr. *saponifier* < Lat. *sapo,* soap.] —*vt.* **1.** To convert (an ester) by saponification.

**2.** To convert (fats) into soap. —*vi.* To undergo saponification. **—sa·pon'i·fi'a·ble** *adj.* **—sa·pon'i·fi'er** *n.*
**sap·o·nin** (săp'ə-nĭn, sə-pō'-) *n.* [Fr. *saponine* < Lat. *sapo,* soap.] A plant glucoside that forms a soapy colloidal solution when mixed and agitated with water, used in making detergents, foaming agents, and emulsifiers.
**sap·o·nite** (săp'ə-nīt') *n.* [Swed. *saponit* < Lat. *sapo,* soap.] An amorphous, hydrous silicate of magnesium occurring as a soaplike mass in the cavities of certain rocks, as diabase.
**sa·por** (sā'pər, -pôr') *n.* [ME < Lat. < *sapere,* to taste.] A quality discernible to the sense of taste : FLAVOR. **—sa'po·rif'ic** (sā'pə-rĭf'ĭk, săp'ə-), **sa'po·rous** (sā'pər-əs, săp'ər-) *adj.*
**sa·po·ta** (sə-pō'tə) *n.* [Sp. *zapota* < Nahuatl *tzapotl.*] The sapodilla.
**sap·pan·wood** (sə-păn'wŏŏd', săp'ăn-, -ən-) *n.* [Malay *sapang,* sappanwood + WOOD.] **1.** A tree, *Caesalpina sappan* of tropical Asia, with wood yielding a red dye. **2.** The wood of the sappanwood.
**sap·per** (săp'ər) *n.* [< SAP².] A military engineer who specializes in sapping and other field fortification activities, esp. one in charge of deployment and detection of mines.
**Sap·phic** (săf'ĭk) *adj.* **1.** Of or relating to the Greek poet Sappho. **2. a.** Designating a verse of dactyls combined with trochees or anapests with iambs, esp. one of 11 syllables. **b.** Designating a stanza of three such verses followed by a measure including a dactyl and a spondee. **c.** Designating an ode made up of such stanzas. —*n.* A Sapphic meter, verse, stanza, or ode.
**sap·phire** (săf'īr') *n.* [ME *saphir* < OFr. *safir* < Lat. *sapphirus* < Gk. *sappheiros,* of Semitic orig.] **1.** A relatively pure form of corundum, esp. a clear deep-blue form used as a gemstone. **2.** A corundum gem. **3.** The deep-blue color of a gem sapphire. —*adj.* Having the deep-blue color of a sapphire.
**sap·phi·rine** (săf'ə-rīn', -rēn', sə-fīr'ĭn) *adj.* Of or resembling sapphire. —*n.* A rare light blue or green aluminum-magnesium silicate mineral.
**sap·py** (săp'ē) *adj.* **-pi·er, -pi·est. 1.** Full of sap : JUICY. **2.** *Slang.* Oversentimental : mawkish. **3.** *Slang.* Silly or foolish : FATUOUS. **—sap'pi·ly** *adv.* **—sap'pi·ness** *n.*
**sapr-** *pref. var. of* SAPRO-.
**sa·pre·mia** *also* **sa·prae·mia** (să-prē'mē-ə) *n.* Septicemia. **—sa·pre'mic** *adj.*
**sapro-** *or* **sapr-** *pref.* [< Gk. *sapros,* rotten.] **1.** Decay : putrefaction <*saprogenic*> **2.** Dead or decaying organic material <*saprophyte*>
**sap·robe** (săp'rōb') *n.* [SAPRO- + Gk. *bios,* life.] An organism deriving its nourishment from nonliving or decaying organic matter. **—sa·pro'bic** (să-prō'bĭk) *adj.* **—sa·pro'bi·cal·ly** *adv.*
**sap·ro·gen·ic** (săp'rə-jĕn'ĭk) *adj.* Of, causing, or produced by putrefaction. **—sap'ro·ge·nic'i·ty** (-jə-nĭs'ĭ-tē) *n.*
**sap·ro·lite** (săp'rə-līt') *n.* Clay, silt, or other rock remnants remaining at the site of disintegration.
**sap·ro·pel** (săp'rə-pĕl') *n.* [SAPRO- + Gk. *pēlos,* mud.] **1.** An aquatic sludge rich in organic matter. **2.** A fluid swamp slime resulting from putrefaction. **—sap'ro·pel'ic** (-pĕl'ĭk, -pē'lĭk) *adj.*
**sa·proph·a·gous** (să-prŏf'ə-gəs) *adj.* Feeding on decaying matter.
**sap·ro·phyte** (săp'rə-fīt') *n.* A plant that derives its nourishment from dead or decaying organic matter. **—sap'ro·phyt'ic** (-fĭt'ĭk) *adj.* **—sap'ro·phyt'ic·al·ly** *adv.*
**sap·ro·zo·ic** (săp'rə-zō'ĭk) *adj.* **1.** Of or designating nutrition by absorption of dissolved organic and inorganic materials, as in protozoans. **2.** Feeding on dead or decaying animal matter.
**sap·sa·go** (săp-sā'gō, săp'sə-gō') *n.* [Alteration of G. *Schabzieger* : *schaben,* to scrape + *Zieger,* whey.] A hard skim-milk cheese colored and flavored with sweet clover.
**sap·suck·er** (săp'sŭk'ər) *n.* Either of two small North American woodpeckers, *Sphyrapicus varius* or *S. thyrsoides,* that drill holes in certain trees in order to drink the sap.
**sap·wood** (săp'wŏŏd') *n.* Newly formed outer wood that lies just inside the cambium of a tree or woody plant and is usu. lighter in color and more active in nutrition than the heartwood.
**sar·a·band** *also* **sar·a·bande** (săr'ə-bănd') *n.* [Fr. *sarabande* < Sp. *zarabanda.*] **1.** A stately 17th- and 18th-cent. court dance in slow triple time. **2.** The music for the saraband.
**Sar·a·cen** (săr'ə-sən) *n.* [ME < OFr. *Saracin* < LLat. *Saracenus* < LGk. *Sarakēnos,* perh. of Ar. orig.] **1.** A member of a pre-Islamic nomadic people of the Syrian-Arabian deserts. **2.** An Arab. **3.** A Moslem, esp. during the Crusades. **—Sar'a·cen'ic** (-sĕn'ĭk) *adj.*
**sa·ran** (sə-răn') *n.* [< *Saran,* a former U.S. trademark.] **1.** A thermoplastic resin derived from vinyl compounds and used in the manufacture of packaging films, fittings, and bristles and as a fiber in various heavy fabrics. **2. Saran.** A trademark for saran outside the United States.
**sa·ra·pe** (sə-rä'pē, -răp'ē) *n. var. of* SERAPE.
**Sar·a·to·ga trunk** (săr'ə-tō'gə) *n.* [After *Saratoga* Springs, New York.] A large traveling trunk with a rounded top.
**sarc-** *pref. var. of* SARCO-.
**sar·casm** (sär'kăz'əm) *n.* [Gk. *sarkasmos* < *sarkazein,* to bite the lips in rage < *sarx,* flesh.] **1.** A mocking or contemptuously ironic remark intended to wound another. **2.** The use of sarcasm.
**sar·cas·tic** (sär-kăs'tĭk) *adj.* [SARC(ASM) + *-astic,* as in *enthusiastic.*] **1.** Expressing sarcasm. **2.** Given to the use of sarcasm. **—sar·cas'ti·cal·ly** *adv.*
☆ ***syns:*** SARCASTIC, ACERBIC, CAUSTIC, MORDANT *adj. core meaning* : bitter, cutting, and contemptuously derisive, esp. in expression <*a sarcastic,* sneering critic><*sarcastic* remarks about the performance>
**sarce·net** (särs'nĭt) *n.* [ME *sarsenet* < AN *sarzinett.*] A fine, soft silk cloth used esp. for trimmings.
**sarco-** *or* **sarc-** *pref.* [Gk. *sarko-* < *sarx,* flesh.] **1.** Flesh <*sarcocarp*> **2.** Striated muscle <*sarcolemma*>
**sar·co·carp** (sär'kə-kärp') *n.* The fleshy pulp surrounding the seed of a drupaceous fruit, as a peach or plum.
**sar·coid** (sär'koid') *adj.* Pertaining to or resembling flesh.
**sar·coid·o·sis** (sär'koi-dō'sĭs) *n.* A disease of unknown origin marked by the formation of granulomatous lesions esp. in the liver, lungs, skin, and lymph nodes.
**sar·co·lem·ma** (sär'kə-lĕm'ə) *n.* [SARCO- + Gk. *lemma,* husk.] A thin membrane surrounding a striated muscle fiber. **—sar'co·lem'mal** *adj.*
**sar·co·ma** (sär-kō'mə) *n., pl.* **-ma·ta** (-mə-tə) *or* **-mas.** A malignant tumor that develops nonepithelial connective tissues. **—sar·co'ma·toid', sar·co'ma·tous** *adj.*
**sar·co·ma·to·sis** (sär-kō'mə-tō'sĭs) *n.* Formation of sarcomatous growths in the body.
**sar·co·mere** (sär'kə-mîr') *n.* A structural segment of a striated muscle fiber.
**sar·coph·a·gi** (sär-kŏf'ə-jī') *n. var. pl. of* SARCOPHAGUS.
**sar·co·phag·ic** (sär'kə-făj'ĭk) *also* **sar·coph·a·gous** (sär-kŏf'ə-gəs) *adj.* Carnivorous.
**sar·coph·a·gus** (sär-kŏf'ə-gəs) *n., pl.* **-gi** (-jī') *or* **-gus·es.** [Lat. < Gk. *sarkophagos,* flesh-eating : *sarx,* flesh + *phagein,* to eat.] A stone coffin.
**sar·co·plasm** (sär'kə-plăz'əm) *n.* The cytoplasm of muscle cells. **—sar'co·plas'mic** *adj.*
**sar·cop·tic mange** (sär-kŏp'tĭk) *n.* [< NLat. *Sarcoptes,* genus name : SARCO- + Gk. *koptein,* to cut.] Mange caused by a mite, *Sarcoptes scabiei,* that burrows in the skin.
**sar·cous** (sär'kəs) *adj.* Relating to or composed of flesh or muscle.
**sard** (särd) *n.* [Fr. *sarde* < Lat. *sarda,* a kind of precious stone, perh. < Gk. *sardion* < *Sardeis,* Sardis, an ancient city in Asia Minor.] A clear or translucent deep orange-red to brownish red chalcedony.
**sar·dine** (sär-dēn') *n.* [ME *sardeyn* < OFr. *sardine* < Lat. *sardina* < *sarda,* a kind of fish.] **1.** A small or half-grown edible herring or related fish of the family Clupeidae, often canned in oil. **2.** Any of various edible small, silvery freshwater or marine fish unrelated to the sardine.
**sar·don·ic** (sär-dŏn'ĭk) *adj.* [Fr. *sardonique* < Gk. *sardonios,* alteration of *sardanios,* bitter, scornful.] Scornfully derisive : SARCASTIC. **—sar·don'i·cal·ly** *adv.* **—sar·don'i·cism** (-ĭ-sĭz'əm) *n.*
**sar·don·yx** (sär-dŏn'ĭks, sär'dn-ĭks') *n.* [ME *sardonix* < Lat. *sardonyx* < Gk. *sardonux* : *sardion,* sard + *onux,* onyx.] An onyx with alternating layers of sard and other minerals, often used for cameos.
**sar·gas·so** (sär-găs'ō) *n.* [Port. *sargaço.*] Gulfweed.
**sar·gas·sum** (sär-găs'əm) *n.* [NLat. *Sargassum,* genus name < SARGASSO.] Gulfweed.
**sarge** (särj) *n. Informal.* Sergeant.
**sa·ri** (sä'rē) *n.* [Hindi *sāṛī* < Skt. *śāṭī.*] An outer garment worn mainly by Hindu women, consisting of a length of lightweight cloth with one end wrapped about the waist to form a long skirt and the other draped over the shoulder or covering the head.
**sar·men·tose** (sär-mĕn'tōs') *adj.* [Lat. *sarmentosus,* full of twigs < *sarmentum,* twigs.] Having long slender stems that root at intervals, as the strawberry.
**sa·rong** (sə-rông', -rŏng') *n.* [Malay.] A skirt consisting of a length of bright-colored cloth wrapped about the waist, worn by both men and women in Malaysia, Indonesia, and the Pacific islands.
**sar·sa·pa·ril·la** (săs'pə-rĭl'ə, särs'-) *n.* [Sp. *zarzaparrilla* : *zarza,* bramble (< Ar. *sharaṣ*) + *parrilla,* dim. of *parra,* vine.] **1. a.** The dried roots of a tropical American plant of the genus *Smilax,* esp. *S. aristolochiaefolia* of Mexico, used as a flavoring. **b.** A soft drink flavored with sarsaparilla. **2.** A North American plant, *Aralia hispida* or *A. nudicaulis,* with clusters of small white flowers.
**sar·to·ri·al** (sär-tôr'ē-əl, -tōr'-) *adj.* [< Lat. *sartor,* tailor. —see SARTORIUS.] Of or pertaining to a tailor, tailoring, or tailored clothing. **—sar·to'ri·al·ly** *adv.*
**sar·to·ri·us** (sär-tôr'ē-əs, -tōr'-) *n.* [NLat. < Lat. *sartor,* tailor < *sartus,* p.part. of *sarcire,* to mend.] A long, flat, narrow thigh muscle, crossing the front of the thigh obliquely from the hip to the inner side of the tibia.
**sash¹** (săsh) *n.* [Ar. *shāsh,* muslin.] An ornamental band worn around the waist or over the shoulder as a symbol of rank.
**sash²** (săsh) *n.* [Alteration of Fr. *châssis,* frame. —see CHASSIS.] **1.** A frame in which the glass panes of a window or door are set. **2.** The framework together with its panes, esp. a sliding section of a window. —*vt.* **sashed, sash·ing, sash·es.** To furnish with a sash.

**sa·shay** (să-shā') [Alteration of CHASSÉ.] *Informal* —*vi.* **shayed, -shay·ing, -shays. 1.** To strut or flounce. **2.** To perform a chassé. —*n.* An excursion : jaunt.

**sa·shi·mi** (sä-shē'mē) *n.* [J.] A Japanese dish of thinly sliced raw fish.

**sas·ka·toon** (săs'kə-to͞on') *n.* [Cree *misāskwatomin,* saskatoon berry.] A shrub, *Amelanchier alnifolia* of northwestern North America, with white flowers and edible dark-purple fruit.

**sas·quatch** *also* **Sas·quatch** (săs'kwăch', -kwăch') *n.* [Of Salish orig.] A large, hairy mammal reputed to exist in the Pacific Northwest and western Canada.

**sass** (săs) [Back-formation < SASSY[1].] *Informal.* —*n.* Impudent, disrespectful speech : BACK TALK. —*vt.* **sassed, sass·ing, sass·es.** To talk impudently to.

**sas·sa·by** (săs'ə-bē) *n., pl.* **-bies.** [Of Bantu orig.] An African antelope, *Damaliscus lunatus,* with curving ridged horns.

**sassaby**
*Approximately 6 feet long including tail*

**sas·sa·fras** (săs'ə-frăs') *n.* [Sp. *sasafrás.*] **1.** A North American tree, *Sassafras albidum,* with irregularly lobed leaves and aromatic bark. **2.** The dried root bark of the sassafras, used for flavoring and as a source of a volatile oil.

**Sas·sa·nid** (săs'ə-nĭd, sə-sä'nĭd, -săn'-) *also* **Sas·sa·ni·an** (sə-sā'nē-ən, să-) *or* **Sas·sa·nide** (săs'ə-nīd') *n.* [Med. Lat. *Sassanidae,* the Sassanids < *Sassanus,* Sassan, grandfather of the founder of the dynasty.] A member of a dynasty of Persian kings ruling from the 3rd through the middle of the 17th cent. A.D. **—Sas'sa·nid** *adj.*

**sass·wood** (săs'wo͝od') *n.* [SASSY[2] + WOOD.] SASSY[2].

**sas·sy[1]** (săs'ē) *adj.* **-si·er, -si·est.** [Alteration of SAUCY.] **1.** Impudent. **2.** Jaunty. **—sas'si·ly** *adv.* **—sas'si·ness** *n.*

**sas·sy[2]** (săs'ē) *n., pl.* **-sies.** [Prob. of African orig.] A tree, *Erythrophloeum guineense* of western Africa, with bark that yields a poison formerly used in trials by ordeal.

**sas·tru·ga** (să-stro͞o'gə) *n.* [R. *zastruga,* groove : *za,* by + *struga,* deep place.] A long wavelike ridge of snow, formed by the winds of the polar plains.

**sat** (săt) *v. p.t. & p.p. of* SIT.

**Sa·tan** (săt'n) *n.* [ME < OE < LLat. < Gk. < Heb. *śāṭān,* devil, adversary < *śāṭan,* he accused.] DEVIL 1.

**sa·tang** (sə-tăng') *n., pl.* **satang.** [Thai *satān.*] —See table at CURRENCY.

**sa·tan·ic** (sə-tăn'ĭk, sā-) *or* **sa·tan·i·cal** (-ĭ-kəl) *adj.* **1.** Of, relating to, or suggestive of Satan or evil. **2.** Extremely evil or vicious : FIENDISH. **—sa·tan'i·cal·ly** *adv.*

**Sa·tan·ism** (săt'n-ĭz'əm) *n.* **1.** The worship of Satan characterized by a mockery of the Christian rites. **2. satanism.** Extreme wickedness. **—Sa'tan·ist** *n.*

**satch·el** (săch'əl) *n.* [ME *sachel* < OFr. < Lat. *saccellus,* dim. of *saccus,* bag. —see SACK[1].] A small bag for books or clothing, often having a shoulder strap.

**satchel charge** *n.* Several blocks of explosives, usu. taped to a board equipped with a looped handle and used esp. by sappers.

**sate[1]** (sāt) *vt.* **sat·ed, sat·ing, sates.** [Prob. alteration of obs. *sade,* to satiate < ME *saden* < OE *sadian.*] **1.** To satisfy or appease (an appetite) fully. **2.** To indulge to excess : GLUT.

**sate[2]** (săt, sāt) *v. Archaic. var. p.t. of* SIT.

**sa·teen** (să-tēn') *n.* [Alteration of SATIN.] A durable cotton fabric with a satinlike finish.

**sat·el·lite** (săt'l-īt') *n.* [OFr. < Lat. *satelles,* attendant.] **1.** *Astron.* A relatively small celestial body orbiting a larger one : MOON. **2.** *Aerospace.* A manmade object designed to orbit a celestial body. **3.** One who attends a powerful dignitary : MINION. **4.** A subservient follower : SYCOPHANT. **5.** A nation politically and economically dominated by another. **6.** An urban or suburban community near a large city.

**satellite station** *n.* A radio or television station that immediately rebroadcasts a received transmission on a different wavelength.

**sa·tem** (sä'təm) *adj.* [Avestan *satəm,* hundred (a word whose initial sound illustrates the sound change).] Designating those Indo-European languages, including the Indo-Iranian, Armenian, Albanian, and Balto-Slavic subfamilies, in which original velar stops became fricatives (as *k* > *s* or *š*) and labiovelar stops became velars (as *kw* > *k*).

**sa·tia·ble** (sā'shə-bəl, -shē-ə-) *adj.* Able to be satiated. **—sa'tia·bil'i·ty** *n.* **—sa'tia·bly** *adv.*

**sa·ti·ate** (sā'shē-āt') *vt.* **-at·ed, -at·ing, -ates.** [Lat. *satiare, satiat-* < *satis,* sufficient.] **1.** To satisfy or appease (an appetite or desire) fully. **2.** SATE 2. —*adj.* (-ĭt). Filled to satisfaction. **—sa'ti·a'tion** *n.*
☆ **syns:** SATIATE, CLOY, GLUT, GORGE, SATE, SURFEIT *v. core meaning* : to satisfy to the fullest or to excess <a large meal that *satiated* our hunger>

**sa·ti·e·ty** (sə-tī'ĭ-tē) *n.* [OFr. *satiete* < Lat. *satietas,* sufficiency < *satis,* sufficient.] The state of being full or glutted : SURFEIT.

**sat·in** (săt'n) *n.* [ME *satyn* < OFr. *satin.*] A smooth fabric, as of silk, nylon, or rayon, with a glossy face and a dull back.

**sat·in·flow·er** (săt'n-flou'ər) *n.* A plant, *Godetia grandiflora,* of California, with red-blotched flowers.

**sat·in·pod** (săt'n-pŏd') *n.* HONESTY 4.

**satin stitch** *n.* An embroidery stitch done in close parallel lines to produce a satinlike finish.

**satin weave** *n.* A basic weave construction with the interlacing of the threads so arranged that the face of the cloth is covered with warp yarn or filling yarn and no twill line is distinguishable.

**sat·in·wood** (săt'n-wo͝od') *n.* **1. a.** A tree, *Chloroxylon swietenia* of southern Asia, with hard, yellowish, close-grained wood. **b.** The wood of the sandalwood. **2. a.** A tree having wood similar to satinwood. **b.** The wood of such a tree.

**sat·in·y** (săt'n-ē) *adj.* Smooth and lustrous like satin.

**sat·ire** (săt'īr') *n.* [OFr. < Lat. *satira,* alteration of *satura,* mixture, poem treating various subjects < fem. of *satur,* sated.] **1. a.** A literary work in which human vice or folly is ridiculed or attacked scornfully. **b.** The branch of literature that comprises such works. **2.** Irony, derision, or caustic wit used to attack or expose folly, vice, or stupidity.

▲ word history: The word *satire* is from Latin *satira,* a variant of *satura,* which denoted a type of long poem that held human follies and vices up to ridicule and scorn. *Satura* originally meant "a full dish," and later "a mixture; a poem treating a variety of subjects." It is derived from *satur,* "full of food," which is related to *satis,* "enough," the source of English *satisfy* and *satiate.*

**sa·tir·i·cal** (sə-tĭr'ĭ-kəl) *or* **sa·tir·ic** (-ĭk) *adj.* Of, pertaining to, or marked by satire. **—sa·tir'i·cal·ly** *adv.*

**sat·i·rist** (săt'ər-ĭst) *n.* One who writes satirical works.

**sat·i·rize** (săt'ə-rīz') *vt.* **-rized, -riz·ing, -riz·es.** To ridicule or attack with satire.

**sat·is·fac·tion** (săt'ĭs-făk'shən) *n.* [ME < OFr. < LLat. < Lat., amends < *satisfacere,* to satisfy.] **1. a.** The fulfillment or gratification of a need, desire, or appetite. **b.** Pleasure derived from such gratification. **2.** Compensation for injury or loss : REPARATION. **3.** A source of fulfillment or gratification.

**sat·is·fac·to·ry** (săt'ĭs-făk'tə-rē) *adj.* Giving the satisfaction needed to meet a demand or requirement : ADEQUATE. **—sat'is·fac'to·ri·ly** *adv.* **—sat'is·fac'to·ri·ness** *n.*

**sat·is·fi·a·ble** (săt'ĭs-fī'ə-bəl) *adj.* Able to be satisfied.

**sat·is·fy** (săt'ĭs-fī') *v.* **-fied, -fy·ing, -fies.** [ME *satisfien* < OFr. *satisfier* < Lat. *satisfacere* : *satis,* sufficient + *facere,* to make.] —*vt.* **1.** To gratify the need, desire, or expectation of. **2.** To fulfill (a need or desire). **3. a.** To free from doubt or question : ASSURE. **b.** To dispel (a doubt). **4. a.** To discharge (e.g., an obligation). **b.** To discharge an obligation to (a creditor). **5.** To comply with the requirements of (a standard or rule). **6.** To make reparation for : REDRESS. **7.** *Math.* To make the left and right sides of an equation equal after substituting equivalent quantities for unknown variables. —*vi.* To give satisfaction. **—sat'is·fi'er** *n.* **—sat'is·fy'ing·ly** *adv.*

**sa·to·ri** (sä-tôr'ē, -tōr'ē, sə-) *n.* [J., insight.] A state of spiritual enlightenment sought by Zen Buddhists.

**sa·trap** (sā'trăp', săt'răp') *n.* [ME *satrape* < OFr. < Lat. *satrapes* < Gk. < OPers. *khshathrapāvā,* protector of the people.] **1.** A provincial governor in ancient Persia. **2.** A subordinate ruler or official.

**sa·tra·py** (sā'trə-pē, -trăp'ē, săt'rə-pē) *n., pl.* **-pies.** The territory or sphere ruled by a satrap.

**sat·u·rant** (săch'ər-ənt) *adj.* Serving to saturate. —*n.* A substance that saturates.

**sat·u·rate** (săch'ə-rāt') *vt.* **-rat·ed, -rat·ing, -rates.** [Lat. *saturare, saturat-,* to fill < *satur,* sated.] **1.** To imbue or impregnate thoroughly <The garden was *saturated* with the scent of flowers.> **2.** To soak, fill, or load to capacity. **3.** *Chem.* **a.** To cause (a solution) to be saturated. **b.** To cause (a compound) to be saturated. —*adj.* (-rĭt). Saturated. **—sat'u·ra·ble** (săch'ər-ə-bəl) *adj.* **—sat'u·ra'tor** *n.*

**sat·u·rat·ed** (săch'ə-rā'tĭd) *adj.* **1.** Unable to hold or contain more : FULL. **2.** Soaked with moisture : DRENCHED. **3.** *Chem.* **a.** Containing the greatest amount of solute that can normally be dissolved at a given temperature. —Used of solutions. **b.** Having all available valence bonds filled. —Used esp. of organic compounds. **4.** *Geol.* Of or relating to minerals that can crystallize from magmas even in the presence of excess silica.

**sat·u·ra·tion** (săch'ə-rā'shən) *n.* **1.** The act or process of saturating or the state of being saturated. **2.** *Physics.* A state of a ferromagnetic substance in which an increase in applied magnetic field strength does not produce an increase in magnetic intensity. **3.** *Chem.* The state of a fully saturated compound or solution. **4.** *Meteorol.* A condition in which air at a specific temperature contains the maximum moisture vapor possible without precipitating, equal to 100% relative humidity. **5.** Degree of difference from a gray of the same lightness or brightness : vividness of hue. **6.** Intensive shelling or bombing in order to achieve complete destruction. **7.** The flooding of a market with the maximum amount of a commodity that consumers can purchase.

**Sat·ur·day** (săt'ər-dē, -dā') *n.* [ME *Saterday* < OE *Sæternesdæg,* transl. of Lat. *dies Saturni,* Saturn's day.] The seventh day of the week.

**Saturday night special** *n. Informal.* A cheap, easily obtainable handgun.

**Sat·urn** (săt'ərn) *n.* [ME *Saturnus* < OE < Lat. *Saturnus.*] **1.** *Rom. Myth.* The god of agriculture. **2.** The sixth planet from the sun, with a diameter of 74,000 miles or 119,000 kilometers, a mass 95 times that of Earth, and an orbital period of 29.5 years at mean distance of approx. 886,000,000 miles or 1,425,000,000 kilometers. **3. Saturn.** *Archaic.* LEAD² 1.

**sat·ur·na·li·a** (săt'ər-nā'lē-ə, -nāl'yə) *pl.n.* [Lat. *saturnalia,* the festival of Saturn < neuter pl. of *Saturnalis,* Saturnian < *Saturnus,* Saturn.] **1. Saturnalia.** The ancient Roman festival of Saturn, beginning on Dec. 17 and lasting seven days. **2.** *(sing. in number).* An unrestrained celebration : ORGY.

**Sa·tur·ni·an** (să-tûr'nē-ən, sə-) *adj.* **1.** Of, relating to, or supposedly influenced astrologically by the planet Saturn. **2.** *Archaic.* Of or relating to the god Saturn or the period of his reign.

**sa·tur·ni·id** (să-tûr'nē-ĭd, sə-) *n.* [NLat. *Saturniidae,* family name < *Saturnia,* type genus < Lat., daughter of Saturn < *Saturnus,* Saturn.] Any of various often large, brightly colored moths of the family Saturniidae. *—adj.* Of or belonging to the Saturniidae.

**sat·ur·nine** (săt'ər-nīn') *adj.* **1.** Having the temperament of one born under the supposed astrological influence of Saturn. **2. a.** Having a melancholy or surly disposition. **b.** Sarcastic. **3. a.** Of or like lead. **b.** Produced by lead absorption. **—sat'ur·nine'ly** *adv.*

**sat·urn·ism** (săt'ər-nĭz'əm) *n.* Lead poisoning.

**Sa·tya·gra·ha** (sə-tyä'grə-hə) *n.* [Skt. *satyāgraha* : *satyam,* truth + *graha,* the act of seizing < *gṛhṇāti,* he seizes.] The policy of nonviolent resistance advocated by Mahatma Gandhi in India as a means of promoting social and political reform.

**sa·tyr** (sā'tər, săt'ər) *n.* [ME < Lat. *satyrus* < Gk. *saturos.*] **1.** *often* **Satyr.** *Gk. Myth.* A woodland god usu. depicted with the pointed ears, legs, and short horns of a goat. **2.** A lecher. **3.** A butterfly of the family Satyridae, usu. having brown wings marked with eyelike spots. **—sa·tyr'ic** (sā-tĭr'ĭk, sə-, să-), **sa·tyr'i·cal** *adj.*

**sauce** (sôs) *n.* [ME < OFr. < Lat. *salsa,* fem. of *salsus,* salted, p.part. of *sallere,* to salt < *sal,* salt.] **1.** A flavorful liquid dressing or relish for food. **2.** Stewed fruit, usu. served with other foods or as a dessert. **3.** Something that adds flavor or piquancy to something else. **4.** *Informal.* Impertinence or disrespect. **5.** *Slang.* Alcoholic liquor. *—vt.* **sauced, sauc·ing, sauc·es. 1.** To season or flavor with sauce. **2.** To add flavor or piquancy to. **3.** *Informal.* To be impertinent or disrespectful to.

**sauce bé·ar·naise** (bā'är-nāz') *n.* [Fr. < *Béarn,* a region in southwestern France.] A hollandaise sauce flavored with tarragon, shallots, and chervil.

**sauce·box** (sôs'bŏks') *n. Informal.* An impertinent person.

**sauce·pan** (sôs'păn') *n.* A deep, usu. straight-sided cooking pan with a handle.

**sau·cer** (sô'sər) *n.* [ME, sauce dish < OFr. *saussier* < *sausse, sauce,* sauce.] **1.** A small shallow dish with a depression in the center for holding a cup. **2.** Something shaped like a saucer.

**sauc·y** (sô'sē) *adj.* **-i·er, -i·est. 1.** Impertinent or disrespectful. **2.** Pert <a *saucy* glance> **—sau'ci·ly** *adv.* **—sau'ci·ness** *n.*

**sau·er·bra·ten** (sour'brät'n) *n.* [G. : *sauer,* sour + *Braten,* roast meat.] A pot roast of beef that is marinated in a mixture of vinegar, water, wine, spices, and aromatic vegetables before cooking.

**sau·er·kraut** (sour'krout') *n.* [G. : *sauer,* sour + *Kraut,* cabbage.] Chopped or shredded cabbage salted and fermented in its own juice.

**sau·ger** (sô'gər) *n.* [Orig. unknown.] A North American freshwater food and game fish, *Stizostedion canadense,* with a spiny dorsal fin.

**Sauk** (sôk) *also* **Sac** (săk, sôk) *n., pl.* **Sauk** or **Sauks** *also* **Sac** or **Sacs.** [Fr., of Algonquian orig.] **1. a.** A tribe of Indians living orig. in Michigan, Wisconsin, and Illinois and now settled in Iowa and Oklahoma. **b.** A member of this tribe. **2.** The Algonquian language of the Sauk.

**sault** (sōō) *n.* [Canadian Fr. < OFr. *saut,* leap < Lat. *saltus* < p.part. of *salire,* to leap.] A waterfall or rapids.

**sau·na** (sou'nə, sô'-) *n.* [Finn.] **1.** A steam bath in which the steam is usu. produced by pouring water over heated rocks. **2.** A room or enclosure for taking a sauna.

**saun·ter** (sôn'tər) *vi.* **-tered, -ter·ing, -ters.** [Prob. < ME *santren,* to muse.] To walk leisurely : STROLL. **—saun'ter** *n.*

**sau·rel** (sôr'əl, sô-rĕl') *n.* [Fr. < NLat. *saurus,* lizard < Gk. *sauros.*] **1.** A marine fish, *Trachurus trachurus* of eastern Atlantic waters. **2.** The jack mackerel.

**sau·ri·an** (sôr'ē-ən) *n.* [< NLat. *Sauria,* suborder name < *saurus,* lizard < Gk. *sauros.*] Any of various reptiles of the suborder Sauria, including the true lizards. *—adj.* Of, belonging to, or characteristic of the Sauria.

**sau·ro·pod** (sôr'ə-pŏd') *n.* [NLat. *Sauropoda,* suborder name : *saurus,* lizard (< Gk. *sauros*) + Gk. *pous,* foot.] A large semiaquatic dinosaur of the suborder Sauropoda, existing during the Jurassic and Cretaceous periods. *—adj.* Of or belonging to the Sauropoda. **—sau·rop'o·dous** (sô-rŏp'ə-dəs) *adj.*

**sauropod**
*Approximately 50 feet long*

**sau·ry** (sôr'ē) *n., pl.* **-ries.** [NLat. *saurus,* lizard < Gk. *sauros.*] Any of various edible marine fishes of the family Scomberesocidae, related to the needlefishes.

**sau·sage** (sô'sĭj) *n.* [ME *sausige* < ONFr. *saussiche* < LLat. *salsicia* < *salsicius,* prepared by salting < *salsus,* salted. —see SAUCE.] Finely chopped and often highly seasoned pork or other meats, encased usu. in a prepared animal intestine and cooked or cured.

**sau·té** (sō-tā', sô-) *vt.* **-téed, -té·ing, -tés.** [Fr., tossed < p.part. of *sauter,* to leap < Lat. *saltare.* —see SALTATION.] To fry lightly in a shallow pan in a small amount of fat. *—n.* Sautéed food.

**sau·terne** or **Sau·terne** (sō-tûrn', sô-) *n.* [Fr. < *Sauternes,* a commune in France.] A delicate, sweet white wine.

**sav·age** (săv'ĭj) *adj.* [ME *sauvage* < OFr. < Lat. *silvaticus,* of the woods, wild < *silva,* forest.] **1.** Not domesticated or cultivated : WILD. **2.** Uncivilized : barbaric. **3.** Ferocious : fierce. **4.** Cruel or merciless : BRUTAL. **5.** Lacking polish or manners : CRUDE. *—n.* **1.** A barbaric or uncivilized person. **2.** A fierce or vicious person. **3.** A crude person. *—vt.* **-aged, -ag·ing, -ag·es. 1.** To attack violently or ferociously. **2.** To make angry or fierce. **—sav'age·ly** *adv.* **—sav'age·ness** *n.*

☆ ***syns:*** SAVAGE, FERAL, WILD *adj. core meaning* : not domesticated, as a beast of prey <a *savage* lion>

**sav·age·ry** (săv'ĭj-rē) *n., pl.* **-ries. 1.** The quality or state of being savage. **2.** A savage act. **3.** Savage behavior or disposition.

**sa·van·na** *also* **sa·van·nah** (sə-văn'ə) *n.* [Sp. *zavana* < Taino *zabana.*] A flat, treeless, tropical or subtropical grassland.

**sa·vant** (să-vänt') *n.* [Fr. < pr.part. of *savoir,* to know < VLat. **sapēre* < Lat. *sapere,* to be wise.] A learned person : SCHOLAR.

**sav·a·rin** (săv'ə-rĭn) *n.* [After Anthelme Brillat-*Savarin* (1755–1826), French gourmet.] A rich bread made with almonds, orange peel, and citron and baked in a ring mold.

**save¹** (sāv) *v.* **saved, sav·ing, saves.** [ME *saven* < OFr. *sauver* < LLat. *salvare* < Lat. *salvus,* safe.] *—vt.* **1.** To rescue from danger, harm, or loss. **2.** To keep in a safe condition : SAFEGUARD. **3.** To prevent the waste or loss of : CONSERVE. **4.** To put aside for future use : STORE. **5.** To treat with care in order to avoid fatigue, wear, or damage : SPARE. **6.** To make unnecessary : AVOID <The short cut *saved* them an extra hour's driving time.> **7. a.** To prevent an opponent from scoring or winning, esp. in hockey. **b.** *Baseball.* To preserve (a victory) by protecting a team's lead. —Used of a relief pitcher. **8.** To deliver from sin : REDEEM. *—vi.* **1.** To avoid waste or expense : ECONOMIZE. **2.** To accumulate money or goods. **3.** To preserve a person or thing from harm or loss. *—n.* **1.** An act that prevents an opponent from scoring in a sport. **2.** *Baseball.* A game in which a relief pitcher preserves a victory by protecting a team's lead. **—sav'a·ble, save'a·ble** *adj.* **—sav'er** *n.*

☆ ***syns:*** **1.** SAVE, DELIVER, RESCUE *v. core meaning* : to free from danger or confinement <commandos who *saved* the hostages> **2.** SAVE, KEEP, LAY AWAY, SALT AWAY *v. core meaning* : to reserve for the future <tried to *save* some money> ***ant:*** consume, spend

**save²** (sāv) *prep.* [ME < OFr. *sauf* < Lat. *salvo,* ablative sing. of *salvus,* safe.] With the exception of : EXCEPT. *—conj.* **1.** Were it not. **2.** *Archaic.* Unless.

**save-all** (sāv'ôl') *n.* **1.** A device that prevents the waste, damage, or loss of something. **2.** A receptacle that captures the waste products of a process for further use in manufacture.

**sav·e·loy** (săv'ə-loi') *n.* [Alteration of obs. Fr. *cervelat* < Ital. *cervellato* < *cervello,* brain < Lat. *cerebellum,* dim. of *cerebrum,* brain.] A smoked and highly seasoned sausage made of pork.

**sav·in** or **sav·ine** (săv′ĭn) *n.* [ME < OE *safine* and OFr. *savine,* both < Lat. *(herba) Sabina,* Sabine (plant).] **1.** A dwarf evergreen shrub, *Juniperus sabina,* with a strong odor. **2.** A shrub or tree related to the savin.

**sav·ing** (sā′vĭng) *n.* **1.** Preservation or rescue from harm, danger, or loss. **2.** Avoidance of overspending : ECONOMY. **3.** A reduction in expenditure or cost. **4. savings.** The amount of money saved. **5.** Something saved. **6.** *Law.* An exception or reservation. *—prep.* With the exception of. *—conj.* Except : save.

**savings account** *n.* An account yielding interest at a bank.

**savings bank** *n.* A bank that receives and invests the savings of individual depositors and pays interest on the deposits.

**savings bond** *n.* A nontransferable registered bond issued by the U.S. Government in denominations of $25 to $1,000.

**sav·ior** (sāv′yər) *n.* [ME *saviour* < OFr. *sauveour* < LLat. *salvator* < *salvare,* to save < *salvus,* safe.] **1.** One who rescues someone or something from danger or harm. **2. Savior.** Christ.

**sav·iour** (sāv′yər) *n. Chiefly Brit. var. of* SAVIOR.

**sa·voir-faire** (săv′wär-fâr′) *n.* [Fr. : *savoir,* to know + *faire,* to do.] The ability to say and do the correct thing : TACT.

**sa·vor** (sā′vər) *n.* [ME *savour* < OFr. < Lat. *sapor* < *sapere,* to taste.] **1.** The taste or smell of something. **2.** A specific taste or smell. **3.** A characteristic property or quality. *—v.* **-vored, -vor·ing, -vors.** *—vi.* **1.** To have a particular savor. **2.** To exhibit a specified quality or characteristic <remarks that *savored* of jealousy> *—vt.* **1.** To give a flavor or aroma to. **2.** To taste or enjoy with gusto : RELISH. **3.** To have or show the savor of. **—sa′vor·er** *n.* **—sa′vor·ous** *adj.*

**sa·vor·y[1]** (sā′və-rē) *adj.* **1.** Appealing to the taste or smell. **2.** Pungent or salty to the taste : not sweet. **3.** Morally attractive : RESPECTABLE. *—n., pl.* **-ies.** A pungent dish, as anchovies on toast, served in Britain at the end of a meal or sometimes as an hors d'oeuvre. **—sa′vor·i·ly** *adv.* **—sa′vor·i·ness** *n.*

**sa·vor·y[2]** (sā′və-rē) *n., pl.* **-ies.** [ME *saverey,* perh. alteration of OE *sætherie* < Lat. *satureia.*] **1.** An aromatic herb, *Satureia hortensis* or *S. montana,* native to the Old World. **2.** The leaves of the savory, used as seasoning.

**sa·vour** (sā′vər) *n. & v. Chiefly Brit. var. of* SAVOR.

**sa·vour·y** (sā′və-rē) *adj. & n. Chiefly Brit. var. of* SAVORY[1].

**sav·vy** (săv′ē) [< Sp. *sabe (usted),* (you) know < *saber,* to know < Lat. *sapere,* to be wise.] *Slang. —vi.* **-vied, -vy·ing, -vies.** To know : comprehend. *—n.* Practical understanding : COMMON SENSE. *—adj.* **-vi·er, -vi·est.** Practical and perceptive.

**saw[1]** (sô) *n.* [ME *sawe* < OE *sagu.*] **1.** A hand-operated or power-driven portable tool, having a thin metal blade or disk with a sharp-toothed edge used to cut hard material, as wood or metal. **2.** A powered disk tool lacking teeth, used to cut metal. **3.** A fixed machine for the operation of a saw or series of saws. *—v.* **sawed, sawed** or **sawn** (sôn), **saw·ing, saws.** *—vt.* **1.** To cut or divide with or as if with a saw. **2.** To produce or shape with or as if with a saw. **3.** To handle with a sawlike motion. *—vi.* **1.** To use a saw. **2.** To cut or be cut with or as if with a saw. **—saw′er** *n.*

**saw[2]** (sô) *n.* [ME *sawe* < OE *sagu,* speech.] A trite or banal saying.

▲ word history: The noun *saw[2],* "a saying," is descended from the same Germanic ancestor as the word *saga,* "a narrative." *Saw* is a native English word whose Old English form was *sagu. Saga* was borrowed from Icelandic in the 18th century as the name of the historical legends of the Scandinavian peoples. In Old Norse and Old Icelandic *saga* basically meant a story or legend transmitted orally; the narratives now called *sagas* were written down several centuries later than the events they recount. Both *saw* and *saga* are related to the verb *say.*

**saw[3]** (sô) *v. p.t. of* SEE[1].

**saw·bones** (sô′bōnz′) *n., pl.* **sawbones** or **saw·bones·es** (-bōn′zĭz). *Slang.* A doctor, esp. a surgeon.

**saw·buck** (sô′bŭk′) *n.* **1.** A sawhorse. **2.** *Slang.* A ten-dollar bill.

**saw·dust** (sô′dŭst′) *n.* Small particles, as of wood, produced by sawing.

**sawed-off** (sôd′ôf′, -ŏf′) *adj.* **1.** Having one end sawed off <a *sawed-off* shotgun> **2.** *Slang.* Short in height.

**saw·fish** (sô′fĭsh′) *n., pl.* **sawfish** or **-fish·es.** A marine fish of the genus *Pristis,* having a bladelike snout with teeth along both sides.

**saw·fly** (sô′flī′) *n.* Any of various destructive insects chiefly of the family Tenthredinidae, the females of which have sawlike ovipositors used for cutting into plant tissue to deposit their eggs.

**saw grass** *n.* A sedge, esp. *Cladium jamaicense,* having leaves with finely toothed margins.

**saw·horse** (sô′hôrs′) *n.* A rack or trestle, esp. one with X-shaped legs, used to support something being sawed.

**saw log** *n.* A log large enough to be sawn into boards.

**saw·mill** (sô′mĭl′) *n.* **1.** A plant or factory where lumber is cut into boards. **2.** A large machine for sawing lumber.

**sawn** (sôn) *v. var. p.p. of* SAW[1].

**saw palmetto** *n.* A low-growing prickly palm, *Serenoa repens* of the southeastern United States.

**saw set** *n.* An instrument used to bend the teeth of a saw slightly outward.

**saw-toothed** (sô′tōōtht′) *adj.* Having teeth or notches like the teeth of a saw : SERRATE.

**saw-whet owl** (sô′hwĕt′, -wĕt′) *n.* [From the resemblance of its call to the sound made in sharpening a saw.] A small brown and white owl, *Aegolius acadicus* of western North America.

**saw·yer** (sô′yər) *n.* [ME *sawier* < *sawen,* to saw < *sawe,* saw.] **1.** One employed at sawing wood. **2.** Any of several longicorn beetles having larvae that bore large holes in living or dead wood.

**sax** (săks) *n. Informal.* A saxophone.

**sax·a·tile** (săk′sə-tĭl′, -tīl) *adj.* [Lat. *saxatilis* < *saxum,* rock.] *Ecol.* Saxicolous.

**sax·horn** (săks′hôrn′) *n.* [After Adolphe *Sax* (1814–1894), its inventor.] Any of a family of valved brass wind instruments that resemble the bugle and have a full, even tone and a wide range.

**sax·ic·o·lous** (săk-sĭk′ə-ləs) *also* **sax·ic·o·line** (-līn′) *adj.* [Lat. *saxum,* rock + -COLOUS.] *Ecol.* Growing on or living among rocks.

**sax·i·frage** (săk′sə-frĭj, -frāj′) *n.* [ME < OFr. < LLat. *saxifraga* < *saxifragus,* rock-breaking (from its being found growing in rock crevices) : Lat. *saxum,* rock + Lat. *frangere,* to break.] A low-growing plant of the genus *Saxifraga* of temperate regions, with leaves that are sometimes arranged in a rosette.

**sax·i·tox·in** (săk′sĭ-tŏk′sĭn) *n.* [NLat. *Saxidomus giganteus,* clam species + TOXIN.] A poisonous substance, $C_{10}H_{17}N_7O_4 \cdot 2HCl$, produced by an organism that causes red tide and found in some species of mollusks.

**Sax·on** (săk′sən) *n.* [ME < OFr. < LLat. *Saxo,* of Germanic orig.] **1.** A member of a West Germanic tribal group that invaded England in the 5th cent. with the Angles and Jutes. **2.** An Englishman as distinguished from an Irishman, Welshman, or Scot. **3.** A native or inhabitant of Saxony. **4. a.** The West Germanic language of any of the Saxon peoples. **b.** The Germanic elements of English as distinguished from the French and Latin elements. **—Sax′on** *adj.*

**Sax·on·ism** (săk′sə-nĭz′əm) *n.* An English word, phrase, or idiom of Anglo-Saxon origin.

**sax·o·ny** *also* **Sax·o·ny** (săk′sə-nē) *n.* **1.** A fine soft woolen fabric, orig. made from the wool of sheep raised in Saxony. **2.** A soft tightly twisted knitting yarn.

**sax·o·phone** (săk′sə-fōn′) *n.* [After Adolphe *Sax* (1814–1894), its inventor.] A wind instrument with a single-reed mouthpiece and a usu. curved conical metal tube. **—sax′o·phon′ist** *n.*

**sax·tu·ba** (săks′tōō′bə, -tyōō′-) *n.* [SAX(HORN) + TUBA.] A large bass saxhorn.

**say** (sā) *v.* **said** (sĕd), **say·ing, says** (sĕz) [ME *sayen* < OE *secgan.*] *—vt.* **1.** To utter aloud : SPEAK. **2.** To express in words. **3.** To state positively : DECLARE. **4.** To recite or repeat <*say* a novena> **5.** To report or maintain : ALLEGE. **6.** To indicate : show <My watch *says* two o'clock.> **7.** To estimate or suppose : ASSUME. *—vi.* To make a statement or express an opinion. *—n.* **1.** A definite assertion or assurance <You have my *say* on that.> **2.** A turn or opportunity to speak. **3.** The right or power to influence a decision : AUTHORITY <had the last *say*> *—adv.* **1.** Approximately <worth, *say,* $20.00> **2.** For instance <a cat, *say* a Siamese> **—that is to say.** In other words.

**say·ing** (sā′ĭng) *n.* A familiar expression or sentiment : ADAGE.

**say-so** (sā′sō′) *n., pl.* **-sos.** *Informal.* **1.** An unsupported statement or assurance. **2.** An authoritative assertion. **3.** The right or authority to make a decision.

**say·yid** (sä′yĭd) *n.* [Ar.] —Used as the equivalent of *lord* or *sir* and as a title of respect for an Islamic dignitary.

**Sb** [Lat. *stibium.*] *symbol for* ANTIMONY.

**'sblood** (zblŭd) *interj.* [Contraction of *God's blood.*] *Archaic.* —Used as an oath.

**Sc** *symbol for* SCANDIUM.

**scab** (skăb) *n.* [ME *scabbe* < ON *skabb.*] **1.** The crustlike exudate that forms on a healing wound. **2.** Scabies or mange in domestic animals or livestock. **3. a.** Any of various plant diseases caused by fungi or bacteria, characterized by crustlike spots on fruit, leaves, or roots. **b.** The spots caused by such a disease. **4. a.** A worker who refuses to join a labor union. **b.** One who works while others are on strike : STRIKEBREAKER. **c.** One who is hired to replace a striking worker. **d.** A mean, contemptible person. *—vi.* **scabbed, scab·bing, scabs. 1.** To become covered with a scab. **2.** To take a job held by a worker on strike.

**scab·bard** (skăb′ərd) *n.* [ME *scauberc* < AN *escaubers* (pl.), of Germanic orig.] A sheath or case, as for a dagger. *—vt.* **-bard·ed, -bard·ing, -bards.** To put into or furnish with a scabbard.

**scab·ble** (skăb′əl) *vt.* **-bled, -bling, -bles.** [ME *scaplen* < OFr. *eschapler,* to dress timber : *es-,* off (< Lat. ex-) + *chapler,* to cut < LLat. *capulare.*] To work or dress (e.g., stone) roughly.

**scab·by** (skăb′ē) *adj.* **-bi·er, -bi·est. 1.** Having, consisting of, or covered with scabs. **2.** Afflicted with scabies. **3.** *Informal.* Low : mean. **—scab′bi·ly** *adv.* **—scab′bi·ness** *n.*

**sca·bies** (skā′bēz) *n., pl.* **scabies.** [Lat., itch < *scabere,* to scratch.] **1.** A contagious skin disease of humans, caused by a mite, *Sarcoptes scabiei,* and characterized by intense itching. **2.** A similar disease in animals, esp. sheep.

**sca·bi·o·sa** (skā'bē-ō'sə, -zə, skăb'ē-) *n.* [NLat. *Scabiosa*, genus name < Med. Lat. *scabiosa (herba)*, (herb) for scabies < Lat. *scabies*, itch < *scabere*, to scratch.] SCABIOUS[2].

**sca·bi·ous[1]** (skā'bē-əs, skăb'ē-) *adj.* **1.** Of, relating to, or like scabies. **2.** Having scabs.

**sca·bi·ous[2]** (skā'bē-əs) *n.* [ME *scabiose* < Med. Lat. *scabiosa (herba)*, (herb) for scabies < Lat. *scabies*, itch < *scabere*, to scratch.] A plant of the genus *Scabiosa*, esp. *S. atropurpurea*, with opposite leaves and variously colored flower heads.

**scab·rous** (skăb'rəs, skā'brəs) *adj.* [Lat. *scabrosus* < *scaber*, scurfy.] **1.** Rough to the touch : scaly or scabby. **2.** Difficult to handle tactfully : THORNY. **3.** Salacious : indecent <a *scabrous* motion picture> **—scab'rous·ly** *adv.* **—scab'rous·ness** *n.*

**scad** (skăd) *n., pl.* **scad** *or* **scads.** [Orig. unknown.] Any of several marine fishes of the family Carangidae, related to the pompanos.

**scads** (skădz) *pl.n.* [Orig. unknown.] *Informal.* A large number or amount : LOTS <*scads* of room in the car>

**scaf·fold** (skăf'əld, -ōld') *n.* [ME < ONFr. *escafaut.*] **1.** A temporary platform used by workers while constructing, painting, or repairing a structure. **2.** A raised wooden framework or platform. **3.** A platform used in the execution of condemned prisoners. *—vt.* **-fold·ed, -fold·ing, -folds. 1.** To furnish or support with a scaffold. **2.** To place on a scaffold.

**scag** (skăg) *n.* [Orig. unknown.] *Slang.* Heroin.

**scagl·io·la** (skăl-yō'lə, -yô'-) *n.* [Ital., dim. of *scaglia*, chip, of Germanic orig.] Plasterwork in imitation of ornamental marble, made of ground gypsum and glue colored with marble or granite dust.

**sca·lade** (skə-lăd', -läd') *also* **sca·la·do** (skə-lā'dō, -lä'-) *n., pl.* **-lades** *also* **-la·dos.** [Ital. *scalada, scalata.* —see ESCALADE.] An escalade.

**scal·age** (skā'lĭj) *n.* **1.** The percentage by which a deduction, as from listed weights or prices of goods, is figured to compensate for shrinkage or other loss. **2.** The estimated amount of lumber in logs being scaled.

**sca·lar** (skā'lər, -lär') *n.* [Lat. *scalaris*, of a staircase < *scalae*, ladder.] **1.** A quantity, as mass, length, time, or temperature, completely specified by a number on an appropriate scale. **2.** A device yielding an output equal to the input multiplied by a constant, as in a linear amplifier. **—sca'lar** *adj.*

**sca·la·re** (skə-lâr'ē, -lär'ē) *n.* [NLat. < Lat. *scalaris*, of a ladder < *scalae*, ladder.] ANGELFISH 2.

**sca·lar·i·form** (skə-lăr'ə-fôrm') *adj.* [Lat. *scalaris*, of a ladder (< *scalae*, ladder) + -FORM.] Ladderlike or having runglike markings, as certain vessels and tissues. **—sca·lar'i·form'ly** *adv.*

**scalar product** *n.* The numerical product of the lengths of two vectors and the cosine of the angle between them.

**scal·a·wag** (skăl'ə-wăg') *also* **scal·ly·wag** (skăl'ē-) *n.* [Orig. unknown.] **1.** *Informal.* A reprobate : rascal. **2.** A white Republican Southerner during Reconstruction.

**scald[1]** (skôld) *v.* **scald·ed, scald·ing, scalds.** [ME *scalden* < ONFr. *escalder* < LLat. *excaldare*, to wash in hot water : Lat. *ex-*, from + Lat. *calidus*, warm.] *—vt.* **1.** To injure or burn with or as if with hot liquid or steam. **2.** To subject to or treat with boiling water, esp. in order to sterilize. **3.** To heat (a liquid) to a temperature just below the boiling point. *—vi.* To be or become scalded. *—n.* **1.** A bodily injury caused by scalding. **2. a.** A discoloration on leaves or fruit, caused by sudden exposure to intense sunlight or the action of gases. **b.** A disease of some cereal grasses caused by a fungus of the genus *Rhynchosporium.*

**scald[2]** (skôld, skäld) *n. var. of* SKALD.

**scald[3]** (skôld) *n. var. of* SCALL.

**scald·ing** (skôl'dĭng) *adj.* **1.** Causing a burning sensation, as from contact with hot liquid. **2.** Boiling <*scalding* hot water> **3.** Scorching : searing. **4.** Harsh : scathing <a *scalding* indictment>

**scale[1]** (skāl) *n.* [ME < OFr. *escale*, shell, of Germanic orig.] **1. a.** A small platelike dermal or epidermal structure that forms the proctective outer covering of fishes, reptiles, and certain mammals. **b.** A similar part, as one of the minute structures that overlap to form the covering on the wings of butterflies and moths. **2.** *Pathol.* A dry, thin flake of epidermis shed from the skin. **3.** A small thin piece. **4.** *Bot.* Any of various thin, often overlapping parts, as one of the rudimentary leaves proctecting the buds of certain trees. **5. a.** A scale insect. **b.** A plant disease or infestation caused by scale insects. **6. a.** A flaky oxide film formed on a metal, as iron, heated to high temperatures. **b.** A flake of rust. **7.** A coating or encrustation formed inside boilers, kettles, and other similar containers after extensive use. *—v.* **scaled, scal·ing, scales.** *—vt.* **1.** To clear or strip of scale or scales. **2.** To remove in layers or scales. **3.** To cover with scales : ENCRUST. **4.** To throw (a thin, flat object) so it soars through the air or skips along the surface of a body of water. *—vi.* **1.** To come off in scales : FLAKE. **2.** To become encrusted.

**scale[2]** (skāl) *n.* [ME, ladder < LLat. *scala* < Lat. *scalae.*] **1. a.** A system of ordered marks at fixed intervals used as a standard of reference in measurement. **b.** An instrument or device bearing such marks. **2. a.** The proportion used in determining the relationship of a representation to that which it represents. **b.** A calibrated line, as on a map or blueprint, that indicates such a proportion. **3.** A progressive classification, as of size, amount, importance, or rank. **4.** A relative level or degree. **5.** *Math.* A system of notation in which the value of numbers is determined by their place relative to the fixed constant of the system <binary *scale*> **6.** *Mus.* An ascending or descending series of tones proceeding by a specified scheme of intervals and varying in pitch arrangement and interval size. *—v.* **scaled, scal·ing, scales.** *—vt.* **1.** To climb up or over : ASCEND. **2.** To produce in accordance with a particular proportion or scale. **3.** To adjust according to a proportion : REGULATE. **4.** To estimate or measure the quantity of lumber in (logs or uncut trees). *—vi.* **1.** To climb : ascend. **2.** To ascend by steps or stages.

**scale[3]** (skāl) *n.* [ME < ON *skāl*, bowl.] **1.** A weighing instrument or machine. **2.** Either of the pans, trays, or dishes of a balance. *—v.* **scaled, scal·ing, scales.** *—vt.* To weigh with scales. *—vi.* To have as a weight, as determined by a scale.

**scale insect** *n.* Any of various small, destructive sucking insects of the family Coccidae, the females of which secrete and remain under waxy scales on plant tissue.

**scale moss** *n.* A leafy liverwort of the order Jungermanniales.

**sca·lene** (skā'lēn', skā-lēn') *adj.* [LLat. *scalenus* < Gk. *skalēnos*, uneven.] Having three unequal sides <a *scalene* triangle>

**scal·er** (skā'lər) *n.* [< SCALE[2].] *Electron.* A circuit that records the aggregate of a specific number of signals that occur too rapidly to be recorded individually.

**scall** (skôl) *also* **scald** (skôld) *n.* [ME < ON *scalli*, a bald head.] A scaly or scabby eruption of the skin or scalp.

**scal·lion** (skăl'yən) *n.* [ME *scaloun* < AN < VLat. **escalonia*, alteration of Lat. *Ascalonia (caepa)*, Ascalonian (onion) < *Ascalo*, Ascalon, a port in Palestine.] **1.** A young onion before the enlargement of the bulb, used chiefly in salads and as a garnish. **2.** An onionlike plant, as a leek or shallot.

**scal·lop** (skŏl'əp, skăl'-) *also* **scol·lop** (skŏl'-) [ME *scalop* < OFr. *escalope*, shell, of Germanic orig.] *—n.* **1.** A free-swimming marine mollusk of the family Pectinidae, with a fan-shaped, fluted bivalve shell. **2.** The edible adductor muscle of a scallop. **3.** A scallop shell or a similarly shaped dish, used for baking and serving seafood or other foods. **4.** One of a series of semicircular curved projections forming an ornamental border. **5.** A thin, boneless slice of meat. *—vt.* **-loped, -lop·ing, -lops. 1.** To design or border (e.g., cloth) with scallops. **2.** To bake in a casserole with milk or a sauce and often with bread crumbs. **3.** To cut (meat) into scallops. **—scal'lop·er** *n.*

**scal·ly·wag** (skăl'ē-wăg') *n. var. of* SCALAWAG.

**sca·lop·pi·ne** *also* **sca·lop·pi·ni** (skăl'ə-pē'nē, skä'lə-) *pl.n.* [Ital., pl. of *scaloppina*, dim. of *scaloppa*, fillet of beef < OFr. *escalope*, shell, of Germanic orig.] Small, thinly sliced pieces of meat, esp. veal, either dredged in flour and fried or sautéed in a sauce of wine or tomatoes and seasonings.

**scalp** (skălp) *n.* [ME, of Scand. orig.] **1.** *Anat.* The skin covering the top of the human head. **2.** A part of this skin with attached hair, formerly cut or torn from an enemy as a battle trophy by certain North American Indians. **3.** A piece of hide from the skull of certain animals, as a wolf, shown as proof of killing in order to collect a bounty. **4.** A victory trophy. **5.** *Informal.* The profit made by a ticket scalper. *—v.* **scalped, scalp·ing, scalps.** *—vt.* **1.** To cut or tear the scalp from. **2.** To deprive of top growth or of an upper layer. **3.** *Informal.* To sell (tickets) at a price higher than the established value. **4.** *Informal.* To buy and sell (e.g., stocks) quickly in order to make many small profits. *—vi. Informal.* To scalp tickets, stocks, or bonds. **—scalp'er** *n.*

**scal·pel** (skăl'pəl, skăl-pĕl') *n.* [Lat. *scalpellum*, dim. of *scalper*, knife < *scalpere*, to cut.] A small straight knife with a thin, sharp blade used in surgery and dissection.

**scalp lock** *n.* A long lock of hair remaining on the top of the shaven head of certain North American Indians.

**scal·y** (skā'lē) *adj.* **-i·er, -i·est. 1.** Having, covered with, or resembling scales. **2.** Shedding scales or flakes : FLAKING. **3.** *Slang.* Despicable : nasty. **—scal'i·ness** *n.*

**scaly anteater** *n.* The pangolin.

**scam** (skăm) *n.* [Orig. unknown.] *Slang.* A fraudulent scheme : SWINDLE.

**scam·mo·ny** (skăm'ə-nē) *n., pl.* **-nies.** [ME *scamonie* < Lat. *scammonea* < Gk. *skammōnia.*] **1.** A plant, *Convolvulus scammonia* of the eastern Mediterranean area, with large roots formerly used as a cathartic. **2.** A preparation made from the scammony.

**scamp[1]** (skămp) *n.* [< obs. *scamp*, to go about idly.] **1.** A rogue : rascal. **2.** A mischievous young person.

**scamp[2]** (skămp) *vt.* **scamped, scamp·ing, scamps.** [Poss. of Scand. orig.] To make or do in a careless, superficial way.

**scam·per** (skăm'pər) *vi.* **-pered, -per·ing, -pers.** [Prob. < Flem. *scamperen*, to decamp < OFr. *escamper* < VLat. **excampare* : Lat. *ex-*, away + Lat. *campus*, field.] To run or go hurriedly or playfully. *—n.* A hasty or playful run or departure.

**scam·pi** (skăm'pē, skäm'-) *n.* [Ital., pl. of *scampo*, a kind of lobster.] A dish of shrimp cooked in a garlic and butter sauce.

**scan** (skăn) *v.* **scanned, scan·ning, scans.** [ME *scannen* < LLat. *scandere* < Lat., to climb.] —*vt.* **1.** To examine closely. **2.** To look over rapidly but thoroughly by moving from one point to another. **3.** To look over or leaf through hastily or casually. **4.** To analyze (verse) into rhythmical patterns. **5.** *Electron.* **a.** To move a finely focused beam of light or electrons in a systematic pattern over (a surface) in order to reproduce or sense and subsequently transmit an image. **b.** To move a radar beam over (a sector of sky) in search of a target. **c.** To search (e.g., a magnetic tape) automatically for specific data. —*vi.* **1.** To analyze verse into rhythmical patterns. **2.** To conform to a metrical pattern. **3.** *Electron.* To be subjected to electronic scanning. —*n.* **1.** An act of scanning. **2.** A range or area of vision. **—scan'na·ble** *adj.* **—scan'ner** *n.*

**scan·dal** (skăn'dl) *n.* [Fr. *scandale* < LLat. *scandalum* < Gk. *skandalon,* trap, scandal.] **1.** An action or circumstance that brings about disgrace or offends public morality. **2.** Malicious talk damaging to the character. **3.** Damage to reputation or character brought about by offensive or grossly improper behavior. **4.** One whose conduct causes disgrace or shame. —*vt.* **-daled, -dal·ing, -dals. 1.** *Archaic.* To defame. **2.** *Obs.* To disgrace.

**scan·dal·ize** (skăn'dl-īz') *vt.* **-ized, -iz·ing, -iz·es. 1.** To shock the propriety or moral sense of. **2.** *Archaic.* To dishonor : disgrace. **—scan'dal·i·za'tion** *n.* **—scan'dal·iz'er** *n.*

**scan·dal·ous** (skăn'dl-əs) *adj.* **1.** Causing scandal : shocking or offensive. **2.** Containing libelous or defamatory information. **—scan'dal·ous·ly** *adv.* **—scan'dal·ous·ness** *n.*

**scandal sheet** *n.* A periodical that prints scandalous stories.

**scan·dent** (skăn'dənt) *adj.* [Lat. *scandens, scandent-,* pr.part. of *scandere,* to climb.] *Bot.* Climbing, as a vine.

**scan·di·a** (skăn'dē-ə) *n.* [< SCANDIUM.] *Chem.* Scandium oxide.

**Scan·di·an** (skăn'dē-ən) *adj.* [< Lat. *Scandia,* Scandinavia.] Scandinavian.

**Scan·di·na·vi·an** (skăn'də-nā'vē-ən, -nāv'yən) *adj.* Of or pertaining to Scandinavia or to its residents, culture, or languages. —*n.* **1.** A native or resident of Scandinavia. **2.** The North Germanic languages.

**scan·di·um** (skăn'dē-əm) *n.* [< Lat. *Scandia,* Scandinavia, where it was discovered.] *Symbol* **Sc** A silvery-white, lightweight metallic element found in various rare minerals; atomic number 21; atomic weight 44.956. **—scan'dic** *adj.*

**scandium oxide** *n.* A white amorphous powder, $Sc_2O_3$, used as a source of scandium and in the manufacture of ceramics.

**scan·ning** (skăn'ĭng) *n.* An electronic or optical technique by which recorded information or images are sensed for later modification, integration, or transmission.

**scanning electron microscope** *n.* An electron microscope that forms a three-dimensional image on a cathode-ray tube by moving a beam of focused electrons across an object and reading both the electrons scattered by the object and the secondary electrons produced by it.

**scan·sion** (skăn'shən) *n.* [LLat. *scansio* < Lat., act of climbing < *scandere,* to climb.] Analysis of verse into rhythmical patterns.

**scan·so·ri·al** (skăn-sôr'ē-əl, -sōr'-) *adj.* [< Lat. *scansorius* < *scandere,* to climb.] Of, related to, or specially adapted for climbing.

**scant** (skănt) *adj.* **-er, -est.** [ME < ON *skamt,* neuter of *skammr,* short.] **1.** Lacking in quantity or amount : INADEQUATE. **2.** Being slightly less than a specific measure <a *scant* two cups> **3.** Not sufficiently supplied <*scant* of common sense> —*vt.* **scant·ed, scant·ing, scants. 1.** To supply with an inadequate portion or allowance : SKIMP. **2.** To limit, as in amount or share : STINT. **3.** To reduce the size or quantity of : CUT DOWN. **4.** To deal with inadequately or give slight attention to : NEGLECT. **—scant'ly** *adv.* **—scant'ness** *n.*

**scant·ling** (skănt'lĭng, -lĭn) *n.* [Alteration of obs. *scantlon,* carpenter's gauge < ME *scantilon* < OFr. *escantillon.*] **1.** A small beam or piece of timber, as an upright in a structural frame. **2.** The dimensions of building materials, as stone or timber, esp. in breadth and thickness. **3.** *often* **scantlings.** *Naut.* The dimensions of the structural parts of a vessel. **4.** A small amount or quantity.

**scant·y** (skăn'tē) *adj.* **-i·er, -i·est. 1.** Barely adequate or sufficient : MEAGER. **2.** Deficient : lacking. **—scant'i·ly** *adv.* **—scant'i·ness** *n.*

**scape[1]** (skāp) *n.* [Lat. *scapus,* stalk.] **1.** *Bot.* A leafless flower stalk arising directly from the crown at ground level. **2.** A part resembling a stalk, as the shaft of a feather or an insect's antenna. **3.** The shaft of a column.

**scape[2]** (skāp) *v.* & *n. Archaic. var. of* ESCAPE.

**-scape** *suff.* [< LANDSCAPE.] Scene : view <*seascape*>

**scape·goat** (skāp'gōt') *n.* [(E)SCAPE + GOAT, as transl. of Heb. *azāzēl,* goat of Azazel, construed as *ēz-ōzēl,* goat that escapes.] **1.** An individual or group bearing blame for others. **2.** A goat sent into the wilderness on the Day of Atonement, symbolically bearing on its head the sins of the children of Israel. —*vt.* **-goat·ed, -goat·ing, -goats.** To make a scapegoat of.

☆ ***syns:*** SCAPEGOAT, FALL GUY, GOAT, PATSY, WHIPPING BOY *n.* *core meaning* : one who is made an object of blame for others <made the senator the *scapegoat* for the party's defeat>

**scape·grace** (skāp'grās') *n.* [SCAPE[2] + GRACE.] An unscrupulous or incorrigible person : REPROBATE.

**scaph·oid** (skăf'oid') *adj.* [NLat. *scaphoides* < Gk. *skaphoeidēs* : *skaphē,* boat + *-eidēs,* -oid.] Boat-shaped. —*n.* The navicular bone.

**scap·o·lite** (skăp'ə-līt') *n.* [Fr. : Lat. *scapus,* stalk + Gk. *lithos,* stone (from the prismatic shape of its crystals).] Any of a group of variously colored mineral silicates composed chiefly of aluminum, calcium, and sodium.

**sca·pose** (skā'pōs') *adj.* Resembling, producing, or consisting of a scape.

**scap·u·la** (skăp'yə-lə) *n., pl.* **-las** *or* **-lae** (-lē') [Lat.] Either of a pair of large, flat, triangular bones that form the back part of the shoulder.

**scap·u·lar** (skăp'yə-lər) *n.* [ME *scapulare* < Med. Lat. *scapulare* < Lat. *scapula,* shoulder.] **1.** A sleeveless outer garment that hangs from the shoulders, worn as part of a monk's habit. **2.** A religious badge or token of devotion consisting of two pieces of cloth joined by strings and worn about the shoulders. **3.** One of the feathers covering the shoulder of a bird. —*adj. Anat.* Of or relating to the shoulder or scapula.

**scap·u·lo·cla·vic·u·lar** (skăp'yə-lō-klə-vĭk'yə-lər) *adj.* Of, relating to, or affecting both the scapula and the clavicle.

**scar[1]** (skär) *n.* [ME < OFr. *escare,* scab < LLat. *eschara* < Gk. *eskhara,* hearth, scab caused by burning.] **1.** A mark left on the skin after a surface injury or wound has healed. **2.** A lingering sign of mental or physical injury or damage. **3.** *Bot.* An indication of a former attachment, as of a leaf to a stem. **4.** A dent, blemish, or disfiguring mark caused by use or contact. —*v.* **scarred, scar·ring, scars.** —*vt.* **1.** To mark with or as if with a scar. **2.** To do lingering damage to. —*vi.* **1.** To form a scar. **2.** To become scarred.

**scar[2]** (skär) *n.* [ME *skerre* < ON *sker,* low reef.] **1.** An isolated rock that juts out. **2.** A bare, rocky place on a steep slope or mountainside.

**scar·ab** (skăr'əb) *n.* [OFr. *scarabee* < Lat. *scarabaeus.*] **1.** A scarabaeid beetle, esp. *Scarabaeus sacer,* considered sacred by the ancient Egyptians. **2.** A ceramic or stone representation of a scarab beetle, formerly used as a talisman and a symbol of the soul.

**scar·a·bae·i** (skăr'ə-bē'ī') *n. var. pl. of* SCARABAEUS.

**scar·a·bae·id** (skăr'ə-bē'ĭd) *n.* [NLat. *Scarabaeidae,* family name < Lat. *scarabaeus,* beetle.] A beetle of the family Scarabaeidae, as the June bug or dung beetle. —*adj.* Of or belonging to the Scarabaeidae.

**scar·a·bae·us** (skăr'ə-bē'əs) *n., pl.* **-bae·us·es** *or* **-bae·i** (-bē'ī') [Lat.] A scarab.

**scar·a·bi·a·sis** (skăr'ə-bī'ə-sĭs) *n.* [SCARAB(AEID) + -IASIS.] Infestation of the intestine with the dung beetle.

**scar·a·boid** (skăr'ə-boid') *adj.* Like or characteristic of a scarabaeid beetle.

**Scar·a·mouch** *also* **Scar·a·mouche** (skăr'ə-mōōsh', -mōōch', -mouch') *n.* [Fr. *Scaramouche* < Ital. *Scaramuccia.*] A stock character in the commedia dell'arte, depicted as a boastful, cowardly braggart or buffoon.

**scarce** (skârs) *adj.* **scarc·er, scarc·est.** [ME *scars* < ONFr. *escars* < VLat. **excarpsus,* p.part. of **excarpere,* alteration of Lat. *excerpere,* to pick out. —see EXCERPT.] **1.** Infrequently appearing or occurring : RARE <a *scarce* mineral> **2.** Not sufficient to meet a demand or need <a *scarce* commodity> **—scarce'ness** *n.*

**scarce·ly** (skârs'lē) *adv.* **1.** By a narrow margin : BARELY. **2.** Almost not : HARDLY. **3.** Certainly not. ***usage:*** *Scarcely* has the force of a negative; therefore it is not properly used with another negative, as in *I could* (not *couldn't*) *scarcely believe my eyes.* A clause after *scarcely* may be introduced by *when* or *before,* but not by *than,* as in *The meeting had scarcely begun when* (or *before* but not *than*) *the fire broke out.*

**scar·ci·ty** (skâr'sĭ-tē) *n., pl.* **-ties. 1.** Lack of sufficient quantity or supply : DEARTH. **2.** Rarity of appearance or occurrence.

**scare** (skâr) *v.* **scared, scar·ing, scares.** [ME *skerren* < ON *skirra* < *skjarr,* timid.] —*vt.* To suddenly frighten : ALARM. —*vi.* To become frightened or alarmed. **—scare up.** *Informal.* To find or gather with considerable effort <finally *scared up* the necessary funds> —*n.* **1.** An instance of sudden fear. **2.** A state of alarm or panic. **—scar'er** *n.* **—scar'ing·ly** *adv.*

**scare·crow** (skâr'krō') *n.* **1.** A roughly fashioned figure of a person used to scare birds away from crops. **2.** Something that is frightening but not dangerous. **3.** An extremely thin or haggard person.

**scare·mon·ger** (skâr'mŭng'gər, -mŏng'-) *n.* One who spreads alarming rumors.

**scarf[1]** (skärf) *n., pl.* **scarfs** *or* **scarves** (skärvz) [ONFr. *escarpe,* sash, sling.] **1.** A square, rectangular, or triangular piece of cloth worn over the head or about the neck or shoulders. **2.** A runner, as for a chest or table. **3.** A sash denoting military or official rank. —*vt.* **scarfed, scarf·ing, scarfs. 1.** To wrap, cover, or decorate with or as if with a scarf. **2.** To wrap around loosely.

**scarf[2]** (skärf) *n., pl.* **scarfs.** [ME *skarf.*] **1.** A joint formed by strapping or bolting two notched pieces of timbers together to make a continuous piece. **2.** The end of a timber notched to form a scarf. —*vt.* **scarfed, scarf·ing, scarfs. 1.** To unite by a scarf joint. **2.** To cut a scarf in.

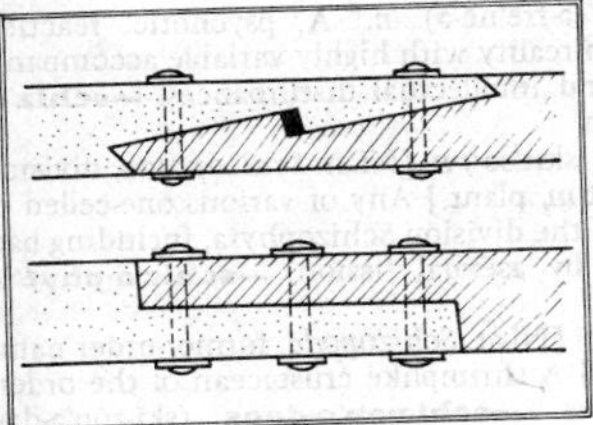

scarf

**scarf·skin** (skärf′skĭn′) *n.* The outer layer of skin, esp. the cuticle.

**scar·i·fi·ca·tor** (skăr′ə-fĭ-kā′tər) *n.* A surgical instrument with several spring-operated lancets used to make superficial cuts in the skin.

**scar·i·fy**[1] (skăr′ə-fī′) *vt.* **-fied, -fy·ing, -fies.** [ME *scarifien* < OFr. *scarifier* < LLat. *scarificare,* alteration of *scarifare* < Gk. *skariphasthai,* to scratch, sketch < *skariphos,* pencil.] **1.** To make superficial cuts in (the skin), as when vaccinating. **2.** To break up and loosen the surface of (topsoil). **3.** To criticize severely : FLAY. **4.** *Bot.* To slit or abrade the outer coat of (seeds) in order to speed germination. **—scar′i·fi·ca′tion** *n.* **—scar′i·fi′er** *n.*

**scar·i·fy**[2] (skâr′ə-fī′) *vt.* **-fied, -fy·ing, -fies.** To scare.

**scar·i·ous** (skâr′ē-əs) *also* **scar·i·ose** (-ōs′) *adj.* [NLat. *scariosus.*] Thin, membranous, and dry <*scarious* bracts>

**scar·la·ti·na** (skär′lə-tē′nə) *n.* [NLat. < Med. Lat. *scarlata,* scarlet.] Scarlet fever. **—scar′la·ti′nal** *adj.*

**scar·la·ti·noid** (skär′lə-tē′noid′) *adj.* [NLat. *scarlatina,* scarlet fever + -OID.] Resembling scarlet fever or its rash.

**scar·let** (skär′lĭt) *n.* [ME < OFr. *escarlate* < Med. Lat. *scarlata,* scarlet cloth < Pers. *sāqalāt,* a kind of rich cloth.] **1.** A strong to vivid red or reddish orange. **2.** Scarlet-colored cloth or clothing. *—adj.* **1.** Having a scarlet color <a face *scarlet* with embarrassment> **2.** Sinful or immoral.

**scarlet fever** *n.* An acute contagious disease caused by a hemolytic streptococcus, usu. affecting children and marked by a scarlet-colored rash and high fever.

**scarlet pimpernel** *n.* Pimpernel.

**scarlet runner** *n.* A climbing bean plant, *Phaseolus coccineus,* native to tropical America, with bright-red flowers and long pods containing edible seeds.

**scarlet sage** *n.* A species of salvia, *Salvia splendens,* with vivid red flowers.

**scarlet tanager** *n.* A New World bird, *Piranga olivacea,* of which the male has bright red plumage with a black tail and wings.

**scarp** (skärp) *n.* [Ital. *scarpa,* perh. of Germanic orig.] **1.** A steep slope : CLIFF. **2.** A steep outer slope of a fortification. *—vt.* **scarped, scarp·ing, scarps.** To carve or make into a scarp.

**scar tissue** *n.* A dense, often hard layer of connective tissue formed over a healing wound or cut.

**scarves** (skärvz) *n. var. pl. of* SCARF[1].

**scar·y** (skâr′ē) *adj.* **-i·er, -i·est. 1.** Causing fear : FRIGHTENING. **2.** Easily frightened.

**scat**[1] (skăt) *vi.* **scat·ted, scat·ting, scats.** [Orig. unknown.] *Informal.* To leave hastily.

**scat**[2] (skăt) *n.* [Perh. imit.] Jazz singing that includes improvised nonsensical syllables. *—vi.* **scat·ted, scat·ting, scats.** To sing scat.

**scat**[3] (skăt) *n.* [Gk. *skōr, skat-,* excrement.] Animal droppings.

**scathe** (skāth) *vt.* **scathed, scath·ing, scathes.** [ME *skathen* < ON *skaða.*] **1.** To damage or injure severely, esp. by fire. **2.** To criticize harshly. *—n.* Damage : injury.

**scath·ing** (skā′thĭng) *adj.* **1.** Extremely harsh or bitter. **2.** Harmful : injurious.

**scato-** *pref.* [Gk. *skato-* < *skōr,* ordure.] Excrement <scatology>

**sca·tol·o·gy** (skă-tŏl′ə-jē, skə-) *n.* **1.** The study of fecal excrement, as in medicine or paleontology. **2. a.** An obsession with excrement or excretory functions. **b.** The psychiatric study of such an obsession. **3.** Interest in or preoccupation with obscenity, as in literature. **—scat′o·log′i·cal** (skăt′l-ŏj′ĭ-kəl), **scat′o·log′ic** *adj.* **—sca·tol′o·gist** *n.*

**scat·ter** (skăt′ər) *v.* **-tered, -ter·ing, -ters.** [ME *scateren.*] *—vt.* **1.** To cause to break up and go in many directions : DISPERSE. **2.** To distribute by or as if by sprinkling or strewing. **3.** *Physics.* To deflect (e.g., radiation or particles). *—vi.* **1.** To break up and go in many directions. **2.** To happen or be present at widely spaced intervals. *—n.* **1.** The act of scattering or state of being scattered. **2.** Something scattered or strewn about. **—scat′ter·er** *n.*

☆ ***syns:*** SCATTER, DISPEL, DISPERSE, DISSIPATE *v. core meaning* : to break up or cause to break up and go in many directions <children *scattering* on the playground><wind that *scattered* the leaves> *ant:* gather

**scat·ter·brain** (skăt′ər-brān′) *n.* A flighty, disorganized person. **—scat′ter·brained′** *adj.*

**scat·ter·good** (skăt′ər-go͝od′) *n.* A spendthrift.

**scat·ter·gun** (skăt′ər-gŭn′) *n.* A shotgun.

**scat·ter·ing** (skăt′ər-ĭng) *n.* **1. a.** The act or process of scattering. **b.** The state of being scattered. **2.** A small, sporadic amount or quantity of something <a *scattering* of boos from the audience> **3.** *Physics.* The dispersal of radiation or a beam of particles into a variety of directions. *—adj.* Occurring or found at irregular intervals. **—scat′ter·ing·ly** *adv.*

**scatter pin** *n.* A small decorative brooch often worn in groups of two or three.

**scatter rug** *n.* A small rug that covers part of a floor.

**scat·ter·shot** (skăt′ər-shŏt′) *adj.* Randomly covering a broad range <*scattershot* artillery fire>

**scat·ter-site** (skăt′ər-sīt′) *adj.* Of, concerning, or designating publicly funded low-income housing units scattered throughout middle-income residential areas.

**scat·ty** (skăt′ē) *adj.* **-ti·er, -ti·est.** [Prob. SCATT(ERBRAIN) + -Y[1].] *Chiefly Brit.* Foolish : crazy.

**scaup** (skôp) *n., pl.* **scaup** *or* **scaups.** [Perh. alteration of Sc. *scalp,* bed of mussels (from its feeding on shellfish).] A diving duck, *Aythya marila* or *A. affinis,* with black and white plumage.

**scav·enge** (skăv′ənj) *v.* **-enged, -eng·ing, -eng·es.** [Back-formation < SCAVENGER.] *—vt.* **1.** To gather and remove dirt or refuse from. **2.** To search through for salvageable material. **3.** To gather (salvageable material) by searching. **4.** To expel (burned gases) from a cylinder of an internal-combustion engine. **5.** *Metallurgy.* To clean and purify (molten metal) by chemically removing impurities. *—vi.* To search through refuse or discarded material for something useful.

**scav·en·ger** (skăv′ən-jər) *n.* [Alteration of obs. *scavager,* street cleaner < ME *skawager,* toll collector < AN *scawager* < *scawage,* a tax on the goods of foreign merchants < Flem. *scawen,* to look at, show.] **1.** An animal feeding on dead or decaying matter. **2.** One who scavenges. **3.** *Chem.* A substance added to a mixture to remove impurities or counteract the unwanted effects of other constituents.

**sce·nar·i·o** (sĭ-nâr′ē-ō′, -när′-, -năr′-) *n., pl.* **-os.** [Ital. < LLat. *scaenarius,* of the stage < Lat. *scaena,* stage. —see SCENE.] **1.** A synopsis of a dramatic or literary plot. **2.** A screenplay. **3.** An outline of a hypothesized or projected chain of events.

**sce·nar·ist** (sĭ-nâr′ĭst, -när′-, -năr′-) *n.* One who writes screenplays.

**scend** *also* **send** (sĕnd) *vi.* **scend·ed, scend·ing, scends** *also* **send·ed, send·ing, sends.** [Prob. alteration of SEND [1].] To rise upward on a wave or swell. *—n.* The upward movement of a ship on a wave or swell.

**scene** (sēn) *n.* [OFr., stage < Lat. *scaena* < Gk. *skēnē,* stage, tent.] **1.** An image or prospect as seen by a viewer : VIEW. **2.** The place where an action or event occurs : LOCALE. **3.** The setting in which the action of a narrative occurs. **4.** A subdivision of an act in a dramatic presentation that takes place in one setting and period of time. **5.** A shot or series of shots in a film that makes up a unit of continuous related action. **6.** The scenery and properties for a dramatic presentation. **7.** *Archaic.* A theater stage. **8.** A real or fictitious episode, esp. when recounted. **9.** A public display of passion or temper <made a *scene* at the restaurant> **10.** *Slang.* **a.** A sphere of activity <the dating *scene*> **b.** A given situation <a bad *scene*> **—behind the scenes. 1.** Backstage. **2.** In private <jockeying for position *behind the scenes*> **—sce′nic** *adj.* **—sce′ni·cal·ly** *adv.*

**scen·er·y** (sē′nə-rē) *n., pl.* **-ies. 1.** A landscape. **2.** The painted backdrops on a theatrical stage.

**scene-steal·er** (sēn′stē′lər) *n.* An actor who attracts the audience's attention from the action or actors meant to be the focus of attention.

**scent** (sĕnt) *n.* [ME *sent* < *senten,* to scent < OFr. *sentir* < Lat. *sentire,* to feel.] **1.** A distinctive or characteristic odor. **2.** A perfume. **3.** An odor left by the passing of an animal. **4.** The trail of a hunted animal or fugitive. **5.** The sense of smell. **6.** An indication or hint of something to come. *—v.* **scent·ed, scent·ing, scents.** *—vt.* **1.** To perceive or recognize by the sense of smell. **2.** To suspect or detect as if by smelling <*scent* trouble ahead> **3.** To perfume. *—vi.* To hunt or track prey by means of the sense of smell. —Used of hounds.

**scent gland** *n.* A gland in many mammals that secretes odorous substances.

**scep·ter** (sĕp′tər) *n.* [ME *sceptre* < OFr. < Lat. *sceptrum* < Gk. *skēptron,* staff.] **1.** A usu. ornamented rod or staff held by a sovereign as an emblem of authority. **2.** Sovereign office or power. *—vt.* **-tered, -ter·ing, -ters.** To invest with royal authority.

**scep·tic** (skĕp′tĭk) *n. var. of* SKEPTIC.

**scep·ti·cal** (skĕp′tĭ-kəl) *adj. var. of* SKEPTICAL.

**scep·ti·cism** (skĕp′tĭ-sĭz′əm) *n. var. of* SKEPTICISM.

**scep·tre** (sĕp′tər) *n.* & *v. Chiefly Brit. var. of* SCEPTER.

**scha·den·freu·de** (shäd′n-froi′də) *n.* [G. : *schaden,* damage + *freude,* joy.] Enjoyment derived from the misfortunes of others.

**schav** (shäv) *n.* [Pol. *szczaw,* sorrel.] A chilled soup of sorrel, onions, lemon juice, eggs, and sugar, served with sour cream.

**sched·ule** (skĕj′o͞ol, -əl; *Brit.* shĕd′yo͞ol) *n.* [ME *sedule,* slip of parchment or paper, note < OFr. *cedule* < LLat. *schedula,* dim. of Lat. *scheda,* papyrus leaf, of Gk. orig.] **1.** A written or printed list of items, usu. in tabular form. **2. a.** A program of future events or appointments. **b.** A program of classes. **3.** A table of departure and arrival times. **4.** A production plan allocating work and specifying deadlines. **5.** A supplemental statement of details appended to a document. —*vt.* **-uled, -ul·ing, -ules. 1.** To enter on a schedule. **2.** To make up a schedule for. **3.** To plan or appoint for a certain date or time.

**schee·lite** (shā'līt') *n.* [After Karl *Scheele* (1742–1786), its discoverer.] A natural form of calcium tungstate, $CaWO_4$, that occurs in igneous rocks and is used as a source of tungsten.

**sche·ma** (skē'mə) *n., pl.* **-ma·ta** (-mə-tə) [Gk. *skhēma, skhēmat-*, form.] A diagrammatic outline or representation.

**sche·mat·ic** (skē-măt'ĭk) *adj.* [NLat. *schematicus* < Gk. *skhēma*, form.] Of, relating to, or in the form of a scheme, schema, or diagram. —*n.* A structural or procedural diagram, esp. of an electrical or mechanical system. **—sche·mat'i·cal·ly** *adv.*

**sche·ma·tism** (skē'mə-tĭz'əm) *n.* The patterned arrangement of constituents within a specific system.

**sche·ma·tize** (skē'mə-tīz') *vt.* **-tized, -tiz·ing, -tiz·es.** [Gk. *skhēmatizein*, to give form to < *skhēma*, form.] To form into a scheme. **—sche'ma·ti·za'tion** *n.*

**scheme** (skēm) *n.* [Lat. *schema*, figure, manner < Gk. *skhēma*, form.] **1.** A systematic plan of action. **2.** An orderly combination of related or successive parts or elements **:** SYSTEM. **3.** A plan, esp. a secret or underhand one **:** PLOT. **4.** A chart, diagram, or outline of a system or object. **5.** A visionary plan. —*v.* **schemed, schem·ing, schemes.** —*vt.* **1.** To contrive a plan or scheme for. **2.** To plot. —*vi.* To make plans, esp. secret or underhand ones. **—schem'er** *n.*

**scher·zan·do** (skĕrt-sän'dō) [Ital., gerund. of *scherzare*, to joke < *scherzo*, joke. —see SCHERZO.] *Mus.* —*adv.* In a light, playful manner. —Used as a direction. —*n., pl.* **-dos.** A passage written for or performed scherzando. **—scher·zan'do** *adj.*

**scher·zo** (skĕr'tsō) *n., pl.* **-zos** *or* **-zi** (-tsē) [Ital., joke, scherzo < MHG *scherz*, joke < *scherzen*, to leap with joy.] *Mus.* A lively movement usu. in 3/4 time.

**Schick test** (shĭk) *n.* [After Bela *Schick* (1877–1967), its deviser.] An intracutaneous skin test of susceptibility to diphtheria.

**Schiff's reagent** (shĭfs) *n.* [After Hugo *Schiff* (1834–1915).] An aqueous solution of rosaniline and sulfurous acid used to indicate the presence of aldehydes.

**schil·ler** (shĭl'ər) *n.* [G.] A metallic luster reflected by certain planes in a mineral grain.

**schil·ling** (shĭl'ĭng) *n.* [G.] —See table at CURRENCY.

**schip·per·ke** (skĭp'ər-kē, -kə) *n.* [Flem., dim. of *schipper*, skipper (from its use as a watchdog on a boat) < MDu. —see SKIPPER[1].] A small dog orig. bred in Belgium, with dense, long black fur.

**schism** (sĭz'əm, skĭz'-) *n.* [ME *scisme* < OFr. < LLat. *schisma* < Gk. *skhisma*, division < *skhizein*, to split.] **1. a.** A separation into factions, esp. a formal division within a Christian church. **b.** The offense of attempting to produce a schism. **2.** A body or sect that creates or participates in a schism. **—schis·mat'ic** (sĭz-măt'ĭk, skĭz-) *adj.* **—schis·mat'i·cal·ly** *adv.*

**schist** (shĭst) *n.* [Fr. *schiste* < Lat. *(lapis) schistos*, fissile (stone) < Gk. *skhistos*, split, divisible < *skhizein*, to split.] A medium- to coarse-grained metamorphic rock made up of laminated, often flaky parallel layers of chiefly micaceous minerals. **—schis'tose'** (shĭs'tōs'), **schis'tous** (shĭs'təs) *adj.*

**schis·to·some** (shĭs'tə-sōm') *n.* [NLat. *Schistosoma*, genus name : Gk. *skhistos*, split + Gk. *soma*, body.] Any of several chiefly tropical trematode worms of the genus *Schistosoma*, many of which are parasitic in the blood of mammals. **—schis'to·som'al** (-sō'məl) *adj.*

**schis·to·so·mi·a·sis** (shĭs'tō-sə-mī'ə-sĭs) *n.* A severe, gen. tropical disease caused by infestation with schistosomes.

**schiz-** *pref. var. of* SCHIZO-.

**schiz·o** (skĭt'sō) *n., pl.* **-os.** *Slang.* A schizoid.

**schizo-** *or* **schiz-** *pref.* [Gk. *skhizo-* < *skhizein*, to split.] **1.** Split **:** cleft <*schizocarp*> **2.** Cleavage **:** fission <*schizogenesis*> **3.** Schizophrenia <*schizoid*>

**schiz·o·carp** (skĭz'ə-kärp', skĭt'sə-) *n.* A dry seed, as of the maple, that splits when mature into two or more closed carpels, each usu. containing one seed. **—schiz'o·car'pous, schiz'o·car'pic** *adj.*

**schiz·o·gen·e·sis** (skĭz'ō-jĕn'ĭ-sĭs, skĭt'sō-) *n. Biol.* Reproduction by fission.

**schi·zog·o·ny** (skĭ-zŏg'ə-nē, skĭt-sŏg'-) *n. Biol.* Reproduction by multiple asexual fission. **—schi·zog'o·nous** *adj.*

**schiz·oid** (skĭt'soid') *adj.* Characteristic of or like schizophrenia. —*n.* A schizophrenic person.

**schiz·o·my·cete** (skĭz'ō-mī'sēt', -mī-sēt', skĭt'sō-) *n.* [NLat. *Schizomycetes*, class name : SCHIZO- + Gk. *mukētes*, pl. of *mukēs*, fungus.] A one-celled microorganism of the class Schizomycetes, which includes the bacteria. **—schiz'o·my·ce'tous** (-mī-sē'təs) *adj.*

**schiz·o·my·co·sis** (skĭz'ō-mī-kō'sĭs, skĭt'sō-) *n.* [SCHIZO(MYCETE) + MYCOSIS.] A disease caused by schizomycetes.

**schiz·ont** (skĭz'ŏnt', skĭt'sŏnt') *n.* A protozoan cell produced by schizogony in the life cycle of a sporozoan.

**schiz·o·phrene** (skĭt'sə-frēn') *n.* [Back-formation < SCHIZOPHRENIA.] One afflicted with schizophrenia.

**schiz·o·phre·ni·a** (skĭt'sə-frē'nē-ə) *n.* A psychotic reaction marked by withdrawal, from reality with highly variable accompanying affective, behavioral, and intellectual disturbances. **—schiz'o·phren'ic** (-frĕn'ĭk) *adj. & n.*

**schiz·o·phyte** (skĭz'ə-fīt', skĭt'sə-) *n.* [NLat. *Schizophyta*, division name : SCHIZO- + Gk. *phuton*, plant.] Any of various one-celled or simple colonial organisms of the division Schizophyta, including bacteria, that usu. reproduce by asexual fission. **—schiz'o·phyt'ic** (-fĭt'ĭk) *adj.*

**schiz·o·pod** (skĭz'ə-pŏd') *n.* [NLat. *Schizopoda*, former order name : SCHIZO- + Gk. *pous*, foot.] A shrimplike crustacean of the orders Euphausiacea or Mysidacea. **—schiz·op'o·dous** (skĭ-zŏp'ə-dəs, skĭt-sŏp'-) *adj.*

**schiz·o·thy·mi·a** (skĭt'sə-thī'mē-ə) *n.* Schizoid behavior that resembles schizophrenia, as in the tendency to withdraw from reality, but remains within the bounds of normality. **—schiz'o·thy'mic** (-thī'mĭk) *adj.*

**schiz·y** (skĭt'sē) *adj.* **-i·er, -i·est.** [Shortening and alteration of SCHIZOID.] *Slang.* Schizoid.

**schle·miel** (shlə-mēl') *n.* [Yiddish *shlumiel*, perh. < Heb. *Shelūmīel*, Shelumiel, a character in the Bible.] *Slang.* A bungler **:** dolt.

**schlep** (shlĕp) [Yiddish *shleppen*, to drag < MLG *slēpen*.] *Slang.* —*v.* **schlepped, schlep·ping, schleps.** —*vt.* To carry clumsily or with difficulty **:** LUG. —*vi.* To move slowly or laboriously. —*n.* **1.** A difficult journey. **2.** A clumsy or stupid person. **—schlep'per** *n.*

**schlie·ren** (shlîr'ən) *pl.n.* [G. < dial. G., pl. of *Shlier*, ulcer < MHG *slier*.] **1.** *Geol.* Irregular tabular bodies that occur as essential components of plutonic rock but differ in texture or composition from the principal mass. **2.** Regions of a transparent medium, as of a flowing gas, that exhibit densities different from that of the bulk of the medium.

**schli·ma·zel** (shlĭ-mä'zəl) *n.* [Yiddish *shlimazel*, bad luck : *shlim*, bad (< MHG *slimp*) + Heb. *mazāl*, luck.] *Slang.* An unlucky or inept person.

**schlock** *also* **shlock** (shlŏk) *n.* [Yiddish < G. *Schlag*, a blow.] *Slang.* Cheap or inferior merchandise. **—schlock'y** *adj.*

**schmaltz** *also* **schmalz** (shmälts) *n.* [Yiddish *shmalts*, melted fat < MHG *smalz* < OHG.] **1.** Highly sentimental art or music. **2.** Maudlin sentimentality. **—schmaltz'y** *adj.*

**schmeer** *also* **schmear** (shmîr) *n.* [Yiddish *shmir* < G. *Schmiere*, grease.] *Slang.* **1.** An aggregate <believed the whole schmeer> **2.** A bribe.

**Schmidt system** (shmĭt) *n.* [After Bernhard *Schmidt* (1879–1935), its inventor.] A system having a concave spherical mirror and a transparent glass plate at its center of curvature, used in reflecting telescopes to offset spherical aberration, coma, and astigmatism.

**schmo** *also* **schmoe** (shmō) *n., pl.* **schmoes.** [Yiddish *shmok* < Slovene *šmok*.] *Slang.* A stupid person.

**schmoose** *also* **schmooze** (shmo͞oz) [Yiddish *shmuesn*, to chat < *shmues*, a chat < Heb. *shemu'ōth*, pl. of *shemu'āh*, rumor < *shāmō'a*, he heard.] *Slang.* —*vi.* **schmoosed, schmoos·ing, schmoos·es** *also* **schmoozed, schmooz·ing, schmooz·es.** To chat idly. —*n.* A chat. **—schmoos'er** *n.*

**schmuck** (shmŭk) *n.* [Yiddish *shmok*, penis, fool < G. *Schmuck*, ornament < MLG *smuck*.] *Slang.* An inept or stupid person.

**schnapps** (shnäps, shnăps) *n., pl.* **schnapps.** [G. *Schnaps* < LG *snaps*, mouthful < *snappen*, to snap < MLG, to speak hastily.] A strong liquor.

**schnau·zer** (shnou'zər, shnout'ser) *n.* [G. < *Schnauze*, snout.] A dog orig. bred in Germany, having a wiry gray coat and a blunt muzzle and ranging in size from fairly small to quite large.

**schnit·zel** (shnĭt'səl) *n.* [G., dim. of *Schnitz*, slice.] A thin veal cutlet sautéed in butter.

**schnook** (shno͝ok) *n.* [Yiddish *shnok*.] A dupe.

**schnor·rer** (shnôr'ər, shnōr'-) *n.* [Yiddish *shnorer* < *shnoren*, to beg < MHG *snurren*, to hum, whir.] *Slang.* One who takes advantage of the generosity of friends **:** SPONGER.

**schnoz·zle** (shnŏz'əl) *also* **schnozz** (shnŏz) *n.* [Prob. alteration of Yiddish *shnoitsl*, dim of *shnoits*, snout < G. *Schnauze*.] *Slang.* The human nose.

**schol·ar** (skŏl'ər) *n.* [ME *scoler* < OFr. *escoler* < LLat. *scholaris*, of a school < Lat. *schola, scola*, school. —see SCHOOL[1].] **1. a.** An erudite person. **b.** A specialist in one of the humanities. **2. a.** One who attends school or studies with a teacher **:** PUPIL. **b.** One considered in respect to his or her aptness at learning. **3.** One who holds or has held a scholarship.

**schol·ar·ly** (skŏl'ər-lē) *adj.* Of, relating to, or characteristic of scholars or scholarship. **—schol'ar·li·ness** *n.*

**schol·ar·ship** (skŏl'ər-shĭp') *n.* **1.** The methods, discipline, and achievements of a scholar. **2.** Knowledge resulting from study and research in a field. **3.** Financial aid awarded to a student.

**scho·las·tic** (skə-lăs'tĭk) *adj.* [Lat. *scholasticus* < Gk. *skholastikos* < *skholazein*, to study < *skholē*, school.] **1.** Of or relating to schools

: ACADEMIC. **2.** *often* **Scholastic.** Of or relating to the medieval Schoolmen. **3.** Pedantic : dogmatic. —*n.* **1.** *often* **Scholastic.** A medieval Schoolman. **2.** A dogmatic person : PEDANT. —**scho·las′ti·cal·ly** *adv.*

**scho·las·ti·cism** (skə-lăs′tĭ-sĭz′əm) *n.* **1.** *often* **Scholasticism.** The dominant theological and philosophical school of the High Middle Ages, based on the authority of the early Church fathers and of Aristotle and his commentators. **2. a.** Close adherence to the methods, traditions, and teachings of a particular sect or school. **b.** A pedantic adherence to scholarly methodology.

**scho·li·a** (skō′lē-ə) *n. var. pl. of* SCHOLIUM.

**scho·li·ast** (skō′lē-ăst′) *n.* [LGk. *skholiastēs* < *skholiazein,* to comment on < Gk. *skholion,* scholium.] A commentator, esp. one who annotated the classical authors.

**scho·li·um** (skō′lē-əm) *n., pl.* **-li·ums** or **-li·a** (-lē-ə) [NLat. < Gk. *skholion,* dim. of *skholē,* lecture, school.] **1.** An explanatory note or commentary, as on a Greek or Latin text. **2.** A note amplifying a proof or process, as in mathematics.

**school¹** (skōōl) *n.* [ME *scole* < OE *scōl* < Med. Lat. *scola* < Lat. < Gk. *skholē.*] **1.** An institution for the instruction of children. **2.** An institution for instruction in a skill or business <a cooking *school*> **3. a.** A college or university. **b.** An institution within or associated with a college or university for instruction in a specialized field <the *school* of agriculture> **c.** The student body of an educational institution. **d.** The building or group of buildings in which instruction is given or in which students work and live. **4.** The process of being instructed formally, esp. instruction comprising a planned curriculum over a number of years. **5.** A session of instruction. **6.** A group of persons, esp. philosophers, artists, or writers, whose thought, work, or style demonstrates a common influence or unifying belief. **7.** A group of people joined together by their manners, customs, or opinions. **8.** Education gained by a set of circumstances or experiences <the *school* of hard knocks> **9.** A division made up of several grades or classes in a private school. **10.** The regulations and drill instructions applying to individuals or units in the military. —*vt.* **schooled, school·ing, schools. 1.** To instruct : educate. **2.** To train : discipline.

**school²** (skōōl) *n.* [ME *scole* < MDu. *schōle,* troop.] A large group of aquatic animals, esp. fish, swimming together : SHOAL. —*vi.* **schooled, school·ing, schools.** To swim in or form into a school.

**school age** *n.* The period during which a child is considered able to attend school and is usu. legally required to do so.

**school bag** *n.* A usu. cloth bag for carrying textbooks and school supplies.

**school board** *n.* A local board that oversees public schools.

**school·book** (skōōl′bŏŏk′) *n.* A textbook for use in school.

**school·boy** (skōōl′boi′) *n.* A boy who attends school.

**school bus** *n.* A publicly or privately owned vehicle used to take schoolchildren to and from school or school-related activities.

**school·child** (skōōl′chīld′) *n.* A child who attends school.

**school committee** *n.* A school board.

**school district** *n.* A limited area within a state designated as the administrative unit of a public-school system.

**school edition** *n.* A simplified, abridged, or emended edition of a book issued esp. for use in schools.

**school·girl** (skōōl′gûrl′) *n.* A girl who attends school.

**school·house** (skōōl′hous′) *n.* A building used as a school.

**school·ing** (skōō′lĭng) *n.* **1.** Formal instruction or training given at school. **2.** Training or instruction gained through experience. **3.** The training of a horse or of a horse and rider in specific techniques.

**school·ma'am** (skōōl′măm′, -măm′) *n. var. of* SCHOOLMARM.

**school·man** (skōōl′mən) *n.* **1.** *often* **Schoolman.** A philosopher or theologian in a medieval university. **2.** A professional educator.

**school·marm** (skōōl′märm′) *also* **school·ma'am** (-măm′, -măm′) *n.* [SCHOOL + dial. *marm,* var. of MA'AM.] *Informal.* A woman who is a teacher, esp. a pedantic or old-fashioned one.

**school·mas·ter** (skōōl′măs′tər) *n.* **1.** A man who is a teacher. **2.** A headmaster of a school. **3.** A reddish-brown edible snapper, *Lutjanus apodus* of the tropical Atlantic and Gulf of Mexico.

**school·mate** (skōōl′māt′) *n.* A school companion.

**school·mis·tress** (skōōl′mĭs′trĭs) *n.* **1.** A woman who is a teacher. **2.** A headmistress of a school.

**school·room** (skōōl′rōōm′, -rŏŏm′) *n.* A classroom.

**school·teach·er** (skōōl′tē′chər) *n.* One who teaches in a school below the college level.

**school·work** (skōōl′wûrk′) *n.* Lessons done at school or assigned as homework.

**school year** *n.* The period of the year constituting a complete annual session of school.

**schoo·ner** (skōō′nər) *n.* [Orig. unknown.] **1.** A ship with two or more fore-and-aft-rigged masts. **2.** A large beer glass, usu. holding a pint or more. **3.** A prairie schooner.

**schorl** (shôrl) *n.* [G. *Schörl.*] Tourmaline, esp. black tourmaline. —**schor·la′ceous** (shôr-lā′shəs) *adj.*

**schot·tische** (shŏt′ĭsh, shŏ-tēsh′) *n.* [G. < *schottisch,* Scottish.] **1.** A round dance in 2/4 time. **2.** Music for the schottische.

**schtick** (shtĭk) *n. var. of* SHTICK.

**schuss** (shōōs, shŏŏs) *vi.* **schussed, schuss·ing, schuss·es.** [G. < OHG *scuz,* shot.] To make a fast straight run in skiing. —*n.* **1.** A straight downhill course for skiing. **2.** The act of schussing.

**schuss·boom·er** (shōōs′bōō′mər, shŏŏs′-) *n. Informal.* A skier who is expert at schussing.

**schwa** (shwä) *n.* [G. < Heb. *shĕwā′.*] **1.** A vowel sound that in English often occurs in an unstressed syllable, as the sound of *a* in *alone* or *e* in *linen.* **2.** The symbol (ə) often used to represent a schwa.

**Schwann cell** (shwän) *n.* [After Theodor *Schwann* (1810–1882).] A cell of the neurilemma of a nerve fiber.

**Schwarz·schild radius** (shwôrts′chīld′, shfärts′shĭld) *n.* [After Karl *Schwarzschild* (d. 1916).] The radius of a collapsing celestial object at which gravitational forces exceed the ability of matter and energy to escape, resulting in a black hole.

**sci·at·ic** (sī-ăt′ĭk) *adj.* [OFr. *sciatique* < LLat. *sciaticus* < Lat. *ischiadicus* < Gk. *iskhiadikos* < *iskhion,* hip.] **1.** Of or relating to the ischium. **2.** Of, relating to, or caused by sciatica.

**sci·at·i·ca** (sī-ăt′ĭ-kə) *n.* [ME < Med. Lat. < LLat., fem. of *sciaticus,* of the hip. —see SCIATIC.] **1.** Neuralgia of the sciatic nerve. **2.** Chronic pain in the hip or thigh.

**sciatic nerve** *n.* A sensory and motor nerve running from the sacral plexus through the pelvis and upper leg.

**sci·ence** (sī′əns) *n.* [ME, knowledge, learning < OFr. < Lat. *scientia* < *sciens,* pr.part. of *scire,* to know.] **1. a.** The observation, identification, description, experimental investigation, and theoretical explanation of natural phenomena. **b.** Such activity restricted to a class of natural phenomena. **c.** Such activity applied to any class of phenomena. **2.** Methodological activity, discipline, or study <culinary *science*> **3.** An activity that appears to require study and method. **4.** Knowledge, esp. that gained through experience.

**science fiction** *n.* Fiction in which actual or potential scientific discoveries and developments form part of the plot.

**sci·en·tial** (sī-ĕn′shəl) *adj.* **1.** Of, relating to, or producing knowledge or science. **2.** Capable : efficient.

**sci·en·tif·ic** (sī′ən-tĭf′ĭk) *adj.* [Med. Lat. *scientificus,* producing knowledge : Lat. *scientia,* knowledge + Lat. *facere,* to make.] Of, pertaining to, using, or based on the methodology of science. —**sci′en·tif′i·cal·ly** *adv.*

**scientific empiricism** *n.* The philosophical view that there are no ultimate differences among the various sciences.

**scientific method** *n.* The principles and processes regarded as characteristic of or needed for scientific investigation, including rules for concept formation, conduct of observations and experiments, and validation of hypotheses by observations or experiments.

**scientific notation** *n.* A method of writing numbers in terms of powers of ten; e.g., the number 11,862 would be represented as $1.1862 \times 10^4$.

**sci·en·tism** (sī′ən-tĭz′əm) *n.* The theory that methods used to investigate the natural sciences should be applied in all fields of inquiry. —**sci′en·tis′tic** *adj.*

**sci·en·tist** (sī′ən-tĭst) *n.* **1.** An expert in one or more sciences. **2. Scientist.** A Christian Scientist.

**sci-fi** (sī′fī′) *n. Informal.* Science fiction.

**scil·i·cet** (sĭl′ĭ-sĕt′, skē′lĭ-kĕt′) *adv.* [Lat., contraction of *scire licet,* it is permitted to know.] That is to say : NAMELY.

**scim·i·tar** (sĭm′ĭ-tər, -tär′) *n.* [Ital. *scimitarra.*] A short, curved Oriental sword with an edge only on its convex side.

**scin·til·la** (sĭn-tĭl′ə) *n.* [Lat., spark.] A tiny amount : TRACE.

**scin·til·late** (sĭn′tl-āt′) *v.* **-lat·ed, -lat·ing, -lates.** [Lat. *scintillare, scintillat-* < *scintilla,* spark.] —*vi.* **1.** To emit sparks : FLASH. **2.** To sparkle or shine. **3.** To twinkle, as a star. **4.** To be lively. —*vt.* To give off (sparks or flashes). —**scin′til·lat′ing·ly** *adv.*

**scin·til·la·tion** (sĭn′tl-ā′shən) *n.* **1.** The act or an instance of scintillating. **2.** A spark : flash. **3.** *Astron.* Rapid change in the light emitted by a celestial body, caused by turbulence in the earth's atmosphere. **4.** *Physics.* A flash of light produced in certain media by absorption of an ionizing particle or photon.

**scintillation counter** *n.* A device that detects and records the number of scintillations produced by ionizing radiation.

**scin·ti·scan** (sĭn′tĭ-skăn′) *n.* [SCINTI(LLATION) + SCAN.] A two-dimensional representation of radiation emitted from a radioisotope introduced into a bodily organ. —**scin′ti·scan′ner** *n.*

**sci·o·lism** (sī′ə-lĭz′əm) *n.* [< Lat. *sciolus,* smatterer, dim. of *scius,* knowing < *scire,* to know.] A pretentious or superficial display of scholarship. —**sci′o·list** *n.* —**sci′o·lis′tic** *adj.*

**sci·on** (sī′ən) *n.* [ME < OFr. *cion,* of Germanic orig.] **1.** A descendant or heir. **2.** A detached living shoot or twig used in grafting.

**sci·re fa·ci·as** (sī′rē fā′shē-əs, fā′shəs) *n.* [< Lat., you should cause (him) to know, a phrase that occurs in the writ.] *Law.* **1.** A writ enjoining the party against whom it is issued to appear and show cause why a judicial record should not be enforced, repealed, or annulled. **2.** A judicial proceeding under a scire facias.

**sci·roc·co** (shə-rŏk′ō, sə-) *n. var. of* SIROCCO.

**scir·rhus** (skĭr′əs, sĭr′-) *n., pl.* **scir·rhi** (skĭr′ī′, sĭr′ī′) *or* **scir·rhus·es.** [NLat. < Gk. *skirrhos, skiros* < *skiros,* hard.] A hard cancerous tumor usu. associated with fibrous connective tissue. **—scir′rhous, scir′rhoid′** *adj.*

**scis·sile** (sĭs′əl, -īl′) *adj.* [Fr. < Lat. *scissilis* < *scindere,* to cut.] Able to be cut or split easily.

**scis·sion** (sĭzh′ən, sĭsh′-) *n.* [Fr. < LLat. *scissio* < Lat. *scindere,* to cut.] The act of cutting or splitting : DIVISION.

**scis·sor** (sĭz′ər) *vt.* **-sored, -sor·ing, -sors.** To cut or cut up with scissors or shears. *—n.* Scissors.

**scis·sors** (sĭz′ərz) *pl.n.* [ME *sisoures* < OFr. *cisoires* < Med. Lat. *cisoria,* pl. of LLat. *cisorium,* cutting instrument < Lat. *caedere,* to cut.] *(sing. or pl. in number).* **1.** A cutting implement having two blades joined by a pin that allows the cutting edges to be opened and closed. **2.** A wrestling hold in which the legs are locked about an opponent's head or body. **3.** A gymnastic exercise in which the movement of the legs resembles the opening and closing of scissors.

**scissors kick** *n.* A swimming kick in which the legs are opened and closed like scissors.

**scis·sor·tail** (sĭz′ər-tāl′) *n.* A long-tailed bird, *Muscivora forficata* of the southwestern United States, Mexico, and Central and South America.

**scissortail**
*11½–15 inches long*

**sclaff** (sklăf) *v.* **sclaffed, sclaff·ing, sclaffs.** [Sc., to strike with a flat surface.] *—vi.* To strike the ground with a golf club behind the ball before hitting it. *—vt.* **1.** To hit (a golf club) against the ground before striking the ball. **2.** To strike (the ground) with a golf club before hitting the ball. *—n.* An act of sclaffing. **—sclaff′er** *n.*

**scler-** *pref. var. of* SCLERO-.

**scle·ra** (sklîr′ə) *n.* [NLat. < Gk. *sklēros,* hard.] The tough, white, fibrous outer layer of tissue covering all of the eyeball except the cornea. **—scle′ral** *adj.*

**scler·e·id** (sklĕr′ē-ĭd) *n.* [SCLERE(NCHYMA) + -ID.] A thick-walled, lignified sclerenchyma cell.

**scle·ren·chy·ma** (sklə-rĕng′kə-mə) *n.* Supportive or protective plant tissue composed of thick-walled, usu. lignified cells. **—scle′·ren·chym′a·tous** (sklĭr′ən-kĭm′ə-təs, -kī′mə-) *adj.*

**scle·rite** (sklîr′īt′) *n.* One of the hard outer plates forming part of the exoskeleton of an arthropod.

**scle·ri·tis** (sklə-rī′tĭs) *n.* Inflammation of the sclera. **—scle·rit′ic** (-rĭt′ĭk) *adj.*

**sclero-** *or* **scler-** *pref.* [< Gk. *sklēros,* hard.] **1.** Hard <*sclerite*> **2.** Hardness <*sclerometer*> **3.** Sclera <*scleritis*>

**scle·ro·der·ma** (sklîr′ə-dûr′mə) *n.* A pathological condition in which the skin becomes thick and hard.

**scle·ro·der·ma·tous** (sklîr′ə-dûr′mə-təs) *adj.* **1.** Pertaining to or afflicted with scleroderma. **2.** *Zool.* Having an outer covering of hard plates or bony scales.

**scle·roid** (sklîr′oid′) *adj. Biol.* Hard or hardened : INDURATED.

**scle·ro·ma** (sklə-rō′mə) *n., pl.* **-ma·ta** (-mə-tə) [NLat. < Gk. *sklērōma,* hardening < *sklēroun,* to harden < *sklēros,* hard.] An abnormal hardening of bodily tissue.

**scle·rom·e·ter** (sklə-rŏm′ĭ-tər) *n.* An instrument that determines relative hardness by measuring of the pressure required to pierce or scratch a substance with a standard diamond stylus.

**scle·ro·pro·tein** (sklîr′ō-prō′tēn′, -tē-ĭn) *n.* Any of a large class of proteins found in skeletal and connective tissue.

**scle·rosed** (sklə-rōzd′, -rōst′) *adj.* [< SCLEROSIS.] **1.** Affected with sclerosis. **2.** Lignified.

**scle·ro·sis** (sklə-rō′sĭs) *n., pl.* **-ses** (-sēz′) [ME *sclirosis* < Med. Lat. < Gk. *sklērōsis* < *sklēroun,* to harden. —see SCLEROMA.] **1. a.** A thickening or hardening of a bodily part, as an artery, esp. from disease or excessive growth of tissue. **b.** A disease marked by sclerosis. **2.** *Bot.* The hardening of an outer cell wall by formation or deposit of lignin.

**scle·ro·ti·a** (sklə-rō′shē-ə, -shə) *n. pl. of* SCLEROTIUM.

**scle·rot·ic** (sklə-rŏt′ĭk) *adj.* **1.** Affected with or marked by sclerosis. **2.** *Anat.* Of or relating to the sclera. *—n.* Sclera.

**scle·ro·ti·um** (sklə-rō′shē-əm, -shəm) *n., pl.* **-ti·a** (-shē-ə, -shə) [NLat. < Gk. *sklērotēs,* hardness < *sklēros,* hard.] A dense, hardened mass of filaments in certain fungi, containing stored food and capable of remaining dormant for long periods.

**scler·o·ti·za·tion** (sklĕr′ə-tĭ-zā′shən) *n.* [Gk. *sklērotēs,* hardness (< *sklēros,* hard) + -IZATION.] The process by which an insect's cuticle is hardened by the cross linkage of chitin protein molecules. **—scler′o·tized′** (sklĕr′ə-tīzd′) *adj.*

**scle·rot·o·my** (sklə-rŏt′ə-mē) *n., pl.* **-mies.** Surgical incision of the sclera.

**scle·rous** (sklîr′əs) *adj.* [Gk. *sklēros,* hard.] Hardened : bony.

**scoff** (skŏf, skôf) *v.* **scoffed, scoff·ing, scoffs.** [ME *scoffen* < *scof,* mockery, poss. of Scand. orig.] *—vt.* To mock at or treat with scorn. *—vi.* To treat or express scornfully : MOCK. *—n.* An expression of derision or scorn. **—scoff′er** *n.* **—scoff′ing·ly** *adv.*

**scoff·law** (skŏf′lô′, skôf′-) *n.* One who habitually violates the law, esp. one who fails to pay parking tickets or traffic fines.

**scold** (skōld) *v.* **scold·ed, scold·ing, scolds.** [ME *scolden,* to rail at < *scolde,* an abusive person, prob. of Scand. orig.] *—vt.* To reprove or criticize severely : BERATE. *—vi.* To criticize openly. *—n.* One who scolds others. **—scold′er** *n.* **—scold′ing·ly** *adv.*

**sco·lex** (skō′lĕks′) *n., pl.* **-li·ces** (-lĭ-sēz′) [NLat. < Gk. *skōlēx,* worm.] The knoblike anterior end of a tapeworm, with suckers or hooklike parts that in the parasitic adult stage attach the tapeworm to its host.

**sco·li·o·sis** (skō′lē-ō′sĭs, skŏl′ē-) *n.* [Gk. *skolios,* crooked + -OSIS.] Abnormal lateral curvature of the spine. **—sco′li·ot′ic** (-ŏt′ĭk) *adj.*

**scol·lop** (skŏl′əp) *n. & v. var. of* SCALLOP.

**scol·o·pen·drid** (skŏl′ə-pĕn′drĭd) *n.* [NLat. *Scolopendridae,* family name < Lat. *scolopendra,* millipede < Gk. *skolopendra.*] An arthropod of the family Scolopendridae, which includes the centipedes. **—scol′o·pen′drine′** (-drīn′, -drĭn) *adj.*

**scom·broid** (skŏm′broid′) *adj.* [NLat. *Scombroidei,* suborder name < Lat. *scomber,* mackerel < Gk. *skombros.*] Of or belonging to the suborder Scombroidei, including marine food fishes such as the mackerel or the tuna. *—n.* A scombroid fish.

**sconce¹** (skŏns) *n.* [Du. *schans* < G. *Schanze.*] A small defensive earthwork or fort.

**sconce²** (skŏns) *n.* [ME < OFr. *esconse,* lantern, hiding place < Med. Lat. *sconsa* < Lat. *absconsus,* p.part. of *abscondere,* to hide away : *ab-,* away + *condere,* to hide.] **1.** A decorative wall bracket for candles or electric lights. **2.** *Informal.* The human head or skull.

**scone** (skōn, skŏn) *n.* [Perh. < Du. *schoonbrood,* fine white bread < MDu. *schoonbroot* : *schoon,* bright + *broot,* bread.] A small biscuit-like pastry or quick bread, baked on a griddle or in an oven.

**scoop** (sko͞op) *n.* [ME < MDu. *schope,* shovel.] **1. a.** A shovellike utensil, usu. having a deep curved bowl and a short handle. **b.** The amount a scoop holds. **2.** A ladle : dipper. **3. a.** A thick-handled utensil for serving balls of soft food, as ice cream, often having a sweeping band in the bowl that is levered by the thumb to free the contents. **b.** A portion of food gathered or served with a scoop. **4.** An implement for bailing water from a boat. **5.** A narrow spoon-shaped instrument for surgical extraction in cavities or cysts. **6.** The bucket or shovel of a steam shovel or dredge. **7.** A scooping action or movement. **8.** *Slang.* An exclusive news story. *—vt.* **scooped, scoop·ing, scoops. 1.** To take up or dip into with or as if with a scoop. **2.** To hollow out by digging : EXCAVATE. **3.** To gather or collect swiftly : GRAB <*scooped* up a handful of peanuts> **4.** *Slang.* To top or outmaneuver (a competitor) in reporting an exclusive news story. **—scoop′er** *n.*

**scoot** (sko͞ot) *vi.* **scoot·ed, scoot·ing, scoots.** [Prob. of Scand. orig.] To go quickly and suddenly. **—scoot** *n.*

**scoot·er** (sko͞o′tər) *n.* [< SCOOT.] **1.** A child's vehicle with a long footboard between two small end wheels that is controlled by an upright steering handle attached to the front wheel. **2.** A motor scooter. **3.** A flat-bottomed sailboat with runners for skimming over water or ice.

**scop** (skŏp) *n.* [OE.] An Anglo-Saxon bard or minstrel.

**scope** (skōp, shōp) *n.* [Ital. *scopo,* aim, purpose < Gk. *skopos,* target, aim.] **1.** The range of one's awareness, thoughts, or actions. **2.** Space or opportunity to function. **3.** The area covered by a given activity or subject. **4.** The length or sweep of a mooring cable. **5.** *Informal.* A viewing instrument, as a periscope or telescope.

**-scope** *suff.* [Lat. *-scopium* < Gk. *-skopion* < *skopein,* to see.] An instrument for viewing or observing <*bronchoscope*>

**sco·pol·a·mine** (skō-pŏl′ə-mēn′, -mĭn) *n.* [G. *Scopolamin* : NLat. *Scopolia,* plant genus (after Giovanni *Scopoli,* 1723–1788) + G. *Amin,* amine.] A thick, syrupy, colorless alkaloid, $C_{17}H_{21}NO_4$, extracted from plants of the nightshade family, esp. henbane, and used as a mydriatic, sedative, and truth serum.

**sco·po·line** (skō′pə-lēn′, -lĭn) *n.* [SCOPOL(AMINE) + -INE.] A crystalline alkaloid, $C_8H_{13}O_2N$, derived from scopolamine for use as a narcotic.

**scop·u·la** (skŏp′yə-lə) *n., pl.* **-lae** (-lē′) [LLat., dim. of Lat. *scopa,* twigs.] A brushlike tuft of hairs, as on certain insects. **—scop′u·late′** (-lāt′) *adj.*

**-scopy** *suff.* [Gk. *-skopia* < *skopein,* to see.] Viewing : seeing : observation <*microscopy*>

**scor·bu·tic** (skôr-byo͞o′tĭk) *adj.* [NLat. *scorbuticus* < LLat.

*scorbutus*, scurvy, perh. of Germanic orig.] Of, relating to, resembling, or afflicted with scurvy. **—scor·bu'ti·cal·ly** *adv.*

**scorch** (skôrch) *v.* **scorched, scorch·ing, scorch·es.** [ME *scorchen*, prob. of Scand. orig.] —*vt.* **1.** To burn slightly so as to change in color, flavor, or texture. **2.** To shivel or parch with intense heat. **3.** To subject to harsh criticism or censure. —*vi.* To become scorched or singed. —*n.* **1.** A slight or surface burn. **2.** Discoloration or damage caused by heat. **3.** Brown spotting on plant leaves due to fungi, heat, or lack of water. **—scorch'ing·ly** *adv.*

**scorched-earth policy** (skôrcht'ûrth') *n.* A military policy of devastating or destroying all land and buildings while advancing or retreating, so as to leave nothing salvageable to the enemy.

**scorch·er** (skôr'chər) *n.* **1.** One that scorches. **2.** *Informal.* A very hot day.

**score** (skôr, skōr) *n.* [ME *scor* < OE **scoru* < ON *skor*, notch, twenty.] **1.** A notch or mark, esp. one made to keep a tally. **2.** A usu. numerical record of a competitive event. **3. a.** The total number of points made by each competitor or side in a contest, either at the end of the contest or at a given stage. **b.** The number of points attributed to a competitor or team. **4.** A result, usu. expressed numerically, of a test or examination <a *score* of 82 on the English exam> **5. a.** An amount due : DEBT. **b.** A grievance that is harbored and requires satisfaction : GRUDGE. **6.** A motive or reason. **7.** A group of 20 items. **8. scores.** Large numbers. **9.** The written form of a musical composition for orchestral or vocal parts, either complete or for a particular instrument or voice. **10.** The music composed for a dramatic presentation. **11.** *Informal.* **a.** The act of gaining an advantage, esp. a large or unexpected one. **b.** The act of buying illicit drugs. **c.** A robbery : heist. —*v.* **scored, scor·ing, scores.** —*vt.* **1.** To mark with lines or notches, esp. in order to keep a record. **2.** To cancel or eliminate by or as if drawing a line or lines through. **3.** To mark the surface of (e.g., a roast or ham) with narrow, usu. parallel cuts. **4. a.** To gain (a point) in a competitive event. **b.** To count or be worth as points. **5.** To achieve : win. **6.** To evaluate and assign a grade to. **7.** *Mus.* **a.** To orchestrate. **b.** To arrange for a specific instrument. **8.** To criticize harshly : BERATE. —*vi.* **1.** To make a point in a game or contest. **2.** To keep a tally of the points made in a game or contest. **3.** *Informal.* To gain a purpose or advantage, esp. a large or unexpected one. **—scor'er** *n.*

**score·board** (skôr'bôrd', skōr'bōrd') *n.* A large board for displaying the score of a game.

**score·card** (skôr'kärd', skōr'-) *n.* **1.** A card used for recording the score of game or competition. **2.** A printed program or card identifying the members of competing teams.

**score·keep·er** (skôr'kē'pər, skōr'-) *n.* An official who records the score during a game or competition.

**score·less** (skôr'lĭs, skōr'-) *adj.* Having no points scored.

**sco·ri·a** (skôr'ē-ə, skōr'-) *n., pl.* **sco·ri·ae** (skôr'ē-ē', skōr'-) [ME, dross < Lat. < Gk. *skōria* < *skōr*, excrement.] **1.** *Geol.* Rough crustlike particles of burnt lava. **2.** *Metallurgy.* The refuse of a smelted metal or ore : SLAG. **—sco'ri·a'ceous** (skôr'ē-ā'shəs, skōr'-) *adj.*

**sco·ri·fy** (skôr'ə-fī', skōr'-) *vt.* **-fied, -fy·ing, -fies.** To separate (an ore) into scoria and a precious metal.

**scorn** (skôrn) *n.* [ME *scorne* < OFr. *escarn*, of Germanic orig.] **1. a.** Contempt for a person or object thought to be hateful or inferior : DISDAIN. **b.** Expression of such an attitude : DERISION. **2.** One treated or spoken of with contempt. —*v.* **scorned, scorn·ing, scorns.** —*vt.* **1.** To consider or treat as contemptible or unworthy. **2.** To reject or refuse with derision : CONTEMN. —*vi.* To express contempt : SCOFF. **—scorn'er** *n.* **scorn'ful** *adj.* **—scorn'ful·ly** *adv.* **—scorn'ful·ness** *n.*

**scor·pae·noid** (skôr-pē'noid') *adj.* [NLat. *Scorpaenoidei*, suborder name < *Scorpaena*, type genus < Lat., a kind of fish < Gk. *skorpaina*, fem. of *skorpios*, a sea fish, scorpion.] Of or belonging to the suborder Scorpaenoidei, which includes the scorpion fishes and rockfishes. —*n.* A scorpaenoid fish.

**Scor·pi·o** (skôr'pē-ō') *n.* [Lat. < *scorpio*, scorpion.] **1.** *var. of* SCORPIUS. **2. a.** The eighth sign of the zodiac. **b.** One born under this sign.

**scor·pi·oid** (skôr'pē-oid') *adj.* [Gk. *skorpioeides*, scorpionlike : *skorpios*, scorpion + *-eidēs*, -oid.] **1.** Pertaining to or resembling a scorpion. **2.** *Bot.* Curved or curled like the tail of a scorpion <a *scorpioid* inflorescence>

**scor·pi·on** (skôr'pē-ən) *n.* [ME *scorpioun* < OFr. *scorpion* < Lat. *scorpio* < Gk. *skorpios.*] **1.** An arachnid of the order Scorpionida of warm, dry regions, having a segmented body and an erectile tail tipped with a venomous sting. **2. Scorpion. a.** The constellation Scorpius. **b.** SCORPIO 2a.

**scorpion fish** *n.* Any of numerous small, often brightly colored marine fishes of the family Scorpaenidae, most species of which have venomous spines in the dorsal fin.

**scorpion fly** *n.* An insect of the order Mecoptera, usu. having in the male a genital structure resembling the sting of a scorpion.

**Scor·pi·us** (skôr'pē-əs) *also* **Scor·pi·o** (-pē-ō') *n.* [Lat. *scorpius*, *scorpio*, scorpion, Scorpius < Gk. *skorpios.*] A constellation in the Southern Hemisphere.

**scot** (scŏt) *n.* [ME, tax, partly < ON *skot*, and partly < OFr. *escot*, of Germanic orig.] Money assessed or paid : LEVY.

**Scot** (skŏt) *n.* [ME *Scottes*, Scotsmen < OE *Scottas*, Scotsmen, Irishmen < LLat. *Scotti.*] **1.** A native or resident of Scotland. **2.** A member of the ancient Gaelic tribe that migrated from Ireland to the northern part of Great Britain in about the 6th cent. A.D.

**scot and lot** (skŏt) *n.* A municipal tax formerly levied in Great Britain on the members of a community according to their ability to pay. **—pay scot and lot.** To pay in full.

**scotch[1]** (skŏch) *vt.* **scotched, scotch·ing, scotch·es.** [ME *scocchen*, perh. < AN *escocher*, to notch : *es-* (intensive < Lat. *ex-*) + OFr. *coche*, notch.] **1.** To cut or score. **2.** To damage or injure so as to make harmless : CRIPPLE. **3.** To put an abrupt end to. —*n.* **1.** A superficial cut or abrasion. **2.** A line drawn on or scratched in the ground, as one used in the game of hopscotch.

**scotch[2]** (skŏch) *n.* [Orig. unknown.] A block or wedge placed behind or under an object to prevent rolling or slipping. —*vt.* **scotched, scotching, scotch·es.** To block (e.g., a wheel) with a wedge or chock.

**Scotch** (skŏch) *n.* [Contraction of SCOTTISH.] **1.** *(pl. in number).* The people of Scotland. **2.** Scots. **3.** Scotch whisky. —*adj.* **1.** Of or pertaining to the people, language, or culture of Scotland. **2.** Close with one's money : THRIFTY.

**Scotch-I·rish** (skŏch'ī'rĭsh) *adj.* Of, pertaining to, or characteristic of the people of northern Ireland who are of Scottish descent, esp. those who emigrated to America.

**Scotch·man** (skŏch'mən) *n.* A Scot.

**Scotch terrier** *n.* A Scottish terrier.

**Scotch verdict** *n.* **1.** *Law.* A verdict indicating only that guilt is not proven, allowed in some criminal cases esp. in Scotland. **2.** An inconclusive judgment or pronouncement.

**Scotch whisky** *n.* A smoky-flavored whiskey distilled in Scotland from malted barley.

**Scotch woodcock** *n.* A dish made of scrambled eggs on toast with anchovies or anchovy paste, served as a savory.

**sco·ter** (skō'tər) *n.* [Orig. unknown.] A dark-colored diving duck of the genera *Oidemia* and *Melanitta*, of northern coastal areas.

**scoter**
*18–20 inches long*

**scot-free** (skŏt'frē') *adj.* Free from obligation, harm, punishment, or penalty.

**scot·o·bi·ot·ic** (skŏt'ō-bī-ŏt'ĭk) *adj.* [Gk. *skotos*, darkness + BIOTIC.] Able to survive in darkness.

**sco·to·ma** (skə-tō'mə) *n., pl.* **-mas** or **-ma·ta** (-mə-tə) [NLat. < Med. Lat., dim sight < Gk. *skotōma*, dizziness < *skotoun*, to darken < *skotos*, darkness.] An area of pathologically diminished vision within the visual field. **—sco·to'ma·tous** *adj.*

**sco·to·pi·a** (skə-tō'pē-ə) *n.* [Gk. *skotos*, darkness + -OPIA.] Ability to see in dim light. **—sco·to·pic** (-tō'pĭk, -tŏp'ĭk) *adj.*

**Scots** (skŏts) *adj.* [ME *scottis*, var. of *scottisc*, Scottish.] Scottish. —*n.* The dialect of English used in Scotland.

**Scots·man** (skŏts'mən) *n.* A Scot.

**Scot·ti·cism** (skŏt'ĭ-sĭz'əm) *n.* A word or expression characteristic of Scottish English.

**Scot·tie** (skŏt'ē) *n.* **1.** A Scotsman. **2.** A Scottish terrier.

**Scot·tish** (skŏt'ĭsh) *adj.* [ME *scottisc* < *Scottes*, Scots. —see SCOT.] Of or characteristic of Scotland or of its people or language. —*n.* **1.** Scots. **2.** The people of Scotland.

**Scottish deerhound** *n.* A deerhound.

**Scottish Gaelic** *n.* The Goidelic language of Scotland.

**Scottish rite** *n.* A ceremonial rite in a Masonic system.

**Scottish terrier** *n.* A terrier orig. bred in Scotland, with a short-legged, heavy-set body, a blunt muzzle, and a dark, wiry coat.

**scoun·drel** (skoun'drəl) *n.* [Orig. unknown.] A rascal : villain. **—scoun'drel·ly** *adj.*

**scour[1]** (skour) *v.* **scoured, scour·ing, scours.** [ME *scouren* < MDu. *scūren* < OFr. *escurer* < LLat. *excurare*, to clean out : *ex-*, out < Lat. + *curare*, to clean < Lat. *cura*, care.] —*vt.* **1. a.** To cleanse or polish by scrubbing vigorously. **b.** To remove by scrubbing. **2.** To remove soil or grease from (e.g., wool fibers) by means of a detergent. **3.** To clean (wheat) before the milling process. **4.** To clear (an area) by removing weeds or other vegetation. **5.** To clear (e.g., a channel) by flushing. —*vi.* **1.** To scrub something in order to clean or polish it.

**2.** To have diarrhea. —Used of livestock. —*n.* **1.** A scouring action or effect. **2.** An area that has been scoured, as by flushing with water. **3.** A cleansing agent for wool. **4. scours** *(sing. or pl. in number).* Diarrhea in livestock.

**scour²** (skour) *v.* **scoured, scour·ing, scours.** [ME *scouren,* prob. of Scand. orig.] —*vt.* **1.** To range over (an area) quickly and energetically. **2.** To search through or over thoroughly <*scoured* the market for bargains> —*vi.* **1.** To range about an area, esp. in a search. **2.** To move quickly : RUN.

**scour·er** (skour′ər) *n.* One that scours.

**scourge** (skûrj) *n.* [ME < AN *escorge* < OFr. *escorgier,* to whip < VLat. **excorrigiare* : Lat. *ex-* (intensive) + Lat. *corrigia,* thong.] **1.** A whip used to inflict punishment or pain. **2.** A means of inflicting harsh punishment or suffering. **3.** A cause of severe and widespread affliction, as pestilence or war. —*vt.* **scourged, scourg·ing, scourg·es. 1.** To flog. **2.** To chastise or punish severely. **3.** To devastate : ravage. **—scourg′er** *n.*

**scouring rush** *n.* A horsetail, esp. *Equisetum hyemale,* with rough-ridged stems formerly used to scour utensils.

**scour·ings** (skour′ĭngz) *pl.n.* **1.** The refuse or material removed by scouring. **2.** The least desirable element : DREGS.

**scouse** (skous) *n.* [Short for LOBSCOUSE.] Lobscouse.

**scout¹** (skout) *v.* **scout·ed, scout·ing, scouts.** [< ME *scoute,* act of watching or spying < OFr. *escoute* < *escouter,* to listen < Lat. *auscultare.*] —*vt.* **1.** To watch or explore carefully in order to obtain information, as about a military enemy : RECONNOITER. **2.** To observe and evaluate (a talented person) for a possible position or role. —*vi.* **1.** To search <*scout* around for a note pad on which to write the message> **2.** To search for talented persons. —*n.* **1. a.** A person, aircraft, or ship sent out to obtain information, esp. in preparation for military action. **b.** The action of reconnoitering. **2.** A lookout : sentinel. **3.** One employed to discover and recruit talent, as in sports or entertainment. **4. a.** A member of the Boy Scouts. **b.** A member of the Girl Scouts. **5.** *Chiefly Brit.* A student's servant at Oxford University. **6.** *Informal.* A fellow <a good *scout*> **—scout′er** *n.*

**scout²** (skout) *v.* **scout·ed, scout·ing, scouts.** [Of Scand. orig.] —*vt.* To reject contemptuously. —*vi.* To scoff <*scouted* at the mere suggestion of political reform>

**scout·ing** (skou′tĭng) *n.* Participation in the activities of the Boy Scouts or Girl Scouts.

**scout·mas·ter** (skout′măs′tər) *n.* The adult leader in charge of a troop of Boy Scouts.

**scow** (skou) *n.* [Du. *schouw.*] A large, rectangular, flat-bottomed boat used mainly to transport freight.

**scowl** (skoul) *v.* **scowled, scowl·ing, scowls.** [ME *scoulen,* of Scand. orig.] —*vi.* To wrinkle or contract the brow in order to show anger or disapproval. —*vt.* To express with a scowl. —*n.* A look of anger or disapproval. **—scowl′er** *n.* **—scowl′ing·ly** *adv.*

**scrab·ble** (skrăb′əl) *v.* **-bled, -bling, -bles.** [MDu. *schrabbelen,* freq. of *schrabben,* to scrape.] —*vi.* **1.** To scratch or grope about frantically with the hands. **2.** To ascend with scrambling, disorderly haste. **3.** To make quick, untidy jottings or markings : SCRIBBLE. —*vt.* **1.** To make or gather by scraping together hastily. **2.** To scribble on or over. —*n.* **1.** The act or an instance of scrabbling. **2.** A scribble.

**scrag** (skrăg) *n.* [ME *cragge,* neck < MDu. *crāghe.*] **1.** A bony or scrawny person or thing. **2.** A piece of lean or bony meat, esp. a neck of mutton. **3.** *Slang.* The human neck. —*vt.* **scragged, scrag·ging, scrags.** *Informal.* To wring the neck of : STRANGLE.

**scrag·gly** (skrăg′lē) *adj.* **-gli·er, -gli·est.** [< SCRAG.] Messy : unkempt.

**scrag·gy** (skrăg′ē) *adj.* **-gi·er, -gi·est. 1.** Jagged : rough. **2.** Bony and lean : SCRAWNY. **—scrag′gi·ly** *adv.* **—scrag′gi·ness** *n.*

**scram** (skrăm) *vi.* **scrammed, scram·ming, scrams.** [Short for SCRAMBLE.] *Slang.* To leave immediately.

**scram·ble** (skrăm′bəl) *v.* **-bled, -bling, -bles.** [Perh. blend of obs. *scamble,* to struggle for, and *cramble,* to crawl.] —*vi.* **1.** To move or ascend rapidly, esp. on the hands and knees. **2.** To struggle or contend frantically. **3.** To take off as fast as possible in response to an enemy air attack. —*vt.* **1.** To mix or throw together quickly or haphazardly. **2.** To cook (eggs) by stirring while frying. **3.** *Electron.* To distort or garble (a signal) in order to make it unintelligible without a special receiver. —*n.* **1.** The act or an instance of scrambling. **2.** A difficult hike or climb over rough terrain. **3.** An undignified scuffle or struggle. **4.** A swift takeoff of military aircraft in response to an enemy attack.

**scrambled eggs** *pl.n.* **1.** Eggs fried with the yolks and whites mixed together. **2.** *Slang.* The gold braid worn on a military cap.

**scram·bler** (skrăm′blər) *n.* **1.** One that scrambles. **2.** An electronic device that distorts or garbles telecommunication signals to make them unintelligible to an eavesdropper.

**scram·jet** (skrăm′jĕt′) *n.* [S(UPERSONIC) + C(OMBUSTION) + RAMJET.] A ramjet airplane engine that burns fuel in the supersonic airstream produced by the plane after reaching supersonic speeds by conventional means.

**scrap¹** (skrăp) *n.* [ME < ON *skrap,* trifles.] **1.** A small section or bit : FRAGMENT. **2. scraps.** Leftover pieces of food. **3.** Discarded waste material, esp. metal suitable for reprocessing. **4. scraps.** Crisp pieces of rendered animal fat : CRACKLINGS. —*vt.* **scrapped, scrap·ping, scraps. 1.** To separate into parts for disposal or salvage. **2.** To abandon or discard as worthless : JUNK.

**scrap²** (skrăp) [Orig. unknown.] *Slang.* —*vi.* **scrapped, scrap·ping, scraps.** To fight, often with the fists. —*n.* A fight or scuffle : BRAWL. **—scrap′per** *n.*

**scrap·book** (skrăp′bo͝ok′) *n.* A book with blank pages for the mounting of pictures or other mementos.

**scrape** (skrāp) *v.* **scraped, scrap·ing, scrapes.** [ME *scrapen* < ON *skrapa.*] —*vt.* **1.** To rub (a surface) with heavy pressure. **2.** To push or draw (a hard or abrasive object) forcefully over a surface. **3.** To remove from a surface by or as if by rubbing. **4.** To clean, smooth, or abrade by rubbing with a sharp or rough instrument. **5.** To injure the surface of by rubbing against something rough or sharp. **6.** To gather together or produce with difficulty <*scrape* up some new information> —*vi.* **1.** To come into sliding, abrasive contact. **2.** To rub or move with a harsh grating noise. **3.** To emit a harsh grating noise. **4.** To practice petty economies : SCRIMP. **5.** To succeed or manage with difficulty <barely *scraped* through> —*n.* **1.** The act or an instance of scraping. **2.** The sound of scraping. **3.** An abrasion on the skin. **4.** *Slang.* **a.** An embarrassing predicament. **b.** A fight : scuffle. **—scrap′er** *n.*

**scrap heap** *n.* **1.** A pile of scrapped metal. **2.** A place for discarding unwanted or worthless material.

**scra·pie** (skrā′pē, skrăp′ē) *n.* [< SCRAPE.] A usu. fatal virus disease of sheep marked by degeneration of the central nervous system.

**scrap·ple** (skrăp′əl) *n.* [Dim. of SCRAP¹.] A mush of pork scraps and cornmeal allowed to set and then sliced and fried.

**scrap·py¹** (skrăp′ē) *adj.* **-pi·er, -pi·est.** Made up of scraps : FRAGMENTARY. **—scrap′pi·ly** *adv.* **—scrap′pi·ness** *n.*

**scrap·py²** (skrăp′ē) *adj.* **-pi·er, -pi·est. 1.** Quarrelsome : contentious. **2.** Full of fighting spirit : AGGRESSIVE. **—scrap′pi·ly** *adv.* **—scrap′pi·ness** *n.*

**scratch** (skrăch) *v.* **scratched, scratch·ing, scratch·es.** [ME *scracchen,* prob. blend of *scratten,* to scratch, and *cracchen,* to scratch < MDu. *cratsen,* to scrape.] —*vt.* **1.** To make a narrow line or mark on (a surface) with a sharp instrument. **2.** To use the nails or claws to dig or scrape at. **3.** To rub or scrape (the skin) to alleviate itching. **4.** To scrape or strike on an abrasive surface. **5.** To write or draw hurriedly. **6.** To remove or cancel (e.g., a word) by or as if by drawing lines through. **7.** To withdraw (an entry) from a contest. —*vi.* **1.** To use the nails or claws to dig, scrape, or injure. **2.** To rub or scrape the skin to alleviate itching. **3.** To make a rough scraping sound. **4.** To gather funds or make a living with difficulty. **5.** To withdraw from a contest. **6.** To make a billiards shot that results in a penalty, as when the cue ball falls into a pocket or jumps the cushion. —*n.* **1. a.** A line or mark made by scratching. **b.** A slight wound. **2.** A hasty scribble. **3.** A harsh sound made by scratching. **4.** The starting line for a race. **5.** One who has been withdrawn from a contest. **6. a.** The act of scratching in billiards. **b.** An accidently successful shot in billiards : FLUKE. **7.** Feed for poultry. **8.** *Slang.* Money. —*adj.* **1.** Done by chance. **2.** Gathered together at random : HAPHAZARD. **3.** Without a golf handicap. **—from scratch.** From the beginning. **—up to scratch.** *Informal.* **1.** Fulfilling the requirements. **2.** In good condition. **—scratch′er** *n.*

**scratch hit** *n. Baseball.* A batted ball counted as a hit although it is not squarely struck or cleanly fielded.

**scratch line** *n.* **1.** A starting line for a race. **2.** A line beyond which a contestant must not step.

**scratch·pad** (skrăch′păd′) *n.* A usu. high-speed internal register used for temporary storage in a computer memory.

**scratch sheet** *n. Slang.* A dope sheet.

**scratch test** *n.* An allergy test made by scratching the skin and applying an allergen to the wound.

**scratch·y** (skrăch′ē) *adj.* **-i·er, -i·est. 1.** Marked by or consisting of scratches. **2.** Making a scratching noise. **3.** Uneven or irregular. **4.** Harsh and irritating. **—scratch′i·ly** *adv.* **—scratch′i·ness** *n.*

**scrawl** (skrôl) *v.* **scrawled, scrawl·ing, scrawls.** [Perh. < obs. *scrawl,* to gesticulate.] —*vt.* To write quickly and often illegibly. —*vi.* To write in a sprawling, irregular manner. —*n.* Irregular, often illegible handwriting. **—scrawl′er** *n.* **—scrawl′y** *adj.*

**scraw·ny** (skrô′nē) *adj.* **-ni·er, -ni·est.** [Orig. unknown.] Thin and bony. **—scraw′ni·ness** *n.*

**screak** (skrēk) *vi.* **screaked, screak·ing, screaks.** [ME *skricken* < ON *skrækja.*] To screech. **—screak** *n.* —*n.* A screech. **—screak′y** *adj.*

**scream** (skrēm) *v.* **screamed, scream·ing, screams.** [ME *scremen,* perh. of Scand. orig.] —*vi.* **1.** To produce a loud piercing cry, as of pain. **2.** To make a loud piercing sound. **3.** To write or speak in an angry, hysterical manner. **4.** To have or produce a startling effect. —*vt.* To utter or say in or as if in a screaming voice. —*n.* **1.** A loud piercing cry or sound. **2.** *Slang.* One that is considered to be extremely funny.

**scream·er** (skrē′mər) *n.* **1.** One that screams. **2.** *Slang.* A large or sensational headline. **3.** *Slang.* Something that evokes screams or laughter. **4.** A large aquatic bird of the family Anhimidae of South America, having a harsh, resonant call.

**scree** (skrē) *n.* [Prob. ult. < ON *skriða*, landslide.] **1.** Small stones or rock debris. **2.** A slope of scree at the base of an incline or cliff.

**screech** (skrēch) *v.* **screeched, screech·ing, screech·es.** [ME *scrichen*, to screech < ON *skrækja*.] —*vi.* **1.** To scream in a shrill, strident voice. **2.** To make a prolonged shrill, grating noise. —*vt.* To utter or say in or as if in a screeching voice. —*n.* A shrill, harsh cry : SHRIEK. **—screech′er** *n.* **—screech′y** *adj.*

**screech owl** *n.* A small owl of the genus *Otus*, esp. *O. asio* of North America, with ear tufts and a quavering, high-pitched call.

**screed** (skrēd) *n.* [ME *screde*, fragment < OE *scrēade*.] **1.** A long, repetitious tirade or piece of writing. **2.** A strip of wood, plaster, or metal placed on a wall or pavement as a guide for the even application of plaster or concrete.

**screen** (skrēn) *n.* [ME *screne* < OFr. *escren* < MDu. *scherm*, shield.] **1.** A movable device, as a panel or hinged framework, designed to divide, conceal, or protect. **2.** Something that serves to divide, conceal, or protect <a dense *screen* of bushes> **3. a.** A coarse sieve used to separate fine particles, as of sand, gravel, or coal, from larger ones. **b.** A system for evaluating and selecting personnel. **4.** A frame fitted with wire or plastic mesh and inserted in a window or door to keep out insects. **5.** The surface on which a picture is projected for viewing. **6.** *Electron.* The phosphorescent surface upon which the image is formed in a cathode-ray tube. **7.** A glass plate crisscrossed with lines, placed in front of the lens of a camera when photographing for halftone reproduction in printing. —*vt.* **screened, screen·ing, screens. 1.** To provide with a screen. **2. a.** To conceal from view. **b.** To protect, guard, or shield. **3. a.** To separate by means of a sieve or screen. **b.** To examine carefully in order to determine suitability <*screened* the applicants for the position> **4.** To show or project (e.g., a motion picture) on a screen. **—screen′er** *n.*

**screen·ing** (skrē′nĭng) *n.* **1. screenings** *(sing. or pl. in number).* Refuse, as waste coal, separated out by a screen. **2.** The wire or plastic meshwork used to make door or window screens. **3.** A presentation of a motion picture.

**screen·land** (skrēn′lănd′) *n.* The motion-picture industry.

**screen·play** (skrēn′plā′) *n.* The script for a motion picture.

**screen test** *n.* A short motion-picture sequence used to judge the talent or suitability of an aspiring actor. **—screen′-test′** *v.* **(-test·ed, -test·ing, -tests).**

**screen·writ·er** (skrēn′rī′tər) *n.* One who writes screenplays.

**screw** (skrōō) *n.* [ME *skrewe* < OFr. *escrove*, female screw, nut < Med. Lat. *scrofa* < Lat., sow.] **1. a.** A cylindrical rod incised with one or more helical or advancing spiral threads. **b.** The tapped collar or socket that receives such a rod. **2.** A metal pin with incised threads and a broad notched head that can be driven by turning with a screwdriver, esp.: **a.** A tapered and sharp-pointed wood screw. **b.** A cylindrical and flat-tipped machine screw. **3.** A device having a helical form, as a corkscrew. **4.** A propeller. **5.** A twist or turn of or as if of a screw. **6.** *Chiefly Brit.* Salary : wages. **7.** *Chiefly Brit.* A small paper packet, as of tobacco or salt. **8.** *Chiefly Brit.* An old broken-down horse : NAG. **9.** *Chiefly Brit.* A crafty bargainer. **10.** *Slang.* **a.** A prison guard. **b.** The turnkey of a jail. —*v.* **screwed, screw·ing, screws.** —*vt.* **1.** To turn or tighten (a screw). **2. a.** To fasten, attach, or tighten by or as if by means of a screw. **b.** To attach (a tapped or threaded fitting or cap) by twisting into place. **c.** To rotate (a part) on a threaded axis. **3.** To twist out of shape : CONTORT. **4.** *Slang.* To take advantage of : CHEAT. —*vi.* **1.** To turn or twist. **2. a.** To become attached by means of screw threads. **b.** To be capable of such attachment. **—screw up.** *Slang.* To make a mess of (an undertaking). **—screw′er** *n.*

**screw·ball** (skrōō′bôl′) *n.* **1.** *Baseball.* A pitched ball that curves in the direction opposite to that of a normal curve ball. **2.** *Slang.* One who is eccentric, unconventionally whimsical, or irrational.

**screw bean** *n.* **1.** A mesquite, *Prosopis pubescens* of the southwestern United States, with small yellowish-white flowers and twisted pods used as fodder. **2.** The pod of the screw bean.

**screw cap** *n.* A cap that screws onto the threaded mouth of a container, as a glass jar or medicine bottle.

**screw·driv·er** (skrōō′drī′vər) *n.* **1.** A tool used to turn screws. **2.** A cocktail of vodka and orange juice.

**screw eye** *n.* A wood screw with an eyelet in place of a head.

**screw jack** *n.* A jackscrew.

**screw pine** *n.* The pandanus.

**screw propeller** *n.* A propeller.

**screw thread** *n.* **1.** The continuous spiral groove on a screw or the inner surface of a nut. **2.** One complete turn of a screw thread.

**screw-up** (skrōō′ŭp′) *n.* *Slang.* **1.** One who screws up : BUNGLER. **2.** A clumsy mistake or disastrous situation.

**screw·worm** (skrōō′wûrm′) *n.* The larva of the screwworm fly.

**screwworm fly** *n.* A blue-green New World fly, *Cochliomyia hominivorax*, that breeds in the living tissue of mammals, having penetrated mainly through open wounds, and whose larvae cause serious injury or death to livestock.

**screw·y** (skrōō′ē) *adj.* **-i·er, -i·est.** *Slang.* **1.** Eccentric : crazy. **2.** Ludicrously odd, unusual, or inappropriate. **—screw′i·ness** *n.*

**scrib·ble** (skrĭb′əl) *v.* **-bled, -bling, -bles.** [ME *scriblen* < Med. Lat. *scribillare*, freq. of Lat. *scribere*, to write.] —*vt.* **1.** To write quickly without regard for legibility or style. **2.** To cover with scribbles, doodles, or meaningless notations. —*vi.* To write or draw in a careless, hurried way. —*n.* **1.** Careless, hurried writing. **2.** Meaningless lines or notations.

**scrib·bler** (skrĭb′lər) *n.* **1.** One who scribbles. **2.** A minor or disreputable author.

**scribe** (skrīb) *n.* [ME < Lat. *scriba* < *scribere*, to write.] **1.** A public secretary or clerk. **2.** A professional copyist of manuscripts and documents. **3.** An author or journalist. **4.** A scriber. —*v.* **scribed, scrib·ing, scribes.** —*vt.* **1.** To mark with a scriber. **2.** To write or inscribe. —*vi.* To work as a scribe. **—scrib′al** *adj.*

**scrib·er** (skrī′bər) *n.* A sharp-pointed tool used to make lines, as on wood, metal, or ceramic.

**scrim** (skrĭm) *n.* [Orig. unknown.] **1.** A durable, loosely woven cotton or linen fabric used for curtains or upholstery lining or in industry. **2.** A transparent fabric used as a theater backdrop or curtain.

**scrim·mage** (skrĭm′ĭj) *n.* [Obs. *scrimish*, alteration of SKIRMISH.] **1. a.** A rough-and-tumble struggle : SCUFFLE. **b.** A skirmish. **2.** *Football.* **a.** The action from the time the ball is snapped until it is out of play. **b.** A team's practice session. **3.** A Rugby scrummage. —*vi.* **-maged, -mag·ing, -mag·es.** *Football.* To engage in a scrimmage.

**scrimp** (skrĭmp) *v.* **scrimped, scrimp·ing, scrimps.** [Perh. of Scand. orig.] —*vi.* To economize sharply. —*vt.* **1.** To be excessively sparing with or of. **2.** To cut or make too small or scanty. **—scrimp′i·ness** *n.* **—scrimp′y** *adj.*

**scrim·shaw** (skrĭm′shô′) *n.* [Orig. unknown.] **1.** The art of carving or incising elaborate designs on whalebone or whale ivory. **2.** A decorative article made by scrimshaw. —*v.* **-shawed, -shaw·ing, -shaws.** —*vt.* To decorate (whalebone or whale ivory) with elaborate carvings or designs. —*vi.* To make scrimshaw.

**scrip**[1] (skrĭp) *n.* [Alteration of SCRIPT.] **1.** A small piece of paper, esp. a short list or schedule. **2.** Paper money issued for temporary emergency use.

**scrip**[2] (skrĭp) *n.* [Short for *subscription receipt*, receipt for a portion of a loan.] **1.** A provisional certificate entitling the holder to a fractional share of stock or of other jointly owned property. **2.** Scrip certificates collectively.

**scrip**[3] (skrĭp) *n.* [ME *scrippe* < Med. Lat. *scrippa*.] *Archaic.* A small wallet, satchel, or bag.

**script** (skrĭpt) *n.* [ME *skript*, a piece of writing < OFr. *escript* < Lat. *scriptum*, neuter p.part. of *scribere*, to write.] **1. a.** Handwriting. **b.** A style of writing with cursive characters. **c.** Alphabet. **2.** A typeface that resembles handwriting. **3.** *Law.* An original document. **4.** The text of a play, broadcast, or motion picture. —*vt.* **script·ed, script·ing, scripts.** To prepare (a text) for or from.

**scrip·to·ri·um** (skrĭp-tôr′ē-əm, -tōr′-) *n., pl.* **-to·ri·ums** or **-to·ri·a** (-tôr′ē-ə, -tōr′-) [Med. Lat. < Lat. *scriptus*, p.part. of *scribere*, to write] A room in a monastery reserved for the copying, writing, or illuminating of manuscripts and records.

**scrip·tur·al** (skrĭp′chər-əl) *adj.* **1.** Of or relating to writing : WRITTEN. **2. Scriptural.** Of, relating to, based on, or found in the Scriptures. **—scrip′tur·al·ly** *adv.*

**Scrip·ture** (skrĭp′chər) *n.* [ME < LLat. *scriptura* < Lat., act of writing < *scriptus*, p.part. of *scribere*, to write.] **1. a.** A sacred writing or book. **b.** A passage from such a writing or book. **2.** *often* **Scriptures.** The Holy Scriptures. **3. scripture.** A statement or text regarded as authoritative.

**script·writ·er** (skrĭpt′rī′tər) *n.* One who writes scripts.

**scriv·en·er** (skrĭv′ə-nər, skrĭv′nər) *n.* [ME *scriveiner* < *scrivein* < OFr. *escrevein* < LLat. *scribanus* < Lat. *scriba*, scribe.] *Archaic.* **1.** A scribe. **2.** A notary public.

**scro·bic·u·late** (skrō-bĭk′yə-lĭt, -lāt′) *adj.* [< Lat. *scrobiculus*, dim. of *scrobis*, trench.] *Biol.* Covered or marked with numerous shallow depressions, grooves, or pits.

**scrod** (skrŏd) *n.* [Poss. < obs. Du. *schrood*, shred.] A young cod or haddock, esp. one split and boned for cooking.

**scrof·u·la** (skrŏf′yə-lə) *n.* [Med. Lat. < Lat. *scrofulae* (pl.), swelling of the glands < *scrofa*, sow.] A rare constitutional disease affecting the tissues of the young and marked by predisposition to tuberculosis, lymphatism, glandular swellings, and respiratory catarrhs.

**scrof·u·lous** (skrŏf′yə-ləs) *adj.* **1.** Relating to, affected with, or resembling scrofula. **2.** Morally degenerate : CORRUPT. **—scrof′u·lous·ly** *adv.* **—scrof′u·lous·ness** *n.*

**scroll** (skrōl) *n.* [ME *scrowle*, alteration of *scrowe* < OFr. *escroue*, strip of parchment, of Germanic orig.] **1.** A roll, as of parchment or papyrus, used esp. for writing. **2. a.** *Archaic.* A piece of writing, as a letter. **b.** A list of names : ROSTER. **3.** An ornament or ornamental design that resembles a partially rolled scroll of paper. **4.** The curved head on an instrument of the violin family.

**scroll saw** *n.* A hand or power tool with a narrow ribbonlike blade used to cut curved or irregular shapes.

**scroll·work** (skrōl′wûrk′) *n.* Decoration with a scroll motif, esp. that which is executed in wood with a scroll saw.
**Scrooge** (skro͞oj) *n.* [After Ebenezer *Scrooge,* a miserly character in *A Christmas Carol* by Charles Dickens (1812–1870).] A petty, mean-tempered miser.
**scro·tum** (skrō′təm) *n., pl.* **-ta** (-tə) *or* **-tums.** [Lat.] The external sac of skin that encloses the testes in most mammals. **—scro′tal** (skrōt′l) *adj.*
**scrounge** (skrounj) *v.* **scrounged, scroung·ing, scroung·es.** [Alteration of dial. *scrunge,* to steal.] *Slang.* —*vt.* **1.** To obtain or gather by foraging or salvaging. **2.** To wheedle : cadge. —*vi.* **1.** To forage about in an effort to acquire something at no cost. **2.** To wheedle. **—scroung′er** *n.*
**scrub¹** (skrŭb) *v.* **scrubbed, scrub·bing, scrubs.** [ME *scrobben,* to currycomb a horse < MDu. *schrobben,* to clean by rubbing.] —*vt.* **1. a.** To rub hard in order to clean. **b.** To remove (dirt or stains) by hard rubbing. **2.** To cleanse (a gas). **3.** *Slang.* To cancel : drop <*scrubbed* their plans due to bad weather> —*vi.* To clean or wash something by hard rubbing. —*n.* An act or instance of scrubbing.
**scrub²** (skrŭb) *n.* [ME, prob. of Scand. orig.] **1. a.** A stunted tree or shrub. **b.** A growth or tract of stunted vegetation. **2.** An undersized or underdeveloped domestic animal. **4.** One who is undersized or insignificant. **5.** A player not on a varsity or first team.
**scrub·ber** (skrŭb′ər) *n.* **1.** One that scrubs. **2.** An air-pollution control device that traps pollutants and cools emissions, as from a smokestack.
**scrub·by** (skrŭb′ē) *adj.* **-bi·er, -bi·est. 1.** Covered with or made up of scrub or underbrush. **2.** Stunted or irregular. **3.** Paltry : shabby. **—scrub′bi·ness** *n.*
**scrub pine** *n.* A small, irregularly shaped pine tree, esp. *Pinus virginiana* of the eastern United States, bearing prickly cones.
**scrub typhus** *n.* An acute infectious disease common in Asia, transmitted by a mite and marked by sudden fever, painful swelling of the lymphatic glands, skin lesions, and skin rash.
**scrub·wom·an** (skrŭb′wo͝om′ən) *n.* A woman who is hired to clean : CHARWOMAN.
**scruff** (skrŭf) *n.* [Orig. unknown.] The back of the neck : NAPE.
**scruf·fy** (skrŭf′ē) *adj.* **-fi·er, -fi·est.** [< obs. *scruff,* scurf.] **1.** Shabby : unkempt. **2.** *Chiefly Brit.* Scaly : scabby.
**scrum** (skrŭm) *Informal.* —*n.* A scrummage. —*vi.* **scrummed, scrum·ming, scrums.** To engage in a scrummage.
**scrum·mage** (skrŭm′ĭj) *n.* [Alteration of SCRIMMAGE.] A Rugby formation in which the two sets of forwards mass together around the ball and, with their heads down, try to shoulder their opponents off the ball so they can kick it to their own team. —*vi.* **-maged, -mag·ing, -mag·es.** To engage in a scrummage. **—scrum′mag·er** *n.*
**scrump·tious** (skrŭmp′shəs) *adj.* [Perh. alteration of SUMPTUOUS.] *Slang.* Delicious : delightful. **—scrump′tious·ly** *adv.*
**scrunch** (skrŭnch, skro͞onch) *v.* **scrunched, scrunch·ing, scrunch·es.** [Alteration of CRUNCH.] —*vt.* **1.** To crush or crunch. **2.** To hunch, as the shoulders. —*vi.* **1.** To hunch. **2.** To move with or produce a crunching sound. —*n.* A crunching sound.
**scru·ple** (skro͞o′pəl) *n.* [OFr. *scrupule* < Lat. *scrupulus,* small stone, scruple < *scrupus,* rough stone.] **1.** A dictate of conscience or ethical principle that inhibits action. **2.** A unit of apothecary weight equal to approx. 1.3 grams or 20 grains. **3.** A tiny part or amount. —*vi.* **-pled, -pling, -ples.** To hesitate because of conscience or principle.
**scru·pu·lous** (skro͞o′pyə-ləs) *adj.* [ME < Lat. *scrupulosus* < *scrupulus,* scruple.] **1.** Having scruples : PRINCIPLED. **2.** Conscientious : exacting. **—scru′pu·los′i·ty** (-lŏs′ĭ-tē), **scru′pu·lous·ness** *n.* **—scru′pu·lous·ly** *adv.*
**scru·ta·ble** (skro͞o′tə-bəl) *adj.* [Med. Lat. *scrutabilis,* searchable < Lat. *scrutari,* to search. —see SCRUTINY.] Capable of being understood : COMPREHENSIBLE.
**scru·ti·nize** (skro͞ot′n-īz′) *vt.* **-nized, -niz·ing, -niz·es.** To inspect or observe carefully and critically. **—scru′ti·niz′er** *n.* **—scru′ti·niz′ing·ly** *adv.*
**scru·ti·ny** (skro͞ot′n-ē) *n., pl.* **-nies.** [Lat. *scrutinium* < *scrutari,* to search, examine < *scruta,* trash.] **1.** Close, careful inspection or study. **2.** Close observation : SURVEILLANCE.
**scu·ba** (sko͞o′bə, skyo͞o′-) *n.* [S(ELF) + C(ONTAINED) + U(NDERWATER) + B(REATHING) + A(PPARATUS).] An apparatus containing compressed air and used for underwater breathing while swimming.
**scuba diver** *n.* One who uses scuba gear in underwater swimming.
**scud** (skŭd) *vi.* **scud·ded, scud·ding, scuds.** [Prob. of Scand. orig.] **1.** To run or move along quickly and easily. **2.** *Naut.* To run before a gale with little or no sail set. —*n.* **1.** The act of scudding. **2. a.** Wind-driven clouds, snow, mist, or rain. **b.** A sudden light shower.
**scu·do** (sko͞o′dō) *n., pl.* **-di** (-dē) [Ital. < Lat. *scutum,* shield.] A monetary unit and gold coin formerly used in Italy and Sicily.
**scuff** (skŭf) *v.* **scuffed, scuff·ing, scuffs.** [Prob. of Scand. orig.] —*vi.* To walk without raising the feet : SHUFFLE. —*vt.* **1.** To scrape with the feet while walking. **2.** To shuffle (the feet), as in embarrassment. **3.** To scrape and roughen the surface of. —*n.* **1.** The act or sound of scuffing. **2.** A mark or worn spot resulting from scuffing. **3.** A flat-soled, backless house slipper.
**scuf·fle¹** (skŭf′əl) *vi.* **-fled, -fling, -fles.** [Prob. of Scand. orig.] **1.** To fight or struggle roughly and confusedly at close quarters. **2. a.** To shuffle. **b.** To walk quickly with shuffling steps. —*n.* A rough, confused struggle at close quarters. **—scuf′fler** *n.*
**scuf·fle²** (skŭf′əl) *n.* [Du. *schoffel,* hoe for weeding.] A hoe that is pushed rather than pulled.
**scull** (skŭl) *n.* [ME *sculle.*] **1.** A long oar mounted on the stern of a boat and twisted from side to side to propel it. **2.** One of a pair of short-handled oars used by a single rower. **3.** A light racing boat for one, two, or four oarsmen. —*v.* **sculled, scull·ing, sculls.** —*vt.* To propel (a boat) with a scull. —*vi.* To use a scull to propel a boat. **—scull′er** *n.*

scull

**scul·ler·y** (skŭl′ə-rē) *n., pl.* **-ies.** [ME < OFr. *escuelerie* < *escuelier,* keeper of dishes < *escuele,* dish < Lat. *scutella,* salver, dim. of *scutra,* platter.] A room adjoining a kitchen where dishwashing and other rough kitchen chores are done.
**scul·lion** (skŭl′yən) *n.* [ME *sculyon,* prob. < OFr. *escovillon,* dishcloth, dim. of *escouvee,* broom < Lat. *scopa.*] *Archaic.* A servant employed to do menial kitchen tasks.
**scul·pin** (skŭl′pĭn) *n., pl.* **-pins** *or* **sculpin.** [Orig. unknown.] **1.** A marine or freshwater fish of the family Cottidae, with a large flattened head and prominent spines. **2.** A scorpion fish, *Scorpaena guttata* of California coastal waters.
**sculpt** (skŭlpt) *vt.* **sculpt·ed, sculpt·ing, sculpts.** [Fr. *sculpter, sculper* < Lat. *sculpere,* to carve.] To sculpture.
**sculp·tor** (skŭlp′tər) *n.* [Lat. < *sculptus,* p.part. of *sculpere,* to carve.] **1.** One who produces sculptures. **2. Sculptor.** A constellation in the Southern Hemisphere.
**sculp·tress** (skŭlp′trĭs) *n.* A woman who sculptures.
**sculp·ture** (skŭlp′chər) *n.* [ME < Lat. *sculptura* < *sculptus,* p.part. of *sculpere,* to carve.] **1.** The art or practice of creating three-dimensional figures or designs, as by chiseling marble, modeling clay, or casting in metal. **2. a.** An artwork created by sculpture. **b.** Such works as a group. **3.** Ridges or other markings, as on a shell, formed by natural processes. —*vt.* **-tured, -tur·ing, -tures. 1.** To fashion into a three-dimensional figure. **2.** To depict in sculpture. **3.** To decorate with sculpture or sculptured form. **4.** To change the shape or contour of naturally, as by erosion. **—sculp′tur·al** *adj.* **—sculp′tur·al·ly** *adv.*
**sculp·tur·esque** (skŭlp′chə-rĕsk′) *adj.* Suggesting or resembling sculpture. **—sculp′tur·esque′ly** *adv.*
**scum** (skŭm) *n.* [ME < MDu. *schūm.*] **1.** A filmy layer of impurities or extraneous material that forms on the surface of a liquid or body of water. **2.** The dross of molten metals. **3.** Refuse or worthless matter. **4.** A vile or worthless person or group of persons. —*v.* **scummed, scum·ming, scums.** —*vt.* To remove the scum from. —*vi.* To form or become covered with scum. **—scum′mer** *n.*
**scum·ble** (skŭm′bəl) *vt.* **-bled, -bling, -bles.** [Freq. of SCUM.] To soften the colors or outlines of (a painting) by applying with a film of opaque or semiopaque color or by rubbing. —*n.* **1.** The effect created by scumbling. **2.** Material used for scumbling.
**scun·ner** (skŭn′ər) *n.* [ME *skunner.*] Extreme dislike : AVERSION.
**scup** (skŭp) *n., pl.* **scup** *or* **scups.** [Narraganset *mishcup.*] A food fish, *Stenotomus chrysops* of western Atlantic waters, resembling the porgy.
**scup·per** (skŭp′ər) *n.* [ME *skopper.*] *Naut.* An opening at deck level in the side of a ship that allows water to run off.
**scup·per·nong** (skŭp′ər-nông′, -nŏng′) *n.* [After the *Scuppernong* River, North Carolina.] **1.** The muscadine grape, esp. a cultivated variety with sweet, yellowish fruit. **2.** A sweet white wine made from scuppernongs.
**scurf** (skûrf) *n.* [ME, of Scand. orig.] **1.** Scaly or flaky dry skin, as dandruff. **2.** A scaly crust covering a surface, esp. of a plant. **—scurf′i·ness** *n.* **—scurf′y** *adj.*
**scur·rile** *also* **scur·ril** (skûr′əl, skŭr′-) *adj.* [Fr. < Lat. *scurrilis,* jeering < *scurra,* buffoon.] *Archaic.* Scurrilous.

**scur·ril·i·ty** (skə-rĭl′ĭ-tē) *n., pl.* **-ties. 1.** The quality of being scurrilous. **2.** A scurrilous statement.

**scur·ri·lous** (skûr′ə-ləs, skŭr′-) *adj.* **1.** Tending to use vulgar or abusive language : FOUL-MOUTHED. **2.** Expressed in or containing vulgar or abusive language. **—scur′ri·lous·ly** *adv.* **—scur′ri·lous·ness** *n.*

**scur·ry** (skûr′ē, skŭr′ē) *vi.* **-ried, -ry·ing, -ries.** [Prob. short for HURRY-SCURRY.] **1.** To move with or as if with light running steps : SCAMPER. **2.** To flurry or swirl about. *—n., pl.* **-ries.** An act or the sound of scurrying.

**scur·vy** (skûr′vē) *n.* [< SCURF.] A disease resulting from a deficiency of vitamin C and marked by spongy and bleeding gums, bleeding under the skin, and extreme weakness. *—adj.* **-vi·er, -vi·est. 1.** Worthless : contemptible. **2.** *Obs.* Scurfy. **—scur′vi·ly** *adv.* **—scur′vi·ness** *n.*

**scurvy grass** *n.* A plant, *Cochlearia officinalis* of northern regions, with bitter foliage that was formerly thought to cure scurvy.

**scut** (skŭt) *n.* [ME, hare.] A stubby erect tail, as of a rabbit.

**scu·tage** (skyōō′tĭj) *n.* [ME < Med. Lat. *scutagium* < Lat. *scutum,* shield.] A feudal tax paid in lieu of military service.

**scutch** (skŭch) *vt.* **scutched, scutch·ing, scutch·es.** [Obs. Fr. *escoucher* < OFr. *escousser* < VLat. **excussare,* freq. of Lat. *excutere,* to shake out : *ex-,* out + *quatere,* to shake.] To separate the valuable fibers of (e.g., flax) from the woody, unusable parts by beating. *—n.* An implement or machine for scutching. **—scutch′er** *n.*

**scutch·eon** (skŭch′ən) *n.* *var. of* ESCUTCHEON.

**scutch grass** *n.* Bermuda grass.

**scu·ta** (skyōō′tə) *n. pl. of* SCUTUM.

**scute** (skyōōt) *n.* [Lat. *scutum,* shield.] *Zool.* A horny, chitinous, or bony external scale or plate.

**scu·tel·la** (skyōō-tĕl′ə) *n. pl. of* SCUTELLUM.

**scu·tel·late** (skyōō-tĕl′ĭt, skyōōt′l-āt′) *also* **scu·tel·lat·ed** (skyōōt′l-ā′tĭd) *adj.* **1.** *Zool.* **a.** Covered with or protected by bony plates or scales. **b.** Having a scutellum. **2.** *Bot.* Shaped like a shield.

**scu·tel·la·tion** (skyōōt′l-ā′shən) *n.* A covering of small scales, as on a bird's leg.

**scu·tel·lum** (skyōō-tĕl′əm) *n., pl.* **-tel·la** (-tĕl′ə) [NLat., dim. of Lat. *scutum,* shield.] **1.** *Zool.* A shieldlike bony plate or scale. **2.** *Bot.* A shield-shaped structure. **—scu·tel′lar** *adj.*

**scut·ter** (skŭt′ər) *vi.* **-tered, -ter·ing, -ters.** [Alteration of SCUTTLE[3].] *Chiefly Brit.* To bustle about : SCUTTLE.

**scut·tle[1]** (skŭt′l) *n.* [ME *skottell* < OFr. *escoutille.*] **1. a.** A small opening or hatch with a movable lid in the deck or hull of a ship. **b.** A similar opening in the roof or floor of a house. **2.** The cover of a scuttle. *—vt.* **-tled, -tling, -tles. 1. a.** To cut or open a hole or holes in (a ship's hull). **b.** To sink (a ship) by this means. **2.** *Informal.* To throw away : discard.

**scut·tle[2]** (skŭt′l) *n.* [ME *scutel,* basket < OE, dish < Lat. *scutella.*] **1.** A metal pail used to carry coal. **2.** A shallow open basket for carrying flowers, vegetables, or grain.

**scut·tle[3]** (skŭt′l) *vi.* **-tled, -tling, -tles.** [Prob. alteration of dial. *scuddle,* freq. of SCUD.] To run quickly : SCURRY. *—n.* A hurried run.

**scut·tle·butt** (skŭt′l-bŭt′) *n.* [SCUTTLE[1] + BUTT[5].] **1.** A ship's drinking fountain. **2.** *Archaic.* A cask on a ship for holding the day's supply of drinking water. **3.** *Slang.* Gossip : rumor.

**scu·tum** (skyōō′təm) *n., pl.* **-ta** (-tə) [Lat., shield.] *Zool.* A bony, calcareous, chitinous, or horny scale or plate.

**scuz·zy** (skŭz′ē) *adj.* **-zi·er, -zi·est.** [Orig. unknown.] *Slang.* Soiled : dirty.

**Scyl·la** (sĭl′ə) *n.* [Lat. < Gk. *Skulla.*] *Gk. Myth.* A female sea monster who devoured sailors who approached too closely. **—between Scylla and Charybdis.** In a situation where avoiding one risk exposes a person to another.

**scy·phis·to·ma** (sī-fĭs′tə-mə) *n., pl.* **-mae** (-mē) *or* **-mas.** [NLat. : Gk. *skuphos,* cup + Gk. *stoma,* mouth.] A larva of a scyphozoan that produces free-swimming medusae.

**scy·pho·zo·an** (sī′fə-zō′ən) *n.* [< NLat. *Scyphozoa,* class name : Gk. *skuphos,* cup + Gk. *zōia,* pl. of *zōion,* animal.] A marine coelenterate of the class Scyphozoa, as the jellyfish, usu. having a highly developed medusoid stage.

**scythe** (sīth) *n.* [ME *sithe* < OE *siðe.*] A tool with a long, curved single-edged blade and a long, bent handle, used for reaping or mowing. *—vt.* **scythed, scyth·ing, scythes.** To cut with or as if with a scythe.

**Scyth·i·an** (sĭth′ē-ən, sĭth′-) *n.* **1.** A member of an ancient nomadic people inhabiting Scythia. **2.** The extinct Iranian language of the Scythians. *—adj.* Of or relating to the Scythians or to their land or language.

**Scyth·o-Dra·vid·i·an** (sĭth′ō-drə-vĭd′ē-ən, sĭth′-) *adj.* [SCYTH(IAN) + DRAVIDIAN.] Of or relating to an ethnic group of northwestern India that has both Iranian and Dravidian characteristics.

**Se** *symbol for* SELENIUM.

**sea** (sē) *n.* [ME *see* < OE *sǣ.*] **1. a.** The body of salt water that covers most of the earth's surface, esp. this body considered as an entity distinct from earth and sky. **b.** An area of water within an ocean. **c.** A relatively large body of salt water that is wholly or partially landlocked. **d.** A relatively large body of fresh water. **2.** *often* **seas.** The ocean's surface considered with regard to its course, flow, swell, or turbulence <a high *sea*><choppy *seas*> **3.** Something that suggests the sea, as in sweep or vastness <a *sea* of humanity> **4.** Seafaring as a way of life. **5.** A lunar mare. **—at sea. 1.** On the open waters of the ocean. **2.** Confused : perplexed.

**sea anchor** *n. Naut.* A drag or float, usu. a conical canvas-covered frame, thrown overboard from a vessel to prevent drifting or maintain a heading into the wind.

**sea anemone** *n.* A marine coelenterate of the class Anthozoa or Actinozoa, with a flexible cylindrical body and petallike tentacles surrounding a central mouth.

**sea bass** *n.* A marine food fish of the genus *Centropristes* and related genera, esp. *C. striatus* of coastal U.S. Atlantic waters.

**sea·bed** (sē′bĕd′) *n.* The bottom of a sea or ocean.

**Sea·bee** (sē′bē′) *n.* [Alteration of *cee bee,* pronunciation of the initial letters of *construction battalion.*] A member of one of the U.S. Navy's construction battalions.

**sea bird** *n.* A bird, as a storm petrel or albatross, that frequents the open waters of the ocean.

**sea biscuit** *n.* Hardtack.

**sea·board** (sē′bôrd′, -bōrd′) *n.* **1.** A seacoast. **2.** The land area near the sea <the Eastern *Seaboard*>

**sea·borne** (sē′bôrn′, -bōrn′) *adj.* **1.** Transported by ship. **2.** Carried on or over the sea.

**sea bread** *n.* Hardtack.

**sea bream** *n.* A marine food fish of the family Sparidae, esp. *Archosargus rhomboidalis* of western Atlantic coastal waters.

**sea breeze** *n.* A cool breeze blowing inland from the sea.

**sea butterfly** *n.* A pteropod.

**sea captain** *n.* The captain of a ship, esp. of a merchant ship.

**sea change** *n.* **1.** A change effected by the sea <"Of his bones are coral made/ Those are pearls that were his eyes/ Nothing of him that doth fade/ But doth suffer a *sea change*" —Shakespeare> **2.** A drastic transformation <"The script suffered considerable *sea changes,* particularly in structure" —Harold Pinter>

**sea·coast** (sē′kōst′) *n.* Land bordering the sea.

**sea cow** *n.* A marine mammal of the order Sirenia, as a dugong or manatee.

**sea cradle** *n.* The chiton.

**sea crayfish** *n.* The spiny lobster.

**sea cucumber** *n.* A cucumber-shaped echinoderm of the class Holothuroidea, with a flexible body and a mouth surrounded by tentacles.

**sea·dog** (sē′dôg′, -dŏg′) *n.* A fogbow.

**sea dog** *n.* **1.** A seal or similar marine mammal. **2.** An experienced sailor.

**sea duck** *n.* A coastal diving duck, as the eider or scoter.

**sea duty** *n.* U.S. Navy duty outside the continental United States.

**sea eagle** *n.* A fish-eating eagle or a similar bird, as the osprey.

**sea elephant** *n.* The elephant seal.

**sea fan** *n.* A fan-shaped coral of the genus *Gorgonia,* esp. *G. flabellum* of coastal waters of Florida and the West Indies.

**sea·far·er** (sē′fâr′ər) *n.* A sailor or mariner.

**sea·far·ing** (sē′fâr′ĭng) *n.* **1.** A sailor's calling. **2.** Traveling by sea. *—adj.* Following a life at sea.

**sea feather** *n.* An anthozoan of the family Pennatulidae, having a featherlike shape.

**sea·floor** (sē′flôr′, -flōr′) *n.* Seabed.

**sea·food** (sē′fōōd′) *n.* Edible fish or shellfish from the sea.

**sea·fowl** (sē′foul′) *n.* **1.** A sea bird. **2.** Sea birds as a group.

**sea front** *n.* A strip of land at the edge of the sea, esp. land desirable for development as a resort.

**sea·girt** (sē′gûrt′) *adj.* Surrounded by the sea.

**sea·go·ing** (sē′gō′ĭng) *adj.* **1.** Made or used for ocean voyages. **2.** Seafaring.

**sea gooseberry** *n.* A marine organism of the genus *Pleurobrachia,* with a round iridescent body.

**sea grape** *n.* A tropical American shrub or small tree, *Coccolobis uvifera,* of sandy beaches, with large, round, leathery leaves and grapelike clusters of hard purplish fruit.

**sea green** *n.* **1.** A medium green or bluish green. **2.** A medium yellow green.

**sea gull** *n.* A gull, esp. one frequenting coastal areas.

**sea holly** *n.* An Old World plant, *Eryngium maritimum,* growing on seashores and having prickly leaves and clusters of small blue or purplish flowers.

**sea horse** *n.* **1.** A small marine fish of the genus *Hippocampus,* swimming erect and having a horselike head, a prehensile tail, and a body covered with bony plates. **2.** A walrus. **3.** A mythical animal, half fish and half horse, ridden by sea gods such as Neptune.

**Sea Island cotton** *n.* [After the *Sea Islands,* an island chain off the coasts of South Carolina, Georgia, and Florida.] A species of cotton, *Gossypium barbadense,* native to tropical America and now widely cultivated for its fine, long-staple fibers.

**sea kale** *n.* A European plant, *Crambe maritima*, with edible shoots and cabbagelike leaves.

**sea king** *n.* A Norse pirate chief of the early Middle Ages.

**seal**[1] (sēl) *n.* [ME *seel* < OFr. < Lat. *sigillum*, dim. of *signum*, sign.] **1. a.** A signet or die with a raised or incised emblem used to stamp an impression on a soft substance, as wax or lead. **b.** The impression made. **c.** The emblem or design itself, belonging solely to the user. **d.** A small disk or wafer of wax, lead, or paper bearing such an imprint and used to close or authenticate a document. **2.** Something, as a commercial hallmark, that serves to authenticate or confirm. **3.** An adhesive agent, as wax or putty, used to close or secure something or prevent seepage of moisture or air. **4.** A device or fluid in a drainpipe that prevents the upward passage of gas. **5.** An airtight or watertight closure. **6.** A small decorative paper sticker. —*vt.* **sealed, seal·ing, seals. 1.** To affix a seal to so as to prove authenticity or attest to a standard, as of accuracy, legal weight, or quality. **2. a.** To close with or as if with a seal. **b.** To close hermetically. **c.** To make fast or fill up, as with plaster or cement. **3.** To grant, certify, or designate under seal or authority. **4.** To fix irrevocably <My fate was *sealed.*> **5.** *Mormon Ch.* To make (e.g., a marriage) binding for life. **—seal off.** To close tightly. **—seal'a·ble** *adj.*

▲ word history: The noun *seal*[1], "a die used to stamp an impression," is related to the word *sign. Sign* is from Latin *signum*, which meant both "a mark" and "a signet ring." *Signum* formed a diminutive noun *sigillum*, which literally meant "little mark" but denoted especially the seal or impression left by a signet ring. *Sigillum* became *seel* in Old French, which was borrowed into English as *seal.*

**seal**[2] (sēl) *n.* [ME *seel* < OE *seolh.*] **1.** Any of various aquatic, carnivorous mammals of the families Phocidae and Otariidae, with a sleek streamlined body and limbs modified into paddlelike flippers. **2.** The pelt or fur of a seal, esp. a fur seal. **3.** Leather made from the hide of a seal. —*vi.* **sealed, seal·ing, seals.** To hunt seals.

**sea lamprey** *n.* A large marine lamprey, *Petromyzon marinus*, sometimes used as food and parasitic to freshwater fish esp. in the Great Lakes.

**sea-lane** (sē'lān') *n.* An established or frequently used sea route.

**seal·ant** (sē'lənt) *n.* A sealing agent.

**sea lavender** *n.* A salt-marsh plant of the genus *Limonium*, with small lavender or pinkish flower clusters.

**sea legs** *pl.n. Informal.* The ability to walk without faltering on board ship, esp. in rough seas.

**seal·er**[1] (sē'lər) *n.* **1.** One that seals. **2.** A substance, as paint or varnish, used to size a surface. **3.** An officer who tests and certifies weights and measures.

**seal·er**[2] (sē'lər) *n.* A person or vessel engaged in seal hunting.

**seal·er·y** (sē'lə-rē) *n., pl.* **-ies. 1.** The process or occupation of hunting seals. **2.** A place where seals are hunted.

**sea lettuce** *n.* A green seaweed of the genus *Ulva*, with thin, irregularly shaped leaflike fronds sometimes used as food.

**sea level** *n.* The level of the ocean's surface, esp. the mean level halfway between high and low tide, used as a standard in measuring land elevation or sea depths.

**sea lily** *n.* Any of various marine crinoids, usu. attached to the ocean floor in deep water, with a long stalk and a flowerlike body.

**sealing wax** *n.* A resinous mixture of shellac and turpentine, soft and fluid when heated and solid when cooled, used to seal letters, batteries, or jars.

**sea lion** *n.* An eared seal of the family Otariidae, esp. *Zalophus californianus* of the Northern Pacific, lacking the valuable underfur of the fur seal.

**seal ring** *n.* A signet ring.

**seal·skin** (sēl'skĭn') *n.* **1.** The pelt or fur of a fur seal, esp. the underfur. **2.** A garment made of sealskin.

**Sea·ly·ham terrier** (sē'lē-hăm', -lē-əm) *n.* [After *Sealyham*, Wales.] A terrier orig. bred in Wales, with a long head, short legs, and a wiry white coat.

**Sealyham terrier**
*Approximately 10½ inches high at shoulder*

**seam** (sēm) *n.* [ME *seem* < OE *sēam.*] **1. a.** A line of junction formed by sewing together two pieces of material along their edges. **b.** A similar line, ridge, or groove formed by fitting or joining together two sections along their edges. **c.** A suture. **d.** A scar. **2.** A line across a surface, as a crack or wrinkle. **3.** A thin stratum or layer, as of coal or rock. —*v.* **seamed, seam·ing, seams.** —*vt.* **1.** To join or fit together with or as if with a seam. **2.** To mark with a seamlike line, as a groove, scar, or wrinkle. **3.** To form ridges in by purling. —*vi.* To develop fissures or furrows : CRACK. **—seam'er** *n.*

**sea-maid·en** (sē'mād'n) *also* **sea-maid** (-mād') *n.* A mermaid or sea nymph.

**sea·man** (sē'mən) *n.* **1.** A mariner or sailor. **2.** An enlisted person in the U.S. Navy or Coast Guard ranking above seaman apprentice and below petty officer.

**seaman apprentice** *n.* An enlisted person in the U.S. Navy or Coast Guard ranking above a seaman recruit and below a seaman.

**seaman recruit** *n.* An enlisted person of the lowest rank in the U.S. Navy or Coast Guard.

**sea·man·ship** (sē'mən-shĭp') *n.* Skill in handling or navigating a boat or ship.

**sea·mark** (sē'märk') *n.* **1.** A landmark visible from the sea, used as a navigational guide. **2.** The mark along a shoreline indicating the extent of high tide.

**sea mew** *n.* A coastal gull, esp. *Larus canus* of Europe.

**sea mile** *n.* A nautical mile.

**sea milkwort** *n.* A small fleshy plant, *Glaux maritima* of shores and brackish marshes, with pink or white flowers.

**seam·less** (sēm'lĭs) *adj.* Lacking seams. **—seam'less·ness** *n.*

**sea·mount** (sē'mount') *n.* A submarine mountain having a summit at least 1,000 feet below sea level.

**sea mouse** *n.* A segmented marine worm of the genus *Aphrodite*, esp. *A. aculeata*, with overlapping scales covered by long hairs.

**seam·ster** (sēm'stər) *n.* [ME *semester* < OE *sēamestre* < *sēam*, seam.] A tailor.

**seam·stress** (sēm'strĭs) *n.* A woman who sews, esp. one who makes her living by sewing.

**seam·y** (sē'mē) *adj.* **-i·er, -i·est. 1.** Having or showing a seam or seams. **2.** Rough and unattractive : SORDID. **—seam'i·ness** *n.*

**sé·ance** (sā'äns', -äɴs') *n.* [Fr. < OFr., a sitting < *seoir*, to sit < Lat. *sedēre.*] **1.** A meeting at which persons attempt to receive spiritualistic messages. **2.** A meeting or session.

**sea oats** *pl.n.* (*sing. or pl. in number*). A tall grass, *Uniola paniculata*, of the southern U.S. coast.

**sea onion** *n.* **1.** A bulbous plant, *Urginea maritima* of the Mediterranean area, yielding a powder used medicinally and as a rat poison. **2.** A small European plant, *Scilla verna*, with sweet-smelling blue flowers.

**sea otter** *n.* A large nearly extinct marine otter, *Enhydra lutris* of northern Pacific coasts, with a soft dark-brown coat.

**sea pen** *n.* [From its resemblance to a quill pen.] Any of various marine anthozoans of the families Stylatulidae and Funiculinidae, resembling and related to the sea feathers.

**sea·plane** (sē'plān') *n.* An aircraft having floats for taking off from or landing on water.

**sea·port** (sē'pôrt', -pōrt') *n.* A harbor, town, or city with facilities for seagoing ships.

**sea power** *n.* **1.** Naval strength. **2.** A nation possessing or wielding great naval strength.

**sea purse** *n.* The purse-shaped egg case produced by skates and certain sharks.

**sea·quake** (sē'kwāk') *n.* An earthquake under the ocean floor.

**sear**[1] (sîr) *v.* **seared, sear·ing, sears.** [ME *seren* < OE *sēarian* < *sēar*, withered.] —*vt.* **1.** To make withered or dried up : SHRIVEL. **2.** To scorch, char, or burn the surface of with or as if with something hot. —*vi.* To become withered or dried up. —*n.* A mark or injury caused by searing.

**sear**[2] (sîr) *n.* [Prob. < OFr. *serre*, lock < *serrer*, to grasp < LLat. *serare*, to bolt < Lat. *sera*, bar, bolt.] The catch in a gunlock that holds the hammer in a halfcocked or fully cocked position.

**sear**[3] (sîr) *adj. var. of* SERE[1].

**sea raven** *n.* A large sculpin, *Hemitripterus americanus* of the western North Atlantic.

**search** (sûrch) *v.* **searched, search·ing, search·es.** [ME *serchen* < OFr. *cerchier* < LLat. *circare*, to go around < Lat. *circus*, circle.] —*vt.* **1.** To examine or look over carefully in order to discover something : EXPLORE. **2.** To examine or investigate carefully : PROBE <*search* one's conscience> **3.** To make a complete check of (a legal document) : SCRUTINIZE <*search* a title> **4. a.** To examine in order to find something lost or concealed. **b.** To examine (a person or the personal effects of) in order to find something lost or concealed. **5.** To come to know : LEARN. —*vi.* To conduct a complete investigation : SEEK <*searching* for answers to the riddle> —*n.* **1.** An act or instance of searching. **2.** The exercise of right of search. **—search'a·ble** *adj.* **—search'er** *n.*

**search·ing** (sûr'chĭng) *adj.* **1.** Examining carefully or completely <a *searching* investigation> **2.** Keenly observant <*searching* insights>

**search·light** (sûrch′līt′) *n.* **1. a.** An apparatus housing a light source and a reflector for projecting a strong beam of approx. parallel rays of light. **b.** The beam of light so projected. **2.** A flashlight.

**search warrant** *n.* A warrant legally authorizing a search.

**sea robin** *n.* A marine fish of the family Triglidae, with a bony head and long pectoral fins with fingerlike rays.

**sea room** *n.* Adequate space for maneuvering a ship at sea.

**sea·scape** (sē′skāp′) *n.* A view or depiction of the sea.

**Sea Scout** *n.* A member of a program that trains Boy and Girl Scouts in seamanship.

**sea serpent** *n.* A legendary large, snakelike marine animal.

**sea·shell** (sē′shĕl′) *n.* The calcareous shell of a marine mollusk or similar marine organism.

**sea·shore** (sē′shôr′, -shōr′) *n.* **1.** Land bordering or near the sea. **2.** *Law.* Ground lying between high-water and low-water marks.

**sea·sick·ness** (sē′sĭk′nĭs) *n.* Nausea or other discomfort caused by the motion of a vessel at sea. **—sea′sick′** *adj.*

**sea·side** (sē′sīd′) *n.* The seashore.

**sea slug** *n.* A marine gastropod of the suborder Nudibranchia, with a colorful shell-less body and fringelike projections.

**sea snake** *n.* A venomous tropical marine snake of the family Hydrophidae, mainly of the Pacific and Indian oceans.

**sea·son** (sē′zən) *n.* [ME *sesoun* < OFr. *seson* < Lat. *satio,* act of sowing < *satus,* p.part. of *serere,* to plant.] **1. a.** One of the four divisions of the year, spring, summer, autumn, and winter, marked by the passage of the sun through an equinox or solstice. **b.** One of the two divisions of the year, rainy and dry, in tropical climates. **2.** A recurrent period marked by certain activities, celebrations, or crops. **3.** A natural, convenient, or appropriate time. **4.** A period of time. *—v.* **-soned, -son·ing, -sons.** *—vt.* **1.** To enhance the flavor of (food) by adding salt, spices, herbs, or other flavorings. **2.** To add interest or zest to <*seasoned* my speech with anecdotes> **3.** To dry (cut wood) until it is usable : CURE. **4.** To make competent through trial and experience <*seasoned* the rookie soldiers> **5.** To accustom : inure. **6.** To moderate or temper. *—vi.* To become usable, competent, or tempered. **—in season. 1.** Available for eating. **2.** Legally available for hunting, fishing, or trapping. **3.** At the right or proper moment : OPPORTUNELY. **4.** In heat. —Used of animals. **—out of season. 1.** Not available for eating <oysters *out of season*> **2.** Not legally available for hunting, fishing, or trapping. **3.** Not at the right or proper moment : INOPPORTUNELY.

**sea·son·a·ble** (sē′zə-nə-bəl, sēz′nə-) *adj.* **1.** Appropriate to the time or season <*seasonable* weather> **2.** Occurring or performed at the proper time : TIMELY. **—sea′son·a·bly** *adv.*

**sea·son·al** (sē′zə-nəl, sēz′nəl) *adj.* Of or dependent on a particular season <*seasonal* jobs at the cannery> **—sea′son·al·ly** *adv.*

**sea·son·er** (sē′zə-nər, sēz′nər) *n.* **1.** One that seasons, esp. one that uses seasonings. **2.** Seasoning.

**sea·son·ing** (sē′zə-nĭng, sēz′nĭng) *n.* **1.** Something, as a spice or herb, used to enhance the flavor of food. **2.** The act or process by which something is seasoned.

**season ticket** *n.* A ticket that can be used during a specified period of time.

**sea spider** *n.* A marine arachnid of the class Pycnogonida, with long legs and a rather small body.

**sea squirt** *n.* [From its habit of squirting water when disturbed.] A sedentary marine animal of the class Ascidiacea, with a transparent sac-shaped body and two siphons.

**sea·strand** (sē′strănd′) *n.* A seashore.

**seat** (sēt) *n.* [ME *sete* < ON *sæti.*] **1.** Something that may be sat upon, as a bench or chair. **2.** A place in which one may sit. **3.** A part of something on which one rests in sitting <the car *seat*> **4. a.** The buttocks. **b.** That part of a garment covering the buttocks. **5. a.** A part that serves as the base of something. **b.** The surface or part upon which another part sits or rests. **6. a.** The place where something is located or based <the *seat* of intelligence> **b.** A center of authority : CAPITAL <the county *seat*> **7.** A place of residence, esp. a large house that is part of an estate. **8.** Membership in a legislature, stock exchange, or other organization, obtained by purchase, appointment, or election. **9.** The manner or way of sitting, as on a horse. *—vt.* **seat·ed, seat·ing, seats. 1. a.** To place in or on a seat. **b.** To cause or assist to sit down <*seat* the guests courteously> **2.** To provide with a particular seat. **3.** To have or provide seats for <We can only *seat* 12 at the table.> **4.** To install in a position of authority or eminence. **5.** To fix firmly in place.

**sea tangle** *n.* A brown seaweed, esp. one of the genus *Laminaria.*

**seat belt** *n.* An adjustable strap or harness that holds one securely in a seat, as in an automobile or aircraft.

**seat·ing** (sē′tĭng) *n.* **1.** The act or an instance of providing or furnishing with a seat or seats. **2.** The arrangement of seats, as in a room or auditorium. **3.** The member or part upon or within which another part is seated. **4.** Material for covering or upholstering seats.

**sea·train** (sē′trān′) *n.* A large seagoing vessel equipped to carry railroad cars.

**sea trout** *n.* **1.** A marine fish of the genus *Cynoscion,* esp. the weakfish. **2.** A trout or similar fish that lives in the sea but migrates to fresh water to spawn.

**sea urchin** *n.* An echinoderm of the class Echinoidea, having a soft body enclosed in a limy, round shell covered with long spines.

**sea wall** *n.* An embankment or wall designed to prevent erosion of a shoreline.

**sea walnut** *n.* A ctenophore of the genus *Mnemiopsis* and related genera, with a translucent, ovoid, ridged body and hairlike cilia.

**sea·ward** (sē′wərd) *adj. & adv.* At or toward the sea. **—sea′wards** (-wərdz) *adv.*

**sea·ware** (sē′wâr′) *n.* [SEA + dial. *ware,* seaweed < ME < OE *wār.*] Sea wrack used as fertilizer.

**sea·wa·ter** (sē′wô′tər, -wŏt′ər) *n.* Water in or coming from the sea.

**sea·way** (sē′wā′) *n.* **1.** A route used at sea. **2.** An inland waterway for ocean shipping. **3.** The headway of a ship. **4.** A rough sea.

**sea·weed** (sē′wēd′) *n.* **1.** Any of numerous marine algae, as the kelp, rockweed, or sea lettuce. **2.** Any of various marine plants.

**sea·wor·thy** (sē′wûr′thē) *adj.* **-thi·er, -thi·est.** Fit to sail. —Used of a ship. **—sea′wor′thi·ness** *n.*

**sea wrack** *n.* Material cast ashore, esp. seaweed.

**se·ba·ceous** (sĭ-bā′shəs) *adj.* [Lat. *sebaceus* < *sebum,* tallow.] *Physiol.* **1.** Of, relating to, or resembling fat or sebum : FATTY. **2.** Secreting fat or sebum.

**sebaceous gland** *n.* A gland in the skin's corium that opens into a hair follicle and produces and secretes sebum.

**se·bac·ic acid** (sĭ-băs′ĭk, -bā′sĭk, sē-) *n.* [< SEBACEOUS.] A white crystalline acid, $C_{10}H_{18}O_4$, used to make various synthetic resins and fibers, plasticizers, and polyester rubbers.

**sebi-** *or* **sebo-** *pref.* [< Lat. *sebum,* tallow.] Fat : sebum <*sebiferous*>

**se·bif·er·ous** (sĭ-bĭf′ər-əs) *also* **se·bip·a·rous** (-bĭp′-). Producing or secreting fatty, oily, or waxy matter : SEBACEOUS.

**sebo-** *pref. var. of* SEBI-.

**seb·or·rhe·a** *also* **seb·or·rhoe·a** (sĕb′ə-rē′ə) *n.* A disease of the sebaceous glands marked by excessive discharge of sebum or an alteration in its quality and resulting in an oily coating, scales, or crusts on the skin. **—seb′or·rhe′ic** *adj.*

**se·bum** (sē′bəm) *n.* [Lat., tallow.] The semifluid discharge of the sebaceous glands.

**sec** (sĕk) *adj.* [Fr.] Dry. —Used of wines, esp. champagne.

**sec·a·lose** (sĕk′ə-lōs′) *n.* [NLat. *Secale,* grass genus (< Lat. *secale,* rye) + -OSE.] A fructose polysaccharide.

**se·cant** (sē′kănt′, -kənt) *n.* [< Lat. *secans, secant-,* cutting, pr.part. of *secare,* to cut.] **1. a.** A straight line that intersects a curve at two or more points. **b.** The straight line drawn from the center through one end of a circular arc and intersecting the tangent to the other end of the arc. **2. a.** The reciprocal of an angle's cosine. **b.** For an acute angle, the ratio of the hypotenuse to the side of a right triangle adjacent to the acute angle.

**sec·co** (sĕk′ō) *n., pl.* **-cos.** [Ital. < Lat. *siccus,* dry.] The art or an example of painting on dry plaster.

**se·cede** (sĭ-sēd′) *vi.* **-ced·ed, -ced·ing, -cedes.** [Lat. *secedere,* to separate : *se-,* apart + *cedere,* to go.] To withdraw formally from membership in an association, organization, or alliance, esp. a political one.

**se·cern** (sĭ-sûrn′) *vt.* **-cerned, -cern·ing, -cerns.** [Lat. *secernere,* to sever : *se-,* apart + *cernere,* to separate.] **1.** To perceive as distinct and separate : DISCRIMINATE. **2.** *Physiol.* To secrete. —Used of a gland or follicle. **—se·cern′ment** *n.*

**se·ces·sion** (sĭ-sĕsh′ən) *n.* [Lat. *secessio* < *secessus,* p.part. of *secedere,* to secede.] **1.** The act or an instance of seceding. **2.** *often* **Secession.** The withdrawal of 11 Southern states from the Federal Union in 1860–61, precipitating the Civil War. **—se·ces′sion·al** *adj.*

**se·ces·sion·ism** (sĭ-sĕsh′ə-nĭz′əm) *n.* The policy of upholding the right of secession. **—se·ces′sion·ist** *n.*

**Seck·el pear** (sĕk′əl, sĭk′-) *n.* [Perh. < the name *Seckel.*] A pear having small, sweet, juicy reddish-brown fruit.

**se·clude** (sĭ-klōōd′) *vt.* **-clud·ed, -clud·ing, -cludes.** [ME *secluden,* to shut off < Lat. *secludere* : *se-,* apart + *claudere,* to shut.] **1.** To set apart or remove from others : ISOLATE. **2.** To screen from view : make hidden.

**se·clud·ed** (sĭ-klōō′dĭd) *adj.* **1.** Removed or far apart from others : SOLITARY. **2.** Screened from view : HIDDEN. **—se·clud′ed·ly** *adv.* **—se·clud′ed·ness** *n.*

**se·clu·sion** (sĭ-klōō′zhən) *n.* [Med. Lat. *seclusio* < Lat. *seclusus,* p.part. of *secludere,* to seclude.] **1.** The act of secluding or the state of being secluded. **2.** A secluded place.

☆ ***syns:*** SECLUSION, RECLUSION, RETIREMENT, SEQUESTRATION *n. core meaning* : the act of secluding or the state of being secluded <sought *seclusion* in the mountains>

**se·clu·sive** (sĭ-klōō′sĭv, -zĭv) *adj.* Of, fond of, or looking for seclusion. **—se·clu′sive·ly** *adv.* **—se·clu′sive·ness** *n.*

**sec·o·bar·bi·tal** (sĕk′ō-bär′bĭ-tôl′) *n.* [*Seconal,* a trademark for secobarbital + BARBITAL.] A barbiturate, $C_{12}H_{18}N_2O_3$, used medically as a sedative and hypnotic in the form of its sodium salt.

**sec·ond[1]** (sĕk′ənd) *n.* [ME *seconde,* one-sixtieth of a minute of arc < OFr. < Med. Lat. *(pars minuta) secunda,* second (small part), fem.

ōō **boot** ou **out** th **thin** *th* **this** ŭ **cut** ûr **urge** y **young**
yōō **abuse** zh **vision** ə **about,** **item,** **edible,** **gallop,** **circus**

of Lat. *secundus*, second, following.] **1.** A unit of time equal to 1/60 of a minute. **2.** *Informal.* A short period of time : MOMENT. **3.** *Math.* A unit of angular measure equal to 1/60 of a minute of arc.

▲ word history: The noun *second¹* is ultimately derived from the same Latin word as the adjective *second²*, meaning "coming next after the first." The Latin word *secundus* was an adjective that meant "following." The use of *secundus* as a noun arose in Medieval Latin. Certain units, as the degrees of a circle, were divided into equal parts, and those parts were further subdivided. The first subdivision was called *pars minuta prima*, or "first small part"; the subdivision of this unit was called *pars minuta secunda*, or "second small part." *Secunda* is the feminine form of *secundus*, agreeing with the feminine noun *pars*. *Secunda* alone was eventually used for the second subdivision of a degree and came into English as *second*.

**sec·ond²** (sĕk′ənd) *adj.* [ME < OFr. < Lat. *secundus*, following.] **1.** Coming next after the first in place, order, rank, time, or quality. **2.** Repeating an original instance : ADDITIONAL <*a second* opportunity> **3.** Inferior to another : SUBORDINATE <*second* to none> **4.** *Mus.* **a.** Having a lower pitch or range. **b.** Singing or playing a part having a lower range. **5.** Having the second-highest ratio. —Used of gears in a sequence. —*n.* **1. a.** The ordinal number that matches the number 2 in a series. **b.** One of two equal parts. **2.** One next in place, order, time, or quality after the first. **3.** *often* **seconds.** An inferior or imperfect article of merchandise. **4.** The official attendant of a contestant in a duel or boxing match. **5.** *Mus.* **a.** The interval between two adjacent tones of the diatonic scale. **b.** A tone separated from another tone by this interval. **c.** A combination of two such tones in notation or in harmony. **d.** The second part, instrument, or voice in a harmonized composition. **6.** An utterance of endorsement. **7.** The forward gears in an automotive transmission having the second-highest ratio. —*vt.* **-ond·ed, -ond·ing, -onds. 1.** To attend (e.g., a boxer) as an aide or assistant. **2.** To promote or encourage : REINFORCE. **3.** To endorse (a motion or nomination) so that discussion or a vote can take place. —*adv.* **1.** In the second place, order, or rank <finished *second*> **2.** But for one other : save one <the *second*-longest river>

**Second Advent** *n.* The Second Coming.

**sec·ond·ar·y** (sĕk′ən-dĕr′ē) *adj.* **1. a.** Of the second rank : not primary. **b.** Inferior. **c.** Minor : lesser. **2.** Derived or resulting from what is primary or original <*a secondary* effect of the accident> **3.** Of or designating the shorter flight feathers along the inner part of the edge of a bird's wing. **4.** *Elect.* Having a current or voltage induced by the magnetic field caused by a current flowing in another coil. —Used of a circuit or coil. **5.** *Chem.* Formed by replacement of two atoms or radicals within a molecule. —Used of a compound. **6.** *Geol.* Resulting from or formed by changes in the pre-existing minerals. **7.** Of or pertaining to a secondary school <*secondary* teaching aids> —*n.*, *pl.* **-ies. 1.** One that acts in an auxiliary, subordinate, or inferior capacity. **2.** One of the shorter flight feathers along the inner part of the edge of a bird's wing. **3.** *Elect.* A circuit or coil with an induced current. **4.** *Astron.* A celestial body that orbits a larger body : SATELLITE. **5.** *Football.* The defensive backfield. **—sec′ond·ar′i·ly** (-dâr′ə-lē) *adv.* **—sec′ond·ar′i·ness** *n.*

**secondary battery** *n. Elect.* A storage battery.

**secondary cell** *n.* A rechargeable electric cell for converting chemical energy into electrical energy by a reversible chemical reaction.

**secondary color** *n.* A color produced by blending approx. equal proportions of two primary colors.

**secondary electron** *n.* An electron produced in secondary emission.

**secondary emission** *n.* Emission of electrons from the surface of a substance bombarded by high-speed electrons or ions.

**secondary offering** *n.* The sale of an extensive block of outstanding stock through dealers but outside of a stock exchange.

**secondary school** *n.* A school higher in level than elementary school that usu. offers general, technical, vocational, or college preparatory curricula.

**secondary sex characteristic** *n.* Any of the genetically transmitted anatomical, physiological, or behavioral characteristics, as abundance of facial hair or breast development, that first appear in humans during puberty and differentiate between the sexes without having a direct reproductive function.

**second base** *n. Baseball.* **1.** The base across the diamond from home plate, to be touched second by a runner. **2.** The position played by a second baseman.

**second baseman** *n. Baseball.* The infielder who is positioned near and usu. to the first-base side of second base.

**second best** *n.* One that is next below the best. **—second best** *adv.*

**sec·ond-best** (sĕk′ənd-bĕst′) *adj.* Next to best.

**second childhood** *n.* Senility : dotage.

**second class** *n.* **1.** Second-class travel accommodations. **2.** Second-class mail.

**sec·ond-class** (sĕk′ənd-klăs′) *adj.* **1.** Of secondary quality status : INFERIOR. **2.** Of or relating to travel accommodations ranking next below the highest or first class. **3.** Of or relating to a class of mail consisting mainly of newspapers and periodicals. —*adv.* By means of second-class mail or second-class travel accommodations.

**Second Coming** *n.* The return of Christ as judge on the last day.

**sec·ond-de·gree burn** (sĕk′ənd-dĭ-grē′) *n.* A burn that blisters the skin.

**Second Empire** *n.* A heavily ornate style of furniture, architecture, and design developed in mid-19th-cent. France.

**second fiddle** *n.* **1.** A secondary or subordinate role. **2.** One who plays a secondary role.

**second generation** *n. Computer Sci.* The period of computer technology that utilized solid-state circuitry and off-line storage and extensively developed software.

**second growth** *n.* Trees that cover a region after the removal of the original growth, as by cutting or fire.

**sec·ond-guess** (sĕk′ənd-gĕs′) *v.* **-guessed, -guess·ing, -guess·es.** —*vt.* **1.** To criticize (a decision) after the outcome is known. **2.** To anticipate the moves of : OUTGUESS. —*vi.* To criticize a decision after its outcome is known. **—sec′ond-guess′er** *n.*

**sec·ond·hand** (sĕk′ənd-hănd′) *adj.* **1.** Previously used or worn by another : not new. **2.** Dealing in previously used merchandise. **3.** Obtained, derived, or borrowed from another <*secondhand* information> —*adv.* In a roundabout way : INDIRECTLY.

**second hand¹** *n.* The hand of a timepiece that marks the seconds.

**second hand²** *n.* An intermediary person or source <heard at *second hand*>

**se·con·di** (sĭ-kôn′dē) *n. pl. of* SECONDO.

**second lieutenant** *n.* An officer in the U.S. Army, Air Force, and Marine Corps of the lowest commissioned grade.

**sec·ond·ly** (sĕk′ənd-lē) *adv.* In the second place : SECOND.

**second mortgage** *n.* A mortgage on property that already carries a mortgage.

**second nature** *n.* Deeply ingrained habits and characteristics.

**se·con·do** (sĭ-kôn′dō) *n.*, *pl.* **-di** (-dē) [Ital. < Lat. *secundus*, second, following.] *Mus.* The second part in a concert piece, esp. the lower part in a piano duet.

**second person** *n.* The form of a pronoun or verb used in referring to the person addressed, as *you* and *may* in *you may not go.*

**sec·ond-rate** (sĕk′ənd-rāt′) *adj.* Inferior or mediocre in quality or value. **—sec′ond-rate′ness** *n.* **—sec′ond-rat′er** *n.*

**second sight** *n.* Clairvoyance.

**sec·ond-sto·ry man** (sĕk′ənd-stôr′ē, -stōr′ē) *n.* A burglar who enters through upstairs windows.

**sec·ond-strike** (sĕk′ənd-strīk′) *adj.* Of or constituting a nuclear-weapons force capable of withstanding nuclear attack and therefore delivering a retaliatory attack.

**sec·ond-string** (sĕk′ənd-strĭng′) *adj.* **1.** Being a substitute, as on a ball team. **2.** Of lesser importance : INFERIOR.

**second thought** *n.* Reconsideration of a previous decision or belief.

**second wind** *n.* Restored energy or strength.

**Second World War** *n.* World War II.

**se·cre·cy** (sē′krĭ-sē) *n.*, *pl.* **-cies.** [Alteration of obs. *secretie* < ME *secretee* < *secret*, secret.] **1.** The condition or quality of being secret or hidden : CONCEALMENT. **2.** The ability to keep secrets.

**se·cret** (sē′krĭt) *adj.* [ME < OFr. < Lat. *secretus*, p.part. of *secernere*, to sever : *se-*, apart + *cernere*, to separate.] **1.** Concealed from public or general knowledge or view. **2.** Reliably close-mouthed : DISCREET. **3.** Operating in a hidden or confidential manner <*a secret* agent> **4.** Not outwardly expressed : INMOST. **5.** Not frequented : SECLUDED. **6.** Known or shared only by the initiated <*secret* rituals> **7.** Beyond ordinary understanding : ESOTERIC. **8.** Designating the security classification above confidential and below top-secret. —*n.* **1.** Something concealed from others or known only to oneself or to a few. **2.** Something beyond ordinary understanding or explanation : MYSTERY. **3.** A method or formula by which a desired end is attained. **4. Secret.** A variable prayer said after the Offertory and before the Preface in the liturgy of the Mass. **—se′cret·ly** *adv.*

☆ ***syns:*** SECRET, CLANDESTINE, COVERT, HUSH-HUSH, UNDERCOVER *adj. core meaning* : purposely concealed from view or knowledge <*secret* intelligence operations>

**se·cre·ta·gogue** (sə-krē′tə-gôg′, -gŏg′) *n.* [SECRET(ION) + -AGOGUE.] An agent that stimulates secretion.

**sec·re·tar·i·at** (sĕk′rĭ-târ′ē-ĭt) *n.* [Fr. *secrétariat* < Med. Lat. *secretariatus* < *secretarius*, secretary.] **1. a.** The department administered by a governmental secretary, esp. for an international organization. **b.** The building or facilities occupied by such a department. **2.** The office or position of a governmental secretary.

**sec·re·tar·y** (sĕk′rĭ-tĕr′ē) *n.*, *pl.* **-ies.** [ME *secretarie*, confidant < Med. Lat. *secretarius*, confidential officer < Lat. *secretus*, secret.] **1.** One employed to handle correspondence, keep files, and do clerical work for an individual or company. **2.** An officer of an organization who is in charge of minutes of meetings and other important documents. **3.** An official presiding over an administrative department of government. **4.** A desk surmounted by a small bookcase. **—sec're·tar'i·al** (-târ'ē-əl) *adj.*

**secretary bird** *n.* A large African bird of prey, *Sagittarius serpentarius,* having long legs and a crest of quills.

**secretary bird**
*Approximately 4½ feet long*

**sec·re·tar·y-gen·er·al** (sĕk'rĭ-tĕr'ē-jĕn'ər-əl) *n., pl.* **sec·re·tar·ies-gen·er·al.** A high-ranking executive officer, as in the United Nations or certain political parties.

**secret ballot** *n.* An Australian ballot.

**se·crete[1]** (sĭ-krēt') *vt.* **-cret·ed, -cret·ing, -cretes.** [Back-formation < SECRETION.] To produce and separate out (a substance) from cells or bodily fluids. **—se·cre'tor** *n.*

**se·crete[2]** (sĭ-krēt') *vt.* **-cret·ed, -cret·ing, -cretes.** [Obs. *secret,* to conceal < SECRET.] To conceal in a hiding place : HIDE.

**se·cre·tin** (sĭ-krēt'n) *n.* [SECRET(ION) + -IN.] A hormone secreted in the duodenum to stimulate the flow of pancreatic juice.

**se·cre·tion** (sĭ-krē'shən) *n.* [Fr. *sécrétion* < Lat. *secretio,* separation < *secernere,* to sever. —see SECERN.] **1.** The act or process of secreting a substance, esp. one that is not a waste, from blood or cells. **2.** A substance so secreted. **—se·cre'tion·ar'y** (-shə-nĕr'ē) *adj.*

**se·cre·tive** (sē'krĭ-tĭv, sĭ-krē'tĭv) *adj.* Practicing or inclined to secrecy. **—se'cre·tive·ly** *adv.* **—se'cre·tive·ness** *n.*

**se·cre·to·ry** (sĭ-krē'tə-rē) *adj.* Relating to or performing the function of secretion.

**secret partner** *n.* A person whose participation in a partnership is unknown to the public.

**secret police** *n.* A police force operating largely in secret and often using terrorism to suppress political opposition.

**secret service** *n.* **1. a.** Intelligence-gathering activities conducted secretly by a government agency. **b.** A government agency engaged in such activities. **2. Secret Service.** A branch of the U.S. Treasury Department charged with suppression of counterfeiting and the protection of the President.

**sect** (sĕkt) *n.* [ME *secte* < OFr. < Lat. *secta,* faction < *sequi,* to follow.] **1.** A group of people that forms a distinct unit within a larger group by virtue of certain refinements or distinctions of belief or practice. **2.** A religious body, esp. one that has separated from a larger denomination. **3.** A small group united by common interests or beliefs.

**-sect** *suff.* [< Lat. *sectus,* p.part. of *secare,* to cut.] **1.** To cut : divide <trisect> **2.** Cut : divided <pinnatisect>

**sec·tar·i·an** (sĕk-târ'ē-ən) *adj.* **1.** Of, relating to, or like a sect. **2.** Adhering to the dogmatic limits of a sect : PARTISAN. **3.** Narrow-minded : parochial. —*n.* **1.** A member of a sect. **2.** One showing bigoted adherence to a factional viewpoint. **—sec·tar'i·an·ism** *n.*

**sec·ta·ry** (sĕk'tə-rē) *n., pl.* **-ries.** [Med. Lat. *sectarius* < Lat. *secta,* sect.] **1.** A sectarian. **2.** One who dissents from an established church, esp. a Protestant nonconformist.

**sec·tile** (sĕk'təl, -tīl') *adj.* [Lat. *sectilis* < *secare,* to cut.] Capable of being cut smoothly by a knife. **—sec·til'i·ty** (-tĭl'ĭ-tē) *n.*

**sec·tion** (sĕk'shən) *n.* [Lat. *sectio,* act of cutting < *sectus,* p.part. of *secare,* to cut.] **1.** A part or piece of something : PORTION. **2.** A subdivision of a written work. **3.** A division of a statute or legal code. **4.** A separate portion of a newspaper. **5.** A distinct area of a town, county, or country. **6.** A land unit of 640 acres or one square mile, equal to 1/36 of a township. **7.** The act or process of separating or cutting, esp. the surgical separation of tissue. **8.** A thin slice, as of tissue, used for microscopic examination. **9.** A segment of a fruit <a grapefruit *section*> **10.** The representation of a solid object as it would appear if cut by an intersecting plane, thus displaying the internal structure. **11.** *Geom.* The set of points formed by the intersection of a solid by a plane. **12. a.** A part of a railroad track maintained by a single crew. **b.** An area in a sleeping car that has an upper and lower berth. **13.** An army tactical unit smaller than a platoon and larger than a squad. **14.** A unit of vessels or aircraft within a military division. **15. a.** The character (§) used in printing to mark the beginning of a section. **b.** This character used as the fourth in a series of reference marks for footnotes. —*vt.* **-tioned, -tion·ing, -tions. 1.** To divide into parts : SEPARATE. **2.** To separate (tissue) surgically. **3.** To shade or crosshatch (part of a drawing) to indicate sections.

**sec·tion·al** (sĕk'shə-nəl) *adj.* **1.** Of, relating to, or like a particular district or region. **2.** Composed of or divided into component sections. —*n.* A piece of furniture having sections that can be used separately or together. **—sec'tion·al·ly** *adv.*

**sec·tion·al·ism** (sĕk'shə-nə-lĭz'əm) *n.* Inordinate attention or devotion to local interests and customs. **—sec'tion·al·ist** *n.*

**sec·tion·al·ize** (sĕk'shə-nə-līz') *vt.* **-ized, -iz·ing, -iz·es.** To divide into sections, esp. into geographic sections. **—sec'tion·al·i·za'tion** *n.*

**Section Eight** *n.* [After *Section VIII,* Army Regulation 615–360, which provided for the discharge of psychopaths and neurotics.] **1.** A U.S. Army discharge based on military unfitness or undesirable character traits. **2.** A soldier discharged under a Section Eight.

**section gang** *n.* A work crew assigned to a section of railroad track.

**section hand** *n.* One who works on a section gang.

**sec·tor** (sĕk'tər, -tôr') *n.* [LLat. < Lat., cutter < *secare,* to cut.] **1.** *Math.* The part of a circle bounded by two radii and one of the intercepted arcs. **2.** A division of a defensive or offensive position for which one military unit is responsible. **3.** A distinct part or division <the agricultural *sector* of the national economy> **4.** *Computer Sci.* A bit or set of bits on a magnetic storage system making up the smallest addressable unit of information. —*vt.* **-tored, -tor·ing, -tors.** To divide into sectors. **—sec·to'ri·al** (-tôr'ē-əl, -tōr'-) *adj.*

**sec·u·lar** (sĕk'yə-lər) *adj.* [ME *seculer* < OFr. < Lat. *saecularis* < *saeculum,* generation, age.] **1.** Worldly rather than spiritual. **2.** Not relating directly to religion or to a religious body <*secular* music> **3.** Relating to or advocating secularism. **4.** Not living in a religious community or bound by monastic restrictions <*secular* clergy> **5.** Happening or observed once in an age or century. **6.** Lasting or continuing from century to century. —*n.* **1.** A secular cleric. **2.** A layperson. **—sec'u·lar'i·ty** (-lăr'ĭ-tē) *n.* **—sec'u·lar·ly** *adv.*

**sec·u·lar·ism** (sĕk'yə-lə-rĭz'əm) *n.* The belief that consideration of the present well-being of mankind should take precedence over religious considerations in civil affairs or public education. **—sec'u·lar·ist** *n.* **—sec'u·lar·is'tic** *adj.*

**sec·u·lar·ize** (sĕk'yə-lə-rīz') *vt.* **-ized, -iz·ing, -iz·es. 1.** To convert from ecclesiastical or religious to civil or lay use, ownership, or control. **2.** To make worldly. **3.** To lift the monastic restrictions from (a member of the clergy). **—sec'u·lar·i·za'tion** *n.*

**se·cund** (sē'kŭnd', sĭ-kŭnd') *adj.* [Lat. *secundus,* following.] *Bot.* Arranged on or facing one side of an axis.

**se·cun·dines** (sē-kŭn'dīnz', sĕk'ən-dīnz') *pl.n.* [ME *secundine* < LLat. *secundinae* < *secundus,* following.] Afterbirth.

**se·cure** (sĭ-kyŏŏr') *adj.* **-cur·er, -cur·est.** [Lat. *securus* : *se-,* without + *cura,* care.] **1.** Free from danger, harm, or risk of loss : SAFE. **2.** Free from fear, anxiety, or doubt : CONFIDENT. **3. a.** Not likely to fail or give way : STABLE. **b.** Well-fastened. **4.** Assured : certain. **5.** *Archaic.* Careless or overconfident. —*vt.* **-cured, -cur·ing, -cures. 1.** To guard from danger, harm, or risk of loss. **2.** To make tight or firm : FASTEN. **3.** To make certain : GUARANTEE. **4.** To make a pledge on (e.g., a loan). **5.** To gain possession of : ACQUIRE. **6.** To bring about : EFFECT. **—se·cur'a·ble** *adj.* **—se·cure'ly** *adv.* **—se·cure'ment** *n.* **—se·cure'ness** *n.* **—se·cur'er** *n.*

**se·cu·ri·ty** (sĭ-kyŏŏr'ĭ-tē) *n., pl.* **-ties.** [ME *securite* < Lat. *securitas* < *securus,* secure.] **1.** Freedom from danger, harm, or risk of loss : SAFETY. **2.** Freedom from doubt, anxiety, or fear : CONFIDENCE. **3.** Something that gives or assures safety. **4.** *Computer Sci.* **a.** The degree to which a program or device is safe from unauthorized use. **b.** Prevention of unauthorized use of a program or device. **5.** Something deposited or given as assurance of the fulfillment of an obligation : PLEDGE. **6.** One who guarantees or assumes the financial obligations of another : SURETY. **7. securities.** Written evidence of ownership or creditorship, esp. stocks or bonds. **8.** Measures adopted to guard against attack or disclosure, as in wartime.

**security blanket** *n.* **1.** A blanket or toy carried by a child to relieve anxiety. **2.** Something that dispels anxiety.

**security guard** *n.* One employed by a private organization to guard a physical plant and maintain order.

**se·dan** (sĭ-dăn') *n.* [Orig. unknown.] **1.** A closed automobile with two or four doors and a front and rear seat. **2.** An enclosed chair for one, with poles front and rear, carried by two men.

**Se·dar·im** (sĭ-dăr'ĭm) *n. var. pl. of* SEDER.

**se·date[1]** (sĭ-dāt') *adj.* [Lat. *sedatus,* p.part. of *sedare,* to settle, calm.] Serene and dignified. **—se·date'ly** *adv.* **—se·date'ness** *n.*

**se·date[2]** (sĭ-dāt') *vt.* **-dat·ed, -dat·ing, -dates.** [Back-formation < SEDATIVE.] To administer a sedative to.

**se·da·tion** (sĭ-dā'shən) *n.* **1.** Administration of a sedative to reduce stress or excitement. **2.** The state induced by a sedative.

**sed·a·tive** (sĕd'ə-tĭv) *adj.* [Fr. *sédatif* < Med. Lat. *sedativus* < Lat. *sedare,* to calm.] Having a calming, soothing, or tranquilizing effect. —*n.* A sedative drug or agent.

**sed·en·tar·y** (sĕd'n-tĕr'ē) *adj.* [Fr. *sédentaire* < Lat. *sedentarius* < *sedens,* pr.part. of *sedēre,* to sit.] **1.** Requiring or marked by much sitting <a *sedentary* desk job> **2.** Used to sitting or to taking little exercise. **3.** Staying in one area : not migratory. **4.** *Zool.* Attached to a surface and not free-moving, as a barnacle. **—sed'en·tar'i·ly** (-târ'ə-lē) *adv.* **—sed'en·tar'i·ness** *n.*

**Se·der** (sā′dər) *n., pl.* **Se·ders** *or* **Se·dar·im** (sĭ-där′ĭm) [Heb. *sēdher,* order.] The feast commemorating the exodus of the Israelites from Egypt, celebrated on the first or first and second evenings of Passover.

**sedge** (sĕj) *n.* [ME *segge* < OE *secg.*] Any of various plants of the family Cyperaceae, resembling grasses but having solid stems.

**se·dil·ia** (sĭ-dēl′yə, -dĭl′-) *pl.n.* [Lat. *sedilia,* pl. of *sedile,* seat < *sedēre,* to sit.] A set of usu. three seats built into the wall on the south side of the choir near the altar in Gothic-style churches for the use of the officiating clergy.

**sed·i·ment** (sĕd′ə-mənt) *n.* [Fr. *sédiment* < Lat. *sedimentum,* act of settling < *sedēre,* to sit, settle.] **1.** Finely divided solid material that settles to the bottom of a liquid : LEES. **2. a.** Such material suspended in water or in the air. **b.** The deposition of such material onto the surface beneath this water or air. **c.** The material so deposited.

**sed·i·men·ta·ry** (sĕd′ə-mĕn′tə-rē, -mĕn′trē) *also* **sed·i·men·tal** (-mĕn′tl) *adj.* **1.** Of, containing, resembling, or derived from sediment. **2.** *Geol.* Of or relating to rocks formed from sediment or from fragments of other rocks deposited in water.

**sed·i·men·ta·tion** (sĕd′ə-mən-tā′shən, -mĕn-) *n.* The act or process of depositing sediment.

**sed·i·men·tol·o·gy** (sĕd′ə-mən-tŏl′ə-jē, -mĕn-) *n.* The science dealing with the examination, description, classification, and origin of sedimentary rock. **—sed′i·men·to·log′ic** (-tə-lŏj′ĭk), **sed′i·men·to·log′i·cal** *adj.* **—sed′i·men·tol′o·gist** *n.*

**se·di·tion** (sĭ-dĭsh′ən) *n.* [ME *sedicioun* < OFr. *sedition* < Lat. *seditio,* separation : *se-,* apart + *itio,* act of going < *itus,* p.part. of *ire,* to go.] **1.** Conduct or language that incites others to rebel against the authority of a state. **2.** Rebellion : insurrection. **—se·di′tion·ist** *n.*

**se·di·tious** (sĭ-dĭsh′əs) *adj.* **1.** Of or constituting sedition. **2.** Engaged in sedition. **—se·di′tious·ly** *adv.* **—se·di′tious·ness** *n.*

**se·duce** (sĭ-do͞os′, -dyo͞os′) *vt.* **-duced, -duc·ing, -duc·es.** [ME *seduisen* < OFr. *seduire, seduis-* < Lat. *seducere,* to lead away : *se-,* apart + *ducere,* to draw.] **1.** To draw (one) away from duty or proper conduct : CORRUPT. **2.** To induce to have sexual intercourse. **3. a.** To beguile or entice into a desired position or state. **b.** To win over : ATTRACT. **—se·duce′a·ble, se·duc′i·ble** *adj.* **—se·duc′er** *n.*

**se·duce·ment** (sĭ-do͞os′mənt, -dyo͞os′-) *n.* Seduction.

**se·duc·tion** (sĭ-dŭk′shən) *n.* **1.** The act or state of seducing or being seduced. **2.** Something that seduces : ENTICEMENT.

**se·duc·tive** (sĭ-dŭk′tĭv) *adj.* Tending to seduce : ALLURING. **—se·duc′tive·ly** *adv.* **—se·duc′tive·ness** *n.*

**se·duc·tress** (sĭ-dŭk′trĭs) *n.* A woman who seduces.

**sed·u·lous** (sĕj′ə-ləs) *adj.* [Lat. *sedulus.*] Diligent or industrious : PAINSTAKING. **—sed′u·lous·ly** *adv.* **—sed′u·lous·ness, se·du′li·ty** (sĭ-do͞o′lĭ-tē, -dyo͞o′-) *n.*

**se·dum** (sē′dəm) *n.* [NLat. *Sedum,* genus name < Lat. *sedum,* houseleek.] Any of various plants of the genus *Sedum,* with thick fleshy leaves.

**see**[1] (sē) *v.* **saw** (sô), **seen** (sēn), **see·ing, sees.** [ME *seen* < OE *sēon.*] *—vt.* **1.** To perceive with the eye. **2.** To perceive or comprehend as if by the sense of sight <*see* with one's fingers> **3.** To have a mental image of : VISUALIZE <could still *see* the town as it once was> **4.** To understand : comprehend <*saw* the point of the story> **5.** To regard : judge. **6.** To believe possible : IMAGINE <I don't *see* you as a wheeler-dealer.> **7.** To foresee. **8.** To know through firsthand experience : UNDERGO. **9.** To be characterized by or bring forth. **10.** To find out : ASCERTAIN. **11.** To refer to : READ <*See* page one of the introduction.> **12.** To take note of. **13.** To meet or be in the company of. **14.** To socialize together often or regularly. **15.** To visit socially or for consultation <*see* an attorney> **16.** To receive socially or for consultation <an attorney who *sees* many clients> **17.** To attend : view. **18.** To escort : attend <*see* my parents off on their trip> **19.** To make certain <*See* that this is taken care of immediately.> **20. a.** To meet (a bet) in card games. **b.** To meet the bet of (another player). *—vi.* **1.** To have the power of sight. **2.** To understand : comprehend. ***usage:*** *See* in the sense of "to understand" sometimes is followed by a clause beginning with *where* in informal speech, as in *I see where you're running for the city council.* Such an informal usage, while permissible in speech, should be avoided in a formal context, and sentences such as the preceding are best expressed as *I see that you are running for the city council.* **3.** To think over : CONSIDER <Let's *see,* where should we go tonight?> **4.** To wait and decide later <We probably will go, but we'll have to *see.*> **5.** To have foresight <"No man can *see* to the end of time" —John F. Kennedy> **6.** To take note. **—see about. 1.** To attend to. **2.** To investigate. **—see through.** To understand the real character or nature of. **—see to.** To attend to.

**see**[2] (sē) *n.* [ME < OFr. *se* < Lat. *sedes,* seat.] **1.** The authority, jurisdiction, position, or official seat of a bishop. **2.** *Obs.* A cathedra.

**see·catch** (sē′kăch′) *n., pl.* **-catch·ie** (-kăch′ē) [R. *sekach*] An adult male Alaskan fur seal.

**seed** (sēd) *n., pl.* **seeds** *or* **seed.** [ME < OE *sǣd.*] **1.** A fertilized and ripened plant ovule having an embryo that is capable of germinating to produce a new plant. **2.** A propagative part of a plant, as a bulb, tuber, or spore. **3.** Seeds as a group. **4.** The seed-bearing stage of a plant. **5.** A beginning or source : ORIGIN. **6.** Offspring : descendants. **7.** Family stock : ANCESTRY. **8.** Sperm : semen. **9.** A young oyster or oysters used for propagating a new oyster bed. **10.** A competitor who has been ranked in a tournament. *—v.* **seed·ed, seed·ing, seeds.** *—vt.* **1.** To plant seeds in : SOW. **2.** To plant in soil. **3.** To remove the seeds from (fruit). **4.** To sprinkle (a cloud) with particles, as of silver iodide, in order to disperse it or induce rain. **5. a.** To arrange (the drawing for positions in a tournament) so that the more skilled contestants meet in the later rounds. **b.** To rank (a contestant) in this manner. *—vi.* **1.** To sow seed. **2.** To go to seed. **—go (***or* **run) to seed. 1.** To progress to the seed-bearing stage. **2.** To become useless or devitalized : DETERIORATE.

**seed·bed** (sēd′bĕd′) *n.* **1.** A plot of soil cultivated for germinating seeds. **2.** An area or source of growth.

**seed·cake** (sēd′kāk′) *n.* A cake or cookie made with aromatic seeds.

**seed coat** *n.* The protective outer covering of a seed.

**seed·er** (sē′dər) *n.* **1.** One that plants seeds. **2.** One that removes the seeds from fruit.

**seed leaf** *n.* COTYLEDON 1.

**seed·ling** (sēd′lĭng) *n.* A young plant grown from a seed.

**seed money** *n.* Money required or given to start a new project.

**seed oyster** *n.* A young oyster, esp. one suitable for transplanting.

**seed pearl** *n.* A tiny, often imperfect pearl.

**seed plant** *n.* A seed-bearing plant.

**seed·pod** (sēd′pŏd′) *n.* POD[1] 1.

**seed stock** *n.* **1.** Seed for planting. **2.** A source of new entities or growth <*a seed stock* of fish in the pool>

**seed·time** (sēd′tīm′) *n.* **1.** A time for sowing seeds. **2.** A time of new growth or development.

**seed vessel** *n. Bot.* A pericarp.

**seed·y** (sē′dē) *adj.* **-i·er, -i·est. 1.** Having many seeds. **2.** Seedlike. **3.** Run-down and shabby : UNKEMPT. **4.** Tired or ill. **5.** Squalid <a *seedy* area of town> **—seed′i·ly** *adv.* **—seed′i·ness** *n.*

**see·ing** (sē′ĭng) *conj.* Considering that.

**Seeing Eye.** A trademark for a dog trained to aid a blind person.

**seek** (sēk) *v.* **sought** (sôt), **seek·ing, seeks.** [ME *seken* < OE *sēcan.*] *—vt.* **1.** To try to find or discover : search for. **2.** To try to obtain or reach. **3.** To go to or toward <Water *seeks* its own level.> **4.** To ask for : REQUEST. **5.** To try : endeavor. **6.** *Obs.* To explore. *—vi.* To make a search or investigation. *—n. Computer Sci.* A serial inspection of stored data in order to find and produce a desired item of data. **—seek′er** *n.*

**seel** (sēl) *vt.* **seeled, seel·ing, seels.** [ME *silen* < OFr. *ciller* < Med. Lat. *cilare* < Lat. *cilium,* eyelid.] To stitch closed the eyes of (a falcon).

**seem** (sēm) *vi.* **seemed, seem·ing, seems.** [ME *semen* < ON *sœma,* to conform to < *sœnt,* fitting.] **1.** To give the impression of being : APPEAR. **2.** To appear to one's own opinion or intellect. **3.** To appear to be evident. **4.** To appear to exist.

**seem·ing** (sē′mĭng) *adj.* Apparent : ostensible. *—n.* External appearance : SEMBLANCE. **—seem′ing·ly** *adv.* **—seem′ing·ness** *n.*

**seem·ly** (sēm′lē) *adj.* **-li·er, -li·est.** [ME *semely* < ON *sœmiligr* < *sœmr,* fitting.] **1.** Conforming to accepted standards of conduct and good taste : APPROPRIATE <*seemly* behavior> **2.** Of pleasant appearance : HANDSOME. *—adv.* In a suitable way. **—seem′li·ness** *n.*

**seen** (sēn) *v. p.p. of* SEE.

**seep** (sēp) *vi.* **seeped, seep·ing, seeps.** [Alteration of dial. *sipe* < ME *sipen* < OE *sipian.*] **1.** To pass slowly through small openings or pores : OOZE. **2.** To enter, leave, or become diffused gradually. *—n.* A place where water or petroleum oozes from the ground to form a pool.

**seep·age** (sē′pĭj) *n.* **1.** The act or process of seeping. **2.** Something that has seeped.

**seer** (sîr, sē′ər) *n.* [ME < *seen,* to see.] **1.** One that sees. **2.** A prophet. **3.** A clairvoyant.

**seer·ess** (sîr′ĭs) *n.* A woman who is a prophet or clairvoyant.

**seer·suck·er** (sîr′sŭk′ər) *n.* [Urdu *sīrsakar* < Pers. *shīr-o-shakar* : *shīr,* milk + *o,* and + *shakar,* sugar < Skt. *śarkarā.*] A lightweight fabric, gen. of cotton or rayon, with a crinkled surface and usu. a striped pattern.

**see·saw** (sē′sô′) *n.* [Redup. of SAW[1].] **1.** A long board balanced on a central fulcrum so that with a person riding on each end, one end goes up as the other goes down. **2.** The act or game of riding a seesaw. **3.** An up-and-down or back-and-forth movement. *—vi.* **-sawed, -saw·ing, -saws. 1.** To play on a seesaw. **2.** To move up and down or back and forth.

**seethe** (sēth) *v.* **seethed, seeth·ing, seethes.** [ME *sethen,* to boil < OE *sēothan.*] *—vi.* **1.** To churn and bubble while or as if while boiling. **2.** To move in confusion : FERMENT. **3.** To be violently agitated or excited. **4.** *Archaic.* To come to a boil. *—vt.* **1.** To steep in liquid. **2.** *Archaic.* To boil. **—seeth** *n.*

**see-through** (sē′thro͞o′) *adj.* Transparent.

**seg·ment** (sĕg′mənt) *n.* [Lat. *segmentum* < *secare,* to cut.] **1.** Any of the parts into which something can be divided : SECTION. **2.** *Math.* A section of a figure cut off by a line or plane, esp.: **a.** The region bounded by a chord and the arc of a curve subtended by the chord. **b.** The part of a curve between any two points on the curve. **c.** The part of a sphere bounded by two parallel planes intersecting or tangent to the sphere. **3.** *Biol.* A distinct subdivision of an organism or part, as a metamere. —*vt.* & *vi.* (sĕg-mĕnt′) **-ment·ed, -ment·ing, -ments.** To divide or become divided into segments. **—seg·men′tar′y** (-mən-tĕr′ē) *adj.*

**seg·men·tal** (sĕg-mĕn′tl) *adj.* **1.** Of or pertaining to segments. **2.** Divided or organized into segments. **—seg·men′tal·ly** *adv.*

**seg·men·ta·tion** (sĕg′mən-tā′shən, -mĕn-) *n.* **1.** Division into segments. **2.** *Biol.* Cleavage.

**segmentation cavity** *n. Biol.* A blastocoel.

**se·gno** (sā′nyō) *n., pl.* **-gnos.** [Ital., sign < Lat. *signum.*] *Mus.* A sign, esp. one marking the beginning or end of a repeat.

**se·go** (sē′gō) *n., pl.* **-gos.** [Of Ute orig.] The succulent edible bulb of the sego lily.

**sego lily** *n.* A plant, *Calochortus nuttallii,* of western North America, with variously colored trumpet-shaped flowers.

**sego lily**

**seg·re·gate** (sĕg′rĭ-gāt′) *v.* **-gat·ed, -gat·ing, -gates.** [Lat. *segregare, segregat-* : *se-,* apart + *grex,* flock.] —*vt.* **1.** To set apart or isolate from others or from a central body or group. **2.** To impose the separation of (a race or class) from the rest of a people. —*vi.* **1.** To become separated from a central body or group. **2.** To practice a policy of racial segregation. **3.** *Genetics.* To undergo genetic segregation. —*adj.* (-gĭt, -gāt′). Set apart : ISOLATED. **—seg′re·ga′tive** *adj.* **—seg′re·ga′tor** *n.*

**seg·re·ga·tion** (sĕg′rĭ-gā′shən) *n.* **1.** The act or process of segregating or the state of being segregated. **2.** The policy or practice of imposing the social separation of races, as in schools, housing, and employment. **3.** *Genetics.* Separation of paired alleles in meiosis.

**seg·re·ga·tion·ist** (sĕg′rĭ-gā′shə-nĭst) *n.* One who advocates or practices a policy of racial segregation.

**se·gue** (sĕg′wā′, sā′gwā′) *vi.* **-gued, -gu·ing, -gues.** [Ital., there follows < *seguire,* to follow < Lat. *sequi.*] To make an immediate transition from one section or theme, as of music, to another.

**se·gui·di·lla** (sĕg′ə-dē′yə, -dēl′yə) *n.* [Sp. < *seguida,* fem. p.part. of *seguir,* to follow < VLat. **sequere* < Lat. *sequi.*] **1.** A Spanish stanza form of four to seven short verses. **2. a.** An energetic Spanish dance. **b.** The music for this dance, in 3/4 time.

**sei·cen·to** (sā-chĕn′tō) *n.* [Ital., short for *milleseicento,* one thousand six hundred.] The 17th cent. period in Italian literature and art.

**seiche** (sāsh, sĕch) *n.* [Dial. Fr.] A wave that oscillates in partially or totally enclosed bodies of water from a few minutes to a few hours, caused by seismic or atmospheric disturbances.

**Seid·litz powder** *also* **Seid·litz powders** (sĕd′lĭts) *n.* [So called from its laxative properties, which are similar to those of the spring water at *Seidlitz,* a village in Bohemia.] A cathartic containing Rochelle salts, sodium bicarbonate, and tartaric acid.

**seign·ior** (sān-yôr′, sān′yôr′) *n.* [ME *seignour* < OFr. *seigneur* < Med. Lat. *senior* < Lat., older, comp. of *senex,* old.] A man of high rank, esp. a feudal lord. **—sei·gnio′ri·al** *adj.*

**seign·ior·age** (sān′yər-ĭj) *n.* [ME *seigneurage* < OFr. < *seigneur,* seignior.] A revenue or profit taken from the minting of coins, usu. the difference between the intrinsic value of the bullion and the face value of the coin.

**seign·io·ry** (sān′yə-rē) *n., pl.* **-ies. 1.** A feudal lord's estate. **2.** The authority and power of a feudal lord.

**seine** (sān) *n.* [ME < OE *segne* < Lat. *sagena* < Gk. *sagēnē.*] A fishing net that hangs vertically in the water by means of weights at the lower edge and floats at the top. —*v.* **seined, sein·ing, seines.** —*vi.* To fish with a seine. —*vt.* To fish for or catch with a seine. **—sein′er** *n.*

**seise** (sēz) *v. var. of* SEIZE 6.

**sei·sin** *also* **sei·zin** (sē′zĭn) *n.* [ME *seisine* < OFr. < *seisir,* to seize.] *Law.* **1.** Possession of land, as a freehold estate. **2. a.** The act of taking possession of land. **b.** Property thus possessed.

**seism** (sī′zəm) *n.* [Gk. *seismos* < *seiein,* to shake.] An earthquake.

**seism-** *pref. var. of* SEISMO-.

**seis·mic** (sīz′mĭk) *adj.* Of, subject to, or caused by an earthquake or earth tremor. **—seis′mi·cal·ly** *adv.* **—seis·mic′i·ty** (-mĭs′ĭ-tē) *n.*

**seis·mism** (sīz′mĭz′əm) *n.* Earthquake phenomena.

**seismo-** *or* **seism-** *pref.* [< Gk. *seismos,* earthquake.] Earthquake <*seismo*graph>

**seis·mo·gram** (sīz′mə-grăm′) *n.* The record of an earth tremor made by a seismograph.

**seis·mo·graph** (sīz′mə-grăf′) *n.* An instrument that automatically detects and records the direction, duration, and strength of a ground vibration, esp. of an earthquake. **—seis·mog′ra·pher** (sīz-mŏg′rə-fər) *n.* **—seis′mo·graph′ic** *adj.* **—seis·mog′ra·phy** *n.*

**seis·mol·o·gy** (sīz-mŏl′ə-jē) *n.* The science that deals with earthquakes and the mechanical properties of the earth. **—seis′mo·log′ic** (-mə-lŏj′ĭk), **seis′mo·log′i·cal** *adj.* **—seis′mo·log′i·cal·ly** *adv.* **—seis·mol′o·gist** *n.*

**seis·mom·e·ter** (sīz-mŏm′ĭ-tər) *n.* A detector of seismic impulses. **—seis′mo·met′ric** (-mə-mĕt′rĭk), **seis′mo·met′ri·cal** *adj.*

**seis·mom·e·try** (sīz-mŏm′ĭ-trē) *n.* The study of earthquakes.

**seis·mo·scope** (sīz′mə-skōp′) *n.* An instrument that indicates only the occurrence or time of occurrence of an earthquake. **—seis′mo·scop′ic** (-skŏp′ĭk) *adj.*

**sei·sor** (sē′zər, -zôr′) *n. var. of* SEIZOR.

**seize** (sēz) *v.* **seized, seiz·ing, seiz·es.** [ME *seisen* < OFr. *seisir,* to take possession, of Germanic orig.] —*vt.* **1.** To take hold of suddenly and forcibly : GRAB. **2.** To grasp mentally : COMPREHEND. **3.** To affect suddenly : OVERWHELM. **4.** To take into custody : ARREST. **5.** To take quick and forcible possession of <*seize* a cache of rifles> **6.** *also* **seise. a.** *Law.* To take into legal custody : CONFISCATE. **b.** *Obs.* To put in possession of a feudal property. **7.** To put to quick advantage, as an opportunity. **8.** *Naut.* To bind with turns of small line. —*vi.* **1.** To lay hold suddenly or forcibly <*seize* on one's chance> **2.** To fuse or cohere with another part due to high pressure or temperature, slowing or stopping further motion. **—seiz′a·ble** *adj.* **—seiz′er** *n.*

**sei·zin** (sē′zĭn) *n. var. of* SEISIN.

**seiz·ing** (sē′zĭng) *n. Naut.* **1.** A binding made with multiple turns of small line. **2.** The line used for seizings.

**seiz·or** *also* **sei·sor** (sē′zər, -zôr′) *n. Law.* One that takes seisin.

**sei·zure** (sē′zhər) *n.* **1.** The act of seizing or the state of being seized. **2.** A sudden, often acute paroxysm, as an epileptic convulsion or heart attack. **3.** A sudden subjective sensation.

**se·la·chi·an** (sĭ-lā′kē-ən) *adj.* [< NLat. *Selachi,* order name < Gk. *selakhē,* pl. of *selakhos,* a cartilaginous fish.] Of or belonging to the order Selachii or Squaliformes, which includes the sharks and rays. —*n.* A member of the Selachii or Squaliformes.

**se·lag·i·nel·la** (sə-lăj′ə-nĕl′ə) *n.* [NLat. *Selaginella,* genus name < Lat. *selago,* a plant resembling the savin.] A fernlike, usu. prostrate plant of the genus *Selaginella,* having scalelike leaves and spores.

**se·lah** (sē′lə, -lä′) *n.* [Heb. *selāh.*] A Hebrew word of unknown meaning that often marks the end of a verse in the Psalms and is thought to be a term indicating a pause or rest.

**sel·dom** (sĕl′dəm) *adv.* [ME *selden* < OE *seldan.*] Not often : RARELY. —*adj. Archaic.* Infrequent : rare. **—sel′dom·ness** *n.*

**se·lect** (sĭ-lĕkt′) *v.* **-lect·ed, -lect·ing, -lects.** [Lat. *seligere, select-* : *se-,* apart + *legere,* to choose.] —*vt.* To pick out from among several : CHOOSE. —*vi.* To make a choice or selection. —*adj. also* **se·lect·ed** (-lĕk′tĭd). **1.** Singled out in preference : CHOSEN. **2.** Of particular value or quality. **—se·lect′** *n.* **—se·lect′ness** *n.*

**se·lect·ee** (sĭ-lĕk′tē′) *n.* One selected, esp. for military service.

**se·lec·tion** (sĭ-lĕk′shən) *n.* **1. a.** The act of selecting or the fact of being selected. **b.** One selected. **2.** A carefully chosen collection. **3.** A literary or musical text chosen for reading or performance. **4.** *Biol.* A process that favors or brings about the survival and perpetuation of one kind of organism in competition with others.

**se·lec·tive** (sĭ-lĕk′tĭv) *adj.* **1.** Of or marked by selection : DISCRIMINATING. **2.** Empowered or tending to select. **3.** *Electron.* Capable of rejecting frequencies other than that selected or tuned. **—se·lec′tive·ly** *adv.* **—se·lec′tive·ness** *n.*

**selective service** *n.* A system for choosing individuals for compulsory military service.

**se·lec·tiv·i·ty** (sĭ-lĕk′tĭv′ĭ-tē, sē′lĕk-) *n.* **1.** The quality or state of being selective. **2.** *Electron.* The degree to which a receiver is selective.

**se·lect·man** (sĭ-lĕkt′măn′, -mən) *n.* One of a board of officers chosen annually in most New England towns to manage local affairs.

**se·lec·tor** (sĭ-lĕk′tər) *n.* One that selects.

**selen-** *pref. var. of* SELENO-.

**sel·e·nate** (sĕl′ə-nāt′) *n.* [SELEN(IC ACID) + -ATE².] A salt or ester of selenic acid.

**se·le·nic** (sə-lē′nĭk, -lĕn′ĭk) *adj.* Of or containing selenium.

**selenic acid** *n.* A corrosive hygroscopic white solid acid with composition $H_2SeO_4$.

**se·le·ni·ous acid** (sĭ-lē′nē-əs) *n.* Selenous acid.

**sel·e·nite** (sĕl′ə-nīt′) *n.* [Lat. *selenites* < Gk. *selēnitēs (lithos),* moon (stone) < *selēnē,* moon, from the belief that it waxed and waned with the moon.] Gypsum in the form of clear colorless crystals.

**se·le·ni·um** (sĭ-lē′nē-əm) *n.* [Gk. *selēnē,* moon + -IUM.] *Symbol*

Se A nonmetallic element resembling sulfur, used as a semiconductor and in xerography; atomic number 34; atomic weight 78.96.

**selenium cell** *n.* A photoconductive cell having an insulated selenium strip between two suitable electrodes.

**seleno-** or **selen-** *pref.* [< Gk. *selenē,* moon.] **1.** Moon <*selenography*> **2.** Selenium <*selenosis*>

**sel·e·nod·e·sy** (sĕl′ə-nŏd′ĭ-sē) *n.* [SELENO- + (GEO)DESY.] The mathematical study of the precise shape and size of the moon.

**sel·e·nog·ra·phy** (sĕl′ə-nŏg′rə-fē) *n.* The study of the physical features of the moon. **—sel′e·nog′ra·pher, sel′e·nog′ra·phist** *n.* **—sel′e·no·graph′ic** (-nə-grăf′ĭk), **sel′e·no·graph′i·cal** *adj.* **—sel′e·no·graph′i·cal·ly** *adv.*

**sel·e·nol·o·gy** (sĕl′ə-nŏl′ə-jē) *n.* The astronomical study of the moon. **—sel′e·no·log′i·cal** (-nə-lŏj′ĭ-kəl) *adj.* **—sel′e·nol′o·gist** *n.*

**sel·e·no·sis** (sĕl′ə-nō′sĭs) *n.* Selenium poisoning.

**se·le·nous acid** (sĭ-lē′nəs) *n.* A clear, colorless crystalline acid, $H_2SeO_3$, used as a chemical reagent.

**self** (sĕlf) *n., pl.* **selves** (sĕlvz) [ME < OE *self* (pronoun).] **1.** The total, essential, or particular being of a person : the individual. **2.** The essential qualities distinguishing one individual from another : INDIVIDUALITY. **3.** Consciousness of one's own being or identity : EGO. **4.** Personal welfare, interests, or advantage <thinking only of *self*> *—pron.* Myself, yourself, himself, or herself <a decent home for *self* and family> *—adj.* **1.** *Obs.* Same or identical. **2.** Of the same character throughout. **3.** Of the same material as the article with which it is used or worn <a robe with a *self* belt>

**self-** *pref.* [ME < OE < *self,* self.] **1.** Oneself : itself <*self*-control> **2.** Automatic : automatically <*self*-loading>

**self-a·ban·doned** (sĕlf′ə-băn′dənd) *adj.* Abandoned by oneself, esp. having yielded totally to one's impulses. **—self′a·ban′don·ment** *n.*

**self-a·base·ment** (sĕlf′ə-bās′mənt) *n.* Humiliation or degradation of oneself, esp. due to feelings of guilt or inferiority.

**self-ab·ne·ga·tion** (sĕlf′ăb′nĭ-gā′shən) *n.* Sacrifice of self-interest for the sake of others or for a principle. **—self′-ab′ne·gat′ing** *adj.*

**self-a·buse** (sĕlf′ə-byo͞os′) *n.* **1.** Abuse of oneself or one's abilities. **2.** Masturbation.

**self-act·ing** (sĕlf′ăk′tĭng) *adj.* Automatic.

**self-ac·tu·al·ize** (sĕlf′ăk′cho͞o-ə-līz′) *vi.* **-ized, -iz·ing, -iz·es.** To achieve one's maximum potential. **—self′-ac′tu·al·i·za′tion** *n.* **—self′-ac′tu·al·iz′er** *n.*

**self-ad·dressed** (sĕlf′ə-drĕst′) *adj.* Addressed to oneself.

**self-ag·gran·dize·ment** (sĕlf′ə-grăn′dĭz-mənt) *n.* Enhancement of one's own importance, power, or reputation. **—self′-ag·gran′diz′ing** (-ə-grăn′dī′zĭng) *adj.*

**self-a·nal·y·sis** (sĕlf′ə-năl′ĭ-sĭs) *n.* A methodical attempt to study and understand one's own personality or emotions. **—self′-an′a·lyt′i·cal** (-ăn′ə-lĭt′ĭ-kəl), **self′-an′a·lyt′ic** *adj.*

**self-an·ni·hi·la·tion** (sĕlf′ə-nī′ə-lā′shən) *n.* **1.** SELF-DESTRUCTION 2. **2.** Loss of self-awareness, as in a mystical state.

**self-ap·point·ed** (sĕlf′ə-poin′tĭd) *adj.* Chosen or designated by oneself rather than by others <a *self-appointed* social critic>

**self-as·sert·ing** (sĕlf′ə-sûr′tĭng) *adj.* **1.** Defending or maintaining one's own personality, rights, or opinions. **2.** Overbearing : arrogant.

**self-as·ser·tion** (sĕlf′ə-sûr′shən) *n.* **1.** Determined advancement or defense of one's own personality, rights, wishes, or opinions. **2.** Arrogance. **—self′-as·ser′tive** *adj.* **—self′-as·ser′tive·ly** *adv.*

**self-as·sured** (sĕlf′ə-sho͝ord′) *adj.* Having or exhibiting confidence in oneself. **—self′-as·sur′ance** *n.*

**self-a·ware** (sĕlf′ə-wâr′) *adj.* Conscious of oneself as an individual entity or personality. **—self′-a·ware′ness** *n.*

**self-cen·tered** (sĕlf′sĕn′tərd) *adj.* Concerned with oneself and one's own interests : SELFISH. **—self′-cen′tered·ly** *adv.* **—self′-cen′tered·ness** *n.*

**self-col·ored** (sĕlf′kŭl′ərd) *adj.* **1.** In the natural or original color. **2.** Of one color only.

**self-com·mand** (sĕlf′kə-mănd′) *n.* Self-control.

**self-com·pla·cent** (sĕlf′kəm-plā′sənt) *adj.* Marked by self-satisfaction. **—self′-com·pla′cen·cy** *n.* **—self′-com·pla′cent·ly** *adv.*

**self-con·cep·tion** (sĕlf′kən-sĕp′shən) *n.* Self-image.

**self-con·cern** (sĕlf′kən-sûrn′) *n.* A selfish or unwholesome regard for oneself. **—self′-con·cerned′** *adj.*

**self-con·fessed** (sĕlf′kən-fĕst′) *adj.* By one's own admission.

**self-con·fi·dence** (sĕlf′kŏn′fĭ-dəns) *n.* Confidence in oneself or one's abilities. **—self′-con′fi·dent** *adj.* **—self′-con′fi·dent·ly** *adv.*

**self-con·scious** (sĕlf′kŏn′shəs) *adj.* **1.** Excessively and usu. uncomfortably aware of one's appearance or manner. **2.** Socially uncomfortable. **3.** Displaying the effects of self-consciousness : STILTED <*self-conscious* prose> **4.** Aware of one's own individuality. **—self′-con′scious·ly** *adv.* **—self′-con′scious·ness** *n.*

**self-con·tained** (sĕlf′kən-tānd′) *adj.* **1.** Having within oneself or itself everything that is necessary : SELF-SUFFICIENT. **2.** Keeping to oneself : RESERVED.

**self-con·tent** (sĕlf′kən-tĕnt′) *adj.* Satisfied with oneself : COMPLACENT. **—self′-con·tent′, self′-con·tent′ment** *n.* **—self′-con·tent′ed** *adj.* **—self′con·tent′ed·ly** (-tĕn′tĭd-lē) *adv.*

**self-con·tra·dic·tion** (sĕlf′kŏn′trə-dĭk′shən) *n.* **1.** The act, state, or fact of contradicting oneself or itself. **2.** An idea or statement that contains contradictory elements. **—self′-con′tra·dic′to·ry** (-dĭk′tə-rē) *adj.*

**self-con·trol** (sĕlf′kən-trōl′) *n.* Control of one's feelings, desires, or actions by one's own will. **—self′-con·trolled′** *adj.*

**self-cor·rect·ing** (sĕlf′kə-rĕk′tĭng) *adj.* Of or comprising a typewriter mechanism that automatically allows for correction of an error <a *self-correcting* key>

**self-crit·i·cal** (sĕlf′krĭt′ĭ-kəl) *adj.* Severely judging one's own faults. **—self′-crit′i·cal·ly** *adv.* **—self′-crit′i·cism** *n.*

**self-de·ceit** (sĕlf′dĭ-sēt′) *n.* Self-deception.

**self-de·ceived** (sĕlf′dĭ-sēvd′) *adj.* Misled by one's own illusion or error.

**self-de·ceiv·ing** (sĕlf′dĭ-sē′vĭng) *adj.* Given to or furthering self-deception.

**self-de·cep·tion** (sĕlf′dĭ-sĕp′shən) *n.* The act of deceiving oneself or the state of being deceived by oneself. **—self′de·cep′tive** *adj.*

**self-de·feat·ing** (sĕlf′dĭ-fē′tĭng) *adj.* Harmful to one's or its own purposes or well-being.

**self-de·fense** (sĕlf′dĭ-fĕns′) *n.* **1.** Defense of oneself when physically attacked. **2.** Defense of what belongs to oneself, as one's reputation or rights. **3.** *Law.* The right to protect oneself against violence or threatened violence with whatever force or means reasonably necessary. **—self′-de·fen′sive** *adj.*

**self-de·ni·al** (sĕlf′dĭ-nī′əl) *n.* Denial of one's own comfort or pleasures. **—self′-de·ny′ing** *adj.* **—self′-de·ny′ing·ly** *adv.*

**self-dep·re·cat·ing** (sĕlf′dĕp′rĭ-kā′tĭng) *adj.* Belittling or detracting from oneself. **—self′-dep′re·cat′ing·ly** *adv.*

**self-dep·re·ca·to·ry** (sĕlf′dĕp′rĭ-kə-tôr′ē, -tōr′ē) *adj.* Self-deprecating <*self-deprecatory* remarks>

**self-de·pre·ci·a·tion** (sĕlf′dĭ-prē′shē-ā′shən) *n.* Belittlement of oneself.

**self-de·struct** (sĕlf′dĭ-strŭkt′) *n.* [Back-formation < SELF-DESTRUCTION.] A mechanism that causes a device to destroy itself. *—vi.* **-struct·ed, -struct·ing, -structs.** To destroy oneself or itself <a warhead that *self-destructs*>

**self-de·struc·tion** (sĕlf′dĭ-strŭk′shən) *n.* **1.** The act or process of destroying oneself. **2.** Suicide. **—self′-de·struc′tive** *adj.* **—self′-de·struc′tive·ly** *adv.* **—self′-de·struc′tive·ness** *n.*

**self-de·ter·mi·na·tion** (sĕlf′dĭ-tûr′mə-nā′shən) *n.* **1.** Freedom to choose one's own fate or course of action without compulsion : FREE WILL. **2.** Freedom of the people of a given area to choose their own political status : INDEPENDENCE.

**self-de·vel·op·ment** (sĕlf′dĭ-vĕl′əp-mənt) *n.* Development of one's potentialities.

**self-de·vo·tion** (sĕlf′dĭ-vō′shən) *n.* Dedication of oneself, esp. to a cause or belief. **—self′-de·vot′ed·ly** (-tĭd-lē) *adv.* **—self′-de·vot′ed·ness** *n.*

**self-di·ges·tion** (sĕlf′dĭ-jĕs′chən, -dī-) *n.* Autolysis.

**self-di·rect·ed** (sĕlf′dĭ-rĕk′tĭd, -dī-) *adj.* Guided or led by oneself, esp. as an independent agent. **—self′-di·rect′ing** *adj.* **—self′-di·rec′tion** *n.*

**self-dis·ci·pline** (sĕlf′dĭs′ə-plĭn) *n.* Training and control of oneself and one's behavior, usu. for personal improvement.

**self-dis·cov·er·y** (sĕlf′dĭ-skŭv′ə-rē) *n.* The act or process of understanding oneself.

**self-dis·trust** (sĕlf′dĭs-trŭst′) *n.* Lack of assurance in oneself. **—self′-dis·trust′ful** *adj.*

**self-doubt** (sĕlf′dout′) *n.* Lack of confidence in oneself. **—self′-doubt′ing** *adj.*

**self-driv·en** (sĕlf′drĭv′ən) *adj.* Driven by oneself or itself : AUTOMOTIVE.

**self-ed·u·cat·ed** (sĕlf′ĕj′ə-kā′tĭd) *adj.* Educated by oneself. **—self′-ed′u·ca′tion** *n.*

**self-ef·fac·ing** (sĕlf′ĭ-fā′sĭng) *adj.* Not drawing attention to oneself : HUMBLE. **—self′-ef·face′ment** *n.*

**self-e·lect·ed** (sĕlf′ĭ-lĕk′tĭd) *adj.* Self-appointed.

**self-em·ployed** (sĕlf′ĕm-ploid′) *adj.* Making a living by working for oneself rather than others. **—self′-em·ploy′ment** *n.*

**self-en·forc·ing** (sĕlf′ĕn-fôr′sĭng, -fōr′-) *adj.* Having within itself the means or a guarantee of its enforcement <a *self-enforcing* executive order>

**self-en·rich·ment** (sĕlf′ĕn-rĭch′mənt) *n.* The act or process of developing or augmenting one's mental faculties or spiritual resources.

**self-es·teem** (sĕlf′ĭ-stēm′) *n.* Satisfaction with oneself.

**self-ev·i·dent** (sĕlf′ĕv′ĭ-dənt) *adj.* Needing no proof or explanation. **—self′-ev′i·dence** *n.* **—self′-ev′i·dent·ly** *adv.*

**self-ex·am·i·na·tion** (sĕlf′ĭg-zăm′ə-nā′shən) *n.* Introspective analysis of one's own thoughts or feelings.

**self-ex·iled** (sĕlf′ĕg′zīld′, -ĕk′sīld′) *adj.* Exiled by one's own decision or will. **—self′-ex′ile′** *n.*

**self-ex·plan·a·to·ry** (sĕlf′ĭk-splăn′ə-tôr′ē, -tōr′ē) *adj.* Requiring no explanation : OBVIOUS.

**self-ex·pres·sion** (sĕlf′ĭk-sprĕsh′ən) *n.* Expression of one's own personality, as through art. **—self′-ex·pres′sive** (-sprĕs′ĭv) *adj.*
**self-fer·til·i·za·tion** (sĕlf′fûr′tl-ĭ-zā′shən) *n.* Fertilization by sperm from the same animal, as in some hermaphrodites, or by pollen from the same flower. **—self′-fer′til·ized′** *adj.*
**self-flag·el·la·tion** (sĕlf′flăj′ə-lā′shən) *n.* Harsh, recriminative criticism of oneself.
**self-for·get·ful** (sĕlf′fər-gĕt′fəl, -fôr-) *adj.* Unselfish : selfless. **—self′-for·get′ful·ly** *adv.* **—self′-for·get′ful·ness** *n.*
**self-ful·fill·ing** (sĕlf′fo͝ol-fĭl′ĭng) *adj.* **1.** Attaining self-fulfillment. **2.** Attaining fulfillment as a result of having been expected or foretold <a *self-fulfilling* prophecy>
**self-ful·fill·ment** (sĕlf′fo͝ol-fĭl′mənt) *n.* Fulfillment of oneself.
**self-giv·en** (sĕlf′gĭv′ən) *adj.* **1.** Emanating or derived from itself. **2.** Self-ordained.
**self-giv·ing** (sĕlf′gĭv′ĭng) *adj.* Marked by self-sacrificing behavior : UNSELFISH.
**self-gov·erned** (sĕlf′gŭv′ərnd) *adj.* **1.** Not controlled by others. **2.** Marked by self-discipline or self-control.
**self-gov·ern·ing** (sĕlf′gŭv′ər-nĭng) *adj.* **1.** Controlling or ruling oneself or itself. **2.** Possessing the right or power of self-government : AUTONOMOUS.
**self-gov·ern·ment** (sĕlf′gŭv′ərn-mənt) *n.* **1.** Political independence : AUTONOMY. **2.** Popular or representative government : DEMOCRACY. **3.** *Archaic.* Self-control.
**self-grat·i·fi·ca·tion** (sĕlf′grăt′ə-fĭ-kā′shən) *n.* The act or process of giving oneself pleasure or gratification.
**self-hard·en·ing** (sĕlf′här′dn-ĭng) *adj.* Of or relating to materials, as certain steels, that harden without special treatment after heating.
**self-heal** (sĕlf′hēl′) *n.* A plant thought to have healing powers, esp. *Prunella vulgaris*, a low-growing plant native to Europe, with violet-blue flower clusters.
**self-help** (sĕlf′hĕlp′) *n.* The act or an example of providing for or improving oneself.
**self·hood** (sĕlf′ho͝od′) *n.* [Transl. of G. *Selbheit.*] **1.** The state of having a distinct identity : INDIVIDUALITY. **2.** The fully developed self : PERSONALITY. **3.** Self-centeredness.
**self-hyp·no·sis** (sĕlf′hĭp-nō′sĭs) *n.* The act or process of hypnotizing oneself.
**self-i·den·ti·fi·ca·tion** (sĕlf′ī-dĕn′tə-fĭ-kā′shən) *n.* Identification of oneself with one that exists outside.
**self-i·den·ti·ty** (sĕlf′ī-dĕn′tĭ-tē) *n.* **1.** Oneness of a thing with itself. **2.** Awareness of oneself as a distinct individual.
**self-im·age** (sĕlf′ĭm′ĭj) *n.* One's idea of oneself or one's status.
**self-im·mo·la·tion** (sĕlf′ĭm′ə-lā′shən) *n.* Sacrifice of oneself, esp. by fire, for a cause or religious ideal.
**self-im·por·tance** (sĕlf′ĭm-pôr′tns) *n.* Excessively high opinion of one's own importance or status : CONCEIT. **—self′-im·por′tant** *adj.* **—self′-im·por′tant·ly** *adv.*
**self-im·posed** (sĕlf′ĭm-pōzd′) *adj.* Voluntarily chosen or endured <*self-imposed* exile>
**self-im·prove·ment** (sĕlf′ĭm-pro͞ov′mənt) *n.* Improvement of oneself or one's condition by one's own efforts.
**self-in·clu·sive** (sĕlf′ĭn-klo͞o′sĭv, -zĭv) *adj.* **1.** Enclosing or including oneself or itself. **2.** Whole or complete in itself.
**self-in·crim·i·na·tion** (sĕlf′ĭn-krĭm′ə-nā′shən) *n.* Incrimination of oneself, esp. by giving evidence or testimony that could lead to criminal prosecution. **—self′-in·crim′i·nat′ing** *adj.* **—self′-in·crim′i·na·to′ry** (-nə-tôr′ē, -tōr′ē) *adj.*
**self-in·duced** (sĕlf′ĭn-do͞ost′, -dyo͞ost′) *adj.* **1.** Induced by oneself or itself. **2.** *Elect.* Produced by self-induction.
**self-in·duc·tion** (sĕlf′ĭn-dŭk′shən) *n.* Generation by a changing current of an electromotive force in the same circuit tending to counteract such change. **—self′-in·duc′tive** *adj.*
**self-in·dul·gence** (sĕlf′ĭn-dŭl′jəns) *n.* Excessive gratification of one's desires. **—self′-in·dul′gent** *adj.* **—self′-in·dul′gent·ly** *adv.*
**self-in·flict·ed** (sĕlf′ĭn-flĭk′tĭd) *adj.* Inflicted or imposed on oneself <a *self-inflicted* injury>
**self-in·struct·ed** (sĕlf′ĭn-strŭk′tĭd) *adj.* Self-taught.
**self-in·struc·tion·al** (sĕlf′ĭn-strŭk′shə-nəl) *adj.* Of, pertaining to, or designed for independent study.
**self-in·sur·ance** (sĕlf′ĭn-sho͝or′əns) *n.* Insurance of oneself or one's possessions by regularly setting aside funds rather than paying premiums on a policy. **—self′-in·sured′** *adj.*
**self-in·ter·est** (sĕlf′ĭn′trĭst, -ĭn′tər-ĭst) *n.* **1.** Personal advantage or interest. **2.** Selfish regard for one's own advantage or interest. **—self′-in′ter·est·ed** *adj.*
**self-in·volved** (sĕlf′ĭn-vŏlvd′) *adj.* Absorbed in one's own interests or activities.
**self·ish** (sĕl′fĭsh) *adj.* **1.** Concerned only or primarily with oneself without regard for others. **2.** Arising from, marked by, or exhibiting selfishness <a *selfish* idea> **—self′ish·ly** *adv.* **—self′ish·ness** *n.*
☆ ***syns:*** SELFISH, SELF-CENTERED, SELF-SEEKING *adj. core meaning* : concerned only with oneself <a greedy, *selfish* person who cared for no one> ***ant:*** altruistic
**self-jus·ti·fy·ing** (sĕlf′jŭs′tə-fī′ĭng) *adj.* **1.** Making excuses for oneself. **2.** Justifying itself automatically <a *self-justifying* typewriter> **—self′-jus′ti·fi·ca′tion** (-jŭs′tə-fĭ-kā′shən) *n.*
**self-knowl·edge** (sĕlf′nŏl′ĭj) *n.* Awareness of one's own nature, abilities, and limitations.
**self·less** (sĕlf′lĭs) *adj.* Concerned about others rather than oneself : UNSELFISH. **—self′less·ly** *adv.* **—self′less·ness** *n.*
**self-lim·it·ing** (sĕlf′lĭm′ĭ-tĭng) *adj.* Limiting oneself or itself. **—self′-lim′i·ta′tion** (-tā′shən) *n.*
**self-liq·ui·dat·ing** (sĕlf′lĭk′wĭ-dā′tĭng) *adj.* **1.** Of, or involving goods that are easily converted into cash. —Used of business transactions. **2.** Yielding a return equal to the sum invested to create or maintain something.
**self-load·ing** (sĕlf′lō′dĭng) *adj.* SEMIAUTOMATIC 2.
**self-love** (sĕlf′lŭv′) *n.* The inclination or desire to promote one's own interests or well-being. **—self′-lov′ing** *adj.*
**self-made** (sĕlf′mād′) *adj.* **1.** Successful through one's own actions <a *self-made* millionaire> **2.** Made by oneself or itself.
**self-mail·er** (sĕlf′mā′lər) *n.* A folder that can be sealed and mailed without being enclosed in an envelope. **—self′-mail′ing** *adj.*
**self-mas·ter·y** (sĕlf′măs′tə-rē) *n.* Self-command.
**self·ness** (sĕlf′nĭs) *n.* **1.** The quality or condition of being self-centered : SELFISHNESS. **2.** Individuality.
**self-ob·ser·va·tion** (sĕlf′ŏb′zər-vā′shən) *n.* **1.** Observation of one's own countenance or appearance. **2.** Examination of one's own feelings or thoughts.
**self-o·pin·ion** (sĕlf′ə-pĭn′yən) *n.* A high or exaggerated opinion of oneself.
**self-o·pin·ion·at·ed** (sĕlf′ə-pĭn′yə-nā′tĭd) *adj.* **1.** Vain : conceited. **2.** Stubbornly insistent on one's own opinion.
**self-or·dained** (sĕlf′ôr-dānd′) *adj.* Ordained by oneself rather than by others <a *self-ordained* arbiter of fashion>
**self-per·cep·tion** (sĕlf′pər-sĕp′shən) *n.* Self-image.
**self-per·pet·u·at·ing** (sĕlf′pər-pĕch′o͞o-ā′tĭng) *adj.* Capable of renewing or perpetuating oneself or itself for an indefinite length of time. **—self′-per·pet′u·a′tion** *n.*
**self-pit·y** (sĕlf′pĭt′ē) *n.* Pity for oneself. **—self′-pit′y·ing** *adj.* **—self′-pit′y·ing·ly** *adv.*
**self-poised** (sĕlf′poizd′) *adj.* **1.** Balanced without need of support. **2.** Being in command of oneself.
**self-pol·li·na·tion** (sĕlf′pŏl′ə-nā′shən) *n.* Pollen transfer from an anther to a stigma of the same flower. **—self′-pol′li·nat′ed** *adj.*
**self-por·trait** (sĕlf′pôr′trĭt, -trāt′, -pōr′-) *n.* A pictorial or verbal portrait of oneself done by oneself.
**self-pos·ses·sion** (sĕlf′pə-zĕsh′ən) *n.* Command of one's faculties, emotions, and behavior : POISE. **—self′-pos·sessed′** *adj.*
**self-pres·er·va·tion** (sĕlf′prĕz′ər-vā′shən) *n.* **1.** Protection of oneself from harm, loss, or destruction. **2.** The instinct for preservation of oneself.
**self-pro·claimed** (sĕlf′prō-klāmd′, -prə-) *adj.* Self-ordained.
**self-pro·pelled** (sĕlf′prə-pĕld′) *adj.* Containing its own means of propulsion <a *self-propelled* vehicle>
**self-pro·tec·tive** (sĕlf′prə-tĕk′tĭv) *adj.* Serving to protect oneself. **—self′-pro·tec′tion** *n.* **—self′-pro·tec′tive·ly** *adv.*
**self-pu·ri·fi·ca·tion** (sĕlf′pyo͝or′ə-fĭ-kā′shən) *n.* **1.** Purification produced by nature. **2.** Purification of oneself.
**self-re·al·i·za·tion** (sĕlf′rē′ə-lĭ-zā′shən) *n.* Complete development or fulfillment of oneself or one's own potential.
**self-re·cord·ing** (sĕlf′rĭ-kôr′dĭng) *adj.* Automatically recording its own functions or operations. —Used of a machine or instrument.
**self-re·crim·i·na·tion** (sĕlf′rĭ-krĭm′ə-nā′shən) *n.* The act or an instance of blaming oneself.
**self-re·flec·tion** (sĕlf′rĭ-flĕk′shən) *n.* Self-examination. **—self′-re·flec′tive** *adj.* **—self′-re·flec′tive·ly** *adv.*
**self-re·gard** (sĕlf′rĭ-gärd′) *n.* **1.** Concern for oneself or one's own interests. **2.** Self-respect.
**self-reg·u·lat·ing** (sĕlf′rĕg′yə-lā′tĭng) *adj.* **1.** Regulating oneself or itself. **2.** Regulating itself automatically. **—self′-reg·u·la′tion** *n.*
**self-re·li·ance** (sĕlf′rĭ-lī′əns) *n.* Reliance on one's own ability, judgment, or resources. **—self′-re·li′ant** *adj.* **—self′-re·li′ant·ly** *adv.*
**self-rep·li·cat·ing** (sĕlf′rĕp′lĭ-kā′tĭng) *adj.* Reproducing itself.
**self-re·proach** (sĕlf′rĭ-prōch′) *n.* The act or an instance of blaming oneself for an error or fault. **—self′-re·proach′ful** *adj.* **—self′-re·proach′ful·ly** *adv.*
**self-re·spect** (sĕlf′rĭ-spĕkt′) *n.* Due respect for oneself and one's character and conduct. **—self′-re·spect′ing** *adj.*
**self-re·straint** (sĕlf′rĭ-strānt′) *n.* Restraint of one's feelings, desires, or inclinations : SELF-CONTROL.
**self-rev·e·la·tion** (sĕlf′rĕv′ə-lā′shən) *n.* Revelation of one's own thoughts, feelings, or ideas, esp. unintentionally. **—self′-re·veal′ing** (-rĭ-vē′lĭng) *adj.*
**self-righ·teous** (sĕlf′rī′chəs) *adj.* Piously or smugly convinced of one's own righteousness : MORALISTIC. **—self′-righ′teous·ly** *adv.* **—self′-righ′teous·ness** *n.*

o͞o **boot** ou **out** th **thin** *th* **this** ŭ **cut** ûr **urge** y **young**
yo͞o **abuse** zh **vision** ə **about, item, edible, gallop, circus**

**self-right·ing** (sĕlf'rī'tĭng) *adj.* Capable of righting itself when overturned <a *self-righting* dinghy>
**self-ris·ing flour** (sĕlf'rī'zĭng) *n.* A commercially produced mixture of flour, salt, and leavening.
**self-rule** (sĕlf'ro͞ol') *n.* Self-government.
**self-sac·ri·fice** (sĕlf'săk'rə-fīs') *n.* Sacrifice of one's own interests or well-being for the sake of others or for a cause or ideal. **—self'-sac'ri·fic'ing** *adj.*
**self·same** (sĕlf'sām') *adj.* [ME *selve same* : *self,* same + *same,* same.] The very same : IDENTICAL. **—self'same'ness** *n.*
**self-sat·is·fac·tion** (sĕlf'săt'ĭs-făk'shən) *n.* Satisfaction with oneself or with one's actions or accomplishments.
**self-seal·ing** (sĕlf'sē'lĭng) *adj.* **1.** Capable of sealing itself, as after being punctured <a *self-sealing* gasket> **2.** Capable of being sealed without moisture <a *self-sealing* envelope>
**self-seek·ing** (sĕlf'sē'kĭng) *adj.* Concerned with or pursuing only one's own interests. **—self'-seek'er** *n.*
**self-serv·ice** (sĕlf'sûr'vĭs) *adj.* Requiring customers or users to help themselves <a *self-service* market><a *self-service* elevator>
**self-serv·ing** (sĕlf'sûr'vĭng) *adj.* Furthering one's own interests, esp. without concern for the needs or interests of others.
**self-slaugh·ter** (sĕlf'slô'tər) *n.* Suicide.
**self-start·er** (sĕlf'stär'tər) *n.* **1.** STARTER 3. **2.** An individual having initiative.
**self-stud·y** (sĕlf'stŭd'ē) *n.* **1.** Study or consideration of oneself. **2.** Independent or self-instructional study.
**self-styled** (sĕlf'stīld') *adj.* As characterized or designated by oneself <a *self-styled* conservative>
**self-suf·fi·cient** (sĕlf'sə-fĭsh'ənt) *also* **self-suf·fic·ing** (-fī'sĭng) *adj.* **1.** Providing for oneself without help : INDEPENDENT. **2.** Overly confident : SMUG. **—self'-suf·fi'cien·cy** *n.*
**self-sup·port** (sĕlf'sə-pôrt', -pōrt') *n.* Support of oneself without the help of others. **—self'-sup·port'ed, self'-sup·port'ing** *adj.*
**self-sus·tain·ing** (sĕlf'sə-stā'nĭng) *adj.* Capable of maintaining oneself or itself independently.
**self-taught** (sĕlf'tôt') *adj.* Educated by oneself without formal instruction or the help of others.
**self-treat·ment** (sĕlf'trēt'mənt) *n.* Treatment of one's own illness without professional assistance.
**self-trust** (sĕlf'trŭst') *n.* Self-confidence.
**self-un·der·stand·ing** (sĕlf'ŭn'dər-stăn'dĭng) *n.* Self-knowledge.
**self-will** (sĕlf'wĭl') *n.* Willfulness, esp. in gratifying one's own wishes : OBSTINACY. **—self'-willed'** *adj.*
**self-wind·ing** (sĕlf'wīn'dĭng) *adj.* Not needing to be wound by hand <a *self-winding* watch>
**Sel·juk** (sĕl'jo͝ok', sĕl-jo͝ok') *n.* [< Turk. *Seljūk,* the eponymous ancestor of the dynasties.] A member of a Turkish dynasty that ruled central and western Asia from the 11th to the 13th cent.
**sell** (sĕl) *v.* **sold** (sōld), **sell·ing, sells.** [ME *sellen* < OE *sellan,* to give.] *—vt.* **1.** To exchange or give up for money or its equivalent. **2.** To offer for sale. **3.** To surrender or give up in exchange for a price or reward. **4.** To promote the sale of <A new ad campaign really *sold* that product.> **5.** To convince of. **6.** *Slang.* To cheat or dupe. *—vi.* **1.** To exchange ownership for money or its equivalent. **2.** To be sold or be on sale. **3.** To attract prospective buyers <an item that *sells* well> **4.** To gain acceptance. **—sell off.** To dispose of by selling, often at a reduced price. **—sell out. 1.** To sell all one's goods or possessions. **2.** *Slang.* To betray <*sold out* to the enemy> *—n.* **1.** The act or an example of selling. **2.** *Slang.* A hoax or swindle. **—sell a bill of goods.** To take unfair advantage of. **—sell down the river.** To betray the true trust or faith of. **—sell short. 1.** To contract for the sale of securities or commodities one expects to own later and at more advantageous terms. **2.** To underestimate the value or worth of. **—sell'a·ble** *adj.*
☆ ***syns:*** SELL, HANDLE, MARKET, MERCHANDISE, RETAIL, VEND *v.* *core meaning* : to offer for sale <doesn't *sell* washing machines> ***ant:*** buy, purchase
**sell·er** (sĕl'ər) *n.* **1.** One who sells : VENDER. **2.** An item that sells in a specified manner <a best *seller*>
**selling climax** *n.* A sharp decline in stock prices on a heavy volume of trading followed by a rally.
**selling point** *n.* An aspect stressed or played up, as in advertising or selling.
**sell·out** (sĕl'out') *n.* **1.** The act or an instance of selling out. **2.** An event for which all the tickets are sold. **3.** *Slang.* One who has betrayed one's principles or a cause.
**sel·syn** (sĕl'sĭn') *n.* [SEL(F) + SYN(CHRONOUS).] A device for the instantaneous transmission and reception, from a generator to a motor, of the angular movement of rotating parts.
**selt·zer** (sĕlt'sər) *n.* [G. *Selterser (Wasser),* (water) of Nieder Selters, a district in West Germany.] **1.** Naturally effervescent spring water. **2.** SODA WATER 1.
**sel·vage** *also* **sel·vedge** (sĕl'vĭj) *n.* [ME, prob. < MDu. *selfegghe* : *self,* self + *egge,* edge.] **1.** The edge of a fabric woven so that it will not ravel. **2.** The edge plate of a lock with a slot for a bolt.
**selves** (sĕlvz) *n. pl. of* SELF.
**se·man·teme** (sĭ-măn'tēm') *n.* [SEMANT(IC) + -EME.] An irreducible linguistic unit of meaning.
**se·man·tic** (sĭ-măn'tĭk) *adj.* [Gk. *sēmantikos,* significant < *sēmainein,* to signify < *sēma,* sign.] **1.** Of or relating to meaning, esp. meaning in language. **2.** Of, relating to, or according to the science of semantics. **—se·man'ti·cal·ly** *adv.*
**se·man·tics** (sĭ-măn'tĭks) *n.* (*sing. in number*). **1.** The study or science of meaning in language forms, esp. with regard to historical changes. **2.** The study of relationships between signs and symbols and what they represent to their interpreters. **—se·man'ti·cist** (-măn'tĭ-sĭst) *n.*
**sem·a·phore** (sĕm'ə-fôr', -fōr') *n.* [Gk. *sēma,* sign + -PHORE.] **1.** A visual signaling apparatus with flags, lights, or mechanically moving arms. **2.** A system for signaling using two flags that are held one in each hand. *—v.* **-phored, -phor·ing, -phores.** *—vt.* To send (information) by semaphore. *—vi.* To signal with a semaphore.
**se·ma·si·ol·o·gy** (sĭ-mā'sē-ŏl'ə-jē, -zē-) *n.* [Gk. *sēmasia,* meaning (< *sēmainein,* to signify < *sēma,* sign) + -LOGY.] SEMANTICS 2. **—se·ma'si·o·log'i·cal** (-ə-lŏj'ĭ-kəl) *adj.* **—se·ma'si·ol'o·gist** *n.*
**se·mat·ic** (sĭ-măt'ĭk) *adj.* [< Gk. *sēma, sēmat-,* sign.] Serving as a warning or signal of danger. —Used esp. of the coloring of certain poisonous or noxious animals.
**sem·bla·ble** (sĕm'blə-bəl) *adj.* [ME < OFr. < *sembler,* to resemble < Lat. *simulare,* to simulate < *similis,* like.] **1.** Resembling : like. **2.** Seeming : apparent. *—n.* One closely resembling something else. **—sem'bla·bly** *adv.*
**sem·blance** (sĕm'bləns) *n.* [ME < OFr. < *semblant,* pr.part. of *sembler,* to resemble. —see SEMBLABLE.] **1.** A superficial or token appearance : SHOW. **2.** A representation : likeness. **3.** The smallest trace : MODICUM <not a *semblance* of common sense>
**se·mé** (sĕ-mā', sə-) *adj.* [Fr., p.part. of *semer,* to sow, scatter < Lat. *seminare* < *semen,* seed.] *Heraldry.* Having a pattern ornamented with small delicate figures, as stars or flowers.
**se·mei·ol·o·gy** (sē'mī-ŏl'ə-jē) *n. var. of* SEMIOLOGY.
**se·mei·ot·ic** (sē'mī-ŏt'ĭk) *also* **se·mei·ot·i·cal** (-ĭ-kəl) *adj. vars. of* SEMIOTIC.
**se·mei·ot·ics** (sē'mī-ŏt'ĭks) *n. var. of* SEMIOTICS.
**se·meme** (sē'mēm') *n.* [Gk. *sēmainein,* to signify (< *sēma,* sign) + -EME.] The meaning of a morpheme.
**se·men** (sē'mən) *n.* [ME < Lat. *semen,* seed.] A thick whitish secretion of the male reproductive organs, the transporting medium for spermatozoa.
**se·mes·ter** (sə-mĕs'tər) *n.* [G. < Lat. *(cursus) se-mestris,* (period) of six months : *sex,* six + *mensis,* month.] One of two divisions, usu. lasting 15 to 18 weeks each, of an academic year.
**sem·i** (sĕm'ē, sĕm'ī) *n. Informal.* A semitrailer.
**semi-** *pref.* [Lat. *semi-,* half.] **1.** Half <*semi*circle> **2.** Partial : partially <*semi*conscious> **3.** Like or having some of the characteristics of <*semi*official><*semi*aquatic> **4.** Occurring two times during <*semi*monthly>
**sem·i·ab·stract** (sĕm'ē-ăb-străkt', -ăb'străkt') *adj.* Of or pertaining to an art form marked by stylized but recognizable subject matter. **—sem'i·ab·strac'tion** *n.*
**sem·i·an·nu·al** (sĕm'ē-ăn'yo͞o-əl, sĕm'ī-) *adj.* Occurring or issued twice a year. **—sem'i·an'nu·al·ly** *adv.*
**sem·i·a·quat·ic** (sĕm'ē-ə-kwŏt'ĭk, -kwăt'-, sĕm'ī-) *adj.* Capable of living or growing in or near water but not entirely aquatic.
**sem·i·ar·id** (sĕm'ē-ăr'ĭd, sĕm'ī-) *adj.* Marked by light annual rainfall and capable of sustaining only short grasses and shrubs. **—sem'i·a·rid'i·ty** (-ə-rĭd'ĭ-tē, -ă-rĭd'-) *n.*
**sem·i·au·to·mat·ic** (sĕm'ē-ô'tə-măt'ĭk, sĕm'ī-) *adj.* **1.** Partially automatic. **2.** Ejecting the shell and loading the succeeding round of ammunition automatically after each shot has been fired. —Used of firearms. *—n.* A semiautomatic firearm.
**sem·i·au·ton·o·mous** (sĕm'ē-ô-tŏn'ə-məs, sĕm'ī-) *adj.* Possessing the powers of self-government within a larger organization or political entity.
**sem·i·breve** (sĕm'ē-brēv', -brĕv', sĕm'ī-) *n. Chiefly Brit. Mus.* A whole note.
**sem·i·cen·ten·ni·al** (sĕm'ē-sĕn-tĕn'ē-əl, sĕm'ī-) *adj.* Occurring every 50 years. *—n.* A 50th anniversary or its celebration.
**sem·i·cir·cle** (sĕm'ĭ-sûr'kəl) *n.* **1.** A half of a circle as divided by a diameter. **2.** Something shaped like a half-circle. **—sem'i·cir'cu·lar** (-kyə-lər) *adj.*
**semicircular canal** *n.* Any of the three tubular and looped structures in the labyrinth of the inner ear, which function together in the maintenance of a sense of balance and orientation.
**sem·i·civ·i·lized** (sĕm'ē-sĭv'ə-līzd', sĕm'ī-) *adj.* Partially civilized.
**sem·i·clas·si·cal** (sĕm'ē-klăs'ĭ-kəl, sĕm'ī-) *adj.* **1.** Of, pertaining to, or being a musical work that in style or form falls between classical and popular music. **2.** Of, pertaining to, or being classical music that has acquired widespread appeal or approval.
**sem·i·co·lon** (sĕm'ĭ-kō'lən) *n.* A punctuation mark (;) indicating a degree of separation greater than the comma but less than the period.
**sem·i·co·ma·tose** (sĕm'ē-kō'mə-tōs', -kŏm'ə-, sĕm'ī-) *adj.* Torpid and disoriented but not in a complete coma.

**sem·i·con·duc·tor** (sĕm′ē-kən-dŭk′tər, sĕm′ī-) *n.* A solid crystalline substance, as germanium or silicon, that has electrical conductivity greater than an insulator but less than a good conductor.
**sem·i·con·scious** (sĕm′ē-kŏn′shəs, sĕm′ī-) *adj.* Partially conscious. **—sem′i·con′scious·ly** *adv.* **—sem′i·con′scious·ness** *n.*
**sem·i·dark·ness** (sĕm′ē-därk′nĭs, sĕm′ī-) *n.* Partial darkness.
**sem·i·des·ert** (sĕm′ē-dĕz′ərt, sĕm′ī-) *n.* An area somewhat like a desert, often located between a desert and a grassland or woodland.
**sem·i·de·tached** (sĕm′ē-dĭ-tăcht′, sĕm′ī-) *adj.* Attached to something else, as another building, on one side only.
**sem·i·di·am·e·ter** (sĕm′ē-dī-ăm′ĭ-tər, sĕm′ī-) *n.* The apparent radius of a celestial body when viewed as a disk from Earth.
**sem·i·di·ur·nal** (sĕm′ē-dī-ûr′nəl, sĕm′ī-) *adj.* **1.** Of, relating to, happening, or performed during a half day. **2.** Occurring or arriving approx. once every 12 hours, as the tides. **3.** Designating the arc described by a celestial body between its meridian passage and its point of rising or setting.
**sem·i·di·vine** (sĕm′ē-dĭ-vīn′, sĕm′ī-) *adj.* Not fully divine but more than mortal, as a Greek demigod.
**sem·i·doc·u·men·ta·ry** (sĕm′ē-dŏk′yə-mĕn′tə-rē, -mĕn′trē, sĕm′ī-) *n.* A television or motion-picture dramatization that incorporates factual details or actual events. **—sem′i·doc′u·men′ta·ry** *adj.*
**sem·i·dome** (sĕm′ē-dōm′, sĕm′ī-) *n.* A roof or domed ceiling over a semicircular space.
**sem·i·el·lip·ti·cal** (sĕm′ē-ĭ-lĭp′tĭ-kəl, sĕm′ī-) *adj.* Formed or shaped like half of an ellipse, esp. when divided along the major axis.
**sem·i·fi·nal** (sĕm′ē-fī′nəl, sĕm′ī-) *adj.* Immediately before the final, as in a series of competitions or examinations. *—n.* (sĕm′ē-fī′nəl, sĕm′ī-). A semifinal match or competition. **—sem′i·fi′nal·ist** *n.*
**sem·i·flu·id** (sĕm′ē-flo͞o′ĭd, sĕm′ī-) *adj.* Intermediate in flow properties between solids and liquids. *—n.* (sĕm′ē-flo͞o′ĭd, sĕm′ī-). A semifluid substance.
**sem·i·for·mal** (sĕm′ē-fôr′məl, sĕm′ī-) *adj.* **1.** Somewhat formal <a *semiformal* party> **2.** Suitable for a somewhat formal occasion <*semiformal* clothes>
**sem·i·group** (sĕm′ē-gro͞op′, sĕm′ī-) *n. Math.* A nonempty set with an associative binary multiplication.
**sem·i·in·de·pend·ent** (sĕm′ē-ĭn′dĭ-pĕn′dənt, sĕm′ī-) *adj.* **1.** Partially independent. **2.** Semiautonomous.
**sem·i·liq·uid** (sĕm′ē-lĭk′wĭd, sĕm′ī-) *adj.* Semifluid : viscous. **—sem′i·liq′uid** *n.*
**sem·i·lit·er·ate** (sĕm′ē-lĭt′ər-ĭt, sĕm′ī-) *adj.* **1.** Having attained an elementary level of reading and writing ability. **2.** Having a limited degree of understanding, as of a technical subject.
**sem·i·log** (sĕm′ē-lôg′, -lŏg′, sĕm′ī-) *adj.* Semilogarithmic.
**sem·i·log·a·rith·mic** (sĕm′ē-lô′gə-rĭth′mĭk, -lŏg′ə-, sĕm′ī-) *adj.* Having one logarithmic and one arithmetic scale <*semilogarithmic* graph paper>
**sem·i·lu·nar** (sĕm′ē-lo͞o′nər, sĕm′ī-) *also* **sem·i·lu·nate** (-lo͞o′nāt′) *adj.* Shaped like a half-moon.
**semilunar bone** *n. Anat.* The lunate bone.
**semilunar valve** *n.* Either of two crescent-shaped valves of the aorta and the pulmonary artery that prevent blood from flowing back into the heart.
**sem·i·lu·nate** (sĕm′ē-lo͞o′nāt′, sĕm′ī-) *adj. var. of* SEMILUNAR.
**sem·i·month·ly** (sĕm′ē-mŭnth′lē, sĕm′ī-) *adj.* Occurring or issued twice a month. *—n., pl.* **-lies.** A semimonthly event or publication. *—adv.* Every half month.
**sem·i·nal** (sĕm′ə-nəl) *adj.* [ME < OFr. < Lat. *seminalis* < *semen,* seed.] **1.** Of, pertaining to, or containing semen or seed. **2.** Of, relating to, or able to originate : CREATIVE. **—sem′i·nal·ly** *adv.*
**sem·i·nar** (sĕm′ə-när′) *n.* [G. < Lat. *seminarium,* seed plot < *semen,* seed.] **1. a.** A small group of advanced college or graduate students engaged in special study or original research under the guidance of a professor. **b.** A course of study so pursued. **c.** A scheduled meeting of such a group. **2.** A meeting for an exchange of ideas in a particular area.
**sem·i·nar·i·an** (sĕm′ə-nâr′ē-ən) *n.* A seminary student.
**sem·i·nar·y** (sĕm′ə-nĕr′ē) *n., pl.* **-ies.** [ME, seed plot < Lat. *seminarium* < *semen,* seed.] **1. a.** A school for training priests, ministers, or rabbis. **b.** A secondary school, esp. a private school for girls. **2.** An area or environment in which something is developed or nurtured.
**sem·i·na·tion** (sĕm′ə-nā′shən) *n.* [Lat. *seminatio,* propagation < *seminatus,* propagation < *seminatus,* p.part. of *seminare,* to sow < *semen,* seed.] Dispersal or production of seed.
**sem·i·nif·er·ous** (sĕm′ə-nĭf′ər-əs) *adj.* [Lat. *semen, semin-,* seed, semen + -FEROUS.] *Biol.* **1.** Carrying or producing semen. **2.** Bearing seed.
**Sem·i·nole** (sĕm′ə-nōl′) *n., pl.* **Seminole** *or* **-noles.** [Creek *simanóli* < Am. Sp. *cimarrón,* wild.] **1. a.** A tribe of Indians, a late Creek offshoot, living orig. in Alabama and later in Florida but now chiefly in Oklahoma. **b.** A member of this tribe. **2.** The Muskhogean language of the Seminole. **—Sem′i·nole′** *adj.*

**sem·i·no·mad** (sĕm′ē-nō′măd′, sĕm′ī-) *n.* One of a people whose living habits are mainly nomadic but who cultivate some crops at a base settlement. **—sem′i·no·mad′ic** (-nō-măd′ĭk) *adj.*
**sem·i·of·fi·cial** (sĕm′ē-ə-fĭsh′əl, sĕm′ī-) *adj.* Having a degree of official authority or sanction. **—sem′i·of·fi′cial·ly** *adv.*
**se·mi·ol·o·gy** (sē′mē-ŏl′ə-jē, -mī-) *also* **se·mei·ol·o·gy** (sē′mī-) *n.* [Gk. *sēmion,* sign + -LOGY.] **1. a.** The science of signs or sign language. **b.** The use of signs in signaling, as with a semaphore. **2.** Symptomatology.
**sem·i·o·paque** (sĕm′ē-ō-pāk′, sĕm′ī-) *adj.* Partially opaque.
**se·mi·ot·ic** (sē′mē-ŏt′ĭk, -mī-) *also* **se·mi·ot·i·cal** (-ĭ-kəl) *or* **se·mei·ot·ic** (sē′mī-) *also* **se·mei·ot·i·cal** (-ĭ-kəl) *adj.* [Gk. *sēmeiōtikos,* observant of signs < *sēmeioun,* to note < *sēmeion,* sign.] **1.** Of or pertaining to semantics. **2.** *Med.* Pertaining to symptomatology.
**se·mi·ot·ics** (sē′mē-ŏt′ĭks, -mī-) *also* **se·mei·ot·ics** (sē′mī-) *n. (sing. in number).* **1.** SEMANTICS 2. **2.** *Med.* Symptomatology. **—se′mi·o·ti′cian** (-ə-tĭsh′ən) *n.*
**sem·i·pal·mate** (sĕm′ē-păl′māt′, -pä′māt′, -päl′māt′, sĕm′ī-) *also* **sem·i·pal·mat·ed** (-mā′tĭd) *adj.* Having limited or reduced webbing between the toes, as some wading birds do.

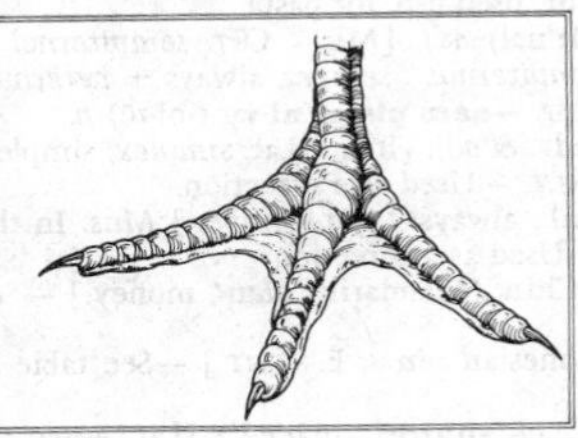
**semipalmate**
*Semipalmate foot of a plover*

**sem·i·par·a·site** (sĕm′ē-păr′ə-sīt′, sĕm′ī-) *n. Biol.* A hemiparasite. **—sem′i·par′a·sit′ic** (-sĭt′ĭk) *adj.*
**sem·i·per·me·a·ble** (sĕm′ē-pûr′mē-ə-bəl, sĕm′ī-) *adj.* **1.** Partially permeable. **2.** Permeable to some molecules in a mixture but not to all. **—sem′i·per′me·a·bil′i·ty** (-bĭl′ĭ-tē) *n.*
**sem·i·po·lit·i·cal** (sĕm′ē-pə-lĭt′ĭ-kəl, sĕm′ī-) *adj.* Political in some respects.
**sem·i·por·ce·lain** (sĕm′ē-pôr′sə-lĭn, -pōr′-, sĕm′ī-) *n.* A glazed ceramic ware resembling porcelain but having little or no translucency.
**sem·i·post·al** (sĕm′ē-pō′stəl, sĕm′ī-) *n.* A postage stamp sold for more than its face value, esp. to raise money for charity.
**sem·i·pre·cious** (sĕm′ē-prĕsh′əs, sĕm′ī-) *adj.* Worth somewhat less than a precious stone.
**sem·i·pri·vate** (sĕm′ē-prī′vĭt, sĕm′ī-) *adj.* Shared with usu. one to three other hospital patients <a *semiprivate* room>
**sem·i·pro** (sĕm′ē-prō′, sĕm′ī-) *adj. Informal.* Semiprofessional. **—sem′i·pro′** *n.*
**sem·i·pro·fes·sion·al** (sĕm′ē-prə-fĕsh′ə-nəl, sĕm′ī-) *adj.* **1.** Taking part in a sport for pay but not on a full-time basis. **2.** Made up of or engaged in by semiprofessional players. *—n.* **1.** A semiprofessional player. **2.** One whose job has some of the characteristics of a profession or of a professional. **—sem′i·pro·fes′sion·al·ly** *adv.*
**sem·i·qua·ver** (sĕm′ē-kwā′vər) *n. Mus. Chiefly Brit.* A sixteenth note.
**sem·i·re·tired** (sĕm′ē-rĭ-tīrd′, sĕm′ī-) *adj.* Working on a part-time basis, esp. due to advanced age. **—sem′i·re·tire′ment** *n.*
**sem·i·rig·id** (sĕm′ē-rĭj′ĭd, sĕm′ī-) *adj.* **1.** Partially rigid. **2.** Having some rigid components.
**sem·i·skilled** (sĕm′ē-skĭld′, sĕm′ī-) *adj.* **1.** Having some skills but not enough to do specialized work. **2.** Calling for only limited skills <a *semiskilled* job>
**sem·i·soft** (sĕm′ē-sôft′, -sŏft′, sĕm′ī-) *adj.* **1.** Of medium softness. **2.** Firm but easily sliced, as a cheese.
**sem·i·sol·id** (sĕm′ē-sŏl′ĭd, sĕm′ī-) *adj.* Intermediate in properties, esp. in rigidity, between solids and liquids. *—n.* (sĕm′ē-sŏl′ĭd, sĕm′ī-). A semisolid substance.
**sem·i·sweet** (sĕm′ē-swēt′, sĕm′ī-) *adj.* Having a slight amount of sweetening <*semisweet* chocolate>
**Sem·ite** (sĕm′īt′) *n.* [NLat. *semita* < LLat. *Sem,* Shem, eponymous ancestor of the Semites < Gk. < Heb. *Shem.*] **1.** One of a Middle Eastern people of Caucasian stock comprising chiefly the Jews and Arabs but in ancient times also including the Babylonians, Assyrians, Canaanites, Arameans, and Phoenicians. **2.** One of the people descended from Shem.
**Se·mit·ic** (sə-mĭt′ĭk) *adj.* **1.** Of or pertaining to the Semites, esp. Jewish or Arabic. **2.** Of, pertaining to, or comprising a subfamily of the Afro-Asiatic language family that includes Arabic, Hebrew, Amharic, and Aramaic. *—n.* **1.** The Semitic subfamily of languages. **2.** One of the Semitic languages.
**Se·mit·ics** (sə-mĭt′ĭks) *n. (sing. in number).* Study of Semitic history, languages, and cultures. **—Se·mit′i·cist** (-ĭ-sĭst) *n.*

**Sem·i·tism** (sĕm′ĭ-tĭz′əm) *n.* **1.** A Semitic word or idiom. **2.** Semitic characteristics, attributes, or customs. **3.** A predisposition or policy in favor of the Jews.

**sem·i·tone** (sĕm′ē-tōn′, sĕm′ī-) *n. Mus.* An interval equal to a half tone in the diatonic scale. **—sem′i·ton′ic** (-tŏn′ĭk) *adj.* **—sem′i·ton′i·cal·ly** *adv.*

**sem·i·trail·er** (sĕm′ē-trā′lər, sĕm′ī-) *n.* A trailer with a set or sets of wheels at the rear only, supported in front by the truck tractor or towing vehicle.

**sem·i·trans·par·ent** (sĕm′ē-trăns-pâr′ənt, -păr′-, sĕm′ī-) *adj.* Partially transparent.

**sem·i·trop·i·cal** (sĕm′ē-trŏp′ĭ-kəl, sĕm′ī-) *adj.* Partly tropical.

**sem·i·vow·el** (sĕm′ĭ-vou′əl) *n.* A letter or vocal sound having the sound of a vowel but used as a consonant, as *w, y,* and *r* : GLIDE.

**sem·i·week·ly** (sĕm′ē-wēk′lē, sĕm′ī-) *adj.* Occurring or issued twice a week. *—n., pl.* **-lies.** A semiweekly event or publication. *—adv.* Twice weekly.

**sem·i·year·ly** (sĕm′ē-yîr′lē, sĕm′ī-) *adj.* Occurring or issued twice a year or once every half year. *—n., pl.* **-lies.** A semiyearly event or publication. *—adv.* Every half year.

**sem·o·li·na** (sĕm′ə-lē′nə) *n.* [Ital. *semolino,* dim. of *semola,* bran < Lat. *simila,* fine flour.] Coarse, gritty particles of wheat left after the finer flour has been sifted out, used esp. for pasta.

**sem·pi·ter·nal** (sĕm′pĭ-tûr′nəl) *adj.* [ME < OFr. *sempiternel* < LLat. *sempiternalis* < Lat. *sempiternus* : *semper,* always + *aeternus,* eternal.] Perpetual : everlasting. **—sem′pi·ter′ni·ty** (-nĭ-tē) *n.*

**sem·pli·ce** (sĕm′plĭ-chā′) *adv. & adj.* [Ital. < Lat. *simplex,* simple.] *Mus.* In a simple way : PLAINLY. —Used as a direction.

**sem·pre** (sĕm′prā) *adv.* [Ital., always < Lat. *semper.*] *Mus.* In the same manner throughout. —Used as a direction.

**sen¹** (sĕn) *n., pl.* **sen.** [J. < Chin. (Mandarin) *qian²,* money.] —See table at CURRENCY.

**sen²** (sĕn) *n., pl.* **sen.** [Indonesian *sén* < E. CENT.] —See table at CURRENCY.

**se·nar·i·us** (sə-nâr′ē-əs) *n., pl.* **-nar·i·i** (-nâr′ē-ī′) [Lat. —see SENARY.] A Greek or Latin verse composed of six feet.

**sen·a·ry** (sĕn′ə-rē) *adj.* [Lat. *senarius* < *seni,* six each < *sex,* six.] **1.** Of or pertaining to the number six. **2.** Having six parts or things.

**sen·ate** (sĕn′ĭt) *n.* [ME *senat* < OFr. < Lat. *senatus* < *senex,* old man.] **1.** An assembly of citizens having the highest legislative functions in a government, esp.: **a. Senate.** The upper house of the U.S. Congress, to which two members are elected from each state. **b.** *often* **Senate.** The upper house in the bicameral legislature of many states in the United States. **c. Senate.** The upper legislative house in such countries as Canada and France. **d.** The highest council of state of the ancient Roman republic and empire. **2.** The building or hall in which a senate meets. **3.** A college or university governing, advisory, or disciplinary body. **—sen′a·tor** *n.* **—sen′a·to′ri·al** (sĕn′ə-tôr′ē-əl, -tōr′-) *adj.* **—sen′a·to′ri·al·ly** *adv.* **—sen′a·tor·ship′** *n.*

**senatorial courtesy** *n.* The custom in the U.S. Senate of refusing to confirm a presidential appointment opposed by both senators from the appointee's state or by the senior senator of the President's party.

**senatorial district** *n.* The voting district represented by a state senator.

**send¹** (sĕnd) *v.* **sent** (sĕnt), **send·ing, sends.** [ME *senden* < OE *sendan.*] *—vt.* **1.** To cause to be conveyed by an intermediary to a destination <*send* books to California by train> **2.** To dispatch, as by a communications medium <*send* a letter><*sent* a telegram> **3. a.** To direct to go on a mission. **b.** To allow or enable to go <*send* a youngster to camp> **c.** To cause to depart : DISMISS. **c.** To direct (one) to a source of information : REFER. **4.** To give off (e.g., heat) : EMIT. **5.** To direct or propel with force : DRIVE <*sent* the ball out of the stadium> **6.** To cause to happen or befall : INFLICT. **7. a.** To put or drive into a particular state or condition <news that *sent* them into a rage> **b.** *Slang.* To transport with delight : THRILL. *—vi.* **1.** To dispatch someone to do an errand or convey a message <*send* out for a pizza> **2.** To dispatch a request or order, esp. by mail <*send* away for the new books> **—send flying.** To scatter in all directions with force or violence. **—send for. 1.** To place an order, esp. by mail. **2.** To request to come by means of a message : SUMMON. **—send up.** *Informal.* To send to jail. **—send′er** *n.*

☆ ***syns:*** SEND, DISPATCH, FORWARD, ROUTE, SHIP, TRANSMIT *v. core meaning* : to cause (something) to be conveyed to a destination <*sent* the package to California> ***ant:*** receive

**send²** (sĕnd) *v. & n. var. of* SCEND.

**sen·dal** (sĕn′dl) *n.* [ME *cendal* < OFr.] A thin silk fabric used in the Middle Ages for fine clothing, church vestments, and banners.

**send·off** (sĕnd′ôf′, -ŏf′) *n.* A demonstration of affection and good wishes for the start of a new undertaking.

**se·ne** (sā′nā) *n., pl.* **sene.** [Samoan < E. CENT.] —See table at CURRENCY.

**Sen·e·ca** (sĕn′ĭ-kə) *n., pl.* **Seneca** *or* **-cas.** [Du. *Sennecaas,* perh. of Algonquian orig.] **1. a.** A tribe of Indians formerly inhabiting western New York. **b.** A member of this tribe. **2.** The Iroquoian language of the Seneca.

**Seneca snakeroot** *n.* A North American plant, *Polygala senega,* bearing a terminal cluster of small white flowers.

**se·nec·ti·tude** (sĭ-nĕk′tĭ-to͞od′, -tyo͞od′) *n.* [Med. Lat. *senectitudo* < Lat. *senectus* < *senex,* old.] Old age.

**sen·e·ga** (sĕn′ĭ-gə) *n.* [Alteration of SENECA.] The dried roots of the Seneca snakeroot, used as an expectorant.

**se·nes·cent** (sĭ-nĕs′ənt) *adj.* [Lat. *senescens, senescent-,* pr.part. of *senescere,* to grow old, inchoative of *senēre,* to be old < *senex,* old.] Aging : elderly. **—se·nes′cence** *n.*

**sen·e·schal** (sĕn′ə-shəl) *n.* [ME < OFr. < Med. Lat. *siniscalcus,* of Germanic orig.] A medieval official in a noble household charged with domestic arrangements and the administration of servants.

**se·nile** (sē′nīl′, sĕn′īl′) *adj.* [OFr. < Lat. *senilis* < *senex,* old.] **1.** Relating to, typical of, or proceeding from old age. **2.** Exhibiting senility. **3.** *Geol.* Worn away almost to the base, as at the end of an erosion cycle. **—se′nile′ly** *adv.*

**senile dementia** *n.* Progressive, premature deterioration of mental faculties and emotional stability in old age.

**se·nil·i·ty** (sĭ-nĭl′ĭ-tē) *n.* **1.** The state of being senile. **2.** Mental and physical deterioration ocurring in old age.

**sen·ior** (sēn′yər) *adj.* [Lat., comp. of *senex,* old.] **1.** Of or designating the older of two, esp. the older of two persons having the same name. **2.** Above others in age, rank, or length of service. **3.** Of or relating to the fourth and last year of high school or college. *—n.* **1.** A senior person. **2.** A fourth-year college or high-school student.

**senior citizen** *n.* One of or over the age of retirement.

**senior high school** *n.* A high school usu. including grades 10, 11, and 12.

**sen·ior·i·ty** (sēn-yôr′ĭ-tē, -yŏr′-) *n.* **1.** The state of being older or higher in rank. **2.** A position of precedence over others by reason of a longer span of service.

**sen·i·ti** (sĕn′ĭ-tē) *n., pl.* **seniti.** [Tongan < E. CENT.] —See table at CURRENCY.

**sen·na** (sĕn′ə) *n.* [NLat. < Ar. *sanā′.*] **1.** Any of various plants of the genus *Cassia,* with compound leaves and usu. yellow flowers. **2.** The dried leaves of *C. angustifolia* or *C. acutifolia,* used as a cathartic.

**sen·net** (sĕn′ĭt) *n.* [Perh. var. of SIGNET.] A call on a trumpet or cornet signaling an entrance or exit on the stage.

**sen·night** (sĕn′īt′) *n.* [ME *seoveniht* < OE *seofon nihta,* seven nights.] *Archaic.* A week.

**sen·nit** (sĕn′ĭt) *n.* [Orig. unknown.] **1.** *Naut.* Cordage formed by braiding several strands of rope fiber or similar material. **2.** Braided straw, grass, or palm leaves used for making hats.

**se·no·pi·a** (sĭ-nō′pē-ə) *n.* [Lat. *senex,* old + -OPIA.] Improvement of near vision that sometimes occurs in the elderly because of swelling of the crystalline lens in incipient cataract.

**se·ñor** (sān-yôr′) *n., pl.* **se·ño·res** (sān-yôr′ās) [Sp. < Med. Lat. *senior,* lord < Lat., senior.] **1.** —Used as the Spanish courtesy title for a man, equivalent to the English *Mr.* or *sir.* **2.** A Spanish or Spanish-speaking man.

**se·ño·ra** (sān-yôr′ə) *n.* [Sp., fem. of *señor,* señor.] **1.** —Used as the Spanish courtesy title for a married woman, equivalent to the English *Mrs.* or *madam.* **2.** A Spanish or Spanish-speaking woman.

**se·ño·res** (sān-yôr′ās) *n. pl. of* SEÑOR.

**se·ño·ri·ta** (sān′yə-rē′tə) *n.* [Sp., dim. of *señora,* señora.] **1.** —Used as the Spanish courtesy title for an unmarried young woman or a girl, equivalent to the English *Miss.* **2.** An unmarried Spanish or Spanish-speaking woman or girl.

**sen·sate** (sĕn′sāt′) *also* **sen·sat·ed** (-sā′tĭd) *adj.* [LLat. *sensatus,* gifted with sense < Lat. *sensus,* sense.] Apprehended through the senses. **—sen′sate′ly** *adv.*

**sen·sa·tion** (sĕn-sā′shən, sən-) *n.* [Med. Lat. *sensatio* < LLat. *sensatus,* gifted with sense < *sensus,* sense.] **1. a.** An awareness associated with stimulation of a sense organ or with a specific bodily condition <the *sensation* of cold> **b.** The faculty to feel or perceive physically. **2.** A condition of heightened interest or emotion <"The anticipation produced in me a *sensation* somewhat between bliss and fear" —James Weldon Johnson> **3. a.** A condition of strong public interest and excitement. **b.** An event or object producing such public excitement.

**sen·sa·tion·al** (sĕn-sā′shə-nəl, sən-) *adj.* **1.** Of or relating to sensation. **2.** Arousing or meant to arouse intense curiosity, interest, or reaction, esp. by exaggerated or lurid details. **3.** Outstanding : spectacular. **—sen·sa′tion·al·ly** *adv.*

**sen·sa·tion·al·ism** (sĕn-sā′shə-nə-lĭz′əm, sən-) *n.* **1. a.** Use of sensational material or methods in order to shock, excite, or arouse curiosity. **b.** Sensational subject matter <journalistic *sensationalism*> **c.** Interest in or the effect of such subject matter. **2.** *Philos.* The theory that sensation is the only source of knowledge. **3.** The ethical doctrine that feeling is the only criterion of good. **—sen·sa′tion·al·ist** *n.* **—sen·sa′tion·al·is′tic** *adj.*

**sense** (sĕns) *n.* [Lat. *sensus,* the faculty of perceiving < p.part. of *sentire,* to feel.] **1. a.** Any of the animal functions of hearing, sight, smell, touch, and taste. **b.** The faculty of self-awareness represented by these functions. **2. senses.** The faculties of sensation as providers of physical gratification and pleasure. **3. a.** Intuitive or acquired ability to estimate or judge <a *sense* of direction> **b.** A capacity to appreciate or understand <a *sense* of humor> **c.** A vague feeling or impression <a *sense* of doom> **d.** Sensate or intellectual recognition or perception : CONSCIOUSNESS <a *sense* of guilt> **4. a.** *often* **senses.** Normal ability to think or reason soundly : JUDGMENT <Come to your *senses.*> **b.** Something sound or reasonable <There's no *sense* in staying here.> **5. a.** Significance : meaning. **b.** Lexical meaning. **c.** The meaning of a word in a particular context. **6.** Viewpoint : consensus <finding out the *sense* of the electorate on nuclear energy> —*vt.* **sensed, sens·ing, sens·es. 1.** To become aware of through the senses : PERCEIVE. **2.** To grasp : comprehend. **3.** To detect automatically <*sense* earthquake activity>

**sense datum** *n.* A basic unanalyzable experience arising from stimulation of a sense organ.

**sense·less** (sĕns′lĭs) *adj.* **1.** Without sense or meaning : PURPOSELESS <a *senseless* assault> **2.** Lacking sense : FOOLISH <a *senseless* proposal> **3.** Unconscious. **—sense′less·ly** *adv.* **—sense′less·ness** *n.*

**sense organ** *n.* A specialized organ or structure, as the eye or ear, that initiates a process of sensory perception when stimulated.

**sense perception** *n.* Perception by the bodily senses.

**sen·si·bil·i·ty** (sĕn′sə-bĭl′ĭ-tē) *n., pl.* **-ties. 1.** The ability to feel or perceive sensations. **2. a.** Keen mental perception <an artistic *sensibility*> **b.** Mental or emotional responsiveness toward something, as the feelings of another. **3.** *often* **sensibilities.** Receptiveness to impression, whether pleasant or unpleasant <"The sufferings of the Cuban people shocked our *sensibilities*" —George F. Kennan> **4.** Refined awareness and appreciation in matters of feeling.

**sen·si·ble** (sĕn′sə-bəl) *adj.* [ME < OFr. < Lat. *sensibilis* < *sensus,* sense.] **1.** Capable of being perceived by the senses or by the mind. **2.** Easily perceived : NOTICEABLE. **3.** Having the ability to feel or perceive. **4.** Having a perception of something : COGNIZANT. **5.** Having or displaying good sense <a *sensible* person><a *sensible* choice> **—sen′si·ble·ness** *n.* **—sen′si·bly** *adv.*

**sen·sil·lum** (sĕn-sĭl′əm) *n., pl.* **-sil·la** (-sĭl′ə) [NLat., dim. of Med. Lat. *sensus,* sense organ < Lat., sense.] An epithelial sense organ composed of one cell or a few cells.

**sen·si·tive** (sĕn′sĭ-tĭv) *adj.* [ME < OFr. *sensitif* < Med. Lat. *sensitivus* < Lat. *sensus,* sense.] **1.** Capable of perceiving with a sense or senses. **2.** Responsive to external conditions or stimulation. **3.** Susceptible to the ideas, emotions, or circumstances of others. **4.** Easily offended : TOUCHY. **5.** Easily irritated <*sensitive* skin> **6.** Readily affected or altered by the action of an agent <film that is *sensitive* to light> **7.** Registering very slight differences or changes <a *sensitive* barometer> **8.** Fluctuating or tending to fluctuate, as stock prices. **9.** Of or concerned with classified information <*sensitive* data> **—sen′si·tive·ly** *adv.* **—sen′si·tive·ness** *n.*

**sensitive plant** *n.* **1.** A woody tropical American plant, *Mimosa pudica,* with leaflets and stems that fold and droop when touched. **2.** A similar plant, as *Cassia nictitans* of eastern North America.

**sen·si·tiv·i·ty** (sĕn′sĭ-tĭv′ĭ-tē) *n., pl.* **-ties. 1.** The quality or state of being sensitive. **2.** Organic or organismic responsiveness to stimulation. **3.** *Electron.* The minimum input signal required to produce a specified output signal. **4.** The degree of response of a film or plate to light, esp. to light of a specified wavelength.

**sensitivity training** *n.* Group therapy in which individuals are taught how to develop a sensitive awareness and understanding of themselves and of their relationships with others.

**sen·si·tize** (sĕn′sĭ-tīz′) *v.* **-tized, -tiz·ing, -tiz·es.** —*vt.* **1.** To make sensitive. **2.** To make (a film or plate) sensitive to light, esp. to light of a specific wavelength. —*vi.* To become sensitive. **—sen′si·ti·za′tion** *n.* **—sen′si·tiz′er** *n.*

**sen·si·tom·e·ter** (sĕn′sĭ-tŏm′ĭ-tər) *n.* [SENSIT(IVE) + -METER.] **1.** A device that measures the sensitivity of photographic film to light. **2.** A device that measures the sensitivity of eyes to light. **—sen′si·to·met′ric** (-tə-mĕt′rĭk) *adj.* **—sen′si·tom′e·try** *n.*

**sen·sor** (sĕn′sər, -sôr′) *n.* [< Lat. *sensus,* sense.] A device, as a photoelectric cell, that detects and responds to a signal or stimulus.

**sen·so·ri·a** (sĕn-sôr′ē-ə, -sōr′-) *n. var. pl. of* SENSORIUM.

**sen·so·ri·al** (sĕn-sôr′ē-əl, -sōr′-) *adj.* Sensory.

**sen·so·ri·mo·tor** (sĕn′sə-rē-mō′tər) *adj.* [SENSOR(Y) + MOTOR.] Of, relating to, or combining sensory and motor functions. —Used of nerves.

**sen·so·ri·neu·ral** (sĕn′sə-rē-no͝or′əl, -nyo͝or′-) *adj.* Of, relating to, or involving the neural aspects of sensory perception.

**sen·so·ri·um** (sĕn-sôr′ē-əm, -sōr′-) *n., pl.* **-so·ri·ums** *or* **-so·ri·a** (-sôr′ē-ə, -sōr′-) [LLat. *sensorium,* organ of sensation < Lat. *sensus,* sense.] **1.** The part of the brain that receives and correlates impressions transmitted to various sensory areas. **2.** The sensory system.

**sen·so·ry** (sĕn′sə-rē) *adj.* **1.** Of or relating to the senses. **2.** Transmitting impulses from sense organs to nerve centers : AFFERENT.

**sensory deprivation** *n.* Stoppage of sensory stimulation, as by prolonged immersion in water, in order to examine physical and esp. psychological reactions.

**sen·su·al** (sĕn′sho͞o-əl) *adj.* **1.** Relating to or affecting a sense or a sense organ. **2. a.** Relating to or preoccupied with the gratification of physical appetites, esp. the sexual appetite. **b.** Suggesting sexuality : VOLUPTUOUS. **c.** Not spiritual or intellectual : PHYSICAL. **d.** Having no moral or spiritual interests : WORLDLY. **3.** Sensory. **—sen′su·al·ly** *adv.* **—sen′su·al·ness** *n.*

**sen·su·al·ism** (sĕn′sho͞o-ə-lĭz′əm) *n.* **1.** Sensuality. **2.** The ethical doctrine that sensual pleasures are the highest good. **3.** *Philos.* SENSATIONALISM 2. **—sen′su·al·ist** *n.* **—sen′su·al·is′tic** *adj.*

**sen·su·al·i·ty** (sĕn′sho͞o-ăl′ĭ-tē) *n.* **1.** The quality or state of being sensual. **2.** Preoccupation with sensual pleasures.

**sen·su·al·ize** (sĕn′sho͞o-ə-līz′) *vt.* **-ized, -iz·ing, -iz·es.** To make sensual. **—sen′su·al·i·za′tion** *n.*

**sen·su·ous** (sĕn′sho͞o-əs) *adj.* **1.** Of, relating to, or arising from the senses. **2.** Appealing to the senses <the *sensuous* aroma of a summer garden> **3.** Greatly appreciative of the pleasures of sensation. **—sen′su·os′i·ty** (-ŏs′ĭ-tē) *n.* **—sen′su·ous·ly** *adv.* **—sen′su·ous·ness** *n.*

**Sen·sur·round** (sĕn′sə-round′). A trademark for a motion-picture sound effect having low-frequency sound signals felt by the audience as vibrations.

**sent** (sĕnt) *v. p.t. & p.p. of* SEND[1].

**sen·tence** (sĕn′təns) *n.* [ME, opinion < OFr. < Lat. *sententia* < *sentire,* to feel.] **1.** A grammatical unit consisting of a word or a group of words that is separate from any other grammatical construction and usu. contains at least one subject with its predicate and a finite verb or verb phrase. **2. a.** A judicial decision, esp. one detailing punishment to be inflicted on a convicted person. **b.** The penalty imposed. **3.** An opinion, esp. a formal one made after deliberation. **4.** *Archaic.* An aphorism. —*vt.* **-tenced, -tenc·ing, -tenc·es.** To pass sentence on (a convicted person). **—sen·ten′tial** (sĕn-tĕn′shəl) *adj.* **—sen·ten′tial·ly** *adv.*

**sen·tenc·er** (sĕn′tən-sər) *n.* One who pronounces sentence.

**sentence stress** *n.* Change in emphasis or vocal stress on the syllables of words within a sentence.

**sen·ten·tia** (sĕn-tĕn′shə, -shē-ə) *n., pl.* **-ti·ae** (-shē-ē′) [Lat. —see SENTENCE.] An adage : aphorism.

**sen·ten·tious** (sĕn-tĕn′shəs) *adj.* [Lat. *sententiosus,* full of meaning < *sententia,* opinion < *sentire,* to feel.] **1.** Concise and forceful in expression : SUCCINCT. **2. a.** Full of aphorisms. **b.** Given to uttering aphorisms. **3.** Abounding in or given to pompous moralizing. **—sen·ten′tious·ly** *adv.* **—sen·ten′tious·ness** *n.*

**sen·tience** (sĕn′shəns, -shē-əns, -tē-əns) *n.* **1.** The quality or condition of being sentient : CONSCIOUSNESS. **2.** Emotion as opposed to perception or thought.

**sen·tient** (sĕn′shənt, -shē-ənt, -tē-ənt) *adj.* [Lat. *sentiens, sentient-,* pr.part. of *sentire,* to feel.] **1.** Capable of feeling : CONSCIOUS. **2.** Experiencing sensation or feeling. —*n.* **1.** One that is sentient. **2.** The human mind. **—sen′ti·ent·ly** *adv.*

**sen·ti·ment** (sĕn′tə-mənt) *n.* [ME *sentement* < OFr. < Med. Lat. *sentimentum* < Lat. *sentire,* to feel.] **1. a.** A general cast of mind regarding something. **b.** A particular view. **2.** An idea, opinion, or attitude based on feeling or emotion rather than reason. **3.** The emotional meaning of a written passage. **4.** Susceptibility to tender, romantic, or nostalgic feeling. **5. a.** Emotion. **b.** Romantic or nostalgic feeling. **6.** Delicate and sensitive feeling, as in art and literature. **7.** A vague awareness or feeling.

☆ ***syns:*** SENTIMENT, ATTITUDE, DISPOSITION, FEELING *n. core meaning* : a general cast of mind with regard to something <anti-Soviet *sentiment* throughout the country>

**sen·ti·men·tal** (sĕn′tə-mĕn′tl) *adj.* **1. a.** Marked by or affected by sentiment. **b.** Extravagantly emotional. **2.** Arising from or colored by emotion rather than reason. **3.** Appealing to the sentiments, esp. to romantic feelings. **—sen′ti·men′tal·ly** *adv.*

**sen·ti·men·tal·ism** (sĕn′tə-mĕn′tl-ĭz′əm) *n.* **1.** A predilection for the sentimental. **2.** A thought or expression characterized by excessive sentiment. **—sen′ti·men′tal·ist** *n.*

**sen·ti·men·tal·i·ty** (sĕn′tə-mĕn-tăl′ĭ-tē) *n., pl.* **-ties. 1.** The quality or state of being excessively or affectedly sentimental. **2.** A sentimental thought or an expression of it.

**sen·ti·men·tal·ize** (sĕn′tə-mĕn′tl-īz′) *v.* **-ized, -iz·ing, -iz·es.** —*vt.* To be sentimental about. —*vi.* To act in a sentimental way. **—sen′ti·men′tal·i·za′tion** *n.*

**sen·ti·nel** (sĕn′tə-nəl) *n.* [Fr. *sentinelle* < Ital. *sentinella,* prob. < *sentire,* to watch < Lat. *sentire,* to feel.] A guard : sentry. —*vt.* **-neled, -nel·ing, -nels** *or* **-nelled, -nel·ling, -nels. 1.** To guard as a sentinel. **2.** To provide with a sentinel. **3.** To post as a sentinel.

**sen·try** (sĕn′trē) *n., pl.* **-tries.** [Perh. alteration of obs. *sentery,* sanctuary.] **1.** A guard, esp. a military one, posted to prevent the passage of unauthorized persons. **2.** The duty of a sentry : WATCH.

**sentry box** *n.* A small enclosure or shelter for a sentry.

**se·pal** (sē′pəl) *n.* [NLat. *sepalum* < *sepa,* sepal < Gk. *skepē,* covering.] One of the usu. green sections forming the calyx of a flower. **—se′paled, sep′a·lous** (sĕp′ə-ləs) *adj.*

**se·pal·oid** (sē′pə-loid′, sĕp′ə-) *also* **se·pal·ine** (-līn′, -lĭn) *adj.* Like or characteristic of a sepal.

**-sepalous** *suff.* [SEPAL + -OUS.] Having a specified kind or number of sepals <gamo*sepalous*>

**sep·a·ra·ble** (sĕp′ər-ə-bəl, sĕp′rə-) *adj.* [OFr. < Lat. *separabilis* < *separare,* to separate.] Capable of being separated. **—sep′a·ra·bil′i·ty** *n.* **—sep′a·ra·bly** *adv.*

**sep·a·rate** (sĕp′ə-rāt′) *v.* **-rat·ed, -rat·ing, -rates.** [ME *separaten* < Lat. *separare* : *se-,* apart + *parare,* to make ready.] —*vt.* **1. a.** To set or keep apart : DISUNITE. **b.** To scatter. **c.** To sort. **2.** To discriminate or differentiate between : DISTINGUISH. **3.** To extract from a mixture or combination : ISOLATE. **4.** To part (a married couple), usu. by decree. **5.** To discharge, as from employment or military service. —*vi.* **1.** To become disunited or severed. **2.** To withdraw : secede <threatening to *separate* from the association> **3.** To part company : DISPERSE. **4.** To discontinue living together as husband and wife. **5.** To become divided into parts or components <Oil and water tend to *separate.*> —*adj.* (sĕp′ər-ĭt, sĕp′rĭt). **1.** Set apart from others : DETACHED. **2.** *Archaic.* Withdrawn from others : ALONE. **3.** Existing by itself : INDEPENDENT. **4.** Not alike : DISSIMILAR. **5.** Not shared : INDIVIDUAL. —*n.* (sĕp′ər-ĭt, sĕp′rĭt). A garment, as a skirt, jacket, or pair of slacks, designed to be worn in various combinations with other garments. **—sep′a·rate·ly** *adv.* **—sep′a·rate·ness** *n.*

☆ ***syns:*** **1.** SEPARATE, BREAK UP, DIVIDE, PART, PARTITION, SECTION, SEGMENT *v. core meaning* : to make a division into parts <*separated* the city into two sectors> ***ant:*** combine **2.** SEPARATE, BREAK, DISJOIN, DIVIDE, DIVORCE, SPLIT *v. core meaning* : to become or cause to become apart from another <families that *separated* over the issue of slavery>

**sep·a·ra·tion** (sĕp′ə-rā′shən) *n.* [ME *separacion* < OFr. < Lat. *separatio* < *separare,* to separate.] **1.** The act or process of separating or the state of being separated. **2.** The site of a division or parting. **3.** A space or interval that separates : GAP. **4. a.** *Law.* An agreement or court decree ending the conjugal relationship of a husband and wife. **b.** Discharge, as from employment or military service.

**sep·a·ra·tion·ist** (sĕp′ə-rā′shə-nĭst) *n.* A separatist.

**sep·a·ra·tist** (sĕp′ər-ə-tĭst, sĕp′rə-tĭst, sĕp′ə-rā′tĭst) *n.* One who withdraws or advocates political or religious separation. **—sep′a·ra·tism′** (-tĭz′əm) *n.* **—sep′a·ra·tis′tic** *adj.*

**sep·a·ra·tive** (sĕp′ə-rā′tĭv, sĕp′ər-ə-tĭv, sĕp′rə-tĭv) *adj.* Tending to separate or cause separation.

**sep·a·ra·tor** (sĕp′ə-rā′tər) *n.* **1.** One that separates. **2.** A device that separates cream from milk.

**Se·phar·di** (sə-fär′dē) *n., pl.* **-dim** (-dĭm) [Modern Heb. *Sĕphāradhī* < *Sĕphārad,* Spain.] A member of the branch of European Jews who settled mainly in Spain, Portugal, and northern Africa. **—Se·phar′dic** (-dĭk) *adj.*

**se·pi·a** (sē′pē-ə) *n.* [Ital. *seppia* < Lat. *sepia* < Gk., cuttlefish.] **1. a.** A dark-brown ink or pigment made from the secretion of the cuttlefish. **b.** A work of art done in sepia. **c.** A photograph with a brown tint. **2.** A dark grayish yellowish brown to dark or moderate olive brown. —*adj.* **1.** Of the color sepia. **2.** Done or made in sepia.

**se·pi·o·lite** (sē′pē-ə-līt′) *n.* [G. *Sepiolith* : Gk. *sēpion,* cuttlebone (< *sēpia,* cuttlefish) + Gk. *lithos,* stone.] *Mineral.* MEERSCHAUM 1.

**se·poy** (sē′poi′) *n.* [Prob. < Port. *sipae* < Urdu *sipāhī* < Pers. < *sipāh,* army.] A regular soldier in some Middle Eastern countries, esp. an Indian soldier serving under British command.

**sep·pu·ku** (sĕp′o͞o-ko͞o) *n.* [J., self-disembowelment : *seppu,* to cut + *ku,* abdomen.] Hara-kiri.

**sep·sis** (sĕp′sĭs) *n., pl.* **-ses** (-sēz′) [Gk. *sēpsis,* putrefaction < *sēpein,* to make rotten.] The presence of disease-causing organisms or their toxins in the blood or tissues.

**sept** (sĕpt) *n.* [Prob. alteration of SECT.] A subdivision of a tribe, esp. in ancient and medieval Ireland : CLAN.

**sep·ta** (sĕp′tə) *n. pl. of* SEPTUM.

**sep·tal** (sĕp′təl) *adj.* Of or relating to a septum.

**sep·tar·i·um** (sĕp-târ′ē-əm) *n., pl.* **-i·a** (-ē-ə) [SEPT(UM) + -ARIUM.] An irregular polygonal system of calcite-filled fissures occurring in some rock concretions. **—sep·tar′i·an** *adj.*

**sep·tate** (sĕp′tāt′) *adj.* Having a septum or septa.

**Sep·tem·ber** (sĕp-tĕm′bər) *n.* [ME *Septembre* < OFr. < Lat. *September,* the seventh month < *septem,* seven.] The ninth month of the year. —See table at CALENDAR.

**Sep·tem·brist** (sĕp-tĕm′brĭst) *n.* A bloodthirsty or terroristic revolutionary.

**sep·te·nar·i·us** (sĕp′tə-nâr′ē-əs) *n., pl.* **-i·i** (-ē-ī′) [Lat. *septenarius,* of seven < *septeni,* seven each < *septem,* seven.] A Greek or Latin verse having seven feet.

**sep·ten·ni·al** (sĕp-tĕn′ē-əl) *adj.* [< Lat. *septennium,* period of seven years < *septennis,* of seven years : *septem,* seven + *annus,* year.] **1.** Occurring every seven years. **2.** Continuing for seven years. —*n.* Something that occurs every seven years. **—sep·ten′ni·al·ly** *adv.*

**sep·ten·tri·on** (sĕp-tĕn′trē-ŏn′, -ən) *n.* [ME *septemtrioun* < OFr. *septentrion* < Lat. *septentriones,* seven plow oxen, the seven principal stars of Ursa Major or Ursa Minor : *septem,* seven + *triones,* pl. of *trio,* plow ox.] *Obs.* The north. **—sep·ten′tri·o·nal** (-trē-ə-nəl) *adj.*

**sep·tet** *also* **sep·tette** (sĕp-tĕt′) *n.* [G. < Lat. *septem,* seven.] **1.** A group of seven. **2.** *Mus.* **a.** A composition for seven voices or instruments. **b.** The musicians performing such a composition.

**sep·tic** (sĕp′tĭk) *adj.* [Lat. *septicus,* putrefying < Gk. *sēptikos* < *sēptos,* rotten < *sēpein,* to make rotten.] **1.** Of, relating to, or having the nature of sepsis. **2.** Causing sepsis : PUTREFACTIVE. —*n.* A septic substance. **—sep·tic′i·ty** (-tĭs′ĭ-tē) *n.*

**sep·ti·ce·mi·a** (sĕp′tĭ-sē′mē-ə) *n.* A systemic disease caused by pathogenic organisms or their toxins in the bloodstream. **—sep′ti·ce′mic** (-mĭk) *adj.*

**sep·ti·ci·dal** (sĕp′tĭ-sīd′l) *adj.* [SEPT(UM) + -CID(E) + -AL.] *Bot.* Splitting along or through the septa. —Used of a seed capsule. **—sep′ti·ci′dal·ly** *adv.*

**septic sore throat** *n.* An often epidemic infection of the throat, caused by hemolytic streptococci and marked by fever and inflammation of the tonsils.

**septic tank** *n.* A tank in which a continuous flow of sewage is decomposed by anaerobic bacteria.

**sep·tif·ra·gal** (sĕp-tĭf′rə-gəl) *adj.* [SEPTUM + Lat. *frangere,* to break.] *Bot.* Marked by the separating of certain parts of the plant from its dividing walls.

**sep·ti·lat·er·al** (sĕp′tə-lăt′ər-əl) *adj.* [Lat. *septem,* seven + LATERAL.] Seven-sided.

**sep·til·lion** (sĕp-tĭl′yən) *n.* [Fr. : *septi-,* seven + *million,* million.] **1.** The cardinal number equal to $10^{24}$. **2.** *Chiefly Brit.* The cardinal number equal to $10^{42}$. **—sep·til′lion** *adj.*

**sep·til·lionth** (sĕp-tĭl′yənth) *n.* **1.** The ordinal number matching the number one septillion in a series. **2.** One of a septillion equal parts. **—sep·til′lionth** *adj. & adv.*

**sep·tu·a·ge·nar·i·an** (sĕp′to͞o-ə-jə-nâr′ē-ən, -tyo͞o-) *n.* [< Lat. *septuagenarius,* of the number seventy < *septuageni,* seventy each < *septuaginta,* seventy < *septem,* seven.] One who is between the ages of 70 and 80. —*adj.* **1.** Being between the ages of 70 and 80. **2.** Of or relating to a septuagenarian.

**Sep·tu·a·ges·i·ma** (sĕp′to͞o-ə-jĕs′ə-mə, -jā′zə-) *n.* [ME < OFr. < LLat. *septuagesima* < Lat. *septuagesimus,* seventieth < *septuaginta,* seventy < *septem,* seven.] The third Sunday before Ash Wednesday.

**Sep·tu·a·gint** (sĕp′to͞o-ə-jĭnt′, sĕp-to͞o′ə-jənt, -tyo͞o′-) *n.* [Lat. *septuaginta,* seventy (from the traditional number of its translators) < *septem,* seven.] A 3rd cent. B.C. Greek translation of the Old Testament.

**sep·tum** (sĕp′təm) *n., pl.* **-ta** (-tə) [Lat. *saeptum,* partition < *saepire,* to enclose < *saepes,* fence.] A thin partition or membrane between two cavities or soft masses of tissue.

**septum pel·lu·ci·dum** (pə-lo͞o′sĭ-dəm) *n.* [NLat., transparent partition.] *Anat.* A thin membrane of nervous tissue forming the medial wall of the lateral ventricles in the brain.

**sep·tu·ple** (sĕp-to͞o′pəl, -tyo͞o′-) *adj.* [LLat. *septuplus,* sevenfold < Lat. *septem,* seven.] **1.** Made up of or containing seven. **2.** Multiplied by seven. —*vt.* **-pled, -pling, -ples.** To multiply by seven.

**sep·ul·cher** (sĕp′əl-kər) *n.* [ME *sepulcre* < OFr. < Lat. *sepulcrum* < *sepultus,* p.part. of *sepelire,* to bury.] **1.** A burial vault. **2.** A container for sacred relics, esp. in an altar. —*vt.* **-chered, -cher·ing, -chers.** To put in a sepulcher : INTER.

**se·pul·chral** (sə-pŭl′krəl, -po͝ol′-) *adj.* **1.** Of or relating to a sepulcher. **2.** Suggestive of the grave : FUNEREAL. **—se·pul′chral·ly** *adv.*

**sep·ul·chre** (sĕp′əl-kər) *n. & v. Chiefly Brit. var. of* SEPULCHER.

**sep·ul·ture** (sĕp′əl-cho͝or′, -chər) *n.* [ME < OFr. < Lat. *sepultura* < *sepultus,* p.part. of *sepelire,* to bury.] *Archaic.* **1.** Interment : burial. **2.** A sepulcher.

**se·qua·cious** (sĭ-kwā′shəs) *adj.* [< Lat. *sequax, sequac-,* pursuing < *sequi,* to follow.] **1.** Following logically and sequentially. **2.** *Archaic.* Apt to follow another : DEPENDENT. **—se·qua′cious·ly** *adv.* **—se·quac′i·ty** (-kwăs′ĭ-tē) *n.*

**se·quel** (sē′kwəl, -kwĕl′) *n.* [ME *sequele* < OFr. *sequelle* < Lat. *sequela* < *sequi,* to follow.] **1.** Something that follows or comes after. **2.** A book, motion picture, or dramatic presentation that continues the narrative of an earlier work. **3.** A consequence or result.

**se·quel·a** (sĭ-kwĕl′ə) *n., pl.* **-quel·ae** (-kwĕl′ē) [Lat., sequel.] Something that follows, esp. a condition arising from a disease.

**se·quence** (sē′kwəns, -kwĕns′) *n.* [LLat. *sequentia* < Lat. *sequens,* pr.part. of *sequi,* to follow.] **1.** A following of one thing after another : SUCCESSION. **2.** An order of succession : ARRANGEMENT. **3.** A related or continuous series. **4.** Three or more playing cards in consecutive order. **5.** A series of single film shots edited so as to constitute a unit : EPISODE. **6.** *Mus.* A melodic or harmonic pattern successively repeated at different pitches with or without a key change. **7.** *Rom. Cath. Ch.* A hymn read between the gradual and the gospel. **8.** *Math.* An ordered set of quantities, as $x$, $2x^2$, $3x^3$, $4x^4$. —*vt.* **-quenced, -quenc·ing, -quenc·es.** To arrange in a sequence.

**se·quenc·er** (sē′kwən-sər, -kwĕn′-) *n. Computer Sci.* A device that sorts cards, data, or programs in a prearranged sequence.

**se·quent** (sē′kwənt) *adj.* [Lat. *sequens, sequent-,* pr.part. of *sequi,* to follow.] **1.** Coming after in order or time : SUBSEQUENT. **2.** Resulting from : CONSEQUENT. —*n.* A result : consequence.

**se·quen·tial** (sĭ-kwĕn′shəl) *adj.* **1.** Forming or marked by a sequence, as of notes or units. **2.** Sequent. **—se·quen′ti·al′i·ty** (-shē-ăl′ĭ-tē) *n.* **—se·quen′tial·ly** *adv.*

**se·ques·ter** (sĭ-kwĕs′tər) *v.* **-tered, -ter·ing, -ters.** [ME *sequestren* < LLat. *sequestrare,* to give up for safekeeping < Lat. *sequester,* depository.] —*vt.* **1.** To set apart or remove : SEGREGATE. **2.** *Law.* To take temporary possession of (property) as security against legal claims. **3.** *Law.* To requisition and confiscate (enemy property). **4.** To cause to seclude oneself. —*vi. Chem.* To undergo sequestration.

**se·ques·tra** (sĭ-kwĕs′trə) *n. pl. of* SEQUESTRUM.

**se·ques·trant** (sĭ-kwĕs′trənt) *n.* A chemical agent that promotes sequestration.

**se·ques·trate** (sē′kwĭ-strāt′, sĕk′wĭ-, sĭ-kwĕs′trāt′) *vt.* **-trat·ed, -trat·ing, -trates.** [LLat. *sequestrare, sequestrat-,* to give up for safekeeping < Lat. *sequester,* depository.] **1.** *Law.* To take possession of : CONFISCATE. **2.** *Archaic.* To sequester.

**se·ques·tra·tion** (sē′kwĭ-strā′shən, sĕk′wĭ-) *n.* [ME *sequestracioun,* excommunication < LLat. *sequestratio,* separation < *sequestrare,* to give up for safekeeping < Lat. *sequester,* depository.] **1.** The act of sequestering or the state of being sequestered. **2.** *Law.* **a.** Seizure of property. **b.** A writ authorizing seizure of property. **3.** *Chem.* The inhibition or stoppage of normal ion behavior by combination with added materials, esp. the prevention of metallic ion precipitation from solution by formation of a coordination complex with a phosphate.

**se·ques·trum** (sĭ-kwĕs′trəm) *n., pl.* **-tra** (-trə) [NLat. < Lat., deposit < *sequester,* depository.] A dead bone fragment that is separated from healthy bone.

**se·quin** (sē′kwĭn) *n.* [Fr. < Ital. *zecchino,* a coin < *zecca,* the mint < Ar. *sikkah,* coin die.] **1.** A small shiny ornamental disk sewn on fabric : SPANGLE. **2.** A gold coin of the Venetian Republic.

**se·quoi·a** (sĭ-kwoi′ə) *n.* [NLat. *Sequoia,* genus name, after *Sequoya* (George Guess), d.1843.] An extremely large evergreen tree of the genus *Sequoia,* which includes the redwood and the giant sequoia.

**se·ra** (sîr′ə) *n. var. pl. of* SERUM.

**sé·rac** (sə-răk′, sā-) *n.* [Fr. < Med. Lat. *seracium,* whey < Lat. *serum.*] A large mass of glacier ice remaining behind in a crevasse after glacial movement or melting.

**se·ra·glio** (sə-răl′yō, -räl′-) *n., pl.* **-glios.** [Ital. *serraglio,* prob. partly < VLat. **serraculum,* enclosure (< Lat. *serare,* to lock up < *sera,* lock), and partly < Turk. *serai,* palace.] **1.** A harem. **2.** A sultan's palace.

**se·ra·pe** *also* **sa·ra·pe** (sə-rä′pē, -răp′ē) *n.* [Mex. Sp. *sarape.*] A Latin-American cloak or poncho made of wool.

**ser·aph** (sĕr′əf) *n., pl.* **-a·phim** (-ə-fĭm) *or* **-aphs.** [Back-formation < *seraphim* (pl.) < ME *seraphin* < OE < LLat. < Heb. *sĕrāphīm,* pl. of *sārāph.*] **1.** An angel with three pairs of fiery wings that guards the throne of God. **2.** An angel of the highest rank in the nine orders of angels. **—se·raph′ic** (sə-răf′ĭk), **se·raph′i·cal** *adj.* **—se·raph′i·cal·ly** *adv.*

**Se·ra·pis** (sə-rā′pĭs) *n. Myth.* An Egyptian god of the netherworld.

**Serb** (sûrb) *n.* [Serbian *Srb.*] A Serbian.

**Ser·bi·an** (sûr′bē-ən) *n.* **1.** A member of a southern Slavic people that is the dominant ethnic group of Serbia and adjacent republics of Yugoslavia. **2.** A Serbo-Croatian. **3.** Serbo-Croatian as spoken in Serbia. —*adj.* Of or pertaining to Serbia or the Serbians.

**Ser·bo-Cro·a·tian** (sûr′bō-krō-ā′shən) *n.* **1.** The Slavic language of the Serbs and the Croats. **2.** A native speaker of Serbo-Croatian. —*adj.* Of or relating to Serbo-Croatian or those who speak it.

**sere**[1] *also* **sear** (sîr) *adj.* [ME < OE *sēar.*] Withered : dry.

**sere**[2] (sîr) *n.* [< SERIES.] A sequence of ecological communities successively occupying an area.

**ser·e·nade** (sĕr′ə-nād′, sĕr′ə-nād′) *n.* [Fr. *sérénade* < Ital. *serenata* < *sereno,* serene < Lat. *serenus.*] **1.** A musical performance that honors or expresses love for someone. **2.** An instrumental composition for a small ensemble, having characteristics of the suite and the sonata. —*v.* **-nad·ed, -nad·ing, -nades.** —*vt.* To perform a serenade for. —*vi.* To perform a serenade. **—ser′e·nad′er** *n.*

**ser·en·dip·i·ty** (sĕr′ən-dĭp′ĭ-tē) *n.* [From its possession by the characters in the Persian fairy tale *The Three Princes of Serendip.*] The faculty of making providential discoveries by accident. **—ser′en·dip′i·tous** *adj.*

**se·rene** (sə-rēn′) *adj.* [Lat. *serenus,* serene, clear.] **1.** Peaceful : tranquil. **2.** Unclouded : clear. **3.** *often* **Serene.** August : exalted. —Used as part of a title for certain royal personages. **—se·rene′ly** *adv.* **—se·rene′ness, se·ren′i·ty** (-rĕn′ĭ-tē) *n.*

**serf** (sûrf) *n.* [ME < OFr. < Lat. *servus,* slave.] **1.** A slave, esp. a member of the lowest feudal class in medieval Europe, owned by a lord and bound to the land. **2.** One in servitude. **—serf′dom** *n.*

**serge** (sûrj) *n.* [ME *sarge* < OFr. < VLat. **sarica* < Lat. *serica,* silks < *sericus,* silken < *Seres,* a people of Eastern Asia.] A twilled cloth of worsted or a blend of worsted and wool.

**ser·geant** (sär′jənt) *n.* [ME *sergeaunte,* a common soldier < OFr. *sergent* < Lat. *serviens,* pr.part. of *servire,* to serve < *servus,* slave.] **1. a.** Any of several ranks of noncommissioned officers in the U.S. Army, Air Force, or Marine Corps. **b.** One holding any of these ranks. **2. a.** The rank of police officer next below a captain, lieutenant, or inspector. **b.** A police officer holding this rank. **3.** A sergeant at arms. **—ser′gean·cy, ser′geant·ship′** *n.*

**sergeant at arms** *n.* An officer appointed to keep order, as at the meetings of a legislative, judicial, or social body.

**sergeant first class** *n.* A noncommissioned officer next below master sergeant in the U.S. Army.

**sergeant fish** *n.* The cobia.

**sergeant major** *n.* **1.** A noncommissioned officer serving as chief administrative assistant of a headquarters unit of the U.S. Army, Air Force, or Marine Corps. **2.** *Chiefly Brit.* A noncommissioned officer of the highest rank. **3.** A tropical Atlantic fish, *Abudefduf saxatilis,* with a flattened body and dark vertical stripes.

**se·ri·al** (sîr′ē-əl) *adj.* **1.** Of, forming, consisting of, or arranged in a series. **2. a.** Published or produced in installments at regular intervals, as a novel or television drama. **b.** Of or relating to such publication or production. **3.** *Mus.* Relating to or based on a 12-tone row. —*n.* A literary or dramatic work published or produced in installments. **—se′ri·al·ly** *adv.*

**se·ri·al·ism** (sîr′ē-ə-lĭz′əm) *n.* **1.** Serial music. **2.** The composition or theory of serial music. **—se′ri·al·ist** *n.*

**se·ri·al·ize** (sîr′ē-ə-līz′) *vt.* **-ized, -iz·ing, -iz·es.** To produce or publish in serial form. **—se′ri·al·i·za′tion** *n.*

**serial number** *n.* A number, one of a series, used for identification, as of a machine.

**se·ri·ate** (sîr′ē-āt′, -ĭt) *adj.* Occurring or organized in a series or in rows. **—se′ri·ate′ly** *adv.*

**se·ri·a·tim** (sîr′ē-ā′tĭm, -ăt′ĭm) *adv.* [Med. Lat. < Lat. *series,* series.] In a series.

**se·ri·ceous** (sĭ-rĭsh′əs) *adj.* [LLat. *sericeus,* silken < Lat. *sericus.* —see SERGE.] **1.** Silky. **2.** *Bot.* Covered with fine, soft, silky hairs.

**ser·i·cin** (sĕr′ĭ-sĭn) *n.* [< Lat. *sericus,* silken.] A gelatinous protein forming on the surface of raw-silk fibers.

**ser·i·cul·ture** (sĕr′ĭ-kŭl′chər) *n.* [Lat. *sericum,* silk (< *Seres,* a people of Eastern Asia) + CULTURE.] The raising of silkworms for the production of raw silk. **—ser′i·cul′tur·al** *adj.* **—ser′i·cul′tur·ist** *n.*

**ser·i·e·ma** (sĕr′ĭ-ē′mə) *n.* [Tupi, crested.] A cranelike South American bird, *Cariama cristata* or *Chunga burmeisteri,* with a tuftlike crest behind the bill.

seriema
*36 inches long*

**se·ries** (sîr′ēz) *n., pl.* **series.** [Lat. *series* < *serere,* to join.] **1.** A number of things or events of the same kind occurring in a row or following one after the other in succession <a *series* of phone calls><a *series* of mishaps> **2.** A group of objects related by a linearly varying morphological or configurational characteristic <the paraffin *series*> **3.** *Math.* The indicated sum of a finite or of a sequentially ordered infinite set of terms. **4.** A sequence of coordinate elements in a sentence. **5.** A sequence of usu. continuously numbered issues or volumes of a publication. **6.** A television or radio program broadcast at regular intervals. **7. a.** A number of games played one after the other by the same opposing teams. **b.** *Baseball.* The World Series. **—in series.** In an arrangement forming a series.

**series circuit** *n.* An electric circuit connected so that current passes through each circuit element in turn without branching.

**se·ries-wound** (sîr′ēz-wound′) *adj.* Of or relating to an electric motor or dynamo in which the armature circuit and the field circuit are connected in series with the external circuit.

**ser·if** (sĕr′ĭf) *n.* [Perh. < Du. *schreef,* line < MDu. *scrēve* < *scriven,* to write < Lat. *scribere.*] A fine line in printing finishing off the main strokes of a letter.

**ser·i·graph** (sĕr′ĭ-grăf′) *n.* [Lat. *sericum,* silk (< *Seres,* a people of Eastern Asia) + -GRAPH.] A silk-screened print. **—se·rig′ra·pher** (sə-rĭg′rə-fər) *n.* **—se·rig′ra·phy** (-fē) *n.*

**ser·in** (sĕr′ĭn) *n.* [Fr.] An Old World finch of the genus *Serinus,* with yellow or yellowish-green plumage.

**ser·ine** (sĕr′ēn′) *n.* [SER(ICIN) + -INE.] An amino acid, $C_3H_7NO_3$, a common component of many proteins.

**se·ri·o·com·ic** (sîr′ē-ō-kŏm′ĭk) *adj.* [SERIO(US) + COMIC.] Partially serious and partially comic. **—se′ri·o·com′i·cal·ly** *adv.*

**se·ri·ous** (sîr′ē-əs) *adj.* [ME *seryous* < OFr. *serieux* < LLat. *seriosus* < Lat. *serius.*] **1.** Grave in quality, character or manner : SOBER. **2.** Said or done earnestly : SINCERE. **3.** Involving important rather than trivial matters : WEIGHTY. **4.** Characterized by much effort or devotion. **5.** Causing worry. **—se′ri·ous·ly** *adv.* **—se′ri·ous·ness** *n.*

**ser·jeant** (sär′jənt) *n. Chiefly Brit. var. of* SERGEANT 2, 3.

**ser·mon** (sûr′mən) *n.* [ME < OFr. < Lat. *sermo,* discourse.] **1.** A discourse delivered during a church service. **2.** An often long-winded and repetitious speech of reproof or exhortation. **—ser·mon′ic** (-mŏn′ĭk), **ser·mon′i·cal** *adj.*

**ser·mon·ize** (sûr′mə-nīz′) *v.* **-ized, -iz·ing, -iz·es.** —*vt.* To deliver a sermon to. —*vi.* To speak as if giving a sermon. **—ser′mon·iz′er** *n.*

**Sermon on the Mount** *n.* A discourse of Jesus, delivered on the Mount of Olives, in the New Testament.

**sero-** *pref.* [< SERUM.] Serum <*sero*therapy>

**se·ro·di·ag·no·sis** (sîr′ō-dī′əg-nō′sĭs, sĕr′ō-) *n.* Diagnosis by means of blood serum reactions. **—se′ro·di′ag·nos′tic** (-nŏs′tĭk) *adj.*

**se·rol·o·gy** (sĭ-rŏl′ə-jē) *n.* The medical science dealing with serums. **—se′ro·log′ic** (sĭr′ə-lŏj′ĭk), **se′ro·log′i·cal** *adj.* **—se·rol′o·gist** *n.*

**se·ro·pu·ru·lent** (sîr′ō-pyŏŏr′ə-lənt, -pyŏŏr′yə-, sĕr′-) *adj.* Made up of serum and pus.

**se·ro·sa** (sĭ-rō′sə, -zə) *n., pl.* **-sas** or **-sae** (-sē′) [NLat., fem. of *serosus,* serous < Lat. *serum,* serum.] A serous membrane. **—se·ro′·sal** (-zəl) *adj.*

**se·ro·ther·a·py** (sîr′ō-thĕr′ə-pē, sĕr′-) *n.* Treatment of disease with a serum or antitoxin.

**se·rot·i·nal** (sĭ-rŏt′n-əl, sĕr′ə-tī′nəl) *adj.* Serotinous.

**se·rot·i·nous** (sĭ-rŏt′n-əs, sĕr′ə-tī′nəs) *adj.* [Lat. *serotinus,* coming late < *sero,* at a late hour < *serus,* late.] *Biol.* Late in maturing or blooming.

**se·ro·to·nin** (sîr′ə-tō′nĭn, sĕr′-) *n.* [SERO- + TON(IC) + -IN.] An organic compound, $C_{10}H_{12}N_2O$, found esp. in the brain, blood serum, and gastric mucosa and capable of raising blood pressure.

**se·ro·type** (sîr′ə-tīp′, sĕr′-) *n.* A group of related microorganisms distinguished by its composition of antigenes.

**se·rous** (sîr′əs) *adj.* Containing, secreting, or like serum.

**serous membrane** *n.* A thin membrane that lines a closed bodily cavity and secretes a serous fluid.

**se·row** (sə-rō′) *n.* [Lepcha *să-ro.*] A goatlike antelope of the genus *Capricornis,* of eastern Asia, with short horns and a dark coat.

**Ser·pens** (sûr′pənz, -pĕnz′) *n.* [Lat. *Serpens* < *serpens,* serpent.] A constellation in the Northern Hemisphere.

**ser·pent** (sûr′pənt) *n.* [ME < OFr. < Lat. *serpens* < pr.part. of *serpere,* to creep.] **1.** A snake. **2.** *often* **Serpent. a.** The creature that tempted Eve in the Garded of Eden. **b.** DEVIL 1. **3.** One who is subtle, sly, or treacherous. **4.** A firework that twists and turns while burning. **5.** *Mus.* A deep-voiced 18th-cent. wind instrument of serpentine shape, approx. eight feet long and made of brass or wood. **6. Serpent.** Serpens.

**ser·pen·tine** (sûr′pən-tēn′, -tīn′) *adj.* [ME < OFr. *serpentin* < LLat. *serpentinus* < Lat. *serpens,* serpent.] **1.** Of or like a serpent in form or movement : SINUOUS. **2.** Subtly sly and tempting. —*n.* (-tēn′) **1.** Something that wins like a snake. **2.** A greenish, brownish, or spotted mineral, $3MgO \cdot 2SiO_2 \cdot 2H_2O$, used as a source of magnesium and a decorative building stone.

**serpent star** *n.* A brittle star.

**ser·pi·go** (sər-pī′gō) *n.* [ME < Med. Lat. < Lat. *serpere,* to creep.] A spreading skin eruption or disease, as ringworm. **—ser·pig′i·nous** (sər-pĭj′ə-nəs) *adj.*

**ser·ran·id** (sə-răn′ĭd, sĕr′ə-nĭd) *adj.* [NLat. *Serranidae,* family name < Lat. *serra,* saw.] Of or belonging to the family Serranidae, which includes the sea basses and groupers. **—ser·ran′id** *n.*

**ser·rate** (sĕr′āt′) *also* **ser·rat·ed** (-ā′tĭd) *adj.* [Lat. *serratus,* saw-shaped < Lat. *serra,* saw.] **1.** Having notched, toothlike projections. **2.** Edged with notched, toothlike projections <*serrate* leaves> **—ser·ra′tion** (sə-rā′shən, sĕ-) *n.*

**ser·ried** (sĕr′ēd) *adj.* [P.part. of obs. *serry,* to close ranks < OFr. *serre,* p.part. of *serrer,* to crowd, grasp. —see SEAR².] Pressed or close together, esp. in rows : CROWDED. **—ser′ried·ly** *adv.*

**ser·ru·late** (sĕr′yə-lĭt, -lāt′, sĕr′ə-) *also* **ser·ru·lat·ed** (-lā′tĭd) *adj.* [NLat. *serrulatus* < Lat. *serrula,* dim. of *serra,* saw.] Having small serrations along the edge.

**ser·tu·lar·i·an** (sûr′chə-lâr′ē-ən, sûr′tl-âr′-) *n.* [< NLat. *Sertularia,* genus name < Lat. *sertula,* dim. of *serta,* garland < *serere,* to entwine.] A colonial hydroid of the genus *Sertularia,* with paired stalkless polyps on a long, branching stem.

**se·rum** (sîr′əm) *n., pl.* **se·rums** or **se·ra** (sîr′ə) [Lat., whey, serum.] **1.** The clear yellowish fluid obtained when whole blood is separated into its solid and liquid components. **2.** The fluid from the tissues of immunized animals, used esp. as an antitoxin. **3.** Watery fluid from animal tissue. **4.** Whey.

**serum albumin** *n.* A protein fraction of blood serum that helps maintain osmotic pressure, used in the treatment of shock.

**serum globulin** *n.* A protein fraction of blood serum mainly containing antibodies.

**serum sickness** *n.* A hypersensitive reaction to the administration of serum, marked by fever, rash, swelling, and enlargement of the lymph nodes.

**ser·val** (sûr′vəl, sər-văl′) *n.* [Fr. < Port. *(lobo) cerval,* deerlike (wolf), lynx < *cervo,* deer < Lat. *cervus.*] A long-legged wild cat, *Felis serval* of Africa, with a black-spotted yellowish coat.

**ser·vant** (sûr′vənt) *n.* [ME < OFr. < pr.part. of *servir,* to serve.] **1.** One privately employed to perform household services. **2.** One publicly employed to perform services, as for a government. **3.** One expressing submission or debt to another <your obedient *servant*> **4.** One that serves another.

**serve** (sûrv) *v.* **served, serv·ing, serves.** [ME *serven* < OFr. *servir* < Lat. *servire* < *servus,* slave.] —*vt.* **1. a.** To work for. **b.** To be a servant to. **2. a.** To prepare and offer (e.g., food)<*serve* lunch> **b.** To set food before (someone) : WAIT ON. **3. a.** To supply goods and services to (customers) <*serving* the public for more than 30 years> **b.** To supply (goods or services) to customers. **c.** To assist (the celebrant) during Mass. **4.** To be of assistance to or promote the interests of : AID <"Both major parties today seek to *serve* the national interest" —John F. Kennedy> **5.** To spend or complete (time) <*serve* six years in the Senate> **6.** To fight or do military duty for. **7.** To honor and obey. **8.** To deal with : REQUITE. **9.** To copulate with. —Used of male animals. **10.** To meet the needs or requirements of : SATISFY <*serve* the purpose> **11.** *Law.* **a.** To deliver or present (a writ or summons). **b.** To deliver such a writ to. **12.** To put (a ball or shuttlecock) in play, as in tennis, badminton, or jai alai. **13.** To bind or whip (a rope) with fine cord or wire. —*vi.* **1.** To work as a servant. **2.** To do a term of duty <*serve* in the army> **3.** To act in a specific capacity <*serve* as a court stenographer> **4.** To be of service or use : FUNCTION <*serve* as a memento of the trip> **5.** To meet requirements or needs : SATISFY <an instrument that will *serve* well> **6.** To wait on table <*serve* at dinner> **7.** To put a ball or shuttlecock into play, as in court games. **8.** To assist the celebrant during Mass. —*n.* The act, manner, or right of serving, as in court games.

**serv·er** (sûr′vər) *n.* **1.** One that serves. **2.** Something, as a tray, used in serving. **3.** An assistant to the celebrant at a Mass. **4.** The player who serves, as in court games.

**serv·ice** (sûr′vĭs) *n.* [ME *servise* < OFr. *service* < Lat. *servitum,* slavery < *servus,* slave.] **1.** The occupation or duties of a servant. **2.** Employment in duties or work for another, esp. for a government <public *service*> **3.** A government branch or department and its employees <the civil *service*> **4. a.** The military forces of a nation. **b.** A branch of the military forces of a nation. **5.** Work or duties performed for a superior. **6.** Work done for others as an occupation or business <a housecleaning *service*> **7.** Installation, maintenance, or repairs done or guaranteed by a dealer or manufacturer. **8.** A facility providing the public with the use of something, as water or transportation. **9.** Acts of devotion to God. **10.** A religious ceremony or rite. **11.** An act of assistance or benefit : FAVOR. **12.** The serving of food or the way in which it is served. **13.** A set of dishes or utensils <a china *service* for eight> **14.** The act, manner, or right of serving in many court games : SERVE. **15.** Copulation with a female. —Used of male animals. **16.** *Law.* The delivery of a writ or summons. **17.** The material, as cord, used to bind or wrap rope. —*vt.* **-iced, -ic·ing, -ic·es. 1.** To adjust, repair, or maintain <*service* a washing machine> **2.** To provide services to. **3.** To make interest payments on (a debt). **4.** To copulate with. —Used of male animals.

**serv·ice·a·ble** (sûr′vĭs-ə-bəl) *adj.* **1.** Ready for service : USABLE. **2.** Capable of giving long service : DURABLE. **—serv′ice·a·bil′i·ty, serv′ice·a·ble·ness** *n.* **—serv′ice·a·bly** *adv.*

**serv·ice·ber·ry** (sûr′vĭs-bĕr′ē, sär′-) *n.* [SERVICE (TREE) + BERRY.] The shadbush or its fruit.

**service break** *n.* A game won on an opponent's serve.

**service cap** *n.* A round, flat-topped military cap with a visor.

**service charge** *n.* An extra charge for a service for which there is already a basic fee.

**service line** *n.* A boundary line, as in tennis or handball, that must not be overstepped in serving.

**serv·ice·man** (sûr′vĭs-măn′, -mən) *n.* **1.** A member of the armed forces. **2.** *also* **service man.** A man whose work is the maintenance and repair of equipment.

**service mark** *n.* A symbol or mark used in the sale or advertising of services to distinguish them from the services of others.

**service station** *n.* **1.** A filling station. **2.** A place where services, esp. repairs, can be obtained.

**serv·ice tree** (sûr′vĭs, sär′-) *n.* [ME *serves,* pl. of *serve,* the service tree < OE *syrfe* < Lat. *sorbus.*] An Old World tree, *Sorbus domestica* or *S. torminalis,* with white flower clusters and edible fruit.

**ser·vi·ette** (sûr′vē-ĕt′) *n.* [Fr. < OFr., towel, napkin < *servir,* to serve.] A table napkin.

**ser·vile** (sûr′vəl, -vīl′) *adj.* [ME < Lat. *servilis* < *servus,* slave.] **1.** Abjectly submissive : SLAVISH. **2.** Of or suitable to a servant or slave <*servile* duties and responsibilities> **—ser′vile·ly** *adv.* **—ser′vile·ness, ser·vil′i·ty** (sər-vĭl′ĭ-tē) *n.*

☆ **syns:** SERVILE, OBSEQUIOUS, SLAVISH, SUBSERVIENT *adj. core meaning* : abjectly submissive <a *servile,* spineless toady>

**serv·ing** (sûr′vĭng) *n.* **1.** The act of one that serves. **2.** A individual portion of food or drink : HELPING.

**ser·vi·tor** (sûr′vĭ-tər, -tôr′) *n.* [ME < OFr. < Lat. *servitor* < *servire,* to serve < *servus,* slave.] A personal servant. **—ser′vi·tor·ship′** *n.*

**ser·vi·tude** (sûr′vĭ-to͞od′, -tyo͞od′) *n.* [ME *servytude* < OFr. < Lat. *servitudo* < *servus,* slave.] **1.** Submission to a master : BONDAGE.

**2.** Forced labor imposed as a punishment for crime <penal *servitude*> **3.** *Law.* A right that grants use of another's property.

**ser·vo** (sûr′vō) *n., pl.* **-vos. 1.** A servomechanism. **2.** A servomotor.

**ser·vo·mech·a·nism** (sûr′vō-mĕk′ə-nĭz′əm) *n.* [SERVO(MOTOR) + MECHANISM.] A feedback system composed of a sensory element, an amplifier, and a servomotor, used in the automatic control of a mechanical device.

**ser·vo·mo·tor** (sûr′vō-mō′tər) *n.* [Fr. *servomoteur* : Lat. *servus,* slave + Fr. *moteur,* motor.] An electric motor or hydraulic piston that powers a servomechanism.

**ses·a·me** (sĕs′ə-mē) *n.* [Lat. *sesamum* < Gk. *sēsamon, sēsamē,* of Semitic orig.] **1.** A plant, *Sesamum indicum* of tropical Asia, yielding small flat seeds used as food and a source of oil. **2.** The seeds of the sesame.

**ses·a·moid** (sĕs′ə-moid′) *adj.* [Gk. *sēsamoeidēs,* shaped like a sesame seed < *sēsamon, sēsamē,* sesame.] Of or designating a small bone, as the kneecap, that develops in a tendon or in the capsule of a joint. **—ses′a·moid′** *n.*

**sesqui-** *pref.* [Lat. : *semis,* half + *-que,* and.] One and a half <*sesquicentennial*>

**ses·qui·cen·ten·ni·al** (sĕs′kwĭ-sĕn-tĕn′ē-əl) *adj.* Occurring every 150 years. *—n.* A 150th anniversary or its celebration.

**ses·qui·pe·da·lian** (sĕs′kwĭ-pĭ-dāl′yən) *also* **ses·quip·e·dal** (sĕ-skwĭp′ĭ-dl) *adj.* [Lat. *sesquipedalis,* of a foot and a half in length : *sesqui-,* one half more + *pes,* foot.] **1.** Long and cumbersome : POLYSYLLABIC. **2.** Tending to use long words. *—n.* **sesquipedalian.** A long, polysyllabic word.

**ses·sile** (sĕs′ĭl′, -əl) *adj.* [Lat. *sessilis,* low, of sitting < *sessus,* p.part. of *sedēre,* to sit.] **1.** *Bot.* Having no stalk and attached directly at the base <*sessile* leaves> **2.** *Zool.* Permanently attached or fixed. **—ses·sil′i·ty** (sĕ-sĭl′ĭ-tē) *n.*

**ses·sion** (sĕsh′ən) *n.* [ME < OFr. < Lat. *sessio,* act of sitting < *sessus,* p.part. of *sedēre,* to sit.] **1. a.** A meeting of a legislative or judicial body. **b.** A series of such meetings. **c.** The term or amount of time taken by such a series of meetings. **2.** The part of a year or of a day during which a school holds classes. **3.** A group of persons gathered for a common purpose or with a common interest <a gossip *session*> **4.** A U.S. court of criminal jurisdiction. **5.** A period of time devoted to a specific activity <a practice *session*> **—ses′sion·al** *adj.* **—ses′sion·al·ly** *adv.*

**ses·terce** (sĕs′tûrs′) *n.* [Lat. *sestertius,* a coin worth two and a half asses : *semis,* half + *tertius,* third.] An ancient Roman silver or bronze coin equivalent to 1/4 denarius.

**ses·ter·tium** (sĕ-stûr′shəm, -shē-əm) *n., pl.* **-tia** (-shə, -shē-ə) [Lat. *(mille) sestertium,* (a thousand) sesterces.] An ancient Roman monetary unit equivalent to 1,000 sesterces.

**ses·tet** (sĕ-stĕt′) *n.* [Ital. *sestetto* < *sesto,* sixth < Lat. *sextus.*] A stanza making up the last six lines of a sonnet.

**ses·ti·na** (sĕ-stē′nə) *n.* [Ital. < *sesto,* sixth < Lat. *sextus.*] An orig. Provençal verse form made up of six six-line stanzas and a three-line envoi that repeats the end words of the first stanza throughout according to a scheme of cruciate retrogradation.

**†set¹** (sĕt) *v.* **set, set·ting, sets.** [ME *setten* < OE *settan.*] **1.** To put in a designated position : PLACE <*set* a book on the desk> **2.** To put into a designated state <*set* the prisoner at liberty> **3.** To put into a secure position : FIX. **4.** To return to a normal and proper state when dislocated or broken <*set* a broken leg> **5. a.** To regulate for proper functioning. **b.** To adjust (a saw) by deflecting the teeth. **c.** To spread open to the wind <*set* the sails> **6.** To adjust according to a standard. **7.** To adjust (an instrument) to a specific point or calibration <*set* a timer> **8.** To arrange properly for use <*set* a place for a dinner guest><*set* the dining room table> **9.** To apply equipment, as curlers and clips, to (hair) in order to style. **10. a.** To arrange (type) into words and sentences preparatory to printing : COMPOSE. **b.** To transpose into type. **11. a.** To compose (music) to suit a given text. **b.** To write (words) to accompany a given melodic line. **12.** To arrange scenery on (a stage). **13.** To prescribe or establish <*set* a precedent> **14.** To prescribe the unfolding of (a dramatic work or scene) in a specific place <a musical *set* in Austria> **15.** To designate as a time for <*set* May 14 as the date of the meeting> **16.** To detail or assign (someone) to a particular task, service, or station <*set* the gardener to planting roses><*set* lookouts around the camp> **17.** To arouse to hostile action <a conflict *setting* brother against brother> **18. a.** To establish as the highest level of performance <*set* a new marathon record> **b.** To establish as a model <*set* a good example for the others> **19. a.** To place in a mounting : MOUNT <*set* a pearl in the ring> **b.** To attach jewels to : STUD <a necklace *set* with diamonds> **20.** To cause to sit. **21. a.** To put (a hen) on eggs in order to hatch them. **b.** To put (eggs) beneath a hen or in an incubator. **22.** To position (oneself) to begin an action, as running a race. **23. a.** To value or consider something at the rate of <*sets* a great store in daily exercise> **b.** To put at a specified amount <*set* bail for $50,000> **c.** To make as an estimate of worth <*set* a high value on human dignity> **24.** To signal the location of (game) by maintaining a fixed attitude. —Used of a hunting dog. **25.** To produce, as after pollination <*set* seed> **26. a.** To prepare (a trap) for catching prey. **b.** To fix (a hook) firmly into a fish's jaw. *—vi.* **1.** To disappear below the horizon <The sun *sets* at six tonight.> **2.** To decline or diminish : WANE. **3.** To sit on eggs. —Used of fowl. **4.** To harden or congeal. **5.** To become returned to a normal state : KNIT <The broken bone *set* quickly.> **6.** To mature or develop, as after pollination. **7.** *Regional.* To sit. **8.** To position oneself in preparation for an action, as running a race. ***usage:*** In most cases *set* is a transitive verb, as in *I set the book down on the table,* whereas *sit* is gen. an intransitive verb, as in *The student sits in the last row.* There are some exceptions, however; we say *The sun sets* (not *sits*) *in the west* and *A hen sets* (or *sits*) *on her eggs.* **—set about.** To begin or start <*set about* weeding the garden> **—set apart. 1.** To put aside for a specific use. **2.** To make noticeable. **—set aside. 1.** To reserve for a special purpose. **2.** To reject or get rid of. **3.** To annul or overrule. **—set at.** To attack or assail. **—set back. 1.** To retard the progress of : HINDER. **2.** *Informal.* To cost <That car *set* me *back* $8,000.> **—set down. 1.** *Informal.* To cause to sit : SEAT <*Set* yourself *down* right here.> **2.** To put in writing : RECORD <*set down* the facts> **3. a.** To regard : consider <Just *set* them *down* as inefficient workers.> **b.** To assign to a cause : ATTRIBUTE <Let's *set* the mistake *down* to hastiness.> **4.** To land (an aircraft). **—set forth. 1.** To present for consideration : PROPOSE. **2.** To express in words <*set forth* my proposal> **—set in. 1.** To insert <*set in* the sleeve of a jacket> **2.** To begin to happen or be apparent <Infection *set in.*> **3.** To move toward the shore. —Used of wind or water. **—set off. 1. a.** To cause to occur. **b.** To cause to explode. **2.** To indicate as being different : DISTINGUISH <an outfit *setting* you *off* from the crowd> **3.** To call attention to by contrast : ACCENTUATE <*set off* a passage with italics> **4.** To leave on a journey <*set off* for California> **—set out. 1.** To begin a serious attempt : UNDERTAKE <*set out* to determine why the scheme had not worked> **2.** To lay out systematically and graphically <*set out* a terrace> **3.** To display for exhibition or sale. **4.** To plant <*set out* seedlings> **5.** To begin a journey <*set out* later than we had planned> **—set to. 1.** To begin working energetically. **2.** To begin fighting. **—set up. 1.** To put in an upright position. **2. a.** To make higher : RAISE. **b.** To raise in authority or power <*set* the general *up* as a dictator> **c.** To claim to be <*set* oneself *up* as an expert in judo> **3.** To put together and erect <*set up* a new machine> **4.** To establish : found <*set up* a day-care center> **5.** To create : cause <*set up* a fuss because I was late> **6.** To establish in business by providing capital or other backing. **7.** *Informal.* **a.** To treat (someone) to drinks. **b.** To pay for (drinks). **8.** *Informal.* To excite or stimulate <The victory really *set* us all *up* for the day.> **9.** To make plans for <*set up* a bank robbery> *—adj.* **1.** Fixed or determined by agreement <a *set* time for lunch> **2.** Established by convention <followed *set* procedures for registering a complaint> **3.** Established deliberately : INTENTIONAL. **4.** Fixed and rigid <grim, *set* eyes> **5.** Unwilling to change <*set* in one's ways> **6.** Ready <We're *set* to go.> *—n.* **1. a.** The act or process of setting. **b.** The condition resulting from setting. **2.** The gradual firming or hardening of a substance, as by cooling. **3.** The deflection of the teeth of a saw. **4.** The manner in which something is positioned <the *set* of your cap> **5.** The carriage or bearing of a part of the body. **6.** A descent below the horizon. **7.** The direction or course of wind or water. **8.** A seedling, slip, or cutting ready to be planted. **9.** The act or an instance of arranging hair by waving and curling it. **—set eyes on.** To catch sight of : SEE. **—set foot in.** To enter. **—set foot on.** To step on. **—set in motion.** To give impetus to. **—set (one's) heart on.** To be determined to do something. **—set (or put) (one's) house in order.** To arrange one's affairs in an orderly manner. **—set (one's) sights on.** To have as a goal <*set* my *sights on* a promotion> **—set (someone) straight.** To inform fully. **—set store by.** To consider valuable or worthwhile.

**set²** (sĕt) *n.* [ME *sette* < OFr. < Lat. *secta,* faction < *sequi,* to follow.] **1.** A group of things of the same kind that belong together and are so used <a *set* of dishes> **2.** A group of persons sharing a common interest or social milieu <the yachting *set*> **3.** A group of books or periodicals published as a unit. **4. a.** A number of couples required for participation in a square dance. **b.** The steps and movements making up a square dance. **5. a.** The scenery constructed for a dramatic performance. **b.** The enclosure in which a motion picture is filmed : SOUND STAGE. **6.** The receiving apparatus assembled to operate a radio or television. **7.** *Math.* A collection of distinct elements <a *set* of negative integers> **8.** A group of games, as in tennis, forming one division or unit of a match.

**se·ta** (sē′tə) *n., pl.* **-tae** (-tē′) [NLat. < Lat. *saeta,* bristle.] *Biol.* A stiff hair, bristle, or bristlelike process. **—se′tal** (sēt′l) *adj.*

**se·ta·ceous** (sĭ-tā′shəs) *adj.* [SET(A) + -ACEOUS.] **1.** Having or consisting of bristles : BRISTLY. **2.** Resembling a bristle or bristles.

**se·tae** (sē′tē′) *n. pl. of* SETA.

**set·back** (sĕt′băk′) *n.* **1.** An unexpected check in progress : REVERSE. **2. a.** A steplike recession, as in a wall. **b.** One of a series of such recessions in the rise of a tall building.

**set back** *n. Football.* An offensive back who lines up behind the quarterback.

**set chisel** *n.* A chisel with a cutting edge on a tapered shaft.

**se·ti·form** (sē′tə-fôrm′) *adj.* Shaped like a seta or bristle.
**set·line** (sĕt′līn′) *n.* A long fishing line to which many smaller lines bearing baited hooks are attached.
**set·off** (sĕt′ôf′, -ŏf′) *n.* **1.** Something, as a decoration, that sets off something else by contrast. **2.** Something that offsets or makes up for something else : COUNTERBALANCE. **3. a.** A counterclaim. **b.** The settlement of a debt by a debtor's establishing such a claim against his or her creditor. **4.** A flat projection, as from a wall : LEDGE.
**se·tose** (sē′tōs′) *adj.* Setaceous.
**set piece** *n.* **1.** A realistic piece of scenery built to stand by itself. **2.** An often brilliantly executed artistic or literary work marked by a formal pattern. **3.** A carefully planned and performed operation, esp. a military one.
**set·screw** (sĕt′skro͞o′) *n.* **1.** A screw, often without a head, used to hold two parts in a position relative to each other without motion. **2.** A screw for regulating the tension of a spring.
**set·tee** (sĕ-tē′) *n.* [Perh. alteration of SETTLE.] **1.** A long wooden bench with a high back. **2.** A small sofa.
**set·ter** (sĕt′ər) *n.* **1.** One that sets. **2.** Any of several breeds of long-haired dogs orig. trained to signal the presence of game by crouching in a set position.
**set theory** *n.* The study of the mathematical properties of sets.
**set·ting** (sĕt′ĭng) *n.* **1. a.** The context and environment in which something occurs. **b.** The time and place in which a literary or a dramatic work is set. **2.** The scenery constructed for a dramatic performance. **3.** Music composed or arranged for a particular text. **4.** A mounting, as for a jewel. **5.** A set of eggs in a hen's nest.
**set·tle** (sĕt′l) *v.* **-tled, -tling, -tles.** [ME *setlen,* to seat < OE *setlan* < *setl,* seat.] —*vt.* **1.** To arrange or fix definitely as desired : put into order. **2.** To put firmly in a desired position or place : ESTABLISH. **3.** To establish as a resident or residents <*settled* my family in Alaska> **4.** To establish residence in <The Spanish *settled* Mexico.> **5.** To establish in a residence, business, or profession. **6.** To restore calm or comfort to. **7. a.** To cause to come to rest, sink, or become compact. **b.** To cause (a liquid) to become clear by forming a sediment. **8.** To make orderly or quiet. **9.** To establish on a permanent basis : STABILIZE. **10. a.** To make compensation for (a claim). **b.** To pay (a debt). **11.** To end (e.g., a dispute) by a final decision or agreement. **12.** To decide (a lawsuit) by mutual agreement of the involved parties without court action <*settled* the case out of court> **13.** *Law.* To secure or assign (property or title) by legal action. —*vi.* **1.** To stop moving and come to rest in one place. **2.** To subside or come down gradually. **3.** To sink and become more compact <The dust *settled.*> **4. a.** To become clear. —Used of liquids. **b.** To be separated from a mixture or solution as a sediment. **5.** To establish one's residence. **6.** To reach a decision : DETERMINE. **7. a.** To compensate for a claim. **b.** To pay a debt. **—settle down. 1.** To start living a more orderly life. **2.** To become less nervous, restless, or unruly. **—settle for.** To accept despite lack of complete satisfaction. —*n.* A long wooden bench with a high back.
**set·tle·ment** (sĕt′l-mənt) *n.* **1.** The act or process of settling. **2. a.** Establishment, as of a person in a business or of people in a new area. **b.** A newly colonized area. **3.** A small community. **4.** An adjustment or understanding reached, as in financial or business proceedings. **5. a.** Transfer of property to provide for the future needs of a person. **b.** Property thus transferred. **6.** A welfare center offering community services in an underprivileged area.
**set·tler** (sĕt′lər) *n.* **1.** One that settles or decides something. **2.** One who settles in a new area.
**set·tlings** (sĕt′lĭngz) *pl.n.* Sediment : dregs.
**set-to** (sĕt′to͞o′) *n., pl.* **-tos.** A brief but usu. heated conflict.
**set·up** (sĕt′ŭp′) *n.* **1.** *Informal.* The manner in which something is arranged or planned. **2.** Bodily carriage : POSTURE. **3.** Physical makeup : PHYSIQUE. **4.** *often* **setups.** *Informal.* The ingredients and mixers, as ice and soda water, needed to serve a variety of alcoholic drinks. **5.** *Slang.* **a.** A contest prearranged to result in an easy or faked victory. **b.** An endeavor that is intentionally made easy. **c.** A hoax or fraud.
**sev·en** (sĕv′ən) *n.* [ME < OE *seofon* : akin to G. *sieben,* Lat. *septem,* Gk. *hepta,* and Skt. *sapta.*] **1.** The cardinal number equal to 6 + 1. **2.** The seventh in a set or sequence. **—sev′en** *adj.* & *pron.*
**sev·en·fold** (sĕv′ən-fōld′) *adj.* **1.** Having seven parts or members. **2.** Having seven times as many or as much. **—sev′en·fold′** *adv.*
**seven seas** *also* **Seven Seas** *pl.n.* All the oceans of the world.
**sev·en·teen** (sĕv′ən-tēn′) *n.* [ME *seventene* < OE *seofontīne* < *seofon,* seven.] **1.** The cardinal number equal to 16 + 1. **2.** The 17th in a set or sequence. **—sev′en·teen′** *adj.* & *pron.*
**sev·en·teenth** (sĕv′ən-tēnth′) *n.* **1.** The ordinal number matching the number 17 in a series. **2.** One of 17 equal parts. **—sev′en·teenth′** *adj.* & *adv.*
**sev·en·teen-year locust** (sĕv′ən-tēn′yîr′) *n.* A cicada, *Magicicada septendecim* of the eastern United States, having a nymphal stage in which it remains underground for 17 or sometimes 13 years.
**sev·enth** (sĕv′ənth) *n.* [ME < *seven,* seven.] **1.** The ordinal number matching the number seven in a series. **2.** One of seven equal parts. **3.** A musical interval including seven diatonic degrees. **—sev′enth** *adj.* & *adv.*
**Sev·enth-Day Adventist** (sĕv′ənth-dā′) *n.* A member of a sect of Adventism noted primarily for its observance of the Sabbath on Saturday.
**seventh heaven** *n.* **1.** The farthest of the concentric spheres containing the stars and comprising the dwelling place of God and the angels in the Moslem and cabalist systems. **2.** A state of great happiness and satisfaction.
**sev·en·ti·eth** (sĕv′ən-tē-ĭth) *n.* **1.** The ordinal number matching the number 70 in a series. **2.** One of 70 equal parts. **—sev′en·ti·eth** *adj.* & *adv.*
**sev·en·ty** (sĕv′ən-tē) *n.* [ME < OE *hundseofontig* : *hund,* hundred + *seofon,* seven + *-tig,* -ty.] The cardinal number equal to 7 × 10. **—sev′en·ty** *adj.* & *pron.*
**sev·en-up** (sĕv′ən-ŭp′) *n.* A card game for two to four persons requiring seven points to win.
**sev·er** (sĕv′ər) *v.* **-ered, -er·ing, -ers.** [ME *severen* < OFr. *severer* < Lat. *separare.* —see SEPARATE.] —*vt.* **1.** To divide or separate into parts. **2.** To cut or break forcibly from a whole. **3.** To break off, as a relationship : DISSOLVE. —*vi.* **1.** To become cut or broken apart. **2.** To separate or go apart : DIVIDE. **—sev′er·a·ble** *adj.*
**sev·er·al** (sĕv′ər-əl, sĕv′rəl) *adj.* [ME *severall,* separate < AN *several* < Med. Lat. *separalis* < Lat. *separ* < *separare,* to separate. —see SEPARATE.] **1.** Being more than two or three but not many <*several* days from now> **2.** Single : distinct <"Pshaw! said I, with an air of carelessness, three *several* times" —Sterne> **3.** Respectively different : VARIOUS <went our *several* ways after graduation> **4.** *Law.* Pertaining separately to each party of a bond or note. —*n.* Several persons or things : FEW. **—sev′er·al·ly** *adv.*
**sev·er·ance** (sĕv′ər-əns, sĕv′rəns) *n.* **1.** The act or process of severing or the state of being severed. **2.** Separation : division.
**severance pay** *n.* A sum of money usu. based on length of employment that an employee is eligible for on termination.
**se·vere** (sə-vîr′) *adj.* **-ver·er, -ver·est.** [OFr. < Lat. *severus.*] **1.** Unsparing and harsh : STRICT. **2.** Corresponding strictly and rigidly to established rule. **3.** Austere or dour : FORBIDDING <a *severe* look on your face> **4.** Extremely plain in style. **5.** Extremely intense <*severe* pain><a *severe* storm> **6.** Extremely difficult : TRYING. **—se·vere′ly** *adv.* **—se·vere′ness, —se·ver′i·ty** (-vĕr′ĭ-tē) *n.*
**Sè·vres** (sĕv′rə) *n.* [After *Sèvres,* France.] A fine, often elaborately decorated French porcelain.
**sew** (sō) *v.* **sewed, sewn** (sōn) *or* **sewed, sew·ing, sews.** [ME *sewen* < OE *seowian.*] —*vt.* **1.** To make, repair, or fasten with a needle and thread <*sew* a skirt><*sew* on a button> **2.** To close, fasten, or attach with stitches <*sew* an incision closed> —*vi.* To work with a needle and thread or with a sewing machine. **—sew up.** *Informal.* **1.** To complete successfully <*sew up* a multimillion dollar business deal> **2.** To control : monopolize.
**sew·age** (so͞o′ĭj) *n.* [SEW(ER) + -AGE.] Liquid and solid waste material carried off with ground water in sewers or drains.
**sew·er**[1] (so͞o′ər) *n.* [ME < OFr. *sewiere* < VLat. **exaquaria* : Lat. *ex-,* out of + Lat. *aqua,* water.] A manmade, usu. underground conduit for carrying off sewage or rainwater.
**sew·er**[2] (so͞o′ər) *n.* [ME < AN *asseour* < OFr. *asseoir,* to seat < Lat. *assidēre,* to sit down : *ad-,* to + *sedēre,* to sit.] A medieval servant who supervised the serving of meals.
**sew·er**[3] (sō′ər) *n.* One that sews.
**sew·er·age** (so͞o′ər-ĭj) *n.* **1.** A system of sewers. **2.** Removal of waste materials by a sewer system. **3.** Sewage.
**sew·ing** (sō′ĭng) *n.* **1.** The act, occupation, or hobby of one who sews. **2.** The article upon which one is working with needle and thread : NEEDLEWORK.
**sewing circle** *n.* A group that meets regularly for the purpose of sewing, often for charitable causes.
**sewing machine** *n.* A machine for sewing, often having additional attachments for special stitching.
**sewn** (sōn) *v. var. p.p. of* SEW.
**sex** (sĕks) *n.* [ME < Lat. *sexus.*] **1. a.** The property or quality by which many living things are classified according to their reproductive functions. **b.** One of the two divisions, either male or female, of this classification. **2.** Males or females as a group. **3. a.** The condition or character of being male or female. **b.** The physiological, functional, and psychological differences that distinguish the male and the female. **4.** The sexual urge or instinct as manifested in behavior. **5.** Sexual intercourse. **6.** The genitalia. —*vt.* **sexed, sex·ing, sex·es.** To determine the sex of (young chickens).
**sex-** *pref.* [Lat. *sex,* six.] Six <*sex*partite>
**sex·a·ge·nar·i·an** (sĕk′sə-jə-nâr′ē-ən) *n.* [< Lat. *sexagenarius,* sexagenary.] One who is between the ages of 60 and 70. —*adj.* **1.** Between the ages of 60 and 70. **2.** Of or relating to a sexagenarian.
**sex·ag·e·nar·y** (sĕk-săj′ə-nĕr′ē) *adj.* [Lat. *sexagenarius* < *sexageni,* sixty each < *sexaginta,* sixty < *sex,* six.] **1.** Relating to or proceeding by sixties. **2.** Sexagenarian. —*n., pl.* **-ies.** A sexagenarian.
**Sex·a·ges·i·ma** (sĕk′sə-jĕs′ə-mə, -jā′zə-) *n.* [LLat. *sexagesima* < Lat.

*sexagesimus,* sixtieth < *sexaginta,* sixty < *sex,* six.] The second Sunday before Lent.
**sex·a·ges·i·mal** (sĕk'sə-jĕs'ə-məl) *adj.* [< Lat. *sexagesimus,* sixty. —see SEXAGESIMA.] Of or based on the number 60.
**sex appeal** *n.* Physical attractiveness that arouses sexual interest in another person.
**sex cell** *n.* A gamete.
**sex·cen·te·nar·y** (sĕk-sĕn'tə-nĕr'ē, sĕk'sĕn-tĕn'ə-rē) *adj.* [< Lat. *sexcenteni,* six hundred each : *sex,* six + *centeni,* a hundred each < *centum,* hundred.] Relating to 600 or to a 600-year period. —*n., pl.* **-ies.** A 600th anniversary or its commemoration.
**sex chromosome** *n.* Either of a pair of chromosomes, usu. designated X or Y, in the germ cells of human beings, most animals, and some plants, that combine to determine the sex of an individual, XX resulting in a female and XY in a male.
**sex·en·ni·al** (sĕk-sĕn'ē-əl) *adj.* [< Lat. *sexennium,* of six years : *sex,* six + *annus,* year.] **1.** Occurring every six years. **2.** Pertaining to or lasting six years. —*n.* An event that occurs every six years. **—sex·en'ni·al·ly** *adv.*
**sex gland** *n.* A gonad.
**sex hormone** *n.* An animal hormone, as estrogen or androgen, that affects the growth or function of the reproductive organs and the development of secondary sex characteristics.
**sex·ism** (sĕk'sĭz'əm) *n.* **1.** Prejudice or discrimination based on sex, esp. against women. **2.** Arbitrary stereotyping of social roles based on gender. **—sex'ist** *adj. & n.*
**sex·less** (sĕks'lĭs) *adj.* **1.** Lacking sexual characteristics : NEUTER. **2.** Arousing or displaying no sexual interest or desire. **—sex'less·ly** *adv.* **—sex'less·ness** *n.*
**sex linkage** *n.* The condition in which a gene responsible for a specific phenotypic trait is located on the X chromosome, resulting in sexually dependent inheritance of the trait.
**sex-linked** (sĕks'lĭngkt') *adj.* **1.** Transmitted by a sex chromosome, esp. an X chromosome. —Used of genes. **2.** Sexually determined. —Used esp. of inherited traits.
**sex·ol·o·gy** (sĕk-sŏl'ə-jē) *n.* The study of human sexual behavior. **—sex'o·log'ic** (-sə-lŏj'ĭk), **sex'o·log'i·cal** *adj.* **—sex·ol'o·gist** *n.*
**sex·par·tite** (sĕks-pär'tīt') *adj.* Composed of or divided into six parts.
**sex·pot** (sĕks'pŏt') *n. Informal.* A sexy person.
**sext** *also* **Sext** (sĕkst) *n.* [ME *sexte* < Lat. *sexta (hora),* sixth (hour) < *sextus,* sixth.] **1.** The fourth of the seven canonical hours. **2.** The time of day set aside for sext, usu. the sixth hour, or noon.
**sex·tan** (sĕks'tən) *n.* [NLat. *sextana (febris),* sextan (fever) < Lat. *sextus,* sixth.] A malarial fever with paroxysms recurring every six days. —*adj.* Happening or recurring every six days.
**Sex·tans** (sĕks'tənz) *n.* [NLat., sextant.] A constellation in the equatorial region of the sky south of Leo.
**sex·tant** (sĕk'stənt) *n.* [NLat. *sextans, sextant-* < Lat., sixth part (from its being graduated in sixths of a circle) < *sextus,* sixth.] **1.** A navigational instrument used to measure the altitudes of celestial bodies. **2. Sextant.** Sextans.
**sex·tet** (sĕk-stĕt') *n.* [Alteration of SESTET.] **1.** *Mus.* **a.** A group of six vocalists or musicians. **b.** A musical composition written for six performers. **2.** A group of six persons or things.
**sex·tile** (sĕk'stīl', -stəl) *adj.* [Lat. *sextilis,* one sixth < *sextus,* sixth.] Designating the position of two celestial bodies when they are 60° apart.
**sex·til·lion** (sĕk-stĭl'yən) *n.* [Fr. : *sex-,* six (< Lat. *sex*) + *million,* million.] **1.** The cardinal number written $10^{21}$. **2.** *Chiefly Brit.* The cardinal number written $10^{36}$. **—sex·til'lion** *adj. & pron.*
**sex·til·lionth** (sĕk-stĭl'yənth) *n.* **1.** The ordinal number matching the number sextillion in a series. **2.** One of sextillion equal parts. **—sex·til'lionth** *adj. & adv.*
**sex·to·dec·i·mo** (sĕk'stō-dĕs'ə-mō') *n., pl.* **-mos.** [Lat. *sextodecimo,* ablative of *sextusdecimus,* one sixteenth : *sextus,* sixth + *decimus,* tenth < *decem,* ten.] **1.** The page size of a book made up of printer's sheets folded into 16 leaves or 32 pages. **2.** A book made up of sextodecimo pages.
**sex·ton** (sĕk'stən) *n.* [ME *segerstone* < AN *segerstaine* < Med. Lat. *sacristanus,* sacristan. —see SACRISTAN.] An employee of a church who is responsible for the maintenance of church property.
**sexton beetle** *n.* The burying beetle.
**sex·tu·ple** (sĕk-stōō'pəl, -styōō'-, -stŭp'əl, sĕk'stŭp'əl) *vt. & vi.* **-pled, -pling, -ples.** [Prob. SEX- + (QUIN)TUPLE.] To multiply or be multiplied by six. —*adj.* **1.** Having six parts : SIXFOLD. **2.** Larger or greater by sixfold. **3.** *Mus.* Having six beats to the measure. —*n.* A number six times larger than another. **—sex·tu'ply** *adv.*
**sex·tu·plet** (sĕk-stŭp'lĭt, -stōō'plĭt, -styōō'-, sĕk'stŭp'lĭt) *n.* [SEXTU(PLE) + (TRI)PLET.] **1.** One of six offspring delivered at one birth. **2. sextuplets.** The six offspring of one birth. **3.** A group of six similar persons or things : SEXTET.
**sex·tu·pli·cate** (sĕk-stōō'plĭ-kĭt, -styōō'-) *adj.* [SEXTU(PLE) + (DU)PLICATE.] **1.** Six times as many or as much : SIXFOLD. **2.** Raised to the sixth power. —*vt.* (-kāt') **-cat·ed, -cat·ing, -cates.** To sextuple. —*n.* (-kĭt). One of six similar things. **—sex·tu'pli·cate·ly** *adv.* **—sex·tu'pli·ca'tion** *n.*
**sex·u·al** (sĕk'shōō-əl) *adj.* [LLat. *sexualis* < Lat. *sexus,* sex.] **1.** Of, relating to, affecting, or typical of sex, the sexes, or the sex organs and their functions. **2.** Implying or symbolizing erotic desires or activity. **3.** Of or involving the union of male and female gametes. **—sex'u·al·ly** *adv.*
**sexual intercourse** *n.* Coitus, esp. between humans.
**sex·u·al·i·ty** (sĕk'shōō-ăl'ĭ-tē) *n.* **1.** The condition of being characterized by sex. **2.** Concern with or interest in sexual activity. **3.** The quality of having a sexual character or potency.
**sex·u·al·ize** (sĕk'shōō-ə-līz') *vt.* **-ized, -iz·ing, -iz·es.** To make sexual in character or quality.
**sex·y** (sĕk'sē) *adj.* **-i·er, -i·est.** Arousing or tending to arouse sexual interest or desire. **—sex'i·ly** *adv.* **—sex'i·ness** *n.*
**Sey·fert galaxy** (sē'fərt, sī'-) *n.* [After Carl K. *Seyfert* (d. 1960).] A spiral galaxy with a small, compact, bright nucleus that displays variable light intensity and emits radio waves.
**sfer·ics** *also* **spher·ics** (sfîr'ĭks, sfĕr'-) *n.* [Short for ATMOSPHERICS.] (*sing. in number*). **1.** The study of atmospherics, esp. by means of electronic detectors. **2.** Atmospherics.
**sfor·zan·do** (sfôrt-sän'dō, -săn'-) *adj. & adv.* [Ital., pr.part of *sforzare,* to use force.] *Mus.* Suddenly and strongly accented. —Used as a direction. —*n., pl.* **-dos** *or* **-di** (-dē). A sforzando tone or chord.
**sgraf·fi·to** (zgrə-fē'tō, skrä-) *n., pl.* **-ti** (-tē) [Ital. < p.part. of *sgraffire,* to scratch.] **1.** Ornamentation made on pottery or ceramic by incising a surface of plaster or glazing to reveal a different color beneath. **2.** Ware decorated with sgraffito.
**sh** (sh) *interj.* —Used to urge silence.
**Sha·ban** *also* **Shaa·ban** (shə-bän') *n.* [Ar. *sha'bān.*] The eighth month of the Moslem year. —See table at CALENDAR.
**Shab·bat** (shə-bät', shä'bəs) *n.* [Heb. *shabbāth,* sabbath.] The Jewish Sabbath.
**shab·by** (shăb'ē) *adj.* **-bi·er, -bi·est.** [< obs. *shab,* scab < ME *schab* < OE *sceabb.*] **1.** Worn-out : ragged. **2.** Wearing worn-out clothing : SEEDY. **3.** Dilapidated : deteriorated. **4.** Contemptible : mean. **—shab'bi·ly** *adv.* **—shab'bi·ness** *n.*
**Sha·bu·oth** (shə-vōō'ōt', -ōth', -əs) *n. var. of* SHAVUOT.
**shack** (shăk) *n.* [Mex. Sp. *jacal* < Nahuatl *xacalli,* adobe hut : *xămitl,* adobe + *calli,* house.] A small crude building : SHANTY.
**shack·le** (shăk'əl) *n.* [ME *schackle* < OE *sceacel,* fetter.] **1.** A metal restraining device, usu. one of a pair, that encircles the ankle or wrist of a prisoner or captive : MANACLE. **2.** A hobble for an animal. **3.** A device used to fasten or couple. **4.** *often* **shackles.** Something that restrains or confines. —*vt.* **-led, -ling, -les. 1.** To put shackles on : FETTER. **2.** To fasten or couple with a shackle. **3.** To restrain, confine, or hamper. **—shack'ler** *n.*
**shack·o** (shăk'ō, shā'kō, shä'-) *n. var. of* SHAKO.
**shad** (shăd) *n., pl.* **shad** *or* **shads.** [ME < OE *sceadd.*] A food fish of the genus *Alosa,* related to the herring, that swims up streams from marine waters to spawn.
**shad·ber·ry** (shăd'bĕr'ē) *n.* The fruit of the shadbush.
**shad·blow** (shăd'blō') *n.* The shadbush.
**shad·bush** (shăd'bŏŏsh') *n.* A North American shrub or tree of the genus *Amelanchier,* with white flowers and edible blue-black or purplish fruit.
**shad·dock** (shăd'ək) *n.* [After Captain *Shaddock,* 17th-cent. English ship commander.] **1.** A tropical tree, *Citrus maxima* or *C. grandis,* related to the grapefruit. **2.** The edible pear-shaped yellow fruit of the shaddock.
**shade** (shād) *n.* [ME *schade* < OE *sceadu.*] **1.** Light reduced in intensity due to interception of the rays : partial darkness. **2.** An area or space of partial darkness. **3.** Cover or shelter from the sun's rays. **4.** A device used to reduce or screen light or heat <window *shades*> **5. shades.** *Slang.* Sunglasses. **6.** Relative obscurity. **7. shades.** Dark shadows gathering at dusk. **8.** The portion of a picture depicting darkness or shadow. **9.** The extent to which a color is mixed with black or is decreasingly illuminated : degree of darkness. **10.** A slight variation or difference : NUANCE. **11.** A small amount : TRACE. **12.** A phantom : ghost. **13. shades.** *Informal.* Reminders : memories. —*vt.* **shad·ed, shad·ing, shades. 1.** To screen from light or heat. **2.** To obscure or darken. **3.** To cause shade in or on. **4. a.** To represent degrees of darkness in. **b.** To produce (gradations of light or color) in. **5.** To alter by slight degrees <*shade* the meaning> **6.** *Informal.* To make a minor reduction in <*shade* prices>
☆ ***syns:*** SHADE, GRADATION, NUANCE *n. core meaning* : a slight variation between nearly identical entities <*shades* of meaning>
**shad·fly** (shăd'flī') *n.* The mayfly.
**shad·ing** (shā'dĭng) *n.* **1.** Protection against light or heat. **2.** Lines or other marks used to fill in outlines of a sketch, engraving, or painting to represent gradations of colors or darkness. **3.** A slight variation, gradation, or difference.
**shad·ow** (shăd'ō) *n.* [ME *schadow* < OE *sceadwe* < *sceadu,* shade, shadow.] **1.** An area that is not or is only partially irradiated or illuminated due to blockage of light by an opaque object. **2.** The rough image cast by an object blocking rays of light. **3.** An imperfect imitation or copy. **4. shadows.** The darkness following sunset.

**5. a.** Gloom : unhappiness. **b.** A cause of gloom or unhappiness. **6.** A shaded area in a picture or photograph. **7.** A mirrored image or reflection. **8.** A phantom : ghost. **9. a.** A detective. **b.** A spy. **10.** A slight indication : PREMONITION. **11.** A vestige : remnant. **12.** An insignificant portion or amount : TRACE. **13.** Shelter : protection. —*vt.* **-owed, -ow·ing, -ows. 1.** To cast a shadow on : SHADE. **2.** To make gloomy or unhappy : CLOUD. **3.** To represent vaguely, mysteriously, or prophetically. **4.** To darken in a painting or drawing. **5.** To follow after, esp. in secret : TRAIL. —*adj.* Unofficial <a *shadow* government of exiled leaders> **—shad'ow·er** *n.*

▲ word history: The meanings of *shade* and *shadow* are distinct enough to obscure the common origin of the words. Both are descended from Old English *sceadu*, which meant both "shade" and "shadow." *Shade* is the direct modern descendent of *sceadu*, which is the nominative singular form. *Shadow* is derived from *sceadwe*, the form of the oblique cases of *sceadu*.

**shad·ow·box** (shăd'ō-bŏks') *vi.* **-boxed, -box·ing, -box·es.** To spar with an imaginary opponent, as for exercise.

**shadow cabinet** *n.* A group of leaders of a parliamentary opposition who will probably hold positions in the official cabinet when their party is returned to power.

**shadow dance** *n.* A dance presented by casting shadows of dancers on a screen.

**shad·ow·graph** (shăd'ō-grăf') *n.* An image produced by throwing a shadow on a screen.

**shadow play** *n.* A play presented by casting shadows of puppets or actors on a screen.

**shad·ow·y** (shăd'ō-ē) *adj.* **-i·er, -i·est. 1.** Relating to or like a shadow. **2.** Full of shadows : DARK. **3.** Vague : indistinct. **4.** SHADY 4. **—shad'ow·i·ness** *n.*

**shad·y** (shā'dē) *adj.* **-i·er, -i·est. 1.** Full of shade : SHADED. **2.** Casting shade. **3.** Quiet, dark, or concealed : HIDDEN. **4.** Of dubious character or honesty : QUESTIONABLE. **—shad'i·ly** *adv.* **—shad'i·ness** *n.*

☆ ***syns:*** SHADY, DOUBTFUL, FISHY, QUESTIONABLE, SUSPECT, SUSPICIOUS *adj. core meaning* : of dubious character <had some *shady* deals with known criminals>

**shaft** (shăft) *n.* [ME < OE *scheaft.*] **1.** The long narrow stem or body of a spear or arrow. **2.** A spear or arrow. **3. a.** A projectile like a spear or arrow. **b.** *Informal.* A derisive comment : BARB. **c.** *Slang.* Harsh or unfair treatment <really gave them the *shaft*> **4.** A ray or beam of light. **5.** The handle of an implement or tool. **6.** The rib of a feather. **7.** *Anat.* **a.** The centersection of a long bone : DIAPHYSIS. **b.** The section of a hair projecting from the surface of the body. **8. a.** A column or obelisk. **b.** The part of a column between the capital and the base. **9.** One of two parallel poles between which an animal is harnessed to a vehicle. **10.** A long, usu. cylindrical bar, esp. one that rotates and transmits power <a drive *shaft*> **11.** A long narrow passage beneath the surface of the ground : TUNNEL. **12.** A vertical passage for an elevator. **13.** An air duct or conduit. —*vt.* **shaft·ed, shaft·ing, shafts. 1.** To provide with a shaft. **2.** *Slang.* To treat harshly or unfairly.

**shaft·ing** (shăf'tĭng) *n.* **1.** A system of shafts, as in a mechanical device, for transmitting motion or power. **2.** Material from which shafts are made.

**shag**[1] (shăg) *n.* [ME *shagge < OE *sceagga*, matted hair.] **1.** A tangle or mass, esp. of rough matted hair. **2. a.** A coarse long nap, as on a woolen cloth. **b.** Cloth having such a nap. **3.** Roughly shredded tobacco. —*vt.* **shagged, shag·ging, shags. 1.** To make shaggy : ROUGHEN. **2. a.** To chase and bring back : FETCH. **b.** *Baseball.* To chase and catch (fly balls) in practice.

**shag**[2] (shăg) *n.* [Orig. unknown.] A 1930's dance step consisting of a hop on each foot in turn. —*vi.* **shagged, shag·ging, shags.** To dance the shag.

**shag**[3] (shăg) *n.* [Perh. from its shaggy crest.] CORMORANT 1.

**shag·bark** (shăg'bärk') *n.* A North American hickory tree, *Carya ovata*, with shaggy bark, compound leaves, and hard-shelled nuts.

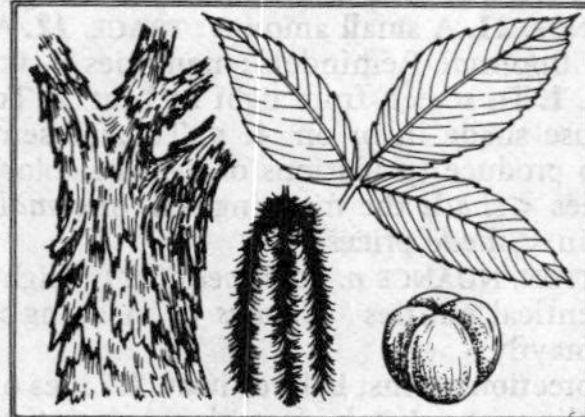

**shagbark**
*The shagbark hickory: detail of its bark, leaves, flowers, and fruit*

**shag·gy** (shăg'ē) *adj.* **-gi·er, -gi·est. 1.** Having, covered with, or like, long rough hair or wool. **2.** Bushy and matted <*shaggy* hair> **3.** Poorly groomed : UNKEMPT. **—shag'gi·ly** *adv.* **—shag'gi·ness** *n.*

**shaggy cap** *n.* The shaggy mane.

**shag·gy-dog story** (shăg'ē-dôg', -dŏg') *n.* A long, drawn-out anecdote ending with an absurd or anticlimactic punch line.

**shaggy mane** *n.* An edible mushroom, *Coprinus comatus*, with shaggy scales covering the cap.

**sha·green** (shə-grēn') *n.* [Fr. *chagrin* < Turk. *sağri*, leather.] **1.** The rough hide of a shark or ray, covered with bony denticles and used as an abrasive and as leather. **2.** An untanned leather with a grainy surface often dyed green. **—sha·green'** *adj.*

**shah** (shä) *n.* [Pers. *shāh*, king.] The title of the former hereditary rulers of Iran.

**shai·tan** (shī-tän', shā-) *n.* [Ar. *shaiṭān* < Heb. *śāṭān* < *śāṭan*, he accused.] **1.** An evil spirit : FIEND.

**shake** (shāk) *v.* **shook** (shŏŏk), **shak·en** (shā'kən), **shak·ing, shakes.** [ME *schaken* < OE *sceacan.*] —*vt.* **1.** To cause to move to and fro with short jerky movements. **2.** To cause to tremble or quiver : VIBRATE. **3.** To cause to stagger or reel. **4.** To remove or displace by jerky movements <*shake* the dust out of the cloth> **5.** To bring to a designated condition by or as if by shaking. **6.** To agitate or disturb : DISCONCERT <was *shaken* by the survivors' reports> **7.** To brandish or wave <*shake* one's fist> **8.** To clasp (hands) in greeting or farewell or as an indication of agreement. **9.** *Mus.* To trill (a note). **10.** To rattle and mix (dice) before throwing. —*vi.* **1.** To move to and fro in short jerky movements. **2.** To tremble, as from cold or in anger. **3.** To totter or waver. **4.** *Mus.* To trill. **5.** To shake hands. **—shake down. 1.** *Informal.* To extort money from. **2.** *Informal.* To make a complete search of. **3.** To subject (a ship or aircraft) to a shakedown test. **—shake off.** To get rid of <*shook off* my pursuers> **—shake up.** *Informal.* To rearrange or reorder drastically. —*n.* **1.** An act of shaking. **2.** A trembling or quivering movement. **3.** *Informal.* An earthquake. **4.** A rock fissure. **5.** A crack in timber caused by wind or frost. **6.** *Slang.* A moment or instant : TRICE <I'll finish this in a *shake.*> **7.** *Mus.* A trill. **8.** A beverage in which the ingredients are blended by shaking <a milk *shake*> **9.** A long rough shingle. **10. shakes.** *Informal.* Uncontrollable trembling, as in a sick person. **11.** *Slang.* Bargain : deal <getting a fair *shake*> **—give (someone) the shake.** *Slang.* To escape from or get rid of. **—no great shakes.** *Slang.* Ordinary : so-so. **—shake a leg.** *Informal.* **1.** To dance. **2.** To rush : hurry. **—shak'a·ble, shake'a·ble** *adj.*

☆ ***syns:*** SHAKE, QUAKE, QUAVER, QUIVER, SHIVER, SHUDDER, TREMBLE, TREMOR *v. core meaning* : to move to and fro in short, jerky movements <*shaking* with fear><ground that *shook* during the earthquake>

**shake·down** (shāk'doun') *n.* **1.** *Informal.* Extortion of money, as by blackmail. **2.** *Informal.* A complete search of a person or place. **3.** A period of appraisal followed by adjustments to improve efficiency or functioning. —*adj.* Designed to test the performance of a new ship or aircraft <a *shakedown* cruise><*shakedown* flights>

**shak·er** (shā'kər) *n.* **1.** One that shakes. **2.** A container used to dispense something by shaking <a pepper *shaker*> **3.** A container used to mix or blend by shaking <a cocktail *shaker*> **4. Shaker.** A member of a religious sect originating in England in 1747, practicing communal living and observing celibacy.

**Shake·spear·e·an** *or* **Shake·spear·i·an** (shāk-spîr'ē-ən) *adj.* Of, relating to, or like Shakespeare, his works, or his style. —*n.* A scholar of Shakespeare or his works.

**Shake·spear·e·an·a** *or* **Shake·spear·i·an·a** (shāk-spîr'ē-ăn'ə, -ä'nə) *n.* A collection of items by or pertaining to Shakespeare.

**Shakespearean sonnet** *n.* The sonnet form perfected by Shakespeare, made up of three quatrains and a final couplet with the rhyme pattern *abab cdcd efef gg* and retaining the break or pause in theme that falls between the octave and sestet in earlier sonnet forms.

**shake·up** (shāk'ŭp') *n.* A complete, often drastic reorganization, as in the personnel of a business or government.

**shaking palsy** *n.* Parkinson's disease.

**shak·o** *also* **shack·o** (shăk'ō, shā'kō, shä'-) *n., pl.* **-os** *or* **-oes.** [Fr. *schako* < Hung. *csákó* < *csákó (süveg)*, pointed (cap) < *csák*, peak < G. *Zacken*, point.] A stiff cylindrical military dress hat with a short visor and a plume.

**Shak·ta** (shäk'tə, säk'-) *n.* [Skt. *śākta* < *śaktiḥ*, Shakti.] One of a Hindu sect worshiping Shakti. **—Shak'tism** *n.* **—Shak'tist** *n.*

**Shak·ti** (shŭk'tē, shäk'-) *n.* [Skt. *śaktiḥ* < *śaknoti*, he is strong.] The wife of the god Shiva, the personification of nature and generative power in Hinduism.

**shak·y** (shā'kē) *adj.* **-i·er, -i·est. 1.** Trembling or quivering : TREMULOUS. **2.** Unsteady or unsound : WEAK <a *shaky* chair> **3.** Unreliable : precarious <a *shaky* confederation> **—shak'i·ly** *adv.* **—shak'i·ness** *n.*

**shale** (shāl) *n.* [Prob. < ME, shell < OE *scealu.*] An easily split sedimentary rock having laminated layers of fine claylike particles.

**shale oil** *n.* A crude oil obtained from oil shale by heating and distillation.

**shall** (shăl) *aux.v.* **should** (shŏŏd) [ME *schal* < OE *sceal.*] **1.** —Used to indicate simple futurity <I *shall* be 40 tomorrow.> **2.** —Used to express: **a.** Determination or promise <You *shall* answer for your misdeeds.> **b.** Inevitability <That day *shall* come.>

c. Command <Students *shall* report weekly to their tutors.> **d.** A directive or requirement <The fine *shall* not exceed $1,000.> **3.** *Archaic.* **a.** To be able to. **b.** To have to : MUST.

**shal·loon** (shə-lo͞on′, shă-) *n.* [Fr. *chalon,* after *Châlons*-sur-Marne, France.] A lightweight twilled wool or worsted fabric, used primarily for coat linings.

**shal·lop** (shăl′əp) *n.* [Fr. *chaloupe* < Du. *sloep,* sloop.] An open boat having oars or sails or both.

**shal·lot** (shə-lŏt′, shăl′ət) *n.* [Obs. Fr. *eschalotte* < OFr. *eschaloigne* <VLat. **escalōnia.* —see SCALLION.] **1.** A plant, *Allium ascalonicum,* related to the onion, grown for its edible bulb that divides into smaller sections. **2.** The mild-flavored bulb of the shallot.

**shal·low** (shăl′ō) *adj.* **-er, -est.** [ME *schalowe.*] **1.** Measuring little from bottom to top or surface. **2.** Lacking depth, as in intellect or meaning. —*n.* A shallow part of a body of water : SHOAL. —*vt.* & *vi.* **-lowed, -low·ing, -lows.** To make or become shallow. **—shal′low·ly** *adv.* **—shal′low·ness** *n.*

**sha·lom** (shä-lōm′, shə-) *interj.* [Heb. *shālōm,* peace.] —Used as a greeting or farewell among Jews.

**sha·lom a·lei·chem** (shô′ləm ə-lā′KHəm, -kəm, shō′-) *interj.* [Heb. *shālōm 'alēkhem,* peace be with you.] —Used as a greeting or farewell among Jews.

**shalt** (shălt) *aux.v. Archaic. 2nd person sing. present tense of* SHALL.

**sham** (shăm) *n.* [Perh. dial. var. of SHAME.] **1.** A spurious imitation : FAKE. **2.** Empty pretense : HYPOCRISY. **3.** One who assumes a false character : IMPOSTOR. **4.** A decorative cover made to simulate an article of household linen and used on top or in place of it <a pillow *sham*> —*adj.* Not genuine : COUNTERFEIT <*sham* modesty> —*v.* **shammed, sham·ming, shams.** —*vt.* To put on the false appearance of : FEIGN. —*vi.* To assume a false appearance or character. **—sham′mer** *n.*

**sha·man** (shä′mən, shā′-) *n.* [R. < Tungus *šaman,* ult. < Skt. *śramaṇas.*] **1.** A priest of shamanism. **2.** A medicine man among some North American Indians.

**sha·man·ism** (shä′mə-nĭz′əm, shā′-) *n.* **1.** A religion practiced by certain native peoples of northern Asia who believe that the good and evil spirits pervading the world can be summoned or heard through inspired priests acting as mediums. **2.** A form of primitive spiritualism, as that practiced by some North American Indians. **—sha′man·ist** *n.* **—sha′man·is′tic** *adj.*

**Sha·mash** (shä′măsh′) *n.* [Akkadian.] The Assyro-Babylonian sun god, regarded as the author of justice and compassion.

**sham·ble** (shăm′bəl) *vi.* **-bled, -bling, -bles.** [< E. *shamble,* awkward, ungainly.] To walk in an ungainly or lazy way, shuffling the feet. **—sham′ble** *n.*

**sham·bles** (shăm′bəlz) *pl.n.* [< dial. *shamble,* a table for selling meat < ME *shamel* < OE *sceamel,* table.] *(sing. in number).* **1.** A condition or example of total chaos or destruction <"The economy was in *shambles*" —W. Bruce Lincoln> **2.** A site of great bloodshed or carnage. **3.** A slaughterhouse. **4.** *Chiefly Brit.* A meat market or butcher shop.

**shame** (shām) *n.* [ME < OE *sceamu.*] **1. a.** A painful feeling brought about by a strong sense of guilt, embarrassment, unworthiness, or disgrace. **b.** Capacity for such a feeling <Have you no *shame*?> **2.** One that brings dishonor, disgrace, or condemnation. **3.** Dishonor or disgrace : IGNOMINY. **4.** A major disappointment. —*vt.* **shamed, sham·ing, shames. 1.** To cause to feel shame. **2.** To bring dishonor or disgrace on. **3.** To force by making ashamed <was *shamed* into making amends> **—put to shame. 1.** To fill with shame : DISGRACE. **2.** To outdo completely : SURPASS.

**shame·faced** (shām′făst′) *adj.* [Alteration of obs. *shamefast,* bashful, ashamed < ME < OE *sceamfæst* : *sceamu,* shame + *fæst,* fixed.] **1.** Showing shame : ASHAMED <a *shamefaced* explanation> **2.** Extremely shy or modest : BASHFUL. **—shame′fac′ed·ly** (-fā′sĭd-lē) *adv.* **—shame′fac′ed·ness** *n.*

**shame·ful** (shām′fəl) *adj.* **1.** Bringing or deserving shame : DISGRACEFUL. **2.** *Archaic.* Full of shame. **—shame′ful·ly** *adv.* **—shame′ful·ness** *n.*

**shame·less** (shām′lĭs) *adj.* **1.** Not subject to shame : BRAZEN. **2.** Displaying a lack of shame <a *shameless* lie> **—shame′less·ly** *adv.* **—shame′less·ness** *n.*

**sham·mes** (shä′məs) *n., pl.* **sham·mo·sim** (shä-mô′sĭm) [Yiddish *shames* < Heb. *shammāsh.*] **1.** A sexton in a synagogue. **2.** The candle used to light the other eight candles of a Chanukah menorah.

**sham·my** (shăm′ē) *n.* & *v. var. of* CHAMOIS.

**sham·poo** (shăm-po͞o′) *n., pl.* **-poos.** [< Hindi *cāpō,* imper. of *cāpna,* to press.] **1.** A preparation of soap or detergent used to wash the hair and scalp. **2.** A cleaning agent for upholstery or rugs. **3.** The act or process of washing with shampoo. —*v.* **-pooed, -poo·ing, -poos.** —*vt.* To wash or clean with shampoo. —*vi.* To wash or clean something with shampoo.

**sham·rock** (shăm′rŏk′) *n.* [Ir. *seamrog,* dim. of *seamar,* clover.] A plant, as a clover or wood sorrel, having compound leaves with three small leaflets.

**sha·mus** (shä′məs, shā′-) *n.* [Perh. var. of SHAMMES.] *Slang.* **1.** A police officer. **2.** A private detective.

**Shan** (shän, shăn) *n., pl.* **Shan** *or* **Shans. 1. a.** One of a group of Mongoloid tribes living in Thailand, Burma, and southern China. **b.** A member of one of these tribes. **2.** The Tai language of the Shan. **—Shan** *adj.*

**shan·dy·gaff** (shăn′dē-găf′) *n.* [Orig. unknown.] A drink of beer or ale mixed with ginger beer, ginger ale, or lemonade.

**shang·hai** (shăng-hī′) *vt.* **-haied, -hai·ing, -hais.** [After *Shanghai,* China, from the former custom of kidnaping sailors to man ships going to China.] **1.** To kidnap (a man) for compulsory shipboard service, esp. after rendering him unconscious. **2.** To compel or induce (someone) to do something, esp. by fraud or force.

**Shang·hai** (shăng-hī′) *n.* [After *Shanghai,* China.] A red and black domestic fowl of a breed said to have been imported from Asia.

**Shan·gri-la** (shăng′grĭ-lä′) *n.* [After *Shangri-La,* the imaginary land in the novel *Lost Horizon* by James Hilton (1900–1954).] An imaginary paradise on earth : UTOPIA.

**shank** (shăngk) *n.* [ME *schank* < OE *sceanca.*] **1.** *Anat.* **a.** The human leg between the knee and ankle. **b.** An analogous part in other vertebrates. **2.** The whole leg of a human being. **3.** A cut of meat from the leg of an animal. **4.** The long narrow part of a nail or pin. **5.** A stalk, stem, or similar part. **6.** The stem of an anchor. **7.** The long shaft of a fishhook. **8.** The part of a tobacco pipe between the stem and bowl. **9.** The shaft of a key. **10.** The narrower section of a spoon's handle. **11. a.** The narrow part of a shoe's sole under the instep. **b.** A piece of material, as of metal, used to reinforce or shape the shank of a shoe. **12.** A projection on the back of a button by which it is attached to cloth. **13. a.** The part of a drill or other tool between the functioning head and the handle. **b.** TANG[1] 4b. **14. a.** The latter or remaining part of a period of time. **b.** The early or best part of a period of time.

**shank·piece** (shăngk′pēs′) *n.* An arch support inserted into the shank of a shoe.

**shan't** *or* **sha'nt** (shănt, shänt). Shall not.

**shan·tey** (shăn′tē) *n. var. of* CHANTEY.

**shan·tung** (shăn-tŭng′) *n.* [After *Shandong* (Shantung), China.] **1.** A heavy, natural silk fabric with a rough, nubby surface. **2.** An imitation of shantung.

**shan·ty[1]** (shăn′tē) *n., pl.* **-ties.** [Prob. Canadian Fr. *chantier* < Fr., timberyard < OFr., gantry < Lat. *cantherius,* rafter.] A roughly built, often ramshackle structure : SHACK

**shan·ty[2]** (shăn′tē) *n. var. of* CHANTEY.

**shan·ty·town** (shăn′tē-toun′) *n.* A town or section of a town consisting of makeshift run-down huts.

**shape** (shāp) *n.* [ME < OE *gesceap,* a creation.] **1. a.** Characteristic surface configuration : FORM. **b.** Something distinguished from its surroundings by its outline. **2.** The contour of one's body : FIGURE. **3. a.** A particular form. **b.** A desirable form <a fabric that doesn't lose its *shape*> **4.** A form or condition in which something may exist or appear : EMBODIMENT <a cake in the *shape* of a turkey> **5.** Assumed or false appearance : GUISE. **6.** A phantom : ghost. **7.** Something, as a mold or pattern, used to impart or determine form. **8.** Proper condition for action, effectiveness, or use <took weeks to get back in *shape* for running> —*vt.* **shaped, shap·ing, shapes. 1.** To give a specific form to : CREATE. **2.** To cause to conform to a specific form or pattern : adapt to fit. **3. a.** To plan : devise. **b.** To embody in a definite form. **—shape up.** *Informal.* **1.** To turn out : DEVELOP. **2.** To improve so as to meet a standard or goal. **—shap′er** *n.*

**shaped charge** *n.* An explosive charge shaped so that its explosive force will be concentrated in a given direction.

**shape·less** (shāp′lĭs) *adj.* **1.** Having no shape : FORMLESS. **2.** Lacking attractive form. **—shape′less·ly** *adv.* **—shape′less·ness** *n.*

**shape·ly** (shāp′lē) *adj.* **-li·er, -li·est.** Having a pleasing shape : well proportioned. **—shape′li·ness** *n.*

**shap·en** (shā′pən) *v. Archaic. var. p.p. of* SHAPE.

**shape·up** (shāp′ŭp′) *n.* A group of longshoremen from which the day's work crew is chosen by a union representative.

**shard** (shärd) *also* **sherd** (shûrd) *n.* [ME *sherd* < OE *sceard.*] **1.** A piece of broken pottery : POTSHERD. **2.** A small piece of a brittle substance, as of glass or metal. **3.** *Zool.* A tough sheath, esp. the outer wing covering of a beetle.

**share[1]** (shâr) *n.* [ME < OE *scearu,* division.] **1.** A part or portion belonging to, distributed to, contributed by, or owed by a person or group. **2.** A fair or full portion <do one's *share* of the housework> **3.** One of the equal parts into which the capital stock of a corporation or company is divided. —*v.* **shared, shar·ing, shares.** —*vt.* **1.** To separate and parcel out in shares : APPORTION. **2.** To take part in, use, or have in common <*share* responsibilities><*share* an office> —*vi.* To have or take a part : PARTICIPATE <*share* in the good time> **—go shares.** To be concerned or partake equally or jointly, as in a financial venture. **—on shares.** With each individual concerned taking a share, usu. equal, of any profit or loss. —Used of an enterprise. **—shar′er** *n.*

**share[2]** (shâr) *n.* [ME < OE *scēar.*] A plowshare.

o͞o **boot** ou **out** th **thin** *th* **this** ŭ **cut** û **urge** y **young**
yo͞o **abuse** zh **vision** ə **a**bout, it**e**m, ed**i**ble, gall**o**p, circ**u**s

**share·crop** (shâr′krŏp′) *vi.* **-cropped, -crop·ping, -crops.** To work as a sharecropper.

**share·crop·per** (shâr′krŏp′ər) *n.* A tenant farmer who gives a share of the crop to the landlord as rent.

**share·hold·er** (shâr′hōl′dər) *n.* A holder of a share or shares of stock : STOCKHOLDER.

**sha·rif** (shə-rēf′) *n. var. of* SHERIF.

**shark** (shärk) *n.* [Orig. unknown.] **1.** Any of numerous chiefly marine fishes of the order Squaliformes or Selachii, which are sometimes large and voracious and have a cartilaginous skeleton and tough skin covered with small toothlike scales. **2.** One who is ruthless, greedy, or dishonest <a loan *shark*> **3.** *Slang.* One who is highly skilled in an activity <a card *shark*> —*vi.* **sharked, shark·ing, sharks.** To live by fraud and trickery.

**shark·skin** (shärk′skĭn′) *n.* **1.** The skin of a shark. **2.** Leather made from a shark's skin. **3.** A rayon and acetate fabric with smooth, somewhat shiny surface.

**shark sucker** *n.* A remora.

**sharp** (shärp) *adj.* **-er, -est.** [ME *scharp* < OE *scearp.*] **1.** Having a thin, keen edge or a fine point and suitable for or capable of cutting or piercing. **2. a.** Having clear, distinct form and detail. **b.** Ending in an edge or point <*sharp* cliffs> **c.** Clearly and distinctly set forth <*sharp* contrasts in behavior> **3.** Abrupt or acute <a *sharp* drop> **4.** Shrewd : astute <a *sharp* mind> **5.** Artful : devious <*sharp* selling practices> **6.** Vigilant : aware. **7.** Brisk : vigorous. **8.** Harsh : acrimonious <*sharp* criticism> **9.** Fierce or impetuous : VIOLENT. **10.** Intense : severe <a *sharp* pain> **11.** Sudden and shrill. **12.** Strongly affecting the senses of smell and taste <a *sharp,* pungent aroma><a *sharp* cheddar> **13.** Made up of hard, angular particles. **14.** *Mus.* **a.** Raised in pitch by a semitone. **b.** Above the proper pitch. **c.** Having the key signature in sharps. **15.** Voiceless. —Used of a consonant. **16.** *Slang.* Attractive or stylish <a *sharp* outfit> —*adv.* **1.** In a sharp manner. **2.** Punctually : exactly <at 3:30 *sharp*> **3.** *Mus.* Above the true or proper pitch. —*n.* **1.** *Mus.* **a.** A musical note or tone raised one semitone above its normal pitch. **b.** A sign (#) indicating this. **2.** A thin sewing needle with a very fine point. **3.** *Informal.* A sharper. —*v.* **sharped, sharp·ing, sharps.** *Mus.* —*vt.* To raise in pitch by a semitone. —*vi.* To play or sing above the proper pitch. **—sharp′ly** *adv.* **—sharp′ness** *n.*

☆ ***syns:*** SHARP, KEEN, WHETTED *adj. core meaning* : having a fine, honed edge <a *sharp* knife> ***ant*** : blunt, dull

**sharp·en** (shär′pən) *vt.* & *vi.* **-ened, -en·ing, -ens.** To make or become sharp or sharper. **—sharp′en·er** *n.*

**sharp·er** (shär′pər) *n.* One that deals dishonestly with others, esp. a gambler who cheats.

**sharp-eyed** (shärp′īd′) *adj.* **1.** Having keen eyesight. **2.** Acutely perceptive or observant.

**sharp·ie** (shär′pē) *n.* [< SHARP.] **1.** A long, narrow, flat-bottomed fishing boat with a centerboard and one or two masts, each rigged with a triangular sail. **2.** *Informal.* One who is alert and quick-witted.

**sharp-set** (shärp′sĕt′) *adj.* **1.** Placed at a sharp angle or presenting a sharp edge. **2.** Having a keen appetite or desire. **—sharp′set′ness** *n.*

**sharp-shinned hawk** (shärp′shĭnd′) *n.* A North American hawk, *Accipiter striatus,* with short rounded wings and a long tail.

**sharp·shoot·er** (shärp′sho͞o′tər) *n.* **1.** An expert marksman. **2. a.** The second military grade of proficiency in the use of rifles and other small arms. **b.** One who has attained this grade of proficiency.

**sharp·shoot·ing** (shärp′sho͞o′tĭng) *n.* **1.** Expert marksmanship. **2.** Correct and often unexpected verbal or written attack.

**sharp-sight·ed** (shärp′sī′tĭd) *adj.* Sharp-eyed.

**sharp-tongued** (shärp′tŭngd′) *adj.* Severe, critical, or sarcastic in speech.

**sharp-wit·ted** (shärp′wĭt′ĭd) *adj.* Acutely perceptive. **—sharp′-wit′ted·ness** *n.*

**shash·lik** *or* **shash·lick** (shäsh-lĭk′, shäsh′lĭk) *n.* [R. *shashlyk,* of Turkic orig.] A dish of marinated cubes of mutton or veal grilled or roasted on a spit, often with slices of eggplant, onion, and tomato : SHISH KEBAB.

**Shas·ta daisy** (shăs′tə) *n.* [After Mt. *Shasta* in California.] A cultivated variety of *Chrysanthemum maximum* with large, white, daisylike flowers.

**shat·ter** (shăt′ər) *v.* **-tered, -ter·ing, -ters.** [ME *schateren* < OE *sceaterian.*] —*vt.* **1.** To cause to break or burst suddenly into pieces, as with a violent blow. **2.** To damage severely : DISABLE. —*vi.* To break into pieces : SMASH. —*n.* **1.** The act of shattering. **2.** *often* **shatters.** A fragmented or splintered condition.

**shatter cone** *n.* A conical fragment of rock that is formed from the high pressure in volcanism or meteorite impact and has striations radiating from the apex.

**shat·ter·proof glass** (shăt′ər-pro͞of′) *n.* Safety glass.

**shave** (shāv) *v.* **shaved, shaved** *or* **shav·en** (shā′vən), **shav·ing, shaves.** [ME *shaven,* to scrape < OE *sceafan.*] **1.** To remove the beard or other body hair from, as with a razor. **2.** To cut (e.g., the beard) at the surface of the skin with a razor. **3.** To trim or mow closely. **4.** To remove thin slices from <*shave* a piece of lumber> **5.** To cut or scrape into thin slices : SHRED <*shave* cheese> **6.** To come close to or graze in passing. **7.** *Informal.* To purchase (a note) at a reduction greater than the legal or customary rate. **8.** *Informal.* To reduce (a price) by a slight margin. —*vi.* To remove a beard or hair with a razor. —*n.* **1.** The act, process, or result of shaving. **2.** A thin slice or scraping. **3.** A tool or implement used for shaving.

**shav·er** (shā′vər) *n.* **1. a.** One that shaves. **b.** A device used for shaving. **2.** *Informal.* A small child, esp. a boy.

**shav·ing** (shā′vĭng) *n.* **1.** A thin strip of material shaved off : SLIVER. **2.** The act of one that shaves.

**Sha·vu·ot** *also* **Sha·bu·oth** (shə-vo͞o′ōt′, -ōth′, -əs) *n.* [Heb. *shābhū′ōth* < *shābhūa′,* week.] A Jewish holiday commemorating the revelation of the Law on Mount Sinai and the celebration of the wheat festival in ancient times.

**shawl** (shôl) *n.* [Pers. *shāl.*] A square or rectangular piece of cloth worn as a covering for the head, neck, and shoulders. —*vt.* **shawled, shawl·ing, shawls.** To cover with or as if with a shawl.

**shawm** (shôm) *n.* [ME *schallemele* < OFr. *chalemel* < VLat. ***calamellus,** dim. of Lat. *calamus,* reed < Gk. *kalamos.*] An early double-reed wind instrument, forerunner of the modern oboe.

**Shaw·nee** (shô-nē′) *n., pl.* **Shawnee** *or* **-nees.** [Obs. *Shawanese* < Delaware *šāonu* < Shawnee *šāwanwa,* a Shawnee.] **1. a.** A tribe of Indians once living in the Tennessee Valley and adjacent areas and now surviving in Oklahoma. **b.** A member of this tribe. **2.** The Algonquian language of the Shawnee.

**Shaw·wal** (shə-wäl′) *n.* [Ar. *Shawwāl.*] The tenth month of the Moslem year. —See table at CALENDAR.

**shay** (shā) *n.* [Back-formation < CHAISE, taken as pl.] *Informal.* A chaise.

**she** (shē) *pron.* [ME *sche,* prob. alteration of OE *hēo,* she, or *sēo,* fem. demonstrative pronoun.] **1.** —Used to represent the female person or animal last mentioned or implied. **2.** —Used traditionally of certain objects and institutions as ships and nations. —*n.* A female animal or person <a *she*-cat><Is the cat a *she?*>

**shea butter** (shē, shā) *n.* A whitish or yellowish fat obtained from the nut of the shea tree, used as food and in the manufacture of soap and candles.

**sheaf** (shēf) *n., pl.* **sheaves** (shēvz) [ME *sheef* < OE *scēaf.*] **1.** A bundle of cut stalks of grain or similar plants bound with straw or twine. **2.** A group of items held or bound together. **3.** A quiver for arrows. —*vt.* **sheafed, sheaf·ing, sheafs.** To bind into a sheaf.

**shear** (shîr) *v.* **sheared, sheared** *or* **shorn** (shôrn, shōrn), **shear·ing, shears.** [ME *scheren* < OE *sceran.*] —*vt.* **1.** To remove (fleece or hair) by cutting or clipping with a sharp implement. **2.** To remove the hair or fleece from. **3.** To cut with or as if with shears <*shearing* a border> **4.** To strip, divest, or deprive of. —*vi.* **1.** To use a cutting tool such as shears. **2.** To move or proceed by or as if by cutting <*shear* through the grain> **3.** *Physics.* To become deformed by forces tending to produce a shearing strain. —*n.* **1.** The act, process, or result of shearing. **2.** Something cut off or removed by shearing. **3.** The act, process, or fact of shearing. —Used to indicate a sheep's age <a two-*shear* ram> **4.** *Physics.* **a.** An applied force or system of forces that tends to produce a shearing strain. **b.** A shearing strain. **—shear′er** *n.*

**shearing strain** *n.* A condition in or deformation of an elastic body caused by forces that tend to produce an opposite but parallel sliding motion of the body's planes.

**shearing stress** *n. Physics.* SHEAR 4a.

**shear legs** *also* **sheer·legs** (shîr′lĕgz′) *n.* A device used to lift heavy weights, having two or more spars joined at the top and spread at the base, the tackle being suspended from the top.

**shear·ling** (shîr′lĭng) *n.* **1.** A year-old sheep that has been shorn once. **2.** The tanned skin of a shearling or of a newly shorn sheep.

**shears** (shîrz) *pl.n.* [ME *schere,* scissors < OE *scēar.*] **1. a.** Large scissors. **b.** An implement or machine that cuts with a scissorlike action. **2.** *also* **sheers.** A shear legs.

**shear stress** *n. Physics.* SHEAR 4a.

**shear·wa·ter** (shîr′wô′tər, -wŏt′ər) *n.* An oceanic bird of the family Procellariidae, esp. of the genus *Puffinus,* having a hooked bill.

**sheat·fish** (shēt′fĭsh′) *n., pl.* **sheatfish** *or* **-fish·es.** [Alteration of obs. *sheathfish* : SHEATH + FISH.] A large freshwater catfish, *Silurus glanis* of Eurasia.

**sheath** (shēth) *n., pl.* **sheaths** (shēthz, shēths) [ME *schethe* < OE *scēath.*] **1.** A case for a blade, as of a knife or sword. **2.** A covering resembling or used like a sheath. **3.** *Biol.* An enveloping structure or part, as the tubular base of a leaf surrounding a stem. **4.** A close-fitting dress. —*vt.* **sheathed, sheath·ing, sheaths.** To sheathe.

**sheath·bill** (shēth′bĭl′) *n.* An Antarctic shore bird, *Chionia alba* or *C. minor,* with white plumage and a horny covering on the base of the bill.

**sheathe** (shēth) *vt.* **sheathed, sheath·ing, sheathes.** **1.** To insert into or provide with a sheath. **2.** To retract (a claw) into a sheath. **3.** To enclose : envelope. **—sheath′er** *n.*

**sheath·ing** (shē′thĭng) *n.* **1.** A layer of material applied to the outer frame of a building to strengthen the structure and serve as a base for an exterior weatherproof cladding. **2.** A covering on the underwater part of a ship's hull that protects it against marine growths.

**sheath knife** *n.* A knife with a fixed blade that fits into a sheath.

**shea tree** (shē, shā) *n.* [Mandekan *si.*] An African tree, *Butyrospermum parkii,* having fruit containing oily seeds that yield shea butter.

**sheave[1]** (shēv) *vt.* **sheaved, sheav·ing, sheaves.** To gather and bind into a sheaf or sheaves.

**sheave[2]** (shēv, shĭv) *n.* [ME *sheve.*] A wheel or disk with a grooved rim, esp. one used as a pulley.

**sheaves** (shēvz) *n. pl. of* SHEAF.

**she·bang** (shə-băng′) *n.* [Orig. unknown.] *Informal.* A situation, organization, contrivance, or collection of facts or things <organized and ran the whole *shebang*>

**She·bat** (shə-bät′, -vät′) *n. var. of* SHEVAT.

**she·been** (shə-bēn′) *n.* [Ir. Gael. *sībīn,* bad ale.] *Chiefly Ir.* An unlicensed drinking establishment.

**shed[1]** (shĕd) *v.* **shed, shed·ding, sheds.** [ME *sheden* < OE *scēadan,* to divide.] —*vt.* **1.** To cause to pour forth <*shed* tears of joy> **2.** To diffuse or give off <*shed* a glow> **3.** To repel without allowing penetration <a treated fabric that *sheds* water> **4.** To lose by a natural process <A snake *sheds* its skin.> —*vi.* **1.** To lose a growth or covering by a natural process. **2.** To pour forth, fall off, or drop out. —*n.* **1.** Something that sheds, esp. an elevation in the earth's surface from which water flows in two directions. **2.** Something that has been shed. **—shed blood.** To take life : KILL.

**shed[2]** (shĕd) *n.* [Alteration of obs. *shadde,* var. of SHADE.] **1.** A small structure, either freestanding or attached to a larger structure, used for storage or shelter. **2.** A large low building often open on one or more sides.

**she'd** (shēd). **1.** She had. **2.** She would.

**shed·der** (shĕd′ər) *n.* One that sheds, as a long-haired animal or a molting lobster.

**shed dormer** *n.* A dormer with a roof that slants in the same direction as the one in which the dormer is located.

**she-dev·il** (shē′dĕv′əl) *n.* An evil or malicious woman.

**sheen** (shēn) *n.* [ME *shene,* beautiful < OE *scīene.*] **1.** Sparkling brightness : SHININESS <the *sheen* of polished silver on the table> **2.** Splendid attire.

**sheep** (shēp) *n., pl.* **sheep.** [ME < OE *scēap.*] **1.** Any of various usu. horned ruminant mammals of the genus *Ovis,* esp. the domesticated species *O. aries* bred for its wool, edible flesh, or skin. **2.** Leather made from the skin of a sheep. **3. a.** A humble and submissive person. **b.** One who is easily influenced or led.

**sheep·ber·ry** (shēp′bĕr′ē) *n.* A North American shrub or tree, *Viburnum lentago,* with white flower clusters and edible blue-black berries.

**sheep·cote** (shēp′kōt′, -kŏt′) *n. Chiefly Brit.* A sheepfold.

**sheep dip** *n.* A liquid disinfectant that destroys vermin in the wool of sheep prior to shearing.

**sheep dog** *also* **sheep·dog** (shēp′dôg′, -dŏg′) *n.* A dog trained to herd and guard sheep.

**sheep·fold** (shēp′fōld′) *n.* A fenced-in enclosure for sheep.

**sheep·herd·er** (shēp′hûr′dər) *n.* One that herds a large flock of sheep : SHEPHERD. **—sheep′herd′ing** *n.*

**sheep·ish** (shē′pĭsh) *adj.* **1.** Embarrassed and apologetic <a *sheepish* look on your face> **2.** Like a sheep in meekness or stupidity. **—sheep′ish·ly** *adv.* **—sheep′ish·ness** *n.*

**sheep ked** (kĕd) *n.* [SHEEP + *ked,* sheep ked, of unknown orig.] Sheep tick.

**sheep laurel** *n.* An evergreen shrub, *Kalmia angustifolia* of eastern North America, with rose-pink flowers and poisonous foliage.

**sheep's eyes** *pl.n.* Shy and usu. amorous glances.

**sheep·shank** (shēp′shăngk′) *n.* A knot for shortening a line.

**sheeps·head** (shēps′hĕd′) *n.* A food fish, *Archosargus probatocephalus* of north Atlantic waters, with dark vertical markings.

**sheep·shear·ing** (shēp′shîr′ĭng) *n.* **1.** The act or process of shearing sheep. **2. a.** The time when sheep are sheared. **b.** The festivities held at this time. **—sheep′shear′er** *n.*

**sheep·skin** (shēp′skĭn′) *n.* **1.** The skin of a sheep either tanned with the fleece left on or in the form of leather or parchment. **2.** A diploma.

**sheep tick** *n.* A wingless fly, *Melophagus ovinus,* that is parasitic to sheep.

**sheer[1]** (shîr) *vi. & vt.* **sheered, sheer·ing, sheers.** [Perh. alteration of SHEAR.] To swerve or cause to swerve from a course. —*n.* **1.** A swerving or veering course. **2.** *Naut.* **a.** The upward curve or amount of upward curve of the fore-and-aft lines of a ship's hull as viewed from the side. **b.** The position in which a ship is placed so that it remains clear of a single bow anchor.

**sheer[2]** (shîr) *adj.* **-er, -est.** [Obs. *shere,* thin, clear < ME *schir,* free from service or guilt, prob. < ON *skærr,* bright, pure.] **1.** Thin, fine, and transparent : DIAPHANOUS <a *sheer* fabric> **2. a.** Undiluted : pure <*sheer* delight> **b.** Unadulterated : unmixed <*sheer* alcohol> **3.** Almost perpendicular : STEEP <*sheer* rock cliffs> —*adv.* Almost perpendicularly. **—sheer′ly** *adv.* **—sheer′ness** *n.*

**sheer·legs** (shîr′lĕgz′) *n. var. of* SHEAR LEGS.

**sheers** (shîrz) *n. var. of* SHEARS 2.

**sheet[1]** (shēt) *n.* [ME *schete,* cloth < OE *scēte.*] **1.** A rectangular piece of cotton or other fabric serving as a basic article of bedding, commonly used in pairs. **2.** A broad, thin, usu. rectangular piece of material, as paper, metal, glass, or wood. **3.** A broad, flat, continuous surface or expanse <a *sheet* of ice> **4.** A newspaper, esp. a tabloid. **5.** *Geol.* A broad, relatively thin deposit or layer of igneous or sedimentary rock. **6.** A large block of unseparated postage stamps printed by a single impression of a plate. —*v.* **sheet·ed, sheet·ing, sheets.** —*vt.* To cover with, enclose in, or provide with a sheet. —*vi.* To flow or fall in a sheet <rain *sheeting* on the roof>

**sheet[2]** (shēt) [ME *schete* < OE *scēata,* corner of a sail.] *Naut.* —*n.* **1.** A rope or chain attached to one or both of the lower corners of a sail, serving to move or extend it. **2. sheets.** The spaces at either end of an open boat in front of and behind the seats. —*vi.* **sheet·ed, sheet·ing, sheets.** To extend in a particular direction. —Used of the sheets of a sail.

**sheet anchor** *n.* [Perh. SHEET[2] + ANCHOR.] **1.** *Naut.* A large extra anchor intended for emergency use. **2.** One that can be turned to in an emergency.

**sheet bend** *n.* A knot in which one rope or piece of yarn is made fast to the bight of another.

**sheet glass** *n.* Molten glass drawn into a wide sheet cut into required lengths after annealing and hardening.

**sheet lightning** *n.* A broad sheetlike illumination, caused by the reflection of a lightning flash.

**sheet music** *n.* Music printed on unbound sheets of paper.

**Sheet·rock** (shēt′rŏk′). A trademark for plasterboard.

**sheik** *also* **sheikh** (shēk) *n.* [Ar. *shaik,* old man < *shākha,* he grew old.] **1.** (*also* shāk). **a.** A Moslem religious official. **b.** A leader of an Arab family, village, or tribe. **2.** *Slang.* A romantically attractive man.

**sheik·dom** (shēk′dəm, shāk′-) *n.* The area ruled by a sheik.

**shek·el** (shĕk′əl) *n.* [Heb. *sheqel* < *shāqal,* he weighed.] **1.**—See table at CURRENCY. **2. a.** Any of several ancient units of weight, esp. a Hebrew unit equal to about a half ounce. **b.** A gold or silver coin equal in weight to one of these units, esp. the chief silver coin of the Hebrews. **3.** *Slang.* **a.** A coin. **b. shekels.** Money.

**She·ki·nah** (shĭ-kē′nə, -KHē′-, -kī′-) *n.* [Heb. *shĕkhīnāh* < *shākhan,* to dwell.] A visible manifestation of the divine presence as described in Jewish theology.

**shel·drake** (shĕl′drāk′) *n.* [ME *sheldedrake* : *sheld-,* variegated + *drake,* drake.] **1.** A large Old World duck of the genus *Tadorna,* esp. *T. tadorna,* with black and white plumage. **2.** The merganser.

**shel·duck** (shĕl′dŭk′) *n.* [SHEL(DRAKE) + DUCK.] SHELDRAKE 1.

**shelf** (shĕlf) *n., pl.* **shelves** (shĕlvz) [ME, prob. < MLG *schelf.*] **1. a.** A flat, usu. rectangular structure of a rigid material fixed at right angles to a wall or other vertical surface and used to hold objects. **b.** The contents or capacity of such a structure. **2.** A structure, as a balcony, that resembles a shelf. **3.** A reef, sandbar, or shoal. **4.** Bedrock. **—on the shelf. 1.** In a state of disuse. **2. a.** Unemployed. **b.** No longer in circulation. **c.** Retired from employment.

**shelf ice** *n.* An extension of glacial ice into coastal waters that is in contact with the bottom near the shore but not toward the outer edge of the shelf.

**shelf life** *n.* The length of time that a product may be stored without deteriorating.

**shell** (shĕl) *n.* [ME < OE *scell.*] **1. a.** The usu. hard outer cover encasing certain organisms. **b.** An outer covering on an egg, fruit, or nut. **c.** The material composing such a covering. **2.** Something resembling or shaped like a shell, esp.: **a.** A framework or exterior, as of a building. **b.** A thin layer of pastry. **c.** A ship's hull. **d.** The external part of the ear. **e.** A long narrow racing boat propelled by oarsmen. **3. a.** A projectile, esp. the hollow tube containing explosives used to propel such a projectile. **b.** A metal or cardboard case containing the charge, primer, and shot fired from a shotgun. **4.** An attitude adopted to mask one's true feelings. **5.** *Physics.* **a.** Any of the set of hypothetical spherical surfaces centered on the nucleus of an atom that contain the orbits of electrons having the same principal quantum number. **b.** Any of a set of groupings of nucleon energy states in a nucleus or of nucleons occupying such states in which the binding energies of states differ from one another by much less than from the binding energies of states in another grouping. **6.** A usu. sleeveless, collarless blouse. —*v.* **shelled, shell·ing, shells.** —*vt.* **1. a.** To remove the shell of : SHUCK. **b.** To remove from a shell. **2.** To separate the kernels of (corn) from the cob. **3.** To fire shells at. **4. a.** To defeat decisively. **b.** *Baseball.* To hit the pitches of hard and with regularity <*shelled* the pitcher for eight runs> —*vi.* **1.** To shed a shell. **2.** To look for or collect shells. **—shell out.** *Informal.* To pay. **—shell′er** *n.* **—shell′y** *adj.*

**she'll** (shēl). **1.** She will. **2.** She shall.

**shel·lac** (shə-lăk′) *n.* [SHEL(L) + LAC[1]] **1.** A purified lac formed into thin yellow or orange flakes, often bleached white and widely used in varnishes, paints, stains, inks, and sealing wax, as a binder, and in phonograph records. **2.** A thin varnish made by dissolving flake shellac in denatured alcohol, used as a wood coating and sealer and for finishing floors. —*v.* **-lacked, -lack·ing, -lacs.** —*vt.* **1.** To apply shellac to. **2.** *Slang.* To defeat decisively. **3.** *Slang.* To batter mercilessly. —*vi.* To apply shellac.

**shell·back** (shĕl′băk′) *n.* A veteran sailor, esp. one who has crossed the equator.

**shell bean** *n.* A bean cultivated for its edible seeds rather than for its pods.

**shell·fire** (shĕl′fīr′) *n.* The firing of shells.

**shell·fish** (shĕl′fĭsh′) *n., pl.* **shellfish** *or* **-fish·es.** An aquatic animal, as a mollusk or crustacean, with a shell or shell-like exoskeleton.

**shell·flow·er** (shĕl′flou′ər) *n.* **1.** A tall plant, *Molucella laevis,* native to Asia, with tiny flowers and conspicuous green calyxes. **2.** A tall plant, *Alpina speciosa,* native to tropical Asia, with showy, variously colored flowers.

**shell game** *n.* **1.** THIMBLERIG 1. **2.** A swindle.

**shell jacket** *n.* **1.** A short tight-fitting military jacket worn buttoned up the front. **2.** A mess jacket.

**shell pink** *n.* A pinkish white to strong yellowish pink, including grayish and light yellowish pinks. **—shell′-pink′** *adj.*

**shell·proof** (shĕl′pro͞of′) *adj.* Able to withstand shellfire.

**shell shock** *n.* **1.** Any of various usu. acute, often hysterical neuroses originating in trauma suffered under fire in modern warfare. **2.** Combat fatigue. **—shell′-shocked′** *adj.*

**shel·ter** (shĕl′tər) *n.* [Orig. unknown.] **1. a.** Something providing cover or protection, as from the weather. **b.** A refuge : haven. **2.** The state of being protected or covered. —*v.* **-tered, -ter·ing, -ters.** —*vt.* To provide protection or cover for. —*vi.* To take cover. **—shel′ter·er** *n.*

**shel·ter·belt** (shĕl′tər-bĕlt′) *n.* A barrier of trees and shrubs that reduces erosion and protects against wind and storms.

**shelter tent** *n.* A small tent usu. composed of two or more pieces of waterproof material.

**shel·tie** *also* **shel·ty** (shĕl′tē) *n., pl.* **-ties.** [Prob. < ON *Hjalti,* Shetlander.] **1.** A Shetland pony. **2.** A Shetland sheepdog.

**shelve** (shĕlv) *v.* **shelved, shelv·ing, shelves.** —*vt.* **1.** To place or arrange on a shelf. **2.** To put aside as if on a shelf. **3.** To cause to retire from service : DISMISS. **4.** To equip with shelves. —*vi.* To slope gradually : INCLINE. **—shelv′er** *n.*

**shelves** (shĕlvz) *n. pl. of* SHELF.

**shelv·ing** (shĕl′vĭng) *n.* **1.** Shelves as a whole. **2.** Material for shelves. **3.** An incline.

**Shem** (shĕm) *n.* [Heb. *Shēm.*] The eldest son of Noah in the Old Testament.

**she·nan·i·gan** (shə-năn′ĭ-gən) *n.* [Orig. unknown.] *Informal.* **1.** *often* **shenanigans.** Mischief. **2.** Treachery : deceit.

**she-oak** (shē′ōk′) *n.* The beefwood.

**she·ol** (shē′ōl′, shē-ōl′) *n.* [Heb. *shə̄ōl.*] **1.** Hell. **2. Sheol.** A place described in the Old Testament as the abode of the dead.

**shep·herd** (shĕp′ərd) *n.* [ME *sheepherde* < OE *scēaphirde* : *scēap,* sheep + *hirde,* herdsman.] **1.** One who herds and tends sheep. **2.** One, as a minister or teacher, who cares for and guides a group of people. —*vt.* **-herd·ed, -herd·ing, -herds.** To herd, tend, or guide as or in the manner of a shepherd.

**shepherd dog** *n.* A dog trained to tend sheep.

**shep·herd·ess** (shĕp′ər-dĭs) *n.* A girl or woman who tends sheep.

**shepherd's pie** *n.* A casserole of cooked meat with gravy, topped by a layer or surrounded by a border of mashed potatoes.

**shep·herd's-purse** (shĕp′ərdz-pûrs′) *n.* [From its pouchlike pods.] A common weed, *Capsella bursa-pastoris,* with small white flowers and flat, heart-shaped fruit.

**Sher·a·ton** (shĕr′ə-tən) *adj.* [After Thomas *Sheraton* (1751–1806), its originator.] Of, relating to, or being a style of English furniture that originated about 1800 and that is marked by straight lines and graceful proportions.

**sher·bet** (shûr′bĭt) *n.* [Turk. < Pers. *sharbat* < Ar. *sharbah,* drink < *shariba,* he drank.] **1.** A sweet-flavored water ice to which milk, egg white, or gelatin has been added. **2.** A beverage of sweetened diluted fruit juice.

**sherd** (shûrd) *n. var. of* SHARD.

**she·rif** *also* **sha·rif** (shə-rēf′) *n.* [Ar. *sharīf,* noble < *sharafa,* he was exalted.] **1.** A descendant of the prophet Mohammed through his daughter Fatima. **2.** The chief magistrate of Mecca. **3.** A Moroccan prince or ruler.

**sher·iff** (shĕr′ĭf) *n.* [ME *schirreff,* the representative of royal authority in a shire < OE *scīrgerēfa* : *scīr,* shire + *gerēfa,* reeve.] **1.** The chief executive of the courts of superior jurisdiction in a U.S. county. **2.** An officer of a shire or county in England, Scotland, and Northern Ireland.

▲ word history: Like many ancient titles, *sheriff* now means different things in different places. In Anglo-Saxon England the sheriff was the chief representative of the king in each county. *Sheriff* was *scīr-gerēfa* in Old English, a compound of *scīr,* "shire," and *gerēfa,* "officer," the ancestor of English *reeve*[1]. In England the sheriff has chiefly ceremonial duties, such as presiding over elections and courts. In Scotland the sheriff is a local judge. In the United States the sheriff became an important officer of law enforcement.

**she·root** (shə-ro͞ot′) *n. var. of* CHEROOT.

**Sher·pa** (shûr′pə) *n., pl.* **Sherpa** *or* **-pas.** A member of a Tibetan people living in northern Nepal.

**sher·ry** (shĕr′ē) *n., pl.* **-ries.** [Alteration of obs. *sherris* (taken as pl.), after *Xeres,* Jerez, Spain.] An amber-colored fortified Spanish wine varying in taste from very dry to sweet.

**Shet·land** (shĕt′lənd) *n.* [After the *Shetland* Islands.] **1.** A fine, loosely twisted yarn made from the wool of sheep raised in the Shetland Islands and used for knitting and weaving. **2.** A garment made of Shetland.

**Shetland pony** *n.* A small, compactly built pony orig. bred in the Shetland Islands.

**Shetland sheepdog** *n.* A dog orig. bred in the Shetland Islands, having a rough coat and resembling a small collie.

**She·vat** (shə-vät′) *also* **She·bat** (-bät′, -vät′) *n.* [Heb. *shĕbhāt.*] The fifth month of the Hebrew year. —See table at CALENDAR.

**shew·bread** (shō′brĕd′) *n. Archaic. var. of* SHOWBREAD.

**Shi·ah** *also* **Shi·a** (shē′ə) *n.* [Ar. *shī'ah,* a following < *shā'a,* he followed.] **1.** The principal minority sect of Islam, composed of the followers of Ali, the cousin and son-in-law of Mohammed, who regard the heirs of Ali as the legitimate successors to the Prophet and reject the other caliphs and the Sunnite legal and political institutions. **2.** A Shiite.

**shi·at·su** (shē-ät′so͞o) *n.* [Short for J. *shiatsuryōhō* : *shi,* finger + *atsu-,* pressure + *ryōhō,* treatment.] A massage in which finger pressure is applied to those areas of the body used in acupuncture.

**shib·ah** (shĭv′ə) *n. var. of* SHIVA.

**shib·bo·leth** (shĭb′ə-lĭth, -lĕth′) *n.* [Heb. *shibbōleth,* an ear of corn, stream, from the use of this word to distinguish Gileadites from Ephraimites, who pronounced it *sibbōleth.*] **1.** A language usage that distinguishes the members of one group or class from another. **2. a.** A catchword : slogan. **b.** A common saying.

**shield** (shēld) *n.* [ME *sheeld* < OE *scield.*] **1.** A piece of armor made of leather, metal, or wood and carried on the forearm. **2.** A means of defense : PROTECTION. **3. a.** An escutcheon. **b.** A police officer's badge. **c.** A decorative emblem often serving to identify. **4.** A steel sheet attached to a gun to protect the gunners from small-arms fire. **5.** *Zool.* A hard outer covering, as a protective plate. **6.** A piece of rubberized or absorbent cloth worn, as at the armpits, to protect a garment from perspiration stains. **7.** *Physics.* A mass of material, as lead or cement, that encloses a nuclear reactor in order to reduce the amount of radiation that escapes into the surrounding area. —*v.* **shield·ed, shield·ing, shields.** —*vt.* **1.** To defend or protect with or as if with a shield : GUARD. **2.** To conceal. —*vi.* To act or function as a shield. **—shield′er** *n.*

**shield law** *n.* A law protecting journalists from being forced to reveal confidential sources of information.

**Shield of David** *n.* The Magen David.

**shiel·ing** (shē′lĭng, -lĭn) *n.* [< Sc. *shiel,* hut < ME *schele,* prob. of Scand. orig.] *Scot.* A shepherd's hut.

**shi·er** (shī′ər) *adj. var. compar. of* SHY[1].

**shies**[1] (shīz) *v. 3rd person sing. present tense of* SHY[1]. *—n. pl. of* SHY[1].

**shies**[2] (shīz) *v. 3rd person sing. present tense of* SHY[2]. *—n. pl. of* SHY[2].

**shi·est** (shī′ĭst) *adj. var. superl. of* SHY[1].

**shift** (shĭft) *v.* **shift·ed, shift·ing, shifts.** [ME *shiften,* to arrange < OE *sciftan.*] —*vt.* **1.** To move from one place or position to another. **2.** To exchange (one thing) for another : SWITCH <*shift* marketing tactics> **3.** To change (gears) in a motor vehicle. **4.** To alter phonetically or as part of a systematic change. —*vi.* **1.** To change position, direction, place, or form. **2. a.** To provide for one's own needs <had to *shift* for myself> **b.** To get along by evasive means. **3.** To change gears, as when driving a motor vehicle. —*n.* **1.** A change from one individual, position, or configuration to another. **2.** A change of direction or form. **3. a.** A group of workers on duty at the same time, as at a mill. **b.** The working period of such a group <worked the day *shift*> **4.** *Mus.* A change of the position of the hand in playing a musical instrument such as the violin. **5. a.** A systematic change of the phonetic or phonemic structure of a language. **b.** Functional shift. **6.** *Football.* A lateral movement of the offensive backfield from one formation to another just prior to putting the ball in play. **7. a.** A loosely fitting dress that hangs straight from the shoulders : CHEMISE. **b.** A woman's undergarment, as a slip. **8. a.** A means to an end : EXPEDIENT. **b.** A stratagem. **—shift′er** *n.*

**shift character** *n.* A data control character that determines the shift of character codes in a message.

**shift·less** (shĭft′lĭs) *adj.* [SHIFT, resourcefulness (obs.) + -LESS.] **1.** Devoid of purpose or ambition : LAZY. **2.** Lacking efficiency. **—shift′less·ly** *adv.* **—shift′less·ness** *n.*

**shift register** *n.* A computer memory device in which data can be moved to the left or to the right.
**shift·y** (shĭfʹtē) *adj.* **-i·er, -i·est. 1.** Crafty : tricky. **2.** Suggesting craft or guile. **3.** Resourceful. **—shiftʹi·ly** *adv.* **—shiftʹi·ness** *n.*
**shi·gel·la** (shĭ-gĕlʹə) *n., pl.* **-gel·lae** (-gĕlʹē) [NLat. *Shigella,* genus name, after Kiyoshi *Shiga* (1870–1957).] Any of various nonmotile rod-shaped bacteria of the genus *Shigella,* including some species that cause dysentery.
**Shi·ism** (shēʹĭzʹəm) *n.* The religion or doctrines of the Shiah.
**Shi·ite** (shēʹītʹ) *n.* A member of the Shiah branch of Islam. **—Shi·it'ic** (-ĭtʹĭk) *adj.*
**shi·ka·ri** (shĭ-kärʹē, -kărʹē) *n.* [Hindi < Pers. *shikārī* < *shikār,* hunting.] A big-game hunting guide.
**shi·ling·i** (shĭ-lĭngʹē) *n., pl.* **shilingi.** [Swahili < E. SHILLING.] —See table at CURRENCY.
**shill** (shĭl) [Orig. unknown.] *Slang.* *—n.* One who works as a decoy, as in a confidence game, by posing as a customer or an innocent bystander. *—vi.* **shilled, shill·ing, shills.** To act as a shill.
**shil·le·lagh** *also* **shil·la·lah** (shə-lāʹlē, -lə) *n.* [After *Shillelagh,* Ireland.] A cudgel, esp. one of oak or blackthorn.
**shil·ling** (shĭlʹĭng) *n.* [ME < OE *scilling.*] **1.** —See table at CURRENCY. **2.** A virgule.
**shil·ly-shal·ly** (shĭlʹē-shălʹē) *vi.* **-lied, -ly·ing, -lies.** [Redup. of the phrase *shall I.*] **1.** To hesitate or waver. **2.** To idle or poke. *—n., pl.* **-lies.** Procrastination : hesitation. **—shilʹly-shalʹly** *adj. & adv.* **—shilʹly-shalʹli·er** *n.*
**shim** (shĭm) *n.* [Orig. unknown.] A thin, often tapered piece of material, as metal, wood, or stone, used as leveler or filler between other materials such as stone or metal or between pieces of furniture and the floor. *—vt.* **shimmed, shim·ming, shims.** To level or fill in by using a shim.
**shim·mer** (shĭmʹər) *vi.* **-mered, -mer·ing, -mers.** [ME *schimeren* < OE *scimerian.*] To shine with a flickering light. **—shimʹmer** *n.* **—shimʹmer·y** *adj.*
**†shim·my** (shĭmʹē) *n., pl.* **-mies.** [Short for *shimmy-shake* < *shimmy,* alteration of CHEMISE.] **1.** A dance popular in the 1920's, marked by rapid shaking of the body. **2.** Abnormal vibration, as in the chassis of a motor vehicle. **3.** *Regional.* A chemise. *—vi.* **-mied, -my·ing, -mies. 1.** To vibrate abnormally. **2.** To shake the body in or as if in dancing the shimmy.
**shin[1]** (shĭn) *n.* [ME *shine* < OE *scinu.*] **1.** *Anat.* **a.** The front part of the leg below the knee and above the ankle. **b.** The tibia. **2.** The lower part of the foreleg in beef cattle as opposed to the upper foreleg or shank. —Used of cuts of meat. *—v.* **shinned, shin·ning, shins.** *—vt.* **1.** To climb (e.g., a rope) by gripping and pulling alternately with the hands and legs. **2.** To kick or hit in the shins. *—vi.* To climb something by shinning.
**shin[2]** (shēn, shĭn) *n.* [Heb. *shīn.*] The 22nd letter of the Hebrew alphabet. —See table at ALPHABET.
**shin·bone** (shĭnʹbōnʹ) *n.* The tibia.
**shin·dig** (shĭnʹdĭgʹ) *n.* [Prob. alteration of SHINDY.] *Slang.* A festive celebration or party.
**shin·dy** (shĭnʹdē) *n., pl.* **-dies.** [Alteration of SHINNY[1].] *Slang.* **1.** An uproar. **2.** A shindig.
**shine** (shīn) *v.* **shone** (shōn) *or* **shined, shin·ing, shines.** [ME *shinen* < OE *scīnan.*] *—vi.* **1.** To emit light : BEAM. **2.** To reflect light : GLINT. **3.** To distinguish oneself in a field or activity : EXCEL. **4.** To become clearly apparent. *—vt.* **1.** To aim or cast the beam or glow of <*Shine* that flashlight over here.> **2.** *p.t. & p.p.* **shined.** To make glossy by polishing. *—n.* **1.** Radiance : brightness. **2.** A shoeshine. **3.** Fair weather <The game will be held, rain or *shine.*> **4. shines.** *Informal.* Foolish pranks. **5.** *Slang.* Moonshine whiskey. **—take a shine to.** To take an immediate liking to.
**shin·er** (shīʹnər) *n.* **1.** One that shines. **2.** *Slang.* A black eye. **3. a.** A small, often silvery North American freshwater fish of the family Cyprinidae, esp. one of the genus *Notropis.* **b.** Any of various other small silvery fishes.
**shin·gle[1]** (shĭngʹgəl) *n.* [ME *schyngle* < Lat. *scindula,* alteration of *scandula.*] **1.** A thin oblong piece of material, as wood or asbestos, laid in overlapping rows to cover the roofs and sides of houses. **2.** *Informal.* A small signboard, as one indicating a lawyer's office <hang out a *shingle*> **3.** A woman's close-cropped haircut. *—vt.* **-gled, -gling, -gles. 1.** To cover (a roof or building) with shingles. **2.** To cut (hair) short and close to the head. **—shinʹgler** *n.*
**shin·gle[2]** (shĭngʹgəl) *n.* [Sc. *chyngill,* perh. of MLG orig.] **1. a.** Beach gravel composed of large smooth pebbles unmixed with finer material. **b.** A stretch of shore covered with coarse, smooth gravel. **2.** Gravel coarser than ordinary gravel. **—shinʹgly** *adj.*
**shin·gles** (shĭngʹgəlz) *pl.n.* [ME *schingles,* by folk ety. < Med. Lat. *cingulus* < Lat., belt < *cingere,* to gird.] *(sing. or pl. in number).* *Pathol.* Herpes zoster.
**shin·leaf** (shĭnʹlēfʹ) *n.* [From the use of its leaves in plasters for sore legs.] A North American woodland plant, *Pyrola elliptica,* with rounded basal leaves and white flowers.

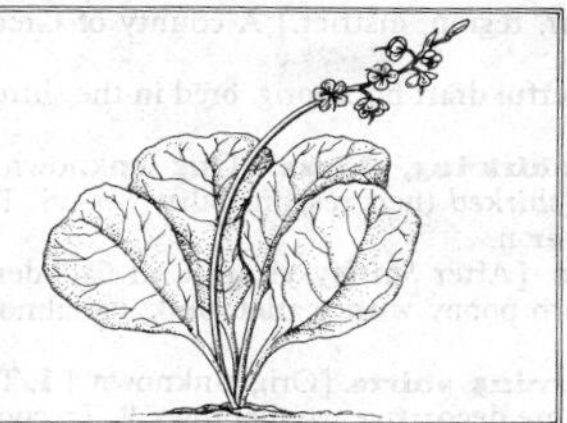
shinleaf

**shin·ny[1]** (shĭnʹē) *n., pl.* **-nies.** [Prob. < the phrase *shin ye,* a cry used in the game.] **1.** A simple form of hockey played by schoolboys. **2.** The curved stick used in shinny.
**shin·ny[2]** (shĭnʹē) *vi.* **-nied, -ny·ing, -nies.** To climb by shinning.
**shin·plas·ter** (shĭnʹplăsʹtər) *n.* [From its resemblance to paper used in plasters for sore legs.] Privately issued paper currency, esp. such currency devalued by lack of backing or by inflation.
**Shin·to** (shĭnʹtō) *n.* [J. *shintō* : *shin,* gods (< Chin. *shen*[2]) + *dō,* way (< Chin. *dao*[4]).] The aboriginal religion of Japan, marked by the veneration of nature spirits and of ancestors. **—Shinʹto·ism** *n.* **—Shinʹto·ist** *n.* **—Shinʹto·isʹtic** *adj.*
**shin·y** (shīʹnē) *adj.* **-i·er, -i·est. 1.** Bright : glistening <a *shiny* coat of paint> **2.** Clear : shining. **—shinʹi·ness** *n.*
**ship** (shĭp) *n.* [ME *schipp* < OE *scip.*] **1.** A rather large vessel adapted for deep-water navigation. **2.** A three-masted sailing vessel with square mainsails on all masts. **3.** *Law.* A vessel intended for marine transportation, without regard to form, rig, or means of propulsion. **4.** A ship's crew. **5.** An aircraft. *—v.* **shipped, ship·ping, ships.** *—vt.* **1.** To place or take on board a ship. **2.** To transport or cause to be transported <*shipped* the goods by air mail to Europe> **3.** To hire for work on a ship. **4.** To take in (water) over the side of a ship. *—vi.* **1. a.** To go aboard a ship. **b.** To travel by ship. **2.** To hire oneself out on a ship.
**-ship** *suff.* [ME < OE *-scipe.*] **1. a.** Quality, state, or condition <scholar*ship*> **b.** One that shows or possesses a quality, state, or condition <town*ship*> **2.** Rank, status, or office <professor*ship*> **3.** Art, skill, or craft <penman*ship*>
**ship biscuit** *n.* Hardtack.
**ship·board** (shĭpʹbôrdʹ, -bōrdʹ) *n.* The side of a ship. **—on shipboard.** On board a ship.
**ship·build·ing** (shĭpʹbĭlʹdĭng) *n.* The business of constructing ships. **—shipʹbuildʹer** *n.*
**ship canal** *n.* A canal deep enough to serve ships.
**ship fever** *n.* Typhus.
**ship·load** (shĭpʹlōdʹ) *n.* The amount a ship carries or is able to carry.
**ship·man** (shĭpʹmən) *n.* *Archaic.* **1.** A sailor. **2.** A shipmaster.
**ship·mas·ter** (shĭpʹmăsʹtər) *n.* The commanding officer of a merchant ship.
**ship·mate** (shĭpʹmātʹ) *n.* A fellow sailor.
**ship·ment** (shĭpʹmənt) *n.* **1.** The act of shipping goods. **2.** The goods shipped.
**ship money** *n.* A tax once levied on English maritime towns and shires to provide revenue for construction of warships.
**ship of the line** *n.* A warship large enough to take a position in the line of battle.
**ship·per** (shĭpʹər) *n.* One that consigns or receives goods for transportation.
**ship·ping** (shĭpʹĭng) *n.* **1.** The act or business of transporting goods. **2.** The body of ships belonging to one port, industry, or country, often referred to in aggregate tonnage. **3.** Passage on a ship.
**shipping clerk** *n.* One employed to manage the shipment or receipt of goods.
**ship-rigged** (shĭpʹrĭgdʹ) *adj.* *Naut.* Rigged with three or more masts and square sails.
**ship·shape** (shĭpʹshāpʹ) *adj.* [Short for obs. *shipshapen,* arranged as a ship should be : SHIP + obs. *shapen,* p.part. of SHAPE.] Tidy <a *shipshape* little apartment> **—shipʹshapeʹ** *adv.*
**ship's papers** *pl.n.* Documents that a ship is required to carry and provide on demand for inspection, as per international law.
**ship·way** (shĭpʹwāʹ) *n.* **1.** The structure supporting a ship during construction or in dry dock. **2.** A ship canal.
**ship·worm** (shĭpʹwûrmʹ) *n.* A wormlike marine mollusk of the genera *Teredo* and *Bankia,* having a rudimentary bivalve shell with which it bores into wood, often doing extensive damage.
**ship·wreck** (shĭpʹrĕkʹ) *n.* [Alteration of obs. *shipwrack* < ME *shipwrak* < OE *scipwræc,* jetsam : *scip,* ship + *wræc,* something driven by the sea.] **1. a.** Destruction of a ship, as by storm or collision. **b.** The remains of a wrecked ship. **2.** Complete ruin or failure. *—vt.* **-wrecked, -wreck·ing, -wrecks. 1.** To cause (a ship or its passengers) to suffer shipwreck. **2.** To ruin utterly.
**ship·wright** (shĭpʹrītʹ) *n.* A carpenter employed in ship construction or maintenance.
**ship·yard** (shĭpʹyärdʹ) *n.* A yard where ships are built or repaired.

**shire** (shīr) *n.* [ME < OE *scīr,* region, district.] A county of Great Britain.
**shire horse** *n.* A large, powerful draft horse orig. bred in the shires or midland region of England.
**shirk** (shûrk) *v.* **shirked, shirk·ing, shirks.** [Orig. unknown.] —*vt.* To avoid discharging <*shirked* their responsibility> —*vi.* To avoid duty or work. **—shirk'er** *n.*
**Shir·ley poppy** (shûr'lē) *n.* [After *Shirley* vicarage in Croyden, England.] A variety of the corn poppy with scarlet, pink, or salmon flowers.
**shirr** (shûr) *vt.* **shirred, shirr·ing, shirrs.** [Orig. unknown.] **1.** To gather (cloth) into three or more decorative parallel rows. **2.** To cook (unshelled eggs) by baking until set. **—shirr** *n.*
**shirt** (shûrt) *n.* [ME *sherte* < OE *scyrte.*] **1.** A garment for the upper part of the body, typically having a collar, sleeves, and a front opening. **2.** An undershirt. **3.** A night shirt. **4.** A protective cloth casing, as for use in shipping perishable goods. **—keep (one's) shirt on.** *Slang.* To remain calm. **—lose (one's) shirt.** *Slang.* To lose all of one's possessions or financial assets.
**shirt·dress** (shûrt'drĕs') *n.* A usu. simple dress styled like a shirt with a collar and buttons down the front.
**shirt·ing** (shûr'tĭng) *n.* Fabric for making shirts.
**shirt-sleeve** (shûrt'slēv') *also* **shirt-sleeves** (-slēvz') *or* **shirt-sleeved** (-slēvd') *adj.* **1. a.** Dressed casually, esp. being without a coat <*shirt-sleeve* office workers> **b.** Calling for the removal of a coat <*shirt-sleeve* weather> **2.** Informal or straightforward <*shirt-sleeve* politics>
**shirt·tail** (shûrt'tāl') *n.* **1.** The part of a shirt that extends below the waist, esp. in the back. **2.** Something small, inadequate, or of little value. **3.** A brief addition at the end of a newspaper article. —*adj.* **1.** Extremely young <*shirttail* kids> **2.** Distantly related <*shirttail* cousins> **3.** Of little value : INADEQUATE.
**shirt·waist** (shûrt'wāst') *n.* **1.** A woman's tailored shirt with details copied from men's shirts. **2.** A woman's dress with the bodice styled like a tailored shirt.
**shirt·y** (shûr'tē) *adj.* **-i·er, -i·est.** Ill-tempered <a *shirty* retort>
**shish ke·bab** *also* **shish ke·bob** *or* **shish ka·bob** (shĭsh' kə-bŏb') *n.* [Turk. *şiş kebabiu* : *şiş,* skewer + *kebap,* roast meat.] A dish of small pieces of meat roasted on skewers and served with condiments.
**shist** (shĭst) *n. var. of* SCHIST.
**shit·tah** (shĭt'ə) *n.* [Heb. *shiṭṭāh.*] A tree, prob. a species of acacia, that was a source of a wood mentioned often in the Bible.
**shit·tim·wood** (shĭt'ĭm-wo͝od') *n.* [Heb. *shiṭṭim,* pl. of *shiṭṭāh,* shittah + WOOD.] The wood of the shittah, used to make the ark of the Hebrew tabernacle.
**shiv** (shĭv) *n.* [Romany *chiv,* blade.] *Slang.* A knife or razor, used esp. as a weapon.
**shiv·a** *also* **shiv·ah** *or* **shib·ah** (shĭv'ə) *n.* [Yiddish < Heb. *shiv'āh,* seven.] A seven-day formal mourning period observed in Judaism after the funeral of a close relative.
**Shi·va** (shē'və, shĭv'ə) *also* **Si·va** (shē'və, sē-, shĭv'ə, sĭv'ə) *n.* [Skt. *śivaḥ* < *śiva-,* gracious.] The Hindu god of destruction and reproduction. **—Shi'va·ism** *n.* **—Shi'va·ist** *n.*
**shiv·ah** (shĭv'ə) *n. var. of* SHIVA.
**shiv·a·ree** (shĭv'ə-rē', shĭv'ə-rē') *n. var. of* CHARIVARI.
**shiv·er**[1] (shĭv'ər) *v.* **-ered, -er·ing, -ers.** [ME *shiveren, chiveren.*] —*vi.* **1.** To shudder from or as if from cold : TREMBLE. **2.** To quiver, as by the force of the wind. —*vt.* To cause (a sail) to flutter in the wind. **—shiv'er** *n.*
**shiv·er**[2] (shĭv'ər) *vi.* & *vt.* **-ered, -er·ing, -ers.** [ME *schiveren* < *schivere,* fragment.] To break or cause to break into splinters.
**shiv·er·y**[1] (shĭv'ə-rē) *adj.* **1.** Trembling, as from cold. **2.** Causing shivers : CHILLING.
**shiv·er·y**[2] (shĭv'ə-rē) *adj.* So brittle as to be easily broken.
**shle·miel** (shlə-mēl') *n. var. of* SCHLEMIEL.
**shlep** (shlĕp) *v. var. of* SCHLEP.
**shlock** (shlŏk) *n. var. of* SCHLOCK.
**shmear** (shmîr) *n. var. of* SCHMEER.
**shmuck** (shmŭk) *n. var. of* SCHMUCK.
**shoal**[1] (shōl) *n.* [ME *schald* < OE *sceald,* shallow.] **1.** A particularly shallow area in a body of water. **2.** A sandy elevation of the bottom of a body of water, constituting a navigational hazard. —*v.* **shoaled, shoal·ing, shoals.** —*vi.* To become shallow. —*vt.* **1.** To make shallow. **2.** To sail into a shallower part of. —*adj.* Having little depth : SHALLOW.
**shoal**[2] (shōl) *n.* [Prob. < MLG *schōle.*] **1.** A large group : CROWD. **2.** A school of marine animals, as fish. —*vi.* **shoaled, shoal·ing, shoals.** To come together in large numbers : THRONG.
**shoat** *also* **shote** (shōt) *n.* [ME *shote.*] A young pig that has just been weaned.
**shock**[1] (shŏk) *n.* [OFr. *choc* < *choquer,* to strike.] **1.** A violent collision or impact. **2. a.** Something, as an event or encounter, that jars the mind or emotions as if with a violent, unexpected blow. **b.** Disturbance of function, equilibrium, or mental faculties caused by such a blow. **3.** A severe offense to one's sense of propriety or decency : OUTRAGE. **4.** *Pathol.* A gen. temporary state of massive physiological reaction to bodily trauma, usu. characterized by marked loss of blood pressure and depression of vital processes. **5.** The sensation and muscular spasm caused by an electric current passing through the body. **6.** Shock therapy. —*v.* **shocked, shock·ing, shocks.** —*vt.* **1.** To strike with great surprise and agitation. **2.** To strike with disgust : OFFEND. **3.** To induce a state of shock in (a person). **4.** To subject (an individual) to an electric shock. —*vi.* To come into contact violently, as in battle.
**shock**[2] (shŏk) *n.* [ME *schock,* perh. of LG orig.] **1.** A number of sheaves of grain stacked upright in a field for drying. **2.** A thick heavy mass <a *shock* of gray hair> —*vt.* **shocked, shock·ing, shocks.** To gather (grain) into shocks.
**shock absorber** *n.* A device for absorbing mechanical shocks, esp. a hydraulically damped coupling for absorbing impulsive forces generated by the contact of automotive wheels with irregular road surfaces.
**shock·er** (shŏk'ər) *n.* Something, as a sensational story or novel, that startles, shocks, or horrifies.
**shock·ing** (shŏk'ĭng) *adj.* **1.** Highly disturbing. **2.** Highly offensive : INDECENT. **3.** Vivid or intense in color or tone <*shocking* pink> **—shock'ing·ly** *adv.*
**shock therapy** *n.* Induction of shock by electric current or drugs, sometimes with convulsions, as a therapy for mental illness.
**shock treatment** *n.* Shock therapy.
**shock troops** *pl.n.* Highly experienced troopers trained to lead attacks.
**shock wave** *n.* **1.** A large-amplitude compression wave, as that produced by an explosion, caused by supersonic motion of a body in a medium. **2.** A violent disruption, disturbance, or reaction <*Shock waves* of protest followed the court's ruling.>
**shod** (shŏd) *v. p.t.* & *p.p. of* SHOE.
**shod·den** (shŏd'n) *v. var. p.p. of* SHOE.
**shod·dy** (shŏd'ē) *n., pl.* **-dies.** [Orig. unknown.] **1. a.** Wool fibers obtained by shredding unfelted woolen or worsted rags or worn garments. **b.** Yarn, fabric, or garments made from or containing such reclaimed wool fibers. **2.** Inferior goods. —*adj.* **-di·er, -di·est.** **1.** Made of or containing inferior material. **2.** Being of poor quality or workmanship. **3.** Dishonest <*shoddy* business practices> **4.** Transparently and cheaply imitative. **5.** Shabby : run-down <*shoddy* housing projects> **—shod'di·ly** *adv.* **—shod'di·ness** *n.*
**shoe** (sho͞o) *n.* [ME *shoo* < OE *scōh.*] **1.** A durable covering for the human foot, esp. one of a matched pair made of leather or similar material reaching about to the ankle and having a rigid sole and heel. **2.** A horseshoe. **3.** A part or device placed at an end, foot, or bottom, esp.: **a.** A strip of metal fitted onto the bottom of a sled runner. **b.** A skid placed under the wheel of a vehicle to retard its motion. **c.** The outer covering, casing, or tread of a pneumatic rubber tire. **4.** The part of a brake that presses against the wheel or drum to retard its motion. **5.** The sliding contact plate on an electric train or streetcar that conducts electricity from the third rail. **6. shoes.** *Informal.* **a.** Status : position. **b. shoes.** Plight <I wouldn't want to be in their *shoes.*> —*vt.* **shod** (shŏd), **shod** *or* **shod·den** (shŏd'n), **shoe·ing, shoes. 1.** To furnish or fit with shoes. **2.** To cover with a wooden or metal guard to protect against wear.
**shoe·bill** (sho͞o'bĭl') *n.* A tall wading bird, *Balaeniceps rex* native to eastern tropical African swampy regions, and having long black legs, a stubby neck, and a large shoelike bill with a hook on the upper mandible.
**shoe·horn** (sho͞o'hôrn') *n.* A curved implement used at the heel to help slip on a shoe. —*vt.* **-horned, -horn·ing, -horns.** To squeeze into an insufficient space <*shoehorned* the car into a tight parking slot>
**shoe·lace** (sho͞o'lās') *n.* A string used for fastening shoes.
**shoe·mak·er** (sho͞o'mā'kər) *n.* A maker or repairer of shoes. **—shoe'mak'ing** *n.*
**shoe·string** (sho͞o'strĭng') *n.* **1.** A shoelace. **2.** A small amount of money or capital <a business that ran on a *shoestring*> —*adj.* **1.** Cut to or in the shape of a shoestring <*shoestring* potatoes> **2.** Marked by or composed of a small amount of money <ran the household on a *shoestring* budget>
**shoe·tree** (sho͞o'trē') *n.* A foot-shaped form inserted into a shoe to preserve its shape.
**sho·far** (shō'fär', -fər) *n., pl.* **sho·fars** *or* **sho·froth** (shō-frōt', -frōth', -frōs') [Heb. *shōphār,* ram's horn.] A trumpet made of a ram's horn, blown for warning, summoning, and ritual purposes by the ancient Hebrews and now sounded in the synagogue at Rosh Hashanah and Yom Kippur.
**sho·gun** (shō'gən) *n.* [J. *shōgun,* general, of Chin. orig.] One of a line of Japanese military leaders who until 1868 exercised absolute rule under the nominal leadership of the emperor.
**sho·gun·ate** (shō'gə-nĭt, -nāt') *n.* The government of a shogun.
**sho·ji** (shō'jē) *n., pl.* **shoji** *or* **-jis.** [J. *shōji* < Chin. (Mandarin) *zhang*[4] *zi*[5].] A translucent paper screen forming a sliding door or partition in a Japanese house.
**shone** (shōn) *v. var. p.t.* & *p.p. of* SHINE.

**shoo** (sho͞o) *interj.* —Used to frighten away animals. —*vt.* **shooed, shoo·ing, shoos.** To frighten away by or as if by crying "shoo."
**shoo·fly** (sho͞o′flī′) *n.* **1.** A child's rocker whose seat is built between two sides cut in the shape of an animal. **2.** *Slang.* An undercover police officer who monitors the honesty and performance of other police officers.
**shoofly pie** *n.* A pie with a molasses and brown sugar filling.
**shoo-in** (sho͞o′ĭn′) *n. Informal.* A sure winner.
**shook**[1] (shŏŏk) *n.* [Orig. unknown.] **1.** A set of parts for assembling a barrel or packing box. **2.** A shock of grain.
**shook**[2] (shŏŏk) *v. p.t. of* SHAKE.
**shook-up** (shŏŏk-ŭp′) *adj. Slang.* Emotionally upset.
**shoot** (sho͞ot) *v.* **shot** (shŏt), **shoot·ing, shoots.** [ME *schoten* < OE *scēotan.*] —*vt.* **1.** To hit, wound, or kill with a missile fired from a weapon. **2.** To fire or let fly (a missile) from a weapon. **3.** To discharge (a weapon). **4. a.** To send forth suddenly, intensely, or swiftly <*shooting* angry glances at me> **b.** To utter (sounds or words) forcefully, quickly, or suddenly <*shot* a retort> **5.** To pass over or through swiftly <*shoot* the rapids> **6.** To cover (country) in hunting for game. **7.** To record (e.g., a movie) on film. **8.** To cause to project or protrude : EXTEND. **9.** To begin to grow or produce. **10.** To pour, empty out, or discharge down or as if down a chute. **11.** To variegate with streaks or threads of a different color <black hair *shot* with gray> **12. a.** To move or propel (a marble or ball) toward its objective. **b.** To score (a point or goal). **c.** To play (golf, craps, or pool). **13.** To slide into or out of a fastening <*shoot* the bolt on the door> **14.** To measure the altitude of, as with a sextant. —*vi.* **1.** To discharge a missile. **2.** To discharge fire : GO OFF. **3.** To move rapidly : DART. **4.** To protrude. **5.** To hunt with a weapon, esp. for sport. **6.** To put forth new growth : GERMINATE. **7.** To take pictures : FILM. **8.** To propel a ball or other object toward a goal in a sport or game. **9.** To take one's turn at play. **—shoot down. 1.** To bring down (e.g., an aircraft) by hitting and damaging with a missile or cannon fire. **2.** *Slang.* To ruin the aspirations of : DISAPPOINT. **3. a.** *Informal.* To put an end to <*shot down* the proposal> **b.** To expose as false <*shot down* my pet theory> **—shoot for** (or **at**). To strive or aim for. **—shoot up. 1.** *Informal.* To grow taller rapidly. **2.** *Slang.* To inject (a narcotic drug) directly into a vein. —*n.* **1.** The movement of something shot. **2. a.** The young growth arising from a germinating seed : SPROUT. **b.** A bud or young leaf on a plant. **3.** A new growing part. **4.** A narrow, swift, or turbulent section of a stream : RAPID. **5.** An inclined channel through which something, such as timber, can be shot : CHUTE. **6.** An organized shooting activity, as a skeet tournament. **7.** *Informal.* A missile launch. **8.** The distance a shot travels : RANGE. **9.** The interval between strokes in rowing. —*interj.* **1.** —Used to express surprise, disbelief, or mild annoyance. **2.** —Used to express readiness to listen. **—shoot from the hip.** To act or speak with no thought of the consequences. **—shoot (one's) bolt.** To exhaust all of one's resources or capabilities. **—shoot the breeze.** To converse idly. **—shoot the works.** To expend all of one's efforts or capital. **—shoot′er** *n.*
**shoot-'em-up** (sho͞ot′əm-ŭp′) *Informal.* A film or television show featuring much physical violence, esp. shooting and killing.
**shooting gallery** *n.* An enclosed target range for firearms practice or competition.
**shooting iron** *n. Informal.* A six-shooter.
**shooting star** *n.* **1.** METEOR 1. **2.** Any of several North American plants of the genus *Dodecatheon,* with nodding flowers having reflexed petals.
**shooting stick** *n.* A stick pointed at one end and opening into a seat at the other, typically used by spectators at races.
**shoot-out** *also* **shoot·out** (sho͞ot′out′) *n.* An armed confrontation in which opponents shoot firearms at one another.
**shoot-up** (sho͞ot′ŭp′) *n. Slang.* An act of shooting up a narcotic drug.
**shop** (shŏp) *n.* [ME *shoppe* < OE *sceoppa,* booth, stall.] **1.** *also* **shoppe** (shŏp). A small retail store or a specialty department in a large store. **2.** An atelier : studio. **3.** A manufacturing or repair facility <a small machine *shop*> **4. a.** A commercial or industrial establishment. **b.** A business establishment. **5.** A home workshop. **6. a.** A schoolroom furnished with machinery and tools for instruction in industrial arts. **b.** Industrial arts as a technical science or course of study. —*v.* **shopped, shop·ping, shops.** —*vi.* **1.** To visit stores in order to inspect and buy goods. **2.** To look for something with the intention of acquiring it. —*vt.* To visit or buy from (a store). **—talk shop.** To talk about one's business.
**shop·keep·er** (shŏp′kē′pər) *n.* One who manages or owns a shop.
**shop·lift·er** (shŏp′lĭf′tər) *n.* One who steals merchandise on display in a store. **—shop′lift′ing** *n.*
**shoppe** (shŏp) *n. var. of* SHOP 1.
**shop·per** (shŏp′ər) *n.* **1.** One that shops. **2.** A commercial agent who compares the goods and prices of competing merchants. **3.** A commercial employee who fills mail or telephone orders. **4.** A usu. free paper carrying mostly advertising and some local news.
**shopping bag** *n.* A strong bag with handles designed for carrying a shopper's purchases.
**shop·ping-bag lady** (shŏp′ĭng-băg′) *n.* A bag lady.
**shopping center** *n.* A group of shops forming a central retail market within a given area.
**shopping mall** *n.* **1.** An urban shopping area limited to pedestrians. **2.** A shopping center with stores facing an enclosed walkway for pedestrians.
**shop steward** *n.* A union member elected to represent the union in its dalings with management.
**shop·talk** (shŏp′tôk′) *n.* Talk concerning one's business.
**shop·worn** (shŏp′wôrn′, -wōrn′) *adj.* **1.** Tarnished, frayed, faded, or otherwise defective from being on display in a store. **2.** Stale due to overuse or familiarity.
**sho·ran** (shôr′ăn′, shōr′-) *n.* [SHO(RT) RA(NGE) N(AVIGATION).] A relatively short-range navigation system by which a ship or aircraft can determine its position with high precision by measuring the times required for a radio signal to reach each of two ground stations of known position and to return.
**shore**[1] (shôr, shōr) *n.* [ME *schore,* prob. of LG orig.] **1.** The land along the edge of an ocean, sea, lake, or river : COAST. **2.** *often* **shores.** Land <my native *shores*>
**shore**[2] (shôr, shōr) *vt.* **shored, shor·ing, shores.** [ME *schoren* < *schore,* prop, prob. of LG orig.] To prop up, as with an inclined timber <*shore* up sagging beams> —*n.* A beam or timber propped against a structure to provide usu. temporary support.
**shore**[3] (shôr, shōr) *v. Archaic. var. p.t. of* SHEAR.
**shore bird** *n.* A bird, as a sandpiper, plover, or snipe, that frequents coastal or inland shores.
**shore·line** (shôr′līn′, shōr′-) *n.* The line marking the edge of a body of water.
**shore patrol** *n.* A detail of the U.S. Navy, Marine Corps, or Coast Guard functioning as military police ashore.
**shore·ward** (shôr′wərd, shōr′-) *also* **shore·wards** (-wərdz) *adv.* Toward the shore.
**shorl** (shôrl) *n. var. of* SCHORL.
**shorn** (shôrn, shōrn) *v. var. p.p. of* SHEAR.
**short** (shôrt) *adj.* **-er, -est.** [ME <OE *sceort.*] **1.** Having little length. **2.** Having little height. **3. a.** Lasting only a short period of time : BRIEF <a *short* rest> **b.** Seeming to pass quickly <finished the job in a few *short* weeks> **4.** Inadequate : insufficient <gas in *short* supply><were *short* on talent> **5.** Lacking in length or amount <a beam *short* three inches> **6.** Lacking in scope. **7.** Succinct : concise <a speech that was *short* and to the point> **8. a.** Not owning the stocks or commodities one is selling. **b.** Relating to or designating a sale of stocks or goods that the seller does not yet own but must produce to meet the terms of a contract. **9.** Not retentive <had a *short* memory> **10.** Rudely brief : ABRUPT. **11.** Containing a large amount of shortening : FLAKY <a *short* pie crust> **12. a.** Indicating a syllable of relatively brief duration. **b.** Unstressed, as in Greek and Latin verse. **13. a.** Designating a particular pronunciation of the letters for the vowel sounds, as the sound of (ă) in *pan* as distinguished from the sound of (ā) in *pane.* **b.** Describing a speech sound of relatively brief duration as opposed to the same or a similar sound of relatively long duration. **14.** *Slang.* Being near the end of a tour of military duty. —*adv.* **1.** Abruptly : quickly <pulled up *short*> **2.** Rudely : curtly. **3.** At a point before a given limit or goal <an arrow that fell *short*> **4.** At a disadvantage <caught *short* by their aggressive response> **5.** Without owning what one is selling <sell *short*> —*n.* **1.** Something short, esp.: **a.** A briefly articulated or unaccented syllable. **b.** A short vowel. **c.** A short sale. **d.** One who sells short. **e. shorts.** Short trousers extending to the knee or above. **f. shorts.** Men's undershorts. **g.** A short subject. **2. shorts.** A by-product of wheat processing, consisting of bran mixed with coarse meal or flour. **3. shorts.** Clippings or trimmings that remain as by-products in manufacturing processes, often used to make an inferior variety of a product. **4.** *Elect.* **a.** A short circuit. **b.** A malfunction caused by a short circuit. **5.** *Baseball.* A shortstop. —*v.* **short·ed, short·ing, shorts.** —*vt.* **1.** To cause a short circuit in. **2.** *Informal.* To give (a person) less than he or she is entitled to : SHORTCHANGE. —*vi.* To short-circuit. **—for short.** As an abbreviation. **—in short.** In summary : BRIEFLY. **—short′ness** *n.*
**short account** *n.* **1.** The account of one who sells short. **2.** The total open short sales of a commodity or security or on the market as a whole.
**short·age** (shôr′tĭj) *n.* A deficiency : insufficiency.
**short·bread** (shôrt′brĕd′) *n.* A kneaded dough of flour, sugar, and butter, rolled thickly, cut into cookies, and baked.
**short·cake** (shôrt′kāk′) *n.* A dessert made of a cake split and filled with fruit and topped with cream.
**short·change** (shôrt′chānj′) *vt.* **-changed, -chang·ing, -chang·es.** *Informal.* **1.** To give (someone) less change than is due. **2.** To swindle, cheat, or trick. **—short′chang′er** *n.*
**short circuit** *n.* An accidentally established low-resistance connection between two points in an electric circuit.
**short-cir·cuit** (shôrt′sûr′kĭt) *v.* **-cuit·ed, -cuit·ing, -cuits.** —*vt.* **1.** To cause to have a short circuit. **2.** *Informal.* To impede the progress of. —*vi.* To become affected with a short circuit.

**short·com·ing** (shôrt′kŭm′ĭng) *n.* A deficiency : flaw.
**short covering** *n.* Purchase of securities, stocks, or commodities so as to close out a short sale.
**short cut** *n.* **1.** A more direct route than the customary one. **2.** A way of saving time or effort.
**short division** *n.* A division of one number by another, usu. no more than two digits, without writing out the remainders.
**short·en** (shôr′tn) *v.* **-ened, -en·ing, -ens.** —*vt.* **1.** To make short or shorter. **2.** *Naut.* To take in (a sail) so that less canvas is exposed to the wind. **3.** To reduce in force, efficacy, or intensity. **4.** To add shortening to (dough) so as to make flaky. —*vi.* To become short or shorter. **—short′en·er** *n.*
**shortened form** *n.* An abbreviated form of a polysyllabic word, as *auto* for *automobile.*
**short·en·ing** (shôr′tn-ĭng, shôrt′nĭng) *n.* **1.** A fat, as butter or lard, used for making cake or pastry light or flaky. **2.** A shortened form, as an abbreviation. **3.** The act of one that shortens.
**short·fall** (shôrt′fôl′) *n.* **1.** A failure to attain a specified amount or level : SHORTAGE. **2.** The amount by which a supply falls short of expectation, need, or demand. **3.** A monetary deficit.
**short fuse** *n.* A quick temper.
**short·hand** (shôrt′hănd′) *n.* **1.** A rapid handwriting system utilizing symbols to represent words, phrases, and letters : STENOGRAPHY. **2.** A system, form, or instance of abbreviated or formulaic reference.
**short-hand·ed** (shôrt′hăn′dĭd) *adj.* Lacking the usual or necessary number of workers, employees, or assistants.
**short·horn** (shôrt′hôrn′) *n.* A beef or dairy cow orig. bred in northern England and having short curved horns.
**short hundredweight** *n.* HUNDREDWEIGHT 1.
**short·leaf pine** (shôrt′lēf′) *n.* A pine, *Pinus echinata,* with short flexible leaves, common in the southern United States.
**short-lived** (shôrt′līvd′, -lĭvd′) *adj.* Living or lasting only a short time.
**short·ly** (shôrt′lē) *adv.* **1.** In a short time : SOON. **2.** In a few words : CONCISELY. **3.** In an abrupt way : CURTLY.
**short order** *n.* Food quickly prepared and served.
**short rib** *n.* A cut of meat consisting of the area between the rib roast and the plate.
**short shrift** *n.* **1.** The short time granted a condemned prisoner for confession before execution. **2. a.** Summary, unsympathetic treatment. **b.** Quick work <made *short shrift* of the committee proceedings>
**short sight** *n.* Myopia.
**short·sight·ed** (shôrt′sī′tĭd) *adj.* **1.** Nearsighted : myopic. **2.** Lacking foresight. **—short′sight′ed·ly** *adv.* **—short′sight′ed·ness** *n.*
**short-spo·ken** (shôrt′spō′kən) *adj.* Short or abrupt in manner or speech : CURT.
**short·stop** (shôrt′stŏp′) *n. Baseball.* **1.** The field position between second and third bases. **2.** The player occupying shortstop.
**short story** *n.* A short piece of prose fiction aiming at unity of characterization, theme, and effect.
**short subject** *n.* A brief film often projected between showings of longer films.
**short-tem·pered** (shôrt′tĕm′pərd) *adj.* Irascible.
**short-term** (shôrt′tûrm′) *adj.* **1.** Lasting a relatively short time. **2.** Payable or reaching maturity within a relatively short time, as a year <*short-term* loans>
**short ton** *n.* TON 1b.
**short wave** *n.* An electromagnetic wave with wavelength in the short-wave region.
**short-wave** (shôrt′wāv′) *adj.* **1.** Having a wavelength of less than approx. 80 meters. **2.** Capable of receiving or transmitting at wavelengths of less than approx. 80 meters.
**short-wind·ed** (shôrt′wĭn′dĭd) *adj.* **1.** Having or characterized by shortness of breath. **2.** Choppy : disconnected.
**Sho·sho·ne** *also* **Sho·sho·ni** (shō-shō′nē) *n., pl.* **Shoshone** *or* **-nes** *also* **Shoshoni** *or* **-nis. 1. a.** A tribe of Indians once inhabiting parts of Nevada, Oregon, Idaho, Utah, Wyoming, and Texas. **b.** A member of this tribe. **2.** The Uto-Aztecan language of the Shoshone.
**Sho·sho·ne·an** (shō-shō′nē-ən) *n.* A group of Uto-Aztecan languages that includes most of the Uto-Aztecan languages found in the United States. **—Sho·sho′ne·an** *adj.*
**Sho·sho·ni** (shō-shō′nē) *n. var. of* SHOSHONE.
**shot**[1] (shŏt) *n.* [ME *schot* < OE *sceot.*] **1.** The discharge of a weapon, as of a gun. **2.** *pl.* **shot.** A projectile, as a bullet or pellet, from a firearm. **3.** *Informal.* **a.** Something, as a throw, hit, or drive in any of several games, that resembles the directed discharge of a weapon in force and carry. **b.** *Baseball.* A home run. **4.** One who shoots <a good *shot*> **5.** The distance over which something is shot : RANGE. **6.** An attempt to hit or land with a missile or rocket <a moon *shot*> **7. a.** An attempt. **b.** A guess. **c.** An opportunity. **8.** The heavy metal ball put for distance in the shot-put. **9.** An explosive charge used in blasting mine shafts. **10. a.** A photograph or one in a series of photographs. **b.** A single cinematic view or take. **11.** A hypodermic injection. **12.** A drink, esp. a jigger, of liquor. **13.** *Naut.* A unit designating chain length, in the United States 15 fathoms, in Great Britain 12½ fathoms. —*vt.* **shot·ted, shot·ting, shots. 1.** To load or weight with shot. **2.** To clean (bottles) by shaking when full of shot. **—like a shot.** Very quickly. **—shot in the arm.** A boost to one's spirits. **—shot in the dark. 1.** A wild, unsubstantiated guess. **2.** An attempt having little chance of success.
**shot**[2] (shŏt) *adj.* **1.** Of changeable or variegated color, as fabric having different colored warp and weft. **2.** *Informal.* Worn-out or ruined.
**shot**[3] (shŏt) *v. p.t. & p.p. of* SHOOT.
**shote** (shōt) *n. var. of* SHOAT.
**shot·gun** (shŏt′gŭn′) *n.* **1.** A shoulder-held firearm that fires multiple pellets through a smooth bore. **2.** *Football.* An offensive formation in which the quarterback lines up a few yards behind the scrimmage line and the other backs play in scattered positions.
**shotgun marriage** *also* **shotgun wedding** *n.* A forced marriage due to pregnancy.
**shot-put** (shŏt′po͝ot′) *n.* **1. a.** An athletic event in which the participants try to throw a heavy ball as far as possible. **b.** The standard ball used in this competition. **2.** A throw. **—shot′-put′ter** *n.*
**shott** (shŏt) *n. var. of* CHOTT.
**shot·ten** (shŏt′n) *adj.* [ME *schotyn,* p.part. of *schoten,* to shoot.] Having recently spawned and therefore being less desirable as food. —Used of fish, esp. herring.
**should** (sho͝od) *aux.v. p.t. of* SHALL. **1.** —Used to express duty or obligation <You *should* write a thank-you note.> **2.** —Used to express probability or expectation <They *should* arrive here soon.> **3.** —Used to express conditionality or contingency <If they *should* fail, then so would I.> **4.** —Used to moderate the directness or bluntness of a statement <I *should* think you would want to come.> ***usage:*** *Should have* is sometimes incorrectly written *should of* by writers who have mistaken the source of the spoken contraction *should've.*
**shoul·der** (shōl′dər) *n.* [ME *shulder* < OE *sculdor.*] **1.** *Anat.* **a.** The part of the human body between the neck and upper arm. **b.** The joint that connects the arm with the trunk. **2.** The part of an animal corresponding to the human shoulder. **3.** *often* **shoulders.** The two shoulders and the area of the back between them <broad *shoulders*> **4.** The forequarter of some animals. **5.** The part of a garment covering the shoulder. **6.** The angle between the face and the flank of a bastion. **7.** The extended flat surface on the body of printing type beyond the letter or character. **8. a.** The edge running on either side of a road. **b.** A shoulderlike slope or projection. —*v.* **-dered, -dering, -ders.** —*vt.* **1.** To place or carry (e.g., a burden) on the shoulders. **2.** To take on : ASSUME <*shouldered* all the blame> **3.** To apply force to with or as if with the shoulder <*shouldered* their way through the throng> —*vi.* To push with the shoulders.
**shoulder bag** *n.* A handbag carried by a strap looped over the shoulder.
**shoulder belt** *n.* An automotive safety belt that is worn diagonally across the body and over the shoulder.
**shoulder blade** *n.* The scapula.
**shoulder girdle** *n.* The pectoral girdle.
**shoulder harness** *n.* A shoulder belt.
**shoulder patch** *n.* An identification patch worn on the upper part of a uniform sleeve.
**should·n't** (sho͝od′nt). Should not.
**shouldst** (sho͝odst) *also* **should·est** (sho͝od′ĭst) *aux.v. Archaic.* 2nd person sing. p.t. of SHALL.
**shout** (shout) *n.* [ME *shoute.*] A loud yell. —*vt. & vi.* **shout·ed, shout·ing, shouts.** To say with or utter a shout. **—shout′er** *n.*
**shouting distance** *n.* A short distance within which one can be heard by another.
**shove** (shŭv) *v.* **shoved, shov·ing, shoves.** [ME *schouven* < OE *scūfan.*] —*vt.* To prod or thrust, often rudely or roughly. —*vi.* To push someone or something often rudely or roughly. **—shove off.** *Informal.* To leave. —*n.* **1.** An act of shoving. **2.** A rude or rough push. **—shov′er** *n.*
**shov·el** (shŭv′əl) *n.* [ME *schovel* < OE *scofl.*] **1.** A tool with a handle and a somewhat flattened scoop for picking up material, as dirt or snow. **2.** A large mechanical device or vehicle for heavy digging or excavation. **3.** A shovelful. —*v.* **-eled, -el·ing, -els** *or* **-elled, -el·ling, -els.** —*vt.* **1.** To move or remove with or as if with a shovel. **2.** To clear or make with a shovel. —*vi.* To dig or work with a shovel.
**shov·el·er** *also* **shov·el·ler** (shŭv′ə-lər, shŭv′lər) *n.* **1.** One that shovels. **2.** A widely distributed duck, *Spatula clypeata* or *Anas clypeata,* whose bill is long and broad.
**shov·el·ful** (shŭv′əl-fo͝ol′) *n.* The amount that a shovel will hold.
**shovel hat** *n.* A stiff, broad-brimmed, low-crowned hat, turned up at the sides and projecting in front, worn by some English clerics.
**shov·el·head** (shŭv′əl-hĕd′) *n.* A shark, *Sphyrna tiburo* of Atlantic and Pacific waters.
**shov·el·nose** (shŭv′əl-nōz′) *n.* A sturgeon, *Scaphirhynchus platorynchus,* of the Mississippi River, with a broad flat snout.
**shov·el-nosed** (shŭv′əl-nōzd′) *adj.* Having a broad flattened snout, bill, or head.
**show** (shō) *v.* **showed, shown** (shōn) *or* **showed, show·ing, shows.** [ME *showen* < OE *scēawian,* to look at.] —*vt.* **1. a.** To make

visible. **b.** To present in public exhibition or competition. **2.** To guide : conduct <*showed* the way> **3.** To point out : demonstrate. **4. a.** To manifest : exhibit <*showed* irritation at the delays> **b.** To indicate : register <The thermostat *shows* 72°F.> **5.** To grant : bestow. **6.** *Law.* To plead : allege <*show* cause> —*vi.* **1.** To be or become visible. **2.** *Informal.* To make an appearance. **3.** To appear : seem. **4.** To be exhibited publicly : RUN <A new movie is *showing* tonight.> **5.** To finish third or better in certain sports, for betting purposes. **—show off.** To behave ostentatiously or conspicuously. **—show up. 1.** To expose or reveal the true nature of. **2.** To be clearly visible. **3.** To put in an appearance : ARRIVE <*showed up* at nine> —*n.* **1.** A display : manifestation <a *show* of power> **2.** A false appearance <a *show* of sympathy> **3.** A striking appearance or display : SPECTACLE. **4.** Pompous or ostentatious display. **5. a.** A public exhibition or entertainment. **b.** A theatrical troupe or company. **6.** *Informal.* An undertaking <tried to run the whole *show*> **7.** Third place or better in certain sports, for betting purposes <win, place, and *show*> **—show (one's) hand. 1.** To display one's cards with the faces up. **2.** To reveal one's intentions or resources. **—show (someone) the door.** *Informal.* To tell (another) to leave.

☆ ***syns:*** SHOW, DEMONSTRATE, DISPLAY, EVIDENCE, EVINCE, EXHIBIT, MANIFEST *v. core meaning* : to make apparent <*showed* prudence by driving carefully>

**show bill** *n.* An advertising poster.

**show biz** (bĭz) *n. Slang.* Show business.

**show·boat** (shō'bōt') *n.* **1.** A river steamboat having a troupe of performers and a theater aboard for entertainment on the river. **2.** SHOWOFF 2.

**show·bread** (shō'brĕd') *n.* [Transl. of G. *Schaubrot,* transl. of Heb. *leḥem pānim,* bread of the Divine Presence.] The 12 loaves of blessed unleavened bread placed every Sabbath in the sanctuary of the Tabernacle by ancient Hebrew priests.

**show business** *n.* The entertainment industry.

**show·case** (shō'kās') *n.* **1.** A display case or cabinet. **2.** A setting in which something may be displayed advantageously. —*vt.* **-cased, -cas·ing, -cas·es.** To display prominently and advantageously.

**show·down** (shō'doun') *n.* **1.** The laying down of the players' hands of cards to determine the winner of the pot in poker. **2.** An event or circumstance that forces an issue to a conclusion.

**show·er¹** (shou'ər) *n.* [ME *shour* < OE *scūr.*] **1. a.** A brief fall of precipitation, as rain, sleet, or hail. **b.** A fall of a group of objects, esp. of a large group, from the sky <a meteor *shower*> **2.** A brief or sudden downpour <a *shower* of dry leaves> **3.** An abundant flow : OUTPOURING <a *shower* of critical comments> **4.** A party held to honor and present gifts to someone <a baby *shower*> <a bridal *shower*> **5.** A shower bath. —*v.* **-ered, -er·ing, -ers.** —*vt.* **1.** To sprinkle : spray. **2.** To bestow abundantly. —*vi.* **1.** To pour down in a shower. **2.** To take a shower bath. **—show'er·y** *adj.*

**show·er²** (shō'ər) *n.* One that shows.

**shower bath** *n.* A bath in which water is sprayed on the bather from an overhead nozzle.

**show·girl** (shō'gûrl') *n.* A chorus girl.

**show·ing** (shō'ĭng) *n.* **1.** Presentation : display. **2.** Performance, as in a competition <a poor *showing* in the golf tournament> **3.** Presentation of evidence, facts, or figures.

**show·man** (shō'mən) *n.* **1.** A theatrical producer. **2.** One having a flair for the dramatic. **—show'man·ship'** *n.*

**shown** (shōn) *v. var. p.p. of* SHOW.

**show·off** (shō'ôf', -ŏf') *n.* **1.** An act of showing off. **2.** One who shows off : EXHIBITIONIST.

**show·piece** (shō'pēs') *n.* Something exhibited as an outstanding example of its kind.

**show place** *also* **show·place** (shō'plās') *n.* A place viewed, admired, and frequented for its excellence or beauty.

**show room** *n.* A room in which merchandise is displayed.

**show·stop·per** (shō'stŏp'ər) *n.* One, as an act or a performer, that evokes so much audience applause that the performance is temporarily interrupted.

**show·y** (shō'ē) *adj.* **-i·er, -i·est. 1.** Making a conspicuous display : STRIKING <*showy* flower clusters> **2.** Displaying brilliance and virtuosity. **3.** Gaudy : flashy. **—show'i·ly** *adv.* **—show'i·ness** *n.*

**shrank** (shrăngk) *v. var. p.t. of* SHRINK.

**shrap·nel** (shrăp'nəl) *n., pl.* **shrapnel.** [After General Henry *Shrapnel* (1761–1842), its inventor.] **1. a.** An artillery shell containing metal balls fused to explode in the air above enemy troops. **b.** The metal balls in such a weapon. **2.** Shell fragments from a high-explosive shell.

**shred** (shrĕd) *n.* [ME *shrede* < OE *scrēade.*] **1.** A long irregular strip cut or torn off. **2.** A small particle. —*vt.* **shred·ded** or **shred, shred·ding, shreds.** To tear or cut into shreds. **—shred'der** *n.*

**shrew** (shrōō) *n.* [ME *shrewe,* villain < OE *scrēawa,* shrewmouse.] **1.** Any of various small, chiefly insectivorous mammals of the family Soricidae, with a long pointed nose and small, often poorly developed eyes. **2.** A scold.

**shrewd** (shrōōd) *adj.* **-er, -est.** [ME *shrewed,* wicked < *shrew,* rascal.] **1.** Having keen insight : ASTUTE. **2.** Artful and cunning. **3.** Sharp : penetrating. **—shrewd'ly** *adv.* **—shrewd'ness** *n.*

☆ ***syns:*** SHREWD, ASTUTE, CAGEY, SLICK *adj. core meaning* : having or showing clever awareness and resourcefulness <a *shrewd* judge of character> <a *shrewd* investment>

**shrew·ish** (shrōō'ĭsh) *adj.* Ill-tempered : nagging <a *shrewish* neighbor> **—shrew'ish·ly** *adv.* **—shrew'ish·ness** *n.*

**shrew mole** *n.* A shrewlike mole of the family Talpidae, esp. *Neurotrichus gibbsi* of western North America, or *Uropsilus soricipes* of eastern Asia.

**shrew·mouse** (shrōō'mous') *n.* SHREW 1.

**shriek** (shrēk) *n.* [< ME *shriken,* to shriek.] **1.** A loud shrill cry : SCREECH. **2.** A sound like a shriek. —*v.* **shrieked, shriek·ing, shrieks.** —*vi.* **1.** To utter a shriek. **2.** To make a sound likened to a shriek. —*vt.* To utter with a shriek. **—shriek'er** *n.*

**shrie·val** (shrē'vəl) *adj.* [< obs. *shrieve,* var. of SHERIFF.] Of or relating to a sheriff. **—shrie'val·ty** *n.*

**shrift** (shrĭft) *n.* [ME < OE *scrift* < *scrīfan,* to shrive < Lat. *scribere,* to write.] *Archaic.* **1.** The act of shriving. **2. a.** Confession to a priest. **b.** Absolution given by a priest.

**shrike** (shrīk) *n.* [Prob. < ME **shrik* < OE *scrīe,* thrush.] A carnivorous bird of the family Laniidae, having a hooked bill and often impaling its prey on sharp-pointed thorns or barbed wire fencing.

**shrike**
*9 inches long*

**shrill** (shrĭl) *adj.* **-er, -est.** [ME *shrille* < *shrillen,* to shriek.] **1.** High-pitched and piercing <a *shrill* voice> **2.** Producing a shrill tone or sound. **3.** Sharp or keen to the senses. —*v.* **shrilled, shrill·ing, shrills.** —*vt.* To utter shrilly. —*vi.* To produce a shrill cry or sound. **—shrill'ness** *n.* **—shril'ly** *adv.*

**shrimp** (shrĭmp) *n., pl.* **shrimp** or **shrimps.** [ME *shrimpe,* perh. of LG orig.] **1. a.** Any of various small, slender-bodied, chiefly marine decapod crustaceans of the suborder Natantia, many species of which are edible. **b.** Any of various crustaceans similar to the shrimp. **2.** *Slang.* One who is small or unimportant. —*vi.* **shrimped, shrimp·ing, shrimps.** To catch or fish for shrimp.

**shrimp·fish** (shrĭmp'fĭsh') *n., pl.* **shrimpfish** or **-fish·es.** Any of various small slender tropical marine fishes of the family Centriscidae, related to the sea horses and pipefish.

**shrimp plant** *n.* A shrubby plant, *Beloperone guttata,* with inconspicuous flowers borne between reddish bracts.

**shrine** (shrīn) *n.* [ME *shrine* < OE *scrīn,* box < Lat. *scrinium,* case for books or papers.] **1.** A receptacle for sacred relics : RELIQUARY. **2.** The tomb of a venerated person, as a saint. **3.** A site hallowed by a venerated object or its associations. —*vt.* **shrined, shrin·ing, shrines.** To enshrine.

**Shrin·er** (shrī'nər) *n.* [After the Ancient Arabic Order of Nobles of the Mystic *Shrine,* their fraternal order.] A member of a U.S. secret fraternal order that is not Masonic but that admits as members only Knights Templars and 32nd-degree Masons.

**shrink** (shrĭngk) *v.* **shrank** (shrăngk) or **shrunk** (shrŭngk), **shrunk** or **shrunk·en** (shrŭng'kən), **shrink·ing, shrinks.** [ME *shrinken* < OE *scrincan.*] —*vi.* **1.** To become constricted from heat, moisture, or cold : CONTRACT. **2.** To become reduced in amount or value : DWINDLE <Their savings quickly *shrank.*> **3.** To draw back : RECOIL. **4.** To be reluctant to do or say something. —*vt.* To cause to shrink. —*n.* **1. a.** The act of shrinking. **b.** Shrinkage. **2.** *Slang.* A psychiatrist or psychologist. **—shrink'a·ble** *adj.* **—shrink'er** *n.*

**shrink·age** (shrĭng'kĭj) *n.* **1.** The process of shrinking. **2.** A reduction in value : DEPRECIATION. **3.** The total weight loss sustained by livestock in shipment to a market. **4.** The amount of a loss caused by shrinkage.

**shrinking violet** *n. Informal.* A shy or retiring person.

**shrink package** *n.* A transparent form-fitting plastic wrapping, esp. of polyethylene or polyvinyl chloride, used to protect a commodity from dust, moisture, and abrasion.

**shrink-pack·age** (shrĭngk'păk'ĭj) *vt.* **-aged, -ag·ing, -ag·es.** To enclose (a commodity) in a shrink package.

**shrink wrap** *n.* A shrink package.

**shrink-wrap** (shrĭngk'răp') *vt.* **-wrapped, -wrap·ping, -wraps.** To shrink-package.

**shrive** (shrīv) *v.* **shrove** (shrōv) or **shrived, shriv·en** (shrĭv'ən) or **shrived, shriv·ing, shrives.** [ME *schriven* < OE *scrīfan* < Lat. *scribere,* to write.] —*vt.* **1.** To hear the confession of and give absolution to (a penitent). **2.** To obtain absolution for (oneself) by confessing and doing penance. *—vi.* **1.** To make or go to confession. **2.** To hear confessions. **—shriv'er** *n.*

▲ word history: The word *shrive* is unique among English verbs because it is the only verb of non-Germanic origin that has a strong conjugation instead of the expected weak one. A strong conjugation in Germanic is one in which the vowel undergoes changes and there is no suffix to mark the past tense. The verb *sing* belongs to the strong category, for its principal parts are *sing, sang,* and *sung.* When Latin *scribere* was borrowed into the original Germanic language, it was conjugated like the ancestor of English *ride,* and its principal parts are correspondingly *shrive, shrove,* and *shriven.* Not until the 14th century did the form *shrived* appear for the past tense, and the past participle form *shrived* did not occur until the 16th century. The noun *shrift* is derived from *shrive* and is not borrowed directly from a Latin word.

**shriv·el** (shrĭv'əl) *v.* **-eled, -el·ing, -els** or **-elled, -el·ling, -els.** [Orig. unknown.] *—vi.* **1.** To shrink and wrinkle, often in drying. **2.** To lose vitality or intensity. *—vt.* To cause to become shriveled.

**shriv·en** (shrĭv'ən) *v. var. p.p. of* SHRIVE.

**Shrop·shire** (shrŏp'shîr', -shər, -shīr') *n.* A large, hornless, black-faced sheep orig. bred in Shropshire, England, and raised for meat and wool.

**shroud** (shroud) *n.* [ME *schrud,* garment < OE *scrūd.*] **1.** A cloth used to wrap the body of a deceased person for burial. **2.** A means of concealment, protection, or screening. **3. a.** *Naut.* One of a set of ropes or wire cables stretched from the masthead to a vessel's sides to support the mast. **b.** A support for a smokestack or comparable structure. **c.** One of the ropes connecting the harness and canopy of a parachute. *—v.* **shroud·ed, shroud·ing, shrouds.** *—vt.* **1.** To wrap (a corpse) in burial clothing. **2.** To screen : hide. **3.** *Archaic.* To shelter : protect. *—vi. Archaic.* To take cover : find shelter.

**shrove** (shrōv) *v. var. p.t. of* SHRIVE.

**Shrove Sunday** *n.* The Sunday before Ash Wednesday.

**Shrove·tide** (shrōv'tīd') *n.* [ME *schroftyde* : *schrof-,* shriving (< *schriven,* to shrive) + *tyde,* time < OE *tīd.*] The three days preceding Ash Wednesday.

**shrub[1]** (shrŭb) *n.* [ME *schrubbe* < OE *scrybb.*] A rather low woody plant having several stems : BUSH.

**shrub[2]** (shrŭb) *n.* [Ar. *shurb,* a drink < *shariba,* to drink.] A beverage made from fruit juice, sugar, and a liquor such as rum or brandy.

**shrub·ber·y** (shrŭb'ə-rē) *n., pl.* **-ies. 1.** Shrubs as a whole. **2.** A group of shrubs.

**shrub·by** (shrŭb'ē) *adj.* **-bi·er, -bi·est. 1.** Composed of, planted with, or covered with shrubs. **2.** Of or like a shrub <*shrubby* trees> **—shrub'bi·ness** *n.*

**shrug** (shrŭg) *v.* **shrugged, shrug·ging, shrugs.** [ME *shruggen.*] *—vt.* To raise (the shoulders) as a gesture esp. of doubt, disdain, or indifference. *—vi.* To raise the shoulders as a gesture esp. of doubt, disdain, or indifference. **—shrug off. 1.** To minimize the importance of. **2.** To get rid of. **3.** To wriggle out of (clothing). *—n.* **1.** The gesture of shrugging. **2.** A woman's short jacket or sweater open down the front. **3.** A fur cape or stole.

**shrunk** (shrŭngk) *v. var. p.t. & p.p. of* SHRINK.

**shrunk·en** (shrŭng'kən) *v. var. p.p. of* SHRINK.

**shtetl** (shtĕt'l, shtāt'l) *n.* [Yiddish < MHG *stetel,* dim. of *stat,* town < OHG, place.] A small Jewish community once common in Eastern Europe.

**shtick** *also* **schtick** (shtĭk) *n.* [Yiddish *shtik* < MHG *stücke,* piece < OHG *stucki.*] *Slang.* **1.** A characteristic talent, attribute, or trait. **2.** A striking portion or detail. **3.** One's method of doing something. **4.** An entertainment routine.

**shuck** (shŭk) *n.* [Orig. unknown.] **1.** An outer covering, as of a pea, an ear of corn, or an oyster. **2.** *Slang.* A deception or sham. *—v.* **shucked, shuck·ing, shucks.** *—vt.* **1.** To remove the outer covering from <*shuck* oysters> **2.** *Informal.* To cast off (e.g., clothing). **3.** *Slang.* To deceive. *—vi. Slang.* To talk or behave deceptively. *—interj.* **shucks.** —Used to express mild disappointment, disgust, or annoyance. **—shuck'er** *n.*

**shud·der** (shŭd'ər) *vi.* **-dered, -der·ing, -ders.** [ME *shoddren.*] **1.** To tremble or shiver convulsively <*shuddered* in fear> **2.** To vibrate : quiver. **—shud'der** *n.* **—shud'der·ing·ly** *adv.*

**shuf·fle** (shŭf'əl) *v.* **-fled, -fling, -fles.** [Prob. < LG *schüffeln,* to walk clumsily.] *—vt.* **1.** To drag (the feet) along the floor or ground while walking : SCUFFLE. **2.** To move (something) from one place to another. **3.** To mix together in a disordered, haphazard fashion. **4.** To put aside or conceal hastily : COVER UP. **5.** To mix together (playing cards, tiles, or dominoes) to change their order of arrangement. *—vi.* **1.** To move with a shuffling gait. **2.** To dance the shuffle. **3.** To shift about from place to place. **4.** To behave in a shifty or deceitful way : EQUIVOCATE. **5.** To mix playing cards, tiles, or dominoes together to change their order of arrangement. *—n.* **1.** A shuffling gait or movement. **2.** A dance in which the feet scrape along the floor at each step. **3.** An evasive or deceitful action : EQUIVOCATION. **4. a.** The act of mixing cards, dominoes, or tiles. **b.** A player's right or turn to do this. **—shuf'fler** *n.*

**shuf·fle·board** (shŭf'əl-bôrd', -bōrd') *n.* [Alteration of obs. *shove-board* : SHOVE + BOARD.] **1.** A game in which disks are pushed or slid along a smooth level surface toward numbered squares with a pronged cue. **2.** The surface on which shuffleboard is played.

**shul** (sho͝ol, sho͞ol) *n.* [Yiddish < MHG *schuol,* school < OHG *scuola* < Lat. *scola.* —see SCHOOL[1].] A synagogue.

**shun** (shŭn) *vt.* **shunned, shun·ning, shuns.** [ME *shunnen* < OE *scunian,* to abhor.] To avoid deliberately and consistently <*shun* one's neighbors> **—shun'ner** *n.*

**shun·pike** (shŭn'pīk') *n.* A side road taken to avoid the tollgates on a major artery. *—vi.* **-piked, -pik·ing, -pikes.** To travel on a shunpike. **—shun'pik'er** *n.*

**shunt** (shŭnt) *n.* [ME *shunten,* to flinch.] **1.** The act or process of moving to an alternate course. **2.** A railroad switch. **3.** *Elect.* A low-resistance connection between two points in an electric circuit that forms an alternative path for part of the current. **4.** *Med.* A surgically created channel that diverts flow, as of blood, from one pathway to another or that allows flow from one region or part of the body to another. *—v.* **shunt·ed, shunt·ing, shunts.** *—vt.* **1.** To divert onto another course <*shunting* traffic around an accident> **2.** To shift or switch (a train or car) from one track to another. **3.** *Elect.* To provide or divert (current) by means of a shunt. **4.** To evade or avoid (e.g., a task) by refusing or putting aside. *—vi.* **1.** To move or turn aside. **2.** *Elect.* To become diverted by a shunt. —Used of a circuit. **3.** To shift one's views or direction. **—shunt'er** *n.*

**shunt-wound** (shŭnt'wound') *adj.* Of or relating to a direct-current motor or generator in which the field coil is connected in parallel with the armature so that the same voltage appears across each.

**shush** (shŭsh) *interj.* —Used to express a request for silence. *—vt.* **shushed, shush·ing, shush·es.** To request silence from by saying "shush."

**shut** (shŭt) *v.* **shut, shut·ting, shuts.** [ME *shutten* < OE *scyttan.*] *—vt.* **1.** To move (e.g., a door) into closed position over or within a conjoined aperture. **2.** To block passage or access to : CLOSE. **3.** To fasten with a lock, catch, or latch. **4.** To prevent or forbid access to. **5.** To close (a business establishment). *—vi.* To move or become moved to a closed position : CLOSE. **—shut down. 1.** To halt the operation of. **2.** To stop operating, esp. automatically. **—shut in.** To prevent egress from. **—shut off. 1.** To halt operation. **2.** To stop operating <a switch that *shuts off* automatically> **3.** To close off <hermits who *shut* themselves *off* from the rest of the world> **—shut out. 1.** To keep from entering. **2.** To prevent (a sports team) from scoring. **—shut up. 1.** To silence (a person). **2.** To stop speaking. *—n.* **1.** The act or time of shutting. **2.** The line of connection between welded pieces of metal.

**shut·down** (shŭt'doun') *n.* A cessation of operations or activity, as work in a factory.

**shut·eye** (shŭt'ī') *n. Slang.* SLEEP 1a, b.

**shut-in** (shŭt'ĭn') *n.* One confined indoors by illness or disability. *—adj.* (shŭt-ĭn'). **1.** Confined indoors, as by illness or disability. **2.** *Psychiat.* Disposed to avoid other people : excessively introverted.

**shut·off** (shŭt'ôf', -ŏf') *n.* **1.** A device that shuts something off. **2.** A stoppage.

**shut·out** (shŭt'out') *n.* **1.** A lockout. **2.** A sports contest in which one side does not score.

**shut·ter** (shŭt'ər) *n.* **1.** One that shuts, esp. a hinged cover or screen for a window, usu. fitted with louvers. **2. shutters.** The movable louvers on a pipe organ, controlled by pedals, that open and close the swell box. **3.** A device that opens and shuts a camera lens to expose a plate or film. *—vt.* **-tered, -ter·ing, -ters.** To furnish or close with shutters.

**shut·ter·bug** (shŭt'ər-bŭg') *n. Informal.* An amateur photographer.

**shut·tle** (shŭt'l) *n.* [ME *schutylle* < OE *scytel,* dart.] **1.** A device used in weaving to carry the woof thread back and forth between the warp threads. **2.** A device for holding the thread in tatting, in netting, and in a sewing machine. **3. a.** Regular travel back and forth over an established, often short route by a vehicle, as an aircraft. **b.** The route used by a vehicle that shuttles. **c.** A vehicle that frequently travels back and forth between points. *—v.* **-tled, -tling, -tles.** *—vi.* To go, move, or travel back and forth by or as if by a shuttle. *—vt.* **1.** To move or cause to move back and forth frequently. **2.** To transport by or as if by a shuttle.

**shut·tle·cock** (shŭt'l-kŏk') *n.* A small rounded piece of cork or rubber with a crown of feathers or plastic, used in badminton. *—vt.* **-cocked, -cock·ing, -cocks.** To throw or send back and forth like a shuttlecock.

**shuttle diplomacy** *n.* Diplomatic negotiations conducted by an official intermediary who travels frequently between the negotiating nations. **—shuttle diplomat** *n.*

**shy[1]** (shī) *adj.* **shi·er** or **shy·er, shi·est** or **shy·est.** [ME *schey* < OE *scēoh.*] **1.** Easily startled : TIMID. **2.** Bashful. **3.** Distrustful : wary. **4.** *Informal.* Not having paid an amount due, as one's ante in poker. **5.** *Informal.* Short : lacking <Twenty-three is one *shy* of two dozen.> *—vi.* **shied, shy·ing, shies. 1.** To move suddenly, as if

ă **pat** ā **pay** âr **care** ä **father** ĕ **pet** ē **be** hw **which** ĭ **pit** ī **tie** îr **pier** ŏ **pot** ō **toe** ô **paw,** for oi **noise** o͝o **took**

startled. **2.** To draw back, as from fear. —*n., pl.* **shies.** A sudden movement, as from fright. **—shy'er** *n.* **—shy'ly** *adv.* **—shy'ness** *n.*

**shy[2]** (shī) *v.* **shied, shy·ing, shies.** [Perh. < SHY.] —*vt.* To throw (something) with a swift sideways motion. —*vi.* To throw something with a swift sideways motion. —*n., pl.* **shies. 1.** A quick throw : FLING. **2.** *Informal.* A gibe : sneer. **3.** *Informal.* An experiment : TRY.

**shy·lock** (shī'lŏk') *n.* **1. Shylock.** The ruthless usurer in Shakespeare's play *The Merchant of Venice.* **2.** A loan shark. —*vi.* **-locked, -lock·ing, -locks.** To lend money at exorbitant interest rates.

**shy·ster** (shī'stər) *n.* [Perh. after *Scheuster,* an unscrupulous 19th-cent. lawyer.] *Slang.* An unethical, unscrupulous practitioner, esp. of the law.

**si** (sē) *n.* [Med. Lat. —see GAMUT.] *Mus.* Ti.

**Si** *symbol for* SILICON.

**si·al** (sī'ăl') *n.* [SI(LICON) + AL(UMINUM).] A layer of rock rich in silica and alumina underlying all continental land masses.

**si·a·lad·en·i·tis** (sī'ə-lăd'n-ī'tĭs) *n.* [Gk. *sialon,* saliva + *adēn,* gland + -ITIS.] Inflammation of a salivary gland.

**si·al·a·gogue** (sī-ăl'ə-gôg', -gŏg') *n.* [Gk. *sialon,* saliva + -AGOGUE.] An agent that increases saliva flow.

**si·al·ic acid** (sī-ăl'ĭk) *n.* [Gk. *sialon,* saliva + -IC.] Any of a group of amino carbohydrates that are found as components of mucoproteins, lipids, and polysaccharides in bacteria and animal tissue.

**si·a·mang** (sē'ə-măng', sē-ăm'əng) *n.* [Malay.] A large black gibbon, *Symphalangus syndactylus* or *Hylobates syndactylus* of Sumatra and the Malay Peninsula, with an inflatable throat sac.

**Si·a·mese** (sī'ə-mēz', -mēs') *adj.* [After *Siam* (Thailand).] **1.** Thai. **2.** Very similar or closely connected : TWIN. **3. siamese.** Connecting two hoses or pipes to a larger hose or pipe. —*n., pl.* **Siamese.** Thai.

**Siamese cat** *n.* A short-haired cat orig. bred in the Orient, having blue eyes and a pale fawn or gray coat with darker ears, face, tail, and feet.

**Siamese fighting fish** *n.* A small, often brightly colored freshwater fish, *Betta splendens* native to tropical Asia, and popular in home aquariums.

**Siamese twin** *n.* [After Chang and Eng (1811–1874), joined Chinese twins born in *Siam* (Thailand).] One of a pair of twins born with their bodies joined together.

**sib** (sĭb) *n.* [ME *sibbe* < OE *sibb.*] **1. a.** A blood relation. **b.** Relatives as a group : KINFOLK. **2.** A sibling. **—sib** *adj.*

**Si·be·ri·an husky** (sī-bîr'ē-ən) *n.* HUSKY[3] 1.

**sib·i·lant** (sĭb'ə-lənt) *adj.* [Lat. *sibilans, sibilant-,* pr.part. of *sibilare,* to hiss.] Of, characterized by, or producing a hissing sound, as the sound of (s) or (sh). —*n.* **1.** A speech sound, as (s), (sh), (z), or (zh), that suggests hissing. **2.** A sibilant consonant. **—sib'i·lance, sib'i·lan·cy** *n.* **—sib'i·lant·ly** *adv.*

**sib·i·late** (sĭb'ə-lāt') *vi. & vt.* **-lat·ed, -lat·ing, -lates.** [Lat. *sibilare, sibilat-,* to hiss.] To utter or pronounce with a hiss. **—sib'i·la'tion** *n.*

**sib·ling** (sĭb'lĭng) *n.* [ME *siblyng* < OE *sibling* < *sibb,* kinsman.] One of two or more persons having one or both parents in common.

**sib·yl** (sĭb'əl) *n.* [ME *Sibile* < OFr. < Lat. *Sibylla* < Gk. *Sibulla.*] **1.** A woman considered to be an oracle or prophetess by the ancient Greeks and Romans. **2.** A woman prophet.

**sib·yl·line** (sĭb'ə-līn', -lēn') *also* **si·byl·ic** *or* **si·byl·lic** (sĭ-bĭl'ĭk) *adj.* **1.** Relating to, coming from, or characteristic of a sibyl. **2.** Prophetic <*sibylline* pronouncements>

**sic[1]** (sĭk) *adv.* [Lat.] Thus : so. —Used in written texts to indicate that a surprising or paradoxical word, phrase, or fact is not a mistake and is to be read as it stands.

**sic[2]** *also* **sick** (sĭk) *vt.* **sicced, sic·cing, sics** *also* **sicked, sick·ing, sicks.** [Dial. var. of SEEK.] **1.** To urge to attack or chase <*sicced* the vicious dog on us> **2.** To set upon or chase.

**sic·ca·tive** (sĭk'ə-tĭv) *n.* [Lat. *siccativus,* drying < *siccare,* to dry < *siccus,* dry.] An additive to paints and some medicines that promotes drying.

**sick[1]** (sĭk) *adj.* **-er, -est.** [ME *sek* < OE *sēoc.*] **1. a.** Afflicted with a physical illness. **b.** Nauseated. **2.** Of or for sick persons <hospital *sick* wards> **3. a.** Mentally ill. **b.** Morbid : unwholesome <*sick* jokes> **c.** Unsound : defective <a *sick* stock market> **4. a.** Distressed : upset. **b.** Disgusted. **c.** Tired : weary <*sick* of this job> **d.** Longing : pining <*sick* for their native land> **5.** Being in need of repairs. —Used of a ship. **6.** Unable to produce a profitable crop. **—sick and tired.** Utterly weary, discouraged, or bored <*sick and tired* of disorganization>

☆ ***syns:*** SICK, ILL, INDISPOSED, UNWELL *adj. core meaning* : not in good physical or mental condition <a *sick* patient> SICK, ILL, and UNWELL are used interchangeably. INDISPOSED refers to minor sickness <Although somewhat *indisposed,* the pianist did not cancel the recital.> ***ant:*** healthy, well

**sick[2]** (sĭk) *v. var. of* SIC[2].

**sick·bay** (sĭk'bā') *n.* **1.** A ship's hospital and dispensary. **2.** A place for the treatment of sick or injured patients.

**sick·bed** (sĭk'bĕd') *n.* A sick person's bed.

**sick call** *n.* **1.** The daily line-up of military personnel requiring medical attention. **2.** The signal announcing sick call.

**sick·ee** (sĭk'ē) *n. var. of* SICKIE.

**sick·en** (sĭk'ən) *vt. & vi.* **-ened, -en·ing, -ens.** To make or become sick. **—sick'en·er** *n.*

**sick·en·ing** (sĭk'ə-nĭng) *adj.* **1.** Producing sickness. **2.** Revolting : disgusting <a *sickening* crime> **—sick'en·ing·ly** *adv.*

**sick headache** *n.* A headache accompanied by nausea.

**sick·ie** *also* **sick·ee** (sĭk'ē) *n. Slang.* An emotionally sick person.

**sick·ish** (sĭk'ĭsh) *adj.* **1.** Slightly ill. **2.** Slightly nauseated. **3.** Slightly revolting. **—sick'ish·ly** *adv.* **—sick'ish·ness** *n.*

**sick·le** (sĭk'əl) *n.* [ME *sikel* < OE *sicol* < Lat. *secula.*] **1.** An implement with a semicircular blade attached to a short handle, used for cutting grain or tall grass. **2.** The cutting mechanism of a reaper or mower. —*vt.* **-led, -ling, -les.** To cut (e.g., grass) with a sickle.

**sick leave** *n.* Paid absence from work allowed an employee because of sickness.

**sick·le·bill** (sĭk'əl-bĭl') *n.* A bird having a sharply curved bill, esp. *Falculea palliata* of Madagascar.

**sickle cell** *n.* An abnormal crescent-shaped red blood cell.

**sickle cell anemia** *n.* A hereditary anemia marked by the presence of oxygen-deficient sickle cells, episodic pain, and leg ulcers.

**sickle feather** *n.* A long curving feather in the tail of a cock.

**sick·le·mi·a** (sĭk'ə-lē'mē-ə) *n.* Presence of sickle cells in blood.

**sick list** *n.* A list of sick personnel, as in the military.

**sick·ly** (sĭk'lē) *adj.* **-li·er, -li·est. 1.** Prone to sickness <a *sickly* child> **2.** Of, caused by, or associated with sickness <a face with a *sickly* pallor> **3.** Conducive to sickness : UNHEALTHFUL <a *sickly,* damp climate> **4.** Nauseating : sickening. **5.** Feeble : weak <a *sickly* handshake> —*vt.* **-lied, -ly·ing, -lies.** To make sickly. **—sick'li·ness** *n.* **—sick'ly** *adv.*

**sick·ness** (sĭk'nĭs) *n.* **1.** Illness. **2.** A disease. **3.** Nausea.

**sick-out** (sĭk'out') *n.* An organized action by employees who claim illness and absent themselves from work, usu. to force the granting of demands or to avoid being penalized for striking formally.

**sick pay** *n.* Wages paid to an employee absent on sick leave.

**sick·room** (sĭk'ro͞om', -ro͝om') *n.* A room that is occupied by a sick person.

**sic pas·sim** (sĭk păs'ĭm) *adv.* [Lat.] Thus everywhere. —Used to indicate that a term or idea is to be found throughout a text.

**sid·dur** (sĭd'ər, -o͞or') *n., pl.* **sid·du·rim** (sĭ-do͞or'ĭm) [Heb. *siddūr,* arrangement < *siddēr,* he arranged.] A Jewish prayer book containing prayers for the various days of the year.

**side** (sīd) *n.* [ME < OE *sīde.*] **1.** *Math.* **a.** A line bounding a plane figure. **b.** A surface bounding a solid figure. **2.** A surface of an object, esp. one joining a top and bottom. **3.** A surface of an object that extends more or less perpendicularly from an observer standing in front <the *side* of a building> **4.** Either of the two surfaces of a flat object, as a piece of paper. **5. a.** The part within an object or area to the left or right of the observer or of its vertical axis. **b.** The left or right half of the trunk of a human or animal body. **6.** The space next to someone or something <stood at my *side*> **7.** One of two or more contrasted parts or places within an area, identified by its location with respect to a center <the north *side* of the park> **8.** An area that is separated from another area by an intervening feature, as a barrier or line <on this *side* of the ocean> **9. a.** One of two or more opposing groups, teams, or sets of opinions <the losing *side*> **b.** A position maintained in a dispute or debate. **10.** A distinct aspect <the gentle *side* of my nature> **11.** Line of descent. **12.** *Chiefly Brit.* Nerve : swagger. —*adj.* **1.** Located on a side <a *side* entrance> **2.** From or to one side : OBLIQUE <a *side* view> **3.** Minor : incidental <*side* interests> **4.** Being in addition to the main part : SUPPLEMENTARY <*side* benefits> —*v.* **sid·ed, sid·ing, sides.** —*vt.* **1.** To be in agreement with : SUPPORT. **2.** To be positioned next to. **3.** To provide sides or siding for <*side* a house> —*vi.* To align oneself with a particular side <*sided* with me in the argument> **—on the side. 1.** In addition to the main part. **2.** In addition to the main occupation or arrangement <took in washing *on the side*> **—side by side.** Next to each other.

**side·arm** (sīd'ärm') *adj. Baseball.* Thrown with or marked by a sweep of the arm between shoulder and hip height <a *sidearm* curve ball> **—side'arm** *adv.*

**side arm** *n.* A small weapon, as a pistol, carried at the side or waist.

**side·band** *also* **side band** (sīd'bănd') *n.* Either of the two bands of frequencies, one just above and one just below a carrier frequency, that result from modulation of a carrier wave.

**side·bar** (sīd'bär') *n.* A short news story that accompanies and presents sidelights of a major news story.

**side·board** (sīd'bôrd', -bōrd') *n.* A piece of dining-room furniture having drawers and shelves for linens and tableware.

**side·burns** (sīd'bûrnz') *pl.n.* [Alteration of BURNSIDES.] Growths of hair down the sides of a man's face in front of the ears, esp. when worn with the rest of the beard shaved off.

**side·car** (sīd'kär') *n.* **1.** A one-wheeled car for a single passenger, attached to the side of a motorcycle. **2.** A cocktail combining brandy, an orange-flavored liqueur, and lemon juice.

o͞o **boot** ou **out** th **thin** *th* **this** ŭ **cut** û **r urge** y **young**
yo͞o abuse zh vision ə **a**bout, it**e**m, edib**i**le, gall**o**p, circ**u**s

**sid·ed** (sīʹdĭd) *adj.* Having sides usu. of a specified number or kind <many-*sided*><marble-*sided*> **—sidʹed·ness** *n.*
**side dish** *n.* A dish served along with a main course.
**side drum** *n.* A snare drum.
**side effect** *n.* A peripheral or secondary effect, esp. an undesirable secondary effect of a drug.
**side-glance** (sīdʹglăns′) *n.* **1.** A glance cast to the side. **2.** An indirect reference : ALLUSION.
**side issue** *n.* A secondary issue.
**side·kick** (sīdʹkĭk′) *n. Slang.* A close friend and associate.
**side·light** (sīdʹlīt′) *n.* **1.** A light coming from the side. **2.** *Naut.* Either of two lights, red to port, green to starboard, shown by ships at night. **3.** Incidental information.
**side·line** *also* **side line** (sīdʹlīn′) *n.* **1. a.** A line along either of the two sides of a playing court or field, marking its limits. **b. sidelines.** The space outside such limits, occupied by spectators and inactive players. **c. sidelines.** The position or viewpoint of nonparticipants in an activity. **2.** A subsidiary line of merchandise. **3.** An activity pursued in addition to one's regular occupation. *—vt.* **-lined, -lin·ing, -lines.** To remove or keep from active participation, as in athletic contests.
**side·lin·er** (sīdʹlī′nər) *n.* A nonparticipant in a game or an activity.
**side·ling** (sīdʹlĭng) *adj.* [ME *sideling* < *side,* side.] **1.** Directed to one side : OBLIQUE. **2.** Inclined : sloping. *—adv.* Obliquely : sideways.
**side·long** (sīdʹlông′, -lŏng′) *adj.* [Alteration of SIDELING.] **1.** Directed to one side <a *sidelong* glance> **2.** Sloping : slanting. *—adv.* **1.** On or toward the side : SIDEWAYS. **2.** Obliquely.
**side·man** (sīdʹmăn′) *n.* A jazz instrumentalist.
**sider-** *pref. var. of* SIDERO-.
**si·de·re·al** (sī-dîrʹē-əl) *adj.* [< Lat. *siderus* < *sidus,* constellation.] **1.** Of, relating to, or concerned with the stars or constellations : STELLAR. **2.** Determined or measured by means of the stars <*sidereal* time> **3.** Relative to the stars.
**sidereal day** *n.* The time required for a complete rotation of the earth, measured as the interval between two successive meridian transits of the vernal equinox, or 23 hours, 56 minutes, 4.09 seconds in units of mean solar time.
**sidereal hour** *n.* A 24th part of a sidereal day.
**sidereal month** *n.* MONTH 4.
**sidereal time** *n.* Time based on the axial and orbital rotation of the earth with reference to the background of stars.
**sidereal year** *n.* The time required for one complete revolution of the earth about the sun, relative to the fixed stars, or 365 days, 6 hours, 9 minutes, 9.54 seconds in units of mean solar time.
**sid·er·ite** (sidʹə-rīt′) *n.* **1.** An impure yellowish-brown iron carbonate mineral. **2.** An iron meteorite. **—sidʹer·itʹic** (-rĭtʹĭk) *adj.*
**sidero-** *or* **sider-** *pref.* [< Gk. *sidēros,* iron.] Iron <*sidero*lite>
**sid·er·o·lite** (sidʹər-ə-līt′) *n.* A meteorite containing iron, nickel, silicon, magnesium, and small amounts of other elements.
**sid·er·o·sil·i·co·sis** (sidʹə-rō-sĭlʹĭ-kōʹsĭs) *n.* Pneumoconiosis caused by excessive inhalation of silica and iron dust.
**sid·er·o·sis** (sidʹə-rōʹsĭs) *n.* Chronic inflammation of the lungs brought about by excessive inhalation of dust containing iron salts or particles.
**sid·er·o·stat** (sidʹər-ə-stăt′) *n.* [Lat. *sidus, sider-,* constellation + -STAT.] An optical system consisting of a rotating clock-driven mirror that reflects light from a celestial body in a relatively fixed direction to a fixed telescope or other bulky astronomical instrument.
**side·sad·dle** (sīdʹsăd′l) *n.* A saddle designed so that a woman may sit with both legs on one side of the horse. **—sideʹsadʹdle** *adv.*
**side show** *n.* **1.** A small show offered in addition to the main attraction, as at a circus. **2.** A diverting subordinate event.
**side·slip** (sīdʹslĭp′) *vi.* **-slipped, -slip·ping, -slips. 1.** To slip to one side. **2.** To slide sideways and downward in skiing. **3.** To fly sideways and downward in an aircraft along the lateral axis to reduce altitude without gaining speed or as the result of excessively deep banking.
**side·spin** (sīdʹspĭn′) *n.* A rotary motion that spins a ball horizontally.
**side·split·ting** (sīdʹsplĭt′ĭng) *adj.* Hilarious <*sidesplitting* jokes> **—sideʹsplitʹting·ly** *adv.*
**side·step** (sīdʹstĕp′) *v.* **-stepped, -step·ping, -steps.** *—vi.* **1.** To step aside. **2.** To dodge an issue or responsibility. *—vt.* **1.** To step out of the way of. **2.** To evade : skirt <*sidestep* an embarrassing question> **—sideʹstepʹper** *n.*
**side step** *n.* **1.** A step to one side. **2.** A step taken sideways.
**side-strad·dle hop** (sīdʹstrăd′l) *n.* JUMPING JACK 2.
**side stroke** *n.* A swimming stroke in which the swimmer swims on one side and thrusts his or her arms forward alternately while performing a scissors kick.
**side·swipe** (sīdʹswīp′) *vt.* **-swiped, -swip·ing, -swipes.** To strike along the side of in passing. **—sideʹswipe′** *n.*
**side·track** (sīdʹtrăk′) *v.* **-tracked, -track·ing, -tracks.** *—vt.* **1.** To switch from a main railroad track to a siding. **2.** To divert from a main issue or course. *—vi.* **1.** To run into a railroad siding. **2.** To deviate from a main issue or course. **—sideʹtrack′** *n.*
**side·walk** (sīdʹwôk′) *n.* A walk for pedestrians along the side of a road.
**sidewalk artist** *n.* An artist who draws pictures, usu. with chalk, on a sidewalk as a way of acquiring money from passers-by.
**sidewalk superintendent** *n. Informal.* A pedestrian who stops to watch construction or demolition work.
**side wall** *n.* A side surface of an automotive tire.
**side·ward** (sīdʹwərd) *adj. & adv.* At or toward one side. **—sideʹwards** (-wərdz) *adv.*
**side·ways** (sīdʹwāz′) *also* **side·way** (-wā′) *or* **side·wise** (-wīz′) *adv.* **1.** From one side. **2.** In a sideward direction : toward one side. **3.** Presenting the side instead of the front or back. *—adj.* Toward or from one side.
**side-wheel** (sīdʹhwēl′, -wēl′) *adj.* Of, relating to, or being a steamboat with a paddle wheel on each side.
**side-wheel·er** (sīdʹhwē′lər, -wē′-) *n.* A side-wheel steamboat.
**side-whis·kers** (sīdʹhwĭs′kərz, -wĭs′-) *pl.n.* Whiskers worn usu. long on the sides of a man's face.
**side·wind·er** (sīdʹwīn′dər) *n.* **1.** A small rattlesnake, *Crotalus cerastes* of the southwestern United States and Mexico, that moves by a distinctive lateral, looping bodily motion. **2.** A powerful blow by the fist delivered from the side. **3.** A short-range supersonic air-to-air missile.
**side·wise** (sīdʹwīz′) *adv. & adj. var. of* SIDEWAYS.
**sid·ing** (sīʹdĭng) *n.* **1.** A short section of railroad track linked to a main track by switches. **2.** Material for surfacing a frame building.
**si·dle** (sīdʹl) *v.* **-dled, -dling, -dles.** [Back-formation < SIDELING.] *—vi. & vt.* To move or cause to move sidewise, esp. furtively or indirectly. **—siʹdle** *n.* **—siʹdling·ly** *adv.*
**siege** (sēj) *n.* [ME *sege* < OFr., seat < VLat. ***sedicum** < Lat. *sedēre,* to be seated.] **1.** Encirclement and blockade of a town or fortress by an army determined to capture it. **2.** A prolonged period, as of illness or distress. **3.** *Obs.* **a.** A seat or throne. **b.** Rank : position. *—vt.* **sieged, sieg·ing, sieg·es.** To besiege.
**Siege Perilous** *n.* A seat at King Arthur's Round Table kept for the knight who was destined to find the Holy Grail and fatal for any other occupant.
**Sieg·fried** (sēgʹfrēd′, sĭgʹ-) *n.* [G. < OHG *Sigifrith* : *sigu,* victory + *fridu,* peace.] A principal character of the *Nibelungenlied.*
**sie·mens** (sēʹmənz) *n.* [After Werner von *Siemens* (1816–1892).] A unit of conductance equal to one ampere per volt.
**si·en·na** (sē-ĕnʹə) *n.* [Short for *terra-sienna* < Ital. *terra di Sienna,* earth of Siena < *Sienna,* Siena, a city in Italy.] **1.** A special clay containing iron and manganese oxides, used as a pigment for oil and water-color painting. **2.** RAW SIENNA 2. **3.** Burnt sienna.
**si·er·o·zem** (sī-ĕrʹə-zĕm′, sē-ĕrʹə-zhôm′) *n.* [R. *serozem* : *seryĭ,* gray + *zemlya,* earth.] Soil found in cool to temperate arid regions that is brownish-gray at the surface with a lighter layer below and is based in a carbonate or hardpan layer.
**si·er·ra** (sē-ĕrʹə) *n.* [Sp. < Lat. *serra,* saw.] **1.** A rugged range of mountains having an irregular or serrated configuration. **2.** Any of several mackerellike fishes of the genus *Scomberomorus,* of tropical seas. **—si·erʹran** *adj.*
**si·es·ta** (sē-ĕsʹtə) *n.* [Sp. < Lat. *sexta (hora),* sixth (hour) < *sextus,* sixth.] A rest or nap, usu. taken after the midday meal.
**sieve** (sĭv) *n.* [ME *sive* < OE *sife.*] A wire mesh utensil used for straining, sifting, ricing, or puréeing. *—vt. & vi.* **sieved, siev·ing, sieves.** To pass through a sieve or to sift something.
**sieve tube** *n.* A series of cells joined end to end, forming a tube through which food is conducted in vascular plants.
**sift** (sĭft) *v.* **sift·ed, sift·ing, sifts.** [ME *siften* < OE *siftan.*] *—vt.* **1.** To put through a straining device, as a sieve, so as to separate the fine from the coarse particles. **2.** To distinguish as if separating with a sieve <*sifted* the candidates for admission> **3.** To apply by scattering with or as if with a sieve. **4.** To examine closely <*sift* all the evidence> *—vi.* **1.** To put something through a sieve. **2.** To pass through or as if through a sieve. **3.** To make a close examination <*sifted* through all the evidence> **—siftʹer** *n.*
**sift·ing** (sĭfʹtĭng) *n. Computer Sci.* An internal sorting technique in which data are displaced to permit the insertion of new data.
**sift·ings** (sĭfʹtĭngz) *pl.n.* Material separated with or as if with a sieve.
**sigh** (sī) *v.* **sighed, sigh·ing, sighs.** [ME *sighen,* alteration of *siken,* to sigh < OE *sīcan.*] *—vi.* **1. a.** To exhale audibly in a long deep breath, as in weariness or relief. **b.** To emit a sound similar to a sigh <willows *sighing* in the wind> **2.** To feel yearning, longing, or grief : MOURN. *—vt.* **1.** To express with or as if with an audible exhalation <*sighed* assent> **2.** *Archaic.* To lament. **—sigh** *n.* **—sighʹer** *n.*
**†sight** (sīt) *n.* [ME < OE *gesihð,* something seen.] **1.** Ability to see. **2.** The act or fact of seeing. **3.** Field of vision. **4.** The foreseeable future : PROSPECT <no end in *sight*> **5.** Something seen. **6.** Something worth seeing <the *sights* of Moscow> **7.** *Informal.* Something unsightly <My hair was a *sight.*> **8. a.** A device used to assist aim by guiding the eye, as on a firearm. **b.** Aim or observation taken with such a device. **9.** An opportunity to observe or inspect. **10.** *Regional.* A large number or quantity <a *sight* of people at the barbecue>

—*vt.* **sight·ed, sight·ing, sights. 1.** To see or observe within one's field of vision <*sight* land> **2.** To observe or take a sight of with an instrument <*sight* a target> **3.** To adjust the sights of (e.g., a rifle). **4.** To take aim with (a firearm). **—out of sight.** *Slang.* Incredible : remarkable. **—sight for sore eyes.** *Informal.* One whose arrival is a cause for joy. **—sight unseen.** Without seeing the object in question <buy a house *sight unseen*>

**sight draft** *n.* A draft or bill payable upon demand or presentation.

**sight·ed** (sī'tĭd) *adj.* **1.** Having sight. **2.** Having eyesight of a specified kind <keen-*sighted*>

**sight gag** *n.* A comic bit or effect that depends on action rather than words.

**sight·less** (sīt'lĭs) *adj.* **1.** Blind. **2.** Invisible. **—sight'less·ly** *adv.* **—sight'less·ness** *n.*

**sight·ly** (sīt'lē) *adj.* **-li·er, -li·est. 1.** Pleasing to the eye : HANDSOME. **2.** Providing a fine view. **—sight'li·ness** *n.*

**sight-read** (sīt'rēd') *v.* **-read** (-rĕd'), **-read·ing, -reads.** —*vt.* To read or perform (e.g., music) without preparation or prior acquaintance. —*vi.* To read or perform something at sight.

**sight rhyme** *n.* An eye rhyme.

**sight·see** (sīt'sē') *vi.* **-saw** (-sô'), **-seen** (-sēn'), **-see·ing, -sees.** To engage in sightseeing. **—sight'se'er** *n.*

**sight·see·ing** (sīt'sē'ĭng) *n.* The act or pastime of touring places of interest. **—sight'see'ing** *adj.*

**sig·il** (sĭj'əl, sĭg'ĭl) *n.* [Lat. *sigillum*, dim. of *signum*, sign.] **1.** A seal or signet. **2.** A sign or image held to be magical.

**sig·ma** (sĭg'mə) *n.* [Gk., of Phoenician orig : akin to Heb. *sāmehk*, samek.] **1.** The 18th letter of the Greek alphabet. —See table at ALPHABET. **2.** *Physics.* Any of three subatomic particles in the baryon family. **—sig'mate'** (-māt') *adj.*

**sig·moid** (sĭg'moid') *also* **sig·moi·dal** (sĭg-moid'l) *adj.* [Gk. *sigmoeidēs* : *sigma*, sigma + *-eidēs*, -oid.] **1.** Shaped like the letter S. **2.** *Physiol.* Of or relating to the sigmoid flexure of the colon.

**sigmoid flexure** *n.* An S-shaped bend in the colon between the descending section and the rectum.

**sign** (sīn) *n.* [ME *signe* < OFr. < Lat. *signum*.] **1.** Something suggesting the presence of a fact, condition, or quality. **2.** A gesture or action used to convey an idea, a desire, data, or a command <gave us the go-ahead *sign*> **3.** A board, poster, or placard displayed in a public place to advertise, impart information, or give directions. **4.** A conventional symbolic device standing for a word, phrase, or operation, as in mathematics or musical notation. **5.** *pl.* **sign.** An indicator, as a spoor or scent, of the presence or trail of an animal <a deer *sign*> **6.** A trace : vestige <not a *sign* of life> **7.** A portentous incident or event : PRESAGE <*signs* of a long hot summer> **8.** A bodily manifestation indicating the presence of a disease or malfunction <Shortness of breath is a *sign* of heart trouble.> **9.** One of the 12 divisions of the zodiac, each named for a constellation and represented by a symbol. —*v.* **signed, sign·ing, signs.** —*vt.* **1.** To affix one's signature to. **2.** To write (one's signature). **3.** To approve or ratify (a document) by affixing a signature or seal <*signed* the bill into law> **4.** To relinquish or transfer title to by signature <*signed* away my claim to the estate> **5.** To express or signify with a sign. **6.** To consecrate with the sign of the cross. —*vi.* **1.** To make a sign or signs : SIGNAL. **2.** To write one's signature. **—sign in.** To record the arrival of by signing. **—sign off.** To stop broadcasting after identifying one's station. **—sign on. 1.** To enlist oneself <I *signed on* as a deck hand.> **2.** To start broadcasting after identifying one's station. **—sign out.** To record the departure of by signing. **—sign up.** To volunteer one's services : ENLIST. **—sign'er** *n.*

☆ ***syns:*** SIGN, EVIDENCE, INDICATION, INDICATOR, MANIFESTATION, MARK, SYMPTOM, TOKEN *n. core meaning* : something visible or evident that gives grounds for believing in the existence of something else <intolerance as a *sign* of bigotry>

**sig·nal** (sĭg'nəl) *n.* [Fr. < OFr. < Med. Lat. *signale* < Lat. *signalis*, of a sign < *signum*, sign.] **1. a.** An indicator, as a mechanical device, functioning as a means of communication. **b.** A message communicated by such means. **2.** Something that incites action <The tax increase was a *signal* for mass protests.> **3.** *Electron.* An impulse or fluctuating electric quantity, as voltage, current, or electric field strength, whose variations represent coded information. **4.** The sound, image, or message transmitted or received in telegraphy, telephony, radio, television, or radar. —*adj.* Extraordinary <a *signal* accomplishment in diplomacy> —*v.* **-naled, -nal·ing, -nals** *or* **-nalled, -nal·ling, -nals.** —*vt.* **1.** To make a signal to. **2.** To communicate by signals. —*vi.* To make a signal. **—sig'nal·er** *n.*

**sig·nal·ize** (sĭg'nə-līz') *vt.* **-ized, -iz·ing, -iz·es. 1.** To render noteworthy. **2.** To draw attention to. **—sig'nal·i·za'tion** *n.*

**sig·nal·ly** (sĭg'nə-lē) *adv.* Conspicuously.

**sig·nal·ment** (sĭg'nəl-mənt) *n.* [Fr. *signalement* < *signaler*, to mark out < *signal*, signal.] A description detailing a person's appearance and features, as for police files.

**sig·na·to·ry** (sĭg'nə-tôr'ē, -tōr'ē) *adj.* [Lat. *signatorius* < *signare*, to mark < *signum*, sign.] Bound by a signed agreement. —*n.*, *pl.* **-ries.** A signer of a document.

**sig·na·ture** (sĭg'nə-chər) *n.* [OFr. < *signer*, to sign < Lat. *signare*, to mark < *signum*, sign.] **1.** The name of one as written by oneself. **2.** A distinctive mark, characteristic, modus operandi, or sound effect indicating identity. **3.** The act of signing one's name. **4.** The part of a physician's prescription containing directions to the patient. **5.** *Mus.* **a.** A sign used to indicate key. **b.** A sign used to indicate tempo. **6. a.** A letter, number, or symbol placed at the bottom of the first page of each form of printed pages of a book as a guide to the proper sequence of the sheets in binding. **b.** A large sheet printed with four or a multiple of four pages that when folded becomes a section of the book.

**sign·board** (sīn'bôrd', -bōrd') *n.* A board that bears a sign, notice, or advertisement.

**sig·net** (sĭg'nĭt) *n.* [ME < OFr., dim. of *signe*, sign.] **1.** A seal, esp. one used on a document. **2.** An impression made with a signet. —*vt.* **-net·ed, -net·ing, -nets.** To mark with a signet.

**signet ring** *n.* A finger ring that bears an engraved signet.

**sig·nif·i·cance** (sĭg-nĭf'ĭ-kəns) *also* **sig·nif·i·can·cy** (-kən-sē) *n.* **1.** The quality or state of being significant. **2.** Import : meaning. **3.** Implied meaning.

**significance level** *n.* Level of significance.

**sig·nif·i·cant** (sĭg-nĭf'ĭ-kənt) *adj.* [Lat. *significans*, *significant-*, pr.part. of *significare*, to signify.] **1.** Having or expressing a meaning : MEANINGFUL. **2.** Having or expressing a covert meaning : SUGGESTIVE <gave me a *significant* look> **3.** Momentous : important <a *significant* news story> **—sig·nif'i·cant·ly** *adv.*

**significant digits** *pl.n. Math.* The digits of the decimal form of a number beginning with the leftmost nonzero digit and extending to the right to include all digits warranted by the accuracy of measuring devices used to obtain the numbers.

**sig·ni·fi·ca·tion** (sĭg'nə-fĭ-kā'shən) *n.* **1.** Intended meaning : SENSE. **2.** The act of signifying : INDICATION.

**sig·nif·i·ca·tive** (sĭg-nĭf'ĭ-kā'tĭv) *adj.* Significant. **—sig·nif'i·ca'tive·ness** *n.*

**sig·ni·fy** (sĭg'nə-fī') *v.* **-fied, -fy·ing, -fies.** [ME *signifien* < OFr. *signifier* < Lat. *significare* : *signum*, sign + *facere*, to make.] —*vt.* **1.** To serve as a sign of : BETOKEN. **2.** To make known : INTIMATE. —*vi.* To have meaning or import. **—sig'ni·fi'er** *n.*

**si·gnior** (sēn-yôr', -yōr') *n.* Signor.

**si·gnio·ry** (sēn'yə-rē) *n. var. of* SIGNORY.

**sign language** *n.* A system of communication by means of hand gestures, used esp. by deaf people.

**sign manual** *n.*, *pl.* **signs manual.** A personal signature, esp. that of a sovereign at the top of a royal decree.

**sign of the cross** *n.* A gesture forming a cross, made in token of faith in Christ or as a blessing.

**si·gnor** (sēn-yôr', -yōr') *n.*, *pl.* **si·gno·ri** (sēn-yôr'ē, -yōr'ē) *or* **si·gnors.** [Ital. *signor*, var. of *signore*.] —Used as a courtesy title for an Italian man, equivalent to the English *Mr.* or *Sir.*

**si·gno·ra** (sēn-yôr'ə, -yōr'ə) *n.*, *pl.* **si·gno·re** (sēn-yôr'ā, -yōr'ā) *or* **si·gno·ras.** [Ital., fem. of *signore*, signore.] —Used as a courtesy title for a married Italian woman, equivalent to the English *Mrs.* or *Madam.*

**si·gno·re** (sēn-yôr'ā, -yōr'ā) *n.*, *pl.* **si·gno·ri** (sēn-yôr'ē, -yōr'ē) [Ital. < Med. Lat. *senior*, lord < Lat., elder. —see SENIOR.] —Used as a courtesy title for an Italian man, equivalent to the English *Mr.* or *Sir.*

**si·gno·ri** (sēn-yôr'ē, -yōr'ē) *n. var. pl. of* SIGNOR.

**si·gno·ri·na** (sēn'yə-rē'nə) *n.*, *pl.* **-ne** (-nā) *or* **-nas.** [Ital., dim. of *signora*, signora.] —Used as a courtesy title for an unmarried Italian woman, equivalent to the English *Miss.*

**si·gno·ry** *or* **si·gnio·ry** (sēn'yə-rē) *n.*, *pl.* **-ries.** [ME *signorie* < OFr. *seigneurie* < *seigneur*, seignior. —see SEIGNIOR.] A seigniory.

**sign·post** (sīn'pōst') *n.* **1.** A post supporting a sign. **2.** Something serving as an indication or guide.

**Sig·urd** (sĭg'ərd) *n.* [ON *Sigurðr.*] *Norse Myth.* A hero who killed the dragon Fafnir.

**Sikh** (sēk) *n.* [Hindi < Skt. *śiṣyaḥ*, pupil < *śikṣati*, he wishes to learn, desiderative of *śaknōti*, he is able.] One who is an adherent of Sikhism. **—Sikh** *adj.*

**Sikh·ism** (sēk'ĭz'əm) *n.* The doctrines and practices of a monotheistic Hindu religious sect founded in the 16th cent.

**si·lage** (sī'lĭj) *n.* Fodder prepared by storing and fermenting green forage plants in a silo.

**sil·ane** (sĭl'ān', sī'lān') *n.* [SIL(ICON) + (METH)ANE.] Any of a group of silicon hydrides with the general formula SiH that are analogous to the paraffin hydrocarbons.

**sild** (sĭld) *n.* [Norw.] A young herring other than a sprat that is processed as a sardine in Norway.

**si·lence** (sī'ləns) *n.* [ME < OFr. < Lat. *silentium* < *silēre*, to be silent.] **1.** The quality or state of being or keeping silent. **2.** Absence of sound : STILLNESS. **3.** A time period without speech or noise. **4.** Refusal or failure to speak out : SECRECY. —*vt.* **-lenced, -lenc·ing, -lenc·es. 1.** To make silent or bring to silence. **2.** To curtail the expression of : SUPPRESS <*silenced* all dissent>

**si·lenc·er** (sī'lən-sər) *n.* One that silences, esp. a device attached to the muzzle of a firearm to muffle the sound of firing.

**si·le·ni** (sī-lē′nī′) *n. pl. of* SILENUS.

**si·lent** (sī′lənt) *adj.* [Lat. *silens, silent-,* pr.part. of *silēre,* to be silent.] **1.** Making no sound or noise : QUIET. **2.** Tending not to speak : TACITURN. **3.** Unable to speak : MUTE. **4.** Refusing to give information or an opinion : SECRETIVE. **5.** Unexpressed : tacit <a *silent* admission of guilt> **6.** Inactive or undisturbed : QUIESCENT <a *silent* volcano> **7.** Having no phonetic value : unpronounced, as the *l* in *solder.* **8.** Having no sound track <an old *silent* film> **—si′lent·ly** *adv.* **—si′lent·ness** *n.*

☆ ***syns:*** SILENT, HUSHED, NOISELESS, QUIET, SOUNDLESS, STILL *adj. core meaning* : marked by, done with, or making no sound or noise <a *silent* reply><a *silent* alarm> ***ant:*** noisy

**silent butler** *n.* A small receptacle with a handle and a hinged cover, used for collecting ashes and crumbs.

**silent partner** *n.* One that makes financial investments in a business but does not participate in its management.

**silent treatment** *n.* The act or an instance of totally disregarding the object of one's contempt or disapproval as a means of expressing one's negative attitude <gave me the *silent treatment*>

**si·le·nus** (sī-lē′nəs) *n., pl.* **-ni** (-nī′) [Lat. < Gk. *silēnos* < *Silēnos,* Silenus.] *Gk. Myth.* Any of various minor woodland deities or spirits and companions of Dionysus.

**Si·le·nus** (sī-lē′nəs). [Lat. < Gk. *Silēnos.*] *Gk. Myth.* A satyr, the foster father of Dionysus.

**si·le·sia** (sī-lē′zhə, -shə) *n.* **1.** A smooth linen fabric first made in Silesia. **2.** A twilled cotton fabric for linings.

**si·lex** (sī′lĕks′) *n.* [Lat., hard stone, flint.] **1.** *Obs.* Silica. **2.** Finely ground tripoli used as an inert paint filler.

**sil·hou·ette** (sĭl′o͞o-ĕt′) *n.* [Fr. < Étienne de *Silhouette* (1709–1767).] **1.** A drawing consisting of the outline of something, esp. a human profile, filled in with a solid color. **2.** An outline of an object that appears dark against a light background <the *silhouette* of a battleship on the horizon> *—vt.* **-et·ted, -et·ting, -ettes.** To cause to be seen as a silhouette : OUTLINE.

**silic-** *pref. var. of* SILICI-.

**sil·i·ca** (sĭl′ĭ-kə) *n.* [NLat. < Lat. *silex,* hard stone, flint.] A white or colorless crystalline compound, $SiO_2$, occurring as quartz, sand, flint, agate, and many other minerals and used to make glass and concrete.

**silica gel** *n.* Amorphous silica resembling white sand, used as a drying and dehumidifying agent, as a catalyst and catalyst carrier, as an anticaking agent in cosmetics, and in chromatography.

**sil·i·cate** (sĭl′ĭ-kāt′, -kĭt) *n.* Any of numerous compounds containing silicon, oxygen, and a metallic or organic radical, occurring in most rocks except limestone and dolomite, and forming the basis of common glass and bricks.

**si·li·ceous** (sĭ-lĭsh′əs) *adj.* [Lat. *siliceus,* of flint < *silex,* flint.] Containing, resembling, relating to, or consisting of silica.

**silici-** *or* **silic-** *pref.* [< SILICON and SILICA.] **1.** Silicon <*silicate*> **2.** Silica <*silicify*>

**si·lic·ic** (sĭ-lĭs′ĭk) *adj.* Relating to, resembling, or derived from silica or silicon.

**silicic acid** *n.* A jellylike substance, $SiO_2.nH_2O$, produced when sodium silicate solution is acidified.

**sil·i·cic·o·lous** (sĭl′ĭ-sĭk′ə-ləs) *adj.* Thriving in soil containing a high silica content.

**sil·i·cide** (sĭl′ĭ-sīd′) *n.* A compound of silicon with another element or radical.

**sil·i·cif·er·ous** (sĭl′ĭ-sĭf′ər-əs) *adj.* Bearing, producing, or in partial combination with silica.

**si·lic·i·fy** (sĭ-lĭs′ə-fī′) *vt.* & *vi.* **-fied, -fy·ing, -fies.** To convert or become converted into silica. **—si·lic′i·fi·ca′tion** *n.*

**sil·i·cle** (sĭl′ĭ-kəl) *n.* [Lat. *silicula,* dim. of *siliqua,* seed pod.] *Bot.* A short flat silique.

**sil·i·con** (sĭl′ĭ-kən, -kŏn′) *n.* [< SILICA.] *Symbol* **Si** A nonmetallic element occurring extensively in the earth's crust in silica and silicates, used in glass, semiconducting devices, concrete, brick, refractories, pottery, and silicones; atomic number 14; atomic weight 28.086.

**silicon carbide** *n.* A bluish-black hard crystalline compound, SiC, used as an abrasive and heat refractory and in single crystals as semiconductors, esp. in high-temperature applications.

**silicon dioxide** *n. Chem.* Silica.

**sil·i·cone** (sĭl′ĭ-kōn′) *n.* Any of a group of semi-inorganic polymers based on the structural unit $R_2SiO$, where R is an organic group, used in adhesives, lubricants, protective coatings, paints, and electrical insulation.

**sil·i·co·sis** (sĭl′ĭ-kō′sĭs) *n.* Fibrosis of the lungs caused by long-term inhalation of silica dust and resulting in a chronic shortness of breath.

**si·lique** (sĭ-lēk′) *n.* [Fr. < Lat. *siliqua,* seed pod.] A long pod divided by a membranous partition and that splits at both seams, characteristic of fruit of the mustards and related plants. **—sil′i·quous** (sĭl′ĭ-kwəs), **sil′i·quose′** (-kwōs′) *adj.*

**silk** (sĭlk) *n.* [ME < OE *sioloc,* prob. of Slav. orig.] **1. a.** The fine lustrous fiber produced by certain insect larvae and spiders, esp. that produced by silkworms. **b.** Thread or fabric made from silk. **c.** A garment made from silk. **2. silks.** The brightly colored identifying garments of a jockey or harness driver. **3.** A silky filamentous material, as that which forms a tuft on an ear of corn. *—vi.* **silked, silk·ing, silks.** To develop silk <corn *silking* in the fields>

**silk cotton** *n.* A silky fiber, as kapok, attached to the seeds of some trees.

**silk-cot·ton tree** (sĭlk′kŏt′n) *n.* Any of several trees of the family Bombacaceae, esp. *Ceiba pentandra,* native to tropical America and cultivated for its leathery fruit containing the silklike fiber kapok.

**silk-cotton tree**

**silk·en** (sĭl′kən) *adj.* **1.** Made of silk. **2.** Suggestive of silk. **3.** Delicately pleasing or caressing <a *silken* voice> **4.** Luxurious.

**silk hat** *n.* A man's silk-covered top hat.

**silk oak** *n.* A tree, *Grevillea robusta,* native to Australia, having divided leaves and showy orange flower clusters.

**silk-screen process** (sĭlk′skrēn′) *n.* A method of producing a stencil in which a design is imposed on a screen of fine fabric, as silk, with blank areas coated with an impermeable substance, and ink being forced through the cloth onto the printing surface. **—silk′-screen′** *v.* **(-screened, -screen·ing, -screens).**

**silk-stock·ing** (sĭlk′stŏk′ĭng) *n.* **1.** A wealthy, aristocratic person. **2.** *Informal.* A member or supporter of the Whig party formed during the early 19th cent. in the United States. *—adj.* Affluent : wealthy <a city's *silk-stocking* district>

**silk tree** *n.* A tree, *Albizzia julibrissin* native to the eastern Mediterranean area, with feathery compound leaves and pinkish flower clusters.

**silk·weed** (sĭlk′wēd′) *n.* MILKWEED 1.

**silk·worm** (sĭlk′wûrm′) *n.* A caterpillar that produces a silk cocoon, esp. the larva of a moth, *Bombyx mori* native to Asia, that spins a cocoon of fine lustrous fiber.

**silk·y** (sĭl′kē) *adj.* **-i·er, -i·est. 1.** Resembling silk. **2.** Made of silk. **3.** Having long, silklike hairs or a silky covering. **4.** Ingratiating : seductive <a *silky* voice> **—silk′i·ly** *adv.* **—silk′i·ness** *n.*

**sill** (sĭl) *n.* [ME *sille* < OE *sylle,* threshold.] **1.** The horizontal member that bears the upright portion of a frame, esp. the base of a window. **2.** *Geol.* A relatively thin sheet of igneous rock intruded between beds of other rock.

**sil·la·bub** (sĭl′ə-bŭb′) *n. var. of* SYLLABUB.

**sil·ly** (sĭl′ē) *adj.* **-li·er, -li·est.** [ME *syly,* defenseless, pitiable, alteration of *sely,* fortunate, holy < OE *gesælig,* blessed.] **1.** Exhibiting a lack of good sense : STUPID. **2.** Frivolous. **3.** Semiconscious : dazed. **—sil′li·ly** (sĭl′ə-lē) *adv.* **—sil′li·ness** *n.*

**si·lo** (sī′lō) *n., pl.* **-los.** [Sp. < Lat. *sirus* < Gk. *siros,* pit for storing grain.] **1. a.** A tall cylindrical structure in which fodder is stored. **b.** A pit dug for the storage of fodder. **2.** An underground missile launch area and shelter. *—vt.* **-loed, -lo·ing, -los.** To store in a silo.

**si·lox·ane** (sĭ-lŏk′sān′, sī-) *n.* [SIL(ICON) + OX(YGEN) + (METH)ANE.] Any of a class of organic or inorganic chemical compounds of silicon, oxygen, and usu. carbon and hydrogen, based on the structural unit $R_2SiO$, where R is $CH_3$, H, $C_2H_5$, or a more complex group.

**silt** (sĭlt) *n.* [ME *cylte,* prob. of Scand. orig.] A sedimentary material composed of fine mineral particles intermediate in size between sand and clay. *—v.* **silt·ed, silt·ing, silts.** *—vi.* To become filled with silt. *—vt.* To fill, cover, or obstruct with silt.

**silt·stone** (sĭlt′stōn′) *n.* Stone made up of hardened silt.

**Sil·u·res** (sĭl′yə-rēz′) *pl.n.* [Lat.] A people described by Tacitus as occupying southwestern Britain at the time of the Roman invasion.

**Si·lu·ri·an** (sĭ-lo͝or′ē-ən, sī-) *adj.* [< SILURES (so called because the Silures lived in the part of Wales where the rocks were first identified).] **1.** Of, belonging to, or designating the geologic time, system of rocks, or sedimentary deposits of the third period of the Paleozoic era, marked by the appearance of land plants. **2.** Of or pertaining to the Silures or their culture. *—n.*The Silurian period or system of deposits.

**si·lu·rid** (sĭ-lo͝or′ĭd, sī-) *adj.* [NLat. *Siluridae,* family name < Lat. *silurus,* a large freshwater fish < Gk. *silouros.*] Of or belonging to the family Siluridae, which includes various freshwater catfishes of Europe and Asia. *—n.* A silurid fish.

**sil·va** also **syl·va** (sĭl′və) *n.* [Lat., forest.] **1.** The trees or forests of a region. **2.** A written work on the trees or forests of a region.
**sil·van** (sĭl′vən) *adj.* & *n. var. of* SYLVAN.
**Sil·va·nus** also **Syl·va·nus** (sĭl-vā′nəs) *n.* [Lat. < *silva,* forest.] *Rom. Myth.* A god of forests, fields, and herding.
**sil·ver** (sĭl′vər) *n.* [ME < OE *siolfor.*] **1.** *Symbol* **Ag** A lustrous white, ductile, malleable metallic element, highly valued for jewelry and tableware, and widely used in coinage, photography, dental and soldering alloys, electrical contacts, and printed circuits; atomic number 47; atomic weight 107.870. **2.** Silver as a commodity or medium of exchange. **3.** Coinage made of silver. **4. a.** Domestic articles, as tableware, made of or plated with silver. **b.** Tableware made of nonprecious metals, as stainless steel. **5.** The color medium gray. **6.** A silver salt, esp. silver nitrate, used to sensitize paper. —*adj.* **1.** Having a lustrous medium-gray color <*silver* whiskers> **2.** Having a bell-like sound. **3.** Eloquently persuasive. **4.** Favoring the adoption of silver as a standard of currency. **5.** Of or designating a 25th anniversary. —*v.* **-vered, -ver·ing, -vers.** —*vt.* **1.** To cover, plate, or adorn with silver or a similar lustrous substance. **2.** To cause to resemble silver. **3.** To coat (photographic paper) with a film of silver nitrate or other silver salt. —*vi.* To become silvery <hair *silvering* with age>
**silver age** *n.* A period of history secondary to that of a golden age <the *silver* age of Russian poetry>
**sil·ver-bell tree** (sĭl′vər-bĕl′) *n.* A tree or shrub of the genus *Halesia,* esp. *H. carolina* of the southeastern United States, having drooping bell-shaped white flowers.
**sil·ver·ber·ry** (sĭl′vər-bĕr′ē) *n.* A North American shrub, *Elaeagnus commutata,* with silvery flowers, leaves, and berries.
**silver bromide** *n.* A pale-yellow crystalline compound, AgBr, that turns black on exposure to light and is used as the light-sensitive component on ordinary photographic films and plates.
**silver certificate** *n.* A paper money bill once issued as legal tender by the U.S. government in representation of deposited silver bullion.
**silver chloride** *n.* A white granular powder, AgCl, that turns dark on exposure to light and is used in photography, photometry, and optics.
**silver cord** *n.* [After *The Silver Cord,* a play by Sidney Howard (1891–1939).] The emotional bond between mother and child.
**sil·ver·fish** (sĭl′vər-fĭsh′) *n., pl.* **silverfish** or **-fish·es. 1.** A fish, as a tarpon, having silvery scales. **2.** A silvery wingless insect, *Lepisma saccharina,* that often causes damage to bookbindings and clothing.
**silver fox** *n.* **1.** A color phase of the North American red fox, *Vulpes fulva,* having black fur tipped with white. **2.** The fur of the silver fox.
**silver iodide** *n.* A pale-yellow, odorless powder, AgI, that darkens on exposure to light and is used in artificial rainmaking, in photography, and as an antiseptic.
**silver nitrate** *n.* A poisonous, colorless crystalline compound, $AgNO_3$, that becomes grayish black when exposed to light in the presence of organic matter and is used in photography, mirror manufacturing, hair dyeing, and silver plating and as an external medicine.
**silver perch** *n.* MADEMOISELLE 3.
**silver plate** *n.* Tableware made of or coated with silver.
**silver protein** *n.* A usu. gelatinous preparation of silver and protein used as an antibacterial agent.
**sil·ver·rod** (sĭl′vər-rŏd′) *n.* A North American plant, *Solidago bicolor,* related to the goldenrods but having white flowers.
**silver salmon** *n.* The coho salmon.
**silver screen** *n.* **1.** A screen for showing films. **2.** MOTION PICTURE 2.
**sil·ver·side** (sĭl′vər-sīd′) also **sil·ver·sides** (-sīdz′) *n.* Any of various marine and freshwater fishes of the family Atherinidae, having a silvery band along each side.
**sil·ver·smith** (sĭl′vər-smĭth′) *n.* One that makes, repairs, or replates silver articles.
**silver spoon** *n.* Inherited wealth.
**silver standard** *n.* A monetary standard under which a specified quantity of silver constitutes the basic unit of currency.
**Silver Star** *n.* A U.S. military decoration awarded for gallantry in action.
**sil·ver-tongued** (sĭl′vər-tŭngd′) *adj.* SILVER 3.
**sil·ver·ware** (sĭl′vər-wâr′) *n.* Tableware made of or plated with silver.
**sil·ver·weed** (sĭl′vər-wēd′) *n.* A plant, *Potentilla anserina,* with yellow flowers and leaves that are silvery beneath.
**sil·ver·y** (sĭl′və-rē) *adj.* **1.** Coated with or containing silver. **2.** Like silver. **3.** Having a clear, ringing sound. **—sil′ver·i·ness** *n.*
**sil·vex** (sĭl′vĕks′) *n.* [Lat. *silva,* forest + EX(TERMINATOR).] A solid, toxic, selective herbicide, $C_9H_7O_3Cl_3$, used chiefly against woody plants.
**sil·vi·chem·i·cal** (sĭl′vĭ-kĕm′ĭ-kəl) *n.* [Lat. *silva,* forest + CHEMICAL.] A chemical derived from wood.
**sil·vic·o·lous** (sĭl-vĭk′ə-ləs) *adj.* [< Lat. *silvicola,* inhabitant of the forest : *silva,* forest + *colere,* to dwell.] Inhabiting forests.
**sil·vi·cul·ture** (sĭl′vĭ-kŭl′chər) *n.* [Fr. : Lat. *silva,* forest + *culture,* culture.] Care and cultivation of forest trees : FORESTRY. **—sil′vi·cul′tur·al** *adj.* **—sil′vi·cul′tur·ist** *n.*
**si·ma** (sī′mə) *n.* [SI(LICA) + MA(GNESIUM).] The lower layer of the earth's outer crust that is rich in silica, iron, and magnesium, and that underlies the sial.
**Sim·chas To·rah** (sĭm′KHəs tôr′ə, tōr′ə) *n.* [Heb. *shimḥath tōrāh,* rejoicing over the Law.] A Jewish holiday celebrated on the 23rd day of Tishri, marking the end of the Feast of Tabernacles.
**Sim·e·on** (sĭm′ē-ən) *n.* [LLat. < Gk. *Symeōn* < Heb. *Shim'ōn* < *shāma',* he heard.] **1.** The second son of Jacob and Leah and ancestor of the tribe of Israel descended from him in the Old Testament. **2.** The man who, upon seeing the infant Jesus, spoke the Nunc Dimittis.
**sim·i·an** (sĭm′ē-ən) *adj.* [< Lat. *simia,* ape < *simus,* snub-nosed < Gk. *simos.*] Relating to, characteristic of, or resembling an ape or monkey. **—sim′i·an** *n.*
**sim·i·lar** (sĭm′ə-lər) *adj.* [Fr. *similaire* < Lat. *similis,* like.] **1.** Resembling though not completely identical. **2.** *Math.* Designating figures having corresponding angles equal and corresponding line segments proportional. **—sim′i·lar′i·ty** *n.* **—sim′i·lar·ly** *adv.*
**sim·i·le** (sĭm′ə-lē) *n.* [Lat., neuter of *similis,* like.] A figure of speech in which two unlike things are compared, often in a phrase introduced by *like* or *as* as in *The soldier was as strong as an ox.*
**si·mil·i·tude** (sĭ-mĭl′ĭ-to͞od′, -tyo͞od′) *n.* [ME < OFr. < Lat. *similitudo* < *similis,* like.] **1.** The quality or state of being similar. **2.** Something closely resembling another : COUNTERPART. **3.** *Archaic.* A simile, allegory, or parable.
**sim·mer** (sĭm′ər) *v.* **-mered, -mer·ing, -mers.** [Alteration of obs. *simper,* to simmer < ME *simperen.*] —*vi.* **1.** To cook gently just at or below the boiling point. **2.** To be filled with barely controlled anger or resentment : SEETHE <*simmered* over the insult> —*vt.* To cook (e.g., food) gently just at or below the boiling point. **—simmer down. 1.** To reduce the liquid volume of by simmering. **2.** To calm down after excitement or anger. **—sim′mer** *n.*
**sim·nel** (sĭm′nəl) *n.* [ME *simenel* < OFr. < Lat. *simila,* fine flour.] *Chiefly Brit.* **1.** A crisp bread of fine wheat flour. **2.** A fruitcake eaten on festive occasions.
**si·mo·le·on** (sĭ-mō′lē-ən) *n.* [Orig. unknown.] *Slang.* A dollar.
**si·mo·ni·ac** (sĭ-mō′nē-ăk′, sĭ-) *n.* One who practices simony. **—si·mo′ni·ac′, si′mo·ni′a·cal** (sī′mə-nī′ə-kəl, sĭm′ə-) *adj.* **—si′mo·ni′a·cal·ly** *adv.*
**Si·mon Le·gree** (sī′mən lə-grē′) *n.* [After *Simon Legree,* a cruel slave dealer in the novel *Uncle Tom's Cabin* by Harriet Beecher Stowe.] A brutal taskmaster.
**si·mon-pure** (sī′mən-pyo͝or′) *adj.* [< the phrase *the real Simon Pure,* after *Simon Pure,* a character impersonated by a rival in the play *A Bold Stroke for a Wife* by Susanna Centlivre.] **1.** Genuinely and utterly pure. **2.** Superficially or hypocritically virtuous.
**si·mo·ny** (sī′mə-nē, sĭm′ə-) *n.* [ME *simonie* < OFr. < LLat. *simonia,* after *Simon* Magus, a sorcerer who tried to buy spiritual powers from the Apostle Peter.] Purchase or sale of ecclesiastical pardons, offices, or emoluments. **—si′mo·nist** *n.*
**si·moom** (sĭ-mo͞om′) also **si·moon** (-mo͞on′) *n.* [Ar. *samūm,* poisonous < *samma,* he poisoned.] A strong, hot, sand-laden wind of the Sahara and Arabian deserts.
**simp** (sĭmp) *n. Slang.* A simpleton.
**sim·pa·ti·co** (sĭm-pä′tĭ-kō′, -păt′ĭ-) *adj.* [Ital. < *simpatia,* sympathy < Lat. *sympathia.* —see SYMPATHY.] **1.** Of like mind or temperament : COMPATIBLE. **2.** Having attractive qualities : PLEASING.
**sim·per** (sĭm′pər) *v.* **-pered, -per·ing, -pers.** [Of Scand. orig.] —*vi.* To smile in a silly way. —*vt.* To utter or express with a simper. **—sim′per** *n.* **—sim′per·er** *n.* **—sim′per·ing·ly** *adv.*
**sim·ple** (sĭm′pəl) *adj.* **-pler, -plest.** [ME < OFr. < Lat. *simplus.*] **1.** Having or composed of only one thing or part. **2.** Not complex : EASY <a *simple* task> **3.** Without additions or modifications : MERE <Just give me a *simple* "yes" or "no."> **4.** Without embellishment <a *simple* suit> **5.** Not elaborate, elegant, or luxurious <a *simple* house> **6.** Unassuming or unpretentious. **7.** Not deceitful : SINCERE. **8.** Humble or lowly <a *simple* peasant> **9.** Ordinary or common <a *simple* head cold> **10.** Unimportant : trivial. **11.** Having or manifesting little sense or intellect : SILLY. **12.** *Biol.* Having no divisions or subdivisions <a *simple* leaf> **13.** *Mus.* Without overtones <a *simple* tone> —*n.* **1.** A single component of a complex, esp. one that is unanalyzable. **2.** A fool. **3.** One of humble birth or condition. **4.** A medicinal plant or the medicine obtained from it.
**simple closed curve** *n.* A Jordan curve.
**simple equation** *n.* A linear equation.
**simple fraction** *n.* A fraction in which both the numerator and the denominator are integers.
**simple fruit** *n.* A fruit, as a pea pod, grape, or almond, that develops from a single pistil.
**simple harmonic motion** *n. Physics.* A periodic motion that may be described as a sinusoidal function of time; specif., the motion of a particle that obeys the equation $x = A\cos(kt + \phi)$, where x is the displacement of the particle from the origin at any time *t,* *A*

is the maximum displacement, $\phi$ is the initial phase or angular displacement at $t = 0$, and $k$ is a constant equal to $2\pi$ times the frequency of the oscillation.

**simple honors** *pl.n.* Three honors in trump or three aces at no-trump held by the same side in bridge.

**simple interest** *n.* Interest paid only on the original principal, not on the interest accrued.

**simple machine** *n.* MACHINE 1b.

**simple microscope** *n.* A microscope having one lens or lens system, as a magnifying glass or hand lens.

**sim·ple-mind·ed** (sĭm′pəl-mīn′dĭd) *adj.* **1.** Not sophisticated : ARTLESS. **2.** Stupid : silly. **3.** Mentally defective. **—sim′ple-mind′ed·ly** *adv.* **—sim′ple-mind′ed·ness** *n.*

**simple pendulum** *n.* PENDULUM 1.

**simple sentence** *n.* A sentence having no coordinate or subordinate clauses, as *The dog growled.*

**Simple Simon** *n.* [After *Simple Simon,* a character in a nursery rhyme.] A fool.

**simple sugar** *n.* A monosaccharide.

**sim·ple·ton** (sĭm′pəl-tən) *n.* [< SIMPLE.] A fool.

**sim·plex** (sĭm′plĕks′) *adj.* [Lat. *simplex,* simple.] Denoting a system of telegraphy in which only one message may be sent in either direction at one time.

**sim·plic·i·ty** (sĭm-plĭs′ĭ-tē) *n., pl.* **-ties.** [ME *symplicite* < OFr. < Lat. *simplicitas* < *simplex,* simple.] **1.** The quality, state, or fact of being simple. **2.** Absence of luxury or showiness. **3.** Absence of affectation or pretense. **4.** Lack of good sense or intelligence : FOOLISHNESS.

**sim·pli·fy** (sĭm′plə-fī′) *vt.* **-fied, -fy·ing, -fies.** [Fr. *simplifier* < Med. Lat. *simplificare* : Lat. *simplus,* simple + *facere,* to make.] To make simple or simpler. **—sim′pli·fi·ca′tion** *n.* **—sim′pli·fi′er** *n.*

**sim·plism** (sĭm′plĭz′əm) *n.* [Fr. *simplisme* < OFr. *simple,* simple.] The tendency to oversimplify an issue by ignoring complexities. **—sim·plis·tic** (sĭm-plĭs′tĭk) *adj.* **—sim·plis′ti·cal·ly** *adv.*

**sim·ply** (sĭm′plē) *adv.* **1.** In a simple way : PLAINLY. **2.** Foolishly. **3.** Merely : only. **4.** Absolutely : altogether <*simply* ludicrous> **5.** Frankly : candidly <You are, quite *simply,* incompetent.>

**simply connected** *adj.* Of, being, or characterized by a mathematical surface that is divided into two separate parts by every simple closed curve within it.

**simply ordered** *adj.* Having any three mathematical elements transitively related and any two elements equal or connected by an asymmetric relationship.

**sim·u·la·cra** (sĭm′yə-lā′krə, -lăk′rə) *n. pl. of* SIMULACRUM.

**sim·u·la·cre** (sĭm′yə-lā′kər, -lăk′ər) *n. Archaic.* A simulacrum.

**sim·u·la·crum** (sĭm′yə-lā′krəm, -lăk′rəm) *n., pl.* **-la·cra** (-lā′krə, -lăk′rə). [Lat. < *simulare,* to simulate < *similis,* like.] **1.** An image or representation. **2.** An unreal or vague semblance.

**sim·u·lar** (sĭm′yə-lər, -lär′) *Archaic.* —*n.* One that simulates : SIMULATOR. —*adj.* Simulated : sham.

**sim·u·late** (sĭm′yə-lāt′) *vt.* **-lat·ed, -lat·ing, -lates.** [Lat. *simulare, simulat-* < *similis,* like.] **1.** To have or take on the appearance, form, or sound of : IMITATE. **2.** To make a pretense of : FEIGN <*simulate* an interest in a conversation> —*adj.* (-lĭt, -lāt′). Simulated. **—sim′u·la′tive** *adj.*

**sim·u·la·tion** (sĭm′yə-lā′shən) *n.* **1.** The act or process of simulating. **2.** An imitation. **3.** A false appearance.

**sim·u·la·tor** (sĭm′yə-lā′tər) *n.* One that simulates, esp. a device that generates test conditions approximating actual or operational conditions.

**si·mul·cast** (sī′məl-kăst′, sĭm′əl-) *vt.* **-cast·ed, -cast·ing, -casts.** [SIMUL(TANEOUS) + (BROAD)CAST.] To broadcast simultaneously, as by FM radio and television. **—si′mul·cast′** *n.*

**si·mul·ta·ne·ous** (sī′məl-tā′nē-əs, sĭm′əl-) *adj.* [Lat. *simul,* at the same time + E. *-aneous,* as in *instantaneous.*] **1.** Occurring, existing, or carried out at the same time. **2.** *Math.* Collectively restricting the values of a set of variables <*simultaneous* equations> **—si′mul·ta′ne·ous·ly** *adv.* **—si′mul·ta′ne·ous·ness, si′mul·ta·ne′i·ty** (-tə-nē′ĭ-tē, -nā′-) *n.*

**sin**[1] (sĭn) *n.* [ME *sinne* < OE *synn.*] **1.** Transgression of a religious or moral law. **2.** Estrangement from God as a result of breaking God's law. **3.** An offense, violation, fault, or error. —*vi.* **sinned, sin·ning, sins. 1.** To violate a religious or moral law. **2.** To commit an offense or violation.

**sin**[2] (sēn, sĭn) *n.* [Heb., var. of *shīn,* the letter shin.] The 21st letter of the Hebrew alphabet. —See table at ALPHABET.

**sin·an·thro·pus** (sī-năn′thrə-pəs, sĭ-, sī′năn-thrō′pəs, sĭn′ăn-) *n.* [NLat. *Sinanthropus,* genus name : SINO- + *anthropus,* human being.] An extinct humanlike primate of the genus *Sinanthropus,* which includes the Peking man.

**sin·a·pism** (sĭn′ə-pĭz′əm) *n.* [Fr. *sinapisme* < LLat. *sinapismus* < Gk. *sinapismos,* use of a mustard plaster < *sinapizein,* to apply a mustard plaster < *sinapi,* mustard.] A mustard plaster.

**since** (sĭns) *adv.* [ME *sinnes,* contraction of *sithenes* < OE *siððan.*] **1.** From then until now or between then and now <left home and hasn't been there *since*> **2.** Before now : AGO <long *since* forgiven> —*prep.* From a specified time in the past <has not been here *since* Monday> —*conj.* **1.** During the time after which <They have not been back *since* they graduated.> **2.** Continuously from the time when <They have not called *since* they left.> **3.** As a result of the fact that : INASMUCH AS <*Since* you're not interested, I won't tell you about it.>

**sin·cere** (sĭn-sîr′) *adj.* **-cer·er, -cer·est.** [Lat. *sincerus.*] **1.** Not feigned or affected : TRUE <*sincere* apologies> **2.** Presenting no false appearance : HONEST <a *sincere* believer> **3.** *Archaic.* Pure : unadulterated. **—sin·cere′ly** *adv.* **—sin·cere′ness** *n.* **—sin·cer′i·ty** (-sĕr′ĭ-tē) *n.*

**sin·ci·put** (sĭn′sə-pət) *n., pl.* **sin·ci·puts** or **sin·cip·i·ta** (sĭn-sĭp′ĭ-tə) [Lat. : *semi-,* half + *caput,* head.] **1.** The upper half of the cranium, esp. the anterior portion above and including the forehead. **2.** The forehead. **—sin·cip′i·tal** (-sĭp′ĭ-tl) *adj.*

**Sin·dhi** (sĭn′dē) *n., pl.* **Sindhi** or **-dhis.** [Ar. *Sindī.*] **1. a.** The predominantly Moslem people of Sind. **b.** A member of this people. **2.** The Indic language of Sind. **—Sin′dhi** *adj.*

**sine** (sīn) *n.* [Med. Lat. *sinus* < Lat., curve.] **1.** The ordinate of the endpoint of an arc of a unit circle centered at the origin of a Cartesian coordinate system, the arc being of length x and measured counterclockwise from the point (1, 0) if x is positive or clockwise if x is negative. **2.** The function of an acute angle that is the ratio of the opposite side to the hypotenuse in a right triangle.

**si·ne·cure** (sī′nĭ-kyŏŏr′, sĭn′ĭ-) *n.* [< Med. Lat. *sine cura,* without cure of souls.] **1.** An ecclesiastical benefice not attached to the spiritual duties of a parish. **2.** A position or office requiring little or no work but providing a salary. **—si′ne·cur·ism** *n.* **—si′ne·cur·ist** *n.*

**sine curve** *n.* The graph of the equation $y = \sin x$.

**si·ne di·e** (sī′nī dī′ē, sĭn′ā dē′ā′) *adv.* [Lat., without a day.] Without a day specified for a future meeting : INDEFINITELY <Parliament was dismissed *sine die.*>

**si·ne qua non** (sĭn′ĭ kwä nōn′, nŏn′, sī′nĭ kwā nŏn′) *n.* [Lat., without which not.] An essential element.

**sin·ew** (sĭn′yōō) *n.* [ME *sinewe* < OE *sinu.*] **1.** A tendon. **2.** Vigorous muscular strength and power. **3.** *often* **sinews.** The source or mainstay of vitality and strength.

**sine wave** *n. Physics.* A waveform with deviation that can be expressed as the sine or cosine of a linear function of time or space or both.

**sin·ew·y** (sĭn′yōō-ē) *adj.* **1.** Consisting of or being like sinew. **2.** Lean and muscular. **3.** Strong : vigorous.

**sin·ful** (sĭn′fəl) *adj.* Characterized by or full of sin : WICKED <*sinful* deeds><*sinful* thoughts> **—sin′ful·ly** *adv.* **—sin′ful·ness** *n.*

**sing** (sĭng) *v.* **sang** (săng), **sung** (sŭng), **sing·ing, sings.** [ME *singen* < OE *singan.*] —*vi.* **1.** To utter a series of sounds or words in musical tones. **2.** To vocalize songs. **3.** To produce or have the effect of melody : LILT. **4.** To produce musical sounds when played <made the violin *sing*> **5.** To make a high whine <The machine *sang.*> **6.** To be filled with a buzzing sound <The great hall *sang* with voices.> **7.** To proclaim or extol something in poetry. **8.** *Slang.* To give information or evidence against someone. —*vt.* **1.** To render in tones with musical inflections of the voice <The messenger *sang* the telegram.> **2.** To produce the musical sound of. **3.** To intone : chant. **4.** To proclaim, esp. in verse <*sang* our praises> **5.** To bring to a specified state by singing <*sang* the child to sleep> **—sing out.** To shout out loudly. —*n.* A gathering of people for group singing. **—sing′a·ble** *adj.*

**sing-a·long** (sĭng′ə-lông′, -lŏng′) *n.* A songfest.

**singe** (sĭnj) *vt.* **singed, singe·ing, sing·es.** [ME *sengen* < OE *sengan.*] **1.** To burn superficially : SCORCH. **2.** To burn the ends of. **3.** To burn off the feathers or bristles of by subjecting briefly to flame. **—singe** *n.* **—sing′er** (sĭn′jər) *n.*

**sing·er** (sĭng′ər) *n.* **1.** A person who sings, esp. a trained or professional vocalist. **2.** A poet. **3.** A songbird.

**Sin·gha·lese** (sĭng′gə-lēz′, -lēs′) *also* **Sin·ha·lese** (sĭn′hə-lēz′, -lēs′) *n., pl.* **Singhalese** *also* **Sinhalese.** [Skt. *Siṁhala,* Sri Lanka + -ESE.] **1.** A people constituting the major portion of the population of Sri Lanka. **2.** The Indic language of the Singhalese that is the chief language of Sri Lanka. —*adj.* Of or relating to the Singhalese or their language.

**sin·gle** (sĭng′gəl) *adj.* [ME *sengle* < OFr. < Lat. *singulus.*] **1.** Unaccompanied by another or others : SOLE <a *single* survivor> **2. a.** Consisting of one part or form <a *single* layer><a *single* standard of quality> **b.** Consisting of one alone <I had but a *single* thought, which was to survive.> **3.** Undivided : unbroken. **4.** Separate from others : DISTINCT. **5.** Designed to accommodate only one <a *single* room> **6. a.** Unmarried <*single* people> **b.** Of or pertaining to celibacy. **7.** *Bot.* Having only one rank or row of petals <a *single* flower> —*n.* **1.** A separate unit : INDIVIDUAL. **2.** An accommodation for one person. **3.** One who is unmarried <a bar for *singles*> **4.** A one-dollar bill. **5. a.** *Baseball.* A one-base hit. **b.** A hit for one run in cricket. **c.** A golf match between two players. **d.** *often* **singles.** A tennis match between two players. —*v.* **-gled, -gling, -gles.** —*vt.* **1.** To choose from among others <*singled* you out for this great honor> **2.** *Baseball.* To cause (a baserunner) to score or advance by making a one-base hit. —*vi. Baseball.* To make a one-base hit. **—sin'gle·ness** *n.*

☆ ***syns:*** **1.** SINGLE, DISCRETE, INDIVIDUAL, SEPARATE, SINGULAR *adj. core meaning* : being a distinct entity <a sentence made up of *single* words> **2.** SINGLE, LONE, ONE, ONLY, PARTICULAR, SOLE, SOLITARY *adj. core meaning* : alone in a given category <a *single* prehistoric monument still standing> **3.** SINGLE, UNMARRIED, UNWED *adj. core meaning* : being without a spouse <*single* parents> ***ant:*** married, wed

**sin·gle-blind** (sĭng'gəl-blīnd') *adj.* Of, pertaining to, or being an experimental procedure in which the experimenters know the composition of the test and control groups but the subjects do not.

**sin·gle-breast·ed** (sĭng'gəl-brĕs'tĭd) *adj.* Closing with a narrow overlap and fastened down the front with a single row of buttons <a *single-breasted* coat>

**single combat** *n.* Combat between two individuals.

**single entry** *n.* A system of bookkeeping in which a business keeps only a single account showing amounts due and amounts owed.

**single file** *n.* A line of individuals, animals, or objects standing or moving one behind the other. **—single file** *adv.*

**sin·gle-foot** (sĭng'gəl-fo͝ot') *n.* A rapid gait of a horse in which each foot strikes the ground separately. —*vi.* **-foot·ed, -foot·ing, -foots.** To go at the single-foot.

**sin·gle-hand·ed** (sĭng'gəl-hăn'dĭd) *adj.* **1.** Working or performed without help : UNASSISTED. **2.** Designed for use with one hand. **3.** Having or using one hand. **—sin'gle-hand'ed·ly** *adv.* **—sin'gle-hand'ed·ness** *n.*

**sin·gle-heart·ed** (sĭng'gəl-här'tĭd) *adj.* Sincere and dedicated. **—sin'gle-heart'ed·ly** *adv.* **—sin'gle-heart'ed·ness** *n.*

**single knot** *n.* An overhand knot.

**sin·gle-mind·ed** (sĭng'gəl-mīn'dĭd) *adj.* **1.** Having one overriding opinion or purpose. **2.** Steadfast. **—sin'gle-mind'ed·ly** *adv.* **—sin'gle-mind'ed·ness** *n.*

**sin·gle-phase** (sĭng'gəl-fāz') *adj.* Producing, carrying, or powered by a single alternating voltage.

**singles bar** *n.* A bar frequented esp. by unmarried people.

**sin·gle-space** (sĭng'gəl-spās') *v.* **-spaced, -spac·ing, -spac·es.** —*vt.* To typewrite (copy) without leaving a blank line between lines. —*vi.* To typewrite copy without line spaces.

**sin·gle·stick** (sĭng'gəl-stĭk') *n.* **1.** A one-handed fencing stick fitted with a hand guard. **2.** The art, sport, or exercise of fencing with a singlestick.

**sin·gle·stick·er** (sĭng'gəl-stĭk'ər) *n. Informal.* A sailboat with one mast : SLOOP.

**sin·glet** (sĭng'glĭt) *n.* **1.** A man's jersey undershirt, usu. without sleeves. **2.** *Physics.* A multiplet with a single member.

**single tax** *n.* A system by which all revenue is derived from a tax on one object, esp. on land.

**sin·gle·ton** (sĭng'gəl-tən) *n.* [< SINGLE.] **1.** A playing card that is the only one of its suit in a player's hand. **2.** An individual separated or distinguished from two or more of its group.

**single-track** (sĭng'gəl-trăk') *adj.* **1.** Having just one track. **2.** Lacking range or flexibility : ONE-TRACK.

**sin·gle·tree** (sĭng'gəl-trē) *n.* A whiffletree.

**sin·gly** (sĭng'glē) *adv.* **1.** Being without help or company : ALONE. **2.** One by one : INDIVIDUALLY.

**sing·song** (sĭng'sông', -sŏng') *n.* **1.** Verse marked by mechanical regularity of rhythm and rhyme. **2.** A monotonously rising and falling speech cadence.

**sing·spiel** (sĭng'spēl', zĭng'shpēl') *n.* [G. : *singen,* to sing + *spiel,* play.] An 18th-cent. German musical comedy featuring folk songs interspersed with dialogue.

**sin·gu·lar** (sĭng'gyə-lər) *adj.* [ME *singuler* < OFr. < Lat. *singularis* < *singulus,* single.] **1. a.** Being only one : INDIVIDUAL. **b.** Deviating strongly from a norm : RARE <a *singular* archaeological find> **2.** Of or being a word form denoting a single person or thing or several considered as a single unit. **3.** *Logic.* Of or relating to the specific : INDIVIDUAL. **4.** Peculiar : eccentric <*singular* behavior> —*n.* **1.** The singular number or a form denoting it. **2.** A word having a singular number. **—sin'gu·lar·ly** *adv.* **—sin'gu·lar·ness** *n.*

**sin·gu·lar·i·ty** (sĭng'gyə-lăr'ĭ-tē) *n., pl.* **-ties. 1.** The quality or state of being singular. **2.** A peculiarity marking one as distinct from others. **3.** Something unusual. **4.** A black hole. **5.** *Math.* A point at which the derivative does not exist for a given function of a random variable but every neighborhood of which contains points for which the derivative exists.

**sin·gu·lar·ize** (sĭng'gyə-lə-rīz') *vt.* **-ized, -iz·ing, -iz·es.** To make conspicuous : DISTINGUISH.

**singular point** *n.* SINGULARITY 5.

**Sin·ha·lese** (sĭn'hə-lēz', -lēs') *n.* & *adj. var. of* SINGHALESE.

**Si·ni·cism** (sī'nĭ-sĭz'əm, sĭn'ĭ-) *n.* [< Med. Lat. *Sinicus,* Chinese < LLat. *Sinae,* the Chinese. —see SINO-.] A custom or trait peculiar to the Chinese.

**Si·ni·cize** (sī'nĭ-sīz', sĭn'ĭ-) *vt.* **-cized, -ciz·ing, -ciz·es.** [< Med. Lat. *Sinicus,* Chinese < LLat. *Sinae.* —see SINO-.] To change or modify by Chinese influence.

**sin·is·ter** (sĭn'ĭ-stər) *adj.* [ME *sinistre* < OFr. < Lat. *sinister,* on the left, unlucky.] **1.** Suggesting or threatening evil <a *sinister* look> **2.** Presaging trouble : OMINOUS <*sinister* clouds> **3.** Situated on the left side : LEFT. **4.** *Heraldry.* Being on the bearer's left and hence on the observer's right. **—sin'is·ter·ly** *adv.* **—sin'is·ter·ness** *n.*

**sin·is·tral** (sĭn'ĭ-strəl, sĭ-nĭs'trəl) *adj.* **1.** Of or facing the left side. **2.** Left-handed. **3.** *Zool.* Designating or relating to a gastropod shell that has its aperture to the left when facing the observer with the apex upward. **—sin'is·tral·ly** *adv.*

**sin·is·trorse** (sĭn'ĭ-strôrs') *adj.* [NLat. *sinistrorsus* < Lat., turned toward the left : *sinister,* left + *versus,* p.part. of *vertere,* to turn.] Growing upward in a spiral that turns from right to left <*sinistrorse* vines> **—sin'is·trorse'ly** *adv.*

**sin·is·trous** (sĭn'ĭ-strəs, sĭ-nĭs'trəs) *adj. Archaic.* Sinister : ill-omened. **—sin'is·trous·ly** *adv.*

**Si·nit·ic** (sī-nĭt'ĭk, sĭ-) *n.* [SIN(O)- + *-itic,* as in *Semitic.*] The branch of Sino-Tibetan that comprises Chinese. **—Si·nit'ic** *adj.*

**sink** (sĭngk) *v.* **sank** (săngk) *or* **sunk** (sŭngk), **sunk, sink·ing, sinks.** [ME *sinken* < OE *sincan.*] —*vi.* **1.** To descend to the bottom. **2.** To move to a lower level, esp. slowly or in stages. **3.** To appear to move downward. **4.** To slope downward : INCLINE. **5.** To pass into a specified condition <*sank* into a deep coma> **6.** To pass into a worsened physical condition <The patient is *sinking* fast.> **7.** To become weaker, quieter, or less forceful <a voice that *sank* to a whisper> **8.** To diminish, as in value <stock prices *sinking*> **9.** To feel great disappointment or discouragement. **10.** To seep : penetrate <water *sinking* into the ground> **11.** To make an impression <The meaning finally *sank* in.> —*vt.* **1.** To cause to descend beneath a surface <*sink* a battleship> **2.** To cause to drop or lower <*sank* the bucket into the well> **3.** To force into the ground <*sink* a piling> **4.** To dig or drill (a mine or well) in the earth. **5.** To make weaker, quieter, or less forceful. **6.** To debase the nature of : DEGRADE. **7.** To suppress : hide. **8.** *Informal.* To defeat, as in a game. **9. a.** To invest. **b.** To invest without any prospect of return. **10.** To pay off (a debt). **11.** To get (a ball) into a hole or basket in a sport. —*n.* **1.** A water basin fixed to a wall or floor and having a drainpipe and gen. a piped supply of water. **2.** A cesspool. **3.** A sinkhole. **4.** A place considered evil and corrupt. **—sink'a·ble** *adj.*

**sink·er** (sĭng'kər) *n.* **1.** One that sinks. **2.** A weight used for sinking fishing lines or nets. **3.** *Slang.* A doughnut.

**sink·hole** (sĭngk'hōl') *n.* A natural depression in a land surface communicating with a subterranean passage, gen. occurring in limestone regions and formed by solution or by collapse of a cavern roof.

**sinking fund** *n.* A fund accumulated to pay off a public or corporate debt.

**sin·less** (sĭn'lĭs) *adj.* Free from sin : INNOCENT <*sinless* babies> **—sin'less·ly** *adv.* **—sin'less·ness** *n.*

**sin·ner** (sĭn'ər) *n.* One who sins.

**Sinn Fein** (shĭn fān') *n.* [Ir. Gael. : *sinn,* we + *fēin,* self.] An Irish political and cultural society founded in about 1905 to promote political and economic independence and the renewal of culture in Ireland.

**Sino-** *pref.* [Fr. < LLat. *Sinae,* the Chinese < Gk. *Sinai* < Ar. *Sīn,* China.] Chinese <*Sinology*>

**si·no·a·tri·al** (sī'nō-ā'trē-əl) *adj.* [SIN(US) + ATRIAL.] Of or relating to the sinoatrial node.

**sinoatrial node** *n.* A small mass of specialized cardiac muscle fibers located in the posterior wall of the right atrium of the heart that generates the initiating impulses of the heartbeat.

**Si·no·logue** *also* **Si·no·log** (sī'nə-lôg', -lŏg', sĭn'ə-) *n.* [Fr., back-formation < *Sinologie,* Sinology.] A student of Sinology.

**Si·nol·o·gy** (sī-nŏl'ə-jē, sĭ-) *n.* [Fr. *Sinologie* < *Sino-,* Sino- + *-logie,* -logy.] Study of Chinese language, literature, or civilization. **—Si'no·log'i·cal** (sī'nə-lŏj'ĭ-kəl, sĭn'ə-) *adj.* **—Si·nol'o·gist** *n.*

**Si·no·phile** (sī'nə-fīl', sĭn'ə-) *n.* One friendly to the Chinese and their interests.

**Si·no-Ti·bet·an** (sī'nō-tĭ-bĕt'n, sĭn'ō-) *n.* A language family that includes the Sinitic and Tibeto-Burman branches. **—Si·no-Ti·bet'an** *adj.*

**sin·ter** (sĭn'tər) *n.* [G., iron dross.] **1.** *Geol.* A chemical sediment or crust, as of porous silica, deposited by a mineral spring. **2.** A mass formed by sintering. —*v.* **-tered, -ter·ing, -ters.** —*vt.* To weld together (e.g., metallic powder) partially and without melting. —*vi.* To form a homogeneous mass by heating without melting.

**sin·u·ate** (sĭn'yo͞o-ĭt, -āt') *also* **sin·u·at·ed** (-ā'tĭd) *adj.* [Lat. *sinuatus,* p.part. of *sinuare,* to bend < *sinus,* curve.] Having a wavy indented margin <large *sinuate* leaves> **—sin'u·ate·ly** *adv.* **—sin'u·a'tion** (-ā'shən) *n.*

**sin·u·ous** (sĭn'yo͞o-əs) *adj.* [Lat. *sinuosus* < *sinus,* curve.] **1.** Marked by many curves or turns : WINDING. **2.** Supple and lithe. **3.** Sinuate. **—sin'u·ous'i·ty** (-ŏs'ĭ-tē) *n.* **—sin'u·ous·ly** *adv.* **—sin'u·ous·ness** *n.*

**si·nus** (sī'nəs) *n.* [Lat. *sinus,* curve, hollow.] **1.** A depression formed by a bending or curving. **2.** *Anat.* **a.** A dilated channel for the passage of chiefly venous blood. **b.** Any of various air-filled cavities in the cranial bones, esp. one communicating with the nostrils. **3.** *Pathol.* A fistula or channel to a suppurating cavity. **4.** *Bot.* A notch or indentation between lobes of a leaf or corolla.

**si·nus·i·tis** (sī′nə-sī′tĭs) *n.* Inflammation of a sinus membrane, esp. in the nasal region.

**si·nu·soid** (sī′nə-soid′, -nyə-) *n.* [Med. Lat. *sinus,* sine < Lat., curve + -OID.] A sine curve. **—si′nu·soi′dal** (-soid′l) *adj.*

**sinusoidal projection** *n.* A map projection in which areas are equal to corresponding areas on a globe, the parallels and the prime meridian being straight lines and the other meridians being increasingly curved outward from the prime meridian.

**Si·on** (sī′ən) *n. var. of* ZION.

**Siou·an** (so͞o′ən) *n.* [SIOU(X) + -AN.] A large North American Indian language family spoken from Lake Michigan to the Rocky Mountains and southward to Arkansas. **—Siouan** *adj.*

**Sioux** (so͞o) *n., pl.* **Sioux** (so͞o, so͞oz) [Fr., short for *Nadowessioux* < Ojibwa *nātowēssiwak,* the Dakota.] **1. a.** Any of the various groups of Indian peoples once occupying parts of the Great Plains in the Dakotas, Minnesota, and Nebraska. **b.** A member of one of these groups. **2.** Any of the languages of the Sioux. **—Sioux** *adj.*

**sip** (sĭp) *v.* **sipped, sip·ping, sips.** [ME *sippen.*] *—vt.* **1.** To imbibe delicately and in small quantities. **2.** To drink from in sips. *—vi.* To imbibe in sips. *—n.* **1.** An act of sipping. **2.** A small quantity sipped. **—sip′per** *n.*

**si·phon** *also* **sy·phon** (sī′fən) *n.* [Fr. < Lat. *sipho* < Gk. *siphōn,* tube.] **1.** A pipe or tube fashioned or deployed in an inverted U shape and filled until atmospheric pressure is sufficient to force a liquid from a reservoir in one end of the tube over a barrier higher than the reservoir and out the other end. **2.** *Zool.* A tubular organ, esp. of aquatic invertebrates as squids or clams, by which water is taken in or expelled. *—v.* **-phoned, -phon·ing, -phons.** *—vt.* To draw off or convey through or as if through a siphon. *—vi.* To pass through a siphon. **—si′phon·al, si·phon′ic** (sī-fŏn′ĭk) *adj.*

**si·phon·o·phore** (sī-fŏn′ə-fôr′, -fōr′, sī′fə-nə-) *n.* [NLat. *Siphonophora,* order name : Lat. *sipho,* siphon + Gk. *pherein,* to bear.] A colonial marine coelenterate of the order Siphonophora, including the Portuguese man-of-war.

**si·phon·o·stele** (sī-fŏn′ə-stēl′, sī′fə-nə-stē′lē) *n.* [SIPHON + STELE.] A vascular tube encircling the pith in the stems of some plants. **—si·phon′o·ste′lic** (-stē′lĭk) *adj.*

**si·phun·cle** (sī′fŭng′kəl) *n.* [Lat. *siphunculus,* dim. of *sipho,* siphon.] **1.** A tubelike structure in the body of a shelled cephalopod, as a chambered nautilus, extending through each chamber of the shell. **2.** A dorsal tube in an aphid, secreting a waxy fluid.

**sir** (sûr) *n.* [ME, var. of *sire.* —see SIRE.] **1.** *often* **Sir.** —Used as a form of respectful address in place of a man's name. **2. Sir.** —Used as an honorific before the given name or the full name of a baronet or a knight. **3.** A high-ranking gentleman.

**sir·dar** (sûr′där′, sər-där′) *n.* [Hindi *sardār* < Pers. : *sar,* head + *-dār,* holder.] One of high rank, esp. in India.

**sire** (sīr) *n.* [ME < OFr. < Lat. *senior,* older, comp. of *senex,* old.] **1. a.** A father. **b.** *Archaic.* A male ancestor : FOREFATHER. **2.** The male parent of an animal, esp. a horse. **3.** *Archaic.* SIR 3. **4.** *Archaic.* A form of address for a superior, used esp. in addressing a king. *—vt.* **sired, sir·ing, sires.** To beget.

**sir·ee** (sə-rē′) *n. var. of* SIRREE.

**si·ren** (sī′rən) *n.* [ME < OFr. *sereine* < LLat. *sirena* < Lat. *Siren* < Gk. *Seirēn.*] **1.** *often* **Siren.** *Gk. Myth.* One of a group of sea nymphs who by their sweet singing lured mariners to destruction on the rocks surrounding their island. **2.** A seductive woman. **3.** A device in which compressed air or steam is driven against a rotating perforated disk to create a loud, penetrating whistle or wailing, as a signal or warning. **4.** Any of several North American amphibians of the family Sirenidae, with an eellike body and no hind limbs.

**si·re·ni·an** (sī-rē′nē-ən) *n.* [< NLat. *Sirenia,* order name < Lat. *Siren,* siren.] A herbivorous aquatic mammal of the order Sirenia, including the manatee and the dugong. **—si·re′ni·an** *adj.*

**siren song** *n.* A deceptively alluring plea or appeal.

**Sir·i·us** (sîr′ē-əs) *n.* [Lat. < Gk. *Seirios* < *seirios,* burning.] A star in the constellation Canis Major, the brightest star in the sky.

**sir·loin** (sûr′loin′) *n.* [OFr. *surlonge* : *sur,* above (< Lat. *super*) + *longe,* loin < Lat. *lumbus.*] A cut of meat, esp. of beef, from the upper part of the loin between the rump and the porterhouse.

**si·roc·co** (sə-rŏk′ō) *also* **sci·roc·co** (shə-) *n., pl.* **-cos.** [Ital. < Ar. *sharq,* east.] **1.** A hot, humid south or southeast wind of southern Italy, Sicily, and the Mediterranean islands, originating in the Sahara as a dry, dusty wind but becoming moist as it passes over the Mediterranean. **2.** A hot or warm southerly wind, esp. one moving toward a low barometric pressure center.

**sir·rah** (sĭr′ə) *n.* [Alteration of SIR.] *Obs.* Mister : fellow. —Used contemptuously as a form of address.

**sir·ree** *also* **sir·ee** (sə-rē′) *n. Informal.* Sir. —Used emphatically after *yes* or *no.*

**sir·up** (sĭr′əp, sûr′-) *n. var. of* SYRUP.

**sir·vente** (sîr-vänt′) *also* **sir·ven·tes** (-vĕn′təs) *n., pl.* **-ventes** (-vänt′, -vänts′) *also* **-ven·tes** (-vĕn′təs) [Fr. < Prov. *sirventes* < *sirvent,* servant < Lat. *serviens,* pr.part. of *servire,* to serve < *servus,* servant.] A form of lyric verse of the Provençal troubadours satirizing political, social, or moral themes.

**sis** (sĭs) *n. Informal.* Sister.

**si·sal** (sī′səl, -zəl) *n.* [Mex. Sp., after *Sisal,* a town in Yucatán.] **1.** A fleshy plant, *Agave sisalana* native to Mexico, cultivated for its large leaves that yield a stiff fiber used for cordage and rope. **2. a.** The fiber of the sisal. **b.** The fiber of certain plants similar or related to the sisal.

**sis·kin** (sĭs′kĭn) *n.* [MDu. *sīseken,* dim. of MLG *sīsek,* of Slav. orig.] A small bird of the family Fringillidae, esp. *Carduelis spinus* of Eurasia, or the pine siskin.

**sis·si·fied** (sĭs′ə-fīd′) *adj.* Of, pertaining to, or having the characteristics of a sissy : EFFEMINATE.

**sis·sy** (sĭs′ē) *n., pl.* **-sies.** [< *sis,* short for SISTER.] **1.** An effeminate boy or man : MILKSOP. **2.** One who is timid or cowardly. **3.** *Informal.* Sister. **—sis′sy** *adj.*

**sissy bar** *n. Informal.* A narrow, inverted U-shaped bar that rises from behind the seat of a motorcycle or bicycle and supports the driver or a passenger.

**sissy bar**
*On a motorcycle*

**sis·ter** (sĭs′tər) *n.* [ME, partly < OE *sweostor,* and partly of Scand. orig.] **1. a.** A woman or girl having the same mother and father as another. **b.** A woman or girl having one parent in common with another. **2.** A woman or girl who shares a common ancestry, allegiance, character, or purpose with another or others, specif.: **a.** A kinswoman. **b.** A fellow member, as of a sorority. **c.** A fellow woman, friend, or companion. **3.** *Informal.* A girl or woman. —Used as a form of address. **4. Sister. a.** A member of a religious order of women : NUN. **b.** —Used as a form of address for a member of a religious order of women. **5.** *Chiefly Brit.* A nurse, esp. the head nurse in a ward. **6.** A thing identified as female and closely related to another thing.

**sis·ter·hood** (sĭs′tər-ho͝od′) *n.* **1.** The state or relationship of being a sister or sisters. **2.** The quality of being sisterly. **3.** A society, esp. a religious society, of women. **4.** Association or unification of women in a common cause <the *sisterhood* of feminists>

**sis·ter-in-law** (sĭs′tər-ĭn-lô′) *n., pl.* **sis·ters-in-law. 1.** The sister of one's husband or wife. **2.** The wife of one's brother. **3.** The wife of the brother of one's spouse.

**sis·ter·ly** (sĭs′tər-lē) *adj.* Having the nature of or befitting a sister or sisters. **—sis′ter·li·ness** *n.* **—sis′ter·ly** *adv.*

**Sis·tine** (sĭs′tēn′, sĭ-stēn′) *also* **Six·tine** (sĭk′stēn′, -stīn′) *adj.* [Ital. *sistino* < NLat. *sixtinus* < *Sixtus,* the name of several popes.] Of or concerning one of the popes named Sixtus.

**sis·trum** (sĭs′trəm) *n., pl.* **-trums** *or* **-tra** (-trə) [ME < Lat. < Gk. *seistron* < *seiein,* to shake.] An ancient Egyptian percussion instrument made up of metal rods or loops attached to a metal frame.

**Sis·y·phus** (sĭs′ə-fəs) *n. Gk. Myth.* A cruel king of Corinth condemned forever to roll a huge stone up a hill in Hades only to have it roll down again on nearing the top. **—Sis′y·phe′an** (-fē′ən) *adj.*

**sit** (sĭt) *v.* **sat** (săt), **sit·ting, sits.** [ME *sitten* < OE *sittan.*] *—vi.* **1.** To assume a position with the torso vertical and the body resting on the buttocks. **2.** To rest with the hindquarters lowered onto a supporting surface. —Used of animals. **3.** To perch. —Used of birds. **4.** To cover eggs for hatching : BROOD. **5.** To be situated : LIE <a tower that *sits* on a mountain> **6.** To pose for a photographer or artist. **7. a.** To occupy a seat as a member of an official body <*sit* in the Senate> **b.** To be in session <The court is *sitting* in Boston today.> **8.** To remain inactive or unused. **9.** To lie or rest in a specified manner <*sitting* idle> **10.** To affect one with or as if with a burden : WEIGH <Official duties *sat* heavily on me.> **11.** To fit, fall, or drape in a given manner <The jacket *sits* perfectly on you.> **12.** To be agreeable to one : PLEASE <an idea that didn't *sit* well with them> **13.** *Chiefly Brit.* To take an examination, as for a degree. **14.** To blow from a certain direction. —Used of the wind. **15.** To baby-sit or keep watch. *—vt.* **1.** To cause to sit : SEAT <*Sit* yourself over there.> **2.** To keep one's seat upon (an animal). **—sit down.** To take a seat. **—sit in. 1.** To attend as an observer <*sat in* on the discussion> **2.** To take part in a sit-in. **—sit on (one's) hands.** To fail to act. **—sit on (**or **upon).** *Informal.* **1.** To suppress <*sat on* the evidence> **2.** To rebuke sharply: REPRIMAND. **—sit out. 1.** To stay until the end of. **2.** To refrain from taking part in <*sit out* a dance> **—sit pretty.** *Informal.* To be in a favorable position. **—sit tight.** *Informal.* To be patient and await developments. **—sit up. 1.** To stay up later than the customary bedtime. **2.** To become suddenly alert <*sit up* and take notice>

**si·tar** (sĭ-tär′) *n.* [Hindi *sitār* < Pers. : *si*, three + *tār*, string.] A Hindu stringed instrument consisting of seasoned gourds and teak and having a track of 20 metal frets with 6 or 7 main playing strings above and 13 sympathetic resonating strings below. **—si·tar′ist** *n.*

**sit·com** *also* **sit-com** (sĭt′kŏm′) *n. Informal.* A situation comedy.

**sit-down** (sĭt′doun′) *n.* **1.** A work stoppage in which the workers refuse to leave their place of employment pending agreement. **2.** An obstruction of normal activity, as of an office, by the act of a large group sitting down to express a grievance or protest. *—adj.* Performed or accomplished while sitting down <a *sit-down* luncheon>

**site** (sīt) *n.* [ME < OFr. < Lat. *situs*, place < p.part. of *sinere*, to allow, put.] **1.** Location. **2.** The setting of an event. *—vt.* **sit·ed, sit·ing, sites.** To locate on a site <*sited* the power plant by the river>

**sith** (sĭth) *conj.* [ME *sithe* < OE *siððan*, since.] *Archaic.* Since. **—sith** *adv. & prep.*

**sit-in** (sĭt′ĭn′) *n.* **1.** A protest demonstration in which participants seat themselves in an appropriate place and refuse to move until their demands are considered or met. **2.** An act of occupying the seats or an area of a segregated establishment to protest racial discrimination.

**si·tol·o·gy** (sī-tŏl′ə-jē) *n.* [Gk. *sitos*, food, grain + -LOGY.] The science of foods, nutrition, and diet.

**si·tos·ter·ol** (sī-tŏs′tə-rôl′, -rōl′, sĭ-) *n.* [Gk. *sitos*, food, grain + STEROL.] Any of a group of sterols that occur in plants and are used in the synthesis of steroid hormones.

**sit·ter** (sĭt′ər) *n.* **1.** One that sits, esp. a baby sitter. **2.** A brooding hen.

**sit·ting** (sĭt′ĭng) *n.* **1.** The act or position of one that sits. **2.** A period during which one is seated and occupied with a single activity. **3.** A session of an official body, as court. **4. a.** A period of incubation. **b.** The number of eggs under a brooding bird.

**sitting duck** *n. Informal.* An easy target or victim.

**sitting room** *n.* A living room.

**sit·u·ate** (sĭch′ōō-āt′) *vt.* **-at·ed, -at·ing, -ates.** [Med. Lat. *situare, situat-*, to put < Lat. *situs*, place.] **1.** To place in a spot: LOCATE. **2.** To place under particular circumstances or in a given condition. *—adj.* (-ĭt, -āt′). *Archaic.* Situated.

**sit·u·a·tion** (sĭch′ōō-ā′shən) *n.* **1. a.** The way in which something is positioned with regard to its surroundings. **b.** The place in which something is situated: LOCATION. **2.** Position with regard to surrounding conditions and attendant circumstances: STATUS. **3.** A combination of circumstances at a given moment: STATE OF AFFAIRS <the international *situation*> **4.** A critical or problematic combination of circumstances. **5.** A position of employment: POST. **—sit′u·a′tion·al** *adj.*

**situation comedy** *n.* A humorous radio or television series with a continuing cast of characters.

**situation ethics** *n. (pl. in number).* A system of ethics based on brotherly love in which acts are morally evaluated within a situational context rather than by application of moral absolutes.

**sit-up** (sĭt′ŭp′) *n.* A physical exercise in which one uses the abdominal muscles to raise the torso from a prone to a sitting position with or without bending the knees and then returns to the original position.

**si·tus** (sī′təs) *n., pl.* **situs.** [Lat., place.] Position, esp. normal position, as of a bodily organ.

**sitz bath** (sĭts) *n.* [Partial transl. of G. *Sitzbad* : *Sitz*, act of sitting + *Bad*, bath.] **1.** A tub in which one bathes in a sitting position. **2.** A therapeutic bath taken in a sitz bath.

**sitz·krieg** (sĭt′skrēg′, zĭt′-) *n.* [G. : *sitz*, act of sitting + *Krieg*, war.] Warfare marked by a lack of aggression or progress.

**sitz·mark** (sĭt′smärk′, zĭt′-) *n.* [Partial transl. of G. *sitzmarke* : *sitz*, act of sitting + *Marke*, mark.] A hollow made in the snow by a skier falling backward.

**Si·va** (shē′və, sē′-, shĭv′ə, sĭv′ə) *n. var. of* SHIVA.

**Si·van** (sĭv′ən) *n.* [Heb. *Sīwān*.] The ninth month of the Hebrew year. —See table at CALENDAR.

**six** (sĭks) *n.* [ME < OE *siex*, akin to G. *sechs*, Lat. *sex*, Gk. *hex*, and Skt. *ṣaṣ*.] **1.** The cardinal number equal to 5 + 1. **2.** The sixth in a set or sequence. **3.** Something having six parts, units, or members, esp. a motor vehicle having six cylinders. **—six** *adj. & pron.*

**six-gun** (sĭks′gŭn′) *n.* A six-shooter.

**six-pack** (sĭks′păk′) *n.* **1.** Six units of a product, esp. six cans or bottles of a beverage sold in a pack. **2.** The contents of a six-pack.

**six·pence** (sĭks′pəns) *n. Chiefly Brit.* **1.** A coin worth six pennies. **2.** The sum of six pennies.

**six·pen·ny** (sĭks′pə-nē) *adj.* **1.** Valued at, selling for, or worth sixpence. **2.** Of little worth: PALTRY. **3.** (sĭks′pĕn′ē). Denoting a size of nails, gen. two inches.

**six-shoot·er** (sĭks′shōō′tər) *n.* A six-chambered revolver.

**six·teen** (sĭk-stēn′) *n.* [ME *sixtene* < OE *sixtyne.*] **1.** The cardinal number equal to 15 + 1. **2.** The sixteenth in a set or sequence. **—six·teen′** *adj. & pron.*

**six·teen·mo** (sĭk-stēn′mō) *n., pl.* **-mos.** Sextodecimo.

**six·teenth** (sĭk-stēnth′) *n.* **1.** The ordinal number matching the number 16 in a series. **2.** One of 16 equal parts. **—six·teenth′** *adj. & adv.*

**sixteenth note** *n. Mus.* A note having $^1/_{16}$ the time value of a whole note.

**sixth** (sĭksth) *n.* **1.** The ordinal number matching the number six in a series. **2.** One of six equal parts. **3.** *Mus.* **a.** An interval of six degrees in a diatonic scale. **b.** A tone separated by this interval from a given tone. **c.** The harmonic combination of two tones separated by this interval. **d.** The sixth tone of a scale: SUBMEDIANT. **—sixth** *adj. & adv.*

**sixth sense** *n.* Perceptive power apparently independent of the five senses: INTUITION.

**six·ti·eth** (sĭk′stē-ĭth) *n.* **1.** The ordinal number matching the number 60 in a series. **2.** One of 60 equal parts. **—six′ti·eth** *adj. & adv.*

**Six·tine** (sĭk′stēn′, -stīn′) *adj. var. of* SISTINE.

**six·ty** (sĭks′tē) *n.* [ME < OE *siextig* : *siex*, six + *-tig*, -ty.] The cardinal number equal to 6 × 10. **—six′ty** *adj. & pron.*

**six·ty-fourth note** (sĭk′stē-fôrth′, -fōrth′) *n. Mus.* A note having $^1/_{64}$ the time value of a whole note.

**siz·a·ble** *also* **size·a·ble** (sī′zə-bəl) *adj.* Having considerable size. **—siz′a·ble·ness** *n.* **—siz′a·bly** *adv.*

**size[1]** (sīz) *n.* [ME *syse* < OFr. *sise* < *assise*, act of sitting. —see ASSIZE.] **1.** Physical proportions, dimensions, magnitude, or extent. **2.** Any of a series of graduated categories of dimension whereby manufactured articles are classified. **3.** Considerable extent, amount, or dimensions. **4.** Moral or mental qualities, rank, or status with reference to relative importance or the capacity to meet given requirements. **5.** The true state of affairs <That's about the *size* of the situation.> *—vt.* **sized, siz·ing, siz·es. 1.** To arrange, classify, or distribute according to size. **2.** To make, cut, or shape in accordance with a required size. **—size up. 1.** To make an estimate, opinion, or judgment of <*sized up* their opponents> **2.** To meet the given specifications or requirements.

**size[2]** (sīz) *n.* [ME *syse.*] Any of several gelatinous or glutinous substances usu. made from glue, wax, or clay and used as a glaze or filler for porous materials such as paper, cloth, or wall surfaces. *—vt.* **sized, siz·ing, siz·es.** To treat or coat with size. **—siz′y** *adj.*

**size·a·ble** (sī′zə-bəl) *adj. var. of* SIZABLE.

**sized** (sīzd) *adj.* Having a particular or specified size <medium-*sized* jackets>

**siz·ing** (sī′zĭng) *n.* **1.** SIZE[2]. **2.** Treatment of a surface or material with size.

**siz·zle** (sĭz′əl) *vi.* **-zled, -zling, -zles.** [Imit.] **1.** To make the hissing sound characteristic of frying fat. **2.** To seethe with anger: BOIL. **3.** To be very hot. —*n.* A hissing sound.

**siz·zler** (sĭz′lər) *n. Informal.* A very hot day.

**skald** *also* **scald** (skôld, skäld) *n.* [ON *skāld.*] An ancient Scandinavian poet: BARD. **—skald′ic** *adj.*

**skat** (skăt) *n.* [G. < Ital. *scarto*, a discarded card < *scartare*, to reject : *s-*, out (< Lat. *ex-*) + *carta*, card < Lat. *charta*, leaf of papyrus. —see CARD[1].] **1.** A card game for 3 persons played with 32 cards, sevens through aces. **2.** A combination of cards occurring in skat.

**skate[1]** (skāt) *n.* [< Du. *schaats* (taken as pl.), stilt, skate < ONFr. *escace*, stilt, perh. of Germanic orig.] **1. a.** A shoe or boot with a bladelike metal runner fixed to its sole, enabling the wearer to glide over ice. **b.** A bladelike metal runner having clamps and straps for attaching it to a shoe or boot. **2.** A roller skate. *—vi.* **skat·ed, skat·ing, skates.** To glide on or as if on skates. **—skat′ing** *n.*

**skate[2]** (skāt) *n.* [ME *scate* < ON *skata.*] A marine fish of the family Rajidae, having a cartilaginous skeleton and a flattened body with the pectoral fins forming winglike lateral extensions.

**skate[3]** (skāt) *n.* [Perh. alteration of dial. *skite*, contemptible person.] A fellow <a good *skate*>

**skate·board** (skāt′bôrd′, -bōrd′) *n.* A short, narrow board having a set of four roller-skate wheels mounted under it.

**skat·er** (skā′tər) *n.* **1.** One who skates. **2.** The water strider.

**skat·ole** (skăt′ōl′, skā′tōl′) *also* **skat·ol** (skăt′ôl′, skā′tōl′) *n.* [Gk. *skōr, skat-*, dung + -OLE.] A white crystalline organic compound, $C_9H_9N$, having a strong fecal odor, found naturally in feces, beets, and coal tar and used as a fixative in perfumery.

**skean** (skēn) *n.* [Ir. Gael. *scian* < OIr. *scīan.*] A double-edged dagger once used in Ireland and Scotland.

**ske·dad·dle** (skĭ-dăd′l) *vi.* **-dled, -dling, -dles.** [Orig. unknown.] *Informal.* To leave hastily: FLEE. **—ske·dad′dler** *n.*

**skeet** (skēt) *n.* [Ult. < ON *skjōta*, to shoot.] Trapshooting in which clay targets are thrown from traps to simulate birds in flight and are shot at from different stations.

**skeg** (skĕg) *n.* [Du. *scheg* < ON *skegg*, beard.] *Naut.* **1.** A timber connecting the keel and the sternpost. **2.** An arm extending to the rear of the keel to support the rudder and protect the propeller. **3.** A series of timbers attached to the stern of a small boat, serving as a keel to keep the boat on course.

**skein** (skān) *n.* [ME *skeyne* < OFr. *escaigne.*] **1. a.** A length of thread or yarn wound in a loose, elongated coil. **b.** Something resembling a skein of thread or yarn <a twisted *skein* of lies> **2.** A flock of birds, as geese, in flight.

**skel·e·ton** (skĕl′ĭ-tn) *n.* [NLat. < Gk., neuter of *skeletos,* dried up.] **1. a.** The internal vertebrate structure composed of bone and cartilage that protects and supports the soft organs, tissues, and parts. **b.** The hard external structure in many invertebrates and certain vertebrates, as turtles : EXOSKELETON. **2.** A supporting structure or framework. **3.** An outline or sketch. **4.** One that is very thin or emaciated. **—skel′e·tal** *adj.* **—skel′e·tal·ly** *adv.*

**skeleton key** *n.* A key with a large portion of the bit filed away so that it can open different locks.

**skep** (skĕp) *n.* [ME, basket, basketful < OE *sceppe* < ON *skeppa,* basket.] A beehive, esp. one of straw.

**skep·tic** *also* **scep·tic** (skĕp′tĭk) *n.* [Lat. *Scepticus,* disciple of Pyrrho < Gk. *Skeptikos* < *skeptesthai,* to examine.] **1.** One who instinctively or consistently doubts, questions, or disagrees with assertions or gen. accepted conclusions. **2.** One inclined to religious skepticism. **3. a.** *often* **Skeptic.** An adherent of a philosophical school of skepticism. **b. Skeptic.** A member of an ancient Greek school of philosophical skepticism, esp. that of Pyrrho of Elis.

**skep·ti·cal** *also* **scep·ti·cal** (skĕp′tĭ-kəl) *adj.* **1.** Instinctively or consistently questioning. **2.** Relating to or characteristic of skeptics or skepticism. **—skep′ti·cal·ly** *adv.*

**skep·ti·cism** *also* **scep·ti·cism** (skĕp′tĭ-sĭz′əm) *n.* **1.** A doubting or questioning attitude : DUBIETY. **2.** The philosophical doctrine that absolute knowledge is impossible and that inquiry must be a process of doubting in order to acquire approximate or relative certainty. **3.** Doubt or disbelief of religious tenets.

**sker·ry** (skĕr′ē) *n., pl.* **-ries.** [Of Scand. orig.] A small rocky reef or island.

**sketch** (skĕch) *n.* [Du. *schets* < Ital. *schizzo* < *schizzare,* to fizz.] **1.** A hasty or undetailed drawing or painting made as a preliminary study. **2.** A brief, incomplete delineation or presentation, as of a book to be completed : OUTLINE. **3. a.** A brief, light, or informal literary composition, as a short story or essay. **b.** *Mus.* A brief composition, esp. for the piano. **c.** A short, often satirical scene or play in a revue or variety show. **4.** *Informal.* One who is amusing. *—v.* **sketched, sketch·ing, sketch·es.** *—vt.* To make a sketch of : OUTLINE. *—vi.* To make a sketch. **—sketch′er** *n.*

☆ **syns:** SKETCH, ACT, SKIT *n. core meaning* : a short theatrical piece within a larger production <a comic *sketch* in a variety show>

**sketch·book** (skĕch′bŏŏk′) *n.* **1.** A pad of paper for sketching. **2.** A book of literary sketches.

**sketch·y** (skĕch′ē) *adj.* **-i·er, -i·est.** **1.** Having the nature of a sketch : presenting only major points or parts. **2. a.** Lacking completeness : INCOMPLETE <only a *sketchy* plan of attack> **b.** Slight : superficial. **—sketch′i·ly** *adv.* **—sketch′i·ness** *n.*

**skew** (skyōō) *v.* **skewed, skew·ing, skews.** [ME *skewen,* to move sideways, escape < ONFr. *eskiuer,* of Germanic orig.] *—vi.* **1.** To take an oblique direction. **2.** To look obliquely. *—vt.* **1.** To turn or place at an angle. **2.** To give a bias to : DISTORT. *—adj.* **1.** Positioned to one side. **2.** Distorted or biased. **3.** Having a part that diverges, as gearing. **4. a.** *Math.* Neither parallel nor intersecting. —Used of straight lines in space. **b.** *Statistics.* Not symmetric about the mean. —Used of distributions. *—n.* An oblique movement, position, or direction. **—skew′ness** *n.*

**skew arch** *n.* An arch having sides not at right angles to the face of its abutments.

**skew·back** (skyōō′băk′) *n.* Either of two inset abutments sloped to support a segmental arch.

**skew·bald** (skyōō′bôld′) *adj.* [Obs. *skewed,* skewbald + BALD.] Having spots or patches of white on a coat of a color other than black <a *skewbald* horse>

**skew·er** (skyōō′ər) *n.* [Var. of dial. *skiver.*] **1.** A long pin used to secure or suspend food during cooking : SPIT. **2.** A pick or rod functioning or shaped similar to a skewer. *—vt.* **-ered, -er·ing, -ers.** To hold together or pierce with or as if with a skewer.

**skew field** *n.* A mathematical field in which commutativity does not hold for multiplication.

**skew lines** *pl.n.* Straight lines that are not in the same plane and do not intersect.

**ski** (skē) *n., pl.* **skis.** [Norw. < ON *skīð.*] **1. a.** One of a pair of long, flat runners of wood, metal, or plastic that curve upward in front and are attached to a boot for traveling over snow. **b.** A water-ski. **2.** Something shaped like a ski that is used as a runner, as on an aircraft. *—v.* **skied, ski·ing, skis.** *—vi.* To travel on skis, esp. as a sport. *—vt.* To travel over on skis. **—ski′er** *n.* **—ski′ing** *n.*

**ski·a·gram** (skī′ə-grăm′) *also* **ski·a·graph** (-grăf′, -gräf′) *n.* [Gk. *skia,* shadow + -GRAM.] A picture or photograph made up of shadows or outlines.

**ski·ag·ra·phy** (skī-ăg′rə-fē) *n.* [SKIA(GRAM) + -GRAPHY.] The art or technique of making skiagrams.

**ski·a·scope** (skī′ə-skōp′) *n.* [Gk. *skia,* shadow + -SCOPE.] A retinoscope.

**ski·as·co·py** (skī-ăs′kə-pē) *n.* Retinoscopy.

**ski·bob** (skē′bŏb′) *n.* A vehicle for gliding downhill over snow consisting of two skis one behind the other on a metal frame, steering handlebars connected to the shorter forward ski, and a low seat attached to the longer rear ski for the rider, who wears small skis for balance. **—ski′bob·ber** *n.*

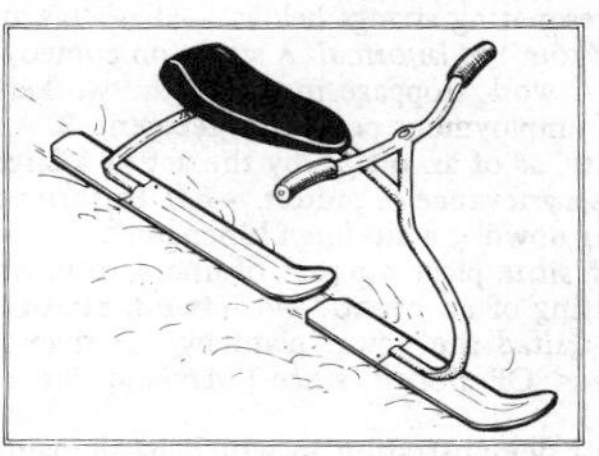
**skibob**

**ski boot** *n.* A stiff padded boot fastened to the foot with strong buckles and locked into place in a ski binding.

**skid** (skĭd) *n.* [Orig. unknown.] **1.** The act of sliding or slipping sideways over a surface. **2. a.** A plank, log, or timber used as a support or as a track for sliding or rolling heavy objects. **b.** A small platform for stacking merchandise to be moved or temporarily stored. **c.** One of several logs or timbers forming a skid road. **3. skids.** *Naut.* A wooden framework attached to the side of a ship to prevent damage, as when unloading. **4.** A shoe or drag applying pressure to a wheel to brake a vehicle. **5.** A runner in the landing gear of certain aircraft. *—v.* **skid·ded, skid·ding, skids.** *—vi.* **1.** To slip or slide sideways while moving because of loss of traction <a car *skidding* on ice> **2.** To slide without revolving <wheels *skidding* on wet, oily pavement> **3.** To move sideways in a turn because of insufficient banking. —Used of an aircraft. *—vt.* **1.** To brake (a wheel) with a skid. **2.** To haul on a skid or skids. **—on the skids.** *Slang.* On a path to destruction or failure.

**skid fin** *n.* An upright auxiliary airfoil placed above the upper wing in biplanes to increase lateral stability.

**skid road** *n.* **1.** A track made of logs laid transversely about five feet apart that is used to haul logs to a loading platform or a mill. **2.** *Slang.* Skid row.

**skid row** *n.* [Alteration of SKID ROAD.] *Slang.* A squalid urban district inhabited by derelicts and vagrants.

**skiff** (skĭf) *n.* [Fr. *esquif* < Ital. *schifo,* of Germanic orig.] *Naut.* A flat-bottomed open boat of shallow draft, having a pointed bow and a square stern and propelled by oars, sail, or motor.

**skif·fle** (skĭf′əl) *n.* [Perh. imit.] Folk or country music played by performers who use unconventional instruments or percussion, as kazoos, washboards, or jugs.

**ski·jor·ing** (skē′jôr′ĭng, -jōr′-) *n.* [Norw. *skikjoring* : *ski,* ski + *kjoring,* driving < *kjore,* to drive < ON *keyra.*] A sport in which a skier is drawn over ice or snow by a vehicle or horse.

**ski jump** *n.* **1.** A jump or leap made by a skier. **2.** A chute prepared for a ski jump.

**skil·ful** (skĭl′fəl) *adj. var. of* SKILLFUL.

**ski lift** *n.* A power-driven conveyor, usu. with attached tow bars, suspended chairs, or gondolas, used to carry skiers to the top of a mountain, slope, or trail.

**skill** (skĭl) *n.* [ME *skile* < ON *skil,* discernment.] **1.** Proficiency, ability, or dexterity. **2.** An art, trade, or technique, esp. one requiring use of the hands or body. **3.** *Obs.* A reason or cause.

**skilled** (skĭld) *adj.* **1.** Possessing or exhibiting skill : EXPERT. **2.** Requiring specialized ability or training <*skilled* labor>

**skil·let** (skĭl′ĭt) *n.* [ME *skelet.*] **1.** A frying pan. **2.** *Chiefly Brit.* A long-handled stewing pan or saucepan occas. having legs.

**skill·ful** *also* **skil·ful** (skĭl′fəl) *adj.* **1.** SKILLED 1. **2.** Requiring skill. **—skill′ful·ly** *adv.* **—skill′ful·ness** *n.*

**skim** (skĭm) *v.* **skimmed, skim·ming, skims.** [ME *skymmen.*] *—vt.* **1.** To remove floating matter from (a liquid). **2.** To remove (floating matter) from a liquid. **3.** To coat or cover with or as if with a thin layer, as of scum. **4. a.** To hurl across and close to the surface of, so as to bounce on water or slide on ice <*skimming* stones across a pool> **b.** To glide or pass quickly and lightly over. **5.** To glance through superficially or quickly <*skimmed* the memorandum> *—vi.* **1.** To pass swiftly and lightly over or near a surface : GLIDE. **2.** To give a quick and superficial reading or consideration : GLANCE <*skimmed* through the magazine> **3.** To become coated with a thin layer. *—n.* **1.** The act of skimming. **2.** Something, as skim milk, that has been skimmed. **3.** A thin layer or film.

**ski mask** *n.* A knitted mask worn esp. by skiers for protection from the cold.
**skim·mer** (skĭm′ər) *n.* **1.** One that skims. **2.** A flat, usu. perforated utensil used in skimming liquids. **3.** A wide-brimmed hat with a flat shallow crown. **4.** A chiefly coastal bird of the genus *Rynchops,* having long narrow wings and a long bill with a longer lower mandible for skimming the water's surface for food.
**skim milk** *n.* Milk from which the cream has been removed.
**skim·ming** (skĭm′ĭng) *n.* **1.** *often* **skimmings.** Matter skimmed off a liquid. **2.** An act of skimming.
**ski·mo·bile** (skē′mō-bēl′, -mə-) *n.* A snowmobile.
**skimp** (skĭmp) *v.* **skimped, skimp·ing, skimps.** [Perh. alteration of SCRIMP.] —*vt.* **1.** To do hastily or carelessly. **2.** To be sparing with : SCRIMP. —*vi.* To be unduly thrifty. **—skimp** *adj.*
**skimp·y** (skĭm′pē) *adj.* **-i·er, -i·est. 1.** Inadequate : scanty <a *skimpy* supper> **2.** Unduly thrifty : NIGGARDLY. **—skimp′i·ly** *adv.* **—skimp′i·ness** *n.*
**skin** (skĭn) *n.* [ME < ON *skinn.*] **1.** The membranous tissue forming the external covering of the animal body : INTEGUMENT. **2.** An animal pelt, esp. the pliable pelt of a small or young animal. **3.** An outer layer like skin, as the rind of fruit or the plating on a ship or rocket. **4.** A liquid container made of animal skin <a wine *skin*> —*v.* **skinned, skin·ning, skins.** —*vt.* **1.** To remove skin from. **2.** To cover with or as if with skin. **3.** To peel off (an outer covering). **4.** *Slang.* To fleece : swindle. **5.** To bruise, cut, or injure the skin or surface of <*skinned* my elbow> —*vi.* **1.** To become covered with or as if with skin. **2.** To pass by or through with little room to spare. **3.** To go hurriedly. **—by the skin of one's teeth.** By the smallest margin. **—get under one's skin. 1.** To anger or irritate. **2.** To be or become an obsession. **—have a thick skin.** To be unperturbed by criticism or insults. **—have a thin skin.** To be easily perturbed by criticism or insults.
**skin-deep** (skĭn′dēp′) *adj.* Devoid of real significance or meaning : SHALLOW. **—skin′-deep′** *adv.*
**skin-dive** (skĭn′dīv′) *vi.* **-dived, -div·ing, -dives.** To take part in skin diving.
**skin diving** *n.* Underwater swimming, exploration, or fishing, often with flippers, a face mask, and a snorkel or scuba equipment. **—skin diver** *n.*
**skin drag** *n.* Skin friction.
**skin effect** *n.* The tendency of electric current density in a conductor carrying alternating current to be greater at the surface than at the center.
**skin·flint** (skĭn′flĭnt′) *n.* A miser.
**skin friction** *n.* Friction caused by air crossing the surface of aircraft or rockets at high speeds.
**skin game** *n.* **1.** A crooked gambling game. **2.** A swindle.
**skin graft** *n.* A surgical graft of skin from one part of the body to another or from one individual to another.
**skin·head** (skĭn′hĕd′) *n. Slang.* A young British working-class tough with close-cropped hair.
**skink** (skĭngk) *n.* [Lat. *scincus* < Gk. *skinkos.*] Any of numerous smooth, shiny lizards of the family Scincidae, with a cylindrical body and short or rudimentary legs.
**skinned** (skĭnd) *adj.* Having a specified kind of skin <The sun often burns fair-*skinned* people.>
**†skin·ner** (skĭn′ər) *n.* **1.** One who flays, dresses, or sells animal skins. **2.** *Western U.S.* A mule driver.
**Skin·ner box** (skĭn′ər) *n.* [After B.F. Skinner (b.1904).] A laboratory apparatus for animal experiments in operant conditioning that usu. contains a bar or lever to be pressed by the animal to gain a reward, as food, or to avoid a painful stimulus, as shock.
**skin·ny** (skĭn′ē) *adj.* **-ni·er, -ni·est. 1.** Of, relating to, or like skin. **2.** Very thin. **—skin′ni·ness** *n.*
**skin·ny-dip** (skĭn′ē-dĭp′) *vi.* **-dipped, -dip·ping, -dips.** *Informal.* To swim in the nude. **—skin′ny-dip′per** *n.* **—skin·ny-dip·ping** *n.*
**skin-pop** (skĭn′pŏp′) *vt.* **-popped, -pop·ping, -pops.** *Slang.* To inject, as a drug, beneath the skin rather than into a vein. **—skin′-pop′ping** *n.*
**skin test** *n.* A test for an allergy or infectious disease, performed by a patch test, scratch test, or an intracutaneous injection of an allergen or extract of the disease-causing organism.
**skin·tight** (skĭn′tīt′) *adj.* Clinging closely to the skin.
**skip** (skĭp) *v.* **skipped, skip·ping, skips.** [ME *skippen.*] —*vi.* **1. a.** To move by springing or hopping first on one foot and then on the other. **b.** To leap lightly about. **2.** To bounce over or be deflected from a surface : RICOCHET. **3. a.** To pass from point to point omitting or disregarding what intervenes <*skipping* over the television channels> **b.** To be promoted in school beyond the next regular class or grade. **4.** *Informal.* To leave hastily : ABSCOND. **5.** To misfire <The engine *skipped* and choked.> —*vt.* **1.** To leap or jump lightly over <*skip* rope> **2.** To pass over, omit, or disregard <*skipped* the minor details> **3.** To cause to ricochet or skim. **4.** To be promoted beyond (the next grade or level). **5. a.** *Informal.* To leave hastily <The deadbeats *skipped* town.> **b.** To fail to attend <*skipped* the early class> —*n.* **1.** A leaping or jumping movement, esp. a gait in which hops and steps alternate. **2.** An omission.
**skip distance** *n.* The smallest separation between a transmitter and a receiver that permits radio signals of a specific frequency to travel from one to the other by reflection from the ionosphere.
**skip·jack** (skĭp′jăk′) *n., pl.* **skipjack** or **-jacks. 1.** Any of several marine food fishes of the genus *Euthynnus,* related to and resembling the tuna. **2.** Any of various fishes, as certain herrings **3.** A small sailboat with a bottom shaped like a flat V and nearly vertical sides.
**ski pole** *n.* A thin pointed pole with a disk above the point, used as an aid by skiers.
**skip·per¹** (skĭp′ər) *n.* [ME *skypper* < MDu. *schipper* < *schip,* ship.] The master of a ship. **—skip′per** *v.* **(-pered, -per·ing, -pers).**
**skip·per²** (skĭp′ər) *n.* **1.** One that skips. **2.** Any of numerous butterflies of the families Hesperiidae and Megathymidae, having a hairy, mothlike body and a darting flight pattern. **3.** A marine fish, esp. a saury, *Cololabis saira* indigenous to Pacific waters.
**skirl** (skûrl) *v.* **skirled, skirl·ing, skirls.** [ME *skrillen,* prob. of Scand. orig.] —*vi.* To produce a shrill, piercing tone. —Used of a bagpipe. —*vt.* To play on the bagpipe. —*n.* **1.** The shrill sound made by a bagpipe. **2.** A shrill piercing sound.
**skir·mish** (skûr′mĭsh) *n.* [ME *skirmisshe* < OFr. *eskermir, eskirmiss-,* to fight with a sword, of Germanic orig.] **1.** A minor encounter in war between small bodies of troops, often as part of larger movements. **2.** A minor or preliminary conflict. —*vi.* **-mished, -mish·ing, -mish·es.** To engage in a skirmish. **—skir′mish·er** *n.*
**skir·ret** (skûr′ĭt) *n.* [ME *skirwhit,* alteration of OFr. *eschervi,* prob. < Ar. *alkarawyā,* caraway.] An Old World plant, *Sium sisarum,* having a sweetish, edible root.
**skirt** (skûrt) *n.* [ME < ON *skyrta,* shirt.] **1.** That part of a garment, as a dress, that hangs from the waist down. **2.** A separate garment hanging from the waist and worn by women and girls. **3. a.** A leather flap hanging from the side of a saddle. **b.** The lower outer section of a rocket. **4.** A border, margin, or outer edge. **5. skirts.** The outskirts, as of a town. **6.** *Slang.* A woman or girl. —*v.* **skirt·ed, skirt·ing, skirts.** —*vt.* **1.** To lie along, form the border of, or surround : BOUND. **2.** To move or pass around rather than across or through. **3.** To evade or elude (as a topic of conversation) by circumlocution. —*vi.* To be near or move along an edge or border.

▲ word history: The connections between England and Scandinavia during the Middle Ages in both peace and war were always close, and the intercourse between them added greatly to the vocabulary of English. Some words borrowed from Scandinavian did not completely supplant the native words with which they were cognate, but existed alongside them. *Shirt* and *skirt* are one such pair. Both are descended from Germanic *skurtjon-,* which became *scyrte* in Old English and *skyrta* in Old Norse. Both words originally meant what *shirt* means today, but after *skyrta* was borrowed into Middle English it came to mean "a lower garment" or "the lower part of a garment." *Shirt* and *skirt* are related to *short,* and the original Germanic word probably denoted a short garment of some kind.

**ski run** *n.* A trail or slope for skiing.
**skit** (skĭt) *n.* [Orig. unknown.] **1.** A short, usu. comic theatrical sketch. **2.** A short humorous or satirical piece of writing.
**ski touring** *n.* Cross-country skiing for pleasure. **—ski tourer** *n.*
**ski tow** *n.* A ski lift in which skiers cling to a continuous rope as they are pulled up a slope.
**skit·ter** (skĭt′ər) *v.* **-tered, -ter·ing, -ters.** [Prob. freq. of dial. *skite,* to run rapidly.] —*vi.* **1.** To skip, glide, or move lightly or rapidly along a surface : FLIT. **2.** To fish by drawing a lure or baited hook over the surface of the water with a skipping movement. —*vt.* To cause to skitter.
**skit·tish** (skĭt′ĭsh) *adj.* [ME.] **1.** Nervously excitable. **2.** Shy, coy, or timid. **3. a.** Extremely lively or frivolous. **b.** Undependable : fickle. **—skit′tish·ly** *adv.* **—skit′tish·ness** *n.*
**skit·tle** (skĭt′l) *n.* [Orig. unknown.] *Chiefly Brit.* **1. skittles.** *(sing. in number).* The game of ninepins, in which a wooden disk or ball is thrown to knock down the pins. **2.** One of the pins used in skittles.
**skive** (skīv) *vt.* **skived, skiv·ing, skives.** [Of Scand. orig.] To shave the surface of (leather or rubber) : PARE.
**skiv·er** (skī′vər) *n.* **1.** A soft, thin leather split off the outside of sheepskin and used for bookbinding. **2.** One that skives. **3.** A cutting device, as a knife, used in skiving.
**Skiv·vies** (skĭv′ēz). A trademark for underwear.
**ski·wear** (skē′wâr′) *n.* Clothing appropriate for skiing.
**skoal** (skōl) *interj.* [Dan. and Norw. *skaal,* cup.] —Used as a drinking toast.
**sku·a** (skyōō′ə) *n.* [NLat. < Faroese *skūvur* < ON *skūfr.*] **1.** A predatory gull-like sea bird, *Catharacta skua,* of northern regions, having brownish plumage. **2.** *Chiefly Brit.* A jaeger.
**skul·dug·ger·y** (skŭl-dŭg′ə-rē) *n. var. of* SKULLDUGGERY.
**skulk** (skŭlk) *vi.* **skulked, skulk·ing, skulks.** [ME *skulken,* of Scand. orig.] **1.** To lie in hiding : LURK. **2.** To move about stealthily. **3.** To evade obligation or work : MALINGER. —*n.* **1.** One who skulks. **2.** A group of foxes. **—skulk′er** *n.*

**skull** (skŭl) *n.* [ME *skulle.*] **1.** The framework of the head of vertebrates, made up of the bones of the brain case and face. **2.** The head, esp. regarded as the seat of thought or intelligence. **3.** A death's-head.
**skull and crossbones** *n.* A representation of a human skull above two long crossed bones, a symbol of death once used by pirates and now used as a warning label on poisons.
**skull·cap** (skŭl'kăp') *n.* **1. a.** A light, close-fitting, brimless cap sometimes worn indoors. **b.** A yarmulke. **2.** A plant of the genus *Scutellaria,* with two-lipped flower clusters.
**skull·dug·ger·y** *also* **skul·dug·ger·y** (skŭl-dŭg'ə-rē) *n., pl.* **-ger·ies.** [Orig. unknown.] Crafty trickery or deception.
**skunk** (skŭngk) *n.* [Massachuset *squnck.*] **1.** A small carnivorous New World mammal of the genus *Mephitis* and related genera, having a bushy tail and black fur with white markings and ejecting a foul-smelling secretion from glands near the anus. **2.** *Slang.* A despicable person. —*vt.* **skunked, skunk·ing, skunks.** *Slang.* **1.** To defeat overwhelmingly, esp. by keeping from scoring. **2.** To cheat, as by failing to pay.
**skunk cabbage** *n.* **1.** An ill-smelling swamp plant, *Symplocarpus foetidus* of eastern North America, having minute flowers enclosed in a mottled greenish or purplish spathe. **2.** A plant, *Lysichitum americanum* of western North America, similar to skunk cabbage.
**skunkweed** (skŭngk'wēd) *n.* Skunk cabbage.
**sky** (skī) *n., pl.* **skies.** [ME < ON *ský,* cloud.] **1.** The upper atmosphere, appearing as a hemisphere above the earth. **2.** The highest level or degree. **3.** The celestial regions. **4.** *often* **skies.** The appearance of the upper atmosphere, esp. with respect to weather <threatening *skies*> —*vt.* **skied, sky·ing, skies. 1.** To hit or throw (e.g., a ball) high in the air. **2.** To hang (e.g., a painting) high up on the wall, above the line of vision, esp. in an exhibition.
**sky blue** *n.* A light to pale blue.
**sky·borne** (skī'bôrn', -bōrn') *adj.* Airborne <*skyborne* pollen>
**sky·cap** (skī'kăp') *n.* [SKY + (RED)CAP.] An airport luggage handler.
**sky·dive** (skī'dīv') *vi.* **-dived, -div·ing, -dives.** To jump from an airplane, performing various maneuvers before pulling the ripcord of a parachute. **—sky'div'er** *n.* **—sky'div'ing** *n.*
**Skye terrier** (skī) *n.* A small terrier orig. bred on the Isle of Skye, having a long low body, short legs, and shaggy hair.
**sky·ey** (skī'ē) *adj.* Of or like the sky.
**sky-high** (skī'hī') *adv.* **1.** To a very high level <debris piled *sky-high*> **2.** In a lavish or enthusiastic way. **3.** In pieces or to pieces : APART <blew the building *sky-high*> —*adj.* **1.** Being high in the air. **2.** Exorbitantly high <*sky-high* prices>
**sky·jack** (skī'jăk') *vt.* **-jacked, -jack·ing, -jacks.** [SKY + (HI)JACK.] To hijack (an aircraft, esp. one in flight) through the use or threat of force. **—sky'jack'er** *n.* **—sky'jack'ing** *n.*
**sky·lark** (skī'lärk') *n.* An Old World bird, *Alauda arvensis,* having brownish plumage and noted for its singing while in flight. —*vi.* **-larked, -lark·ing, -larks.** To frolic.
**sky·light** (skī'līt') *n.* An overhead window.
**sky·line** (skī'līn') *n.* **1.** The line along which the surface of the earth and sky appear to meet : HORIZON. **2.** The outline of a group of buildings or a mountain range seen against the sky.
**sky·lounge** (skī'lounj') *n.* A vehicle that picks up passengers and then is carried by a helicopter between a downtown terminal and an airport.
**sky marshal** *n.* An armed federal law-enforcement officer assigned to prevent skyjackings.
**sky pilot** *n. Slang.* A chaplain.
**sky·rock·et** (skī'rŏk'ĭt) *n.* A firework that rises high into the air where it explodes in a brilliant cascade of sparks. —*vi. & vt.* **-et·ed, -et·ing, -ets.** To rise or cause to rise rapidly and suddenly, as in quantity, position, or reputation.
**sky·sail** (skī'səl, -sāl') *n. Naut.* A small square sail above the royal in a square-rigged vessel.
**sky·scrap·er** (skī'skrā'pər) *n.* An exceptionally tall building.
**sky·walk** (skī'wôk') *n.* An elevated usu. enclosed walkway between two buildings.
**sky·ward** (skī'wərd) *adj. & adv.* At or toward the sky. **—sky'wards** *adv.*
**sky·way** (skī'wā') *n.* **1.** An airline route : AIR LANE. **2.** An elevated highway.
**sky·writ·ing** (skī'rī'tĭng) *n.* **1.** The process of writing in the sky by releasing a visible vapor from an aircraft. **2.** The letters or words that are formed in skywriting. **—sky'writ'er** *n.*
**slab[1]** (slăb) *n.* [ME *slabbe.*] **1.** A broad, flat, rather thick piece, as of cake, stone, or cheese. **2.** An outside piece cut from a log when squaring it for lumber. **3.** *Baseball.* The pitcher's rubber. —*vt.* **slabbed, slab·bing, slabs. 1.** To make or shape into a slab. **2.** To cover or pave with slabs. **3.** To dress (a log) by cutting slabs.
**slab[2]** (slăb) *adj.* [Prob. of Scand. orig.] *Archaic.* Viscid. —Used in the phrase *thick and slab.*
**slab-sid·ed** (slăb'sī'dĭd) *adj. Informal.* **1.** Having flat sides. **2.** Tall and slim : LANKY.
**slack[1]** (slăk) *adj.* [ME *slak* < OE *slæc.*] **1.** Lacking liveliness or motion : SLUGGISH. **2.** Not busy <a *slack* selling season> **3.** Not tense or taut : LOOSE <a *slack* anchor line> **4.** Lacking firmness <a *slack* handshake> **5.** Lacking in diligence : NEGLIGENT <a *slack* worker> **6.** Flowing or blowing with little speed. —Used of the wind or tide. —*v.* **slacked, slack·ing, slacks.** —*vt.* **1.** To slacken. **2.** To be remiss about. **3.** To slake (lime). —*vi.* To be or become slack. **—slack off.** To decrease in activity or intensity : ABATE <The wind *slacked off.*> —*n.* **1.** A loose or slack part of something, as a rope or sail. **2.** Lack of tension : LOOSENESS. **3.** A period of little activity : LULL. **4. a.** Cessation of movement in a current of air or water. **b.** An area of still water. **5. slacks.** Separate trousers not part of a suit. **—slacks, slack'ly** *adv.* **—slack'ness** *n.*
**slack[2]** (slăk) *n.* [ME *sleck.*] A mixture of coal fragments, coal dust, and dirt that remains after screening coal.
**slack[3]** (slăk) *n.* [ME *slak* < ON *slakki.*] *Chiefly Brit.* **1.** A small dell. **2.** A bog : morass.
**slack-baked** (slăk'bākt') *adj.* **1.** Not perfectly baked : UNDERDONE <*slack-baked* bread> **2.** Imperfectly made.
**slack·en** (slăk'ən) *v.* **-ened, -en·ing, -ens.** —*vt.* **1.** To make slower <*slacken* one's pace> **2.** To make less vigorous, intense, firm, or severe. **3.** To reduce the tension or tautness of : LOOSEN. —*vi.* **1.** To slow down. **2.** To become less energetic, active, firm, or strict. **3.** To become less tense or taut : LOOSEN.
**slack·er** (slăk'ər) *n.* One who shirks work or duty, esp. one who evades military service in time of war.
**slack water** *n.* **1.** The period at high or low tide when there is no visible flow of water. **2.** Still water in a sea or river.
**slag** (slăg) *n.* [MLG *slagge.*] **1.** The vitreous mass left as a residue by the smelting of metallic ore. **2.** SCORIA 1. —*vt. & vi.* **slagged, slag·ging, slags.** To change into slag or form slag. **—slag'gy** *adj.*
**slain** (slān) *v. p.p. of* SLAY.
**slake** (slāk) *v.* **slaked, slak·ing, slakes.** [ME *slaken,* to abate < OE *slacian* < *slæc,* slack, sluggish.] —*vt.* **1.** To quench or satisfy <*slaked* my thirst with lemonade> **2.** To lessen the activity or force of : MODERATE <*slaking* their anger> **3.** To refresh by wetting or moistening. **4.** To combine (lime) chemically with water or moist air. —*vi.* To undergo a slaking process.
**sla·lom** (slä'ləm) *n.* [Norw. : *slad,* sloping + *lom,* path.] **1.** Skiing in a zigzag course. **2.** A ski race along a zigzag course, laid out with flag-marked poles. **—sla'lom** *v.* **(-lomed, -lom·ing, -loms).**
**slam[1]** (slăm) *v.* **slammed, slam·ming, slams.** [Perh. of Scand. orig.] —*vt.* **1.** To shut forcefully and loudly <*slammed* the door> **2.** To put, throw, or move forcefully so as to produce a loud noise <*slam* one's fist on a desk> **3.** To hit or strike with great force. **4.** *Slang.* To criticize harshly. —*vi.* **1.** To close or swing into place with force so as to produce a loud noise. **2.** To hit something with force : CRASH. —*n.* **1. a.** A forceful movement producing a loud noise. **b.** A loud noise produced by a forceful movement. **2.** *Slang.* Harsh or devastating criticism.
**slam[2]** (slăm) *n.* [Orig. unknown.] The winning of all the tricks or all but one during the play of one hand in bridge and other whist-derived card games.
**slam-bang** (slăm'băng') *adv.* **1.** Swiftly and noisily. **2.** Recklessly.
**slam·mer** (slăm'ər) *n.* [< SLAM[1].] *Slang.* A jail.
**slan·der** (slăn'dər) *n.* [ME *slaundre* < OFr. *esclandre* < Lat. *scandalum,* scandal < Gk. *skandalon,* trap.] **1.** *Law.* Utterance of defamatory statements injurious to the reputation or well-being of a person. **2.** A malicious report or statement. —*v.* **-dered, -der·ing, -ders.** —*vt.* To utter damaging reports about. —*vi.* To spread slander. **—slan'der·er** *n.* **—slan'der·ous** *adj.* **—slan'der·ous·ly** *adv.*
**slang** (slăng) *n.* [Orig. unknown.] **1.** The nonstandard vocabulary of a given culture or subculture, consisting typically of arbitrary and often ephemeral coinages and figures of speech marked by spontaneity and occas. by raciness. **2.** Language peculiar to a group : ARGOT. **—slang'i·ly** *adv.* **—slang'i·ness** *n.* **—slang'y** *adj.*
**slant** (slănt) *v.* **slant·ed, slant·ing, slants.** [Alteration of obs. *slent* < ME *slenten,* to lie aslant, of Scand. orig.] —*vt.* **1.** To impart an oblique direction to. **2.** To present so as to conform to a particular bias <*slanted* the story in favor of the government> —*vi.* To incline obliquely. —*n.* **1. a.** A sloping direction, plane, or course : INCLINE. **b.** Slope. **2.** A biased point of view. **—slant'ing·ly** *adv.*
**slant·ways** (slănt'wāz) *adv.* Slantwise.
**slant·wise** (slănt'wīz') *adv.* At a slant or slope : OBLIQUELY. —*adj.* Slanting : oblique.
**slap** (slăp) *n.* [LG *slapp.*] **1. a.** A smacking blow made with a flat object, as the open hand. **b.** The sound of a smacking blow. **2.** An injury, as to one's pride. —*v.* **slapped, slap·ping, slaps.** —*vt.* **1.** To strike with a flat object, as the open hand. **2.** To criticize sharply. **3.** To put with a slapping sound <*slapped* a dollar on the counter> —*vi.* To strike or beat with the force and sound of a slap. **—slap down. 1.** To forbid and prevent one from acting in a specific way by means of a sharp blow or emphatic censure. **2.** To put a sudden end to : SUPPRESS. —*adv. Informal.* Directly and forcefully. **—slap'per** *n.*
**slap·dash** (slăp'dăsh') *adj.* Hasty or careless. **—slap'dash'** *adv.*
**slap·hap·py** (slăp'hăp'ē) *adj.* **-pi·er, -pi·est.** *Slang.* **1.** Dazed, silly, or incoherent from or as if from blows to the head. **2.** Happy-go-lucky.

**slap·jack** (slăp′jăk′) *n.* [SLAP + (FLAP)JACK.] A flapjack.
**slap shot** *n.* A shot in ice hockey that consists of a swinging stroke.
**slap·stick** (slăp′stĭk′) *n.* **1.** Comedy marked by many chases, collisions, crude practical jokes, and similar boisterous actions. **2.** A paddle designed to make a loud whacking sound, once used by performers in farces.
**slash** (slăsh) *v.* **slashed, slash·ing, slash·es.** [ME *slashen.*] —*vt.* **1.** To form by cutting with violent sweeping strokes. **2.** To lash violently with sweeping strokes. **3.** To make a gash or gashes in. **4.** To cut a slit or slits in. **5.** To criticize sharply. **6.** To curtail drastically <*slash* gas prices> —*vi.* **1.** To make violent and sweeping strokes with or as if with a sharp instrument. **2.** To cut one's way with such strokes <We *slashed* through the dense thickets.> —*n.* **1.** A sweeping stroke made with a sharp instrument. **2.** An injury, as a cut, made by a sweeping stroke : SLIT. **3.** An ornamental slit in a fabric or garment. **4.** Branches and other residue left on a forest floor after the cutting of timber. **5.** *often* **slashes.** Wet or swampy ground overgrown with bushes and trees. **6.** A virgule. —**slash′er** *n.*
**slash·ing** (slăsh′ĭng) *adj.* **1.** Bitingly critical or satiric <the critic's *slashing* wit> **2.** Dashing : pelting <a *slashing* rainstorm> **3.** Brilliant : intense <*slashing* colors>
**slash pine** *n.* Any of several pine trees of the southeastern United States and adjacent regions that grow in swampy coastal areas.
**slat** (slăt) *n.* [ME *sclat* < OFr. *esclat,* splinter.] **1.** A narrow strip of metal, plastic, or wood, as in a Venetian blind. **2.** A movable auxiliary airfoil running along the leading edge of the wing of an aircraft. **3. slats.** *Slang.* The ribs. —*vt.* **slat·ted, slat·ting, slats.** To provide or make with slats.
**†slatch** (slăch) *n.* [Var. of SLACK[1].] *New England.* **1.** A momentary lull between breaking waves, favorable for launching a boat. **2.** A lull in a high windstorm.
**slate** (slāt) *n.* [ME *sclate* < OFr. *esclate,* splinter, fem. of *esclat.*] **1.** A fine-grained metamorphic rock that splits into thin, smooth-surfaced layers. **2. a.** A piece of slate cut for use as roofing material or a writing surface. **b.** A writing tablet made of material similar to slate. **3.** A record of past performance or activity <started with a clean *slate*> **4.** A list of the candidates of a political party running for offices. **5.** A dark gray to bluish gray, to dark bluish or dark purplish gray. —*vt.* **slat·ed, slat·ing, slates. 1.** To cover (e.g., a roof) with slate. **2.** To put on a list of candidates. **3.** To designate or destine <was *slated* to retire in June>
**slate black** *n.* A purplish black. —**slate′-black′** (slāt′blăk′) *adj.*
**slate blue** (slāt′blo͞o′) *n.* A grayish blue to dark bluish gray. —**slate′-blue′** *adj.*
**slat·er** (slā′tər) *n.* **1.** One employed to lay slate roofs. **2.** A small isopod crustacean, as a sow bug.
**slath·er** (slăth′ər) *vt.* **-ered, -er·ing, -ers.** [Orig. unknown.] *Informal.* **1.** To use great amounts of : LAVISH. **2. a.** To spread thickly with. **b.** To spread thickly on. —*n. often* **slathers.** *Slang.* A great amount <had *slathers* of money>
**slat·ing** (slā′tĭng) *n.* **1.** The act, process, or occupation of laying slates. **2.** Slates as a whole.
**slat·tern** (slăt′ərn) *n.* [Perh. < dial. *slattering,* slovenly, pr.part. of dial. *slatter,* to slop.] An untidy woman.
**slat·tern·ly** (slăt′ərn-lē) *adj.* **1.** Slovenly : untidy <a *slatternly* tenement> **2.** Characteristic of or befitting a slattern <*slatternly* old dresses> —**slat′tern·li·ness** *n.*
**slaugh·ter** (slô′tər) *n.* [ME, of Scand. orig.] **1.** The killing of animals for food. **2.** The killing of a large number of persons : CARNAGE. —*vt.* **-tered, -ter·ing, -ters. 1.** To kill (animals) for food : BUTCHER. **2. a.** To kill (persons) in large numbers. **b.** To kill violently or brutally. —**slaugh′ter·er** *n.* —**slaugh′ter·ous** *adj.*
**slaugh·ter·house** (slô′tər-hous′) *n.* **1.** A place where animals are butchered. **2.** A scene of carnage.
**Slav** (släv) *n.* [ME *Sclave* < Med. Lat. *Sclavus* < LGk. *Sklabos.*] A member of one of the Slavic-speaking peoples of eastern Europe.
**slave** (slāv) *n.* [ME *sclave* < OFr. *esclave* < Med. Lat. *sclavus* < *Sclavus,* Slav.] **1.** One bound in servitude to a person or household as an instrument of labor. **2.** One who is submissive or subject to a person or influence. **3.** An extremely hard worker. **4.** A machine or component controlled by another machine or component. —*vi.* **slaved, slav·ing, slaves.** To work like a slave.
**slave driver** *n.* **1.** A severely exacting employer or supervisor. **2.** An overseer of slaves.
**slav·er[1]** (slăv′ər) *vi.* **-ered, -er·ing, -ers.** [ME *slaveren,* prob. < ON *slafra.*] **1.** To slobber. **2.** To fawn : drivel. —*n.* **1.** Saliva drooling from the mouth. **2.** Slobbering flattery or drivel.
**slav·er[2]** (slā′vər) *n.* **1.** A ship engaged in slave traffic. **2.** One who traffics in slaves.
**slav·er·y** (slā′və-rē, slāv′rē) *n., pl.* **-ies. 1.** Bondage to a master or household. **2.** A mode of production in which slaves constitute the principal work force. **3.** Subjugation or addiction to a specified influence <the *slavery* of narcotics> **4.** Hard work and subjection <wage *slavery*>
**slave state** *n.* **1. Slave State.** One of the 15 states of the Union in which slavery was legal before the Civil War. **2.** A country under totalitarian rule.
**slave trade** *n.* Traffic in slaves.
**Slav·ic** (slä′vĭk) *adj.* Of or relating to the Slavs or their languages. —*n.* A branch of the Indo-European language family including Bulgarian, Byelorussian, Czech, Polish, Russian, Serbo-Croatian, Slovak, Slovene, and Ukrainian.
**slav·ish** (slā′vĭsh) *adj.* **1.** Relating to or characteristic of a slave : SERVILE <*slavish* devotion> **2.** Relating to or characteristic of the institution of slavery : OPPRESSIVE. **3.** Showing no originality : blindly imitative <a *slavish* copy of an original novel> —**slav′ish·ly** *adv.* —**slav′ish·ness** *n.*
**slav·oc·ra·cy** (slā-vŏk′rə-sē) *n., pl.* **-cies.** The power structure formed by the advocates of slavery in the United States before the Civil War. —**slav′o·crat′** (slā′və-krăt′) *n.* —**slav′o·crat′ic** *adj.*
**Sla·von·ic** (slə-vŏn′ĭk) *n.* Slavic. —**Sla·von′ic** *adj.*
**Slav·o·phile** (släv′ə-fīl′) *also* **Slav·o·phil** (-fĭl′) *n.* **1.** An admirer of the Slavs. **2.** An advocate of the supremacy of Slavic and esp. Russian, culture. —**Sla·voph′i·lism** (slə-vŏf′ə-lĭz′əm) *n.*
**slaw** (slô) *n.* Coleslaw.
**slay** (slā) *vt.* **slew** (slo͞o), **slain** (slān), **slay·ing, slays.** [ME *sleen* < OE *slēan.*] **1.** To kill violently or deliberately. **2.** *Slang.* To overwhelm, as with laughter. —**slay′er** *n.*
**sleave** (slēv) *vt.* **sleaved, sleav·ing, sleaves.** [ME **sleven* < OE *slǣfan,* to cut.] To separate, as a twisted mass of threads. —*n.* **1.** A tangled or knotted thread or ravel. **2.** A thin thread.
**sleaze** (slēz) *n.* [Back-formation < SLEAZY.] *Informal.* A sleazy quality, state, or appearance.
**slea·zy** (slē′zē) *adj.* **-zi·er, -zi·est.** [Orig. unknown.] **1.** Thin and loosely woven : FLIMSY <a coat with a *sleazy* lining> **2.** Made of low-quality materials : CHEAP. **3.** Vulgar : disreputable <*sleazy* nightclubs> —**slea′zi·ly** *adv.* —**slea′zi·ness** *n.*
**sled** (slĕd) *n.* [ME *sledde* < MLG.] **1.** A vehicle mounted on runners, used for carrying people or loads over ice and snow. **2.** A light frame on runners, used by children for coasting over snow or ice. —*v.* **sled·ded, sled·ding, sleds.** —*vt.* To carry on or convey by a sled. —*vi.* To ride or use a sled. —**sled′der** *n.*
**sled·ding** (slĕd′ĭng) *n.* **1.** The act of using a sled for hauling, transportation, or sport. **2.** The conditions under which one may use a sled. **3.** *Informal.* Progress or existence : GOING <tough economic *sledding*>
**sledge** (slĕj) *n.* [MDu. *sleedse.*] A vehicle on low runners drawn by horses, dogs, or other work animals and used for transporting loads across ice and snow. —*vt.* & *vi.* **sledged, sledg·ing, sledg·es.** To convey or travel on a sledge.
**sledge·ham·mer** (slĕj′hăm′ər) *n.* [ME *slegge,* sledgehammer (< OE *slecg*) + HAMMER.] A long heavy hammer, often wielded with both hands, used for driving wedges and posts and for other heavy work. —*vt.* **-mered, -mer·ing, -mers.** To strike with or as if with a sledgehammer. —*adj.* Ruthlessly severe : CRUSHING.
**sleek** (slēk) *adj.* **-er, -est.** [Var. of SLICK.] **1.** Smooth and lustrous as if polished : GLOSSY <*sleek* hair> **2.** Well-groomed and neatly tailored. **3.** Healthy or well-fed : THRIVING <*sleek* cattle> **4.** Polished or smooth in behavior : SLICK. —*vt.* **sleeked, sleek·ing, sleeks. 1.** To make lustrous or smooth : POLISH. **2.** To gloss over : CONCEAL. —**sleek′ly** *adv.* —**sleek′ness** *n.*
**sleep** (slēp) *n.* [ME *slepe* < OE *slǣp.*] **1. a.** A natural, periodically recurring physiological state of rest, marked by relative physical and nervous inactivity, unconsciousness, and lessened responsiveness to external stimuli. **b.** A period of physiological rest. **c.** A condition of inactivity, such as unconsciousness, dormancy, hibernation, or death, that resembles physiological sleep. **2.** *Bot.* The folding together of leaves or petals at night or in the absence of light. —*v.* **slept** (slĕpt), **sleep·ing, sleeps.** —*vi.* **1.** To be in the state of sleep or to fall asleep. **2.** To be in a condition resembling sleep, as hibernation, dormancy, or death. —*vt.* **1.** To pass or get rid of by sleeping <*slept* away the day><went home to *sleep* off a hangover> **2.** To provide with accommodations for sleeping. —**sleep in. 1.** To sleep at one's place of employment. **2. a.** To oversleep. **b.** To sleep late on purpose. —**sleep on it.** To consider something overnight before deciding what to do. —**sleep out. 1.** To sleep outside. **2.** To sleep at one's own home, not at one's place of employment. **3.** To sleep away from one's home.
**sleep·er** (slē′pər) *n.* **1.** One that sleeps. **2.** A sleeping car. **3.** Any of various usu. small marine and freshwater fishes of the family Eleotridae. **4.** *Football.* An offensive player stationed in an obscure field position with the hope that he will not be noticed by the other team until after he has performed his function in the play. **5.** *Chiefly Brit.* A heavy beam supporting rails in a railroad track : CROSSTIE. **6.** *Informal.* One that achieves unexpected recognition or success, as a race horse, book, or marketed product. **7.** An earmarked unbranded calf.
**sleep-in** (slēp′ĭn′) *adj.* Being one that lives where employed <a *sleep-in* maid>
**sleeping bag** *n.* A large, warmly lined, usu. zippered bag in which a person may sleep outdoors.
**sleeping car** *n.* A railroad car having accommodations to sleep passengers.

**sleeping pill** *n.* A sedative, esp. a barbiturate, in the form of a pill or capsule used to relieve insomnia.
**sleeping sickness** *n.* **1.** An often fatal endemic infectious disease of human beings and animals in tropical Africa, caused by either of two protozoans of the genus *Trypanosoma,* transmitted by the tsetse fly, and marked by fever and lethargy. **2.** *Pathol.* Encephalitis lethargica.
**sleep·less** (slēp'lĭs) *adj.* **1. a.** Lacking sleep <a *sleepless* night> **b.** Unable to sleep. **2.** Constantly alert or active <*sleepless* vigilance> **—sleep'less·ly** *adv.* **—sleep'less·ness** *n.*
**sleep·walk·ing** (slēp'wô'kĭng) *n.* Somnambulism. **—sleep'walk'·er** *n.*
**sleep·wear** (slēp'wâr') *n.* Nightclothes.
**sleep·y** (slē'pē) *adj.* **-i·er, -i·est. 1.** Ready for or requiring sleep : DROWSY. **2.** Dull : inactive. **3.** Inducing sleep. **4.** Quiet <a *sleepy* little town> **—sleep'i·ly** *adv.* **—sleep'i·ness** *n.*
**sleep·y·head** (slē'pē-hĕd') *n. Informal.* A sleepy person.
**sleet** (slēt) *n.* [ME *slete.*] **1.** Precipitation composed of gen. transparent frozen or partially frozen raindrops. **2.** A mixture of rain and snow or hail. **3.** A thin icy coating that forms when rain or sleet freezes, as on trees or streets. *—vi.* **sleet·ed, sleet·ing, sleets.** To shower sleet. **—sleet'y** *adj.*
**sleeve** (slēv) *n.* [ME *sleve* < OE *slēf.*] **1.** The part of a garment covering all or a part of the arm. **2.** An encasement into which an object fits <a phonograph record *sleeve*> *—vt.* **sleeved, sleev·ing, sleeves.** To furnish or fit with a sleeve or sleeves. **—up (one's) sleeve.** Hidden but ready to be used. **—sleeve'less** *adj.*
**sleeve coupling** *n.* A thin steel cylinder uniting two lengths of shafting or pipe.
**sleigh** (slā) *n.* [Du. *slee,* alteration of *slede* < MDu. *slēde.*] A light vehicle mounted on runners for use on snow or ice, having one or more seats and usu. drawn by a horse. *—vi.* **sleighed, sleigh·ing, sleighs.** To ride in or drive a sleigh. **—sleigh'er** *n.*
**sleight** (slīt) *n.* [ME < ON *slœgð* < *slœgr,* sly.] **1.** Skill : dexterity. **2.** A clever or skillful trick or deception : STRATAGEM.
**sleight of hand** *n.* **1.** Tricks or feats performed by jugglers or magicians so quickly that their manner of execution cannot be detected : LEGERDEMAIN. **2.** Skill in performing sleight of hand. **3.** The performance of sleight of hand.
**slen·der** (slĕn'dər) *adj.* **-er, -est.** [ME *sclendre.*] **1.** Having little width in proportion to the height or length : gracefully slim. **2.** Spare or small in amount or extent : MEAGER <*slender* wages> **3.** Having little force or justification : FEEBLE <only a *slender* chance for success> **—slen'der·ly** *adv.* **—slen'der·ness** *n.*
**slen·der·ize** (slĕn'də-rīz') *v.* **-ized, -iz·ing, -iz·es.** *—vi.* To become slender or more slender. *—vt.* **1.** To make slender or slim. **2.** To cause to appear slender.
**slept** (slĕpt) *v. p.t. & p.p. of* SLEEP.
**sleuth** (sloo͞th) *n.* [Short for SLEUTHHOUND.] **1.** *Informal.* A detective. **2.** SLEUTHHOUND 1. *—v.* **sleuthed, sleuth·ing, sleuths.** *—vt.* To track or follow. *—vi.* To act as a detective.
**sleuth·hound** (sloo͞th'hound') *n.* [ME : *sleuth,* animal track (< ON *slōð*) + *hound,* hound.] **1.** A dog, as a bloodhound, used for tracking or pursuing. **2.** A detective.
**slew[1]** *also* **slue** (sloo͞) *n.* [Ir. Gael. *sluag* < OIr. *slúag.*] *Informal.* A large amount or number : LOT <a whole *slew* of complaints>
**slew[2]** (sloo͞) *v. p.t. of* SLAY.
**slew[3]** (sloo͞) *n. var. of* SLOUGH[1].
**slew[4]** (sloo͞) *v. & n. var. of* SLUE[1]
**slice** (slīs) *n.* [ME *sclice,* splinter < OFr. *esclice* < *esclicer,* to splinter, of Germanic orig.] **1.** A broad thin piece cut from a larger object. **2.** A portion : share <a *slice* of the profits> **3. a.** A knife with a broad, thin, flexible blade, used for cutting and serving food. **b.** A similar implement for spreading printing ink. **4. a.** A stroke that causes a ball to curve off course to the right or, if the player is left-handed, to the left. **b.** The course followed by such a ball. *—v.* **sliced, slic·ing, slic·es.** *—vt.* **1.** To cut or divide into slices. **2.** To cut or remove from a larger piece <*slice* off a piece of ham> **3.** To cut through or across with or as if with a knife. **4.** To divide into portions or shares. **5.** To spread, work at, or clear away with a bladed tool, as a slice bar. **6.** To hit (a ball) with a slice. *—vi.* **1.** To move like a knife <The PT boat *sliced* through the water.> **2.** To hit a ball with a slice. **—slice'a·ble** *adj.* **—slic'er** *n.*
**slice bar** *n.* An iron tool with a flat, broad end, used for loosening and clearing out clinkers from furnace grates.
**slice-of-life** (slīs'əv-līf') *adj.* Of, relating to, or being a dramatic work that accurately portrays a segment of actual life experience.
**slick** (slĭk) *adj.* [ME *slike.*] **1.** Smooth, glossy, and slippery <roads *slick* with oil> **2.** Deftly executed : ADROIT. **3.** Wily : shrewd. **4.** Superficially attractive but without depth or sound quality <a *slick* writing style> *—n.* **1.** A smooth or slippery surface or area. **2.** An implement for making a surface slick, esp. a chisel used for smoothing and polishing. **3.** *Informal.* A magazine printed on glossy, high-quality paper, featuring articles and fiction of popular appeal but little literary merit. **4.** A racing automobile tire with a smooth tread. *—vt.* **slicked, slick·ing, slicks. 1.** To make smooth, glossy, or oily. **2.** *Informal.* To make neat, trim, or tidy.
**slick·en·side** (slĭk'ən-sīd') *n.* [Dial. *slicken,* glossy < SLICK + SIDE.] A polished and striated rock surface caused by one rock mass sliding over another in a fault plane.
**slick·er** (slĭk'ər) *n.* **1.** A glossy raincoat. **2.** A tool for dressing hides. **3.** *Informal.* A swindler. **4.** *Informal.* A person with stylish clothing and manners <a city *slicker*>
**slid·den** *v. Archaic. var. p.p. of* SLIDE.
**slide** (slīd) *v.* **slid** (slĭd), **slid·ing, slides.** [ME *sliden* < OE *slīdan.*] *—vi.* **1.** To move over a surface while maintaining smooth, continuous contact. **2.** To coast on a slippery surface, such as ice or snow. **3.** To pass quietly and smoothly : GLIDE <a cat burglar *sliding* from room to room> **4.** To go unattended <Let the problem *slide.*> **5.** To lose one's balance or intended direction on a slippery surface. **6.** To move downward <Prices suddenly began to *slide.*> **7.** To move accidentally out of place : SLIP <The tray *slid* out of my hand.> **8.** To return to a less favorable or less worthy state. **9.** *Baseball.* To drop down and skid, usu. feet first, into a base to avoid being tagged out. *—vt.* To cause to slide or slip. *—n.* **1.** A sliding action or movement. **2.** A smooth surface for sliding, usu. inclined <a water *slide*> **3.** A playground apparatus for children to slide on, typically consisting of a smooth chute mounted by way of a ladder. **4.** A part that operates by sliding, as the U-shaped section of tube on a trombone that is moved to produce various tones. **5.** An image on a transparent plate for projection on a screen. **6.** A small glass plate for mounting specimens to be examined under a microscope. **7.** An avalanche. **8.** *Mus.* **a.** A portamento. **b.** An ornamentation consisting of two grace notes approaching the main note. **—slid'er** *n.*
☆ ***syns:*** SLIDE, COAST, DRIFT, GLIDE, SLIP *v. core meaning* : to pass smoothly, quietly, and effortlessly on or as if on a slippery surface <skaters *sliding* across the ice><fish *sliding* under the clear water>
**slide rule** *n.* A device consisting of two logarithmically scaled rules mounted to slide along each other so that multiplication, division, and sometimes more complex computations may be reduced to the mechanical equivalent of addition or subtraction.
**slide valve** *n.* A valve that slides back and forth over ports in the cylinder wall of a steam engine, permitting the intake and outflow of steam to move the piston.
**sliding scale** *n.* A scale in which indicated prices, taxes, or wages vary in accordance with another factor, as wages with the cost-of-living index or medical charges with a patient's income.
**sli·er** (slī'ər) *adj. var. compar. of* SLY.
**sli·est** (slī'ĭst) *adj. var. compar. of* SLY.
**slight** (slīt) *adj.* **-er, -est.** [ME, slender, of Scand. orig.] **1.** Small in size, degree, or amount. **2.** Of small importance : TRIFLING. **3.** Slender or frail : DELICATE. *—vt.* **slight·ed, slight·ing, slights. 1.** To treat with discourteous reserve or inattention. **2.** To do negligently or thoughtlessly : SHIRK. **3.** To give insufficient consideration to. *—n.* An act of pointed disrespect or discourtesy. **—slight'ness** *n.*
**slight·ing** (slī'tĭng) *adj.* Being or conveying a slight : DISRESPECTFUL. **—slight'ing·ly** *adv.*
**slight·ly** (slīt'lē) *adv.* **1.** Carelessly. **2.** To a small degree or extent.
**slim** (slĭm) *adj.* **slim·mer, slim·mest.** [Du., inferior, < MDu. *slimp,* bad.] **1.** Small in girth or thickness in proportion to height or length : SLENDER. **2.** Small in quality or amount : SCANT. *—vt. & vi.* **slimmed, slim·ming, slims.** To make or become slim or thin. **—slim'ly** *adv.* **—slim'ness** *n.*
**slime** (slīm) *n.* [ME < OE *slīm.*] **1.** A thick, sticky, slippery substance. **2.** A mucous substance secreted by certain animals, as fish or slugs. *—vt.* **slimed, slim·ing, slimes. 1.** To smear with slime. **2.** To remove slime from, as from fish to be canned.
**slime fungus** *n.* Slime mold.
**slime mold** *n.* Any of various fungi of the class Myxomycetes, having a vegetative body consisting of a slimy, motile, multinucleate mass of protoplasm.
**slim·nas·tics** (slĭm-năs'tĭks) *n.* [SLIM + (GYM)NASTICS.] (*sing. or pl. in number*). Physical exercises designed to facilitate weight loss.
**slim·sy** (slĭm'zē) *also* **slimp·sy** (slĭmp'sē) *adj.* **-si·er, -si·est.** *Informal.* [Blend of SLIM and FLIMSY.] Frail and flimsy.
**slim·y** (slī'mē) *adj.* **-i·er, -i·est. 1.** Made up of or like slime : VISCOUS. **2.** Covered with or exuding slime. **3.** Vile : foul. **—slim'i·ly** *adv.* **—slim'i·ness** *n.*
**sling[1]** (slĭng) *n.* [ME.] **1. a.** A weapon composed of a looped strap in which a stone is whirled and then let fly. **b.** A slingshot. **2.** A looped rope, strap, or chain for supporting, cradling, or hoisting, esp.: **a.** A strap over the heel to hold a shoe in place. **b.** A strap for carrying a rifle over the shoulder. **c.** *Naut.* A rope or chain for supporting a yard. **d.** A band suspended from the neck to support an injured arm or hand. **3.** The act of slinging. *—vt.* **slung** (slŭng), **sling·ing, slings. 1.** To hurl from or as if from a sling : FLING. **2.** To place or carry in a sling. **3.** To raise or lower by means of a sling. **4.** To let swing loosely or freely. **—sling'er** *n.*
**sling[2]** (slĭng) *n.* [Orig. unknown.] A drink of brandy, whiskey, or gin, sweetened and usu. lemon-flavored.

**sling·shot** (slĭng′shŏt′) *n.* A Y-shaped stick with an elastic strap attached to the prongs, used for flinging small stones.

**slink** (slĭngk) *v.* **slunk** (slŭngk), **slink·ing, slinks.** [ME *slynken* < OE *slincan.*] —*vi.* To move quietly and furtively : SNEAK. —*vt.* To give birth to prematurely. —Used esp. of cows. —*n.* An animal, esp. a calf, born prematurely. —*adj.* That has been born prematurely. **—slink′ing·ly** *adv.*

**slink·y** (sling′kē) *adj.* **-i·er, -i·est. 1.** Stealthy : furtive. **2.** *Informal.* Graceful, sinuous, and sleek <a *slinky* gown>

**slip¹** (slĭp) *v.* **slipped, slip·ping, slips.** [ME *slippen,* prob. < MLG.] —*vi.* **1. a.** To move smoothly and quietly : GLIDE. **b.** To move stealthily. **2.** To pass gradually, easily, or imperceptibly <The years *slipped* by.> **3. a.** To slide unexpectedly and by accident. **b.** To slide out of place <The gear *slipped.*> **c.** To escape, as from a grip <The dog *slipped* out of its collar.> **4.** To get away completely. **5.** *Informal.* To decline in physical or mental ability, strength, or keenness. **6.** *Informal.* To decline : FALL OFF <Your work is *slipping.*> **7.** To fall behind a scheduled production rate. **8.** To fall into fault or error. —*vt.* **1.** To cause to move in a smooth, easy, or sliding motion <*slipped* the bolt into place> **2.** To place or insert smoothly and quietly. **3.** To put on or remove (clothing) quickly or easily <*slip* on a jacket><*slip* off one's boots> **4.** To get loose from. **5.** To bring forth (young) prematurely. —Used of animals. **6.** To unleash or free (a dog or hawk) to pursue game. **7.** To unfasten <*slip* the knot> **8.** To dislocate (a bone). **9.** To pass (a knitting stitch) from one needle to another without knitting it. —*n.* **1.** The act of slipping or sliding. **2.** An accident or mishap, esp. a falling down. **3. a.** An error in conduct or thinking : DEVIATION. **b.** A slight error or oversight, as in speech or writing <a *slip* of the tongue> **4. a.** A docking place for a ship : PIER. **b.** A space for a ship between two docks or wharves. **c.** A slipway. **5.** The difference between a vessel's actual speed through water and the speed at which the vessel would move if the screw were propelling against a solid. **6.** A woman's undergarment, serving as a lining for a dress. **7.** A pillowcase. **8.** *Geol.* **a.** A smooth crack at which rock strata have moved on each other. **b.** A small fault. **9.** The difference between optimal and actual output in a mechanical device. **10.** Movement between two parts where none should exist, as between a pulley and belt. **11.** The sliding movement of an aircraft in certain attitudes. **—give (someone) the slip.** *Slang.* To escape the pursuit of. **—let slip.** To say inadvertently. **—slip one over on.** *Informal.* To hoodwink or trick.

**slip²** (slĭp) *n.* [ME *slippe,* scion.] **1.** A part of a plant cut or broken off for grafting or planting. **2.** A long narrow piece : STRIP. **3.** A youthful, slender person. **4.** A small piece of paper, esp. a small form <a sales *slip*> **5.** A narrow church pew. —*vt.* **slipped, slip·ping, slips.** To make a slip from (a plant or plant part).

**slip³** (slĭp) *n.* [ME *slyppe,* slime < OE *slypa,* a soft mass.] Thinned potter's clay for decorating or coating ceramics.

**slip·case** (slĭp′kās′) *n.* An open-ended protective box for a book.

**slip·cov·er** (slĭp′kŭv′ər) *n.* A fitted, removable cover for upholstered furniture. —*vt.* **-ered, -er·ing, -ers.** To provide with a slipcover <*slipcover* a sofa>

**slip·knot** (slĭp′nŏt′) *n.* **1.** A knot made with a loop so that it slips easily along the rope or cord around which it is tied. **2.** A knot made so that it can readily be untied by pulling one free end.

**slip-on** (slĭp′ŏn′, -ôn′) *n.* A garment easily donned or removed.

**slip·o·ver** (slĭp′ō′vər) *n.* A garment that can be donned or removed over the head.

**slip·page** (slĭp′ĭj) *n.* **1. a.** A slipping. **b.** The amount or extent of slipping. **2.** Loss of motion or power due to slipping.

**slipped disk** *n.* Herniation of an intervertebral disk resulting in back pain or sciatica.

**slip·per** (slĭp′ər) *n.* A light, low shoe slipped on and off easily.

**slip·per·wort** (slĭp′ər-wûrt′, -wôrt′) *n.* The calceolaria.

**slip·per·y** (slĭp′ə-rē) *adj.* **-i·er, -i·est.** [Alteration of obs. *slipper,* slippery < ME < OE *slipor.*] **1.** Causing or tending to cause sliding or slipping. **2.** Tending to slip, as from one's grasp <a *slippery* bar of soap> **3.** Elusive : evasive. **—slip′per·i·ly** *adv.* **—slip′per·i·ness** *n.*

**slippery elm** *n.* A tree, *Ulmus rubra* of eastern North America, having twigs and leaves with a mucilaginous, aromatic juice once used medicinally.

**slip ring** *n.* A metal ring mounted on a rotating machine part to provide a continuous electrical connection through brushes on stationary contacts.

**slip-sheet** (slĭp′shēt′) *n.* A blank sheet of paper slipped between newly printed sheets to prevent offsetting. —*vt.* **-sheet·ed, -sheet·ing, -sheets.** To insert blank sheets between (printed sheets).

**slip·shod** (slĭp′shŏd′) *adj.* **1.** Poorly made or done : CARELESS <*slipshod* work> **2.** Slovenly : shabby.

**slip·slop** (slĭp′slŏp′) *n.* [Redup. of SLOP.] **1.** Trivial conversation or writing. **2.** *Archaic.* Unappetizing liquid or watery food : SLOPS.

**slip·stitch** (slĭp′stĭch′) *n.* A stitch used wherever stitching must be invisible on the right side of a garment, as on hems and facings, made by picking up one or two threads of fabric and then loosely catching the needle in the hem edge.

**slip·stream** (slĭp′strēm′) *n.* **1.** The turbulent flow of air driven backward by the propeller or propellers of an aircraft. **2.** The region of reduced air pressure and forward suction produced by and just behind a fast-moving ground vehicle. —*vi.* **-streamed, -stream·ing, -streams.** To drive in the slipstream of a ground vehicle.

**slip-up** (slĭp′ŭp′) *n.* *Informal.* An error or oversight.

**slip·way** (slĭp′wā′) *n.* A sloping incline leading down to the water on which ships are built or repaired.

**slit** (slĭt) *n.* [ME *slitte.*] A long narrow cut, tear, or incision. —*vt.* **slit, slit·ting, slits. 1.** To make a long narrow incision in. **2.** To cut lengthwise into strips : SPLIT.

**slith·er** (slĭth′ər) *v.* **-ered, -er·ing, -ers.** [ME *sliddren* < OE *slidrian,* freq. of *slīdan,* to slide.] —*vi.* **1.** To slip and slide, as on a loose or uneven surface. **2.** To move sinuously. —*vt.* To cause to slither or glide. —*n.* A slithering movement. **—slith′er·y** *adj.*

☆ ***syns:*** SLITHER, SNAKE, UNDULATE *v. core meaning* : to move sinuously <A large worm *slithered* through the grass.>

**slit trench** *n.* A narrow, shallow trench dug during combat for the protection of a single soldier or a small group.

**sliv·er** (slĭv′ər) *n.* [ME *slifere* < *slyven,* to split.] **1.** A slender piece cut, split, or broken off : SPLINTER. **2.** (*also* slī′vər). A continuous strand of loose wool, flax, or cotton, ready for drawing and twisting. —*vt.* & *vi.* **-ered, -er·ing, -ers.** To split or become split into slivers.

**sliv·o·vitz** (slĭv′ə-vĭts) *n.* [Serbo-Croatian *šljivovica* < *šljiva,* plum.] A dry, colorless plum brandy.

**slob** (slŏb) *n.* [Ir. Gael. *slab,* mud.] *Informal.* An obnoxious, crude, or slovenly person.

**slob·ber** (slŏb′ər) *v.* **-bered, -ber·ing, -bers.** [ME *sloberen.*] —*vi.* **1.** To allow saliva to dribble from the mouth : SLAVER. **2.** To spill (liquid or food) from the mouth while eating or drinking. **3.** To indulge in mawkish, sentimental expression. —*vt.* To wet or smear with or as if with saliva or food dribbled from the mouth. —*n.* **1.** Saliva or liquid running from the mouth. **2.** Mawkish, sentimental expression : DRIVEL. **—slob′ber·er** *n.* **—slob′ber·y** *adj.*

**sloe** (slō) *n.* [ME *sloo* < OE *slā.*] **1.** The blackthorn. **2.** The tart, blue-black, plumlike fruit of the sloe.

**sloe-eyed** (slō′īd′) *adj.* Having slanted eyes.

**sloe gin** *n.* A liqueur having a gin base, flavored with fresh sloes.

**slog** (slŏg) *v.* **slogged, slog·ging, slogs.** [Orig. unknown.] —*vt.* To strike with heavy blows, as in boxing. —*vi.* **1.** To walk with a slow, plodding gait. **2.** To work diligently for long hours. —*n.* **1.** Long hard work. **2.** A long exhausting march. **—slog′ger** *n.*

**slo·gan** (slō′gən) *n.* [Sc. *slogorne,* battle cry < Gael. *sluagh-ghairm* : *sluagh,* host + *gairm,* shout.] **1.** A phrase expressing the aims or nature of an enterprise or organization : MOTTO. **2.** A battle cry of the Scottish clans. **3.** A catch phrase used in advertising or promotion.

**slo·gan·eer** (slō′gə-nîr′) *n.* One that invents or uses slogans. **—slo′gan·eer′** *v.* **(-eered, -eer·ing, -eers).**

**slo·gan·ize** (slō′gə-nīz′) *vt.* **-ized, -iz·ing, -iz·es.** To express in the form of slogans.

**sloop** (slo͞op) *n.* [Du. *sloep.*] *Naut.* A single-masted, fore-and-aft-rigged sailing boat with a short standing bowsprit or none at all and a single headsail set from the forestay.

**sloop of war** *n.* A small armed vessel larger than a gunboat, carrying guns on one deck only.

**slop¹** (slŏp) *n.* [ME *sloppe,* a muddy place.] **1.** Spilled or splashed liquid. **2.** Soft slush or mud. **3.** Unappetizing watery food. **4.** *often* **slops.** Waste food used to feed pigs or other animals : SWILL. **5.** *often* **slops.** Mash remaining after the process of alcohol distillation. **6.** *often* **slops.** Human excrement. **7.** Repulsively effusive expression. —*v.* **slopped, slop·ping, slops.** —*vi.* **1.** To spill or splash, as a liquid. **2.** To spill over. **3.** To gush with excessive sentimentality. **4.** To plod or tramp awkwardly as if walking through mud. —*vt.* **1.** To spill (liquid). **2.** To spill liquid upon. **3.** To dish out or serve unappetizingly or clumsily. **4.** To feed slops to (animals).

**slop²** (slŏp) *n.* [ME *sloppe,* a kind of garment.] **1. slops.** Garments and bedding issued to sailors from a ship's stores. **2. slops.** Short, full trousers or breeches worn in the 16th cent. **3.** A loose outer garment, as a smock or overalls. **4. slops.** *Chiefly Brit.* Cheap ready-made garments.

**slope** (slōp) *v.* **sloped, slop·ing, slopes.** [< ME *slope,* sloping.] —*vi.* **1.** To incline upward or downward. **2.** To ascend or descend on a slanting course. —*vt.* To cause to slope. —*n.* **1.** An inclined line, surface, plane, position, or direction. **2.** A stretch of ground forming a natural or artificial incline <ski *slopes*> **3. a.** Deviation from the horizontal. **b.** The amount or degree of deviation from the horizontal. **4.** *Math.* The rate at which an ordinate of a point of a line on a coordinate plane changes with respect to a change in the abscissa. **5.** *Math.* The slope of the line tangent to a plane curve at a given point. **—slop′er** *n.* **—slop′ing·ly** *adv.*

**slo-pitch** (slō′pĭch′) *n. var. of* SLOW-PITCH.

**slop·py** (slŏp′ē) *adj.* **-pi·er, -pi·est. 1.** Of, like, or covered with slop : MUDDY <*sloppy* ground> **2.** Watery and unappetizing <a *sloppy* stew> **3.** Spotted or splashed with liquid or slop. **4.** *Informal.* Untidy : messy <a *sloppy* room> **5.** *Informal.* SLIPSHOD 1. **6.** *Informal.* Oversentimental : gushy. **—slop′pi·ly** *adv.* **—slop′pi·ness** *n.*

**sloppy joe** (jō) *n.* Ground cooked meat in a usu. spicy sauce served on a bun.

**slosh** (slŏsh) *v.* **sloshed, slosh·ing, slosh·es.** [Alteration of SLUSH.] —*vt.* **1.** To stir or splash (a liquid). **2.** To agitate in a liquid <*slosh* clothes in bleach> —*vi.* To splash or flounder in a liquid, as water. —*n.* **1.** Slush. **2.** The sound of splashing liquid. **—slosh'y** *adj.*

**sloshed** (slŏsht) *adj.* [< p.part. of SLOSH.] *Slang.* Drunk.

**slot**[1] (slŏt) *n.* [ME, indentation running down the middle of the breast < OFr. *esclot.*] **1.** A long narrow groove, opening, or notch, as for receiving coins in a vending machine. **2.** A gap between a main and an auxiliary airfoil to provide space for airflow and facilitate the smooth passage of air over the wing. **3.** *Informal.* A suitable position or niche. —*vt.* **slot·ted, slot·ting, slots. 1.** To cut or make a slot or slots in. **2.** *Informal.* To put into or assign to a slot.

**slot**[2] (slŏt) *n.* [OFr. *esclot*, horse's hoofprint, perh. < ON *slōð*, track.] The trail of an animal, esp. a deer.

**slot car** *n.* An electric toy racing car that fits into a slotted track and is controlled by a rheostat held by the operator.

**sloth** (slôth, slōth, slŏth) *n.* [ME *slowth* < *slow*, slow < OE *slāw.*] **1.** Aversion to exertion or work : LAZINESS. **2.** A slow-moving arboreal mammal of the family Bradypodidae of tropical America, esp.: **a.** A member of the genus *Bradypus*, having three long-clawed toes on each foot. **b.** A member of the genus *Choloepus*, having two toes on the forefeet and three on the hind feet. **3.** A company of bears.

**sloth bear** *n.* A bear, *Melursus ursinus* of south-central Asia, having a long snout and dark shaggy hair.

**sloth·ful** (slôth'fəl, slōth'-, slŏth'-) *adj.* Lazy. **—sloth'ful·ly** *adv.* **—sloth'ful·ness** *n.*

**slot machine** *n.* A vending or gambling machine operated by the insertion of coins into a slot.

**slot racing** *n.* The game of racing slot cars. **—slot racer** *n.*

**slouch** (slouch) *v.* **slouched, slouch·ing, slouch·es.** [Orig. unknown.] —*vi.* **1.** To sit, stand, or walk with an awkward, drooping posture. **2.** To droop or hang carelessly, as a hat. —*vt.* To cause to droop. —*n.* **1.** A drooping posture. **2.** An awkward, lazy, or inept person. **—slouch'i·ly** *adv.* **—slouch'i·ness** *n.* **—slouch'y** *adj.*

**slouch hat** *n.* A soft hat with a broad flexible brim.

**slough**[1] (slo͞o, slou) *also* **slew** (slo͞o) *n.* [ME *slogh* < OE *slōh.*] **1.** A depression or hollow, usu. filled with deep mud or mire. **2.** *also* **slue** (slo͞o). A stagnant swamp, marsh, bog, or pond, esp. as part of a bayou, inlet, or backwater. **3.** Deep despair or moral degradation.

**slough**[2] (slŭf) *n.* [ME *slughe.*] **1.** The dead outer skin shed by a snake or amphibian. **2.** *Med.* Dead tissue separated from a living structure. **3.** An outer layer or covering that is shed. —*v.* **sloughed, slough·ing, sloughs.** —*vi.* **1.** To be cast off or shed. **2.** To shed a slough. **3.** *Med.* To separate from surrounding tissue. —Used of dead tissue. —*vt.* To discard as undesirable. **—slough'y** *adj.*

**Slo·vak** (slō'väk', -văk') *also* **Slo·va·ki·an** (slō-vä'kē-ən, -văk'ē-ən) *n.* [Slovak *Slovák.*] **1.** A member of a Slavic people living in Slovakia. **2.** The Slavic language of the Slovaks. —*adj.* Of or relating to Slovakia, the Slovaks, or their language.

**slov·en** (slŭv'ən) *n.* [ME *sloveyn.*] A slovenly person.

**Slo·vene** (slō'vēn') *n.* **1.** A native or inhabitant of Slovenia. **2.** The Slavic language of Slovenia. **—Slo'vene'** *adj.*

**slov·en·ly** (slŭv'ən-lē) *adj.* Careless in personal appearance or work. **—slov'en·ly** *adv.*

**slow** (slō) *adj.* **-er, -est.** [ME < OE *slāw.*] **1. a.** Not moving or capable of moving quickly <a *slow* tugboat> **b.** Marked by a retarded tempo <a *slow* love song> **2. a.** Taking a long time <the *slow* process of editing a manuscript> **b.** Taking more than the usual time <*slow* workers><*slow* progress in negotiations> **3. a.** Registering a time or rate behind or below the correct one <a *slow* watch> **b.** Not on time : TARDY <The subway is *slow* today.> **4.** Not prompt or willing <*slow* to accept our invitation> **5.** Sluggish : inactive <Business is *slow.*> **6.** *Informal.* Lacking liveliness or interest : BORING <a *slow* cocktail party> **7.** Not quick to understand : OBTUSE <a *slow* learner> **8.** Only moderately warm : LOW <a *slow* oven> —*adv.* **1.** So as to fall behind <The watch runs *slow.*> **2.** At a low speed <Go *slow!*> ***usage:*** *Slow* often occurs as an adverb in speech and informal writing, as in *Drive slow! Slow* is also the established form in certain fixed expressions, such as *My watch is running slow.* Otherwise, *slowly* is preferable in formal contexts. —*v.* **slowed, slow·ing, slows.** —*vt.* **1.** To make slow or slower. **2.** To delay. —*vi.* To become slow or slower. **—slow'ly** *adv.* **—slow'ness** *n.*

☆ ***syns:*** SLOW, DILATORY, LAGGARD, POKY, TARDY *adj. core meaning* : moving at a pace less than usual or desired <a lane for *slow* traffic><*slow* progress toward peace> ***ant:*** fast

**slow burn** *n.* A gradual accumulation of anger.

**slow·down** (slō'doun') *n.* A slackening of pace, esp. an intentional slowing down of production.

**slow-foot·ed** (slō'fo͝ot'ĭd) *adj.* Advancing at a tediously slow pace.

**slow match** *n.* A slow-burning fuse used to set off explosives.

**slow motion** *n.* A motion-picture technique in which the action as projected is slower than the original action.

**slow neutron** *n.* A neutron in thermal equilibrium with the surrounding medium, esp. one produced by fission, slowed by a moderator, and having an average speed of approx. 2,200 meters per second.

**slow-pitch** *also* **slo-pitch** (slō'pĭch') *n.* A softball game with ten players to a team and in which a pitch must travel in an arc from three to ten feet high in order to be legal.

**slow·poke** (slō'pōk') *n.* *Informal.* One who moves or performs very slowly.

**slow virus** *n.* Any of a group of animal viruses, such as the one that causes multiple sclerosis, marked by prolonged incubation periods in which the virus remains dormant in the body prior to the development of symptoms.

**slow-wit·ted** (slō'wĭt'ĭd) *adj.* Obtuse : dull. **—slow'wit'ted·ly** *adv.* **—slow'wit'ted·ness** *n.*

**slow·worm** (slō'wûrm') *n.* [ME *slowurm* < OE *slāwyrm.*] A limbless European lizard, *Anguis fragilis*, with a smooth snakelike body.

**slowworm**
*2–3 feet long*

**sloyd** (sloid) *n.* [Swed. *slöjd*, skill, skilled labor.] A system of manual training developed in Sweden, based on woodcarving as an exercise in the use of tools.

**slub** (slŭb) *vt.* **slubbed, slub·bing, slubs.** [Orig. unknown.] To draw out and twist (a sliver of silk or other textile fiber) in preparation for spinning. —*n.* **1.** A soft thick nub in yarn that is either an imperfection or has been purposely set for a desired effect. **2.** A slightly twisted roll of fiber.

**sludge** (slŭj) *n.* [Perh. alteration of dial. *slutch*, mire.] **1.** Mud, mire, or ooze covering the ground or forming a deposit, as on a river bed. **2.** Slushy matter, as that precipitated by sewage treatment. **3.** Finely broken or half-formed ice on a body of water. **—sludg'y** *adj.*

**slue**[1] *also* **slew** (slo͞o) *v.* **slued, slu·ing, slues** *also* **slewed, slew·ing, slews.** [Orig. unknown.] —*vt.* **1.** To turn or twist (something) sideways. **2.** To twist (a mast or boom) around on its axis. —*vi.* To turn, twist, move, or skid to the side. —*n.* **1.** The act of sluing. **2.** The position to which something has slued.

**slue**[2] *n.* *var. of* SLEW[1].

**slue**[3] *n.* *var. of* SLOUGH[1] 2.

**slug**[1] (slŭg) *n.* [Prob. < SLUG[2].] **1.** A round bullet larger than buckshot. **2.** *Informal.* A shot of liquor. **3.** A small metal disk for use in a slot machine, esp. one used illegally. **4.** A lump of metal or glass ready for processing. **5. a.** A strip of type metal, less than type-high and thicker than a lead, used for spacing. **b.** A line of cast type in a single strip of metal. **c.** A compositor's type line of identifying marks or instructions, inserted temporarily in copy. **6.** *Physics.* The unit of mass that is accelerated at the rate of one foot per second per second when acted upon by a force of one pound weight. —*vt.* **slugged, slug·ging, slugs.** To add slugs to in printing.

**slug**[2] (slŭg) *n.* [ME *slugge*, sluggard, prob. of Scand. orig.] **1.** A terrestrial gastropod mollusk of the genus *Limex* and related genera, with an elongated body and no external shell. **2.** The smooth soft larva of an insect, as the sawfly. **3.** *Informal.* A sluggard.

**slug**[3] (slŭg) *vt.* **slugged, slug·ging, slugs.** [Perh. < SLUG[1].] To strike heavily, esp. with the fist. —*n.* A hard, heavy blow, as with the fist or a baseball bat.

**slug·a·bed** (slŭg'ə-bĕd') *n.* A lazy person inclined to stay in bed.

**slug·fest** (slŭg'fĕst') *n.* *Slang.* **1.** A vicious fistfight. **2.** *Baseball.* A game in which many hits and runs are scored.

**slug·gard** (slŭg'ərd) *n.* [ME *sluggart*, prob. < *sluggen*, to be lazy, prob. of Scand. orig.] A lazy, indolent person. **—slug'gard, slug'gard·ly** *adj.* **—slug'gard·ness** *n.*

**slug·ger** (slŭg'ər) *n.* **1.** A fighter who swings out with the fists. **2.** *Baseball.* A hard hitter.

**slug·gish** (slŭg'ĭsh) *adj.* [ME, prob. < *sluggen*, to be lazy. —see SLUGGARD.] **1.** Displaying little movement or activity : SLOW. **2.** Lacking alertness, vigor, or energy : DULL. **3.** Slow to perform or respond to stimulation <a *sluggish* old car> **—slug'gish·ly** *adv.* **—slug'gish·ness** *n.*

**sluice** (slo͞os) *n.* [ME *scluse* < OFr. *excluse* < Lat. *exclusa*, fem. p.part. of *excludere*, to shut out. —see EXCLUDE.] **1. a.** An artificial channel for conducting water with a valve or gate to regulate the flow. **b.** The body of water so regulated or held back. **2.** The valve or gate used in a sluice. **3.** An artificial channel for carrying off excess water. **4.** A long inclined trough, as for carrying logs or separating gold ore. —*v.* **sluiced, sluic·ing, sluic·es.** —*vt.* **1. a.** To flood or drench by a sluice. **b.** To wash with a sudden flow of water : FLUSH. **2.** To draw off or let out by a sluice. **3.** To send (logs) down a sluice. —*vi.* To flow out from or as if from a sluice.

**slum** (slŭm) *n.* [Orig. unknown.] *often* **slums.** A squalid, heavily populated urban area. —*vi.* **slummed, slum·ming, slums.** To visit a slum, esp. from curiosity or for amusement <go *slumming*>

**slum·ber** (slŭm'bər) *v.* **-bered, -ber·ing, -bers.** [ME *slumeren*, freq. of *slumen*, to doze < *slume*, sleep < OE *sluma*.] —*vi.* **1.** To sleep. **2.** To be inactive or dormant. —*vt.* To pass (time) in sleep. —*n.* **1.** Sleep. **2.** Inactivity or dormancy. **—slum'ber·er** *n.* **—slum'ber·ing·ly** *adv.*

**slum·ber·ous** (slŭm'bər-əs) *also* **slum·ber·y** (-bə-rē) *or* **slum·brous** (-brəs) *adj.* **1.** Sleepy : drowsy. **2. a.** Like sleep. **b.** Quiet : tranquil. **3.** Inducing sleep : SOPORIFIC. **—slum'ber·ous·ly** *adv.* **—slum'ber·ous·ness** *n.*

**slumber party** *n.* An overnight party in which teen-age girls wear nightclothes, socialize, and sometimes sleep.

**slum·gul·lion** (slŭm'gŭl'yən) *n.* [Perh. dial. *slum*, slime + dial. *gullion*, mud.] A watery meat stew.

**slum·lord** (slŭm'lôrd') *n. Informal.* A landlord of slum property.

**slump** (slŭmp) *vi.* **slumped, slump·ing, slumps.** [Prob. of Scand. orig.] **1.** To fall or sink suddenly, as into a bog or through a crust of snow or ice. **2.** To decline suddenly : SINK <Sales *slumped*.> **3.** To slide down suddenly. **4.** To droop, as in sitting or standing : SLOUCH. —*n.* A sudden decline, as in interest or activity.

**slung** (slŭng) *v.* *p.t.* & *p.p.* of SLING[1].

**slung·shot** (slŭng'shŏt') *n.* A small, heavy weight attached to a thong, used as a weapon.

**slunk** (slŭngk) *v.* *p.t.* & *p.p.* of SLINK.

**slur** (slûr) *vt.* **slurred, slur·ring, slurs.** [< ME *sloor*, mud.] **1.** To treat without due consideration : make light of. **2.** To pronounce indistinctly. **3.** To speak slightingly of : DISPARAGE. **4.** *Mus.* **a.** To glide over (a series of notes) smoothly without a break. **b.** To mark with a slur. **5.** To blur or smear (print). —*n.* **1.** A disparaging remark : ASPERSION. **2.** A slurred utterance or sound. **3.** *Mus.* **a.** A curved line connecting notes on a score to indicate that they are to be played or sung legato. **b.** A passage played or sung in this manner. **4.** A smeared or blurred impression in printing.

**slurb** (slûrb) *n. Informal.* [SL(OVENLY) + (SUB)URB.] A poorly planned and constructed suburban housing development.

**slurp** (slûrp) *v.* **slurped, slurp·ing, slurps.** [Du. *slurpen*.] —*vt.* To eat or drink noisily. —*vi.* To eat or drink something noisily.

**slur·ry** (slûr'ē) *n., pl.* **-ries.** [ME *slory*, perh. < *sloor*, mud.] A thin mixture of a liquid, esp. water, and any of several finely divided substances, as cement, plaster of Paris, or clay particles.

**slush** (slŭsh) *n.* [Perh. of Scand. orig.] **1.** Partially melted ice or snow. **2.** Soft mud : MIRE. **3.** Refuse grease or fat from a ship's galley. **4.** A greasy compound used as a lubricant for machinery. **5.** Maudlin expression. —*v.* **slushed, slush·ing, slush·es.** —*vt.* **1.** To daub (machinery) with slush. **2.** To fill (joints in masonry) with mortar. **3.** To wash down (a deck) by dashing water on. **4.** To splash or soak with slush or mud. —*vi.* **1.** To move through slush. **2.** To make a splashing sound. **—slush'i·ness** *n.* **—slush'y** *adv.*

**slush fund** *n.* **1.** A fund raised for undesignated purposes, esp.: **a.** A fund used by a group, as office workers, for entertainment. **b.** A fund raised by a group for corrupt practices, as bribery or graft. **2.** Money once raised by the sale of garbage from a warship to buy small items of luxury for the crew.

**slut** (slŭt) *n.* [ME *slutte*.] **1.** A slovenly, dirty woman : SLATTERN. **2. a.** An immoral woman. **b.** A prostitute. **3.** A brazen girl. **4.** A female dog. **—slut'tish** *adj.* **—slut'tish·ly** *adv.* **—slut'tish·ness** *n.*

**sly** (slī) *adj.* **sli·er, sli·est** also **sly·er, sly·est.** [ME *sleih* < ON *slœgr*, clever.] **1.** Stealthily clever : CRAFTY. **2.** Secretive : underhand. **3.** Playfully mischievous : ROGUISH <a *sly* wink> **—sly'ly** *adv.* **—sly'ness** *n.*

☆ ***syns:*** SLY, FURTIVE, SECRETIVE, SURREPTITIOUS *adj. core meaning* : stealthily clever and underhand <a series of *sly* political maneuvers>

**sly·boots** (slī'bo͞ots') *n. Informal.* A sly person.

**slype** (slīp) *n.* [Orig. unknown.] A covered passage, esp. one between the transept and chapter house of a cathedral.

**Sm** *symbol for* SAMARIUM.

**smack[1]** (smăk) *v.* **smacked, smack·ing, smacks.** [Prob. of MLG orig.] —*vt.* **1.** To make a sound by pressing together the lips and pulling them apart quickly. **2.** To kiss noisily. **3.** To strike heartily and noisily. —*vi.* To make or give a smack. —*n.* **1.** The loud sharp sound of smacking. **2.** A noisy kiss. **3.** A sharp blow or slap. —*adv.* **1.** With a smack <fell *smack* on my head> **2.** Directly <a plan that went *smack* against all regulations>

**smack[2]** (smăk) *n.* [ME < OE *smæc*.] **1. a.** A distinctive flavor. **b.** A suggestion : trace. **2.** A small amount : SMATTERING. —*vi.* **smacked, smack·ing, smacks. 1.** To have a distinctive flavor. **2.** To give an indication : SUGGEST <This *smacks* of illegality.>

**smack[3]** (smăk) *n.* [Du. *smak*.] *Naut.* A sloop-rigged boat used chiefly in fishing, esp. to transport the catch to a market.

**smack[4]** (smăk) *n.* [Orig. unknown.] *Slang.* Heroin.

**smack-dab** (smăk'dăb') *adv. Slang.* SMACK[1] 2.

**smack·er** (smăk'ər) *n.* **1.** A loud kiss. **2.** A resounding blow. **3.** *Slang.* A dollar.

**smack·ing** (smăk'ĭng) *adj.* Brisk : vigorous <a *smacking* breeze>

**small** (smôl) *adj.* **-er, -est.** [ME < OE *smæl*.] **1.** Having relatively little size or slight dimensions. **2.** Of limited importance or significance : TRIVIAL <a *small* matter> **3.** Limited in scope or degree <*small* farm operations> **4.** Lacking position, influence, or status : MINOR. **5.** Modest : unpretentious. **6.** Not fully grown. **7.** Petty <*small* minds> **8.** Humiliated. **9.** Diluted : weak. —Used of alcoholic beverages. **10.** Lacking strength <a *small* cry> —*adv.* **1.** In small pieces <cut up *small*> **2.** Softly. **3.** In a small way. —*n.* **1.** Something smaller than the rest <the *small* of my back> **2. smalls. a.** Small things as a whole. **b.** *Chiefly Brit.* Smallclothes. **—small'ness** *n.*

**small arm** *n.* A hand-held firearm.

**small beer** *n.* **1.** Weak or inferior beer. **2.** Trivia.

**small calorie** *n.* CALORIE 1.

**small capital** *n.* A smaller letter having the form of a capital letter.

**small change** *n.* **1.** Coins of low denomination. **2.** Something of little value or significance.

**small-claims court** (smôl'klāmz') *n.* A court established for the simplified and efficient handling of small claims on debts.

**small·clothes** (smôl'klōthz', -klōz') *pl.n.* **1.** Men's close-fitting knee breeches worn in the 18th cent. **2.** *Chiefly Brit.* Small items of clothing, as underclothes or handkerchiefs.

**small fry** *n.* **1.** Young or small fish. **2.** Small children. **3.** Trivia.

**small hours** *pl.n.* The early postmidnight hours.

**small intestine** *n.* The part of the intestine between the oulet of the stomach and the large intestine.

**small-mind·ed** (smôl'mīn'dĭd) *adj.* **1.** Selfish or narrow in attitude. **2.** Petty or selfish. **—small'-mind'ed·ly** *adv.* **—small'mind'ed·ness** *n.*

**small·mouth bass** (smôl'mouth') *n.* A North American freshwater food and game fish, *Micropterus dolomieui*.

**small potatoes** *pl.n. Informal.* Trivia.

**small·pox** (smôl'pŏks') *n.* An acute, highly infectious disease caused by a virus and initially characterized by chills, high fever, headache, and backache, with subsequent widespread eruption of pimples that eventually blister, suppurate, and form pockmarks.

**small-scale** (smôl'skāl') *adj.* **1.** Limited in scope or extent <a *small-scale* plan> **2.** Created on a small scale.

**small talk** *n.* Casual or trivial conversation.

**small-time** (smôl'tīm') *adj. Informal.* Insignificant or unimportant : MINOR <a *smalltime* actor> **—small'tim'er** *n.*

**smalt** (smôlt) *n.* [Fr. < Ital. *smalto*, of Germanic orig.] A deep-blue paint and ceramic pigment produced by pulverizing a glass made of silica, potash, and cobalt oxide.

**smalt·ite** (smôl'tīt') *also* **smalt·ine** (smôl'tĭn, -tēn') *n.* [Alteration of *smaltine* < Fr. < *smalt*, smalt.] A white to silver-gray ore of nickel and cobalt, essentially $(Co,Ni)As_2$.

**sma·rag·dite** (smə-răg'dīt') *n.* [Fr. < Lat. *smaragdus*, a kind of precious stone < Gk. *smaragdos*.] A thin, foliated, light-green amphibole mineral.

**smarm·y** (smär'mē) *adj.* **-i·er, -i·est.** [< dial. *smarm*, to gush, slobber.] **1.** Sleek. **2.** Gushingly or unctuously flattering <"He is a *smarmy*, obsequious sort" —*New Yorker*> **—smarm·i·ness** *n.*

**smart** (smärt) *adj.* **-er, -est.** [ME, causing pain < OE *smeart*.] **1 a.** Characterized by sharp, quick thought : INTELLIGENT. **b.** Amusingly clever : WITTY <a *smart* rejoinder> **c.** Impertinent <a *smart* answer> **2.** Sharp and quick in movement : ENERGETIC <a *smart* pace> **3.** Shrewd in dealings and intelligent : CANNY <a *smart* business person> **4.** Fashionable : elegant <a *smart* suit><a *smart* hotel> **5.** Of, relating to, or being a device that imitates human intelligence. —*vi.* **smart·ed, smart·ing, smarts. 1. a.** To cause a sharp, usu. superficial stinging pain, as an acrid liquid or a slap. **b.** To be the source of such a pain, as a wound. **c.** To feel such a pain. **2.** To suffer acutely, as from mental distress, wounded feelings, or remorse <*smarted* from the rebuke> **3.** To suffer or pay a heavy penalty. —*n.* **1.** Sharp mental or physical pain. **2. smarts.** *Slang.* Intelligence : expertise. **—smart'ly** *adv.* **—smart'ness** *n.*

**smart al·eck** (ăl'ĭk) *n.* [SMART + *Aleck*, nickname for *Alexander*.] *Informal.* An obnoxiously self-assertive, arrogant person. **—smart'-al'eck·y** *adj.*

**smart bomb** *n.* A bomb guided to its target by radio waves or a laser beam.

**smart copier** *n.* A machine that has the capability of copying, printing, and collating documents, and through electronic hardware and software, can also store and transmit them to other terminals.

**smart·en** (smär'tn) *v.* **-ened, -en·ing, -ens.** —*vt.* **1.** To improve in appearance or style. **2.** To make quicker <*smarten* one's pace> —*vi.* To make oneself smart or smarter.

**smart money** *n.* **1.** Compensation beyond the value of actual harm, awarded by jury in cases of gross negligence or willful misconduct. **2.** A bet or bets placed by experienced gamblers or those having privileged information.

**smart·weed** (smärt'wēd') *n.* A marsh plant of the genus *Polygonum* or *Persicaria,* having small, densely clustered pink, white, or green flowers.

**smar·ty-pants** (smär'tē-pănts') *n. (sing. in number) Informal.* A smart aleck.

**smash** (smăsh) *v.* **smashed, smash·ing, smash·es.** [Perh. blend of SMACK and CRASH.] —*vt.* **1.** To break into pieces suddenly, noisily, and violently : SHATTER. **2. a.** To throw or dash (something) violently so as to shatter or crush. **b.** To strike with a heavy blow : BATTER. **3.** To hit (a ball or shuttlecock) in a violent overhand stroke. **4.** To crush or destroy utterly <*smashed* the rebellion> —*vi.* **1.** To move or be moved suddenly, noisily, and violently. **2.** To break into pieces, as from a violent blow or collision. **3.** *Sports.* To hit a ball or shuttlecock in a violent overhand stroke. **4.** To be crushed or destroyed. **5.** To go bankrupt. —*n.* **1. a.** The act or sound of smashing. **b.** The state of having been smashed. **2. a.** Utter defeat or destruction : RUIN. **b.** Financial failure : BANKRUPTCY. **3.** A collision : crash. **4. a.** A drink of mint, sugar, soda water, and alcoholic liquor, usu. brandy. **b.** A soft drink made of crushed fruit. **5.** A violent overhand stroke as in tennis or badminton. **6.** *Informal.* A resounding success. —*adj. Informal.* Of, relating to, or being a resounding success. **—smash** *adv.* **—smash'er** *n.*

☆ ***syns:*** SMASH, BANG, CRASH, SLAM, WHAM *n. core meaning* : a forceful movement causing a loud noise <hit the wall with a *smash*>

**smashed** (smăsht) *adj. Slang.* Intoxicated.

**smash·ing** (smăsh'ĭng) *adj. Informal.* Extraordinarily impressive or fine <a *smashing* time at the party>

**smash·up** (smăsh'ŭp') *n.* **1.** Utter collapse or defeat. **2.** A serious collision between vehicles.

**smat·ter** (smăt'ər) *v.* **-tered, -ter·ing, -ters.** [ME *smateren.*] —*vt.* **1.** To speak (a language) without fluency. **2.** To study or treat superficially. —*vi.* To prattle. **—smat'ter** *n.* **—smat'ter·er** *n.*

**smat·ter·ing** (smăt'ər-ĭng) *n.* **1.** Superficial or piecemeal knowledge. **2.** A small scattered number or amount <a *smattering* of raindrops>

**smaze** (smāz) *n.* [SM(OKE) + (H)AZE.] A relatively dry atmospheric mixture of smoke and haze.

**smear** (smîr) *v.* **smeared, smear·ing, smears.** [ME *smeren,* to anoint < OE *smerian.*] —*vt.* **1.** To spread or daub with a sticky, greasy, or dirty substance. **2.** To stain by or as if by spreading or daubing with a sticky, greasy, or dirty substance. **3.** To stain or attempt to destroy the reputation of : VILIFY <political enemies *smearing* our name> **4.** *Slang.* To defeat utterly : SMASH. —*vi.* To be or become stained or dirtied. —*n.* **1.** A mark made by smearing : SPOT. **2.** A substance that is to be spread on a surface. **3.** A substance placed on a slide for microscopic examination. **4.** An attempt to destroy a reputation : SLANDER.

**smear·case** (smîr'kās') *n.* [G. *Schmierkäse* : *schmieren,* to smear + *Käse,* cheese < Lat. *caseus.*] Cottage cheese.

**smear word** *n.* An abusive or disparaging word or phrase intended to denigrate a person, group, or race.

**smear·y** (smîr'ē) *adj.* **-i·er, -i·est. 1.** That is smeared. **2.** Tending to smear or soil. **—smear'i·ness** *n.*

**smell** (smĕl) *v.* **smelled** *or* **smelt** (smĕlt), **smell·ing, smells.** [ME *smellen.*] —*vt.* **1.** To perceive the scent of (something) by means of the olfactory nerves. **2.** To sense the presence of by or as if by the olfactory nerves : DETECT <We *smelled* trouble ahead.> —*vi.* **1. a.** To use the sense of smell. **b.** To perceive the scent of something. **2.** To have or emit an odor. **3.** To be suggestive <an abandoned mine that *smells* of terror> **4.** To have or emit an unpleasant odor : STINK. **5.** To appear to be dishonest or to suggest evil. —*n.* **1.** The sense by which odors are perceived. **2.** The quality of something that may be perceived by the olfactory sense : ODOR. **3.** An act or instance of smelling. **4.** A distinctive quality : AURA <the *smell* of victory> **—smell a rat.** *Slang.* To suspect that something is wrong.

**smelling salts** *pl.n. (sing. or pl. in number).* Any of several preparations based on spirits of ammonia, sniffed as a restorative.

**smell·y** (smĕl'ē) *adj.* **-i·er, -i·est.** *Informal.* Having a bad odor.

☆ ***syns:*** SMELLY, FETID, FOUL, MALODOROUS, REEKING, STINKING *adj. core meaning* : having an offensive odor <*smelly* chemical wastes> ***ant:*** fragrant

**smelt**[1] (smĕlt) *v.* **smelt·ed, smelt·ing, smelts.** [MLG *smelten.*] —*vt.* To melt or fuse (ores), separating the metallic constituents. —*vi.* To melt or fuse. —Used of ores.

**smelt**[2] (smĕlt) *n., pl.* **smelts** *or* **smelt.** [ME < OE.] A small silvery marine or freshwater food fish of the family Osmeridae, esp. *Osmerus mordax* of North America and *O. eperlanus* of Europe.

**smelt**[3] (smĕlt) *v. var. p.t. & p.p. of* SMELL.

**smelt·er** (smĕl'tər) *n.* **1. a.** A device for smelting. **b.** *also* **smelt·er·y** (smĕl'tə-rē). An establishment for smelting. **2.** One engaged in the smelting industry.

**smew** (smyōō) *n.* [Orig. unknown.] A small, crested Old World duck, *Mergus albellus,* with a narrow bill and white and black plumage in the male.

**smid·gen** *also* **smid·geon** *or* **smid·gin** (smĭj'ən) *n.* [Prob. alteration of dial. *smitch,* particle.] *Informal.* A minute quantity or portion : BIT.

**smi·lax** (smī'lăks') *n.* [NLat. *Smilax,* genus name < Lat. *smilax,* bindweed < Gk.] **1.** A plant of the genus *Smilax,* including climbing vines such as the catbrier. **2.** A vine, *Asparagus asparagoides,* with glossy foliage.

**smile** (smīl) *n.* [< ME *smilen,* to smile.] **1.** A facial expression characterized by an upward curving of the corners of the mouth and expressing pleasure, amusement, or derision. **2.** A pleasant or favorable disposition. —*v.* **smiled, smil·ing, smiles.** —*vi.* **1.** To have or form a smile. **2.** To express or appear to express approval or beneficence. —*vt.* **1.** To express with a smile. **2.** To effect or accomplish with or as if with a smile. **—smil'er** *n.* **—smil'ing·ly** *adv.* **—smil'ing·ness** *n.*

☆ ***syns:*** SMILE, BEAM, GRIN *v. core meaning* : to curl the lips upward in amusement, pleasure, or happiness <*smiled* and said hello> ***ant:*** frown

**smirch** (smûrch) *vt.* **smirched, smirch·ing, smirch·es.** [ME *smorchen.*] **1.** To soil, stain, or dirty with a smearing agent. **2.** To dishonor : defame. **—smirch** *n.*

**smirk** (smûrk) *vi.* **smirked, smirk·ing, smirks.** [ME *smirken* < OE *smearcian,* to smile.] To smile in a self-conscious, knowing, or simpering way. **—smirk** *n.* **—smirk'er** *n.* **—smirk'ing·ly** *adv.*

**smite** (smīt) *v.* **smote** (smōt), **smit·ten** (smĭt'n) *or* **smote, smit·ing, smites.** [ME *smiten* < OE *smītan.*] —*vt.* **1. a.** To inflict a heavy blow on with or as if with the hand, a tool, or a weapon. **b.** To drive or strike (e.g., a weapon) forcefully onto or into something else. **2.** To attack, damage, or destroy by or as if by blows. **3. a.** To afflict <*smitten* with disease> **b.** To chasten or chastise. **4.** To affect sharply with deep feeling <*smitten* with love> —*vi.* To strike or beat. **—smit'er** *n.*

**smith** (smĭth) *n.* [ME < OE *smiđ.*] **1.** A metalworker, esp. one who works metal when it is hot and malleable. —Often used in combination <silver*smith*><gold*smith*> **2.** A blacksmith.

**smith·er·eens** (smĭth'ə-rēnz') *pl.n.* [Ir. Gael. *smidirīn,* dim. of *smiodar,* small fragment.] *Informal.* Fragments : bits <broke the dish into *smithereens*>

**smith·er·y** (smĭth'ə-rē) *n., pl.* **-ies. 1.** The occupation or craft of a smith. **2.** SMITHY 1.

**smith·son·ite** (smĭth'sə-nīt') *n.* [After James *Smithson* (1765–1829).] A white or yellow-to-brown mineral, chiefly $ZnCO_3$, used as a source of zinc.

**smith·y** (smĭth'ē, smĭth'ē) *n., pl.* **-ies.** [ME < ON *smiđja.*] **1.** A blacksmith's shop. **2.** A blacksmith.

**smit·ten** (smĭt'n) *v. var. p.p. of* SMITE.

**smock** (smŏk) *n.* [ME *smok,* woman's undergarment < OE *smoc.*] A loose coatlike outer garment, often worn to protect the clothes while working. —*vt.* **smocked, smock·ing, smocks. 1.** To clothe in a smock. **2.** To decorate (fabric) with smocking.

**smock·ing** (smŏk'ĭng) *n.* Needlework decoration of small regularly spaced gathers stitched into a honeycomb pattern.

**smog** (smŏg, smôg) *n.* [SM(OKE) + (F)OG.] Fog polluted and mixed with smoke. **—smog'gy** *adj.* **—smog'less** *adj.*

**smoke** (smōk) *n.* [ME < OE *smoca.*] **1.** Vapor made up of small particles of carbonaceous matter in the air, resulting chiefly from incomplete combustion of organic material, such as wood or coal. **2.** A suspension of particles in a gaseous medium. **3.** A cloud of fine particles. **4.** Insubstantial, unreal, or transitory matter. **5.** The act or duration of smoking tobacco. **6.** *Informal.* Tobacco that can be smoked, esp. a cigarette. **7.** A substance used in warfare to produce a smoke screen. **8.** A pale to grayish blue to bluish gray or dark gray. —*v.* **smoked, smok·ing, smokes.** —*vi.* **1.** To emit smoke or a smokelike substance. **2.** To emit too much smoke. **3.** To draw in and exhale smoke from a cigarette, cigar, or pipe. —*vt.* **1.** To draw in and exhale the smoke of (e.g., tobacco). **2.** To preserve (meat or fish) by exposure to the aromatic smoke of burning hardwood, usu. after pickling in salt or brine. **3.** To fumigate (e.g., a house). **4.** To expose (glass) to smoke in order to darken or change its color. **—smoke out. 1.** To force out of a place of hiding or concealment by or as if by the use of smoke. **2.** To detect and bring to public view : EXPOSE.

**smoke·chas·er** (smōk'chā'sər) *n.* A forest firefighter, esp. one whose light equipment permits quick arrival at a fire.

**smoke detector** *n.* An alarm device that automatically detects the presence of smoke.

**smoke·house** (smōk'hous') *n.* A structure in which meat or fish is cured with smoke.

**smoke·jack** (smōk'jăk') *n.* A device for turning a roasting spit in a chimney, activated by the current of rising gases.

**smoke·jump·er** (smōk'jŭm'pər) *n.* A firefighter who parachutes into a forest fire.

**smoke·less** (smōk'lĭs) *adj.* Emitting little or no smoke.

**smokeless powder** *n.* A propellant charge composed chiefly of nitrocellulose, which produces little or no smoke, used in projectiles and small artillery rockets.

**smok·er** (smō'kər) *n.* **1.** One that smokes. **2.** A railroad car in which smoking is permitted. **3.** An informal social gathering for men.

**smoke screen** *n.* **1.** A mass of dense artificial smoke used to conceal military areas or operations from an enemy. **2.** An action or statement used to conceal actual plans or intentions.

**smoke·stack** (smōk'stăk') *n.* A large chimney or vertical pipe through which combustion vapors, gases, and smoke are discharged.

**smoke tree** *n.* [From the resemblance of the flower clusters to puffs of smoke.] A tree, *Cotinus obovatus* of the southern United States, or *C. coggygria* of Eurasia, having plumelike clusters of small yellowish flowers.

**Smok·ey** (smō'kē) *n.* [From the resemblance of some state troopers' hats to that of *Smokey* the Bear, an animal who warns against fires in U.S. Forest Service posters.] *Slang.* A law-enforcement officer on highway patrol.

**smoking car** *n.* SMOKER 2.

**smoking gun** *n. Informal.* Indisputable evidence or proof, esp. of a crime.

**smoking jacket** *n.* An evening jacket, often made of a fine fabric, elaborately trimmed, and usu. worn at home.

**smoking room** *n.* A room, as in a hotel or private club, set aside for smokers.

**smok·ing-room** (smō'kĭng-ro͞om', -ro͝om') *adj.* Marked by indecency : OBSCENE <*smoking-room* humor>

**smok·y** (smō'kē) *adj.* **-i·er, -i·est. 1.** Emitting smoke profusely. **2.** Mixed or filled with smoke. **3.** Resembling smoke. **4.** Discolored or soiled with or as if with smoke. **5.** Tasting of smoke. **—smok'i·ly** *adv.* **—smok'i·ness** *n.*

**smoky quartz** *n.* Cairngorm.

**smol·der** *also* **smoul·der** (smōl'dər) *vi.* **-dered, -der·ing, -ders.** [ME *smolderen* < *smolder,* smolder.] **1.** To burn with little smoke and no flame. **2.** To exist in a suppressed state. **3.** To manifest repressed anger or hatred. **—smol'der, smoul'der** *n.*

**smolt** (smōlt) *n.* [ME.] A young salmon at the stage at which it migrates from fresh water to the sea.

**smooch** (smo͞och) *n.* [Perh. imit. of the sound of a kiss.] *Slang.* A kiss. *—vi.* **smooched, smooch·ing, smooch·es.** *Slang.* To kiss.

**smooth** (smo͞oth) *adj.* **-er, -est.** [ME *smothe* < OE *smōð.*] **1.** Having a surface free from roughness, irregularities, or projections : EVEN. **2.** Fine in texture. **3.** Even in consistency <a *smooth* pudding> **4.** Having an even or gentle motion <a *smooth* ride> **5.** Having no difficulties or obstructions <a *smooth* sales operation> **6.** Serene. **7.** Bland <a *smooth* wine> **8.** Artfully suave : INGRATIATING <*smooth* talker> *—v.* **smoothed, smooth·ing, smoothes.** *—vt.* **1.** To make (a surface) even, level, or unwrinkled. **2.** To rid of obstructions, hindrances, or difficulties. **3.** To make calm : SOOTHE. **4.** To make less harsh or crude : REFINE. *—vi.* To become smooth. *—n.* **1.** A smooth part or surface. **2.** An act of smoothing. **—smooth'er** *n.* **—smooth'ly** *adv.* **—smooth'ness** *n.*

**smooth·bore** *also* **smooth bore** (smo͞oth'bôr', -bōr') *adj.* Having no rifling within the barrel <a *smoothbore* firearm> **—smooth'bore'** *n.*

**smooth breathing** *n.* The symbol (') written over some initial vowels in classical Greek to indicate that they are not aspirated.

**smooth·en** (smo͞o'thən) *vt. & vi.* **-ened, -en·ing, -ens.** To make or become smooth.

**smooth·ie** *also* **smooth·y** (smo͞o'thē) *n. Slang.* An assured, artfully ingratiating person.

**smooth muscle** *n.* The unstriated involuntary muscle of the internal organs, as of the intestine, bladder, and blood vessels, excluding the heart.

**smooth-tongued** (smo͞oth'tŭngd') *adj.* Speaking in a pleasing, flattering way.

**smooth·y** (smo͞o'thē) *n. var. of* SMOOTHIE.

**smor·gas·bord** (smôr'gəs-bôrd', -bōrd') *n.* [Swed. *smörgåsbord* : *smörgås,* sandwich + *bord,* table.] A buffet meal featuring a varied number of dishes.

**smote** (smōt) *v. p.t. & var. p.p. of* SMITE.

**smoth·er** (smŭth'ər) *v.* **-ered, -er·ing, -ers.** [ME *smotheren* < *smorther,* dense smoke.] *—vt.* **1. a.** To suffocate (another). **b.** To deprive (a fire) of the oxygen necessary for combustion. **2.** To conceal : suppress <*smothered* the true facts> **3.** To cover (a foodstuff) thickly with another foodstuff <*smother* chicken in sauce> **4.** To lavish a given emotion on (someone) <*smothered* me with their love> *—vi.* **1. a.** To suffocate. **b.** To be extinguished. **2.** To be concealed or suppressed. **3.** To be lavished with a given emotion. *—n.* Something, as a dense cloud of smoke, that smothers. **—smoth'er·y** *adj.*

**smoul·der** (smōl'dər) *v. & n. var. of* SMOLDER.

**smudge** (smŭj) *v.* **smudged, smudg·ing, smudg·es.** [ME *smogen.*] *—vt.* **1.** To make dirty. **2.** To smear : blur. **3.** To fill (a planted area) with dense smoke from a smudge pot in order to prevent damage from insects or frost. *—vi.* **1.** To smear, as with dirt, soot, or ink. **2.** To be smudged. *—n.* **1.** A blotch : smear. **2.** A smoky fire used as a protection against insects or frost. **—smudg'i·ly** *adv.* **—smudg'i·ness** *n.* **—smudg'y** *adj.*

**smudge pot** *n.* A receptacle in which a smoky fuel, as oil, is burned, so as to protect an orchard from insects or frost.

**smug** (smŭg) *adj.* **smug·ger, smug·gest.** [Prob. < LG *smuck,* neat < MLG < *smucken,* to adorn.] Self-righteously complacent. **—smug'ly** *adv.* **—smug'ness** *n.*

**smug·gle** (smŭg'əl) *v.* **-gled, -gling, -gles.** [LG *smuggeln.*] *—vt.* **1.** To import or export without paying lawful customs charges or duties. **2.** To bring in or take out illicitly or secretly. *—vi.* To engage in smuggling. **—smug'gler** *n.*

☆ ***syns:*** SMUGGLE, BOOTLEG, CONTRABAND, RUN *v. core meaning* : to import or export secretly and illegally <*smuggled* cocaine from South America>

**smut** (smŭt) *n.* [Alteration of *smot,* stain < ME *smotten,* to besmirch.] **1. a.** A particle of dirt. **b.** A smudge made by soot, smoke, or dirt. **2.** Obscenity. **3. a.** Any of various plant diseases caused by fungi of the order Ustilaginales that result in the formation of black powdery masses of spores on the affected parts. **b.** A fungus causing a plant disease. *—v.* **smut·ted, smut·ting, smuts.** *—vt.* **1.** To blacken or smudge, as with smoke or grime. **2.** To affect (a plant) with smut. **3.** To free (e.g., grain) from smut. **4.** To make obscene. *—vi.* **1.** To emit smut. **2.** To be or become blackened or smudged. **3.** To become affected with smut, as a plant. **—smut'ti·ly** *adv.* **—smut'ti·ness** *n.* **—smut'ty** *adj.*

**smutch** (smŭch) *vt.* **smutched, smutch·ing, smutch·es.** [Perh. alteration of SMUDGE.] To soil or stain. **—smutch** *n.* **—smutch'y** *adj.*

**Sn** [< Lat. *stannum,* tin.] *symbol for* TIN.

**snack** (snăk) *n.* [ME *snake,* a bite < *snaken,* to bite.] **1.** A hurried or light meal. **2.** Food eaten between meals. *—vi.* **snacked, snack·ing, snacks.** To eat a hurried or light meal.

**snack bar** *n.* A lunch counter where light meals are served.

**snaf·fle** (snăf'əl) *n.* [Orig. unknown.] A bit for a horse, consisting of two bars jointed at the center. *—vt.* **-fled, -fling, -fles.** To put on or control with a snaffle.

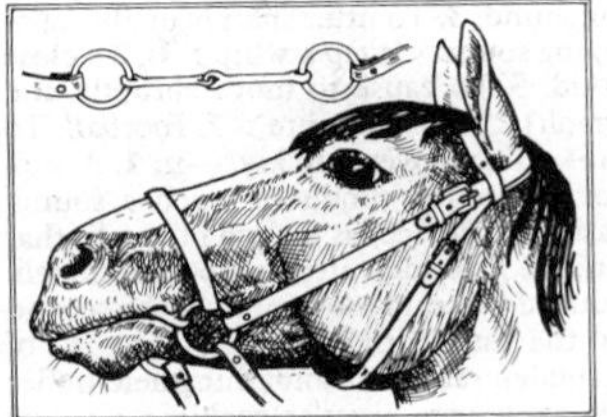
snaffle

**sna·fu** (snă-fo͞o') [S(ITUATION) + N(ORMAL) + A(LL) + F(OULED) + U(P).] *Slang. —adj.* In a state of complete confusion. *—vt.* **-fued, -fu·ing, -fus.** To make chaotic or confused. **—sna·fu'** *n.*

**snag** (snăg) *n.* [Of Scand. orig.] **1.** A sharp, rough, or jagged protuberance. **2.** A tree or a part of a tree that protrudes above the surface in a body of water. **3.** A snaggletooth. **4.** A break, pull, or tear in a fabric. **5.** An unforeseen or hidden obstacle. *—v.* **snagged, snag·ging, snags.** *—vt.* **1.** To hinder, break, tear, or destroy by or as if by a snag. **2.** To free of snags. **3.** *Informal.* To catch unexpectedly and quickly. *—vi.* To be damaged by a snag. **—snag'gy** *adj.*

**snag·gle·tooth** (snăg'əl-to͞oth') *n.* [Dial. *snaggled,* irregular (< SNAG) + TOOTH.] A broken or unaligned tooth.

**snail** (snāl) *n.* [ME < OE *snægl.*] **1.** Any of numerous aquatic or terrestrial mollusks of the class Gastropoda, with a spiral, coiled shell, a broad retractile foot, and a distinct head. **2.** A slow-moving, lazy, or sluggish person.

**snail fever** *n.* Schistosomiasis.

**snail-paced** (snāl'pāst') *adj.* Moving very slowly.

**snake** (snāk) *n.* [ME < OE *snaca.*] **1.** Any of various scaly, legless, occas. venomous reptiles of the suborder Serpentes, having a long, tapering, cylindrical body. **2. Snake.** HYDRA 3. **3.** A treacherous person. **4.** A long, highly flexible metal wire used for cleaning drains. **5.** The concept of fixing the value of currencies to each other within defined parameters, which, when graphed visually, show these currencies remaining parallel in value to each other as a unit despite fluctuations with other currencies. *—v.* **snaked, snak·ing, snakes.** *—vt.* **1.** To drag or pull lengthwise, esp. to drag with a rope or chain. **2.** To pull with quick jerks. **3.** To move like a snake. *—vi.* To move with a snakelike motion : CRAWL.

**snake·bird** (snāk'bûrd') *n.* A long-necked, long-billed bird of the genus *Anhinga,* as the water turkey.

**snake·bite** (snāk'bīt') *n.* **1.** The bite of a snake. **2.** Poisoning resulting from the bite of a venomous snake.

**snake charmer** *n.* One who uses rhythmic music and bodily movements to control snakes.
**snake dance** *n.* **1.** A dance performed as part of a biennial religious ceremony of the Hopi Indians, in which the dancers carry live rattlesnakes in their mouths. **2.** A procession of persons who join hands and move forward in a zigzag line.
**snake fence** *n.* A worm fence.
**snake·head** (snāk′hĕd′) *n.* The turtlehead.
**snake in the grass** *n.* SNAKE 3.
**snake·mouth** (snāk′mouth′) *n.* An orchid, *Pogonia ophioglossoides* of eastern North America, having a solitary rose-purple flower with a fringed lip.
**snake oil** *n.* A worthless preparation fraudulently peddled as a cure for any and all ills.
**snake pit** *n. Slang.* **1.** A disorderly, chaotic place. **2.** A mental institution.
**snake plant** *n.* Any of several tropical Old World plants of the genus *Sansevieria*, having narrow, rigid, often mottled leaves.
**snake·root** (snāk′ro͞ot′, -ro͝ot′) *n.* A plant whose roots were reputed to cure snakebite, esp. a plant of the genus *Eupatorium* or the genus *Rauwolfia*.
**snake·skin** (snāk′skĭn′) *n.* The skin of a snake, esp. when prepared as leather.
**snake·stone** (snāk′stōn′) *n.* **1.** A small stone or piece of porous substance reputed to cure snakebite. **2.** A whetstone.
**snake·weed** (snāk′wēd′) *n.* Any of various plants reputed to have the power of curing snakebite.
**snak·y** (snā′kē) *adj.* **-i·er, -i·est. 1.** Pertaining to or characteristic of snakes. **2.** Having the form or movement of a snake : SERPENTINE. **3.** Overrun with snakes <*snaky* jungles> **4.** Treacherous. **—snak′i·ly** *adv.* **—snak′i·ness** *n.*
**snap** (snăp) *v.* **snapped, snap·ping, snaps.** [MLG *snappen*, to seize.] —*vi.* **1.** To make a brisk, sharp cracking sound. **2.** To break suddenly with a brisk, sharp cracking sound. **3.** To give way abruptly under pressure or tension. **4.** To bring the jaws briskly together, often with a clicking sound : BITE. **5.** To snatch or grasp suddenly and eagerly <*snap* at a chance to go to Paris> **6.** To speak sharply or abruptly. **7.** To move swiftly and smartly <*snapped* to attention> **8.** To flash or appear to flash light : SPARKLE <eyes *snapping* with anger> **9.** To open or close with a click <The lock *snapped* shut.> —*vt.* **1.** To snatch at with or as if with the teeth : BITE. **2.** To come apart or break with a snapping sound. **3.** To utter sharply or abruptly. **4. a.** To cause to emit a snapping sound <*snap* a whip> **b.** To close or latch with a snapping sound. **5.** To cause to move abruptly and smartly. **6.** To take (a photograph) <*snap* a picture> **7.** *Football.* To center (the ball). **—snap back.** To recover quickly. —*n.* **1.** A sudden, sharp cracking sound or the action producing such a sound. **2.** A sudden breaking. **3.** A fastening device, as a clasp or catch, that operates with a snapping sound. **4.** A sudden attempt to bite, snatch, or grasp. **5. a.** The sound produced by rapid movement of the second finger from the thumb tip to the base of the thumb. **b.** The act of producing this sound. **6.** The sudden release of something held under pressure or tension. **7.** A thin, crisp, usu. circular cookie <a ginger *snap*> **8.** *Informal.* Briskness, liveliness, or energy. **9.** A brief spell of brisk, cold weather. **10.** *Informal.* An effortless task <thought that this job would be a *snap*> **11. a.** A snapshot. **b.** The taking of a snapshot. **12.** A snap bean. **13.** *Football.* The passing of the ball from the center to a back that initiates each play. —*adj.* **1.** Made or carried out on the spur of the moment <a *snap* judgment> **2.** Fastening with a snap. **3.** *Informal.* Simple : easy. **—snap** *adv.*
☆ ***syns:*** SNAP, BARK, SNARL *v. core meaning* : to speak abruptly and sharply <*snapped* at the child for misbehaving>
**snap bean** *n.* A bean, as the string bean, cultivated for its crisp, edible pods.
**snap-brim** (snăp′brĭm′) *n.* A hat having a flexible brim, usu. turned down in front and up at the back.
**snap·drag·on** (snăp′drăg′ən) *n.* [From the imagined resemblance of the flowers to the mouth of a dragon.] A plant of the genus *Antirrhinum*, esp. a widely cultivated species, *A. majus* of the Mediterranean region, with showy, two-lipped, variously colored flower clusters.
**snap·per** (snăp′ər) *n.* **1.** One that snaps. **2.** *pl.* **snapper** *or* **-pers.** Any of numerous widely distributed marine fishes of the family Lutjanidae, many of which are valued as food and game fishes. **3.** A snapping turtle.
**snapping beetle** *n.* The click beetle.
**snapping turtle** *n.* A New World freshwater turtle of the family Chelydridae, esp. *Chelydra serpentina* or *Macrochelys temmincki* of North America, with a rough shell and powerful hooked jaws.
**snap·pish** (snăp′ĭsh) *adj.* **1.** Liable to snap or bite. **2.** Liable to speak curtly. **—snap′pish·ly** *adv.* **—snap′pish·ness** *n.*
**snap·py** (snăp′ē) *adj.* **-pi·er, -pi·est. 1.** *Informal.* Lively : energetic. **2.** *Informal.* Smart or chic. **3.** Snappish. **—snap′pi·ly** *adv.* **—snap′pi·ness** *n.*
**snap roll** *n.* An aerial maneuver in which an aircraft is put through a sharp roll of 360° about its longitudinal axis.
**snap·shoot** (snăp′sho͞ot′) *vt.* **-shot** (shŏt), **-shoot·ing, -shoots.** [Back-formation < SNAPSHOT.] To take a snapshot of.
**snap·shot** (snăp′shŏt′) *n.* A picture taken with a small hand-held camera.
**snare¹** (snâr) *n.* [ME < OE *sneare* < ON *snara*.] **1.** A trapping device, often made up of a noose, used for capturing birds and small animals. **2.** Something that entangles the unwary. **3.** A surgical instrument with a wire loop controlled by a mechanism in the handle, used to remove growths, as tumors or polyps. —*vt.* **snared, snar·ing, snares. 1.** To trap with a snare. **2.** To entrap (someone). **—snar′er** *n.*
**snare²** (snâr) *n.* [Prob. < Du. *snaar*, string.] **1.** Any of the wires or cords stretched across the lower skin of a snare drum to increase reverberation. **2.** A snare drum.
**snare drum** *n.* A small double-headed drum with a snare or snares stretched across the bottom head.
**snarl¹** (snärl) *v.* **snarled, snarl·ing, snarls.** [Freq. of obs. *snar*, to growl.] —*vi* **1.** To growl viciously while baring the teeth <a *snarling* dog> **2.** To speak harshly, angrily, or threateningly. —*vt.* To utter with harshness, anger, or threats. —*n.* **1.** A vicious growl. **2.** A vicious utterance. **—snarl′er** *n.* **—snarl′ing·ly** *adv.* **—snarl′y** *adj.*
**snarl²** (snärl) *n.* [ME *snarle*, trap < *snare*. —see SNARE¹.] **1.** A tangled mass <a *snarl* of yarn> **2.** A confused, complex, or tangled situation. —*v.* **snarled, snarl·ing, snarls.** —*vi.* To become confused or tangled. —*vt.* **1.** To tangle (e.g., hair). **2.** To confuse : complicate. **—snarl′er** *n.* **—snarl′y** *adj.*
**snatch** (snăch) *v.* **snatched, snatch·ing, snatch·es.** [ME *snacchen*, to snap at.] —*vt.* **1.** To grasp hastily, eagerly, or suddenly. **2.** To seize illicitly <kidnapers who had *snatched* a child> —*vi.* To make grasping or seizing motions. —*n.* **1.** An act of snatching. **2.** A brief period of time <"At the end we preferred to travel all night, sleeping in *snatches*" —T.S. Eliot> **3.** A fragment <heard only a *snatch* of the conversation> **4.** *Slang.* A kidnaping. **5.** A lift in weightlifting in which the weight is raised in one uninterrupted movement from the floor to an overhead position. **—snatch′er** *n.*
**snatch block** *n. Naut.* A block that can be opened on one side to receive the looped part of a rope.
**snatch·y** (snăch′ē) *adj.* **-i·er, -i·est.** Intermittent : disconnected.
**snaz·zy** (snăz′ē) *adj.* **-zi·er, -zi·est.** [Perh. a blend of SNAPPY and JAZZY.] *Slang.* Fashionable : flashy.
**sneak** (snēk) *v.* **sneaked, sneak·ing, sneaks.** [Orig. unknown.] —*vi.* **1.** To move quietly and stealthily. **2.** To behave in a cowardly or servile way. —*vt.* To move, give, take, or put in a quiet, stealthy way. —*n.* **1.** A stealthy, cowardly person. **2.** An instance of sneaking.
☆ ***syns:*** SNEAK, CREEP, GLIDE, LURK, PROWL, SKULK, SLIDE, SLINK, SLIP, STEAL *v. core meaning* : to move silently and furtively <*sneaked* through the fence and past the guards>
**sneak·er** (snē′kər) *n.* **1.** One that sneaks. **2.** A sports shoe usu. made of canvas and having a soft rubber sole.
**sneak·ing** (snē′kĭng) *adj.* **1.** Stealthy : furtive. **2.** Unavowed : secret <a *sneaking* fondness for violence>. **3.** Gradually growing <a *sneaking* suspicion> **—sneak′ing·ly** *adv.*
**sneak preview** *n.* A single public showing of a film prior to its general release.
**sneak thief** *n.* A burglar who enters without breaking in.
**sneak·y** (snē′kē) *adj.* **-i·er, -i·est.** Furtive : surreptitious <*sneaky* behavior> **—sneak′i·ly** *adv.* **—sneak′i·ness** *n.*
**sneer** (snîr) *n.* [Perh. of LG orig.] **1.** A scornful facial expression marked by a slight raising of one corner of the upper lip. **2.** A contemptuous sound or statement. —*v.* **sneered, sneer·ing, sneers.** —*vt.* To utter with or as if with a sneer. —*vi.* **1.** To take on a scornful, contemptuous, or derisive facial expression. **2.** To speak scornfully, contemptuously, or derisively. **—sneer′er** *n.* **—sneer′ful** *adj.* **—sneer′ing·ly** *adv.*
**sneeze** (snēz) *vi.* **sneezed, sneez·ing, sneez·es.** [ME *snesen*, alteration of *fnesen* < OE *fnēosan*.] To expel air forcibly from the mouth and nose in an explosive, spasmodic involuntary action resulting from irritation of the nasal mucosa. **—sneeze** *n.* **—sneez′er** *n.* **—sneez′y** *adj.*
▲ word history: The word *sneeze* is descended from Old English *fnēosan* by a combination of circumstances that caused the replacement of initial *f* by *s*. The most important factor in the change was the similarity in appearance of *f* and *s* in some medieval manuscripts and printed books. Initial *s* looked like an *f* without the crossbar, so it was very easy to misread one letter for the other. Contributing to the confusion was the rarity of the cluster *fn-* at the beginning of a word in English. It was uncommon even in Old English, and had been all but eliminated in Middle English times. On top of all this, the sounds represented by *f* and *s* were very similar and this would make the spelling change more acceptable. The original verb *fnēosan* is related to Greek *pneuma*, "breath," from which *pneumatic* is derived.
**sneeze·weed** (snēz′wēd′) *n.* **1.** Any of several North American plants of the genus *Helenium*, with rayed, yellow flowers. **2.** The sneezewort.

**sneeze·wort** (snēz'wûrt', -wôrt') *n.* A plant, *Achillea ptarmica* native to Europe, with white flower clusters.

**snell** (snĕl) *n.* [Orig. unknown.] A length of fine threadlike material, as monofilament or gut, that connects a fishhook to a heavier line : LEADER.

**snib** (snĭb) *vt.* **snibbed, snib·bing, snibs.** [Orig. unknown.] *Chiefly Brit.* To latch (a door).

**snick·er** (snĭk'ər) *vi.* **-ered, -er·ing, -ers.** [Imit.] To utter a stifled laugh. **—snick'er** *n.* **—snick'er·ing·ly** *adv.*

**snick·er·snee** (snĭk'ər-snē') *n.* [Alteration of obs. *stick or snee,* to cut and thrust in fighting with a knife : Du. *steken,* to stab + Du. *snijden,* to cut.] **1.** A knife similar to a sword. **2.** *Archaic.* The act of fighting with knives.

**snide** (snīd) *adj.* **snid·er, snid·est.** [Orig. unknown.] Derogatory in a malicious, superior way. **—snide'ly** *adv.* **—snide'ness** *n.*

**sniff** (snĭf) *v.* **sniffed, sniff·ing, sniffs.** [ME *sniffen.*] *—vi.* **1.** To inhale a short audible breath through the nose. **2.** To indicate doubt, ridicule, or contempt by or as if by sniffing <*sniffed* at their bad manners> **3.** To savor an odor by sniffing. *—vt.* **1.** To inhale forcibly through the nose. **2.** To smell or try to smell by sniffing. **3.** To detect by or as if by sniffing <*sniffed* unpleasantness ahead> *—n.* **1. a.** An instance of sniffing. **b.** The sound of sniffing. **2.** Something sniffed or perceived by sniffing : WHIFF. **—sniff'er** *n.*

**snif·fle** (snĭf'əl) *vi.* **-fled, -fling, -fles.** [Freq. of SNIFF.] **1.** To breathe audibly through a congested nose. **2.** To whimper lightly with spasmodic sniffing. *—n.* **1.** An act or sound of sniffling. **2. sniffles.** *Informal.* A condition, as a head cold, accompanied by nasal discharge.

**sniff·y** (snĭf'ē) *adj.* **-i·er, -i·est.** *Informal.* Disposed to exhibiting arrogance or contempt : HAUGHTY.

**snif·ter** (snĭf'tər) *n.* [Dial. *snifter,* sniff < ME *snifteren,* to sniff.] **1.** A pear-shaped goblet with a narrow top, used in serving aromatic liquors, as brandy. **2.** *Slang.* A small portion of liquor.

**snig·ger** (snĭg'ər) *n.* [Alteration of SNICKER.] A snicker. *—vi.* **-gered, -ger·ing, -gers.** To snicker.

**snip** (snĭp) *v.* **snipped, snip·ping, snips.** [LG *snippen,* to snap.] *—vt.* To cut, clip, or separate in a short, quick stroke with scissors or shears. *—vi.* To cut or clip with short, quick strokes. *—n.* **1.** An instance of snipping or the sound made by snipping. **2. a.** A small cut made with scissors or shears. **b.** A small piece cut or clipped off. **3.** *Informal.* **a.** One that is small or slight. **b.** A small, mischievous, or annoying person. **4. snips.** Small hand shears for cutting sheet metal. **5.** *Slang.* SNAP 10.

**snipe** (snīp) *n., pl.* **snipe** or **snipes.** [ME *snype,* prob. of Scand. orig.] **1. a.** A long-billed wading bird of the genus *Capella,* esp. the common, widely distributed species *C. gallinago.* **b.** A bird similar or related to the snipe. **2.** A gunshot fired from a concealed place. *—vi.* **sniped, snip·ing, snipes. 1.** To shoot at individuals from a concealed place. **2.** To shoot snipe. **3.** To make malicious, underhand remarks or attacks <a politician who *sniped* at all opponents>

**snip·er** (snī'pər) *n.* **1.** A skilled military rifleman detailed to spot and pick off enemy troops from a concealed place. **2.** One who shoots at others from a concealed place.

**snip·pet** (snĭp'ĭt) *n.* **1.** A tidbit <*snippets* of gossip> **2.** *Informal.* A small or mischievous person.

**snip·pet·y** (snĭp'ĭ-tē) *adj.* **-i·er, -i·est. 1.** Made up of snippets. **2.** *Informal.* Snippy.

**snip·py** (snĭp'ē) *adj.* **-pi·er, -pi·est.** *Informal.* **1.** Impertinently curt or snappish. **2.** Fragmentary.

**snit** (snĭt) *n.* [Orig. unknown.] *Slang.* Agitation or irritation <They are in a *snit* over the delay.>

**snitch** (snĭch) *v.* **snitched, snitch·ing, snitch·es.** [Orig. unknown.] *Slang* *—vt.* To steal (something of little or no value) : PILFER. *—vi.* To turn informer <*snitched* on one's colleagues> *—n.* **1.** A thief. **2.** An informer. **—snitch'er** *n.*

**sniv·el** (snĭv'əl) *vi.* **-eled, -el·ing, -els** or **-elled, -el·ling, -els.** [ME *snyvelen,* to run at the nose, of OE orig.] **1.** To cry with sniffling. **2.** To complain or whine tearfully. **3.** To run at the nose. **4.** To sniffle. *—n.* **1.** An act of sniffling or sniveling. **2.** Nasal mucus. **—sniv'el·er** *n.*

**snob** (snŏb) *n.* [Obs. *snob,* person of the lower classes < dial. *snob,* cobbler.] **1.** One who is convinced of and flaunts one's social superiority. **2.** One who despises one's inferiors and whose condescension arises from social or intellectual pretension.

**snob·ber·y** (snŏb'ə-rē) *n., pl.* **-ies.** Snobbish behavior.

**snob·bish** (snŏb'ĭsh) *adj.* Of, characteristic of, befitting, or resembling a snob. **—snob'bish·ly** *adv.* **—snob'bish·ness** *n.*

☆ ***syns:*** SNOBBISH, ELITIST, HIGH-HAT, SNOOTY, UPPITY *adj. core meaning* : characteristic of or resembling a snob <the *snobbish* nouveau riche>

**snob·bism** (snŏb'ĭz'əm) *n.* Snobbery.

**snood** (snŏŏd) *n.* [ME **snood* < OE *snōd.*] **1.** A small netlike cap worn by women to hold the hair in place. **2.** A headband once worn in Scotland by young unmarried girls. *—vt.* **snood·ed, snood·ing, snoods.** To hold (the hair) in place with a snood.

**snook** (snŏŏk, snŏŏk) *n., pl.* **snook** or **snooks.** [Du. *snoek,* pike < MDu. *snoec.*] A chiefly marine fish of the family Centropomidae, esp. *Centropomus undecimalis* of warm Atlantic waters.

**snook·er** (snŏŏk'ər) *n.* [Orig. unknown.] A pocket billiards game in which 15 red and 6 nonred balls are used.

**snoop** (snōōp) [Du. *snoepen,* to eat on the sly.] *Informal.* *—vi.* **snooped, snoop·ing, snoops.** To pry into the business of others. *—n.* One who snoops. **—snoop'er** *n.*

☆ ***syns:*** SNOOP, NOSE (around), POKE, PRY *v. core meaning* : to look into or inquire about curiously, inquisitively, or in a meddlesome fashion <*snooping* through a neighbor's garage>

**snoop·y** (snōō'pē) *adj.* **-i·er, -i·est.** *Informal.* Tending to snoop. **—snoop'i·ly** *adv.* **—snoop'i·ness** *n.*

**snoot** (snōōt) [ME *snute,* snout.] *Slang.* **1.** A snout or nose. **2.** A snob.

**snoot·y** (snōō'tē) *adj.* **-i·er, -i·est.** *Slang.* **1.** Snobbish. **2.** Exclusive : high-class. **—snoot'i·ly** *adv.* **—snoot'i·ness** *n.*

**snooze** (snōōz) [Orig. unknown.] *Informal.* *—vi.* **snoozed, snooz·ing, snooz·es** To take a light nap : DOZE. **—snooze** *n.*

**snore** (snôr, snōr) *vi.* **snored, snor·ing, snores.** [ME *snoren,* to snort.] To breathe through the nose and mouth while sleeping, making snorting noises caused by vibration of the soft palate. *—n.* **1.** An act or instance of snoring. **2.** The noise produced by snoring. **—snor'er** *n.*

**snor·kel** (snôr'kəl) *n.* [G. *Schnorchel,* snorkel, snout < *schnarchen,* to snore.] **1.** A retractable vertical tube in a submarine that contains air-intake and exhaust pipes for the engines and for ventilation and that permits extended submergence at periscope depth. **2.** A breathing apparatus used by skin divers, composed of a long tube held in the mouth. *—vi.* **-keled, -kel·ing, -kels.** To dive using a skin-diving snorkel. **—snor'kel·er** *n.*

**snort** (snôrt) *n.* [< ME *snorten,* to snort.] **1. a.** A rough, noisy sound made by breathing forcefully through the nostrils. **b.** A sound like a snort <the *snort* of a steam engine> **2.** *Slang.* **a.** A drink of liquor, esp. when swallowed in one gulp. **b.** Inhalation of a drug, as cocaine. **c.** The liquor or drug swallowed or inhaled. *—v.* **snort·ed, snort·ing, snorts.** *—vi.* **1.** To breathe noisily and forcefully through the nostrils. **2.** To make an abrupt noise expressing scorn, ridicule, or contempt. **3.** *Slang.* To inhale a drug. *—vt.* **1.** To express with a snort <*snorted* their disapproval> **2.** *Slang.* To inhale (a drug). **—snort'er** *n.*

**snot** (snŏt) *n.* [ME < OE *gesnot.*] *Slang.* **1.** Nasal mucus : PHLEGM. **2.** An untrustworthy, malicious person.

**snot·ty** (snŏt'ē) *adj.* **-ti·er, -ti·est.** *Slang.* **1.** Dirtied with nasal mucus. **2.** Mean and nasty <*snotty* comments>.

**snout** (snout) *n.* [ME *snute,* prob. of MLG orig.] **1. a.** The projecting nose, jaws, or anterior facial part of an animal's muzzle. **b.** A similar prolongation of the anterior portion of the head in certain insects, as the weevils. **2.** A spout or nozzle likened to a snout. **3.** *Slang.* The human nose.

**snout beetle** *n.* Any of numerous weevils of the family Curculionidae, with the front of the head elongated to form a snout.

**snow** (snō) *n.* [ME < OE *snāw.*] **1.** Solid precipitation in the form of variously shaped white or translucent ice crystals originating in the upper atmosphere as frozen particles of water vapor. **2. a.** Something like snow. **b.** The white specks on a television screen resulting from weak reception. **c.** *Slang.* Cocaine. **d.** *Slang.* Heroin. **3.** A falling of snow : SNOWSTORM. *—v.* **snowed, snow·ing, snows.** *—vi.* To fall as or in snow. *—vt.* **1.** To cover, shut off, or close off with snow <were *snowed* in all winter> **2.** *Slang.* To overwhelm with insincere talk, esp. with flattery. **—snow under. 1.** To overwhelm <was *snowed* under with extra work> **2.** To defeat decisively.

**snow·ball** (snō'bôl') *n.* **1.** A mass of soft wet snow packed into a ball that can be thrown. **2.** A plant or shrub having rounded clusters of white flowers, esp. a cultivated variety of *Viburnum opulus.* *—v.* **-balled, -ball·ing, -balls.** *—vi.* **1.** To throw snowballs. **2.** To grow rapidly in importance, significance, or size <a problem that had *snowballed*> *—vt.* **1.** To throw snowballs at. **2.** To cause to grow or increase rapidly.

**snow·bell** (snō'bĕl') *n.* A shrub, *Styrax grandifolia* or *S. americana* of the southeastern United States, with white bell-shaped flowers.

**snow·ber·ry** (snō'bĕr'ē) *n.* A shrub of the genus *Symphoricarpos,* esp. *S. albus,* with small pinkish flowers and white berries.

**snow·bird** (snō'bûrd') *n.* A bird, as the junco, that is seen under snowy conditions.

**snow blindness** *n.* Conjunctivitis and deteriorated vision that is caused by sunlight reflected from snow or ice. **—snow'-blind', snow'blind'ed** *adj.*

**snow·blink** (snō'blĭngk') *n.* A white glare reflected from snowfields.

**snow·bound** (snō'bound') *adj.* Confined in one place by heavy snow.

**snow bunting** *n.* A bird, *Plectrophenax nivalis* of northern regions, with predominantly white winter plumage.

**snow·bush** (snō'bŏŏsh') *also* **snow·brush** (-brŭsh') *n.* A shrub,

ōō **boot** ou **out** th **thin** *th* **this** ŭ **cut** ûr **urge** y **young**
yōō **abuse** zh **vision** ə **about,** it**e**m, edib**i**le, gall**o**p, circ**u**s

*Ceanothus velutinus* of western North America, with large white flower clusters.

**snow·cap** (snō′kăp′) *n.* A cap of snow. **—snow′capped′** *adj.*

**snow·drift** (snō′drĭft′) *n.* Snow piled up by the wind.

**snow·drop** (snō′drŏp′) *n.* Any of several bulbous plants of the genus *Galanthus,* native to Eurasia, with solitary nodding white flowers that bloom in early spring.

**snowdrop tree** *n.* The silverbell tree.

**snow·fall** (snō′fôl′) *n.* **1.** A fall of snow. **2.** The amount of snow that falls during a specific period or in a given locale.

**snow fence** *n.* Temporary fencing of thin upright slats, used near roads and walks to prevent snow from drifting onto them.

**snow·flake** (snō′flāk′) *n.* **1.** A single crystal of snow. **2.** Any of several bulbous plants of the genus *Leucojum,* native to Europe, with white or whitish flowers.

**snow goose** *n.* A goose, *Chen hyperborea,* having white plumage with black wing tips that breeds in northern regions.

**snow job** *n. Slang.* An attempt to deceive, overwhelm, or persuade with insincere talk, esp. flattery.

**snow leopard** *n.* A large feline mammal, *Uncia uncia* of the highlands of central Asia, having long, thick, whitish fur with dark markings.

**snow lily** *n.* The fawn lily.

**snow line** *n.* **1.** The lower altitudinal boundary of a snow-covered area, esp. of one that is perennially covered, as the snowcap of a mountain. **2.** The fluctuating latitudinal boundaries around the polar regions marking the extent of snow cover.

**snow maker** *n.* A device that makes snow artificially, as on a ski slope. **—snow′mak·ing′** *n.* & *adj.*

**snow·man** (snō′măn) *n.* The figure of a person, made of snow.

**snow·mo·bile** (snō′mō-bēl′) *n.* A small vehicle with skilike runners in front and tanklike treads, used for driving in or traveling on snow. **—snow′mo·bil′er** *n.* **—snow′mo·bil′ing** *n.*

**snow-on-the-moun·tain** (snō′ŏn-*th*ə-moun′tən, -ôn-) *n.* A widely cultivated plant, *Euphorbia marginata* of central North America, with white-margined leaves and showy white bracts.

**snow pellets** *n.* Graupel.

**snow plant** *n.* A saprophytic plant, *Sarcodes sanguinea* of the mountains of western North America, with a fleshy, scaly, reddish stalk and scarlet flowers.

**snow·plow** (snō′plou′) *n.* **1.** A plowlike device or vehicle used for snow removal. **2.** A maneuver in skiing in which the tips of the skis are brought together so as to slow down or stop. **—snow′plow′** *v.* **(-plowed, -plow·ing, -plows).**

**snow·shed** (snō′shĕd′) *n.* Roofing built over portions of a railroad track to protect them from snowslides.

**snow·shoe** (snō′sho͞o′) *n.* A racket-shaped frame containing interlaced leather strips that can be attached to the foot to facilitate walking in deep snow. **—snow′shoe′** *v.* **(-shoed, -shoe·ing, -shoes).**

**snowshoe rabbit** *n.* A hare, *Lepus americanus* of northern North America, with large heavily furred feet and fur that is white in winter and brown in summer.

**snow·slide** snō′slīd′ *n.* An avalanche of snow.

**snow·storm** (snō′stôrm′) *n.* A storm marked by heavy snowfall and often high winds.

**snow·suit** (snō′so͞ot′) *n.* A child's zippered winter coverall.

**snow tire** *n.* A tire with a deep tread that gives added traction on snow.

**snow-white** (snō′hwīt′, -wīt′) *adj.* That is white as snow.

**snow·y** (snō′ē) *adj.* **-i·er, -i·est. 1. a.** Full of or covered with snow. **b.** Subject to snow <a *snowy* climate> **2.** Like snow. **—snow′i·ly** *adv.* **—snow′i·ness** *n.*

**snub** (snŭb) *vt.* **snubbed, snub·bing, snubs.** [ME *snubben,* to rebuke < ON *snubba.*] **1.** To slight by ignoring or behaving coldly toward. **2.** To dismiss, turn down, or frustrate the expectations of. **3. a.** To check suddenly the movement of (a running rope or cable) by turning it about a post. **b.** To secure (e.g., a vessel) in this way. *—n.* **1.** A deliberate slight. **2.** A sudden checking, as of a rope or cable running out. **—snub′ber** *n.*

**snub-nosed** (snŭb′nōzd′) *adj.* **1.** Having a short, turned-up nose. **2.** Having a very short barrel <a *snub-nosed* revolver>

**snuck** (snŭk) *v. Nonstandard. var. p.t. & p.p. of* SNEAK <"He ran up huge hotel bills and then *snuck* out without paying" —Professor George Stade>

**snuff¹** (snŭf) *v.* **snuffed, snuff·ing, snuffs.** [Prob. < MDu. *snuffen.*] *—vt.* **1.** To inhale through the nose : SNIFF. **2.** To sense or examine by smelling. *—vi.* To sniff : inhale. *—n.* An act of snuffing or the sound made in snuffing.

**snuff²** (snŭf) *n.* [ME *snoffe.*] The charred part of a candlewick. *—vt.* **snuffed, snuff·ing, snuffs. 1.** To cut off the charred part of (a candlewick). **2.** To extinguish <*snuff* out a candle> **3.** To put a sudden end to : DESTROY.

**snuff³** (snŭf) *n.* [Du. *snuf,* short for *snuftabak* < MDu. *snuffen,* to sniff.] **1. a.** A mixture of finely pulverized tobacco that can be inhaled. **b.** A quantity of snuff inhaled at a single time : PINCH. **2.** A powdery substance, as a medicine, taken by inhaling. *—vi.* **snuffed, snuff·ing, snuffs.** To use or inhale snuff. **—up to snuff.** *Informal.* **1.** Normal in health. **2.** Up to standard.

**snuff·box** (snŭf′bŏks′) *n.* A small box for carrying snuff.

**snuff·er¹** (snŭf′ər) *n.* One who uses snuff.

**snuff·er²** (snŭf′ər) *n.* **1.** One that snuffs out candles. **2. snuffers.** An instrument similar to a pair of shears that is used for cutting the snuff from or for extinguishing candles.

**snuf·fle** (snŭf′əl) *v.* **-fled, -fling, -fles.** [Prob. < LG *snuffelen.*] *—vi.* **1.** To breathe noisily, as through a blocked nose. **2.** To sniff. **3.** To talk nasally : WHINE. *—vt.* To utter (something) in a snuffling tone. *—n.* **1.** An act of snuffling or the sound made in snuffling. **2. snuffles.** *Informal.* The sniffles. **—snuf′fler** *n.*

**snug¹** (snŭg) *adj.* **snug·ger, snug·gest.** [Perh. of Scand. orig.] **1.** Comfortably sheltered : COZY. **2.** Small but well-arranged <a *snug* little room> **3. a.** Closely secured and well-built : COMPACT <a *snug* little sports car> **b.** Seaworthy. **c.** Close-fitting <a *snug* winter jacket> *—v.* **snugged, snug·ging, snugs.** *—vt.* To make snug or secure. *—vi.* To snuggle. **—snug, snug′ly** *adv.* **—snug′ness** *n.*

**snug²** *n.* [Short for SNUGGERY.] *Chiefly Brit.* A small private room in a pub.

**snug·ger·y** (snŭg′ə-rē) *n., pl.* **-ies.** *Chiefly Brit.* A snug position or place.

**snug·gle** (snŭg′əl) *v.* **-gled, -gling, -gles.** [Freq. of SNUG.] *—vi.* To lie or press close together. *—vt.* To draw close or hold closely, as for comfort : HUG.

☆ ***syns:*** SNUGGLE, CUDDLE, NESTLE, NUZZLE *v. core meaning* : to lie or press close together, usu. with another <*snuggled* together in the sleeping bag>

**so¹** (sō) *adv.* [ME < OE *swā.*] **1.** In the state or manner indicated or expressed : THUS <Why do you say *so?*> **2.** To the amount or degree expressed or understood : to such an extent <They were *so* tired that they fell.> **3.** To a great extent : to such an evident degree <But the idea is *so* absurd.> **4.** Because of the reason given : CONSEQUENTLY <They were tired and *so* they fell.> **5.** Approximately <The ticket costs $17 or *so.*> **6.** In the same way : LIKEWISE <You were late and *so* was I.> **7.** Then : apparently. —Used in expressing astonishment, disapproval, or sarcasm <*So* you think you've got problems!> **8.** In truth : INDEED <"You aren't right." "I am *so!*"> *—adj.* True : factual <I wouldn't have said it if it weren't *so.*> *—conj.* With the result that <They failed to appear, *so* we began without them.> ***usage:*** In formal contexts the conjunction *so* is preferably followed by *that* when it introduces a clause stating the purpose of or the reason for an action, as in *I usually stay late so that I can avoid the rush-hour traffic.* However, when used to introduce a clause stating a result or consequence, *so* gen. stands alone, as in *The traffic was unusually heavy so I delayed my departure for an hour.* *—pron.* Such as has already been suggested or specified : the same <became a loyal friend and remained *so*> *—interj.* —Used to express surprise or understanding. **—so as to.** In order to <Mail your package early *so as to* ensure its timely arrival.> **so that.** In order that <I stopped *so that* you could catch up with me.>

**so²** (sō) *n. Mus. var. of* SOL.

**soak** (sōk) *v.* **soaked, soak·ing, soaks.** [ME *soken* < OE *socian.*] *—vt.* **1. a.** To wet thoroughly by or as if by placing in liquid. **b.** To immerse in liquid for a time. **2.** To absorb (liquid) through pores or interstices. **3.** *Informal.* To take in mentally, esp. eagerly <*soaked* up the juicy gossip> **4.** *Informal.* **a.** To drink (liquor) to excess. **b.** To make (a person) drunk. **5.** *Slang.* To overcharge (a person) for something. *—vi.* **1.** To be immersed until thoroughly saturated with a liquid. **2.** To penetrate or permeate deeply. **3.** *Slang.* To drink excessively. *—n.* **1.** The act or process of soaking or the state of being soaked. **2.** Liquid in which something may be soaked. **3.** *Slang.* A drunkard. **—soak′er** *n.*

**soak·age** (sō′kĭj) *n.* **1.** The act of soaking or the state of being soaked. **2.** The amount of liquid that soaks into, through, or out of an object.

**so-and-so** (sō′ən-sō′) *n., pl.* **-sos. 1.** One that is unnamed or unspecified. **2.** *Informal.* An unpleasant person.

**soap** (sōp) *n.* [ME *sope* < OE *sāpe.*] **1.** A cleansing agent manufactured in bars, granules, flakes, or liquid, made from a mixture of the sodium salts of various fatty acids of natural oils and fats. **2.** A metallic salt of a fatty acid, as of aluminum or iron. **3.** *Slang.* Money, esp. that used for bribery. **4.** *Slang.* A soap opera. *—vt.* **soaped, soap·ing, soaps. 1.** To treat or cover with or as if with soap. **2.** *Slang.* To bribe. **—no soap.** *Slang.* **1.** Not permissible or possible. **2.** Futile.

**soap·bark** (sōp′bärk′) *n.* **1. a.** A tree, *Quillaja saponaria* of western South America, with bark used as soap and as a source of saponin. **b.** The bark of the soapbark. **2.** A tree or shrub with bark similar to that of the soapbark.

**soap·ber·ry** (sōp′bĕr′ē) *n.* **1.** Any of various chiefly tropical New World trees of the genus *Sapindus,* bearing pulpy fruit that lathers like soap. **2.** The fruit of a soapberry tree.

**soap·box** *also* **soap box** (sōp′bŏks′) *n.* **1.** A carton in which soap is packed. **2.** A temporary platform used while making an impromptu or nonofficial public speech.

ă **pat** ā **pay** âr **care** ä **father** ĕ **pet** ē **be** hw **wh**ich ĭ **pit** ī **tie** îr **pier** ŏ **pot** ō **toe** ô **paw, for** oi **noise** o͝o **took**

**Soapbox Derby** *n.* A service mark for a downhill race using children's homemade racing cars without motors or pedals.
**soap bubble** *n.* **1.** A bubble formed from soapy water. **2.** Something beautiful but insubstantial, illusory, or transient.
**soap opera** *n.* [From its orig. having been sponsored by soap companies.] A radio or television serial drama characterized by stock characters and situations, sentimentality, and melodrama.
**soap plant** *n.* **1.** A plant, *Chlorogalum pomeridianum* of California, with small white flowers and a bulbous root once used as soap. **2.** Any of various plants whose parts are used as soap.
**soap·stone** (sōp'stōn') *n.* [From its soapy texture.] Steatite.
**soap·suds** (sōp'sŭdz') *pl.n.* Suds from soapy water.
**soap·wort** (sōp'wûrt', -wôrt') *n.* [From its yielding a soapy substance when the leaves are bruised.] The bouncing Bet.
**soap·y** (sō'pē) *adj.* **-i·er, -i·est. 1.** Containing or composed of soap. **2.** Covered with soap. **3.** Relating to or like soap. **4.** *Slang.* Unctuous. **—soap'i·ly** *adv.* **—soap'i·ness** *n.*
**soar** (sôr, sōr) *vi.* **soared, soar·ing, soars.** [ME *soren* < OFr. *esorer* < VLat. **exaurare* : Lat *ex-*, out of + Lat. *aura*, air < Gk., breeze.] **1.** To rise, fly, or glide high and apparently with little effort. **2.** To climb swiftly or powerfully. **3.** To glide in an aircraft while maintaining altitude. **4.** To rise suddenly above the normal or accustomed level <Our spirits *soared.*> —*n.* **1.** The act of soaring. **2.** The altitude or scope attained in soaring. **—soar'er** *n.* **—soar'ing·ly** *adv.*
**soar·ing** (sôr'ĭng, sōr'-) *n.* The act of gliding while maintaining altitude, esp. the sport of flying a heavier-than-air craft by utilizing ascending air currents.
**so·a·ve** (sō-ä'vā) *n.* [Ital. < Lat. *suavis*, sweet.] A dry white Italian table wine.
**sob** (sŏb) *v.* **sobbed, sob·bing, sobs.** [ME *sobben.*] —*vi.* **1.** To weep aloud with convulsive gasping : CRY. **2.** To make a sound resembling that of sobbing. —*vt.* **1.** To utter with sobs. **2.** To put or bring (oneself) into a specified condition by sobbing <*sob* oneself to sleep> **—sob** *n.* **—sob'bing·ly** *adv.*
**so·ber** (sō'bər) *adj.* **-er, -est.** [ME < OFr. *sobre* < Lat. *sobrius.*] **1.** Habitually abstemious in the use of liquor : TEMPERATE. **2.** Not intoxicated. **3.** Straightforward : serious. **4.** Plain or subdued <*sober* attire for church> **5.** Devoid of frivolity, exaggeration, or speculative imagination <gave a *sober* assessment of the problem> **6.** Self-controlled and sane : REASONABLE. —*vt.* & *vi.* **-bered, -ber·ing, -bers.** To make or become sober. **—so'ber·ly** *adv.* **—so'ber·ness** *n.*
**so·ber-sid·ed** (sō'bər-sī'dĭd) *adj.* SOBER 3, 5.
**so·ber·sides** (sō'bər-sīdz') *n.* (*sing. or pl. in number*). A sober-sided person.
**so·bri·e·ty** (sō-brī'ĭ-tē) *n.* **1.** Seriousness or gravity in bearing, manner, or treatment : SOLEMNITY. **2.** Absence of alcoholic intoxication.
**so·bri·quet** (sō'brĭ-kā', sō'brĭ-kā') *also* **sou·bri·quet** (so͞o'brĭ-kā', so͞o'brĭ-kā') *n.* [Fr.] **1.** An affectionate or humorous nickname. **2.** An assumed name.
**sob sister** *n.* **1.** A journalist employed as a writer or editor of sob stories. **2.** A sentimental, ineffective do-gooder.
**sob story** *n.* **1.** A tale of personal misfortune intended to arouse pity. **2.** A maudlin plea given as an explanation or rationalization.
**soc·age** (sŏk'ĭj, sō'kĭj) *n.* [ME *sokage* < *soke*, soke.] Feudal tenure of land by a tenant who was not a knight, in return for agricultural or other nonmilitary services or for payment of rent in money. **—soc'ag·er** *n.*
**so-called** (sō'kôld') *adj.* **1.** Commonly called <a *so-called* economic summit> **2.** Incorrectly or falsely termed <a *so-called* liberal>
**soc·cer** (sŏk'ər) *n.* [Shortening and alteration of *association football.*] A game played on a rectangular field with net goals at either end in which 2 teams of 11 players each maneuver a round ball mainly by kicking or butting or by using any part of the body except the arms and hands in attempts to score goals.
▲ word history: Soccer, which is probably the most popular team sport everywhere except in the United States, was invented in England. The word *soccer* is derived from *association.* The official name of the game is *Association Football,* that is, football as played under the rules of the Football Association of England, which was founded in 1863.
**so·cia·ble** (sō'shə-bəl) *adj.* [OFr. < Lat. *sociabilis* < *sociare*, to share < *socius*, partner.] **1.** Pleasant, friendly, and affable. **2.** Providing occasion for conversation and conviviality. —*n.* A social. **—so'cia·bil'i·ty** (-bĭl'ĭ-tē) *n.* **—so'cia·ble·ness** *n.* **—so'cia·bly** *adv.*
**so·cial** (sō'shəl) *adj.* [Lat. *socialis*, of companionship < *socius*, partner.] **1. a.** Living together in communities. **b.** Of or relating to communal living. **c.** Of or relating to society. **2.** Living in an organized group or similar close aggregate <*social* bees> **3.** Involving allies or members of a confederacy. **4.** Of or relating to the upper classes. **5.** Enjoying the company of others : SOCIABLE. **6.** Intended for convivial activities. **7.** Of, relating to, or occupied with welfare work. —*n.* An informal social gathering.
**social climber** *n.* One who strives to be accepted in fashionable society.
**social contract** *n.* An agreement among the members of an organized society or between the governed and the government defining and limiting the rights and duties of each.
**social disease** *n.* **1.** Venereal disease. **2.** A disease having its highest incidence among social classes predisposed to it by a given set of adverse living or working conditions.
**so·cial·ism** (sō'shə-lĭz'əm) *n.* **1. a.** A social system in which the producers possess political power and the means of producing and distributing goods. **b.** The theory or practice of those who support such a social system. **2.** Construction of the material base for Marxist-Leninist communism under the dictatorship of the proletariat.
**so·cial·ist** (sō'shə-lĭst) *n.* **1.** An advocate of socialism. **2.** A member of a socialist party. —*adj.* **1.** Of, promoting, or practicing socialism. **2. Socialist.** Of, belonging to, or designating a socialist party.
**so·cial·is·tic** (sō'shə-lĭs'tĭk) *adj.* Of, advocating, or tending toward socialism. **—so'cial·is'ti·cal·ly** *adv.*
**socialist party** *n.* A political party advocating socialism to be achieved by democratic process.
**socialist realism** *n.* A Marxist aesthetic doctrine that promotes the development of social consciousness through the didactic use of literature, art, and music.
**so·cial·ite** (sō'shə-līt') *n.* A member of fashionable society.
**so·ci·al·i·ty** (sō'shē-ăl'ĭ-tē) *n., pl.* **-ties. 1. a.** The quality or state of being sociable. **b.** An instance of sociableness. **2.** A tendency to form communities and societies.
**so·cial·ize** (sō'shə-līz') *v.* **-ized, -iz·ing, -iz·es.** —*vt.* **1.** To place under government or group ownership or control. **2.** To fit for companionship with others. **3.** To convert or adapt to the needs of society. —*vi.* To participate in social activities. **—so'cial·i·za'tion** *n.* **—so'cial·iz'er** *n.*
**socialized medicine** *n.* Provision of medical and hospital care for patients at nominal cost by means of government regulation of health services and subsidies derived from taxation.
**so·cial·ly** (sō'shə-lē) *adv.* **1.** In a social way. **2.** With regard to society. **3.** By society <*socially* accepted behavior>
**social register** *n.* A directory listing prominent members of a community.
**social science** *n.* The study of society and of individual relationships in and to society, gen. held to include sociology, psychology, anthropology, economics, political science, and history.
**social secretary** *n.* A personal secretary who handles social correspondence and appointments.
**social security** *n.* **1.** A government program providing economic assistance to persons faced with unemployment, disability, or old age, financed by assessment of employers and employees. **2.** Economic assistance provided by social security.
**social service** *n.* **1.** Organized efforts to advance human welfare : SOCIAL WORK. **2.** *often* **social services.** Welfare services, as free school lunches, provided by a government for its needy citizens.
**social studies** *pl.n.* A course of study including geography, history, government, and sociology, taught in secondary and elementary schools.
**social work** *n.* Welfare work usu. involving casework. **—social worker** *n.*
**so·ci·e·tal** (sə-sī'ĭ-təl) *adj.* Of or relating to the structure, organization, or functioning of society. **—so·ci'e·tal·ly** *adv.*
**so·ci·e·ty** (sə-sī'ĭ-tē) *n., pl.* **-ties.** [OFr. *societe* < Lat. *societas*, fellowship < *socius*, partner.] **1. a.** The totality of social relationships among human beings. **b.** A group of human beings broadly distinguished from other groups by mutual interests, participation in characteristic relationships, shared institutions, and a common culture. **c.** The institutions and culture of a distinct self-perpetuating group. **2. a.** The rich, privileged, and fashionable social class. **b.** The socially dominant members of a community. **3.** Companionship : company <enjoyed the *society* of their neighbors> **4.** *Biol.* A colony or community of organisms, usu. of the same species.
**Society of Friends** *n.* A Christian sect, founded in about 1650 in England, that rejects ritual sacraments, a formal creed, the priesthood, and violence.
**Society of Jesus** *n.* The Jesuits.
**So·cin·i·an** (sō-sĭn'ē-ən) *n.* [NLat. *Socinianus*, after Laelius *Socinus* and Faustus *Socinus*, 16th-cent. Italian theologians.] An adherent of a 16th-cent. Italian sect holding unitarian views, including denial of the divinity of Jesus. **—So·cin'i·an** *adj.* **—So·cin'i·an·ism** *n.*
**socio-** *pref.* [Fr. < Lat. *socius*, fellow.] **1.** Society <*sociometry*> **2.** Social <*socioeconomic*>
**so·ci·o·cul·tur·al** (sō'sē-ō-kŭl'chər-əl, -shē-) *adj.* Being both social and cultural. **—so'ci·o·cul'tur·al·ly** *adv.*
**so·ci·o·ec·o·nom·ic** (sō'sē-ō-ĕk'ə-nŏm'ĭk, -ē'kə-, -shē-) *adj.* Being both social and economic.
**so·ci·o·lin·guis·tics** (sō'sē-ō-lĭng-gwĭs'tĭks) *n.* (*sing. in number*). The study of linguistic behavior as influenced by social and cultural factors. **—so·ci·o·lin·guist** *n.* **—so·ci·o·lin·guis·tic** *adj.*
**so·ci·ol·o·gy** (sō'sē-ŏl'ə-jē, -shē-) *n.* [Fr. *sociologie* : *socio-*, socio- + *-logie*, -logy.] **1.** The study of human social behavior, esp. the study of the origins, organization, institutions, and development of human

society. **2.** Analysis of a social institution or societal segment as a self-contained entity or in relation to society as a whole. **—so'ci·o·log'ic** (-ə-lŏj'ĭk), **so'ci·o·log'i·cal** *adj.* **—so'ci·o·log'i·cal·ly** *adv.* **—so'ci·ol'o·gist** *n.*

**so·ci·om·e·try** (sō'sē-ŏm'ĭ-trē, -shē-) *n.* Quantitative study of interpersonal relationships in populations, esp. study and measurement of preferences.

**so·ci·o·path** (sō'sē-ə-păth', -shē-) *n.* A person manifesting asocial or antisocial behavior or character traits. **—so'ci·o·path'ic** *adj.*

**so·ci·o·po·li·ti·cal** (sō'sē-ō-pə-lĭt'ĭ-kəl, -shē-) *adj.* Being both social and political.

**so·ci·o·re·lig·ious** (sō'sē-ō-rĭ-lĭj'əs, -shē-) *adj.* Being both social and religious.

**sock[1]** (sŏk) *n.* [ME *socke* < OE *socc*, a kind of light shoe < Lat. *soccus*.] **1.** *pl.* **socks** or **sox.** A short stocking reaching a point between the ankle and the knee. **2. a.** A light shoe worn by comic actors in ancient Greek and Roman plays. **b.** Comic drama : COMEDY. **3.** A windsock. *—vt.* **socked, sock·ing, socks.** To provide with socks. **—sock away.** *Informal.* To put away (money) : STASH. **—sock in.** To close to air traffic <Fog *socked in* the airport.>

**sock[2]** (sŏk) [Orig. unknown.] *Slang.* *—v.* **socked, sock·ing, socks.** *—vt.* To hit or strike forcefully : PUNCH. *—vi.* To deliver a forceful blow. **—sock** *n.*

**sock·dol·a·ger** *also* **sock·dol·o·ger** (sŏk-dŏl'ə-jər) *n.* [Orig. unknown.] *Slang.* **1.** A conclusive blow or remark. **2.** Something outstanding of its kind.

**sock·et** (sŏk'ĭt) *n.* [ME *soket* < AN, spearhead, dim. of OFr. *soc*, plowshare, prob. of Celtic orig.] **1.** An opening into which an inserted part is designed to fit <an electric *socket*> **2.** *Anat.* **a.** The hollow part of a joint that receives the end of a bone. **b.** A hollow into which a part, as the eye, fits. *—vt.* **-et·ed, -et·ing, -ets.** To furnish with or insert into a socket.

**socket wrench** *n.* A usu. bar-shaped wrench with a removable socket that fits over a nut or bolt.

**sock·eye salmon** (sŏk'ī') *n.* [By folk ety. < Salish *sukkegh.*] A salmon, *Oncorhynchus nerka* of northern Pacific coastal waters, that is a commercially valuable food fish.

**sock·o** (sŏk'ō) *adj.* [< SOCK[2].] *Slang.* Very impressive, successful, or popular <The comedian gave a *socko* performance.>

**so·cle** (sō'kəl) *n.* [Fr. < Ital. *zoccolo*, wooden shoe < Lat. *socculus*, dim. of *soccus*, a kind of light shoe.] **1.** A plain square block higher than a plinth, serving as a pedestal for sculpture, a vase, or a column. **2.** A plain plinth supporting a wall.

**sod** (sŏd) *n.* [ME < MLG *sode.*] **1.** An area of grass-covered surface soil held together by matted roots : TURF. **2.** The ground, esp. when covered with grass. *—vt.* **sod·ded, sod·ding, sods.** To cover (an area) with sod.

**so·da** (sō'də) *n.* [Med. Lat., barilla.] **1. a.** Any of various forms of sodium carbonate. **b.** Chemically combined sodium. **2.** Carbonated water or a soft drink containing it. **3.** A refreshment made from carbonated water, ice cream, and usu. a flavoring. **4.** The card turned face up at the beginning of faro.

**soda ash** *n.* Crude anhydrous sodium carbonate used esp. as an industrial chemical.

**soda biscuit** *n.* **1.** A breadlike biscuit leavened with baking soda. **2.** A soda cracker.

**soda cracker** *n.* A thin, usu. square or rectangular cracker leavened slightly with baking soda.

**soda fountain** *n.* **1.** An apparatus with faucets for dispensing soda water. **2.** A counter equipped for preparing and serving soft drinks, ice-cream dishes, or sandwiches.

**soda jerk** *n.* [Short for *soda jerker.*] *Slang.* One who works behind a soda fountain.

**soda lime** *n.* A mixture of calcium hydroxide and sodium or potassium hydroxide that is used as a drying agent and carbon dioxide absorbent.

**so·da·list** (sō'də-lĭst, sō-dăl'ĭst) *n.* A member of a sodality.

**so·da·lite** (sō'də-līt') *n.* A blue-white vitreous mineral, essentially $Na_4Al_3Si_3O_{12}Cl$, found in igneous rocks.

**so·dal·i·ty** (sō-dăl'ĭ-tē) *n., pl.* **-ties.** [Lat. *sodalitas*, fellowship < *sodalis*, fellow.] **1.** A society or association, esp. a devotional or charitable society in the Roman Catholic Church. **2.** Fellowship.

**so·da·mide** (sō'də-mīd') *n.* Sodium amide.

**soda niter** *n.* Sodium nitrate.

**soda pop** *n.* *Informal.* A soft drink : SODA.

**so·da-pop wine** (sō'də-pŏp') *n.* A pop wine.

**soda water** *n.* **1.** Effervescent water charged under pressure with purified carbon dioxide gas, used as a beverage or mixer. **2.** A solution of water, sodium bicarbonate, and acid.

**sod·den** (sŏd'n) *adj.* [ME *soden*, boiled, p.part. of *sethen*, to boil. —see SEETHE.] **1.** Utterly soaked. **2.** Soggy and heavy from improper cooking : DOUGHY. **3.** Bloated, esp. from drink. **4.** Unimaginative : torpid <*sodden* prose> *—vt. & vi.* **-dened, -den·ing, -dens.** To make or become sodden. **—sod'den·ly** *adv.* **—sod'den·ness** *n.*

**so·di·um** (sō'dē-əm) *n.* [SOD(A) + -IUM.] *Symbol* **Na** A soft, light, extremely malleable silver-white metallic element, used esp. in the production of a wide variety of industrially important compounds; atomic number 11; atomic weight 22.99.

**sodium ammonium phosphate** *n.* A colorless, odorless crystalline compound, $NaNH_4HPO_4{\cdot}4H_2O$, used as an analytical reagent.

**sodium benzoate** *n.* The sodium salt of benzoic acid, $C_6H_5COONa$, used as a food preservative, antiseptic, and intermediate in dye making and in production of pharmaceuticals.

**sodium bicarbonate** *n.* A white crystalline compound, $NaHCO_3$, with a slightly alkaline taste, used to make effervescent salts and beverages, artificial mineral water, baking soda and pharmaceuticals and in manufacturing fire extinguishers.

**sodium borate** *n.* A crystalline compound, $Na_2B_4O_7{\cdot}10H_2O$, used to make glass, detergents, and pharmaceuticals.

**sodium carbonate** *n.* **1.** A white powdery compound, $Na_2CO_3$, used to make sodium bicarbonate, sodium nitrate, glass, ceramics, detergents, and soap. **2.** A hydrated carbonate of sodium, as sal soda.

**sodium chlorate** *n.* A colorless crystalline compound, $NaClO_3$, used as a bleaching and oxidizing agent and in explosives.

**sodium chloride** *n.* A colorless crystalline compound, NaCl, used to make chemicals and also utilized as a food preservative and seasoning.

**sodium cyanide** *n.* A poisonous white crystalline compound, NaCN, used in the extraction of gold and silver from ores and in dye manufacture.

**sodium cyclamate** *n.* A soluble white crystalline powder, $C_6H_{11}NHSO_3Na$, 30 times as sweet as sugar and once a major constituent of low-calorie sweetening agents.

**sodium dichromate** *n.* A red-orange crystalline compound, $Na_2Cr_2O_7{\cdot}2H_2O$, used as an oxidizing agent.

**sodium glutamate** *n.* A white crystalline compound, $COOH(CH_2)_2CH(NH_2)COONa$, having a meatlike taste, used in cooking.

**sodium hydrosulfite** *n.* A yellowish powder, $Na_2S_2O_4$, used as a bleaching and reducing agent.

**sodium hydroxide** *n.* A strongly alkaline compound, NaOH, used in making chemicals and soaps and in petroleum refining.

**sodium hypochlorite** *n.* An unstable salt, NaOCl, usu. stored in solution and used as a fungicide and an oxidizing bleach.

**sodium hyposulfite** *n.* **1.** Sodium hydrosulfite. **2.** Sodium thiosulfate.

**sodium nitrate** *n.* A white crystalline compound, $NaNO_3$, used in solid rocket propellants and in making explosives and fertilizer.

**sodium perborate** *n.* A white odorless crystalline compound, $NaBO_2{\cdot}H_2O_2{\cdot}3H_2O$, used as a mild alkaline oxidizing agent in dentifrices, as a topical antiseptic and deodorant, and as an industrial reagent.

**sodium peroxide** *n.* A yellowish-white powder, $Na_2O_2$, used industrially as an oxidizing and bleaching agent and medically as a germicide, antiseptic, and disinfectant.

**sodium phosphate** *n.* Any of the three sodium salts of phosphoric acid, $NaH_2PO_4$, $Na_2HPO_4$, and $Na_3PO_4$, used in industry, pharmaceutical manufacturing, medicine, and chemistry.

**sodium propionate** *n.* A clear crystalline compound, $C_2H_5COONa$, capable of retarding the growth of molds and bacteria and used to prevent food spoilage.

**sodium silicate** *n.* Any of various water-soluble silicate glass compounds used as a preservative for eggs, in plaster and cement, and in purification and refining processes.

**sodium sulfate** *n.* A white crystalline compound, $Na_2SO_4$, used in making paper, glass, dyes, and pharmaceuticals.

**sodium sulfide** *n.* A hygroscopic yellow compound, $Na_2S$, used as a metal ore reagent and in photography, engraving, and printing.

**sodium sulfite** *n.* A white crystalline or powdered compound, $Na_2SO_3$, used in preserving foods, silvering mirrors, developing photographs, and manufacturing dyes.

**sodium thiosulfate** *n.* A white translucent crystalline compound, $Na_2S_2O_3{\cdot}5H_2O$, used as a photographic fixing agent and as a bleach.

**so·di·um-va·por lamp** (sō'dē-əm-vā'pər) *n.* An electric lamp containing a small amount of sodium and neon gas, used in generating yellow light for illuminating streets and highways.

**Sod·om** or **sod·om** (sŏd'əm) *n.* [After *Sodom*, a wicked city in ancient Palestine.] A place notorious for vice and corruption.

**so·ev·er** (sō-ĕv'ər) *adv.* In any way : at all <"Space to breathe, how short *soever*" —B. Jonson>

**so·fa** (sō'fə) *n.* [Ar. *sufah*, dais.] A long upholstered seat having arms and a back.

**sofa bed** *n.* A sofa whose seat unfolds to form a bed.

**so·far** (sō'fär') *n.* [SO(UND) F(IXING) A(ND) R(ANGING).] A system for detecting and locating underwater explosions propagated over long distances through deep ocean layers, used to find survivors lost at sea.

**sof·fit** (sŏf'ĭt) *n.* [Fr. *soffite* < Ital. *soffito* < VLat. *suffictus < Lat. *suffixus*, p.part. of *suffigere*, to fasten beneath. —see SUFFIX.] The underside of a structural component as a beam, arch, staircase, or cornice.

**soft** (sôft, sŏft) *adj.* **-er, -est.** [ME, pleasant, calm < OE *sōfte.*] **1. a.** Easily cut, worked, or molded. **b.** Yielding readily to pressure or weight. **2.** Out of condition : FLABBY <*soft* muscles> **3.** Smooth or fine to the touch <*soft* pile> **4. a.** Not loud, harsh, or irritating : LOW <a *soft*, gentle voice> **b.** Not glaring or brilliant : SUBDUED <*soft* colors> **5.** Not sharply drawn or delineated <*soft* charcoal shading> **6.** Balmy : mild <a *soft* summer breeze> **7. a.** Gentle in disposition : TENDER. **b.** Affectionate <*soft* glances> **c.** Attracted or emotionally involved. **d.** Not stern : LENIENT. **e.** Lacking strength of character : WEAK. **f.** *Informal.* Feeble-minded <*soft* in the head> **g.** Gradually declining in trend : not firm <a *soft* economy> **8. a.** *Informal.* Easy : cushy <a *soft* job> **b.** Based on conciliation or negotiation rather than on threats or power plays <took a *soft* line toward the opposition> **9.** Of or relating to a paper currency. **10.** Having low dissolved mineral content. **11. a.** Sibilant rather than guttural, as *c* in *certain* and *g* in *gem.* **b.** Voiced and weakly articulated <*soft* consonants> **c.** Palatalized, as certain consonants in Slavic languages. **12.** Occurring under such circumstances and at such a speed as to preclude destructive impact <The jet made a *soft* touchdown.> **13.** Unprotected against enemy attack <*soft* launching sites> —*n.* A soft object or part. —*adv.* In a soft way : GENTLY. **—soft'ly** *adv.* **—soft'ness** *n.*

**soft·ball** (sôft'bôl', sŏft'-) *n.* **1.** A variation of baseball played on a smaller diamond with a larger, softer ball pitched underhand. **2.** The ball used in softball.

**soft-boiled** (sôft'boild', sŏft'-) *adj.* **1.** Boiled in the shell to a soft consistency <*soft-boiled* eggs> **2.** *Informal.* **a.** SOFT 7d. **b.** Sentimental.

**soft·bound** (sôft'bound', sŏft'-) *adj.* Not bound between hard covers <*softbound* books>

**soft clam** *n.* The soft-shell clam.

**soft coal** *n.* Bituminous coal.

**soft-core** (sôft'kôr', -kōr', sŏft'-) *adj.* **1.** Being less explicit than hard-core material in depicting or describing sexual activity <*soft-core* pornography> **2.** Moderate <a *soft-core* sports fan>

**soft drink** *n.* A nonalcoholic usu. carbonated beverage.

**soft drug** *n.* A nonaddictive drug, as marijuana, considered less damaging to the health than a hard drug.

**soft·en** (sô'fən, sŏf'ən) *v.* **-ened, -en·ing, -ens.** —*vt.* **1.** To make soft or softer. **2.** To weaken the strength or resistance of by preliminary harassment. —*vi.* To become soft or softer. **—soft'en·er** *n.*

**soft-finned** (sôft'fĭnd', sŏft'-) *adj. Zool.* Having fins supported by flexible cartilaginous rays.

**soft goods** *pl.n.* Dry goods.

**soft hail** *n.* Graupel.

**soft·head** (sôft'hĕd', sŏft'-) *n.* A feeble-minded person.

**soft·head·ed** (sôft'hĕd'ĭd, sŏft'-) *adj.* Lacking judgment, common sense, or firmness. **—soft'head'ed·ly** *adv.* **—soft'head'ed·ness** *n.*

**soft·heart·ed** (sôft'här'tĭd, sŏft'-) *adj.* Easily moved : TENDER. **—soft'heart'ed·ly** *adv.* **—soft'heart'ed·ness** *n.*

**soft-land** (sôft'lănd', sŏft'-) *vi.* & *vt.* **-land·ed, -land·ing, -lands.** To make or cause to make a soft landing.

**soft landing** *n.* The landing of a space vehicle on a celestial body in such a way as to prevent damage or destruction.

**soft-lin·er** (sôft'lī'nər, sŏft'-) *n.* A proponent of moderation or flexibility, esp. on political issues.

**soft palate** *n.* The movable fold, consisting of muscular fibers enclosed in mucous membrane, that is suspended from the rear of the hard palate and closes off the nasal cavity from the oral cavity during swallowing or sucking.

**soft paste** *also* **soft-paste** (sôft'pāst', sŏft'-) *n.* A ceramic containing frit and refined clay.

**soft pedal** *n.* A pedal used to mute tone, as on a piano.

**soft-ped·al** (sôft'pĕd'l, sŏft'-) *vt.* **-aled, -al·ing, -als** or **-alled, -al·ling, -als. 1.** *Mus.* To soften or mute the tone of by depressing the soft pedal. **2.** *Informal.* To make less obvious : PLAY DOWN.

**soft rock** *n.* Rock 'n' roll marked by the predominance of melody and minimal use of electronic modulations.

**soft roe** *n.* The spermatozoa or testes of a fish : MILT.

**soft sculpture** *n.* A sculpture made of pliant materials, as cloth or foam rubber.

**soft sell** *n. Informal.* A low-pressure way of selling or advertising.

**soft-shell** (sôft'shĕl', sŏft'-) *also* **soft-shelled** (-shĕld') *adj.* Having a soft, brittle, or unhardened shell.

**soft-shell clam** *n.* A common edible clam, *Mya arenaria,* whose shell is thin and elongated.

**soft-shell crab** *n.* A marine crab before its shell has hardened after molting, esp. the edible species, *Callinectes sapidus* of eastern North America, in this stage.

**soft-shelled turtle** *n.* Any of various freshwater turtles of the family Trionychidae, having a flat carapace covered with leathery skin.

**soft-shoe** (sôft'shōō', sŏft'-) *n.* Tap dancing performed without metal taps on the shoes.

**soft shoulder** *n.* A border of soft earth along the edge of a road.

**soft soap** *n.* **1.** A fluid or semifluid soap. **2.** *Informal.* Cajolery.

**soft-soap** (sôft'sōp', sŏft'-) *vt.* **-soaped, -soap·ing, -soaps.** *Informal.* To cajole. **—soft'-soap'er** *n.*

**soft-spo·ken** (sôft'spō'kən, sŏft'-) *adj.* **1.** Speaking with a soft voice. **2.** Smooth and ingratiating.

**soft spot** *n.* **1.** A tender, sentimental feeling. **2.** Either of the points of juncture of the sagittal and lambdoid or sagittal and coronal sutures in an infant's skull.

**soft touch** *n.* One easily persuaded or taken advantage of.

**soft·ware** (sôft'wâr', sŏft'-) *n. Computer Sci.* **1.** Written or printed data, as programs, routines, and symbolic languages, requisite to computer operations. **2.** Documents containing information on computer operation and maintenance.

**soft water** *n.* Water containing little or no dissolved salts of calcium or magnesium, esp. water containing less than 85.5 parts per million of calcium carbonate.

**soft·wood** (sôft'wo͝od', sŏft'-) *n.* **1.** The wood of a coniferous tree. **2.** A coniferous tree.

**soft·y** (sôf'tē, sŏf'-) *n., pl.* **-ies.** *Informal.* **1.** A soft touch. **2.** A person who finds it difficult to punish or be strict.

**sog·gy** (sŏg'ē, sô'gē) *adj.* **-gi·er, -gi·est.** [< dial. *sog,* to soak.] **1.** Saturated with moisture : SOAKED. **2.** Lacking spirit : DULL. **3.** Very moist, humid, and warm. **—sog'gi·ly** *adv.* **—sog'gi·ness** *n.*

**soi-di·sant** (swä'dē-zäN') *adj.* [Fr.] Self-styled <a *soi-disant* pillar of society>

**soi·gné** *also* **soi·gnée** (swän-yā') *adj.* [Fr. < p.part. of *soigner,* to take care of < Med. Lat. *soniare.*] **1.** Exhibiting sophisticated elegance : FASHIONABLE <a *soigné* little hotel><a *soigné* dinner dress> **2.** Well-groomed : polished.

**soil**[1] (soil) *n.* [ME < AN, a piece of ground < Lat. *solium,* seat.] **1.** The top layer of the earth suitable for the growth of plants. **2.** A specific type of earth <rich *soil*> **3.** Country : land <one's native *soil*> **4.** Agricultural life <relinquished city life for the *soil*> **5.** A place or condition favorable to growth.

**soil**[2] (soil) *v.* **soiled, soil·ing, soils.** [ME *soilen* < OFr. *souiller* < VLat. **suculare* < Lat. *suculus,* dim. of *sus,* pig.] —*vt.* **1.** To make dirty. **2.** To tarnish (e.g., one's reputation), by deeds or words. **3.** To corrupt : defile. **4.** To dirty with excrement. —*vi.* To become dirty, stained, or tarnished. —*n.* **1. a.** The state of being soiled. **b.** A stain. **2.** Filth, sewage, or refuse matter. **3.** Manure, esp. human excrement, used as fertilizer.

**soil**[3] (soil) *vt.* **soiled, soil·ing, soils.** [Orig. unknown.] **1.** To feed (livestock) with soilage. **2.** To purge (livestock) by feeding with green food.

**soil·age** (soi'lĭj) *n.* Green crops cut for feeding penned livestock.

**soil pipe** *n.* A drainpipe for carrying off plumbing wastes, esp. from a toilet.

**soil·ure** (soil'yər) *n.* **1.** The act of soiling or the state of being soiled. **2.** A stain, blot, or smudge.

**soi·ree** *also* **soi·rée** (swä-rā') *n.* [Fr. *soirée* < *soir,* evening < Lat. *serus,* late.] An evening reception or party.

**so·journ** (sō'jûrn', sō-jûrn') *vi.* **-journed, -journ·ing, -journs.** [ME *sojournen* < OFr. *sojorner* < VLat. **subdiurnare* : Lat. *sub-,* under + LLat. *diurnum,* day < Lat. *diurnus,* daily < *dies,* day.] To stay or reside temporarily. **—so'journ'** *n.* **—so'journ'er** *n.*

**soke** (sōk) *n.* [ME < Med. Lat. *soca* < OE *socn,* act of seeking.] **1.** The right of local jurisdiction in early English law, gen. one of the feudal rights of lordship. **2.** The district over which soke jurisdiction was exercised.

**sol**[1] (sōl) *also* **so** (sō) *n.* [ME < Med. Lat. —see GAMUT.] *Mus.* **1.** The syllable used to represent the fifth tone of a diatonic scale. **2.** The tone G.

**sol**[2] (sŏl) *n.* [ME < OFr. < Lat. *solidus,* solidus.] **1.** A former monetary unit of France, equal to 12 deniers. **2.** An old French coin worth 12 deniers.

**sol**[3] (sōl) *n., pl.* **so·les** (sō'lās) [Sp. < Lat. *sol,* sun.] —See table at CURRENCY.

**sol**[4] (sôl, sōl) *n.* [Short for HYDROSOL.] A liquid colloidal dispersion.

**Sol** (sŏl, sōl) *n.* [ME < Lat.] The sun.

**so·la**[1] (sō'lə) *n. var. pl. of* solum.

**so·la**[2] (sō'lə) *adj. feminine of* SOLUS.

**sol·ace** (sŏl'ĭs) *also* **sol·ace·ment** (-mənt) *n.* [ME *solas* < OFr. < Lat. *solacium* < *solari,* to console.] **1.** Comfort in sorrow, distress, or misfortune : CONSOLATION. **2.** A source of comfort or consolation. —*vt.* **-aced, -ac·ing, -ac·es. 1.** To give comfort, cheer, or consolation to. **2.** To assuage <*solace* one's grief in work> **—sol'ac·er** *n.*

**so·la·nine** (sō'lə-nēn', -nĭn) *n.* [Fr. < Lat. *solanum,* nightshade < *sol,* sun.] A bitter poisonous alkaloid, $C_{45}H_{73}NO_{15}$, derived from potato sprouts, tomatoes, and nightshade, once used to treat epilepsy.

**so·lar** (sō'lər) *adj.* [ME < Lat. *solaris* < *sol,* sun.] **1.** Of, relating to, or proceeding from the sun <*solar* rays> **2.** Using or operated by energy derived from the sun <a passive *solar* heating system> **3.** Determined or measured with respect to the sun.

**solar battery** *n.* A system comprised of a large number of connected solar cells.

**solar cell** *n.* A semiconductor device that converts solar energy into electric energy, used chiefly in space vehicles as a power supply.

**solar collector** *n.* A device for absorbing solar radiation to be used in producing electricity or in heating buildings or water.
**solar constant** *n.* The amount of solar radiation perpendicularly impinging on a surface of unit area at a distance of one astronomical unit from the sun in a unit interval of time, having an average value of 1.94 calories per minute per square centimeter.
**solar day** *n.* The interval between two successive meridian passages of the sun.
**solar flare** *n.* A temporary burst of solar gases from a small area of the sun's surface, a source of intense radiation.
**solar furnace** *n.* A parabolic reflector that focuses solar radiation at a point to obtain temperatures as high as 4,000°C.
**solar house** *n.* A house constructed with large quantities of heat-absorbing material behind large glass areas, intended to replace or supplement conventional heating systems.
**so·lar·i·a** (sō-lâr'ē-ə) *n. var. pl. of* SOLARIUM.
**so·lar·im·e·ter** (sō'lə-rĭm'ĭ-tər) *n.* An instrument for measuring the flux of solar radiation through a surface.
**so·lar·i·um** (sō-lâr'ē-əm) *n., pl.* **-i·a** (-ē-ə) *or* **-i·ums.** [Lat., terrace < *sol,* sun.] A glassed-in room, gallery, or porch exposed to the sun.
**so·lar·ize** (sō'lə-rīz') *v.* **-ized, -iz·ing, -iz·es.** —*vt.* **1.** To affect by exposure to the sun's rays. **2.** To convert (a building) to solar heat. —*vi.* To be overexposed. —Used of photographic film. **—so'lar·i·za'tion** *n.*
**solar month** *n.* MONTH 6.
**solar panel** *n.* A panel of connected solar cells.
**solar plexus** *n.* [From its radially branching ganglia.] **1.** The large network of sympathetic nerves and ganglia located in the peritoneal cavity behind the stomach and having branching tracts that supply nerves to the abdominal viscera. **2.** *Informal.* The pit of the stomach.
**solar system** *n.* The sun together with the nine planets and all other celestial bodies that orbit it.
**solar wind** *n.* Plasma ejected at high speeds from the sun's surface.
**solar year** *n.* A tropical year.
**so·la·ti·um** (sō-lā'shē-əm) *n., pl.* **-ti·a** (-shē-ə) [LLat., solace < Lat. *solari,* to console.] *Law.* Compensation for suffering, loss, or feelings injured.
**sold** *v. p.t. & p.p. of* SELL.
**sol·dan** (sŏl'dən, sōl'-) *also* **sou·dan** (sōōd'n) *n.* [ME < OFr. < Ar. *sulṭān.* —see SULTAN.] *Archaic.* A sultan.
**sol·der** (sŏd'ər) *n.* [ME *soudur* < OFr. *soudure* < *souder,* to solder < Lat. *solidare,* to make solid < *solidus,* solid.] **1.** Any of various fusible, usu. tin and lead alloys used to join metallic parts when applied in the melted state to the solid metal. **2.** Something that joins or cements. —*v.* **-dered, -der·ing, -ders.** —*vt.* To function as a bond between : JOIN. —*vi.* **1.** To unite or repair objects with solder. **2.** To be joined by or as if by solder. **—sol'der·er** *n.*

▲ word history: The letter *l* in the word *solder* has a long history. The ancestor of *solder* is Latin *solidare,* "to make solid, to fasten together." The Latin word descended into Old French as *souder,* meaning "to join with melted metal"; loss of *l* in certain situations was a normal development in Old French. An *l* was reinserted in the Old French word by those who knew its derivation from Latin, and the variant *solder* existed alongside *souder.* These forms were both borrowed into English, as *solder* and *soder* respectively, each being pronounced as it was spelled. The spelling *soder* eventually was displaced by *solder.* In British English both pronunciations still exist, but in American English the pronunciation without *l* has completely prevailed.

**sol·dier** (sōl'jər) *n.* [ME *soudeour,* mercenary < OFr. *soudier* < *soude,* pay < Lat. *solidus,* solidus.] **1.** One who serves in an armed force, esp. an army. **2.** An enlisted person. **3.** An active, loyal, and militant follower. **4.** A sexually undeveloped form of certain ants and termites whose jaws are specialized to serve as fighting weapons. —*vi.* **-diered, -dier·ing, -diers. 1.** To be or serve as a soldier. **2.** To make a pretense of working in order to escape punishment.
**sol·dier·ly** (sōl'jər-lē) *adj.* Of, relating to, or befitting a soldier.
**soldier of fortune** *n.* One who will serve in any army for personal gain, love of adventure, or pleasure.
**soldiers' home** *n.* A government-funded institution for the care of armed forces veterans.
**sol·dier·y** (sōl'jə-rē) *n., pl.* **-ies. 1.** Soldiers as a group. **2.** A body of soldiers. **3.** The military profession.
**sold-out** (sōld'out') *adj.* Having all tickets or accommodations completely sold, esp. ahead of time <a *sold-out* rock concert>
**sole**[1] (sōl) *n.* [ME < OFr. *solea,* sandal < *solum,* bottom.] **1.** The undersurface of the foot. **2.** The undersurface of a shoe or boot. **3.** The part on which an object rests while standing, esp.: **a.** The bottom surface of a plow. **b.** The bottom surface of the head of a golf club. —*vt.* **soled, sol·ing, soles. 1.** To furnish (a shoe or boot) with a sole. **2.** To put the sole of (a golf club) on the ground, as in preparing to make a stroke.
**sole**[2] (sōl) *adj.* [ME, alone < OFr. *sol* < Lat. *solus.*] **1.** Being the only one : SINGLE <My *sole* purpose was to pass the course.> **2.** Of or relating to only one individual or group : EXCLUSIVE <The appeals court has the *sole* right to decide.> **3.** *Law.* Single or unmarried.
**sole**[3] (sōl) *n., pl.* **sole** *or* **soles.** [ME < OFr. < Lat. *solea,* sandal, flatfish < *solum,* bottom.] **1.** Any of various chiefly marine flatfishes of the family Soleidae, related to and resembling the flounders, esp. any of several European species, as *Solea solea,* valued as a source of food. **2.** A flatfish.
**sol·e·cism** (sŏl'ĭ-sĭz'əm, sō'lĭ-) *n.* [Lat. *soloecismus* < Gk. *soloikismos* < *soloikos,* speaking incorrectly, after *Soloi,* an Athenian colony in Cilicia where a substandard dialect was spoken.] **1.** A nonstandard word usage or grammatical construction. **2.** A violation of etiquette. **3.** Something that deviates from the normal, proper, or gen. accepted order. **—sol'e·cist** *n.* **—sol'e·cis'tic** *adj.*
**sole·ly** (sōl'lē, sō'lē) *adv.* **1.** Alone : singly <*solely* responsible for the errors> **2.** Exclusively <did it *solely* for love>
**sol·emn** (sŏl'əm) *adj.* [ME *solempne* < OFr. < Lat. *sollemnis,* established, customary.] **1.** Deeply earnest : GRAVE <a *solemn* tone of voice> **2.** Impressive and serious <a *solemn* occasion> **3.** Performed with great ceremony <a *solemn* High Mass> **4.** Invoking the force of religion : SACRED <*solemn* vows> **5.** Gloomy. **—so·lem'ni·ty** (sə-lĕm'nĭ-tē) *n.* **—sol'emn·ly** *adv.* **—sol'emn·ness** *n.*
**sol·em·nize** (sŏl'əm-nīz') *vt.* **-nized, -niz·ing, -niz·es. 1.** To celebrate or observe (e.g., a religious event) with formal ceremonies or rites. **2.** To perform with formal ceremony. **3.** To make serious or grave. **—sol'em·ni·za'tion** *n.*
**so·le·noid** (sō'lə-noid') *n.* [Fr. *solénoïde* < Gk. *sōlēnoeidēs,* pipe-shaped : *sōlēn,* pipe + *-eidēs,* -oid.] **1.** A cylindrical coil of insulated wire in which an axial magnetic field is established by a flow of electric current. **2.** An assembly often used as a switch, composed of a coil and a metal core free to slide along the coil axis under the influence of the magnetic field. **—so'le·noi'dal** (-noid'l) *adj.* **—so'le·noi'dal·ly** *adv.*
**sole·plate** (sōl'plāt') *n.* The undersurface of a clothes iron.
**sole·print** (sōl'prĭnt') *n.* **1.** A print of the sole of the foot. **2.** A soleprint made for identification, as of an infant.
**so·les** (sō'lās) *n. pl. of* SOL[3].
**sol-fa** (sōl-fä') *n.* [SOL[1] + FA.] *Mus.* **1.** The set of syllables *do, re, mi, fa, sol, la,* and *ti,* that represent the tones of the scale. **2.** The use of the sol-fa syllables. —*vi. & vt.* **-faed, -fa·ing, -fas.** To use the sol-fa syllables or to sing (a song) using the sol-fa syllables.
**sol·fa·ta·ra** (sōl'fə-tär'ə) *n.* [Ital., sulfurous volcano < *solfo,* sulfur < Lat. *sulfur.*] A volcanic fissure emitting sulfurous vapors, steam, and occas. hot mud. **—sol'fa·ta'ric** *adj.*
**sol·fège** (sŏl-fĕzh') *n.* [Fr. < Ital. *solfeggio.*] Solfeggio.
**sol·feg·gio** (sŏl-fĕj'ē-ō', -fĕj'ō) *n.* [Ital. < *sol-fa,* sol-fa.] *Mus.* **1.** Use of the sol-fa syllables to note the tones of the scale : SOLMIZATION. **2.** A singing exercise in which the sol-fa syllables are used.
**sol·fe·ri·no** (sŏl'fə-rē'nō) *n.* [After *Solferino,* Italy, from the discovery of a dye of this color in the same year that a battle was fought there.] A moderate purplish red. **—sol'fe·ri'no** *adj.*
**so·lic·it** (sə-lĭs'ĭt) *v.* **-it·ed, -it·ing, -its.** [ME *soliciten,* to disturb < OFr. *solliciter* < Lat. *sollicitare* < *sollicitus,* solicitous.] —*vt.* **1.** To try to obtain by entreaty, persuasion, or formal application <*solicit* new votes> **2.** To petition persistently : IMPORTUNE <*solicited* us for donations> **3.** To entice into evil or illegal action. **4.** To approach with an offer of sexual services. —*vi.* To make solicitation or petition for something desired. **—so·lic'i·ta'tion** *n.*
**so·lic·i·tor** (sə-lĭs'ĭ-tər) *n.* **1.** One that solicits, esp. one that seeks contributions or trade. **2.** The chief law officer of a city, town, or government department. **3.** *Chiefly Brit.* A lawyer who is not a member of the bar and who may be heard only in the lower courts.
**solicitor general** *n., pl.* **solicitors general.** A law officer primarily assisting an attorney general.
**so·lic·i·tous** (sə-lĭs'ĭ-təs) *adj.* [Lat. *sollicitus,* troubled : *sollus,* entire + *citus,* p.part. of *ciere,* to move.] **1.** Expressing or exhibiting solicitude : CONCERNED <a *solicitous* inquiry about a friend's health> **2.** Full of desire : EAGER. **3.** Anxious <*solicitous* about their children's future> **4.** Very careful : METICULOUS <*solicitous* in editorial matters> **—so·lic'i·tous·ly** *adv.* **—so·lic'i·tous·ness** *n.*
**so·lic·i·tude** (sə-lĭs'ĭ-tōōd', -tyōōd') *n.* **1. a.** The state or quality of being solicitous. **b.** An attitude of solicitous concern. **2.** *often* **solicitudes.** A cause of anxiety or concern.
**sol·id** (sŏl'ĭd) *adj.* [ME *solide,* not hollow < OFr. < Lat. *solidus.*] **1.** Having a definite configuration and volume. **2.** Not hollowed out <a *solid* block of wood> **3.** Being the same substance or color throughout <a *solid* green car> **4.** Of or relating to three-dimensional geometric figures or bodies. **5.** Being without gaps or breaks : CONTINUOUS <a *solid* line of spectators> **6.** Constructed firmly and well <a *solid* foundation> **7.** Substantial : hearty <a *solid* meal> **8.** Reliable : sound <*solid* evidence> **9.** Financially sound. **10.** Upstanding <a good, *solid* citizen> **11.** Written without a hyphen or space <The word "software" is a *solid* compound.> **12.** Set without leads between the lines in printing. **13.** Acting together : UNANIMOUS. —*n.* **1.** A neither liquid nor gaseous substance. **2.** A three-dimensional geometric figure. **—sol'id·ly** *adv.* **—sol'id·ness** *n.*
**solid angle** *n.* An angle subtended at a point by a surface, meas-

ured in steradians with respect to the area delimited on the unit sphere centered on that point by the locus of points of intersection of the sphere with the lines joining the point to the perimeter of the surface.

**sol·i·dar·i·ty** (sŏl′ĭ-dăr′ĭ-tē) *n.* A union of interests, sympathies, or purposes among the members of a group : FELLOWSHIP.

**solid geometry** *n.* The geometry of three-dimensional figures and surfaces.

**sol·i·di** (sŏl′ĭ-dī′) *n. pl. of* SOLIDUS.

**so·lid·i·fy** (sə-lĭd′ə-fī′) *vt. & vi.* **-fied, -fy·ing, -fies. 1.** To make or become solid or united. **—so·lid′i·fi·ca′tion** *n.*

**so·lid·i·ty** (sə-lĭd′ĭ-tē) *n.* **1.** The state or property of being solid. **2.** Something solid.

**solid of revolution** *n.* A volume generated by the rotation of a plane figure about an axis in its plane.

**solid propellant** *n.* A rocket propellant in solid form, combining both fuel and oxidizer in the form of a compact, cohesive grain.

**solid solution** *n. Chem.* A homogeneous crystalline structure in which one or more types of atoms or molecules may be partly substituted for the original atoms and molecules without changing the structure.

**sol·id-state** (sŏl′ĭd-stāt′) *adj.* **1.** Relating to or characteristic of the physical properties of solid materials, esp. to the electromagnetic, thermodynamic, and structural properties of crystalline solids. **2.** Based on or composed chiefly or exclusively of semiconducting materials, components, and related devices.

**sol·i·dus** (sŏl′ĭ-dəs) *n., pl.* **-di** (-dī′) [ME < Lat. < *solidus,* solid.] **1.** An ancient Roman coin used until the fall of the Byzantine Empire. **2.** A virgule.

**so·lil·o·quize** (sə-lĭl′ə-kwīz′) *vi. & vt.* **-quized, -quiz·ing, -quiz·es.** To utter or put into a soliloquy. **—so·lil′o·quist, so·lil′o·quiz′er** *n.*

**so·lil·o·quy** (sə-lĭl′ə-kwē) *n., pl.* **-quies.** [LLat. : Lat. *solus,* alone + Lat. *loqui,* to speak.] **1.** Literary or dramatic discourse in which a character talks to himself or herself or reveals his or her thoughts in the form of a monologue without addressing a listener. **2.** The act of speaking to oneself.

**sol·ip·sism** (sŏl′ĭp-sĭz′əm, sō′lĭp-) *n.* [Lat. *solus,* alone + Lat. *ipse,* self + -ISM.] *Philos.* **1.** The theory that the self is the only thing that can be known and verified. **2.** The theory or view that the self is the only reality. **—sol′ip·sist** *n.* **—sol′ip·sis′tic** *adj.*

**sol·i·taire** (sŏl′ĭ-târ′) *n.* [Fr. < OFr., solitary < Lat. *solitarius* < *solus,* alone.] **1.** A gem, as a diamond, set alone. **2.** Any of a number of card games played by one person.

**sol·i·tar·y** (sŏl′ĭ-tĕr′ē) *adj.* [ME < Lat. *solitarius* < *solus,* alone.] **1.** Existing, living, or acting without others <a *solitary* passenger> **2.** Occurring, carried out, or made alone <a *solitary* evening> **3.** Remote : secluded <a *solitary* mountain retreat> **4.** Standing alone : SOLE <a *solitary* pine standing on the hill> *—n., pl.* **-ies. 1.** One who lives alone : RECLUSE. **2.** *Informal.* Solitary confinement. **—sol′i·tar′i·ly** (-târ′ə-lē) *adv.* **—sol′i·tar′i·ness** *n.*

**solitary confinement** *n.* Confinement of a prisoner in a cell in which he or she is isolated from all others.

**sol·i·tude** (sŏl′ĭ-to͞od′, -tyo͞od′) *n.* [ME < OFr. < Lat. *solitudo* < *solus,* alone.] **1.** The quality or state of being alone or remote from others. **2.** A lonely or secluded place.

☆ ***syns:*** SOLITUDE, ISOLATION *n. core meaning* : the quality or state of being alone <preferred *solitude* to crowded streets>

**sol·i·tud·i·nar·i·an** (sŏl′ĭ-to͞od′n-âr′ē-ən, -tyo͞od′-) *n.* [Lat. *solitudo, solitudin-,* solitude + -ARIAN.] A recluse.

**sol·ler·et** (sŏl′ə-rĕt′) *n.* [OFr., dim. of *soller,* shoe.] A steel shoe of overlapping plates, worn as part of medieval armor.

**sol·mi·za·tion** (sŏl′mĭ-zā′shən) *n.* [Fr. *solmisation* < *solmiser,* to sol-fa.] *Mus.* The act or a system of using syllables, as *do, re,* and *mi,* to represent the tones of the scale.

**so·lo** (sō′lō) *n., pl.* **-los.** [Ital. < Lat. *solus,* alone.] **1.** An accompanied or unaccompanied musical composition or passage for a single voice or instrument. **2.** Something performed by one person. **3.** A card game in which one player singly opposes others. *—adj.* **1.** Composed, arranged for, or performed by a single voice or instrument. **2.** Made or carried out by one person. *—adv.* Unaccompanied : alone. *—vi.* **-loed, -lo·ing, -los.**To perform alone, esp. to fly an aircraft without a companion or instructor.

**so·lo·ist** (sō′lō-ĭst) *n.* A solo performer.

**Solomon's seal** *n.* [After *Solomon,* 10th-cent. king of Israel.] **1.** A six-pointed star or hexagram held to possess mystical powers. **2.** Any of several plants of the genus *Polygonatum,* with paired, drooping, greenish or yellowish flowers.

**so·lon** (sō′lən, -lŏn′) *n.* [After *Solon,* Athenian statesman of the 7th to 6th cent. B.C.] **1.** A wise lawgiver. **2.** A member of a legislature.

**so long** *interj. Informal.* —Used to express farewell.

**sol·stice** (sŏl′stĭs, sōl′-, sôl′-) *n.* [ME < OFr. < Lat. *solstitium* : *sol,* sun + *sistere,* to stand.] **1.** *Astron.* Either of two times of the year when the sun has no apparent northward or southward motion, at the most northern or most southern point of the ecliptic; the summer solstice, when the sun is in the zenith at the tropic of Cancer, occurs about Jun. 22, and the winter solstice, when it is over the tropic of Capricorn, occurs about Dec. 22. **2.** A highest point : ZENITH. **—sol·sti′tial** (-stĭsh′əl) *adj.*

**sol·u·bi·lize** (sŏl′yə-bə-līz′) *vt.* **-lized, -liz·ing, -liz·es.** To make (substances such as fats and lipids, which are not appreciably soluble under standard conditions) soluble in water by the action of a detergent or similar agent.

**sol·u·ble** (sŏl′yə-bəl) *adj.* [ME < OFr. < LLat. *solubilis* < *solvere,* to loosen.] **1.** Capable of being dissolved. **2.** Capable of being explained or solved. **—sol′u·bil′i·ty** (-bĭl′ĭ-tē) *n.* **—sol′u·ble·ness** *n.* **—sol′u·bly** *adv.*

**soluble glass** *n.* Sodium silicate.

**so·lum** (sō′ləm) *n., pl.* **-la** (-lə) *or* **-lums.** [NLat. < Lat., foundation.] The surface layers of a soil profile in which topsoil formation occurs.

**so·lus** (sō′ləs) *adj. & adv.* [Lat., alone.] By oneself : ALONE. —Used as a stage direction.

**sol·ute** (sŏl′yo͞ot′) *n.* [< Lat. *solutus,* p.part. of *solvere,* to loosen.] A substance dissolved in another substance, usu. the component of a solution present in the lesser amount. **—sol′ute′** *adj.*

**so·lu·tion** (sə-lo͞o′shən) *n.* [ME < OFr. < Lat. *solutio* < *solutus,* p.part. of *solvere,* to loosen.] **1. a.** A spontaneously forming homogeneous mixture of two or more substances, retaining its constitution in subdivision to molecular volumes, displaying no settling, and having various possible proportions of the constituents, which may be solids, liquids, gases, or intercombinations. **b.** Formation of such a mixture. **c.** The state of being dissolved. **2. a.** The method or process of solving a problem. **b.** The answer to or disposition of a problem. **3.** *Law.* Payment or satisfaction of a claim or debt. **4.** The act of separating or breaking up : DISSOLUTION.

**So·lu·tre·an** *also* **So·lu·tri·an** (sə-lo͞o′trē-ən) *adj.* [After *Solutré,* France.] Of or relating to the Old World Upper Paleolithic culture that succeeded the Aurignacian and was marked by improved flint implements and stylized symbolic forms of art.

**sol·va·tion** (sŏl-vā′shən, sôl-) *n.* [SOLV(ENT) + -ATION.] Any of a class of chemical reactions, such as formation of hydrated copper sulfate in aqueous solution, in which solute and solvent molecules combine with relatively weak covalent bonds.

**Sol·vay process** (sŏl′vā′) *n.* [After Ernest *Solvay* (1838–1932).] A process used to produce large quantities of sodium bicarbonate from salt, ammonia, carbon dioxide, and limestone.

**solve** (sŏlv, sôlv) *vt.* **solved, solv·ing, solves.** [ME *solven,* to loosen < Lat. *solvere.*] **1.** To find a solution to. **2.** To work out a correct solution to (a problem). **—solv′a·bil′i·ty, solv′a·ble·ness** *n.* **—solv′a·ble** *adj.* **—solv′er** *n.*

**sol·vent** (sŏl′vənt, sôl′-) *adj.* [Lat. *solvens, solvent-,* pr.part. of *solvere,* to loosen.] **1.** Able to meet one's financial obligations. **2.** Capable of dissolving another substance. *—n.* **1.** *Chem.* **a.** The component of a solution that is present in excess or that undergoes no change of state. **b.** A liquid capable of dissolving another substance. **2.** Something that solves. **—sol′ven·cy** *n.*

**sol·vol·y·sis** (sŏl-vŏl′ĭ-sĭs, sôl-) *n.* [SOLV(ENT) + -LYSIS.] Any of a class of ionic chemical reactions, as hydrolysis, in which solute and solvent react and alter the acidity or relative ionic concentrations of the solution. **—sol′vo·lyt′ic** (-və-lĭt′ĭk) *adj.*

**so·ma** (sō′mə) *n., pl.* **-ma·ta** (-mə-tə) *or* **-mas.** [NLat. < Gk. *sōma,* body.] *Biol.* The physical entity of an organism, exclusive of the germ cells.

**So·ma·li** (sō-mä′lē) *n., pl.* **Somali** *or* **-lis. 1.** A member of one of a group of Hamitic tribes of Somaliland. **2.** The Cushitic language of the Somali.

**so many** *adj.* **1.** Forming an unspecified number <issued *so many* regulations each year> **2.** Forming a pack or group <fought like *so many* tigers>

**so·ma·ta** (sō′mə-tə) *n. var. pl. of* SOMA.

**so·mat·ic** (sō-măt′ĭk) *adj.* [Gk. *sōmatikos* < *sōma,* body.] **1.** Of or relating to the body : PHYSICAL. **2.** Of or relating to the wall of the body cavity. **3.** Of or relating to somatoplasm. **—so·mat′i·cal·ly** *adv.*

**somatic cell** *n.* A bodily cell other than a germ cell.

**somato-** *pref.* [< Gk. *sōma, sōmat-,* body.] **1.** Body <*somato*logy> **2.** Soma <*somato*plasm>

**so·mat·o·gen·ic** (sō-măt′ə-jĕn′ĭk) *also* **so·mat·o·ge·net·ic** (-jə-nĕt′ĭk) *adj.* Arising within the body in response to environment.

**so·ma·tol·o·gy** (sō′mə-tŏl′ə-jē) *n.* **1.** Physiological and anatomical study of the body. **2.** Physical anthropology. **—so′ma·to·log′i·cal** (sō′mə-tə-lŏj′ĭ-kəl, sō-măt′ə-) *adj.*

**so·mat·o·plasm** (sō-măt′ə-plăz′əm) *n.* **1.** The entirety of specialized protoplasm, other than germ plasm, that constitutes the body. **2.** The protoplasm of a somatic cell.

**so·mat·o·pleure** (sō-măt′ə-plo͝or′) *n.* [NLat. *somatopleura* : SOMATO- + Gk. *pleura,* side.] A complex sheet of embryonic cells in certain vertebrates, formed by association of part of the mesoderm with the ectoderm and developing as the internal body wall. **—so·mat′o·pleu′ric** (-plo͝or′ĭk) *adj.*

**so·mat·o·type** (sō-măt′ə-tīp′) *n.* The morphological type of a human body : PHYSIQUE. **—so·mat′o·typ′ic** (-tĭp′ĭk) *adj.*

**som·ber** (sŏm′bər) *adj.* [Fr. *sombre.*] **1.** Gloomy : dark. **2.** Dark in color. **3.** Melancholy <a *somber* frame of mind> **—som′ber·ly** *adv.* **—som′ber·ness** *n.*

**som·bre** (sŏm′bər) *adj. Chiefly Brit. var. of* SOMBER.

**som·bre·ro** (sŏm-brâr′ō, səm-) *n., pl.* **-ros.** [Sp. < *sombra*, shade, of Lat. orig.] A large hat with a broad brim and tall crown, worn esp. in Mexico and the American southwest.

**som·brous** (sŏm′brəs) *adj. Archaic.* Somber.

**some** (sŭm) *adj.* [ME < OE *sum*, a certain one.] **1.** Being an unspecified quantity or number <*some* salt><*some* students><*some* sugar> **2.** Unknown or unidentified by name <*Some* student called.> **3.** *Logic.* Being part and perhaps all of a class. **4.** *Informal.* Remarkable <You are *some* athlete!> *—pron.* **1.** An indefinite or unspecified number or quantity. **2.** An indefinite additional quantity <did the assigned tasks and then *some*> *—adv.* **1.** Approximately : about <*some* 50 employees> **2.** *Informal.* To a certain extent : SOMEWHAT.

**-some¹** *suff.* [ME *-som* < OE *-sum.*] Characterized by a specified quality, condition, or action <bother*some*>

**-some²** *suff.* [ME *-sum* < *sum*, some.] A group of a specified number of members <three*some*>

**-some³** *suff.* [< Gk *sōma*, body.] **1.** Body <centro*some*> **2.** Chromosome <mono*some*>

**some·bod·y** (sŭm′bŏd′ē, -bə-dē) *pron.* A person unspecified or unknown : SOMEONE. *—n., pl.* **-ies.** *Informal.* An important person <You think you're a real *somebody.*>

**some·day** (sŭm′dā′) *adv.* At an unspecified future time.

**some·how** (sŭm′hou′) *adv.* In a way not specified, understood, or known <got the job done *somehow*>

**some·one** (sŭm′wŭn′, -wən) *pron.* Somebody. *—n. Informal.* An important person.

**some·place** (sŭm′plās′) *adv.* Somewhere.

**som·er·sault** *also* **sum·mer·sault** (sŭm′ər-sôlt′) [OFr. *sombresault*, var. of *sobresault* : *sobre-*, above (< Lat. *supra*) + *sault*, leap (< Lat. *saltus* < *salire*, to leap).] *—n.* **1.** An acrobatic stunt in which the body rolls in a complete circle, head over heels. **2.** A complete reversal, as of opinions. *—vi.* **-sault·ed, -sault·ing, -saults.** To perform a somersault.

**som·er·set** *also* **sum·mer·set** (sŭm′ər-sĕt′) *n.* A somersault. *—vi.* **-set·ted, -set·ting, -sets.** To somersault.

**some·thing** (sŭm′thĭng) *pron.* A thing undetermined or unspecified. *—n. Informal.* One that is remarkable or important. *—adv.* **1.** A little : SOMEWHAT <looks *something* like our cat> **2.** *Informal.* To an extreme degree <complains *something* fierce> **—something else.** *Informal.* One that is special or remarkable <Their new beach house is *something else.*> **something of.** To some extent <The director is *something of* an eccentric.>

**some·time** (sŭm′tīm′) *adv.* **1.** At an unstated or indefinite time <I'll see you *sometime* tomorrow.> **2.** At an indefinite future time <Let's have lunch *sometime.*> **3.** Sometimes. **4.** *Archaic.* Formerly. *—adj.* **1.** Having been such at a prior time : FORMER <a *sometime* secretary> **2.** *Nonstandard.* Occasional.

**some·times** (sŭm′tīmz′) *adv.* **1.** Now and then : at times <*sometimes* goes to New York> **2.** *Obs.* At a prior time : FORMERLY.

**some·way** (sŭm′wā′) *also* **some·ways** (-wāz′) *adv.* In some way or another : SOMEHOW.

**some·what** (sŭm′hwŏt′, -wŏt′, -hwət, -wət) *adv.* To some extent : RATHER. *—pron. Archaic.* Something.

**some·where** (sŭm′hwâr′, -wâr′) *adv.* **1.** At, in, or to an unspecified or unknown place <found the sock *somewhere* in the closet> **2.** To a place or state of further development or progress <We're finally getting *somewhere.*> **3.** Approximately <*somewhere* about halfway down the road> *—n.* An unknown or unspecified place.

**some·wheres** (sŭm′hwârz′, -wârz′) *adv. Informal.* Somewhere.

**so·mite** (sō′mīt′) *n.* [Gk. *sōma*, body + -ITE¹.] **1.** *Zool.* A metamere. **2.** A segmental mass of mesoderm in the vertebrate embryo, occurring in pairs along the notochord.

**som·me·lier** (sŭm′əl-yā′) *n.* [Fr. < OFr., officer in charge of provisions, pack-animal driver < *somme*, pack < LLat. *sagma* < Gk., pack-saddle.] A wine steward in a restaurant.

**somn-** *pref. var. of* SOMNI-.

**som·nam·bu·late** (sŏm-năm′byə-lāt′) *vi.* **-lat·ed, -lat·ing, -lates.** To walk in one's sleep. **—som·nam′bu·lar** (-lər) *adj.*

**som·nam·bu·lism** (sŏm-năm′byə-lĭz′əm) *n.* Walking while one is asleep. **—som·nam′bu·list** *n.* **—som·nam′bu·lis′tic** *adj.*

**somni-** *or* **somn-** *pref.* [< Lat. *somnus*, sleep.] Sleep <*somnam*bulate>

**som·ni·fa·cient** (sŏm′nə-fā′shənt) *adj.* Tending to produce sleep. **—som′ni·fa′cient** *n.*

**som·nif·er·ous** (sŏm-nĭf′ər-əs) *also* **som·nif·ic** (-nĭf′ĭk) *adj.* Inducing sleep. **—som·nif′er·ous·ly** *adv.*

**som·no·lence** (sŏm′nə-ləns) *n.* Sleepiness : drowsiness.

**som·no·lent** (sŏm′nə-lənt) *adj.* [ME *sompnolent* < OFr. < Lat. *somnolentus* < *somnus*, sleep.] **1.** Sleepy : drowsy. **2.** Inducing or tending to induce sleep : SOPORIFIC. **—som′no·lent·ly** *adv.*

**so much** *adv.* By the amount or degree indicated or expressed <If we win the award, *so much* the better for us.> *—adj.* —Used as an intensive <The pompous speech was *so much* baloney.> *—pron.* **1.** Something unspecified <charged *so much* a yard> **2.** Everything that can be said or done.

**so much as** *adv.* Even <wouldn't *so much as* glance at me>

**son** (sŭn) *n.* [ME < OE *sunu.*] **1.** A male offspring. **2.** A male descendant. **3. a.** An adopted male child. **b.** A son-in-law. **4.** A man or boy strongly influenced by or associated with a place, cause, race, or school. **5.** A young man. —Used as a familiar term of address.

**so·nance** (sō′nəns) *n.* Sound.

**so·nant** (sō′nənt) *n.* [Lat. *sonans, sonant-*, pr.part. of *sonare*, to sound.] **1.** A voiced speech sound. **2.** A syllabic consonant. **—so′nant** *adj.*

**so·nar** (sō′När′) *n.* [SO(UND) NA(VIGATION) R(ANGING).] **1.** A system using transmitted and reflected acoustic waves to detect and locate submerged objects. **2.** An apparatus, as in a submarine, using acoustic waves to detect submerged objects.

**so·na·ta** (sə-nä′tə) *n.* [Ital. < fem. p.part. of *sonare*, to sound < Lat. *sonare.*] An instrumental musical composition, as for the piano, consisting of three or four independent movements varying in key, mood, and tempo.

**sonata form** *n.* A musical form of three sections, the exposition, development, and recapitulation, often followed by a coda.

**son·a·ti·na** (sŏn′ə-tē′nə) *n.* [Ital., dim. of *sonata*, sonata.] A sonata with shorter movements than the typical sonata.

**sone** (sōn) *n.* [< Lat. *sonus*, a sound.] A subjective unit of loudness, equal to the loudness of a pure tone having a frequency of 1,000 hertz at 40 decibels above the listener's threshold of audibility.

**son et lu·mière** (sôN′ ā lüm-yâr′) *n.* [Fr.: *son*, sound + *et*, and + *lumière*, light.] A dramatic spectacle using special light and sound effects, often held at a historic site.

**song** (sông, sŏng) *n.* [ME < OE *sang.*] **1.** A brief musical composition written or adapted for singing. **2. a.** The act of singing. **b.** The art of singing. **3.** A melodious utterance, as a bird call. **4. a.** Poetry : verse. **b.** A lyric poem. **—for a song.** At a low price.

**song and dance** *n.* **1.** A theatrical performance that combines singing and dancing. **2.** *Slang.* **a.** An excessively elaborate attempt to explain or justify. **b.** An elaborate story or explanation intended to mislead or deceive.

**song·bird** (sông′bûrd′, sŏng′-) *n.* A bird, esp. one of the suborder Passeres, having a melodious song or call.

**song·fest** (sông′fĕst′, sŏng′-) *n.* A casual gathering for group singing.

**song·ful** (sông′fəl, sŏng′-) *adj.* Melodious. **—song′ful·ly** *adv.*

**Song of Solomon** *n.* —See table at BIBLE.

**Song of Songs** *n.* —See table at BIBLE.

**song·smith** (sông′smĭth′, sŏng′-) *n.* A songwriter.

**song sparrow** *n.* A North American songbird, *Melospiza melodia*, with streaked brownish plumage.

**song·ster** (sông′stər, sŏng′-) *n.* **1.** A singer. **2.** A songwriter.

**song thrush** *n.* An Old World songbird, *Turdus philomelos*, with brown upper plumage and a spotted breast.

**song thrush**
*9 inches long*

**song·writ·er** (sông′rī′tər, sŏng′-) *n.* One who writes lyrics or composes tunes for songs.

**son·ic** (sŏn′ĭk) *adj.* [< Lat. *sonus*, a sound.] **1.** Of or relating to audible sound <a *sonic* wave> **2.** Having a speed approaching or being that of sound in air, approx. 738 miles per hour at sea level.

**sonic barrier** *n.* The sudden sharp increase in aerodynamic drag experienced by aircraft approaching the speed of sound.

**sonic boom** *n.* A loud transient explosive sound caused by the shock wave preceding an aircraft traveling at supersonic speeds.

**son-in-law** (sŭn′ĭn-lô′) *n., pl.* **sons-in-law.** The husband of one's daughter.

**son·net** (sŏn′ĭt) *n.* [Fr. < Ital. *sonetto* < OProv. *sonet*, dim. of *son*, song < Lat. *sonus*, a sound.] **1.** A fixed 14-line verse form of Italian origin whose lines are typically 5-foot iambics rhyming according to a prescribed scheme. **2.** A poem written in this pattern.

**son·net·eer** (sŏn′ĭ-tîr′) *n.* **1.** A composer of sonnets. **2.** An inferior poet.

**son·ny** (sŭn′ē) *n., pl.* **-nies.** [Dim. of SON.] A little boy. —Used as a familiar form of address.

**son of a gun** *n. Informal.* A fellow.

**son of God** *n.* Christ.
**so·no·rant** (sə-nôr′ənt, -nōr′-, sŏn′ər-) *n.* [SONOR(OUS) + -ANT.] A voiced consonant regarded as a syllabic sound, as the last sound in the word *sudden.*
**so·nor·i·ty** (sə-nôr′ĭ-tē, -nŏr′-) *n., pl.* **-ties. 1.** The quality or state of being sonorous. **2.** A sound.
**so·no·rous** (sə-nôr′əs, -nōr′-, sŏn′ər-) *adj.* [Lat. *sonorus* < *sonor,* sound < *sonare,* to sound.] **1.** Having or producing sound. **2.** Having or producing a full, rich, or deep sound. **3.** Impressive : grandiloquent. **—so·no′rous·ly** *adv.* **—so·no′rous·ness** *n.*
**soo·chong** (sōō′chông′, -shông′) *n. var. of* SOUCHONG.
**soon** (sōōn) *adv.* **-er, -est.** [ME *sone* < OE *sōna,* immediately.] **1.** In the near future. **2.** Without hesitation : PROMPTLY <called as *soon* as possible> **3.** Before the usual or appointed time : EARLY. **4.** With willingness : READILY <I'd as *soon* stop right now.> **5.** *Obs.* Immediately. **—no sooner than.** As soon as <*No sooner* was the frost off the ground *than* the work began.> ***usage:*** As a comparative adverb, *no sooner* shoud be followed by *than,* and not by *when,* as in *No sooner had they arrived than the phone rang.* **—sooner or later.** Sometime : eventually.
**soon·er** (sōō′nər) *n.* [< SOON.] *Slang* **1.** One who settled homestead land in the early West before it was officially made available, so as to have first choice of location. **2. Sooner.** A native or resident of Oklahoma.
**soot** (sŏŏt, sŭt, sōōt) *n.* [ME < OE *sōt.*] A fine dispersion of black particles, mainly carbon, generated by incomplete combustion of coal, oil, wood, or other fuels. —*vt.* **soot·ed, soot·ing, soots.** To cover or smudge with soot.
**sooth** (sōōth) *adj.* [ME < OE *sōð.*] *Archaic.* **1.** Real : true. **2.** Soft : sweet. —*n. Archaic.* Truth : reality. **—sooth′ly** *adv.*
**soothe** (sōōth) *v.* **soothed, sooth·ing, soothes.** [ME *sothen,* to verify < OE *sōðian* < *sōð,* truth.] —*vt.* **1.** To placate or calm. **2.** To ease the pain of. —*vi.* To bring comfort, composure, or relief. **—sooth′er** *n.*
**sooth·fast** (sōōth′făst′) *adj.* [ME *sothfast* < OE *sōðfæst* : *sōð,* truth + *fæst,* fixed, fast.] *Archaic.* **1.** Truthful : honest. **2.** True : real.
**sooth·ing** (sōō′thĭng) *adj.* Tending to soothe <a soft, *soothing* tone of voice> **—sooth′ing·ly** *adv.* **—sooth′ing·ness** *n.*
**sooth·say** (sōōth′sā′) *vi.* **-said** (-sĕd′), **-say·ing, -says** (-sĕz′) [Back-formation < SOOTHSAYER.] To foretell future events.
**sooth·say·er** (sōōth′sā′ər) *n.* One who claims to foretell events or predict the future : SEER.
**sooth·say·ing** (sōōth′sā′ĭng) *n.* **1.** The art or practice of foretelling events. **2.** A prediction or prophecy.
**soot·y** (sŏŏt′ē, sŭt′ē, sōō′tē) *adj.* **-i·er, -i·est. 1.** Covered with soot. **2.** Of or generating soot. **3.** Dark like soot. **—soot′i·ness** *n.*
**sooty grouse** *n.* The blue grouse.
**sooty mold** *n.* A black fungus of the genus *Capnodium,* growing on plants in the droppings of sucking insects such as aphids.
**sop** (sŏp) *v.* **sopped, sop·ping, sops.** [< ME *soppe,* dipped bread < OE *sopp.*] —*vt.* **1.** To dip, soak, or drench in a liquid : SATURATE. **2.** To take up by absorption <*sop* up water with rags> —*vi.* To be or become utterly soaked or saturated. —*n.* **1.** A piece of food soaked or dipped in a liquid. **2. a.** Something yielded to soothe or placate. **b.** A bribe.
**soph·ism** (sŏf′ĭz′əm) *n.* [ME *sophime* < OFr. < Lat. *sophisma* < Gk., clever device < *sophos,* clever.] **1.** A plausible but fallacious argument. **2.** Deceptive or fallacious argumentation.
**soph·ist** (sŏf′ĭst) *n.* [Lat. *sophistēs* < Gk. *'sophistēs,* expert < *sophizesthai,* to play tricks < *sophos,* clever.] **1. Sophist. a.** A member of a pre-Socratic school of philosophy in ancient Greece. **b.** Any of a class of later Greek teachers of rhetoric and philosophy known for their overly subtle, often misleading arguments. **2.** A scholar or thinker, esp. one skillful in devious argumentation.
**so·phis·tic** (sə-fĭs′tĭk) or **so·phis·ti·cal** (-tĭ-kəl) *adj.* **1.** Of, relating to, or characteristic of sophists. **2.** Specious : fallacious. **—so·phis′ti·cal·ly** *adv.*
**so·phis·ti·cate** (sə-fĭs′tĭ-kāt′) *v.* **-cat·ed, -cat·ing, -cates.** [Med. Lat. *sophisticare, sophisticat-* < Lat. *sophisticus,* sophistic < Gk. *sophistikos* < *sophistēs,* sophist.] —*vt.* **1.** To cause to become less natural, esp. to make more worldly. **2.** To corrupt or pervert : ADULTERATE. **3.** To make more complex or inclusive : REFINE. —*vi.* To use sophistry. —*n.* (-kĭt). A sophisticated person. **—so·phis′ti·ca′tion** *n.* **—so·phis′ti·ca′tor** *n.*
**so·phis·ti·cat·ed** (sə-fĭs′tĭ-kā′tĭd) *adj.* **1.** Having acquired worldly knowledge or refinement. **2.** Quite complex <the latest and most *sophisticated* computer technology> **3.** Suitable for or appealing to the tastes of sophisticates <a *sophisticated* novel> **—so·phis′ti·cat′ed·ly** *adv.*
☆ ***syns:*** SOPHISTICATED, COSMOPOLITAN, WORLDLY, WORLDLY-WISE *adj. core meaning* : experienced in the ways of the world <a *sophisticated* diplomat experienced in negotiations> ***ant:*** naive, unsophisticated
**soph·is·try** (sŏf′ĭ-strē) *n., pl.* **-tries. 1.** A plausible but misleading or fallacious argument. **2.** Plausible but fallacious argumentation.
**soph·o·more** (sŏf′ə-môr′, -mōr′) *n.* [Gk. *sophos,* wise + Gk. *mōros,* foolish.] **1.** A second-year student in an American college or high school. **2.** One in the second year of an endeavor.
**soph·o·mor·ic** (sŏf′ə-môr′ĭk, -mōr′-, -mŏr′-) *adj.* **1.** Of or characteristic of a sophomore. **2.** Immature and overconfident. **—soph′o·mor′i·cal·ly** *adv.*
**So·phy** (sō′fē) *n., pl.* **-phies.** [Pers. *Safi,* surname of a Persian dynasty (1500–1736) < Ar. *Ṣafiuddin,* purity of religion.] A title formerly given to kings of Persia.
**so·por** (sō′pər, -pôr′) *n.* [Lat., sleep.] An abnormally deep sleep.
**sop·o·rif·er·ous** (sŏp′ə-rĭf′ər-əs, sō′pə-) *adj.* SOPORIFIC 1. **—sop′o·rif′er·ous·ly** *adv.* **—sop′o·rif′er·ous·ness** *n.*
**sop·o·rif·ic** (sŏp′ə-rĭf′ĭk, sō′pə-) *adj.* **1.** Inducing or tending to induce sleep. **2.** Drowsy. —*n.* A sleep-inducing drug.
**sop·ping** (sŏp′ĭng) *adj.* SOPPY 1.
**sop·py** (sŏp′ē) *adj.* **-pi·er, -pi·est. 1.** Utterly soaked : SOPPING. **2.** *Slang.* Overly sentimental : MAWKISH.
**so·pra·ni·no** (sō′prə-nē′nō, sŏp′rə-) *n., pl.* **-nos.** [Ital., dim. of *soprano,* soprano.] A musical instrument, as a recorder, that is higher in pitch than the soprano of its family.
**so·pran·o** (sə-prăn′ō, -prä′nō) *n., pl.* **-os.** [Ital. < *sopra,* above < Lat. *supra.*] **1.** The highest singing voice of a woman or young boy. **2.** A singer with a soprano voice. **3.** A part written in the range of the soprano voice. **4.** The tonal range characteristic of a soprano.
**so·ra** (sôr′ə, sōr′ə) *n.* [Prob. of American Indian orig.] A North American marsh bird, *Porzana carolina,* with grayish-brown plumage.
**sorb**[1] (sôrb) *vt.* **sorbed, sorb·ing, sorbs.** [Back-formation < ADSORB and ABSORB.] To take up and hold, as by absorption or adsorption.
**sorb**[2] (sôrb) *n.* [Fr. *sorbe* < Lat. *sorbus,* service tree.] **1.** An Old World tree of the genus *Sorbus* or related genera, as the service tree or the rowan. **2.** The fruit of the sorb.
**Sorb** (sôrb) *n.* [G. *Sorbe.*] A Wend.
**Sor·bi·an** (sôr′bē-ən) *n.* [< SORB.] **1.** A Wend. **2.** Wendish. **—Sor′bi·an** *adj.*
**sor·bic acid** (sôr′bĭk) *n.* [< SORB[2].] A white crystalline solid, $C_6H_8O_2$, found in the unripe berries of the mountain ash and also synthesized, used as a fungicide and a food preservative.
**sor·bose** (sôr′bōs′) *n.* [SORB[2] + -OSE.] A white sweetish crystalline sugar, $C_6H_{12}O_6$, used in making ascorbic acid.
**sor·cer·er** (sôr′sər-ər) *n.* [ME *sorser* < OFr. *sorcier* < VLat. **sortiarius* < Lat. *sors,* lot, chance.] One who practices sorcery.
**sor·cer·ess** (sôr′sər-ĭs) *n.* A woman who practices sorcery.
**sor·cer·y** (sôr′sə-rē) *n.* [ME *sorcerie* < OFr. < *sorcier,* sorcerer.] Use of supernatural power over others through the assistance of evil spirits : NECROMANCY. **—sor′cer·ous** *adj.* **—sor′cer·ous·ly** *adv.*
**sord** (sôrd) *n.* [ME *sorde* < *sorden,* to rise up in flight < OFr. *sordre* < Lat. *surgere,* to rise.] A flight of mallards.
**sor·did** (sôr′dĭd) *adj.* [Fr. *sordide* < Lat. *sordidus* < *sordēre,* to be dirty.] **1.** Dirty or filthy : FOUL. **2.** Squalid : wretched <*sordid* slums> **3.** Morally degraded : BASE. **4.** Exceedingly avaricious : GRASPING. **—sor′did·ly** *adv.* **—sor′did·ness** *n.*
☆ ***syns:*** SORDID, BASE, CONTEMPTIBLE, DESPICABLE, LOW, SQUALID, VILE *adj. core meaning* : having or proceeding from low moral standards <a *sordid* affair that resulted in a scandal>
**sor·di·no** (sôr-dē′nō) *n., pl.* **-ni** (-nē) [Ital. < *sordo,* deaf, mute < Lat. *surdus.*] MUTE 3.
**sore** (sôr, sōr) *adj.* **sor·er, sor·est.** [ME < OE *sār.*] **1.** Painful to the touch : TENDER <a *sore* toe> **2.** Experiencing physical pain <*sore* all over> **3.** Causing distress, sorrow, or misery : GRIEVOUS <*sore* need> **4.** Causing embarrassment or irritation <a *sore* subject> **5.** Full of distress : SORROWFUL. **6.** *Informal.* Angry : offended <Don't get *sore.*> —*adv. Archaic.* Sorely. —*n.* **1.** An open skin lesion, wound, or ulcer. **2.** A cause of pain, distress, or irritation. **—sore′ness** *n.*
**sore·head** (sôr′hĕd′, sōr′-) *n. Slang.* One easily offended or angered.
**sore·ly** (sôr′lē, sōr′-) *adv.* **1.** Painfully : grievously. **2.** Extremely : greatly <a skill that is *sorely* needed>
**sore throat** *n.* Inflammation of the tonsils, pharynx, or larynx marked by pain in swallowing.
**sor·gho** (sôr′gō) *n. var. of* SORGO.
**sor·ghum** (sôr′gəm) *n.* [NLat. < Ital. *sorgo.*] **1.** An Old World grass, *Sorghum vulgare,* several varieties of which are cultivated as grain and forage or as a source of syrup. **2.** Syrup made from the juice of the sorghum.
**sor·go** *also* **sor·gho** (sôr′gō) *n., pl.* **-gos** *also* **-ghos.** [Ital.] A sorghum, esp. *Sorghum vulgare saccharatum,* cultivated as a source of syrup.
**so·ri** (sôr′ī′, sōr′ī′) *n. pl. of* SORUS.
**sor·i·cine** (sôr′ĭ-sīn′, sōr′-, sŏr′-) *adj.* [Lat. *soricinus* < *sorex,* shrew.] Of or belonging to the family Soricidae, including the shrews.
**so·ri·tes** (sə-rī′tēz) *n., pl.* **sorites.** [Lat. *sorites* < Gk. *sōreitēs* < *sōros,* heap.] *Logic.* A form of argument in which a series of incomplete syllogisms is so arranged that the predicate of each premise

forms the subject of the next until the subject of the first is joined with the predicate of the last in the conclusion.

**so·ro·ral** (sə-rôr′əl, -rōr′-) *adj.* [< Lat. *soror,* sister.] Of, relating to, or like a sister : SISTERLY.

**so·ro·rate** (sə-rôr′ĭt, -rōr′-) *n.* [< Lat. *soror,* sister.] The custom of marriage of a man to his wife's sister or sisters, usu. after the wife has died or proved sterile.

**so·ror·i·cide** (sə-rôr′ĭ-sīd′, -rōr′-) *n.* [Lat. *soror,* sister + -CIDE.] **1.** The murder of one's sister. **2.** One who murders one's own sister. **—so·ror′i·cid′al** (-sīd′l) *adj.*

**so·ror·i·ty** (sə-rôr′ĭ-tē, -rŏr′-) *n., pl.* **-ties.** [Med. Lat. *sororitas* < Lat. *soror,* sister.] A social or civic club for women, esp. one at a college.

**sorp·tion** (sôrp′shən) *n.* [Back-formation < ABSORPTION and ADSORPTION.] The process of sorbing or the state of being sorbed. **—sorp′tive** *adj.*

**sor·rel[1]** (sôr′əl, sŏr′-) *n.* [ME *sorel* < OFr. *surele* < *sur,* sour, of Germanic orig.] **1.** A plant of the genus *Rumex,* having acid-flavored leaves occas. used as salad greens, esp. *R. acetosella,* a widely naturalized Eurasian species. **2.** Any of various plants of the genus *Oxalis.*

**sor·rel[2]** (sôr′əl, sŏr′-) *n.* [< ME *sorelle,* sorrel-colored < OFr. *sorel* < *sor,* red-brown.] **1.** A brownish orange to light brown. **2.** A sorrel colored horse.

**sorrel tree** *n.* Sourwood.

**sor·row** (sŏr′ō, sôr′ō) *n.* [ME *sorow* < OE *sorg.*] **1.** Mental anguish or pain caused by injury, loss, or despair. **2.** A cause of sorrow : MISFORTUNE. **3.** Expression of sorrow. *—vi.* **-rowed, -row·ing, -rows.** To feel or display sorrow. **—sor′row·er** *n.*

**sor·row·ful** (sŏr′ō-fəl, -ə-fəl, sôr′-) *adj.* **1.** Feeling or causing sorrow. **2.** Characterized by or expressing sorrow. **—sor′row·ful·ly** *adv.* **—sor′row·ful·ness** *n.*

☆ ***syns:*** SORROWFUL, DOLEFUL, DOLOROUS, LUGUBRIOUS, MOURNFUL, PLAINTIVE, RUEFUL, WOEFUL *adj. core meaning* : full of or expressive of sorrow <*sorrowful* wailing at the funeral><*sorrowful* faces of refugees> ***ant:*** gay

**sor·ry** (sŏr′ē, sôr′ē) *adj.* **-ri·er, -ri·est.** [ME *sory* < OE *sārig,* painful < *sār,* sore.] **1.** Feeling or showing sympathy, pity, or regret <I'm *sorry* I was rude.> **2.** Worthless or inferior : PALTRY <a *sorry* excuse> **3.** Causing sorrow, grief, or misfortune : GRIEVOUS <a *sorry* state of affairs> **—sor′ri·ly** *adv.* **—sor′ri·ness** *n.*

☆ ***syns:*** SORRY, COMPUNCTIOUS, CONTRITE, PENITENT, REGRETFUL, REMORSEFUL, REPENTANT *adj. core meaning* : feeling or expressing regret <were *sorry* for their misdeeds>

**sort** (sôrt) *n.* [ME < OFr. *sorte,* prob. < Lat. *sors,* lot.] **1.** A class that is defined by the common attribute or attributes possessed by all its members : KIND. **2.** Character or nature : TYPE <a friendly *sort* of person> **3.** A way of acting or behaving. **4.** *often* **sorts.** One of the characters in a font of printing type. *—vt.* **sort·ed, sort·ing, sorts. 1.** To arrange according to class, kind, or size : CLASSIFY. **2.** To set apart from others <*sort* out the wheat from the chaff> **3.** To clarify by going over mentally <tried to *sort* out the problems> **—after a sort.** In a haphazard or imperfect way. **—of sorts. 1.** Of a mediocre or inferior kind. **2.** Of one kind or another. **—out of sorts.** *Informal.* **1.** Slightly ill. **2.** Irritable : cross. **—sort of.** *Informal.* Somewhat : rather <*sort of* dumb> **—sort′a·ble** *adj.* **—sort′er** *n.*

**sor·tie** (sôr′tē, sôr-tē′) *n.* [Fr. < OFr. *sortir,* to go out.] **1.** An armed attack made from a place surrounded by enemy forces. **2.** A flight of an aircraft on a combat mission. *—vi.* **-tied, -tie·ing, -ties.** To go on a sortie.

**sor·ti·lege** (sôr′tl-ĭj) *n.* [ME < OFr. < Med. Lat. *sortilegium* < *sortilegus,* diviner : Lat. *sors,* lot + Lat. *legere,* to read.] **1.** The act or practice of predicting the future by drawing lots. **2.** Witchcraft.

**so·rus** (sôr′əs, sōr′-) *n., pl.* **so·ri** (sôr′ī′, sōr′ī′) [NLat. < Gk. *sōros,* heap.] **1.** A cluster of spore cases borne by ferns on the undersides of the fronds. **2.** A structure in certain fungi and lichens similar to a sorus.

**S O S** (ĕs′ō-ĕs′, ĕs′ō-ĕs′) *n.* **1.** The letters represented by the signal • • • — — — • • •, used internationally as a distress signal, esp. by ships and aircraft. **2.** A call for help.

**so-so** (sō′sō′) *adj.* Mediocre : passable. *—adv.* Tolerably : passably.

**so·ste·nu·to** (sō′stə-nōō′tō, sô′-) [Ital., p.part. of *sostenere,* to sustain < Lat. *sustinēre.* —see SUSTAIN.] *Mus. —adv.* In a sustained or prolonged manner. —Used as a direction. *—n., pl.* **-tos** *or* **-ti** (-tē). A passage played or sung sostenuto. **—so′ste·nu′to** *adj.*

**sot** (sŏt) *n.* [ME, fool < OE *sott.*] A chronic drunkard.

**so·te·ri·ol·o·gy** (sō-tîr′ē-ŏl′ə-jē) *n.* [Gk. *sōtērion,* deliverance (< *sōtēr,* savior < *saos,* safe) + -LOGY.] The doctrine of salvation as effected by Christ. **—so·te′ri·o·log′ic** (-ə-lŏj′ĭk), **so·te′ri·o·log′i·cal** *adj.*

**So·thic** (sō′thĭk, sŏth′ĭk) *adj.* [< Gk. *Sōthis,* the star Sirius.] **1.** Of, relating to, or deriving from the name of Sothis. **2.** Designating the ancient Egyptian year, consisting of 365¼ days. **3.** Designating a cycle consisting of 1,460 years of 365 days in the ancient Egyptian calendar.

**So·this** (sō′thĭs) *n.* [Gk. *Sōthis.*] Sirius.

**So·tho** (sō′tō) *n.* **1.** A group of Bantu languages spoken in southern Africa. **2.** Any of the Sotho languages.

**so·tol** (sō′tōl′) *n.* [Mex. Sp. *sotole* < Nahuatl *tzotolli.*] Any of several tall woody plants of the genus *Dasylirion,* of the southwestern United States and adjacent Mexico, with prickly-margined leaves and a large cluster of whitish flowers.

**sot·tish** (sŏt′ĭsh) *adj.* **1.** Resembling a drunk. **2.** Stupid : dull. **—sot′tish·ly** *adv.* **—sot′tish·ness** *n.*

**sot·to vo·ce** (sŏt′ō vō′chē) *adv.* [Ital., under the voice.] **1.** Softly, so as not to be overheard. **2.** *Mus.* Very softly. —Used as a direction.

**sou** (sōō) *n.* [Fr. < obs. *sol* < Lat. *solidus,* solidus.] A five-centime piece once in circulation in France.

**sou·a·ri nut** (sōō-är′ē) *n.* [Fr. *saouari* < Galibi *sawarra.*] **1.** A South American tree, *Caryocar nuciferum,* yielding nuts used as food and a source of cooking oil. **2.** The nut of the souari nut tree.

**sou·brette** (sōō-brĕt′) *n.* [Fr. < Prov. *soubreto,* fem. of *soubret,* conceited < *soubra,* to leave aside < OProv. *sobras,* to be excessive < Lat. *superare* < *super,* above.] **1. a.** A saucy, coquettish, and intriguing lady's maid in comedies or comic opera. **b.** An actress or singer taking such a part. **2.** A flirtatious or frivolous young woman.

**sou·bri·quet** (sōō′brĭ-kā′, sōō′brĭ-kā′) *n. var. of* SOBRIQUET.

**sou·chong** *also* **soo·chong** (sōō′chông′, -shông′) *n.* [Chin. (Mandarin) *xiao[3] zhong[3]* : *xiao[3],* small + *zhong[3],* kind.] A black tea native to China and adjacent regions.

**sou·dan** (sōōd′n) *n. var. of* SOLDAN.

**souf·flé** (sōō-flā′) *n.* [Fr. < p.part. of *souffler,* to puff up < Lat. *sufflare* : *sub-,* under + *flare,* to blow.] A light, fluffy baked dish made with egg yolks and beaten egg whites combined with other ingredients and served as a main dish or sweetened as a dessert. *—adj.* Made light and puffy by beating and baking or cooking. **—souf·fléd′** *adj.*

**sough** (sŭf, sou) *vi.* **soughed, sough·ing, soughs.** [ME *swoghen* < OE *swōgan.*] To make a soft rustling or murmuring sound. *—n.* A soft murmuring or rustling sound, as of the wind or a gentle surf.

**sought** (sôt) *v. p.t. & p.p. of* SEEK.

**soul** (sōl) *n.* [ME < OE *sāwol.*] **1.** The animating and vital principle in the human being, credited with the faculties of thought, action, and emotion and often conceived as an immaterial entity. **2.** The spiritual nature of the human being, regarded as immortal, separable from the body at death, and susceptible to happiness or misery in a future state. **3.** The disembodied spirit of a dead human being : SHADE. **4. Soul.** *Christian Science.* GOD 1c. **5.** A human being <didn't see a *soul* I knew> **6.** The central or integral part <You are the *soul* of the movement.> **7.** A person considered as the perfect embodiment of an intangible quality : PERSONIFICATION <I will be the very *soul* of discretion.> **8.** A person's emotional or moral nature. **9. a.** An awareness of and pride in the physical and cultural aspects of the African heritage. **b.** A strong, deeply felt emotion conveyed by a performer or artist.

**soul brother** *n. Slang.* A fellow black man.

**soul food** *n.* Food, as ham hocks and collard greens, traditionally eaten by southern American blacks.

**soul·ful** (sōl′fəl) *adj.* Full of or expressing deep feeling. **—soul′ful·ly** *adv.* **—soul′ful·ness** *n.*

**soul kiss** *n.* A French kiss.

**soul·less** (sōl′lĭs) *adj.* Lacking sensitivity or the capacity for deep feeling. **—soul′less·ly** *adv.* **—soul′less·ness** *n.*

**soul mate** *n.* One of two persons compatible with each other in disposition, viewpoint, or sensitivity.

**soul music** *n.* Music developed by American blacks, combining elements of gospel music and rhythm and blues.

**soul-search·ing** (sōl′sûr′chĭng) *n.* A penetrating examination of one's motives, convictions, and emotional attitudes.

**soul sister** *n. Slang.* A fellow black woman.

**sound[1]** (sound) *n.* [ME *soun* < OFr. *son* < Lat. *sonus.*] **1. a.** A vibratory disturbance in the pressure and density of a fluid or in the elastic strain in a solid, with frequency in the approximate range between 20 and 20,000 hertz, capable of being detected by the organs of hearing. **b.** A disturbance of any frequency. **c.** The sensation stimulated in the organs of hearing by such a disturbance. **d.** Such sensations collectively. **2.** A distinctive noise <a metallic *sound*> **3.** The distance over which something can be heard. **4. a.** An articulation made by the vocal apparatus. **b.** The distinctive character of such an articulation <The words "bear" and "bare" have the same *sound.*> **5.** A mental impression : IMPLICATION <didn't like the *sound* of the invitation> **6.** Auditory material recorded, as for a film. **7.** Meaningless noise. **8.** *Archaic.* Rumor : report. *—v.* **sound·ed, sound·ing, sounds.** *—vi.* **1.** To produce or emit a sound. **2.** To present a particular impression <That idea *sounds* sensible.> *—vt.* **1.** To cause to produce or emit a sound <*sounded* the alarm> **2.** To summon, announce, or signal by a sound <*sound* a warning> **3.** To pronounce : articulate <*sound* a vowel> **4.** To make known : PROCLAIM. **5.** To examine (a bodily organ or part) by causing it to emit sound : AUSCULTATE. **—sound off. 1.** To count cadence when marching in military formation. **2.** To express one's views vigorously <*sounded off* about high taxes>

**sound[2]** (sound) *adj.* **-er, -est.** [ME *sund* < OE *gesund.*] **1.** Free from defect, decay, or damage. **2.** Free from injury or disease : HEALTHY.

3. Having a firm basis : UNSHAKABLE. **4.** Financially secure <a *sound* economy> **5. a.** Based on valid reasoning : SENSIBLE <a *sound* suggestion> **b.** Founded on thorough knowledge or experience <*sound* data> **6.** Thorough : complete <a *sound* spanking> **7.** Deep and unbroken : UNDISTURBED <a *sound* sleep> **8.** Free from moral defect : UPRIGHT. **9.** Deserving confidence : TRUSTWORTHY. **10.** Marked by or showing common sense and good judgment : LEVELHEADED. **11.** Compatible with an accepted point of view : CONSERVATIVE. **12.** *Law.* Valid : legal. **—sound'ly** *adv.* **—sound'ness** *n.*

☆ ***syns:*** **1.** SOUND, FIRM, SECURE, SOLID, STABLE, STURDY, SURE *adj. core meaning* : not easily moved or shaken <a *sound* structural support> **2.** SOUND, COGENT, SOLID, VALID *adj. core meaning* : based on good judgment, reasoning, or evidence <a *sound* proposal>

**sound[3]** (sound) *n.* [ME, partly from OE *sund,* swimming, sea, and partly < ON *sund,* strait, channel < *sundr,* asunder.] **1.** A long, relatively wide body of water, larger than a strait or a channel, connecting larger bodies of water. **2.** A long, wide ocean inlet. **3.** The air bladder of a fish.

**sound[4]** (sound) *v.* **sound·ed, sound·ing, sounds.** [ME *sounden* < OFr. *sonder* < *sonde,* sounding line, prob. of Germanic orig.] *—vt.* **1.** To measure the depth of (water), esp. by a weighted line : FATHOM. **2.** To attempt to learn the attitudes or opinions of <*sounded* out the voters' feelings on gun control> **3.** To probe (a bodily cavity) with a sound. *—vi.* **1.** To measure depth. **2.** To dive swiftly downward. —Used of a whale or fish. **3.** To look into : INVESTIGATE. *—n.* An instrument for examining bodily cavities. **—sound'a·ble** *adj.*

**sound barrier** *n.* The sonic barrier.

**sound·board** (sound'bôrd', -bōrd') *n.* SOUNDING BOARD 1.

**sound box** *n.* A hollow chamber in the body of a musical instrument, as a violin or cello, that intensifies the resonance of the tone.

**sound camera** *n.* A motion-picture camera equipped to record sound and image synchronously.

**sound effects** *pl.n.* Imitative sounds, as of thunder or an explosion, produced artificially for theatrical effects.

**sound·er** (soun'dər) *n.* One that sounds, esp. a device for making soundings of the sea.

**sound·ing[1]** (soun'dĭng) *n.* **1.** The act of one that sounds. **2.** An environmental probe for scientific observation. **3. a.** A measured depth of water. **b.** *often* **soundings.** Water shallow enough for depth measurements to be taken by a hand line.

**sound·ing[2]** (soun'dĭng) *adj.* **1.** Emitting a full sound : RESONANT. **2.** Noisy but with little significance : HIGH-SOUNDING.

**sounding board** *n.* **1.** A thin board that forms the upper part of the resonant chamber in a musical instrument, as a violin or piano, and that serves to increase resonance. **2.** A structure suspended behind or over a podium or platform to reflect the speaker's voice to the audience. **3.** A person or group whose reactions to an idea, opinion, or viewpoint will serve as a measure of its effectiveness or acceptability. **4.** A device or means serving to spread or popularize an idea or viewpoint.

**sounding lead** *n.* The metal weight at the end of a sounding line.

**sounding line** *n.* A line marked at intervals of fathoms and weighted at one end, used to determine the water depth.

**sounding rocket** *n.* A rocket used to make observations anywhere within the earth's atmosphere.

**sound·less** (sound'lĭs) *adj.* Having or making no sound. **—sound'less·ly** *adv.* **—sound'less·ness** *n.*

**sound pollution** *n.* Noise pollution.

**sound·proof** (sound'pro͞of') *adj.* Impenetrable by audible sound. *—vt.* **-proofed, -proof·ing, -proofs.** To make soundproof.

**sound ranging** *n.* Electronic location of a sound source, as of enemy weapons, by checking time intervals indicated by microphones of known position.

**sound stage** *n.* A usu. soundproof room or studio for producing motion pictures.

**sound·track** (sound'trăk') *n.* **1.** The narrow strip at one side of a motion-picture film that carries the sound recording. **2. a.** Music accompanying a motion picture. **b.** A commercial phonograph record or tape of such music.

**sound truck** *n.* A vehicle, as a truck, having one or more loudspeakers, usu. on top, for area broadcasting.

**sound wave** *n.* A wave of sound.

**soup** (so͞op) *n.* [Fr. *soupe* < OFr., broth, of Germanic orig.] **1.** A liquid food prepared from meat, fish, or vegetable stock with other ingredients added and often containing pieces of solid food. **2.** *Slang.* Something resembling the consistency of soup, esp.: **a.** Dense fog. **b.** Nitroglycerine. **3.** A chaotic or unfortunate situation. **—in the soup.** *Slang.* In trouble. **—soup up.** *Slang.* To add horsepower or greater speed potential to (an engine or vehicle).

**soup·çon** (so͞op-sôN', so͞op'sŏn') *n.* [Fr. < VLat. **suspectio* < Lat. *suspectus,* p.part. of *suspicere,* to suspect. —see SUSPECT.] A tiny amount : TRACE.

**soup du jour** (so͞op' də zho͝or') *n.* [Fr., soup of the day.] A soup featured by a restaurant on a given day.

**soup kitchen** *n.* A place where food is offered at very low cost or free of charge to the needy.

**soup·spoon** (so͞op'spo͞on') *n.* A rather large spoon for eating soup.

**soup·y** (so͞o'pē) *adj.* **-i·er, -i·est. 1.** Like soup, as in consistency or appearance. **2.** Foggy. **3.** *Informal.* Sentimental.

**sour** (sour) *adj.* **-er, -est.** [ME < OE *sūr.*] **1.** Having the sharp, tart, or tangy taste characteristic of that produced by acids. **2.** Made acid or rancid by fermentation. **3.** Having the taste or smell characteristic of fermentation or rancidity. **4.** Bad-tempered and morose : PEEVISH <a *sour* disposition> **5.** Below the expected or usual ability or quality : BAD <Their pitching was *sour.*> **6.** Designating soil that is excessively acid and damaging to crops. **7.** Containing an excess of sulfur compounds. —Used of gasoline. *—n.* **1.** The sensation of sour taste. **2.** Something sour. **3.** A mixed drink made esp. with whiskey, lemon or lime juice, sugar, and occas. soda water. *—vt. & vi.* **soured, sour·ing, sours. 1.** To make or become sour. **2.** To make or become disagreeable or disenchanted. **—sour'ly** *adv.* **—sour'ness** *n.*

**sour·ball** (sour'bôl') *n.* A round piece of hard, tart candy.

**source** (sôrs, sōrs) *n.* [ME *sourse* < OFr. < fem. p.part. of *sourdre,* to rise < Lat. *surgere.* —see SURGE.] **1.** A point of origin. **2.** A body of water, as a spring or lake at which a stream or river originates. **3.** One that causes, creates, or initiates : MAKER. **4.** One that supplies information <a reporter's confidential *sources*> **5.** A record, as a book or document supplying primary or firsthand information.

**source book** *n.* **1.** A primary document, as of history, literature, or religion, on which secondary writings are based. **2.** A collection of source books.

**sour cherry** *n.* **1.** A tree, *Prunus cerasus,* native to Eurasia, having white flowers and tart red fruit. **2.** The edible fruit of the sour cherry.

**sour cream** *n.* **1.** Cream that has soured naturally by the action of lactic-acid bacteria, used in baking. **2.** A smooth, thick, artificially soured cream, used as an ingredient in soups, salads, and meat dishes.

**sour·dine** (so͞or-dēn') *n.* [Fr. < Ital. *sordina* < *sordo,* deaf, mute < Lat. *surdus.*] **1.** An obsolete double-reed instrument with a soft tone. **2.** A mute, esp. one for a violin. **3.** A stop on an organ producing a low, soft, muted tone.

**sour·dough** (sour'dō') *n.* **1.** Sour fermented dough used as leaven in making bread. **2.** *Slang.* An old-time settler or prospector, esp. in Alaska and northwestern Canada.

**sour gum** *n.* A tree, *Nyssa sylvatica* of eastern North America, with glossy, somewhat leathery leaves and soft wood.

**sour mash** *n.* **1.** A mixture of new mash and mash from a preceding run used to distill certain malt whiskeys. **2.** Whiskey distilled from sour mash.

**sour·puss** (sour'po͝os') *n.* [SOUR + PUSS[2].] *Slang.* A habitually sullen or gloomy person.

**sour salt** *n.* Crystals of citric acid used in cooking.

**sour·sop** (sour'sŏp') *n.* **1.** A tropical American tree, *Annona muricata,* yielding spiny fruit with tart, edible pulp. **2.** The fruit of the soursop.

**sour·wood** (sour'wo͝od') *n.* [So called from its sour-tasting leaves.] A tree, *Oxydendrum arboreum* of the southeastern United States, with clusters of small white flowers.

**sou·sa·phone** (so͞o'zə-fōn', -sə-) *n.* [After John Philip *Sousa* (1854–1932).] A large brass wind instrument similar to the tuba, having a flaring bell.

**souse[1]** (sous) *v.* **soused, sous·ing, sous·es.** [ME *sousen,* to pickle < *souse,* pickled meat < OFr. *sous,* of Germanic orig.] *—vt.* **1.** To plunge in a liquid. **2.** To make soaking wet. **3.** To steep in a mixture, as in pickling. **4.** *Slang.* To make drunk. *—vi.* To become immersed or soaking wet. *—n.* **1.** The act or process of sousing. **2. a.** Food steeped in pickle, esp. pork trimmings. **b.** The liquid used in pickling : BRINE. **3.** *Slang.* A drunkard.

▲ word history: The original meaning of the verb *souse[1]* was "to pickle" and of the noun, "pickled meat." Various liquids can be used to pickle or preserve food, but a common one is brine, which is very highly salted water. The etymology of *souse[1]* reflects this fact, for the word is ultimately derived from the Germanic stem *sult–,* a variant of *salto–,* the direct ancestor of English *salt.* The Germanic stem, or a word derived from it, was borrowed into Old French as *sous,* and this form entered Middle English and became the modern word *souse.*

**souse[2]** (sous) *v.* **soused, sous·ing, sous·es.** [ME *souce,* swooping motion, perh. alteration of *sourse,* source. —see SOURCE.] *Archaic.* *—vt.* To pounce on : ATTACK. *—vi.* To swoop down, as an attacking hawk does. *—n. Obs.* A swooping motion of attack.

**sou·tache** (so͞o-tăsh') *n.* [Fr. < Hung. *sujtá.*] A narrow flat braid in a herringbone pattern, used for trimming and embroidery.

**sou·tane** (so͞o-tän', -tăn') *n.* [Fr. < Ital. *sottana* < *sotto,* under < Lat. *subtus* < *sub.*] A cassock worn by a Roman Catholic priest.

**south** (south) *n.* [ME < OE *sūð.*] **1. a.** The direction along a meridian to the right of an observer facing in the direction of the earth's rotation : the direction to the right of sunrise. **b.** The cardinal point on the mariner's compass 180° clockwise from north. **c.** An area or region lying in this direction. **2.** *often* **South. a.** The southern part of the earth. **b.** The southern part of a region or country. **3. South.**

The southern part of the United States, esp. the states that fought for the Confederacy in the Civil War. —*adj.* **1.** Moving toward or situated in the south. **2.** Coming from or originating in the south <a hot *south* wind> **3. South.** That officially designates the southern part of a continent, country, or other geographic area <*South* America> —*adv.* In, from, or toward the south.

**south·bound** (south′bound′) *adj.* Going toward the south.

**south by east** *n.* The direction or point on the mariner's compass halfway between due south and south-southeast that is 168°45′ east of due north. —*adv.* & *adj.* Toward or from a south by east direction or point.

**south by west** *n.* The direction or point on the mariner's compass halfway between due south and south-southwest that is 168°45′ west of due north. —*adv.* & *adj.* Toward or from a south by west direction or point.

**South·down** (south′doun′) *n.* [After the *South Downs,* a range of hills in southeastern England.] Any of a breed of small hornless sheep of English origin, with dense, short, fine-textured wool.

**south·east** (south-ēst′, sou-ēst′) *n.* **1.** The direction or point on the mariner's compass halfway between south and east that is 135° east of due north. **2.** An area or region lying in the southeast. —*adj.* **1.** Situated toward, facing, or in the southeast. **2.** Coming from or originating in the southeast <a *southeast* wind> —*adv.* In, from, or toward the southeast. **—south·east′ern** *adj.*

**southeast by east** *n.* The direction or point on the mariner's compass halfway between southeast and east-southeast that is 123° 45′ east of due north. —*adv.* & *adj.* Toward or from a southeast by east direction or point.

**southeast by south** *n.* The direction or point on the mariner's compass halfway between southeast and south-southeast that is 146° 15′ east of due north. —*adv.* & *adj.* Toward or from a southeast by south direction or point.

**south·east·er** (south-ē′stər, sou-ē′-) *n.* **1.** A strong southeast wind. **2.** A storm having southeast winds.

**south·east·er·ly** (south-ē′stər-lē, sou-ē′-) *adj.* **1.** Toward or in the southeast. **2.** From the southeast. **—south·east′er·ly** *adv.*

**south·east·ward** (south-ēst′wərd, sou-ēst′-) *adj.* & *adv.* At or toward the southeast. —*n.* **1.** A direction or point toward the southeast. **2.** A region or part situated in or toward the southeast. **—south·east′ward·ly** *adj.* & *adv.* **—south·east·wards** *adv.*

**south·er** (sou′thər) *n.* A strong wind coming from the south.

**south·er·ly** (sŭth′ər-lē) *adj.* **1.** Situated toward the south. **2.** Coming from the south <*southerly* winds> —*n., pl.* **-lies.** A storm or wind from the south. **—south′er·ly** *adv.*

**south·ern** (sŭth′ərn) *adj.* [ME *southerne* < OE *sūðerne.*] **1.** Situated toward, in, or facing the south. **2.** Coming from the south <*southern* breezes> **3.** Native to or growing in the south. **4.** *often* **Southern.** Of, relating to, or characteristic of southern regions or the South. **5.** Being south of the equator.

**Southern Cross** *n.* Crux.

**Southern Crown** *n.* Corona Australis.

**south·ern·er** (sŭth′ər-nər) *n.* **1.** A native or inhabitant of the south. **2.** *often* **Southerner.** A native or inhabitant of the southern United States.

**Southern Hemisphere** *n.* The half of the earth that is south of the equator.

**South·ern·ism** (sŭth′ər-nĭz′əm) *n.* **1.** An expression or a pronunciation characteristic of the southern United States. **2.** A trait, attitude, or practice characteristic of the South or Southerners, esp. in the United States.

**southern lights** *pl.n.* The aurora australis.

**south·ern·most** (sŭth′ərn-mōst′) *adj.* Farthest south.

**south·ern·wood** (sŭth′ərn-wo͝od′) *n.* An aromatic woody plant, *Artemisia abrotanum* native to Europe, with finely divided grayish foliage.

**south·ing** (sou′thĭng) *n.* **1.** The difference in latitude between two positions as a result of a movement to the south. **2.** Progress toward the south.

**south·land** *or* **South·land** (south′lănd′, -lənd) *n.* A region in the south of an area or a country.

**south·paw** (south′pô′) *n. Slang.* **1.** A left-handed person. **2.** A left-handed baseball pitcher.

**South Pole** *n.* **1.** The southern end of the earth's axis of rotation. **2.** The celestial zenith of the heavens as viewed from the south terrestrial pole. **3. south pole.** The south-seeking magnetic pole of a magnet.

**south·ron** (sŭth′rən) *n.* [ME < *southerne,* southern.] **1.** *often* **Southron.** *Chiefly Scot.* One who lives in the south. —Used esp. by the Scots of the English. **2.** A native or inhabitant of the American South. —Used by the Confederate side in the Civil War. —*adj. Chiefly Scot.* Southern.

**south-south·east** (south′south-ēst′, sou′sou-ēst′) *n.* The direction or point on the mariner's compass halfway between due south and southeast that is 157° 30′ east of due north. —*adj.* Situated toward, facing, or in the south-southeast. —*adv.* In, from, or toward the south-southeast.

**south-south·west** (south′south-wĕst′, sou′sou-wĕst′) *n.* The direction or point on the mariner's compass halfway between due south and southwest that is 157° 30′ west of due north. —*adj.* Situated toward, facing, or in the south-southwest. —*adv.* In, from, or toward the south-southwest.

**south·ward** (south′wərd, sŭth′ərd) *adj.* & *adv.* At or toward the south. —*n.* **1.** A direction toward the south. **2.** A region situated in or toward the south. **—south′ward·ly** *adj.* & *adv.* **—south′wards** *adv.*

**south·west** (south-wĕst′, sou-wĕst′) *n.* **1. a.** The direction or point on the mariner's compass halfway between south and west that is 135° west of due north. **b.** An area or region lying in this direction. **2. Southwest.** A region of the southwestern United States gen. including New Mexico, Arizona, Texas, California, Nevada, Utah, and Colorado. —*adj.* **1.** To, toward, of, facing, or in the southwest. **2.** Coming from or originating in the southwest <a *southwest* wind> —*adv.* In, from, or toward the southwest. **—south·west′ern** *adj.*

**southwest by south** *n.* The direction or point on the mariner's compass halfway between southwest and south-southwest that is 146°15′ west of due north. —*adv.* & *adj.* Toward or from a southwest by south direction or point.

**southwest by west** *n.* The direction or point on the mariner's compass halfway between southwest and west-southwest that is 123° 45′ west of due north. —*adv.* & *adj.* Toward or from a southwest by west direction or point.

**south·west·er** (south-wĕs′tər, sou-wĕs′-) *n.* **1. a.** A storm from the southwest. **b.** A strong wind from the southwest. **2.** A waterproof hat of material such as plastic, oilskin, or canvas with a broad brim behind to protect the neck.

**south·west·er·ly** (south-wĕs′tər-lē, sou-wĕs′-) *adj.* **1.** Toward or in the southwest. **2.** From the southwest. **—south·west′er·ly** *adv.*

**south·west·ward** (south-wĕst′wərd) *adj.* & *adv.* At or toward the southwest. —*n.* **1.** A direction or point toward the southwest. **2.** A region or part situated in or toward the southwest. **—south·west′ward·ly** *adj.* & *adv.* **—south·west′wards** *adv.*

**sou·ve·nir** (so͞o′və-nîr′, so͞o′və-nîr′) *n.* [Fr., memory < *souvenir,* to recall < Lat. *subvenire,* to come to mind : *sub-,* under + *venire,* to come.] A token of remembrance : MEMENTO.

**sov·er·eign** (sŏv′ər-ĭn, sŏv′rĭn) *n.* [ME *souverein* < OFr. < VLat. **superanus* < Lat. *super,* above.] **1.** The chief of state in a monarchy. **2.** A gold coin formerly used in Great Britain. —*adj.* **1.** Paramount : supreme. **2.** Having supreme rank or power. **3.** Self-governing : independent <a *sovereign* state> **4. a.** Superlative in strength or efficacy <a *sovereign* remedy> **b.** Unmitigated : utter <*sovereign* contempt> **—sov′er·eign·ly** *adv.*

**sov·er·eign·ty** (sŏv′ər-ĭn-tē, sŏv′rĭn-) *n., pl.* **-ties. 1.** Supremacy of authority or rule as exercised by a sovereign or a sovereign state. **2.** Royal rank, power, or authority. **3.** Total independence and self-government. **4.** A territory existing as an independent state.

**so·vi·et** (sō′vē-ĕt′, -ĭt, sŏv′ē-) *n.* [R. *sovet,* council.] **1.** One of the popularly elected local, regional, and national legislative assemblies in the Soviet Union. **2. Soviets.** The government and people of the Soviet Union. —*adj.* **1.** *often* **Soviet.** Of or relating to the Union of Soviet Socialist Republics. **2.** Of or relating to a soviet.

**so·vi·et·ize** *also* **So·vi·et·ize** (sō′vē-ĭ-tīz′, sŏv′ē-) *vt.* **-ized, -iz·ing, -iz·es. 1.** To cause to come under Soviet control. **2.** To cause to conform to Soviet political, social, and cultural policy. **—so′vi·et·i·za′tion** *n.*

**sov·khoz** (sŏf-kôz′, -KHôz′) *n.* [R., short for *sovetskoe khozyaĭstvo,* soviet farm.] A state-owned farm in the U.S.S.R. whose workers are paid wages.

**sow¹** (sō) *v.* **sowed, sown** (sōn) *or* **sowed, sow·ing, sows.** [ME *sowen* < OE *sāwan.*] —*vt.* **1.** To scatter (seed) over the ground for growing. **2.** To impregnate (a growing medium) with seed. **3.** To propagate : disseminate <*sow* discontent among the workers> **4.** To strew or cover with something. —*vi.* To scatter seed for growing. **—sow′er** *n.*

**sow²** (sou) *n.* [ME < OE *sugu.*] **1.** An adult female hog. **2. a.** A channel that conducts molten iron to the molds in a pig bed. **b.** The mass of metal solidified in such a channel or mold.

**sow·bel·ly** (sou′bĕl′ē) *n. Informal.* Salt pork.

**sow·bread** (sou′brĕd′) *n.* The cyclamen.

**sow bug** *n.* [From its piglike shape.] Any of various small terrestrial crustaceans, chiefly of the genera *Oniscus* and *Porcellio,* usu. found under logs or stones and having oval, segmented bodies.

**sown** (sōn) *v. var. p.p. of* SOW¹.

**sow thistle** *n.* A plant of the genus *Sonchus,* esp. *S. oleraceus,* native to Europe, with prickly leaves and yellow flowers.

**sox** (sŏks) *n. var. pl. of* SOCK¹ 1.

**soy** (soi) *n.* [J. *shō-yu* < Chin. (Mandarin) *jiang⁴ you²* : *jiang⁴,* soy paste + *you²,* sauce.] **1.** The soybean. **2.** A salty brown liquid condiment made by fermenting soybeans in brine.

**soy·a** (soi′ə) *n.* [Du. *soja* < J. *shō-yu,* soy.] SOYBEAN 2.

**soy·bean** (soi′bēn′) *n.* **1.** A leguminous Asiatic plant, *Glycine max,* widely cultivated for forage and soil improvement and for its nutritious, edible seeds. **2.** The seed of the soybean.

**soy·milk** (soi′mĭlk′) *n.* A milk substitute made from soybeans and often supplemented with vitamins.

**†spa** (spä) *n.* [After *Spa*, a resort town in Belgium.] **1.** A mineral spring. **2.** A resort area with mineral springs. **3.** A chic hotel or resort. **4.** *Regional.* SODA FOUNTAIN 2.

**space** (spās) *n.* [ME, distance < OFr. *espace* < Lat. *spatium.*] **1. a.** A set of elements or points satisfying specified geometric postulates <non-Euclidean *space*> **b.** The infinite extension of the three-dimensional field of everyday life. **2.** The expanse in which the solar system, stars, and galaxies exist : UNIVERSE. **3.** A blank area <the *spaces* between individual words> **4.** An area provided for a specific purpose <a parking *space*> **5.** Reserved or available accommodation on a public transportation vehicle. **6. a.** A period of time. **b.** A little while. **7.** *Informal.* Sufficient freedom from external pressure to develop or explore one's needs, interests, and individuality. **8.** *Mus.* One of the intervals between the lines of a staff. **9.** One of the blank pieces of type or other means used for separating words or characters in printing. **10.** One of the intervals during the telegraphic transmission of a message when the key is open or not in contact. **11.** Broadcast time or areas in printed material available esp. for use by advertisers. —*vt.* **spaced, spac·ing, spac·es. 1.** To organize or arrange with spaces between. **2.** To separate or keep apart. **—spac′er** *n.*

**space biology** *n.* Exobiology.

**space·borne** (spās′bôrn′) *adj.* Operating in or involving equipment operating in outer space <*spaceborne* satellites>

**space capsule** *n. Aerospace.* CAPSULE 5.

**space charge** *n.* An electric charge in a vacuum or region of low gas pressure, as in a vacuum tube, that is carried by a stream of electrons or ions.

**space·craft** (spās′krăft′) *n., pl.* **spacecraft.** A vehicle designed to be launched into space.

**spaced** (spāst) *adj. Slang.* Spaced-out.

**spaced-out** (spāst′out′) *adj. Slang.* Stupefied from or as if from a drug : DOPEY.

**space flight** *n.* Flight beyond the earth's atmosphere.

**space lattice** *n.* Any of the 14 possible geometric arrangements of points at which the atoms of a crystal may occur.

**space·less** (spās′lĭs) *adj.* Having no limits or boundaries.

**space·man** (spās′măn′, -mən) *n.* **1.** One who travels in outer space. **2.** A visitor from outer space.

**space medicine** *n.* The science of the biological, physiological, and psychological effects of space flight upon human beings.

**space·port** (spās′pôrt′, -pōrt′) *n.* An installation for testing and launching spacecraft.

**space probe** *n.* A spacecraft carrying instruments designed to explore the physical properties of outer space or of celestial bodies other than Earth.

**space science** *n.* **1.** Any of several scientific disciplines, such as exobiology, that study phenomena occurring in the upper atmosphere, in space, or on celestial bodies other than Earth. **2.** A discipline related to or dealing with the problems of space flight.

**space·ship** or **space ship** (spās′shĭp′) *n.* A spacecraft.

**space shuttle** *n.* A space vehicle designed to transport astronauts between Earth and an orbiting space station.

**space sickness** *n.* Any of various ailments resulting from manned space flight.

**space station** *n.* A large manned satellite designed for permanent orbit around Earth and used for scientific research and military reconnaissance or as an assembly point for long-range spacecraft.

**space suit** *n.* A protective pressure suit having an independent air supply and other devices designed to permit the wearer relatively free movement in space.

**space-time** (spās′tīm′) *n.* The four-dimensional continuum of one temporal and three spatial coordinates in which any event or physical object is located.

**space vehicle** *n.* A spacecraft.

**space walk** *n.* An excursion by an astronaut outside a spacecraft in space : EXTRAVEHICULAR ACTIVITY. **—space walk** *v.* **(walked, walk·ing, walks) —space walker** *n.*

**space·ward** (spās′wərd) *adv.* Toward outer space.

**space writer** *n.* A writer, as a journalist, paid according to the amount of space his or her material occupies in print.

**spac·ey** (spā′sē) *adj. var. of* SPACY.

**spa·cial** (spā′shəl) *adj. var. of* SPATIAL.

**spac·ing** (spā′sĭng) *n.* **1. a.** The act or result of arranging by spaces. **b.** A system of or allowance for intervals. **2.** A space or spaces, as in printed matter.

**spa·cious** (spā′shəs) *adj.* **1.** Offering or containing much space or room : EXTENSIVE <a *spacious* dining room> **2.** Vast in scope or range <a *spacious* view> **—spa′cious·ly** *adv.* **—spa′cious·ness** *n.*

**Spack·le** (spăk′əl). A trademark for a powder to be mixed with water or a ready-to-use plastic paste designed to fill cracks and holes in plaster before painting or papering.

**spac·y** or **spac·ey** (spā′sē) *adj.* **-i·er, -i·est.** *Slang.* **1.** Spaced-out. **2.** Eccentric : offbeat.

**spade[1]** (spād) *n.* [ME < OE *spadu.*] **1.** A sturdy digging tool having a thick handle and a heavy, flat iron blade that can be pressed into the ground with the foot. **2.** A digging or cutting tool like a spade. —*vt.* **spad·ed, spad·ing, spades.** To dig or cut with a spade. **—call a spade a spade.** To speak frankly and truly. **—spad′er** *n.*

**spade[2]** (spād) *n.* [Ital. *spada*, broadsword < Lat. *spatha* < Gk. *spathē*, broad blade.] **1.** A black figure shaped like an inverted heart with a short stalk at the bottom, on a playing card. **2. spades.** The suit of cards identified by a spade. **3.** A card bearing a spade.

▲ word history: The word *spade*[1], "a digging tool," has no etymological relation to *spade*[2], "a suit of playing cards." *Spade*[1] comes directly from Old English *spadu*, and *spade*[2] is from Italian *spada*, "sword." On Italian playing cards a broad-bladed sword was used as the symbol of a suit; this suit was called "spades" in English. Because this word *spade* was confused with the native English word, on English playing cards the symbol was reshaped to resemble a digging tool.

**spade·fish** (spād′fĭsh′) *n., pl.* **spadefish** or **-fish·es.** [From its shape.] A marine food fish of the family Ephippidae, esp. *Chaetodipterus faber* of the Atlantic, or *C. zonatus* of the Pacific.

**spade·work** (spād′wûrk′) *n.* **1.** Work requiring a spade. **2.** Preparation required for a project or activity.

**spa·dix** (spā′dĭks) *n., pl.* **spa·di·ces** (spā′dĭ-sēz′) [Lat., broken-off palm branch < Gk.] A clublike spike bearing minute flowers, usu. enclosed within a sheathlike spathe, as in the calla and the jack-in-the-pulpit.

**spadix**
*The A. spadix and B. spathe of a jack-in-the-pulpit*

**spa·ghet·ti** (spə-gĕt′ē) *n.* [Ital., pl. dim. of *spago*, string.] **1.** A pasta made into long, solid strings and cooked by boiling. **2.** *Elect.* A slender tube of insulating material into which bare wire is inserted, esp. in radio circuits.

**spa·ghet·ti·ni** (spăg′ĭ-tē′nē) *n.* [Ital., dim. of *spaghetti*, spaghetti.] Pasta thinner than spaghetti but not as thin as vermicelli.

**spaghetti Western** *n.* A low-budget Western film made by the Italian film industry.

**spa·gyr·ic** (spə-jĭr′ĭk) *also* **spa·gyr·i·cal** (-ĭ-kəl) *adj.* [NLat. *spagiricus.*] Relating to or resembling alchemy : ALCHEMICAL.

**spake** (spāk) *v. Archaic. var. p.t. of* SPEAK.

**spall** (spôl) *n.* [ME *spalle.*] A flake, chip, or fragment from a piece of stone or ore. —*v.* **spalled, spall·ing, spalls.** —*vt.* To break up into flakes, chips, or fragments. —*vi.* To chip or crumble.

**spal·la·tion** (spô-lā′shən) *n.* A nuclear reaction in which many particles are ejected from an atomic nucleus by an incident particle of sufficiently high energy.

**Spam** (spăm). A trademark for spiced pork products, luncheon meat, and deviled luncheon-meat spread.

**span[1]** (spăn) *n.* [ME, unit of measurement < OE *spann.*] **1.** The extent of space between two points or extremities, as of a bridge or roof : BREADTH. **2.** The distance between the tips of an aircraft's wings. **3.** The section between two intermediate supports of a bridge. **4.** Something, as a railroad trestle or bridge, that spans. **5.** The distance from the tip of the thumb to the tip of the little finger when the hand is fully extended, formerly used as a unit of measure equal to about nine inches. **6.** A time period <a life *span* of 90 years> —*vt.* **spanned, span·ning, spans. 1.** To measure by or as if by the fully extended hand. **2.** To encircle with the hand or hands in or as if in measuring. **3.** To extend across <a film career that *spanned* 50 years> **4.** To form a span over.

**span[2]** (spăn) *vt.* **spanned, span·ning, spans.** [MDu. *spannen.*] To bind or fetter. —*n.* **1.** *Naut.* A stretch of rope made fast at either end. **2.** A pair of animals, as oxen, matched in size, strength, or color.

**span[3]** (spăn) *v. Archaic. var. p.t. & p.p. of* SPIN.

**span·drel** *also* **span·dril** (spăn′drəl) *n.* [ME *spaundrell* < AN *spaundre* < OFr. *espandre*, to spread out < Lat. *expandere.* —see EXPAND.] **1.** The triangular space between the left or right exterior curve of an arch and the rectangular framework surrounding it. **2.** The space between two arches and a horizontal molding or cornice above them.

**spang** (spăng) *adv.* [Orig. unknown.] Squarely : precisely.

**span·gle** (spăng′gəl) *n.* [ME *spangele*, dim. of *spange*, perh. < MDu., buckle.] **1.** A small, often circular, decorative piece of sparkling metal or plastic sewn esp. on garments. **2.** A small sparkling object, drop, or spot <*spangles* of sunshine> —*v.* **-gled, -gling, -gles.** —*vt.* To adorn or cause to sparkle by covering with or as if with spangles. —*vi.* To sparkle. **—span'gly** *adj.*

**Span·iard** (spăn'yərd) *n.* A native or inhabitant of Spain.

**span·iel** (spăn'yəl) *n.* [ME *spanyel* < OFr. *espaignol,* Spaniard, Spanish dog < Lat. *Hispaniolus,* Spanish < *Hispania,* Spain.] **1.** Any of several breeds of small to medium-sized dogs, usu. having drooping ears, short legs, and a silky, wavy coat. **2.** A servile or docile person.

**Span·ish** (spăn'ĭsh) *adj.* Of or relating to Spain, its inhabitants, or their language or culture. —*n.* **1.** The Romance language of Spain and most of Central and South America. **2.** The inhabitants of Spain.

**Span·ish-A·mer·i·can** (spăn'ĭsh-ə-mĕr'ĭ-kən) *adj.* **1.** Of or relating to the countries or people of Spanish America. **2.** Of or relating to people of Spanish descent residing in the United States. —*n.* **1.** A native or inhabitant of a Spanish-American country. **2.** A person of Spanish descent who lives in the United States.

**Spanish bayonet** *n.* **1.** A New World plant of the genus *Yucca,* esp. *Y. aloifolia,* with a tall woody stem, stiff pointed leaves, and a large cluster of white flowers. **2.** A plant, *Y. filamentosa,* similar to the Spanish bayonet.

**Spanish cedar** *n.* **1.** A tropical American tree of the genus *Cedrela,* esp. *C. odorata,* with reddish, aromatic wood used for cabinetwork and cigar boxes. **2.** The wood of the Spanish cedar.

**Spanish chestnut** *n.* **1.** A tree, *Castanea sativa* of the Mediterranean area, yielding edible nuts. **2.** The nut of the Spanish chestnut.

**Spanish fly** *n.* **1.** A European blister beetle, *Lytta vesicatoria.* **2.** Cantharides.

**Spanish mackerel** *n.* A marine food fish of the genus *Scomberomorus,* esp. a commercially important species, *S. maculatus* of American Atlantic coastal waters.

**Spanish moss** *n.* An epiphytic plant, *Tillandsia usneoides,* growing on trees of the southeastern United States and tropical America, with gray threadlike stems drooping in long, densely matted clusters.

**Spanish needles** *pl.n. (sing. or pl. in number).* A North American plant, *Bidens bipinnata,* with yellowish flowers and slender barbed fruit.

**Spanish omelet** *n.* An omelet served with an often spicy sauce of tomatoes, onions, and peppers.

**Spanish onion** *n.* A mild-flavored, yellow-skinned onion, prob. derived from *Allium fistulosum.*

**Spanish paprika** *n.* A mild seasoning made from pimientos.

**Spanish rice** *n.* A dish of rice cooked with tomatoes, spices, chopped onions, olives, and green peppers.

**spank** (spăngk) *v.* **spanked, spank·ing, spanks.** [Perh. imit.] —*vt.* To slap on the buttocks with a flat object or with the open hand as punishment. —*vi.* To move briskly or spiritedly. **—spank** *n.*

**spank·er** (spăng'kər) *n.* **1.** One that spanks. **2.** *Naut.* A quadrilateral gaff sail set abaft the after mast of a square-rigged sailing ship.

**spank·ing** (spăng'kĭng) *adj.* [Orig. unknown.] **1.** *Informal.* Exceptional : remarkable. **2.** Brisk and fresh <a *spanking* breeze> —*adv.* Very <a car that was *spanking* clean>

**span·ner** (spăn'ər) *n.* [G., winding tool < *spannen,* to stretch.] **1.** One that spans. **2.** *Chiefly Brit.* A wrench. **3.** A measuring worm.

**†span-new** (spăn'no͞o', -nyo͞o') *adj.* [ME *spannewe,* partial transl. of ON *spānnȳr* : *spānn,* chip + *nȳr,* new.] *Regional.* Entirely new.

**span·worm** (spăn'wûrm') *n.* [< SPAN[2].] A measuring worm.

**spar**[1] (spär) *n.* [ME *sparre,* rafter < ON *sperra,* beam.] **1.** *Naut.* A wooden or metal pole, as a mast, boom, yard, or bowsprit, used to support rigging. **2.** A usu. metal pole used as part of a crane or derrick. **3.** A principal structural member in an airplane wing that runs from tip to tip or from root to tip. —*vt.* **sparred, spar·ring, spars. 1.** To furnish with spars. **2.** *Archaic.* To fasten with a bolt.

**spar**[2] (spär) *vi.* **sparred, spar·ring, spars.** [ME *sparren,* to thrust or strike rapidly.] **1.** To box, esp. for practice. **2.** To bandy words about in argument : DISPUTE. **3.** To fight by striking with the feet and spurs. —Used of cocks. —*n.* **1.** An act of sparring. **2.** A boxing match.

**spar**[3] (spär) *n.* [MLG.] A nonmetallic, readily cleavable mineral with a vitreous luster, as feldspar.

**Spar** *also* **SPAR** (spär) *n.* [Contraction of Lat. *semper paratus,* always prepared, the motto of the U.S. Coast Guard.] A member of the women's reserve of the U.S. Coast Guard.

**spare** (spâr) *v.* **spared, spar·ing, spares.** [ME *sparen,* to leave unharmed < OE *sparian.*] —*vt.* **1. a.** To treat leniently or mercifully. **b.** To refrain from harming or destroying. **2.** To save or relieve from undergoing or doing something <I *spared* myself the trouble of the long drive.> **3.** To use with restraint <Don't *spare* the sugar.> **4.** To give or grant out of one's resources : AFFORD <Can you *spare* 30 minutes of your time?> —*vi.* **1.** To be frugal. **2. a.** To be lenient or merciful. **b.** To refrain or forbear. —*adj.* **spar·er, spar·est. 1. a.** Kept in reserve <a *spare* wheel> **b.** In excess of what is needed : EXTRA <*spare* cash> **c.** Unoccupied <*spare* time> **2. a.** Economical : meager <a *spare* budget> **b.** Thin : lean <a tall, *spare* farmer> —*n.* **1.** A replacement, as a tire, reserved for future need. **2. a.** The act of knocking down all ten pins with two successive rolls of a bowling ball. **b.** The score made by doing this. **—to spare.** In addition to what is needed <I paid my bills and had money *to spare.*> **—spare'ly** *adv.* **—spare'ness** *n.* **—spar'er** *n.*

**spare·ribs** (spâr'rĭbz') *pl.n.* [By folk ety. < MLG *ribbespēr* : *ribbe,* rib + *spēr,* spit.] A cut of pork consisting of the ribs with most of the meat trimmed off.

**sparge** (spärj) *vt.* **sparged, sparg·ing, sparg·es.** [OFr. *espargier* < Lat. *spargere.*] **1.** To spray or sprinkle. **2.** To introduce air or gas into (a liquid) and thereby agitate it. **—sparge** *n.*

**spar·id** (spăr'ĭd, spâr'-) *adj.* [NLat. *Sparidae,* family name < *Sparus,* type genus < Lat., a kind of fish < Gk. *sparos.*] Of or belonging to the family Sparidae, which includes the porgies and similar fishes. —*n.* A member of the Sparidae.

**spar·ing** (spâr'ĭng) *adj.* **1.** Frugal : thrifty. **2.** Lenient : forbearing. **—spar'ing·ly** *adv.* **—spar'ing·ness** *n.*

**spark**[1] (spärk) *n.* [ME *sparke* < OE *spærca.*] **1.** An incandescent particle, esp.: **a.** One thrown off from a burning substance. **b.** One resulting from friction. **c.** One remaining in an otherwise extinguished fire : EMBER. **2.** A shining particle, as of metal. **3. a.** A flash of light, esp. a flash caused by electric discharge. **b.** A short pulse of electric current. **4.** A trace or suggestion, as: **a.** A quality or feeling with latent potential <a *spark* of musical genius> **b.** A vital, animating, or activating factor <the *spark* of revolution> **5. sparks** *(sing. in number). Informal.* A ship's radio operator. **6.** *Elect.* **a.** The luminous phenomenon resulting from a disruptive discharge through an insulating material. **b.** The discharge itself. —*v.* **sparked, spark·ing, sparks.** —*vi.* **1.** To emit a spark or sparks. **2.** To operate correctly. —Used of the ignition system of an internal-combustion engine. —*vt.* **1.** To put in motion : ACTIVATE. **2.** To rouse to action. **—spark'er** *n.*

**spark**[2] (spärk) *n.* [Perh. of Scand. orig.] **1.** A young dandy or gallant. **2.** A lover or a suitor. —*v.* **sparked, spark·ing, sparks.** —*vt.* To court or woo. —*vi.* To play the suitor. **—spark'er** *n.*

**spark arrester** *n.* **1.** A device to keep sparks from escaping, as at a chimney opening. **2.** A device to control electric sparking at a point where a circuit is made or broken.

**spark chamber** *n.* A device consisting of electrically charged parallel metal plates in a chamber filled with inert gas, used to detect and measure charged subatomic particles as they pass from one plate to another, leaving a trail of sparks.

**spark coil** *n.* An induction coil used to produce a spark, as in an internal-combustion engine.

**spark gap** *n.* A gap in an otherwise complete electric circuit across which a discharge occurs at some prescribed voltage.

**spark generator** *n.* SPARK TRANSMITTER 1.

**sparking plug** *n. Chiefly Brit. var. of* SPARK PLUG 1.

**spar·kle** (spär'kəl) *vi.* **-kled, -kling, -kles.** [ME *sparklen,* freq. of *sparken,* to spark.] **1.** To emit sparks. **2.** To emit or reflect flashes of light : GLITTER <gems *sparkling* in the candlelight> **3.** To shine with animation. **4.** To flash with wit. **5.** To release gas bubbles : EFFERVESCE. —*n.* **1.** A small spark or gleaming particle. **2.** A glittering quality. **3.** Vivacity : animation. **4.** Effervescence.

**spar·kler** (spär'klər) *n.* **1.** One that sparkles. **2.** A firework that burns slowly and gives off a shower of sparks. **3.** *Informal.* A diamond.

**sparkling wine** *n.* An effervescent wine, as champagne, produced by fermentation in the bottle.

**spark plug** *n.* **1.** A device inserted in the head of an internal-combustion-engine cylinder that ignites the fuel mixture by way of an electric spark. **2.** *Informal.* One who infuses life or energy into an undertaking.

**spark·plug** (spärk'plŭg') *vt.* **-plugged, -plug·ging, -plugs.** *Informal.* To inspire or energize (e.g., an undertaking).

**spark transmitter** *n.* **1.** A source of alternating current that derives its output from the discharge of a condenser across a spark gap. **2.** A now obsolete radio transmitter using a discharge across a spark gap to create a signal.

**spark·y** (spär'kē) *adj.* Lively : animated.

**spar·ling** (spär'lĭng) *n.* [ME *sperlinge* < OFr. *esperlinge,* of Germanic orig.] **1.** The European smelt. **2.** A young herring.

**sparring partner** *n.* One who serves as a boxer's opponent during practice.

**spar·row** (spăr'ō) *n.* [ME *sparowe* < OE *spearwe.*] **1.** Any of various small New World birds of the genera *Spizella, Zonotrichia, Melospiza,* and other closely related genera within the family Fringillidae, with grayish or brownish feathers. **2.** A bird, as the common house sparrow, that is similar or related to the sparrow.

**†spar·row·grass** (spăr'ə-grăs', -grəs) *n.* [By folk ety. < ASPARAGUS.] *Regional.* Asparagus.

**sparrow hawk** *n.* **1.** A small North American falcon, *Falco sparverius,* that preys on small birds and animals. **2.** A European hawk, *Accipiter nisus,* similar to the sparrow hawk.

**sparse** (spärs) *adj.* **spars·er, spars·est.** [Lat. *sparsus,* p.part. of *spargere,* to scatter.] Occurring and esp. growing or settled at widely spaced intervals <a *sparse* population> **—sparse'ly** *adv.* **—sparse'ness, spar'si·ty** (spär'sĭ-tē) *n.*

**Spar·tan** (spär'tn) *adj.* **1.** Of or relating to the ancient Greek city-state Sparta or its people. **2.** Marked by self-discipline : AUSTERE <*Spartan* eating habits> —*n.* **1.** A citizen of Sparta. **2.** One of Spartan character. **—Spar'tan·ism** *n.*

**spar varnish** *n.* A waterproof varnish.

**spasm** (spăz'əm) *n.* [ME *spasme* < OFr. < Lat. *spasmus* < Gk. *spasmos* < *span,* to pull.] **1.** A sudden, involuntary muscular contraction. **2.** A sudden burst of energy, activity, or emotion.

**spas·mod·ic** (spăz-mŏd'ĭk) *adj.* [NLat. *spasmodicus* < Gk. *spasmodikos* < *spasmos,* spasm.] **1.** Relating to, affected by, or of the nature of a spasm : CONVULSIVE. **2.** Occurring intermittently : FITFUL <*spasmodic* shelling> **3.** Given to sudden outbursts of energy or feeling. **—spas·mod'i·cal·ly** *adv.*

**spas·tic** (spăs'tĭk) *adj.* [Lat. *spasticus* < Gk. *spastikos* < *span,* to pull.] Of, relating to, or marked by spasms. —*n.* **1.** One suffering from muscular spasms. **2.** One afflicted with spastic paralysis. **—spas'ti·cal·ly** *adv.* **—spas·tic'i·ty** (spă-stĭs'ĭ-tē) *n.*

**spastic paralysis** *n.* A chronic pathological condition involving exaggerated tendon reflexes and muscular spasms accompanying sclerosis of the spinal cord.

**spat¹** (spăt) *v. var. p.t. & p.p. of* SPIT¹.

**spat²** (spăt) *n., pl.* **spat** *or* **spats.** [Orig. unknown.] An oyster or similar bivalve mollusk in the larval stage, esp. when it settles to the bottom and begins to develop a shell. —*vi.* **spat·ted, spat·ting, spats.** To spawn. —Used of oysters and similar mollusks.

**spat³** (spăt) *n.* [Short for E. *spatterdash* : SPATTER + DASH.] *often* **spats.** A gaiter covering the shoe upper and the ankle and fastening under the shoe with a strap.

**spat⁴** (spăt) *n.* [Orig. unknown.] **1.** A brief petty quarrel. **2.** *Informal.* A slap or smack. **3.** A spattering sound. —*v.* **spat·ted, spat·ting, spats.** —*vi.* **1.** To engage in a brief petty quarrel. **2.** To strike with a light spattering sound : SLAP. —*vt. Informal.* To slap.

**spate** (spāt) *n.* [ME.] **1.** A sudden flood, rush, or outpouring <a *spate* of criticism> **2.** *Chiefly Brit.* **a.** A flash flood. **b.** A freshet caused by a downpour of rain or melting of snow. **c.** A sudden heavy rainfall.

**spathe** (spāth) *n.* [Lat. *spatha,* broad-sword < Gk. *spathē,* broad blade.] *Bot.* A leaflike organ that encloses or spreads from the base of the spadix of certain plants, as the jack-in-the-pulpit or the calla.

**spath·ic** (spăth'ĭk) *adj.* [< G. *spath, spat,* spar.] Having good cleavage. —Used of minerals.

**spa·tial** *also* **spa·cial** (spā'shəl) *adj.* [< Lat. *spatium,* space.] Of, relating to, involving, or having the nature of space. **—spa'ti·al'i·ty** (spā'shē-ăl'ĭ-tē) *n.* **—spa'tial·ly** *adv.*

**spa·ti·o·tem·po·ral** (spā'shē-ō-tĕm'pər-əl) *adj.* [Lat. *spatium,* space + TEMPORAL.] **1.** Of, relating to, or existing in both space and time. **2.** Of or relating to space-time. **—spa'ti·o·tem'po·ral·ly** *adv.*

**spat·ter** (spăt'ər) *v.* **-tered, -ter·ing, -ters.** [Perh. of LG orig.] —*vt.* **1.** To scatter (a liquid) in drops or small splashes. **2.** To spot, splash, or soil <shoes *spattered* with mud> **3.** To sully the reputation of : DEFAME. —*vi.* **1.** To spit off drops or small splashes : SPLATTER. **2.** To fall with a splash or a splashing sound. —*n.* **1.** The act of spattering. **2.** A spattering sound. **3.** A drop or splash of something spattered <*spatters* of paint>

**spat·ter·dock** (spăt'ər-dŏk') *n.* [SPATTER + DOCK⁴.] An aquatic plant, *Nuphar advena* of eastern North America, with broad leaves and globe-shaped yellow flowers.

**spat·u·la** (spăch'ə-lə) *n.* [Lat., dim. of *spatha,* broadsword < Gk. *spathē,* broad blade.] **1.** A small implement having a broad, flat, flexible blade that is used esp. to mix, spread, or lift soft material. **2.** An implement, as a small wooden paddle, for pressing down the tongue. **—spat'u·lar** *adj.*

**spat·u·late** (spăch'ə-lĭt) *adj.* Shaped like a spatula.

**spav·in** (spăv'ĭn) *n.* [ME *spaveyne* < OFr. *espavin.*] Either of two diseases affecting the hock joint of horses: **a.** Bog spavin, an infusion of lymph that enlarges the joint. **b.** Bone spavin, a bony deposit that stiffens the joint. **—spav'ined** *adj.*

**spawn** (spôn) *n.* [ME *spawne* < *spawnen,* to spawn < AN *espaundre* < OFr. *espandre,* to spread < Lat. *expandere.* —see EXPAND.] **1.** The eggs of aquatic animals such as bivalve mollusks, fishes, and amphibians. **2.** Offspring occurring in numbers : BROOD. **3.** One who is regarded as the issue of a parent or family <the *spawn* of a madman> **4.** A product : outcome. **5.** Fragments of mycelia used to start a mushroom culture. —*v.* **spawned, spawn·ing, spawns.** —*vi.* **1.** To deposit eggs : produce spawn. **2.** To produce offspring in numbers like spawn. —*vt.* **1.** To produce (spawn). **2.** To give birth to. **3.** To give rise to : ENGENDER <oppression that *spawned* revolution> **4.** To bring forth : PRODUCE. **5.** To plant with mycelia.

**spay** (spā) *vt.* **spayed, spay·ing, spays.** [ME *spayen* < OFr. *espeer,* to cut with a sword < *espee,* sword < Lat. *spatha.* —see SPATHE.] To excise the ovaries of (a female animal).

**speak** (spēk) *v.* **spoke** (spōk), **spo·ken** (spō'kən), **speak·ing, speaks.** [ME *speken* < OE *specan.*] —*vi.* **1.** To utter words with ordinary speech modulation : TALK. **2. a.** To express oneself. **b.** To be on good terms <They are no longer *speaking.*> **3.** To deliver a lecture <*spoke* at graduation> **4.** To convey a message <Actions *speak* louder than words.> **5.** To be expressive. **6. a.** To produce a characteristic sound <The drums *spoke.*> **b.** To give off a sound on firing <The artillery *spoke.*> **7.** To make communicative sounds. —*vt.* **1.** To articulate in a speaking voice. **2.** To converse in or be able to converse in (a language) <*speak* Russian> **3. a.** To express aloud : TELL <*spoke* the truth> **b.** To express in writing. **4.** To show to be : REVEAL. **5.** *Naut.* To hail and communicate with (another vessel) at sea. **—so to speak.** In a manner of speaking. **—speak down to.** To speak condescendingly to <never *spoke down to* the students> **—speak (one's) mind.** To express one's opinion. **—speak out.** To talk freely and fearlessly. **—speak up.** To speak fearlessly and unhesitatingly. **—to speak of.** Worthy of mention. **—speak'a·ble** *adj.*

**speak·eas·y** (spēk'ē'zē) *n., pl.* **-ies.** A place for the illegal sale of alcoholic drinks, as during U.S. Prohibition.

**speak·er** (spē'kər) *n.* **1. a.** One who speaks. **b.** A spokesperson. **2.** One who delivers a public speech. **3.** *often* **Speaker.** The presiding officer of a legislative assembly. **4.** A loudspeaker.

**speak·ing** (spē'kĭng) *adj.* **1.** Expressive or telling : ELOQUENT. **2.** True to life : STRIKING <a *speaking* likeness>

**speaking tube** *n.* A tube for speaking from one room or building to another.

**spear** (spîr) *n.* [ME *spere* < OE.] **1.** A weapon with a long sharply pointed shaft. **2.** A shaft with a sharp point and barbs for spearing fish. **3.** A spearman. **4.** A slender stalk, as of asparagus. —*v.* **speared, spear·ing, spears.** —*vt.* **1.** To pierce with or as if with a spear. **2.** To catch with a thrust of the arm <*speared* the football> —*vi.* **1.** To stab with or as if with a spear. **2.** To sprout like a spear. **—spear'er** *n.*

**spear·fish** (spîr'fĭsh') *n., pl.* **spearfish** *or* **-fish·es.** A large marine game fish, *Tetrapturus angustirostris* or *T. belone,* whose upper jaw is elongated into a spearlike projection.

**spear·head** (spîr'hĕd') *n.* **1.** The sharpened head of a spear. **2. a.** The leading forces in a military drive. **b.** The driving force in an action or endeavor. —*vt.* **-head·ed, -head·ing, -heads.** To be the leader of (e.g., an endeavor).

**spear·man** (spîr'mən) *n.* A soldier armed with a spear.

**spear·mint** (spîr'mĭnt') *n.* An aromatic plant, *Mentha spicata* native to Europe, with clusters of small purplish flowers and yielding an oil widely used as flavoring.

**spear·wort** (spîr'wûrt', -wôrt') *n.* A plant related to the buttercups, esp. *Ranunculus flammula* native to Eurasia, with lance-shaped leaves and yellow flowers.

**spe·cial** (spĕsh'əl) *adj.* [ME < OFr. *especial* < Lat. *specialis* < *species,* kind. —see SPECIES.] **1.** Surpassing the usual : EXCEPTIONAL <a *special* event> **2. a.** Distinct among others of a kind : SINGULAR. **b.** Primary <one's *special* concern> **3.** Peculiar to a specific person or thing : PARTICULAR <my own *special* chair> **4. a.** Having a limited or specific function, application, or scope <played a *special* role in the negotiations> **b.** Arranged for a particular occasion or purpose <a *special* session of the legislature> **5.** Esteemed : close <*special* friends> **6.** Additional : extra <a *special* weekend bus> —*n.* **1.** Something arranged, issued, or appropriated to a particular service or occasion <caught the commuter *special*> **2.** A featured attraction, as a reduced price <a *special* on steak> **3.** A single television production featuring a specific work, a given topic, or a particular performer. **—spe'cial·ly** *adv.*

☆ ***syns:*** SPECIAL, ESPECIAL, INDIVIDUAL, PARTICULAR, SPECIFIC *adj. core meaning* : of, relating to, or intended for a distinctive thing or group <a *special* exterior paint><*special* credentials>

**special act** *n.* A legislative act applicable only to a particular person or area.

**special court-martial** *n., pl.* **special courts-martial.** A court-martial consisting of at least three officers for trying intermediate offenses.

**special delivery** *n.* Delivery of mail, for an additional charge, by a special messenger rather than by scheduled delivery.

**special effects** *pl.n.* Visual or sound effects added to a film or taped television show during processing.

**Special Forces** *pl.n.* A division of the U.S. Army composed of soldiers specially trained in guerrilla fighting.

**special handling** *n.* The handling of fourth-class or parcel-post mail as first-class mail for an extra charge.

**spe·cial·ism** (spĕsh'ə-lĭz'əm) *n.* **1.** Confinement or limitation to a given field of study or occupation. **2.** A field of specialization.

**spe·cial·ist** (spĕsh'ə-lĭst) *n.* **1. a.** One who has devoted oneself to a particular branch of study or research. **b.** A physician certified to practice in a specified field. **2.** Any of several enlisted ranks in the U.S. Army that correspond to those of corporal through sergeant first class. **—spe'cial·is'tic** *adj.*

**spe·ci·al·i·ty** (spĕsh'ē-ăl'ĭ-tē) *n., pl.* **-ties.** **1.** A distinguishing feature or mark. **2. specialities.** Special points of consideration : PARTICULARS. **3.** *Chiefly Brit.* A specialty.

**spe·cial·ize** (spĕsh'ə-līz') *v.* **-ized, -iz·ing, -iz·es.** —*vi.* **1.** To train or employ oneself in a special study or activity. **2.** *Biol.* To develop so as to become adapted to a specific environment or function. **3.** To concentrate on a particular activity or product <The store *specializes* in ski equipment.> —*vt.* **1.** To make specific mention of : PAR-

TICULARIZE. **2.** To give a particular character or function to. **3.** *Biol.* To adapt by specialization. **4.** To specify the payee in endorsing (a check). **—spe'cial·i·za'tion** (-lĭ-zā'shən) *n.*

**special pleading** *n.* **1.** *Law.* Assertion of new or special matter to offset the opposing party's allegations as an alternative to direct denial. **2.** A presentation of an argument that emphasizes only a favorable or a single aspect of the question at issue.

**special relativity** *n.* The physical theory of space and time developed by Albert Einstein that is based on the postulates that all the laws of physics are equally valid in all nonaccelerated frames of reference and that light is propagated rectilinearly in all directions at a constant speed and that has as consequences the relativistic mass increase of rapidly moving objects, the Lorentz contraction, time dilatation, and the principle of mass-energy equivalence.

**special session** *n.* An extraordinary session of a legislature or court.

**special theory of relativity** *n.* Special relativity.

**spe·cial·ty** (spĕsh'əl-tē) *n., pl.* **-ties. 1.** A special occupation, pursuit, aptitude, or skill. **2.** A branch of medicine, as pediatrics, to which a physician confines his or her practice. **3.** A special feature or characteristic : PECULIARITY. **4.** The quality or state of being special. **5.** *Law.* A special contract or agreement, esp. a deed kept under seal.

**spe·ci·a·tion** (spē'shē-ā'shən, -sē-) *n.* [SPECI(ES) + -ATION.] The evolutionary process by which new species are formed.

**spe·cie** (spē'shē, -sē) *n.* [Lat. *(in) specie,* (in) kind < ablative of *speciēs.* —see SPECIES.] Coined money : COIN. **—in specie. 1.** In coin. **2.** In the same kind or shape : in kind.

**spe·cies** (spē'shēz, -sēz) *n., pl.* **species. 1.** *Biol.* **a.** A fundamental taxonomic classification category, ranking after a genus and consisting of organisms capable of interbreeding. **b.** An organism belonging to such a category, represented in taxonomic nomenclature by a Latin adjective or epithet following a genus name. **2.** *Logic.* A class of individuals or objects grouped by virtue of their common attributes and assigned a common name : a division subordinate to a genus. **3.** Kind : variety. **4.** *Obs.* An outward appearance. **5.** *Rom. Cath. Ch.* **a.** The outward appearance or form of the Eucharistic elements that is retained after their consecration. **b.** Either of the consecrated elements of the Eucharist. **6.** *Obs.* Specie.

**spe·cif·ic** (spĭ-sĭf'ĭk) *adj.* [Med. Lat. *specificus* < Lat. *species,* kind, species.] **1.** Set forth explicitly : DEFINITE. **2.** Relating to, characterizing, or distinguishing a species. **3.** Special, distinctive, or unique <customs *specific* to these islands> **4.** Intended for, applying to, or acting on a given thing. **5.** Denoting a disease produced by a particular microorganism or condition. **6. a.** Denoting a customs charge levied on merchandise by unit or weight rather than according to value. **b.** Denoting a commodity rate applicable to the transportation of a single commodity between named points. —*n.* **1.** Something, as a quality, statement, or attribute, that is specific. **2.** A remedy for a particular disorder. **—spe·cif'i·cal·ly** *adv.* **—spec'i·fic'i·ty** (spĕs'ə-fĭs'ĭ-tē) *n.*

**spec·i·fi·ca·tion** (spĕs'ə-fĭ-kā'shən) *n.* **1.** An act of specifying. **2. a. specifications.** A detailed and exact statement of particulars, esp. a statement prescribing materials, dimensions, and workmanship for something to be built, installed, or manufactured. **b.** A single item or article that has been specified. **3.** An exact written description of an invention by an applicant for a patent.

**specific gravity** *n.* The ratio of the mass of a solid or liquid to the mass of an equal volume of distilled water at 4°C or of a gas to an equal volume of air or hydrogen under prescribed temperature and pressure conditions.

**specific heat** *n.* **1.** The ratio of the amount of heat required to raise the temperature of a unit mass of a substance by one unit of temperature to the amount of heat required to raise the temperature of a similar mass of a reference material, usu. water, by the same amount. **2.** The amount of heat, measured in calories, required to raise the temperature of one gram of a substance by one centigrade degree.

**specific impulse** *n.* A performance measure for rocket propellants equal to units of thrust per unit weight of propellant consumed per unit time.

**specific performance** *n. Law.* Performance of a legal contract as specified in its terms.

**specific resistance** *n. Elect.* RESISTIVITY 2.

**specific thrust** *n.* Specific impulse.

**spec·i·fy** (spĕs'ə-fī') *vt.* **-fied, -fy·ing, -fies.** [ME *specifien* < OFr. *specifier* < Med. Lat. *specificare* < *specificus,* specific.] **1.** To state explicitly <*specified* our requirements> **2.** To include in a specification. **—spec'i·fi'a·ble** (-fī'ə-bəl) *adj.*

**spec·i·men** (spĕs'ə-mən) *n.* [Lat., example < *specere,* to look at.] **1.** An individual, item, or part taken as representative of an entire set or whole : SAMPLE. **2.** A sample, as of tissue, blood, or urine, used for analysis and medical diagnosis.

**spe·cious** (spē'shəs) *adj.* [ME, attractive < Lat. *speciosus* < *species,* appearance < *specere,* to look at.] **1.** Deceptively attractive. **2.** Having the ring of truth or plausibility but actually fallacious. **—spe'cious·ly** *adv.* **—spe'cious·ness** *n.*

**speck** (spĕk) *n.* [ME *specke* < OE *specca.*] **1.** A small spot, mark, or discoloration. **2.** A very small bit : PARTICLE <not a *speck* of truth to the rumor> —*vt.* **specked, speck·ing, specks.** To mark with specks.

**speck·le** (spĕk'əl) *n.* [ME *spakle,* perh. of MLG orig.] A speck or small spot, esp. a natural dot of color on skin, plumage, or foliage. —*vt.* **-led, -ling, -les.** To mark or cover with or as if with speckles.

**speck·led** (spĕk'əld) *adj.* **1.** Dotted or covered with speckles, esp. flecked with small spots of contrasting color. **2.** Of a mixed character : MOTLEY.

**speckled trout** *n.* The brook trout.

**specs** (spĕks) *pl.n. Informal.* **1.** *also* **specks.** Eyeglasses. **2.** SPECIFICATION 2a.

**spec·ta·cle** (spĕk'tə-kəl) *n.* [ME < OFr. < Lat. *spectaculum* < *spectare,* to watch, freq. of *specere,* to look at.] **1.** A public performance or display. **2. a.** An object of interest. **b.** A regrettable public display, as of bad behavior <made a *spectacle* of yourself> **3. a.** Something seen or able to be seen. **b.** The sight of something. **4. spectacles. a.** GLASSES 4b. **b.** Something resembling eyeglasses, as in shape or function.

**spec·ta·cled** (spĕk'tə-kəld) *adj.* **1.** Wearing spectacles. **2.** Having markings that resemble spectacles. —Used of animals.

**spec·tac·u·lar** (spĕk-tăk'yə-lər) *adj.* Like a spectacle : SENSATIONAL <a *spectacular* sunset> —*n.* A single, unusually long or lavish theatrical production. **—spec·tac'u·lar'i·ty** (-lăr'ĭ-tē) *n.* **—spec·tac'u·lar·ly** *adv.*

**spec·ta·tor** (spĕk'tā'tər) *n.* [Lat. *spectator* < *spectare,* to watch. —see SPECTACLE.] An observer of an event.

**spec·ter** (spĕk'tər) *n.* [Fr. *spectre* < Lat. *spectrum,* appearance < *specere,* to look at.] **1.** A phantom. **2.** A threatening or haunting possibility <the awful *specter* of global war>

**spec·tra** (spĕk'trə) *n. var. pl. of* SPECTRUM.

**spec·tral** (spĕk'trəl) *adj.* **1.** Resembling a specter : GHOSTLY. **2.** Of, relating to, or produced by a spectrum. **—spec·tral'i·ty** (-trăl'ĭ-tē), **spec'tral·ness** *n.* **—spec'tral·ly** *adv.*

**spectral line** *n.* An isolated peak of intensity in a spectrum, esp. one of the visible dispersed images of the slit through which light enters the collimator of a spectroscope, generated by light of a single wavelength.

**spec·tre** (spĕk'tər) *n. Chiefly Brit. var. of* SPECTER.

**spectro-** *pref.* [< SPECTRUM.] Spectrum <*spectro*graph>

**spec·tro·gram** (spĕk'trə-grăm') *n.* A diagram or photograph of a spectrum.

**spec·tro·graph** (spĕk'trə-grăf') *n.* **1.** A spectroscope equipped to photograph spectra. **2.** A spectrogram. **—spec'tro·graph'ic** *adj.* **—spec'tro·graph'i·cal·ly** *adv.* **—spec·trog'ra·phy** (-trŏg'rə-fē) *n.*

**spec·tro·he·li·o·gram** (spĕk'trō-hē'lē-ə-grăm') *n.* A photograph of the sun taken in a narrow wavelength band centered on a selected wavelength.

**spec·tro·he·li·o·graph** (spĕk'trō-hē'lē-ə-grăf') *n.* An instrument for making spectroheliograms. **—spec'tro·he'li·o·graph'ic** *adj.* **—spec'tro·he'li·og'ra·phy** (-ŏg'rə-fē) *n.*

**spec·tro·he·li·o·scope** (spĕk'trō-hē'lē-ə-skōp') *n.* A device for observing solar radiation. **—spec'tro·he'li·o·scop'ic** (-skŏp'ĭk) *adj.*

**spec·trom·e·ter** (spĕk-trŏm'ĭ-tər) *n.* [SPECTRO(SCOPE) + -METER.] A spectroscope with scales for measuring the positions of spectral lines. **—spec'tro·met'ric** (-trə-mĕt'rĭk) *adj.* **—spec·trom'e·try** *n.*

**spec·tro·pho·tom·e·ter** (spĕk'trō-fō-tŏm'ĭ-tər) *n. Physics.* An instrument for determining the distribution of energy in a spectrum of luminous radiation. **—spec'tro·pho'to·met'ric** (-fō'tə-mĕt'rĭk) *adj.* **—spec'tro·pho·tom'e·try** *n.*

**spec·tro·scope** (spĕk'trə-skōp') *n.* An instrument for resolving and observing or recording spectra. **—spec'tro·scop'ic** (-skŏp'ĭk), **spec'tro·scop'i·cal** *adj.* **—spec'tro·scop'i·cal·ly** *adv.*

**spectroscopic analysis** *n.* Analysis of a spectrum in order to determine characteristics of its source.

**spec·tros·co·py** (spĕk-trŏs'kə-pē) *n.* The study of spectra, esp. the experimental observation of optical spectra. **—spec·tros'co·pist** *n.*

**spec·trum** (spĕk'trəm) *n., pl.* **-tra** (-trə) *or* **-trums.** [Lat., appearance < *specere,* to look at.] **1.** *Physics.* Distribution of a characteristic of a physical system or phenomenon, esp.: **a.** Distribution of energy emitted by a radiant source, as by an incandescent body, arranged in order of wavelengths. **b.** Distribution of atomic or subatomic particles in a system, as in a magnetically resolved molecular beam, arranged in order of masses. **c.** A graphic or photographic representation of such a distribution. **2. a.** A range of values of a quantity or set of related quantities. **b.** A broad sequence or range of related qualities, ideas, or activities <the whole *spectrum* of human thought and accomplishments>

**spec·u·la** (spĕk'yə-lə) *n. var. pl. of* SPECULUM.

**spec·u·lar** (spĕk'yə-lər) *adj.* Of, resembling, or produced by a mirror or speculum.

**spec·u·late** (spĕk'yə-lāt') *vi.* **-lat·ed, -lat·ing, -lates.** [Lat. *speculari, speculat-,* to observe < *specula,* watchtower < *specere,* to look at.] **1.** To meditate on a given subject : REFLECT. **2.** To engage in the buying or selling of a commodity with an element of risk on the chance of profit.

**spec·u·la·tion** (spĕk'yə-lā'shən) *n.* **1. a.** The act of speculating. **b.** Profound contemplation. **c.** An opinion, theory, or conclusion reached by speculating. **2. a.** Engagement in risky business deals on the chance of large, quick profits. **b.** An instance of speculating.

**spec·u·la·tive** (spĕk'yə-lə-tĭv, -lā'-) *adj.* **1.** Of, marked by, or based on contemplative speculation. **2. a.** Given to speculation or conjecture. **b.** Spent in speculation. **3. a.** Engaging in, given to, or involving financial speculation. **b.** Involving chance : RISKY <*speculative* business ventures> **—spec'u·la·tive·ly** *adv.* **—spec'u·la·tive·ness** *n.*

**spec·u·la·tor** (spĕk'yə-lā'tər) *n.* One that speculates.

**spec·u·lum** (spĕk'yə-ləm) *n., pl.* **-la** (-lə) or **-lums.** [Lat., mirror < *specere,* to look at.] **1.** A mirror or polished metal plate used as a reflector in optical instruments. **2.** An instrument for dilating the opening of a body cavity for medical examination. **3.** *Biol.* **a.** A bright, often iridescent patch of color on the wings of certain birds, esp. ducks. **b.** A transparent spot in the wings of some butterflies or moths.

**sped** (spĕd) *v. var. p.t. & p.p. of* SPEED.

**speech** (spēch) *n.* [ME *speche* < OE *spæc, spræc.*] **1. a.** The faculty or act of speaking. **b.** The faculty or act of expressing or describing thoughts, feelings, or perceptions by the articulation of words. **2.** Something spoken : UTTERANCE. **3.** Vocal communication : CONVERSATION. **4. a.** A public address : TALK. **b.** A printed copy of a public address. **5.** One's habitual manner of speaking. **6.** The language or dialect of a region or nation. **7.** The sounding of a musical instrument. **8.** The study of oral communication, speech sounds, and vocal physiology. **9.** *Archaic.* Rumor.

☆ **syns: 1.** SPEECH, DISCOURSE, TALK, UTTERANCE, VOICE *n. core meaning* : the faculty, act, or product of speaking <a sore throat that made *speech* difficult><expressions occurring often in *speech*> **2.** SPEECH, LECTURE, ORATION, TALK *n. core meaning* : a formal oral communication to an audience <a *speech* at graduation>

**speech community** *n.* All speakers of a particular language or dialect, whether located in one area or scattered.

**speech·i·fy** (spē'chə-fī') *vi.* **-fied, -fy·ing, -fies.** To make a speech. **—speech'i·fi'er** *n.*

**speech·less** (spēch'lĭs) *adj.* **1.** Lacking the faculty of speech : DUMB. **2.** Temporarily unable to speak. **3.** Refraining from speech : SILENT. **4.** Unexpressed or inexpressible in words <*speechless* horror> **—speech'less·ly** *adv.* **—speech'less·ness** *n.*

**speech·mak·er** (spēch'mā'kər) *n.* One who gives a speech. **—speech'mak'ing** *n.*

**speed** (spēd) *n.* [ME *spede* < OE *spēd,* success.] **1.** *Math. & Physics.* The rate or a measure of the rate of motion, esp.: **a.** Distance traveled divided by the time of travel. **b.** The limit of this quotient as the time of travel becomes vanishingly small : the first derivative of distance with respect to time. **c.** Magnitude of a velocity. **2.** Swiftness of action. **3.** The act or state of moving rapidly : RAPIDITY. **4.** A transmission gear or set of gears in a motor vehicle. **5. a.** A numerical expression of the sensitivity of a photographic film, plate, or paper to light. **b.** The capacity of a lens to accumulate light at an appropriate aperture. **c.** The length of time required or permitted for a camera shutter to open and admit light. **6.** *Slang.* Methamphetamine. **7.** *Archaic.* Prosperity : luck. *—v.* **sped** (spĕd) *or* **speed·ed, speed·ing, speeds.** *—vt.* **1. a.** To hasten. **b.** To send with haste. **2. a.** To increase the speed or rate of : ACCELERATE <*speed* up a process> **b.** To set the speed of (a machine). **3.** To drive (a motor vehicle) at a high or illegal rate of speed. **4. a.** To wish Godspeed to. **b.** *Archaic.* To help to succeed or prosper : AID. **c.** To further, promote, or expedite (e.g., a legal action). *—vi.* **1.** To go or move rapidly. **2. a.** To drive fast. **b.** To exceed a traffic speed limit. **3.** To pass quickly. **4.** To move, perform, or happen at a faster rate : ACCELERATE. **5.** *Obs.* **a.** To prove successful. **b.** To get along in a specified manner : FARE.

**speed·ball** (spēd'bôl') *n. Slang.* An intravenous dose of cocaine and heroin.

**speed·boat** (spēd'bōt') *n.* A fast motorboat.

**speed·boat·ing** (spēd'bō'tĭng) *n.* The act or sport of driving a speedboat. **—speed'boat'er** *n.*

**speed·er** (spē'dər) *n.* One that speeds, esp. a driver who exceeds a legal or safe speed.

**speed freak** *n. Slang.* A habitual user of amphetamines and esp. of methamphetamine.

**speed·ing** (spē'dĭng) *adj.* Moving with speed. *—n.* The act of driving esp. a motor vehicle faster than the legal limit.

**speed limit** *n.* The maximum speed legally permitted on a given stretch of road.

**speed·om·e·ter** (spĭ-dŏm'ĭ-tər, spē-) *n.* **1.** An automotive instrument for indicating speed. **2.** An odometer.

**speed-read** (spēd'rēd') *vi.* **-read** (-rĕd'), **-read·ing, -reads.** To practice or engage in speed-reading. **—speed'read'er** *n.*

**speed-read·ing** (spēd'rē'dĭng) *n.* A method of reading rapidly by assimilating several words or phrases at a glance or by skimming.

**speed shop** *n.* An automotive shop catering to hot rodders.

**speed·ster** (spēd'stər) *n.* **1.** A speeder. **2.** A fast car.

**speed trap** *n.* Deployment of concealed police officers or electronic devices on a road to apprehend speeders.

**speed·up** (spēd'ŭp') *n.* Acceleration of production without increase in pay.

**speed·way** (spēd'wā') *n.* **1.** A course for automobile racing. **2.** A multi-lane road for fast-moving traffic : EXPRESSWAY.

**speed·well** (spēd'wĕl') *n.* Any of various plants of the genus *Veronica,* with small, usu. blue flower clusters.

**speed·y** (spē'dē) *adj.* **-i·er, -i·est. 1.** Marked by rapid motion : SWIFT. **2.** Accomplished or arrived at without delay : PROMPT. **—speed'i·ly** *adv.* **—speed'i·ness** *n.*

**speiss** (spīs) *n.* [G. *Speise,* food < OHG *spīsa* < Med. Lat. *spesa,* provisions < Lat. *expensa,* p.part. of *expendere,* to pay out.—see EXPEND.] A basic arsenic or antimony compound of iron, often with nickel, copper, or other metals, having a metallic luster and a strong tendency to crystallize, produced during the smelting of various ores.

**spe·le·ol·o·gy** (spē'lē-ŏl'ə-jē) *n.* [Lat. *speleum,* cave (< Gk. *spēlaion*) + -LOGY.] **1.** The study of the physical, geologic, and biological aspects of caves. **2.** Cave exploration. **—spe'le·o·log'i·cal** (-ə-lŏj'ĭ-kəl) *adj.* **—spe'le·ol'o·gist** *n.*

**spell**[1] (spĕl) *v.* **spelled** *or* **spelt** (spĕlt), **spell·ing, spells.** [ME *spellen,* to read letter by letter < OFr. *espelir,* of Germanic orig.] *—vt.* **1.** To name or write in order the letters constituting (a word or part of a word). **2.** To be the letters of (a word). **3.** To mean : signify <a policy that *spells* disaster> *—vi.* To form a word or words correctly by means of letters. **—spell down.** To defeat in a spelldown. **—spell out. 1.** To make perfectly clear. **2.** To comprehend by study.

**spell**[2] (spĕl) *n.* [ME, discourse < OE, story.] **1.** An incantational word or formula. **2.** Compelling attraction : FASCINATION. **3.** A bewitched state : TRANCE. *—vt.* **spelled, spell·ing, spells.** To put under a spell.

▲ <u>word history:</u> The noun *spell*[2], "an incantation," and the verb *spell*[1] "to give the letters of a word in proper order," are etymologically related. Both are derived from a root that basically means "a sound" and "to speak aloud." The noun *spell*[2] comes from Old English *spell,* which had a much wider meaning than its modern descendant; it was used with the senses "discourse," "narration," "sermon," and "story." Old English *spell* also appears as the second element of *gospel,* literally "good tidings." The meaning "incantation" did not appear until late medieval or early modern times. The verb *spell*[1] is descended from Middle English *spellen,* which meant "to read letter by letter." *Spellen* is a borrowing of Old French *espeller,* "to interpret, to explain," but the French word is ultimately descended from the same root as the English noun *spell*[2].

**spell**[3] (spĕl) *n.* [< ME *spelen,* to spare < OE *spelian,* to represent.] **1.** A short yet indefinite time period. **2.** *Informal.* A period of weather of a given kind <a wet *spell*> **3.** A period of work : SHIFT. **4.** *Informal.* **a.** A period, as of irritability. **b.** A sudden bout of illness. **5.** *Informal.* A short distance. *—v.* **spelled, spell·ing, spells.** *—vt.* **1.** To relieve (another) from work temporarily by taking a turn. **2.** To allow to rest a while. *—vi.* To rest for a time from an activity.

☆ **syns:** SPELL, ACCESS, ATTACK, FIT, SEIZURE *n. core meaning* : a sudden, often acute manifestation of a disease <experienced frequent *spells* of indigestion>

**spell·bind** (spĕl'bīnd') *vt.* **-bound** (-bound'), **-bind·ing, -binds.** To hold under or as if under a spell : ENCHANT.

**spell·bind·er** (spĕl'bīn'dər) *n.* One, as a speaker or novel, that holds others spellbound.

**spell·bound** (spĕl'bound') *adj.* Fascinated : entranced.

**spell·down** (spĕl'doun') *n.* A contest in which competitors are eliminated as they fail to spell a given word correctly.

**spell·er** (spĕl'ər) *n.* **1.** One who spells words. **2.** An elementary spelling textbook.

**spell·ing** (spĕl'ĭng) *n.* **1. a.** Formation of words with letters in an accepted order : ORTHOGRAPHY. **b.** The art or study of orthography. **2.** The way in which a word is spelled.

**spelling bee** *n.* A spelldown.

**spelt**[1] (spĕlt) *n.* [ME < OE < LLat. *spelta,* of Germanic orig.] A hardy wheat, *Triticum spelta,* grown mostly in Europe.

**spelt**[2] (spĕlt) *v. var. p.t. & p.p. of* SPELL[1].

**spel·ter** (spĕl'tər) *n.* [Prob. of MLG orig.] Zinc, esp. in ingots, slabs, or plates.

**spe·lunk·er** (spĭ-lŭng'kər, spē'lŭng'-) *n.* [< obs. *spelunk,* cave < ME < Lat. *spelunca* < Gk. *spēlunx.*] An explorer of caves : SPELEOLOGIST. **—spe'lunk'ing** *n.*

**spen·cer**[1] (spĕn'sər) *n.* [Perh. < the name *Spencer.*] *Naut.* A trysail.

**spen·cer**[2] (spĕn'sər) *n.* [After George John *Spencer* (1758–1834), 2nd Earl Spencer.] **1.** A short double-breasted overcoat worn by men in the early 19th cent. **2.** A close-fitting, waist-length jacket worn by women.

**Spen·ce·ri·an**[1] (spĕn-sîr'ē-ən) *adj.* Of or relating to Herbert Spencer or his philosophy. *—n.* A follower of Herbert Spencer.

**Spen·ce·ri·an**[2] (spĕn-sîr'ē-ən) *adj.* [After Platt Rogers *Spencer*

(1800–1864), its inventor.] Of or relating to an ornate style of penmanship employing rounded letters slanted to the right.

**Spen·cer·ism** (spĕn′sə-rĭz′əm) *also* **Spen·ce·ri·an·ism** (spĕn-sîr′ē-ə-nĭz′əm) *n.* The system of logical positivism developed by Herbert Spencer, setting forth the idea that evolution is the passage from the simple, indefinite, and incoherent to the complex, definite, and coherent.

**spend** (spĕnd) *v.* **spent** (spĕnt), **spend·ing, spends.** [ME *spenden,* partly < OE *spendan* (< Lat. *expendere,* to expend) and partly < OFr. *despendre,* to dispend.] —*vt.* **1.** To use up : EXPEND <*spent* hours studying Latin> **2.** To pay out (money). **3.** To deprive of force or strength : WEAR OUT <The hurricane finally *spent* itself.> **4.** To pass (time) in a specified manner or place <*spent* the winter in Florida> **5. a.** To throw away : SQUANDER <*spent* our youth on decadent pursuits> **b.** To sacrifice. —*vi.* **1.** To pay out or expend money. **2.** *Obs.* To be exhausted or consumed. **—spend′a·ble** *adj.* **—spend′er** *n.*

**spending money** *n.* Cash for small personal needs.

**spend·thrift** (spĕnd′thrĭft′) *n.* [SPEND + THRIFT, accumulated wealth (obs.).] One who squanders money. —*adj.* Extravagant : wasteful.

**Spen·se·ri·an** (spĕn-sîr′ē-ən) *adj.* Of, relating to, or resembling Edmund Spenser, his poetry, or his style. —*n.* A scholar of Spenser's poetry or life.

**Spenserian sonnet** *n.* [After Edmund *Spenser* (1552–1599).] A sonnet form comprising three interlocking quatrains and a couplet with the rhyme pattern *abab bcbc cdcd ee.*

**Spenserian stanza** *n.* A stanza consisting of eight lines of iambic pentameter and a final Alexandrine, rhymed *ababbcbcc,* used by Edmund Spenser in *The Faerie Queene.*

**spent** (spĕnt) *adj.* [P.part. of SPEND.] **1.** Used up : CONSUMED. **2.** Having come to an end. **3.** Depleted of energy, force, or strength. **4.** *Naut.* Of or relating to a vessel at the end of a voyage, with fuel, stores, and water consumed and cargo discharged.

**sperm[1]** (spûrm) *n., pl.* **sperm** *or* **sperms.** [ME *sperme,* semen < OFr. *esperme* < LLat. *sperma* < Gk.] **1.** Spermatozoon. **2.** The male fluid of fertilization : SEMEN. **—sperm′ous** *adj.*

**sperm[2]** (spûrm) *n.* [Short for SPERMACETI.] A substance, as spermaceti, associated with the sperm whale.

**sperm-** *pref. var. of* SPERMI-.

**-sperm** *suff.* [< Gk. *sperma,* seed.] Seed <endo*sperm*>

**sperma-** *pref. var. of* SPERMI-.

**sper·ma·ce·ti** (spûr′mə-sē′tē, -sĕt′ē) *n.* [ME < Med. Lat. *sperma-ceti* : Lat. *sperma,* semen (< Gk.) + Lat. *ceti,* genitive of *cetus,* whale (< Gk. *kētos).*] A white, waxy substance consisting of various esters of fatty acids, obtained from the head of the sperm whale and used for making candles, ointments, and cosmetics.

**sper·ma·ry** (spûr′mə-rē) *n., pl.* **-ries.** [NLat. *spermarium* < LLat. *sperma,* semen.] An organ in which male gametes are formed, esp. in invertebrate animals.

**spermat-** *pref. var. of* SPERMATO-.

**sper·ma·the·ca** (spûr′mə-thē′kə) *n.* [LLat. *sperma,* semen + THECA.] A receptacle for storing spermatozoa in certain female invertebrates, esp. insects. **—sper′ma·the′cal** *adj.*

**sper·ma·ti·a** (spər-mā′shē-ə, -shə) *n. pl. of* SPERMATIUM.

**sper·mat·ic** (spər-măt′ĭk) *adj.* **1.** Of, relating to, or resembling sperm : SPERMOUS. **2.** Of or relating to a spermary.

**spermatic cord** *n.* A cordlike structure consisting of the vas deferens and its arteries, veins, nerves, and lymphatic vessels that passes from the abdominal cavity through the inguinal canal down into the scrotum to the back of the testicle.

**spermatic fluid** *n.* Semen.

**sper·ma·tid** (spûr′mə-tĭd) *n.* One of four haploid cells formed during mitosis in the male that develop into spermatozoa without further division.

**sper·ma·ti·um** (spər-mā′shē-əm, -shəm) *n., pl.* **-ti·a** (-shē-ə, -shə) [NLat. < Gk. *spermation,* dim. of *sperma,* semen.] A nonmotile sporelike structure in red algae and certain fungi, gen. functioning as a male gamete. **—sper·ma′tial** (-shəl) *adj.*

**spermato-** *or* **spermat-** *pref.* [< Gk. *sperma, spermat-,* seed.] **1.** Seed <*spermato*phyte> **2. a.** Sperm <*spermat*ic> **b.** Spermatozoon <*spermato*phore>

**sper·mat·o·cide** (spər-măt′ə-sīd′) *n.* A sperm-killing agent. **—sper·mat′o·cid′al** (-sīd′l) *adj.*

**sper·mat·o·cyte** (spər-măt′ə-sīt′) *n.* A diploid cell that is converted by meiotic division into four spermatids.

**sper·mat·o·gen·e·sis** (spər-măt′ə-jĕn′ĭ-sĭs) *n.* Generation of sperm by male meiosis and spermiogenesis. **—sper·mat′o·ge·net′ic** (-jə-nĕt′ĭk), **sper·mat′o·gen′ic** (-jĕn′ĭk) *adj.*

**sper·mat·o·go·ni·um** (spər-măt′ə-gō′nē-əm) *n., pl.* **-ni·a** (-nē-ə). Any of the cells of the gonads in male animals that are the progenitors of primary spermatocytes. **—sper·mat′o·go′ni·al** *adj.*

**sper·ma·toid** (spûr′mə-toid′) *adj.* Resembling sperm.

**sper·mat·o·phore** (spər-măt′ə-fôr′, -fōr′) *n.* An extruded mass or capsule of spermatozoa in certain invertebrates and primitive vertebrates. **—sper′ma·toph′o·ral** (spûr′mə-tŏf′ər-əl) *adj.*

**sper·mat·o·phyte** (spər-măt′ə-fīt′) *n.* Any of a group of plants of the division Spermatophyta, including all seed-bearing plants. **—sper·mat′o·phyt′ic** (-fĭt′ĭk) *adj.*

**sper·mat·or·rhe·a** *also* **sper·mat·or·rhoe·a** (spər-măt′ə-rē′ə) *n.* Involuntary seminal discharge without orgasm.

**sper·mat·o·zo·a** (spər-măt′ə-zō′ə, spûr′mə-tə-) *n. pl. of* SPERMATOZOON.

**sper·mat·o·zo·id** (spər-măt′ə-zō′ĭd, spûr′mə-tə-) *n.* [SPERMATOZO(ON) + -ID.] A ciliated male gamete produced in an antheridium. —*adj.* Resembling a spermatozoon.

**sper·mat·o·zo·on** (spər-măt′ə-zō′ŏn′, -ən, spûr′mə-tə-) *n., pl.* **-zo·a** (-zō′ə). A usu. long nucleated cell with a thin, motile tail that is the fertilizing gamete of a male animal. **—sper·mat′o·zo′al, sper·mat′o·zo′an, sper·mat′o·zo′ic** *adj.*

**spermi-** *or* **sperma-** *or* **spermo-** *or* **sperm-** *pref.* [Gk. *sperm-, spermo-* < *sperma,* seed.] **1.** Seed <*spermo*phile> **2.** Sperm <*sperm*ine>

**sper·mi·cide** (spûr′mĭ-sīd′) *n.* A sperm-killing agent : SPERMATOCIDE. **—sper′mi·cid′al** (-sīd′l) *adj.*

**sper·mine** (spûr′mēn′) *n.* A crystalline compound, $C_{10}H_{26}N_4$, found as a phosphate in semen, yeast, and ox pancreas.

**sper·mi·o·gen·e·sis** (spûr′mē-ō-jĕn′ĭ-sĭs) *n.* [NLat. *spermium,* spermatozoon, prob. < Lat. *sperma,* semen + -GENESIS.] Transformation of a spermatid into a spermatozoon.

**spermo-** *pref. var. of* SPERMI-.

**sper·mo·go·ni·um** (spûr′mə-gō′nē-əm) *n., pl.* **-ni·a** (-nē-ə). *Bot.* A hollow structure in which spermatia are formed.

**sperm oil** *n.* A yellow waxy oil obtained chiefly from the head of the sperm whale and used as an industrial lubricant.

**sperm·o·phile** (spûr′mə-fīl′) *n.* Any of various North American ground squirrels of the genus *Citellus* or *Spermophilus.*

**sperm whale** *n.* A whale, *Physeter catodon,* having a very large head with cavities containing sperm oil and spermaceti and a long, narrow, toothed lower jaw.

**sper·ry·lite** (spĕr′ĭ-līt′) *n.* [After F.L. *Sperry,* 19th-cent. Canadian mineralogist.] A white crystalline platinum mineral, $PtAs_2$.

**spes·sar·tite** (spĕs′ər-tīt′) *also* **spes·sar·tine** (-tēn′) *n.* [Fr., after *Spessart,* a hilly area in West Germany.] A mineral silicate of manganese and aluminum, usu. containing some iron.

**spew** (spyōō) *v.* **spewed, spew·ing, spews.** [ME *spewen* < OE *spiwan.*] —*vt.* **1.** To cast out through the mouth. **2.** To force out in a stream : EJECT <a volcano *spewing* flames, dust, and molten lava> —*vi.* To vomit. **—spew** *n.*

**sphag·num** (sfăg′nəm) *n.* [NLat. < Lat. *sphagnos,* a kind of moss < Gk.] Any of various pale or ashy mosses of the genus *Sphagnum,* the decomposed remains of which form peat. **—sphag′nous** *adj.*

**sphal·er·ite** (sfăl′ə-rīt′) *n.* [G. *Sphalerit* < Gk. *sphaleros,* slippery < *sphallein,* to trip.] A yellow, brown, black, or red zinc ore, essentially ZnS with some cadmium, iron, and manganese.

**sphen-** *pref. var. of* SPHENO-.

**sphene** (sfēn) *n.* [Fr. *sphène* < Gk. *sphēn,* wedge.] A titanium ore, mostly $CaTiSiO_5$, occas. used as a gemstone.

**sphe·nic** (sfē′nĭk) *adj.* Shaped like a wedge.

**spheno-** *or* **sphen-** *pref.* [< Gk. *sphēn,* wedge.] Wedge : wedge-shaped <*spheno*don><*sphen*oid>

**sphe·no·don** (sfē′nə-dŏn′, sfĕn′ə-) *n.* The tuatara.

**sphe·no·gram** (sfē′nə-gram′, sfĕn′ə-) *n.* A cuneiform character.

**sphe·noid** (sfē′noid′) *n.* The sphenoid bone. —*adj.* **1.** Wedge-shaped. **2.** Of or relating to the sphenoid bone. **—sphe·noi′dal** (-noid′l) *adj.*

**sphenoid bone** *n.* A compound bone with winglike processes, situated at the base of the skull.

**spher-** *pref. var. of* SPHERO-.

**spher·al** (sfîr′əl) *adj.* **1.** SPHERICAL 1, 2. **2.** Symmetric.

**sphere** (sfîr) *n.* [ME *spere* < OFr. *espere* < Lat. *sphaera,* ball < Gk. *sphaira.*] **1.** *Math.* A three-dimensional surface all points of which are equidistant from a fixed point. **2.** A spherical figure or object. **3.** A celestial body, as a planet or star. **4.** The sky, appearing as a hemisphere to an observer. **5.** Any of a series of concentric, transparent, revolving globes that together were once thought to contain the moon, sun, planets, and stars. **6.** The extent of one's knowledge, interests, or social position. **7.** An area of influence, power or control : DOMAIN. —*vt.* **sphered, spher·ing, spheres. 1.** To form into a sphere. **2.** To put in or within a sphere. **3.** To surround or encompass. **—sphe·ric′i·ty** (sfĭ-rĭs′ĭ-tē) *n.*

**sphere of influence** *n.* A territorial area over which political or economic influence is wielded by one nation.

**spher·i·cal** (sfîr′ĭ-kəl, sfĕr′-) *also* **spher·ic** (sfîr′ĭk, sfĕr′-) *adj.* **1.** Shaped like a sphere : GLOBULAR. **2.** Of or pertaining to a sphere. **3.** Of or pertaining to celestial bodies. **—spher′i·cal·ly** *adv.* **—spher′i·cal·ness** *n.*

**spherical aberration** *n.* An optical defect of refracting and reflecting spherical surfaces in which light rays from one axial point, incident on the surface at different distances from the optical axis, do not come to a common focus.

**spherical angle** *n.* The angle formed at the intersection of the arcs of two great circles.

**spherical astronomy** *n.* The branch of astronomy dealing with positions on the celestial sphere.

**spher·i·cal-co·or·di·nate system** (sfîr'ĭ-kəl-kō-ôr'dn-ĭt, -āt', sfĕr'-) *n.* A three-dimensional system for locating points in space by means of a radius vector and two angles measured from the center of a sphere with respect to two arbitrary, fixed, perpendicular directions.

**spherical excess** *n.* The difference between the sum of the angles of a spherical triangle and the sum of the angles of a plane triangle.

**spherical geometry** *n.* The geometry of circles, angles, and figures on the surface of a sphere.

**spherical polygon** *n.* A part of a spherical surface that is bounded by arcs of three or more great circles.

**spherical triangle** *n.* A triangle the three sides of which are arcs of great circles.

**spherical trigonometry** *n.* A modified form of trigonometry applied to spherical triangles.

**spher·ics** (sfîr'ĭks, sfĕr'-) *n. (sing. in number).* **1. a.** Spherical geometry. **b.** Spherical trigonometry. **2.** *var. of* SFERICS.

**sphero-** or **spher-** *pref.* [Lat. *sphaero-* < Gk. *sphairo-*, sphere.] Sphere <*spherometer*>

**sphe·roid** (sfîr'oid', sfĕr'-) *n.* An ellipsoid generated by revolving an ellipse around one of its axes. **—sphe·roi'dal** (-oid'l), **sphe·roi'dic** (-oi'dĭk) *adj.* **—sphe·roi'dal·ly** *adv.* **—sphe'roi·dic'i·ty** (-oi-dĭs'ĭ-tē) *n.*

**sphe·rom·e·ter** (sfĭ-rŏm'ĭ-tər) *n.* An instrument for measuring the curvature of a surface.

**spher·ule** (sfîr'ōōl, -yōōl, sfĕr'-) *n.* [LLat. *sphaerula*, dim. of Lat. *sphaera*, ball < Gk. *sphaira*.] A miniature sphere : GLOBULE. **—spher'u·lar** (sfîr'yə-lər, sfĕr'-) *adj.*

**spher·u·lite** (sfîr'yə-līt', -ə-līt', sfĕr'-) *n.* [SPHERUL(E) + -ITE.] A small, usu. spheroid crystalline body having a radiating structure and found in obsidian and other silicic lava flows. **—spher'u·lit'ic** (-lĭt'ĭk) *adj.*

**spher·y** (sfîr'ē) *adj.* **-i·er, -i·est. 1.** Of or relating to the celestial spheres. **2.** Resembling a heavenly body : STARLIKE.

**sphinc·ter** (sfĭngk'tər) *n.* [LLat. < Gk. *sphinkter* < *sphingein*, to bind tight.] A ringlike muscle that normally maintains constriction of a bodily passage or orifice and relaxes as required by normal physiological functioning. **—sphinc'ter·al** *adj.*

**sphinx** (sfĭngks) *n., pl.* **sphinx·es** or **sphin·ges** (sfĭn'jēz') [ME *spynx* < Lat. *sphinx* < Gk.] **1.** *Myth.* An Egyptian figure having the body of a lion and the head of a man, ram, or hawk. **2.** *Gk. Myth.* A winged monster having the head of a woman and the body of a lion that destroyed all who could not answer its riddle. **3.** An enigmatic or mysterious person.

**sphinx moth** *n.* The hawk moth.

**sphra·gis·tics** (sfrə-jĭs'tĭks) *n.* [Fr. *sphragistique* < LGk. *sphragistikos* < Gk. *sphragis*, seal.] *(sing. in number).* Study of seals and signets.

**sphygm-** *pref. var. of* SPHYGMO-.

**sphyg·mic** (sfĭg'mĭk) *adj.* [Gk. *sphugmikos* < *sphugmos*, pulsation < *sphuzein*, to throb.] *Physiol.* Of or relating to the pulse.

**sphygmo-** or **sphygm-** *pref.* [< Gk. *sphugmos*, pulsation.] Pulse <*sphygmograph*>

**sphyg·mo·gram** (sfĭg'mə-grăm') *n.* The record or tracing produced by a sphygmograph.

**sphyg·mo·graph** (sfĭg'mə-grăf') *n.* An instrument for graphically recording the character and variations of the arterial pulse. **—sphyg'mo·graph'ic** *adj.* **—sphyg·mog'ra·phy** (-mŏg'rə-fē) *n.*

**sphyg·moid** (sfĭg'moid') *adj. Physiol.* Resembling a pulse.

**sphyg·mo·ma·nom·e·ter** (sfĭg'mō-mə-nŏm'ĭ-tər) *also* **sphyg·mom·e·ter** (sfĭg-mŏm'ĭ-tər) *n.* An instrument for measuring arterial blood pressure. **—sphyg'mo·man'o·met'ric** (-măn'ə-mĕt'rĭk) *adj.* **—sphyg'mo·man'o·met'ri·cal·ly** *adv.* **—sphyg'mo·ma·nom'e·try** *n.*

**spi·ca** (spī'kə) *n., pl.* **-cae** (-kē') or **-cas.** [Lat., ear of grain.] A bandage applied in overlapping opposite spirals to immobilize a digit or limb.

**Spi·ca** (spī'kə) *n.* [Lat. < *spica*, ear of grain.] The brightest star in the constellation Virgo, 212 light-years from Earth.

**spi·cae** (spī'kē') *n. var. pl. of* SPICA.

**spic-and-span** (spĭk'ən-spăn') *adj. var. of* SPICK-AND-SPAN.

**spi·cate** (spī'kāt') *adj.* [Lat. *spicatus*, p.part. of *spicare*, to provide with spikes < *spica*, ear of grain.] *Bot.* Borne in or forming a spike.

**spic·ca·to** (spĭ-kä'tō) [Ital., p.part. of *spiccare*, to separate.] *n., pl.* **-tos.** *Mus.* A technique of bowing in which the bow is made to bounce slightly from the string. **—spic·ca'to** *adj.*

**spice** (spīs) *n.* [ME < OFr. *espice* < LLat. *species*, wares, spices < Lat., kind < *specere*, to look at.] **1. a.** A pungently aromatic vegetable substance, as cinnamon or nutmeg, used to flavor foods or beverages. **b.** Such substances as a group. **2.** Something that adds zest or flavor. **3.** A pungent aroma : PERFUME. *—vt.* **spiced, spic·ing, spic·es. 1.** To season with spices. **2.** To add zest or flavor to.

**spice·ber·ry** (spīs'bĕr'ē) *n.* A plant or shrub having spicy berries, as the wintergreen.

**spice·bush** (spīs'bŏŏsh') *n.* **1.** An aromatic shrub, *Lindera benzoin* of eastern North America, having small, early-blooming yellow flower clusters. **2.** An aromatic shrub, *Calycanthus occidentalis* of California, having fragrant, brownish flowers.

**spic·er·y** (spī'sə-rē) *n., pl.* **-ies. 1.** SPICE 1b. **2.** The aromatic or pungent quality of spices. **3.** *Obs.* A place where spices are stored.

**spick-and-span** *also* **spic-and-span** (spĭk'ən-spăn') *adj.* [Short for obs. *spick and spannew* : *spick*, spike + SPANNEW.] **1.** Neat and clean : SPOTLESS. **2.** Brand-new.

**spic·u·la**[1] (spĭk'yə-lə) *n. pl. of* SPICULUM.

**spic·u·la**[2] (spĭk'yə-lə) *n. var. of* SPICULE.

**spic·ule** (spĭk'yōōl) *also* **spic·u·la** (-yə-lə) *n., pl.* **-ules** *also* **-u·lae** (-yə-lē') [Lat. *spiculum*, spiculum.] A small needlelike structure or part, such as one of the silicate or calcium carbonate processes supporting the soft tissue of certain invertebrates, esp. sponges. **—spic'u·lar** (-yə-lər), **spic'u·late** (-yə-lĭt, -lāt') *adj.*

**spic·u·lum** (spĭk'yə-ləm) *n., pl.* **-la** (-lə) [Lat. *spiculum*, dim. of *spica*, point.] A needlelike structure, as a spicule.

**spic·y** (spī'sē) *adj.* **-i·er, -i·est. 1.** Having the characteristics of spice, as flavor and aroma. **2.** Piquant : zesty. **3.** Slightly scandalous : RISQUÉ <*spicy* anecdotes> **—spic'i·ly** *adv.* **—spic'i·ness** *n.*

**spi·der** (spī'dər) *n.* [ME *spither* < OE *spīðra*.] **1.** Any of numerous arachnids of the order Araneae, having eight legs, a body divided into a cephalothorax and an abdomen, and several spinnerets that produce silk used to make nests, cocoons, or webs for trapping insects. **2.** One considered similar to a spider, as in appearance, character, or movement. **3.** A cast-iron frying pan with a long handle. **4.** A trivet.

**spider crab** *n.* A crab, as one of the genus *Libinia* or *Macrocheira*, with long legs and a small body.

**spi·der·flow·er** (spī'dər-flou'ər) *n.* The cleome.

**spider lily** *n.* A chiefly tropical American plant of the genus *Hymenocallis*, with narrow leaves and white flower clusters.

**spider monkey** *n.* Any of several tropical American monkeys of the genus *Ateles*, with long legs and a long prehensile tail.

**spi·der·wort** (spī'dər-wûrt', -wôrt') *n.* A New World plant of the genus *Tradescantia*, esp. *T. virginiana*, having three-petaled blue or purple flowers.

**spi·der·y** (spī'də-rē) *adj.* **1.** Like a spider. **2. a.** Like a spider's legs. **b.** Like a spider's web. **3.** Infested with spiders.

**spied** (spīd) *v. p.t. & p.p. of* SPY.

**spie·gel** (spē'gəl) *n.* [Short for SPIEGELEISEN.] Spiegeleisen.

**spie·gel·ei·sen** (spē'gə-lī'zən) *n.* [G. *Spiegel*, mirror (< Lat. *speculum*) + *Eisen*, iron.] An alloy of iron containing approx. 15% manganese and small amounts of carbon and silicon, used in the Bessemer process.

**spiel** (spēl) [G., play.] *Slang —n.* A lengthy, usu. extravagant speech or argument intended to persuade. *—v.* **spieled, spiel·ing, spiels.** *—vi.* To talk at length or extravagantly. *—vt.* To say at length or extravagantly. **—spiel'er** *n.*

**spies** (spīz) *v. 3rd person sing. present tense of* SPY. *—n. pl. of* SPY.

**spif·fy** (spĭf'ē) *adj.* **-fi·er, -fi·est.** [Orig. unknown.] *Slang.* Chic : stylish. **—spif'fi·ness** *n.*

**spig·ot** (spĭg'ət) *n.* [ME.] **1.** A faucet. **2.** The vent plug of a cask. **3.** A wooden faucet placed in the bunghole of a cask.

**spike**[1] (spīk) *n.* [ME *spyk*, perh. of Scand. or MLG. orig.] **1. a.** A long, thick, sharp-pointed piece of wood or metal. **b.** A heavy nail. **2. a.** A sharp-pointed projection along the top of a fence or wall. **b.** One of several sharp metal projections set in the sole or sole and heel of an athletic shoe for grip. **c. spikes.** A pair of athletic shoes having spikes. **3. spikes.** Shoes with spike heels. **4.** A young deer's unbranched antler. **5.** A small young mackerel. **6. a.** The act of driving a volleyball at a sharp angle into the opponent's court by jumping near the net and hitting the ball down hard from above. **b.** *Informal.* The act of excitedly slamming a football to the ground after scoring a touchdown or making a big play. *—vt.* **spiked, spik·ing, spikes. 1.** To secure or provide with a spike. **2.** To impale, pierce, or injure with a spike. **3.** To render (a muzzle-loading gun) useless by driving a spike into the vent. **4.** To put an end to : BLOCK <*spiked* that scheme> **5.** *Slang.* To add alcoholic liquor to. **6.** To drive (a volleyball or football) in a spike.

**spike**[2] (spīk) *n.* [ME *spik* < Lat. *spica*, ear of grain.] **1.** An ear of grain. **2.** *Bot.* A usu. elongated inflorescence with stalkless or nearly stalkless flowers arranged along an axis.

**spike heel** *n.* A very high thin heel used on women's shoes.

**spike lavender** *n.* An aromatic plant, *Lavandula latifolia* of southern Europe, yielding an oil similar to that of true lavender.

**spike·let** (spīk'lĭt) *n.* A small or secondary spike, esp. one of those forming the inflorescence of grasses or similar plants.

**spike·nard** (spīk'närd') *n.* [ME < Med. Lat. *spica nardi*, spike of a nard.] **1.** An aromatic plant, *Nardostachys jatamansi* of India, with rose-purple flowers. **2.** A costly ointment of antiquity, prob. prepared

from the spikenard. **3.** A North American plant, *Aralia racemosa,* having small greenish flowers and an aromatic root.
**spik·y** (spī′kē) *adj.* **-i·er, -i·est.** Having a projecting sharp point. **—spik′i·ly** *adv.* **—spik′i·ness** *n.*
**spile** (spīl) *n.* [MLG *spīle,* wooden peg.] **1.** A foundation post : PILE. **2.** A wooden plug : BUNG. **3.** A spigot used in taking sap from a tree. —*vt.* **spiled, spil·ing, spiles.** To support, plug, or tap with a spile.
**spill[1]** (spĭl) *v.* **spilled** *or* **spilt** (spĭlt), **spill·ing, spills.** [ME *spillen,* to kill < OE *spillan.*] —*vt.* **1.** To cause or allow (a substance) to run or fall out of a container. **2.** To shed (blood). **3.** To let the wind out of (a sail). **4.** To cause to fall <The horse *spilled* its rider.> **5.** *Informal.* To divulge <*spilled* the story to the press> —*vi.* **1.** To run or fall out of a container. **2.** To come to the ground suddenly and involuntarily. **3.** To pour out or spread beyond limits <The angry mob *spilled* onto the streets.> —*n.* **1.** An act of spilling. **2.** An amount spilled. **3.** A fall, as from a horse. **—spill the beans.** *Informal.* To divulge all. **—spill′er** *n.*
**spill[2]** (spĭl) *n.* [MLG *spīle,* wooden peg.] **1.** A piece of wood or rolled paper used to light a fire. **2.** A small peg used as a plug.
**spill·age** (spĭl′ĭj) *n.* SPILL[1] 1, 2.
**spil·li·kin** (spĭl′ĭ-kĭn) *n.* [Perh. < obs. Du. *spelleken,* small peg.] **1.** A jackstraw. **2. spillikins.** Jackstraws.
**spill·way** (spĭl′wā′) *n.* A channel for an overflow of water, as from a reservoir.
**spilt** (spĭlt) *v. var. p.t. & p.p. of* SPILL[1].
**spilth** (spĭlth) *n.* [< SPILL.] Spillage.
**spin** (spĭn) *v.* **spun** (spŭn), **spin·ning, spins.** [ME *spinnen* < OE *spinnan.*] —*vt.* **1. a.** To draw out and twist (fibers) into thread. **b.** To form (thread or yarn) in this manner. **2.** To form (e.g., a web) by extruding viscous filaments. **3.** To make or produce by or as if by drawing out and twisting. **4.** To prolong or extend. **5.** To relate, esp. imaginatively <*spin* yarns> **6.** To cause to rotate swiftly : TWIRL. —*vi.* **1.** To make thread or yarn by drawing out and twisting fibers. **2.** To extrude viscous filaments, forming thread <spiders *spinning* their webs> **3.** To rotate rapidly : WHIRL. **4.** To have the sensation of turning fast in circles : REEL. **5.** To ride or drive rapidly. **6.** To fish with spinning tackle. **—spin off.** To derive (e.g., a product) from something larger and more or less unrelated. —*n.* **1.** The act of spinning. **2.** A swift whirling motion. **3.** Mental confusion. **4.** *Informal.* A short drive in a vehicle. **5.** The flight condition of an aircraft in a nose-down, spiraling, stalled descent. **6.** *Physics.* **a.** The intrinsic angular momentum of a subatomic particle. **b.** The total angular momentum of an atomic nucleus. **c.** A nonnegative integral or half-integral quantum number that specifies the value of such momenta in units of Planck's constant divided by $2\pi$.
☆ ***syns:*** SPIN, REEL, SWIM, SWIRL, WHIRL *v. core meaning* : to have the sensation of turning in circles <Too many drinks made our heads *spin.*>
**spin·ach** (spĭn′ĭch) *n.* [OFr. *espinache* < OSp. *espinaca* < Ar. *isfānākh.*] **1.** A widely cultivated plant, *Spinacia oleracea* native to Asia, having succulent edible leaves. **2.** The edible leaves of the spinach.
**spi·nal** (spī′nəl) *adj.* **1.** Of, relating to, or located near the spine or spinal cord : VERTEBRAL. **2.** Resembling a spine or spinous part. —*n.* A spinal anesthetic. **—spi′nal·ly** *adv.*
**spinal anesthesia** *n.* Partial or complete anesthesia produced by infusing an anesthetic substance into the spinal canal.
**spinal canal** *n.* The canal formed by the successive openings in the vertebrae through which the spinal cord and its membranes pass.
**spinal column** *n.* The columnar assemblage of articulated vertebrae extending from the cranium to the coccyx or the end of the tail, encasing the spinal cord and forming the supporting axis of the body : BACKBONE.
**spinal cord** *n.* The part of the central nervous system contained within the spinal canal and continuous at its cranial end with the medulla oblongata.
**spinal meningitis** *n. Pathol.* Cerebrospinal meningitis.
**spin·dle** (spĭn′dl) *n.* [ME *spindel* < OE *spinel.*] **1. a.** A notched stick for spinning fibers into thread by hand. **b.** A pin or rod holding a bobbin or spool on which thread is wound on a spinning wheel or spinning machine. **2.** A slender mechanical part that revolves or serves as an axis for a larger revolving part, as in a lock or an axle. **3.** *Biol.* The axis between cytoplasm centers, along which the chromosomes are distributed in mitosis. —*v.* **-dled, -dling, -dles.** —*vt.* To impale or perforate on the spike of a spindle. —*vi.* To grow into a thin, elongated, or weakly form.
**spindle tree** *n.* [So called because the wood is often used to make spindles.] A shrub or tree of the genus *Euonymus,* many species of which have brightly colored fruit.
**spin·dling** (spĭnd′lĭng) *adj.* Spindly.
**spin·dly** (spĭnd′lē) *adj.* **-dli·er, -dli·est. 1.** Slender and elongated. **2.** Of weak growth.
**spin·drift** (spĭn′drĭft′) *n.* [Var. of SPOONDRIFT.] Wind-blown sea spray.
**spine** (spīn) *n.* [ME < OFr. *espine* < Lat. *spina.*] **1.** The vertebrate spinal column. **2.** *Zool.* Any of various pointed projections, processes, or appendages of animals. **3.** *Bot.* A sharp-pointed, usu. woody process arising from the stem of a plant : THORN. **4.** The hinged back of a book. **5.** An object resembling a spine.

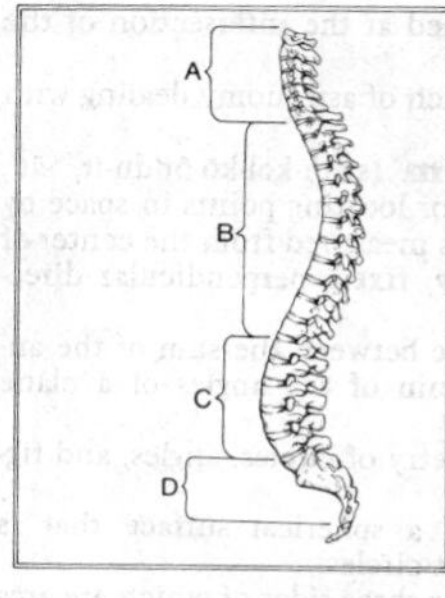


**spine**
*The human spine:*
*A. cervical vertebrae,*
*B. thoracic vertebrae,*
*C. lumbar vertebrae,*
*D. sacral vertebrae*

**spi·nel** *also* **spi·nelle** (spĭ-nĕl′) *n.* [Ital. *spinella,* dim. of *spina,* thorn (from its sharply pointed crystals) < Lat. *spina.*] Any of several hard white, orange, red, green, blue, or black minerals with composition $MgAl_2O_4$, the red variety being valued as a gem.
**spine·less** (spīn′lĭs) *adj.* **1.** Lacking a vertebral column. **2.** Having no spiny processes. **3.** Lacking courage or will power. **—spine′less·ly** *adv.* **—spine′less·ness** *n.*
**spi·nelle** (spĭ-nĕl′) *var. of* SPINEL.
**spi·nes·cent** (spī-nĕs′ənt) *adj.* [LLat. *spinescens, spinescent-,* pr.part. of *spinescere,* to become thorny < Lat. *spina,* thorn.] *Biol.* **1.** Having a spine. **2.** Having or tending toward the form of a spine. **—spi·nes′cence** *n.*
**spin·et** (spĭn′ĭt) *n.* [Obs. Fr. *espinette* < Ital. *spinetta.*] **1.** A small, compact upright piano. **2.** A small harpsichord with a single keyboard.
**spi·nif·er·ous** (spī-nĭf′ər-əs) *adj.* [< Lat. *spinifer* < *spina,* thorn.] SPINY 1.
**spi·ni·fex** (spī′nə-fĕks′) *n.* [NLat. *Spinifex,* genus name : Lat. *spina,* thorn + Lat. *facere,* to make.] An Australian grass, chiefly of the genus *Spinifex,* that grows in arid regions and has spiny leaves or seeds.
**spin·na·ker** (spĭn′ə-kər) *n.* [Orig. unknown.] *Naut.* A large triangular sail set on a spar that swings out opposite the mainsail, used on racing yachts when running before the wind.
**spin·ner** (spĭn′ər) *n.* **1.** One that spins. **2.** An angler's lure that spins rapidly. **3.** A fairing fitted over the hub of the propeller in some aircraft. **4.** A device composed of a dial and an arrow that is spun to indicate the next move in a board game.
**spin·ner·et** (spĭn′ə-rĕt′) *n.* **1.** *Zool.* A posterior structure in spiders and certain insect larvae, containing passages through which silky filaments are secreted. **2.** A device for making rayon, nylon, and other synthetic fibers, consisting of a plate pierced with holes through which plastic material is extruded in filaments.
**spin·ner·y** (spĭn′ə-rē) *n., pl.* **-ies.** A spinning mill.
**spin·ney** (spĭn′ē) *n., pl.* **-neys.** [OFr. *espinei,* thicket < Lat. *spinetum,* thorn hedge < *spina,* thorn.] *Chiefly Brit.* A small grove : COPSE.
**spin·ning** (spĭn′ĭng) *n.* The process of making fibrous material into yarn or thread.
**spinning frame** *n.* A machine that draws and twists fibers into yarn and winds it on spindles.
**spinning jenny** *n.* An early spinning machine that had several spindles.
**spinning wheel** *n.* A device for making yarn or thread, consisting of a foot- or hand-driven wheel and a single spindle.
**spin-off** (spĭn′ôf′, -ŏf′) *n.* **1.** Something, as a product, derived from something larger and more or less unrelated : BY-PRODUCT. **2.** Something derived from an earlier work, esp. a television show starring a character who had a popular minor role in an earlier show.
**spi·nose** (spī′nōs′) *adj.* [Lat. *spinosus* < *spina,* thorn.] SPINY 1 <a *spinose* plant> **—spi′nose′ly** *adv.* **—spi·nos′i·ty** (-nŏs′ĭ-tē) *n.*
**spi·no·tec·tal** (spī′nō-tĕk′təl) *adj.* Of or relating to the spinal cord and the tectum.
**spi·nous** (spī′nəs) *adj.* **1.** Like a spine or thorn. **2.** Having spines or similar projections : SPINY.
**spinous process** *n.* The rearward projection from the arch of a vertebra that with those of the other vertebrae forms the spine.
**spin·ster** (spĭn′stər) *n.* [ME *spinnester* < *spinnen,* to spin.] **1.** A woman who has remained single beyond the conventional age for marrying. **2.** A single woman. **3.** A woman whose occupation is spinning. **—spin′ster·hood′** *n.* **—spin′ster·ish** *adj.*
**spin·thar·i·scope** (spĭn-thăr′ĭ-skōp′) *n.* [Gk. *spintharis,* spark + SCOPE.] A device for observing individual scintillations produced by ionizing radiation, consisting of a tube with a magnifying lens at one end and a phosphorescent screen and a speck of radioactive salt at the other. **—spin·thar′i·scop′ic** (-skŏp′ĭk) *adj.*

**spin-the-bot·tle** (spĭn′*th*ə-bŏt′l) *n.* A game in which a spinning bottle is used to determine one's partner, as for kissing.
**spi·nule** (spĭn′yo͞ol) *n.* [Lat. *spinula,* dim. of *spina,* thorn.] *Bot.* A small spine.
**spi·nu·lose** (spĭn′yə-lōs′) *also* **spi·nu·lous** (spĭ′nyə-ləs) *adj.* **1.** Having spinules. **2.** Shaped like a spinule.
**spin wave** *n.* A sinusoidal wave of quantized energy propagated through a substance as a result of shifts in atomic magnetic fields as a response to outside stimuli.
**spin·y** (spī′nē) *adj.* **-i·er, -i·est. 1.** Bearing or covered with spines, thorns, or similar stiff projections. **2.** Shaped like a spine. **3.** Difficult : troublesome <*spiny* problems> **—spin′i·ness** *n.*
**spiny anteater** *n.* The echidna.
**spin·y-finned** (spī′nē-fĭnd′) *adj.* Having fins supported by sharp, spiny, inflexible rays.
**spin·y-head·ed worm** (spī′nē-hĕd′ĭd) *n.* Any of various worms of the phylum Acanthocephala, endoparasitic to vertebrates and characterized by an anterior cylindrical retractile proboscis bearing many rows of hooks.
**spiny lobster** *n.* Any of various edible marine decapod crustaceans of the family Palinuridae, having a spiny carapace and lacking the large pincers characteristic of true lobsters.
**spin·y-rayed** (spī′nē-rād′) *adj.* Spiny-finned.
**spir·a·cle** (spĭr′ə-kəl, spī′rə-) *n.* [Lat. *spiraculum,* breathing hole < *spirare,* to breathe.] **1.** *Zool.* A respiratory aperture, esp.: **a.** Any of several tracheal openings in the exoskeleton of an insect or spider. **b.** A small respiratory opening behind the eye of fishes, as sharks, rays, and skates. **c.** The blowhole of a cetacean. **2.** *Geol.* A small volcanic vent formed by gases on a lava flow. **3.** An aperture through which air is admitted and expelled. **—spi·rac′u·lar** (spī-răk′yə-lər, spĭ-) *adj.*
**spi·rae·a** (spī-rē′ə) *n. var. of* SPIREA.
**spi·ral** (spī′rəl) *n.* [Med. Lat. *spiralis* < Lat. *spira,* coil < Gk. *speira.*] **1.** The locus in a plane of a point moving around a fixed center at a monotonically increasing or decreasing distance from the center. **2. a.** The three-dimensional locus of a point moving parallel to and about a central axis at a constant or continuously varying distance : HELIX. **b.** Something shaped like such a curve <*spirals* of smoke> **3.** The course or flight path of an object rotating on its longitudinal axis. **4.** A continuously accelerating increase or decrease <the wage-price *spiral*> —*adj.* **1.** Of or resembling a spiral. **2.** Coiling in a constantly changing plane : HELICAL. **3.** Circling around to form a series of constantly changing planes. —*v.* **-raled, -ral·ing, -rals** *or* **-ralled, -ral·ling, -rals.** —*vi.* **1.** To take a spiral form or course. **2.** To rise or fall with steady acceleration. —*vt.* To cause to take a spiral form or course. **—spi·ral′i·ty** (spī-răl′ĭ-tē) *n.* **—spi′ral·ly** *adv.*
**spiral binding** *n.* A binding for notebooks and booklets in which a cylindrical spiral is passed through a row of punched holes at the edge of each sheet.
**spiral galaxy** *n.* A galaxy having a spiral structure with spiral arms consisting mainly of gas, dust, and stars.
**spi·rant** (spī′rənt) *n.* [Lat. *spirans, spirant-,* pr.part. of *spirare,* to breathe.] A fricative. **—spi′rant** *adj.*
**spire¹** (spīr) *n.* [ME < OE *spīr.*] **1.** A top part tapering upward : PINNACLE. **2.** A structure, as a steeple, that tapers to a point at the top. **3.** A slender tapering part, as a newly sprouting blade of grass. —*v.* **spired, spir·ing, spires.** —*vt.* To furnish with a spire. —*vi.* To rise taperingly.
**spire²** (spīr) *n.* [Fr. < Lat. *spira,* coil < Gk. *speira.*] **1.** A spiral, esp. a single turn of a spiral : WHORL. **2.** *Zool.* The area farthest from the aperture and nearest the apex on a coiled gastropod shell.
**spi·re·a** *also* **spi·rae·a** (spī-rē′ə) *n.* [Lat. *spiraea,* meadowsweet < Gk. *speiraia* < *speira,* coil.] A plant or shrub of the genus *Spiraea,* including the bridal wreath, hardhack, and meadowsweet, with small white or pink flower clusters.
**spi·reme** (spī′rēm′) *also* **spi·rem** (-rēm′) *n.* [G. *Spirem* < Gk. *speirēma,* coil < *speira.*] *Biol.* **1.** The tangle of filaments that appears at the beginning of prophase in meiosis or mitosis. **2.** One of the filaments appearing in meiosis or mitosis.
**spi·rif·er·ous** (spī-rĭf′ər-əs) *adj.* [SPIR(E)² + -FEROUS.] Having a spiral structure or spiral parts.
**spi·ril·lum** (spī-rĭl′əm) *n., pl.* **-ril·la** (-rĭl′ə) [NLat. *Spirillum,* genus name, dim. of Lat. *spira,* coil < Gk. *speira.*] A flagellated aerobic bacteria of the genus *Spirillum,* with an elongated spiral form.
**spir·it** (spĭr′ĭt) *n.* [ME < AN < Lat. *spiritus,* breath < *spirare,* to breathe.] **1.** The vital principle or animating force traditionally believed to be within living beings. **2.** The soul, considered as departing from the body of a person at death. **3. Spirit.** The Holy Ghost. **4. Spirit.** *Christian Science.* GOD 1c. **5.** A supernatural being. **6. a.** The part of a human being associated with the mind and feelings as distinguished from the physical body. **b.** A person's essential nature. **7.** A person as characterized by a stated quality <a bold *spirit*> **8.** A specific inclination or tendency. **9. spirits.** An emotional state. **10.** A particular emotional state characterized by vigor and animation. **11.** Strong loyalty or dedication. **12.** The predominant mood of an occasion or period <the *spirit* of 1776> **13.** The real sense or significance of something <Heed the *spirit* of the law.> **14.** *often* **spirits.** An alcohol solution of an essential or volatile substance. **15. spirits.** An alcoholic beverage. —*vt.* **-it·ed, -it·ing, -its. 1.** To carry off mysteriously or secretly <kidnapers who *spirited* the child off> **2.** To impart courage, animation, or determination to : INSPIRIT.
☆ ***syns:*** SPIRIT, BRIO, DASH, ÉLAN, ESPRIT, LIVELINESS, PEP *n. core meaning* : a lively, emphatic, eager quality or manner <worked with extraordinary *spirit*>
**spir·it·ed** (spĭr′ĭ-tĭd) *adj.* **1.** Full of or marked by animation, vigor, or courage <a *spirited* discussion> **2.** Having a specified mood or nature <low-*spirited*> **—spir′it·ed·ly** *adv.* **—spir′it·ed·ness** *n.*
**spir·it·ism** (spĭr′ĭ-tĭz′əm) *n.* SPIRITUALISM 1a. **—spir′it·ist** *n.* **—spir′it·is′tic** *adj.*
**spirit lamp** *n.* A lamp using alcohol or other liquid fuel.
**spir·it·less** (spĭr′ĭt-lĭs) *adj.* Devoid of energy or enthusiasm : LISTLESS. **—spir′it·less·ly** *adv.* **—spir′it·less·ness** *n.*
**spirit level** *n.* LEVEL 7a.
**spirit of ammonia** *n. var. of* SPIRITS OF AMMONIA.
**spirit of turpentine** *n. var. of* SPIRITS OF TURPENTINE.
**spirit of wine** *n. var. of* SPIRITS OF WINE.
**spir·it·ous** (spĭr′ĭ-təs) *adj.* **1.** Spirituous. **2.** *Archaic.* Refined : pure.
**spirit rapping** *n.* Communication by rapping, held to be produced by spirits of the dead, as at a séance.
**spirits of ammonia** *also* **spirit of ammonia** *n.* A colorless aromatic solution made from ammonium carbonate, ammonia water, alcohol, and water, with small amounts of various aromatic agents, used as a remedy for faintness.
**spirits of turpentine** *also* **spirit of turpentine** *n.* TURPENTINE 1.
**spirits of wine** *also* **spirit of wine** *n.* Rectified ethyl alcohol.
**spir·i·tu·al** (spĭr′ĭ-cho͞o-əl) *adj.* **1.** Of, relating to, consisting of, or having the nature of spirit. **2.** Of, concerned with, or affecting the soul. **3.** Of, from, or relating to God : DEIFIC. **4.** Of or belonging to a church or religion : SACRED. **5.** Relating to or having the nature of spirits : SUPERNATURAL. —*n.* **1. a.** A religious folk song of black American origin. **b.** A work composed in imitation of a spiritual. **2.** *often* **spirituals.** Religious, spiritual, or ecclesiastical matters. **—spir′i·tu·al·ly** *adv.* **—spir′i·tu·al·ness** *n.*
**spiritual bouquet** *n.* A card sent by a Roman Catholic indicating that certain devotional acts will be undertaken on behalf of a person.
**spir·i·tu·al·ism** (spĭr′ĭ-cho͞o-ə-lĭz′əm) *n.* **1. a.** The belief that the dead communicate with the living, usu. through a medium. **b.** The practices or doctrines of those holding such a belief. **2.** A philosophy, doctrine, or religion emphasizing the spiritual rather than the material. **—spir′i·tu·al·ist** *n.* **—spir′i·tu·al·is′tic** *adj.*
**spir·i·tu·al·i·ty** (spĭr′ĭ-cho͞o-ăl′ĭ-tē) *n., pl.* **-ties. 1.** The state, quality, or fact of being spiritual. **2.** The clergy. **3.** *often* **spiritualities.** Something, as property or revenue, belonging to the church or to a cleric.
**spir·i·tu·al·ize** (spĭr′ĭ-cho͞o-ə-līz′) *vt.* **-ized, -iz·ing, -iz·es. 1.** To impart a spiritual nature to. **2.** To invest with or treat as having a spiritual sense or meaning. **—spir′i·tu·al·i·za′tion** *n.* **—spir′i·tu·al·iz′er** *n.*
**spir·i·tu·al·ty** (spĭr′ĭ-cho͞o-əl-tē) *n., pl.* **-ties.** Spirituality.
**spir·i·tu·el** *also* **spir·i·tu·elle** (spĭr′ĭ-cho͞o-ĕl′, spē′rē-to͞o-ĕl′, -tü-) *adj.* [Fr., spiritual.] Having or evidencing a refined mind.
**spir·i·tu·ous** (spĭr′ĭ-cho͞o-əs) *adj.* Having the nature of or containing alcohol. **—spir′i·tu·os′i·ty** (-ŏs′ĭ-tē), **spir′i·tu·ous·ness** *n.*
**spiro-** *pref.* [< Lat. *spirare,* to breathe.] Respiration <*spiro*meter>
**spi·ro·chete** *also* **spi·ro·chaete** (spī′rə-kēt′) *n.* [NLat. *Spirochaeta,* genus name : Lat. *spira,* coil + Lat. *chaeta,* bristle < Gk. *khaitē,* long hair.] Any of various slender, nonflagellated, twisted microorganisms of the order Spirochaetales, many of which are pathogenic, causing syphilis, relapsing fever, yaws, and other diseases. **—spi′ro·chet′al** (-kēt′l) *adj.*
**spi·ro·che·to·sis** (spī′rə-kē-tō′sĭs) *n.* [SPIROCHET(E) + -OSIS.] A disease, as syphilis, caused by a spirochete.
**spi·ro·graph** (spī′rə-grăf′) *n.* An instrument for registering the depth and rapidity of respiratory movements. **—spi′ro·graph′ic** *adj.* **—spi′ro·graph′i·cal·ly** *adv.* **—spi·rog′ra·phy** (spī-rŏg′rə-fē) *n.*
**spi·ro·gy·ra** (spī′rə-jī′rə) *n.* [NLat. *Spirogyra,* genus name : Lat. *spira,* coil + Gk. *guros,* ring.] Any of various green, filamentous freshwater algae of the genus *Spirogyra,* having chloroplasts in spirally twisted bands.
**spi·roid** (spī′roid′) *adj.* Like a spiral.
**spi·rom·e·ter** (spī-rŏm′ĭ-tər) *n.* An instrument for measuring the volume of air entering and leaving the lungs. **—spi′ro·met′ric** (-rə-mĕt′rĭk) *adj.* **—spi·rom′e·try** *n.*
**spi·ro·no·lac·tone** (spə-rō′nō-lăk′tōn′, spī-rŏn′ə-) *n.* [SPIR(O)- + -*no-* (of unknown orig.) + LACTONE.] A steroid, $C_{24}H_{32}O_4S$, used medically as a diuretic.
**spirt** (spûrt) *n.* & *v. Chiefly Brit. var. of* SPURT.
**spir·u·la** (spĭr′yə-lə, spĭr′ə-) *n., pl.* **-lae** (-lē′) [NLat. *Spirula,* genus name, dim. of Lat. *spira,* coil < Gk. *speira.*] A small cephalopod

mollusk of the genus *Spirula,* with a spirally coiled, partitioned internal shell.
**spit**[1] (spĭt) *n.* [ME *spitten* < OE *spittan.*] **1.** Expectorated saliva : SPITTLE. **2.** The act of expectorating. **3.** Something, as the frothy secretion of certain insects, that is felt to resemble saliva. **4.** A brief, scattered fall of rain or snow. —*v.* **spat** (spăt) or **spit, spit·ting, spits.** —*vt.* **1.** To eject from the mouth. **2.** To eject as if by spitting <*spat* out an order> —*vi.* **1.** To expectorate. **2.** To express contempt or animosity by or as if by spitting. **3.** To make a hissing or sputtering noise. **4.** To rain or snow in light, scattered drops or flakes.
**spit**[2] (spĭt) *n.* [ME < OE *spitu.*] **1.** A slender, pointed rod on which meat is skewered for broiling. **2.** A narrow point of land extending into a body of water. —*vt.* **spit·ted, spit·ting, spits.** To skewer on or as if on a spit.
**spit·al** (spĭt'l) *n.* [ME *spitel* < Med. Lat. *hospitale.* —see HOSPITAL.] A hospital, esp. one for contagious diseases.
**spit·ball** (spĭt'bôl') *n.* **1.** Paper chewed and shaped into a lump for use as a projectile. **2.** *Baseball.* An illegal pitch in which the ball is moistened on one side with spit.
**spit curl** *n.* [From the use of saliva to fix the curl.] A spiral curl pressed flat against the cheek or forehead.
**spite** (spīt) *n.* [ME, outrage, insult, ill will < OFr. *despite.* —see DESPITE.] **1.** Malicious ill will prompting an urge to hurt or humiliate. **2.** An instance of malicious feeling. —*vt.* **spit·ed, spit·ing, spites. 1. a.** To show spite toward. **b.** To vent spite upon. **2. a.** To fill with spite. **b.** To annoy : irritate. **—in spite of.** Regardless of.
**spite·ful** (spīt'fəl) *adj.* Filled with, caused by, or displaying spite : MALICIOUS. **—spite'ful·ly** *adv.* **—spite'ful·ness** *n.*
**spit·fire** (spĭt'fīr') *n.* A quick-tempered, excitable person.
**spit·ter** (spĭt'ər) *n.* **1.** One that spits. **2.** *Baseball.* SPITBALL 2. **3.** A young deer with unbranched horns : PRICKET.
**spitting image** *n.* [Alteration of *spit and image* < *spit,* an exact likeness.] A perfect likeness or counterpart.
**spit·tle** (spĭt'l) *n.* [ME *spyttle,* alteration of *spatel* < OE *spatl.*] **1.** SPIT[1] 1. **2.** The frothy liquid secreted by spittlebugs.
**spit·tle·bug** (spĭt'l-bŭg') *n.* An insect of the family Cercopidae, the nymphs of which form frothy masses of liquid on plant stems.
**spit·toon** (spĭ-to͞on') *n.* [SPIT + *-oon,* as in *balloon.*] A bowl-shaped, usu. metal vessel into which one may spit.
**spitz** (spĭts) *n.* [G. < *spitz,* pointed.] A dog orig. bred in Germany, with a long, thick, usu. white coat and a tail curled over the back.
**spiv** (spĭv) *n.* [Dial. *spiff,* dandy.] *Chiefly Brit.* **1.** One who is usu. unemployed and who lives by one's wits. **2.** A shirker of work or responsibility : SLACKER.
**splanch·nic** (splăngk'nĭk) *adj.* [Gk. *splankhnikos,* of the bowels < *splankhna,* inward parts.] Of or relating to the viscera : VISCERAL <a *splanchnic* nerve>
**splash** (splăsh) *v.* **splashed, splash·ing, splash·es.** [Alteration of PLASH.] —*vt.* **1.** To scatter or dash (a liquid) about in flying masses. **2. a.** To dash liquid on. **b.** To wet or soil by splashing. **3.** To cause to splash. —*vi.* **1. a.** To cause a liquid to fly in scattered masses. **b.** To fall into or move through liquid with this effect. **2.** To move, spill, or fly about in scattered masses. **—splash down.** To land on water <The spacecraft *splashed down* in the Pacific.> —*n.* **1.** The act or sound of splashing. **2.** A flying mass of liquid. **3.** A marking produced by or as if by scattered liquid <great *splash* of light> **4.** A great though often short-lived impression : STIR. **—splash'er** *n.*
**splash·board** (splăsh'bôrd', -bōrd') *n.* **1.** A structure that protects a vehicle from splashes of mud. **2.** A screen on a boat to keep water from splashing on the deck. **3.** A board for closing a spillway or sluice.
**splash·down** (splăsh'doun') *n.* The landing of a spacecraft or missile on water.
**splash·y** (splăsh'ē) *adj.* **-i·er, -i·est. 1.** Making or apt to make splashes. **2.** Covered with splashes of color. **3.** Ostentatious : showy. **—splash'i·ly** *adv.* **—splash'i·ness** *n.*
**splat**[1] (splăt) *n.* [Orig. unknown.] A slat of wood, as one in the middle of a chair back.
**splat**[2] (splăt) *n.* [Imit.] A slapping noise. **—splat** *adv.*
**splat·ter** (splăt'ər) *v.* **-tered, -ter·ing, -ters.** [Blend of SPLASH and SPATTER.] —*vt.* To spatter (something), esp. to soil with splashes of liquid. —*vi.* To spatter, esp. to move or fall so as to cause heavy splashes. **—splat'ter** *n.*
**splay** (splā) *adj.* [< ME *splayen,* to spread out, short for *displayen.* —see DISPLAY.] **1.** Turned or spread out. **2.** Clumsy : awkward. —*n.* **1.** Expansion : spread. **2.** An oblique slope given to the sides of an opening in a wall so that the opening is wider at one face than at the other. —*v.* **splayed, splay·ing, splays.** —*vt.* **1.** To spread (e.g., the limbs) out or apart, esp. in a clumsy way. **2.** To make slanting or sloping : BEVEL. **3.** To dislocate (a bone). —Used of an animal. —*vi.* **1.** To be spread out or apart. **2.** To slope or slant.
**splay·foot** (splā'fo͝ot') *n.* **1.** A physical deformity marked by abnormally flat and turned-out feet. **2.** A foot abnormally flat and turned-out : FLATFOOT. **—splay'foot'ed** *adj.*
**spleen** (splēn) *n.* [ME *splene* < OFr. *esplen* < Lat. *splen* < Gk *splēn.*] **1. a.** A visceral organ in human beings composed of a white pulp of lymphatic nodules and tissue and a red pulp of venous tissue, functioning as a blood filter and to store blood. **b.** A homologous organ or tissue in other vertebrates. **2.** *Obs.* **a.** The seat of emotions or passions. **b.** A whim : caprice. **3.** *Archaic.* Melancholy. **4.** Ill temper. **—spleen'y** *adj.*
**spleen·wort** (splēn'wûrt', -wôrt') *n.* [So called because it was thought to cure spleen disorders.] Any of various ferns of the genus *Asplenium,* with featherlike, often evergreen fronds.
**splen-** *pref. var. of* SPLENO-.
**splen·dent** (splĕn'dənt) *adj.* [ME < Lat. *splendens,* pr.part. of *splendēre,* to shine.] **1.** Shining : brilliant <*splendent* luster> **2.** Conspicuously illustrious <the *splendent* genius of Einstein>
**splen·did** (splĕn'dĭd) *adj.* [Fr. *splendide* < Lat. *splendidus* < *splendēre,* to shine.] **1.** Brilliant with color or light : RADIANT. **2.** Imposing : magnificent. **3.** Glorious : illustrious. **4.** Admirable for boldness or purity : TRANSCENDENT. **5.** Exceptionally good or satisfying <a *splendid* feast> **—splen'did·ly** *adv.* **—splen'did·ness** *n.*
**splen·dif·er·ous** (splĕn-dĭf'ər-əs) *adj.* [ME < Med. Lat. *splendiferus* : Lat. *splendor,* splendor + *ferre,* to bear.] Splendid.
**splen·dor** (splĕn'dər) *n.* [ME *splendure* < OFr. *splendeur* < Lat. *splendor* < *splendēre,* to shine.] **1.** Great luster or light : BRILLIANCE. **2. a.** Magnificent display : GRANDEUR. **b.** Something magnificent. **3.** Glory : illustriousness. **—splen'dor·ous, splen'drous** (splĕn'drəs) *adj.*
**splen·dour** (splĕn'dər) *n. Chiefly Brit. var. of* SPLENDOR.
**sple·nec·to·my** (splĭ-nĕk'tə-mē) *n., pl.* **-mies.** Surgical removal of the spleen.
**sple·net·ic** (splĭ-nĕt'ĭk) *also* **sple·net·i·cal** (-ĭ-kəl) *adj.* [LLat. *spleneticus* < Lat. *splen,* spleen <Gk.] **1.** Of or relating to the spleen. **2.** Ill-humored and irritable. —*n.* An ill-humored person. **—sple·net'i·cal·ly** *adv.*
**splen·ic** (splĕn'ĭk) *adj.* Of, in, near, or relating to the spleen.
**sple·ni·i** (splē'nē-ī') *n. pl. of* SPLENIUS.
**sple·ni·tis** (splĭ-nī'tĭs) *n.* Inflammation of the spleen.
**sple·ni·us** (splē'nē-əs) *n., pl.* **-ni·i** (-nē-ī') [NLat. < Lat. *splenius,* patch, plaster < Gk. *splēnion* < *splēn,* spleen.] Either of two muscles of the back of the neck, extending from the vertebral column to the skull, that rotate and extend the head and neck. **—sple'ni·al** *adj.*
**spleno-** or **splen-** *pref.* [Gk. *splēno-* < *splēn,* spleen.] Spleen <*splenitis*>
**sple·no·meg·a·ly** (splē'nō-mĕg'ə-lē, splĕn'ō-) *n.* Abnormal enlargement of the spleen.
**splice** (splīs) *vt.* **spliced, splic·ing, splic·es.** [MDu. *splissen.*] **1. a.** To join (e.g., film) at the ends. **b.** To join (ropes) by intertwining strands. **2.** To join (pieces of wood) by overlapping and binding at the ends. **3.** *Informal.* To unite in marriage. —*n.* **1.** A joint made by splicing. **2.** A place where parts have been spliced. **—splic'er** *n.*
**spline** (splīn) *n.* [Orig. unknown.] **1. a.** Any of a series of projections on a shaft that fit into slots on a corresponding shaft, enabling both to rotate together. **b.** The groove or slot for such a projection. **2.** A flexible piece of material, as wood, hard rubber, or metal, used in drawing curves. **3.** A wooden or metal strip : SLAT.
**splint** (splĭnt) *n.* [ME < MLG *splinte.*] **1.** A thin piece split off from a larger piece : SPLINTER. **2.** A rigid device for preventing movement of a joint or of the ends of a fractured bone. **3.** A thin flexible wooden strip, as one used in basketmaking and chair caning. **4.** A plate or strip of metal. **5.** A bony enlargement of the cannon bone or splint bone of a horse. —*vt.* **splint·ed, splint·ing, splints.** To support or restrict with or as if with a splint.
**splint bone** *n.* Either of two small metacarpal or metatarsal bones in horses or related animals.
**splin·ter** (splĭn'tər) *n.* [ME < MDu.] **1.** A sharp slender piece, as of wood, bone, glass, or metal, split or broken off from a main body. **2.** A group, as a religious sect or a political faction, that has broken away from a parent group. —*v.* **-tered, -ter·ing, -ters.** —*vi.* & *vt.* To split or cause to split into sharp slender pieces. **—splin'ter·y** *adj.*
**split** (splĭt) *v.* **split, split·ting, splits.** [Du. *splitten* < MDu.] —*vt.* **1.** To divide cleanly or sharply, esp. into lengthwise sections or into two parts of approx. equal size. **2.** To break, burst, or rip apart with force : REND. **3.** To separate (persons or groups) : DISUNITE. **4.** To divide and share <*split* dessert> **5.** To separate (e.g., leather) into layers. **6.** To mark (a ballot or vote) in favor of candidates from different parties. **7.** To win half the games of (a series or double-header). **8.** *Slang.* To depart from : LEAVE <Let's *split* the party early.> —*vi.* **1.** To become separated into parts, esp. to divide lengthwise. **2.** To become broken or ripped apart, esp. from internal pressure. **3.** To part company due to discord or disagreement <The couple *split.*> **4.** To divide or share something with others. **5.** *Slang.* To depart. —*n.* **1.** The act or result of splitting. **2.** A breach or rupture in a group. **3.** SPLINTER 1. **4.** Something divided and portioned out : SHARE. **5.** A split strip of flexible wood used in basketmaking. **6. a.** A bottle of an alcoholic or carbonated beverage half the usual size, usu. about six ounces <a *split* of champagne> **b.** A drink of half the usual quantity. **c.** A half pint. **7.** A dessert of sliced fruit, ice cream, and toppings. **8.** *often* **splits.** An acrobatic feat in which the legs are stretched out in opposite directions at right angles to the

trunk. **9.** A single thickness of a split hide. **10.** An arrangement of bowling pins left standing after the first bowl with one or more intermediate pins knocked down. *—adj.* **1.** Separated or divided. **2.** Fissured longitudinally : CLEFT. **3.** Quoted in 16ths rather than in 8ths. —Used of stocks. **—split hairs.** To make trivial distinctions : QUIBBLE. **—split'ter** *n.*

**split infinitive** *n.* An infinitive verb form with usu. an adverbial element interposed between *to* and the verb form, as in *to carefully examine.*

**split-lev·el** (splĭt'lĕv'əl) *adj.* Having the floor levels of adjoining rooms separated by about a half story <a *split-level* house>

**split personality** *n.* Hysteria in which the afflicted person manifests two or more relatively distinct personalities.

**split rail** *n.* A fence rail split lengthwise from a log.

**split second** *n.* An instant : flash.

**split shift** *n.* A working shift divided into several time periods, as mornings and evenings, with a break of several hours between.

**split ticket** *n.* **1.** A ballot cast for candidates of two or more political parties. **2.** A ballot including the names of candidates of more than one party.

**split·ting** (splĭt'ĭng) *adj.* Very severe <a *splitting* headache>

**splotch** (splŏch) *n.* [Perh. blend of SPOT and BLOTCH.] An irregularly shaped stain, spot, or discoloration. *—vt.* **splotched, splotch·ing, splotch·es.** To mark with a splotch or splotches. **—splotch'i·ness** *n.* **—splotch'y** *adj.*

**splurge** (splûrj) *v.* **splurged, splurg·ing, splurg·es.** [Orig. unknown.] *—vi.* **1.** To indulge in an extravagant luxury or expense. **2.** To be ostentatious or showy. *—vt.* To spend extravagantly or wastefully. *—n.* **1.** Extravagant display. **2.** An expensive indulgence : SPREE <a shopping *splurge*> **—splurg'y** *adj.*

**splut·ter** (splŭt'ər) *v.* **-tered, -ter·ing, -ters.** [Perh. alteration of SPUTTER.] *—vi.* **1.** To make a spitting sound. **2.** To speak incoherently, as when confused or angry. *—vt.* To utter or express hastily and incoherently <*spluttered* their objections> *—n.* A spluttering sound. **—splut'ter·er** *n.*

**Spode** (spōd). A trademark for porcelain or chinaware of fine quality.

**spod·u·mene** (spŏj'ə-mēn') *n.* [Fr. *spodumène* < Gk. *spodoumenos,* pr.part. of *spodousthai,* to be burned to ashes < *spodos,* wood ashes.] A greenish to pinkish or lilac mineral, essentially $LiAlSi_2O_6$, used as a source of lithium and in transparent varieties as a gemstone.

**spoil** (spoil) *v.* **spoiled** or **spoilt** (spoilt), **spoil·ing, spoils.** [ME *spoilen,* to plunder < OFr. *espoillier* < Lat. *spoliare* < *spolium,* booty.] *—vt.* **1.** To impair the value or quality of : DAMAGE. **2.** To impair the perfection of : FLAW. **3.** To disrupt. **4.** To overindulge or overpraise <*spoiled* the baby> **5.** *Obs.* **a.** To plunder. **b.** To take by force. *—vi.* **1.** To become tainted, rotten, or otherwise unfit for use : DECAY. **2.** *Obs.* To pillage. **—spoil for.** To be eager for <*spoiled for* a fight> *—n.* **1. spoils.** Goods or property seized from a victim after a conflict. **2. spoils.** Incidental benefits reaped by a winner, esp. political patronage enjoyed by a successful party or candidate. **3.** *Archaic.* The act of plundering. **4.** An object of plunder. **5.** Refuse removed from an excavation.

**spoil·age** (spoi'lĭj) *n.* **1.** The state or process of becoming spoiled. **2. a.** Something spoiled. **b.** The degree to which something has been spoiled.

**spoil·er** (spoi'lər) *n.* **1.** One who seizes spoils. **2.** Something causing spoilage. **3.** A long, narrow hinged plate on the upper surface of an aircraft wing, whose position affects the lift. **4.** A candidate for office whose chances of victory are slight but who may acquire enough votes to prevent a leading candidate from winning.

**spoil·sport** (spoil'spôrt', -spōrt') *n.* One who behaves in such a way as to mar the pleasure of others.

**spoils system** *n.* The practice after an election of rewarding loyal supporters of the winning candidates and party with appointive public offices.

**spoilt** (spoilt) *v. var. p.t. & p.p. of* SPOIL.

**spoke[1]** (spōk) *n.* [ME < OE *spāca.*] **1.** A rod or brace connecting the hub and the rim of a wheel. **2.** A handle that projects from the rim of a ship's steering wheel. **3.** A rod or stick that may be inserted into a wheel to prevent it from turning. **4.** A rung of a ladder. *—vt.* **spoked, spok·ing, spokes. 1.** To equip with spokes. **2.** To impede (a wheel) by inserting a rod.

**spoke[2]** (spōk) *v.* **1.** *p.t. of* SPEAK. **2.** *Archaic. var. p.p. of* SPEAK.

**spo·ken** (spō'kən) *adj.* [P.part. of SPEAK.] **1.** Expressed orally : UTTERED. **2.** Using speech in a specified manner <a soft-*spoken* person who never offended anyone>

**spokes·man** (spōks'mən) *n.* [Prob. < *spoke,* obs. p.part. of SPEAK + MAN.] One who speaks on behalf of another or others.

**spokes·per·son** (spōks'pûr'sən) *n.* A spokesman or a spokeswoman.

**spokes·wom·an** (spōks'wo͝om'ən) *n.* [SPOKES(MAN) + WOMAN.] A woman who speaks on behalf of another or others.

**spo·li·a·tion** (spō'lē-ā'shən) *n.* [ME *spoliacioun* < Lat. *spoliatio* < *spoliare,* to despoil. —see SPOIL.] **1.** The act of plundering, esp. the seizure of neutral vessels at sea by a belligerent power in wartime. **2.** *Law.* Intentional alteration or destruction of a document. **—spo'li·a'tor** *n.*

**spon·da·ic** (spŏn-dā'ĭk) *adj.* [Fr. *spondaïque* < Lat. *spondaicus* < Gk. *spondeiakos* < *spondeios,* spondee.] Of, relating to, or consisting of spondees.

**spon·dee** (spŏn'dē') *n.* [ME *sponde* < OFr. *spondee* < Lat. *spondeum* < Gk. *spondeios* < *spondē,* libation, from its use in songs performed at libations.] A metrical foot consisting of two long syllables or two stressed ones.

**spon·dy·li·tis** (spŏn'də-lī'tĭs) *n.* [Gk. *spondulos,* vertebra + -ITIS.] Inflammation of one or more of the vertebrae.

**sponge** (spŭnj) *n.* [ME < OE < Lat. *spongia* < Gk. *sphongos.*] **1. a.** Any of numerous primitive, chiefly marine animals of the phylum Porifera, with a porous skeleton composed of fibrous material or siliceous or calcareous spicules and often forming irregularly shaped colonies attached to an underwater surface. **b.** The light, fibrous, absorbent connective structure of some of these organisms, used for bathing or cleaning. **2.** A substance, as rubber, cellulose, or plastic, that has spongelike qualities. **3.** A gauze pad for absorbing blood and other fluids, as in surgery. **4.** Dough that is leavened or is being leavened. **5.** Sponge cake. **6.** A sponge bath. **7. a.** *Informal.* A glutton. **b.** *Slang.* A drunkard. **8.** One who habitually depends on others for maintenance or for support. *—v.* **sponged, spong·ing, spong·es.** *—vt.* **1.** To moisten, wipe, or clean with or as if with a sponge. **2.** To wipe out : ERASE. **3.** *Informal.* To obtain free <*sponge* money> *—vi.* **1.** To fish for sponges. **2.** *Informal.* To live by relying on the generosity of others. **—throw (or toss) in the sponge.** *Informal.* To abandon an effort : GIVE UP.

**sponge bath** *n.* A washing of the body with a sponge or cloth and without immersion.

**sponge cake** *n.* A very light porous cake made of flour, sugar, beaten eggs, and flavoring and containing no shortening.

**sponge mushroom** *n.* The morel.

**spong·er** (spŭn'jər) *n.* **1.** A gatherer of sponges. **2.** SPONGE 8.

**sponge rubber** *n.* A soft, porous rubber used in toys, cushions, gaskets, and weather stripping and as a vibration dampener.

**spon·gin** (spŭn'jĭn) *n.* [G. < Lat. *spongia,* sponge.] The fibrous material forming the skeletal structure of sponges.

**spon·gi·o·blast** (spŭn'jĭ-ə-blăst') *n.* [Lat. *spongia,* sponge + -BLAST.] Embryonic epithelial cells giving rise to the neuroglia cells.

**spon·gi·o·cyte** (spŭn'jĭ-ə-sīt') *n.* [Lat. *spongia,* sponge + -CYTE.] A neuroglia cell.

**spon·go·coel** (spŏng'gə-sēl') *n.* [SPONG(E) + -COEL.] *Zool.* The central cavity of a sponge that opens to the outside by way of the osculum.

**spong·y** (spŭn'jē) *adj.* **-i·er, -i·est.** Like a sponge <*spongy* soil> **—spong'i·ness** *n.*

**spon·son** (spŏn'sən) *n.* [Orig. unknown.] **1.** A structure that projects from the side of a boat or ship, esp. a gun platform. **2.** A short, curved, air-filled projection on a seaplane hull that provides stability in the water.

**spon·sor** (spŏn'sər) *n.* [Lat. < *spondēre,* to pledge.] **1.** One who assumes responsibility for a person or group during instruction, apprenticeship, or probation. **2.** One who vouches for the suitability of a candidate for admission. **3.** A legislator who proposes and urges adoption of a bill. **4.** One who presents a candidate for baptism or confirmation : GODPARENT. **5.** A business that pays for a television or radio program, usu. in return for advertising time. *—vt.* **-sored, -sor·ing, -sors.** To serve as a sponsor for. **—spon·so'ri·al** (-sôr'ē-əl, -sōr'-) *adj.* **—spon'sor·ship'** *n.*

**spon·ta·ne·ous** (spŏn-tā'nē-əs) *adj.* [LLat. *spontaneus* < Lat. *sponte,* voluntarily.] **1.** Occurring without apparent external cause : SELF-GENERATED. **2.** Impulsive : unpremeditated. **3.** Unconstrained and unstudied in behavior. **4.** Growing without cultivation or human labor : INDIGENOUS. **—spo'ta·ne'i·ty** spŏn'tə-nē'ĭ-tē, -nā'-) *n.* **—spon·ta'ne·ous·ly** *adv.* **—spon·ta'ne·ous·ness** *n.*

☆ ***syns:*** SPONTANEOUS, AUTOMATIC, IMPULSIVE, INSTINCTIVE, REFLEX, UNPREMEDITATED *adj. core meaning* : acting or happening without apparent forethought, prompting, or planning <embraced in a *spontaneous* gesture of affection> ***ant:*** premeditated

**spontaneous abortion** *n.* MISCARRIAGE 2.

**spontaneous combustion** *n.* Ignition in a thermally isolated substance, as in oily rags or hay, caused by a localized heat-increasing reaction between the oxidant and the fuel.

**spontaneous generation** *n. Biol.* Abiogenesis.

**spon·toon** (spŏn-to͞on') *n.* [Fr. *sponton* < Ital. *spuntone* < *spuntare,* to blunt : *s-,* off (< Lat. *dis-*) + *punto,* point < Lat. *punctum* < *pungere,* to pierce.] A short pike carried by 18th-cent. infantry officers.

**spoof** (spo͞of) *n.* [< *Spoof,* a trademark for a card game characterized by nonsense and hoaxing.] **1.** Nonsense. **2.** A hoax. **3.** Gentle satirical imitation. *—vt.* **spoofed, spoof·ing, spoofs. 1.** To deceive. **2.** To satirize gently.

**spook** (spo͞ok) [Du.] *Informal.* —*n.* **1.** A ghost. **2.** A secret agent : SPY. —*vt.* **spooked, spook·ing, spooks. 1.** To haunt. **2.** To frighten, esp. to startle (cattle).

**spook·y** (spo͞o'kē) *adj.* **-i·er, -i·est.** *Informal.* **1.** Ghostly. **2.** Easily startled. **—spook'i·ly** *adv.* **—spook'i·ness** *n.*

**spool** (spo͞ol) *n.* [ME *spole* < OFr. *espole* < MDu. *spoele.*] **1.** A cylinder on which wire, thread, or string is wound. **2.** The amount of material wound on a spool. **3.** A reel for magnetic paper or plastic tape. **4.** Something similar to a spool. —*vt.* **spooled, spool·ing, spools.** To wind on a spool.

**spool·ing** (spo͞o'lĭng) *n. Computer Sci.* A procedure that involves storing information temporarily on a file while awaiting further processing.

**spoon** (spo͞on) *n.* [ME < OE *spōn,* chip of wood.] **1.** A utensil consisting of a small shallow bowl on a handle, used in preparing, serving, or eating food. **2.** A device similar to a spoon, esp.: **a.** A shiny, curved metallic fishing lure. **b.** A paddle or oar with a curved blade. **3.** The three wood golf club. —*v.* **spooned, spoon·ing, spoons.** —*vt.* **1.** To lift, scoop up, or carry with or as if with a spoon. **2.** To shove or scoop (a ball) into the air, as in some games. —*vi.* **1.** To fish with a spoon lure. **2.** To give a ball an upward scoop in some games. **3.** *Informal.* To kiss or caress.

**spoon·bill** (spo͞on'bĭl') *n.* **1.** Any of several long-legged wading birds of the subfamily Plataleinae, having a long flat bill with a broadly spatulate tip. **2.** Any of various broad-billed ducks, as the shoveler. **3.** The paddlefish.

**spoon bread** *n.* A soft light bread made with corn meal, eggs, and milk, baked in a casserole, and served with a spoon.

**spoon·drift** (spo͞on'drĭft') *n.* [Obs. *spoon,* to drive back and forth + DRIFT.] Spindrift.

**spoon·er·ism** (spo͞o'nə-rĭz'əm) *n.* [After William A. *Spooner* (1844–1930).] An unintentional transposition of sounds of two or more words, as *Let me sew you to your sheet* for *Let me show you to your seat.*

**spoon·ey** (spo͞o'nē) *adj. var. of* SPOONY.

**spoon-fed** (spo͞on'fĕd') *adj.* **1.** Fed with a spoon. **2.** Overindulged. **3.** Given no chance to act or think independently.

**spoon·ful** (spo͞on'fo͝ol') *n., pl.* **-fuls.** The amount a spoon holds.

**spoon·y** *also* **spoon·ey** (spo͞o'nē) *adj.* **-i·er, -i·est. 1.** Enamored in a silly or sentimental way. **2.** Feebly sentimental.

**spoor** (spo͝or) *n.* [Afr. < MDu.] The track or trail of an animal, esp. a wild animal. —*v.* **spoored, spoor·ing, spoors.** —*vt.* To track by following a spoor. —*vi.* To track an animal by its spoor.

**spor-** *pref. var. of* SPORO-.

**spo·rad·ic** (spə-răd'ĭk, spô-) *also* **spo·rad·i·cal** (-ĭ-kəl) *adj.* [Med. Lat. *sporadicus* < Gk. *sporadikos,* isolated < *sporas,* scattered.] **1.** Occurring at irregular intervals : having no pattern or order. **2.** Appearing singly or at widely scattered localities, as a plant. **3.** Occurring in isolated instances <*sporadic* static on the radio> **—spo·rad'i·cal·ly** *adv.* **—spo·rad'i·cal·ness** *n.*

**spo·ran·gi·a** (spə-răn'jē-ə) *n. pl. of* SPORANGIUM.

**spo·ran·gi·o·phore** (spə-răn'jē-ə-fôr', -fōr') *n.* [SPORANGI(UM) + -PHORE.] A specialized branch or filament bearing sporangia.

**spo·ran·gi·um** (spə-răn'jē-əm) *n., pl.* **-gi·a** (-jē-ə) [NLat. : SPOR(O)- + Gk. *angeion,* vessel.] A spore-bearing structure in certain plants, as fungi, mosses, and ferns. **—spo·ran'gi·al** *adj.*

**spore** (spôr, spōr) *n.* [NLat. *spora* < Gk., seed.] **1.** An asexual, usu. single-celled reproductive organ characteristic of nonflowering plants such as fungi, mosses, or ferns. **2.** A microorganism, as a bacterium, in a dormant or resting state. —*vi.* **spored, spor·ing, spores.** To produce spores. **—spo·ra'ceous** (spə-rā'shəs, spô-) *adj.*

**spore case** *n.* A structure containing spores : SPORANGIUM.

**spo·ri·cide** (spôr'ĭ-sīd', spōr'-) *n.* A spore-killing agent. **—spo'ri·cid'al** (-sīd'l) *adj.*

**spo·rif·er·ous** (spə-rĭf'ər-əs, spô-, spō-) *adj.* Generating spores.

**sporo-** *or* **spor-** *pref.* [< NLat. *spora,* spore < Gk., seed.] Spore <*sporocyte*>

**spo·ro·carp** (spôr'ə-kärp', spōr'-) *n.* A multicellular structure in which spores are formed.

**spo·ro·cyst** (spôr'ə-sĭst', spōr'-) *n.* **1.** A resting cell that produces asexual plant spores. **2.** A protective case containing spores of certain protozoans. **3.** A saclike larval stage in many trematode worms.

**spo·ro·cyte** (spôr'ə-sīt', spōr'-) *n.* A cell that produces haploid spores during meiosis.

**spo·ro·gen·e·sis** (spôr'ə-jĕn'ĭ-sĭs, spōr'-) *n.* Production or formation of spores. **—spo'ro·gen'ic** (-jĕn'ĭk), **spo·rog'e·nous** (spə-rŏj'ə-nəs, spô-) *adj.*

**spo·ro·go·ni·um** (spôr'ə-gō'nē-əm, spōr'-) *n., pl.* **-ni·a** (-nē-ə). A structure in mosses that generates asexual spores.

**spo·rog·o·ny** (spə-rŏg'ə-nē, spô-) *n.* Production of spores resulting from sexual fusion of gametes prior to multiple fission, characteristic of certain protozoans. **—spo'ro·gon'ic** (spôr'ə-gŏn'ĭk, spōr'-), **spo·rog'o·nous** (spə-rŏg'ə-nəs, spô-) *adj.*

**spo·ro·phore** (spôr'ə-fôr', spōr'ə-fōr') *n.* A spore-bearing structure, esp. in fungi.

**spo·ro·phyll** (spôr'ə-fĭl', spōr'-) *n.* A leaf or leaflike organ that bears spores.

**spo·ro·phyte** (spôr'ə-fīt', spōr'-) *n.* The spore-producing phase in plants that reproduce by metagenesis. **—spo'ro·phyt'ic** (-fĭt'ĭk) *adj.*

**spo·ro·pol·len·in** (spôr'ə-pŏl'ə-nĭn, spōr'-) *n.* A polymer that comprises the exine of spores and pollen grains.

**-sporous** *suff.* [< SPOR(E) + -OUS.] Having a specified number or kind of spores <hetero*sporous*>

**spo·ro·zo·an** (spôr'ə-zō'ən, spōr'-) *n., pl.* **-zo·a** (-zō'ə) [< NLat. *Sporozoa,* class name : SPORO- + Gk. *zōia,* pl. of *zōion,* animal.] Any of numerous parasitic protozoans of the class Sporozoa, many of which have complex reproductive processes. **—spo'ro·zo'an** *adj.*

**spo·ro·zo·ite** (spôr'ə-zō'īt', spōr'-) *n.* [SPOROZO(A) + -ITE.] A sporozoan that has been released from a spore and is ready to penetrate a new host cell.

**spor·ran** (spôr'ən, spŏr'-) *n.* [Sc. Gael. *sporan* < LLat. *bursa,* bag < Gk., leather.] A leather or fur pouch worn by Scottish Highlanders at the front of the kilt.

sporran

**sport** (spôrt, spōrt) *n.* [ME *sporten,* to amuse < *disporten.* —see DISPORT.] **1.** An active pastime : RECREATION. **2.** A specific diversion, usu. involving physical exercise and having a set form and body of rules : GAME. **3.** Mockery : jest. **4.** One known for the manner of one's acceptance of the rules of a game or of a difficult situation <a good *sport*> **5.** *Informal.* One who lives a gay, extravagant life. **6.** *Genetics.* An organism that shows a marked change from the parent stock : MUTATION. **7.** *Archaic.* Amorous dalliance. —*v.* **sport·ed, sport·ing, sports.** —*vi.* **1.** To play or frolic. **2.** To trifle or joke. **3.** To mutate. —*vt.* To display or show off <*sported* a large yellow hat> —*adj.* Of, relating to, or appropriate for sports. **—in sport.** In jest. **—sport'ful** *adj.* **—sport'ful·ly** *adv.* **—sport'ful·ness** *n.*

**sport·ing** (spôr'tĭng, spōr'-) *adj.* **1.** Appropriate for or used in sports. **2.** Marked by sportsmanship. **3.** Of or having to do with gambling. **—sport'ing·ly** *adv.*

**sporting chance** *n. Informal.* An even chance for success.

**spor·tive** (spôr'tĭv, spōr'-) *adj.* **1.** Playful : frolicsome. **2.** Relating to or interested in sports. **3.** *Obs.* Amorous : wanton. **—spor'tive·ly** *adv.* **—spor'tive·ness** *n.*

**sports car** *n.* A car equipped for racing, esp. an aerodynamically shaped one- or two-passenger vehicle having a low center of gravity, and steering and suspension designed for precise control at high speeds on curving roads.

**sports·cast** (spôrts'kăst', spōrts'-) *n.* [SPORTS + (BROAD)CAST.] A television or radio broadcast of a sports event or of sports news. **—sports'cast'er** *n.*

**sports·man** (spôrts'mən, spōrts'-) *n.* **1.** A man active in sports. **2.** One who abides by the rules of a contest and accepts victory or defeat graciously. **—sports'man·ly** *adj.*

**sports·man·ship** (spôrts'mən-shĭp', spōrts'-) *n.* The qualities and conduct befitting a sportsman or sportswoman.

**sports medicine** *n.* Medicine dealing with the diseases and injuries resulting from sports participation.

**sports·wear** (spôrts'wâr', spōrts'-) *n.* Comfortable, casual clothes.

**sports·wom·an** (spôrts'wo͝om'ən, spōrts'-) *n.* A woman active in sports.

**sports·writ·er** (spôrts'rī'tər, spōrts'-) *n.* A writer about sports.

**sport·y** (spôr'tē, spōr'-) *adj.* **-i·er, -i·est.** *Informal.* **1.** Appropriate to sport or participation in sports. **2.** Casual in style <*sporty* pants> **3.** Carefree : gay. **—sport'i·ly** *adv.* **—sport'i·ness** *n.*

**spor·u·late** (spôr'yə-lāt', spōr'-) *vi.* **-lat·ed, -lat·ing, -lates.** [< NLat. *sporula,* small spore, dim. of *spora,* spore.] To produce or release spores. **—spor'u·la'tion** *n.*

**spot** (spŏt) *n.* [ME.] **1.** A specific place with relatively small and definite limits. **2. a.** A mark on a surface differing sharply in color from the surroundings, esp. a stain. **b.** A mark on a playing card indicating its value <a two-*spot*> **3.** A position : location. **4.** *Informal.* A troublesome situation. **5.** A personal defect or injury, as in one's reputation. **6.** *pl.* **spots** *or* **spot.** An edible marine fish, *Leiostomus xanthurus* of North American waters, with a dark spot above each pectoral fin. **7.** *Chiefly Brit.* A small amount : BIT <a *spot* of tea> **8.** *Informal.* A spotlight. **9.** A short presentation or commercial on radio or television between major programs, esp. by a local station on a network broadcast. —*v.* **spot·ted, spot·ting, spots.** —*vt.* **1.** To cause a spot or spots to appear on, esp.: **a.** To dirty with spots. **b.** To decorate with spots. **2.** To place in a specific location. **3.** To discern. **4.** To yield as a handicap in a sport <*spotted* their opponents 14 points> —*vi.* **1.** To become marked with spots. **2.** To make a stain : DISCOLOR. **3.** To locate targets from the air during combat or training missions. —*adj.* **1.** Made, paid, or delivered at once <*spot* cash> **2.** Presented between major radio or television programs <a *spot* commercial> **—hit the spot.** *Informal.* To be exactly what is needed. **—on the spot. 1.** Without delay or movement : at once. **2.** At the scene of action. **3.** Under pressure. **—spot'ta·ble** *adj.*

**spot check** *n.* An inspection or investigation carried out at random or in a limited way.

**spot-check** (spŏt'chĕk') *v.* **-checked, -check·ing, -checks.** *vt.* & *vi.* To subject to or make a spot check.

**spot·less** (spŏt'lĭs) *adj.* **1.** Entirely clean. **2.** Free from blemish : IMPECCABLE. **—spot'less·ly** *adv.* **—spot'less·ness** *n.*

**spot·light** (spŏt'līt') *n.* **1. a.** A strong beam of light that illuminates only a small area, used esp. to focus attention on a stage performer. **b.** A lamp that produces such a light. **2.** Public notoriety. **3.** An artificial light source with a strongly focused beam, as on a car. —*vt.* **-light·ed** or **-lit** (-lĭt), **-light·ing, -lights. 1.** To illuminate with a spotlight. **2.** To focus attention on.

**spot price** *n.* A commodity's market price.

**spot·ted** (spŏt'ĭd) *adj.* Marked or stained with spots.

**spotted cranesbill** *n.* The wild geranium.

**spotted fever** *n.* **1.** An often fatal infectious disease, as typhus or Rocky Mountain spotted fever, caused by *Rickettsiae,* that is transmitted by ticks and mites and is characterized by skin eruptions. **2.** An epidemic form of cerebrospinal meningitis.

**spotted sandpiper** *n.* A small brownish-gray North American shore bird, *Actitis macularia.*

**spot·ter** (spŏt'ər) *n.* **1.** One that applies spots. **2.** One that looks for, locates, and reports something, esp.: **a.** A military or civil-defense lookout. **b.** *Informal.* A person hired to detect dishonest acts by employees, as in a bank. **3. a.** One who identifies players on the field, as for a television or radio announcer <a football *spotter*> **b.** One who is responsible for watching and guarding a performer during practice to prevent injury, as in gymnastics or water-skiing. **4.** One employed by a dry cleaner to remove spots.

**spot·ty** (spŏt'ē) *adj.* **-ti·er, -ti·est. 1.** Having or marked with spots : SPOTTED. **2.** Inconsistent : uneven <*spotty* work> **—spot'ti·ly** *adv.* **—spot'ti·ness** *n.*

**spou·sal** (spou'zəl, -səl) *adj.* Of or relating to marriage : NUPTIAL. —*n. often* **spousals.** Marriage : nuptials.

**spouse** (spous, spouz) *n.* [ME < OFr. *espous* < Lat. *sponsus,* betrothal < p.part. of *spondēre,* to pledge.] A marriage partner. —*vt.* (spouz, spous) **spoused, spous·ing, spous·es.** *Archaic.* To marry.

**spout** (spout) *v.* **spout·ed, spout·ing, spouts.** [ME *spouten.*] —*vi.* **1.** To gush forth in a rapid stream or in spurts. **2.** To discharge a liquid continuously or in spurts. **3.** *Informal.* To speak volubly and tediously. —*vt.* **1.** To cause to flow or spurt out. **2.** To utter pompously and volubly. **3.** *Chiefly Brit.* To pawn. —*n.* **1.** A tube, mouth, or pipe through which liquid is released. **2.** A continuous stream of liquid. **3.** *Chiefly Brit.* A pawnbroker's shop. **—spout'er** *n.*

**sprach·ge·fühl** (shpräKH'gə-fül') *n.* [G.] An ear for the idiomatically correct or appropriate language.

**sprag** (sprăg) *n.* [Perh. of Scand. orig.] **1. a.** A piece of wood or metal wedged beneath a wheel or between spokes to keep a vehicle from rolling. **b.** A pointed stake lowered at an angle into the ground from a vehicle to prevent movement. **2.** A prop to support a mine roof.

**sprain** (sprān) *n.* [Orig. unknown.] **1.** A wrenching or laceration of the ligaments of a joint. **2.** The condition resulting from a sprain, characterized by pain, swelling, and disablement of the joint. —*vt.* **sprained, sprain·ing, sprains.** To cause a sprain in (a muscle or joint).

**sprang** (sprăng) *v. var. p.t. of* SPRING.

**sprat** (sprăt) *n.* [Alteration of ME *sprotte* < OE *sprott.*] **1.** A small marine food fish, *Clupea sprattus* of northeastern Atlantic waters. **2.** A fish, as a young herring, similar to the sprat.

**sprawl** (sprôl) *v.* **sprawled, sprawl·ing, sprawls.** [ME *sprawlen* < OE *sprēawlian.*] —*vi.* **1.** To sit or lie with the body and limbs spread out awkwardly. **2.** To spread out in a straggling or disordered fashion <filthy slums *sprawling* toward the river> —*vt.* To cause to spread out in a straggling or disordered way. —*n.* **1.** A sprawling posture or position. **2.** Haphazard growth or extension outward, esp. that resulting from new housing on the outskirts of a city. **—sprawl'er** *n.*

**spray¹** (sprā) *n.* [< MDu. *sprayen,* to sprinkle.] **1.** Liquid, as water, moving in a mass of dispersed droplets <sea *spray*> **2. a.** A fine jet of liquid discharged from a pressurized container. **b.** A pressurized container : ATOMIZER. **c.** A commercial product, as paint and insecticide, dispensed from a container in this manner. —*v.* **sprayed, spray·ing, sprays.** —*vt.* **1.** To disperse (a liquid) in a mass or jet of droplets. **2.** To apply a spray to (a surface). —*vi.* **1.** To discharge sprays of liquid. **2.** To move in the form of a spray. **—spray'er** *n.*

**spray²** (sprā) *n.* [ME.] **1.** A small branch yielding buds, flowers, or berries. **2.** Something resembling a spray.

**spray gun** *n.* A gunlike device for applying sprays.

**spread** (sprĕd) *v.* **spread, spread·ing, spreads.** [ME *spreden* < OE *sprǣdan.*] —*vt.* **1.** To open to a fuller extent or width : STRETCH. **2.** To move farther apart <*spread* one's fingers> **3. a.** To distribute over a surface in a layer : APPLY. **b.** To cover with a layer <*spread* a wall with paint> **4.** To distribute widely <The hurricane *spread* destruction.> **5.** To cause to become widely known : DISSEMINATE <*spread* the news> **6. a.** To prepare (a table) for eating. **b.** To arrange (food or a meal) on a table. —*vi.* **1.** To be extended or enlarged. **2.** To become widely distributed. **3.** To become known or prevalent over a wide area <The bad news *spread* fast.> **4.** To become distributed in a layer. **5.** To become separated. —*n.* **1. a.** An act of spreading. **b.** Dissemination, as of news. **2. a.** An open area of land : EXPANSE. **b.** A ranch or farmland. **3.** The extent or limit to which something is or can be spread : RANGE. **4.** A cloth covering for a piece of furniture, as a bed or table. **5.** *Informal.* An abundant meal laid out on a table. **6.** A food to be layered on bread or crackers. **7. a.** Facing pages of a magazine or newspaper with related matter extending across the fold. **b.** A story or advertisement running across two or more columns. **8.** A difference, as between two figures or totals.

☆ ***syns:*** **1.** SPREAD, EXPAND, OPEN, OUTSTRETCH, UNFOLD *v. core meaning* : to move or arrange so as to cover a larger area <*spread* the blanket on the grass> **2.** SPREAD, DIFFUSE, DISPERSE, RADIATE, SCATTER *v. core meaning* : to extend over a wide area <toxic fumes that *spread* quickly>

**spread eagle** *n.* **1. a.** The figure of an eagle with wings and legs spread. **b.** The emblem on the Great Seal of the United States. **2.** A posture or design resembling a spread eagle.

**spread-ea·gle** (sprĕd'ē'gəl) *adj.* **1.** With the arms and legs stretched out. **2.** *Informal.* Full of patriotic or jingoistic rhetoric. —*v.* **-gled, -gling, -gles.** —*vt.* To place in a spread-eagle position, esp. as a way of punishment. —*vi.* To make a grandiloquent, patriotic speech.

**spread·er** (sprĕd'ər) *n.* One that spreads, esp.: **a.** A butter knife. **b.** A farm implement for scattering fertilizer or seed. **c.** A device, as a bar, for keeping wires or stays apart.

**spree** (sprē) *n.* [Perh. alteration of Sc. *spreath,* cattle raid < Ir. Gael. *sprēidh* < Lat. *praeda,* booty.] **1.** A lively outing. **2.** A drinking bout. **3.** Overindulgence in an activity <a shopping *spree*>

**spri·er** (sprī'ər) *adj. var. comp. of* SPRY.

**spri·est** (sprī'ĭst) *adj. var. superl. of* SPRY.

**sprig** (sprĭg) *n.* [ME *sprigge.*] **1. a.** A small shoot or twig of a plant. **b.** An ornament in this shape. **2.** A small brad without a head. **3.** A young immature person. —*vt.* **sprigged, sprig·ging, sprigs. 1.** To decorate with a design of sprigs. **2.** To remove a sprig or sprigs from (e.g., a tree). **3.** To fasten with a small headless brad. **—sprig'ger** *n.*

**spright** (sprīt) *n. var. of* SPRITE.

**spright·ly** (sprīt'lē) *adj.* **-li·er, -li·est.** Buoyantly animated. —*adv.* With buoyant animation : GAILY. **—spright'li·ness** *n.*

**spring** (sprĭng) *v.* **sprang** (sprăng) or **sprung** (sprŭng), **sprung, spring·ing, springs.** [ME *springen* < OE *springan.*] —*vi.* **1.** To move upward or forward in a single quick motion. **2.** To emerge or appear suddenly. **3.** To shift position suddenly <The trap door *sprang* shut.> **4.** To arise from a source : DEVELOP. **5.** To become bent, warped, or cracked. —Used of wood. **6.** To come loose or move out of place, as a machine part. —*vt.* **1.** To cause to leap, dart, or emerge suddenly. **2.** To jump over : VAULT. **3.** To release from a checked or inoperative position : ACTUATE <The raccoon *sprang* the box trap.> **4.** To cause to warp, bend, or crack, as by force. **5.** To present unexpectedly <*spring* a trick question on me> **6.** *Slang.* To cause to be released, esp. from prison. —*n.* **1.** An elastic device, as a coil of wire, that regains its original shape after being compressed or extended. **2.** An actuating force : IMPETUS. **3.** Elasticity : resilience. **4. a.** The act of springing. **b.** The distance covered by a leap. **5.** A flock of teal. **6.** A usu. rapid return to normal shape after removal of stress : RECOIL. **7.** A natural fountain or flow of water. **8.** Source : origin. **9.** The season between winter and summer. **10.** A warping, bending, or cracking, as that caused by excessive force.

☆ ***syns:*** SPRING, HOP, LOPE, SKIP, SKITTER, TRIP *v. core meaning* : to bound lightly <rabbits *springing* across the field>

**spring beauty** *n.* A plant of the genus *Claytonia,* esp. *C. virginica,* of eastern North America, having narrow leaves and white or pinkish flowers.

**spring·board** (sprĭng'bôrd', -bōrd') *n.* **1.** A flexible board mounted on a fulcrum with one end secured, used by gymnasts to gain momentum at the start of an exercise. **2.** A diving board. **3.** A starting place.

**spring·bok** (sprĭng'bŏk') *n., pl.* **springbok** or **-boks.** [Afr. : *spring,* to leap up + *bok,* male deer.] A small brown and white gazelle, *Antidorcas marsupialis* of southern Africa, capable of leaping high into the air.

**spring chicken** *n.* **1.** A young chicken, esp. one from two to ten months old, with tender meat. **2.** *Slang.* A young or naive person.
**spring-clean·ing** (sprĭng'-klē'nĭng) *n.* Extensive cleaning, esp. of a house after winter.
**springe** (sprĭnj) *n.* [ME.] **1.** A device for snaring small game, made by attaching a noose to a branch under tension. **2.** A trap : snare.
**†spring·er** (sprĭng'ər) *n.* **1.** One that springs. **2.** A springer spaniel. **3.** *Western U.S.* A cow about to give birth. **4. a.** The impost of an arch. **b.** The bottom stone of an arch resting on the impost.
**springer spaniel** *n.* A dog orig. bred in England or Wales, having drooping ears and a silky brown and white or black and white coat.
**spring fever** *n.* The feelings of languor, rejuvenation, or yearning that may affect people at the beginning of spring.
**Spring·field rifle** (sprĭng'fēld') *n.* [After *Springfield,* Massachusetts, where it was first made.] A magazine-fed breechloading bolt-action .30-caliber U.S. Army rifle.
**spring·form pan** (sprĭng'fôrm') *n.* A baking pan with an upright removable rim fastened to the bottom with a spring.
**spring·halt** (sprĭng'hôlt') *n.* [Alteration of STRINGHALT.] A stringhalt.
**spring·house** (sprĭng'hous') *n.* A small structure built over a spring and used to keep food cool.
**spring·let** (sprĭng'lĭt) *n.* A small spring of water : RILL.
**spring lock** *n.* A lock in which the bolt shoots automatically by a spring.
**spring peeper** *n.* A small brownish tree frog, *Hyla crucifer* of eastern North America, with a shrill high-pitched call.
**spring roll** *n.* An egg roll.
**spring·tail** (sprĭng'tāl') *n.* Any of various small wingless insects of the order Collembola, with abdominal appendages acting as springs to catapult them through the air.
**spring tide** *n.* **1.** The tide gen. having the greatest rise and fall, occurring at or shortly after the new moon or the full moon when the sun, moon, and earth are approx. aligned. **2.** A great flood or rush, as of emotion.
**spring·time** (sprĭng'tīm') *n.* SPRING 9.
**spring·wood** (sprĭng'wo͝od') *n.* Young, usu. soft wood that lies directly beneath the bark and develops in early spring.
**spring·y** (sprĭng'ē) *adj.* **-i·er, -i·est. 1.** Resilient : elastic. **2.** Abounding with freshwater springs. **—spring'i·ly** *adv.* **—spring'i·ness** *n.*
**sprin·kle** (sprĭng'kəl) *v.* **-kled, -kling, -kles.** [ME *sprenklen,* perh. of MLG orig.] —*vt.* **1.** To release or scatter in drops or small particles. **2.** To scatter drops or particles on. —*vi.* **1.** To scatter small drops or particles. **2.** To fall or rain in small or infrequent drops. —*n.* **1.** An act of sprinkling. **2.** A light rainfall. **3.** A small amount <a *sprinkle* of chopped parsley>
☆ ***syns:*** SPRINKLE, BESPRINKLE, DUST, POWDER *v. core meaning* : to scatter or release in small particles or drops <*sprinkled* chocolate bits on the sundae>
**sprin·kler** (sprĭng'klər) *n.* **1.** One that sprinkles, esp.: **a.** An outlet on a sprinkler system. **b.** A device attached to the end of a water hose, used for sprinkling water on lawns. **2.** A sprinkler system.
**sprinkler system** *n.* A fire-extinguishing system made up of a network of water pipes equipped to release water automatically when temperatures rise above a predetermined limit.
**sprin·kling** (sprĭng'klĭng) *n.* **1.** Something sprinkled. **2.** A small amount or quantity <a *sprinkling* of applause> **3.** A small quantity that is tossed or distributed sparsely.
**sprint** (sprĭnt) *n.* [Of Scand. orig.] A short race run at top speed. —*vi.* **sprint·ed, sprint·ing, sprints.** To run at top speed. **—sprint'er** *n.*
**sprint car** *n.* A medium-sized car with a large powerful engine that is run in dirt track races.
**sprit** (sprĭt) *n.* [ME *sprytt* < OE *sprēot,* pole.] *Naut.* **1.** A pole extending diagonally across a fore-and-aft sail from the lower part of the mast to the peak of the sail. **2.** A bowsprit.
**sprite** *also* **spright** (sprīt) *n.* [ME *spreit* < OFr. *esprit* < Lat. *spiritus.* —see SPIRIT.] **1. a.** A small or elusive supernatural being. **b.** An elflike person. **2. a.** A ghost. **b.** *Archaic.* A soul.
**sprit·sail** (sprĭt'səl, -sāl') *n. Naut.* A sail extended by a sprit.
**sprock·et** (sprŏk'ĭt) *n.* [Orig. unknown.] Any of various toothlike projections arranged on a wheel rim to engage the links of a chain.
**sprocket wheel** *n.* A wheel rimmed with sprockets, used to engage the links of a chain in a pulley or drive system.
**sprout** (sprout) *v.* **sprout·ed, sprout·ing, sprouts.** [ME *spruten* < OE *sprūtan.*] —*vi.* **1.** To begin to grow and give off shoots or buds. **2.** To grow or develop quickly. —*vt.* To cause to grow. —*n.* **1.** A young plant growth. **2.** Something resembling a sprout. **3. sprouts.** Brussels sprouts.
**spruce¹** (spro͞os) *n.* [Short for obs. *Spruce fir,* Prussian fir < *Spruce,* Prussia < ME *Sprewse,* alteration of *Pruce* < OFr. < Med. Lat. *Prussia.*] **1. a.** Any of various coniferous evergreen trees of the genus *Picea,* with needlelike foliage, drooping cones, and soft wood often used for paper pulp. **b.** Any of various trees similar or related to the spruce. **c.** The wood of any of these trees. **2.** A grayish or dark grayish to greenish black.
**spruce²** (spro͞os) *adj.* **spruc·er, spruc·est.** [Perh. < obs. *Spruce leather,* Prussian leather. —see SPRUCE¹.] Neat, trim, or dapper. —*vt.* & *vi.* **spruced, spruc·ing, spruc·es.** To make or become spruce. **—spruce'ly** *adv.* **—spruce'ness** *n.*
**spruce pine** *n.* The black spruce.
**sprue** (spro͞o) *n.* [Du. *spruw* < MDu. *sprouwe.*] A chronic, chiefly tropical disease marked by diarrhea, emaciation, and anemia.
**sprung** (sprŭng) *v. p.p.* & *var. p.t. of* SPRING.
**sprung rhythm** *n.* [Coined by Gerard Manley Hopkins (1844–1889).] A forcefully accentual verse rhythm in which a stressed syllable is followed by an irregular number of unstressed or slack syllables to form a foot having a metrical value equal to that of the other feet in the line.
**spry** (sprī) *adj.* **spri·er** *or* **spry·er, spri·est** *or* **spry·est.** [Perh. of Scand. orig.] Briskly active. **—spry'ly** *adv.* **—spry'ness** *n.*
**spud** (spŭd) *n.* [ME *spudde,* short knife.] **1.** A sharp tool resembling a spade for rooting or digging out weeds. **2.** *Slang.* A potato. —*vt.* **spud·ded, spud·ding, spuds.** To remove (e.g., weeds) with a spud.
**spue** (spyo͞o) *v.* & *n. Obs. var. of* SPEW.
**spume** (spyo͞om) *n.* [ME < OFr. *espume* < Lat. *spuma.*] Foam or froth on a liquid. —*vi.* **spumed, spum·ing, spumes.** To froth or foam. **—spu'mous, spum'y** *adj.*
**spu·mo·ne** *also* **spu·mo·ni** (spo͞o-mō'nē) *n.* [Ital. < *spuma,* foam < Lat.] An Italian frozen dessert of ice cream containing fruit, nuts, or candies.
**spun** (spŭn) *v. p.t.* & *p.p. of* SPIN.
**spun glass** *n.* **1.** Fiber glass. **2.** Fine blown glass having delicate, often spiral threading or filigree.
**spunk** (spŭngk) *n.* [Sc. Gael. *spong,* tinder < Lat. *spongia,* sponge.] **1.** Tinder, as punk or touchwood. **2.** *Informal.* Pluck : spirit. **—spunk'i·ly** *adv.* **—spunk'i·ness** *n.* **—spunk'y** *adj.*
**spun silk** *n.* A yarn made from short-fibered silk.
**spun sugar** *n.* Sugar threaded into a candylike fluff.
**spun yarn** *n.* A lightweight line made of several rope yarns loosely wound together, used for seizings on board ship.
**spur** (spûr) *n.* [ME *spure* < OE *spura.*] **1.** One of a pair of spikes or spiked wheels attached to a rider's heels and used to urge a horse forward. **2.** An incentive. **3.** A spurlike attachment or projection, as: **a.** A spinelike process on the leg of some birds. **b.** A climbing iron : CRAMPON. **c.** The gaff attached to the leg of a gamecock. **d.** A short or stunted tree branch. **e.** Ergot growing on rye. **4.** A lateral ridge projecting from a mountain or mountain range. **5.** An oblique reinforcing prop or stay of timber or masonry. **6.** *Bot.* A tubular extension of the corolla or calyx of a flower, as in a columbine or larkspur. **7.** A spur track. —*v.* **spurred, spur·ring, spurs.** —*vt.* **1.** To urge (a horse) on by the use of spurs. **2.** To incite : prompt. —*vi.* To ride quickly on horseback by using spurs. **—on the spur of the moment.** On a sudden impulse.
**spurge** (spûrj) *n.* [ME < OFr. *espurge* < *espurgier,* to purge (from its use as a purgative) < Lat. *expurgare.* —see EXPURGATE.] Any of various chiefly tropical plants of the genus *Euphorbia,* having milky juice and small flowers surrounded by showy bracts in some species.
**spur gear** *n.* A gear with teeth radially arrayed on the rim parallel to its axis.
**spurge laurel** *n.* A low-growing shrub, *Daphne laureola* of southern Europe, with glossy evergreen leaves and small yellowish-green flowers.
**spu·ri·ous** (spyo͝or'ē-əs) *adj.* [LLat. *spurius,* false < Lat., illegitimate.] **1.** Lacking validity or authenticity : FALSE. **2.** Being a forgery or interpolation. **3.** Illegitimate : bastard. **4.** *Bot.* Similar in appearance but unlike in function or structure. **—spu'ri·ous·ly** *adv.* **—spu'ri·ous·ness** *n.*
**spurn** (spûrn) *v.* **spurned, spurn·ing, spurns.** [ME *spurnen* < OE *spurnan.*] —*vt.* **1.** To refuse or reject disdainfully : SCORN. **2.** *Archaic.* **a.** To kick disdainfully. **b.** To trample on. —*vi.* To refuse something disdainfully. —*n.* **1.** A disdainful refusal. **2.** *Archaic.* A kick. **—spurn'er** *n.*
**spurred** (spûrd) *adj.* **1.** Wearing spurs. **2.** Having a spur or spurs.
**spur·rey** (spûr'ē) *n. var. of* SPURRY.
**spur·ri·er** (spûr'ē-ər) *n.* [ME *sporior* < *spore, spure,* spur.] A maker of spurs.
**spur·ry** *also* **spur·rey** (spûr'ē) *n., pl.* **-ries** *also* **-reys.** [Du. *spurrie* < MDu. *speurie,* prob. < Med. Lat. *spergula,* prob. < Lat. *spargere,* to scatter.] A weedy, low-growing plant of the genera *Spergula* or *Spergularia,* esp. *Spergula arvensis* native to Europe, with small white flowers and whorled leaves.
**spurt** (spûrt) *n.* [ME *sprutten,* to sprout < OE *spryttan.*] **1.** A sudden forcible gush. **2.** A sudden short burst of energy or activity. —*v.* **spurt·ed, spurt·ing, spurts.** —*vi.* To gush forth. —*vt.* To force out in a gush.
**spur track** *n.* A short side railroad track that connects with the main track.
**spu·ta** (spyo͞o'tə) *n. pl. of* SPUTUM.
**sput·nik** (spo͝ot'nĭk, spŭt'-, spo͞ot'-) *n.* [R. *sputnik (zemlyi),* fellow traveler (of Earth).] An artificial earth satellite launched by the U.S.S.R., esp. the first, launched on Oct. 4, 1957.

**sput·ter** (spŭt'ər) *v.* **-tered, -ter·ing, -ters.** [Prob. of LG orig.] *—vi.* **1. a.** To spit out small particles in short bursts, often with corresponding sounds or noises. **b.** To make the sporadic coughing noise characteristic of sputtering. **2.** To speak hastily or confusedly. *—vt.* **1.** To spit out (e.g., saliva) in short bursts. **2.** To utter hastily or confusedly. *—n.* **1.** An act of sputtering. **2.** The sound of sputtering. **3.** The particles emitted during sputtering. **4.** Hasty, confused utterances. **—sput'ter·er** *n.*

**spu·tum** (spyōō'təm) *n., pl.* **-ta** (-tə) [Lat. *sputum* < *sputus,* p.part. of *spuere,* to spit.] **1.** Expectorated saliva : SPITTLE. **2.** Expectorated matter, including saliva, substances from the respiratory tract, and foreign material.

**spy** (spī) *n., pl.* **spies.** [ME *spie* < OFr. *espie* < *espier,* to watch, of Germanic orig.] **1.** A clandestine agent employed by a state to obtain intelligence relating to its actual or potential enemies at home or abroad. **2.** One who secretly watches another or others. **3.** The act of watching secretly or covertly. *—v.* **spied, spy·ing, spies.** *—vt.* **1.** To keep under surveillance with hostile intent. **2.** To catch sight of : SEE <*spied* an old friend> *—vi.* **1. a.** To observe secretly and closely. **b.** To engage in espionage. **2.** To investigate <*spying* into the neighbors' activities>

**spy·glass** (spī'glăs') *n.* A small telescope.

**squab** (skwŏb) *n.* [Prob. of Scand. orig.] **1.** A young, unfledged pigeon. **2.** A short, fat person. **3.** A soft cushion. **4.** A couch : sofa. *—adj.* **1.** Short and broad : SQUAT. **2.** Newly hatched or unfledged.

**squab·ble** (skwŏb'əl) *vi.* **-bled, -bling, -bles.** [Prob. of Scand. orig.] To engage in a trivial quarrel. **—squab'ble** *n.* **—squab'bler** *n.*

**squab·by** (skwŏb'ē) *adj.* **-bi·er, -bi·est.** SQUAB 1.

**squad** (skwŏd) *n.* [OFr. *esquadre* < OSp. *escuadra* and OItal. *squadra,* both < VLat. **exquadrare,* to make square. —see SQUARE.] **1.** A small group of persons organized for a specific purpose. **2.** The smallest unit of military personnel. **3.** An athletic team.

**squad car** *n.* A police patrol car.

**squad·ron** (skwŏd'rən) *n.* [Ital. *squadrone* < *squadra,* squad. —see SQUAD.] **1.** A group of naval vessels constituting two or more divisions of a fleet. **2.** An armored cavalry unit consisting of two to four troops, a headquarters, and auxiliary units. **3.** The basic tactical air force unit, subordinate to a group and consisting of two or more flights. **4.** A multitude : legion <a *squadron* of gnats>

**squad room** *n.* A room in a police station where officers assemble, as for assignment or briefing.

**squa·lene** (skwā'lēn') *n.* [< NLat. *Squalus,* shark genus (from its occurrence in the liver oil of sharks) < Lat. *squalus,* a sea fish + -ENE.] A natural unsaturated aliphatic hydrocarbon, $C_{30}H_{50}$, found in human sebum and other fatty deposits, that is an intermediate in the biosynthesis of cholesterol and is used in biochemical research.

**squal·id** (skwŏl'ĭd) *adj.* [Lat. *squalidus* < *squalere,* to be filthy < *squalus,* filthy.] **1.** Dirty or wretched in appearance <a *squalid* shack> **2.** Morally repulsive : SORDID <a *squalid* tale> **—squa·lid'i·ty** (skwŏ-lĭd'ĭ-tē), **squal'id·ness** *n.* **—squal'id·ly** *adv.*

**squall**[1] (skwôl) *n.* [Prob. of Scand. orig.] A loud, harsh outcry. *—vi.* **squalled, squall·ing, squalls.** To scream or cry loudly and harshly. **—squall'er** *n.*

**squall**[2] (skwôl) *n.* [Prob. of Scand. orig.] **1.** A brief, sudden, and violent windstorm, often accompanied by rain or snow. **2.** *Informal.* A commotion : disturbance. *—vi.* **squalled, squall·ing, squalls.** To blow strongly for a brief time.

**squall line** *n.* A zone of squalls and other violent weather changes that marks replacement of a warm air current by cold air.

**squall·y** (skwô'lē) *adj.* **-i·er, -i·est. 1.** Marked by squalls : GUSTY. **2.** *Informal.* Characterized by disturbance or trouble.

**squal·or** (skwŏl'ər) *n.* [Lat.] The quality or state of being squalid.

**squa·ma** (skwā'mə, skwä'-) *n., pl.* **-mae** (-mē') [Lat.] **1.** A scale or scalelike structure. **2.** A thin plate of bone. **—squa'mate'** (-māt') *adj.*

**squa·ma·tion** (skwə-mā'shən) *n.* **1.** The state of being scaly. **2.** An arrangement of scales, as on a fish.

**Squa·mish** (skwä'mĭsh) *n. var. of* SUQUAMISH.

**squa·mo·sal** (skwə-mō'səl, -zəl) *adj.* [< Lat. *squamosus,* squamous.] Of or relating to the squamous area of the temporal bone. *—n.* A squamosal bone.

**squa·mous** (skwā'məs, skwä'-) *also* **squa·mose** (-mōs') *adj.* [Lat. *squamosus* < *squama,* scale.] **1.** Covered with or formed of scales : SCALY. **2.** Suggestive of a scale or scales. **—squa'mous·ly** *adv.* **—squa'mous·ness** *n.*

**squamous cell** *n.* A flat scaly epithelial cell.

**squamous epithelium** *n.* Single-layered epithelium composed of flat scaly cells.

**squa·mu·lose** (skwā'myə-lōs', skwä'-) *adj.* [< NLat. *squamula,* dim. of Lat. *squama,* scale.] Having or made up of minute scales.

**squan·der** (skwŏn'dər) *vt.* **-dered, -der·ing, -ders.** [Orig. unknown.] **1.** To spend extravagantly or wastefully : DISSIPATE. **2.** *Obs.* To scatter. *—n.* Extravagant, wasteful expenditure : PRODIGALITY. **—squan'der·er** *n.* **—squan'der·ing·ly** *adv.*

**square** (skwâr) *n.* [ME < OFr. *esquare* < VLat. **exquadra* < **exquadrare* : Lat. *ex-* (intensive) + *quadrare,* to square < *quadrus,* a square.] **1.** A rectangle with four equal sides. **2.** Something with an equal-sided rectangular shape. **3.** A T-shaped or L-shaped instrument for drawing or testing right angles. **4.** The product of a number or quantity multiplied by itself. **5.** One of the quadrilateral spaces dividing a checkerboard. **6. a.** An open, often quadrilateral area at the intersection of two or more streets. **b.** A rectangular space enclosed by streets and occupied by buildings. **7.** *Slang.* A rigidly conventional or unsophisticated person. *—adj.* **squar·er, squar·est. 1.** Having four equal sides and four right angles. **2.** Forming a right angle. **3. a.** Expressed in units measuring area <*square* yards> **b.** Having a specified length in each of two equal dimensions. **4.** *Naut.* Set at right angles to the mast and keel, as the yards of a square-rigged ship. **5. a.** Of somewhat quadrate dimensions <a *square* building> **b.** Marked by blocklike solidity or sturdiness. **6.** Honest : direct. **7.** Just : equitable. **8.** Paid up : SETTLED. **9.** Even or tied in golf. **10.** *Slang.* Rigidly conventional : STRAIGHT. *—v.* **squared, squar·ing, squares.** *—vt.* **1.** To cut to a square or rectangular shape. **2.** To test for conformity to a desired plane, straight line, or right angle. **3.** To test by comparison. **4.** To bring into conformity or agreement. **5.** To set straight or at right angles <*square* one's cap> **6.** To bring into balance : SETTLE <*squared* the debt> **7.** To even the score of in golf : TIE. **8.** To raise (a quantity or number) to the second power. **9.** To find a square equal in area to (the area of a given figure). *—vi.* **1.** To be at right angles. **2.** To agree or conform : BALANCE. **—square away. 1.** To square the yards of a sailing vessel. **2.** To put away or in order. **—square off.** To take on a fighting posture. *—adv.* **1.** At right angles. **2.** In a square shape. **3.** Solidly. **4.** Directly : straight. **5.** In an honest way. **—square'ly** *adv.* **—square'ness** *n.*

**square bracket** *n.* BRACKET 4a.

**square dance** *n.* **1.** A dance in which sets of four couples form squares. **2.** Any of various group dances of English rural origin.

**square-dance** (skwâr'dăns') *vi.* **-danced, -danc·ing, -danc·es.** To perform a square dance. **—square'-danc'er** *n.*

**square knot** *n.* A common double knot with the loose ends parallel to the standing parts.

**square matrix** *n.* *Math.* A matrix with equal numbers of rows and columns.

**square measure** *n.* A system of units used in measuring area.

**square rig** *n.* *Naut.* A sailing-ship rig with sails of rectangular cut set approx. at right angles to the keel line from horizontal yards. **—square'-rigged'** *adj.*

**square-rig·ger** (skwâr'rĭg'ər) *n.* *Naut.* A square-rigged vessel.

**square root** *n.* A divisor of a quantity that when squared gives the quantity.

**square sail** *n.* *Naut.* A four-sided sail that is bent to a yard set athwart the mast.

**squar·rose** (skwăr'ōs', skwär'-) *adj.* [Lat. *squarrosus,* scabby.] **1.** *Biol.* Having rough or spreading scalelike processes. **2.** *Bot.* Spreading or recurved at the tip <*squarrose* bracts>

**squash**[1] (skwŏsh, skwôsh) *n.* [Short for obs. *isquoutersquash* < Massachuset *askōōtasquash.*] **1.** Any of various plants of the genus *Cucurbita,* having fleshy edible fruit with a hard rind. **2.** The fruit of a squash, used as a vegetable.

**squash**[2] (skwŏsh, skwôsh) *v.* **squashed, squash·ing, squash·es.** [OFr. *esquasshe* < VLat. **exquassare* : Lat. *ex-* (intensive) + *quassare,* to shatter, freq. of *quatere,* to shake.] *—vt.* **1.** To squeeze, beat, or flatten to a pulp : CRUSH. **2.** To suppress : quash <*squash* an uprising> **3.** To silence (a person), as with harsh words. *—vi.* **1.** To be flattened or crushed. **2.** To move with a squelching sound. *—n.* **1.** The impact or sound of a soft body dropping against a surface. **2.** The sound of water being squeezed out, as from spongy ground or wet shoes. **3.** A crush <a *squash* of people> **4.** *Chiefly Brit.* A citrus-based soft drink. **5. a.** A game played in a walled court with a racket and a hard rubber ball. **b.** A similar game played with an inflated rubber ball. *—adv.* With a squashing sound. **—squash'er** *n.*

**squash**[2]

**squash bug** *n.* A blackish North American insect, *Anasa tristis,* destructive to crops, as squash and pumpkins.

**squash·y** (skwŏ'shē, skwô'shē) *adj.* **-i·er, -i·est. 1.** Easily squashed. **2.** Soft and pulpy : OVERRIPE. **3.** Boggy : squishy <*squashy* soil> **—squash'i·ly** *adv.* **—squash'i·ness** *n.*

**squat** (skwŏt) *v.* **squat·ted, squat·ting, squats.** [ME *squatten* < OFr. *esquatir,* to crush : *es-* (intensive < Lat. *ex-*) + *quatir,* to press flat < VLat. **coactire* < Lat. *cogere,* to compress (*co-,* together + *agere,* to drive).] —*vi.* **1.** To sit on one's heels. **2.** To settle on unoccupied land without legal claim. **3.** To occupy a given piece of public land so as to acquire title to it. **4.** To crouch close to the ground. —*vt.* **1.** To put (oneself) in a crouching posture. **2.** To occupy as a squatter. —*adj.* **squat·ter, squat·test. 1.** Seated in a squatting position. **2.** Short and thick. —*n.* **1. a.** A squatting or crouching posture. **b.** The act of squatting or crouching. **2.** The lair of a hare. **3.** The land occupied by a squatter. **—squat'ter** *n.*

**squaw** (skwô) *n.* [Massachuset *squa.*] A North American Indian woman.

**squaw·fish** (skwô'fĭsh') *n., pl.* **squawfish** or **-fish·es.** Any of several large freshwater fishes of the genus *Ptychocheilus* of western North America.

**squawk** (skwôk) *v.* **squawked, squawk·ing, squawks.** [Perh. blend of SQUALL and SQUEAK.] —*vi.* **1.** To utter a harsh scream : SCREECH. **2.** To make a loud, angry protest. —*vt.* To utter with or as if with a squawk. —*n.* **1.** A loud screech : SQUALL. **2.** A protest. **—squawk'er** *n.*

**squaw·root** (skwô'ro͞ot', -ro͝ot') *n.* A plant, *Conopholis americana* of eastern North America, that has yellowish flowers and a stem covered with brownish scales and is parasitic on the roots of certain trees, as oaks.

**squeak** (skwēk) *v.* **squeaked, squeak·ing, squeaks.** [ME *squeken.*] —*vi.* **1.** To utter or make a brief thin, shrill cry or sound. **2.** To pass or win by a slight margin <*squeaked* through the test> **3.** *Slang.* To turn informer. —*vt.* To utter in a squeaky voice. **—squeak** *n.* **—squeak'er** *n.*

**squeak·y** (skwē'kē) *adj.* **-i·er, -i·est. 1.** Marked by squeaking tones. **2.** Tending to squeak. **—squeak'i·ly** *adv.* **—squeak'i·ness** *n.*

**squeal** (skwēl) *v.* **squealed, squeal·ing, squeals.** [ME *squelen.*] —*vi.* **1.** To utter or produce a loud, shrill cry or sound. **2.** *Slang.* To betray a friend or a secret : SNITCH. —*vt.* To utter or produce with a squeal. **—squeal** *n.* **—squeal'er** *n.*

**squea·mish** (skwē'mĭsh) *adj.* [ME *squaymisch,* alteration of AN *escoymous.*] **1. a.** Easily sickened or nauseated. **b.** Affected with nausea. **2.** Easily offended or disgusted. **3.** Excessively fastidious : OVERSENSITIVE. **—squea'mish·ly** *adv.* **—squea'mish·ness** *n.*

**squee·gee** (skwē'jē') *n.* [Perh. < E. *squeege,* to press, alteration of SQUEEZE.] **1.** A T-shaped implement having a crosspiece edged with rubber or leather, used to remove water from a surface, as a window. **2.** A roller used in printing and photography. —*vt.* **-geed, -gee·ing, -gees.** To wipe or smooth with a squeegee.

**squeeze** (skwēz) *v.* **squeezed, squeez·ing, squeez·es.** [Alteration of obs. *quease,* to press < ME *queysen* < OE *cwȳsan.*] —*vt.* **1.** To press hard on or together : COMPRESS. **2.** To exert pressure on, as by way of extracting liquid <*squeeze* a lemon> **3.** To extract from by applying pressure <*squeeze* juice from an orange> **4.** To extract dishonestly : EXTORT. **5.** To obtain room or passage for by pressure : CRAM. **6.** To oppress with burdensome exactions <a middle class *squeezed* by inflation and high taxes> **7.** To force (an opponent) in bridge to discard a potentially winning card. —*vi.* **1.** To give way under pressure. **2.** To exert pressure. **3.** To force one's way, as through a crowd. **—squeeze off.** To fire (a round of bullets) by pulling the trigger of a gun. —*n.* **1.** An act or instance of compressing. **2.** A handclasp or brief embrace. **3.** A group crowded together. **4. a.** An amount squeezed out of something. **b.** A minor ingredient : PINCH. **5. a.** *Informal.* A squeeze play. **b.** Financial pressure caused by shortages or narrowing economic margins. **6.** A forced discard of a potentially winning card in bridge. **—squeez'er** *n.*

**squeeze play** *n.* **1.** *Baseball.* A play in which the batter attempts to bunt so that a runner on third base may score. **2.** *Informal.* Pressure exerted to obtain a concession or achieve a goal.

**squelch** (skwĕlch) *v.* **squelched, squelch·ing, squelch·es.** [Imit.] —*vt.* **1.** To crush by or as if by trampling : SQUASH. **2.** To put down or silence, as with a harsh remark. **3.** To cause to make a squishing sound. —*vi.* To make or move with a splashing, squashing, or sucking sound. —*n.* **1.** A squishing sound. **2.** A harsh reply. **3.** An electric circuit that cuts off a radio receiver when the signal is too weak for reception of anything but noise. **—squelch'er** *n.*

**sque·teague** (skwĭ-tēg') *n., pl.* **squeteague.** [Prob. of Algonquian orig.] **1.** The weakfish. **2.** A fish related to the weakfish.

**squib** (skwĭb) *n.* [Prob. imit.] **1. a.** A firecracker. **b.** A broken firecracker that burns but does not explode. **2.** A brief, sometimes witty literary effort, as a lampoon. —*v.* **squibbed, squib·bing, squibs.** —*vi.* To write squibs. —*vt.* **1.** To write squibs about. **2.** *Football.* To kick (the ball) low on a kickoff, making it difficult to field and return.

**squib kick** *n.* *Football.* A kickoff in which the ball is kicked low so that it will bounce along the ground, making it difficult to field and return.

**squid** (skwĭd) *n., pl.* **squids** or **squid.** [Orig. unknown.] Any of various marine cephalopod mollusks of the genera *Loligo, Rossia,* and related genera, having a usu. elongated body, ten arms surrounding the mouth, a vestigial internal shell, and a pair of triangular or rounded fins.

**squig·gle** (skwĭg'əl) *n.* [Blend of SQUIRM and WRIGGLE.] A small wiggly mark or scrawl. —*vi.* **-gled, -gling, -gles. 1.** To squirm and wriggle. **2.** To write quickly : SCRIBBLE. **—squig'gly** *adj.*

**squill** (skwĭl) *n.* [ME < Lat. *squilla* < Gk. *skilla.*] **1.** A bulbous plant of the genus *Scilla,* native to Eurasia, having narrow leaves and small, bell-shaped blue, white, or pink flowers. **2. a.** The sea onion. **b.** The dried inner scales of the bulbs of the squill, used as rat poison and formerly as a cardiac stimulant, expectorant, and diuretic.

**squil·la** (skwĭl'ə) *n., pl.* **squil·las** or **squil·lae** (skwĭl'ē') [NLat. *Squilla,* type genus < Lat. *squilla,* shrimp.] Any of various burrowing marine crustaceans of the order Stomatopoda, having a pair of jointed grasping appendages.

**squinch** (skwĭnch) *n.* [Shortening and alteration of *scuncheon,* an architectural detail < OFr. *escoinson* : *es-,* out (< LLat. *ex-*) + *coin,* angle, wedge. —see COIN.] A quarter-spherical segment of masonry vaulting or corbeling used across the upper inside corners of a square tower as the transition to a circular or octagonal superstructure.

**squint** (skwĭnt) *v.* **squint·ed, squint·ing, squints.** [Short for ASQUINT.] —*vi.* **1.** To look with partly closed eyes. **2.** To look to the side. **3.** To suffer from strabismus. **4.** To have an indirect or implicit tendency. —*vt.* **1.** To cause to squint. **2.** To close (the eyes) partly. —*n.* **1.** An act of squinting. **2.** An inclination or tendency. **3.** Strabismus. —*adj.* **1.** Looking obliquely or askance. **2.** SQUINT-EYED 1. **—squint'er** *n.* **—squint'ing·ly** *adv.*

**squint-eyed** (skwĭnt'īd') *adj.* **1.** Having strabismus. **2.** Looking with narrowed or squinting eyes. **3.** Looking askance : BIASED.

**squir·ar·chy** (skwīr'är'kē) *n. var. of* SQUIREARCHY.

**squire** (skwīr) *n.* [ME *squier* < OFr. *esquier.* —see ESQUIRE.] **1.** A young feudal nobleman attendant on a knight and ranked next below a knight. **2.** An English country gentleman. **3.** A local dignitary, as a judge. **4.** A man who attends or escorts a woman : GALLANT. —*vt.* **squired, squir·ing, squires.** To attend as a squire or escort.

**squire·ar·chy** or **squir·ar·chy** (skwīr'är'kē) *n., pl.* **-chies. 1.** Squires as a group. **2.** Government by landed proprietors.

**squirm** (skwûrm) *vi.* **squirmed, squirm·ing, squirms.** [Perh. imit.] **1.** To twist about in a wriggling, snakelike motion : WRITHE. **2.** To feel or show humiliation or embarrassment. —*n.* **1.** An act of squirming. **2.** A squirming motion. **—squirm'er** *n.* **—squirm'y** *adj.*

**squir·rel** (skwûr'əl, skwŭr'-) *n.* [ME *squirel* < AN *esquirel* < VLat. **scuriolus* < Lat. *sciurus* < Gk. *skiouros* : *skia,* shadow + *oura,* tail.] **1.** Any of various arboreal rodents of the genus *Sciurus* and related genera, usu. with gray or reddish-brown fur and a long, flexible, bushy tail. **2.** An animal of the family Sciuridae, as the ground squirrel or the flying squirrel. **3.** The fur of a squirrel.

**squirrel corn** *n.* A low-growing North American plant, *Dicentra canadensis,* with finely divided leaves, cream-colored flowers, and tubers resembling grains of corn.

**squir·rel·fish** (skwûr'əl-fĭsh', skwŭr'-) *n., pl.* **squirrelfish** or **-fish·es.** A fish of the genus *Holocentrus* or related genera, of warm marine waters, with large eyes and a usu. reddish body.

**squir·rel·ly** (skwûr'ə-lē, skwŭr'-) *adj. Slang.* Eccentric.

**squirrel monkey** *n.* A tropical American monkey, *Saimiri sciureus* or *S. örstedii,* with short, thick fur and a long, nonprehensile tail.

**squirt** (skwûrt) *v.* **squirt·ed, squirt·ing, squirts.** [ME *squirten.*] —*vi.* **1.** To be ejected in a swift thin stream. **2.** To eject a swift thin stream. —*vt.* **1.** To eject (liquid) in a swift thin stream. **2.** To wet with liquid ejected in a swift thin stream. —*n.* **1.** The act of squirting. **2.** A device for squirting. **3.** A stream squirted. **4.** *Informal.* **a.** An impudent youngster. **b.** A child. **—squirt'er** *n.*

**squirt gun** *n.* A toy designed to squirt a stream of water.

**squirting cucumber** *n.* A hairy vine, *Ecballium elaterium* of the Mediterranean region, having fruit that when ripe discharges its seeds and juice explosively.

**squish** (skwĭsh) *v.* **squished, squish·ing, squish·es.** [Alteration of SQUASH.] *Informal.* —*vt.* To squash noisily. —*vi.* To emit a sound like that of soft mud being compressed. **—squish** *n.* **—squish'y** *adj.*

**Sr** *symbol for* STRONTIUM.

**sRNA** *n.* [S(OLUBLE) RNA.] Transfer RNA.

**-st** *suff. var. of* -EST².

**stab** (stăb) *v.* **stabbed, stab·bing, stabs.** [< ME *stabbe,* a stab wound.] —*vt.* **1.** To pierce or wound with or as if with a pointed object. **2.** To plunge (a weapon) into a body. —*vi.* **1.** To lunge with or as if with a pointed weapon. **2.** To inflict a wound by stabbing. —*n.* **1.** A thrust made with a pointed instrument or weapon. **2.** A wound inflicted by or as if by stabbing. **3.** An attempt : effort. **—stab'ber** *n.*

**sta·bile** (stā'bĭl, -bīl') *adj.* [Lat. *stabilis,* stable.] Immobile : unchangeable. —*n.* (-bēl'). An abstract, usu. sheet metal sculpture with no mobile parts.

**sta·bil·i·ty** (stə-bĭl'ĭ-tē) *n., pl.* **-ties. 1.** Resistance to sudden change, dislodgment, or overthrow. **2. a.** Constancy of character or purpose : STEADFASTNESS. **b.** Reliability : dependability. **3.** A vow committing a friar to a single monastery for life.

**sta·bi·lize** (stā′bə-līz′) *v.* **-lized, -liz·ing, -liz·es.** —*vt.* **1.** To make stable. **2.** To maintain the stability of. —*vi.* To become stable. **—sta′bi·li·za′tion** *n.*

**sta·bi·liz·er** (stā′bə-lī′zər) *n.* **1.** One that stabilizes. **2.** *Naut.* A gyroscopic device used to prevent excessive rolling of a vessel in a rough sea. **3.** An airfoil for stabilizing an aircraft in flight. **4.** *Chem.* An additive that prevents or keeps a solution, suspension, or system resistant to chemical change.

**sta·ble¹** (stā′bəl) *adj.* **-bler, -blest.** [ME < OFr. *estable* < Lat. *stabilis.*] **1. a.** Resisting sudden change of position or condition. **b.** Maintaining equilibrium : SELF-RESTORING. **2.** *Physics.* Having no known mode of decay : indefinitely long-lived. —Used of atomic particles. **3.** Unchanging and permanent : ENDURING. **4.** *Chem.* Not easily decomposed or otherwise modified chemically. **5. a.** Consistently dependable. **b.** Not aberrant or irrational in character or personality <a *stable* person> **—sta′ble·ness** *n.* **—sta′bly** *adv.*

**sta·ble²** (stā′bəl) *n.* [ME < OFr. *estable* < Lat. *stabulum.*] **1. a.** A building for sheltering and feeding domestic animals, esp. horses and cattle. **b.** The animals kept in a stable. **2.** All of the racehorses belonging to a particular owner. **3.** Personnel employed to keep and train a collection of racehorses. **4.** A group managed by a single individual or authority <a *stable* of authors> —*v.* **-bled, -bling, -bles.** —*vt.* To put or keep in a stable. —*vi.* To live in or as if in a stable.

**sta·bling** (stā′blĭng) *n.* **1.** Stables as a whole. **2.** Accommodations for animals in a stable.

**stab·lish** (stăb′lĭsh) *v. Archaic. var. of* ESTABLISH.

**stac·ca·to** (stə-kä′tō) *adj.* [Ital., p.part. of *staccare,* to detach < OFr. *destachier.* —see DETACH.] **1.** *Mus.* Crisply detached : DISCONNECTED <a *staccato* passage> **2.** Composed of short, distinct parts or sounds <the *staccato* hammering of a woodpecker> **—stac·ca·to** *n.* **—stac·ca′to** *adv.*

**stack** (stăk) *n.* [ME < ON *stakkr.*] **1.** A large, usu. conical pile of straw or hay left outdoors for storage. **2.** An orderly mass or pile. **3.** *Computer Sci.* A section of memory and its associated registers for temporary storage, operating on the principle that the information put in late will be the first to be retrieved. **4.** A group of three or more unslung rifles with their butts downward so as to form a cone. **5. a.** A chimney : flue. **b.** A group of chimneys. **6.** A vertical exhaust pipe. **7.** A series of enclosed bookshelves. **8. stacks.** The area of a library in which most of the books are shelved. **9.** An English measure of coal or cut wood, equal to 108 cubic feet. **10.** *Informal.* A large quantity. —*vt.* **stacked, stack·ing, stacks. 1.** To pile up in a stack. **2.** To load with stacks of material. **3.** To prearrange the order of (e.g., playing cards) in order to cheat. **—stack up.** To bear comparison. **—stack′er** *n.*

**stacked** (stăkt) *adj. Slang.* Having a very shapely, voluptuous body. —Used of a woman.

**stacked heel** *n.* A shoe heel made of several layers of material.

**stack·up** (stăk′ŭp′) *n.* A group of airplanes awaiting landing instructions, arranged at different levels over a airport.

**stac·te** (stăk′tē) *n.* [Lat. *stacte* < Gk. *staktē* < *staktos,* oozing < *stazein,* to ooze.] A spice used by the ancient Jews to make incense.

**stad·dle** (stăd′l) *n.* [ME *stathel* < OE *staðol.*] A foundation, esp. a base for stacking hay or straw.

**stad·hold·er** (stăd′hōl′dər) *n.* [Partial transl. of Du. *stadhouder* : *stad,* place + *houder,* holder.] **1.** A governor or viceroy in a province of the Netherlands. **2.** The chief magistrate of the Netherlands.

**sta·di·a** (stā′dē-ə) *n.* [Ital., prob. < Lat., pl. of *stadium,* a unit of length. —see STADIUM.] **1. a.** A method of surveying distances using a telescopic instrument with two parallel lines that are used to intercept intervals on a calibrated rod, the intervals being proportional to the intervening distance. **b.** The graduated rod used for this purpose. **2.** The parallel lines in a telescope. **3.** *pl. of* STADIUM.

**sta·di·um** (stā′dē-əm) *n., pl.* **-di·a** (-dē-ə) [ME, unit of length < Lat. < Gk. *stadion,* alteration of *spadion,* racetrack < *span,* to pull.] **1.** A usu. open structure of ancient Greece with tiers of seats for spectators having a track on which footraces were held. **2.** An ancient Greek measure of distance, based on the length of the course in a stadium and equal to about 607 feet or 185 kilometers. **3.** *pl.* **-di·ums.** A large, often open structure for holding athletic events. **4.** A stage in the progress of a disease.

**stadium coat** *n.* A casual medium-length winter coat.

**staff¹** (stăf) *n.* [ME *staf* < OE *stæf.*] *pl.* **staffs** *or* **staves** (stāvz). **1.** A pole, rod, or stick used esp. as: **a.** An aid in walking or climbing. **b.** A weapon : CUDGEL. **c.** A device for displaying a flag. **d.** A symbol of authority. **2.** *pl.* **staffs.** A graduated stick for testing or measuring, as in surveying. **3.** *pl.* **staffs. a.** A group of assistants who aid an executive or direction. **b.** A group of military or naval officers under a commanding officer who do not participate in combat and who lack the authority to command. **c.** The personnel carrying out a specific enterprise <the nursing *staff* of a hospital> **4.** *Mus.* The set of horizontal lines and their intermediate spaces upon which notes are written or printed. —*vt.* **staffed, staff·ing, staffs. 1.** To furnish with a staff. **2.** To serve on the staff of.

**staff²** (stăf) *n.* [Prob. < G. *staffen,* to adorn.] A building material composed of plaster and fiber, similar to stucco in appearance, and used as a wall covering over the skeleton of temporary buildings.

**staff·er** (stăf′ər) *n. Informal.* A staff member.

**staff of life** *n.* A food, esp. bread, that is a staple.

**Staf·ford·shire terrier** (stăf′ərd-shîr′) *n.* [After *Staffordshire,* a county in England.] A terrier orig. bred in England, with a short, glossy, variously colored coat and widely set forelegs.

**staff sergeant** *n.* **1.** A noncommissioned army officer who ranks above a sergeant and below a sergeant first class. **2.** A noncommissioned air force officer who ranks above an airman first class and below a technical sergeant. **3.** A noncommissioned marine corps officer who ranks above a sergeant and below a gunnery sergeant.

**staff tree** *n.* A dicotyledonous shrubby plant of the genus *Celastrus,* which includes the bittersweet.

**stag** (stăg) *n.* [ME < OE *stagga.*] **1.** The adult male of various deer, esp. the red deer. **2.** An animal castrated after sexual maturity. **3.** A man who attends a social function without a woman. **4.** A social function for men only. —*adj.* For or attended by men only <a *stag* party> —*adv.* As a single man unaccompanied by a woman <go *stag* to a dance> —*vi.* **stagged, stag·ging, stags.** To attend a social function without a woman.

**stag beetle** *n.* A large beetle of the family Lucanidae, with long, powerful, antlerlike mandibles.

**stage** (stāj) *n.* [ME < OFr. *estage* < VLat. ***staticum** < Lat. *stare,* to stand.] **1.** A raised and level floor or platform. **2.** A viewing platform on a microscope. **3.** A scaffold for workers. **4. a.** A raised platform on which theatrical performances are presented. **b.** An area in which actors perform. **c.** The acting profession. **d.** The theater as a professional activity. **5.** The setting for an event. **6.** A resting place on a journey, providing overnight accommodations. **7.** The distance between stopping places on a journey. **8.** A stagecoach. **9.** A level or story of a building. **10.** A level, as of the surface of a river, in relation to a particular point <flood *stage*> **11.** A level, degree, or period of time in the course of a process or activity : STEP. **12.** One of two or more successive propulsion units of a rocket vehicle that fires after the preceding one has been jettisoned. **13.** *Geol.* A subdivision in the classification of stratified rocks, ranking just below a series and representing rock formed during a chronological age. **14.** *Electron.* An element or group of elements in a complex arrangement of parts, esp. a single tube or transistor and its accessory components in an amplifier. —*v.* **staged, stag·ing, stag·es.** —*vt.* **1.** To exhibit, present, or perform on or as if on a stage <*stage* a talent show> **2.** To produce or direct (a theatrical performance).

**stage·coach** (stāj′kōch′) *n.* A four-wheeled horse-drawn vehicle once used to transport mail, parcels, and passengers.

**stage·craft** (stāj′krăft′) *n.* Skill in the use of theatrical techniques or devices.

**stage fright** *n.* Fright or nervousness brought on by the prospect of performing or speaking before an audience.

**stage·hand** (stāj′hănd′) *n.* One who works backstage in a theater.

**stage-man·age** (stāj′măn′ĭj) *vt.* **-aged, -ag·ing, -ag·es. 1.** To act as manager of the stage and actors for (a theatrical production). **2.** To supervise (e.g., a political campaign) from behind the scenes. **—stage management** *n.* **—stage manager** *n.*

**stag·er** (stā′jər) *n.* A very experienced person : VETERAN.

**stage-struck** (stāj′strŭk′) *adj.* Entranced by the glamour of the stage or by the hope of taking up acting as a career.

**stage whisper** *n.* **1.** The conventional whisper of an actor, intended to be heard by the audience but supposedly not audible to the other performers. **2.** A whisper that can be overheard.

**stag·ey** (stā′jē) *adj. var. of* STAGY.

**stag·fla·tion** (stăg-flā′shən) *n.* [STAG(NATION) + (IN)FLATION.] A condition in which a high rate of price and wage inflation is accompanied by stagnant consumer demand and a high rate of unemployment. **—stag·fla′tion·ar′y** (-shə-nĕr′ē) *adj.*

**stag·ger** (stăg′ər) *v.* **-gered, -ger·ing, -gers.** [Alteration of ME *stakeren* < ON *stakra,* freq. of *staka,* to push.] —*vi.* **1.** To stand or proceed unsteadily : TOTTER. **2.** To lose strength or confidence. —*vt.* **1.** To cause to sway. **2.** To overwhelm with emotion or surprise <I was *staggered* by the news.> **3.** To place on alternating sides of a midline in a usu. regular pattern <*staggered* seating> **4.** To arrange in alternating or overlapping time periods <*stagger* workers' breaks> —*n.* **1.** An act of staggering. **2.** A staggered pattern, arrangement, or order. **3. staggers** (*sing. in number*) A disease of the animal nervous system, esp. a cerebrospinal disease of horses marked by loss of coordination and staggering. **—stag′ger·er** *n.*

**stag·ger·bush** (stăg′ər-bo͝osh′) *n.* A shrub, *Lyonia mariana* of the eastern United States, having poisonous foliage.

**stag·ger·ing** (stăg′ər-ĭng) *adj.* Overwhelming <*staggering* medical bills> **—stag′ger·ing·ly** *adv.*

☆ ***syns:*** STAGGERING, MIND-BLOWING, MIND-BOGGLING, OVERWHELMING, SPECTACULAR, STUNNING *adj. core meaning* : of such a character as to overwhelm <the *staggering* costs of the war>

**stag·horn fern** (stăg′hôrn′) *n.* A tropical epiphytic fern of the genus *Platycerium,* with large divided fronds that resemble antlers.

**stag·hound** (stăg′hound′) *n.* A dog, such as a deerhound, that was once used to hunt deer.

**stag·ing** (stā′jĭng) *n.* **1.** A temporary platform : SCAFFOLD. **2. a.** Forward movement of troops and war materiel in stages. **b.** Assemblage of troops in transit at a given location. **3.** The process of producing and directing a theatrical work. **4.** The jettisoning of a stage of a multistage rocket. **5. a.** Operation of stagecoaches as a business. **b.** Travel by stagecoach.

**staging area** *n.* A place where armed forces and military supplies are assembled before deployment.

**stag·nant** (stăg′nənt) *adj.* [Lat. *stagnans, stagnant-,* pr.part. of *stagnare,* to be stagnant < *stagnum,* swamp.] **1.** Not moving or flowing : MOTIONLESS <*stagnant* water> **2.** Foul from standing still : STALE <*stagnant* air> **3.** Sluggish <a *stagnant* period for sales> **—stag′nan·cy** *n.* **—stag′nant·ly** *adv.*

**stag·nate** (stăg′nāt′) *vi.* **-nat·ed, -nat·ing, -nates.** [Lat. *stagnare, stagnat-* < *stagnum,* swamp.] **1.** To be or become stagnant. **2.** To fail to progress or develop. **—stag·na′tion** *n.*

**St. Ag·nes′ Eve** (sānt ăg′nĭs) *n. var. of* SAINT AGNES' EVE.

**stag·y** *also* **stag·ey** (stā′jē) *adj.* **-i·er, -i·est.** Theatrical in character or quality, esp. affected. **—stag′i·ly** *adv.* **—stag′i·ness** *n.*

**staid** (stād) *adj.* [< obs. *staid,* p.part. of STAY.] Marked by a sedate and usu. grave reserve : SOBER. **—staid′ly** *adv.* **—staid′ness** *n.*

**stain** (stān) *v.* **stained, stain·ing, stains.** [ME *steynen,* partly < OFr. *desteindre,* to deprive of color (Lat. *dis-,* apart + *tingere,* to dye), and partly < ON *steina,* to paint.] *—vt.* **1.** To discolor : soil. **2.** To taint : corrupt. **3.** To color with dye. **4.** To treat with a reagent or dye to facilitate microscopic examination. *—vi.* To make or receive discolorations. *—n.* **1.** A spot or smudge of foreign matter. **2.** A blemish on one's moral character or reputation. **3.** A substance applied esp. to wood that penetrates the surface and imparts color. **4.** A colored solution for staining microscopic specimens. **—stain′a·ble** *adj.* **—stain′er** *n.*

**stained glass** *n.* Glass that is colored as by fusing colored metallic oxides onto the glass, or by painting and baking transparent colors on the glass surface.

**stain·less** (stān′lĭs) *adj.* **1.** Having no stain or blemish. **2.** Resistant to stain or corrosion. **—stain′less·ly** *adv.*

**stainless steel** *n.* A steel alloyed with chromium so as to inhibit the corrosion, oxidation, or rusting resulting from exposure of ordinary steel to moisture.

**stair** (stâr) *n.* [ME < OE *stǣger.*] **1. stairs.** A series or flight of steps : STAIRCASE. **2.** One of a flight of steps.

**stair·case** (stâr′kās′) *n.* A flight or series of flights of steps and a supporting structure connecting separate levels.

**stair·way** (stâr′wā′) *n.* A staircase.

**stair·well** (stâr′wĕl′) *n.* A vertical shaft around which a staircase has been built.

**stake** (stāk) *n.* [ME < OE *staca.*] **1.** A sharpened piece, esp. of wood or metal for driving into the ground, as a marker, fence pole, or tent peg. **2. a.** A vertical post to which an offender is bound for execution by burning. **b.** Execution by burning at the stake. **3.** A vertical post secured at the edge of a platform, as on a truck. **4.** *Mormon Ch.* A territorial division consisting of a group of wards under the jurisdiction of a president. **5. a.** Money or property risked in a wager or game of chance. **b.** The prize awarded the winner of a contest or race : PURSE. **6.** A race in which a reward or prize is offered to the winner. **7.** A share in an enterprise. **8.** A grubstake. *—vt.* **staked, stak·ing, stakes. 1.** To indicate the boundaries of with or as if with stakes. **2.** To attach or support with a stake. **3.** To tie to a stake. **4.** To gamble or risk : HAZARD. **5.** To provide working capital for : FINANCE. **—at stake.** In jeopardy. **—stake out. 1.** To assign (a police officer) to conduct surveillance of a given area. **2.** To establish a stakeout on.

**stake·out** (stāk′out′) *n.* Police surveillance.

**Sta·kha·nov·ite** (stə-kä′nə-vīt′) *n.* [After Alexei *Stakhanov* (1905–1977).] A Soviet worker whose diligence and zeal have earned governmental recognition.

**sta·lac·tite** (stə-lăk′tīt′, stăl′ək-) *n.* [NLat. *stalactites* < Gk. *stalaktos,* dripping < *stalassein,* to drip.] A deposit that projects down from the roof of a cavern due to the dripping of mineral-rich water. **—sta·lac′ti·form′** *adj.* **—stal′ac·tit′ic** (stăl′ăk-tĭt′ĭk, stə-lăk′-) *adj.*

**sta·lag** (stä′läg′, stăl′ăg′) *n.* [G., short for *Stammlager,* base camp.] A German prisoner-of-war camp for enlisted personnel.

**sta·lag·mite** (stə-lăg′mīt′, stăl′əg-) *n.* [NLat. *stalagmites* < Gk. *stalagma,* a drop < *stalagmos,* dripping < *stalassein,* to drip.] A deposit that projects upward from the floor of a cavern as a result of the dripping of mineral-rich water. **—stal′ag·mit′ic** (stăl′əg-mĭt′ĭk, stə-lăg′-) *adj.*

**stale¹** (stāl) *adj.* **stal·er, stal·est.** [ME, well-aged (as beer).] **1.** Having lost freshness, effervescence, or palatability. **2.** Lacking originality or spontaneity <a *stale* joke> **3.** Impaired in efficacy or strength. **4.** Having lost legal efficacy or force through lack of exercise or action. *—vt. & vi.* **staled, stal·ing, stales.** To make or become stale. **—stale′ly** *adv.* **—stale′ness** *n.*

☆ ***syns:*** STALE, FLAT, TIRED *adj. core meaning* : lacking freshness or effectiveness through age or overuse <*stale* ideas><*stale* bread> STALE has broad literal and figurative application <*stale* crackers><*stale* news> FLAT suggests a lack or loss of sparkle, either literal or metaphorical <*flat* champagne><*flat* prose> TIRED refers to what is worn out or hackneyed <a *tired* metaphor>

**stale²** (stāl) *vi.* **staled, stal·ing, stales.** [ME *stalen.*] To urinate. —Used esp. of horses and camels. **—stale** *n.*

**stale·mate** (stāl′māt′) *n.* [Obs. *stale,* stalemate (< ME < AN *estale*) + MATE².] **1.** A position in chess in which only the king can move but only into check. **2.** A situation in which further action by opponents is impossible : DEADLOCK. *—vt.* **-mat·ed, -mat·ing, -mates.** To bring into a stalemate.

**Sta·lin·ism** (stä′lə-nĭz′əm) *n.* The bureaucratic and authoritarian exercise of state power and mechanistic application of Marxist-Leninist principles characteristic of the period of the leadership of Joseph Stalin in the Soviet Union. **—Sta′lin·ist** *n.* **—Sta′lin·ize** *v.* **(-ized, -iz·ing, -iz·es).**

**stalk¹** (stôk) *n.* [ME.] **1. a.** A stem or main axis of a herbaceous plant. **b.** A stem or similar structure that supports a plant part such as a flower, flower cluster, or leaf. **2.** A slender or elongated support. **—stalked** (stôkt) *adj.* **—stalk′less** *adj.* **—stalk′y** *adj.*

**stalk²** (stôk) *v.* **stalked, stalk·ing, stalks.** [ME *stalken* < OE *(be)stealcian,* to walk softly.] *—vi.* **1.** To walk with a stiff, haughty, or angry gait. **2.** To move threateningly or menacingly. **3.** To track game. *—vt.* **1.** To pursue by tracking. **2.** To go through (a region) in pursuit of game or other quarry. **—stalk′er** *n.*

**stalk·ing-horse** (stô′kĭng-hôrs′) *n.* **1. a.** A horse trained to conceal a hunter stalking game. **b.** A representation of a horse, used for similar concealment. **2.** A concealment for one's true purpose : DECOY. **3.** A sham candidate put forward to conceal the candidacy of another or to divide the opposition.

**stall¹** (stôl) *n.* [ME < OE *steall,* cattle stall.] **1.** A compartment for a domestic animal in a barn or shed. **2.** A small compartment. **3. a.** An enclosed seat in a church chancel. **b.** A church pew. **4.** *Chiefly Brit.* A theater seat in the front part of the orchestra. **5.** A parking space for a car. **6.** A protective sheath for a finger or toe. **7.** A sudden loss in the power or effectiveness of an engine. **8.** An interruption of airflow causing a loss of lift and a tendency to drop in an aircraft or airfoil. *—v.* **stalled, stall·ing, stalls.** *—vt.* **1.** To lodge (an animal) in a stall. **2.** To maintain (an animal) in a stall for fattening. **3.** To check the motion or progress of. **4.** To cause (an engine) accidentally to stop running. **5.** To cause (an airplane) to go into a stall. *—vi.* **1.** To live or be lodged in a stall. **2.** To stick fast in mud or snow. **3.** To come to a standstill. **4.** To stop operating because of mechanical failure. —Used of an engine. **5.** To experience a stall as a result of a loss in forward flying speed.

**stall²** (stôl) *n.* [AN *estal,* decoy bird, of Germanic orig.] A stratagem employed to delay or gain time. *—v.* **stalled, stall·ing, stalls** *—vi.* To gain time by means of a stratagem or ruse : DELAY. *—vt.* To use delaying tactics on.

**stall-feed** (stôl′fēd′) *vt.* **-fed** (-fĕd′), **-feed·ing, -feeds.** To lodge and feed (an animal) in a stall for the purpose of fattening.

**stal·lion** (stăl′yən) *n.* [ME *stalyone* < OFr. *estalon,* of Germanic orig.] An uncastrated adult male horse.

**stal·wart** (stôl′wərt) *adj.* [ME, alteration of *stalworth* < OE *stǣlwierthe,* serviceable.] **1.** Having physical strength : ROBUST. **2.** Resolute : uncompromising. *—n.* **1.** A physically and morally strong person. **2.** An active supporter of an organization or cause. **—stal′wart·ly** *adv.* **—stal′wart·ness** *n.*

**sta·men** (stā′mən) *n., pl.* **sta·mens** *or* **sta·mi·na** (stā′mə-nə, stăm′ə-) [Lat. *stamen, stamin-,* thread.] The pollen-producing reproductive organ of a flower, usu. consisting of a filament and an anther.

**stamin-** *pref.* [Lat. *stamen, stamin-,* thread.] Stamen <*staminate*>

**stam·i·na¹** (stăm′ə-nə) *n.* [Lat., pl. of *stamen,* thread.] The strength required to resist or withstand disease, fatigue, or hardship : ENDURANCE. **—stam·i·nal** *adj.*

**sta·mi·na²** (stā′mə-nə, stăm′ə-) *n. var. pl. of* STAMEN.

**sta·mi·nal** (stā′mə-nəl, stăm′ə-) *adj.* Relating to a stamen.

**sta·mi·nate** (stā′mə-nĭt, -nāt′, stăm′ə-) *adj. Bot.* **1.** Having a stamen or stamens. **2.** Bearing stamens but no pistils.

**stam·i·node** (stā′mə-nōd′, stăm′ə-) *also* **stam·i·no·di·um** (stā′mə-nō′dē-əm, stăm′ə-) *n., pl.* **-nodes** *also* **-no·di·a** (-nō′dē-ə) [NLat. *staminodium* < Lat. *stamen,* thread.] *Bot.* A sterile, functionless stamen.

**sta·mi·no·dy** (stā′mə-nō′dē, stăm′ə-) *n.* [STAMIN- + Gk. *-ōdēs,* like.] Transformation of a floral organ into a stamen.

**stam·mel** (stăm′əl) *n.* [Prob. alteration of *stamin* < ME *stamyn* < Lat. *stamineus,* consisting of threads < *stamen,* thread.] **1.** *Obs.* A coarse, usu. red woolen cloth. **2.** *Archaic.* The red color of stammel.

**stam·mer** (stăm′ər) *v.* **-mered, -mer·ing, -mers.** [ME *stameren* < OE *stamerian.*] *—vi.* To make involuntary pauses or syllabic repetitions while speaking. *—vt.* To utter or say with a stammer. *—n.* The habit of stammering or an instance of it. **—stam′mer·er** *n.* **—stam′mer·ing·ly** *adv.*

**stamp** (stămp) *v.* **stamped, stamp·ing, stamps.** [ME *stampen.*] *—vt.* **1.** To bring down (the foot) forcibly. **2.** To bring the foot down on forcibly. **3.** To bring into a specified condition by or as if by thrusting downward forcibly with the foot. **4.** To form or cut out by application of a mold, form, or die. **5.** To imprint or impress with a mark, design, or seal. **6.** To impress forcibly or permanently. **7.** To affix a stamp to. **8.** To identify, characterize, or reveal <*stamped* them traitors> —*vi.* **1.** To thrust the foot forcibly downward. **2.** To walk with forcible, heavy steps. —*n.* **1.** An act of stamping. **2. a.** An implement or device used to impress, cut out, or shape something to which it is applied. **b.** The impression or shape thus formed. **3.** A mark, design, or seal, whose impression indicates ownership, approval, or completion. **4. a.** A small piece of paper that is sold by a government for affixing to an article to be mailed : POSTAGE STAMP. **b.** A similar piece of paper issued for a specific purpose <trading *stamps*> **5.** An identifying or characterizing mark or impression. **6.** Characteristic nature or quality.

**stam·pede** (stăm-pēd′) *n.* [Sp. *estampida,* uproar < Prov. < *estampier,* to stamp, of Germanic orig.] **1.** A sudden headlong rush of startled animals. **2.** A sudden headlong rush of a crowd of people. **3.** A precipitous mass movement. —*v.* **-ped·ed, -ped·ing, -pedes.** —*vt.* **1.** To cause (animals) to move in a headlong rush. **2.** To cause (a group of people) to act on a single, common impulse. —*vi.* **1.** To move in a headlong rush. **2.** To act on a single, common impulse. **—stam·ped′er** *n.*

**stamping ground** *n.* **1.** One's customary environment. **2.** A favorite gathering place.

**stamp mill** *n.* **1.** A machine that crushes ore. **2.** A building in which ore is crushed.

**stance** (stăns) *n.* [OFr. *estance,* position < Ital. *stanza* < VLat. **stantia* < Lat. *stare,* to stand.] **1. a.** The attitude or position of a standing person or animal. **b.** The position assumed by an athlete or sportsman directly preparatory to action. **2.** An emotional or intellectual position.

**stanch**[1] *also* **staunch** (stônch, stänch) *vt.* **stanched, stanch·ing, stanch·es** *also* **staunched, staunch·ing, staunch·es.** [ME *stanchen* < OFr. *estanchier* < VLat. **stanticare* < Lat. *stans,* pr.part. of *stare,* to stand.] **1.** To stop or check the flow of. **2.** To check the flow of blood from (a wound). **3.** To stop or prevent the progress of. **—stanch′er** *n.*

**stanch**[2] (stônch, stänch) *adj. var. of* STAUNCH[1].

**stan·chion** (stăn′chən, -shən) *n.* [ME *stanchon* < OFr. *estanchon* < *estance,* prop. —see STANCE.] **1.** An upright pole, post, or support. **2.** One of the vertical posts used to secure cattle in a stall. —*vt.* **-chioned, -chion·ing, -chions. 1.** To equip with stanchions. **2.** To confine (cattle) in stanchions.

**stand** (stănd) *v.* **stood** (sto͝od), **stand·ing, stands.** [ME *standen* < OE *standan.*] —*vi.* **1. a.** To take or maintain an upright position on the feet. **b.** To be placed in or maintain an erect position. **c.** To grow in a vertical direction. **2.** To assume a standing position in a manner specified <*stand* straight> **3.** To point or range in hunting. **4.** To be equal to a specified height when erect <*stands* five feet tall> **5.** To remain valid, intact, or unchanged <The rule still *stands* on the books.> **6.** To have a specified position, expectation, or opportunity <*stand* to gain> **7.** To be situated or placed. **8.** To be in a specified class or degree : RANK <*stood* second in the class> **9.** To remain in a stationary position. **10. a.** To remain without movement, flow, or disturbance. **b.** To stagnate. **11.** To take up or keep to an attitude, conviction, or course <*stand* fast> **12.** *Chiefly Brit.* To be a candidate for public office. **13.** To take or hold a particular course or direction on the water <a ship *standing* to windward> —*vt.* **1.** To cause to stand : place upright. **2.** To encounter : meet <*stand* battle> **3. a.** To resist : withstand. **b.** To tolerate : endure <can't *stand* the tension> **4.** To be subjected to : UNDERGO <*stand* trial> **5.** *Informal.* To pick up the check for : TREAT <*stand* a friend to a drink> **—stand for.** To represent : symbolize. **—stand in.** To act as a stand-in. **—stand off. 1.** To maintain a distance from. **2.** To fail in or deny compliance or agreement. **3.** To put off : evade. **4.** To take or maintain a course away from shore. **—stand out.** To refuse compliance. **—stand up. 1.** To assume a standing position : RISE. **2.** To prove valid, satisfactory, or durable. **3.** *Informal.* To fail to keep an appointment with someone <My date *stood* me *up.*> —*n.* **1.** The act of standing. **2.** A halt. **3.** Cessation of work or activity : STANDSTILL. **4.** A stop on a performance tour <was booked for a series of one-night *stands*> **5.** The place where one stands. **6.** A small booth or stall for goods for sale. **7.** A space reserved for taxis. **8.** A desperate or decisive halt for defense or resistance. **9.** A position one is prepared to defend <take a *stand*> **10. stands.** The bleachers at a playing field or stadium. **11.** A witness stand. **12.** A rack or framework for holding various articles <a music *stand*> **13.** A growth of plants or trees <a *stand* of maples> **—stand a chance.** To have a chance of gaining or accomplishing. **—stand one's ground.** To hold one's position. **—stand to reason.** To be consistent with reason. **—stand′er** *n.*

**stand-a·lone** (stănd′ə-lōn′) *adj.* Of, relating to, or being a device that can function independently <a *stand-alone* word processing terminal>

**stan·dard** (stăn′dərd) *n.* [ME < OFr. *estandard,* rallying place, of Germanic orig.] **1.** A flag, banner, or ensign, specif.: **a.** The ensign of a chief of state, nation, or city. **b.** A long tapering flag bearing the distinctive device of a person or corporation. **c.** The colors of a mounted or motorized military unit. **2. a.** An accepted measure of comparison for quantitative or qualitative value : CRITERION. **b.** An object that under specified conditions defines, represents, or records the magnitude of a unit **3.** The proportion by weight of gold or silver to alloy metal that has been set for use in coinage. **4.** The commodity used as the basis for a monetary system. **5.** A pedestal, stand, or base. **6.** *Bot.* **a.** The large upper petal of the flower of a pea or related plant. **b.** One of the narrow upright petals of an iris. **7.** A shrub or plant trained to grow with a single stem of limited height. —*adj.* **1. a.** Serving as a standard of measurement or value. **b.** Commonly used and accepted as an authority <a *standard* text in the field of economics> **c.** Of average but acceptable quality. **d.** Of normal or prescribed size or quantity. **2.** Conforming to an established norm of educated usage in language.

**stan·dard-bear·er** (stăn′dərd-bâr′ər) *n.* **1.** One that bears the colors of a military unit. **2.** One in the forefront of a movement.

**stan·dard·bred** (stăn′dərd-brĕd′) *n.* One of an American breed of horses developed for harness racing.

**standard candle** *n.* A candela.

**standard deviation** *n. Statistics.* **1.** The square root of the variance. **2.** A statistic used as a measure of dispersion in a distribution, the square root of the arithmetic average of the squares of the deviations from the mean.

**standard gauge** *n.* **1.** A railroad track that is 56½ inches wide. **2.** A railroad or railroad car that meets standard gauge specification.

**stan·dard·ize** (stăn′dər-dīz′) *vt.* **-ized, -iz·ing, -iz·es.** To cause to be in agreement with a standard. **—stan′dard·i·za′tion** *n.*

**standard of living** *n.* A measure of the goods and services affordable by and available to a person or a country.

**standard operating procedure** *n.* An established procedure to be followed in a given situation.

**standard time** *n.* The time in any of 24 time zones, usu. the mean solar time at the central meridian of each zone.

**stand·by** (stănd′bī′) *n., pl.* **-bys. 1.** One that can always be depended on. **2.** A favorite or frequent choice. **3. a.** One kept in readiness to serve as a substitute. **b.** Readiness to serve as a substitute <was on *standby* for a flight to London>

**stand·ee** (stăn-dē′) *n.* An occupant of standing room.

**stand-in** (stănd′ĭn′) *n.* **1.** One who substitutes for an actor during lights and camera adjustments. **2.** A substitute.

**stand·ing** (stăn′dĭng) *n.* **1.** The act or position of one that stands. **2.** Standing room. **3. a.** Status with respect to achievement or reputation. **b.** High reputation : ESTEEM <a person of *standing* in the community> **4.** Length of time : DURATION. —*adj.* **1.** Remaining upright : ERECT. **2.** Made or performed from an upright position <*standing* jumps> **3.** Permanent and unchanging <a *standing* order> **4.** Not movable : STATIONARY. **5.** Not flowing or circulating : STAGNANT.

**standing army** *n.* A permanent army of paid soldiers.

**standing crop** *n.* The total amount of living organisms in a specific area at a given time.

**standing room** *n.* Space in which to stand, as at a performance for which all seats are filled.

**standing wave** *n.* A wave in which the amplitude of the resultant of a transmitted and a reflected wave is stationary in time and in which some of the energy of the transmitted wave is absorbed by the reflecting boundary.

**stand·off** (stănd′ôf′, -ŏf′) *n.* A standoff insulator.

**stand-off** (stănd′ôf′, -ŏf′) *n.* **1.** A tie in a race or contest : DRAW. **2.** A neutralizing or counterbalancing effect.

**standoff insulator** *n.* An insulator used to support a conductor a specified distance from a surface.

**stand·off·ish** (stănd-ô′fĭsh, -ŏf′ĭsh) *adj.* Unsociable : aloof.

**stand oil** *n.* A drying oil, as linseed, tung, or soya, heated until thickened and used in oil enamel paints.

**stand·out** (stănd′out′) *n.* One that is outstanding.

**stand·pipe** (stănd′pīp′) *n.* A large vertical pipe into which water is pumped in order to produce a desired pressure.

**stand·point** (stănd′point′) *n.* [Transl. of G. *Standpunkt.*] A position from which things are considered or judged : POINT OF VIEW.

**St. An·drew's cross** (sānt′ ăn′dro͞oz) *n.* [After the apostle *St. Andrew* (d. ca. A.D. 60).] **1.** A cross shaped like the letter X. **2.** A shrubby New World plant, *Ascyrum hypericoides,* with four-petaled yellow flowers.

**stand·still** (stănd′stĭl′) *n.* A halt <The work on the project came to a *standstill.*>

**stand·up** or **stand-up** (stănd′ŭp′) *adj.* **1.** Erect : upright. **2.** Taken standing <a *standup* supper> **3.** Of or designating a performance staged without costume, props, or assisting persons <a *standup* comedian>

**Stan·ford-Bi·net scale** (stăn′fərd-bĭ-nā′) *n.* [After *Stanford* University, California.] A revision of the Binet-Simon scale used in one form or another since 1916.

**stang** (stăng) *v. Obs. var. p.t. of* STING.

**stan·hope** (stăn′hōp′, stăn′əp) *n.* [After the Rev. Fitzroy *Stanhope* (1787–1864).] A light carriage having one seat and two or four wheels.

**stank** (stăngk) *v. var. p.t. of* STINK.

**stan·nic** (stăn′ĭk) *adj.* [Prob. < Fr. *stannique* < LLat. *stannum,* tin < Lat., an alloy of silver and lead.] Of, relating to, or containing tin, esp. with valence 4.

**stannic chloride** *n.* A colorless caustic liquid, $Na_2SnCl_6 \cdot H_2O$, made from tin treated with chlorine and used in making textiles, sensitized papers, and perfumes.

**stan·nite** (stăn′īt′) *n.* [G. *Stannit* < LLat. *stannum,* tin. —see STANNIC.] A lustrous, gray to black mineral consisting chiefly of $Cu_2FeSnS$.

**stan·nous** (stăn′əs) *adj.* [< LLat. *stannum,* tin.] Of or relating to tin, esp. with valence 2.

**stannous fluoride** *n.* A white powder, $SnF_2$, used to fluoridate toothpaste.

**St. An·tho·ny's fire** (sānt′ ăn′thə-nēz) *n.* [After *St. Anthony* (d. ca. A.D. 350).] Erysipelas.

**stan·za** (stăn′zə) *n.* [Ital.—see STANCE.] One of the divisions of a poem, consisting of two or more lines usu. characterized by a common and repeated pattern of meter, rhyme, and number of lines. **—stan·za′ic** (-zā′ĭk) *adj.*

**sta·pe·dec·to·my** (stā′pĭ-dĕk′tə-mē, -pē-) *n., pl.* **-mies.** [NLat. *stapes, staped-,* stapes + -ECTOMY.] Surgical removal of the stapes.

**sta·pes** (stā′pēz) *n., pl.* **stapes** *or* **sta·pe·des** (stā′pĭ-dēz′) [NLat. *stapes, staped-* < Med. Lat., stirrup.] A small bone of the inner ear, shaped somewhat like a stirrup. **—sta·pe′di·al** (stā-pē′dē-əl) *adj.*

**staph** (stăf) *n.* Staphylococcus.

**staphylo-** *pref.* [NLat. < Gk. *staphulē,* bunch of grapes.] **1.** Cluster : similar to a cluster <*staphylo*coccus> **2.** Uvula <*staphylo*plasty>

**staph·y·lo·coc·cus** (stăf′ə-lō-kŏk′əs) *n., pl.* **-coc·ci** (-kŏk′sī′, -kŏk′ī′). Any of various Gram-positive, spherical parasitic bacteria of the genus *Staphylococcus,* causing boils, septicemia, and other infections. **—staph′y·lo·coc′cal** (-kŏk′əl), **staph′y·lo·coc′cic** (-kŏk′sĭk, -kŏk′ĭk) *adj.*

**staph·y·lo·plas·ty** (stăf′ə-lō-plăs′tē) *n.* Corrective surgery of the uvula and the soft palate. **—staph′y·lo·plas′tic** *adj.*

**staph·y·lor·rha·phy** *also* **staph·y·lor·a·phy** (stăf′ə-lôr′ə-fē) *n.* [STAPHYLO- + Gk. *rhaptein,* to sew.] Correction of a cleft palate or divided uvula by plastic surgery.

**sta·ple¹** (stā′pəl) *n.* [ME, market town < OFr. *estaple* < MDu. *stapel,* emporium.] **1.** A major commodity grown or produced in a region. **2.** A major trade item in steady demand. **3.** A major element or feature. **4.** Raw material. **5.** Cotton, wool, or flax fiber graded with respect to length and fineness. *—vt.* **-pled, -pling, -ples.** To grade (fibers) according to length and fineness.

**sta·ple²** (stā′pəl) *n.* [ME *stapel* < OE *stapol,* post.] **1.** A U-shaped metal loop with pointed ends that are both driven into a surface to hold a bolt, hook, or hasp, or to hold wiring in place. **2.** A thin U-shaped piece of wire used as a fastening for materials such as paper or cloth. *—vt.* **-pled, -pling, -ples.** To fasten with a staple.

**sta·pler¹** (stā′plər) *n.* One who deals in staple goods or staple fibers.

**sta·pler²** (stā′plər) *n.* A device for binding material together with staples.

**star** (stär) *n.* [ME *sterre* < OE *steorra.*] **1.** *Astron.* A self-luminous, self-containing mass of gas in which the energy generated by nuclear reactions in the interior is balanced by the outflow of energy to the surface, and in which inward-directed gravitational forces and the outward-directed gas and radiation pressures are in balance. **2.** Any of the celestial bodies that can be seen at night from Earth as usu. twinkling points of light. **3.** Something held to resemble a star. **4.** A symbol or graphic design with five or more radiating points. **5.** A performer widely acknowledged as outstanding. **6.** ASTERISK 1. **7.** Something that resembles a star. **8. stars.** The constellations of the zodiac believed to influence personal destiny. **9. stars.** The future : DESTINY. *—v.* **starred, star·ring, stars.** *—vt.* **1. a.** To ornament with stars. **b.** To award or mark with a star for excellence. **2.** To mark with an asterisk. **3.** To present or feature (a performer) in a leading role. *—vi.* **1.** To play the leading role in a theatrical production <*starring* on Broadway> **2.** To perform in an outstanding manner. **—star′less** *adj.* **—star′like** *adj.*

**star anise** *n.* **1.** An east Asian aromatic tree, *Illicium verum* bearing purple-red flowers and starlike clusters of anise-scented fruit. **2.** The fruit of the star anise.

star anise

**star apple** *n.* **1.** A tropical American tree, *Chrysophyllum cainito,* bearing smooth-skinned, greenish-purple fruit. **2.** The edible fruit of the star apple.

**star·board** (stär′bərd) *n.* [ME *sterbord* < OE *stēorbord* : *stēor,* rudder + *bord,* side of a ship.] The right-hand side of a ship or aircraft as one faces forward. *—adj.* Located on the right-hand side. *—adv.* To or toward the right-hand side.

**starch** (stärch) *n.* [ME *starche* < *starchen,* to stiffen, of OE orig.] **1.** A naturally abundant nutrient carbohydrate, $(C_6H_{10}O_5)$, found chiefly in the seeds, fruits, tubers, roots, and stem pith of plants, such as corn, potatoes, or rice, varying widely in appearance but commonly prepared as a white, amorphous, tasteless powder. **2.** A substance used to stiffen fabrics. **3.** Foods with a high starch content. **4.** Stiff, formal behavior. **5.** Vigor : mettle. *—vt.* **starched, starch·ing, starch·es.** To stiffen with starch.

**Star Chamber** *n.* [So called because the ceiling of the original courtroom was decorated with stars.] **1.** A 15th- to 17th-cent. English court of judges appointed by the Crown to sit in closed session on cases involving state security. **2. star chamber.** A court or group marked by harsh or arbitrary procedures.

**star-cham·ber** (stär′chām′bər) *adj.* [< STAR CHAMBER.] Marked by harsh or arbitrary procedures.

**starch·y** (stär′chē) *adj.* **-i·er, -i·est. 1.** Of or like starch. **2.** Containing starch. **3.** Stiffened with starch. **4.** *Informal.* Stiff : formal. **—starch′i·ly** *adv.* **—starch′i·ness** *n.*

**star·dom** (stär′dəm) *n.* **1.** The status of an acknowledged star. **2.** Stars as a group.

**stare** (stâr) *v.* **stared, star·ing, stares.** [ME *staren* < OE *starian.*] *—vi.* **1.** To look with a steady, often wide-eyed gaze. **2.** *Chiefly Brit.* To stand out conspicuously. **3.** To bristle, as hair or feathers. *—vt.* To look at intently. **—stare down.** To cause to waver or give in by or as if by staring. *—n.* An intent gaze. **—star′er** *n.*

**sta·rets** (stär′yəts) *n., pl.* **star·tsy** (stärt′sē) [R. < *staryĭ,* old.] A respected spiritual adviser, often an Eastern Orthodox monk or religious hermit.

**star facet** *n.* One of the eight small triangular facets in the crown of a brilliant.

**star·fish** (stär′fĭsh′) *n., pl.* **starfish** *or* **-fish·es.** Any of various marine echinoderms of the class Asteroidea, having a radially symmetrical form with five arms extending from a central disk.

**star·flow·er** (stär′flou′ər) *n.* **1.** Any of several small North American plants of the genus *Trientalis,* with white starlike flowers. **2.** Any of several plants with starlike flowers.

**star·gaze** (stär′gāz′) *vi.* **-gazed, -gaz·ing, -gaz·es. 1.** To gaze at the stars. **2.** To daydream.

**star·gaz·er** (stär′gā′zər) *n.* **1. a.** *Informal.* An astronomer. **b.** An astrologer. **2.** Any of various marine bottom-dwelling fishes of the families Uranoscopidae and Dactyloscopidae, with eyes on the top of the head.

**star grass** *n.* **1.** Any of various plants of the genus *Hypoxis,* with grasslike leaves and star-shaped flowers. **2.** A plant, as the colicroot, that is similar to the star grass.

**stark** (stärk) *adj.* **-er, -est.** [ME < OE *stearc,* hard, severe.] **1.** Bare : blunt <the *stark* truth> **2.** Complete : absolute <*stark* poverty> **3.** Harsh in appearance : GRIM <*stark* cliffs> *—adv.* Utterly : entirely <*stark* naked><*stark* raving mad> **—stark′ly** *adv.* **—stark′ness** *n.*

**stark·ers** (stär′kərz) *adj. Chiefly Brit.* Completely naked.

**star·let** (stär′lĭt) *n.* **1.** A small star. **2.** A young film actress.

**star·light** (stär′līt′) *n.* The light produced by the stars.

**star·ling¹** (stär′lĭng) *n.* [ME < OE *stærlinc* : *stær,* starling + *-linc,* -ling.] An Old World bird of the family Sturnidae, with dark, often iridescent plumage, esp. *Sturnus vulgaris,* widely naturalized in North America.

**star·ling²** (stär′lĭng) *n.* [Perh. alteration of ME *stadelinge* < ME *stathel,* foundation < OE *staðol.*] A protective structure of pilings surrounding a pier of a bridge.

**star·lit** (stär′lĭt′) *adj.* Illuminated by starlight.

**star-nosed mole** (stär′nōzd′) *n.* A mole, *Condylura cristata* of eastern North America, with 22 small fleshy tentacles encircling the end of its nose.

**star-of-Beth·le·hem** (stär′əv-bĕth′lĭ-hĕm′) *n.* [After the star that guided the Magi to Bethlehem.] **1.** A European plant, *Ornithogalum umbellatum,* with narrow leaves and a cluster of star-shaped white flowers. **2.** Any of several plants that are similar or related to the star-of-Bethlehem.

**Star of David** *n.* The Magen David.

**star·quake** (stär′kwāk′) *n.* A seismological occurrence on a star.

**star·ry** (stär′ē) *adj.* **-ri·er, -ri·est. 1.** Marked or set with stars or starlike objects. **2.** Shining or glittering like stars. **3.** Star-shaped. **4.** Lighted by stars : STARLIT. **5.** Of, relating to, or from the stars : STELLAR. **—star′ri·ness** *n.*

**star·ry-eyed** (stär'ē-īd') *adj.* **1.** Naively enthusiastic. **2.** Marked by idealism : VISIONARY.

**Stars and Bars** *n. (sing. or pl. in number).* The first Confederate flag.

**Stars and Stripes** *n. (sing. or pl. in number).* The flag of the United States.

**star sapphire** *n.* A sapphire whose polished convex surface exhibits a starlike figure produces by reflected light.

**star shell** *n.* An artillery shell that explodes in midair with a shower of lights.

**Star-Span·gled Banner** (stär'spăng'gəld) *n.* **1.** The U.S. flag. **2.** The U.S. national anthem.

**start** (stärt) *v.* **start·ed, start·ing, starts.** [ME *sterten* < OE *styrtan,* to leap up.] —*vi.* **1.** To begin an activity or movement : SET OUT. **2.** To have a beginning : COMMENCE. **3.** To move suddenly or involuntarily <*started* at the least noise> **4.** To come quickly into view, life, or activity : spring forth. **5.** To be in the line-up for a contest. **6.** To protrude or bulge. **7.** To become loosened or disengaged. —*vt.* **1.** To begin. **2.** To set into motion, operation, or activity <*started* the car> **3.** To introduce : originate <*start* a fad> **4. a.** To cause to enter into a race or game. **b.** To put (a player) into the starting line-up for a game. **5.** To found : establish <*start* a business> **6.** To tend to the development of <*start* seedlings> **7.** To rouse from a hiding place or lair : FLUSH. **8.** To cause to become loosened. —*n.* **1.** A beginning : commencement. **2.** A startled movement. **3.** A dislocated or loosened part. **4.** A place or time of beginning. **5. a.** A starting line for a race. **b.** A signal to begin a race. **6.** A position of advantage over another : LEAD. **7.** An opportunity for a beginning <a fresh *start*> **—start something.** *Informal.* To cause trouble. **—to start with.** At the beginning : INITIALLY.

**start·er** (stär'tər) *n.* **1.** One that starts. **2.** A worker who supervises the departure of vehicles. **3.** A device for starting an internal-combustion engine. **4. a.** One who signals the start of a race. **b.** One that starts in a race or game.

**star thistle** *n.* A plant of the genus *Centaurea,* esp. *C. calcitrapa,* native to Eurasia, with spiny purplish flower heads.

**star·tle** (stär'tl) *v.* **-tled, -tling, -tles.** [ME *stertlen* < OE *steartlian,* to kick.] —*vt.* **1.** To cause to make a quick involuntary movement or start. **2.** To alarm or surprise. —*vi.* To become startled. —*n.* A sudden mild shock. **—star'tling·ly** *adv.* **—star'tling·ness** *n.*

☆ ***syns:*** STARTLE, JOLT, SHOCK *v. core meaning* : to cause to experience a sudden, momentary shock <The exploding firecrackers *startled* us.>

**star tracker** *n.* A telescopic instrument, used chiefly on rockets, that provides a guidance reference by remaining fixed on a celestial body.

**star·tsy** (stärt'sē) *n. pl. of* STARETS.

**starve** (stärv) *v.* **starved, starv·ing, starves.** [ME *sterven,* to die < OE *steorfan.*] —*vi.* **1.** To suffer or die from extreme or prolonged lack of food. **2.** To suffer from deprivation <*starved* for companionship> **3.** *Informal.* To be hungry. **4.** *Archaic.* To suffer or die from cold. —*vt.* **1.** To cause to starve. **2.** To bring to a specified state by starving. **—star·va'tion** (stär-vā'shən) *n.*

▲ word history: The verb *starve* is descended from Old English *steorfan,* which meant simply "to die." Only in modern times did the verb develop more specific meanings. In standard English *starve* became restricted to meaning "to die from lack of food." In the northern English dialects *starve* also acquired the sense of "to die of cold," but this usage no longer occurs in the standard language.

**starve·ling** (stärv'lĭng) *n.* One that is emaciated from or as if from starvation.

**star·wort** (stär'wûrt', -wôrt') *n.* Any of various plants with star-shaped flowers.

**stash** (stăsh) *vt.* **stashed, stash·ing, stash·es.** [Orig. unknown.] To hide or store away in a secret place. —*n.* **1.** A cache of money or valuables. **2.** Something hidden away.

**sta·sis** (stā'sĭs) *n., pl.* **-ses** (-sēz') [NLat. < Gk., standstill.] **1.** *Pathol.* The slowing down or stopping of the flow of a bodily fluid, esp. of blood. **2.** Balance among various forces : MOTIONLESSNESS.

**-stasis** *suff.* [< Gk. *stasis,* standstill.] **1.** Slowing : stoppage <bacteriostasis> **2.** Stable state <homeo*stasis*>

**-stat** *suff.* [NLat. *-stata* < Gk. *-statēs,* one that causes to stand.] **1.** Something that stabilizes <rheo*stat*> **2.** A device for reflecting something specified in a constant direction <helio*stat*>

**state** (stāt) *n.* [ME < OFr. *estat* < Lat. *status,* position, p.part. of *stare,* to stand.] **1.** A condition or mode of existence. **2.** A stage or condition of structure, growth, or development <a dormant *state*> **3.** A mental or emotional condition. **4.** *Informal.* An excited or disturbed condition. **5.** *Physics.* The condition of a physical system as specified by a set of appropriate macroscopic or quantum variables <the proton *state* of the nucleon> **6.** A social position or rank : ESTATE. **7.** Ceremony : pomp <dined in *state* at the embassy> **8.** The sovereign power of political entity. **9.** A particular form of government <a fascist *state*> **10.** A body politic, esp. one constituting a nation <the *states* of east Asia> **11.** One of the more or less internally autonomous units of a federal government <the United *States* of America> —*vt.* **stat·ed, stat·ing, states.** To put into words : DECLARE. **—state'hood** *n.*

**State attorney** *or* **State's attorney** *n.* A state prosecuting attorney.

**state·craft** (stāt'krăft') *n.* The art of leading a country.

**state house** *also* **State House** *n.* A building in which a state legislature holds sessions : state capitol.

**state·ly** (stāt'lē) *adj.* **-li·er, -li·est.** [ME *statly,* suitable to a person of rank < *state,* state, rank.] **1.** Dignified : formal. **2.** Majestic : lofty. —*adv.* In a ceremonious or imposing way. **—state'li·ness** *n.*

**state·ment** (stāt'mənt) *n.* **1.** The act of stating or declaring. **2.** Something stated : DECLARATION. **3.** *Law.* A formal pleading. **4.** An abstract of a commercial or financial account showing an amount due : BILL. **5.** A monthly report sent to a debtor or bank depositor. **6.** *Computer Sci.* An elementary instruction in the source language of a computer.

**state of affairs** *n.* The present situation.

**state of the art** *n.* The highest level of development, as of a device, technique, or science, achieved at a given time.

**state prison** *n.* A prison maintained by a state for the persons convicted of serious crimes, such as felonies.

**state·room** (stāt'rōōm') *n.* A private compartment with sleeping accommodations on a ship or train.

**state's evidence** *also* **State's evidence** *n.* **1.** Evidence for the prosecution in U.S. state or federal trials. **2.** One who gives evidence for the prosecution in criminal proceedings.

**States-Gen·er·al** (stāts'jĕn'ər-əl) *n.* [Transl. of Fr. *états généreaux.*] **1.** A legislative assembly of representatives from the estates of the nation, as opposed to a provincial assembly. **2.** The legislative assembly in France before the Revolution.

**state·side** (stāt'sīd') *adj.* & *adv.* Of, to, or in the continental United States.

**states·man** (stāts'mən) *n.* **1.** A national or international government leader. **2.** A political leader considered above partisan politics. **—states'man·like', states'man·ly** *adj.* **—states'man·ship'** *n.*

**States' rights** *also* **State rights** *pl.n.* All rights that are not granted to the federal government nor denied to the states by the U.S. Constitution.

**state university** *n.* A university funded and operated as part of the public education system of a state.

**state·wide** (stāt'wīd') *adj.* & *adv.* Taking place throughout a state.

**stat·ic** (stăt'ĭk) *adj.* [NLat. *staticus* < Gk. *statikos,* causing to stand < *statos,* standing.] **1. a.** Having no motion : at rest. **b.** Marked by the absence of motion or progress. **2.** *Elect.* Of, pertaining to, or producing stationary charges : ELECTROSTATIC. **3.** Of, relating to, or produced by random radio noise. —*n.* **1.** Random noise in a receiver or specks on a television screen produced by atmospheric disturbances. **2.** *Slang.* **a.** Back talk. **b.** Interference : obstruction. **—stat'i·cal** *adj.* **—stat'i·cal·ly** *adv.*

**static dump** *n.* A printout of the contents of a computer memory consistently performed at a particular point in the machine's run.

**stat·i·ce** (stăt'ĭ-sē') *n.* [NLat. *Statice,* genus name < Lat., an astringent plant < Gk. *statikē* < fem. of *statikos,* causing to stand, astringent. —see STATIC.] Sea lavender.

**static electricity** *n.* **1.** An accumulation of electric charge on an insulated body. **2.** Electric discharge resulting from the accumulation of electric charge on an insulated body.

**static memory** *n.* A computer memory with no moving parts.

**static routine** *n.* A computer subroutine with the addresses of the operands involved as the only parameters.

**stat·ics** (stăt'ĭks) *n. (sing. in number).* Equilibrium mechanics of stationary bodies.

**static tube** *n.* A specialized tube used to measure the static pressure in a stream of fluid.

**sta·tion** (stā'shən) *n.* [ME *stacioun* < OFr. *estation* < Lat. *statio* < *status,* p.part. of *stare,* to stand.] **1.** The place or position where one stands or is assigned to stand : POST. **2.** The place from which a service is provided or operations are directed <a fire *station*> **3.** A stopping place along a route, as for refueling or for taking on passengers : DEPOT. **4.** Social position : RANK. **5.** An establishment equipped for observation and study <a radar *station*> **6.** An establishment equipped for radio or television transmission. **7.** An input or output point along a communications system. —*vt.* **-tioned, -tion·ing, -tions.** To assign to a position.

☆ ***syns:*** STATION, ASSIGN, POST, SET *v. core meaning* : to appoint and send to a particular place for duty <patrols *stationed* along the boundaries of the Korean DMZ>

**sta·tion·ar·y** (stā'shə-nĕr'ē) *adj.* [ME *stacionarye* < Lat. *stationarius* < *statio,* station.] **1. a.** Not moving. **b.** Not capable of being moved : FIXED. **2.** Unchanging <a *stationary* sound> —*n., pl.* **-ries.** One that is stationary.

**stationary front** *n.* A transition zone between two nearly stationary air masses of different density.

**stationary orbit** *n.* Synchronous orbit.

**stationary satellite** *n.* An artificial satellite in a synchronous orbit.

**stationary wave** *n.* A standing wave.

**station break** *n.* A pause in a broadcast program to allow for identification of the network or station.

**sta·tion·er** (stā'shə-nər) *n.* [ME *staciouner* < Med. Lat. *stationarius,* shopkeeper < *statio,* shop < Lat., station.] **1.** A seller of stationery. **2.** *Obs.* **a.** A publisher. **b.** A bookseller.

**sta·tion·er·y** (stā'shə-nĕr'ē) *n.* **1.** Writing paper and envelopes. **2.** Writing or typing materials.

**station house** *n.* **1.** A police station. **2.** A fire station.

**sta·tion·mas·ter** (stā'shən-măs'tər) *n.* An official in charge of a railroad station.

**Stations of the Cross** *pl.n.* **1.** A devotion that consists of meditation before each of the images or representations set up usu. in a church to commemorate 14 events in the passion of Jesus. **2.** The 14 images representing the events of the passion of Christ.

**station wagon** *n.* An automobile having an extended interior, third seat or luggage platform, and a tailgate.

**sta·tis·tic** (stə-tĭs'tĭk) *n.* [Back-formation < STATISTICS.] **1.** A numerical datum. **2.** An estimate of a parameter, as of the population mean or variance, obtained from a sample. **3.** A random variable that takes on the characteristics of a statistic.

**sta·tis·ti·cal** (stə-tĭs'tĭ-kəl) *adj.* Of, relating to, or using statistics or the principles of statistics. **—sta·tis'ti·cal·ly** *adv.*

**stat·is·ti·cian** (stăt-ĭ-stĭsh'ən) *n.* **1.** A specialist in statistics. **2.** One who compiles of statistical data.

**sta·tis·tics** (stə-tĭs'tĭks) *n.* [G. *Statistik,* political science < NLat. *statisticus,* of state affairs < Lat. *status,* state. —see STATE.] **1.** *(sing. in number).* The mathematics of the collection, organization, and interpretation of numerical data. **2.** *(pl. in number).* A collection of numerical data.

**stato-** *pref.* [< Gk. *statos,* standing, placed.] **1.** Resting : remaining <*stato*blast> **2.** Equilibrium : balance <*stato*cyst>

**stat·o·blast** (stăt'ə-blăst') *n.* An asexually produced encapsulated bud of a freshwater bryozoan from which new individuals develop after the parent colony has disintegrated.

**stat·o·cyst** (stăt'ə-sĭst') *n.* A small organ of balance in many invertebrates, consisting of a fluid-filled sac containing statoliths that help indicate position when the animal moves.

**stat·o·lith** (stăt'l-ĭth') *n.* A small movable concretion of calcium carbonate found in statocysts.

**sta·tor** (stā'tər) *n.* [Lat., one that stands < *status,* p.part of *stare,* to stand.] The stationary part of a machine, such as a motor, dynamo, or turbine, about which a rotor turns.

**stat·o·scope** (stăt'ə-skōp') *n.* **1.** A barometer for recording small variations in atmospheric pressure. **2.** A device for indicating small changes in an airplane's altitude.

**stat·u·ar·y** (stăch'o͞o-ĕr'ē) *n., pl.* **-ies.** [Partly < Lat. *statuaria,* art of making statues, and partly < Lat. *statuarius,* sculptor, both < *statuarius,* of a statue < *statua,* statue.] **1.** Statues collectively. **2.** A sculptor. **3.** The art of making statues.

**stat·ue** (stăch'o͞o) *n.* [ME < OFr. < Lat. *statua* < *statuere,* to set up. —see STATUTE.] A form or likeness sculpted, modeled, carved, or cast in material such as stone, clay, wood, or bronze.

**stat·u·esque** (stăch'o͞o-ĕsk') *adj.* Like a statue, esp. in size, grace, or dignity : STATELY. **—stat'u·esque'ly** *adv.*

**stat·u·ette** (stăch'o͞o-ĕt') *n.* A small statue.

**stat·ure** (stăch'ər) *n.* [ME < OFr. < Lat. *statura* < *status,* p.part. of *stare,* to stand.] **1.** The natural height of a human or animal body in an upright position. **2.** A level achieved : STATUS.

**stat·us** (stā'təs, stăt'əs) *n.* [Lat., condition, p.part of *stare,* to stand.] **1.** The legal character or condition of a person or thing <the *status* of a minor> **2.** A stage of progress or development. **3. a.** Relative position in a ranked group or in a social system <the high *status* of physicians> **b.** High relative position <a job with *status*> **4.** A state of affairs : SITUATION.

**sta·tus quo** (stā'təs kwō', stăt'əs) *n.* [Lat., state in which.] The existing condition : STATE OF AFFAIRS.

**status word** *n.* A computer storage location which provides data to restore an interrupted program.

**stat·u·ta·ble** (stăch'ə-tə-bəl) *adj.* **1.** Enacted, regulated, or authorized by statute : STATUTORY. **2.** Legally punishable : recognized by statute <a *statutable* offense>

**stat·ute** (stăch'o͞ot) *n.* [ME < OFr. *estatut* < LLat. *statutum* < Lat., *statutus,* p.part. of *statuere,* to set up < *status,* p.part of *stare,* to stand.] **1.** A law enacted by the legislative assembly of a nation or state. **2.** A decree or edict. **3.** An established law or rule, esp. of a corporation.

**statute law** *n.* A law established by legislative enactment.

**statute mile** *n.* MILE 1.

**statute of limitations** *n.* *Law.* A statute setting a time limit on legal action in certain cases.

**stat·u·to·ry** (stăch'ə-tôr'ē, -tōr'ē) *adj.* **1.** Of or relating to a statute. **2.** Enacted, regulated, or authorized by statute.

**statutory offense** *n.* A legal offense declared by statute.

**statutory rape** *n.* Sexual intercourse with a girl who has not reached the statutory age of consent.

**staunch[1]** (stônch, stänch) *also* **stanch** (stänch, stănch) *adj.* **-er, -est.** [ME *staunche,* watertight < OFr. *estanche* < *estanchier,* to stanch. —see STANCH.] **1.** Firm and steadfast : TRUE. **2.** Having a strong construction or constitution. ***usage:*** The adjective form is commonly spelled *staunch,* but the verb form is more commonly spelled *stanch.* **—staunch'ly** *adv.* **—staunch'ness** *n.*

**staunch[2]** (stônch, stänch) *v. var. of* STANCH[1].

**stau·ro·lite** (stôr'ə-līt') *n.* [Fr. < Gk. *stauros,* cross.] A brownish-black mineral, chiefly $FeAl_4Si_2O_{10}(OH)_2$, often with crossed intergrown crystals, occas. used as a gem. **—stau'ro·lit'ic** (-lĭt'ĭk) *adj.*

**stave** (stāv) *n.* [Back-formation < *staves,* pl. of STAFF.] **1.** A narrow strip of wood forming part of the sides of a container, such as a barrel or tub. **2.** A rung of a ladder or chair. **3.** A staff. **4.** A musical staff. **5.** A set of verses : STANZA. —*v.* **staved** or **stove** (stōv), **stav·ing, staves.** —*vt.* **1.** To break in or puncture the staves of. **2.** To break or smash a hole in. **3.** To crush or smash inward. —*vi.* To be or become crushed in. **—stave off.** To keep or ward hold off.

**staves** (stāvz) *n. var. pl. of* STAFF[1] 1, 2, 4.

**staves·a·cre** (stāvz'ā'kər) *n.* [By folk ety. < ME *staphisagre* < Lat. *staphis agria* < Gk., wild raisin.] **1.** A larkspur, *Delphinium staphisagria* of southern Europe, with greenish-white flowers. **2.** The poisonous seeds of the stavesacre, formerly used externally as a parasiticide.

**stay[1]** (stā) *v.* **stayed, stay·ing, stays.** [ME *steyen,* to halt < OFr. *ester,* to stop < Lat. *stare,* to stand.] —*vi.* **1.** To remain in a given place or condition <*stay* home><couldn't *stay* interested in the subject> **2.** To sojourn as a guest or lodger <*stayed* with as at our country house> **3.** To stop moving : CEASE. **4.** To wait : PAUSE. **5.** To hold on : ENDURE. **6.** To keep up in a race or contest. **7.** To meet a bet in poker without raising it. —*vt.* **1.** To stop or halt : CHECK. **2.** To postpone : delay. **3.** To delay or stop the effect of (e.g., an order) by legal action or mandate. **4.** To satisfy temporarily <*stayed* my hunger with a snack> **5.** *Obs.* To wait for : AWAIT. —*n.* **1.** The act of halting : CHECK. **2.** The act of coming to a halt. **3.** A brief period of residence or visiting. **4.** Suspension or postponement of a legal action <a *stay* of execution>

☆ ***syns:*** STAY, SOJOURN, VISIT *v. core meaning* : to remain as a guest or lodger <*stayed* with friends>

**stay[2]** (stā) *vt.* **stayed, stay·ing, stays.** [OFr. *estayer,* to support < *estaie,* a support, of Germanic orig.] **1.** To support or prop up. **2.** To sustain mentally or spiritually. **3.** To rest or fix on for support. —*n.* **1.** A support : brace. **2.** A strip of bone, plastic, or metal, used to stiffen a garment. **3. stays.** A corset.

**stay[3]** (stā) *n.* [ME < OE *stæg.*] **1.** A heavy rope or cable that serves as a brace or support for a mast or spar. **2.** A rope used to guide or brace something. —*v.* **stayed, stay·ing, stays.** —*vt.* **1.** To brace with a stay. **2.** To put (a ship) on the opposite tack. —*vi.* To come about.

**staying power** *n.* Stamina : endurance.

**stay-in strike** (stā'ĭn') *n.* A job action that consists of a slowdown or work stoppage by employees who remain at their work place.

**stay·sail** (stā'səl, -sāl') *n.* *Naut.* A triangular sail hoisted on a stay.

**St. Ber·nard** (sānt' bər-närd') *n.* The Saint Bernard.

**stead** (stĕd) *n.* [ME *stede* < OE.] The place or position of another <My friend went to the meeting in my *stead.*> —*vt.* **stead·ed, stead·ing, steads.** To be of advantage to : BENEFIT.

**stead·fast** *also* **sted·fast** (stĕd'făst', -fəst) *adj.* [ME *stedefast* < OE *stedefæst* : *stede,* place + *fæst,* fixed, fast.] **1.** Fixed or unchanging : STEADY. **2.** Firmly loyal or constant. **—stead'fast'ly** *adv.* **—stead'fast'ness** *n.*

**stead·y** (stĕd'ē) *adj.* **-i·er, -i·est. 1.** Firm in position or place : STABLE. **2.** Direct and unfaltering : SURE. **3.** Continuous in movement, quality, or pace <a slow, *steady* trot> **4.** Not easily excited or upset <*steady* nerves> **5.** Reliable : dependable. **6.** Marked by temperance : SOBER. —*v.* **stead·ied, stead·y·ing, stead·ies.** *vt.* & *vi.* To make or become steady. —*n., pl.* **-ies.** *Slang.* The person one dates regularly and exclusively. **—stead'i·er** *n.* **—stead'i·ly** *adv.* **—stead'i·ness** *n.*

**steady state** *n.* A stable condition that does not change over time or in which change in one direction is continually balanced by change in another.

**stead·y-state theory** (stĕd'ē-stāt') *n.* A cosmological theory that assumes that the large-scale view of the universe is independent of the position of the observer in space and time and that the expansion of the universe, required on other grounds, is compensated for by the continuous creation of matter.

**steak** (stāk) *n.* [ME *steyke* < ON *steik.*] **1.** A piece of meat, esp. beef, typically cut in a thick slice across the muscle grain. **2.** A thick slice of a large fish cut across the body. **3.** A patty of ground meat that has been prepared like a steak.

**steak house** *n.* A restaurant specializing in beefsteak dishes.

**steak knife** *n.* A table knife with a sharp, occas. serrated blade.

**steak tar·tare** (tär'tär') *n.* [STEAK + Fr. *tartare,* Tartar.] A dish of raw ground beef mixed with onion, seasoning, and raw egg.

**steal** (stēl) *v.* **stole** (stōl), **sto·len** (stō'lən), **steal·ing, steals.** [ME *stelen* < OE *stelan.*] —*vt.* **1.** To take (the property of another) without right or permission. **2.** To get or accomplish secretly or artfully. **3.** To move stealthily or gradually. **4.** *Baseball.* To gain (another base) without the ball being batted, by running during the delivery of a pitch. —*vi.* **1.** To commit theft. **2.** To move, happen, or go by stealthily or unobtrusively. **3.** *Baseball.* To steal a base. —*n.* **1.** The act of stealing : THEFT. **2.** *Baseball.* The act of stealing a base. **3.** *Slang.* A bargain. **—steal (one's) thunder.** To take or use as one's own something invented by another. **—steal'er** *n.*

**stealth** (stĕlth) *n.* [ME *stelth.*] **1.** The act of moving or proceeding covertly. **2.** Furtiveness : covertness. **3.** *Archaic.* The act of stealing.

**stealth·y** (stĕl'thē) *adj.* **-i·er, -i·est.** Marked by stealth. **—stealth'i·ly** *adv.* **—stealth'i·ness** *n.*

**steam** (stēm) *n.* [ME *steme,* vapor < OE *stēam.*] **1. a.** The vapor phase of water. **b.** The mist of cooling water vapor. **2.** Steam heating. **3. a.** Power : energy. **b.** Emotional or physical tension <wanted to let off *steam* after a hard day> —*v.* **steamed, steam·ing, steams.** —*vi.* **1.** To produce or emit steam. **2.** To rise up as steam. **3.** To become misted or covered with steam. **4.** To move by steam power. **5.** *Informal.* To become very angry : FUME. —*vt.* To expose to steam, esp. so as to cook.

**steam bath** *n.* **1.** A bath that involves exposure to steam. **2.** A room or building equipped to provide steam baths.

**steam beer** *n.* A highly effervescent beer of western U.S. origin.

**steam·boat** (stēm'bōt') *n.* A steamship.

**steam boiler** *n.* A closed tank in which water is converted into steam under pressure.

**steam chest** *n.* A compartment in a steam engine through which steam is delivered from the boiler to a cylinder.

**steam engine** *n.* An engine for converting the heat energy of pressurized steam into mechanical energy, esp. one in which a piston in a closed cylinder is driven by steam.

**steam·er** (stē'mər) *n.* **1.** A steamship. **2.** A container in which something is steamed. **3.** A soft-shell clam.

**steamer rug** *n.* A blanket used esp. by shipboard passengers while sitting in deck chairs.

**steamer trunk** *n.* A trunk orig. designed for storing under the bunk of a steamship cabin.

**steam·fit·ter** (stēm'fĭt'ər) *n.* One whose occupation is the installation and repair of heating, ventilating, refrigerating, or air-conditioning systems.

**steam heating** *n.* A heating system in which the steam generated in a boiler is piped to radiators.

**steam iron** *n.* An iron that holds and heats water to be emitted as steam on the article that is being pressed.

**steam·rol·ler** *also* **steam roller** (stēm'rō'lər) *n.* **1.** A vehicle equipped with a heavy roller for smoothing road surfaces. **2.** A ruthless or irresistible force. —*v.* **-lered, -ler·ing, -lers.** —*vt.* **1.** To smooth or level with a steamroller. **2.** To overwhelm by means of great force. —*vi.* To move or proceed with overwhelming or crushing force.

**steam·ship** (stēm'shĭp') *n.* A vessel propelled by one or more steam-driven screws or propellers.

**steam shovel** *n.* A steam-driven machine for digging.

**steam table** *n.* A table in which containers of cooked food are kept warm by steam or hot water circulating below.

**steam turbine** *n.* A turbine operated by highly pressurized steam directed against or through vanes on a rotor.

**steam·y** (stē'mē) *adj.* **-i·er, -i·est.** **1.** Filled with or emitting steam <a *steamy* kitchen> **2.** *Slang.* Erotic <a *steamy* novel> **—steam'i·ly** *adv.* **—steam'i·ness** *n.*

**ste·ap·sin** (stē-ăp'sĭn) *n.* [Gk. *stear,* tallow + (PE)PSIN.] An enzyme of pancreatic juice that catalyzes the hydrolysis of fats to fatty acids and glycerol.

**ste·a·rate** (stē'ə-rāt', stîr'āt') *n.* [STEAR(IC) + -ATE.] A salt or ester of stearic acid.

**ste·ar·ic** (stē-ăr'ĭk, stîr'ĭk) *adj.* [Fr. *stéarique* < Gk. *stear,* tallow.] Of, relating to, or similar to stearin or fat.

**stearic acid** *n.* A colorless, odorless, waxlike fatty acid, $CH_3(CH_2)_{16}COOH$, occurring in natural animal and vegetable fats.

**ste·a·rin** (stē'ər-ĭn, stîr'ĭn) *n.* [Fr. *stéarine* < Gk. *stear,* tallow.] **1.** A colorless, odorless, tasteless ester of glycerol and stearic acid, $C_3H_5(C_{18}H_{35}O_2)_3$, used in making soap and candles and for textile sizing. **2.** Stearic acid. **3.** The solid form of fat.

**ste·a·rop·tene** (stē'ə-rŏp'tēn') *n.* [STEAR(IC) + Gk. *ptēnos,* flying.] The portion of a natural essential oil that separates out as a white crystalline solid on cooling or standing.

**ste·a·tite** (stē'ə-tīt') *n.* [Lat. *steatitis,* a precious stone < Gk. < *stear,* tallow.] A massive white-to-green talc used in paints, ceramics, and insulation. **—ste'a·tit'ic** (-tĭt'ĭk) *adj.*

**steato-** *pref.* [Gk. < *stear,* tallow.] Fat <*steato*lysis>

**ste·a·tol·y·sis** (stē'ə-tŏl'ĭ-sĭs) *n.* Digestive emulsification of fats prior to assimilation.

**ste·at·o·pyg·i·a** (stē-ăt'ə-pĭj'ē-ə, -pī'jē-ə) *n.* [STEATO- + Gk. *pugē,* rump.] Excessive accumulation of fat on the buttocks. **—ste'at·o·pyg'ic** (-pĭj'ĭk, -pī'jĭk), **ste'a·to·py'gous** (-pī'gəs) *adj.*

**ste·at·or·rhe·a** (stē'ə-tə-rē'ə) *also* **ste·at·or·rhoe·a** (stē-ăt'ə-rē'ə) *n.* **1.** Overaction of the sebaceous glands. **2.** Excessive discharge of fat in the feces.

**sted·fast** (stĕd'făst', -fəst) *adj. var. of* STEADFAST.

**steed** (stēd) *n.* [ME *stede* < OE *stēda,* stallion.] A spirited horse.

**steel** (stēl) *n.* [ME *stel* < OE *stȳle.*] **1.** Any of various gen. hard, strong, durable, malleable alloys of iron and carbon, usu. containing between 0.2–1.5% carbon, often with other constituents, and widely used as a structural material. **2.** A hard unflinching character or quality. **3.** Something made of steel. **4.** A knife sharpener consisting of a handled steel rod. **5.** A slender strip or band of steel used for stiffening. **6.** A dark gray to purplish gray. —*adj.* Of the color steel. —*vt.* **steeled, steel·ing, steels.** **1.** To cover, plate, edge, or point with steel. **2.** To make hard, strong, or obdurate : STRENGTHEN <I *steeled* myself for disappointment.>

**steel band** *n.* A musical band of Trinidadian origin, composed chiefly of percussion instruments fashioned from oil drums.

**steel blue** *n.* **1.** A medium grayish blue. **2.** One of several blue colors taken on by steel while being tempered.

**steel engraving** *n.* **1.** The art or process of engraving on a steel plate. **2.** An impression produced with an engraved steel plate.

**steel guitar** *n.* A Hawaiian guitar.

**steel·head** (stēl'hĕd') *n.* The rainbow trout when occurring in marine waters or large inland lakes.

**steel-trap** (stēl'trăp') *adj.* Very quick and keen : TRENCHANT <a *steel-trap* mind>

**steel wool** *n.* Fine fibers of steel woven or matted together to form an abrasive for cleaning, smoothing, or polishing.

**steel·work** (stēl'wûrk') *n.* **1.** Something made of steel. **2. steelworks.** A plant where steel is made. **—steel'work'er** *n.*

**steel·y** (stē'lē) *adj.* **-i·er, -i·est.** **1.** Made of steel. **2.** Like steel <a *steely* glance> **—steel'i·ness** *n.*

**steel·yard** (stēl'yärd') *n.* [STEEL + YARD (rod).] A balance consisting of a scaled arm suspended off center, a hook at the shorter end on which to hang the object being weighed, and a counterbalance at the longer end that can be moved to find the weight.

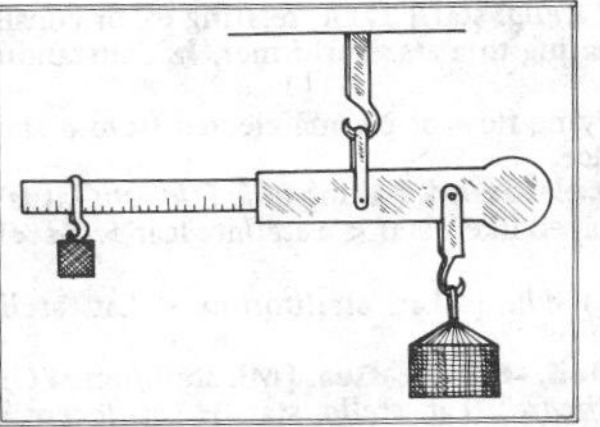
steelyard

**steen·bok** (stēn'bŏk', stān'-) *also* **stein·bok** (stīn'bŏk') *n.* [Afr. < MDu. *steenboc,* ibex : *steen,* stone + *boc,* buck.] An African antelope, *Raphicerus campestris,* with a brownish coat and short, pointed horns in the male.

**steep¹** (stēp) *adj.* **-er, -est.** [ME *stepe* < OE *stēap,* lofty.] **1.** Sharply inclined : PRECIPITOUS. **2.** Rising or falling rapidly or precipitously <a *steep* rise in prices> **3. a.** Excessive : stiff <a *steep* fine> **b.** Ambitious : difficult —*n.* A precipitous slope. **—steep'ly** *adv.* **—steep'ness** *n.*

**steep²** (stēp) *v.* **steeped, steep·ing, steeps.** [ME *stepen.*] —*vt.* **1.** To soak in liquid so as to cleanse, soften, or extract a given property from. **2.** To infuse or subject thoroughly to <*steeped* the students in African culture> **3.** To make thoroughly wet : SATURATE. —*vi.* To undergo a soaking in liquid. —*n.* **1.** The act of steeping or the state of being steeped. **2.** A liquid, bath, or solution in which something is steeped. **—steep'er** *n.*

**steep·en** (stē'pən) *vt.* & *vi.* **-ened, -en·ing, -ens.** To make or become steeper.

**stee·ple** (stē'pəl) *n.* [ME *stepel* < OE *stēpel,* tall tower.] **1.** A tall tower forming the superstructure of a building, as a church or temple, and usu. surmounted by a spire. **2.** A spire.

**stee·ple·bush** (stē'pəl-bo͝osh') *n.* The hardhack.

**stee·ple·chase** (stē'pəl-chās') *n.* [From the use of church steeples as goals.] A horse race across open country or over an obstacle course. **—stee'ple·chas'er** *n.*

**stee·ple·jack** (stē'pəl-jăk') *n.* A worker on steeples or other very high structures.

**steer¹** (stîr) *v.* **steered, steer·ing, steers.** [ME *steren* < OE *stīeran.*] —*vt.* **1.** To guide by a device such as a rudder, paddle, or wheel. **2. a.** To direct the course of. **b.** To maneuver (a person) into a place or course of action. —*vi.* **1.** To guide a vessel or vehicle. **2.** To follow or move in a set course. **3.** To be capable of being steered or guided <a car that *steers* easily> —*n.* A piece of advice. **—steer'a·ble** *adj.* **—steer'er** *n.*

**steer²** (stîr) *n.* [ME < OE *stēor.*] A young ox castrated before sexual maturity and raised for beef.

**steer·age** (stîr′ĭj) *n.* **1.** The act or practice of steering. **2.** A ship's steering mechanism. **3.** The section of a passenger ship, orig. near the rudder, providing the cheapest passenger accommodations.
**steer·age·way** (stîr′ĭj-wā′) *n.* The minimum rate of motion required for the helm of a ship or boat to have effect.
**steering committee** *n.* A committee that sets agendas and schedules business, as for a legislative body.
**steering gear** *n.* The mechanism by which dispositions of the steering controls of a vehicle are transferred to the part that interacts with the external medium.
**steering wheel** *n.* A wheel that controls steering.
**steers·man** (stîrz′mən) *n.* A helmsman.
**steeve**[1] (stēv) *n.* [ME *steven,* to stow < OFr. *estiver* < Sp. *estibar,* to cram < Lat. *stipare.*] A derrick or spar with a block at one end, used for stowing cargo. —*vt.* **steeved, steev·ing, steeves.** To pack or stow (cargo) in the hold of a ship.
**steeve**[2] (stēv) [Orig. unknown.] —*n. Naut.* The angle formed by the bowsprit and the horizon or the keel. —*v.* **steeved, steev·ing, steeves.** —*vt.* To incline (a bowsprit) upward at an angle with the horizon or the keel. —*vi.* To have an upward inclination. —Used of a bowsprit.
**steg·o·don** *also* **steg·o·dont** (stĕg′ə-dŏn′, -dŏnt′) *n.* [NLat. *Stegodon,* genus name : Gk. *stegos,* roof (< *stegein,* to cover) + Gk. *odous, odont-,* tooth.] An extinct elephantlike mammal of the genus *Stegodon* and of related genera, of the Pliocene to Pleistocene epochs.
**steg·o·saur** (stĕg′ə-sôr′) *also* **steg·o·sau·rus** (stĕg′ə-sôr′əs) *n.* [NLat. *Stegosaurus,* genus name : Gk. *stegos,* roof (< *stegein,* to cover) + *osaurs,* lizard.] An herbivorous dinosaur of the genus *Stegosaurus* and of related genera, of the Triassic to the Cretaceous periods, that had a double row of upright bony plates along the back.
**stein** (stīn) *n.* [G., prob. short for *Steingut,* stoneware : *Stein,* stone + *Gut,* goods.] A usu. one-pint mug, esp. for beer.
**stein·bok** (stīn′bŏk′) *n. var. of* STEENBOK.
**ste·le** (stē′lē) *n., pl.* **-les** *or* **-lae** (-lē) [Gk. *stēlē,* pillar.] **1.** An upright stone or slab with an inscribed or sculptured surface, used as a monument or as a commemorative tablet in the face of a building. **2.** *Bot.* The central core of vascular tissue in a plant stem. **—ste′lar** (-lər) *adj.*
**stel·lar** (stĕl′ər) *adj.* [Lat. *stella,* star.] **1.** Of, relating to, or consisting of stars. **2. a.** Of or relating to a star performer. **b.** Outstanding <a *stellar* performance>
**stellar wind** *n.* The varying flow of plasma ejected from a star's surface into interstellar space.
**stel·late** (stĕl′āt′) *also* **stel·lat·ed** (-ā′tĭd) *adj.* [Lat. *stellatus* < *stella,* star.] Arranged or shaped like a star <a *stellate* leaf> **—stel′late·ly** *adv.*
**stel·li·form** (stĕl′ə-fôrm′) *adj.* [NLat. *stelliformis* < Lat. *stella,* star.] Stellate.
**stel·li·fy** (stĕl′ə-fī′) *vt.* **-fied, -fy·ing, -fies.** [ME *stellifien* < OFr. *stellifier* < Med. Lat. *stellificare* : Lat. *stella,* star + Lat. *facere,* to make.] To transform into a star.
**stel·lu·lar** (stĕl′yə-lər) *adj.* [< LLat. *stellula,* dim. of Lat. *stella,* star.] **1.** Having the form of a small star. **2.** Bespangled with small stars.
**St. El·mo's fire** (sānt′ ĕl′mōz) *n.* Saint Elmo's fire.
**stem**[1] (stĕm) *n.* [ME < OE *stefn,* prow.] **1. a.** The main ascending axis of a plant : a stalk or trunk. **b.** A slender stalk supporting or connecting another plant part, as a leaf or flower. **2.** A banana stalk yielding several bunches of bananas. **3.** A connecting or supporting part, esp.: **a.** The tube of a tobacco pipe. **b.** The slender upright support of a wine goblet. **c.** The small projecting shaft with an expanded crown by which a watch is wound. **d.** The rounded rod in the center of a lock about which a key fits and is turned. **e.** The shaft of a feather or hair. **f.** The upright stroke of a typeface or letter. **g.** The vertical line extending from the head of a musical note. **4.** The main line of genealogical descent. **5.** The main part of a word to which affixes are added. **6.** The curved upright beam at the fore of a vessel into which the hull timbers are scarfed to form the prow. **7.** The tubular glass structure mounting the filament or electrodes in an incandescent bulb or vacuum tube. —*v.* **stemmed, stem·ming, stems.** —*vt.* **1.** To remove the stem of. **2.** To provide with a stem. **3.** To make headway against. —*vi.* To derive from or originate in. **—from stem to stern.** From one end to another. **—stem′less** *adj.*
**stem**[2] (stĕm) *v.* **stemmed, stem·ming, stems.** [ME *stemmen* < ON *stemma.*] —*vt.* **1.** To stop or hold back by or as if by damming : STANCH. **2.** To plug or tamp (e.g., a blast hole). **3.** To point (skis) inward. —*vi.* To point skis inward in order to slow down or turn.
**stem cell** *n.* An unspecialized cell that gives rise to a specific specialized cell, as a blood cell.
**stem·ma** (stĕm′ə) *n., pl.* **stem·ma·ta** (stĕm′ə-tə) *or* **stem·mas.** [Lat., garland < Gk. *stephein,* to encircle.] **1.** An ancient Roman scroll recording the genealogy of a family : FAMILY TREE. **2.** A diagram showing the relationships of the manuscripts of a literary work.
**stemmed** (stĕmd) *adj.* **1.** Having the stems removed. **2.** Provided with a stem <long-*stemmed* roses>
**stem·mer** (stĕm′ər) *n.* One that removes stems, as from fruit or tobacco.
**stem rust** *n.* A rust disease affecting the stem of a plant.
**stem·son** (stĕm′sən) *n.* [STEM (prow) + (KEEL)SON.] *Naut.* A piece of supporting timber bolted to the stem and keelson at their junction near the bow of a wooden vessel.
**stem turn** *n.* A skiing turn made by stemming the downhill ski and placing one's weight upon it while bringing the other ski into a parallel position.
**stem·ware** (stĕm′wâr′) *n.* Glassware mounted on a stem.
**stem-wind·er** (stĕm′wīn′dər) *n.* A stem-winding watch.
**stem-wind·ing** (stĕm′wīn′dĭng) *adj.* Wound by turning an expanded crown on the stem.
**stench** (stĕnch) *n.* [ME < OE *stenc,* odor.] A strong foul odor : FUG.
**sten·cil** (stĕn′səl) *n.* [< ME *stanselen,* to adorn with bright colors < OFr. *estenceler* < *estencele,* spark < VLat. **stincilla,* alteration of Lat. *scintilla,* spark.] **1.** A sheet of plastic, cardboard, or other material in which a desired lettering or design has been cut so that ink or paint applied to the sheet will reproduce the pattern on the surface beneath. **2.** The lettering or design produced by stencil. —*vt.* **-ciled, -cil·ing, -cils** *or* **-cilled, -cil·ling, -cils. 1.** To mark with a stencil. **2.** To make by stencil. **—sten′cil·er** *n.*
**stencil paper** *n.* Strong tissue-thin paper for making stencils.
**sten·o** (stĕn′ō) *n., pl.* **-os. 1.** A stenographer. **2.** Stenography.
**steno-** *pref.* [< Gk. *stenos,* narrow.] Narrow : small <*steno*topic>
**sten·o·bath·ic** (stĕn′ə-băth′ĭk) *adj.* Of or relating to an organism able to live only within a narrow range of water depths. **—sten′o·bath′** *n.*
**sten·o·graph** (stĕn′ə-grăf′) *n.* [Back-formation < STENOGRAPHY.] **1.** A keyboard machine for reproducing letters in a shorthand system. **2.** A character in shorthand. —*vt.* **-graphed, -graph·ing, -graphs.** To write in shorthand.
**ste·nog·ra·pher** (stə-nŏg′rə-fər) *n.* One skilled in shorthand, esp. one hired to take and transcribe dictation.
**ste·nog·ra·phy** (stə-nŏg′rə-fē) *n.* **1.** The art or process of writing in shorthand. **2.** Material in shorthand. **—sten′o·graph′ic** (stĕn′ə-grăf′ĭk), **sten′o·graph′i·cal** *adj.* **—sten′o·graph′i·cal·ly** *adv.*
**sten·o·ha·line** (stĕn′ə-hā′līn, -hăl′īn) *adj.* Of or relating to an organism able to live only within a narrow range of water salinity.
**ste·noph·a·gous** (stə-nŏf′ə-gəs) *adj.* Feeding on a single kind or limited range of food.
**ste·nosed** (stə-nōzd′, -nōst′) *adj.* [STENOS(IS) + -ED.] Marked by stenosis.
**ste·no·sis** (stə-nō′sĭs) *n.* [NLat. < Gk. *stenōsis,* a narrowing < *stenoun,* to narrow < *stenos,* narrow.] Constriction of a passage or duct. **—ste·not′ic** (-nŏt′ĭk) *adj.*
**sten·o·ther·mal** (stĕn′ə-thûr′məl) *adj.* Of or relating to organisms adapted to living only within a limited range of temperature.
**sten·o·top·ic** (stĕn′ə-tŏp′ĭk) *adj.* [STENO- + Gk *topos,* place.] Having narrow limits of adaptation to environmental conditions.
**sten·o·type** (stĕn′ə-tīp′) *n.* [(STENO)GRAPHY + TYPE.] **1.** A symbol or combination of symbols representing a sound, word, or phrase, esp. in shorthand. **2.** A keyboard machine used to record dictation by a phonetic system.
**sten·tor** (stĕn′tôr′) *n.* [NLat. *Stentor,* genus name, after *Stentor,* a Greek herald. —see STENTORIAN.] Any of several trumpet-shaped aquatic microorganisms of the genus *Stentor,* with cilia around the oral cavity.
**sten·to·ri·an** (stĕn-tôr′ē-ən, -tōr′-) *adj.* [After *Stentor,* a loud-voiced Greek herald in the *Iliad,* a Homeric poem.] Very loud <a senator who spoke with a *stentorian* voice>
**step** (stĕp) *n.* [ME < OE *stæp.*] **1. a.** The single complete movement of raising one foot and putting it down in another spot, as in walking. **b.** Manner of walking : GAIT. **c.** A fixed pace or rhythm, as in marching. **d.** The sound of a footstep. **e.** A footprint. **2. a.** The distance traversed by moving one foot ahead of the other. **b.** A very short distance <Our house is just a *step* away.> **c. steps.** Course : path <followed in their parents' *steps*> **3. a.** A rest for the foot in ascending or descending. **b. steps.** Stairs. **4. a.** One of a series of actions or measures undertaken to reach a goal. **b.** A stage in a process. **5.** A degree in progress or a grade or rank in a scale <a *step* ahead of our competitors> **6.** *Mus.* The interval that separates two successive tones of a scale. **7.** *Computer Sci.* A single instructor in a computer sequence. **8.** *Naut.* The block in which the heel of a mast is fixed. —*v.* **stepped, step·ping, steps.** —*vi.* **1.** To put or press the foot <*step* on the accelerator> **2.** To move or shift slightly by taking a step or two <*step* forward> **3.** To walk a short distance to a specified place or in a specified direction <*step* over to the counter> **4.** To move with the feet in a particular way <Let's *step* lively!> **5.** To move into a new situation by or as if by taking a single step <*stepped* into a life of hardship> **6.** To treat with arrogant indifference <always *stepping* on people> —*vt.* **1.** To put or set (the foot) down. **2.** To measure by pacing <*step* off five yards> **3.** To furnish with steps. **4.** To cause a computer to execute a single instruction. **5.** *Naut.* To place (a mast) in its step. **—in step. 1.** Moving in rhythm. **2.** *Informal.* In conformity with one's environment <stayed *in step* with the times> **—out of step. 1.** Not in step <troops

marching *out of step*> **2.** Not in conformity with one's environment <*out of step* with the times> **—step by step.** By degrees. **—step down. 1.** To resign from a high office. **2.** To reduce, esp. in stages <*stepping down* the electric current> **—step in. 1.** To enter into an activity or situation. **2.** To intervene. **—step on it.** *Informal.* To go faster: HURRY. **—step out. 1.** To walk briskly. **2.** To go outside for a short time <*stepped out* for a minute> **3.** *Informal.* To go out for a special evening of entertainment. **4.** To withdraw: quit. **—step up. 1.** To increase, esp. in stages <*step up* the electric current> **2.** To make oneself known. **—watch (one's) step.** To proceed carefully.

**step-** *pref.* [ME < OE *stēop-.*] Related by remarriage rather than by blood <*stepparent*>

**step·broth·er** (stĕp′brŭth′ər) *n.* The son of one's stepparent by a previous marriage.

**step·child** (stĕp′chīld′) *n.* The child of one's spouse by a previous marriage.

**step dance** *n.* A dance in which the performer emphasizes certain steps, as clogging or tapping, rather than body position or gesture.

**step·daugh·ter** (stĕp′dô′tər) *n.* The daughter of one's spouse by a previous marriage.

**step-down** (stĕp′doun′) *adj.* Decreasing in stages <a *step-down* gear> —*n.* A reduction in amount or size.

**step-down transformer** *n.* A transformer that has a greater number of turns in the primary winding than in the secondary, used to transform high voltage to low voltage.

**step·fa·ther** (stĕp′fä′thər) *n.* The husband of one's mother by a later marriage.

**steph·a·no·tis** (stĕf′ə-nō′tĭs) *n.* [NLat. *Stephanotis,* genus name < Gk. *stephanōtis,* deserving a crown < *stephanos,* crown, wreath < *stephein,* to crown.] A woody climbing plant of the genus *Stephanotis,* cultivated for its fragrant, showy, white flowers.

**step-in** (stĕp′ĭn′) *adj.* Donned by stepping into <*step-in* tights> —*n.* **1. step-ins.** Panties with wide legs. **2.** A step-in garment.

**step·lad·der** (stĕp′lăd′ər) *n.* A portable ladder with a hinged supporting frame and usu. topped with a small platform.

**step·moth·er** (stĕp′mŭth′ər) *n.* The wife of one's father by a later marriage.

**step·par·ent** (stĕp′pâr′ənt, -păr′-) *n.* A stepfather or a stepmother.

**steppe** (stĕp) *n.* [R. *step′.*] A vast grass-covered semiarid plain, as found in southeastern Europe and Siberia.

**stepped-up** (stĕpt′ŭp′) *adj.* Increased in pace or intensity: HEIGHTENED <a *stepped-up* election campaign>

**step·per** (stĕp′ər) *n.* **1.** One that steps, esp. in a spirited way. **2.** *Slang.* A dancer.

**step·ping-off place** (stĕp′ĭng-ôf′, -ŏf′) *n.* **1.** The last stop on an outbound rail or bus line. **2.** A point from which one leaves for unfamiliar regions.

**step·ping·stone** (stĕp′ĭng-stōn′) *n.* **1.** A stone providing a place to step, as in crossing a yard. **2.** A position for advancement toward a goal <a job that is a *steppingstone* to top management>

**step rocket** *n.* A multistage rocket.

**step·sis·ter** (stĕp′sĭs′tər) *n.* The daughter of one's stepparent by a previous marriage.

**step·son** (stĕp′sŭn′) *n.* The son of one's spouse by a previous marriage.

**step turn** *n.* A skiing turn made by lifting a ski, putting it down again pointed in the direction of the turn, and placing one's weight on it while bringing the other ski into parallel position.

**step-up** (stĕp′ŭp′) *adj.* Increasing in or by steps. —*n.* An increase in size, amount, or activity.

**step-up transformer** *n.* A transformer that has fewer turns in the primary winding than in the secondary, used to transform low voltage to high voltage.

**step·wise** (stĕp′wīz′) *adj.* **1.** Marked by a gradual progression as if step by step. **2.** Moving from one musical tone to an adjacent one. **—step′wise′** *adv.*

**-ster** *suff.* [ME < OE *-estre.*] **1.** One that is associated with, participates in, makes, or does <*songster*> **2.** One that is <*youngster*>

**ste·ra·di·an** (stĭ-rā′dē-ən) *n.* [STE(REO)- + RADIAN.] A unit of measure equal to the solid angle subtended at the center of a sphere by an area equal to the radius squared on the surface of the sphere <The total solid angle of a sphere is $4\pi$ *steradians.*>

**ster·co·ra·ceous** (stûr′kə-rā′shəs) *also* **ster·co·rous** (stûr′kər-əs) *adj.* [Lat. *stercus,* ′*stercor-,* dung + -ACEOUS.] Made up of or relating to excrement.

**stere** (stîr) *n.* [Fr. *stère* < Gk. *stereos,* solid, hard.] A unit of volume equal to one cubic meter.

**ster·e·o** (stĕr′ē-ō′, stîr′-) *n., pl.* **-os. 1. a.** A stereophonic sound reproduction system. **b.** Stereophonic sound. **2.** A stereotype. **3.** A stereoscopic system or photograph.

**stereo-** *pref.* [< Gk. *stereos,* solid.] **1.** Solid: solid body <*stereotropism*> **2.** Three-dimensional <*stereoscope*>

**ster·e·o·bate** (stĕr′ē-ō-bāt′, stîr′-) *n.* [Lat. *stereobata* < Gk. *stereobatēs* : *stereos,* solid + *bainein,* to go.] **1.** A stylobate. **2.** The foundation of a stone building.

**ster·e·o·chem·is·try** (stĕr′ē-ō-kĕm′ĭ-strē, stîr′-) *n.* The chemical study of spatial arrangements of atoms in molecules and of the effects of these arrangements on the molecule's properties. **—ster′e·o·chem′i·cal** *adj.*

**ster·e·o·chro·my** (stĕr′ē-ō-krō′mē, stîr′-) *n.* Mural painting in which pigments are mixed with water glass. **—ster′e·o·chrome′** *n.* **—ster′e·o·chro′mic** *adj.* **—ster′e·o·chro′mi·cal·ly** *adv.*

**ster·e·o·gram** (stĕr′ē-ō-grăm′, stîr′-) *n.* **1.** A picture or diagram designed to give the impression of solidity. **2.** A stereograph.

**ster·e·o·graph** (stĕr′ē-ō-grăf′, stîr′-) *n.* Two stereoscopic pictures or one picture with two superposed stereoscopic images, intended to create a three-dimensional effect when viewed through a stereoscope. —*vt.* **-graphed, -graph·ing, -graphs.** To make (a stereographic picture).

**ster·e·og·ra·phy** (stĕr′ē-ŏg′rə-fē, stîr′-) *n.* **1.** The art or technique of depicting solid bodies on a plane surface. **2.** Photography involving the use of stereoscopic equipment. **—ster′e·o·graph′ic** (-ə-grăf′ĭk), **ster′e·o·graph′i·cal** *adj.* **—ster′e·o·graph′i·cal·ly** *adv.*

**ster·e·o·i·so·mer** (stĕr′ē-ō-ī′sə-mər, stîr′-) *n.* ISOMER 1c.

**ster·e·o·i·som·er·ism** (stĕr′ē-ō-ī-sŏm′ə-rĭz′əm, stîr′-) *n.* Isomerism caused by differences in the spatial arrangement of atoms in a molecule. **—ster′e·o·i′so·mer′ic** (-ī′-sə-mĕr′ĭk) *adj.*

**ster·e·ol·o·gy** (stĕr′ē-ŏl′ə-jē) *n.* The study of three-dimensional properties of objects or matter usu. observed two-dimensionally. **—ster′e·o·log′ic** (-ə-lŏj′ĭk), **ster′e·o·log′i·cal** *adj.* **—ster′e·o·log′i·cal·ly** *adv.* **—ster′e·ol′o·gist** *n.*

**ster·e·o·mi·cro·scope** (stĕr′ē-ō-mī′krə-skōp′, stîr′-) *n.* A microscope optically equipped for stereoscopic viewing. **—ster′e·o·mi′cro·scop′ic** *adj.*

**ster·e·o·phon·ic** (stĕr′ē-ō-fŏn′ĭk, stîr′-) *adj.* Of or used in a sound reproduction system employing two or more separate channels to give a more natural sound distribution. **—ster′e·o·phon′i·cal·ly** *adv.* **—ster′e·oph′on·y** (-ē-ŏf′ə-nē) *n.*

**ster·e·op·sis** (stĕr′ē-ŏp′sĭs, stîr′-) *n.* Stereoscopic vision.

**ster·e·op·ti·con** (stĕr′ē-ŏp′tĭ-kŏn′, stîr′-) *n.* [NLat. STEREO- + Greek *optikon,* neuter of *optikos,* optic.] A magic lantern, esp. one made double so as to produce dissolving views.

**ster·e·o·scope** (stĕr′ē-ə-skōp′, stîr′-) *n.* An optical instrument used to impart a three-dimensional effect to two photographs of the same scene taken at slightly different angles and viewed through two eyepieces.

**ster·e·o·scop·ic** (stĕr′ē-ə-skŏp′ĭk, stîr′-) *adj.* **1.** Of or relating to stereoscopy, esp. three-dimensional. **2.** Of or relating to a stereoscope. **—ster′e·o·scop′i·cal·ly** *adv.*

**ster·e·os·co·py** (stĕr′ē-ŏs′kə-pē, stîr′-) *n.* **1.** The viewing of objects as three-dimensional. **2.** The technique of making or using stereoscopes and stereoscopic slides. **—ster′e·os′co·pist** *n.*

**ster·e·o·tax·is** (stĕr′ē-ə-tăk′sĭs, stîr′-) *also* **ster·e·o·tax·y** (stĕr′ē-ə-tăk′sē, stîr′-) *n.* Thigmotaxis. **—ster′e·o·tac′tic** (-tăk′tĭk), **ster′e·o·tac′ti·cal** *adj.* **—ster′e·o·tac′ti·cal·ly** *adv.*

**ster·e·ot·ro·pism** (stĕr′ē-ŏt′rə-pĭz′əm, stîr′-) *n.* Thigmotropism. **—ster′e·o·trop′ic** (-ē-ə-trŏp′ĭk) *adj.*

**ster·e·o·type** (stĕr′ē-ə-tīp′, stîr′-) *n.* [Fr. *stéréotype* : *stéréo-,* stereo- + *-type,* -type.] **1.** A conventional, formulaic, usu. oversimplified opinion, conception, or belief. **2.** One, as a person, group, event, or issue that is thought to typify or conform to an unvarying pattern or manner, lacking any individuality. **3.** A metal printing plate cast from a matrix that is molded from a raised printing surface, as type. —*vt.* **-typed, -typ·ing, -types. 1.** To develop a fixed, idea about. **2.** To make a stereotype of. **3.** To print from a stereotype. **—ster′e·o·typ′er** *n.* **—ster′e·o·typ′ic** (-tĭp′ĭk), **ster′e·o·typ′i·cal** *adj.*

**ster·e·o·typed** (stĕr′ē-ə-tīpt′, stîr′-) *adj.* **1.** Printed or reproduced from stereotype plates. **2.** Conventional, formulaic, and oversimplified <*stereotyped* characterizations>

**ster·e·o·ty·py** (stĕr′ē-ə-tī′pē, stîr′-) *n.* **1.** The art or process of making stereotype plates. **2.** Excessive repetition or lack of variation in movements, ideas, or speech patterns.

**ster·e·o·vi·sion** (stĕr′ē-ō-vĭzh′ən, stîr′-) *n.* Visual perception of or exhibition in three dimensions.

**ster·ic** (stĕr′ĭk, stîr′-) *adj.* [STER(EO)- + -IC.] Of or relating to the spatial arrangement of atoms in a molecule. **—ster′i·cal·ly** *adv.*

**ste·rig·ma** (stə-rĭg′mə) *n., pl.* **-ma·ta.** [NLat. < Gk. *stērigma,* support < *stērizein,* to support.] A slender projection of the basidium of some fungi which bears a basidiospore.

**ster·il·ant** (stĕr′ə-lənt) *n.* A sterilizing agent.

**ster·ile** (stĕr′əl, -īl′) *adj.* [OFr. < Lat. *sterilis,* unfruitful.] **1.** Incapable of sexual reproduction: INFERTILE. **2.** Capable of producing little or no vegetation: UNFRUITFUL. **3.** Free from microorganisms. **4.** Lacking imagination or vitality <*sterile* writing> **5.** Lacking power to function productively or effectively. **—ster′ile·ly** *adv.* **—ster′ile·ness, ste·ril′i·ty** (stə-rĭl′ĭ-tē) *n.*

☆ **syns:** STERILE, SANITARY, SANITIZED, STERILIZED *adj.* core *meaning*: free from microorganisms <*sterile* surgical instruments>

**ster·il·ize** (stĕr′ə-līz′) *vt.* **-ized, -iz·ing, -iz·es. 1.** To make sterile. **2.** To place (gold) in safekeeping so as not to affect the supply of money or credit. **—ster′il·i·za′tion** (-lĭ-zā′shən) *n.* **—ster′il·iz′er** *n.*

ōō **boot** ou **out** th **thin** *th* **this** ŭ **cut** ûr **urge** y **young**
yōō **abuse** zh **vision** ə **about,** **item,** **edible,** **gallop,** **circus**

**ster·let** (stûr′lĭt) *n.* [R. *sterlyad′.*] A sturgeon, *Acipenser ruthenus,* of the Black Sea and adjacent waters, used as a source of caviar.

**ster·ling** (stûr′lĭng) *n.* [ME, silver penny.] **1.** British money, esp. the pound as the basic monetary unit of the United Kingdom. **2.** British coinage of silver or gold, having as a standard of fineness 0.500 for silver and 0.91666 for gold. **3. a.** Sterling silver. **b.** Articles, as flatware or hollowware, made of sterling silver. —*adj.* **1.** Consisting of or relating to sterling or British money. **2.** Made of sterling silver. **3.** Of the highest quality <a person of *sterling* character>

**sterling silver** *n.* **1.** An alloy of 92.5% silver with copper or another metal. **2.** Objects made of sterling silver.

**stern**[1] (stûrn) *adj.* **-er, -est.** [ME *sterne* < OE *styrne.*] **1.** Firm or unyielding : INFLEXIBLE <*stern* military discipline> **2.** Grave : austere <a *stern* judge> **3.** Grim, gloomy, or forbidding <*stern* and massive cliffs> **4.** Inexorable : relentless <*stern* economic demands>

**stern**[2] (stûrn) *n.* [ME *sterne,* perh. of Scand. orig.] **1.** The rear part of a ship or boat. **2.** A rear part or section.

**ster·na** (stûr′nə) *n. var. pl. of* STERNUM.

**ster·nal** (stûr′nəl) *adj.* [NLat. *sternalis* < *sternum,* sternum.] Of, near, or relating to the sternum.

**stern chaser** *n.* A gun or cannon mounted on the stern of a ship for firing at a pursuing vessel.

**stern·fore·most** (stûrn′fôr′mōst′, -fōr′-) *adv.* With the stern foremost.

**stern·most** (stûrn′mōst′) *adj.* Farthest astern.

**ster·no·cos·tal** (stûr′nō-kŏs′təl) *adj.* [STERN(UM) + Lat. *costa,* rib + -AL.] Of or relating to both the sternum and the ribs.

**stern·post** (stûrn′pōst′) *n.* The principal upright post at the stern of a vessel, usu. serving to support the rudder.

**stern sheets** *pl.n.* The stern area of an open boat.

**stern·son** (stûrn′sən) *n.* [STERN + (KEEL)SON.] A bar of metal or wood set between the keelson and the sternpost to fortify the joint.

**ster·num** (stûr′nəm) *n., pl.* **-nums** *or* **-na** (-nə) [NLat. < Gk. *sternon.*] A long flat bone articulating with the cartilages of and forming the midventral support of most of the ribs in tetrapod vertebrates, and also of the collarbone in human beings and some other vertebrates.

**ster·nu·ta·tion** (stûr′nyə-tā′shən) *n.* [Lat. *sternutatio, sternutation-* < *sternutare,* freq. of *sternuere,* to sneeze.] **1.** The act of sneezing. **2.** A sneeze.

**ster·nu·ta·tor** (stûr′nyə-tā′tər) *n.* A substance that irritates the nasal and respiratory passages and causes coughing, sneezing, lachrimation, and sometimes vomiting.

**ster·nu·ta·to·ry** (stûr-nyōō′tə-tôr′ē, -tōr′ē, -nōō′-) *also* **ster·nu·ta·tive** (stûr′nyə-tā′tĭv) *adj.* Causing or tending to cause sneezing. —*n., pl.* **-ries.** A sternutatory substance, as pepper.

**stern·ward** (stûrn′wərd) *adj. & adv.* At or in the stern : ASTERN. **—stern′wards** *adv.*

**stern·way** (stûrn′wā′) *n.* The backward movement of a vessel.

**stern-wheel·er** (stûrn′hwē′lər, -wē′lər) *n.* A steamboat propelled by a paddle wheel at the stern.

**ster·oid** (stîr′oid′, stĕr′-) *n.* [STER(OL) + -OID.] Any of numerous naturally occurring, fat-soluble organic compounds having a 17-carbon-atom ring as a basis, and including the sterols and bile acids, many hormones, certain natural drugs such as digitalis compounds, and the precursors of certain vitamins.

**ste·roid·o·gen·e·sis** (stĭ-roi′də-jĕn′ĭ-sĭs, stîr′oi-, stĕr′-) *n.* Production of steroids. **—ste·roid′o·gen′ic** *adj.*

**ster·ol** (stîr′ôl′, -ōl′, stĕr′-) *n.* [Short for CHOLESTEROL.] Any of a group of predominantly unsaturated solid alcohols of the steroid group, as cholesterol and ergosterol, occurring in the fatty tissues of plants and animals.

**Ster·o·pe** (stĕr′ə-pē′) *n.* [Gk. < *asteropē,* lightning.] **1.** *Gk. Myth.* One of the seven Pleiades. **2.** A star in the constellation Pleiades.

**ster·son knee** (stûr′sən) *n.* A sternson.

**ster·tor** (stûr′tər) *n.* [NLat. < Lat. *stertere,* to snore.] A heavy snoring sound in respiration. **—ster′to·rous** (stûr′tər-əs) *adj.*

**stet** (stĕt) *n.* [Lat., let it stand.] A printer's direction that a letter, word, or other matter marked for omission or correction is to be retained. —*vt.* **stet·ted, stet·ting, stets.** To nullify a correction or omission previously made in (printed matter) by underlining with dots and writing the word *stet* in the margin.

**steth·o·scope** (stĕth′ə-skōp′) *n.* [Fr. *stéthoscope* : Gk. *stēthos,* chest + Fr. *-scope,* -scope.] An instrument for listening to internal bodily sounds. **—steth′o·scop′ic** (-skŏp′ĭk), **steth′o·scop′i·cal** *adj.* **—steth′o·scop′i·cal·ly** *adv.* **—ste·thos′co·py** (stĕ-thŏs′kə-pē) *n.*

**Stet·son** (stĕt′sən). A trademark for a hat having a high crown and wide brim.

**ste·ve·dore** (stē′və-dôr′, -dōr′) *n.* [Sp. *estibador* < *estivar,* to stow < Lat. *stipare,* to pack.] A worker who loads or unloads ships. —*v.* **-dored, -dor·ing, -dores.** —*vt.* To load or unload the cargo of (a ship). —*vi.* To load or unload a ship.

**stevedore's knot** *also* **stevedore knot** *n.* A knot used to prevent a line from coming out of a pulley.

**stew** (stōō, styōō) *v.* **stewed, stew·ing, stews.** [ME *stewen* < OFr. *estuver,* to bathe in hot water.] —*vt.* To cook (food) by simmering or boiling slowly. —*vi.* **1.** To be cooked by boiling slowly or simmering. **2.** *Informal.* To suffer with oppressive heat : SWELTER. **3.** *Informal.* To worry : fret. —*n.* **1.** A dish cooked by stewing, esp. a mix of meat or fish and vegetables with stock. **2.** *Informal.* Mental agitation. **3.** *often* **stews.** *Archaic.* A brothel.

**stew·ard** (stōō′ərd, styōō′-) *n.* [ME < OE *stigweard* : *stig,* hall + *weard,* keeper.] **1.** One who manages another's property, finances, or other affairs. **2.** One in charge of the household affairs of a large estate, club, hotel, or resort. **3.** A ship's officer in charge of provisions and dining arrangements. **4.** An attendant on a ship or aircraft. **5.** A shop steward. —*v.* **-ard·ed, -ard·ing, -ards.** —*vt.* To serve as steward of : MANAGE. —*vi.* To serve as a steward.

**stew·ard·ess** (stōō′ər-dĭs, styōō′-) *n.* A woman who works as a steward, esp. one who works as a flight attendant.

**stewed** (stōōd, styōōd) *adj.* **1.** Cooked by stewing <*stewed* tomatoes> **2.** *Slang.* Drunk.

**Sthe·no** (sthē′nō) *n.* [Gk. *Sthenō* < *sthenos,* strong.] *Gk. Myth.* One of the three Gorgons.

**stib·ine** (stĭb′ēn′) *n.* [< Lat. *stibium,* antimony.] A colorless flammable poisonous gas, $SbH_3$, often used as a fumigant.

**stib·nite** (stĭb′nīt′) *n.* [Fr. *stibine,* stibnite (< Lat *stibium,* antimony < Gk. *stibi,* of Egypt. orig.) + -ITE.] A lead-gray mineral, $Sb_2S_3$, occas. containing silver and gold, and the chief source of antimony.

**stich** (stĭk) *n.* [Gk. *stikhos.*] A line of verse.

**stich·ic** (stĭk′ĭk) *adj.* Of or relating to verse composed in homogeneous and recurrent lines, as in recitative poetry.

**sti·chom·e·try** (stĭ-kŏm′ĭ-trē) *n.* [Gk. *stikhos,* stich + -METRY.] Division of a prose piece into lines whose lengths correspond to the natural divisions of sense or to the natural cadences, as in manuscripts written before the adoption of punctuation. **—stich′o·met′ric** (stĭk′ə-mĕt′rĭk) *adj.*

**stich·o·myth·i·a** (stĭk′ə-mĭth′ē-ə) *also* **sti·chom·y·thy** (stĭ-kŏm′ə-thē) *n.* [Gk. *stikhomuthia* < *stikhomuthein,* to speak in alternating lines : *stikhos,* stich + *muthos,* speech.] An ancient Greek arrangement of dialogue in drama, poetry, and disputation in which single lines of verse are spoken by alternate speakers. **—stich′o·myth′ic** *adj.*

**stick** (stĭk) *n.* [ME *stykke* < OE *sticca.*] **1.** A long slender piece of wood, esp.: **a.** A branch or stem cut from a shrub or tree. **b.** A piece of wood, as a tree branch, used for fuel, cut for lumber, or shaped for a given purpose. **c.** A staff, wand, or rod. **d.** A sticklike implement used in a game or sport. **2.** A cane or walking stick. **3.** Something slender and often cylindrical <a *stick* of dynamite> **4.** *Slang.* A marijuana cigarette. **5.** An aircraft control that operates the elevators and ailerons. **6.** *Naut.* A mast or a part of a mast. **7. a.** A composing stick in printing. **b.** The type contents of a composing stick. **8.** A group of bombs released to fall in a straight row across a target. **9.** A timber tree. **10.** A poke, thrust, or stab with a stick or similar object. **11.** The state or power of adhering <a glue with lots of *stick*> **12. sticks.** *Informal.* **a.** A remote area : BACKWOODS. **b.** A dull or unsophisticated city or town <hated living in the *sticks*> **13.** *Informal.* A stiff, spiritless, or boring person. **14.** *Archaic.* A difficulty or obstacle : DELAY. —*v.* **stuck** (stŭk), **stick·ing, sticks.** —*vt.* **1.** To puncture or penetrate with a pointed instrument. **2.** To kill by piercing. **3.** To thrust or push (a pointed instrument) into or through an object. **4.** To fasten into place by forcing an end or point into something <*stick* a hook on the wall> **5.** To fasten or attach with or as if with pins or nails. **6.** To fasten or attach with an adhesive material. **7.** To cover or decorate with objects piercing the surface. **8.** To fix, impale, or transfix on a pointed object <*stick* an onion on a skewer> **9.** To put, thrust, or poke into a specified place or position. **10.** To detain : delay. **11.** *p.t. & p.p.* **sticked.** To prop (a vine or other plant) with sticks or brush on which to grow. **12.** *p.t. & p.p.* **sticked.** To set (printing type) in a composing stick. **13.** *Informal.* To confuse, baffle, or puzzle <Sometimes even simple questions *stick* me.> **14.** To cover or smear with a sticky substance. **15.** To put blame or responsibility on <*stuck* me with the debts> **16.** *Slang.* To defraud or cheat. —*vi.* **1.** To be or become fixed or embedded in place by having the point thrust in. **2.** To become or remain attached or in close association by or as if by adhesion : CLING <*stick* together> **3. a.** To remain firm, determined, or resolute <*stuck* to their opinion> **b.** To remain loyal or faithful <*stick* by a friend in trouble> **4.** To persist, endure, or persevere. **5.** To hesitate or scruple <*sticks* at nothing> **6. a.** To be at or come to a standstill and be unable to proceed <*stuck* in the mud> **b.** To become jammed or obstructed <The window *stuck* tight.> **7.** To project, extend, or protrude. **—be stuck on.** *Informal.* To be very fond of. **—stick around.** *Informal.* To remain : linger. **—stick it to.** *Slang.* To treat severely or unfairly. **—stick one's neck out.** *Informal.* To voluntarily make oneself vulnerable. **—stick out.** To be prominent. **—stick to one's knitting.** *Informal.* To mind one's own business. **—stick to (one's) ribs.** *Informal.* To be substantial or filling <a meal that *sticks to your ribs*> **—stick up.** To rob, esp. at gunpoint.

**stick·ball** (stĭk′bôl′) *n.* Baseball played with a rubber ball and a stick or a broom handle for a bat.

ă **pat** ā **pay** âr **care** ä **father** ĕ **pet** ē **be** hw **which** ĭ **pit**
ī **tie** îr **pier** ŏ **pot** ō **toe** ô **paw, for** oi **noise** ōō **took**

**stick·er** (stĭk′ər) *n.* **1.** One that sticks. **2.** An adhesive label or patch. **3.** A tenacious or persistent person. **4.** A prickle, thorn, or barb.
**stick figure** *n.* A picture of a human or animal figure depicting the head as a circle and the rest of the body as a combination of straight lines.
**sticking plaster** *n.* Adhesive tape.
**sticking point** *n.* An issue creating or likely to create an impasse.
**stick insect** *n.* An insect of the family Phasmidae, resembling a stick or twig, as the walking stick.
**stick-in-the-mud** (stĭk′ĭn-*th*ə-mŭd′) *n. Informal.* One lacking initiative, imagination, or enthusiasm.
**stick·le** (stĭk′əl) *vi.* **-led, -ling, -les.** [ME *stightlen,* to contend, freq. of *stighten,* to arrange < OE *stihtian.*] **1.** To argue stubbornly, esp. about trifling matters. **2.** To have or raise objections : SCRUPLE.
**stick·le·back** (stĭk′əl-băk′) *n.* [ME *stykylbak* : OE *sticel,* prick + ME *bak,* back.] Any of various small freshwater and marine fishes of the family Gasterosteidae, with erectile spines along the back.

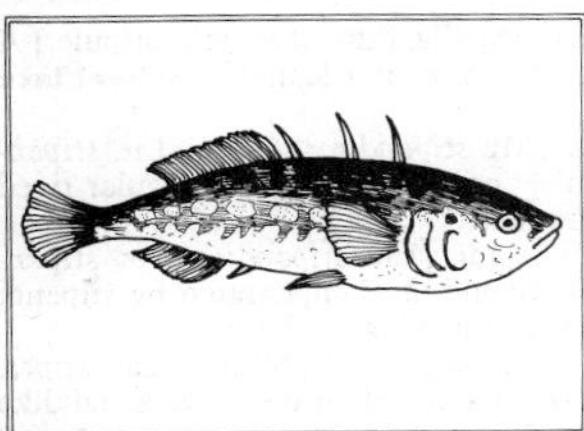
**stickleback**
*3 inches long*

**stick·ler** (stĭk′lər) *n.* **1.** One who insists on something unyieldingly <a *stickler* for accuracy> **2.** Something difficult or puzzling.
**stick·pin** (stĭk′pĭn′) *n.* A decorative pin worn on a necktie or lapel.
**stick·seed** (stĭk′sēd′) *n.* Any of various plants of the genus *Lappula,* with small prickly fruits that cling to clothing or fur.
**stick shift** *n.* An automotive gearshift operated by hand.
**stick·tight** (stĭk′tīt′) *n.* A plant, as the bur marigold, having barbed clinging seeds or fruit.
**stick-to-it·ive·ness** (stĭk-to͞o′ĭ-tĭv-nĭs) *n. Informal.* Unwavering pertinacity : PERSEVERANCE.
**stick·up** (stĭk′ŭp′) *n. Slang.* A robbery, esp. at gunpoint.
**stick·weed** (stĭk′wēd′) *n.* A plant having clinging seeds or fruit.
**stick·y** (stĭk′ē) *adj.* **-i·er, -i·est. 1.** Having the property of adhering or sticking to a surface : ADHESIVE. **2.** Covered with an adhesive agent. **3.** Warm and humid : MUGGY. **4.** *Informal.* Painful or difficult <a *sticky* problem> **—stick′i·ly** *adv.* **—stick′i·ness** *n.*
☆ ***syns:*** STICKY, HUMID, MUGGY, SOGGY *adj. core meaning* : very damp and warm <a breezeless, *sticky* summer day>
**sticky wicket** *n. Informal.* A problem or situation that is embarrassing or difficult.
**stied** (stīd) *v. p.t. & p.p. of* STY[1].
**sties[1]** (stīz) *n. pl. of* STY[1]. *—v. 3rd person sing. present tense of* STY[1].
**sties[2]** (stīz) *n. var. pl. of* STY[2].
**stiff** (stĭf) *adj.* **-er, -est.** [ME *stiffe* < OE *stīf.*] **1.** Difficult to bend or stretch : RIGID <a *stiff* collar> **2.** Not moving or operating easily <a *stiff* leg joint> **3.** Tightly drawn : TAUT. **4.** Rigidly or excessively formal, awkward, or constrained <a *stiff* farewell> **5.** Not liquid, loose, or fluid : THICK. **6.** Firm in purpose or resistance : STUBBORN. **7.** Having a strong, swift force or movement <battled the *stiff* current> **8.** Potent <a *stiff* drink of whiskey> **9.** Difficult, laborious, or arduous <a *stiff* climb> **10.** Harsh : severe <*stiff* punishment> **11.** Excessively high <paid a *stiff* price> **12.** *Naut.* Not heeling over much in spite of great wind or the press of the sail. *—adv.* **1.** In a stiff manner. **2.** Completely : totally <scared *stiff*> *—n. Slang.* **1.** A corpse. **2.** An overformal or constrained person. **3.** A drunk. **4.** A person <working *stiffs*> **5.** A hobo : tramp. **6.** One who tips poorly. **—stiff′ly** *adv.* **—stiff′ness** *n.*
☆ ***syns:*** STIFF, CARDBOARD, STARCHY, STILTED, WOODEN *adj. core meaning* : so rigidly constrained, formal, or awkward as to lack all grace and spontaneity <a *stiff,* uncomfortable interview>
**stiff-arm** (stĭf′ärm′) *vt.* **-armed, -arm·ing, -arms.** *Football.* To straight-arm.
**stiff·en** (stĭf′ən) *vt. & vi.* **-ened, -en·ing, -ens.** To make or become stiff or stiffer. **—stiff′en·er** *n.*
**stiff-necked** (stĭf′nĕkt′) *adj.* Stubborn : unyielding.
**sti·fle[1]** (stī′fəl) *v.* **-fled, -fling, -fles.** [ME *stufflen.*] *—vt.* **1.** To kill by preventing respiration : SUFFOCATE. **2.** To interrupt or cut off (e.g., the voice). **3.** To hold back : SUPPRESS <*stifle* a yawn> *—vi.* **1.** To die of suffocation. **2.** To feel suffocated by or as if by close confinement. **—sti′fler** *n.* **—sti′fling·ly** *adv.*

**sti·fle[2]** (stī′fəl) *n.* [ME.] The joint of the hind leg analogous to the human knee in certain quadrupeds, as the horse.
**stig·ma** (stĭg′mə) *n., pl.* **stig·ma·ta** (stĭg-mä′tə, stĭg′mə-) *or* **stig·mas.** [Lat. < Gk., tattoo mark < *stizein,* to prick.] **1. a.** A mark or token of infamy, disgrace, or reproach. **b.** *Archaic.* A mark burned into the skin of a criminal or slave : BRAND. **2.** A small mark, as a scar or birthmark. **3.** *Med.* **a.** A spot on the skin that bleeds as a symptom of hysteria. **b.** A mark indicative of a history of disease or abnormality. **4.** *Biol.* A small mark, spot, or pore, as the respiratory spiracle of an insect or an eyespot in certain algae. **5.** *pl.* **-mas.** *Bot.* The apex of the pistil of a flower, on which pollen is deposited at pollination. **6. stigmata.** Marks or sores corresponding to and resembling the crucifixion wounds of Jesus, sometimes occurring during religious ecstasy or hysteria. **—stig′mal** *adj.*
**stig·mas·ter·ol** (stĭg-măs′tə-rôl′, -rōl′) *n.* [NLat. *(Physo)stigma,* Calabar bean genus + STEROL.] A sterol, $C_{29}H_{48}O$, obtained from soybeans or Calabar beans.
**stig·ma·ta** (stĭg-mä′tə, stĭg′mə-) *n. pl. of* STIGMA.
**stig·mat·ic** (stĭg-măt′ĭk) *adj.* **1.** Relating to, resembling, or having a stigma or stigmata. **2.** Anastigmatic. *—n.* One marked with religious stigmata. **—stig·mat′i·cal·ly** *adv.*
**stig·ma·tism** (stĭg′mə-tĭz′əm) *n.* **1.** The state of being affected by stigmata. **2.** The state of a refracting or reflecting system that focuses at a point light rays from an off-axis point.
**stig·ma·tist** (stĭg′mə-tĭst) *n.* A stigmatic.
**stig·ma·tize** (stĭg′mə-tīz′) *vt.* **-tized, -tiz·ing, -tiz·es.** [Med Lat. *stigmatizare,* to brand < Gk. *stigmatizein,* to mark < *stigma,* tattoo mark < *stizein,* to prick.] **1.** To brand as disgraceful or ignominious. **2.** To mark with stigmata. **3.** To cause stigmata to appear on. **—stig′ma·ti·za′tion** *n.* **—stig′ma·tiz′er** *n.*
**stil·bene** (stĭl′bēn′) *n.* [Gk. *stilbos,* shining < *stilbein,* to shimmer + -ENE.] A colorless or yellowish crystalline compound, $C_{14}H_{12}$, used in making dyes and optical bleaches and as a phosphor.
**stil·bes·trol** (stĭl-bĕs′trôl′, -trōl′) *n.* [STILB(ENE) + ESTR(US) + -OL.] Diethylstilbestrol.
**stil·bite** (stĭl′bīt′) *n.* [Fr. < Gk. *stilbos,* shining < *stilbein,* to shimmer.] A white or yellow lustrous zeolite mineral, essentially $(Ca,Na)_2Al_2Si_7O_{18}\cdot7H_2O$.
**stile[1]** (stīl) *n.* [ME < OE *stigel.*] **1.** A series of steps for crossing a fence or wall. **2.** A turnstile.
**stile[2]** (stīl) *n.* [Prob. < Du. *stijl,* doorpost.] A vertical member of a panel or frame, as in a door or window sash.
**sti·let·to** (stĭ-lĕt′ō) *n., pl.* **-tos** *or* **-toes.** [Ital., dim. of *stilo,* dagger < Lat. *stilus,* stylus.] **1. a.** A small dagger with a slender, tapering blade. **b.** Something shaped like a stiletto. **2.** A small sharp-pointed instrument used for making eyelet holes in needlework.
**stiletto heel** *n.* A high heel on women's shoes that is thinner than a spike heel.
**†still[1]** (stĭl) *adj.* **-er, -est.** [ME < OE.] **1.** Quiet : silent. **2.** Hushed : subdued. **3.** Devoid of movement : being at rest. **4.** Free from disturbance, commotion, or agitation : TRANQUIL. **5.** Free from a noticeable current <a *still* river> **6.** Not carbonated : lacking effervescence <a *still* wine> **7.** Of or relating to a single or static photograph as opposed to a motion picture. *—n.* **1.** Quiet : silence <the *still* of the night> **2.** A still photograph, esp. one from a scene of a film used for advertising. **3.** A still-life picture. *—adv.* **1.** Without movement : MOTIONLESSLY <stood *still*> **2.** Up to or at the time indicated : YET <*still* alive> **3.** In increasing amount or degree <*still* further criticism> **4.** All the same : NEVERTHELESS. **5.** *Archaic & Regional.* Always : constantly. *—v.* **stilled, still·ing, stills.** *—vt.* **1.** To make still. **2.** To make quiet : SILENCE. **3.** To make motionless. **4.** To allay : calm <*stilled* our fears>. *—vi.* To become still. **—still′ness** *n.*
**still[2]** (stĭl) *n.* [< ME *stillen,* to distill < *distillen.* —see DISTILL.] **1.** An apparatus for distilling liquids, esp. alcohols, composed of a vessel in which the substance is vaporized by heat and a cooling device in which the vapor is condensed. **2.** A distillery.
**still alarm** *n.* A fire alarm transmitted by means, such as the telephone, other than by sounding the conventional signal apparatus.
**still·birth** (stĭl′bûrth′) *n.* **1.** The birth of a dead child or fetus. **2.** A child or fetus born dead.
**still·born** (stĭl′bôrn′) *adj.* Dead at birth.
**still hunt** *n.* The hunting of game by stalking or ambushing.
**still-hunt** (stĭl′hŭnt′) *v.* **-hunt·ed, -hunt·ing, -hunts.** *—vt.* To pursue (game) stealthily. *—vi.* To engage in a still hunt.
**still life** *n., pl.* **still lifes. 1.** Representation of inanimate objects, as flowers or fruit, in painting or photography. **2.** A painting or photograph of inanimate objects.
**still·man** (stĭl′măn′) *n.* **1.** One who owns or manages a still or distillery. **2.** An operator of a distillation apparatus, as in an oil refinery.
**still·y** (stĭl′ē) *adj.* **-i·er, -i·est.** Quiet : calm. **—stil′ly** *adv.*
**Stil·son** (stĭl′sən). A trademark for a monkey wrench with serrated jaws.
**stilt** (stĭlt) *n.* [ME *stilte.*] **1.** Either of a pair of long slender poles each equipped with a raised footrest to permit walking elevated above the ground. **2.** A tall post or pillar used as a support, as for a dock or building. **3.** *pl.* **stilts** *or* **stilt. a.** A long-legged wading bird, *Himantopus mexicanus* or *H. himantopus,* with black and white plumage and a long, slender bill. **b.** A related bird, *Cladorhyncus*

*leucocephala,* of Australia. —*vt.* **stilt·ed, stilt·ing, stilts.** To place or raise on stilts.

**stilt·ed** (stĭl′tĭd) *adj.* **1.** Artificially formal : STIFF <*stilted* language> **2.** Having some vertical length between the impost and the beginning of the curve. —Used of an arch. **—stilt′ed·ly** *adv.* **—stilt′ed·ness** *n.*

**Stil·ton cheese** (stĭl′tn) *n.* [After *Stilton,* a parish in Huntingdon, England.] A rich waxy cheese having a blue-green mold and a wrinkled rind.

**stim·u·lant** (stĭm′yə-lənt) *n.* **1.** An agent that temporarily arouses or accelerates physiological or organic activity. **2.** STIMULUS 3. **3.** An alcoholic beverage.

**stim·u·late** (stĭm′yə-lāt′) *v.* **-lat·ed, -lat·ing, -lates.** [Lat. *stimulare, stimulat-,* to goad on < *stimulus,* goad.] —*vt.* To rouse to activity or heightened action, as by goading : EXCITE. —*vi.* To act or serve as a stimulant or stimulus. **—stim′u·lat′er, stim′u·la′tor** *n.* **—stim′u·la′tion** *n.* **—stim′u·la′tive** (-lā′tĭv), **stim′u·la·to′ry** (-lə-tôr′ē, -tōr′ē) *adj.*

**stim·u·lus** (stĭm′yə-ləs) *n., pl.* **-li** (-lī′) [Lat., goad.] **1.** Something causing or viewed as causing a response. **2.** An agent, action, or state that elicits or accelerates a physiological or psychological activity. **3.** Something that incites or rouses to action : INCENTIVE <literature as a *stimulus* to the imagination>

☆ ***syns:*** STIMULUS, CATALYST, IMPETUS, IMPULSE, INCENTIVE, MOTIVATION, SPUR, STIMULANT *n. core meaning* : something that causes and encourages a given response <free enterprise as a *stimulant* to the economy>

**sting** (stĭng) *v.* **stung** (stŭng), **sting·ing, stings.** [ME *stingen* < OE *stingan.*] —*vt.* **1.** To pierce or wound painfully with or as if with a sharp-pointed structure or organ, as that of certain insects. **2.** To cause to feel a sharp, smarting pain by or as if by pricking with a sharp point. **3.** To cause to suffer keenly in the mind or feelings <The angry words *stung* me bitterly.> **4.** To spur on by or as if by sharp irritation. **5.** *Slang.* To cheat or overcharge. —*vi.* **1.** To have, use, or wound with or as if with a sharp-pointed structure or organ, as that of certain insects. **2.** To cause or feel a sharp, smarting pain. —*n.* **1.** An act of stinging. **2.** A wound or pain caused by or as if by stinging. **3.** A sharp, piercing organ or part, often ejecting a venomous secretion, as the modified ovipositor of a bee or wasp or the spine of certain fishes. **4.** A stinging power, quality, or capacity. **5.** Keen stimulus or incitement : GOAD. **6.** *Informal.* A complicated confidence game planned and executed carefully, esp. one executed by undercover agents to catch criminals. **—sting′ing·ly** *adv.*

☆ ***syns:*** STING, BITE, BURN, SMART *v. core meaning* : to feel or cause to feel a sensation of heat or discomfort <smoke that made my eyes *sting*>

**sting·a·ree** (stĭng′ə-rē) *n.* [Alteration of STINGRAY.] The stingray.

**sting·er** (stĭng′ər) *n.* **1.** One that stings, esp. an insult that wounds another. **2.** A stinging organ or part. **3.** A cocktail of crème de menthe and brandy.

**stinging cell** *n.* A cnidoblast.

**sting·ray** (stĭng′rā′) *n.* Any of various rays of the family Dasyatidae, having a whiplike tail armed with a venomous spine capable of inflicting severe injury.

**stin·gy** (stĭn′jē) *adj.* **-gi·er, -gi·est.** [Obs. *stingy,* stinging < STING.] **1.** Giving or spending reluctantly or unwillingly <a *stingy* person> **2.** Meager <a *stingy* supper> **—stin′gi·ly** *adv.* **—stin′gi·ness** *n.*

☆ ***syns:*** STINGY, CHEAP, CLOSE, MEAN, MISERLY, NIGGARDLY, PARSIMONIOUS, PENURIOUS, TIGHT, TIGHTFISTED *adj. core meaning* : reluctant to give or spend <too *stingy* to pay the employees well> ***ant:*** generous

**stink** (stĭngk) *v.* **stank** (stăngk) *or* **stunk** (stŭngk), **stunk, stink·ing, stinks.** [ME *stinken* < OE *stincan.*] —*vi.* **1.** To emit a strong foul odor. **2. a.** To be highly offensive. **b.** To be in very bad repute. **3.** *Slang.* To have a quality to an extreme or offensive degree <an act that *stinks* of treason> **4.** *Slang.* To be of a very low quality. —*vt.* To cause to stink. —*n.* A strong offensive odor. **—make (or raise) a stink.** *Slang.* To make a great fuss.

**stink·bug** (stĭngk′bŭg′) *n.* Any of numerous insects of the family Pentatomidae, having a broad flattened body and emitting a foul odor.

**stink·er** (stĭng′kər) *n.* **1.** One that stinks. **2.** *Slang.* A contemptible, disgusting person. **3.** *Slang.* Something very difficult <The exam was a real *stinker.*>

**stink·horn** (stĭngk′hôrn′) *n.* A foul-smelling fungus of the order Phallales, as *Phallus impudicus* or *P. ravenelii,* having a thick cylindrical stalk and a narrow cap.

**stink·ing** (stĭng′kĭng) *adj.* **1.** Foul smelling : FETID. **2.** *Slang.* Very drunk. —*adv. Slang.* To an offensive or extreme degree <"proceeded to get *stinking* drunk" —James Jones> **—stink′ing·ly** *adv.* **—stink′ing·ness** *n.*

**stinking chamomile** *n.* The mayweed.

**stink·pot** (stĭngk′pŏt′) *n.* **1.** An earthenware jar containing combustibles emitting a suffocating smoke, formerly used in warfare. **2.** *Slang.* A mean, despicable person. **3.** A musk turtle, *Sternotherus odoratus,* of eastern North America.

**stink stone** *also* **stink·stone** (stĭngk′stōn′) *n.* A limestone that emits a bad odor when struck or rubbed.

**stink·weed** (stĭngk′wēd′) *n.* Any of various plants having flowers or foliage emitting an unpleasant smell.

**stink·wood** (stĭngk′wo͝od′) *n.* **1. a.** A tree, *Ocotea bullata* of southern Africa, with foul-smelling wood. **b.** The hard, heavy wood of this tree, used in cabinetwork. **2.** Any of several trees having wood with a foul odor.

**stint**[1] (stĭnt) *v.* **stint·ed, stint·ing, stints.** [ME *stinten,* to cease < OE *styntan,* to blunt.] —*vt.* **1.** To restrict or limit, as in amount or number. **2.** *Archaic.* To cause to stop. —*vi.* **1.** To subsist on a meager allowance. **2.** *Archaic.* To stop or desist. —*n.* **1.** A fixed amount or share of work to be performed within a given time period. **2.** A limitation : restriction. **—stint′er** *n.*

☆ ***syns:*** STINT, PINCH, SCRAPE, SCRIMP, SKIMP *v. core meaning* : to be severely sparing in order to economize <had to *stint* on expenditures in redecorating the office>

**stint**[2] (stĭnt) *n.* [ME *stynt.*] Any of several small sandpipers of the genera *Erolia* or *Calidris,* of northern regions.

**stipe** (stīp) *n.* [Fr. < Lat. *stipes,* post.] *Bot.* A stalk or stalklike structure, as the stemlike support of the cap of a mushroom or the main stem of a fern frond.

**sti·pel** (stī′pəl, stī-pĕl′) *n.* [NLat. *stipella,* dim. of *stipula,* stipule.] A minute or secondary stipule at the base of a leaflet. **—sti·pel′late** (stī-pĕl′ĭt, stī′pə-lāt′) *adj.*

**sti·pend** (stī′pĕnd′, -pənd) *n.* [ME *stipendie* < OFr. < Lat. *stipendium,* tax : *stips,* contribution + *pendere,* to pay.] A regular fixed payment, as a salary or an allowance.

**sti·pen·di·ar·y** (stī-pĕn′dē-ĕr′ē) *adj.* [Lat. *stipendiarius* < *stipendium,* stipend.] **1.** Receiving a stipend. **2.** Compensated by stipend. —*n., pl.* **-ies.** One who receives a stipend, as a cleric.

**sti·pes** (stī′pēz′) *n., pl.* **stip·i·tes** (stĭp′ĭ-tēz′) [NLat. < Lat. *stipes,* post.] **1.** The basal segment of the maxilla of an insect. **2.** A stalklike support or structure. **—sti′pi·form′** (-pə-fôrm′), **stip′i·ti·form′** (stĭp′ĭ-tə-fôrm′) *adj.*

**stip·i·tate** (stĭp′ĭ-tāt′) *adj.* [< Lat. *stipes, stipit-,* post.] Having or supported on a stipe.

**stip·i·tes** (stĭp′ĭ-tēz′) *n. pl. of* STIPES.

**stip·ple** (stĭp′əl) *vt.* **-pled, -pling, -ples.** [Du. *stippelen,* freq. of *stippen,* to speckle < *stip,* dot.] **1.** To draw, engrave, or paint in dots or short strokes. **2.** To apply (e.g., paint) in dots or short strokes. **3.** To dot, fleck, or speckle <a field *stippled* with purple flowers> —*n.* **1.** The method of drawing, engraving, or painting by stippling. **2.** The effect produced by stippling. **—stip′pler** *n.*

**stip·u·lar** (stĭp′yə-lər) *adj.* Of, relating to, or like stipules.

**stip·u·late**[1] (stĭp′yə-lāt′) *v.* **-lat·ed -lat·ing, -lates.** [Lat. *stipulari, stipulat-,* to bargain.] —*vt.* **1.** To specify as a condition of an agreement : require by contract. **2.** To guarantee in an agreement. —*vi.* **1.** To make an express demand or provision in an agreement. **2.** To make an agreement. **—stip′u·la′tion** *n.* **—stip′u·la′tor** *n.* **—stip′u·la·to′ry** (-lə-tôr′ē, -tōr′ē) *adj.*

☆ ***syns:*** STIPULATE, DETAIL, PARTICULARIZE, SPECIFY *v. core meaning* : to make specific, as a condition or requirement <a contract that *stipulates* the obligations of all parties>

**stip·u·late**[2] (stĭp′yə-lĭt) *adj.* Having stipules.

**stip·ule** (stĭp′yo͞ol) *n.* [NLat. *stipula* < Lat., stalk.] One of the usu. small, paired leaflike appendages at the base of a leaf or leafstalk in certain plants. **—stip′u·lar** *adj.* **—stip′uled′** *adj.*

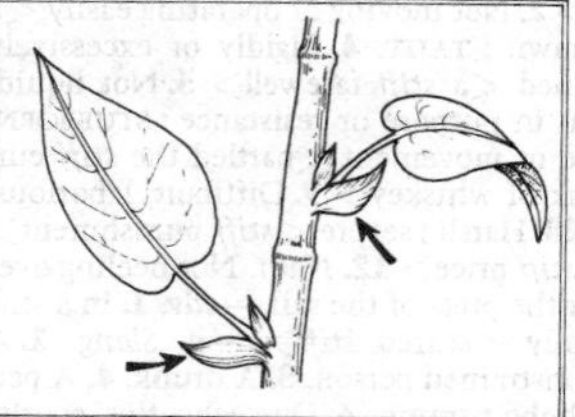

stipule

**stir**[1] (stûr) *v.* **stirred, stir·ring, stirs.** [ME *stiren* < OE *styrian,* to excite.] —*vt.* **1.** To pass an implement through (e.g., a liquid) in circular motions so as to mix or cool the contents. **2.** To alter the placement of slightly : DISARRANGE <The breeze *stirred* my hair.> **3.** To move briskly or vigorously. **4.** To rouse (someone), as from sleep. **5.** To incite, provoke, or instigate. **6.** To excite the emotions of. —*vi.* **1.** To change position slightly, as while sleeping. **2.** To move about actively : VENTURE. **3.** To take place : HAPPEN. **4.** To be capable of being stirred. —*n.* **1.** A stirring movement. **2.** A disturbance : commotion. **3.** An excited reaction : FERMENT. **—stir′rer** *n.*

**stir**[2] (stûr) *n.* [Orig. unknown.] *Slang.* Prison.

**stir crazy** *adj. Slang.* Upset from long confinement in or as if in prison.

**stir-fry** (stûr′frī′) *vt.* **-fried, -fry·ing, -fries.** To fry quickly in a small amount of oil over high heat while stirring continuously.

**stirk** (stûrk) *n.* [ME < OE *stirc.*] A yearling heifer or bullock.

**stirps** (stûrps) *n., pl.* **stir·pes** (stûr′pēz, -pāz′) [Lat., stem, lineage.] **1.** A line of descendants of common ancestry : STOCK. **2.** *Law.* One from whom a family is descended.

**stir·ring** (stûr′ĭng) *adj.* **1.** Rousing : exciting <a *stirring* oration> **2.** Lively : active. **—stir′ring·ly** *adv.*

**stir·rup** (stûr′əp, stĭr′-) *n.* [ME *stirope* < OE *stigrāp.*] **1.** A flat-based loop or ring hung from either side of a horse's saddle to support the rider's foot in mounting and riding. **2.** A device or part shaped like a stirrup in which something is supported. **3.** *Naut.* A rope on a ship hanging from a yard and having an eye at the end through which a footrope is passed for support.

**stir·rup bone** *n.* The stapes.

**stir·rup-cup** (stûr′əp-kŭp′, stĭr′-) *n.* A farewell drink, esp. for a rider mounted to depart.

**stirrup leather** *n.* The strap used to fasten a stirrup to a saddle.

**stish·ov·ite** (stĭsh′ə-vīt′) *n.* [After S.M. *Stishov,* 20th-cent. Russian mineralogist.] A dense tetragonal form of silicon dioxide that is a polymorph of quartz and that is formed under great pressure.

**stitch** (stĭch) *n.* [ME *stiche* < OE *stice,* sting.] **1.** A single complete movement of a threaded needle in sewing or surgical suturing. **2. a.** A single loop of yarn around an implement such as a knitting needle. **b.** The link, loop, or knot made in this way. **3.** A style of arranging the threads in sewing, knitting, or crocheting <a purl *stitch*> **4.** A sudden sharp pain in the side. **5.** *Informal.* An article of clothing <not a *stitch* on> **6.** *Informal.* The least part : BIT <wouldn't do a *stitch* of work> **7.** A ridge between two furrows. *—v.* **stitched, stitch·ing, stitch·es.** *—vt.* **1.** To fasten or join with or as if with stitches. **2.** To decorate with or as if with stitches. **3.** To fasten with staples. *—vi.* To make stitches : SEW. **—in stitches.** *Informal.* Uncontrollably laughing. **—stitch′er** *n.*

**stitch·wort** (stĭch′wûrt′, -wôrt′) *n.* Any of several low-growing plants of the genus *Stellaria,* having small, white, star-shaped flowers.

**stith·y** (stĭth′ē, stĭth′ē) *n., pl.* **-ies.** [ME *stethy* < ON *steđi.*] **1.** An anvil. **2.** A forge or smithy.

**sti·ver** (stī′vər) *n.* [Du. *stuiver* < MDu. *stuyver.*] **1.** An obsolete Dutch coin worth 1/20 of a guilder. **2.** Something of little value.

**St. John's bread** (sānt′jŏnz′) *n.* [After *St. John* the Baptist, who lived on honey and locusts (prob. locust beans, or carob) while preaching in the desert.] The long blackish, sugary, edible pod of the carob.

**St. Johns·wort** (sānt jŏnz′wûrt′, -wôrt′) *n.* [So called because it was gathered on *St. John's* Eve to ward off evil spirits.] Any of various plants or shrubs of the genus *Hypericum,* having yellow flowers.

**sto·a** (stō′ə) *n., pl.* **sto·ae** (stō′ē′) *or* **sto·as.** [Gk., porch.] An ancient Greek covered walk or colonnade, usu. having columns on one side and a wall on the other.

**stoat** (stōt) *n., pl.* **stoats** *or* **stoat.** [ME *stote.*] *Chiefly Brit.* The ermine, esp. when in its brown color phase.

**sto·chas·tic** (stō-kăs′tĭk) *adj.* [Gk. *stokhiastikos* < *stokhazesthai,* to guess at < *stokhos,* aim.] **1.** Of, denoting, or marked by conjecture : CONJECTURAL. **2. a.** Involving or containing a random variable <*stochastic* calculus> **b.** Involving probability or chance. **—sto·chas′ti·cal·ly** *adv.*

**stock** (stŏk) *n.* [ME *stok* < OE *stocc,* tree trunk.] **1.** A supply gathered for future use : STORE. **2.** The total merchandise kept on hand by a merchant or commercial establishment. **3.** All the animals kept or raised on a farm : LIVESTOCK. **4. a.** The capital or fund that a corporation raises through the sale of shares entitling the holder to dividends and to other ownership rights. **b.** The number of shares possessed by each stockholder. **c.** A certificate showing ownership of a stated number of shares. **d.** The part of a tally or record of account formerly given to a creditor. **e.** A debt symbolized by a tally. **5.** The trunk or main stem of a plant as opposed to the branches and roots. **6. a.** A plant or stem onto which a graft is made. **b.** A plant from which cuttings and slips are taken. **7. a.** The original progenitor of a family line. **b.** Ancestry or lineage : ANTECEDENTS. **c.** The type from which a group of animals or plants has descended. **d.** A race, family, or other related group of animals or plants. **e.** A major division of humankind, as an ethnic group. **f.** A group of related languages. **g.** A group of related families of languages. **8.** The raw material from which something is made. **9.** The broth from boiled meat or fish, used as a base in preparing soup, gravy, or sauces. **10. a.** A main upright part, esp. a supporting structure. **b. stocks.** The timber frame supporting a ship during construction. **c.** A frame in which a horse or other animal is held for shoeing or for veterinary treatment. **11. stocks.** A former instrument of punishment, consisting of a heavy timber frame with holes for confining the ankles and occas. the wrists. **12.** *Naut.* A crosspiece at the end of an anchor's shank. **13.** The wooden block from which a bell is suspended. **14. a.** The rear handle of a firearm, to which the barrel and the firing and loading mechanisms are attached. **b.** The long mooring beam of field gun carriages. **15.** A handle, as of a whip or fishing rod. **16.** The frame of a plow, to which the share, handles, colter, and other parts are fastened. **17. a.** A theatrical stock company. **b.** The repertoire of such a company. **c.** A theater or theatrical activity, esp. outside of a main theatrical center. **18.** An Old World plant of the genus *Mathiola,* esp. *M. incana,* widely cultivated for its showy, variously colored flower clusters. **19.** The portion of a pack of cards or group of dominoes not dealt out but drawn from during a game. **20.** *Geol.* A body of intrusive igneous rock of which less than 40 square miles is exposed. **21. a.** An assessment : estimate <took *stock* of the situation> **b.** Personal reputation or status <The manager's *stock* with the employees was falling.> **c.** Confidence or credence <put no *stock* in the rumor> **22.** A broad scarf worn around the neck. *—v.* **stocked, stock·ing, stocks.** *—vt.* **1.** To provide (e.g., a store) with stock : SUPPLY. **2.** To keep for future sale or use. **3.** To provide (e.g., a rifle) with a stock. **4.** *Obs.* To put (someone) in the stocks as a punishment. *—vi.* **1.** To gather and store a supply of something <*stock* up on frozen peas> **2.** To put forth or sprout new shoots. *—adj.* **1.** Kept in stock <a *stock* item> **2.** Commonplace : ordinary <a *stock* response> **3.** Employed in dealing with or caring for stock or merchandise. **4. a.** Of or relating to the raising of livestock. **b.** Used for breeding <a *stock* mare> **5.** Of or relating to a stock company or its repertoire. **—in stock.** On hand for sale or use. **—out of stock.** Not available for sale or use. **—stock in trade.** One's resources for any purpose <Flattery appears to be your *stock in trade.*>

**stock·ade** (stŏ-kād′) *n.* [Fr. *estacade* < Sp. *estacada* < *estaca,* stake, of Germanic orig.] **1.** A defensive barrier made of strong posts or timbers driven upright side by side in the ground. **2.** A fenced or enclosed area, esp. one used for protection or imprisonment. *—vt.* **-ad·ed, -ad·ing, -ades.** To fortify, protect, encircle, or imprison with a stockade <POW's *stockaded* into a compound>

**stock·breed·ing** (stŏk′brē′dĭng) *n.* The raising of livestock. **—stock′breed′er** *n.*

**stock·bro·ker** (stŏk′brō′kər) *n.* One who acts as an agent in the buying and selling of stocks. **—stock′bro′ker·age** *n.*

**stock car** *n.* **1.** A car of a standard make modified for racing. **2.** A railroad car carrying livestock.

**stock certificate** *n.* A certificate establishing ownership of a stated number of shares in a company's stock.

**stock company** *n.* **1.** A company whose capital is divided into shares. **2.** A group of actors and technicians attached to a single theater and performing in repertory.

**stock dove** *n.* [Prob. so called from its living in hollow tree trunks.] A common Old World bird, *Columba oenas,* with grayish plumage.

**stock exchange** *n.* **1.** A place where stocks, bonds, or other securities are bought and sold. **2.** An association of stockbrokers who meet to buy and sell stocks and bonds according to fixed regulations.

**stock·fish** (stŏk′fĭsh′) *n., pl.* **stockfish** *or* **-fish·es.** A fish, as cod or haddock, cured by being split and air-dried without salt.

**stock·hold·er** (stŏk′hōl′dər) *n.* One who owns a share or shares of stock in a company.

**stock·i·net** *also* **stock·i·nette** (stŏk′ə-nĕt′) *n.* [Alteration of *stocking net.*] An elastic knitted fabric used esp. in making bandages or undergarments.

**stock·ing** (stŏk′ĭng) *n.* [Obs. *stock,* to cover with hose < obs. *stock,* a stocking.] **1.** A close-fitting, usu. knitted covering for the foot and leg. **2.** An object resembling a stocking.

**stocking cap** *n.* A knitted cap for casual winter wear that has a long cone-shaped tail usu. with a pom-pom attached.

**stock·job·ber** (stŏk′jŏb′ər) *n.* **1.** *Chiefly Brit.* A stock exchange operator who deals only with brokers and not with the public. **2.** A stockbroker. **—stock′job′ber·y** *n.*

**stock·man** (stŏk′mən) *n.* **1.** One who owns or raises livestock. **2.** One who is in charge of livestock or works on a stock farm. **3.** An employee in a stockroom or warehouse.

**stock market** *n.* **1.** A stock exchange. **2.** The business transacted at a stock exchange. **3.** The prices offered for stocks and bonds in general.

**stock·pile** *also* **stock pile** (stŏk′pīl′) *n.* A carefully maintained supply stored for future use. *—vt.* **-piled, -pil·ing, -piles.** To accumulate a stockpile of. **—stock′pil′er** *n.*

**stock·pot** (stŏk′pŏt′) *n.* **1.** A pot for preparing soup stock. **2.** A rich supply.

**stock·room** *also* **stock room** (stŏk′ro͞om′, -ro͝om′) *n.* A room in which a store of materials is kept.

**stock-still** (stŏk′stĭl′) *adj.* Very still : MOTIONLESS.

**stock·tak·ing** (stŏk′tā′kĭng) *n.* **1.** The inventorying of merchandise or supplies on hand. **2.** The act of evaluating a situation at a given point.

**stock·y** (stŏk′ē) *adj.* **-i·er, -i·est. 1.** Solidly built : STURDY. **2.** Plump : chubby. **—stock′i·ly** *adv.* **—stock′i·ness** *n.*

**stock·yard** (stŏk′yärd′) *n.* A large enclosed yard, usu. with pens or stables, in which livestock is temporarily kept until slaughtered or shipped.

**stodg·y** (stŏj′ē) *adj.* **-i·er, -i·est.** [< E. *stodge,* thick filling food < *stodge,* to cram.] **1. a.** Dull and commonplace. **b.** Pompous : stuffy.

2. Heavy, starchy, and hard to digest <a *stodgy* dinner> **3.** Solidly built : STOCKY. **—stodg'i·ly** *adv.* **—stodg'i·ness** *n.*

**sto·gy** or **sto·gie** (stō'gē) *n., pl.* **-gies.** [After *Conestoga,* Pennsylvania.] **1.** A long, thin, inexpensive cigar. **2.** A rough, heavy shoe or boot.

**sto·ic** (stō'ĭk) *also* **sto·i·cal** *adj.* [< Lat. *Stoicus,* a Stoic < Gk. *Stōikos* < *stoa* (*Poikilē*), (Painted) Porch, where Zeno taught.] Apparently indifferent to or unaffected by pleasure or pain : IMPASSIVE <"*stoic* resignation in the face of hunger" —John F. Kennedy> —*n.* **stoic. 1.** A stoic person. **2. Stoic.** A member of a Greek school of philosophy, founded by Zeno about 308 B.C., holding that human beings should be free from passion and calmly accept all occurrences as the unavoidable result of divine will. **—sto'i·cal·ly** *adv.* **—sto'i·cal·ness** *n.*

**stoi·chi·om·e·try** (stoi'kē-ŏm'ĭ-trē) *n.* [Gk. *stoicheion,* element + -METRY.] Methodology and technology by which quantities of reactants and products in chemical reactions are determined. **—stoi'chi·o·met'ric** (-ō-mĕt'rĭk) *adj.* **—stoi'chi·o·met'ri·cal·ly** *adv.*

**sto·i·cism** (stō'ĭ-sĭz'əm) *n.* **1.** Indifference to pleasure or pain : IMPASSIVENESS. **2. Stoicism.** The philosophical doctrines of the Stoics.

**stoke** (stōk) *v.* **stoked, stok·ing, stokes.** [Back-formation < STOKER.] —*vt.* **1.** To stir up and feed amply, as a fire. **2.** To feed fuel to and tend (a furnace). —*vi.* To feed fuel to and tend a furnace.

**stoke·hold** (stōk'hōld') *n. Naut.* The compartment into which a ship's furnaces or boilers open.

**stoke·hole** (stōk'hōl') *n.* [Transl. of Du. *stookgat.*] **1.** The space about the opening in a furnace or boiler. **2.** *Naut.* A stokehold.

**stok·er** (stō'kər) *n.* [Du. < *stoken,* to poke.] **1.** One who feeds fuel to and tends a furnace, as a fireman on a locomotive. **2.** A mechanical device for feeding coal to a furnace.

**stole¹** (stōl) *n.* [ME, long robe < OE *stol* < Lat. *stola* < Gk. *stolē,* garment.] **1.** A long usu. embroidered silk or linen scarf worn over the left shoulder by deacons and over both shoulders by priests and bishops while officiating. **2.** A long cloth or fur scarf worn about a woman's shoulders. **3.** A long robe worn by ancient Roman matrons.

**stole²** (stōl) *v. p.t. of* STEAL.

**sto·len** (stō'lən) *v. p.p. of* STEAL.

**stol·id** (stŏl'ĭd) *adj.* [Lat. *stolidus,* stupid.] Feeling or exhibiting little emotion : IMPASSIVE. **—sto·lid'i·ty** (stŏ-lĭd'ĭ-tē, stə-), **stol'id·ness** *n.* **—stol'id·ly** *adv.*

**stol·len** (stō'lən) *n., pl.* **-len** *or* **-lens.** [G.] A rich yeast bread containing raisins, citron, and chopped nutmeats.

**sto·lon** (stō'lŏn', -lən) *n.* [Lat. *stolo, stolon-,* branch.] **1.** *Bot.* A stem growing along or under the ground and taking root at the nodes or apex to form new plants. **2.** *Zool.* A stemlike structure of certain colonial organisms from which new individuals bud. **—sto'lon·ate'** (-lə-nāt') *adj.*

**sto·lon·if·er·ous** (stō'lə-nĭf'ər-əs) *adj.* Yielding or forming stolons. **—sto'lon·if'er·ous·ly** *adv.*

**sto·ma** (stō'mə) *n., pl.* **-ma·ta** (-mə-tə) *or* **-mas.** [NLat. < Gk., mouth.] **1.** *Bot.* One of the minute pores in the epidermis of a leaf or stem through which gases and water vapor pass. **2.** *Anat.* **a.** A small aperture in the surface of a membrane. **b.** A tiny opening in the surface of the peritoneum thought to be for the passage of fluid into the lymphatic vessels. **3.** *Zool.* A mouthlike opening, as the oral cavity of a nematode.

**stom·ach** (stŭm'ək) *n.* [ME *stomak* < OFr. *stomaque* < Lat. *stomachus* < Gk. *stomakhos* < *stoma,* mouth.] **1. a.** The enlarged saclike portion of the alimentary canal, one of the principal organs of digestion, located in vertebrates between the esophagus and the small intestine. **b.** A similar digestive structure of many invertebrates. **2.** The abdomen : belly. **3.** An appetite for food. **4.** A desire or inclination. **5.** *Obs.* Courage : spirit. **6.** *Obs.* Pride. —*vt.* **-ached, -ach·ing, -achs. 1.** To bear : tolerate <couldn't *stomach* that insult> **2.** *Obs.* To resent.

**stom·ach·ache** (stŭm'ək-āk') *n.* Abdominal pain.

**stom·ach·er** (stŭm'ə-kər) *n.* A heavily embroidered or jeweled garment formerly worn over the chest and stomach, esp. by women.

**sto·mach·ic** (stə-măk'ĭk) *adj.* **1.** Of or relating to the stomach : GASTRIC. **2.** Beneficial to or stimulating digestion in the stomach. —*n.* An agent that strengthens or stimulates the stomach. **—sto·mach'i·cal·ly** *adv.*

**stomach pump** *n.* A suction pump with a flexible tube inserted into the stomach through the mouth and esophagus to empty the contents of the stomach in an emergency, as in poisoning.

**stomach worm** *n.* A parasitic nematode worm that infests the stomachs of animals, esp. *Haemonchus contortus,* a parasite of sheep and other ruminants.

**stomat-** *pref. var. of* STOMATO-.

**sto·ma·ta** (stō'mə-tə) *n. pl. of* STOMA.

**sto·ma·tal** (stō'mə-təl) *adj.* Of, relating to, or having a stoma.

**sto·mat·ic** (stō-măt'ĭk) *adj.* **1.** Of or pertaining to the mouth. **2.** Stomatal.

**sto·ma·ti·tis** (stō'mə-tī'tĭs) *n.* Inflammation of the mucous tissue of the mouth.

**stomato-** or **stomat-** *pref.* [< Gk. *stoma, stomat-,* mouth.] Mouth : stoma <*stomatitis*>

**sto·ma·tol·o·gy** (stō'mə-tŏl'ə-jē) *n.* Medical study of the physiology and pathology of the mouth. **—sto'ma·to·log'i·cal** (-tə-lŏj'ĭ-kəl), **sto'ma·to·log'ic** *adj.* **—sto'ma·tol'o·gist** *n.*

**sto·mat·o·pod** (stō-măt'ə-pŏd') *n.* [NLat. *Stomatopoda,* order name : Gk. *stoma,* mouth + Gk. *pous,* foot.] Any of various marine crustaceans of the order Stomatopoda, including the squilla.

**sto·ma·tous** (stō'mə-təs) *adj.* Stomatal.

**sto·mo·de·um** *also* **sto·mo·dae·um** (stō'mə-dē'əm) *n., pl.* **-de·a** *also* **-dae·a** (-dē'ə) [NLat. : Gk. *stoma,* mouth + Gk. *hodaios,* on the way < *hodos,* road.] An embryonic oral cavity. **—sto'mo·de'al** *adj.*

**stomp** (stŏmp, stômp) *v.* **stomped, stomp·ing, stomps.** [Var. of STAMP.] —*vt.* To trample heavily or violently on. —*vi.* To trample heavily or violently. —*n.* **1.** A dance involving a rhythmical and heavy step. **2.** The jazz music for the stomp.

**-stomy** *suff.* [< Gk. *stoma,* opening, mouth.] A surgical operation in which an artificial opening is made into a specified organ or part <colostomy>

**stone** (stōn) *n.* [ME < OE *stān.*] **1. a.** Concreted earthy or mineral matter : ROCK. **b.** This material used for construction. **2.** A small piece of rock. **3.** Rock shaped or finished for a specific purpose, esp.: **a.** A gravestone. **b.** A grindstone, millstone, or whetstone. **c.** A milestone or boundary stone. **4.** A gem. **5.** Something like a stone in shape or hardness, as a hailstone. **6.** *Bot.* The hard covering enclosing the kernel in certain fruits, as the cherry or plum. **7.** *Pathol.* A mineral concretion in a hollow organ, as in the kidney. **8.** *pl.* **stone.** A unit of weight in Britain, 6.36 kilograms or 14 pounds avoirdupois. **9.** A table with a smooth surface on which page forms are composed. —*vt.* **stoned, ston·ing, stones. 1.** To hurl or throw stones at, esp. to kill with stones. **2.** To remove the stones or pits from. **3.** To furnish, fit, pave, or line with stones. **4.** To rub on or with a stone in order to polish or sharpen. **5.** *Obs.* To make hard or indifferent. **—ston'er** *n.*

**Stone Age** *n.* The earliest known period of human culture, marked by the use of stone tools.

**stone-blind** (stōn'blīnd') *adj.* Totally blind.

**stone-broke** (stōn'brōk') *adj.* Totally broke : PENNILESS.

**stone cell** *n.* A nearly isometric sclereid found in certain fruits.

**stone·chat** (stōn'chăt') *n.* A small Old World bird, *Saxicola torquata,* with dark plumage.

**stone·crop** (stōn'krŏp') *n.* **1.** Any of various plants of the genus *Sedum,* with fleshy leaves and variously colored flowers. **2.** Any of various plants related to the stonecrop.

**stone·cut·ter** (stōn'kŭt'ər) *n.* One that cuts stone. **—stone'cut'ting** *n.*

**stoned** (stōnd) *adj. Slang.* **1.** Intoxicated : drunk. **2.** Being under the influence of a mind-altering drug, as marijuana.

**stone-deaf** (stōn'dĕf') *adj.* Totally deaf.

**stone·fish** (stōn'fĭsh') *n., pl.* **stonefish** *or* **-fish·es.** Any of several tropical marine fishes of the family Scorpaenidae, with spines that eject a deadly venom.

**stone·fly** (stōn'flī') *n.* Any of numerous winged insects of the order Plecoptera, found on banks of streams and used as fishing bait both in the larval and adult stage.

**stone fruit** *n.* A drupe.

**stone-ground** (stōn'ground') *adj.* Ground in a buhrstone mill <*stone-ground* flour>

**stone lily** *n.* A fossil crinoid.

**stone marten** *n.* **1.** A Eurasian mammal, *Martes foina,* having brown fur with lighter underfur. **2.** The fur of the stone marten.

**stone·ma·son** (stōn'mā'sən) *n.* One who prepares and lays stones in building. **—stone'ma'son·ry** *n.*

**stone mint** *n.* A North American plant, *Cunila origanoides,* with small purplish or white flower clusters.

**stone's throw** *n.* A short distance <just a *stone's throw* away>

**stone·wall** (stōn'wôl') *v.* **-walled, -wall·ing, -walls.** —*vi.* **1.** To play defensively rather than trying to score in cricket. **2.** *Informal.* **a.** To engage in delaying tactics : STALL <"*Stonewalling* for a time in order to close the missile gap" —James Reston> **b.** To refuse to answer or cooperate. —*vt. Informal.* To refuse to answer or cooperate with <"I want you to *stonewall* it, let them plead the Fifth Amendment. . ." —Richard M. Nixon> **—stone'wall'er** *n.*

**stone·ware** (stōn'wâr') *n.* A heavy, nonporous pottery.

**stone·work** (stōn'wûrk') *n.* **1.** The process or technique of working in stone. **2.** Stone masonry. **—stone'work'er** *n.*

**stone·wort** (stōn'wûrt', -wôrt') *n.* Any of various green algae of the genus *Chara* that grow submerged in fresh or brackish water and are often encrusted with deposits of calcium carbonate.

**ston·y** *also* **ston·ey** (stō'nē) *adj.* **-i·er, -i·est. 1.** Covered with or full of stones. **2.** Like stone, as in hardness. **3.** Unemotional : hardhearted. **4.** Rigid : impassive <a *stony* stare> **5.** Emotionally numbing. **—ston'i·ly** *adv.* **—ston'i·ness** *n.*

**ston·y·heart·ed** (stō'nē-här'tĭd) *adj.* STONY 3. **—ston'y·heart'ed·ness** *n.*

**stood** (sto͝od) *v. p.t. & p.p. of* STAND.

**stooge** (sto͞oj) *n.* [Orig. unknown.] **1.** The straight man to a comedian. **2.** One who allows oneself to be used for another's profit : PUPPET. **3.** STOOL PIGEON 2, 3. *—vi.* **stooged, stoog·ing, stoog·es.** To be or behave as a stooge.

**stool** (sto͞ol) *n.* [ME *stol* < OE *stōl.*] **1.** A backless and armless single seat supported on legs or a pedestal. **2.** A low bench or support for the feet or knees in sitting or kneeling, as a footrest. **3.** A toilet seat : PRIVY. **4. a.** A bowel movement. **b.** Fecal matter. **5. a.** A stump or rootstock producing shoots or suckers. **b.** A shoot or growth from a stump or rootstock. *—vi.* **stooled, stool·ing, stools. 1.** To send up shoots or suckers. **2.** To defecate. **3.** *Slang.* To function or behave as a stool pigeon.

**stool·ie** (sto͞o′lē) *n.* STOOL PIGEON 3.

**stool pigeon** *n.* [From the practice of tying decoy pigeons to a stool.] **1.** A pigeon used as a decoy. **2.** *Slang.* A person functioning as a decoy. **3.** *Slang.* An informer, esp. a spy for the police.

**stoop**[1] (sto͞op) *v.* **stooped, stoop·ing, stoops.** [ME *stupen* < OE *stūpian.*] *—vi.* **1.** To bend forward and down from the waist or the middle of the back. **2.** To walk or stand with the head and upper back bent forward. **3.** To bend or sag downward. **4.** To lower oneself : CONDESCEND. **5.** To yield : submit. **6.** To swoop down, as a bird in pursuing its prey. *—vt.* **1.** To bend (the head or body) forward and down. **2.** To debase : humble. *—n.* **1.** An act of stooping. **2.** An habitual bending forward of the head and upper back. **3.** A lowering of oneself : CONDESCENSION. **4.** A descent, as of a bird of prey.

**stoop**[2] (sto͞op) *n.* [Du. *stoep,* front verandah.] A small porch, platform, or staircase leading to the entrance of a building.

**stoop**[3] (sto͞op) *n. var. of* STOUP.

**stoop·ball** (sto͞op′bôl′) *n.* A game patterned on baseball in which a player throws a ball against a stoop or wall and then runs to base.

**stop** (stŏp) *v.* **stopped, stop·ping, stops.** [ME *stoppen* < OE *stoppian* < LLat. *stuppare* < Lat. *stuppa,* tow, broken flax < Gk. *stuppē.*] *—vt.* **1.** To close (an opening) by filling in, covering, or plugging up. **2.** To constrict (an opening). **3.** To obstruct or block passage on (e.g., a road). **4.** To prevent the passage or flow of. **5.** To cause to halt, cease, or desist. **6.** To desist from : CEASE <*stop* talking> **7.** To order a bank to withhold payment of <*stopped* the check> **8.** To cause (e.g., a motor) to cease operation or function : HALT. **9. a.** To press down (a string on a stringed instrument) on the fingerboard to produce a desired pitch. **b.** To close (a hole on a wind instrument) with the finger in sounding a desired pitch. *—vi.* **1.** To cease moving, progressing, acting, or operating. **2.** To put an end to what one is doing : CEASE. **3.** To interrupt a journey for a brief visit or stay <*stopped* at the pub> *—n.* **1. a.** An act of stopping. **b.** The state of being stopped : CESSATION. **2.** A finish : end. **3.** A stay or visit, as during a trip. **4.** A place stopped at <a trolley *stop*> **5.** A device or means that obstructs, blocks, or plugs up. **6.** An order given to a bank to withhold payment on a check. **7. a.** A machine part that stops or regulates movement. **b.** A perforated screen or diaphragm that limits the effective aperture of a lens, producing an image of improved definition but lowered intensity. **8.** A punctuation mark, esp. a period. **9.** *Mus.* **a.** The act of stopping a string or hole on a musical instrument. **b.** A hole on a wind instrument. **c.** A fret on a stringed instrument. **d.** A key for closing the hole on a wind instrument. **10.** *Mus.* **a.** A tuned set of pipes, as in an organ. **b.** A knob, key, or pull that regulates such a set of pipes. **11.** *Naut.* A line used for securing something temporarily <a sail *stop*> **12.** A consonant, as English *p, t,* or *k,* marked by an articulation in which the air passage is completely closed. **13.** The depression between the muzzle and top of the skull of a dog. *—adj.* Of, relating to, or being of use at the end of an operation or activity <a *stop* code>

☆ ***syns:*** **1.** STOP, ARREST, CEASE, CHECK, DISCONTINUE, HALT, STAY *v. core meaning* : to prevent the occurrence or continuation of <*stopped* the execution of the prisoner><told us to *stop* the noise> ***ant:*** start **2.** STOP, CEASE, DESIST, DISCONTINUE, HALT, LAY OFF, LEAVE OFF, QUIT *v. core meaning* : to come to a cessation <snow that finally *stopped*><a guard who yelled for us to *stop*>

**stop·cock** (stŏp′kŏk′) *n.* A valve that regulates the flow of fluid through a pipe : FAUCET.

**stope** (stōp) *n.* [Perh. < LG step.] An excavation in the form of steps made by the mining of ore from steeply inclined or vertical veins. *—vt. & vi.* **stoped, stop·ing, stopes.** To remove (ore) from or mine by means of a stope.

**stop·gap** (stŏp′găp′) *n.* An improvised substitute : temporary expedient.

**stop·light** (stŏp′līt′) *n.* **1.** A traffic signal. **2.** A light on the rear of a vehicle activated when the brakes are applied.

**stop order** *n.* An order to a broker to buy or sell a stock when it reaches a specified level of decline or gain.

**stop·o·ver** (stŏp′ō′vər) *n.* **1.** An interruption in the course of a journey for stopping at a certain place. **2.** A place visited briefly during a journey.

**stop·page** (stŏp′ĭj) *n.* The act of stopping or the state of being stopped.

**stop payment** *n.* An order to one's bank not to honor a check.

**stop·per** (stŏp′ər) *n.* **1.** A device, as a cork or plug, inserted to close an opening. **2.** One that causes something to stop. **3.** *Computer Sci.* The topmost memory location in a device or system. *—vt.* **-pered, -per·ing, -pers.** To close with or as if with a stopper.

**stop·ple** (stŏp′əl) *n.* [ME *stoppell* < *stoppen,* to stop.] A stopper : plug. *—vt.* **-pled, -pling, -ples.** To close with a stopple.

**stop sign** *n.* A traffic sign that orders traffic to stop.

**stop street** *n.* A street intersection at which a vehicle must come to a complete stop before entering a through street.

**stop·watch** (stŏp′wŏch′) *n.* A timepiece that can be instantly started and stopped by pushing a button.

**stor·age** (stôr′ĭj, stōr′-) *n.* **1. a.** The act of storing goods. **b.** The state of being stored. **c.** A space for storing goods. **d.** The price charged for storing goods. **2.** Recharging of a storage battery. **3.** *Computer Sci.* The part of a computer that stores information for subsequent use or retrieval.

**storage battery** *n.* A group of reversible or rechargeable secondary cells acting as a unit.

**storage cell** *n.* **1.** A secondary cell. **2.** *Computer Sci.* An elementary unit of storage.

**sto·rax** (stôr′ăks′, stōr′-) *n.* [ME < Lat., alteration of *styrax* < Gk. *sturax,* perh. of Semitic orig.] **1.** Any of various trees of the genus *Styrax,* some of which yield an aromatic resin. **2.** An aromatic resin obtained from a storax tree. **3.** A brownish aromatic resin used in perfume and medicine and obtained from a tree of the genus *Liquidambar,* esp. *L. orientalis,* of Asia Minor.

**store** (stôr, stōr) *n.* [ME *stor* < OFr. *estor* < *estorer,* to build < Lat. *instaurare,* to restore.] **1.** A place where merchandise is offered for sale : SHOP. **2.** A supply reserved for future use. **3. stores.** Supplies, esp. of food, clothing, or arms. **4.** A place where commodities are kept. **5.** A great number or quantity : ABUNDANCE. *—vt.* **stored, stor·ing, stores. 1.** To reserve or put away for future use. **2.** To fill, supply, or stock. **3.** To deposit or receive in a storehouse or warehouse for safekeeping. **—in store.** Forthcoming <trouble *in store* for us> **—set store by.** To regard with esteem : VALUE <*sets* great *store by* honesty>

**store-bought** (stôr′bôt′, stōr′-) *adj. Informal.* Manufactured and purchased at retail <*store-bought* shoes>

**store cheese** *n.* Cheddar cheese.

**store·front** (stôr′frŭnt′, stōr′-) *n.* **1.** The side of a store facing a street. **2.** A room or suite in a store building at street level <a clinic in a *storefront*> **—store′front′** *adj.*

**store·house** (stôr′hous′, stōr′-) *n.* **1.** A building in which goods are stored : WAREHOUSE. **2.** An abundant source or supply <a veritable *storehouse* of knowledge>

**store·keep·er** (stôr′kē′pər, stōr′-) *n.* **1.** One who keeps a retail store or shop : SHOPKEEPER. **2.** One in charge of receiving or distributing stores or supplies, as military or naval supplies.

**store·room** (stôr′ro͞om′, -ro͝om′, stōr′-) *n.* A room in which items are stored.

**sto·rey** (stôr′ē, stōr′ē) *n. var. of* STORY[2].

**sto·ried**[1] (stôr′ēd, stōr′-) *adj.* **1.** Celebrated in history or story <"the *storied* infamies of the Emperor Tiberius on the Isle of Capri" —George Marye> **2.** Ornamented with designs representing historical or legendary scenes <*storied* tapestry>

**sto·ried**[2] *also* **sto·reyed** (stôr′ēd, stōr′-) *adj.* Having or consisting of a specified number of stories <a four-*storied* house>

**stork** (stôrk) *n.* [ME < OE *storc.*] Any of various large wading birds of the family Ciconiidae, chiefly of warm regions, having long legs and a long straight bill.

**stork's-bill** (stôrks′bĭl′) *n.* Any of various plants of the genus *Erodium,* having fruit with a narrow, beaklike point.

**storm** (stôrm) *n.* [ME < OE.] **1.** An atmospheric disturbance manifested in strong winds accompanied by rain, snow, or other precipitation and often by thunder and lightning. **2.** *Meteorol.* A wind ranging from 64 to 72 miles per hour. **3.** A heavy shower of objects <a *storm* of bullets> **4.** A strong or violent emotional outburst. **5.** A violent disturbance, as in political, social, or domestic affairs. **6.** A sudden, violent attack on a fortified place. *—v.* **stormed, storm·ing, storms.** *—vi.* **1. a.** To blow forcefully. **b.** To rain, snow, hail, or sleet. **2.** To be very angry : rant and rave. **3.** To move or rush violently, tumultuously, or angrily <*stormed* out of the meeting> *—vt.* To capture or try to capture by a violent, sudden attack <*stormed* the firebase>

**storm·bound** (stôrm′bound′) *adj.* Delayed, confined, or cut off from communication by a storm.

**storm cellar** *n.* A cyclone cellar.

**storm center** *n.* **1.** The central area covered by a storm, esp. the point of lowest barometric pressure within a storm. **2.** A center of trouble, disturbance, or argument.

**storm door** *n.* An outer door added for protection against inclement weather.

**storm petrel** *n.* A small sea bird of the family Hydrobatidae, esp. *Hydrobates pelagicus,* of the North Atlantic and the Mediterranean.

o͞o **boot** ou **out** th **thin** *th* **this** ŭ **cut** ûr **urge** y **young**
yo͞o **abuse** zh **vision** ə **about, item, edible, gallop, circus**

**storm trooper** *n.* **1.** A member of the Nazi militia noted for violence and brutality. **2.** One who behaves like a Nazi storm trooper.

**storm window** *n.* An outer window attached over the usual window to protect against inclement weather.

**storm·y** (stôr′mē) *adj.* **-i·er, -i·est. 1.** Subject to, marked by, or affected by storms : TEMPESTUOUS. **2.** Characterized by violent emotions, passions, speech, or actions <a *stormy* meeting> **—storm′i·ly** *adv.* **—storm′i·ness** *n.*

**stormy petrel** *n.* **1.** The storm petrel. **2.** One who brings discord or appears at the onset of trouble : REBEL.

**sto·ry[1]** (stôr′ē, stōr′ē) *n., pl.* **-ries.** [ME *storie* < OFr. *estorie* < Lat. *historia.* —see HISTORY.] **1.** Narration of an event or series of events, either true or fictitious. **2.** A usu. fictional prose or verse narrative intended to interest or amuse : TALE. **3.** A short story. **4.** The plot of a narrative or dramatic work. **5.** A report, statement, or allegation of facts. **6. a.** A news article or broadcast. **b.** The material for such an article or broadcast. **7.** An anecdote. **8.** A lie. **9.** Romantic legend or tradition. —*vt.* **-ried, -ry·ing, -ries. 1.** To decorate with scenes representing historical or legendary events. **2.** *Archaic.* To tell as a story.

▲ word history: The word *story* is a doublet of *history,* since both are derived from the Latin word *historia. Story* was borrowed from Old French *estorie,* but *history* was taken directly from Latin. At first *story* and *history* were synonymous, both signifying a narrative, such as a legend or biography, which might contain either factual or mythological elements, or both. The two words gradually diverged so that *story* now includes fictional plots and narratives, even lies, while history includes the records of actual events as well as the systematic study of the past.

**sto·ry[2]** *also* **sto·rey** (stôr′ē, stōr′ē) *n., pl.* **-ries** *also* **-reys.** [ME < Med. Lat. *historia* (prob. from painted windows or sculpture on the front of buildings) < Lat., history. —see HISTORY.] **1.** A complete horizontal division of a building, making up the area between two adjacent levels. **2.** The set of rooms on the same level of a building.

**sto·ry·book** (stôr′ē-bo͝ok′, stōr′-) *n.* A book containing a collection of stories, usu. for children. —*adj.* Happening in or suggestive of the style of a storybook : ROMANTIC <a *storybook* wedding>

**story line** *n.* The plot of a story or a dramatic work.

**sto·ry·tell·er** (stôr′ē-tĕl′ər, stōr′-) *n.* **1.** One who tells or writes stories. **2.** *Informal.* One who tells lies : FIBBER.

**stoss** (stōs, stôs, shtōs) *adj.* [< G. *stossen,* to push < OHG *stōzan.*] Facing the direction from which a glacier moves. —Used of a rock or slope in its path.

**sto·tin·ki** (stō-tĭng′kə) *n., pl.* **stotinki.** [Bulgarian.] —See table at CURRENCY.

**stound** (stound) *n.* [ME < OE *stund.*] *Obs.* A short time : WHILE.

**stoup** *also* **stoop** (sto͞op) *n.* [ME *stoup,* bucket < ON *staup.*] **1.** A font for holy water at the entrance of a church. **2.** *Scot.* A bucket or pail. **3.** A drinking vessel, as a cup or tankard.

**stout** (stout) *adj.* **-er, -est.** [ME < OFr. *estout,* of Germanic orig.] **1.** Determined, bold, or brave. **2.** Physically strong : STURDY. **3.** Strong in structure or substance : SUBSTANTIAL. **4.** Corpulent. **5.** Forceful : powerful. **6.** Staunch : firm. —*n.* **1. a.** A stout person. **b.** A garment size for a stout person. **2.** A strong, very dark beer or ale. **—stout′ly** *adv.* **—stout′ness** *n.*

**stout·en** (stout′n) *vt.* & *vi.* **-ened, -en·ing, -ens.** To make or become stout.

**stout·heart·ed** (stout′här′tĭd) *adj.* Brave : courageous. **—stout′heart′ed·ly** *adv.* **—stout′heart′ed·ness** *n.*

**stove[1]** (stōv) *n.* [ME, heated room < MLG.] **1.** An apparatus in which electricity or a fuel is used to furnish heat, as for cooking. **2.** A device for providing heat. **3.** A kiln. **4.** A hothouse.

**stove[2]** (stōv) *v. var. p.t.* & *p.p. of* STAVE.

**stove·pipe** (stōv′pīp′) *n.* **1.** A pipe, usu. of thin sheet iron, used to conduct smoke or fumes from a stove into a chimney flue. **2.** A man's tall silk hat.

**sto·ver** (stō′vər) *n.* [ME, provisions < Norman Fr. *estovers* < OFr. *estovier,* to be necessary < Lat. *est opus,* it is necessary.] The dried stalks and leaves of a cereal crop, used as fodder after the grain has been harvested.

**stow** (stō) *vt.* **stowed, stow·ing, stows.** [ME *stowen* < *stow,* place < OE *stōw.*] **1.** To arrange, place, or store away, esp. in a neat, compact way. **2.** To fill by packing tightly. **3.** *Slang.* To cease : stop <*stow* the complaints> **4.** *Obs.* To provide lodging for : QUARTER. **5.** *Slang.* To eat (food) greedily. **—stow away.** To be a stowaway.

**stow·age** (stō′ĭj) *n.* **1. a.** The act, manner, or process of stowing. **b.** The state of being stowed. **2.** Space or room for storage. **3.** Stored goods. **4.** A charge for storing goods.

**stow·a·way** (stō′ə-wā′) *n.* One who hides aboard a ship, aircraft, or other vehicle so as to obtain free passage.

**stra·bis·mus** (strə-bĭz′məs) *n.* [NLat. < Gk. *strabismos,* condition of squinting < *strabizein,* to squint < *strabos,* squinting.] A defect in which one eye cannot focus with the other on an objective due to imbalance of the eye muscles. **—stra·bis′mal, stra·bis′mic** *adj.*

**stra·bot·o·my** (strə-bŏt′ə-mē) *n., pl.* **-mies.** [Gk. *strabos,* squinting + -TOMY.] Cutting of an ocular muscle or tendon to correct strabismus.

**strad·dle** (străd′l) *v.* **-dled, -dling, -dles.** [< STRIDE.] —*vt.* **1.** To sit astride of. **2.** To appear to favor both sides of (an issue). **3.** To fire shots behind and in front of (a target) so as to determine the range. —*vi.* **1.** To sit astride. **2.** To be wide apart : SPRAWL. **3.** To appear to favor both sides of an issue. —*n.* **1.** The act or posture of sitting astride. **2.** An equivocal or noncommittal position. **3.** The privileged option of either delivering or buying stocks at a specified price within a stated time period. **—strad′dler** *n.*

**Strad·i·var·i·us** (străd′ə-vâr′ē-əs, -văr′-) *n.* A stringed instrument, as a violin, made in the workshop of Antonio Stradivari.

**strafe** (strāf) *vt.* **strafed, straf·ing, strafes.** [< G. *(Gott) strafe (England),* (God) punish (England), a common World War I salutation.] To attack (e.g., ground troops) with machine-gun fire from low-flying aircraft. **—strafe** *n.*

**strag·gle** (străg′əl) *vi.* **-gled, -gling, -gles.** [ME *straglen.*] **1.** To fall behind. **2.** To proceed or spread out in a scattered or irregular group. **—strag′gler** *n.*

**strag·gly** (străg′lē) *adj.* **-gli·er, -gli·est.** Spread out irregularly <*straggly* long hair>

**straight** (strāt) *adj.* **-er, -est.** [ME < *strecchen,* to stretch.] **1.** Extending continuously in the same direction without curving <a *straight* line> **2.** Having no waves or bends <soft, *straight* hair> **3.** Erect : upright. **4.** Candid and direct <a *straight* reply> **5.** Uninterrupted : unbroken <talked for two *straight* hours> **6.** Made up of five cards constituting a sequence in poker. **7.** Politically undeviating <voted a *straight* party line> **8.** Being in the correct sequence. **9.** Upright : honorable. **10.** *Slang.* **a.** Conventional, conservative, or law-abiding. **b.** Heterosexual. **c.** SQUARE 10. **11.** *Slang.* Not being under the influence of alcohol or drugs. **12.** Not deviating from the normal or strict form. **13.** Undiluted or unmixed <*straight* vodka> **14.** Orderly and neat. **15.** Sold without discount regardless of the amount purchased. —*adv.* **1.** In a straight line : DIRECTLY. **2.** In an erect posture : UPRIGHT. **3.** Without detour or delay <went *straight* to the airport> **4.** Without circumlocution : CANDIDLY. **5.** Honestly : virtuously. **6.** Continuously. —*n.* **1.** The straight part of a racecourse between the winning post and the last turn. **2.** A straight line. **3.** A straight part, piece, or position. **4. a.** A numerical sequence of five cards of various suits in poker. **b.** A hand containing a straight. **5.** *Slang.* **a.** A heterosexual. **b.** A conventional person. **6.** *Slang.* A nonuser of illegal drugs. **—straight′ly** *adv.* **—straight′ness** *n.*

**straight and narrow** *n.* Proper conduct and moral integrity <kept strictly to the *straight and narrow* in our dealings>

**straight angle** *n.* An angle of 180°.

**straight-arm** (strāt′ärm′) *vt.* **-armed, -arm·ing, -arms.** *Football.* To ward off (a tackler) by holding the arm out straight.

**straight arrow** *n. Informal.* STRAIGHT 5b.

**straight-a·way** (strāt′ə-wā′) *adj.* **1.** Extending in a straight line or course without a curve or turn. **2.** Unhesitating : immediate. —*n.* (strāt′ə-wā′). A straight course, stretch, or track. —*adv.* (strāt′ə-wā′). At once : IMMEDIATELY.

**straight chain** *n.* An open linear organic molecular structure with no side chains.

**straight·edge** (strāt′ĕj′) *n.* A rigid flat rectangular bar with a straight edge for drawing straight lines. **—straight′edged′** *adj.*

**straight·en** (strāt′n) *vt.* & *vi.* **-ened, -en·ing, -ens.** To make or become straight. **—straighten out. 1.** To put to rights or restore order to : RECTIFY. **2.** To reform or improve <*straighten out* a juvenile delinquent. **—straight′en·er** *n.*

**straight face** *n.* A face betraying no emotion. **—straight′faced′** (strāt′fāst′) *adj.*

**straight·for·ward** (strāt-fôr′wərd) *adj.* **1.** Moving in a straight course : DIRECT. **2.** Honest : frank. —*adv.* In a straightforward course or manner. **—straight·for′ward·ly** *adv.* **—straight·for′ward·ness** *n.* **—straight·for′wards** (-wərdz) *adv.*

**straight·jack·et** (strāt′jăk′ĭt) *n.* & *v. var. of* STRAITJACKET.

**straight-leg** (strāt′lĕg′) *adj.* Being pants with legs having essentially the same diameter at the top as at the bottom.

**straight-line** (strāt′līn′) *adj.* **1.** Lying in a straight line. **2.** Relating to a device whose linkage produces or copies motion in straight lines. **3.** Designating a mode of amortization by equal payments at stated intervals over a given period of time.

**straight man** *n.* An actor serving as a foil for a comedian.

**straight off** *adv.* Straightaway.

**straight-out** (strāt′out′) *adj.* **1.** Straightforward : blunt <gave the salesperson a *straight-out* "no"> **2.** Complete : unmitigated <a *straight-out* computational error>

**straight razor** *n.* A razor consisting of a blade hinged to a handle into which it slips when not in use.

**straight ticket** *n.* A ballot cast for all the candidates of one party.

**straight·way** (strāt′wā′, -wā′) *adv.* Straightaway.

**strain[1]** (strān) *v.* **strained, strain·ing, strains.** [ME *streynen* < OFr. *estreindere,* to bind tightly < Lat. *stringere.*] —*vt.* **1.** To pull, draw, or stretch tight. **2.** To tax or exert to the utmost. **3.** To injure or impair by overuse or overexertion : WRENCH. **4.** To stretch or force beyond the usual, legitimate, or proper limit <*strain* credibility> **5.** To alter (the relations between the parts of a structure or shape) by applying external force: DEFORM. **6.** To pass (a substance) through a filtering agent. **7.** To remove or draw off by filtration. **8.** To embrace or clasp tightly: HUG. —*vi.* **1.** To strive hard. **2.** To be or become wrenched or twisted. **3.** To be subjected to great stress. **4.** To pull violently or forcibly. **5.** To stretch or exert one's muscles or nerves to the utmost. **6.** To filter, trickle, or ooze. **7.** To be very hesitant: BALK. —*n.* **1.** The act of straining or the state of being strained. **2.** Great effort, exertion, or tension. **3.** An injury resulting from excessive effort or use <a muscle *strain*> **4.** *Physics.* A deformation made by stress. **5.** Great or excessive pressure or stress on one's emotions or resources <suffered *strain* from overwork>

**strain[2]** (strān) *n.* [ME *strene* < OE *strēon.*] **1.** The collective descendants of a common ancestor. **2.** Any of the various lines of ancestry united in an individual or family: LINEAGE. **3.** *Biol.* A group of organisms of the same species, having distinctive characteristics but not usu. thought of as a separate breed or variety. **4.** A kind: sort. **5. a.** An inborn or inherited tendency or character. **b.** A streak: trace. **6.** The tone or tenor of a verbal utterance. **7.** *often* **strains.** A musical passage: TUNE. **8.** A poetic passage. **9.** A flow of eloquent or impassioned language.

**strained** (strānd) *adj.* **1.** Passed through a strainer <*strained* apples> **2.** Marked by or done with excessive effort: FORCED <a *strained* smile> **3.** Antagonized to the point of conflict <*strained* diplomatic relations>

**strain·er** (strā'nər) *n.* **1.** A device used to separate liquids from solids. **2.** An apparatus for tightening, stretching, or strengthening.

**strain gauge** *n.* A device in which mechanical motion, as of a thin wire or piezoelectric crystal, is converted into an electric variation that is used as a sensitive measure of strain.

**straining beam** *n.* A horizontal tie beam connecting two queen posts in a roof truss.

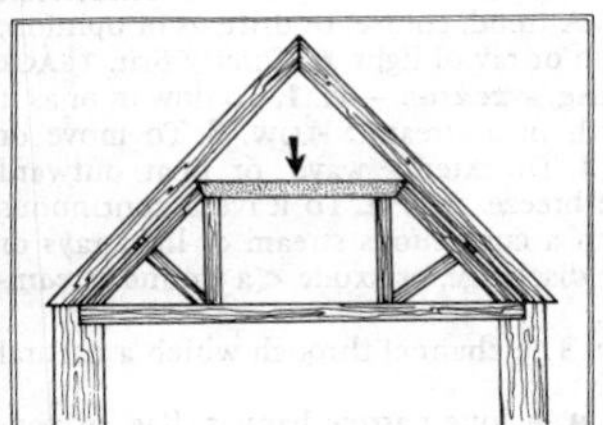

**straining beam**

**strain·om·e·ter** (strā-nŏm'ĭ-tər) *n.* An extensometer.

**strait** (strāt) *n.* [ME *streit* < OFr. *estreit* < Lat. *strictus,* narrow < p.part. of *stringere,* to bind tightly.] **1.** *often* **straits.** A narrow passage of water joining two larger bodies of water. **2.** *often* **straits.** A situation of difficulty, perplexity, distress, or need <desperate financial *straits*> —*adj. Archaic.* **1.** Narrow or constricted. **2.** Affording little space or room. **3.** Strict, rigid, or righteous.

**strait·en** (strāt'n) *vt.* **-ened, -en·ing, -ens. 1.** To make narrow: LIMIT. **2.** To put into hardship, esp. financial hardship.

**strait·jack·et** *also* **straight·jack·et** (strāt'jăk'ĭt) —*n.* **1.** A long-sleeved jacketlike garment used to bind the arms tightly against the body as a means of restraining a violent patient or prisoner. **2.** A tight restriction <an economic *straitjacket*> —*vt.* **-et·ed, -et·ing, -ets.** To restrict or restrain by or as if by confining in a straitjacket.

**strait-laced** (strāt'lāst') *adj.* **1.** Too strict in behavior, morality, or opinions. **2.** *Archaic.* Having or wearing a tightly laced garment. **—strait'-lac'ed·ly** (-lā'sĭd-lē, -lāst'lē) *adv.* **—strait'-lac'ed·ness** *n.*

**strake** (strāk) *n.* [ME.] *Naut.* A single continuous line of planking or metal plating extending on a vessel's hull from stem to stern.

**stra·mo·ni·um** (strə-mō'nē-əm) *n.* [NLat.] **1.** The jimsonweed. **2.** The dried poisonous leaves of the jimsonweed, used in treating asthma.

**strand[1]** (strănd) *n.* [ME < OE.] Land bordering a body of water: BEACH. —*v.* **strand·ed, strand·ing, strands.** —*vt.* **1.** To drive or run ashore or aground. **2.** To bring into or leave in a difficult or helpless position <was *stranded* in a snowstorm> —*vi.* **1.** To be driven or run ashore or aground. **2.** To be brought into or left in a difficult or helpless position.

**strand[2]** (strănd) *n.* [ME *strond.*] **1.** Fibers or filaments twisted together so as to form a cable, rope, thread, or yarn. **2.** A single filament, as a fiber or thread <a *strand* of hair> **3.** Something, as a string of pearls, that is plaited or twisted into a ropelike length. —*vt.* **strand·ed, strand·ing, strands. 1.** To make or form (e.g., a rope) by twisting strands together. **2.** To break a strand of (e.g., a rope).

**strand line** *also* **strand·line** (strănd'līn') *n.* A shore line, esp. one marking an earlier and higher water level.

**strange** (strānj) *adj.* **strang·er, strang·est.** [ME *straunge* < OFr. *estrange* < Lat. *extraneus* < *extra,* outside.] **1.** Previously unknown. **2.** Notably out of the ordinary. **3.** Unusual. **4.** Not of one's own or a particular locality, environment, or kind: EXOTIC. **5.** *Archaic.* Alien: foreign. **6.** Inexperienced: unacquainted <*strange* to their new responsibilities> —*adv.* In a strange way. **—strange'ly** *adv.*

☆ ***syns:* 1.** STRANGE, NEW, UNACCUSTOMED, UNFAMILIAR *adj. core meaning*: previously unknown <*strange* faces> **2.** STRANGE, ECCENTRIC, ODD, PECULIAR, QUAINT, QUEER, UNUSUAL *adj. core meaning*: notably out of the ordinary <*strange* behavior><a *strange* animal> ***ant:*** familiar

**strange·ness** (strānj'nĭs) *n.* **1.** The quality of being strange. **2.** *Physics.* A quantum number equal to hypercharge minus baryon number, indicating the possible transformations of an elementary particle upon strong interaction with another elementary particle.

**strange particle** *n.* An unstable elementary particle created in high-energy particle collisions with a short life and a strangeness quantum number other than zero.

**strang·er** (strān'jər) *n.* [ME *strangere* < OFr. *estrangier* < Lat. *extraneus,* strange < *extra,* outside.] **1.** One who is neither friend nor acquaintance. **2.** A newcomer, foreigner, or outsider. **3.** One unaccustomed to or unacquainted with something specified: NOVICE <a *stranger* to our local customs> **4.** A visitor: guest. **5.** *Law.* One neither privy nor party to a title, act, or contract.

**stran·gle** (străng'gəl) *v.* **-gled, -gling, -gles.** [ME *stranglen* < OFr. *estrangler* < Lat. *strangulare* < Gk. *strangalan* < *strangalē,* halter.] —*vt.* **1. a.** To kill by squeezing the throat so as to choke or suffocate: throttle. **b.** To cut off the oxygen supply of: SMOTHER. **2.** To repress, suppress, or stifle <*strangle* a sneeze> **3.** To inhibit the growth or action of: RESTRICT. —*vi.* To suffer or die from suffocation or strangulation: CHOKE. **—stran'gler** *n.*

**strangle hold** *n.* **1.** An illegal wrestling hold used to choke an opponent. **2.** A force, influence, or act that restricts or suppresses freedom or progress.

**stran·gles** (străng'gəlz) *pl.n.* [ME *strangle* (sing.), strangulation < *stranglen,* to strangle.] *(sing. in number).* An infectious disease of horses and related animals, caused by the bacterium *Streptococcus equi* and marked by nasal inflammation and oral abscesses.

**stran·gu·late** (străng'gyə-lāt') *v.* **-lat·ed, -lat·ing, -lates.** [Lat. *strangulare, strangulat-,* to strangle.] —*vt.* **1.** To strangle. **2.** *Pathol.* To compress, constrict, or obstruct so as to cut off the flow of blood or other fluid. —*vi.* To be or become strangled or constricted. **—stran'gu·la'tion** (-lā'shən) *n.*

**stran·gu·ry** (străng'gyə-rē, -gyŏŏr'ē) *n.* [ME < Lat. *stranguria* < Gk. *strangouria*: *stranx,* drop + *-ouria,* -uria.] Slow, painful urination accompanied by spasms of the urethra and bladder.

**strap** (străp) *n.* [Alteration of STROP.] **1. a.** A long, narrow strip of pliant material, as leather. **b.** Such a strip fitted with a buckle or similar fastener. **2.** A flat thin metal or plastic band used for fastening or clamping objects together or into position. **3.** A narrow band formed into a loop for grasping with the hand. **4.** A razor strop. **5.** A strip of leather used in flogging. —*vt.* **strapped, strap·ping, straps. 1.** To fasten or secure with a strap. **2.** To beat with a strap. **3.** To sharpen (e.g., a razor).

**strap·hang·er** (străp'hăng'ər) *n.* A passenger, as on a bus or subway, who grips a hanging strap for support.

**strap·less** (străp'lĭs) *adj.* **1.** Without a strap or straps. **2.** Of, relating to, or being a dress or undergarment having no straps and leaving the shoulders bare. —*n.* A strapless garment.

**strap·pa·do** (stră-pā'dō, -pä'-) *n., pl.* **-does.** [Fr. *strapade* < Ital. *strappata* < *strappare,* to drag, prob. of Germanic orig.] **1.** Torture consisting of hoisting the tied-up victim by means of a pulley and then dropping the victim halfway down to the ground. **2.** The apparatus employed in strappado.

**strapped** (străpt) *adj.* [< STRAP.] *Informal.* Having little or no money.

**strap·per** (străp'ər) *n.* A tall sturdy person.

**strap·ping** (străp'ĭng) *adj.* Tall and sturdy.

**strass** (străs) *n.* [G. or < Fr., after Joseph *Strasser,* an 18th-cent. German jeweler.] PASTE[1] 7b.

**stra·ta** (strā'tə, străt'ə) *n. pl. of* STRATUM.

**strat·a·gem** (străt'ə-jəm) *n.* [Fr. *stratagème* < Lat. *strategema* < Gk. *stratēgēma* < *stratēgein,* to be a general < *stratēgos,* general: *stratos,* army + *agein,* to lead.] **1.** A military maneuver intended to surprise or deceive an enemy. **2.** A deception.

**stra·te·gic** (strə-tē'jĭk) *also* **stra·te·gi·cal** (-jĭ-kəl) *adj.* **1.** Of or relating to strategy. **2. a.** Important or essential in relation to strategy <a *strategic* retreat> **b.** Essential to the effective conduct of war <*strategic* supplies> **c.** Designed to destroy the military potential of an enemy <*strategic* bombing of military targets> **—stra·te'gi·cal·ly** *adv.*

**stra·te·gics** (strə-tē'jĭks) *n. (sing. in number).* The art of military strategy.

**strat·e·gy** (străt'ə-jē) *n., pl.* **-gies.** [Fr. *stratégie* < Gk. *stratēgos,* general. —see STRATAGEM.] **1.** The science or art of military command as applied to the overall planning and conduct of large-scale combat operations. **2.** A plan of action resulting from the practice of strategy. **3.** The art or skill of using stratagems esp. in politics and business. **—strat′e·gist** (-jĭst) *n.*

**strath** (străth) *n.* [Sc. Gael. *srath.*] *Scot.* A wide, flat river valley.

**stra·ti** (strā′tī′, străt′ī′) *n. pl. of* STRATUS.

**strati-** *pref.* [< STRATUM.] Stratum <*strati*form>

**stra·tic·u·late** (stră-tĭk′yə-lĭt) *adj.* [< STRATUM.] Having thin strata. **—stra·tic′u·la′tion** *n.*

**strat·i·fi·ca·tion** (străt′ə-fĭ-kā′shən) *n.* **1.** The act or process of stratifying. **2.** A stratified configuration.

**stratified charge engine** *n.* An internal-combustion engine that runs on a lean mixture of fuel by means of a divided ignition cylinder that burns rich fuel in a small chamber near the spark plug and a very lean mixture throughout the rest of the cylinder.

**strat·i·form** (străt′ə-fôrm′) *adj.* Having the form of strata.

**strat·i·fy** (străt′ə-fī′) *v.* **-fied, -fy·ing, -fies.** [Fr. *stratifier* < NLat. *stratificare* : STRATUM + Lat. *facere*, to make.] *—vt.* **1.** To form, arrange, or deposit in strata. **2.** To preserve (seeds) by placing between layers of moist sand or similar material. *—vi.* **1.** To form strata. **2.** To develop different levels, as of social class, privilege, or status.

**stra·tig·ra·phy** (strə-tĭg′rə-fē) *n.* The study of rock strata, esp. of their distribution, deposition, and age. **—strat′i·graph′ic** (străt′ĭ-grăf′ĭk), **strat′i·graph′i·cal** *adj.* **—strat′i·graph′i·cal·ly** *adv.*

**stra·toc·ra·cy** (strə-tŏk′rə-sē) *n., pl.* **-cies.** [Gk. *stratos*, army + -CRACY.] Government by the military, esp. the army. **—strat′o·crat′ic** (străt′ə-krăt′ĭk) *adj.*

**stra·to·cu·mu·lus** (strā′tō-kyōō′myə-ləs, străt′ō-) *n., pl.* **-li** (-lī′) [STRAT(US) + CUMULUS.] A low-lying cloud occurring in extensive horizontal layers with massive, rounded summits.

**strat·o·sphere** (străt′ə-sfîr′) *n.* [Fr. *stratosphère* : NLat. *stratum*, stratum + *sphère*, sphere < Lat. *sphaera* < Gk. *sphaira.*] The relatively isothermal part of the atmosphere above the troposphere and below the mesosphere. **—strat′o·spher′ic** (-sfîr′ĭk, -sfĕr′-) *adj.*

**strat·o·vol·ca·no** (străt′ō-vŏl-kā′nō, strā′tō-) *n.* [STRAT(UM) + VOLCANO.] A volcano composed of lava and ash deposited in alternating conical layers.

**stra·tum** (strā′təm, străt′əm) *n., pl.* **-ta** (-tə) *or* **-tums.** [NLat. < Lat. *stratum*, a covering < *sternere*, to spread.] **1.** A horizontal layer of a material, esp. one of several parallel layers arranged one on top of another. **2.** *Geol.* **a.** A layer of rock having the same composition throughout. **b.** A formation containing a number of layers of rock of the same kind. **3.** A social level composed of people of similar social, cultural, or economic status. **—stra′tal** *adj.*

**stra·tus** (strā′təs, străt′əs) *n., pl.* **stra·ti** (strā′tī′, străt′ī′) [< Lat. p.part. of *sternere*, to spread.] A low-altitude cloud typically resembling a horizontal layer of fog.

**straw** (strô) *n.* [ME < OE *strēaw.*] **1. a.** Stalks of grain after thrashing, used as bedding and food for animals and for weaving or braiding. **b.** A single stalk of straw. **2.** A slender tube used for sucking up a liquid. **3. a.** Something of little value or importance. **b.** Something with too little substance to provide support in a crisis. *—vt.* **strawed, straw·ing, straws. 1.** To cover (e.g., a surface) with straw : STREW. **2.** To provide with straw. *—adj.* **1.** Of the color of straw : YELLOWISH. **2.** Having little or no value or substance : UNIMPORTANT. **—straw in the wind.** A hint of something to come.

**straw·ber·ry** (strô′bĕr′ē) *n.* [ME < OE *strēawberige* : *strēaw*, straw + *berige*, berry.] **1.** Any of various low-growing plants of the genus *Fragaria*, with white flowers and red, fleshy, edible fruit. **2.** The fruit of the strawberry.

**strawberry bass** *n.* The black crappie.

**strawberry blite** *n.* A weedy plant, *Chenopodium capitatum* of northern regions, with tiny petalless flowers and red berrylike fruit.

**strawberry bush** *n.* A North American shrub, *Euonymus americanus*, with showy pinkish fruit and inconspicuous flowers.

**strawberry mark** *n.* A small reddish birthmark.

**strawberry roan** *n.* A horse having reddish hair mixed with white.

**strawberry shrub** *n.* Any of several North American shrubs of the genus *Calycanthus*, with aromatic reddish-brown flowers.

**strawberry tomato** *n.* **1.** A North American plant, *Physalis pruinosa*, with yellow flowers and edible yellowish fruit enclosed in a husk. **2.** The fruit of the strawberry tomato.

**strawberry tree** *n.* A tree, *Arbutus unedo*, native to southern Europe, with strawberrylike fruit and evergreen leaves.

**straw·board** (strô′bôrd′, -bōrd′) *n.* A coarse yellow cardboard made of straw pulp.

**straw boss** *n. Informal.* A worker who acts as a boss or assistant foreman in addition to regular duties.

**straw·flow·er** (strô′flou′ər) *n.* An Australian plant, *Helichrysum bracteatum*, having flowers with showy, variously colored bracts that retain their color when dried.

**straw-hat** (strô′hăt′) *adj.* [From the fashion of wearing straw hats during the summer.] Of or relating to summer theater in suburban or resort areas.

**straw man** *n.* **1.** A bundle of straw made into the likeness of a human being and often used as a scarecrow. **2.** One set up as cover for a questionable enterprise. **3.** An argument or opponent set up so as to be easily refuted or defeated.

**straw vote** *n.* An unofficial vote or poll indicating the trend of political opinion.

**straw·worm** (strô′wûrm′) *n.* The destructive larva of a fly, *Harmolita grandis* of western North America, that infests grain stalks.

**stray** (strā) *vi.* **strayed, stray·ing, strays.** [ME *straien* < OFr. *estraier* < VLat. **estragare* : Lat. *extra-*, outside + Lat. *vagari*, to wander < *vagus*, wandering.] **1. a.** To wander from a given place or group or beyond established limits : ROAM. **b.** To become lost. **2.** To rove, wander about, or meander. **3.** To deviate from a course regarded as right or moral. **4.** To deviate from the subject matter at hand : DIGRESS. *—n.* One that has strayed, esp. a domestic animal lost or at large. *—adj.* **1. a.** Straying or having strayed. **b.** Lost. **2.** Scattered or separate <a few *stray* crumbs on the mat> **—stray′er** *n.*

**streak** (strēk) *n.* [ME < OE *strica.*] **1.** A line, mark, smear, or band differentiated by color or texture from its surroundings. **2.** A slight trace or tendency : TRAIT <a gentle *streak*> **3.** *Informal.* A brief stretch of time : RUN <a *streak* of bad luck> **4.** *Mineral.* The color of the powder of a mineral, used as a distinguishing characteristic. *—v.* **streaked, streak·ing, streaks.** *—vt.* To mark with streaks. *—vi.* **1.** To form streaks. **2.** To be or become streaked. **3. a.** To move at high speed : RUSH <antelope *streaking* across the plain> **b.** *Slang.* To run unclothed through a public place as a prank. **—streak′er** *n.*

**streak·ing** (strē′kĭng) *n.* **1.** Creation of a streaked effect by the chemical lightening of several strands of hair. **2.** *Slang.* The act of running unclothed through a public place as a prank.

**streak·y** (strē′kē) *adj.* **-i·er, -i·est. 1.** Marked with, characterized by, or occurring in streaks. **2.** Variable or uneven in character or quality. **—streak′i·ly** *adv.* **—streak′i·ness** *n.*

**stream** (strēm) *n.* [ME < OE *strēam.*] **1. a.** A body of running water, esp. one moving over the earth's surface in a channel or bed, as a brook, rivulet, or river. **b.** A steady current in such a body of water. **2.** A steady current of a fluid. **3.** A steady flow or succession <a *stream* of invectives> **4.** A trend, course, or drift, as of opinion, thought, or history. **5.** A beam or ray of light. **6.** *Chiefly Brit.* TRACK 5. *—v.* **streamed, stream·ing, streams.** *—vi.* **1.** To flow in or as if in a stream. **2.** To pour forth in a stream : FLOW. **3.** To move or proceed in large numbers. **4.** To extend, wave, or float outward <The flags *streamed* in the breeze.> **5. a.** To leave a continuous trail of light. **b.** To give forth a continuous stream of light rays or beams : SHINE. *—vt.* To emit, discharge, or exude <a wound *streaming* blood> **—stream′y** *adj.*

**stream·bed** (strēm′bĕd′) *n.* The channel through which a natural stream of water runs or once ran.

**stream·er** (strē′mər) *n.* **1. a.** A long narrow banner, flag, or pennant. **b.** A long narrow strip of material <a car decorated with paper *streamers*> **2.** A shaft or ray of light extending upward from the horizon. **3.** A newspaper headline that runs across a full page.

**stream·let** (strēm′lĭt) *n.* A small stream.

**stream·line** (strēm′līn′) *n.* **1.** A fluid line having the property that the tangent at every point on the line is aligned with the fluid's local velocity. **2.** The path of one particle in a flowing fluid. **3.** A contour of a body constructed so as to offer minimum resistance to a fluid flow. *—vt.* **-lined, -lin·ing, -lines. 1.** To build or design in a streamlined configuration. **2.** To improve the efficiency or look of : MODERNIZE.

**stream·lined** (strēm′līnd′) *adj.* **1.** Designed or configured to offer resistance to fluid flow. **2.** Improved in minimal efficiency or look : MODERNIZED <*streamlined* methods>

**stream·lin·er** (strēm′lī′nər) *n.* Something streamlined, esp. a streamlined passenger train.

**stream of consciousness** *n.* **1.** *Psychol.* The conscious experience of an individual regarded as a continuous rather than a discrete series of events. **2.** A fictional narrative technique by which a character's feelings and thoughts are recorded.

**street** (strēt) *n.* [ME < OE *strǣt* < LLat. *strata* < Lat. *sternere*, to cover.] **1. a.** A public thoroughfare in a city or town, usu. including the sidewalks lining one or both sides. **b.** Such a thoroughfare for vehicles. **2.** The people who live, work, or habitually gather in or along a street <The whole *street* gathered for a block party.> **3. Street.** A district, as Wall Street in New York City, that is identified with a particular profession. **4.** The streets of a large city thought of as the scene of crime, poverty, or dereliction. **—on (or in) the street. 1.** Out of work : IDLE. **2.** At liberty : out of prison.

**street Arab** *n.* A homeless or neglected child roaming the streets.

**street·car** (strēt′kär′) *n.* A public passenger car operated on rails along a regular route, usu. through city streets.

**street·light** (strēt′līt′) *n.* One of a series of lights usu. attached to tall poles, spaced at intervals along a public street or roadway and illuminated automatically from dusk to dawn.

**street theater** *n.* Dramatization of social and political issues, usu. presented outside, as on the street or in a park.

**street·walk·er** (strēt′wô′kər) *n.* A prostitute who solicits in the streets. **—street′walk′ing** *n.*

**street·wise** (strēt′wīz′) *adj.* **1.** Experienced in dealing with inner-city dwellers <a *streetwise* social worker> **2.** Capable of surviving and satisfying one's wants and needs on the streets of a large city.

**strength** (strĕngkth, strĕngth) *n.* [ME < OE *strengðu.*] **1.** The quality, state, or property of being strong : physical power. **2. a.** Power to withstand strain, force, or stress : TOUGHNESS. **b.** Power to sustain or resist attack : IMPREGNABILITY. **3.** Legal, intellectual, or moral force. **4. a.** A source of power or force. **b.** One viewed as the embodiment of protective or supportive power : STAY. **5.** Firm will or character : moral courage or power. **6.** Effective or binding force : EFFICACY <the *strength* of a well thought out argument> **7.** The power or capability of generating a reaction or effect : operative potency <the *strength* of a vise> **8.** Degree of concentration, distillation, or saturation : POTENCY. **9.** Vehemence or intensity, as of emotion, language, or action. **10. a.** Numerical force or supportive personnel measured as to concentration. **b.** Military force in numbers of personnel or materiel <a battalion at half *strength*> **11.** A continuous rising tendency in prices. **12.** Power derived from the value of playing cards held. **—on the strength of.** On the basis of.

**strength·en** (strĕngk′thən, strĕng′-) *vt.* & *vi.* **-ened, -en·ing, -ens.** To make or become strong or stronger. **—strength′en·er** *n.*

**stren·u·ous** (strĕn′yōō-əs) *adj.* [Lat. *strenuus.*] **1.** Requiring or marked by great energy, effort, or exertion. **2.** Vigorously active : ENERGETIC. **—stren′u·os′i·ty** (-ŏs′ĭtē), **stren′u·ous·ness** *n.* **—stren′u·ous·ly** *adv.*

**strep** (strĕp) *adj.* Streptococcal.

**strep throat** (strĕp) *n.* Septic sore throat.

**strepto-** *pref.* [< Gk. *streptos,* twisted < *strephein,* to turn.] **1.** Twisted : twisted chain <*strepto*coccus> **2.** Streptococcus <*strepto*lysin>

**strep·to·ba·cil·lus** (strĕp′tō-bə-sĭl′əs) *n.* Any of various Gram-negative, rod-shaped, often pathogenic bacteria of the genus *Streptobacillus,* occurring in chains.

**strep·to·coc·cus** (strĕp′tə-kŏk′əs) *n., pl.* **-coc·ci** (-kŏk′sī′, -kŏk′ī′) [NLat. *Streptococcus,* genus name.] Any of various round to ovoid pathogenic bacteria of the genus *Streptococcus,* occurring in pairs or chains. **—strep′to·coc′cal, strep′to·coc′cic** (-kŏk′sĭk, -kŏk′ĭk) *adj.*

**strep·to·kin·ase** (strĕp′tō-kĭn′ās′, -āz′, -nās′, -kī′nāz′) *n.* A proteolytic enzyme derived from hemolytic streptococci, capable of dissolving fibrin, and used to dissolve blood clots.

**strep·to·ly·sin** (strĕp′tə-lī′sĭn) *n.* An antigenic hemolysin derived from some strains of streptococci.

**strep·to·my·ces** (strĕp′tə-mī′sēz′) *n.* [NLat. *Streptomyces,* genus name : STREPTO- + Gk. *mukēs,* fungus.] Any of various actinomycetes of the genus *Streptomyces,* including some strains that produce antibiotics.

**strep·to·my·cin** (strĕp′tə-mī′sĭn) *n.* [< STREPTOMYCES.] An antibiotic, $C_{21}H_{39}N_7O_{12}$, produced from mold cultures of bacteria of the genus *Streptomyces* and used medicinally to combat various Gram-positive and Gram-negative bacteria and tuberculosis.

**strep·to·thri·cin** (strĕp′tə-thrī′sĭn, -thrĭs′ĭn) *n.* [< NLat. *Streptothrix,* genus of bacteria : Gk. *streptos,* twisted + Gk. *thrix,* hair.] An antibiotic isolated from a soil fungus, *Streptomyces lavendulae,* and active against both Gram-positive and Gram-negative bacteria and some fungi.

**stress** (strĕs) *n.* [ME *stresse,* hardship < *distresse* OFr. *destresse.* —see DISTRESS.] **1.** Importance, significance, or emphasis <put great *stress* on economic issues> **2. a.** The relative force with which a sound or syllable is spoken. **b.** The emphasis placed on the sound or syllable spoken loudest in a given word or phrase. **3. a.** The relative emphasis given a syllable or word in accordance with a metrical pattern. **b.** A syllable receiving a strong relative emphasis. **4.** *Mus.* ACCENT 7. **5.** *Physics.* An applied force or system of forces that tends to strain or deform a body. **6.** Mental, emotional, or physical tension, strain, or distress. *—vt.* **stressed, stress·ing, stress·es. 1.** To place emphasis on <*stressed* the importance of accuracy> **2.** To subject to pressure or strain. **3.** To subject to mechanical pressure or force. **4.** To construct so as to withstand a specified stress.

**STRESS** (strĕs) *n.* [STR(UCTURAL) + E(NGINEERING) + S(YSTEMS) + S(OLVER).] A computer language designed for use in solving structural analysis problems in civil engineering.

**stres·sor** (strĕs′ər) *n.* An agent causing stress.

**stretch** (strĕch) *v.* **stretched, stretch·ing, stretch·es.** [ME *strecchen* < OE *streccan.*] *—vt.* **1.** To widen, lengthen, or distend by pulling. **2.** To cause to extend from one place to another or across a given space <*stretch* lines from one telephone pole to another> **3.** To make taut : TIGHTEN. **4.** To extend or reach forth <*stretch* out one's hand> **5.** To extend (oneself or one's extremities) at full length. **6.** To flex the muscles of <*stretch* one's legs> **7.** To exert to the utmost : STRAIN <*stretch* every nerve to win> **8.** To wrench or strain (e.g., a muscle). **9.** *Informal.* To fell by a blow <The boxer was *stretched* in the first round.> **10.** To put to torture on the rack. **11. a.** To extend or broaden to fulfill a greater function <*stretched* the budget> **b.** To increase the quantity of by admixture or dilution <*stretch* a meal by thinning the soup> **12. a.** To extend the limits of <*stretch* the rules> **b.** To strain, esp. to unreasonable or questionable limits <*stretch* the truth> **13.** To prolong <*stretch* out an argument> *—vi.* **1.** To become widened, lengthened, or distended. **2.** To extend over a distance or area or in a certain direction <"On both sides of us *stretched* the wet plain" —Hemingway> **3.** To lie down at full length. **4.** To flex or extend one's limbs or muscles. **5.** To extend over a given period of time <"This story *stretches* over a whole generation" —William Golding> *—n.* **1.** The act of stretching or the state of being stretched. **2.** The extent to which something can be stretched : ELASTICITY. **3.** A continuous or unbroken length, area, or expanse <an empty *stretch* of road> **4.** A straight section of a racecourse or track, esp. the section leading to the finish line. **5. a.** A continuous time period. **b.** *Slang.* A prison term. <a ten-year *stretch*> **c.** *Informal.* The last stage of an event, period, or process. **—stretch′a·ble** *adj.* **—stretch′y** *adj.*

**stretch·er** (strĕch′ər) *n.* **1.** One that stretches. **2.** A litter, usu. of canvas stretched over a frame, used to transport the sick, wounded, or dead. **3.** A device for stretching and shaping, such as the wooden framework on which canvas is stretched for an oil painting. **4. a.** A usu. horizontal tie beam or brace that supports or extends a framework. **b.** A brick or stone laid parallel to the face of a wall.

**stretch·er-bear·er** (strĕch′ər-bâr′ər) *n.* One who helps carry a stretcher or litter.

**stretch-out** (strĕch′out′) *n.* An increase in the labor required of industrial workers without a commensurate pay increase.

**stretch receptor** *n.* A proprioceptor in a muscle or tendon that is stimulated by a stretch.

**stret·to** (strĕt′ō) *n., pl.* **stret·ti** (strĕt′ē) *or* **stret·tos.** [Ital. < Lat. *strictus,* strict.] *Mus.* **1.** A close succession or overlapping of voices in a fugue, esp. in the final section. **2.** A final section, as of an oratorio, performed with an acceleration in tempo to produce a climax.

**streu·sel** (strōō′zəl, stroi′-) *n.* [G. < MHG *ströusel,* something strewn < *ströuwen,* to sprinkle < OHG *strowwen.*] A crumblike topping for coffee cakes and rich breads, consisting of flour, sugar, butter, cinnamon, and sometimes chopped nutmeats.

**strew** (strōō) *vt.* **strewn** (strōōn) *or* **strewed, strew·ing, strews.** [ME *strewen* < OE *strewian.*] **1.** To scatter here and there. **2.** To cover (a surface) with things scattered <"Italy . . . was *strewn* thick with the remains of Roman buildings" —Bernard Berenson> **3.** To be or become dispersed over (a surface).

**strewn field** *n.* An area abundant with tektites.

**stri·a** (strī′ə) *n., pl.* **stri·ae** (strī′ē′) [Lat.] **1.** A thin, narrow groove or channel. **2.** A thin line or band, esp. one of several that are parallel or close together.

**stri·ate** (strī′āt′) *also* **stri·at·ed** (-ā′tĭd) *adj.* [Lat. *striatus,* p.part. of *striare,* to make furrows < *stria,* furrow.] **1.** Marked with striae : striped, grooved, or ridged. **2.** Consisting of a stria or striae. *—vt.* **-at·ed, -at·ing, -ates.** To mark with striae.

**striated muscle** *n.* Skeletal, voluntary, and cardiac muscle, having transverse striations of the fibers.

**stri·a·tion** (strī-ā′shən) *n.* **1.** The state of being striated or having striae. **2.** The form taken by striae. **3.** A stria.

**strick·en** (strĭk′ən) *adj.* [P.part. of STRIKE.] **1.** Struck or wounded, as by a projectile. **2.** Afflicted with something overwhelming, as strong emotion or trouble. **3.** Having the contents made even with the top of a measuring device or container : LEVEL <a *stricken* measure of sugar>

**strick·le** (strĭk′əl) *n.* [ME *strikelle* < OE *stricel.*] **1.** An instrument for leveling off grain or other material in a measure. **2.** A foundry tool for shaping a mold in sand or loam. **3.** A tool for sharpening scythes. *—vt.* **-led, -ling, -les.** To apply a strickle to.

**strict** (strĭkt) *adj.* **-er, -est.** [Lat. *strictus* < p.part. of *stringere,* to bind tightly.] **1.** Exact : precise <a *strict* translation of the document> **2.** Complete : absolute <kept in *strict* confidence> **3.** Kept within specific and narrow limits <a *strict* construction of the Constitution> **4.** Imposing exacting discipline : not permissive <a *strict* parent> **5.** Rigorously maintained or enforced : STRINGENT <*strict* rules> **6.** Rigidly conforming : DEVOUT. **7.** *Bot.* Stiff, narrow, and upright. **—strict′ly** *adv.* **—strict′ness** *n.*

**stric·ture** (strĭk′chər) *n.* [ME < Lat. *strictura* < *stringere,* to bind tightly.] **1.** A restraint, limit, or restriction. **2.** An adverse remark or criticism : CENSURE. **3.** *Pathol.* An abnormal narrowing of a duct or passage.

**stride** (strīd) *v.* **strode** (strōd), **strid·den** (strĭd′n), **strid·ing, strides.** [ME *striden* < OE *strīdan.*] *—vi.* **1.** To walk with long steps, esp. vigorously. **2.** To take a single long step, as in passing over an obstruction. *—vt.* **1.** To stride on, along, or through. **2.** To be astride of : STRADDLE. *—n.* **1.** An act of striding. **2. a.** A single long step. **b.** The distance traveled in such a step. **3. a.** A single coordinated movement of the four legs of an animal, as a horse, completed when the legs are returned to their initial relative position. **b.** The distance traveled in a stride. **4.** *often* **strides.** A step forward : ADVANCE <great *strides* in the space program> **—take in (one's) stride.** To handle or accept without disruption of normal routine. **—strid′er** *n.*

☆ ***syns:*** STRIDE, MARCH, STALK *v. core meaning* : to walk with long steps, esp. in a vigorous way <*strode* into the office and picked up the telephone>

**stri·dent** (strīd′nt) *adj.* [Lat. *stridens, strident-,* pr.part. of *stridēre,* to make harsh sounds.] Harsh, grating, and loud : SHRILL. **—stri′dence, stri′den·cy** *n.* **—stri′dent·ly** *adv.*

**stride piano** *n.* Jazz pianism in which the melody is played by the right hand while a single note is played by the left hand in alternation with a chord that is an octave, or more than an octave, higher.

**stri·dor** (strī′dər, -dôr′) *n.* [Lat. < *stridēre,* to make harsh sounds.] **1.** A strident sound. **2.** *Pathol.* A harsh, high-pitched sound in inhalation or exhalation.

**strid·u·late** (strĭj′ə-lāt′) *vi.* **-lat·ed, -lat·ing, -lates.** [< Lat. *stridulus,* stridulous.] To produce a shrill grating sound by rubbing body parts together. —Used esp. of certain insects. **—strid′u·la′tion** *n.* **—strid′u·la·to′ry** (-lə-tôr′ē, -tōr′ē) *adj.*

**strid·u·lous** (strĭj′ə-ləs) *adj.* [Lat. *stridulus* < *stridēre,* to make harsh sounds.] Making or marked by a strident sound.

**strife** (strīf) *n.* [ME *strif* < OFr. *estrif* < *estriver,* to fight, of Germanic orig.] **1.** Heated, often violent conflict : bitter dissension. **2.** A struggle between rivals : CONTENTION. **3.** *Archaic.* Earnest endeavor.

**strig·il** (strĭj′əl) *n.* [Lat. *strigilis* < *stringere,* to touch lightly.] An ancient Greek and Roman instrument for scraping the skin after a bath.

**stri·gose** (strī′gōs′) *adj.* [NLat. *strigosus* < Lat. *striga,* furrow.] **1.** Marked with fine, close-set grooves or streaks. **2.** *Bot.* Having stiff, closely pressed hairs or bristles.

**strike** (strīk) *v.* **struck** (strŭk), **struck** or **strick·en** (strĭk′ən), **strik·ing, strikes.** [ME *striken* < OE *strīcan,* to stroke.] *—vt.* **1. a.** To hit sharply, as with the fist, hand, or a weapon. **b.** To inflict (a blow). **2. a.** To collide with <*struck* the chair with my knee> **b.** To move into violent contact <*struck* my knee against the chair> **3.** To assault : attack. **4.** To afflict suddenly with a disease or impairment. **5.** To wound by biting. **6.** To hook (a fish) that has taken the bait. **7.** To form by stamping, printing, or punching <*strike* a commemorative coin> **8.** To produce by hitting a device, as a key on a musical instrument or a typewriter <*strike* a C sharp> **9.** To indicate by a percussive sound <The clock *struck* ten.> **10. a.** To produce by friction. **b.** To produce flame, light, or a spark from by friction. **11.** To eliminate : expunge <*struck* that testimony from the record> **12.** To come upon : DISCOVER <*struck* gold> **13.** To fall upon : REACH <A blinding light *struck* my face.> **14.** To impress abruptly or freshly, causing an immediate response <That *strikes* me as a good idea.> **15.** To occur or appear to <The thought of going now suddenly *struck* me.> **16.** To cause (an emotion) to penetrate deeply <The specter of war *struck* terror into our hearts.> **17. a.** To make or conclude (a bargain). **b.** To achieve (e.g. a balance) by careful weighing. **18.** To fall into or assume (e.g., a pose). **19.** *Naut.* **a.** To haul down (a mast or sail). **b.** To lower (a flag or sail) in salute or surrender. **c.** To lower (cargo) into a hold. **20.** To remove (theatrical properties) from the stage. **21.** To remove or pack up <*strike* the tents> **22.** To undertake a strike against (an employer). **23.** To smooth or level : STRICKLE. **24.** To send out or down (e.g., roots). *—vi.* **1.** To deal a blow or blows with or as if with the fist or a weapon : HIT. **2.** To aim a stroke or blow. **3.** To make contact suddenly or violently : COLLIDE. **4.** To begin an attack. **5.** To pierce : penetrate. **6.** To take bait. —Used esp. of a fish. **7. a.** To make a percussive sound. **b.** To be indicated by sounds <The hour has *struck.*> **8.** To become ignited. **9.** To come suddenly or unexpectedly. **10.** To fall : impinge. **11.** To proceed, esp. in a new direction : SET OUT. **12.** To engage in a strike against an employer. **—strike out. 1.** To begin an action. **2.** To undertake enthusiastically. **3.** *Baseball.* **a.** To pitch three strikes to a batter, putting him or her out. **b.** To be struck out. **—strike up. 1.** To start to play or sound vigorously <*Strike up* the orchestra.> **2.** To initiate or begin <*strike up* a conversation> *—n.* **1.** An act or gesture of striking. **2.** An attack, esp. a military air attack on a single group of targets. **3. a.** A cessation of work by employees in support of demands made upon their employer <a *strike* for higher pay> **b.** A temporary stoppage of normal activity undertaken as a protest. **4.** A sudden achievement or valuable discovery, as of a precious mineral. **5. a.** A taking of bait by a fish. **b.** A pull on a fishing line indicating a strike. **6.** A quantity of coins or medals struck at the same time. **7. a.** *Baseball.* A pitched ball counted against the batter, typically one swung at and missed, fouled off, or judged to have passed through the strike zone. **b.** A perfectly thrown ball. **8.** The knocking down of all the pins in bowling with the first bowl of a frame. **9.** *Geol.* The direction of a horizontal line in the plane of an inclined structural feature such as a rock bed or vein. **10.** A strickle.

**strike·bound** (strīk′bound′) *adj.* Immobilized, closed, or slowed down by a labor strike.

**strike·break·er** (strīk′brā′kər) *n.* One who works or provides an employer with workers during a strike : SCAB. **—strike′break′ing** *n.*

**strike·out** (strīk′out′) *n.* *Baseball.* An act or instance of striking out.

**strik·er** (strī′kər) *n.* **1.** One that strikes. **2.** An employee who is on strike against his or her employer. **3.** A device, as the clapper in a bell, that strikes. **4. a.** A harpoon. **b.** A harpooner. **5.** An enlisted person in training for a specified naval technical rating.

**strike zone** *n.* *Baseball.* The area over home plate through which a pitch must pass to be called a strike, defined as being between the batter's armpits and knees.

**strik·ing** (strī′kĭng) *adj.* Immediately or vividly impressive <a *striking* resemblance> **—strik′ing·ly** *adv.* **—strik′ing·ness** *n.*

**striking price** *n.* The price at which a put or call option may be exercised.

**string** (strĭng) *n.* [ME < OE *streng.*] **1.** A usu. fiber cord for fastening, tying, or lacing. **2.** Something shaped into a long thin line <*strings* of thin spaghetti> **3.** A set of objects threaded together <a *string* of pearls> **4.** A series of related events, acts, or items arranged or falling in a line <a *string* of defeats> **5.** *Computer Sci.* A set of data arranged in ascending or descending sequence according to a key within the data. **6.** *Informal.* A set of animals, esp. racehorses, belonging to a single owner : STABLE. **7.** A group of players constituting a ranked team within a team <made the first *string*> **8.** *Mus.* **a.** A cord stretched across the sounding board of an instrument that is struck, plucked, or bowed to produce tones. **b. strings.** Instruments having such strings, esp. the instruments of the violin family. **9. a.** A stringboard. **b.** A stringcourse. **10.** The balk line in billiards. **11.** *often* **strings.** A limiting, often hidden condition or proviso <an offer with no *strings* attached> *—v.* **strung** (strŭng), **string·ing, strings.** *—vt.* **1.** To fit or furnish with a string or strings <*string* a balalaika> **2.** To thread on a string <*string* beads> **3.** To arrange in a string or series. **4.** To fasten, tie, or hang with a string or strings. **5.** To stretch out : EXTEND <*string* a wire across a garden> **6.** To strip (vegetables) of strings. *—vi.* **1.** To form strings or become stringlike. **2.** To extend or progress in a line or succession. **—pull strings.** To use one's influence, often in secret, to obtain an advantage. **—string along. 1.** To keep (another) waiting or dangling. **2.** To cheat : deceive. **—string up.** *Informal.* To hang (someone).

**string bass** *n.* A double bass.

**string bean** *n.* **1.** A bushy or climbing plant, *Phaseolus vulgaris,* cultivated for its narrow, green, edible pods. **2.** The green pod of a bean cooked whole or broken into sections that retain the beans. **3.** *Slang.* A tall thin person.

**string·board** (strĭng′bôrd′, -bōrd′) *n.* A board running along the side of a staircase to support or cover the ends of the steps.

**string·course** (strĭng′kôrs′, -kōrs′) *n.* A horizontal molding set in the face of a building as a design element.

**stringed instrument** *n.* A musical instrument played by plucking, striking, or bowing taut strings.

**strin·gen·do** (strĭn-jĕn′dō) *adj. & adv.* [Ital. < *stringere,* to tighten < Lat.] *Mus.* Played with an accelerating tempo. —Used as a direction.

**strin·gent** (strĭn′jənt) *adj.* [Lat. *stringens, stringent-,* pr.part. of *stringere,* to bind tightly.] **1.** Imposing strict standards : SEVERE <*stringent* quality control procedures> **2.** Constricted : tight <a *stringent* time limit> **3.** Marked by scarcity of money, credit restrictions, or other financial strain <*stringent* governmental economic policies> **—strin′gen·cy** *n.* **—strin′gent·ly** *adv.*

**string·er** (strĭng′ər) *n.* **1.** One that strings. **2. a.** A long heavy horizontal timber used to connect or support other members. **b.** A stringboard. **3.** A lengthwise timber used to support rails. **4.** A member of a specified string or squad on a team <a first-*stringer*> **5.** A part-time or free-lance news correspondent for the electronic or print media.

**string·halt** (strĭng′hôlt′) *n.* [STRING, tendon (obs.) + HALT[2].] Lameness in a horse, accompanied by spasmodic movements in the hind legs.

**string quartet** *n.* **1.** A quartet of musicians playing stringed instruments, usu. including a first and second violinist, a violist, and a cellist. **2.** A musical composition for a string quartet.

**string tie** *n.* A narrow necktie, usu. tied in a bow.

**string·y** (strĭng′ē) *adj.* **-i·er, -i·est. 1.** Resembling, forming, or consisting of a string or strings. **2.** Slender and sinewy : WIRY. **—string′i·ly** *adv.* **—string′i·ness** *n.*

**strip[1]** (strĭp) *v.* **stripped, strip·ping, strips.** [ME *stripen* < OE *strīpan.*] *—vt.* **1. a.** To remove the clothing or covering from. **b.** To remove (clothing or covering). **2.** To deprive of honors, rank, or functions : DIVEST <*stripped* the generals of their rank> **3.** To remove all excess detail from <a news story *stripped* to the bare facts> **4.** To remove the leaves from the stalks of. —Used esp. of tobacco. **5.** To dismantle (something) piece by piece. **6.** To damage or break the threads or teeth of (e.g., a screw). **7.** To milk (a milk-giving creature). **8.** To rob : despoil. *—vi.* **1. a.** To undress completely. **b.** To perform a striptease. **2.** To fall away or be removed : PEEL. *—n.* A striptease.

**strip[2]** (strĭp) *n.* [Perh. alteration of STRIPE[1].] **1.** A long narrow piece, usu. of uniform width. **2.** A comic strip. **3.** An airstrip. *—vt.* **stripped, strip·ping, strips.** To cut or tear into strips.

**strip-crop·ping** (strĭp'krŏp'ĭng) *n.* The growing of a cultivated crop, as cotton, and a sod-forming crop, as alfalfa, in alternating strips following the contour of the land so as to minimize erosion.

**stripe[1]** (strīp) *n.* [Poss. < MDu. *stripe.*] **1. a.** A long narrow band distinguished, as by color or texture, from the surrounding material or surface. **b.** A fabric having such a band or bands. **c. stripes.** A garment of such fabric, esp. a prisoner's uniform. **2.** A strip of cloth or braid worn on a uniform to indicate rank, awards received, or length of service : CHEVRON. **3.** Sort : kind <"All Fascists are not of one mind, one *stripe*" —Lillian Hellman> —*vt.* **striped, strip·ing, stripes.** To mark with a stripe or stripes.

**stripe[2]** (strīp) *n.* [ME.] A stroke or blow, as with a whip.

**striped** (strīpt, strī'pĭd) *adj.* Having a stripe or stripes.

**striped bass** *n.* A food and game fish, *Roccus saxatilis,* of North American coastal waters, with dark longitudinal stripes along its sides.

**striped maple** *n.* The moosewood.

**strip·er** (strī'pər) *n.* **1.** *Slang.* A member of the armed forces or crew member of a commercial aircraft who wears stripes designating rank or length of service <a three-*striper*> **2.** *Informal.* A striped bass.

**strip·film** (strĭp'fĭlm') *n.* A filmstrip.

**strip·ling** (strĭp'lĭng) *n.* [ME.] An adolescent youth.

**strip mine** *n.* An open mine, esp. a coal mine, whose seams or outcrops run close to ground level and are exposed by the removal of topsoil and overburden. —*vt.* **strip-mine, -mined, -min·ing, -mines.** To mine (an ore) from a strip mine.

**strip·per** (strĭp'ər) *n.* **1.** One that strips. **2.** *Slang.* A striptease artist.

**strip poker** *n.* A poker game in which the losing players in each hand must remove an article of clothing.

**strip·tease** *also* **strip tease** (strĭp'tēz') *n.* An entertainment featuring a person who slowly removes clothing usu. to a musical accompaniment. **—strip'teas'er** *n.*

**strip·y** (strī'pē) *adj.* **-i·er, -i·est.** Resembling or marked with stripes.

**strive** (strīv) *vi.* **strove** (strōv), **striv·en** (strĭv'ən) or **strived, striv·ing, strives.** [ME *striven* < OFr. *estriver,* of Germanic orig.] **1.** To exert energy or effort. **2.** To struggle : contend. **—striv'er** *n.*

**strobe** (strōb) *n.* **1.** A stroboscope. **2.** A strobe light.

**strobe light** *n.* A flash lamp that produces high-intensity short-duration light pulses by electric discharge in a gas.

**stro·bi·la** (strō-bī'lə) *n., pl.* **-lae** (-lē') [NLat. < Gk. *strobilē,* twisted plug of lint < *strobilos,* pine cone.] A part or structure as the main body part of a tapeworm or the polyp stage in certain jellyfish, that buds to form a series of segments. **—stro·bi'lar** *adj.*

**stro·bi·la·ceous** (strō'bə-lā'shəs) *adj.* Of or like a strobile : CONE-LIKE.

**stro·bi·la·tion** (strō'bə-lā'shən) *n.* Asexual reproduction by division into body segments, as in tapeworms and jellyfish.

**stro·bile** (strō'bīl', -bəl) *also* **stro·bi·lus** (strō-bī'ləs) *n., pl.* **-biles** *or* **-bi·li** (-bī'lī') [NLat. *strobilus* < Gk. *strobilos,* pine cone.] A fruiting structure characterized by rows of overlapping scales, as a pine cone or the fruit of the hop.

**strob·o·scope** (strō'bə-skōp') *n.* [Gk. *strobos,* a whirling round + -SCOPE.] An instrument used to view, calibrate, balance, or otherwise adjust moving, rotating, or vibrating objects by making them appear stationary, esp. with pulsed illumination or mechanical devices that intermittently interrupt observation. **—stro'bo·scop'ic** (-skŏp'ĭk) *adj.* **—strob'o·scop'i·cal·ly** *adv.*

**stro·bo·tron** (strō'bə-trŏn') *n.* [STROBO(SCOPE) + -TRON.] A gas-filled cathode tube that produces bright flashes of light for a stroboscope.

**strode** (strōd) *v. p.t. of* STRIDE.

**stroke** (strōk) *n.* [ME.] **1.** An impact : strike <the heavy *stroke* of an ax> **2.** An act of striking. **3. a.** The striking of a bell or gong. **b.** The sound so produced. **c.** The time so indicated <the *stroke* of noon> **4.** A sudden occurrence having a powerful immediate effect for good or ill <a *stroke* of bad luck> **5. a.** The sudden severe onset of a malady, as apoplexy. **b.** Apoplexy. **6.** An inspired or effective idea or act <a *stroke* of genius> **7. a.** A single completed movement of the limbs and body, as in swimming or rowing. **b.** The rate or manner of executing such a movement. **8. a.** The member of a rowing crew who sits nearest the coxswain or the stern and sets the tempo for the oarsmen. **b.** The position this crew member occupies. **9. a.** A movement of the upper torso and arms for striking a ball, as in golf or tennis. **b.** The manner of executing such a movement. **10.** Any of a series of movements of a piston from one end of the limit of its motion to the other. **11. a.** A single mark made by a marking implement, as a pen. **b.** The act of making such a mark. **c.** A printed line in a graphic character resembling such a mark. **12.** A single deft touch, as in literary composition. **13.** A light caressing movement, as of the hand. —*vt.* **stroked, strok·ing, strokes.** **1. a.** To rub lightly with or as if with the hand or something held in the hand : CARESS. **b.** *Slang.* To flatter (another). **2.** To set the pace for (a rowing crew).

**stroll** (strōl) *v.* **strolled, stroll·ing, strolls.** [Prob. < dial. G. *strollen.*] —*vi.* To go for a leisurely walk <*stroll* on the beach> —*vt.* To walk through or on at a leisurely pace <*stroll* the beach at dusk> —*n.* A leisurely walk.

☆ ***syns:*** STROLL, AMBLE, MOSEY, RAMBLE, SAUNTER, WANDER *v.* *core meaning* : to walk at a leisurely pace <*strolled* through the park>

**stroll·er** (strō'lər) *n.* **1.** One that strolls. **2.** A strolling player. **3.** A vagabond. **4.** A light four-wheeled chair used for transporting small children.

**stro·ma** (strō'mə) *n., pl.* **-ma·ta** (-mə-tə) [NLat. < Lat., covering < Gk. *strōma* < *stornunai,* to spread.] The tissue framework of an anatomical structure, as an organ or gland. **—stro·mat'ic** (-măt'ĭk) *adj.*

**stro·mat·o·lite** (strō-măt'l-īt') *n.* [Gk. *strōma, strōmat-,* bed covering + -LITE.] A sedimentary fossil composed of laminated layers of algal origin. **—stro·mat'o·lit'ic** (-măt'l-ĭt'ĭk) *adj.*

**strong** (strông) *adj.* **-er, -est.** [ME < OE *strang.*] **1.** Possessing great physical strength. **2.** Being in good or sound health : ROBUST. **3.** Financially sound. **4.** Having force of character, will, morality, or intelligence. **5.** Possessing or exhibiting ability or achievement in a given field <*strong* in science> **6.** Capable of enduring : SOLID <a *strong* foundation> **7.** Capable of being defended <the army's *strong* right flank> **8.** Having a specified number of units or members <50 soldiers *strong*>> **9.** Having force of motion or action <a *strong* current> **10. a.** Persuasive, effective, and cogent <a *strong defensive* argument> **b.** Forceful and pointed : EMPHATIC <a *strong* protest> **11.** Extreme : drastic <took *strong* measures> **12.** Capable of exerting authority effectively <*strong* management> **13.** Having force of conviction or feeling <*strong* faith> **14.** Intense <*strong* emotions> **15. a.** Having an intense effect on the senses <a *strong* odor> **b.** Containing a high percentage of alcohol <*strong* punch> **16.** Having a high degree of saturation <a *strong* color> **17.** Designating those verbs in Germanic languages that form a past tense other than by means of a dental suffix; e.g., *fly, flew; sing, sang.* —*adv.* In a strong, powerful, or vigorous way : FORCEFULLY. **—strong'ly** *adv.*

**strong-arm** (strông'ärm') *Informal.* —*adj.* Using physical force or coercion <*strong-arm* enforcement tactics> —*vt.* **-armed, -arm·ing, -arms.** To use physical force or coercion against.

**strong·box** (strông'bŏks') *n.* A stout box or safe in which valuables are deposited.

**strong force** *n. Physics.* Strong interaction.

**strong·hold** (strông'hōld') *n.* **1.** A fortress. **2.** An area dominated or occupied by a special group.

**strong interaction** *n. Physics.* Fundamental interaction between elementary particles causing protons and neutrons to bind together in the atomic nucleus.

**strong-mind·ed** (strông'mīn'dĭd) *adj.* Having a strong will. **—strong'-mind'ed·ly** *adv.* **—strong'-mind'ed·ness** *n.*

**strong room** *n.* A secure fireproof room designed for the safekeeping of money or valuables.

**strong suit** *n.* Long suit.

**stron·gyle** *also* **stron·gyl** (strŏn'jĭl', -jəl) *n.* [NLat. *Strongylus,* type genus < Gk. *strongulos,* compact.] Any of various nematode worms of the family Strongylidae, often parasitic in the gastrointestinal tract of mammals, esp. horses.

**stron·gy·lo·sis** (strŏn'jə-lō'sĭs) *n.* Infestation with strongyles.

**stron·ti·an·ite** (strŏn'chē-ə-nīt', -tē-ə-nīt') *n.* [*Strontian,* var. of STRONTIUM + -ITE.] A gray to yellowish-green strontium ore, essentially $SrCO_3$.

**stron·ti·um** (strŏn'chē-əm, -tē-əm) *n.* [After *Strontian,* Scotland.] *Symbol* **Sr** A soft, silvery, easily oxidized metallic element used in pyrotechnic compounds and various alloys; atomic number 38; atomic weight 87.62. **—stron'tic** (-tĭk) *adj.*

**strontium 90** *n.* The strontium isotope with mass 90, having a half-life of 28 years, used for its high-energy beta emission in nuclear electric power sources and constituting a radiation hazard in fallout.

**strop** (strŏp) *n.* [ME *stroppe,* band of leather < MLG *strop* < Lat. *stroppus* < Gk. *strophion.*] A flexible leather or canvas strip for sharpening a razor. —*vt.* **stropped, strop·ping, strops.** To sharpen (a razor) on a strop.

**stro·phan·thin** (strō-făn'thĭn) *n.* [< NLat. *Strophanthus,* genus of tropical trees or vines : Gk. *strophos,* twisted + Gk. *anthos,* flower.] A toxic glycoside or mix of glycosides, used as a cardiac tonic.

**stro·phe** (strō'fē) *n.* [Gk. *strophē,* movement of the chorus < *strephein,* to turn.] **1. a.** A stanza, esp. the first of a pair of stanzas of alternating form on which the structure of a poem is based. **b.** A rhythmic system constituting a section of a poem, typically consisting of a series of asymmetric lines. **2.** The first division of the triad constituting a section of a Pindaric ode. **3. a.** The movement of the chorus in classical Greek drama while turning from one side of the orchestra to the other. **b.** The part of a choral ode sung while this movement is executed. **—stro'phic** (strō'fĭk, strŏf'ĭk) *adj.*

**stro·phoid** (strō'foid') *n.* [Gk. *strophos,* twisted (< *strephein,* to twist) + -OID.] A plane curve generated by a point that maintains a distance from the y-axis along a straight line equal to the y-intercept.

**stroph·u·lus** (strŏf′yə-ləs) *n.* [NLat. < Gk. *strophos*, twisted cord < *strephein*, to turn.] A disease, esp. common among children, sometimes associated with intestinal disturbances and marked by a papular skin eruption.

**strove** (strōv) *v. var. p.t. of* STRIVE.

**struck** (strŭk) *adj.* [P.part. of STRIKE.] Affected by or shut down due to a labor strike.

**struck jury** *n. Law.* A jury selected from an original panel of 48 members from which each party strikes off names until the list is reduced to 12.

**struck measure** *n.* A dry measure having the contents leveled off and not heaped.

**struc·tur·al** (strŭk′chər-əl) *adj.* **1.** Of, pertaining to, having, or marked by structure. **2.** Used in or needed for construction. **3.** *Geol.* Of or pertaining to the structure of rocks and other aspects of the earth's crust. **4.** *Biol.* Of or pertaining to organic structure : MORPHOLOGICAL. **—struc′tur·al·ly** *adv.*

**structural formula** *n.* A chemical formula representing the configuration of atoms and bonds in a molecule.

**structural gene** *n.* A gene determining the amino acid sequence of a protein.

**struc·tur·al·ize** (strŭk′chər-ə-līz′) *vt.* **-ized, -iz·ing, -iz·es.** To incorporate into a structure. **—struc′tur·al·i·za′tion** *n.*

**structural steel** *n.* Steel shaped for construction purposes.

**struc·ture** (strŭk′chər) *n.* [ME < OFr. < Lat. *structura* < *struere*, to construct.] **1.** Something made up of a number of parts held or put together in a specific way. **2.** The manner in which parts are arranged or combined to form a whole. **3.** Interrelation of parts in a complex entity. **4.** Relatively intricate or extensive organization. **5.** Something constructed, esp. a building or part. *—vt.* **-tured, -tur·ing, -tures.** To give form or arrangement to.

**struc·tured** (strŭk′chərd) *adj.* **1.** Highly organized <a *structured* society> **2.** *Psychol.* Having a limited number of correct or nearly correct answers. —Used of a test.

**stru·del** (strōōd′l, shtrōōd′l) *n.* [G. < MHG, whirlpool.] A pastry of fruit or cheese rolled up in a thin sheet of dough and baked.

**strug·gle** (strŭg′əl) *v.* **-gled, -gling, -gles.** [ME *struglen.*] *—vi.* **1.** To exert muscular energy, as in opposition to a material force or mass. **2.** To be vigorously involved with a task, problem, or undertaking. **3.** To make a strenuous effort : STRIVE <*struggling* to be civil> **4. a.** To contend against. **b.** To compete with. **5.** To progress or penetrate with difficulty. *—vt.* To move or place with an effort <*struggle* a large suitcase into a car> *—n.* **1.** An act of struggling. **2.** Strenuous effort. **3.** Combat : strife. **—strug′gler** *n.* **—strug′gling·ly** *adv.*

**strum** (strŭm) *v.* **strummed, strum·ming, strums.** [Perh. blend of STRING and THRUM¹.] *—vt.* To play idly on or as if on (a stringed musical instrument) by plucking the strings with the fingers. *—vi.* To play an instrument by strumming. **—strum** *n.* **—strum′mer** *n.*

**stru·ma** (strōō′mə) *n., pl.* **-mae** (-mē′) *or* **-mas.** [Lat.] **1.** *Pathol.* **a.** Scrofula. **b.** Goiter. **2.** *Bot.* A cushionlike swelling at the base of a moss capsule. **—stru·mat′ic** (-măt′ĭk), **stru′mose′** (-mōs′), **stru′·mous** (-məs) *adj.*

**strum·pet** (strŭm′pĭt) *n.* [ME.] A whore.

**strung** (strŭng) *v. p.t. & p.p. of* STRING.

**strung-out** (strŭng′out′) *adj. Slang.* **1. a.** Addicted to a drug. **b.** Stupefied from ingestion of a drug. **2.** Emotionally or physically debilitated from or as if from long-term drug use.

**strut** (strŭt) *v.* **strut·ted, strut·ting, struts.** [ME *strouten*, to stand out < OE *strūtian*, to stand out stiffly.] *—vi.* To walk pompously : SWAGGER. *—vt.* To brace with a supporting bar or rod. *—n.* **1.** A pompous, self-important gait. **2.** A bar or rod used to brace a structure against forces applied from the side. **—strut′ter** *n.* **—strut′ting·ly** *adv.*

**stru·thi·ous** (strōō′thē-əs, -thē-əs) *adj.* [< LLat. *struthio*, ostrich < Gk. *strouthion* < *strouthos.*] Of, relating to, or resembling the ostrich or a related bird.

**strych·nine** (strĭk′nīn′, -nĭn, -nēn′) *n.* [Fr. < NLat., *Strychnos*, genus of tropical trees and vines < Lat. *strychnos*, a kind of nightshade < Gk. *strukhnos.*] A highly toxic white crystalline alkaloid, $C_{21}H_{22}N_2O_2$, derived from nux vomica and related plants, used as a poison for rodents and other pests and medicinally as a stimulant for the central nervous system.

**strych·nin·ism** (strĭk′nĭ-nĭz′əm, -nī-, -nē-) *n.* Chronic strychnine poisoning.

**stub** (stŭb) *n.* [ME *stubbe* < OE *stybb.*] **1.** The usu. short end remaining after something has been removed or used up <a *stub* of chalk> **2. a.** The part of a check or receipt retained as a record. **b.** The part of a ticket returned as a voucher of payment. *—vt.* **stubbed, stub·bing, stubs. 1.** To pull up (weeds) by the roots. **2.** To clear (a field) of weeds. **3.** To strike (one's toe or foot) against an object. **4.** To snuff out (a cigarette or cigar butt) by crushing.

**stub·ble** (stŭb′əl) *n.* [ME *stuble* < OFr. *estuble* < Lat. *stupla*, var. of *stipula*, straw.] **1.** The short stiff stalks of a grain or hay crop remaining on a field after the crop has been harvested. **2.** Something like stubble, esp. a short stiff growth of beard. **—stub′bly** *adj.*

**stub·born** (stŭb′ərn) *adj.* [ME *stuborn.*] **1.** Firmly, often unduly determined in will or purpose : OBSTINATE. **2.** Marked by perseverance : PERSISTENT. **3. a.** Difficult to handle or work : RESISTANT <*stubborn* soil> **b.** Difficult to cure or alleviate <a *stubborn* cough> **—stub′born·ly** *adv.* **—stub′born·ness** *n.*

**stub·by** (stŭb′ē) *adj.* **-bi·er, -bi·est. 1.** Of the nature of a stub : short and thick. **2.** Covered with or made of stubs. **3.** Short and bristly. **—stub′bi·ly** *adv.* **—stub′bi·ness** *n.*

**stuc·co** (stŭk′ō) *n., pl.* **-coes** *or* **-cos.** [Ital., of Germanic orig.] **1.** A durable finish for exterior walls, applied wet and usu. consisting of cement, sand, and lime. **2.** A fine plaster for interior wall ornamentation, as moldings. **3.** A plaster or cement finish for interior walls. *—vt.* **-coed, -co·ing, -coes** *or* **-cos.** To finish or decorate with stucco.

**stuck** (stŭk) *v. p.t. & p.p. of* STICK.

**stuck-up** (stŭk′ŭp′) *adj. Informal.* Snobbish : conceited.

**stud¹** (stŭd) *n.* [ME *stode* < OE *studu.*] **1.** An upright post in the framework of a wall for supporting sheets of lath or wallboard. **2.** A small knob, nail head, or rivet fixed in and slightly projecting from a surface. **3. a.** A small ornamental button mounted on a short post for insertion through an eyelet, as on a dress shirt. **b.** A buttonlike, usu. pierced earring. **4.** A protruding pin or peg in machinery. **5.** A metal crosspiece used as a brace in a link, as in a chain cable. *—vt.* **stud·ded, stud·ding, studs. 1.** To provide with or construct with a stud or studs. **2.** To set with a stud or studs <*stud* a bracelet with diamonds> **3.** To be dotted about on, esp. ornamentally <Poppies *studded* the field.>

**stud²** (stŭd) *n.* [ME *stod* < OE *stōd.*] **1. a.** A group of animals, esp. horses, kept for breeding. **b.** A stable or farm where they are kept. **2.** A male animal, as a stallion, kept for breeding. **3.** Stud poker. **4.** *Slang.* A virile man. **—at stud.** Available or offered for breeding.

**stud·book** (stŭd′bŏŏk′) *n.* A book registering the pedigrees of thoroughbred animals, esp. horses.

**stud·ding** (stŭd′ĭng) *n.* **1. a.** The wood framework of a wall or partition. **b.** Lumber cut for studs. **2.** Something with which a surface is studded.

**stud·ding·sail** (stŭn′səl, stŭd′ĭng-sāl′) *also* **stun·sail** *or* **stun·s'l** (stŭn′səl) *n.* [Orig. unknown.] *Naut.* A narrow rectangular sail set from extensions of the yards of square-rigged ships.

**studdingsail**

**stu·dent** (stōōd′nt, styōōd′-) *n.* [ME < Lat. *studens*, pr.part. *studēre*, to study.] **1.** One who attends a school, college, or university. **2. a.** One who makes a study of something <a *student* of medicine> **b.** An attentive observer <a *student* of world affairs>

**student teacher** *n.* A college student who practices teaching under supervision.

**student union** *n.* A building on a college campus with facilities for social and organizational activities.

**stud·fish** (stŭd′fĭsh′) *n., pl.* **studfish** *or* **-fish·es.** A small brightly colored freshwater fish, *Fundulus catenatus* or *F. stellifer*, of the southeastern United States.

**stud·horse** *also* **stud horse** (stŭd′hôrs′) *n.* A stallion.

**stud·ied** (stŭd′ēd) *adj.* **1. a.** Prepared or considered with care <a *studied* effect> **b.** Devoid of spontaneity : CONTRIVED <a *studied* smile> **2.** Knowledgeable. **—stud′ied·ly** *adv.* **—stud′ied·ness** *n.*

**stu·di·o** (stōō′dē-ō, styōō′-) *n., pl.* **-os.** [Ital. < Lat. *studium*, eagerness.] **1.** An artist's workroom. **2.** A photographer's establishment. **3.** An establishment where an art is studied or taught <a dance *studio*> **4. a.** A room or building for motion-picture, television, or radio productions. **b.** A room or building where tapes and records are recorded.

**studio apartment** *n.* A small apartment with one main living space, a small kitchen, and a bathroom.

**studio couch** *n.* A couch that can be made to serve as a double bed by sliding the frame of a cot from beneath it.

**stu·di·ous** (stōō′dē-əs, styōō′-) *adj.* [ME < Lat. *studiosus* < *studium*, eagerness.] **1.** Devoted to study. **2.** Diligent : earnest <a *studious* effort> **3.** Giving or evincing careful attention : HEEDFUL <*studious* of one's appearance> **4.** Deliberate. **5.** Conducive to study. **—stu′di·ous·ly** *adv.* **—stu′di·ous·ness** *n.*

☆ ***syns:*** STUDIOUS, BOOKISH, SCHOLARLY *adj. core meaning* : devoted to study <a *studious*, yet popular young person>

ă **pat** ā **pay** âr **care** ä **father** ĕ **pet** ē **be** hw **which** ĭ **pit** ī **tie** îr **pier** ŏ **pot** ō **toe** ô **paw, for** oi **noise** ŏŏ **took**

**stud poker** *n.* Poker in which the first round of cards, and often the last, is dealt face down and the others face up.
**stud·work** (stŭd'wûrk') *n.* **1.** Work ornamented or covered with studs. **2.** The framework supporting a wall or partition.
**stud·y** (stŭd'ē) *n., pl.* **-ies.** [ME *studie* < OFr. *estudie* < Lat. *studium* < *studēre,* to study.] **1.** The act or process of studying: the pursuit of knowledge, as by reading, observation, or research. **2.** Attentive scrutiny. **3.** A branch of knowledge. **4. studies.** A branch or department of learning <graduate *studies*> **5. a.** A work (e.g., a thesis) resulting from studious endeavor. **b.** A literary work on a subject. **c.** A preliminary sketch, as for a work of art. **6.** A musical composition designed as a technical exercise. **7.** Mental absorption <in a deep *study*> **8.** A room intended or equipped for studying or writing. **9. a.** One who memorizes something, esp. a performer with reference to his or her ability to memorize a part. **b.** Memorization of a theatrical part. —*v.* **stud·ied, stud·y·ing, stud·ies.** —*vt.* **1.** To apply one's mind purposefully to the acquisition of knowledge or understanding of (a subject) <*study* a science> **2.** To read carefully <*study* a text> **3.** To memorize. **4.** To take (a course) at a school. **5.** To inquire into: INVESTIGATE <*study* the mood of the country> **6.** To examine closely: SCRUTINIZE <*study* a blueprint> **7.** To give careful thought to: CONTEMPLATE <*study* a move in chess> —*vi.* **1.** To apply oneself to learning, esp. by reading. **2.** To pursue a course of study. **3.** To ponder: reflect.
**study hall** *n.* **1.** A schoolroom reserved for study. **2.** A school period set aside for study.
**stuff** (stŭf) *n.* [ME < OFr. *estoffe* < *estoffer,* to equip.] **1.** The material out of which something is made: SUBSTANCE. **2.** The most central and material part: ESSENCE <the *stuff* heroes are made of> **3.** Unspecified material <Put your *stuff* over there.> **4.** *Informal.* Household or personal articles as a whole: BELONGINGS. **5.** Worthless material. **6. a.** *Slang.* Specific talk or actions <Don't give me that *stuff* about being bored.> **b.** Special capability <athletes who showed their *stuff* and won the game> **7.** *Chiefly Brit.* Woven material, esp. woolens. **8.** *Slang.* Money: cash. **9.** *Slang.* A habit-forming drug, esp. heroin. —*v.* **stuffed, stuff·ing, stuffs.** —*vt.* **1. a.** To pack tightly: CRAM <*stuff* a bag with trash> **b.** To block (a passage): PLUG <*stuff* a leak with plaster> **2. a.** To fill with an appropriate stuffing <*stuff* a pillow> **b.** To fill (an animal skin) to restore the natural form. **3.** To cram with food <*stuffed* myself at dinner> **4.** To fill (the mind) <Their heads are *stuffed* with silly ideas.> **5.** To put fraudulent votes into (a ballot box). **6.** To apply a preservative and softening agent to (leather). —*vi.* To overeat: gorge. **—stuff'er** *n.*
**stuffed derma** *n.* DERMA[2].
**stuffed shirt** *n. Informal.* A pretentious, pompous person.
**stuff·ing** (stŭf'ĭng) *n.* Material used to stuff or fill, esp.: **a.** Padding for cushions and upholstered furniture. **b.** Food put in the cavity of meat or vegetables.
**stuff shot** *n. Basketball.* A dunk shot.
**stuff·y** (stŭf'ē) *adj.* **-i·er, -i·est.** **1.** Lacking sufficient ventilation. **2.** Having the respiratory passages blocked <a *stuffy* nose> **3.** *Informal.* **a.** Dull <a *stuffy* party> **b.** Strait-laced. **—stuff'i·ly** *adv.* **—stuff'i·ness** *n.*
**stull** (stŭl) *n.* [Prob. < G. *Stollen,* support < OHG *stollo.*] **1.** A prop supporting the roof of a mine opening. **2.** A platform braced against the sides of a working area in a mine.
**stul·ti·fy** (stŭl'tə-fī') *vt.* **-fied, -fy·ing, -fies.** [LLat. *stultificare,* to make foolish: Lat. *stultus,* foolish + Lat. *facere,* to make.] **1.** To make useless or ineffectual: CRIPPLE. **2. a.** To cause to appear stupid, inconsistent, or ridiculous. **b.** To make torpid or dull. **3.** *Law.* To allege or prove insane and so not legally responsible. **—stul'ti·fi·ca'tion** *n.* **—stul'ti·fi'er** *n.*
**stum** (stŭm) *n.* [Du. *stom* < *stom,* dumb < MDu.] **1.** MUST[3]. **2.** Vapid wine renewed by an admixture of stum. —*vt.* **stummed, stum·ming, stums.** To ferment (vapid wine) by adding stum.
**stum·ble** (stŭm'bəl) *v.* **-bled, -bling, -bles.** [ME *stumblen,* prob. of Scand. orig.] —*vi.* **1. a.** To miss one's step and trip in walking or running. **b.** To act or speak clumsily, unsteadily, or falteringly: FLOUNDER <*stumble* through a lecture> **2.** To make a mistake: BLUNDER. **3.** To fall into evil ways: ERR. **4.** To come upon unexpectedly or accidentally <*stumbled* on the evidence> —*vt.* To cause to stumble. —*n.* **1.** An act of stumbling. **2.** A mistake, blunder, or sin. **—stum'bler** *n.* **—stum'bling·ly** *adv.*
**stum·ble·bum** (stŭm'bəl-bŭm') *n. Slang.* **1.** A blundering or inept person. **2.** A punch-drunk or second-rate prizefighter.
**stumbling block** *n.* An obstacle or impediment, esp. one that hinders progress.
**stump** (stŭmp) *n.* [ME *stumpe* < MLG *stump.*] **1.** The part of a tree trunk left protruding from the ground after the tree has fallen or been felled. **2.** A part, as of a branch, limb, or tooth, remaining after the main part has been cut away, broken off, or worn down. **3. a. stumps.** *Informal.* The legs. **b.** A prosthetic leg. **4.** A short, thickset person. **5.** A heavy footfall. **6.** A place or an occasion used for political or campaign oratory. **7.** A short, pointed roll of leather or paper or wad of rubber for rubbing on a charcoal or pencil drawing to shade or soften it. **8.** One of the three upright sticks in a wicket. —*vt.* **stumped, stump·ing, stumps.** **1.** To reduce to a stump. **2.** To clear stumps from (land). **3.** To stub (a toe or foot). **4.** To traverse (a district) making political speeches. **5.** To shade (a drawing) with a stump. **6.** *Informal.* To puzzle: baffle <a problem that *stumped* me> **—stump'er** *n.* **—stump'i·ness** *n.* **—stump'y** *adj.*
**stump·age** (stŭm'pĭj) *n.* **1.** Standing timber regarded as a commodity. **2.** The value of standing timber. **3.** The right to cut standing timber.
**stun** (stŭn) *vt.* **stunned, stun·ning, stuns.** [ME *stonen* < OFr. *estoner* < VLat. **extonare*: Lat. *ex-* (intensive) + Lat. *tonare,* to thunder.] **1.** To render senseless, as by a blow or a loud noise. **2.** To stupefy, as with emotional impact. **—stun** *n.*
**stung** (stŭng) *v. p.t. & p.p. of* STING.
**stunk** (stŭngk) *v. p.p. & var. p.t. of* STINK.
**stun·ner** (stŭn'ər) *n.* **1.** One that stuns. **2.** *Informal.* An exceptionally good-looking person.
**stun·ning** (stŭn'ĭng) *adj.* **1.** Causing or capable of causing loss of consciousness or emotional shock. **2.** Strikingly attractive in appearance. **—stun'ning·ly** *adv.*
**stun·sail** or **stun·s'l** (stŭn'səl) *n. vars. of* STUDDINGSAIL.
**stunt[1]** (stŭnt) *vt.* **stunt·ed, stunt·ing, stunts.** [Prob. < dial. *stont,* short in duration, of Scand. orig.] To check the growth or development of. —*n.* **1.** One that stunts. **2.** One that is stunted. **3.** A disease of plants that causes dwarfing. **—stunt'ed·ness** *n.*
**stunt[2]** (stŭnt) *n.* [Orig. unknown.] **1.** A feat displaying unusual strength, skill, or daring. **2.** Something unusual done for publicity. —*vi.* **stunt·ed, stunt·ing, stunts.** To perform a stunt.
**stunt box** *n.* An electronic device designed to control the nonprinting functions of a communications terminal.
**stunt man** *n.* A man who substitutes for an actor in scenes requiring physical prowess or involving physical risk.
**stunt person** *n.* A stunt man or a stunt woman.
**stunt woman** *n.* A woman who substitutes for an actress in scenes requiring physical prowess or involving physical risk.
**stu·pa** (sto͞o'pə) *n.* [Skt. *stūpaḥ,* summit.] TOPE[3].
**stupe** (sto͞op, styo͞op) *n.* [ME < Lat. *stuppa,* tow < Gk. *stuppē.*] A hot medicated compress.
**stu·pe·fa·cient** (sto͞o'pə-fā'shənt, styo͞o'-) *adj.* [Lat. *stupefaciens, stupefacient-,* pr.part. of *stupefacere,* to stupefy.] Inducing stupor. —*n.* A drug that induces stupor.
**stu·pe·fac·tion** (sto͞o'pə-făk'shən, styo͞o'-) *n.* **1.** The act of stupefying or the state of being stupefied. **2.** Astonishment or consternation.
**stu·pe·fac·tive** (sto͞o'pə-făk'tĭv, styo͞o'-) *adj. & n.* Stupefacient.
**stu·pe·fy** (sto͞o'pə-fī', styo͞o'-) *vt.* **-fied, -fy·ing, -fies.** [OFr. *stupefier* < Lat. *stupefacere*: *stupēre,* to be stunned + *facere,* to make.] **1.** To put into a stupor. **2.** To astonish: amaze. **—stu'pe·fi'er** *n.*
**stu·pen·dous** (sto͞o-pĕn'dəs, styo͞o-) *adj.* [Lat. *stupendus,* gerund. of *stupēre,* to be stunned.] **1.** Astounding in volume, degree, or size: TREMENDOUS. **2.** Causing astonishment. **—stu·pen'dous·ly** *adv.* **—stu·pen'dous·ness** *n.*
**stu·pid** (sto͞o'pĭd, styo͞o'-) *adj.* **-er, -est.** [Fr. *stupide* < Lat. *stupidus* < *stupēre,* to be stunned.] **1. a.** Slow to comprehend. **b.** Lacking intelligence. **2.** Dazed: stunned. **3.** *Informal.* Pointless: worthless <a *stupid* idea> —*n. Informal.* A stupid person. **—stu'pid·ly** *adv.* **—stu'pid·ness** *n.*
☆ ***syns:*** STUPID, DENSE, DUMB, OBTUSE, THICK *adj. core meaning*: lacking intelligence <a *stupid* person><a *stupid* interpretation> ***ant***: intelligent
**stu·pid·i·ty** (sto͞o-pĭd'ĭ-tē, styo͞o-) *n., pl.* **-ties.** **1.** The quality or state of being stupid. **2.** A stupid act, remark, or idea.
**stu·por** (sto͞o'pər, styo͞o'-) *n.* [ME < Lat. *stupēre,* to be stunned.] **1.** Reduced sensibility: TORPOR. **2.** Mental confusion: DAZE. **—stu'por·ous** *adj.*
**stur·dy** (stûr'dē) *adj.* **-di·er, -di·est.** [ME < OFr. *estourdi,* stunned, p.part. of *estourir,* to stun.] **1.** Strongly built: SUBSTANTIAL. **2.** Stalwart: robust. **3.** Vigorous: lusty. —*n.* The gid. **—stur'di·ly** *adv.* **—stur'di·ness** *n.*
**stur·geon** (stûr'jən) *n.* [ME < OFr. *estourgeon,* of Germanic orig.] Any of various large freshwater and marine fishes of the family Acipenseridae, of the Northern Hemisphere, having edible flesh and roe valued as a source of caviar.
**Sturm und Drang** (shto͝orm' o͝ont drăng') *n.* [G., storm and stress, after *Sturm und Drang,* a drama written by Friedrich Maximilian von Klinger (1752–1831).] **1.** A late 18th-cent. German romantic literary movement, the works of which usu. depicted the impulsive man struggling against conventional society. **2.** Turmoil: ferment.
**stut·ter** (stŭt'ər) *v.* **-tered, -ter·ing, -ters.** [Freq. of dial. E. *stut,* to stutter < ME *stutten.*] —*vi.* To speak with a spasmodic hesitation, prolongation, or repetition of sounds. —*vt.* To utter or say with or as if with a stutter. —*n.* The act or habit of stuttering. **—stut'ter·er** *n.* **—stut'ter·ing·ly** *adv.*
**St. Vitus' dance** *also* **St. Vi·tus's dance** (sānt' vī'tə-sĭz) *n.* [After *St. Vitus,* a 3rd-cent. martyr.] *Pathol.* Chorea.
**sty[1]** (stī) *n., pl.* **sties.** [ME < OE *stig.*] **1.** An enclosure for pigs. **2.** A

very dirty place. —*v.* **stied, sty·ing, sties.** —*vt.* & *vi.* To confine to or live in a sty.

**sty²** (stī) *n., pl.* **sties.** [Short for dial. *styan* < ME **styan* < OE *stīgend* < *stīgan*, to rise.] Inflammation of one or more sebaceous glands of an eyelid.

**styg·i·an** *also* **Styg·i·an** (stĭj′ē-ən) *adj.* [Lat. *Stygius* < Gk. *Stugios* < *Stux*, Styx.] **1.** Of or relating to the river Styx. **2. a.** Dark and gloomy. **b.** Infernal : hellish.

**styl–** *pref. var. of* STYLO-.

**sty·lar** (stī′lər, -lär′) *adj.* **1.** Of, relating to, or resembling a stylus. **2.** *Biol.* Of or relating to a style.

**sty·late** (stī′lāt′) *adj.* Having a style or styles.

**style** (stīl) *n.* [ME < OFr. < Lat. *stilus.*] **1.** The manner in which something is done, expressed, or performed <a *style* of writing> **2.** The combination of distinctive literary and artistic features of expression, execution, or performance characterizing a particular person, group, school, or era. **3.** Sort : type <a modern *style* of furniture> **4.** A quality of imagination and individuality expressed in one's actions and tastes. **5. a.** A comfortable, elegant mode of existence <lived in *style*> **b.** A specified mode of living <a relaxed, informal *style*> **6. a.** The fashion of the moment, esp. of dress : VOGUE <out of *style*> **b.** A particular fashion <the *style* of the 1940's> **7.** A customary manner of presenting printed material, including usage, punctuation, spelling, typography, and arrangement. **8.** Form of address : TITLE. **9.** A slender, pointed writing instrument used by the ancients on wax tablets. **10.** An implement for etching or engraving. **11.** A phonograph needle. **12.** The gnomon of a sundial. **13.** *Bot.* The usu. slender part of a pistil, rising from the ovary and tipped by the stigma. **14.** *Zool.* A slender, tubular, or bristlelike process. **15.** *Obs.* A pen. **16.** STYLET 2. —*vt.* **styled, styl·ing, styles. 1.** To call or name : DESIGNATE. **2.** To make consistent with rules of style <*style* a book> **3.** To give a particular style to <*style* hair> **—styl′er** *n.*

☆ ***syns:*** STYLE, FASHION, MANNER, TONE, VEIN *n. core meaning* : a distinctive way of expressing oneself <an unconventional journalistic *style*>

**style book** *n.* A manual giving rules and examples of usage, punctuation, and typography, used esp. in the preparation of copy for publication.

**sty·let** (stī-lĕt′, stī′lĭt) *n.* [Fr. < Ital. *stiletto*, stiletto. —see STILETTO.] **1.** A slender, pointed instrument or weapon, as a stiletto. **2.** A slender surgical probe. **3.** *Zool.* A small, stiff, needlelike process in some invertebrates.

**sty·li** (stī′lī′) *n. var. pl. of* STYLUS.

**styli–** *pref. var. of* STYLO-.

**sty·li·form** (stī′lə-fôrm′) *adj.* Shaped like a style.

**styl·ish** (stī′lĭsh) *adj.* Conforming to the current fashion : MODISH. **—styl′ish·ly** *adv.* **—styl′ish·ness** *n.*

**styl·ist** (stī′lĭst) *n.* **1.** A speaker or writer who cultivates an artful literary style. **2.** A designer of or consultant on styles in decorating, dress, or beauty.

**sty·lis·tic** (stī-lĭs′tĭk) *adj.* Of or pertaining to style, esp. literary style. **—sty·lis′ti·cal·ly** *adv.*

**sty·lite** (stī′līt′) *n.* [LGk. *stulitēs* < Gk. *stulos*, pillar.] An early Christian ascetic who lived unsheltered on the top of a high pillar. **—sty·lit′ic** (-lĭt′ĭk) *adj.* **—sty′lit·ism** (stī′lī′tĭz-əm) *n.*

**styl·ize** (stī′līz′) *vt.* **-ized, -iz·ing, -iz·es. 1.** To restrict or conform to a particular style. **2.** To represent conventionally : CONVENTIONALIZE <"An air of fastidious, *stylized* melancholy" —Elizabeth Bowen> **—styl′i·za′tion** *n.* **—styl′iz′er** *n.*

**stylo–** *or* **styli–** *or* **styl–** *pref.* [< Lat. *stilus*, stake, stem, style.] Style <*stylo*graph>

**sty·lo·bate** (stī′lə-bāt′) *n.* [Lat. *stylobata* < Gk. *stulobatēs* : *stulos*, pillar + *bainein*, to walk.] The immediate foundation of a row of classical columns.

**sty·log·ra·phy** (stī-lŏg′rə-fē) *n.* The art or a method of etching, engraving, or writing with a style. **—sty′lo·graph′ic** (-lə-grăf′ĭk), **sty′lo·graph′i·cal** *adj.*

**sty·loid** (stī′loid′) *adj.* Slender and pointed.

**sty·lo·lite** (stī′lə-līt′) *n.* [Gk. *stulos*, pillar + -LITE.] A small columnar rock development in limestone and other calcareous rocks that is at right angles to the bed and has striated sides.

**sty·lo·po·di·um** (stī′lə-pō′dē-əm) *n., pl.* **-di·a** (-dē-ə). An enlargement at the base of the style of certain flowers.

**sty·lus** (stī′ləs) *n., pl.* **-lus·es** *or* **-li** (-lī′) [Lat. *stilus.*] **1.** A sharp, pointed instrument for writing, marking, or engraving. **2.** A phonograph needle. **3.** A sharp, pointed tool for cutting record grooves.

**sty·mie** *also* **sty·my** (stī′mē) [Orig. unknown.] —*vt.* **-mied, -mie·ing, -mies** *also* **-mied, -my·ing, -mies.** To block, impede, or thwart. —*n.* **1.** An obstacle : obstruction. **2.** A situation in golf in which an opponent's ball obstructs the line of play of one's own ball on the putting green.

**styp·sis** (stĭp′sĭs) *n.* [LLat. < Gk. *stupsis* < *stuphein*, to contract.] The action or application of a styptic.

**styp·tic** (stĭp′tĭk) *adj.* [ME *stiptik* < LLat. *stypticus* < Gk. *stuptikos* < *stuphein*, to contract.] **1.** Contracting the tissues or blood vessels : ASTRINGENT. **2.** Tending to check bleeding. —*n.* A styptic drug or substance. **—styp·tic′i·ty** (-tĭs′ĭ-tē) *n.*

**styptic pencil** *n.* A short medicated stick, often of alum, applied to a slight cut to check bleeding.

**sty·rax** (stī′răks) *n.* STORAX 2, 3.

**sty·rene** (stī′rēn′) *n.* [< Lat. *styrax*, storax. —see STORAX.] A colorless oily liquid, $C_8H_8$, the monomer for polystyrene.

**Sty·ro·foam** (stī′rə-fōm′). A trademark for a light, resilient polystyrene plastic.

**Styx** (stĭks) *n.* [Lat. < Gk. *Stux.*] *Gk. Myth.* A river in Hades, across which the souls of the dead were ferried.

**su·a·ble** (sōō′ə-bəl) *adj.* Liable to a court suit. **—su′a·bil′i·ty** *n.*

**sua·sion** (swā′zhən) *n.* [ME < Lat. *suasio* < *suadēre*, to persuade.] Persuasion <moral *suasion*> **—sua′sive** (-sĭv, -zĭv) *adj.* **—sua′sive·ly** *adv.* **—sua′sive·ness** *n.*

**suave** (swäv) *adj.* [OFr., agreeable < Lat. *suavis.*] Smoothly or blandly gracious : URBANE. **—suave′ly** *adv.* **—suave′ness, suav′i·ty** (swä′vĭ-tē) *n.*

**†sub¹** (sŭb) *n. Informal.* **1.** A submarine. **2.** *Regional.* HERO 5.

**sub²** (sŭb) *Informal.* —*n.* A substitute. —*vi.* **subbed, sub·bing, subs.** To act as a substitute.

**sub–** *pref.* [Lat. < *sub*, under, below.] **1.** Below : under : beneath <*sub*soil> **2. a.** Subordinate : secondary <*sub*plot> **b.** Subdivision <*sub*region> **3.** Less than completely or normally : nearly : almost <*sub*human>

**sub·ab·dom·i·nal** (sŭb′ăb-dŏm′ə-nəl) *adj.* Located or occurring below the abdomen.

**sub·a·cute** (sŭb′ə-kyōōt′) *adj.* Somewhat acute. —Used of a disease. **—sub′a·cute′ly** *adv.*

**sub·ad·dress** (sŭb′ə-drĕs′) *n. Computer Sci.* A section of an input/output device accessible through an order code.

**sub·aer·i·al** (sŭb′âr′ē-əl) *adj.* Located or occurring on or near the surface of the earth.

**sub·al·pine** (sŭb′ăl′pīn′) *adj.* **1.** Of or relating to regions at or near the foot of the Alps. **2.** Of, designating, or growing or living in mountainous regions just below the timberline.

**sub·al·tern** (sŭb′ôl′tərn, sŭb′əl-tûrn′) *adj.* [LLat. *subalternus* : Lat. *sub-*, below + Lat. *alternus*, alternate < *alter*, other.] **1.** Lower in position or rank : SECONDARY. **2.** *Chiefly Brit.* Holding a military rank just below that of captain. **3.** *Logic.* In the relation of a particular proposition to a universal with the same subject, predicate, and quality. —*n.* **1.** A subordinate. **2.** *Chiefly Brit.* A subaltern officer. **3.** *Logic.* A subaltern proposition.

**sub·al·ter·nate** (sŭb′ôl′tər-nĭt) *adj.* **1.** Subordinate. **2.** Arranged in an alternating pattern but tending to become opposite. —Used of leaves. **—sub·al′ter·na′tion** *n.*

**sub·ant·arc·tic** (sŭb′ănt-ärk′tĭk, -är′tĭk) *adj.* Of or like regions just north of the Antarctic Circle.

**sub·a·pi·cal** (sŭb′ā′pĭ-kəl) *adj.* Located below or near an apex. **—sub′a′pi·cal·ly** *adv.*

**sub·a·que·ous** (sŭb′ā′kwē-əs, -ăk′wē-) *adj.* **1.** Formed or adapted for underwater operation : SUBMARINE. **2.** Found or taking place under water.

**sub·arc·tic** (sŭb′ärk′tĭk, -är′tĭk) *adj.* Of or resembling regions just south of the Arctic Circle.

**sub·ar·id** (sŭb′ăr′ĭd) *adj.* Somewhat arid.

**sub·a·tom·ic** (sŭb′ə-tŏm′ĭk) *adj.* **1.** Of or relating to the constituents of the atom. **2.** Having dimensions or participating in reactions characteristic of the constituents of the atom.

**sub·au·di·tion** (sŭb′ô-dĭsh′ən) *n.* [LLat. *subauditio*, supplying a missing word < *subaudire*, to supply an omitted word : Lat. *sub-*, below + Lat. *audire*, to hear.] **1.** The act of understanding and mentally supplying a word or thought that has been implied but not expressed. **2.** A word or thought supplied by subaudition.

**sub·base** (sŭb′bās′) *n.* The lowermost front strip or molding of a baseboard.

**sub·base·ment** (sŭb′bās′mənt) *n.* A story or floor beneath a main basement of a building.

**sub·bass** (sŭb′bās′) *n. Mus.* A pedal stop on an organ that produces the lowest tones, having 16 or 32 feet.

**sub·cal·i·ber** (sŭb′kăl′ə-bər) *adj.* **1.** Being of a smaller caliber than the barrel of the gun from which it was fired. —Used of a projectile. **2.** Of or relating to a subcaliber projectile.

**sub·car·ri·er** (sŭb′kăr′ē-ər) *n.* A section of a transmitted wave used to modify the information-carrying section of the wave.

**sub·car·ti·lag·i·nous** (sŭb′kär-tə-lăj′ə-nəs) *adj.* **1.** Located beneath a cartilage. **2.** Partly cartilaginous.

**sub·ce·les·tial** (sŭb′sĭ-lĕs′chəl) *adj.* **1.** Lower than celestial : TERRESTRIAL. **2.** Mundane.

**sub·chas·er** (sŭb′chā′sər) *n.* A submarine chaser.

**sub·class** (sŭb′klăs′) *n.* **1.** A subdivision of a class. **2.** A taxonomic category ranking between a class and an order.

**sub·cla·vi·an** (sŭb′klā′vē-ən) *adj.* [NLat. *subclavius* : Lat. *sub-*, below + Lat. *clavis*, key.] *Anat.* **1.** Situated beneath the clavicle. **2.** Of or pertaining to a subclavian part. **3.** Of or pertaining to the subclavian artery or vein. —*n.* A subclavian structure, as a nerve or muscle.

**subclavian artery** *n.* A short part of a major artery originating under the clavicle and continuous with the axillary artery extending to the upper extremities or forelimbs.

**subclavian vein** *n.* A part of a major vein of the upper extremities or forelimbs continuous with the axillary vein and situated beneath the clavicle.

**sub·cli·max** (sŭb'klī'măks') *n.* A stage in the ecological succession of a plant or animal community immediately preceding a climax, and often persisting because of the effects of fire, flood, or other conditions. **—sub'cli·mac'tic** (-klī-măk'tĭk) *adj.*

**sub·clin·i·cal** (sŭb-klĭn'ĭ-kəl) *adj.* Of or relating to a disease or condition in which no characteristic symptoms are manifested. **—sub·clin'i·cal·ly** *adv.*

**sub·com·mit·tee** (sŭb'kə-mĭt'ē) *n.* A subordinate committee composed of members appointed from the main committee.

**sub·com·pact** (sŭb'kŏm'păkt') *n.* A car smaller in size than a compact.

**sub·con·scious** (sŭb'kŏn'shəs) *adj.* **1.** Occurring without conscious perception on the part of the individual <a *subconscious* desire to die> **2.** Not wholly conscious. **—sub'con'scious** *n.* **—sub'con'scious·ly** *adv.* **—sub'con'scious·ness** *n.*

**sub·con·ti·nent** (sŭb'kŏn'tə-nənt) *n.* A large land mass, as India, that is separate to some degree but still part of a continent. **—sub'con'ti·nent'al** *adj.*

**sub·con·tract** (sŭb'kŏn'trăkt') *n.* A contract that assigns some of the obligations of a prior contract to another party. *—v.* (sŭb'kŏn'-trăkt', sŭb'kən-trăkt') **-tract·ed, -tract·ing, -tracts.** *—vt.* To make a subcontract for. *—vi.* To make a subcontract.

**sub·con·trac·tor** (sŭb'kŏn'trăk'tər, sŭb'kən-trăk'tər) *n.* One that enters into a subcontract and assumes some of the obligations of the primary contractor.

**sub·cool** (səb'kōōl') *vt.* & *vi.* **-cooled, -cool·ing, -cools.** To supercool or to become super-cooled.

**sub·cor·tex** (sŭb'kôr'tĕks) *n., pl.* **-ti·ces** (-tĭ-sēz'). The portion of the brain immediately below the cerebral cortex. **—sub'cor'ti·cal** (-tĭ-kəl) *adj.* **—sub'cor'ti·cal·ly** *adv.*

**sub·cul·ture** (sŭb'kŭl'chər) *n.* **1.** One culture of microorganisms derived from another. **2.** A cultural subgroup differentiated by status, ethnic background, residence, religion, or other factors that functionally unify the group and act collectively on each member. **—sub'cul'tur·al** *adj.*

**sub·cu·ta·ne·ous** (sŭb'kyōō-tā'nē-əs) *adj.* Located or introduced just beneath the skin <*subcutaneous* nerves><*subcutaneous* injections> **—sub'cu·ta'ne·ous·ly** *adv.*

**sub·cu·tis** (sŭb-kyōō'tĭs) *n.* A layer of connective tissue beneath the dermis.

**sub·dea·con** (sŭb'dē'kən) *n.* [ME < LLat. *subdiaconus,* partial transl. of LGk. *hupodiakonos* : Gk. *hupo-,* below + Gk. *diakonos,* attendant.] **1.** A cleric ranking just below a deacon. **2.** A cleric who acts as assistant to the deacon at High Mass.

**sub·deb** (sŭb'dĕb') *n. Informal.* A subdebutante.

**sub·deb·u·tante** (sŭb'dĕb'yə-tänt') *n.* **1.** A teen-ager approaching her debut. **2.** A young woman in her middle teens.

**sub·di·ac·o·nate** (sŭb'dī-ăk'ə-nĭt) *n.* [LLat. *subdiaconatus* < *subdiaconus,* subdeacon.] The office, order, or rank of subdeacon. **—sub'di·ac'o·nal** *adj.*

**sub·di·vide** (sŭb'dĭ-vīd', sŭb'dĭ-vīd') *v.* **-vid·ed, -vid·ing, -vides.** [ME *subdividen* < LLat. *subdividere* : Lat. *sub-,* secondary + Lat. *dividere,* to divide.] *—vt.* **1.** To divide a part or parts of into smaller parts. **2.** To divide into a number of parts, esp. to divide (land) into lots. *—vi.* To form into subdivisions. **—sub'di·vid'a·ble** *adj.* **—sub'di·vid'er** *n.*

**sub·di·vi·sion** (sŭb'dĭ-vĭzh'ən, sŭb'dĭ-vĭzh'ən) *n.* **1.** The act or process of subdividing. **2.** A subdivided part. **3.** An area consisting of subdivided lots. **—sub'di·vi'sion·al** *adj.*

**sub·dom·i·nant** (sŭb'dŏm'ə-nənt) *n. Mus.* The fourth tone of a diatonic scale, next below the dominant.

**sub·duc·tion** (səb-dŭk'shən) *n.* [LLat. *subductio,* act of taking away < Lat. *subducere,* to withdraw : *sub-,* under + *ducere,* to lead.] A geological process in which one edge of crustal plate descends below another. **—sub·duct'** *v.* **(-duct·ed, -duct·ing, -ducts).**

**sub·due** (səb-dōō', -dyōō') *vt.* **-dued, -du·ing, -dues.** [ME *subduen* < OFr. *suduire,* to seduce < Lat. *subducere,* to withdraw : *sub-,* away + *ducere,* to lead.] **1.** To conquer and subjugate : VANQUISH. **2.** To quiet or bring under control by physical force or persuasion. **3.** To make less intense or prominent : TONE DOWN. **4.** To bring (land) under cultivation. **—sub·du'a·ble** *adj.* **—sub·du'er** *n.*

**sub·dur·al** (səb-dŏŏr'əl, -dyŏŏr'-) *adj.* Located or occurring beneath the dura mater.

**sub·e·qua·to·ri·al** (sŭb'ē-kwə-tôr'ē-əl, -tōr'-, -ĕk-wə-) *adj.* Belonging to a region adjacent to the equatorial area.

**su·ber·ic acid** (sōō-bĕr'ĭk) *n.* [Fr. *subérique* (< Lat. *suber,* cork) + ACID.] A colorless crystalline dibasic acid, $C_8H_{14}O_4$, used in drug synthesis and plastics manufacture.

**su·ber·in** (sōō'bər-ĭn) *n.* [Fr. *subérine* < Lat. *suber,* cork.] A waxy waterproof substance present in the cell walls of cork tissue in plants.

**su·ber·i·za·tion** (sōō'bər-ĭ-zā'shən) *n.* Formation of suberin in the walls of plant cells, thus converting them into cork tissue.

**su·ber·ize** (sōō'bə-rīz') *vt.* **-ized, -iz·ing, -iz·es.** [< Lat. *suber,* cork.] To cause to undergo suberization.

**su·ber·ose** (sōō'bə-rōs') *also* **su·ber·ous** (-bər-əs) *adj.* [NLat. *suberosus* < Lat. *suber,* cork.] Of, relating to, or resembling cork or cork tissue.

**sub·fam·i·ly** (sŭb'făm'ə-lē) *n., pl.* **-lies. 1.** *Biol.* A taxonomic category ranking between a family and a genus. **2.** A division of languages below a family and above a branch.

**sub·field** (sŭb'fēld') *n.* **1.** A mathematical field that is a subject of another field. **2.** A subdivision of a field of study.

**sub·freez·ing** (sŭb-frē'zĭng) *adj.* Below freezing.

**sub·ge·nus** (sŭb'jē'nəs) *n., pl.* **-gen·e·ra** (-jĕn'ər-ə). *Biol.* An occas. used taxonomic category ranking between a genus and a species. **—sub'ge·ner'ic** (-jə-nĕr'ĭk) *adj.*

**sub·gla·cial** (sŭb'glā'shəl) *adj.* Deposited or formed beneath a glacier. **—sub'gla'cial·ly** *adv.*

**sub·group** (sŭb'grōōp') *n.* **1.** A distinct group within a group. **2.** *Math.* A nonempty subset of a group. **3.** A subordinate group. **4.** *Biol.* A taxonomic division of an order.

**sub·gum** (sŭb'gŭm') *n.* [Cantonese *shap kam,* mixture.] A Chinese dish of mixed vegetables.

**sub·har·mon·ic** (sŭb'här-mŏn'ĭk) *adj.* Of, relating to, or being a wave with a frequency that is a fraction of a fundamental.

**sub·head** (sŭb'hĕd') *n.* **1.** *also* **sub·head·ing** (-hĕd'ĭng). The heading of a subdivision of a printed text. **2.** A subordinate heading.

**sub·hu·man** (sŭb'hyōō'mən) *adj.* **1.** Below the human race in evolutionary development. **2.** Not fully human. **—sub'hu'man** *n.*

**sub·in·dex** (sŭb'ĭn'dĕks) *n., pl.* **-di·ces** (-dĭ-sēz'). *Math.* A subscript.

**sub·in·feu·date** (sŭb'ĭn-fyōō'dāt') *also* **sub·in·feud** (-fyōōd') *vt.* **-dat·ed, -dat·ing, -dates** *also* **-feud·ed, -feud·ing, -feuds.** To lease (lands) by subinfeudation.

**sub·in·feu·da·tion** (sŭb'ĭn-fyōō-dā'shən) *n.* **1.** The sublease of a portion of a feudal estate by a vassal to a subtenant who pays fealty to the vassal. **2.** The lands leased in subinfeudation. **—sub'in·feu'da·to'ry** (-fyōō'də-tôr'ē, -tōr'ē) *adj.*

**sub·ir·ri·gate** (sŭb'ĭr'ĭ-gāt') *vt.* **-gat·ed, -gat·ing, -gates.** To irrigate from beneath, as by underground pipes. **—sub'ir·ri·ga'tion** *n.*

**su·bi·to** (sōō'bē-tō') *adv.* [Ital. < Lat., unexpectedly < *subire,* to come secretly : *sub-,* below + *ire,* to come.] *Mus.* Quickly : suddenly. —Used as a direction.

**sub·ja·cent** (sŭb'jā'sənt) *adj.* [Lat. *subjacens, subjacent-,* pr.part. of *subjacēre,* to lie beneath : *sub-,* beneath + *jacēre,* to lie] **1.** Located beneath or below : UNDERLYING. **2.** Lying at a lower level but not directly beneath. **—sub'ja'cen·cy** *n.*

**sub·ject** (sŭb'jĭkt) *adj.* [ME < OFr. *subget* < Lat. *subjectus* < *subicere,* to subject : *sub-,* below + *jacere,* to throw.] **1.** Being under the authority, control, or power of another <*subject* to the law> **2.** Disposed : prone <*subject* to headaches> **3.** Liable to incur or receive <comments *subject* to misinterpretation> **4.** Contingent <*subject* to approval> *—n.* **1.** One under the authority, control, or power of another, esp. one owing allegiance to a government or ruler <a *subject* of the British Crown> **2. a.** A topic of interest. **b.** The primary theme of an artistic work. **c.** A theme of a musical composition, esp. a fugue. **3.** A course or area of study. **4.** A basis for action : CAUSE. **5. a.** One that experiences or is subjected to something <the *subject* of ridicule> **b.** One that is the object of clinical or pathological study. **6.** A word or phrase in a sentence that denotes the doer of the action, the receiver of the action in passive constructions, or that which is described or identified. **7.** *Logic.* The term of a proposition about which something is affirmed or denied. **8.** *Philos.* **a.** The essential nature or substance of something as distinguished from its attributes. **b.** The mind or thinking part as distinguished from the object of thought. *—vt.* (səb-jĕkt') **-ject·ed, -ject·ing, -jects. 1.** To submit for consideration. **2.** To submit to the authority of. **3.** To render liable to something : EXPOSE <*subjected* to infection> **4.** To cause to experience <*subjected* to torture> **5.** To subjugate : subdue. **—sub·jec'tion** (səb-jĕk'shən) *n.*

**sub·jec·tive** (səb-jĕk'tĭv) *adj.* **1. a.** Of, produced by, or resulting from an individual's mind or state of mind <*subjective* opinions> **b.** Particular to a given individual : PERSONAL <a *subjective* experience> **2.** Moodily introspective. **3.** Existing only in the mind : ILLUSORY. **4.** *Psychol.* Existing only within the experiencer's mind and incapable of external verification. **5.** *Med.* Designating a symptom or condition perceived by the patient and not by the examiner. **6.** Expressing or bringing into prominence the individuality of the artist or author. **7.** Designating the nominative case. **8.** Having to do with the real nature of something : ESSENTIAL. **—sub·jec'tive·ly** *adv.* **—sub·jec'tive·ness, sub'jec·tiv'i·ty** (sŭb'jĕk-tĭv'ĭ-tē) *n.*

**sub·jec·tiv·ism** (səb-jĕk'tə-vĭz'əm) *n.* **1.** The quality of being subjective. **2. a.** The doctrine that all knowledge is restricted to the conscious self and its sensory states. **b.** A theory or doctrine that emphasizes the subjective elements in experience. **3.** The theory that individual conscience is the only valid standard of moral judgment. **—sub·jec'tiv·ist** *n.* **—sub·jec'tiv·is'tic** *adj.*

**sub·jec·tiv·ize** (səb-jĕk'tə-vīz') *vt.* **-ized, -iz·ing, -iz·es.** To make subjective. **—sub·jec'tiv·i·za'tion** *n.*

**subject matter** *n.* Matter treated in a written work or speech : THEME.

**sub·join** (səb-join') *vt.* **-joined, -join·ing, -joins.** [OFr. *subjoindre* < Lat. *subjungere* : *sub-*, under + *jungere*, to join.] To add at the end : APPEND.

**sub·join·der** (səb-join'dər) *n.* [SUBJOIN + *-der*, as in *rejoinder*.] Something subjoined.

**sub ju·di·ce** (sŭb jōō'dĭ-sē', yōō'dĭ-kā') *adv.* [Lat.] *Law.* Under judicial deliberation.

**sub·ju·gate** (sŭb'jə-gāt') *vt.* **-gat·ed, -gat·ing, -gates.** [ME *subjugaten* < Lat. *subjugare* : *sub-*, under + *jugum*, yoke.] **1.** To bring under the dominion of another : SUBDUE. **2.** To make subservient : ENSLAVE. **—sub'ju·ga'tion** *n.* **—sub'ju·ga'tor** *n.*

▲ word history: The word *subjugate* is borrowed from Latin *subjugare*, literally "to bring under the yoke." An ancient Latin custom required that when an army was completely defeated, the survivors were forced to pass under a symbolic yoke made of two upright spears and a third used as a crossbar. The Romans regarded subjugation as the worst possible humiliation.

**sub·junc·tion** (səb-jŭngk'shən) *n.* [LLat. *subjunctio* < *subjungere*, to subjoin.] **1.** The act of subjoining or the state of being subjoined. **2.** Something subjoined.

**sub·junc·tive** (səb-jŭngk'tĭv) *adj.* [LLat. *subjunctivus* < Lat. *subjungere*, to subordinate, subjoin.] Designating a verb form or set of forms used in English to express a contingent or hypothetical action. —*n.* **1.** The subjunctive mood. **2.** A subjunctive construction.

**sub·king·dom** (sŭb'kĭng'dəm) *n. Biol.* A former taxonomic category constituting a major division of a kingdom.

**sub·lease** (sŭb'lēs') *vt.* **-leased, -leas·ing, -leas·es. 1.** To sublet (property). **2.** To rent (property) under a sublease. —*n.* (sŭb'lēs'). A lease of property granted by a lessee.

**sub·let** (sŭb'lĕt') *vt.* **-let, -let·ting, -lets. 1.** To rent (property one holds by lease) to another. **2.** To subcontract (work). —*n.* (sŭb'lĕt'). *Informal.* Property, esp. an apartment, rented by a tenant to another party.

**sub·le·thal** (sŭb-lē'thəl) *adj.* Being less than lethal <a *sublethal* dosage> **—sub·le'thal·ly** *adv.*

**sub·lev·el** (sŭb'lĕv-əl) *n.* A level beneath or lower than another level.

**sub·li·mate** (sŭb'lə-māt') *v.* **-mat·ed, -mat·ing, -mates.** [Lat. *sublimare, sublimat-*, to raise < *sublimis*, uplifted.] —*vt.* **1.** *Chem.* To cause (a solid or a gas) to change state without becoming a liquid. **2.** *Psychol.* To modify the natural expression of (an instinctual impulse) in a socially acceptable manner. —*vi.* To transform directly from the solid to the gaseous state or from the gaseous to the solid state without becoming a liquid. **—sub'li·ma'tion** *n.*

**sub·lime** (sə-blīm') *adj.* [Lat. *sublimis*, uplifted.] **1.** Noble : majestic. **2. a.** Of high spiritual, moral, or intellectual worth. **b.** Not to be excelled : SUPREME. **3.** Inspiring awe : IMPRESSIVE. **4.** *Obs.* Lofty in bearing or appearance : HAUGHTY. **5.** *Archaic.* Raised aloft : set high. —*n.* **1.** Something sublime. **2.** An ultimate example. —*v.* **-limed, -lim·ing, -limes.** —*vt.* **1.** To render sublime. **2.** *Chem.* To cause to sublimate. —*vi. Chem.* To sublimate. **—sub·lime'ly** *adv.* **—sub·lime'ness, sub·lim'i·ty** (sə-blĭm'ĭ-tē) *n.*

**sub·lim·i·nal** (sŭb-lĭm'ə-nəl) *adj.* [SUB- + Lat. *limen, limin-*, threshold.] *Psychol.* **1.** Below the threshold of conscious perception. —Used of stimuli. **2.** Inadequate to produce conscious awareness. **—sub·lim'i·nal·ly** *adv.*

**sub·lin·gual** (sŭb'lĭng'gwəl) *adj.* Situated beneath or on the underside of the tongue.

**sub·lit·to·ral** (sŭb'lĭt'ər-əl) *adj.* **1.** Near the seashore. **2.** Shallow and lying between the shoreline and the edge of the continental shelf or ranging in depth to about 50 fathoms.

**sub·lu·na·ry** (sŭb'lōō'nə-rē, sŭb'lōō-nĕr'ē) *also* **sub·lu·nar** (-lōō'nər) *adj.* [LLat. *sublunaris* : Lat. *sub-*, beneath + Lat. *luna*, moon.] **1.** Situated beneath the moon. **2.** Of this world : EARTHLY.

**sub·lux·a·tion** (sŭb'lŭk-sā'shən) *n.* Incomplete dislocation of a bone in a joint.

**sub·ma·chine gun** (sŭb'mə-shēn') *n.* A lightweight automatic or semiautomatic gun fired from the shoulder or hip.

**sub·man·dib·u·lar** (sŭb'măn-dĭb'yə-lər) *adj.* Submaxillary.

**sub·mar·gin·al** (sŭb'mär'jə-nəl) *adj.* **1.** Beneath a margin. **2.** Of low productivity : INFERTILE.

**sub·ma·rine** (sŭb'mə-rēn', sŭb'mə-rēn') *adj.* Being or operating beneath the surface of the water : UNDERSEA. —*n.* **1.** A ship capable of operating submerged. **2.** *Slang.* HERO 5.

**submarine chaser** *n.* A small fast boat equipped to pursue and attack submarines.

**sub·ma·rin·er** (sŭb'mə-rē'nər, sŭb'măr'ə-nər) *n.* A crew member of a submarine.

**sub·max·il·lar·y** (sŭb'măk'sə-lĕr'ē) *adj.* **1.** Of or pertaining to the lower jaw. **2.** Situated beneath the maxilla. —*n., pl.* **-ies.** An anatomical part, as a gland or nerve, situated beneath the maxilla.

**sub·me·di·ant** (sŭb'mē'dē-ənt) *n. Mus.* The sixth tone of a diatonic scale.

**sub·merge** (səb-mûrj') *v.* **-merged, -merg·ing, -merg·es.** [Lat. *submergere* : *sub-*, under + *mergere*, to plunge.] —*vt.* **1.** To place under water. **2.** To cover with water : INUNDATE <fields *submerged* by flood waters> **3.** To hide from view : OBSCURE <meanings *submerged* in legal jargon> —*vi.* To go under or as if under water. **—sub·mer'gence** *n.*

**sub·mer·gi·ble** (səb-mûr'jə-bəl) *adj.* Able to be plunged into or to remain under water. **—sub·mer'gi·bil'i·ty** *n.*

**sub·merse** (səb-mûrs') *vt.* **-mersed, -mers·ing, -mers·es.** [Lat. *submergere, submers-*, to submerge.] To submerge. **—sub·mer'sion** (-mûr'zhən, -shən) *n.*

**sub·mers·i·ble** (səb-mûr'sə-bəl) *adj.* Submergible. —*n.* A vessel capable of operating or remaining under water.

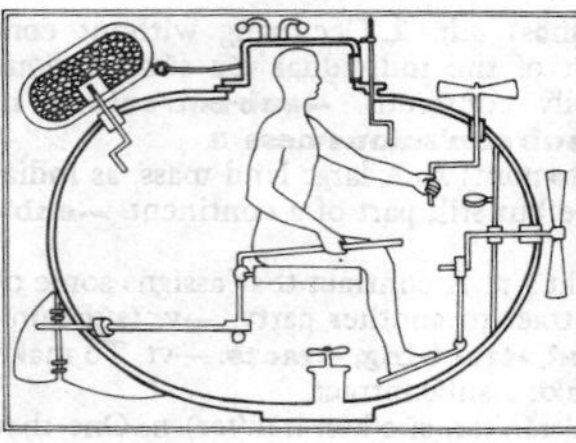

**submersible**
*A drawing of the original submersible, "Turtle"*

**sub·mi·cro·scop·ic** (sŭb'mī-krə-skŏp'ĭk) *adj.* Too small to be resolved by an optical microscope. **—sub'mi·cro·scop'i·cal·ly** *adv.*

**sub·min·i·a·ture** (sŭb'mĭn'ē-ə-chŏŏr', -chər) *adj.* Smaller than miniature : extremely small.

**sub·min·i·a·tur·ize** (sŭb'mĭn'ē-ə-chə-rīz') *vt.* **-ized, -iz·ing, -iz·es.** To make subminiature <*subminiaturized* electronic equipment> **—sub'min'i·a·tur·i·za'tion** *n.*

**sub·miss** (səb-mĭs') *adj.* [Lat. *submissus* < p.part. of *submittere*, to set under. —see SUBMIT.] *Archaic.* **1.** Submissive. **2.** Soft in tone.

**sub·mis·sion** (səb-mĭsh'ən) *n.* **1. a.** The act of submitting to the power of another. **b.** The state of having submitted. **2.** Meek compliance. **3. a.** The act of submitting something for consideration. **b.** Something thus submitted.

**sub·mis·sive** (səb-mĭs'ĭv) *adj.* Meekly compliant : DOCILE. **—sub·mis'sive·ly** *adv.* **—sub·mis'sive·ness** *n.*

**sub·mit** (səb-mĭt') *v.* **-mit·ted, -mit·ting, -mits.** [ME *submitten* < Lat. *submittere*, to set under : *sub-*, under + *mittere*, to cause to go.] —*vt.* **1.** To surrender or yield (oneself) to the will or authority of another. **2.** To subject to a condition or process. **3.** To commit (something) to the consideration or judgment of another. **4.** To offer as a proposition or contention <I *submit* that the answer is incorrect.> —*vi.* **1.** To yield to the opinion or authority of another : GIVE IN. **2.** To allow oneself to be subjected : ACQUIESCE. **—sub·mit'tal** *n.* **—sub·mit'ter** *n.*

**sub·mon·tane** (sŭb'mŏn'tān', -mŏn-tān') *adj.* [LLat. *submontanus* : Lat. *sub-*, under + Lat. *montanus*, mountainous < *mons*, mountain.] Located under or at the base of a mountain or mountain range.

**sub·mu·co·sa** (sŭb'myōō-kō'sə) *n.* A layer of loose connective tissue beneath a mucous membrane. **—sub'mu·co'sal** *adj.* **—sub'mu·co'sal·ly** *adv.*

**sub·mul·ti·ple** (sŭb'mŭl'tə-pəl) *n.* A number that is an exact divisor of another number.

**sub·net** (sŭb'nĕt') *n.* A system of interconnections within a communications system that allows the component parts to communicate directly with each other.

**sub·nor·mal** (sŭb'nôr'məl) *adj.* Less than normal : below the average <*subnormal* temperatures> —*n.* One who is subnormal, as in intelligence or coordination. **—sub'nor·mal'i·ty** (-nôr-măl'ĭ-tē) *n.* **—sub·nor'mal·ly** *adv.*

**sub·o·ce·an·ic** (sŭb'ō-shē-ăn'ĭk) *adj.* Formed, situated, or taking place beneath the ocean or the ocean bed.

**sub·or·bit·al** (sŭb-ôr'bĭ-tl) *adj.* **1.** Of or for less than one orbit of the earth <*suborbital* space flights> **2.** Located below the eye or its orbit.

**sub·or·der** (sŭb'ôr'dər) *n.* **1.** *Biol.* A taxonomic category ranking after an order and before a family. **2.** A subdivision of an order <a *suborder* of deacons>

**sub·or·di·nate** (sə-bôr'də-nĭt) *adj.* [ME *subordinat* < Med. Lat. *subordinatus*, p.part. of *subordinare*, to put in a lower rank : Lat. *sub-*, below + Lat. *ordinare*, to set in order < *ordo*, order.] **1.** Belonging to a class or rank lower than another. **2.** Subject to the control or authority of another. —*n.* One subordinate to another. —*vt.* (sə-bôr'də-nāt') **-nat·ed, -nat·ing, -nates. 1.** To put in a lower rank or class. **2.** To make subservient. **—sub·or'di·nate·ly** *adv.* **—sub·or'di·nate·ness, sub·or'di·na'tion** *n.* **—sub·or'di·na'tive** *adj.*

☆ ***syns:*** SUBORDINATE, INFERIOR, JUNIOR, SUBALTERN, UNDERLING *n. core meaning* : one of a lower class or rank than another <supervisors who recognized the worth of their *subordinates*> ***ant:*** superior

**subordinate clause** *n.* A dependent clause.

**subordinate conjunction** *n.* A conjunction, as *that, who, which,* and *where,* that introduces a dependent clause.

**sub·or·di·na·tion·ism** (sə-bôr'də-nā'shə-nĭz'əm) *n.* The doctrine that the second and third persons of the Trinity are subordinate to the first person. **—sub·or'di·na'tion·ist** *n.*

**sub·orn** (sə-bôrn') *vt.* **-orned, -orn·ing, -orns.** [Lat. *subornare* : *sub-,* secretly + *ornare,* to equip.] **1. a.** To induce to commit an unlawful act. **b.** To induce to commit perjury <*suborn* a witness> **2.** To procure (perjured testimony). **—sub'or·na'tion** (sŭb'ôr-nā'shən) *n.* **—sub·orn'er** *n.*

**sub·ox·ide** (sŭb'ŏk'sīd') *n.* An oxide containing a relatively small amount of oxygen.

**sub·phy·lum** (sŭb'fī'ləm) *n., pl.* **-la** (-lə). *Biol.* A taxonomic category ranking between a phylum and a class.

**sub·plot** (sŭb'plŏt') *n.* A subordinate literary plot.

**sub·poe·na** (sə-pē'nə) *n.* [ME *suppena* < Lat. *sub poena,* under a penalty.] A legal writ requiring the recipient to appear in court to testify. *—vt.* **-naed, -na·ing, -nas.** To serve with a subpoena.

**sub·prin·ci·pal** (sŭb'prĭn'sə-pəl) *n.* **1.** An assistant school principal. **2.** An auxiliary or bracing rafter in a frame. **3.** *Mus.* An open diapason subbass in an organ.

**sub·pro·gram** (sŭb'prō'grăm, -grəm) *n.* A computer program contained within another program that operates semi-independently of the encasing program.

**sub·re·gion** (sŭb'rē'jən) *n.* A subdivision of a region, esp. of an ecological region. **—sub're'gion·al** *adj.*

**sub·rep·tion** (sŭb-rĕp'shən) *n.* [Lat. *subreptio,* theft < *subrepere,* to take away secretly : *sub-,* secretly + *rapere,* to take away.] **1.** Calculated misrepresentation through concealment of the facts. **2.** An inference drawn from such a misrepresentation. **—sub'rep·ti'tious** (sŭb'rĕp-tĭsh'əs) *adj.*

**sub·ring** (sŭb'rĭng') *n.* A subset of a mathematical ring that is itself a ring.

**sub·ro·gate** (sŭb'rō-gāt') *vt.* **-gat·ed, -gat·ing, -gates.** [Lat. *subrogare, subrogat-* : *sub-,* instead of + *rogare,* to ask.] To substitute (one person) for another. **—sub'ro·ga'tion** (-gā'shən) *n.*

**sub ro·sa** (sŭb'rō'zə) *adv.* [Lat., under the rose (from the practice of hanging a rose over a meeting as a symbol of secrecy).] In secret : PRIVATELY <negotiations held *sub rosa*>

**sub-ro·sa** (səb-rō'zä) *adj.* Intended to be secret, private, or confidential <*sub-rosa* negotiations>

**sub·rou·tine** (sŭb'rōō-tēn') *n.* A set of computer instructions that performs a specific task for a main routine, requiring direction back to the proper place in the main routine on completion of the task.

**sub·scap·u·lar** (sŭb'skăp'yə-lər) *adj. Anat.* Situated below or on the underside of the scapula. *—n.* A subscapular part, as an artery or nerve.

**sub·scribe** (səb-skrīb') *v.* **-scribed, -scrib·ing, -scribes.** [ME *subscriben* < Lat. *subscribere* : *sub-,* under + *scribere,* to write.] *—vt.* **1.** To sign (one's name) at the end of a document. **2.** To sign one's name to in attestation, testimony, or consent <*subscribe* a will> **3.** To pledge or contribute (a sum of money). *—vi.* **1.** To sign one's name. **2.** To affix one's signature to a document as a witness or to show consent. **3.** To express concurrence or approval : ASSENT <*subscribe* to a particular belief> **4.** To promise to contribute money <*subscribe* to a charity> **5.** To contract to receive and pay for a certain number of issues of a publication. **—sub·scrib'er** *n.*

**sub·script** (sŭb'skrĭpt') *adj.* [Lat. *subscriptus,* p.part. of *subscribere,* to subscribe.] Written beneath. *—n.* A distinguishing character or symbol written directly beneath or next to and slightly below a letter or number.

**sub·scrip·tion** (səb-skrĭp'shən) *n.* **1.** A written signature, as on a document. **2.** A purchase made by signed order, as for a periodical for a specified period of time or for a series of performances. **3.** Acceptance, as of articles of faith, demonstrated by the signing of one's name. **4. a.** The raising of money from subscribers. **b.** A sum of money so raised. **—sub·scrip'tive** *adj.* **—sub·scrip'tive·ly** *adv.*

**sub·se·quence** (sŭb'sĭ-kwĕns', -kwəns) *n.* **1.** Something subsequent : SEQUEL. **2.** The quality or fact of being subsequent. **3.** *Math.* A sequence contained in another sequence.

**sub·se·quent** (sŭb'sĭ-kwĕnt', -kwənt) *adj.* [ME < OFr. *subsequent* < Lat. *subsequens,* pr.part. of *subsequi,* to follow close after : *sub-,* after + *sequi,* to follow.] Coming after in time, order, or place. **—sub'se·quent'ly** *adv.* **—sub'se·quent'ness** *n.*

**sub·sere** (sŭb'sîr') *n. Ecol.* A secondary series of communities that succeeds an interrupted climax community.

**sub·serve** (səb-sûrv') *vt.* **-served, -serv·ing, -serves.** [Lat. *subservire* : *sub-,* under + *servire,* to serve.] To promote (an end).

**sub·ser·vi·ent** (səb-sûr'vē-ənt) *adj.* [Lat. *subserviens, subservient-,* pr.part. of *subservire,* to subserve.] **1.** Obsequious : servile. **2.** Subordinate in function or capacity. **3.** Useful as a means or instrument : serving to promote an end. **—sub·ser'vi·ence, sub·ser'vi·en·cy** *n.* **—sub·ser'vi·ent·ly** *adv.*

**sub·set** (sŭb'sĕt') *n.* A mathematical set contained within a set.

**sub·shell** (sŭb'shĕl') *n.* Any of the orbitals making up the electron shell of an atom.

**sub·shrub** (sŭb'shrŭb') *n.* **1.** A herbaceous plant with a woody lower stem. **2.** A low shrub : UNDERSHRUB.

**sub·side** (səb-sīd') *vi.* **-sid·ed, -sid·ing, -sides.** [Lat. *subsidere* : *sub-,* down + *sidere,* to settle.] **1.** To sink to a lower or normal level. **2.** To sink to the bottom, as a sediment : SETTLE. **3.** To become less agitated : ABATE <The storm *subsided.*> **—sub·si'dence** *n.*

**sub·sid·i·ar·y** (səb-sĭd'ē-ĕr'ē) *adj.* [Lat. *subsidiarius* < *subsidium,* support < *subsidere,* to subside.] **1.** Serving to supplement or assist : AUXILIARY. **2.** Secondary in importance : SUBORDINATE. **3.** Of, relating to, or of the nature of a subsidy. *—n., pl.* **-ies. 1.** One that is subsidiary. **2.** An affiliate of a larger company : SUBSIDIARY COMPANY. **3.** *Mus.* A theme subordinate to a main theme or subject. **—sub·sid'i·ar'i·ly** *adv.*

☆ ***syns:*** SUBSIDIARY, AFFILIATE, BRANCH, DIVISION, SUBSIDIARY COMPANY *n. core meaning* : a local or auxiliary unit of a parent company controlled by the parent company <worked for a *subsidiary* of a multinational corporation>

**subsidiary company** *n.* A company with over half of its stock owned by another company.

**sub·si·dize** (sŭb'sĭ-dīz') *vt.* **-dized, -diz·ing, -diz·es. 1.** To support or assist with a subsidy. **2.** To obtain the assistance of by granting a subsidy. **—sub'si·di·za'tion** *n.* **—sub'si·diz'er** *n.*

**sub·si·dy** (sŭb'sĭ-dē) *n., pl.* **-dies.** [ME *subsidie* < AN < Lat. *subsidium,* support. —see SUBSIDIARY.] **1.** Financial assistance granted by a government to an individual or a private business. **2.** Financial assistance given by one person or government to another. **3.** Money formerly granted to the British monarch by Parliament.

**sub·sist** (səb-sĭst') *v.* **-sist·ed, -sist·ing, -sists.** [Lat. *subsistere,* to stand up : *sub-,* up + *sistere,* to stand.] *—vt.* **1. a.** To exist. **b.** To remain in existence. **2.** To maintain life : LIVE <*subsisted* on a sparse diet> **3.** *Philos.* To be logically conceivable. *—vt.* To support or maintain with provisions. **—sub·sist'er** *n.*

**sub·sis·tence** (səb-sĭs'təns) *n.* **1.** The act or state of subsisting. **2.** A means of subsisting : SUSTENANCE. **3.** Something that has real or substantial existence. **4.** Hypostasis. **—sub·sis'tent** *adj.*

**sub·soil** (sŭb'soil') *n.* The layer or bed of earth beneath the surface soil. *—vt.* **-soiled, -soil·ing, -soils.** To plow or turn up the subsoil of. **—sub'soil'er** *n.*

**sub·so·lar** (sŭb'sō'lər) *adj.* **1.** Situated directly beneath the sun. **2.** Located between the tropics : EQUATORIAL.

**sub·son·ic** (sŭb'sŏn'ĭk) *adj.* **1.** Of less than audible frequency. **2.** Having a speed less than that of sound in a designated medium.

**sub·spe·cies** (sŭb'spē'shēz, -sēz) *n., pl.* **subspecies.** A subdivision of a taxonomic species, usu. based on geographic distribution. **—sub'spe·cif'ic** (-spĭ-sĭf'ĭk) *adj.*

**sub·stage** (sŭb'stāj') *n.* The part of a microscope located below the stage by which attachments are held in place.

**sub·stance** (sŭb'stəns) *n.* [ME < OFr. < Lat. *substantia* < *substans,* pr.part. of *substare,* to be present : *sub,* under + *stare,* to stand.] **1. a.** That which has mass, occupies space, and can be perceived by the senses : MATTER. **b.** A material of a particular kind or constitution. **2.** The most central and material part : ESSENCE. **3.** Solid, substantial quality or character <a plan with *substance*> **4.** Physical density : BODY <Air has little *substance.*> **5.** Material possessions : WEALTH <a person of *substance*>

**sub·stan·dard** (sŭb'stăn'dərd) *adj.* **1.** Being below standard, as in quality. **2.** Considered unacceptable usage by the educated members of a speech community.

**sub·stan·tial** (səb-stăn'shəl) *adj.* [ME *substancial* < LLat. *substantialis* < Lat. *substantia,* substance.] **1.** Of, relating to, or having substance : MATERIAL. **2.** Not imaginary : REAL. **3.** Solidly built : STRONG. **4.** Ample <a *substantial* dinner> **5.** Being of considerable importance, value, degree, amount, or extent <lost the election by a *substantial* margin> **6.** Possessing wealth or property : WELL-TO-DO. *—pl.n.* **substantials.** Substantial things. **—sub·stan'ti·al'i·ty** (-shē-ăl'ĭ-tē), **sub·stan'tial·ness** *n.* **—sub·stan'tial·ly** *adv.*

**sub·stan·ti·ate** (səb-stăn'shē-āt') *vt.* **-at·ed, -at·ing, -ates.** [Med. Lat. *substantiare, substantiat-* < Lat. *substantia,* substance.] **1.** To support and verify with proof or evidence <*substantiate* previous testimony> **2. a.** To give material form to : EMBODY. **b.** To make firm or solid. **3.** To give substance or reality to. : make real or actual. **—sub·stan'ti·a'tion** *n.*

**sub·stan·ti·val** (sŭb'stən-tī'vəl) *adj.* Of, relating to, or of the nature of a grammatical substantive. **—sub'stan·ti'val·ly** *adv.*

**sub·stan·tive** (sŭb'stən-tĭv) *adj.* [ME *substantif* < OFr. < LLat. *substantivus* < Lat. *substantia,* substance.] **1.** SUBSTANTIAL 5. **2.** Independent in function or existence : not subordinate. **3.** Not imaginary : ACTUAL. **4.** Of or relating to the essence or substance :

ESSENTIAL <*substantive* data> **5.** Having a solid basis : FIRM. **6.** Expressing or denoting existence; e.g., the verb *to be.* **7.** Denoting a noun or noun equivalent. —*n.* A word or group of words functioning as a noun. **—sub′stan·tive·ly** *adv.* **—sub′stan·tive·ness** *n.*

**sub·sta·tion** (sŭb′stā′shən) *n.* A subsidiary or branch station, as of a power plant.

**sub·stit·u·ent** (səb-stĭch′o͞o-ənt) *n.* [Lat. *substituens, substituent-*, pr. part. of *substituere,* to substitute.] An atom, radical, or group substituted for another in a compound. —*adj.* **—sub·stit′u·ent** *adj.*

**sub·sti·tute** (sŭb′stĭ-to͞ot′, -tyo͞ot′) *n.* [< Lat. *substitutus,* p.part. of *substituere,* to substitute : *sub-,* in place of + *statuere,* to cause to stand.] **1.** One that takes the place of another : FILL-IN. **2.** A word or construction used in place of another word or construction. —*v.* **-tut·ed, -tut·ing, -tutes.** —*vt.* To put or use in place of another. —*vi.* To take the place of another <"Only art can *substitute* for nature" —Leonard Bernstein> **—sub′sti·tut′a·bil′i·ty** *n.* **—sub′sti·tut′a·ble** *adj.* **—sub′sti·tu′tion** (-to͞o′shən, -tyo͞o′-) *n.*

☆ ***syns:*** SUBSTITUTE, ALTERNATE, FILL-IN, PINCH HITTER, REPLACEMENT, STAND-IN, SUB, SURROGATE *n. core meaning* : one that takes the place of another <hired a *substitute* for the sick teacher>

**sub·sti·tu·tive** (sŭb′stĭ-to͞o′tĭv, -tyo͞o′-) *adj.* Serving or capable of serving as a substitute.

**sub·strate** (sŭb′strāt′) *n.* [< SUBSTRATUM.] **1.** The material or substance on which an enzyme acts. **2.** *Biol.* A surface on which a plant or animal grows or is attached. **3.** A substratum.

**sub·stra·tum** (sŭb′strā′təm, -străt′əm) *n., pl.* **-stra·ta** (-strā′tə, -străt′ə) *or* **-stra·tums.** [Med. Lat. < Lat. *substratus,* p.part. of *substernere,* to lay under : *sub-,* under + *sternere,* to spread out.] **1. a.** An underlying layer. **b.** A layer of earth beneath the surface soil : SUBSOIL. **2.** The foundation : groundwork. **3.** The material upon which another material is coated or fabricated. **4.** *Philos.* The characterless substance that supports attributes of reality. **5.** *Biol.* A substrate. **—sub′stra′tive** *adj.*

**sub·struc·tion** (sŭb′strŭk′shən) *n.* [Lat. *substructio* < *substruere,* to build beneath : *sub-,* beneath + *struere,* to build.] A foundation : substructure. **—sub′struc′tion·al** *adj.*

**sub·struc·ture** (sŭb′strŭk′chər) *n.* **1.** A supporting structural part : FOUNDATION. **2.** The earth bank or bed supporting railroad tracks. **—sub′struc′tur·al** *adj.*

**sub·sume** (səb-so͞om′) *vt.* **-sumed, -sum·ing, -sumes.** [Med. Lat. *subsumere* : Lat. *sub-,* from below + Lat. *sumere,* to take up.] **1.** To include within a broader class, group, category, or order. **2.** To show (e.g., an idea) to be covered by a broad principle or rule. **—sub·sum′a·ble** *adj.*

**sub·sump·tion** (səb-sŭmp′shən) *n.* [Med. Lat. *subsumptio,* a subsuming < *subsumere,* to subsume.] **1. a.** An act or instance of subsuming. **b.** Something subsumed. **2.** *Logic.* The minor premise of a syllogism. **—sub·sump′tive** *adj.*

**sub·tem·per·ate** (sŭb-tĕm′pər-ĭt, -tĕm′prĭt) *adj.* Of, relating to, or occurring within the colder regions of the Temperate Zones.

**sub·ten·ant** (sŭb-tĕn′ənt) *n.* One that rents property, as land or a house, from a tenant. **—sub′ten′an·cy** *n.*

**sub·tend** (səb-tĕnd′) *vt.* **-tend·ed, -tend·ing, -tends.** [Lat. *subtendere,* to extend underneath : *sub-,* beneath + *tendere,* to extend.] **1.** *Math.* To be opposite to and delimit <The side of a triangle *subtends* the opposite angle.> **2.** To underlie so as to enclose or surround <flowers *subtended* by leafy bracts>

**sub·ter·fuge** (sŭb′tər-fyo͞oj′) *n.* [Fr. < LLat. *subterfugium* < Lat. *subterfugere,* to escape : *subter,* secretly + *fugere,* to flee.] **1.** A deceptive stratagem or device. **2.** Deception by artifice so as to conceal, evade, or escape.

**sub·ter·ra·ne·an** (sŭb′tə-rā′nē-ən) *adj.* [Lat. *subterraneus* : *sub-,* under + *terra,* earth.] **1.** Situated or operating beneath the earth's surface : UNDERGROUND. **2.** Hidden : secret <*subterranean* plots> **—sub′ter·ra′ne·an·ly** *adv.*

**sub·ter·res·tri·al** (sŭb′tə-rĕs′trē-əl) *adj.* SUBTERRANEAN 1. —*n.* An animal living underground.

**sub·tile** (sŭt′l, sŭb′təl) *adj.* **-til·er, -til·est.** Subtle. **—sub′tile·ly** *adv.* **—sub·til′i·ty** (səb-tĭl′ĭ-tē), **sub′tile·ness, sub′til·ty** *n.*

**sub·ti·lin** (sŭb′tə-lĭn) *n.* [NLat. *subtilis,* specific epithet of *Bacillus subtilis* + -IN.] An antibiotic obtained from the bacterium *Bacillus subtilis* that is active against Gram-positive microorganisms.

**sub·til·ize** (sŭt′l-īz′, sŭb′tə-līz′) *v.* **-ized, -iz·ing, -iz·es.** [Med. Lat. *subtilizare* < Lat. *subtilis,* subtle.] —*vt.* To render subtle. —*vi.* To argue or discuss with subtlety. **—sub′til·i·za′tion** *n.*

**sub·ti·tle** (sŭb′tīt′l) *n.* **1.** A secondary, usu. explanatory title, as of a literary work. **2. a.** A printed translation of the dialogue of a foreign-language movie shown at the bottom of the screen. **b.** A printed narration or portion of dialogue flashed on the screen between the scenes of a silent movie.

**sub·tle** (sŭt′l) *adj.* **-tler, -tlest.** [ME *subtil* < OFr. *sotil* < Lat. *subtilis.*] **1. a.** So slight as to be difficult to detect or analyze : ELUSIVE. **b.** Not immediately clear : ABSTRUSE. **2.** Capable of making fine distinctions <a *subtle* mind> **3. a.** Skillful or ingenious : CLEVER. **b.** Marked by craft or slyness : DEVIOUS. **c.** Operating in a hidden and usu. injurious way : INSIDIOUS <*subtle* demoralization of the company's staff> **—sub′tle·ness** *n.* **—sub′tly** *adv.*

**sub·tle·ty** (sŭt′l-tē) *n., pl.* **-ties. 1.** The quality or state of being subtle. **2.** Something subtle, esp. a fine distinction.

**sub·ton·ic** (sŭb′tŏn′ĭk) *n. Mus.* The seventh tone of a diatonic scale, immediately below the tonic.

**sub·to·pi·a** (sŭb-tō′pē-ə) *n.* [SUB(URB) + (U)TOPIA.] *Chiefly Brit.* A city's suburbs. **—sub·to′pi·an** *adj.*

**sub·top·ic** (sŭb′tŏp′ĭk) *n.* One of the divisions into which a main topic may be divided.

**sub·tor·rid** (sŭb′tôr′ĭd, -tŏr′-) *adj.* Subtropical.

**sub·to·tal** (sŭb-tōt′l) *adj.* Less than total : INCOMPLETE. —*n.* (sŭb′tōt′l). The total of part of a series of numbers. —*v.* (sŭb′tōt′l) **-taled, -tal·ing, -tals** *also* **-talled, -tal·ling, -tals.** —*vt.* To total part of (a series of numbers). —*vi.* To arrive at a subtotal.

**sub·tract** (səb-trăkt′) *v.* **-tract·ed, -tract·ing, -tracts.** [Lat. *substahere, substract-* : *sub-,* away + *trahere,* to draw.] —*vt.* To take away : DEDUCT. —*vi.* To perform the arithmetic operation of subtraction. **—sub·tract′er** *n.*

**sub·trac·tion** (səb-trăk′shən) *n.* **1.** The act or process of subtracting : DEDUCTION. **2.** The arithmetic operation of finding a quantity that when added to one of two quantities produces the other.

**sub·trac·tive** (səb-trăk′tĭv) *adj.* **1.** Producing or involving subtraction. **2.** Designating a color produced by light passing through more than one colorant, each of which inhibits certain wavelengths, as in mixtures of pigments. **3.** Designating a photographic process that produces a positive image by superposing or mixing substances that selectively absorb colored light.

**sub·tra·hend** (sŭb′trə-hĕnd′) *n.* [< Lat. *subtrahendum,* neuter gerund. of *subtrahere,* to subtract.] A quantity or number to be subtracted from another.

**sub·trop·i·cal** (sŭb-trŏp′ĭ-kəl) *adj.* Of, pertaining to, or being the geographic areas adjacent to the tropics.

**sub·trop·ics** (sŭb-trŏp′ĭks) *pl.n.* Subtropical regions.

**su·bu·late** (so͞o′byə-lĭt, -lāt′, sŭb′yə-) *adj. Biol.* [NLat. *subulatus* < Lat. *subula,* awl.] *Biol.* Tapering to a point <*subulate* leaves>

**sub·um·brel·la** (sŭb′ŭm-brĕl′ə) *n. Zool.* The concave undersurface of the body of a jellyfish.

**sub·urb** (sŭb′ûrb′) *n.* [ME < OFr. *suburbe* < Lat. *suburbium* : *sub-,* close to + *urbs,* city.] **1.** A usu. residential area or community outlying a city. **2. suburbs.** The usu. residential region around a large city : ENVIRONS.

**sub·ur·ban** (sə-bûr′bən) *adj.* **1.** Of, relating to, or characteristic of a suburb. **2.** Located or living in a suburb. **3.** Of or relating to the life-style of those living in the suburbs. —*n.* A suburbanite.

**sub·ur·ban·ite** (sə-bûr′bə-nīt′) *n.* One who lives in a suburb.

**sub·ur·ban·ize** (sə-bûr′bə-nīz′) *vt.* **-ized, -iz·ing, -izes.** To impart a suburban character to. **—sub·ur′ban·i·za′tion** *n.*

**sub·ur·bi·a** (sə-bûr′bē-ə) *n.* **1.** Suburbs. **2. a.** Suburbanites as a group. **b.** Suburbanites as a cultural class.

**sub·ven·tion** (səb-vĕn′shən) *n.* [ME *subvencioun* < OFr. *subvention* < LLat. *subventio* < Lat. *subvenire,* to come to help : *sub-,* up + *venire,* to come.] **1.** Provision of help or support. **2.** A grant of financial aid, esp. an endowment, as that given by a government to an institution for research. **—sub·ven′tion·ar′y** *adj.*

**sub·ver·sion** (səb-vûr′zhən, -shən) *n.* [ME *subversioun* < OFr. *subversion* < LLat. *subversio* < Lat. *subvertere,* to subvert.] **1.** The act of subverting or the state of being subverted. **2.** *Obs.* A cause of overthrow or ruin. **—sub·ver′sion·ar′y** *adj.*

**sub·ver·sive** (səb-vûr′sĭv, -zĭv) *adj.* Intended or serving to subvert <*subversive* political activity> —*n.* One who advocates or is regarded as advocating subversive means or policies. **—sub·ver′sive·ly** *adv.* **—sub·ver′sive·ness** *n.*

**sub·vert** (səb-vûrt′) *vt.* **-vert·ed, -vert·ing, -verts.** [ME *subverten* < OFr. *subvertir* < Lat. *subvertere* : *sub-,* from below + *vertere,* to turn.] **1.** To destroy completely : RUIN. **2.** To undermine the character, morals, or allegiance of. **3.** To overthrow completely <"economic assistance . . . must *subvert* the existing . . . feudal or tribal order" —Henry A. Kissinger> **—sub·vert′er** *n.*

**sub·way** (sŭb′wā′) *n.* **1. a.** An underground urban railroad, usu. operated by electricity. **b.** A passage for such a railroad. **2.** An underground passage, as for a water main.

**Su·ca·ryl** (so͞o′kə-rĭl′). A trademark for either of two compounds used as low-calorie sweeteners.

**suc·ceed** (sək-sēd′) *v.* **-ceed·ed, -ceed·ing, -ceeds.** [ME *succeden* < OFr. *succeder* < Lat. *succedere* : *sub-,* after + *cedere,* to go.] —*vi.* **1.** To follow next after in time or succession, esp. to replace another in an office or position <*succeeded* to the Presidency> **2.** To accomplish something desired or intended <*succeeded* in having my own way> **3.** *Obs.* To devolve upon a person by way of inheritance. —*vt.* **1.** To come after in time or order. **2.** To follow in office : REPLACE. **—suc·ce′dent** (sək-sēd′nt) *adj.* **—suc·ceed′er** *n.*

**suc·cès d'es·time** (sük-sĕ′ dĕs-tēm′) *n.* [Fr. : *succès,* success + *de,* of + *estime,* esteem.] **1.** An artistic work receiving critical acclaim, often without achieving popular success. **2.** Critical acclaim.

**suc·cès fou** (sük-sĕ′ fo͞o′) *n.* [Fr.] A wild success.

**suc·cess** (sək-sĕs′) *n.* [Lat. *successus* < p.part. of *succedere,* to succeed.] **1.** The gaining of something desired, planned, or attempted. **2. a.** The gaining of fame or prosperity. **b.** The extent of such gain. **3.** One that is successful. **4.** *Obs.* A result or outcome.

**suc·cess·ful** (sək-sĕs′fəl) *adj.* **1.** Having a favorable outcome. **2.** Having achieved success <a *successful* lawyer> **—suc·cess′ful·ly** *adv.* **—suc·cess′ful·ness** *n.*

**suc·ces·sion** (sək-sĕsh′ən) *n.* **1.** The act or process of following in order or sequence. **2.** A group arranged or following in order : SEQUENCE. **3. a.** The sequence in which one person after another succeeds to a title, office, throne, or estate. **b.** The right of a person or line of persons to so succeed. **c.** The person or line vested with such a right. **4. a.** The act or process of succeeding to the rights or duties of another. **b.** The act or process of becoming entitled as a legal beneficiary to the property of a deceased person. **5.** *Ecol.* The gradual and orderly process of ecosystem development brought about by changes in species populations that culminates in the production of a climax characteristic of a particular geographic region. **—suc·ces′sion·al** *adj.* **—suc·ces′sion·al·ly** *adv.*

**suc·ces·sive** (sək-sĕs′ĭv) *adj.* **1.** Following in uninterrupted order or sequence. **2.** Of, marked by, or involving succession. **—suc·ces′sive·ly** *adv.* **—suc·ces′sive·ness** *n.*

**successive approximation** *n.* A method for estimating the value of an unknown quantity by repeated comparison to a sequence of known quantities.

**suc·ces·sor** (sək-sĕs′ər) *n.* One that succeeds another.

**suc·cinct** (sək-sĭngkt′) *adj.* [Lat. *succinctus* < p.part. of *succingere,* to gird from below : *sub-,* below + *cingere,* to gird.] **1.** Expressed clearly in few words : CONCISE. **2.** Marked by brevity and clarity in speech or writing. **3.** *Archaic.* Encircled as if by a girdle. **—suc·cinct′ly** *adv.* **—suc·cinct′ness** *n.*

**suc·cin·ic acid** (sək-sĭn′ĭk) *n.* [Fr. *succinique* < Lat. *succinum,* amber.] A colorless crystalline compound, $C_4H_6O_4$, occurring naturally in amber and synthesized for use in pharmaceuticals and perfumes.

**suc·cor** (sŭk′ər) *n.* [ME *sucurs* (pl.) < OFr. *secors* < Med. Lat. *succursus* < Lat. *succurrere,* to be useful for : *sub-,* up + *currere,* to run.] **1.** Assistance in time of distress : RELIEF. **2.** One that provides assistance or relief. *—vt.* **-cored, -cor·ing, -cors.** To give assistance to in time of distress. **—suc′cor·a·ble** *adj.* **—suc′cor·er** *n.*

**suc·co·ry** (sŭk′ə-rē) *n., pl.* **-ries.** [Alteration of ME *cicoree.*] CHICORY 2.

**suc·co·tash** (sŭk′ə-tăsh′) *n.* [Narraganset *msíckquatash.*] A mixture of cooked corn, lima beans, and tomatoes.

**Suc·coth** *also* **Suk·koth** (so͝ok′ōt, -əs) *n.* [Heb. *sukkōth* < *sukkāh,* tabernacle.] A Jewish harvest festival beginning on the eve of the 15th of Tishri and celebrated for nine days.

**suc·cour** (sŭk′ər) *n. Chiefly Brit. var. of* SUCCOR.

**suc·cu·bus** (sŭk′yə-bəs) *also* **suc·cu·ba** (sŭk′yə-bə) *n., pl.* **-bus·es** or **-bi** (-bī′, -bē′) *also* **-bae** (-bē′, -bī) [Med. Lat. < LLat. *succuba,* prostitute < Lat. *succubare,* to lie under : *sub-,* under + *cubare,* to lie down.] **1.** A woman demon supposed to descend upon and have sexual intercourse with a man while he sleeps. **2.** An evil spirit.

**suc·cu·lent** (sŭk′yə-lənt) *adj.* [Lat. *succulentus* < *succus,* juice.] **1.** Full of sap or juice : JUICY. **2.** *Bot.* Having thick, fleshy leaves or stems that conserve moisture. **3.** Having desirable qualities, as vitality or richness. *—n.* A succulent plant. **—suc′cu·lence, suc′cu·len·cy** *n.* **—suc′cu·lent·ly** *adv.*

**suc·cumb** (sə-kŭm′) *vi.* **-cumbed, -cumb·ing, -cumbs.** [ME *succomben* < OFr. *succomber* < Lat. *succumbere.*] **1.** To yield to an overpowering force or overwhelming desire. **2.** To die <The patient *succumed* to cancer.>

**such** (sŭch) *adj.* [ME < OE *swylc.*] **1.** Of this or that kind <could not do *such* work> **2.** Being the same as something implied but left undefined <*Such* people are never happy.> **3.** Of so great or extreme a degree or quality <*such* luck> *—adv.* **1.** To such a degree <*such* good writing><*such* a bad job> **2.** Especially : very <has been in *such* poor health> *—pron.* **1.** Such a person or persons or thing or things. **2.** One implied or indicated <*Such* are the vicissitudes of life.> **3.** The like <pots, pans, and *such*> **—as such. 1.** As being the person or thing implied or mentioned previously <An executive *as such* must make decisions.> **2.** In itself or by itself <Money *as such* seldom brings happiness.> **—such as. 1.** For example. **2.** Of the stated or implied kind or degree : LIKE <an idea *such as* that one>

**such and such** *adj.* Not yet specified : UNDETERMINED <They agreed to meet at *such and such* a restaurant.>

**such·like** (sŭch′līk′) *adj.* Of a similar kind : LIKE. *—pron.* Persons or things of such a kind.

**suck** (sŭk) *v.* **sucked, suck·ing, sucks.** [ME *souken* < OE *sūcan.*] *—vt.* **1.** To draw (liquid) into the mouth by inhalation. **2. a.** To draw in by establishing a partial vacuum. **b.** To draw in by or as if by a current in a fluid. **3.** To draw nourishment through or from. **4.** To hold, moisten, or maneuver (e.g., a piece of candy) in the mouth. *—vi.* **1.** To draw in by or as if by suction. **2.** To draw nourishment : SUCKLE. **3.** To make a sucking sound. **—suck in.** *Slang.* To take advantage of : CHEAT. *—n.* **1.** The act of sucking. **2.** Suction. **3.** Something drawn in by sucking.

**suck·er** (sŭk′ər) *n.* **1.** One that sucks. **2.** *Informal.* One easily deceived : DUPE. **3.** A lollipop. **4. a.** A piston or piston valve, as in a suction pump or syringe. **b.** A tube or pipe, as a siphon, through which a liquid is sucked. **5.** Any of numerous chiefly North American freshwater fishes of the family Catostomidae, having a thick-lipped mouth adapted for feeding by suction. **6.** A structure or part adapted for clinging by suction. **7.** *Bot.* A secondary shoot arising from the base of a tree trunk or from the lower part of certain shrubs. *—v.* **-ered, -er·ing, -ers.** *—vt.* **1.** To strip suckers or shoots from <*sucker* tomato plants> **2.** *Informal.* To deceive or trick : DUPE. *—vi. Bot.* To send out suckers or shoots.

**suck·er·fish** (sŭk′ər-fĭsh′) *n., pl.* **suckerfish** or **-fish·es.** The remora.

**suck·ing** (sŭk′ĭng) *adj.* Not yet weaned.

**sucking louse** *n.* Any of various small wingless insects of the order Anoplura with mouth parts adapted for sucking and piercing.

**suck·le** (sŭk′əl) *v.* **-led, -ling, -les.** [Prob. back-formation < SUCKLING.] *—vt.* **1.** To cause or allow to take milk at the breast or udder : NURSE. **2.** To take in as sustenance. **3.** To bring up : FOSTER. *—vi.* To suck at the breast. **—suck′ler** *n.*

**suck·ling** (sŭk′lĭng) *n.* [ME *suklinge.*] A young mammal not yet weaned.

**su·crase** (so͞o′krās′, -krāz′) *n.* [Fr. *sucre,* sugar + -ASE.] *Chem.* Invertase.

**su·cre** (so͞o′krā) *n.* [Sp., after Antonio José de *Sucre* (1795–1830).] —See table at CURRENCY.

**su·crose** (so͞o′krōs′) *n.* [Fr. *sucre,* sugar + -OSE.] A crystalline disaccharide carbohydrate, $C_{12}H_{22}O_{11}$, obtained chiefly from sugar cane and sugar beet, and widely used as a sweetener, preservative, and in making plastics and cellulose.

**suc·tion** (sŭk′shən) *n.* [LLat. *suctio* < Lat. *sugere,* to suck.] **1.** The act or process of sucking. **2.** A force causing a fluid or solid to be drawn into an interior space or to adhere to a surface due to the difference between external and internal pressures. *—adj.* **1.** Creating suction. **2.** Operating by suction.

**suction pump** *n.* A pump for drawing up a liquid by suction produced via a piston drawn through a cylinder.

**suction stop** *n.* CLICK 3.

**suc·to·ri·al** (sŭk-tôr′ē-əl, -tōr′-) *adj.* [NLat. *suctorius* < Lat. *sugere,* to suck.] **1.** Adapted for sucking or clinging by suction <a *suctorial* organ> **2.** Having suctorial organs or parts.

**Su·dan·ic** (so͞o-dăn′ĭk) *n.* The non-Bantu, non-Hamitic languages of the Sudan. **—Su·dan′ic** *adj.*

**su·da·to·ri·um** (so͞o′də-tôr′ē-əm, -tōr′-) *n., pl.* **-to·ri·a** (-tôr′ē-ə, -tōr′ē-ə) [Lat. < *sudatorius,* sweating < *sudare,* to sweat.] A hot-air room used for sweat baths.

**su·da·to·ry** (so͞o′də-tôr′ē, -tōr′ē) *adj.* Sudorific. *—n., pl.* **-ries. 1.** A sudatorium. **2.** A sudorific.

**sudd** (sŭd) *n.* [Ar.] Floating masses of vegetation that often obstruct navigation on the White Nile.

**sud·den** (sŭd′n) *adj.* [ME < OFr. *sodein* < Lat. *subitaneus* < *subitus,* sudden < p.part. of *subire,* to approach stealthily : *sub-,* secretly + *ire,* to go.] **1.** Taking place without warning <a *sudden* attack> **2.** Hasty : abrupt <a *sudden,* unannounced departure> **3.** Brought about in a short time. **—all of a sudden.** Very quickly and unexpectedly. **—sud′den·ly** *adv.* **—sud′den·ness** *n.*

**sudden death** *n.* **1.** A death not preceded by any condition that would appear fatal. **2. a.** A game played to break a tie. **b.** Extra minutes of play added to a tied game, with the winning team being the first team to score.

**sudden infant death syndrome** *n.* The unexpected death of an apparently healthy infant that usu. occurs during the first four months of life while the infant is sleeping.

**su·dor·if·er·ous** (so͞o′də-rĭf′ər-əs) *adj.* [LLat. *sudoriferus* : Lat. *sudor,* sweat + Lat. *ferre,* to carry.] Producing or secreting sweat.

**su·dor·if·ic** (so͞o′də-rĭf′ĭk) *adj.* [NLat. *sudorificus* < Lat. *sudor,* sweat.] Causing or increasing sweat. *—n.* A sweat-inducing agent.

**Su·dra** (so͞o′drə) *n.* [Skt. *śūdraḥ.*] **1.** The lowest of the major Hindu castes, the members of which were orig. menials but are now largely artisans and laborers. **2.** A member of the Sudra caste.

**suds** (sŭdz) *pl.n.* [Poss. MDu. *sudse,* marsh.] **1.** Soapy water. **2.** Foam : lather. **3.** *Slang.* Beer.

**suds·y** (sŭd′zē) *adj.* **-i·er, -i·est.** Resembling or filled with suds.

**sue** (so͞o) *v.* **sued, su·ing, sues.** [ME *sewen* < AN *suer* < Lat. *sequi,* to follow.] *—vt.* **1.** To make a petition to : BESEECH. **2.** *Law.* **a.** To petition (a court) for redress of grievances or recovery of a right. **b.** To institute legal proceedings against (a person) for redress of grievances. **c.** To carry (an action) through to a final decision. **3.** To court : woo. *—vi.* **1.** To institute legal proceedings. **2.** To make an appeal or entreaty. **3.** To woo. **—su′er** *n.*

**suede** *also* **suède** (swād) *n.* [Fr. *suède* < *Suède,* Sweden.] **1.** Leather with a soft napped surface. **2.** Fabric made to look like suede.

**su·et** (sōō'ĭt) *n.* [ME *sewet* < AN **sewet,* dim. of *sue,* tallow < Lat. *sebum.*] The hard fatty tissues around the kidneys of cattle and sheep, used in cooking and making tallow.

**suf·fer** (sŭf'ər) *v.* **-fered, -fer·ing, -fers.** [ME *sufferen* < AN *suffrir* < Lat. *sufferre* : *sub-,* from below + *ferre,* to bear.] —*vi.* **1.** To feel, experience, or endure pain or distress. **2.** To sustain loss, injury, harm, or punishment. **3.** To appear at a disadvantage <This house *suffers* by comparison with others.> —*vt.* **1.** To undergo or sustain (pain, injury, or unpleasantness). **2.** To experience <*suffer* a change in staffing> **3.** To endure : bear <cannot *suffer* this job> **4.** To permit : allow <would not *suffer* their foul behaviour> **—suf'fer·er** *n.* **—suf'fer·ing·ly** *adv.*

**suf·fer·a·ble** (sŭf'ər-ə-bəl, sŭf'rə-) *adj.* Capable of being suffered, endured, or permitted : TOLERABLE. **—suf'fer·a·bly** *adv.*

**suf·fer·ance** (sŭf'ər-əns, sŭf'rəns) *n.* [ME < OFr. < LLat. *sufferentia* < Lat. *sufferre,* to suffer.] **1.** Capacity to tolerate pain or distress. **2.** Sanction or permission implied or given by failure to prohibit : tacit assent. **3.** Suffering : misery. **4.** Patient endurance.

**suf·fer·ing** (sŭf'ər-ĭng, sŭf'rĭng) *n.* **1.** The act or state of one that suffers. **2.** Something that causes pain or distress.

**suf·fice** (sə-fīs') *v.* **-ficed, -fic·ing, -fic·es.** [ME *suffisen* < OFr. *suffire, suffis-* < Lat. *sufficere* : *sub-,* under + *facere,* to make.] —*vi.* **1.** To meet present needs or requirements : be sufficient to <These clothes will *suffice* until next week.> <*Suffice* it to say that I feel wronged.> **2.** To be equal to a specified task <No words will *suffice* to convey our sorrow.> —*vt.* To be enough or sufficient for. **—suf·fic'er** *n.*

**suf·fi·cien·cy** (sə-fĭsh'ən-sē) *n.* **1.** The quality or state of being sufficient. **2.** An adequate amount or quantity.

**suf·fi·cient** (sə-fĭsh'ənt) *adj.* [ME < OFr. < Lat. *sufficiens,* pr.part. of *sufficere,* to suffice.] **1.** That is enough : ADEQUATE <*sufficient* water> **2.** *Archaic.* Competent : qualified. **—suf·fi'cient·ly** *adv.*

**suf·fix** (sŭf'ĭks) *n.* [< Lat. *suffixus,* p.part. of *suffigere,* to affix : *sub-,* secondary + *figere,* to fix.] An affix appended to the end of a word or stem, serving to form a new word or functioning as an inflectional ending, as *-ness* in *gentleness, -ing* in *walking,* or *-s* in *sits.* —*vt.* **-fixed, -fix·ing, -fix·es.** To add as a suffix. **—suf'fix·al** *adj.* **—suf·fix'ion** (sə-fĭk'shən) *n.*

**suf·fo·cate** (sŭf'ə-kāt') *v.* **-cat·ed, -cat·ing, -cates.** [Lat. *suffocare, suffocat-* : *sub-,* under + *fauces,* throat.] —*vt.* **1.** To kill or destroy by cutting off the oxygen supply : ASPHYXIATE. **2.** To cause discomfort by or as if by cutting off the supply of air. **3.** To suppress the development, imagination, or creativity of : STIFLE. —*vi.* **1.** To die from suffocation. **2.** To be stifled : SMOTHER. **—suf'fo·ca'tion** *n.* **—suf'fo·ca'tive** *adj.*

**Suf·folk** (sŭf'ək) *n.* [After *Suffolk* County, England.] **1.** Any of an English breed of hornless sheep producing high-quality mutton. **2.** Any of a breed of English draft horses with short legs and a thickset, heavy body.

**Suffolk**
*A Suffolk sheep, 3 feet high at shoulder*

**suf·fra·gan** (sŭf'rə-gən) *n.* [ME < OFr. < Med. Lat. *suffraganeus* < Lat. *suffragium,* suffrage.] **1.** A bishop elected or appointed as an assistant to the bishop or ordinary of a diocese, having administrative and episcopal responsibilities but no jurisdictional functions. **2.** A bishop considered subordinate to his archbishop or metropolitan. —*adj.* Of, being, or relating to a suffragan : AUXILIARY. **—suf'fra·gan·ship'** *n.*

**suf·frage** (sŭf'rĭj) *n.* [Partly < ME, intercessory prayer (< OFr. < Med. Lat. *suffragium* < Lat., vote), and partly < Lat. *suffragium,* vote.] **1.** A vote cast in deciding a disputed question or in electing a person to office. **2. a.** Right or privilege of voting : FRANCHISE. **b.** Exercise of voting rights. **3.** A short intercessory prayer.

**suf·fra·gette** (sŭf'rə-jĕt) *n.* A woman who is an advocate of suffrage for women. **—suf'fra·get'tism** *n.*

**suf·fra·gist** (sŭf'rə-jĭst) *n.* An advocate of the extension of political voting rights, esp. to women.

**suf·fru·tes·cent** (sŭf'rōō-tĕs'ənt) *also* **suf·fru·ti·cose** (sŭf-rōō'tĭ-kōs') *adj.* [NLat. *suffrutescens, suffrutescent-* : Lat. *sub-,* under + NLat. *frutescens,* frutescent < Lat. *frutex,* shrub.] *Bot.* Having a woody stem or base.

**suf·fuse** (sə-fyōōz') *vt.* **-fused, -fus·ing, -fus·es.** [Lat. *suffundere, suffus-* : *sub-,* below + *fundere,* to pour.] To spread through or over, as with liquid, color, or light. **—suf·fu'sion** *n.* **—suf·fu'sive** (sə-fyōō'sĭv, -zĭv) *adj.*

**Su·fi** (sōō'fē) *n.* [Ar. *sūfīy* < *sūf,* wool.] A member of a Moslem mystic sect. **—Su'fic** (-fĭk), **Su·fis'tic** (-fĭs'tĭk) *adj.*

**Su·fism** (sōō'fĭz'əm) *n.* A sect of Islamic mysticism, dating from the 8th cent. A.D. and developed chiefly in Persia.

**sug·ar** (shŏŏg'ər) *n.* [ME *sugre* < OFr. *sukere* < OItal. *zucchero* < Med. Lat. *succarum* < Ar. *sukkar* < Pers. *shakar* < Skt. *śarkarā.*] **1.** Sucrose. **2.** Any of a class of water-soluble crystalline carbohydrates, including sucrose and lactose, having a sweet taste. **3.** A specified amount of sugar, as a cube. **4.** *Slang.* Sweetheart. —Used as a term of endearment. —*v.* **-ared, -ar·ing, -ars.** —*vt.* **1.** To coat, sweeten, or cover with sugar. **2.** SUGARCOAT 2. —*vi.* To form sugar : GRANULATE.

**sugar apple** *n.* The sweetsop.

**sugar beet** *n.* A form of the common beet, *Beta vulgaris,* with white roots from which sugar is obtained.

**sug·ar·ber·ry** (shŏŏg'ər-bĕr'ē) *n.* The hackberry.

**sugar bush** *n.* A grove of sugar maples used as a source of maple syrup or maple sugar.

**sugar cane** *n.* A tall grass, *Saccharum officinarum,* native to the East Indies, with thick tough stems that are one of the chief commercial sources of sugar.

**sug·ar-coat** (shŏŏg'ər-kōt') *vt.* **-coat·ed, -coat·ing, -coats. 1.** To coat with sugar <*sugar-coat* a pill> **2.** To cause to seem more appealing or pleasant.

**sugar corn** *n.* Sweet corn.

**sug·ar-cured** (shŏŏg'ər-kyŏŏrd') *adj.* Cured with a mixture of sugar, salt, and nitrate <*sugar-cured* ham>

**sugar daddy** *n. Slang.* A wealthy, usu. older man who gives expensive gifts to a young woman in return for her sexual favors or companionship.

**sug·ar·house** (shŏŏg'ər-hous') *n.* A sugar refinery or processing plant, esp. a building in which maple sap is boiled down to yield maple syrup and maple sugar.

**sugaring off** *n.* **1.** The boiling down of maple sap to yield maple syrup and maple sugar. **2.** An informal social gathering in which the guests help make maple sugar.

**sugar loaf** *n.* **1.** A conical loaf of pure concentrated sugar. **2.** Something shaped like a loaf of sugar. **—sug'ar-loaf'** (shŏŏg'ər-lōf') *adj.*

**sugar maple** *n.* A maple tree, *Acer saccharum* of eastern North America, having hard variously grained wood used in cabinetmaking and sap that is the source of maple syrup and maple sugar.

**sugar of lead** *n.* Lead acetate.

**sugar of milk** *n.* Lactose.

**sugar orchard** *n.* Sugar bush.

**sugar pine** *n.* A tall evergreen timber tree, *Pinus lambertiana,* of the Pacific coast of North America.

**sug·ar·plum** (shŏŏg'ər-plŭm') *n.* A small piece of sugary candy.

**sug·ar·y** (shŏŏg'ə-rē) *adj.* **-i·er, -i·est. 1.** Made of sugar. **2.** Tasting like or resembling sugar. **3.** Deceitfully or cloyingly sweet : mawkishly sentimental. **—sug'ar·i·ness** *n.*

**sug·gest** (səg-jĕst', sə-jĕst') *vt.* **-gest·ed, -gest·ing, -gests.** [Lat. *suggerere, suggest-* : *sub-,* up + *gerere,* to carry.] **1.** To offer for action or consideration : PROPOSE. **2. a.** To call to mind by association or logic : EVOKE <a cave that *suggests* a dark tomb> **b.** To serve as or provide a motive for <Such behavior *suggests* harsh punishment.> **3.** To make evident indirectly. **—sug·gest'er** *n.*

☆ ***syns:*** SUGGEST, HINT, IMPLY, INSINUATE, INTIMATE *v. core meaning* : to make evident indirectly <actions that *suggested* jealousy> SUGGEST in this context usu. refers to a process in which something is called to mind by an association of ideas <a cavern that *suggests* a cathedral> IMPLY refers to something suggested by logical necessity <life *implying* birth, growth, and death> HINT refers to indirect expression containing pointed clues <*hinted* that it was time to leave> INTIMATE applies to veiled expression that may be the result of discretion or reserve <*intimated* that there was trouble ahead> INSINUATE refers to communicating something, usu. unpleasant, in a covert way <*insinuated* that I was dishonest> ***ant:*** express

**sug·gest·i·ble** (səg-jĕs'tə-bəl, sə-jĕs'-) *adj.* Readily influenced by or susceptible to suggestion. **—sug·gest'i·bil'i·ty** (-bĭl'ĭ-tē, sə-jĕs'-) *n.*

**sug·ges·tion** (səg-jĕs'chən, sə-jĕs'-) *n.* **1.** An act of suggesting. **2.** Something suggested. **3.** The sequential thought process by which one idea or concept leads to another. **4. a.** The psychological process by which an idea is induced in or adopted without argument, command, or coercion. **b.** An idea or response so induced. **5.** A trace : hint <not a *suggestion* of scandal>

**sug·ges·tive** (səg-jĕs'tĭv, sə-jĕs'-) *adj.* **1. a.** Tending to suggest thoughts or ideas. **b.** Conveying a suggestion or hint : INDICATIVE. **2.** Tending to suggest something considered improper or indecent. **—sug·ges'tive·ly** *adv.* **—sug·ges'tive·ness** *n.*

☆ ***syns:*** SUGGESTIVE, ALLUSIVE, EVOCATIVE, IMPRESSIONISTIC, REMINISCENT *adj. core meaning* : tending to bring something, as a memory, mood, or image, subtly or indirectly to mind <music *suggestive* of the chaos of war>

**su·i·cid·al** (sōō′ĭ-sīd′l) *adj.* **1.** Relating to, involving, or tending toward suicide. **2.** Dangerous to oneself or to one's interests : SELF-DESTRUCTIVE. **—su′i·cid′al·ly** *adv.*

**su·i·cide** (sōō′ĭ-sīd′) *n.* [Lat. *sui*, of oneself + -CIDE.] **1.** An act or instance of intentionally killing oneself. **2.** Destruction or ruin of one's own interests. **3.** One who commits suicide.

**su·i·cid·ol·o·gy** (sōō′ĭ-sī-dŏl′ə-jē) *n.* The study of suicide, suicidal behavior, and suicide prevention. **—su′i·cid·ol′o·gist** *n.*

**su·i ge·ne·ris** (sōō′ī′ jĕn′ər-ĭs, sōō′ē) *adj.* [< Lat., of its own kind.] Unique : individual.

**su·i ju·ris** (sōō′ī′ jŏŏr′ĭs, sōō′ē) *adj.* [Lat.] *Law.* Capable of managing one's own affairs.

**su·int** (sōō′ĭnt, swĭnt) *n.* [Fr. < OFr. < *suer*, to sweat < Lat. *sudare.*] A natural grease formed from dried perspiration found in the fleece of sheep, used as a source of potash.

**suit** (sōōt) *n.* [ME < AN *suite* < OFr. *sieute* < VLat. **sequita* < **sequere* < Lat. *sequi*, to follow.] **1.** A set of garments consisting of a coat and pants or skirt that match in color or fabric. **2.** A group of things united into a set or series by sharing a common form or function. **3.** Any of the four sets of playing cards, each with similar spots, constituting a deck. **4. a.** Attendance required of a vassal at his feudal lord's court or manor. **b.** *Law.* A court proceeding to recover a right or claim. **5.** An act or instance of courting a woman. —*v.* **suit·ed, suit·ing, suits.** —*vt.* **1.** To fulfill the requirements of : ACCOMMODATE <This house does not *suit* our needs.> **2.** To make appropriate or suitable : ADAPT <We can *suit* the product to your specifications.> **3.** To please : satisfy <This movie doesn't *suit* me.> **4.** To provide with clothing : DRESS. —*vi.* To be suitable or acceptable. **—follow suit. 1.** To play a card of the same suit as the one led. **2.** To follow another's example.

**suit·a·ble** (sōō′tə-bəl) *adj.* That is appropriate to a given purpose. **—suit′a·bil′i·ty, suit′a·ble·ness** *n.* **—suit′a·bly** *adv.*

**suit·case** (sōōt′kās′) *n.* A usu. rectangular piece of luggage for carrying clothing and other personal belongings.

**suite** (swēt) *n.* [Fr. < OFr. *sieute.* —see SUIT.] **1.** A staff or train of attendants or followers : RETINUE. **2.** A succession of related things intended to be used together. **3.** A series of connected rooms functioning as a living unit. **4.** (swēt, sōōt). A set of matching furniture <a bedroom *suite*> **5.** *Mus.* An instrumental composition consisting of a succession of dances in the same or related keys.

**suit·ing** (sōō′tĭng) *n.* Fabric from which suits are made.

**suit·or** (sōō′tər) *n.* [ME < AN *seutor*, follower < Lat. *secutor* < *sequi*, to follow.] **1.** One who petitions or requests. **2.** A party that sues in a court of law : PLAINTIFF. **3.** A man who courts a woman.

**su·ki·ya·ki** (sōō′kē-yä′kē, skē-yä′kē) *n.* [J.] A Japanese dish of sliced meat, vegetables, and seasoning fried together.

**Suk·koth** (sōōk′ōt, -əs) *n. var. of* SUCCOTH.

**sul·cate** (sŭl′kāt′) *adj.* [Lat. *sulcatus*, p.part. of *sulcare*, to furrow < *sulcus*, furrow.] *Biol.* Having narrow longitudinal indentations.

**sul·cus** (sŭl′kəs) *n., pl.* **-ci** (-kī′, -sī′) [Lat.] **1.** A deep narrow furrow or groove. **2.** *Anat.* Any of the narrow fissures separating adjacent cerebral convolutions. **—sul′cal** *adj.*

**sulf-** *or* **sulfo-** *pref.* [< SULFUR.] Sulfur <*sulf*ate>

**sul·fa·di·a·zine** (sŭl′fə-dī′ə-zēn′) *n.* [SULFA (DRUG) + DIAZINE.] An antibacterial sulfa drug, $C_{10}H_{10}O_2N_4S$, used in the treatment of meningitis and other infections.

**sul·fa drug** (sŭl′fə) *n.* [SULFA(NILAMIDE) + DRUG.] Any of a group of synthetic organic compounds, as sulfadiazine, chemically similar to sulfonamide and capable of inhibiting bacterial growth and activity.

**sul·fa·nil·a·mide** (sŭl′fə-nĭl′ə-mīd′, -mĭd) *n.* [SULF- + ANIL(INE) + AMIDE.] A white, odorless crystalline sulfonamide, $C_6H_8N_2SO_2$, used in the treatment of various bacterial infections.

**sul·fa·tase** (sŭl′fə-tās′) *n.* [SULFAT(E) + -ASE.] Any of various esterases that catalyze the hydrolysis of sulfuric esters.

**sul·fate** (sŭl′fāt′) *n.* [Fr. < Lat. *sulfur*, sulfur.] A chemical compound containing the bivalent group $SO_4$. —*v.* **-fat·ed, -fat·ing, -fates.** —*vt.* **1.** To treat or react with sulfuric acid or a sulfate. **2.** *Elect.* To cause lead sulfate to accumulate on (the plates of a lead-acid storage battery). —*vi.* To become sulfated.

**sul·fide** (sŭl′fīd′) *n.* A compound of bivalent sulfur with an electropositive element or group, esp. a binary compound of sulfur with a metal.

**sul·fi·nyl** (sŭl′fə-nĭl′) *n.* [SULF- + -IN + -YL.] The bivalent group SO.

**sul·fite** (sŭl′fīt′) *n.* A salt or ester of sulfurous acid. **—sul·fit′ic** (sŭl-fĭt′ĭk) *adj.*

**sulfo-** *pref. var. of* SULF-.

**sulfon-** *pref.* [< SULFONE.] **1.** Sulfonic <*sulfon*amide> **2.** Sulfonyl <*sulfon*methane>

**sul·fon·a·mide** (sŭl-fŏn′ə-mīd′, -mĭd) *n.* Any of a group of organic sulfur compounds having the general formula $RSO_2NH_2$.

**sul·fo·nate** (sŭl′fə-nāt′) *n.* A compound in which a hydrogen atom is replaced by the sulfonic acid group $SO_2OH$. —*vt.* **-nat·ed, -nat·ing, -nates. 1.** To introduce into (an organic compound) one or more sulfonic acid groups. **2.** To treat with sulfonic acid.

**sul·fone** (sŭl′fōn′) *n.* Any of various organic sulfur compounds having a sulfonyl group attached to two carbon atoms, esp. such a compound used to treat leprosy or tuberculosis.

**sul·fon·ic** (sŭl-fŏn′ĭk) *adj.* Of or pertaining to the chemical group $SO_3H$.

**sul·fon·ic acid** (sŭl-fŏn′ĭk, -fō′nĭk) *n.* Any of several organic acids containing one or more sulfonic groups, $SO_2OH$.

**sul·fo·ni·um** (sŭl-fō′nē-əm) *n.* [SULF- + (AMM)ONIUM.] The univalent cation $H_3S$.

**sul·fon·meth·ane** (sŭl′fōn-mĕth′ān′, -fōn-) *n.* A colorless crystalline or powdered compound, $C_7H_{16}S_2O_4$, used medicinally as a hypnotic.

**sul·fo·nyl** (sŭl′fə-nĭl′) *n.* The bivalent radical $SO_2$.

**sulf·ox·ide** (sŭl-fŏk′sīd′) *n.* Any of various organic compounds containing a sulfinyl group.

**sul·fur** *also* **sul·phur** (sŭl′fər) [ME *sulphre* < Lat. *sulfur.*] —*n. Symbol* **S** A pale-yellow nonmetallic element used in gunpowder, rubber vulcanization, the making of insecticides and pharmaceuticals, and in preparing industrial chemicals; atomic number 16; atomic weight 32.064. —*vt.* **-fured, -fur·ing, -furs** *also* **-phured, -phur·ing, -phurs.** To treat with sulfur or a compound of sulfur.

**sul·fu·rate** (sŭl′fə-rāt′, -fyə-) *vt.* **-rat·ed, -rat·ing, -rates.** To treat or react with sulfur. **—sul′fu·ra′tion** *n.*

**sulfur bacterium** *n.* A bacterium able to oxidize sulfur compounds.

**sulfur dioxide** *n.* A colorless, very irritating gas or liquid, $SO_2$, used esp. in manufacturing sulfuric acid.

**sul·fu·re·ous** (sŭl-fyŏŏr′ē-əs) *adj.* Of or relating to sulfur : SULFUROUS.

**sul·fu·ret** (sŭl′fə-rĕt′, -fyə-) *vt.* **-ret·ed, -ret·ing, -rets** *or* **-ret·ted, -ret·ting, -rets.** [< NLat. *sulfuretum*, sulfide < Lat. *sulfur*, sulfur.] To sulfurize. —*n.* A sulfide.

**sul·fu·ric** (sŭl-fyŏŏr′ĭk) *adj.* Of, pertaining to, or containing sulfur, esp. with valence 6.

**sulfuric acid** *n.* A highly corrosive, dense oily liquid, $H_2SO_4$, used to make chemicals and materials including fertilizers, paints, detergents, and explosives.

**sul·fu·rize** (sŭl′fə-rīz′, -fyə-) *vt.* **-ized, -iz·ing, -iz·es. 1.** To treat or impregnate with sulfur : SULFURET. **2.** To bleach or fumigate with sulfur or sulfur dioxide. **—sul′fur·i·za′tion** *n.*

**sul·fur·ous** (sŭl′fər-əs, -fyər-, sŭl-fyŏŏr′əs) *adj.* **1.** Of, pertaining to, derived from, or containing sulfur, esp. in its lower valence. **2.** Characteristic of or emanating from burning sulfur. **3.** *also* **sul·phur·ous.** Of or suggesting the fires of hell : HELLISH.

**sulfurous acid** *n.* A colorless solution of sulfur dioxide in water, $H_2SO_3$, characterized by a suffocating odor, used as a bleaching agent, preservative, and disinfectant.

**sulfur trioxide** *n.* A corrosive compound, $SO_3$, having three solid forms that may coexist in a given sample, used in the sulfonation of organic compounds.

**sul·fur·y** (sŭl′fə-rē) *adj.* **1.** Like or suggesting sulfur. **2.** *also* **sul·phur·y.** SULFUROUS 3.

**sul·fur·yl** (sŭl′fə-rĭl′, -fyə-) *n.* Sulfonyl.

**sulfuryl chloride** *n.* A colorless liquid, $SO_2Cl_2$, with a pungent odor, used as a chlorinating and dehydrating agent and in making pharmaceuticals, dyestuffs, and poison gases.

**sulk** (sŭlk) *vi.* **sulked, sulk·ing, sulks.** [Back-formation < SULKY.] To be sullenly withdrawn or aloof, as in silent resentment or protest. —*n.* A mood or exhibition of sulking.

☆ **syns:** SULK, MOPE, POUT *v. core meaning* : to be sullenly withdrawn or aloof, as in silent resentment or protest <*sulked* because they were not invited to the party>

**sulk·y**[1] (sŭl′kē) *adj.* **-i·er, -i·est.** [Orig. unknown.] **1.** Sullenly withdrawn or aloof. **2.** Dismal : gloomy <*sulky* winter weather> **—sulk′i·ly** *adv.* **—sulk′i·ness** *n.*

**sulk·y**[2] (sŭl′kē) *n., pl.* **-ies.** [< SULKY[1] (from its having a single seat).] A light two-wheeled vehicle holding a single passenger and drawn by one horse.

**sul·lage** (sŭl′ĭj) *n.* [Perh. < OFr. *souiller*, to soil. —see SOIL[2].] **1.** Silt deposited by a current of water. **2.** Waste materials : SEWAGE.

**sul·len** (sŭl′ən) *adj.* **-er, -est.** [ME *solein* < OFr. *sol*, alone < Lat. *solus.*] **1.** SULKY[1] 1. **2.** SULKY[1] 2. **3.** Sluggish : slow <a *sullen* march> **—sul′len·ly** *adv.* **—sul′len·ness** *n.*

**sul·ly** (sŭl′ē) *vt.* **-lied, -ly·ing, -lies.** [Prob. < OFr. *souiller*, to soil. —see SOIL[2].] **1.** To spoil the luster or cleanness of : SOIL. **2.** To defile : taint <a scandal that *sullied* a hitherto illustrious family name> —*n., pl.* **-lies.** *Archaic.* Something that sullies : STAIN.

**sul·phur**[1] (sŭl′fər) *n.* [< SULPHUR[2].] A butterfly of the genus *Colias* or related genera, having yellow or orange wings marked with black.

**sul·phur**[2] (sŭl′fər) *n.* & *v. var. of* SULFUR.

**sul·phur-bot·tom** (sŭl′fər-bŏt′əm) *n.* The blue whale.

**sulphur butterfly** *n.* SULPHUR[1].

**sul·tan** (sŭl′tən) *n.* [OFr. < Med. Lat. *sultanus* < Ar. *sulṭān.*] The ruler of a Moslem country, esp. of the former Ottoman Empire. **—sul′tan·ate** *n.*

**sul·tan·a** (sŭl-tăn'ə, -tä'nə) *n.* [Ital., fem. of *sultano,* sultan < Ar. *sulṭān.*] **1.** The wife, mother, sister, or daughter of a sultan. **2.** The mistress of a sultan, king, or prince. **3.** A small yellow seedless raisin of a kind orig. produced in Asia Minor.

**sul·try** (sŭl'trē) *adj.* **-tri·er, -tri·est.** [< obs. *sulter,* to swelter, var. of SWELTER.] **1.** Quite hot and humid. **2.** Extremely hot : TORRID. **3.** Sensual : voluptuous <a *sultry* torch singer> **—sul'tri·ly** *adv.* **—sul'tri·ness** *n.*

**Su·lu** (so͞o'lo͞o) *n., pl.* **Su·lus** or **Sulu.** [Sulu *sulug,* current.] **1. a.** A Moro people of the Sulu Archipelago. **b.** A member of the Sulus. **2.** The Austronesian language of the Sulus. **—Su'lu·an** *adj.* & *n.*

**sum** (sŭm) *n.* [ME *summe* < OFr. < Lat. *summa* < *summus,* highest.] **1.** The total obtained as a result of adding. **2.** The whole amount, quantity, or number : AGGREGATE <the *sum* of modern scientific knowledge> **3.** An amount of money <a trifling *sum*> **4.** An arithmetic problem. **5.** A summary : gist. *—vt.* **summed, sum·ming, sums.** To add. **—sum up.** To summarize.

**su·mac** *also* **su·mach** (so͞o'măk', sho͞o'-) *n.* [ME < OFr. < Ar. *summāq.*] A shrub or small tree of the genus *Rhus,* having compound leaves and small greenish flower clusters followed by usu. hairy red fruits, some species of which cause an acute itching rash on contact.

**Su·me·ri·an** (so͞o-mîr'ē-ən, -mĕr'-) *adj.* Of or relating to ancient Sumer, its people, culture, or language. *—n.* **1.** A member of an ancient Babylonian people, prob. of non-Semitic origin, who established one of the earliest historic civilizations in Sumer in the fourth millennium B.C. **2.** The language of the Sumerians, of no known linguistic affiliation.

**sum·ma cum lau·de** (so͝om'ə ko͝om lou'də) *adv.* [Lat.] With the greatest praise <graduated *summa cum laude*>

**sum·ma·rize** (sŭm'ə-rīz') *vt.* **-rized, -riz·ing, -riz·es.** To make a summary of. **—sum'ma·ri·za'tion** *n.* **—sum'ma·ri'zer** *n.*

**sum·ma·ry** (sŭm'ə-rē) *adj.* [ME < Med. Lat. *summarius* < Lat. *summa,* sum.] **1.** Presenting the substance in a condensed form : CONCISE. **2.** Performed or meted out speedily and unceremoniously <*summary* punishment> *—n., pl.* **-ries.** A condensation of the substance of a larger work : ABSTRACT. **—sum·mar'i·ly** (sə-mĕr'ə-lē) *adv.* **—sum'ma·ri·ness** *n.*

**summary court-martial** *n.* A court-martial for trying minor offenses, consisting of one officer.

**sum·ma·tion** (sə-mā'shən) *n.* [< Med. Lat *summare,* to sum up < Lat. *summa,* sum.] **1.** The act or process of adding or totaling. **2.** A sum : aggregate. **3.** A concluding statement containing a summary of principal points, esp. of a case before a court of law.

**sum·mer[1]** (sŭm'ər) *n.* [ME *sumer* < OE *sumor.*] **1.** The usu. warmest season of the year occurring between spring and autumn. **2.** A period considered as a time of fruition, fulfillment, happiness, or beauty. **3.** A year <a person of 35 *summers*> *—v.* **-mered, -mer·ing, -mers.** *—vt.* To lodge or keep during the summer. *—vi.* To pass the summer. **—sum'mer·ly** *adj.* & *adv.*

**sum·mer[2]** (sŭm'ər) *n.* [ME < Norman Fr. *sumer,* beam, pack animal < VLat. **saumarius* < LLat. *sagmarius,* packhorse < *sagma,* packsaddle. —see SUMPTER.] **1.** A heavy horizontal timber serving as a supporting beam, esp. for the floor above. **2.** A lintel. **3.** A large heavy stone usu. set on the top of a column or pilaster to support an arch or lintel.

**summer cypress** *n.* A plant, *Kochia scoparia,* native to Eurasia, with dense foliage that turns bright red.

**sum·mer·house** (sŭm'ər-hous') *n.* A small roofed house in a garden or park providing shade : GAZEBO.

**sum·mer·sault** (sŭm'ər-sôlt') *n.* & *v. var. of* SOMERSAULT.

**summer savory** *n.* A European herb, *Satureja hortensis,* used as a seasoning.

**summer school** *n.* A summer academic session.

**summer solstice** *n. Astron.* A solstice.

**summer squash** *n.* A squash, as the crookneck or the cymling, eaten shortly after being picked rather than kept for storage.

**summer stock** *n.* The theatrical productions of stock companies presented during the summer.

**sum·mer·time** (sŭm'ər-tīm') *n.* SUMMER[1] 1.

**sum·mer·wood** (sŭm'ər-wo͝od') *n.* Wood that develops during the latter part of the growing season and is harder and less porous than springwood.

**sum·mer·y** (sŭm'ə-rē) *adj.* Relating to or resembling summer.

**sum·mit** (sŭm'ĭt) *n.* [ME *somette* < OFr. *sommette,* dim. of *som,* top < Lat. *summus,* highest.] **1.** The highest point, as of a mountain. **2.** The highest degree of achievement or status : ACME. **3. a.** The highest level, as of government. **b.** A summit conference.

**summit conference** *n.* A conference, esp. of the highest ranking officials of a government or governments.

**sum·mon** (sŭm'ən) *vt.* **-moned, -mon·ing, -mons.** [ME *somonen* < OFr. *somondre* < VLat. **summonere* < Lat. *summonēre,* to remind privately : *sub-,* secretly + *monēre,* to warn.] **1.** To call together : CONVENE. **2.** To request to appear : send for. **3.** To order to appear in court by the issuance of a summons. **4.** To order to do a specific act. **5.** To call forth : EVOKE <managed to *summon* a smile> **—sum'mon·er** (sŭm'ə-nər) *n.*

**sum·mons** (sŭm'ənz) *n., pl.* **-mons·es.** [ME *somones* < OFr. *somonse,* p.part. of *somondre,* to summon.] **1.** A call or order to appear or perform a specified act. **2.** *Law.* **a.** A notice summoning a defendant to report to a court. **b.** A notice issued to a person summoning him or her to report to court as a juror or witness. *—vt.* **-monsed, -mons·ing, -mons·es.** To serve a court summons to.

**sum·mum bo·num** (so͝om'əm bō'nəm) *n.* [Lat.] The highest or supreme good.

**su·mo** (so͞o'mō) *n.* [J. *sumō.*] Stylized Japanese wrestling in which a fighter loses if forced from the ring or if any part of his body except the soles of his feet touches the ground.

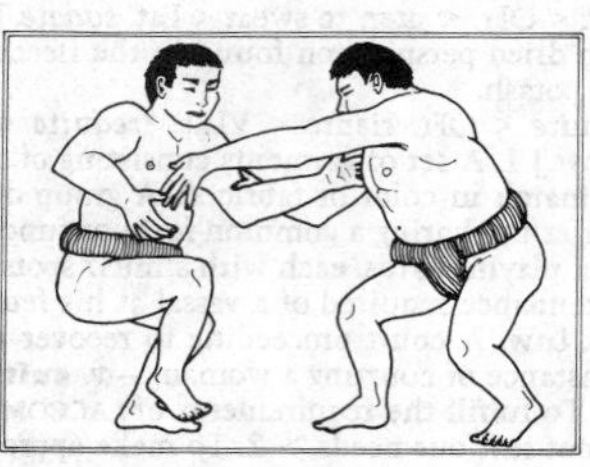

**sumo**

**sump** (sŭmp) *n.* [ME *sompe,* marsh < MLG *sump.*] **1. a.** A low land area receiving drainage. **b.** A cesspool. **2.** A hole at the lowest point of a mine shaft into which water is drained in order to be pumped out. **3.** The crankcase of an internal-combustion engine.

**sump·ter** (sŭmp'tər) *n.* [ME, driver of a packhorse < OFr. *sometier* < VLat. **saumatarius* < LLat. *sagma,* packsaddle < Gk. < *sattein,* to pack.] A pack animal, as a mule or horse.

**sump·tu·ar·y** (sŭmp'cho͞o-ĕr'ē) *adj.* [Lat. *sumptuarius* < *sumptus,* expense < *sumere,* to spend.] **1.** Regulating or limiting expenses. **2.** Attempting to regulate personal behavior on moral or religious grounds.

**sump·tu·ous** (sŭmp'cho͞o-əs) *adj.* [ME < OFr. *sumptueux* < Lat. *sumptuosus* < *sumptus,* expense. —see SUMPTUARY.] Of a size or magnificence suggesting great expense : LAVISH <a *sumptuous* palace> **—sump'tu·ous·ly** *adv.* **—sump'tu·ous·ness** *n.*

**sun** (sŭn) *n.* [ME *sonne* < OE *sunne.*] **1.** The central star of the solar system, having a mean distance from Earth of 93 million miles or approx. 150 million kilometers, a diameter of 864,000 miles or approx. 1,390,000 kilometers, and a mass about 330,000 times that of Earth. **2.** A star that is the center of a planetary system. **3.** The radiant energy, esp. heat and visible light, emitted by the sun : SUNSHINE. *—v.* **sunned, sun·ning, suns.** *—vt.* **1.** To expose to the sun's rays. **2.** To dry, warm, or tan in the sun. *—vi.* To bask in the sun. **—place in the sun.** A dominant or favorable position or situation. **—under the sun.** On earth : in the world <the best food *under the sun*>

**sun·bath** (sŭn'băth', -bäth') *n.* Exposure of the body to sun.

**sunbathe** (sŭn'bāth') *vi.* **-bathed, -bath·ing, -bathes.** To expose the body to the sun : SUN. **—sun'bath'er** (-bā'thər) *n.*

**sun·beam** (sŭn'bēm') *n.* A ray of sunlight.

**sun·belt** *or* **Sun·belt** (sŭn'bĕlt') *n.* The U.S. southern and southwestern states.

**sun·bird** (sŭn'bûrd') *n.* Any of various small, tropical Old World birds of the family Nectariniidae, with a slender downward-curving bill and often brightly colored plumage in the male.

**sun bittern** *n.* A cranelike tropical American bird, *Eurypyga helias,* having mottled brownish plumage and often spreading its wings and tail in a showy display.

**sun·bon·net** (sŭn'bŏn'ĭt) *n.* A wide-brimmed bonnet with a flap at the back to protect the neck from the sun.

**sun·bow** (sŭn'bō') *n.* A rainbowlike display of colors resulting from refraction of sunlight through a spray of water or mist.

**sun·burn** (sŭn'bûrn') *n.* Inflammation or blistering of the skin caused by overexposure to direct sunlight. *—vt.* & *vi.* **-burned** *or* **-burnt** (-bûrnt'), **-burn·ing, -burns.** To afflict with or be afflicted with sunburn.

**sun·burst** (sŭn'bûrst') *n.* **1.** A sudden burst of sunlight, as through broken clouds. **2. a.** A pattern or design consisting of a central disk with radiating spires projecting like sunbeams. **b.** A jeweled brooch with such a design.

**sun·dae** (sŭn'dē, -dā') *n.* [Orig. unknown.] A dish of ice cream with a topping including syrup, fruits, nuts, or whipped cream.

**sun dance** *n.* A ritual dance performed by the North American Plains Indians at the summer solstice.

**Sun·day** (sŭn'dē, -dā') *n.* [ME *Soneday* < OE *Sūnnandæg,* transl. of Lat. *dies solis,* day of the sun.] The first day of the week and the Christian Sabbath.

**Sunday school** *n.* **1.** A school, gen. affiliated with a church, that

offers religious instruction for children on Sundays. **2.** The teachers and pupils of a Sunday school.
**sun deck** *n.* A deck for sunbathing.
**sun·der** (sŭn′dər) *v.* **-dered, -der·ing, -ders.** [ME *sundren* < OE *sundrian.*] —*vt.* To break (something) apart : SEVER. —*vi.* To break into parts. **—sun′der·ance** *n.*
**sun·dew** (sŭn′do͞o, -dyo͞o) *n.* [Transl. of Lat. *ros solis.*] Any of several insectivorous plants of the genus *Drosera,* growing in wet ground and having leaves covered with sticky hairs.
**sun·di·al** (sŭn′dī′əl) *n.* An instrument indicating local apparent solar time by measuring the hour angle of the sun with a style that casts a shadow on a calibrated dial.
**sun disk** *n.* An ancient Near Eastern symbol composed of a disk set between outspread wings, representing the sun god.
**sun·dog** (sŭn′dôg′, -dŏg′) *n.* **1.** A parhelion. **2.** A small halo or rainbow near the horizon just off the parhelic circle.
**sun·down** (sŭn′doun′) *n.* SUNSET 1.
**sun·down·er** (sŭn′dou′nər) *n.* **1.** *Austral.* A tramp : vagrant. **2.** *Chiefly Brit.* A drink of liquor taken at sunset.
**sun·dries** (sŭn′drēz) *pl.n.* [< SUNDRY.] Miscellaneous articles too small or numerous to be specified.
**sun·drops** (sŭn′drŏps′) *pl.n.* [So called because the flowers remain open during the hours of sunlight.] *(sing. or pl. in number).* Any of several New World plants of the genus *Oenothera,* with four-petaled yellow flowers.
**sun·dry** (sŭn′drē) *adj.* [ME *sundri* < OE *syndrig,* separate.] Various : miscellaneous <*sundry* cosmetic items>
**sun·fish** (sŭn′fĭsh′) *n., pl.* **sunfish** or **-fish·es.** [From its roundish body and bright colors.] **1.** Any of various small North American freshwater fishes of the family Centrarchidae, with laterally compressed, often brightly colored bodies. **2.** Any of several large marine fishes of the family Molidae, esp. the ocean sunfish.
**sun·flow·er** (sŭn′flou′ər) *n.* **1.** A plant of the genus *Helianthus,* esp. *H. annuus,* with tall coarse stems and large yellow-rayed flowers that yield edible seeds rich in oil. **2.** A brilliant yellow to strong or vivid orange yellow. **—sun′flow′er** *adj.*
**sung** (sŭng) *v. p.p. & var. p.t. of* SING.
**sun·glass** (sŭn′glăs′) *n.* A burning glass.
**sun·glass·es** (sŭn′glăs′ĭz) *pl.n.* Eyeglasses with tinted or polarizing lenses to protect the eyes from the glare of the sun.
**sun·glow** (sŭn′glō′) *n.* A rose or yellow glow in the sky preceding sunrise or following sunset.
**sun god** *n.* A god personifying the sun.
**sunk** (sŭngk) *v. p.p. & var. p.t. of* SINK.
**sunk·en** (sŭng′kən) *adj.* [P.part. of SINK.] **1.** Depressed, fallen in, or hollowed <*sunken* cheeks> **2.** Situated below the surface of the water or ground <a *sunken* schooner><a *sunken* garden>
**sun lamp** *n.* **1.** A lamp that radiates over a wide range of the spectrum from ultraviolet to infrared and is used in therapeutic and cosmetic treatments. **2.** A high-intensity lamp with parabolic mirrors, used in photography.
**sun·less** (sŭn′lĭs) *adj.* **1.** Without sunlight. **2.** Cheerless : gloomy. **—sun′less·ness** *n.*
**sun·light** (sŭn′līt′) *n.* SUNSHINE 1.
**sun·lit** (sŭn′lĭt′) *adj.* Illuminated by the sun.
**sunn** (sŭn) *n.* [Hindi *san* < Skt. *sāṇa,* hempen.] **1.** A plant, *Crotalaria juncea* of tropical Asia, with yellow flower clusters. **2.** A tough fiber from the stems of sunn, used for cordage.
**Sun·na** *also* **Sun·nah** (so͝on′ə) *n.* [Ar. *sunnah.*] The body of traditional Moslem law, observed by the orthodox Moslems and based on the practices and teachings of Mohammed.
**Sun·ni** (so͝on′ē) *n.* [Ar. *sunnīy* < *sunnah,* Sunna.] The great branch of Islam following orthodox tradition and accepting the first four caliphs as rightful successors of Mohammed.
**Sun·nite** (so͝on′īt′) *n.* [< SUNNI.] A Moslem of the Sunni.
**sun·ny** (sŭn′ē) *adj.* **-ni·er, -ni·est. 1.** Exposed to or abounding in sunshine <a *sunny* porch> **2.** Cheerful <*sunny* smiles> **—sun′ni·ly** *adv.* **—sun′ni·ness** *n.*
**sun·ny-side up** (sŭn′ē-sīd′ ŭp′) *adj.* Fried only on one side<eggs *sunny-side up*>
**sun·rise** (sŭn′rīz′) *n.* **1.** The event or time of the daily first appearance of the sun above the eastern horizon. **2.** An outset or emergence <the *sunrise* of a new generation>
**sun·roof** (sŭn′ro͞of′, -ro͝of′) *n.* A roof on a car with a panel that can be slid back or lifted up.
**sun·scald** (sŭn′skôld′) *n.* An injury to woody plants marked by localized death of plant tissues and caused by excessive sun in summer and by the combined effects of sun and low temperatures in winter.
**sun·screen** (sŭn′skrēn′) *n.* A substance, often a cream or lotion, used to protect the skin from the sun's damaging ultraviolet rays. **—sun′screen′ing** *adj.*
**sun·seek·er** (sŭn′sē′kər) *n.* **1.** One who travels to a warm sunny region. **2.** A photoelectric navigational device on a spacecraft or artificial satellite that maintains a constant fix on the sun.
**sun·set** (sŭn′sĕt′) *n.* **1.** The event or time of the daily disappearance of the sun below the western horizon. **2.** A decline or final phase <the *sunset* of the Russian empire>
**sun·shade** (sŭn′shād′) *n.* Something, as an awning or a billed cap, used or worn as a protection from the sun's rays.
**sun·shine** (sŭn′shīn′) *n.* **1.** The sun's light or its direct rays. **2. a.** Happiness : cheerfulness. **b.** A source of happiness or cheerfulness. **—sun′shin′y** *adj.*
**sun·spot** (sŭn′spŏt′) *n.* Any of the relatively dark spots that appear in groups on the surface of the sun, that have an approximate 11-year cycle, and are associated with strong magnetic fields.
**sun·stone** (sŭn′stōn′) *n.* Aventurine.
**sun·stroke** (sŭn′strōk′) *n.* Heat stroke caused by direct exposure to the sun and marked by a rise in temperature, convulsions, and coma.
**sun·tan** (sŭn′tăn′) *n.* A tan color on the skin caused by exposure to the sun. **—sun′tanned′** *adj.*
**sun·up** (sŭn′ŭp′) *n.* SUNRISE 1.
**sun·ward** (sŭn′wərd) *adj. & adv.* At or toward the sun. **—sun′wards** (-wərdz) *adv.*
**sun·wise** (sŭn′wīz′) *adv.* From left to right, like the sun's course as viewed in the Northern Hemisphere.
**sup**[1] (sŭp) *v.* **supped, sup·ping, sups.** [ME *soupen* < OE *sūpan.*] —*vt.* To take (a liquid) into the mouth by sips. —*vi.* To take liquid into the mouth in small amounts. —*n.* A small mouthful of liquid.
**sup**[2] (sŭp) *vi.* **supped, sup·ping, sups.** [ME *soupen* < OFr. *souper* < *soupe,* soup. —see SOUP.] To eat the evening meal.
**supe** (so͞op) *n.* [Short for SUPERNUMERARY.] *Slang.* A supernumerary actor : EXTRA.
**su·per** (so͞o′pər) *n.* **1.** *Informal.* A superintendent of an apartment or office building. **2.** *Informal.* An extra person, esp. a super. **3.** An article or product of superior size or quality. **4.** A thin starched cotton mesh for reinforcing books. —*adj. Slang.* Ideal : first-rate <a *super* way to end an evening> —*vt.* **-pered, -per·ing, -pers.** To reinforce or strengthen (a book) with super.
**super-** *pref.* [< Lat. *super,* over, above.] **1.** Above : over : upon <*superimpose*> **2.** Superior in size, quality, number, or degree <*superfine*> **3. a.** Exceeding a norm <*supersaturate*> **b.** Excessive in degree or intensity <*supersubtle*> **c.** Containing a specified ingredient in an unusually high proportion <*superphosphate*> **4.** More inclusive than a specified category <*superorder*>

▲ word history: The prefix *super-* is from Latin *super,* "above." The comparative and superlative forms of Latin *super* also appear in English. The comparative is *superior,* "higher," which also meant "greater" and "better"; the English word *superior* also has these meanings. The superlative is *supremus,* which meant "highest" as well as "last, ultimate, final." English *supreme* preserves the sense "highest" in its figurative use. Latin *super* is descended from the Indo-European form *uper,* "over, above," which descended into English as *over.* The English prefix *hyper-* is a borrowing of *hyper,* "over, above," the Greek descendent of the same Indo-European form.

**su·per·a·ble** (so͞o′pər-ə-bəl) *adj.* [Lat. *superabilis* < *superare,* to overcome < *super,* over.] Capable of being surmounted or overcome. **—su′per·a·ble·ness** *n.* **—su′per·a·bly** *adv.*
**su·per·a·bound** (so͞o′pər-ə-bound′) *vi.* **-bound·ed, -bound·ing, -bounds.** [ME *superabounden* < LLat. *superabundare* : Lat. *super-,* excessively + Lat. *abundare,* to overflow. —see ABOUND.] To be unusually or excessively abundant.
**su·per·a·bun·dant** (so͞o′pər-ə-bŭn′dənt) *adj.* [ME < LLat. *superabundans,* pr.part. of *superabundare,* to superabound.] Abundant to excess. **—su′per·a·bun′dance** *n.* **—su′per·a·bun′dant·ly** *adv.*
**su·per·al·loy** (so͞o′pər-ăl′oi) *n.* A complex, extremely strong, temperature-resistant alloy.
**su·per·an·nu·ate** (so͞o′pər-ăn′yo͞o-āt′) *vt.* **-at·ed, -at·ing, -ates.** [Back-formation from SUPERANNUATED.] **1.** To allow to retire on a pension due to age or infirmity. **2.** To discard or set aside as old-fashioned or obsolete.
**su·per·an·nu·at·ed** (so͞o′pər-ăn′yo͞o-ā′tĭd) *adj.* [Med. Lat. *superannuatus,* p.part. of *superannuari,* to be too old : Lat. *super-,* over + Lat. *annus,* year.] **1.** Retired or ineffective because of advanced age. **2.** Obsolete : antiquated.
**su·perb** (so͞o-pûrb′) *adj.* [Lat. *superbus,* proud < *super,* over.] **1.** Of the highest quality <a *superb* painting> **2.** Majestic : imposing. **3.** Rich : luxurious. **—su·perb′ly** *adv.* **—su·perb′ness** *n.*
**super band** *n.* The radio frequency range from 216 to 600 megahertz, used chiefly for citizens band and cable television transmissions.
**su·per·cal·en·der** (so͞o′pər-kăl′ən-dər) *n.* A calender with a number of rollers for giving a high gloss or finish to paper. —*vt.* **-dered, -der·ing, -ders.** To give high gloss or finish to.
**su·per·car·go** (so͞o′pər-kär′gō) *n., pl.* **-goes** or **-gos.** [Sp. *sobrecargo* : *sobre-,* over (< Lat. *super-*) + *cargo,* cargo. —see CARGO.] A merchant marine officer in charge of a cargo and its sale and purchase.
**su·per·charge** (so͞o′pər-chärj′) *vt.* **-charged, -charg·ing, -charg·es. 1.** To increase the power of (e.g., an engine) as by fitting it with

a supercharger. **2.** To charge or load excessively : OVERLOAD. —*n.* An extra charge.

**su·per·charg·er** (so͞o′pər-chär′jər) *n.* A compressor, usu. driven by the engine, for supplying air under high pressure to the cylinders of an internal-combustion engine.

**su·per·cil·i·ar·y** (so͞o′pər-sĭl′ē-ĕr′ē) *adj.* [NLat. *superciliaris* < Lat. *supercilium*, eyebrow. —see SUPERCILIOUS.] **1.** Of or relating to the eyebrow. **2.** Situated over the eyebrow.

**su·per·cil·i·ous** (so͞o′pər-sĭl′ē-əs) *adj.* [Lat. *superciliosus* < *supercilium*, pride, eyebrow : *super-*, above + *cilium*, eyelid.] Haughtily disdainful. **—su′per·cil′i·ous·ly** *adv.* **—su′per·cil′i·ous·ness** *n.*

**su·per·cil·i·um** (so͞o′pər-sĭl′ē-əm) *n.* [Lat. —see SUPERCILIOUS.] The eyebrow.

**su·per·cit·y** (so͞o′pər-sĭt′ē) *n.* A megalopolis.

**su·per·class** (so͞o′pər-klăs′) *n. Biol.* A taxonomic category ranking between a phylum and a class.

**su·per·clus·ter** (so͞o′pər-klŭs′tər) *n.* A large group of galaxies relatively close to each other.

**su·per·co·lum·nar** (so͞o′pər-kə-lŭm′nər) *adj.* **1.** Having one order of columns above another. **2.** Situated above a colonnade or column.

**su·per·con·duc·tiv·i·ty** (so͞o′pər-kŏn′dŭk-tĭv′ĭ-tē) *n., pl.* **-ties.** The flow of electric current without resistance in certain metals and alloys at temperatures near absolute zero. **—su′per·con·duc′tive** *adj.* **—su′per·con·duc′tor** *n.*

**su·per·con·ti·nent** (so͞o′pər-kŏn′tə-nənt) *n.* A protocontinent.

**su·per·cool** (so͞o′pər-ko͞ol′) *v.* **-cooled, -cool·ing, -cools.** —*vt.* To cool (a liquid) below a transition temperature without the transition occurring, esp. to cool below the freezing point without solidification. —*vi.* To become supercooled.

**su·per·cur·rent** (so͞o′pər-kûr′ənt, -kŭr′-) *n.* An electrical current flowing through a superconductor.

**su·per·dom·i·nant** (so͞o′pər-dŏm′ə-nənt) *n. Mus.* Submediant.

**su·per·du·per** (so͞o′pər-do͞o′pər) *adj.* [Redup. of SUPER.] *Slang.* Great : wonderful <a *superduper* vacation>

**su·per·e·go** (so͞o′pər-ē′gō, -ĕg′ō) *n. Psychoanal.* The division of the psyche that develops by the incorporation of the perceived moral standards of the community, is mainly unconscious, and includes the conscience.

**su·per·em·i·nent** (so͞o′pər-ĕm′ə-nənt) *adj.* [LLat. *supereminens, supereminent-* < Lat., pr.part. of *superemin̄ere*, to rise above : *super-*, above + *emin̄ere*, to stand out.] Supremely eminent. **—su′per·em′i·nence** *n.* **—su′per·em′i·nent·ly** *adv.*

**su·per·er·o·gate** (so͞o′pər-ĕr′ə-gāt′) *vi.* **-gat·ed, -gat·ing, -gates.** [LLat. *supererogare, supererogat-*, to spend more : Lat. *super-*, above + Lat. *erogare*, to spend (*ex-*, out + *rogare*, to ask).] To do more than is required or expected. **—su′per·er′o·ga′tion** (-gā′shən) *n.*

**su·per·e·rog·a·to·ry** (so͞o′pər-ĭ-rŏg′ə-tôr′ē, -tōr′ē) *also* **su·per·e·rog·a·tive** (-tĭv) *adj.* **1.** Performed or observed beyond the degree required or expected. **2.** Superfluous <*supererogatory* personnel>

**su·per·fam·i·ly** (so͞o′pər-făm′ə-lē) *n., pl.* **-lies.** A taxonomic category ranking between an order or its subdivisions and a family.

**su·per·fec·ta** (so͞o′pər-fĕk′tə) *n.* [Blend of SUPER- and PERFECTA.] A method of betting in which the bettor must pick the first four finishers of a race in the correct sequence in order to win.

**su·per·fe·cun·da·tion** (so͞o′pər-fē′kən-dā′shən, -fĕk′ən-) *n.* Impregnation of more than one ovum within a single menstrual cycle by separate acts of coitus, esp. by different males.

**su·per·fe·tate** (so͞o′pər-fē′tāt′) *vi.* **-tat·ed, -tat·ing, -tates.** [Lat. *superfetare, superfetat-* : *super-*, over + *fetare*, to breed < *fetus*, offspring.] To conceive when a fetus is already present in the uterus.

**su·per·fe·ta·tion** (so͞o′pər-fē-tā′shən) *n.* The presence of fetuses of different ages resulting from the fertilization and development of two or more ova liberated at different periods of ovulation in the same uterus.

**su·per·fi·cial** (so͞o′pər-fĭsh′əl) *adj.* [ME < LLat. *superficialis* < Lat. *superficies*, surface. —see SUPERFICIES.] **1.** Of, affecting, or being on or near the surface <a *superficial* cut> **2.** Concerned only with what is apparent or obvious : SHALLOW. **3. a.** Apparent rather than actual or substantial. **b.** Trivial : insignificant. **—su′per·fi′ci·al′i·ty** (-fĭsh′ē-ăl′ĭ-tē), **su′per·fi′cial·ness** *n.* **—su′per·fi′cial·ly** *adv.*

☆ ***syns:*** SUPERFICIAL, CURSORY, SHALLOW, SKETCHY, SKIN-DEEP, UNCRITICAL *adj. core meaning* : lacking intellectual depth or thoroughness <a *superficial* report on a critical issue>

**su·per·fi·cies** (so͞o′pər-fĭsh′ēz, -fĭsh′ē-ēz′) *n., pl.* **superficies.** [Lat. : *super-*, above + *facies*, face.] **1.** The surface of an area or body. **2.** External appearance or aspect.

**su·per·fine** (so͞o′pər-fīn′) *adj.* **1.** Of exceptional quality or refinement. **2.** Overdelicate or refined. **3.** Of extra fine texture <*superfine* sugar> **—su′per·fine′ness** *n.*

**su·per·flu·id** (so͞o′pər-flo͞o′ĭd) *n.* A fluid, as an electric current or a form of helium, exhibiting a frictionless flow at temperatures close to absolute zero. **—su′per·flu·id′i·ty** (-flo͞o-ĭd′ĭ-tē) *n.*

**su·per·flu·i·ty** (so͞o′pər-flo͞o′ĭ-tē) *n., pl.* **-ties. 1.** The quality or state of being superfluous. **2.** Something superfluous. **3.** Overabundance.

**su·per·flu·ous** (so͞o-pûr′flo͞o-əs) *adj.* [ME < Lat. *superfluus* < *superfluere*, to overflow : *super-*, over + *fluere*, to flow.] Being beyond what is sufficient or required : EXTRA. **—su·per′flu·ous·ly** *adv.* **—su·per′flu·ous·ness** *n.*

**su·per·gal·ax·y** (so͞o′pər-găl′ək-sē) *n., pl.* **-ies.** An exceptionally large group of galaxies.

**su·per·gi·ant** (so͞o′pər-jī′ənt) *n.* A very large star gen. of sufficient rotational speed to flatten from a spherical shape.

**su·per·graph·ics** (so͞o′pər-grăf′ĭks) *n. (sing. or pl. in number).* Brightly colored and simply designed graphic shapes of billboard proportions.

**su·per·heat** (so͞o′pər-hēt′) *vt.* **-heat·ed, -heat·ing, -heats. 1.** To heat excessively : OVERHEAT. **2.** To heat (steam or other vapor not in contact with its own liquid) beyond its saturation point at a given pressure. **3.** To heat (a liquid) above its boiling point at a given pressure without causing vaporization. —*n.* (so͞o′pər-hēt′). **1.** The amount that a vapor is superheated. **2.** The heat imparted during superheating. **—su′per·heat′er** *n.*

**su·per·het·er·o·dyne** (so͞o′pər-hĕt′ər-ə-dīn′) *adj.* [SUPER(SONIC) + HETERODYNE.] Indicating or relating to a form of radio reception in which the frequency of an incoming radio signal is converted to an intermediate frequency, by mixing with a locally generated signal, to facilitate amplification and the rejection of unwanted signals. —*n.* A superheterodyne radio receiver.

**su·per·high frequency** (so͞o′pər-hī′) *n.* A radio frequency between 3,000 and 30,000 megahertz.

**su·per·high·way** (so͞o′pər-hī′wā′) *n.* A broad arterial highway, as an expressway, intended for high-speed vehicular traffic.

**su·per·hu·man** (so͞o′pər-hyo͞o′mən) *adj.* **1.** Being above or beyond the human : DIVINE. **2.** Being beyond normal human ability, power, or experience <"soldiers driven mad by *superhuman* misery" —John Reed> **—su′per·hu·man′i·ty** (-măn′ĭ-tē) *n.* **—su′per·hu′man·ly** *adv.*

**su·per·im·pose** (so͞o′pər-ĭm-pōz′) *vt.* **-posed, -pos·ing, -pos·es.** To place on or over something else. **—su′per·im·pos′a·ble** *adj.* **—su′per·im′po·si′tion** (-ĭm′pə-zĭsh′ən) *n.*

**su·per·in·cum·bent** (so͞o′pər-ĭn-kŭm′bənt) *adj.* [Lat. *superincumbens, superincumbent-*, pr.part. of *superincumbere*, to lie on top of : *super-*, above + *incumbere*, to lie down. —see INCUMBENT.] Lying or resting on or above something else. **—su′per·in·cum′bence, su′per·in·cum′ben·cy** *n.*

**su·per·in·duce** (so͞o′pər-ĭn-do͞os′, -dyo͞os′) *vt.* **-duced, -duc·ing, -duc·es.** [LLat. *superinducere*, to bring upon : Lat. *super-*, over + Lat. *inducere*, to lead in. —see INDUCE.] To introduce as an addition to an existing effect or condition. **—su′per·in·duc′tion** *n.*

**su·per·in·tend** (so͞o′pər-ĭn-tĕnd′, so͞o′prĭn-) *vt.* **-tend·ed, -tend·ing, -tends.** [LLat. *superintendere* : Lat. *super-* over + Lat. *intendere*, to direct one's attention to. —see INTEND.] To exercise supervision over : be in charge of. **—su′per·in·ten′dence** *n.*

**su·per·in·ten·dent** (so͞o′pər-ĭn-tĕn′dənt, so͞o′prĭn-) *n.* One who is authorized to supervise or direct others. **—su′per·in·ten′dent** *adj.*

**su·pe·ri·or** (so͞o-pîr′ē-ər) *adj.* [ME < OFr. < Lat., comp. of *superus*, upper < *super*, over.] **1.** Higher in rank, station, or authority than others <a *superior* army officer> **2.** Of a higher nature or kind than others. **3.** Of great value or excellence : EXTRAORDINARY <*superior* work> **4.** Greater in number or amount than others. **5.** Affecting an attitude of disdain or conceit : HAUGHTY <a *superior* glance> **6.** Above being affected or influenced : indifferent or immune <"Trust magnates were *superior* to law" —Gustavus Myers> **7.** Located higher than another : UPPER. **8.** *Bot.* Located above and not in contact with the calyx and corolla. —Used of an ovary. **9.** Set above the main line of printed type. **10.** *Logic.* Of wider or more comprehensive application : GENERIC. —Used of a term or proposition. —*n.* **1.** One who surpasses another in rank or quality. **2.** The head of a religious community. **3.** A superior character or letter in printing. **—su·pe′ri·or′i·ty** (-ôr′ĭ-tē, -ŏr′-) *n.* **—su·pe′ri·or·ly** *adv.*

**superior conjunction** *n.* The position of a celestial body when it is on the opposite side of the sun from Earth.

**superior court** *n.* A court of general jurisdiction, above the inferior courts and below the higher courts of appeal.

**superiority complex** *n.* **1.** A feeling of being superior to others. **2.** A psychological defense in which feelings of superiority counter feelings of inferiority.

**superior planet** *n.* A planet whose mean distance from the sun is greater than that of Earth.

**su·per·ja·cent** (so͞o′pər-jā′sənt) *adj.* [Lat. *superjacens, superjacent-*, pr.part. of *superjacēre*, to lie over : *super-*, over + *jacēre*, to lie.] Resting or lying immediately above or upon something else.

**su·per·jet** (so͞o′pər-jĕt′) *n.* A supersonic jet aircraft.

**su·per·la·tive** (so͞o-pûr′lə-tĭv) *adj.* [ME *superlatyf* < OFr. < LLat. *superlativus* < Lat. *superlatus*, excessive : *super-*, over + *latus*, p.part. of *ferre*, to carry.] **1.** Highest in order, quality, or degree. **2.** Excessive or exaggerated <*superlative* compliments> **3.** Expressing or involving the extreme degree of comparison of an adjective or adverb. —*n.* **1.** Something of the greatest excellence. **2.** The highest degree : ACME. **3. a.** The superlative degree. **b.** An adjective or adverb expressing the superlative degree. **—su·per'la·tive·ly** *adv.*

**su·per·lu·na·ry** (sōō'pər-lōō'nə-rē) *also* **su·per·lu·nar** (-nər) *adj.* Located beyond the moon.

**su·per·man** (sōō'pər-măn') *n.* **1.** One apparently having more than human powers. **2.** An ideal superior man who, according to Nietzsche, represents the goal of human evolution because of his exercise of creative power and his ability to forgo transient pleasure.

**su·per·mar·ket** (sōō'pər-mär'kĭt) *n.* A large self-service retail market that sells food and household goods.

**su·per·mol·e·cule** (sōō'pər-mŏl'ĭ-kyōōl') *n.* Macromolecule.

**su·per·nal** (sōō-pûr'nəl) *adj.* [ME < OFr. < Lat. *supernus* < *super,* over.] **1.** Celestial : heavenly. **2.** Of, coming from, or being in the sky or high above. **—su·per'nal·ly** *adv.*

**su·per·na·tant** (sōō'pər-nāt'nt) *adj.* [Lat. *supernatans, supernatant-,* pr.part. of *supernatare,* to float : *super-,* above + *natare,* to swim.] Floating on the surface. **—su'per·na'tant** *n.*

**su·per·nate** (sōō'pər-nāt') *n.* [Short for SUPERNATANT.] The clear supernatant fluid over a sediment or precipitate.

**su·per·nat·u·ral** (sōō'pər-năch'ər-əl) *adj.* **1.** Of or relating to existence outside the natural world. **2.** Attributed to a power that seems to violate or go beyond natural laws : MIRACULOUS. **3.** Of or relating to a deity. —*n.* One that is supernatural. **—su'per·nat'u·ral·ly** *adv.* **—su'per·nat'u·ral·ness** *n.*

**su·per·nat·u·ral·ism** (sōō'pər-năch'ər-ə-lĭz'əm) *n.* **1.** The quality of being supernatural. **2.** Belief in a supernatural agency that intervenes in the course of natural laws. **—su'per·nat'u·ral·ist** *n.* & *adj.* **—su'per·nat'u·ral·is'tic** *adj.*

**su·per·nor·mal** (sōō'pər-nôr'məl) *adj.* **1.** Greatly exceeding the average or normal. **2.** Paranormal. **—su'per·nor·mal'i·ty** *n.*

**su·per·no·va** (sōō'pər-nō'və) *n., pl.* **-vae** (-vē) *or* **-vas.** A rare celestial phenomenon involving the explosion of most of the material in a star, resulting in an extremely bright short-lived object that emits vast amounts of energy.

**su·per·nu·mer·ar·y** (sōō'pər-nōō'mə-rĕr'ē, -nyōō'-) *adj.* [LLat. *supernumerarius* : Lat. *super-,* above + Lat. *numerus,* number.] **1.** Exceeding a fixed, prescribed, or standard number : EXTRA. **2.** Being in excess of the required or desired number : SUPERFLUOUS. —*n., pl.* **-ies. 1.** One that is in excess of the regular, necessary, or usual number. **2.** A performer without a speaking part.

**su·per·or·der** (sōō'pər-ôr'dər) *n.* A taxonomic category ranking between a class or one of its subdivisions and an order.

**su·per·o·vu·la·tion** (sōō'pər-ō'vyə-lā'shən, -ŏv'yə-) *n.* Production of a large number of ova at one time.

**su·per·phos·phate** (sōō'pər-fŏs'fāt') *n.* **1.** An acid phosphate. **2.** A fertilizer made by the action of sulfuric acid on phosphate rock that consists mainly of tribasic calcium phosphate, to form a mixture of gypsum and monobasic calcium phosphate.

**su·per·phys·i·cal** (sōō'pər-fĭz'ĭ-kəl) *adj.* Exceeding the known laws of physics.

**su·per·pose** (sōō'pər-pōz') *vt.* **-posed, -pos·ing, -pos·es. 1.** To set or place over or above something else. **2.** To place (one geometric figure) over another so that all like parts coincide.

**su·per·pow·er** (sōō'pər-pou'ər) *n.* A powerful, influential nation, esp. one dominating its satellites and allies in an international power bloc <the U.S. and the U.S.S.R. as *superpowers*>

**su·per·sat·u·rate** (sōō'pər-săch'ə-rāt') *vt.* **-rat·ed, -rat·ing, -rates.** To cause (a chemical solution) to be more highly concentrated than is normally possible under given temperature and pressure conditions. **—su'per·sat'u·ra'tion** *n.*

**su·per·scribe** (sōō'pər-skrīb') *vt.* **-scribed, -scrib·ing, -scribes.** [Lat. *superscribere,* to write over : *super-,* over + *scribere,* to write.] **1.** To write (e.g., a character) on the outside or upper part of. **2.** To write (e.g., a name) on the top or outside of : ADDRESS. **—su'per·scrip'tion** *n.*

**su·per·script** (sōō'pər-skrĭpt') *n.* [< Lat. *superscriptus,* p.part. of *superscribere,* to write over. —see SUPERSCRIBE.] A figure, symbol, or letter printed or written above and immediately to one side of another <3 is the *superscript* in $x^3$> **—su'per·script** *adj.*

**su·per·sede** (sōō'pər-sēd') *vt.* **-sed·ed, -sed·ing, -sedes.** [ME *superceden,* to postpone < OFr. *superceder* < Lat. *supersedēre,* to refrain from : *super-,* above + *sedēre,* to sit.] **1.** To replace : supplant. **2.** To cause to be set aside or replaced by another. **—su'per·sed'er** *n.* **—su'per·ses'sion** (-sĕsh'ən) *n.*

**su·per·se·de·as** (sōō'pər-sē'dē-əs) *n.* [Med. Lat. < Lat., you must desist, the first word of the writ.] *Law.* A writ issued to stay a legal proceedings.

**su·per·se·dure** (sōō'pər-sē'jər) *n.* **1.** The act or process of superseding. **2.** Replacement of a queen bee by a superior or younger one.

**su·per·sen·si·ble** (sōō'pər-sĕn'sə-bəl) *adj.* Being beyond or outside perception by the senses. **—su'per·sen'si·bly** *adv.*

**su·per·son·ic** (sōō'pər-sŏn'ĭk) *adj.* Having, caused by, related to, or traveling at a speed greater than the speed of sound in a specified medium. **—su'per·son'i·cal·ly** *adv.*

**su·per·son·ics** (sōō'pər-sŏn'ĭks) *n. (sing. in number).* The study of phenomena produced by the motion of a body through a medium at velocities greater than that of sound.

**supersonic transport** *n.* A large transport aircraft engineered to fly at supersonic speeds.

**su·per·star** (sōō'pər-stär') *n.* A star, as in films, music, or sports, who has great popular appeal and whose talents are considered exceptional.

**su·per·sti·tion** (sōō'pər-stĭsh'ən) *n.* [ME *superstición* < OFr. *superstition* < Lat. *superstitio* < *superstare,* to stand over : *super-,* over + *stare,* to stand.] **1. a.** A belief, practice, or rite held in spite of evidence to the contrary, resulting from ignorance of the laws of nature or from faith in magic or chance. **b.** A fearful or abject state of mind resulting from such ignorance or irrationality. **2.** A body of superstitious beliefs.

**su·per·sti·tious** (sōō'pər-stĭsh'əs) *adj.* **1.** Having superstitions. **2.** Of, marked by, or proceeding from superstition. **—su'per·sti'tious·ly** *adv.* **—su'per·sti'tious·ness** *n.*

**su·per·stra·tum** (sōō'pər-strā'təm, -străt'əm) *n., pl.* **-stra·ta** (-strā'tə, -străt'ə). One stratum superimposed on another.

**su·per·struc·ture** (sōō'pər-strŭk'chər) *n.* **1.** A structure built on top of another structure. **2.** That part of a structure, as a building, built above the foundation. **3.** The rails, sleepers, and other parts of a railway. **4.** The parts of a ship's structure above the main deck. **—su'per·struc'tur·al** *adj.*

**su·per·sub·tle** (sōō'pər-sŭt'l) *adj.* Subtle to an extreme degree.

**su·per·ton·ic** (sōō'pər-tŏn'ĭk) *n. Mus.* The second tone of a diatonic scale.

**su·per·vene** (sōō'pər-vēn') *vi.* **-vened, -ven·ing, -venes.** [Lat. *supervenire* : *super-,* in addition to + *venire,* to come.] **1.** To occur or come as an extraneous, additional, or unexpected element or factor. **2.** To take place immediately after : ENSUE. **—su'per·ven'ient** (-vēn'yənt) *adj.* **—su'per·ven'tion** (-vĕn'shən) *n.*

**su·per·vise** (sōō'pər-vīz') *vt.* **-vised, -vis·ing, -vis·es.** [Med. Lat. *supervidēre, supervis-,* to look over : Lat. *super-,* over + Lat. *vidēre,* to see.] To direct and watch over the work and performance of. **—su'per·vi'sion** (-vĭzh'ən) *n.*

☆ ***syns:*** SUPERVISE, BOSS, OVERSEE, SUPERINTEND *v. core meaning* : to direct and watch over the work and performance of others <*supervised* a team of investigators>

**su·per·vi·sor** (sōō'pər-vī'zər) *n.* **1.** One who supervises. **2.** An elected administrative officer in some U.S. counties and townships. **3.** One in charge of a department or unit, as in a governmental agency or a school system. **—su'per·vi'so·ry** (-vī'zə-rē) *adj.*

**su·pi·nate** (sōō'pə-nāt') *v.* **-nat·ed, -nat·ing, -nates.** [Lat. *supinare, supinat-,* to bend backward < *supinus,* backward.] —*vt.* To rotate (e.g., the hand) so that the palm faces upward. —*vi.* To assume a position with the palm upward. **—su'pi·na'tion** *n.*

**su·pi·na·tor** (sōō'pə-nā'tər) *n.* A muscle in the forearm that makes supination possible.

**su·pine¹** (sōō-pīn', sōō'pīn') *adj.* [Lat. *supinus.*] **1.** Lying on the back with the face upward. **2.** Having the palm facing upward. —Used of the hand. **3.** Indisposed to act : PASSIVE. **—su·pine'ly** *adv.* **—su·pine'ness** *n.*

**su·pine²** (sōō'pīn') *n.* [LLat. *supinum* < Lat. *supinus,* backward.] A Latin verbal noun having an accusative in *-um* and an ablative in *-ū.*

**sup·per** (sŭp'ər) *n.* [ME *suppere* < OFr. *souper.* —see SUP².] **1. a.** An evening meal. **b.** A light evening meal. **2.** A social affair, as a dance, during which supper is served.

**supper club** *n.* A nightclub.

**sup·plant** (sə-plănt') *vt.* **-plant·ed, -plant·ing, -plants.** [ME *supplanten* < *supplanter* < Lat. *supplantare,* to trip up : *sub-,* from below + *planta,* sole of the foot.] SUPERSEDE 1. **—sup·plant'er** *n.*

**sup·ple** (sŭp'əl) *adj.* **-pler, -plest.** [ME *souple* < OFr. < Lat. *supplex,* bending at the knees, humble.] **1.** Capable of being bent : PLIANT. **2.** Moving and bending with agility : LIMBER. **3.** Readily adaptable to influences or change. —*v.* **-pled, -pling, -ples.** —*vt.* **1.** To make supple. **2.** To make amiable, friendly, and agreeable. —*vi.* To become supple. **—sup'ple·ness** *n.* **—sup'ply, sup'ple·ly** *adv.*

**sup·ple·ment** (sŭp'lə-mənt) *n.* [ME < Lat. *supplementum* < *supplēre,* to complete. —see SUPPLY.] **1.** Something added to complete a thing, offset a deficiency, or strengthen the whole. **2.** A section added to a book or document to provide additional data, or to correct errors. **3.** A separate section treating a special subject that is inserted into a periodical, as a newspaper. **4.** The angle or arc that when added to a given angle or arc makes 180° or a semicircle. —*vt.* (sŭp'lə-mĕnt') **-ment·ed, -ment·ing, -ments.** To provide or form a supplement to. **—sup'ple·men'ta·ry** (-tə-rē, -trē), **sup'ple·men'tal** *adj.* **—sup'ple·men·ta'tion** (-mĕn-tā'shən) *n.*

**supplementary angle** *n.* SUPPLEMENT 4.

**sup·pli·ant** (sŭp'lē-ənt) *adj.* [ME < OFr., pr.part. of *supplier,* to entreat < Lat. *supplicare.* —see SUPPLICATE.] Expressing supplication : BESEECHING. —*n.* A supplicant. **—sup'pli·ant·ly** *adv.*

**sup·pli·cant** (sŭp'lĭ-kənt) *n.* One who supplicates. —*adj.* That supplicates <a *supplicant* pilgrim>

**sup·pli·cate** (sŭp'lĭ-kāt') *v.* **-cat·ed, -cat·ing, -cates.** [ME *supplicaten* < Lat. *supplicare,* to kneel down : *sub-,* down + *plicare,* to fold up.] —*vt.* **1.** To beg for humbly or earnestly, as by praying.

**2.** To make a humble request of : BESEECH. —*vi.* To make a humble and earnest request : BEG. **—sup′pli·ca′tion** *n.* **—sup′pli·ca·to′ry** (-kə-tôr′ē, -tōr′ē) *adj.*

**sup·ply** (sə-plī′) *v.* **-plied, -ply·ing, -plies.** [ME *supplen* < OFr. *soupleer* < Lat. *supplēre* : *sub-*, from below + *plēre*, to fill.] —*vt.* **1.** To make available for use : PROVIDE <*supplied* new uniforms> **2.** To equip with <*supplied* the troops with new uniforms> **3.** To fill sufficiently : SATISFY <*supply* a need> **4.** To make up for (e.g., a deficiency) : compensate for. **5.** To serve temporarily as a substitute in (e.g., a church) <*supplied* another priest's pulpit> —*vi.* To fill a position as a substitute. —*n., pl.* **-plies. 1.** The act of supplying. **2.** Something supplied. **3.** An amount available or sufficient for a given use : STOCK. **4.** *often* **supplies.** Materials or provisions stored and dispensed when needed. **5.** The amount of a commodity available for meeting a demand or for purchase at a given price. **6.** A cleric serving as a temporary substitute for another pastor. **—sup·pli′er** *n.*

**sup·ply-side** (sə-plī′sīd′) *adj.* Of, relating to, or supporting supply-side economics. **—sup·ply′-sid′er** *n.*

**supply-side economics** *n. (sing. in number).* A theory holding that large Federal tax cuts and other incentives for business will stimulate capital investments in new plants and equipment, spur demand for goods and services, and will be offset by the new tax revenues generated during the resultant period of increased economic activity.

**sup·port** (sə-pôrt′, -pōrt′) *vt.* **-port·ed, -port·ing, -ports.** [ME *supporten* < OFr. *supporter* < Lat. *supportare*, to carry : *sub-*, from below + *portare*, to carry.] **1.** To carry the weight of, esp. from below. **2.** To maintain in position so as to keep from falling, sinking, or slipping. **3.** To be able to bear : WITHSTAND. **4.** To keep from failing or yielding during stress. **5.** To provide for, by supplying with money or necessities <*support* a large family>. **6.** To furnish corroborating evidence for <*support* a witness's testimony> **7.** To aid the cause of by approving, favoring, or advocating <*support* a political candidate> **8.** To endure : tolerate. **9. a.** To act (a part or role). **b.** To act in a secondary or subordinate role to (a leading performer). —*n.* **1.** The act of supporting or the state of being supported. **2.** One that supports. **3.** A means of maintenance or subsistence. **—sup·port′a·ble** *adj.* **—sup·port′a·bly** *adv.*

**sup·port·er** (sə-pôr′tər, -pōr′-) *n.* **1.** One that supports. **2.** One who promotes or advocates : ADHERENT. **3.** An athletic supporter.

**sup·por·tive** (sə-pôr′tĭv, -pōr′-) *adj.* Providing support.

**sup·pose** (sə-pōz′) *v.* **-posed, -pos·ing, -pos·es.** [ME *supposen* < OFr. *supposer* < Med. Lat. *supponere* < Lat., to put under : *sub-*, under + *ponere*, to place.] —*vt.* **1.** To assume to be real or true for the sake of an argument or explanation. **2. a.** To believe, esp. on uncertain or tentative grounds. **b.** To consider to be likely. **3.** To imply as an antecedent condition : PRESUPPOSE. **4.** To consider as a suggestion <*Suppose* we vacation together.> —*vi.* To imagine : conjecture. **—sup·pos′a·ble** (-pō′zə-bəl) *adj.* **—sup·pos′a·bly** *adv.*

**sup·posed** (sə-pōzd′, -pō′zĭd) *adj.* **1.** Regarded as true, real, or genuine, esp. on dubious grounds. **2.** Intended <a drug that is *supposed* to cure arthritis> **3. a.** Required <You are *supposed* to take the test.> **b.** Permitted <You are not *supposed* to smoke here.> **—sup·pos′ed·ly** (-pō′zĭd-lē) *adv.*

**sup·pos·ing** (sə-pō′zĭng) *conj.* Assuming that <*Supposing* the rumor is true, what do we do now?>

**sup·po·si·tion** (sŭp′ə-zĭsh′ən) *n.* **1.** The act of supposing. **2.** Something supposed <testimony based on *supposition*> **—sup′po·si′tion·al** *adj.* **—sup′po·si′tion·al·ly** *adv.*

**sup·pos·i·ti·tious** (sə-pŏz′ĭ-tĭsh′əs) *adj.* [Lat. *suppositicius* < *supponere*, to substitute. —see SUPPOSE.] **1.** Substituted with intent to defraud : COUNTERFEIT. **2.** Hypothetical. **—sup·pos′i·ti′tious·ly** *adv.* **—sup·pos′i·ti′tious·ness** *n.*

**sup·pos·i·tive** (sə-pŏz′ĭ-tĭv) *adj.* Of the nature of, based on, or involving supposition. —*n.* A conjunction, such as *if* or *providing*, that introduces a supposition. **—sup·pos′i·tive·ly** *adv.*

**sup·pos·i·to·ry** (sə-pŏz′ĭ-tôr′ē, -tōr′ē) *n., pl.* **-ries.** [Med. Lat. *suppositorium* < LLat. *suppositorius*, placed under < *supponere*, to put under. —see SUPPOSE.] A solid medication designed to melt within a body cavity other than in the mouth.

**sup·press** (sə-prĕs′) *vt.* **-pressed, -press·ing, -press·es.** [ME *suppressen* < Lat. *supprimere* : *sub-*, down + *premere*, to press.] **1.** To put an end to forcibly : QUASH. **2.** To curtail or prohibit the activities of <*suppress* political opposition> **3. a.** To keep from being revealed, published, or circulated <*suppress* a news story> **b.** To keep from conscious awareness. **4.** To hold back (e.g., an impulse) : CHECK. **5.** To reduce the incidence or severity of (e.g., a hemorrhage). **—sup·press′er, sup·pres′sor** (-prĕs′ər) *n.* **—sup·press′i·ble** *adj.* **—sup·pres′sive** *adj.*

**sup·pres·sion** (sə-prĕsh′ən) *n.* **1.** The act of suppressing or the state of being suppressed. **2.** *Psychoanal.* Conscious exclusion of painful desires or thoughts from awareness.

**sup·pres·sor** (sə-prĕs′ər) *n.* **1.** A gene that completely reduces the phenotypic expression of a mutant gene. **2.** The grid between screen and plate in a pentode.

**sup·pu·rate** (sŭp′yə-rāt′) *vi.* **-rat·ed, -rat·ing, -rates.** [Lat. *suppurare, suppurat-* : *sub-*, under + *pus*, pus.] To form or discharge pus <an abscess that *suppurated*>

**sup·pu·ra·tion** (sŭp′yə-rā′shən) *n.* **1.** Formation or discharge of pus. **2.** Pus. **—sup′pu·ra′tive** *adj.*

**supra-** *pref.* [Lat. < *supra*, above, beyond, earlier.] **1.** Above : over : on top of <*supra*renal> **2.** Greater than : transcending <*supra*molecular> **3.** Earlier than <*supra*lapsarian>

**su·pra·glot·tal** (sōō′prə-glŏt′l) *adj.* **1.** Being above or anterior to the glottis. **2.** Designating a phone or phoneme produced by the speech organs anterior to the glottis.

**su·pra·lap·sar·i·an** (sōō′prə-lăp-sâr′ē-ən) *n.* [SUPRA- + Lat. *lapsus*, fall + -ARIAN.] A Calvinist who believes that God's determination of the elect preceded the fall of man from grace and that the fall itself had been predestined. **—su′pra·lap·sar′i·an** *adj.* **—su′pra·lap·sar′i·an·ism** *n.*

**su·pra·lim·i·nal** (sōō′prə-lĭm′ə-nəl) *adj.* Being above the threshold of consciousness or sensation. **—su′pra·lim′i·nal·ly** *adv.*

**su·pra·mo·lec·u·lar** (sōō′prə-mə-lĕk′yə-lər) *adj.* **1.** Made up of more than one molecule. **2.** Of greater complexity than a molecule.

**su·pra·or·bi·tal** (sōō′prə-ôr′bĭ-təl) *adj.* Sited above the orbit of the eye.

**su·pra·re·nal** (sōō′prə-rē′nəl) *adj.* Located on or above the kidney. —*n.* **1.** A suprarenal part. **2.** An adrenal gland.

**suprarenal gland** *n.* An adrenal gland.

**su·pra·scap·u·lar** (sōō′prə-skăp′yə-lər) *adj.* That is sited above the scapula.

**su·prem·a·cist** (sōō-prĕm′ə-sĭst) *n.* One believing that a certain group is or should be supreme.

**su·prem·a·cy** (sōō-prĕm′ə-sē) *n., pl.* **-cies. 1.** The quality or state of being supreme. **2.** Supreme power or authority.

**su·preme** (sōō-prēm′) *adj.* [Lat. *supremus*, superl. of *superus*, upper < *super*, over.] **1.** Highest in power, authority, or rank : PARAMOUNT. **2.** Greatest in importance, degree, significance, character, or achievement. **3.** Ultimate : final <the *supreme* sacrifice> **—su·preme′ly** *adv.* **—su·preme′ness** *n.*

**Supreme Being** *n.* GOD 1a.

**Supreme Court** *n.* **1.** The highest federal court in the United States, consisting of nine justices and having jurisdiction over all other courts in the nation. **2. supreme court.** The highest court in most U.S. states.

**Supreme Soviet** *n.* The legislature of the Soviet Union, consisting of two houses one of which represents the overall population and the other, the constituent republics.

**su·pre·mo** (sōō-prē′mō, sə-) *n., pl.* **-mos.** [Sp. and Ital. < *supremo*, supreme < Lat. *supremus*.] *Chiefly Brit.* One who is highest in authority or command.

**Su·qua·mish** (sə-kwä′mĭsh) *also* **Squa·mish** (skwä′-) *n., pl.* **Suquamish** *also* **Squamish** or **-mish·es. 1. a.** A tribe of Indians of the northwestern Pacific coast, west of Puget Sound. **b.** A member of this tribe. **2.** The Salish language of the Suquamish.

**sur-** *pref.* [ME < OFr. < Lat. *super-*. —see SUPER-.] **1.** Over : above : upon <*sur*print> **2.** Additional <*sur*tax>

**su·ra** (sŏŏr′ə) *n.* [Ar. *sūrah*.] One of the main sections of the Koran.

**su·rah** (sŏŏr′ə) *n.* [Fr. *surat*, after *Surat*, India, where it was originally made.] A soft twilled fabric of silk or of a silk and rayon blend.

**su·ral** (sŏŏr′əl) *adj.* [NLat. *suralis* < Lat. *sura*, calf of the leg.] Of or pertaining to the calf of the leg.

**sur·base** (sûr′bās′) *n.* A molding or border above the base of a structure such as a baseboard.

**sur·based** (sûr′bāst′) *adj.* **1.** Having a surbase. **2.** Relating to an arch with a rise less than half its span.

**sur·cease** (sûr′sēs′, sər-sēs′) [ME *sursesen* < OFr. *sursesoir*, to refrain < Lat. *supersedēre*. —see SUPERSEDE.] —*vt.* & *vi.* **-ceased, -ceas·ing, -ceas·es.** To bring or come to an end : STOP. —*n.* Cessation : end.

**sur·charge** (sûr′chärj′) *n.* [< ME *surchargen*, to overcharge < OFr. *surcharger* : *sur-*, excessively + *chargier*, to charge. —see CHARGE.] **1.** An additional sum added to the usual charge. **2.** An overcharge, esp. when unlawful. **3.** An additional excessive burden : OVERLOAD. **4. a.** A new value or denomination overprinted on a postage or revenue stamp. **b.** The stamp to which a new value has been applied. **5.** *Law.* The act of surcharging. —*vt.* **-charged, -charg·ing, -charg·es. 1.** To overcharge (a person). **2.** To place an excessive burden on : OVERLOAD. **3.** To fill beyond usual capacity : OVERFILL. **4.** To print a surcharge on (a postage or revenue stamp). **5.** *Law.* To show an omission of a credit in (an account).

**sur·cin·gle** (sûr′sĭng′gəl) *n.* [ME *sursengle* < OFr. *surcengle* : *sur-*, over (< Lat. *super-*) + *cengle*, belt < Lat. *cingula* < *cingere*, to gird.] **1.** A girth that binds a saddle, pack, or blanket to the body of a horse. **2.** The belt on a cassock. —*vt.* **-gled, -gling, -gles.** To bind or fasten with a surcingle.

**sur·coat** (sûr′kōt′) *n.* [ME *surcote* < OFr. : *sur-*, over + *cote*, coat. —see COAT.] **1.** A loose outer coat or gown. **2.** A tunic worn in the Middle Ages by a knight over his armor.

**sur·cu·lose** (sûr′kyə-lōs′) *adj.* [Lat. *surculosus*, woody < *surculus*, dim. of *surus*, branch.] *Bot.* Producing suckers.

**surd** (sûrd) *n.* [< Lat. *surdus*, deaf, trans. of Ar. (*jadhr*) *asạm*, deaf (root), transl. of Gk. *alogos*, speechless, irrational.] **1.** A sum, as $\sqrt{2} + \sqrt{3}$, containing one or more irrational roots of numbers. **2.** A voiceless speech sound. —*adj.* Voiceless.

**sure** (shŏŏr) *adj.* **sur·er, sur·est.** [ME < OFr. *sur* < Lat. *securus*, safe. —see SECURE.] **1.** Being beyond doubt or dispute : CERTAIN <*sure* proof> **2.** Unhesitating : firm <a *sure* conviction> **3.** Certain in expectations : CONFIDENT <*sure* of winning> **4. a.** Bound to happen : INEVITABLE <*sure* disaster> **b.** Having one's course inevitably directed : DESTINED <*sure* to inherit the throne> **5.** Steady <a *sure* hand on the rudder> **6.** Trustworthy : reliable. **7.** *Obs.* Free from danger or harm : SAFE. —*adv. Informal.* Surely : certainly. **—for sure.** Certainly : unquestionably <We'll lose *for sure.*> **—make sure.** To establish without doubt. **—to be sure.** Indeed : certainly. **—sure′ness** *n.*

**sure-fire** (shŏŏr′fīr′) *adj. Informal.* That is bound to be successful or perform as expected.

**sure-foot·ed** (shŏŏr′fŏŏt′ĭd) *adj.* Not apt to stumble or fall. **—sure′foot′ed·ly** *adv.* **—sure′foot′ed·ness** *n.*

**sure·ly** (shŏŏr′lē) *adv.* **1.** With confidence : UNHESITATINGLY. **2.** Undoubtedly : certainly <You *surely* can't be serious.> **3.** Without fail.

**sur·e·ty** (shŏŏr′ĭ-tē) *n., pl.* **-ties.** [ME *surte* < OFr. < Lat. *securitas* < *securus*, sure. —see SECURE.] **1.** The state of being sure, esp. of oneself : SELF-ASSURANCE. **2.** Something sure : CERTAINTY. **3.** A formal pledge made to secure against loss, damage, or default : GUARANTEE. **4.** One who has contracted to be responsible for another, esp. one who assumes responsibilities or debts in the event of default. **—sur′e·ty·ship′** *n.*

**surf** (sûrf) *n.* [Orig. unknown.] The waves of the sea as they break upon a reef or shore. —*vi.* **surfed, surf·ing, surfs.** To engage in surfing. **—surf′y** *adj.*

**sur·face** (sûr′fəs) *n.* [Fr. : *sur-*, above + *face*, face < OFr. —see FACE.] **1. a.** The exterior face of an object. **b.** A material layer constituting such an exterior face. **2.** *Math.* **a.** The boundary of a three-dimensional figure. **b.** The two-dimensional locus of points located in three-dimensional space whose height $z$ above each point $(x,y)$ of a region of a coordinate plane is specified by a function $f(x,y)$ of two arguments. **3.** Superficial or outward appearance. **4.** An airfoil. —*adj.* **1.** Relating to, on, or at a surface <*surface* algae> **2.** Intended to operate or be carried on land <*surface* troops> <*surface* mail> **3. a.** Superficial. **b.** Merely apparent as opposed to real. —*v.* **-faced, -fac·ing, -fac·es.** —*vt.* **1.** To form the surface of, as by smoothing or leveling <*surface* a road> **2.** To provide with a surface. —*vi.* **1.** To rise to the surface. **2.** To emerge after concealment <New evidence *surfaced.*>

**sur·face-ac·tive** (sûr′fəs-ăk′tĭv) *adj.* Designating a substance capable of reducing the surface tension of a liquid in which it is dissolved <*surface-active* reagents>

**sur·face-ef·fect ship** (sûr′fəs-ĭ-fĕkt′) *n.* A ground-effect vehicle that operates over water.

**surface of revolution** *n.* A surface generated by revolving a plane curve about an axis in its plane.

**surface plate** *n.* A planometer.

**surface tension** *n.* A property of liquids arising from unbalanced molecular cohesive forces at or near the surface, as a result of which the surface tends to contract and has properties resembling those of a stretched elastic membrane.

**sur·face-to-air missile** (sûr′fəs-tŏŏ-âr′) *n.* A guided missile launched from the ground against an airborne target.

**sur·fac·tant** (sər-făk′tənt, sûr′făk′-) *n.* [SURF(ACE) + ACT(IVE) + -ANT.] A surface-active substance.

**surf and turf** *n. Informal.* Seafood and beefsteak served as the main course of a meal, as in a restaurant.

**surf·bird** (sûrf′bûrd′) *n.* A shore bird, *Aphriza virgata*, of the Pacific coast of North and South America, with dark spotted plumage.

**surf·board** (sûrf′bôrd′, -bōrd′) *n.* A long, narrow, somewhat rounded board for surfing.

**surf·boat** (sûrf′bōt′) *n.* A strong seaworthy boat that can be launched or landed in heavy surf.

**surf·cast·ing** (sûrf′kăs′tĭng) *n.* The sport of fishing from shore, by casting one's line into the surf. **—surf′cast′er** *n.*

**sur·feit** (sûr′fĭt) *v.* **-feit·ed, -feit·ing, -feits.** [ME, excess < OFr. < *surfaire*, to overdo : *sur-*, excessively + *faire*, to do < Lat. *facere*.] —*vt.* To feed or supply to fullness or excess : SATIATE. —*vi. Archaic.* To overindulge. —*n.* **1. a.** Overindulgence in food or drink. **b.** The result of such overindulgence : SATIETY. **2.** An excessive amount. **—sur′feit·er** *n.*

**surf·er** (sûr′fər) *n.* One who engages in the sport of surfing.

**surfer's knee** *n.* Surfer's knobs.

**surfer's knobs** *pl.n.* Tumorlike overgrowths of connective tissue just below the knees, on the tops of the feet, and often on the toes, found in surfers who paddle in a kneeling position.

**surfer's knot** *n.* Surfer's knobs.

**sur·fi·cial** (sər-fĭsh′əl) *adj.* [SURF(ACE) + (SUPERF)ICIAL.] Of, relating to, or taking place on the earth's surface.

**surf·ing** (sûr′fĭng) *n.* The sport of riding the crests of waves on a surfboard.

**surf·perch** (sûrf′pûrch′) *n., pl.* **surfperch** *or* **-perch·es.** Any of various viviparous marine fishes of the family Embiotocidae, of North American Pacific coastal waters.

**surge** (sûrj) *v.* **surged, surg·ing, surg·es.** [Prob. < OFr. *sourgir* < Lat. *surgere*, to rise : *sub-*, from below + *regere*, to lead.] —*vi.* **1.** To move in a heavy, violent, swelling manner in or as if in waves. **2.** To roll or be tossed about on waves, as a boat. **3.** To increase suddenly. **4.** *Naut.* To slip around a windlass. —Used of a rope. —*vt.* To loosen or slacken (a cable) gradually. —*n.* **1.** A heavy, violent, or swelling motion. **2. a.** A wave, ground swell, or billow. **b.** Such waves as a whole. **3.** A sudden onrush <a *surge* of anger> **4.** A sudden transient increase in electric current. **5.** *Naut.* The part of a windlass into which the cable surges.

**sur·geon** (sûr′jən) *n.* [ME *surgien* < Norman Fr., short for OFr. *serurgien* < *serurgie*, surgery.] A physician specializing in surgery. **—sur′geon·cy** *n.*

**sur·geon·fish** (sûr′jən-fĭsh′) *n., pl.* **surgeonfish** *or* **-fish·es.** [From its lancetlike spines, which resemble surgeons' instruments.] Any of various bright-colored tropical marine fishes of the family Acanthuridae, with a sharp erectile spine near the base of the tail.

**surgeon general** *n., pl.* **surgeons general.** The chief medical officer of an armed force or a public health service.

**surgeon's knot** *n.* Any of several knots used in surgery for tying ligatures or stitching incisions.

**sur·ger·y** (sûr′jə-rē) *n., pl.* **-ies.** [ME *surgerie* < OFr. < *serurgie*, *cerurgie* < Lat. *chirurgia* < Gk. *kheirourgia* < *kheirurgos*, working by hand : *kheir*, hand + *ergon*, work.] **1.** Medical diagnosis and treatment of injury, deformity, and disease by manual and instrumental operations. **2.** An operating room or laboratory of a surgeon or of a hospital's surgical staff. **3.** The skill or work of a surgeon. **4.** *Chiefly Brit.* A physician's office.

**sur·gi·cal** (sûr′jĭ-kəl) *adj.* [< SURGEON.] **1. a.** Of, relating to, or characteristic of surgeons or surgery. **b.** Used in surgery. **c.** Resulting from or occurring after surgery <*surgical* complications> **2.** Held to resemble surgery, esp. in precision <*surgical* air strikes> **—sur′gi·cal·ly** *adv.*

**su·ri·cate** (sŏŏr′ĭ-kāt′) *n.* [Fr., of African orig.] A small gregarious burrowing mammal, *Suricata suricatta* of southern Africa, with grayish fur and a long tail.

**sur·ly** (sûr′lē) *adj.* **-li·er, -li·est.** [Obs. *sirly*, masterful < SIR.] **1.** Sullenly ill-tempered. **2.** *Obs.* Arrogant : domineering. **—sur′li·ly** *adv.* **—sur′li·ness** *n.*

**sur·mise** (sər-mīz′) *v.* **-mised, -mis·ing, -mis·es.** [ME *surmysen*, to accuse < OFr. *surmettre*, *surmis-* < Med. Lat. *supermittere* < Lat., to throw on : *super-*, over + *mittere*, to put.] —*vt.* To infer (something) without conclusive evidence. —*vi.* To make a conjecture. —*n.* An idea or opinion based on inconclusive evidence : CONJECTURE. **—sur·mis′er** *n.*

**sur·mount** (sər-mount′) *vt.* **-mount·ed, -mount·ing, -mounts.** [ME *surmonten*, to excel < OFr. *surmonter* : *sur-*, above + *monter*, to mount. —see MOUNT[1].] **1.** To overcome (e.g., an obstacle) : CONQUER. **2.** To climb to the top of. **3. a.** To place something above : TOP. **b.** To be above or on top of <The office tower *surmounts* the street.> **4.** *Obs.* To exceed in amount. **—sur·mount′a·ble** *adj.* **—sur·mount′er** *n.*

**sur·mul·let** (sər-mŭl′ĭt, sûr′mŭl′-) *n., pl.* **surmullet** *or* **-lets.** [Fr. *surmulet* < OFr. *sormulet* : prob. *sor*, reddish brown (of Germanic orig.) + *mulet*, mullet. —see MULLET.] The goatfish.

**sur·name** (sûr′nām′) *n.* [ME : *sur-*, sur- + *name*, name.] **1.** One's family name. **2.** A nickname or epithet added to one's name, as *the Great* in *Catherine the Great.* —*vt.* **-named, -nam·ing, -names.** To give a surname to.

**sur·pass** (sər-păs′) *vt.* **-passed, -pass·ing, -pass·es.** [OFr. *surpasser* : *sur-*, over + *passer*, pass. —see PASS.] **1.** To exceed the limit, powers, or extent of : TRANSCEND. **2.** To be superior to.

☆ ***syns:*** SURPASS, EXCEED, EXCEL, OUTDO, OUTSHINE, OUTSTRIP, PASS, TOP, TRANSCEND *v. core meaning* : to be greater or better than <*surpassed* last month's production quota>

**sur·pass·ing** (sər-păs′ĭng) *adj.* That surpasses the usual or average : EXCEPTIONAL <*surpassing* splendor> **—sur·pass′ing·ly** *adv.*

**sur·plice** (sûr′plĭs) *n.* [ME *surplis* < Norman Fr. *surpliz* < OFr. *sourpeliz* < Med. Lat. *superpellicium* : Lat. *super-*, over + *pellicium*, fur coat < Lat. *pellicius*, made of skin < *pellis*, skin.] A loose-fitting white gown with wide sleeves, worn over a cassock by some clerics.

**sur·plus** (sûr′pləs, -plŭs′) *adj.* [ME < OFr. < Med. Lat. *superplus* : Lat. *super-*, over + Lat. *plus*, more.] Being more than what is needed or required. —*n.* **1.** An amount or quantity in excess of what is needed. **2.** Total assets minus the sum of all liabilities. **3.** Excess of a corporation's net assets over the face value of its capital stock. **4.** Excess of receipts over expenditures.

**sur·plus·age** (sûr′plə-sĭj) *n.* **1.** SURPLUS 1. **2.** An excess of words : VERBIAGE. **3.** Irrelevant matter in a legal pleading.

**surplus value** *n.* The difference, in the Marxian analysis of capitalism, between the value of the product produced by labor and the actual price of labor as paid out in wages.

**sur·print** (sûr′prĭnt′) *vt.* **-print·ed, -print·ing, -prints. 1.** To overprint. **2.** To superimpose (a second negative) upon a previously printed image of the first negative. **—sur′print′** *n.*

**sur·pris·al** (sər-prī′zəl) *n.* SURPRISE 1.

**sur·prise** *also* **sur·prize** (sər-prīz′) *vt.* **-prised, -pris·ing, -pris·es** *also* **-prized, -priz·ing, -priz·es.** [ME *surprysen,* to overcome < OFr. *surprendre, surpris-* : *sur-,* over + *prendre,* to take < Lat. *praehendere,* to seize.] **1.** To encounter suddenly or unexpectedly : catch unawares. **2.** To attack or capture suddenly and with no warning. **3.** To cause to feel wonder or astonishment. **4. a.** To cause (someone) to do or say something unintended. **b.** To elicit or detect through surprise. —*n.* **1.** The act of surprising or the state of being surprised. **2.** Something that surprises. **—sur·pris′er** *n.*

**sur·re·al** (sə-rē′əl) *adj.* [Back-formation < SURREALISM.] **1.** Having qualities attributed to surrealism. **2.** Grotesque : bizarre <*surreal* management techniques>

**sur·re·al·ism** (sə-rē′ə-lĭz′əm) *n.* [Fr. *surréalisme* : *sur-,* beyond + *réalisme,* realism < *réel,* real < OFr. *reel* < LLat. *realis* < Lat. *res,* thing.] A 20th-cent. literary and artistic movement that attempts to express the workings of the subconscious by fantastic imagery and irrational juxtaposition of subject matter. **—sur·re′al·ist** *n.* **—sur·re′al·is′tic** (-lĭs′tĭk) *adj.* **—sur·re′al·is′ti·cal·ly** *adv.*

**sur·re·but·ter** (sûr′rĭ-bŭt′ər) *also* **sur·re·but·tal** (-bŭt′l) *n.* *Law.* A plaintiff's reply to a defendant's rebutter.

**sur·re·join·der** (sûr′rĭ-join′dər) *n.* *Law.* A plaintiff's reply to a defendant's rejoinder.

**sur·ren·der** (sə-rĕn′dər) *v.* **-dered, -der·ing, -ders.** [ME *sorendren* < OFr. *surrendre* : *sur-,* over + *rendre,* to deliver. —see RENDER.] —*vt.* **1.** To relinquish possession or control of to another as a result of demand or compulsion. **2.** To give up in favor of another. **3.** To give up or give back something that has been granted <*surrender* a contractual right> **4.** To give up or abandon <*surrender* all hope of success> **5.** To give over or resign (oneself) to something, as to an emotion. —*vi.* To give oneself up, as to an enemy. —*n.* **1.** An act or an instance of surrendering. **2.** Delivery of a prisoner, fugitive from justice, or principal in a legal suit into legal custody.

**surrender value** *n.* The value of an insurance policy either to the owner or to the beneficiary upon its expiration.

**sur·rep·ti·tious** (sûr′əp-tĭsh′əs) *adj.* [Lat. *surrepticius* < *surripere,* to take away secretly : *sub-,* secretly + *rapere,* to seize.] Made, done, acquired by clandestine or stealthy means. **—sur′rep·ti′tious·ly** *adv.* **—sur′rep·ti′tious·ness** *n.*

**sur·rey** (sûr′ē, sŭr′ē) *n., pl.* **-reys.** [Short for *Surrey cart,* after *Surrey,* a county in England where it was first built.] A four-wheeled horse-drawn carriage with two seats.

**sur·ro·gate** (sûr′ə-gĭt, -gāt′, sŭr′-) *n.* [< Lat. *surrogare,* to substitute < *subrogare.* —see SUBROGATE.] **1.** One taking the place of another : SUBSTITUTE. **2.** A judge in some U.S. states having jurisdiction over the probate of wills and the settlement of estates. —*adj.* Substitute <a *surrogate* parent> —*vt.* (-gāt′) **-gat·ed, -gat·ing, -gates. 1.** To put in the place of another, esp. as a successor : REPLACE. **2.** To appoint (another) as a replacement for oneself.

**sur·round** (sə-round′) *vt.* **-round·ed, -round·ing, -rounds.** [ME *sourrounden,* to inundate < OFr. *suronder* < LLat. *superundare* : Lat. *super-,* over + Lat. *undare,* to rise in waves < *unda,* wave.] **1.** To encircle on all sides of simultaneously : GIRD. **2.** To cause to be surrounded by something. **3.** To confine on all sides so as to prevent escape or cut off outside communication.

☆ ***syns:*** SURROUND, CIRCLE, COMPASS, ENCIRCLE, ENCLOSE, GIRD, HEM IN, RING *v. core meaning* : to shut in on all sides <*surrounded* by water>

**sur·round·ings** (sə-roun′dĭngz) *pl.n.* The external circumstances, conditions, and objects that affect existence and development : ENVIRONMENT.

**sur·sum cor·da** (so͝or′səm kôr′də) *n.* [Lat., (lift) up (your) hearts.] **1.** *often* **Sursum Corda.** An ecclesiastical versicle offering praise and thanksgiving to God. **2.** Something instilling courage or fervor.

**sur·tax** (sûr′tăks′) *n.* **1.** An additional tax. **2.** A graduated income tax added to the normal income tax levied on the amount by which one's net income exceeds a certain sum. —*vt.* **-taxed, -tax·ing, -tax·es.** To levy a surtax on.

**sur·veil** (sər-văl′) *vt.* **-veilled, -veil·ling, -veils.** [Back-formation < SURVEILLANCE.] To keep under surveillance.

**sur·veil·lance** (sər-vā′ləns) *n.* **1.** Close observation, esp. of one under suspicion. **2.** The act of observing or the state of being observed.

**sur·veil·lant** (sər-vā′lənt) *adj.* [Fr., pr.part. of *surveiller,* to watch over : *sur-,* over + *veiller,* to watch < Lat. *vigilare* < *vigil,* watchful.] Exercising surveillance. —*n.* One that exercises surveillance.

**sur·vey** (sər-vā′, sûr′vā′) *v.* **-veyed, -vey·ing, -veys.** [ME *surveyen* < OFr. *surveeir* < Med. Lat. *supervidēre,* to look over : Lat. *super-,* over + Lat. *vidēre,* to look.] —*vt.* **1.** To examine in a comprehensive way : APPRAISE. **2.** To inspect carefully : SCRUTINIZE. **3.** To determine the boundaries, area, or elevations of (land or structures on the earth's surface) by means of measuring angles and distances, using the techniques of geometry and trigonometry. —*vi.* To make a survey. —*n.* (sûr′vā′) *pl.* **-veys. 1.** A detailed inspection or investigation. **2.** A comprehensive view. **3. a.** The process of surveying. **b.** A report on or map of something surveyed. **—sur·vey′or** *n.*

**survey course** *n.* A course consisting of an overview of a broad topic or field of knowledge.

**sur·vey·ing** (sər-vā′ĭng) *n.* Measurement of dimensional relationships, as of horizontal distances, elevations, directions, and angles, on the earth's surface esp. for use in locating property boundaries, construction layout, and mapmaking.

**surveyor's level** *n.* A level having a telescope and attached spirit level mounted on a tripod and rotating around a vertical axis.

**sur·viv·a·ble** (sər-vī′və-bəl) *adj.* Having the capability to survive or to be survived. **—sur·viv′a·bil′i·ty** *n.*

**sur·viv·al** (sər-vī′vəl) *n.* **1. a.** The act or process of surviving. **b.** The state of having survived. **2.** Something, as an ancient belief, that has survived.

**survival of the fittest** *n.* Natural selection conceived of as a struggle in which only those organisms best adapted to existing conditions survive.

**survival value** *n.* Usefulness in the struggle for survival <"emotional activation most clearly has *survival value*" —Norman L. Munn>

**sur·viv·ance** (sər-vī′vəns) *n.* Survival.

**sur·vive** (sər-vīv′) *v.* **-vived, -viv·ing, -vives.** [ME *surviven* < Norman Fr. *survivre* < OFr. *sourvivre* < LLat. *supervivere* : Lat. *super-,* over + Lat. *vivere,* to live.] —*vi.* To remain alive or in existence. —*vt.* **1.** To live longer than : OUTLIVE <*survive* one's spouse> **2.** To persist through <plants *surviving* a frost> **—sur·vi′vor** *n.*

**sur·vi·vor·ship** (sər-vī′vər-shĭp′) *n.* **1.** *Law.* The right of one who survives a partner or joint owner to the entire ownership of something previously owned jointly. **2.** The state of being a survivor.

**Su·san B. An·tho·ny Day** (so͞o′zən-bē-ăn′thə-nē) *n.* Feb. 15 observed in honor of the birthday of Susan B. Anthony.

**Su·san·na** (so͞o-zăn′ə) *n.* [Heb. *Shōshannāh* < *shōshannāh,* lily.] —See table at BIBLE.

**sus·cep·tance** (sə-sĕp′təns) *n.* [SUSCEPT(IBILITY) + (CONDUCT)ANCE.] *Elect.* The imaginary part of the complex representation of admittance.

**sus·cep·ti·bil·i·ty** (sə-sĕp′tə-bĭl′ĭ-tē) *n., pl.* **-ties. 1.** The quality or state of being susceptible. **2.** The capacity to be affected by deep emotions or strong feelings : SENSITIVITY. **3. susceptibilities.** Feelings : sensibilities.

**sus·cep·ti·ble** (sə-sĕp′tə-bəl) *adj.* [Med. Lat. *susceptibilis,* receivable < Lat. *suscipere,* to receive : *sub-,* from below + *capere,* to take.] **1.** Easily affected or influenced <*susceptible* to gentle persuasion> **2.** Likely to be stricken with or by <*susceptible* to fevers> **3.** Particularly sensitive and impressionable. **4.** That can be affected with : PERMITTING <testimony *susceptible* of proof> **—sus·cep′ti·ble·ness** *n.* **—sus·cep′ti·bly** *adv.*

**sus·cep·tive** (sə-sĕp′tĭv) *adj.* **1.** Receptive. **2.** Susceptible. **—sus·cep′tive·ness, sus·cep′tiv′i·ty** (sə-sĕp′tĭv′ĭ-tē) *n.*

**su·shi** (so͞o′shē) *n.* [J.] A Japanese dish consisting of thin slices of fresh raw fish or seaweed wrapped around a cake of cooked rice.

**sus·pect** (sə-spĕkt′) *v.* **-pect·ed, -pect·ing, -pects.** [Lat. *suspectare,* freq. of *suspicere,* to watch : *sub-,* from below + *specere,* to look at.] —*vt.* **1.** To suppose to be true or probable : IMAGINE. **2.** To have doubts about : DISTRUST <*suspected* their motives> **3.** To think (a person) guilty without proof <They *suspect* the doctor of murder.> —*vi.* To have suspicion. —*n.* (sŭs′pĕkt′). One who is suspected of committing a crime. —*adj.* (sŭs′pĕkt′, sə-spĕkt′). That is open to or regarded with suspicion <*suspect* behavior>

**sus·pend** (sə-spĕnd′) *v.* **-pend·ed, -pend·ing, -pends.** [ME *suspenden* < OFr. *suspendre* < Lat. *suspendere,* to hang up : *sub-,* from below + *pendere,* to hang.] —*vt.* **1.** To bar for a period from a privilege, office, or position, usu. as a punishment <*suspend* a lawyer from the bar> **2.** To cause to stop for a period <*suspend* stock market trading> **3. a.** To hold in abeyance : DEFER <*suspend* judgment> **b.** To render temporarily ineffective <*suspend* the regulations> **4.** To hang so as to allow free movement. **5.** To support or keep from falling without apparent attachment, as by buoyancy <*suspend* oneself in the water> —*vi.* **1.** To cease for a period : DELAY. **2.** To fail to make payments or fulfill obligations.

☆ ***syns:*** SUSPEND, BREAK OFF, CEASE, DISCONTINUE, INTERRUPT, TERMINATE *v. core meaning* : to cause to stop for a period <*suspended* the negotiations indefinitely>

**suspended animation** *n.* A dormant condition resembling death and induced by reversible cessation of the vital functions.

**sus·pend·er** (sə-spĕn′dər) *n.* **1.** One that suspends. **2. suspenders.** A pair of often elastic straps, worn over the shoulders to support trousers. **3.** *Chiefly Brit.* A garter.

**sus·pense** (sə-spĕns') *n.* [ME < Norman Fr. < OFr. < *suspendre,* to suspend.] **1.** The quality or state of being undecided, uncertain, or doubtful. **2. a.** Anxiety or apprehension resulting from an uncertain, undecided, or mysterious situation. **b.** Agreeable stimulation or excitement as to dénouement <a novel of *suspense* and intrigue> **3.** The state of being suspended. **—sus·pense'ful** *adj.*

**suspense account** *n.* A temporary account in which entries of credits or charges are made until their proper disposition can be determined.

**sus·pen·sion** (sə-spĕn'shən) *n.* **1.** The act of suspending or the state of being suspended, esp.: **a.** Temporary abrogation or deferment. **b.** Debarment, as from office or privilege. **c.** Postponement of judgment, opinion, or decision. **2.** *Mus.* **a.** Prolongation of one or more tones of a chord into a following chord to as to create a temporary dissonance. **b.** The tone prolonged by suspension. **3.** A device from which a mechanical part is suspended. **4.** The system of springs that protects the chassis of a motor vehicle from shocks transmitted through the wheels. **5.** *Chem.* A relatively coarse, noncolloidal dispersion of solid particles in a liquid.

**suspension bridge** *n.* A bridge having the roadway suspended from cables that are usu. supported by towers.

**suspension points** *pl.n.* A series of dots, usu. three, used to indicate the omission of a word or words from a written text.

**sus·pen·sive** (sə-spĕn'sĭv) *adj.* **1.** Serving or tending to suspend something temporarily. **2.** Marked by or causing suspense. **—sus·pen'sive·ly** *adv.* **—sus·pen'sive·ness** *n.*

**sus·pen·soid** (sə-spĕn'soid') *n.* [SUSPENS(ION) + (COLL)OID.] A colloid solution with solid dispersed particles.

**sus·pen·sor** (sə-spĕn'sər) *n.* [Med. Lat., one that suspends < Lat. *suspendere,* to suspend. —see SUSPEND.] **1.** *Bot.* A cell or cellular structure developed from a zygote in seed-bearing plants and connecting the embryo to the embryo sac. **2.** A suspensory.

**sus·pen·so·ry** (sə-spĕn'sə-rē) *adj.* **1.** Supporting or suspending <a *suspensory* bandage> **2.** Delaying completion. *—n., pl.* **-ries. 1.** A support or truss. **2.** An athletic supporter.

**suspensory ligament** *n.* A ligament supporting an organ or bodily part.

**sus·pi·cion** (sə-spĭsh'ən) *n.* [ME *suspecioun* < Norman Fr., var. of OFr. *sospecon* < Med. Lat. *suspectio* < *suspicere,* to watch. —see SUSPECT.] **1.** The act or an instance of suspecting a wrong, crime, or guilt, without sufficient evidence or proof. **2.** The condition of being suspected, esp. of wrongdoing. **3.** Uncertainty : doubt. **4.** A minute amount : TRACE <a *suspicion* of garlic> *—vt.* **-cioned, -cion·ing, -cions.** *Nonstandard.* To suspect. **—sus·pi'cion·al** *adj.*

**sus·pi·cious** (sə-spĭsh'əs) *adj.* **1.** Arousing or apt to arouse suspicion : QUESTIONABLE <*suspicious* activities> **2.** Tending to suspect : DISTRUSTFUL <a *suspicious* mind> **3.** Exhibiting suspicion <a *suspicious* glance> **—sus·pi'cious·ly** *adv.* **—sus·pi'cious·ness** *n.*

**sus·pire** (sə-spīr') *vi.* **-pired, -pir·ing, -pires.** [ME *suspiren* < Lat. *suspirare,* to sigh : *sub,* from below + *spirare,* to breathe.] **1.** To breathe. **2.** To sigh. **—sus'pi·ra'tion** (sŭs'pə-rā'shən) *n.*

**Sus·sex spaniel** (sŭs'ĭks) *n.* A dog orig. bred in Sussex, England, and having long ears, short legs, and a silky golden-brown coat.

**sus·tain** (sə-stān') *vt.* **-tained, -tain·ing, -tains.** [ME *susteynen* < Norman Fr. *sustein* < OFr. *sustenir* < Lat. *sustinēre,* to hold up : *sub-,* from below + *tenēre,* to hold.] **1.** To keep in existence : MAINTAIN. **2.** To provide with nourishment or sustenance. **3.** To support from below : PROP <columns *sustaining* the arches> **4.** To support the spirits, vitality, or resolution of : ENCOURAGE. **5.** To endure or withstand : BEAR UP <*sustain* hardships> **6.** To experience or suffer (loss or injury). **7.** To affirm the validity or justice of <The judge *sustained* the objection.> **8.** To corroborate : confirm. **—sus·tain'a·ble** *adj.* **—sus·tain'er** *n.* **—sus·tain'ment** *n.*

☆ ***syns:*** SUSTAIN, BOLSTER, BUOY (up), PROP, SUPPORT, UPHOLD *v. core meaning* : to keep from yielding or failing during stress or difficulty <strong convictions that *sustained* us during the crisis>

**sustaining program** *n.* A radio or television program supported by the station or network on which it appears and having no commercial announcements.

**sus·te·nance** (sŭs'tə-nəns) *n.* [ME < Norman Fr. *sustenaunce* < OFr. *soustenance* < *sustenir,* to sustain.] **1.** The act of sustaining or the state of being sustained. **2.** The sustaining of life or health : MAINTENANCE. **3.** Something, esp. food, that sustains life or health. **4.** Means of livelihood.

**sus·ten·tac·u·lar** (sŭs'tən-tăk'yə-lər, -tĕn-) *adj.* [< Lat. *sustentaculum,* a support < *sustentare,* to support, freq. of *sustinēre,* to sustain.] *Anat.* Serving to support.

**sus·ten·ta·tion** (sŭs'tən-tā'shən, -tĕn-) *n.* [ME *sustentacion,* maintenance < Norman Fr. < OFr. < Lat. *sustentatio* < *sustentare,* to support. —see SUSTENTACULAR.] **1.** Something that sustains : SUPPORT. **2.** Sustenance. **—sus'ten·ta'tive** *adj.*

**Su·su** (sōō'sōō) *n., pl.* **Susu** or **Su·sus. 1.** A West African people residing in Guinea and the Sudan and along the northern border of Sierra Leone. **2.** A member of the Susu. **3.** The Mande language of the Susu.

**su·sur·ra·tion** (sōō'sə-rā'shən) *also* **su·sur·rus** (sōō-sûr'əs, -sŭr'-) *n.* [ME *susurracyoun* < LLat. *susurratio* < Lat. *susurrare,* to whisper < *susurrus,* whisper.] A soft whispering or rustling sound : MURMUR <*susurration* of leaves in the wind> **—su·sur'rant** (sōō-sûr'ənt, -sŭr'-), **su·sur'rous** (-sûr'əs, -sŭr'-) *adj.*

**sut·ler** (sŭt'lər) *n.* [MDu. *soetler,* of Germanic orig.] A follower of an army camp who peddled provisions to the troops.

**su·tra** (sōō'trə) *n.* [Skt. *sūtram.*] **1.** Any of various aphoristic doctrinal summaries in Hinduism, produced gen. between 500 and 200 B.C. and later incorporated into Hindu and Buddhist literature. **2.** A scriptural narrative, esp. a text traditionally regarded as a discourse of the Buddha.

**sut·tee** (sŭ-tē', sŭt'ē') *n.* [Skt. *satī,* virtuous woman, fem. of *sat,* true, pr.part. of *asti,* he is.] **1.** The act or practice of a Hindu widow allowing herself to be cremated on her husband's funeral pyre. **2.** A widow cremated by suttee.

**su·ture** (sōō'chər) *n.* [OFr. < Lat. *sutura* < *suere,* to sew.] **1. a.** The process of joining two surfaces or edges together along a line by or as if by sewing. **b.** The material, as thread, gut, or wire, that is used in this procedure. **2.** *Anat.* The line of junction or an immovable joint between two bones, esp. of the skull. **3.** *Biol.* A seamlike joint or line of articulation, as the line of dehiscence in a seed or fruit or the spiral seam marking the junction of whorls of a gastropod shell. *—vt.* **-tured, -tur·ing, -tures.** To join by sutures. **—su'tur·al** *adj.* **—su'tur·al·ly** *adv.*

**su·ze·rain** (sōō'zər-ən, -zə-rān') *n.* [Fr. : prob. *sus,* up (< Lat. *sursum,* upward) + *souverain,* sovereign.] **1.** A feudal lord to whom fealty was due. **2.** A nation that controls another nation in international affairs but allows it domestic sovereignty.

**su·ze·rain·ty** (sōō'zər-ən-tē, -zə-rān'tē) *n., pl.* **-ties.** The position, power, or domain of a suzerain.

**svelte** (svĕlt) *adj.* **svelt·er, svelt·est.** [Fr. < Ital. *svelto* < *svellere,* to stretch out < VLat. ***exvellere** < Lat. *evellere* : *ex-,* out + *vellere,* to pull.] Slender or graceful : SLIM. **—svelte'ly** *adv.* **—svelte'ness** *n.*

**swab** *also* **swob** (swŏb) [Prob. < MDu. *swabbe,* mop.] *—n.* **1.** A small piece of absorbent material attached to the end of a stick or wire and used for cleansing or applying medicine. **2.** Matter, as a specimen of mucus, removed with a swab. **3.** A mop for cleaning floors or decks. **4.** *Slang.* A sailor. **5.** A lout. *—vt.* **swabbed, swab·bing, swabs** *also* **swobbed, swob·bing, swobs. 1.** To use a swab on. **2.** To clean with or as if with a swab.

**swab·bie** *also* **swab·by** (swŏb'ē) *n., pl.* **-bies.** *Slang.* A sailor.

**swad·dle** (swŏd'l) *vt.* **-dled, -dling, -dles.** [ME *swadlen,* alteration of *swedlen, swethlen* < *swethel,* swaddling clothes < OE *swæðel.*] **1.** To wrap or bind in bandages : SWATHE. **2.** To wrap (a baby) in swaddling clothes. **3.** To restrain or restrict. *—n.* Cloth used for swaddling.

**swaddling clothes** *pl.n.* **1.** Strips of cloth wrapped around a newborn infant to hold its legs and arms still. **2.** Rigid restrictions imposed on the immature.

**swag** (swăg) *n.* [Prob. of Scand. orig.] **1.** An ornamental hanging, as a valance or wreath, draped in a curve between two points. **2.** *Slang.* Stolen money or property : LOOT. **3.** *Austral.* A pack or bundle holding a swagman's personal belongings. *—vi.* **swagged, swag·ging, swags.** To lurch or sway.

**swage** (swāj) *n.* [ME, ornamental border < OFr. *souage.*] **1.** A tool for bending or shaping cold metal. **2.** A stamp or die for marking or shaping metal with a hammer. **3.** A swage block. *—vt.* **swaged, swag·ing, swag·es.** To bend or shape by or as if by using a swage.

**swage block** *n.* A metal block with holes or grooves for shaping metal objects.

**swag·ger** (swăg'ər) *v.* **-gered, -ger·ing, -gers.** [Prob. of Scand. orig.] *—vi.* **1.** To walk or conduct oneself with an insolent air : STRUT. **2.** To boast : brag. *—vt.* To influence by swaggering : BULLY. *—n.* **1.** Swaggering gait or movement. **2.** Boastful expression : BRAGGADOCIO. **—swag'ger·er** *n.* **—swag'ger·ing·ly** *adv.*

**swagger stick** *n.* A short metal-tipped cane carried esp. by some military officers.

**swag·man** (swăg'măn') *n. Austral.* An itinerant worker.

**Swa·hi·li** (swä-hē'lē) *n., pl.* **Swahili** or **-lis.** [Swahili, belonging to the coasts < Ar. *sawāḥil,* pl. of *sāḥil,* coast.] **1.** A Bantu language widely used as a lingua franca in eastern and central Africa. **2.** One of the original speakers of Swahili, a Bantu people of Zanzibar and the neighboring mainland. **—Swa·hi'li·an** *adj.*

**swain** (swān) *n.* [ME *swayn* < ON *sveinn,* boy.] **1.** A country youth, esp. a young shepherd. **2.** A male suitor. **—swain'ish** *adj.*

**swale** (swāl) *n.* [Perh. < ME, shade.] A low tract of marshy land.

**swal·low**[1] (swŏl'ō) *v.* **-lowed, -low·ing, -lows.** [ME *swalowen* < OE *swelgan.*] *—vt.* **1.** To cause (e.g., food) to pass through the mouth and throat into the stomach. **2.** To destroy or consume as if by ingestion : DEVOUR <a building that was *swallowed* up in flames> **3. a.** To bear humbly : TOLERATE <had to *swallow* the insult> **b.** *Slang.* To believe unquestioningly. **4. a.** To refrain from expressing : SUPPRESS <*swallow* one's anger> **b.** To take back : retract <*swallow* one's words> *—vi.* To perform the act of swallowing.

—*n.* **1.** The act of swallowing. **2.** An amount swallowed. **3.** *Naut.* The channel through which a rope runs in a block or a mooring chock. **—swal′low·er** *n.*

**swal·low²** (swŏl′ō) *n.* [ME *swalowe* < OE *swealwe.*] **1.** Any of various birds of the family Hirundinidae, with long, pointed wings and a usu. notched or forked tail. **2.** A bird, as a swift, that is similar to the swallow.

**swal·low·tail** (swŏl′ō-tāl′) *n.* **1. a.** The deeply forked tail of a swallow. **b.** Something thought to be similar to the tail of a swallow. **2.** *Informal.* A swallow-tailed coat. **3.** Any of various butterflies of the family Papilionidae, usu. having a taillike extension at the end of each hind wing.

**swal·low-tailed** (swŏl′ō-tāld′) *adj.* **1.** Having a deeply forked tail. —Used of various birds. **2.** Like of the tail of a swallow <a *swallow-tailed* kite>

**swallow-tailed coat** *n.* A man's black coat worn for formal daytime occasions and having a long rounded and split tail.

**swal·low·wort** (swŏl′ō-wûrt′, -wôrt′) *n.* [From the shape of its pod.] **1.** CELANDINE 1. **2.** A vine of the genus *Cynanchum,* native to Europe, esp. *C. nigrum,* with small brownish-purple flower clusters.

**swam** (swăm) *v. p.t. of* SWIM.

**swa·mi** (swä′mē) *n.* [Hindi *svāmī,* master < Skt. *svāmin.*] **1.** Lord : master. —Used as a Hindu title of respect. **2.** A Hindu religious teacher. **3.** A mystic : yogi. **4.** A very learned man: PUNDIT

**swamp** (swŏmp, swômp) *n.* [Perh. of LG orig.] A lowland region saturated with water : MARSH. —*v.* **swamped, swamp·ing, swamps.** —*vt.* **1.** To drench in or cover with or as if with water. **2.** To inundate or burden : OVERWHELM <*swamped* with projects> **3.** To fill or sink (a ship) with water. —*vi.* To become full of water or sink, as a ship.

**swamp boat** *n.* A flat-bottomed boat powered by an airplane propeller above the stern and used in swamps or shallow waters.

**swamp·er** (swŏm′pər, swôm′-) *n.* **1.** One who lives in or near a swamp. **2.** One who clears a swamp or forest. **3. a.** A menial helper, as in a restaurant. **b.** A truck driver's assistant.

**swamp fever** *n.* **1.** MALARIA 1. **2.** A viral disease in horses, marked by progressive anemia, a staggering gait, and fever.

**swamp·land** (swŏmp′lănd′, swômp′-) *n.* Swampy land : MARSHLAND.

**swamp·y** (swŏm′pē, swôm′-) *adj.* **-i·er, -i·est.** Of, relating to, or like a swamp : MARSHY. **—swamp′i·ness** *n.*

**swan** (swŏn) *n.* [ME < OE.] **1.** A large aquatic bird, chiefly of the genera *Cygnus* or *Olor,* having webbed feet, a long slender neck, and usu. white plumage. **2. Swan.** Cygnus.

**swan dive** *n.* A dive performed with the back arched, the legs straight together, and the arms stretched out from the sides.

**swank** (swăngk) *adj.* **-er, -est.** [Perh. < MHG *swanken,* to swing.] **1.** Imposingly elegant or fashionable : GRAND. **2.** Ostentatious : pretentious. —*n.* **1.** Elegance : smartness. **2.** Swaggering behavior. —*vi.* **swanked, swank·ing, swanks.** To act ostentatiously or pretentiously : SWAGGER.

**swank·y** (swăng′kē) *adj.* **-i·er, -i·est.** Swank. **—swank′i·ly** *adv.* **—swank′i·ness** *n.*

**swan·ner·y** (swŏn′ə-rē) *n., pl.* **-ies.** A place where swans are raised and bred.

**Swan River daisy** *n.* An Australian plant, *Brachycome iberidifolia,* cultivated for its blue or white flower heads.

**swan's-down** *also* **swans·down** (swŏnz′doun′) *n.* **1.** The soft down of a swan. **2.** A soft woolen fabric used esp. for baby clothes. **3.** Canton flannel.

**swan·skin** (swŏn′skĭn′) *n.* **1.** The skin of a swan with the feathers attached. **2.** A soft-napped flannel or cotton fabric.

**swan song** *n.* [Transl. of G. *Schwanenlied.*] **1.** The legendary last utterance or song of a dying swan, held to be of great sweetness. **2.** A farewell or final appearance, action, or work.

**swap** *also* **swop** (swŏp) [Obs. *swap,* to strike hands in closing a bargain < ME *swappen,* to hit.] *Informal* —*v.* **swapped, swap·ping, swaps** *also* **swopped, swop·ping, swops.** —*vi* To trade one thing for another. —*vt.* To exchange. —*n.* An exchange of one thing for another. **—swap′per** *n.*

**sward** (swôrd) *also* **swarth** (swôrth) *n.* [ME < OE *sweard,* skin.] **1.** Land covered with grassy turf. **2.** A lawn or meadow.

**sware** (swâr) *v. Archaic. var. p.t. of* SWEAR.

**swarm¹** (swôrm) *n.* [ME, group of bees < OE *swearm.*] **1.** A large number of small organisms, as insects, esp. when in motion. **2.** A group of bees with a queen bee in migration to establish a new colony. **3.** A moving mass of persons or animals <A *swarm* of fans surrounded the star.> —*v.* **swarmed, swarm·ing, swarms.** —*vi.* **1. a.** To move or emerge in a swarm. **b.** To leave a hive as a swarm. —Used of bees. **2.** To move or gather in large numbers. **3.** To be overrun : TEEM <a garbage can *swarming* with flies> —*vt.* To fill with a crowd : THRONG. **—swarm′er** *n.*

**swarm²** (swôrm) *vt. & vi.* **swarmed, swarm·ing, swarms.** [Orig. unknown.] To climb by gripping with the arms and legs.

**swarm spore** *n.* A zoospore.

**swart** (swôrt) *adj.* [ME *swarte* < OE *sweart.*] Swarthy. **—swart′ness** *n.*

**swarth** (swôrth) *n. var. of* SWARD.

**swar·thy** (swôr′thē) *adj.* **-thi·er, -thi·est.** [Alteration of obs. *swarty* < SWART.] Having a dark color or complexion. **—swar′thi·ly** *adv.* **—swar′thi·ness** *n.*

**swash** (swŏsh, swôsh) *n.* [Prob. imit.] **1. a.** A splash of liquid. **b.** The sound of a splash of liquid. **2.** A narrow channel through which tides flow. **3.** A bar over which waves wash freely. **4. a.** Swagger : bluster. **b.** A swaggering, blustering person. —*v.* **swashed, swash·ing, swash·es.** —*vi.* **1.** To strike, move, or wash with a splashing sound. **2.** To swagger. —*vt.* **1.** To splash (a liquid). **2.** To splash a liquid against.

**swash·buck·ler** (swŏsh′bŭk′lər, swôsh′-) *n.* [From the striking of bucklers in fighting.] **1.** A flamboyant adventurer or swordsman. **2.** A drama or fictional work about a swashbuckler. **—swash′buck′ling** *adj.*

**swash letter** *n.* [Orig. unknown.] An ornamental italic letter with elaborate, flowing flourishes and tails.

**swas·ti·ka** (swŏs′tĭ-kə) *n.* [Skt. *svastikaḥ,* a sign of good luck < *svasti,* success : *su-,* good + *asti,* it is.] **1.** An ancient cosmic or religious symbol formed by a Greek cross with the ends of the arms bent at right angles in either a clockwise or a counterclockwise direction. **2.** The emblem of Nazi Germany and of various neo-Nazi groups.

**swat** (swŏt) *vt.* **swat·ted, swat·ting, swats.** [Alteration of SQUAT, to squash (obs.).] To hit with a sharp blow : SLAP. —*n.* A sharp blow.

**swatch** (swŏch) *n.* [Orig. unknown.] A sample cut from a piece of fabric.

**swath** (swŏth, swôth) *also* **swathe** (swŏth, swôth) *n.* [ME *swathe* < OE *swæð,* track.] **1. a.** The width of a scythe stroke or a mowing-machine blade. **b.** A path of this width made in mowing. **c.** The mown grass or grain lying on such a path. **2.** Something held to be like a swath. **—cut a (wide) swath.** To create a great stir, impression, or display.

**swathe¹** (swŏth, swôth, swāth) *vt.* **swathed, swath·ing, swathes.** [ME *swathen* < OE *swaðian.*] **1.** To wrap with or as if with bandages. **2.** To enfold or constrict. —*n.* A wrapping, binding, or bandage. **—swath′er** *n.*

**swathe²** (swŏth, swôth) *n. var. of* SWATH.

**swat·ter** (swŏt′ər) *n.* **1.** One that swats. **2.** A fly swatter. **3.** *Baseball.* A hard-hitting batter.

**sway** (swā) *v.* **swayed, sway·ing, sways.** [ME *sweyen.*] —*vi.* **1.** To move back and forth with a swinging motion : OSCILLATE. **2.** To bend to one side : VEER. **3. a.** To tend toward change, as in opinion or feeling. **b.** To fluctuate, as in viewpoint. —*vt.* **1.** To cause to move back and forth. **2.** To cause to bend to one side. **3.** *Naut.* To swing into position, as a yard or a mast. **4. a.** To divert : deflect. **b.** To exert influence on or control over <a speech that *swayed* the voters> **5.** *Archaic.* **a.** To rule or govern. **b.** To wield, as a weapon. —*n.* **1.** The act of moving from side to side with a swinging motion. **2.** Power : influence. **3.** Dominion : control. **—sway′er** *n.* **—sway′ing·ly** *adv.*

☆ ***syns:*** SWAY, TEETER, TOTTER, WAVER, WEAVE, WOBBLE *v. core meaning* : to move back and forth or from side to side, as if about to fall <*swayed* dizzily at the top of the stairs>

**sway·back** (swā′băk′) *n.* An excessive inward or downward curvature of the spine. **—sway′backed′** *adj.*

**Swa·zi** (swä′zē) *n., pl.* **Swazi** *or* **-zis.** A tribesman of the Bantu people of Swaziland.

**swear** (swâr) *v.* **swore** (swôr, swōr), **sworn** (swôrn, swōrn), **swear·ing, swears.** [ME *swerien* < OE *swerian.*] —*vi.* **1.** To make a solemn declaration <*swore* that they had told the truth> **2.** To make a solemn promise : VOW. **3.** To use oaths : CURSE. **4.** *Law.* To give testimony or evidence under oath. —*vt.* **1.** To declare or affirm solemnly. **2.** To promise or pledge with a solemn oath : VOW <*swore* their eternal love> **3.** To utter or bind oneself to (an oath). **4.** To administer a legal oath to. **5.** To say or affirm earnestly and with conviction. **—swear by.** To have full confidence in. **—swear in.** To administer a legal or official oath to <*swear in* the mayor> **—swear off.** *Informal.* To pledge to renounce or give up <*swore off* cigarettes> **—swear′er** *n.*

**swear·word** (swâr′wûrd′) *n.* An obscene or blasphemous term.

**sweat** (swĕt) *v.* **sweat·ed** *or* **sweat, sweat·ing, sweats.** [ME *sweten* < OE *swǣtan.*] —*vi.* **1.** To exude perspiration through the pores in the skin : PERSPIRE. **2.** To exude in droplets, as moisture from certain cheeses or sap from a tree. **3.** To condense atmospheric moisture. **4. a.** To release moisture, as hay in the swath. **b.** To ferment, as tobacco during curing. **5.** *Informal.* To work hard and long <*sweated* over the exam> **6.** *Informal.* To suffer much, as for a misdeed. **7.** *Informal.* To fret or worry. —*vt.* **1.** To excrete (moisture) through a porous surface. **2.** To gather and condense (moisture) on a surface. **3.** To cause to perspire. **4.** To make damp or wet with perspiration. **5.** To cause to work excessively : OVERWORK. **6.** To overwork, esp. at low pay. **7.** *Informal.* **a.** To interrogate (someone) under duress <*sweated* the suspect for 12 hours> **b.** To extract (information) from someone under duress. **—sweat out.** *Slang.* **1.** To endure anxiously. **2.** To await (something) anxiously. —*n.* **1.** The product of the sweat glands. **2.** Condensation of moisture in the form of droplets on a surface. **3. a.** The process of sweating. **b.** The state of being sweated. **4.** Strenuous, exhaustive labor. **5.** Exercise given a horse prior to a race. **6.** *Informal.* An anxious, fretful condition : IMPATIENCE. **—no sweat.** *Slang.* Easily done or handled. **—sweat blood.** *Slang.* **1.** To work strenuously. **2.** To worry intensely. **—sweat'i·ly** *adv.* **—sweat'i·ness** *n.* **—sweat'y** *adj.*

**sweat·band** (swĕt'bănd') *n.* **1.** A fabric or leather band sewn inside the crown of a hat as protection against sweat. **2.** A band of material around the forehead or wrist to absorb sweat.

**sweat·box** (swĕt'bŏks') *n.* **1.** A box in which something, as hide or fruit, is fermented by sweating. **2.** A confined place where one sweats, esp.: **a.** An interrogation room. **b.** A punishment cell in a prison.

**sweat·er** (swĕt'ər) *n.* **1.** One that sweats, esp. profusely. **2.** An agent, esp. a sudorific, that induces sweating. **3.** A garment, as a jacket or pullover, made esp. of knit, crocheted, or woven wool.

**sweat gland** *n.* Any of the numerous small, tubular glands that in humans are found nearly everywhere in the skin and that secrete perspiration externally through pores.

**sweat pants** *pl.n.* Cotton jersey pants usu. having a drawstring or elasticized waist and elasticized cuffs, worn esp. for exercising.

**sweat shirt** *n.* A usu. long-sleeved cotton jersey pullover.

**sweat·shop** (swĕt'shŏp') *n.* A business establishment where employees work under bad conditions for long hours and low wages.

**sweat suit** *n.* A two-piece outfit consisting of cotton jersey pants tightly fitted at the ankles and waist and a sweat shirt.

**swede** (swēd) *n.* [From its introduction from Sweden.] A vegetable, the rutabaga.

**Swede** (swēd) *n.* [< MLG.] **1.** A native or inhabitant of Sweden. **2.** One of Swedish descent.

**Swed·ish** (swē'dĭsh) *adj.* Of or relating to Sweden, its people, or its language. —*n.* The North Germanic language of Sweden.

**Swedish massage** *n.* A system of massage and exercises for the muscles and joints.

**Swedish turnip** *n.* The rutabaga.

**sweep** (swēp) *v.* **swept** (swĕpt), **sweep·ing, sweeps.** [ME *swepen.*] —*vt.* **1.** To clear the surface of with or as if with a broom or brush. **2.** To clear away (e.g., dirt) with or as if with a broom or brush. **3.** To clear (a space) with or as if with a broom. **4.** To touch or brush lightly with or as if with a trailing garment <Low branches *swept* the ground.> **5. a.** To move or convey with a flowing motion, as by water or wind. **b.** To affect or unbalance emotionally <a love that *swept* me off my feet> **6.** To remove or destroy with forceful movement <flood waters that *swept* the village away> **7.** To range throughout with speed, violence, or intensity <Plague *swept* Europe.> **8.** To traverse, as when searching <Searchlights *swept* the compound.> **9.** To drag the bottom of (a body of water). **10. a.** To win all the stages of (a game or contest) <a team that *swept* the World Series> **b.** To win overwhelmingly in (a contest or election) —*vi.* **1.** To clean or clear a surface with or as if with a broom or brush. **2. a.** To move, surge, or flow with smooth and steady force <frigid winds *sweeping* over the steppes> **b.** To move swiftly or majestically. **3.** To trail, as a garment. **4.** To extend gracefully or majestically <The white cliffs *sweep* down to the sea.> —*n.* **1.** The act of sweeping. **2.** The motion of sweeping <a *sweep* of one's arm> **3.** The range or scope encompassed by sweeping <the *sweep* of an antiaircraft gun> **4.** Reach : extent <the great *sweep* of the English language> **5.** A curve or contour <the *sweep* of the model's long hair> **6.** One who sweeps, esp. a chimney sweep. **7. sweeps.** Sweepings. **8. a.** The winning of all stages of a game or contest. **b.** An overwhelming victory or success. **9.** A long oar used to propel a boat. **10.** A long pole attached to a pivot and with a bucket at one end, used to raise water from a well. **11.** *Informal.* A sweepstakes. **12.** *Electron.* The steady motion of an electron beam across a cathode-ray tube. **—sweep'er** *n.*

**sweep·back** (swēp'băk') *n.* The backward slant of the leading edge of an airfoil.

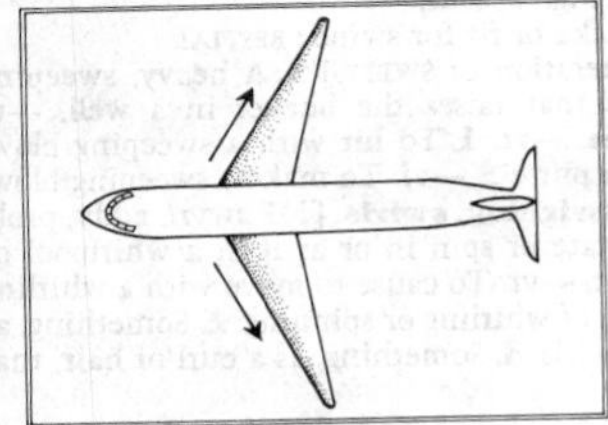

**sweepback**

**sweep·ing** (swē'pĭng) *adj.* **1.** Influencing or ranging over a great area <*sweeping* political changes> **2.** Curving in form or motion <a *sweeping* gesture of disdain> —*n.* **1.** The act or occupation of one that sweeps. **2. sweepings.** Things that are swept up : REFUSE. **—sweep'ing·ly** *adv.*

**sweep·stakes** (swēp'stāks') *also* **sweep·stake** (-stāk') *n., pl.* **sweepstakes. 1.** A lottery in which the participants' contributions form a fund to be awarded as a prize to the winner or winners. **2.** An event or contest, esp. a horse race, the result of which determines the winner of a sweepstakes. **3.** The prize won in a sweepstakes.

**sweet** (swēt) *adj.* **-er, -est.** [ME *swete* < OE *swēte.*] **1. a.** Having a sugary taste. **b.** Containing or derived from sugar. **2.** Pleasing to the feelings, senses, or mind : GRATIFYING <*sweet* success> **3.** Pleasing in disposition : LOVABLE <a *sweet* child> **4.** Not saline <*sweet* water> **5.** Not spoiled, sour, or decaying : FRESH <This milk is still *sweet.*> **6.** Devoid of acid. **7.** *Mus.* Designating jazz characterized by adherence to a melodic line and a time signature. **8.** Low in sulfur content <*sweet* oil> —*n.* **1.** The quality of being sweet : SWEETNESS. **2.** Something that is sweet to the taste or contains sugar. **3.** A candy, preserve, or confection. **4.** *Chiefly Brit.* Something relatively sweet served as a dessert. **5.** A dear or beloved person. **—sweet on.** In love with. **—sweet'ly** *adv.* **—sweet'ness** *n.*

**sweet alyssum** *n.* A widely cultivated plant, *Lobularia maritima,* native to the Mediterranean region, having small, fragrant white or purplish flower clusters.

**sweet-and-sour** (swēt'n-sour') *adj.* Flavored with a sauce made of sugar and vinegar and often fruit <*sweet-and-sour* pork>

**sweet basil** *n.* BASIL 1.

**sweet bay** *n.* A small tree, *Magnolia virginiana* of the southeastern United States, with large, fragrant white flowers.

**sweet·bread** (swēt'brĕd') *n.* The thymus gland of a young animal, esp. a calf, used for food.

**sweet·bri·er** *also* **sweet·bri·ar** (swēt'brī'ər) *n.* A rose, *Rosa eglanteria,* native to Europe, with prickly stems, fragrant leaves, and pink flowers.

**sweet cherry** *n.* **1.** A widely cultivated Eurasian tree, *Prunus avium,* with white flowers and sweet, edible fruit. **2.** The fruit of the sweet cherry.

**sweet chocolate** *n.* Chocolate to which sugar has been added.

**sweet cicely** *n.* **1.** A plant of the genus *Osmorhiza,* with aromatic roots, compound leaves, and small white flower clusters. **2.** An aromatic European plant, *Myrrhis odorata,* with compound leaves and small white flower clusters.

**sweet cider** *n.* Unfermented cider.

**sweet clover** *n.* The melilot.

**sweet corn** *n.* A variety of corn, *Zea mays rugosa,* having kernels that are sweet when young and that is the common edible corn.

**sweet·en** (swēt'n) *v.* **-ened, -en·ing, -ens.** —*vt.* **1.** To make sweet or sweeter by or as if by addition of sugar. **2.** To make more pleasurable or gratifying. **3.** To make bearable : ALLEVIATE. **4.** *Informal.* To increase the value of (collateral for a loan) by adding more securities. **5.** To increase the value of (an unwon poker pot) by adding stakes before reopening. —*vi.* To become sweet. **—sweet'en·er** *n.*

**sweet·en·ing** (swēt'n-ĭng) *n.* **1.** The act or process of making sweet. **2.** Something that sweetens.

**sweet fern** *n.* An aromatic shrub, *Myrica asplenifolia* or *Comptonia peregrina* of eastern North America, with narrow, shallowly lobed, fernlike foliage.

**sweet flag** *n.* A plant, *Acorus calamus,* growing in moist places and having bladelike leaves, minute greenish flowers, and aromatic roots.

**sweet gale** *n.* A swamp shrub, *Myrica gale* of northern regions, having aromatic resinous leaves.

**sweet gum** *n.* **1.** A New World tree, *Liquidambar styraciflua,* having sharply lobed leaves, prickly ball-like fruit clusters, and wood used to make furniture. **2.** The aromatic resin obtained from the sweet gum.

**sweet·heart** (swēt'härt') *n.* **1.** One who loves and is loved by another. **2.** A lovable person.

**sweetheart deal** *n.* A deal arranged by collusion between union officers and an employer, the terms of which are disadvantageous to union members.

**swee·tie** (swē'tē) *n.* *Informal.* Sweetheart : dear.

**sweet·ing** (swē'tĭng) *n.* **1.** A sweet apple. **2.** *Archaic.* A sweetheart.

**sweet marjoram** *n.* MARJORAM 1.

**sweet·meat** (swēt'mēt') *n.* Candy or another sweet delicacy.

**sweet pea** *n.* A climbing plant, *Lathyrus odoratus,* native to southern Europe, cultivated for its fragrant, variously colored flowers.

**sweet pepper** *n.* The bell pepper.

**sweet pepperbush** *n.* A North American shrub, *Clethra alnifolia,* growing in moist ground and having small, fragrant white flower clusters.

**sweet potato** *n.* **1.** A tropical American vine, *Ipomoea batatas,* cultivated for its thick, orange-colored, edible root. **2.** The root of the sweet potato, eaten cooked as a vegetable. **3.** *Informal.* The ocarina.

**sweet·shop** (swēt'shŏp') *n.* *Chiefly Brit.* A candy store.

**sweet·sop** (swēt'sŏp') *n.* **1.** A tropical American tree, *Annona* *squamosa,* having yellowish-green fruit with sweet, edible pulp **2.** The fruit of the sweetsop.

**sweet sorghum** *n.* Sorgo.

**sweet spot** *n. Informal.* The area around the center of mass of a racket, club, or bat that provides the greatest power when hitting a ball.

**sweet sultan** *n.* An Old World plant, *Centaurea moschata,* cultivated for its variously colored flowers.

**sweet talk** *n.* Flattery.

**sweet-talk** (swēt'tôk') *v.* **-talked, -talk·ing, -talks.** —*vt.* To coax or cajole with flattery. —*vi.* To use flattery.

**sweet tooth** *n. Informal.* Inordinate craving for sweets.

**sweet William** *n.* A widely cultivated Eurasian plant, *Dianthus barbatus,* having flat, dense, varicolored flowered clusters.

**swell** (swĕl) *v.* **swelled, swelled** *or* **swol·len** (swō'lən), **swell·ing, swells.** [ME *swellen* < OE *swellan.*] —*vi.* **1.** To increase in size or volume due to internal pressure : EXPAND. **2. a.** To increase in force, size, number, or degree <Membership in the union *swelled.*> **b.** To grow in loudness or intensity, as a sound. **3.** To bulge out, as a sail : PROTRUDE. **4.** To rise in billows above the surrounding level, as clouds. **5. a.** To be or become filled with an emotion <*swelled* with pride> **b.** To behave pompously or self-importantly. —*vt.* **1.** To cause to increase in size, volume, number, degree, or intensity. **2.** To fill with emotion. —*n.* **1.** The act or process of swelling or the state of being swollen. **2.** A swollen part. **3.** A long ocean wave that moves continuously without breaking. **4.** A rounded hill. **5.** *Informal.* One who is fashionably dressed or prominent in fashionable society. **6.** *Mus.* **a.** A crescendo followed by a gradual diminuendo. **b.** The sign indicating this. **c.** A device on some instruments, as an organ or harpsichord, for regulating volume. —*adj.* **-er, -est.** *Informal.* **1.** Fashionably elegant : STYLISH. **2.** Fine : excellent <"It was *swell* out, just cool enough" —James T. Farrell>

**swell box** *n.* A chamber housing one or more sets of organ pipes and having shutters that can be opened or shut to regulate the volume of tone.

**swelled head** *n.* Undue high opinion of oneself : CONCEIT.

**swell·fish** (swĕl'fĭsh') *n., pl.* **swellfish** *or* **-fish·es.** PUFFER 2.

**swell·head** (swĕl'hĕd') *n.* A conceited person. **—swell'head'ed** *adj.* **—swell'head'ed·ness** *n.*

**swell·ing** (swĕl'ĭng) *n.* **1.** The state of being swollen. **2.** Something swollen, esp. an abnormally swollen or protuberant bodily part.

**swel·ter** (swĕl'tər) *v.* **-tered, -ter·ing, -ters.** [ME *swelteren,* freq. of *swelten,* to faint from heat < OE *sweltan,* to die.] —*vi.* To be affected by oppressive heat. —*vt.* **1.** To affect with oppressive heat. **2.** *Archaic.* To exude. —*n.* Oppressive heat.

**swel·ter·ing** (swĕl'tər-ĭng) *adj.* **1.** Oppressively hot. **2.** Suffering from oppressive heat. **—swel'ter·ing·ly** *adv.*

**swel·try** (swĕl'trē) *adj.* **-tri·er, -tri·est.** Sweltering.

**swept** (swĕpt) *v. p.t. & p.p. of* SWEEP.

**swept·back** (swĕpt'băk') *adj.* Angled rearward from the points of attachment. —Used esp. of aircraft wings.

**swerve** (swûrv) *vt. & vi.* **swerved, swerv·ing, swerves.** [ME *swerven* < OE *sweorfan,* to rub.] To turn aside suddenly from a straight course : VEER. —*n.* The act of swerving.

**swift** (swĭft) *adj.* **-er, -est.** [ME < OE.] **1.** Moving or able to move fast. **2.** Coming, happening, or accomplished quickly <a *swift* kick> **3.** Quick to act or react : PROMPT <*swift* to respond> —*adv.* Quickly <*swift*-running rivers> —*n.* **1. a.** A cylinder on a carding machine. **b.** A reel for holding yarn as it is wound off. **2.** Any of various dark-colored birds of the family Apodidae, with long, narrow wings and a relatively short tail. **3.** Any of various small, fast-moving North American lizards of the genera *Sceloporus* and *Uta.* **—swift'ly** *adv.* **—swift'ness** *n.*

**swig** (swĭg) [Orig. unknown.] *Informal.* —*n.* A large swallow of a liquid : GULP. —*vt. & vi.* **swigged, swig·ging, swigs.** To drink in swigs : GULP. **—swig'ger** *n.*

**swill** (swĭl) *v.* **swilled, swill·ing, swills.** [ME *swilen* < OE *swilian,* to wash out.] —*vt.* **1.** To drink greedily or eagerly. **2.** To flood with water, as for washing. **3.** To feed (animals) with slop. —*vi.* To eat or drink greedily or eagerly. —*n.* **1.** A mixture of liquid and solid food, as table scraps, fed to animals, esp. pigs. **2.** Refuse : garbage. **3.** A deep draft of liquor. **—swill'er** *n.*

**swim** (swĭm) *v.* **swam** (swăm), **swum** (swŭm), **swim·ming, swims.** [ME *swimmen* < OE *swimman.*] —*vi.* **1.** To propel oneself through water by means of movements of the body. **2.** To move as though gliding through water. **3.** To float on water. **4.** To be immersed <veal *swimming* in sauce> **5.** To feel faint or giddy <Champagne makes my head *swim.*> **6.** To appear to spin hazily. —*vt.* **1.** To propel oneself through or across (a body of water) by swimming. **2.** To cause to swim or float on a body of water. —*n.* **1.** An act or instance of swimming. **2.** A period of swimming. **3.** Dizziness. **—in the swim.** *Informal.* Participating in what is current or fashionable. **—swim against the stream.** To move in opposition to a popular trend. **—swim'mer** *n.*

**swim bladder** *n.* AIR BLADDER 1.

**swim·mer·et** (swĭm'ə-rĕt', swĭm'ə-rĕt') *n.* One of the paired abdominal appendages of certain aquatic crustaceans, as shrimps, lobsters, and isopods, that function as organs of respiration or locomotion.

**swimmer's ear** *n.* An external otitis often seen in people who swim for considerable periods of time or fail to dry completely their ear canals afterward.

**swimmer's itch** *n.* A dermatitis caused by the penetration of the skin by cercariae of certain schistosomes.

**swim·ming·ly** (swĭm'ĭng-lē) *adv.* Easily and successfully <The project went *swimmingly.*>

**swimming pool** *n.* A pool constructed for swimming.

**swim·suit** (swĭm'sōōt') *n.* A garment worn for swimming.

**swin·dle** (swĭn'dl) *v.* **-dled, -dling, -dles.** [Back-formation < *swindler* < G. *Schwindler,* dizzy person < *schwindeln,* to be dizzy < OHG *swintilōn,* freq. of *swintan,* to vanish.] —*vt.* **1.** To defraud or cheat (a victim) of money or property. **2.** To obtain (e.g., money) by fraud. —*vi.* To practice fraud as a way of obtaining money. —*n.* An act or instance of swindling : FRAUD. **—swin'dler** *n.*

**swine** (swīn) *n., pl.* **swine.** [ME < OE *swīn.*] **1.** An ungulate mammal of the family Suidae, including the pig, hog, and boar. **2.** A contemptible, vicious person.

**swine·herd** (swīn'hûrd') *n.* One who tends or keeps swine.

**swine·pox** (swīn'pŏks') *n.* A disease of domesticated swine caused by a virus similar to that causing cowpox and smallpox and characterized by skin lesions.

**swing** (swĭng) *v.* **swung, swing·ing, swings.** [ME *swingen* < OE *swingan,* to flog.] —*vi* **1.** To move rhythmically back and forth suspended or as if suspended from above. **2.** To try to strike a ball with a sweeping motion of the arm. **3.** To move laterally or in a curve <The car *swung* over to the curb.> **4.** To turn in place, as on a hinge. **5.** To move from one attitude, position, or emotion to another : VACILLATE. **6.** *Slang.* To be executed by hanging. **7. a.** *Mus.* To have a compulsive rhythm. **b.** To play a piece with a compulsive rhythm. **8.** *Slang.* To be spirited and up-to-date. —*vt.* **1.** To cause to move back and forth. **2.** To cause to move in a broad arc <*swing* a baseball bat> **3.** To move with a sweeping motion <*swing* one's legs> **4.** To hang or suspend (something) so that it can move freely. **5. a.** To suspend on hinges <*swing* a blind> **b.** To cause to turn on hinges <*swing* the gate open> **6.** To cause to move from one attitude, position, or emotion to another. **7.** *Slang.* To manipulate or manage successfully <was unable to *swing* the deal> **8.** To perform (popular music) in the style of swing. —*n.* **1.** An act of swinging, esp.: **a.** A rhythmic back-and-forth movement. **b.** A single movement or series of movements in one direction. **2.** The space covered while swinging <The giant pendulum's *swing* is 40 feet.> **3.** The way in which one swings something, as a baseball bat or golf club. **4.** Freedom and scope of movement or action. **5. a.** A swaying, graceful motion. **b.** A sweep or swoop <the *swing* of a kite across the sky> **6.** A seat suspended from above, on which one may ride back and forth in an arc. **7.** *Mus.* **a.** An innovation in popular dance music developed about 1935 and based on jazz but using a larger band and simpler harmonic and rhythmic patterns. **b.** The rhythmic quality of this music. —*adj.* Relating to or performing swing. **—in full swing.** At full speed or intensity <a party *in full swing*>

**swing-by** (swĭng'bī') *n., pl.* **-bys.** An interplanetary mission in which a space vehicle utilizes planetary gravitation for course changes.

**swinge** (swĭnj) *vt.* **swinged, swinge·ing, swing·es.** [ME *swengen,* to shake < OE *swengan.*] *Archaic.* To strike or beat : SWIPE. **—swing'er** (swĭn'jər) *n.*

**swing·er** (swĭng'ər) *n.* **1.** One that swings. **2.** *Slang.* One who actively seeks excitement and pleasure and moves with the latest trends.

**swing·ing** (swĭng'ĭng) *adj.* Spirited and up-to-date.

**swin·gle·tree** (swĭng'gəl-trē) *n.* [< E. *swingle,* wooden instrument < ME < MDu. *swinghel.*] A whiffletree.

**swing·man** (swĭng'mən) *n.* A team member who has the ability to play effectively in two different positions, as both forward and guard in basketball.

**swing shift** *n. Informal.* A factory work shift between the day and night shifts, lasting from about 4 P.M. to midnight.

**swing-wing** (swĭng'wĭng') *adj.* Of or being an aircraft with wings constructed to allow the outer portion to fold back along the fusilage to produce streamlining at high speeds.

**swin·ish** (swī'nĭsh) *adj.* Like or fit for swine : BESTIAL.

**swipe** (swīp) *n.* [Perh. alteration of SWEEP.] **1.** A heavy, sweeping blow. **2.** A lever, esp. one that raises the bucket in a well. —*v.* **swiped, swip·ing, swipes.** —*vt.* **1.** To hit with a sweeping blow. **2.** *Slang.* To steal <*swipe* a purse> —*vi.* To make a sweeping blow.

**swirl** (swûrl) *v.* **swirled, swirl·ing, swirls.** [ME *swyrl,* eddy, prob. of LG orig.] —*vi.* **1.** To rotate or spin in or as if in a whirlpool or eddy. **2.** To be dizzy or faint. —*vt.* To cause to move with a whirling motion. —*n.* **1.** The motion of whirling or spinning. **2.** Something, as a whirlpool or eddy, that swirls. **3.** Something, as a curl of hair, that is swirled. **—swirl'y** *adj.*

**swish** (swĭsh) *v.* **swished, swish·ing, swish·es.** [Imit.] —*vi.* **1.** To move with a whistle or hiss. **2.** To rustle, as certain fabrics. —*vt.* **1.** To cause to make a swishing sound or movement. **2.** To chastise with a rod. —*n.* **1. a.** A sharp whistling or rustling sound <the *swish* of a sword> **b.** A movement making such a sound. **2. a.** A rod for flogging. **b.** A stroke made with such a rod. —*adj. Slang. Chiefly Brit.* Fashionable.

**swish·y** (swĭsh'ē) *adj.* **-i·er, -i·est. 1.** Creating a swishing sound. **2.** *Slang.* Effeminate.

**Swiss** (swĭs) *adj.* [OFr. *Suisse* < MHG *Swizer* < *Swiz*, Switzerland.] Of, relating to, or characteristic of Switzerland, its residents, or its culture. —*n., pl.* **Swiss. 1.** A native or resident of Switzerland. **2.** One of Swiss descent. **3.** A crisp, sheer cotton cloth used for curtains or light garments.

**Swiss chard** *n.* Chard.

**Swiss cheese** *n.* A firm white or pale-yellow cheese with many large holes, orig. produced in Switzerland.

**Swiss steak** *n.* A round or shoulder steak pounded with flour, braised, and usu. served with a seasoned sauce.

**switch** (swĭch) *n.* [Perh. < MDu. *swijch*, twig.] **1.** A slender flexible rod or stick, used esp. for whipping. **2.** The bushy tip of the tail of an animal <a cow's *switch*> **3.** A thick bunch of real or synthetic hair used in a coiffure. **4.** A flailing or lashing, as with a slender rod. **5.** *Elect.* A device for breaking or opening an electrical circuit or for diverting current from one conductor to another. **6.** A device consisting of two sections of railroad track and the accompanying apparatus, used to transfer rolling stock from one track to another. **7. a.** The act or process of operating a switching device. **b.** The result achieved by such an act. **8.** A shift, as of opinion or attention. —*v.* **switched, switch·ing, switch·es.** —*vt.* **1.** To whip with or as if with a switch. **2.** To jerk or swish abruptly. **3.** To shift, transfer, change, or divert <*switch* the conversation to a neutral topic> **4.** To exchange <*switch* sides> **5.** To connect, disconnect, or divert (an electric current) by operating a switch. **6. a.** To cause (an electric current or appliance) to begin operation <*switch* on the lights> **b.** To cause (an electric current or appliance) to cease operation <*switch* off the lights> **7.** To move (rolling stock) from one track to another : SHUNT. —*vi.* **1.** To shift or change <*switch* from car to motorcycle> **2.** To be shifted or changed. **—switch'er** *n.*

**switch·back** (swĭch'băk') *n.* **1.** A road, roadbed, or trail that ascends a steep incline in a winding course. **2.** *Chiefly Brit.* A roller coaster.

**switch·blade knife** (swĭch'blād') *n.* A pocket knife having a spring-operated blade that unsheathes when a release on the handle is pressed.

**switch·board** (swĭch'bôrd', -bōrd') *n.* One or more panels accommodating control switches, indicators, and other apparatus for operating electric circuits.

**switch engine** *n.* A railroad engine used to switch cars.

**switch·er·oo** (swĭch'ə-ro͞o') *n.* [Alteration of SWITCH.] *Slang.* An unexpected reversal or change.

**switch hitter** *n. Baseball.* An ambidextrous batter.

**switch·man** (swĭch'mən) *n.* One who operates railroad switches.

**switch·yard** (swĭch'yärd') *n.* An area where railroad cars are switched and trains assembled.

**Swit·zer** (swĭt'sər) *n.* [MHG *Swizer.* —see SWISS.] A Swiss.

**swiv·el** (swĭv'əl) *n.* [ME *swyvel.*] **1.** A fastening, as a link or pivot, so designed that it permits the free turning of attached parts. **2.** A pivoted support that allows an attached object, as a chair or gun, to turn in a horizontal plane. **3.** A cannon that turns on a pivot. —*v.* **-eled, -el·ing, -els** or **-elled, -el·ling, -els.** —*vt.* **1.** To turn or rotate on or as if on a swivel. **2.** To secure, fit, or support with a swivel. —*vi.* To turn on or as if on a swivel <*swivel* in one's chair>

**swivel chair** *n.* A chair that swivels on its base.

**swiv·el-hipped** (swĭv'əl-hĭpt') *adj.* Marked by or moving with exaggerated swinging of the hips.

**swiv·et** (swĭv'ĭt) *n.* [Orig. unknown.] *Informal.* A state of extreme distress or discomposure.

**swiz·zle** (swĭz'əl) *n.* [Orig. unknown.] Any of various tall mixed drinks, usu. made with rum.

**swizzle stick** *n.* A rod for stirring mixed drinks.

**swob** (swŏb) *n.* & *v. var. of* SWAB.

**swol·len** (swō'lən) *v. var. p.p. of* SWELL.

**swoon** (swo͞on) *vi.* **swooned, swoon·ing, swoons.** [ME *swounen*, prob. < OE *swōgan*, to suffocate.] To faint. —*n.* A fainting spell.

**swoop** (swo͞op) *v.* **swooped, swoop·ing, swoops.** [ME *swopen*, to sweep along < OE *swāpan*, to sweep.] —*vi.* To make a sudden sweeping movement <The hawk *swooped* upon its prey.> —*vt.* To take or snatch suddenly : SCOOP. —*n.* The act of swooping.

**swoosh** (swo͞osh, swo͝osh) *v.* **swooshed, swoosh·ing, swoosh·es.** [Imit.] —*vi.* **1.** To make a rushing sound. **2.** To flow or swirl copiously. —*vt.* To emit or carry with a rushing sound.

**swop** (swŏp) *v.* & *n. var. of* SWAP.

**sword** (sôrd) *n.* [ME < OE *sweord.*] **1.** A weapon with a long blade for thrusting or cutting. **2.** An instrument of death, combat, or destruction. **3. a.** Use of force, as in war. **b.** Power or jurisdiction. **4.** Something resembling a sword. **—at swords' points.** Ready for combat : ANTAGONISTIC **—cross swords. 1.** To fight. **2.** To quarrel violently. **—put to the sword.** To kill with a sword.

**sword bayonet** *n.* A short swordlike bayonet.

**sword·bill** (sôrd'bĭl') *n.* A hummingbird, *Ensifera ensifera* of tropical South America, with a very long, slender bill.

**sword cane** *n.* A cane designed to hide a sword or dagger.

**sword dance** *n.* A dance performed with swords, esp. one performed around swords laid on the ground.

**sword·fish** (sôrd'fĭsh') *n., pl.* **swordfish** or **-fish·es.** A large marine game and food fish, *Xiphias gladius,* with a long swordlike extension of the upper jaw.

**sword grass** *n.* Any of various grasses or grasslike plants with pointed bladelike leaves.

**sword·knot** (sôrd'nŏt') *n.* A decorative loop or tassle attached to the hilt of a sword.

**Sword of Damocles** *n.* [After *Damocles*, courtier of Dionysius I, tyrant of Syracuse, who was forced to sit under a sword suspended by a hair to demonstrate the precariousness of a king's fortune.] An impending disaster.

**sword·play** (sôrd'plā') *n.* Use of a sword, as in combat or fencing. **—sword'play'er** *n.*

**swords·man** (sôrdz'mən) *n.* **1.** A skilled user of sword. **2.** One armed with a sword. **—swords'man·ship'** *n.*

**sword·tail** (sôrd'tāl') *n.* A small, brightly colored freshwater fish, *Xiphophorus helleri* of Central America, that has a long, tapering extension of the caudal fin in the male and is popular in home aquariums.

**swordtail**
*Approximately 3 inches long*

**swore** (swôr) *v. p.t. of* SWEAR.

**sworn** (swôrn) *v. p.p. of* SWEAR.

**swounds** or **swouns** *interj. var. of* ZOUNDS.

**swum** (swŭm) *v. p.p. of* SWIM.

**swung** (swŭng) *v. p.t.* & *p.p. of* SWING.

**swung dash** *n.* A character ~ used in printing to save space by standing for all or part of a previously spelled out word.

**syb·a·rite** also **Syb·a·rite** (sĭb'ə-rīt') *n.* [Lat. *Sybarita*, native of Sybaris < Gk. *Subaritēs* < *Subaris*, Sybaris, Italy (from the notorious luxury of the inhabitants of Sybaris).] One fond of pleasure and luxury : VOLUPTUARY. **—syb'a·rit'ic** (-rĭt'ĭk), **syb'a·rit'i·cal** *adj.* **—syb'a·rit'i·cal·ly** *adv.*

**syc·a·mine** (sĭk'ə-mīn, -mĭn) *n.* [Lat. *sycaminus* < Gk. *sukaminos*, of Semitic orig.] A tree mentioned in the New Testament, thought to be a species of mulberry.

**syc·a·more** (sĭk'ə-môr', -mōr') *n.* [ME *sicamour*, a kind of fig tree < OFr. *sicamor* < Lat. *sycomorus* < Gk. *sukomoros.*] **1.** A deciduous tree, *Platanus occidentalis* of eastern North America, with lobed leaves, ball-like seed clusters, and bark that often flakes off in large patches. **2.** A Eurasian tree, *Acer pseudoplatanus*, resembling the maple. **3.** A tree, *Ficus sycomorus* of northeastern Africa and adjacent Asia, related to the fig and mentioned in the Bible.

**syce** (sīs) *n.* [Hindi *sā'is* < Ar. < *sāsa*, to administer.] A groom for horses, esp. in India.

**sy·cee** (sī'sē') *n.* [Chin. (Cantonese) *sai⁴ si¹*, fine silk (so called because the pure silver can be spun into fine threads).] Lumps of pure silver bearing the stamp of a banker or assayer and formerly used in China as money.

**syc·o·more** (sĭk'ə-môr', -mōr') *n. Obs. var. of* SYCAMORE 3.

**sy·co·ni·um** (sī-kō'nē-əm) *n., pl.* **-ni·a** (-nē-ə) [NLat. < Gk. *sukon*, fig.] The fleshy multiple fruit of the fig, consisting primarily of the enlarged floral receptacle.

**syc·o·phan·cy** (sĭk'ə-fən-sē) *n., pl.* **-cies.** The act, practice, or behavior of a sycophant : servile flattery.

**syc·o·phant** (sĭk'ə-fənt) *n.* [Lat. *sycophanta* < Gk. *sukophantēs*, informer : *sukon*, fig + *-phantēs* < *phainein*, to show.] One who attempts to progress or win favor by flattering influential people. **—syc'o·phan'tic** (-făn'tĭk), **syc'o·phan'ti·cal** *adj.* **—syc'o·phan'ti·cal·ly** *adv.*

**sy·co·sis** (sī-kō'sĭs) *n.* [Gk. *sukōsis*, ulcer resembling a fig < *sukon*, fig.] Chronic inflammation of the hair follicles, esp. of the beard and scalp.

**sy·e·nite** (sī′ə-nīt′) *n.* [Lat. *Syenites (lapis),* (stone) of Syene < *Syene,* Syene, ancient city in Egypt < Gk. *Suēnē.*] An igneous rock composed primarily of alkali feldspar together with other minerals, as hornblende. **—sy′e·nit′ic** (-nĭt′ĭk) *adj.*

**sy·li** (sē′lē) *n.* [Native word in Guinea.] —See table at CURRENCY.

**syl·la·bar·y** (sĭl′ə-bĕr′ē) *n., pl.* **-ies.** [NLat. *syllabarium* < Lat. *syllaba,* syllable. —see SYLLABLE.] A list of syllables, esp. a list or set of written characters, each one representing a syllable.

**syl·la·bi** (sĭl′ə-bī′) *n. var. pl. of* SYLLABUS.

**syl·lab·ic** (sĭ-lăb′ĭk) *adj.* [Med. Lat. *syllabicus* < Gk. *sullabikos* < *sullabē,* syllable.] **1.** Of, relating to, or composed of a syllable or syllables. **2.** Designating a consonant that forms a syllable without a vowel, as the *l* in *riddle* (rĭd′l). **3.** Pronouncing every syllable distinctly. **4.** Designating a verse form based on the number of syllables per line rather than on the arrangement of accents or quantities. *—n.* A syllabic sound. **—syl·lab′i·cal·ly** *adv.*

**syl·lab·i·cate** (sĭ-lăb′ə-kāt′) *also* **syl·lab·i·fy** (-fī′) *vt.* **-cat·ed, -cat·ing, -cates** *also* **-fied, -fy·ing, -fies.** To form or divide into syllables. **—syl·lab′i·ca′tion, syl·lab′i·fi·ca′tion** *n.*

**syl·la·bism** (sĭl′ə-bĭz′əm) *n.* [< Lat. *syllaba,* syllable. —see SYLLABLE.] **1.** Use of written characters representing syllables. **2.** Division into syllables.

**syl·la·bize** (sĭl′ə-bīz′) *vt.* **-bized, -biz·ing, -biz·es.** [Med. Lat. *syllabizare* < Gk. *sullabizein* < *sullabē,* syllable.] To syllabicate.

**syl·la·ble** (sĭl′ə-bəl) *n.* [ME *sillable* < Norman Fr. < OFr. *sillabe* < Lat. *syllaba* < Gk. *sullabē* < *sullambanein,* to combine in pronunciation : *sun-,* together + *lambanein,* to take.] **1.** A unit of spoken language consisting of a single uninterrupted sound formed by a vowel or diphthong alone, of a syllabic consonant alone, or of either with one or more consonants. **2.** One or more letters or phonetic symbols written or printed to approximate a spoken syllable. **3.** The slightest bit : JOT. *—vt.* **-bled, -bling, -bles.** To pronounce (e.g., a line of poetry) in syllables.

**syl·la·bub** *also* **sil·la·bub** (sĭl′ə-bŭb′) *n.* [Orig. unknown.] A beverage or a dessert consisting of wine or liquor mixed with sweetened milk or cream and beaten to a froth.

**syl·la·bus** (sĭl′ə-bəs) *n., pl.* **-bus·es** or **-bi** (-bī′) [Med. Lat., prob. alteration of *syllabos,* alteration of Lat. *sittybas* < Gk. *sittubas,* accusative pl. of *sittuba,* title slip.] **1.** An outline of the main points of a text, speech, or course of study. **2.** *Law.* A short statement preceding a report on an adjudged case and containing the court rulings on the legal points involved.

**syl·lep·sis** (sĭ-lĕp′sĭs) *n., pl.* **-ses** (-sēz′) [LLat. < Gk. *sullēpsis* : *sun-,* together + *lēpsis,* a taking < *lambanein,* to take.] A construction in which one word seems to be in the same relation to two or more words but in fact is not; e.g., *We lost our coats and our tempers.* **—syl·lep′tic** *adj.*

**syl·lo·gism** (sĭl′ə-jĭz′əm) *n.* [ME *silogisme* < OFr. < Lat. *syllogismus* < Gk. *sullogismos* < *sullogizesthai,* to infer : *sun-,* with + *logizesthai,* to reason < *logos,* reason.] **1.** *Logic.* Deductive reasoning consisting of a major premise, a minor premise, and a conclusion; e.g., *All creatures are foolish* (major premise); *Smith is a creature* (minor premise); *therefore, Smith is foolish* (conclusion). **2.** Reasoning from the general to the specific : DEDUCTION. **3.** Subtle or specious reasoning. **—syl′lo·gis′tic, syl′lo·gis′ti·cal** *adj.* **—syl′lo·gis′ti·cal·ly** *adv.*

**syl·lo·gis·tics** (sĭl′ə-jĭs′tĭks) *n.* *(used with a sing. verb).* **1.** The branch of logic dealing with syllogisms. **2.** The art of reasoning by syllogism.

**syl·lo·gize** (sĭl′ə-jīz′) *v.* **-gized, -giz·ing, -giz·es.** *—vi.* To reason or argue by syllogisms. *—vt.* To deduce by syllogism. **—syl′lo·gi·za′tion** *n.* **—syl′lo·giz′er** *n.*

**sylph** (sĭlf) *n.* [NLat. *sylphus.*] **1.** Any of a class of elemental beings without souls that were supposed to inhabit the air. **2.** A slim, graceful girl or woman.

**sylph·id** (sĭl′fĭd) *n.* [Fr. *sylphide* < *sylphe,* sylph < NLat. *sylphus.*] A young or diminutive sylph. *—adj.* Relating to or like a sylph.

**syl·va** (sĭl′və) *n. var. of* SILVA.

**syl·van** *also* **sil·van** (sĭl′vən) *adj.* [Med. Lat. *silvanus* < Lat. *silva,* forest.] **1.** Relating to or characteristic of forests. **2.** Found or living in a forest. **3.** Abounding in trees : WOODED. *—n.* One that lives in or frequents the woods.

**syl·van·ite** (sĭl′və-nīt′) *n.* [Fr. after *Transylvania,* Rumania, where it was first found.] A pale brass-yellow to silver-white gold and silver ore, chiefly (Au, Ag)$Te_2$.

**Syl·va·nus** (sĭl-vā′nəs) *n. var. of* SILVANUS.

**syl·vat·ic** (sĭl-văt′ĭk) *adj.* [Lat. *silvaticus,* of the forest, wild < *silva,* forest.] **1.** Affecting wild animals <a *sylvatic* disease> **2.** Sylvan.

**syl·vite** (sĭl′vīt′) *also* **syl·vine** (-vēn′) or **syl·vin·ite** (-vĭn-īt′) *n.* [Fr., alteration of *sylvine* < NLat. *(sal digestivus) Sylvii,* (digestive salt) of Sylvius.] A colorless vitreous potassium chloride mineral used as a major source of potassium compounds.

**sym-** *pref. var. of* SYN-.

**sym·bi·ont** (sĭm′bē-ŏnt′, -bī-) *n.* [Gk. *sumbiōn, sumbiont-,* pr.part. of *sumbioun,* to live together. —see SYMBIOSIS.] An organism in a symbiotic relationship. **—sym′bi·on′tic** *adj.*

**sym·bi·o·sis** (sĭm′bē-ō′sĭs, -bī-) *n.* [Gk. *sumbiōsis,* companionship < *sumbioun,* to live together : *sun-,* together + *bios,* life.] *Biol.* The relationship of two or more different organisms in a close association that may be but is not necessarily of benefit to each. **—sym′bi·ot′ic** (-ŏt′ĭk), **sym′bi·ot′i·cal** (-ŏt′ĭ-kəl) *adj.* **—sym′bi·ot′i·cal·ly** *adv.*

**sym·bi·ote** (sĭm′bē-ōt′, -bī-) *n.* A symbiont.

**sym·bol** (sĭm′bəl) *n.* [Lat. *symbolum* < Gk. *sumbolon,* token for identification (by comparison with a counterpart) < *sumballein,* to compare : *sun-,* together + *ballein,* to throw.] **1.** Something representing something else by association, resemblance, or convention, esp. a material object representing something abstract. **2.** A printed or written sign used to represent an operation, element, quantity, quality, or relation, as in mathematics, statistics, or music. *—vt.* **-boled, -bol·ing, -bols.** To symbolize.

**sym·bol·ic** (sĭm-bŏl′ĭk) *also* **sym·bol·i·cal** (-ĭ-kəl) *adj.* **1.** Of, relating to, or expressed by a symbol or symbols. **2.** Functioning as a symbol. **3.** Marked by the use of symbolism, as a work of art. **—sym·bol′i·cal·ly** *adv.* **—sym·bol′i·cal·ness** *n.*

**symbolic address** *n. Computer Sci.* An identification of particular data without regard to the location of the data within the device.

**symbolic language** *n. Computer Sci.* A programming language designed for ease of use by human operators rather than computers.

**symbolic logic** *n.* A treatment of formal logic in which a calculus or system of symbols is used to represent quantities and relationships.

**sym·bol·ism** (sĭm′bə-lĭz′əm) *n.* **1.** The practice of representing things by means of symbols or of attributing symbolic meanings or significance to objects, events, or relationships. **2.** A system of symbols or representations. **3.** A symbolic meaning or representation. **4.** The revelation or suggestion of intangible conditions or truths by artistic invention.

**sym·bol·ist** (sĭm′bə-lĭst) *n.* **1.** One who uses symbols or symbolism. **2. a.** One who interprets or represents conditions or truths by the use of symbolism. **b.** Any of a group of chiefly French artists and writers of the late 19th cent. who expressed their ideas and emotions through symbols. **—sym′bol·is′tic, sym′bol·is′ti·cal** *adj.*

**sym·bol·ize** (sĭm′bə-līz′) *v.* **-ized, -iz·ing, -iz·es.** *—vt.* **1.** To be or serve as a symbol of. **2.** To represent or identify by a symbol or symbols. *—vi.* To use symbols. **—sym′bol·i·za′tion** *n.*

**sym·bol·o·gy** (sĭm-bŏl′ə-jē) *n.* **1.** Study or interpretation of symbols or symbolism. **2.** The art or practice of symbolic expression.

**sym·met·al·ism** (sĭm-mĕt′l-ĭz′əm) *n.* A system of coinage in which a unit of currency consists of a combination of two or more metals in fixed proportions.

**sym·met·ric** (sĭ-mĕt′rĭk) *also* **sym·met·ri·cal** (-rĭ-kəl) *adj.* Of or exhibiting symmetry. **—sym·met′ri·cal·ly** *adv.*

**symmetric group** *n.* A permutation group comprised of all possible permutations of a given number of items.

**symmetric matrix** *n.* A matrix that is its own transpose.

**sym·met·rics** (sə-mĕt′rĭks) *n.* [Blend of SYMMETRICAL and METRICS.] *(sing. in number).* An epigrammatic verse form invented (1948) by David McCord, consisting of a quatrain beginning and ending with the same word in which the second rhyming word of the first couplet is repeated at the beginning of the second couplet, the pair forming a third line when printed, being centered between the couplets (of which they are a part) with a colon spaced equally left and right to divide them.

**sym·me·trize** (sĭm′ĭ-trīz′) *vt.* **-trized, -triz·ing, -triz·es.** To make symmetrical. **—sym′me·tri·za′tion** *n.*

**sym·me·try** (sĭm′ĭ-trē) *n., pl.* **-tries.** [Obs. Fr. *symmetrie* < Lat. *symmetria* < Gk. *summetria* < *summetros,* of like measure : *sun-,* like + *metron,* measure.] **1.** A relationship of characteristic correspondence, equivalence, or identity among constituents of a system or between different systems <electric charge *symmetry*> **2.** Correspondence of form and arrangement of parts on opposite sides of a boundary, as a plane or line or around a point or axis. **3.** Structural or functional independence of direction : ISOTROPY. **4.** Beauty resulting from balanced or harmonious arrangement.

**sym·pa·thec·to·my** (sĭm′pə-thĕk′tə-mē) *n., pl.* **-mies.** [SYMPATH(ETIC) + -ECTOMY.] Surgical removal of a part of a sympathetic nerve or a number of sympathetic ganglia.

**sym·pa·thet·ic** (sĭm′pə-thĕt′ĭk) *adj.* [Gk. *sumpathētikos* < *sumpatheia,* sympathy.] **1.** Of, feeling, expressing, or resulting from sympathy. **2.** Exhibiting agreement or approval : FAVORABLE <*sympathetic* to our proposal> **3.** Being in accord with one's disposition or mood <a *sympathetic* family relationship> **4.** Relating to or affecting the sympathetic nervous system. **—sym′pa·thet′i·cal·ly** *adv.*

**sympathetic ink** *n.* Invisible ink.

**sympathetic nervous system** *n.* The part of the autonomic nervous system that contains adrenergic fibers and tends to depress secretion, decrease smooth-muscle tone and contractility, and cause vascular contraction.

**sym·pa·thin** (sĭm′pə-thĭn) *n.* [SYMPATH(ETIC) + -IN.] A hormone resembling epinephrine, believed to be formed in the muscle cells by sympathetic nerve impulses.

**sym·pa·thize** (sĭm′pə-thīz′) *vi.* **-thized, -thiz·ing, -thiz·es.**

1. To feel or express compassion : COMMISERATE. 2. To share or understand another's feelings or ideas. 3. *Obs.* To agree in quality or disposition : CORRESPOND. **—sym'pa·thiz'er** *n.* **—sym'pa·thiz'ing·ly** *adv.*

**sym·pa·tho·lyt·ic** (sĭm'pə-thō-lĭt'ĭk) *adj.* [SYMPATH(ETIC) + -LYTIC.] Of or relating to an agent that opposes the activity of the sympathetic nervous system.

**sym·pa·tho·mi·met·ic** (sĭm'pə-thō-mĭ-mĕt'ĭk, -mī-) *adj.* [SYMPATH(ETIC) + MIMETIC.] Of or relating to an agent that stimulates the sympathetic nervous system.

**sym·pa·thy** (sĭm'pə-thē) *n., pl.* **-thies.** [Lat. *sympathia* < Gk. *sumpatheia* < *sumpathēs*, affected by like feelings : *sun-*, like + *pathos*, emotion.] **1. a.** Mutual affinity between individuals in which whatever affects one correspondingly affects the other. **b.** Mutual understanding or affection arising from this affinity. **2. a.** The act of or capacity for sharing or understanding the feelings of another person. **b.** Pity or sorrow for the distress of another : COMMISERATION. **3.** Agreement : accord <I am not in *sympathy* with your motives.> **4.** A feeling of loyalty : ALLEGIANCE.

**sympathy strike** *n.* A strike by a group of workers for the purpose of supporting a cause or another group of strikers.

**sym·pat·ric** (sĭm-păt'rĭk) *adj.* [SYM- + Gk. *patra*, fatherland (< *patēr*, father) + -IC.] Occupying the same or overlapping geographic areas without interbreeding.—Used of populations of closely related species. **—sym·pat'ri·cal·ly** *adv.*

**sym·pet·al·ous** (sĭm-pĕt'l-əs) *adj.* Gamopetalous.

**sym·phon·ic** (sĭm-fŏn'ĭk) *adj.* **1.** Relating to or having the character or form of a symphony. **2.** Harmonious in sound.

**symphonic poem** *n.* Program music based on an extramusical theme in a single, extended movement for symphony orchestra and typical mainly of the late 19th cent.

**sym·pho·ni·ous** (sĭm-fō'nē-əs) *adj.* Agreeing, esp. with regard to sound : HARMONIOUS. **—sym·pho'ni·ous·ly** *adv.*

**sym·pho·nist** (sĭm'fə-nĭst) *n.* A composer of symphonies.

**sym·pho·ny** (sĭm'fə-nē) *n., pl.* **-nies.** [ME *symphonie*, harmony < OFr. < Lat. *symphonia* < Gk. *sumphōnia* < *sumphōnos*, harmonious : *sun-*, together + *phōnē*, sound.] **1.** *Mus.* **a.** A usu. long sonata for orchestra, consisting of four related movements. **b.** An instrumental passage in a vocal or choral composition. **c.** An instrumental overture or interlude, as in early opera. **2. a.** A symphony orchestra. **b.** An orchestral concert. **3.** Harmony, esp. of sound or color. **4.** Something marked by a harmonious combination of elements <a *symphony* of muted colors>

**symphony orchestra** *n.* A large orchestra composed of string, wind, and percussion sections, designed for playing symphonic works.

**sym·phy·sis** (sĭm'fĭ-sĭs) *n., pl.* **-ses** (-sēz') [Gk. *sumphusis* < *sumphuein*, to cause to grow together : *sun-*, together + *phuein*, to cause to grow.] **1.** Synarthrosis. **2.** Coalescence of similar parts or organs. **—sym'phy·se'al** (sĭm'fĭ-sē'əl), **sym·phys'i·al** (sĭm-fĭz'ē-əl) *adj.*

**sym·po·di·um** (sĭm-pō'dē-əm) *n., pl.* **-dia** (-dē-ə) [NLat. : SYM- + Gk. *podion*, base < *pous*, foot.] *Bot.* A primary axis that develops from a series of short lateral branches and has a zigzag or irregular form. **—sym·po'di·al** *adj.* **—sym·po'di·al·ly** *adv.*

**sym·po·si·a** (sĭm-pō'zē-ə) *n. var. pl. of* SYMPOSIUM.

**sym·po·si·ac** (sĭm-pō'zē-ăk') *adj.* Of, of the nature of, appropriate to, or occurring at a symposium. *—n. Archaic.* A meeting or conference : SYMPOSIUM.

**sym·po·si·arch** (sĭm-pō'zē-ärk') *n.* [Gk. *sumposiarkhos* : *sumposion*, symposium + *arkhein*, to rule.] **1.** The master or director of an ancient Greek symposium. **2.** A toastmaster.

**sym·po·si·um** (sĭm-pō'zē-əm) *n., pl.* **-si·ums** or **-si·a** (-zē-ə) [Lat. < Gk. *sumposion*, drinking party : *sun-*, together + *posis*, drink.] **1.** A meeting or conference for discussion of some topic. **2.** A collection of writings on a topic, as in a magazine. **3.** A convivial meeting for drinking, music, and intellectual discussion in ancient Greece.

**symp·tom** (sĭm'təm, sĭmp'-) *n.* [LLat. *symptoma* < Gk. *sumptōma* < *sumpiptein*, to coincide : *sun-*, together + *piptein*, to fall.] **1.** A phenomenon or circumstance considered as an indication or characteristic of a condition or event. **2.** *Med.* A phenomenon experienced by an individual as a departure from normal function, sensation, or appearance, gen. indicating disease or disorder. **—symp'to·mat'ic** (-tə-măt'ĭk) *adj.* **—symp'to·mat'i·cal·ly** *adv.*

**symp·to·ma·tol·o·gy** (sĭm'tə-mə-tŏl'ə-jē, sĭmp'-) *n.* [NLat. *symptomatologia* : Gk. *sumptōma*, symptom + *-logia*, -logy.] **1.** The medical science of disease symptoms. **2.** The complex of symptoms of a disease.

**syn-** or **sym-** *pref.* [Gk. *sun-* < *sun*, together.] **1. a.** Together : with <*synecology*> **b.** United <*syncarp*> **2. a.** Same : similar <*sympatric*> **b.** At the same time <*synesthesia*>

**syn·aer·e·sis** (sĭ-nĕr'ĭ-sĭs) *n. var. of* SYNERESIS.

**syn·aes·the·sia** (sĭn'ĭs-thē'zhə) *n. var. of* SYNESTHESIA.

**syn·a·gogue** *also* **syn·a·gog** (sĭn'ə-gŏg') *n.* [ME *synagoge* < OFr. *sinagoge* < Lat. *synagoga* < Gk. *sunagōgē* < *sunagein*, to bring together : *sun-*, together + *agein*, to lead.] **1.** A building in which Jewish worship and religious instruction take place. **2.** A congregation of Jews for worship or religious study. **3.** The Jewish religion as organized or typified in local congregations. **—syn'a·gog'al** (-gŏg'əl), **syn'a·gog'i·cal** (-gŏj'ĭ-kəl) *adj.*

**syn·a·le·pha** *also* **syn·a·loe·pha** (sĭn'ə-lē'fə) *n.* [NLat. < Gk. *sunaloiphē* < *sunaleiphein*, to unite two syllables : *sun-*, together + *aleiphein*, to smear.] The blending into one syllable of two successive vowels of adjacent syllables; e.g., *th' elite* for *the elite*.

**syn·apse** (sĭn'ăps', sĭ-năps') *n.* [Gk. *sunapsis*, point of contact < *sunaptein*, to join together : *sun-*, together + *haptein*, to fasten.] The point at which a nerve impulse is transmitted from an axon of one neuron to the dendrite of another. *—vi.* **-apsed, -aps·ing, -aps·es.** To form a synapse.

**syn·ap·sis** (sĭ-năp'sĭs) *n., pl.* **-ses** (-sēz') [NLat. < *sunapsis*, point of contact. —see SYNAPSE.] **1.** *Biol.* Fusion of similar paternal and maternal chromosome pairs during meiosis. **2.** Synapse. **—syn·ap'tic** *adj.* **—syn·ap'ti·cal·ly** *adv.*

**syn·ap·ti·ne·mal complex** *also* **syn·ap·to·ne·mal complex** (sĭ-năp'tə-nĕm'əl) *n.* [SYNAPTI(C) + Gk. *nēma*, thread + -AL.] A ribbonlike structure consisting of three protein components that extends across the region of synapsing chromosomes in the first stage of meiosis.

**syn·ar·thro·di·a** (sĭn'är-thrō'dē-ə) *n., pl.* **-di·ae** (-dē-ē') [SYN- + Gk. *arthrōdia*, a kind of articulation < *arthron*, joint.] Synarthrosis. **—syn'ar·thro'di·al** *adj.* **—syn'ar·thro'di·al·ly** *adv.*

**syn·ar·thro·sis** (sĭn'är-thrō'sĭs) *n., pl.* **-ses** (-sēz) [Gk. *sunarthrōsis* : *sun-*, together + *arthrōsis*, articulation < *arthron*, a joint.] *Anat.* A form of bone articulation in which the bones are rigidly joined without an intervening cavity.

**sync** or **synch** (sĭngk) *Informal.* *—n.* **1.** Synchronization. **2.** Synchronism. *—vi. & vt.* **synced, sync·ing, syncs** or **synched, synch·ing, synchs.** To synchronize.

**syn·carp** (sĭn'kärp') *n.* A fleshy fruit composed of the fruits of several flowers or several carpels of a single flower.

**syn·car·pous** (sĭn-kär'pəs) *adj.* Having or composed of united carpels. **—syn'car'py** (sĭn'kär'pē) *n.*

**synch** (sĭngk) *n. & v. var. of* SYNC.

**syn·chon·dro·sis** (sĭng'kŏn-drō'sĭs, sĭn'-) *n.* [NLat. < Gk. *sunkhondrōsis*, cartilaginous joint : *sun-*, together + *khondros*, cartilage + *-ōsis*, -osis.] A synarthrosis in which the surfaces of the bones are connected by cartilage.

**syn·chro** (sĭng'krō, sĭn'-) *n., pl.* **-chros.** [Short for SYNCHRONOUS.] A selsyn.

**synchro-** *pref.* [< SYNCHRONIZED.] Synchronized : synchronous <*synchrotron*>

**syn·chro·cy·clo·tron** (sĭng'krō-sī'klə-trŏn', sĭn'-) *n.* A proton and positive ion accelerator, the chief components and configuration of which are similar to those of a cyclotron and in which the phase of the accelerating potential is synchronized with the frequency of the accelerated particles by frequency modulation to compensate for relativistic increases in particle mass at high speeds.

**syn·chro·flash** (sĭng'krō-flăsh', sĭn'-) *n.* A device on a camera that synchronizes the peak of a flash created by a flash lamp with the opening of the shutter. **—syn'chro·flash'** *adj.*

**syn·chro·mesh** (sĭng'krō-mĕsh', sĭn'-) *n.* **1.** An automotive gearshift system in which the gears are synchronized at the same speeds before engaging to effect a smooth change. **2.** A gear in such a system. **—syn'chro·mesh'** *adj.*

**syn·chro·ne·i·ty** (sĭng'krə-nē'ə-tē, sĭn'-, -nā'-) *n.* The state of being synchronous.

**syn·chron·ic** (sĭn-krŏn'ĭk, sĭng-) *adj.* **1.** Synchronous. **2. a.** Descriptive. **b.** Studying the events of a particular time or era without consideration of historical data. **—syn·chron'i·cal·ly** *adv.*

**synchronic linguistics** *n.* Descriptive linguistics.

**syn·chro·nism** (sĭng'krə-nĭz'əm, sĭn'-) *n.* **1.** Synchroneity. **2.** A chronological listing of historical personages or events so as to indicate parallel existence or occurrence. **3.** Representation in the same artwork of two or more events that occurred at different times. **—syn'chro·nis'tic, syn'chro·nis'ti·cal** *adj.* **—syn'chro·nis'ti·cal·ly** *adv.*

**syn·chro·nize** (sĭng'krə-nīz', sĭn'-) *v.* **-nized, -niz·ing, -niz·es.** [< SYNCHRONOUS.] *—vi.* **1.** To take place at the same time. **2.** To operate in unison. *—vt.* **1.** To cause to operate with exact coincidence in time or rate <Let's *synchronize* our ... hes.> **2.** To arrange (historical events) so as to indicate parallel existence or occurrence. **3.** To cause (sound effects or dialogue) to coincide with an action. **—syn'chro·ni·za'tion** *n.*

**syn·chro·niz·er** (sĭng'krə-nī'zər, sĭn'-) *n.* **1.** One that synchronizes. **2.** *Computer Sci.* A storage device that compensates for a difference in the rate at which information is processed between two or more devices.

**syn·chro·nous** (sĭng'krə-nəs, sĭn'-) *adj.* [LLat. *synchronos* < Gk. *sunkhronos* : *sun-*, same + *khronos*, time.] **1.** Happening at the same time. **2.** Moving or operating at the same rate. **3. a.** Having identical periods. **b.** Having identical period and phase. **—syn'chro·nous·ly** *adv.* **—syn'chro·nous·ness** *n.*

**synchronous motor** *n.* A motor having a speed directly proportional to the frequency of the electric current that operates it.

**synchronous orbit** *n.* An orbit having a period the same as the period of axial rotation of the earth and so oriented that any body in it maintains a position over one point on the earth's surface.

**syn·chro·ny** (sĭng′krə-nē, sĭn′-) *n., pl.* **-nies.** [< SYNCHRONOUS.] A synchronous occurrence, arrangement, or movement.

**syn·chro·tron** (sĭng′krə-trŏn′, sĭn′-) *n.* An accelerator in which charged particles are accelerated around a fixed circular path by a radio-frequency potential and held to the path by a time-varying magnetic field.

**synchrotron radiation** *n.* Electromagnetic radiation emitted by high-energy particles when accelerated to relativistic speeds by a magnetic field.

**syn·cli·nal** (sĭn-klī′nəl) *adj.* [SYN- + Gk. *klinein,* to lean.] **1.** Sloping downward from opposite directions to meet in a common point. **2.** *Geol.* Relating to, formed by, or forming a syncline. —*n.* A syncline.

**syn·cline** (sĭn′klīn′) *n.* [Back-formation < SYNCLINAL.] A low troughlike area in bedrock, in which rocks incline together from opposite sides.

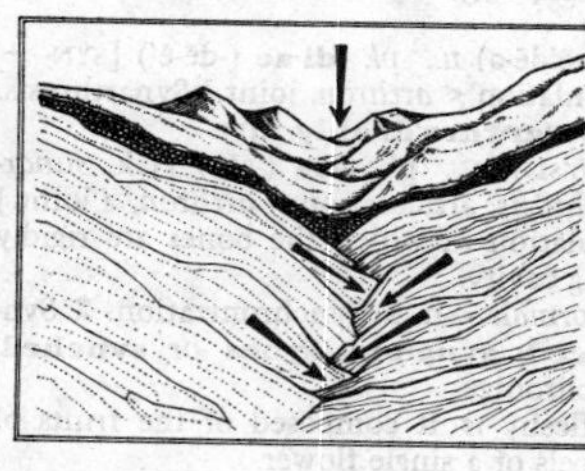

**syncline**

**syn·com** (sĭn′kŏm′) *n.* [SYN(CHRONOUS) + COM(MUNICATION).] A communications satellite moving in a synchronous orbit.

**syn·co·pate** (sĭng′kə-pāt′, sĭn′-) *vt.* **-pat·ed, -pat·ing, -pates.** [Med. Lat. *syncopare, syncopat-* < LLat. *syncope,* syncope.] **1. a.** To shorten (a word) by syncope. **b.** To drop (a letter or sound) from the spelling or pronunciation of a word. **2.** To modify (musical rhythm) by syncopation. **—syn′co·pa′tor** *n.*

**syn·co·pa·tion** (sĭng′kə-pā′shən, sĭn′-) *n.* **1.** The act of syncopating or state of being syncopated. **2.** Something syncopated. **3.** *Mus.* A shift of accent in a passage or composition that occurs when a normally weak beat is stressed. **4.** SYNCOPE 1.

**syn·co·pe** (sĭng′kə-pē, sĭn′-) *n.* [LLat. < Gk. *sunkopē* < *sunkoptein,* to cut short : *sun-,* together + *koptein,* to strike.] **1.** The shortening of a word by dropping a sound, letter, or syllable from the middle of the word; e.g., *bos'n* for *boatswain.* **2.** *Pathol.* A brief loss of consciousness caused by transient anemia : SWOON. **—syn′co·pal** *adj.*

**syn·cre·tism** (sĭng′krĭ-tĭz′əm, sĭn′-) *n.* [Gk. *sunkrētismos,* union < *sunkrētizein,* to unite (in the manner of the Cretan cities) : *sun-,* together + *Krēs,* Cretan.] **1.** The attempt or tendency to combine or reconcile differing philosophical or religious beliefs. **2.** Fusion of two or more orig. different inflectional forms into a single form. **—syn′cre·tist** *n.* **—syn′cre·tis′tic** *adj.*

**syn·cre·tize** (sĭng′krĭ-tīz′, sĭn′-) *v.* **-tized, -tiz·ing, -tiz·es.** [Gk. *sunkrētizein.* —see SYNCRETISM.] —*vt.* To reconcile or attempt to reconcile (e.g., differing beliefs). —*vi.* To combine differing beliefs.

**syn·cy·ti·um** (sĭn-sĭsh′ē-əm) *n., pl.* **-cy·ti·a** (-sĭsh′ē-ə) [NLat. : SYN- + Gk. *kutos,* hollow vessels.] A protoplasmic mass with many nuclei but no clear cell boundaries. **—syn·cy′ti·al** *adj.*

**syn·dac·tyl** or **syn·dac·tyle** (sĭn-dăk′təl) *also* **syn·dac·ty·lous** (-tə-ləs) *adj.* [Fr. *syndactyle* : Gk. *sun-,* together + Gk. *daktulos,* finger.] *Biol.* Having two or more wholly or partially fused digits. —*n.* A syndactyl animal. **—syn·dac′tyl·ism** *n.*

**syn·des·mo·sis** (sĭn′dĕz-mō′sĭs, -dĕs-) *n.* [NLat. < Gk. *sundesmos,* ligament < *sundein,* to bind together. —see SYNDETIC.] Articulation of bones by ligaments. **—syn′des·mot′ic** (-mŏt′ĭk) *adj.*

**syn·det·ic** (sĭn-dĕt′ĭk) *adj.* [Gk. *sundetikos* < *sundetos,* bound together < *sundein,* to bind together : *sun-,* together + *dein,* to bind.] **1.** Serving to connect, as a conjunction : CONJUNCTIVE. **2.** Linked by a conjunction. **—syn·det′i·cal·ly** *adv.*

**syn·dic** (sĭn′dĭk) *n.* [Fr. < LLat. *syndicus* < Gk. *sundikos,* public advocate : *sun-,* with + *dikē,* court case.] **1.** One appointed to represent an organization, as a corporation or university, in business transactions : BUSINESS AGENT. **2.** A government official, as a civil magistrate. **—syn′di·cal** *adj.*

**syn·di·cal·ism** (sĭn′dĭ-kə-lĭz′əm) *n.* [Fr. *syndicalisme* < *chambre syndicale,* trade union.] A radical political movement that advocates bringing industry and government under the control of labor unions by use of direct action, as general strikes and sabotage. **—syn′di·cal·ist** *n.* **—syn′di·cal·is′tic** *adj.*

**syn·di·cate** (sĭn′dĭ-kĭt) *n.* [Fr. *syndicat* < *syndic,* syndic.] **1. a.** An association of people authorized to transact business requiring a heavy outlay of capital. **b.** An association of people formed to carry out an enterprise, esp. a group of organized crime figures working to control vice, drug traffic, and gambling. **2.** An agency that sells articles for publication in a number of newspapers or periodicals simultaneously. **3.** The office, position, or jurisdiction of a syndic or body of syndics. —*v.* (sĭn′dĭ-kāt′) **-cat·ed, -cat·ing, -cates.** —*vt.* **1.** To organize into a syndicate. **2.** To sell (e.g., an article) through a syndicate for publication. —*vi.* To organize a syndicate.

**syn·drome** (sĭn′drōm′) *n.* [Gk. *sundromē,* concurrence of symptoms : *sun-,* together + *dramein,* to run.] **1.** A group of signs and symptoms that collectively indicate or characterize a disease, psychological disorder, or other abnormal condition. **2. a.** A complex of symptoms indicating the existence of an undesirable condition or quality. **b.** A distinctive or characteristic behavior pattern. **—syn·drom′ic** (-drō′mĭk, -drŏm′ĭk) *adj.*

**syn·ec·do·che** (sĭ-nĕk′də-kē) *n.* [Lat. < Gk. *sunekdokhē* < *sunekdekhesthai,* to take with : *sun-,* with + *ekdekhesthai,* to understand (*ek-,* out of + *dekhesthai,* to take).] A figure of speech by which a more inclusive term is used for a less inclusive term or vice versa; e.g., *head* for *cattle* or *the law* for *a police officer.* **—syn′ec·doch′ic** (sĭn′ĕk-dŏk′ĭk), **syn′ec·doch′i·cal** *adj.*

**syn·e·cious** (sĭ-nē′shəs) *adj. var. of* SYNOECIOUS.

**syn·e·col·o·gy** (sĭn′ĭ-kŏl′ə-jē) *n.* The study of the environmental interrelationships among communities of organisms. **—syn′e·co·log′ic** (-kə-lŏj′ĭk), **syn′e·co·log′i·cal** *adj.*

**syn·er·e·sis** *also* **syn·aer·e·sis** (sĭ-nĕr′ĭ-sĭs) *n., pl.* **-ses** (-sēz′) [LLat. *synaeresis* < Gk. *sunairesis* < *sunairein,* to draw together : *sun-,* together + *hairein,* to take.] **1.** The drawing together into one syllable of two consecutive vowels ordinarily pronounced separately. **2.** *Chem.* Exudation of the liquid component of a gel.

**syn·er·gid** (sĭ-nûr′jĭd, sĭn′ər-) *n.* [NLat. *synergida* < Gk. *sunergos,* working together. —see SYNERGISM.] One of two small cells lying near the egg in the mature embryo of a seed plant.

**syn·er·gism** (sĭn′ər-jĭz′əm) *n.* [NLat. *synergismus* < Gk. *sunergos,* working together : *sun-,* with + *ergon,* work.] **1.** The action of two or more substances, organs, or organisms to achieve an effect of which each is individually incapable. **2.** The theological doctrine that regeneration is effected by a combination of human will and divine grace. **—syn′er·get′ic** (-jĕt′ĭk), **syn·er′gic** (sĭ-nûr′jĭk) *adj.*

**syn·er·gist** (sĭn′ər-jĭst) *n.* **1.** An adherent of theological synergism. **2.** A synergetic organ or substance. **—syn′er·gis′tic, syn′er·gis′ti·cal** *adj.* **—syn′er·gis′ti·cal·ly** *adv.*

**syn·er·gy** (sĭn′ər-jē) *n., pl.* **-gies.** SYNERGISM 1.

**syn·e·sis** (sĭn′ə-sĭs) *n.* [Gk. *sunesis,* understanding < *sunienai,* to understand : *sun-,* together + *hienai,* to send.] A construction in which a form differs in number but agrees in meaning with the word governing it; e.g., *If anyone arrives, tell them to wait.*

**syn·es·the·sia** *also* **syn·aes·the·sia** (sĭn′ĭs-thē′zhə) *n.* [SYN- + (AN)ESTHESIA.] A phenomenon in which one type of stimulation evokes the sensation of another, as the hearing of a sound resulting in the sensation of the visualization of a color. **—syn′es·thet′ic** (-thĕt′ĭk) *adj.*

**syn·e·ze·sis** (sĭn′ĭ-zē′sĭs) *n. var. of* SYNIZESIS.

**syn·fu·el** (sĭn′fyo͞o′əl) *n.* [SYN(THETIC) + FUEL.] A liquid, gaseous, or solid hydrocarbon fuel derived from naturally occurring fossil fuels as coal, shale, or tar sand.

**syn·ga·my** (sĭng′gə-mē) *n.* Fusion of two gametes. **—syn·gam′ic** (sĭn-găm′ĭk), **syn′ga·mous** (sĭng′gə-məs) *adj.*

**syn·gen·e·sis** (sĭn-jĕn′ĭ-sĭs) *n.* Sexual reproduction. **—syn′ge·net′ic** (-jə-nĕt′ĭk) *adj.*

**syn·i·ze·sis** *also* **syn·e·ze·sis** (sĭn′ĭ-zē′sĭs) *n., pl.* **-ses** (-sēz′) [LLat. *synizesis* < Gk. *sunizēsis* < *sunizein,* to collapse : *sun-,* together + *hizein,* to settle down.] **1.** Contraction of two syllables into one by joining in pronunciation two adjacent vowels. **2.** *Biol.* The phase of meiosis in which the chromatin contracts into a mass at one side of the nucleus.

**syn·kar·y·on** (sĭn-kăr′ē-ŏn′, -ē-ən) *n.* [SYN- + Gk. *karuon,* nut.] The nucleus of a fertilized egg immediately after fusion of the male and female nuclei. **—syn·kar′y·on′ic** (-ŏn′ĭk) *adj.*

**syn·od** (sĭn′əd) *n.* [ME < LLat. *synodus* < Gk. *sunodos,* meeting : *sun-,* together + *hodos,* road.] **1.** A council of churches or church officials. **2.** A lay council or assembly. **—syn′od·al** (sĭn′ə-dəl) *adj.*

**syn·od·i·cal** (sĭ-nŏd′ĭ-kəl) *also* **syn·od·ic** (-nŏd′ĭk) *adj.* **1.** Relating to or having the nature of a synod. **2.** Relating to the conjunction of celestial bodies, esp. the interval between two successive conjunctions of a planet or the moon with the sun. **—syn·od′i·cal·ly** *adv.*

**synodic month** *n.* MONTH 5.

**syn·oe·cious** *also* **syn·e·cious** (sĭ-nē′shəs) *adj.* [SYN- + (MON)OECIOUS.] *Bot.* Having male and female organs in the same structure.

**syn·o·nym** (sĭn′ə-nĭm′) *n.* [ME *sinonyme* < Lat. *synonymum* < Gk. *sunōnumon* < *sunōnumos,* synonymous.] **1.** A word having a meaning that is the same or nearly the same as that of another word in the same language. **2.** A word or expression accepted as a figurative or symbolic substitute for another word or expression. **3.** *Biol.* A taxonomic name of an organism that is equivalent to or has been superseded by another designation. **—syn'o·nym'ic** (-nĭm'ĭk), **syn'o·nym'i·cal** *adj.* **—syn'o·nym'i·ty** (-nĭm'ĭ-tē) *n.*

**syn·on·y·mist** (sĭ-nŏn'ə-mĭst) *n.* One who studies or discriminates synonyms.

**syn·on·y·mize** (sĭ-nŏn'ə-mīz') *vt.* **-mized, -miz·ing, -miz·es.** To provide or analyze the synonyms of (a word).

**syn·on·y·mous** (sĭ-nŏn'ə-məs) *adj.* [Med. Lat. *synonymus* < Gk. *sunōnumos* : *sun-*, same + *onoma* name.] Expressing the same or nearly the same meaning as another word. **—syn·on'y·mous·ly** *adv.*

**syn·on·y·my** (sĭ-nŏn'ə-mē) *n., pl.* **-mies. 1.** The quality or state of being synonymous. **2.** Study and classification of synonyms. **3.** A list, book, or system of synonyms. **4.** A chronological list or record of the scientific names applied to a species and its subdivisions.

**syn·op·sis** (sĭ-nŏp'sĭs) *n., pl.* **-ses** (-sēz') [LLat. < Gk. *sunopsis*, general view : *sun-*, together + *opsis*, view.] A brief outline or statement of a topic : ABSTRACT.

**syn·op·size** (sĭ-nŏp'sīz') *vt.* **-sized, -siz·ing, -siz·es.** [LGk. *sunopsizein* < Gk. *sunopsis*, general view. —see SYNOPSIS.] To give or write a synopsis of (a topic).

**syn·op·tic** (sĭ-nŏp'tĭk) *also* **syn·op·ti·cal** (-tĭ-kəl) *adj.* **1.** Of or being a synopsis. **2.** Presenting an account from the same viewpoint. —Used esp. of the first three Gospels of the New Testament, which correspond closely. **3.** Involving or presenting data on atmospheric and weather conditions over a broad area at a single given time. **—syn·op'ti·cal·ly** *adv.*

**syn·os·to·sis** (sĭn'ŏs-tō'sĭs) *n., pl.* **-ses** (-sēz') [SYN- + Gk. *osteon*, bone + -OSIS.] Fusion of two bones. **—syn'os·tot'ic** (-tŏt'ĭk) *adj.*

**syn·o·vi·a** (sĭ-nō'vē-ə) *n.* [NLat.] A clear, viscid lubricating fluid secreted by membranes in joint cavities, sheaths of tendons, and bursae. **—syn·o'vi·al** *adj.*

**sy·no·vi·tis** (sĭ'nə-vī'tĭs) *n.* [SYNOV(IAL MEMBRANE) + -ITIS.] Inflammation of a synovial membrane.

**syn·sep·al·ous** (sĭn-sĕp'ə-ləs) *adj.* Gamosepalous.

**syn·tac·tics** (sĭn-tăk'tĭks) *n.* [< SYNTACTIC.] *(sing. or pl. in number).* The branch of semiotics that deals with the formal properties of signs and symbols.

**syn·tax** (sĭn'tăks') *n.* [Fr. *syntaxe* < LLat. *syntaxis* < Gk. *suntaxis* < *suntassein*, to combine : *sun-*, together + *tassein*, to arrange.] **1. a.** The way in which terms are combined to form phrases and sentences. **b.** The branch of grammar dealing with the formation of phrases and sentences. **2.** *Computer Sci.* The rules governing the construction of a machine language. **—syn·tac'tic** (-tăk'tĭk), **syn·tac'ti·cal** *adj.* **—syn·tac'ti·cal·ly** *adv.*

**syn·the·sis** (sĭn'thĭ-sĭs) *n., pl.* **-ses** (-sēz') [Lat. < Gk. *sunthesis* < *suntithenai*, to put together : *sun-*, together + *tithenai*, to put.] **1. a.** Fusion of separate elements or substances to form a coherent whole. **b.** The whole so formed. **2.** *Chem.* Formation of a compound from its constituents. **3.** *Philos.* **a.** Reasoning from the general to the particular : logical deduction. **b.** The combination of thesis and antithesis in the dialectical process, producing a new and higher form of being. **—syn'the·sist** *n.*

**synthesis gas** *n.* A synthetic fuel produced by controlled combustion of coal in the presence of water vapor.

**syn·the·size** (sĭn'thĭ-sīz') *also* **syn·the·tize** (-tīz') *v.* **-sized, -siz·ing, -siz·es** *also* **-tized, -tiz·ing, -tiz·es.** —*vt.* **1.** To combine so as to form a new, complex product. **2.** To make by combining separate elements. —*vi.* To form a synthesis.

**syn·the·siz·er** (sĭn'thĭ-sī'zər) *n.* **1.** One that synthesizes. **2.** A machine having a simple keyboard and using solid-state circuitry to duplicate the sounds of musical instruments, often up to 12 instruments simultaneously.

**syn·thet·ic** (sĭn-thĕt'ĭk) *also* **syn·thet·i·cal** (-ĭ-kəl) *adj.* [Gk. *sunthetikos*, component < *suntithenai*, to put together. —see SYNTHESIS.] **1.** Relating to, involving, or having the nature of a synthesis. **2.** *Chem.* Produced by synthesis, esp. not of natural origin : MANMADE. **3.** Not genuine : ARTIFICIAL <*synthetic* charm> **4.** Denoting a language, as Latin or Russian, that uses inflectional affixes to express syntactic relationships. —*n.* **synthetic.** A synthetic chemical compound or material. **—syn·thet'i·cal·ly** *adv.*

**synthetic division** *n.* A method of dividing a polynomial by another, when the second is of first order, by writing only the coefficients of the terms and changing the sign of the constant term in the divisor.

**syn·tro·phism** (sĭn-trō'fĭz'əm) *n.* An ecological relationship in which microorganisms are mutually dependent upon one another for nutritional requirements.

**sy·pher** (sī'fər) *vt.* **-phered, -pher·ing, -phers.** [Alteration of CIPHER.] To overlap and even (chamfered or beveled plank edges) so as to form a flush surface.

**syphil-** *pref. var. of* SYPHILO-.

**syph·i·lis** (sĭf'ə-lĭs) *n.* [NLat., alteration of *Syphilus*, protagonist of a poem by Girolamo Francastoro (1483–1553) in which he is represented as the first victim of the disease.] A chronic infectious venereal disease caused by a spirochete, *Treponema pallidum*, transmitted by direct, usu. sexual contact and progressing through three stages respectively characterized by local formation of chancres, ulcerous skin eruptions, and systemic infection leading to general paresis. **—syph'i·lit'ic** (-lĭt'ĭk) *adj.* & *n.*

**syphilo-** *or* **syphil-** *pref.* [< SYPHILIS.] Syphilis <*syphiloma*>

**syph·i·loid** (sĭf'ə-loid') *adj.* Characteristic of syphilis.

**syph·i·lol·o·gy** (sĭf'ə-lŏl'ə-jē) *n.* The sum of knowledge concerning the origin, nature, course, complications, and treatment of syphilis. **—syph'i·lol'o·gist** *n.*

**syph·i·lo·ma** (sĭf'ə-lō'mə) *n., pl.* **-mas** *or* **-ma·ta** (-mə-tə). A lesion formed in an advanced stage of syphilis : GUMMA. **—syph'i·lom'a·tous** (-lŏm'ə-təs) *adj.*

**sy·phon** (sī'fən) *n.* & *v. var. of* SIPHON.

**Syr·ette** (sĭ-rĕt') A trademark for a collapsible tube having an attached hypodermic needle containing a single dose of medicine.

**Syr·i·ac** (sîr'ē-ăk') *n.* An ancient Aramaic language spoken in Syria from the 3rd to the 13th cent. A.D. that survives as the liturgical language of several eastern Christian churches.

**Syr·i·an** (sîr'ē-ən) *adj.* Of or relating to Syria, its people, or its culture. —*n.* **1.** A native or inhabitant of Syria. **2.** A member of a Christian church using the Syriac language.

**sy·rin·ga** (sə-rĭng'gə) *n.* [NLat. < Gk. *surinx*, shepherd's pipe (from the use of its hollow stems to make pipes).] MOCK ORANGE 1.

**sy·ringe** (sə-rĭnj', sîr'ĭnj) *n.* [ME *syryng* < Med. Lat. *syringa* < Gk. *surinx*, shepherd's pipe.] **1.** A medical instrument for injecting fluids into the body or drawing them out of it. **2.** A hypodermic syringe.

**sy·rin·go·my·e·li·a** (sə-rĭng'gō-mī-ē'lē-ə) *n.* [NLat. : Gk. *surinx*, spinal cavity + Gk. *muelos*, marrow < *mus*, muscle, mouse.] A chronic disease of the spinal cord marked by the presence of liquid-filled cavities and leading to spasticity and sensory disturbances.

**syr·inx** (sîr'ĭngks) *n., pl.* **sy·rin·ges** (sə-rĭn'jēz', -rĭng'gēz') *or* **syr·inx·es.** [Lat. < Gk. *surinx.*] **1.** A panpipe. **2.** *Zool.* The vocal organ of a bird, made up of thin vibrating muscles at or near the division of the trachea. **—sy·rin'ge·al** (sə-rĭn'jē-əl) *adj.*

**syr·phid** (sûr'fĭd) *n.* [NLat. *Syrphidae*, family name < Gk. *surphos*, gnat.] Any of numerous flies of the family Syrphidae, many of which have a form or coloration mimicking that of bees or wasps. —*adj.* Of or belonging to the Syrphidae.

**syr·phus fly** (sûr'fəs) *n.* [NLat. *Syrphus*, fly genus < Gk. *surphos*, gnat.] Syrphid.

**syr·up** *also* **sir·up** (sĭr'əp, sûr'-) *n.* [ME *sirop* < OFr. < Med. Lat. *siropus* < Ar. *sharāb* < *shariba*, he drank.] **1.** A thick, sweet, sticky liquid, composed of a sugar base, natural or artificial flavorings, and water. **2.** The juice of a fruit or plant boiled with sugar until thick and sticky. **—syr'up·y** *adj.*

**sys·sar·co·sis** (sĭs'är-kō'sĭs) *n.* [Gk. *sussarkōsis*, a being overgrown with flesh < *sussarkousthai*, to be overgrown with flesh : *sun-*, with + *sarkousthai*, passive of *sarkoun*, to cover with flesh < *sarx*, flesh.] Union of bones, as the hyoid bone and lower jaw, by muscle.

**sys·tal·tic** (sĭ-stôl'tĭk, -stăl'-) *adj.* [LLat. *systalticus* < Gk. *sustaltikos* < *sustellein*, to contract : *sun-*, together + *stellein*, to make compact.] Alternately contracting and expanding, as the heart : PULSATING.

**sys·tem** (sĭs'təm) *n.* [LLat. *systema*, *systemat-* < Gk. *sustēma* < *sunistanai*, to combine : *sun-*, together + *histanai*, to make stand.] **1.** A group of interrelated, interacting, or interdependent constituents forming a complex whole. **2.** A functionally related group of elements, esp.: **a.** The human body regarded as a functional physiological unit. **b.** A group of physiologically complementary organs or parts <the nervous *system*> **c.** A group of interacting mechanical or electrical components. **d.** A network of structures and channels, as for communications, travel, or distribution <a broadcasting *system*><a rail *system*> **3.** A structurally or anatomically related group of parts or elements. **4.** A set of interrelated ideas or principles. **5.** A social, economic, or political organizational form <the capitalist *system*> **6.** A naturally occurring group of objects or phenomena <the solar *system*> **7.** A set of objects or phenomena grouped together for classification or analysis. **8.** Harmonious, orderly interaction. **9.** A method : procedure. **10.** Organized society : ESTABLISHMENT <You can't beat the *system.*>

**sys·tem·at·ic** (sĭs'tə-măt'ĭk) *also* **sys·tem·at·i·cal** (-ĭ-kəl) *adj.* **1.** Of, marked by, based on, or making up a system. **2.** Carried on in a step-by-step procedure. **3.** Purposefully regular : METHODICAL. **4.** Of or relating to taxonomic classification. **—sys'tem·at'i·cal·ly** *adv.*

**sys·tem·at·ics** (sĭs'tə-măt'ĭks) *n.* *(sing. in number).* Classification of organisms into an orderly system indicating natural relationships.

**sys·tem·a·tism** (sĭs'tə-mə-tĭz'əm, sĭ-stĕm'ə-) *n.* **1.** The practice of classifying or systematizing. **2.** Adherence to a system.

**sys·tem·a·tist** (sĭs'tə-mə-tĭst, sĭ-stĕm'ə-) *n.* **1.** One who formulates or adheres to a system. **2.** A taxonomist.

**sys·tem·a·tize** (sĭs'tə-mə-tīz') *vt.* **-tized, -tiz·ing, -tiz·es.** To formulate into or reduce to a system <amass and *systematize* knowledge> **—sys'tem·a·ti·za'tion** *n.* **—sys'tem·a·tiz'er** *n.*

**sys·tem·ic** (sĭ-stĕm'ĭk) *adj.* **1.** Of or relating to a system. **2.** Of, relating to, or affecting the entire body. **—sys·tem'i·cal·ly** *adv.*

**sys·tem·ize** (sĭs'tə-mīz') *vt.* **-ized, -iz·ing, -iz·es.** To systematize. **—sys'tem·i·za'tion** *n.* **—sys'tem·iz'er** *n.*

**systems analysis** *n.* **1.** The study of an activity by mathematical means to determine its desired end and the most efficient method of obtaining it. **2.** The act, process, or profession of systems analysis. **—systems analyst** *n.*

**sys·to·le** (sĭs′tə-lē) *n.* [Gk. *sustolē,* contraction < *sustellein,* to contract. —see SYSTALTIC.] Rhythmic contraction of the heart by which blood is driven through the aorta and pulmonary artery after each dilation or diastole. **—sys·tol′ic** (-tŏl′ĭk) *adj.*

**syz·y·gy** (sĭz′ə-jē) *n., pl.* **-gies.** [LLat. *syzygia* < Gk. *suzugia,* conjunction < *suzugos,* paired : *sun-,* together + *zugon,* yoke.] **1.** *Astron.* **a.** Either of two points in the orbit of a celestial body where the body is in opposition to or in conjunction with the sun. **b.** Either of two points in the orbit of the moon when the moon lies in a straight line with the sun and the earth. **c.** The configuration of the sun, the moon, and the earth lying in a straight line. **2.** Combination of two feet into a single metrical unit in classical prosody.

# T t

**t** *or* **T** (tē) *n., pl.* **t's** *or* **T's. 1.** The 20th letter of the English alphabet. **2.** A speech sound represented by the letter *t.* **3.** The 20th in a series. **4.** Something shaped like the letter T. **—to a T.** Perfectly : precisely <fits the role *to a T*>

**Ta** *symbol for* TANTALUM.

**Taal** (täl) *n.* [Du. *taal,* speech < MDu. *tāle.*] Afrikaans.

**tab** (tăb) *n.* [Orig. unknown.] **1.** A projection, flap, or short strip attached to an object to facilitate opening, handling, or identification. **2.** A small, usu. decorative tongue or flap on a garment. **3.** A small auxiliary control surface attached to a larger one to help stabilize an airplane. **4.** *Informal.* A bill, as for a restaurant meal. **5.** A tabulator, as on a typewriter. *—vt.* **tabbed, tab·bing, tabs.** To supply with a tab or tabs. **—keep tabs on.** To account for : WATCH.

**ta·ba·nid** (tə-bā′nĭd, -băn′ĭd) *n.* [NLat. *Tabanidae,* family name < Lat. *tabanus,* horsefly.] Any of various blood-sucking flies of the family Tabanidae, including the horseflies. **—ta·ba′nid** *adj.*

**tab·ard** (tăb′ərd) *n.* [ME < OFr. *tabart.*] **1.** A short heavy cape of coarse cloth once worn outdoors. **2. a.** A tunic or capelike garment worn by a knight over his armor and emblazoned with his coat of arms. **b.** A similar garment worn by a herald and bearing his lord's coat of arms. **3.** An embroidered pennant attached to a trumpet.

tabard

**tab·a·ret** (tăb′ə-rĕt′) *n.* [Prob. < TABBY.] A strong upholstery fabric having alternating stripes of satin and moiré.

**Ta·bas·co** (tə-băs′kō). A trademark for a spicy-hot sauce made from a strong-flavored red pepper.

**tab·bou·leh** (tə-bo͞o′lə) *also* **ta·boo·ley** (-lē) *n.* [Ar. *tabbūla.*] A Lebanese salad made with bulgur wheat, oil, scallions, tomatoes, and parsley.

**tab·by** (tăb′ē) *n., pl.* **-bies.** [Fr. *tabis* < Med. Lat. *attabi* < Ar. *'attābī,* after *Al-'attābīya,* a suburb of Baghdad, Iraq.] **1.** A rich watered silk. **2.** A plain weave fabric. **3. a.** A gray or tawny striped domestic cat. **b.** A domestic cat, esp. a female. **4.** An old maid. **5.** A prying woman : GOSSIP. *—adj.* **1.** Having light and dark striped markings <a *tabby* cat> **2.** Made of or resembling watered silk.

**tab·er·na·cle** (tăb′ər-năk′əl) *n.* [ME < OFr. < LLat. *tabernaculum* < Lat., tent, dim. of *taberna,* hut.] **1.** *often* **Tabernacle. a.** The portable sanctuary in which the Jews carried the Ark of the Covenant through the desert. **b.** The Jewish temple. **2.** *often* **Tabernacle.** A box or case on a church altar containing the consecrated host and wine of the Eucharist. **3. a.** A place of worship distinguished from a church. **b.** TEMPLE[1] 4. **4.** A niche for a relic or statue. **5.** *Naut.* A boxlike support in which the heel of a mast is stepped. *—v.* **-cled, -cling, -cles.** *—vt.* To enshrine. *—vi.* To dwell temporarily. **—tab′er·nac′u·lar** (-năk′yə-lər) *adj.*

**ta·bes** (tā′bēz) *n., pl.* **tabes.** [Lat.] **1.** Progressive bodily wasting or emaciation. **2.** Tabes dorsalis. **—ta·bet′ic** *adj.*

**tabes dor·sa·lis** (dôr-sā′lĭs, -săl′ĭs, -sä′lĭs) *n.* [NLat., dorsal tabes.] A syphilitic disease resulting in a hardening of the dorsal columns of the spinal cord and marked by shooting pains, unsteadiness, and loss of coordination in voluntary movements.

**tab·la** (tŭb′lə) *n.* [Hindi < Ar. *ṭabl.*] A small hand drum of India.

**tab·la·ture** (tăb′lə-cho͝or′) *n.* [OFr. < Med. Lat. *tabulatus,* tablet < Lat. *tabula.*] **1.** *Mus.* An obsolete system of notation using letters and symbols to indicate playing directions rather than tones. **2.** An engraved tablet or surface.

**ta·ble** (tā′bəl) *n.* [ME < OFr. < Lat. *tabula,* board.] **1.** A piece of furniture supported by one or more vertical legs and having a flat horizontal surface. **2.** The objects laid out on a table for a meal. **3.** The food and drink served at meals : FARE. **4.** The group of people assembled around a table, as for a meal. **5.** *often* **tables.** A gaming table, as for faro, roulette, or dice. **6. a.** Either of the leaves of a backgammon board. **b. tables.** *Obs.* The game of backgammon. **7.** A tableland or plateau. **8. a.** A flat facet cut across the top of a precious stone. **b.** A stone cut in this way. **9.** *Mus.* The belly of a stringed instrument. **10. a.** A raised or sunken rectangular panel on a wall. **b.** A raised horizontal surface or continuous band on an exterior wall : STRINGCOURSE. **11.** *Geol.* A horizontal rock stratum. **12.** A part of the palm framed by four lines. **13.** An orderly arrangement of data, esp. one in which the data are arranged in columns and rows in a rectangular form. **14.** An abbreviated list, as of contents : SYNOPSIS. **15.** A slab or tablet, as of stone, bearing an inscription or device. **16. tables.** A system of laws or decrees : CODE <the *tables* of Moses> *—vt.* **-bled, -bling, -bles. 1.** To put on a table. **2.** To postpone consideration of (e.g., a piece of legislation) : SHELVE. **3.** To enter in a list or table : TABULATE. **—on the table.** Postponed to a later date. **—turn the tables.** To reverse a situation and gain the upper hand. **—under the table. 1.** In secret. **2.** Into a completely intoxicated state <could drink you *under the table* anytime>

**tab·leau** (tăb′lō′, tă-blō′) *n., pl.* **tab·leaux** *or* **tab·leaus** (tăb′lōz′, tă-blōz′) [Fr. < OFr. *tablel,* dim. of *table,* surface prepared for painting. —see TABLE.] **1.** A vivid description <The film is a *tableau* of circus life.> **2.** A striking incidental scene, as of a picturesque group of people. **3.** An interlude during a scene when all the performers on stage freeze in position and then resume action. **4.** A tableau vivant.

**tab·leau vi·vant** (tă-blō′ vē-väɴ′) *n., pl.* **tab·leaux vi·vants** (tă-blō′ vēväɴ′) [Fr. : *tableau,* tableau + *vivant,* living.] A scene presented on stage by costumed performers who remain silent and motionless as if in a picture.

**ta·ble·cloth** (tā′bəl-klôth′, -klŏth′) *n.* A cloth to cover a table.

**ta·ble d'hôte** (tä′bəl dōt′) *n., pl.* **ta·bles d'hôte** (tä′bəl dōt′) [Fr. : *table,* table + *de,* of + *hôte,* host.] **1.** A communal table for all guests at a hotel or restaurant. **2.** A full-course meal served at a fixed price in a restaurant or hotel.

**ta·ble-hop** (tā′bəl-hŏp′) *vi.* **-hopped, -hop·ping, -hops.** To go from table to table greeting friends, as in a restaurant. **—ta′ble-hop′per** *n.*

**ta·ble·land** (tā′bəl-lănd′) *n.* A flat elevated region : MESA.

**table linen** *n.* Tablecloths and napkins.

**table salt** *n.* **1.** A refined mixture of salts, chiefly sodium chloride, used in cooking and as a seasoning. **2.** Sodium chloride.

**ta·ble·spoon** (tā′bəl-spo͞on′) *n.* **1.** A large spoon for eating soups and serving food. **2.** A household cooking measure equal to 3 teaspoons, 1/2 fluid ounce, or approx. 15 milliliters. **3.** Tablespoonful.

**ta·ble·spoon·ful** (tā′bəl-spo͞on′fo͝ol′) *n., pl.* **-fuls.** The amount a tablespoon holds.

**table sugar** *n.* Sucrose.

**tab·let** (tăb′lĭt) *n.* [ME *tablette* < OFr. *tablete,* dim. of *table,* table.]

1. A slab or plaque, as of stone or ivory, with a surface intended for or bearing an inscription. **2. a.** A thin sheet, as of clay or ivory, used as a writing surface. **b.** A set of such leaves fastened together, as in a book. **3.** A pad consisting of sheets of writing paper glued together along one edge. **4.** A small flat cake of a prepared substance, as soap. **5.** A small flat pellet of oral medication. *—vt.* **-let·ed, -let·ing, -lets. 1.** To inscribe on a tablet. **2.** To form into a tablet.

**table talk** *n.* Casual mealtime conversation.

**table tennis** *n.* A game similar to lawn tennis, played on a table with wooden paddles and a small plastic ball.

**ta·ble·ware** (tā′bəl-wâr′) *n.* The dishes, glassware, and silverware used in setting a table for a meal.

**table wine** *n.* An unfortified wine served with a meal.

**tab·loid** (tăb′loid′) *n.* [TABL(ET) + -OID.] A small-format newspaper presenting the news in condensed form, usu. with illustrated, often sensational material.

**ta·boo** *also* **ta·bu** (tə-bo͞o′, tă-) [Tongan *tabu.*] *—n., pl.* **-boos** *also* **-bus. 1. a.** A prohibition excluding something from use, approach, or mention because of its sacred and inviolable nature. **b.** An object, word, or act protected by a taboo. **2.** A ban or inhibition attached to something by social custom or emotional aversion. **3.** Belief in or conformity to religious or social prohibitions. **4.** A proscription devised and observed by a group for its own protection. *—adj.* Excluded or forbidden from use, approach, or mention <a *taboo* subject> *—vt.* **-booed, -boo·ing, -boos** *also* **-bued, -bu·ing, -bus.** To exclude from use, approach, or mention.

**ta·bor** *also* **ta·bour** (tā′bər) *n.* [ME *tabur* < OFr.] A small drum played by a fifer to accompany his or her fife.

**tab·o·ret** *also* **tab·ou·ret** (tăb′ə-rĕt′, -rā′) *n.* [Fr. *tabouret,* dim. of OFr. *tabur,* tabor.] **1.** A low stool without a back or arms. **2.** A low stand or cabinet. **3.** An embroidery frame.

**ta·bour** (tā′bər) *n. var. of* TABOR.

**tab·ou·ret** (tăb′ə-rĕt′, -rā′) *n. var. of* TABORET.

**ta·bu** (tə-bo͞o′, tă-) *n., adj., & v. var. of* TABOO.

**tab·u·lar** (tăb′yə-lər) *adj.* [Lat. *tabularis,* of boards < *tabula,* board.] **1.** Having a plane surface : FLAT. **2.** Organized as a table or list. **3.** Calculated by means of a table. **—tab′u·lar·ly** *adv.*

**tab·u·la ra·sa** (tăb′yə-lə rä′sə, rä′zə) *n.* [Lat., erased tablet.] The mind before it receives the impressions gained from experience, esp. the unformed featureless mind in the philosophy of Locke.

**tab·u·lar·ize** (tăb′yə-lə-rīz′) *vt.* **-ized, -iz·ing, -iz·es.** TABULATE 1. **—tab′u·lar·i·za′tion** *n.*

**tab·u·late** (tăb′yə-lāt′) *vt.* **-lat·ed, -lat·ing, -lates.** [< Lat. *tabula,* writing tablet.] **1.** To condense and list in tabular form. **2.** To cut or form with a plane surface. *—adj.* (tăb′yə-lĭt, -lāt′). Having a plane surface. **—tab′u·la′tion** *n.*

**tab·u·la·tor** (tăb′yə-lā′tər) *n.* **1.** One who makes tabulations. **2.** A machine into which data can be fed for tabulation. **3.** A typewriter mechanism by means of which automatic stops or margins for columns can be set.

**tac·a·ma·hac** (tăk′ə-mə-hăk′) *n.* [Sp. *tacamahaca* < Nahuatl *tecamaca.*] **1.** An aromatic resinous substance used in ointments and incenses. **2.** The balsam poplar.

**ta·cet** (tā′sĭt, tăs′ĭt, tä′kĕt′) *v.* [Lat., it is silent < *tacēre,* to be silent.] *Mus.* Be silent. —Used as a direction.

**tache** (tăch) *n.* [ME < OFr., of Germanic orig.] *Archaic.* A buckle or clasp.

**tach·i·na fly** (tăk′ə-nə) *n.* [NLat. *Tachina,* fly genus < Gk. *takhinos,* swift < *takhos,* speed.] Any of several bristly, usu. grayish flies of the family Tachinidae whose larvae live as parasites within the bodies of other insects.

**tach·i·nid** (tăk′ə-nĭd′) *n.* [NLat. *Tachinidae,* family name < *Tachina,* fly genus. —see TACHINA FLY.] Tachina fly. **—tach′i·nid** *adj.*

**tach·ism** (tăsh′ĭz′əm) *n.* [Fr. *tachisme* < *tache,* stain < OFr., of Germanic orig.] Action painting. **—tach′ist** *n.*

**ta·chis·to·scope** (tă-kĭs′tə-skōp′) *n.* [Gk. *takhistos,* superl. of *takhus,* swift + -SCOPE.] A device projecting transient images onto a screen to test visual perception. **—ta·chis′to·scop′ic** (-skŏp′ĭk) *adj.* **—ta·chis′to·scop′i·cal·ly** *adv.*

**ta·chom·e·ter** (tă-kŏm′ĭ-tər) *n.* [Gk. *takhos,* speed + -METER.] An instrument for determining speed, esp. the rotational speed of a shaft. **—tach′o·met′ric** (tăk′ə-mĕt′rĭk) *adj.* **—ta·chom′e·try** *n.*

**tachy-** *pref.* [< Gk. *takhus,* swift.] Accelerated : rapid <*tachy*meter><*tachy*cardia>

**tach·y·ar·rhyth·mi·a** (tăk′ē-ə-rĭth′mē-ə) *n.* Excessively rapid heartbeat accompanied by arrhythmia.

**tach·y·car·di·a** (tăk′ĭ-kär′dē-ə) *n.* [TACHY- + Gk. *kardia,* heart.] Excessively rapid heartbeat.

**ta·chyg·ra·phy** (tă-kĭg′rə-fē) *n.* The art or practice of rapid writing or shorthand, esp. ancient Greek and Roman stenography.

**tach·y·lyte** *also* **tach·y·lite** (tăk′ə-līt′) *n.* [G. *Tachylyt* : Gk. *takhos* + Gk. *lutos,* soluble < *luein,* to dissolve.] A black glassy basaltic rock. **—tach′y·lyt′ic** (-lĭt′ĭk) *adj.*

**ta·chym·e·ter** (tă-kĭm′ĭ-tər) *n.* A surveying device for rapid measurement of distances, elevations, and bearings. **—ta·chym′e·try** *n.*

**tach·y·on** (tăk′ē-ŏn′) *n.* A hypothetical particle that travels faster than the speed of light. **—tach′y·on′ic** *adj.*

**tach·yp·ne·a** (tăk′ĭp-nē′ə) *n.* [TACHY- + Gk. *pnoiē,* breathing < *pnein,* to breathe.] Excessively rapid respiration.

**tac·it** (tăs′ĭt) *adj.* [Lat. *tacitus,* silent < p.part. of *tacēre,* to be silent.] **1.** Expressed nonverbally : not spoken <Your frown was a *tacit* warning.> **2.** Implied by or inferred from actions or statements <*tacit* agreement> **3.** *Archaic.* Not speaking : SILENT. **—tac′it·ly** *adv.* **—tac′it·ness** *n.*

**tac·i·turn** (tăs′ĭ-tûrn′) *adj.* [Fr. *taciturne* < *taciturnus* < *tacitus,* silent. —see TACIT.] Uncommunicative : laconic. **—tac′i·tur′ni·ty** (-tûr′nĭ-tē) *n.* **—tac′i·turn·ly** *adv.*

**tack¹** (tăk) *n.* [ME *tak,* something that attaches < OFr. *tache,* of Germanic orig.] **1.** A short light nail with a sharp point and a flat head. **2.** *Naut.* **a.** A rope for holding down the weather clew of a course. **b.** A rope for hauling the outer lower corner of a studdingsail to the boom. **c.** The part of a sail to which a tack is fastened, as the weather clew of a course. **d.** The lower forward corner of a fore-and-aft sail. **3.** *Naut.* **a.** The position of a vessel relative to the trim of its sails. **b.** The act of changing from one tack to another. **c.** The distance or leg sailed between changes of tack. **4. a.** A course of action intended to minimize opposition to the achievement of an objective. **b.** An approach, esp. one of a series of changing approaches. **5.** A large loose stitch made as a temporary binding or as a marker. **6.** Stickiness, as of a newly painted surface. **7.** Gear, as saddles and bridles, for use on saddle horses. *—v.* **tacked, tack·ing, tacks.** *—vt.* **1.** To fasten or attach with or as if with a tack. **2.** To fasten or mark (e.g., cloth or a seam) with a loose basting stitch. **3.** To put together loosely and arbitrarily <*tacked* some ideas together into a proposal> **4.** To add as an extra item : APPEND <*tacked* $20 onto the bill for extra labor> **5.** To bring (a sailboat) into the wind in order to change tack. *—vi.* **1. a.** To change the direction or course of a vessel. **b.** To change tack <The ship *tacked* to starboard.> **2.** To modify one's course of action <I *tacked* in mid-sentence and tried another line of argument.>

**tack²** (tăk) *n.* [Orig. unknown.] Food, esp. coarse foodstuffs.

**tack hammer** *n.* A light hammer for driving tacks.

**tack·le** (tăk′əl) *n.* [ME *takel.*] **1.** Equipment used in a sport or occupation, esp. in fishing : GEAR. **2.** (tăk′əl, tā′kəl). *Naut.* **a.** A system of ropes and blocks for raising and lowering weights of rigging and pulleys for applying tension. **b.** A rope and its pulley. **3.** *Football.* **a.** Either of two line players in football positioned between the guard and the end. **b.** The position of a tackle. **c.** The act of stopping another player by seizing and bringing him down. *—v.* **-led, -ling, -les.** *—vt.* **1.** To come to grips with (e.g., an opponent or problem) in order to surmount or overcome. **2.** *Football.* To seize and bring down (another player). **3.** To harness (a horse). *—vi. Football.* To tackle an opponent. **—tack′ler** *n.*

**tack·ling** (tăk′lĭng) *n.* TACKLE 1.

**tack room** *n.* A room in a stable for storing and maintaining tack, as saddles and bridles.

**tack·y¹** (tăk′ē) *adj.* **-i·er, -i·est.** [< TACK¹.] Somewhat adhesive : STICKY. **—tack′i·ness** *n.*

**tack·y²** (tăk′ē) *adj.* **-i·er, -i·est.** [< *tacky,* an inferior horse.] *Informal.* **1.** Marked by neglect and disrepair : SHABBY. **2. a.** Lacking style : DOWDY <*tacky* clothes> **b.** Lacking good taste or manners : VULGAR <*tacky* jokes><a *tacky* remark> **—tack′i·ly** *adv.* **—tack′i·ness** *n.*

**ta·co** (tä′kō) *n., pl.* **-cos.** [Mex. Sp. < Sp., roll < *atacar,* to plug < Ital. *attacare,* to attack, of Germanic orig.] A tortilla wrapped around a filling, as of ground meat or cheese.

**tac·o·nite** (tăk′ə-nīt′) *n.* [After the *Taconic* Mountains, New York and western New England.] A fine-grained sedimentary rock of magnetite, hematite, and quartz, mined as a low-grade iron ore.

**tact** (tăkt) *n.* [Fr. < Lat. *tactus,* touch < p.part. of *tangere,* to touch.] **1.** The ability to appreciate the delicacy of a situation and to do or say the kindest or most fitting thing. **2.** *Archaic.* The sense of touch.
☆ ***syns:*** TACT, ADDRESS, DIPLOMACY, SAVOIR-FAIRE, TACTFULNESS *n. core meaning* : the ability to say and do the right thing at the right time <used *tact* in dealing with their many clients> ***ant:*** tactlessness

**tact·ful** (tăkt′fəl) *adj.* Possessing or showing tact : DISCREET <a *tactful* salesperson><a *tactful* suggestion> **—tact′ful·ly** *adv.* **—tact′ful·ness** *n.*

**tac·tic** (tăk′tĭk) *n.* An expedient for achieving a goal : MANEUVER. *—adj.* Of or relating to order or arrangement.

**tac·ti·cal** (tăk′tĭ-kəl) *adj.* **1.** Of or relating to tactics. **2.** Marked by adroitness in maneuvering.

**tac·ti·cian** (tăk-tĭsh′ən) *n.* **1.** One skilled in the planning and execution of military tactics. **2.** A clever maneuverer.

**tac·tics** (tăk′tĭks) *n.* [NLat. *tactica* < Gk. *taktika* < neuter pl. of *taktikos,* of order < *taktos,* arranged < *tassein,* to arrange.] (*sing. in number*). **1.** The technique or science of securing the objectives designated by strategy, esp. the art of deploying and directing troops, ships, and aircraft in coordinated maneuvers against an enemy. **2.** The skill or art of using available means to achieve an end.

**tac·tile** (tăk′təl, -tīl′) *adj.* [Lat. *tactilis* < *tangere,* to touch.] **1.** Perceptible to the touch : TANGIBLE. **2.** Used for feeling <a *tactile* organ> **3.** Of, relating to, or proceeding from the sense of touch <a *tactile* reflex> **—tac·til′i·ty** (-tĭl′ĭ-tē) *n.*

**tac·tion** (tăk′shən) *n.* [Lat. *tactio, taction-* < *tangere,* to touch.] The act of touching : CONTACT.

**tact·less** (tăkt′lĭs) *adj.* Lacking tact <a *tactless* remark><a *tactless* person> **—tact′less·ly** *adv.* **—tact′less·ness** *n.*

**tac·to·re·cep·tor** (tăk′tō-rĭ-sĕp′tər) *n.* [Lat. *tactus,* touch + RECEPTOR.] A receptor that responds to touch.

**tac·tu·al** (tăk′cho͞o-əl) *adj.* [< Lat. *tactus,* touch.] TACTILE 3. **—tac′tu·al·ly** *adv.*

**tad** (tăd) *n.* [Perh. < dial. *tad,* toad < ME *tode.*] *Informal.* **1.** A small boy. **2.** A small amount or degree : BIT.

**tad·pole** (tăd′pōl′) *n.* [ME *taddepol* : *tode,* toad (< OE *tādige*) + *pol,* head (< MLG *poll*).] The aquatic larval stage of a frog or toad, with a tail and external gills that disappear as the limbs develop and the adult stage is reached.

**tae kwon do** (tī kwŏn′dō) *n.* [Korean.] A Korean form of karate.

**tael** (tāl) *n.* [Port. < Malay *tahil,* prob. < Hindi *tolā,* a weight < Skt. *tulā.*] **1.** Any of various units of weight used in eastern Asia, roughly equivalent to 38 grams or $1^1/_3$ ounces. **2.** A Chinese monetary unit once in use, equivalent in value to 38 grams or $1^1/_3$ ounces of standard silver.

**tae·ni·a** *also* **te·ni·a** (tē′nē-ə) *n., pl.* **-ni·ae** (-nē-ē′) *or* **-ni·as.** [Lat. < Gk. *tainia.*] **1.** A narrow ribbon or band for the hair worn in ancient Greece. **2.** A band in the Doric order separating the frieze from the architrave. **3.** A ribbonlike anatomical structure. **4.** A flatworm of the genus *Taenia,* which includes many tapeworms.

**tae·ni·a·cide** (tē′nē-ə-sīd′) *n. var. of* TENIACIDE.

**tae·ni·a·sis** (tē-nī′ə-sĭs) *n. var. of* TENIASIS.

**taf·fe·ta** (tăf′ĭ-tə) *n.* [ME < OFr. *taffetas* < OItal. *taffettà* < Turk. *tafta* < Pers. *tāftah,* woven < *tāftan,* to weave.] A smooth, crisp fabric with a slight sheen, made of various fibers and used esp. for women's garments. **—taf′fe·ta** *adj.*

**taffeta weave** *n.* Plain weave.

**taf·fi·a** (tăf′ē-ə) *n. var. of* TAFIA.

**taff·rail** (tăf′rāl′, -rəl) *n.* [Alteration of *tafferel* < Du. *taffereel,* dim. of *tafel,* panel < MDu. *tāvele* < OFr. *tablel* < Lat. *tabula,* board.] *Naut.* **1.** The rail around the stern of a vessel. **2.** The flat upper part of the stern of a vessel, made of often richly carved wood.

**taffrail log** *n.* A screw log.

**taf·fy** (tăf′ē) *n., pl* **-fies.** [Orig. unknown.] **1.** A sweet chewy candy of molasses or brown sugar boiled until very thick and pulled until glossy and firm. **2.** *Informal.* Wheedling flattery.

**taffy pull** *n.* A social gathering at which taffy is made.

**taf·i·a** *also* **taf·fi·a** (tăf′ē-ə) *n.* [Native word in the West Indies.] A cheap West Indian rum distilled from molasses and refuse sugar.

**tag[1]** (tăg) *n.* [ME *tagge,* dangling piece of cloth on a garment, poss. of Scand. orig.] **1.** A strip, as of paper, metal, or plastic, hung from a wearer's neck or attached to something to identify, classify, or label <a price *tag*> **2.** The metal or plastic tip at the end of a shoelace. **3.** The contrastingly colored tip of an animal's tail. **4.** A bright piece of feather, floss, or tinsel surrounding the shank of the hook on a fishing fly. **5. a.** A matted dirty lock of wool. **b.** A loose lock of hair. **6.** A rag or tatter. **7.** A fragment <overheard *tags* of a conversation> **8.** An ornamental flourish at the end of a signature. **9. a.** A brief quotation used in speaking for added effect. **b.** A cliché, saw, or similar short conventional idea used in speaking as an embellishment <*tags* of wit and wisdom> **10. a.** The refrain or last lines of a song or poem. **b.** The closing lines of a speech in a play : CUE. **11.** A designation or epithet <the *tag* of new kid at school> *—v.* **tagged, tag·ging, tags.** *—vt.* **1.** To label, identify, or recognize with or as if with a tag. **2.** To ticket (an automobile and its driver) for a traffic or parking violation. **3.** To charge with a crime or an infraction of a rule or law <was *tagged* for illegal passing> **4.** To add as an appendage to, as a bill : TACK[1] 4. **5.** To follow closely. **6.** To cut the tags from (a sheep). *—vi.* To follow along after : ACCOMPANY.

**tag[2]** (tăg) *n.* [Orig. unknown.] **1.** A children's game in which one player pursues the others until able to touch one of them, who then becomes the pursuer. **2. a.** *Baseball.* The act of putting another player out by touching that player with the ball when he or she is not on base. **b.** The act of touching a player in touch football. *—vt.* **tagged, tag·ging, tags. 1.** To touch (another player) in the game of tag. **2. a.** *Baseball.* To touch (a runner) with the ball in order to put him or her out. **b.** To touch (the runner) in touch football as a substitute for tackling. **—tag up.** *Baseball.* To return to and touch a base with one foot before running to the next base after a fielder has caught a flyball.

**Ta·ga·log** (tə-gä′lôg′) *n., pl.* **Tagalog** *or* **-logs.** [Tagalog : *taga,* belonging to + *ilog,* river.] **1.** A member of a people native to the Philippines and inhabiting Manila and its adjacent provinces. **2.** The Austronesian language of the Tagalog.

**tag·a·long** (tăg′ə-lông′, -lŏng′) *n.* A persistent follower of another.

**tag day** *n.* A day on which collectors for a charitable fund solicit contributions, giving each contributor a tag.

**tag·ger** (tăg′ər) *n.* **1.** One that tags, esp. the pursuer in the game of tag. **2. taggers.** A very thin sheet iron, usu. plated with tin.

**tag line** *n.* **1.** An ending line, as in a play or joke, that serves to make a point. **2.** An often repeated phrase associated with an individual, organization, or commercial product.

**tag sale** *n.* A garage sale.

**Ta·hi·tian** (tə-hē′shən) *adj.* Of or relating to Tahiti or its people or language. *—n.* **1.** A native or inhabitant of Tahiti. **2.** The Polynesian language of Tahiti.

**Ta·hi·ti orange** (tə-hē′tē) *n.* The Otaheite orange.

**tahr** (tär) *n.* [Nepalese *thār.*] A goatlike mammal of the genus *Hemitragus* of mountainous regions of Asia, with a shaggy coat and curved horns.

**tah·sil·dar** *also* **tah·seel·dar** (tə-sēl′där′) *n.* [Urdu *taḥsīldār* < Pers. < Ar. : *taḥsīl,* collection + Pers. *-dār,* having.] A district official in India in charge of revenues and taxation.

**Tai** (tī) *n.* A family of languages spoken in Southeast Asia and southern China that includes Thai, Lao, and Shan. *—adj.* Of or relating to the Tai language family.

**tai chi** (tī′ chē′, jē′) *or* **tai chi chuan** (chwän′, cho͞o-än′) *n.* [Chin. (Mandarin) *tai*$^4$ *ji*$^2$ *quan*$^2$ : *tai*$^4$, highest + *ji*$^2$, reach + *quan*$^2$, boxing.] A Chinese system of physical exercises designed esp. for meditation and for the development of self-discipline and a sense of harmonious well-being.

**tai·ga** (tī′gə) *n.* [R. *taĭga.*] The subarctic evergreen forest of Siberia and of similar regions elsewhere in Eurasia and North America.

**tail[1]** (tāl) *n.* [ME < OE *tægel.*] **1.** The posterior part of an animal, esp. when elongated and extending beyond the main part or trunk of the body. **2.** The bottom, rear, or hindmost part. **3.** The rear end of a vehicle. **4. a.** The rear portion of an aircraft's fuselage. **b.** An assembly of stabilizing planes and control surfaces in this region. **5.** The vaned rear portion of a missile or bomb. **6.** An appendage to the rear or bottom <the *tail* of a kite> **7.** A braid of hair : PIGTAIL. **8.** Something that follows or takes the last place <the *tail* of the trip> **9.** A retinue of followers. **10.** The end of a line of persons or things. **11.** The short closing line of certain stanzas. **12.** Refuse or dross remaining from a process such as distilling or milling. **13.** The bottom margin of a page. **14. tails.** The reverse of a coin <heads or *tails*> **15.** *Informal.* The trail of a fleeing person or animal <The police were on the suspect's *tail.*> **16.** *Informal.* An agent assigned to follow and report on someone's movements and actions. **17. tails. a.** A man's formal evening suit. **b.** A swallow-tailed coat. *—v.* **tailed, tail·ing, tails.** *—vt.* **1.** To provide with a tail <*tail* a kite> **2.** To deprive of a tail : DOCK. **3.** To serve as the concluding part of <The clowns *tailed* the parade.> **4.** To connect (objects often dissimilar or incongruous) by or as if by the tail or end <*tail* two theories together> **5.** To set one end of (a beam, board, or brick) into a wall. **6.** *Informal.* To follow and keep under surveillance. *—vi.* **1.** To become lengthened or spaced when moving in a line <The patrol *tailed* out in pairs.> **2.** To be inserted at one end, as a beam or floor timber. **3.** *Informal.* To follow. **4.** *Naut.* **a.** To go aground with the stern foremost. **b.** To be pointed in some direction with the stern when riding at anchor or on a mooring <was *tailing* into the wind> **—tail down.** To ease a heavy load down a steep slope. **—tail off (**or **away).** To diminish gradually : SUBSIDE <The fireworks *tailed off* into darkness.>

**tail[2]** (tāl). [ME *taille* < OFr., division < *tailler,* to cut. —see TAILOR.] *Law.* *—n.* Limitation of the inheritance of an estate to a particular person or persons. *—adj.* In tail <a *tail* estate>

**tail·back** (tāl′băk′) *n.* *Football.* The offensive back lining up farthest from the line of scrimmage.

**tail beam** *n.* TAILPIECE 3.

**tail·board** (tāl′bôrd′, -bōrd′) *n.* TAILGATE 2.

**tail·bone** (tāl′bōn′) *n.* The coccyx.

**tail end** *n.* **1.** The hindmost part. **2.** The very end : CONCLUSION.

**tail·gate** (tāl′gāt′) *n.* **1.** One of the pair of gates downstream in a canal lock. **2.** A hinged board or closure at the rear of a vehicle, as a station wagon, that can be lowered during loading and unloading. *—v.* **-gat·ed, -gat·ing, -gates.** *—vt.* To drive so closely behind (another vehicle) that one cannot stop or swerve in an emergency. *—vi.* To follow another vehicle too closely.

**tail-heav·y** (tāl′hĕv′ē) *adj.* **-i·er, -i·est.** Too weighty at the rear either from overloading or from poor design and construction.

**tail·ing** (tā′lĭng) *n.* **1. tailings.** Refuse or dross remaining after processes such as milling, distilling, or mining. **2.** The portion of a tailed beam, brick, or board inside a wall.

**tail lamp** *n.* A taillight.

**taille** (tā′yə, tāl) *n.* [Fr. < OFr., division < *tailler,* to cut. —see TAILOR.] A form of direct royal taxation levied in France before 1789 on nonprivileged subjects and lands and that tended to weigh most heavily on the peasants.

**tail·light** (tāl′līt′) *n.* A red light or one of a pair mounted on the rear end of a vehicle.

**tai·lor** (tā′lər) *n.* [ME < AN *taillour* < OFr. *tailler,* to cut < VLat. **taliare* **talliare* < Lat. *talea,* a cutting.] One who makes, repairs, and alters garments. —*v.* **-lored, -lor·ing, -lors.** —*vt.* **1.** To produce (a garment). **2.** To outfit (someone) with clothes. **3.** To make, alter, or adapt for a given end <a speech *tailored* to an educated audience> —*vi.* To work as a tailor.

**tai·lor·bird** (tā′lər-bûrd′) *n.* An Old World tropical bird of the genus *Orthotomus* that uses plant fibers to stitch leaves together in making its nest.

**tai·lored** (tā′lərd) *adj.* **1.** Made by a tailor : CUSTOM-MADE. **2.** Simple, trim, or severe in line or design <a neat, *tailored* suit>

**tai·lor-made** (tā′lər-mād′) *adj.* **1.** Made by a tailor. **2.** Perfectly fitted to a condition, preference, or purpose : made or as if made to order <a *tailor-made* job> —*n.* A garment made by a tailor.

**tailor's chalk** *n.* A thin piece of hard chalk used in tailoring for making temporary alteration marks on garments.

**tail·piece** (tāl′pēs′) *n.* **1.** A piece forming an end to something : APPENDAGE. **2.** An ornamental engraving or design at the end of a chapter or at the bottom of a page. **3.** A beam tailed into a wall. **4.** *Mus.* A triangular piece of ebony to which the lower ends of the strings of a violin or cello are attached.

**tail pipe** *n.* The pipe through which exhaust gases from an engine are discharged.

**tail·race** (tāl′rās′) *n.* **1.** The part of a millrace below the water wheel through which the spent water flows. **2.** A channel for floating away mine tailings and refuse.

**tail·skid** (tāl′skĭd′) *n.* A skid attached to the rear underside of certain aircraft to act as a runner.

**tail·spin** (tāl′spĭn′) *n.* **1.** The descent of an aircraft in a spin, marked by the rapid spiral movement of the tail section. **2.** An emotional collapse.

**tail·stock** (tāl′stŏk′) *n.* The adjustable stock of a lathe supporting the spindle containing the dead center.

**tail wind** *n.* A wind blowing in the same direction as that of the course of a vehicle.

**tain** (tān) *n.* [Fr. < *étain,* tin.] **1.** A type of paper-thin tin plate. **2.** Tinfoil used as a backing for mirrors.

**Tai·no** (tī′nō) *n., pl.* **Taino** or **-nos.** [Sp., of American Indian orig.] **1.** An extinct aboriginal Arawakan Indian people of the West Indies. **2.** The language of the Taino.

**taint** (tānt) *v.* **taint·ed, taint·ing, taints.** [Partly < ME *tainten,* to color (< AN *teinter* < *teint,* p.part. of OFr. *teindre,* to color < Lat. *tingere,* to dye), and partly < ME *taynten,* to convict (< OFr. *ataint,* p.part of *ataindre,* to attain —see ATTAIN).] —*vt.* **1.** To cast aspersions on <The reporter's reputation was *tainted* by the scandal.> **2.** To expose to contagion : infect with or as if with a disease. **3.** To make rotten or poisonous. **4.** To corrupt morally. —*vi.* To become discolored : ROT. —*n.* **1.** A moral defect considered as a stain or spot. **2.** An infecting touch, influence, or tinge.

☆ ***syns:*** TAINT, BEFOUL, BLACKEN, DIRTY, SMEAR, SULLY, TARNISH *v. core meaning* : to cast aspersions on <a good name now *tainted* by rumors of improprieties>

▲ word history: *Taint* is a doublet of *tint,* since they both have the same source but entered English by different routes. *Tint,* whose earlier form is *tinct,* is a direct borrowing of Latin *tinctus,* "a dyeing," from *tingere,* "to dip, to dye." *Tingere* is also the source of *tinge.* In Old French *tinctus* became *taint* or *teint,* "a dye, tint, color," which was borrowed into English as *taint.* In English there was another word *taint,* a shortened form of *attaint,* a legal term denoting the forfeiture of rights suffered by a condemned criminal. The confusion of the two words spelled *taint* led to the development of the sense "moral stain" for the word that originally meant just "color."

**taj** (täzh, täj) *n.* [Ar. *tāj* < Pers.] A tall conical cap worn by Moslems as a headdress of distinction.

**taj**

**ta·ka** (tä′kə) *n.* [Bengali *ṭākā* < Skt. *ṭaṅkaḥ,* coin.] —See table at CURRENCY.

**ta·ka·he** (tə-kī′) *n.* [Maori.] A nearly extinct flightless bird, *Notornis mantelli* of New Zealand, with a large bill and brightly colored plumage.

**take** (tāk) *v.* **took** (to͝ok), **tak·en** (tā′kən), **tak·ing, takes.** [ME *taken* < OE *tacan* < ON *taka.*] —*vt.* **1.** To get into one's possession by force, skill, or artifice, esp.: **a.** To capture physically : SEIZE <*take* an enemy stronghold> **b.** To kill, snare, or trap (e.g., fish or game). **c.** To acquire in a game or competition : WIN <*take* your opponent's knight> **d.** To seize authoritatively : CONFISCATE. **e.** To catch (a ball in play), esp. in baseball <*took* it on the fly> **2.** To grasp with the hands : GRIP <*take* the child's hand> **3.** To be affected with : CONTRACT <I had *taken* the flu.> **4.** To encounter or catch in a particular situation <*took* me by surprise> **5.** To deal a blow to : HIT <I *took* my opponent a sharp jab to the ribs.> **6.** To affect favorably : CHARM <completely *taken* by the kitten> **7.** To receive (e.g., air, food, or drink) into the body <*take* a deep breath><*took* a sip of tea> **8.** To expose one's body to (e.g., healthful or pleasurable treatment) <*take* some sun> **9.** To bring or receive into a particular relation, association, or connection <*take* a new member into the club> **10.** To have sexual intercourse with. **11.** To accept and place under one's care or keeping. **12.** To appropriate for one's own or another's use or benefit : BUY <We always *take* season tickets.> **13. a.** To assume for oneself <*take* credit> **b.** To charge or oblige oneself with the fulfillment of (e.g., a task or duty) : commit oneself to <*took* chairmanship of the committee> **c.** To pledge one's obedience to : impose (a vow or promise) upon oneself. **d.** To subject oneself to <We *took* extra time to do the job properly.> **e.** To adopt or accept for one's own <*take* the unpopular side of an issue> **f.** To put forth or adopt as a point of argument, defense, or discussion <Your interpretation of the poem is well *taken.*> **g.** To require or have as a proper or fitting accompaniment <Intransitive verbs *take* no direct object.> **14.** To obtain through competition <*took* first place> **15.** To defeat <Our team *took* our rivals three to one.> **16. a.** To select <*take* any card> **b.** To choose for one's own use <*took* a rented car> **c.** To use, as in operating <This camera *takes* 35mm. film.> **d.** To use as a means of conveyance or transportation <*take* a plane to Chicago> **e.** To use as a means of safety or refuge <*take* shelter from the storm> **17.** To assume occupancy of <*take* a seat> **18.** To have as a necessity for something : REQUIRE <It *takes* determination to achieve success.> **19.** To obtain from a source : DERIVE <The book *takes* its title from the Bible.> **20.** To obtain through particular procedures, as through measurement <*take* one's temperature> **21. a.** To put down in writing : WRITE <*take* dictation> **b.** To put down an image, likeness, or representation of by or as if by drawing, painting, or photography <*take* a picture of the family> **22. a.** To accept (something owed, offered, or given) either reluctantly or willingly <*took* the advice with a grain of salt> **b.** To submit to : ENDURE <didn't *take* the punishment very well> **c.** To withstand <The dam *took* the heavy flood waters.> **d.** To accept or believe as true <I'll *take* your word.> **e.** To follow (e.g., advice, a suggestion, or a lead). **f.** To accept or deal with <I *take* things in stride.> **g.** To consider in a particular relation or from a particular viewpoint <*take* the bitter with the sweet> **23.** To impose upon oneself : UNDERTAKE <*take* precautions><*take* the responsibility> **24. a.** To allow to come in : ADMIT <The boat *took* a lot of water.> **b.** To provide room for : ACCOMMODATE <We can *take* 40 campers.> **c.** To become saturated or impregnated with (e.g., dye). **25. a.** To understand or interpret <*took* my question as an insult> **b.** To consider : assume <*take* the matter as settled> **c.** To consider to be equal to : RECKON <We *take* their number at a thousand.> **d.** To perceive or feel : EXPERIENCE <*take* offense> **26.** To carry along or cause to go with one to another place <Remember to *take* your umbrella.> **27.** To convey to another place <This bus *takes* you to New York.> **28.** To remove or obtain by removing <*took* the coat from the rack><The dentist *took* two molars.> **29.** To cause to die : DESTROY <The blight *took* these tomatoes.> **30.** To subtract <*take* 15 from 30> **31.** To commit oneself to the study of : enroll in <*take* a computer course> **32.** To swindle <*taken* by a con artist> **33.** *Baseball.* To refrain from swinging at (a pitched ball). —*vi.* **1.** To acquire possession. **2.** To engage or mesh, as gears. **3.** To start growing : GERMINATE <Have the seeds *taken?*> **4.** To have the intended effect : WORK <The inoculation apparently *took.*> **5.** To gain popularity or favor <The TV series didn't *take* and was canceled.> **6.** To detract <smokestacks that *take* from the view> **7.** To become <I *took* sick.> **—take back.** To retract something stated or written. **—take down. 1.** To bring from a higher to a lower position. **2.** To dismantle : take apart <*take down* the tent> **3.** To lower the self-esteem of <*took* me *down* a peg or two> **4.** To put down in writing. **—take in. 1.** To grant admittance to : receive as a guest or employee. **2.** To reduce in size <*take in* the waist on a pair of pants> **3.** To include or comprise. **4.** To understand. **5.** To deceive or swindle. **6.** To look at thoroughly : VIEW <*took in* the sights> **7.** To accept (work) to be done at home for pay <*took in* washing> **8.** To convey (a prisoner) to a police station. **—take off. 1.** To remove, as clothing <*take* one's hat *off*> **2.** To release <*took* the brake *off*> **3.** To deduct as a discount <*took* 20% *off* the bill> **4.** To carry off or away. **5.** *Slang.* **a.** To leave <*took off* in a hurry> **b.** To achieve wide use or popularity. **6.** To become airborne <The plane *took off* on time.> **7.** To discontinue <*took off* the commuter special> **8.** To withhold service due, as from one's work <*took* three days *off* to go fishing> **—take on. 1.** To undertake or begin to handle <*took on* extra responsibilities> **2.** To hire <*took on* more workers> **3.** To oppose in competition <a chess player who *took on* all comers> **4.** *Informal.* To display violent or passionate emotion <Don't *take on* so!> **5.** To acquire (e.g., an appearance) as or as if one's own <*took on* the look of a conservative> **—take out. 1.** To remove <*took* the splinter *out*> **2.** To secure (e.g., a license) by application to an authority. **3.** *Informal.* To escort, as a date <*took* me *out* to dinner> **4.** To give vent to <*took out* my frustration on them> **5.** To obtain as an equivalent in a different form <*took out* the money owed in services> **6.** *Informal.* To begin a course : SET OUT <*took out* after them> **—take over.** To assume the control or management of. **—take up. 1.** To raise or lift <*take up* a shovel> **2.** To reduce in size : shorten or tighten <*take up* a hem> **3.** To pay off an outstanding debt, mortgage, or note. **4.** To accept as offered, as an option, a bet, or a challenge. **5.** To begin again : RESUME <*take up* where we left off> **6.** To use up, consume, or occupy <The extra duties *took up* most of my time.> **7. a.** To develop an interest in or begin to practice <*take up* mountain climbing><*take up* smoking> **b.** To deal with <Let's *take up* each problem, one at a time.> **c.** To assume <*took up* a friendly attitude> **d.** To absorb or adsorb <crops *taking up* nutrients> **e.** To enter into (a profession or business) <*took up* engineering> *—n.* **1. a.** The act or process of taking. **b.** That which is taken. **2.** The number of fish, game birds, or animals caught or killed at one time. **3.** A quantity collected at one time, esp. the amount of profit or receipts taken on a business venture. **4.** *Slang.* The amount of money collected as admission to a sporting event : GATE. **5.** The uninterrupted running of a motion-picture or television camera or set of recording equipment in filming a movie or television program or cutting a record. **6. a.** A scene filmed or televised without interrupting the run of the camera. **b.** A recording made in a single session. **7. a.** A physical reaction, as a rash, indicating a successful vaccination. **b.** A successful graft. **8.** *Slang.* An attempt or try <found the solution on the third *take*> **—on the take. 1.** Seeking to take advantage of another. **2.** Seeking to take or taking bribes or illegal income. **—take a bath.** To experience serious financial loss. **—take account of.** To take into account. **—take advantage of. 1.** To use to one's advantage : derive profit from. **2.** To impose upon (another) to his or her detriment : EXPLOIT. **—take after. 1.** To follow as an example. **2.** To resemble in character, temperament, or appearance <You *take after* your parents.> **—take amiss.** To be offended through misunderstanding of <*took* my comments *amiss*> **—take apart. 1.** To disassemble or divide into parts. **2.** To dissect or analyze (e.g., an object or theory), usu. to discover hidden weaknesses or flaws. **3.** *Informal.* To tear into violently : beat up. **—take a powder.** To leave quickly. **—take care.** To be careful. **—take care of.** To assume responsibility for the maintenance, support, or treatment of. **—take charge.** To assume command or control. **—take effect. 1.** To become operative, as under regulation or law <The curfew *takes effect* at midnight.> **2.** To have the intended effect, as a drug. **—take five** (or **ten**). To take a short rest or break. **—take for. 1.** To consider or suppose to be : regard as <Don't *take* me *for* a fool.> **2.** To consider mistakenly <*took* them *for* dead> **—take for granted. 1.** To think of as true, real, or forthcoming. **2.** To underestimate the value of <*took* the teachers *for granted*> **—take heart.** To be confident or courageous. **—take hold. 1.** To seize, as by grasping. **2.** To become established <The newly planted vines *took hold.*> **—take into account.** To make allowances for. **—take into consideration.** To take into account. **—take in vain.** To use a name, esp. a sacred one, profanely or blasphemously. **—take issue.** To assume an opposing position : DISAGREE. **—take it. 1.** To understand : assume <As I *take it,* you haven't accepted the job.> **2.** *Informal.* To endure harsh treatment. **—take it lying down.** *Informal.* To submit passively to harsh treatment. **—take it on the chin.** *Informal.* To endure punishment, suffering, or defeat. **—take it or leave it.** To accept or reject unconditionally. **—take it out on.** *Informal.* To abuse another in venting one's own anger. **—take notice of.** To pay attention to. **—take (one's) time.** To act slowly or at one's leisure. **—take part.** To participate in : JOIN. **—take place. 1.** To have as a locality. **2.** To happen. **—take root. 1.** To become fixed or established. **2.** To become rooted. **—take shape.** To take on a distinctive form. **—take sides.** To associate with and support a particular faction, group, or cause. **—take stock. 1.** To take an inventory. **2.** To make an estimate or appraisal, as of resources. **—take stock in.** To trust, believe in, or attach importance to. **—take the cake.** To come in first in or as if in a contest. **—take the count. 1.** To be defeated. **2.** To be counted out in boxing. **—take the floor.** To rise to deliver a formal speech, as to an assembly. **—take to. 1.** To have recourse to <*took to* the mountains> **2.** To develop as a habit or steady practice <*take to* drink> **3.** To become fond of or attached to. **—take to task.** To reprimand severely. **—take up for.** To support (e.g., a person or a group) in an argument. **—take up with.** To begin to associate with.

**take·down** (tāk′doun′) *adj.* Capable of being taken down or apart <a *takedown* rifle> *—n.* **1.** A takedown article. **2.** The mechanism allowing an article to be easily taken down. **3.** *Informal.* The act or an instance of humiliating someone.

**take-home pay** (tāk′hōm′) *n.* The amount of one's salary remaining after various deductions, as income taxes and insurance, have been withheld.

**take-in** (tāk′ĭn′) *n.* An act or instance of deception : SWINDLE.

**tak·en** (tā′kən) *v. p.p. of* TAKE.

**take·off** (tāk′ôf′, -ŏf′) *n.* **1.** The act of rising in flight. —Used of an aircraft or rocket. **2.** The point or place from which one takes off. **3.** *Informal.* An amusing imitative caricature or burlesque.

**take-out** (tāk′out′) *adj.* **1.** Intended to be consumed away from the premises <*take-out* pizza> **2.** Dealing with or designed for take-out products <a *take-out* counter><*take-out* cups> **—take′-out** *n.*

**take·o·ver** *also* **take-o·ver** (tāk′ō′vər) *n.* The act or an instance of assuming control or management of or responsibility for, esp. the forcible seizure of power, as in a nation or political organization. **—take′o′ver** *adj.*

**tak·er** (tā′kər) *n.* One who accepts or takes up something, as a wager or purchase.

**take-up** (tāk′ŭp′) *n.* **1.** A device for reducing slack or taking up lost motion, as in a loom. **2.** The act of taking or tightening up.

**ta·kin** (tä′kēn′) *n.* [Of Tibeto-Burman orig.] A goatlike mammal, *Budorcas taxicolor* of the mountains of central Asia, with a shaggy coat and backward-pointing horns.

**tak·ing** (tā′kĭng) *adj.* **1.** Capturing the interest : WINNING <a *taking* smile> **2.** Contagious. —Used of an infectious disease. *—n.* **1.** The act of one that takes. **2.** Something taken, as a catch of fish. **3. takings.** Monetary receipts.

**ta·la** (tä′lə) *n.* [Samoan < E. DOLLAR.] —See table at CURRENCY.

**tal·a·poin** (tăl′ə-poin′) *n.* [Fr. < Port. *talapão,* monk < Mon *tala pōi.*] A small African monkey, *Miopithecus talapoin* or *Cercopithecus talapoin,* with greenish fur and a long tail.

**ta·lar·i·a** (tə-lâr′ē-ə) *pl.n.* [Lat. < *talaris,* of the ankles < *talus,* ankle.] Winged sandals such as those represented in Greco-Roman painting and sculpture.

**talc** (tălk) *n.* [OFr. *talc* < Med. Lat. *talcum* < Ar. *ṭalq* < Pers. *talk.*] A fine-grained white, greenish, or gray mineral, essentially $Mg_3Si_4O_{10}(OH)_2$, having a soft soapy texture and used in talcum and face powder, as a paper coating, and as a filler for paint and plastics. *—vt.* **talcked, talck·ing, talcs** *or* **talced, talc·ing, talcs.** To apply talc to (e.g., a photographic plate).

**talc·ose** (tăl′kōs′) *also* **talc·ous** (tăl′kəs) *or* **talck·y** (tăl′kē) *adj.* Made of or containing talc.

**tal·cum** (tăl′kəm) *n.* [Med. Lat., talc.] **1.** Soapstone or talc. **2.** Talcum powder.

**talcum powder** *n.* A fine, often perfumed body powder made from purified talc.

**tale** (tāl) *n.* [ME < OE *talu.*] **1.** A recital of events <I told them my *tale* of woe.> **2.** A malicious story : GOSSIP. **3.** A deliberate lie : FALSEHOOD. **4.** A diverting or edifying narrative of real or imaginary events. **5.** *Archaic.* A tally or reckoning : TOTAL.

**tale·bear·er** (tāl′bâr′ər) *n.* One who spreads malicious stories or gossip. **—tale′bear′ing** *adj.* & *n.*

**tal·ent** (tăl′ənt) *n.* [ME < OE *talente,* a unit of money < Lat. *talentum* < Gk. *talanton.*] **1.** A mental or physical aptitude : natural or acquired ability. **2. a.** Superior natural endowment or ability. **b.** A person or group of persons having such ability <The company makes good use of its *talent.*> **3.** A variable unit of weight and money used in ancient Greece, Rome, and the Middle East. **—tal′ent·ed** *adj.*

▲ word history: The word *talent* is a borrowing of Greek *talanton,* "an amount of money." The meaning "ability, aptitude," comes from the metaphorical use of *talanton* in the parable recorded in Matthew 25:14–30. This parable tells how a master entrusted money to each of his three servants in his absence. The servant with five talents and the one with two talents both doubled their money and were rewarded on their master's return. The servant who had been given one talent buried the money and returned only the original sum, and for this he was reproached. The parable has been interpreted to mean that everyone has a duty to improve the natural gifts and abilities that God has given, and failure to do so would result in eternal punishment.

**talent scout** *n.* An agent sent on tour in search of talented people, as for acting, sports, or business.

**talent show** *n.* A show featuring amateur performers whose talent may win recognition or a special award.

**ta·ler** *also* **tha·ler** (tä′lər) *n., pl.* **taler** *or* **-lers** *also* **thaler** *or* **-lers.** [G. —see DOLLAR.] Any of numerous silver coins used as currency in certain Germanic countries between the 15th and 19th cent.

**ta·les** (tā′lēz) *n., pl.* **tales.** [ME < Med. Lat. *tales de circumstantibus,* such (persons) from those standing about (a phrase used in the writ).] *Law.* **1.** A group of persons summoned to fill vacancies on a jury that has become deficient in number. **2.** The writ allowing for a summons of jurors.

**tales·man** (tālz′mən, tā′lēz-) *n. Law.* One summoned under a writ of tales.

**tale·tell·er** (tāl′tĕl′ər) *n.* **1.** One who tells stories : STORYTELLER. **2.** A talebearer. **—tale′tell′ing** *adj.* & *n.*

**ta·li** (tā′lī′) *n. pl. of* TALUS¹.

**tal·i·on** (tăl′ē-ən) *n.* [ME *talioun* < OFr. *talion* < Lat. *talio.*] A punishment identical to the offense, as death for a murderer.

**tal·i·ped** (tăl′ə-pĕd′) *adj.* Afflicted with clubfoot. **—tal′i·ped′** *n.*

**tal·i·pes** (tăl′ə-pēz′) *n.* [NLat. *talipes, taliped-* : Lat. *talus,* ankle + Lat. *pes,* foot.] Clubfoot.

**tal·i·pot** (tăl′ə-pŏt′) *n.* [Bengali *tālipōt.*] A tall palm tree, *Corypha umbraculifera* of tropical Asia, with a spreading crown of large fan-like leaves.

**tal·is·man** (tăl′ĭs-mən, tăl′ĭz-) *n., pl.* **-mans.** [Fr. *talisman* or Sp. *talismán* or Ital. *talismano,* all < Ar. *ṭilsām* < LGk. *telesma* < Gk., consecration < *telein,* to consecrate < *telos,* result.] **1.** An object marked with magical signs and held to confer on its bearer supernatural powers or protection. **2.** Something having apparently magical power.

**tal·is·man·ic** (tăl′ĭs-măn′ĭk, tăl′ĭz-) *also* **tal·is·man·i·cal** (-ĭ-kəl) *adj.* **1.** Of or relating to talismans <*talismanic* formulas> **2.** Possessing magical power <a *talismanic* amulet>

**talk** (tôk) *v.* **talked, talk·ing, talks.** [ME *talken.*] *—vt.* **1.** To articulate (words) in speech : UTTER. **2.** To articulate (something) in words <*talk* sedition> **3.** To speak of or discuss (something) <*talk* money> **4.** To speak (an idiom) <*talks* Yiddish> **5.** To spend (a period of time) by or as if by talking <*talked* the afternoon away> *—vi.* **1.** To converse by means of spoken language <We *talked* for days.> **2.** To articulate words <The child can *talk.*> **3.** To imitate the sounds of human speech <The parrot *talks.*> **4.** To manifest one's thoughts other than by articulate language <*talk* with one's hands> **5.** To express one's thoughts in writing <*talks* about Paris in the essay> **6.** To parley or negotiate with someone <Let's *talk* before acting.> **7.** To chatter incessantly <did nothing but *talk*> **8.** To gossip <If we do that, people will *talk.*> **9.** To consult or confer with someone <I *talked* with the doctor.> **10.** To yield under coercion wanted information <Has the prisoner *talked* yet?> **—talk around. 1.** To persuade <*talked* them *around* to my viewpoint> **2.** To speak indirectly about something. **—talk at. 1.** To say something intended for a listener without addressing the listener directly <*talked at* me in the debate> **2.** To address someone without regard to his or her response <just *talks at* people, never with them> **—talk back. 1.** To make an impertinent reply <taught not to *talk back* to their parents> **2.** To make a belligerent response <heavy artillery *talking back*> **—talk down. 1.** To depreciate <*talked down* the importance of money> **2.** To address (someone) with insulting condescension <*talked down* to the students> **3.** To silence (someone) <I could *talk* you *down* with one word.> **—talk out. 1.** To discuss a matter exhaustively <Let's *talk out* our problems.> **2.** To speak clearly or loudly. **3.** To exhaust someone by talking. **4.** *Chiefly Brit.* To filibuster proposed legislation. **—talk over. 1.** To win someone over by persuasion <*talked* you *over* to my side> **2.** To discuss a subject <*talked* the matter *over*> **—talk up. 1.** To propagandize in favor of <*talked* the candidate *up*><*talked up* the new product> **2.** To speak up impertinently or defiantly, esp. to a superior. *—n.* **1.** Articulation of ideas in conversation <skilled at making diverting and intelligent *talk*> **2.** An informal speech <give a short *talk*> **3.** Hearsay, rumor, or speculation <There is *talk* of a recession.> **4.** A subject of conversation <the *talk* of the town> **5.** A conference or negotiation <peace *talks*> **6.** Slang or jargon <prison *talk*><street *talk*> **7.** Empty speech <much *talk* and no action> **8.** Manner of speech <high-sounding *talk*> **9.** Something, as the sounds of animals, resembling talk <whale *talk*> **—talk big.** *Slang.* To brag. **—talk sense.** To speak rationally and logically. **—talk (someone) into.** To persuade. **—talk (someone) out of.** To dissuade. **—talk through (one's) hat.** To speak irrationally or illogically. **—talk turkey.** To speak forthrightly.

**talk·a·thon** (tôk′ə-thŏn′) *n.* [TALK + (MAR)ATHON.] A prolonged session of discussions, speech-making, or debate.

**talk·a·tive** (tô′kə-tĭv) *adj.* Inclined to talk : GARRULOUS. **—talk′a·tive·ly** *adv.* **—talk′a·tive·ness** *n.*

☆ ***syns:*** TALKATIVE, CHATTY, GABBY, GARRULOUS, LOQUACIOUS, TALKY *adj. core meaning* : inclined to talk <a *talkative,* tiring person>

**talk·er** (tô′kər) *n.* One who talks, esp. loquaciously.

**talk·ie** (tô′kē) *n. Informal.* A motion-picture film with a sound track.

**talking book** *n.* A phonograph record or a tape-recorded cassette of a reading of a book, designed for use by the blind.

**talking point** *n.* A persuasive point that helps to support an argument.

**talk·ing-to** (tô′kĭng-too͞′) *n., pl.* **-tos.** *Informal.* A scolding.

**talk show** *n.* **1.** A radio or television show in which usu. noted individuals participate in discussions or are interviewed. **2.** A radio or television show in which one or more hosts answer calls from listeners or viewers.

**talk·y** (tô′kē) *adj.* **-i·er, -i·est. 1.** Talkative. **2.** Containing excessive talk <a *talky,* boring film>

**tall** (tôl) *adj.* **-er, -est.** [ME, brave < OE *getæl,* swift.] **1. a.** Having greater than ordinary height <a *tall* person> **b.** Having considerable height <*tall* pines> **2.** Having a stated height <a hedge three feet *tall*> **3.** *Informal.* Boastful or fanciful <*tall* talk> **4.** *Archaic.* Excellent : comely : fine. **5.** Long <a *tall* glass for iced coffee> **6.** Impressively great or difficult <a *tall* order to fill> *—adv.* With proud bearing : STRAIGHT <stand *tall*>

☆ ***syns:*** TALL, HIGH, LOFTY, TOWERING *adj. core meaning* : extending to a great height <the *tall* redwood trees> ***ant:*** short

**tal·lage** (tăl′ĭj) *n.* [ME *taillage* < OFr., feudal fee < *tailler,* to cut. —see TAILOR.] An occasional tax levied by the Anglo-Norman kings on crown lands and royal towns. *—vt.* **-laged, -lag·ing, -lag·es.** To levy a tax on.

**tall·boy** (tôl′boi′) *n. Chiefly Brit.* A highboy.

**tall drink** *n.* A drink served in a tall glass and consisting typically of a liquor base with any of various diluents and flavorings.

**tal·lith** (tä′lĭs, -lĭth) *n., pl.* **tal·lith·im** (tä′lĭ-sēm′, -thēm′) [Heb. *ṭallīth.*] A fringed prayer shawl with bands of black or blue, worn during worship by Orthodox and Conservative Jewish men.

tallith

**tall oil** (täl, tôl) *n.* [Partial transl. of G. *Tallöl* < partial transl. of Swed. *tallolja* : *tall,* pine (< ON *thöll,* young pine tree) + *olja,* oil.] An oily resinous liquid composed of a mixture of rosin acids and fatty acids derived as a by-product in the treatment of pine pulp and used in soaps, emulsions, and lubricants.

**tal·low** (tăl′ō) *n.* [ME *talow.*] **1.** A mixture of the whitish, tasteless solid or hard fat obtained from parts of the bodies of cattle, sheep, or horses, and used in foods or to make candles, leather dressing, soap, and lubricants. **2.** A fat similar to tallow. *—vt.* **-lowed, -low·ing, -lows. 1.** To smear or cover with tallow. **2.** To fatten (animals) in order to obtain tallow. **—tal′low·y** *adj.*

**tal·ly** (tăl′ē) *n., pl.* **-lies.** [ME *taly* < Med. Lat. *talea* < Lat., stick.] **1.** A stick on which notches are made to keep a count or score. **2.** The reckoning or score kept on a piece of wood, gunstock, blackboard, or scorecard. **3.** A mark used in recording a number of acts or objects, usu. in series of five, consisting of four vertical lines canceled diagonally or horizontally by a fifth line. **4.** An identification label used in gardens and greenhouses. **5.** A part very similar to another part, as in shape or function. **6.** A metal plate attached to a ship's machinery and bearing instructions for its use. *—v.* **-lied, -ly·ing, -lies.** *—vt.* **1.** To record by making a mark. **2.** To count or reckon. **3.** To label with a tally. **4.** To cause to correspond or agree. *—vi.* **1.** To be alike : AGREE. **2.** To keep score or a reckoning.

**tal·ly·ho** (tăl′ē-hō′) [Prob. < Fr. *taïaut* < OFr. *thialau.*] *interj.* —Used to urge hounds in fox hunting. *—v.* **-hoed, -ho·ing, -hos.** *—vt.* To excite (hounds) on a fox hunt by shouting "tallyho" when the fox is sighted. *—vi.* To shout "tallyho" as a hunting cry. *—n., pl.* **-hos. 1.** The cry of "tallyho." **2.** A pleasure coach pulled by four horses.

**tal·ly·man** (tăl′ē-mən) *n.* **1.** A recorder or scorekeeper. **2.** *Chiefly Brit.* A merchant whose customers pay by the week according to a simple reckoning, without regular bookkeeping and billing.

**Tal·mi gold** (tăl′mē) *n.* [G. *Talmigold.*] A composite metal made of gold and brass, used in making jewelry.

**Tal·mud** (täl′mo͝od′, tăl′məd) *n.* [Heb. *talmūd.*] The collection of ancient Rabbinic writings composed of the Mishnah and the Gemara, forming the basis of religious authority for traditional Judaism. **—Tal·mu′dic** (tăl-mo͞o′dĭk, -myo͞o′-, täl-), **Tal·mu′di·cal** *adj.* **—Tal′mud·ist** (täl′mo͝od-ĭst, tăl′məd-) *n.*

**tal·on** (tăl′ən) *n.* [ME < OFr., heel < Lat. *talus,* ankle.] **1. a.** The claw of a bird of prey. **b.** The claw of a predatory animal. **2.** Something similar to or like a claw. **3.** The part of a lock which the key presses in order to shoot the bolt. **4.** The part of the deck of cards in certain card games left on the table after the deal. **—tal′oned** *adj.*

**ta·lus¹** (tā′ləs) *n., pl.* **-li** (-lī′) [NLat. < Lat., ankle.] **1.** A tarsal bone that articulates with the tibia and fibula to form the anklebone. **2.** The ankle.

**ta·lus²** (tā'ləs) *n., pl.* **-lus·es.** [OFr., sloping side of an earthwork.] **1.** A slope formed by an accumulation of debris. **2.** A sloping mass of debris accumulated at the base of a cliff.

**tam** (tăm) *n.* A tam-o'-shanter.

**ta·ma·le** (tə-mä'lē) *n.* [Mex. Sp. *tamales,* pl. of *tamal* < Nahuatl *tamalli.*] An often highly seasoned Mexican dish made of fried chopped meat and crushed peppers rolled in cornmeal dough, wrapped in corn husks, and steamed.

**ta·man·du·a** (tə-măn'dōō-ə) *n.* [Port. *tamanduá* < Tupi.] A chiefly arboreal anteater, *Tamandua tetradactyla* of tropical America, having a dense furry coat.

**tam·a·rack** (tăm'ə-răk') *n.* [Of Algonquian orig.] A North American larch tree, esp. *Larix laricina,* with short deciduous needles.

**tam·a·rau** *also* **ta·ma·rao** (tăm'ə-rou') *n.* [Tagalog *tamaráw.*] A small short-horned buffalo, *Anoa mindorensis* of the island of Mindoro in the Philippines.

**tam·a·rin** (tăm'ə-rĭn, -răn') *n.* [Fr. < Galibi.] A small long-tailed monkey of the genus *Saguinus* of tropical South America.

**tam·a·rind** (tăm'ə-rĭnd') *n.* [Med. Lat. *tamarindus* < Ar. *tamr hindī* : *tamr,* date + *hindī,* of India.] **1.** A tropical Old World tree, *Tamarindus indica,* having compound leaves and red-striped yellow flowers. **2.** The fruit of the tamarind.

**tam·a·risk** (tăm'ə-rĭsk') *n.* [ME *tamarisc* < LLat. *tamariscus* < Lat. *tamarix.*] Any of numerous shrubs or small trees of the genus *Tamarix* native to Eurasia, with small scalelike leaves and pink flower clusters.

**tam·bac** *or* **tam·bak** (tŏm'băk) *n. vars. of* TOMBAC.

**tam·ba·la** (täm-bä'lə) *n.* [Native word in Malawi.] —See table at CURRENCY.

**tam·bour** (tăm'bŏŏr', tăm-bŏŏr') *n.* [ME < OFr.] **1. a.** A drum. **b.** A drummer. **2. a.** A small wooden embroidery frame composed of two concentric hoops between which fabric is stretched. **b.** Embroidery made on a tambour. **3.** A rolling front or top for a desk, consisting of narrow strips of wood glued to canvas. **4. a.** The wall of a circular building surrounded with columns. **b.** The vertical section of a cupola. —*v.* **-boured, -bour·ing, -bours.** —*vt.* To do (embroidery) on a tambour. —*vi.* To embroider at a tambour frame.

**tam·bou·ra** *also* **tam·bu·ra** (tŭm-bŏŏr'ə) *n.* [Pers. *ṭanbūr.*] An unfretted lute of India, used as a harmonic drone.

**tam·bou·rin** (tăm'bŏŏ-rĭn, tăN-bŏŏ-răN') *n.* [Prov. *tamborin,* dim. of *tambor,* var. of OFr. *tambour,* tambour.] **1. a.** A long narrow drum used in Provence. **b.** One who plays the tambourin. **2.** A dance in lively two-beat rhythm, accompanied by the tambourin.

**tam·bou·rine** (tăm'bə-rēn') *n.* [OFr. *tambourin,* dim. of *tambour,* tambour.] A musical instrument consisting of a small drumhead with jingling disks fitted into the rim that is carried and shaken with one hand and struck with the other.

**tam·bu·ra** (tŭm-bŏŏr'ə) *n. var. of* TAMBOURA.

**tame** (tām) *adj.* **tam·er, tam·est.** [ME < OE *tam.*] **1.** Brought from wildness into a tractable or domesticated state. **2.** Naturally gentle or unafraid : not timid <"The sea otter is gentle and relatively *tame*" —Peter Matthiessen> **3.** Submissive : docile <a *tame* child> **4.** Insipid : flat <a *tame* New Year's Eve party> **5.** Sluggish : inactive <a *tame* river> —*vt.* **tamed, tam·ing, tames. 1.** To make tractable : DOMESTICATE. **2.** To subdue or curb. **3.** To tone down : SOFTEN. **—tam'a·ble, tame'a·ble** *adj.*

**Tam·il** (tăm'əl, tŭm'-) *n.* [Tamil.] **1.** A member of a Dravidian race of southern India and Sri Lanka. **2.** The Dravidian language of the Tamil. —*adj.* Of or relating to the Tamil.

**Tam·muz** *also* **Tham·muz** (tä'mŏŏz') *n.* [Heb. *Tammūz* < Babylonian *Du'uzu,* the name of a god.] The tenth month in the Hebrew year. —See table at CALENDAR.

**tam-o'-shan·ter** (tăm'ə-shăn'tər) *n.* [After the hero of *Tam o'Shanter,* a poem by Robert Burns (1759–96).] A tight-fitting Scottish cap or braided bonnet, occas. having a pompom, tassel, or feather in the center.

**tamp** (tămp) *vt.* **tamped, tamp·ing, tamps.** [Prob. back-formation < TAMPION.] **1.** To pack down tightly by a succession of blows or taps. **2.** To pack clay, sand, or dirt into (a drill hole) above an explosive.

**tam·per¹** (tăm'pər) *v.* **-pered, -per·ing, -pers.** [Alteration of TEMPER.] —*vi.* **1.** To interfere in a harmful way <*tampering* with a delicate mechanism> **2.** To meddle foolishly or rashly <*tamper* with someone's emotions> **3.** To bring about an improper situation or condition by clandestine means <*tamper* with a jury><*tamper* with a contract> —*vt.* To alter improperly. **—tam'per·er** *n.*

☆ ***syns:*** TAMPER, FOOL, MEDDLE, MESS, MONKEY, TINKER *v. core meaning* : to handle something idly, ignorantly, or destructively <instructions not to *tamper* with the radio>

**tamp·er²** (tăm'pər) *n.* **1.** One that tamps. **2.** A neutron reflector in an atomic bomb that also delays the expansion of the exploding material, making possible a longer-lasting, more energetic, and more efficient explosion.

**Tam·pi·co hemp** (tăm-pē'kō) *n.* [After *Tampico,* Mex.] PITA² 1.

**tam·pi·on** (tăm'pē-ən) *also* **tom·pi·on** (tŏm'-) *n.* [ME *tampyne,* plug < OFr. *tampon,* of Germanic orig.] A cover or plug for the muzzle of a gun or cannon to keep out dust and moisture.

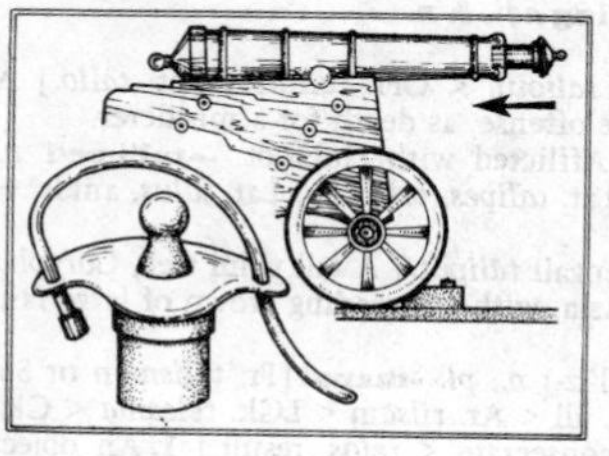

**tampion**

**tam·pon** (tăm'pŏn') *n.* [Fr. < OFr., of Germanic orig.] A plug of absorbent material inserted into a bodily cavity or wound to stop a flow of blood or absorb secretions. —*vt.* **-poned, -pon·ing, -pons.** To plug or stop with a tampon.

**tam-tam** (tŭm'tŭm', tăm'tăm') *n.* [Hindi *ṭamṭam.*] **1.** One of a set of tuned gongs used in a gamelan orchestra. **2.** A tom-tom.

**tan** (tăn) *v.* **tanned, tan·ning, tans.** [ME *tannen* < OFr. *tanner* < Med. Lat. *tannare* < *tannum,* tanbark, prob. of Celt. orig.] —*vt.* **1.** To convert (hide) into leather, as by treating with tannin. **2.** To make brown by exposure to sun. **3.** *Informal.* To thrash or spank. —*vi.* To become brown or tawny from exposure to sun. —*n.* **1.** A light or moderate yellowish brown to brownish orange. **2.** The brown color sun rays impart to the skin. **3.** Tanbark. **4.** Tannin or a solution derived from it. —*adj.* **tan·ner, tan·nest. 1.** Being of the color tan. **2.** Having a suntan. **3.** Used in or relating to tanning.

**Tan** (tăn) *n., pl.* **Tan** *or* **Tans.** The Tanka.

**tan·a·ger** (tăn'ĭ-jər) *n.* [NLat. *tanagra* < Port. *tangará* < Tupi.] A small New World bird of the family Thraupidae, often with brightly colored plumage in the male.

**tan·bark** (tăn'bärk') *n.* **1.** The bark of various trees, used as a source of tannin. **2.** Shredded bark from which the tannin has been extracted, used to cover a surface, as a circus arena or a racetrack.

**tan·dem** (tăn'dəm) *n.* [< Lat., at last, at length. ] **1.** A two-wheeled carriage drawn by horses harnessed one behind the other. **2.** A team of carriage horses harnessed in single file. **3.** A tandem bicycle. **4.** An arrangement of two or more persons or objects placed one behind the other. —*adv.* One behind the other <ride *tandem*> **—in tandem. 1.** In a tandem order. **2.** In a tandem relationship. **—tan'dem** *adj.*

**tandem bicycle** *n.* A bicycle for two or more persons sitting tandem.

**tang¹** (tăng) *n.* [ME *tonge,* of Scand. orig.] **1.** A sharp, distinctive flavor, taste, or odor. **2.** A distinctive quality. **3.** A trace, hint, or smattering. **4. a.** A sharp point, shank, tongue, or prong. **b.** A projection by which a tool, as a chisel or knife, is attached to its handle or stock. —*vt.* **tanged, tang·ing, tangs.** To furnish with a tang. **—tang'y** *adj.*

**tang²** (tăng) *n.* [Imit.] A loud ringing sound : TWANG. —*vi.* & *vt.* **tanged, tang·ing, tangs.** To twang or cause to twang.

**tan·ge·lo** (tăn'jə-lō') *n., pl.* **-los.** [Blend of TANGERINE and POMELO.] **1.** A hybrid citrus tree that is a cross between certain varieties of grapefruit and tangerine. **2.** The fruit of the tangelo, with an acid orange pulp.

**tan·gent** (tăn'jənt) *adj.* [Lat. *tangens, tangent-,* pr.part. of *tangere,* to touch.] **1. a.** Making contact at a single point or along a line. **b.** Touching but not intersecting. **2.** Diverging from an original purpose or the matter at hand. —*n.* **1.** A line, curve, or surface touching but not intersecting another line, curve, or surface. **2.** The trigonometric function of an acute angle in a right triangle that is the ratio of the length of the side opposite the angle to the length of the side adjacent to the angle. **3.** A sudden digression or change of course <The minister went off on a *tangent* while giving the sermon.> **—tan'gen·ce** (-jəns), **tan'gen·cy** (tăn'jən-sē) *n.*

**tan·gen·tial** (tăn-jĕn'shəl) *also* **tan·gen·tal** (-jĕn'tl) *adj.* **1.** Of, relating to, or moving along or in the direction of a tangent. **2.** Merely touching or slightly connected. **3.** Superficially relevant : DIVERGENT <*tangential* remarks> **—tan·gen'ti·al'i·ty** (-shē-ăl'ĭ-tē) *n.* **—tan·gen'tial·ly** *adv.*

**tangent plane** *n.* The plane containing all the lines tangent to a specified point on a surface.

**tan·ger·ine** (tăn'jə-rēn', tăn'jə-rēn') *n.* [< Fr. *Tanger,* Tangier, Morocco.] **1. a.** A widely cultivated citrus tree, *Citrus nobilis deliciosa,* yielding edible fruit with an easily peeled deep-orange skin and sweet juicy pulp. **b.** The fruit of the tangerine. **2.** A strong reddish orange to strong or vivid orange. **—tan'ger·ine'** *adj.*

**tan·gi·ble** (tăn'jə-bəl) *adj.* [LLat. *tangibilis* < Lat. *tangere,* to touch.] **1. a.** Discernible by the touch or capable of being touched.

**b.** Capable of being treated as fact : REAL <*tangible* evidence> **2.** Capable of being understood or realized <a *tangible* benefit> **3.** *Law.* Capable of being valued monetarily, as land or securities <*tangible* property> —*n.* **1.** Something palpable or concrete. **2. tangibles.** Material assets. **—tan'gi·bil'i·ty, tan'gi·ble·ness** *n.* **—tan'gi·bly** *adv.*

**tan·gle[1]** (tăng'gəl) *v.* **-gled, -gling, -gles.** [ME *tangilen,* to involve in embarrassment.] —*vt.* **1.** To mix together or intertwine in a confused mass : SNARL. **2.** To involve in awkward or hampering complications : ENTANGLE. **3.** To trap or ensnare. —*vi.* To be or become entangled. —*n.* **1.** A confused intertwined mass. **2.** A confused or jumbled condition. **3.** Bewilderment. **4.** *Informal.* An argument. **—tangle with.** *Informal.* To come to blows with <*tangled with* local street gangs> **—tan'gly** *adj.*

**tan·gle[2]** (tăng'gəl) *n.* [Of Scand. orig.] A large seaweed of the genus *Laminaria.*

**tan·go** (tăng'gō) *n., pl.* **-gos.** [Am. Sp., poss. of African orig.] **1.** A Latin-American ballroom dance in 2/4 or 4/4 time. **2.** The music for the tango. —*vi.* **-goed, -go·ing, -gos.** To dance the tango.

**tang·o·re·cep·tor** (tăng'gō-rĭ-sĕp'tər) *n.* [Lat. *tangere,* to touch + RECEPTOR.] A cutaneous receptor that responds to touch and pressure.

**tan·gram** (tăng'grəm) *n.* [Perh. Chin. *tang*[2], a Chinese dynasty + -GRAM.] A Chinese puzzle made of a square cut into five triangles, a square, and a rhomboid, to be reassembled into different figures.

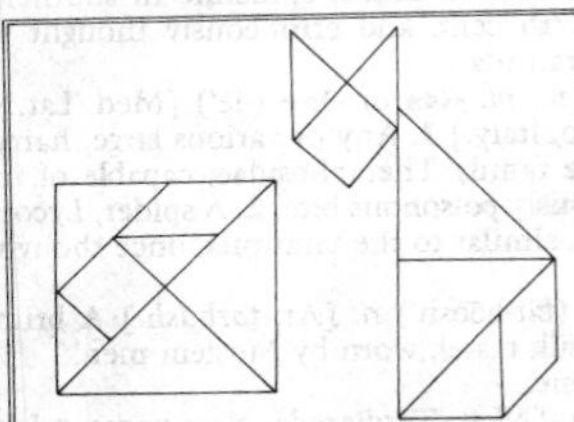

**tangram**

**tan·ist** (tăn'ĭst, thô'nĭst) *n.* [Ir. Gael. *tānaiste* < OIr. *tānaise.*] The heir apparent to an ancient Celtic chief, elected during the chief's lifetime.

**tan·ist·ry** (tăn'ĭ-strē, thô'nĭ-) *n.* The system of electing a tanist.

**tank** (tăngk) *n.* [Port. *tanque,* pond < Lat. *stagnum.*] **1.** A large container for holding or storing gases or fluids. **2.** The amount a tank holds. **3.** A pool, pond, reservoir, or cistern, esp. one for drinking water or irrigation. **4.** A heavily armored closed combat vehicle mounted with cannon and guns and that moves on caterpillar treads. **5.** *Slang.* A jail or jail cell. —*vt.* **tanked, tank·ing, tanks.** To place, store, or process in a tank. **—tank up.** *Slang.* **1.** To drink alcoholic beverages to excess. **2.** To be or become intoxicated.

**tan·ka[1]** (tăng'kə) *n.* [J.] A Japanese verse form in five lines, the first and third composed of five syllables and the rest of seven.

**tan·ka[2]** (tăng'kə) *n.* [Tibetan *taṅka.*] A Tibetan religious painting, usu. mounted on fabric.

**Tan·ka** (tăng'kä) *n., pl.* **Tanka** or **-kas.** [Cantonese.] A people in southern China and Hong Kong who live on small boats clustered in colonies.

**tank·age** (tăng'kĭj) *n.* **1. a.** The capacity or contents of a tank. **b.** The act or process of putting or storing in a tank. **c.** The fee for tankage. **2.** Animal residues left after rendering fat in a slaughterhouse and used for fertilizer or feed.

**tank·ard** (tăng'kərd) *n.* [ME.] A large drinking cup having a single handle and often a hinged cover, esp. a tall silver or pewter mug.

**tanked** (tăngkt) *adj. Slang.* Drunk.

**tank·er** (tăng'kər) *n.* **1.** A ship, plane, or truck constructed to transport liquids, as oil, in bulk. **2.** A tank crew member.

**tank suit** *n.* A simply designed one-piece swimsuit with shoulder straps.

**tank top** *n.* [From its resemblance to a TANK SUIT.] A casual sleeveless shirt with usu. wide shoulder straps and no front opening.

**tank town** *n.* [So called because trains would stop there only to replenish water.] A small or unimportant town.

**tank trailer** *n.* A truck trailer equipped as a tanker.

**tan·nage** (tăn'ĭj) *n.* **1.** The act, process, or skill of tanning. **2.** Something tanned.

**tan·nate** (tăn'āt') *n.* [TANN(IN) + -ATE.] A salt or ester of tannic acid.

**tan·ner[1]** (tăn'ər) *n.* One who tans hides.

**tan·ner[2]** (tăn'ər) *n.* [Orig. unknown.] *Chiefly Brit. Slang.* A sixpence.

**tan·nery** (tăn'ə-rē) *n., pl.* **-ies.** A place where hides are tanned.

**tan·nic** (tăn'ĭk) *adj.* [TANN(IN) + IC.] Relating to or derived from tannin.

**tannic acid** *n.* A lustrous yellowish to light-brown amorphous, powdered, flaked, or spongy mass having the approximate composition $C_{76}H_{52}O_{46}$, obtained from the bark and fruit of many plants and used in tanning, as a mordant, to clarify wine and beer, and as an astringent and styptic.

**tan·nif·er·ous** (tăn-ĭf'ər-əs) *adj.* Having or yielding tannin.

**tan·nin** (tăn'ĭn) *n.* [Fr. < *tanner,* to tan < OFr.] **1.** Tannic acid. **2.** A chemical substance capable of promoting tanning.

**tan·ning** (tăn'ĭng) *n.* **1.** The art or process of making leather from rawhides. **2.** The browning of the skin by exposure to sun and weather. **3.** *Informal.* A beating, whipping, or spanking.

**Ta·no·an** (tä'nō-ən) *n.* [< *Tano,* an American Indian people.] An American Indian language family of New Mexico and northeastern Arizona. **—Ta'no·an** *adj.*

**tan·rec** (tăn'rĕk') *n. var. of* TENREC.

**tan rot** *n.* A disease of the strawberry plant caused by the fungus *Pezizella lythri* and marked by the formation of tan indentations on the surface of the fruit.

**tan·sy** (tăn'zē) *n., pl.* **-sies.** [ME < OFr. *tanesie* < Med. Lat. *athanasia* < Gk., immortality : *a-,* without + *thanatos,* death.] A plant of the genus *Tanacetum,* esp. *T. vulgare* native to the Old World, with buttonlike yellow flower clusters and aromatic, pungent juice occas. used medicinally and as a flavoring.

**tan·tal·ic** (tăn-tăl'ĭk) *adj.* Of, relating to, or containing tantalum.

**tan·ta·lite** (tăn'tə-līt') *n.* [Swed. *tantalit* < *tantalum,* tantalum.] A black to red-brown mineral, essentially $(Fe,Mn)(Ta,Nb)_2O_6$, distinguished from columbite by the predominance of tantalum over niobium and used as an ore of both elements.

**tan·ta·lize** (tăn'tə-līz') *vt.* **-lized, -liz·ing, -liz·es.** [After *Tantalus.*] To excite (another) by exposing something desirable while keeping it out of reach : TEASE. **—tan'ta·li·za'tion** *n.* **—tan'ta·liz'er** *n.* **—tan'ta·liz'ing·ly** *adv.*

☆ ***syns:*** TANTALIZE, BAIT, TEASE, TORMENT *v. core meaning* : to excite (another) by exposing something desirable while keeping it out of reach <*tantalized* the hungry dog with pieces of raw meat>

▲ word history: The religion of ancient Greece and Rome was swept away by Christianity, but its gods and heroes live on. *Tantalize,* for example, is derived from *Tantalus,* a legendary king of Lydia. He stole the food of the gods and gave it to mortals, a crime punishable by death. Tantalus, however, could not die, because he had eaten the divine food and become immortal. The gods instead condemned him to suffer eternal hunger and thirst in the presence of food and drink that remained just out of his reach. The verb *tantalize* means basically "to suffer the torments of Tantalus," although in a much milder form.

**tan·ta·lum** (tăn'tə-ləm) *n.* [After *Tantalus,* from its non-absorbent quality.] *Symbol* **Ta** A very hard, heavy gray metallic element used to make electric-light-bulb filaments, lightning arresters, nuclear reactor parts, and some surgical instruments; atomic number 73; atomic weight 180.948.

**Tan·ta·lus** (tăn'tə-ləs) *n.* [Lat. < Gk. *Tantalos.*] **1.** *Gk. Myth.* A king who for his crimes was condemned in Hades to stand in water that receded when he tried to drink, and with fruit hanging above him that receded when he reached for it. **2. tantalus.** A locked-up stand for displaying decanters.

**tan·ta·mount** (tăn'tə-mount') *adj.* [< obs. *tantamount,* an equivalent < AN *tant amunter,* to amount to as much.] Equivalent in significance, effect, or value <a statement *tantamount* to a confession of murder>

**tan·ta·ra** (tăn-tăr'ə, tär'ə) *n.* [Imit.] **1. a.** A trumpet or horn fanfare. **b.** A sound resembling a tantara. **2.** A hunting cry.

**tan·tiv·y** (tăn-tĭv'ē) *adv.* [Orig. unknown.] At full gallop : at top speed. —*n., pl.* **-ies. 1.** A hunting cry. **2.** Full gallop : top speed.

**tan·tra** (tŭn'trə) *n.* [Skt. *tantram* < *tanoti,* he weaves.] One of a class of Hindu or Buddhist religious writings concerned with mysticism and magic. **—tan'tric** (-trĭk) *adj.*

**tan·trum** (tăn'trəm) *n.* [Orig. unknown.] A fit of temper.

**tan·yard** (tăn'yärd') *n.* A section in a tannery for the tanning vats.

**tan·zan·ite** (tăn'zə-nīt') *n.* [After *Tanzania.*] A hydrated calcium aluminum silicate mineral, exhibiting blue, violet, or greenish color depending on polarization of incident light, and used as a gem.

**Tao·ism** (tou'ĭz'əm, dou'-) *n.* [< Chin. (Mandarin) *dao*[4], way.] A principal Chinese philosophy and religion based on the teachings of Lao-tse in the 6th cent. B.C. **—Tao'ist** *n.* **—Tao·is'tic** *adj.*

**tap[1]** (tăp) *v.* **tapped, tap·ping, taps.** [ME *tappen,* poss. < OFr. *taper.*] —*vt.* **1.** To strike gently with a light blow or series of blows <*tapped* me on the shoulder> **2.** To give a light rap with <*tap* a pen on a desk> **3.** To produce with a succession of light blows <*tap* out a fast rhythm> **4. a.** To repair (shoe heels or toes) by applying a tap. **b.** To attach metal taps to. —*vi.* **1.** To deliver a light gentle blow or series of blows. **2.** To walk making light clicks. —*n.* **1. a.** A gentle blow. **b.** The sound made by such a blow. **2. a.** A thin layer of material, as leather, applied to a worn-down shoe. **b.** A metal plate attached to the toe or heel of a shoe, as for tap-dancing.

**tap[2]** (tăp) *n.* [ME < OE *tæppa.*] **1.** A valve and spout for regulating delivery of a fluid at the end of a pipe. **2.** A plug for a bunghole : SPIGOT. **3. a.** Liquor drawn from a tap. **b.** Liquor of a particular brew, cask, or quality. **4.** *Med.* Removal of bodily fluid <a spinal *tap*> **5.** A tool for cutting an internal screw thread. **6.** A makeshift terminal in an electric circuit. —*vt.* **tapped, tap·ping, taps. 1.** To provide with a tap or spigot. **2.** To pierce so as to draw off liquid <*tap* maple trees> **3.** To draw (liquid) from a container. **4.** *Med.* To withdraw fluid from (a bodily cavity). **5.** To make a connection with or open outlets from <*tap* a gas main> **6. a.** To wiretap (a telephone). **b.** To establish an electric connection in (a power line), as to divert current secretly. **7.** To cut screw threads in (a collar or socket). **8.** *Informal.* To ask (someone) for money. **—on tap. 1.** In a tapped cask and ready to be drawn <beer *on tap*> **2.** Available for immediate use <extra workers *on tap*>

**ta·pa** (tä′pə, tăp′ə) *n.* [Marquesan and Tahitian.] **1.** The inner bark of the paper mulberry. **2.** A paperlike cloth made in the Pacific islands by pounding tapa or similar bark.

**tap dance** *n.* A dance in which the rhythm is sounded out by the clicking heels and toes of a dancer's shoes. **—tap dancer** *n.*

**tap-dance** (tăp′dăns′) *vi.* **-danced, -danc·ing, -danc·es.** To perform a tap dance.

**tape** (tāp) *n.* [ME < OE *tæppe.*] **1.** A narrow strip of strong woven fabric, as that used in bookbinding or sewing. **2.** A narrow, flexible, continuous strip of material, esp.: **a.** Adhesive tape. **b.** Magnetic tape. **c.** A tape measure. **3.** A string stretched across the finish line of a racetrack to be broken by the winner. **4.** A tape recording. —*v.* **taped, tap·ing, tapes.** —*vt.* **1. a.** To fasten, strengthen, or wrap with tape. **b.** To bind together (sections of a book) by applying strips of tape. **2.** To measure with a tape measure. **3.** To record on magnetic tape. —*vi.* **1.** To measure. **2.** To record something on magnetic tape.

**tape cartridge** *n.* **1.** A cartridge containing an endless loop of magnetic tape and designed for automatic use on insertion into a tape recorder or player designed to receive it. **2.** CASSETTE 2a.

**tape deck** *n.* A tape recorder and player with no built-in amplifiers or speakers, used as a component in a high-fidelity sound system.

**tape grass** *n.* An aquatic plant, *Vallisneria spiralis,* with long grasslike submerged leaves.

**tape·line** (tāp′līn′) *n.* A tape measure.

**tape measure** *n.* A tape marked off in a linear scale, as inches or centimeters, for taking measurements.

**tape player** *n.* A self-contained machine for playing back recorded magnetic tapes.

**ta·per** (tā′pər) *n.* [ME < OE *tapor,* poss. < Lat. *papyrus,* papyrus.] **1.** A small or very slender candle. **2.** A long wax-coated wick used to light gas lamps or candles. **3.** Something that gives off a feeble light. **4.** A gradual decrease in width or thickness of an elongated object. —*v.* **-pered, -per·ing, -pers.** —*vi.* **1.** To become gradually thinner or narrower toward one end. **2.** To become gradually smaller or less <The thunderstorm *tapered* off.> —*vt.* **1.** To make narrower or thinner at one end. **2.** To diminish or make smaller gradually. —*adj.* Gradually decreasing in size toward a point. **—ta′per·ing·ly** *adv.*

**tape-re·cord** (tāp′rĭ-kôrd′) *vt.* **-cord·ed, -cord·ing, -cords.** To record on magnetic tape.

**tape recorder** *n.* A machine used for recording sound on magnetic tape and usu. for playing back the sound so recorded.

**tape recording** *n.* **1. a.** Magnetized tape on which sound has been recorded. **b.** The sound recorded on a magnetic tape. **2.** The act of recording on magnetic tape.

**tap·es·try** (tăp′ĭ-strē) *n., pl.* **-tries.** [ME *tapstery* < OFr. *tapisserie* < *tappisser,* to cover with carpet < *tapis,* carpet < Gk. *tapēs.*] **1.** A heavy cloth woven with rich, complex, often varicolored designs or scenes, usu. hung on walls for decoration and sometimes used to cover furniture. **2.** Something resembling a tapestry, as in complexity or elegance of design. —*vt.* **-tried, -try·ing, -tries. 1.** To hang or decorate with tapestry. **2.** To make, weave, or depict in a tapestry.

**ta·pe·tum** (tə-pē′təm) *n., pl.* **-ta** (-tə) [NLat. < Lat. *tapete,* carpet < Gk. *tapēs.*] **1.** *Bot.* A layer of nutritive cells within the sporangium of ferns and related plants or within the anther of seed plants. **2.** *Anat.* A membranous region or layer, esp. in the choroid coat or retina. **3.** A stratum of fibers of the corpus callosum.

**tape·worm** (tāp′wûrm′) *n.* Any of various ribbonlike, often very long flatworms of the class Cestoda, that are parasitic in the intestines of vertebrates, including humans.

**tap house** *n.* A tavern or bar.

**tap·i·o·ca** (tăp′ē-ō′kə) *n.* [Port. and Sp., both < Guarani *tipiog.*] A beady starch obtained from the root of the cassava, used for puddings and as a thickening agent in cooking.

**ta·pir** (tā′pər, tə-pîr′) *n.* [NLat. *Tapirus,* genus name < Tupi *tapira,* tapir.] An ungulate mammal of the genus *Tapirus* of tropical America or southern Asia, with a heavy body, short legs, and a fleshy proboscis.

**tap·per** (tăp′ər) *n.* One that taps.

**tap·pet** (tăp′ĭt) *n.* [< TAP[1].] A projecting arm or lever that moves or is moved by contact with another part, usu. to communicate a certain motion, as between a driving mechanism and a valve.

**tap·pit-hen** (tăp′ĭt-hĕn′) *n.* [Sc. *tappit,* crested + HEN.] *Scot.* **1.** A crested hen. **2.** A large mug with a knobbed lid.

**tap·room** (tăp′ro͞om′, -ro͝om′) *n.* A bar or barroom.

**tap·root** (tăp′ro͞ot′, -ro͝ot′) *n.* The main root of a plant, usu. stouter than the lateral roots and growing straight downward from the stem.

**taps** (tăps) *pl.n.* [Perh. alteration of obs. *taptoo,* tattoo. —see TATTOO[1].] *(sing. in number).* A bugle call or a drum signal sounded at night as an order to put out lights and at military funerals and memorial services.

**tap·ster** (tăp′stər) *n.* A person who draws and serves liquor for customers.

**Ta·pu·ya** (tä-po͞o′yə) *n., pl.* **Tapuya** or **-yas.** [Tupi *Tapua.*] A Tapuyan Indian.

**Ta·pu·yan** (tä-po͞o′yən) *n.* A South American Indian linguistic stock of Brazil. **—Ta·pu′yan** *adj.*

**tar[1]** (tär) *n.* [ME *taar* < OE *teru.*] **1.** A dark, oily, viscid mixture, consisting mainly of hydrocarbons, produced by the destructive distillation of organic substances such as wood, coal, or peat. **2.** Coal tar. —*vt.* **tarred, tar·ring, tars.** To cover with tar. **—tar and feather. 1.** To punish (someone) by covering with tar and feathers. **2.** *Informal.* To criticize severely **:** EXCORIATE.

**tar[2]** (tär) *n.* [Short for TARPAULIN.] *Informal.* A sailor.

**Tar·a·ca·hi·tian** (tăr′ə-kə-hē′shən) *adj.* [Blend of *Tarahumara* and *Cahita,* two peoples of Mexico.] Of, relating to, or constituting a language family of the Uto-Aztecan group.

**tar·a·did·dle** (tăr′ə-dĭd′l) *var. of* TARRADIDDLE.

**tar·an·tel·la** (tăr′ən-tĕl′ə) *n.* [Ital., after *Taranto,* Italy.] **1.** A lively, whirling southern Italian dance once thought to be a remedy for tarantism. **2.** The music for the tarantella, in 6/8 time.

**tar·an·tism** (tăr′ən-tĭz′əm) *n.* [After *Taranto,* Italy.] A malady marked by an uncontrollable urge to dance, epidemic in southern Italy from the 15th to the 17th cent. and erroneously thought to result from the bite of the tarantula.

**ta·ran·tu·la** (tə-răn′chə-lə) *n., pl.* **-las** or **-lae** (-lē′) [Med. Lat. < OItal. *tarantola,* after *Taranto,* Italy.] **1.** Any of various large, hairy, chiefly tropical spiders of the family Theraphosidae, capable of inflicting a painful but not seriously poisonous bite. **2.** A spider, *Lycosa tarentula* of southern Europe, similar to the tarantula, once thought to cause tarantism.

**tar·boosh** *also* **tar·bush** (tär-bo͞osh′) *n.* [Ar. *ṭarbūsh.*] A brimless, usu. red felt cap with a silk tassel, worn by Moslem men.

**tar camphor** *n.* Naphthalene.

**tar·di·grade** (tär′dĭ-grād′) *n.* [NLat. *Tardigrada,* class name < Lat. *tardigradus,* slow-moving : *tardus,* slow + *gradi,* to go.] Any of various slow-moving minute arthropods of the class Tardigrada, with eight legs and living in water or damp moss. —*adj.* **1.** Of or belonging to the Tardigrada. **2.** Moving slowly.

**tar·dy** (tär′dē) *adj.* **-di·er, -di·est.** [ME *tardyve,* slow < OFr. *tardif* < Lat. *tardus.*] **1.** Happening, arriving, or acting later than expected or scheduled. **2.** Moving slowly. **—tar′di·ly** *adv.* **—tar′di·ness** *n.*

**tare[1]** (târ) *n.* [ME.] **1.** The common vetch, *Vicia sativa.* **2.** Any of several weedy plants that grow in grain fields. **3. tares.** An undesirable or bad element that endangers the well-being of what is good or beneficial.

**tare[2]** (târ) *n.* [ME < OFr. < OSp. *tara* < Ar. *ṭarḥah,* that which is thrown away < *ṭaraḥa,* he rejected.] **1.** The weight of a container or wrapper deducted from the gross weight to obtain net weight. **2.** A deduction from gross weight made to allow for the weight of a container. **3.** *Chem.* A counterbalance, esp. an empty vessel used to counterbalance the weight of a similar container. —*vt.* **tared, tar·ing, tares.** To determine, allow for, or indicate the tare of.

**targe** (tärj) *n.* [ME < OFr. —see TARGET.] *Archaic.* A light shield or buckler.

**tar·get** (tär′gĭt) *n.* [ME, small targe < OFr. *targette,* dim. of *targe,* light shield, of Germanic orig.] **1.** An object, as a padded disc, with a marked surface, that is shot at to test accuracy in rifle or archery practice. **2.** Something aimed or fired at. **3. a.** An object of attack or criticism. **b.** Something to be acted on with a view to transforming it. **4.** A desired goal. **5.** A railroad signal that indicates the position of a switch by its color, position, and shape. **6.** The sliding sight on a surveyor's leveling rod. **7.** A small round shield. **8. a.** A structure in a camera tube with a storage surface that is scanned by an electron beam to generate a signal output current similar to the charge-density pattern stored on the surface. **b.** A usu. metal part in an x-ray tube on which a beam of electrons is focused and from which x-rays are emitted. —*vt.* **-get·ed, -get·ing, -gets. 1.** To make a target of. **2.** To aim at or for. **3.** To establish as a goal.

**target date** *n.* A date established for the completion of a project or the launch of an operation.

**Tar·gum** (tär′go͞om′, -go͞om′) *n.* [Heb. *targūm,* interpretation < *targēm,* to interpret.] Any of several Aramaic translations or paraphrasings of the Old Testament.

**Tar Heel** or **Tar·heel** (tär′hēl′) *n.* A native or resident of North Carolina.

**tar·iff** (tăr′ĭf) *n.* [Ital. *tariffa* < Ar. *ta'rīf,* notification < *'arafa,* he made known.] **1.** A list or system of duties imposed by a government on imported or exported goods. **2.** A duty imposed by a government on imported or exported goods. **3.** A schedule of prices or fees. —*vt.* **-iffed, -iff·ing, -iffs.** To fix a duty or price on.

**tar·la·tan** *also* **tar·le·tan** (tär′lə-tən) *n.* [Fr. *tarlatane.*] A thin, stiffly starched open-weave muslin fabric.
**tar·mac** (tär′măk′) *n.* [Orig. a trademark.] **1.** A bituminous substance used as a binder in paving. **2.** A tarmacadam pavement, runway, or road.
**tar·mac·ad·am** (tär′mə-kăd′əm) *n.* A pavement of layered crushed stone with a tar binder pressed to a smooth surface.
**tarn** (tärn) *n.* [ME *tarne,* of Scand. orig.] A small mountain lake.
**†tar·nal** (tär′nəl) *adj.* & *adv.* [Alteration of ETERNAL.] *Regional.* Damned. **—tar′nal·ly** *adv.*
**†tar·na·tion** (tär-nā′shən) *n.* & *interj. Regional.* Damnation.
**tar·nish** (tär′nĭsh) *v.* **-nished, -nish·ing, -nish·es.** [OFr. *ternir, terniss-,* to dull.] —*vt.* **1.** To dull the luster of : discolor, esp. by exposure to air or dirt. **2.** To spoil or detract from : TAINT <gossip that *tarnished* my reputation> —*vi.* **1.** To lose luster. **2.** To diminish or become tainted. —*n.* **1.** The state of being tarnished. **2.** A changed or discolored luster. **3.** The state of being spoiled or tainted. **—tar′nish·a·ble** *adj.*
**ta·ro** (tär′ō, tăr′ō) *n., pl.* **-ros.** [Of Polynesian orig.] **1.** A widely cultivated tropical plant, *Colocasia esculenta,* with broad leaves and a large, starchy, edible rootstock. **2.** The rootstock of the taro.
**tar·ok** *also* **tar·oc** (tăr′ək) *n.* [Ital. *tarrochi,* pl. of *tarocco,* tarot.] A card game developed in Italy in the 14th cent., played with a 78-card pack consisting of four suits plus the 22 tarot cards as trumps.
**tar·ot** (tăr′ō) *n.* [OFr. < OItal. *tarocco.*] **1.** Any of a set of 22 playing cards consisting of a joker plus 21 cards depicting vices, virtues, and elemental forces, used in fortunetelling and as trump in tarok games. **2. tarots.** Tarok.
**tarp** (tärp) *n.* [Short for TARPAULIN.] A tarpaulin.
**tar·pa·per** (tär′pā′pər) *n.* Heavy paper impregnated or coated with tar, used as a waterproof protective material in building.
**tar·pau·lin** (tär-pô′lĭn, tär′pə-lĭn) *n.* [Obs. *tarpawling* : TAR¹ + PALL¹.] **1.** Waterproof material, as canvas, used to cover and protect things from moisture. **2.** A sheet of tarpaulin.
**tar·pon** (tär′pən) *n., pl.* **tarpon** *or* **-pons.** [Orig. unknown.] A fish of the family Elopidae or Megalopidae, esp. a large silvery game fish, *Megalops atlantica* of Atlantic coastal waters.

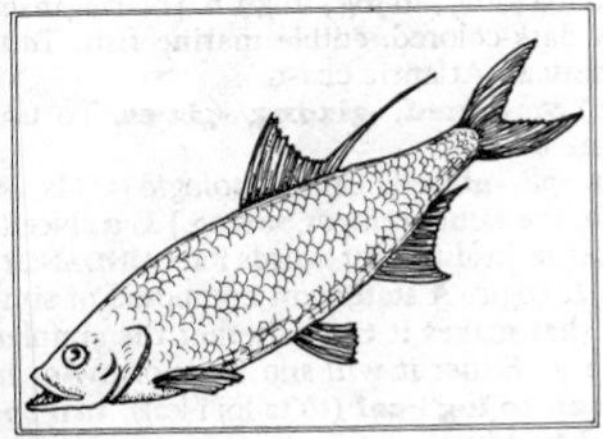
**tarpon**
*Size varies greatly, some to 6½ feet long*

**tar·ra·did·dle** *also* **tar·a·did·dle** (tăr′ə-dĭd′l) *n.* [Orig. unknown.] *Informal.* A petty falsehood : FIB.
**tar·ra·gon** (tăr′ə-gŏn′, -gən) *n.* [Med. Lat. *tragonia* < Med. Gk. *tarkhōn,* prob. < Ar. *ṭarkhūn.*] **1.** An aromatic herb, *Artemisia dracunculus,* native to Eurasia. **2.** The leaves of the tarragon, used as seasoning.
**tar·ri·ance** (tăr′ē-əns) *n. Archaic.* **1.** The act of tarrying. **2.** A temporary stay : SOJOURN.
**tar·ry¹** (tăr′ē) *v.* **-ried, -ry·ing, -ries.** [ME *tarien.*] —*vi.* **1.** To delay or be late in going or coming : LINGER. **2.** To wait. **3.** To remain or stay temporarily, as in a place : SOJOURN. —*vt. Archaic.* To await. **—tar′ri·er** *n.* **—tar′ry** *n.*
**tar·ry²** (tär′ē) *adj.* **-ri·er, -ri·est.** Of, suggesting, or coated with tar.
**tar·sal** (tär′səl) *adj.* [NLat. *tarsalis* < Gk. *tarsos,* ankle.] **1.** Of, relating to, or situated near the tarsus of the foot. **2.** Of or relating to the tarsus of the eyelid.
**tarsal gland** *n.* Any of the branched sebaceous glands located in the tarsus of the eyelid.
**tarsal plate** *n.* TARSUS 2.
**tar·si** (tär′sī′) *n. pl. of* TARSUS.
**tar·si·er** (tär′sē-ər, -sē-ā′) *n.* [Fr. < *tarse,* tarsus < NLat. *tarsus.*] Any of several small nocturnal primates of the genus *Tarsius* of the East Indies, with a long tail and large round eyes.
**tar·so·met·a·tar·sus** (tär′sō-mĕt′ə-tär′səs) *n., pl.* **-si** (-sī′). A compound bone between the tibia and the toes of a bird's leg, formed by fusion of the tarsal and metatarsal bones.
**tar·sus** (tär′səs) *n., pl.* **-si** (-sī′) [NLat. < Gk. *tarsos,* ankle.] **1. a.** The section of the vertebrate foot between the leg and the metatarsus. **b.** The seven bones composing this section. **2.** A fibrous plate supporting and shaping the edge of the eyelid. **3.** *Zool.* **a.** The tarsometatarsus. **b.** The distal segmented structure on the leg of an insect or an arachnid.

**tart¹** (tärt) *adj.* **-er, -est.** [ME < OE *teart,* severe.] **1.** Having a sharp pungent taste. **2.** Sharp or bitter in tone or meaning : CUTTING. **—tart′ly** *adv.* **—tart′ness** *n.*
**tart²** (tärt) *n.* [ME *tarte* < OFr.] **1.** A small open pie with a sweet filling. **2.** A loose woman : PROSTITUTE. **—tart up.** *Chiefly Brit.* To dress up or decorate in a tawdry garish way.
**tar·tan¹** (tär′tn) *n.* [Poss. < OFr. *tertaine,* linsey-woolsey.] **1. a.** Any of many textile patterns consisting of stripes of varying widths and colors crossed at right angles against a solid background, each forming a distinctive design worn by the members of a Scottish clan. **b.** A twilled wool fabric or garment having a tartan pattern. **2.** A plaid fabric. **—tar′tan** *adj.*
**tar·tan²** (tär′tn, tär-tăn′) *n.* [Fr. *tartane* < OItal. *tatana.*] A small single-masted Mediterranean ship with a large lateen sail.
**tar·tar** (tär′tər) *n.* [ME *tartre* < OFr. < Med. Lat. *tartarum* < Med. Gk. *tartaron.*] **1.** A reddish acid compound, chiefly potassium bitartrate, found in the juice of grapes and deposited on the sides of casks during wine-making. **2.** A hard yellowish deposit on the teeth, consisting of organic secretions and food particles deposited in various salts, as calcium carbonate.
**Tar·tar** (tär′tər) *also* **Ta·tar** (tä′tər) *n.* [ME *Tartre* < OFr. *Tartare* < Med. Lat. *Tartarus,* alteration of Pers. *Tātār,* of Turkic orig.] **1.** A member of any of the Mongolian peoples of central Asia who invaded western Asia and eastern Europe in the 13th cent. **2.** A descendant of these Mongolian peoples. **3.** *often* **tartar** *or* **tatar.** A violent or ferocious person. **4.** Any of the Turkic languages of the Tartars.
**tartar emetic** *n.* A poisonous crystalline compound, $KSbOC_4H_4O_6 \cdot 1/2H_2O$, used in the medical treatment of amoebiasis.
**tar·tar·e·ous** (tär-târ′ē-əs) *adj.* Consisting of or similar to tartar.
**tar·tare steak** (tär′tər) *n.* Steak tartare.
**tar·tar·ic** (tär-tăr′ĭk) *adj.* Of, pertaining to, or derived from tartar or tartaric acid.
**tartaric acid** *n.* Any of four isomeric crystalline organic compounds, $C_4H_6O_6$, used in the manufacture of cream of tartar, as a sequestrant, in tanning, and in effervescent beverages, baking powders, and photographic chemicals.
**tar·tar·ize** (tär′tə-rīz′) *vt.* **-ized, -iz·ing, -iz·es.** To treat, impregnate, or combine with tartar, tartar emetic, or cream of tartar. **—tar′tar·i·za′tion** *n.*
**tar·tar·ous** (tär′tər-əs) *adj.* Made up of, derived from, or containing tartar.
**tar·tar sauce** *also* **tar·tare sauce** (tär′tər) *n.* Mayonnaise mixed with chopped onion, olives, pickles, and capers and served as a sauce with fish.
**Tar·ta·rus** (tär′tər-əs) *n.* [Lat. < Gk. *Tartaros.*] **1.** *Gk. Myth.* The abysmal regions below Hades where the Titans were confined. **2.** A hellish place. **—Tar·tar′e·an** (-târ′ē-ən) *adj.*
**tar·trate** (tär′trāt′) *n.* A salt or ester of tartaric acid.
**tar·trat·ed** (tär′trā′tĭd) *adj.* Containing, mixed with, or derived from tartaric acid.
**tar·tuffe** *also* **tar·tufe** (tär-to͝of′, -to͞of′) *n.* [After the protagonist of *Tartuffe,* a play by Molière (1622–1673).] **1.** One who affects religious piety. **2.** A hypocrite.
**tar·weed** (tär′wēd′) *n.* **1.** Any of several strong-smelling, resinous western American plants of the genus *Madia,* with rayed, yellow flowers. **2.** Any of several plants related or similar to the tarweed.
**task** (tăsk) *n.* [ME *taske,* imposed work < ONFr. *tasque* < Med. Lat. *tasca* < *taxare,* to tax.] **1.** A piece of work assigned or carried out as part of one's duties. **2.** A difficult, tedious undertaking. **3.** A function to be performed : OBJECTIVE. —*vt.* **tasked, task·ing, tasks. 1.** To assign a task to. **2.** To overburden with labor : TAX. **—take to task.** To reprimand.
☆ ***syns:*** TASK, CHORE, DUTY, JOB, STINT *n. core meaning* : a piece of work that has been assigned <editorial *tasks* such as copy-editing and proofreading>
**task force** *n.* A temporary grouping of forces and resources, esp. of military units, for the accomplishment of an objective.
**task·mas·ter** (tăsk′măs′tər) *n.* One who imposes work, esp. heavy or burdensome labor.
**Tas·ma·ni·an devil** (tăz-mā′nē-ən, -mān′yən) *n.* A burrowing carnivorous marsupial, *Sarcophilus harrisii* of Tasmania, with a blackish coat and a long, almost hairless tail.

**Tasmanian devil**
*Approximately 3 feet long*

**Tasmanian wolf** *n.* The thylacine.
**tasse** (tăs) *also* **tas·set** (tăs′ĭt) *n.* [Poss. < OFr., pouch.] One of a

series of jointed overlapping metal splints hanging from the corselet, used as armor for the lower trunk and thighs.

**tas·sel** (tăs′əl) *n.* [ME < OFr. < VLat. **tascellus* < Lat. *taxillus,* small die.] **1.** A bunch of loose cords or threads bound at one end and hanging free at the other, used as an ornament, as on curtains or clothing. **2.** Something, as the pollen-bearing inflorescence of a corn plant, that resembles a tassel. —*v.* **-seled, -sel·ing, -sels** *or* **-selled, -sel·ling, -sels.** —*vt.* To fringe or decorate with tassels. —*vi.* To put forth a tassellike inflorescence <corn *tasseling* in the garden>

**tas·set** (tăs′ĭt) *n. var. of* TASSE.

**taste** (tāst) *v.* **tast·ed, tast·ing, tastes.** [ME *tasten* < OFr. *taster* < VLat. **taxitare,* freq. of Lat. *taxare,* to touch.] —*vt.* **1.** To ascertain the flavor of by taking into the mouth <*tasted* the soup and then seasoned it> **2.** To eat or drink a small quantity of. **3.** To experience, esp. for the first time <finally *tasted* victory> **4.** *Archaic.* To appreciate : enjoy. —*vi.* **1.** To ascertain flavors in the mouth. **2.** To have a distinct flavor <The medicine *tastes* bitter.> **3.** To eat or drink a small amount. **4.** To have an experience : PARTAKE. —*n.* **1. a.** The sense that ascertains the sweet, sour, salty, and bitter qualities of substances coming in contact with the taste buds on the tongue. **b.** This sense in combination with the senses of smell and touch, which together receive a sensation of a substance in the mouth. **2. a.** The sensation of sweet, sour, salty, or bitter qualities produced by or as if by a substance placed in the mouth. **b.** The unified sensation produced by any of these qualities plus a distinct smell and texture : FLAVOR. **3.** The act of tasting. **4.** A small quantity tasted or eaten. **5.** A limited or first experience : SAMPLE <recruits who got a *taste* of war> **6.** A personal preference or liking <a *taste* for the mysterious> **7. a.** The faculty of discerning what is aesthetically excellent or appropriate. **b.** A manner indicative of the quality of such discernment <decorated with superb *taste*> **8. a.** The sense of what is proper, fitting, or least likely to offend in a given social situation. **b.** A manner indicative of the quality of this sense. **9.** *Obs.* The act of testing : TRIAL. **—tast′a·ble** *adj.*

**taste bud** *n.* Any of numerous spherical or ovoid nests of cells that are distributed over the tongue and embedded in the epithelium, consisting of gustatory cells and supporting cells and constituting the end organs of the sense of taste.

**taste·ful** (tāst′fəl) *adj.* **1.** Displaying good taste. **2.** Tasty. **—taste′ful·ly** *adv.* **—taste′ful·ness** *n.*

**taste·less** (tāst′lĭs) *adj.* **1.** Displaying poor taste. **2.** Lacking flavor : INSIPID. **—taste′less·ly** *adv.* **—taste′less·ness** *n.*

**tast·er** (tā′stər) *n.* **1.** One who tastes, esp. one who samples a food or beverage for quality. **2.** A device or implement used in tasting.

**tast·y** (tā′stē) *adj.* **-i·er, -i·est. 1.** Having a pleasing flavor : SAVORY. **2.** Having good taste : TASTEFUL. **—tast′i·ly** *adv.* **—tast′i·ness** *n.*

**tat** (tăt) *vi.* & *vt.* **tat·ted, tat·ting, tats.** [Prob. back-formation < TATTING.] To make or produce by tatting. **—tat′ter** *n.*

**ta·ta·mi** (tä-tä′mē, tə-) *n.* [J.] Straw matting used as a floor covering esp. in Japan.

**Ta·tar** (tä′tər) *n. var. of* TARTAR.

**†ta·ter** (tā′tər) *n.* [Shortening and alteration of POTATO.] *Regional.* A potato.

**tat·ter** (tăt′ər) *n.* [ME *tater,* of Scand. orig.] **1.** A torn and hanging piece of cloth : SHRED. **2. tatters.** Torn and ragged clothing : RAGS. —*vt.* & *vi.* **-tered, -ter·ing, -ters.** To make or become ragged.

**tat·ter·de·mal·ion** (tăt′ər-dĭ-māl′yən, -mā′lē-ən) *n.* [Orig. unknown.] One wearing ragged or tattered clothing : RAGAMUFFIN. —*adj.* Being in a very shabby, tattered state.

☆ ***syns:*** TATTERDEMALION, RAGGED, RAGGEDY, TATTERED *adj. core meaning* : being in a very shabby condition <the *tatterdemalion* attire of an old beggar>

**tat·tered** (tăt′ərd) *adj.* **1.** Torn into shreds or tatters : RAGGED. **2.** Dressed in ragged clothes.

**tat·ter·sall** *also* **Tat·ter·sall** (tăt′ər-sôl′, -səl) [After *Tattersall's* horse market, London, England.] —*adj.* Having a pattern of dark lines forming squares on a solid, gen. light background. —*n.* **1.** A pattern of dark lines forming squares on a light background. **2.** Fabric woven in a tattersall pattern.

**tat·ting** (tăt′ĭng) *n.* [Orig. unknown.] **1.** Handmade lace worked by looping and knotting a single strand of heavy-duty thread on a small hand shuttle. **2.** The act or art of making tatting.

**tat·tle** (tăt′l) *v.* **-tled, -tling, -tles.** [ME *tattlen,* to stammer, of Flem. orig.] —*vi.* **1.** To reveal the plans or activities of another : GOSSIP. **2.** To chatter aimlessly. —*vt.* To reveal through gossiping. —*n.* **1.** Aimless chatter : PRATTLE. **2.** Gossip. **3.** A tattletale.

**tat·tler** (tăt′lər) *n.* **1.** One who tattles. **2.** A shore bird related to and resembling the sandpiper, esp. one of the genus *Heteroscelus* of Pacific coastal areas.

**tat·tle·tale** (tăt′l-tāl′) *n.* One who tattles on others : INFORMER. —*adj.* Revealing : telltale.

**tat·too¹** (tă-to͞o′) *n., pl.* **-toos.** [Obs. *taptoo* < Du. *taptoe* : *tap,* spigot, tap + *toe,* shut.] **1.** A signal sounded on a drum or bugle to summon military personnel to their quarters at night. **2.** A display of military exercises offered as evening entertainment. **3.** A continuous even drumming or rapping <the *tattoo* of rain on the roof> —*v.* **-tooed, -too·ing, -toos.** —*vi.* To beat out an even rhythm, as with the fingers. —*vt.* To beat or tap rhythmically on. **—tat·too′er** *n.*

**tat·too²** (tă-to͞o′) *n., pl.* **-toos.** [Of Polynesian orig.] A permanent design or mark made on the skin by pricking it and ingraining in it an indelible pigment or by raising scars on it. —*vt.* **-tooed, -too·ing, -toos. 1.** To mark (the skin) with a tattoo. **2.** To form (a tattoo) on the skin. **—tat·too′er** *n.*

**tat·ty** (tăt′ē) *adj.* **-ti·er, -ti·est.** [Orig. unknown.] Shabby or frayed.

**tau** (tou, tô) *n.* [Gk., of Phoenician orig.; akin to Heb. *tāw,* tav.] The 19th letter of the Greek alphabet. —See table at ALPHABET.

**tau cross** *n.* A cross shaped like a T.

**taught** (tôt) *v. p.t.* & *p.p. of* TEACH.

**taunt¹** (tônt) *vt.* **taunt·ed, taunt·ing, taunts.** [Prob. < OFr. *tant pour tant,* so much for so much.] **1.** To deride or reproach contemptuously : MOCK. **2.** To incite (someone) to action by taunting. —*n.* A scornful remark : JEER. **—taunt′er** *n.* **—taunt′ing·ly** *adv.*

**taunt²** (tônt) *adj.* [Orig. unknown.] *Naut.* Unusually tall. —Used of masts.

**taupe** (tōp) *n.* [Fr., mole < Lat. *talpa.*] A brownish gray to a dark yellowish brown. **—taupe** *adj.*

**tau·rine¹** (tôr′īn′) *adj.* [Lat. *taurinus* < *taurus,* bull < Gk. *tauros.*] Of or pertaining to a bull.

**tau·rine²** (tôr′ēn′) *n.* [Gk. *tauros,* bull (from having been obtained first from ox bile) + -INE.] A crystalline amino acid, $C_2H_7NO_3S$, found in bile.

**tau·ro·cho·lic acid** (tôr′ō-kō′lĭk, -kŏl′ĭk) *n.* [Gk. *tauros,* bull (from having been obtained first from ox bile) + CHOLIC ACID.] A crystalline acid, $C_{26}H_{45}NO_7S$, occurring as a constituent of bile.

**Tau·rus** (tôr′əs) *n.* [ME < Lat., bull < Gk. *tauros.*] **1.** A constellation in the Northern Hemisphere. **2. a.** The second sign of the zodiac. **b.** One born under this sign.

**taut** (tôt) *adj.* **-er, -est.** [ME *toght,* distended.] **1.** Drawn or pulled tight : not slack. **2.** Strained or tense, as nerves. **3.** Kept in proper condition <runs a *taut* ship> **—taut′ly** *adv.* **—taut′ness** *n.*

**taut-** *pref. var. of* TAUTO-.

**taut·en** (tôt′n) *vt.* & *vi.* **-ened, -en·ing, -ens.** To make or become taut.

**tauto-** *or* **taut-** *pref.* [< Gk. *tauto,* the same, contraction of *to auto.*] Same : identical <*tautomerism*>

**tau·tog** *also* **tau·taug** (tô′tôg′, -tŏg′, tô-tôg′, -tŏg′) *n.* [Narraganset *tautauog,* plural of *taut.*] A dark-colored, edible marine fish, *Tautoga onitis* of the North American Atlantic coast.

**tau·tol·o·gize** (tô-tŏl′ə-jīz′) *vi.* **-gized, -giz·ing, -giz·es.** To use tautology. **—tau·tol′o·gist** *n.*

**tau·tol·o·gy** (tô-tŏl′ə-jē) *n., pl.* **-gies.** [LLat. *tautologia* < Gk. < *tautologos,* redundant : *tauto,* the same + *logos,* saying.] **1. a.** Needless repetition of the same sense in different words : REDUNDANCY. **b.** An instance of tautology. **2.** *Logic.* A statement composed of simpler statements in a fashion that makes it true whether the simpler statements are true or false; e.g., *Either it will snow tomorrow or it will not snow tomorrow.* **—tau′to·log′i·cal** (tô′tə-lŏj′ĭ-kəl), **tau′to·log′ic** *adj.* **—tau′to·log′i·cal·ly** *adv.*

**tau·tom·er·ism** (tô-tŏm′ə-rĭz′əm) *n.* [TAUTO- + (ISO)MERISM.] Chemical isomerism marked by relatively easy interconversion of isomeric forms in equilibrium. **—tau′to·mer** (tô′tə-mər) *n.* **—tau′to·mer′ic** (tô′tə-mĕr′ĭk) *adj.*

**tau·to·nym** (tô′tə-nĭm′) *n.* A taxonomic designation, as *Gorilla gorilla,* commonly used in zoology but no longer in botany, in which the genus and species names are the same. **—tau′to·nym′ic, tau·ton′y·mous** (tô-tŏn′ə-məs) *adj.* **—tau·ton′y·my** *n.*

**tav** *also* **taw** (täf, tôf) *n.* [Heb. *tāw.*] The 23rd letter of the Hebrew alphabet. —See table at ALPHABET.

**tav·ern** (tăv′ərn) *n.* [ME *taverne* < OFr. < Lat. *taberna.*] **1.** An establishment licensed to sell liquor and beer to be drunk on the premises : BAR. **2.** INN 1.

**taw¹** (tô) *vt.* **tawed, taw·ing, taws.** [ME *tawen* < OE *tawian.*] To convert (skin) into white leather by mineral tanning, as with alum and salt. **—taw′er** *n.*

**taw²** (tô) *n.* [Orig. unknown.] **1.** A large fancy marble used for shooting. **2.** The line from which a player shoots in marbles. **3.** A game of marbles. —*vi.* **tawed, taw·ing, taws.** To shoot a marble.

**taw³** (tô) *n. var. of* TAV.

**taw·dry** (tô′drē) *adj.* **-dri·er, -dri·est.** [< *tawdry lace,* lace necktie, after *St. Audrey* (d. 679).] Gaudy and cheap, esp. in appearance. —*n.* Vulgarly ornamental finery. **—taw′dri·ly** *adv.* **—taw′dri·ness** *n.*

▲ word history: The word *tawdry* is an alteration of the name *Saint Audrey,* an Anglo-Saxon princess who died of a throat ailment. A fair was held annually in her honor at which cloth neckbands or neckties, called *tawdry laces,* were sold. These must have been showy but inexpensive souvenirs, because the word *tawdry* became an adjective in its own right meaning "gaudy and cheap."

**taw·ny** (tô′nē) *n.* [ME < AN *tauné* < OFr. *tane* < *tan,* tan < Med. Lat. *tannum,* tanbark, prob. of Celt. orig.] A light brown to brownish orange. **—taw′ny** *adj.*

**tax** (tăks) *n.* [< ME *taxen,* to tax < OFr. *taxer* < Med. Lat. *taxare* < Lat., to handle, freq. of *tangere,* to touch.] **1.** A contribution for the support of a government required of persons, groups, or businesses within the domain of that government. **2.** A due or fee levied on the members of an organization to meet expenses. **3.** A burdensome or excessive demand : STRAIN <a *tax* on my patience> —*vt.* **taxed, tax·ing, tax·es. 1.** To place a tax on (income, property, or goods). **2.** To exact a tax from. **3.** *Law.* To assess (e.g., court costs). **4.** To make difficult or excessive demands on. **5.** To make a charge against : ACCUSE <*taxed* them with irresponsible behavior> **—tax'a·bil'i·ty, tax'a·ble·ness** *n.* **—tax'a·ble** *adj.* **—tax'er** *n.*

**tax-** *pref. var. of* TAXO-.

**ta·xa** (tăk'sə) *n. pl. of* TAXON.

**tax·a·tion** (tăk-sā'shən) *n.* **1. a.** Imposition of taxes. **b.** The fact of being taxed. **2.** An assessed amount of tax.

**tax-de·duct·i·ble** (tăks'dĭ-dŭk'tə-bəl) *adj.* Exempt from inclusion in one's taxable income <a *tax-deductible* contribution>

**tax·eme** (tăk'sēm') *n.* [TAX- + (PHON)EME.] A minimal linguistic feature, as the order or stress of words in a compound or phonemes in a word.

**tax·es** (tăk'sēz') *n. pl. of* TAXIS.

**tax evasion** *n.* Intentional avoidance of tax payment usu. by inaccurate declaration of taxable income.

**tax-ex·empt** (tăks'ĭg-zĕmpt') *adj.* Exempt from taxation, as the capital or income of a philanthropic organization.

**tax·i** (tăk'sē) *n., pl.* **-is** *or* **-ies.** [Short for TAXICAB.] A taxicab. —*v.* **tax·ied, taxi·ing** *or* **taxy·ing, tax·ies** or **tax·is.** —*vi.* **1.** To be transported by taxi. **2.** To move slowly on a runway or on the surface of the water before takeoff or after landing. —*vt.* **1.** To transport by taxi. **2.** To cause (an aircraft) to taxi.

**taxi-** *pref. var. of* TAXO-.

**tax·i·cab** (tăk'sē-kăb') *n.* [TAXI(METER) + CAB.] An automobile that carries passengers for a fare, usu. calculated by a taximeter.

**taxi dancer** *n.* A woman employed, as by a dance hall or nightclub, to dance with the patrons for a fee.

**tax·i·der·my** (tăk'sĭ-dûr'mē) *n.* The art or operation of preparing, stuffing, and mounting the skins of dead birds and animals for exhibition in a lifelike state. **—tax'i·der'mal, tax'i·der'mic** *adj.* **—tax'i·der'mist** *n.*

**tax·i·me·ter** (tăk'sē-mē'tər) *n.* [Fr. *taximètre* : *taxe,* charge (< OFr. *taxer,* to tax) + *-metre,* -meter.] An instrument in a taxicab for computing fares automatically.

**tax·ing** (tăk'sĭng) *adj.* Causing fatigue : BURDENSOME.

**tax·is** (tăk'sĭs) *n., pl.* **tax·es** (tăk'sēz'). [Gk., arrangement < *tattein,* to arrange.] **1.** *Biol.* Responsive movement of an organism toward or away from an external stimulus. **2.** Movement of an organ, as in a dislocation, into normal position by manipulation.

**-taxis** *suff.* [< Gk. *taxis,* arrangement < *tattein,* to arrange.] **1.** Order : arrangement <homo*taxis*> **2.** Taxis : responsive movement <chemo*taxis*>

**taxi stand** *n.* A reserved area where taxicabs park while waiting to be hired.

**taxo-** or **taxi-** or **tax-** *pref.* [< Gk. *taxis,* arrangement < *tattein,* to arrange.] Order : arrangement <*taxi*dermy>

**tax·on** (tăk'sŏn') *n., pl.* **ta·xa** (tăk'sə) [NLat., back-formation < TAXONOMY.] *Biol.* A group of organisms sharing common characteristics in varying degrees of distinction that constitute one of the categories in taxonomic classification, such as a phylum, order, family, genus, or species.

**tax·on·o·my** (tăk-sŏn'ə-mē) *n.* [Fr. *taxonomie* : Gk. *taxis,* arrangement + *-nomie,* -nomy.] **1.** The science, laws, or principles of classification. **2.** *Biol.* The theory, principles, and process of classifying organisms in categories. **—tax'o·nom'ic** (tăk'sə-nŏm'ĭk), **tax'o·nom'i·cal** *adj.* **—tax'o·nom'i·cal·ly** *adv.* **—tax·on'o·mist** *n.*

**tax·pay·er** (tăks'pā'ər) *n.* One that pays taxes.

**tax shelter** *n.* A financial operation, as the use of special depletion allowances, that reduces taxes on current earnings. **—tax'-shel'tered** *adj.*

**-taxy** *suff.* [Gk. *-taxia* < *tattein,* to arrange.] -TAXIS <antho*taxy*>

**Ta·yg·e·ta** (tā-ĭj'ĭ-tə) *n.* [Lat. *Taygete* < Gk. *Taugetē.*] **1.** *Gk. Myth.* One of the Pleiades. **2.** One of the six visible stars in the Pleiades cluster.

**Tay-Sachs disease** (tā'săks') *n.* [After Warren *Tay* (1843–1927) and Bernard *Sachs* (1858–1944).] A hereditary disease in which the absence of a specific enzyme causes the accumulation of certain lipids in nerve and brain cells resulting in mental retardation, convulsions, blindness, and death usu. in childhood.

**Tb** *symbol for* TERBIUM.

**T-bar lift** (tē'bär') *n.* A ski lift consisting of a bar suspended like an inverted T against which skiers lean while being towed uphill.

**T-bone** (tē'bōn') *n.* A thick porterhouse steak taken from the small end of the loin and containing a T-shaped bone.

**Tc** *symbol for* TECHNETIUM.

**T cell** *n.* A lymphocyte influenced by the thymus that functions in the defense against intracellular pathogens such as viruses and tubercle bacilli.

**Te** *symbol for* TELLURIUM.

**tea** (tē) *n.* [Du. *thee* < Malay *teh* < Chin. (Amoy) *te.*] **1.** A shrub, *Thea sinensis* or *Camellia sinensis* of eastern Asia, with fragrant white flowers and evergreen leaves. **2.** The dried leaves of the tea plant, prepared and used to make a hot beverage. **3.** An aromatic, slightly bitter beverage made by steeping tea leaves in boiling water. **4.** A beverage made, as by steeping the leaves of certain plants or by extracting an infusion esp. from beef. **5.** A plant whose leaves are used to make a tealike infusion. **6. a.** *Chiefly Brit.* An afternoon refreshment consisting usu. of sandwiches and cakes served with tea. **b.** A social gathering at which tea is taken. **7.** *Chiefly Brit.* High tea. **8.** *Slang.* Marijuana.

**tea bag** *n.* A small porous sack holding enough tea leaves to make an individual serving of tea.

**tea ball** *n.* A small perforated metal ball for holding tea leaves that are to be steeped in hot water.

**tea·ber·ry** (tē'bĕr'ē) *n., pl.* **-ries.** Wintergreen.

**tea biscuit** *n.* A plain cookie or biscuit often served with tea.

**tea·cart** (tē'kärt') *n.* A tea wagon.

**teach** (tēch) *v.* **taught** (tôt), **teach·ing, teach·es.** [ME *techen* < OE *tǣcan.*] —*vt.* **1.** To impart knowledge or skill to : INSTRUCT. **2.** To provide knowledge of : instruct in. **3.** To cause to learn by example or experience <The accident *taught* me to be more careful.> **4.** To advocate or preach <*teach* kindness and tolerance> —*vi.* To give instruction, esp. as an occupation. ***usage:*** Although some grammarians object to the use of *teach* as a transitive verb when the object denotes an institution of learning, as in *I teach elementary school,* this usage has wide currency at all levels, and should be regarded as completely acceptable.

☆ ***syns:*** TEACH, INSTRUCT, TRAIN, TUTOR *v. core meaning* : to impart knowledge or skill to <*teach* schoolchildren> TEACH is the most widely applicable, while INSTRUCT usu. suggests methodical direction in a specific subject or area <*instructing* students in English literature> TUTOR usu. refers to private instruction of one student or a small group <*tutoring* children in mathematics after school> TRAIN gen. implies concentration on particular skills intended to fit one for a desired role <*training* young musicians><a flight school that *trains* fighter pilots>

**teach·a·ble** (tē'chə-bəl) *adj.* **1.** Capable of being taught. **2.** Receptive to learning. **—teach'a·bil'i·ty, teach'a·ble·ness** *n.* **—teach'a·bly** *adv.*

**teach·er** (tē'chər) *n.* One who teaches, esp. one hired to teach.

**teachers college** *also* **teachers' college** *n.* A college with a special curriculum for training teachers.

**teacher's pet** *n.* **1.** A student in special favor with a teacher. **2.** One who has gained favor with an authority.

**teach-in** (tēch'ĭn') *n.* An extended session, as on a college or university campus, for lectures and discussions on an important and usu. controversial issue.

**teach·ing** (tē'chĭng) *n.* **1.** The work or occupation of teachers. **2.** A doctrine or precept <religious *teachings*>

**teaching fellow** *n.* A graduate student in a college or university who is awarded a fellowship that provides him or her with financial aid in exchange for teaching duties. **—teaching fellowship** *n.*

**teaching hospital** *n.* A hospital closely associated with a medical school and that serves as a practical educational setting for medical students, interns, and residents.

**teaching machine** *n.* A device designed to teach by presenting the student with a planned sequence of statements and questions and providing an immediate response to his or her answers.

**tea·cup** (tē'kŭp') *n.* A small cup for serving tea.

**tea·cup·ful** (tē'kŭp'fo͝ol') *n., pl.* **-fuls.** The amount a teacup will hold.

**tea dance** *n.* A late-afternoon dance.

**tea garden** *n.* A public garden where tea and light refreshments may be consumed.

**tea·house** (tē'hous') *n.* A public establishment serving tea and other refreshments.

**teak** (tēk) *n.* [Port. *teca* < Malayalam *tēkka.*] **1. a.** A tall evergreen tree, *Tectona grandis* of southeastern Asia, with hard, heavy, durable wood. **b.** The yellowish-brown hard wood of the teak, used for furniture and in shipbuilding. **2.** An olive gray or dark olive, to grayish yellowish brown or grayish to moderate brown. **—teak** *adj.*

**tea·ket·tle** (tē'kĕt'l) *n.* A kettle, usu. with a spout, used for boiling water for tea.

**teak·wood** (tēk'wo͝od') *n.* TEAK 1b.

**teal** (tēl) *n., pl.* **teal** or **teals.** [ME *tele.*] **1.** A small, widely distributed river duck, esp. one of the genus *Anas,* with bright plumage. **2.** A moderate or dark bluish green to greenish blue. **—teal** *adj.*

**team** (tēm) *n.* [ME< OE *tēam.*] **1. a.** Two or more draft animals used to pull a vehicle or a piece of farm machinery. **b.** A vehicle together with the animal or animals harnessed to it. **2.** A group of animals exhibited or performing together, as horses at an equestrian show. **3.** A group on the same side, as in a game. **4.** A group organized to work together <a *team* of scientists> **5.** A brood or flock. **6.** *Obs.*

Offspring: lineage. —*v.* **teamed, team·ing, teams.** —*vt.* **1.** To harness or join together so as to form a team. **2.** To haul or transport with a draft team. —*vi.* **1.** To form a team. **2.** To drive a team or truck.

**team·mate** (tēm′māt′) *n.* A fellow member of a team.

**team play** *n.* **1.** Collective play participated in by team members. **2.** Mutual cooperative effort. **—team player** *n.*

**team·ster** (tēm′stər) *n.* **1.** One who drives a team. **2.** A truck driver.

**team teaching** *n.* A method of classroom instruction in which several teachers are jointly responsible for teaching a single group of students. **—team-teach** (tēm′tēch′) *v.* **(-taught, -teach·ing, -teach·es).**

**team·work** (tēm′wûrk′) *n.* Cooperative effort by the members of a group or team to achieve a common goal.

**tea·pot** (tē′pŏt′) *n.* A covered pot with a spout in which tea is steeped and from which it is served.

**tea·poy** (tē′poi′) *n.* [Hindi *tipāī* : *tīn,* three + Pers. *pāī,* foot.] **1.** A small table for holding a tea service. **2.** A small decorative three-legged table.

**tear[1]** (târ) *v.* **tore** (tôr, tōr), **torn** (tôrn, tōrn), **tear·ing, tears.** [ME *teren* < OE *teran.*] —*vt.* **1.** To pull apart or into pieces : REND. **2.** To make (an opening) by ripping. **3.** To lacerate (e.g., the skin). **4.** To separate forcefully : WRENCH. **5.** To divide or disunite <*torn* between going and staying> —*vi.* **1.** To become torn. **2.** To move with heedless speed : rush headlong. **—tear around. 1.** To move about in excited, often angry haste. **2.** To lead a wild life. **—tear away.** To remove (as oneself) unwillingly or reluctantly. **—tear down. 1.** To demolish <*tear down* old buildings> **2.** To take apart : DISASSEMBLE <*tear down* an engine> **3.** To denigrate or vilify. **—tear into.** To attack with great violence or vigor. **—tear off.** To produce hurriedly <*tear off* a memo> **—tear up.** To tear to pieces. —*n.* **1.** An act of tearing. **2.** The result of tearing. **3.** A great rush : HURRY. **4.** *Slang.* A carousal or spree.

**tear[2]** (tîr) *n.* [ME *tere* < OE *tēar.*] **1.** A drop of the clear saline liquid that is secreted by the lachrymal gland of the eye and lubricates the surface between the eyeball and the eyelid. **2.** A drop of liquid or hardened fluid. **3. tears.** The act of weeping <The story left me in *tears.*> —*vi.* **teared, tear·ing, tears.** To fill with tears.

**tear·down** (târ′down′) *n.* The act or process of taking apart or demolishing.

**tear·drop** (tîr′drŏp′) *n.* **1.** A single tear. **2.** An object, as an earring, shaped like a tear.

**tear·ful** (tîr′fəl) *adj.* **1.** Filled with or accompanied by tears. **2.** So piteous as to excite tears. **—tear′ful·ly** *adv.* **—tear′ful·ness** *n.*

☆ ***syns:*** TEARFUL, LACHRYMOSE, TEARY, WEEPING, WEEPY *adj.* *core meaning* : filled with or accompanied by tears <a sad, *tearful* farewell>

**tear gas** (tîr) *n.* An agent that on dispersal, usu. from grenades or projectiles, irritates the eyes and causes blinding tears.

**tear-jerk·er** (tîr′jûr′kər) *n. Slang.* A pathetic story, drama, or performance apt to make one weep.

**tea·room** (tē′ro͞om′, -ro͝om′) *n.* An establishment serving tea and other refreshments.

**tea rose** *n.* **1.** Any of several cultivated roses derived from *Rosa odorata,* with fragrant yellowish or pink flowers. **2.** A pale to strong yellowish pink. **—tea′-rose′** *adj.*

**tear sheet** (târ) *n.* A page torn from a publication and used chiefly as evidence to an advertiser of the insertion of an advertisement.

**tease** (tēz) *v.* **teased, teas·ing, teas·es.** [ME *tesen,* to comb apart < OE *tǣsan.*] —*vt.* **1.** To annoy or vex. **2.** To make fun of playfully. **3.** To tantalize. **4. a.** To coax. **b.** To gain by persistent coaxing. **5.** To cut (e.g., tissue) into pieces for examination. **6.** To disentangle and dress the fibers of (e.g., wool). **7.** To raise the nap of (cloth) by dressing, as with a fuller's teasel. **8.** To ruffle (the hair) by combing from the ends toward the scalp for a bouffant effect. —*vi.* To annoy or make fun of someone persistently. —*n.* **1.** The act of teasing. **2.** One that teases, as: **a.** One given to playful mocking. **b.** A coquettish woman. **c.** A preliminary remark or action designed to stimulate the curiosity. **—teas′er** *n.* **—teas′ing·ly** *adv.*

**tea·sel** (tē′zəl) *n.* [ME *tesel* < OE *tǣsel.*] **1.** Any of several plants of the genus *Dipsacus,* native to the Old World, with thistlelike flowers surrounded by prickly bracts. **2. a.** The bristly flower head of *D. fullonum,* used to produce a napped surface on fabrics. **b.** A wire device used similarly. —*vt.* **-seled** *or* **-selled, -sel·ing** *or* **-sel·ling, -sels.** To produce a napped surface on (a fabric).

**tea service** *n.* The articles, such as cups and a teapot, used in serving tea.

**tea·shop** (tē′shŏp′) *n.* **1.** A tearoom. **2.** *Chiefly Brit.* A lunchroom.

**tea·spoon** (tē′spo͞on′) *n.* **1.** A small spoon used esp. with tea, coffee, and desserts. **2.** A household cooking measure equal to approx. 13 milliliters or 1/3 tablespoon.

**tea·spoon·ful** (tē′spo͞on′fo͝ol′) *n., pl.* **-fuls.** The amount a teaspoon holds.

**teat** (tēt, tĭt) *n.* [ME *tete* < OFr., of Germanic orig.] A mammary gland or nipple. **—teat′ed** *adj.*

**tea wagon** *n.* A small table on wheels for serving tea or holding dishes.

**tech·ne·ti·um** (tĕk-nē′shē-əm, -shəm) *n.* [< Gk. *tekhnētos,* artificial < *teknasthai,* to make by art < *tekhnē,* art.] *Symbol* **Tc** A silvery-gray, radioactive metallic element, used as a tracer and to eliminate corrosion in steel; atomic number 43; longest-lived isotope Tc 97.

**tech·nic** (tĕk′nĭk) *n.* [<Gk. *tekhnikos,* of art < *tekhnē,* art.] **1. technics.** *(sing. or pl. in number).* The theory, principles, or study of an art or process. **2. technics.** *(sing. or pl. in number).* Technical details, rules, or methods. **3.** *var. of* TECHNIQUE 2. —*adj.* Technical.

**tech·ni·cal** (tĕk′nĭ-kəl) *adj.* [< Gk. *tekhnikos,* of art < *tekhnē,* art.] **1.** Of, relating to, or derived from technique. **2.** Specialized <a *technical* school> **3. a.** Theoretical or abstract <a *technical* analysis> **b.** Scientific. **4.** According to principle, esp. formal rather than practical <a *technical* advantage> **5.** Industrial and mechanical : TECHNOLOGICAL. **6.** Designating or relating to a stock market in which prices are determined or affected by internal manipulation and speculation. **—tech′ni·cal·ly** *adv.* **—tech′ni·cal·ness** *n.*

**tech·ni·cal·i·ty** (tĕk′nĭ-kăl′ĭ-tē) *n., pl.* **-ties. 1.** The quality or state of being technical. **2.** Something meaningful or relevant only to a specialist <was acquitted on a legal *technicality*>

**technical knockout** *n.* A victory in boxing, with immediate termination of the match, awarded by the referee when it appears that one fighter is too badly beaten to continue.

**tech·ni·cian** (tĕk-nĭsh′ən) *n.* An expert in a technique, as: **a.** One whose occupation requires training in a specific technical process <an x-ray *technician*> **b.** One who is noted for skill in an artistic or intellectual technique.

**Tech·ni·col·or** (tĕk′nĭ-kŭl′ər). A trademark for a motion-picture color process.

**tech·nique** (tĕk-nēk′) *n.* [Fr. < *technique,* technical < Gk. *tekhnikos.*] **1.** The systematic procedure by which a complex or scientific task is accomplished. **2.** *also* **tech·nic** (tĕk′nĭk). The degree of skill or command of fundamentals exhibited in a performance.

**tech·noc·ra·cy** (tĕk-nŏk′rə-sē) *n., pl.* **-cies.** [Gk. *tekhnē,* skill + -CRACY.] A government and social system controlled by scientific technicians. **—tech′no·crat′** (-nə-krăt′) *n.* **—tech′no·crat′ic** *adj.*

**tech·no·log·i·cal** (tĕk′nə-lŏj′ĭ-kəl) *also* **tech·no·log·ic** (-lŏj′ĭk) *adj.* **1.** Relating to or involving technology, esp. scientific technology. **2.** Resulting from or affected by scientific and industrial progress. **—tech′no·log′i·cal·ly** *adv.*

**tech·nol·o·gize** (tĕk-nŏl′ə-jīz′) *vt.* **-gized, -giz·ing, -giz·es.** To make technological.

**tech·nol·o·gy** (tĕk-nŏl′ə-jē) *n., pl.* **-gies.** [Gk. *tekhnē,* skill + -LOGY.] **1. a.** The application of science esp. to industrial or commercial objectives. **b.** The whole body of methods and materials used to achieve such objectives. **2.** The body of knowledge available to a civilization that is of use in fashioning implements, practicing manual arts and skills, and extracting or collecting materials. **—tech·nol′o·gist** *n.*

**tech·no·struc·ture** (tĕk′nō-strŭk′chər) *n.* [TECHNO(LOGY) + STRUCTURE.] **1.** A large-scale corporate system. **2.** A network of skilled professionals who control a technostructure.

**tech·y** (tĕch′ē) *adj. var. of* TETCHY.

**tec·ton·ic** (tĕk-tŏn′ĭk) *adj.* [LLat. *tectonicus* < Gk. *tektonikos* < *tektōn,* builder.] **1.** Relating to construction. **2.** Architectural. **3.** *Geol.* Relating to, causing, or resulting from structural deformation in the earth's crust. **—tec·ton′i·cal·ly** *adv.*

**tec·ton·ics** (tĕk-tŏn′ĭks) *n. (sing. in number).* **1.** The art or science of construction, esp. of large buildings. **2.** The geology of the earth's structural deformation.

**tec·ton·ism** (tĕk′tə-nĭz′əm) *n.* [TECTON(IC) + -ISM.] Diastrophism.

**tec·trix** (tĕk′trĭks) *n., pl.* **-tri·ces** (-trĭ-sēz′) [NLat. < fem. of Lat. *tector,* plasterer < *tegere,* to cover.] *often* **tectrices.** One of the coverts of a bird's wing.

**tec·tum** (tĕk′təm) *n., pl.* **-ta** (-tə) [NLat. < Lat., roof < neuter p.part. of *tegere,* to cover.] A rooflike bodily structure, esp. the dorsal part of the midbrain. **—tec′tal** *adj.*

**ted** (tĕd) *vt.* **ted·ded, ted·ding, teds.** [ME *tedden.*] To strew or spread (e.g., newly mown grass) for drying. **—ted′der** *n.*

**ted·dy** (tĕd′ē) *n., pl.* **-dies.** [Orig. unknown.] A woman's undergarment combining a camisole top and loose-fitting panties.

**teddy bear** *also* **Teddy bear** *n.* [After *Teddy,* nickname of Theodore Roosevelt (1858–1919), who was depicted in a cartoon sparing the life of a bear cub.] A child's toy bear, usu. stuffed with soft material and covered with furlike plush.

**Teddy boy** *n.* [< *Teddy,* nickname for *Edward.*] A tough British youth wearing a modified style of Edwardian clothes.

**Te De·um** (tā′ dā′əm, tē′ dē′əm) *n., pl.* **Te De·ums.** [< Lat. *Te Deum (laudamus),* You, God, (we praise), the opening words of the hymn.] A hymn of praise sung as part of a liturgy.

**te·di·ous** (tē′dē-əs) *adj.* [ME < LLat. *taediosus* < Lat. *taedium,* tedium.] **1.** Tiresome or boring due to extreme length or slowness : WEARISOME <a *tedious* lecture> **2.** *Obs.* Moving or progressing very slowly. **—te'di·ous·ly** *adv.* **—te'di·ous·ness** *n.*

**te·di·um** (tē'dē-əm) *n.* [Lat. *taedium* < *taedēre,* to weary.] The quality or state of being tedious.

**tee¹** (tē) *n.* The letter *t.*

**tee²** (tē) *n.* [Orig. unknown.] **1.** A small peg with a concave top for holding a golf ball for an initial drive. **2.** The designated area from which a golfer makes a first stroke. *—vt.* **teed, tee·ing, tees.** To place (a golf ball) on a tee. **—tee off. 1.** To drive a golf ball from the tee. **2.** *Slang.* To begin <They *teed off* the fund-raising campaign with a cocktail party.> **3.** *Slang.* To make or become angry. **—tee up.** To place or set up a golf ball for driving.

**tee³** (tē) *n.* [Orig. unknown.] A mark aimed at in some games, as curling or quoits. **—to a tee.** Perfectly : exactly <The suit fits *to a tee.*>

**teem¹** (tēm) *v.* **teemed, teem·ing, teems.** [ME *temen,* to give birth to < OE *tīeman.*] *—vi.* **1.** To be full of : SWARM. **2.** *Obs.* To produce young : BEAR. *—vt. Archaic.* To give birth to. **—teem'er** *n.*

☆ **syns:** TEEM, ABOUND, BRISTLE, CRAWL, FLOW, PULLULATE, SWARM *v. core meaning* : to be so full of as to overflow with <a drop of water *teeming* with microorganisms>

**teem²** (tēm) *vt.* **teemed, teem·ing, teems.** [ME *temen* < ON *tōma.*] To pour out or empty <*teemed* molten ore into a huge mold>

**teen¹** (tēn) *n.* A teen-ager. *—adj.* Teen-age.

**teen²** (tēn) *n.* [ME *tene* < OE *tēona.*] *Archaic.* Injury : grief.

**teen-age** (tēn'āj') *also* **teen-aged** (-ājd') *adj.* Of, relating to, or descriptive of those aged 13 through 19.

**teen-ag·er** (tēn'ā'jər) *n.* One between the ages of 13 and 19.

**teens** (tēnz) *pl.n.* **1.** The numbers 13 through 19. **2.** The years of one's age between 13 and 19.

**tee·ny** (tē'nē) *also* **teen·sy** (tēn'sē) *adj.* **-ni·er, -ni·est** *also* **-si·er, -si·est.** [Alteration of TINY.] Tiny.

**teen·y·bop·per** (tē'nē-bŏp'ər) *n.* [TEEN¹ + -Y¹ + BOP¹ + -ER¹.] *Slang.* A usu. young teen-ager who follows the latest fad or craze, as in dress or music.

**tee·pee** (tē'pē) *n. var. of* TEPEE.

**tee shirt** *n. var. of* T-SHIRT.

**tee·ter** (tē'tər) *v.* **-tered, -ter·ing, -ters.** [ME *titeren,* prob. < ON *titra,* to shake.] *—vi.* **1.** To walk or move unsteadily : TOTTER. **2.** To vacillate. *—vt.* To cause to teeter or vacillate. *—n.* **1.** A seesaw. **2.** A teetering or seesaw motion.

**teeter board** *n.* A seesaw.

**tee·ter-tot·ter** (tē'tər-tŏt'ər) *n.* A seesaw.

**teeth** (tēth) *n. pl. of* TOOTH.

**teethe** (tēth) *vi.* **teethed, teeth·ing, teethes.** [ME *tethen* < *teeth,* pl. of *tooth,* tooth. —see TOOTH.] To cut teeth.

**teething ring** *n.* A ring of hard rubber or plastic upon which a teething baby can bite.

**tee·to·tal·er** *or* **tee·to·tal·ler** (tē'tōt'l-ər) *also* **tee·to·tal·ist** (-ĭst) *n.* [TEE¹ (pronunciation of the first letter in *total*) + *total (abstinence)* + -ER¹.] One who entirely abstains from alcoholic liquors. **—tee'to'tal** *adj.* **—tee'to'tal·ism** *n.*

▲ word history: The word *teetotaler* is derived from the phrase *teetotal abstainer,* which was coined by temperance workers in the 1830's. The syllable *tee-* is simply a spelling of the pronunciation of the letter T, which is the reduplication of the initial letter of *total. Teetotal* was therefore a catchy way of saying "absolutely total."

**tee·to·tum** (tē'tō'təm) *n.* [TEE¹ + Lat. *totum,* neuter sing. of *totus,* all (from the inscription of a *T* on the top meaning 'take all').] A top, usu. having four lettered sides, used to play various games of chance.

**Tef·lon** (tĕf'lŏn'). A trademark for a waxy, opaque material, polytetrafluoroethylene, used as a coating on cooking utensils and in industrial applications to prevent sticking.

**teg·men** (tĕg'mən) *n., pl.* **-mi·na** (-mə-nə) [NLat. < Lat., covering < *tegere,* to cover.] *Biol.* A covering or integument, as the tough, leathery forewing of certain insects or the inner coat of a seed.

**teg·men·tum** (tĕg-mĕn'təm) *n.* [NLat. < Lat., covering < *tegere,* to cover.] *Biol.* Tegmen. **—teg·men'tal** *adj.*

**te·gua** (tā'gwä, tā'wä) *n.* [Native word in Mexico.] An ankle-high moccasin worn by Mexicans and Indians.

**teg·u·lar** (tĕg'yə-lər) *also* **teg·u·lat·ed** (-lā'tĭd) *adj.* [< Lat. *tegula,* tile < *tegere,* to cover.] Relating to or resembling a tile. **—teg'u·lar·ly** *adv.*

**teg·u·ment** (tĕg'yə-mənt) *n.* [ME < Lat. *tegumentum* < *tegere,* to cover.] An outer covering : INTEGUMENT. **—teg'u·men'ta·ry** (-mĕn'tə-rē, -mĕn'trē), **teg'u·men'tal** (-mĕn'tl) *adj.*

**Te·huel·che** (tā-wĕl'chā) *n., pl.* **Tehuelche** *or* **-ches.** [Of Araucanian orig.] One of a nomadic Indian people of southern Argentina, virtually exterminated by the Spanish colonists. **—Te·huel'che·an** (-chē-ən) *adj.*

**teig·lach** (tāg'läKH', tĭg'-) *pl.n.* [Yiddish *teyglekh* < dim. of *teyg,* dough < MHG *teig* < OHG *teic.*] A confection consisting of bits of dough cooked briefly in a mixture of honey, brown sugar, and nuts, then cooled and rolled into balls.

**tek·tite** (tĕk'tīt') *n.* [Gk. *tēktos,* molten (< *tēkein,* to melt) + -ITE.] Any of numerous dark brown to green glass objects, gen. small and rounded, believed to be of extraterrestrial origin, found chiefly in Czechoslovakia, Australia, Indonesia, the Philippines, Texas, and Georgia, and having a largely silica composition with various oxides. **—tek·tit'ic** *adj.*

**tel-¹** *pref. var. of* TELE-.

**tel-²** *pref. var. of* TELO-.

**tel·aes·the·sia** (tĕl'ĭs-thē'zhə) *n. var. of* TELESTHESIA.

**tel·a·mon** (tĕl'ə-mŏn') *n., pl.* **tel·a·mon·es** (tĕl'ə-mō'nēz) [Lat. < Gk. *telamōn,* bearer.] A male figure used as a supporting pillar.

**tel·an·gi·ec·ta·sia** (tĕl-ăn'jē-ĕk-tā'zhə) *also* **tel·an·gi·ec·ta·sis** (-ĕk'tə-sĭs) *n.* [NLat. : TEL(O)- + Gk. *angos,* vessel + Gk. *ektasis,* expansion < *ekteinein,* to stretch out (*ek-,* out + *teinein,* to stretch).] A chronic dilation of capillaries of the blood vascular system causing dark-red blotches on the skin. **—tel·an'gi·ec·tat'ic** (-tăt'ĭk) *adj.*

**tele-** *or* **tel-** *pref.* [< Gk. *tēle,* at a distance.] **1.** Distance : distant <*tel*esthesia> **2. a.** Telegraph <*tele*gram> **b.** Television <*tele*cast>

**tel·e·cast** (tĕl'ə-kăst') *v.* **-cast** *or* **-cast·ed, -cast·ing, -casts.** *—vi.* To broadcast by television. *—vt.* To broadcast (a program) by television. *—n.* A television broadcast. **—tel'e·cast'er** *n.*

**tel·e·com·mu·ni·ca·tion** (tĕl'ə-kə-myōō'nĭ-kā'shən) *n.* **1.** *often* **telecommunications** *(sing. in number).* The science and technology of communication by electronic transmission of impulses, as by telegraphy, cable, telephony, radio, or television. **2.** A message transmitted by means of telecommunications.

**tel·e·course** (tĕl'ə-kôrs', -kōrs') *n.* A course of televised lectures, as offered by a university.

**tel·e·du** (tĕl'ə-dōō') *n.* [Malay *tēledu.*] A brownish-black carnivorous mammal, *Mydaus javanensis* of the East Indies, that can emit an offensive odor.

**tel·e·film** (tĕl'ə-fĭlm) *n.* A motion picture produced for television.

**tel·e·gen·ic** (tĕl'ə-jĕn'ĭk) *adj.* Presenting a pleasing appearance on television.

**te·leg·o·ny** (tə-lĕg'ə-nē) *n.* The supposed influence of one sire on offspring sired by subsequent males on the same female. **—tel'e·gon'ic** (tĕl'ə-gŏn'ĭk), **te·leg'o·nous** *adj.*

**tel·e·gram** (tĕl'ə-grăm') *n.* A message transmitted by telegraph.

**tel·e·graph** (tĕl'ə-grăf') *n.* **1.** A communication system that transmits and receives simple unmodulated electric impulses, esp. one in which the transmission and reception stations are directly connected by wires. **2.** A telegram. *—v.* **-graphed, -graph·ing, -graphs.** *—vt.* **1.** To transmit (a message) by telegraph. **2.** To send or convey a message to by telegraph. **3.** To make known, as an intended action, in advance or unwittingly. *—vi.* To send or transmit a telegram. **—te·leg'ra·pher** (tə-lĕg'rə-fər), **te·leg'ra·phist** *n.* **—te·leg'ra·phy** *n.*

**tel·e·graph·ic** (tĕl'ə-grăf'ĭk) *also* **tel·e·graph·i·cal** (-ĭ-kəl) *adj.* **1.** Relating to or transmitted by telegraph. **2.** Brief or concise like a telegram. **—tel'e·graph'i·cal·ly** *adv.*

**telegraph plant** *n.* A tropical Asiatic plant, *Desmodium motorium* or *D. gyrans,* with trifoliolate compound leaves, of which the lateral leaflets move or rotate.

**Tel·e·gu** *also* **Tel·u·gu** (tĕl'ə-gōō') *n., pl.* **Telegu** *also* **Telugu** *or* **-gus.** [Native word in India.] **1.** A Dravidian language spoken in Andhra Pradesh, India. **2.** A member of a Dravidian people who speak Telegu. **—Tel'e·gu** *adj.*

**tel·e·ki·ne·sis** (tĕl'ə-kĭ-nē'sĭs, -kī-) *n.* Movement of objects by scientifically inexplicable means, as by the exercise of mystical powers. **—tel'e·ki·net'ic** (-nĕt'ĭk) *adj.* **—tel'e·ki·net'i·cal·ly** *adv.*

**Te·lem·a·chus** (tə-lĕm'ə-kəs) *n.* [Lat. < Gk. *Tēlemakhos.*] *Gk. Myth.* The son of Odysseus and Penelope who helped his father kill Penelope's suitors.

**tel·e·mark** *also* **Tel·e·mark** (tĕl'ə-märk') *n.* [Norw., after *Telemark,* a region in Norway.] A turn or stop in skiing performed by shifting the weight forward on the ski that will be on the outside of the turn and pulling its tip gradually inward.

**tel·e·me·ter** (tĕl'ə-mē'tər, tə-lĕm'ĭ-tər) *n.* A measuring device used in telemetry. *—vt.* (tĕl'ə-mē'tər) **-tered, -ter·ing, -ters.** To measure and transmit (data) automatically from a distant source, as from a spacecraft or electric power grid, to a receiving station for recording or display. **—tel'e·met'ric** (tĕl'ə-mĕt'rĭk), **tel'e·met'ri·cal** *adj.* **—tel'e·met'ri·cal·ly** *adv.*

**te·lem·e·try** (tə-lĕm'ĭ-trē) *n.* The science and technology of automatic data measurement and transmission, as by wire or radio, from remote sources, such as space vehicles, to a receiving station for recording and analysis. **—tel'e·met'ric** *adj.* **—tel'e·met'ri·cal·ly** *adv.*

**tel·en·ceph·a·lon** (tĕl'ĕn-sĕf'ə-lŏn', -lən) *n. Anat.* The anterior portion of the forebrain, including the cerebral cortex and related parts. **—tel'en·ce·phal'ic** (-sə-făl'ĭk) *adj.*

**tel·e·ol·o·gy** (tĕl'ē-ŏl'ə-jē, tē'lē-) *n., pl.* **-gies. 1.** The philosophical study of design or purpose in natural phenomena. **2.** The use of ultimate purpose or design as a means of explaining natural phenomena. **—tel'e·o·log'i·cal** (-ə-lŏj'ĭ-kəl), **tel'e·o·log'ic** *adj.* **—tel'e·o·log'i·cal·ly** *adv.* **—tel'e·ol'o·gist** *n.*

**tel·e·ost** (tĕl'ē-ŏst', tē'lē-) *also* **tel·e·os·te·an** (tĕl'ē-ŏs'tē-ən, tē'lē-) [< NLat. *Teleostei,* group name (Gk. *teleos,* complete + *osteon,* bone) and NLat. *Teleostomi,* group name (Gk. *teleos,* complete + *stoma,* mouth).] *—adj.* Of or belonging to the Teleostei or Teleostomi, a group consisting of fishes with bony skeletons and rayed fins. *—n.* A teleost fish.

**te·lep·a·thy** (tə-lĕp'ə-thē) *n.* Communication through means other than the senses, as by the exercise of mystical powers. **—tel'e·path'ic** (tĕl'ə-păth'ĭk) *adj.* **—tel'e·path'i·cal·ly** *adv.* **—te·lep'a·thist** *n.*

**tel·e·phone** (tĕl'ə-fōn') *n.* An instrument that directly modulates carrier waves with voice or other acoustic source signals to be transmitted to remote locations and that directly reconverts received waves into audible signals, esp. such an instrument connected to others by wire. *—v.* **-phoned, -phon·ing, -phones.** *—vt.* **1.** To communicate with by telephone. **2.** To call (someone) on the telephone. **3.** To transmit by telephone. *—vi.* To communicate by telephone. **—tel'e·phon'er** *n.*

**telephone book** *n.* A directory containing the names of telephone subscribers with their addresses and telephone numbers.

**telephone booth** *n.* A small enclosure for a public telephone.

**telephone exchange** *n.* A central switching system that establishes connections between individual telephones.

**telephone receiver** *n.* The part of a telephone in which incoming electrical impulses are converted into sound.

**tel·e·phon·ic** (tĕl'ə-fŏn'ĭk) *adj.* **1.** Of or relating to a telephone. **2.** Transmitted by a telephone. **—tel'e·phon'i·cal·ly** *adv.*

**te·leph·o·ny** (tə-lĕf'ə-nē) *n.* **1.** Electrical transmission of sound between distant stations, esp. by radio or telephone. **2.** The technology and manufacture of telephone equipment.

**tel·e·pho·to** (tĕl'ə-fō'tō) *adj.* **1.** Of or relating to a photographic lens or lens system used to produce a large image of a distant object. **2.** TELEPHOTOGRAPH 1.

**tel·e·pho·to·graph** (tĕl'ə-fō'tə-grăf') *n.* **1.** A photograph made with a telephoto lens. **2.** A photograph transmitted and reproduced by telephotography. *—vt.* **-graphed, -graph·ing, -graphs. 1.** To photograph with a telephoto lens. **2.** To transmit by telephotography.

**tel·e·pho·tog·ra·phy** (tĕl'ə-fə-tŏg'rə-fē) *n.* **1.** The technique or process of photographing distant objects, using a telephoto lens on a camera. **2.** The process or technique of transmitting charts, pictures, and photographs over a distance. **—tel'e·pho'to·graph'ic** (-fō'tə-grăf'ĭk) *adj.*

**tel·e·play** (tĕl'ə-plā') *n.* A play written or adapted for television.

**tel·e·print·er** (tĕl'ə-prĭn'tər) *n.* A teletypewriter.

**tel·e·proc·ess·ing** (tĕl'ə-prŏs'ĕs'ĭng, -prō'sĕs'-) *n.* Computer service by means of terminals remote from the central computer.

**Tel·e·Promp·Ter** (tĕl'ə-prŏmp'tər). A trademark for a device used in television to show an actor or speaker an enlarged line-by-line reproduction of a script.

**tel·e·ran** (tĕl'ə-răn') *n.* [Orig. a trademark.] An air-traffic control system in which the image of a ground-based radar unit is televised to aircraft in the vicinity so that a pilot may see his or her position in relation to other aircraft.

**tel·e·scope** (tĕl'ə-skōp') *n.* [NLat. *telescopium* or Ital. *telescopio,* both < Gk. *tēleskopos,* far seeing : *tēle,* at a distance + *skopos,* watcher.] An instrument for collecting and examining electromagnetic radiation, esp.: **1.** An arrangement of lenses or mirrors or both that gathers visible light, allowing direct observation or photographic recording of distant objects. **2.** A device, as a radio telescope, used to detect and observe distant objects by their emission, transmission, reflection, or other interaction with invisible radiation. *—v.* **-scoped, -scop·ing, -scopes.** *—vt.* **1.** To cause to slide inward or outward in overlapping cylindrical sections. **2.** To make shorter or more precise : CONDENSE. *—vi.* To slide inward or outward in or as if in overlapping cylindrical sections.

**tel·e·scop·ic** (tĕl'ə-skŏp'ĭk) *adj.* **1.** Of or relating to a telescope. **2.** Seen or obtained by means of a telescope <*telescopic* data> **3.** Visible only by means of a telescope <a *telescopic* binary star> **4.** Able to discern distant objects <*telescopic* vision> **5.** Extensible or compressible by or as if by the successive sliding of overlapping concentric tubular sections. **—tel'e·scop'i·cal·ly** *adv.*

**Tel·e·sco·pi·um** (tĕl'ə-skō'pē-əm) *n.* [NLat. < *telescopium,* telescope.] A constellation in the Southern Hemisphere.

**te·les·co·py** (tə-lĕs'kə-pē) *n.* The art or study of making and operating telescopes. **—te·les'co·pist** *n.*

**tel·e·ster·e·o·scope** (tĕl'ə-stĕr'ē-ə-skōp', -stîr'-) *n.* A binocular telescope for stereoscopic viewing of distant objects.

**tel·es·the·sia** *also* **tel·aes·the·sia** (tĕl'ĭs-thē'zhə) *n.* Perception of or response to distant stimuli by extrasensory means. **—tel'es·thet'ic** (-thĕt'ĭk) *adj.*

**tel·e·text** (tĕl'ə-tĕkst') *n.* An electronic communication system in which printed information is broadcast by television signal to sets equipped with a decoder.

**tel·e·ther·mo·scope** (tĕl'ə-thûr'mə-skōp') *n.* An apparatus for indicating or recording the temperatures of remote or inaccessible locations.

**tel·e·thon** (tĕl'ə-thŏn') *n.* [TELE- + (MARA)THON.] A long continuous television program, usu. to raise funds for charity.

**tel·e·tran·scrip·tion** (tĕl'ə-trăn-skrĭp'shən) *n.* Transcription of television programs by means of a kinescope or videotape.

**Tel·e·type** (tĕl'ə-tīp'). A trademark for a teletypewriter.

**tel·e·type·writ·er** (tĕl'ə-tīp'rī'tər) *n.* An electromechanical typewriter that either transmits or receives messages coded in electrical signals carried by telegraph or telephone wires.

**te·leu·to·spore** (tə-lōō'tə-spôr', -spōr') *n.* [Gk. *teleutē,* termination (< *telos,* end) + SPORE.] A teliospore. **—te·leu'to·spor'ic** *adj.*

**tel·e·vise** (tĕl'ə-vīz') *vt.* & *vi.* **-vised, -vis·ing, -vis·es.** [Back-formation < TELEVISION.] To broadcast by television.

**tel·e·vi·sion** (tĕl'ə-vĭzh'ən) *n.* [Fr. *télévision* : *télé-,* tele- + *vision,* vision.] **1.** Transmission of visual images of moving and stationary objects, gen. with accompanying sound, as electromagnetic waves and the reconversion of received waves into visual images. **2. a.** An electronic device that receives electromagnetic waves and displays the reconverted images on a screen. **b.** The integrated audible and visible content of the electromagnetic waves received and converted by such an apparatus. **c.** A televison receiving set. **3. a.** The television broadcast industry. **b.** Television as a communication medium.

**tel·e·vi·sor** (tĕl'ə-vī'zər) *n.* A television transmitter.

**tel·ex** (tĕl'ĕks') *n.* [TEL(ETYPEWRITER) + EX(CHANGE).] **1.** A communication system consisting of teletypewriters connected to a telephonic network to send and receive signals. **2.** A message sent or received by telex. *—vt.* **-exed, -ex·ing, -ex·es.** To send (a message) by telex.

**tel·ic** (tĕl'ĭk, tē'lĭk) *adj.* [Gk. *telikos,* final < *telos,* end.] Directed or tending toward a goal : PURPOSEFUL.

**te·li·o·spore** (tē'lē-ə-spôr', -spōr') *n.* [TELI(UM) + SPORE.] A thick-walled, blackish resting spore of rusts and smuts, from which the basidium arises. **—te'li·o·spor'ic** *adj.*

**te·li·um** (tē'lē-əm) *n., pl.* **-li·a** (-lē-ə) [NLat. < Gk. *teleios,* complete < *telos,* end.] A pustulelike blackish structure formed on the tissue of a plant infected by a rust fungus, and generating teliospores. **—te'li·al** *adj.*

**tell** (tĕl) *v.* **told** (tōld), **tell·ing, tells.** [ME *tellen* < OE *tellan.*] *—vt.* **1.** To give a detailed account of : NARRATE. **2.** To communicate by speech or writing : express with words. **3.** To make known to : INFORM <*Tell* me the facts.> **4.** To make known : REVEAL <*tell* fortunes> **5.** To command : ORDER <You never do what I *tell* you.> **6.** To assure <I *tell* you, they're honest people.> **7.** To discover by observation : IDENTIFY. **8.** To say (a rosary). *—vi.* **1.** To give an account, enumeration, or description. **2.** To give evidence or indication. **3.** To have an effect or impact <In this business every mistake *tells.*> **—tell off. 1.** To count and set apart, esp. aloud. **2.** *Informal.* To rebuke severely. **—tell on.** *Informal.* To tattle on. **—tell'a·ble** *adj.*

**tell·er** (tĕl'ər) *n.* **1.** One who tells. **2.** A bank employee who receives and pays out money. **3.** One appointed to count votes in a legislative assembly. **—tell'er·ship'** *n.*

**tell·ing** (tĕl'ĭng) *adj.* **1.** Having force or effect : STRIKING <a *telling* defense of the accused> **2.** Full of special meaning : REVEALING <a *telling* example of their generosity> **—tell'ing·ly** *adv.*

**tell·tale** (tĕl'tāl') *n.* **1.** One who informs on another : TATTLETALE. **2.** Something that indicates or reveals information : SIGN. **3.** A device that indicates or registers information, esp.: **a.** A time clock. **b.** A device indicating the position of a ship's rudder. **c.** A row of strips hung above a railroad track to warn an approaching train of a low clearance ahead. **4.** A metal strip, 2 or 2½ feet high, across the bottom of the front wall of a racquets or squash court above which the ball must be hit.

**tellur-** *pref. var. of* TELLURO-.

**tel·lu·ri·an** (tĕ-lŏŏr'ē-ən) *adj.* Of, relating to, or inhabiting the earth. *—n.* **1.** An inhabitant of the earth : TERRESTRIAL. **2.** *var. of* TELLURION.

**tel·lu·ric** (tĕ-lŏŏr'ĭk) *adj.* **1.** Of or pertaining to the earth : TERRESTRIAL. **2.** Derived from or containing tellurium, esp. with valence 6.

**telluric acid** *n.* A white crystalline inorganic acid, $H_6TeO_6$, used as a chemical reagent.

**tel·lu·ride** (tĕl'yə-rīd') *n.* A binary compound of tellurium.

**tel·lu·ri·on** (tĕ-lŏŏr'ē-ŏn') *also* **tel·lu·ri·an** (-ən) *n.* An instrument that shows how the movement of the earth on its axis and around the sun causes day and night and the seasons.

**tel·lu·ri·um** (tĕ-lŏŏr'ē-əm) *n. Symbol* **Te** A brittle silvery-white metallic element, used to alloy stainless steel and lead, in ceramics, and, in the form of bismuth telluride, in thermoelectric devices; atomic number 52; atomic weight 127.60.

**telluro-** or **tellur-** *pref.* [< Lat. *tellus, tellur-,* earth.] **1.** Earth <*tellurian*> **2.** Tellurium <*tellurous*>

**tel·lu·rom·e·ter** (tĕl'yə-rŏm'ĭ-tər) *n.* A surveying instrument that measures distance by means of microwaves.

**tel·lu·rous** (tĕl'yər-əs, tĕ-lŏŏr'əs) *adj.* Of, pertaining to, or derived from tellurium, esp. with valence 4.

**tel·ly** (tĕl′ē) *n., pl.* **-lies.** *Chiefly Brit.* TELEVISION 2C.
**telo-** *or* **tel-** *pref.* [< Gk. *telos,* end.] End <*telophase*>
**tel·o·cen·tric** (tĕl′ə-sĕn′trĭk, tē′lə-) *adj.* Of or relating to a chromosome whose centromere is terminally located.
**tel·o·lec·i·thal** (tĕl′ə-lĕs′ə-thəl, tē′lə-) *adj.* Of or relating to an ovum in which the yolk is concentrated at one end.
**tel·o·mere** (tĕl′ə-mîr′, tē′lə-) *n.* A centromere located in a terminal position on a chromosome.
**tel·o·phase** (tĕl′ə-fāz′, tē′lə-) *n.* The final phase of mitosis, in which the chromosomes of daughter cells are grouped in new nuclei.
**tel·pher** (tĕl′fər) *n.* [TEL(E)- + Gk. *pherein,* to carry.] **1.** A light transportation car suspended from overhead wire cables, usu. propelled by an electric motor. **2.** A transportation system using telphers. —*vt.* **-phered, -pher·ing, -phers.** To transport by telpher.
**tel·son** (tĕl′sən) *n.* [Gk., limit.] A terminal structure of the posterior section of certain arthropods, as the middle lobe of the tail fin of a lobster or shrimp or the sting of a scorpion.
**Tel·star** (tĕl′stär′) *n.* One of two privately financed, low-altitude, active communications satellites launched by the U.S. Government in 1962 and 1963, and used commercially to transmit television pictures and telephone messages.
**Tel·u·gu** (tĕl′ə-gōō′) *n. & adj. var. of* TELEGU.
**tem·blor** (tĕm′blər, -blôr′) *n.* [Sp. < *temblar,* to shake.] An earthquake.
**tem·er·ar·i·ous** (tĕm′ə-râr′ē-əs) *adj.* [Lat. *temerarius* < *temere,* rashly.] Arrogantly or recklessly daring. **—tem′er·ar′i·ous·ly** *adv.* **—tem′er·ar′i·ous·ness** *n.*
**te·mer·i·ty** (tə-mĕr′ĭ-tē) *n.* [ME *temeryte* < Lat. *temeritas* < *temere,* rashly.] Excessive, arrogant self-confidence.
☆ ***syns:*** TEMERITY, CHUTZPAH, GALL, PRESUMPTION *n. core meaning* : excessive, arrogant self-confidence <had the *temerity* to come to dinner uninvited>
**temp** (tĕmp) *n.* [Short for *temporary worker.*] *Informal.* A temporary worker, as in an office.
**tem·peh** (tĕm′pā′) *n.* [Indonesian *tĕmpe.*] A high-protein food of Indonesian origin made from partially cooked fermented soybeans.
**tem·per** (tĕm′pər) *v.* **-pered, -per·ing, -pers.** [ME *temperen* < OE *temprian* < Lat. *temperare.*] —*vt.* **1.** To modify by the addition of an agent or quality : MODERATE <"*temper* its doctrinaire logic with a little practical wisdom" —Robert Houghwout Jackson> **2.** To bring to a specified physical condition, as consistency, texture, or hardness, by or as if by blending, admixing, or kneading. **3.** To harden, strengthen, or toughen (metal) by application of heat or by alternate heating and cooling. **4. a.** To attune. **b.** *Mus.* To adjust (the pitch of an instrument) to a temperament. —*vi.* To be or become tempered. —*n.* **1.** A state of mind or emotions : DISPOSITION <an even *temper*> **2.** Calmness of mind or emotions : COMPOSURE <tried not to lose my *temper*> **3. a.** A tendency to become easily angry or upset <a quick *temper*> **b.** An outburst of rage <a fit of *temper*> **4. a.** The state of being tempered. **b.** The degree of hardness and elasticity of a metal, chiefly steel, as a result of tempering. **5.** An agent or substance added to something to alter it. **6.** *Obs.* The character or constitution of a human being according to medieval physiology, as determined by the mixture within him of the four humors. **7.** *Archaic.* A compromise between extremes : a middle course. **—tem′per·a·bil′i·ty** *n.* **—tem′per·a·ble** *adj.* **—tem′per·er** *n.*
**tem·per·a** (tĕm′pər-ə) *n.* [Ital. < *temperare,* to mingle < Lat.] **1.** A painting medium in which pigment is mixed with water-soluble glutinous materials, as size or egg yolk. **2.** Painting done with tempera.
**tem·per·a·ment** (tĕm′prə-mənt, tĕm′pər-ə-) *n.* [ME < Lat. *temperamentum* < *temperare,* to temper.] **1. a.** The manner of thinking, behaving, or reacting characteristic of a specific individual <a high-strung *temperament*> **b.** *Obs.* The distinguishing mental and physical characteristics that established the constitution of a human being according to medieval physiology, caused by the dominance of one of the four humors. **2.** Excessive irritability or sensitiveness. **3.** *Mus.* Equal temperament.
**tem·per·a·men·tal** (tĕm′prə-mĕn′tl, tĕm′pər-ə-) *adj.* **1.** Relating to, caused by, or endowed with temperament or temper. **2. a.** Overly sensitive or irritable : MOODY. **b.** Unpredictable in performance <a *temperamental* copy machine> **—tem′per·a·men′tal·ly** *adv.*
**tem·per·ance** (tĕm′pər-əns, tĕm′prəns) *n.* **1.** The quality or state of being temperate : moderation or self-restraint. **2.** Total abstinence from alcoholic liquors.
**tem·per·ate** (tĕm′pər-ĭt, tĕm′prĭt) *adj.* [ME < Lat. *temperatus* < p.part. of *temperare,* to temper.] **1.** Exercising moderation and self-restraint. **2.** Moderate : tempered <a *temperate* handling of the dispute> **3.** Neither hot nor cold in climate : mild.
☆ ***syns:*** TEMPERATE, ABSTEMIOUS, ABSTENTIOUS, ABSTINENT, CONTINENT, SOBER *adj. core meaning* : exercising moderation and self-restraint in appetites and behavior <a *temperate* person who did nothing to excess> ***ant:*** intemperate
**Temperate Zone** *n.* Either of two middle latitude zones of the earth, the North Temperate Zone and the South Temperate Zone, lying between 23½° and 66½°north and south.
**tem·per·a·ture** (tĕm′pər-ə-chŏŏr′, tĕm′prə-) *n.* [Lat. *temperatura,* composition < *temperare,* to mix.] **1. a.** Degree of hotness or coldness of a body or environment. **b.** A specific degree of hotness or coldness as indicated on or referred to a standard scale : a scalar quantity that is independent of the size of the system and that determines the direction of heat flow between any two systems in thermal contact. **2.** An abnormally high temperature usu. caused by illness : FEVER.
**temperature gradient** *n.* The rate of change of temperature with displacement in a given direction from a given reference point.
**tem·pered** (tĕm′pərd) *adj.* **1.** Having a specified temper or disposition <hot-*tempered*> **2.** *Mus.* Tuned to temperament. —Used of a scale, interval, semitone, or intonation. **3.** Moderated by the admixture of another substance, quality, or factor <delivered a *tempered* admonishment> **4.** Having the required degree of hardness or elasticity <*tempered* steel>
**tem·pest** (tĕm′pĭst) *n.* [ME < OFr. *tempeste* < Lat. *tempestas* < *tempus,* time.] **1.** A violent windstorm, often accompanied by rain, snow, or hail. **2.** Furious commotion : UPROAR. —*vt.* **-pest·ed, -pest·ing, -pests.** To disturb or agitate violently.
**tem·pes·tu·ous** (tĕm-pĕs′chōō-əs) *adj.* [LLat. *tempestuosus* < *tempestas,* tempest.] **1.** Relating to or characteristic of a tempest. **2.** Tumultuous : stormy <a *tempestuous* relationship> **—tem·pes′tu·ous·ly** *adv.* **—tem·pes′tu·ous·ness** *n.*
**tem·pi** (tĕm′pē) *n. var. pl. of* TEMPO.
**Tem·plar** (tĕm′plər) *n.* [ME < AN *templer,* var. of OFr. *templier* < Med. Lat. *templarius* < Lat. *templum,* temple.] **1.** A Knight Templar. **2. templar.** *Chiefly Brit.* A lawyer or law student having chambers in the Temple in London.
**tem·plate** *also* **tem·plet** (tĕm′plĭt) *n.* [Prob. < Fr. *templet,* dim. of OFr. *temple,* temple, device in a loom.] **1.** A pattern or gauge, as a thin metal plate with a cut pattern, used as a guide in making something accurately, as in woodworking. **2.** A piece of stone or timber used to distribute weight or pressure, as over a door frame. **3.** *Biol.* A molecule, as DNA, that serves as a model for the synthesis of a macromolecule, as RNA.
**tem·ple**[1] (tĕm′pəl) *n.* [ME, partly < OE *tempel,* and partly < OFr. *temple,* both < Lat. *templum.*] **1.** A place or building dedicated to the worship or the presence of a deity. **2. Temple.** Any of three successive buildings in ancient Jerusalem dedicated to the worship of Jehovah. **3.** *Informal.* A synagogue. **4.** *Mormon Ch.* A building in which the sacred ordinances are administered. **5.** Something held to contain a divine presence. **6.** The headquarters of a fraternal order, esp. that of the Knights Templar. **7.** A place or building reserved for a special purpose <a *temple* of learning><a *temple* of art> **8. Temple.** Either of the two Inns of Court in London housing England's major law societies, and once used by the Knights Templar.
**tem·ple**[2] (tĕm′pəl) *n.* [ME < OFr. < Lat. *tempora,* pl. of *tempus,* temple of the head.] The flat region on either side of the forehead.
**tem·ple**[3] (tĕm′pəl) *n.* [ME *tempylle* < OFr. *temple,* poss. < Lat. *templum,* small piece of wood.] A device in a loom that keeps the cloth stretched to the proper width during weaving.
**tem·plet** (tĕm′plet) *n. var. of* TEMPLATE.
**tem·po** (tĕm′pō) *n., pl.* **-pos** *or* **-pi** (-pē) [Ital. < Lat. *tempus,* time.] **1.** *Mus.* Relative speed at which a composition is to be played, as indicated by a descriptive or metronomic direction to the performer. **2.** A characteristic rate or rhythm of activity : PACE <"the *tempo* and the feeling of modern life" —Robert L. Heilbroner>
**tem·po·ral**[1] (tĕm′pər-əl, tĕm′prəl) *adj.* [ME < Lat. *temporalis* < *tempus,* time.] **1.** Relating to, concerned with, or limited by time. **2.** Relating to or concerned with worldly affairs. **3.** Lasting briefly : EPHEMERAL <the *temporal* dreams of youth> **4.** Civil, secular, or lay. **5.** Expressing time <a *temporal* adverb> **—tem′po·ral·ly** *adv.*
**tem·po·ral**[2] (tĕm′pər-əl, tĕm′prəl) *adj.* [LLat. *temporalis* < Lat. *tempora,* pl. of *tempus,* temple.] Of, relating to, or near the temples of the skull.
**temporal bone** *n.* Either of two complex, three-part bones forming the sides and base of the skull.
**tem·po·ral·i·ty** (tĕm′pə-răl′ĭ-tē) *n., pl.* **-ties. 1.** The state of being short-lived or temporary. **2. temporalities.** Temporal possessions, esp. of the church or clergy.
**tem·po·rar·y** (tĕm′pə-rĕr′ē) *adj.* [Lat. *temporarius* < *tempus,* time.] Lasting, used, or enjoyed for a limited time : IMPERMANENT. —*n., pl.* **-ies.** *Informal.* One that serves for a limited time, esp. an office worker. **—tem′po·rar′i·ly** *adv.* **—tem′po·rar′i·ness** *n.*
**tem·po·rize** (tĕm′pə-rīz′) *vi.* **-rized, -riz·ing, -riz·es.** [OFr. *temporiser,* to pass one's time < Med. Lat. *temporizare* < Lat. *tempus,* time.] **1.** To act evasively in order to gain time, avoid argument, or postpone a decision. **2. a.** To behave appropriately under the circumstances. **b.** To yield to current conditions : COMPROMISE. **—tem′po·ri·za′tion** *n.*
**tempt** (tĕmpt) *vt.* **tempt·ed, tempt·ing, tempts.** [ME *tempten* < OFr. *tempter* < Lat. *temptare,* to feel, try.] **1.** To entice (someone) to commit an unwise or immoral act, esp. by a promise of reward. **2.** To be inviting or attractive to <The fresh strawberries *tempted* me.> **3.** To provoke or to risk provoking, as fate. **4.** To incline or

dispose strongly <was *tempted* to quit the job> **—tempt'a·ble** *adj.* **—tempt'er** *n.* **—tempt'ress** (tĕmp'trĭs) *n.*

☆ ***syns:*** TEMPT, ALLURE, ENTICE, INVEIGLE, LURE, SEDUCE *v. core meaning* : to beguile or draw into a wrong or foolish course of action <was *tempted* into accepting bribes>

**temp·ta·tion** (tĕmp-tā'shən) *n.* **1.** The act of tempting or the state of being tempted. **2.** One that tempts or entices.

**tempt·ing** (tĕmp'tĭng) *adj.* Enticing : alluring. **—tempt'ing·ly** *adv.* **—tempt'ing·ness** *n.*

**tem·pu·ra** (tĕm'po͝o-rə, tĕm-po͝or'ə) *n.* [J.] A Japanese dish of vegetables and shrimp or other seafood dipped in batter and deep-fried.

**ten** (tĕn) *n.* [ME < OE *tīen*; akin to G. *zehn*, Lat. *decem*, Gk. *deka*, Skt. *daśa*.] **1.** The cardinal number equal to 9 + 1. **2.** The tenth in a set or sequence. **3.** Something having ten parts, units, or members. **4.** A playing card marked with ten spots. **5.** A ten-dollar bill. **—ten** *adj.* & *pron.*

**ten·a·ble** (tĕn'ə-bəl) *adj.* [Fr. < OFr. < *tenir*, to hold < Lat. *tenēre*.] **1.** Capable of being defended or sustained : LOGICAL <a *tenable* hypothesis> **2.** Defensible from armed assault <a *tenable* outpost> **—ten'a·bil'i·ty, ten'a·ble·ness** *n.* **—ten'a·bly** *adv.*

**ten·ace** (tĕn'ās', tĕ-nās', tĕn'ĭs) *n.* [Fr. < Sp. *tenaza* < Lat. *tenax*, tenacious.] A combination of two high cards, as the king and jack, held in a player's hand, esp. in bridge and whist.

**te·na·cious** (tə-nā'shəs) *adj.* [< Lat. *tenax, tenac-*, holding fast < *tenēre*, to hold.] **1.** Holding or tending to hold firmly : PERSISTENT <people *tenacious* of their opinions> **2. a.** Holding together firmly, as wood or metal : COHESIVE. **b.** Clinging to another object or surface : ADHESIVE <*tenacious* burs> **3.** Tending to retain : RETENTIVE <a *tenacious* memory> **—te·na'cious·ly** *adv.* **—te·na'cious·ness** *n.* **—te·nac'i·ty** (-năs'ĭ-tē) *n.*

**te·nac·u·lum** (tə-năk'yə-ləm) *n., pl.* **-la** (-lə) [NLat. < LLat., holder < Lat. *tenēre*, to hold.] A slender, hooked, long-handled instrument for lifting and holding parts, as blood vessels, during surgery.

**ten·an·cy** (tĕn'ən-sē) *n., pl.* **-cies.** [< TENANT.] *Law.* **1.** Possession or occupancy of lands or tenements by title, under a lease, or on payment of rent. **2.** The period of a tenant's occupancy or possession. **3.** A habitation held or occupied by a tenant.

**ten·ant** (tĕn'ənt) *n.* [ME < OFr. < pr.part. of *tenir*, to hold < Lat. *tenēre*.] **1.** One who pays rent to use or occupy land, a building, or other property owned by another. **2.** An occupant, inhabitant, or dweller in a place. **3.** *Law.* One who holds or possesses lands, tenements, and sometimes personal property by any kind of right or title. *—v.* **-ant·ed, -ant·ing, -ants.** *—vt.* To hold as a tenant : OCCUPY. *—vi.* To be a tenant.

**tenant farmer** *n.* One who farms land owned by another and pays rent in cash or in kind.

**ten·ant·ry** (tĕn'ən-trē) *n., pl.* **-ries. 1.** Tenants as a group. **2.** The state of being a tenant.

**ten-cent store** (tĕn'sĕnt') *n.* A five-and-ten.

**tench** (tĕnch) *n., pl.* **tench** or **tench·es.** [ME *tenche* < OFr. < LLat. *tinca*.] An edible Eurasian freshwater fish, *Tinca tinca*, with small scales and two barbels near the mouth.

**Ten Commandments** *pl.n.* The ten injunctions given by God to Moses on Mount Sinai, serving as the basis of Mosaic Law.

**tend[1]** (tĕnd) *vi.* **tend·ed, tend·ing, tends.** [ME *tenden* < OFr. *tendre* < Lat. *tendere*.] **1.** To extend or move in a certain direction <Our course *tended* toward the north.> **2.** To be likely <These things *tend* to work themselves out.> **3.** To be disposed or inclined <*tends* toward self-indulgence>

**tend[2]** (tĕnd) *v.* **tend·ed, tend·ing, tends.** [ME *tenden*, short for *attenden*, to attend. —see ATTEND.] *—vt.* **1.** To minister to the needs of : look after <*tend* a baby> **2.** To serve at <*tend* bar> *—vi.* **1.** To serve or wait. **2.** *Informal.* To apply one's attention <*Tend* to your own concerns.>

**ten·den·cy** (tĕn'dən-sē) *n., pl.* **-cies.** [Med. Lat. *tendentia* < Lat. *tendens*, pr.part. of *tendere*, to tend.] **1.** A demonstrated inclination to think, act, or behave in a certain way : PROPENSITY <a *tendency* to lie> **2.** The purposeful trend of something communicated : AIM.

**ten·den·tious** *also* **ten·den·cious** (tĕn-dĕn'shəs) *adj.* [TENDENCY.] Written or said to promote a cause. **—ten·den'tious·ly** *adv.* **—ten·den'tious·ness** *n.*

**ten·der[1]** (tĕn'dər) *adj.* **-er, -est.** [ME < OFr. *tendre* < Lat. *tener*.] **1. a.** Easily crushed or bruised : FRAGILE <a *tender* petal> **b.** Easily chewed or cut <*tender* beef> **c.** Having a delicate quality <*tender* music> **2.** Young and vulnerable <of *tender* age> **3.** Physically frail : WEAK. **4.** Sensitive to frost or severe cold <*tender* green leaves> **5. a.** Easily hurt : SENSITIVE <*tender* skin> **b.** Painful : sore <a *tender* bruise> **6. a.** Gentle and solicitous <a *tender* parent> **b.** Expressing gentle emotions : LOVING <a *tender* glance> **c.** Given to sympathy or sentimentality : SOFT <a *tender* heart> **7. a.** Wary and protective <*tender* of our reputation> **b.** Scrupulous : chary <*tender* of making promises I can't keep> **8.** *Naut.* Likely to lean under sail : CRANK. *—vt.* **-dered, -der·ing, -ders. 1.** To make tender. **2.** *Archaic.* To treat with tender regard. **—ten'der·ly** *adv.* **—ten'der·ness** *n.*

**ten·der[2]** (tĕn'dər) *n.* [< OFr. *tendre*, to offer < Lat. *tendere*, to hold forth.] **1.** A formal offer, as: **a.** *Law.* An offer of money or service in payment of an obligation. **b.** A written offer to contract goods or services at a specified rate or cost : BID. **2.** Something tendered, esp. money <legal *tender*> *—vt.* **-dered, -der·ing, -ders.** To offer formally <*tendered* my letter of resignation> **—ten'der·er** *n.*

**tend·er[3]** (tĕn'dər) *n.* **1.** One who tends something. **2.** A vessel that tends another vessel or vessels, esp. one that ferries supplies between ship and shore. **3.** A railroad car attached to the rear of a locomotive and designed to carry fuel and water.

**ten·der·foot** (tĕn'dər-fo͝ot') *n., pl.* **-foots** or **-feet. 1.** A newcomer not yet hardened to rough outdoor life : GREENHORN. **2.** One who is inexperienced : NOVICE. **3.** A beginner in the ranks of the Boy Scouts.

**ten·der·heart·ed** (tĕn'dər-här'tĭd) *adj.* Easily affected by another's distress : COMPASSIONATE. **—ten'der·heart'ed·ly** *adv.* **—ten'der·heart'ed·ness** *n.*

**ten·der·ize** (tĕn'də-rīz') *vt.* **-ized, -iz·ing, -iz·es.** To make (meat) tender, as by pounding, marinating, or applying a tenderizer. **—ten'der·i·za'tion** *n.*

**ten·der·iz·er** (tĕn'də-rī'zər) *n.* A substance applied to meat to make it tender.

**ten·der·loin** (tĕn'dər-loin') *n.* [Sense 2, after the *Tenderloin*, an area of New York City (from the easy income it affords a corrupt police officer).] **1.** The tenderest part of a loin of beef, pork, etc. **2.** A city district notorious for vice and graft.

**ten·der-mind·ed** (tĕn'dər-mīn'dĭd) *adj.* Resisting harsh facts : inclined to idealism.

**ten·di·ni·tis** (tĕn'də-nī'tĭs) *n.* [NLat. *tendo, tendin-*, tendon (< Med. Lat. *tendo*) + -ITIS.] Inflammation of a tendon.

**ten·di·nous** (tĕn'də-nəs) *adj.* [NLat. *tendinosus* < *tendo, tendin-*, tendon < Med. Lat. *tendo*.] **1.** Of, having, or resembling a tendon. **2.** Made up of tendons : SINEWY.

**ten·don** (tĕn'dən) *n.* [Med. Lat. *tendo, tendon-*, < Lat. *tendere*, to stretch.] A band of tough inelastic fibrous tissue connecting a muscle with its bony attachment.

**ten·do·ni·tis** (tĕn'də-nī'tĭs) *n.* Tendinitis.

**tendon of Achilles** *n.* The Achilles' tendon.

**ten·dril** (tĕn'drəl) *n.* [OFr. *tendrillon* < *tendron*, young shoot < Lat. *tener*, tender.] **1.** A long, slender, coiling extension, as of a stem, serving as an organ of attachment for certain climbing plants. **2.** Something, as a ringlet of hair, resembling a tendril.

**Ten·e·brae** (tĕn'ə-brā', -brē') *pl.n.* [Med. Lat. < Lat. *tenebrae*, darkness.] (*sing. or pl. in number*). *Rom. Cath. Ch.* The office of matins and lauds sung on the last three days of Holy Week, with a ceremony of candles.

**ten·e·brif·ic** (tĕn'ə-brĭf'ĭk) *adj.* [Lat. *tenebrae*, darkness + -FIC.] **1.** Serving to darken or obscure. **2.** Dark and gloomy.

**te·neb·ri·o·nid** (tə-nĕb'rē-ə-nĭd', tĕn'ə-brī'-) *n.* [< NLat. *Tenebrionidae*, family name < *Tenebrio*, type genus < Lat. *tenebrio*, one who avoids light < *tenebrae*, darkness.] A dark-colored herbivorous beetle of the family Tenebrionidae. **—te·neb'ri·o·nid'** *adj.*

**ten·e·brous** (tĕn'ə-brəs) *also* **te·neb·ri·ous** (tə-nĕb'rē-əs) *adj.* [< Lat. *tenebrae*, darkness.] TENEBRIFIC 2. **—ten'e·bros'i·ty** (-brŏs'ĭ-tē) *n.*

**ten·e·ment** (tĕn'ə-mənt) *n.* [ME, house < OFr. < Med. Lat. *tenementum* < Lat. *tenēre*, to hold.] **1.** A building to live in, esp. one intended for rent : RESIDENCE. **2.** A run-down low-rental apartment building whose facilities and maintenance barely meet minimum standards. **3.** *Chiefly Brit.* A room or an apartment leased to a tenant. **4.** *Law.* Permanent property, as land, rents, or franchises, that may be held by one person for another. **—ten'e·men'tal** (-mĕn'tl), **ten'e·men'ta·ry** (-mĕn'tə-rē) *adj.*

**te·nes·mus** (tə-nĕz'məs) *n.* [Med. Lat., var. of Lat. *tenesmos* < Gk. *teinesmos* < *teinein*, to strain.] A painfully urgent yet ineffectual attempt to defecate or urinate.

**ten·et** (tĕn'ĭt) *n.* [< Lat., he holds < *tenēre*, to hold.] An opinion, doctrine, or principle considered as being true by an individual or esp. by an organization.

**ten·fold** (tĕn'fōld') *adj.* **1.** Composed of ten parts or members. **2.** Ten times as great or as many. *—adv.* Ten times in extent or number.

**ten-gal·lon hat** (tĕn'găl'ən) *n.* A felt hat having an extremely tall crown and wide brim.

**te·ni·a** (tē'nē-ə) *n. var. of* TAENIA.

**te·ni·a·cide** *also* **tae·ni·a·cide** (tē'nē-ə-sīd') *n.* [TENIA + -CIDE.] An agent that destroys tapeworms.

**te·ni·a·sis** *also* **tae·ni·a·sis** (tē-nī'ə-sĭs) *n.* [TEN(IA) + -IASIS.] Infestation with tapeworms.

**ten·nis** (tĕn'ĭs) *n.* [ME *tenetz*, court tennis, prob. < OFr. *tenez*, imper. of *tenir*, to receive.] **1.** A game played with rackets and a light ball by two players or two pairs of players on a court divided by a net. **2.** Lawn tennis. **3.** Court tennis.

**tennis shoe** *n.* SNEAKER 2.

**teno-** *pref.* [< Gk. *tenōn*, tendon.] Tendon <*tenotomy*>

**ten·on** (tĕn'ən) *n.* [ME < OFr. < *tenir*, to hold < Lat. *tenēre*.] A projection on the end of a piece of wood shaped for insertion into a mortise. —*vt.* **-oned, -on·ing, -ons. 1.** To furnish with a tenon. **2.** To connect with a tenon.

**ten·or** (tĕn′ər) *n.* [ME < OFr. < Lat., uninterrupted course < *tenēre,* to continue.] **1. a.** The flow of meaning apparent in something communicated. **b.** General sense : PURPORT. **2. a.** *Law.* The exact meaning or actual wording of a document as distinct from its effect. **b.** An exact copy or transcript of a document. **3.** *Mus.* **a.** The highest natural adult male voice. **b.** A part for a tenor. **c.** One who sings tenor.

**te·nor·rha·phy** (tĕ-nôr′ə-fē) *n., pl.* **-phies.** [TENO- + Gk. *rhaphē,* suture.] Surgical union of divided tendons with sutures.

**ten·o·syn·o·vi·tis** (tĕn′ō-sĭn′ō-vī′tĭs) *n.* [TENO- + SYNOV(IA) + -ITIS.] Inflammation of a tendon sheath.

**te·not·o·my** (tĕ-nŏt′ə-mē) *n., pl.* **-mies.** Surgical division of a tendon for the relief of deformities caused by the shortening of a muscle.

**ten·pence** (tĕn′pəns) *n. Chiefly Brit.* A sum of money equal to ten pennies.

**ten·pen·ny** (tĕn′pĕn′ē, -pə-nē) *adj. Chiefly Brit.* Costing or valued at tenpence.

**tenpenny nail** *n.* [From its orig. price per hundred.] A nail three inches long.

**ten·pin** (tĕn′pĭn′) *n.* **1.** A bowling pin used in playing tenpins. **2. tenpins** *(sing. in number).* The game of bowling.

**ten·rec** (tĕn′rĕk′) *also* **tan·rec** (tăn′-) *n.* [Fr. < Malagasy *tàndraka.*] Any of various insectivorous, often hedgehoglike mammals of the family Tenrecidae of Madagascar and adjacent islands.

**tense**[1] (tĕns) *adj.* **tens·er, tens·est.** [Lat. *tensus* < p.part. of *tendere,* to stretch out.] **1.** Tightly stretched : STRAINED <*tense* muscles> **2.** Being in a state of mental or emotional tension. **3.** Nerve-racking or suspenseful <spent a *tense* night at the hospital> **4.** Enunciated with taut muscles, as the consonant *t.* —*vt. & vi.* **tensed, tens·ing, tens·es.** To make or become tense. **—ten′si·ty** (tĕn′sĭ-tē) *n.*

**tense**[2] (tĕns) *n.* [ME *tens* < OFr., time < Lat. *tempus.*] **1.** One of the inflected forms in the conjugation of a verb that indicates the time as well as the continuance or completion of the action or state. **2.** A set of tense forms indicating a particular time.

**ten·sile** (tĕn′səl, -sīl′) *adj.* [NLat. *tensilis* < Lat. *tensus,* stretched out < p.part. of *tendere,* to stretch.] **1.** Of or relating to tension. **2.** Able to be stretched or extended : DUCTILE. **—ten·sil′i·ty** (tĕn-sĭl′-ə-tē) *n.*

**tensile strength** *n.* Resistance of a material to a force tending to tear it apart.

**ten·sim·e·ter** (tĕn-sĭm′ĭ-tər) *n.* [TENSI(ON) + -METER.] An apparatus for measuring differences in vapor pressure.

**ten·si·om·e·ter** (tĕn′sē-ŏm′ĭ-tər) *n.* [TENSIO(N) + -METER.] **1.** An instrument for measuring tensile strength. **2.** A torsion-balance apparatus for measuring the surface tension of a liquid. **—ten′si·o·met′ric** *adj.* **—ten′si·om′e·try** *n.*

**ten·sion** (tĕn′shən) *n.* [OFr. < LLat. *tensio* < Lat. *tendere,* to stretch.] **1.** The act or process of stretching or the state of being stretched. **2. a.** A force tending to stretch or elongate something. **b.** The measure of such a force <a *tension* of 60 pounds> **3. a.** Mental, emotional, or nervous strain. **b.** Barely controlled hostility between persons or groups. **c.** Uneasy suspense. **4.** A device for regulating tautness, esp. a device regulating the tautness of thread on a sewing machine. **5.** *Elect.* Voltage or potential : ELECTROMOTIVE FORCE. —*vt.* **-sioned, -sion·ing, -sions.** To make taut. **—ten′sion·al** *adj.*

**ten·sive** (tĕn′sĭv) *adj.* Of, relating to, or causing tension.

**ten·sor** (tĕn′sər, -sôr′) *n.* **1.** *Anat.* A muscle that functions to stretch a part. **2.** *Math.* An element of an abstract system used to denote position determined within the context of more than one coordinate system, a special case of which is a vector that is determined in a single coordinate system. **—ten·so′ri·al** (-sôr′ē-əl, -sōr′-) *adj.*

**Ten·sor lamp** (tĕn′sər, -sôr′). A trademark for a high-intensity electric lamp.

**ten-strike** (tĕn′strīk′) *n.* **1.** A strike in tenpins. **2.** *Informal.* A remarkably successful stroke or accomplishment.

**tent**[1] (tĕnt) *n.* [ME < OFr. *tente* < VLat. **tenta* < fem. p.part. of Lat. *tendere,* stretch out.] **1.** A portable shelter of canvas, plastic, or skins stretched over a supporting framework of poles with ropes and pegs. **2.** Something suggesting a tent in construction or outline. —*v.* **tent·ed, tent·ing, tents.** —*vi.* To encamp in a tent. —*vt.* **1.** To form a tent over. **2.** To put up in tents.

**tent**[2] (tĕnt) *n.* [ME *tente* < OFr. < *tenter,* to probe < Lat. *tentare,* to feel.] A small, usu. lint or gauze roll or plug placed in a wound or orifice to keep it open. —*vt.* **tent·ed, tent·ing, tents.** To keep (a wound or cut) open with a tent.

**tent**[3] (tĕnt) *vt.* **tent·ed, tent·ing, tents.** [ME *tenten* < *tent,* attention, short for *attent* < OFr. *attente* < *attendre,* to attend < Lat. *attendere.* —see ATTEND.] *Scot.* **1.** To pay heed to. **2.** To wait on : ATTEND.

**ten·ta·cle** (tĕn′tə-kəl) *n.* [NLat. *tentaculum* < Lat. *tentare,* to touch.] **1.** *Zool.* An elongated, flexible, unsegmented protrusion, as one of those surrounding the mouth or oral cavity of the squid. **2.** *Bot.* One of the hairs on the leaves of insectivorous plants, as the sundew. **3.** Something resembling a tentacle, esp. in grasping or holding ability. **—ten·tac′u·lar** (-tăk′yə-lər) *adj.*

**tent·age** (tĕn′tĭj) *n.* **1.** Tents as a whole. **2.** A supply of tents available for accommodation. **3.** Tent equipment.

**ten·ta·tive** (tĕn′tə-tĭv) *adj.* [Med. Lat. *tentativus* < Lat. *tentare,* to try.] **1.** Experimental : provisional. **2.** Uncertain <a *tentative* smile> —*n.* An experiment or preliminary undertaking. **—ten′ta·tive·ly** *adv.* **—ten′ta·tive·ness** *n.*

**tent caterpillar** *n.* A destructive caterpillar, esp. the hairy larva of a North American moth *Malacosoma americanum,* living in colonies in tentlike webs constructed in deciduous trees.

**tent·ed** (tĕn′tĭd) *adj.* **1.** Covered with tents. **2.** Sheltered in or by tents. **3.** Suggestive of a tent.

**ten·ter** (tĕn′tər) *n.* [ME *teyntur.*] **1.** A framework on which milled cloth is stretched for drying without shrinkage. **2.** *Obs.* A tenterhook. —*vt.* **-tered, -ter·ing, -ters.** To stretch (cloth) on a tenter.

**ten·ter·hook** (tĕn′tər-ho͝ok′) *n.* A hooked nail for securing cloth on a tenter. **—on tenterhooks.** In a state of tension, suspense, or anxiety <students *on tenterhooks* before an exam>

**tenth** (tĕnth) *n.* [ME *tenthe* (< *ten,* ten), alteration of *tethe* < OE *teoða.*] **1.** The ordinal number matching the number ten in a series. **2.** One of ten equal parts. **—tenth** *adj. & adv.*

**tent stitch** *n.* A short diagonal embroidery stitch that forms close, even, parallel rows to fill in a pattern or a background.

**ten·u·is** (tĕn′yo͞o-ĭs) *n., pl.* **-u·es** (-yo͞o-ēz′) [NLat. (transl. of Gk. *psilos*) < Lat., thin.] A voiceless stop in Greek.

**te·nu·i·ty** (tĕ-no͞o′ĭ-tē, -nyo͞o′-) *n.* [Lat. *tenuitas,* thinness < *tenuis,* thin.] The quality or state of being tenuous.

**ten·u·ous** (tĕn′yo͞o-əs) *adj.* [Lat. *tenuis.*] **1.** Having a thin or slender form. **2.** Thin in consistency : DILUTE. **3.** Having little substance : FLIMSY. **—ten′u·ous·ly** *adv.* **—ten′u·ous·ness** *n.*

☆ ***syns:*** TENUOUS, FEEBLE, FLIMSY, INSUBSTANTIAL, UNSUBSTANTIAL *adj. core meaning* : having little substance or significance and not firmly grounded <a *tenuous* argument>

**ten·ure** (tĕn′yər, -yo͝or′) *n.* [ME < OFr. < *tenir,* to hold < Lat. *tenēre.*] **1.** The fact or condition of holding something, as real estate or an office. **2.** The terms under which something is held. **3. a.** The period of holding something. **b.** Permanence of position, often granted an employee after a specified number of years <academic *tenure*> **—ten′ured** *adj.* **—ten·u′ri·al** (tĕn-yo͝or′ē-əl) *adj.* **—ten·u′ri·al·ly** *adv.*

☆ ***syns:*** TENURE, INCUMBENCY, OCCUPANCY, OCCUPATION *n. core meaning* : the holding of something, such as an official position <a three-year *tenure* in office>

**ten·ured** (tĕn′yərd, -yo͝ord′) *adj.* Having academic tenure.

**te·nu·to** (tā-no͞o′tō) *adv. & adj.* [Ital. < p.part. of *tenere,* to hold < Lat. *tenēre.*] *Mus.* So as to be held for the full time value : SUSTAINED.

**te·o·cal·li** (tē′ə-kăl′ē) *n., pl.* **-lis.** [Nahuatl : *teotl,* sacred + *calli,* house.] **1.** An ancient Mexican and Central American temple usu. built on a truncated pyramidal mound. **2.** The mound on which a teocalli is built.

**te·o·sin·te** (tē′ə-sĭn′tē, tā′ō-) *n.* [Mex. Sp. < Nahuatl *teocentli* : *teotl,* sacred + *centli,* dried ear of corn.] A tall Central American grass, *Euchlaena mexicana,* related to corn and cultivated for fodder.

**te·pal** (tē′pəl, tĕp′əl) *n.* [Fr. *tépale.*] *Bot.* A division of the perianth of a flower having petals and sepals that are almost indistinguishable.

**tep·a·ry bean** (tĕp′ə-rē) *n.* [Orig. unknown.] **1.** A vine, *Phaseolus acutifolius latifolius* of the southwestern United States and adjacent Mexico, bearing edible beans. **2.** The bean borne by the tepary bean.

**te·pee** *also* **tee·pee** or **ti·pi** (tē′pē) *n.* [Dakota *tipi.*] A cone-shaped tent of skins or bark used by North American Indians, esp. the Plains Indians.

**tep·e·fy** (tĕp′ə-fī′) *vt. & vi.* **-fied, -fy·ing, -fies.** [Lat. *tepefacere* : *tepēre,* to be tepid + *facere,* to make.] To make or become tepid. **—tep′e·fac′tion** (-făk′shən) *n.*

**tep·id** (tĕp′ĭd) *adj.* [Lat. *tepidus* < *tepēre,* to be lukewarm.] **1.** Moderately warm : LUKEWARM. **2.** Lacking real interest, enthusiasm, or involvement <a *tepid* welcome> **—te·pid′i·ty** (tĕ-pĭd′ĭ-tē), **tep′id·ness** *n.* **—tep′id·ly** *adv.*

☆ ***syns:*** TEPID, HALFHEARTED, LUKEWARM *adj. core meaning* : lacking real warmth, interest, enthusiasm, or involvement <a *tepid* review of the play>

**te·qui·la** (tə-kē′lə) *n.* [Mex. Sp., after *Tequila,* Mexico.] An alcoholic liquor distilled from a Central American century plant, *Agave tequilana.*

**tera-** *pref.* [< Gk. *teras,* monster.] One trillion ($10^{12}$) <*terahertz*>

**ter·a·hertz** (tĕr′ə-hûrts′) *n.* One trillion ($10^{12}$) hertz.

**ter·a·ohm** (tĕr′ə-ōm′) *n.* One trillion ($10^{12}$) ohms.

**ter·aph** (tĕr′əf) *n., pl.* **ter·a·phim** (tĕr′ə-fĭm) [< Heb. *tĕrāphīm,* idols.] An image of a Semitic household idol.

**ter·a·to·car·ci·no·ma** (tĕr′ə-tō-kär′sə-nō′mə) *n.* [Gk. *teras, terat-,* monster + CARCINOMA.] A carcinomatous teratoma.

**te·rat·o·gen** (tə-răt′ə-jən, tĕr′ə-tə-) *n.* [Gk. *teras, terat-,* monster + -GEN.] A teratogenic agent.

**ter·a·to·gen·ic** (tĕr′ə-tə-jĕn′ĭk) *adj.* Causing fetal malformations or monstrosities. **—ter′a·to·ge·nic′i·ty** (-jə-nĭs′ĭ-tē) *n.*

**ter·a·toid** (tĕr′ə-toid′) *adj.* [Gk. *teras, terat-*, monster + -OID.] Like a monster : MONSTROUS.

**ter·a·tol·o·gy** (tĕr′ə-tŏl′ə-jē) *n.* [Gk. *teras, terat-*, monster + -LOGY.] Study of production, development, anatomy, and classification of monsters. **—ter′a·to·log′i·cal** (-tə-lŏj′ĭ-kəl) *adj.*

**ter·a·to·ma** (tĕr′ə-tō′mə) *n., pl.* **-mas** *or* **-ma·ta** (-mə-tə) [Gk. *teras, terat-*, monster + -OMA.] A tumor consisting of different types of tissue, caused by development of independent germ cells. **—ter′a·to′ma·tous** (-tō′mə-təs) *adj.*

**ter·bi·um** (tûr′bē-əm) *n.* [After *Ytterby*, Sweden.] *Symbol* **Tb** A soft silvery-gray metallic rare-earth element, used in electronics and as a laser material; atomic number 65; atomic weight 158.924.

**terbium metal** *n.* Any of several rare-earth metals separable from other metals as a group and including europium, terbium, and gadolinium.

**terce** (tûrs) *n. var. of* TIERCE.

**ter·cel** (tûr′səl) *also* **tier·cel** (tîr′səl) *n.* [ME < OFr. < VLat. **tertiolus*, dim. of Lat. *tertius*, third.] A male hawk used in falconry.

**ter·cen·ten·a·ry** (tûr′sĕn-tĕn′ə-rē, tər-sĕn′tə-nĕr′ē) *n., pl.* **-ries.** [Lat. *ter*, thrice + CENTENARY.] A 300th anniversary or its celebration. —*adj.* Of or relating to a time span of 300 years or to a 300th anniversary.

**ter·cen·ten·ni·al** (tûr′sĕn-tĕn′ē-əl) *n.* & *adj.* Tercentenary.

**ter·cet** (tûr′sĭt) *n.* [Ital. *terzetto*, dim. of *terzo*, third < Lat. *tertius.*] **1.** A poetic triplet of lines that rhyme or are connected with adjacent rhymes. **2.** *Mus.* TRIPLET 4.

**ter·e·bene** (tĕr′ə-bēn′) *n.* [Fr. *térébène* < *térébinthe*, terebinth < OFr. *terebinte.*] A mixture of terpenes prepared from oil of turpentine, used as an antiseptic and expectorant.

**te·reb·ic acid** (tə-rĕb′ĭk, -rē′bĭk) *n.* [TEREB(INTH) + -IC + ACID.] A white crystalline compound, $C_7H_{10}O_4$, produced by the action of nitric acid on turpentine.

**ter·e·binth** (tĕr′ə-bĭnth′) *n.* [ME *terebinthe* < OFr. *terebinte* < Lat. *terebinthus* < Gk. *terebinthos.*] A small tree, *Pistachia terebinthus* of the Mediterranean region, that yields a resinous liquid.

**ter·e·bin·thine** (tĕr′ə-bĭn′thĭn, -thīn′) *also* **ter·e·bin·thic** (-thĭk) *adj.* **1.** Of or relating to the terebinth. **2.** Relating to, consisting of, or resembling turpentine.

**te·re·do** (tə-rē′dō, -rā′dō) *n., pl.* **-dos.** [NLat. *Teredo*, mollusk genus < Lat. *teredo*, a kind of worm < Gk. *terēdōn.*] A shipworm.

**te·rete** (tē-rēt′) *adj.* [Lat. *teres, teret-*, rounded.] *Bot.* Cylindrical but usu. tapering slightly at both ends, circular in cross section, and smooth-surfaced <*terete* pods>

**Te·reus** (tē′rōōs′, tîr′ē-əs) *n.* [Lat. < Gk. *Tēreus.*] *Gk. Myth.* A king of Thrace who raped Philomela and who was changed into a hoopoe.

**ter·ga** (tûr′gə) *n. pl. of* TERGUM.

**ter·gal** (tûr′gəl) *adj.* Of or relating to the tergum : DORSAL.

**ter·giv·er·sate** (tər-jĭv′ər-sāt′, tûr′jĭ-vər-) *vi.* **-sat·ed, -sat·ing, -sates.** [Lat. *tergiversari, tergiversat-* : *tergum*, a back + *versare*, to whirl.] **1.** To use ambiguities or evasions. **2.** To change sides : APOSTATIZE. **—ter·giv′er·sa′tion** *n.* **—ter′gi·ver·sa′tor** (-sā′tər) *n.*

**ter·gum** (tûr′gəm) *n., pl.* **-ga** (-gə) [NLat. < Lat., back.] *Zool.* The upper or dorsal surface, esp. of a body segment of an insect.

**ter·i·ya·ki** (tĕr′ē-yä′kē) *n.* [J. : *teri*, sunshine, flame + *yaki*, to broil.] A Japanese dish of skewered and broiled marinated slices of meat or shellfish.

**term** (tûrm) *n.* [ME *terme* < OFr. < Lat. *terminus*, boundary.] **1. a.** A limited time period during which something lasts. **b.** The time during which a court is in session. **c.** A division of a school year. **2. a.** A point of time beginning or ending a period. **b.** A deadline, as for making a payment. **c.** The end of a normal gestation period <carried the baby to *term*> **3.** *Law.* **a.** A fixed time period during which an estate may be held. **b.** The estate to be granted. **c.** A time period allowed a debtor to meet an obligation. **4. a.** A word having an explicit meaning, esp. one peculiar to a particular group or activity <complex legal *terms*> **b. terms.** Language or a mode of expression used <refused in no uncertain *terms*> **5. a.** A stipulation or condition that defines the nature and limits of an agreement <the *terms* of a treaty> **b. terms.** The relation between two persons or groups : FOOTING <on speaking *terms* with the neighbors> **6.** *Math.* **a.** Each of the quantities composing a ratio or a fraction or forming a series. **b.** Each of the quantities connected by addition or subtraction signs in an equation : MEMBER. **7.** *Logic.* Each of the two concepts being compared or related in a proposition. **8.** A post or stone marking a boundary, esp. a square and tapering pillar adorned with a head and upper torso. —*vt.* **termed, term·ing, terms.** To designate or call <*termed* them cowards> **—come to terms.** To reach an agreement. **—in terms of.** With regard to. **—term′ly** *adv.*

**ter·ma·gant** (tûr′mə-gənt) *n.* [ME *Termagaunt*, imaginary Moslem deity portrayed as a violent and overbearing character in medieval mystery plays.] A scolding, quarrelsome woman : SHREW. —*adj.* Abusive : shrewish.

**term·er** (tûr′mər) *n.* One serving a specified term <a second *termer* in prison>

**ter·mi·na·ble** (tûr′mə-nə-bəl) *adj.* **1.** Capable of being terminated. **2.** Terminating after a specified date <*terminable* insurance> **—ter′mi·na·bil′i·ty, ter′mi·na·ble·ness** *n.* **—ter′mi·na·bly** *adv.*

**ter·mi·nal** (tûr′mə-nəl) *adj.* [Lat. *terminalis* < *terminus*, boundary.] **1.** Of, relating to, situated at, or forming an end or boundary. **2.** *Biol.* Growing or appearing at the end, as of a stem, branch, or stalk. **3.** Relating to or occurring at the end of a section or series : FINAL <made some *terminal* remarks> **4.** Relating to or occurring in a term or each term <*terminal* payments> **5.** Ending in death : FATAL <a *terminal* illness> —*n.* **1.** A terminating point, limit, or part. **2.** An ornamental figure or object, as a finial of a lamp, situated at the end of another object. **3.** *Elect.* **a.** A position in an electric circuit or device at which an electric connection is normally established or broken. **b.** A passive conductor at such a position used to facilitate the connection. **4. a.** Either end of a transportation line, as a railroad : TERMINUS. **b.** A station at such a point or at a major junction on such a line. **c.** A town at the end of a carrier line. **5.** *Computer Sci.* An instrument through which data or information can enter or leave a computer. **—ter′mi·nal·ly** *adv.*

**terminal leave** *n.* Final leave equal to accumulated unused leave granted to a member of the armed forces just prior to separation or discharge.

**ter·mi·nate** (tûr′mə-nāt′) *v.* **-nat·ed, -nat·ing, -nates.** [Lat. *terminare, terminat-* < *terminus*, end.] —*vt.* **1.** To bring to an end or a halt. **2.** To occur at or form the end of : CONCLUDE. **3.** To discontinue the employment of. —*vi.* **1.** To come to an end. **2.** To have as an end or result.

**ter·mi·na·tion** (tûr′mə-nā′shən) *n.* **1.** The act of terminating or the state of being terminated. **2.** Spatial or temporal end. **3.** A result : outcome. **4.** The end of a word, as an inflectional ending, suffix, or final morpheme. **—ter′mi·na′tion·al** *adj.*

**ter·mi·na·tive** (tûr′mə-nā′tĭv) *adj.* Serving, designed, or tending to terminate : CONCLUSIVE. **—ter′mi·na′tive·ly** *adv.*

**ter·mi·na·tor** (tûr′mə-nā′tər) *n.* **1.** One that terminates. **2.** The dividing line between the bright and shaded regions of the disk of the moon or an inner planet.

**ter·mi·ni** (tûr′mə-nī′) *n. var. pl. of* TERMINUS.

**ter·mi·nol·o·gy** (tûr′mə-nŏl′ə-jē) *n., pl.* **-gies.** [Med. Lat. *terminus*, expression (< Lat., limit) + -LOGY.] **1.** The vocabulary of technical terms and usages appropriate to a particular field, subject, science, or art : NOMENCLATURE. **2.** The study of nomenclature. **—ter′mi·no·log′i·cal** (-nə-lŏj′ĭ-kəl) *adj.* **—ter′mi·no·log′i·cal·ly** *adv.* **—ter′mi·nol′o·gist** *n.*

**term insurance** *n.* Insurance for a designated period providing coverage for losses to the insured during that period but becoming void upon its expiration.

**ter·mi·nus** (tûr′mə-nəs) *n., pl.* **-nus·es** *or* **-ni** (-nī′) [Lat.] **1.** The final point : END. **2.** A terminal on a transportation line or the town in which it is located. **3. a.** A border or boundary. **b.** A stone or post marking a terminus.

**ter·mite** (tûr′mīt′) *n.* [Lat. *termes, termit-*, a wood-eating worm.] Any of numerous superficially antlike social insects of the order Isoptera, many species of which feed on wood and are highly destructive to living trees and wooden structures.

**term·less** (tûrm′lĭs) *adj.* **1.** Having no bounds or limits : without end. **2.** Unconditional.

**term paper** *n.* A lengthy report or essay required of a student usu. on a topic drawn from the subject matter of a course of study.

**tern**[1] (tûrn) *n.* [Of Scand. orig.] Any of various sea birds of the genus *Sterna* and related genera, related to and resembling the gulls but smaller and with a forked tail.

**tern**[2] (tûrn) *n.* [Lat. *terni*, three each < *ter*, thrice.] **1.** A set of three, esp. a combination of three numbers that wins a lottery prize. **2.** A three-masted schooner.

**ter·na·ry** (tûr′nə-rē) *adj.* [ME < Lat. *ternarius* < *terni*, three each < *ter*, thrice.] **1.** Made up of three or arranged in threes. **2.** *Math.* **a.** Having the base three. **b.** Involving three variables. —*n., pl.* **-ries.** A set or group of three.

**ter·nate** (tûr′nāt′, -nĭt) *adj.* [NLat. *ternatus* < Med. Lat., p.part. of *ternare*, to multiply by three < Lat. *terni*, three each < *ter*, thrice.] Arranged in or consisting of sets or groups of three, as a compound leaf with three leaflets. **—ter′nate′ly** *adv.*

**terne·plate** (tûrn′plāt′) *n.* [Prob. Fr. *terne*, dull (< OFr. < *ternir*, to tarnish) + PLATE.] Sheet iron or steel plated with an alloy of three or four parts of lead to one part of tin, used as a roofing material.

**ter·pene** (tûr′pēn′) *n.* [Obs. *terpentin*, turpentine + -ENE.] Any of various unsaturated hydrocarbons, $C_{10}H_{16}$, found in essential oils and oleoresins of plants such as conifers and used in organic syntheses. **—ter·pen′ic** *adj.* **—ter′pe·noid′** *adj.* & *n.*

**ter·pin·e·ol** (tər-pĭn′ē-ōl′, -ōl′) *n.* [TERP(ENE) + -INE + -OL.] Any of three isomeric alcohols, $C_{10}H_{17}OH$, occurring naturally in the essential oils of certain plants and used as a solvent, in perfumes, soaps, and medicine.

**ter·pol·y·mer** (tər-pŏl′ə-mər) *n.* [Lat. *ter*, three times + POLYMER.] A polymer made up of three distinct monomers.

**Terp·sich·o·re** (tûrp-sĭk′ə-rē) *n.* [Gk. *Terpsikhorē* : *terpein,* to delight + *khoros,* dance.] *Gk. Myth.* The Muse of dancing and choral singing.

**terp·si·cho·re·an** (tûrp′sĭk-ə-rē′ən, tûrp′sĭ-kôr′ē-ən, -kōr′-) *adj.* [< TERPSICHORE.] Of or relating to dancing. —*n.* A dancer.

**ter·ra al·ba** (tĕr′ə ăl′bə, ôl′bə) *n.* [NLat. < Lat., white earth.] **1.** Finely ground gypsum used in making paper, paints, and plastics. **2.** Kaolin.

**ter·race** (tĕr′ĭs) *n.* [OFr. < Lat. *terra,* earth.] **1. a.** An open colonnaded platform, as a promenade. **b.** A platform extending outdoors from a floor of a house or apartment building. **2.** An open, often paved area adjacent to a house serving as an outdoor living area : PATIO. **3.** A raised bank of earth having vertical or sloping sides and a flat top. **4.** A narrow, flat stretch of ground, often having a steep slope facing a river or sea. **5. a.** A row of buildings erected on raised ground or on a sloping site. **b.** A section of row houses. **6. a.** A narrow strip of landscaped earth in the middle of a street. **b.** A street, esp. one divided by such a terrace. —*vt.* **-raced, -rac·ing, -rac·es.** To make into or supply with a terrace.

**ter·ra cot·ta** (tĕr′ə kŏt′ə) *n.* [Ital. : *terra,* earth + *cotta,* baked.] **1. a.** A hard, semifired, waterproof ceramic clay used in pottery and building construction. **b.** Ceramic wares made of terra cotta. **2.** A brownish orange color. **—ter′ra-cot′ta** *adj.*

**ter·ra fir·ma** (tĕr′ə fûr′mə) *n.* [Lat.] Dry land : solid ground.

**ter·rain** (tə-rān′, tĕ-) *n.* [Fr. < OFr. < Lat. *terrenus,* terrene.] **1.** A tract of land : GROUND. **2.** The physical character of land : TOPOGRAPHY. **3.** A particular geographic area : REGION. **4.** *var. of* TERRANE.

**Ter·ra·my·cin** (tĕr′ə-mī′sĭn). A trademark for oxytetracycline.

**Ter·ran** (tĕr′ən) *n.* [< Lat. *terra,* earth.] An inhabitant of the earth.

**ter·rane** *also* **ter·rain** (tə-rān′, tĕ-) *n.* [Alteration of TERRAIN.] **1.** A series of related rock formations. **2.** An area having a preponderance of a particular rock or rock formations.

**ter·ra·pin** (tĕr′ə-pĭn) *n.* [Of Algonquian orig.] **1.** An aquatic North American turtle of the genus *Malaclemys* and related genera. **2.** *Chiefly Brit.* A semiterrestrial freshwater turtle.

**ter·ra·que·ous** (tĕr-ā′kwē-əs, -ăk′wē-) *adj.* [Lat. *terra,* earth + AQUEOUS.] Made up of both land and water.

**ter·rar·i·um** (tə-râr′ē-əm) *n., pl.* **-i·ums** or **-i·a** (-ē-ə) [Lat. *terra,* earth + -ARIUM.] A closed container or small enclosure in which small plants are grown or small animals, as turtles or lizards, are kept.

**ter·raz·zo** (tə-răt′sō, -răz′ō) *n.* [Ital.] A flooring material of stone chips, as marble, set in mortar and polished.

**ter·rene** (tĕ-rēn′, tĕr′ēn′) *adj.* [ME < Lat. *terrenus* < *terra,* earth.] Of or relating to the earth : EARTHLY.

**ter·re·plein** (tĕr′ə-plān′) *n.* [OFr. *terreplein* < OItal. *terrapieno* < *terrapienare,* to fill with earth : *terra,* earth (< Lat.) + *pieno,* full (< Lat. *plenus*).] A horizontal platform behind a parapet where heavy guns are mounted.

**ter·res·tri·al** (tə-rĕs′trē-əl) *adj.* [ME < Lat. *terrestris* < *terra,* earth.] **1.** Of or relating to the earth or its inhabitants. **2.** Having a worldly, mundane character or quality. **3.** Of, relating to, or made up of land. **4.** *Biol.* Growing or living on land. —*n.* An inhabitant of the earth. **—ter·res′tri·al·ly** *adv.* **—ter·res′tri·al·ness** *n.*

**ter·ret** (tĕr′ĭt) *n.* [ME *teret* < OFr. *toret,* dim. of *tor,* a round.] **1.** One of the metal rings on a harness through which the reins pass. **2.** A ring on an animal's collar, used for attaching a leash.

**terre-verte** (tĕr′vĕrt′) *n.* [Fr. : *terre,* earth + *verte,* green.] An olive-green pigment used by artists and usu. made from glauconite.

**ter·ri·ble** (tĕr′ə-bəl) *adj.* [ME < OFr. < Lat. *terribilis* < *terrēre,* to frighten.] **1.** Causing fear or alarm : DREADFUL <a *terrible* scream> **2.** Extremely formidable <the *terrible* responsibilities of a head of state> **3.** Extreme <paid a *terrible* price in winning the war> **4. a.** Disagreeable or unpleasant <had a *terrible* time finding a parking space> **b.** Markedly objectionable <*terrible* hypocrisy> **—ter′ri·ble·ness** *n.* **—ter′ri·bly** *adv.*

**ter·ric·o·lous** (tĕ-rĭk′ə-ləs) *adj.* [< Lat. *terricola,* earth-dweller : *terra,* earth + *colere,* to dwell.] *Biol.* Living on or in the ground.

**ter·ri·er** (tĕr′ē-ər) *n.* [Fr. *terrier,* burrow < *terre,* earth < Lat. *terra.*] Any of various usu. small, active dogs orig. bred for hunting burrowing animals.

**ter·rif·ic** (tə-rĭf′ĭk) *adj.* [Lat. *terrificus* : *terrēre,* to frighten + *facere,* to make.] **1.** Causing great fear : TERRIFYING <a *terrific* shriek> **2.** Very bad or unpleasant <a *terrific* stomachache> **3.** Very good or fine : SPLENDID <a *terrific* skater> **4.** Causing awe or amazement : ASTOUNDING <*terrific* speed> **—ter·rif′i·cal·ly** *adv.*

**ter·ri·fy** (tĕr′ə-fī′) *vt.* **-fied, -fy·ing, -fies.** [Lat. *terrificare* < *terrificus,* terrific.] **1.** To fill with terror : ALARM. **2.** To intimidate <was *terrified* by the prospect of the examination>

**ter·rig·e·nous** (tĕ-rĭj′ə-nəs) *adj.* [Lat. *terrigena,* earth-born : *terra,* earth + *gignere,* to bring forth.] Derived from the land, esp. by erosion. —Used primarily of sediments.

**ter·ri·to·ri·al** (tĕr′ĭ-tôr′ē-əl, -tōr′-) *adj.* **1.** Of or relating to a territory or to its jurisdictional powers. **2.** Relating or restricted to a particular territory : REGIONAL. **3. Territorial.** Organized for national or home defense. —*n.* **Territorial.** A member of a territorial army. **—ter′ri·to′ri·al·ly** *adv.*

**ter·ri·to·ri·al·ism** (tĕr′ĭ-tôr′ē-ə-lĭz′əm, -tōr′-) *n.* **1.** A social system that invests state authority and influence in the landowners. **2.** A system of church government based on primacy of civil power. **—ter′ri·to′ri·al·ist** *n.*

**ter·ri·to·ri·al·i·ty** (tĕr′ĭ-tôr′ē-ăl′ĭ-tē, -tōr′-) *n., pl.* **-ties. 1.** The status of a territory. **2.** A behavior pattern in animals consisting of the occupation and defense of a territory.

**ter·ri·to·ri·al·ize** (tĕr′ĭ-tôr′ē-ə-līz′, -tōr′-) *vt.* **-ized, -iz·ing, -iz·es. 1.** To add to by the acquisition of territory. **2.** To reduce to the status of a territory. **3.** To distribute among territories. **—ter′ri·to′ri·al·i·za′tion** *n.*

**territorial waters** *pl.n.* Inland and coastal waters under the jurisdiction of a state, esp. the ocean waters within three miles of the shoreline.

**ter·ri·to·ry** (tĕr′ĭ-tôr′ē, -tōr′ē) *n., pl.* **-ries.** [ME < Lat. *territorium* < *terra,* land.] **1.** An area of land : REGION. **2.** The land and waters under the jurisdiction of a state, nation, or sovereign. **3. Territory. a.** A part of the United States that is not admitted as a state, that is administered by a governor, and that has its own legislature. **b.** A semi-autonomous geographic region, as a colonial possession, that is dependent on an external government. **4.** The area for which one is responsible as representative or agent <a sales representative's *territory*> **5.** The area of a field defended by a sports team. **6.** *Biol.* An area inhabited by an individual animal or a mating pair or group of animals, and often vigorously defended against intruders. **7.** A sphere of interest : PROVINCE <This county is Democratic *territory.*>

**ter·ror** (tĕr′ər) *n.* [ME *terrour* < OFr. < Lat. *terror* < *terrēre,* to frighten.] **1.** Intense, overwhelming fear. **2.** Something, as a terrifying object or event, that instills intense fear. **3.** Ability to instill intense fear. **4.** Violence promoted by a group to achieve or maintain supremacy. **5.** *Informal.* One that is annoying or difficult to manage, esp. a child : NUISANCE.

**ter·ror·ism** (tĕr′ə-rĭz′əm) *n.* Systematic use of violence, terror, and intimidation to achieve an end. **—ter′ror·ist** *n.* **—ter′ror·is′tic** *adj.*

**ter·ror·ize** (tĕr′ə-rīz′) *vt.* **-ized, -iz·ing, -iz·es. 1.** To fill or overwhelm with terror : TERRIFY. **2.** To coerce by intimidation or fear. **—ter′ror·i·za′tion** *n.* **—ter′ror·iz′er** *n.*

**ter·ry** (tĕr′ē) *n., pl.* **-ries.** [Orig. unknown.] **1.** Any of the uncut loops forming the pile of a fabric. **2.** *also* **terry cloth.** A pile fabric with uncut loops on both sides, used for articles such as bath towels and bathrobes.

**terse** (tûrs) *adj.* **ters·er, ters·est.** [Lat. *tersus,* p.part. of *tergēre,* to cleanse.] Free of superfluity : CONCISE <a *terse* summation of the case> **—terse′ly** *adv.* **—terse′ness** *n.*

**ter·tian** (tûr′shən) *adj.* [ME, tertian fever < Lat. *tertianus,* of the third < *tertius,* third.] Recurring every other day or, when considered inclusively, every third day. —*n. Pathol.* Tertian fever.

**tertian fever** *n. Pathol.* Malaria caused by the invasion of *Plasmodium vivax* into new red blood cells, marked by a 48-hour life cycle in the human body with a recurrence of fever paroxysms at the end of each such period.

**ter·ti·ar·y** (tûr′shē-ĕr′ē) *adj.* [Lat. *tertiarius* < *tertius,* third.] **1.** Being third in position, order, degree, or rank. **2.** Of, relating to, or designating the short flight feathers nearest the body on the inner edge of a bird's wing. **3.** *Chem.* **a.** Of or relating to salts of acids containing three replaceable hydrogen atoms. **b.** Of or relating to organic compounds in which a group, as an alcohol or amine, is bound to three nonelementary radicals. **4.** Of or relating to the third order of a monastic system. **5. Tertiary.** Of, belonging to, or designating the geologic time, system of rocks, and sedimentary deposits of the first period of the Cenozoic era, extending from the Cretaceous period of the Mesozoic era to the Quaternary period of the Cenozoic era, marked by the appearance of modern flora and of apes and other large mammals. —*n., pl.* **-ies. 1.** A tertiary feather. **2.** A member of a tertiary order of a monastic system. **3. Tertiary.** The Tertiary period or system of deposits.

**tertiary color** *n.* A color resulting from the mixture of two secondary colors.

**ter·ti·um quid** (tûr′shē-əm kwĭd′) *n.* [LLat.] An element or factor intermediate between two groups.

**ter·va·lent** (tər-vā′lənt, tûr′vā′-) *adj. var. of* TRIVALENT.

**ter·za ri·ma** (tĕr′tsə rē′mə) *n., pl.* **ter·ze ri·me** (tĕr′tsā rē′mā) [Ital. : *terza,* third + *rima,* rhyme.] A verse form consisting of a series of triplets having 10-syllable or 11-syllable lines of which the middle line of one triplet rhymes with the first and third lines of the following triplet.

**tes·la** (tĕs′lə) *n.* [After Nikola *Tesla* (1857–1943).] The unit of magnetic flux density in the International System, equal to one weber per square meter.

**tesla coil** *n.* A transformer with an air-core primary and a capacitor-tuned secondary, used as a source of high-frequency power, as for x-ray tubes.

**tes·sel·late** (tĕs′ə-lāt′) *vt.* **-lat·ed, -lat·ing, -lates.** [Lat. *tessellatus,* of small square stones < *tessella,* small cube, dim. of *tessera,* a square.] To form into a mosaic pattern, as by using small squares of stone or glass. **—tes′sel·la′tion** *n.*

**tes·ser·a** (tĕs′ər-ə) *n., pl.* **tes·ser·ae** (tĕs′ə-rē′) [Lat., a square < Gk. *tesseres,* four.] One of the small squares of stone or glass used in making mosaic patterns.

**test¹** (tĕst) *n.* [ME, cupel < OFr., pot < Lat. *testum.*] **1.** A means employed to examine, try, or prove. **2.** A series of questions or problems designed to determine knowledge or intelligence. **3.** A standard or criterion. **4.** *Chem.* **a.** A physical or chemical reaction by which a substance may be detected or its properties ascertained. **b.** The reagent used in such determination. **c.** A positive result obtained. **5.** A cupel. —*v.* **test·ed, test·ing, tests.** —*vt.* **1.** To subject to a test : EXAMINE. **2. a.** To ascertain the presence or properties of (a substance). **b.** To assay (metal) in a cupel. —*vi.* **1.** To undergo a test. **2.** To achieve as a score or rating through testing. **3.** To exhibit certain properties under test conditions. **4.** To administer a test in order to diagnose <*test* for acid content>

☆ ***syns:*** TEST, EXPERIMENT, EXPERIMENTATION, TRIAL *n. core meaning* : an operation employed to resolve an uncertainty <ran *tests* for dioxin contamination>

**test²** (tĕst) *n.* [Lat. *testa,* shell.] A hard external covering of invertebrates, as that of certain insects.

**tes·ta** (tĕs′tə) *n., pl.* **-tae** (-tē′) [Lat., shell.] The often thick or hard outer coat of a seed.

**tes·ta·cean** (tĕs-tā′shən) *n.* [< NLat. *Testacea,* order name < Lat. *testaceus,* covered with a shell < *testa,* brick, tile.] A rhizopod of the order Testacea containing a shell. **—tes·ta′cean** *adj.*

**tes·ta·ceous** (tĕs-tā′shəs) *adj.* [Lat. *testaceus* < *testa,* shell.] **1.** *Biol.* Of, relating to, or having a shell or shell-like outer covering. **2.** Having the reddish-brown or brownish-yellow color of bricks.

**tes·ta·cy** (tĕs′tə-sē) *n. Law.* The state of being testate.

**tes·tae** (tĕs′tē′) *n. pl. of* TESTA.

**tes·ta·ment** (tĕs′tə-mənt) *n.* [ME < Lat. *testamentum* < *testari,* to make a will < *testis,* witness.] **1.** *Law.* **a.** A written document providing for the disposition of one's personal property after death. **b.** WILL¹ 9. **2. a.** A tangible proof or tribute that testifies to or serves as evidence. **b.** A statement of belief or conviction : CREDO. **3. a.** *Archaic.* A covenant between man and God. **b. Testament.** Either of the two main divisions of the Bible.

**tes·tate** (tĕs′tāt′) *adj.* [ME < Lat. *testatus,* p.part. of *testari,* to make one's will.] Having made a legally valid will before death.

**tes·ta·tor** (tĕs′tā′tər, tĕs-tā′tər) *n.* [ME *testatour* < AN < Lat. *testator* < *testari,* to make one's will.] One who has made a legally valid will before death.

**tes·ta·trix** (tĕs-tā′trĭks) *n.* [Lat., fem. of *testator,* testator.] A woman who has made a legally valid will before death.

**test case** *n.* A legal action whose outcome is likely to set a precedent or test the constitutionality of a statute.

**test·cross** (tĕst′krôs′, -krŏs′) *n.* A genetic cross between an individual having a dominant trait with a homozygous recessive individual to determine if the genotype of the dominant individual is homozygous or heterozygous. —*vt.* **-crossed, -cross·ing, -cross·es.** To subject to a testcross.

**test-drive** (tĕst′drīv′) *vt.* **-drove** (-drōv), **-driv·en** (-drĭv′ən), **-driv·ing, -drives.** To drive (e.g., a new car) so as to evaluate its performance.

**tes·ter¹** (tĕs′tər) *n.* [ME < Med. Lat. *testerium* < LLat. *testa,* skull < Lat., shell.] A canopy over a bed.

**tes·ter²** (tĕs′tər) *n.* [Alteration of TESTON.] TESTON 2.

**test·er³** (tĕs′tər) *n.* One that tests.

**tes·tes** (tĕs′tēz′) *n. pl. of* TESTIS.

**tes·ti·cle** (tĕs′tĭ-kəl) *n.* [ME *testicule* < Lat. *testiculus,* dim. of *testis,* testis.] A testis.

**tes·tic·u·late** (tĕs-tĭk′yə-lĭt) *also* **tes·tic·u·lar** (-lər) *adj.* Shaped like a testicle : OVOID.

**tes·ti·fy** (tĕs′tə-fī′) *v.* **-fied, -fy·ing, -fies.** [ME *testifien* < Lat. *testificari* : *testis,* witness + *facere,* to make.] —*vi.* **1.** To make a declaration of truth or fact under oath. **2.** To make a serious statement in support of an argument, position, or asserted fact. **3.** *Law.* To submit testimony : bear witness. **4.** To serve as witness or evidence. —*vt.* **1.** To provide evidence for : bear witness to. **2.** To state or affirm under oath. **3.** *Archaic.* To declare publicly. **—tes′ti·fi·ca′tion** *n.* **—tes′ti·fi′er** *n.*

**tes·ti·mo·ni·al** (tĕs′tə-mō′nē-əl) *n.* [ME < OFr. < LLat. *testimonialis,* of evidence < Lat. *testimonium,* testimony.] **1.** A formal or written statement testifying to a particular truth or fact. **2.** A written affirmation of another's worth or character. **3.** Something given as a tribute for one's service or achievement. **—tes′ti·mo′ni·al** *adj.*

☆ ***syns:*** TESTIMONIAL, SALUTE, SALVO, TRIBUTE *n. core meaning* : a formal token of appreciation and admiration for a person's high achievements <a plaque as a *testimonial* to your public service>

**tes·ti·mo·ny** (tĕs′tə-mō′nē) *n., pl.* **-nies.** [ME < Lat. *testimonium* < *testari,* to testify.] **1.** A declaration or affirmation of truth or fact, as that given before a court. **2.** Evidence in support of a fact or assertion : PROOF. **3.** The collective written and spoken testimony offered in a legal case. **4.** A public declaration regarding a religious experience. **5.** *often* **Testimony. a.** The law of Moses, inscribed on the tablets of stone. **b.** The ark containing these tablets.

**tes·tis** (tĕs′tĭs) *n., pl.* **-tes** (-tēz′) [Lat.] The male reproductive gland, the source of spermatozoa and of the androgens, usu. paired in an external scrotum in man and certain other mammals.

**tes·ton** (tĕs′tŏn′) *also* **tes·toon** (tĕs-to͞on′) *n.* [OFr. < OItal. *testone,* dim. of *testa,* head < LLat., skull < Lat., shell.] A coin with the image of a head on one side, esp.: **1.** A 16th-cent. French silver coin. **2.** An English coin of Henry VIII.

**tes·tos·ter·one** (tĕs-tŏs′tə-rōn′) *n.* [TEST(IS) + STER(OL) + -ONE.] A male sex hormone, $C_{19}H_{28}O_2$, produced in the testes and controlling secondary sex characteristics.

**test paper** *n.* **1.** A paper saturated with a reagent, as litmus, used in making chemical tests. **2.** A paper or booklet containing a student's work for an examination.

**test pilot** *n.* A pilot who flies new or experimental aircraft to test them for conformity to planned standards.

**test stand** *n.* A facility for static test firing of rocket engines to ascertain performance characteristics.

**test tube** *n.* A cylindrical clear glass tube usu. open at one end and rounded at the other, used in laboratory experimentation.

**test-tube baby** (tĕst′to͞ob′) *n.* A baby conceived outside the womb through fertilization of an egg removed from the mother.

**tes·tu·di·nate** (tĕs-to͞od′n-ĭt, -āt′, -tyo͞od′-) *adj.* [< NLat. *Testudinata,* order name < Lat. *testudo,* tortoise. —see TESTUDO.] Of or relating to a tortoise or turtle. **—tes·tu′di·nate′** *n.*

**tes·tu·do** (tĕs-to͞o′dō, -tyo͞o′-) *n., pl.* **-dos.** [Lat. < *testa,* shell.] An ancient Roman siege device consisting of a movable arched screen protecting the besiegers' approach to a wall.

**tes·ty** (tĕs′tē) *adj.* **-ti·er, -ti·est.** [ME *testif,* headstrong < AN < OFr. *teste,* head.] **1.** Having an irritable disposition : ILL-TEMPERED. **2.** Marked by irritability, impatience, or exasperation <a *testy* reply> **—tes′ti·ly** *adv.* **—tes′ti·ness** *n.*

**Tet** (tĕt) *n.* [Vietnamese *tết,* of Chin. orig.] The lunar New Year as celebrated in Southeast Asia.

**te·tan·ic** (tĕ-tăn′ĭk) *adj.* **1.** Of or relating to tetanus. **2.** Of or relating to tetany. **—te·tan′i·cal·ly** *adv.*

**tet·a·nize** (tĕt′n-īz′) *vt.* **-nized, -niz·ing, -niz·es.** To produce or induce tetanus in. **—tet′a·ni·za′tion** *n.*

**tet·a·nus** (tĕt′n-əs) *n.* [ME *tetane* < Lat. *tetanus* < Gk. *tetanos* < *teinein,* to stretch.] **1.** An acute, often fatal infectious disease caused by a bacillus, *Clostridium tetani,* that gen. enters the body through wounds, marked by rigidity and spasmodic contraction of the voluntary muscles. **2.** Continuous muscular contraction caused by reaction to rapidly repeated stimuli. **—tet′a·nal** (tĕt′n-əl) *adj.*

**tet·a·ny** (tĕt′n-ē) *n.* [< TETANUS.] An abnormal condition, occurring chiefly in children and young adults, marked by periodic painful muscular spasms caused by faulty calcium metabolism.

**tetch·y** *also* **tech·y** (tĕch′ē) *adj.* **-i·er, -i·est.** [Orig. unknown.] Easily annoyed : TOUCHY. **—tetch′i·ly** *adv.* **—tetch′i·ness** *n.*

**tête-à-tête** (tāt′ə-tāt′) *adv.* [Fr : *tête,* head + *à,* to + *tête,* head.] Together without the intrusion of a third person : in intimate privacy <spent the evening *tête-à-tête*> —*n.* **1.** A private conversation between two people. **2.** A sofa for two, esp. an S-shaped one allowing the occupants to face each other. —*adj.* Intimate : private.

**tête-bêche** (tĕt′bĕsh′) *adj.* [Fr., head to foot : *tête,* head + OFr. *bechevet,* head to foot.] Of, relating to, or characteristic of postage stamps printed upside-down in relation to one another.

**teth** (tĕt, tĕs) *n.* [Heb. *ṭēth.*] The ninth letter in the Hebrew alphabet. —See table at ALPHABET.

**teth·er** (tĕth′ər) *n.* [ME *tethir* < ON *tjōðr.*] **1.** A rope, chain, or halter for an animal, allowing it a short radius in which to move about. **2.** The range or scope of one's abilities or resources. —*vt.* **-ered, -er·ing, -ers.** To restrict with or as if with a tether. **—at the end of (one's) tether.** At the limit of one's endurance or resources.

**teth·er·ball** (tĕth′ər-bôl′) *n.* A game played by two people with paddles and a ball hung by a cord from an upright post, the objective being to wind the cord around the post.

**Te·thys** (tē′thĭs) *n.* [Gk. *Tēthus.*] *Gk. Myth.* A Titaness and sea goddess who was both sister and wife of Oceanus.

**tetr-** *pref. var. of* TETRA-.

**tet·ra** (tĕt′rə) *n.* [Short for NLat. *Tetragonopterus,* former genus name : LLat. *tetragonum,* tetragon + Gk. *pteron,* wing.] A small colorful tropical freshwater fish of the family Characidae, popular in home aquariums.

**tetra-** *or* **tetr-** *pref.* [Gk.] **1.** Four <*tetr*ode> **2.** Containing four of a specified kind of atom, radical, or group <*tetra*chloride>

**tet·ra·ba·sic** (tĕt′rə-bā′sĭk) *adj.* **1.** Containing four replaceable hydrogen atoms in a molecule. —Used of acids. **2.** Containing four univalent basic atoms or radicals. —Used of bases or salts. **—tet′ra·ba·sic′i·ty** (-sĭs′ĭ-tē) *n.*

**tet·ra·caine hydrochloride** (tĕt′rə-kān′) *n.* A crystalline compound, $C_{15}H_{24}N_2O_2 \cdot HCl$, used medically as a local anesthetic.

**tet·ra·chlo·ride** (tĕt′rə-klôr′īd′, -klōr′-) *n.* A chemical compound containing four chlorine atoms per molecule.

**tet·ra·chord** (tĕt′rə-kôrd′) *n.* [Gk. *tetrakhordon* < *tetrakhordos,* four-stringed : *tetra-,* four + *khordē,* string.] *Mus.* A series of four diatonic tones encompassing the interval of a perfect fourth. **—tet′ra·chor′dal** *adj.*

**te·trac·id** (tĕ-trăs′ĭd) *adj.* **1.** Capable of reacting with four molecules of a monobasic acid. —Used of a base. **2.** Containing four replaceable hydrogen atoms. —Used of an acid or acid salt. —*n.* An acid containing four replaceable hydrogen atoms.

**tet·ra·cy·cline** (tĕt′rə-sī′klēn′) *n.* [TETRA- + CYCL(IC) + -INE².] A yellow crystalline compound, $C_{22}H_{24}N_2O_8$, synthesized or derived from certain microorganisms of the genus *Streptomyces* and used as an antibiotic.

**tet·rad** (tĕt′răd′) *n.* [Gk. *tetras, tetrad-,* group of four < *tetra-,* four.] **1.** A group or set of four. **2.** A tetravalent atom, radical, or element. **3.** *Biol.* **a.** A group of four chromatids formed at meiosis by synapsis of two chromatids from each of a pair of homologous chromosomes. **b.** A body formed of four cells, as pollen grains from one mother cell.

**te·trad·y·mite** (tĕ-trăd′ə-mīt′) *n.* [G. *Tetradymit* < LGk. *tetradumos* : Gk. *tetra-,* four + *didumos,* double.] A steel-gray bismuth ore, chiefly $Bi_2Te_2S$.

**tet·ra·dy·na·mous** (tĕt′rə-dī′nə-məs) *adj.* [TETRA- + Gk. *dynamis,* strength + -OUS.] *Bot.* Having six stamens, two of which are shorter than the others.

**tet·ra·eth·yl lead** (tĕt′rə-ĕth′əl) *also* **tet·ra·eth·yl·lead** (-lĕd′) *n.* An oily, colorless, poisonous liquid, $Pb(C_2H_5)_4$, used in gasoline for internal-combustion engines as an antiknock agent.

**tet·ra·gon** (tĕt′rə-gŏn′) *n.* [LLat. < Gk. *tetragonon* : *tetra-,* four + *-gonon,* -gon.] A four-sided polygon : QUADRILATERAL. **—te·trag′o·nal** (tĕ-trăg′ə-nəl) *adj.* **—te·trag′o·nal·ly** *adv.*

**Tet·ra·gram·ma·ton** (tĕt′rə-grăm′ə-tŏn′) *n.* [ME *Tetragramaton* < Gk. *tetragrammaton,* four-letter word : *tetra-,* four + *gramma,* letter.] The four Hebrew letters usu. transliterated as YHWH or JHVH, for Yahweh or Jehovah, and used as a Biblical proper name for God.

**tet·ra·he·dra** (tĕt′rə-hē′drə) *n. pl. of* TETRAHEDRON.

**tet·ra·he·dral** (tĕt′rə-hē′drəl) *adj.* **1.** Relating to a tetrahedron. **2.** Having four faces. **—tet′ra·he′dral·ly** *adv.*

**tet·ra·he·drite** (tĕt′rə-hē′drīt′) *n.* [G. *Tetraëdrit* < Gk. *tetraedros,* four-faced. —see TETRAHEDRON.] A grayish-black copper ore, essentially $(CuFe)_{12}Sb_4S_{13}$, often containing other elements and occas. used as an ore of silver.

**tet·ra·he·dron** (tĕt′rə-hē′drən) *n., pl.* **-drons** *or* **-dra** (-drə) [Gk. *tetraedron* < *tetraedros,* four-faced : *tetra-,* four + *hedra,* face.] A four-faced polyhedron.

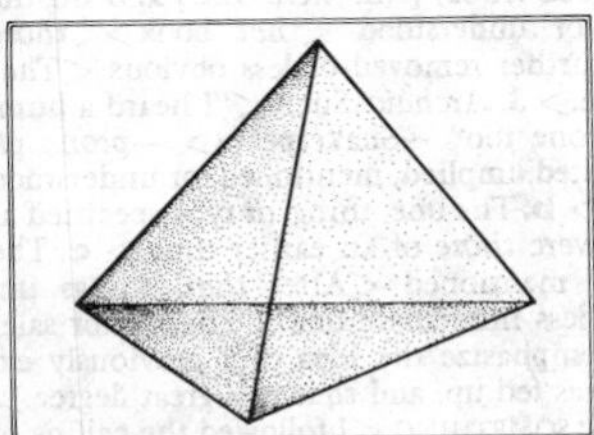
**tetrahedron**

**tet·ra·hy·dro·can·nab·i·nol** (tĕt′rə-hī′drə-kə-năb′ə-nôl′, -nōl′) *n.* A compound found in cannabis or made synthetically that is the primary intoxicant in marijuana.

**tet·ra·hy·drox·y** (tĕt′rə-hī-drŏk′sē) *adj.* Having four hydroxyl groups in a molecule.

**te·tral·o·gy** (tĕ-trăl′ə-jē, -trŏl′-) *n., pl.* **-gies.** [Gk. *tetralogia* : *tetra-,* four + *-logia,* -logy.] **1.** A series of four dramas, three tragic, one satiric, performed at the festivals dedicated to Dionysus in ancient Athens. **2.** A series of four related dramas, operas, or literary works. **3.** *Med.* A complex of four symptoms.

**tetralogy of Fallot** *n.* Fallot's tetralogy.

**tet·ra·mer** (tĕt′rə-mər) *n.* A polymer consisting of four identical monomers. **—tet′ra·mer′ic** (-mĕr′ĭk) *adj.*

**te·tram·er·ous** (tĕ-trăm′ər-əs) *adj.* [NLat. *tetramerus* < Gk. *tetramerēs* : *tetra-,* four + *meros,* part.] **1.** Having or consisting of four similar parts. **2.** *Bot.* Having flower parts, as sepals, petals, and stamens, in sets of four. **—te·tram′er·ism** *n.*

**te·tram·e·ter** (tĕ-trăm′ĭ-tər) *n.* [LLat. *tetrametrus* < Gk. *tetrametros* (adj.) : *tetra-,* tetra- + *-metron,* measure.] **1.** A line of verse consisting of four metrical feet. **2.** A verse consisting of tetrameters. **—te·tram′e·ter** *adj.*

**tet·ra·ploid** (tĕt′rə-ploid′) *Genetics.* —*adj.* Having four haploid sets of chromosomes. —*n.* A tetraploid individual.

**tet·ra·pod** (tĕt′rə-pŏd′) *adj.* [NLat. *tetrapodus* < Gk. *tetrapous,* four-footed : *tetra-,* four + *pous,* foot.] Having four feet, legs, or leglike appendages. **—tet′ra·pod′** *n.*

**te·trap·ter·ous** (tĕ-trăp′tər-əs) *adj.* [Gk. *tetrapteros* : *tetra-,* four + *pteron,* wing.] Having four wings <a *tetrapterous* insect>

**tet·rarch** (tĕt′rärk′, tē′trärk′) *n.* [ME < LLat. *tetrarcha* < Lat. *tetrarches* < Gk. *tetrarkhēs* : *tetra-,* four + *arkein,* to rule.] **1.** A governor of one of the four divisions of a country or province, esp. under the ancient Roman Empire. **2. a.** A subordinate ruler. **b.** One of four joint rulers. **3.** The commander of a subdivision of a phalanx in ancient Greece. **—te·trar′chic** (tĕ-trär′kĭk, tē-) *adj.*

**tet·rar·chy** (tĕt′rär′kē, tē′trär′-) *also* **tet·rar·chate** (-kāt′, -kĭt) *n., pl.* **-chies** *also* **-chates.** **1.** The area or jurisdiction of a tetrarch. **2. a.** Joint rule by four governors. **b.** Four governors ruling jointly.

**tet·ra·spore** (tĕt′rə-spôr′, -spōr′) *n.* One of four spores produced in a group from a sporangium, as in red algae. **—tet′ra·spor′ic** *adj.*

**te·tras·ti·chous** (tĕ-trăs′tĭ-kəs) *adj.* [NLat. *tetrastichus* < Gk. *tetrastikhos,* having four rows : *tetra-,* four + *stikhos,* row.] Arranged in four vertical rows, as leaves or flowers on a stalk.

**tet·ra·tom·ic** (tĕt′rə-tŏm′ĭk) *adj.* **1.** Having four atoms per molecule. **2.** Having four replaceable univalent atoms or radicals.

**tet·ra·va·lent** (tĕt′rə-vā′lənt) *adj. Chem.* Having valence 4.

**Tet·raz·zi·ni** (tĕt′rə-zē′nē) *adj.* [After Luisa *Tetrazzini* (1871–1940).] Made with noodles, mushrooms, and almonds in a cream sauce topped with cheese <turkey *Tetrazzini*>

**tet·rode** (tĕt′rōd′) *n.* [TETR(A)- + -ODE.] A four-element electron tube with a cathode, a control grid, a screen grid, and an anode.

**te·trox·ide** (tĕ-trŏk′sīd′) *n.* A chemical compound containing four oxygen atoms per molecule.

**tet·ryl** (tĕt′rəl) *n.* A yellow crystalline explosive, $C_7H_5N_5O_8$, used chiefly as a primer or detonator.

**tet·ter** (tĕt′ər) *n.* [ME *teter* < OE.] A skin disease, as psoriasis, herpes, and esp. eczema, characterized by itching and eruptions.

**Teu·ton** (to͞ot′n, tyo͞ot′n) *n.* [Lat. *Teutoni.*] **1. Teutons.** An ancient people, prob. of Germanic or Celtic origin, who lived in Jutland until about 100 B.C. **2.** One of the peoples speaking a Germanic language, esp. a German.

**Teu·ton·ic** (to͞o-tŏn′ĭk, tyo͞o-) *adj.* [Lat. *Teutonicus* < *Teutoni,* Teutons.] **1.** Of or pertaining to the Teutons. **2.** Of or pertaining to the Germanic languages. —*n.* Germanic.

**Teu·ton·ism** (to͞ot′n-ĭz′əm, tyo͞ot′-) *also* **Teu·ton·i·cism** (to͞o-tŏn′ĭ-sĭz′əm, tyo͞o-) *n.* **1.** A German practice or idiom. **2.** German character or civilization. **—Teu′ton·ist** *n.*

**Teu·ton·ize** (to͞ot′n-īz′, tyo͞ot′-) *vt.* **-ized, -iz·ing, -iz·es.** To give a German quality to. **—Teu′ton·i·za′tion** *n.*

**Te·vet** *also* **Te·bet** *or* **Te·beth** (tā′vəs, tā-vāt′) *n.* [Heb. *tēbhēth* < Akkadian *ṭebētu.*] The fourth month of the Hebrew year. —See table at CALENDAR.

**Te·wa** (tā′wə, tē′wə) *n., pl.* **Tewa** *or* **-was.** **1. a.** An Indian tribe of New Mexico and northeastern Arizona. **b.** A member of this tribe. **2.** The Tanoan language of the Tewa.

**tex·as** (tĕk′səs) *n.* [After *Texas.*] The structure on a river steamboat containing the pilothouse and the officers' quarters.

**Tex·as fever** *n.* An infectious disease of cattle and related animals, transmitted by ticks and caused by a parasitic microorganism, *Babesia bigemina.*

**Texas leaguer** *n.* [After *Texas League,* a baseball minor league.] *Baseball.* A fly ball that drops between the infielder and the outfielder for a hit.

**Texas Ranger** *n.* **1.** A member of the Texas mounted police force. **2.** A member of a band of men orig. organized in Texas to fight Indians and maintain order.

**Texas tower** *n.* [After *Texas.*] An offshore radar tower.

**text** (tĕkst) *n.* [ME *texte* < OFr. < Med. Lat. *textus* < Lat. < *texere,* to construct.] **1. a.** The words or wording of something written or printed. **b.** The words of a speech appearing in print. **2.** The formal content of a printed work. **3.** An author's original wording as opposed to a translation, revision, or condensation. **4.** A topic or theme. **5.** A reference used as the starting point of a discussion. **6.** A textbook.

**text·book** (tĕkst′bo͝ok′) *n.* A book used as a standard work for the formal study of a particular subject.

**text edition** *n.* An edition of a book designed esp. for use in schools.

**tex·tile** (tĕks′tīl′, -təl) *n.* [Lat. < *textilis,* woven < *textus,* p.part. of *texere,* to weave.] **1.** Fabric, esp. one woven or knitted. **2.** Fiber or yarn for weaving or knitting into fabric.

**tex·tu·al** (tĕks′cho͞o-əl) *adj.* **1.** Of, relating to, or contained in a text. **2.** Based on or conforming to a text. **3.** Being word for word : LITERAL. **—tex′tu·al·ly** *adv.*

**textual criticism** *n.* **1.** A study of a written work that attempts to establish the original text. **2.** Literary criticism stressing scholarly study and analysis of the text.

**tex·tu·al·ism** (tĕks′cho͞o-ə-lĭz′əm) *n.* **1.** Strict adherence to a text, esp. of the Scriptures. **2.** Textual criticism, esp. of the Scriptures. **—tex′tu·al·ist** *n.*

**tex·tu·ar·y** (tĕks′cho͞o-ĕr′ē) *adj.* Of, relating to, or contained in a text : TEXTUAL. —*n., pl.* **-ies.** A specialist in Scriptural studies.

**tex·ture** (tĕks′chər) *n.* [Lat. *textura* < *textum,* that which is woven < *tegere,* to weave.] **1. a.** The appearance of a fabric resulting from the woven arrangement of its fibers or yarns. **b.** A surface appearance suggesting the weave of a fabric. **2.** A grainy, fibrous, woven, or dimensional quality <Brick walls give a room *texture.*> **3.** Representation of the structure of a surface as distinct from color or form. **4.** Composition or structure of a substance : GRAIN. **5.** Distinctive or identifying character <the *texture* of urban life> **—tex′tur·al** *adj.* **—tex′tur·al·ly** *adv.*

**tex·tured** (tĕks′chərd) *adj.* **1.** Having a particular kind of texture <a rough-*textured* tweed> **2.** Having marked texture <a *textured* wall of old brick>

**tex·tus re·cep·tus** (tĕks′təs rĭ-sĕp′təs) *n.* [Lat.] Received text, esp. the received text of the Greek New Testament.

**T-group** (tē′gro͞op′) *n.* [T(RAINING) GROUP.] A group of individuals, as in a corporation, who seek to improve awareness and sensitivity in interpersonal relations during sessions with a trained leader.

**Th** *symbol for* THORIUM.

**-th**[1] *suff. var. of* -ETH[1].

**-th**[2] *suff.* [ME < OE *-ðu,* n. suffix.] **1.** Act : process <spil*th*> **2.** State : quality <dear*th*>

**-th**[3] *suff.* [ME *-the* < OE *-ða, -ðe.*] —Used to form ordinal numbers <million*th*>

**Thai** (tī) *n., pl.* **Thai. 1. a.** A native or citizen of Thailand. **b.** A member of the predominant ethnic group of Thailand, a people with both Mongoloid and Indonesian characteristics. **2.** The Tai language that is the official language of Thailand. *—adj. also* **Tai.** Of or relating to Thailand, its people, or its language.

**thal·a·men·ceph·a·lon** (thăl′ə-mĕn-sĕf′ə-lŏn′) *n.* [THALAM(US) + ENCEPHALON.] *Anat.* The diencephalon.

**thal·a·mus** (thăl′ə-məs) *n., pl.* **-mi** (-mī′) [NLat. < Gk. *thalamos,* inner chamber.] **1.** *Anat.* A large ovoid mass of gray matter that relays sensory stimuli to the cerebral cortex and acts in integrative and nonspecific functions. **2.** *Bot.* The receptacle of a flower. **—tha·lam′ic** (thə-lăm′ĭk) *adj.* **—tha·lam′i·cal·ly** *adv.*

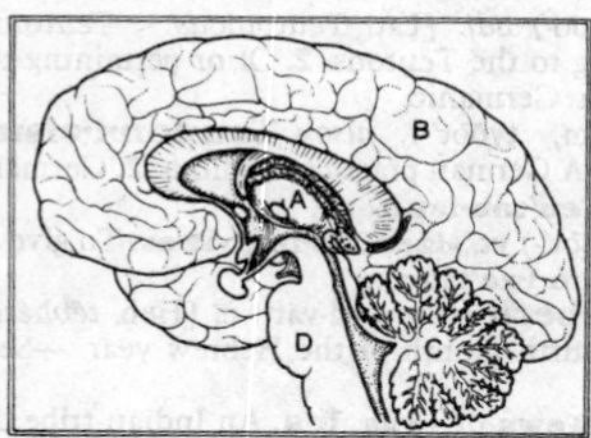


**thalamus**
*A. thalamus, B. cerebrum, C. cerebellum, D. medulla oblongata*

**thal·as·se·mi·a** (thăl′ə-sē′mē-ə) *n.* [Gk. *thalassa,* sea + -EMIA.] An inherited form of anemia that results from a faulty synthesis of hemoglobin. **—thal′as·se′mic** *adj.*

**tha·las·sic** (thə-lăs′ĭk) *adj.* [Fr. *thalassique* < Gk. *thalassa,* sea.] **1.** Of or relating to oceans or deep seas. **2.** Of, relating to, or situated about inland seas.

**thal·as·soc·ra·cy** (thăl′ə-sŏk′rə-sē) *n., pl.* **-cies.** [Gk. *thalassokratia* : *thalassa,* sea + *-kratia,* -cracy.] Supremacy on the seas. **—tha·las′so·crat′** (thə-lăs′ə-krăt′) *n.*

**tha·ler** (tä′lər) *n. var. of* TALER.

**Tha·li·a** (thə-lī′ə, thā′lē-ə) *n.* [Gk. *Thaleia* < *thallein,* to bloom.] *Gk. Myth.* **1.** The Muse of comedy and pastoral poetry. **2.** One of the three Graces.

**tha·lid·o·mide** (thə-lĭd′ə-mīd′) *n.* [(PH)THAL(IC ACID) + (IM)ID(E) + (I)MIDE.] A sedative and hypnotic drug, $C_{13}H_{10}N_2O_4$, withdrawn from sale because of association with fetal abnormalities.

**thall-** *pref. var. of* THALLO-.

**thal·li** (thăl′ī′) *n. var. pl. of* THALLUS.

**thal·lic** (thăl′ĭk) *adj.* Of, relating to, or containing thallium, esp. with valence 3.

**thal·li·um** (thăl′ē-əm) *n.* [THALL(O)- (from its green spectral line) + -IUM.] *Symbol* **Tl** A soft, malleable, highly toxic metallic element, used in rodent and ant poisons and low-melting glass; atomic number 81; atomic weight 204.37.

**thallo-** or **thall-** *pref.* [< Gk. *thallos,* young green shoot.] **1. a.** Young green shoot <*thall*ium> **b.** Thallus <*thall*oid> **2.** Thallium <*thall*ous>

**thal·loid** (thăl′oid′) *also* **thal·loi·dal** (thə-loid′l) *adj.* Of, resembling, or constituting a thallus.

**thal·lo·phyte** (thăl′ə-fīt′) *n.* A plant or plantlike organism of the division or subkingdom Thallophyta, which includes the algae, fungi, and bacteria. **—thal′lo·phyt′ic** (-fĭt′ĭk) *adj.*

**thal·lous** (thăl′əs) *adj.* Of, relating to, or containing thallium, esp. with valence 1.

**thal·lus** (thăl′əs) *n., pl.* **thal·li** (thăl′ī′) or **-lus·es.** [NLat. < Lat., green stalk < Gk. *thallos* < *thallein,* to sprout.] The undifferentiated stemless, rootless, leafless plant body characteristic of thallophytes.

**Tham·muz** (tä′mo͞oz′) *n. var. of* TAMMUZ.

**than** (thăn, thən) *conj.* [ME < OE *ðanne.*] **1.** —Used to introduce the second element or clause of an unequal comparison <I am a safer driver *than* you.> **2.** —Used to introduce the rejected alternative in statements of preference <I would rather swim *than* jog.> *—prep.* In comparison with <admired no one more *than* my parents>

**than·age** (thā′nĭj) *n.* **1.** The rank, jurisdiction, or office of a thane. **2.** The land held by a thane.

**than·a·top·sis** (thăn′ə-tŏp′sĭs) *n.* [Gk. *thanatos,* death + -OPSIS.] A meditation on death.

**Than·a·tos** (thăn′ə-tōs′) *n.* [Gk.] **1.** Death personified or as a philosophical notion. **2. thanatos.** An instinct toward self-destruction : DEATH WISH. **—than′a·tot′ic** (-tŏt′ĭk) *adj.*

**thane** (thān) *n.* [ME < OE *ðegn.*] **1. a.** A freeman granted land by the king in return for military service in Anglo-Saxon England. **b.** A man ranking above an ordinary freeman and below a nobleman in Anglo-Saxon England. **2.** A Scottish feudal lord or baron. **—thane′ship′** *n.*

**thank** (thăngk) *vt.* **thanked, thank·ing, thanks.** [ME *thanken* < OE *ðancian.*] **1.** To express gratitude to. **2.** To hold responsible <had only myself to *thank* for the accident>

**thank·ful** (thăngk′fəl) *adj.* **1.** Grateful. **2.** Expressing thanks. **—thank′ful·ly** *adv.* **—thank′ful·ness** *n.*

**thank·less** (thăngk′lĭs) *adj.* **1.** Not feeling or showing gratitude : UNGRATEFUL. **2.** Not apt to be appreciated. **—thank′less·ly** *adv.* **—thank′less·ness** *n.*

☆ ***syns:*** THANKLESS, UNGRATEFUL, UNTHANKFUL *adj. core meaning* : not apt to be appreciated <the often *thankless* job of a police officer> ***ant:*** appreciated

**thanks** (thăngks) *pl.n.* **1.** An acknowledgment of a favor, gift, or benefit. **2.** An expression of gratitude <give *thanks*> *—interj.* —Used to express thanks. **—no thanks to.** Without the benefit of help from <I finally found the answer, *no thanks to* you.> **—thanks to. 1.** Thanks be given to. **2.** On account of : because of <We were late for the ceremony, *thanks to* the traffic.>

**thanks·giv·ing** (thăngks-gĭv′ĭng) *n.* **1.** An expression of gratitude, esp. to God. **2. Thanksgiving.** Thanksgiving Day.

**Thanksgiving Day** *n.* A national holiday set apart for giving thanks to God, celebrated on the fourth Thursday of Nov. in the United States and on the second Monday of Oct. in Canada.

**thank·wor·thy** (thăngk′wûr′thē) *adj.* **-thi·er, -thi·est.** Worthy of thanks.

**thank-you** (thăngk′yo͞o′) *n.* An expression of gratitude.

**thank-you-ma'am** (thăngk′yo͞o-măm′) *n.* A bump or hollow in a road.

**that** (thăt, thət) *adj., pl.* **those** (thōz) [ME < OE *ðæt.*] **1.** Being the one singled out, implied, or understood <*that* book><*those* clothes> **2.** Being the one further removed or less obvious <*That* route is shorter than this one.> **3.** *Archaic.* Such <"I heard a humming, /And *that* a strange one too" —Shakespeare> *—pron., pl.* **those. 1. a.** The one designated, implied, mentioned, or understood <What breed of dog is *that?*> **b.** The one, thing, or type specified as follows <The relics found were *those* of an earlier time.> **c.** The event, action, or time just mentioned <After *that,* things improved.> **2.** The further or less immediate one <*That* is for sale; this is not.> **3.** —Used to emphasize the idea of a previously expressed word or phrase <I was fed up, and *that* to a great degree.> **4. a.** The one, kind, or thing : SOMETHING <I followed the calling of *that* I loved.><What's *that* you say?> **b. those.** Some persons <*those* who decided to stay> **5.** —Used as a relative pronoun to introduce a clause, esp. a restrictive clause <the car *that* has the broken windshield> **6. a.** In, on, by, or with which <each summer *that* the concerts are performed> **b.** In accordance with <I never met them, *that* I know of.> —Used after a negative. *—adv.* **1.** To such an extent or degree <Is your illness *that* serious?> **2.** To a high degree <shouldn't take your job *that* lightly> *—conj.* **1.** —Used to introduce a noun clause that is chiefly the subject or object of a verb or a predicate nominative <heard *that* you were leaving> **2.** —Used to introduce a subordinate clause stating a fact, wish, reason, or cause <We thought *that* they were lost.><I hoped *that* you would arrive on time.> **3. a.** —Used to introduce an anticipated subordinate clause following the expletive *it* occurring as subject of the verb <It is likely *that* I will reject the proposal.> **b.** —Used to introduce a subordinate clause modifying an adverb or adverbial expression <can stop anywhere *that* serves breakfast> **c.** —Used to introduce a subordinate clause that is joined to an adjective or noun as a complement <I was sure *that* you were wrong.> **4.** —Used to introduce an elliptical exclamation of desire <Oh, *that* we were home!> **—all that. 1.** All of the kind specified <beauty, brains, and *all that*> **2. a.** More of the same type <a store selling nails, hammers, saws, and *all that*> **b.** To the degree indicated <It is really not as bad as *all that.*> **—at that. 1.** In addition : BESIDES <merchandise of good quality and inexpensive *at that*> **2.** Regardless of what has been said or implied.

**thatch** (thăch) *n.* [ME *thacche* < *thacchen,* to thatch < OE *ðeccan,* to cover.] **1.** Plant stalks or foliage, as reeds or palm fronds, used for roofing. **2.** Something, as a thick growth of hair on the head, that resembles thatch. —*vt.* **thatched, thatch·ing, thatch·es.** To cover with or as if with thatch. **—thatch'er** *n.* **—thatch'y** *adj.*

**thau·ma·tol·o·gy** (thô'mə-tŏl'ə-jē) *n., pl.* **-gies.** [Gk. *thauma, thaumat-,* wonder + -LOGY.] The study of or a discourse on miracles.

**thau·ma·turge** (thô'mə-tûrj') *also* **thau·ma·tur·gist** (-tûr'-jĭst) *n.* [Med. Lat. *thaumaturgus* < Gk. *thaumatourgos* : *thauma,* wonder + *-ergos,* working < *ergon,* work.] A performer of miracles or magic feats.

**thau·ma·tur·gy** (thô'mə-tûr'jē) *n.* The working of miracles or wonders. **—thau'ma·tur'gic, thau'ma·tur'gi·cal** *adj.*

**thaw** (thô) *v.* **thawed, thaw·ing, thaws.** [ME *thawen* < OE *ðawian.*] —*vi.* **1.** To change from a frozen solid to a liquid by gradual warming. **2.** To lose stiffness, numbness, or impermeability by being warmed. **3.** To become warm enough for snow and ice to melt. **4.** To become less reserved : RELAX. —*vt.* To melt (a frozen solid) by gradual warming. —*n.* **1.** The process of thawing. **2.** A period of warm weather during which ice and snow melt. **3.** A relaxation of reserve, restraints, or tensions.

**THC** *n.* Tetrahydrocannabinol.

**the**[1] (thē *before a vowel;* thə *before a consonant*) *def.art.* [ME < OE *ðe,* alteration of *se,* masc. nominative sing. demonstrative adj.] **1. a.** —Used before singular or plural nouns and noun phrases that denote particular persons or things <read *the* newspaper> **b.** —Used before a noun, and gen. stressed, emphasizing one of a group or type as the most outstanding or prominent <*the* social event of the summer season> **c.** —Used before a title or rank or office designating its holder <*The* Prince of Wales><*The* President of the United States><*the* Reverend John Smith> **d.** —Used before nouns that designate natural phenomena or points of the compass <*The* weather looks stormy.><A wind is coming from *the* north.> **e.** —Used as the equivalent of a possessive adjective before names of some parts of the body <took me by *the* hand><an infection of *the* foot> **f.** —Used before a noun specifying a particular human endeavor <*the* law><*the* television industry><*the* stage> **g.** —Used before a proper name, as of a monument or ship <*the* Alamo><*the* Titanic> **h.** —Used before the plural form of a numeral denoting a specific decade or period in one's life <rural life in *the* twenties> **2.** —Used before a singular noun indicating that the noun is generic <*The* bald eagle is an endangered species.> **3. a.** —Used before an adjective extending it to signify a class and giving it the function of a noun <"*The* rich were dull and they drank too much" —Ernest Hemingway> **b.** —Used before an absolute adjective <*the* best we could find> **4.** —Used before a present participle, signifying the action in the abstract <*the* weaving of rugs> **5.** —Used before a noun with the force of *per* <ten dollars *the* box>

**the**[2] (thē *before a vowel;* thə *before a consonant*) *adv.* [ME < OE *ðy,* instrumental case of *the* and *that.*] **1.** To that extent : by that much <*the* sooner *the* better> **2.** Beyond any other <enjoyed reading mysteries *the* most>

**the-** *pref. var. of* THEO-.

**the·an·throp·ic** (thē'ăn-thrŏp'ĭk) *also* **the·an·throp·i·cal** (-ĭ-kəl) *adj.* [< LGk. *theanthrōpos,* god-man : Gk. *theos,* god + Gk. *anthrōpos,* man.] Both divine and human in nature or quality.

**the·an·thro·pism** (thē-ăn'thrə-pĭz'əm) *n.* **1.** Attribution of human traits to God : ANTHROPOMORPHISM. **2.** The theological doctrine of the union of human and divine natures in Christ. **—the·an'thro·pist** *n.*

**the·ar·chy** (thē'är'kē) *n., pl.* **-chies.** [LGk. *thearkhia* : Gk. *theos,* god + Gk. *-arkhia,* -archy.] **1.** Government or rule by a god : THEOCRACY. **2.** A hierarchy of gods.

**the·a·ter** *also* **the·a·tre** (thē'ə-tər) *n.* [ME *theatre* < OFr. < Lat. *theatrum* < Gk. *theatron* < *theasthai,* to watch.] **1.** A building, room, or outdoor structure for the presentation of dramatic performances, as plays or motion pictures. **2.** A room often with tiers of seats used for lectures or demonstrations : AUDITORIUM. **3. a.** Dramatic literature or its performance <the *theater* of Shakespeare> **b.** The milieu of actors and playwrights. **c.** The quality or effectiveness of a theatrical production <This play is good *theater.*> **4.** The audience gathered for a dramatic performance. **5. a.** A place that is the setting for dramatic events. **b.** A large geographic area in which military operations are carried on.

**the·a·ter·go·er** (thē'ə-tər-gō'ər) *n.* A person who often attends the theater.

**the·a·ter-in-the-round** (thē'ə-tər-ĭn-thə-round') *n., pl.* **theaters-in-the-round.** An arena theater.

**theater of the absurd** *n.* Dramatic literature that deals with the absurd.

**the·a·tre** (thē'ə-tər) *n. var. of* THEATER.

**the·at·ri·cal** (thē-ăt'rĭ-kəl) *also* **the·at·ric** (-ăt'rĭk) *adj.* **1.** Of, pertaining to, or suitable for the theater or dramatic performance. **2.** Characterized by exaggerated self-display and affectedly dramatic behavior : HISTRIONIC. —*n. often* **theatricals. 1.** Stage performances, esp. by amateurs. **2.** THEATRICS 2. **—the·at'ri·cal'i·ty, the·at'ri·cal·ness** *n.* **—the·at'ri·cal·ly** *adv.*

**the·at·ri·cal·ism** (thē-ăt'rĭ-kə-lĭz'əm) *n.* Theatrical manner or style : SHOWINESS.

**the·at·ri·cal·ize** (thē-ăt'rĭ-kə-līz') *vt.* **-ized, -iz·ing, -iz·es. 1.** To make theatrical : DRAMATIZE. **2.** To display in a showy way. **—the·at'ri·cal'i·za'tion** *n.*

**the·at·rics** (thē-ăt'rĭks) *n.* **1.** *(sing. in number).* The art of the theater. **2.** *(pl. in number).* Theatrical mannerisms calculated for effect.

☆ ***syns:*** THEATRICS, DRAMATICS, HISTRIONICS, MELODRAMATICS, THEATRICALS *n. core meaning* : overly emotional, exaggerated behavior calculated for effect <Let's spare the *theatrics* and address the issue squarely.>

**the·ba·ine** (thē'bə-ēn', thĭ-bā'ĭn) *n.* [Gk. *Thēbai,* Thebes + -INE.] A poisonous alkaloid, $C_{19}H_{21}NO_3$, obtained from opium.

**the·be** (thā'bā) *n.* [Native word in Botswana.] —See table at CURRENCY.

**the·ca** (thē'kə) *n., pl.* **-cae** (-sē', -kē') [NLat. < Lat., case < Gk. *thēkē.*] A case or sheath, as the spore case of a moss capsule or the outer covering of the pupa of certain insects. **—the'cal** (-kəl) *adj.*

**the·cate** (thē'kāt') *adj.* Having a theca : encased or sheathed.

**thee** (thē) *pron.* [ME < OE *ðe,* dative and accusative of *ðū,* thou.] *Archaic.* **1.** —Used: **a.** As the direct object of a verb. **b.** As the indirect object of a verb. **c.** As the object of a preposition. **2.** —Used in the nominative as well as the objective case, esp. by members of the Society of Friends.

**thee·lin** (thē'lĭn) *n.* [Gk. *thēlus,* female + -IN.] Estrone.

**thee·lol** (thē'lôl', -lōl') *n.* [THEEL(IN) + -OL.] Estriol.

**theft** (thĕft) *n.* [ME < OE *ðiefð.*] **1.** The act or an instance of stealing : LARCENY. **2.** *Obs.* That which is stolen.

**their** (thâr) *adj.* [ME < ON *ðeira.*] **1.** —Used to indicate possession or the agent or recipient of an action <*their* car><investing *their* earnings><suffered *their* first setback> **2.** *Informal.* His, hers, or its <Does everyone have *their* papers?>

**theirs** (thârz) *pron.* [ME < *their,* their.] *(sing. or pl. in number).* **1.** That or those belonging to them <The decision ought to be *theirs.*><Mine is here, and *theirs* are at the bus station.> **2.** *Informal.* His or hers <brought her tapes and assumed everybody else would bring *theirs*>

**the·ism** (thē'ĭz'əm) *n.* Belief in the existence of a god or gods, esp. belief in a personal God as creator and ruler of the world. **—the'ist** *n.* **—the·is'tic, the·is'ti·cal** *adj.* **—the·is'ti·cal·ly** *adv.*

**them** (thĕm, thəm) *pron.* [ME, partly < ON *ðeim,* and partly < OE *ðem.*] *The objective case of* THEY. —Used: **a.** As the direct object of a verb <The mayor accompanied *them.*> **b.** As the indirect object of a verb <I offered *them* a vacation in Europe.> **c.** As the object of a preposition <It was decided by *them.*>

**the·mat·ic** (thĭ-măt'ĭk) *adj.* [Gk. *thematikos* < *thema, themat-,* theme.] **1.** Of, constituting, or relating to a theme. **2.** Constituting the stem of a word. **—the·mat'i·cal·ly** *adv.*

**theme** (thēm) *n.* [ME < Lat. *thema* < Gk., proposition < *tithenai,* to place.] **1.** A topic of discourse or discussion. **2.** An idea, point of view, or perception embodied and expanded upon in a work of art. **3.** A short composition assigned to a student as a writing exercise. **4.** A principal melody in a musical composition. **5.** STEM[1] 5.

**theme song** *n.* **1.** A melody or song recurring throughout a dramatic performance and often intended to convey a mood. **2.** A song identified with a performer, group, or radio or television program.

**them·selves** (thĕm-sĕlvz', thəm-) *pron.* **1.** Those ones identical with them. —Used: **a.** Reflexively as the direct or indirect object of a verb or the object of a preposition <prepared *themselves* for the trip><gave *themselves* adequate time><were left by *themselves*> **b.** For emphasis <They *themselves* were discouraged.> **2.** Their normal or healthy state or condition <The crew members were *themselves* again after the storm passed.>

**then** (thĕn) *adv.* [ME < OE *ðenne.*] **1.** At that time in the past <We were happier *then.*> **2.** Next in time, space, or order : immediately afterward <played a last round of bridge and *then* went home> **3.** —Used after *but* to make noticeable or to balance a preceding statement <lost the election, but *then* never really expected to win> **4.** In that case : ACCORDINGLY <If you're late, *then* you'd better hurry.> **5.** In addition : moreover <*Then* there's the expense to consider.> **6.** As it appears <The plans, *then,* are confirmed.> **7.** Consequently <If x equals 3 and *y* equals 2, *then* x plus *y* equals 5.> —*n.* A specific time or moment <Until *then* let's not worry about it.> —*adj.* Being so at that time <the *then* governor> **—and then some.** With considerably more in addition <It took all my courage *and then some.*>

**the·nar** (thē'när') *n.* [Gk.] The fleshy mass on the palm of the hand at the base of the thumb. **—the'nar'** *adj.*

**thence** (thĕns, thĕns) *adv.* [ME *thannes* < *thanne,* from there < OE *ðanon.*] **1.** From that place : from there. **2.** From that time : THENCEFORTH. **3.** From that circumstance or source : THEREFROM.

**thence·forth** (thĕns-fôrth', -fōrth', thĕns-) *adv.* From that time forward : THEREAFTER.

**thence·for·ward** (thĕns-fôr′wərd, thĕns-) *also* **thence·for·wards** (-wərdz) *adv.* **1.** Thenceforth. **2.** From that time or place onward.

**theo-** *or* **the-** *pref.* [Gk. *theos*, god.] God <theomorphism>

**the·o·bro·mine** (thē′ō-brō′mēn′) *n.* [NLat. *Theobroma*, genus of trees (Gk. *theos*, god + *brōma*, food) + -INE.] A bitter, colorless alkaloid, $C_7H_8N_4O_2$, found in chocolate products, derived chiefly from the cacao bean, and used as a diuretic and a nerve stimulant.

**the·o·cen·tric** (thē′ō-sĕn′trĭk) *adj.* Centering on God as the prime concern.

**the·oc·ra·cy** (thē-ŏk′rə-sē) *n., pl.* **-cies.** [Gk. *theokratia* : *theos*, god + *-kratia*, -cracy.] **1.** Government by a god regarded as the ruling power or by priests or officials claiming divine sanction. **2.** A state governed by a theocracy.

**the·o·crat** (thē′ə-krăt′) *n.* **1.** A ruler of a theocracy. **2.** A believer in theocracy. **—the′o·crat′ic, the′o·crat′i·cal** *adj.* **—the′o·crat′i·cal·ly** *adv.*

**the·od·i·cy** (thē-ŏd′ĭ-sē) *n., pl.* **-cies.** [After *Théodicée*, a work by Gottfried Wilhelm von Leibnitz (1646–1716) : Gk. *theos*, god + Gk. *dikē*, order, right.] A vindication of divine justice in the face of the existence of evil.

**the·od·o·lite** (thē-ŏd′ə-līt′) *n.* [Orig. unknown.] A surveying instrument for measuring horizontal and vertical angles with a small telescope that can move in horizontal and vertical planes. **—the·od′o·lit′ic** *adj.*

**the·og·o·ny** (thē-ŏg′ə-nē) *n., pl.* **-nies.** [Gk. *theogonia* : *theos*, god + *-gonia*, -gony.] A recitation of the origin and genealogy of the gods, esp. as in ancient epic poetry. **—the′o·gon′ic** (-ə-gŏn′ĭk) *adj.* **—the·og′o·nist** *n.*

**the·o·lo·gi·an** (thē′ə-lō′jən) *n.* A specialist in theology.

**the·ol·o·gize** (thē-ŏl′ə-jīz′) *v.* **-gized, -giz·ing, -giz·es.** —*vt.* To make theological in form or significance. —*vi.* To speculate about theology. **—the·ol′o·giz′er** *n.*

**the·ol·o·gy** (thē-ŏl′ə-jē) *n., pl.* **-gies.** [ME *theologie* < OFr. < Lat. *theologia* < Gk. : *theos*, god + *-logia*, -logy.] **1.** The study of the nature of God and religious truth, esp. by an organized religious community. **2.** An organized, often formalized body of opinions concerning God and man's relationship to God. **3.** A course of specialized religious study usu. at a college or seminary. **—the′o·log′i·cal, the′o·log′ic** *adj.* **—the′o·log′i·cal·ly** *adv.*

**the·om·a·chy** (thē-ŏm′ə-kē) *n., pl.* **-chies.** [Gk. *theomakhia* : *theos*, god + *makhia*, fighting < *makhē*, battle.] Strife or battle among gods, as in the Homeric poems.

**the·o·mor·phism** (thē′ō-môr′fĭz′əm) *n.* The depiction or conception of man as having the form of a god. **—the′o·mor′phic** *adj.*

**the·oph·a·ny** (thē-ŏf′ə-nē) *n., pl.* **-nies.** [Med. Lat. *theophania* < LGk. *theophaneia* : Gk. *theos*, god + Gk. *phainein*, to show.] An appearance of a god to a human beings : divine manifestation.

**the·oph·yl·line** (thē-ŏf′ə-lĭn, thē′ō-fĭl′ēn′) *n.* [THEO(BROMINE) + PHYLL(O)- + -INE.] A colorless crystalline alkaloid, $C_7H_8N_4O_2.H_2O$, derived from tea leaves and also made synthetically, used as a diuretic and cardiac stimulant.

**the·or·bo** (thē-ôr′bō) *n., pl.* **-bos.** [Ital. *tiorba*.] A 17th-cent. lute having two sets of strings and an S-shaped neck with two sets of pegs, one set above and somewhat to the side of the other.

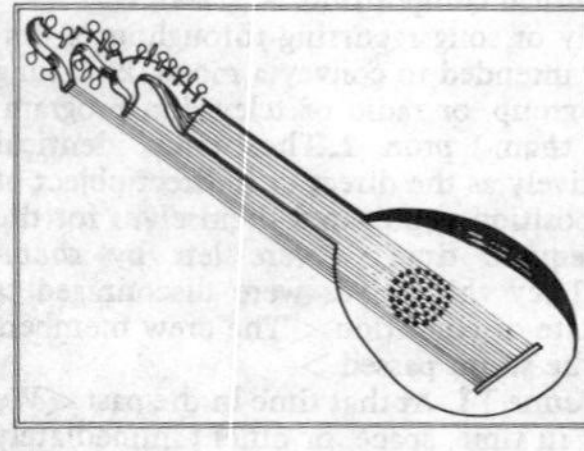

**theorbo**

**the·o·rem** (thē′ər-əm, thîr′əm) *n.* [LLat. *theorema* < Gk. *theōrēma* < *theōrein*, to look at < *theōros*, spectator. —see THEORY.] **1.** An idea demonstrably true or assumed to be so. **2.** *Math.* **a.** A proposition provable on the basis of explicit assumptions. **b.** A proven proposition.

**the·o·ret·i·cal** (thē′ə-rĕt′ĭ-kəl) *also* **the·o·ret·ic** (-rĕt′ĭk) *adj.* [LLat. *theoreticus* < Gk. *theōrētikos*, contemplative < *theōrētos*, observable < *theōrein*, to look at. —see THEOREM.] **1.** Of, relating to, or based on theory. **2.** Lacking verification or practical application : restricted to theory <*theoretical* physics> **3.** Tending to theorize : SPECULATIVE. **—the′o·ret′i·cal·ly** *adv.*

**the·o·re·ti·cian** (thē′ər-ə-tĭsh′ən) *n.* One who formulates, studies, or is expert in the theory of a science or art.

**the·o·ret·ics** (thē′ə-rĕt′ĭks) *n. (sing. in number).* The theoretical aspect of a science or art.

**the·o·rist** (thē′ər-ĭst) *n.* A theoretician.

**the·o·rize** (thē′ə-rīz′) *vi.* **-rized, -riz·ing, -riz·es. 1.** To formulate or analyze theories. **2.** To analyze by way of theory. **3.** To speculate. **—the′o·ri·za′tion** *n.* **—the′o·riz′er** *n.*

**the·o·ry** (thē′ə-rē, thîr′ē) *n., pl.* **-ries.** [LLat. *theoria* < Gk. *theōria* < *theōros*, spectator < *theasthai*, to observe.] **1. a.** Systematically organized knowledge applicable in a relatively wide variety of circumstances, esp. a system of assumptions, accepted principles, and rules of procedure devised to analyze, predict, or otherwise explain the nature or behavior of a specified set of phenomena. **b.** Such knowledge or such a system distinguished from experiment or practice. **2.** Abstract reasoning : SPECULATION. **3.** An assumption or guess based on limited knowledge or information : HYPOTHESIS.

**theory of games** *n.* Game theory.

**the·os·o·phy** (thē-ŏs′ə-fē) *n., pl.* **-phies.** [Med. Lat. *theosophia* < LGk. : Gk. *theos*, god + Gk. *sophia*, wisdom.] **1.** Religious philosophy or speculation based on mystical insight into the nature of God and divine teachings. **2.** *often* **Theosophy.** The doctrines and beliefs of a modern religious sect, the Theosophical Society, incorporating aspects of Buddhism and Brahmanism. **—the′o·soph′ic** (-ə-sŏf′ĭk), **the′o·soph′i·cal** *adj.* **—the′o·soph′i·cal·ly** *adv.* **—the·os′o·phist** *n.*

**ther·a·peu·tic** (thĕr′ə-pyōō′tĭk) *also* **ther·a·peu·ti·cal** (-tĭ-kəl) *adj.* [Gk. *therapeutikos* < *therapeutēs*, one who administers < *therapeuein*, to treat medically. —see THERAPY.] **1.** Having healing powers. **2.** Of or relating to therapeutics. **—ther′a·peu′ti·cal·ly** *adv.*

**ther·a·peu·tics** (thĕr′ə-pyōō′tĭks) *n. (sing. in number).* Medical treatment of disease. **—ther′a·peu′tist** *n.*

**ther·a·pist** (thĕr′ə-pĭst) *n.* A specialist in therapy.

**the·rap·sid** (thə-răp′sĭd) *n.* [< NLat. *Therapsida*, order name < Gk. *theraps*, attendant.] Any of various reptiles of the order Therapsida of the Permian and Triassic periods that are thought to be direct ancestors of mammals. **—the·rap′sid** *adj.*

**ther·a·py** (thĕr′ə-pē) *n., pl.* **-pies.** [NLat. *therapia* < Gk. *therapeia* < *therapeuein*, to treat medically < *theraps*, attendant.] **1.** Treatment of illness or disability. **2.** Healing power or quality. **3.** Psychotherapy.

**Ther·a·va·da** (thĕr′ə-vä′də) *n.* [Pali *theravāda* : *thera*, elder (< Skt. *sthavira-*, old) + *vāda*, doctrine (< Skt. *vādaḥ*, speech, doctrine).] Hinayana.

**there** (thâr) *adv.* [ME < OE *ðǣr*.] **1.** At or in that place <Stand over *there*.> **2.** To, into, or toward that place <wouldn't take the bus *there*> **3.** At a point of action or time <Stop *there* before you make a mess.> **4.** In that matter <I disagree with you *there*.> —*pron.* **1.** —Used to introduce a clause or sentence <*There* are many tales about the place.> **2.** —Used in place of a name <Hello *there*.> —*adj.* **1.** —Used as an intensive <That person *there* might help us.> ***usage:*** In some dialects *there* may occur together with *that* or *those* among the modifiers before a noun, as in *that there red book*. This usage is inappropriate in formal English. **2.** Present <You were still *there* when I left.> —*n.* That place or point <turned the car around and went on from *there*> —*interj.* —Used to express emotion, as relief, satisfaction, or consolation <*There*, now I can have some rest!>

**there·a·bouts** (thâr′ə-bouts′) *also* **there·a·bout** (-bout′) *adv.* **1.** Near that number or degree <was 21, or *thereabouts*> **2.** Near that place or time <lived in New Mexico, or *thereabouts*>

**there·af·ter** (thâr-ăf′tər) *adv.* From a specified time onward <an apprentice for three years, an electrician *thereafter*>

**there·a·gainst** (thâr′ə-gĕnst′) *adv.* In opposition to.

**there·at** (thâr-ăt′) *adv.* **1.** At a specified place. **2.** At such time or occasion.

**there·by** (thâr-bī′) *adv.* **1.** Through or with the agency of. **2.** In a specified connection or relation wherein.

**there·for** (thâr-fôr′) *adv. Archaic.* For that, this, or it.

**there·fore** (thâr′fôr′, -fōr′) *adv.* For that reason : HENCE <Your information is inaccurate and your conclusion *therefore* wrong.>

**there·from** (thâr-frŭm′, -frŏm′) *adv.* Coming from that time, location, or thing.

**there·in** (thâr-ĭn′) *adv.* **1.** In that place. **2.** In that circumstance or respect <*Therein* lies the problem.>

**there·in·af·ter** (thâr′ĭn-ăf′tər) *adv.* In a later or subsequent portion, as of a speech or book.

**ther·e·min** (thĕr′ə-mĭn) *n.* [After Leo *Theremin* (b. 1896), its inventor.] An electronic consolelike musical instrument often used for high tremolo effects.

**there·of** (thâr-ŭv′, -ŏv′) *adv.* **1.** Of or concerning this, that, or it. **2.** From or because of a stated cause or origin : THEREFROM.

**there·on** (thâr-ŏn′, -ôn′) *adv.* **1.** On or upon this, that, or it. **2.** Directly following that.

**there·to** (thâr-tōō′) *adv.* **1.** Thereunto <affixed my seal *thereto*> **2.** *Archaic.* In addition to that : FURTHERMORE.

**there·to·fore** (thâr′tə-fôr′, -fōr′) *adv.* Until or prior to a specified time : before that <a *theretofore* little known musician>

**there·un·der** (thâr-ŭn′dər) *adv.* Under this, that, or it.

**there·un·to** (thâr′ŭn-tōō′) *adv.* To that, this, or it.

**there·up·on** (thâr′ə-pŏn′, -pôn′) *adv.* **1.** Upon this, that, or it. **2.** Directly following that. **3.** In consequence of that : THEREFORE.
**there·with** (thâr-wĭth′, -wĭth′) *adv.* **1.** With that, this, or it. **2.** In addition to that. **3.** Immediately thereafter.
**there·with·al** (thâr′wĭth-ôl′, -wĭth-) *adv.* **1.** With all that, this, or it : BESIDES. **2.** *Obs.* With that, this, or it : THEREWITH.
**the·ri·o·mor·phic** (thîr′ē-ə-môr′fĭk) *also* **the·ri·o·mor·phous** (-fəs) *adj.* [Gk. *thērion,* beast, dim. of *thēr,* beast + -MORPHIC.] Having the form of a beast.
**therm** (thûrm) *n.* [Gk. *thermē,* heat.] A unit of heat equal to: **a.** One hundred thousand British thermal units. **b.** One thousand large calories. **c.** The large calorie. **d.** The small calorie.
**therm-** *pref. var. of* THERMO-.
**-therm** *suff.* [< Gk. *thermē,* heat.] An animal having a specified kind of body temperature <poikilo*therm*>
**ther·mal** (thûr′məl) *also* **ther·mic** (-mĭk) *adj.* [< Gk. *thermē,* heat.] **1.** Of, relating to, utilizing, producing, or caused by heat. **2.** Designed to help retain body heat <*thermal* underwear> —*n.* A rising current of warm air. **—ther′mal·ly** *adv.*
**thermal neutron** *n.* A slow neutron.
**therm·i·on** (thûr′mī′ən) *n.* An electrically charged particle or ion emitted by a conducting material at high temperatures. **—therm′i·on′ic** (-mī-ŏn′ĭk) *adj.*
**thermionic current** *n.* A flow of thermions.
**thermionic emission** *n.* Emission of thermions from a conducting material at high temperatures.
**therm·i·on·ics** (thûr′mī-ŏn′ĭks) *n. (sing. in number).* The physics of thermionic phenomena.
**thermionic tube** *n.* An electron tube in which the source of electrons is a heated electrode.
**therm·is·tor** (thûr′mĭs′tər) *n.* [THERM(AL) + (RES)ISTOR.] A resistor composed of semiconductors having resistance that varies rapidly and predictably with temperature.
**thermo-** *or* **therm-** *pref.* [< Gk. *thermē,* heat.] **1.** Heat <*thermo*chemistry> **2.** Thermoelectric <*thermo*junction>
**ther·mo·chem·is·try** (thûr′mō-kĕm′ĭ-strē) *n.* The chemistry of heat and heat-associated chemical phenomena. **—ther′mo·chem′i·cal** (-kĕm′ĭ-kəl) *adj.* **—ther′mo·chem′ist** *n.*
**ther·mo·cline** (thûr′mə-klīn′) *n.* The region in a thermally stratified body of water, as a lake, in which the temperature decrease with depth is greater than that of the water above and below it.
**ther·mo·co·ag·u·la·tion** (thûr′mō-kō-ăg′yə-lā′shən) *n.* Use of heat produced by high frequency currents to bring about the localized destruction of tissues.
**ther·mo·cou·ple** (thûr′mə-kŭp′əl) *n.* A thermoelectric device for measuring temperatures accurately, esp. one composed of two dissimilar metals joined so that a potential difference generated between the points of contact is a measure of the temperature difference between the points.
**ther·mo·dur·ic** (thûr′mō-dŏŏr′ĭk, -dyŏŏr′-) *adj.* [THERMO- + Lat. *durare,* to last + -IC.] Of or relating to microorganisms that can survive high temperatures.
**ther·mo·dy·nam·ics** (thûr′mō-dī-năm′ĭks) *n. (sing. in number).* The physics of the interrelationships between heat and other energy forms. **—ther′mo·dy·nam′ic** *adj.* **—ther′mo·dy·nam′i·cal·ly** *adv.*
**ther·mo·e·lec·tric** (thûr′mō-ĭ-lĕk′trĭk) *also* **ther·mo·e·lec·tri·cal** (-trĭ-kəl) *adj.* Characteristic of or resulting from electrical phenomena occurring in conjunction with a heat flow. **—ther′mo·e·lec′tri·cal·ly** *adv.*
**ther·mo·e·lec·tric·i·ty** (thûr′mō-ĭ-lĕk-trĭs′ĭ-tē) *n.* Electricity generated by a heat flow, as in a thermocouple.
**ther·mo·e·lec·tron** (thûr′mō-ĭ-lĕk′trŏn′) *n.* An electron emitted by a material at high temperatures.
**ther·mo·gram** (thûr′mə-grăm′) *n.* A record made by a thermograph.
**ther·mo·graph** (thûr′mə-grăf′) *n.* A thermometer that records the temperature it indicates.
**ther·mog·ra·phy** (thər-mŏg′rə-fē) *n.* **1.** Production of raised lettering, as on stationery or calling cards, by transferring the inked lines on a plate to the paper by pressure and suction. **2.** A diagnostic technique for measuring blood flow by determining the variations in heat emitted from the body. **—ther′mo·graph′ic** *adj.* **—ther′mo·graph′i·cal·ly** *adv.*
**ther·mo·junc·tion** (thûr′mō-jŭngk′shən) *n.* A point of contact between two dissimilar metals at which a thermoelectric current is generated.
**ther·mo·la·bile** (thûr′mō-lā′bĭl′, -bəl) *adj.* Subject to destruction, to decomposition, or to great change by moderate heating. —Used esp. of certain biochemicals. **—ther′mo·la·bil′i·ty** *n.*
**ther·mo·lu·mi·nes·cence** (thûr′mō-lōō′mə-nĕs′əns) *n.* A phenomenon in which certain minerals release previously absorbed radiation when moderate heat is applied. **—ther′mo·lu′mi·nes′cent** *adj.*
**ther·mol·y·sis** (thər-mŏl′ĭ-sĭs) *n.* **1.** *Physiol.* Loss of heat from the body. **2.** *Chem.* Dissociation or decomposition of compounds by heat. **—ther·mo·lyt′ic** *adj.*
**ther·mom·e·ter** (thər-mŏm′ĭ-tər) *n.* An instrument for temperature measurement, esp. one having a graduated glass tube with a bulb containing a liquid, typically mercury, that expands and rises in the tube as the temperature increases.
**ther·mom·e·try** (thər-mŏm′ĭ-trē) *n.* **1.** Measurement of temperature. **2.** The technology of temperature measurement. **—ther′mo·met′ric** (thûr′mō-mĕt′rĭk) *adj.*
**ther·mo·mo·tor** (thûr′mō-mō′tər) *n.* An engine operated by heat.
**ther·mo·nu·cle·ar** (thûr′mō-nōō′klē-ər, -nyōō′-) *adj.* **1.** Of, relating to, or derived from the fusion of atomic nuclei at high temperatures. **2.** Of or relating to atomic weapons based on fusion.
**ther·mo·pe·ri·od·ism** (thûr′mō-pîr′ē-ə-dĭz′əm) *also* **ther·mo·pe·ri·o·dic·i·ty** (-dĭs′ĭ-tē) *n.* The effect of the rhythmic fluctuation of temperature upon an organism, including responses corresponding to thermal changes due to alternation of day and night.
**ther·mo·phil·ic** (thûr′mə-fĭl′ĭk) *adj. Biol.* Requiring high temperatures for normal development, as certain bacteria. **—ther′mo·phile′** (-fīl′) *n.*
**ther·mo·pile** (thûr′mə-pīl′) *n.* [THERMO- + PILE[1].] A device to measure temperature, consisting of a number of thermocouples connected in series.
**ther·mo·plas·tic** (thûr′mə-plăs′tĭk) *adj.* Becoming soft when heated and hard when cooled. **—ther′mo·plas′tic** *n.* **—ther′mo·plas·tic′i·ty** (-plă-stĭs′ĭ-tē) *n.*
**ther·mo·re·cep·tor** (thûr′mō-rĭ-sĕp′tər) *n.* A sensory receptor that responds to heat and cold.
**ther·mo·reg·u·la·tion** (thûr′mō-rĕg′yə-lā′shən) *n.* Maintenance of a constant internal body temperature regardless of environmental temperature. **—ther′mo·reg′u·la·to′ry** (-rĕg′yə-lə-tôr′ē, -tōr′ē) *adj.*
**Ther·mos bottle** (thûr′məs). A trademark for a vacuum flask.
**ther·mo·set·ting** (thûr′mō-sĕt′ĭng) *adj.* Permanently hardening or solidifying on being heated <*thermosetting* resins>
**ther·mo·sphere** (thûr′mə-sfîr′) *n.* The outermost shell of the atmosphere between the mesosphere and outer space, where temperatures increase steadily with altitude. **—ther′mo·spher′ic** *adj.*
**ther·mo·sta·ble** (thûr′mō-stā′bəl) *also* **ther·mo·sta·bile** (-bəl, -bīl′) *adj.* Unaffected by high temperatures. **—ther′mo·sta·bil′i·ty** (-stə-bĭl′ĭ-tē) *n.*
**ther·mo·stat** (thûr′mə-stăt′) *n.* A device that automatically responds to temperature changes and activates switches controlling equipment as furnaces, refrigerators, and air conditioners. **—ther′mo·stat′ic** *adj.*
**ther·mo·tax·is** (thûr′mə-tăk′sĭs) *n.* **1.** Movement of a living organism in response to heat. **2.** Normal regulation or adjustment of body temperature. **—ther′mo·tac′tic** (-tăk′tĭk) *adj.*
**ther·mo·ther·a·py** (thûr′mō-thĕr′ə-pē) *n.* Therapy by heat.
**ther·mot·ro·pism** (thər-mŏt′rə-pĭz′əm) *n. Biol.* Growth or movement of organisms, as plants, in response to heat. **—ther′mo·trop′ic** (thûr′mə-trŏp′ĭk) *adj.*
**-thermy** *suff.* [NLat. *-thermia* < Gk. *thermē,* heat.] Heat <dia*thermy*>
**the·ro·pod** (thîr′ə-pŏd′) *n.* [NLat. *Theropoda,* suborder name : Gk. *thēr,* beast + Gk. *pous,* foot.] Any of various carnivorous dinosaurs of the suborder Theropoda, of the Jurassic and Cretaceous periods, having small forelimbs. **—the·rop′o·dan** (thĭ-rŏp′ə-dən) *adj.* & *n.*
**the·sau·rus** (thĭ-sôr′əs) *n., pl.* **-sau·ri** (-sôr′ī′) *or* **-sau·rus·es.** [Lat., collection < Gk. *thēsauros,* treasure.] **1.** A book of selected words or concepts, as a specialized vocabulary of a given field, as medicine or music. **2.** A book of synonyms.

▲ word history: A thesaurus is literally a treasury of words and information, for the word itself comes from Greek *thesauros,* "treasure." The Romans borrowed the Greek word, and Latin *thesaurus* developed into Old French *tresor,* which was borrowed into English as *treasure.* The Latin word *thesaurus* was used in titles of English books in early modern times, but it did not become a generic term for a particular kind of book until the 19th century.

**these** (thēz) *pron.* & *adj. pl. of* THIS.
**The·se·us** (thē′sē-əs, -syōōs′) *n. Gk. Myth.* A hero and king of Athens who slew the Minotaur and conquered the Amazons and married Phaedra, their queen. **—The·se′an** (thĭ-sē′ən) *adj.*
**the·sis** (thē′sĭs) *n., pl.* **-ses** (-sēz′) [Lat. < Gk. < *tithenai,* to place.] **1. a.** A proposition maintained, as by a degree candidate, by argument. **b.** A hypothetical proposition, esp. one put forth for the sake of argument or one to be accepted without proof. **2.** A dissertation advancing an original point of view as a result of research, esp. as a requirement for an academic degree. **3.** The first stage of dialectic. **4.** The unstressed part of a foot in prosody. **5.** *Mus.* The accented section of a measure : DOWNBEAT.
**Thes·pi·an** (thĕs′pē-ən) *adj.* **1.** Of or relating to Thespis. **2.** *often* **thespian.** Of or relating to drama : DRAMATIC. —*n. also* **thes·pi·an.** An actor or actress.
**Thes·sa·lo·ni·ans** (thĕs′ə-lō′nē-ənz) *pl.n. (sing. in number).* —See table at BIBLE.
**the·ta** (thā′tə, thē′-) *n.* [Gk. *thēta,* of Phoenician orig.; akin to Heb.

*ṭēth*, teth.] The eighth letter in the Greek alphabet. —See table at ALPHABET.

**thet·ic** (thĕt'ĭk, thē'tĭk) *also* **thet·i·cal** (thĕt'ĭ-kəl, thē'tĭ-) *adj.* [Gk. *thetikos*, fit for placing < *thetos*, placed < *tithenai*, to place.] **1.** Beginning with, constituting, or pertaining to the thesis in prosody. **2.** Presented dogmatically. **—thet'i·cal·ly** *adv.*

**The·tis** (thē'tĭs) *n.* [Gk.] *Gk. Myth.* One of the Nereids, the wife of Peleus and mother of Achilles.

**the·ur·gy** (thē'ûr-jē) *n., pl.* **-gies.** [LLat. *theurgia* < Gk. *theourgia*, sorcery : *theos*, god + *-ergos*, working.] **1.** Divine or supernatural intervention in human affairs. **2.** Performance of miracles with supernatural assistance. **3.** Magic supposedly performed with the aid of beneficent spirits, as practiced by Neo-Platonists. **—the·ur'gic** (thē-ûr'jĭk), **the·ur'gi·cal** *adj.* **—the·ur'gi·cal·ly** *adv.* **—the'ur·gist** *n.*

**thew** (thyōō) *n.* [ME, good quality < OE *ðēaw*, a characteristic.] **1.** A well-developed sinew or muscle. **2. thews.** Muscular power or strength. **—thew'y** *adj.*

**they** (*th*ā) *pron.* [ME < ON *ðeir*, masc. pl. demonstrative and personal pronoun.] **1.** Those ones <We had a dog and cat, but *they* fought all the time.> **2.** People in general <*They* said we couldn't do it.><a lawyer as tough as *they* come> **3.** *Informal.* He or she <Every person has legal rights but *they* don't always know them.>

**they'd** (*th*ād). **1.** They had. **2.** They would.

**they'll** (*th*āl). They will.

**they're** (*th*âr). They are.

**they've** (*th*āv). They have.

**thi-** *pref. var. of* THIO-.

**thi·a·ben·da·zole** (thī'ə-bĕn'də-zōl') *n.* [THIA(ZOLE) + BENZ(O)- + (IMI)D(E) + AZOLE.] A white compound, $C_{10}H_7N_3S$, used medically as an antifungal agent and as a drug to control parasitic roundworms.

**thi·a·mine** (thī'ə-mĭn, -mēn') *also* **thi·a·min** (-mĭn) *n.* [Alteration of *thiamin* : THI(O)- + (VIT)AMIN.] A B-complex vitamin, $C_{12}H_{17}ClN_4OS$, produced synthetically and occurring naturally in the bran coat of grains, in yeast, and in meat, that is essential for carbohydrate metabolism, maintenance of normal neural activity, and the prevention of beriberi.

**thi·a·zine** (thī'ə-zēn') *n.* Any of a class of organic chemical compounds containing a ring composed of one sulfur atom, one nitrogen atom, and four carbon atoms.

**thi·a·zole** (thī'ə-zōl') *n.* **1.** A pale-yellow or colorless liquid, $C_3H_3NS$, containing a five-member ring composed of a nitrogen atom, a sulfur atom, and three carbon atoms, used in making dyes and fungicides. **2.** Any of various thiazole derivatives.

**thick** (thĭk) *adj.* **-er, -est.** [ME *thicke* < OE *ðicce.*] **1. a.** Relatively great in depth or in extent from one surface to the opposite <*thick* timber> **b.** Measuring in this dimension <three inches *thick*> **2.** Heavy in build : THICKSET. **3.** Having compactly arranged parts : DENSE <a *thick* forest> **4.** Having a heavy or viscous consistency <*thick* pea soup> **5.** Lacking fresh air : STUFFY. **6.** Having a great number of : TEEMING <a room *thick* with flies> **7.** Impenetrable by the eyes <a *thick* fog> **8.** Indistinctly articulated, as speech. **9.** Readily apparent : MARKED <a *thick* brogue> **10.** *Informal.* Lacking mental agility : STUPID. **11.** *Informal.* Being on intimate terms <*thick* friends> **12.** *Informal.* Going beyond a limit : EXCESSIVE. *—adv.* So as to be thick <Slice it *thick.*> *—n.* **1.** The thickest part <the *thick* of one's thumb> **2.** The most active or intense part <in the *thick* of the fighting> **—through thick and thin.** Through both good and bad times. **—thick'ly** *adv.*

**thick·en** (thĭk'ən) *v.* **-ened, -en·ing, -ens.** *—vt.* **1.** To make thick or thicker. **2.** To make more intense, intricate, or complex. *—vi.* To become thickened. **—thick'en·er** *n.*

**thick·en·ing** (thĭk'ə-nĭng) *n.* **1.** The act or process of making or becoming thick. **2.** Material, as flour, used to thicken liquid. **3.** A thickened part.

**thick·et** (thĭk'ĭt) *n.* [ME **thikket* < OE *ðiccet* < *ðicce*, thick.] **1.** A dense growth of shrubs or underbrush : COPSE. **2.** Something suggesting a thicket in impenetrability or thickness <"the *thicket* of unreality which stands between us and the facts of life" —Daniel J. Boorstin>

**thick·head** (thĭk'hĕd') *n.* A stupid person. **—thick'head'ed** *adj.*

**thick·ness** (thĭk'nĭs) *n.* **1.** The quality or condition of being thick. **2.** The dimension between two of an object's opposite surfaces, usu. the dimension of smallest measure. **3.** A layer, sheet, stratum, or ply <a double *thickness* of cloth>

**thick·set** (thĭk'sĕt') *adj.* **1.** Having a solid, stocky body : STOUT. **2.** Positioned or placed closely together <*thickset* trees>

**thick-skinned** (thĭk'skĭnd') *adj.* **1.** Having a thick skin. **2.** Not easily offended or hurt : INSENSITIVE.

**thick-wit·ted** (thĭk'wĭt'ĭd) *adj.* Lacking intelligence : STUPID.

**thief** (thēf) *n., pl.* **thieves** (thēvz) [ME *theef* < OE *ðēof.*] One who steals.

**thieve** (thēv) *v.* **thieved, thiev·ing, thieves.** [< THIEF.] *—vt.* To take by theft : STEAL. *—vi.* To commit theft.

**thiev·er·y** (thē'və-rē) *n., pl.* **-ies.** The act or practice of thieving.

**thieves** (thēvz) *n. pl. of* THIEF.

**thiev·ish** (thē'vĭsh) *adj.* **1.** Given to thieving. **2.** Of, similar to, or characteristic of a thief : FURTIVE.

**thigh** (thī) *n.* [ME < OE *ðēoh.*] **1. a.** The section of the human leg between the hip and the knee. **b.** A homologous structure in animals. **2.** The femur of an insect's leg.

**thigh·bone** (thī'bōn') *n.* FEMUR 1a.

**thig·mo·tax·is** (thĭg'mə-tăk'sĭs) *n.* [Gk. *thigma*, touch (< *thinganein*, to touch) + -TAXIS.] Movement of an organism in response to a direct tactile stimulus. **—thig'mo·tac'tic** (-tăk'tĭk) *adj.* **—thig'mo·tac'ti·cal·ly** *adv.*

**thig·mot·ro·pism** (thĭg-mŏt'rə-pĭz'əm) *n.* [Gk. *thigma*, touch (< *thinganein*, to touch) + -TROPISM.] Response or motion of an organism to direct contact with a surface or object.

**thill** (thĭl) *n.* [ME *thille.*] Either of the two long shafts between which an animal is fastened when pulling a wagon.

**thim·ble** (thĭm'bəl) *n.* [ME *thymbyl*, leather sheath for fingers < OE *ðymel* < *ðūma*, thumb.] **1.** A small metal, ceramic, or plastic cup worn to protect the finger that pushes the sewing needle. **2.** Any of various tubular sockets or sleeves in machinery. **3.** *Naut.* **a.** A metal ring fitted in an eye of a sail to prevent chafing. **b.** A metal ring around which a rope splice is passed.

**thim·ble·ber·ry** (thĭm'bəl-bĕr'ē) *n., pl.* **-ries.** A raspberry or a related plant having thimble-shaped fruit, esp. *Rubus parviflora* of western and central North America.

**thimbleberry**

**thim·ble·ful** (thĭm'bəl-fŏŏl') *n.* A very small quantity.

**thim·ble·rig** (thĭm'bəl-rĭg') *n.* **1.** A gambling game, usu. a swindle, in which the operator shuffles three inverted thimbles or shells, under one of which he or she has placed a marker, and spectators bet on the location of the marker. **2.** One who operates a thimblerig. *—vt.* **-rigged, -rig·ging, -rigs.** To swindle with or as if with a thimblerig. **—thim'ble·rig'ger** *n.*

**thim·ble·weed** (thĭm'bəl-wēd') *n.* **1.** Any of several North American plants of the genus *Anemone*, having thimblelike, cylindrical fruiting heads. **2.** A coneflower.

**thi·mer·o·sal** (thī-mĕr'ə-săl') *n.* [THI(O)- + MER(CURY) + SAL(ICYLATE).] A crystalline powder, $C_9H_9HgNaO_2S$, used as an antiseptic for surface tissues.

**thin** (thĭn) *adj.* **thin·ner, thin·nest.** [ME *thinn* < OE *ðynne.*] **1. a.** Having a relatively small distance between opposite sides or surfaces. **b.** Not great in diameter or cross section : FINE <*thin* wire> **2.** Lean or slender in physique. **3. a.** Not dense or concentrated : SPARSE <*thin* vegetation> **b.** More rarefied than normal <*thin* air> **4. a.** Flowing easily : not viscous <a *thin* oil> **b.** Watery <*thin* soup> **5.** Sparsely supplied or provided : SCANTY <*thin* earnings> **6.** Lacking force or substance : FLIMSY <a *thin* attempt at humor> **7.** Lacking fullness or resonance <The piano had a *thin* sound.> **8.** Lacking radiance or intensity <*thin* light> **9.** Not having enough photographic density or contrast to make satisfactory prints. *—adv.* So as to be thin <Slice it *thin.*> *—vt. & vi.* **thinned, thin·ning, thins.** To make or become thin or thinner. **—thin'ly** *adv.* **—thin'ness** *n.*

☆ **syns:** THIN, BONY, LANKY, LEAN, SCRAWNY, SKINNY, SLENDER, SPARE, TWIGGY *adj. core meaning* : having little flesh or fat on the body <*thin* POW's> *ant:* fat, fleshy

**thine** (*th*īn) *pron.* [ME *thin* < OE *ðīn.*] *Archaic.* **1.** Belonging to thee. —Used predicatively. **2.** The one or ones that belong to thee. —Used substantively. **3.** —Used instead of *thy* before an initial vowel or *h* <*thine* enemy>

**thing** (thĭng) *n.* [ME < OE *ðing.*] **1.** An entity, idea, or quality perceived, known, or considered to have a separate existence. **2.** Real or concrete substance of an entity as opposed to its appearances or to the name, word, or symbol denoting it. **3.** An entity existing in space or time. **4.** An inanimate object. **5. a.** A creature <the poor little *thing*> **b.** An individual. **6. a.** That which can be possessed or owned. **b. things.** Possessions : belongings <unpacked my *things*> **c.** An article of clothing <couldn't find a *thing* to wear> **7. things.** The equipment needed for an activity or a special purpose <Where are my fishing *things*?> **8.** An object or entity that cannot or need not be named <How do you use this *thing*?> **9. a.** An act, deed, or work <tried to do the right *things*> **b.** The result of work or activity. **10.** A thought, notion, or utterance <What a rude *thing* to say!> **11.** A piece of information <didn't know a *thing* about your plans> **12.** A means to an end <just the *thing* to attract customers> **13.** A matter of concern <many *things* on my mind> **14.** A turn of events : CIRCUMSTANCE <The earthquake was a terrible *thing.*> **15. a. things.** General state of affairs : CONDITIONS <*Things* are steadily improving.> **b.** A particular state of affairs : SITUATION <Let's deal with this *thing* promptly.> **16.** A persistent illogical feeling : OBSESSION <has a *thing* about spiders> **17. the thing.** The current custom or latest fashion : RAGE. **18.** *Slang.* An activity uniquely suitable and satisfying <Let me do my own *thing.*> **19. a.** A specified point or item : DETAIL <worries about every little *thing*> **b.** Matter or objects of a specified kind <avoids eating sweet *things*> **—first thing.** Right away. **—see** (or **hear**) **things.** To have hallucinations. **—sure thing.** *Informal.* **1.** A certainty <Your election is a *sure thing.*> **2.** Of course : CERTAINLY <*Sure thing,* I'll be there!>

▲ word history: The word *thing* is derived from a Germanic ancestor that has no relatives in the other branches of Indo-European. The Germanic word was very important, however, and it appears in some form in all the Germanic languages. The Gothic word, recorded in the mid-5th century, meant "an appointed time," which is probably close to the basic meaning of the original word. In Old English *thing* meant "an assembly," and "a legal matter"; these senses are also found for the word in many of the other languages. These Old English meanings of *thing* are now obsolete, but the related senses "a matter of concern" and "a circumstance" survive. The sense "an entity" has always been a meaning of the English word.

**thing·a·ma·bob** (thĭng'ə-mə-bŏb') *n.* [< THING.] *Informal.* A thingamajig.

**thing·a·ma·jig** *also* **thing·um·a·jig** (thĭng'ə-mə-jĭg') *n.* [< THING.] *Informal.* An unnamed item difficult to classify.

**T-hinge** (tē'hĭnj') *n.* A hinge whose two parts shape the letter T.

**thing-in-itself** (thĭng'ĭn-ĭt-sĕlf') *n., pl.* **things-in-them·selves** (thĭngz'ĭn-*th*ĕm-sĕlvz') [Transl. of G. *Ding an sich.*] An ultimate metaphysical reality conceived by Kant as beyond the perception of human senses and thought : NOUMENON.

**think** (thĭngk) *v.* **thought** (thôt), **think·ing, thinks.** [ME *thenken* < OE *ðencan.*] *—vt.* **1.** To have or formulate in the mind. **2. a.** To reason about or reflect on : PONDER <*Think* how complex language is.><*Think* the problem through.> **b.** To decide by thinking <*thinking* what to do> **3.** To judge or regard <I *think* it only reasonable.> **4.** To believe or suppose <Do you *think* people will care?> **5.** To hope or expect <*thought* I'd make enough money but didn't> **6.** To remember <I can't *think* where I saw it.> **7.** To imagine <didn't *think* so many people would come> **8.** To devise or evolve : INVENT <*thought* up a way to beat the system> **9.** To bring into a given state by mental preoccupation <*thought* myself into a panic> **10.** To concentrate one's thoughts on <does nothing but *think* profits> *—vi.* **1.** To exercise the power of reason. **2.** To weigh or consider the idea <They are *thinking* of marrying.> **3.** To recall a thought or image <*thought* of my childhood> **4.** To consider or regard <I *think* of myself as honest.> **5.** To have care or consideration <*Think* first of the family.> **6.** To dispose the mind in a given way <*Think* positively.> **—think better of.** To decide against after reconsidering. **—think much of.** To consider satisfactory : APPROVE <didn't *think much of* my new suit> **—think nothing of.** To regard as routine or usual. **—think twice.** To consider something carefully.

**think·a·ble** (thĭng'kə-bəl) *adj.* **1.** Capable of being thought about : conceivable. **2.** Capable of happening or being done : POSSIBLE. **—think'a·bly** *adv.*

**think·er** (thĭng'kər) *n.* **1.** One who devotes time to thought or meditation. **2.** One who thinks or reasons in a given way <a methodical *thinker*>

**think·ing** (thĭng'kĭng) *n.* **1.** Thought. **2.** A way of reasoning : JUDGMENT <not to my *thinking* a workable solution> *—adj.* Marked by the ability to think : RATIONAL.

**thinking cap** *n.* A state in which one thinks, esp. carefully <put on one's *thinking cap*>

**think piece** *n.* A newspaper article consisting of news analysis, background material, and personal opinions.

**think tank** *n.* A group or institution organized for intensive research and problem-solving, esp. in technology or strategic politics.

**thin·ner** (thĭn'ər) *n.* A liquid, as turpentine, mixed with paint to reduce viscosity.

**thin-skinned** (thĭn'skĭnd') *adj.* **1.** Having a thin skin or rind. **2.** Easily offended or hurt : OVERSENSITIVE.

**thio-** *or* **thi-** *pref.* [Gk. *theio-* < *theion,* sulfur.] Containing sulfur <*thiourea*>

**thi·o·car·ba·mide** (thī'ō-kär'bə-mīd') *n.* Thiourea.

**thi·o·cy·a·nate** (thī'ō-sī'ə-nāt') *n.* A salt or ester of thiocyanic acid.

**thi·o·cy·an·ic acid** (thī'ō-sī-ăn'ĭk) *n.* An unstable colorless liquid, HSCN, used in the form of esters as an insecticide.

**thi·ol** (thī'ôl', -ōl') *n.* Mercaptan.

**thion-** *pref.* [< Gk. *theion,* sulfur.] Sulfur <*thionic*>

**thi·on·ic** (thī-ŏn'ĭk) *adj.* Of, relating to, containing, or derived from sulfur.

**thi·o·nyl** (thī'ə-nĭl') *n.* Sulfinyl.

**thi·o·pen·tal sodium** (thī-ō-pĕn'tăl', -tôl') *n.* [THIO- + PENT(OBARBIT)AL.] A yellowish-white hygroscopic powder, $C_{11}H_{17}N_2O_2SNa$, injected intravenously as a general anesthetic.

**thi·o·phene** (thī'ə-fēn') *n.* [THIO- + PH(ENO)- + -ENE.] A colorless liquid, $C_4H_4S$, used as a solvent.

**thi·o·sul·fate** (thī'ō-sŭl'fāt') *n.* A salt of thiosulfuric acid.

**thi·o·sul·fu·ric acid** (thī'ō-sŭl-fyo͝or'ĭk) *n.* An acid, $H_2S_2O_3$, formed by the replacement of an oxygen atom by a sulfur atom in sulfuric acid, known only in solution or by its salts and esters.

**thi·o·ur·a·cil** (thī'ō-yo͝or'ə-sĭl') *n.* [THIO- + UR(O)- + AC(ETIC) + E. *-il,* a substance related to another substance.] A white crystalline compound, $C_4H_4N_2OS$, used medically to depress thyroxin synthesis.

**thi·o·u·re·a** (thī'ō-yo͞o-rē'ə) *n.* A white, lustrous crystalline compound, $(NH_2)_2CS$, used in photography, photocopying paper, and various organic syntheses.

**third** (thûrd) *n.* [ME *thridd* (adj.) < OE *ðirdda.*] **1.** The ordinal number that matches the number three in a series. **2.** One of three equal parts. **3.** One sixtieth of a second, as a measure of time or of the arc of an angle. **4.** *Mus.* **a.** An interval of three degrees in a diatonic scale. **b.** A tone separated by three degrees from a given tone, esp. the third tone of a scale. **5.** The third forward gear in the transmission of a motor vehicle. **6.** *Baseball.* Third base. **7. thirds.** Merchandise whose quality is below the standard set for seconds. **—third** *adj. & adv.*

▲ word history: The ordinal *third* and its corresponding cardinal, *three,* exhibit what was once a very common linguistic process in English, that of metathesis. *Metathesis* literally means "change of position." In the case of *third,* the *r* sound changed its position with respect to the vowel. The Indo-European root from which both *three* and *third* are derived is *trei-. Three* preserves the original position of *r; third* represents a change. The usual form of the ordinal in Old English was *thridda,* although the form *thirdda* was also used. Both the forms *thrid* and *third* occurred throughout the Middle English period, and *thrid* did not die out until modern times. Metathesis of *r* is exhibited by other pairs of related words, among them *burn*¹/*brand* and *work*/*wrought.*

**third base** *n. Baseball.* **1.** The third base to be reached by a runner. **2.** The position played by the third baseman.

**third baseman** *n. Baseball.* The infielder stationed near third base.

**third class** *n.* **1.** A class of mail in the U.S. postal system including all printed matter, except newspapers and magazines, that weighs less than 16 ounces and is unsealed. **2.** Accommodations, as on a ship or train, of the third and usu. lowest order of price and luxury. **—third'-class'** *adj. & adv.*

**third degree** *n.* Physical or mental torture to obtain information or a confession from a prisoner.

**third-degree burn** (thûrd'dĭ-grē') *n.* A severe burn in which the epidermis is destroyed and sensitive nerve endings are exposed.

**third dimension** *n.* **1.** Depth or thickness. **2.** The quality of seeming real or lifelike. **—third'-di·men'sion·al** (thûrd'dĭ-mĕn'shə-nəl) *adj.*

**third·hand** (thûrd'hănd') *adj.* **1.** Acquired from or through two intermediate sources <a *thirdhand* report of the corporate takeover> **2. a.** That has been previously used by two other owners. **b.** Dealing in thirdhand merchandise.

**third house** *n.* [From its role with respect to the two houses of which many legislatures consist.] A legislative lobby.

**third·ly** (thûrd'lē) *adv.* In the third place, rank, or order.

**Third Order** *n.* A confraternity of lay people associated with a religious order of the Roman Catholic Church.

**third party** *n.* A political party organized as opposition to the existing parties in a two-party system.

**third person** *n.* **1. a.** A set of grammatical forms used in referring to a person or thing other than the speaker or the one addressed. **b.** A grammatical form belonging to such a set. **2.** Reference of a grammatical form to a person or thing other than the speaker or the one addressed.

**third rail** *n.* The rail through which the current runs to power the train on an electric railway.

**third-rate** (thûrd'rāt') *adj.* Being of third quality or value, esp. of less value than second-rate.

**third-stream** (thûrd'strēm') *adj.* Of, pertaining to, or being music that blends classical music with jazz improvisation.

**Third World** *also* **third world** *n.* **1.** Underdeveloped or developing countries, esp. those not allied with the Communist or non-Communist blocs. **2.** Minority groups as a whole within a larger prevailing culture. **—Third Worlder** *n.*

**thirst** (thûrst) *n.* [ME < OE *ðurst.*] **1. a.** A sensation of dryness in the mouth related to a need or desire to drink. **b.** The desire to drink. **2.** An insistent desire : CRAVING. —*vi.* **thirst·ed, thirst·ing, thirsts. 1.** To feel a need to drink. **2.** To have a strong craving : YEARN. —**thirst'er** *n.*

**thirst·y** (thûr'stē) *adj.* **-i·er, -i·est. 1.** Desiring to drink. **2.** Arid : parched. **3.** Craving something. —**thirst'i·ly** *adv.* —**thirst'i·ness** *n.*

**thir·teen** (thûr-tēn') *n.* [ME *thrittene* < OE *ðrēotīne.*] **1.** The cardinal number equal to 12 + 1. **2.** The 13th in a set or sequence. **3.** Something having 13 parts, units, or members. —**thir·teen'** *adj.* & *pron.*

**thir·teenth** (thûr-tēnth') *n.* **1.** The ordinal number matching the number 13 in a series. **2.** One of 13 equal parts. —**thir·teenth'** *adj.* & *adv.*

**thir·ti·eth** (thûr'tē-ĭth) *n.* **1.** The ordinal number matching the number 30 in a series. **2.** One of 30 equal parts. —**thir'ti·eth** *adj.* & *adv.*

**thir·ty** (thûr'tē) *n., pl.* **-ties.** [ME *thritty* < OE *ðrītig.*] **1.** The cardinal number equal to 3 × 10. **2.** An indication of the end of a news story, usu. written 30. **3.** The second point scored by one side in tennis. —**thir'ty** *adj.* & *pron.*

**thir·ty-sec·ond note** (thûr'tē-sĕk'ənd) *n.* A musical note with a time value equivalent to 1/32 of a whole note.

**thir·ty-two·mo** (thûr'tē-to͞o'mō) *n., pl.* **-mos. 1.** The page size (3½ by 5½ inches) that results when a printers' sheet is folded into 32 equal sections. **2.** A book composed of pages of thirty-twomos.

**this** (thĭs) *pron., pl.* **these** (thēz) [ME < OE *ðis* (nominative and accusative neuter sing.).] **1. a.** The person or thing present, nearby, or just mentioned <*This* is my cat.><*These* are my plants.> **b.** What is about to be said <Try not to laugh when you hear *this.*> **c.** The present occasion or time <thought I'd be here before *this*> **2.** The one nearer than another or the one compared with the other <*This* is mine and that is yours.> ***usage:*** In referring to a thought expressed earlier, either *this* or *that* may be used, as in *The window was open; this* (or *that*) *in itself was suspicious.* When the referent is to follow, only *this* is used, as in *This* (not *that*) *is what I want: a permanent position and a private office.* —*adj., pl.* **these. 1.** Being just mentioned or present in space, time, or thought <I left early *this* morning.> **2.** Being nearer than another or compared with another <*this* wall and that wall> **3.** Being about to be stated or described <Just wait till you hear *this* story.> —*adv.* To this extent : SO <never felt *this* tired>

**this·tle** (thĭs'əl) *n.* [ME *thistel* < OE *ðistel.*] **1.** Any of numerous weedy plants, chiefly of the genera *Cirsium, Carduus,* or *Onopordum,* with prickly leaves and usu. purplish flowers surrounded by prickly bracts. **2.** A plant similar or related to the thistle.

**thistle butterfly** *n.* The painted lady.

**this·tle·down** (thĭs'əl-doun') *n.* The silky down attached to the seeds of a thistle.

**thith·er** (thĭth'ər, *thĭth'-*) *adv.* [ME <OE *ðider.*] **1.** To or toward that place or direction : THERE <running hither and *thither*> **2.** *Archaic.* To or toward that end or result. —*adj.* Located or being on the more distant side : FARTHER <the *thither* side of the lake>

**thith·er·to** (thĭth'ər-to͞o', *thĭth'-*) *adv.* Up to that time : until then.

**thith·er·ward** (thĭth'ər-wərd, *thĭth'-*) *adv.* THITHER 1.

**thix·ot·ro·py** (thĭk-sŏt'rə-pē) *n.* [Gk. *thixis,* touch (< *thinganien,* to touch) + -TROPY.] The property exhibited by certain gels of liquefying when stirred or shaken and returning to the hardened state upon standing. —**thix'o·trop'ic** (thĭk'sə-trŏp'ĭk) *adj.*

**tho** (thō) *conj.* & *adv. Informal.* Though.

**†thole** (thōl) *vt.* **tholed, thol·ing, tholes.** [ME *tholen* < OE *ðolian.*] *Regional.* To endure or bear : UNDERGO.

**thole pin** (thōl) *n.* [ME *tholle* < OE *ðol.*] *Naut.* A wooden peg set in pairs in the gunwale of a boat to serve as an oarlock.

**Tho·mism** (tō'mĭz'əm) *n.* The theological and philosophical system of Saint Thomas Aquinas, which later became the basis of scholasticism. —**Tho'mist** (tō'mĭst) *n.* —**Tho·mis'tic** (tō-mĭs'tĭk) *adj.*

**Thomp·son submachine gun** (tŏmp'sən) *n.* [After John *Thompson* (d. 1940), its co-inventor.] A .45-caliber submachine gun.

**thong** (thông, thŏng) *n.* [ME < OE *ðwong.*] **1.** A narrow strip of material, as leather, used for binding or lashing. **2.** A whiplash of plaited leather or cord. **3.** A sandal held on the foot by a thong that fits between the toes and is connected to a strap usu. passing over the top or around the sides of the foot.

**Thor** (thôr) *n.* [ON *Ðórr.*] *Norse Myth.* The god of thunder.

**tho·ra·ces** (thôr'ə-sēz', thōr'-) *n. var. pl. of* THORAX.

**tho·rac·ic** (thə-răs'ĭk) *adj.* Of, pertaining to, or situated in or near the thorax. —**tho·rac'i·cal·ly** *adv.*

**thoracic duct** *n.* The main duct of the lymphatic system, ascending along the spinal cord and discharging into the venous system.

**tho·ra·cot·o·my** (thôr'ə-kŏt'ə-mē, thōr'-) *n., pl.* **-mies.** [Lat. *thorax, thorac-,* thorax + -TOMY.] Surgical incision of the chest wall.

**tho·rax** (thôr'ăks', thōr'-) *n., pl.* **tho·rax·es** or **tho·ra·ces** (thôr'ə-sēz', thōr'-) [ME < Lat. < Gk. *thōrax.*] **1.** *Anat.* The part of the human body between the neck and the diaphragm, partially encased by the ribs : CHEST. **2.** A part in animals that corresponds to the thorax. **3.** The second or middle region of the body of an arthropod, in insects bearing the true legs and wings.

**tho·ri·a** (thôr'ē-ə, thōr'-) *n.* [< THORIUM.] Thorium dioxide.

**tho·ric** (thôr'ĭk, thŏr'-, thōr'-) *adj.* [THOR(IUM) + -IC.] Of, relating to, or containing thorium.

**tho·rite** (thôr'īt', thōr'-) *n.* [THOR(IUM) + -ITE[1].] A vitreous brownish-yellow to black thorium ore, essentially $ThSiO_4$.

**tho·ri·um** (thôr'ē-əm, thōr'-) *n.* [After *Thor.*] *Symbol* **Th** A silvery-white metallic element used in magnesium alloys; atomic number 90; atomic weight 232.038.

**thorium dioxide** *n.* A heavy white powder, $ThO_2$, used mainly in ceramics, gas mantles, and nuclear fuels.

**thorn** (thôrn) *n.* [ME < OE *ðorn.*] **1.** *Bot.* A modified branch in the form of a sharp woody spine. **2.** A shrub, tree, or woody plant bearing thorns. **3.** A sharp spiny protuberance on an animal. **4.** One that causes sharp pain, irritation, or discomfort. **5.** The runic letter **þ** orig. representing either sound of the Modern English *th,* as in *the* and *thin,* used in Old English and Middle English manuscripts.

**thorn apple** *n.* A plant of the genus *Datura,* esp. jimsonweed.

**thorn·back** (thôrn'băk') *n.* A ray, *Raja clavata* of European waters or *Platyrhinoidis triseriata* of Pacific waters, with a spiny back.

**thorn·y** (thôr'nē) *adj.* **-i·er, -i·est. 1.** Full of or covered with thorns. **2.** Shaped like a thorn : SPINY. **3.** Painfully controversial : VEXATIOUS <a *thorny* issue> —**thorn'i·ness** *n.*

**tho·ron** (thôr'ŏn', thōr'-) *n.* A radioactive isotope of radon having a half-life of 54.5 seconds and produced by disintegration of thorium.

**thor·ough** (thûr'ō) *adj.* [ME *thorow,* thorough, through < OE *ðuruh,* through.] **1.** Exhaustively complete <a *thorough* investigation> **2.** Painstakingly accurate or careful <*thorough* research> **3.** Completely satisfactory : THOROUGHGOING <a *thorough* pleasure> —**thor'ough·ly** *adv.* —**thor'ough·ness** *n.*

**thor·ough·bass** (thûr'ō-bās', thûr'ə-) *n.* A continuo.

**thorough brace** *n.* One of several leather bands passed from front to back of a carriage, supporting it and serving as a spring. —**thor'ough-braced'** *adj.*

**thor·ough·bred** (thûr'ō-brĕd', thûr'ə-) *n.* **1.** A purebred or pedigreed animal. **2. Thoroughbred.** Any of a breed of horse originating from a cross of Arabian stallions with English mares. **3.** A well-bred individual. —*adj.* **1.** Bred of pure stock : PUREBRED. **2. Thoroughbred.** Relating or belonging to the Thoroughbred breed of horses. **3.** Of good upbringing : WELL-BRED.

☆ ***syns:*** THOROUGHBRED, FULL-BLOODED, HIGHBRED, PUREBLOOD, PUREBRED *adj. core meaning* : of pure breeding stock <a *thoroughbred* mare>

**thor·ough·fare** (thûr'ō-fâr', thûr'ə-) *n.* [ME *thurghfare* : *thurgh,* through + *fare,* road.] **1.** A main road or public highway. **2. a.** A place of passage from one location to another. **b.** Right to such passage. **3.** A heavily traveled passage, as a waterway, strait, or channel.

**thor·ough·go·ing** (thûr'ō-gō'ĭng, thûr'ə-) *adj.* **1.** Very thorough : COMPLETE. **2.** Being without exception : ABSOLUTE.

**thor·ough·paced** (thûr'ō-pāst', thûr'ə-) *adj.* **1.** Trained in all paces or gaits, as a horse. **2.** Thoroughgoing : complete.

**thor·ough·pin** (thûr'ō-pĭn', thûr'ə-) *n.* [THOROUGH, through (obs.) + PIN.] An abnormal swelling on either side of the hock joint of horses and related animals.

**thor·ough·wort** (thûr'ō-wûrt', -wôrt', thûr'ə-) *n.* The boneset.

**thorp** (thôrp) *n.* [ME < OE *ðorp.*] *Obs.* A hamlet.

**those** (thōz) *adj.* & *pron. pl. of* THAT.

**thou[1]** (thou) *pron.* [ME < OE *ðū.*] —Used to indicate the one that is spoken to esp. in a literary or ecclesiastical context.

**thou[2]** (thou) *n. Slang.* A thousand.

**though** (thō) *conj.* [ME, of Scand. orig.] **1.** Despite the fact that : ALTHOUGH <continued to search, *though* I knew it was useless> **2.** Conceding or supposing that : even if <*Though* they may not succeed, they won't give up.> —*adv.* However : nevertheless <We can expect some rain, *though* not this afternoon.>

**thought** (thôt) *n.* [ME < OE *geðōht.*] **1.** The act or process of thinking : COGITATION. **2.** A product of thinking : IDEA. **3.** The intellectual activity or production of a particular time or social class <ancient Greek *thought*> **4.** Attention or heedful regard <didn't give much *thought* to what you said> **5.** A plan of action : INTENTION <had no *thought* of quitting> **6.** A small amount or slight degree <You could be a *thought* more considerate.>

**thought·ful** (thôt'fəl) *adj.* **1.** Occupied with thought : CONTEMPLATIVE. **2.** Well thought out and considered <a *thoughtful* essay> **3.** Showing considerate regard for others. —**thought'ful·ly** *adv.* —**thought'ful·ness** *n.*

**thought·less** (thôt'lĭs) *adj.* **1. a.** Not giving sufficient thought : CARELESS. **b.** Marked by a reckless disregard : RASH. **2.** Devoid of consideration for others : INCONSIDERATE. —**thought'less·ly** *adv.* —**thought'less·ness** *n.*

**thought reading** *n.* MIND READING 1.

**thou·sand** (thou'zənd) *n.* [ME < OE *ðūsend.*] The cardinal number equal to 10 × 100 or $10^3$. —**thou'sand** *adj.* & *pron.*

**Thousand Island dressing** *n.* [Perh. after the *Thousand Is-*

*lands,* islands in the St. Lawrence River.] A salad dressing made with mayonnaise, chili sauce, and seasonings.

**thou·sandth** (thou′zəndth, -zənth) *n.* **1.** The ordinal number matching the number 1,000 in a series. **2.** One of 1,000 equal parts. **—thou′sandth** *adj.* & *adv.*

**Thra·cian** (thrā′shən) *adj.* Of or relating to Thrace or its people. —*n.* **1.** A native or inhabitant of Thrace. **2.** The Indo-European language of the ancient Thracians.

**thrall** (thrôl) *n.* [ME < OE *ðrǣl* < ON *ðræll.*] **1. a.** One who is held in bondage, as a serf or slave. **b.** One who is intellectually or morally enslaved. **2.** Bondage : servitude. —*vt.* **thralled, thrall·ing, thralls.** *Archaic.* To enslave. **—thrall′dom, thral′dom** (-dəm) *n.*

**thrash** (thrăsh) *v.* **thrashed, thrash·ing, thrash·es.** [Alteration of THRESH.] —*vt.* **1.** To beat with or as if with a whip or stick : FLOG. **2.** To swing or strike in a way similar to the action of a flail <The crocodile *thrashed* its tail.> **3.** To defeat utterly : VANQUISH. **4.** To thresh (grain). **5.** To sail (a boat) against opposing winds or tides. —*vi.* **1.** To move violently or wildly. **2.** To strike or flail. **3.** To thresh grain. **4.** To sail against opposing tides or winds. **—thrash out.** To discuss fully. —*n.* **1.** The act of thrashing. **2.** A swimming kick in the backstroke and crawl. **—thrash′er** *n.*

☆ ***syns:*** THRASH, FLAIL, THRESH *v. core meaning* : to swing about or strike at wildly <The captured animal *thrashed* about in the net.>

**thrash·er** (thrăsh′ər) *n.* [Perh. alteration of THRUSH.] Any of various New World songbirds of the genus *Toxostoma,* with a long tail, a long, curved beak, and, in several species, a spotted breast.

**thrash·ing** (thrăsh′ĭng) *n.* A severe beating.

**thra·son·i·cal** (thrā-sŏn′ĭ-kəl, thrə-) *adj.* [After *Thraso,* a character in the play *Eunuchus* by Terence (185–159 B.C.).] Given to bragging : BOASTFUL. **—thra·son′i·cal·ly** *adv.*

**thread** (thrĕd) *n.* [ME < OE *ðrǣd.*] **1. a.** A fine cord of a fibrous material, as cotton or flax, constructed from two or more filaments twisted together and used in needlework and the weaving of cloth. **b.** A piece of thread. **2.** A strand, fiber, or filament of natural or manufactured material. **3.** Something suggesting the fineness or thinness of thread <a *thread* of smoke> **4.** Something suggesting the continuousness of thread <lost the *thread* of my thought> **5.** A helical or spiral ridge on a screw, nut, or bolt. **6. threads.** *Slang.* Clothes. —*v.* **thread·ed, thread·ing, threads.** —*vt.* **1. a.** To pass one end of a thread through the eye of (e.g., a needle). **b.** To pass (something) through in the manner of a thread <*thread* the wire through the opening> **c.** To pass a tape or film into or through a device <*thread* a motion-picture projector> **2.** To connect by running a thread through : STRING <*thread* popcorn> **3.** To make one's way cautiously through <*threading* dark alleys> **4.** To occur throughout : PERVADE <dark cloth *threaded* with gold> **5.** To machine a thread on (a screw, nut, or bolt). —*vi.* **1.** To make one's way cautiously. **2.** To proceed by a winding course. **3.** To form a thread when dropped from a spoon, as boiling sugar syrup. **—thread′er** *n.*

☆ ***syns:*** THREAD, FIBER, FIBRIL, FILAMENT *n. core meaning* : a very fine, continuous strand <polyester *thread*>

**thread·bare** (thrĕd′bâr′) *adj.* **1.** Having the nap worn down so that the filling or warp threads show through. **2.** Wearing shabby, old clothing. **3.** Hackneyed : trite <a *threadbare* excuse>

**thread·fin** (thrĕd′fĭn′) *n.* A chiefly tropical marine fish of the family Polynemidae, with threadlike rays extending from the lower part of the pectoral fin.

**thread mark** *n.* A marking made in paper currency by a threading of colored silk fibers to make counterfeiting difficult.

**thread·worm** (thrĕd′wûrm′) *n.* A threadlike nematode worm, esp. the pinworm.

**thread·y** (thrĕd′ē) *adj.* **-i·er, -i·est. 1.** Consisting of or resembling thread : FILAMENTOUS. **2.** Capable of forming or tending to form threads, as a syrupy liquid : VISCID. **3.** *Med.* Weak and shallow <a *thready* pulse> **4.** Lacking resonance : THIN <a *thready* voice> **—thread′i·ness** *n.*

**threat** (thrĕt) *n.* [ME < OE *ðrēat.*] **1.** An expression of an intention to inflict something harmful. **2.** An indication of impending danger or harm. **3.** One regarded as a possible danger : MENACE. —*vt.* **threat·ed, threat·ing, threats.** *Archaic.* To threaten.

**threat·en** (thrĕt′n) *v.* **-ened, -en·ing, -ens.** —*vt.* **1.** To express a threat against. **2.** To serve as a threat to : ENDANGER. **3.** To give signs or warning of : PORTEND. **4.** To announce as possible <*threatened* to move out of town> —*vi.* **1.** To express or use threats. **2.** To indicate danger or harm. **—threat′en·er** *n.* **—threat′en·ing·ly** *adv.*

☆ ***syns:*** THREATEN, FOREBODE, FOREWARN *v. core meaning* : to give warning signs of (impending peril) <economic indicators *threatening* recession>

**three** (thrē) *n.* [ME < OE *ðrī;* akin to G. *drei,* Lat. *tres,* Gk. *treis,* Skt. *tri.*] **1.** The cardinal number equal to 2 + 1. **2.** The third in a set or sequence. **3.** Something having three parts, units, or members. **—three** *adj.* & *pron.*

**three-bag·ger** (thrē′băg′ər) *n.* *Baseball.* A three-base hit.

**three-base hit** (thrē′bās′) *n.* *Baseball.* A base hit that allows the batter to reach third base without being put out : TRIPLE.

**three-card mon·te** (thrē′kärd mŏn′tē) *n.* A gambling game in which each player is dealt and shown three cards, which are then placed face down on the table, the players betting they can identify a particular card.

**three-col·or** (thrē′kŭl′ər) *adj.* Designating a color printing or photographic process in which three primary colors are transferred by three different plates or filters to a surface, reproducing all the colors of the subject matter.

**three-D** *or* **3-D** (thrē′dē′) *adj.* Three-dimensional. —*n.* A three-dimensional medium, display, or performance, esp. a cinematic or graphic display in three dimensions.

**three-deck·er** (thrē′dĕk′ər) *n.* **1.** A ship having three decks, esp. one of a class of sail-powered warships with guns on three decks. **2.** Something with three layers, esp. a sandwich having three slices of bread.

**three-di·men·sion·al** (thrē′dĭ-mĕn′shə-nəl) *adj.* **1.** Of, relating to, having, or existing in three dimensions. **2.** Having or appearing to have extension in depth.

**three·fold** (thrē′fōld′) *adj.* **1.** Having or consisting of three parts. **2.** Three times as many or as much : TREBLE. **—three′fold′** *adv.*

**three-gait·ed** (thrē′gā′tĭd) *adj.* Trained in the walk, trot, and canter <a *three-gaited* horse>

**Three Graces** *pl.n.* The Graces.

**three-leg·ged race** (thrē′lĕg′ĭd, -lĕgd′) *n.* A race in which partners run with their near legs tied together.

**three-mile limit** (thrē′mīl′) *n.* *Law.* The outer limit of the area extending three miles out to sea from the coast of a land that constitutes that land's territorial waters.

**three·pence** (thrĕp′əns, thrĭp′-, thrŭp′-) *n., pl.* **-pence** *or* **-penc·es.** *Chiefly Brit.* **1.** A coin worth three pennies. **2.** The sum of three pennies.

**three·pen·ny** (thrĕp′ə-nē, thrĭp′-, thrŭp′-) *adj.* *Chiefly Brit.* **1.** Worth or priced at threepence. **2.** Of very little worth.

**three-piece** (thrē′pēs′) *adj.* Made in or consisting of three parts or pieces <a *three-piece* suit>

**three-ply** (thrē′plī′) *adj.* Consisting of three layers or strands.

**three-point landing** (thrē′point′) *n.* An aircraft landing in which the tailskid or tail wheel and the two forward wheels all touch the ground simultaneously.

**three-quar·ter** (thrē′kwôr′tər) *adj.* Relating to, consisting of, or extending to three-fourths of the usual full length of something.

**three-quarter binding** *n.* A type of bookbinding in which the leather or fabric covering the spine extends onto the covers for one third of their width.

**three-ring circus** (thrē′rĭng′) *n.* **1.** A circus having simultaneous performances in three separate rings. **2.** A situation marked by confusing, engrossing, or amusing activity.

**three R's** *pl.n.* [From the phrase *reading, 'riting,* and *'rithmetic,* alteration of *reading, writing,* and *arithmetic.*] Reading, writing, and arithmetic, regarded as the fundamentals of elementary education.

**three·score** (thrē′skôr′, -skōr′) *adj.* Three times twenty : SIXTY. **—three′score′** *n.*

**three·some** (thrē′səm) *adj.* Consisting of or performed by three. —*n.* **1.** A group of three persons. **2.** An activity involving three persons, esp. a golf match in which one player competes against two others who alternate their play.

**three-square** (thrē′skwâr′) *adj.* Having an equilateral triangular cross section <a *three-square* file>

**three wood** *n.* A wooden golf club with more loft than a brassie.

**threm·ma·tol·o·gy** (thrĕm′ə-tŏl′ə-jē) *n.* [Gk. *thremma, thremmat-,* nursling + -LOGY.] The scientific breeding of domestic plants and animals.

**thren·o·dy** (thrĕn′ə-dē) *n., pl.* **-dies.** [Gk. *thrēnōidia* : *thrēnos,* lament + *ōidē,* song.] A poem or song of lamentation. **—thre·no′di·al** (thrə-nō′dē-əl), **thre·nod′ic** (-nŏd′ĭk) *adj.* **—thren′o·dist** *n.*

**thre·o·nine** (thrē′ə-nēn′) *n.* [Orig. unknown.] An essential component of human nutrition that is a colorless crystalline amino acid, $C_4H_9NO_3$, derived from the hydrolysis of protein.

**thresh** (thrĕsh) *v.* **threshed, thresh·ing, thresh·es.** [ME *thresshen* < OE *ðrescan.*] —*vt.* **1. a.** To beat the stems and husks of (grain or cereal plants) with a machine or flail to separate the grain or seeds from the straw. **b.** To separate (grain or seed) by threshing. **2.** To discuss or go over (e.g., an issue) repeatedly. **3.** To beat severely : THRASH. —*vi.* **1.** To thresh grain. **2.** To thrash about : TOSS.

**thresh·er** (thrĕsh′ər) *n.* **1.** One that threshes. **2.** A threshing machine. **3.** A shark of the genus *Alopias,* having a tail with a long whiplike upper lobe.

**threshing machine** *n.* A farm machine used in threshing grain or seed plants.

**thresh·old** (thrĕsh′ōld′, -hōld′) *n.* [ME *thresshold* < OE *ðerscold.*] **1.** The piece of wood or stone placed beneath a door : DOORSILL. **2.** An entrance or doorway. **3.** A place or point of beginning : OUTSET. **4.** The intensity below which a mental or physical stimulus cannot be perceived and can produce no response <a low *threshold* of pain>

**threw** (thrōō) *v.* *p.t. of* THROW.

**thrice** (thrīs) *adv.* [ME *thries,* adv. genitive of *thrie* < OE *ðriga.*] **1.** Three times. **2.** In a threefold quantity or degree. **3.** *Archaic.* Extremely : greatly.

**thrift** (thrĭft) *n.* [ME, prosperity < ON *ðrift* < *ðrīfask,* to thrive.] **1.** Wise economy in the management of resources, esp. of money. **2.** Vigorous growth of living things such as plants. **3.** A densely tufted, chiefly European plant of the genus *Armeria,* esp. *A. maritima,* with rounded pink flower clusters.

**thrift·less** (thrĭft'lĭs) *adj.* **1.** Lacking usefulness or value. **2.** Careless in handling money : WASTEFUL.

**thrift shop** *n.* A shop that sells used articles and esp. clothing, often to benefit a charitable organization.

**thrift·y** (thrĭf'tē) *adj.* **-i·er, -i·est. 1.** Practicing thrift : ECONOMICAL. **2.** Industrious and thriving : PROSPEROUS. **3.** Growing vigorously, as a plant. **—thrift'i·ly** *adv.* **—thrift'i·ness** *n.*

**thrill** (thrĭl) *v.* **thrilled, thrill·ing, thrills.** [ME *thrillen,* var. of *thirlen,* to pierce < OE *ðyrlian* < *ðyrel,* hole < *ðurh,* through.] *—vt.* **1.** To cause to feel a sudden intense emotion : excite greatly. **2.** To give great pleasure to : DELIGHT. **3.** To cause to quiver, tremble, or vibrate. *—vi.* **1.** To feel a sudden quiver of emotion. **2.** To quiver, tremble, or vibrate. *—n.* **1. a.** A quivering or trembling caused by sudden emotion. **b.** Something that produces such excitement. **2.** *Pathol.* A slight vibration that accompanies a cardiac or vascular murmur. **—thrill'ing·ly** *adv.*

**thrill·er** (thrĭl'ər) *n.* **1.** One that thrills. **2.** *Informal.* A sensational or suspenseful book, story, or motion picture.

**thrips** (thrĭps) *n., pl.* **thrips.** [Lat., woodworm < Gk.] Any of various small, often wingless insects of the order Thysanoptera, many of which are destructive to plants.

**thrive** (thrīv) *vi.* **throve** (thrōv) *or* **thrived, thrived** *or* **thriv·en** (thrĭv'ən), **thriv·ing, thrives.** [ME *thriven* < ON *ðrīfask,* reflexive of *ðrīfa,* to seize.] **1.** To make steady progress : PROSPER. **2.** To grow vigorously : FLOURISH. **—thriv'er** *n.*

**throat** (thrōt) *n.* [ME *throte* < OE *ðrote.*] **1.** *Anat.* **a.** The portion of the digestive tract that lies between the rear of the mouth and the esophagus and includes the fauces and the pharynx. **b.** The anterior part of the neck. **2.** *Bot.* The outer, expanded part of a tubular corolla. **3.** A narrow passage or part suggesting the human throat <the *throat* of a horn><the *throat* of a chimney> *—vt.* **throat·ed, throat·ing, throats.** To pronounce with a harsh or guttural voice.

**throat·latch** (thrōt'lăch') *n.* A strap passing under the neck of a horse for holding a bridle or halter in place.

**throat·y** (thrō'tē) *adj.* **-i·er, -i·est.** Uttered or sounding as if uttered deep in the throat. **—throat'i·ly** *adv.* **—throat'i·ness** *n.*

**throb** (thrŏb) *vi.* **throbbed, throb·bing, throbs.** [ME *throbben.*] **1.** To beat rapidly or violently : POUND <hearts throbbing with excitement> **2.** To vibrate, pulsate, or sound with a steady, pronounced rhythm <boat engines *throbbing*> *—n.* **1.** The act of throbbing. **2.** A beat, palpitation, or vibration. **—throb'bing·ly** *adv.*

**throe** (thrō) *n.* [ME *throwe* < OE *ðrawe.*] **1.** A severe pang or spasm of pain, as in childbirth. **2. throes.** Agonizing struggle or effort.

☆ ***syns:*** THROE, CONVULSION, PAROXYSM *n. core meaning* : a condition of anguished struggle and disorder <a country in the *throes* of revolution>

**thromb-** *pref. var. of* THROMBO-.

**throm·bi** (thrŏm'bī') *n. pl. of* THROMBUS.

**throm·bin** (thrŏm'bĭn) *n.* An enzyme in blood that facilitates blood clotting by reacting with fibrinogen to form fibrin.

**thrombo-** *or* **thromb-** *pref.* [< Gk. *thrombos,* clot.] Blood clot : blood clotting <*thrombo*plastic>

**throm·bo·cyte** (thrŏm'bə-sīt') *n.* A blood platelet. **—throm'bo·cyt'ic** (-sĭt'ĭk) *adj.*

**throm·bo·cy·to·pe·ni·a** (thrŏm'bə-sī'tə-pē'nē-ə) *n.* A condition marked by an abnormal decrease in the number of blood platelets. **—throm'bo·cy'to·pe'nic** *adj.*

**throm·bo·em·bo·lism** (thrŏm'bō-ĕm'bə-lĭz'əm) *n.* Occlusion of a blood vessel by a thrombus dislodged from a vein.

**throm·bo·phle·bi·tis** (thrŏm'bō-flĭ-bī'tĭs) *n.* Inflammation of a vein with the formation of a thrombus.

**throm·bo·plas·tic** (thrŏm'bō-plăs'tĭk) *adj.* **1.** Causing or promoting blood clotting. **2.** Of or relating to thromboplastin. **—throm'bo·plas'ti·cal·ly** *adv.*

**throm·bo·plas·tin** (thrŏm'bō-plăs'tĭn) *n.* A protein complex essential for thrombin formation and blood clotting.

**throm·bo·sis** (thrŏm-bō'sĭs) *n., pl.* **-ses** (-sēz') [NLat. < Gk. *thrombōsis,* a clotting < *thrombousthai,* to clot < *thrombos,* clot.] Formation, presence, or development of a thrombus.

**throm·bus** (thrŏm'bəs) *n., pl.* **-bi** (-bī') [NLat. < Gk. *thrombos,* clot.] A blood clot occluding a vessel or formed in a heart cavity.

**throne** (thrōn) *n.* [ME *thron* < OFr. *trone* < Lat. *thronus* < Gk. *thronos.*] **1.** The chair occupied by an exalted personage such as a sovereign or bishop on state or ceremonial occasions. **2. a.** The occupant of a throne. **b.** Sovereign rank or power : SOVEREIGNTY. **3. thrones.** The third of the nine orders of angels. *—vt. & vi.* **throned, thron·ing, thrones.** To enthrone or occupy a throne.

**throng** (thrông, thrŏng) *n.* [ME < OE *ðrang.*] A large group of people or things gathered or crowded closely together : MULTITUDE. *—v.* **thronged, throng·ing, throngs.** *—vt.* **1.** To crowd into : FILL. **2.** To press in on. *—vi.* To gather, press, or move in a throng.

**thros·tle** (thrŏs'əl) *n.* [ME < OE *ðrostle.*] **1.** Any of various Old World thrushes. **2.** A machine once used for spinning fibers such as cotton or wool.

**throt·tle¹** (thrŏt'l) *n.* [Perh. dim. of obs. *throte,* throat < ME.] **1. a.** A valve in an internal-combustion engine that regulates the amount of vaporized fuel entering the cylinders. **b.** A valve in a steam engine regulating the amount of steam. **c.** A lever or pedal controlling either of these valves. **2. a.** THROAT 1. **b.** The windpipe.

**throt·tle²** (thrŏt'l) *vt.* **-tled, -tling, -tles.** [ME *throttlen,* to strangle, perh. < *throte,* throat.] **1. a.** To regulate the flow of (fuel) in an engine. **b.** To regulate the speed of (an engine) with a throttle. **2.** To choke or strangle. **3.** To keep from being published or transmitted : SUPPRESS. **—throt'tler** *n.*

**throt·tle·hold** (thrŏt'l-hōld') *n.* STRANGLE HOLD 2.

**through** (thrōō) *prep.* [ME < OE *ðurh.*] **1.** In one side and out the opposite or another side of <went *through* the tunnel> **2.** In the midst of <a walk *through* the rows of corn> **3.** By way of <went out *through* the back door> **4.** By the means or agency of <got an antique *through* a dealer> **5.** Here and there in : AROUND <a bicycle tour *through* Holland> **6.** From the beginning to the end of <slept all *through* the concert> **7.** At or to the end of <We are *through* the initial testing period.> **8.** Without stopping for <drove *through* a red light> **9.** Because of <failed *through* lack of practice> *—adv.* **1.** From one end or side to another or opposite end or side. **2.** From beginning to end : COMPLETELY. **3.** To a conclusion or accomplishment. **4.** Out into the open. *—adj.* **1.** Passing or extending from one end, side, or surface to another <a *through* beam> **2.** Allowing direct or continuous passage <a *through* route> **3.** Affording transportation to a destination with few or no stops and no transfers <a *through* bus> **4.** No longer effective, capable, or valued : WASHED-UP. **5.** At completion : FINISHED <We are *through* with the project.> **—through and through. 1.** In every part of : THROUGHOUT. **2.** In every aspect : COMPLETELY.

**through·ly** (thrōō'lē) *adv. Archaic.* Thoroughly.

**through·out** (thrōō-out') *prep.* In, to, through, or during every part of <*throughout* the world> *—adv.* **1.** In or through all parts : EVERYWHERE. **2.** During the entire time or extent.

**through·put** (thrōō'po͝ot') *n.* Output or production, as of a computer program, over a period of time.

**through street** *n.* A street on which traffic is permitted to move without having to stop, as for traffic entering from intersecting streets.

**through·way** (thrōō'wā') *n. var. of* THRUWAY.

**throve** (thrōv) *v. var. p.t. of* THRIVE.

**throw** (thrō) *v.* **threw** (thrōō), **thrown** (thrōn), **throw·ing, throws.** [ME *throwen* < OE *ðrāwan,* to twist.] *—vt.* **1.** To propel through the air with a swift arm motion : HURL <*threw* a big rock> **2.** To discharge into the air <a howitzer that *throws* a shell 20 miles> **3.** To hurl with great force, as in anger <I *threw* myself at the bully.> **4.** To hurl to the floor or ground <The horse *threw* its rider.> **5.** To perplex or mislead <The phony clue *threw* me.> **6.** To put on or take off hastily or carelessly <*throw* on a jacket> **7.** To put abruptly or forcibly into a specified condition <*threw* us into turmoil> **8.** To form on a potter's wheel <*throw* a vase> **9.** To twist (fibers) into thread. **10. a.** To roll (dice). **b.** To roll (a particular combination) with dice. **c.** To discard or play (a card). **11.** To cast <The streetlight *threw* shadows across the yard.> **12.** To bear (young), as cows or horses. **13.** *Slang.* To arrange or give (e.g., a party). **14.** To move (a controlling lever or switch). **15.** *Informal.* To lose (a contest) on purpose. **16.** To abandon oneself to <I *threw* a fit when they wrecked my car.> **17.** To commit (oneself) esp. for leniency or support <*threw* myself on the mercy of the court> **18.** To deliver (a punch), as in boxing <*threw* a left hook> *—vi.* To hurl, fling, or cast an object. **—throw away. 1. a.** To get rid of as useless. **b.** To discard <*threw away* two aces> **2. a.** To fail to take advantage of <*threw away* a chance to travel> **b.** To waste or use foolishly <*threw away* the family fortune> **—throw back. 1.** To hinder or check the progress of. **2.** To revert to an earlier type or stage in one's past. **3.** To cause to depend: make reliant. **—throw in. 1.** To engage (e.g., a clutch). **2.** To add (an extra amount) with no additional charge. **3.** To insert or introduce into a course of something <*threw in* some nasty remarks> **—throw off. 1.** To cast out or reject. **2.** To give off or emit. **3.** To rid oneself of <*threw off* my cold> **—throw out. 1.** To emit <searchlights *throwing out* powerful beams> **2.** To reject or discard <*threw out* the original plan> **3.** To get rid of as useless <*threw out* all the junk> **4.** To offer, as a suggestion. **5.** To disengage (a clutch or gears). **6.** *Baseball.* To put out (a base runner) by throwing the ball to the player guarding the base to which he or she is running. **7.** To eject from a place, esp. in an abrupt or unexpected manner <*threw* the troublemaker *out*> **—throw over. 1.** To overturn. **2.** To abandon. **—throw up. 1.** To abandon or relinquish. **2.** To vomit. **3.** To construct hurriedly. **4.** To refer to something repeatedly and often reproachfully <always *throwing up* my past to me> *—n.* **1.** The act of throwing. **2.** The distance, height, or direction of something thrown <a low *throw*> **3. a.** A roll or cast of dice. **b.** The combination of numbers so obtained. **4.** A venture or chance. **5.** A technique used to throw an opponent in wrestling. **6. a.** A light blanket or coverlet, as an afghan. **b.** A scarf or shawl. **7. a.** The length of the radius of a circle described by a machine part, as a crank or cam. **b.** The maximum displacement of a machine part moved by a crank, cam, or similar part. **8.** *Geol.* **a.** The amount of vertical displacement of a fault. **b.** The vertical component of the net slip. **—throw in the sponge (or towel).** To give up in a contest or undertaking: admit defeat <finally *threw in the sponge* after 15 years of marriage> **—throw (one's) weight around.** To use power or authority, esp. in an excessive or heavy-handed way. **—throw up (one's) hands.** To admit defeat. **—throw'er** *n.*

☆ ***syns:*** THROW, CAST, EMIT, PROJECT, RADIATE, SHED *v. core meaning*: to send out heat, light, or energy <the moon *throwing* ghostly light over the moors>

**throw·a·way** (thrō'ə-wā') *n.* A free handbill distributed on the street. *—adj.* **1.** Designed or intended to be discarded after use. **2.** Written or delivered in a low-key or offhand manner <*throwaway* comedy>

**throw·back** (thrō'băk') *n.* **1.** A reversion to a former type or ancestral characteristic. **2.** ATAVISM 2.

**thrown** (thrōn) *v. p.p. of* THROW.

**throw rug** *n.* A scatter rug.

**thru** (thrōō) *prep. adv. & adj. Informal.* Through.

**thrum**[1] (thrŭm) *v.* **thrummed, thrum·ming, thrums.** [Imit.] *—vt.* **1.** To play (a stringed instrument) idly or monotonously. **2.** To recite or repeat in a monotone. *—vi.* **1.** To strum idly on a stringed instrument. **2.** To speak in a monotone: DRONE. **—thrum** *n.*

**thrum**[2] (thrŭm) *n.* [ME < OE (*tunge*) *ðrum*, ligament (of the tongue).] **1. a.** The fringe of warp threads left on a loom after the cloth has been cut off. **b.** One of these threads. **2.** A loose end, fringe, or tuft of thread. **3. thrums.** *Naut.* Short bits of rope yarn inserted into canvas to roughen the surface. *—vt.* **thrummed, thrum·ming, thrums. 1.** To cover or trim with thrums: FRINGE. **2.** *Naut.* To sew thrums in (canvas).

**thrush**[1] (thrŭsh) *n.* [ME *thrushe* < OE *ðrȳsce.*] **1.** Any of various songbirds of the family Turdidae, with brownish upper plumage and a spotted breast. **2.** A bird related or similar to the thrush.

**thrush**[2] (thrŭsh) *n.* [Prob. of Scand. orig.] **1.** An oral infection with a fungus, *Candida albicans,* characterized by white eruptions in the mouth. **2.** A suppurative infection of a horse's foot caused by standing in a wet dirty stall.

**thrust** (thrŭst) *v.* **thrust, thrust·ing, thrusts.** [ME *thrusten* < ON *ðrȳsta.*] *—vt.* **1. a.** To push or drive quickly and forcibly <*thrust* the ax into the tree> **b.** To stab or pierce, as with a dagger. **2.** To force (oneself or another) into a specified condition or situation <was *thrust* into a position of awesome responsibility> **3.** To interject, as questions or comments. *—vi.* **1.** To shove into something: PUSH. **2.** To pierce or stab with or as if with a pointed weapon. **3.** To force one's way. *—n.* **1.** A forceful shove or push: LUNGE. **2. a.** A driving force or pressure. **b.** The forward-directed force developed in a jet or rocket engine as a reaction to the rearward ejection of fuel gases at high velocities. **3.** A stab. **4.** General direction or tendency <The whole *thrust* of the campaign was to raise funds.> **5.** Outward or lateral stress in a structure, as an arch. **—thrust'er** *n.*

**thrust fault** *n. Geol.* A reverse fault having a low angle of inclination in relation to the horizontal plane.

**thru·way** *also* **through·way** (thrōō'wā') *n.* An expressway.

**thud** (thŭd) *n.* [Perh. < ME *thudden,* to strike with a weapon < OE *ðyddan.*] **1.** A dull sound, as that of a heavy object striking a solid surface. **2.** A blow or fall causing a thud. *—vi.* **thud·ded, thud·ding, thuds.** To make a thud.

**thug** (thŭg) *n.* [Hindi *ṭhag* < Skt. *sthagaḥ,* a cheat < *sthagati,* he conceals.] **1.** One of a band of professional assassins once active in northern India. **2.** A brutal ruffian or gangster. **—thug'ger·y** *n.* **—thug'gish** *adj.*

☆ ***syns:*** THUG, APE, GOON, GORILLA, HOOD, HOODLUM, HOOLIGAN, RUFFIAN *n. core meaning*: a person who treats others violently and roughly, esp. for hire <*thugs* who intimidated the witness>

**thu·ja** (thōō'jə, thyōō'-) *n.* [NLat. *Thuja,* arborvitae genus < Med. Lat. *thuja,* cedar < Gk. *thuia.*] ARBORVITAE 1.

**Thu·le** (thōō'lē, thyōō'-) *n.* [Lat. < Gk. *Thoulē.*] The most northerly region of the ancient habitable world.

**thu·li·um** (thōō'lē-əm, thyōō'-) *n.* [< THULE.] *Symbol* **Tm** A bright silvery rare-earth element, one isotope of which is used in small portable medical x-ray units; atomic number 69; atomic weight 168.934.

**thumb** (thŭm) *n.* [ME < OE *ðūma.*] **1. a.** The short first digit of the human hand, opposable to each of the other four digits. **b.** A corresponding digit in other animals, esp. primates. **c.** The part of a glove or mitten covering the thumb. **2.** An ovolo. *—v.* **thumbed, thumb·ing, thumbs.** *—vt.* **1.** To disarrange, soil, or impair by careless or frequent handling. **2.** *Informal.* To solicit (a ride) from a passing vehicle by signaling with the thumb. *—vi.* To hitchhike. **—thumb (one's) nose.** To express scorn or derision by or as if by placing the thumb on the nose and wiggling the fingers. **—thumb through.** To browse rapidly through (the pages of a publication).

**thumb·hole** (thŭm'hōl') *n.* The hole on a wind instrument that is opened or closed with the thumb.

**thumb index** *n.* A series of rounded labeled indentations cut into the front edge of a reference book to indicate separate sections.

**thumb-in·dex** (thŭm'ĭn'dĕks) *vt.* **-dexed, -dex·ing, -dex·es.** To supply with a thumb index.

**thumb·nail** (thŭm'nāl') *n.* The nail of the thumb. *—adj.* **1.** Of the size of a thumbnail. **2.** Brief <a *thumbnail* sketch>

**thumb·nut** (thŭm'nŭt') *n.* A wing nut.

**thumb·print** (thŭm'prĭnt') *n.* A print made by the thumb.

**thumb·screw** (thŭm'skrōō') *n.* **1.** A screw so designed that it can be turned with the thumb and fingers. **2.** An instrument of torture used to compress the thumb.

**thumb·tack** (thŭm'tăk') *n.* A tack with a smooth, rounded head that can be pressed into place with the thumb. *—vt.* **-tacked, -tack·ing, -tacks.** To affix with a thumbtack.

**thump** (thŭmp) *n.* [Imit.] **1.** A blow with a blunt instrument. **2.** The muffled sound produced by or as if by a blow with a blunt instrument: THUD. *—v.* **thumped, thump·ing, thumps.** *—vt.* **1.** To beat with or as if with a blunt instrument so as to produce a thud. **2.** To beat thoroughly: DRUB. *—vi.* **1.** To hit or fall in such a way as to produce a thump: POUND. **2.** To walk with heavy steps. **3.** To throb audibly. **—thump'er** *n.*

**thump·ing** (thŭm'pĭng) *adj. Informal.* **1.** Very large or great. **2.** Thoroughly enjoyable. **—thump'ing·ly** *adv.*

**thun·der** (thŭn'dər) *n.* [ME < OE *ðunor.*] **1.** The sound emitted by rapidly expanding gases along the path of the electrical discharge of lightning. **2.** A sound similar to thunder. *—v.* **-dered, -der·ing, -ders.** *—vi.* **1.** To produce thunder. **2.** To produce sounds like thunder. **3.** To utter loud, vociferous remarks or threats. *—vt.* To express violently, commandingly, or angrily: ROAR. **—thun'der·er** *n.*

**thun·der·bird** (thŭn'dər-bûrd') *n.* Thunder, lightning, and rain personified as a huge bird in the mythology of some North American Indians.

**thun·der·bolt** (thŭn'dər-bōlt') *n.* **1.** The discharge of lightning that accompanies thunder. **2.** A flash of lightning imagined as a bolt or dart hurled from the heavens. **3.** One that acts with sudden and destructive fury.

**thun·der·clap** (thŭn'dər-klăp') *n.* **1.** A single sharp crash of thunder. **2.** Something similar to a thunderclap in violence, as a startling or shocking piece of news.

**thun·der·cloud** (thŭn'dər-kloud') *n.* **1.** A large, dark cloud charged with electricity and producing thunder and lightning: CUMULONIMBUS. **2.** Something dreadful or menacing.

**thun·der·head** (thŭn'dər-hĕd') *n.* The swollen upper portion of a thundercloud, often associated with the coming of a thunderstorm.

**thunder lizard** *n.* A brontosaur.

**thun·der·ous** (thŭn'dər-əs) *adj.* **1.** Producing thunder or a similar sound. **2.** Loud and unrestrained <*thunderous* cheers> **—thun'der·ous·ly** *adv.*

**thun·der·show·er** (thŭn'dər-shou'ər) *n.* A brief rainstorm accompanied by thunder and lightning.

**thun·der·stone** (thŭn'dər-stōn') *n.* **1.** A mineral concretion, as a belemnite, formerly considered to be a thunderbolt. **2.** *Archaic.* A flash of lightning conceived as a stone: THUNDERBOLT.

**thun·der·storm** (thŭn'dər-stôrm') *n.* An electrical storm accompanied by heavy rain.

**thun·der·struck** (thŭn'dər-strŭk') *adj.* Struck with sudden astonishment or amazement.

**thu·ri·ble** (thŏŏr'ə-bəl) *n.* [ME *thoryble* < Lat. *thuribulum* < *thus,* incense < Gk. *thuos* < *thuein,* to sacrifice.] A censer.

**thu·ri·fer** (thŏŏr'ə-fər) *n.* [Lat., incense-bearing : *thus,* incense + *-fer,* -fer.] An acolyte who carries a thurible.

**Thu·rin·gi·an** (thōō-rĭn'jē-ən, -jən) *adj.* Of or relating to Thuringia or its people. —*n.* **1.** One of an ancient tribe inhabiting central Germany until the 6th cent. A.D. **2.** A native or inhabitant of Thuringia.

**Thurs·day** (thûrz'dē, -dā') *n.* [ME < OE *Đunresdæg* : *Đunres*, Thor's + *dæg*, day.] The fifth day of the week.

▲ word history: The English name of the fifth day of the week is actually a translation of its Latin name, *dies Jovis*, "Jupiter's day." The Old English form of *Thursday* was *Thunresdæg*; *thunres* is the genitive of *thunor*, the ancestor of the modern word *thunder*. *Thunor* was also the name of a Germanic thunder god; the Norse form of *Thunor* is *Thor*. Since thunder was one of Jupiter's attributes, the two gods were associated and Thor's, or Thunor's, name was given to Jupiter's day.

**thus** (thŭs) *adv.* [ME < OE *đus*.] **1.** In this way <wore my hair *thus*> **2.** To a stated degree or extent <*thus* far> **3.** Therefore : consequently.

**thus·ly** (thŭs'lē) *adv.* Thus.

**thwack** (thwăk) *vt.* **thwacked, thwack·ing, thwacks.** [Imit.] To hit with something flat : WHACK. **—thwack** *n.*

**thwart** (thwôrt) *vt.* **thwart·ed, thwart·ing, thwarts.** [ME *thwerten* < *thwert*, across < ON *đvert*, neuter of *đverr*, transverse.] **1.** To prevent from taking place : FRUSTRATE. **2.** To challenge, oppose, or offend : ANTAGONIZE. —*n.* A seat across a boat for an oarsman. —*adj.* **1.** Extending, lying, or passing across something : TRANSVERSE. **2.** Perverse. —*adv.* & *prep. Archaic.* Athwart : across. **—thwart'er** *n.* **—thwart'ly** *adv.*

**thy** (thī) *pron.* [ME < OE *đīn*.] *Archaic.* —Used attributively to indicate possession, agency, or reception of an action by the person or persons spoken to.

**thy·la·cine** (thī'lə-sīn') *n.* [NLat. *Thylacinus*, genus name < Gk. *thulakos*, sack.] A wolflike marsupial, *Thylacinus cynocephalus* of forest areas of Tasmania, with dark transverse bands across its back.

**thyme** (tīm, thīm) *n.* [ME < OFr. *thym* < Lat. *thymum* < Gk. *thumon*.] **1.** An aromatic herb or low shrub of the genus *Thymus*, esp. *T. vulgaris* of southern Europe, with small purplish flowers. **2.** The leaves of the thyme, used as seasoning.

**thy·mec·to·my** (thī-mĕk'tə-mē) *n.*, *pl.* **-mies.** [THYM(US) + -ECTOMY.] Surgical removal of the thymus.

**-thymia** *suff.* [NLat. < Gk. *thumos*, mind, soul.] State or condition of mind <schizo*thymia*>

**thy·mic¹** (tī'mĭk, thī'-) *adj.* Of or relating to thyme.

**thy·mic²** (thī'mĭk) *adj.* Of or relating to the thymus.

**thy·mi·dine** (thī'mĭ-dēn') *n.* [THYM(INE) + -ID(E) + -INE².] A nucleoside, $C_{10}H_{14}N_2O_5$, composed of thymine and deoxyribose.

**thy·mine** (thī'mēn') *n.* [THYM(US) + -INE².] A pyrimidine base, $C_5H_6N_2O_2$, that is an essential constituent of deoxyribonucleic acid.

**thy·mo·cyte** (thī'mə-sīt') *n.* [THYM(US) + -CYTE.] A lymphocyte derived from the thymus.

**thy·mol** (thī'môl', -mōl') *n.* A white, crystalline, aromatic compound, $C_{10}H_{14}O$, derived from thyme oil and other oils and used as an antiseptic, in perfumery, and as a preservative.

**thy·mus** (thī'məs) *n.* [Gk. *thumos*.] A ductless glandlike structure situated just behind the top of the sternum that plays some part in building resistance to disease, reaching its maximum development during early childhood but usu. vestigial in adults.

**thyro-** *or* **thyr-** *pref.* [< THYROID.] Thyroid <*thyro*xin>

**thy·ro·cal·ci·to·nin** (thī'rō-kăl'sĭ-tō'nĭn) *n.* Calcitonin.

**thy·roid** (thī'roid') *adj.* [Gk. *thureoeidēs* : *thureos*, oblong shield + *-eidēs*, -oid.] Of or relating to the thyroid gland or the thyroid cartilage. —*n.* **1.** The thyroid gland. **2.** The thyroid cartilage. **3.** A dried and powdered preparation of the thyroid gland of certain domestic animals that is used in the treatment of hypothyroid conditions.

**thyroid cartilage** *n.* The largest cartilage of the larynx, with two broad processes that join anteriorly to form the Adam's apple.

**thy·roid·ec·to·my** (thī'roi-dĕk'tə-mē) *n.*, *pl.* **-mies.** Surgical removal of the thyroid gland.

**thyroid gland** *n.* A two-lobed endocrine gland found in all vertebrates, located in front of and on either side of the trachea in humans, and producing the hormone thyroxin.

**thy·roid·i·tis** (thī'roi-dī'tĭs) *n.* Inflammation of the thyroid gland.

**thyroid stimulating hormone** *n.* Thyrotropin.

**thy·ro·tox·i·co·sis** (thī'rō-tŏk'sĭ-kō'sĭs) *n.* Poisoning from hyperthyroidism.

**thy·ro·tro·pin** (thī'rə-trō'pĭn) *also* **thy·ro·tro·phin** (-fĭn) *n.* A hormone of the anterior pituitary that stimulates and regulates the development and secretion of the thyroid gland hormone.

**thy·rox·in** (thī-rŏk'sĭn) *also* **thy·rox·ine** (-sēn', -sĭn) *n.* [THYR(O)- + OX(Y)- + -IN.] An iodine-containing hormone, $C_{15}H_{11}I_4NO_4$, produced by the thyroid gland to regulate metabolism and made synthetically for treatment of thyroid disorders.

**thyrse** (thûrs) *n.* [NLat. *thyrsus* < Lat., thyrsus.] A branched flower cluster, as of the lilac, whose main axis does not terminate in a flower.

**thyr·si** (thûr'sī') *n. pl. of* THYRSUS.

**thyr·soid** (thûr'soid') *also* **thyr·soid·al** (thûr-soid'l) *adj.* Shaped like a thyrse.

**thyr·sus** (thûr'səs) *n.*, *pl.* **-si** (-sī') [Lat. < Gk. *thursos*.] **1.** A staff tipped with a pine cone and twined with ivy, represented as carried by Dionysius, Dionysian revelers, or satyrs. **2.** *Bot.* A thyrse.

**thyrsus**

**thy·self** (thī-sĕlf') *pron. Archaic.* Yourself. —Used as the reflexive or emphatic form of *thee* or *thou*.

**ti¹** (tē) *n.* [Alteration of SI.] *Mus.* A syllable representing the seventh tone of the diatonic scale in solmization.

**ti²** (tē) *n.* [Tahitian and Maori.] A tree or shrub of the genus *Cordyline* of tropical Asia and adjacent Pacific regions, esp. *C. australis*, with a terminal tuft of long narrow leaves.

**Ti** *symbol for* TITANIUM.

**ti·ar·a** (tē-ăr'ə, -âr'ə, -är'ə) *n.* [Lat., turban < Gk.] **1.** The pope's triple crown. **2.** An ornamental crownlike headpiece, often decorated with jewels, worn by noblewomen and royalty on formal occasions.

**Ti·bet·an** (tĭ-bĕt'n) *adj.* Of or relating to Tibet, its people, or their language or culture. —*n.* **1.** One of the Mongoloid people of Tibet. **2.** The Tibeto-Burman language of Tibet.

**Ti·bet·o-Bur·man** (tĭ-bĕt'ō-bûr'mən) *n.* A branch of the Sino-Tibetan language family that includes Tibetan and Burmese. **—Ti·bet'o-Bur'man** *adj.*

**tib·i·a** (tĭb'ē-ə) *n.*, *pl.* **-i·ae** (-ē-ē') *or* **-i·as.** [Lat.] **1. a.** The inner and larger of the two bones of the lower human leg from the knee to the ankle. **b.** A homologous bone in animals. **2.** The fourth division of an insect's leg, between the femur and the tarsi. **3.** An ancient flute orig. made from an animal's leg bone. **—tib'i·al** *adj.*

**tic** (tĭk) *n.* [Fr.] A habitual spasmodic muscular contraction, usu. of the face or extremities.

**tic dou·lou·reux** (tĭk' dōō'lōō-rōō') *n.* [Fr. : *tic*, tic + *douloureux*, painful.] Trigeminal neuralgia.

**tick¹** (tĭk) *n.* [ME *tek*.] **1.** The recurring sharp clicking sound made by a machine, esp. a clock. **2.** *Chiefly Brit.* A moment. **3.** A light mark used to check off or point out an item. —*v.* **ticked, tick·ing, ticks.** —*vi.* **1.** To emit recurring sharp, clicking sounds, as a clock. **2.** To behave in a characteristic way, as if by means of a motivating mechanism <What makes you *tick*?> —*vt.* **1.** To count or record by means of ticks <The meter *ticked* off the taxi fare.> **2.** To mark or check off (a listed item) with a tick. **—tick off.** *Slang.* To make angry or annoyed.

**tick²** (tĭk) *n.* [ME *teke*.] **1.** Any of numerous bloodsucking parasitic arachnids of the family Ixodidae within the order Acarina, many of which transmit infectious diseases. **2.** Any of various usu. wingless, louselike insects of the family Hippoboscidae that are parasitic on animals such as sheep and goats.

**tick³** (tĭk) *n.* [ME *tikke*, prob. < MDu. *tīke* < Lat. *theca*, case < Gk. *thēkē*.] **1. a.** The cloth case of a mattress or pillow. **b.** A light mattress without inner springs. **2.** Ticking.

**tick⁴** (tĭk) *n.* [Short for TICKET.] *Chiefly Brit.* Credit or a credit account <on *tick*>

**tick-borne** (tĭk'bôrn', -bōrn') *adj.* Transmitted by ticks.

**tick·er** (tĭk'ər) *n.* **1. a.** A telegraphic instrument that receives and records stock-market quotations on a paper tape. **b.** A device that records similar information by electronic means rather than paper tape. **2.** *Slang.* A watch. **3.** *Slang.* The human heart.

**ticker tape** *n.* The paper strip on which a telegraphic ticker prints.

**tick·et** (tĭk'ĭt) *n.* [OFr. *estiquet*, short document < *estiquier*, to stick < MDu. *steken*.] **1.** A paper slip or card indicating that its holder has paid for or is entitled to a specified service, right, or consideration. **2.** A certifying document, esp. a captain's or pilot's license. **3.** An identifying or descriptive tag attached esp. to merchandise : LABEL. **4.** A list of candidates proposed or endorsed by a political party : SLATE. **5.** A legal summons, esp. for a traffic violation. **6.** *Informal.* The suitable thing <A trip to the ocean would be just the *ticket* for you.> —*vt.* **-et·ed, -et·ing, -ets. 1.** To provide with a ticket for admission or passage. **2.** To attach a tag to : LABEL. **3.** To designate for a specified use or end : DESTINE. **4.** To serve (an offender) with a legal summons.

**ticket scalper** *n.* A profiteer who buys up desirable admission tickets and resells them at higher prices.

**tick fever** *n.* Rocky Mountain spotted fever.

**tick·ing** (tĭk'ĭng) *n.* A strong, tightly woven cotton or linen fabric used esp. to make mattress and pillow coverings.

**tick·le** (tĭk'əl) *v.* **-led, -ling, -les.** [ME *tikelen,* perh. freq. of *ticken,* to touch lightly.] —*vt.* **1.** To touch (the body) lightly so as to cause laughter or twitching movements. **2. a.** To tease or excite pleasurably : TITILLATE. **b.** To fill with mirth or pleasure : DELIGHT. —*vi.* To feel or cause an itching or a tingling sensation. —*n.* **1.** The act of tickling. **2.** A tickling sensation. **—tickle pink.** *Informal.* To please greatly : DELIGHT <was *tickled pink* by the invitation>

**tick·ler** (tĭk'lər) *n.* **1.** One that tickles. **2.** A memorandum book or file to aid the memory.

**tick·lish** (tĭk'lĭsh) *adj.* **1.** Sensitive to tickling. **2.** Easily offended or upset : TOUCHY. **3.** Requiring skillful or tactful handling <a *ticklish* diplomatic situation> **—tick'lish·ly** *adv.* **—tick'lish·ness** *n.*

**tick·seed** (tĭk'sēd') *n.* [TICK[2] (from its seed's shape) + SEED.] The coreopsis.

**tick·tack** *also* **tic-tac** (tĭk'tăk') *n.* [Imit.] **1.** A steady ticking sound, as of a clock. **2.** A prankster's device for tapping on a door or window from a distance.

**tick·tack·toe** *also* **tick-tack-toe** (tĭk'tăk'tō') *n.* [From the sounds of the original game, in which players dropped pencils on a slate.] A game played by two persons, each trying to make a line of three X's or three O's in a boxlike figure with nine spaces.

**tick·tock** (tĭk'tŏk') *n.* [Imit.] The ticking sound made by a clock.

**tick trefoil** *n.* [< TICK[2] (from the way its pods adhere to animals).] Any of various plants of the genus *Desmodium,* having compound leaves with three leaflets, small purplish or white flower clusters, and jointed seed pods with sticky, easily separable segments.

**tick·y-tack·y** (tĭk'ē-tăk'ē) *n.* [Coined by Malvina Reynolds (b. 1900).] Shoddy material, as for buildings.

**tic-tac** (tĭk'tăk') *n. var. of* TICKTACK.

**tid·al** (tīd'l) *adj.* **1.** Having, related to, or affected by <a *tidal* river> **2.** Depending on or scheduled by the time of high tide <a *tidal* ship> **—tid'al·ly** *adv.*

**tidal wave** *n.* **1.** An unusual rise or incursion of water along the seashore, as from a storm or a combination of wind and spring tide. **2.** A tsunami. **3.** An overwhelming manifestation, as of sentiment, desire, or public opinion.

**tid·bit** (tĭd'bĭt') *also* **tit·bit** (tĭt'-) *n.* [Perh. dial. *tid,* tender + BIT[1].] A choice morsel, as of food or gossip.

**tid·dly·winks** (tĭd'lē-wĭngks') *also* **tid·dle·dy·winks** (tĭd'l-dē-) *n.* [Perh. dial. *tiddly,* little + WINK.] *(sing. in number).* A game in which players try to snap small disks into a cup by pressing them on the edge with a larger disk.

**tide[1]** (tīd) *n.* [ME *tid* < OE *tīd,* season.] **1. a.** Periodic variation in the surface level of the oceans and of bays, gulfs, inlets, and tidal regions of rivers, caused by gravitational attraction of the sun and moon, with the lunar effect being the more powerful. **b.** A specific occurrence of a tide. **c.** The waters in a tide. **2.** Stress exerted on a body or part of a body by the gravitational attraction of another <atmospheric *tide*> **3.** Something that fluctuates like the waters of the tide <the rising *tide* of rebellion> **4.** A time or season <even*tide*><Christmas*tide*> **5.** *Archaic.* A favorable occasion : OPPORTUNITY. —*v.* **tid·ed, tid·ing, tides.** —*vi.* **1.** To rise and fall like the tide. **2.** To drift or ride with the tide. —*vt.* To carry along with or as if with the tide. **—tide over.** To support through a difficult period <loaned me $20 to *tide* me *over* till payday>

▲ word history: The words *time* and *tide*[1] are related and were once synonymous, both meaning "an interval of time." This sense for *tide*[1] is archaic except in compounds like *eventide* and *Yuletide.* The usual sense of *tide*[1], "the periodic variation in the level of the earth's waters," is a development of the original meaning of the word, since the tides rise and fall at predictable times of the day.

**tide[2]** (tīd) *vi.* **tid·ed, tid·ing, tides.** [ME *tiden* < OE *tīdan.*] *Archaic.* To betide : befall.

**tide·land** (tīd'lănd') *n.* Coastal land under water during high tide.

**tide·mark** (tīd'märk') *n.* A line or artificial indicator marking the high-water or low-water limit of the tides.

**tide·rip** (tīd'rĭp') *n.* A rip current.

**tide·wait·er** (tīd'wā'tər) *n.* A customs officer who boards incoming ships at a harbor.

**tide·wa·ter** (tīd'wô'tər, -wŏt'ər) *n.* **1.** Water that inundates land at flood tide. **2.** Water affected by the tides, esp. tidal streams. **3.** Low coastal land drained by tidal streams.

**tide·way** (tīd'wā') *n.* A channel in which a tidal current runs.

**tid·ings** (tī'dĭngz) *pl.n.* [Pl. of obs. *tiding,* event < ME, perh. < ON *tiðendi,* events < *tiðr,* occurring.] News <good *tidings*>

**ti·dy** (tī'dē) *adj.* **-di·er, -di·est.** [ME *tidi,* healthy < *tid,* time < OE *tīd.*] **1.** Orderly and neat. **2.** *Informal.* **a.** Adequate : satisfactory <a *tidy* answer> **b.** Substantial : considerable <a *tidy* sum> —*v.* **-died, -dy·ing, -dies.** —*vt.* To put in order <*tidied* up the living room> —*vi.* To make things tidy <*tidied* up after dinner> —*n., pl.* **-dies.** A fancy protective covering for the arms or headrest of a chair. **—ti'di·ly** *adv.* **—ti'di·ness** *n.*

**ti·dy·tips** (tī'dē-tĭps') *n.* *(sing. or pl. in number).* A plant, *Layia elegans* of California, having daisylike flowers with yellow, white-tipped rays.

**tie** (tī) *v.* **tied, ty·ing, ties.** [ME *tyen* < OE *tīgan.*] —*vt.* **1.** To fasten or secure with or as if with a cord, rope, or strap. **2.** To fasten by drawing together the parts or sides of and knotting with strings or laces <*tied* my shoes> **3. a.** To make (a knot or bow) by tying. **b.** To put a knot or bow in <*tie* a necktie> **4.** To confine or restrict as if with cord <*tied* to my family> **5.** To bring together closely : UNITE. **6.** To equal (an opponent or an opponent's score) in a contest. **7.** *Mus.* To join (notes) by a tie. —*vi.* **1.** To be fastened with strings. **2.** To achieve equal scores in a contest. **—tie in.** To have a connection with : COORDINATE. **—tie up. 1.** To keep occupied or inaccessible <I was *tied up* in a meeting.><The phone was *tied up* for an hour.> **2.** To impede the progress of <The accident *tied up* traffic.> —*n.* **1.** A means by which something is tied, as a cord or string. **2.** Something that unites : BOND <marital *ties*> **3.** A necktie. **4.** A beam or rod that joins parts and gives support. **5.** One of the timbers laid across a railroad bed to support the tracks. **6. a.** An equality of scores, votes, or performance in a contest. **b.** A contest resulting in a tie : DRAW. **7.** *Mus.* A curved line above or below two notes of the same pitch, indicating that the tone is to be sustained for their combined duration. **—tie one on.** *Slang.* To get drunk <students who *tied one on* after the football game> **—tie the knot. 1.** To get married. **2.** To perform a marriage ceremony.

☆ **syns:** TIE, BIND, KNOT, SECURE *v. core meaning* : to fasten with or as if with a cord or string <*tied* the package with twine><*tied* my shoelaces> ***ant:*** untie

**tie·back** (tī'băk') *n.* **1.** A decorative loop of fabric, cord, or metal for parting and draping curtains to the sides. **2. tiebacks.** A pair of curtains meant to be tied back at about midlength.

**tie beam** *n.* A horizontal beam connecting the rafters in a roof.

**tie clasp** *n.* An ornamental device holding the ends of a necktie to a shirt front.

**tie-dye** (tī'dī') *n.* A method of dyeing fabric by tying parts of it to prevent absorption of the dye, creating a streaked or mottled pattern. **—tie-dye** *vt.* **(-dyed, -dye·ing, -dyes).**

**tie-in** (tī'ĭn') *n.* A connection or relation, as between a television program and a book or other product derived from or based on it.

**tie line** *n.* **1.** A communication link between extensions of a private telephone system. **2.** A connection between major systems, as electrical power.

**tier[1]** (tîr) *n.* [OFr. *tire,* rank.] One of a series of rows placed one above another. —*vt.* & *vi.* **tiered, tier·ing, tiers.** To arrange or rise in tiers.

**ti·er[2]** (tī'ər) *n.* One that ties.

**tierce** (tîrs) *n.* [ME < OFr. < fem. of *tiers,* third < Lat. *tertium.*] **1.** *also* **terce** (tûrs). **a.** The third of the seven canonical hours. **b.** The time of day set aside for this prayer, usu. the third hour after sunrise. **2.** A former measure of liquid capacity, equal to a third of a pipe, or 42 gallons. **3.** A sequence of three cards of the same suit. **4.** The third position from which a parry or thrust can be made in fencing. **5.** *Mus.* An interval of a third.

**tier·cel** (tîr'səl) *n. var. of* TERCEL.

**tier table** (tîr) *n.* A table with two or more successive tops.

**tie tack** *n.* A short pin with a decorative head for attaching a tie to a shirt front by means of a snap or chain.

**tie-up** (tī'ŭp') *n.* A temporary stoppage or suspension of activity.

**tiff** (tĭf) *n.* [Orig. unknown.] **1.** A fit of irritation. **2.** A petty quarrel. —*vi.* **tiffed, tiff·ing, tiffs.** To quarrel.

**tif·fa·ny** (tĭf'ə-nē) *n., pl.* **-nies.** [Prob. < OFr. *tiphanie,* Epiphany < Med. Lat. *theophania,* theophany. —see THEOPHANY.] A thin, transparent gauze of silk or cotton muslin.

**Tiffany glass** *n.* [After Louis C. *Tiffany* (1848–1933).] Stained or iridescent glass of a kind popular in the early 1900's for decorative objects or lamps.

**tif·fin** (tĭf'ĭn) *n.* [Short for obs. *tiffing,* gerund. of *tiff,* to sip.] *Chiefly Brit.* Luncheon.

**ti·ger** (tī'gər) *n.* [ME *tigre* < OFr. < Lat. *tigris* < Gk.] **1. a.** A large carnivorous feline mammal, *Panthera tigris* of Asia, having a tawny coat with transverse black stripes. **b.** Any of various other similar felines. **2.** A fierce, aggressive, or audacious person. **—ti'ger·ish** *adj.*

**tiger beetle** *n.* Any of numerous active, often varicolored beetles of the family Cicindelidae, chiefly of warm, sandy regions.

**tiger cat** *n.* Any of various small felines resembling the tiger in either appearance or behavior.

**ti·ger-eye** (tī'gər-ī') *also* **ti·ger's-eye** (tī'gərz-) *n.* A semiprecious, yellow-brown chatoyant gemstone made of silicified crocidolite.

**tiger lily** *n.* A plant, *Lilium tigrinum* native to Asia, having large black-spotted reddish-orange flowers with reflexed petals.

**tiger moth** *n.* Any of numerous, often brightly colored moths of the family Arctiidae, having wings marked with spots or lines.

**tiger salamander** *n.* A salamander, *Ambystoma tigrinum,* with distinctive yellowish markings, that is widely distributed over most of North America.

**ti·ger's-eye** (tī'gərz-ī') *n. var. of* TIGER-EYE.

**†tight** (tīt) *adj.* **-er, -est.** [ME, var. of *thight,* dense, of Scand. orig.] **1. a.** Of such close construction, texture, or organization as to be impermeable, esp. by water or air. **b.** Closely reasoned or worded. **2.** Held, fastened, or closed securely. **3.** Compressed, leaving few or no intervening spaces : COMPACT. **4.** Drawn out to the fullest extent : TAUT. **5.** Cramped : constrained <*tight* living quarters> **6.** Snug, often uncomfortably so <a *tight* fit> **7.** Constricted <a *tight* feeling in the chest> **8.** Stingy. **9. a.** Difficult to obtain <*tight* money> **b.** Affected by scarcity <a *tight* market> **10.** Difficult to deal with or escape from <a *tight* spot> **11.** Barely profitable <a *tight* bargain> **12.** Closely contested <a *tight* race> **13.** *Regional.* Neat and trim. **14.** *Slang.* Drunk. **15.** *Slang.* Friendly : compatible. —Usu. used with *with.* —*adv.* **-er, -est. 1.** Firmly : securely <hold *tight*> **2.** Soundly <sleep *tight*> **—sit tight.** To watch and wait. **—tight′ly** *adv.* **—tight′ness** *n.*

☆ ***syns:*** TIGHT, TAUT, TENSE *adj. core meaning* : stretched to the fullest extent <a *tight* bowstring> ***ant:*** loose, slack

**tight·en** (tīt′n) *vt. & vi.* **-ened, -en·ing, -ens.** To make or become tight or tighter. **—tight′en·er** *n.*

**tight end** *n. Football.* The offensive player at the right end of the team in the modern T-formation and who is positioned close to the adjoining tackle.

**tight·fist·ed** (tīt′fĭs′tĭd) *adj.* Stingy.

**tight·lipped** (tīt′lĭpt′) *adj.* **1.** Having the lips pressed together. **2.** Saying little : RETICENT.

**tight·rope** (tīt′rōp′) *n.* **1.** A tightly stretched, usu. wire rope on which acrobats perform high above the ground. **2.** A precarious situation <a diplomatic *tightrope*>

**tights** (tīts) *pl.n.* A snug stretchable garment covering the body from the waist or neck down, worn by acrobats and dancers and also designed for general wear.

**tight·wad** (tīt′wŏd′) *n. Slang.* A miser.

**tig·lic acid** (tĭg′lĭk) *n.* [NLat. *tiglium,* specific epithet of *Croton tiglium,* perh. < Gk. *tilos,* liquid feces.] A thick, syrupy poisonous liquid, $C_5H_8O_2$, derived from croton oil, having a spicy odor and used in perfumery and flavoring agents.

**ti·glon** (tī′glən) *also* **ti·gon** (tī′gən) *n.* [Blend of TIGER and LION.] The hybrid offspring of a male tiger and a female lion.

**Ti·gré** (tē-grā′) *n.* A Semitic language of northern Ethiopia.

**ti·gress** (tī′grĭs) *n.* **1.** A female tiger. **2.** A fierce, aggressive, or audacious woman.

**Ti·gri·nya** (tə-grēn′yə) *n.* A Semitic language of northern Ethiopia.

**tike** (tīk) *n. var. of* TYKE.

**ti·ki** (tē′kē) *n.* [Maori.] **1. Tiki.** A male figure in Polynesian mythology, occas. identified as the first man. **2.** A wood or stone image of a Polynesian god. **3.** A Maori figurine representing an ancestor, often intricately carved from greenstone and worn about the neck as a talisman.

**til** (tĭl) *n.* [Hindi < Skt. *tilaḥ.*] The sesame plant, esp. as used in India as a source of food and oil.

**til·bur·y** (tĭl′bĕr′ē, -bə-rē) *n., pl.* **-ies.** [After *Tilbury,* a 19th-cent. London coach builder.] A light open two-wheeled carriage that seats two persons.

**til·de** (tĭl′də) *n.* [Sp. < Lat. *titulus,* superscription.] The diacritical mark (˜) placed over the letter *n* in Spanish to indicate the palatal nasal sound (ny), as in *cañon,* or over a vowel in Portuguese to indicate nasalization, as in *lã, pão.*

**tile** (tīl) *n.* [ME < OE *tigele* < Lat. *tegula* < *tegere,* to cover.] **1.** A thin, flat, or convex slab of material, as baked clay or plastic, laid in rows to cover walls, floors, and roofs. **2.** A short length of pipe made of clay or concrete, used in sewers and drains. **3.** A hollow fired clay or concrete block used for building walls. **4.** Tiles as a whole. **5.** A marked playing piece, as in mahjong. —*vt.* **tiled, til·ing, tiles.** To cover or provide with tiles. **—til′er** *n.*

**tile·fish** (tīl′fĭsh′) *n., pl.* **tilefish** or **-fish·es.** [*Tile-,* short for NLat. *Lopholatilus,* genus name + FISH.] A marine food fish of the family Branchiostegidae, esp. *Lopholatilus chamaeleonticeps* of deep Atlantic waters, with varicolored markings.

**till¹** (tĭl) *vt.* **tilled, till·ing, tills.** [ME *tilien* < OE *tilian,* to labor.] To prepare (land) for the raising of crops by plowing, harrowing, and fertilizing. **—till′a·ble** *adj.*

**till²** (tĭl) *prep.* [ME < ON, to.] Until. —*conj.* **1.** Until. **2.** Before or unless.

**till³** (tĭl) *n.* [ME *tylle.*] A drawer or small compartment for money, esp. in a store.

**till⁴** (tĭl) *n.* [Orig. unknown.] Glacial drift composed of an unconsolidated, heterogeneous mixture of clay, sand, gravel, and boulders.

**till·age** (tĭl′ĭj) *n.* **1.** Cultivation of land. **2.** Tilled land.

**til·land·si·a** (tĭ-lănd′zē-ə) *n.* [NLat., genus name, after Elias *Tillands* (1640–1693).] Any of various usu. epiphytic plants of the genus *Tillandsia,* as Spanish moss, of tropical and subtropical America.

**till·er¹** (tĭl′ər) *n.* One that tills land.

**til·ler²** (tĭl′ər) *n.* [ME *tiler,* stock of a crossbow < OFr. *telier,* weaver's beam < Med. Lat. *telarium* < Lat. *tela.*] A lever used to turn a rudder and steer a boat.

**til·ler³** (tĭl′ər) *n.* [ME **tiller* < OE *telgor.*] A shoot, esp. one that sprouts from the base of a grass. —*vi.* **-lered, -ler·ing, -lers.** To send forth tillers.

**tilt¹** (tĭlt) *v.* **tilt·ed, tilt·ing, tilts.** [ME *tylten,* to cause to fall, perh. of Scand. orig.] —*vt.* **1.** To cause to slope, as by raising one end : INCLINE. **2. a.** To aim or thrust (a lance) in a joust. **b.** To charge against (an opponent). **3.** To forge with a tilt hammer. —*vi.* **1.** To slope. **2.** To joust. **3.** To quarrel. —*n.* **1. a.** An inclination from the horizontal or vertical : SLANT. **b.** A sloping surface, as of the ground. **2.** The act of tilting. **3. a.** A medieval sport in which two mounted knights with lances charged together and attempted to unhorse one another. **b.** A thrust or blow with a lance. **4.** A verbal duel. **5.** A tilt hammer. **—at full tilt.** At full speed.

**tilt²** (tĭlt) *n.* [ME *telte,* tent < OE *teld.*] A canopy or awning for a boat, wagon, or cart. —*vt.* **tilt·ed, tilt·ing, tilts.** To cover with a tilt.

**tilth** (tĭlth) *n.* [ME < OE *tilð* < *tilian,* to labor.] **1.** TILLAGE 1. **2.** TILLAGE 2.

**tilt hammer** *n.* A heavy forge hammer with a pivoted lever by which it is tilted up and then allowed to drop.

**tilt·yard** (tĭlt′yärd′) *n.* An enclosed yard for tilting contests.

**tim·bal** *also* **tym·bal** (tĭm′bəl) *n.* [Fr. *timbale,* var. of obs. *tamballe* < OSp. *atabal,* small drum < Ar. *aṭ-ṭabl,* drum.] A kettledrum.

**tim·bale** (tĭm′bəl, tĭm-băl′, tăm-) *n.* [Fr., mold, timbal.] **1.** A custardlike dish of cheese, chicken, fish, or vegetables baked in a drum-shaped pastry mold. **2.** The pastry mold in which a timbale is baked.

**tim·ber** (tĭm′bər) *n.* [ME < OE.] **1.** Trees or wooded land regarded as a source of wood. **2. a.** Wood as a building material : LUMBER. **b.** A dressed piece of wood, esp. a beam in a structure. **c.** A rib in a ship's frame. **3.** Material <You're executive *timber.*> —*vt.* **-bered, -ber·ing, -bers.** To support or shore up with timbers. —*interj.* —Used to warn of a falling tree.

**tim·bered** (tĭm′bərd) *adj.* **1. a.** Built of or covered with timber. **b.** Built with exposed timbers. **2.** Wooded.

**tim·ber·head** (tĭm′bər-hĕd′) *n. Naut.* A timber end that projects above a deck and is used as a bollard.

**timber hitch** *n. Naut.* A knot used for fastening a rope around a spar or log to be hoisted or towed.

**tim·ber·land** (tĭm′bər-lănd′) *n.* Forested land considered of commercial value.

**tim·ber·line** *also* **timber line** (tĭm′bər-līn′) *n.* The limit of altitude in mountainous regions beyond which trees do not grow.

**timber right** *n.* A claim to the trees on property belonging to another.

**timber wolf** *n.* A grayish or whitish wolf, *Canis lupus* of forested northern regions.

**tim·ber·work** (tĭm′bər-wûrk′) *n.* The part of a structure made with timbers, such as the framework of a house or boat.

**tim·bre** (tăm′bər, tĭm′-) *n.* [Fr. < OFr., timbrel < Med. Gk. *timbanon* < Gk. *tumpanum.*] The quality of a sound distinguishing it from other sounds of the same pitch and volume, esp. the distinctive tone of a musical instrument, a voice, or a voiced speech sound.

**tim·brel** (tĭm′brəl) *n.* [ME *timbre* < OFr. —see TIMBRE.] An ancient percussion instrument similar to a tambourine.

**time** (tīm) *n.* [ME < OE *tīma,* interval between events.] **1. a.** A nonspatial continuum in which events occur in apparently irreversible succession from the past through the present to the future. **b.** An interval separating two points on this continuum, measured essentially by selecting a regularly recurring event, as the sunrise, and counting the number of its occurrences during the interval : DURATION. **c.** A number, as of years, days, or minutes, representing such an interval. **d.** A similar number representing a specific point, as the present, as reckoned from an arbitrary past point on the continuum. **e.** A system by which such intervals are measured or such numbers are reckoned <standard *time*><solar *time*> **f.** A unit of time as indicated by a timepiece or calendar <What *time* is it?> **2.** *often* **times.** An interval, esp. a span of years, characterized by similar events, conditions, or phenomena : ERA <a *time* of prosperity> **3.** A suitable or opportune moment or season. **4.** A moment or period designated, as by custom, for a given activity <harvest *time*><bed*time*> **5.** An appointed or fated moment, esp. of death <died before their *time*> **6.** One of several instances. **7.** An occasion. **8.** *Informal.* A prison sentence. **9. a.** The customary period of work <hired for part *time*> **b.** The period spent working. **10.** The rate of speed of a measured activity <marching in double *time*> **11.** The characteristic beat of musical rhythm <four-four *time*> —*adj.* **1.** Of or pertaining to time. **2.** Designed so as to operate at a particular moment <a *time* fuse> **3.** Payable on a future date or dates. **4.** Of or pertaining to installment buying. —*vt.* **timed, tim·ing, times. 1.** To set the time for (an event or occasion). **2.** To adjust to keep accurate time. **3.** To regulate or adjust for the orderly sequence of movements or events <*timed* the jump perfectly> **4.** To record the speed or duration of. **5.** To set or maintain the tempo, speed, or duration of. **—against time.** With a quickly approaching time limit. **—at one time. 1.** Simultaneously. **2.** At a period or moment in the past. **—at the same time.** Nonetheless. **—at times.** On occasion. **—behind the times.** Out-of-date. **—for the**

**time being.** Temporarily. **—from time to time.** Once in a while. **—gain time.** To run too fast. —Used of a timepiece. **—high time.** Long overdue. **—in good time. 1.** In a reasonable length of time. **2.** When or before due. **3.** Quickly. **—in no time.** Almost instantly. **—in time. 1.** Before a time limit expires. **2.** Within an indefinite amount of passing time. **3.** In proper tempo. **—keep time. 1.** To indicate the correct time. —Used of a timepiece. **2.** To maintain the tempo or rhythm. **—lose time. 1.** To run too slowly. —Used of a timepiece. **2.** To delay advancement. **—on time. 1.** According to schedule. **2.** By paying in installments.

☆ ***syns:* 1.** TIME, HITCH, STRETCH *n. core meaning* **:** a term of service, as in the military <did their *time* in the army> **2.** TIME, PERIOD, SEASON *n. core meaning* **:** a span designated for a given activity <harvest *time*>

**time and a half** *n.* A rate of pay that is one and a half times the regular rate, as for overtime work.

**time and motion study** *n.* Analysis of the efficiency with which an industrial operation is performed.

**time bill** *n.* A bill of exchange payable at an indicated future time.

**time bomb** *n.* A bomb with a detonating mechanism that can be set for a particular time.

**time capsule** *n.* A sealed container preserving articles and records of contemporary culture for examination by scientists and scholars of the distant future.

**time·card** (tīm′kärd′) *n.* A card, either maintained by an employee or stamped by a time clock, recording the employee's arrival and departure time each day.

**time clock** *n.* A clock that records the arrival and departure times of employees, usu. by punching timecards.

**time deposit** *n.* A bank deposit that cannot be withdrawn before a date specified at the time of deposit.

**time dilatation** *also* **time dilation** *n.* The relativistic slowing of a clock that moves with respect to a stationary observer.

**time exposure** *n.* **1.** A photographic exposure made for a relatively long duration. **2.** An image made by time exposure.

**time-hon·ored** (tīm′ŏn′ərd) *adj.* Respected or adhered to because of age or age-old observance <*time-honored* customs>

**time immemorial** *n.* **1.** Time long past. **2.** *Law.* Time antedating legal records.

**time·keep·er** (tīm′kē′pər) *n.* **1.** A timepiece. **2.** The person who keeps track of time, as in a sports event. **3.** A railroad dispatcher.

**time-lapse** (tīm′lăps′) *adj.* Of or using a motion-picture technique for filming a naturally slow process, as the unfolding of a leaf, by photographing it at intervals so that the continuous projection of the frames gives an accelerated view of it.

**time·less** (tīm′lĭs) *adj.* **1.** Having no limit or end **:** ETERNAL. **2.** Unaffected by time **:** AGELESS. **3.** *Obs.* Untimely. **—time′less·ly** *adv.* **—time′less·ness** *n.*

**time loan** *n.* A loan to be paid within or by a specified time.

**time lock** *n.* A lock set to open at a specific time.

**time·ly** (tīm′lē) *adj.* **-li·er, -li·est. 1.** Occurring at a suitable or opportune time. **2.** *Archaic.* Early **:** premature. —*adv.* **1.** Opportunely. **2.** *Archaic.* Early **:** soon. **—time′li·ness** *n.*

**time machine** *n.* A machine or device that in theory permits travel into the future and the past.

**time money** *n.* A time loan.

**time note** *n.* A promissory note payable at a specified date.

**time·ous** (tī′məs) *adj. Scot.* Timely. **—time′ous·ly** *adv.*

**time-out** *also* **time out** (tīm′out′) *n.* **1.** A brief suspension of play at the request of a sports team for rest or consultation. **2. a.** A short break from work or play. **b.** A short time of isolation from a group, as for misbehavior.

**time·piece** (tīm′pēs′) *n.* An instrument that measures, registers, or records time.

**tim·er** (tī′mər) *n.* **1.** TIMEKEEPER 2. **2.** A timepiece, esp. one used for measuring intervals of time. **3.** A switch or regulator that controls or activates another mechanism at fixed intervals.

**time reversal** *n.* A mathematical operation representing a transformation from a given physical system undergoing a given sequence of events to a system in which the exact reverse sequence of events is undergone.

**times** (tīmz) *prep.* Multiplied by <Six *times* two is twelve.>

**time·sav·ing** (tīm′sā′vĭng) *adj.* Serving to save time, as through an efficient method. **—time′sav′er** *n.*

**time·serv·er** (tīm′sûr′vər) *n.* One who conforms to the prevailing ways and opinions of one's time or condition for personal advantage **:** OPPORTUNIST. **—time′serv′ing** *adj.* & *n.*

**time-shar·ing** (tīm′shâr′ĭng) *n.* **1.** A technique permitting many users simultaneous access to a central computer through remote terminals. **2.** The joint ownership or lease of vacation property through which the principals occupy the property individually for set periods of time. **—time′-share′** *v.* **(-shared, -shar·ing, -shares).**

**time sheet** *n.* A sheet that records the number of hours worked by an employee during a pay period.

**time signature** *n. Mus.* A symbol, commonly in the form of a numerical fraction, placed on a staff to indicate the meter.

**times sign** *n.* The symbol (×) used to indicate multiplication.

**time study** *n.* Time and motion study.

**time·ta·ble** (tīm′tā′bəl) *n.* A schedule listing the times at which certain events, as arrivals and departures at a transportation station, are expected to take place.

**time-test·ed** (tīm′tĕs′tĭd) *adj.* Proved effective over a long period of time <a *time-tested* remedy>

**time warp** *n.* Discontinuity or distortion thought to occur in the flow of time.

**time·work** (tīm′wûrk′) *n.* Work paid for in specified time units, as by the hour. **—time′work′er** *n.*

**time·worn** (tīm′wôrn′, -wōrn′) *adj.* **1.** Exhibiting the effects of long wear or use. **2.** Having been used too often **:** HACKNEYED.

**time zone** *n.* Any of the 24 longitudinal divisions of the earth's surface in which a standard time is kept, the primary division being that bisected by the Greenwich meridian; each zone is 15° of longitude in width, with local variations, and observes a clock time one hour earlier than the zone immediately to the east.

**tim·id** (tĭm′ĭd) *adj.* **-er, -est.** [Lat. *timidus* < *timēre,* to fear.] **1.** Shrinking from difficult or dangerous circumstances **:** FEARFUL. **2.** Shrinking from public attention **:** SHY. **—ti·mid′i·ty** (tə-mĭd′ĭ-tē), **tim′id·ness** *n.* **—tim′id·ly** *adv.*

**tim·ing** (tī′mĭng) *n.* The art or operation of regulating occurrence, pace, or coordination to achieve the most desirable effects.

**ti·moc·ra·cy** (tī-mŏk′rə-sē) *n., pl.* **-cies.** [OFr. *tymocracie* < Med. Lat. *timocratia* < Gk. *timokratia* : *timē,* honor, value + *-kratia,* -cracy.] **1.** A state described by Plato as being governed on principles of honor and military glory. **2.** An Aristotelian state in which civic honor or political power is proportional to the property one owns. **—ti′mo·crat′ic** (tī′mə-krăt′ĭk) *adj.*

**tim·or·ous** (tĭm′ər-əs) *adj.* [ME *tymerous* < OFr. *timoureus* < Med. Lat. *timorosus* < Lat. *timor,* fear < *timēre,* to fear.] Apprehensive **:** timid. **—tim′or·ous·ly** *adv.* **—tim′or·ous·ness** *n.*

**tim·o·thy** (tĭm′ə-thē) *n.* [Prob. after *Timothy* Hanson, an 18th-cent. American farmer, who reportedly took the grass from New York to the Carolinas.] A grass, *Phleum pratense* native to Eurasia, with narrow cylindrical flower spikes, and widely cultivated for hay.

**Tim·o·thy** (tĭm′ə-thē) *n.* —See table at BIBLE.

**tim·pa·ni** *also* **tym·pa·ni** (tĭm′pə-nē) *pl.n.* [Ital., pl. of *timpano,* kettledrum < Lat. *tympanum,* drum. —see TYMPANUM.] A set of kettledrums. **—tim′pa·nist** *n.*

**tim·pa·num** (tĭm′pə-nəm) *n. var. of* TYMPANUM.

**tin** (tĭn) *n.* [ME < OE.] **1.** *Symbol* **Sn** A malleable, silvery metallic element used to coat other metals to prevent corrosion and in numerous alloys, as soft solder, pewter, type metal, and bronze; atomic number 50; atomic weight 118.69. **2.** Tin plate. **3.** A tin container or box. **4.** *Chiefly Brit.* A can for preserved foodstuffs. —*vt.* **tinned, tin·ning, tins. 1.** To plate or coat with tin. **2.** *Chiefly Brit.* To preserve or pack in tins **:** CAN.

**tin·a·mou** (tĭn′ə-mo͞o′) *n.* [Fr. < Galibi *tinamu.*] A chickenlike bird of the family Tinamidae of Central and South America.

**tin·cal** (tĭng′kəl) *n.* [Malay *tingkal* < Skt. *ṭankanaḥ.*] Crude borax.

**tin can** *n.* **1.** A container of tin-plated sheet steel used esp. for preserving food. **2.** *Slang.* DESTROYER 2.

**tinct** (tĭngkt) [Lat. *tinctus,* a dyeing < p.part. of *tingere,* to dye.] *Archaic.* —*n.* A color or tint. —*adj.* Tinged.

**tinc·to·ri·al** (tĭngk-tôr′ē-əl, -tōr′-) *adj.* [Lat. *tinctorius* < *tingere,* to dye.] Pertaining to the processes of dyeing or coloring. **—tinc·to′ri·al·ly** *adv.*

**tinc·ture** (tĭngk′chər) *n.* [ME < Lat. *tinctura,* a dyeing < *tingere,* to dye.] **1.** A dyeing substance **:** PIGMENT. **2.** An imparted color **:** TINT. **3.** A quality that colors, pervades, or distinguishes. **4.** A vestige or trace. **5.** A component of a substance extracted by means of a solvent. **6.** An alcohol solution of a nonvolatile medicine <*tincture* of iodine> **7.** A heraldic metal, color, or fur. —*vt.* **-tured, -tur·ing, -tures. 1.** To tinge with a color **:** TINT. **2.** To infuse, as with a quality **:** IMPREGNATE.

**tin·der** (tĭn′dər) *n.* [ME < OE *tynder.*] Readily combustible material, as dry twigs, used to kindle fires.

**tin·der·box** (tĭn′dər-bŏks′) *n.* **1.** A metal box for holding tinder. **2.** A potentially explosive place or condition.

**tine** (tīn) *n.* [ME *tyne* < OE *tind.*] **1.** A branch of a deer's antlers. **2.** A prong on an implement, as a fork.

**tin·e·a** (tĭn′ē-ə) *n.* [Lat., a gnawing worm.] A fungous skin disease, as ringworm. **—tin′e·al** *adj.*

**tinea cap·i·tis** (kăp′ĭ-tĭs) *n.* [NLat., worm of the head.] A fungous infection of the scalp.

**tin ear** *n.* **1.** A cauliflower ear. **2.** An insensitive ear <has a *tin ear* for music>

**tin·foil** *also* **tin foil** (tĭn′foil′) *n.* A thin, pliable sheet of aluminum or of tin-lead alloy, used as a protective wrapping.

**ting** (tĭng) *n.* [ME *tyngen,* to cause to ring.] A single light metallic sound, as of a small bell. —*vi.* **tinged** (tĭngd), **ting·ing, tings.** To emit a light metallic sound.

**tinge** (tĭnj) *vt.* **tinged** (tĭnjd), **tinge·ing** or **ting·ing, ting·es.** [ME *tyngen* < Lat. *tingere.*] **1.** To apply a trace of color to **:** TINT.

2. To affect slightly, as with a contrasting quality <humor *tinged* with pathos> —*n.* **1.** A faint trace of a color incorporated or added. **2.** A slight admixture : TRACE.

**tin·gle** (tĭng'gəl) *v.* **-gled, -gling, -gles.** [ME *tinglen.*] —*vi.* **1.** To have a prickling, stinging sensation, as from cold, a sharp slap, or excitement. **2.** To cause a prickling, stinging sensation. —*vt.* To cause to tingle. **—tin'gle** *n.* **—tin'gler** *n.* **—tin'gly** *adj.*

**tin·horn** (tĭn'hôrn') *n. Slang.* A petty braggart, esp. a gambler, who pretends to be wealthy.

**tin·ker** (tĭng'kər) *n.* [ME *tinkere.*] **1.** A traveling mender of metal household utensils. **2.** One who enjoys repairing and experimenting with machine parts. **3.** One who is a clumsy worker : BUNGLER. —*v.* **-kered, -ker·ing, -kers.** —*vi.* **1.** To work as a tinker. **2.** To toy with machine parts experimentally. —*vt.* To mend as a tinker.

**tinker's damn** *also* **tinker's dam** *n.* [Prob. from the cursing attributed to tinkers.] *Slang.* Something of the smallest value <not worth a *tinker's damn*>

**Tin·ker·toy** (tĭng'kər-toi'). A trademark for a construction toy consisting of pieces that fit together.

**tin·kle** (tĭng'kəl) *v.* **-kled, -kling, -kles.** [ME *tynclen,* freq. of *tynken,* to emit a brief metallic sound.] —*vi.* To make light metallic sounds, as those of a small bell. —*vt.* **1.** To cause to tinkle. **2.** To signal or call by tinkling. **—tin'kle** *n.* **—tin'kly** *adj.*

**tin lizzie** (lĭz'ē) *n.* [< *Lizzie,* a nickname for Elizabeth.] *Slang.* A cheap or dilapidated car.

**tin·ner** (tĭn'ər) *n.* **1.** A tin miner. **2.** A tinsmith.

**tin·ni·tus** (tĭn'ĭ-təs) *n.* [Lat. < p.part. of *tinnire,* to ring.] A sound in the ears, as buzzing, ringing, or whistling, caused by a defect in the auditory nerve.

**tin·ny** (tĭn'ē) *adj.* **-ni·er, -ni·est.** **1.** Of, containing, or yielding tin. **2.** Cheaply shiny and attractive. **3.** Having a thin metallic sound. **4.** Tasting or smelling of tin, as food from a tin can. **—tin'ni·ly** *adv.* **—tin'ni·ness** *n.*

**Tin Pan Alley** *also* **tin-pan alley** (tĭn'păn') *n.* [From the use of *tin pans* for drums.] **1.** A district associated with musicians, composers, and publishers of popular music. **2.** The publishers and composers of popular music as a group.

**tin plate** *n.* Thin sheet iron or steel coated with tin.

**tin-plate** (tĭn'plāt') *vt.* **-plat·ed, -plat·ing, -plates.** To coat with tin. **—tin'-plat'er** *n.*

**tin·sel** (tĭn'səl) *n.* [< OFr. *estincele,* adorned with metallic thread < *estinceler,* to sparkle < *estencele,* spark. —see STENCIL.] **1.** Very thin sheets, strips, or threads of a glittering material used as a decoration. **2.** Something superficially sparkling or showy but worthless. —*adj.* **1.** Made of or decorated with tinsel. **2.** Gaudy and showy but worthless. —*vt.* **-seled, -sel·ing, -sels** or **-selled, -sel·ling, -sels.** **1.** To decorate with or as if with tinsel. **2.** To give a false sparkle to.

**tin·smith** (tĭn'smĭth') *n.* One that makes and repairs tin articles.

**tin·stone** (tĭn'stōn') *n.* Cassiterite.

**tint** (tĭnt) *n.* [Alteration of TINCT.] **1.** A shade of a color, esp. a pale or delicate variation : TINGE. **2.** A gradation of a color made by adding white to it to lessen its saturation. **3.** A slight coloration : HUE. **4.** A barely detectable degree : TRACE. **5.** A shaded effect in engraving produced by hatching. **6.** A panel of color on which matter in another color, as an illustration, may be printed. **7.** A dye for the hair. —*vt.* & *vi.* **tint·ed, tint·ing, tints.** To give a tint to or take on a tint.

**tin·tin·nab·u·la** (tĭn'tĭ-năb'yə-lə) *n. pl. of* TINTINNABULUM.

**tin·tin·nab·u·lar** (tĭn'tĭ-năb'yə-lər) *also* **tin·tin·nab·u·lar·y** (-lĕr'ē) or **tin·tin·nab·u·lous** (-ləs) *adj.* [< TINTINNABULUM.] Of or relating to bells or the ringing of bells.

**tin·tin·nab·u·la·tion** (tĭn'tĭ-năb'yə-lā'shən) *n.* [< TINTINNABULUM.] The ringing of bells.

**tin·tin·nab·u·lum** (tĭn'tĭ-năb'yə-ləm) *n., pl.* **-la** (-lə) [Lat. < *tintinnare,* to jingle, redup. of *tinnire,* to ring.] A small tinkling bell.

**tin·type** (tĭn'tīp') *n.* FERROTYPE 1.

**tin·work** (tĭn'wûrk') *n.* **1.** Work in tin. **2. tinworks** *(sing.* or *pl. in number).* A place where tin is smelted and rolled.

**ti·ny** (tī'nē) *adj.* **-ni·er, -ni·est.** [Alteration of ME *tine.*] Very small : MINUTE. **—ti'ni·ness** *n.*

**-tion** *suff.* [ME *-cioun* < OFr. *-tion* < Lat. *-tio.*] Action : process <adsorption>

**tip¹** (tĭp) *n.* [ME.] **1.** The end of an object, esp. of a pointed or projecting object. **2.** A piece or attachment, as a cap or ferrule, designed to be fitted to the end of an object. —*vt.* **tipped, tip·ping, tips.** **1.** To furnish with a tip. **2.** To cover, decorate, or remove the tip of. **3.** To attach (an insert) in a book by gluing along the binding edge. **4.** To dye strands or ends of hair, as of furs, in order to blend or improve appearance. **—tip of the iceberg.** The most obvious or superficial manifestation.

**tip²** (tĭp) *v.* **tipped, tip·ping, tips.** [ME *tipen.*] —*vt.* **1.** To knock over or upset : TOPPLE. **2.** To move to a slanting position : TILT. **3.** To touch or raise (one's hat) in greeting. —*vi.* **1.** To topple over : OVERTURN. **2.** To become tilted : SLANT. **3.** *Chiefly Brit.* To empty by overturning : DUMP. —*n.* **1.** A tilt, slant, or incline. **2.** *Chiefly Brit.* An area or place for dumping something, as refuse from a mine.

**tip³** (tĭp) *vt.* **tipped, tip·ping, tips.** [ME *tippen.*] **1.** To strike gently : TAP. **2.** *Baseball.* To hit (the ball) with the side of the bat so that it glances off. —*n.* A light blow : TAP.

**tip⁴** (tĭp) *n.* [< slang *tip,* to give.] **1.** A small sum of money given as an acknowledgment of services rendered : GRATUITY. **2. a.** Advance or inside information given as a guide to action. **b.** A helpful hint. —*v.* **tipped, tip·ping, tips.** —*vt.* **1.** To give a gratuity to. **2.** To provide advance or inside information to. —*vi.* To give a tip or tips. **—tip (one's) hand.** To reveal one's intentions. **—tip'per** *n.*

**tip·cart** (tĭp'kärt') *n.* A cart having a body that can be tilted to facilitate unloading.

**ti·pi** (tē'pē) *n. var. of* TEPEE.

**tip-off** (tĭp'ôf', -ŏf') *n. Informal.* **1.** An item of advance or inside information. **2.** *Basketball.* The act or practice of beginning a game or overtime period with a jump ball.

**tip·pet** (tĭp'ĭt) *n.* [ME *tipet.*] **1.** A covering for the shoulders, as of fur, with long ends that hang in front. **2.** A long stole worn by Anglican clerics. **3.** A long hanging part, as of a sleeve, hood, or cape.

**tip·ple¹** (tĭp'əl) *v.* **-pled, -pling, -ples.** [Back-formation < obs. *tippler,* drunkard < ME *tipeler,* bartender.] —*vi.* To drink alcoholic liquor, esp. habitually. —*vt.* To drink (alcoholic liquor), esp. habitually. —*n.* Alcoholic liquor. **—tip'pler** *n.*

**tip·ple²** (tĭp'əl) *n.* [< dial. *tipple,* to overturn, freq. of TIP².] An apparatus for unloading freight cars by tipping them, or the place where this is done.

**tip·staff** (tĭp'stăf') *n., pl.* **-staves** (-stāvz', -stăvz') or **-staffs.** [Alteration of *tipped staff.*] **1.** A staff with a metal tip, carried as a sign of office. **2.** An officer, as a bailiff or constable, who carries a tipstaff.

**tip·ster** (tĭp'stər) *n. Informal.* A seller of inside information, as to bettors or speculators.

**tip·sy** (tĭp'sē) *adj.* **-si·er, -si·est.** [< TIP².] **1.** Slightly drunk. **2.** Likely to tip over. **—tip'si·ly** *adv.* **—tip'si·ness** *n.*

**tip·toe** (tĭp'tō') *vi.* **-toed, -toe·ing, -toes.** To walk stealthily on or as if on the tips of one's toes. —*n.* The tip of a toe. —*adj.* **1.** Standing or walking on or as if on the tips of one's toes. **2.** Stealthy : wary. —*adv.* On or as if on tiptoe.

**tip·top** (tĭp'tŏp') *n.* **1.** The highest point : SUMMIT. **2.** The highest degree of quality or excellence. —*adj.* Excellent : first-rate. —*adv.* At the highest point of excellence.

**ti·rade** (tī'răd', tī-răd') *n.* [Fr. < OItal. *tirata,* volley < p.part. of *tirare,* to draw.] A long angry or violent speech, usu. of censure or denunciation.

☆***syns:*** TIRADE, DIATRIBE, FULMINATION, HARANGUE, JEREMIAD, OBLOQUY, PHILIPPIC *n. core meaning* : a long, violent, or blustering speech, usu. of censure or denunciation <delivered a *tirade* against pornography>

**tire¹** (tīr) *v.* **tired, tir·ing, tires.** [ME *tyren* < OE *tyrian.*] —*vi.* **1.** To grow weary. **2.** To grow bored or impatient : lose interest <*tired* of watching TV> —*vt.* **1.** To make weary : FATIGUE. **2.** To exhaust the interest or patience of : BORE. **—tire out.** To fatigue or exhaust.

**tire²** (tīr) *n.* [ME *tyre,* curved metal plates for wheels, prob. < *tyr,* attire.] **1.** A covering for a wheel, usu. made of rubber reinforced with cords of nylon, fiber glass, or other material, and filled with compressed air. **2.** A hoop of metal or rubber fitted around a wheel.

**tire³** (tīr) [ME *tiren,* short for *attiren,* to attire. —see ATTIRE.] *Archaic.* —*vt.* **tired, tir·ing, tires.** To adorn or attire. —*n.* **1.** Attire. **2.** A headband or headdress.

**tired** (tīrd) *adj.* **1. a.** Marked by weariness : FATIGUED. **b.** Impatient : bored. **2.** Lacking freshness : HACKNEYED <a *tired* joke> **—tired'ly** *adv.* **—tired'ness** *n.*

**tire·less** (tīr'lĭs) *adj.* Untiring : indefatigable. **—tire'less·ly** *adv.* **—tire'less·ness** *n.*

**tire·some** (tīr'səm) *adj.* Causing fatigue or boredom. **—tire'some·ly** *adv.* **—tire'some·ness** *n.*

**tire·wom·an** (tīr'wo͝om'ən) *n.* **1.** A dressing assistant, as in a theater. **2.** *Archaic.* A lady's maid.

**ti·ro** (tī'rō) *n. var. of* TYRO.

**'tis** (tĭz). *Archaic.* It is.

**ti·sane** (tĭ-zăn', -zän') *n.* [ME *tysan* < OFr. *tisane* < Lat. *ptisana.* —see PTISAN.] An infusion of herbs or a similar preparation drunk as a beverage or for its mildly medicinal effect.

**Tish·ri** (tĭsh'rē) *n.* [Heb. *Tishrī* < Akkadian *Tashrītu* < *shurru,* to begin.] The first month of the civil year in the Hebrew calendar. —See table at CALENDAR.

**Ti·siph·o·ne** (tĭ-sĭf'ə-nē) *n.* [Lat. < Gk. *Tisiphonē.*] *Gk. Myth.* One of the three Furies.

**tis·sue** (tĭsh'o͞o) *n.* [ME *tyssu,* a rich kind of cloth < OFr. *tissu* < p.part. of *tistre,* to weave < Lat. *texere.*] **1.** *Biol.* **a.** An aggregation of morphologically and functionally similar cells. **b.** Cellular matter regarded as a collective entity. **2.** A soft, absorbent piece of paper, gen. composed of two thin layers and used as a disposable handkerchief or towel. **3.** *also* **tissue paper.** Thin, nearly translucent paper used for packing, wrapping, or protecting delicate articles. **4.** A fine sheer cloth, as gauze. **5.** An interwoven or interrelated number of things : NETWORK <"The text is a *tissue* of mocking echoes" —Richard Kain> **—tis'su·lar** *adj.*

**tissue culture** *n.* In vitro preparation and growth of tissue cells in culture media.

**tit** (tĭt) *n.* [Short for TITMOUSE.] **1.** Any of various small Old World birds of the family Paridae, related to and resembling the New World chickadees. **2.** A bird similar or related to the tit.

**Ti·tan** (tīt'n) *n.* [Gk.] **1.** *Gk. Myth.* One of a family of giants, the children of Uranus and Gaea, who sought to rule heaven and were overthrown and supplanted by the family of Zeus. **2. titan.** A person of colossal size, strength, or achievement <a *titan* of industry> **3.** *Astron.* Saturn's largest satellite.

**ti·tan·ate** (tīt'n-āt') *n.* A salt of titanic acid.

**Ti·tan·ess** (tīt'n-ĭs) *n. Gk. Myth.* A female Titan.

**Ti·ta·ni·a** (tĭ-tā'nē-ə, -tăn'yə, tī-) *n.* The queen of the fairies and wife of Oberon in medieval folklore.

**ti·tan·ic[1]** (tī-tăn'ĭk) *adj.* **1. a.** Having great stature or enormous strength : HUGE. **b.** Enormous in scope, power, or influence. **2. Titanic.** Of or relating to the Titans. **—ti·tan'i·cal·ly** *adv.*

**ti·tan·ic[2]** (tī-tăn'ĭk, -tā'nĭk, tĭ-) *adj.* Relating to or containing titanium, esp. with valence 4.

**titanic acid** *n.* **1.** A white, powdered inorganic acid, $H_2TiO_3$, derived from an acid solution of titanates and used as a mordant. **2.** Titanium dioxide.

**ti·tan·if·er·ous** (tīt'n-ĭf'ər-əs) *adj.* Containing or yielding titanium.

**Ti·tan·ism** (tīt'n-ĭz'əm) *n.* Defiance of and revolt against the established order or authority.

**ti·tan·ite** (tīt'n-īt') *n.* [G. *Titanit.*] *Mineral.* Sphene.

**ti·ta·ni·um** (tī-tā'nē-əm, tĭ-) *n.* [NLat. < Gk. *Titan,* Titan.] *Symbol* **Ti** A strong, low-density, highly corrosion-resistant, lustrous white metallic element used in alloys requiring low weight, strength, and high-temperature stability; atomic number 22; atomic weight 47.90.

**titanium dioxide** *n.* A white powder, $TiO_2$, used as an exceptionally opaque white pigment.

**titanium white** *n.* Titanium dioxide used as a paint pigment with great covering power and durability.

**ti·tan·o·there** (tī-tăn'ə-thîr') *n.* [NLat. *Titanotherium* : Gk. *Titan,* Titan + Gk. *thērion,* little beast < *thēr,* beast.] An extinct herbivorous mammal of the genus *Brontotherium* and related genera, of the Eocene and Oligocene epochs, resembling the rhinoceros.

**ti·tan·ous** (tī-tăn'əs, -tā'nəs, tĭ-, tīt'n-) *adj.* Relating to or containing titanium, esp. with valence 3.

**tit·bit** (tĭt'bĭt') *n. var. of* TIDBIT.

**ti·ter** *also* **ti·tre** (tī'tər) *n.* [Fr. < OFr., title.] **1.** The concentration of a substance in solution or the strength of such a substance determined by titration. **2.** The minimum volume needed to cause a particular result in titration.

**tit for tat** *n.* [Alteration of *tip for tap.*] Repayment in kind.

**tithe** (tīth) *n.* [ME < OE *tēotha.*] **1.** A tenth part of one's annual income, either in kind or money, contributed voluntarily for charity or due as a tax for the support of the clergy or church. **2.** A tax or assessment of one tenth. **3. a.** A tenth part. **b.** A very small part. *—v.* **tithed, tith·ing, tithes.** *—vt.* **1.** To contribute or pay a tenth part of (one's annual income). **2.** To levy a tithe on. *—vi.* To pay a tithe. **—tith'a·ble** (tī'thə-bəl) *adj.* **—tith'er** (tī'thər) *n.*

**tith·ing** (tī'thĭng) *n.* **1.** The act of levying or paying tithes. **2.** A tithe. **3.** An administrative division consisting of ten householders in the old English system of frankpledge.

**ti·ti[1]** (tī'tī', tē'tē') *n.* [Orig. unknown.] A New World shrub of the genus *Cyrilla* and related genera, esp. *C. racemiflora* of warm swamps, with leathery leaves, white flower clusters, and yellow fruit.

**ti·ti[2]** (tē-tē') *n.* [Sp., perh. of Tupian orig.] A small, long-tailed South American monkey of the genus *Callicebus.*

**titi[2]**
*2–4 feet including tail*

**ti·tian** (tĭsh'ən) *n.* [After *Titian* (1477–1576), from his frequent use of the color in his paintings.] A brownish orange. **—ti'tian** *adj.*

**tit·il·late** (tĭt'l-āt') *vt.* **-lat·ed, -lat·ing, -lates.** [Lat. *titillare, titillat-,* to tickle.] **1.** To stimulate by touching or tickling lightly. **2.** To excite agreeably. **—tit'il·lat'ing·ly** *adv.* **—tit'il·la'tion** *n.* **—tit'il·la'tive** *adj.*

**tit·lark** (tĭt'lärk') *n.* [TIT(MOUSE) + LARK.] The pipit.

**ti·tle** (tīt'l) *n.* [ME < OFr. < Lat. *titulus.*] **1.** An identifying name given to a book, play, film, musical composition, or work of art. **2.** A general or descriptive heading, as of a book chapter. **3. a.** Written matter included in a movie or in a television show to give credits. **b.** The subtitle in a movie. **4. a.** The heading that names a legal document or statute. **b.** The heading or caption of a legal document in a court proceeding. **5.** A division of a law book, declaration, or bill, gen. larger than a section or article. **6.** *Law.* **a.** The coincidence of all the elements that constitute the fullest legal right to control and dispose of property or a claim. **b.** The aggregate evidence that gives rise to a legal right of possession or control. **c.** The evidence of such means. **d.** The instrument constituting this evidence, as a deed. **7.** Something that provides ground for or justifies a claim. **8.** A formal appellation attached to a person or family by virtue of office, rank, hereditary privilege, noble birth, attainment, or as a mark of respect. **9.** A descriptive appellation : EPITHET. **10.** A championship in sports. **11. a.** A source of income or area of work required of a candidate for ordination in the Church of England. **b.** A Roman Catholic church in or near Rome having a cardinal for its nominal head. *—vt.* **-tled, -tling, -tles.** To give a title to : confer a name on.

**ti·tled** (tīt'ld) *adj.* Having a title, esp. of nobility.

**ti·tle·hold·er** (tīt'l-hōl'dər) *n.* One that holds a title, esp. for a championship.

**title page** *n.* A page at the front of a book giving the complete title, the names of the author and publisher, and the place of publication.

**tit·mouse** (tĭt'mous') *n., pl.* **-mice** (-mīs') [By folk ety. < ME *titmose* : *tit- (of Scand. orig.) + *mose,* titmouse < OE *māse,* a kind of bird.] **1.** Any of several small North American birds of the genus *Parus,* having grayish plumage and a pointed crest. **2.** TIT 1.

**Ti·to·ism** (tē'tō-ĭz'əm) *n.* **1.** The Communist policies and practices associated with the late Marshal Tito of Yugoslavia. **2.** The assertion by a Communist state of its national interests independently of and often in opposition to Soviet policy.

**ti·trant** (tī'trənt) *n.* A reagent used in titration.

**ti·trate** (tī'trāt') *v.* **-trat·ed, -trat·ing, -trates.** [Fr. *titrer* < *titre,* titer.] *—vt.* To determine the concentration of (a solution) by titration. *—vi.* To perform titration. **—ti'trat'a·ble** *adj.* **—ti'tra·tor** *n.*

**ti·tra·tion** (tī-trā'shən) *n.* Determination of the concentration of a substance in solution by adding to it a standard reagent of known concentration in carefully measured amounts until a reaction of definite and known proportion is completed, as shown by a color change or by electrical measurement, followed by calculation of the unknown concentration.

**ti·tre** (tī'tər) *n. var. of* TITER.

**ti·tri·met·ric** (tī'trə-mĕt'rĭk) *adj.* [TITR(ATION) + -METRIC.] Utilizing titration. **—ti'tri·met'ri·cal·ly** *adv.*

**tit·ter** (tĭt'ər) *vi.* **-tered, -ter·ing, -ters.** [Imit.] To utter a suppressed, nervous giggle. **—tit'ter** *n.* **—tit'ter·er** *n.* **—tit'ter·ing·ly** *adv.*

**tit·tle** (tĭt'l) *n.* [ME *titel* < Med. Lat. *titulus* < Lat., title.] **1.** A small diacritical mark, as an accent, vowel mark, or dot over an *i.* **2.** The tiniest bit : IOTA.

▲ word history: The word *tittle* is a doublet of *title,* since both are derived from Latin *titulus.* The Latin word originally meant "label, title," but in Medieval Latin it also meant "a small diacritical mark." The meaning "the tiniest bit" is derived directly from the sense of the Medieval Latin word. It attained currency in English from the translation of Matthew 5:18, which is rendered in the King James Version as "For verily I say unto you, Till heaven and earth pass, one jot or one tittle shall in no wise pass from the law, till all be fulfilled."

**tit·tle-tat·tle** (tĭt'l-tăt'l) *n.* [Redup. of TATTLE.] Petty gossip. *—vi.* **-tled, -tling, -tles.** To engage in gossip.

**tit·tup** (tĭt'əp) *vi.* **-tuped, -tup·ing, -tups** *or* **-tupped, -tup·ping, -tups.** [Imit. of the sound of a horse's hoofs.] To move in a lively, capering way : PRANCE. **—tit'tup** *n.*

**tit·u·ba·tion** (tĭch'ə-bā'shən) *n.* [Lat. *titubatio,* a staggering < *titubare,* to stagger.] A stumbling or staggering gait characteristic of some nervous disorders.

**tit·u·lar** (tĭch'ə-lər) *adj.* [< Lat. *titulus,* title.] **1.** Relating to, having the nature of, or constituting a title. **2.** Existing as such in name only : NOMINAL <the *titular* head of the department> **3. a.** Having a title. **b.** Related to or coming from a title, as honors. **4.** Having been derived from a title <the *titular* role in a Shakespearean play> **5.** Of or designating one of the ancient churches in or near Rome from which a cardinal takes his title. *—n. also* **tit·u·lar·y** (tĭch'ə-lĕr'ē), *pl.* **-ies.** One holding a title. **—tit'u·lar·ly** *adv.*

**Ti·tus** (tī'təs) *n.* —See table at BIBLE.

**Ti·u** (tē'o͞o) *n.* [OE *Tīw.*] *Norse Myth.* The Germanic god of war and the sky.

**tiz·zy** (tĭz'ē) *n., pl.* **-zies.** [Orig. unknown.] *Slang.* A state of nervous confusion or excitement, esp. over trivia.

**Tl** *symbol for* THALLIUM.

**Tlin·git** (tlĭng'gĭt, tlĭng'ĭt) *n., pl.* **Tlingit** *or* **-gits. 1. a.** A group of Indian seafaring tribes inhabiting the coastal areas of southern Alaska and northern British Columbia. **b.** A member of any of these tribes. **2.** A language family of the Na-dene phylum consisting only of the language of the Tlingit.

**T lymphocyte** *n.* T cell.

o͞o **boot** ou **out** th **thin** *th* **this** ŭ **cut** ûr **urge** y **young**
yo͞o **abuse** zh **vision** ə **about,** it**e**m, edib**le,** gall**o**p, circ**u**s

**Tm** *symbol for* THULIUM.
**tme·sis** (tmē′sĭs, mē′-) *n.* [LLat. < Gk. *tmēsis,* a cutting < *temnein,* to cut.] Separation of the parts of a compound word by one or more intervening words; e.g., *where I go ever* instead of *wherever I go.*
**TNT** (tē′ĕn-tē′) *n.* [T(RI)N(ITRO)T(OLUENE).] Trinitrotoluene.
**to** (to͞o; tə *when unstressed*) *prep.* [ME < OE *tō.*] **1.** In a direction toward <went *to* town><turned *to* us and spoke><going back *to* the original plan> **2. a.** Reaching as far as <cut *to* the quick><rotten *to* the core> **b.** To the degree or extent of <beaten *to* death><defended the fortress *to* the last soldier> **c.** With the resultant condition of <torn *to* bits> **3.** Toward a specified state <the Communist rise *to* power> **4.** In contact with : AGAINST <the children's faces pressed *to* the glass> **5.** In front of <stood face *to* face> **6.** —Used to indicate possession <I have the belt *to* this coat.> **7.** Concerning <deaf *to* our pleas> **8.** In a given relationship with <at right angles *to* the line> **9.** As an accompaniment or complement of <danced *to* the music> **10.** In respect of <the secret *to* my success> **11.** Composing <two pints *to* a quart> **12.** In accord with <not *to* my taste><*to* the best of my knowledge> **13.** In comparison with <a score of ten *to* eight><a book superior *to* your others> **14. a.** Before <The time is five *to* ten.> **b.** Until <worked from eight *to* four> **15. a.** For the purpose of <came *to* my assistance><went out *to* dinner> **b.** In honor of <a toast *to* you> **16.** —Used before a verb to indicate the infinitive <I'd like *to* go.> —Also used alone when the infinitive is understood <Leave if you want *to.*> **17. a.** —Used to indicate the relationship of a verb with its complement <refer *to* a thesaurus ><refer me *to* a thesaurus> **b.** —Used with a reflexive pronoun to indicate exclusivity or separateness <had the train all *to* ourselves> —*adv.* **1.** In a direction toward <ran *to* and fro> **2.** Into a shut or closed position <pushed the gate *to*> **3.** Into a state of consciousness <brought the patient *to*> **4.** Into a state of action <sat down for dinner and fell *to*> **5.** *Naut.* Into the wind. —Used of a sailing vessel.
**toad** (tōd) *n.* [ME *tode* < OE *tādige.*] **1.** Any of numerous tailless amphibians chiefly of the family Bufonidae, related to and resembling the frogs but more terrestrial and having rougher, drier skin. **2.** The horned toad. **3.** A loathsome person.
**toad·eat·er** (tōd′ē′tər) *n.* A toady.
**toad·fish** (tōd′fĭsh′) *n., pl.* **toadfish** or **-fish·es.** Any of various bottom-dwelling, chiefly marine fishes of the family Batrachoididae, with a broad, flattened head and a wide mouth.
**toad·flax** (tōd′flăks′) *n.* A plant of the genus *Linaria,* with narrow leaves and spurred two-lipped flowers, esp. the butter-and-eggs.
**toad·stool** (tōd′sto͞ol′) *n.* [ME *tadestole.*] A poisonous or inedible fungus with an umbrella-shaped fruiting body.
**toad·y** (tō′dē) *n., pl.* **-ies.** [< TOADEATER.] An obsequious flatterer : SYCOPHANT. —*vt.* & *vi.* **-ied, -y·ing, -ies.** To be a toady to or behave as a toady.
**to and fro** *adv.* Back and forth <a pendulum swinging *to and fro*> **—to′-and-fro′** (to͞o′ən-frō′) *adj.* & *n.*
**toast**[1] (tōst) *v.* **toast·ed, toast·ing, toasts.** [ME *tosten* < OFr. *toster* < Lat. *torrēre,* to bake.] —*vt.* **1.** To heat and brown (e.g., bread) by placing it in a toaster or close to a fire. **2.** To warm thoroughly, as before a fire <*toast* one's hands> —*vi.* To become toasted. —*n.* Sliced bread that has been heated and browned.
**toast**[2] (tōst) *n.* [< TOAST[1] (from the use of spiced toast to flavor drinks).] **1.** The act of drinking to the health of or in honor of another. **2.** The one honored by a toast. **3.** One receiving much attention or acclaim <the *toast* of Hollywood> —*v.* **toast·ed, toast·ing, toasts.** —*vt.* To drink to the health or honor of. —*vi.* To propose or drink a toast.
**toast·er** (tō′stər) *n.* A device for toasting bread, esp. by exposure to electrically heated coils.
**toast·mas·ter** (tōst′măs′tər) *n.* One who offers toasts and introduces speakers at a banquet.
**toast·mis·tress** (tōst′mĭs′trĭs) *n.* A woman who offers toasts and introduces speakers at a banquet.
**toast·y** (tō′stē) *adj.* **-i·er, -i·est.** Cozily warm.
**to·bac·co** (tə-băk′ō) *n., pl.* **-cos** or **-coes.** [Sp. *tabaco,* of Caribbean orig.] **1.** A plant of the genus *Nicotiana,* esp. *N. tabacum* native to tropical America, widely cultivated for its leaves, which are used primarily for smoking. **2.** The leaves of cultivated tobacco, dried and processed chiefly for use in cigarettes, cigars, or snuff or for smoking in pipes. **3.** Products made from tobacco. **4.** The habit of smoking tobacco. **5.** A tobacco crop.
**tobacco mosaic** *n.* Any of several viral diseases of tobacco and nightshade marked by mottled leaves.
**to·bac·co·nist** (tə-băk′ə-nĭst) *n.* A dealer in tobacco.
**to-be** (to͞o-bē′) *adj.* That is to be : FUTURE <a parent-*to-be*>
**To·bit** (tō′bĭt) *n.* [Gk. *Tōbit* < Heb. *Ṭobhiyyāh* : *ṭōbh,* good + *yāh,* God.] —See table at BIBLE.
**to·bog·gan** (tə-bŏg′ən) *n.* [Canadian Fr. *tobagan* < Micmac *tobākan.*] A long, narrow, runnerless sled made of thin boards curled upward in front. —*vi.* **-ganed, -gan·ing, -gans. 1.** To coast or ride on a toboggan. **2.** To decline rapidly <Sales *tobogganed.*> **—to·bog′gan·er, to·bog′gan·ist** *n.*
**to·by** *also* **To·by** (tō′bē) *n., pl.* **-bies.** [After *Toby,* a nickname for *Tobias.*] A drinking mug usu. shaped in the form of a stout man wearing a large three-cornered hat.
**toc·ca·ta** (tə-kä′tə) *n.* [Ital. < fem. p.part. of *toccare,* to touch.] A musical composition, usu. for the organ or other keyboard instrument, written in free style with elaborate runs and harmonies.
**To·char·i·an** *also* **To·khar·i·an** (tō-kâr′ē-ən, -kär′-, -kăr′-) *n.* [Lat. *Tochari* < Gk. *Tokharoi.*] **1.** A member of a people of possible European origin with an advanced culture, living in Asia until about the 10th cent. **2.** An Indo-European language of central Asia attested in documents of the 7th cent.
**to·coph·er·ol** (tō-kŏf′ə-rôl′, -rōl′) *n.* [Gk. *tokos,* offspring (< *tiktein,* to beget) + Gk. *pherein,* to carry + -OL.] Any of a group of four chemically related compounds, differing slightly in structure, that together make up vitamin E.
**toc·sin** (tŏk′sĭn) *n.* [OFr. *toquassen* < OProv. *tocasenh* : *tocar,* to strike + *senh,* bell < Lat. *signum,* sign.] **1.** An alarm sounded on a bell. **2.** A warning <ignored the unmistakable *tocsins* of war>
**tod** (tŏd) *n.* [ME.] *Chiefly Brit.* **1.** A unit of weight used esp. for wool, equivalent to about 28 pounds. **2.** A bushy clump, as of ivy.
**to·day** *also* **to-day** (tə-dā′) [ME < OE *tō dæg.*] —*adv.* **1.** During or on the present day. **2.** During or at the present time. —*n.* The present day, time, or age.
**tod·dle** (tŏd′l) *vi.* **-dled, -dling, -dles.** [Orig. unknown.] To walk with short, unsteady steps. —*n.* A slow, unsteady gait.
**tod·dler** (tŏd′lər) *n.* **1.** A child who is beginning to learn how to walk. **2.** A clothing size for children between the ages of approx. one and three years.
**tod·dy** (tŏd′ē) *n., pl.* **-dies.** [Hindi *tāṛī,* sap of palm < *tāṛ,* palm < Skt. *tālaḥ.*] **1.** A drink of brandy or other liquor mixed with hot water, sugar, and spices. **2. a.** The sweet sap of a tropical Asian palm tree, esp. *Caryota urens,* used as a beverage. **b.** A liquor fermented from this sap.
**to-do** (tə-do͞o′) *n., pl.* **-dos** (-do͞oz′). *Informal.* Commotion : fuss.
**to·dy** (tō′dē) *n., pl.* **-dies.** [Fr. *todier* < Lat. *todus,* a kind of small bird.] Any of various small colorful birds of the family Todidae, of the West Indies.

**tody**
*3¾ inches long*

**toe** (tō) *n.* [ME < OE *tā.*] **1.** One of the digits of the foot, esp. of a vertebrate. **2.** The part of a shoe, sock, or boot covering the toes. **3. a.** The base or lower tip, as the end of the head of a golf club. **b.** Something resembling a toe. —*v.* **toed, toe·ing, toes.** —*vt.* **1.** To touch, kick, or reach with the toe. **2.** To drive (a golf ball) with the toe of the club. **3. a.** To drive (a spike or nail) at an oblique angle. **b.** To fasten with nails or spikes driven obliquely. —*vi.* To move with the toes pointed in a specified direction <You *toe* out.> **—on (one's) toes.** Prepared to act : ALERT. **—step on (someone's) toes.** To hurt, offend, or encroach upon the feelings or province of (another). **—toe the mark (***or* **line). 1.** To touch a mark or line with the toe or hands in readiness for the start of a race or competition. **2.** To obey rules conscientiously : CONFORM.
**toe·a** (toi′ə) *n., pl.* **toea.** [Prob. Pidgin English < E. DOLLAR.] —See table at CURRENCY.
**toe·cap** (tō′kăp′) *n.* A reinforced covering for the toe of a boot or shoe.
**toed** (tōd) *adj.* **1.** Having a specified number or kind of toes <a two-*toed* sloth> **2. a.** Driven obliquely <a *toed* nail> **b.** Secured by obliquely driven nails <a *toed* beam>
**toe dance** *n.* A dance performed on the toes. **—toe dancer** *n.*
**toe·hold** (tō′hōld′) *n.* **1.** A small indentation on which the toe can find support in climbing. **2.** An advantage useful for future progress <finally achieved a *toehold* on the executive ladder> **3.** A wrestling hold in which one competitor wrenches the other's foot.
**toe·nail** (tō′nāl′) *n.* **1.** The nail on a toe. **2.** An obliquely driven nail, as used to join vertical and horizontal beams. —*vt.* **-nailed, -nail·ing, -nails.** To secure (beams) with obliquely driven nails.
**toff** (tŏf) *n.* [Prob. var. of TUFT, a gold tassel worn by titled students at Oxford and Cambridge.] *Chiefly Brit.* A dandy : fop.
**tof·fee** (tŏf′ē, tô′fē) *n.* [Alteration of TAFFY.] A hard chewy candy made of brown sugar and butter.

**toft** (tôft, tŏft) *n.* [ME < OE < ON *topt.*] *Chiefly Brit.* **1.** A homestead along with its arable land. **2.** A hillock.
**to·fu** (tō'fōō) *n.* [J. *tōfu.*] Bean curd.
**tog** (tŏg, tôg) [Obs. *togeman* < Fr. *toge,* cloak < Lat. *toga,* garment, toga.] *Informal.* —*n.* **1.** A coat or cloak. **2. togs.** Clothes <jogging *togs*> —*vt.* **togged, tog·ging, togs.** To clothe or dress <*togged* themselves out in high boots>
**to·ga** (tō'gə) *n.* [Lat. < *tegere,* to cover.] **1.** A loose one-piece outer garment worn in public by Roman citizens. **2.** A loose robe or gown characteristic of a profession or office. **—to'gaed** (tō'gəd) *adj.*
**toga vi·ri·lis** (və-rē'lĭs, -rĭl'ĭs) *n.* [Lat., toga of a man.] A white toga symbolizing manhood that Roman boys were allowed to wear at age 15.
**to·geth·er** (tə-gĕth'ər) *adv.* [ME < OE *tōgædere.*] **1.** In or into a single mass, group, or place <gathered *together* for dinner> **2.** In relationship, one to another: mutually or reciprocally <couldn't get along *together*> **3.** Regarded collectively <I am worth more than all of you *together.*> **4.** At once : SIMULTANEOUSLY <The alarms sounded *together.*> **5.** In accord or harmony <stood *together* on the issue> **6.** *Informal.* In an effective, coherent condition <Get yourself *together* and stop complaining.> —*adj. Slang.* **1.** Being in tune with what is going on: HIP <a *together* person> **2.** Unified and performing effectively. **—get** (or **put**) **it (all) together.** *Slang.* To unify, harmonize, and integrate one's resources so as to perform with top effectiveness. **—together with.** In addition to. ***usage:*** *Together with* sometimes follows the subject of a sentence or clause to introduce an additional element. In such a circumstance, however, the number of the verb is governed only by the subject and does not change. Therefore it is correct to write *The ambassador* (singular), *together with two aides, is* (singular) *going to attend the conference.* The same principle applies to the use of expressions such as *along with, as well as, besides, in addition to,* and *plus: Common sense as well as tact is necessary for the job.* **—to·geth'er·ness** *n.*
**tog·gery** (tŏg'ə-rē, tô'gə-) *n., pl.* **-ies.** *Informal.* **1.** Clothing. **2.** A clothing store.
**tog·gle** (tŏg'əl) *n.* [Orig. unknown.] **1.** A pin, rod, or crosspiece fitted or inserted into a loop in a rope, chain, or strap to prevent slipping, to tighten, or to hold an attached object. **2.** An ornamental crosspiece or button, as of wood or bone, inserted into a loop, as of rope, as a closure or fastening. **3.** A device with a toggle joint. —*vt.* **-gled, -gling, -gles.** To furnish or fasten with a toggle.
**toggle bolt** *n.* A fastener made up of a threaded bolt and a mated toggle.
**toggle joint** *n.* An elbowlike joint made up of two arms pivoted together so that force applied to the pivot point to straighten the joint produces an outward force at the ends of each arm.
**toggle switch** *n.* A switch in which a projecting lever employing a toggle joint with a spring is used to open or close an electric circuit.
**togue** (tōg) *n.* [Canadian Fr.] The lake trout.
**toil¹** (toil) *vi.* **toiled, toil·ing, toils.** [ME *toilen* < AN *toiler,* to strive < OFr. *toillier,* to stir up < Lat. *tudiclare,* to stir about < *tudicula,* a machine for bruising olives < *tudes,* hammer.] **1.** To labor untiringly and continuously <*toiled* in the fields> **2.** To advance with difficulty <*toiling* over the alpine paths> —*n.* **1.** Exhausting labor. **2.** *Obs.* Strife.
**toil²** (toil) *n.* [OFr. *toile,* web < Lat. *tela.*] **1.** *often* **toils.** A snare or entanglement : ENTRAPMENT <caught up in the *toils* of despair> **2.** A net for trapping game.
**toile** (twäl) *n.* [Fr. < OFr., cloth. —see TOIL².] **1.** A sheer linen fabric. **2.** Fine cretonne printed in a single color.
**toi·let** (toi'lĭt) *n.* [Fr. *toilette* < OFr., cloth cover for a dressing table, dim. of *toile,* cloth. —see TOIL².] **1. a.** An apparatus consisting of a porcelain bowl fitted with a hinged seat and a flushing device, used for disposal of bodily wastes. **b.** A room or booth containing such an apparatus. **2.** TOILETTE 1. **3.** *Archaic.* A dressing table. **4.** Attire.
**toilet paper** *n.* Thin absorbent paper, usu. in rolls, used for cleansing the body after defecation or urination.
**toi·let·ry** (toi'lĭ-trē) *n., pl.* **-ries.** An article or cosmetic used in dressing or grooming.
**toi·lette** (twä-lĕt') *n.* [Fr. —see TOILET.] **1.** The act or process of dressing or grooming oneself. **2.** One's dress or style of dress. **3.** A gown or costume.
**toilet training** *n.* The training of a child to use the toilet for defecation and urination.
**toilet water** *n.* A scented liquid weaker than perfume and stronger than cologne.
**toil·some** (toil'səm) *adj.* Marked by or requiring toil. **—toil'some·ly** *adv.* **—toil'some·ness** *n.*
**to-ing and fro-ing** (tōō'ĭng ən frō'ĭng) *n., pl.* **to-ings and fro-ings.** Movement back and forth <much *to-ing and fro-ing* in the convention hall>
**to·ka·mak** (tō'kə-măk', tŏk'ə-) *n.* [R.] A small doughnut-shaped nuclear reactor in which a plasma is heated and confined by electric and magnetic fields.
**To·kay** (tō-kā') *n.* **1.** A grape orig. grown near Tokay, Hungary. **2.** A wine made from Tokay grapes.
**toke** (tōk) *n.* [Orig. unknown.] *Slang.* A puff of marijuana.
**to·ken** (tō'kən) *n.* [ME < OE *tācen.*] **1.** An indication or representation of a fact, event, or emotion : SIGN <Wedding rings are *tokens* of love.> **2.** A symbol or evidence of authority, validity, or identity <the scepter as a *token* of royalty> **3.** A keepsake : souvenir. **4.** A piece of stamped metal used as a substitute for currency. —*vt.* **-kened, -ken·ing, -kens.** To symbolize or betoken. —*adj.* **1.** Performed or done as an indication or pledge <a *token* payment> **2. a.** Perfunctory <a *token* gesture of friendship> **b.** Merely symbolic <a *token* minority on the board of directors> **—by the same token.** In like manner. **—in token of.** As an indication of.
**to·ken·ism** (tō'kə-nĭz'əm) *n.* The policy of making merely a perfunctory effort or symbolic gesture toward the accomplishment of a goal, as racial integration or eradication of sexism.
**To·khar·i·an** (tō-kâr'ē-ən, -kär'-, -kăr'-) *n. var. of* TOCHARIAN.
**to·la** (tō'lə, tō-lä') *n.* [Hindi *tolā* < Skt. *tulā,* weight.] A unit of weight used in India, equal to the weight of one silver rupee, 11.7 grams, or 180 troy grains.
**tol·booth** *also* **toll·booth** (tōl'bōōth') *n.* [ME (Scotland) *tolbothe,* town hall containing customs offices and prison cells : *tol,* toll + *bothe,* booth.] *Scot.* A prison : jail.
**tol·bu·ta·mide** (tŏl-byōō'tə-mīd') *n.* [TOL(U) + BUT- + AMIDE.] A white powder, $C_{12}H_{18}N_2O_3S$, used in the treatment of diabetes.
**told** (tōld) *v. p.t.* & *p.p. of* TELL.
**tole** *also* **tôle** (tōl) *n.* [Fr. *tôle,* sheet metal < Lat. *tabula,* board.] Lacquered or enameled 18th-cent. metalware, usu. gilded and elaborately painted.
**To·le·do** *also* **to·le·do** (tə-lē'dō) *n., pl.* **-dos.** A fine-tempered sword or steel sword blade made in Toledo, Spain.
**tol·er·a·ble** (tŏl'ər-ə-bəl) *adj.* **1.** Able to be tolerated. **2.** Allowable : permissible <a *tolerable* number of errors> **3.** Fair : passable. **—tol'er·a·bil'i·ty, tol'er·a·ble·ness** *n.* **—tol'er·a·bly** *adv.*
**tol·er·ance** (tŏl'ər-əns) *n.* **1.** Recognition of and respect for the opinions, practices, or behavior of others. **2. a.** The amount of variation from a standard that is allowed. **b.** The permissible deviation from a specified value of a structural dimension. **3.** Capacity to withstand pain or hardship. **4. a.** Physiological resistance to poison. **b.** Capacity to absorb a drug continuously or in large dosages without adverse effect.
**tol·er·ant** (tŏl'ər-ənt) *adj.* [Lat. *tolerans, tolerant-,* pr.part. of *tolerare,* to tolerate.] **1.** Inclined to recognize and respect the beliefs, practices, or traits of others : FORBEARING. **2.** Capable of withstanding or enduring an adverse environmental condition <plants *tolerant* of extreme cold> **—tol'er·ant·ly** *adv.*
**tol·er·ate** (tŏl'ə-rāt') *vt.* **-at·ed, -at·ing, -ates.** [Lat. *tolerare, tolerat-,* to bear.] **1.** To neither forbid nor prevent : PERMIT. **2.** To recognize and respect (the rights, opinions, or practices of others). **3.** To put up with : ENDURE <could not *tolerate* rudeness> **4.** *Med.* To have tolerance for (a drug or poison). **—tol'er·a'tive** *adj.* **—tol'er·a'tor** *n.*
**tol·er·a·tion** (tŏl'ə-rā'shən) *n.* **1.** Tolerance. **2.** Official recognition of the rights of persons and groups to hold dissenting views, esp. with regard to religion.
**tol·i·dine** (tŏl'ĭ-dēn') *n.* [TOL(UENE) + -ID(E) + -INE.] Any of several isomeric bases, $C_{14}H_{16}N_2$, derived from toluene, one of which is used as a reagent for gold and for chlorine in water.
**toll¹** (tōl) *n.* [ME *tol* < OE *toll* < Med. Lat. *toloneum* < LLat. *teloneum,* tollbooth < Gk. *telōnion* < *telōnēs,* tax collector < *telos,* tax.] **1.** A fixed fee or tax for a privilege, esp. for passage across a bridge or along a road. **2.** A charge for a service, as a long-distance telephone call. **3.** The extent or amount of loss or destruction, as of life, health, or property, caused by a disaster <The death *toll* from the storm mounted.> —*vt.* **tolled, toll·ing, tolls.** To exact as a toll.
☆ ***syns:*** TOLL, CHARGE, EXACTION, FEE *n. core meaning* : a fixed amount of money charged for a privilege or service <a *toll* for using the expressway>
**toll²** (tōl) *v.* **tolled, toll·ing, tolls.** [ME *tollen,* to ring an alarm, perh. < *tollen,* to pull.] —*vt.* **1.** To sound (a large bell) slowly and at regular intervals. **2.** To summon or announce by tolling. —*vi.* To sound in slowly repeated single tones. —*n.* **1.** An act of tolling. **2.** The sound of a tolling bell.
**toll·booth¹** (tōl'bōōth') *n.* A booth at a tollgate where a toll is collected.
**toll·booth²** (tōl'bōōth') *n. Scot. var. of* TOLBOOTH.
**toll call** *n.* A telephone call for which a higher rate is charged than that standard for a local call.
**toll collector** *n.* One employed to collect toll payments.
**toll·er** (tō'lər) *n.* **1.** One who tolls a bell. **2.** A bell for tolling.
**toll·gate** (tōl'gāt') *n.* A gate barring passage to a tunnel, road, or bridge until a toll is collected.
**toll·house** (tōl'hous') *n.* **1.** A house occupied by the toll collector adjoining a tollgate. **2.** TOLLBOOTH¹.
**toll line** *n.* A long-distance telephone line or circuit.

**Tol·tec** (tŏl′tĕk′, tōl′-) *n.* [Sp. *Tolteca* < Nahuatl *tolecatl,* artisan.] One of an ancient Nahuatl people of central and southern Mexico whose culture flourished in about 1000 A.D. —*adj. also* **Tol·tec·an** (tŏl-tĕk′ən, tōl-). Of or relating to the Toltecs or their culture.

**to·lu** (tə-lo͞o′) *n.* [Sp. *tolú,* after Santiago de *Tolú,* Colombia.] The balsam of Tolu.

**tol·u·ate** (tŏl′yo͞o-āt′) *n.* [TOLU(IC ACID) + -ATE.] A salt or ester of toluic acid.

**tol·u·ene** (tŏl′yo͞o-ēn′) *n.* [TOLU (from which it was orig. obtained) + -ENE.] A colorless flammable liquid, $CH_3C_6H_5$, derived from coal tar or petroleum and used in aviation and other high-octane fuels, in dyestuffs, explosives, and as a solvent for gums and lacquers.

**to·lu·ic acid** (tə-lo͞o′ĭk) *n.* [TOLU(ENE) + -IC.] Any of four isomeric acids, $C_8H_8O_2$, derived from toluene.

**to·lu·i·dine** (tə-lo͞o′ĭ-dēn′) *n.* [TOLU(ENE) + -ID(E) + -INE.] Any of three isomeric compounds, $C_7H_9N$, used to make dyes.

**tol·u·ol** (tŏl′yo͞o-ôl′, -ōl′) *n.* Commercial-grade toluene.

**tol·yl** (tŏl′əl) *n.* [TOL(U) + -YL.] The univalent organic radical $CH_3C_6H_4$.

**tom** (tŏm) *n.* [*Tom,* nickname for *Thomas.*] The male of various animals, esp. a male cat or turkey.

**tom·a·hawk** (tŏm′ə-hôk′) *n.* [Algonquian (Virginia) *tamahaac.*] **1.** A light ax once used as a tool or weapon by North American Indians. **2.** An implement similar to a tomahawk. —*vt.* **-hawked, -hawk·ing, -hawks.** To strike with a tomahawk.

**to·mal·ley** (tə-măl′ē, tŏm′ăl′ē) *n., pl.* **-leys.** [Of Cariban orig.] The liver of a lobster, esteemed as a culinary delicacy.

**Tom and Jer·ry** (tŏm′ ən jĕr′ē) *n.* [After Corinthian *Tom and Jerry* Hawthorn, characters in the novel *Life in London* by Pierce Egan (1772–1849).] A hot drink consisting of rum, a beaten egg, milk or water, sugar, and spices.

**to·ma·to** (tə-mā′tō, -mä′-) *n., pl.* **-toes.** [Alteration of Sp. *tomate* < Nahuatl *tomatl.*] **1.** A plant, *Lycopersicon esculentum,* native to South America, widely cultivated for its edible, fleshy, usu. red fruit. **2.** The fruit of the tomato.

**tomb** (to͞om) *n.* [ME < OFr. *tombe* < Lat. *tumba* < Gk. *tumbos.*] **1.** A vault or chamber for burying the dead. **2.** A place of burial. **3.** A monument commemorating the dead.

**tom·bac** *also* **tam·bac** *or* **tom·back** *or* **tam·bak** (tŏm′băk′) *n.* [Fr. < Du. *tombak* < Malay *tĕmbaga.*] An alloy of copper and zinc, used in making cheap jewelry.

**tom·boy** (tŏm′boi′) *n.* A young girl who behaves like a boy. —**tom′boy′ish** *adj.*

**tomb·stone** (to͞om′stōn′) *n.* A gravestone.

**tom·cat** (tŏm′kăt′) *n.* A male cat.

**tom·cod** (tŏm′kŏd′) *n., pl.* **tomcod** *or* **-cods.** An edible marine fish, *Microgadus tomcod* of the North American Atlantic, or *M. proximus* of the northern Pacific, related to and resembling the cod.

**Tom Col·lins** (tŏm′ kŏl′ĭnz) *n.* [*Tom,* a kind of gin + *Collins,* a name.] A beverage of gin, lemon or lime juice, carbonated water, and sugar.

**Tom, Dick, and Har·ry** (tŏm′ dĭk′ ən hăr′ē) *n.* Anybody : everyone <Every *Tom, Dick, and Harry* came to the reception.>

**tome** (tōm) *n.* [OFr. < Lat. *tomus* < Gk. *tomos* < *temnein,* to cut.] **1.** One of the books in a work of several volumes. **2.** A large or scholarly book.

**-tome** *suff.* [NLat. *-tomus* < Gk. *-tomos,* a cutting < *temnein,* to cut.] **1.** Part : area : segment <dermatome> **2.** Cutting instrument <microtome>

**to·men·tose** (tō-mĕn′tōs′, tō′mən-tōs′) *adj.* [NLat. *tomentosus* < Lat. *tomentum,* cushion stuffing.] *Biol.* Covered with short, dense, matted hairs.

**to·men·tum** (tō-mĕn′təm) *n., pl.* **-ta** (-tə) [NLat. < Lat., cushion stuffing.] **1.** *Anat.* A network of very small blood vessels passing between the pia mater and cerebral cortex. **2.** *Biol.* A covering of closely matted woolly hairs.

**tom·fool** (tŏm′fo͞ol′) *n.* [ME *Thome Fole.*] A stupid, foolish person : BLOCKHEAD. —**tom′fool′** *adj.* —**tom·fool′er·y** *n.*

**tom·my** (tŏm′ē) *n., pl.* **-mies.** [*Tommy,* nickname for *Thomas.*] *Chiefly Brit.* **1.** A loaf or piece of bread. **2.** Food : provisions. **3.** *often* **Tommy.** A Tommy Atkins.

**Tommy At·kins** (ăt′kĭnz) *n.* [From the use of the name on sample forms.] A British soldier.

**Tommy gun** *n. Informal.* A Thompson submachine gun.

**tom·my·rot** (tŏm′ē-rŏt′) *n.* [Dial. *tommy,* fool + ROT.] *Informal.* Nonsense.

**to·mog·ra·phy** (tō-mŏg′rə-fē) *n.* [Gk. *tomos,* section (< *temnein,* to cut) + -GRAPHY.] A method for making x-ray pictures of a predetermined plane section of a solid object by blurring out the images of other planes. —**to′mo·gram′** (tō′mə-grăm′) *n.* —**to′mo·graph′** (-grăf′) *n.*

**to·mor·row** (tə-môr′ō, -mŏr′ō) *n.* [ME *to morow* < OE *tō morgenne,* in the morning.] **1.** The day following today. **2.** The near future. —*adv.* On or for the day following today.

**tom·pi·on** (tŏm′pē-ən) *n. var. of* TAMPION.

**Tom Thumb** *n.* **1.** A tiny hero of English folklore. **2.** A midget.

**tom·tit** (tŏm′tĭt′) *n.* A small bird, as a tit.

**tom-tom** (tŏm′tŏm′) *n.* [Hindi *ṭamṭam.*] **1.** Any of various usu. long narrow small-headed drums beaten with the hands. **2.** A gong with a metal disk struck by a felt-covered hammer or stick. **3.** A monotonous rhythmical drumbeat or sound similar to it.

**-tomy** *suff.* [NLat. *-tomia* < Gk. < *tomos,* cutting, sharp < *temnein,* to cut.] Act of cutting : incision <gastro*tomy*>

**ton** (tŭn) *n.* [ME *toun,* a measure of weight < OE *tunne,* large cask.] **1. a.** A unit of weight equal to 1.016 metric ton, 2,240 pounds, or 1016.06 kilograms. **b.** A unit of weight equal to .907 metric ton, 2,000 pounds, or 907.20 kilograms. **c.** A metric ton. **2.** A unit of capacity for cargo in maritime shipping, usu. estimated at 40 cubic feet. **3.** A unit of inside capacity of a ship equal to 100 cubic feet. **4.** *Informal.* A very large amount <a *ton* of homework>

**to·nal** (tō′nəl) *adj.* Of or relating to a tone, tones, or tonality. —**to′nal·ly** *adv.*

**to·nal·i·ty** (tō-năl′ĭ-tē) *n., pl.* **-ties. 1.** *Mus.* **a.** A system of seven tones built on a tonic key. **b.** Arrangement of all the tones and chords of a musical composition in relation to a tonic. **2.** The scheme or interrelation of the color tones in a painting.

**tone** (tōn) *n.* [ME < Lat. *tonus* < Gk. *tonos.*] **1. a.** A sound of distinct pitch, quality, or duration. **b.** The quality or character of sound <the clear *tones* of a horn> **2.** *Mus.* **a.** The interval of a major second. **b.** The characteristic quality or timbre of a specific instrument or voice. **3.** The pitch of a word used to determine its meaning or to distinguish differences in meaning. **4.** The particular or relative pitch of a word, phrase, or sentence. **5.** Manner of expression <a hectoring *tone* of voice> **6.** General quality, effect, or atmosphere <a room decorated in an elegant *tone*><an antagonistic *tone* in the discussion> **7. a.** A color or shade of color. **b.** Quality of color. **8.** *Physiol.* **a.** The tension in resting muscles. **b.** Normal firmness of tissue. —*v.* **toned, ton·ing, tones.** —*vt.* **1.** To give a particular tone or inflection to. **2.** To soften or change the color of (e.g., a painting or photographic negative). **3.** To sound monotonously : INTONE. —*vi.* **1.** To assume a given color quality. **2.** To harmonize in color. —**tone down.** To make less vivid, harsh, or violent : MODERATE <Let's *tone down* the rhetoric.> —**tone up.** To make or become brighter or more vigorous.

☆ ***syns:*** TONE, TIMBRE, TONALITY *n. core meaning* : a sound of distinct pitch and quality <the *tone* of a French horn>

**tone arm** *n.* The pivoted arm of a record player that holds the cartridge and stylus.

**tone color** *n.* The timbre of a singing voice or instrument.

**tone language** *n.* A language that distinguishes meanings among words of similar form by variations in pitch and tone.

**tone·less** (tōn′lĭs) *adj.* **1.** Lacking tone. **2.** Devoid of vitality : LISTLESS. —**tone′less·ly** *adv.* —**tone′less·ness** *n.*

**tone poem** *n.* A symphonic poem.

**tong**[1] (tông, tŏng) *vt.* **tonged, tong·ing, tongs.** [Back-formation < TONGS.] To seize or manipulate with tongs.

**tong**[2] (tông, tŏng) *n.* [Cantonese, assembly hall.] **1.** A Chinese association, clan, or fraternity. **2.** A Chinese secret society in the United States, at one time believed to control criminal activity among Chinese Americans.

**Ton·gan** (tŏng′gən, tŏng′ən) *n.* A Polynesian language spoken in Tonga.

**tongs** (tôngz, tŏngz) *pl.n.* [ME < OE *tong.*] *(sing. or pl. in number).* A grasping implement consisting of two arms joined at one end by a pivot or hinge.

**tongue** (tŭng) *n.* [ME < OE *tunge.*] **1.** The movable, fleshy muscular organ in the mouth that functions in tasting, speech, and as an aid in chewing and swallowing. **2.** The tongue of an animal, as a cow, used as food. **3.** A spoken language or dialect <one's native *tongue*> **4.** Style or quality of utterance <your sharp *tongue*> **5.** The flap of material under the laces or buckles of a shoe. **6.** A spit of land : PROMONTORY. **7.** Something resembling a tongue, as in shape <*tongues* of flame> **8.** A bell clapper. **9.** The harnessing pole attached to the front axle of a horse-drawn vehicle. **10.** A protruding strip along the edge of a board that fits into a matching groove on the edge of another board. —*v.* **tongued, tongu·ing, tongues.** —*vt.* **1.** To separate or articulate (musical notes played on a wind instrument) by shutting off air with the tongue. **2.** To lick or touch with the tongue. **3. a.** To provide (a board) with a tongue. **b.** To join by a tongue and groove. **4.** *Archaic.* To scold. —*vi.* **1.** To articulate notes on a wind instrument. **2.** To project, as a promontory. —**hold (one's) tongue.** To be or keep silent. —**on the tip of (one's) tongue.** On the verge of being recalled or expressed <a name *on the tip of* my *tongue*>

**tongue and groove** *n.* A joint made by fitting a tongue on the edge of a board into a matching groove on another board.

**tongue·fish** (tŭng′fĭsh′) *n., pl.* **tonguefish** *or* **-fish·es.** [From its tongue-shaped body.] Any of various marine flatfishes of the family Cynoglossidae, with the posterior tapering to a point.

**tongue-in-cheek** (tŭng′ĭn-chēk′) *adj.* Marked by irony, insincerity, or facetious exaggeration.

**tongue-lash·ing** (tŭng′lăsh′ĭng) *n. Informal.* A harsh scolding.

**tongue-tie** (tŭng'tī') *n.* Restricted mobility of the tongue resulting from abnormal shortness of the frenum. —*vt.* **-tied, -ty·ing, -ties.** To make tongue-tied.

**tongue-tied** (tŭng'tīd') *adj.* **1.** Speechless, as from shyness, embarrassment, or astonishment. **2.** Afflicted with tongue-tie.

**tongue twister** *n.* **1.** A word or words difficult to articulate rapidly, usu. due to a succession of similar consonantal sounds. **2.** A word or expression difficult to pronounce.

**-tonia** *suff.* [NLat. < *tonus.* —see TONE.] Degree or state of tonicity <myo*tonia*>

**†ton·ic** (tŏn'ĭk) *n.* [< Gk. *tonikos,* capable of extension < *tonos,* tone.] **1.** An invigorating, refreshing, or restorative agent. **2.** A medicine that restores or increases bodily tone. **3.** *Mus.* The first note of a diatonic scale : KEYNOTE. **4. a.** Quinine water. **b.** *Regional.* A flavored carbonated beverage. **5. a.** A tonic accent. **b.** *Obs.* A voiced sound. —*adj.* **1.** Producing or stimulating mental, physical, or emotional vigor. **2.** *Mus.* Of or based on the tonic. **3. a.** Stressed, as a syllable : ACCENTED. **b.** *Obs.* Voiced. **4.** *Physiol.* Of or relating to tissue or muscular tension. **—ton'i·cal·ly** *adv.*

**tonic accent** *n.* A stress produced by rising pitch as distinguished from increased volume.

**to·nic·i·ty** (tō-nĭs'ĭ-tē) *n.* **1.** Normal functional readiness in bodily tissues. **2.** Active resistance to stretching in muscles.

**tonic sol-fa** *n. Mus.* A system of notation that is based on key relationships and that replaces usual staff notation with solmization syllables or their abbreviations.

**to·night** (tə-nīt') *adv.* [ME *to night* < OE *tō niht,* at night.] On or during the present or coming night. —*n.* This night or the night of this day.

**ton·ka bean** (tŏng'kə) *n.* [Perh. < Galibi *tonka.*] **1.** A South American tree of the genus *Dipteryx,* esp. *D. odorata,* with seeds that yield the fragrant compound coumarin. **2.** The seed of the tonka bean tree.

**ton·nage** (tŭn'ĭj) *n.* **1.** The number of tons of water a ship displaces afloat. **2.** The capacity of a merchant ship in units of 100 cubic feet. **3.** A duty or charge per ton on cargo, as at a port or canal. **4.** The total shipping of a country or port, figured in tons, with reference to carrying capacity. **5.** Weight that is measured in tons.

**to·nom·e·ter** (tō-nŏm'ĭ-tər) *n.* [Gk. *tonos,* tension + -METER.] **1.** An instrument for measuring fluid or vapor pressure. **2.** *Mus.* An instrument or device, as a graduated set of tuning forks, for determining the pitch or vibration rate of tones. **3.** An instrument for measuring hydrostatic pressure within the eyeball. **—to'no·met'ric** (tō'nə-mĕt'rĭk) *adj.* **—to·nom'e·try** *n.*

**to·no·plast** (tō'nə-plăst') *n.* [Gk. *tonos,* tension + -PLAST.] The cytoplasmic membrane surrounding a vacuole of a plant cell.

**ton·sil** (tŏn'səl) *n.* [Lat. *tonsillae,* tonsils.] A mass of lymphoid tissue, esp. either of two such masses, embedded in the lateral walls of the aperture between the mouth and pharynx. **—ton'sil·lar** (tŏn'sə-lər) *adj.*

**tonsill-** *pref. var. of* TONSILLO-.

**ton·sil·lec·to·my** (tŏn'sə-lĕk'tə-mē) *n., pl.* **-mies.** Surgical removal of the tonsils.

**ton·sil·li·tis** (tŏn'sə-lī'tĭs) *n.* Inflammation of the tonsils. **—ton'sil·lit'ic** (-lĭt'ĭk) *adj.*

**tonsillo-** *or* **tonsill-** *pref.* [< Lat. *tonsillae,* tonsils.] Tonsil <*tonsill*ectomy>

**ton·sil·lot·o·my** (tŏn'sə-lŏt'ə-mē) *n., pl.* **-mies.** Surgical incision of a tonsil.

**ton·so·ri·al** (tŏn-sôr'ē-əl, -sōr'-) *adj.* [Lat. *tonsorius* < *tonsor,* barber < *tondēre,* to shear.] Of or relating to a barber or to barbering.

**ton·sure** (tŏn'shər) *n.* [ME < Med. Lat. *tonsura* < Lat. *tondēre,* to shear.] **1.** The act of shaving the crown of the head, esp. prior to becoming a priest or a monk. **2.** The part of a priest's or monk's head that has been shaved. —*vt.* **-sured, -sur·ing, -sures.** To shave the head of.

**ton·tine** (tŏn'tēn', tŏn-tēn') *n.* [Fr., after Lorenzo *Tonti* (1635–1690).] **1.** An annuity or insurance plan whereby a group of participants hold shares in a common fund with right of survivorship, each participant's share being increased as one of the other dies, and the final survivor receiving the whole. **2.** Each member's share of a tontine. **3.** The subscribers to a tontine as a group.

**to·nus** (tō'nəs) *n.* [Lat., tone.] Tonicity.

**ton·y** (tō'nē) *adj.* **-i·er, -i·est.** Elegant or exclusive in manner or quality <a *tony* neighborhood>

**To·ny** (tō'nē) *n.* [After *Tony,* nickname of *Antoinette* Perry (1888–1946).] An annual award for outstanding theatrical achievement.

**too** (tōō) *adv.* [ME *to* < OE *tō.*] **1.** As well : ALSO <I'm coming *too.*> **2.** More than enough : EXCESSIVELY <You study *too* much.> **3.** Very : extremely <They're only *too* willing to help.> **4.** *Informal.* Indeed : so <You will *too* sit down!>

**took** (tŏŏk) *v. p.t. of* TAKE.

ōō **boot** ou **out** th **thin** *th* **this** ŭ **cut** ûr **urge** y **young**
yōō **abuse** zh **vision** ə **about,** it**e**m, edib**i**le, gall**o**p, circ**u**s

**tool** (tōōl) *n.* [ME < OE *tōl.*] **1.** A hand-held implement, as a hammer, saw, or drill, used in accomplishing work. **2. a.** A machine, as a lathe, for cutting and shaping mechanical parts. **b.** The cutting part of such a machine. **3.** Something used in the performance of an operation : INSTRUMENT <"modern democracies have the fiscal and monetary *tools* . . . to end chronic slumps and galloping inflations" —Paul A. Samuelson> **4.** Something regarded as necessary to the performance of one's occupational or professional tasks <Words are the *tools* of my trade.> **5. a.** One utilized to carry out the designs of another : DUPE. **b.** One manipulated or used by another : CAT'S-PAW. **6. a.** A bookbinder's hand stamp. **b.** A design impressed on a book cover by this means. —*v.* **tooled, tool·ing, tools.** —*vt.* **1.** To form, work, or decorate with a tool. **2.** To ornament (a book cover) with a bookbinder's tool. **3.** *Informal.* To drive (a vehicle). —*vi.* **1.** To work with a tool. **2.** *Informal.* To travel in a vehicle <*tooled* along at 70 mph> **—tool up.** To prepare an industry or a factory for production by providing tools and machinery suitable for a given job.

**tool·box** (tōōl'bŏks') *n.* A case for carrying or storing hand tools.

**tool·ing** (tōō'lĭng) *n.* **1.** Work or ornamentation done with tools, esp. stamped or gilded designs on leather. **2.** Provision of machinery to a factory in preparation for production.

**tool·ma·ker** (tōōl'mā'kər) *n.* A master machinist skilled in making tools and parts.

**toon** (tōōn) *n.* [Hindi *tūn* < Skt.] **1.** A tall tree, *Cedrela toona* or *Toona ciliata,* of tropical Asia and Australia, with reddish aromatic wood. **2.** The wood of the toon.

**toot** (tōōt) *v.* **toot·ed, toot·ing, toots.** [Prob. imit.] —*vi.* **1.** To sound a horn or whistle in short blasts. **2.** To make the sound of a horn or whistle blown in short blasts. —*vt.* **1.** To blow or sound (a horn or whistle). **2.** To sound (a blast or series of blasts) on a horn or whistle. **—toot** *n.* **—toot'er** *n.*

**tooth** (tōōth) *n., pl.* **teeth** (tēth) [ME < OE *tōð.*] **1. a.** One of a set of hard bonelike structures rooted in sockets in the jaws of most vertebrates, typically composed of a core of soft pulp surrounded by a layer of hard dentine that is coated with cement or enamel at the crown, used to seize, hold, or masticate. **b.** A similar structure in invertebrates, as one of the pointed denticles or ridges on the exoskeleton of an arthropod or the shell of a mollusk. **2.** A projecting part resembling a tooth, as on a comb, gear, or saw. **3.** A small notched projection along a margin, esp. of a leaf. **4. teeth. a.** Something highly injurious or destructive in its concentrated force <the *teeth* of the storm> **b.** Effective means of enforcement : MUSCLE <regulations with *teeth* to them> **5.** Taste or appetite <a sweet *tooth*> —*v.* (tōōth, tōōth) **toothed, tooth·ing, tooths.** —*vt.* **1.** To furnish (e.g., a tool) with teeth. **2.** To make a jagged edge on. —*vi.* To become interlocked : MESH. **—get (one's) teeth into.** To become actively involved in. **—in the teeth of. 1.** Directly and forcefully in contact with <sailing *in the teeth of* a hurricane> **2.** In defiance of. **—put teeth into.** To make (e.g., a law) effective or forceful. **—show (one's) teeth.** To express a readiness to fight. **—to the teeth.** Lacking nothing <bandits armed *to the teeth*>

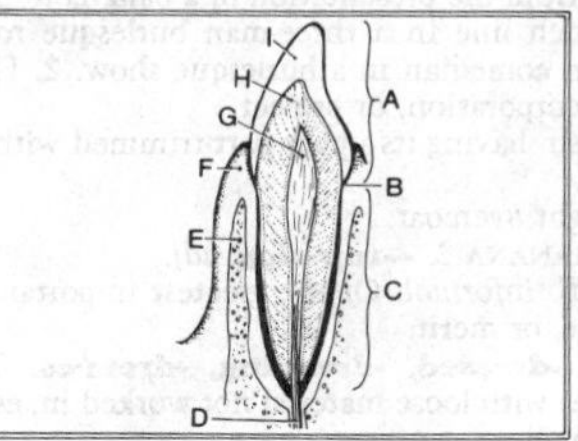


**tooth**
*Cross section of a tooth showing: A. crown, B. neck, C. root, D. root canal, E. bone, F. gum, G. pulp, H. dentine, I. enamel*

**tooth·ache** (tōōth'āk') *n.* An aching pain in or near a tooth.

**toothache tree** *n.* The prickly ash.

**tooth and nail** *adv.* With great ferocity.

**tooth·brush** (tōōth'brŭsh') *n.* A small brush for cleaning teeth.

**toothed** (tōōtht, tōōthd) *adj.* **1.** Having teeth. **2.** Having a certain number or kind of teeth <saw-*toothed*>

**toothed whale** *n.* Any of various whales of the suborder Odontoceti with numerous conical teeth.

**tooth·less** (tōōth'lĭs) *adj.* **1.** Lacking teeth. **2.** Lacking force <*toothless* rules> **—tooth'less·ly** *adv.* **—tooth'less·ness** *n.*

**tooth·paste** (tōōth'pāst') *n.* A paste for cleaning teeth.

**tooth·pick** (tōōth'pĭk') *n.* A small piece of wood or plastic for removing food particles from between the teeth.

**tooth·pow·der** (tōōth'pou'dər) *n.* A powder for cleaning teeth.

**tooth shell** *n.* Any of various burrowing marine mollusks of the class Scaphopoda, with a long, tapering, slightly curved tubular shell.

**tooth·some** (tōōth'səm) *adj.* **1.** Extremely tasty : DELICIOUS <a *toothsome* morsel of cake> **2.** Pleasant : attractive <a *toothsome* business offer> **3.** Sexually attractive. **—tooth'some·ly** *adv.* **—tooth'some·ness** *n.*

**tooth·wort** (tōōth'wûrt', -wôrt') *n.* **1.** A plant of the genus *Dentaria,* as the crinkleroot. **2.** A parasitic European plant, *Lathraea* *squamaria,* having scaly cream-colored or pink stems and pinkish flowers.

**tooth·y** (to͞o'thē) *adj.* **-i·er, -i·est.** Having prominent teeth. **—tooth'i·ly** *adv.*

**too·tle** (to͞ot'l) *vi.* **-tled, -tling, -tles.** [Freq. of TOOT.] To toot softly and repeatedly, as on a flute. **—too'tle** *n.*

**toots** (to͝ots) *n.* [Orig. unknown.] *Slang.* Dear : sweetheart.

**toot·sy** *also* **toot·sie** (to͝ot'sē) *n., pl.* **-sies.** [Alteration of *footsy* < FOOT.] *Informal.* One's foot.

**top¹** (tŏp) *n.* [ME < OE.] **1.** The highest part, point, surface, or end. **2.** The crown of the human head. **3.** The part of a plant, as a rutabaga, above the ground. **4.** A part, as a lid or cap, that covers or forms the uppermost section of something. **5.** *Naut.* A platform enclosing the head of each mast of a sailing ship, to which the topmast rigging is attached. **6. a.** A stroke, as in tennis or golf, that lands above the center of a ball, giving it forward spin. **b.** A forward spin on a ball resulting from such a stroke. **7.** The highest degree, pitch, or point : ZENITH. **8. a.** The highest position or rank <at the *top* of the legal profession> **b.** One in this position. **9.** The highest card or cards in a suit or a hand. **10.** The best part : PICK. **11.** The earliest part : BEGINNING <the *top* of the eighth inning> *—v.* **topped, top·ping, tops.** *—vt.* **1.** To form, furnish with, or serve as a top. **2.** To reach the top of. **3.** To go over the top of. **4.** To surpass or exceed. **5.** To be at the head of <I *topped* my class.> **6.** To remove the uppermost part from : CROP <*topped* all of the peach trees> **7. a.** To strike the upper part of (a ball), giving it forward spin. **b.** To make (a stroke) in this way. *—vi.* To make a finish, end, or conclusion. **—blow (one's) top.** *Slang.* **1.** To lose one's temper. **2.** To lose one's mind. **—off the top of (one's) head.** In an impromptu way <provided sales projections *off the top of* my *head*> **—on top. 1.** At the highest point. **2.** In a dominant, controlling, or successful position. **—on top of. 1.** On or at the uppermost part or side of. **2.** *Informal.* **a.** In control of. **b.** Fully informed about <The President is *on top of* developments in the Middle East.> **3.** In addition to : BESIDES. **4.** Following closely upon or immediately after. **—on top of the world.** In a position of great happiness or success. **—over the top. 1.** Over the breastwork, as an attack in trench warfare. **2.** Surpassing a goal or quota. **—top off.** To finish up. **—top out. 1.** To put the framework for the top story on (a building). **2.** *Informal.* To give up one's career just as one becomes highly successful.

**top²** (tŏp) *n.* [ME *topp* < OE *top.*] A toy consisting of a symmetrical rigid body spun on a pointed end about the axis of symmetry.

**top-** *pref. var. of* TOPO-.

**Top 40** *n.* The 40 most popular phonograph records during a specified time period.

**to·paz** (tō'păz') *n.* [ME *topace* < OFr. < Lat. *topazus* < Gk. *topazos.*] **1.** A colorless, blue, yellow, brown, or pink aluminum silicate mineral, often found in association with granitic rocks and valued as a gemstone. **2.** A yellow gemstone, esp. a yellow sapphire or corundum. **3.** A light-yellow quartz. **4.** A colorful South American hummingbird, *Topaza pyra* or *T. pella.*

**top banana** *n.* [So called from the presentation of a banana to the comedian who has the punch line in a three-man burlesque routine.] *Informal.* **1.** The main comedian in a burlesque show. **2.** The head person, as of a group, corporation, or project.

**top boot** *n.* A high boot usu. having its upper part trimmed with a contrasting color or texture.

**top·coat** (tŏp'kōt') *n.* A light overcoat.

**top dog** *n. Informal.* TOP BANANA 2. **—top'-dog'** *adj.*

**top-drawer** (tŏp'drôr') *adj. Informal.* Of the greatest importance or the highest rank, privilege, or merit.

**top-dress** (tŏp'drĕs') *vt.* **-dressed, -dress·ing, -dress·es.** To cover (land or a road surface) with loose material not worked in, esp. to cover (farmland) with fertilizer.

**top dressing** *n.* **1.** A cover of fertilizer, as manure, spread on soil without being plowed under. **2.** Loose gravel on top of a road.

**tope¹** (tōp) *v.* **toped, top·ing, topes.** [Prob. < obs. *tope,* interjection used in proposing a toast.] *—vt.* To drink (alcoholic liquors) habitually and excessively. *—vi.* To drink to excess habitually.

**tope²** (tōp) *n.* [Orig. unknown.] A small shark, esp. one of the genus *Galeorhinus.*

**tope³** (tōp) *n.* [Hindi *tōp.*] A dome-shaped Buddhist shrine topped with a cupola.

**to·pee** (tō-pē', tō'pē) *n. var. of* TOPI.

**top·er** (tō'pər) *n.* A drunkard.

**top·flight** (tŏp'flīt') *adj.* First-rate <a *topflight* performance>

**top·full** *also* **top·ful** (tŏp'fo͝ol') *adj.* Full to the brim.

**top·gal·lant** (tə-găl'ənt, tŏp-) *adj. Naut.* Designating the mast above the topmast, its sails, or its rigging.

**top-ham·per** *also* **top hamper** (tŏp'hăm'pər) *n.* **1.** *Naut.* Rigging not immediately necessary and stored either aloft or on the upper decks. **2.** Cumbersome, unnecessary, or meaningless matter.

**top hat** *n.* A man's hat with a narrow brim and a tall cylindrical crown, usu. made of silk.

**top-heav·y** (tŏp'hĕv'ē) *adj.* **-i·er, -i·est. 1.** That is likely to topple because of being overloaded at the top. **2.** Overcapitalized. **—top'-heav'i·ness** *n.*

**To·phet** (tō'fĕt', -fĭt) *n.* [ME < Heb. *tōpheth.*] **1.** A shrine near Gehenna where human sacrifices were made. **2. a.** Hell. **b.** A hellish place.

**to·phus** (tō'fəs) *n., pl.* **-phi** (-fī') [Lat., tufa.] **1.** *Pathol.* A urate deposit found in tissue, as cartilage, around the joints. **2.** A concretion of mineral salts and organic matter deposited on the surface of the teeth.

**to·pi** *also* **to·pee** (tō-pē', tō'pē) *n., pl.* **-pis** *also* **-pees.** [Hindi *topī,* hat.] A pith helmet.

**to·pi·ar·y** (tō'pē-ĕr'ē) *adj.* [Lat. *topiarius* < *topia,* ornamental gardening < Gk. *topia,* pl. of *topion,* field, dim. of *topos,* place.] Of or characterized by the clipping or trimming of live shrubs or trees into decorative shapes, as those of animals or birds. *—n., pl.* **-ies. 1.** Topiary work or art. **2.** A topiary garden.

**top·ic** (tŏp'ĭk) *n.* [Obs. *topic,* rhetorical argument < *Topics,* a work by Aristotle < Lat. *Topica* < Gk. *Topika,* neuter pl. of *topikos,* of a place < *topos,* place.] **1.** A subject discussed in a speech, essay, thesis, or part of a discourse : THEME. **2.** A subject of discussion or conversation. **3.** A subdivision of a theme, thesis, or outline.

**top·i·cal** (tŏp'ĭ-kəl) *adj.* [Gk. *topikos* < *topos,* place.] **1.** Of or belonging to a particular location or place : LOCAL. **2.** Of current interest : CONTEMPORARY <*topical* issues> **3.** *Med.* Of or applied to an isolated part of the body. **4.** Of or relating to a particular topic or topics. **—top'i·cal'i·ty** (-kăl'ĭ-tē) *n.* **—top'i·cal·ly** *adv.*

**topic sentence** *n.* The sentence within a paragraph that states the main thought, often placed at the beginning.

**top kick** *n. Slang.* A first sergeant.

**top·knot** (tŏp'nŏt') *n.* **1.** A crest or knot of hair or feathers on the crown of the head. **2.** A decorative ribbon or bow worn as a headdress.

**top·less** (tŏp'lĭs) *adj.* **1.** Having no top <*topless* jars> **2.** So high as to appear to extend out of sight <the *topless* mountain peaks>

**top·loft·y** (tŏp'lôf'tē, -lŏf'tē) *adj.* **-i·er, -i·est.** *Informal.* Pretentious : haughty. **—top'loft'i·ness** *n.*

**top·mast** (tŏp'məst, -măst') *n. Naut.* The mast below the topgallant mast in a square-rigged ship and next above the lower mast in a fore-and-aft-rigged ship.

**top·min·now** (tŏp'mĭn'ō) *n.* [So called because it swims near the surface of water.] **1.** Any of several small New World freshwater fishes of the genus *Fundulus,* related to the killifishes. **2.** Any of various small viviparous New World fishes of the family Poeciliidae found in fresh or brackish waters.

**top·most** (tŏp'mōst') *adj.* Being the highest : UPPERMOST.

**top·notch** (tŏp'nŏch') *adj. Informal.* First-rate <a *topnotch* unit of soldiers>

**topo-** *or* **top-** *pref.* [< Gk. *topos,* place.] Place : region <*topony my*>

**to·pog·ra·pher** (tə-pŏg'rə-fər, tō-) *n.* **1.** One skilled in topography. **2.** One who describes and maps the surface features of geographic regions.

**to·pog·ra·phy** (tə-pŏg'rə-fē, tō-) *n., pl.* **-phies.** [ME *topographie* < LLat. *topographia* < Gk. < *topographein,* to describe a place : *topos,* place + *graphein,* to write.] **1.** Detailed, precise description of a place or region. **2.** The technique of graphically representing the exact physical features of a place or region on a map. **3.** The physical features of a place or region. **4.** Surveying of the features of a region or place. **—top'o·graph'** (tŏp'ə-grăf', tō'pə-) *n.* **—top'o·graph'ic** (-ə-grăf'ĭk), **top'o·graph'i·cal** *adj.* **—top'o·graph'i·cal·ly** *adv.*

**to·pol·o·gy** (tə-pŏl'ə-jē, tō-) *n., pl.* **-gies. 1.** Topographical study of a given place in relation to its history. **2.** Anatomy of specific physiological areas. **3.** *Math.* Study of the properties of geometric configurations invariant under transformation by continuous mappings. **—top'o·log'ic** (tŏp'ə-lŏj'ĭk, tō'pə-), **top'o·log'i·cal** *adj.* **—top'o·log'i·cal·ly** *adv.* **—to·pol'o·gist** *n.*

**top·o·nym** (tŏp'ə-nĭm', tō'pə-) *n.* [Back-formation < TOPONYMY.] A name that is derived from a place or region. **—top'o·nym'ic, top'o·nym'i·cal** *adj.*

**to·pon·y·my** (tə-pŏn'ə-mē, tō-) *n., pl.* **-mies. 1.** *Anat.* Nomenclature with respect to a bodily region rather than to organs or structures. **2.** Study of place names.

**top·o·type** (tŏp'ə-tīp', tō'pə-) *n. Biol.* A specimen of an organism taken from the area typical for that species.

**top·per** (tŏp'ər) *n.* **1.** One that removes tops <a carrot *topper*> **2.** A woman's short lightweight topcoat. **3.** *Slang.* A top hat. **4.** *Slang* One that outdoes what has been said before, esp. a bantering remark **5.** One found at or on the top.

**top·ping** (tŏp'ĭng) *n.* **1.** A sauce, frosting, or garnish for food. **2.** A part or layer that forms the top. **3. toppings.** The cropped parts of plants or trees after pruning. *—adj.* **1.** Very highly thought of : OUTSTANDING. **2.** *Chiefly Brit.* First-rate : excellent.

**top·ple** (tŏp'əl) *v.* **-pled, -pling, -ples.** [Freq. of TOP¹.] *—vt.* To push over : OVERTURN. *—vi.* **1.** To totter and fall. **2.** To lean over as if about to fall.

**top round** *n.* A cut of meat, as steak, taken from the inner section of a round of beef.
**tops** (tŏps) *adj. Slang.* First-rate : excellent <*tops* in your field>
**top·sail** (tŏp′səl, -sāl′) *n. Naut.* **1.** A square sail set above the lowest sail on the mast of a square-rigged ship. **2.** A triangular or square sail set above the gaff of a lower sail on a fore-and-aft-rigged ship.
**topsail schooner** *n.* A schooner carrying two or more square topsails on the foremast.
**top-se·cret** (tŏp′sē′krĭt) *adj.* Designating documents or data of the highest level of security classification.
**top sergeant** *n. Informal.* A first sergeant.
**top·side** (tŏp′sīd′) *n.* **1.** *often* **topsides.** The upper parts of a ship above the main deck. **2.** The highest position of authority. —*adv.* & *adj.* **1.** On or to the upper parts of a ship : on deck. **2.** In a position of authority.
**top·sid·er** (tŏp′sī′der) *n.* One at the highest level of authority.
**Top-Sid·er** (tŏp′sī′der). A trademark for a soft leather or canvas shoe with a rubber sole.
**top·soil** (tŏp′soil′) *n.* The surface layer of soil. —*vt.* **-soiled, -soil·ing, -soils.** To remove the surface layer of soil from (land).
**top·stitch** (tŏp′stĭch′) *vt.* **-stitched, -stitch·ing, -stitch·es.** To sew a line of stitching close to the seam or edge of (a garment) on the right side of the fabric.
**top·sy-tur·vy** (tŏp′sē-tûr′vē) *adv.* [Prob. TOP[1] + obs. *terve,* to overturn.] **1.** With the top down and the bottom up : UPSIDE-DOWN. **2.** In utter confusion or disorder. —*adj.* Being in a confused or disordered state. —*n.* Confusion : disorder. **—top′sy-tur′vi·ly** *adv.* **—top′sy-tur′vi·ness** *n.*
**toque** (tōk) *n.* [Fr. < Sp. *toca.*] **1.** A woman's small brimless, close-fitting hat. **2.** A plumed velvet cap with a full crown and small rolled brim, worn by 16th-cent. French people.
**tor** (tôr) *n.* [ME < OE *torr.*] A high rock or pile of rocks on the top of a hill.
**to·rah** *also* **To·rah** (tôr′ə, tōr′ə) *n.* [Heb. *tōrāh* < *hārāh,* he taught.] **1.** The complete body of Jewish religious law and learning including sacred literature and oral tradition. **2.** A scroll or scrolls of parchment containing the Pentateuch, used in a synagogue during services.
**tor·bern·ite** (tôr′bər-nīt′) *n.* [G. *Torbernit,* after *Torbern* O. Bergman (1735–1784).] An emerald- or grass-green hydrous crystalline phosphate of uranium and copper.
**torch** (tôrch) *n.* [ME *torche* < OFr. < Lat. *torquēre,* to twist.] **1.** A portable light consisting of flaming material wound about the end of a stick of wood : FLAMBEAU. **2.** A portable apparatus producing a very hot flame by the combustion of gases, and used in welding and construction. **3.** Something that illuminates, enlightens, or guides. **4.** *Chiefly Brit.* A flashlight. **5.** *Slang.* An arsonist. —*vt.* **torched, torch·ing, torch·es.** *Slang.* To set fire to <*torched* the building> **—carry a** (or **the**) **torch for.** To love (someone) who does not reciprocate.
**torch·bear·er** (tôrch′bâr′ər) *n.* **1.** One who carries a torch. **2.** One who imparts knowledge, truth, or inspiration to others.
**tor·chon lace** (tôr′shŏn′) *n.* [Fr. *torchon,* duster < OFr., twisted straw < *torche,* torch.] A lace of coarse linen or cotton thread twisted in simple geometric patterns.
**torch song** *n.* A sentimental popular song, typically one in which the singer laments a lost or unrequited love. **—torch singer** *n.*
**torch·wood** (tôrch′wo͝od′) *n.* **1.** A tropical American tree of the genus *Amyris,* esp. *A. balsamifera,* having resinous wood that burns with a torchlike flame. **2.** The wood of the torchwood.
**tore**[1] (tôr, tōr) *v. p.t. of* TEAR[1].
**tore**[2] (tôr, tōr) *n.* [Fr. < Lat. *torus.*] TORUS 4.
**tor·e·a·dor** (tôr′ē-ə-dôr′) *n.* [Sp. < *toreado,* p.part. of *torear,* to fight bulls < *toro,* bull < Lat. *taurus.*] A bullfighter.
**to·re·ro** (tə-râr′ō) *n., pl.* **-ros.** [Sp. < LLat. *taurarius* < Lat. *taurus,* bull.] A matador or one of his team.
**to·reu·tics** (tə-ro͞o′tĭks) *n.* [< Gk. *toreutikos,* of metal work < *toreuein,* to work in relief < *toreus,* a boring tool.] *(sing. in number).* The art of working esp. metal by the use of embossing and chasing to form minute detailed reliefs. **—to·reu′tic** *adj.*
**to·ri** (tôr′ī′, tōr′ī′) *n. pl. of* TORUS.
**tor·ic** (tôr′ĭk, tŏr′-) *adj.* Of, relating to, or shaped like a torus or a part of a torus.
**to·ri·i** (tôr′ē-ē′, tōr′-) *n., pl.* **torii.** [J. : *tori,* bird + *iru-,* to dwell.] The gateway of a Shinto temple, consisting of two uprights with a concave lintel above a straight crosspiece.
**tor·ment** (tôr′mĕnt′) *n.* [ME < OFr. < Lat. *tormentum* < *torquēre,* to twist.] **1.** Extreme physical pain or mental anguish. **2.** A source of harassment, annoyance, or pain. **3.** Torture inflicted on prisoners being interrogated. —*vt.* (tôr-mĕnt′, tôr′mĕnt′) **-ment·ed, -ment·ing, -ments. 1.** To cause to undergo extreme physical pain or mental anguish. **2.** To upset or agitate greatly. **3.** To annoy, pester, or harass. **—tor·ment′ing·ly** *adv.*

**tor·men·til** (tôr′mən-tĭl′) *n.* [ME *tormentille* < Med. Lat. *tormentilla.*] A Eurasian plant, *Potentilla tormentilla* or *P. erecta,* with yellow flowers and astringent roots.
**tor·men·tor** *also* **tor·ment·er** (tôr-mĕn′tər, tôr′mĕn′tər) *n.* **1.** One that torments. **2.** A hanging at each side of a stage directly behind the proscenium to block the wing area and sidelights from the audience. **3.** A sound-absorbent screen used on a film set to prevent echo.
**torn** (tôrn, tōrn) *v. p.p. of* TEAR[1].
**tor·na·do** (tôr-nā′dō) *n., pl.* **-does** or **-dos.** [Alteration of Sp. *tronada,* thunderstorm < *tronar,* to thunder < Lat. *tonare.*] **1.** A rotating column of air usu. accompanied by a funnel-shaped downward extension of a cumulonimbus cloud and having a vortex several hundred yards in diameter whirling destructively at speeds of up to 300 miles per hour. **2.** A violent squall accompanying a thunderstorm in West Africa and nearby Atlantic waters. **3.** A violently destructive whirlwind. **—tor·na′dic** (-nā′dĭk, -năd′ĭk) *adj.*
**to·roid** (tôr′oid′, tōr′-) *n.* [TOR(US) + -OID.] **1.** *Math.* **a.** A surface generated by a closed curve rotating about, but not intersecting or containing, an axis in its own plane. **b.** A solid having such a surface. **2.** An object shaped like a toroid. **—to·roi′dal** (tô-roid′l) *adj.*
**to·rose** (tôr′ōs′, tōr′-) *adj.* [Lat. *torosus,* knotty < *torus,* knot.] Cylindrical in shape and having ridges or swellings.
**tor·pe·do** (tôr-pē′dō) *n., pl.* **-does.** [NLat. *Torpedo,* genus of fish that give electric shocks < Lat. *torpedo,* electric ray < *torpēre,* to be astounded.] **1.** A cigar-shaped, self-propelled underwater projectile launched from a ship, submarine, or aircraft, and designed to detonate on contact with or in the vicinity of a target. **2.** A submarine explosive device, esp. a submarine mine. **3.** A small explosive placed on a railroad track that is fired by the weight of the train to sound a warning of an approaching hazard. **4.** An explosive fired in an oil or gas well to begin or increase the flow. **5.** A small firework made up of gravel wrapped in tissue paper with a percussion cap that explodes when thrown against a hard surface. **6.** Any of several cartilaginous fishes of the genus *Torpedo,* related to the skates and rays. **7.** A professional assassin or thug. **8.** HERO 5. —*vt.* **-doed, -do·ing, -does.** To attack, explode, or destroy with or as if with a torpedo.
▲ word history: The original sense of *torpedo* in English was "electric ray," a fish that produces an electric charge. The word was borrowed from Latin *torpedo,* which also denoted the same fish but which basically meant "numbness." *Torpedo* is related to *torpidus,* "numb," the Latin source of English *torpid.* The word *torpedo* was first applied to drifting underwater mines in the early 19th century; self-propelled torpedos were a later invention.
**torpedo boat** *n.* A fast, thinly plated boat carrying heavy machine guns and fitted with torpedo tubes.
**tor·pe·do-boat destroyer** (tôr-pē′dō-bōt′) *n.* A fast vessel, larger and more heavily armed than a torpedo boat, designed to destroy the latter, but often serving the same purpose.
**torpedo tube** *n.* The torpedo-launch tube esp. in the hull of a submarine.
**tor·pid** (tôr′pĭd) *adj.* [Lat. *torpidus* < *torpēre,* to be sluggish.] **1.** Having been deprived of the power of motion or feeling : BENUMBED. **2.** Dormant : hibernating. **3.** Lethargic : apathetic. **—tor·pid′i·ty** (-pĭd′ĭ-tē) *n.* **—tor′pid·ly** *adv.*
**tor·por** (tôr′pər) *n.* [Lat. < *torpēre,* to be numb.] **1.** Mental or physical inactivity or insensibility : SLUGGISHNESS. **2.** Lethargy : apathy. **—tor′po·rif′ic** (-pə-rĭf′ĭk) *adj.*
**tor·quate** (tôr′kwāt′) *adj.* [Lat. *torquatus,* having a collar < *torques,* collar. —see TORQUE[2].] *Zool.* Having a ringlike or collarlike band or marking about the neck.
**torque**[1] (tôrk) *n.* [< Lat. *torquēre,* to twist.] **1.** The moment of a force, a measure of its tendency to produce torsion and rotation about an axis, equal to the vector product of the radius vector from the axis of rotation to the point of application of the force by the force applied. **2.** A turning or twisting force. —*vt.* **torqued, torqu·ing, torques.** To impart torque to. **—torqu′er** *n.*
**torque**[2] (tôrk) *n.* [Fr. < Lat. *torques* < *torquēre,* to twist.] A collar, necklace, or armband made of a strip of twisted metal, worn by the ancient Gauls, Germans, and Britons.
**torque converter** *n.* A mechanical or hydraulic device for changing the ratio of torque to speed between the input and output shafts of a mechanism.
**tor·ques** (tôr′kwēz′) *n.* [Lat., collar. —see TORQUE[2].] *Zool.* A band of feathers, hair, or coloration around the neck.
**torr** (tôr) *n., pl.* **torr.** [After Evangelista *Torricelli* (1608–1647).] A unit of pressure equal to $1.316 \times 10^{-3}$ atmosphere.
**tor·rent** (tôr′ənt, tŏr′-) *n.* [Fr. < Ital. *torrente* < Lat. *torrens* < pr.part. of *torrēre,* to burn.] **1.** A swift turbulent stream. **2.** A raging flood : DELUGE. **3.** A turbulent flow <*torrents* of criticism><*torrents* of telegrams>
**tor·ren·tial** (tô-rĕn′shəl, tə-) *adj.* **1.** Of or relating to a torrent <*torrential* spring rains> **2.** Resembling a torrent : WILD <*torrential* insults> **3.** Resulting from a torrent or torrents <*torrential* erosion in the hills> **—tor·ren′tial·ly** *adv.*
**tor·rid** (tôr′ĭd, tŏr′-) *adj.* [Lat. *torridus* < *torrēre,* to parch.] **1.** Dried by the heat of the sun : PARCHED. **2.** Scorching : burning <the *torrid*

desert sun> **3.** Passionate : ardent <a *torrid* love affair> **—tor·rid'i·ty** (tô-rĭd'ĭ-tē), **tor'rid·ness** *n.* **—tor'rid·ly** *adv.*

**Torrid Zone** *n.* The region of the earth's surface between the tropics of Cancer and Capricorn.

**tor·sade** (tôr-säd', -sād') *n.* [Fr. < obs. *tors,* twisted < LLat. *torsus* < p.part. of Lat. *torquēre,* to twist.] A decorative hat trimming made of twisted ribbon or cord.

**tor·si** (tôr'sē') *n. var. pl. of* TORSO.

**tor·sion** (tôr'shən) *n.* [LLat. *torsio* < *torsus,* twisted. —see TORSADE.] **1.** The act of twisting or turning or the state of being twisted or turned. **2.** Stress produced when one end of an object is twisted in one direction and the other end is held motionless or twisted in the opposite direction. **—tor'sion·al** *adj.* **—tor'sion·al·ly** *adv.*

**torsion balance** *n.* An instrument with which small forces, as of electricity or magnetism, are measured by means of the torsion they produce in a wire or slender rod.

**torsion bar** *n.* A part of an automotive suspension consisting of a bar that twists to maintain stability.

**tor·so** (tôr'sō) *n., pl.* **-sos** *or* **-si** (-sē') [Ital., trunk of a statue < Lat. *thyrsus,* stalk. —see THYRSUS.] **1.** TRUNK 2a. **2.** A statue of the human trunk, esp. with the head and limbs truncated. **3.** An unfinished or truncated thing.

**tort** (tôrt) *n.* [ME, injury < OFr. < Med. Lat. *tortum* < Lat., neuter p.part. of *torquēre,* to twist.] *Law.* A wrongful act, damage, or injury done willfully, negligently, or in circumstances involving strict liability, but not involving breach of contract, for which a civil suit can be brought.

**torte** (tôrt, tôr'tə) *n., pl.* **tortes** *or* **tor·ten** (tôr'tn) [G., perh. < Ital. *torta,* cake < LLat. *torta,* a kind of bread.] A rich cake made with many eggs and little flour and usu. containing chopped nuts.

**tor·tel·li·ni** (tôr'tl-ē'nē) *n.* [Ital., dim. of *tortelli,* a kind of pasta < LLat. *torta,* a kind of bread.] Small stuffed pasta dumplings.

**tor·ti·col·lis** (tôr'tĭ-kŏl'ĭs) *n.* [Lat. *tortus,* twisted (< *torquēre,* to twist) + *collum,* neck.] Contracted neck muscles producing an unnatural position of the head. **—tor'ti·col'lar** (-kŏl'ər) *adj.*

**tor·til·la** (tôr-tē'yə) *n.* [Mex. Sp. < Sp., omelet < Sp. *torta,* cake < LLat. *torta,* a kind of bread.] A round thin unleavened bread, usu. made from cornmeal or flour and served hot with toppings of ground meat or cheese.

**tor·toise** (tôr'tĭs) *n.* [ME *tortuce* < OFr. *tortue.*] **1. a.** A terrestrial turtle, esp. one of the family Testudinidae, with thick scaly limbs and a high rounded carapace. **b.** *Chiefly Brit.* A terrestrial or freshwater chelonian. **2.** One that moves or acts slowly.

**tor·toise·shell** *also* **tor·toise-shell** *or* **tortoise shell** (tôr'tĭs-shĕl') *n.* **1.** The mottled, horny, translucent brownish covering of the carapace of a sea turtle, esp. the hawksbill, used to make combs, jewelry, etc. **2.** A domestic cat having fur with brown, black, and yellowish markings. **3.** Any of several butterflies, chiefly of the genus *Nymphalis,* having wings with orange, black, and brown markings. **—tor'toise·shell'** *adj.*

**tor·tu·os·i·ty** (tôr'chōō-ŏs'ĭ-tē) *n., pl.* **-ties. 1.** The state of being tortuous or crooked. **2.** A bent or twisted part, passage, or thing.

**tor·tu·ous** (tôr'chōō-əs) *adj.* [ME < OFr. < Lat. *tortuosus* < *tortus,* a twisting < p.part of *torquēre,* to twist.] **1.** Marked by or having repeated turns or bends : TWISTING <a *tortuous* road through the Alps> **2.** Not straightforward : DEVIOUS <a *tortuous* criminal plot> **3.** Extremely involved : COMPLEX <*tortuous* judicial procedures> **—tor'tu·ous·ly** *adv.* **—tor'tu·ous·ness** *n.*

**tor·ture** (tôr'chər) *n.* [OFr. < LLat. *tortura* < Lat. *torquēre,* to twist.] **1. a.** Infliction of severe physical pain as punishment or coercion. **b.** The state of being tortured. **2.** Mental anguish. **3.** A cause of pain or anguish. *—vt.* **-tured, -tur·ing, -tures. 1.** To subject to torture. **2.** To afflict with great physical or mental pain. **3.** To twist or turn abnormally : DISTORT. **—tor'tur·er** *n.* **—tor'tur·ous** *adj.* **—tor'tur·ous·ly** *adv.*

**to·rus** (tôr'əs, tōr'-) *n., pl.* **to·ri** (tôr'ī', tōr'ī') [Lat., bulge.] **1.** A large semicircular convex molding located at the base of a classical column. **2.** *Anat.* A bulging or rounded projection or swelling. **3.** *Biol.* A moundlike or rounded structure, as the receptacle of a flower. **4.** *Math.* A toroid generated by a circle.

**To·ry** (tôr'ē, tōr'ē) *n., pl.* **-ries.** [Ir. Gael. *tōraidhe,* robber < OIr. *tōir,* pursuit.] **1.** A member of a British political party founded in 1689 that was the opposition party to the Whigs and has been known as the Conservative Party since about 1832. **2.** An American who favored the English side during the American Revolution. **3.** *often* **tory.** A member of a Conservative Party, as in Canada. **—To'ry** *adj.* **—To'ry·ism** *n.*

**toss** (tôs, tŏs) *v.* **tossed, toss·ing, toss·es.** [Poss. of Scand. orig.] *—vt.* **1.** To throw or heave continuously about <*tossed* by the waves> **2.** To throw lightly with or as if with the hand or hands <*toss* a ball> **3.** *Informal.* To discuss informally : bandy about. **4.** To move or lift (the head) with a sudden motion. **5.** To disturb or agitate. **6.** To throw to the ground. **7. a.** To flip (coins) so as to decide a matter. **b.** To flip coins with. **8.** To mix (a salad) lightly so as to cover with dressing. *—vi.* **1.** To be thrown here and there <small craft *tossing* in stormy waters> **2.** To move oneself about vigorously and from side to side <*toss* in one's sleep> **3.** To flip a coin. **—toss down.** To drink in one draft by suddenly tilting <We *tossed down* one glass of beer after another.> **—toss off. 1.** To drink up in one draft. **2.** To do, finish, or accomplish easily and casually <I *tossed off* a few jokes.> *—n.* **1.** The act of tossing or the state of being tossed. **2.** The distance something can be tossed. **3.** A rapid movement or lift, as of the head. **4.** An even chance. **—toss'er** *n.*

☆ ***syns:*** TOSS, HEAVE, PITCH, ROCK, ROLL *v. core meaning* : to move vigorously from side to side or up and down <a small boat *tossing* in heavy seas>

**toss·pot** (tôs'pŏt', tŏs'-) *n.* A drunkard.

**toss·up** (tôs'ŭp', tŏs'-) *n. Informal.* **1.** The flip of a coin to decide an issue. **2.** An even chance.

**tot¹** (tŏt) *n.* [Orig. unknown.] **1.** A small child. **2.** A small amount, as of liquor <a *tot* of brandy>

**tot²** (tŏt) *vt.* **tot·ted, tot·ting, tots.** To total <*totted* up the bill>

**to·tal** (tōt'l) *n.* [ME, whole < OFr. < Med. Lat. *totalis* < Lat. *tōtus.*] **1.** The quantity or amount reached by addition : SUM. **2.** A whole quantity : ENTIRETY. *—adj.* **1.** Being or relating to the whole : ENTIRE. **2.** Complete : absolute <a *total* flop> *—v.* **-taled, -tal·ing, -tals** *or* **-talled, -tal·ling, -tals.** *—vt.* **1.** To determine the sum or total of. **2.** To equal a total of <The number of victims *totals* 200.> **3.** *Slang.* To demolish (a vehicle) completely <*totaled* my motorcycle> *—vi.* To add up : AMOUNT <It *totals* to $100.> **—to'tal·ly** *adv.*

**total eclipse** *n.* An eclipse during which the entire surface of one celestial body is obscured by another.

**to·tal·i·tar·i·an** (tō-tăl'ĭ-târ'ē-ən) *adj.* [TOTAL + (AUTHOR)ITARIAN.] Having or exercising complete political power and control and brooking no political opposition. **—to·tal'i·tar'i·an** *n.* **—to·tal'i·tar'i·an·ism** *n.*

**to·tal·i·ty** (tō-tăl'ĭ-tē) *n., pl.* **-ties. 1.** The state of being total. **2.** An aggregate amount : SUM. **3.** The state of a total eclipse.

**to·tal·i·za·tor** (tōt'l-ĭ-zā'tər) *n.* A machine for computing and showing totals, esp. a pari-mutuel machine showing the total number and amounts of bets at a racetrack.

**to·tal·ize** (tōt'l-īz') *vt.* **-ized, -iz·ing, -iz·es.** To make or combine into a total. **—to'tal·i·za'tion** *n.*

**to·tal·iz·er** (tōt'l-ī'zər) *n.* **1.** A pari-mutuel machine. **2.** An adding machine.

**to·ta·quine** (tō'tə-kwīn', -kwēn', -kwĭn) *n.* [TOTA(L) + Sp. *quina,* cinchona bark.] A powdered, yellowish, bitter mix of quinine and alkaloids from cinchona bark, used as an antimalarial.

**tote¹** (tōt) [Orig. unknown.] *Informal.* *—vt.* **tot·ed, tot·ing, totes. 1.** To haul, esp. on the back or in the arms : LUG. **2.** To have on one's person : PACK <*toting* guns> *—n.* **1.** A load : burden. **2.** A tote bag. **—tot'er** *n.*

**tote²** (tōt) *n.* [Short for TOTALIZATOR.] *Informal.* TOTALIZER 1.

**tote bag** *n. Informal.* A large handbag or shopping bag.

**to·tem** (tō'təm) *n.* [Ojibwa *nintōtēm.*] **1. a.** An animal, plant, or natural object serving among some primitive peoples as the emblem of a clan or family by virtue of an asserted ancestral relationship. **b.** A representation of this being. **2.** A social group having a common totemic affiliation. **3.** A venerated symbol or emblem. **—to·tem'ic** (-tĕm'ĭk) *adj.*

**to·tem·ism** (tō'tə-mĭz'əm) *n.* **1.** Belief in kinship through common totemic affiliation or identification of an individual or group with a totem. **2.** The primitive kinship system of which totemism is a reflection. **—to'tem·ist** *n.* **—to'tem·is'tic** *adj.*

**totem pole** *n.* **1.** A post carved and painted with a series of totemic symbols and erected in front of a dwelling, as among some Indian peoples of the northwestern coast of North America. **2.** *Slang.* A hierarchy.

**toth·er** *or* **t'oth·er** (tŭth'ər) *pron. & adj.* [ME < *thet other,* that other.] *Informal.* The other.

**to·ti·pal·mate** (tō'tĭ-păl'māt') *adj.* [Lat. *totus,* whole + PALMATE.] Having webs connecting each of the four anterior toes, as in pelicans and gannets.

**to·tip·o·ten·cy** (tō-tĭp'ə-tən-sē, tō'tĭ-pōt'n-sē) *also* **to·tip·o·tence** (tō-tĭp'ə-təns, tō'tĭ-pōt'ns) *n.* [Lat. *totus,* whole + POTENCY.] The ability of a cell, as an egg, to generate unlike cells and thus to form a new individual or part. **—to·tip'o·tent** *adj.*

**tot·ter** (tŏt'ər) *vi.* **-tered, -ter·ing, -ters.** [ME *toteren* < MDu. *touter,* to swing.] **1. a.** To sway as if about to fall. **b.** To appear about to collapse <a government *tottering* on the brink of collapse> **2.** To walk unsteadily. **3.** To vacillate <*tottered* between capitulation and resistance> *—n.* The act or state of tottering. **—tot'ter·er** *n.* **—tot'ter·y** *adj.*

**tou·can** (tōō'kăn', -kän', tōō-kăn', -kän') *n.* [Fr. < Port. *tucano* < Tupi *tucana.*] Any of various tropical American birds of the family Ramphastidae, with brightly colored plumage and a very large bill.

**touch** (tŭch) *v.* **touched, touch·ing, touch·es.** [ME *touchen* < OFr. *tochier.*] *—vt.* **1.** To cause or permit a bodily part, esp. the hand or fingers, to come into contact with so as to feel. **2. a.** To bring something into contact with <*touched* the metal plate with a wire> **b.** To bring (one thing) into contact with something else <*touch* a wire to the metal plate> **3.** To cause (one thing) to be in contact with something else <*touch* a button and activate a switch> **4.** To tap or nudge lightly. **5.** To lay hands on in violence <I never *touched* you!> **6.** To eat or drink : TASTE <You didn't *touch* your dinner.> **7.** To disturb by handling <didn't *touch* anything at the scene of the crime> **8. a.** To adjoin : border. **b.** *Math.* To be tangent to. **9.** To measure up to : EQUAL. **10.** To deal with as a subject : treat of. **11.** To be relevant to : CONCERN <a problem that *touches* all of society> **12.** To have an emotional effect on <Your grief *touched* me deeply.> **13.** To injure or spoil slightly. **14.** To color slightly : TINGE. **15.** To draw with light strokes. **16.** To change or improve by adding fine lines or strokes. **17.** To strike or pluck the keys or strings of (a musical instrument). **18.** To play (a musical piece). **19.** To stamp (tested metal). **20.** *Slang.* To wheedle a loan from <*touched* me for $20> —*vi.* **1.** To touch someone or something. **2.** To be or come into contact. **—touch down.** To land or make contact with a landing surface, as an aircraft or spacecraft. **—touch off. 1.** To cause to explode : FIRE. **2.** To initiate (e.g., a chain of events) : TRIGGER <a riot that *touched off* a revolution> **—touch on (*or* upon). 1.** To deal with (a subject) in passing. **2.** To pertain to : CONCERN. **3.** To approach being : verge on <acts that *touch on* sedition> **—touch up.** To improve by making minor changes or additions. —*n.* **1.** An act or instance of touching. **2.** The physiological sense by which external objects or forces are perceived through contact with the body. **3.** A sensation experienced in touching something with a characteristic texture. **4.** A mild tap or shove. **5.** A discernible mark or effect left by contact with something. **6.** A subtle effect brought about by a small change or addition. **7.** A suggestion : hint <a *touch* of scandal> **8.** A mild attack <a *touch* of indigestion> **9.** A tiny amount : TRACE <a *touch* of garlic> **10. a.** A manner or technique of striking the keys of a keyboard instrument, as a piano or typewriter. **b.** The resistance to being struck by the fingers characteristic of a keyboard. **11.** A characteristic manner of doing things. **12.** A facility : knack. **13.** The state of being in contact or communication <Let's keep in *touch*.> **14.** The official stamp indicating the quality of a metal product. **15.** *Slang.* **a.** An approach to wheedle a loan. **b.** A sum of money borrowed. **c.** One liable to be the victim of an approach for a loan <a soft *touch*><an easy *touch*> **16.** The area just outside the sidelines in Rugby football and soccer. **—touch'a·ble** *adj.* **—touch'a·ble·ness** *n.* **—touch'er** *n.*

**touch-and-go** (tŭch'ən-gō') *adj.* **1.** Of insecure future : UNCERTAIN. **2.** Performed without attention : CASUAL.

**touch·back** (tŭch'băk') *n. Football.* A play in which a player recovers and touches the ball to the ground behind his own goal line after it has been propelled into or beyond the end zone by a player on the opposing team.

**touch·down** (tŭch'doun') *n.* **1.** *Football.* A play worth six points, accomplished by being in possession of the ball when it is declared dead on or behind the opponent's goal line. **2.** The contact, or moment of contact, of an aircraft or spacecraft with the landing surface.

**tou·ché** (tōō-shā') [Fr. < p.part. of *toucher,* to touch (from a fencing term indicating the success of an opponent's attack).] *interj.* —Used to express concession to an opponent for a point well made, as in an argument.

**touched** (tŭcht) *adj.* **1.** Emotionally moved. **2.** Somewhat crazy.

**touch football** *n.* Football played on an improvised field and without protective equipment, involving the substitution of touching for tackling.

**touch·hole** (tŭch'hōl') *n.* The opening in early firearms and cannons through which the powder was ignited.

**touch·ing** (tŭch'ĭng) *adj.* Eliciting a tender reaction. —*prep. Archaic.* Concerning. **—touch'ing·ly** *adv.* **—touch'ing·ness** *n.*

**touch·line** (tŭch'līn') *n.* Either of the sidelines bordering the playing field in Rugby.

**touch-me-not** (tŭch'mē-nŏt') *n.* [From the bursting of ripe seed pods when touched.] A plant of the genus *Impatiens,* esp. the jewelweed.

**touch paper** *n.* [From its use to ignite gunpowder.] Paper impregnated with saltpeter so that it burns slowly and without a flame.

**touch·stone** (tŭch'stōn') *n.* **1.** A hard black stone, as jasper or basalt, once used to test the quality of gold or silver by comparing the streak left on the stone by one of these metals with that of a standard alloy. **2.** A criterion <"The *touchstone* of an art is its precision" —Ezra Pound>

**touch-type** (tŭch'tīp') *vi.* **-typed, -typ·ing, -types.** To type without having to look at the keyboard, the fingers having been trained to locate the keys by position.

**touch·up** (tŭch'ŭp') *n.* The act or process of finishing or improving by small alterations and additions.

**touch·wood** (tŭch'wŏŏd') *n.* [From its being easy to ignite.] Decayed wood or similar material used as tinder : PUNK.

**touch·y** (tŭch'ē) *adj.* **-i·er, -i·est. 1.** Easily taking offense : OVERSENSITIVE. **2.** Requiring tact or skill <a *touchy* marital situation> **3.** Sensitive to touch. —Used of a bodily part. **4.** Easily ignited : FLAMMABLE. **—touch'i·ly** *adv.* **—touch'i·ness** *n.*

**tough** (tŭf) *adj.* **-er, -est.** [ME < OE *tōh.*] **1.** So strong and resilient as to withstand great strain without tearing or breaking. **2.** Hard to chew or cut <a *tough* steak> **3.** Physically hardy : RUGGED <a *tough* paratrooper> **4.** Severe : harsh <a *tough* winter> **5.** Aggressive : pugnacious. **6.** Demanding or troubling : DIFFICULT <a *tough* problem> **7.** Strong-minded : resolute <a *tough* hoodlum> **8.** Vicious. **9.** *Informal.* Unfortunate. **10.** *Slang.* Fine : great. —*n.* A thug <attacked by city *toughs*> **—tough it out.** *Slang.* To remain unyielding in adverse circumstances <*toughed it out* and refused to resign> **—tough'ly** *adv.* **—tough'ness** *n.*

**tough·en** (tŭf'ən) *vt.* & *vi.* **-ened, -en·ing, -ens.** To make or become tough. **—tough'en·er** *n.*

**tough·ie** (tŭf'ē) *n. Informal.* **1.** A hoodlum : tough. **2.** A difficult problem.

**tough-mind·ed** (tŭf'mīn'dĭd) *adj.* Not sentimental or afraid. **—tough'-mind'ed·ly** *adv.* **—tough'-mind'ed·ness** *n.*

**tou·pee** (tōō-pā') *n.* [Fr. *toupet,* tuft of hair < OFr. *toup.*] **1.** A partial wig or hair piece worn to cover a bald spot. **2.** A curl or lock of hair worn during the 18th cent. as a topknot on a periwig.

**tour** (tŏŏr) *n.* [ME, one's turn < OFr., turn < Lat. *tornus,* lathe. —see TURN.] **1.** A comprehensive trip including visits to points of interest <a *tour* of Europe> **2.** A group organized for a comprehensive trip or for a shorter sightseeing excursion. **3.** A brief trip to or through a place for the purpose of viewing it <a *tour* of the great houses of England> **4.** A journey to fulfill a round of engagements in several places <a orchestral concert *tour*> **5.** A shift, as in a factory. **6.** A period of duty at a single place or job. —*v.* **toured, tour·ing, tours.** —*vi.* To go on a tour. —*vt.* **1.** To make a tour of. **2.** To present (a theatrical performance) on a tour.

**tou·ra·co** *also* **tu·ra·co** (tŏŏr'ə-kō') *n., pl.* **-cos.** [Fr.] Any of various African birds of the family Musophagidae, many of which have brightly colored plumage.

**tour·bil·lion** (tŏŏr-bĭl'yən) *n.* [ME *turbilloun* < OFr. *torbeillon* < Lat. *turbo.*] **1. a.** A whirlwind. **b.** A vortex, as of a whirlwind or whirlpool. **2.** A skyrocket having a spiral flight pattern.

**tour de force** (tŏŏr' də fôrs', fōrs') *n.* [Fr.] A feat of strength or virtuosity <a pianistic *tour de force*>

**tour·ing** (tŏŏr'ĭng) *n.* **1.** Participation in a tour. **2.** Ski touring.

**touring car** *n.* A large open car seating five or more people, popular in the 1920's.

**tour·ism** (tŏŏr'ĭz'əm) *n.* **1.** Traveling for pleasure. **2.** The business of providing tours and services for tourists.

**tour·ist** (tŏŏr'ĭst) *n.* One who travels for pleasure. **—tou·ris'tic** (tōō-rĭs'tĭk) *adj.*

**tourist class** *n.* A grade of travel accommodations less luxurious than first class or cabin class.

**tourist trap** *n.* A place, as a shop or resort area, that offers goods and services to tourists at inflated prices.

**tour·ma·line** *also* **tur·ma·line** (tŏŏr'mə-lĭn, -lēn') *n.* [Singhalese *toramalli,* carnelian.] A complex crystalline silicate containing aluminum and boron, used in electronic instrumentation and, esp. in its green, clear, and blue varieties, as a gemstone.

**tour·na·ment** (tŏŏr'nə-mənt, tûr'-) *n.* [ME *tournement,* a medieval sport < OFr. *torneiement* < *torneier,* to tourney.] **1.** A contest involving a number of competitors who vie against each other in a series of elimination games or trials. **2.** A medieval martial sport in which two groups of mounted and armored contestants fought against each other with blunted lances or swords.

**tour·ne·dos** (tŏŏr'nə-dō') *n., pl.* **tour·ne·dos** (-dō', -dōz') [Fr. : *tourner,* to turn + *dos,* back < Lat. *dorsum.*] A fillet of beef cut from the tenderloin, often bound in bacon for cooking.

**tour·ney** (tŏŏr'nē, tûr'-) *vi.* **-neyed, -ney·ing, -neys.** [ME *torneyen* < OFr. *torneier* < Lat. *tornus,* lathe. —see TURN.] To compete in a tournament. —*n., pl.* **-neys.** A tournament.

**tour·ni·quet** (tŏŏr'nĭ-kĭt, tûr'-) *n.* [Fr. < *tourner,* to turn < OFr. —see TURN.] A device used to stop temporarily the flow of blood through a large artery in a limb, esp. a cloth band tightened around a limb, often over a pad placed to focus pressure on the artery.

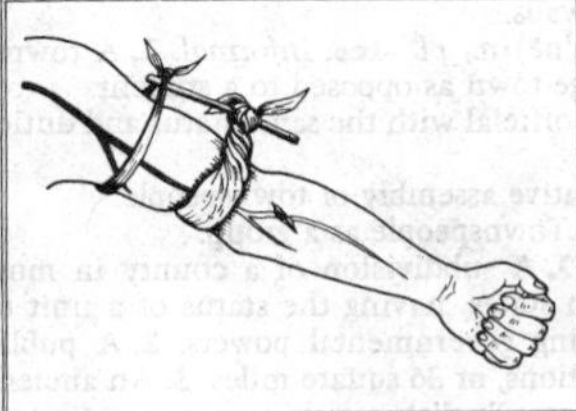

**tourniquet**

**tou·sle** *also* **tou·zle** (tou'zəl) [ME *touselen,* freq. of *tousen,* to pull roughly.] —*vt.* **sled, -sling, -sles** *also* **-zled, -zling, -zles.** To disarrange or rumple : DISHEVEL. —*n.* A disheveled mass, as of hair.

**tout** (tout) [ME *tuten,* to peer.] *Informal.* —*v.* **tout·ed, tout·ing, touts.** —*vi.* **1.** To solicit customers, votes, or patronage, esp. in a brazen way. **2.** To obtain and deal in horseracing information. —*vt.* **1.** To solicit : importune. **2.** To obtain or sell information on (a racing horse or stable) for the guidance of bettors. **3.** To publicize as being of great worth <a restaurant highly *touted* by the food critics> —*n.* **1.** One who obtains information on racehorses and their prospects and sells it to bettors. **2.** One who solicits customers persistently or brazenly. **—tout'er** *n.*

**to·va·rish** *or* **to·va·rich** (tə-vär'ĭsh, -ĭch) *n.* [R.] A comrade.

**tow[1]** (tō) *vt.* **towed, tow·ing, tows.** [ME *towen* < OE *togian.*] To draw or pull along behind by a chain or line. —*n.* **1.** An act of towing or the state of being towed. **2.** Something, as a barge or car, that is towed. **3.** Something, as a tugboat, that tows. **4.** A rope or cable used in towing <a ski *tow*> **—in tow.** Under one's control <a teacher with students *in tow*>

**tow[2]** (tō) *n.* [ME, poss. < OE *tow-,* spinning.] Coarse broken flax or hemp fiber prepared for spinning.

**tow·age** (tō'ĭj) *n.* **1.** The act or service of towing. **2.** A towing charge.

**to·ward** (tôrd, tōrd, tə-wôrd') *also* **to·wards** (tôrdz, tōrdz, tə-wôrdz') *prep.* [ME < OE *tōweard* : *tō,* to + *-weard,* -ward.] **1.** In the direction of <driving *toward* town> **2.** In a position facing <had your back *toward* me> **3.** Somewhat before in time <It began to sleet *toward* morning.> **4.** With regard to <I don't like your attitude *toward* them.> **5.** In furtherance or partial fulfillment of <paid only $20 *toward* the bill> **6.** By way of achieving <efforts *toward* reconciliation> —*adj.* **toward** (tôrd, tōrd). **1.** Favorable. **2.** Being in progress or imminent. **3.** Tractable : docile.

**to·ward·ly** (tôrd'lē, tōrd'-) *adj. Archaic.* **1.** Promising. **2.** Advantageous : favorable. **—to'ward·li·ness** *n.*

**tow-a·way zone** (tō'ə-wā') *n.* A no-parking zone from which cars may be legally towed.

**tow·boat** (tō'bōt') *n.* A tugboat.

**tow·el** (tou'əl) *n.* [ME < OFr. *toaille,* of Germanic orig.] A piece of absorbent cloth or paper used for wiping or drying. —*v.* **-eled, -el·ing, -els** *or* **-elled, -el·ling, -els.** —*vt.* To wipe or rub dry with a towel. —*vi.* To dry oneself with a towel.

**tow·el·ette** (tou'ə-lĕt') *n.* A small, usu. moist piece of material used for cleansing, as of the hands or face.

**tow·el·ing** (tou'ə-lĭng) *n.* A fabric of cotton or linen for making towels.

**tow·er** (tou'ər) *n.* [ME *tour* < OE *torr* and OFr. *tor,* both < Lat. *turris* < Gk. *tursis.*] **1. a.** An extremely tall building : SKYSCRAPER. **b.** An extremely tall part of a building. **2.** A tall framework or structure used for observation, signaling, or pumping. —*vi.* **-ered, -er·ing, -ers. 1.** To rise to a conspicuous height : LOOM. **2.** To fly directly upward before swooping or falling. —Used of certain birds.

**tow·er·ing** (tou'ər-ĭng) *adj.* **1.** Of imposing height. **2.** Outstanding : pre-eminent <*towering* scientific achievements> **3.** Awesomely intense <a *towering* rage>

**tow·head** (tō'hĕd') *n.* **1.** A head of white-blond hair. **2.** One with a towhead. **—tow'head'ed** *adj.*

**tow·hee** (tō'hē, tō-hē') *n.* [Imit. of the song of some of these birds.] A North American bird of the genera *Pipilo* or *Chlorura,* esp. *P. erythrophthalmus,* with black, white, and rust-colored plumage in the male.

**tow·line** (tō'līn') *n.* A line, cable, or chain used in towing a vessel or vehicle.

**town** (toun) *n.* [ME < OE *tūn,* hamlet.] **1.** An often incorporated population center larger than a village and smaller than a city. **2.** *Informal.* A city. **3.** *Chiefly Brit.* A rural village that has a market or fair periodically. **4.** The commercial district or center of an area. **5.** The residents of a town <The *town* protested.> **—go to town.** *Slang.* To go all out. **—on the town.** *Slang.* On a spree.

**town clerk** *n.* A public official who keeps the records of a town.

**town crier** *n.* One formerly employed by a town to walk the streets, ringing a bell and proclaiming announcements.

**town hall** *n.* The building that contains the offices of the public officials of a town and houses the town council and courts.

**town house** *n.* **1.** A city residence. **2.** One of a row of houses connected by common side walls.

**town·ie** *also* **town·y** (tou'nē) *n., pl.* **-ies.** *Informal.* **1.** A townsman. **2.** A resident of a college town as opposed to a student.

**town manager** *n.* A town official with the same status and duties as a city manager.

**town meeting** *n.* A legislative assembly of townspeople.

**towns·folk** (tounz'fōk') *n.* Townspeople as a group.

**town·ship** (toun'shĭp') *n.* **1.** A subdivision of a county in most northeastern and midwestern states, having the status of a unit of local government with varying governmental powers. **2.** A public land surveying unit of 36 sections, or 36 square miles. **3.** An ancient administrative division of a large English parish.

**towns·man** (tounz'mən) *n.* **1.** A resident of a town. **2.** A fellow resident of one's town.

**towns·peo·ple** (tounz'pē'pəl) *pl.n.* The inhabitants or citizens of a town or city.

**towns·wom·an** (tounz'wŏŏm'ən) *n.* **1.** A woman resident of a town. **2.** A woman resident of one's town.

**town·y** (tou'nē) *n. var. of* TOWNIE.

**tow·path** (tō'păth', -päth') *n.* A path along a canal or river used by animals towing boats.

**tow truck** *n.* WRECKER 2a.

**tox-** *pref. var. of* TOXI-.

**tox·a·phene** (tŏk'sə-fēn') *n.* [Blend of TOXI- and CAMPHENE.] A toxic solid compound, $C_{10}H_{10}Cl_8$, used as an insecticide.

**tox·e·mi·a** (tŏk-sē'mē-ə) *n.* A condition in which toxins produced by cells at a local source of infection or derived from the growth of microorganisms are contained in the blood. **—tox·e'mic** *adj.*

**toxi-** *or* **toxo-** *or* **tox-** *pref.* [< Lat. *toxicum.* —see TOXIC.] Poison : poisonous <*tox*albumin>

**tox·ic** (tŏk'sĭk) *adj.* [LLat. *toxicus* < Lat. *toxicum,* poison < Gk. *toxikon,* poison for arrows < *toxikos,* of a bow < *toxon,* bow.] **1.** Of or relating to a toxin. **2.** Harmful, destructive, or deadly <*toxic* wastes><*toxic* fumes> **—tox'i·cal·ly** *adv.*

**toxic-** *pref. var. of* TOXICO-.

**tox·i·cant** (tŏk'sĭ-kənt) *n.* [Med. Lat. *toxicans, toxicant-,* pr.part. of *toxicare,* to poison < Lat. *toxicum,* poison. —see TOXIC.] A poisonous agent : POISON. **—tox'i·cant** *adj.*

**tox·ic·i·ty** (tŏk-sĭs'ĭ-tē) *n., pl.* **-ties. 1.** The quality or state of being toxic. **2.** The degree to which a poison is toxic.

**toxico-** *or* **toxic-** *pref.* [Lat. *toxicum.* —see TOXIC.] Poison <*toxico*sis>

**tox·i·co·gen·ic** (tŏk'sĭ-kō-jĕn'ĭk) *adj.* **1.** Producing poison or toxic substances. **2.** Derived from toxic matter.

**tox·i·col·o·gy** (tŏk'sĭ-kŏl'ə-jē) *n.* The study of the nature, effects, and detection of poisons and the treatment of poisoning. **—tox'i·co·log'i·cal** (-kə-lŏj'ĭ-kəl) *adj.* **—tox'i·co·log'i·cal·ly** *adv.* **—tox'i·col'o·gist** *n.*

**tox·i·co·sis** (tŏk'sĭ-kō'sĭs) *n., pl.* **-ses** (-sēz'). A pathological condition resulting from poisoning.

**tox·i·gen·ic** (tŏk'sə-jĕn'ĭk) *adj.* Producing toxins. **—tox'i·ge·nic'i·ty** (-jə-nĭs'ĭ-tē) *n.*

**tox·in** (tŏk'sĭn) *n.* A poisonous substance, having a protein structure, that is secreted by certain organisms and is capable of causing toxicosis when introduced into the body tissues but is also capable of inducing a counteragent or an antitoxin.

**tox·in-an·ti·tox·in** (tŏk'sĭn-ăn'tĭ-tŏk'sĭn) *n.* A mixture of a toxin and its antitoxin with a slight excess of toxin, once used as a vaccine.

**toxo-** *pref. var. of* TOXI-.

**tox·oid** (tŏk'soid') *n.* A toxin that has lost toxicity but has retained the capacity to stimulate the production of or combine with antitoxins, used in immunization.

**tox·o·plas·ma** (tŏk'sə-plăz'mə) *n.* [< NLat. *Toxoplasma,* genus name : TOXO- + Lat. *plasma,* plasma.] Any of various microorganisms of the genus *Toxoplasma,* including some vertebrate pathogens.

**tox·o·plas·mo·sis** (tŏk'sō-plăz-mō'sĭs) *n.* [TOXOPLASMA + -OSIS.] A disease caused by infection with a microorganism, *Toxoplasma gondii,* and marked by lesions in the brain and eye, esp. in the case of infants.

**toy** (toi) *n.* [ME *toye,* amorous play.] **1.** An object for children to play with. **2.** Something of little importance : TRIFLE. **3.** A small ornament : BAUBLE. **4.** A diminutive thing or person. **5.** A dog of a very small breed or one much smaller than is characteristic of its breed. **6.** *Scot.* A loose covering for the head, once worn by women. —*vi.* **toyed, toy·ing, toys.** To amuse oneself idly : TRIFLE.

**toy·on** (toi'ŏn') *n.* [Am. Sp.] An evergreen shrub, *Heteromeles arbutifolia* or *Photinia arbutifolia,* of the Pacific coast of southern North America, with fragrant white flower clusters and red, berry-like fruit.

**tra·be·at·ed** (trā'bē-ā'tĭd) *also* **tra·be·ate** (-bē-ĭt, -āt') *adj.* [< Lat. *trabs,* beam.] Having horizontal beams or lintels rather than arches. **—tra'be·a'tion** *n.*

**tra·bec·u·la** (trə-bĕk'yə-lə) *n., pl.* **-lae** (-lē') [Lat., dim. of *trabs,* beam.] **1.** A small supporting beam or bar. **2.** *Anat.* Any of the supporting strands of connective tissue projecting into an organ and being part of the framework of that organ. **3.** *Bot.* A transverse rod-like or platelike structure, often extending across a cavity. **—tra·bec'u·lar** *adj.*

**trace[1]** (trās) *n.* [ME, track < OFr. < *tracier,* to make one's way < Lat. *tractus,* a drawing < p.part. of *trahere,* to draw.] **1.** A visible mark or sign of a person or thing formerly present. **2.** A barely perceptible indication : TOUCH. **3. a.** A very small amount <a *trace* of smoke> **b.** A constituent, as a chemical compound or element, present in less than standard quantities. **4.** A path through a wilderness that has been beaten out by passing animals or people. **5.** *Archaic.* A route followed. **6.** A line drawn by a recording instrument, as a cardiograph. **7.** *Math.* **a.** The point at which a line, or the curve in which a surface, intersects a coordinate plane. **b.** The sum of the elements of

ă pat ā pay âr care ä father ĕ pet ē be hw which ĭ pit
ī tie îr pier ŏ pot ō toe ô paw, for oi noise ŏŏ took

the principal diagonal of a matrix. —*v.* **traced, trac·ing, trac·es.** —*vt.* **1.** To follow the trail of. **2.** To ascertain the successive stages in the development or progress of <*trace* the development of a culture> **3.** To locate or discover (e.g., a cause) by researching evidence. **4.** To delineate or sketch (a figure). **5.** To imprint (a design) on something. **6.** To form (letters) with special concentration or care. **7.** To copy by following lines seen through a sheet of transparent paper. **8.** To make a design or series of markings on (a surface). **9.** To record (a variable), as on a graph. —*vi.* **1.** To make one's way. **2.** To have origins. **—trace'a·bil'i·ty, trace'a·ble·ness** *n.* **—trace'a·ble** *adj.* **—trace'a·bly** *adv.*

☆ ***syns:*** TRACE, RELIC, REMAINS, VESTIGE *n. core meaning* **:** a mark or remnant that indicates the former presence of something or someone <*traces* of an old Indian village>

**trace[2]** (trās) *n.* [ME *trais* (pl.) < OFr., pl. of *trait,* strap < Lat. *tractus,* a hauling < p.part. of *trahere,* to haul.] **1.** One of two side straps or chains connecting a harnessed draft animal to the vehicle it pulls. **2.** A bar or rod, hinged at either end to another part, that transfers movement from one machine part to another.

**trace element** *n.* A chemical element occurring in minute quantities in a substance.

**trac·er** (trā'sər) *n.* **1.** One employed to locate missing goods or people. **2.** An inquiry organized to trace missing goods or people. **3.** A drawing instrument. **4.** A tracer bullet. **5.** An identifiable substance, as a dye or radioactive isotope, that can be followed through the course of a mechanical or biological process, providing information on the pattern of events in the process or on the redistribution of the parts or elements involved.

**tracer bullet** *n.* A bullet that leaves a luminous or smoky trail.

**trac·er·y** (trā'sə-rē) *n., pl.* **-ies.** [< TRACE[1].] Ornamental work of interlaced and ramified lines, esp. the lacy openwork in a Gothic window.

**trache-** *pref. var. of* TRACHEO-.

**tra·che·a** (trā'kē-ə) *n., pl.* **-che·ae** (-kē-ē') *or* **-che·as.** [ME *trache* < Med. Lat. *trachea* < LLat. *trachia* < Gk. (*artēria*) *trakheia,* rough (artery) < fem. of *trakhus,* rough.] **1.** *Anat.* A thin-walled cartilaginous and membranous tube descending from the larynx to the bronchi and carrying air to the lungs. **2.** *Zool.* One of the internal respiratory tubes of insects and certain other terrestrial arthropods. **3.** *Bot.* One of the tubular conductive vessels in the xylem of plants. **—tra'che·al** *adj.*

**tra·che·id** (trā'kē-ĭd, -kēd') *n.* One of the elongated tapering supporting and conductive cells in woody tissue. **—tra·che'i·dal** (trā-kē'ĭ-dl, -kēd'l) *adj.*

**tra·che·i·tis** (trā'kē-ī'tĭs) *n.* Inflammation of the trachea.

**tracheo-** *or* **trache-** *pref.* [NLat. < Med. Lat. *trachea.* —see TRACHEA.] Trachea <*tracheid*>

**tra·che·o·e·soph·a·ge·al** (trā'kē-ō'ĭ-sŏf'ə-jē'əl) *adj.* Relating to the trachea and the esophagus.

**tra·che·o·phyte** (trā'kē-ə-fīt') *n.* [< NLat. *Tracheophyta,* division name : TRACHEO- + Gk. *phuta,* pl. of *phuton,* plant.] Any of various plants of the division Tracheophyta that includes all vascular plants characterized by their specialized conducting system of xylem and phloem.

**tra·che·ot·o·my** (trā'kē-ŏt'ə-mē) *n., pl.* **-mies.** Surgical incision into the trachea via the neck.

**tra·cho·ma** (trə-kō'mə) *n.* [Gk. *trakhōma* < *trakhus,* rough.] A contagious viral disease of the conjunctiva of the eye characterized by inflammation, hypertrophy, and granules of adenoid tissue. **—tra·cho'ma·tous** (-kō'mə-təs) *adj.*

**tra·chyte** (trā'kīt', trăk'īt') *n.* [Fr. < Gk. *trakhus,* rough.] A light-colored igneous rock composed mainly of alkalic feldspar. **—tra·chyt'ic** (trə-kĭt'ĭk) *adj.*

**trac·ing** (trā'sĭng) *n.* **1.** A reproduction made by superimposing a transparent sheet and tracing the original upon it. **2.** A graphic record made by a recording instrument, as a cardiograph.

**track** (trăk) *n.* [ME *trak* < OFr. *trac,* perh. of Germanic orig.] **1. a.** A mark left by a passing person, animal, or thing. **b.** The path, route, or course indicated by such marks <*tracks* left in the desert by long-gone armies> **2.** A course of action **:** METHOD. **3. a.** A road or course laid out for running or racing. **b.** Track events. **c.** Track and field. **4.** A rail or set of parallel rails upon which a train or trolley runs. **5.** One of several courses of study to which students are assigned in tracking. —*v.* **tracked, track·ing, tracks.** —*vt.* **1.** To follow the traces or footprints of **:** TRAIL. **2.** To pursue successfully <finally *tracked* down the thief> **3.** To move over or along **:** TRAVERSE. **4.** To deposit as footprints (matter carried on the shoes) <*tracked* grime into the house> **5. a.** To monitor the course of (e.g., aircraft), as by radar. **b.** *Informal.* To observe (something) carefully <*tracking* the corporation's profits> **6.** To equip with a track. **7.** To assign to a curricular track. —*vi.* **1.** To keep a constant distance apart. —Used of a pair of wheels. **2.** To be in alignment. **3.** To pursue a track. **—in (one's) tracks.** Precisely where one is standing. **—track'a·ble** *adj.* **—track'er** *n.*

o͞o **boot** ou **out** th **thin** *th* **this** ŭ **cut** ûr **urge** y **young**
yo͞o **abuse** zh **vision** ə **about,** item, edible, gallop, circus

**track·age** (trăk'ĭj) *n.* **1.** Railway tracks. **2. a.** The right of one railroad company to use the track system of another. **b.** The charge for trackage.

**track and field** *n.* Athletic events performed on a running track and the field associated with it.

**track events** *pl.n.* The running events at a track meet.

**track·ing** (trăk'ĭng) *n.* Homogeneous grouping of students into any of several study courses according to the students' intelligence or ability levels.

**tracking station** *n.* An observation station for maintaining radar or radio contact with an object in the atmosphere or in space.

**track·less** (trăk'lĭs) *adj.* **1.** Not running on tracks or rails. **2.** Unmarked by trails or paths <a *trackless* jungle>

**trackless trolley** *n.* A trolley bus.

**track·man** (trăk'mən) *n.* A worker employed to repair or inspect railroad tracks.

**track meet** *n.* A track and field competition.

**track record** *n.* A record of performance <a company's *track record*>

**track·side** (trăk'sīd') *adj.* Of, pertaining to, or situated in the area near a track.

**track·suit** (trăk'so͞ot') *n.* A loose-fitting outfit usu. consisting of a jacket and pants worn while exercising.

**track·walk·er** (trăk'wô'kər) *n.* A trackman.

**tract[1]** (trăkt) *n.* [Lat. *tractus* < p.part. of *trahere,* to draw.] **1.** A stretch of land. **2.** *Anat.* **a.** A system of organs and tissues that together perform a specialized function <the digestive *tract*><the respiratory *tract*> **b.** A bundle of nerve fibers with a common origin, termination, and function. **3.** *Archaic.* A stretch or lapse of time.

**tract[2]** (trăkt) *n.* [ME *tracte* < Lat. *tractatus* < p.part. of *tractare,* to discuss < *trahere,* to draw.] A paper or pamphlet containing a declaration or appeal, esp. one distributed by a special interest group.

**tract[3]** (trăkt) *n.* [ME *tracte* < Med. Lat. *tractus* < Lat., a drawing out (from singing without a break in one's voice). —see TRACT[1].] *Rom. Cath. Ch.* The verses from Scripture sung during Lent or on Ember days after the gradual in the Mass.

**trac·ta·ble** (trăk'tə-bəl) *adj.* [Lat. *tractabilis* < *tractare,* to manage, freq. of *trahere,* to draw.] **1.** Easily controlled or managed **:** GOVERNABLE <a *tractable* child> **2.** Easily handled or worked **:** MALLEABLE. **—trac'ta·bil'i·ty, trac'ta·ble·ness** *n.* **—trac'ta·bly** *adv.*

**Trac·tar·i·an·ism** (trăk-târ'ē-ə-nĭz'əm) *n.* The religious principles and views of the founders of the Oxford movement, published in a series of 90 pamphlets entitled *Tracts for the Times,* printed at Oxford, England (1833–41). **—Trac·tar'i·an** *adj.* & *n.*

**trac·tate** (trăk'tāt') *n.* [Lat. *tractatus,* tract.] A dissertation.

**tract house** *n.* One of numerous houses of similar or complementary design constructed on a tract of land. **—tract housing** *n.*

**trac·tile** (trăk'təl, -tīl') *adj.* [LLat. *tractilis* < Lat. *trahere,* to draw.] Capable of being drawn out in length **:** DUCTILE <*tractile* metals> **—trac·til'i·ty** (-tĭl'ĭ-tē) *n.*

**trac·tion** (trăk'shən) *n.* [Med. Lat. *tractio* < Lat. *trahere,* to pull.] **1.** The act of pulling, as a load over a surface by motor power. **2.** The state of being pulled. **3.** Adhesive friction, as of a wheel on a track. **4.** The pulling power of a railroad engine. **—trac'tion·al** *adj.*

**trac·tive** (trăk'tĭv) *adj.* [< Lat. *trahere, tract-,* to draw.] Exerting traction.

**trac·tor** (trăk'tər) *n.* [< Lat. *trahere, tract-,* to draw.] **1.** A gasoline- or diesel-powered vehicle with large heavily treaded tires, used in farming for pulling machinery. **2.** A truck with a cab and no body, used for pulling large vehicles such as trailers. **3.** An aircraft having a propeller mounted forward of the supporting surfaces.

**trade** (trād) *n.* [ME *trad,* habit < MLG *trade,* track.] **1.** An occupation requiring skilled labor **:** CRAFT <the *trade* of a plumber> **2.** The business of buying and selling commodities **:** COMMERCE. **3.** The people working in or associated with a specified industry or business. **4.** The customers of a specified industry or business. **5.** An instance of buying or selling **:** TRANSACTION. **6.** Exchange of one thing for another. **7. trades.** The trade winds. —*v.* **trad·ed, trad·ing, trades.** —*vi.* **1.** To engage in buying and selling for profit. **2.** To make an exchange of one thing for another. **3.** To shop regularly at a given store <always *traded* at local markets> —*vt.* **1.** To give in exchange for something else. **2.** To buy and sell (e.g., stock). **3.** To pass back and forth <We *traded* anecdotes.> **—trade in.** To give (an old or used item) as partial payment on a new purchase. **—trade on.** To put to advantage **:** UTILIZE <*traded on* looks and charm> **—trade up.** To trade an item in, as one less valuable for one more valuable. **—trad'a·ble** *adj.*

**trade acceptance** *n.* A bill of exchange for the amount of a purchase drawn by the seller on the purchaser, bearing the purchaser's signature and specifying time and place of payment.

**trade book** *n.* A book published for distribution to the general public through booksellers.

**trade discount** *n.* A discount on the list price granted by a manufacturer or wholesaler to buyers in the same trade.

**trade edition** *n.* A trade book or a trade version of a book <a *trade edition* of a school dictionary>

**trade-in** (trād′ĭn′) *n.* **1.** Merchandise accepted as partial payment for a new purchase. **2.** A transaction involving a trade-in.

**trade language** *n.* A language, as pidgin, common to peoples of diverse speech that is used esp. for communication in commercial trade.

**trade magazine** *n.* A magazine published regularly by a business or industry to give pertinent news and developments.

**trade·mark** (trād′märk′) *n.* **1.** A name, symbol, or other device identifying a product, officially registered and legally restricted to the use of the owner or manufacturer. **2.** A distinctive sign by which a person or thing comes to be known. —*vt.* **-marked, -mark·ing, -marks. 1.** To label (a product) with a trademark. **2.** To register as a trademark.

**trade name** *n.* **1.** The name by which a commodity, service, or process is known to the trade. **2.** The name under which a business firm operates.

**trade·off** *also* **trade-off** (trād′ôf′, -ŏf′) *n.* Exchange of one thing in return for another, esp. relinquishment of something desirable, as a benefit or advantage, for one regarded as more desirable. **—trade′-off** *adj.*

**trad·er** (trā′dər) *n.* **1.** One that trades : DEALER. **2.** A ship employed in foreign trade. **3.** A member of a stock exchange who trades for himself or herself and not as a broker for customers.

**trade rat** *n.* PACK RAT 1.

**trade route** *n.* A sea lane used by trading ships.

**trade school** *n.* A secondary school providing instruction in skilled trades : VOCATIONAL SCHOOL.

**trade secret** *n.* A secret method, formula, or device that gives one an advantage over one's competitors.

**trades·man** (trādz′mən) *n.* **1.** One engaged in the retail trade, esp. a shopkeeper : DEALER. **2.** A skilled worker.

**trade union** *n.* A labor union, esp. one whose membership is limited to people in the same trade. **—trade unionism** *n.* **—trade unionist** *n.*

**trade wind** *n.* A consistent system of winds occupying most of the tropics, being the major component of the general circulation of the atmosphere, blowing northeasterly in the Northern Hemisphere and southeasterly in the Southern Hemisphere.

**trading card** *n.* A picture card or playing card with a design on the back, collected and traded esp. by children.

**trading post** *n.* A store in a sparsely settled area established by traders to barter supplies for local products.

**trading stamp** *n.* A stamp given by a retailer to a buyer for a purchase of a specified amount and intended to be redeemed in quantity for merchandise.

**tra·di·tion** (trə-dĭsh′ən) *n.* [ME *tradicion* < OFr. < Lat. *traditio* < *tradere,* to hand down : *trans,* over + *dare,* to give.] **1.** Transmittal of elements of a culture from one generation to another, esp. by oral communication. **2. a.** A mode of thought or behavior passed from one generation to another. **b.** Customs and usages transmitted from one generation to another and viewed as a coherent body of precedents influencing the present : HERITAGE. **3.** A body of unwritten religious precepts. **4.** A time-honored practice or a set of such practices. **5.** *Law.* Transfer of property to another. **—tra·di′tion·al** *adj.* **—tra·di′tion·al·ize′** *v.* **(-ized, -iz·ing, -iz·es). —tra·di′tion·al·ly** *adv.*

**tra·di·tion·al·ism** (trə-dĭsh′ə-nə-lĭz′əm) *n.* **1.** Adherence to tradition, esp. strict adherence to religious heritage. **2.** A philosophical system holding that all knowledge is derived from original divine revelation and is transmitted by tradition. **—tra·di′tion·al·ist** *n.* **—tra·di′tion·al·is′tic** *adj.*

**trad·i·tor** (trăd′ĭ-tər) *n., pl.* **-to·res** (trăd′ĭ-tôr′ēz, -tōr′-) [ME *traditour* < Lat. *traditor,* betrayer < *tradere,* to betray. —see TRADITION.] One of the early Christians who betrayed fellow Christians during the Roman persecutions.

**tra·duce** (trə-dōōs′, -dyōōs′) *vt.* **-duced, -duc·ing, -duc·es.** [Lat. *traducere,* to dishonor : *trans,* across + *ducere,* to lead.] **1.** To speak falsely of : MISREPRESENT. **2.** To betray : violate <*traduce* one's principles> **—tra·duce′ment** *n.* **—tra·duc′er** *n.* **—tra·duc′i·ble** *adj.* **—tra·duc′ing·ly** *adv.*

**tra·du·cian·ism** (trə-dōō′shə-nĭz′əm, -dyōō′-) *n.* [Med. Lat. *traducianus,* believer in traducianism < *tradux,* inheritance < Lat., shoot for propagation < *traducere,* to lead across. —see TRADUCE.] Belief that the soul is inherited from the parents along with the body. **—tra·du′cian·ist** *n.* **—tra·du′cian·is′tic** *adj.*

**traf·fic** (trăf′ĭk) *n.* [OFr. *traffique* < OItal. *traffico* < *trafficare,* to trade.] **1. a.** Commercial exchange of goods : TRADE <heavy *traffic* in steel> **b.** Illegal or improper commercial activity <drug *traffic*> **2. a.** The business of moving passengers and cargo through a transportation system. **b.** The amount of cargo or number of passengers conveyed in a transportation system. **3. a.** Passage of persons, vehicles, or messages through transportation routes. **b.** The amount, as of vehicles, in transit <heavy *traffic* on the expressways> **4.** Dealings between groups or individuals. —*v.* **-ficked, -fick·ing, -fics.** —*vi.* To carry on traffic. —*vt.* **1.** To travel over <roads that are heavily *trafficked*> **2.** To trade or barter. **—traf′fick·er** *n.*

**traffic circle** *n.* A circular one-way road at a junction of streets, facilitating uninterrupted traffic.

**traffic island** *n.* A raised area over which vehicles may not pass, placed at a junction of streets, or between opposing traffic lanes.

**traffic light** *n.* A signal that flashes a red, green, or yellow warning light to direct vehicular and pedestrian traffic to stop, go, or proceed with caution.

**trag·a·canth** (trăg′ə-kănth′, trăj′-) *n.* [Lat. *tragacantha* < Gk. *tragakantha* : *tragos,* goat + *akantha,* thorn.] **1.** A thorny shrub of the genus *Astragalus,* esp. *A. gummifer* of southwestern Asia, yielding a gum used in pharmacy, adhesives, and textile printing. **2.** The gum of the tragacanth.

**tra·ge·di·an** (trə-jē′dē-ən) *n.* [ME *tragedyen,* prob. < OFr. *tragediane* < *tragedie,* tragedy.] **1.** A writer of tragedies. **2.** A performer of tragic roles.

**tra·ge·di·enne** (trə-jē′dē-ĕn′) *n.* [Fr., fem. of *tragédien,* tragedian < OFr. *tragediane.*] An actress who performs tragic roles.

**trag·e·dy** (trăj′ĭ-dē) *n., pl.* **-dies.** [ME *tragedie* < OFr. < Lat. *tragoedia* < Gk. *tragōidia* : *tragos,* goat + *ōidē,* song.] **1.** A dramatic or literary work depicting a protagonist engaged in a morally significant struggle ending in ruin or utter disappointment, specif.: **a.** A classical verse drama in which a noble protagonist is ruined as a result of an extreme quality that is both his greatness and his downfall. **b.** A Renaissance or modern drama like the classical model in its depiction of terrible struggle and calamity but freer in style and choice of protagonist. **c.** A drama or narrative that seriously treats of disastrous events and has an unhappy yet meaningful end. **d.** The literary genre of tragic dramatic works. **2.** A dramatic, disastrous event, esp. one of moral significance <the personal *tragedy* of a disgraced President> **3.** A tragic element or aspect.

**tra·gi** (trā′gī′, -jī′) *n. pl. of* TRAGUS.

**trag·ic** (trăj′ĭk) *also* **trag·i·cal** (trăj′ĭ-kəl) *adj.* [OFr. *tragique* < Lat. *tragicus* < Gk. *tragikos* < *tragōidia,* tragedy.] **1.** Relating to, being in the style of, or having the nature of tragedy. **2.** Writing or performing in tragedy <a *tragic* playwright> **3.** Having the characteristics of tragedy : DISASTROUS <a *tragic* air crash> **—trag′i·cal·ly** *adv.* **—trag′i·cal·ness** *n.*

**tragic flaw** *n.* A flaw in the character of the protagonist of a dramatic or literary tragedy that causes his ruin.

**tragic irony** *n.* IRONY 3.

**trag·i·com·e·dy** (trăj′ĭ-kŏm′ĭ-dē) *n., pl.* **-dies.** [OFr. *tragicomédie* < LLat. *tragicōmoedia* < Lat. *tragicocomoedia* : *tragicus,* tragic + *comoedia,* comedy. —see COMEDY.] A drama containing elements of both tragedy and comedy. **—trag′i·com′ic** (-kŏm′ĭk), **trag′i·com′i·cal** *adj.* **—trag′i·com′i·cal·ly** *adv.*

**trag·o·pan** (trăg′ə-păn′) *n.* [NLat. *Tragopan,* genus name < Lat. *tragopan,* a fabulous bird < Gk. : *tragos,* goat + *Pan,* Pan.] Any of several Asian pheasants of the genus *Tragopan,* the male of which has bright plumage and two hornlike appendages on the head.

**tragopan**
*Approximately 28 inches long*

**tra·gus** (trā′gəs) *n., pl.* **-gi** (-gī′, -jī′) [NLat. < Gk. *tragos,* a part of the ear.] **1.** The projection of skin-covered cartilage in front of the meatus of the external ear. **2.** Any of the hairs growing at the entrance to the meatus of the external ear.

**trail** (trāl) *v.* **trailed, trail·ing, trails.** [ME *trailen,* prob. < ONFr. *trailler,* to tow < VLat. **tragulare,* to drag < Lat. *tragula,* dragnet < *trahere,* to pull.] —*vt.* **1.** To permit to drag or stream behind, as along the ground. **2.** To drag (e.g., the body) heavily or wearily. **3. a.** To follow the traces or scent of, as in hunting : TRACK <dogs *trailing* deer> **b.** To follow in the footsteps of (another). **4. a.** To follow slowly. **b.** To lag behind (an opponent). —*vi.* **1.** To drag or be dragged along, brushing the ground. **2.** To extend, grow, or droop over a surface <grape vines *trailing* over an arbor> **3.** To drift in a tenuous stream <smoke *trailing* from a cigarette> **4.** To become gradually fainter : DWINDLE <voices *trailing* off in the distance> **5. a.** To walk with dragging steps : TRUDGE. **b.** To fall behind in competition : LAG. —*n.* **1.** Something that hangs long and loose <*trails* of ribbons> **2.** Something that is drawn along or follows behind : TRAIN. **3.** The part of a gun carriage that rests or slides on the ground. **4. a.** A mark, trace, course, or path left by a moving body <white jet *trails*> **b.** Scent : track <the *trail* of a wolf> **c.** A beaten track or blazed path, as through wilderness. **d.** A chain of consequences <a *trail* of regrets> **5.** The act of trailing.

**trail bike** *n.* A small motorcycle designed not for highway use but for cross-country, off-road riding.

**trail·blaz·er** (trāl′blā′zər) *n.* **1.** One that blazes a trail. **2.** A pioneer in a specific field of endeavor.

**trail boss** *n.* The person in charge of a cattle drive in the West.

**trail·break·er** (trāl′brā′kər) *n.* A trailblazer.

**trail·er** (trā′lər) *n.* **1.** One that trails. **2.** A large transport vehicle designed to be hauled by a truck or tractor. **3.** A furnished van drawn by a truck or car and used as a house or office. **4. a.** A short filmed advertisement for a motion picture. **b.** A short, blank strip of film at the end of a reel. —*vt.* & *vi.* **-ered, -er·ing, -ers.** To transport or be transported by a trailer <*trailered* the snowmobiles to the truck> **—trail′er·a·ble** *adj.*

**trailer camp** *n.* A campsite for house trailers.

**trailing arbutus** *n.* A low-growing plant, *Epigaea repens* of eastern North America, with evergreen leaves and fragrant pink or white flower clusters.

**trailing edge** *n.* The rearmost edge, esp. of an airfoil.

**train** (trān) *n.* [ME *trayne* < OFr. *train* < *trainer,* to drag < Lat. *trahere.*] **1.** A part of a long gown that trails behind the wearer <a bridal *train*> **2.** A staff of attendants : RETINUE <courtiers in the monarch's *train*> **3.** A service unit of troops, vehicles, and equipment following and supporting an army <a supply *train*> **4.** A long line of moving persons, animals, or vehicles. **5.** A string of connected railroad cars. **6.** An orderly series of related thoughts or events : SEQUENCE. **7.** A set of linked mechanical parts. **8.** A string of gunpowder that acts as a fuse for exploding a charge. —*v.* **trained, train·ing, trains.** —*vt.* **1.** To coach in or accustom to a mode of behavior or performance. **2.** To make proficient with special instruction and practice. **3.** To prepare physically, as with a regimen <*train* a sprint swimmer> **4.** To cause (a plant or one's hair) to take a desired course or shape, as by manipulating. **5.** To focus or direct : AIM <*Train* your sights on the hilltop.> **6.** To draw, drag, or trail. —*vi.* To give or undergo a course of training. **—train′a·ble** *adj.*

**train·bear·er** (trān′bâr′ər) *n.* An attendant who holds up the train of a robe or gown, as in a procession.

**train·ee** (trā-nē′) *n.* One who is undergoing training.

**train·ee·ship** (trā-nē′shĭp′) *n.* The status or position of a trainee.

**train·er** (trā′nər) *n.* **1.** One who trains, esp. one who coaches athletes, racehorses, or show animals. **2.** A device used in training. **3.** A naval gun crew member who trains cannons horizontally.

**train·ing** (trā′nĭng) *n.* **1.** The act, process, or routine of one who trains. **2.** The state of being trained.

**training school** *n.* **1.** A school providing practical vocational and technical instruction. **2.** A detention house for juvenile delinquents that provides vocational training.

**training table** *n.* A table, as in a mess hall, providing carefully planned meals for athletes undergoing training.

**train·load** (trān′lōd′) *n.* The load that a train can carry.

**train·man** (trān′mən) *n.* A crew member on a railroad train.

**train·mas·ter** (trān′măs′tər) *n.* A railroad supervisor of a division of a rail line.

**train oil** *n.* [ME *trane,* train oil < MLG *trān.*] Oil obtained esp. from the blubber of a whale.

**traipse** (trāps) *vi.* **traipsed, traips·ing, traips·es.** [Orig. unknown.] *Informal.* To walk about idly or intrusively.

**trait** (trāt) *n.* [Fr. < OFr., stroke < Lat. *tractus,* a drawing. —see TRACT[1].] **1. a.** A distinguishing feature, as of one's character. **b.** A characteristic that is inherited. **2. a.** A stroke with or as if with a pencil. **b.** A touch : trace.

**trai·tor** (trā′tər) *n.* [ME < OFr. *traitre* < Lat. *traditor* < *tradere,* to betray. —see TRADITION.] One who betrays one's country, a cause, or a trust, esp. one who commits treason.

**trai·tor·ous** (trā′tər-əs) *adj.* **1.** Having the character of a traitor : DISLOYAL. **2.** Constituting treason. **—trai′tor·ous·ly** *adv.* **—trai′tor·ous·ness** *n.*

**trai·tress** (trā′trĭs) *or* **trai·tor·ess** (trā′tər-ĭs) *n.* A woman who betrays her country, a cause, or a trust, esp. one who commits treason.

**tra·ject** (trə-jĕkt′) *vt.* **-ject·ed, -ject·ing, -jects.** [Lat. *trajicere, traject-,* to throw across : *trans,* across + *jacere,* to throw.] To transmit <*traject* light rays through a prism> **—tra·jec′tion** *n.*

**tra·jec·to·ry** (trə-jĕk′tə-rē) *n., pl.* **-ries.** [Med. Lat. *trajectorius* < Lat. *trajicere,* to throw across. —see TRAJECT.] **1.** The path of a moving body or particle, esp. such a path in three dimensions. **2.** *Math.* A curve that cuts all of a given family of curves or surfaces at the same angle.

**tram[1]** (trăm) *n.* [Dial., shaft of a barrow, prob. < MLG *trame,* beam.] **1.** *Chiefly Brit.* **a.** A streetcar. **b.** A tramway. **c.** A cable car. **2.** A four-wheeled, open box-shaped wagon or iron car run on tracks in a coal mine. —*vt.* **trammed, tram·ming, trams.** To move or transport in a tram.

**tram[2]** (trăm) *n.* [Short for TRAMMEL.] **1.** TRAMMEL 5. **2.** Accurate mechanical adjustment. —*vt.* **trammed, tram·ming, trams.** To adjust or align (mechanical parts) with a trammel.

**tram·car** (trăm′kär′) *n.* **1.** *Chiefly Brit.* A streetcar. **2.** TRAM[1] 2.

**tram·line** (trăm′līn′) *n.* *Chiefly Brit.* A streetcar line.

**tram·mel** (trăm′əl) *n.* [ME *tramale,* a kind of net < OFr. *tramail* < LLat. *tremaculum* : Lat. *tres,* three + Lat. *macula,* mesh.] **1.** A shackle for teaching a horse to amble. **2.** *often* **trammels.** A restriction on free activity or movement. **3.** A vertically set fishing net of three layers, consisting of a finely meshed net between two nets of coarse mesh. **4. a.** An instrument for describing ellipses. **b.** The pivoted beam of a beam compass. **5.** An instrument for gauging and adjusting machine parts. **6.** An arrangement of links and a hook in a fireplace for raising or lowering a kettle. —*vt.* **-meled, -mel·ing, -mels** *or* **-melled, -mel·ling, -mels. 1.** To hinder, restrict, or confine. **2.** To entrap : enmesh. **—tram′mel·er** *n.*

**tra·mon·tane** (trə-mŏn′tān′, trăm′ən-tān′) *adj.* [Ital. *tramontano* < Lat. *transmontanus* : *trans,* beyond + *montanus,* mountainous.] **1.** Dwelling beyond or coming from the far side of the mountains, esp. the Alps as viewed from Italy. **2.** Sweeping down from the mountains <a *tramontane* wind> —*n.* **1.** One who lives beyond the mountains : OUTSIDER. **2.** A north or cold wind in Italy.

**tramp** (trămp) *v.* **tramped, tramp·ing, tramps.** [ME *trampen.*] —*vi.* **1.** To walk with a heavy, firm step : TRUDGE. **2. a.** To go on foot : HIKE. **b.** To wander about aimlessly. —*vt.* **1.** To traverse on foot <*tramp* the woods> **2.** To tread down : TRAMPLE <*tramp* down snow> —*n.* **1. a.** A heavy footfall. **b.** The sound made by heavy walking or marching <the *tramp* of hobnailed boots> **2.** A walking trip : HIKE. **3.** One who travels aimlessly about on foot, doing odd jobs or begging for a living : VAGRANT. **4.** A prostitute. **5.** A cargo vessel that has no regular schedule but takes on freight wherever it may be found and discharges it wherever required. **6.** A metal plate attached to the sole of a shoe for protection, as when spading ground. **—tramp′er** *n.*

**tram·ple** (trăm′pəl) *v.* **-pled, -pling, -ples.** [ME *tramplen,* freq. of *trampen,* to tramp.] —*vt.* **1.** To beat down with the feet so as to bruise, crush, or destroy. **2.** To treat ruthlessly or harshly. —*vi.* To tread heavily or contemptuously. **—tram′ple** *n.* **—tram′pler** *n.*

**tram·po·line** (trăm′pə-lēn′, -lĭn) *n.* [Sp. *trampolín* < Ital. *trampolino* < *trampoli,* stilts, of Germanic orig.] A sheet of strong taut canvas attached with springs to a metal frame and used for acrobatic tumbling. **—tram′po·lin′er, tram′po·lin′ist** *n.*

**tram·way** (trăm′wā′) *n.* *Chiefly Brit.* **1. a.** A street track or railway for trams. **b.** A streetcar line. **2.** A cable or system of cables for a cablecar.

**trance** (trăns) *n.* [ME *traunce* < OFr. *transe* < *transir,* to depart < Lat. *transire,* to go across. —see TRANSIENT.] **1.** A hypnotic, cataleptic, or ecstatic state. **2.** Detachment from one's physical surroundings, as in contemplation or daydreaming : REVERIE. **3.** A dazed state, as between sleeping and waking : STUPOR. —*vt.* **tranced, tranc·ing, tranc·es.** To put into a trance.

☆ ***syns:*** TRANCE, ABSTRACTION, MUSE, REVERIE, STUDY *n.* *core meaning* : the condition of being so lost in thought as to be unaware of one's surroundings <sat in a *trance* gazing out the window>

**tran·quil** (trăng′kwəl, trăn′-) *adj.* [Lat. *tranquillus.*] **1.** Free from disturbance or agitation : SERENE <a *tranquil* life in the country> **2.** Steady : even <a *tranquil* flame> **—tran′quil·ly** *adv.* **—tran′quil·ness** *n.*

**tran·quil·ize** *also* **tran·quil·lise** (trăng′kwə-līz′, trăn′-) *vt.* & *vi.* **-ized, -iz·ing, -iz·es** *also* **-lised, -lis·ing, -lis·es.** To make or become tranquil. **—tran′quil·i·za′tion** *n.*

**tran·quil·iz·er** (trăng′kwə-lī′zər, trăn′-) *n.* **1.** Something, as music or liquor, that tranquilizes. **2.** A calmative drug.

**tran·quil·li·ty** *or* **tran·quil·i·ty** (trăn-kwĭl′ĭ-tē, trăng-) *n.* The quality or state of being tranquil : SERENITY.

☆ ***syns:*** TRANQUILLITY, CALM, PLACIDITY, QUIET *n.* *core meaning* : absence of motion or disturbance <the *tranquillity* of a glen hidden in an evergreen forest>

**tran·quil·lize** (trăng′kwə-līz′, trăn′-) *v.* *var. of* TRANQUILIZE.

**trans-** *pref.* [< Lat. *trans,* beyond, through.] **1.** Across : on the other side : beyond <*trans*polar> **2.** Through <*trans*cutaneous> **3.** Change : transfer <*trans*literate>

**trans·act** (trăn-săkt′, -zăkt′) *v.* **-act·ed, -act·ing, -acts.** [Lat. *transigere, transact-,* to carry through : *trans,* through + *agere,* to drive.] —*vt.* To do, perform, carry out, manage, or conduct (e.g., business). —*vi.* To do business. **—trans·ac′tor** *n.*

**trans·ac·tion** (trăn-săk′shən, -zăk′-) *n.* **1.** The act of transacting or the fact of being transacted. **2.** Something transacted. **3. transactions.** The proceedings, as of a convention. **—trans·ac′tion·al** *adj.*

**transactional analysis** *n.* A psychotherapeutic system that seeks to analyze intrapsychic conflict and interpersonal interactions so as to afford insight and facilitate constructive communication.

**trans·al·pine** (trăns-ăl′pīn′, trănz-) *adj.* Relating to, living on, or coming from the northern side of the Alps.

**trans·am·i·nase** (trăns-ăm′ə-nās′, -nāz′, trănz-) *n.* Any of a group of enzymes that catalyze transamination.

**trans·am·i·na·tion** (trăns-ăm′ə-nā′shən, trănz-) *n.* **1.** Transfer of an amino group from one chemical compound to another. **2.** Transposition of an amino group within a chemical compound.

**trans·at·lan·tic** (trăns′ət-lăn′tĭk, trănz′-) *adj.* **1.** Situated on the other side of the Atlantic. **2.** Spanning or crossing the Atlantic <*transatlantic* flights>

**trans·ceiv·er** (trăn-sē′vər, -zē′-) *n.* [TRANS(MITTER) + (RE)CEIVER.] A module composed of a radio receiver and transmitter.

**tran·scend** (trăn-sĕnd′) *v.* **-scend·ed, -scend·ing, -scends.** [ME *transcenden* < OFr. *transcendre* < Lat. *transcendere* : *trans,* over + *scandere,* to climb.] —*vt.* **1. a.** To pass beyond (a human limit) <feelings that *transcend* understanding> **b.** To exist above and independent of (material experience or the universe). **2.** To rise above or across. —*vi.* To surpass.

**tran·scen·dent** (trăn-sĕn′dənt) *adj.* **1.** That surpasses all others of the same kind : PRE-EMINENT. **2. a.** *Philos.* Transcending the Aristotelian categories. **b.** Designating knowledge that is beyond the limits of experience in Kant's theory of knowledge. **3.** Above and independent of the material universe. —Used of the Deity. **—tran·scen′dence, tran·scen′den·cy** *n.* **—tran·scen′dent·ly** *adv.*

**tran·scen·den·tal** (trăn′sĕn-dĕn′tl, -sən-) *adj.* **1.** *Philos.* **a.** Concerned with the a priori basis of knowledge : minimizing the importance or denying the reality of sense experience. **b.** Asserting a fundamental irrationality or supernatural element in experience. **2.** Rising above common thought or ideas : MYSTICAL. **3.** *Math.* **a.** Incapable of being determined by any combination of a finite number of equations with rational integral coefficients. **b.** Not expressible as an integer or quotient of integers. —Used of numbers, esp. nonrepeating infinite decimals. **—tran′scen·den′tal·ly** *adv.*

**tran·scen·den·tal·ism** (trăn′sĕn-dĕn′tl-ĭz′əm, -sən-) *n.* **1.** *Philos.* **a.** The belief that knowledge of reality is derived from intuitive sources rather than from objective experience. **b.** A doctrine based on this belief, as the philosophy of Kant. **2.** The quality or state of being transcendental. **—tran′scen·den′tal·ist** *n.*

**transcendental meditation** *n.* A meditational technique in which deep mental and physical relaxation is achieved esp. through the use of a mantra.

**trans·con·ti·nen·tal** (trăns′kŏn-tə-nĕn′tl) *adj.* Spanning or crossing a continent <*transcontinental* flights>

**tran·scribe** (trăn-skrīb′) *vt.* **-scribed, -scrib·ing, -scribes.** [Lat. *transcribere* : *trans,* across + *scribere,* to write.] **1. a.** To write or type a copy of : write out fully, as from shorthand notes or via an electronic recording medium <*transcribe* a memo> **b.** To transfer (data) from one recording and storing system to another. **2.** To adapt or arrange (a musical composition) for a voice or instrument other than the original. **3.** To record, usu. on tape, for broadcasting at a later date <*transcribe* a talk show> **4.** To represent (speech sounds) by phonetic symbols. **5.** *Genetics.* To cause (DNA) to undergo transcription. **—tran·scrib′a·ble** *adj.* **—tran·scrib′er** *n.*

**tran·script** (trăn′skrĭpt′) *n.* [ME < OFr. *transcrit* < Lat. *transcriptum* < neuter p.part. of *transcribere,* to transcribe.] **1.** Transcribed matter, esp. a written, typewritten, or printed copy, as of a legal or academic record. **2.** *Genetics.* The RNA sequence produced by transcription.

**tran·scrip·tion** (trăn-skrĭp′shən) *n.* **1.** The act or process of transcribing. **2.** Something transcribed, esp.: **a.** An adaptation of a musical composition. **b.** A recorded radio or television program. **3.** *Genetics.* The process by which a messenger RNA molecule is synthesized from a DNA molecule template that results in the transfer of genetic information from the DNA to the messenger RNA. **—tran·scrip′tion·al** *adj.* **—tran·scrip′tion·al·ly** *adv.*

**trans·cul·tu·ra·tion** (trăns′kŭl-chə-rā′shən) *n.* Cultural change induced by the introduction of elements of a foreign culture.

**trans·duc·er** (trăns-dōō′sər, -dyōō′-, trănz-) *n.* [< Lat. *transducere,* to transfer : *trans,* across + *ducere,* to lead.] A substance or device, as a piezoelectric crystal or a photoelectric cell, that converts input energy of one form into output energy of another. **—trans·duce′** *v.* **(-duced, -duc·ing, -duc·es).**

**trans·duc·tion** (trăns-dŭk′shən, trănz-) *n.* [Lat. *transductio,* transfer < *transducere,* to transfer. —see TRANSDUCER.] Transfer of genetic material by bacteriophage from one bacterium to another.

**tran·sect** (trăn-sĕkt′) *vt.* **-sect·ed, -sect·ing, -sects.** To divide by cutting transversely. **—tran·sec′tion** *n.*

**tran·sept** (trăn′sĕpt′) *n.* [TRANS- + Lat. *saeptum,* partition. —see SEPTUM.] Either of the two lateral arms of a cruciform church.

**trans·fer** (trăns-fûr′, trăns′fûr′) *v.* **-ferred, -fer·ring, -fers.** [ME *transferren* < OFr. *transferer* < Lat. *transferre* : *trans,* across + *ferre,* to carry.] —*vt.* **1.** To carry, remove, or shift from one person, position, or place to another. **2.** To convey or make over the possession or legal title of (e.g., property) to another. **3.** To shift (e.g., a design) from one surface to another. —*vi.* **1.** To move oneself from one location, job, or academic institution to another. **2.** To change from one motor carrier to another. —*n.* (trăns′fûr′). **1.** *also* **trans·fer·al** (trăns-fûr′əl). Conveyance or removal of a thing from one person or place to another. **2. a.** *also* **transferal.** One that has or has been transferred, as a student enrolled in a new school. **b.** A design conveyed or to be conveyed from one surface to another. **3. a.** A ticket entitling a passenger to change from one motor carrier to another. **b.** A place where such changes are required or permitted. **4.** *also* **transferal.** *Law.* **a.** Conveyance of title or property from one party to another. **b.** The document effecting such conveyance. **—trans·fer′a·bil′i·ty** *n.* **—trans·fer′a·ble** *adj.* **—trans·fer′rer** *n.*

**trans·fer·al** *also* **trans·fer·ral** (trăns-fûr′əl) *n.* **1.** A transfer. **2.** *Psychoanal.* TRANSFERENCE 2.

**trans·fer·ase** (trăns′fə-rās′, -rāz′) *n.* Any of various enzymes that catalyze the transfer of atoms or groups of atoms from one molecule to another.

**trans·fer·ee** (trăns′fə-rē′) *n.* **1.** *Law.* One to whom a transfer of title or property is made. **2.** One who is transferred.

**trans·fer·ence** (trăns-fûr′əns, trăns′fər-əns) *n.* **1.** The act or process of transferring or the state of being transferred. **2.** *Psychoanal.* The process in and by which a person's feelings, thoughts, and wishes shift from one person to another, esp. this process in psychoanalysis with the analyst made the object of the shift. **—trans′fer·en′tial** (trăns′fə-rĕn′shəl) *adj.*

**trans·fer·or** (trăns′fə-rôr′) *n.* *Law.* One who effects a transfer of title or property.

**trans·fer·rin** (trăns-fĕr′ĭn) *n.* A blood globulin that can combine reversibly with and transport iron ions in the body.

**transfer RNA** *n.* A ribonucleic acid that acts as a carrier in the transport of a specific amino acid to the ribosomal site where a protein molecule is being synthesized.

**trans·fig·u·ra·tion** (trăns-fĭg′yə-rā′shən) *n.* **1.** Radical transformation of figure or appearance : METAMORPHOSIS. **2. Transfiguration. a.** The sudden emanation of radiance from Jesus' person that occurred on the mountain. **b.** The Christian commemoration of this, observed on Aug. 6.

**trans·fig·ure** (trăns-fĭg′yər) *vt.* **-ured, -ur·ing, -ures.** [ME *transfiguren* < Lat. *transfigurare* : *trans,* beyond + *figura,* figure.] **1.** To transform the figure or outward appearance of <a face *transfigured* with pain> **2.** To exalt : glorify. **—trans·fig′ure·ment** *n.*

**trans·fi·nite** (trăns-fī′nīt′) *adj.* Surpassing the finite.

**transfinite number** *n.* A cardinal or ordinal number that is not an integer.

**trans·fix** (trăns-fĭks′) *vt.* **-fixed, -fix·ing, -fix·es.** [Lat. *transfigere, transfix-* : *trans,* through + *figere,* to pierce.] **1.** To pierce with or as if with a pointed weapon : IMPALE. **2.** To make motionless, as with terror, amazement, or awe. **—trans·fix′ion** (-fĭk′shən) *n.*

**trans·form** (trăns-fôrm′) *v.* **-formed, -form·ing, -forms.** [ME *transformen* < Lat. *transformare* : *trans,* across + *forma,* shape.] —*vt.* **1.** To alter markedly the appearance or form of <*transform* a liquid into gas> **2.** To change the nature, function, or condition of : CONVERT. **3.** To subject to a mathematical transformation. **4.** To subject to a linguistic transformation. **5.** *Elect.* To subject to the action of a transformer. —*vi.* To undergo a transformation. —*n.* (trăns′fôrm′). **1.** The result, esp. a mathematical quantity, of a transformation. **2.** TRANSFORMATION 3b. **—trans·form′a·ble** *adj.*

**trans·for·ma·tion** (trăns′fər-mā′shən, -fôr-) *n.* **1. a.** An act or instance of transforming or the state of being transformed. **b.** Something transformed. **2.** *Math.* **a.** Replacement of the variables in an algebraic expression by their values in terms of another set of variables. **b.** A mapping of one space onto another or onto itself. **3. a.** The process of converting a syntactic construction into a semantically equivalent construction according to the rules shown to generate the syntax of the language. **b.** A construction derived by such transformation. **—trans·for′ma·tive** (-fôr′mə-tĭv) *adj.*

**trans·for·ma·tion·al grammar** (trăns′fər-mā′shə-nəl, -fôr-) *n.* A grammar that accounts for the constructions of a language by linguistic transformations and phrase structures, esp. generative transformational grammar.

**trans·form·er** (trăns-fôr′mər) *n.* One that transforms, esp. a device that employs mutual induction to convert variations of electric current in a primary circuit into variations of current and voltage in a secondary circuit.

**trans·fuse** (trăns-fyōōz′) *vt.* **-fused, -fus·ing, -fus·es.** [ME *transfusen,* to transmit < Lat. *transfundere,* to pour out : *trans,* across + *fundere,* to pour.] **1.** To transfer (liquid) by pouring from one container into another. **2.** To permeate, instill, or imbue. **3.** *Med.* To administer a transfusion of or to. **—trans·fus′er** *n.* **—trans·fus′i·ble** *adj.* **—trans·fu′sive** (-fyōō′sĭv, -zĭv) *adj.*

**trans·fu·sion** (trăns-fyōō′zhən) *n.* **1.** The act or process of transfusing. **2.** *Med.* Direct injection of whole blood or plasma into the blood stream. **—trans·fu′sion·al** *adj.*

**trans·gress** (trăns-grĕs′, trănz-) *v.* **-gressed, -gress·ing, -gress·es.** [Lat. *transgredi, transgress-,* to step across : *trans,* across + *gradi,* to step.] —*vt.* **1.** To breach (a limit or boundary). **2.** To act in violation of (e.g., the law). —*vi.* To trespass : sin. **—trans·gress′i·ble** *adj.* **—trans·gres′sion** (-grĕsh′ən, trănz-) *n.* **—trans·gres′sive** *adj.* **—trans·gres′sive·ly** *adv.* **—trans·gres′sor** *n.*

**tran·ship** (trăn-shĭp′, trăns-) *v. var. of* TRANSSHIP.

**trans·hu·mance** (trăns-hyōō′məns, trănz-) *n.* [Fr. < *transhumer,* to move livestock seasonally < Sp. *transhumar* : Lat. *trans,* across +

Lat. *humus,* ground.] Seasonal movement of livestock and herders to different grazing grounds. **—trans·hu'mant** *adj.* & *n.*

**tran·sient** (trăn'shənt, -zhənt, -zē-ənt) *adj.* [Lat. *transiens, transeunt-,* pr.part. of *transire,* to go over : *trans,* over + *ire,* to go.] **1.** Lasting only a short time : TRANSITORY <a *transient* civilization now extinct> **2. a.** Passing through from one place to another <*transient* farm laborers> **b.** Intended for the use of temporary residents <*transient* billets> **3.** *Physics.* Decaying with time, esp. as a simple exponential function of time. *—n.* **1.** One that is transient <*transients* at a hotel> **2.** *Physics.* A transient phenomenon or property, esp. a transient electric current. **—tran·sience** (trăn'shəns, -zhəns, -zē-əns), **tran·sien·cy** (-shən-sē, -zhən-sē, -zē-ən-sē) *n.* **—tran'sient·ly** *adv.*

**trans·il·lu·mi·na·tion** (trăns'ĭ-lōō'mə-nā'shən, trănz'-) *n. Med.* Examination of a bodily part or organ by passing a light through its walls. **—trans'il·lu'mi·nate'** (-lōō'mə-nāt') *v.* **(-nat·ed, -nat·ing, -nates). —trans'il·lu'mi·na'tor** *n.*

**tran·sis·tor** (trăn-zĭs'tər, -sĭs'-) *n.* [TRANS(FER) + (RES)ISTOR.] **1.** A three-terminal semiconductor device for amplification, switching, and detection, typically containing two rectifying junctions and operating so that the current between one pair of terminals controls the current between the other pair, one terminal being common to input and output. **2.** A radio fitted with transistors.

**tran·sis·tor·ize** (trăn-zĭs'tə-rīz', -sĭs'-) *vt.* **-ized, -iz·ing, -iz·es.** To fit (an electronic circuit or device) with transistors. **—tran·sis'tor·i·za'tion** *n.*

**transistor radio** *n.* TRANSISTOR 2.

**tran·sit** (trăn'sĭt, -zĭt) *n.* [Lat. *transitus* < p.part. of *transire,* to go across. —see TRANSIENT.] **1. a.** Passage over, across, or through. **b.** Conveyance of goods or persons from one place to another, esp. on a local public transport system. **2.** A transition or change, esp. from one life to another at death. **3.** *Astron.* **a.** The passage of a celestial body across the observer's meridian. **b.** The passage of a smaller celestial body across the disk of a larger celestial body. **4.** A surveying instrument that measures horizontal and vertical angles. *—v.* **-sit·ed, -sit·ing, -sits.** *—vt.* **1.** To pass over, across, or through. **2.** To revolve (the telescope of a surveying transit) about the horizontal transverse axis for direction reversal. *—vi. Astron.* To make a transit. **—in transit.** In the course of transit.

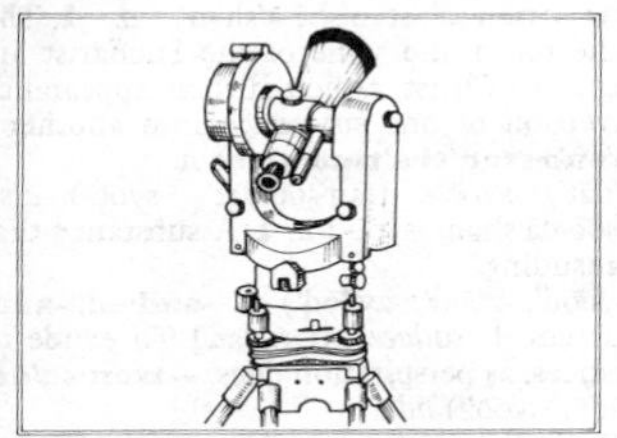

**transit**

**tran·si·tion** (trăn-zĭsh'ən, -sĭsh'-) *n.* **1.** An act, process, or instance of changing from one state, form, activity, or place to another. **2.** Passage from one subject to another. **3.** *Mus.* **a.** A modulation, esp. a brief one. **b.** A passage connecting two themes. **—tran·si'tion·al, tran·si'tion·ar'y** (-ə-nĕr'ē) *adj.* **—tran·si'tion·al·ly** *adv.*

**transition element** *n.* **1.** Any of the elements that serve as transitional links between the most and the least electropositive in a series of elements, and that are marked by high melting points, densities, magnetic moments, multiple valences, and the ability to form stable complex ions. **2.** Any of the elements in which an inner electron shell rather than an outer shell is only partially filled, gen. taken to include elements 21–29, 38–46, and 71–78.

**transition metal** *n.* A transition element.

**tran·si·tive** (trăn'sĭ-tĭv, -zĭ-) *adj.* [LLat. *transitivus* < *transitio,* transition < *transire,* to go over. —see TRANSIENT.] **1.** Expressing an action carried from the subject to the object and requiring a direct object to complete the meaning. —Used of a verb. **2.** Marked by or effecting transition. *—n.* A transitive verb. **—tran'si·tive·ly** *adv.* **—tran'si·tive·ness, tran'si·tiv'i·ty** (-tĭv'ĭ-tē) *n.*

**tran·si·to·ry** (trăn'sĭ-tôr'ē, -tōr'ē, trăn'zĭ-) *adj.* [ME *transitorie* < AN < LLat. *transitorius* < Lat., having a passageway < *transitus,* transit. —see TRANSIT.] TRANSIENT 1. **—tran'si·to'ri·ly** *adv.* **—tran'si·to'ri·ness** *n.*

☆ ***syns:*** TRANSITORY, MOMENTARY, PASSING, SHORT-LIVED, TEMPORARY, TRANSIENT *adj. core meaning* : lasting or existing only for a short time <the *transitory* Arctic summer><*transitory* popularity> ***ant:*** enduring, lasting

**trans·late** (trăns-lāt', trănz-, trăns'lāt', trănz'-) *v.* **-lat·ed, -lat·ing, -lates.** [ME *translaten* < Lat. *transferre, translat-* : *trans,* across + *ferre,* to carry.] *—vt.* **1.** To express in another language, while systematically retaining the original sense. **2.** To put into simpler terms : EXPLAIN <*translate* complex legal documents into lay people's language> **3.** To change from one form or style to another : CONVERT <*translate* ideas into reality> **4.** To transfer (a bishop) to another see. **5.** To forward or retransmit (a telegraphic message). **6.** To convey to heaven without natural death. **7.** *Physics.* To subject (a body) to translation. **8.** *Archaic.* To transport : enrapture. **9.** *Genetics.* To subject (a genetic code) to translation during protein synthesis. *—vi.* **1. a.** To make a translation. **b.** To work as a translator. **2.** To admit of translation <a Russian proverb that does not *translate* easily> **3.** *Aerospace.* To move from one place to another in space by means of reaction power. **—trans·lat'a·bil'i·ty, trans·lat'a·ble·ness** *n.* **—trans·lat'a·ble** *adj.*

**trans·la·tion** (trăns-lā'shən, trănz-) *n.* **1.** The act or process of translating, esp. from one language to another, or the state of being translated. **2.** A translated version of a text. **3.** *Physics.* Motion of a body in which every point of the body moves parallel to and the same distance as every other point of the body. **4.** *Genetics.* The process by which the genetic information in a messenger RNA molecule directs the linear sequence of amino acids in a protein molecule during protein synthesis on a ribosome. **—trans·la'tion·al** *adj.*

**trans·la·tor** (trăns-lā'tər, trănz-, trăns'lā'tər, trănz'-) *n.* **1.** One who translates, esp. one hired to translate written works. **2.** An interpreter. **—trans'la·to'ri·al** (trăns'lə-tôr'ē-əl, -tōr'-, trănz'-) *adj.*

**trans·lit·er·ate** (trăns-lĭt'ə-rāt', trănz-) *vt.* **-at·ed, -at·ing, -ates.** [TRANS- + Lat. *littera,* letter + -ATE[1].] To represent (letters or words) in the corresponding characters of another alphabet. **—trans·lit'er·a'tion** *n.*

**trans·lo·cate** (trăns'lō-kāt', trănz'-, trăns-lō'-, trănz-) *vt.* **-cat·ed, -cat·ing, -cates.** To cause to change from one position to another : DISPLACE.

**trans·lo·ca·tion** (trăns'lō-kā'shən, trănz'-) *n.* **1.** A change in location. **2.** *Genetics.* A chromosomal aberration in which different nonhomologous genes are interchanged.

**trans·lu·cent** (trăns-lōō'sənt, trănz-) *adj.* [Lat. *translucens, translucent-,* pr.part. of *translucēre,* to shine through : *trans,* through + *lucēre,* to shine.] Admitting and diffusing light so that objects beyond cannot be clearly perceived. **—trans·lu'cence, trans·lu'cen·cy** *n.* **—trans·lu'cent·ly** *adv.*

**trans·ma·rine** (trăns'mə-rēn', trănz'-) *adj.* [Lat. *transmarinus* : *trans,* across + *mare,* sea.] **1.** Crossing the sea <*transmarine* shipments> **2.** Located beyond or coming from across the sea <*transmarine* peoples>

**trans·mem·brane** (trăns-mĕm'brān', trănz-) *adj.* Occurring or passing across a membrane.

**trans·mi·grant** (trăns-mī'grənt, trănz-) *n.* **1.** One who transmigrates. **2.** An immigrant passing through a country while traveling to the country of destination.

**trans·mi·grate** (trăns-mī'grāt', trănz-, trăns'mī'-, trănz'-) *vi.* **-grat·ed, -grat·ing, -grates.** [Lat. *transmigrare, transmigrat-* : *trans,* across + *migrare,* to migrate.] **1.** To migrate : emigrate. **2.** To pass into another body after death. —Used of the soul. **—trans·mi'gra'tor** *n.* **—trans·mi'gra·to'ry** (-mī'grə-tôr'ē, -tōr'ē) *adj.*

**trans·mi·gra·tion** (trăns'mī-grā'shən, trănz'-) *n.* **1.** The act or process of transmigrating. **2.** Passage of a soul into another body after death : METEMPSYCHOSIS. **—trans'mi·gra'tion·ism** *n.*

**trans·mis·si·ble** (trăns-mĭs'ə-bəl, trănz-) *adj.* Able to undergo transmission <*transmissible* viruses> **—trans·mis'si·bil'i·ty** *n.*

**trans·mis·sion** (trăns-mĭsh'ən, trănz-) *n.* [Lat. *transmissio,* a sending across < *transmittere,* to transmit.] **1.** The act or process of transmitting or the state of being transmitted. **2.** Something, as a voice or message, that is transmitted. **3. a.** An automotive assembly of gears and associated parts by which power is transmitted from the engine to a drive shaft. **b.** A system of gears. **4.** Passage of modulated carrier waves from a transmitter. **—trans·mis'sive** (-mĭs'ĭv) *adj.* **—trans'mis·siv'i·ty** (-sĭv'ĭ-tē) *n.*

**trans·mis·som·e·ter** (trăns'mĭ-sŏm'ĭ-tər, trănz'-) *n.* [TRANSMIS-S(ION) + -METER.] A device for measuring the transmission of light through a medium. **—trans'mis·som'e·try** *n.*

**trans·mit** (trăns-mĭt', trănz-) *v.* **-mit·ted, -mit·ting, -mits.** [ME *transmitten* < Lat. *transmittere* : *trans,* across + *mittere,* to send.] *—vt.* **1.** To convey or dispatch from one person, thing, or place to another. **2.** To cause to spread <*transmit* a contagious disease> **3.** To convey to others by heredity. **4.** *Electron.* To send (a signal), as by wire or radio. **5.** *Physics.* To cause (a disturbance) to propagate through a medium. **6.** To convey (force or energy) from one part of a mechanism to another. *—vi.* To send out a signal. **—trans·mit'ta·ble** *adj.* **—trans·mit'tal** (trăns-mĭt'l, trănz-) *n.*

**trans·mit·tance** (trăns-mĭt'ns, trănz-) *n.* **1.** TRANSMISSION 1. **2.** *Physics.* The ratio of the radiant energy transmitted to the total radiant energy incident on a given body.

**trans·mit·ter** (trăns-mĭt'ər, trănz-) *n.* One that transmits, as: **a.** A telegraphic sending instrument. **b.** The part of a telephone that converts the incident sounds into electrical impulses that are themselves conveyed to a remote receiver. **c.** Electronic equipment that generates and amplifies a carrier wave, modulates it with a meaningful signal, as derived from speech or other sources, and radiates the resulting signal from an antenna.

**trans·mog·ri·fy** (trăns-mŏg′rə-fī′, trănz-) *vt.* **-fied, -fy·ing, -fies.** [Orig. unknown.] To change into a different, esp. fantastic or bizarre, shape or form. **—trans·mog′ri·fi·ca′tion** *n.*

**trans·mon·tane** (trăns-mŏn′tān′, trănz-, trăns′mŏn-tān′, trănz′-) *adj.* [Lat. *transmontanus.* —see TRAMONTANE.] TRAMONTANE 1.

**trans·moun·tain** (trăns-moun′tən) *adj.* Extending over or through a mountain <a *transmountain* highway><a *transmountain* tunnel>

**trans·mu·ta·tion** (trăns′myōō-tā′shən, trănz′-) *n.* **1.** The act of transmuting or the state of being transmuted. **2.** Alchemical conversion of base metals into gold or silver. **3.** *Physics.* Transformation of one element into another by one or a series of nuclear reactions. **—trans′mu·ta′tion·al, trans·mut′a·tive** (-myōō′tə-tĭv) *adj.*

**trans·mute** (trăns-myōōt′, trănz-) *vt.* **-mut·ed, -mut·ing, -mutes.** [ME *transmuten* < Lat. *transmutare* : *trans,* across + *mutare,* to change.] To change from one nature, form, substance, or state into another : TRANSFORM. **—trans·mut′a·bil′i·ty, trans·mut′a·ble·ness** *n.* **—trans·mut′a·ble** *adj.* **—trans·mut′a·bly** *adv.* **—trans·mut′er** *n.*

**trans·na·tion·al** (trăns-năsh′ə-nəl, trănz-) *adj.* Transcending national boundaries <*transnational* issues>

**trans·o·ce·an·ic** (trăns′ō-shē-ăn′ĭk, trănz′-) *adj.* **1.** Located beyond or on the other side of the ocean <*transoceanic* cities> **2.** Spanning or crossing the ocean <*transoceanic* flights>

**tran·som** (trăn′səm) *n.* [ME *traunson,* prob. < Lat. *transtrum* < *trans,* across.] **1. a.** A small hinged window above another window or a door. **b.** The horizontal crosspiece to which such a window is hinged. **2.** A horizontal piece of wood or stone in a window that serves to divide it. **3.** *Naut.* **a.** A transverse beam affixed to the sternpost of a wooden ship and forming part of the stern. **b.** The aftermost transverse structural member including the floor, frame, and beam assembly at the sternpost of a steel ship. **c.** The stern of a square-sterned boat when it is a structural member. **4.** The horizontal beam on a cross or gallows. **—tran′somed** *adj.*

**tran·son·ic** (trăn-sŏn′ĭk) *adj.* [TRANS- + (SUPER)SONIC.] Of or relating to aerodynamic flow or flight conditions at speeds close to the speed of sound.

**trans·pa·cif·ic** (trăns′pə-sĭf′ĭk) *adj.* **1.** Crossing the Pacific Ocean. **2.** Located across or beyond the Pacific Ocean.

**trans·par·en·cy** (trăns-pâr′ən-sē, -păr′-) *n., pl.* **-cies. 1.** *also* **trans·par·ence** (-pâr′əns, -păr′-). The quality or state of being transparent. **2.** A transparent object, esp. a photographic slide.

**trans·par·ent** (trăns-pâr′ənt, -păr′-) *adj.* [ME < OFr. < Med. Lat. *transparens,* pr.part. of *transparēre,* to be seen through : Lat. *trans,* through + Lat. *parēre,* to show.] **1.** Capable of transmitting light so that objects or images beyond can be clearly perceived. **2.** Permeable to electromagnetic radiation of specified frequencies, as to visible light or radio waves. **3.** So fine or delicate in texture that objects may be easily seen on the other side : DIAPHANOUS. **4. a.** Easily detected : OBVIOUS <*transparent* lies> **b.** Readily understandable : CLEAR <a *transparent* explanation> **5.** Without guile : CANDID. **6.** *Obs.* Luminous. **—trans·par′ent·ly** *adv.* **—trans·par′ent·ness** *n.*

**trans·per·son·al** (trăns-pûr′sə-nəl) *adj.* Transcending the personal or individual.

**trans·pierce** (trăns-pîrs′) *vt.* **-pierced, -pierc·ing, -pierc·es.** To penetrate or pierce.

**tran·spi·ra·tion** (trăn′spə-rā′shən) *n.* The act or process of transpiring, esp. through the stomata of plants or skin pores.

**tran·spire** (trăn-spīr′) *v.* **-spired, -spir·ing, -spires.** [Fr. *transpirer* : Lat. *trans,* across + Lat. *spirare,* to breathe.] *—vt.* To give off (vapor containing waste products) through the pores of the skin or the stomata of plant tissue. *—vi.* **1.** To give off vapor containing waste products through animal or plant pores. **2.** To be revealed : come to light. **3.** To happen : occur.

**trans·pla·cen·tal** (trăns′plə-sĕn′tl) *adj.* Passing through the placenta. **—trans′pla·cen′tal·ly** *adv.*

**trans·plant** (trăns-plănt′) *v.* **-plant·ed, -plant·ing, -plants.** [ME *transplaunten* < LLat. *transplantare* : Lat. *trans,* across + Lat. *plantare,* to plant.] *—vt.* **1.** To uproot and replant (a growing plant). **2.** To transfer from one residence or place to another : RELOCATE. **3.** *Med.* To transfer (tissue or an organ) from one body, or body part, to another. *—vi.* **1.** To engage in transplanting. **2.** To survive transplanting. *—n.* (trăns′plănt′). **1.** The act or process of transplanting. **2.** Something transplanted. **—trans·plant′a·bil′i·ty** *n.* **—trans·plant′a·ble** *adj.* **—trans′plan·ta′tion** *n.* **—trans·plant′er** *n.*

**trans·po·lar** (trăns-pō′lər) *adj.* Extending across or crossing over either of the geographic polar regions.

**tran·spond·er** (trăn-spŏn′dər) *n.* [TRAN(SMITTER) + (RE)SPONDER.] A radio or radar receiver-transmitter activated for transmission by reception of a predetermined signal.

**trans·pon·tine** (trăns-pŏn′tīn′) *adj.* **1.** Located across or beyond a bridge. **2.** Like or characteristic of melodramas once popular in London theaters located south of the Thames River.

**trans·port** (trăns-pôrt′, -pōrt′, trăns′pôrt′, -pōrt′) *vt.* **-port·ed, -port·ing, -ports.** [ME *transporten* < OFr. *transporter* < Lat. *transportare* : *trans,* across + *portare,* to carry.] **1.** To convey from one place to another. **2.** To affect with strong emotion : ENRAPTURE. **3.** To send abroad to a penal colony. *—n.* (trăns′pôrt′, -pōrt′). **1.** The act of transporting : CONVEYANCE. **2.** The state of being moved by emotion : RAPTURE. **3.** A ship or aircraft used to transport troops or military equipment. **4.** A vehicle, as an aircraft, used to transport mail, freight, or passengers. **—trans·port′a·bil′i·ty** *n.* **—trans·port′a·ble** *adj.* **—trans·port′er** *n.* **—trans·por′tive** *adj.*

**trans·por·ta·tion** (trăns′pər-tā′shən) *n.* **1.** The act of transporting or the state of being transported. **2. a.** A means of transport : CONVEYANCE. **b.** The business of transporting goods, materials, or passengers. **3.** A charge for transporting : FARE. **4.** Deportation to a penal colony. **—trans′por·ta′tion·al** *adj.*

**trans·pose** (trăns-pōz′) *v.* **-posed, -pos·ing, -pos·es.** [ME *transposen,* to transform < OFr. *transposer* < Lat. *transponere* : *trans,* across + *ponere,* to place.] *—vt.* **1.** To reverse the order or place of : INTERCHANGE. **2.** To move into a different position or order. **3.** *Math.* To move (a term) from one side of an algebraic equation to the other side, reversing the sign to maintain equality. **4.** *Mus.* To write or perform (a composition) in a key other than the original or given key. **5.** To alter in nature or form : TRANSFORM. *—vi.* **1.** *Mus.* To write or perform music in a different key. **2.** To admit of being transposed. **—trans·pos′a·ble** *adj.* **—trans·pos′er** *n.*

**trans·po·si·tion** (trăns′pə-zĭsh′ən) *n.* **1.** The act of transposing or the state of being transposed. **2.** Something transposed. **—trans′po·si′tion·al** *adj.*

**trans·sex·u·al** (trăns-sĕk′shōō-əl) *n.* **1.** One predisposed to become a member of the opposite sex. **2.** One whose sex has been changed externally by surgery and by hormone injections. **—trans·sex′u·al** *adj.* **—trans·sex′u·al·ism, trans·sex′u·al′i·ty** (-ăl′ĭ-tē) *n.*

**trans·ship** (trăns-shĭp′) *also* **tran·ship** (trăn-shĭp′, trăns-) *v.* **-shipped, -ship·ping, -ships.** *—vt.* To transfer from one vessel or vehicle to another for reshipment. *—vi.* To transfer cargo from one vessel or vehicle to another. **—trans·ship′ment** *n.*

**trans·tho·rac·ic** (trăns′thə-răs′ĭk) *adj.* Extending across or effected by way of the thoracic cavity. **—trans′tho·rac′i·cal·ly** *adv.*

**tran·sub·stan·ti·ate** (trăn′səb-stăn′shē-āt′) *vt.* **-at·ed, -at·ing, -ates.** [Med. Lat. *transubstantiare, transubstantiat-* : Lat. *trans,* beyond + Lat. *substantia,* substance.] **1.** To change (one substance) into another : TRANSMUTE. **2.** To change the substance of (the Eucharistic bread and wine) into the presence of Christ.

**tran·sub·stan·ti·a·tion** (trăn′səb-stăn′shē-ā′shən) *n.* **1.** The theological doctrine that the bread and wine of the Eucharist are transformed into the presence of Christ, although their appearance remains the same. **2.** Conversion of one substance into another : TRANSFORMATION. **—tran′sub·stan′ti·a′tion·al·ist** *n.*

**tran·su·date** (trăn-sōō′dāt′, -syōō′-, trăn′sōō-dāt′, -syōō-) *also* **tran·su·da·tion** (trăn′sōō-dā′shən, -syōō-) *n.* **1.** A substance that transudes. **2.** The act of transuding.

**tran·sude** (trăn-sōōd′, -syōōd′, -zōōd′, -zyōōd′) *vi.* **-sud·ed, -sud·ing, -sudes.** [Lat. *trans,* across + *sudare,* to sweat.] To exude or pass through pores or interstices, as perspiration does. **—tran·su′da·to′ry** (trăn-sōō′də-tôr′ē, -tōr′ē, -syōō′-) *adj.*

**trans·u·ran·ic** (trăns′yōō-răn′ĭk, -rā′nĭk, trănz′-) *also* **trans·u·ra·ni·um** (-rā′nē-əm) *adj.* [TRANS- + URAN(IUM) + -IC.] Having an atomic number greater than 92.

**trans·val·u·ate** (trăns-văl′yōō-āt′) *vt.* **-at·ed, -at·ing, -ates.** To transvalue.

**trans·val·ue** (trăns-văl′yōō, trănz-) *vt.* **-ued, -u·ing, -ues.** To evaluate by a new standard or principle, esp. by one varying from conventional standards. **—trans·val′u·a·tion** *n.*

**trans·ver·sal** (trăns-vûr′səl, trănz-) *adj.* Transverse. *—n.* A line that intersects a system of lines.

**trans·verse** (trăns-vûrs′, trănz-, trăns′vûrs′, trănz′-) *adj.* [Lat. *transversus* < p.part. of *transvertere,* to direct across : *trans,* across + *vertere,* to turn.] Situated or lying across : CROSSWISE. *—n.* (trăns′vûrs′, trănz′-). Something, as a part or beam, that is transverse. **—trans·verse′ly** *adv.* **—trans·verse′ness** *n.*

☆ ***syns:*** TRANSVERSE, CROSSING, CROSSWISE, THWART, TRANSVERSAL, TRAVERSE *adj. core meaning* : situated or lying across <enormous *transverse* arches in the cathedral> ***ant:*** longitudinal

**transverse colon** *n.* The section of the colon that lies across the upper part of the abdominal cavity.

**transverse process** *n.* A process projecting laterally from the side of a vertebra.

**trap¹** (trăp) *n.* [ME < OE *træppe.*] **1.** A device, as a net or a clamp-like apparatus that springs shut suddenly, for catching and holding animals. **2.** A stratagem or device utilized in betraying, tricking, or exposing an unsuspecting victim. **3. a.** A receptacle, as a grease trap, used for collecting waste materials. **b.** A device for sealing a passage against the escape of gases, esp. a bend in a drainpipe that prevents backup of sewer gas. **4.** A device that hurls clay pigeons, balls, or disks into the air to be shot at. **5.** A land hazard or bunker on a golf course. **6.** A light two-wheeled carriage with springs. **7.** A trap door.

**8. traps.** Percussion instruments, as snare drums, cymbals, or bells. **9.** *Slang.* The human mouth. **10. traps.** A measured length of roadway over which electronic timers register the speed of a racing vehicle. —*v.* **trapped, trap·ping, traps.** —*vt.* **1.** To catch in or as if in a trap : ENSNARE. **2.** To place in a confining or embarrassing position. **3.** To seal off (gases) by a trap. **4.** To equip (a drain) with a trap. —*vi.* **1.** To set traps. **2.** To trap fur-bearing animals, esp. as a business.

**trap[2]** (trăp) *n.* [ME *trap,* trapping < OFr. *drap,* cloth.] *often* **traps.** *Informal.* One's personal belongings or household goods. —*vt.* **trapped, trap·ping, traps.** To furnish with trappings.

**trap[3]** (trăp) *n.* [Swed. *trapp* < *trappa,* step < MLG *trappe.*] A dark, fine-grained igneous rock used esp. in road building.

**tra·pan** (trə-păn′) *v. var. of* TREPAN[2].

**trap door** *n.* A hinged or sliding door in a floor, roof, or ceiling.

**trap-door spider** (trăp′dôr′, -dōr′) *n.* A spider of the family Ctenizidae that constructs a silk-lined burrow concealed by a hinged lid.

**tra·peze** (tră-pēz′, trə-) *n.* [Fr. *trapèze* < LLat. *trapezium,* trapezium. —see TRAPEZIUM.] A short horizontal bar suspended from the ends of two parallel ropes, used for acrobatic stunts or exercises.

**trapeze artist** *n.* A performer on the trapeze.

**tra·pe·zi·um** (trə-pē′zē-əm) *n., pl.* **-zi·ums** *or* **-zi·a** (-zē-ə) [LLat. < Gk. *trapezion,* dim. of *trapeza,* table : *tra-,* four + *peza,* foot.] **1. a.** A quadrilateral with no parallel sides. **b.** *Chiefly Brit.* TRAPEZOID 1. **2.** A bone in the wrist at the base of the thumb.

**tra·pe·zi·us** (trə-pē′zē-əs) *n.* [NLat. < LLat. *trapezium,* trapezium (from the shape of the muscles paired).] Either of two large flat muscles running from the base of the occiput to the middle of the back that support the head and shoulders and enable one to raise them.

**tra·pe·zo·he·dron** (trə-pē′zō-hē′drən, trăp′ə-zō-) *n., pl.* **-drons** *or* **-dra** (-drə) [TRAPEZ(IUM) + -HEDRON.] Any of several forms of crystal with trapeziums as faces.

**trap·e·zoid** (trăp′ĭ-zoid′) *n.* [NLat. *trapezoides* < Gk. *trapezoeidēs,* trapezium-shaped: *trapeza,* table + *-eidēs,* -oid.] **1.** A quadrilateral with two parallel sides. **2.** A small wrist bone near the base of the index finger. **—trap′e·zoid′, trap′e·zoi′dal** (-zoid′l) *adj.*

**trap gun** *n.* A shotgun designed for trapshooting.

**trap house** *n.* The enclosure housing the spring traps that hurl clay pigeons, balls, or disks into the air in trapshooting and skeet.

**trap·light** (trăp′līt′) *n.* A device using a light to trap insects.

**trap·per** (trăp′ər) *n.* One who traps animals for their furs.

**trap·ping** (trăp′ĭng) *n.* **1.** *often* **trappings.** An ornamental covering or harness for a horse : CAPARISON. **2. trappings. a.** Articles of dress or adornment. **b.** Outward signs or indications <has all the *trappings* of a tin-horned dictator>

**Trap·pist** (trăp′ĭst) *n.* A member of a branch of a Cistercian order of monks established in 1664 in La Trappe, Normandy, characterized by austerity and a vow of absolute silence. **—Trap′pist** *adj.*

**trap·shoot·ing** (trăp′shōō′tĭng) *n.* The sport of shooting at clay pigeons. **—trap′shoot′er** *n.*

**tra·pun·to** (trə-pōōn′tō, -pŏŏn′-) *n., pl.* **-tos.** [Ital. < *trapungere,* to embroider : Lat. *trans,* across + Lat. *pungere,* to prick.] Quilting having a high-relief effect made by outlining the design with running stitches and then filling it with cotton.

**trash** (trăsh) *n.* [Orig. unknown.] **1.** Worthless, discarded material : REFUSE. **2. a.** Empty expressions or ideas. **b.** Worthless literary or artistic matter. **3.** Something broken off or removed to be discarded, esp. plant trimmings. **4.** One considered ignorant or contemptible. —*vt.* **trashed, trash·ing, trash·es. 1.** To cut off leaves or branches from, esp. to lop off the outer leaves from (growing sugar cane). **2.** To throw away : DISCARD. **3.** *Slang.* **a.** To vandalize <a store *trashed* by hoodlums> **b.** To smash <*trash* store windows> **c.** To criticize harshly <*trashed* the producer's latest play>

**trash·y** (trăsh′ē) *adj.* **-i·er, -i·est.** Resembling trash : INFERIOR. **—trash′i·ly** *adv.* **—trash′i·ness** *n.*

**trass** (trăs) *n.* [Du. *tras.*] A light-colored tuff used in making hydraulic cement.

**trau·ma** (trou′mə, trô′-) *n., pl.* **-mas** *or* **-ma·ta** (-mə-tə) [Gk.] **1.** *Pathol.* A wound, esp. one caused by sudden physical injury. **2.** *Psychiat.* An emotional shock that creates substantial and lasting damage to the psychological development of the individual, gen. leading to neurosis. **3.** Something that severely jars the mind or emotions. **—trau·mat′ic** (-măt′ĭk) *adj.* **—trau·mat′i·cal·ly** *adv.*

☆ **syns:** TRAUMA, BLOW, JOLT, SHOCK *n. core meaning* : something that severely jars the mind or emotions <the national *trauma* of the Vietnam War>

**trau·ma·tism** (trou′mə-tĭz′əm, trô′-) *n.* **1.** A wound produced by injury : TRAUMA. **2.** Development or an instance of trauma.

**trau·ma·tize** (trou′mə-tīz′, trô′-) *vt.* **-tized, -tiz·ing, -tiz·es. 1.** To wound or injure. **2.** To damage the psychological development of (an individual).

**tra·vail** (trə-vāl′, trăv′āl′) *n.* [ME < OFr. < *travailler,* to work hard < LLat. *tripalium,* instrument of torture < Lat. *tripalis,* having three stakes : *ter,* three + *palus,* stake.] **1.** Strenuous physical or mental exertion. **2.** Tribulation : agony. **3.** The labor of childbirth. —*vi.* **-vailed, -vail·ing, -vails. 1.** To labor strenuously : TOIL. **2.** To undergo the labor of childbirth.

**trave** (trāv) *n.* [ME < OFr. < Lat. *trabs.*] **1.** A crossbeam. **2.** A section, as of a ceiling, formed by crossbeams.

**trav·el** (trăv′əl) *v.* **-eled, -el·ing, -els** *or* **-elled, -el·ling, -els.** [ME *travelen* < OFr. *travailler,* to travail. —see TRAVAIL.] —*vi.* **1.** To move from one place to another : JOURNEY. **2.** To journey from one place to another as a traveling sales representative. **3.** To be transmitted, as light. **4.** To keep company : ASSOCIATE <*travels* in a fast circle> **5.** To admit of being transported <Some perishables *travel* poorly.> **6.** *Informal.* To move swiftly. **7.** *Basketball.* To walk or run illegally while holding the ball. —*vt.* To pass over or through <*travel* the byroads of England> —*n.* **1.** The act or process of traveling. **2. travels. a.** A series of journeys. **b.** A written account of a series of journeys. **3.** Activity or traffic along a route or through a given point.

▲ word history: The hardships of making a journey in earlier times is reflected in the etymological identity of the words *travel* and *travail.* Both are derived from Old French *travailler,* which originally meant "to torment, to trouble," and later came to mean "to be troubled, to be in pain, to work hard." *Travailler* was borrowed into English as *travail,* which at first had the same meanings as the Old French word but which later came to mean "to toil, to make a difficult journey," and simply "to journey." *Travel* was originally a variant of *travail,* but it has now become a separate word used exclusively in the sense "to journey."

**travel agency** *n.* An agency that arranges for travel itineraries, tickets, and accommodations. **—travel agent** *n.*

**travel bureau** *n.* A travel agency.

**trav·eled** *also* **trav·elled** (trăv′əld) *adj.* **1.** Having journeyed widely. **2.** Much frequented by travelers <a heavily *traveled* path>

**trav·el·er** *also* **trav·el·ler** (trăv′əl-ər, trăv′lər) *n.* **1.** One who travels. **2.** *Chiefly Brit.* A traveling sales representative. **3.** *Naut.* **a.** A metal ring that moves freely back and forth on a rope, rod, or spar. **b.** The rope, rod, or spar on which such a ring moves.

**traveler's check** *n.* An internationally redeemable draft purchasable from a bank, express company, or travel agency, in various denominations, valid only with the holder's own endorsement against his or her original signature.

**trav·el·er's-joy** (trăv′əl-ərz-joi′, trăv′lərz-) *n.* A climbing vine of the genus *Clematis,* esp. *C. vitalba* of the Old World, with white flower clusters.

**traveling salesman** *n.* A sales representative who solicits business orders or sells merchandise through personal dealings with potential customers within a given territory.

**trav·e·logue** *also* **trav·e·log** (trăv′ə-lôg′, -lŏg′) *n.* **1.** A lecture illustrated by travel slides or films. **2.** A narrated film about travels.

**tra·verse** (trə-vûrs′, trăv′ərs) *v.* **-versed, -vers·ing, -vers·es.** [ME *traversen* < OFr. *traverser* < LLat. *traversare* < Lat. *transvertere,* to direct across. —see TRANSVERSE.] —*vt.* **1.** To pass across, over, or through. **2.** To cross and recross. **3. a.** To go up, down, or across (e.g., a hill) at an angle. **b.** To ski across rather than down (a hill). **4.** To swivel (e.g., a gun) laterally. **5.** To extend across : CROSS. **6.** To examine carefully. **7.** To go counter to : THWART. **8.** *Law.* **a.** To deny formally (an allegation of fact by the opposition) in a suit. **b.** To join issue upon (an indictment). **9.** To make a traverse survey of. **10.** *Naut.* To brace (a yard) fore and aft. —*vi.* **1.** To pass along, across, or back and forth. **2.** To turn laterally : SWIVEL. **3.** To descend a slope in a zigzag way, as in skiing. **4.** To glide or pressure one's fencing blade toward the hilt of the opponent's foil. —*n.* (trăv′ərs, trə-vûrs′). **1. a.** The act of traversing. **b.** A route or path across. **2.** Something lying across something else, esp.: **a.** A transversal. **b.** A structural crosspiece : TRANSOM. **c.** A gallery, deck, or loft crossing from one side of a building to the other. **d.** A railing, curtain, or screen. **e.** A defensive barrier across a rampart or trench, as a bank of earth thrown up for protection from enfilade fire. **3.** An obstruction : obstacle. **4. a.** *Naut.* The zigzag route of a vessel forced by contrary winds to sail on different courses. **b.** The zigzag course of a skier moving down a steep slope. **5.** The horizontal swivel of a mounted gun. **6. a.** A lateral movement, as of a lathe tool across a piece of work. **b.** A mechanical part that moves in this way. **7.** A line established by sighting in surveying a tract of land. **8.** *Law.* Formal denial of an allegation of fact in a suit. —*adj.* **tra·verse** (trăv′ərs, trə-vûrs′). **1.** Extending or lying across : TRANSVERSE. **2.** Relating to the installation or operation of draperies that can be drawn <*traverse* rods> **—tra·vers′a·ble** *adj.* **—tra·vers′al** (trə-vûr′səl) *n.* **—tra·vers′er** *n.*

**trav·er·tine** (trăv′ər-tēn′, -tĭn) *n.* [Ital. *travertino* < Lat. *(lapis) tiburtinus,* (stone) of Tibur, an ancient Italian city.] **1.** A light-colored, porous calcite, $CaCO_3$, deposited from solution in ground or surface waters and forming the stalactites and stalagmites of caverns. **2.** A compact creamy-colored calcium carbonate, used as a facing material in construction.

**trav·es·ty** (trăv′ĭ-stē) *n., pl.* **-ties.** [Fr. *travesti,* p.part. of *travestir,* to take on someone's habits < OItal. *travestire,* to disguise : *tra-,* trans- + *vestire,* to dress < Lat. < *vestis,* garment.] **1.** A grotesque imitation with intent to ridicule. **2.** A broad and grotesque parody on a lofty work or theme. **3.** A grotesque, debased, highly inferior imitation <a *travesty* of the Mass> —*vt.* **-tied, -ty·ing, -ties.** To make a travesty on or of.

**tra·vois** (trə-voi′, trăv′oi′) *or* **tra·voise** (trə-voiz′, trăv′oiz′) *n., pl.* **tra·vois** (trə-voiz′, trăv′oiz′) *or* **tra·vois·es** (trə-voi′zĭz, trăv′oi′zĭz) [Canadian Fr.] A primitive sledge once used by Plains Indians, consisting of a platform or netting supported by two long trailing poles, the forward ends of which were fastened to a dog or horse.

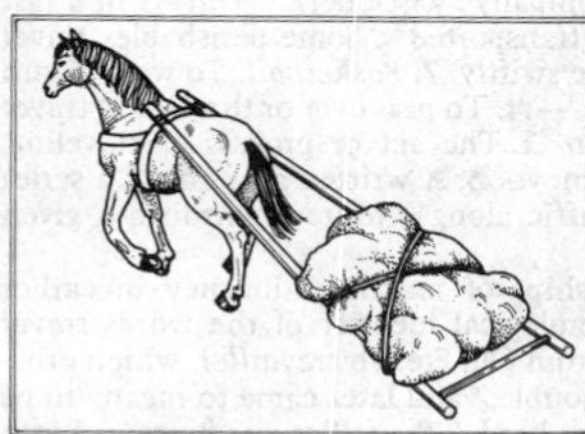
travois

**trawl** (trôl) *n.* [ME *trawelle,* perh. < MDu. *tragel,* dragnet.] **1.** A large tapered and flattened or conical fishing net towed along the sea bottom. **2.** A setline. —*v.* **trawled, trawl·ing, trawls.** —*vt.* To catch (fish) with a trawl. —*vi.* **1.** To fish with a trawl net or line. **2.** To troll.

**trawl·er** (trô′lər) *n.* **1.** A boat for trawling. **2.** One who trawls.

**tray** (trā) *n.* [ME < OE *trēg.*] A shallow flat receptacle with a raised edge or rim, used for carrying, holding, or displaying articles.

**treach·er·ous** (trĕch′ər-əs) *adj.* **1.** That betrays a trust : TRAITOROUS. **2. a.** Undependable : unreliable <a *treacherous* ladder> **b.** Dangerous <*treacherous* whirlpools> **—treach′er·ous·ly** *adv.* **—treach′er·ous·ness** *n.*

**treach·er·y** (trĕch′ə-rē) *n., pl.* **-ies.** [ME *trecherie* < OFr. *trecherie* < *trichier,* to trick.] **1.** Willful betrayal of confidence or trust : PERFIDY. **2.** An act or instance of betraying confidence or trust.

**trea·cle** (trē′kəl) *n.* [ME, antidote for poison < OFr. *triacle* < Lat. *theriaca* < Gk. *thēriakē* < *thērion,* poisonous beast, dim. of *thēr,* beast.] **1.** Cloying expression or sentiment. **2.** *Chiefly Brit.* Molasses. **3.** A medicinal compound once used as an antidote for poison. **—trea′cly** (-klē) *adj.*

**tread** (trĕd) *v.* **trod** (trŏd), **trod·den** (trŏd′n) *or* **trod, tread·ing, treads.** [ME *treden* < OE *tredan.*] —*vt.* **1.** To walk over, on, or along. **2.** To press beneath the foot : TRAMPLE. **3.** To put down harshly or cruelly : CRUSH. **4.** To make (e.g., a path) by walking or trampling. **5.** To execute (e.g., a step or measure) by walking or dancing. **6.** To copulate with. —Used of male birds. —*vi.* **1.** To go on foot. **2.** To trample so as to crush or injure. **3.** To copulate. —Used of birds. —*n.* **1. a.** The act, manner, or sound of treading. **b.** An instance of treading. **2.** The horizontal section of a step in a staircase. **3.** The portion of a wheel that makes contact with the ground or rails. **4.** The grooved face of a tire. **5.** The part of a shoe sole that touches the ground. **—tread water.** To keep one's head above water while in an upright position by moving the feet up and down. **—tread′less** *adj.*

**trea·dle** (trĕd′l) *n.* [ME *tredel* < OE, step of a stair < *tredan,* to tread.] A foot pedal for activating a circular drive, as in a potter's wheel or sewing machine. —*vi.* **-led, -ling, -les.** To work a treadle. **—tread′ler** (trĕd′lər) *n.*

**tread·mill** (trĕd′mĭl′) *n.* **1. a.** A mechanism operated by people walking on the moving steps of a wheel, or treading an endless sloping belt. **b.** A similar device operated by animals. **2.** A monotonous task.

**trea·son** (trē′zən) *n.* [ME < AN *treson* < Lat. *traditio,* surrender. —see TRADITION.] **1.** Violation of allegiance toward one's country or sovereign, esp. the betrayal of one's own country by waging war against it or by consciously and purposely acting to aid its enemies. **2.** Betrayal of confidence or trust. **—trea′son·a·ble** (trē′zə-nə-bəl) *adj.* **—trea′son·a·ble·ness** *n.* **—trea′son·a·bly** *adv.*

**trea·son·ous** (trē′zə-nəs) *adj.* Pertaining to or involving treason. **—trea′son·ous·ly** *adv.*

**treas·ure** (trĕzh′ər) *n.* [ME *tresure* < OFr. *tresor* < Lat. *thesaurus* < Gk. *thēsauros.*] **1.** Accumulated or hidden wealth in the form of valuables, as money or jewels. **2.** One regarded as esp. precious or valuable. —*vt.* **-ured, -ur·ing, -ures. 1.** To accumulate and save for future use. **2.** To value highly. **—treas′ur·a·ble** *adj.*

**treas·ur·er** (trĕzh′ər-ər) *n.* [ME *tresurer* < AN *tresorer* < OFr. *tresor,* treasure.] One who is in charge of funds or revenues, esp. a financial officer for a government, company, or society. **—treas′ur·er·ship′** *n.*

**treas·ure-trove** (trĕzh′ər-trōv′) *n.* [AN *tresor trove,* found treasure.] **1. a.** Unclaimed treasure found hidden. **b.** *Law.* Silver or gold, as in the form of bullion, plate, or money, found hidden, the ownership of which is unknown. **2.** TROVE 1.

**treas·ur·y** (trĕzh′ə-rē) *n., pl.* **-ies.** [ME *tresory* < OFr. *tresorie* < *tresor,* treasure.] **1.** A place where treasure is stored. **2.** A place where private or public funds are received, kept, managed, and disbursed. **3.** Public funds or revenues. **4. a.** A collection of valuables. **b.** A collection of literary or artistic treasures <a *treasury* of medieval verse> **5. Treasury.** The executive department of a government having charge of the collection, management, and expenditure of public revenue.

**treasury note** *n.* A note or bill issued by the U.S. Treasury as legal tender for all debts.

**treat** (trēt) *v.* **treat·ed, treat·ing, treats.** [ME *treten* < AN *treter* < Lat. *tractare,* freq. of *trahere,* to draw.] —*vt.* **1.** To have to do with or behave in a specified manner toward <*treat* all employees fairly> **2.** To consider or regard in a certain way <*treated* me as a member of the family> **3.** To deal with in speech or writing <an article *treating* English antiques> **4.** To represent or deal with in a specified style or manner, as in art or literature <*treat* a subject allegorically> **5.** To entertain (another) at one's own expense <*treated* me to dinner> **6.** To subject to an action, process, or change, esp.: **a.** To give medical aid to <*treat* sick patients> **b.** To subject to a chemical or physical process or application. —*vi.* **1.** To deal with a subject or topic in speech, writing, or thought <The essay *treats* of medieval chivalry.> **2.** To pay for another's entertainment or food. **3.** To negotiate : bargain. —*n.* **1.** Something, as one's food or entertainment, paid for by another person. **2.** Provision of a treat, esp. in return for something else. **3.** A special delight <A trip to Paris was a real *treat.*> **—treat′a·ble** *adj.* **—treat′er** *n.*

**trea·tise** (trē′tĭs) *n.* [ME *treatis* < AN *tretis* < *treter,* to treat.] **1.** A formal written account treating a subject systematically and in detail. **2.** *Obs.* A tale : narrative.

**treat·ment** (trēt′mənt) *n.* **1.** The act or manner of treating. **2.** Medical application of remedies so as to effect a cure : THERAPY.

**trea·ty** (trē′tē) *n., pl.* **-ties.** [ME *tretee* < AN *trete* < Lat. *tractus,* discussion < *tractare,* to handle. —see TREAT.] **1. a.** A formal agreement between two or more nations. **b.** A document in which such an agreement is set down. **2.** A contract or agreement. **3.** *Obs.* Negotiation in order to reach an agreement. **4.** *Obs.* An entreaty.

☆ ***syns:*** TREATY, ACCORD, AGREEMENT, CONCORD, CONVENTION, PACT *n. core meaning* : a formal, usu. written settlement between nations <a nuclear arms control *treaty*>

**treaty port** *n.* A port once kept open for foreign trade according to the terms of a treaty, esp. in China, Korea, and Japan.

**treb·le** (trĕb′əl) *adj.* [ME < OFr. < Lat. *triplus.*] **1.** Triple : threefold. **2.** *Mus.* Of, having, or performing the highest part, voice, or range. **3.** High-pitched : shrill. —*n.* **1.** *Mus.* **a.** The highest part, voice, instrument, or range : SOPRANO. **b.** A singer or player that performs this part. **2.** A high, shrill voice or sound. —*vt. & vi.* **-led, -ling, -les.** To make or become triple. **—treb′le·ness** *n.* **—treb′ly** (trĕb′lē) *adv.*

**treble clef** *n. Mus.* A symbol centered on the second line of the staff to indicate the position of G above middle C.

**treb·u·chet** (trĕb′yə-shĕt′) *also* **treb·uc·ket** (trĕb′ə-kĕt′) *n.* [ME < OFr. < *trebucher,* to overthrow : *tre-,* trans- + *buc,* trunk of the body.] A medieval catapult for hurling large stones at an enemy.

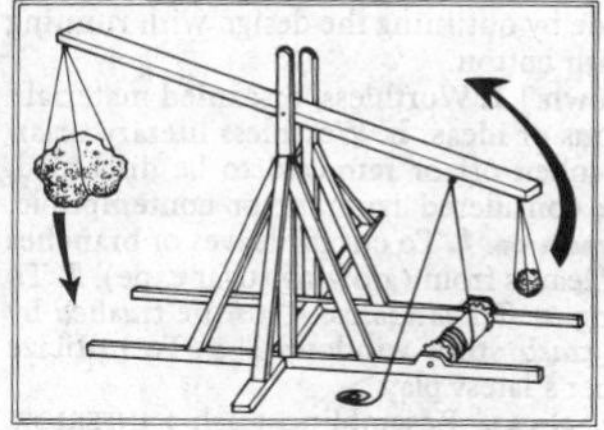
trebuchet

**tre·cen·to** (trā-chĕn′tō) *n.* [Ital., short for *milletrecento,* one thousand three hundred.] The 14th cent. period of art and literature.

**tree** (trē) *n.* [ME < OE *treow.*] **1.** A usu. tall, woody plant having comparatively great height and a single trunk. **2.** A plant, as a shrub, resembling a tree in form or size. **3.** A wooden beam, post, stake, or bar used as a part of a structure or framework. **4.** *Archaic.* A gallows. **5.** *often* **Tree.** *Archaic.* The cross on which Jesus was crucified. **6.** A saddletree. **7.** Something similar to a tree <a clothes *tree*> **8.** A diagram showing a family genealogy. —*vt.* **treed, tree·ing, trees. 1.** To force to climb a tree in evasion of pursuit <a cat *treed* by dogs> **2.** *Informal.* To force into a difficult position : CORNER. **3.** To stretch (shoes) on a shoetree. **—up a tree.** *Informal.* In a difficult position from which escape is difficult if not impossible. **—tree′like** *adj.*

**tree farm** *n.* Forested land on which trees are grown commercially.

**tree fern** *n.* A treelike tropical fern, esp. one of the family Cyatheaceae, having a woody trunklike stem and a terminal crown of large divided fronds.

**tree frog** *n.* A small arboreal frog of the genus *Hyla* and related genera, with long toes ending in adhesive disks.

**tree house** *n.* A structure built by or for children in the branches of a tree.

**tree line** *n.* **1.** The limit of northern or southern latitude beyond which trees cannot grow except as stunted forms. **2.** A timberline.

**tree·nail** or **tre·nail** (trēʹnālʹ, trĕnʹəl, trŭnʹəl) *also* **trun·nel** (trŭnʹəl) *n.* A wooden peg that swells when wet, used esp. in shipbuilding to fasten timbers.

**tree of heaven** *n.* The ailanthus.

**tree of knowledge** *n.* The tree in the Garden of Eden whose forbidden fruit Adam and Eve ate, causing loss of innocence.

**tree of life** *n.* **1.** ARBORVITAE 1. **2.** A tree in the Garden of Eden whose fruit, if eaten, gave human beings immortality.

**tree poppy** *n.* A shrub, *Dendromecon rigidum* of southern California, with evergreen foliage and showy, golden flowers.

**tree surgery** *n.* Professional treatment of diseased or damaged trees by filling cavities, pruning, or bracing weak limbs. **—tree surgeon** *n.*

**tree toad** *n.* A tree frog.

**tree·top** (trēʹtŏpʹ) *n.* The top of a tree.

**tref** (trāf) *adj.* [Yiddish *treyf* < Heb. *tĕrēphāh* < *tāraph,* he tore.] Unclean and unfit for human consumption according to Jewish dietary law.

**tre·foil** (trēʹfoilʹ, trĕfʹoilʹ) *n.* [ME < AN *trifoil* < Lat. *trifolium* : *ter,* three + *folium,* leaf.] **1.** Any of various plants of the genera *Trifolium, Lotus,* and related genera, having compound leaves with three leaflets. **2.** Something, as an ornament, symbol, or architectural form, that has the stylized appearance of a trifoliate leaf.

**tre·ha·la** (trĭ-häʹlə) *n.* [Turk. *tīqāla* < Pers. *tīghāl.*] A sugarlike, edible substance obtained from the pupal case of an Old World beetle, *Larinus maculatus.*

**tre·ha·lase** (trĭ-häʹlāsʹ) *n.* An enzyme that catalyzes the hydrolysis of trehalose.

**tre·ha·lose** (trĭ-häʹlōsʹ, -lōzʹ) *n.* A sweet-tasting crystalline disaccharide, $C_{12}H_{22}O_{11}{\cdot}2H_2O$, found in trehala and in many fungi that store it instead of starch.

**treil·lage** (trĕ-yäzhʹ, trāʹlĭj) *n.* [Fr. < OFr. *treille,* bower supported by trelliswork < Lat. *trichila,* bower.] Latticework, esp. a trellis for climbing vines.

**trek** (trĕk) *vi.* **trekked, trek·king, treks.** [Afr., to travel by ox wagon < Du. *trekken,* to travel.] **1.** To make a slow arduous journey. **2.** To travel by ox wagon in South Africa. *—n.* **1.** A journey or leg of a journey, esp. when slow and difficult. **2.** A migration. **3.** A journey by ox wagon in South Africa. **—trekʹker** *n.*

**trel·lis** (trĕlʹĭs) *n.* [ME *trelis* < OFr. < Lat. *trilix,* woven with three threads: *ter,* three + *licium,* thread.] **1.** A frame supporting open latticework, used for training vines and other climbing plants. **2.** An arch made with a trellis : ARBOR. *—vt.* **-lised, -lis·ing, -lis·es. 1.** To provide with a trellis, esp. to train (a climbing plant) on a trellis. **2.** To make in the form of a trellis : INTERWEAVE.

**trel·lis·work** (trĕlʹĭs-wûrkʹ) *n.* Latticework.

**trem·a·tode** (trĕmʹə-tōdʹ) *n.* [NLat. *Trematoda,* class name < Gk. *trēmatōdēs,* having holes < *trēma,* hole.] Any of numerous parasitic flatworms of the class Trematoda, with a thick outer cuticle, and one or more suckers for attaching to host tissue. *—adj.* Of or belonging to the Trematoda.

**trem·a·to·di·a·sis** (trĕmʹə-tō-dīʹə-sĭs) *n.* Infestation with trematodes.

**trem·ble** (trĕmʹbəl) *vi.* **-bled, -bling, -bles.** [ME *tremblen* < OFr. *trembler* < Lat. *tremulus,* tremulous < *tremere,* to tremble.] **1.** To shake involuntarily, as from fear, cold, or illness : QUIVER. **2.** To feel or express fear or anxiety <I *tremble* to think what they'll do next.> **3.** To vibrate : oscillate <leaves *trembling* in the strong wind> *—n.* **1.** The act or state of trembling. **2.** *often* **trembles.** A convulsive fit of trembling. **3. trembles** *(sing. in number). Med.* **a.** A viral encephalomyelitis of sheep. **b.** Poisoning of domestic animals, esp. cattle and sheep, caused by eating white snakeroot. **—tremʹbler** *n.* **—tremʹbling·ly** *adv.* **—tremʹbly** *adj.*

**tre·men·dous** (trĭ-mĕnʹdəs) *adj.* [Lat. *tremendus,* gerund. of *tremere,* to tremble.] **1.** Fearful : terrible <a *tremendous* family tragedy> **2. a.** Particularly large : ENORMOUS. **b.** *Informal.* Marvelous : wonderful <had a *tremendous* vacation> **—tre·menʹdous·ly** *adv.* **—tre·menʹdous·ness** *n.*

**trem·o·lite** (trĕmʹə-lītʹ) *n.* [Fr. *trémolite,* after *Tremola,* valley in Switzerland.] A white to dark-gray calcium magnesium amphibole, $Ca_2Mg_5Si_8O_{22}(OH)_2$, usu. occurring in aggregates, used as a substitute for asbestos and in paints and ceramics.

**trem·o·lo** (trĕmʹə-lōʹ) *n., pl.* **-los.** [Ital. < Lat. *tremulus,* tremulous.] *Mus.* **1. a.** A tremulous effect effected by the rapid repetition of a single tone. **b.** A similar effect caused by rapid alternation of two tones. **2.** A device on an organ for producing a tremolo. **3.** A vibrato in singing, used for emotional effect or symptomatic of poor vocal control.

**trem·or** (trĕmʹər) *n.* [ME, a trembling < OFr. *tremour* < Lat. *tremor* < *tremere,* to tremble.] **1.** A quick vibratory or shaking movement <measured earth *tremors*> **2.** Involuntary trembling of the body. **3.** A nervous quiver : THRILL. **4.** Nervous agitation or tension <all in a *tremor* before the game> **5. a.** A feeling of uncertainty or insecurity. **b.** A cause of uncertainty or insecurity. **6.** A tremulous sound : QUAVER.

**trem·u·lous** (trĕmʹyə-ləs) *adj.* [Lat. *tremulus* < *tremere,* to tremble.] **1.** Vibrating : quivering. **2.** Timid. **—tremʹu·lous·ly** *adv.* **—tremʹu·lous·ness** *n.*

**tre·nail** (trēʹnālʹ, trĕnʹəl, trŭnʹəl) *n. var. of* TREENAIL.

**trench** (trĕnch) *n.* [ME *trenche* < OFr. < *trenchier,* to cut < Lat. *truncare* < *truncus,* trunk.] **1.** A deep furrow. **2. a.** A ditch. **b.** A long, narrow, crooked ditch embanked with its own soil and used for protection and concealment in warfare. *—v.* **trenched, trench·ing, trench·es.** *—vt.* **1.** To cut or dig a trench in. **2.** To fortify with a trench. **3.** To put into a trench. **4. a.** To cut. **b.** To slash, sever, slice, or gash by cutting. *—vi.* **1.** To dig a trench or ditch. **2.** To cut, carve, or slash. **3.** To verge or encroach <*trenched* upon my land>

**trench·ant** (trĕnʹchənt) *adj.* [ME < OFr., cutting < *trenchier,* to cut. —see TRENCH.] **1.** Forcefully effective : VIGOROUS <*trenchant* arguments> **2. a.** Very perceptive : INCISIVE <a *trenchant* analysis of the problem> **b.** Caustic <*trenchant* wit> **3.** Distinct : clear-cut <*trenchant* patterns> **—trenchʹan·cy** *n.* **—trenchʹant·ly** *adv.*

**trench coat** *n.* A loose-fitting, belted raincoat with numerous pockets and flaps, suggestive of a military style.

**trench·er¹** (trĕnʹchər) *n.* [ME *trenchur* < AN *trenchour* < OFr. *trenchier,* to cut. —see TRENCH.] **1.** A wooden board or plate on which food is cut or served. **2.** One that carves meat.

**trench·er²** (trĕnʹchər) *n.* One that digs trenches.

**trench·er·man** (trĕnʹchər-mən) *n.* **1.** A hearty eater. **2.** *Archaic.* One who frequents another's table : HANGER-ON.

**trench fever** *n.* An acute infectious relapsing fever caused by a microorganism, *Rickettsia quintana,* and carried by a louse, *Pediculus humanus.*

**trench foot** *n.* A condition of the feet resembling frostbite, often afflicting soldiers, and caused by prolonged exposure to cold and dampness.

**trench mouth** *n.* A gingivitis marked by pain, foul odor, and formation of a gray film over the diseased area.

**trend** (trĕnd) *n.* [ME *trenden,* to roll < OE *trendan.*] **1.** A general inclination or tendency : DRIFT <a *trend* to smaller cars> **2.** A direction of movement : COURSE. *—vi.* **trend·ed, trend·ing, trends. 1.** To extend, bend, turn, or move in a specified direction <The prevailing wind *trends* west-northwest.> **2.** To have a general tendency : TEND <a society *trending* to socialism>

**trend·set·ter** (trĕndʹsĕtʹər) *n.* One that sets a trend.

**trend·y** (trĕnʹdē) *adj.* **-i·er, -i·est.** *Informal.* Of or being in accord with the latest fashion or fad <*trendy* attire> **—trendʹi·ly** *adv.* **—trendʹi·ness** *n.*

**trente et qua·rante** (träɴtʹ āʹ kä-räɴtʹ) *n.* [Fr. : *trente,* thirty + *et,* and + *quarante,* forty.] Rouge et noir.

**tre·pan¹** (trĭ-pănʹ) *n.* [ME *trepane,* surgical crown saw < Med. Lat. *trepanum* < Gk. *trupanon,* borer < *trupan,* to pierce < *trupē,* hole.] **1.** A rock-boring tool for sinking mine shafts. **2.** A trephine. *—vt.* **-panned, -pan·ning, -pans. 1.** To bore (a mine shaft) with a trepan. **2.** *Med.* To trephine. **—trepʹa·naʹtion** (trĕpʹə-nāʹshən) *n.*

**tre·pan²** (trĭ-pănʹ) *also* **tra·pan** (trə-pănʹ) [Orig. unknown.] *Archaic.* *—vt.* **-panned, -pan·ning, -pans.** To trap or ensnare. *—n.* **1.** A prankster : trickster. **2.** A trick : stratagem.

**tre·pang** (trĭ-păngʹ) *n.* [Malay *tĕripang.*] **1.** Any of several sea cucumbers of the genus *Holothuria,* of the southern Pacific and Indian oceans. **2.** The eviscerated, dried, or smoked body of the trepang, used in the Orient as an ingredient in soup.

**tre·phine** (trĭ-fīnʹ) *n.* [< Lat. *tres fines,* three ends.] A surgical instrument with circular sawlike edges, used for excising disks of bone, usu. from the skull. *—vt.* **-phined, -phin·ing, -phines.** To operate on with a trephine. **—trephʹi·naʹtion** (trĕfʹə-nāʹshən) *n.*

**trep·id** (trĕpʹĭd) *adj.* [Lat. *trepidus,* anxious.] Timid.

**trep·i·da·tion** (trĕpʹĭ-dāʹshən) *n.* [Lat. *trepidatio* < *trepidare,* to be in a state of confusion < *trepidus,* anxious.] **1.** Great alarm or dread <viewed the snake with *trepidation*> **2.** A quivering motion.

**trep·o·ne·ma** (trĕpʹə-nēʹmə) *n., pl.* **-ma·ta** (-mə-tə) *or* **-mas.** [NLat. *Treponema, Treponemat-,* treponeme genus.] A treponeme. **—trepʹo·neʹmal, trepʹo·nemʹa·tous** (-nĕmʹə-təs) *adj.*

**trep·o·ne·ma·to·sis** (trĕpʹə-nēʹmə-tōʹsĭs) *n., pl.* **-to·ses.** [TREPONEMAT(A) + -OSIS.] Infestation with treponemes.

**trep·o·neme** (trĕpʹə-nēmʹ) *n.* [NLat. *Treponema,* genus name : Gk. *trepein,* to turn + Gk. *nēma,* thread.] Any of a group of spirochetes of the genus *Treponema,* including those that cause syphilis and yaws.

**tres·pass** (trĕsʹpəs, -păsʹ) *vi.* **-passed, -pass·ing, -pass·es.** [ME *trespassen* < OFr. *trespasser* < Med. Lat. *transpassare* : Lat. *trans,* across + Med. Lat. *passare,* to pass < Lat. *passus,* step.] **1.** To commit a sin or offense : TRANSGRESS. **2.** To infringe upon the privacy, time, or attention of another. **3.** *Law.* To invade the property, rights, or person of another without his or her consent and with the actual or implied commission of violence, esp. to enter onto another's land illegally. —*n.* (trĕs′păs′, -pəs). **1.** Transgression of a law or duty. **2.** A transgression against another. **3.** *Law.* **a.** The act of trespassing. **b.** A legal suit brought for trespassing. **—tres′pass·er** *n.*

**tress** (trĕs) *n.* [ME *tresse* < OFr.] **1.** A lock of hair, esp. a long lock of a woman's hair. **2.** A plait or braid of hair. **3. tresses.** A woman's long hair, esp. when unbound.

**tres·tle** (trĕs′əl) *n.* [ME *trestel* < OFr. < Lat. *transtrum,* beam.] **1.** A horizontal beam or bar held up by two pairs of divergent legs and used as a support, as for a table. **2.** A framework composed of vertical, slanted supports and horizontal crosspieces holding up a bridge.

**trestle table** *n.* A table that is supported by trestles, esp. one having two or three supporting trestles connected by a longitudinal bar rather than legs.

**tres·tle·tree** (trĕs′əl-trē′) *n. Naut.* One of a pair of horizontal beams set into a masthead to support the crosstrees.

**tres·tle·work** (trĕs′əl-wûrk′) *n.* A trestle or system of trestles, as that supporting a bridge.

**trews** (tro͞oz) *pl.n.* [Sc. Gael. *triubhas.*] *(sing. in number).* Close-fitting trousers, usu. of tartan fabric.

**trey** (trā) *n.* [ME *treye* < OFr. *treie* < Lat. *tres,* three.] A card, die, or domino with three pips : THREE.

**tri-** *pref.* [ME < Lat. (< *tres,* three) and Gk. (< *treis,* three).] **1.** Three <*tri*lobate> **2. a.** Occurring at intervals of three <*tri*monthly> **b.** Occurring three times during <*tri*weekly>

**tri·a·ble** (trī′ə-bəl) *adj.* **1.** Capable of being tried or tested. **2.** *Law.* Subject to judicial examination. **—tri′a·ble·ness** *n.*

**tri·ac·id** (trī-ăs′ĭd) *adj.* **1.** Capable of reacting with three molecules of a monobasic acid. —Used of a base. **2.** Containing three replaceable hydrogen atoms. —Used of an acid or an acid salt. —*n.* An acid containing three replaceable hydrogen atoms.

**tri·ad** (trī′ăd′, -əd) *n.* [LLat. *trias, triad-* < Gk. < *treis,* three.] **1.** A group of three persons or things. **2.** *Mus.* A chord of three tones, esp. one built on a given root tone plus a major or minor third and a perfect fifth. **—tri·ad′ic** (trī-ăd′ĭk) *adj.*

**tri·age** (trē-äzh′) *n.* [Fr. < *trier,* to sort < OFr.] **1.** The screening and classification of wounded, sick, or injured patients during war or another disaster to determine priority needs and thereby ensure the most efficient use of medical and surgical manpower, equipment, and facilities. **2.** A system used to allocate a scarce commodity, as food, only to those capable of deriving the greatest benefit from it.

**tri·al** (trī′əl, trīl) *n.* [AN < OFr. *trier,* to try.] **1.** *Law.* Examination of evidence and applicable law by a competent tribunal to determine the issue of specified charges or claims. **2. a.** An operation or process intended to resolve an uncertainty : EXPERIMENT. **b.** A single complete instance of such experimentation, esp. as part of a series of tests. **3.** An effort or attempt. **4.** Pain or anguish caused by a difficult situation or condition. **5.** A source of vexation or distress that tests patience or endurance <As a child I was a *trial* to my parents.> —*adj.* **1.** Of or relating to a trial. **2.** Made, done, used, or performed during the course of a trial or trials.

**trial and error** *n.* An empirical method of establishing a satisfactory solution to a problem for which there is no existing or conveniently applicable theory, consisting of repeating experimental trials of various hypotheses until error is sufficiently reduced or eliminated.

**trial balance** *n.* A bookkeeping statement of all the open debit and credit items in a double-entry ledger made to test their equality.

**trial balloon** *n.* [From the use of balloons to test weather conditions.] A preliminary action or statement intended to test public reaction.

**trial jury** *n.* A petit jury.

**tri·a·logue** (trī′ə-lôg′, -lŏg′) *n.* [TRI- + (DI)ALOGUE.] A discourse or scene in which three individuals participate.

**trial run** *n.* An experimental test <We took a *trial run* in the new automobile.>

**tri·am·cin·o·lone** (trī′ăm-sĭn′ə-lōn′) *n.* [TRI- + AM(YL) + E. *cinene,* a terpene + E. *prednisolone,* a corticoid.] A white crystalline compound, $C_{21}H_{27}FO_6$, used in photographic developing and in the medical treatment of respiratory disorders.

**tri·an·gle** (trī′ăng′gəl) *n.* [ME < OFr. or < Lat. *triangulum* < *triangulus,* three-angled : *tri-* tri- + *angulus,* angle.] **1.** A plane geometric figure with three angles and three sides. **2.** Something shaped like a triangle. **3.** A flat, three-sided drawing or drafting guide, used to draw straight lines at specified angles. **4.** *Mus.* A percussion instrument composed of a piece of metal in the shape of a triangle open at one angle. **5.** A relationship among three persons, two of whom are in love with the third.

**tri·an·gu·lar** (trī-ăng′gyə-lər) *adj.* **1.** Of, relating to, or shaped like a triangle. **2.** Having a triangle for a base. **3.** Relating to, involving, or made up of three interrelated entities, as three persons or objects. **—tri·an′gu·lar′i·ty** (-lăr′ĭ-tē) *n.* **—tri·an′gu·lar·ly** *adv.*

**tri·an·gu·late** (trī-ăng′gyə-lāt′) *vt.* **-lat·ed, -lat·ing, -lates. 1.** To divide into triangles. **2.** To survey by triangulation. **3.** To make triangular. **4.** To measure by using trigonometry. —*adj.* (trī-ăng′gyə-lĭt). **1.** TRIANGULAR 1. **2.** Made up of or marked with triangles. **—tri·an′gu·late′ly** *adv.*

**tri·an·gu·la·tion** (trī-ăng′gyə-lā′shən) *n.* **1. a.** A surveying technique in which a region is divided into a series of triangular elements based on a line of known length so that accurate measurements of distances and directions may be made by the application of trigonometry. **b.** The network of triangles so laid out. **2.** Location of an unknown point, as in navigation, by forming a triangle having the unknown point and two known points as the vertices.

**Tri·an·gu·lum** (trī-ăng′gyə-ləm) *n.* [< Lat. *triangulum,* triangle.] A constellation in the Northern Hemisphere.

**Triangulum Aus·tra·le** (ô-strā′lē) *n.* [NLat.] A constellation in the Southern Hemisphere.

**tri·ar·chy** (trī′är′kē) *n., pl.* **-chies.** [Gk. *triarkhia* : *tri-,* tri- + *-arkhia,* -archy.] **1.** Government by three leaders : TRIUMVIRATE. **2.** A country governed by a triarchy.

**Tri·as·sic** (trī-ăs′ĭk) *adj.* [< LLat. *trias,* triad (from the subdivision of this period into three parts).] Of, belonging to, or designating the geologic time, system of rocks, and sedimentary deposits of the first period of the Mesozoic era, after the Permian period of the Paleozoic era and before the Jurassic period of the Mesozoic era. —*n.* The Triassic period or system of deposits.

**tri·a·tom·ic** (trī′ə-tŏm′ĭk) *adj.* **1.** Having three atoms per molecule. **2.** Having three replaceable atoms or radicals.

**tri·ax·i·al** (trī-ăk′sē-əl) *adj.* Involving or having three axes. **—tri·ax′i·al′i·ty** (-ăl′ĭ-tē) *n.*

**tri·a·zine** (trī′ə-zēn′, trī-ăz′ēn′) *n.* **1.** Any of three isomeric compounds, $C_3H_3N_3$, each composed of three carbon and three nitrogen atoms in a six-membered ring. **2.** A compound derived from triazine.

**tri·a·zole** (trī′ə-zōl′, trī-ăz′ōl′) *n.* Any of several compounds with composition $C_2H_3N_3$ having a five-membered ring of two carbon atoms and three nitrogen atoms.

**trib·al** (trī′bəl) *adj.* Of the nature of or relating to a tribe <*tribal* rites> **—trib′al·ly** *adv.*

**trib·al·ism** (trī′bə-lĭz′əm) *n.* **1.** Organization, culture, or beliefs of a tribe. **2.** Tribal spirit.

**tri·ba·sic** (trī-bā′sĭk) *adj.* **1.** Containing three replaceable hydrogen atoms per molecule. —Used of acids. **2.** Containing three univalent basic atoms or radicals per molecule. —Used of bases or salts.

**tribe** (trīb) *n.* [ME < OFr. *tribu* or < Lat. *tribus.*] **1.** A system of social organization comprising several local villages, bands, districts, lineages, or other groups and sharing a common ancestry, culture, language, and name. **2.** A political, ethnic, or ancestral division of ancient states and cultures, esp.: **a.** Any of the three divisions of the ancient Romans. **b.** Any of the 12 divisions of ancient Israel. **c.** An ancient Greek phyle. **3.** A group of persons sharing a common occupation, interest, or habit. **4.** *Informal.* A large family. **5.** *Biol.* A taxonomic category occas. placed between a family and a genus.

**tribes·man** (trībz′mən) *n.* A member of a tribe.

**tri·bo·e·lec·tric·i·ty** (trī′bō-ĭ-lĕk-trĭs′ĭ-tē, -ē′lĕk-, trĭb′ō-) *n.* [Gk. *tribein,* to rub + ELECTRICITY.] An electrical charge produced by friction between two objects. **—tri′bo·e·lec′tric** *adj.*

**tri·bol·o·gy** (trī-bŏl′ə-jē, trĭb-ŏl′-) *n.* [Gk. *tribein,* to rub + -LOGY.] The science of the mechanisms of friction, lubrication, and wear of interacting surfaces that are in relative motion. **—tri′bo·log′i·cal** (trī′bə-lŏj′ĭ-kəl, trĭb′ə-) *adj.* **—tri·bol′o·gist** *n.*

**tri·brach** (trī′brăk′) *n.* [Lat. *tribrachys* < Gk. *tribrakhus* : *tri-,* tri- + *brakhus,* short.] A metrical foot in prosody consisting of three short or unstressed syllables.

**tri·bro·mo·eth·a·nol** (trī-brō′mō-ĕth′ə-nôl′, -nōl′) *n.* A white crystalline compound, $CBr_3CH_2OH$, with a slight aromatic odor and taste, used as a basal anesthetic.

**trib·u·la·tion** (trĭb′yə-lā′shən) *n.* [ME *tribulacioun* < OFr. < LLat. *tribulatio* < *tribulare,* to oppress < Lat. *tribulum,* threshing-sledge.] **1.** Great trial, affliction, or distress. **2.** A cause of great distress.

**tri·bu·nal** (trī-byo͞o′nəl, trĭ-) *n.* [Lat. < *tribunus,* tribune.] **1.** A court of justice <a military *tribunal*> **2.** The platform or seat on which a judge or other presiding court officer sits. **3.** One empowered to determine or judge.

**trib·u·nate** (trĭb′yə-nāt′, trĭ-byo͞o′nĭt) *n.* The rank, office, dignity, or authority of a tribune.

**trib·une**[1] (trĭb′yo͞on′, trĭ-byo͞on′) *n.* [ME < Lat. *tribunus* < *tribus,* tribe.] **1.** An ancient Roman official chosen by the plebs to protect their rights against the patricians. **2.** A champion of the people. **—trib′u·nar′y** (trĭb′yə-nĕr′ē) *adj.*

**trib·une**[2] (trĭb′yo͞on′, trĭ-byo͞on′) *n.* [Fr. < OItal. *tribuna* < Med. Lat. *tribuna,* var. of Lat. *tribunal.*] A raised platform or dais from which a speaker addresses an audience.

**trib·u·tar·y** (trĭb′yə-tĕr′ē) *adj.* [ME *tributarye,* of paying tribute < Lat. *tributarius* < *tributum,* tribute. —see TRIBUTE.] **1.** Making additions or providing supplies : CONTRIBUTORY. **2.** Of the nature of tribute <*tributary* payments> **3.** Paying tribute <*tributary* colonies> —*n., pl.* **-ies. 1.** A river or stream flowing into a larger river or stream. **2.** A payer of tribute.

**trib·ute** (trĭb′yo͞ot) *n.* [ME *tribut* < Lat. *tributum* < neuter p.part. of *tribuere,* to assign < *tribus,* tribe.] **1.** A gift or testimonial expressing gratitude, respect, or admiration. **2. a.** Money or other valuables paid by one ruler or nation to another as acknowledgment of submission or as the price for protection by that nation. **b.** A payment made for protection <shopkeepers paying *tribute* to local gangs> **3. a.** A payment or tax given by a feudal vassal to his overlord. **b.** The obligation involved in such a payment.

**tri·car·box·yl·ic** (trī′kär-bŏk-sĭl′ĭk) *adj.* Containing three carboxyl groups.

**tricarboxylic acid cycle** *n.* The Krebs cycle.

**trice** (trīs) *n.* [ME *tryse* < *trisen,* to hoist < MDu.] A very short time <I'll be there in a *trice.*> —*vt.* **triced, tric·ing, tric·es.** *Naut.* To hoist and secure (e.g., a sail) **:** LASH.

**tri·cen·ten·ni·al** (trī′sĕn-tĕn′ē-əl) *adj.* Tercentenary. —*n.* A tercentenary event or celebration.

**tri·ceps** (trī′sĕps′) *n.* [< Lat., three-headed : *tri-,* tri- + *caput,* head.] A large three-headed muscle running along the back of the upper arm and functioning to extend the forearm.

**tri·cer·a·tops** (trī-sĕr′ə-tŏps′) *n.* [NLat. *Triceratops,* genus name : *tri-,* tri- + Gk. *keras,* horn + *ops,* face.] A horned herbivorous dinosaur of the genus *Triceratops* of the Cretaceous period, having a bony plate covering the neck.

**trich-** *pref. var. of* TRICHO-.

**tri·chi·a·sis** (trĭ-kī′ə-sĭs) *n.* [LLat. < Gk. *trikhiasis* < *trikhian,* to be hairy.] A condition of ingrowing hairs about an orifice, esp. ingrowing eyelashes.

**tri·chi·na** (trĭ-kī′nə) *n., pl.* **-nae** (-nē) *or* **-nas.** [NLat. < Gk. *trikhinos,* hairy < *thrix,* hair.] A parasitic nematode worm, *Trichinella spiralis,* infesting the intestines of mammals, and having larvae that move through the blood vessels and become encysted in muscles.

**trich·i·nize** (trĭk′ə-nīz′) *vt.* **-nized, -niz·ing, -niz·es.** To infest with trichinae. **—trich′i·ni·za′tion** *n.*

**trich·i·no·sis** (trĭk′ə-nō′sĭs) *n.* A disease caused by eating inadequately cooked pork containing trichinae, and marked by intestinal disorders, fever, muscular swelling, pain, and insomnia.

**tri·chi·nous** (trĭ-kī′nəs, trĭk′ə-nəs) *adj.* **1.** Containing trichinae. **2.** Of or pertaining to trichinae or trichinosis.

**trich·ite** (trĭk′īt′) *n.* [G. *Trichit* < Gk. *thrix,* hair.] A small needle-shaped filament or crystal. **—tri·chit′ic** (trĭ-kĭt′ĭk) *adj.*

**tri·chlor·fon** (trī-klôr′fŏn′, -klōr′-) *n.* [TRI- + CHLOR(O)- + *-fon* (< PHOSPHONATE).] A colorless crystalline compound, $C_4H_8Cl_3O_4P$, used as an agricultural insecticide.

**tri·chlo·ride** (trī-klôr′īd′, -klōr′-) *also* **tri·chlo·rid** (-klôr′ĭd, -klōr′-) *n.* A compound having three chlorine atoms per molecule.

**tri·chlo·ro·a·ce·tic acid** (trī-klôr′ō-ə-sē′tĭk, -klōr′-) *n.* A colorless, deliquescent, corrosive, crystalline compound, $CCl_3COOH$, used as a herbicide and topically as an astringent and antiseptic.

**tri·chlo·ro·eth·yl·ene** (trī-klôr′ō-ĕth′ə-lēn′, -klōr′-) *also* **tri·chlor·eth·yl·ene** (trī′klôr-ĕth′ə-lēn′, -klōr-) *n.* A heavy, colorless, toxic liquid, $CHCl{:}CCl_2$, used to degrease metals, as an extraction solvent for oils and waxes, as a refrigerant, in dry cleaning, and as a fumigant.

**tricho-** *or* **trich-** *pref.* [Gk. *trikho-* < *thrix,* hair.] Hair **:** thread **:** filament <*tricho*cyst>

**trich·o·cyst** (trĭk′ə-sĭst′) *n.* One of the minute capsulelike bodies in the outer cytoplasm of certain protozoans, capable of ejecting a threadlike or bristlelike extension. **—trich′o·cys′tic** *adj.*

**trich·o·gyne** (trĭk′ə-jīn′, -gīn′) *n.* A receptive filament of the female reproductive structure of certain fungi or algae.

**trich·oid** (trĭk′oid′, trī′koid′) *adj.* [Gk. *trikhoeidēs* : *trikho-,* tricho- + *-oeidēs,* -oid.] Like hair.

**trich·ome** (trĭk′ōm′, trī′kōm′) *n.* [G. *Trichom* < Gk. *trikhoma,* growth of hair < *trikhoun,* to cover with hair < *thrix,* hair.] A hairlike or bristlelike outgrowth, as from the epidermis of a plant. **—tri·chom′ic** (trĭ-kŏm′ĭk, -kō′mĭk, trī-) *adj.*

**trich·o·mo·nad** (trĭk′ə-mō′năd′) *n.* [NLat. *Trichomonas, Trichomonad-,* genus name : TRICHO- + LLat. *monas,* unit. —see MONAD.] Any of various flagellate protozoans of the genus *Trichomonas,* occurring in the digestive and urogenital tracts of vertebrates. **—trich′o·mo·nad′al, trich′o·mon′al** *adj.*

**trich·o·mo·ni·a·sis** (trĭk′ə-mə-nī′ə-sĭs) *n., pl.* **-ses** (-sēz′). **1.** A vaginal infection caused by a protozoan, *Trichomonas vaginalis,* and resulting in a persistent discharge, inflammation, and discomfort. **2.** An infection caused by trichomonads.

**tri·chop·ter·an** (trĭ-kŏp′tər-ən) *n.* [NLat. *Trichoptera,* order name.] An insect of the order Trichoptera, including the caddis fly.

**tri·cho·sis** (trĭ-kō′sĭs) *n.* [NLat. < Gk. *trikhōsis,* growth of hair < *trikhoun,* to cover with hair. —see TRICHOME.] Disease of the hair.

**tri·chot·o·my** (trī-kŏt′ə-mē) *n., pl.* **-mies.** [Gk. *trikhia,* in three parts + -TOMY.] Division into three parts, esp. the theological division of human beings into body, soul, and spirit. **—tri·chot′o·mous** *adj.* **—tri·chot′o·mous·ly** *adv.*

**-trichous** *suff.* [Gk. *-trikhos* < *thrix,* hair.] Having a specified kind of hair or hairlike part <*peri*trichous>

**tri·chro·ism** (trī′krō-ĭz′əm) *n.* [Gk. *trikhroos,* three-colored : *tri-,* three + *khrōs,* color.] The property possessed by certain minerals of exhibiting three different colors when illuminated by white light and viewed from three different directions. **—tri·chro′ic** *adj.*

**tri·chro·mat·ic** (trī′krō-măt′ĭk) *also* **tri·chrome** (trī′krōm′) *or* **tri·chro·mic** (trī-krō′mĭk) *adj.* **1.** Of, pertaining to, or composed of three colors, as in photography or printing. **2.** Having visual perception of the three primary colors, as in normal vision. **—tri·chro′ma·tism** (trī-krō′mə-tĭz′əm) *n.*

**trich·u·ri·a·sis** (trĭk′yə-rī′ə-sĭs) *n., pl.* **-ses.** [NLat. *Trichuris,* genus name (TRICH(O)- + Gk. *oura,* tail) + -IASIS.] Infestation of the large intestine with whipworms of the genus *Trichuris.*

**trick** (trĭk) *n.* [ME *trik* < ONFr. *trique* < *trikier,* to deceive.] **1.** An indirect, often deceptive or fraudulent means of achieving an end. **2.** A mischievous act **:** PRANK. **3.** A stupid, disgraceful, or childish act. **4.** A peculiar trait **:** MANNERISM <"Mimicry is the *trick* by which a moth comes to look like a wasp" —Marston Bates> **5.** A special skill **:** KNACK <Is there a *trick* to working this puzzle?> **6.** A feat of magic or legerdemain. **7.** A difficult, dexterous, or clever act designed to amuse or impress. **8.** All the cards played in a single round. **9.** A period or turn of duty **:** SHIFT. **10.** *Slang.* A prostitute's customer. —*v.* **tricked, trick·ing, tricks.** —*vt.* **1.** To cheat or deceive. **2.** To adorn <*tricked* out in feathers and rhinestones> —*vi.* To practice deception or trickery. —*adj.* Weak, defective, or liable to fail <a *trick* elbow> **—do** (*or* **turn**) **the trick.** To bring about a desired result. **—trick′er** *n.*

**trick·er·y** (trĭk′ə-rē) *n., pl.* **-ies.** Deception by stratagem.

**trick·ish** (trĭk′ĭsh) *adj.* Marked by or apt to use tricks or trickery. **—trick′ish·ly** *adv.* **—trick′ish·ness** *n.*

**trick·le** (trĭk′əl) *v.* **-led, -ling, -les.** [ME *triklen.*] —*vi.* **1.** To fall or flow in drops or droplets or in a thin stream. **2.** To move or advance slowly or bit by bit <The crowd *trickled* into the hall.> —*vt.* To cause to trickle. —*n.* **1.** The act or condition of trickling. **2.** A slow, small, or irregular quantity that moves, proceeds, or occurs intermittently.

**trick·le-down** (trĭk′əl-doun′) *adj.* Of or relating to the trickle-down theory of economics.

**trickle-down theory** *n.* An economic theory that financial benefits accorded to big-business enterprises will in turn pass down to smaller businesses and consumers.

**trick·ster** (trĭk′stər) *n.* One that plays tricks **:** SWINDLER.

**trick·y** (trĭk′ē) *adj.* **-i·er, -i·est. 1.** Given to or marked by trickery **:** WILY. **2.** Requiring skill or caution <a *tricky* pastry recipe> **—trick′i·ly** *adv.* **—trick′i·ness** *n.*

**tri·clin·ic** (trī-klĭn′ĭk) *adj.* Having three unequal axes intersecting at oblique angles. —Used of certain crystals.

**tri·clin·i·um** (trī-klĭn′ē-əm) *n., pl.* **-i·a** (-ē-ə) [Lat. < Gk. *triklinion,* dim. of *triklinos,* room with three couches : *tri-,* tri- + *klinē,* couch.] **1.** A couch surrounding three sides of a table, used by the ancient Romans for reclining at meals. **2.** A room containing a triclinium.

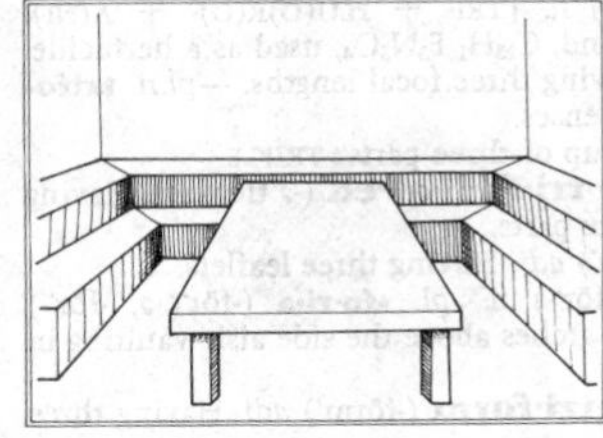

**triclinium**

**tri·col·or** (trī′kŭl′ər) *n.* **1.** A three-color flag. **2. Tricolor.** The French flag. **—tri′col′ored** *adj.*

**tri·corn** *also* **tri·corne** (trī′kôrn′) *n.* [Fr. *tricorne* or < Lat. *tricornis,* three-horned : *tri-,* tri + *cornu,* horn.] A hat whose brim is turned up on three sides.

**tri·cor·nered** (trī′kôr′nərd) *adj.* Having three corners.

**tri·cos·tate** (trī-kŏs′tāt′) *adj.* Having three costae or riblike ridges.

**tri·cot** (trē′kō) *n.* [Fr. < *tricoter,* to knit.] **1.** A plain, warp-knitted cloth. **2.** A soft ribbed cloth of wool or a wool blend.

**tric·o·tine** (trĭk′ə-tēn′, trē′kə-) *n.* [Fr. < *tricot,* tricot.] A sturdy worsted fabric with a double twill.

**tri·cot·y·le·do·nous** (trī′kŏ-tə-lēd′n-əs) *adj.* Having three cotyledons <*tricotyledonous* plants>

**tri·crot·ic** (trī-krŏt′ĭk) *adj.* [Gk. *trikrotos,* having a triple beat : *tri-,* tri- + *krotein,* to beat.] *Med.* Having three waves or elevations to one beat of the pulse. **—tri′cro·tism** (trī′krə-tĭz′əm) *n.*

**tri·cus·pid** (trī-kŭs′pĭd) *also* **tri·cus·pi·dal** (-pĭ-dəl) *adj.* [Lat. *tricuspis, tricuspid-,* having three points : *tri-,* tri- + *cuspis,* point.] **1.** Having three points or cusps, as a molar tooth. **2.** Of or relating to the tricuspid heart valve. —*n.* A tricuspid organ or part, esp. a tooth.

**tricuspid valve** *n.* The three-segmented heart valve that prevents the blood from flowing back from the right ventricle into the right atrium.

**tri·cy·cle** (trī′sĭk′əl, -sĭ-kəl) *n.* A three-wheeled vehicle usu. propelled by pedals.

**tri·dac·tyl** (trī-dăk′təl) *also* **tri·dac·ty·lous** (-tə-ləs) *adj.* [Gk. *tridaktulos,* three-fingered : *tri-,* tri- + *daktulos,* finger.] Having three toes, claws, or similar parts.

**tri·dent** (trīd′nt) *n.* [Lat. *tridens, trident-* : *tri-,* tri- + *dens,* tooth.] A long, three-pronged fork or weapon, esp. the three-pronged spear carried by Neptune or Poseidon. **—tri·den′tate′** (trī-dĕn′tāt′) *adj.*

**Tri·den·tine** (trī-dĕn′tīn′, -tēn′) *adj.* [Med. Lat. *Tridentinus* < *Tridentum,* Trent, a city in Italy.] Of or pertaining to a council held by the Roman Catholic Church in Trent, Italy, from 1545 to 1563.

**tri·di·men·sion·al** (trī′dĭ-mĕn′shə-nəl) *adj.* Of, relating to, or having three dimensions.

**tri·e·cious** (trī-ē′shəs) *adj. var. of* TRIOECIOUS.

**tried** (trīd) *adj.* [< p.part. of TRY.] **1.** Thoroughly tested and proved to be good or reliable. **2.** Made to undergo trials or distress <a much-*tried* teacher>

**tried and true** *adj.* TRIED 1.

**tri·en·ni·al** (trī-ĕn′ē-əl) *adj.* [Lat. *triennis* < *triennium,* triennium.] **1.** Happening every third year. **2.** Lasting three years. *—n.* **1.** A third anniversary. **2.** An event, as a celebration, occurring every three years. **—tri·en′ni·al·ly** *adv.*

**tri·en·ni·um** (trī-ĕn′ē-əm) *n., pl.* **-ni·ums** or **-ni·a** (-ē-ə) [Lat. : *tri-,* tri- + *annus,* year.] A three-year period.

**tri·er·arch** (trī′ə-rärk′) *n.* [Lat. *trierarchus* < Gk. *triērarkhos* : *triērēs,* trireme + *-arkhos,* -arch.] **1.** The captain of a Greek trireme. **2.** An Athenian who equipped and maintained a trireme as a part of his civic duties.

**tri·er·ar·chy** (trī′ə-rär′kē) *n., pl.* **-chies. 1.** The authority or office of a trireme commander. **2.** The ancient Athenian system whereby individual citizens equipped and maintained triremes as a part of their public duty.

**tri·fa·cial** (trī-fā′shəl) *adj.* Trigeminal.

**tri·fec·ta** (trī-fĕk′tə) *n.* [TRI- + (PER)FECTA.] A system of betting in which the bettor must pick the first three winners in the correct sequence.

**tri·fid** (trī′fĭd′) *adj.* [Lat. *trifidus,* split into three : *tri-,* tri- + *findere,* to split.] Divided or cleft into three narrow parts or lobes.

**tri·fle** (trī′fəl) *n.* [ME < OFr. *trufle,* trickery.] **1.** Something unimportant or of little value. **2.** A small amount : JOT. **3.** A dessert consisting of sponge cake spread with jam, soaked in wine, sprinkled with crushed macaroons, and topped with custard and whipped cream. **4. a.** A moderately hard pewter. **b. trifles.** Utensils made from moderately hard pewter. *—v.* **-fled, -fling, -fles.** *—vi.* **1.** To deal with something as if it were of little importance or value. **2.** To act, perform, or speak with little seriousness or purpose : JEST. **3.** To play or toy with something <*trifle* with someone's affections> *—vt.* To waste (e.g., time). **—a trifle.** Somewhat <*a trifle* miserly> **—tri′fler** (trī′flər) *n.*

**tri·fling** (trī′flĭng) *adj.* **1.** Of little importance : INSIGNIFICANT. **2.** Marked by frivolity or idleness. **—tri′fling·ly** *adv.*

**tri·flu·ra·lin** (trī-flo͝or′ə-lĭn) *n.* [TRI- + FLU(O)R(O)- + A(NI)LIN(E).] A crystalline compound, $C_{13}H_{16}F_3N_3O_4$, used as a herbicide.

**tri·fo·cal** (trī-fō′kəl) *adj.* Having three focal lengths. *—pl.n.* **trifocals.** Eyeglasses with trifocal lenses.

**tri·fold** (trī′fōld′) *adj.* Made up of three parts : TRIPLE.

**tri·fo·li·ate** (trī-fō′lē-ĭt) *also* **tri·fo·li·at·ed** (-ā′tĭd) *adj.* Having three leaves, leaflets, or leaflike parts.

**tri·fo·li·o·late** (trī-fō′lē-ə-lāt′) *adj.* Having three leaflets.

**tri·fo·ri·um** (trī-fôr′ē-əm, -fōr′-) *n., pl.* **-fo·ri·a** (-fôr′ē-ə, -fōr′-) [Orig. unknown.] A gallery of arches above the side aisle vaulting in a church nave.

**tri·formed** (trī′fôrmd′) *also* **tri·form** (-fôrm′) *adj.* Having three different forms or parts.

**tri·fur·cate** (trī-fûr′kĭt, -kāt′, trī′fər-kāt′) *also* **tri·fur·cat·ed** (trī′fər-kā′tĭd) *adj.* That has three forks or branches <*trifurcate* tree limbs> **—tri′fur·ca′tion** *n.*

**trig**[1] (trĭg) *adj.* [ME, true < ON *tryggr.*] **1.** Neat and trim : TIDY. **2.** Being in good condition. *—vt.* **trigged, trig·ging, trigs.** To make neat or trim, esp. in dress. **—trig′ly** *adv.* **—trig′ness** *n.*

**trig**[2] (trĭg) *vt.* **trigged, trig·ging, trigs.** [Perh. of Scand. orig.] **1.** To stop (a wheel) from rolling, as with a wedge. **2.** To prop up : SUPPORT. *—n.* A braking device, as a wedge.

**trig**[3] (trĭg) *n.* Trigonometry.

**tri·gem·i·nal** (trī-jĕm′ə-nəl) *adj.* Relating to the trigeminus.

**trigeminal neuralgia** *n.* An intensely painful inflammation of the facial area around the trigeminal nerve.

**tri·gem·i·nus** (trī-jĕm′ə-nəs) *n., pl.* **-ni** (-nī′) [NLat. < Lat. threefold : *tri-,* tri- + *geminus,* twin.] The chief facial sensory nerve and the motor nerve of the masticatory muscles.

**trig·ger** (trĭg′ər) *n.* [Du. *trekker* < MDu. *trecker* < *trecken,* to pull.] **1.** The lever pressed by the finger in discharging a firearm. **2.** A device used to release or activate a mechanism. **3.** An event that precipitates other events. *—vt.* **-gered, -ger·ing, -gers.** To initiate : set off <heated words that *triggered* a family quarrel>

**trig·ger·fish** (trĭg′ər-fĭsh′) *n., pl.* **triggerfish** or **-fish·es.** Any of various brightly colored fishes of the family Balistidae, of warm coastal seas, having a sharp erectile dorsal spine.

**trig·ger-hap·py** (trĭg′ər-hăp′ē) *adj. Slang.* Apt to react violently at the slightest provocation.

**trig·ger·man** (trĭg′ər-mən) *n.* An underworld gunman who shoots the victim in a planned murder.

**tri·glyc·er·ide** (trī-glĭs′ə-rīd′) *n.* An ester of three fatty acids and glycerol.

**tri·glyph** (trī′glĭf′) *adj.* [Lat. *triglyphus* < Gk. *trigluphos* : *tri-,* tri- + *gluphē,* carving < *gluphein,* to carve.] An ornament in a Doric frieze composed of a projecting block with three parallel vertical channels on the face. **—tri·glyph′ic** *adj.*

**tri·gon** (trī′gŏn′) *n.* [Lat. *trigonum* < Gk. *trigōnon* < *trigōnos,* triangular.] **1.** An ancient Greek or Roman triangular lyre or harp. **2.** TRIPLICITY 3. **3.** *Obs.* A triangle.

**trigonometric function** *n.* A function of an angle expressed as the ratio of two of the sides of a right triangle that contains the angle.

**trig·o·nom·e·try** (trĭg′ə-nŏm′ĭ-trē) *n.* [Gk. *trigōnon,* triangle + Gk. *-metria,* -metry.] The study of the properties and applications of trigonometric functions. **—trig′o·no·met′ric** (-nə-mĕt′rĭk), **trig′o·no·met′ri·cal** *adj.* **—trig′o·no·met′ri·cal·ly** *adv.*

**tri·he·dral** (trī-hē′drəl) *adj.* Formed by the plane surfaces of a trihedron. *—n.* A trihedron.

**tri·he·dron** (trī-hē′drən) *n., pl.* **-drons** or **-dra** (-drə). A figure formed by the intersection of three noncoplanar lines.

**tri·hy·brid** (trī-hī′brĭd) *n. Genetics.* An individual that is heterozygous for three pairs of genes.

**tri·lat·er·al** (trī-lăt′ər-əl) *adj.* [Lat. *trilaterus* : *tri-,* tri- + *latus,* side.] Having three sides. **—tri·lat′er·al·ly** *adv.*

**tril·by** (trĭl′bē) *n., pl.* **-bies.** [So called because such a hat was worn in the original London stage production of the novel *Trilby* by George du Maurier (1834–1896).] *Chiefly Brit.* A soft felt hat with a deeply creased crown.

**tri·lin·e·ar** (trī-lĭn′ē-ər) *adj.* Pertaining to, having, or bounded by three lines.

**tri·lin·gual** (trī-lĭng′gwəl) *adj.* Having, expressed in, or able to use three languages.

**tri·lit·er·al** (trī-lĭt′ər-əl) *adj.* Composed of three letters. —Used chiefly of consonantal roots in Semitic languages. *—n.* A three-letter word or word element.

**trill** (trĭl) *n.* [Ital. *trillo* < *trillare,* to trill.] **1.** A tremulous sound : WARBLE. **2.** *Mus.* **a.** The rapid alternation of two tones either a whole or a half tone apart. **b.** A vibrato. **3. a.** A rapid vibration of one speech organ against another, as of the tongue against the alveolar ridge in Spanish *rr.* **b.** A speech sound pronounced with such a vibration. *—v.* **trilled, trill·ing, trills.** *—vt.* **1.** To sound, sing, or play with a trill. **2.** To articulate (a sound) with a trill. *—vi.* To produce or emit a trill.

**tril·lion** (trĭl′yən) *n.* [Fr. : *tri-,* third power + *(m)illion,* million.] **1.** The cardinal number equal to $10^{12}$. **2.** *Chiefly Brit.* The cardinal number equal to $10^{18}$. **—tril′lion** *adj.*

**tril·lionth** (trĭl′yənth) *n.* **1.** The ordinal number matching the number one trillion in a series. **2.** One of a trillion equal parts. **—tril′lionth** *adj. & adv.*

**tril·li·um** (trĭl′ē-əm) *n.* [NLat. *Trillium,* genus name < Swed. *trilling,* triplet (from its three leaves).] Any of various plants of the genus *Trillium* of North America and eastern Asia, usu. having a single whorl of three leaves, and a variously colored, three-petaled flower.

**tri·lo·bate** (trī-lō′bāt′) *or* **tri·lo·bat·ed** (-lō′bā′tĭd) *also* **tri·lobed** (trī′lōbd′) *adj.* Having three lobes <a *trilobate* leaf>

**tri·lo·bite** (trī′lə-bīt′) *n.* [NLat. *Trilobites,* division name < Gk. *trilobos,* three-lobed : *tri-,* tri- + *lobos,* lobe.] Any of numerous extinct marine arthropods of the class Trilobita of the Paleozoic era, having a segmented exoskeleton divided by grooves or furrows into three longitudinal lobes. **—tri′lo·bit′ic** (-bĭt′ĭk) *adj.*

**tri·loc·u·lar** (trī-lŏk′yə-lər) *adj.* Having three chamberlike cavities or divisions.

**tril·o·gy** (trĭl′ə-jē) *n., pl.* **-gies.** [Gk. *trilogia* : *tri,* tri- + *-logia,* -logy.] A group of three literary or dramatic works that are related in subject matter or theme.

**trim** (trĭm) *v.* **trimmed, trim·ming, trims.** [Perh. < ME **trimmen* < OE *trymman,* to arrange.] *—vt.* **1.** To make neat by clipping, smoothing, or pruning. **2.** To remove (excess) by cutting <*trimmed* the budget> **3.** To ornament : decorate <*trim* a Christmas tree> **4.** *Informal.* **a.** To defeat soundly : THRASH. **b.** To cheat. **5.** *Naut.* **a.** To adjust (the sails and yards) for proper wind reception. **b.** To balance (a ship) by shifting the cargo or contents. **6.** To balance (an aircraft) in flight by regulating the control surfaces and tabs. **7.** To furnish or equip. *—vi.* **1.** *Naut.* **a.** To be in or retain equilibrium. **b.** To make sails and yards ready for sailing. **2. a.** To maintain cautious neutrality. **b.** To fashion one's views for momentary popularity or advantage. *—n.* **1.** State of order, arrangement, or appearance : CONDITION <in good *trim*> **2. a.** Exterior ornamentation, as moldings or framework. **b.** Ornamentation, as for clothing. **3.** WINDOW-

DRESSING 1b. **4.** Rejected or excised material. **5.** Personal quality : CHARACTER. **6.** *Naut.* **a.** The readiness of a vessel for sailing with regard to ballast, sails, and yards. **b.** The balance of a ship. **c.** The difference between the draft at the bow and at the stern. **7.** The position of an aircraft relative to its horizontal axis. *—adj.* **trim·mer, trim·mest. 1.** In good or neat order. **2.** Having lines, edges, or forms of neat and pleasing simplicity <a *trim* figure><a *trim* sailboat> *—adv.* In a trim way. **—trim'ly** *adv.* **—trim'ness** *n.*

**tri·ma·ran** (trī'mə-răn') *n.* [TRI- + (CATA)MARAN.] A fast sailboat equipped with three parallel hulls.

**tri·mer** (trī'mər) *n.* [TRI- + Gk. *meros,* part.] A polymeric compound made up of three identical monomeric molecules. **—tri·mer'ic** (-mĕr'ĭk) *adj.*

**trim·er·ous** (trĭm'ər-əs) *adj.* [NLat. *trimerus* < Gk. *trimerēs* : *tri-*, tri- + *meros,* part.] **1.** Having three similar segments or parts. **2.** *Bot.* Having flower parts, as petals, sepals, and stamens, in sets of three. **—trim'er·ism** *n.*

**tri·mes·ter** (trī-mĕs'tər, trī'mĕs'tər) *n.* [Fr. *trimestre* < Lat. *trimestris,* of three months : *tri-*, tri- + *mensis,* month.] **1.** A stage or period of three months. **2.** One of three equal academic terms in some colleges and universities. **—tri·mes'tral** (-trəl), **tri·mes'tri·al** (-trē-əl) *adj.*

**trim·e·ter** (trĭm'ĭ-tər) *n.* [Lat. *trimetrus* < Gk. *trimetros* : *tri-*, tri- + *metros,* meter.] A verse line of three metrical feet or three prosodic units. **—tri·met'ric** (trī-mĕt'rĭk), **tri·met'ri·cal** (-rĭ-kəl) *adj.*

**tri·meth·a·di·one** (trī-mĕth'ə-dī'ōn') *n.* [TRI- + METH(YL) + DI- + -ONE.] A granular crystalline substance, $C_6H_9NO_3$, used in treating epilepsy.

**tri·met·ro·gon** (trī-mĕt'rə-gŏn') *n.* [TRI- + Gk. *metron,* measure + -GON.] A system of aerial photography in which one vertical and two oblique photographs are simultaneously taken for use in topographic mapping.

**trim·mer** (trĭm'ər) *n.* **1.** One that trims, esp. a device for trimming lumber. **2.** One who changes his or her opinions to suit the needs of the moment.

**trim·ming** (trĭm'ĭng) *n.* **1.** Material added as decoration or ornament. **2. trimmings.** Accessories <turkey with all the *trimmings*> **3. trimmings.** Scraps removed when an object is trimmed. **4.** *Informal.* A sound defeat or beating.

**tri·mo·lec·u·lar** (trī'mə-lĕk'yə-lər) *adj.* Relating to or formed from three molecules.

**tri·month·ly** (trī-mŭnth'lē) *adj.* Performed, occurring, or appearing every three months. **—tri·month'ly** *adv.*

**tri·morph** (trī'môrf') *n.* [Back-formation < TRIMORPHIC.] **1.** A substance that occurs in three distinct forms. **2.** One of the three forms in which a trimorphic substance occurs.

**tri·mor·phic** (trī-môr'fĭk) *also* **tri·mor·phous** (-fəs) *adj.* [Gk. *trimorphos,* having three forms : *tri-*, tri- + *morphē,* shape.] **1.** *Biol.* Having or occurring in three differing forms. **2.** *Chem.* Crystallizing in three distinct forms. **—tri·mor'phi·cal·ly** *adv.* **—tri·mor'phism** *n.*

**Tri·mur·ti** (trĭ-mŏŏr'tē) *n.* [Skt. *trimūrtiḥ* : *tri-*, three + *murtiḥ,* form.] The triad of Hindu gods including Brahma, Vishnu, and Shiva.

**tri·nal** (trī'nəl) *adj.* [Lat. *trinalis* < Lat. *trinus,* trine.] Consisting of three parts : THREEFOLD.

**tri·na·ry** (trī'nə-rē) *adj.* [LLat. *trinarius* < Lat. *trinus,* trine.] Consisting of three parts or proceeding by threes : TERNARY.

**trine** (trīn) *adj.* [ME < OFr. < Lat. *trinus* < *tres,* three.] **1.** Threefold : triple. **2. a.** Situated in astrological trine. **b.** Of or pertaining to a favorable positioning of two planets. *—n.* **1.** A group of three. **2.** The astrological aspect of two planets when 120° apart.

**Tri·nil man** (trē'nĭl') *n.* [After *Trinil,* Indonesia.] The pithecanthropus.

**Trin·i·tar·i·an** (trĭn'ĭ-târ'ē-ən) *adj.* **1.** Describing or pertaining to the Trinity. **2.** Believing or professing belief in the Trinity or the religious doctrine of the Trinity. **3. trinitarian.** Having three parts, members, or facets. *—n.* **1.** A believer in the religious doctrine of the Trinity. **2.** A member of a religious teaching and nursing order for men, founded in 1198. **—Trin'i·tar'i·an·ism** *n.*

**tri·ni·tro·ben·zene** (trī-nī'trō-bĕn'zēn', -bĕn-zēn') *n.* A yellow crystalline compound, $C_6H_3N_3O_6$, derived from trinitrotoluene and used as an explosive.

**tri·ni·tro·cre·sol** (trī-nī'trō-krē'sôl', -sōl') *n.* A yellow crystalline compound, $C_7H_5N_3O_7$, used in high explosives.

**tri·ni·tro·glyc·er·in** (trī-nī'trō-glĭs'ər-ĭn) *n. Chem.* Nitroglycerin.

**tri·ni·tro·phe·nol** (trī-nī'trō-fē'nōl', -nôl') *n. Chem.* Picric acid.

**tri·ni·tro·tol·u·ene** (trī-nī'trō-tŏl'yōō-ēn') *n.* A yellow crystalline compound, $C_7H_5N_3O_6$, used chiefly as a high explosive.

**tri·ni·tro·tol·u·ol** (trī-nī'trō-tŏl'yōō-ôl', -ōl') *n.* Trinitrotoluene.

**trin·i·ty** (trĭn'ĭ-tē) *n., pl.* **-ties.** [ME *trinite* < OFr. < Lat. *trinitas* < *trinus,* trine.] **1.** A group of three closely related members. **2. Trinity.** The union of three divine figures, the Father, Son, and Holy Ghost, in one Godhead. **3. Trinity.** Trinity Sunday.

**Trinity Sunday** *n.* The first Sunday after Pentecost, dedicated to the Trinity.

**trin·ket** (trĭng'kĭt) *n.* [Orig. unknown.] **1.** A small ornament. **2.** A trivial thing : TRIFLE.

**tri·no·mi·al** (trī-nō'mē-əl) *adj.* [TRI- + (BI)NOMIAL.] **1.** Composed of three terms or names, as a taxonomic designation. **2.** *Math.* Having three algebraic terms connected by plus or minus signs. *—n.* **1.** *Math.* A trinomial algebraic expression. **2.** A three-part taxonomic designation indicating genus, species, and subspecies or variety.

**tri·nu·cle·o·tide** (trī-nōō'klē-ə-tīd', -nyōō'-) *n.* A triplet of nucleotides : CODON.

**tri·o** (trē'ō) *n., pl.* **-os.** [Fr., composition for three voices < Ital. < *tre,* three < Lat. *tres.*] **1.** A group or set of three. **2.** *Mus.* **a.** A composition for three performers. **b.** The group performing a trio. **c.** The middle section of a minuet or scherzo, a march, or of various dance forms.

**tri·ode** (trī'ōd') *n.* A highly evacuated electron tube containing an anode, a cathode, and a control grid.

**tri·oe·cious** *also* **tri·e·cious** (trī-ē'shəs) *adj.* [< NLat. *Trioecia,* former order name : TRI- + Gk. *oikia,* dwelling < *oikos,* house.] *Bot.* Having male, female, and bisexual flowers borne on separate plants. **—tri·oe'cious·ly** *adv.*

**tri·ol** (trī'ôl', -ōl') *n.* A chemical compound containing three hydroxyl groups.

**tri·o·let** (trē'ə-lĭt, trī'-, trē'ə-lā') *n.* [Fr., dim. of *trio,* trio. —see TRIO.] A poem or stanza of eight lines with a rhyme scheme *ABaAabAB* in which the fourth and seventh lines are the same as the first, and the eighth line is the same as the second.

**tri·ose** (trī'ōs') *n.* One of a group of monosaccharides that contain three carbon atoms.

**tri·ox·ide** (trī-ŏk'sīd') *also* **tri·ox·id** (-ŏk'sĭd) *n.* A chemical compound having three oxygen atoms per molecule.

**trip** (trĭp) *n.* [ME, maneuver to cause someone to fall < *trippen,* to move nimbly < OFr. *tripper* < MDu. *trippen,* to hop.] **1.** Travel from one place to another : JOURNEY. **2.** *Slang.* **a.** A hallucinatory experience induced by a psychedelic drug. **b.** An intense, exciting experience. **3.** *Slang.* **a.** A usu. temporary yet absorbing interest <on a clean air *trip*> **b.** A certain lifestyle <did the whole rich executive *trip*> **4.** A light or nimble tread. **5.** A stumble : fall. **6.** A maneuver causing one to stumble or fall. **7.** A mistake. **8. a.** A device, as a pawl, for triggering a mechanism. **b.** The action of such a device. *—v.* **tripped, trip·ping, trips.** *—vi.* **1.** To stumble. **2.** To move nimbly with light, rapid steps : SKIP. **3.** To make a mistake. **4.** To be released, as a tooth on an escapement wheel in a watch. **5.** To make a trip. **6.** *Slang.* To have a drug-induced hallucination <*tripped* out on acid> *—vt.* **1.** To cause to stumble or fall. **2.** To trap or catch in an error. **3.** *Archaic.* To perform (a dance) nimbly. **4.** To release (a catch, trigger, or switch), setting a mechanism into operation. **5.** *Naut.* **a.** To raise (an anchor) from the bottom. **b.** To tip or turn (a yardarm) into a position for lowering. **c.** To lift (an upper mast) in order to remove the fid before lowering. **—trip the light fantastic.** To dance. **—trip'per** *n.*

**tri·pal·mi·tin** (trī-păl'mĭ-tĭn) *n.* Palmitin.

**tri·par·tite** (trī-pär'tīt') *adj.* **1.** Of or divided into three parts. **2.** Pertaining to or carried out by three parties <a *tripartite* treaty>

**tri·par·ti·tion** (trī'pär-tĭsh'ən) *n.* Division into three parts or among three parties <*tripartition* of conquered territories>

**tripe** (trīp) *n.* [ME < OFr.] **1.** The light-colored rubbery lining of the stomach of ruminants, as cattle, used as food. **2.** *Informal.* Something worthless : RUBBISH.

**tri·ped·al** (trī-pĕd'l) *adj.* [Lat. *tripedalis* : *tri-*, three + *pes, ped-*, foot.] Having three legs or feet : TRIPODAL.

**tri·pep·tide** (trī-pĕp'tīd') *n.* A peptide with three amino acids.

**tri·pet·al·ous** (trī-pĕt'l-əs) *adj. Bot.* Having three petals.

**trip hammer** *also* **trip·ham·mer** *or* **trip-ham·mer** (trĭp'-hăm'ər) *n.* A heavy power-operated hammer lifted by a cam or lever and then dropped.

**tri·phen·yl·meth·ane** (trī'fĕn'əl-mĕth'ān', -fē'nəl-) *n.* A colorless crystalline hydrocarbon, $(C_6H_5)_3CH$, from which synthetic dyes are derived by substitution.

**tri·phib·i·an** (trī-fĭb'ē-ən) *adj.* [TRI- + (AM)PHIBIAN.] Designed to operate on land, water, or in the air. *—n.* A triphibian aircraft.

**tri·phos·phate** (trī-fŏs'fāt') *n.* A salt or ester containing three phosphate groups.

**triph·thong** (trĭf'thông', -thŏng', trĭp'-) *n.* [TRI- + (DI)PHTHONG.] A compound vowel sound resulting from the succession of three simple sounds and functioning as a unit. **—triph·thon'gal** (trĭf-thông'əl, -thŏng'əl, trĭp-) *adj.*

**triph·y·lite** (trĭf'ə-līt') *also* **triph·y·line** (-lēn') *n.* [TRI- + Gk. *phulon,* tribe + -ITE.] Any of a vitreous, bluish-gray mineral series of lithium, iron, and manganese phosphates.

**tri·pin·nate** (trī-pĭn'āt') *adj. Bot.* Divided into leaflets subdivided into smaller, further subdivided leaflets. **—tri·pin'nate'ly** *adv.*

**tri·plane** (trī'plān') *n.* An airplane with three wings placed one above the other.

**tri·ple** (trĭp'əl) *adj.* [OFr. < Lat. *triplus* < Gk. *triplous.*] **1.** Having three parts : THREEFOLD. **2.** Three times as many or as much. **3.** Repeated three times. **4.** *Mus.* Marked by three beats in a measure. *—n.*

**1.** A number or quantity three times as great as another. **2.** A group or set of three : TRIAD. **3.** *Baseball.* A three-base hit. **4.** A trifecta. —*v.* **-pled, -pling, -ples.** —*vt.* To make three times as great in number or amount. —*vi.* **1.** To be or become three times as great in number or amount. **2.** *Baseball.* To make a three-base hit.

**Triple Crown** *n.* **1.** An unofficial championship title attained by a horse that wins the three traditional races for a specified category. **2.** *Baseball.* An unofficial championship title achieved by a player who is at the head of a league in batting average, home runs, and runs batted in.

**tri·ple-head·er** (trĭp′əl-hĕd′ər) *n.* A sports contest featuring three contests in a row.

**triple measure** *n.* Triple time.

**triple play** *n. Baseball.* A defensive play in which three putouts, on two base runners and the batter, are executed during one turn at bat, thereby ending suddenly the offensive threat and the inning.

**triple point** *n.* A point on a phrase diagram that represents a set of conditions in which the liquid, gaseous, and solid phases of a substance such as water can exist in a state of equilibrium.

**trip·let** (trĭp′lĭt) *n.* [TRIPL(E) + (DOUBL)ET.] **1.** A group or set of three of one kind. **2.** One of three offspring born at one birth. **3.** A group of three lines of verse. **4.** *Mus.* A group of three notes having the time value of two notes of the same kind. **5.** *Physics.* A multiplet with three components.

**tri·ple·tail** (trĭp′əl-tāl′) *n.* A chiefly marine fish of the family Lobotidae, esp. *Lobotes surinamensis,* having prominent dorsal and anal fins resembling extra tails.

**triple time** *n.* A musical time or rhythm having three beats to the measure, with the accent on the first beat.

**tri·plex** (trĭp′lĕks′, trī′plĕks′) *adj.* [Lat.] Consisting of three parts : THREEFOLD <*triplex* windows> —*n.* Something triplex, as a building containing three apartments.

**trip·li·cate** (trĭp′lĭ-kĭt) *n.* [ME < Lat. *triplicatus,* p.part. of *triplicare,* to triple < *triplex,* threefold.] **1.** One of a set of three identical things. **2.** Three identical copies <typed in *triplicate*> —*vt.* (trĭp′lĭ-kāt′) **-cat·ed, -cat·ing, -cates. 1.** To make threefold. **2.** To make three identical copies of. **—trip′li·cate·ly** *adv.* **—trip′li·ca′tion** *n.*

**tri·plic·i·ty** (trĭ-plĭs′ĭ-tē, trī-) *n., pl.* **-ties.** [ME, three signs of zodiac < LLat. *triplicitas,* triplicity < *triplex,* triplex.] **1.** The quality or state of being triple. **2.** TRIPLE 2. **3.** One of four groups of the zodiac, each consisting of three astrological signs.

**trip·lo·blas·tic** (trĭp′lō-blăs′tĭk) *adj.* [Lat. *triplus,* triple + -BLASTIC.] Having three germ layers.

**trip·loid** (trĭp′loid′) *adj.* [TRIPL(E) + (HAPL)OID.] Having three haploid sets of chromosomes in each nucleus. **—trip′loid** *n.*

**tri·pod** (trī′pŏd′) *n.* [Lat. *tripus, tripod-,* tripod < Gk. *tripous,* three-footed : *tri-,* tri- + *pous,* foot.] **1.** A three-legged object, as a caldron, stool, or table. **2.** An adjustable three-legged stand, as for supporting a transit or camera. **—trip′o·dal** (trĭp′ə-dəl, trī′pŏd′l) *adj.*

**trip·o·li** (trĭp′ə-lē) *n.* [After *Tripoli,* Libya.] A porous, lightweight, siliceous rock of various colors, derived from weathering of chert and siliceous limestone.

**tri·pos** (trī′pŏs′) *n.* [Alteration of Lat. *tripus,* tripod (from the stool upon which a candidate sat to dispute humorously with other candidates).] Any of the examinations for the B.A. degree with honors at Cambridge University in England.

**trip·pet** (trĭp′ĭt) *n.* [ME *tripet,* piece of wood used in a game < *trippen,* to trip. —see TRIP.] A cam or projection in a mechanism designed to strike another part at regular intervals.

**trip·ping·ly** (trĭp′ĭng-lē) *adv.* Lightly and easily : FLUENTLY.

**trip·tane** (trĭp′tān′) *n.* [Short for *trimethylbutane.*] A colorless liquid antiknock additive, $C_7H_{16}$, used in aviation fuels.

**trip·tych** (trĭp′tĭk) *n.* [< Gk. *triptukhos,* threefold : *tri-,* tri- + *ptukhē,* fold < *ptussein,* to fold.] **1.** An ancient Roman hinged writing tablet made up of three leaves. **2.** A work of art composed of three hinged or folding panels, esp. one with a religious theme.

**tri·reme** (trī′rēm′) *n.* [Lat. *triremis* : *tri-,* tri- + *remus,* oar.] An ancient Greek or Roman galley or warship with three banks of oars on each side.

**tri·sac·cha·ride** (trī-săk′ə-rīd′, -rĭd) *n.* A carbohydrate that upon hydrolysis yields three monosaccharides.

**tri·sect** (trī′sĕkt′, trī-sĕkt′) *vt.* **-sect·ed, -sect·ing, -sects.** To divide into three equal parts. **—tri′sec′tion** (trī′sĕk′shən, trī-sĕk′-) *n.* **—tri′sec′tor** (trī′sĕk′tər, trī-sĕk′-) *n.*

**tri·sep·al·ous** (trī-sĕp′ə-ləs) *adj.* Having three sepals.

**tri·skel·i·on** (trī-skĕl′ē-ən, trĭs-kĕl′-) *also* **tri·skele** (trī′skēl′, trĭs′kēl′) *n., pl.* **tri·skel·i·a** (trī-skĕl′ē-ə, trĭs-kĕl′-) *also* **tri·skeles.** [NLat. < Gk. *triskelēs,* three-legged : *tri-,* tri- + *skelos,* leg.] A usu. symbolic figure composed of three curved lines or branches or three stylized human arms or legs that radiate from a common center.

**tris·mus** (trĭz′məs) *n.* [NLat. < Gk. *trismos,* a grinding.] Lockjaw. **—tris′mic** *adj.*

**tris·oc·ta·he·dron** (trĭs-ŏk′tə-hē′drən) *n., pl.* **-drons** or **-dra** (-drə) [Gk. *tris,* thrice + OCTAHEDRON.] *Math.* **1.** A solid figure having 24 congruent triangular faces and an octahedron as a base. **2.** A trapezohedron. **—tris·oc′ta·he′dral** *adj.*

**tri·so·di·um** (trī-sō′dē-əm) *adj.* Containing three sodium atoms.

**tri·so·mic** (trī-sō′mĭk) *adj.* [TRI- + (CHROMO)SOM(E) + -IC.] Having at least one triploid chromosome in an otherwise diploid set. **—tri′some′** (trī′sōm′) *n.* **—tri·so′my** *n.*

**Tris·tan** (trĭs′tən, -tän′, -tăn′) *n.* A prince in Arthurian legend who fell in love with the Irish princess Iseult and died with her.

**triste** (trēst) *adj.* [ME < OFr. < Lat. *tristis.*] Sorrowful : sad.

**tri·ste·a·rin** (trī-stē′ə-rĭn, -stîr′ĭn) *n.* Stearin.

**trist·ful** (trĭst′fəl) *adj.* [ME *trist,* sad (< OFr. *triste* < Lat. *tristis*) + -FUL.] *Archaic.* Triste. **—trist′ful·ly** *adv.*

**tris·tich** (trĭs′tĭk) *n.* [TRI- + (DI)STICH.] A stanza or strophic unit consisting of three lines.

**Tris·tram** (trĭs′trəm) *n.* Tristan.

**tri·sul·fide** (trī-sŭl′fīd′) *also* **tri·sul·phide** or **tri·sul·fid** or **tri·sul·phid** (-fĭd) *n.* A sulfide having three sulfur atoms per molecule.

**tri·syl·la·ble** (trī′sĭl′ə-bəl) *n.* A word of three syllables. **—tri′syl·lab′ic** (-sĭ-lăb′ĭk), **tri′syl·lab′i·cal** *adj.* **—tri′syl·lab′i·cal·ly** *adv.*

**tri·tan·o·pi·a** (trī′tə-nō′pē-ə) *n.* [Gk. *tritos,* a third + Gk. *anopia,* blindness.] A rare visual defect involving an inability to distinguish the color blue.

**trite** (trīt) *adj.* **trit·er, trit·est.** [Lat. *tritus* < *terere,* to wear out.] Devoid of freshness or appeal due to overuse. **—trite′ly** *adv.* **—trite′ness** *n.*

☆ **syns:** TRITE, BANAL, COMMONPLACE, HACKNEYED, PLATTITUDINOUS, TIMEWORN, WORN-OUT *adj. core meaning* : devoid of freshness or appeal due to overuse <sophomoric writing full of *trite* expressions> *ant:* fresh, original

**tri·the·ism** (trī′thē-ĭz′əm) *n.* The belief that the Father, Son, and Holy Ghost are three separate and distinct gods. **—tri′the·ist** *n.* **—tri′the·is′tic, tri′the·is′ti·cal** *adj.*

**trit·i·um** (trĭt′ē-əm, trĭsh′ē-) *n.* [NLat. < Gk. *tritos,* third.] A rare radioactive hydrogen isotope with atomic mass 3 and half-life 12.5 years, prepared artificially for use as a tracer and as a constituent of hydrogen bombs.

**tri·ton¹** (trīt′n) *n.* [Lat. < Gk. *Tritōn.*] **1. Triton.** *Gk. Myth.* A god of the sea, son of Poseidon and Amphitrite, portrayed as having the head and trunk of a man and the tail of a fish. **2.** A chiefly tropical marine gastropod mollusk of the genus *Cymatium* and related genera, with a pointed, spirally twisted, often colorfully marked shell.

**Triton¹**

**tri·ton²** (trī′tŏn′) *n.* [TRIT(IUM) + -ON².] The nucleus of a tritium atom composed of two neutrons and one proton.

**tri·tone** (trī′tōn′) *n.* [Med. Lat. *tritonus* < Gk. *tritonos,* having three tones : *tri-,* tri- + *tonos,* tone.] *Mus.* An interval of three whole tones.

**trit·u·rate** (trĭch′ə-rāt′) *vt.* **-rat·ed, -rat·ing, -rates.** [LLat. *triturare, triturat-,* to thresh < Lat. *tritura,* a threshing < *tritor,* grinder < *terere,* to rub.] To rub, grind, crush, or pound into fine particles : PULVERIZE. —*n.* (trĭch′ər-ĭt). A triturated substance, esp. a powdered drug. **—trit′u·ra·ble** (trĭch′ər-ə-bəl) *adj.* **—tri′u·ra′tion** *n.* **—trit′u·ra′tor** *n.*

**tri·umph** (trī′əmf) *vi.* **-umphed, -umph·ing, -umphs.** [Lat. *triumphare* < *triumphus,* triumph.] **1.** To be victorious : WIN. **2.** To be jubilant over a victory : EXULT. **3.** To receive honors upon return from a victory in ancient Rome. —*n.* **1.** An instance or the fact of being victorious : SUCCESS. **2.** Jubilant exultation derived from victory. **3.** A public celebration in ancient Rome to welcome a returning victorious commander and his army. **4.** *Obs.* A public celebration or spectacular pageant. **—tri′umph·er** *n.*

**tri·um·phal** (trī-ŭm′fəl) *adj.* **1.** Relating to or characteristic of a triumph. **2.** Celebrating, honoring, or commemorating a triumph.

**tri·um·phant** (trī-ŭm′fənt) *adj.* **1.** Exulting in success or victory. **2.** Victorious : conquering <the return of a *triumphant* army> **3.** *Archaic.* Triumphal. **4.** *Obs.* Splendid. **—tri·um′phant·ly** *adv.*

**tri·um·vir** (trī-ŭm′vər) *n., pl.* **-virs** or **-vi·ri** (-və-rī′) [Lat., back-formation < *triumviri,* board of three : *tres,* three + *vir,* man.] One of three men sharing public administration or civil authority, as in ancient Rome. **—tri·um′vi·ral** *adj.*

**tri·um·vi·rate** (trī-ŭm′vər-ĭt) *n.* [Lat. *triumviratus* < *triumviri,* board of three. —see TRIUMVIR.] **1.** Government by triumvirs.

**2.** The office or term of a triumvir. **3.** A body or group of triumvirs. **4.** A group or association of three.

**tri·une** (trī'yo͞on') *adj.* [TRI- + Lat. *unus*, one.] Being three in one. —Used esp. of the Trinity. —*n.* A trinity.

**tri·u·ni·ty** (trī-yo͞o'nĭ-tē) *n., pl.* **-ties.** TRINITY 1.

**tri·va·lent** (trī-vā'lənt) *also* **ter·va·lent** (tər-vā'lənt, tûr'vā'lənt) *adj.* Having valence 3. **—tri·va'lence, tri·va'len·cy** *n.*

**tri·valve** (trī'vălv') *adj.* Having three valves.

**triv·et** (trĭv'ĭt) *n.* [ME *trevet*, prob. < OE **trefet* < Lat. *tripes*, three-footed : *tri-*, tri- + *pes*, foot.] **1.** A three-legged stand. **2.** A metal stand with short feet, used under a hot dish on a table.

**triv·i·a¹** (trĭv'ē-ə) *pl.n.* [NLat., back-formation < Lat. *trivialis*, trivial.] *(sing. or pl. in number).* Insignificant or superfluous matters : TRIFLES.

☆ ***syns:*** TRIVIA, MINUTIAE, TRIFLES, TRIVIALITY *n. core meaning* : unimportant matters or concerns <was bored by the *trivia* of their suburban lives>

**triv·i·a²** (trĭv'ē-ə) *n. pl. of* TRIVIUM.

**triv·i·al** (trĭv'ē-əl) *adj.* [Lat. *trivialis*, ordinary < *trivium*, public square. —see TRIVIUM.] **1.** Relatively insignificant : UNIMPORTANT. **2.** Commonplace : ordinary. **3.** Concerned with or involving trivia. **—triv'i·al'i·ty** (-ăl'ĭ-tē) *n.* **—triv'i·al·ly** *adv.*

**triv·i·al·ize** (trĭv'ē-ə-līz') *vt.* **-ized, -iz·ing, -iz·es.** To make trivial. **—triv'i·al·i·za'tion** *n.*

**trivial name** *n.* **1.** The term in taxonomic nomenclature that follows the genus name and designates the species. **2.** A common name as opposed to a taxonomic designation.

**triv·i·um** (trĭv'ē-əm) *n., pl.* **-i·a** (-ē-ə) [Med. Lat. < Lat., public square : *tri-*, tri- + *via*, road.] The lower division of the seven liberal arts in medieval schools, consisting of grammar, logic, and rhetoric.

**tri·week·ly** (trī-wēk'lē) *adj.* **1.** Occurring, done, or appearing three times a week. **2.** Occurring, done, or appearing every three weeks. —*adv.* **1.** Three times a week. **2.** Every three weeks. —*n., pl.* **-lies.** A periodical published on a triweekly basis.

**-trix** *suff.* [ME < Lat., fem. of *-tor*, n. suffix.] **1.** A woman associated with a specified thing <aviatrix> **2.** A geometric point, line, or surface <directrix>

**tRNA** (tē'är-ĕn'ā') *n.* Transfer RNA.

**tro·car** (trō'kär') *n.* [Fr. *trocart* : *trois*, three + *carre*, side of an instrument.] A sharp-pointed surgical instrument, used with a cannula to puncture a body cavity for fluid aspiration.

**tro·cha·ic** (trō-kā'ĭk) *adj.* [Fr. *trochaïque* or < Lat. *trochaicus* < Gk. *trokhaikos* < *trokhaios*, trochee.] Of, relating to, or composed of trochees. **—tro·cha'ic** *n.*

**tro·chal** (trō'kəl) *adj.* [< Gk. *trokhos*, wheel.] Wheel-shaped.

**tro·chan·ter** (trō-kăn'tər) *n.* [Gk. *trokhantēr* < *trekhein*, to run.] **1.** Any of several bony processes on the upper part of the femur of many vertebrates. **2.** The second proximal segment of an insect's leg. **—tro·chan'ter·al, tro·chan·ter'ic** *adj.*

**tro·che** (trō'kē) *n.* [ME *trocis* < LLat. *trochiscus* < Gk. *trokhiskos*, dim. of *trokhos*, wheel < *trekhein*, to run.] A small, usu. round medicinal lozenge.

**tro·chee** (trō'kē) *n.* [Fr. *trochée* < Lat. *trochaeus* < Gk. *trokhaios* < *trokhos*, a running < *trekhein*, to run.] A metrical foot in prosody consisting of one long or stressed syllable followed by one short or unstressed syllable.

**troch·le·a** (trŏk'lē-ə) *n., pl.* **-le·ae** (-lē-ē') [Lat., system of pulleys < Gk. *trokhalia.*] *Anat.* A structure resembling a pulley, esp. the part of the distal end of the humerus that articulates with the ulna.

**troch·le·ar** (trŏk'lē-ər) *adj.* **1.** *Anat.* Of, resembling, or situated near a trochlea. **2.** *Anat.* Of or relating to the trochlear, or fourth cranial nerve. **3.** *Bot.* Shaped like a pulley.

**tro·choid** (trō'koid', trŏk'oid') *also* **tro·choi·dal** (trō-koid'l, trŏk-oid'l) *adj.* [Gk. *trokhoeidēs*, wheellike : *trokhos*, wheel + *-eidēs*, -oid.] Capable of or exhibiting rotation about a central axis. —*n.* The plane locus of a point on the radius or on an extension of the radius of a circle, as the circle rolls along a fixed straight line. **—tro·choi'dal·ly** *adv.*

**troch·o·phore** (trŏk'ə-fôr', -fōr') *n.* [Gk. *trokhos*, wheel (< *trekhein*, to run) + -PHORE.] The small aquatic larva of various invertebrates, including certain mollusks and annelids.

**trod** (trŏd) *v. p.t. & var. p.p. of* TREAD.

**trod·den** (trŏd'n) *v. p.p. of* TREAD.

**trof·fer** (trŏf'ər, trô'fər) *n.* [Alteration of TROUGH.] An inverted trough suspended from a ceiling as a fixture for fluorescent lights.

**trog·lo·dyte** (trŏg'lə-dīt') *n.* [< Lat. *Troglodytae*, cave dwellers < Gk. *Troglodutai* : *troglos*, cave + *duein*, to enter.] **1.** A prehistoric cave dweller. **2.** One felt to be like a cave dweller, as in reclusiveness or brutishness. **—trog'lo·dyt'ic** (-dĭt'ĭk), **trog'lo·dyt'i·cal** *adj.*

**tro·gon** (trō'gŏn') *n.* [NLat. *Trogonidae*, family name < Gk. *trōgōn*, pr.part. of *trogein*, to gnaw.] A colorful tropical bird of the family Trogonidae, as the quetzal.

**troi·ka** (troi'kə) *n.* [R. *troyka* < *troje*, three.] **1.** A Russian carriage drawn by a team of three horses abreast. **2.** A triumvirate.

**Troi·lus** (troi'ləs, trō'ə-ləs) *n.* A son of Priam of Troy, depicted as Cressida's lover in medieval romance.

**Tro·jan** (trō'jən) *n.* [ME *Troyan* < Lat. *Troianus* < *Troia*, Troy < Gk. *Trōias*, the plain of Troy < *Trōs*, the mythical founder of Troy.] **1.** A native or inhabitant of ancient Troy. **2.** A courageous, determined, or energetic person. **—Tro'jan** *adj.*

**Trojan horse** *n.* **1.** The hollow wooden horse in which, according to legend, Greeks hid and gained entrance to Troy, later opening the gates to their army. **2.** A subversive group or device placed within enemy ranks.

**Trojan War** *n.* The prehistoric ten-year war waged against Troy by the confederated Greeks, caused by the abduction of the Spartan queen Helen by the Trojan prince Paris, and resulting in the burning and destruction of Troy.

**troll¹** (trōl) *v.* **trolled, troll·ing, trolls.** [ME *trollen*, to wander about.] —*vt.* **1.** To fish for by running a baited line behind a slowly moving boat. **2.** To trail (a baited line) in fishing. **3.** To sing in succession the parts of (e.g., a round). **4.** To sing heartily <*troll* a carol> **5.** To roll or revolve. —*vi.* **1.** To fish by running a baited line behind a moving boat. **2.** To sing gaily or heartily. **3.** To roll or spin around. **4.** To wander about : RAMBLE. —*n.* **1.** A vocal composition in successive parts : ROUND. **2.** The act of trolling for fish. **3.** A lure, as a spoon or spinner, used for trolling. **—troll'er** *n.*

**troll²** (trōl) *n.* [Norw. < ON.] A supernatural creature of Scandinavian folklore variously portrayed as a dwarf or a giant living in caves or under bridges.

**trol·ley** *also* **trol·ly** (trŏl'ē) [Prob. < TROLL¹.] —*n., pl.* **-leys** *also* **-lies. 1.** A streetcar. **2.** A wheeled carriage, cage, or basket suspended from and traveling on an overhead track. **3.** A device that collects electric current from an underground conductor, an overhead wire, or a third rail, and transmits it to the motor of an electric vehicle. **4.** A small truck or car operating on a track and used in a mine, quarry, or factory for carrying materials back and forth. **5.** *Chiefly Brit.* A cart. —*v.* **-leyed, -ley·ing, -leys** *also* **-lied, -ly·ing, -lies.** —*vt.* To convey by trolley. —*vi.* To travel by trolley.

**trolley bus** *n.* An electric bus that does not run on tracks and is powered by electricity from an overhead wire.

**trolley car** *n.* A streetcar.

**trol·lop** (trŏl'əp) *n.* [Perh. < TROLL¹.] **1.** A slovenly woman : SLATTERN. **2.** A sexually promiscuous woman : STRUMPET.

**trom·bic·u·li·a·sis** (trŏm-bĭk'yə-lī'ə-sĭs) *also* **trom·bic·u·lo·sis** (-lō'sĭs) *or* **trom·bi·di·a·sis** (trŏm'bĭ-dī'ə-sĭs) *n.* [NLat. *Trombicula*, genus of mites + -IASIS.] Infestation with chiggers.

**trom·bone** (trŏm-bōn', trəm-, trŏm'bōn') *n.* [Ital., aug. of *tromba*, trumpet, of Germanic orig.] A brass musical instrument composed of a long cylindrical tube bent upon itself twice and ending in a bell-shaped mouth. **—trom·bon'ist** *n.*

**trom·mel** (trŏm'əl) *n.* [G., drum < MHG *trummel* < *trumme.*] A revolving sieve shaped like a cylinder and used for screening or sizing rock and ore.

**tromp** (trŏmp) *v.* **tromped, tromp·ing, tromps.** [Alteration of TRAMP.] *Informal.* —*vi.* To walk noisily and heavily : TRAMP. —*vt.* **1.** To trample underfoot. **2.** To defeat soundly : TROUNCE.

**trompe** (trŏmp) *n.* [Fr.] An apparatus in which water falling through a perforated pipe entrains air into and down the pipe to produce an air blast for a furnace or forge.

**trompe l'oeil** (trômp'loi') *n.* [Fr. : *tromper*, to deceive + *le*, the + *oeil*, eye.] **1.** A style of painting that creates an illusion of photographic reality. **2.** A trompe l'oeil painting or effect.

**-tron** *suff.* [Gk., n. suffix.] **1.** Vacuum tube <dynatron> **2.** Device for manipulating subatomic particles <betatron>

**tro·na** (trō'nə) *n.* [Swed.] A natural vitreous gray or white mineral, $Na_2CO_3 \cdot NaHCO_3 \cdot 2H_2O$, used as a source of sodium compounds.

**troop** (tro͞op) *n.* [OFr. *trope*, prob. of Germanic orig.] **1.** A group of people, animals, or things. **2.** A group of soldiers. **3. troops.** Military units : SOLDIERS. **4.** A unit of at least five Boy Scouts or Girl Scouts under the guidance of an adult leader. **5.** A great many. —*vi.* **trooped, troop·ing, troops. 1.** To advance or go as a throng. **2.** To assemble or move in crowds. **3.** To consort : associate <*troops* with a bad crowd>

**troop carrier** *n.* A transport aircraft for deploying troops.

**troop·er** (tro͞o'pər) *n.* **1. a.** A cavalry soldier. **b.** A cavalry horse. **c.** A soldier. **2.** A mounted police officer. **3.** A state police officer.

**troop·ship** (tro͞op'shĭp') *n.* A ship for transporting troops.

**troost·ite** (tro͞o'stīt') *n.* [After Gerald *Troost* (1776–1850).] A reddish crystalline mineral, a variety of willemite, in which the zinc is partly replaced by manganese.

**trop-** *pref. var. of* TROPO-.

**trope** (trōp) *n.* [Lat. *tropus* < Gk. *tropos*, manner.] **1.** Figurative use of a word or expression : FIGURE OF SPEECH. **2.** A word or phrase interpolated as an embellishment in the sung parts of certain medieval liturgies. **—trop'i·cal** *adj.*

**troph-** *pref. var. of* TROPHO-.

**troph·al·lax·is** (trŏf'ə-lăk'sĭs, trō'fə-) *n.* [TROPH(O)- + Gk. *allaxis*, exchange < *allassein*, to exchange < *allos*, other.] Exchange of food substances between organisms, esp. among social insects.

**-trophic** *suff.* [< -TROPHY.] Of, relating to, or marked by a specified kind of nutrition <polytrophic>

**trophic level** *n. Ecol.* A feeding stratum in a food chain of an ecosystem characterized by organisms that occupy a similar functional position in the ecosystem.

**tropho-** *or* **troph-** *pref.* [< Gk. *trophē,* food < *trephein,* to nourish.] Nutrition : nutritive <*tropho*blast>

**tro·pho·blast** (trō′fə-blăst′) *n.* The outermost layer of cells of the morula that attaches the fertilized ovum to the uterine wall and acts as a nutritive pathway. **—tro′pho·blas′tic** *adj.*

**tro·pho·derm** (trō′fə-dûrm′) *n.* Trophoblast.

**tro·pho·zo·ite** (trō′fə-zō′īt′) *n.* A protozoan of the class Sporozoa in the active stage.

**tro·phy** (trō′fē) *n., pl.* **-phies.** [OFr. < Lat. *trophaeum,* monument to victory < Gk. *tropaion* < *tropē,* a putting to flight.] **1.** An object won or received as a symbol of achievement or victory, often mounted or preserved as a memento. **2.** A monument customarily erected in classical antiquity to commemorate an enemy's defeat. **3.** An architectural ornament depicting a group of weapons or armor.

**-trophy** *suff.* [NLat. *-trophia* < Gk. < *trophē,* food.] Nutrition : growth <hyper*trophy*>

**trop·ic** (trŏp′ĭk) *n.* [ME *tropik,* solstice < LLat. *tropicus* < Gk. *tropikos* < *tropē,* turn.] **1.** *Astron.* Either of two circles on the celestial sphere parallel to and at an angular distance of 23° 27′ from the equator that are the limits of the apparent northern and southern passages of the sun. **2. a.** Either of the two corresponding parallels of latitude on the earth that constitute the boundaries of the Torrid Zone. **b. Tropics.** The region of the earth's surface lying between these latitudes. *—adj.* Of or pertaining to the Tropics : TROPICAL.

**-tropic** *suff.* [< Gk. *tropos,* turn.] **1.** Turning in response to a specified stimulus <helio*tropic*> **2.** Acting on something specified <gonado*tropic*>

**trop·i·cal** (trŏp′ĭ-kəl) *adj.* **1.** Of, occurring in, or characteristic of the Tropics. **2.** Hot and humid : TORRID. **—trop′i·cal·ly** *adv.*

**tropical cyclone** *n.* A very low pressure area 80 to 160 kilometers or 50 to 100 miles in radius that originates in tropical regions and is frequently characterized by winds of hurricane strength circulating around a calm eye in the center.

**tropical fish** *n.* Any of various small or brightly colored fishes native to tropical waters and often kept in home aquariums.

**tropical storm** *n.* A tropical cyclone having winds ranging from approx. 48 to 121 kilometers or 30 to 75 miles per hour.

**tropical year** *n.* The time interval between two successive passages of the sun through the vernal equinox : the calendar year, or 365.2422 mean solar days.

**trop·ic·bird** (trŏp′ĭk-bûrd′) *n.* Any of several predominantly white sea birds of the genus *Phaethon,* of warm regions, with a pair of long slender projecting tail feathers.

**tropic of Cancer** *n.* The parallel of latitude 23° 27′ north of the equator, the northern boundary of the Torrid Zone, and the most northerly latitude at which the sun reaches an altitude of 90°.

**tropic of Capricorn** *n.* The parallel of latitude 23° 27′ south of the equator, the southern boundary of the Torrid Zone, and the most southerly latitude at which the sun reaches an altitude of 90°.

**tro·pine** (trō′pēn′, -pĭn) *also* **tro·pin** (-pĭn) *n.* [< ATROPINE.] A white, crystalline, poisonous alkaloid, $C_8H_{15}NO$, having a tobacco odor and used as a medicine.

**tro·pism** (trō′pĭz′əm) *n.* [< -TROPISM.] Responsive growth movement of an organism toward or away from an external stimulus. **—tro′pic** *adj.*

**-tropism** *suff.* [< Gk. *tropos,* turn.] Tropism <photo*tropism*>

**tropo-** *or* **trop-** *pref.* [< Gk. *tropos,* turn.] **1.** Turning : change <*tropo*sphere> **2.** Tropism <*tropo*taxis>

**tro·pol·o·gy** (trō-pŏl′ə-jē) *n., pl.* **-gies.** [LLat. *tropologia* < LGk. : Gk. *tropos,* trope + Gk. *-logia,* -logy.] Biblical interpretation that stresses the morally edifying sense of tropes in Scripture. **—tro′po·log′ic** (trō′pə-lŏj′ĭk, trŏp′ə-), **tro′po·log′i·cal** *adj.* **—tro′po·log′i·cal·ly** *adv.*

**tro·po·pause** (trō′pə-pôz′, trŏp′ə-) *n.* The boundary between the upper troposphere and the lower stratosphere that varies in altitude from approx. 8 kilometers or 5 miles at the poles to approx. 18 kilometers or 11 miles at the equator.

**tro·po·phyte** (trō′pə-fīt′, trŏp′ə-) *n.* A plant adapted to climatic conditions in which periods of heavy rainfall alternate with periods of drought. **—tro′po·phyt′ic** (-fĭt′ĭk) *adj.*

**tro·po·sphere** (trō′pə-sfîr′, trŏp′ə-) *n.* The lowest atmospheric region between the earth's surface and the tropopause, marked by decreasing temperature with increasing altitude. **—tro′po·spher′ic** *adj.*

**-tropous** *suff.* [Gk. *-tropos,* of turning < *trepein,* to turn.] Turning in a specified way or from a specified stimulus <amphi*tropous*>

**-tropy** *suff.* [Gk. *-tropia* < *-tropos,* -tropous.] The state of turning in a specified way or from a specified stimulus <thixo*tropy*>

**trot** (trŏt) *n.* [ME < OFr. < *troter,* to trot, of Germanic orig.] **1.** A gait of a four-footed animal, between a walk and a run in speed, in which diagonal pairs of legs move forward together. **2.** A person's gait, faster than a walk : JOG. **3.** A race for trotters. **4.** *Informal.* A literal translation of a foreign text, used esp. by students. **5.** A toddler. **6.** An old woman : CRONE. *—v.* **trot·ted, trot·ting, trots.** *—vi.* **1.** To go or move at a trot. **2.** To move rapidly : HURRY. *—vt.* To cause to move at a trot. **—trot out.** *Informal.* To bring out and show for inspection or admiration <*trot out* one's jewelry>

**troth** (trôth, trŏth, trōth) *n.* [ME *trothe* < OE *trēowð,* truth.] **1.** Good faith : FIDELITY. **2. a.** One's pledged fidelity. **b.** Betrothal. *—vt.* **trothed, troth·ing, troths.** To betroth or pledge.

**troth·plight** (trôth′plīt′, trŏth′-, trōth′-) *Archaic. —n.* A betrothal. *—vt.* **-plight·ed, -plight·ing, -plights.** To betroth.

**trot·line** (trŏt′līn′) *n.* [Perh. < TROT.] A setline.

**Trots·ky·ism** (trŏt′skē-ĭz′əm) *n.* The theories of Communism advocated by Leon Trotsky and his followers, who argued for worldwide revolution and bitterly opposed Stalin and his leadership. **—Trots′ky·ist, Trots′ky·ite′** *n.*

**trot·ter** (trŏt′ər) *n.* **1.** A horse that trots, esp. one trained for harness racing. **2.** *Informal.* A foot, esp. the foot of a pig or sheep prepared as food. **3.** *Informal.* An energetic person.

**trou·ba·dour** (trōō′bə-dôr′, -dōr′, -dŏŏr′) *n.* [Fr. < OProv. *trobador* < *trobar,* to compose, perh. < VLat. **tropare* < Lat. *tropus,* trope.] **1.** One of a class of 12th- and 13th-cent. lyric poets attached to the courts of Provence and northern Italy, who composed songs in complex metrical forms. **2.** A strolling minstrel.

**trou·ble** (trŭb′əl) *n.* [ME < OFr. < *troubler,* to trouble < Lat. *turbidare* < *turbidus,* confused < *turba,* turmoil.] **1.** Distress, affliction, danger, or need <in *trouble* with the authorities> **2.** A cause or source of distress, affliction, danger, or need. **3.** Pains : exertion <went to a lot of *trouble* to give us a good time> **4.** Pain, disease, or malfunction <heart *trouble*><car *trouble*> *—v.* **-bled, -bling, -bles.** *—vt.* **1.** To stir up : agitate. **2.** To afflict with discomfort or pain <My knee is *troubling* me.> **3.** To cause confusion or distress in : PERTURB. **4.** To inconvenience : bother <May I *trouble* you to speak more softly?> *—vi.* To take pains <*troubled* over every aspect of the schedule> **—trou′bler** *n.* **—trou′bling·ly** *adv.*

**trou·ble·mak·er** (trŭb′əl-mā′kər) *n.* A maker of trouble.

**trou·ble·shoot·er** (trŭb′əl-shōō′tər) *n.* **1.** A worker who locates and eliminates sources of trouble, as in mechanical operations. **2.** A mediator skilled in settling disputes, esp. diplomatic or political ones. **—trou′ble·shoot′** *v.* **(-shot, -shoot·ing, -shoots).**

**trou·ble·some** (trŭb′əl-səm) *adj.* **1.** Causing trouble or anxiety : WORRISOME. **2.** Difficult to treat, manage, or cope with : TRYING. **—trou′ble·some·ly** *adv.* **—trou′ble·some·ness** *n.*

☆ ***syns:*** TROUBLESOME, MEAN, PESKY, VEXATIOUS, WICKED *adj. core meaning* : hard to treat, manage, or cope with <bothered by a *troublesome* cold><*troublesome* marital problems>

**trou·blous** (trŭb′ləs) *adj.* **1. a.** Full of trouble. **b.** Unsettled : uneasy : troubled. **2.** TROUBLESOME 1.

**trou-de-loup** (trōō′də-lōō′) *n., pl.* **trous-de-loup** (trōō′də-lōō′) [Fr. : *trou,* hole + *de,* of + *loup,* wolf.] Any of a series of conical pits having pointed stakes set upright in their centers, once used as an obstacle to enemy cavalry.

**trough** (trôf, trŏf) *n.* [ME < OE *trog.*] **1.** A long narrow, gen. shallow receptacle, esp. one for holding water or feed for animals. **2.** A gutter below the eaves of a roof. **3.** A long narrow depression, as between waves or ridges. **4.** A low point in a business cycle or on a statistical graph. **5.** *Meteorol.* An elongated region of low atmospheric pressure, often associated with a front.

**trounce** (trouns) *vt.* **trounced, trounc·ing, trounc·es.** [Orig. unknown.] **1.** To thrash : beat. **2.** To defeat decisively : WHIP.

**troupe** (trōōp) *n.* [Fr., troop.] A company, esp. of theatrical or dramatic performers. *—vi.* **trouped, troup·ing, troupes.** To tour with a theatrical company.

**troup·er** (trōō′pər) *n.* **1.** A member of a theatrical company. **2.** A veteran performer.

**trou·pi·al** (trōō′pē-əl) *n.* [Fr. *troupiale* < *troupe,* flock. —see TROOP.] A tropical American bird of the genus *Icterus,* related to the oriole and the New World blackbird, esp. *I. icterus,* with orange and black plumage.

**trou·ser** (trou′zər) *adj.* [Back-formation < TROUSERS.] Being of, designed for, or found on trousers <*trouser* cuffs>

**trou·sers** *also* **trow·sers** (trou′zərz) *pl.n.* [< obs. *trouse* < Sc. Gael. *triubhas.*] An outer garment for covering the body from the waist to the ankles, divided into sections to fit each leg separately.

**trous·seau** (trōō′sō, trōō-sō′) *n., pl.* **-seaux** (-sōz, -sōz′) *or* **-seaus.** [Fr. < OFr., dim. of *trousse,* bundle. —see TRUSS.] The personal possessions, as clothing and linens, that a bride assembles for her marriage.

**trout** (trout) *n., pl.* **trout** *or* **trouts.** [ME *troute* < OE *trūht* < LLat. *tructa* < Gk. *trōktēs,* a kind of seafish with sharp teeth < *trōgein,* to gnaw.] **1. a.** A freshwater or anadromous food or game fish of the genera *Salvelinus* or *Salmo,* usu. having a speckled body. **b.** A fish similar to the trout. **2.** *Chiefly Brit.* A silly old woman.

**trout lily** *n.* [From its spotted leaves.] The dogtooth violet.

**trou·vère** (trōō-vâr′) *also* **trou·veur** (-vûr′) *n.* [Fr. < OFr. *trovere* < *trover,* to compose, perh. < VLat. **tropare* < Lat. *tropus,* trope.] Any of a school of poets flourishing in 12th- and 13th-cent. northern

France who composed chiefly narrative works, as the chansons de geste.

**trove** (trōv) *n.* [(TREASURE-)TROVE.] **1.** Something valuable discovered or found : FIND. **2.** A collection of usu. valuable objects <a *trove* of English porcelain>

▲ word history: In origin the word *trove* is *trové,* the past participle of the Old French verb *trover,* "to find." The word entered English in the phrase *tresor trové,* literally "found treasure." The English form of *tresor trové* is *treasure-trove. Trove,* a shortened form of *treasure-trove,* is used synonymously with that phrase but it has also developed a new meaning of "storehouse, treasury."

**tro·ver** (trō'vər) *n.* [< OFr., to find. —see TROUVÈRE.] *Law.* A common-law action to recover damages for property illegally withheld or wrongfully converted to use by another.

**trow** (trō) *vi.* **trowed, trow·ing, trows.** [ME *trowen* < OE *trēowian.*] *Archaic.* To think : suppose.

**trow·el** (trou'əl) *n.* [ME *trowell* < OFr. *truelle* < LLat. *truella* < Lat. *trulla,* dim. of *trua,* ladle.] **1.** A flat-bladed hand tool for spreading, leveling, or shaping substances, as cement or mortar. **2.** A small implement with a pointed, scoop-shaped blade used for digging, as in setting plants. —*vt.* **-eled, -el·ing, -els** or **-elled, -el·ling, -els.** To spread, level, form, dig, or scoop with a trowel. **—trow'el·er** *n.*

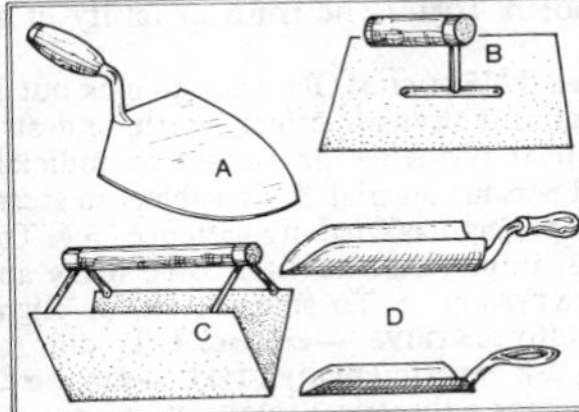


**trowel**
*Four types of trowels: A. brick, B. plaster, C. corner, and D. garden*

**trow·sers** (trou'zərz) *pl.n. var. of* TROUSERS.

**troy** (troi) *adj.* [ME, after *Troyes,* France.] Of or expressed in troy weight.

**troy weight** *n.* A system of units of weight in which the grain is the same as in the avoirdupois system and the pound contains 12 ounces, 240 pennyweights, or 5,760 grains.

**tru·ant** (trōō'ənt) *n.* [ME, beggar < OFr., of Celt. orig.] **1.** One who is absent without permission, esp. from school. **2.** One who shirks duty or work. —*adj.* **1.** Absent without permission, esp. from school. **2.** Idle, lazy, and neglectful. —*vi.* **-ant·ed, -ant·ing, -ants.** To be truant. **—tru'an·cy** (trōō'ən-sē), **tru'ant·ry** (-ən-trē) *n.*

**truant officer** *n.* An officer who investigates unauthorized absences from school.

**truce** (trōōs) *n.* [ME *trewes,* pl. of *trewe,* truce < OE *trēow,* pledge.] **1.** A temporary suspension or cessation of hostilities by agreement of the opposing sides : ARMISTICE. **2.** A respite from a disagreeable state of affairs. **—truce** *v.* **(truced, truc·ing, truc·es).**

**truck[1]** (trŭk) *n.* [Prob. short for TRUCKLE.] **1.** A heavy automotive vehicle used for transporting loads. **2.** A two-wheeled barrow for moving heavy objects by hand. **3.** A sometimes motorized wheeled platform for carrying loads in a warehouse or freight yard. **4.** *Chiefly Brit.* A railroad freight car without a top. **5.** One of the swiveling frames of wheels under each end of a railroad or trolley car. —*v.* **trucked, truck·ing, trucks.** —*vt.* To transport by truck. —*vi.* **1.** To carry goods by truck. **2.** To drive a truck.

**truck[2]** (trŭk) *v.* **trucked, truck·ing, trucks.** [ME *trukken* < OFr. *troquer.*] —*vt.* **1.** To exchange : barter. **2.** To peddle. —*vi.* To have dealings or commerce : TRAFFIC. —*n.* **1.** Trade goods. **2.** Garden produce raised for the market. **3.** *Informal.* Worthless articles : RUBBISH. **4.** Barter : exchange. **5.** *Informal.* Dealings : business <have no *truck* with mobsters>

**truck·age** (trŭk'ĭj) *n.* **1.** Transport of goods by truck. **2.** A charge for transportation by truck.

**truck·er** (trŭk'ər) *n.* **1.** A truck driver. **2.** One engaged in trucking goods.

**truck farm** *n.* A farm producing vegetables for the market. **—truck farmer** *n.* **—truck farming** *n.*

**truck·ing** (trŭk'ĭng) *n.* Transport of goods by truck.

**truck·le** (trŭk'əl) *n.* [ME *trocle,* pulley < AN < Lat. *trochlea,* system of pulleys. —see TROCHLEA.] **1.** A small wheel or roller : CASTER. **2.** A trundle bed. —*vi.* **-led, -ling, -les.** To be obsequiously servile. **—truck'ler** *n.*

**truck·load** (trŭk'lōd') *n.* The load that a truck carries.

**truck·man** (trŭk'mən) *n.* **1.** A truck driver. **2.** A crew member of a hook-and-ladder fire truck.

**truck system** *n.* Payment of wages in goods rather than by money.

**truc·u·lent** (trŭk'yə-lənt) *adj.* [Lat. *truculentus* < *trux,* fierce.] **1.** Savagely cruel : FIERCE. **2.** Vitriolic : scathing. **3.** Inclined to fight : PUGNACIOUS <a *truculent* bully> **—truc'u·lence, truc'u·len·cy** *n.* **—truc'u·lent·ly** *adv.*

**trudge** (trŭj) *vi.* **trudged, trudg·ing, trudg·es.** [Orig. unknown.] To walk in a heavy-footed, labored way : PLOD. —*n.* A long, laborious walk. **—trudg'er** *n.*

**trudg·en** *also* **trudg·eon** (trŭj'ən) *n.* [After John *Trudgen* (1852–1902).] A swimming stroke in which a double overarm movement is combined with a scissors kick.

**true** (trōō) *adj.* **tru·er, tru·est.** [ME < OE *trēowe,* loyal.] **1.** Being consistent with reality or fact. **2.** Conforming exactly to a rule, standard, or pattern. **3.** Reliable : accurate <a *true* prophecy> **4.** Genuine : real <a *true* artist> **5.** Faithful : loyal. **6.** *Archaic.* Honorable : upright. **7.** Sincerely expressed or felt : UNFEIGNED <*true* sympathy> **8.** Fundamental : essential <the criminal's *true* motive> **9.** Legitimate : RIGHTFUL <the *true* heir to the throne> **10.** Shaped or fitted accurately. **11.** Placed, delivered, or thrown accurately. **12.** Quick and exact in sensing and responding. **13.** Determined with reference to the earth's axis, not the magnetic poles <*true* south> **14.** Conforming to the definitive criteria of a natural group <The horseshoe crab is not a *true* crab.> —*adv.* **1.** Truthfully. **2.** Unswervingly : precisely <aimed *true* and hit the target> **3.** So as to conform to a type, standard, or pattern. —*vt.* **trued, tru·ing** or **true·ing, trues.** To adjust or fit so as to conform with a standard. —*n.* **1.** Truth. **2.** Proper alignment or adjustment <out of *true*> **—true'ness** *n.*

☆ ***syns:*** **1.** TRUE, RIGHTFUL *adj. core meaning* : being so legitimately <the *true* heir> ***ant:*** false **2.** TRUE, GENUINE, REAL, SINCERE *adj. core meaning* : devoid of hypocrisy or pretense <*true* grief>

**true bill** *n. Law.* A bill of indictment endorsed by a grand jury.

**true·blue** *also* **true blue** (trōō'blōō') *n.* [From an association of blue with constancy.] A truly loyal person. **—true'-blue'** *adj.*

**true·born** (trōō'bôrn') *adj.* Authentically or genuinely such by birth.

**true-false test** (trōō'fôls') *n.* A test in which statements are to be marked either true or false.

**true-life** (trōō'līf') *adj.* True to life.

**true·love** (trōō'lŭv') *n.* One's beloved : SWEETHEART.

**true lovers' knot** *n.* A stylized knot, gen. a bowknot, used as a symbol of love.

**true·pen·ny** (trōō'pĕn'ē) *n., pl.* **-nies.** [From an association with a genuine coin.] A trustworthy person.

**true rhyme** *n.* Perfect rhyme.

**true rib** *n.* Any of the ribs, esp. in humans any of the upper seven, that are attached to the sternum by a costal cartilage.

**truf·fle** (trŭf'əl) *n.* [OFr. *truffe* < OProv. *trufa* < VLat. **tufera* < Lat. *tuber.*] Any of various fleshy edible subterranean fungi, chiefly of the genus *Tuber.*

**tru·ism** (trōō'ĭz'əm) *n.* An obvious truth. **—tru·is'tic** (trōō-ĭs'tĭk) *adj.*

**trull** (trŭl) *n.* [Perh. < G. *Trulle.*] A prostitute.

**tru·ly** (trōō'lē) *adv.* **1.** Sincerely : genuinely <*truly* sorry> **2.** Truthfully : accurately. **3.** Indeed <*truly* ugly> **4.** Properly.

**trump[1]** (trŭmp) *n.* [Alteration of TRIUMPH.] **1. a.** *often* **trumps.** A suit in card games which outranks all other suits for the duration of a hand. **b.** A card of such a suit. **2.** A key resource to be used at the opportune moment. **3.** *Informal.* One who is reliable or admirable. —*v.* **trumped, trump·ing, trumps.** —*vt.* To take (a card or trick) with a trump. —*vi.* To play a trump card. **—trump up.** To devise fraudulently <*trumped up* false charges>

**trump[2]** (trŭmp) *n.* [ME *trompe* < OFr.] A trumpet.

**trump·er·y** (trŭm'pə-rē) *n., pl.* **-ies.** [ME *trompery,* deceit < OFr. *tromperie* < *tromper,* to deceive.] **1.** Showy, worthless finery. **2.** Nonsense. **3.** Fraud : deception.

**trum·pet** (trŭm'pĭt) *n.* [ME *trumpette* < OFr. *trompette,* dim. of *tromp,* trumpet.] **1.** A soprano brass wind instrument composed of a long metal tube looped once and ending in a flared bell. **2.** Something shaped like or sounding like a trumpet. **3.** An organ stop producing a tone like that of a trumpet. **4.** A resounding call, as that of the elephant. —*v.* **-pet·ed, -pet·ing, -pets.** —*vi.* **1.** To play a trumpet. **2.** To emit a resounding call. —*vt.* To sound or proclaim loudly.

**trumpet creeper** *n.* A woody vine, *Campsis radicans* of the eastern United States, with compound leaves and trumpet-shaped reddish-orange flowers.

**trum·pet·er** (trŭm'pĭt-ər) *n.* **1.** A trumpet player. **2.** One who announces something, as on a trumpet : HERALD. **3.** Any of several large birds of the genus *Psophia* of tropical South America, having a loud resonant call. **4.** The trumpeter swan.

**trumpeter swan** *n.* A large white swan, *Olor buccinator* of western North America, with a loud buglelike call.

**trumpet honeysuckle** *n.* A vine, *Lonicera sempervirens* of the eastern United States, with reddish tubular flowers.

**trumpet vine** *n.* Trumpet creeper.

**trun·cate** (trŭng'kāt') *vt.* **-cat·ed, -cat·ing, -cates.** [Lat. *truncare, truncat-* < *truncus,* trunk.] **1.** To shorten by or as if by cutting off. **2.** To replace (the edge of a crystal) with a plane face. —*adj.*

1. Appearing to terminate abruptly, as a leaf or a coiled gastropod shell that lacks a spire. 2. Truncated. **—trun′cate′ly** *adv.* **—trun·ca′tion** *n.*

**trun·ca·ted** (trŭng′kā′tĭd) *adj.* Having the apex cut off and replaced by a plane, esp. one parallel to the base. —Used of a cone or pyramid.

**trun·cheon** (trŭn′chən) *n.* [ME *tronchon,* club < OFr. < *truncus,* trunk.] **1.** A staff carried as a symbol of office or authority : BATON. **2.** A short stick carried by police officers : BILLY. **3.** A heavy club : CUDGEL. **4.** *Obs.* A thick cutting from a plant, as for grafting. —*vt.* **-cheoned, -cheon·ing, -cheons.** To beat with a truncheon.

**trun·dle** (trŭn′dl) *n.* [Var. of dial. *trendle,* wheel < ME *trendel* < OE, circle.] **1.** A small wheel or roller. **2.** The motion or noise of rolling. **3.** A trundle bed. **4.** A low-wheeled cart : DOLLY. —*v.* **-dled, -dling, -dles.** —*vt.* **1.** To push or propel on wheels or rollers. **2.** To spin : twirl. —*vi.* To move along by or as if by rolling. **—trun′dler** *n.*

**trundle bed** *n.* A low bed on casters that can be rolled under another bed when not in use.

**trunk** (trŭngk) *n.* [ME *troncke* < OFr. *tronc* < Lat. *truncus.*] **1.** The principal woody axis of a tree. **2. a.** The human body excluding the head and limbs : TORSO. **b.** A part of an organism, as the thorax of an insect, analogous to the human torso. **3.** A main body, apart from tributaries or appendages. **4.** A trunk line. **5.** A large packing case that clasps shut, used as luggage or for storage. **6.** A covered compartment for luggage and storage, gen. at the rear of a car. **7.** A proboscis, esp. that of an elephant. **8.** A chute or conduit. **9.** *Naut.* A shaft connecting two or more decks. **10.** *Naut.* The housing for a vessel's centerboard. **11.** *Naut.* Any of certain structures projecting above part of a main deck, as: **a.** A covering over a ship's hatches. **b.** An expansion chamber on a tanker. **c.** A cabin on a small boat. **12.** The shaft of a column. **13. trunks.** Men's shorts worn for athletics, as swimming or running.

**trunk·fish** (trŭngk′fĭsh′) *n., pl.* **trunkfish** *or* **-fish·es.** Any of various tropical marine fishes of the family Ostraciidae, with boxlike armor enclosing the body.

**trunk hose** *pl.n.* [Perh. < obs. *trunk,* to truncate.] Short, ballooning breeches extending from the waist to midthigh, worn by 16th- and 17th-cent. men.

**trunk line** *n.* **1.** A direct line between two telephone switchboards. **2.** The main line of a transportation system.

**trun·nel** (trŭn′əl) *n. var. of* TREENAIL.

**trun·nion** (trŭn′yən) *n.* [Fr. *trognon,* stump.] A pin or gudgeon, esp. either of two small cylindrical projections on a cannon forming an axis on which it pivots.

**truss** (trŭs) *n.* [ME *trusse,* bundle < OFr. *trousse* < *trousser,* to truss.] **1.** *Med.* A supportive device worn to prevent enlargement of a hernia or the return of a reduced hernia. **2.** A rigid framework designed to support a structure. **3.** BRACKET 1. **4.** Matter gathered into a bundle : PACK. **5.** *Naut.* An iron fitting by which a lower yard is secured to a mast. **6.** A compact flower cluster at the end of a stalk. —*vt.* **trussed, truss·ing, truss·es. 1.** To tie up or bind tightly. **2.** To bind or skewer the wings or legs of (a fowl) before cooking. **3.** To brace or support with a truss.

**truss bridge** *n.* A bridge supported by trusses.

**trust** (trŭst) *n.* [ME *truste,* perh. < ON *traust,* confidence.] **1.** Total confidence in the integrity, ability, and good character of another. **2.** One in whom confidence is placed. **3.** Custody : care. **4.** Something committed into the care of another : CHARGE. **5.** The condition and resulting obligation of having confidence placed in one <officials who had violated their public *trust*> **6.** Reliance on something in the future : HOPE. **7.** Reliance on the intention and ability of a purchaser to pay in the future : CREDIT. **8.** *Law.* **a.** A legal title to property held by one party for the benefit of another. **b.** The confidence reposed in a trustee in giving him or her legal title to property to administer for another, and his or her obligation with respect to the property and the beneficiary. **c.** The property so held. **9.** A combination of firms or corporations for the purpose of reducing competition and controlling prices throughout a business or industry <tried to break the great steel *trusts*> —*v.* **trust·ed, trust·ing, trusts.** —*vi.* **1.** To depend : rely. **2.** To be confident : HOPE. **3.** To sell on credit. —*vt.* **1.** To have confidence in. **2.** To expect with assurance : ASSUME <I *trust* you will not be late.> **3.** To believe <I *trust* what you say.> **4.** To place in the care of another : ENTRUST. **5.** To rely or depend on confidently <Shall I *trust* them with the keys?> **6.** To extend credit to. **—in trust.** In the possession or care of a trustee <an estate held *in trust*> **—trust′er** *n.*

**trust·bust·er** (trŭst′bŭs′tər) *n. Informal.* One who seeks to prosecute or dissolve business trusts.

**trust company** *n.* A commercial bank that manages trusts.

**trus·tee** (trŭs-tē′) *n.* **1.** One, as a bank, that holds legal title to property in order to administer it for a beneficiary. **2.** An elected or appointed member of a board empowered to manage the funds and direct the policy of an institution. **3.** A country having the responsibility of supervising a trust territory. —*v.* **-teed, -tee·ing, -tees.** —*vt.* To place (property) in the care of a trustee. —*vi.* To serve or function as a trustee.

**trus·tee·ship** (trŭs-tē′shĭp′) *n.* **1.** The position or function of a trustee. **2. a.** Administration of a territory by a country or countries so commissioned by the United Nations. **b.** A trust territory.

**trust·ful** (trŭst′fəl) *adj.* Full of trust <a *trustful* little child> **—trust′ful·ly** *adv.* **—trust′ful·ness** *n.*

**trust fund** *n.* Property, esp. money and securities, that are held or settled in trust.

**trust territory** *n.* A colony or territory placed under the administration of a country or countries by commission of the United Nations.

**trust·wor·thy** (trŭst′wûr′thē) *adj.* **-thi·er, -thi·est.** Worthy of trust : RELIABLE. **—trust′wor′thi·ly** *adv.* **—trust′wor′thi·ness** *n.*

**trust·y** (trŭs′tē) *adj.* **-i·er, -i·est.** Trustworthy. —*n., pl.* **-ies.** A trusted person, esp. a convict deemed worthy of trust and therefore granted special privileges. **—trust′i·ly** *adv.* **—trust′i·ness** *n.*

**truth** (trōōth) *n., pl.* **truths** (trōōthz, trōōths) [ME *trewthe,* fidelity < OE *trēowð.*] **1.** Conformity to fact or actuality. **2.** Fidelity to an original or standard. **3.** Reality : actuality. **4.** A statement proven to be or accepted as true. **5.** Sincerity : integrity. **6. Truth.** *Christian Science.* GOD 1c.

**truth·ful** (trōōth′fəl) *adj.* **1.** Telling the truth. **2.** Conforming to reality. **—truth′ful·ly** *adv.* **—truth′ful·ness** *n.*

**truth serum** *n.* Any of various hypnotic drugs that are assumed to cause a subject under interrogation to talk uninhibitedly.

**truth-val·ue** (trōōth′văl′yōō) *n. Logic.* The truth or falsity of a proposition.

**try** (trī) *v.* **tried, try·ing, tries.** [ME *trien* < OFr. *trier,* to pick out.] —*vt.* **1.** To test in order to determine strength, effect, worth, or desirability. **2. a.** To examine or hear (evidence or a case) by judicial process. **b.** To put (an accused person) on trial. **3.** To subject to great strain or hardship : TAX <This project has *tried* my patience.> **4.** To melt (e.g., lard) to separate out impurities : RENDER. **5.** To make an effort to do or accomplish : ATTEMPT. **6.** To smooth, fit, or align accurately. —*vi.* To make an effort : STRIVE. **—try on. 1.** To don (a garment) to test the fit. **2.** To use experimentally : TEST. **—try out.** To undergo a qualifying test, as for a theatrical role. —*n., pl.* **tries.** An effort : attempt. **—try one's hand.** To attempt to do something for the first time <*tried my hand* at waterskiing>

**try·ing** (trī′ĭng) *adj.* Causing strain, hardship, or distress.

**try·out** (trī′out′) *n.* A test to determine the qualifications of applicants, as for a theatrical role.

**try·pan·o·some** (trĭ-păn′ə-sōm′) *n.* [NLat. *Trypanosoma,* genus name : Gk. *trupanon,* auger + Gk. *sōma,* body.] Any of various parasitic protozoans of the genus *Trypanosoma,* that are transmitted to the vertebrate blood stream by certain insects, and often cause diseases such as sleeping sickness. **—try·pan′o·som′ic** *adj.*

**try·pan·o·so·mi·a·sis** (trĭ-păn′ə-sō-mī′ə-sĭs) *n., pl.* **-ses** (-sēz′). A disease caused by a trypanosome.

**try·pars·a·mide** (trĭ-pär′sə-mīd′) *n.* [TRYP(ARSAMIDE) + ARS(ENIC) + AMIDE.] A white crystalline powder, $C_8H_{10}AsN_2O_4Na \cdot 1/2H_2O$, used in treating spirochetal and trypanosomic diseases.

**tryp·sin** (trĭp′sĭn) *n.* [Perh. Gk. *tripsis,* a rubbing (from its having been first obtained by rubbing a pancreas with glycerin) + -IN.] One of the proteolytic enzymes of the pancreatic juice, active in the digestive processes. **—tryp′tic** (-tĭk) *adj.*

**tryp·sin·o·gen** (trĭp-sĭn′ə-jən) *n.* The substance produced by the pancreas that is converted into trypsin when acted upon by certain enzymes.

**tryp·to·phan** (trĭp′tə-făn′) *also* **tryp·to·phane** (-fān′) *n.* [TRYPT(IC) + -PHAN(E).] An amino acid, $C_{11}H_{12}N_2O_2$, produced in the digestive process and essential in human nutrition.

**try·sail** (trī′səl, -sāl′) *n.* [TRY, to lie to in a storm (obs.) + SAIL.] *Naut.* A small fore-and-aft sail hoisted abaft the foremast and mainmast in a storm to keep a ship's bow to the wind.

**try square** *n.* A ruled metal straightedge set at right angles to a wooden straight piece, used esp. by carpenters to measure and mark square work.

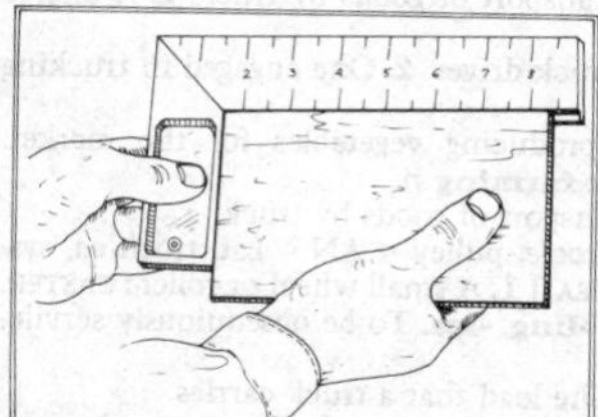

try square

**tryst** (trĭst) *n.* [ME *trist* < OFr. *triste,* an appointed station in hunting.] **1.** An agreement to meet at a certain time and place <a lovers'

ă pat ā pay âr care ä father ĕ pet ē be hw which ĭ pit
ī tie îr pier ŏ pot ō toe ô paw, for oi noise ŏŏ took

*tryst*> **2.** A meeting or meeting place that has been agreed on. —*vi.* **tryst·ed, tryst·ing, trysts.** To keep a tryst. **—tryst'er** *n.*

**tsa·de** (tsä'də, -dē) *n. var. of* SADE.

**tsar** (tsär) *n. var. of* CZAR.

**tset·se disease** (tsĕt'sē, tsē'tsē) *n.* Nagana.

**tset·se fly** *also* **tzet·ze fly** (tsĕt'sē, tsē'tsē) *n.* [Tswana *tsetse.*] A bloodsucking African fly of the genus *Glossina,* often carrying and transmitting pathogenic trypanosomes to humans and livestock.

**Tshi** (chwē, chē) *n. var. of* TWI.

**T-shirt** *also* **tee shirt** (tē'shûrt') *n.* [From its being shaped like the letter *T.*] **1.** A short-sleeved or sleeveless collarless undershirt. **2.** An outer shirt of a design similar to the T-shirt.

**T-square** (tē'skwâr') *n.* A rule having a short, sometimes sliding, perpendicular crosspiece at one end, used by drafters for establishing and drawing parallel lines.

**tsu·na·mi** (tsōō-nä'mē) *n.* [J. : *tsu,* port + *nami,* wave.] A huge ocean wave caused by an underwater earthquake or a volcanic eruption.

**tsu·tsu·ga·mu·shi disease** (tsōō'tsōō-gə-mōō'shē) *n.* [J. *tsutsugamushi* : *tsutsuga,* illness + *mushi,* tick.] Scrub typhus.

**Tswa·na** (tswä'nə, sä'-) *n.* **1.** A Bantu people of southern Africa, living chiefly in Botswana. **2.** The Sotho language of the Tswanas.

**Tua·reg** (twä'rĕg') *n., pl.* **Tuareg** *or* **-regs.** [Ar. *Tawāriq.*] A member of one of the tall, nomadic, Hamitic-speaking peoples who occupy western and central Sahara and an area along the Niger and who have adopted the Moslem religion.

**tu·a·ta·ra** (tōō'ə-tär'ə) *n.* [Maori *tuatāra.*] A large spiny reptile, *Sphenodon punctatus* of New Zealand, the only surviving representative of the order Rhynchocephalia.

**tub** (tŭb) *n.* [ME *tubbe* < MDu.] **1. a.** A round, open, flat-bottomed vessel, usu. wider than it is tall, used for packing, storing, or washing. **b.** The amount held by a tub. **c.** The contents of a tub. **2. a.** A bathtub. **b.** *Informal.* A bath taken in a bathtub. **3.** *Informal.* A clumsy, slow boat. **4. a.** A bucket for conveying ore or coal up a mine shaft. **b.** A coal car used in a mine. —*v.* **tubbed, tub·bing, tubs.** —*vt.* **1.** To pack or store in a tub. **2.** To wash or bathe in a tub. —*vi.* To take a bath. **—tub'ba·ble** *adj.* **—tub'ber** *n.* **—tub'ful** *n.*

**tu·ba** (tōō'bə, tyōō'-) *n.* [Ital. < Lat., trumpet.] **1.** A large brass musical wind instrument with valves and a bass pitch. **2.** A reed stop in an organ, having eight-foot pitch.

**tu·bal** (tōō'bəl, tyōō'-) *adj.* Of, relating to, or taking place in a tube, esp. the Fallopian tube <a *tubal* ligation> <a *tubal* pregnancy>

**tu·bate** (tōō'bāt', tyōō'-) *adj.* Forming or having a tube.

**tub·by** (tŭb'ē) *adj.* **-bi·er, -bi·est. 1.** Short and fat. **2.** Lacking resonance. —Used of a musical instrument. **—tub'bi·ness** *n.*

**tube** (tōōb, tyōōb) *n.* [Fr. or < Lat. *tubus.*] **1. a.** A hollow cylinder, esp. one conveying a fluid or functioning as a passage. **b.** An organic structure so shaped or so functioning : DUCT. **2.** A small, flexible cylindrical container sealed at one end and having a screw cap at the other, used for pigments, toothpaste, etc. **3.** The cylindrical part of a wind instrument. **4. a.** An electron tube. **b.** A vacuum tube. **5.** *Bot.* The lower joined part of a gamopetalous corolla or a gamosepalous calyx. **6.** *Chiefly Brit.* SUBWAY 1. **7.** A tunnel. **8.** A flexible airtight cylinder inserted into the casing of a pneumatic tire for holding air under pressure. **9.** *Informal.* **a.** Television. **b.** A television set. —*vt.* **tubed, tub·ing, tubes. 1.** To provide with a tube or insert a tube in. **2.** To place in or enclose in a tube. **—down the tubes** (*or* **tube**). *Slang.* Into a state of failure or ruin.

**tube foot** *n.* One of the numerous external, fluid-filled muscular tubes of echinoderms, as the starfish, serving chiefly as organs of locomotion.

**tube·less tire** (tōōb'lĭs, tyōōb'-) *n.* A pneumatic tire in which the air is held in the assembly of the casing and rim without an inner tube.

**tu·ber** (tōō'bər, tyōō'-) *n.* [Lat., lump.] **1.** *Bot.* A swollen, usu. underground stem, as the potato, bearing buds from which new plant shoots arise. **2.** *Anat.* A swelling.

**tu·ber·cle** (tōō'bər-kəl, tyōō'-) *n.* [Lat. *tuberculum,* dim. of *tuber,* lump.] **1.** A small rounded prominence or process, as a wartlike excrescence on the roots of some leguminous plants or a knoblike process in the skin or on a bone. **2.** *Pathol.* **a.** A nodule or swelling. **b.** A tubercular lesion.

**tubercle bacillus** *n.* A rod-shaped bacterium, *Mycobacterium tuberculosis,* that causes tuberculosis.

**tu·ber·cu·lar** (tōō-bûr'kyə-lər, tyōō-) *adj.* **1.** TUBERCULATE 1. **2.** Of, pertaining to, or afflicted with tuberculosis. —*n.* One having tuberculosis.

**tu·ber·cu·late** (tōō-bûr'kyə-lĭt, tyōō-) *also* **tu·ber·cu·lat·ed** (-lā'tĭd) *adj.* **1.** Of, pertaining to, or covered with tubercles. **2.** TUBERCULAR 2. **—tu·ber'cu·late·ly** *adv.* **—tu·ber'cu·la'tion** *n.*

**tu·ber·cu·lin** (tōō-bûr'kyə-lĭn, tyōō-) *n.* [Lat. *tuberculum,* tubercle + -IN.] A substance derived from cultures of tubercle bacilli and used in diagnosing and treating tuberculosis.

**tuberculin test** *n.* A test for determining past or present infection with the tubercle bacillus and based on hypersensitivity to tuberculin.

**tu·ber·cu·loid** (tōō-bûr'kyə-loid', tyōō-) *adj.* **1.** Suggestive of tuberculosis. **2.** Resembling a tubercle.

**tu·ber·cu·lo·sis** (tōō-bûr'kyə-lō'sĭs, tyōō-) *n.* [Lat. *tuberculum,* tubercle + -OSIS.] **1.** A communicable disease of humans and animals caused by a microorganism, *Mycobacterium tuberculosis,* and manifesting itself in lesions of the lung, bone, and other bodily parts. **2.** Tuberculosis of the lungs.

**tu·ber·cu·lous** (tōō-bûr'kyə-ləs, tyōō-) *adj.* [NLat. *tuberculosus* < Lat. *tuberculum,* tubercle.] **1.** TUBERCULAR 2. **2.** Of, afflicted with, or caused by tubercles. **—tu·ber'cu·lous·ly** *adv.*

**tube·rose[1]** (tōōb'rōz', tyōōb'-, tōō'bə-rōz', -rōs', tyōō'-) *n.* [NLat. *tuberosa,* specific epithet < fem. of Lat. *tuberosus,* full of lumps.] A tuberous plant, *Polianthes tuberosa,* native to Mexico, cultivated for its fragrant white flowers.

**tu·ber·ose[2]** (tōō'bə-rōs', tyōō'-) *adj. var. of* TUBEROUS.

**tu·ber·os·i·ty** (tōō'bə-rŏs'ĭ-tē, tyōō'-) *n., pl.* **-ties.** A protuberance, esp. one at the end of a bone for the attachment of a muscle or tendon.

**tu·ber·ous** (tōō'bər-əs, tyōō'-) *also* **tu·ber·ose** (-bə-rōs') *adj.* [Lat. *tuberosus,* full of lumps < *tuber,* lump.] *Bot.* **1.** Producing or bearing tubers. **2.** Resembling a tuber <*tuberous* roots>

**tu·bi·fex** (tōō'bə-fĕks', tyōō'-) *n., pl.* **tubifex** *or* **-fex·es.** [NLat. *Tubifex,* genus name : Lat. *tubus,* tube + Lat. *facere,* to make.] Any of various small, slender, reddish freshwater worms of the genus *Tubifex,* often used as food for tropical aquarium fish.

**tub·ing** (tōō'bĭng, tyōō'-) *n.* **1. a.** Tubes as a whole. **b.** A system of tubes. **2.** A length of tube. **3.** Tubular fabric, such as that used in making pillowcases.

**tu·bu·lar** (tōō'byə-lər, tyōō'-) *adj.* **1.** Of, relating to, or having the form of a tube. **2.** Constituting or consisting of tubes. **—tu'bu·lar'i·ty** (-lăr'ĭ-tē) *n.* **—tu'bu·lar·ly** *adv.*

**tu·bu·late** (tōō'byə-lĭt, -lāt', tyōō'-) *also* **tu·bu·lat·ed** (-lā'tĭd) *adj.* [Lat. *tubulatus* < *tubulus,* dim. of *tubus,* tube.] **1.** Formed into or suggestive of a tube : TUBULAR. **2.** Having a tube. **—tu'bu·la'tion** *n.* **—tu'bu·la'tor** *n.*

**tu·bule** (tōō'byōōl, tyōō'-) *n.* [Lat. *tubulus,* dim. of *tubus,* tube.] A very small tube.

**tu·bu·lif·er·ous** (tōō'byə-lĭf'ər-əs, tyōō'-) *adj.* Having or consisting of tubules.

**tu·bu·li·flo·rous** (tōō'byə-lə-flôr'əs, -flōr'-, tyōō'-) *adj.* Having flowers or florets with tubular corollas.

**tu·bu·lous** (tōō'byə-ləs, tyōō'-) *adj.* [< Lat. *tubulus,* dim. of *tubus,* tube.] **1.** Tubular. **2.** Consisting of tubes or having tubular parts. **—tu'bu·lous·ly** *adv.*

**Tu·ca·na** (tōō-kā'nə, -kä'-, tyōō-) *n.* [Tupi *tucana,* toucan.] A constellation in the Southern Hemisphere.

**tu·chun** (dōō'jün', -jōōn') *n., pl.* **-chuns** *or* **tuchun.** [Chin. (Mandarin) *du[1] jun[1]* : *du[1],* to supervise + *jun[1],* army.] **1.** A Chinese provincial military governor. **2.** A Chinese warlord.

**tuck[1]** (tŭk) *v.* **tucked, tuck·ing, tucks.** [ME *tukken* < OE *tūcian,* to torment.] —*vt.* **1.** To make one or more folds in. **2.** To gather up and push in the lower end of so as to secure tightly <*tuck* one's shirt into one's pants> **3. a.** To put into a snug spot. **b.** To put in an out-of-the-way place <a cabin *tucked* into a forest> **c.** To store in a safe spot : SAVE <*tuck* away money> **4.** To draw in : CONTRACT. —*vi.* To make tucks. —*n.* **1.** A flattened, often very narrow pleat or fold stitched in place. **2.** The act of tucking. **3.** *Naut.* The part of a ship's hull under the stern where the ends of the bottom planks meet. **4.** *Chiefly Brit.* Food, esp. sweets and pastry. **5. a.** A bodily position used in sports, as diving, in which the knees are bent, the thighs are drawn close to the chest, and the hands are clasped around the shins. **b.** A skiing position in which the skier squats while holding the poles parallel to the ground and under the arms.

**tuck[2]** (tŭk) *n.* [< ME *tukken,* to beat a drum < ONFr. *toquer,* to strike.] A beat, esp. on a drum.

**tuck[3]** (tŭk) *n.* [OFr. *étoc,* stick, of Germanic orig.] *Archaic.* A slender sword : RAPIER.

**tuck[4]** (tŭk) *n.* [Orig. unknown.] Vigor : energy.

**tuck·a·hoe** (tŭk'ə-hō') *n.* [Algonquian *taccaho.*] Any of various plants or plant parts used by American Indians as food, esp. the edible rootstocks of certain arums or the sclerotium of certain fungi.

**tuck·er[1]** (tŭk'ər) *n.* **1.** One that tucks, esp. a sewing-machine attachment that makes tucks. **2.** A piece of linen or frill of lace once worn around the neckline of a dress.

**tuck·er[2]** (tŭk'ər) *vt.* **-ered, -er·ing, -ers.** [< TUCK[1].] *Informal.* To make tired : EXHAUST <all *tuckered* out>

**tuck·er-bag** (tŭk'ər-băg') *n. Austral.* A bag for carrying food, used by a traveler in the bush or by a swagman.

**tuck·et** (tŭk'ĭt) *n.* [ME *tuk* < *tukken,* to beat a drum. —see TUCK[2].] A trumpet fanfare.

**tuck pointing** *n.* The pointing of grooved mortar joints with a thin ridge of fine lime mortar or putty.

**tuck-shop** (tŭk'shŏp') *n. Chiefly Brit.* A confectionery.

**-tude** *suff.* [OFr. < Lat. *-tudo.*] A condition, state, or quality <exac*titude*>

**Tu·dor** (to͞o'dər, tyo͞o'-) *adj.* **1.** Of or relating to the royal house that ruled England from 1485 through 1603. **2. a.** Of, relating to, or characteristic of the architecture of the Tudor period. **b.** Of, relating to, or characteristic of an architectural style derived from the Tudor period, having exposed beams as a typical feature.

**Tues·day** (to͞oz'dē, -dā', tyo͞oz'-) *n.* [ME *Tuesdai* < OE *Tīwesdæg,* Tiu's day.] The third day of the week, following Monday and preceding Wednesday.

**tu·fa** (to͞o'fə, tyo͞o'-) *n.* [Obs. Ital. < Lat. *tofus.*] **1.** Calcareous and siliceous rock deposits of springs, lakes, or ground water. **2.** Tuff. **—tu·fa'ceous** (-fā'shəs) *adj.*

**tuff** (tŭf) *n.* [OFr. *tuf* < OItal. *tufo,* tufa.] A rock made up of compacted volcanic ash varying in size from fine sand to coarse gravel. **—tuff·a'ceous** (tŭ-fā'shəs) *adj.*

**tuf·fet** (tŭf'ĭt) *n.* [Alteration of TUFT.] **1.** A clump of grass. **2.** A low seat.

**tu·fo·li** (to͞o-fō'lē, tyo͞o-) *n.* [Sicilian, pl. of *tufolo,* duct < LLat. *tubulus,* dim. of Lat. *tubus,* tube.] A large macaroni shell.

**tuft** (tŭft) *n.* [ME, prob. < OFr. *tofe.*] **1.** A short cluster of elongated strands, as of yarn, hair, or grass, attached at the base or growing close together. **2.** A dense clump, esp. of bushes or trees. **3.** A goatee. *—v.* **tuft·ed, tuft·ing, tufts.** *—vt.* **1.** To furnish or decorate with a tuft. **2.** To pass threads through the layers of (e.g., a quilt or mattress), securing the thread ends with a button or knot in the depressions created. *—vi.* **1.** To form or separate into tufts. **2.** To grow in a tuft or tufts. **—tuft'er** *n.* **—tuft'y** *adj.*

**tug** (tŭg) *v.* **tugged, tug·ging, tugs.** [ME *tuggen.*] *—vt.* **1.** To pull and strain at vigorously. **2.** To move by pulling or straining : DRAG. **3.** To tow by tugboat. *—vi.* **1.** To pull hard <*tug* at one's boots> **2.** To toil or struggle : STRAIN. **3.** To vie : contend. *—n.* **1.** A strong pull or pulling force <the *tug* of the waves> **2.** A contest : struggle. **3.** A tugboat. **4.** A rope, chain, or strap used in hauling, esp. a harness trace. **—tug'ger** *n.*

**tug·boat** (tŭg'bōt') *n.* A small powerful boat designed for towing larger vessels.

**tug of war** *n.* **1.** A contest of strength in which two teams tug on opposite ends of a rope, each trying to pull the other across a dividing line. **2.** A struggle for supremacy.

**tu·grik** (to͞o'grĭk) *n.* [Mongolian *dughurik.*] —See table at CURRENCY.

**tuille** (twēl) *n.* [ME *toile* < OFr. *tieule* < Lat. *tugula,* tile < *tegere,* to cover.] A steel thigh protector used in medieval armor.

**tu·i·tion** (to͞o-ĭsh'ən, tyo͞o-) *n.* [ME *tuicion,* protection < OFr. < Lat. *tuitio* < *tueri,* to protect.] **1.** A fee for instruction, esp. at an institution of learning. **2.** Instruction. **3.** *Archaic.* Guardianship. **—tu·i'tion·al, tu·i'tion·ar'y** (-ə-nĕr'ē) *adj.*

**tu·la·re·mi·a** (to͞o'lə-rē'mē-ə, tyo͞o'-) *n.* [NLat., after *Tulare* county, California.] An infectious disease caused by the bacterium *Pasteurella tularensis,* transmitted from infected rodents to humans by insect vectors or by handling infected animals and marked by fever and swelling of the lymph nodes. **—tu'la·re'mic** *adj.*

**tu·le** (to͞o'lē) *n.* [Sp. < Nahuatl *tollin,* reed.] Any of several bulrushes of the genus *Scirpus,* growing in marshes of the southwestern United States.

**tu·lip** (to͞o'lĭp, tyo͞o'-) *n.* [NLat. *Tulipa,* genus name < Turk. *tülibend,* turban < Pers. *dulband.*] **1.** Any of several bulbous plants of the genus *Tulipa,* native to Asia, cultivated for their showy, variously colored flowers. **2.** The flower of the tulip.

**tulip poplar** *n.* The tulip tree.

**tulip tree** *n.* A tall deciduous tree, *Liriodendron tulipifera,* with large tuliplike green and orange flowers and yellowish soft wood.

**tu·lip·wood** (to͞o'lĭp-wo͝od', tyo͞o'-) *n.* **1.** The wood of the tulip tree. **2.** The irregularly striped ornamental wood of a tree related or similar to the tulipwood, esp. that of *Dalbergia variabilis* of tropical South America.

**tulle** (to͞ol) *n.* [Fr., after *Tulle,* France.] A fine, often starched net.

**tum·ble** (tŭm'bəl) *v.* **-bled, -bling, -bles.** [ME *tumblen,* freq. of *tumben,* to dance < OE *tumbian.*] *—vi.* **1.** To perform acrobatic feats such as somersaults. **2. a.** To fall or roll end over end. **b.** To spill or roll out in confusion or disorder. **c.** To pitch headlong : FALL. **d.** To proceed haphazardly. **3. a.** To topple, as from power : FALL. **b.** To collapse. **c.** To drop <Stock prices *tumbled.*> **4.** To come upon accidentally <*tumbled* on a fine country inn> **5.** *Slang.* To come to a sudden understanding. *—vt.* **1.** To cause to fall. **2.** To put, spill, or toss haphazardly. **3.** To toss or whirl in a drum, tumbler, or tumbling box. *—n.* **1.** An act of tumbling : FALL. **2.** Confusion : disorder.

**tum·ble·bug** (tŭm'bəl-bŭg') *n.* Any of various beetles of the family Scarabaeidae that roll up balls of dung to protect their eggs and serve as food for the newly hatched larvae.

**tum·ble-down** (tŭm'bəl-doun') *adj.* Rickety and dilapidated.

**tum·bler** (tŭm'blər) *n.* **1.** One that tumbles, esp. a gymnast or an acrobat. **2. a.** A drinking glass, orig. with a rounded bottom. **b.** A flat-bottomed glass without a handle, foot, or stem. **c.** The contents of a drinking glass. **3.** A toy constructed with a rounded, weighted base so that it can rock over and then right itself. **4.** One of a breed of domestic pigeons that somersault in flight. **5.** A piece in a gunlock that forces the hammer forward via action of the mainspring. **6.** The part in a lock that releases the bolt when moved by a key. **7. a.** The drum in a clothes dryer. **b.** A tumbling box. **8. a.** A projecting piece on a revolving or rocking part in a mechanism that transmits motion to the part it engages. **b.** The rocking frame that moves a gear into place in a transmission, as in a car. **—tum'bler·ful'** *n.*

**tum·ble·weed** (tŭm'bəl-wēd') *n.* A densely branched New World plant, chiefly of the genus *Amaranthus,* that when withered breaks off and is rolled about by the wind, esp. *A. albus* of western prairies.

**tum·bling** (tŭm'blĭng) *n.* The sport, skill, or practice of gymnastic falling, rolling, or somersaulting.

**tumbling box** *n.* A revolving drum in which objects are dried, reduced in size, polished, or cleaned.

**tum·brel** *or* **tum·bril** (tŭm'brəl) *n.* [ME *tumberell* < OFr. *tomberel* < *tomber,* to let fall, of Germanic orig.] **1.** A two-wheeled cart, esp. a farmer's cart that can be tilted. **2.** A crude cart used to carry condemned prisoners to their executions, as during the French Revolution.

**tumbrel**

**tu·me·fa·cient** (to͞o'mə-fā'shənt, tyo͞o'-) *adj.* [Lat. *tumefaciens, tumefacient-,* pr.part. of *tumefacere,* to tumefy : *tumēre,* to swell + *facere,* to make.] Producing or tending to produce tumefaction.

**tu·me·fac·tion** (to͞o'mə-făk'shən, tyo͞o'-) *n.* [OFr. < Lat. *tumefacere,* to tumefy. —see TUMEFACIENT.] **1. a.** The act or process of swelling. **b.** TUMESCENCE 1b. **2.** TUMESCENCE 2. **—tu'me·fac'tive** *adj.*

**tu·me·fy** (to͞o'mə-fī', tyo͞o'-) *vi.* & *vt.* **-fied, -fy·ing, -fies.** [OFr. *tumefier* < Lat. *tumēre.*] To swell or cause to swell.

**tu·mes·cence** (to͞o-mĕs'əns, tyo͞o-) *n.* **1. a.** A swelling. **b.** A swollen condition. **2.** A swollen part or organ.

**tu·mes·cent** (to͞o-mĕs'ənt, tyo͞o-) *adj.* [Lat. *tumescens, tumescent-,* pr.part. of *tumescere,* to begin to swell < *tumēre,* to swell.] Somewhat swollen.

**tu·mid** (to͞o'mĭd, tyo͞o'-) *adj.* [Lat. *tumidus* < *tumēre,* to swell.] **1.** Swollen : distended. —Used of a bodily part or organ. **2.** Bulging in shape : PROTUBERANT. **3.** Bombast : overblown <*tumid* political rhetoric> **—tu·mid'i·ty** (-mĭd'ĭ-tē), **tu'mid·ness** *n.* **—tu'mid·ly** *adv.*

**tum·my** (tŭm'ē) *n., pl.* **-mies.** [Alteration of STOMACH.] *Informal.* The abdomen : stomach.

**tu·mor** (to͞o'mər, tyo͞o'-) *n.* [Lat. *tumor* < *tumēre,* to swell.] **1.** A circumscribed noninflammatory growth arising from existing tissue but growing independently of the normal rate or structural development of such tissue and serving no physiological function. **2.** A swollen part. **—tu'mor·al, tu'mor·ous** *adj.*

**tu·mor·i·gen·e·sis** (to͞o'mər-ə-jĕn'ĭ-sĭs, tyo͞o'-) *n. Path.* Formation of tumors.

**tu·mor·i·gen·ic** (to͞o'mər-ə-jĕn'ĭk, tyo͞o'-) *adj.* Causing tumors. **—tu'mor·i·ge·nic'i·ty** (-jə-nĭs'ĭ-tē) *n.*

**tump·line** (tŭmp'līn') *n.* [*Tump,* tumpline (perh. of Algonquian orig.) + LINE.] A strap slung across the forehead or the chest to support a load carried on the back.

**tu·mu·lar** (to͞o'myə-lər, tyo͞o'-) *adj.* Relating to or having the shape of a tumulus.

**tu·mu·li** (to͞o'myə-lī', tyo͞o'-) *n. pl. of* TUMULUS.

**tu·mu·lose** (to͞o'myə-lōs', tyo͞o'-) *also* **tu·mu·lous** (-ləs) *adj.* [Lat. *tumulosus* < *tumulus,* mound.] Having many small hills or mounds. **—tu'mu·los'i·ty** (-lŏs'ĭ-tē) *n.*

**tu·mult** (to͞o'mŭlt', tyo͞o'-) *n.* [ME *tumulte* < Lat. *tumultus.*] **1.** Commotion and noise produced by a very large crowd. **2. a.** A great disturbance. **b.** A tempestuous uprising : RIOT. **3.** Mental or emotional agitation.

**tu·mul·tu·ar·y** (to͞o-mŭl'cho͞o-ĕr'ē, tyo͞o-) *adj.* [Lat. *tumultuarius* < *tumultus,* commotion.] Characterized or attended by haste, confusion, and disorder.

**tu·mul·tu·ous** (to͞o-mŭl'cho͞o-əs, tyo͞o-) *adj.* **1.** Marked by tumult. **2.** Causing tumult. **3.** Confusedly or violently agitated. **—tu·mul'tu·ous·ly** *adv.* **—tu·mul'tu·ous·ness** *n.*

**tu·mu·lus** (to͞o'myə-ləs, tyo͞o'-) *n., pl.* **-li** (-lī') [Lat.] An ancient grave mound : BARROW.

**tun** (tŭn) *n.* [ME < OE *tunne,* poss. of Celt. orig.] **1.** A large cask for liquids, esp. wine. **2.** A measure of liquid capacity, esp. one equivalent to approx. 954 liters or 252 gallons.

**tu·na**[1] (to͞o′nə, tyo͞o′-) *n., pl.* **tuna** or **-nas.** [Ult. < Lat. *thunnus.* —see TUNNY.] **1. a.** An often large marine food fish of the genus *Thunnus* and related genera, commercially important as a source of canned fish. **b.** A fish, as the bonito, related to the tuna. **2.** The edible flesh of tuna.

**tu·na**[2] (to͞o′nə, tyo͞o′-) *n.* [Sp. < Taino.] **1.** A tropical American cactus of the genus *Opuntia,* including the prickly pear, esp. *O. tuna,* bearing edible red fruit. **2.** The edible fruit of the tuna.

**tun·a·ble** *also* **tune·a·ble** (to͞o′nə-bəl, tyo͞o′-) *adj.* **1.** Able to be tuned. **2.** *Archaic.* Tuneful. **—tun′a·ble·ness** *n.* **—tun′a·bly** *adv.*

**tuna fish** *n.* TUNA[1] 2.

**tun·dra** (tŭn′drə) *n.* [R. < Lapp.] A treeless area that is located between the ice cap and the tree line of arctic regions, that has a permanently frozen subsoil and supports low-growing vegetation such as lichens, mosses, and stunted shrubs.

**tune** (to͞on, tyo͞on) *n.* [ME, var. of *tone,* tone. —see TONE.] **1.** A melody, esp. a simple one. **2. a.** Correct musical pitch. **b.** Proper adjustment for pitch <a violin out of *tune*> **3. a.** Agreement in pitch <play in *tune* with the violin> **b.** Concord : harmony <in *tune* with modern times> **c.** *Archaic.* Frame of mind : DISPOSITION. **4.** *Electron.* Adjustment of a receiver or circuit for maximum response to a given signal or frequency. **5.** *Obs.* A musical tone. —*v.* **tuned, tun·ing, tunes.** —*vt.* **1.** To put into proper musical pitch. **2.** To adjust so as to bring into harmony <*tune* oneself to life in arctic regions> **3.** To adjust (e.g., an engine) for maximum performance. **4.** *Archaic.* To utter musically : SING. —*vi.* To become attuned. **—change (one's) tune.** To change one's approach or attitude. **—to the tune of.** To the sum of <a fine *to the tune of* $50> **—tune in. 1.** To adjust a radio or television receiver to receive signals at a given frequency. **2.** *Slang.* To make or become aware or responsive. **—tune out. 1.** To adjust a radio receiver so as not to receive a given signal. **2.** *Slang.* **a.** To disassociate oneself from one's environment. **b.** To become unresponsive to : IGNORE <*tuned out* the children's bickering> **—tune up. 1.** To adjust a musical instrument to a desired pitch or key. **2.** To adjust a machine so as to put it into proper condition. **3.** To prepare oneself for a specified activity.

**tune·a·ble** (to͞o′nə-bəl, tyo͞o′-) *adj. var. of* TUNABLE.

**tuned-in** (to͞ond′ĭn′, tyo͞ond′-) *adj. Slang.* Highly aware of and responsive to one's environment or to current trends.

**tune·ful** (to͞on′fəl, tyo͞on′-) *adj.* **1.** Full of tune : MELODIOUS. **2.** Producing musical sounds. **—tune′ful·ly** *adv.* **—tune′ful·ness** *n.*

**tune·less** (to͞on′lĭs, tyo͞on′-) *adj.* **1.** Deficient in melody. **2.** Producing no music : SILENT. **—tune′less·ly** *adv.* **—tune′less·ness** *n.*

**tun·er** (to͞o′nər, tyo͞o′-) *n.* **1.** One that tunes <a piano *tuner*> **2.** A device for tuning, esp. an electronic device or circuit used to select signals at a specific radio frequency for amplification and conversion to sound.

**tune-up** (to͞on′ŭp′, tyo͞on′-) *n.* **1.** Adjustment of a motor or engine for maximum efficiency. **2.** An engine warm-up.

**tung oil** (tŭng) *n.* A yellow or brownish oil obtained from the seeds of the tung tree and used as a drying agent in varnishes and paints and for waterproofing.

**tung-oil tree** (tŭng′oil′) *n.* A tung tree.

**tung·state** (tŭng′stāt′) *n.* [TUNG(STEN) + -ATE.] A chemical compound derived from tungstic acid and containing tungsten with valence 6.

**tung·sten** (tŭng′stən) *n.* [Swed. : *tung,* heavy (< ON *þungr*) + *sten,* stone (< ON *steinn*).] *Symbol* **W** A hard, brittle, corrosion-resistant gray to white metallic element used in high-temperature structural materials and electrical elements, notably lamp filaments requiring thermally compatible glass-to-metal seals; atomic number 74; atomic weight 183.85. **—tung·sten′ic** (-stĕn′ĭk) *adj.*

**tungsten carbide** *n.* A very hard, fine gray powder used in tools, dies, wear-resistant machine parts, and abrasives.

**tungsten lamp** *n.* An incandescent electric lamp with a tungsten filament.

**tungsten steel** *n.* A very hard, heat-resistant steel containing tungsten.

**tung·stic** (tŭng′stĭk) *adj.* Of, relating to, or containing tungsten, esp. with valence 6.

**tungstic acid** *n.* A yellow powder, $H_2WO_4$, used in textiles and plastics.

**tung·stite** (tŭng′stīt′) *n.* A yellow or yellowish-green mineral, essentially $WO_3$, often occurring with tungsten ores.

**tung tree** (tŭng) *n.* [Chin. (Mandarin) *tong*[2], tung tree + TREE.] An Asian tree of the genus *Aleurites,* esp. *A. fordii,* cultivated for its seeds that yield a valuable drying oil.

**Tun·gus** (to͝ong-go͞oz′, tŭn-) *n., pl.* **Tungus** or **-gus·es.** [R.] **1.** A Mongoloid people inhabiting eastern Siberia. **2.** The Tungusic language of the Tungus.

**Tun·gus·ic** (to͝ong-go͞o′zĭk, tŭn-) *n.* A subfamily of the Altaic language family spoken in eastern Siberia and northern Manchuria that includes Tungus and Manchu. —*adj.* Of or relating to the Tungus peoples or to Tungusic.

**tu·nic** (to͞o′nĭk, tyo͞o′-) *n.* [Lat. *tunica,* of Semitic orig.] **1. a.** A loose-fitting sleeved or sleeveless garment extending to the knees and worn esp. by ancient Greeks and Romans. **b.** A medieval surcoat. **2. a.** A long plain close-fitting military jacket, usu. with a high stiff collar. **b.** A long plain sleeved or sleeveless blouse worn over a skirt. **c.** A short pleated and belted dress worn by women for some sports. **3.** *Anat.* A coat or layer enveloping an organ or part. **4.** *Bot.* A membranous outer covering, as of a seed. **5.** A tunicle.

**tu·ni·ca** (to͞o′nĭ-kə, tyo͞o′-) *n., pl.* **-cae** (-kē′, -sē′) [NLat. < Lat., tunic.] An enclosing membrane or layer of tissue.

**tu·ni·cate** (to͞o′nĭ-kĭt, -kāt′, tyo͞o′-) *n.* [< Lat. *tunicare, tunicat-,* to clothe with a tunic < *tunica,* tunic.] Any of various chordate marine animals of the subphylum Urochordata or Tunicata, having a cylindrical or globular body enclosed in a tough outer covering, and including the sea squirts and salps. —*adj.* **1.** Of or relating to the tunicates. **2.** *Anat.* Having a tunic. **3.** *Bot.* Having concentric layers, as the bulb of an onion.

**tu·ni·cle** (to͞o′nĭ-kəl, tyo͞o′-) *n.* [ME < Lat. *tunicula,* dim. of *tunica,* tunic.] A short vestment worn by a subdeacon over the alb or by a bishop or cardinal with the dalmatic.

**tuning fork** *n.* A small two-pronged metal device that when struck makes a sound of fixed pitch that is used as a reference, as in tuning musical instruments.

**Tu·ni·sian** (to͞o-nē′zhən, -nĭzh′ən, tyo͞o-) *adj.* Of or relating to Tunisia, Tunis, or their inhabitants. —*n.* A native or inhabitant of Tunisia or Tunis.

**tun·nel** (tŭn′əl) *n.* [ME *tonel,* tubular net < OFr. < *ton,* tun.] **1.** A passage under the ground or under the water. **2.** A passage through or under a barrier. **3.** *Obs.* A main chimney flue. —*v.* **-neled, -nel·ing, -nels** or **-nelled, -nel·ling, -nels.** —*vt.* **1.** To make a tunnel through or under. **2.** To shape or dig in the form of a tunnel. —*vi.* To make a tunnel. **—tun′nel·er, tun′nel·ler** *n.*

**tunnel vision** *n.* **1.** A constricted visual field in which peripheral perception is eliminated. **2.** Narrow-mindedness.

**tun·ny** (tŭn′ē) *n., pl.* **tun·nies** or **tunny.** [OItal. *tonno* < OProv. *ton* < Lat. *thynnus* < Gk. *thunnos.*] TUNA[1] 1a.

**tup** (tŭp) *n.* [ME *tup.*] **1.** *Chiefly Brit.* A male sheep : RAM. **2.** A heavy metal body, esp. the head of a power hammer. —*v.* **tupped, tup·ping, tups.** —*vt.* To copulate with (a ewe). —Used of a ram. —*vi.* To copulate with a ewe. —Used of a ram.

**tu·pe·lo** (to͞o′pə-lō′, tyo͞o′-) *n., pl.* **-los.** [Creek *ito opilwa* : *ito,* tree + *opilwa,* swamp.] **1.** A tree of the genus *Nyssa,* esp. *N. aquatica* of the southeastern United States, having soft, light wood. **2.** The wood of a tupelo.

**Tu·pi** (to͞o′pē, to͞o-pē′) *n., pl.* **Tupi** or **-pis. 1.** A member of any of a group of peoples living along the coast of Brazil, in the Amazon River valley, and in Paraguay. **2.** The Tupian language of the Tupi.

**Tu·pi·an** (to͞o′pē-ən, to͞o-pē′-) *adj.* Of or relating to the Tupi. —*n.* A subdivision of Tupi-Guarani that includes Tupi.

**Tu·pi-Gua·ra·ni** (to͞o-pē′gwär-ə-nē′, to͞o′pē-) *n.* A language family widely spread throughout the Amazon River valley, coastal Brazil, and northeastern South America. **—Tu·pi′-Gua·ra·ni′, Tu·pi′-Gua·ra·ni′an** *adj.*

**tup·pence** (tŭp′əns) *n. Chiefly Brit. var. of* TWOPENCE.

**tup·pen·ny** (tŭp′nē) *n. Chiefly Brit. var. of* TWOPENNY.

**tuque** (to͞ok, tyo͞ok) *n.* [Canadian Fr. < Fr. *toque,* toque < Sp. *toca.*] A knitted woolen cap shaped like a cylindrical bag with tapered ends that is worn with one end tucked into the other.

**tu quo·que** (to͞o kwō′kwē, kō′-, tyo͞o) *n.* [< Lat., you also.] A retort accusing an accuser of a similar offense or similar behavior.

**tu·ra·co** (to͝or′ə-kō′) *n. var. of* TOURACO.

**Tu·ra·ni·an** (to͞o-rā′nē-ən, -rä′-, tyo͞o-) *adj.* [< Pers. *Tūrān,* a region of central Asia.] Of or relating to the Ural-Altaic languages or to the peoples who speak them. —*n.* **1.** Ural-Altaic. **2.** A member of any of the peoples who speak languages of the Ural-Altaic group.

**tur·ban** (tûr′bən) *n.* [OFr. *turbant* < OItal. *turbante* < Turk. *tülibend* < Pers. *dulband.*] **1.** A headdress of Moslem origin, consisting of a cap attached to a long scarf of linen, cotton, or silk, wound around the head. **2.** A woman's close-fitting hat resembling a turban.

**tur·ba·ry** (tûr′bə-rē) *n., pl.* **-ries.** [ME *turbarye* < AN *turberie* < Med. Lat. *turbaria* < *turb,* peat, of Germanic orig.] **1.** A peat bog. **2.** *Law.* The right to dig peat or turf on someone else's ground in Great Britain.

**tur·bel·lar·i·an** (tûr′bə-lâr′ē-ən) *n.* [NLat. *Turbellaria,* class name < Lat. *turbella,* bustle < *turba,* turmoil.] Any of various chiefly aquatic ciliate flatworms of the class Turbellaria. —*adj.* Of or belonging to the Turbellaria.

**tur·bid** (tûr′bĭd) *adj.* [Lat. *turbidus,* disordered < *turba,* turmoil.] **1.** Having suspended or stirred up particles or sediment. **2.** Heavy, dark, or dense, as smoke or fog. **3.** Being in turmoil : MUDDLED <*turbid* feelings> **—tur′bid·ly** *adv.* **—tur′bid·ness, tur·bid′i·ty** (-bĭd′ĭ-tē) *n.*

☆ ***syns:*** TURBID, MUDDY, ROILED, ROILY *adj. core meaning* :

having suspended or stirred up particles or sediment <*turbid* river water> ***ant:*** clear

**tur·bi·dim·e·ter** (tûr′bĭ-dĭm′ĭ-tər) *n.* An instrument for measuring the scattering of a light beam through a solution containing suspended particulate matter. **—tur′bi·di·met′ric** (-də-mĕt′rĭk) *adj.* **—tur′bi·di·met′ri·cal·ly** *adv.* **—tur′bi·dim′e·try** *n.*

**tur·bi·nal** (tûr′bə-nəl) [Lat. *turbo, turbin-,* spinning top.] *Anat.* —*adj.* Shaped like a cone resting on its apex. —*n.* A turbinate bone.

**tur·bi·nate** (tûr′bə-nĭt, -nāt′) *also* **tur·bi·nat·ed** (-nā′tĭd) *adj.* [Lat. *turbinatus* < *turbo,* spinning top.] **1.** Shaped like a top. **2.** Spinning like a top. **3.** *Zool.* Spiral and decreasing sharply in diameter from base to apex <*turbinate* shells> **4.** *Anat.* Designating a small curved bone that extends horizontally along the lateral wall of the nasal passage. **—tur′bi·na′tion** *n.*

**tur·bine** (tûr′bĭn, -bīn′) *n.* [Fr. < Lat. *turbo,* spinning top.] A machine in which the kinetic energy of a moving fluid is converted to mechanical power by the impulse or reaction of the fluid with a series of buckets, paddles, or blades arrayed about the circumference of a wheel or cylinder.

**tur·bit** (tûr′bĭt) *n.* [Orig. unknown.] A domestic pigeon having a small crested head and a ruffled breast.

**turbo–** *pref.* [< TURBINE.] **1.** Turbine <*turbo*charger> **2.** Driven by a turbine <*turbo*jet>

**tur·bo·charg·er** (tûr′bō-chär′jər) *n.* A device that uses the exhaust gas of an internal-combustion engine to drive a turbine that in turn drives a supercharger attached to the engine.

**tur·bo·fan** (tûr′bō-făn′) *n.* **1.** A turbojet engine in which a fan supplements the total thrust by forcing air diverted from the main engine directly into the hot turbine exhaust. **2.** An aircraft equipped with a turbofan.

**tur·bo·jet** (tûr′bō-jĕt′) *n.* **1.** A jet engine having a turbine-driven compressor and developing thrust from the exhaust of hot gases. **2.** An aircraft equipped with a turbojet.

**tur·bo·prop** (tûr′bō-prŏp′) *n.* [Short for *turbopropeller.*] **1.** A turbojet engine used to drive an external propeller. **2.** An aircraft equipped with a turboprop.

**tur·bo·ram·jet** (tûr′bō-răm′jĕt′) *n.* **1.** A turbojet engine that at high speeds compresses air taken in as a ramjet and increases exhaust velocities with an afterburner. **2.** An aircraft that is equipped with a turboramjet.

**tur·bo·su·per·charg·er** (tûr′bō-sōō′pər-chär′jər) *n.* A supercharger that uses an exhaust-driven turbine to maintain air-intake pressure in high-altitude aircraft.

**tur·bot** (tûr′bət) *n., pl.* **turbot** *or* **-bots.** [ME < OFr. *torbout.*] **1.** A European flatfish, *Psetta maxima* or *Scophthalmus maximus,* valued as food. **2.** A flatfish similar or related to the turbot.

**tur·bu·la·tor** (tûr′byə-lā′tər) *n.* [< TURBULENT.] A device designed to cause turbulence in fluids.

**tur·bu·lent** (tûr′byə-lənt) *adj.* [Lat. *turbulentus* < *turba,* turmoil.] **1.** Violently agitated or disturbed : TUMULTUOUS <*turbulent* rapids> **2.** Having a chaotic or restless character or tendency <*turbulent* politics> **3.** Causing unrest or disturbance : UNRULY. **—tur′bu·lence** *n.* **—tur′bu·lent·ly** *adv.*

☆ ***syns:*** TURBULENT, STORMY, TEMPESTUOUS, TUMULTUOUS *adj. core meaning* : marked by unrest or disturbance <*a turbulent* era fraught with civil disobedience>

**turbulent flow** *n.* Motion of a fluid having local velocities and pressures that fluctuate randomly.

**Tur·co·man** (tûr′kə-mən) *adj. & n. var. of* TURKOMAN.

**tu·reen** (tōō-rēn′, tyōō-) *n.* [Fr. *terrine* < OFr. < *terrin,* earthen < Lat. *terra,* earth.] A broad deep dish, usu. with a cover, for serving liquid foods, as soups.

**turf** (tûrf) *n.* [ME < OE.] **1.** A surface layer of soil containing a dense growth of grass and its matted roots : SOD. **2.** A piece cut from a layer of earth or sod. **3.** A piece of peat burned for fuel. **4.** *Slang.* **a.** The area claimed by a juvenile gang as its personal territory **b.** An indefinite geographic area : TERRITORY <a county that was Democratic *turf*> **5. a.** A racetrack. **b.** The sport or business of racing horses. **—turf′y** *adj.*

**tur·ges·cence** (tûr-jĕs′əns)) *n.* [< Lat. *turgescens,* pr.part. of *turgescere,* to begin to swell < *turgēre,* to be swollen.] **1.** The process of swelling or the state of being swollen. **2.** Self-importance : pomposity. **—tur·ges′cent** *adj.*

**tur·gid** (tûr′jĭd) *adj.* [Lat. *turgidus* < *turgēre,* to be swollen.] **1.** Swollen or distended : BLOATED <*turgid* legs> **2.** Excessively ornate in style or language : GRANDILOQUENT <*turgid* prose> **—tur·gid′i·ty** (tûr-jĭd′ĭ-tē), **tur′gid·ness** *n.* **—tur′gid·ly** *adv.*

**tur·gor** (tûr′gər, -gôr′) *n.* [LLat. < Lat. *turgēre,* to be swollen.] **1.** The state of being turgid. **2.** *Biol.* Normal fullness or tension produced by the fluid content of blood vessels, capillaries, and plant or animal cells.

**Turk** (tûrk) *n.* [ME < OFr. *Turc* < Med. Lat. *Turcus* < Turk. *Türk.*] **1.** A native or inhabitant of Turkey. **2.** A speaker of a Turkic language. **3.** A Moslem. **4.** A brutal or tyrannical person.

**tur·key** (tûr′kē) *n., pl.* **-keys.** [After *Turkey,* from a confusion with the guinea fowl imported from Turkish territory.] **1. a.** A large, widely domesticated North American bird, *Meleagris gallopavo,* with brownish plumage and a bare, wattled head and neck. **b.** A related bird, *Agriocharis ocellata,* of Mexico and Central America. **2.** *Slang.* A failure, esp. a failed theatrical production. **3.** *Slang.* One regarded as being inept or undesirable.

▲ word history: The bird commonly known as the *turkey* and familiar as the centerpiece of the Thanksgiving feast is a native of the New World. It acquired the name of an Old World country as a result of two different mistakes. The name *turkey,* or *turkey cock,* was originally applied to an African bird now known as the *guinea fowl,* which was believed to have originated in Turkey. When the Europeans came upon the American turkey, they thought it was the same bird as the African guinea fowl, and so gave it the name *turkey,* although the two species are quite distinct.

**turkey buzzard** *n.* A New World vulture, *Cathartes aura,* having dark plumage and a bare red head and neck similar to that of the turkey.

**turkey cock** *n.* **1.** A male turkey. **2.** A conceited person.

**Turkey red** *n.* A moderate red.

**turkey trot** *n.* A ragtime dance characterized by a springy walk with the feet well apart and a swinging up-and-down movement of the shoulders.

**turkey vulture** *n.* A turkey buzzard.

**Tur·ki** (tûr′kē) *adj.* [Pers. *turkī* < *Turk,* Turk < Turk. *Türk.*] **1.** Of or relating to Turkic. **2.** Of or pertaining to the Turks, esp. those speaking an Eastern Turkic language. —*n.* **1.** One of the Turkic languages. **2.** A member of a people speaking a Turkic language.

**Turk·ic** (tûr′kĭk) *n.* A subfamily of the Altaic language family that includes Turkish. —*adj.* **1.** Of or pertaining to the Turks. **2.** Of or relating to Turkic.

**Turk·ish** (tûr′kĭsh) *adj.* **1.** Of or pertaining to Turkey or the Turks. **2.** Of or relating to the Turkic language of Turkey. —*n.* The Turkic language of Turkey.

**Turkish bath** *n.* **1.** A steam bath that induces heavy perspiration and is followed by a shower and massage. **2.** An establishment equipped with Turkish bath facilities.

**Turkish coffee** *n.* A sweetened brew of pulverized coffee.

**Turkish delight** *n.* A candy usu. made of jellylike cubes covered with powdered sugar.

**Turkish towel** *n.* A towel with a nap of thick uncut pile.

**Turk·ism** (tûr′kĭz′əm) *n.* The cultural, religious, or social system of the Turks.

**Tur·ko·man** *also* **Tur·co·man** (tûr′kə-mən) *n., pl.* **-mans.** [Med. Lat. *Turcomannus* < Pers. *Turkuman* < *Turkmān,* like a Turk < *Turk,* Turk < Turk. *Türk.*] **1.** Any of a once nomadic people inhabiting the Turkmen, Uzbek, Kazakh, and Kara-Kalpak republics of the U.S.S.R. **2.** The Turkic language of the Turkomans. —*adj.* **1.** Of or pertaining to Turkoman. **2.** Of or pertaining to the Turkomans.

**Turk's-cap lily** (tûrks′kăp′) *n.* **1.** A North American lily, *Lilium superbum,* with spotted orange-red flowers. **2.** The martagon.

**Turk's-head** (tûrks′hĕd′) *n. Naut.* A turban-shaped knot made on a rope with a piece of smaller rope.

**Turk's-head**

**tur·ma·line** (tōōr′mə-lĭn, -lēn′) *n. var. of* TOURMALINE.

**tur·mer·ic** (tûr′mər-ĭk) *n.* [OFr. *terre mérite* < Med. Lat. *terra merita* : Lat. *terra,* earth + Lat. *merita,* deserved.] **1.** A plant, *Curcuma longa* of India, with yellow flowers and an aromatic rootstock. **2.** The powdered rootstock of the turmeric, used as a condiment and a yellow dye. **3.** A plant with roots similar to those of the turmeric.

**turmeric paper** *n.* Paper saturated with turmeric and used as an indicator of the presence of alkalis, which turn the paper brown, or of boric acid, which turns it red-brown.

**tur·moil** (tûr′moil′) *n.* Utter confusion and agitation.

**turn** (tûrn) *v.* **turned, turn·ing, turns.** [ME *turnen* < OE *tyrnan* and OFr. *tourner, torner,* both < Lat. *tornare,* to turn in a lathe < *tornus,* lathe < Gk. *tornos.*] —*vt.* **1.** To cause to move around a central point or axis : ROTATE <*turned* the knob> **2.** To change the position of by rotating <*turn* a house plant frequently> **3.** To control or alter the operation of (e.g., a mechanical device) by use of a rotating movement <*turn* the dial to the left on the radio> **4.** To perform by rotating or revolving <*turn* somersaults> **5. a.** To reverse the position of so that the underside becomes the upperside <*turn* chicken on a spit><*turn* a page> **b.** To spade or plow (soil) to bring the undersoil to the surface. **c.** To reverse and resew the material of (e.g., a collar). **6. a.** To produce a rounded shape in (e.g., wood) by using a cutting tool. **b.** To produce a rounded form in by any means <*turn* a heel in knitting a sock> **c.** To shape : form <*turn* a vase on a potter's wheel> **d.** To render distinctive, artistic, or graceful in form <*turn* a phrase> **7. a.** To change the position of by traversing an arc of a circle : PIVOT <*turned* my chair toward the podium> **b.** To injure by twisting <*turned* my ankle> **c.** To make nauseated <a sight that *turned* my stomach> **8.** To change the direction of <*turn* a car to the left> **9. a.** To divert : deflect <*turn* a stampede> **b.** To reverse the course of : cause to retreat <managed to *turn* the enemy's advance> **10.** To make a course around or about <*turn* a corner> **11.** To change the intention or content of by persuasion or influence <The closing argument *turned* one juror's thinking.> **12.** To change the order or disposition of : UNSETTLE. **13.** To set in a specified way or direction : POINT <*turned* the guns north> **14.** To focus <*turn* one's gaze to the sky> **15.** To devote or apply (e.g., oneself) to something <*turn* oneself to music> **16.** To become, reach, or surpass (a certain age, time, or amount) <The price had *turned* $100 dollars by the last bid.> **17.** To make antagonistic <a civil war that *turned* one family against another> **18.** To cause to go in a given direction : DIRECT <They *turned* their way back.> **19.** To send away or drive out, often forcibly <*turn* a drunk out of a bar> **20.** To pour, let fall, or otherwise release (contents) from a container <*turned* the dough onto the floured board> **21.** To make sour : FERMENT <Lack of refrigeration *turned* the milk.> **22.** To cause the color of to change <Autumn *turns* the foliage.> **23.** To transform into something different <*turn* an old piece of lace into a lovely collar> **24.** To convert <*turn* one's talents into extra cash> **25.** To cause to take on a specified character, nature, or appearance. **26. a.** To fold, bend, or curve (something). **b.** To make a bend or curve in <could *turn* a bar of steel> **c.** To blunt or dull (the edge of a cutting instrument). **27.** To keep in circulation <*turned* a lot of merchandise during the holidays> **28.** To obtain by buying and selling <*turn* a profit> —*vi.* **1.** To rotate or revolve around an axis or center <wheels *turning* rapidly> **2.** To have a revolving or whirling sensation, esp. due to dizziness. **3.** To roll from side to side or back and forth <tossed and *turned* all night> **4. a.** To operate a lathe. **b.** To be formed on a lathe. **5.** To direct one's way or course. **6.** To reverse one's way, course, or direction. **7.** To have a specific reaction or effect, esp. when adverse. **8.** To become hostile or antagonistic <My friends have *turned* against me.> **9.** To attack suddenly and viciously with no apparent motive <The dog *turned* on the children.> **10.** To channel one's attention, interest, or thought toward or away from something. **11.** To convert from one religion to another. **12.** To switch one's loyalty from one side or party to another. **13.** To have recourse to a person or thing for help, support, or information. **14.** To apply oneself to a given activity. **15.** To depend on something for success or failure : RELY <The entire project *turned* on the ability of one person.> **16.** To become transformed : CHANGE. **17.** To change color <The maple leaves have *turned.*> **18.** To be saleable <These new books will *turn* easily.> **19.** To become dull or blunt after bending back. —Used of the edge of a cutting instrument. **—turn away. 1.** To send away : DISMISS <*turned* the sales representative *away*> **2.** To avert : deflect <*turned away* all criticism> **3.** To start to leave. **—turn back. 1. a.** To stop going forward. **b.** To move in a reverse direction. **2.** To halt the advance of <*turned* the attacking enemy *back*> **3.** To fold back. **—turn down. 1.** To diminish the speed, volume, intensity, or flow of. **2.** *Informal.* To reject or refuse, as a person, advice, or a suggestion. **3.** To fold or be capable of folding down <*turn* a collar *down*><a collar that *turns down*> **—turn in. 1.** To hand in <*turn in* an income-tax return> **2.** To inform on (another) : BETRAY. **3.** To produce <*turns in* good work> **4.** To bend inward <toes that *turn in*> **5.** *Informal.* To go to bed <*turned in* at eleven> **—turn off. 1.** To stop the operation, activity, or flow of : SHUT OFF. **2.** *Slang.* **a.** To affect with dislike, displeasure, or revulsion <Your behavior *turns* me *off.*> **b.** To affect with boredom. **c.** To lose or cause to lose interest : WITHDRAW <kids *turning off* to school> **3.** To divert : deflect. **4.** To leave a path or road at one point and enter another <*turned off* at the first exit> **—turn on. 1.** To cause to begin the operation, activity, or flow of <*turn on* the light bulb><*turn on* the charm> **2.** *Slang.* **a.** To smoke or ingest a drug for the purpose of experiencing a heightened sensual response. **b.** To be or cause to become interested or pleasurably excited or stimulated <Surfing *turns* them *on.*> **—turn out. 1.** To shut off <*turn* a light *out*> **2.** To assemble, as for a public event or entertainment <thousands *turning out* for the speech> **3.** To produce by a given process : MAKE <an assembly line *turning out* cars> **4.** To be found to be, as after experience or trial <The machine *turned out* to be in perfect repair.> **5.** To end up : RESULT <The dessert *turned out* beautifully.> **6.** To equip : outfit. **7.** *Informal.* To get out of bed. **8.** To evict : expel. **—turn over. 1.** To reverse in position. **2.** To shift the position of, as by rolling from one side to the other. **3.** To rotate <The engine won't *turn over* in cold weather.> **4.** To think about : CONSIDER. **5. a.** To transfer to another. **b.** To give up. **6.** To do business to the extent or amount of <*turn over* two million dollars a year> **—turn to. 1.** To begin work on. **2.** To refer to, as for information or support. **—turn up. 1. a.** To find <*turned up* the missing papers> **b.** To be found <The papers will *turn up.*> **2.** To make an appearance : ARRIVE <*turned up* late> **3.** To happen unexpectedly <Something *turned up* and I was unable to go.> **4.** To be evident <a name that *turns up* in gossip columns> —*n.* **1.** The act of turning or the state of being turned. **2.** A change of direction, motion, or position <a left *turn*> **3.** A deviation, as in a trend <a strange *turn* of events> **4.** A point of change in time <at the *turn* of the nineteenth century> **5. a.** A chance to do or perform <took a *turn* at the wheel of the boat> **b.** One of a series of such opportunities accorded individuals in succession or in scheduled order <waiting my *turn* at bat> **6.** A period of participation <a *turn* at creative writing> **7.** A characteristic mood, style, or habit <a devious *turn* of mind> **8.** A propensity or adeptness <a *turn* for gymnastics> **9.** Movement or development in a given direction <took a *turn* for the worse> **10.** An act or deed having a specified effect <do a friend a good *turn*> **11.** Advantage : purpose <It served my *turn* to remain silent.> **12.** A short excursion <a *turn* in the park> **13.** A distortion in shape. **14.** The state of being twisted or wound. **15. a.** A winding of one thing about another. **b.** A single wind or convolution, as of wire upon a spool. **16.** *Mus.* A figure or ornament consisting of four or more notes in rapid succession and including in addition to the principal note the one a degree above and the one a degree below it. **17.** An attack, as of illness or severe nervousness : SPELL. **18.** *Informal.* A momentary shock or scare <The news gave me quite a *turn.*> **19. a.** A brief theatrical act. **b.** A performer in such an act. **c.** A histrionic performance. **20. a.** A stock market transaction involving both a sale and a purchase. **b.** A similar commercial transaction. **—at every turn.** In every place and at every moment. **—by turns.** One after another : ALTERNATELY. **—in turn.** In the proper sequence or order. **—out of turn. 1.** Not in the proper sequence or order. **2.** At an inappropriate time or in an inappropriate way. **—take turns.** To take part or do in order, one after another. **—to a turn.** To a precise degree : PERFECTLY <a roast done *to a turn*> **—turn a blind eye.** To refuse to see <*turned a blind eye* to the scandal> **—turn a deaf ear.** To refuse to listen to or hear <*turned a deaf ear* to their protests> **—turn a hair.** To become afraid or upset <didn't *turn a hair* during the holdup> **—turn loose. 1.** To set free : RELEASE. **2.** To fire off : DISCHARGE <*turned loose* the artillery> **—turn (one's) back on. 1.** To deny : reject. **2.** To abandon : forsake. **—turn (one's) hand.** To apply oneself to a task. **—turn (one's) head. 1.** To become infatuated. **2.** To be egotistical and conceited. **—turn over a new leaf.** To change for the better. **—turn tail.** To run away. **—turn the other cheek.** To respond to insult or injury passively and patiently. **—turn the scales.** To shift the balance of power or influence. **—turn the tables.** To reverse the fortunes of two contending parties. **—turn the trick.** To accomplish a desired goal or end. **—turn up (one's) nose.** To regard disdainfully or scornfully.

**†turn·a·bout** (tûrn'ə-bout') *n.* **1.** The act of turning about and facing or moving in the opposite direction. **2.** A shift or change in opinion, loyalty, or allegiance. **3.** *Regional.* A dance or party to which girls invite boys.

**turn·a·round** (tûrn'ə-round') *n.* **1.** A space, as in a driveway, permitting a vehicle to turn around. **2.** The time required to load, unload, and service a vehicle. **3.** An act or instance of reversing the course or direction of <achieved an economic *turnaround*>

**turn·buck·le** (tûrn'bŭk'əl) *n.* A metal coupling, used for tightening a rod or wire rope, having an oblong piece internally threaded at both ends into which a threaded rod is screwed.

**turn·coat** (tûrn'kōt') *n.* A traitorous defector.

**turn·down** (tûrn'doun') *n.* **1. a.** A rejection. **b.** Something rejected. **2.** A downturn.

**turned-on** (tûrnd'ŏn',-ôn') *adj. Slang.* **1.** Highly aware of and responsive to what is fashionable and up-to-date. **2.** Pleasantly excited or stimulated.

**turn·er¹** (tûr'nər) *n.* One that turns, esp. one who works a lathe.

**tur·ner²** (tûr'nər) *n.* [G. < *turnen*, to do gymnastics < OHG *turnēn*, to turn < Lat. *tornare*, to turn in a lathe. —see TURN.] A tumbler or gymnast, esp. a member of a turnverein.

**turn·er·y** (tûr'nə-rē) *n., pl.* **-ies.** The work or workshop of a lathe operator.

**turn·ing** (tûr'nĭng) *n.* **1.** The act or course of one that turns. **2.** Deviation from a straight course : TURN. **3.** The shaping of metal or wood on a lathe.

**turning point** *n.* **1.** A decisive moment <reached a *turning point* in my career> **2.** *Math.* A maximum or minimum point on a curve.

**tur·nip** (tûr'nĭp) *n.* [Perh. TURN (from its rounded shape) + dial. *nepe*, turnip < ME < OE *næp* < Lat. *napus.*] **1.** A widely cultivated Old World plant, *Brassica rapa*, with a large edible root. **2.** The root of the turnip. **3.** A plant similar or related to the turnip. **4.** A large round pocket watch.

**turnip cabbage** *n.* Kohlrabi.

**turn·key** (tûrn′kē′) *n., pl.* **-keys.** A jailer.

**turn·off** (tûrn′ôf′, -ŏf′) *n.* **1.** A branch of a road leading from a main thoroughfare, esp. an exit on a highway. **2.** An act or instance of turning off. **3.** *Slang.* One that is distasteful or uninteresting.

**turn-on** (tûrn′ŏn′, -ôn′) *n. Slang.* A cause of pleasure or excitement.

**turn·out** (tûrn′out′) *n.* **1.** The act of turning out. **2.** The number of people attending an event or performance. **3.** The number of things produced : OUTPUT. **4.** *Chiefly Brit.* **a.** A labor strike. **b.** A striking worker. **5.** An array of equipment : OUTFIT. **6.** An outfit of a carriage with its horse or horses : EQUIPAGE. **7.** A railroad siding. **8.** A widening in a highway to allow vehicles to pass.

**turn·o·ver** (tûrn′ō′vər) *n.* **1. a.** The act of turning over. **b.** An upset or overthrow. **2.** An abrupt change : REVERSAL. **3.** A small pastry made by covering one half of a piece of dough with fruit, preserves, or other filling and turning the other half over on top. **4.** The number of times a particular stock of goods is sold and restocked during a given time. **5.** The amount of business transacted during a given time. **6.** The number of shares of stock sold on the market during a given time. **7.** The amount of capital loaned on call during a given time. **8. a.** The number of workers hired by an establishment to replace those who have left. **b.** The ratio of this number to the number of employed workers. —*adj.* Capable of being turned or folded down or over <a *turnover* collar>

**turn·pike** (tûrn′pīk′) *n.* [ME *turnepike,* spiked barrier : *turnen,* to turn + *pike,* pike.] **1.** An expressway or wide highway with tollgates. **2.** A tollgate.

**turn·sole** (tûrn′sōl′) *n.* [ME *turnesole,* purple dye obtained from the plant < OFr. *tournesol* < OItal. *tornasole,* heliotrope : *tornare,* to turn (< Lat. *tornare*) + *sole,* sun (< Lat. *sol*).] A plant, as the heliotrope, that moves or is thought to move in response to the sun.

**turn·spit** (tûrn′spĭt′) *n.* **1.** One who turns a roasting spit. **2.** A dog once used in a treadmill to turn a roasting spit.

**turn·stile** (tûrn′stīl′) *n.* **1.** A device, typically consisting of several horizontal arms supported by and radially projecting from a central vertical post, that is used for controlling passage from one public area to another. **2.** A structure similar to a turnstile that permits the passage of persons but not of horses or cattle.

**turn·stone** (tûrn′stōn′) *n.* [From its method of finding food.] A wading bird, *Arenaria interpres* with reddish and white plumage, or *A. melanocephala* with black and white plumage.

**turn·ta·ble** (tûrn′tā′bəl) *n.* **1.** A circular, usu. horizontal rotating platform equipped with a railway track, used for turning locomotives, as in a roundhouse. **2. a.** The circular horizontal rotating platform of a phonograph on which a record is placed. **b.** A phonograph exclusive of amplifying circuitry and speakers. **3.** A rotating platform or disk, as on a microscope.

**turn·up** (tûrn′ŭp′) *n.* Something, as a trouser cuff, that is turned up or that turns up. —*adj.* Turned up or capable of being turned up <a *turnup* collar>

**turn·ver·ein** (tûrn′və-rīn′, tŏŏrn′-) *n.* [G. : *turnen,* to do gymnastics + *Verein,* club < *vereinen,* to unite.] A club of gymnasts.

**tur·pen·tine** (tûr′pən-tīn′) *n.* [ME *terpentin,* resin of the terebinth < OFr. *terbentine* < Lat. *terebinthina* < *terebinthus,* terebinth.] **1.** A thin volatile essential oil, $C_{10}H_{16}$, obtained, as by steam distillation, from the wood or the exudate of pine trees and used as a paint thinner, solvent, and medicinally as a liniment. **2.** The sticky mixture of resin and volatile oil from which turpentine is distilled. **3.** A brownish-yellow resinous liquid obtained from the terebinth. —*vt.* **-tined, -tin·ing, -tines. 1.** To apply turpentine to or mix turpentine with. **2.** To extract turpentine from (a tree). **—tur′pen·tin′ic** (-tĭn′ĭk), **tur′pen·tin′ous** (-tĭn′əs) *adj.*

**tur·peth** (tûr′pĭth) *n.* [ME *turbit,* purgative < OFr. < Med. Lat. *turbitum* < Ar. *turbid.*] **1.** A vine, *Ipomoea turpethum* or *Operculina turpethum* of tropical Asia and Australia, with roots yielding a resinous substance used medicinally as a cathartic. **2.** The root of the turpeth.

**tur·pi·tude** (tûr′pĭ-tōōd′, -tyōōd′) *n.* [Lat. *turpitudo* < *turpis,* shameful.] **1.** Depravity <moral *turpitude*> **2.** A base act.

**tur·quoise** (tûr′kwoiz′, -koiz′) *n.* [ME *turkeis* < OFr. *turqueise* < *turqueis,* Turkish < *Turc,* Turk.] **1.** A blue to blue-green mineral of aluminum and copper, chiefly $CuAl_6(PO_4)_4(OH)_8 \cdot 4H_2O$, valued in its polished blue form as a gemstone. **2.** A light to brilliant bluish green. **—tur′quoise′** *adj.*

**tur·ret** (tûr′ĭt) *n.* [ME *turet* < OFr. *tourete,* dim. of *tour,* tower < Lat. *turris* < Gk. *tursis.*] **1.** A small ornamented tower or tower-shaped projection on a building. **2. a.** A low, heavily armored, usu. horizontally rotating structure containing mounted guns and their gun crew, as on a tank or warship. **b.** A transparent domelike gunner's enclosure projecting from the fuselage of a military combat aircraft. **3.** A tall wooden structure mounted on wheels and used by attacking troops in ancient warfare to scale the walls of an enemy fortress. **4.** An attachment for a lathe consisting of a rotating cylindrical block holding cutting tools.

**tur·ret·ed** (tûr′ĭ-tĭd) *adj.* **1.** Equipped with a turret or turrets. **2.** Shaped like a turret.

**tur·tle¹** (tûr′tl) *n.* [Perh. < Fr. *tortue.*] **1.** Any of various reptiles of the order Chelonia, having toothless horny jaws and the body enclosed in a bony or leathery shell into which the head, legs, and tail can be retracted. **2.** *Chiefly Brit.* A marine chelonian. —*vi.* **-tled, -tling, -tles.** To hunt for turtles, esp. as an occupation.

**tur·tle²** (tûr′tl) *n.* [ME < OE < Lat. *turtur.*] *Archaic.* A turtledove.

**tur·tle³** (tûr′tl) *n.* A turtleneck.

**tur·tle·dove** (tûr′tl-dŭv′) *n.* **1.** A slender European dove, *Streptopelia turtur,* having a white-edged tail and a soft purring voice. **2.** The mourning dove.

**tur·tle·head** (tûr′tl-hĕd′) *n.* A plant of the genus *Chelone,* esp. *C. glabra* of eastern North America, with pink or white flowers.

**tur·tle·neck** (tûr′tl-nĕk′) *n.* **1.** A high, turned-down collar that fits closely about the neck. **2.** A garment, as a sweater, having a turtleneck.

**turves** (tûrvz) *n. Archaic. var. pl. of* TURF.

**Tus·can** (tŭs′kən) *adj.* [Lat. *Tuscanus* < *Tuscus,* Etruscan.] **1.** Of or relating to Tuscany or to its people. **2.** Of or relating to the Tuscan architectural order. —*n.* **1.** A native or inhabitant of Tuscany. **2. a.** Any of the dialects of Italian spoken in Tuscany. **b.** The standard literary form of Italian.

**Tuscan order** *n.* A classical architectural order similar to Roman Doric, but having an unfluted shaft with a simplified base, capital, and entablature.

**Tus·ca·ro·ra** (tŭs′kə-rôr′ə, -rōr′ə) *n., pl.* **Tuscarora** or **-ras.** [Tuscarora *Skarūrę*.] **1. a.** A tribe of Indians once living in North Carolina and now living in New York and Ontario. **b.** A member of this tribe. **2.** The Iroquoian language of the Tuscarora.

**tu·sche** (tōōsh′ə) *n.* [G., back-formation < *tuschen,* to lay on colors < Fr. *toucher* < OFr. *tochier,* to touch.] A black substance used for drawing in lithography and as a resist in etching and silk-screen work.

**tush¹** (tŭsh) *interj.* [ME *tussch.*] —Used to express mild reproof, disapproval, or admonition.

**tush²** (tŭsh) *n.* [ME *tusche* < OE *tūsc.*] A tusk. —*vt.* **tushed, tush·ing, tush·es.** To tusk.

**tusk** (tŭsk) *n.* [ME < OE *tūsc.*] **1.** An elongated, pointed tooth, usu. one of a pair, extending outside of the mouth in animals such as the walrus, elephant, or wild boar. **2.** A long projecting tooth or toothlike part. —*vt.* **tusked, tusk·ing, tusks.** To dig or gore with the tusks or a tusk. **—tusked** (tŭskt) *adj.*

**tusk·er** (tŭs′kər) *n.* An animal, as a wild boar, that has tusks.

**tusk shell** *n.* A tooth shell.

**tus·sah** (tŭs′ə, tŭs′ô′) *also* **tus·sore** (tŭs′ôr′, -ōr′) *n.* [Hindi *tasar* < Skt. *tasaram,* shuttle.] **1.** An undomesticated Asian silkworm, *Antheraea paphia,* that produces a coarse brownish or yellowish silk. **2.** The silk produced by the tussah or a fabric woven from it.

**tus·sis** (tŭs′ĭs) *n.* [Lat.] A cough. **—tus′sive** (tŭs′ĭv) *adj.*

**tus·sle** (tŭs′əl) *vi.* **-sled, -sling, -sles.** [ME *tussillen,* freq. of *tousen,* to pull roughly.] To struggle : scuffle <children *tussling* on the rug> —*n.* **1.** A rough-and-tumble scuffle. **2.** An argument : dispute.

**tus·sock** (tŭs′ək) *n.* [Orig. unknown.] **1.** A clump or tuft, as of growing grass. **2.** A tuft of hair or feathers. **—tus′sock·y** *adj.*

**tussock moth** *n.* Any of various moths of the family Lymantriidae, having hairy caterpillars often destructive to deciduous trees.

**tut** (tŭt) *interj.* —Used to express mild annoyance, impatience, or reproof.

**tu·tee** (tōō-tē′, tyōō-) *n.* [TUT(OR) + -EE¹.] One who is tutored.

**tu·te·lage** (tōōt′l-ĭj, tyōōt′-) *n.* [Lat. *tutela* < *tueri,* to guard.] **1.** The act or capacity of a tutor. **2.** The state of being under the control or guidance of a guardian or tutor. **3.** The function or capacity of a guardian.

**tu·te·lar·y** (tōōt′l-ĕr′ē, tyōōt′-) *also* **tu·te·lar** (tōōt′l-ər, -är′, tyōōt′-) [< Lat. *tutelarius,* guardianship < *tutelaris,* tutelar < *tutela,* tutelage < *tueri,* to guard.] —*adj.* **1.** Being or serving as a guardian or protector <*tutelary* gods> **2.** Of or pertaining to a guardian or guardianship. —*n., pl.* **-lar·ies** *also* **-lars.** One having tutelary powers.

**tu·tor** (tōō′tər, tyōō′-) *n.* [ME *tutour* < OFr. < Lat. *tutor* < *tueri,* to guard.] **1. a.** A private instructor. **b.** A provider of additional, specialized, or remedial instruction <a *math* tutor> **2.** A college teacher or teaching assistant ranking below an instructor. **3.** A graduate responsible for the special supervision of an undergraduate at some British universities. **4.** *Law.* The guardian of a minor and of his or her property. —*v.* **-tored, -tor·ing, -tors.** —*vt.* **1.** To instruct privately. **2.** To have the guardianship, tutelage, or care of. —*vi.* **1.** To function as a tutor. **2.** To be instructed by or study under a tutor. **—tu·to′ri·al** (tōō-tôr′ē-əl, -tōr′-, tyōō-) *adj.* **—tu′tor·ship′** *n.*

**tutorial system** *n.* An instructional system in which college tutors are responsible for the special supervision of students individually or in small groups.

**tut·ti** (tōō′tē) [Ital., pl. of *tutto,* all < Lat. *totus.*] *Mus.* —*adj.* & *adv.* All. —Used as a direction to indicate that all performers are to take part. **—tut′ti** *n.*

**tut·ti-frut·ti** (to͞o′tē-fro͞o′tē) *n.* [Ital. : *tutti*, all + *frutti*, fruits.] **1.** A confection, esp. ice cream, containing various chopped candied fruits. **2.** A flavoring simulating the flavor of many fruits. **—tut′ti-frut′ti** *adj.*

**tut·ty** (tŭt′ē) *n.* [ME < OFr. < Ar. *tūtiyā*.] An impure zinc oxide obtained as a sublimate from the flues of zinc-smelting furnaces and used as a polishing powder.

**tu·tu** (to͞o′to͞o) *n.* [Fr.] A very short ballet skirt made of many layers of gathered sheer fabric, as tulle.

**tux·e·do** (tŭk-sē′dō) *n.*, *pl.* **-dos** *or* **-does.** [After *Tuxedo* Park, New York.] **1.** A usu. dark jacket with satin or grosgrain lapels worn for formal or semiformal occasions. **2.** A complete outfit including a tuxedo jacket, black trousers with a stripe down the side, and a black bow tie.

**tu·yère** (twē-yâr′) *n.* [Fr. < OFr. *tuyere* < *tuyau*, pipe.] The opening, as a pipe or nozzle, through which air is forced into a blast furnace or forge to facilitate combustion.

**TV** (tē′vē′) *n.*, *pl.* **TVs** or **TV's.** Television.

**TV Dinner.** A trademark for a packaged frozen meal that only needs to be heated before serving.

**twad·dle** (twŏd′l) *vi.* **-dled, -dling, -dles.** [Prob. var. of dial. *twattle*.] To talk foolishly. **—twad′dle** *n.* **—twad′dler** *n.*

**twain** (twān) *adj.* [ME *twayne* < OE *twēgen*.] *Archaic.* Two. *—n.* **1.** A set of two <"Oh, East is East, and West is West, and never the *twain* shall meet" —Kipling> **2.** The two-fathom mark on a sounding line used on riverboats.

**twang** (twăng) *v.* **twanged, twang·ing, twangs.** [Imit.] *—vt.* **1.** To emit a sharp vibrating sound, as the string of a musical instrument when plucked. **2.** To resound with a sharp vibrating sound. *—vt.* **1.** To cause to make a sharp vibrating sound. **2.** To utter with an excessively nasal tone of voice. *—n.* **1.** A sharp vibrating sound, as that of a plucked string. **2.** An excessively nasal tone of voice. **—twang′y** *adj.*

**tway·blade** (twā′blād′) *n.* [Obs. *tway*, two + BLADE.] Any of various small terrestrial orchids of the genera *Liparis* and *Listera*, with two basal leaves and a terminal greenish or purplish flower cluster.

**tweak** (twēk) *vt.* **tweaked, tweak·ing, tweaks.** [Prob. var. of dial. *twick* < ME *twikken* < OE *twiccian*.] To pinch, pluck, or twist sharply. **—tweak** *n.* **—tweak′y** *adj.*

**tweed** (twēd) *n.* [Alteration of Sc. *tweel*, twill < ME *twyl*.] **1.** A coarse, rugged, often nubby woolen fabric made in various twill weaves and used mainly for casual suits and coats. **2. tweeds.** Clothing made of tweed.

**twee·dle·dum and twee·dle·dee** (twēd′l-dŭm′ ən twēd′l-dē′) *n.* [After *Tweedledum* and *Tweedledee*, proverbial rival fiddlers, imit. of low and high musical notes.] Two that resemble each other so closely as to be virtually indistinguishable.

**tweed·y** (twē′dē) *adj.* **-i·er, -i·est. 1.** Made of or like tweed. **2. a.** Wearing tweeds. **b.** *Informal.* Suggestive of casual, informal taste and lifestyle <a *tweedy*, preppy look> **—tweed′i·ness** *n.*

**'tween** (twēn) *prep.* Between.

**tweet** (twēt) *vi.* **tweet·ed, tweet·ing, tweets.** [Imit.] To utter a weak chirping sound. **—tweet** *n.*

**tweet·er** (twē′tər) *n.* A loudspeaker designed to reproduce high-pitched sounds in a high-fidelity audio system.

**tweeze** (twēz) *vt.* **tweezed, tweez·ing, tweez·es.** [Back-formation < TWEEZERS.] To handle or extract with tweezers.

**tweez·ers** (twē′zərz) *pl.n.* [< obs. *tweeze*, a case of small instruments < Fr. *étui*, étui. —see ÉTUI.] A small usu. metal pincerlike tool used for plucking or handling small objects.

**twelfth** (twĕlfth) *n.* [ME *twelfthe* < OE *twelfta*.] **1.** The ordinal number matching the number 12 in a series. **2.** One of 12 equal parts. **3.** *Mus.* **a.** A 12-degree interval in a diatonic scale. **b.** A tone 12 degrees below or above a given tone. **—twelfth** *adj.* & *adv.*

**Twelfth-day** (twĕlfth′dā′) *n.* Jan. 6, the day of Epiphany, 12 days after Christmas.

**Twelfth-night** (twĕlfth′nīt′) *n.* The evening of Jan. 5, before Twelfth-day.

**Twelfth·tide** (twĕlfth′tīd′) *n.* Epiphany.

**twelve** (twĕlv) *n.* [ME < OE *twelf*.] **1.** The cardinal number equal to 11 + 1. **2.** The 12th in a set or sequence. **—twelve** *adj.* & *pron.*

▲ word history: The word *twelve*, like the word *eleven*, was formed in an unusual way. *Twelve* is derived from the Germanic form *twalif-*, a compound of *twa*, "two," and *lif-*, a form derived from an Indo-European root meaning "to leave." *Twelve* therefore means "two left over after counting to ten." In contrast, most of the other Indo-European languages formed the word for "twelve" from the words for "two" and "ten."

**twelve·mo** (twĕlv′mō′) *n.*, *pl.* **-mos.** DUODECIMO 2.

**twelve·month** (twĕlv′mŭnth′) *n.* A year.

**twelve-tone** (twĕlv′tōn′) *adj.* *Mus.* Relating to, consisting of, or based on atonal arrangement of the 12 chromatic tones.

**twen·ti·eth** (twĕn′tē-ĭth) *n.* **1.** The ordinal number matching the number 20 in a series. **2.** One of 20 equal parts. **—twen′ti·eth** *adj.* & *adv.*

**twen·ty** (twĕn′tē) *n.*, *pl.* **-ties.** [ME < OE *twēntig*.] The cardinal number equal to 2 × 10. **—twen′ty** *adj.* & *pron.*

**twen·ty-one** (twĕn′tē-wŭn′) *n.* BLACKJACK[1] 3.

**twen·ty-twen·ty** *or* **20/20** (twĕn′tē-twĕn′tē) *adj.* [From a method of testing vision by reading charts at a distance of 20 feet.] Having normal visual acuity.

**twerp** *also* **twirp** (twûrp) *n.* [Orig. unknown.] *Slang.* A silly contemptible person : TWIT.

**Twi** *also* **Tshi** (chwē, chē) *n.* A western African language spoken esp. by the Ashanti.

**twi·bil** *also* **twi·bill** (twī′bĭl′) *n.* [ME < OE : *twi-*, two + *bil*, billhook.] **1.** A battle-ax with two cutting edges. **2.** A mattock with one arm like an ax and the other like an adz.

**twice** (twīs) *adv.* [ME < OE *twiga*.] **1.** In two cases or on two occasions. **2.** In doubled degree or amount.

**twice-laid** (twīs′lād′) *adj.* Made from strands of old or used rope.

**twice-told** (twīs′tōld′) *adj.* Very familiar due to repeated telling <*twice-told* tales>

**twid·dle** (twĭd′l) *v.* **-dled, -dling, -dles.** [Poss. a blend of TWIRL and FIDDLE.] *—vt.* To turn over or around idly or lightly : fiddle with. *—vi.* **1.** To trifle with something. **2.** To be busy about trifles. **3.** To twirl or rotate without purpose. **—twiddle one's thumbs. 1.** To twirl one's thumbs idly around each other. **2.** To be idle. **—twid′dle** *n.* **—twid′dler** *n.*

**twig[1]** (twĭg) *n.* [ME < OE *twigge*.] A small branch or slender shoot, as of a tree.

**twig[2]** (twĭg) *v.* **twigged, twig·ging, twigs.** [Ir. Gael. *tuigim*, I understand.] *Chiefly Brit.* *—vt.* **1.** To observe or watch : NOTICE. **2.** To understand. *—vi.* To be aware of the situation : UNDERSTAND.

**twig[3]** (twĭg) *n.* [Orig. unknown.] *Chiefly Brit.* The current style.

**twig·gen** (twĭg′ən) *adj.* Made of twigs : WICKER.

**twig·gy** (twĭg′ē) *adj.* **-gi·er, -gi·est. 1.** Like a twig : SLENDER. **2.** Abounding in twigs.

**twi·light** (twī′līt′) *n.* [ME *twylyghte*.] **1. a.** The time interval during which the sun is below the horizon at an angle less than any of several standard angular distances. **b.** Illumination of the atmosphere esp. after a sunset. **2.** A dim or faint illumination. **3.** A decline following growth, glory, or success <in the *twilight* of one's life> **—twi′light′** *adj.*

**twilight sleep** *n.* An amnesiac condition characterized by the absence of sensibility to pain without loss of consciousness that is induced by an injection of morphine and scopolamine administered during labor in childbirth.

**twill** (twĭl) *n.* [ME *twyl* < OE *twilic*.] **1.** A fabric with diagonal parallel ribs. **2.** The weave used to produce twill. *—vt.* **twilled, twill·ing, twills.** To weave (cloth) so as to produce a pattern of twill.

**twilled** (twĭld) *adj.* Woven so as to have diagonal parallel ribs.

**twin** (twĭn) *n.* [ME < OE *twinn*, twofold.] **1.** One of two offspring born at the same birth. **2.** One of two identical or similar persons, animals, or things : COUNTERPART. **3. Twins.** Gemini. **4. twins.** Two interwoven crystals in which unlike faces are parallel. *—adj.* **1.** Being two or one of two offspring born at the same birth. **2.** Being one of two identical or similar persons, animals, or things <a *twin* bed> **3.** Consisting of two identical or similar related or connected parts. *—v.* **twinned, twin·ning, twins.** *—vi.* **1.** To give birth to twins. **2.** *Archaic.* To be one of twin offspring. **3.** To be paired or coupled. *—vt.* **1.** To pair or couple. **2.** To provide a match or counterpart to.

**twin·ber·ry** (twĭn′bĕr′ē) *n.* The partridgeberry.

**twine** (twīn) *v.* **twined, twin·ing, twines.** [ME *twinen* < *twin*, strong string < OE *twīn*.] *—vt.* **1.** To twist together, as threads : INTERTWINE. **2.** To form by twisting, intertwining, or interlacing. **3.** To encircle or coil about. **4.** To cause to be encircled. *—vi.* **1.** To become twisted, interlaced, or interwoven. **2.** To go in a winding course <a stream *twining* through the forest> *—n.* **1.** A strong string or cord formed of two or more threads twisted together. **2.** Something formed by twining <a *twine* of bread dough> **3.** A tangle : knot. **—twin′er** *n.*

**twin·flow·er** (twĭn′flou′ər) *n.* A creeping evergreen plant, *Linnaea borealis* of northern regions, with roundish evergreen leaves and paired, bell-shaped, pinkish flowers.

**twinge** (twĭnj) *n.* [ME *twengen*, to pinch < OE *twengan*.] **1.** A sudden sharp physical pain. **2.** Mental or emotional pain <a *twinge* of remorse> *—v.* **twinged, twing·ing, twing·es.** *—vt.* **1.** To cause to feel a sharp pain. **2.** *Obs.* To tweak. *—vi.* To feel a twinge.

**twi·night** (twī′nīt′) *adj.* [TWI(LIGHT) + NIGHT.] *Baseball.* Designating a double-header in which the first game begins in late afternoon.

**twin·kle** (twĭng′kəl) *vi.* **-kled, -kling, -kles.** [ME *twynklen* < OE *twinclian*.] **1.** To shine with slight, intermittent gleams : GLIMMER. **2.** To be bright or sparkling. **3.** To blink : wink. *—n.* **1.** A slight, intermittent gleam : GLIMMER. **2.** A sparkle of merriment or delight in the eye. **3.** TWINKLING 3. **—twin′kler** *n.*

**twin·kling** (twĭng′klĭng) *n.* **1.** An act of blinking. **2.** A blink or twinkle. **3.** The time it takes to blink once : INSTANT.

o͞o **boot** ou **out** th **thin** *th* **this** ŭ **cut** û **urge** y **young**
yo͞o **abuse** zh **vision** ə **about**, **item**, **edible**, **gallop**, **circus**

**twin-leaf** (twĭn′lēf′) *n., pl.* **-leaves** (-lēvz′). A woodland plant, *Jeffersonia diphylla* of eastern North America, with deeply cleft leaves and a single white flower.
**twinned** (twĭnd) *adj.* **1.** Born at a single birth. **2.** Paired or coupled with an identical or similar thing. **3.** Formed of crystals by the process of twinning.
**twin·ning** (twĭn′ĭng) *n.* **1.** Bearing of twins. **2.** A pairing or union of two similar or identical things. **3.** Formation of twin crystals.
**twin-screw** (twĭn′skro͞o′) *adj. Naut.* Having two propellers, one on either side of the keel, that usu. revolve in opposite directions.
**twin-size** (twĭn′sīz′) *adj.* Pertaining to or being a bed that is 39 inches by 75 inches in dimension <*twin-size* sheets>
**twirl** (twûrl) *v.* **twirled, twirl·ing, twirls.** [Orig. unknown.] —*vt.* **1.** To rotate briskly : SPIN. **2.** To twist or wind around <*twirl* thread on a spindle> **3.** *Baseball.* To pitch. —*vi.* **1.** To spin around rapidly, suddenly, or repeatedly. **2.** To whirl suddenly : make an about-face. —*n.* **1.** A quick spinning or twisting. **2.** Something twirled : TWIST <a *twirl* of cotton candy> **—twirl′er** *n.*
**twirp** (twûrp) *n. var. of* TWERP.
**twist** (twĭst) *v.* **twist·ed, twist·ing, twists.** [ME *twisten.*] —*vt.* **1. a.** To entwine (two or more threads) in order to make a single strand. **b.** To form in this manner <*twist* a length of line> **2.** To wind or coil (e.g., vines or rope) about an object or structure. **3.** To interlock or interlace <*twist* flowers in one's hair> **4.** To impart a coiling or spiral shape to. **5. a.** To turn or open by turning. **b.** To pull, break, or snap by turning <*twist* off a dead branch> **6.** To wrench or sprain <*twist* one's ankle> **7.** To alter the normal appearance of : CONTORT <*twist* one's mouth into grimace> **8.** To distort the intended meaning of <*twist* someone's words> —*vi.* **1.** To be or become twisted. **2.** To move or progress in a winding course : MEANDER <a brook *twisting* through the field> **3.** To squirm : writhe <*twist* with pain> **4.** To rotate : revolve. **5.** To dance the twist. **6.** To move so as to face in another direction. —*n.* **1.** Something twisted or formed by twisting, esp.: **a.** A length of yarn, cord, or thread, esp. a strong silk thread used chiefly to bind the edges of buttonholes. **b.** Tobacco leaves processed into a rope or roll. **c.** A bakery product, as bread, for which the dough was twisted before being baked. **d.** A sliver of citrus peel twisted over or dropped into a beverage to impart flavor. **2.** The act of twisting or the state of being twisted : ROTATION. **3.** A spinning motion given to a ball when thrown or struck in a specific way. **4. a.** The state of being twisted into a spiral. **b.** The degree or angle of stress caused by such twisting. **5.** A sprain or wrench, as of a muscle. **6.** An unexpected departure from a pattern <a *twist* of fate><a story with a quirky *twist*> **7.** A distortion, as of the face : CONTORTION. **8.** A personal eccentricity <a strange *twist* of character> **9.** A dance characterized by vigorous arm and hip motions. **—twist′a·bil′i·ty** *n.* **—twist′a·ble** *adj.* **—twist′ing·ly** *adv.*
**twist drill** *n.* A drill with deep helical grooves along the shank from the point.
**twist·er** (twĭs′tər) *n.* **1.** One that twists. **2.** A ball thrown or batted with a twist. **3.** *Informal.* **a.** A cyclone. **b.** A tornado.
**twit** (twĭt) *vt.* **twit·ted, twit·ting, twits.** [ME *atwiten* < OE *ætwītan* : *æt*, at + *wītan*, to reproach.] To taunt, ridicule, or tease, esp. for embarrassing mistakes or faults. —*n.* **1.** The act of twitting. **2.** A reproach, gibe, or taunt. **3.** *Chiefly Brit.* A silly or foolish person.
**†twitch** (twĭch) *v.* **twitched, twitch·ing, twitch·es.** [ME *twicchen.*] —*vt.* To draw, pull, or move suddenly and sharply : JERK. —*vi.* **1.** To move jerkily or spasmodically. **2.** To ache sharply from time to time : TWINGE. —*n.* **1.** A sudden involuntary or spasmodic muscular movement <a *twitch* in the eye> **2.** A sudden tug. **3.** *Western U.S.* A looped cord used to restrain a horse by tightening it around the animal's upper lip. **—twitch′ing·ly** *adv.*
**twitch grass** *n.* Couch grass.
**twit·ter¹** (twĭt′ər) *v.* **-tered, -ter·ing, -ters.** [ME *twiteren.*] —*vi.* **1.** To utter a succession of light chirping or tremulous sounds : CHIRRUP <birds *twittering*> **2.** To titter. **3.** To tremble with nervous agitation or excitement. —*vt.* To utter or say with a twitter. —*n.* **1.** The light chirping sounds made by certain birds. **2.** Light tremulous speech or laughter. **3.** Agitation or excitement : FLUTTER. **—twit′ter·er** *n.* **—twit′ter·y** *adj.*
**twit·ter²** (twĭt′ər) *n.* One who twits.
**twixt** *also* **′twixt** (twĭkst) *prep.* Betwixt.
**two** (to͞o) *n.* [ME < OE *twā* : akin to G. *zwei*, Lat. *duo*, Gk. *duo*, Skt. *dva.*] **1.** The cardinal number equal to 1 + 1. **2.** The second in a set or sequence. **3.** Something having two parts, units, or members, esp. a playing card, die, or domino with two pips. **—two** *adj. & pron.*
**two-base hit** (to͞o′bās′) *n. Baseball.* A hit enabling the batter to reach second base : DOUBLE.
**two-bit** (to͞o′bĭt′) *adj. Slang.* Worth very little.
**two bits** *pl.n. Informal.* **1.** Twenty-five cents. **2.** A petty sum.
**two-by-four** (to͞o′bī-fôr′, -fōr′, to͞o′bə-) *adj.* **1.** Measuring two by four inches, or in the same ratio in other units. **2.** *Informal.* Small in size or area <a *two-by-four* apartment> —*n.* A length of lumber measuring 1⅝ inches in thickness and 3⅜ inches in width.
**two cents worth** *n.* An opinion <got in my *two cents worth*>
**two-di·men·sion·al** (to͞o′dĭ-mĕn′shə-nəl) *adj.* **1.** Having only two dimensions : FLAT. **2.** Limited in range or depth <*two-dimensional* characters in the novel>
**two-edged** (to͞o′ĕjd′) *adj.* **1.** Having a keen edge on both sides, as a razor or sword blade. **2.** Having two contrasting effects, meanings, or interpretations.
**two-faced** (to͞o′fāst′) *adj.* **1.** Having two faces or surfaces. **2.** Hypocritical : deceitful. **—two′-fac′ed·ly** (to͞o′fā′sĭd-lē, -fāst′lē) *adv.* **—two′-fac′ed·ness** *n.*
**two-fer** *also* **two·fer** (to͞o′fər) *n.* [Shortening and alteration of *two for the price of one.*] *Informal.* **1.** A special offer of two tickets, as for a play or show, for the price of one. **2.** A cheap discounted item.
**two-fisted** (to͞o′fĭs′tĭd) *adj. Informal.* Aggressive : vigorous <a *two-fisted* vodka drinker>
**two·fold** (to͞o′fōld′, -fōld′) *adj.* **1.** Having two components. **2.** Having twice as much or twice as many : DOUBLE. —*adv.* Two times as much or as many : DOUBLY.
**two-hand·ed** (to͞o′hăn′dĭd) *adj.* **1.** Requiring the use of two hands at once. **2.** Made to be operated by two people. **3.** Able to use both hands with equal facility : AMBIDEXTROUS. **4.** Having two hands.
**two iron** *n.* A midiron.
**two-mast·er** (to͞o′măs′tər) *n.* A sailing vessel rigged with two masts.
**two-name** (to͞o′nām′) *adj.* Relating to or designating commercial paper bearing the signatures of two persons liable to the obligation.
**two·pence** (tŭp′əns) *n., pl.* **twopence** *or* **-penc·es.** *Chiefly Brit.* **1.** Two pennies regarded as a monetary unit. **2. a.** A silver coin worth two pennies, since 1662 minted only for distribution on Maundy Thursday. **b.** A copper coin of this value minted during the reign of George III. **3.** A very small amount : BIT <didn't care *twopence* about that problem>
**two·pen·ny** (tŭp′ə-nē, to͞o′pĕn′ē) *adj.* **1.** Worth or costing twopence <*twopenny* candy> **2.** Cheap : worthless.
**two-phase** (to͞o′fāz′) *adj.* Relating to two alternating electrical currents with phases at 90°.
**two-piece** (to͞o′pēs′) *adj.* Made in or of two parts or pieces, as a clothing ensemble. **—two′-piece′** *n.*
**two-ply** (to͞o′plī′) *adj.* **1.** Made of two interwoven layers. **2.** Consisting of two thicknesses or strands <*two-ply* yarn>
**two·some** (to͞o′səm) *n.* **1.** Two persons or things together : PAIR. **2.** A game, as a round of golf, played by two people.
**two-spot** (to͞o′spŏt′) *n.* **1.** A playing card bearing two spots or pips : DEUCE. **2.** *Slang.* **a.** A two-dollar bill. **b.** Two dollars.
**two-step** (to͞o′stĕp′) *n.* **1.** A ballroom dance in 2/4 time and marked by long, sliding steps. **2.** The music for a two-step.
**two-time** (to͞o′tīm′) *vt.* **-timed, -tim·ing, -times.** *Slang.* To be unfaithful or deceitful to (a loved one). **—two′-tim′er** *n.*
**two-tone** (to͞o′tōn′) *or* **two-toned** (-tōnd′) *adj.* Consisting of two colors or two shades of a single color.
**two-way** (to͞o′wā′) *adj.* **1.** Affording passage to vehicular traffic in two directions <a *two-way* street> **2.** Permitting communication in two directions, as a telephone connection. **3. a.** Expressive of or involving mutual action, relationship, or responsibility. **b.** Involving two participants, as a treaty. **4.** Permitting flow in either of two directions <a *two-way* valve>
**-ty** *suff.* [ME *-te* < OFr. < Lat. *-tas.*] Condition : quality <real*ty*>
**ty·coon** (tī-ko͞on′) *n.* [J. *taikun*, title of a shogun, of Chin. orig.] **1.** A wealthy, powerful businessman or industrialist : MAGNATE. **2.** A title once applied to the Japanese shogun.
**tyke** *also* **tike** (tīk) *n.* [ME, mongrel < ON *tīk*, bitch.] **1.** *Informal.* A small child, esp. a mischievous one. **2.** A mongrel : cur. **3.** *Scot.* A mean or uncouth fellow : BOOR.
**ty·lo·sin** (tī′lə-sĭn′) *n.* [Orig. unknown.] An antibiotic, $C_{45}H_{77}NO_{17}$, obtained from the actinomycete *Streptomyces fradiae* and used as an antibacterial drug in veterinary medicine.
**tym·bal** (tĭm′bəl) *n. var. of* TIMBAL.
**tym·pan** (tĭm′pən) *n.* [ME *timpan*, drum < OE *timpana* < Lat. *tympanum* < Gk. *tumpanon.*] **1.** Paper or cloth padding placed over a printing press platen to provide support for the sheet being printed. **2.** TYMPANUM 3. **3.** A tightly stretched sheet or membrane, as on the head of a drum.
**tym·pa·na** (tĭm′pə-nə) *n. var. pl. of* TYMPANUM.
**tym·pa·ni** (tĭm′pə-nē) *n. var. of* TIMPANI.
**tym·pan·ic** (tĭm-păn′ĭk) *adj.* [< TYMPANUM.] **1.** Pertaining to or resembling a drum. **2.** *also* **tym·pa·nal** (tĭm′pə-nəl). *Anat.* Of or pertaining to the tympanum.
**tympanic bone** *n.* The part of the temporal bone of the skull that partially encloses the auditory canal and supports the tympanic membrane.
**tympanic membrane** *n.* The thin, semitransparent, oval membrane separating the middle and external ear.
**tym·pa·nist** (tĭm′pə-nĭst) *n.* [Lat. *tympanista* < Gk. *tumpanistēs* < *tumpanizein*, to beat a drum < *tumpanon*, drum.] The musician

who plays the kettledrums and other percussion instruments in an orchestra.

**tym·pa·ni·tes** (tĭm′pə-nī′tēz) *n.* [ME < LLat. *tympanites* < Gk. *tumpanitēs* < *tumpanon,* drum.] Distention of the abdomen caused by accumulation of gas or air in the intestine or peritoneal cavity. **—tym′pa·nit′ic** *adj.*

**tym·pa·num** *also* **tim·pa·num** (tĭm′pə-nəm) *n., pl.* **-na** (-nə) *or* **-nums.** [Med. Lat. < Lat., drum < Gk. *tumpanon.*] **1. a.** The middle ear. **b.** The eardrum. **2.** *Zool.* A membranous external auditory structure, as in certain insects. **3. a.** The recessed, ornamental space or panel enclosed by the cornices of a triangular pediment. **b.** A similar space between an arch and the lintel of a portal. **4.** A telephone diaphragm.

**tympanum**

**tym·pa·ny** (tĭm′pə-nē) *n., pl.* **-nies.** [Med. Lat. *tympanias,* tympanites < Gk. *tumpanias* < *tumpanon,* drum.] **1.** Inflated manner or style : BOMBAST. **2.** A low-pitched resonance obtained by percussion.

**typ·al** (tī′pəl) *adj.* Relating to or serving as a type : TYPICAL.

**type** (tīp) *n.* [LLat. < Lat., figure < Gk. *tupos,* impression.] **1.** A group of persons or things that share common traits or characteristics distinguishing them as an identifiable group or class : CATEGORY. **2.** One having the features of a group or class. **3.** An example or model : EMBODIMENT. **4.** *Informal.* One regarded as exemplifying a particular profession, rank, or social group <a group of rich executive *types*> **5.** A figure, representation, or symbol of something to come, as an event in the Old Testament that foreshadows another in the New Testament. **6. a.** A taxonomic designation, as the name of a species or genus, used as the basis of ascription to or characterization of the next highest taxonomic category. **b.** A specimen or sample used as the basis of description of a species. **7. a.** A small block of metal or wood bearing a raised letter or character on the upper end, that, when inked and pressed upon paper, leaves a printed impression. **b.** Such pieces collectively. **8.** Printed or typewritten characters : PRINT. **9.** A pattern, design, or image impressed or stamped upon the face of a coin. —*v.* **typed, typ·ing, types.** —*vt.* **1.** To write (something) with a typewriter : TYPEWRITE. **2.** To determine the type of (a blood sample). **3.** To classify according to a particular type. **4.** To represent : typify. **5.** To prefigure. —*vi.* To write with a typewriter : TYPEWRITE.

**type·cast** (tīp′kăst′) *vt.* **-cast, -cast·ing, -casts. 1.** To cast in an acting role akin or natural to one's own personality or fitted to one's physical appearance. **2.** To assign (a theatrical performer) repeatedly to the same kind of part.

**type·face** (tīp′fās′) *n.* **1. a.** The surface of a body of printing type that makes the impression. **b.** The impression made. **2.** The size or style of the letter or character on the type. **3.** The full range of type of the same design.

**type foundry** *n.* A factory where type metal is cast. **—type founder** *n.*

**type genus** *n.* The name of a taxonomic genus designated as representative of the family to which it belongs.

**type-high** (tīp′hī′) *adj.* Being as high as the standard height of printing type, 0.9186 of an inch.

**type metal** *n.* An alloy consisting chiefly of tin, lead, and antimony, used for making metal printing type.

**type·script** (tīp′skrĭpt′) *n.* **1.** A typewritten copy, as of a book. **2.** Typewritten matter.

**type·set·ter** (tīp′sĕt′ər) *n.* One that sets printing type : COMPOSITOR. **—type′set′ting** *n.*

**type-site** (tīp′sīt′) *n.* An archaeological site regarded as definitively characteristic of a particular culture and often supplying the culture with its name.

**type species** *n.* The name of a taxonomic species that is designated as representative of the genus to which it belongs.

**type specimen** *n.* The individual specimen used as a basis for determining the characteristics of a species.

**type·write** (tīp′rīt′) *vt.* & *vi.* **-wrote** (-rōt′), **-writ·ten** (-rĭt′n), **-writ·ing, -writes.** [Back-formation < TYPEWRITER.] To write (something) with a typewriter or to write with a typewriter.

**type·writ·er** (tīp′rī′tər) *n.* **1.** A keyboard machine that prints characters and numerals esp. by means of a set of metal hammers bearing raised, inked type that strike the paper when actuated by keys. **2.** *Archaic.* A typist. **3.** A type style like that of typewritten copy.

**type·writ·ing** (tīp′rī′tĭng) *n.* **1.** The act, process, or skill of using a typewriter. **2.** Copy produced by typewriting.

**typh·lo·sole** (tĭf′lə-sōl′) *n.* [Gk. *tuphlos,* blind + Gk. *sōlēn,* pipe.] *Zool.* A longitudinal fold of the dorsal intestinal wall in some invertebrates that serves to increase the absorptive and digestive surface of the intestine.

**ty·pho·gen·ic** (tī′fə-jĕn′ĭk) *adj.* Causing typhus.

**ty·phoid** (tī′foid′) *n.* Typhoid fever. **—ty′phoid** *adj.*

**typhoid fever** *n.* An acute, highly infectious disease caused by the typhoid bacillus, *Salmonella typhosa,* transmitted by contaminated food or water and characterized by red rashes, high fever, bronchitis, and intestinal hemorrhaging.

**Typhoid Mary** *n.* [After *Mary* Mallon (d. 1938), a carrier of typhoid.] One from whom something undesirable or deadly spreads to others.

**Ty·phon** (tī′fŏn′) *n.* [Gk. *Tuphōn.*] *Gk. Myth.* A monster called by Hesiod the son of Typhoeus.

**ty·phoon** (tī-fo͞on′) *n.* [Cantonese *tai fung.*] A severe tropical hurricane occurring in the western Pacific or the China Sea.

**ty·phus** (tī′fəs) *n.* [NLat. < Gk. *tuphos,* stupor arising from a fever < *tuphein,* to make smoke.] Any of several forms of an infectious disease caused by microorganisms of the genus *Rickettsia,* esp. when flea-borne as in endemic typhus, louse-borne as in epidemic typhus, or mite-borne as in scrub typhus, and gen. characterized by severe headache, sustained elevated fever, depression, delirium, and red rashes. **—ty′phous** (-fəs) *adj.*

**typhus fever** *n.* Typhus.

**typ·i·cal** (tĭp′ĭ-kəl) *also* **typ·ic** (-ĭk) *adj.* [LLat. *typicalis* < *typicus* < Gk. *tupikos,* impressionable < *tupos,* impression.] **1.** Exhibiting the traits or characteristics peculiar to its kind, class, or group <a *typical* urban community> **2.** Of or relating to a representative specimen : DISTINCTIVE. **3.** Conforming to a type, as a species. **4.** Of the nature of, constituting, or serving as a type : EMBLEMATIC. **—typ′i·cal·ly** *adv.* **—typ′i·cal·ness, typ′i·cal′i·ty** *n.*

**typ·i·fy** (tĭp′ə-fī′) *vt.* **-fied, -fy·ing, -fies.** [TYP(E) + -FY.] **1.** To serve as a typical example of. **2.** To represent by an image, form, or model : SYMBOLIZE. **—typ′i·fi·ca′tion** *n.* **—typ′i·fi′er** *n.*

**typ·ist** (tī′pĭst) *n.* An operater of a typewriter.

**ty·po** (tī′pō) *n., pl.* **-os.** *Informal.* A typographical error.

**ty·pog·ra·pher** (tī-pŏg′rə-fər) *n.* A typesetter.

**typographical error** *n.* A mistake in printing or typing.

**ty·pog·ra·phy** (tī-pŏg′rə-fē) *n., pl.* **-phies.** [Med. Lat. *typographia* : Gk. *tupos,* impression + *-graphia,* -graphy.] **1. a.** Composition of printed material from movable type. **b.** The art and technique of typography. **2.** Arrangement and appearance of printed matter. **—ty′po·graph′i·cal** (tī′pə-grăf′ĭ-kəl), **ty′po·graph′ic** *adj.* **—ty′po·graph′i·cal·ly** *adv.*

**ty·pol·o·gy** (tī-pŏl′ə-jē) *n., pl.* **-gies. 1.** The study of types, as in a systematic classification. **2.** A theory or doctrine of types, as in scriptural studies. **—ty′po·log′i·cal** (tī′pə-lŏj′ĭ-kəl) *adj.* **—ty′po·log′i·cal·ly** *adv.* **—ty·pol′o·gist** *n.*

**Tyr** (tîr) *n.* [ON *Týr.*] *Norse Myth.* A god of war, son of Odin.

**ty·ra·mine** (tī′rə-mēn′) *n.* [TYR(OSINE) + AMINE.] A colorless crystalline amine, $C_8H_{11}NO$, found in mistletoe, putrefied animal tissue, certain cheeses, and ergot and also produced synthetically, used in medicine.

**ty·ran·ni·cal** (tĭ-răn′ĭ-kəl, tī-) *also* **ty·ran·nic** (-răn′ĭk) *adj.* Of, relating to, or of the nature of a tyrant : DESPOTIC. **—ty·ran′ni·cal·ly** *adv.* **—ty·ran′ni·cal·ness** *n.*

**tyr·an·nize** (tĭr′ə-nīz′) *v.* **-nized, -niz·ing, -niz·es.** [OFr. *tyranniser* < *tyran,* tyrant.] —*vi.* **1.** To exercise absolute power, esp. arbitrarily. **2.** To rule as a tyrant. —*vt.* To treat tyrannically : OPPRESS. **—tyr′an·niz′er** *n.* **—tyr′an·niz′ing·ly** *adv.*

**ty·ran·no·saur** (tĭ-răn′ə-sôr′, tī-) *also* **ty·ran·no·saur·us** (tĭ-răn′ə-sôr′əs, tī-) *n.* [NLat. *Tyrannasaurus,* genus name : Gk. *turannos,* tyrant + Gk. *sauros,* lizard.] A large carnivorous dinosaur of the genus *Tyrannosaurus,* of the Cretaceous period, with a large head and small forelimbs.

**tyr·an·nous** (tĭr′ə-nəs) *adj.* Tyrannical. **—tyr′an·nous·ly** *adv.*

**tyr·an·ny** (tĭr′ə-nē) *n., pl.* **-nies.** [ME < OFr. *tyrannie* < LLat. *tyrannia* < Gk. *turannia* < *turannos,* tyrant.] **1.** A government in which a single ruler is vested with absolute power. **2.** The office, authority, or jurisdiction of an absolute ruler. **3.** Absolute power, esp. when exercised unjustly or cruelly. **4. a.** Arbitrary use of absolute power. **b.** A tyrannical act. **5.** Extreme severity or harshness : RIGOR.

**ty·rant** (tī′rənt) *n.* [ME < OFr. < Lat. *tyrannus* < Gk. *turannos.*] **1.** A ruler who exercises power in a harsh, cruel manner : OPPRESSOR. **2.** A tyrannical or despotic person <a corporate *tyrant*> **3.** An absolute ruler, esp. in ancient Greece.

**tyre** (tīr) *n. Chiefly Brit. var. of* TIRE[2].

**Tyr·i·an purple** (tĭr′ē-ən) *n.* [After *Tyre,* ancient capital of Phoenicia, famous for its purple dyes.] A reddish dyestuff obtained from the bodies of certain mollusks of the genus *Murex* and highly prized in ancient times.

**ty·ro** *also* **ti·ro** (tī'rō) *n., pl.* **-ros.** [Med. Lat., squire < Lat. *tiro,* recruit.] An inexperienced person : BEGINNER.

**ty·ro·ci·dine** (tī'rə-sīd'n) *n.* [TYRO(THRICIN) + (GRAMI)CID(IN) + -INE².] A polypeptide antibiotic produced by the soil microorganism *Bacillus brevis.*

**Ty·ro·le·an** (tĭ-rō'lē-ən, tī-) *n. & adj.* Tyrolese.

**Tyr·o·lese** (tĭr'ə-lēz', -lēs', tī'rə-) *n., pl.* **Tyrolese.** A native or inhabitant of Tyrol. **—Tyr'o·lese'** *adj.*

**ty·ros·i·nase** (tī-rŏs'ə-nās', -nāz') *n.* A copper-containing enzyme of plant and animal tissues that catalyzes the production of melanin from tyrosine, as in the blackening of a potato exposed to air.

**ty·ro·sine** (tī'rə-sēn') *n.* [Gk. *turos,* cheese + -INE².] A white crystalline amino acid, $C_9H_{11}NO_3$, derived from the hydrolysis of protein, used as a growth factor in nutrition and as a dietary supplement.

**ty·ro·thri·cin** (tī'rō-thrī'sĭn) *n.* [< NLat. *Tyrothrix, Tyrothric-,* former bacteria genus name : Gk. *turos,* cheese + Gk. *thrix,* hair.] A grayish to brown mixture of antibiotics obtained from soil and bacteria cultures, esp. *Bacillus brevis,* used topically in treating infections caused by Gram-positive bacteria.

**tzar** (tsär) *n. var. of* CZAR.

**tzet·ze fly** (tsĕt'sē, tsē'sē) *n. var. of* TSETSE FLY.

# Uu

**u** *or* **U** (yo͞o) *n., pl.* **u's** *or* **U's. 1.** The 21st letter of the English alphabet. **2.** A speech sound represented by the letter *u.* **3.** The 21st in a series. **4. U** A grade indicating that a student's work is unsatisfactory. **5.** Something shaped like the letter U.

**U¹** *symbol for* URANIUM.

**U²** (yo͞o) *adj.* [U(PPER CLASS).] *Informal.* Characteristic of or appropriate to the upper class, as in language usage and tastes.

**u·biq·ui·tous** (yo͞o-bĭk'wĭ-təs) *adj.* Being or seeming to be everywhere at the same time : OMNIPRESENT <"plodded through the shadows fruitlessly like an *ubiquitous* spook" —Joseph Heller> **—u·biq'ui·tous·ly** *adv.* **—u·biq'ui·tous·ness** *n.*

**u·biq·ui·ty** (yo͞o-bĭk'wĭ-tē) *n.* [NLat. *ubiquitas* < Lat. *ubique,* everywhere : *ubi,* where + *-que,* generalizing particle.] Existence everywhere at the same time : OMNIPRESENCE.

**U-boat** (yo͞o'bōt') *n.* [Transl. of G. *U-boot,* short for *Unterseeboot* : *unter,* under + *See,* sea + *Boot,* boat.] A German submarine.

**U-bolt** (yo͞o'bōlt') *n.* A bolt shaped like the letter U, fitted with threads and a nut at each end.

**UBV photometry** (yo͞o'bē-vē') *n.* [U(LTRAVIOLET) + B(LUE) + V(ISUAL).] A system of photometry used to obtain stellar magnitudes by comparing observed magnitudes to a standard sequence of stars.

**ud·der** (ŭd'ər) *n.* [ME < OE *ūder.*] A baglike mammary organ with two or more teats, as in cows, sheep, and goats.

**u·do** (o͞o'dō) *n.* [J.] A Japanese plant, *Aralia cordata,* whose young shoots are blanched and eaten as a vegetable.

**UFO** (yo͞o'ĕf-ō') *n., pl.* **UFOs** *or* **UFO's.** An unidentified flying object.

**u·fol·o·gy** (yo͞o-fŏl'ə-jē) *n.* [UFO + -LOGY.] Study of unidentified flying objects. **—u·fol'o·gist** (yo͞o-fŏl'ə-jĭst) *n.*

**U·ga·rit·ic** (o͞o'gə-rĭt'ĭk, yo͞o'-) *n.* The Semitic language of the ancient city of Ugarit. **—U'ga·rit'ic** *adj.*

**ugh** (ŭg, ŭk) *interj.* —Used to express horror or disgust.

**ug·li** (ŭg'lē) *n., pl.* **-lis** *or* **-lies.** [Poss. < UGLY, from the appearance of its wrinkled skin.] A citrus fruit indigenous to Jamaica, produced by crossing a grapefruit and a tangerine and having a loose, wrinkled yellowish rind.

**ug·li·fy** (ŭg'lə-fī') *vt.* **-fied, -fy·ing, -fies.** To make ugly. **—ug'li·fi·ca'tion** *n.*

**ug·ly** (ŭg'lē) *adj.* **-li·er, -li·est.** [ME, frightful < ON *uggligr* < *uggr,* fear.] **1.** Displeasing to the eye : very unattractive. **2.** Repulsive or offensive : OBJECTIONABLE. **3.** Morally reprehensible <an *ugly* habit><an *ugly* crime> **4.** Threatening or ominous <*ugly* seas><an *ugly* turn of events> **5.** Cross or disagreeable <an *ugly* mood> **—ug'li·ly** *adv.* **—ug'li·ness** *n.*

☆ ***syns:*** UGLY, HIDEOUS, UNSIGHTLY *adj. core meaning* : displeasing to the eye <an *ugly* Halloween mask><a *hideous* gash><an *unsightly* scar> UGLY and HIDEOUS, the stronger term, can also describe what is emotionally, morally, or otherwise offensive <the *ugly* details of the argument><a *hideous* murder> ***ant:*** beautiful

**ugly duckling** *n.* [< *The Ugly Duckling,* story by Hans Christian Andersen (1805–1875).] One considered ugly or unpromising but having the potential of becoming beautiful or admirable.

**U·gri·an** (o͞o'grē-ən, yo͞o'-) *n.* [OR *Ugrin',* of Turkic orig.] **1.** A member of a group of Finno-Ugric peoples of western Siberia and Hungary, including the Magyars. **2.** Ugric. **—U'gri·an** *adj.*

**U·gric** (o͞o'grĭk, yo͞o'-) *n.* The branch of the Finno-Ugric subfamily of languages that includes Hungarian. **—U'gric** *adj.*

**ug·some** (ŭg'səm) *adj.* [ME : *uggen,* to fear (< ON *ugga*) + *-some,* -some.] *Archaic.* Loathsome : disgusting.

**uh** (ŭ) *interj.* —Used to express hesitation or uncertainty.

**uh-huh** (ə-hŭ') *interj. Informal.* —Used to express the affirmative.

**uh·lan** *also* **u·lan** (o͞o'län', yo͞o'lən) *n.* [G. < Pol. < Turk. *uğlan,* youth.] One of a body of mounted lancers that formed part of the Polish and, later, German armies.

**Ui·gur** *also* **Ui·ghur** (wē'go͝or) *n.* [Uigur.] **1.** One of a Turkic people dominant in Mongolia and eastern Turkestan from the 8th to the 12th cent., now inhabiting northwestern China. **2.** The Turkic language of the Uigurs. **—Ui·gu'ri·an** (wē-go͝or'ē-ən), **Ui·gu'ric** *adj.*

**u·in·ta·ite** (yo͞o-ĭn'tə-īt') *n.* [After the *Uinta,* mountains in Utah.] Gilsonite.

**uit·land·er** (oit'lăn'dər, īt'-) *n.* [Afr. < MDu. *utelander* < *utelant,* foreign land : *ute,* out + *land,* land.] **1.** A foreigner in South Africa. **2. Uitlander.** A native of Great Britain living in the former republics of the Orange Free State or Transvaal.

**u·kase** (yo͞o-kās', -kāz', yo͞o'kās', -kāz') *n.* [Fr. < R. *ukaz,* decree < *ukazat',* to order.] **1.** A proclamation of a czar having the force of law in imperial Russia. **2.** An authoritative order : EDICT.

**U·krain·i·an** (yo͞o-krā'nē-ən) *n.* [Ukrainian *Ukrayina* < OR *Ukraina.*] **1.** A native or resident of the Ukraine. **2.** The Slavic language of the Ukrainians. **—U·krain'i·an** *adj.*

**u·ku·le·le** (yo͞o'kə-lā'lē, o͞o'kə-) *n.* [Hawaiian *'ukulele* : *'uku,* flea + *lele,* jumping.] A small four-stringed Hawaiian guitar.

**u·lan** (o͞o'län', yo͞o'lən) *n. var. of* UHLAN.

**-ular** *suff.* [Lat. *-ularis* < *-ulus,* -ule.] Of, relating to, or resembling <tub*ular*>

**ul·cer** (ŭl'sər) *n.* [ME < OFr. *ulcere* < Lat. *ulcus.*] **1.** An inflammatory, often suppurating lesion on the skin or an internal mucous surface of the body, as in the duodenum, resulting in necrosis of the tissue. **2.** A continuing source of corruption.

**ul·cer·ate** (ŭl'sə-rāt') *v.* **-at·ed, -at·ing, -ates.** —*vi.* To form an ulcer. —*vt.* To cause an ulcer in. **—ul'cer·a'tion** (-rā'shən) *n.* **—ul'cer·a'tive** (-rā'tĭv, -sər-ə-tĭv) *adj.*

**ul·cer·o·gen·ic** (ŭl'sə-rō-jĕn'ĭk) *adj.* Tending to cause ulcers.

**ul·cer·ous** (ŭl'sər-əs) *adj.* Characterized by or afflicted with ulcer.

**-ule** *suff.* [Fr. < Lat. *-ulus,* dim. suffix.] Small one <valv*ule*>

**u·lex·ite** (yo͞o'lĭk-sīt', yo͞o-lĕk'-) *n.* [After George L. *Ulex* (d. 1883).] A white mineral, $NaCaB_5O_9{\cdot}8H_2O$, that forms rounded masses of very fine acicular crystals.

**ul·lage** (ŭl'ĭj) *n.* [ME *oylage* < OFr. *ouillage* < *ouiller,* to fill up a cask < *oeil,* eye, bunghole < Lat. *oculus,* eye.] The amount of liquid lost from a container during shipment or storage.

**ul·na** (ŭl'nə) *n., pl.* **-nae** (-nē) *or* **-nas.** [NLat. < Lat., elbow.] *Anat.* **1.** The larger bone of the human forearm, extending from the elbow to the wrist on the side opposite to the thumb. **2.** A homologous bone in the vertebrate forelimb. **—ul'nar** *adj.*

**u·lot·ri·chous** (yo͞o-lŏt'rĭ-kəs) *adj.* [< Gk. *oulothrix, oulotrikh-* : *oulos,* woolly + *thrix,* hair.] Having woolly or wiry hair. **—u·lot'ri·chy** *n.*

**ul·ster** (ŭl'stər) *n.* [After *Ulster,* Ireland.] A long loose overcoat made of heavy rugged fabric.

**ul·te·ri·or** (ŭl-tîr'ē-ər) *adj.* [Lat., farther, comp. of *ulter,* on the other side.] **1.** Lying beyond or outside the area of immediate interest. **2.** Lying beyond what is evident or avowed, esp. concealed so as to deceive <*ulterior* motives> **3.** Occurring later : SUBSEQUENT.

**ul·ti·ma** (ŭl′tə-mə) *n.* [Lat., fem. of *ultimus,* last. —see ULTIMATE.] The last syllable of a word.

**ul·ti·ma·ta** (ŭl′tə-mā′tə, -mä′tə) *n. var. pl. of* ULTIMATUM.

**ul·ti·mate** (ŭl′tə-mĭt) *adj.* [Med. Lat. *ultimatus,* p.part. of *ultimare,* to come to an end < Lat. *ultimus,* last, superl. of *ulter,* on the other side.] **1.** Representing the farthest possible extent of analysis or division into parts <an *ultimate* particle> **2.** Fundamental : elemental <the *ultimate* rules of nature> **3.** Greatest in size or significance : MAXIMUM <the *ultimate* authority> **4. a.** Most remote in time or space : FARTHEST. **b.** Final, as in a series or progression <Space is the *ultimate* frontier.> **c.** Eventual <planning for *ultimate* retirement> **d.** Utmost : extreme <the *ultimate* penalty> —*n.* **1.** The basic or fundamental fact. **2.** The final point : CONCLUSION. **3.** The greatest extreme <the *ultimate* in political chicanery> **—ul′ti·mate·ly** *adv.* **—ul′ti·mate·ness** *n.*

**ultima Thu·le** (thōō′lē) *n.* [Lat., farthest Thule.] **1.** The northernmost region of the habitable world as conceived by ancient geographers. **2.** The most remote goal or ideal attainable.

**ul·ti·ma·tum** (ŭl′tə-mā′təm, -mä′təm) *n., pl.* **-tums** *or* **-ta** (-tə) [NLat. < Med. Lat., neuter of *ultimatus,* last. —see ULTIMATE.] A proposal or statement of terms, esp. one that expresses or implies the threat of serious penalties if the terms are not accepted.

**ul·ti·mo** (ŭl′tə-mō′) *adv.* [Lat. *ultimo (mense),* in the last (month) < *ultimus,* last. —see ULTIMATE.] In or of the month before the present.

**ul·tra** (ŭl′trə) *adj.* [< ULTRA-.] Going beyond the limit, as in adherence to a belief or fashion : EXTREME. —*n.* An extremist.

**ultra-** *pref.* [Lat. < *ultra,* beyond < *ulter,* on the other side < *uls,* beyond.] **1.** Beyond : on the other side of <*ultra*violet> **2.** Beyond the range, scope, or limit of <*ultra*sonic> **3.** Beyond the normal or proper degree : EXCESSIVELY <*ultra*conservative>

**ul·tra·cen·tri·fuge** (ŭl′trə-sĕn′trə-fyōōj′) *n.* A convection-free high-velocity centrifuge used in the separation of colloidal or submicroscopic particles. **—ul′tra·cen·trif′u·gal** (-trĭf′yə-gəl, -trĭf′ə-gəl) *adj.* **—ul′tra·cen′tri·fu·ga′tion** (-fyōō-gā′shən) *n.*

**ul·tra·con·ser·va·tive** (ŭl′trə-kən-sûr′və-tĭv) *adj.* Conservative to an extreme, esp. in political beliefs : REACTIONARY. —*n.* One who is extremely conservative. **—ul′tra·con·ser′va·tism** *n.*

**ul·tra·fiche** (ŭl′trə-fēsh′) *n.* A microfiche on which material is reduced by a factor of 100 or more.

**ul·tra·fil·tra·tion** (ŭl′trə-fĭl-trā′shən) *n.* Filtration of a collodial substance through a semipermeable medium that allows only the passage of small molecules.

**ul·tra·high** (ŭl′trə-hī′) *adj.* Exceedingly high.

**ultrahigh frequency** *n.* A band of radio frequencies from 300 to 3,000 megacycles per second.

**ul·tra·ism** (ŭl′trə-ĭz′əm) *n.* Extremism, esp. in politics or government : RADICALISM. **—ul′tra·ist** *n.*

**ul·tra·lib·er·al** (ŭl′trə-lĭb′ər-əl, -lĭb′rəl) *adj.* Liberal to an extreme, esp. in political beliefs : RADICAL. —*n.* One who is extremely liberal. **—ul′tra·lib′er·al·ism** *n.*

**ul·tra·ma·rine** (ŭl′trə-mə-rēn′) *n.* [< Med. Lat. *ultramarinus,* from beyond the sea : Lat. *ultra,* beyond + Lat. *marinus,* of the sea < *mare,* sea.] **1. a.** A blue pigment made from powdered lapis lazuli. **b.** A similar pigment made from other substances. **2.** A vivid or strong blue to purplish blue. —*adj.* **1.** Having a deep-blue purplish color. **2.** Of or from a place beyond the sea.

**ul·tra·mi·cro·fiche** (ŭl′trə-mī′krō-fēsh′) *n.* Ultrafiche.

**ul·tra·mi·crom·e·ter** (ŭl′trə-mī-krŏm′ĭ-tər) *n.* An extremely accurate micrometer.

**ul·tra·mi·cro·scope** (ŭl′trə-mī′krə-skōp′) *n.* A microscope with high-intensity illumination used to study very minute objects, as colloidal particles, by means of their diffraction system that appears as a bright spot against a black background. **—ul′tra·mi·cros′co·py** (-mī-krŏs′kə-pē) *n.*

**ul·tra·mi·cro·scop·ic** (ŭl′trə-mī′krə-skŏp′ĭk) *adj.* **1.** Too small to be seen with an ordinary microscope. **2.** Of or pertaining to an ultramicroscope. **—ul′tra·mi′cro·scop′i·cal·ly** *adv.*

**ul·tra·mi·cro·tome** (ŭl′trə-mī′krə-tōm′) *n.* A microtome for cutting very thin sections of material for use in electron microscopy. **—ul′tra·mi·crot′o·my** (-mī-krŏt′ə-mē) *n.*

**ul·tra·mil·i·tant** (ŭl′trə-mĭl′ĭ-tnt) *adj.* Militant to an extreme. —*n.* One who is extremely militant.

**ul·tra·min·i·a·ture** (ŭl′trə-mĭn′ē-ə-chŏŏr′, -mĭn′ə-, -chər) *adj.* Subminiature. **—ul′tra·min′i·a·tur·i·za′tion** *n.*

**ul·tra·mod·ern** (ŭl′trə-mŏd′ərn) *adj.* Extremely modern or advanced in ideas or style. **—ul′tra·mod′ern·ism** *n.* **—ul′tra·mod′ern·ist** *n.* **—ul′tra·mod′ern·is′tic** *adj.*

**ul·tra·mon·tane** (ŭl′trə-mŏn′tān′, -mŏn-tān′) *adj.* [Med. Lat. *ultramontanus* : Lat. *ultra,* beyond + Lat. *montanus,* of mountains < *mons,* mountain.] **1.** Of or pertaining to peoples or regions lying beyond the mountains, esp. the Alps. **2.** Favoring the supremacy of the papal court over national or diocesan authority in the Roman Catholic Church. **3.** Relating to or supporting the doctrine of papal supremacy. —*n.* **1.** One living beyond the mountains, esp. south of the Alps. **2.** *often* **Ultramontane.** A Roman Catholic who advocates support of papal policy in ecclesiastical and political matters.

**ul·tra·mon·ta·nism** (ŭl′trə-mŏn′tə-nĭz′əm) *n. often* **Ultramontanism.** The policy that absolute authority in the Roman Catholic Church should be vested in the pope.

**ul·tra·mun·dane** (ŭl′trə-mŭn′dān′, -mŭn-dān′) *adj.* [LLat. *ultramundanus* : Lat. *ultra,* beyond + Lat. *mundanus,* of the world < *mundus,* world.] Extending or being beyond the world or the limits of the universe.

**ul·tra·na·tion·al·ism** (ŭl′trə-năsh′ə-nə-lĭz′əm) *n.* Extreme nationalism, esp. when opposed to international cooperation. **—ul′tra·na′tion·al** *adj.* **—ul′tra·na′tion·al·ist** *n.* **—ul′tra·na′tion·al·is′tic** *adj.*

**ul·tra·pure** (ŭl′trə-pyŏŏr′) *adj.* Of exceeding purity <an *ultrapure* metal>

**ul·tra·son·ic** (ŭl′trə-sŏn′ĭk) *adj.* Relating to acoustic frequencies above the range audible to the human ear, or above approx. 20,000 cycles per second. **—ul′tra·son′i·cal·ly** *adv.*

**ul·tra·son·ics** (ŭl′trə-sŏn′ĭks) *n.(sing. in number).* **1.** The branch of science dealing with ultrasonic sound. **2.** A technology using ultrasonic sound, as for medical therapy.

**ul·tra·so·nog·ra·phy** (ŭl′trə-sə-nŏg′rə-fē) *n.* [ULTRASON(IC) + -GRAPHY.] Diagnostic use of ultrasonic waves to visualize internal body structures. **—ul′tra·son′o·graph′ic** (-sŏn′ə-grăf′ĭk, -sō′nə-) *adj.*

**ul·tra·so·phis·ti·cat·ed** (ŭl′trə-sə-fĭs′tĭ-kā′tĭd) *adj.* Extremely sophisticated.

**ul·tra·sound** (ŭl′trə-sound′) *n.* **1.** Ultrasonic sound. **2.** The use of ultrasonic sound to outline the shape of various bodily organs and tissues for diagnostic or therapeutic purposes.

**ul·tra·thin** (ŭl′trə-thĭn′) *adj.* Very thin.

**ul·tra·vi·o·let** (ŭl′trə-vī′ə-lĭt) *adj.* Of or relating to the range of radiation wavelengths from approx. 4,000 angstroms, just beyond the violet in the visible spectrum, to approx. 40 angstroms, on the border of the x-ray region. —*n.* Ultraviolet radiation.

**ultraviolet lamp** *n.* A mercury-vapor lamp producing ultraviolet light.

**ul·tra·vi·rus** (ŭl′trə-vī′rəs) *n.* A virus small enough to pass through the finest bacterial filter.

**ul·u·lant** (ŭl′yə-lənt) *adj.* Howling : wailing.

**ul·u·late** (ŭl′yə-lāt′) *vi.* **-lat·ed, -lat·ing, -lates.** [Lat. *ululare, ululat-.*] To howl, wail, or lament loudly. **—ul′u·la′tion** *n.*

**U·lys·ses** (yōō-lĭs′ēz′) *n.* [Lat., alteration of *Ulixes.*] Odysseus.

**u·man·gite** (ōō-măn′gīt′, -măng′-) *n.* [After Sierra de *Umango,* a province in Argentina.] A dark red mineral, $Cu_3Se_2$, consisting of copper selenide.

**um·bel** (ŭm′bəl) *n.* [NLat. *umbella* < Lat., umbrella < dim. of *umbra,* shadow.] *Bot.* A flat-topped or rounded flower cluster in which the individual flower stalks spring from about the same point in an axis, as in the carrot, parsley, and related plants.

**um·bel·late** (ŭm′bə-lāt′, ŭm-bĕl′ĭt) *adj.* Having, forming, or characteristic of an umbel.

**um·bel·lif·er·ous** (ŭm′bə-lĭf′ər-əs) *adj.* [< NLat. *umbellifer* : *umbella,* umbel + Lat. *ferre,* to bear.] Bearing umbels.

**um·bel·lule** (ŭm′bəl-yōōl′, ŭm-bĕl′yōōl′) *n.* [NLat. *umbellula,* dim. of *umbella,* umbel.] *Bot.* One of the smaller secondary umbels forming a compound umbel.

**um·ber** (ŭm′bər) *n.* [Prob. < obs. *umber,* shade < ME < OFr. *umbre* < Lat. *umbra.*] **1.** A natural brown earth containing chiefly iron and manganese oxides and used as pigment in both its raw and burnt states. **2.** Any of the shades of brown produced by umber in its various states. —*adj.* **1.** Of or related to umber. **2.** Brownish in hue. —*vt.* **-bered, -ber·ing, -bers.** To darken with or as if with umber.

**um·bil·i·cal** (ŭm-bĭl′ĭ-kəl) *adj.* **1.** Of, relating to, or like an umbilicus. **2.** Pertaining to or located near the central area of the abdomen. —*n. Aerospace.* UMBILICAL CORD 2.

**umbilical cord** *n.* **1.** *Anat.* A flexible, cordlike structure connecting the fetus at the navel with the placenta and containing two umbilical arteries and one vein that nourish the fetus and remove its wastes. **2.** *Aerospace.* **a.** An external electrical line or fluid tube supplying a rocket before launch. **b.** A line attached to a spacecraft that supplies an astronaut working outside the vehicle with air and usu. a means of communication.

**um·bil·i·cate** (ŭm-bĭl′ĭ-kĭt) *also* **um·bil·i·cat·ed** (-kā′tĭd) *adj.* **1.** Having a central mark or depression suggestive of a navel. **2.** Having an umbilicus. **—um·bil′i·ca′tion** *n.*

**um·bil·i·cus** (ŭm-bĭl′ĭ-kəs, ŭm′bə-lī′kəs) *n., pl.* **-ci** (-sī′) [Lat.] **1.** The navel. **2.** *Biol.* A small opening or depression similar to a navel, as the hollow at the base of the shell of some gastropod mollusks or one of the openings in the shaft of a feather.

**um·bo** (ŭm′bō) *n., pl.* **um·bo·nes** (ŭm-bō′nēz) *or* **um·bos.** [Lat.] **1.** The boss or knob at the center of a shield. **2.** *Biol.* A knoblike protuberance similar to an umbo, as a prominence near the hinge of a bivalve shell. **3.** *Anat.* A slight projection at the center of the outer surface of the tympanic membrane of the ear. **—um′bo·nal** (ŭm′bə-nəl, ŭm-bō′nəl), **um·bon′ic** (ŭm-bŏn′ĭk) *adj.*

**um·bo·nate** (ŭm′bə-nāt′, ŭm-bō′nĭt) *adj.* Having or resembling a knob or knoblike protuberance.

ōō **boot** ou **out** th **thin** *th* **this** ŭ **cut** ûr **urge** y **young**
yōō **abuse** zh **vision** ə **about,** it**e**m, edib**i**le, gall**o**p, circ**u**s

**um·bra** (ŭm′brə) *n., pl.* **-brae** (-brē) [Lat., shadow.] **1.** A dark area, esp. the blackest part of a shadow. **2.** *Astron.* **a.** The shadow region over an area of the earth where a solar eclipse is total. **b.** The darkest region of a sunspot. **3.** A shade : ghost.

**um·brage** (ŭm′brĭj) *n.* [ME, shade < OFr. < Lat. *umbraticum,* neuter of *umbraticus,* of shade < Lat. *umbra,* shadow.] **1.** Offense : resentment <took *umbrage* at my remark> **2.** *Archaic.* **a.** Something that affords shade. **b.** Shadow or shade. **3.** A vague indication : HINT.

**um·bra·geous** (ŭm-brā′jəs) *adj.* **1.** Affording or creating shade : SHADY. **2.** Easily offended : IRRITABLE. **—um·bra′geous·ly** *adv.* **—um·bra′geous·ness** *n.*

**um·brel·la** (ŭm-brĕl′ə) *n.* [Ital. *ombrella,* dim. of *ombra,* shade < Lat. *umbra.*] **1.** A portable, collapsible device for protection against rain and snow consisting of a fabric canopy mounted on a sliding framework of ribs radiating from a central rod. **2.** Something that covers or protects, as military aircraft shielding ground operations. **3.** Something that encompasses or covers many different elements or groups. **4.** *Zool.* The contractile, gelatinous, rounded mass constituting the major part of the body of most jellyfishes. *—vt.* **-laed, -la·ing, -las.** To protect, cover, or furnish with an umbrella.

**umbrella bird** *n.* A tropical American bird of the genus *Cephalopterus,* esp. *C. ornatus,* having a retractile black crest and a long feathered wattle.

**umbrella bird**
*18 inches long*

**umbrella leaf** *n.* A plant, *Diphylleia cymosa* of the southeastern United States, with a broad rounded basal leaf and a terminal white flower cluster.

**umbrella palm** *n.* A palm tree, *Hedyscepe canterburyana* of the South Pacific, cultivated for its drooping feathery foliage.

**umbrella tree** *n.* **1.** A tree of the genus *Magnolia* of the southeastern United States, esp. *M. tripetala,* having large leaves in umbrellalike clusters at the ends of the branches. **2.** An Australian tree, *Schefflera actinophylla,* having compound leaves and widely cultivated in its smaller forms as a house plant.

**Um·bri·an** (ŭm′brē-ən) *adj.* Of or relating to Umbria, its people, or their language. *—n.* **1.** A native or resident of ancient or modern Umbria. **2.** The Italic language of ancient Umbria.

**u·mi·ak** (o͞o′mē-ăk′) *n.* [Eskimo.] A large open Eskimo boat made of skins stretched on a wooden frame, usu. propelled by paddles.

**um·laut** (o͝om′lout′) *n.* [G. : *um-,* around (< MHG *umb-* < OHG *umbi-*) + *Laut,* sound (< MHG *lut* < OHG *hlūt*).] **1. a.** A change in a vowel sound caused by partial assimilation to a vowel or semivowel occurring in the following syllable. **b.** A vowel sound changed in this manner. **2.** The diacritical mark (¨) placed over a vowel to indicate an umlaut, esp. in German. *—vt.* **-laut·ed, -laut·ing, -lauts. 1.** To modify by umlaut. **2.** To write or print (a vowel) with an umlaut.

**ump** (ŭmp) *n.* UMPIRE 1. *—vi.* **umped, ump·ing, umps.** To serve as an umpire.

**um·pir·age** (ŭm′pīr′ĭj) *n.* **1.** The position, function, or authority of an umpire. **2.** A ruling or decision of an umpire.

**um·pire** (ŭm′pīr′) *n.* [ME *(an) oumpere,* alteration of *(a) noumpere,* mediator < OFr. *nomper* : *non,* not (< Lat.) + *per,* equal < Lat. *par.*] **1.** One appointed to rule on sports plays, esp. in baseball. **2.** One selected or empowered to settle a dispute or deadlock : ARBITRATOR. **3.** A judge. *—v.* **-pired, -pir·ing, -pires.** *—vt.* To act as umpire for. *—vi.* To act as umpire.

**ump·teen** (ŭmp′tēn′, ŭm′-) *adj.* [Slang *umpty,* dash in Morse code + *-teen,* as in *thirteen.*] *Informal.* Too many to be counted : INNUMERABLE <*umpteen* cats><*umpteen* guests> **—ump′teenth′** *adj.*

**un-**[1] *pref.* [ME < OE.] **1.** Not <*unhappy*> **2.** Opposite of : contrary to <*unrest*>

**un-**[2] *pref.* [ME < OE *on-,* alteration of *ond-, and-,* against.] **1.** To reverse or undo a specified action <*unbind*> **2. a.** To deprive of or remove a specified thing <*unfrock*> **b.** To release, free, or remove from <*unyoke*> **3.** —Used as an intensive <*unloose*>

**un·a·bashed** (ŭn′ə-băsht′) *adj.* **1.** Not embarrassed or disconcerted. **2.** Not disguised <*unabashed* feelings> **—un′a·bash′ed·ly** (-băsh′ĭd-lē) *adv.*

**un·a·bat·ed** (ŭn′ə-bā′tĭd) *adj.* Not diminished in force or intensity <*unabated* fury> **—un′a·bat′ed·ly** *adv.*

**un·a·ble** (ŭn-ā′bəl) *adj.* **1.** Lacking the necessary power, authority, or means : not able. **2.** Lacking mental capability : INCOMPETENT.

**un·a·bridged** (ŭn′ə-brĭjd′) *adj.* Having the full original content : not condensed. —Used of books and documents.

**un·ac·cent·ed** (ŭn-ăk′sĕn-tĭd) *adj.* **1.** Having no diacritical mark. —Used of a word, syllable, or letter. **2.** Having weak or no stress in pronunciation.

**un·ac·cept·a·ble** (ŭn′ăk-sĕp′tə-bəl) *adj.* Not worthy of acceptance : UNSATISFACTORY <*unacceptable* work> **—un′ac·cept′a·bil′i·ty** *n.* **—un′ac·cept′a·bly** *adv.*

**un·ac·com·mo·dat·ed** (ŭn′ə-kŏm′ə-dā′tĭd) *adj.* **1.** Not adapted or accommodated. **2.** Lacking accommodations.

**un·ac·com·pa·nied** (ŭn′ə-kŭm′pə-nēd) *adj.* **1.** Being without a companion. **2.** *Mus.* Having no instrumental accompaniment.

**un·ac·com·plished** (ŭn′ə-kŏm′plĭsht) *adj.* **1.** Not completed or carried out : UNFINISHED. **2.** Lacking special skills or abilities, as in the social graces : UNPOLISHED.

**un·ac·count·a·ble** (ŭn′ə-koun′tə-bəl) *adj.* **1.** Not capable of being explained or accounted for : INEXPLICABLE. **2.** Not to be held to account : not responsible. **—un′ac·count′a·bil′i·ty, un′ac·count′a·ble·ness** *n.* **—un′ac·count′a·bly** *adv.*

**un·ac·cus·tomed** (ŭn′ə-kŭs′təmd) *adj.* **1.** Not used to : not habituated <*unaccustomed* to public speaking> **2.** Not familiar or customary <*unaccustomed* luxury>

**u·na cor·da** (o͞o′nə kôr′də) *adj.* & *adv.* [Ital., one string.] *Mus.* With the soft pedal of the piano depressed. —Used as a direction.

**un·a·dorned** (ŭn′ə-dôrnd′) *adj.* Lacking adornment : PLAIN.

**un·a·dul·ter·at·ed** (ŭn′ə-dŭl′tə-rā′tĭd) *adj.* Not mingled or diluted with extraneous matter : PURE.

**un·ad·vised** (ŭn′əd-vīzd′) *adj.* **1.** Not informed or counseled. **2.** Not prudent or discreet : ILL-ADVISED. **—un′ad·vis′ed·ly** (-vī′zĭd-lē) *adv.* **—un′ad·vis′ed·ness** *n.*

**un·af·fect·ed** (ŭn′ə-fĕk′tĭd) *adj.* **1.** Not changed, modified, or affected. **2.** Free from affectation and pretense : NATURAL. **—un′af·fect′ed·ly** *adv.* **—un′af·fect′ed·ness** *n.*

**un·a·fraid** (ŭn′ə-frād′) *adj.* Devoid of fear.

**un·al·ien·a·ble** (ŭn-āl′yə-nə-bəl, -ā′lē-ə-) *adj.* Inalienable.

**un·a·ligned** (ŭn′ə-līnd′) *adj.* Nonaligned.

**un·al·loyed** (ŭn′ə-loid′) *adj.* **1.** Not mixed with other metals : PURE. **2.** Complete : unqualified <an *unalloyed* triumph>

**un·al·ter·a·ble** (ŭn-ôl′tər-ə-bəl) *adj.* Not capable of being altered <an *unalterable* decision> **—un·al′ter·a·bil′i·ty, un·al′ter·a·ble·ness** *n.* **—un·al′ter·a·bly** *adv.*

**un·am·big·u·ous** (ŭn′ăm-bĭg′yo͞o-əs) *adj.* Not ambiguous or uncertain : CLEAR. **—un′am·big′u·ous·ly** *adv.*

**un-A·mer·i·can** (ŭn′ə-mĕr′ĭ-kən) *adj.* Considered contrary to the institutions or principles of the United States.

**un·a·neled** (ŭn′ə-nēld′) *adj.* *Archaic.* Not having received extreme unction.

**u·na·nim·i·ty** (yo͞o′nə-nĭm′ĭ-tē) *n.* The state of being unanimous.

**u·nan·i·mous** (yo͞o-năn′ə-məs) *adj.* [Lat. *unanimus* : *unus,* one + *animus,* mind.] **1.** Sharing the same opinions or views : being in complete harmony or accord. **2.** Based on or marked by complete assent or agreement. **—u·nan′i·mous·ly** *adv.* **—u·nan′i·mous·ness** *n.*

**un·an·swer·a·ble** (ŭn-ăn′sər-ə-bəl) *adj.* **1.** Impossible to answer. **2.** Not to be refuted : CONCLUSIVE. **—un·an′swer·a·bil′i·ty** *n.* **—un·an′swer·a·bly** *adv.*

**un·an·tic·i·pat·ed** (ŭn′ăn-tĭs′ə-pā′tĭd) *adj.* Not expected : UNFORESEEN. **—un′an·tic′i·pat′ed·ly** *adv.*

**un·ap·peal·a·ble** (ŭn′ə-pē′lə-bəl) *adj.* Not subject to appeal.

**un·ap·pe·tiz·ing** (ŭn-ăp′ĭ-tī′zĭng) *adj.* Not appetizing or attractive. **—un·ap′pe·tiz′ing·ly** *adv.*

**un·ap·proach·a·ble** (ŭn′ə-prō′chə-bəl) *adj.* **1.** Not friendly : ALOOF. **2.** Not accessible : INAPPROACHABLE. **—un′ap·proach′a·bil′i·ty, un′ap·proach′a·ble·ness** *n.* **—un′ap·proach′a·bly** *adv.*

**un·ap·pro·pri·at·ed** (ŭn′ə-prō′prē-ā′tĭd) *adj.* **1.** Not designated for a specific use <*unappropriated* funds> **2.** Not possessed by or formally assigned to a specific person or organization.

**un·arm** (ŭn-ärm′) *vt.* **-armed, -arm·ing, -arms.** To divest of arms : DISARM.

**un·armed** (ŭn-ärmd′) *adj.* **1.** Lacking weapons or armor : DEFENSELESS. **2.** *Biol.* Having no thorns or spines.

**un·ar·tic·u·lat·ed** (ŭn′är-tĭk′yə-lā′tĭd) *adj.* Not articulated, esp. not carefully or thoroughly thought out.

**un·asked** (ŭn-ăskt′, -äskt′) *adj.* **1.** Not asked. **2.** Not invited <*unasked* guests> **3.** Not requested <*unasked* confessions>

**un·as·sail·a·ble** (ŭn′ə-sā′lə-bəl) *adj.* **1.** Not capable of being disputed or disproven : UNDENIABLE. **2.** Not capable of being attacked or seized successfully : IMPREGNABLE. **—un′as·sail′a·bil′i·ty, un′as·sail′a·ble·ness** *n.* **—un′as·sail′a·bly** *adv.*

**un·as·ser·tive** (ŭn′ə-sûr′tĭv) *adj.* Not assertive : RESERVED.

**un·as·sist·ed** (ŭn′ə-sĭs′tĭd) *adj.* **1.** Not assisted or aided. **2.** *Baseball.* Designating a play handled by only one fielder.

**un·as·sum·ing** (ŭn′ə-so͞o′mĭng) *adj.* Free of pretense or ostentation : MODEST. **—un′as·sum′ing·ly** *adv.* **—un′as·sum′ing·ness** *n.*

**un'ab·bre'vi·at'ed** *adj.*
**un'ac·cred'it·ed** *adj.*
**un'ac·knowl'edged** *adj.*
**un'ac·quaint'ed** *adj.*
**un·ad'ver·tised'** *adj.*
**un'af·fil'i·at'ed** *adj.*
**un·aid'ed** *adj.*
**un'a·larmed'** *adj.*
**un'a·like'** *adj.*
**un'an·nounced'** *adj.*
**un·an'swered** *adj.*
**un'ap·par'ent** *adj.*
**un'ap·peal'ing** *adj.*
**un'ap·peased'** *adj.*
**un'ap·pre'ci·at'ed** *adj.*
**un'ap·pre'cia·tive** *adj.*
**un'ar·tis'tic** *adj.*
**un'a·shamed'** *adj.*
**un'as·signed'** *adj.*
**un'as·sim'i·lat'ed** *adj.*
**un'as·so'ci·at'ed** *adj.*
**un'at·tain'a·ble** *adj.*
**un'at·tempt'ed** *adj.*
**un'at·tend'ed** *adj.*
**un'at·trac'tive** *adj.*
**un'at·trib'ut·ed** *adj.*
**un·au'thor·ized'** *adj.*
**un'a·vail'a·ble** *adj.*
**un'a·venged'** *adj.*
**un'a·vowed'** *adj.*
**un'a·ward'ed** *adj.*
**un·bap'tized'** *adj.*
**un·bleached'** *adj.*
**un·blem'ished** *adj.*
**un·block'** *v.*
**un·brand'ed** *adj.*
**un·break'a·ble** *adj.*
**un·bridge'a·ble** *adj.*
**un·burned'** *adj.*
**un·cap'i·tal·ized'** *adj.*
**un·car'bo·nat'ed** *adj.*
**un·cared'-for** *adj.*
**un·car'pet·ed** *adj.*
**un·cashed'** *adj.*
**un·cat'a·logued'** *adj.*
**un·caught'** *adj.*
**un·cen'sored** *adj.*
**un·chal'lenged** *adj.*
**un·changed'** *adj.*
**un·chang'ing** *adj.*
**un·chap'er·oned'** *adj.*
**un'char·ac·ter·is'tic** *adj.*
**un'char·ac·ter·is'ti·cal·ly** *adv.*
**un·chas'tened** *adj.*
**un·checked'** *adj.*
**un·cho'sen** *adj.*
**un·chris'tened** *adj.*
**un·claimed'** *adj.*
**un·cleaned'** *adj.*
**un·cleared'** *adj.*
**un·cloud'ed** *adj.*
**un·clut'tered** *adj.*
**un'col·lect'ed** *adj.*
**un·col'ored** *adj.*
**un·combed'** *adj.*
**un'com·bined'** *adj.*
**un·com'fort·ed** *adj.*
**un·com'pen·sat'ed** *adj.*
**un'com·plet'ed** *adj.*
**un'com·pre·hend'ing** *adj.*
**un'con·cealed'** *adj.*
**un'con·fined'** *adj.*
**un'con·firmed'** *adj.*
**un·con'quered** *adj.*
**un·con'se·crat'ed** *adj.*
**un'con·sol'i·dat'ed** *adj.*
**un'con·sumed'** *adj.*
**un'con·tam'i·nat'ed** *adj.*
**un'con·tend'ed** *adj.*
**un'con·test'ed** *adj.*
**un'con·vinced'** *adj.*
**un'con·vinc'ing** *adj.*
**un·cooked'** *adj.*
**un'co·op'er·a·tive** *adj.*
**un'co·or'di·nat'ed** *adj.*
**un'cor·rect'ed** *adj.*
**un·cor're·lat'ed** *adj.*
**un'cor·rob'o·rat'ed** *adj.*
**un'cor·rupt'ed** *adj.*
**un·count'a·ble** *adj.*
**un·cred'it·ed** *adj.*
**un·crowd'ed** *adj.*
**un·crowned'** *adj.*
**un·cul'ti·vat'ed** *adj.*
**un·cul'tured** *adj.*
**un·curbed'** *adj.*
**un·cured'** *adj.*
**un·dam'aged** *adj.*
**un·dat'ed** *adj.*
**un'de·ci'pher·a·ble** *adj.*
**un'de·clared'** *adj.*
**un'de·feat'ed** *adj.*
**un'de·fend'ed** *adj.*
**un'de·fined'** *adj.*
**un'de·mand'ing** *adj.*
**un'dem·o·crat'ic** *adj.*
**un'de·pend'a·ble** *adj.*
**un'de·served'** *adj.*
**un'de·serv'ing** *adj.*
**un'de·tect'ed** *adj.*
**un'de·terred'** *adj.*
**un'de·vel'oped** *adj.*
**un'dif·fer·en'ti·at'ed** *adj.*
**un'di·lut'ed** *adj.*
**un'di·min'ished** *adj.*
**un'dis·cern'ing** *adj.*
**un·dis'ci·plined** *adj.*
**un'dis·closed'** *adj.*
**un'dis·cov'ered'** *adj.*
**un'dis·guised'** *adj.*
**un'dis·mayed'** *adj.*
**un'dis·put'ed** *adj.*
**un'dis·solved'** *adj.*
**un'di·vid'ed** *adj.*
**un'dra·mat'ic** *adj.*
**un·drink'a·ble** *adj.*
**un·eat'a·ble** *adj.*
**un·eat'en** *adj.*
**un'e·co·nom'i·cal** *adj.*
**un·ed'i·ble** *adj.*
**un·ed'i·fy'ing** *adj.*
**un'en·closed'** *adj.*
**un'en·cum'bered** *adj.*
**un·end'ing** *adj.*
**un'en·force'a·ble** *adj.*
**un'en·forced'** *adj.*
**un'en·gaged'** *adj.*
**un'en·joy'a·ble** *adj.*
**un'en·light'ened** *adj.*
**un'en·light'en·ing** *adj.*
**un'en·thu'si·as'tic** *adj.*
**un·en'ti·tled** *adj.*
**un·en'vi·a·ble** *adj.*
**un'e·quipped'** *adj.*
**un·eth'i·cal** *adj.*
**un'ex·ag'ger·at'ed** *adj.*
**un'ex·celled'** *adj.*
**un'ex·change'a·ble** *adj.*
**un'ex·cit'ed** *adj.*
**un'ex·cit'ing** *adj.*
**un'ex·pired'** *adj.*
**un'ex·plained'** *adj.*
**un'ex·plored'** *adj.*
**un'ex·posed'** *adj.*
**un'ex·pressed'** *adj.*
**un·ex'pur·gat'ed** *adj.*
**un·fal'ter·ing** *adj.*
**un·fath'omed** *adj.*
**un·fea'si·ble** *adj.*
**un·fed'** *adj.*
**un·fem'i·nine** *adj.*
**un·fenced'** *adj.*
**un·fer'til·ized'** *adj.*
**un·filled'** *adj.*
**un·fil'tered** *adj.*
**un·flag'ging** *adj.*
**un·fla'vored** *adj.*
**un·forced'** *adj.*
**un'fore·see'a·ble** *adj.*
**un'for·giv'a·ble** *adj.*
**un'for·giv'ing** *adj.*
**un·for'mu·lat'ed** *adj.*
**un·framed'** *adj.*
**un'ful·filled'** *adj.*
**un·fur'nished** *adj.*
**un·gen'tle·man·ly** *adj.*
**un·gird'** *v.*
**un·gov'erned** *adj.*
**un·grace'ful** *adj.*
**un·grad'ed** *adj.*
**un·guid'ed** *adj.*
**un·ham'pered** *adj.*
**un·harmed'** *adj.*
**un·heat'ed** *adj.*
**un·heed'ed** *adj.*
**un·her'ald·ed** *adj.*
**un'he·ro'ic** *adj.*
**un·hin'dered** *adj.*
**un·hur'ried** *adj.*
**un·hurt'** *adj.*
**un'hy·gi·en'ic** *adj.*
**un'i·den'ti·fied'** *adj.*
**un'id·i·o·mat'ic** *adj.*
**un'i·mag'i·na·ble** *adj.*
**un'i·mag'i·na·tive** *adj.*
**un'im·paired'** *adj.*
**un'im·pres'sive** *adj.*
**un'in·formed'** *adj.*
**un'in·hab'i·ta·ble** *adj.*
**un'in·sured'** *adj.*
**un'in·tend'ed** *adj.*
**un·in'ter·est·ing** *adj.*
**un'in·vit'ed** *adj.*
**un'in·vit'ing** *adj.*
**un·knowl'edge·a·ble** *adj.*
**un·la'beled** *adj.*
**un·lik'a·ble** *adj.*
**un·lined'** *adj.*
**un·lit'** *adj.*
**un·liv'a·ble** *adj.*
**un·lov'a·ble** *adj.*
**un·loved'** *adj.*
**un·man'age·a·ble** *adj.*
**un·mas'cu·line** *adj.*
**un·matched'** *adj.*
**un·meas'ured** *adj.*
**un·melt'ed** *adj.*
**un·men'tioned** *adj.*
**un·mer'it·ed** *adj.*
**un·mixed'** *adj.*
**un'mo·lest'ed** *adj.*
**un·mov'a·ble** *adj.*
**un·moved'** *adj.*
**un·mu'si·cal** *adj.*
**un·nam'a·ble** *adj.*
**un·named'** *adj.*
**un·nav'i·ga·ble** *adj.*
**un·no'ticed** *adj.*
**un'ob·jec'tion·a·ble** *adj.*
**un'ob·served'** *adj.*
**un'ob·struct'ed** *adj.*
**un'ob·tain'a·ble** *adj.*
**un·o'pened** *adj.*
**un'op·posed'** *adj.*
**un·owned'** *adj.*
**un·par'don·a·ble** *adj.*
**un·pas'teur·ized'** *adj.*
**un'pa·tri·ot'ic** *adj.*
**un·paved'** *adj.*
**un'per·turbed'** *adj.*
**un·pile'** *v.*
**un·planned'** *adj.*
**un·plant'ed** *adj.*
**un·plowed'** *adj.*
**un'po·et'ic** *adj.*
**un·pol'ished** *adj.*
**un'pol·lut'ed** *adj.*
**un·posed'** *adj.*
**un'pre·sent'a·ble** *adj.*
**un·pressed'** *adj.*
**un'pre·vent'a·ble** *adj.*
**un·proc'essed** *adj.*
**un·prom'is·ing** *adj.*
**un·prompt'ed** *adj.*
**un'pro·nounced'** *adj.*
**un'pro·pi'tious** *adj.*
**un'pro·tect'ed** *adj.*
**un'prov'en** *adj.*
**un·pub'lished** *adj.*
**un·pun'ished** *adj.*
**un·quench'a·ble** *adj.*
**un·ques'tion·ing** *adj.*
**un're·al·is'tic** *adj.*
**un're·al·is'ti·cal·ly** *adv.*
**un·re'al·ized'** *adj.*
**un're·cord'ed** *adj.*
**un're·deem'a·ble** *adj.*
**un're·fined'** *adj.*
**un're·gard'ed** *adj.*
**un·reg'is·tered** *adj.*
**un·reg'u·lat'ed** *adj.*
**un're·lat'ed** *adj.*
**un're·pen'tant** *adj.*
**un're·port'ed** *adj.*
**un·rep're·sent'a·tive** *adj.*
**un're·quit'ed** *adj.*
**un're·sist'ing** *adj.*
**un're·solved'** *adj.*
**un're·strict'ed** *adj.*
**un're·ward'ing** *adj.*
**un·rhymed'** *adj.*
**un'ro·man'tic** *adj.*
**un·sat'is·fied'** *adj.*
**un·scent'ed** *adj.*
**un·sched'uled** *adj.*
**un·schol'ar·ly** *adj.*
**un·see'ing** *adj.*
**un'sen·ti·men'tal** *adj.*
**un·serv'ice·a·ble** *adj.*
**un·shad'ed** *adj.*
**un·shaved'** *adj.*
**un·shorn'** *adj.*
**un·signed'** *adj.*
**un·sink'a·ble** *adj.*
**un·soiled'** *adj.*
**un·sold'** *adj.*
**un·sol'dier·ly** *adj.*
**un'so·lic'it·ed** *adj.*
**un·solv'a·ble** *adj.*

**un·at·tached** (ŭn′ə-tăcht′) *adj.* **1.** Not attached or joined, as to surrounding tissue. **2. a.** Not committed to or associated with any person, group, or organization. **b.** Not engaged or married. **3.** *Law.* Not possessed or seized as security.

**un·at·test·ed** (ŭn′ə-tĕs′tĭd) *adj.* Not attested.

**u·nau** (yōō′nô, ōō′nou) *n.* [Fr. < Tupi *unáu.*] A two-toed sloth of the genus *Choloepus.*

**un·a·vail·ing** (ŭn′ə-vā′lĭng) *adj.* Useless : futile. **—un′a·vail′ing·ly** *adv.* **—un′a·vail′ing·ness** *n.*

**u·na vo·ce** (yōō′nə vō′sē) *adj. & adv.* [Lat.] With one voice.

**un·a·void·a·ble** (ŭn′ə-voi′də-bəl) *adj.* Not able to be avoided : INEVITABLE. **—un′a·void′a·bil′i·ty, un′a·void′a·ble·ness** *n.* **—un′a·void′a·bly** *adv.*

**un·a·ware** (ŭn′ə-wâr′) *adj.* Not aware or cognizant. *—adv.* Unawares.

**un·a·wares** (ŭn′ə-wârz′) *adv.* **1.** By surprise : UNEXPECTEDLY. **2.** Without forethought or plan <went to the big city all *unawares*>

**un·backed** (ŭn-băkt′) *adj.* **1.** Lacking support or backing. **2.** Not having a back. **3.** Never ridden <an *unbacked* horse>

**un·bal·ance** (ŭn-băl′əns) *vt.* **-anced, -anc·ing, -anc·es. 1.** To upset the balance or equilibrium of. **2.** To derange (the mind). **—un·bal′ance** *n.*

**un·bal·anced** (ŭn-băl′ənst) *adj.* **1.** Not balanced or properly balanced. **2. a.** Mentally deranged. **b.** Not of sound judgment. **3.** Not adjusted so that debit and credit correspond <an *unbalanced* bank account>

**un·bal·last·ed** (ŭn-băl′ə-stĭd) *adj.* **1.** Not stabilized or properly stabilized by ballast. **2.** Unsteady : wavering.

**un·bar** (ŭn-bär′) *vt. & vi.* **-barred, -bar·ring, -bars.** To remove a bar or bars from.

**un·bat·ed** (ŭn-bā′tĭd) *adj.* **1.** Unabated. **2.** *Archaic.* Not blunted by a guard on the tip.

**un·bear·a·ble** (ŭn-bâr′ə-bəl) *adj.* Not endurable <*unbearable* anguish> **—un·bear′a·bly** *adv.*

**un·beat·a·ble** (ŭn-bē′tə-bəl) *adj.* Impossible to defeat or surpass. **—un·beat′a·bly** *adv.*

**un·beat·en** (ŭn-bēt′n) *adj.* **1.** Not defeated or surpassed. **2.** Untrod. **3.** Not beaten or pounded.

**un·be·com·ing** (ŭn′bĭ-kŭm′ĭng) *adj.* **1.** Not appropriate, attractive, or flattering <an *unbecoming* color> **2.** Not seemly : IMPROPER <*unbecoming* behavior> **—un′be·com′ing·ly** *adv.*

**un·be·got·ten** (ŭn′bĭ-gŏt′n) *adj.* **1.** Not yet born. **2.** Self-existent : eternal.

**un·be·known** (ŭn′bĭ-nōn′) *adj.* [UN-[1] + obs. *beknown,* known < ME *beknowen,* p.part. of *beknowen,* to get to know < OE *becnawan.*] Occurring or existing without one's knowledge : UNKNOWN.

**un·be·knownst** (ŭn′bĭ-nōnst′) *adj.* [UNBEKNOWN + *-st,* as in *amongst.*] Unbeknown. *—adv.* Without one's knowledge <had been ill for years, *unbeknownst* to anyone>

**un·be·lief** (ŭn′bĭ-lēf′) *n.* Lack of belief, esp. in religious matters.

**un·be·liev·a·ble** (ŭn′bĭ-lē′və-bəl) *adj.* Not to be believed : INCREDIBLE. **—un′be·liev′a·bly** *adv.*

**un·be·liev·er** (ŭn′bĭ-lēv′-ər) *n.* One who does not believe, esp. one who habitually or instinctively doubts or questions.

☆ ***syns:*** UNBELIEVER, DOUBTER, DOUBTING THOMAS, SKEPTIC *n.* *core meaning* : one who habitually or instinctively doubts or questions <an *unbeliever* in matters of religion> ***ant:*** believer

**un·be·liev·ing** (ŭn′bĭ-lē′vĭng) *adj.* Not believing : DOUBTING <The accident happened right before our *unbelieving* eyes.>

**un·bend** (ŭn-bĕnd′) *v.* **-bent** (-bĕnt′), **-bend·ing, -bends.** *—vt.* **1.** To free from tension, as the mind : RELAX. **2.** To release (e.g., a bow) from flexure or tension. **3.** *Naut.* To loosen or untie (a rope or sail). **4.** To straighten (something crooked or bent). *—vi.* **1.** To become less tense : UNWIND. **2.** To become less strict. **3.** To become straight.

**un·bend·ing** (ŭn-bĕn′dĭng) *adj.* **1.** Inflexible : unyielding <an *unbending* will> **2.** Extremely reserved : ALOOF.

**un·bi·ased** *also* **un·bi·assed** (ŭn-bī′əst) *adj.* Having no bias or prejudice : IMPARTIAL. **—un·bi′ased·ly** *adv.* **—un·bi′ased·ness** *n.*

**un·bid·den** (ŭn-bĭd′n) *also* **un·bid** (-bĭd′) *adj.* Not invited or asked <came to the meeting *unbidden*>

**un·bind** (ŭn-bīnd′) *vt.* **-bound** (-bound′), **-bind·ing, -binds. 1.** To untie or unfasten, as wrappings or bindings. **2.** To release from restraints or bonds : FREE.

**un·blenched** (ŭn-blĕncht′) *adj.* Undaunted.

**un·blessed** *also* **un·blest** (ŭn-blĕst′) *adj.* **1.** Deprived of a blessing. **2.** Unholy : evil.

**un·blink·ing** (ŭn-blĭng′kĭng) *adj.* **1.** Not blinking. **2.** Not showing visible emotion. **3.** Fearless in facing reality <an *unblinking* self-analysis> **—un·blink′ing·ly** *adv.*

**un·blush·ing** (ŭn-blŭsh′ĭng) *adj.* **1.** Lacking shame or remorse. **2.** Not blushing. **—un·blush′ing·ly** *adv.*

**un·bod·ied** (ŭn-bŏd′ēd) *adj.* **1.** Being without body or form : INCORPOREAL. **2.** Disembodied <*unbodied* spirits>

**un·bolt** (ŭn-bōlt′) *vt.* **-bolt·ed, -bolt·ing, -bolts.** To release the bolts of (e.g., a door) : UNLOCK.

**un·bolt·ed**[1] (ŭn-bōl′tĭd) *adj.* Not fastened with a bolt.

**un·bolt·ed**[2] (ŭn-bōl′tĭd) *adj.* Not sifted, as flour.

**un·born** (ŭn-bôrn′) *adj.* **1.** Not yet born <an *unborn* child> **2.** Having not yet appeared : FUTURE.

**un·bos·om** (ŭn-bŏŏz′əm, -bōō′zəm) *v.* **-omed, -om·ing, -oms.** *—vt.* **1.** To confide (one's thoughts or feelings) : DISCLOSE. **2.** To unburden (oneself) of troublesome thoughts or feelings. *—vi.* To reveal one's thoughts or feelings. **—un·bos′om·er** *n.*

**un·bound** (ŭn-bound′) *adj.* **1.** Not bound, as a book. **2.** Freed from bonds or restraints : RELEASED.

**un·bound·ed** (ŭn-boun′dĭd) *adj.* **1.** Having no boundaries or limits. **2.** Not kept within bounds : UNRESTRAINED <*unbounded* glee> **—un·bound′ed·ly** *adv.* **—un·bound′ed·ness** *n.*

**un·bowed** (ŭn-boud′) *adj.* **1.** Not bowed : UNBENT. **2.** Not subdued : UNYIELDING <"My head is bloody but *unbowed*" —W.E. Henley>

**un·brace** (ŭn-brās′) *vt.* **-braced, -brac·ing, -brac·es. 1.** To set free by removing braces or bands. **2.** To release from tension : RELAX. **3.** To weaken.

**un·breath·a·ble** (ŭn-brē′*th*ə-bəl) *adj.* Not fit for breathing.

**un·bred** (ŭn-brĕd′) *adj.* **1.** Not taught or trained. **2.** Not yet bred, as a heifer. **3.** *Obs.* Ill-bred.

**un·bri·dle** (ŭn-brīd′l) *vt.* **-dled, -dling, -dles. 1.** To free from a bridle. **2.** To free from restraint.

**un·bri·dled** (ŭn-brīd′ld) *adj.* **1.** Not wearing or fitted with a bridle. **2.** Unrestrained : uncontrolled <*unbridled* fury>

**un·bro·ken** (ŭn-brō′kən) *adj.* **1.** Not broken or tampered with : INTACT. **2.** Not breached or violated. **3.** Uninterrupted : continuous. **4.** Not broken to harness : UNTAMED. **5.** Not disordered or disorganized. **—un·bro′ken·ly** *adv.* **—un·bro′ken·ness** *n.*

**un·buck·le** (ŭn-bŭk′əl) *v.* **-led, -ling, -les.** *—vt.* To undo or loosen the buckle or buckles of. *—vi.* **1.** To undo buckles. **2.** *Informal.* To relax.

**un·bur·den** (ŭn-bûr′dn) *vt.* **-dened, -den·ing, -dens. 1.** To free from a burden. **2.** To relieve (e.g., one's mind) of fears or concerns. **3.** To cast off, as fears.

**un·but·ton** (ŭn-bŭt′n) *v.* **-toned, -ton·ing, -tons.** *—vt.* **1.** To unfasten the buttons of. **2.** To free (a button) from a buttonhole. **3.** To open as if by undoing buttons <*unbutton* the hatches> *—vi.* To undo buttons.

**un·caged** (ŭn-kājd′) *adj.* **1.** Not confined in or as if in a cage. **2.** Released from a cage or similar restraint.

**un·cal·cu·lat·ed** (ŭn-kăl′kyə-lā′tĭd) *adj.* Not thought out in advance.

---

**un·solved′** *adj.*
**un·sort′ed** *adj.*
**un·spec′i·fied′** *adj.*
**un·spoiled′** *adj.*
**un·sports′man·like′** *adj.*
**un·stained′** *adj.*
**un·stat′ed** *adj.*
**un′sub·stan′ti·at′ed** *adj.*
**un·suit′ed** *adj.*
**un·sul′lied** *adj.*
**un·su′per·vised′** *adj.*
**un′sup·port′ed** *adj.*
**un′sup·pressed′** *adj.*
**un·sure′** *adj.*
**un′sur·passed′** *adj.*
**un′sus·pi′cious** *adj.*
**un·sweet′ened** *adj.*
**un·swerv′ing** *adj.*
**un′sym·pa·thet′ic** *adj.*
**un′sys·te·mat′ic** *adj.*
**un′sys·te·mat′i·cal·ly** *adv.*
**un·taint′ed** *adj.*
**un·tal′ent·ed** *adj.*
**un·tamed′** *adj.*
**un·tar′nished** *adj.*
**un·taxed′** *adj.*
**un·teach′a·ble** *adj.*
**un·test′ed** *adj.*
**un·thought′ful** *adj.*
**un·touched′** *adj.*
**un·trace′a·ble** *adj.*
**un·trained′** *adj.*
**un·tram′meled** *adj.*
**un·trimmed′** *adj.*
**un·trod′** *adj.*
**un·trou′bled** *adj.*
**un·trust′wor′thy** *adj.*
**un·us′a·ble** *adj.*
**un·var′ied** *adj.*
**un·var′y·ing** *adj.*
**un·ver′i·fi′a·ble** *adj.*
**un·ver′i·fied′** *adj.*
**un·versed′** *adj.*
**un·vis′it·ed** *adj.*
**un·want′ed** *adj.*
**un·watched′** *adj.*
**un·wa′ver·ing** *adj.*
**un·weaned′** *adj.*
**un·wear′a·ble** *adj.*
**un·wed′** *adj.*
**un·wel′come** *adj.*
**un·wom′an·ly** *adj.*
**un·work′a·ble** *adj.*
**un·wor′ried** *adj.*
**un·wound′ed** *adj.*
**un·wrin′kled** *adj.*

**un·cal·cu·lat·ing** (ŭn-kăl′kyə-lā′tĭng) *adj.* Not using or involving calculation.
**un·called-for** (ŭn-kôld′fôr′) *adj.* **1.** Not requested or required : UNNECESSARY. **2.** Being out of place and unjustified <*uncalled-for* curtness>
**un·can·ny** (ŭn-kăn′ē) *adj.* **-ni·er, -ni·est. 1.** Exciting wonder and fear : INEXPLICABLE. **2.** So keen or perceptive as to seem supernatural <*uncanny* aim> **—un·can′ni·ly** *adv.* **—un·can′ni·ness** *n.*
**un·cap** (ŭn-kăp′) *vt.* **-capped, -cap·ping, -caps.** To remove the cap or covering of.
**un·caused** (ŭn-kôzd′) *adj.* Existing without having been caused.
**un·ceas·ing** (ŭn-sē′sĭng) *adj.* Never ceasing : CONTINUOUS. **—un·ceas′ing·ly** *adv.*
**un·cel·e·brat·ed** (ŭn-sĕl′ə-brā′tĭd) *adj.* **1.** Not formally or officially honored. **2.** Not famous or well-known : OBSCURE.
**un·cer·e·mo·ni·ous** (ŭn-sĕr′ə-mō′nē-əs) *adj.* **1.** Not ceremonious : INFORMAL. **2.** Lacking due courtesy or formality : rudely abrupt. **—un·cer′e·mo′ni·ous·ly** *adv.* **—un·cer′e·mo′ni·ous·ness** *n.*
**un·cer·tain** (ŭn-sûr′tn) *adj.* **1.** Not known or established : DOUBTFUL <an *uncertain* effect> **2.** Not determined : UNDECIDED <*uncertain* plans> **3.** Not having sure knowledge <*uncertain* about the time> **4.** Subject to change : VARIABLE <*uncertain* weather> **5.** Unsteady : fitful <*uncertain* light> **—un·cer′tain·ly** *adv.* **—un·cer′tain·ness** *n.*
**un·cer·tain·ty** (ŭn-sûr′tn-tē) *n., pl.* **-ties. 1.** The state of being in doubt. **2.** Something uncertain.
**uncertainty principle** *n.* A principle in quantum mechanics that increasing the accuracy of measurement of one observable quantity increases the uncertainty with which other quantities may be known.
**un·chain** (ŭn-chān′) *vt.* **-chained, -chain·ing, -chains.** To release from or as if from a chain or bond : set free.
**un·change·a·ble** (ŭn-chān′jə-bəl) *adj.* Not capable of being altered : IMMUTABLE. **—un·change′a·bil′i·ty, un·change′a·ble·ness** *n.* **—un·change′a·bly** *adv.*
**un·charged** (ŭn-chärjd′) *adj.* **1.** Not loaded. —Used of weapons. **2.** *Law.* Not formally accused. **3.** Lacking electric charge.
**un·char·i·ta·ble** (ŭn-chăr′ĭ-tə-bəl) *adj.* Lacking in charity, as in judging another : HARSH. **—un·char′i·ta·ble·ness** *n.* **—un·char′i·ta·bly** *adv.*
**un·chart·ed** (ŭn-chär′tĭd) *adj.* Not charted on a map or plan : UNKNOWN <*uncharted* seas>
**un·chaste** (ŭn-chāst′) *adj.* Not chaste or pure. **—un·chaste′ly** *adv.* **—un·chaste′ness, un·chas′ti·ty** (-chăs′tĭ-tē) *n.*
**un·chris·tian** (ŭn-krĭs′chən) *adj.* **1.** Not Christian. **2.** Not in accordance with the Christian spirit. **3.** Uncivilized.
**un·church** (ŭn-chûrch′) *vt.* **-churched, -church·ing, -church·es. 1.** To expel from a church or from church membership : EXCOMMUNICATE. **2.** To deprive (a congregation or sect) of the status of a church.
**un·ci** (ŭn′sī′) *n. pl. of* UNCUS.
**un·cial** *also* **Un·cial** (ŭn′shəl, -sē-əl) *adj.* [LLat. *uncialis,* inch-high < Lat. *uncia,* a twelfth part, ounce, inch.] Of or relating to a style of writing characterized by somewhat rounded capital letters and found esp. in Greek and Latin manuscripts of the 4th to the 8th cent. —*n.* **1.** The uncial style or hand. **2.** An uncial letter.

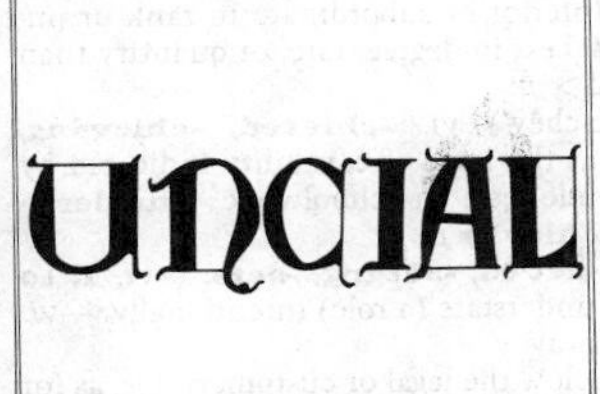

uncial

**un·ci·form** (ŭn′sə-fôrm′) *adj.* [NLat. *unciformis* : Lat. *uncus,* hook + Lat. *forma,* shape.] Hook-shaped.
**un·ci·nar·i·a** (ŭn′sə-nâr′ē-ə) *n.* [NLat. *Uncinaria,* hookworm genus < Lat. *uncinus,* barb < *uncus,* hook.] Hookworm.
**un·ci·nate** (ŭn′sə-nāt′, -nĭt) *adj.* [Lat. *uncinatus* < *uncinus,* barb < *uncus,* hook.] Hooked at the tip.
**un·ci·nus** (ŭn-sī′nəs) *n., pl.* **-ni** (-nī′) [NLat. < Lat., barb < *uncus,* hook.] A small hooklike structure, as one of the setae of certain annelid worms.
**un·cir·cum·cised** (ŭn-sûr′kəm-sīzd′) *adj.* **1.** Not circumcised. **2.** Spiritually unpure : HEATHEN. **—un·cir′cum·ci′sion** (-sĭzh′ən) *n.*
**un·civ·il** (ŭn-sĭv′əl) *adj.* **1.** Discourteous : rude. **2.** Uncivilized. **—un·civ′il·ly** *adv.*

ōō **boot** ou **out** th **thin** *th* **this** ŭ **cut** ûr **urge** y **young**
yōō ab**use** zh vi**sion** ə **a**bout, it**e**m, ed**i**ble, gall**o**p, circ**u**s

**un·civ·i·lized** (ŭn-sĭv′ə-līzd′) *adj.* Not civilized : BARBAROUS.
**un·clad** (ŭn-klăd′) *adj.* Not wearing clothes : NAKED.
**un·clasp** (ŭn-klăsp′) *v.* **-clasped, -clasp·ing, -clasps.** —*vt.* **1.** To loosen the clasp of. **2.** To release from a clasp or grasp. —*vi.* **1.** To become unfastened. **2.** To release or relax a clasp or grasp : let go.
**un·clas·si·fied** (ŭn-klăs′ə-fīd′) *adj.* **1.** Not placed or included in a class or category. **2.** Not restricted for security purposes, as a document <*unclassified* weapons manuals>
**un·cle** (ŭng′kəl) *n.* [ME < OFr. *oncle* < Lat. *avunculus,* maternal uncle.] **1. a.** The brother of one's mother or father. **b.** The husband of one's aunt. **2.** A form of respectful address to an older man, used esp. by children. **3.** One who counsels. **4. Uncle.** Uncle Sam. —*interj. Slang.* —Used to express surrender <cry *uncle*>
**un·clean** (ŭn-klēn′) *adj.* **-er, -est. 1.** Not clean : DIRTY. **2.** Morally defiled : UNCHASTE. **3.** Ceremonially impure.
**un·clean·ly**[1] (ŭn-klĕn′lē) *adj.* **-li·er, -li·est.** Unclean. **—un·clean′li·ness** *n.*
**un·clean·ly**[2] (ŭn-klēn′lē) *adv.* In an unclean manner.
**un·clear** (ŭn-klîr′) *adj.* **-er, -est.** Not clearly defined : VAGUE.
**un·clench** (ŭn-klĕnch′) *v.* **-clenched, -clench·ing, -clench·es.** —*vt.* To loosen from a clenched position, as a fist. —*vi.* To become unclenched.
**Uncle Sam** (săm) *n.* [< *U.S.,* abbr. of *United States.*] **1.** The U.S. Government, as personified by a tall thin man with white chin whiskers and wearing a blue swallow-tailed coat, red-and-white-striped trousers, and a tall hat with a band of stars. **2.** The American nation or its people.
**Uncle Tom** (tŏm) *n.* [After *Uncle Tom,* a slave in *Uncle Tom's Cabin,* a novel by Harriet Beecher Stowe (1811–1896).] A black who is held to be humiliatingly subservient or deferential to whites. —*vi.* **Uncle Tommed, Uncle Tom·ming, Uncle Toms.** To act like an Uncle Tom. **—Uncle Tom′ism** *n.*
**un·cloak** (ŭn-klōk′) *vt.* **-cloaked, -cloak·ing, -cloaks. 1.** To remove a cloak or cover from. **2.** To expose : reveal.
**un·clog** (ŭn-klŏg′) *vt.* **-clogged, -clog·ging, -clogs.** To rid of a blockage or a difficulty.
**un·close** (ŭn-klōz′) *v.* **-closed, -clos·ing, -clos·es.** —*vt.* To open or disclose. —*vi.* To become opened or disclosed.
**un·clothe** (ŭn-klōth′) *vt.* **-clothed, -cloth·ing, -clothes.** To remove the clothing or cover from : STRIP.
**un·co** (ŭng′kō) [ME *unkow,* var. of *uncouth,* strange. —see UNCOUTH.] *Scot.* —*adj.* So unusual as to be surprising : UNCANNY. —*n., pl.* **-cos. 1.** An unusual or amazing person. **2.** A stranger. **3. uncos.** News. —*adv.* Extremely : remarkably.
**un·coil** (ŭn-koil′) *v.* **-coiled, -coil·ing, -coils.** —*vt.* To unwind : untwist. —*vi.* To become unwound or untwisted.
**un·coined** (ŭn-koind′) *adj.* **1.** Not minted. **2.** Not artificial or counterfeit.
**un·com·fort·a·ble** (ŭn-kŭm′fər-tə-bəl, -kŭmf′tə-bəl) *adj.* **1.** Experiencing discomfort. **2.** Causing or involving discomfort. **—un·com′fort·a·ble·ness** *n.* **—un·com′fort·a·bly** *adv.*
**un·com·mer·cial** (ŭn′kə-mûr′shəl) *adj.* **1.** Not engaged in or involving commerce <an *uncommercial* zone of the city> **2.** Not in accordance with the principles of commerce.
**un·com·mit·ted** (ŭn′kə-mĭt′ĭd) *adj.* Not pledged to a specific cause or course of action.
**un·com·mon** (ŭn-kŏm′ən) *adj.* **-er, -est. 1.** Not common : UNUSUAL. **2.** Far beyond the usual, normal, or customary : REMARKABLE. **—un·com′mon·ly** *adv.* **—un·com′mon·ness** *n.*
☆ ***syns:*** UNCOMMON, EXCEPTIONAL, EXTRAORDINARY, RARE, REMARKABLE, SINGULAR, UNUSUAL *adj. core meaning* : far beyond what is usual, normal, or customary <*uncommon* intelligence><an *uncommon* problem> ***ant:*** common
**un·com·mu·ni·ca·ble** (ŭn′kə-myōō′nĭ-kə-bəl) *adj.* Incommunicable.
**un·com·mu·ni·ca·tive** (ŭn′kə-myōō′nĭ-kā′tĭv, -kə-tĭv) *adj.* Not inclined to be communicative : TACITURN. **—un′com·mu′ni·ca′tive·ly** *adv.* **—un′com·mu′ni·ca′tive·ness** *n.*
**un·com·plain·ing** (ŭn′kəm-plā′nĭng) *adj.* Not complaining. **—un′com·plain′ing·ly** *adv.*
**un·com·pli·cat·ed** (ŭn-kŏm′plĭ-kā′tĭd) *adj.* **1.** Not complex : SIMPLE. **2.** Not complicated by something extraneous, esp. not involving medical complications.
**un·com·pli·men·ta·ry** (ŭn′kŏm-plə-mĕn′tə-rē, -mĕn′trē) *adj.* Not complimentary : DEROGATORY.
**un·com·pro·mis·ing** (ŭn-kŏm′prə-mī′zĭng) *adj.* Not granting concessions : INFLEXIBLE. **—un·com′pro·mis′ing·ly** *adv.*
**un·con·ceiv·a·ble** (ŭn′kən-sē′və-bəl) *adj.* Inconceivable.
**un·con·cern** (ŭn′kən-sûrn′) *n.* **1.** Lack of interest : INDIFFERENCE. **2.** Lack of worry or anxiety.
**un·con·cerned** (ŭn′kən-sûrnd′) *adj.* **1.** Not interested or involved : INDIFFERENT. **2.** Not anxious or apprehensive. **—un′con·cern′ed·ly** (-sûr′nĭd-lē) *adv.* **—un′con·cern′ed·ness** (-sûr′nĭd-nĭs) *n.*
**un·con·di·tion·al** (ŭn′kən-dĭsh′ə-nəl) *adj.* Being without conditions or limitations : ABSOLUTE. **—un′con·di′tion·al·ly** *adv.*
**un·con·di·tioned** (ŭn′kən-dĭsh′ənd) *adj.* **1.** Unconditional. **2.** *Psychol.* **a.** Not resulting from or dependent on conditioning <an

*unconditioned* response> **b.** Evoking an unconditioned response <an *unconditioned* stimulus>

**un·con·form·a·ble** (ŭn′kən-fôr′mə-bəl) *adj.* **1.** Not conforming or capable of conforming. **2.** *Geol.* Showing unconformity. **—un′con·form′a·bil′i·ty, un′con·form′a·ble·ness** *n.* **—un′con·form′a·bly** *adv.*

**un·con·form·i·ty** (ŭn′kən-fôr′mĭ-tē) *n., pl.* **-ties. 1.** Lack of conformity : NONCONFORMITY. **2.** *Geol.* **a.** An eroded space. **b.** A space caused by lack of deposit that separates younger strata from older rocks.

**un·con·gen·ial** (ŭn′kən-jēn′yəl) *adj.* **1.** Not compatible <*uncongenial* partners> **2. a.** Not suitable or appropriate <a climate *uncongenial* to citrus trees> **b.** Not agreeable or pleasing <*uncongenial* chores> **—un′con·ge′ni·al′ity** (-jē′nē-ăl′ĭ-tē) *n.*

**un·con·nect·ed** (ŭn′kə-nĕk′tĭd) *adj.* **1.** Not joined or connected. **2.** Not coherent : DISCONNECTED <*unconnected* speech> **—un′con·nect′ed·ly** *adv.* **—un′con·nect′ed·ness** *n.*

**un·con·quer·a·ble** (ŭn-kŏng′kər-ə-bəl) *adj.* Incapable of being conquered. **—un·con′quer·a·bly** *adv.*

**un·con·scion·a·ble** (ŭn-kŏn′shə-nə-bəl) *adj.* **1.** Not restrained by conscience : UNSCRUPULOUS. **2.** Beyond prudence or reason : EXCESSIVE. **—un·con′scion·a·ble·ness** *n.* **—un·con′scion·a·bly** *adv.*

**un·con·scious** (ŭn-kŏn′shəs) *adj.* **1.** Without conscious thought or feeling, esp. without psychological awareness and hence not capable of being consciously scrutinized <*unconscious* resentment><*unconscious* anxieties> **2. a.** Having lost, esp. temporarily, the capacity for sensory perception. **b.** Temporarily lacking full awareness. **3.** Being without conscious control : INVOLUNTARY. —*n.* The division of the psyche not subject to direct conscious observation but inferred from its effects on conscious processes and behavior. **—un·con′scious·ly** *adv.* **—un·con′scious·ness** *n.*

**un·con·sid·ered** (ŭn′kən-sĭd′ərd) *adj.* Not reasoned or considered : RASH <an *unconsidered* response>

**un·con·sti·tu·tion·al** (ŭn′kŏn-stĭ-to͞o′shə-nəl, -tyo͞o′-) *adj.* Not in accord with the principles set forth in the constitution of a nation. **—un′con·sti·tu′tion·al′i·ty** *n.* **—un′con·sti·tu′tion·al·ly** *adv.*

**un·con·trol·la·ble** (ŭn′kən-trō′lə-bəl) *adj.* Not able to be controlled or governed. **—un′con·trol′la·bil′i·ty, un′con·trol′la·ble·ness** *n.* **—un′con·trol′la·bly** *adv.*

**un·con·trolled** (ŭn′kən-trōld′) *adj.* Not under control or governance. **—un′con·trolled′ness** *n.*

**un·con·ven·tion·al** (ŭn′kən-vĕn′shə-nəl) *adj.* Not adhering to convention : unusual or out of the ordinary. **—un′con·ven′tion·al′i·ty** *n.* **—un′con·ven′tion·al·ly** *adv.*

**un·cool** (ŭn-ko͞ol′) *adj. Slang.* **1.** Lacking assurance, self-control, or sophistication. **2.** Not in accord with the customs or requirements of a specific group.

**un·cork** (ŭn-kôrk′) *vt.* **-corked, -cork·ing, -corks. 1.** To draw the cork from. **2.** To free from a sealed or constrained state.

**un·count·ed** (ŭn-koun′tĭd) *adj.* **1.** Not counted. **2.** Impossible to count : INNUMERABLE.

**un·cou·ple** (ŭn-kŭp′əl) *v.* **-led, -ling, -les.** —*vt.* **1.** To disconnect <*uncouple* train cars> **2.** To set loose or release from a couple. —*vi.* To come or break loose. **—un·cou′pler** *n.*

**un·couth** (ŭn-ko͞oth′) *adj.* [ME, unknown, strange < OE *uncuð* : *un-*, not + *cuð*, known.] **1.** Crude or unrefined, esp. in language or manners. **2.** Awkward or ungraceful. **3.** *Archaic.* Foreign : unfamiliar. **—un·couth′ly** *adv.* **—un·couth′ness** *n.*

▲ word history: The history of the word *uncouth* reveals the propensity of human beings to react with hostility and aversion to something unknown or strange. *Uncouth* is the descendent of Old English *uncūth*, which meant simply "unknown, unfamiliar." It is formed from the negative prefix *un–* and *–cūth*, the past participle of *cunnan*, "to know, to be able." The meaning "unfamiliar, strange," for *uncouth* eventually developed into the meaning "odd, awkward," which further developed into "crude, unrefined," the ordinary modern sense of the word. The original unsuffixed form *couth* meaning "known, familiar," did exist at one time but it is now obsolete. A new adjective *couth*, which is not historically continuous with Old English *cūth*, has very recently been formed as an antonym of *uncouth* in its most recently developed sense.

**un·cov·e·nant·ed** (ŭn-kŭv′ə-nən-tĭd) *adj.* **1.** Not bound by a covenant. **2.** Not promised or guaranteed by a covenant.

**un·cov·er** (ŭn-kŭv′ər) *v.* **-ered, -er·ing, -ers.** —*vt.* **1.** To remove the cover from. **2.** To lay bare : DISCLOSE. **3.** To remove the hat from, as in respect or reverence. —*vi.* **1.** To remove a cover. **2.** To bare the head in respect or reverence.

**un·cov·ered** (ŭn-kŭv′ərd) *adj.* **1.** Having no cover or protection. **2.** Not covered by insurance or collateral security. **3.** Bareheaded.

**un·cre·at·ed** (ŭn′krē-ā′tĭd) *adj.* **1.** Not yet existing. **2.** Existing of itself : UNCAUSED.

**un·crit·i·cal** (ŭn-krĭt′ĭ-kəl) *adj.* **1.** Not discriminating. **2.** Not using critical standards or methods of evaluation.

**un·cross** (ŭn-krôs′, -krŏs′) *vt.* **-crossed, -cross·ing, -cross·es.** To move (e.g., one's legs) from a crossed position.

**unc·tion** (ŭngk′shən) *n.* [ME < Lat. *unctio* < *unguere*, to anoint.] **1.** The act of anointing as part of a religious, ceremonial, or healing ritual. **2.** An ointment or oil used for anointing. **3.** Something that serves to soothe : BALM. **4.** Affected or exaggerated earnestness, esp. in language.

**unc·tu·ous** (ŭngk′cho͞o-əs) *adj.* [ME < Med. Lat. *unctuosus* < Lat. *unctum*, ointment < *unguere*, to anoint.] **1.** Having the characteristics of oil or ointment : GREASY. **2.** Containing or composed of oil or fat. **3.** Abundant in organic materials <*unctuous* soil> **4.** Marked by affected, exaggerated, or insincere earnestness : offensively smooth or suave <*unctuous* flattery> **—unc′tu·ous·ly** *adv.* **—unc′tu·ous·ness, unc′tu·os′i·ty** (-ŏs′ĭ-tē) *n.*

**un·cus** (ŭng′kəs) *n., pl.* **un·ci** (ŭn′sī′) [NLat. < Lat., hook.] *Biol.* A part or process shaped like a hook.

**un·cut** (ŭn-kŭt′) *adj.* **1.** Not cut. **2.** Not having the page edges slit or trimmed <an *uncut* book> **3.** Not cut to a specific shape. —Used of a gemstone. **4.** Not shortened : UNABRIDGED.

**un·damped** (ŭn-dămpt′) *adj.* **1.** Not tending toward a state of rest : not damped. —Used of oscillations. **2.** Not diminished or discouraged <*undamped* fervor>

**un·daunt·a·ble** (ŭn-dôn′tə-bəl, -dän′-) *adj.* Not capable of being discouraged or disheartened.

**un·daunt·ed** (ŭn-dôn′tĭd, -dän′-) *adj.* Not discouraged or disheartened : RESOLUTE. **—un·daunt′ed·ly** *adv.* **—un·daunt′ed·ness** *n.*

**un·de·ceive** (ŭn′dĭ-sēv′) *vt.* **-ceived, -ceiv·ing, -ceives.** To free from deception or illusion.

**un·de·cid·ed** (ŭn′dĭ-sī′dĭd) *adj.* **1.** Not yet settled or determined. **2.** Not having reached a decision. **—un′de·cid′ed·ly** *adv.* **—un′de·cid′ed·ness** *n.*

**un·decked**[1] (ŭn-dĕkt′) *adj.* Not decorated or unornamented.

**un·decked**[2] (ŭn-dĕkt′) *adj. Naut.* Having no deck. —Used of a ship.

**un·de·mon·stra·tive** (ŭn′dĭ-mŏn′strə-tĭv) *adj.* Not given to expressions of feeling : RESERVED. **—un′de·mon′stra·tive·ly** *adv.* **—un′de·mon′stra·tive·ness** *n.*

**un·de·ni·a·ble** (ŭn′dĭ-nī′ə-bəl) *adj.* **1.** Not able to be denied : IRREFUTABLE. **2.** Unquestionably good <*undeniable* credit references> **—un′de·ni′a·ble·ness** *n.* **—un′de·ni′a·bly** *adv.*

**un·der** (ŭn′dər) *prep.* [ME < OE.] **1.** In a lower position or place than <a sign *under* a portrait> **2.** Beneath the surface of <*under* the snow> **3.** Beneath the assumed surface or guise of <entered the country *under* a false name> **4.** Less than : smaller than. **5.** Less than the required amount or degree of <*under* drinking age> **6.** Inferior to in rank or status. **7.** Subject to the authority, rule, or control of <*under* a tyrannical junta> **8.** Subject to the supervision, instruction, or influence of <*under* parental guidance> **9.** Undergoing or receiving the effects of <*under* neonatal care> **10.** Subject to the restraint or obligation of <a star *under* contract to the studio> **11.** Within the group or classification of <listed *under* chemistry> **12.** In the process of <a plan *under* discussion> **13.** In view of <*under* those conditions> **14.** With the authorization of <*under* the monarch's seal> **15.** Sowed or planted with <an acre *under* rye> —*adv.* **1.** In or into a place below or beneath. **2.** In or into a subordinate or inferior condition or position. **3.** So as to be covered or enveloped by. **4.** So as to be less than the required amount or degree. —*adj.* **1.** Located or situated on a lower level or beneath something else <the *under* parts of a machine> **2.** Lower in rank, power, or authority : SUBORDINATE. **3.** Less than required or customary <an *under* dose of antibiotics>

**under-** *pref.* [ME < OE < *under*, under.] **1.** Beneath or below in position <*under*ground> **2.** Inferior or subordinate in rank or importance <*under*secretary> **3.** Less in degree, rate, or quantity than normal or proper <*under*sized>

**un·der·a·chieve** (ŭn′dər-ə-chēv′) *vi.* **-chieved, -chiev·ing, -chieves.** To fail to perform at the level of capability indicated by tests of intelligence and aptitude, esp. in schoolwork. **—un′der·a·chieve′ment** *n.* **—un′der·a·chiev′er** *n.*

**un·der·act** (ŭn′dər-ăkt′) *v.* **-act·ed, -act·ing, -acts.** —*vt.* **1.** To perform (a role) weakly. **2.** To understate (a role) intentionally. —*vi.* To perform in an understated way.

**un·der·age** (ŭn′dər-āj′) *adj.* Below the legal or customary age, as for drinking or voting.

**un·der·arm** (ŭn′dər-ärm′) *adj.* **1.** Located, placed, or used under the arm. **2.** Executed with the hand kept below the level of the shoulder <an *underarm* pitch> —*adv.* With an underarm motion or delivery. —*n.* The armpit.

**un·der·bel·ly** (ŭn′dər-bĕl′ē) *n., pl.* **-lies. 1.** The lower part or underside of an animal's body. **2.** The vulnerable or weak part <"the soft *underbelly* of Europe" —Winston Churchill>

**un·der·bid** (ŭn′dər-bĭd′) *v.* **-bid, -bid·ding, -bids.** —*vt.* **1.** To bid lower than (a competitor). **2.** To bid less than the full value of (one's hand) in bridge. —*vi.* To bid unnecessarily low. **—un′der·bid′der** *n.*

**un·der·bod·y** (ŭn′dər-bŏd′ē) *n.* **1.** UNDERBELLY 1. **2.** The under parts of the body of a motor vehicle.

**un·der·bred** (ŭn′dər-brĕd′) *adj.* **1.** Badly brought up : ILL-BRED. **2.** Not purebred.

**un·der·brush** (ŭn′dər-brŭsh′) *n.* Small trees, shrubs, or similar plants growing beneath the taller trees in a forest.

**un·der·car·riage** (ŭn′dər-kăr′ĭj) *n.* **1.** The supporting framework of a vehicle. **2.** The landing gear of an aircraft.

**un·der·charge** (ŭn′dər-chärj′) *vt.* **-charged, -charg·ing, -charg·es. 1.** To charge (someone) less than is customary or required. **2.** To load (a firearm) with an insufficient charge. —*n.* (ŭn′dər-chärj′). An insufficient or improper charge.

**un·der·class** (ŭn′dər-klăs′) *n.* The lowest stratum of society, usu. composed of the disadvantaged.

**un·der·class·man** (ŭn′dər-klăs′mən) *n.* A freshman or sophomore at a secondary school or college.

**un·der·clothes** (ŭn′dər-klōz′, -klōthz′) *pl.n.* Clothes worn next to the skin : UNDERWEAR.

**un·der·cloth·ing** (ŭn′der-klō′thĭng) *n.* Underclothes.

**un·der·coat** (ŭn′dər-kōt′) *n.* **1.** A coat worn beneath another coat. **2.** A growth of short hairs or fur concealed by the longer outer hairs of an animal's coat. **3.** *also* **un·der·coat·ing** (-kō′tĭng). **a.** A coat of sealing material applied to a surface as a base for the topcoat. **b.** A tarlike substance applied to the underside of a motor vehicle to prevent rust. —*vt.* **-coat·ed, -coat·ing, -coats.** To apply an undercoat to <*undercoat* a new truck>

**un·der·cool** (ŭn′dər-kōōl′) *vt.* **-cooled, -cool·ing, -cools.** To supercool.

**un·der·cov·er** (ŭn′dər-kŭv′ər) *adj.* **1.** Carried out in secret <an *undercover* investigation> **2.** Engaged or employed in spying or secret investigation <an *undercover* agent>

**un·der·croft** (ŭn′dər-krôft′, -krŏft′) *n.* [ME *under croft* : *under,* under + *croft,* crypt < Med. Lat. *crupta* < Lat. *crypta.* —see CRYPT.] A crypt, esp. one under a church.

**un·der·cur·rent** (ŭn′dər-kûr′ənt) *n.* **1.** A current, as of air or water, below another current or beneath a surface. **2.** An underlying tendency or feeling often at odds with what is superficially evident : INTIMATION <an *undercurrent* of hostility>

**un·der·cut** (ŭn′dər-kŭt′) *v.* **-cut, -cut·ting, -cuts.** —*vt.* **1.** To make a cut under or below. **2.** To create an overhang by cutting material away from, as in carving. **3.** To sell at lower prices or work for lower wages than (a competitor). **4.** To diminish or destroy the province or effectiveness of : UNDERMINE. **5. a.** To impart backspin to (a ball) by striking downward as well as forward, as in golf and baseball. **b.** To cut or slice (a ball) with an underarm stroke, as in tennis. —*vi.* To undercut someone or something. —*n.* (ŭn′dər-kŭt′). **1. a.** A cut made in the under part to remove material. **b.** The material so removed. **2.** A notch cut in the base of a tree to direct its fall and insure a clean break. **3.** *Chiefly Brit.* The tenderloin or fillet of beef. **4. a.** A backspin given to a ball, as in golf. **b.** A cut or slice made with an underarm stroke, as in tennis.

**un·der·de·vel·oped** (ŭn′dər-dĭ-vĕl′əpt) *adj.* **1.** Not adequately or normally developed. **2.** Not kept in a developing solution long enough to produce a normal degree of contrast <an *underdeveloped* photograph> **3.** Industrially or economically backward <*underdeveloped* countries>

**un·der·do** (ŭn′dər-dōō′) *vt.* **-did** (-dĭd′), **-done** (-dŭn′), **-do·ing, -does** (-dŭz′). To do to an insufficient degree.

**un·der·dog** (ŭn′dər-dôg′, -dŏg′) *n.* **1.** One expected to lose a contest or struggle, as in sports or politics. **2.** One at a disadvantage, as because of discrimination.

**un·der·done** (ŭn′dər-dŭn′) *adj.* Not sufficiently cooked.

**un·der·draw·ers** (ŭn′dər-drôrz′) *pl.n.* Shorts or briefs worn as undergarments, esp. those for a man.

**un·der·dress** (ŭn′dər-drĕs′) *n.* **1.** Apparel worn beneath outer garments. **2.** An outer garment worn under another as part of a costume or suit, as a dress beneath a tunic. —*vi.* (ŭn′dər-drĕs′). To dress too informally for the occasion. **—un′der·dressed′** *adj.*

**un·der·drive** (ŭn′dər-drīv′) *n.* A gearing device causing the output drive shaft to rotate at a slower rate than the engine input shaft.

**un·der·ed·u·cat·ed** (ŭn′dər-ĕj′ə-kā′tĭd) *adj.* Poorly or insufficiently educated. **—un′der·ed′u·ca′tion** *n.*

**un·der·em·pha·size** (ŭn′dər-ĕm′fə-sīz′) *vt.* **-sized, -siz·ing, -siz·es.** To fail to give enough emphasis to. **—un′der·em′pha·sis** (-sĭs) *n.*

**un·der·em·ployed** (ŭn′dər-ĕm-ploid′) *adj.* Partially or inadequately employed, esp. employed at a low-paying job for which one is overqualified. **—un′der·em·ploy′ment** *n.*

**un·der·es·ti·mate** (ŭn′dər-ĕs′tə-māt′) *vt.* **-mat·ed, -mat·ing, -mates.** To make too low an estimate of the quantity, degree, or worth of. —*n.* (ŭn′dər-ĕs′tə-mĭt). An estimate that is or proves to be too low. **—un′der·es′ti·ma′tion** *n.*

**un·der·ex·pose** (ŭn′dər-ĭk-spōz′) *vt.* **-posed, -pos·ing, -pos·es.** To expose (film) to light for too short a time to produce normal image contrast. **—un′der·ex·po′sure** (-ĭk-spō′zhər) *n.*

**un·der·feed** (ŭn′dər-fēd′) *vt.* **-fed** (-fĕd′), **-feed·ing, -feeds. 1.** To feed insufficiently. **2.** To feed (an engine) with fuel from below.

**un·der·flow** (ŭn′dər-flō′) *n.* A data-processing error arising when a computed quantity is a smaller number than the device can display.

**un·der·foot** (ŭn′dər-fŏŏt′) *adv.* **1.** Below or at one's feet <soft grass *underfoot*> **2.** In the way <cats *underfoot* in the kitchen>

**un·der·fund** (ŭn′dər-fŭnd′) *vt.* **-fund·ed, -fund·ing, -funds.** To provide insufficient funding for.

**un·der·fur** (ŭn′dər-fûr′) *n.* The dense soft fur beneath the coarse outer hairs of certain mammals.

**un·der·gar·ment** (ŭn′dər-gär′mənt) *n.* A garment worn under outer garments, esp. one worn next to the skin.

**un·der·gird** (ŭn′dər-gûrd′) *vt.* **-gird·ed** *or* **-girt** (-gûrt′), **-gird·ing, -girds.** To gird, support, or strengthen from beneath.

**un·der·glaze** (ŭn′dər-glāz′) *n.* Coloring applied to pottery before it is glazed.

**un·der·go** (ŭn′dər-gō′) *vt.* **-went** (-wĕnt′), **-gone** (-gôn′, -gŏn′), **-go·ing, -goes** (-gōz′) [ME *undergon* : *under,* under + *gon,* go.] **1.** To be subjected to : EXPERIENCE <*undergo* treatment> **2.** To endure : suffer.

**un·der·grad·u·ate** (ŭn′dər-grăj′ōō-ĭt) *n.* A college or university student who has not yet received a degree. —*adj.* **1.** Of, relating to, or characteristic of undergraduates. **2.** Having undergraduate status.

**un·der·ground** (ŭn′dər-ground′) *adj.* **1.** Situated, occurring, or operating below the surface of the earth <*underground* chambers><*underground* nuclear testing> **2.** Conducted in secret : CLANDESTINE <*underground* resistance to a dictator> **3.** Involved in secret or illegal activity <*underground* dealers in art> **4.** Of or relating to an avant-garde movement or its films, publications, and art, usu. privately produced and of special appeal and often concerned with social or artistic experiment. —*n.* **1.** A secret, often nationalist organization set up to resist or overthrow a government in power, such as an occupying military government. **2.** *Chiefly Brit.* A subway system. **3.** An avant-garde movement or publication. —*adv.* (ŭn′dər-ground′). **1.** Below the surface of the earth. **2.** In secret : STEALTHILY. —*vt.* **-ground·ed, -ground·ing, -grounds.** To situate under the ground, as telephone lines.

**Underground Railroad** *n.* A secret network operating in the United States before 1861 to help fugitive slaves reach sanctuary in the free states or Canada.

**un·der·grown** (ŭn′dər-grōn′) *adj.* Not fully grown : PUNY.

**un·der·growth** (ŭn′dər-grōth′) *n.* **1. a.** Low plants, saplings, and shrubs growing beneath the trees in a forest. **b.** Something resembling this, as a dog's undercoat. **2.** The state of being undergrown.

**un·der·hand** (ŭn′dər-hănd′) *adj.* **1.** Done in a treacherous or deceitful way : SNEAKY. **2.** UNDERARM 2. —*adv.* **1.** With an underhand movement. **2.** Slyly : secretly.

**un·der·hand·ed** (ŭn′dər-hăn′dĭd) *adj.* **1.** Underhand. **2.** Lacking the required number of workers or players : SHORT-HANDED. **—un′der·hand′ed·ly** *adv.* **—un′der·hand′ed·ness** *n.*

**un·der·hung** (ŭn′dər-hŭng′) *adj.* **1. a.** Protruding from beneath <an *underhung* jaw> **b.** Supported by or lying over something that projects. **2.** Resting on a supporting track, as a sliding door. **3.** Underslung, as a vehicle.

**un·der·kill** (ŭn′dər-kĭl′) *n.* Insufficient force to defeat an enemy.

**un·der·laid** (ŭn′dər-lād′) *adj.* **1.** Placed or laid underneath. **2.** Supported or raised by something underneath : having an underlay.

**un·der·lay** (ŭn′dər-lā′) *vt.* **-laid, -lay·ing, -lays. 1.** To put (one thing) under another. **2.** To provide with a base or sublining. **3.** To raise with an underlay in printing. —*n.* (ŭn′dər-lā′). **1.** Something underlaid, as felt under a carpet. **2.** Paper or other material inserted under type or cuts to raise the level of the printing bed.

**un·der·let** (ŭn′dər-lĕt′) *vt.* **-let, -let·ting, -lets. 1.** To lease for less than the proper value. **2.** To sublet.

**un·der·lie** (ŭn′dər-lī′) *vt.* **-lay** (-lā′), **-lain** (-lān′), **-ly·ing, -lies. 1.** To be located under or below. **2.** To be at the basis of <Many factors *underlie* inflation.> **3.** To have a prior financial claim over <Dividends for preferred stock *underlie* those of common stock.>

**un·der·line** (ŭn′dər-līn′, ŭn′dər-līn′) *vt.* **-lined, -lin·ing, -lines. 1.** To draw a line under, esp. so as to emphasize. **2.** To place emphasis on : STRESS. —*n.* (ŭn′dər-līn′). A line under something, as a symbol, word, or phrase, to indicate emphasis or italic type.

**un·der·ling** (ŭn′dər-lĭng) *n.* A subordinate or inferior.

**un·der·lin·ing** (ŭn′dər-lī′nĭng) *n.* **1.** The act of drawing a line under. **2.** Emphasis. **3.** A lining for a garment.

**un·der·lip** (ŭn′dər-lĭp′) *n.* The lower lip.

**un·der·ly·ing** (ŭn′dər-lī′ĭng) *adj.* **1.** Lying under or beneath <*underlying* strata> **2.** Basic : fundamental <an *underlying* truth> **3.** Implicit : hidden <an *underlying* significance> **4.** Taking precedence : PRIOR <an *underlying* financial claim>

**un·der·manned** (ŭn′dər-mănd′) *adj.* Understaffed.

**un·der·mine** (ŭn′dər-mīn′) *vt.* **-mined, -min·ing, -mines. 1.** To dig a mine or tunnel beneath. **2.** To weaken by wearing away the supporting base <Water *undermined* the house's foundation.> **3.** To weaken, injure, or impair, often by degrees <Poor eating habits *undermine* one's health.>

**un·der·mod·u·late** (ŭn′dər-mŏj′ə-lāt′) *vt.* **-lat·ed, -lat·ing, -lates.** To utilize less of a sound reproduction or transmission device than optimally possible. **—un′der·mod′u·la′tion** *n.*

ōō **boot** ou **out** th **thin** *th* **this** ŭ **cut** ûr **urge** y **young** yōō **abuse** zh **vision** ə **about,** it**em,** ed**i**ble, gall**o**p, circ**us**

**un·der·most** (ŭn′dər-mōst′) *adj. & adv.* Lowest in position, rank, or place.

**un·der·neath** (ŭn′dər-nēth′) *adv.* [ME *undernethe* < OE *underneoðan* : *under,* under + *neoðan,* below.] **1.** In a place beneath : BELOW. **2.** On the lower face or underside. —*prep.* **1.** Under : below. **2.** Under the power or control of. —*adj.* Lower : under. —*n.* The part or side below or under.

**un·der·nour·ish** (ŭn′dər-nûr′ĭsh) *vt.* **-ished, -ish·ing, -ish·es.** To provide with insufficient quantity or quality of nourishment to sustain proper health and growth. **—un′der·nour′ish·ment** *n.*

**un·der·nu·tri·tion** (ŭn′dər-nōō-trĭsh′ən, -nyōō-) *n.* Inadequate nutrition due to undernourishment or poor assimilation of food.

**un·der·pants** (ŭn′dər-pănts′) *pl.n.* Pants or briefs worn as underwear.

**un·der·pass** (ŭn′dər-păs′) *n.* A passage underneath something, esp. a section of road that passes under a highway or railroad.

**un·der·pay** (ŭn′dər-pā′) *vt.* **-paid, -pay·ing, -pays.** To pay less than is required or deserved.

**un·der·pin** (ŭn′dər-pĭn′) *vt.* **-pinned, -pin·ning, -pins. 1.** To support from below, as with props, girders, or masonry. **2.** To corroborate or substantiate <*underpin* a theory with scientific proof>

**un·der·pin·ning** (ŭn′dər-pĭn′ĭng) *n.* **1.** Material or masonry used to support a wall or other structure. **2.** *often* **underpinnings.** Something serving as a support or foundation. **3.** *often* **underpinnings.** *Informal.* The legs.

**un·der·play** (ŭn′dər-plā′, ŭn′dər-plā′) *v.* **-played, -play·ing, -plays.** —*vt.* To play (a role or scene) subtly or with restraint. —*vi.* To act subtly or with restraint.

**un·der·pop·u·lat·ed** (ŭn′dər-pŏp′yə-lā′tĭd) *adj.* Lacking the normal or required population density. **—un′der·pop·u·la′tion** *n.*

**un·der·price** (ŭn′dər-prīs′) *vt.* **-priced, -pric·ing, -pric·es. 1.** To price lower than the real value. **2.** To undercut in price <*underprice* a competitor>

**un·der·priv·i·leged** (ŭn′dər-prĭv′ə-lĭjd) *adj.* Deprived of the opportunities and advantages enjoyed by other members of society.

**un·der·pro·duc·tion** (ŭn′dər-prə-dŭk′shən) *n.* **1.** Production below full capacity. **2.** Production below demand. **—un′der·pro·duc′tive** *adj.*

**un·der·proof** (ŭn′dər-prōōf′) *adj.* Having a smaller proportion of alcohol than proof spirit.

**un·der·prop** (ŭn′dər-prŏp′) *vt.* **-propped, -prop·ping, -props.** To prop from below : SUPPORT.

**un·der·quote** (ŭn′dər-kwōt′) *vt.* **-quot·ed, -quot·ing, -quotes. 1.** To quote (goods) at a price lower than the official list or market price. **2.** To quote a lower price than (a competitor).

**un·der·rate** (ŭn′dər-rāt′) *vt.* **-rat·ed, -rat·ing, -rates.** To rate too low : UNDERESTIMATE.

**un·der·re·port** (ŭn′dər-rĭ-pôrt′, -pōrt′) *vt.* **-port·ed, -port·ing, -ports.** To report (e.g., income) to be less than the facts indicate.

**un·der·run** (ŭn′dər-rŭn′) *vt.* **-ran** (-răn′), **-run, -run·ning, -runs. 1.** To run, pass, or go beneath. **2.** *Naut.* To pass under (a line or cable) in a boat for inspection or repairs.

**un·der·score** (ŭn′dər-skôr′, -skōr′) *vt.* **-scored, -scor·ing, -scores. 1.** To draw a line under : UNDERLINE. **2.** To emphasize or stress. —*n.* A line drawn under writing to indicate emphasis or italic type.

**un·der·sea** (ŭn′dər-sē′) *adj.* Relating to, existing, or created for use beneath the surface of the sea. —*adv.* (ŭn′dər-sē′) *also* **un·der·seas** (-sēz′). Beneath the surface of the sea.

**un·der·sec·re·tar·y** (ŭn′dər-sĕk′rə-tĕr′ē) *n., pl.* **-ies.** An official directly subordinate to a cabinet member.

**un·der·sell** (ŭn′dər-sĕl′) *vt.* **-sold** (-sōld′), **-sell·ing, -sells. 1.** To sell goods for a lower price than (a competitor). **2.** To sell at a price less than the actual value.

**un·der·set** (ŭn′dər-sĕt′) *n.* An ocean undercurrent.

**un·der·sexed** (ŭn′dər-sĕkst′) *adj.* Having less sexual desire than normal.

**un·der·shirt** (ŭn′dər-shûrt′) *n.* A usu. short-sleeved undergarment worn next to the skin under a shirt.

**un·der·shoot** (ŭn′dər-shōōt′) *v.* **-shot** (-shŏt′), **-shoot·ing, -shoots.** —*vt.* **1.** To shoot a missile short of (a target). **2.** To land an aircraft short of (a landing area). —*vi.* To undershoot a target or a landing area.

**un·der·shorts** (ŭn′dər-shôrts′) *pl.n.* Underdrawers.

**un·der·shot** (ŭn′dər-shŏt′) *adj.* **1.** Driven by water passing below, as a water wheel. **2.** Projecting from below <an *undershot* jaw>

**un·der·shrub** (ŭn′dər-shrŭb′) *n.* A low-growing shrub.

**un·der·side** (ŭn′dər-sīd′) *n.* The side or surface underneath.

**un·der·sign** (ŭn′dər-sīn′) *vt.* **-signed, -sign·ing, -signs.** To sign one's name at the end of (a letter or document).

**un·der·signed** (ŭn′dər-sīnd′) *n., pl.* **undersigned.** One whose name is signed at the end of a document <The *undersigned* have agreed to the terms.>

**un·der·sized** (ŭn′dər-sīzd′) *also* **un·der·size** (-sīz′) *adj.* Of subnormal or insufficient size.

**un·der·skirt** (ŭn′dər-skûrt′) *n.* A skirt worn under another, as a petticoat.

**un·der·sleeve** (ŭn′dər-slēv′) *n.* **1.** A sleeve worn under another. **2.** An ornamental sleeve worn under another, designed to extend below or show through slashes in the outer sleeve.

**un·der·slung** (ŭn′dər-slŭng′) *adj.* **1.** Having springs attached to the axles from below. —Used of a vehicle. **2.** Low to the ground : SQUAT.

**un·der·soil** (ŭn′dər-soil′) *n.* Soil below the ground surface.

**un·der·spin** (ŭn′dər-spĭn′) *n.* A backspin.

**un·der·staffed** (ŭn′dər-stăft′) *adj.* Lacking sufficient personnel.

**un·der·stand** (ŭn′dər-stănd′) *v.* **-stood** (-stŏŏd′), **-stand·ing, -stands.** —*vt.* [ME *understanden* < OE *understandan* : *under,* under + *standan,* to stand.] **1.** To perceive and comprehend the nature and significance of <"I don't pretend to *understand* the Universe—it's a great deal bigger than I am" —Carlyle> **2.** To know thoroughly through close contact with or long experience of <*understand* teenagers><*understand* politics> **3. a.** To grasp what is intended or expressed by (another). **b.** To comprehend the meaning of (a language) <*understands* Latin> **4.** To know and be tolerant or sympathetic toward <I *understand* your position even though I disagree.> **5.** To learn indirectly, as by hearsay : ASSUME <I *understand* there will be a strike.> **6.** To conclude : infer <Am I to *understand* that you are leaving town?> **7.** To accept as an agreed fact <It is *understood* that the fee will be waived.> —*vi.* **1.** To have understanding, knowledge, or comprehension. **2.** To learn indirectly or at secondhand : GATHER <I *understand* the partners are no longer speaking to each other.> **—un′der·stand′a·bil′i·ty** *n.* **—un′der·stand′a·ble** *adj.* **—un′der·stand′a·bly** *adv.*

**un·der·stand·ing** (ŭn′dər-stăn′dĭng) *n.* **1.** The quality or state of one who understands : COMPREHENSION. **2.** The faculty by which one understands : INTELLIGENCE. **3.** Individual judgment or interpretation : OPINION. **4. a.** A compact implicit between two or more persons or groups. **b.** The matter implicit in such a compact. **5.** Reconciliation of differences : AGREEMENT <reached an *understanding*> —*adj.* **1.** Having or marked by comprehension, good sense, or discernment. **2.** Compassionate and sympathetic. **—un′der·stand′ing·ly** *adv.*

**un·der·state** (ŭn′dər-stāt′) *v.* **-stat·ed, -stat·ing, -states.** —*vt.* **1.** To state with less completeness or truth than seems warranted by the facts <*understated* the complex geopolitical problems> **2.** To express with restraint, esp. ironically or for dramatic impact. —*vi.* To make an understatement.

**un·der·state·ment** (ŭn′dər-stāt′mənt) *n.* **1.** A disclosure or statement that is less than complete. **2.** Intentional lack of emphasis in expression, as in irony.

**un·der·stood** (ŭn′dər-stŏŏd′) *adj.* **1.** Agreed upon. **2.** Implied but not expressed. **3.** Comprehended.

**un·der·stra·tum** (ŭn′dər-strā′təm, -străt′əm) *n., pl.* **-stra·ta** (-strā′tə, -străt′ə) *or* **-stra·tums.** A substratum.

**un·der·stud·y** (ŭn′dər-stŭd′ē) *v.* **-ied, -y·ing, -ies.** —*vt.* **1.** To study (a role) so as to be able to replace the regular performer when required. **2.** To act as an understudy to. —*vi.* To understudy a role or actor. —*n., pl.* **-ies. 1.** A performer who understudies. **2.** One trained to do the work of another.

**un·der·sur·face** (ŭn′dər-sûr′fəs) *n.* An underside.

**un·der·take** (ŭn′dər-tāk′) *v.* **-took** (-tŏŏk′), **-tak·en, -tak·ing, -takes.** [ME *undertaken* : *under,* under + *taken,* to take.] —*vt.* **1.** To take upon oneself : decide or agree to do <*undertake* an assignment> **2.** To pledge or commit oneself to : CONTRACT. **3.** *Obs.* To accept combat with : TAKE ON. —*vi. Archaic.* To make oneself responsible.

**un·der·tak·er** (ŭn′dər-tā′kər) *n.* **1.** One who undertakes a task or job, esp. an entrepreneur. **2.** (ŭn′dər-tā′kər). One whose business it is to arrange for the burial or cremation of the dead and to assist at funeral rites : MORTICIAN.

▲ word history: The subject of death has always inspired euphemism; the word *undertaker* is one example. Derived from the verb *undertake,* "to take upon oneself," *undertaker* originally denoted one who undertakes any kind of task. Around 1700, during a period of general refinement of speech and manners, the word *undertaker* was applied specifically to those who undertake to prepare the dead for the grave.

**un·der·tak·ing** (ŭn′dər-tā′kĭng) *n.* **1.** A task or assignment undertaken : VENTURE. **2.** A guaranty, engagement, or promise. **3.** The profession or duties of an undertaker.

**un·der-the-count·er** (ŭn′dər-*thə*-koun′tər) *adj.* Transacted, given, or sold secretly or illicitly <*under-the-counter* drugs>

**un·der-the-ta·ble** (ŭn′dər-*thə*-tā′bəl) *adj.* Secret and usu. illegal or unethical <*under-the-table* bribes>

**un·der·tint** (ŭn′dər-tĭnt′) *n.* A slight or subtle tint.

**un·der·tone** (ŭn′dər-tōn′) *n.* **1.** A tone of low pitch or volume, esp. of spoken sound. **2. a.** A pale or subdued color. **b.** A color applied under or seen through another color. **3.** An underlying or implied tendency or meaning.

**un·der·tow** (ŭn′dər-tō′) *n.* The seaward pull of receding waves breaking on a shore.

ă **pat** ā **pay** âr **care** ä **father** ĕ **pet** ē **be** hw **which** ĭ **pit**
ī **tie** îr **pier** ŏ **pot** ō **toe** ô **paw, for** oi **noise** ŏŏ **took**

**un·der·trick** (ŭn′dər-trĭk′) *n.* A card trick, esp. in bridge, the loss of which prevents a declarer from making his or her contract.

**un·der·trump** (ŭn′dər-trŭmp′) *vi.* **-trumped, -trump·ing, -trumps.** To play a trump lower than another card player's trump when trump has not been led.

**un·der·val·ue** (ŭn′dər-văl′yo͞o) *vt.* **-ued, -u·ing, -ues. 1.** To assign too low a value to <*undervalue* an antique> **2.** To have too little regard or esteem for <*undervalue* a long-standing friendship> **—un′der·val′u·a′tion** *n.*

**un·der·vest** (ŭn′dər-vĕst′) *n. Chiefly Brit.* An undershirt.

**un·der·wa·ter** (ŭn′dər-wô′tər, -wŏt′ər) *adj.* Occurring, used, or performed beneath the surface of the water. **—un′der·wa′ter** *adv.*

**under way** *adv.* **1.** Put in motion or operation. **2.** Already commenced or initiated. **3.** *Naut.* Not anchored and not moored to a fixed object : in motion.

**un·der·wear** (ŭn′dər-wâr′) *n.* Clothing worn under the outer clothes and next to the skin.

**un·der·weight** (ŭn′dər-wāt′) *adj.* Weighing less than is normal, healthy, or required. *—n.* Insufficiency of weight.

**un·der·whelm** (ŭn′dər-hwĕlm′, -wĕlm′) *vt.* **-whelmed, -whelm·ing, -whelms.** *Informal.* To fail to excite or impress.

**un·der·wing** (ŭn′dər-wĭng′) *n.* **1.** One of a pair of hind wings partially or wholly covered by the forewings, as in certain moths. **2.** Any of various moths of the genus *Calocala,* characterized by brightly colored underwings.

**underwing**
*Of a moth*

**un·der·wood** (ŭn′dər-wo͝od′) *n.* Underbrush.

**un·der·world** (ŭn′dər-wûrld′) *n.* **1.** A region or realm beneath the earth. **2.** The opposite side of the earth : ANTIPODES. **3.** *Gk. & Rom. Myth.* The world of the dead, held to be below the world of the living : HADES. **4.** A hidden sphere of society engaged in organized crime and vice. **5.** *Archaic.* The world beneath the heavens : EARTH.

**un·der·write** (ŭn′dər-rīt′) *v.* **-wrote** (-rōt′), **-writ·ten** (-rĭt′n), **-writ·ing, -writes.** *—vt.* **1. a.** To write under or at the end of. **b.** To subscribe to, esp. to sign or endorse (a document). **2.** To assume financial responsibility for <*underwrite* a concert series> **3. a.** To sign (an insurance policy) so as to assume liability in case of specified losses. **b.** To insure. **c.** To insure against losses to the extent of (a given amount). **4.** To agree to buy (stock not yet offered publicly) at a fixed time and price. *—vi.* To act as an underwriter, esp. to issue an insurance policy.

**un·der·writ·er** (ŭn′dər-rī′tər) *n.* **1. a.** One engaged in an insurance business. **b.** An insurance agent who assesses the risk of enrolling an applicant for coverage or a policy. **2.** One that guarantees the purchase of a full issue of stocks or bonds.

**undescended testicle** *n.* A testicle that has remained within the inguinal canal and has not descended to the scrotum.

**un·de·serv·ed·ly** (ŭn′dĭ-zûr′vĭd-lē) *adv.* Unfairly or unjustifiably.

**un·de·sign·ing** (ŭn′dĭ-zī′nĭng) *adj.* Having no ulterior motives : STRAIGHTFORWARD.

**un·de·sir·a·ble** (ŭn′dĭ-zīr′ə-bəl) *adj.* Not desirable. *—n.* An undesirable person. **—un′de·sir′a·bil′i·ty** *n.* **—un′de·sir′a·bly** *adv.*

**undesirable discharge** *n.* Formal discharge from military service under conditions less than honorable.

**un·de·ter·mined** (ŭn′dĭ-tûr′mĭnd) *adj.* **1.** Not yet determined or decided. **2.** Not specifically known or ascertained.

**un·dies** (ŭn′dēz) *pl.n. Informal.* Women's or girls' underwear.

**un·dig·ni·fied** (ŭn-dĭg′nə-fīd′) *adj.* Not dignified.

**un·dine** (ŭn-dēn′, ŭn′dēn′) *n.* [NLat. *undina* < Lat. *unda,* wave.] A water nymph who, according to Paracelsus, could earn a soul by marrying a mortal and bearing his child.

**un·dip·lo·mat·ic** (ŭn-dĭp′lə-măt′ĭk) *adj.* Not diplomatic or tactful. **—un·dip′lo·mat′i·cal·ly** *adv.*

**un·di·rect·ed** (ŭn′dĭ-rĕk′tĭd, -dī-) *adj.* Without plan or purpose : not guided <*undirected* studies>

**un·dis·crim·i·nat·ing** (ŭn′dĭs-krĭm′ə-nā′tĭng) *adj.* **1.** Indiscriminate. **2.** Lacking sensitivity, taste, or judgment.

**un·dis·posed** (ŭn′dĭs-pōzd′) *adj.* **1.** Not settled, removed, or resolved. **2.** Disinclined : unwilling.

**un·dis·tin·guished** (ŭn′dĭ-stĭng′gwĭsht) *adj.* Not distinguished.

**un·dis·turbed** (ŭn′dĭ-stûrbd′) *adj.* Not disturbed.

**un·do** (ŭn-do͞o′) *v.* **-did** (-dĭd′), **-done** (-dŭn′), **-do·ing, -does** (-dŭz′) [ME *undon* < OE *undōn* : *un-,* un- + *don,* to do.] *—vt.* **1.** To reverse or erase (something done) : ANNUL <no way to *undo* the mistakes of the past> **2.** To untie, disassemble, or loosen <*undo* a ribbon> **3.** To open or unwrap <*undo* a parcel> **4.** *Obs.* To solve or interpret : UNRAVEL. **5. a.** To cause the ruin or downfall of : DESTROY. **b.** To throw into confusion : UNSETTLE. *—vi.* To come open or undone. **—un·do′er** *n.*

**un·dock** (ŭn-dŏk′) *vt.* **-docked, -dock·ing, -docks.** To uncouple.

**un·do·ing** (ŭn-do͞o′ĭng) *n.* **1.** A reversal or annulment of an accomplishment. **2.** An act of unfastening or loosening. **3. a.** An act of bringing to ruin. **b.** A cause of ruin : DOWNFALL <Pride was my *undoing.*>

**un·doubt·ed** (ŭn-dou′tĭd) *adj.* Beyond question : UNDISPUTED. **—un·doubt′ed·ly** *adv.*

**un·draw** (ŭn-drô′) *vt.* **-drew** (-dro͞o′), **-drawn** (-drôn′), **-draw·ing, -draws.** To draw to one side, as a curtain.

**un·dress** (ŭn-drĕs′) *v.* **-dressed, -dress·ing, -dress·es.** *—vt.* **1.** To remove the clothing or covering of. **2.** To remove the dressing from (e.g., a wound). *—vi.* To take off one's clothing. *—n.* **1.** Informal attire. **2.** Nakedness : nudity.

☆ ***syns:*** UNDRESS, DISROBE, STRIP *v. core meaning* : to take off one's clothing <*undressed* and went to bed> ***ant:*** dress

**un·dressed** (ŭn-drĕst′) *adj.* **1. a.** Naked. **b.** Not fully dressed. **2.** Not specially treated or processed <*undressed* leather>

**un·due** (ŭn-do͞o′, -dyo͞o′) *adj.* **1.** Exceeding the appropriate or normal : EXCESSIVE <*undue* remorse> **2.** Not just, proper, or legal <*undue* exercise of power> **3.** Not yet payable or due, as a bill.

**un·du·lant** (ŭn′jə-lənt, ŭn′dyə-, ŭn′də-) *adj.* Resembling waves in occurrence, appearance, or motion.

**undulant fever** *n.* A remittent fever caused by bacteria of the genus *Brucella,* contracted by contact with infected animals or consumption of their meat or milk and marked by weakness, anemia, and painful joints.

**un·du·late** (ŭn′jə-lāt′, ŭn′dyə-, ŭn′də-) *v.* **-lat·ed, -lat·ing, -lates.** [< Lat. *undulatus,* having wave-like markings < *unda,* wave.] *—vt.* **1.** To cause to move in a smooth wavelike motion. **2.** To give a wavelike appearance to. *—vi.* **1.** To move in waves or with a wavelike or sinuous motion : RIPPLE. **2.** To have a wavelike appearance. *—adj.* (ŭn′jə-lĭt, -lāt′, ŭn′dyə-, ŭn′də-) *also* **un·du·lat·ed** (-lā′tĭd). Having a wavy outline or appearance <leaves with *undulate* margins>

**un·du·la·tion** (ŭn′jə-lā′shən, ŭn′dyə-, ŭn′də-) *n.* **1.** A wavelike rising and falling or movement back and forth. **2.** A wavy form, outline, or appearance. **3.** One of a series of waves or wavelike segments. **—un′du·la·to′ry** (-lə-tôr′ē, -tōr′ē) *adj.*

**un·du·ly** (ŭn-do͞o′lē, -dyo͞o′-) *adv.* **1.** Excessively : immoderately <*unduly* anxious> **2.** Improperly : wrongfully <*unduly* suspicious of other people's motives>

**un·du·ti·ful** (ŭn-do͞o′tĭ-fəl, -dyo͞o′-) *adj.* Lacking a sense of duty. **—un·du′ti·ful·ly** *adv.* **—un·du′ti·ful·ness** *n.*

**un·dy·ing** (ŭn-dī′ĭng) *adj.* Endless : eternal.

**un·earned** (ŭn-ûrnd′) *adj.* **1.** Not gained by work or service <*unearned* income> **2.** Not deserved <*unearned* praise> **3.** Not yet earned <*unearned* interest>

**unearned increment** *n.* An increase in property value resulting from factors independent of the efforts of the owner, such as a general rise in demand for land.

**un·earth** (ŭn-ûrth′) *vt.* **-earthed, -earth·ing, -earths. 1.** To bring up out of the earth : dig up. **2.** To bring to public notice : UNCOVER.

**un·earth·ly** (ŭn-ûrth′lē) *adj.* **-li·er, -li·est. 1.** Not of this earth : SUPERNATURAL. **2.** Frighteningly weird and unaccountable : UNNATURAL <an *unearthly* howl> **3.** Ridiculous : absurd <awakened at an *unearthly* hour> **—un·earth′li·ness** *n.*

**un·eas·y** (ŭn-ē′zē) *adj.* **-i·er, -i·est. 1.** Lacking ease, comfort, or a sense of security. **2.** Affording no ease or reassurance <an *uneasy* quiet> **3.** Awkward or constrained <*uneasy* with foreigners> **—un·ease′, un·eas′i·ness** *n.* **—un·eas′i·ly** *adv.*

**un·ed·it·ed** (ŭn-ĕd′ĭ-tĭd) *adj.* **1.** Not edited or revised. **2.** Not adapted for a special audience or purpose. **3.** Still to be edited.

**un·ed·u·cat·ed** (ŭn-ĕj′ə-kā′tĭd) *adj.* Not educated.

**un·e·mo·tion·al** (ŭn′ĭ-mō′shə-nəl) *adj.* **1.** Not easily stirred or moved. **2.** Involving little or no emotion. **—un′e·mo′tion·al·ly** *adv.*

**un·em·ploy·a·ble** (ŭn′ĭm-ploi′ə-bəl) *adj.* Not suitable for employment. **—un′em·ploy′a·ble** *n.*

**un·em·ployed** (ŭn′ĭm-ploid′) *adj.* **1.** Not employed : out of work. **2.** Not being used : IDLE. *—n.* One who does not have a job. **—un′em·ploy′ment** *n.*

**unemployment compensation** *n.* Money provided for unemployed workers by the social security system, usu. paid weekly and for a set time period.

**un·en·dur·a·ble** (ŭn′ən-do͝or′ə-bel, -dyo͝or′-) *adj.* Unbearable. **—un′en·dur′a·bly** *adv.*

**un-Eng·lish** (ŭn-ĭng′glĭsh) *adj.* **1.** Not having the characteristics of English. **2.** Not in agreement with standard English usage.

**un·e·qual** (ŭn-ē′kwəl) *adj.* **1.** Not the same in any measurable aspect, as extent or quantity <*unequal* treatment> **2.** Not the same as another in rank or social position. **3.** Consisting of ill-matched opponents <an *unequal* chess game> **4.** Having unbalanced sides or parts : ASYMMETRIC. **5.** Not even or consistent : VARIABLE <an *unequal* performance> **6.** Not adequate : INSUFFICIENT <talents *unequal* to the role> **7.** *Obs.* Not fair or equitable. —*n.* One that is not the equal of another. **—un·e′qual·ly** *adv.*

**un·e·qualed** (ŭn-ē′kwəld) *adj. also* **un·e·qualled** *adj.* Not matched or paralleled by others of its kind : UNRIVALED.

**un·e·quiv·o·cal** (ŭn′ĭ-kwĭv′ə-kəl) *adj.* Admitting of no doubt or misunderstanding : CLEAR. **—un′e·quiv′o·cal·ly** *adv.*

**un·err·ing** (ŭn-ûr′ĭng, -ĕr′ĭng) *adj.* Committing no mistakes : consistently accurate. **—un·err′ing·ly** *adv.*

**un·es·sen·tial** (ŭn′ĭ-sĕn′shəl) *adj.* Not necessary : DISPENSABLE. —*n.* A nonessential.

**un·e·ven** (ŭn′ē′vən) *adj.* **-er, -est. 1.** Not equal, as in size or quality. **2.** Not uniform or consistent <*uneven* teeth><an *uneven* concert> **3.** Not level or smooth <the *uneven* surface of a stone wall> **4.** Not straight or parallel <*uneven* margins> **5.** Not fair or equitable <an *uneven* contest> **6.** Designating an odd number. **—un·e′ven·ly** *adv.* **—un·e′ven·ness** *n.*

**un·e·vent·ful** (ŭn′ĭ-vĕnt′fəl) *adj.* Lacking in significant events : without incident. **—un′e·vent′ful·ly** *adv.* **—un′e·vent′ful·ness** *n.*

**un·ex·am·pled** (ŭn′ĭg-zăm′pəld) *adj.* Without precedent : UNPARALLELED <*unexampled* generosity>

**un·ex·cep·tion·a·ble** (ŭn′ĭk-sĕp′shə-nə-bəl) *adj.* Beyond the least reasonable objection : IRREPROACHABLE <*unexceptionable* judicial conduct> **—un′ex·cep′tion·a·ble·ness** *n.* **—un′ex·cep′tion·a·bly** *adv.*

**un·ex·cep·tion·al** (ŭn′ĭk-sĕp′shə-nəl) *adj.* **1.** Not varying from a norm : USUAL. **2.** Not subject to exceptions : ABSOLUTE. **—un′ex·cep′tion·al·ly** *adv.*

**un·ex·pect·ed** (ŭn′ĭk-spĕk′tĭd) *adj.* Not expected : UNFORESEEN. **—un′ex·pect′ed·ly** *adv.* **—un′ex·pect′ed·ness** *n.*

**un·ex·ploit·ed** (ŭn′ĕk-sploi′tĭd) *adj.* Not exploited or developed.

**un·ex·pres·sive** (ŭn′ĭk-sprĕs′ĭv) *adj.* **1.** Not conveying the meaning intended or the emotion felt. **2.** *Obs.* Inexpressible. **—un′ex·pres′sive·ly** *adv.* **—un′ex·pres′sive·ness** *n.*

**un·fad·ing** (ŭn-fā′dĭng) *adj.* **1.** Retaining color or freshness. **2.** Retaining value or usefulness. **—un·fad′ing·ly** *adv.*

**un·fail·ing** (ŭn-fā′lĭng) *adj.* **1.** Never giving out : INEXHAUSTIBLE <an *unfailing* supply> **2.** Constant : unflagging <*unfailing* optimism> **3.** Incapable of error : INFALLIBLE. **—un·fail′ing·ly** *adv.* **—un·fail′ing·ness** *n.*

**un·fair** (ŭn-fâr′) *adj.* **-er, -est. 1.** Not fair : BIASED. **2.** Contrary to laws or conventions, esp. in commerce : UNETHICAL. **—un·fair′ly** *adv.* **—un·fair′ness** *n.*

☆ ***syns:*** UNFAIR, INEQUITABLE, UNEQUAL, UNJUST *adj. core meaning* : not fair, right, or just <*unfair* housing laws> ***ant:*** fair, just

**un·faith** (ŭn-fāth′) *n.* An absence of faith : DISBELIEF.

**un·faith·ful** (ŭn-fāth′fəl) *adj.* **1.** Not true to a pledge or obligation : DISLOYAL. **2.** Not constant to a sexual partner, esp. guilty of adultery. **3.** Not justly representing or reflecting the original : INACCURATE. **4.** *Obs.* Lacking religious faith. **—un·faith′ful·ly** *adv.* **—un·faith′ful·ness** *n.*

**un·fa·mil·iar** (ŭn′fə-mĭl′yər) *adj.* **1.** Not previously known or experienced : STRANGE <an *unfamiliar* food> **2.** Not well acquainted : not conversant <*unfamiliar* with modern art> **—un′fa·mil·iar′i·ty** (-mĭl-yăr′ĭ-tē, -mĭl′ē-ăr′ə-tē) *n.* **—un′fa·mil′iar·ly** *adv.*

**un·fash·ion·a·ble** (ŭn-făsh′ə-nə-bəl) *adj.* **1.** Not currently fashionable. **2.** Not socially approved <an *unfashionable* section of town> **—un·fash′ion·a·bly** *adv.*

**un·fas·ten** (ŭn-făs′ən) *v.* **-tened, -ten·ing, -tens.** —*vt.* To separate the connected parts of. —*vi.* To become loosened or separated.

**un·fa·thered** (ŭn-fä′thərd) *adj.* **1.** Having no father : FATHERLESS. **2.** Illegitimate. **3.** Of uncertain origin or authenticity.

**un·fath·om·a·ble** (ŭn-făth′ə-mə-bəl) *adj.* **1.** Incapable of being understood. **2.** Incapable of being measured.

**un·fa·vor·a·ble** (ŭn-fā′vər-ə-bəl, -fā′vrə-bəl) *adj.* **1.** Not propitious <*unfavorable* signs> **2.** Negative : adverse <*unfavorable* reports> **3.** Undesirable : disadvantageous <an *unfavorable* rate of exchange> **4.** Not pleasing. **—un′fa′vor·a·ble·ness** *n.* **—un′fa′vor·a·bly** *adv.*

**un·feel·ing** (ŭn-fē′lĭng) *adj.* **1.** Having no feeling or sensation : INSENTIENT. **2.** Having no feelings of sympathy or kindness : CALLOUS. **—un·feel′ing·ly** *adv.* **—un·feel′ing·ness** *n.*

**un·feigned** (ŭn′fānd′) *adj.* Not simulated : GENUINE. **—un·feign′ed·ly** (ŭn-fā′nĭd-lē) *adv.*

**un·fet·ter** (ŭn-fĕt′ər) *vt.* **-tered, -ter·ing, -ters. 1.** To free from fetters. **2.** To liberate <*unfettered* from convention>

**un·fin·ished** (ŭn-fĭn′ĭsht) *adj.* **1.** Not brought to an end : INCOMPLETE <*unfinished* business> **2.** Having received no final finish or special processing : NATURAL <*unfinished* wood>

**un·fit** (ŭn-fĭt′) *adj.* **1.** Not meant or adapted for a given purpose : UNSUITABLE <*unfit* for human consumption> **2.** Not qualified : INCOMPETENT <an *unfit* teacher> **3.** Not in good physical or mental health. —*vt.* **-fit·ted, -fit·ting, -fits.** To cause to be unsuited or unqualified : DISQUALIFY. **—un·fit′ly** *adv.* **—un·fit′ness** *n.*

**un·fix** (ŭn-fĭks′) *vt.* **-fixed, -fix·ing, -fix·es. 1.** To detach from what secures : UNFASTEN. **2.** To unsettle : disturb.

**un·flap·pa·ble** (ŭn-flăp′ə-bəl) *adj.* Not easily upset or excited. **—un·flap′pa·bil′i·ty** *n.* **—un·flap′pa·bly** *adv.*

**un·flat·ter·ing** (ŭn-flăt′ər-ĭng) *adj.* Not flattering : UNFAVORABLE <an *unflattering* review> **—un·flat′ter·ing·ly** *adv.*

**un·fledged** (ŭn-flĕjd′) *adj.* **1.** Not yet having the flight feathers necessary for flight. **2.** Inexperienced, immature, or untried.

**un·flinch·ing** (ŭn-flĭn′chĭng) *adj.* Not showing fear or indecision : RESOLUTE. **—un·flinch′ing·ly** *adv.* **—un·flinch′ing·ness** *n.*

**un·fo·cused** *also* **un·fo·cussed** (ŭn-fō′kəst) *adj.* **1.** Not brought into focus. **2.** Not centered on something specific.

**un·fold** (ŭn-fōld′) *v.* **-fold·ed, -fold·ing, -folds.** —*vt.* **1.** To open and spread out (something folded) : EXTEND. **2.** To remove the coverings from : disclose to view. **3.** To reveal gradually in words : make known. —*vi.* **1.** To become spread out : open out. **2.** To develop <as the tale *unfolds*> **3.** To be revealed gradually to the view or understanding. **—un·fold′ment** *n.*

**un·fore·seen** (ŭn′fər-sēn′) *adj.* Not anticipated in advance : UNEXPECTED <*unforeseen* problems>

**un·for·get·ta·ble** (ŭn′fər-gĕt′ə-bəl) *adj.* Lastingly set in the memory : MEMORABLE. **—un′for·get′ta·bil′i·ty, un′for·get′ta·ble·ness** *n.* **—un′for·get′ta·bly** *adv.*

**un·for·mat·ted** (ŭn-fôr′măt′ĭd) *adj. Computer Sci.* Designating input or output data that have not been edited before display.

**un·formed** (ŭn-fôrmd′) *adj.* **1.** Having no definite shape or structure : SHAPELESS. **2.** Not yet developed to maturity : IMMATURE. **3.** Not yet given a physical existence : UNCREATED.

**un·for·tu·nate** (ŭn-fôr′chə-nĭt) *adj.* **1.** Characterized by lack of good fortune. **2.** Causing misfortune : DISASTROUS. **3.** Marked by inappropriateness <an *unfortunate* lack of tact> —*n.* A victim of bad luck. **—un·for′tu·nate·ly** *adv.* **—un·for′tu·nate·ness** *n.*

☆ ***syns:*** **1.** UNFORTUNATE, HAPLESS, ILL-FATED, LUCKLESS, UNHAPPY, UNLUCKY *adj. core meaning* : involving or undergoing chance misfortune <an *unfortunate* turn of events> ***ant:*** fortunate **2.** UNFORTUNATE, AWKWARD, INAPPROPRIATE, INFELICITOUS, UNHAPPY *adj. core meaning* : marked by inappropriateness, esp. in expression <an *unfortunate* remark> ***ant:*** appropriate, felicitous

**un·found·ed** (ŭn-foun′dĭd) *adj.* **1.** Not yet established. **2.** Not based on fact or sound observation : GROUNDLESS <*unfounded* fears> **—un·found′ed·ly** *adv.* **—un·found′ed·ness** *n.*

**un·fre·quent·ed** (ŭn′frē-kwĕn′tĭd, ŭn-frē′kwən-tĭd) *adj.* Receiving few or no visitors : UNPATRONIZED <an *unfrequented* spot>

**un·friend·ed** (ŭn-frĕn′dĭd) *adj.* Having no friends.

**un·friend·ly** (ŭn-frĕnd′lē) *adj.* **-li·er, -li·est. 1.** Feeling or exhibiting hostility. **2.** Indicating a bad prospect : UNFAVORABLE <an *unfriendly* climate> **—un·friend′li·ness** *n.*

☆ ***syns:*** UNFRIENDLY, HOSTILE *adj. core meaning* : feeling or exhibiting hostility <*unfriendly* nations><an *unfriendly* look> ***ant:*** friendly

**un·frock** (ŭn-frŏk′) *vt.* **-frocked, -frock·ing, -frocks. 1.** To strip of priestly privileges and functions. **2.** To deprive of the right to practice a profession.

**un·fruit·ful** (ŭn-fro͞ot′fəl) *adj.* **1.** Not bearing fruit or offspring : BARREN. **2.** Not productive of good or valuable results : UNPROFITABLE. **—un·fruit′ful·ly** *adv.* **—un·fruit′ful·ness** *n.*

**un·fund·ed** (ŭn-fŭn′dĭd) *adj.* **1.** Not funded, as a floating debt. **2.** Not supplied with funds <an *unfunded* service>

**un·furl** (ŭn-fûrl′) *v.* **-furled, -furl·ing, -furls.** —*vt.* To spread or open out : UNROLL. —*vi.* To become spread or opened out.

**un·fuss·y** (ŭn-fŭs′ē) *adj.* **1.** Not particular or concerned, as with details. **2.** Not cluttered or complicated, as with extraneous matters.

**un·gain·ly** (ŭn-gān′lē) *adj.* **-li·er, -li·est. 1.** Being without grace or ease of movement : CLUMSY <an *ungainly* youth><*ungainly* language> **2.** Difficult to use or move : UNWIELDY <*ungainly* furniture> **—un·gain′li·ness** *n.*

**un·gen·er·os·i·ty** (ŭn-jen′ə-rŏs′ĭ-tē) *n.* The quality or state of being ungenerous : STINGINESS.

**un·gen·er·ous** (ŭn-jĕn′ər-əs) *adj.* **1.** Not generous : STINGY. **2.** UNKIND 1. **—un·gen′er·ous·ly** *adv.*

**un·girt** (ŭn-gûrt′) *adj.* **1.** Having the belt or girdle removed or loosened. **2.** Loose or free : SLACK.

**un·glue** (ŭn-glo͞o′) *vt.* **-glued, -glu·ing, -glues.** To separate by or as if by dissolving a glue or other adhesive.

**un·glued** (ŭn-glo͞od′) *adj.* Loosened or separated. **—come unglued.** *Slang.* To suffer a loss of composure : become disordered or upset.

**un·god·ly** (ŭn-gŏd′lē) *adj.* **-li·er, -li·est. 1.** Not revering God : IMPIOUS. **2.** Sinful : wicked. **3.** Outrageous : unconscionable <woke us at an *ungodly* hour> **—un·god′li·ness** *n.*

**un·gov·ern·a·ble** (ŭn-gŭv′ər-nə-bəl) *adj.* Not capable of being governed or controlled <an *ungovernable* temper> **—un·gov′ern·a·ble·ness** *n.* **—un·gov′ern·a·bly** *adv.*

**un·gra·cious** (ŭn-grā′shəs) *adj.* **1.** Lacking social grace or graciousness : RUDE. **2.** Not pleasant or acceptable : DISAGREEABLE. **3.** *Archaic.* Evil : wicked. **—un·gra′cious·ly** *adv.* **—un·gra′cious·ness** *n.*

**un·gram·mat·i·cal** (ŭn′grə-măt′ĭ-kəl) *adj.* Not in accord with the rules of grammar.

**un·grate·ful** (ŭn-grāt′fəl) *adj.* **1.** Not feeling or exhibiting gratitude, thanks, or appreciation. **2.** Not agreeable or pleasant : DISTASTEFUL. **—un·grate′ful·ly** *adv.* **—un·grate′ful·ness** *n.*

**un·gual** (ŭng′gwəl) *adj.* [< Lat. *unguis,* nail.] *Zool.* Of, resembling, or bearing a hoof, nail, or claw.

**un·guard·ed** (ŭn-gär′dĭd) *adj.* **1.** Having no guard or protection : VULNERABLE. **2.** Without discretion : INCAUTIOUS <an *unguarded* comment> **—un·guard′ed·ly** *adv.* **—un·guard′ed·ness** *n.*

**un·guent** (ŭng′gwənt) *n.* [ME < Lat. *unguentum* < *unguere,* to anoint.] A soothing or healing salve : OINTMENT.

**un·guic·u·late** (ŭng-gwĭk′yə-lĭt, -lāt′) *adj.* [NLat. *unguiculatus* < Lat. *unguiculus,* fingernail, dim. of *unguis,* nail.] **1.** *Zool.* Having nails or claws. **2.** *Bot.* Having a claw-shaped base, as a petal. —*n.* A mammal having nails or claws.

**un·guis** (ŭng′gwĭs) *n., pl.* **-gues** (-gwēz) [Lat., nail.] A nail, claw, hoof, or clawlike structure.

**un·gu·late** (ŭng′gyə-lĭt, -lāt′) *adj.* [LLat. *ungulatus* < *ungula,* hoof, dim. of *unguis,* nail.] **1.** Having hoofs. **2.** Of or belonging to the former order Ungulata, now divided into the orders Perissodactyla and Artiodactyla, and including all hoofed mammals, such as horses, cattle, swine, deer, and elephants. —*n.* An ungulate mammal.

**un·hal·low** (ŭn-hăl′ō) *vt.* **-lowed, -low·ing, -lows.** *Archaic.* To profane : desecrate.

**un·hal·lowed** (ŭn-hăl′ōd) *adj.* **1.** Not hallowed or consecrated : UNHOLY <*unhallowed* ground> **2. a.** Impious : irreligious. **b.** Not conforming to accepted moral standards : IMMORAL.

**un·hand** (ŭn-hănd′) *vt.* **-hand·ed, -hand·ing, -hands.** To let go one's hand from.

**un·hand·some** (ŭn′hăn′səm) *adj.* **1.** Not attractive or beautiful : HOMELY. **2.** Lacking in courtesy or good taste : UNGRACIOUS. **—un·hand′some·ly** *adv.* **—un·hand′some·ness** *n.*

**un·hand·y** (ŭn-hăn′dē) *adj.* **-i·er, -i·est. 1.** Difficult to handle or manage : UNWIELDY. **2.** Having little or no manual skill or dexterity. **—un·hand′i·ly** *adv.* **—un·hand′i·ness** *n.*

**un·hap·py** (ŭn-hăp′ē) *adj.* **-pi·er, -pi·est. 1.** Not happy or joyful : SAD. **2.** Bringing misfortune : UNLUCKY <an *unhappy* event> **3.** Not suitable : INAPPROPRIATE <an *unhappy* choice of friends> **—un·hap′pi·ly** *adv.* **—un·hap′pi·ness** *n.*

**un·har·ness** (ŭn′här′nĭs) *vt.* **-nessed, -ness·ing, -ness·es. 1.** To remove the harness from. **2.** To release or liberate.

**un·health·y** (ŭn-hĕl′thē) *adj.* **-i·er, -i·est. 1.** In a state of ill health : SICK. **2.** Resulting from or symptomatic of ill health <an *unhealthy* pallor> **3.** Conducive to poor health : UNWHOLESOME <an *unhealthy* diet of junk food> **4.** Morally harmful : CORRUPTIVE. **5.** Risky : dangerous. **—un·health′i·ly** *adv.* **—un·health′i·ness** *n.*

**un·heard** (ŭn-hûrd′) *adj.* **1.** Not heard. **2.** Not given a hearing : not listened to. **3.** *Archaic.* Unheard of.

**un·heard-of** (ŭn-hûrd′ŭv′, -ŏv′) *adj.* **1.** Not previously known. **2.** Unprecedented <an *unheard-of* price>

**un·hes·i·tat·ing** (ŭn-hĕz′ĭ-tā′tĭng) *adj.* **1.** Prompt : ready. **2.** Unfaltering : steadfast. **—un·hes′i·tat′ing·ly** *adv.*

**un·hinge** (ŭn-hĭnj′) *vt.* **-hinged, -hing·ing, -hing·es. 1.** To remove from hinges. **2.** To remove the hinges from. **3.** To disrupt or derange.

**un·hip** (ŭn-hĭp′) *adj. Slang.* Unaware of the latest fashions, ideas, or developments.

**un·hitch** (ŭn-hĭch′) *vt.* **-hitched, -hitch·ing, -hitch·es.** To release from or as if from a hitch : UNFASTEN.

**un·ho·ly** (ŭn′hō′lē) *adj.* **-li·er, -li·est. 1.** Not hallowed or consecrated. **2.** Wicked : immoral. **3.** *Informal.* Outrageous <an *unholy* risk> **—un·ho′li·ly** *adv.* **—un·ho′li·ness** *n.*

**un·hook** (ŭn-ho͝ok′) *vt.* **-hooked, -hook·ing, -hooks. 1.** To release or remove from a hook. **2.** To unfasten the hooks of.

**un·hoped** (ŭn-hōpt′) *adj. Archaic.* Unhoped-for.

**un·hoped-for** (ŭn-hōpt′fôr′) *adj.* Not hoped for or expected.

**un·horse** (ŭn-hôrs′) *vt.* **-horsed, -hors·ing, -hors·es. 1.** To cause to fall from a horse. **2.** To overthrow or upset.

**un·hou·seled** (ŭn-hou′zəld) *adj. Archaic.* Not having received the Eucharist.

**uni-** *pref.* [Lat. < *unus,* one.] Single : one <*unicycle*>

**U·ni·at** *also* **U·ni·ate** (yo͞o′nē-ăt′) *n.* [R. *uniyat* < Pol. *uniat* < *unja,* union < LLat. *unio* < *unus,* one.] A member of a Uniat Church. **—U′ni·at′, U′ni·ate′** *adj.*

**Uniat Church** *also* **Uniate Church** *n.* An Eastern Christian church that recognizes the supremacy of the pope but retains its own distinctive liturgy.

**u·ni·ax·i·al** (yo͞o′nē-ăk′sē-əl) *adj.* **1.** Having only one axis. **2.** Of or along a single axis.

**u·ni·cam·er·al** (yo͞o′nĭ-kăm′ər-əl) *adj.* Having or consisting of a single legislative chamber. **—u′ni·cam′er·al·ly** *adv.*

**UNICEF** (yo͞o′nĭ-sĕf′) *n.* United Nations International Children's Emergency Fund.

**u·ni·cel·lu·lar** (yo͞o′nĭ-sĕl′yə-lər) *adj.* Having or consisting of a single cell <a *unicellular* organism>

**u·ni·col·or** (yo͞o′nĭ-kŭl′ər) *adj.* Monochromatic.

**u·ni·corn** (yo͞o′nĭ-kôrn′) *n.* [ME < OFr. < Lat. *unicornis,* having one horn : *unus,* one + *cornu,* horn.] **1.** A legendary creature usu. represented as a horse with a single spiraled horn projecting from its forehead and often with a goat's beard and a lion's tail. **2. Unicorn.** *Astron.* The constellation Monoceros.

**unicorn plant** *n.* A plant, *Proboscidea louisiana* of the southern United States, having yellowish flowers mottled with purple and a beaked woody pod.

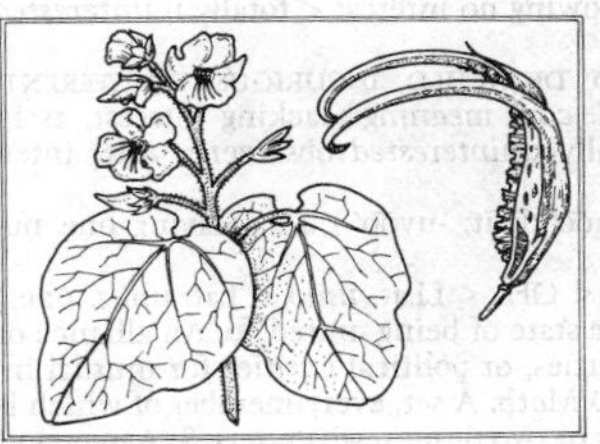
**unicorn plant**

**u·ni·cos·tate** (yo͞o′nĭ-kŏs′tāt′) *adj.* Having a single main costa, rib, or riblike part.

**u·ni·cy·cle** (yo͞o′nĭ-sī′kəl) *n.* A vehicle consisting of a frame mounted over a single wheel and usu. propelled by pedals. **—u′ni·cy′clist** *n.*

**unidentified flying object** *n.* Any of various objects of unknown nature and origin reportedly seen in the air.

**u·ni·di·rec·tion·al** (yo͞o′nĭ-dĭ-rĕk′shən-əl, -dī-rĕk′shən-əl) *adj.* Having, operating, or moving in one direction only.

**u·ni·fi·a·ble** (yo͞o′nə-fī′ə-bəl) *adj.* Capable of being unified.

**unified field theory** *n.* A physical theory that combines the treatment of two or more types of fields in order to deduce previously unrecognized interrelationships, esp. such a theory unifying the theories of nuclear, electromagnetic, and gravitational forces.

**u·ni·fi·lar** (yo͞o′nə-fī′lər) *adj.* Having or utilizing only one filament, as a thread or wire.

**u·ni·fo·li·ate** (yo͞o′nə-fō′lē-ĭt, -āt′) *adj.* Having a single leaf.

**u·ni·fo·li·o·late** (yo͞o′nĭ-fō′lē-ə-lāt′) *adj.* Structurally compound, but having a single leaflet and often a winglike extension along the leafstalk.

**u·ni·form** (yo͞o′nə-fôrm′) *adj.* [OFr. *uniforme* < Lat. *uniformis* : *unus,* one + *forma,* shape.] **1. a.** Always the same : UNVARYING <*uniform* quality> **b.** Being without variation or fluctuation : CONSISTENT <*uniform* speed> **2.** Being the same as another or others : IDENTICAL <"Language was not *uniform* throughout the country but fell into dialects" —Kemp Malone> **3.** Consistent in appearance <*uniform* rows of houses> —*n.* **1.** A distinctive outfit intended to identify those who wear it as members of a specific group. **2.** Any characteristic style of dress. —*vt.* **-formed, -form·ing, -forms. 1.** To make uniform. **2.** To provide or dress with a uniform. **—u′ni·for′mi·ty** (-fôr′mĭ-tē), **u′ni·form′ness** *n.* **—u′ni·form′ly** *adv.*

**u·ni·for·mi·tar·i·an·ism** (yo͞o′nə-fôr′mĭ-târ′ē-ə-nĭz′əm) *n.* The theory that all geological phenomena are the result of existing forces that have operated uniformly from the origin of the earth to the present time. **—u′ni·for′mi·tar′i·an** *adj.* & *n.*

**u·ni·fy** (yo͞o′nə-fī′) *v.* **-fied, -fy·ing, -fies.** [OFr. *unifier* < LLat. *unificare* : Lat. *unus,* one + Lat. *facere,* to make.] —*vt.* To make into a unit : CONSOLIDATE. —*vi.* To become unified. **—u′ni·fi·ca′tion** *n.* **—u′ni·fi′er** *n.*

**u·ni·lat·er·al** (yo͞o′nə-lăt′ər-əl) *adj.* **1.** Of, relating to, involving, or affecting only one side. **2.** Obligating only one of two or more parties, nations, or persons <a *unilateral* treaty> **3.** Emphasizing or recognizing only one side of a subject. **4.** Having only one side. **5.** Tracing the lineage of one parent only <a *unilateral* genealogy> **—u′ni·lat′er·al·ly** *adv.*

**u·ni·lin·e·ar** (yo͞o′nĭ-lĭn′ē-ər) *adj.* Developing in a progressive sequence usu. from the primitive to the advanced, as in a culture.

**u·ni·lin·gual** (yo͞o′nĭ-lĭng′gwəl) *adj.* Written in or making use of one language only.

**u·ni·loc·u·lar** (yo͞o′nə-lŏk′yə-lər) *adj. Bot.* Having a single compartment or chamber.

**un·im·pas·sioned** (ŭn′ĭm-păsh′ənd) *adj.* Dispassionate.

**un·im·peach·a·ble** (ŭn′ĭm-pē′chə-bəl) *adj.* Being beyond doubt or reproach : IMPECCABLE. **—un′im·peach′a·bly** *adv.*

**un·im·por·tant** (ŭn′ĭm-pôr′tnt) *adj.* Not important. **—un′im·por′tance** *n.*

**un·im·proved** (ŭn′ĭm-pro͞ovd′) *adj.* **1.** Not improved or bettered. **2.** Not made use of or put to advantage. **3.** Not built on or cultivated so as to increase in value. —Used of land.

**un·in·form·a·tive** (ŭn'ĭn-fôr'mə-tĭv) *adj.* Not informative. **—un'in·form'a·tive·ly** *adv.*

**un·in·hab·it·ed** (ŭn'ĭn-hăb'ĭ-tĭd) *adj.* Having no inhabitants.

**un·in·hib·it·ed** (ŭn'ĭn-hĭb'ĭ-tĭd) *adj.* **1.** Not restrained or held back <*uninhibited* joy> **2.** Free from social or moral constraints <*uninhibited* nudity> **—un'in·hib'it·ed·ly** *adv.* **—un'in·hib'it·ed·ness** *n.*

**un·in·i·ti·ate** (ŭn'ĭ-nĭsh'ē-ĭt) *adj.* Not experienced.

**un·in·spired** (ŭn'ĭn-spīrd') *adj.* Having no intellectual, emotional, or spiritual excitement : DULL <an *uninspired* sermon>

**un·in·tel·li·gent** (ŭn'ĭn-tĕl'ə-jənt) *adj.* Lacking intelligence. **—un'in·tel'li·gent·ly** *adv.*

**un·in·ter·est·ed** (ŭn-ĭn'trĭs-tĭd, -ĭn'tə-rĕs'tĭd) *adj.* **1. a.** Being without an interest <*uninterested* parties> **b.** Not having a financial interest. **2.** Having or showing no interest <totally *uninterested* in the lecture>

☆ ***syns:*** UNINTERESTED, DETACHED, INCURIOUS, INDIFFERENT, REMOTE, UNCONCERNED *adj. core meaning* **:** lacking interest, as in one's surroundings <a totally *uninterested* observer> ***ant:*** interested

**u·ni·nu·cle·ate** (yōō'nĭ-nōō'klē-ĭt, -nyōō'-) *adj.* Having one nucleus.

**un·ion** (yōōn'yən) *n.* [ME < OFr. < LLat. *unio* < Lat. *unus,* one.] **1. a.** An act of uniting or the state of being united. **b.** An alliance or confederation of persons, parties, or political entities for mutual interest or benefit. **2.** *Symbol* **U** *Math.* A set, every member of which is an element of one or another of two or more given sets. **3.** Agreement resulting from an alliance : CONCORD. **4. a.** Matrimony : marriage. **b.** Sexual intercourse. **5. a.** A combination of parishes for joint administration of relief for the poor in Britain. **b.** A workhouse maintained by such a union. **6.** A labor union. **7.** Any of various coupling devices for connecting pipes or machine parts. **8.** A device on a flag or ensign, occupying the upper inner corner or the entire field, that signifies the union of two or more sovereignties. **9. Union. a.** An organization at a college or university that provides facilities for recreation. **b.** A building housing such facilities. **10. Union.** The United States of America, esp. during the Civil War.

**union catalog** *n.* A library catalog combining in alphabetical sequence the contents of several catalogs or libraries.

**un·ion·ism** (yōōn'yə-nĭz'əm) *n.* **1.** The principle or theory of forming a union. **2.** The principles, theory, or system of a union, esp. a trade union. **3. Unionism.** Loyalty to the federal union of the United States, esp. during the Civil War. **—un'ion·ist** *n.*

**un·ion·ize** (yōōn'yə-nīz') *v.* **-ized, -iz·ing, -iz·es.** —*vt.* **1.** To organize into a labor union. **2.** To cause to join a labor union. —*vi.* To organize or join a labor union. **—un'ion·i·za'tion** *n.*

**union jack** *n.* **1.** A flag consisting entirely of a union. **2. Union Jack.** The flag of the United Kingdom.

**union label** *n.* An identifying mark attached to a product indicating it has been produced by members of a trade union.

**union shop** *n.* A business or industrial establishment in which the employer is required by contract to employ only union members but may hire nonmembers who agree to join the union within a specified time.

**union suit** *n.* A one-piece undergarment combining shirt and pants.

**u·nip·a·rous** (yōō-nĭp'ər-əs) *adj.* **1.** Producing only one offspring at a time. **2.** Having produced only one offspring. **3.** *Bot.* Forming a single axis at each branching, as certain flower clusters.

**u·ni·per·son·al** (yōō'nĭ-pûr'sə-nəl) *adj.* Existing as or manifested in the form of only one person <a *unipersonal* spirit>

**u·ni·pla·nar** (yōō'nĭ-plā'nər, -när') *adj.* Lying in one plane.

**u·ni·po·lar** (yōō'nĭ-pō'lər) *adj.* Of, having, or produced by a single magnetic or electric pole.

**u·nique** (yōō-nēk') *adj.* [Fr. < Lat. *unicus,* sole < *unus,* one.] **1.** Being the only one of its kind : SOLE. **2.** Being without equal or rival. **3.** *Informal.* Unusual. **—u·nique'ly** *adv.* **—u·nique'ness** *n.*

☆ ***syns:*** UNIQUE, INCOMPARABLE, MATCHLESS, PEERLESS, UNEQUALED, UNPARALLELED, UNRIVALED *adj. core meaning* **:** being without equal or rival <an artist with *unique* creativity>

**u·ni·sex** (yōō'nĭ-sĕks') *n.* Elimination or absence of sexual distinctions, esp. in dress. —*adj.* **1.** Not distinguishable on the basis of sex <*unisex* hair styles> **2.** Designed for or suitable to both sexes <*unisex* clothing>

**u·ni·sex·u·al** (yōō'nĭ-sĕk'shōō-əl) *adj.* **1.** Of only one sex. **2.** Having only one type of sexual organ. **3.** *Bot.* Having either stamens or pistils but not both. **4.** Unisex. **—u'ni·sex'u·al'i·ty** *n.* **—u'ni·sex'u·al·ly** *adv.*

**u·ni·son** (yōō'nĭ-sən, -zən) *n.* [OFr. < Med. Lat. *unisonus,* in unison : Lat. *unus,* one + Lat. *sonus,* sound.] **1. a.** Identity of musical pitch : the interval of a perfect prime. **b.** The combination of musical parts at the same pitch or in octaves. **2.** Agreement : concord. **—in unison. 1.** In complete agreement <Our ideas are *in unison* with yours.> **2.** Simultaneously <They recited the vows *in unison.*>

**u·nit** (yōō'nĭt) *n.* [Back-formation from UNITY.] **1.** An individual, group, structure, or other entity regarded as an elementary structural or functional constituent of a whole. **2.** A group regarded as a distinct entity within a larger group. **3. a.** A mechanical part or module. **b.** An entire apparatus or the equipment that performs a specific function. **4.** A precisely specified quantity in terms of which the magnitudes of other quantities of the same kind can be stated. **5.** A fixed amount of scholastic work used as a basis for calculating academic credits, usu. measured in terms of classroom or laboratory time. **6.** The number immediately to the left of the decimal point in the Arabic numeral system.

**U·ni·tar·i·an** (yōō'nĭ-târ'ē-ən) *n.* [< NLat. *unitarius* < Lat. *unitas,* unity < *unus,* one.] **1.** A monotheist who rejects the doctrine of the Trinity. **2.** A member of a Christian denomination that rejects the doctrine of the Trinity and emphasizes freedom and tolerance in religious belief and the autonomy of each congregation. **—U'ni·tar'i·an** *adj.* **—U'ni·tar'i·an·ism** *n.*

**u·ni·tar·y** (yōō'nĭ-tĕr'ē) *adj.* **1.** Of or pertaining to a unit or units. **2.** Having the character of a unit : WHOLE. **3.** Based on or characterized by unity.

**u·nite** (yōō-nīt') *v.* **-nit·ed, -nit·ing, -nites.** [ME *uniten* < LLat. *unire* < Lat. *unus,* one.] —*vt.* **1.** To bring together so as to form a whole. **2.** To bring together by a common interest, attitude, or action <students *united* by a love of learning> **3.** To join (a couple) in marriage. **4.** To cause to adhere. **5.** To have or exhibit (e.g., qualities) in combination <*unites* intellect and wit> —*vi.* **1.** To become or seem to become joined, formed, or combined into a unit. **2.** To join and act together in a common purpose or endeavor. **3.** To be or become bound together by adhesion.

**United Church of Christ** *n.* A Protestant denomination founded in 1957 by a merger of the Congregational Christian Church and the Evangelical and Reformed Church.

**United Methodist Church** *n.* A Protestant church formed in 1968 by the union of the Methodist Church and the Evangelical United Brethren.

**United Nations** *pl.n. (sing. or pl. in number).* An international organization composed of most of the countries of the world, formed in 1945 to promote peace, security, and economic development.

**United Nations Trust Territory** *n.* A trust territory.

**United States** *pl.n. (sing. or pl. in number).* A federation of states, esp. one forming a nation within a specified geographical area <a proposal for a *United States* of Africa>

**u·ni·tive** (yōō'nĭ-tĭv, yōō-nī'-) *adj.* Tending to promote unity or serving to unite.

**u·nit·ize** (yōō'nĭ-tīz') *vt.* **-ized, -iz·ing, -iz·es.** To separate, classify, or package in discrete units.

**unit pricing** *n.* The pricing of goods, usu. foodstuff, on the basis of cost per unit of measure.

**unit rule** *n.* A rule in a Democratic Party national convention that a state's entire vote must go to the candidate preferred by the majority of that state's delegates.

**u·ni·ty** (yōō'nĭ-tē) *n., pl.* **-ties.** [ME *unite* < OFr. <Lat. *unitas* < *unus,* one.] **1.** The state of being one : SINGLENESS. **2.** The quality or state of accord or agreement : CONCORD. **3. a.** Combination or arrangement of parts into a whole : UNIFICATION. **b.** The unified entity thus formed. **4. a.** An ordering of all elements in a work of art or literature so that each contributes to a unified aesthetic effect. **b.** The effect thus produced. **5.** Singleness or constancy of purpose or action : CONTINUITY. **6.** *Math.* **a.** The number l. **b.** An element $I$ in a groupoid satisfying $x{\cdot}I = x = I{\cdot}x$ for each x in the groupoid. **7.** One of three principles of dramatic composition derived from Aristotle's *Poetics,* based on unity of time, action, and place, and stating that a drama should have but one plot, the action of which should be contained within one day and confined to one locality.

**u·ni·va·lent** (yōō'nĭ-vā'lənt) *adj. Chem.* **1.** Having valence 1. **2.** Having only one valence. —*n. Genetics.* An unpaired chromosome.

**u·ni·valve** (yōō'nĭ-vălv') *n.* **1.** A mollusk, esp. a gastropod, having a single shell. **2.** The shell of a univalve mollusk. **—u'ni·valve'** *adj.*

**u·ni·ver·sal** (yōō'nə-vûr'səl) *adj.* **1.** Including, extending to, or affecting the entire world or all within the world : WORLDWIDE <a *universal* drought><*universal* hunger> **2.** Relating to, involving, or affecting all the members of a class or group <the *universal* concerns of parenthood> **3.** Applicable or common to all uses, situations, or conditions <a *universal* language> **4.** Of or relating to the universe or cosmos : COSMIC. **5.** Comprehensively broad in subject matter. **6.** Adapted or adjustable to many sizes or mechanical uses. **7.** *Logic.* Predicable of all the members of a class or genus denoted by the subject. —Used of a proposition. —*n.* **1.** *Logic.* **a.** A universal proposition. **b.** A general or abstract concept or term considered absolute or axiomatic. **2.** A general or widely held principle, concept, or notion. **3.** A trait or pattern of behavior characteristic of all the members of a particular culture or of all human beings. **—u'ni·ver'sal·ly** *adv.* **—u'ni·ver'sal·ness** *n.*

**universal coupling** *n.* A universal joint.

**universal donor** *n.* One having blood type O.

**u·ni·ver·sal·ism** (yōō'nə-vûr'sə-lĭz'əm) *n.* **1. Universalism.** The theological doctrine of universal salvation. **2.** Universality.

**U·ni·ver·sal·ist** (yōō'nə-vûr'sə-lĭst) *n.* One who believes that salvation is extended to all humankind, esp. a member of a Christian denomination that adheres to this doctrine. **—U'ni·ver'sal·ist** *adj.*
**u·ni·ver·sal·i·ty** (yōō'nə-vər-săl'ĭ-tē) *n., pl.* **-ties. 1.** The quality, fact, or condition of being universal. **2.** Unbounded versatility of range or comprehension.
**u·ni·ver·sal·ize** (yōō'nə-vûr'sə-līz') *vt.* **-ized, -iz·ing, -iz·es.** To make universal : GENERALIZE. **—u'ni·ver'sal·i·za'tion** *n.*
**universal joint** *n.* A joint or coupling that allows parts of a machine not collinear with each other limited freedom of movement in any direction while transmitting rotary motion.

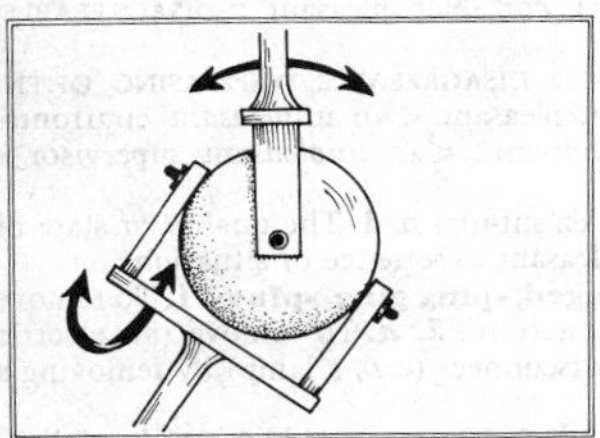
universal joint

**Universal Product Code** *n.* A series of vertical bars of varying widths printed on consumer product packages and used esp. for computerized inventory control.
**universal set** *n.* A mathematical set containing all elements of the variety under consideration.
**universal time** *n.* Greenwich mean time.
**u·ni·verse** (yōō'nə-vûrs') *n.* [ME < OFr. *univers* < Lat. *Universum,* neuter of *universus,* whole : *unus,* one + *versus,* p.part. of *vertere,* to turn.] **1.** All existing things, including the earth, the heavens, the galaxies, and all therein, regarded as a whole. **2. a.** The earth together with all its creatures. **b.** All humankind. **3.** A distinct sphere or realm, as of the imagination, that exists as an independent unit. **4.** *Logic.* The universe of discourse.
**universe of discourse** *n. Logic.* A class containing all the entities referred to in a discourse or argument.
**u·ni·ver·si·ty** (yōō'nə-vûr'sĭ-tē) *n., pl.* **-ties.** [ME *universite* < OFr. < Med. Lat. *universitas* < LLat., a society < Lat., the whole < *universus,* whole. —see UNIVERSE.] **1.** An institution of higher learning, having facilities for teaching and research and comprising an undergraduate division that awards bachelor's degrees and graduate and professional schools that award master's degrees and doctorates. **2.** The buildings and grounds of a university. **3.** The students and faculty of a university.
**u·ni·vo·cal** (yōō-nĭv'ə-kəl) *adj.* [LLat. *univocus* : Lat. *unus,* one + Lat. *vox,* voice.] Having only one meaning. **—u·niv'o·cal·ly** *adv.*
**un·just** (ŭn-jŭst') *adj.* **1.** Violating principles of justice or fairness : UNFAIR. **2.** *Archaic.* Faithless : dishonest. **—un·just'ly** *adv.* **—un·just'ness** *n.*
**un·kempt** (ŭn-kĕmpt') *adj.* [UN- + *kempt,* p.part. of dial. *kemb,* to comb < ME *kemben* < OE *cemban.*] **1. a.** Uncombed <*unkempt* hair> **b.** Not neat or orderly <an *unkempt* room> **2.** *Archaic.* Unpolished : rude.
**un·ken·nel** (ŭn-kĕn'əl) *vt.* **-neled, -nel·ing, -nels** *or* **-nelled, -nel·ling, -nels. 1. a.** To drive from a lair or den. **b.** To loose from a kennel. **2.** To bring out into the open : DISCLOSE.
**un·kind** (ŭn-kīnd') *adj.* **-er, -est. 1.** Lacking in kindness or consideration <an *unkind* criticism> **2.** Cruel : harsh <*unkind* seas> **—un·kind'ly** *adv.* **—un·kind'ness** *n.*
**un·kind·ly** (ŭn-kīnd'lē) *adj.* **-li·er, -li·est.** Unkind. *—adv.* In an unkind way. **—un·kind'li·ness** *n.*
**un·kink** (ŭn-kĭngk') *v.* **-kinked, -kink·ing, -kinks.** *—vt.* To free from kinks : make straight. *—vi.* To become relaxed.
**un·knit** (ŭn-nĭt') *v.* **-knit** *or* **-knit·ted, -knit·ting, -knits.** *—vt.* To unravel or undo (something knit or tied). *—vi.* To become undone.
**un·know·a·ble** (ŭn-nō'ə-bəl) *adj.* Transcending the limits of human experience or understanding. **—un·know'a·ble** *n.* **—un·know'a·ble·ness** *n.* **—un·know'a·bly** *adv.*
**un·know·ing** (ŭn-nō'ĭng) *adj.* Not knowing : UNAWARE. **—un·know'ing·ly** *adv.*
**un·known** (ŭn-nōn') *adj.* **1.** Not known : UNFAMILIAR <a place *unknown* to most people> **2. a.** Not disclosed or identified <a disease of *unknown* origin> **b.** Not determined or verified <a gem of *unknown* value> *—n.* **1.** One that is unknown. **2.** *Math.* **a.** A quantity of unknown numerical value. **b.** The symbol for this quantity.
**un·la·bored** (ŭn-lā'bərd) *adj.* **1.** Done with or requiring little effort : EFFORTLESS. **2.** Not tilled : UNCULTIVATED.

**un·lace** (ŭn-lās') *vt.* **-laced, -lac·ing, -lac·es. 1. a.** To loosen or undo the lacing or laces of. **b.** To remove or loosen the clothing of. **2.** *Obs.* To disgrace.
**un·lade** (ŭn-lād') *v.* **-lad·ed, -lad·ing, -lades.** *—vt.* **1.** To unload (cargo) from a ship. **2.** To unload (a ship). *—vi.* To discharge cargo.
**un·lash** (ŭn-lăsh') *vt.* **-lashed, -lash·ing, -lash·es.** To untie or loosen the lashing of.
**un·latch** (ŭn-lăch') *v.* **-latched, -latch·ing, -latch·es.** *—vt.* To unfasten or open by releasing a latch. *—vi.* To become unfastened or opened.
**un·law·ful** (ŭn-lô'fəl) *adj.* **1.** Not lawful : ILLEGAL. **2.** Illegitimate <*unlawful* offspring> **3.** Immoral : illicit. **—un·law'ful·ly** *adv.* **—un·law'ful·ness** *n.*
**un·lay** (ŭn-lā') *v.* **-laid** (-lād'), **-lay·ing, -lays.** *Naut. —vt.* To untwist the strands of (a rope). *—vi.* To untwist.
**un·lead** (ŭn-lĕd') *vt.* **-lead·ed, -lead·ing, -leads. 1.** To remove the lead from. **2.** To extricate the leads from between (lines of printing type).
**un·lead·ed** (ŭn-lĕd'ĭd) *adj.* **1.** Not containing lead <*unleaded* gasoline> **2.** Not spaced or separated with lead, as lines of printing type.
**un·learn** (ŭn-lûrn') *vt.* **-learned, -learn·ing, -learns. 1.** To put (something learned) out of the mind : FORGET. **2.** To get rid of the effect or habit of <vowed to *unlearn* smoking>
**un·learn·ed** (ŭn-lûr'nĭd) *adj.* **1.** Not educated : ignorant or illiterate. **2.** Not versed in a specified discipline <*unlearned* in Latin> **3.** (ŭn-lûrnd'). Not acquired by training or study <an *unlearned* response> **—un·learn'ed·ly** (-lûr'nĭd-lē) *adv.*
**un·leash** (ŭn-lēsh') *vt.* **-leashed, -leash·ing, -leash·es.** To release or loose from or as if from a leash <*unleashed* vicious dogs><*unleash* hydrogen bombs>
**un·leav·ened** (ŭn-lĕv'ənd) *adj.* Baked without leavening.
**un·less** (ŭn-lĕs') *conj.* [ME *unlesse,* alteration of *onlesse* : *on,* on + *lesse,* less.] Except on the condition that or under the circumstances that. *—prep.* Except for : EXCEPT.
**un·let·tered** (ŭn-lĕt'ərd) *adj.* **1. a.** Not educated. **b.** Illiterate. **2.** Devoid of lettering.
**un·li·censed** (ŭn-lī'sənst) *adj.* **1.** Having no license. **2.** Unauthorized. **3.** Unrestrained.
**un·like** (ŭn'līk') *adj.* **1.** Bearing no resemblance : not like. **2.** Not equal, as in strength. *—prep.* **1.** Different from <a book *unlike* any other> **2.** Not typical of <It's *unlike* them not to call.> **—un·like'ness** *n.*
**un·like·li·hood** (ŭn-līk'lē-hŏŏd') *n.* The state of being unlikely : IMPROBABILITY.
**un·like·ly** (ŭn-līk'lē) *adj.* **-li·er, -li·est. 1.** Not likely : IMPROBABLE. **2.** Likely to fail <an *unlikely* alliance> **—un·like'li·ness** *n.*
☆ **syns:** UNLIKELY, DOUBTFUL, IMPROBABLE, QUESTIONABLE *adj. core meaning* : showing little or no likelihood of happening or being true <an *unlikely* story><an *unlikely* alibi> ***ant:*** likely
**un·lim·ber** (ŭn-lĭm'bər) *v.* **-bered, -ber·ing, -bers.** *—vt.* **1.** To detach (a gun or caisson) from its limber. **2.** To make ready for action <We *unlimbered* out throats and sang.> *—vi.* To prepare for action.
**un·lim·it·ed** (ŭn-lĭm'ĭ-tĭd) *adj.* Having no limits, bounds, or qualifications. **—un·lim'it·ed·ly** *adv.* **—un·lim'it·ed·ness** *n.*
**un·link** (ŭn-lĭngk') *vt.* **-linked, -link·ing, -links.** To disconnect the links of : UNFASTEN. *—vi.* To become detached.
**un·list·ed** (ŭn-lĭs'tĭd) *adj.* **1.** Not appearing on a list <an *unlisted* phone number> **2.** Designating securities not listed on a stock exchange.
**un·live** (ŭn-lĭv') *vt.* **-lived, -liv·ing, -lives.** To live in such a manner as to undo the effects of : REVERSE.
**un·load** (ŭn-lōd') *v.* **-load·ed, -load·ing, -loads.** *—vt.* **1. a.** To remove the load or cargo from. **b.** To discharge (a load or cargo). **2. a.** To relieve of something oppressive : UNBURDEN. **b.** To give vent to : pour forth <*unloaded* our misgivings> **3.** To remove the charge from (a firearm). **4.** To dispose of, esp. by selling in great quantity : DUMP. *—vi.* To discharge a cargo or other burden. **—un·load'er** *n.*
**un·lock** (ŭn-lŏk') *v.* **-locked, -lock·ing, -locks.** *—vt.* **1. a.** To undo (a lock) by turning a key or a corresponding part. **b.** To undo the lock of. **2.** To make accessible : OPEN <*unlocked* the child's mind> **3.** To set free : RELEASE <*unlock* pent-up emotions> **4.** To provide a key to <*unlock* a mystery> *—vi.* To become unfastened, loosened, or freed from restraints or controls.
**un·looked-for** (ŭn-lŏŏkt'fôr') *adj.* Not looked for or expected : UNFORESEEN <*unlooked-for* good fortune>
**un·loose** (ŭn-lōōs') *vt.* **-loosed, -loos·ing, -loos·es.** [ME *unloosen* : *un-,* un- + *loosen,* to loosen < *loos,* loose.] **1.** To let loose or set free : RELEASE <*unloose* one's feelings> **2.** To relax, as a hold on something <*unloose* one's grip on a rope>
**un·loos·en** (ŭn-lōō'sən) *vt.* **-ened, -en·ing, -ens.** To unloose.
**un·love·ly** (ŭn-lŭv'lē) *adj.* **-li·er, -li·est.** Disagreeable : unpleasant.
**un·luck·y** (ŭn-lŭk'ē) *adj.* **-i·er, -i·est. 1.** Subjected to or marked by misfortune. **2.** Forecasting bad luck : INAUSPICIOUS. **3.** Not producing the desired outcome : DISAPPOINTING <an *unlucky* attempt at humor> **—un·luck'i·ly** *adv.* **—un·luck'i·ness** *n.*

**un·make** (ŭn-māk′) *vt.* **-made** (-mād′), **-mak·ing, -makes. 1.** To deprive of position, rank, or authority : DEPOSE. **2.** To undo the making of : DESTROY. **3.** To alter the characteristics or nature of.

**un·man** (ŭn-măn′) *vt.* **-manned, -man·ning, -mans. 1.** To cause to lose courage. **2.** To deprive of virility : EMASCULATE.

**un·man·ly** (ŭn-măn′lē) *adj.* **-li·er, -li·est. 1. a.** Dishonorable : degrading. **b.** Cowardly. **2.** Effeminate. **—un·man′li·ness** *n.*

**un·manned** (ŭn-mănd′) *adj.* **1.** Having no crew <an *unmanned* space mission> **2.** *Obs.* Untrained. —Used of a hawk.

**un·man·nered** (ŭn-măn′ərd) *adj.* **1.** Lacking in good manners : RUDE. **2.** Not affected or artificial : NATURAL.

**un·man·ner·ly** (ŭn-măn′ər-lē) *adj.* UNMANNERED 1. **—un·man′ner·li·ness** *n.*

**un·mar·ried** (ŭn-măr′ēd) *adj.* Not married.

**un·marked** (ŭn-märkt′) *adj.* **1.** Not bearing a mark. **2.** Not observed or noticed <Our departure was *unmarked.*>

**un·mask** (ŭn-măsk′, -mäsk′) *v.* **-masked, -mask·ing, -masks.** —*vt.* **1.** To remove a mask from. **2.** To disclose the true character of : EXPOSE. —*vi.* To remove one's mask.

**un·mean·ing** (ŭn-mē′nĭng) *adj.* **1.** Having no meaning : SENSELESS. **2.** Showing no expression : VACANT. **—un·mean′ing·ly** *adv.*

**un·meant** (ŭn-mĕnt′) *adj.* Not meant or intentional.

**un·meet** (ŭn-mēt′) *adj.* Improper : unseemly.

**un·men·tion·a·ble** (ŭn-mĕn′shə-nə-bəl) *adj.* **1.** Not fit to be mentioned. **2.** Unspeakable. —*n.* **1.** One that is not to be or is not fit to be mentioned. **2. unmentionables.** Underwear. **—un·men′tion·a·ble·ness** *n.* **—un·men′tion·a·bly** *adv.*

**un·mer·ci·ful** (ŭn-mûr′sĭ-fəl) *adj.* **1.** Having no mercy. **2.** Excessive <*unmerciful* folly> **—un·mer′ci·ful·ly** *adv.* **—un·mer′ci·ful·ness** *n.*

**un·mind·ful** (ŭn-mīnd′fəl) *adj.* Inattentive or oblivious <*unmindful* of the late hour> **—un·mind′ful·ly** *adv.* **—un·mind′ful·ness** *n.*

**un·mis·tak·a·ble** (ŭn′mĭ-stā′kə-bəl) *adj.* Obvious : evident <*unmistakable* signs of life> **—un′mis·tak′a·bly** *adv.*

**un·mit·i·gat·ed** (ŭn-mĭt′ĭ-gā′tĭd) *adj.* **1.** Not eased or lessened : UNRELIEVED <*unmitigated* grief> **2.** Unqualified : absolute <an *unmitigated* scoundrel> **—un·mit′i·gat′ed·ly** *adv.*

**un·mod·u·lat·ed** (ŭn-mŏj′ə-lā′tĭd) *adj.* Not modulated.

**un·moor** (ŭn-mŏŏr′) *v.* **-moored, -moor·ing, -moors.** —*vt.* **1.** To release from or as if from moorings. **2.** *Naut.* To release (a ship) from all but one anchor. —*vi. Naut.* To cast off moorings.

**un·mor·al** (ŭn-môr′əl, -mŏr′əl) *adj.* Having no moral quality or characteristic : AMORAL. **—un·mor′al·ly** *adv.*

**un·mor·tise** (ŭn-môr′tĭs) *vt.* **-tised, -tis·ing, -tis·es.** To loosen or separate (something mortised).

**un·muf·fle** (ŭn-mŭf′əl) *v.* **-fled, -fling, -fles.** —*vt.* To free from something that muffles. —*vi.* To cast off something that muffles.

**un·my·e·lin·at·ed** (ŭn-mī′ə-lĭ-nā′tĭd) *adj.* Lacking a myelin sheath.

**un·nat·u·ral** (ŭn-năch′ər-əl) *adj.* **1.** Not conforming to natural law. **2.** Inconsistent with the usual pattern or custom. **3.** Deviating from the accepted or natural modes of behavior <an *unnatural* parent> **4.** Contrived or constrained : ARTIFICIAL <an *unnatural* politeness> **5.** Outrageously contrary to natural feelings : INHUMAN. **—un·nat′u·ral·ly** *adv.* **—un·nat′u·ral·ness** *n.*

**un·nec·es·sar·y** (ŭn-nĕs′ĭ-sĕr′ē) *adj.* Not necessary : NEEDLESS. **—un·nec′es·sar′i·ly** (-sâr′ə-lē) *adv.*

**un·nerve** (ŭn-nûrv′) *vt.* **-nerved, -nerv·ing, -nerves.** To deprive of courage, vigor, or composure : make weak or ineffectual.

**un·no·tice·a·ble** (ŭn-nō′tĭ-sə-bəl) *adj.* Not readily noticeable.

**un·num·bered** (ŭn-nŭm′bərd) *adj.* **1.** Innumerable : countless. **2.** Not marked with an identifying number.

**un·ob·tru·sive** (ŭn′əb-trōō′sĭv) *adj.* Unnoticeable : inconspicuous. **—un′ob·tru′sive·ly** *adv.* **—un′ob·tru′sive·ness** *n.*

**un·oc·cu·pied** (ŭn-ŏk′yə-pīd′) *adj.* **1.** Having no occupants or inhabitants : VACANT. **2.** Unemployed : idle.

**un·of·fi·cial** (ŭn′ə-fĭsh′əl) *adj.* **1.** Not official. **2.** Not acting officially. **—un′of·fi′cial·ly** *adv.*

**un·or·gan·ized** (ŭn-ôr′gə-nīzd′) *adj.* **1.** Lacking order, system, or unity. **2.** Having no organic qualities : INORGANIC. **3.** Not unionized.

**un·o·rig·i·nal** (ŭn′ə-rĭj′ə-nəl) *adj.* Lacking originality : TRITE.

**un·or·tho·dox** (ŭn-ôr′thə-dŏks′) *adj.* Breaking with convention or tradition. **—un·or′tho·dox′ly** *adv.* **—un·or′tho·dox′y** *n.*

**un·pack** (ŭn-păk′) *v.* **-packed, -pack·ing, -packs.** —*vt.* **1.** To remove the contents of <*unpack* a suitcase> **2.** To remove from a container or from packaging <*unpack* a gift> **3.** To remove a pack from (a pack animal). —*vi.* To unpack objects from a container.

**un·paged** (ŭn-pājd′) *adj.* Having no page numbers.

**un·paid** (ŭn-pād′) *adj.* **1.** Not paid <an *unpaid* bill> **2.** Not salaried <an *unpaid* position>

**un·pal·at·a·ble** (ŭn-păl′ə-tə-bəl) *adj.* **1.** Unpleasant to the taste. **2.** Disagreeable : distasteful <*unpalatable* facts> **—un′pal·at·a·bil′i·ty** *n.*

☆ ***syns:*** UNPALATABLE, DISTASTEFUL, UNAPPETIZING, UNSAVORY *adj. core meaning* : so unpleasant in flavor as to be inedible <a tough, *unpalatable* steak> ***ant:*** palatable

**un·par·al·leled** (ŭn-păr′ə-lĕld′) *adj.* Unequaled.

**un·par·lia·men·ta·ry** (ŭn′pär-lə-mĕn′tə-rē, -mĕn′trē) *adj.* Contrary to parliamentary procedure.

**un·peg** (ŭn-pĕg′) *vt.* **-pegged, -peg·ging, -pegs. 1.** To remove a peg from. **2.** To unfasten.

**un·peo·ple** (ŭn-pē′pəl) *vt.* **-pled, -pling, -ples.** To depopulate.

**un·peo·pled** (ŭn-pē′pəld) *adj.* Uninhabited.

**un·per·fo·rat·ed** (ŭn-pûr′fə-rā′tĭd) *adj.* Imperforate.

**un·per·son** (ŭn′pûr′sən) *n.* A nonperson.

**un·pick** (ŭn-pĭk′) *vt.* **-picked, -pick·ing, -picks.** To undo (sewing) by removing stitches <*unpick* a seam>

**un·pin** (ŭn-pĭn′) *vt.* **-pinned, -pin·ning, -pins. 1.** To remove a pin from. **2.** To unfasten or free by or as if by removing pins. **b.** To free.

**un·pleas·ant** (ŭn-plĕz′ənt) *adj.* Not pleasant : DISAGREEABLE. **—un·pleas′ant·ly** *adv.*

☆ ***syns:*** UNPLEASANT, BAD, DISAGREEABLE, DISPLEASING, OFFENSIVE *adj. core meaning* : not pleasant <an *unpleasant* confrontation><an *unpleasant* assignment><an *unpleasant* supervisor> ***ant:*** pleasant

**un·pleas·ant·ness** (ŭn-plĕz′ənt-nĭs) *n.* **1.** The quality or state of being unpleasant. **2.** An unpleasant experience or situation.

**un·plug** (ŭn-plŭg′) *vt.* **-plugged, -plug·ging, -plugs. 1.** To remove a plug, stopper, or obstruction from. **2. a.** To remove (an electric plug) from an outlet. **b.** To disconnect (e.g., a lamp) by removing a plug from an outlet.

**un·plumbed** (ŭn-plŭmd′) *adj.* **1.** Not measured with a plumb line. **2.** Not explored in depth or meaning.

**un·po·lit·i·cal** (ŭn′pə-lĭt′ĭ-kəl) *adj.* Not political.

**un·polled** (ŭn-pōld′) *adj.* **1.** Not interviewed in a poll. **2.** Not registered at the polls.

**un·pop·u·lar** (ŭn-pŏp′yə-lər) *adj.* Lacking general approval or acceptance. **—un′pop·u·lar′i·ty** (-lăr′ĭ-tē) *n.*

**un·prac·ticed** (ŭn-prăk′tĭst) *adj.* **1.** Not yet tried or familiar. **2.** Not experienced or skilled.

**un·prec·e·dent·ed** (ŭn′prĕs′ĭ-dĕn′tĭd) *adj.* Being without precedent : EXAMPLED. **—un·prec′e·dent′ed·ly** *adv.*

**un·pre·dict·a·ble** (ŭn′prĭ-dĭk′tə-bəl) *adj.* Not predictable. **—un′pre·dict′a·bil′i·ty** *n.* **—un′pre·dict′a·bly** *adv.*

**un·prej·u·diced** (ŭn-prĕj′ə-dĭst) *adj.* Free from prejudice : IMPARTIAL <an *unprejudiced* opinion>

**un·pre·med·i·tat·ed** (ŭn′prĭ-mĕd′ĭ-tā′tĭd) *adj.* Not premeditated. **—un′pre·med′i·tat′ed·ly** *adv.*

**un·pre·pared** (ŭn-prĭ-pârd′) *adj.* **1.** Not prepared or equipped, as for war. **2.** Having no previous preparation : IMPROMPTU <an *unprepared* lecture> **—un′pre·par′ed·ly** (-pâr′ĭd-lē) *adv.* **—un′pre·par′ed·ness** (-pâr′ĭd-nĭs, -pârd′nĭs) *n.*

**un·pre·pos·sess·ing** (ŭn′prē-pə-zĕs′ĭng) *adj.* Failing to impress favorably : NONDESCRIPT. **—un′pre·pos·sess′ing·ly** *adv.*

**un·pre·tend·ing** (ŭn′prĭ-tĕn′dĭng) *adj.* Unpretentious.

**un·pre·ten·tious** (ŭn′prĭ-tĕn′shəs) *adj.* Lacking pretention or affectation : MODEST. **—un′pre·ten′tious·ly** *adv.* **—un′pre·ten′tious·ness** *n.*

**un·priced** (ŭn-prīst′) *adj.* Having no price assigned.

**un·prin·ci·pled** (ŭn-prĭn′sə-pəld) *adj.* Lacking moral principles : UNSCRUPULOUS <*unprincipled* actions>

**un·print·a·ble** (ŭn-prĭn′tə-bəl) *adj.* Improper or unfit for publication, as for legal or social reasons.

**un·pro·duc·tive** (ŭn′prə-dŭk′tĭv) *adj.* **1.** Not productive. **2.** Adding nothing to exchangeable value <*unproductive* consumption of fuel> **—un′pro·duc′tive·ly** *adv.* **—un′pro·duc′tive·ness** *n.*

**un·pro·fes·sion·al** (ŭn′prə-fĕsh′ə-nəl) *adj.* **1.** Not belonging to or connected with a profession. **b.** Not a qualified member of a professional group. **2.** Not conforming to the standards of a profession <*unprofessional* conduct> **3.** Lacking professional competence : AMATEURISH. **—un′pro·fes′sion·al·ly** *adv.*

**un·prof·it·a·ble** (ŭn-prŏf′ĭ-tə-bəl) *adj.* **1.** Not profitable. **2.** Serving no purpose : USELESS.

**un·pro·nounce·a·ble** (ŭn′prə-noun′sə-bəl) *adj.* **1.** Difficult to pronounce correctly. **2.** Unfit to be mentioned.

**un·pro·vid·ed** (ŭn′prə-vī′dĭd) *adj.* Not supplied, furnished, or equipped. **—un′pro·vid′ed·ly** *adv.*

**un·pro·voked** (ŭn′prə-vōkt′) *adj.* Not provoked or prompted <an *unprovoked* assault>

**un·qual·i·fied** (ŭn-kwŏl′ə-fīd′) *adj.* **1.** Lacking the proper or required qualifications <an *unqualified* tutor> **2.** Not modified by reservations or restrictions : ABSOLUTE <an *unqualified* success>

**un·ques·tion·a·ble** (ŭn-kwĕs′chə-nə-bəl) *adj.* Not open to doubt or dispute : CERTAIN. **—un·ques′tion·a·bil′i·ty, un·ques′tion·a·ble·ness** *n.* **—un·ques′tion·a·bly** *adv.*

**un·ques·tioned** (ŭn-kwĕs′chənd) *adj.* **1.** Not subjected to questioning or inquiry. **2.** Not able to be questioned or disputed : INDISPUTABLE. **3.** Not doubted.

**un·qui·et** (ŭn-kwī′ĭt) *adj.* **-er, -est. 1.** Mentally or emotionally uneasy : ANXIOUS. **2.** Marked by unrest or disorder : TURBULENT. **—un·qui′et·ly** *adv.* **—un·qui′et·ness** *n.*

**un·quote** (ŭn-kwōt′) *n.* —Used by a speaker to indicate the end of a quotation.
**un·rav·el** (ŭn-răv′əl) *v.* **-eled, -el·ing, -els** *or* **-elled, -el·ling, -els.** —*vt.* **1. a.** To undo the knitted or woven fabric of. **b.** To separate (entangled threads). **2.** To separate and clarify the elements of (something intricate or obscure) <*unravel* a plot> —*vi.* To become unraveled.
**un·read** (ŭn-rĕd′) *adj.* **1.** Not read or studied. **2.** Having read little : ignorant or unlearned.
**un·read·a·ble** (ŭn-rē′də-bəl) *adj.* **1.** Not legible. **2.** Too dull to be worth reading. **3.** Too obscure to understand : INCOMPREHENSIBLE. **4.** Unsuitable for reading. **—un·read′a·bil′i·ty** *n.*
**un·read·y** (ŭn-rĕd′ē) *adj.* **-i·er, -i·est. 1.** Not ready or prepared, as for use. **2.** Slow to see or respond : not prompt or alert. **—un·read′i·ly** *adv.* **—un·read′i·ness** *n.*
**un·re·al** (ŭn-rē′əl, -rēl′) *adj.* **1.** Not real or substantial : ILLUSORY. **2.** *Slang.* Excellent : fantastic. **3.** *Slang.* Unbelievable <Their behavior was *unreal.*>
**un·re·al·i·ty** (ŭn′rē-ăl′ĭ-tē) *n., pl.* **-ties. 1.** The quality or state of being unreal. **2.** Something unreal, insubstantial, or imaginary. **3.** Incompetence in dealing with reality : IMPRACTICALITY.
**un·rea·son** (ŭn-rē′zən) *n.* **1.** Lack of reason or sanity : IRRATIONALITY. **2.** Nonsense : absurdity.
**un·rea·son·a·ble** (ŭn-rē′zə-nə-bəl) *adj.* **1.** Not governed by reason. **2.** Going beyond reasonable limits : IMMODERATE <*unreasonable* demands> **—un·rea′son·a·ble·ness** *n.* **—un·rea′son·a·bly** *adv.*
**un·rea·son·ing** (ŭn-rē′zə-nĭng) *adj.* Not governed by reason or good judgment <*unreasoning* terror> **—un·rea′son·ing·ly** *adv.*
**un·reck·on·a·ble** (ŭn-rĕk′ə-nə-bəl) *adj.* Incalculable.
**un·re·con·struct·ed** (ŭn′rē-kən-strŭk′tĭd) *adj.* Not reconciled or receptive to social and economic change.
**un·reel** (ŭn-rēl′) *v.* **-reeled, -reel·ing, -reels.** —*vt.* To unwind from or as if from a reel. —*vi.* To unwind.
**un·reeve** (ŭn-rēv′) *v.* **-reeved** *or* **-rove** (-rōv′), **-reeved** *or* **-ro·ven** (-rō′vən), **-reev·ing, -reeves.** *Naut.* —*vt.* To withdraw (e.g., a rope) from an opening such as a block or thimble. —*vi.* **1.** To become unreeved. **2.** To unreeve a rope.
**un·re·flec·tive** (ŭn′rĭ-flĕk′tĭv) *adj.* Not reflective : HEEDLESS. **—un′re·flec′tive·ly** *adv.*
**un·re·gen·er·ate** (ŭn′rĭ-jĕn′ər-ĭt) *adj.* **1.** Not regenerated or repentant. **2.** Not reformed or converted : UNRECONSTRUCTED. **3.** Stubborn. **—un′re·gen′er·a·cy** (-ə-sē) *n.* **—un′re·gen′er·ate·ly** *adv.*
**un·re·hearsed** (ŭn′rĭ-hûrst′) *adj.* Not rehearsed.
**un·re·lent·ing** (ŭn′rĭ-lĕn′tĭng) *adj.* **1.** Not yielding : INEXORABLE. **2.** Not diminishing in intensity, speed, or effort <*unrelenting* pursuit of money and power>
**un·re·li·a·ble** (ŭn′rĭ-lī′ə-bəl) *adj.* Not reliable. **—un′re·li′a·bil′i·ty, un′re·li′a·ble·ness** *n.* **—un′re·li′a·bly** *adv.*
☆ **syns:** UNRELIABLE, UNDEPENDABLE, UNTRUSTWORTHY *adj. core meaning* : not to be depended on <*unreliable* workers> <an *unreliable* car> **ant:** dependable, reliable, trustworthy
**un·re·li·gious** (ŭn′rĭ-lĭj′əs) *adj.* **1.** Irreligious. **2.** Having no connection with religion.
**un·re·mark·a·ble** (ŭn′rĭ-mär′kə-bəl) *adj.* Lacking distinction : ORDINARY. **—un′re·mark′a·bly** *adv.*
**un·re·marked** (ŭn′rĭ-märkt′) *adj.* Not noticed.
**un·re·mit·ting** (ŭn′rĭ-mĭt′ĭng) *adj.* Never slackening : INCESSANT. **—un′re·mit′ting·ly** *adv.* **—un′re·mit′ting·ness** *n.*
**un·re·proved** (ŭn′rĭ-prōōvd′) *adj.* Not censured <an *unreproved* misdeed>
**un·re·serve** (ŭn′rĭ-zûrv′) *n.* Frankness of manner : CANDOR.
**un·re·served** (ŭn′rĭ-zûrvd′) *adj.* **1.** Not set aside for a particular person or use <*unreserved* seating> **2.** Given without reservation : UNQUALIFIED <*unreserved* admiration> **3.** Not reserved in manner : CANDID. **—un′re·serv′ed·ly** (-zûr′vĭd-lē) *adv.*
**un·re·spon·sive** (ŭn′rĭ-spŏn′sĭv) *adj.* Not responsive. **—un′re·spon′sive·ly** *adv.* **—un′re·spon′sive·ness** *n.*
**un·rest** (ŭn-rĕst′) *n.* Uneasiness : disquiet <social *unrest*>
**un·re·strained** (ŭn′rĭ-strānd′) *adj.* **1. a.** Unchecked <*unrestrained* growth> **b.** Not given to restraint. **2.** Not constrained : NATURAL. **—un′re·strain′ed·ly** (-strā′nĭd-lē) *adv.*
**un·rid·dle** (ŭn-rĭd′l) *vt.* **-dled, -dling, -dles.** To solve the riddle of : SOLVE.
**un·ri·fled** (ŭn-rī′fəld) *adj.* Having a smooth bore, as a gun.
**un·rig** (ŭn-rĭg′) *vt.* **-rigged, -rig·ging, -rigs.** *Naut.* To strip (a vessel) of rigging.
**un·right·eous** (ŭn-rī′chəs) *adj.* **1.** Not righteous : WICKED <an *unrighteous* person> **2.** Not right or fair : UNJUST <an *unrighteous* law> **—un·right′eous·ly** *adv.* **—un·right′eous·ness** *n.*
**un·rip** (ŭn-rĭp′) *vt.* **-ripped, -rip·ping, -rips.** To rip open.
**un·ripe** (ŭn-rīp′) *adj.* **-rip·er, -rip·est. 1.** Not ripe : IMMATURE. **2.** Not yet ready or developed. **—un·ripe′ness** *n.*
**un·ri·valed** (ŭn-rī′vəld) *adj.* Unequaled : peerless.
**un·roll** (ŭn-rōl′) *v.* **-rolled, -roll·ing, -rolls.** —*vt.* **1.** To unwind and open out (something rolled up). **2.** To unfold : reveal. —*vi.* To become unrolled.
**un·root** (ŭn-rōōt′, -rŏŏt′) *vt.* **-root·ed, -root·ing, -roots.** To uproot <*unroot* a tree>
**un·round** (ŭn-round′) *vt.* **-round·ed, -round·ing, -rounds.** To pronounce (a sound) with the lips in a flattened or neutral position.
**un·ruf·fled** (ŭn-rŭf′əld) *adj.* **1.** Not agitated or upset : CALM. **2.** Not ruffled : SMOOTH.
**un·ru·ly** (ŭn-rōō′lē) *adj.* **-li·er, -li·est.** [ME *unreuly* : *un-*, un- + *reuly*, easy to govern < *reule*, rule.] Difficult or impossible to govern or control. **—un·ru′li·ness** *n.*
**un·sad·dle** (ŭn-săd′l) *v.* **-dled, -dling, -dles.** —*vt.* **1.** To remove the saddle from. **2.** To throw from the saddle : UNHORSE. —*vi.* To remove the saddle from a horse.
**un·safe** (ŭn-sāf′) *adj.* Not safe : DANGEROUS.
**un·said** (ŭn-sĕd′) *v. p.t. & p.p. of* UNSAY.
**un·san·i·tar·y** (ŭn-săn′ĭ-tĕr′ē) *adj.* Not sanitary.
**un·sat·is·fac·to·ry** (ŭn-săt′ĭs-făk′tə-rē) *adj.* Not satisfactory. **—un·sat′is·fac′to·ri·ly** *adv.* **—un·sat′is·fac′to·ri·ness** *n.*
**un·sat·u·rate** (ŭn-săch′ə-rĭt) *n.* An unsaturated compound.
**un·sat·u·rat·ed** (ŭn-săch′ə-rā′tĭd) *adj.* **1.** Of or relating to a compound, esp. of carbon, containing atoms that share more than one valence bond. **2.** Capable of dissolving more of a solute at a given temperature.
**un·saved** (ŭn-sāvd′) *adj.* Not saved, esp. not redeemed from sin.
**un·sa·vor·y** (ŭn-sā′və-rē) *adj.* **1. a.** Not savory : TASTELESS. **b.** Unpleasant in taste or smell. **2.** Distasteful or disagreeable <an *unsavory* task> **3.** Morally offensive <a legislator with an *unsavory* past> **—un·sa′vor·i·ly** *adv.* **—un·sa′vor·i·ness** *n.*
**un·say** (ŭn-sā′) *vt.* **-said** (-sĕd′), **-say·ing, -says.** To retract (something said).
**un·scathed** (ŭn-skāthd′) *adj.* Unharmed : uninjured.
**un·schooled** (ŭn-skōōld′) *adj.* **1.** Not schooled or trained. **2.** Not the result of training : NATURAL <*unschooled* dexterity>
**un·sci·en·tif·ic** (ŭn′sī-ən-tĭf′ĭk) *adj.* **1.** Not conforming to the principles or requirements of science. **2.** Not knowledgeable about science or scientific method. **—un′sci·en·tif′i·cal·ly** *adv.*
**un·scram·ble** (ŭn-skrăm′bəl) *vt.* **-bled, -bling, -bles. 1.** To disentangle : STRAIGHTEN OUT. **2.** To restore (a scrambled message) to intelligible form. **—un·scram′bler** *n.*
**un·screw** (ŭn-skrōō′) *v.* **-screwed, -screw·ing, -screws.** —*vt.* **1.** To take out the screw or screws from. **2.** To loosen, remove, or detach by rotating. —*vi.* To become or allow to become unscrewed.
**un·scru·pu·lous** (ŭn-skrōō′pyə-ləs) *adj.* Lacking a sense of what is right or honorable : devoid of scruples. **—un·scru′pu·lous·ly** *adv.* **—un·scru′pu·lous·ness** *n.*
**un·seal** (ŭn-sēl′) *vt.* **-sealed, -seal·ing, -seals.** To break or remove the seal of : OPEN <*unseal* a letter>
**un·seam** (ŭn-sēm′) *vt.* **-seamed, -seam·ing, -seams.** To undo the seam or seams of <*unseam* a dress>
**un·search·a·ble** (ŭn-sûr′chə-bəl) *adj.* Not capable of being explored or investigated : INSCRUTABLE.
**un·sea·son·a·ble** (ŭn-sē′zə-nə-bəl) *adj.* **1.** Not suitable to or usual for the season <*unseasonable* attire> **2.** Not characteristic of the time of year <*unseasonable* weather> **3.** Poorly timed : INOPPORTUNE <an *unseasonable* request> **—un·sea′son·a·ble·ness** *n.* **—un·sea′son·a·bly** *adv.*
**un·sea·soned** (ŭn-sē′zənd) *adj.* **1.** Not made savory with seasoning. **2.** Inadequately aged or seasoned <*unseasoned* wood> **3.** Inexperienced <an *unseasoned* teacher>
**un·seat** (ŭn-sēt′) *vt.* **-seat·ed, -seat·ing, -seats. 1.** To remove from a seat, esp. from a saddle. **2.** To dislodge from a position or office <*unseated* our senator in the last election>
**un·seem·ly** (ŭn-sēm′lē) *adj.* **-li·er, -li·est.** Not conforming to what is proper or in good taste : INDECOROUS. **—un·seem′li·ness** *n.* **—un·seem′ly** *adv.*
**un·seen** (ŭn-sēn′) *adj.* Not directly evident : INVISIBLE.
**un·seg·re·gat·ed** (ŭn-sĕg′rĭ-gā′tĭd) *adj.* Not segregated, esp. not racially segregated.
**un·se·lect·ed** (ŭn′sĭ-lĕk′tĭd) *adj.* **1.** Chosen randomly.
**un·se·lec·tive** (ŭn′sĭ-lĕk′tĭv) *adj.* **1.** Not selective : INDISCRIMINATE. **2.** Marked by random selection.
**un·sel·fish** (ŭn-sĕl′fĭsh) *adj.* Not selfish : GENEROUS. **—un·sel′fish·ly** *adv.* **—un·sel′fish·ness** *n.*
**un·set** (ŭn′sĕt′) *adj.* **1.** Not yet firm, stiff, or solidified <*unset* gelatin> **2.** Unmounted in a setting <an *unset* diamond>
**un·set·tle** (ŭn-sĕt′l) *v.* **-tled, -tling, -tles.** —*vt.* **1.** To displace from a settled condition : DISRUPT. **2.** To make uneasy : DISTURB. —*vi.* To become unsettled. **—un·set′tle·ment** *n.*
**un·set·tled** (ŭn-sĕt′ld) *adj.* **1.** Not orderly or stable : DISORDERED <*unsettled* times> **2.** Liable to vary : UNCERTAIN <*unsettled* weather> **3.** Not determined or resolved <an *unsettled* case> **4.** Not paid or adjusted <an *unsettled* account> **5.** Unpopulated. **6.** Not fixed or established. **—un·set′tled·ness** *n.*
**un·sex** (ŭn-sĕks′) *vt.* **-sexed, -sex·ing, -sex·es. 1.** To deprive of the appropriate sexual attributes. **2.** To render impotent : CASTRATE.

**un·shack·le** (ŭn-shăk′əl) *vt.* **-led, -ling, -les.** To free from shackles or restraints.

**un·shak·a·ble** (ŭn-shā′kə-bəl) *adj.* Not capable of being shaken : FIRM <*unshakable* faith> **—un·shak′a·bly** *adv.*

**un·shaped** (ŭn-shāpt′) *adj.* **1.** Not shaped or formed. **2.** Imperfectly shaped or formed.

**un·shap·en** (ŭn-shā′pən) *adj.* Unshaped.

**un·sheathe** (ŭn-shēth′) *vt.* **-sheathed, -sheath·ing, -sheathes.** To draw from or as if from a sheath or scabbard.

**un·shell** (ŭn-shĕl′) *vt.* **-shelled, -shell·ing, -shells.** To remove from or as if from a shell.

**un·ship** (ŭn-shĭp′) *v.* **-shipped, -ship·ping, -ships.** *—vt.* **1.** To unload from a ship : DISCHARGE. **2.** To remove (e.g., an oar or tiller) from its proper place. *—vi.* To become detached or removable.

**un·shod** (ŭn-shŏd′) *adj.* Not having or wearing shoes.

**un·sight·ed** (ŭn-sī′tĭd) *adj.* **1.** Not seen or examined. **2.** Not equipped with a sight for aiming. **3.** Not able to see.

**un·sight·ly** (ŭn-sīt′lē) *adj.* **-li·er, -li·est.** Unpleasant or offensive to look at <an *unsightly* rash> **—un·sight′li·ness** *n.*

**un·skilled** (ŭn-skĭld′) *adj.* **1.** Lacking skill or technical training <*unskilled* labor> **2.** Requiring no skill or training <*unskilled* jobs> **3.** Showing no skill : CRUDE <*unskilled* drawings>

**un·skill·ful** (ŭn-skĭl′fəl) *adj.* **1.** Lacking in skill or proficiency : INEXPERT <*unskillful* painters> **2.** *Obs.* Ignorant. **—un·skill′ful·ly** *adv.* **—un·skill′ful·ness** *n.*

**un·sling** (ŭn-slĭng′) *vt.* **-slung** (-slŭng′), **-sling·ing, -slings. 1.** To remove from a sling or a slung position. **2.** *Naut.* To remove the slings of (e.g., a yard).

**un·snap** (ŭn-snăp′) *vt.* **-snapped, -snap·ping, -snaps.** To undo the snaps of : UNFASTEN.

**un·snarl** (ŭn-snärl′) *vt.* **-snarled, -snarl·ing, -snarls.** To free of snarls : DISENTANGLE.

**un·so·cia·ble** (ŭn′sō′shə-bəl) *adj.* **1.** Not inclined to seek the company of others : RESERVED. **2.** Not conducive to social exchange <an *unsociable* atmosphere> **—un·so′cia·bil′i·ty, un·so′cia·ble·ness** *n.* **—un·so′cia·bly** *adv.*

**un·so·cial** (ŭn-sō′shəl) *adj.* Unsociable. **—un·so′cial·ly** *adv.*

**un·so·phis·ti·cat·ed** (ŭn′sə-fĭs′tĭ-kā′tĭd) *adj.* **1.** Not complex : SIMPLE. **2.** Free of artificiality : ARTLESS. **3.** Not changed or adulterated : PURE. **—un′so·phis′ti·cat′ed·ly** *adv.* **—un′so·phis′ti·cat′ed·ness** *n.* **—un′so·phis′ti·ca′tion** *n.*

**un·sought** (ŭn-sôt′, ŭn′sôt′) *adj.* Not looked for or requested.

**un·sound** (ŭn-sound′) *adj.* **-er, -est. 1.** Not dependably strong or solid. **2.** Not physically or mentally healthy. **3.** Not true or logically valid : FALLACIOUS. **—un·sound′ly** *adv.* **—un·sound′ness** *n.*

**un·spar·ing** (ŭn-spâr′ĭng) *adj.* **1.** Not sparing or frugal. **2.** Unmerciful : severe. **—un·spar′ing·ly** *adv.* **—un·spar′ing·ness** *n.*

**un·speak** (ŭn-spēk′) *vt.* **-spoke** (-spōk′), **-spo·ken** (-spō′kən), **-speak·ing, -speaks.** *Obs.* To retract : unsay.

**un·speak·a·ble** (ŭn-spē′kə-bəl) *adj.* **1.** Incapable of being expressed or described. **2.** Inexpressibly bad or evil. **3.** Not fit to be spoken. **—un·speak′a·ble·ness** *n.* **—un·speak′a·bly** *adv.*

☆ ***syns:*** **1.** UNSPEAKABLE, INDESCRIBABLE, INEXPRESSIBLE, UNUTTERABLE *adj. core meaning* : that cannot be described <*unspeakable* happiness> **2.** UNSPEAKABLE, ABOMINABLE, FRIGHTFUL, REVOLTING, SHOCKING, SICKENING *adj. core meaning* : too awful to be described <*unspeakable* atrocities> **3.** UNSPEAKABLE, UNMENTIONABLE, UNUTTERABLE *adj. core meaning* : unfit to be spoken or mentioned <*unspeakable* words of abuse>

**un·spe·cial·ized** (ŭn-spĕsh′ə-līzd′) *adj.* Having no special function : without specialty or specialization.

**un·sphere** (ŭn-sfîr′) *vt.* **-sphered, -spher·ing, -spheres.** To remove from its sphere, as a planet.

**un·spot·ted** (ŭn-spŏt′ĭd) *adj.* **1.** Not spotted. **2.** Morally unblemished. **—un·spot′ted·ness** *n.*

**un·sta·ble** (ŭn-stā′bəl) *adj.* **-bler, -blest. 1. a.** Tending strongly to change or fall <an *unstable* government> **b.** Not constant : FLUCTUATING. **2. a.** Of fickle temperament : FLIGHTY. **b.** Emotionally maladjusted. **3.** Not firmly fixed : UNSTEADY <*unstable* ground> **4.** *Chem.* **a.** Decomposing readily. **b.** Highly or violently reactive. **5.** *Physics.* **a.** Decaying with relatively short lifetime. —Used of subatomic particles. **b.** Radioactive. **—un·sta′ble·ness** *n.* **—un·sta′bly** *adv.*

**un·stead·y** (ŭn-stĕd′ē) *adj.* **-i·er, -i·est. 1.** Not steady or firm : SHAKY <an *unsteady* gait> **2.** Fluctuating : inconstant <an *unsteady* breeze> **3.** Wavering : uneven <an *unsteady* stock market> *—vt.* **-ied, -y·ing, -ies.** To cause to become unsteady. **—un·stead′i·ly** *adv.* **—un·stead′i·ness** *n.*

**un·steel** (ŭn-stēl′) *vt.* **-steeled, -steel·ing, -steels.** To soften : disarm.

**un·step** (ŭn-stĕp′) *vt.* **-stepped, -step·ping, -steps.** *Naut.* To remove (a mast) from a step.

**un·stick** (ŭn-stĭk′) *vt.* **-stuck** (-stŭk′), **-stick·ing, -sticks.** To free from being stuck.

**un·stint·ing** (ŭn-stĭn′tĭng) *adj.* Generous : unsparing. **—un·stint′ing·ly** *adv.*

**un·stop** (ŭn-stŏp′) *vt.* **-stopped, -stop·ping, -stops. 1.** To remove a stopper or stop from. **2.** To remove an obstruction from : OPEN.

**un·stop·pa·ble** (ŭn-stŏp′ə-bəl) *adj.* Not capable of being stopped. **—un·stop′pa·bly** *adv.*

**un·stopped** (ŭn-stŏpt′) *adj.* **1.** Not stopped. **2.** Capable of being prolonged, as the consonants *z* and *l.*

**un·strap** (ŭn-străp′) *vt.* **-strapped, -strap·ping, -straps.** To remove or loosen a strap of.

**un·strat·i·fied** (ŭn-străt′ə-fīd′) *adj.* Lacking definite layers.

**un·stressed** (ŭn-strĕst′) *adj.* **1.** Not stressed or having the weakest stress <an *unstressed* syllable> **2.** Not emphasized.

**un·stri·at·ed** (ŭn-strī′ā′tĭd) *adj.* Lacking striations.

**un·string** (ŭn-strĭng′) *vt.* **-strung** (-strŭng′), **-string·ing, -strings. 1.** To remove from a string. **2.** To loosen or remove the strings of. **3.** To weaken the nerves of : UNNERVE <*unstrung* by the defeat>

**un·struc·tured** (ŭn-strŭk′chərd) *adj.* **1.** Lacking structure. **2.** *Psychol.* **a.** Having no intrinsic or objective meaning : meaningful by subjective interpretation only <*unstructured* inkblot tests> **b.** Not regulated or regimented <an *unstructured* environment>

**un·strung** (ŭn-strŭng′) *adj.* **1.** Having a string or strings loosened or removed. **2.** Emotionally upset : UNNERVED.

**un·stud·ied** (ŭn-stŭd′ēd) *adj.* **1.** Not contrived for effect : NATURAL <an *unstudied* innocence> **2.** Not acquired by instruction.

**un·sub·stan·tial** (ŭn′səb-stăn′shəl) *adj.* **1.** Lacking material substance : INSUBSTANTIAL <*unsubstantial* meals> **2.** Lacking firmness or strength : FLIMSY. **3.** Lacking basis in fact <an *unsubstantial* argument> **—un′sub·stan′ti·al′i·ty** *n.* **—un′sub·stan′tial·ly** *adv.*

**un·suc·cess·ful** (ŭn′sək-sĕs′fəl) *adj.* Not succeeding. **—un′suc·cess′ful·ly** *adv.* **—un′suc·cess′ful·ness** *n.*

**un·suit·a·ble** (ŭn-sōō′tə-bəl) *adj.* Not suitable : INAPPROPRIATE. **—un·suit′a·bil′i·ty, un·suit′a·ble·ness** *n.* **—un·suit′a·bly** *adv.*

☆ ***syns:*** UNSUITABLE, IMPROPER, INAPPROPRIATE, INAPT, MALAPROPOS, UNBECOMING, UNFIT, UNSEEMLY *adj. core meaning* : not suited to the circumstances <*unsuitable* attire for the office><*unsuitable* behavior in public> ***ant:*** suitable

**un·sung** (ŭn-sŭng′) *adj.* **1.** Not sung. **2.** Not honored or celebrated, as in song or story <*unsung* heroes>

**un·sus·pect·ed** (ŭn′sə-spĕk′tĭd) *adj.* **1.** Not arousing suspicion. **2.** Not known or thought to exist. **—un′sus·pect′ed·ly** *adv.*

**un·sus·pect·ing** (ŭn′sə-spĕk′tĭng) *adj.* Not suspicious : TRUSTING. **—un′sus·pect′ing·ly** *adv.*

**un·swathe** (ŭn-swŏth′, -swôth′, -swāth′) *vt.* **-swathed, -swath·ing, -swathes.** To remove the swathes or bindings from.

**un·swear** (ŭn-swâr′) *v.* **-swore** (-swôr′, -swōr′), **-sworn** (-swôrn′, -swōrn′), **-swear·ing, -swears.** *Archaic. —vt.* To retract (an oath), often by swearing another oath. *—vi.* To recant something sworn.

**un·sym·met·ri·cal** (ŭn′sĭ-mĕt′rĭ-kəl) *adj.* Asymmetric. **—un′sym·met′ri·cal·ly** *adv.*

**un·tan·gle** (ŭn-tăng′gəl) *vt.* **-gled, -gling, -gles. 1.** To free from a tangle : DISENTANGLE. **2.** To clear up : RESOLVE.

**un·tapped** (ŭn-tăpt′) *adj.* **1.** Not having been tapped <an *untapped* cask> **2.** Not utilized <*untapped* reserves>

**un·taught** (ŭn-tôt′) *adj.* **1.** Not instructed : IGNORANT. **2.** Not acquired by instruction : NATURAL <*untaught* charm>

**un·teach** (ŭn′tēch′) *vt.* **-taught** (-tôt′), **-teach·ing, -teach·es. 1.** To cause to forget or unlearn something. **2.** To negate (what has been taught) with contradictory information.

**un·ten·a·ble** (ŭn-tĕn′ə-bəl) *adj.* **1.** Not capable of being defended or maintained <was in an *untenable* position due to public outrage> **2.** Not capable of being occupied. **—un·ten′a·bil′i·ty, un·ten′a·ble·ness** *n.* **—un·ten′a·bly** *adv.*

**un·thank·ful** (ŭn-thăngk′fəl) *adj.* **1.** Not thankful : UNGRATEFUL. **2.** Not appreciated : THANKLESS. **—un·thank′ful·ly** *adv.* **—un·thank′ful·ness** *n.*

**un·think** (ŭn-thĭngk′) *vt.* **-thought** (-thôt′), **-think·ing, -thinks.** To dismiss from the mind : DISREGARD.

**un·think·a·ble** (ŭn-thĭng′kə-bəl) *adj.* **1.** Impossible to imagine : INCONCEIVABLE. **2.** Not to be considered : out of the question. **3.** Contrary to what is reasonable or probable. **—un·think′a·bly** *adv.*

**un·think·ing** (ŭn′thĭng′kĭng) *adj.* **1.** Not heedful or considerate : THOUGHTLESS. **2.** Showing lack of thought : INADVERTENT. **3.** Incapable of the power of thought. **—un·think′ing·ly** *adv.* **—un·think′ing·ness** *n.*

**un·thread** (ŭn-thrĕd′) *vt.* **-thread·ed, -thread·ing, -threads. 1.** To draw out the thread from. **2.** To find one's way out of (e.g., a labyrinth).

**un·throne** (ŭn-thrōn′) *vt.* **-throned, -thron·ing, -thrones.** To depose from or as if from a throne : DETHRONE.

**un·ti·dy** (ŭn-tī′dē) *adj.* **-di·er, -di·est. 1.** Not neat and tidy <an *untidy* closet> **2.** Lacking orderliness or organization <an *untidy* mind> **—un·ti′di·ly** *adv.* **—un·ti′di·ness** *n.*

**un·tie** (ŭn-tī′) *v.* **-tied, -ty·ing, -ties.** *—vt.* **1.** To loosen or undo (a knot or something knotted). **2.** To free from something that binds or restrains. **3.** To straighten out (e.g., difficulties) : RESOLVE. *—vi.* To become untied.

**un·til** (ŭn-tĭl′) *prep.* [ME : *un-*, till + *til*, till.] **1.** Up to the time of <danced *until* dawn> **2.** Before a specified time <We can't leave *until* Monday.> **3.** *Chiefly Scot.* Unto : to. —*conj.* **1.** Up to the time that <We worked *until* it was dark.> **2.** Before <You can't leave *until* your homework is finished.> **3.** To the point or extent that <talked *until* I was hoarse>

**un·time·ly** (ŭn-tīm′lē) *adj.* **-li·er, -li·est. 1.** Occurring or done at an inappropriate time : INOPPORTUNE. **2.** Occurring too soon : PREMATURE <an *untimely* death> —*adv.* **1.** Inopportunely. **2.** Prematurely. **—un·time′li·ness** *n.*

**un·tir·ing** (ŭn-tīr′ĭng) *adj.* **1.** Not tiring : INDEFATIGABLE <*untiring* volunteers> **2.** Not ceasing despite fatigue or frustration : TIRELESS <*untiring* efforts for peace> **—un·tir′ing·ly** *adv.*

**un·ti·tled** (ŭn-tīt′ld) *adj.* **1.** Having no right or claim. **2.** Having no title <an *untitled* poem><*untitled* nobility>

**un·to** (ŭn′to͞o) *prep.* [ME : *un-*, till + *to*, to.] To.

**un·told** (ŭn-tōld′) *adj.* **1.** Not told or revealed. **2.** Being beyond description or enumeration <*untold* anguish><*untold* wealth>

**un·touch·a·ble** (ŭn-tŭch′ə-bəl) *adj.* **1.** Not to be touched. **2.** Lying out of reach : UNOBTAINABLE. **3.** Being beyond criticism, impeachment, or attack. **4.** Loathsome or disagreeable to the touch. —*n. often* **Untouchable.** A member of the lowest caste in India, with whom physical contact was considered defiling by Hindus of higher castes. **—un·touch′a·bil′i·ty** *n.*

**un·to·ward** (ŭn-tôrd′, -tōrd′) *adj.* **1.** Not favorable : INAUSPICIOUS. **2.** Hard to guide or control : REFRACTORY. **3.** Improper <*untoward* behavior> **4.** *Archaic.* Awkward. **—un·to′ward·ly** *adv.* **—un·to′ward·ness** *n.*

**un·trav·eled** (ŭn-trăv′əld) *adj.* **1.** Not traversed <an *untraveled* road> **2. a.** Not having traveled. **b.** Provincial : narrow-minded.

**un·tread** (ŭn-trĕd′) *vt. Archaic.* **-trod** (-trŏd′), **-trod·den** (-trŏd′n) or **-trod, -tread·ing, -treads.** To retrace (one's course).

**un·tried** (ŭn-trīd′) *adj.* **1.** Not attempted, tested, or proved <*untried* theories> **2.** Not tried in a court of law.

**un·true** (ŭn-tro͞o′) *adj.* **-tru·er, -tru·est. 1.** Contrary to fact : FALSE. **2.** Deviating from a standard of correctness : INEXACT. **3.** Not faithful : DISLOYAL. **—un·tru′ly** *adv.*

**un·truss** (ŭn-trŭs′) *v.* **-trussed, -truss·ing, -truss·es.** *Archaic.* —*vt.* **1.** To unfasten : undo. **2.** To undress. —*vi.* To remove one's clothes, esp. one's breeches.

**un·truth** (ŭn-tro͞oth′) *n.* **1.** Something untrue : LIE. **2.** The state of being false : lack of truth. **3.** *Archaic.* Unfaithfulness.

**un·truth·ful** (ŭn-tro͞oth′fəl) *adj.* **1.** Contrary to truth. **2.** Given to falsehood. **—un·truth′ful·ly** *adv.* **—un·truth′ful·ness** *n.*

**un·tu·tored** (ŭn-to͞o′tərd, -tyo͞o′-) *adj.* **1.** Having no formal education or instruction <an *untutored* computer whiz> **2.** Unsophisticated : naive.

**un·twine** (ŭn-twīn′) *v.* **-twined, -twin·ing, -twines.** —*vt.* **1.** To loosen or separate, as strands of twisted fiber. **2.** To disentangle. —*vi.* To become untwined.

**un·twist** (ŭn-twĭst′) *v.* **-twist·ed, -twist·ing, -twists.** —*vt.* To loosen or separate (material twisted together) by turning in the opposite direction : UNWIND. —*vi.* To become untwisted.

**un·used** (ŭn-yo͞ozd′) *adj.* **1.** Not in use or put to use. **2.** Never having been used. **3.** (ŭn-yo͞ost′). Not accustomed <*unused* to late hours>

**un·u·su·al** (ŭn-yo͞o′zho͞o-əl) *adj.* Not usual, common, or ordinary. **—un·u′su·al·ly** *adv.* **—un·u′su·al·ness** *n.*

**un·ut·ter·a·ble** (ŭn-ŭt′ər-ə-bəl) *adj.* **1.** Not capable of being expressed : INEXPRESSIBLE <*unutterable* bliss> **2.** Not capable of being pronounced. **—un·ut′ter·a·ble·ness** *n.* **—un·ut′ter·a·bly** *adv.*

**un·val·ued** (ŭn-văl′yo͞od) *adj.* **1.** Not prized or valued : UNAPPRECIATED. **2.** Not appraised, as a gem. **3.** *Obs.* Invaluable.

**un·var·nished** (ŭn′vär′nĭsht) *adj.* **1.** Not varnished <an *unvarnished* floor> **2.** Not embellished or disguised : UNADORNED <the *unvarnished* truth>

**un·veil** (ŭn-vāl′) *v.* **-veiled, -veil·ing, -veils.** —*vt.* **1.** To remove a veil or covering from. **2.** To disclose : reveal. —*vi.* To reveal oneself.

**un·voice** (ŭn-vois′) *vt.* **-voiced, -voic·ing, -voic·es.** To prononce (a voiced speech sound) without vibrating the vocal cords : DEVOICE.

**un·voiced** (ŭn-voist′) *adj.* **1.** Not expressed or uttered. **2.** Uttered without vibrating the vocal cords : VOICELESS.

**un·war·rant·a·ble** (ŭn-wôr′ən-tə-bəl, -wŏr′-) *adj.* Not justifiable : INEXCUSABLE. **—un·war′rant·a·bly** *adv.*

**un·war·rant·ed** (ŭn-wôr′ən-tĭd, -wŏr′-) *adj.* Having no justification : GROUNDLESS <*unwarranted* suspicions>

**un·war·y** (ŭn-wâr′ē) *adj.* **-i·er, -i·est.** Not alert to potential danger or deception : UNGUARDED. **—un·war′i·ly** *adv.* **—un·war′i·ness** *n.*

**un·washed** (ŭn-wŏsht′, -wôsht′) *adj.* **1.** Not washed : UNCLEAN. **2.** Plebeian : common <the *unwashed* masses>

**un·wea·ried** (ŭn-wîr′ēd) *adj.* **1.** Not tired : FRESH. **2.** Never wearying : TIRELESS. **—un·wea′ried·ly** *adv.*

**un·well** (ŭn-wĕl′) *adj.* Not well : ILL.

**un·wept** (ŭn-wĕpt′) *adj.* **1.** Not mourned or wept for <an *unwept* death> **2.** Not shed <*unwept* tears>

**un·whole·some** (ŭn-hōl′səm) *adj.* **1.** Injurious to physical, mental, or moral health. **2.** Suggestive of disease or degeneracy <an *unwholesome* fascination with pain> **3.** Offensive : loathsome. **—un·whole′some·ly** *adv.* **—un·whole′some·ness** *n.*

**un·wield·y** (ŭn-wēl′dē) *adj.* **-i·er, -i·est. 1.** Difficult to carry, handle, or manage, as because of bulk or shape. **2.** Awkward : ungainly.

**un·willed** (ŭn-wĭld′) *adj.* Involuntary : spontaneous.

**un·will·ing** (ŭn-wĭl′ĭng) *adj.* **1.** Reluctant : loath <*unwilling* to go> **2.** Done, given, or said reluctantly <an *unwilling* promise> **—un·will′ing·ly** *adv.* **—un·will′ing·ness** *n.*

**un·wind** (ŭn-wīnd′) *v.* **-wound** (-wound′), **-wind·ing, -winds.** —*vt.* **1.** To reverse the winding or twisting direction of. **2.** To separate the tangled parts of. **3.** To free from tension : RELAX. —*vi.* **1.** To become unwound. **2.** To become relaxed.

**un·wise** (ŭn-wīz′) *adj.* **-wis·er, -wis·est.** Lacking wisdom or good judgment : foolish or imprudent. **—un·wise′ly** *adv.*

**un·wish** (ŭn-wĭsh′) *vt.* **-wished, -wish·ing, -wish·es. 1.** To retract a wish for. **2.** *Obs.* To wish out of existence.

**un·wit·ting** (ŭn-wĭt′ĭng) *adj.* [ME : *un-*, un- + *witting*, p.part. of *witen*, to know < OE *witan*.] **1.** Not knowing : UNAWARE <an *unwitting* accomplice> **2.** Not intended : UNINTENTIONAL <an *unwitting* neglect> **—un·wit′ting·ly** *adv.*

**un·wont·ed** (ŭn-wôn′tĭd, -wōn′-, -wŭn′-) *adj.* **1.** Not usual or habitual : RARE <*unwonted* generosity> **2.** *Archaic.* Not accustomed. **—un·wont′ed·ly** *adv.* **—un·wont′ed·ness** *n.*

**un·world·ly** (ŭn-wûrld′lē) *adj.* **-li·er, -li·est. 1.** Not of this world : SPIRITUAL. **2.** Not concerned with worldly matters. **3.** Not worldly-wise : NAIVE. **—un·world′li·ness** *n.*

**un·worn** (ŭn-wôrn′, -wōrn′) *adj.* **1.** Not worn out or worn away. **2.** Not worn before : NEW. **3.** Not stale or overused : FRESH.

**un·wor·thy** (ŭn-wûr′thē) *adj.* **-thi·er, -thi·est. 1.** Not deserving <*unworthy* of consideration> **2.** Not suiting or befitting <language *unworthy* of a teacher> **3.** Lacking value or merit : WORTHLESS. **4.** Vile : base. **—un·wor′thi·ly** *adv.* **—un·wor′thi·ness** *n.*

**un·wrap** (ŭn-răp′) *v.* **-wrapped, -wrap·ping, -wraps.** —*vt.* To remove the wrappings from : OPEN. —*vi.* To become unwrapped.

**un·writ·ten** (ŭn-rĭt′n) *adj.* **1.** Not written or recorded : ORAL. **2.** Not formulated in writing : effective through custom or tradition <*unwritten* rules of social conduct> **3.** Not written upon : BLANK.

**unwritten law** *n.* A law or code of morality, conduct, or procedure whose authority lies chiefly in custom or general usage rather than in a specific enactment.

**un·yield·ing** (ŭn-yēl′dĭng) *adj.* **1.** Not bending : INFLEXIBLE. **2.** Stubbornly resistant : OBDURATE. **—un·yield′ing·ly** *adv.* **—un·yield′ing·ness** *n.*

**un·yoke** (ŭn-yōk′) *v.* **-yoked, -yok·ing, -yokes.** —*vt.* **1.** To release from or as if from a yoke. **2.** To separate or disjoin. —*vi.* **1.** To remove a yoke. **2.** *Archaic.* To stop working.

**un·zip** (ŭn-zĭp′) *v.* **-zipped, -zip·ping, -zips.** —*vt.* To open (a zipper or something held together by a zipper). —*vi.* To become unzipped.

**up** (ŭp) *adv.* [ME *up*, upward and *uppe*, on high, both < OE.] **1.** From a lower to a higher position. **2.** In or toward a higher position <glancing *up*> **3.** From a reclining to an upright position <The child sat *up* in bed.> **4. a.** Above a surface <dolphins coming *up* for air> **b.** Above the horizon. **5.** Into view, existence, or consideration <bring a subject *up* for discussion> **6.** In or toward a position regarded as higher, as on a chart, scale, or map. **7.** To or at a higher price. **8.** So as to advance, increase, or improve <coming *up* in society> **9.** With or to a greater intensity, pitch, or volume. **10.** Into a state of excitement or turbulence. **11.** So as to detach or unearth <pulling *up* wire grass> **12.** —Used as an intensifier of the action of a verb <typed *up* a list> **13.** Into pieces : APART <tore it *up*> **14.** *Naut.* To windward. **15.** Each : apiece <the score was 3 *up*> **16.** Completely : entirely <fastened *up* my jacket> —*adj.* **1.** High or relatively high. **2. a.** Standing : erect. **b.** Out of bed. **3.** Moving or directed upward <an *up* escalator> **4. a.** Actively functioning : HEALTHY <*up* and around> **b.** *Slang.* Happily excited : EUPHORIC. **5.** Rising toward the flood level. **6.** Marked by agitation or acceleration <The tides are *up*.> **7.** *Informal.* Taking place <What's *up*?> **8.** Being considered <an old contract that is *up* for renewal> **9.** Running as a candidate. **10.** On trial : CHARGED. **11.** Finished : over <The time is *up*.> **12.** *Informal.* Well-informed <I am not *up* on sports.> **13.** Being ahead of the opponent <*up* two holes in a golf match> **14.** *Baseball.* At bat. **15.** As a bet : at stake. **16.** *Naut.* Bound for a specified place. —*prep.* **1.** From a lower to or toward a higher point on. **2.** Toward or at a point farther along <*up* the avenue> **3.** In a direction toward the source of <*up* the Seine> **4.** Against <*up* the wind> —*n.* **1.** An upward slope. **2.** An upward trend or movement. **3.** *Slang.* Excitement : euphoria. —*v.* **upped, up·ping, ups.** —*vt.* **1.** To increase or improve. **2.** To raise to a higher level, esp. to promote to a higher position. —*vi.* **1.** To get up : RISE. **2.** *Informal.* To act suddenly or unexpectedly <*upped* and left> **—on the up and up.** *Slang.* Open and honest. **—up against.** Confronted with : FACING. **—up to. 1.** Occupied with, esp. devising and scheming <children who are *up to* mischief> **2.** Primed and prepared for

<Are you *up to* going?> **3.** Dependent upon <The success of this project is *up to* us.>

**up-** *pref.* [ME < OE.] **1.** Up : upward <*upheave*> **2.** Upper <*upland*>

**up-and-com·ing** (ŭp′ən-kŭm′ĭng) *adj.* Marked for future success : PROMISING <an *up-and-coming* author>

**up-and-down** (ŭp′ən-doun′) *adj.* **1.** Having an alternating upward and downward movement or surface. **2.** Vertical.

**U·pan·i·shad** (o͞o-păn′ə-shăd′) *n.* [Skt. *upaniṣad.*] Any of a group of philosophical treatises elaborating on the earlier Vedas and contributing to ancient Hindu theology. **—U·pan′i·shad′ic** *adj.*

**u·pas** (yo͞o′pəs) *n.* [Malay (*pōhun*) *upas,* poison (tree).] **1.** A tropical Asian tree, *Antiaris toxicaria,* yielding a milky juice used as an arrow poison. **2.** The poison obtained from the upas or similar trees or plants.

**upas**

**up·beat** (ŭp′bēt′) *n. Mus.* **1.** An unaccented beat, esp. the last beat of a measure. **2.** Upswing. *—adj.* Optimistic : cheerful.

**up-bow** (ŭp′bō′) *n. Mus.* A stroke performed on a stringed instrument in which the bow is moved across the strings from its tip to its heel.

**up·braid** (ŭp-brād′) *vt.* **-braid·ed, -braid·ing, -braids.** [ME *upbreyden* < OE *ūpbrēdan* : *up,* up + *bregdan,* to throw, turn.] To scold or criticize sharply. **—up·braid′er** *n.* **—up·braid′ing·ly** *adv.*

▲ word history: *Upbraid* is a compound formed in Old English of the prefix *up–,* "up," and the verb *bregdan,* which had a variety of meanings derived from the idea of "turning." The compound *upbregdan* in Old English meant "to turn up a matter against someone as a reproach." In later times the word came to mean simply "to reproach."

**up·bring·ing** (ŭp′brĭng′ĭng) *n.* The care and training received during childhood : REARING.

**up·build** (ŭp-bĭld′) *vt.* **-built** (-bĭlt′), **-build·ing, -builds.** To enlarge or enhance : BUILD UP. **—up·build′er** *n.*

**up·cast** (ŭp′kăst′) *adj.* Directed or thrown upward. *—n.* **1.** Something cast upward. **2.** A ventilating shaft, as in a mine.

**up·chuck** (ŭp′chŭk′) *v.* **-chucked, -chuck·ing, -chucks.** *Slang.* *—vi.* To vomit. *—vt.* To vomit (stomach contents).

**up·com·ing** (ŭp′kŭm′ĭng) *adj.* Forthcoming : anticipated.

**up·coun·try** (ŭp′kŭn′trē) *n.* The interior region of a country. *—adj.* (ŭp′kŭn′trē). Of, located in, or coming from the upcountry. *—adv.* (ŭp-kŭn′trē). In, to, or toward the upcountry.

**up·date** (ŭp-dāt′) *vt.* **-dat·ed, -dat·ing, -dates.** To bring up to date <*update* a news story> *—n.* (ŭp′dāt′). **1.** Information that updates. **2.** An act or instance of updating.

**up·draft** (ŭp′drăft′, -dräft′) *n.* An upward current of air.

**up·end** (ŭp-ĕnd′) *v.* **-end·ed, -end·ing, -ends.** *—vt.* **1.** To set on end. **2.** To overturn or defeat. *—vi.* To be upended.

**up-front** (ŭp′frŭnt′) *adj.* **1.** Straightforward : frank <an *up-front* discussion of grievances> **2.** Required in advance <*up-front* cash>

**up·grade** (ŭp′grād′) *vt.* **-grad·ed, -grad·ing, -grades. 1.** To raise to a higher grade, as in value, quality, rank, or importance. **2.** To improve the quality of (livestock) by selective breeding. *—n.* An upward incline. *—adj.* Uphill. *—adv.* Up an incline. **—on the upgrade.** Improving or progressing.

**up·growth** (ŭp′grōth′) *n.* **1.** The process of growing upward. **2.** Upward growth or development.

**up·heav·al** (ŭp-hē′vəl) *n.* **1.** The process or an instance of being heaved upward. **2.** A profound and violent disruption or charge <"The psychic *upheaval* caused by war" —Wallace Fowlie> **3.** *Geol.* A lifting up of the earth's crust by movement of stratified or other rocks.

**up·heave** (ŭp-hēv′) *v.* **-heaved, -heav·ing, -heaves.** *—vt.* To heave upward. *—vi.* To be lifted or thrust upward.

**up·hill** (ŭp′hĭl′) *adj.* **1.** Going up a hill or slope. **2.** Prolonged and laborious <an *uphill* struggle> *—n.* (ŭp′hĭl′). An upward slope or incline. *—adv.* (ŭp′hĭl′). **1.** To or toward higher ground : UPWARD. **2.** Against adversity : with difficulty.

**up·hold** (ŭp-hōld′) *vt.* **-held** (-hĕld′), **-hold·ing, -holds.** [ME *upholden* : *up,* up + *holden,* to hold.] **1.** To hold aloft : RAISE. **2.** To prevent from falling or sinking : SUPPORT. **3.** To defend or affirm in the face of a challenge <*uphold* one's rights to privacy and freedom of speech> **—up·hold′er** *n.*

**up·hol·ster** (ŭp-hōl′stər) *vt.* **-stered, -ster·ing, -sters.** [Back-formation from UPHOLSTERER.] To provide (furniture) with stuffing, springs, cushions, and covering fabric.

▲ word history: The word *upholster* has a tortuous history which begins with the simple English verb *uphold,* "to support." To this verb the noun suffixes *–ster* and *–er* were added, producing *upholdsterer* or *upholsterer,* which originally meant "a repairer of stuffed furniture." The verb *upholster* was formed from *upholsterer* by deleting the suffix *–er.* Although the noun *upholsterer* is recorded from the 17th century, the verb *upholster* is not recorded until after 1900.

**up·hol·ster·er** (ŭp-hōl′stər-ər) *n.* [< obs. *upholster* < ME *upholdester* < *upholden,* to repair.] One that upholsters furniture.

**up·hol·ster·y** (ŭp-hōl′stə-rē, -strē) *n., pl.* **-ies. 1.** The fabrics and other materials used in upholstering. **2.** The business of upholstering.

**up·keep** (ŭp′kēp′) *n.* **1.** Maintenance in good condition or repair. **2.** The cost of proper maintenance.

**up·land** (ŭp′lənd, -lănd′) *n.* **1.** The higher parts of a region or tract of land. **2.** Inland country : UPCOUNTRY. **—up′land** *adj.* **—up′land·er** (ŭp′lən-dər, -lăn′dər) *n.*

**upland cotton** *n.* A plant, *Gossypium hirsutum* native to tropical America and widely cultivated for its fiber.

**upland plover** *n.* A brownish, long-necked New World bird, *Bartramia longicauda,* found in fields and prairies.

**up·lift** (ŭp-lĭft′) *vt.* **-lift·ed, -lift·ing, -lifts. 1.** To raise up or aloft : ELEVATE. **2.** To raise to a higher social, moral, or intellectual level. **3.** To raise to spiritual or emotional heights : EXALT. *—adj.* (ŭp′lĭft′). Uplifted. *—n.* (ŭp′lĭft′). **1.** The act, process, or result of lifting up. **2.** A movement to improve social, moral, or intellectual standards. **3.** Moral or spiritual elevation. **4.** An agent or influence causing upward movement or lifting. **5.** *Geol.* An upheaval. **—up·lift′er** *n.*

**up·man·ship** (ŭp′mən-shĭp′) *n.* One-upmanship.

**up·most** (ŭp′mōst′) *adj.* Uppermost.

**up·on** (ə-pŏn′, ə-pôn′) *prep.* [ME : *up,* up + *on,* on.] On.

**up·per** (ŭp′ər) *adj.* **1.** Higher in place, position, or rank. **2. a.** Situated on higher ground. **b.** Lying farther inland. **c.** Northern. **3. Upper.** Being a later division of the geological and archaeological period named. *—n.* **1.** That part of a shoe or boot above the sole. **2.** An upper berth, as on a train. **3. uppers.** The upper teeth or a set of upper dentures. **4.** *Slang.* **a.** A drug used as a stimulant, esp. an amphetamine. **b.** Something that causes a feeling of well-being or euphoria. **—on one's uppers.** *Informal.* Impoverished.

**upper atmosphere** *n.* The atmosphere that extends above the troposphere.

**upper bound** *n. Math.* A number that is not exceeded by any number in a given set.

**Upper Carboniferous** *adj.* & *n. Geol.* Pennsylvanian.

**upper case** *n.* The case of printing type containing the capital letters and special characters.

**up·per-case** (ŭp′ər-kās′) *adj.* Relating to or printed in capital letters : CAPITAL. *—vt.* **-cased, -cas·ing, -cas·es.** To print in upper-case letters.

**upper class** *n.* The upper and usu. highest social class of a society.

**up·per-class** (ŭp′ər-klăs′) *adj.* Of, belonging to, or typical of the upper class.

**up·per·class·man** (ŭp′ər-klăs′mən) *n.* A student in the junior or senior class of a secondary school or college.

**upper crust** *n. Informal.* The highest social class or group.

**up·per·cut** (ŭp′ər-kŭt′) *n.* A short swinging blow directed upward, as to a boxing opponent's chin. **—up′per·cut′** *v.* **(-cut, -cut·ting, -cuts).**

**upper hand** *n.* The controlling position : ADVANTAGE.

**Upper House** or **upper house** *n.* The branch of a bicameral legislature that is smaller and less broadly representative of the population, as the U.S. Senate.

**up·per·most** (ŭp′ər-mōst′) *adj.* Highest in position, place, rank, or influence. *—adv.* In the first or highest rank, position, or place : FIRST.

**up·pish** (ŭp′ĭsh) *adj.* Uppity. **—up′pish·ly** *adv.* **—up′pish·ness** *n.*

**up·pi·ty** (ŭp′ĭ-tē) *adj.* [< UP.] *Informal.* Snobbish or arrogant.

**up·raise** (ŭp-rāz′) *vt.* **-raised, -rais·ing, -rais·es.** To raise or lift up : ELEVATE.

**up·rear** (ŭp-rîr′) *v.* **-reared, -rear·ing, -rears.** *—vt.* **1.** To raise or lift up. **2.** To build or erect. *—vi.* To be raised up : RISE.

**up·right** (ŭp′rīt′) *adj.* [ME < OE *upriht* : *up,* up + *riht,* right.] **1. a.** In a vertical position or direction. **b.** Erect in posture or carriage. **2.** Morally respectable : HONORABLE. *—adv.* Vertically <walk *upright*> *—n.* **1.** The state of being vertical. **2.** Something standing upright, as a beam. **3.** An upright piano. **—up′right′ly** *adv.* **—up′right′ness** *n.*

**upright piano** *n.* A piano having the strings mounted vertically in a rectangular case with the keyboard at a right angle to the case.

**up·rise** (ŭp-rīz′) *vi.* **-rose** (-rōz′), **-ris·en** (-rĭz′ən), **-ris·ing, -ris·es. 1.** To get up or stand up : RISE. **2.** To move or incline upward : ASCEND. **3.** To rise into view, esp. from below the horizon. **4.** To increase in size : SWELL. —*n.* (ŭp′rīz′). **1.** The act or process of rising up. **2.** An upward slope.

**up·ris·ing** (ŭp′rī′zĭng) *n.* **1.** A revolt or insurrection. **2.** An act of rising or rising up. **3.** An upward slope.

**up·riv·er** (ŭp′rĭv′ər) *adj.* & *adv.* Toward or near the source of a river : in the direction opposite to that of the flow of water. —*n.* A region lying upriver.

**up·roar** (ŭp′rôr′, -rōr′) *n.* [By folk ety. < Du. *oproer* < MDu. : *op,* up + *roer,* motion.] **1.** A state of noisy agitation and confusion : TUMULT. **2.** Intense public controversy and excitement.

☆ ***syns:*** UPROAR, BROUHAHA, SENSATION, STIR, TO-DO *n. core meaning* : intense public controversy and excitement <the *uproar* created by the President's resignation>

**up·roar·i·ous** (ŭp-rôr′ē-əs, -rōr′-) *adj.* **1.** Characterized by uproar. **2.** Loud and full : BOISTEROUS <*uproarious* laughter> **3.** Causing hearty laughter : HILARIOUS <an *uproarious* movie> **—up·roar′i·ous·ly** *adv.* **—up·roar′i·ous·ness** *n.*

**up·root** (ŭp-ro͞ot′, -ro͝ot′) *vt.* **-root·ed, -root·ing, -roots. 1.** To pull up by the roots <*uproot* a tree> **2.** To destroy or remove completely : ERADICATE <*uproot* crime> **3.** To force to leave an accustomed or native habitat <*uprooted* by the war> **—up·root′er** *n.*

**ups and downs** *pl.n.* Alternating periods of good and bad fortune or spirits.

**up·set** (ŭp-sĕt′) *v.* **-set, -set·ting, -sets.** [ME *upsetten,* to set up : *up,* up + *setten,* to set.] —*vt.* **1.** To overturn or capsize <*upset* a canoe> **2.** To disturb the normal functioning, order, or course of <*upset* the stomach> **3.** To distress mentally or emotionally. **4.** To defeat unexpectedly. **5.** To shorten and thicken (e.g., an iron bar) by hammering on the end : SWAGE. —*vi.* **1.** To become overturned. **2.** To become distressed. —*n.* (ŭp′sĕt′). **1.** An act of upsetting or the condition of being upset. **2.** A disturbance, disorder, or agitation. **3.** A contest or game in which the favorite is defeated. **4. a.** A tool used for upsetting : SWAGE. **b.** An upset part or piece. —*adj.* (ŭp-sĕt′). **1.** Overturned : capsized. **2.** Disordered : disturbed. **3.** Distressed : distraught. **—up·set′ter** *n.*

**upset price** *n.* The lowest price at which merchandise or property will be auctioned or sold at public sale.

**up·shot** (ŭp′shŏt′) *n.* [Obs. *upshot,* the last shot in an archery contest.] The final issue or result : OUTCOME.

**up·side-down** (ŭp′sīd-doun′) *adj.* [Alteration of ME *up so down,* up as if down.] **1.** Overturned so that the upper side is down. **2.** Totally disordered or confused : TOPSY-TURVY. —*adv. also* **upside down.** Topsy-turvy.

**upside-down cake** *n.* A cake baked with a layer of sliced fruit at the bottom and served with the fruit side up.

**up·si·lon** (ŭp′sə-lŏn′, yo͞op′sə-lŏn′) *n.* [Late Gk., *u psilon,* simple u.] The 20th letter of the Greek alphabet. —See table at ALPHABET.

▲ word history: The word *upsilon,* the name of the 20th letter of the Greek alphabet, literally means "bare," "naked," or "simple" *u.* The letter was given this name in Late Greek times to distinguish it from the diphthong *oi.* In Late Greek *u* and *oi* were pronounced the same, and to call the vowel simply *u* without the qualifying adjective *psilon* would have been unclear.

**up·spring** (ŭp-sprĭng′) *vi.* **-sprang** (-sprăng′) or **-sprung** (-sprŭng′), **-sprung, -spring·ing, -springs. 1.** To spring up, as from the soil. **2.** To come into being : ARISE.

**up·stage** (ŭp′stāj′) *adj.* **1.** Relating to, involving, or located at the back of the stage. **2.** Haughty : aloof. —*adv.* Toward, on, or at the back of the stage. —*vt.* (ŭp-stāj′) **-staged, -stag·ing, -stag·es. 1.** To distract attention from (another performer) by moving upstage and forcing him or her to face away from the audience. **2.** To steal the show from. **3.** To treat haughtily.

**up·stairs** (ŭp′stârz′) *adv.* **1.** On or to an upper floor. **2.** To or at a higher level <moved *upstairs* to management> **3.** *Informal.* In the head <not quite right *upstairs*> —*adj.* (ŭp′stârz′). Of or to an upper floor or floors <an *upstairs* maid> —*n.* (ŭp′stârz′). *(sing. or pl. in number).* A floor or story above ground level. **—kick upstairs.** *Informal.* To dispose of by promotion to a higher but less effectual position.

**up·stand·ing** (ŭp-stăn′dĭng, ŭp′stan′dĭng) *adj.* **1.** Standing erect or upright. **2.** Morally upright : HONORABLE.

**up·start** (ŭp′stärt′) *n.* **1.** One of humble origin who has risen suddenly or recently to wealth, power, or prestige, esp. one who assumes an arrogant attitude : PARVENU. **2.** One having an exaggerated sense of importance. —*adj.* **1.** Suddenly raised to a position of consequence. **2.** Self-important : presumptuous. —*vi.* (ŭp-stärt′) **-start·ed, -start·ing, -starts.** To spring or start up suddenly.

**up·state** (ŭp′stāt′) *adj.* & *adv.* Of, at, in, or toward that part of a state lying inland or farther north of a large city. —*n.* An upstate region. **—up′stat′er** *n.*

**up·stream** (ŭp′strēm′) *adv.* **1.** In, at, or toward the source of a stream. **2.** Against the current of a stream.

**up·stroke** (ŭp′strōk′) *n.* An upward stroke, as of a brush.

**up·surge** (ŭp-sûrj′) *vi.* **-surged, -surg·ing, -surg·es.** To surge up. —*n.* (ŭp′sûrj′). A rapid upward swell or rise.

**up·sweep** (ŭp′swēp′) *n.* **1.** A curve or sweep upward. **2.** A hairdo smoothed upward in the back and piled on top of the head. —*vt.* **-swept** (-swĕpt′), **-sweep·ing, -sweeps.** To sweep upward.

**up·swing** (ŭp′swĭng′) *n.* **1.** An upward swing or trend. **2.** An increase, as in movement or activity <an *upswing* in sales>

**up·take** (ŭp′tāk′) *n.* **1.** A passage for drawing up smoke or air. **2.** Understanding : comprehension <quick on the *uptake*> **3.** An act or instance of taking up or absorbing.

**up·tem·po** (ŭp′tĕm′pō) *n.* A fast-paced tempo, as in jazz.

**up·throw** (ŭp′thrō′) *n.* **1.** A throwing upward. **2.** *Geol.* An upward displacement of rock on one side of a fault.

**up·tick** (ŭp′tĭk′) *n.* A rise or increase, as in the price of a stock market security.

**up·tight** *also* **up tight** (ŭp′tīt′) *adj. Slang.* **1.** Tense : nervous. **2.** Financially pressed. **3.** Outraged : angry. **4.** Adhering rigidly to convention : INFLEXIBLE.

**up·time** (ŭp′tīm′) *n.* The time during which a device, as a computer, is functioning or available for use.

**up-to-date** (ŭp′tə-dāt′) *adj.* **1.** Most recent : CURRENT. **2.** Using the latest information, technology, or methods. **3.** Keeping abreast of the newest styles and ideas : MODERN. **—up′-to-date′ness** *n.*

**up-to-the-min·ute** (ŭp′tə-thə-mĭn′ĭt) *adj.* Marked by or including the very latest information.

**up·town** (ŭp′toun′) *adv.* In or toward the upper part of a town or city. —*n.* The upper part of a town or city. **—up′town′** *adj.*

**up·turn** (ŭp′tûrn′, ŭp-tûrn′) *v.* **-turned, -turn·ing, -turns.** —*vt.* **1.** To turn up or over, as soil. **2.** To upset : overturn. **3.** To direct upward. —*vi.* To turn up or upward. —*n.* (ŭp′tûrn′). An upward movement, curve, or trend.

**up·ward** (ŭp′wərd) *also* **up·wards** (-wərdz) *adv.* [ME < OE *ūpweard, up,* up + *-weard,* -ward.] **1.** In, to, or toward a higher place, level, or position. **2.** To or toward the source, origin, or interior. **3.** Toward the head or upper parts of the body. **4.** Toward a higher amount, degree, or rank <Prices moved *upward.*> **5.** Toward a later time or age <from childhood *upward*> **6.** Toward something greater or better. —*adj.* **upward.** Directed toward a higher place or position <*upward* mobility> **—upwards of.** More than : in excess of <invited *upwards of* 100 people> **—up′ward·ly** *adv.* **—up′ward·ness** *n.*

**up·wind** (ŭp′wĭnd′) *adv.* In or toward the direction from which the wind is blowing. **—up′wind′** *adj.*

**ur-**[1] *pref. var. of* URO-[1].

**ur-**[2] *pref. var. of* URO-[2].

**u·ra·cil** (yo͝or′ə-sĭl) *n.* [UR(EA) + AC(ETIC) + -IL(E).] A pyrimidine, $C_4H_4N_2O_2$, a constituent of ribonucleic acids.

**u·rae·mi·a** (yo͞o-rē′mē-ə) *n. var. of* UREMIA.

**u·rae·us** (yo͞o-rē′əs) *n.* [NLat. < LGk. *ouraios,* cobra, of Egyptian orig.] The figure of the sacred serpent, depicted on the headdress of ancient Egyptian rulers and deities as an emblem of sovereignty.

**uraeus**

**U·ral-Al·ta·ic** (yo͝or′əl-ăl-tā′ĭk) *n.* A hypothetical language group that comprises the Uralic and Altaic language families. **—U·ral-Al·ta′ic** *adj.*

**U·ral·ic** (yo͞o-răl′ĭk) *also* **U·ra·li·an** (yo͞o-rā′lē-ən) *n.* A language family comprising the Finno-Ugric and Samoyedic subfamilies. **—U·ral′ic, U·ra′li·an** *adj.*

**uran-** *pref. var. of* URANO-.

**u·ra·ni·a** (yo͞o-rā′nē-ə, -rān′yə) *n.* [NLat. < URANIUM.] Uranium dioxide.

**U·ra·ni·a** (yo͞o-rā′nē-ə, -rān′yə) *n.* [Lat. < Gk. *Ourania* < *ouranos,* heaven.] *Gk. Myth.* The Muse of astronomy.

**u·ran·ic** (yo͞o-răn′ĭk, -rā′nĭk) *adj.* [Sense 1 : < Gk. *ouranos,* heaven. Sense 2 : < URANIUM.] **1.** Of or pertaining to the heavens : CELESTIAL. **2.** *Chem.* Of, relating to, or derived from uranium, esp. with valence higher than in comparable uranous compounds.

**u·ra·ni·nite** (yo͞o-rā′nə-nīt′) *n.* [G. *Uranin,* uraninite (< NLat. *uranium,* uranium) + -ITE[1].] A complex brownish-black mineral, the chief ore of uranium, chiefly $UO_2$ partially oxidized to $UO_3$, containing variable amounts of radium, lead, thorium, rare-earth metals, helium, argon, and nitrogen.

**u·ra·ni·um** (yo͞o-rā′nē-əm) *n.* [After URANUS.] *Symbol* **U** A heavy silvery-white radioactive metallic element used in research, nuclear fuels, and nuclear weapons; atomic number 92; atomic weight 238.03.

**uranium 235** *n.* The uranium isotope with mass number 235 and half-life $7.13 \times 10^8$ years, fissionable with slow neutrons and capable in a critical mass of sustaining a chain reaction that can proceed explosively with appropriate mechanical arrangements.

**uranium 238** *n.* The most common isotope of uranium, having mass number 238 and half-life $4.51 \times 10^9$ years, nonfissionable but irradiated with neutrons to produce fissionable plutonium 239.

**uranium dioxide** *n.* A highly toxic, black, crystalline powder, $UO_2$, once used in ceramic glazes and gas mantles, now used chiefly to pack nuclear fuel rods.

**uranium enrichment** *n.* A chemical process performed on natural uranium to increase the ratio of uranium 235 to uranium 238, used in fission technology.

**uranium trioxide** *n.* A radioactive orange powder, $UO_3$, used for uranium refining and as a coloring agent in ceramics.

**urano-** *or* **uran-** *pref.* [< URANIUM.] Uranium <*uranyl*>

**u·ra·nous** (yo͞o-rā′nəs, yo͝or′ə-nəs) *adj. Chem.* Of or relating to uranium, esp. with valence lower than in comparable uranic compounds.

**U·ra·nus** (yo͝or′ə-nəs, yo͞o-rā′nəs) *n.* [Lat. < Gk. *Ouranos* < *ouranos*, heaven.] **1.** *Gk. Myth.* The earliest supreme god, a personification of the sky, son and consort of Gaea and father of the Cyclopes and Titans. **2.** The seventh planet from the sun, revolving about it every 84.02 years at a distance of approx. 1,790 million miles or 2,880 million kilometers, with an equatorial diameter of approx. 30,000 miles or 48,270 kilometers, a mass 14.6 times that of Earth, and five satellites.

**u·ra·nyl** (yo͝or′ə-nĭl, yo͞o-rā′nəl) *n.* The divalent radical $UO_2$.

**u·rase** (yo͝or′ās′, -āz′) *n. var. of* UREASE.

**u·rate** (yo͝or′āt′) *n.* [UR(IC ACID) + -ATE².] A salt of uric acid.

**ur·ban** (ûr′bən) *adj.* [Lat. *urbanus* < *urbs*, city.] **1.** Of, relating to, or constituting a city. **2.** Characteristic of the city or city life.

**urban district** *n.* An administrative district of England, Wales, and Northern Ireland, usu. composed of several densely populated communities, resembling a borough but lacking a borough charter.

**ur·bane** (ûr-bān′) *adj.* [Fr. *urbain* < Lat. *urbanus*, of a city < *urbs*, city.] Polished and elegant in manner or style <a witty and *urbane* novel> **—ur·bane′ly** *adv.*

▲ word history: *Urban* and *urbane* are both derived from Latin *urbanus*, "belonging to a city," and they were once synonymous in meaning. *Urbane* was borrowed first, from Old French *urbain*, and it preserves the French pattern of stress. After *urban* was borrowed directly from Latin *urbanus*, *urbane* developed the more specialized sense of "refined, polite, elegant." These desirable qualities were considered to be characteristic of urban rather than country folk.

**ur·ban·ism** (ûr′bə-nĭz′əm) *n.* **1.** The culture or way of life of city dwellers. **2.** The condition of being or becoming urbanized.

**ur·ban·ite** (ûr′bə-nīt′) *n.* A city dweller.

**ur·ban·i·ty** (ûr-băn′ĭ-tē) *n., pl.* **-ties. 1.** Refinement and elegance : polished courtesy. **2. urbanities.** Courtesies : civilities.

**ur·ban·ize** (ûr′bə-nīz′) *v.t.* **-ized, -iz·ing, -iz·es.** To make urban in character. **—ur′ban·i·za′tion** *n.*

**ur·ban·ol·o·gist** (ûr′bə-nŏl′ə-jĭst) *n.* A specialist in urban problems. **—ur′ban·ol′o·gy** *n.*

**urban renewal** *n.* Rehabilitation of slum neighborhoods in urban areas, as by replacement or renovation of substandard buildings and facilities.

**urban sprawl** *n.* Gradual spread of urban dwellings, businesses, and industry to relatively undeveloped land near a city.

**ur·ce·o·late** (ûr-sē′ə-lĭt, ûr′sē-ə-lāt′) *adj.* [NLat. *urceolatus* < Lat. *urceolus*, dim. of *urceus*, jug.] Urn-shaped <an *urceolate* corolla>

**ur·chin** (ûr′chĭn) *n.* [ME *urchone*, hedgehog < OFr. *herichon* < Lat. *ericius* < *er*.] **1.** A small mischievous youngster : SCAMP. **2.** A sea urchin. **3.** *Archaic.* A hedgehog.

**Ur·du** (o͝or′do͞o, ûr′-) *n.* [Hindi, short for *zabān-i-urdū*, language of the camp.] An Indic language that is an official literary language of Pakistan and is widely used in India, esp. by Moslems.

**-ure** *suff.* [ME < OFr. < Lat. *-ura*.] **1.** Act : process <*erasure*> **2. a.** Function : office <*judicature*> **b.** Body performing a function <*legislature*>

**u·re·a** (yo͞o-rē′ə) *n.* [NLat. < Fr. *urée* < *urine*, urine < OFr. < Lat. *urina*.] A white crystalline or powdery compound, $CO(NH_2)_2$, found in mammalian urine and other body fluids, synthesized from ammonia and carbon dioxide, and used as fertilizer, in animal feed, and in the synthesis of plastics and resins.

**u·re·a-for·mal·de·hyde resin** (yo͞o-rē′ə-fôr-măl′də-hīd′) *n.* Any of various thermosetting resins made by combining urea and formaldehyde and widely used to make molded household and mechanical objects.

**u·re·ase** (yo͝or′ē-ās′, -āz′) *also* **u·rase** (yo͝or′ās′, -āz′) *n.* [URE(A) + -ASE.] An enzyme occurring in urine, jack beans, soy beans, and as a secretion of certain microorganisms, used to determine the urea content of blood and urine.

**u·re·da·stage** (yo͞o-rē′də-stāj′) *n.* [URED(INIA) + STAGE.] The stage of a rust fungus in which uredinia are produced.

**u·re·din·i·o·spore** (yo͝or′ə-dĭn′ē-ə-spôr′, -spōr′) *n.* [UREDINI(UM) + SPORE.] *var. of* UREDOSPORE.

**u·re·din·i·um** (yo͝or′ə-dĭn′ē-əm) *also* **u·re·di·um** (yo͞o-rē′dē-əm) *n., pl.* **-din·i·a** (-dĭn′ē-ə) *also* **-di·a** (-dē-ə) [NLat. < Lat. *uredo*, blight < *urere*, to burn.] A reddish pustulelike structure formed on the tissue of a plant infected by a rust fungus, and having hyphae that produce uredospores.

**u·re·do** (yo͞o-rē′dō) *n.* [Lat., burning itch < *urere*, to burn.] *Pathol.* Urticaria.

**u·re·do·spore** (yo͞o-rē′də-spôr′, -spōr′) *also* **u·re·din·i·o·spore** (yo͝or′ə-dĭn′ē-ə-spôr′, -spōr′) *n.* [URED(INIUM) + SPORE.] A reddish spore produced in the uredinium of a rust fungus that spreads to and infects other plants.

**u·re·ide** (yo͝or′ē-īd′) *n.* [URE(A) + -IDE.] *Chem.* A derivative of urea.

**u·re·mi·a** (yo͞o-rē′mē-ə) *also* **u·rae·mi·a** *n.* **1.** Excess urea in the blood. **2.** A toxic condition usu. accompanying severe kidney disease and marked by headache, nausea, vomiting, and coma.

**u·re·o·tel·ic** (yo͞o-rē′ə-tĕl′ĭk, yo͝or′ē-ō-) *adj.* Excreting unneeded nitrogen in the form of urea. **—u·re′o·tel′ism** (yo͞o-rē′ə-tĕl′ĭz′əm, yo͝or′ē-ŏt′l-ĭz′əm) *n.*

**u·re·ter** (yo͝or′ĭ-tər, yo͞o-rē′tər) *n.* [Gk. *ourētēr* < *ourein*, to urinate < *ouron*, urine.] A long, narrow duct that conveys urine from the kidneys to the bladder or cloaca.

**u·re·thane** (yo͝or′ə-thān′) *n.* [UR(O)-¹ + ETH(YL) + -ANE.] **1.** A colorless crystalline or white granular compound, $C_3H_7NO_2$, used in palliative treatment for leukemia and as a solvent. **2.** Any of several esters, other than the ethyl ester, of carbamic acid.

**u·re·thra** (yo͞o-rē′thrə) *n., pl.* **-thras** *or* **-thrae** (-thrē) [LLat. *urethra* < Gk. *ourethra* < *ourein*, to urinate < *ouron*, urine.] The canal through which urine is discharged from the bladder in most mammals and which serves as the male genital duct. **—u·re′thral** *adj.*

**u·re·thri·tis** (yo͝or′ĭ-thrī′tĭs) *n.* Inflammation of the urethra.

**u·re·thro·scope** (yo͞o-rē′thrə-skōp′) *n.* An instrument for examining the interior of the urethra.

**u·ret·ic** (yo͞o-rĕt′ĭk) *adj.* [LLat. *ureticus* < Gk. *ourētikos* < *ourein*, to urinate < *ouron*, urine.] Of or pertaining to urine : URINARY.

**urge** (ûrj) *v.* **urged, urg·ing, urg·es.** [Lat. *urgere*.] *—vt.* **1.** To drive forward or onward forcefully : IMPEL. **2.** To entreat strongly and repeatedly : EXHORT <The crowd was *urged* to disband.> **3.** To advocate or press earnestly or persistently <*urge* caution in driving> **4.** To stimulate : excite. **5.** To persuade, force, or otherwise move to a course of action <*urged* by hunger> *—vi.* **1.** To present a forceful argument, claim, or case. **2.** To exert an impelling force : push vigorously. *—n.* **1.** The act or process of urging. **2.** An irresistible or impelling force, influence, or instinct <"There is a human *urge* to clarify, rationalize, justify" —Leonard Bernstein>

☆ ***syns:*** URGE, EXHORT, PRESS, PROD, PROMPT *v. core meaning:* to impel to action <*urged* them to report the crime>

**ur·gen·cy** (ûr′jən-sē) *n., pl.* **-cies. 1.** The condition of being urgent : pressing importance. **2.** A compelling or insistent need.

**ur·gent** (ûr′jənt) *adj.* [ME < OFr. < Lat. *urgens*, pr.part. of *urgere*, to urge.] **1.** Compelling immediate action : PRESSING <an *urgent* situation> **2.** Earnestly insistent or importunate <*urgent* pleas> **3.** Conveying a sense of pressing importance <an *urgent* call>

**-urgy** *suff.* [NLat. *-urgia* < Gk. *-ourgos*, worker < *ergon*, work.] Technique or process for working with <*zymurgy*>

**-uria** *suff.* [NLat. < Gk. *-ouria* < *ouron*, urine.] **1.** The condition of having a specified substance in the urine <*aciduria*> **2.** The condition of having a specified kind of urine <*polyuria*>

**U·ri·ah** (yo͞o-rī′ə) *n.* [Heb. *ūriyāh*.] Bathsheba's husband, a Hittite officer in the Israelite army, whose death in battle was arranged by David so that he could marry Bathsheba.

**u·ric** (yo͝or′ĭk) *adj.* Pertaining to, contained in, or obtained from urine.

**uric acid** *n.* A white crystalline compound, $C_5H_4N_4O_3$, the end product of purine metabolism in humans and other primates, birds, terrestrial reptiles, and most insects.

**u·ri·co·sur·ic** (yo͝or′ĭ-kə-so͝or′ĭk) *adj.* [URIC + -O- + *-s-* (connective element) + URIC.] Promoting excretion of uric acid in the urine.

**u·ri·co·tel·ic** (yo͝or′ĭ-kō-tĕl′ĭk) *adj.* Excreting unneeded nitrogen in the form of uric acid. **—u′ri·co·tel′ism** (-tĕl′ĭz′əm, -kŏt′l-) *n.*

**u·ri·dine** (yo͝or′ĭ-dēn′) *n.* A white, odorless powder, $C_9H_{12}N_2O_6$, that is the nucleoside of uracil, important in carbohydrate metabolism and used in biochemical experiments.

**U·ri·el** (yo͝or′ē-əl) *n.* [Heb. *ūrī′ēl*.] One of the archangels.

**U·rim and Thum·mim** (yo͝or′ĭm ən thŭm′ĭm) *pl.n.* Objects carried inside the breastplate of the chief priests of ancient Israel and used as oracular media to divine the will of God.

**urin-** *pref. var. of* URINO-.

**u·ri·nal** (yo͝or′ə-nəl) *n.* [ME, chamber pot < OFr. < LLat. < *urina*, urine.] **1. a.** An upright wall fixture used by men for urinating. **b.** A room or other enclosure containing such a fixture. **2.** A receptacle for urine used by a bedridden patient.

**u·ri·nal·y·sis** (yo͝or'ə-năl'ĭ-sĭs) *n.* [URIN(O)- + (AN)ALYSIS.] Chemical analysis of urine.

**u·ri·nar·y** (yo͝or'ə-nĕr'ē) *adj.* Of or pertaining to urine, its production, function, or excretion.

**urinary bladder** *n.* A muscular membranous sac situated in the anterior part of the pelvic cavity and serving as a urine reservoir prior to excretion.

**urinary calculus** *n. Pathol.* A solid concretion of mineral and organic substances in the urinary tract.

**u·ri·nate** (yo͝or'ə-nāt') *vi.* **-nat·ed, -nat·ing, -nates.** [Med. Lat. *urinare, urinat-* < Lat. *urina,* urine.] To discharge urine.

**u·rine** (yo͝or'ĭn) *n.* [ME < OFr. < Lat. *urina.*] The fluid and dissolved substances secreted by the kidneys, stored in the bladder, and discharged from the body through the urethra.

**u·ri·nif·er·ous** (yo͝or'ə-nĭf'ər-əs) *adj.* Conveying urine.

**urino-** *or* **urin-** *pref.* [< Lat. *urina,* urine.] Urine <*urinalysis*>

**u·ri·no·gen·i·tal** (yo͝or'ə-nō-jĕn'ĭ-təl) *n. var. of* UROGENITAL.

**u·ri·nous** (yo͝or'ə-nəs) *also* **u·ri·nose** (-nōs') *adj.* Of, resembling, or containing urine.

**urn** (ûrn) *n.* [ME *urne* < Lat. *urna.*] **1.** A vase of varying size and shape, usu. having a footed base or pedestal, esp. an ornamental vase for holding the ashes of the dead. **2.** A closed metal vessel having a spigot and used for making or serving tea or coffee. **3.** *Bot.* The spore-bearing part of a moss capsule.

**uro-¹** *or* **ur-** *pref.* [NLat. < Gk. *ouro-* < *ouron,* urine.] **1.** Urine <*uric*> **2.** Urinary tract <*urology*> **3.** Urea <*urethane*>

**uro-²** *or* **ur-** *pref.* [NLat. < Gk. *oura,* tail.] Tail <*urochord*>

**u·ro·chord** (yo͝or'ə-kôrd') *n.* [URO-² + CHORD.] *Zool.* A notochord limited to the caudal region, as in tunicates.

**u·ro·chrome** (yo͝or'ə-krōm') *n.* The pigment responsible for the normal yellow color of urine.

**u·ro·gen·i·tal** (yo͝or'ō-jĕn'ĭ-təl) *also* **u·ri·no·gen·i·tal** (yo͝or'ə-nō-) *adj.* Of, relating to, or involving the urinary and genital organs and their functions : GENITOURINARY.

**u·ro·ki·nase** (yo͝or'ō-kī'nās) *n.* An enzyme that is found in human urine and is used for dissolving intravascular blood clots.

**u·ro·lith** (yo͝or'ə-lĭth') *n.* A urinary calculus. **—u'ro·lith'ic** *adj.*

**u·ro·lith·i·a·sis** (yo͝or'ə-lĭ-thī'ə-sĭs) *n.* Formation or presence of urinary calculi.

**u·rol·o·gy** (yo͞o-rŏl'ə-jē) *n.* The branch of medicine concerned with the physiology and pathology of the urogenital tract.

**-uronic** *suff.* [< Gk. *ouron,* urine.] Connected with urine <hyal*uronic*>

**u·ro·pod** (yo͝or'ə-pŏd') *n.* [URO-² + -POD.] One of a pair of posterior abdominal appendages of certain crustaceans, as the lobster or shrimp.

**u·ro·py·gi·al gland** (yo͝or'ə-pī'jē-əl, -pĭj'ē-) *n.* An oil-secreting gland at the base of a bird's tail.

**u·ro·py·gi·um** (yo͝or'ə-pī'jē-əm, -pĭj'ē-) *n.* [NLat. < Gk. *ouropygion* : *oura,* tail + *pugē,* rump.] The posterior part of a bird's body, from which the tail feathers grow. **—u'ro·py'gi·al** *adj.*

**u·ros·co·py** (yo͞o-rŏs'kə-pē) *n., pl.* **-pies.** Microscopic examination of urine.

**-urous** *suff.* [NLat. *-urus* < Gk. *-ouros* < *oura,* tail.] Having a specified kind of tail <an*urous*>

**Ur·sa Major** (ûr'sə) *n.* [Lat., the greater bear.] A constellation in the Northern Hemisphere.

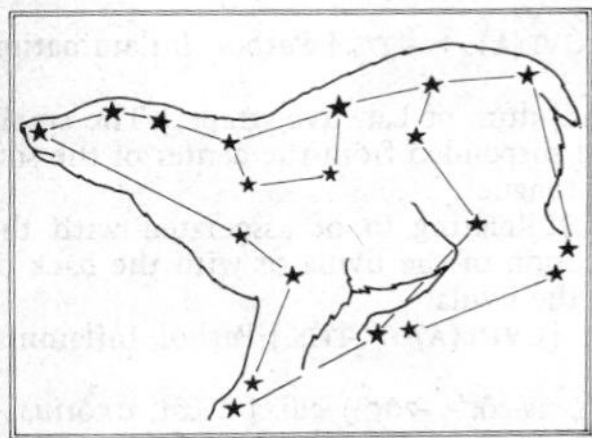

Ursa Major

**Ursa Minor** *n.* [Lat., the lesser bear.] A constellation having the shape of a ladle with Polaris at the tip of its handle.

**ur·sine** (ûr'sīn') *adj.* [Lat. *ursinus* < *ursus,* bear.] Of or characteristic of a bear.

**Ur·spra·che** (o͞or'shprä'кнə) *n.* [G. : *ur-,* original (< OHG *ur,* out of) + *Sprache,* language < OHG *sprāhha,* speech.] A parent language reconstructed from the evidence of later languages.

**Ur·su·line** (ûr'sə-lĭn, -līn', -lēn', ûr'syə-) *n.* [After St. *Ursula.*] A member of an order of nuns of the Roman Catholic Church, founded in Italy in the early 16th cent. and devoted to the education of girls. *—adj.* Of or belonging to the Ursuline.

**ur·ti·cant** (ûr'tĭ-kənt) *adj.* Causing itching or stinging. *—n.* A substance that causes itching or stinging.

**ur·ti·car·i·a** (ûr'tĭ-kâr'ē-ə) *n.* [NLat. < Lat. *urtica,* nettle.] A skin condition marked by intensely itching wheals and usu. caused by allergic reactions to internal or external agents.

**ur·ti·cate** (ûr'tĭ-kāt') *vi.* **-cat·ed, -cat·ing, -cates.** [Med. Lat. *urticare, urticat-* < Lat. *urtica,* nettle.] To sting or whip with or as if with nettles. *—adj.* (ûr'tĭ-kĭt, -kāt'). Characterized by the presence of itching or stinging wheals.

**ur·ti·ca·tion** (ûr'tĭ-kā'shən) *n.* **1.** A lashing with nettles once used to treat a paralyzed bodily part. **2.** The sensation of having been stung by nettles. **3.** Urticaria.

**u·rus** (yo͝or'əs) *n.* [Lat., of Germanic orig.] An extinct bovine mammal, *Bos primigenius* of northern Africa, Europe, and western Asia, thought to be the forerunner of domestic cattle.

**u·ru·shi·ol** (o͞o-ro͞o'shē-ôl', -ōl') *n.* [J. *urushi,* lacquer + -OL.] A toxic substance present in the resin of plants of the genus *Rhus,* which includes poison ivy and the lacquer tree, *R. verniciflua,* from which a black Japanese lacquer is obtained.

**us** (ŭs) *pron.* [ME < OE *ūs.*] *The objective case of* WE. —Used: **a.** As the direct object of a verb <The film impressed *us.*> **b.** As the indirect object of a verb <The singer gave *us* an autograph.> **c.** As the object of a preposition <The singer gave an autograph to *us.*>

**us·a·ble** *also* **use·a·ble** (yo͞o'zə-bəl) *adj.* **1.** Capable of being used. **2.** Fit for use. **—us'a·bil'i·ty, us'a·ble·ness** *n.* **—us'a·bly** *adv.*

**us·age** (yo͞o'sĭj, -zĭj) *n.* [ME < OFr. < *user,* to use. —see USE.] **1.** The act or manner of using or treating <computer *usage*> **2.** Customary and accepted practice or procedure. **3.** The customary way in which the elements of a language are used in speech or writing <modern English *usage*> **4.** A particular expression in speech or writing <a nonstandard *usage*>

**us·ance** (yo͞o'zəns) *n.* [ME, usage < OFr. < VLat **usantia* < **usare,* to use. —see USE.] **1.** The length of time, established by custom and varying between countries, that is allowed for payment of a foreign bill of exchange. **2.** *Obs.* Use. **3.** *Obs.* Usage : custom. **4.** *Obs.* Interest paid on money.

**Us·beg** (o͞os'bĕg', ŭs'-) *or* **Us·bek** (-bĕk') *n. vars. of* UZBEK.

**use** (yo͞oz) *v.* **used, us·ing, us·es.** [ME *usen* < OFr. *user* < VLat. **usare* < Lat. *uti.*] *—vt.* **1.** To bring or put into service or action : EMPLOY <*use* a pen><*use* your imagination> **2.** To put to some purpose : avail oneself of <*use* the bus to get to work> **3.** To conduct oneself toward : TREAT <*use* someone unkindly> **4.** *Informal.* To exploit for one's own advantage or gain <*used* people to get ahead> **5.** To take or partake of regularly, as tobacco, alcohol, or drugs. *—vi.* (yo͞os, yo͞ost). —Used in the past tense with an infinitive to indicate a former state, habitual practice, or custom <Doctors *used* to make house calls.> **—use up.** To consume completely : EXHAUST <*used up* all the paint> *—n.* (yo͞os). **1. a.** The act of using or putting to a purpose <the *use* of a car> **b.** The condition or fact of being used. **2.** The manner of using <the proper *use* of a dictionary> **3. a.** The permission, privilege, or benefit of using something <have *use* of the library> **b.** The power or ability to use something <regained the *use* of both eyes> **4.** The need or occasion to use or employ <Do you still have *use* for a tutor?> **5.** The quality of being suitable or adaptable to an end : USEFULNESS. **6.** The goal, object, or purpose for which something is used. **7.** Accustomed or usual procedure : habitual practice. **8.** *Law.* **a.** The enjoyment of property, as by occupying or exercising it. **b.** The benefit or profit of lands and tenements of which the legal title and possession are vested in another who holds them in trust for the beneficiary. **c.** The arrangement establishing the equitable right to such benefits and profits. **9.** The distinctive form of ritual or liturgy practiced in a particular church, ecclesiastical district, or community. **10.** *Obs.* Usual occurrence or experience.

**use·a·ble** (yo͞o'zə-bəl) *adj. var. of* USABLE.

**used** (yo͞ozd) *adj.* Not new : SECONDHAND <*used* clothing>

**use·ful** (yo͞os'fəl) *adj.* Capable of being used advantageously : SERVICEABLE. **—use'ful·ly** *adv.* **—use'ful·ness** *n.*

**use·less** (yo͞os'lĭs) *adj.* **1.** Having no practical purpose or use <a *useless* device> **2.** Of no avail : FUTILE <*useless* to complain> **—use'less·ly** *adv.* **—use'less·ness** *n.*

**us·er** (yo͞o'zər) *n.* **1.** One that uses. **2.** *Law.* Exercise or enjoyment of a right or property. **3.** *Slang.* A drug addict.

**ush·er** (ŭsh'ər) *n.* [ME < AN *usser* < OFr. *ussier* < Med. Lat. *ustiarius* < Lat. *ostiarius* < *ostium,* entrance.] **1.** One who serves as official doorkeeper, as in a courtroom or legislative chamber. **2.** One who escorts people to their seats, as in a theater or church. **3.** A male attendant at a wedding. **4.** An official whose duty it is to make introductions between strangers or to walk before a person of rank in a procession. **5.** *Archaic.* An assistant teacher in a school. *—v.* **-ered, -er·ing, -ers.** *—vt.* **1.** To serve as an usher to : ESCORT. **2.** To lead or conduct : cause to enter. **3.** To serve to introduce or inaugurate <a party to *usher* in the new year> *—vi.* To act as an usher.

**ush·er·ette** (ŭsh'ə-rĕt') *n.* A woman employed to escort people to their seats, as in a theater.

**us·ne·a** (ŭs′nē-ə, ŭz′-) *n.* [< NLat. *Usnea,* genus name < Ar. *ushnah,* moss.] Any of various widely distributed lichens of the genus *Usnea,* characterized by a gray pendulous thallus.

**us·que·baugh** (ŭs′kwĭ-bô′, -bä′) *n.* [Sc. Gaelic and Ir. Gaelic *uisge beatha,* water of life.] *Ir. & Scot.* Whiskey.

**u·su·al** (yo͞o′zho͞o-əl) *adj.* [ME < OFr. < LLat. *usualis* < Lat. *usus,* use < p.part. of *uti,* to use.] **1.** Commonly encountered, experienced, observed, or used : ORDINARY <not the *usual* college student> **2.** Habitual or customary : NORMAL <the *usual* routine> **—as usual.** In the usual manner <grumpy *as usual*> **—u′su·al·ly** *adv.* **—u′su·al·ness** *n.*

**u·su·fruct** (yo͞o′zə-frŭkt′, -sə-) *n.* [Lat. *ususfructus* : *usus,* use (< p.part. of *uti,* to use) + *fructus,* enjoyment < *frui,* to enjoy.] *Law.* The right to utilize and enjoy the profits and advantages of something belonging to another so long as the property is not damaged or altered.

**u·su·fruc·tu·ar·y** (yo͞o′zə-frŭk′cho͞o-ĕr′ē, -sə-) *n., pl.* **-ies.** *Law.* One who holds property by usufruct. **—u′su·fruc′tu·ar′y** *adj.*

**u·su·rer** (yo͞o′zhər-ər) *n.* [ME < AN < Med. Lat. *usuarius* < Lat. *usura,* usury. —see USURY.] One who lends money at an exorbitant or illegal rate of interest.

**u·su·ri·ous** (yo͞o-zho͝or′ē-əs) *adj.* **1.** Practicing usury. **2.** Of or constituting usury <a *usurious* rate of interest> **—u·su′ri·ous·ly** *adv.* **—u·su′ri·ous·ness** *n.*

**u·surp** (yo͞o-sûrp′, -zûrp′) *v.* **-surped, -surp·ing, -surps.** [ME *usurpen* < OFr. *usurper* < Lat. *usurpare,* to make use of.] *—vt.* **1.** To seize and hold, as the power, position, or rights of another, by force and without right or authority <*usurped* the throne> **2.** To take over or occupy physically, as territory or possessions. *—vi.* To seize another's power, rights, or possessions illegally. **—u·surp′er** *n.* **—u·surp′ing·ly** *adv.*

**u·sur·pa·tion** (yo͞o′sər-pā′shən, -zər-) *n.* **1.** An act of usurping, esp. illegal seizure of royal sovereignty. **2.** *Law.* Illegal encroachment on or exercise of authority or privilege belonging to another.

**u·su·ry** (yo͞o′zhə-rē) *n., pl.* **-ries.** [ME < AN *usurie < Med. Lat. *usuria* < Lat. *usura* < *usus,* use. —see USUAL.] **1. a.** The act or practice of lending money at an exorbitant or illegal rate of interest. **b.** An exorbitant or illegal rate of interest. **2.** The act or practice of lending money at any rate of interest. **3.** *Archaic.* Interest paid on a loan.

**ut** (ŭt, o͞ot) *n.* [Med. Lat. —see GAMUT.] *Mus.* A syllable representing the tone *C,* otherwise represented by *do,* in the French system of solmization.

**Ute** (yo͞ot) *n., pl.* **Ute** *or* **Utes.** [Ute *Yuta.*] **1. a.** A tribe of North American Indians once inhabiting Utah, Colorado, and New Mexico and now living on reservations in Utah and Colorado. **b.** A member of this tribe. **2.** The Uto-Aztecan language of the Ute.

**u·ten·sil** (yo͞o-tĕn′səl) *n.* [ME *utensele* < OFr. *utensile* < Lat. *utensilia,* utensils, neuter pl. of *utensilis,* fit for use < *uti,* to use.] **1.** An implement or vessel used domestically, as in a kitchen <cooking *utensils*> **2.** Any useful tool or vessel.

**u·teri** (yo͞o′tə-rī′) *n. pl. of* UTERUS.

**u·ter·ine** (yo͞o′tər-ĭn, -tə-rīn′) *adj.* [LLat. *uterinus* < *uterus,* uterus.] **1.** Of or pertaining to the uterus. **2.** Having the same mother but different fathers <*uterine* siblings>

**u·ter·us** (yo͞o′tər-əs) *n., pl.* **u·teri** (yo͞o′tə-rī′) [Lat.] **1.** A pear-shaped muscular organ of gestation situated in the pelvic cavity of female mammals and that receives and holds the fertilized ovum during development of the fetus and is the principal agent in its expulsion at birth. **2.** A part of the female reproductive tract in many invertebrates that is similar to the uterus, serving as a repository for storage or development of eggs or embryos.

**U·ther** (yo͞o′thər) *or* **Uther Pen·drag·on** (pĕn-drăg′ən) *n.* A legendary king of Britain and father of King Arthur.

**u·tile** (yo͞ot′l, yo͞o′tīl′) *adj.* [ME < OFr. < Lat. *utilis* < *uti,* to use.] Useful.

**u·til·i·tar·i·an** (yo͞o-tĭl′ĭ-târ′ē-ən) *adj.* [UTILIT(Y) + -ARIAN.] **1.** Relating to or based on utility. **2.** Stressing the value of practical over aesthetic qualities. **3.** Intended or made for utility <*utilitarian* clothing> **4.** Believing in or advocating utilitarianism. *—n.* An advocate or adherent of utilitarianism.

**u·til·i·tar·i·an·ism** (yo͞o-tĭl′ĭ-târ′ē-ə-nĭz′əm) *n.* **1.** *Philos.* The doctrine that considers utility as the criterion of action and the useful as good or worthwhile. **2.** The ethical theory proposed by Jeremy Bentham and John Stuart Mill that all moral, social, or political action should be directed toward achieving the greatest good for the greatest number of people.

**u·til·i·ty** (yo͞o-tĭl′ĭ-tē) *n., pl.* **-ties.** [ME *utilite* < OFr. < Lat. *utilitas* < *utilis,* useful < *uti,* to use.] **1.** The quality or state of being useful : USEFULNESS. **2.** A useful article or device. **3.** A service provided to the public, as electricity, water, or transportation. *—adj.* **1.** Designed primarily for practical use <a *utility* car> **2.** Of the lowest U.S. Government grade of meat.

**utility man** *n.* **1.** A member of a theatrical cast who can play any of the smaller roles on short notice. **2.** A reserve player capable of playing several positions in a sport, esp. baseball. **3.** A worker expected to serve in several capacities.

**u·til·ize** (yo͞ot′l-īz′) *vt.* **-ized, -iz·ing, -iz·es.** [Fr. *utiliser* < Ital. *utilizzare* < *utile,* useful < Lat. *utilis* < *uti,* to use.] To put to use. **—u′til·iz′a·ble** *adj.* **—u′til·i·za′tion** *n.* **—u′til·iz′er** *n.*

**ut·most** (ŭt′mōst′) *adj.* [ME < OE *ūtmest* : *ūt,* out + *-mest,* -most.] **1.** Being or situated at the farthest limit or point : most extreme. **2.** Of the highest or greatest degree, amount, or intensity <the *utmost* secrecy> *—n.* The greatest possible amount, degree, or extent <the *utmost* in luxury><tried their *utmost*>

**U·to-Az·tec·an** (yo͞o′tō-ăz′tĕk′ən) *n.* [UTE + AZTEC.] **1.** A language phylum of North and Central America that includes Ute, Hopi, Nahuatl, and Shoshone. **2. a.** A tribe speaking a Uto-Aztecan language. **b.** A member of such a tribe. *—adj.* Of or relating to the Uto-Aztecans or to the languages spoken by them.

**u·to·pi·a** (yo͞o-tō′pē-ə) *n.* [NLat. : Gk. *ou,* not + Gk. *topos,* place.] **1. Utopia.** An imaginary island of ideal perfection described in Sir Thomas More's *Utopia* (1516). **2. Utopia.** An ideally perfect place, esp. in its socio-political aspects. **2.** An impractical, idealistic concept for social and political reform.

**u·to·pi·an** (yo͞o-tō′pē-ən) *adj.* **1. Utopian.** Of, relating to, or having the characteristics of Utopia. **2.** Based on or involving the ideal of a perfect society. **3.** Given to visionary and impractical theories of social reform. *—n.* A zealous but impractical reformer.

**u·to·pi·an·ism** (yo͞o-tō′pē-ə-nĭz′əm) *n.* The ideals or principles of a utopian : idealistic and impractical social theory.

**utopian socialism** *n.* A belief that propertied groups will peacefully and voluntarily relinquish their holdings in order to achieve common ownership of the means of production.

**u·tri·cle** (yo͞o′trĭ-kəl) *also* **u·tric·u·lus** (yo͞o-trĭk′yə-ləs) *n., pl.* **u·tri·cles** *also* **u·tric·u·li** (-lī′) [Fr. < Lat. *utriculus,* dim. of *uter,* leather bottle.] **1.** A small, delicate membranous sac connecting with the semicircular canals of the inner ear and functioning in the maintenance of bodily equilibrium and coordination. **2.** *Bot.* A small bladderlike one-seeded fruit. **—u·tric′u·lar** (yo͞o-trĭk′yə-lər) *adj.*

**u·tric·u·lus** (yo͞o-trĭk′yə-ləs) *n. var. of* UTRICLE.

**ut·ter¹** (ŭt′ər) *vt.* **-tered, -ter·ing, -ters.** [ME *utteren* < MDu. *ūteren.*] **1.** To express audibly <*utter* a cry> **2.** To express by means of speech : PRONOUNCE. **3. a.** To put (counterfeit money or a forgery) into circulation. **b.** To deliver (something counterfeit) to another. **4.** *Obs.* To sell (goods). **—ut′ter·a·ble** *adj.* **—ut′ter·er** *n.*

**ut·ter²** (ŭt′ər) *adj.* [ME < OE *ūtera,* outer < *ūt,* out.] Completely such, without qualification or exception.

☆ ***syns:*** UTTER, ALL-OUT, ARRANT, COMPLETE, CONSUMMATE, FLAT, OUT-AND-OUT, OUTRIGHT, POSITIVE, PURE, SHEER, THOROUGH, TOTAL, UNMITIGATED, UNQUALIFIED *adj. core meaning* : completely such, without qualification or exception <an *utter* fool><*utter* chaos>

**ut·ter·ance¹** (ŭt′ər-əns) *n.* **1. a.** The act of expressing vocally. **b.** The power of speaking. **2.** Something uttered or expressed, esp. a verbal or written statement.

**ut·ter·ance²** (ŭt′ər-əns) *n.* [ME < OFr. *outrance* < *outrer,* to go beyond limits < VLat. *ultrare < Lat. *ultra,* beyond.] The uttermost end or extremity : bitter end <fight to the *utterance*>

**ut·ter·ly** (ŭt′ər-lē) *adv.* Completely : absolutely <*utterly* mad>

**ut·ter·most** (ŭt′ər-mōst′) *adj.* [ME : *utter,* outer + *-most,* -most.] **1.** Utmost. **2.** Outermost. *—n.* Utmost.

**U-turn** (yo͞o′tûrn′) *n.* A turn, as by a vehicle, completely reversing the direction of travel.

**u·va·rov·ite** (yo͞o-vär′ə-vīt′, o͞o-) *n.* [G. *Uvarovit,* after Count Sergei S. *Uvarov* (1785–1855).] An emerald-green garnet, $Ca_3Cr_2(SiO_4)_3$, found in chromium deposits.

**u·ve·a** (yo͞o′vē-ə) *n.* [Med. Lat. < Lat. *uva,* grape.] The pigmented vascular layer of the eye comprising the iris, ciliary body, and choroid. **—u′ve·al** *adj.*

**u·ve·i·tis** (yo͞o′vē-ī′tĭs) *n.* [UVE(A) + -ITIS.] *Pathol.* Inflammation of the uvea.

**u·vu·la** (yo͞o′vyə-lə) *n.* [LLat., dim. of Lat. *uva,* grape.] The small, conical, fleshy mass of tissue suspended from the center of the soft palate above the back of the tongue.

**u·vu·lar** (yo͞o′vyə-lər) *adj.* **1.** Relating to or associated with the uvula. **2.** Articulated by vibration of the uvula or with the back of the tongue near or touching the uvula.

**u·vu·li·tis** (yo͞o′vyə-lī′tĭs) *n.* [UVUL(A) + -ITIS.] *Pathol.* Inflammation of the uvula.

**ux·o·ri·al** (ŭk-sôr′ē-əl, -sōr′-, ŭg-zôr′-, -zōr′-) *adj.* [< Lat. *uxorius* < *uxor,* wife.] Of, pertaining to, or characteristic of a wife.

**ux·o·ri·cide** (ŭk-sôr′ĭ-sīd′, -sōr′-, ŭg-zôr′-, ŭg-zōr′-) *n.* [Med. Lat. *uxoricidium* : Lat. *uxor,* wife + Lat. *-cidium,* killing < *caedere,* to kill.] **1.** Murder of a wife by her husband. **2.** A man who kills his wife.

**ux·o·ri·ous** (ŭk-sôr′ē-əs, -sōr′-, ŭg-zôr′-, -zōr′-) *adj.* [Lat. *uxorius* < *uxor,* wife.] Excessively submissive or devoted to one's wife. **—ux·o′ri·ous·ly** *adv.* **—ux·o′ri·ous·ness** *n.*

**Uz·bek** (o͝oz′bĕk′, ŭz′-) *also* **Uz·beg** (-bĕg′) *or* **Us·bek** (o͝os′bĕk′, ŭs′-) *or* **Us·beg** (-bĕg′). **1.** A member of a group of Turkic people inhabiting the Uzbek S.S.R. **2.** The Turkic language of the Uzbeks.

ă **pat** ā **pay** âr **care** ä **father** ĕ **pet** ē **be** hw **which** ĭ **pit**
ī **tie** îr **pier** ŏ **pot** ō **toe** ô **paw, for** oi **noise** o͝o **took**

# Vv

**v** or **V** (vē) *n.*, *pl.* **v's** or **V's. 1.** The 22nd letter of the English alphabet. **2.** A speech sound represented by the letter *v*. **3.** The 22nd in a series. **4.** Something shaped like the letter V. **5. V** The Roman numeral for five.

**V** *symbol for* VANADIUM.

**V-1** (vē'wŭn') *n.* [G. *Vergeltungswaffe eins*, retaliation weapon (number) one.] ROBOT BOMB 1.

**V-2** (vē'too') *n.* [G. *Vergeltungswaffe zwei*, retaliation weapon (number) two.] A long-range liquid-fuel rocket used by the Germans as a World War II ballistic missile.

**va·can·cy** (vā'kən-sē) *n.*, *pl.* **-cies. 1.** The state of being vacant or unoccupied. **2.** Empty space. **3.** An unfilled or unoccupied position, office, or accommodation. **4.** Inanity. **5.** *Archaic.* A period of idle leisure.

**va·cant** (vā'kənt) *adj.* [ME < OFr. < Lat. *vacans*, pr.part. of *vacare*, to be empty.] **1.** Containing nothing : EMPTY. **2.** Being without an incumbent or occupant <a *vacant* position in the department> **3.** Not put to use or occupied <a weedy *vacant* lot> **4.** *Law.* Not claimed, as by an heir <a *vacant* estate> **5. a.** Lacking knowledge or intelligence. **b.** Expressionless : blank <looked at me with a *vacant* stare> **6.** Not filled with activity <*vacant* early morning hours> **—va'cant·ly** *adv.* **—va'cant·ness** *n.*

**va·cate** (vā'kāt', vā-kāt') *v.* **-cat·ed, -cat·ing, -cates.** [Lat. *vacare*, *vacat-*, to be empty.] *—vt.* **1. a.** To cease to occupy or hold : GIVE UP. **b.** To empty of occupants or incumbents. **2.** *Law.* To make void : ANNUL. *—vi.* To leave a job, office, or lodging.

**va·ca·tion** (vā-kā'shən, və-) *n.* [ME *vacacioun* < OFr. *vacation* < Lat. *vacatio*, freedom from occupation < *vacare*, to be at leisure.] **1.** A period of time for pleasure, rest, or relaxation, esp. one with pay granted to an employee. **2. a.** A holiday. **b.** A fixed holiday period, esp. one during which a school, court, or business suspends activities. **3.** *Archaic.* An act or instance of vacating. *—vi.* **-tioned, -tion·ing, -tions.** To take or spend a vacation. **—va·ca'tion·er, va·ca'tion·ist** *n.*

**va·ca·tion·land** (vā-kā'shən-lănd') *n.* A place having special attractions for vacationers.

**vac·ci·nal** (văk'sə-nəl, văk-sē'-) *adj.* Of or pertaining to vaccine or vaccination.

**vac·ci·nate** (văk'sə-nāt') *v.* **-nat·ed, -nat·ing, -nates.** [< VACCINE.] *—vt.* To inoculate with a vaccine so as to produce immunity to a disease. *—vi.* To give a vaccination. **—vac'ci·na'tor** *n.*

**vac·ci·na·tion** (văk'sə-nā'shən) *n.* **1.** Inoculation with a vaccine so as to protect against a given disease. **2.** A scar on the skin caused by vaccinating.

**vac·cine** (văk-sēn', văk'sēn') *n.* [< Lat. *vaccinus*, of cows < *vacca*, cow.] **1.** A suspension of attenuated or killed microorganisms, as of viruses or bacteria, incapable of inducing severe infection but capable when inoculated of counteracting the unmodified species. **2.** A vaccine prepared from the cowpox virus and inoculated against smallpox.

▲ word history: The first vaccine was prepared from the virus that causes cowpox, as the derivation of the word *vaccine* from Latin *vacca*, "cow," might suggest. Cowpox is a mild disease of cows that can be caught by human beings. In the late 18th century, Dr. Edward Jenner discovered that someone who has had cowpox is almost always immune to smallpox, a related, but much more serious, disease. Jenner invented a method of inoculating human beings with the cowpox virus; he called the inoculating agent "vaccine." Since that time many other diseases have been prevented by inoculation, and the word *vaccine* has been applied to the inoculating agent in all such cases.

**vac·cin·i·a** (văk-sĭn'ē-ə) *n.* [NLat. < Lat. *vaccinus*, of cows.] Cowpox. **—vac·cin'i·al** *adj.*

**vac·il·lant** (văs'ə-lənt) *adj.* Vacillating.

**vac·il·late** (văs'ə-lāt') *vi.* **-lat·ed, -lat·ing, -lates.** [Lat. *vacillare*, *vacillat-*, to waver.] **1.** To sway from side to side : OSCILLATE. **2.** To swing indecisively from one course of action or opinion to another. **—vac'il·lat'ing·ly** *adv.* **—vac'il·la'tion** *n.* **—vac'il·la'tor** *n.*

**vac·il·la·to·ry** (văs'ə-lə-tôr'ē, -tōr'ē) *adj.* Tending to waver : IRRESOLUTE <a *vacillatory* foreign policy>

**vac·u·a** (văk'yoo-ə) *n. var. pl. of* VACUUM.

**va·cu·i·ty** (vă-kyoo'ĭ-tē, və-) *n.*, *pl.* **-ties.** [OFr. *vacuite* < Lat. *vacuitas* < *vacuus*, empty. —see VACUUM.] **1.** Complete absence of matter : EMPTINESS. **2.** An empty space : VACUUM. **3.** Total lack of ideas. **4.** Absence of meaningful occupation : IDLENESS. **5.** The quality or fact of being devoid of something specified <a *vacuity* of good taste> **6.** Something, esp. a remark, that is inane or pointless.

**vac·u·o·lat·ed** (văk'yoo-ō-lā'tĭd) *also* **vac·u·o·late** (-lāt', -lĭt) *adj.* Containing a vacuole or vacuoles.

**vac·u·ole** (văk'yoo-ōl') *n.* [Fr. < Lat. *vacuum*, vacuum. —see VACUUM.] A small cavity in cell protoplasm. **—vac'u·o'lar** (-ō'lər, -lär') *adj.* **—vac'u·o·la'tion** *n.*

**vac·u·ous** (văk'yoo-əs) *adj.* [Lat. *vacuus*, empty. —see VACUUM.] **1.** Devoid of matter : EMPTY. **2. a.** Lacking intelligence : STUPID. **b.** Devoid of meaning or substance : INANE <a *vacuous* remark> **3.** Lacking meaningful purpose or occupation : IDLE. **—vac'u·ous·ly** *adv.* **—vac'u·ous·ness** *n.*

**vac·u·um** (văk'yoo-əm, -yoom, -yəm) *n.*, *pl.* **-u·ums** or **-u·a** (-yoo-ə) [Lat., neuter of *vacuus*, empty < *vacare*, to be empty.] **1. a.** Absence of matter. **b.** A space empty of matter. **c.** A space relatively empty of matter. **2.** A state of emptiness : VOID. **3.** A state of being sealed off from external or environmental influences : ISOLATION. **4.** *pl.* **vacuums.** A vacuum cleaner. *—vt.* & *vi.* **-umed, -um·ing, -ums.** To clean with or use a vacuum cleaner.

**vacuum bottle** *n.* A bottle or flask having a vacuum between its inner and outer walls, designed to maintain the desired temperature of the contents.

**vacuum casting** *n.* The casting of metals under a vacuum.

**vacuum cleaner** *n.* An electrical appliance that cleans surfaces by suction.

**vacuum drying** *n.* Removal of liquid material from a solution or mixture under reduced air pressure, resulting in drying at a lower temperature than required at full pressure.

**vacuum gauge** *n.* A device for ascertaining the pressure in a partial vacuum.

**vac·u·um-packed** (văk'yoo-əm-păkt', văk'yoom-, văk'yəm-) *adj.* Packed in a container with little or no air.

**vacuum pump** *n.* **1.** A pump used to evacuate an enclosure. **2.** PULSOMETER 1.

**vacuum tube** *n.* An electron tube having an internal vacuum sufficiently high to permit electrons to move with low interaction with any remaining gas molecules.

**va·de me·cum** (vā'dē mē'kəm, vä'dē mā'-) *n.*, *pl.* **vade me·cums.** [Lat., go with me.] **1.** Something useful that one constantly carries about. **2.** A book, as a guidebook, for ready reference.

**va·dose** (vā'dōs') *adj.* [Lat. *vadosus*, shallow < *vadum*, a shallow, ford.] Pertaining to or being water located in the zone of aeration in the earth's crust above the ground water level.

**vag·a·bond** (văg'ə-bŏnd') *n.* [ME *vagabonde* < OFr. *vagabond* < Lat. *vagabundus*, wandering < *vagari*, to wander < *vagus*, wandering.] **1.** A homeless person who moves from place to place. **2.** TRAMP 3. *—adj.* **1.** Of, relating to, or characteristic of a wanderer : NOMADIC. **2.** Aimless : drifting. **3.** Irregular in course or behavior : UNPREDICTABLE. *—vi.* **-bond·ed, -bond·ing, -bonds.** To lead the life of a vagabond. **—vag'a·bond'age** *n.* **—vag'a·bond'ism** *n.*

**va·gal** (vā'gəl) *adj.* Of, relating to, or mediated by the vagus nerve. **—va'gal·ly** *adv.*

**va·ga·ry** (vā'gə-rē, və-gâr'ē) *n.*, *pl.* **-ries.** [< Lat. *vagari*, to wander < Lat. *vagus*, wandering.] A flight of fancy.

**va·gi** (vā'gī', -jī') *n. pl. of* VAGUS.

**va·gil·i·ty** (və-jĭl'ĭ-tē, vă-) *n.* [< obs. *vagile*, free to move about < Lat. *vagus*, wandering.] An organism's capacity or tendency to become widely dispersed.

**va·gi·na** (və-jī'nə) *n.*, *pl.* **-nas** or **-nae** (-nē) [Lat., sheath.] **1.** *Anat.* **a.** The passage leading from the external genital orifice to the uterus in female mammals. **b.** A similar structure in some invertebrates. **2.** *Biol.* A sheathlike structure or part, as that formed by the base of a leaf enclosing a stem.

**vag·i·nal** (văj'ə-nəl) *adj.* **1.** Of or pertaining to the vagina. **2.** Relating to or resembling a sheath. **—vag'i·nal·ly** *adv.*

**vag·i·nate** (văj'ə-nĭt, -nāt') *also* **vag·i·nat·ed** (-nā'tĭd) *adj.* Forming or enclosed in a sheath.

**vag·i·nec·to·my** (văj'ə-nĕk'tə-mē) *n.*, *pl.* **-mies.** [VAGIN(A) + -ECTOMY.] **1.** Surgical excision of all or part of the vagina. **2.** Surgical excision of the serous membrane covering the testis and epididymus.

**vag·i·nis·mus** (văj′ə-nĭz′məs) *n.* [NLat. : VAGIN(A) + *-ismus,* -ism.] Painful contractional spasm of the vagina.

**vag·i·ni·tis** (văj′ə-nī′tĭs) *n.* [VAGIN(A) + -ITIS.] Inflammation of the vagina.

**va·got·o·my** (vā-gŏt′ə-mē) *n., pl.* **-mies.** [VAG(US) + -TOMY.] Surgical division of the lower thoracic or upper abdominal fibers of the vagus nerve, used to diminish acid secretion of the stomach and control a duodenal ulcer.

**va·go·to·ni·a** (vā′gə-tō′nē-ə) *n.* [VAG(US) + -TONIA.] Pathological overactivity of the vagus nerve. **—va′go·ton′ic** (-tŏn′ĭk) *adj.*

**va·go·tro·pic** (vā′gə-trō′pĭk) *adj.* [VAG(US) + -TROPIC.] Affecting or acting on the vagus nerve. —Used chiefly of drugs.

**va·gran·cy** (vā′grən-sē) *n., pl.* **-cies. 1.** The state of being a vagrant. **2.** The conduct or lifestyle of a vagrant. **3.** Mental wandering. **4.** *Law.* The offense of being a vagrant.

**va·grant** (vā′grənt) *n.* [ME *vagraunt,* prob. < OFr. *wacrant,* pr.part. of *wacrer,* to wander, of Germanic orig.] **1.** One who wanders from place to place without a permanent home or job. **2.** A wanderer. **3.** One, as a drunkard, who lives on the streets and is considered a public nuisance. —*adj.* **1.** Wandering from place to place and lacking a job or income. **2.** Wayward : unrestrained. **3.** Moving in a random way. **—va′grant·ly** *adv.*

**vague** (vāg) *adj.* **vagu·er, vagu·est.** [OFr. < Lat. *vagus.*] **1.** Not clearly expressed or outlined <*vague* directions> **2.** Not thinking or expressing oneself clearly <*vague* about their future plans> **3.** Lacking definite shape, form, or character : INDISTINCT <saw *vague* outlines through the smog> **4.** Ambiguous <*vague* promises> **5.** Indistinctly felt, perceived, understood, or recalled : HAZY <a *vague* recollection of the event> **—vague′ly** *adv.* **—vague′ness** *n.*

☆ ***syns:*** VAGUE, CLOUDY, FOGGY, FUZZY, HAZY, INDEFINITE, INDISTINCT, MISTY, UNCLEAR *adj. core meaning* : not clearly perceived or perceptible <a *vague* form in the mist><a *vague* memory of the accident> ***ant:*** clear

**va·gus** (vā′gəs) *n., pl.* **-gi** (-gī′, -jī′) [NLat. *vagus (nervus),* wandering (nerve).] The tenth and longest of the cranial nerves, passing through the neck and thorax into the abdomen and supplying sensation to part of the ear, the larynx, and the pharynx, motor impulses to the vocal-cord muscles, and motor and secretory impulses to the abdominal and thoracic viscera.

**vagus nerve** *n.* The vagus.

**va·hi·ne** (vä-hē′nē, -nā′) *n. var. of* WAHINE.

**vail[1]** (vāl) *v.* **vailed, vail·ing, vails.** [ME *valen* < *avalen* < OFr. *avaler* < *aval,* downward < Lat. *ad vallem,* to the valley.] *Archaic.* —*vt.* **1.** To lower (e.g., a banner). **2.** To doff (one's hat) as an indication of respect or submission. —*vi.* **1.** To descend. **2.** To doff one's hat.

**vail[2]** (vāl) *n. Obs. var. of* VEIL.

**vain** (vān) *adj.* **-er, -est.** [ME < OFr. < Lat. *vanus.*] **1.** Not yielding the desired outcome : FRUITLESS <a *vain* attempt to escape> **2.** Lacking substance or worth <*vain* chitchat> **3.** Overly proud of one's appearance or accomplishments. **4.** *Archaic.* Foolish. **—in vain. 1.** To no avail <All of our work was *in vain.*> **2.** Irreverently <took God's name *in vain*> **—vain′ly** *adv.* **—vain′ness** *n.*

☆ ***syns:*** VAIN, CONCEITED, NARCISSISTIC *adj. core meaning* : unduly preoccupied with one's own appearance <a *vain,* brittle society person>

**vain·glo·ri·ous** (vān-glôr′ē-əs, -glōr′-) *adj.* Boastful. **—vain·glo′ri·ous·ly** *adv.* **—vain·glo′ri·ous·ness** *n.*

**vain·glo·ry** (vān′glôr′ē, -glōr′ē, vān-glôr′ē, -glōr′ē) *n., pl.* **-ries.** [ME *veyn glory* < OFr. *vaine glorie* < Lat. *vana gloria,* empty pride.] **1.** Boastful and unwarranted pride in one's accomplishments or qualities. **2.** Vain, ostentatious display.

**vair** (vâr) *n.* [ME *vaire* < OFr. *vair* < Lat. *varius,* variegated.] **1.** A fur, prob. squirrel, much used in medieval times to line and trim robes. **2.** A heraldic representation of fur.

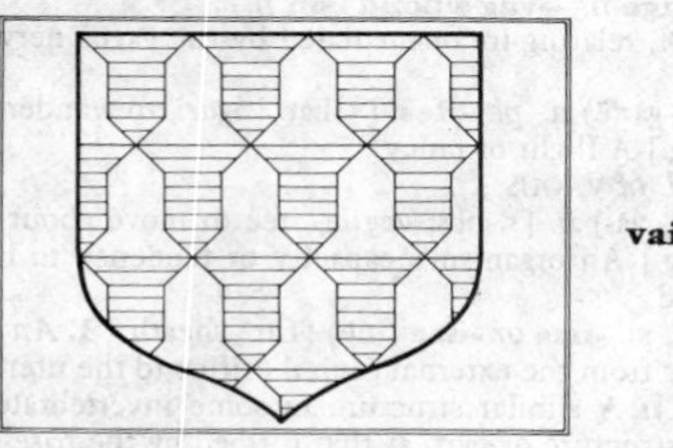

**vair**

**Vaish·na·va** (vīsh′nə-və) *n.* [Skt. *vaiṣṇava-,* relating to Vishnu < *Viṣṇuḥ,* Vishnu.] A member of a Hindu sect that worships Vishnu. **—Vaish′na·vism** (-vĭz′əm) *n.*

**Vais·ya** (vī′shə, vīsh′yə) *n.* [Skt. *vaiśyaḥ,* settler < *viś,* house.] **1.** A Hindu caste orig. composed of farmers and herders but now largely made up of merchants and business people. **2.** A member of the Vaisya caste.

**val·ance** (văl′əns, vā′ləns) *n.* [ME *valaunce.*] **1.** An ornamental drapery hung across a top edge, as of a bed, table, or canopy. **2.** A short drapery, decorative board, or metal strip mounted esp. across the top of a window to hide structural fixtures. —*vt.* **-anced, -anc·ing, -anc·es.** To supply with a valance.

**vale** (vāl) *n.* [ME < OFr. *val* < Lat. *valles.*] A valley.

**val·e·dic·tion** (văl′ĭ-dĭk′shən) *n.* [< Lat. *valedicere,* to say farewell : *vale,* farewell + *dicere,* to say.] **1.** An act of bidding farewell. **2.** A speech or statement made as a farewell, esp. at a high-school commencement.

**val·e·dic·to·ri·an** (văl′ĭ-dĭk-tôr′ē-ən, -tōr′-) *n.* A student, usu. ranking highest in a graduating class, who delivers the farewell speech at a commencement ceremony.

**val·e·dic·to·ry** (văl′ĭ-dĭk′tə-rē) *adj.* Relating to or by way of a farewell. —*n., pl.* **-ries.** A farewell speech, esp. one delivered by a valedictorian at commencement exercises.

**va·lence** (vā′ləns) *also* **va·len·cy** (-lən-sē) *n., pl.* **-lenc·es** *also* **-len·cies.** [LLat. *valentia,* capacity < Lat. *valens,* pr.part. of *valēre,* to be strong.] **1.** *Chem.* **a.** The capacity of an atom or group of atoms to combine in specific proportions with other atoms or groups of atoms. **b.** An integer, often one of several for any given element, used to represent this capacity in terms of an arbitrary assignment of 1 to an atom or group capable of forming a single bond with chlorine and of –1 to an atom or group capable of forming a single bond with hydrogen. **2.** The capacity of something to unite, react, or interact with something else.

**valence bond** *n.* A covalent bond.

**valence electron** *n.* An electron in an outer or next outer shell of an atom that can participate in forming chemical bonds with other atoms.

**valence shell** *n.* A shell of an atom that contains the valence electrons.

**Va·len·ci·ennes** (və-lĕn′sē-ĕn′, -ĕnz′, văl′ən-sē-) *n.* A fine lace with a floral pattern orig. made in Valenciennes, France.

**-valent** *suff.* [< VALENCE.] Having a specified valence or valences <poly*valent*>

**val·en·tine** (văl′ən-tīn′) *n.* **1. a.** A usu. sentimental greeting card sent to one's sweetheart on Saint Valentine's Day. **b.** A greeting or gift sent to one's sweetheart on Saint Valentine's Day. **2.** One singled out as one's sweetheart on Saint Valentine's Day.

**Valentine's Day** *or* **Valentines Day** *n.* Saint Valentine's Day.

**va·le·ri·an** (və-lîr′ē-ən) *n.* [ME < OFr. *valeriane* < Med. Lat. *valeriana,* prob. < fem. of Lat. *Valerianus,* of Valeria, Roman province where the plant originated.] **1.** A plant of the genus *Valeriana,* having small white or pinkish flower clusters, esp. *V. officinalis,* native to Eurasia. **2.** The dried roots of a valerian, *V. officinalis,* used as a sedative.

**va·le·ric acid** (və-lîr′ĭk, -lĕr′-) *n.* [< VALERIAN, from its occurrence in the plant's root.] A colorless liquid, $C_5H_{10}O_2$, used in flavorings, perfumes, plasticizers, and pharmaceuticals.

**val·et** (văl′ĭt, văl′ā′, vă-lā′) *n.* [Fr. < OFr. *vaslet,* servant < Med. Lat. **vassellitus,* dim. of *vassus,* vassal, of Celt. orig.] **1.** A man's personal attendant. **2.** A hotel employee who performs personal services for patrons. —*v.* **-et·ed, -et·ing, -ets.** —*vt.* To act as a personal servant to : ATTEND. —*vi.* To work as a valet.

**val·e·tu·di·nar·i·an** (văl′ĭ-tōōd′n-âr′ē-ən, -tyōōd′-) *n.* [Lat. *valetudinarius* < *valetudo,* state of health < *valēre,* to be well.] A weak or sickly individual, esp. one constantly and morbidly concerned with health matters. **—val′e·tu′di·nar′i·an·ism** *n.*

**val·e·tu·di·nar·y** (văl′ĭ-tōōd′n-ĕr′ē, -tyōōd′-) *adj.* Of, pertaining to, or typical of a valetudinarian. —*n., pl.* **-ies.** A valetudinarian.

**val·gus** (văl′gəs) *n., pl.* **-gus·es.** [< Lat., bowlegged.] *Pathol.* A knock-kneed person. **—val′goid′** (-goid′) *adj.*

**Val·hal·la** (văl-hăl′ə) *n.* [ON *Valhǫll* : *valr,* the slain + *hǫll,* hall.] *Norse Myth.* The great hall of immortality in which the souls of warriors slain heroically were received by Odin and enshrined.

**val·iant** (văl′yənt) *adj.* [ME *valiaunt* < OFr. *vaillant* < Lat. *valens, valent-,* pr.part. of *valēre,* to be strong.] **1.** Having or exhibiting valor : COURAGEOUS. **2.** Characterized by or performed with valor <*valiant* feats in combat> —*n.* A valiant person. **—val′iance, val′ian·cy, val′iant·ness** *n.* **—val′iant·ly** *adv.*

**val·id** (văl′ĭd) *adj.* [Fr. *valide* < OFr. < Lat. *validus,* strong < *valēre,* to be strong.] **1.** Well-grounded <a *valid* concern for quality> **2.** Producing the desired results : EFFICACIOUS <*valid* remedies> **3.** Legally sound and effective : INCONTESTABLE <*valid* claims> **4.** *Logic.* **a.** Containing premises from which the conclusion may logically be derived. **b.** Correctly inferred or deduced from a premise <a *valid* conclusion> **5.** *Archaic.* Of sound health : ROBUST. **—va·lid′i·ty** (və-lĭd′ĭ-tē) *n.* **—val′id·ly** *adv.* **—val′id·ness** *n.*

**val·i·date** (văl′ĭ-dāt′) *vt.* **-dat·ed, -dat·ing, -dates. 1.** To declare or make legally valid. **2.** To mark with an indication of official sanction <*validate* a parking ticket> **3.** To substantiate : verify <Experiments *validated* the new theory.> **—val′i·da′tion** *n.*

**val·ine** (văl′ēn′, vā′lēn′) *n.* [VAL(ERIC ACID) + -INE.] A crystalline amino acid, $C_5H_{11}NO_2$, required for normal human growth.
**val·in·o·my·cin** (văl′ə-nō-mī′sĭn) *n.* [VALIN(E) + -MYCIN.] An antibiotic, $C_{54}H_{90}N_6H_{18}$, that is produced by the bacterium *Streptomyces fulvissimus.*
**va·lise** (və-lēs′) *n.* [Fr. < It. *valigia.*] A small piece of hand luggage.
**Val·i·um** (văl′ē-əm). A trademark for the tranquilizing drug diazepam.
**Val·kyr·ie** (văl-kîr′ē, -kī′rē) *n.* [ON *Valkyrja,* the chooser of the slain.] *Norse Myth.* Any of Odin's handmaidens who chose the heroes to be slain in battle and then conducted their souls to Valhalla.
**val·la·tion** (vă-lā′shən) *n.* [LLat. *vallatio* < Lat. *vallare,* to surround with a rampart < *vallum,* rampart < *vallus,* stake.] **1.** An earthen wall for military defense : RAMPART. **2.** The art or process of planning or erecting earth fortifications. **—val′la·to′ry** (văl′ə-tôr′ē, -tōr′ē) *adj.*
**val·lec·u·la** (vă-lĕk′yə-lə, və-) *n., pl.* **-lae** (-lē′) [LLat., dim. of Lat. *valles,* valley.] *Biol.* A shallow groove, depression, or furrow. **—val·lec′u·lar** (-lər), **val·lec′u·late** (-lĭt, -lāt′) *adj.*
**val·ley** (văl′ē) *n., pl.* **-leys.** [ME *valey* < OFr. *valee* < Lat. *valles.*] **1.** An elongated lowland between mountain ranges, hills, or other uplands, often having a river or stream running along the bottom. **2.** Extensive land irrigated or drained by a river system. **3.** A depression or hollow like a valley, as the point at which the two slopes of a roof meet.
**va·lo·ni·a** (və-lō′nē-ə, -lōn′yə) *n.* [Ital. *vallonia* < Mod. Gk. *balania,* pl. of *balani,* acorn < Gk. *balanos.*] An extract from the dried acorn cups of an oak tree, *Quercus aegilops* of eastern Europe and Asia Minor, used chiefly in tanning and dyeing.
**val·or** (văl′ər) *n.* [ME *valour* < OFr. < Med. Lat. *valor* < Lat. *valēre,* to be worth.] Courage : bravery.
**val·or·ize** (văl′ə-rīz′) *vt.* **-ized, -iz·ing, -iz·es.** [Port. *valorizar* < *valor,* value < Med. Lat. —see VALOR.] To establish and maintain the price of (a commodity) by government action. **—val′or·i·za′tion** *n.*
**val·or·ous** (văl′ər-əs) *adj.* Valiant. **—val′or·ous·ly** *adv.* **—val′or·ous·ness** *n.*
**val·our** (văl′ər) *n. Chiefly Brit. var. of* VALOR.
**val·u·a·ble** (văl′yōō-ə-bəl, văl′yə-) *adj.* **1.** Of high monetary or material value <*valuable* jewelry> **2.** Of great importance, utility, or service <*valuable* marketing data> **3.** Having admirable or esteemed qualities or characteristics <a *valuable* colleague> *—n. often* **valuables.** A personal possession having high monetary or material value. **—val′u·a·ble·ness** *n.* **—val′u·a·bly** *adv.*
☆ ***syns:*** VALUABLE, INVALUABLE, PRECIOUS, PRICELESS *adj. core meaning* : of great value <a *valuable* diamond> ***ant:*** worthless
**val·u·ate** (văl′yōō-āt′) *vt.* **-at·ed, -at·ing, -ates.** [Back-formation < VALUATION.] To set a value for : APPRAISE.
**val·u·a·tion** (văl′yōō-ā′shən) *n.* **1.** An act or process of assessing value or price : APPRAISAL. **2.** Assessed value or price. **3.** An estimation of worth, merit, or character. **—val′u·a′tion·al** *adj.*
**val·u·a·tor** (văl′yōō-ā′tər) *n.* One who estimates values : APPRAISER.
**val·ue** (văl′yōō) *n.* [ME *valew* < OFr. *value* < *valoir,* to be worth < Lat. *valēre.*] **1.** An amount regarded as a suitable equivalent for something else, esp. a fair price or return for goods or services. **2.** Monetary or material worth <the fluctuating *value* of silver> **3.** Worth in usefulness or importance to the possessor <the *value* of a college education> **4.** A principle, standard, or quality regarded as worthwhile or desirable <traditional moral *values*> **5.** Precise meaning or import, as of a term. **6.** *Math.* An assigned or calculated numerical quantity. **7.** *Mus.* Relative duration of a tone or rest. **8.** The relative darkness or lightness of a color. **9.** The sound quality of a letter or diphthong. **10.** One of a series of specified values <issued a postage stamp of new *value*> *—vt.* **-ued, -u·ing, -ues. 1.** To determine or estimate the worth or value of : APPRAISE. **2.** To regard highly : ESTEEM <*valued* my colleagues' opinions> **3.** To rate according to relative estimate of worth or desirability : EVALUATE <*valued* money over all else> **4.** To assign a value to (e.g., a unit of currency). **—val′u·er** *n.*
**val·ue-ad·ded tax** (văl′yōō-ăd′ĭd) *n.* A tax on the estimated market value added to a product or material at each stage of its manufacture or distribution, ultimately passed on to the consumer.
**val·ued** (văl′yōōd) *adj.* Highly esteemed.
**valued policy** *n.* An insurance policy requiring the insurer to pay the insured the full face value of the policy in the event of total loss, regardless of the actual value of the lost property.
**value judgment** *n.* A judgment that assigns a value, as to an object or action.
**val·ue·less** (văl′yōō-lĭs) *adj.* Having no value : WORTHLESS.
**val·vate** (văl′vāt′) *adj.* **1.** Having valvelike parts. **2.** *Bot.* Meeting at the edges without overlapping, as petals.
**valve** (vălv) *n.* [ME, leaf of a door < Lat. *valva.*] **1.** *Anat.* A membranous structure in a hollow organ or passage, as in an artery or vein, that retards or prevents the return flow of a bodily fluid. **2. a.** A device that regulates the flow of gases, liquids, or loose materials through a structure, as a pipe, or through an aperture by opening, closing, or obstructing a port or passageway. **b.** The movable control element of such a device. **c.** *Mus.* A device in a brass wind instrument that allows change in pitch by a rapid varying of the air column in a tube. **3.** *Biol.* **a.** One of the paired hinged shells of many mollusks and of brachiopods. **b.** A similar paired part, as of the cell wall of a diatom. **4.** *Bot.* **a.** One of the sections into which a seed pod or other dehiscent fruit splits. **b.** A lidlike covering of an anther. **5.** *Chiefly Brit.* An electron tube or vacuum tube. **6.** *Archaic.* Either half of a double or folding door. *—vt.* **valved, valv·ing, valves. 1.** To equip with a valve. **2.** To control by a valve.
**valve-in-head engine** (vălv′ĭn-hĕd′) *n.* An internal-combustion engine having the inlet and exhaust valves in the cylinder head.
**val·vu·la** (văl′vyə-lə) *n. var. of* VALVULE.
**val·vu·lar** (văl′vyə-lər) *adj.* Relating to, having, or operating by means of valves or valvelike parts.
**val·vule** (văl′vyōōl′) *also* **val·vu·la** (văl′vyə-lə) *n., pl.* **-vules** *also* **-vu·lae** (-vyə-lē′). A small valve or valvelike structure.
**val·vu·li·tis** (văl′vyə-lī′tĭs) *n.* Inflammation of a valve, esp. of a cardiac valve.
**vam·brace** (văm′brās′) *n.* [ME *vambras* < AN *vauntbras* < OFr. *avauntbras* : *avaunt,* before + *bras,* arm.] Armor for protecting the forearm. **—vam′braced′** *adj.*
**va·moose** (vă-mōōs′, və-) *vi.* **-moosed, -moos·ing, -moos·es.** [< Sp. *vamos,* let's go < Lat. *vadamus,* 1st person pl. subjunctive of *vadere,* to go.] *Slang.* To leave hurriedly.
**vamp**[1] (vămp) *n.* [ME *vampe,* sock < OFr. *avantpie* : *avant,* before (< Lat. *abante,* from before) + *pie,* foot (< Lat. *pes*).] **1.** The part of a shoe or boot covering the instep and occas. extending over the toe. **2.** An improvised musical accompaniment. *—v.* **vamped, vamp·ing, vamps.** *—vt.* **1.** To provide (a shoe or boot) with a new vamp. **2.** To patch up (something old). **3.** *Mus.* To improvise (e.g., an accompaniment) for a solo. *—vi. Mus.* To improvise a vamp. **—vamp′er** *n.*
**vamp**[2] (vămp) [Short for VAMPIRE.] *Informal. —n.* An unscrupulous woman who seduces or exploits men. *—v.* **vamped, vamp·ing, vamps.** *—vt.* To seduce or exploit (a man) in the manner of a vamp. *—vi.* To play the vamp. **—vamp′ish** *adj.*
**vam·pire** (văm′pīr′) *n.* [Fr. < G. *Vampir,* of Slav. orig.] **1.** A reanimated corpse held to rise from the grave at night to suck the blood of sleeping persons. **2.** One who preys on others, as: **a.** An extortionist. **b.** A woman who uses sexuality to exploit men. **3. a.** Any of various tropical American bats of the family Desmodontidae that feed on the blood of living mammals. **b.** Any of various other bats, as those of the family Megadermatidae, erroneously believed to feed on blood. **—vam·pir′ic** (văm-pĭr′ĭk) *adj.*
**vam·pir·ism** (văm′pīr-ĭz′əm) *n.* **1.** Belief in vampires. **2.** The practice or actions of a vampire.
**van**[1] (văn) *n.* [Short for CARAVAN.] **1.** A covered or enclosed truck for transporting goods or livestock. **2.** *Chiefly Brit.* A closed railroad car for carrying baggage or freight.
**van**[2] (văn) *n.* [Short for VANGUARD.] The vanguard : forefront.
**van**[3] (văn) *n.* [ME < OE *fann* and OFr. *van,* both < Lat. *vannus.*] **1.** A wing. **2.** *Archaic.* A winnowing device, as a fan.
**van·a·date** (văn′ə-dāt′) *n.* [< VANADIUM.] A salt or an ester of a vanadic acid.
**va·na·dic acid** (və-nā′dĭk, -năd′ĭk) *n.* **1.** An acid containing a vanadate group, esp. $HVO_3$, $H_3VO_4$, or $H_4V_2O_7$, not existing in a pure state. **2.** Vanadium pentoxide.
**va·na·di·nite** (və-năd′n-īt′, -năd′-, văn′ə-dē′nīt′) *n.* [VANAD(IUM) + -IN + -ITE[1].] A deep ruby-red or yellow to brown vanadium and lead ore, essentially $Pb_5(VO_4)_3Cl$.
**va·na·di·um** (və-nā′dē-əm) *n.* [< ON *Vanadīs,* the goddess Freya.] *Symbol* **V** A bright white soft ductile metallic element, used in rust-resistant high-speed tools, as a carbon stabilizer in some steels, and as a catalyst; atomic number 23; atomic weight 50.942.
**vanadium pentoxide** *n.* A yellow to red crystalline powder, $V_2O_5$, used as a catalyst in various organic reactions and as a starting material for other vanadium salts.
**vanadium steel** *n.* Steel alloyed with vanadium for added strength, hardness, and high-temperature stability.
**Van Al·len belt** (văn ăl′ən) *n.* [After James A. *Van Allen* (b. 1914).] Either of two zones of high-intensity particulate radiation trapped in the earth's magnetic field and surrounding the planet, beginning at an altitude of approx. 800 kilometers and extending several tens of thousands of kilometers into space.
**van·co·my·cin** (văng′kə-mī′sĭn, văn′kə-) *n.* [*vanco-* (of unknown orig.) + -MYCIN.] An antibiotic produced by the bacterium *Streptomyces orientalis* that is effective against staphylococci and spirochetes.
**Van·dal** (văn′dl) *n.* [Lat. *Vandalus,* of Germanic orig.] **1.** A member of a Germanic people that overran Gaul, Spain, and northern Africa in the 4th and 5th cent. A.D. and sacked Rome in A.D. 455. **2. vandal.** One who willfully or maliciously defaces or destroys public or private property. **—Van·dal′ic** (văn-dăl′ĭk) *adj.*
**van·dal·ism** (văn′dl-ĭz′əm) *n.* Willful or malicious destruction of public or private property. **—van′dal·is′tic** *adj.*

**van·dal·ize** (văn'dl-īz') *vt.* **-ized, -iz·ing, -iz·es.** To destroy or deface (public or private property) willfully or maliciously. **—van'dal·i·za'tion** *n.*
**Van de Graaff generator** (văn' də grăf') *n.* [After Robert J. *Van de Graaff* (b.1901).] An electrostatic generator in which an electric charge is either removed from or transferred to a large hollow spherical electrode by a rapidly moving belt, in some configurations producing potentials over a million volts, and used with an accelera-tion tube as an electron or ion accelerator.
**van der Waals force** (văn' dər wôlz') *n.* [After Johannes D. *van der Waals* (1837–1923).] A force between nonpolar molecules caused by a temporary change in dipole moment arising from a brief shift of orbital electrons to one side of one molecule, causing a similar shift in adjacent molecules, with resulting polarization and attraction.
**Van·dyke** (văn-dīk') *n.* **1.** A Vandyke beard. **2.** A Vandyke collar. **3. a.** A V-shaped point that is part of a decorative border. **b.** A border composed of such points.
**Vandyke beard** *n.* [After Sir Anthony *Vandyke* (1599–1641).] A short pointed beard.

**Vandyke beard**

**Vandyke brown** *n.* [After Sir Anthony *Vandyke* (1599–1641), from its frequent use in his paintings.] A moderate to grayish brown. **—Van·dyke'-brown'** *adj.*
**Vandyke collar** *n.* [After Sir Anthony *Vandyke* (1599–1641).] A large linen or lace collar with a deeply indented or scalloped edge.
**vane** (vān) *n.* [ME < OE *fana,* flag.] **1.** A device that pivots on an elevated object, as a rooftop or spire, to indicate wind direction. **2.** One of several usu. relatively thin, rigid, flat, or occas. curved surfaces radially mounted along an axis that is turned by or used to turn a fluid. **3.** The flattened weblike part of a feather consisting of a series of barbs on either side of the shaft. **4. a.** The movable target on a leveling rod. **b.** A sight on a quadrant or compass. **5.** A metal guidance or stabilizing fin attached to the tail of a missile, as a bomb.
**vang** (văng) *n.* [Du., a catch < *vangen,* to catch.] *Naut.* A guy rope running from the peak of a gaff or derrick to the deck.
**van·guard** (văn'gärd') *n.* [ME *vandgard* < *avaunt garde* < OFr. *avant-garde* : *avant,* before + *garde,* guard < *garder,* to guard.] **1.** An army or fleet's foremost position. **2. a.** The foremost or leading position in a movement or trend <the *vanguard* in literary criticism> **b.** Those occupying a foremost position.
**va·nil·la** (və-nĭl'ə) *n.* [Sp. *vainilla,* dim. of *vaina,* sheath < Lat. *vagina* (from the shape of its seed pods).] **1.** A tropical American orchid of the genus *Vanilla,* esp. *V. planifolia,* cultivated for its long narrow seed pods from which a flavoring agent is extracted. **2.** The aromatic seed pod of a vanilla. **3.** A flavoring extract prepared from the seed pods of a vanilla or produced synthetically.
**vanilla bean** *n.* VANILLA 2.
**va·nil·lic** (və-nĭl'ĭk) *adj.* Of, pertaining to, or derived from vanilla or vanillin.
**va·nil·lin** (və-nĭl'ĭn, văn'ə-lĭn) *n.* [VANILL(A) + -IN.] A white or yellowish crystalline compound, $C_8H_8O_3$, found in vanilla beans and certain balsams and resins and used in perfumery, flavoring, and pharmaceuticals.
**Va·nir** (vä'nîr') *pl.n.* [ON.] *Norse Myth.* An early race of gods who dwelt with the Aesir in Asgard.
**van·ish** (văn'ĭsh) *vi.* **-ished, -ish·ing, -ish·es.** [ME *vanisshen* < OFr. *esvanir, esvaniss-* < VLat. **exvanire,* alteration of Lat. *evanescere* : *e-,* ex- + *vanescere,* to vanish < *vanus,* empty.] **1.** To disappear, esp. quickly or in an unexplained way. **2.** To pass out of existence. **3.** *Math.* To become zero. —Used of a function or variable. **—van'ish·er** *n.*
**vanishing cream** *n.* A cosmetic preparation containing less oil than cold cream, used as a powder base and night cream.
**vanishing point** *n.* **1.** A point in a drawing at which parallel lines drawn in perspective converge or seem to converge. **2.** A point at which a thing disappears or ceases to exist.
**van·i·ty** (văn'ĭ-tē) *n., pl.* **-ties.** [ME *vanite* < OFr. < Lat. *vanitas* < *vanus,* empty.] **1.** The quality or state of being vain. **2.** Excessive pride in one's appearance or accomplishments : CONCEIT. **3.** Lack of usefulness, worth, or effect : WORTHLESSNESS. **4. a.** Something vain, futile, or worthless. **b.** Something about which one is vain or conceited. **5.** A vanity case. **6.** A dressing table.
**vanity case** *n.* **1.** A woman's compact. **2.** A small handbag or case for carrying cosmetics or toiletries.
**Vanity Fair** *also* **vanity fair** *n.* [< *Vanity-Fair,* the fair in *Pilgrim's Progress* by John Bunyan (1628–1688).] A place of ostentation, or empty idle amusement and frivolity.
**vanity plate** *n.* A motor vehicular license plate having a combination of letters or numbers specially selected by the purchaser.
**vanity press** *n.* A press that publishes a book at the author's expense.
**van·quish** (văng'kwĭsh, văn'-) *vt.* **-quished, -quish·ing, -quish·es.** [ME *vaynquysshen* < OFr. *vainquir, vainquiss-* < Lat. *vincere.*] **1. a.** To defeat in battle. **b.** To defeat in a contest, conflict, or competition. **2.** To overcome or subdue (e.g., an emotion) : SUPPRESS <Success *vanquished* all of our fears.> **—van'quish·a·ble** *adj.* **—van'quish·er** *n.* **—van'quish·ment** *n.*
**van·tage** (văn'tĭj) *n.* [ME, short for OFr. *avantage,* advantage.] **1.** An advantage in a competition or conflict : SUPERIORITY. **2.** Something, as a strategic position, that provides superiority or advantage. **3.** ADVANTAGE 4.
**van·ward** (văn'wərd) *adj.* Located in the forefront : ADVANCED. *—adv.* Toward or to the forefront : FORWARD.
**vap·id** (văp'ĭd, vā'pĭd) *adj.* [Lat. *vapidus.*] Lacking liveliness, zest, or interest : FLAT <a *vapid* little play> **—va·pid'i·ty** (vă-pĭd'ĭ-tē, vă-, və-), **vap'id·ness** *n.* **—vap'id·ly** *adv.*
**va·por** (vā'pər) *n.* [ME *vapour* < OFr. < Lat. *vapor.*] **1.** Barely visible or cloudy diffused matter, as mist, fumes, or smoke, suspended in the air. **2. a.** The state of a substance that exists below its critical temperature and that may be liquefied by applying sufficient pressure. **b.** The gaseous state of a substance that, under ordinary conditions, is liquid or solid. **3. a.** The vaporized form of a substance for use in industrial, military, or medical processes. **b.** A mixture of a vapor and air, as the explosive gasoline-air mixture burned in an internal-combustion engine. **4.** *Archaic.* **a.** Something insubstantial or fleeting. **b.** A fantastic or foolish idea. **5. vapors. a.** *Archaic.* Exhalations within a bodily organ, esp. the stomach, held to affect the mental or physical condition. **b.** Hysteria or emotional depression. *—v.* **-pored, -por·ing, -pors.** *—vt.* To vaporize. *—vi.* **1.** To emit vapor. **2.** To evaporate. **3.** To boast. **—va'por·er** *n.*
**va·por·es·cence** (vā'pə-rĕs'əns) *n.* Formation of vapor.
**va·por·if·ic** (vā'pə-rĭf'ĭk) *adj.* [VAPOR + -FIC.] **1.** Producing or turning to vapor. **2.** Of the nature of vapor : VAPOROUS.
**va·por·ing** (vā'pər-ĭng) *n.* Boastful or bombastic talk or behavior.
**va·por·ish** (vā'pər-ĭsh) *adj.* **1.** Like or resembling vapor. **2.** *Archaic.* Inclined to spells of hysteria or low spirits. **—va'por·ish·ness** *n.*
**va·por·ize** (vā'pə-rīz') *vt. & vi.* **-ized, -iz·ing, -iz·es.** To convert or be converted into vapor. **—va'por·iz'a·ble** *adj.* **—va'por·i·za'tion** (vā'pər-ĭ-zā'shən) *n.*
**va·por·iz·er** (vā'pə-rī'zər) *n.* One that vaporizes, esp. a device used to vaporize medicine for inhalation.
**vapor lock** *n.* A pocket of vaporized gasoline in the fuel line of an internal-combustion engine that obstructs normal flow of fuel.
**va·por·ous** (vā'pər-əs) *adj.* **1.** Relating to or like vapor. **2. a.** Producing vapors : VOLATILE. **b.** Giving off or full of vapors. **3.** Insubstantial, vague, or ethereal. **4.** Extravagantly fanciful : HIGH-FLOWN. **—va·por·os'i·ty** (vā'pə-rŏs'ĭ-tē), **va'por·ous·ness** *n.* **—va'por·ous·ly** *adv.*
**vapor pressure** *n.* The pressure exerted by a vapor in equilibrium with its solid or liquid phase.
**vapor trail** *n.* A contrail.
**va·por·y** (vā'pə-rē) *adj.* Vaporous.
**va·pour** (vā'pər) *n. & v. Chiefly Brit. var. of* VAPOR.
**†va·que·ro** (vä-kâr'ō) *n., pl.* **-ros.** [Sp. < *vaca,* cow < Lat. *vacca.*] *Southwestern U.S.* A cowboy.
**va·ra** (vär'ə) *n.* [Sp. and Port. *vara,* rod, both < Lat. *vara,* forked pole < *varus,* bent.] **1.** A Spanish, Portuguese, and Latin American unit of linear measure varying from about 81 to 109 centimeters, or 32 to 43 inches. **2.** A square vara.
**va·rac·tor** (və-răk'tər, vä-) *n.* [VAR(YING) + (RE)ACTOR.] A semiconductor in which the capacitance is sensitive to the applied voltage at the boundary of the semiconductor material and an insulator.
**vari-** *pref. var. of* VARIO-.
**var·i·a** (vâr'ē-ə, văr'-) *n.* [Lat., neuter pl. of *varius,* various.] A miscellany, esp. of literary works.
**var·i·a·ble** (vâr'ē-ə-bəl, văr'-) *adj.* **1. a.** Tending or apt to vary. **b.** Fickle : inconstant. **2.** *Biol.* Tending to deviate from an established type : ABERRANT. **3.** *Math.* Having no fixed quantitative value. *—n.* **1.** Something that varies or is prone to variation. **2.** *Astron.* A variable star. **3.** *Math.* **a.** A quantity capable of assuming any of a set of values. **b.** A symbol representing such a quantity. **—var'i·a·bil'i·ty, var'i·a·ble·ness** *n.* **—var'i·a·bly** *adv.*
**variable cost** *n.* A cost fluctuating directly with output changes.
**variable field** *n. Computer Sci.* A set of adjacent columns on a punchcard that can be varied in length as to need.
**variable logic** *n. Computer Sci.* An internal machine logic that can be changed to match programming formats.
**variable star** *n.* A star whose brightness varies due to internal changes or periodic eclipsing of component stars.

**var·i·ance** (vâr′ē-əns, văr′-) *n.* **1. a.** An act of varying. **b.** The quality or state of being variant or variable : VARIATION. **c.** A difference between what is expected and what actually takes place. **2.** A difference of opinion : DISSENSION. **3.** *Law.* **a.** A discrepancy between two statements or documents in a legal proceeding. **b.** The license to engage in an act contrary to a usual rule <approved the zoning *variance*> **4.** *Statistics.* The mean of the squares of the variations from the mean of a frequency distribution. **5.** *Chem.* The number of thermodynamic variables required to specify a state of equilibrium of a system, given by the phase rule. **—at variance.** Conflicting : differing.

**var·i·ant** (vâr′ē-ənt, văr′-) *adj.* [ME < OFr. < Lat. *varians, variant-*, pr.part. of *variare,* to vary.] **1.** Having or exhibiting variation. **2.** Tending or apt to vary : VARIABLE. **3.** Deviating from a standard, usu. by only a slight difference. —*n.* Something that differs in form only slightly from something else <a spelling *variant*>

**var·i·ate** (vâr′ē-ĭt, -āt′, văr′-) *n.* [< Lat. *variatus,* p.part. of *variare,* to vary.] **1.** VARIABLE 1. **2.** *Statistics.* A random variable with a numerical value defined on a given sample space.

**var·i·a·tion** (vâr′ē-ā′shən, văr′-) *n.* **1. a.** The act, process, or result of varying. **b.** The state or fact of being varied. **2.** Extent or degree of varying <a *variation* of 20 pounds in weight> **3.** Magnetic declination. **4.** Something slightly different from another of the same type. **5.** *Biol.* Marked difference or deviation from characteristic form, function, or structure. **6.** *Math.* A function that relates the values of one variable to those of other variables. **7.** *Mus.* **a.** A form that is an altered version of a given theme, diverging from it by melodic ornamentation and by changes in harmony, rhythm, or key. **b.** One of a series of forms based on a single theme. **8.** A solo dance, esp. one forming part of a larger work. **—var′i·a′tion·al** *adj.*

**varic-** *pref. var. of* VARICO-.

**var·i·cel·la** (văr′ĭ-sĕl′ə) *n.* [NLat. < *variola,* variola.] Chicken pox. **—var′i·cel′loid′** (-sĕl′oid′) *adj.*

**var·i·ces** (văr′ĭ-sēz′) *n. pl. of* VARIX.

**varico-** *or* **varic-** *pref.* [< Lat. *varix, varic-,* varix.] Varix : varicose vein <*varicosis*>

**var·i·co·cele** (văr′ĭ-kō-sēl′) *n.* [VARICO- + -CELE¹.] A varicose condition of veins of the spermatic cord or the ovaries, forming a soft tumor.

**var·i·col·ored** (vâr′ĭ-kŭl′ərd, văr′-) *adj.* VARIEGATED 1.

**var·i·cose** (văr′ĭ-kōs′) *adj.* [Lat. *varicosus* < *varix,* swollen vein.] **1.** Designating blood or lymph vessels that are abnormally dilated, knotted, and tortuous. **2.** That causes unusual swelling.

**var·i·co·sis** (văr′ĭ-kō′sĭs) *n.* [VARIC(O)- + -OSIS.] VARICOSITY 1.

**var·i·cos·i·ty** (văr′ĭ-kŏs′ĭ-tē) *n., pl.* **-ties. 1.** The state of being varicose. **2. a.** A varicose distention or swelling. **b.** The condition of having varicose veins.

**var·i·cot·o·my** (văr′ĭ-kŏt′ə-mē) *n., pl.* **-mies.** [VARICO- + -TOMY.] Subcutaneous incision to cure varicose veins.

**var·ied** (vâr′ēd, văr′-) *adj.* **1.** Characterized by variety. **2.** Modified : altered. **3.** VARIEGATED 1. **—var′ied·ly** *adv.*

**varied thrush** *n.* A bird, *Ixoreus naevius* of western North America, similar to the robin but having a black transverse stripe on its breast.

**var·i·e·gate** (vâr′ē-ĭ-gāt′, vâr′ĭ-gāt′, văr′-) *vt.* **-gat·ed, -gat·ing, -gates.** [Lat. *variegare, variegat-* < Lat. *varius,* various.] **1.** To change the appearance of, esp. by marking with different colors : STREAK. **2.** To give variety to. **—var′i·e·ga′tion** (-ĭ-gā′shən, vâr′ĭ-gā′-, văr′-) *n.* **—var′i·e·ga′tor** *n.*

**var·i·e·gat·ed** (vâr′ē-ĭ-gā′tĭd, vâr′ĭ-gā′-, văr′-) *adj.* **1.** Having streaks, marks, or patches of a different color or colors. **2.** Distinguished or marked by variety : DIVERSIFIED.

**var·i·er** (vâr′ē-ər, văr′-) *n.* One that varies.

**va·ri·e·tal** (və-rī′ĭ-tl) *adj.* [< VARIETY.] Of, indicating, or characterizing a variety, esp. a biological variety. **—va·ri′e·tal·ly** *adv.*

**va·ri·e·ty** (və-rī′ĭ-tē) *n., pl.* **-ties.** [OFr. *variete* < Lat. *varietas* < *varius,* various.] **1.** The quality or state of being various or varied : DIVERSITY. **2.** A number of varied things, esp. of a particular group : ASSORTMENT <a *variety* of meats> **3.** A group set off from other groups by a particular characteristic or set of characteristics. **4.** *Biol.* **a.** A taxonomic category forming a subdivision of a species and consisting of naturally occurring or selectively bred individuals with varying characteristics. **b.** An organism, esp. a plant, belonging to such a category. **5.** A variety show.

**variety meat** *n.* Meat, as liver or sweetbreads, that has been taken from a part other than skeletal muscles or that has been processed, as sausage.

**variety show** *n.* A theatrical entertainment consisting of successive unrelated acts, as songs, dances, and comedy skits.

**variety store** *n.* A retail store selling varied goods.

**var·i·form** (vâr′ə-fôrm′, văr′-) *adj.* [VARI(O)- + -FORM.] Having a variety or diversity of forms.

**vario-** *or* **vari-** *pref.* [< Lat. *varius,* speckled.] Variety : difference : variation <*variometer*>

**va·ri·o·la** (və-rī′ə-lə, vâr′ē-ō′lə, văr′-) *n.* [NLat. < Med. Lat., pustule < Lat. *varius,* speckled.] Smallpox.

**var·i·o·late** (vâr′ē-ə-lāt′, văr′-) *adj.* Having pustules or marks resembling those of smallpox. —*vt.* **-lat·ed, -lat·ing, -lates.** To inoculate with smallpox vaccine.

**var·i·o·lite** (vâr′ē-ə-līt′, văr′-) *n.* A basic rock having a pockmarked appearance due to many rounded, white spherules embedded in it.

**var·i·o·loid** (vâr′ē-ə-loid′, văr′-, və-rī′ə-loid′) *n.* A mild smallpox in persons who have previously been vaccinated or who have had the disease before.

**var·i·om·e·ter** (vâr′ē-ŏm′ĭ-tər, văr′-) *n.* A variable inductor for measuring variations in terrestrial magnetism.

**var·i·o·rum** (vâr′ē-ôr′əm, -ōr′-, văr′-) *n.* [< Lat. *(editio cum notis) variorum,* (edition with the notes) of various persons.] **1.** An edition of the works of an author, with notes by scholars or editors. **2.** An edition containing various versions of a text. **—var′i·o′rum** *adj.*

**var·i·ous** (vâr′ē-əs, văr′-) *adj.* [Lat. *varius.*] **1. a.** Of diverse kinds <for *various* purposes> **b.** Unlike : different. **2.** More than one : SEVERAL <studied *various* books on antiques> **3.** Versatile : many-sided <people of *various* talents> **4.** Having a variegated nature or appearance. **5.** Being an individual or separate member of a class or group <The *various* accounts all differed.> **6.** *Archaic.* Changeable : variable. ***usage:*** The use of *various* as a collective noun followed by *of* (*various of the members*) is unacceptable in standard English. **—var′i·ous·ly** *adv.* **—var′i·ous·ness** *n.*

**var·i·sized** (vâr′ĭ-sīzd′, văr′-) *adj.* Of varying sizes.

**var·ix** (văr′ĭks) *n., pl.* **var·i·ces** (văr′ĭ-sēz′) [Lat., swollen vein.] **1.** An abnormally twisted and dilated vein, artery, or lymph vessel. **2.** One of the longitudinal ridges indicating a resting stage in the development of the lip of a gastropod shell.

**var·let** (vär′lĭt) *n.* [ME < OFr., var. of *vaslet.* —see VALET.] *Archaic.* **1.** An attendant or servant. **2.** A knight's page. **3.** A rascal : knave.

**var·let·ry** (vär′lĭ-trē) *n. Archaic.* A crowd of attendants or menials, esp. when disorderly : RABBLE.

**var·mint** (vär′mĭnt) *n.* [Var. of VERMIN.] *Informal.* **1.** A bird or animal regarded as undesirable or troublesome. **2.** An obnoxious, contemptible person.

**var·nish** (vär′nĭsh) *n.* [ME *vernysshe* < OFr. *vernis* < Med. Lat. *veronix,* sandarac resin, prob. < Gk. *Berenikē,* Berenice, a city in Cyrenaica.] **1.** An oil-based paint with a solvent and an oxidizing or an evaporating binder in it for coating a surface with a hard, glossy, thin film. **2. a.** The smooth coating or gloss produced by applying varnish. **b.** Something resembling or suggestive of varnish. **3.** A deceptively attractive external appearance. —*vt.* **-nished, -nish·ing, -nish·es. 1.** To cover with varnish. **2.** To give a smooth and glossy finish to. **3.** To give a deceptively attractive appearance to. **—var′nish·er** *n.*

**varnish tree** *n.* Any of several trees with milky juice for making varnish.

**var·si·ty** (vär′sĭ-tē) *n., pl.* **-ties.** [Alteration of UNIVERSITY.] **1.** The principal team representing a university, college, or school in sports or other competitions. **2.** *Chiefly Brit.* A university.

**Var·u·na** (vär′ə-nə) *n.* [Skt. *Varuṇaḥ.*] *Myth.* The Vedic god of the skies and seas.

**var·us** (vâr′əs, văr′-) *n., pl.* **-us·es.** [< Lat., crooked.] Abnormal positioning of a leg or foot bone.

**varve** (värv) *n.* [Swed. *varv,* layer < *varva,* to bend < ON *hverfa.*] **1.** A sedimentary layer deposited in one year. **2.** A pair of distinct layers of sediment, indicating seasonal deposits.

**var·y** (vâr′ē, văr′ē) *v.* **-ied, -y·ing, -ies.** [ME *varien,* to undergo change < OFr. *varier* < Lat. *variare* < *varius,* various.] —*vt.* **1.** To make or produce changes in attributes or characteristics : MODIFY. **2.** To make varied <a song recital that was *varied* for all tastes> **3.** To express in a different manner <*vary* the speed> —*vi.* **1.** To undergo or exhibit change <Precipitation *varied* throughout the year.> **2.** To be different : DEVIATE <*vary* from customary styles of dress> **3.** To undergo successive or alternate changes in qualities or attributes. **—var′y·ing·ly** *adv.*

**varying hare** *n.* The snowshoe rabbit.

**vas** (văs) *n., pl.* **va·sa** (vā′zə) [NLat. < Lat., vessel.] An organic vessel or duct.

**vas-** *pref. var. of* VASO-.

**va·sa** (vā′zə) *n. pl. of* VAS.

**vas·cu·la** (văs′kyə-lə) *n. pl. of* VASCULUM.

**vas·cu·lar** (văs′kyə-lər) *adj.* [< Lat. *vasculum,* dim. of *vas,* vessel.] **1.** *Biol.* Of, typified by, or having vessels for transmission or circulation of plant or animal fluids as blood, lymph, or sap. **2.** Marked by vigor and ardor : PASSIONATE. **—vas′cu·lar′i·ty** (-lăr′ĭ-tē) *n.*

**vascular bundle** *n.* A strand of supportive and conductive plant tissue composed essentially of xylem and phloem.

**vas·cu·lar·i·za·tion** (văs′kyə-lər-ĭ-zā′shən) *n.* New blood-vessel formation.

**vascular plant** *n.* Any of various plants of the division Tracheophyta, which includes the ferns and seed-bearing plants typified by a system of specialized conductive and supportive tissue.

**vascular tissue** *n.* Plant tissue made up of vascular bundles.

**vas·cu·la·ture** (văs′kyə-lə-chŏŏr′, -chər) *n.* Arrangement of blood vessels in the body or in a bodily organ or part.

**vas·cu·lum** (văs′kyə-ləm) *n., pl.* **-la** (-lə) [Lat., small vessel. —see VASCULAR.] A small box or case for carrying newly gathered plant specimens.
**vas def·er·ens** (văs′ dĕf′ər-ənz, -ə-rĕnz′) *n., pl.* **va·sa def·er·en·ti·a** (vā′zə dĕf′ə-rĕn′shē-ə) [NLat., deferent vessel.] The vertebrate duct that carries sperm from the epididymal duct to the ejaculatory duct.
**vase** (vās, vāz, väz) *n.* [Fr. < Lat. *vas,* vessel.] An open container for holding flowers or for decoration.
**va·sec·to·my** (və-sĕk′tə-mē, vā-zĕk′-) *n., pl.* **-mies.** Surgical excision of a part of the vas deferens, used as a method of sterilization.
**Vas·e·line** (văs′ə-lēn′). A trademark for a petroleum jelly used chiefly as a vehicle for external applications of medicinal agents and as a protective coating for metal surfaces.
**vaso-** *or* **vas-** *pref.* [< Lat. *vas,* vessel.] **1.** Blood vessel <*vaso*constriction> **2.** Vas deferens <*vas*ectomy>
**va·so·ac·tive** (vā′zō-ăk′tĭv) *adj.* Affecting blood vessels. **—va′so·ac·tiv′i·ty** (-tĭv′ĭ-tē) *n.*
**va·so·con·stric·tion** (vā′zō-kən-strĭk′shən) *n.* Constriction of a blood vessel. **—vas′o·con·stric′tive** *adj.*
**va·so·con·stric·tor** (vā′zō-kən-strĭk′tər) *n.* An agent, as a nerve or a drug, causing vasoconstriction.
**va·so·dil·a·ta·tion** (vā′zō-dĭl′ə-tā′shən, -dī′lə-) *also* **va·so·di·la·tion** (-dī-lā′shən, -dĭ-) *n.* Dilatation of a blood vessel.
**va·so·di·la·tor** (vā′zō-dī-lā′tər, -dĭ-) *n.* An agent, as a nerve or drug, causing vasodilatation.
**va·so·mo·tor** (vā′zō-mō′tər) *adj.* Causing or controlling vasoconstriction or vasodilatation.
**va·so·pres·sin** (vā′zō-prĕs′ĭn) *n.* [Orig. a trademark.] A hormone secreted by the posterior lobe of the pituitary gland that has an antidiuretic and pressor effect.
**va·so·pres·sor** (vā′zō-prĕs′ər) *adj.* Causing a rise in blood pressure. —*n.* An agent that causes blood pressure to rise.
**vas·sal** (văs′əl) *n.* [ME < OFr. < Med. Lat. *vassallus* < *vassus,* of Celt. orig.] **1.** One who held land from a feudal lord and received protection in return for homage and allegiance. **2.** A subordinate or dependent. **3.** A bondman.
**vas·sal·age** (văs′ə-lĭj) *n.* **1.** The state of being a vassal. **2.** The service, homage, and fealty demanded of a vassal. **3.** A position of subordination or subjection : SERVITUDE.
**vast** (văst) *adj.* **-er, -est.** [Lat. *vastus.*] **1.** Very great in size, number, amount, or quantity <the *vast* stars> **2.** Very great in area or extent : IMMENSE <the *vast* oceans> **3.** Very great in degree or intensity <*vast* pride> —*n. Archaic.* An immense space. **—vast′ly** *adv.* **—vast′ness** *n.*
**vast·y** (văs′tē) *adj.* **-i·er, -i·est.** *Archaic.* Vast.
**vat** (văt) *n.* [ME < OE *fæt.*] A large vessel, as a tub, cistern, or barrel, for storing or holding fluids. —*vt.* **vat·ted, vat·ting, vats.** To put into or treat in a vat.
**vat dye** *n.* Any of a series of dyes that create a fast color by impregnating the fiber with a reduced soluble form that is then oxidized to an insoluble form. **—vat′-dyed′** (văt′dīd′) *adj.*
**vat·ic** (văt′ĭk) *also* **vat·i·cal** (-ĭ-kəl) *adj.* [< Lat. *vates,* seer.] Of or having the nature of a prophet : ORACULAR.
**Vat·i·can** (văt′ĭ-kən) *n.* [Fr. < Lat. *Vaticanus,* the Vatican Hill.] **1.** The official residence of the pope in Vatican City, Italy. **2.** The papal government.
**va·tic·i·nal** (vă-tĭs′ə-nəl, və-) *adj.* Prophetic.
**va·tic·i·nate** (vă-tĭs′ə-nāt′, və-) *v.* **-nat·ed, -nat·ing, -nates.** [Lat. *vaticinari, vaticinat-* < *vates,* seer.] —*vt.* To prophesy : foretell. —*vi.* To be a prophet. **—va·tic′i·na′tor** *n.*
**va·tic·i·na·tion** (vă-tĭs′ə-nā′shən, və-) *n.* **1.** An act of prophesying. **2.** A prophecy.
**vaude·ville** (vôd′vĭl′, vōd′-, vô′də-) *n.* [Fr. < OFr. *vaudevire,* short for *chanson du Vau de Vire,* song of Vau de Vire, a region in Normandy.] **1. a.** Stage entertainment offering a variety of short acts such as slapstick comedy, song-and-dance routines, and juggling or acrobatic performances. **b.** A theatrical performance of this kind : VARIETY SHOW. **2.** A light comic play often including songs, pantomime, and dances. **3.** A popular, often satirical song.
**vaude·vil·lian** (vôd-vĭl′yən, vōd-, vô′də-) *n.* One who works in vaudeville, esp. as a performer. **—vaude·vil′lian** *adj.*
**Vau·dois** (vō-dwä′) *pl.n.* [Fr. < Med. Lat. *Waldenses.* —see WALDENSES.] The Waldenses.
**vault¹** (vôlt) *n.* [ME *vaute* < OFr. < Lat. *voluta,* fem. p.part. of *volvere,* to roll.] **1. a.** An arched structure, usu. of stone, brick, or concrete, forming a ceiling or roof. **b.** An arched covering, as the sky, resembling a vault. **2.** A room or space having arched walls and ceiling, esp. when underground, as a cellar or storeroom. **3.** A room or compartment for safe storage of valuables <a bank *vault*> **4.** A burial chamber, esp. when underground. **5.** *Anat.* An arched anatomical part. —*vt.* **vault·ed, vault·ing, vaults. 1.** To construct or supply with an arched ceiling. **2.** To build in the shape of a vault.
**vault²** (vôlt) *v.* **vault·ed, vault·ing, vaults.** [OFr. *volter* < OItal. *voltare* < VLat. **volvitare,* freq. of Lat. *volvere,* to turn.] —*vt.* To jump or leap over, esp. with the aid of a support, as the hands or a pole. —*vi.* **1.** To jump or leap, esp. by using the hands or a pole. **2.** To do something as if by leaping suddenly or vigorously <*vaulted* into a position of leadership> **—vault** *n.* **—vault′er** *n.*
**vault·ing¹** (vôl′tĭng) *n.* Something vaulted or arched.
**vault·ing²** (vôl′tĭng) *adj.* **1.** Leaping upward or over. **2.** Reaching too far : EXAGGERATED <*vaulting* drive for fame> **3.** Used in leaping over <new *vaulting* poles>
**vaunt** (vônt, vŏnt) *v.* **vaunt·ed, vaunt·ing, vaunts.** [ME *vaunten* < OFr. *vanter* < LLat. *vanitare,* to talk frivolously < Lat. *vanus,* empty.] —*vt.* To describe boastfully. —*vi.* To boast : brag. —*n.* **1.** A boastful remark. **2.** Speech of lavish self-praise. **—vaunt′er** *n.* **—vaunt′ing·ly** *adv.*
**vaunt-cour·i·er** (vônt′kŏŏr′ē-ər, -kûr′-, -kŭr′-, vŏnt′-) *n.* [Short for OFr. *avant-courier* : *avant,* in front + *courier,* courier.] **1.** *Obs.* A member of an army's advance guard. **2.** One sent in advance, as a herald.
**vav** (väv, vôv) *n.* [Heb. *wāw.*] The sixth letter of the Hebrew alphabet. —See table at ALPHABET.
**vav·a·sor** *also* **vav·a·sour** (văv′ə-sôr′, -sōr′, -sŏŏr′) *n.* [ME *vavasour* < OFr. < Med. Lat. *vavassor,* poss. contraction of *vassus vassorum,* vassal of vassals.] A feudal tenant who ranked directly under a baron or peer.
**V-day** (vē′dā′) *n.* [V(ICTORY) DAY.] A day of victory, as at the end of a war.
**-'ve.** Have <I'*ve* been invited.>
**Ve·a·dar** (vā′ä-där′, vā′ə-) *n.* [Heb. *va'adhar,* and Adar.] An extra month of the Hebrew year, with 29 days, added in leap years following the regular month of Adar.
**veal** (vēl) *n.* [ME *veel* < OFr. < Lat. *vitellus,* dim. of *vitulus,* calf.] **1.** The meat of a calf. **2.** *also* **veal·er** (vē′lər). A calf raised to be slaughtered for food.
**vec·tor** (vĕk′tər) *n.* [Lat., carrier < *vehere,* to carry.] **1.** *Math.* **a.** A quantity completely specified by a magnitude and a direction. **b.** A one-dimensional array. **c.** An element of a vector space. **2.** *Pathol.* An organism that carries pathogens from one host to another. **3.** A force or influence. **—vec·to′ri·al** (vĕk-tôr′ē-əl, -tōr′-) *adj.*
**vector product** *n.* A vector, *C,* that has magnitude equal to the product of the magnitudes of two vectors, *A* and *B,* and the sine of the angle between *A* and *B,* and that is perpendicular to the plane of *A* and *B* and in a right-handed coordinate system directed so that a right-handed rotation about *C* carries *A* into *B* through an angle not greater than 180 degrees.
**vector space** *n.* A set of elements of vectors that are commutative under addition, unchanged after multiplication by a field multiplicative identity, and commutative, closed, and distributive under the multiplicative operation of the field.
**Ve·da** (vā′də, vē′-) *n.* [Skt. *vedaḥ,* sacred knowledge, Veda.] Any of the oldest sacred writings of Hinduism, including the psalms, incantations, hymns, and formulas of worship incorporated in four collections.
**Ve·dan·ta** (vā-dän′tə, -dăn′-, və-) *n.* [Skt. *vedantaḥ,* essence of the Veda.] The system of Hindu philosophy that further develops the implications in the Upanishads that all reality is one principle, Brahman, and teaches that the believer's goal is to transcend the limitations of self-identity and realize unity with Brahman. **—Ve·dan′tic** *adj.* **—Ve·dan′tism** *n.* **—Ve·dan′tist** *n.*
**V-E Day** (vē′ē′) *n.* [V(ICTORY IN) E(UROPE) DAY.] May 8, 1945, the day of victory for the Allied forces in Europe during World War II.
**Ved·da** *also* **Ved·dah** (vĕd′ə) *n.* [Singhalese, hunter.] One of a small dark-skinned wavy-haired aboriginal people of Sri Lanka.
**ve·dette** *also* **vi·dette** (vĭ-dĕt′) *n.* [Fr. < Ital. *vedetta,* alteration of *veletta* < Sp. *vela,* watch < *velar,* to watch < Lat. *vigilare,* to watch through the night.] **1.** A mounted sentinel posted in advance of an outpost. **2.** A small scouting boat for observing and reporting on an enemy naval force.
**Ve·dic** (vā′dĭk, vē′-) *adj.* Of or relating to the Veda or Vedas, the language in which they are written, or the Hindu culture that produced them.
**vee** (vē) *n.* The letter *v.*
**vee·na** (vē′nə) *n. var. of* VINA.
**veep** (vēp) *n.* [Pronunciation of *V.P.,* abbr. of *vice president.*] *Slang.* **1.** A vice president. **2. Veep.** The Vice President of the United States.
**veer¹** (vîr) *v.* **veered, veer·ing, veers.** [OFr. *virer,* poss. of Celt. orig.] —*vi.* **1.** To turn aside from a course, direction, or purpose : SWERVE. **2.** To shift in direction by a clockwise motion. —Used of the wind. **3.** *Naut.* To change the direction of a ship by turning away from the direction of the wind. —*vt.* **1.** To alter the direction of : TURN. **2.** *Naut.* To change the course of (a ship) by turning away from the direction of the wind. —*n.* A change in direction : SWERVE.
**veer²** (vîr) *vt.* **veered, veer·ing, veers.** [ME *veren* < MDu. *vieren.*] *Naut.* To let out or release (e.g., an anchor chain).
**vee·ry** (vîr′ē) *n., pl.* **-ries.** [Poss. imit. of its song.] A thrush, *Hylocichla fuscescens* of the New World, with a reddish-brown back and an indistinctly spotted breast.
**Ve·ga** (vē′gə, vā′-) *n.* [Med. Lat. < Ar. *(al nasr) al wāqi',* the falling (vulture).] The brightest star in the constellation Lyra.

**veg·an·ism** (vĕj′ə-nĭz′əm) *n.* [Alteration of VEGETARIANISM.] A type of vegetarianism in which there is no consumption of animal food or dairy products and no use of products derived from animals, as leather or soap. **—veg′an** (vĕj′ən, -ăn′) *n.*
**veg·e·ta·ble** (vĕj′tə-bəl, vĕj′ĭ-tə-) *n.* [< ME, vegetative < Med Lat. *vegetabilis* < LLat., enlivening < Lat. *vegetare,* to enliven < *vegetus,* lively < *vegēre,* to be lively.] **1. a.** A plant, as the beet or spinach, raised for an edible part, as the root, stem, leaf, or flower. **b.** The edible part of such a plant. **2.** A member of the vegetable kingdom. **3.** One who leads a dull, passive, or merely physical existence. *—adj.* **1.** Of, relating to, or derived from a plant or plants. **2. a.** Like or resembling a vegetable, as in passivity or dullness of existence. **b.** Boundlessly growing or multiplying.
**vegetable ivory** *n.* A hard, ivorylike material obtained from the ivory nut and used in making small objects, as buttons.
**vegetable kingdom** *n.* The category of living organisms that comprises all plants.
**vegetable marrow** *n. Chiefly Brit.* An edible squash bearing very large elongated greenish fruit.
**vegetable oil** *n.* Any of various oils extracted from plants, used in food products and industrially.
**vegetable oyster** *n.* SALSIFY 2.
**vegetable silk** *n.* Silky fiber from the seed pods of certain plants.
**vegetable sponge** *n.* Loofa.
**vegetable tallow** *n.* Any of various waxy fats extracted from certain plants, as the bayberry, and used to make soap and candles.
**vegetable wax** *n.* A waxy substance of plant origin, as that derived from certain palm trees.
**veg·e·tal** (vĕj′ĭ-tl) *adj.* [Fr. < Med. Lat. *vegetalis* < Lat. *vegetare,* to enliven. —see VEGETABLE.] **1.** Of, pertaining to, or typical of a plant or plants. **2.** Relating to growth instead of to sexual reproduction : VEGETATIVE.
**veg·e·tar·i·an** (vĕj′ĭ-târ′ē-ən) *n.* [VEGET(ABLE) + -ARIAN.] **1.** A practitioner or advocate of vegetarianism. **2.** A herbivore. *—adj.* **1.** Of, relating to, practicing, or recommending vegetarianism. **2.** Composed mainly of vegetables and vegetable products <a *vegetarian* meal>
**veg·e·tar·i·an·ism** (vĕj′ĭ-târ′ē-ə-nĭz′əm) *n.* The practice of or belief in eating a diet made up chiefly of vegetables, grains, fruits, nuts, seeds, and occas. dairy products, as milk or cheese.
**veg·e·tate** (vĕj′ĭ-tāt′) *vi.* **-tat·ed, -tat·ing, -tates.** [Lat. *vegetare,* to enliven. —see VEGETABLE.] **1.** To sprout or grow as a plant does. **2.** *Pathol.* To grow or spread abnormally. **3.** To lead a dull, passive, or merely physical existence.
**veg·e·ta·tion** (vĕj′ĭ-tā′shən) *n.* **1.** The act or process of vegetating. **2.** The plants of an area or region. **3.** *Pathol.* An abnormal growth on the body. **—veg′e·ta′tion·al** *adj.*
**veg·e·ta·tive** (vĕj′ĭ-tā′tĭv) *also* **veg·e·tive** (vĕj′ĭ-tĭv) *adj.* **1.** Of, relating to, or typical of plants or plant growth. **2.** *Biol.* **a.** Of, relating to, or capable of growth. **b.** Of, relating to, or functioning in processes such as growth or nutrition instead of sexual reproduction. **c.** Of or relating to asexual reproduction, as fission or budding.
**veg·gies** *or* **veg·ies** (vĕj′ēz) *pl.n. Informal.* Vegetables.
**ve·he·ment** (vē′ə-mənt) *adj.* [OFr. < Lat. *vehemens.*] **1.** Marked by vigorous expression or profound emotion, passion, or conviction : FERVID <*vehement* urgings> **2.** Characterized by or full of vigor or energy : STRONG. **—ve′he·mence, ve′he·men·cy** *n.* **—ve′he·ment·ly** *adv.*
**ve·hi·cle** (vē′ĭ-kəl) *n.* [Fr. *véhicule* < Lat. *vehiculum* < *vehere,* to carry.] **1.** A device, as a motor vehicle or a piece of mechanized equipment, for transporting passengers, goods, or apparatus : CONVEYANCE. **2.** A medium, as a novel or a painting, through which something is conveyed, transmitted, expressed, or accomplished. **3.** A play, role, or piece of music for displaying the particular talents of a performer or company. **4.** A substance without therapeutic value used as the medium in which active medicines are administered. **5.** A substance, as oil, in which paint pigments are mixed for application. **—ve·hic′u·lar** (vē-hĭk′yə-lər) *adj.*
**veil** (vāl) *n.* [ME *veile* < Norman Fr. < Lat. *vela,* pl. of *velum.*] **1. a.** A piece of often transparent and wide-meshed cloth worn by women over the head, shoulders, and often part of the face for concealment or protection or as a token of modesty. **b.** A length of netting fastened to a woman's hat or headdress for decoration, hanging in front of all or part of the face. **2. a.** The part of a nun's headdress that frames the face and falls over the shoulders. **b.** The life or vows of a nun <took the *veil*> **3. a.** A piece of light fabric hung to conceal or separate what is behind it : CURTAIN. **b.** Something that conceals, separates, or screens like a curtain <a *veil* of secrecy> **4.** *Biol.* A membranous covering, as that partially or wholly enclosing the developing fruiting body of certain mushrooms : VELUM. *—vt.* **veiled, veil·ing, veils.** To cover, conceal, mask, or disguise with or as if with a veil.
**vein** (vān) *n.* [ME *veine* < OFr. < Lat. *vena.*] **1. a.** *Anat.* A vessel that transports blood toward the heart. **b.** A blood vessel. **2.** *Bot.* One of the vascular bundles that form the branching framework and support of a leaf. **3.** *Zool.* One of the chitinous, usu. longitudinal ribs that stiffen and support the wing of an insect. **4.** *Geol.* A regularly shaped and lengthy deposit of an ore : LODE. **5.** A long wavy strip of color, as in wood or marble. **6.** A fissure, crack, or cleft. **7.** A pervading quality or character : STREAK <"all through the interminable narrative there ran a *vein* of impressive earnestness" —Mark Twain> **8.** A passing or temporary mood or humor <discussion in a lighter *vein*> *—vt.* **veined, vein·ing, veins. 1.** To supply or fill with veins. **2.** To mark or decorate with veins. **—vein′al** *adj.*
**veined** (vānd) *adj.* Displaying veins or markings that suggest veins.
**vein·ing** (vā′nĭng) *n.* Venation.
**vein·let** (vān′lĭt) *n.* A small or secondary vein, as of an insect's wing.
**vein·stone** (vān′stōn′) *n.* Mineral matter in a vein exclusive of the ore : GANGUE.
**vein·y** (vā′nē) *adj.* **-i·er, -i·est. 1.** Displaying or full of veins. **2.** Veined.
**ve·la** (vē′lə) *n. pl. of* VELUM.
**ve·la·men** (və-lā′mən) *n., pl.* **ve·lam·i·na** (və-lăm′ə-nə) [Lat., covering < *velare,* to cover < *velum,* a covering.] **1.** *Anat.* A membranous covering or integument : VELUM. **2.** *Bot.* The thick, spongy outer covering of the aerial roots of epiphytic orchids and certain other plants, capable of absorbing atmospheric moisture. **—vel′a·men′·tous** (vĕl′ə-mĕn′təs) *adj.*
**ve·lar** (vē′lər) *adj.* [Lat. *velaris,* of a curtain < *velum,* curtain.] **1. a.** Of or relating to a velum. **b.** Using the soft palate. **2.** Formed with the back of the tongue on or near the soft palate, as (g) in *good* and (k) in *cup.* *—n.* A velar sound.
**ve·lar·ize** (vē′lə-rīz′) *vt.* **-ized, -iz·ing, -iz·es.** To articulate (a sound) by retracting the back of the tongue toward the soft palate. **—ve′lar·i·za′tion** *n.*
**ve·late** (vē′lāt′, -lĭt) *adj.* [Lat. *velatus,* p.part. of *velare,* to cover < *velum,* a covering.] *Biol.* Having or covered by a velum or veil.
**Vel·cro** (vĕl′krō′). A trademark for a fastening tape used esp. for cloth products.
**veldt** *also* **veld** (vĕlt, fĕlt) *n.* [Afrikaans *veld* < MDu., field.] *So. Afr.* An open grazing area.
**ve·li·ger** (vē′lə-jər, vĕl′ə-) *n.* [NLat. : VELUM + Lat. *gerere,* to bear.] A larval stage of a mollusk, in which the ciliated velum is present.
**vel·le·i·ty** (vĕ-lē′ĭ-tē, və-) *n., pl.* **-ties.** [NLat. *velleitas* < Lat. *velle,* to want.] **1.** The lowest level of volition. **2.** A mere wish unaccompanied by action or effort to obtain it.
**vel·lum** (vĕl′əm) *n.* [ME *velim* < OFr. *velin,* pertaining to a calf < *veel,* calf, veal.] **1.** A fine parchment made from the skins of calf, lamb, or kid and used for the pages and binding of fine books. **2.** A work written or printed on vellum. **3.** A heavy off-white fine-quality paper similar to vellum.
**ve·lo·ce** (vā-lō′chā) *adv.* [Ital. < Lat. *velox,* swift.] *Mus.* Rapidly. —Used as a direction.
**ve·lo·cim·e·ter** (vē′lō-sĭm′ĭ-tər, vĕl′ō-) *n.* [VELOCI(TY) + -METER.] A device for measuring the speed of sound in water.
**ve·loc·i·pede** (və-lŏs′ə-pēd′) *n.* [Fr. *vélocipède* : *véloci-,* swift (< Lat. *velox, veloci-*) + *-pède,* -ped.] **1.** An early bicycle propelled by pushing the feet along the ground while straddling the vehicle. **2.** Any of several early bicycles with pedals fastened to the front wheel. **3.** A tricycle.

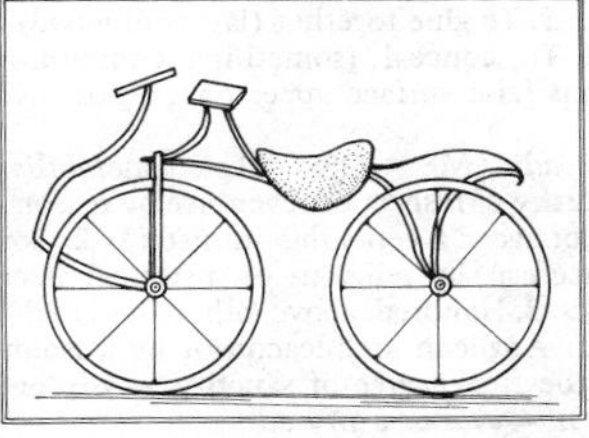
**velocipede**

**ve·loc·i·ty** (və-lŏs′ĭ-tē) *n., pl.* **-ties.** [Fr. *vélocité* < Lat. *velocitas* < *velox,* swift.] **1.** Rapidity or speed. **2.** *Physics.* A vector quantity whose magnitude is a body's speed and whose direction is the body's direction of motion. **3.** The rate of rapidity or action.
**ve·lour** *or* **ve·lours** (və-lo͝or′) *n., pl.* **-lours** (-lo͝orz′) [Fr. *velours,* velvet < OFr. *velous* < Lat. *villosus,* hairy < *villus,* shaggy hair.] **1.** A closely napped velvetlike fabric, used mainly for clothing and upholstery. **2.** A felt resembling velvet, used in making hats.
**ve·lou·té** (və-lo͞o-tā′) *n.* [Fr. < *velouter,* to give the appearance of velvet < *velours,* velvet. —see VELOUR.] A white sauce prepared with flour, butter, and chicken or veal stock.
**ve·lum** (vē′ləm) *n., pl.* **-la** (-lə) [NLat. < Lat., veil.] **1.** *Biol.* A covering or partition of thin membranous tissue, as the veil of a mushroom. **2.** *Anat.* The soft palate. **3.** *Zool.* A ciliated swimming organ that develops in certain larval stages of many marine gastropod mollusks.

o͞o **boot** ou **out** th **thin** *th* **this** ŭ **cut** ûr **urge** y **young**
yo͞o **abuse** zh **vision** ə **about,** it**e**m, edib**le**, gall**o**p, circ**u**s

**ve·lure** (vĕ-lo͝or′, vĕl′yər) *n.* [Fr. *velours.* —see VELOUR.] *Obs.* **1.** Velvet. **2.** A velvetlike fabric.

**ve·lu·ti·nous** (və-lo͞ot′n-əs) *adj.* [NLat. *velutinus* < Med. Lat. *velutum,* velvet < *villutus,* shaggy. —see VELVET.] Coated with dense, soft, silky hairs : VELVETY.

**vel·vet** (vĕl′vĭt) *n.* [ME *veluet* < OFr. *velute* < *velu,* shaggy < Med. Lat. *villutus* < Lat. *villus,* shaggy hair.] **1. a.** A fabric made usu. of silk or a synthetic fiber, as rayon or nylon, and having a smooth dense pile and a plain back. **b.** Something suggesting or similar to velvet. **2.** Smoothness : softness. **3.** The soft coating on the newly developing antlers of deer and related animals. **—vel′vet·y** *adj.*

**velvet ant** *n.* Any of various wasps of the family Mutillidae, with a dense, hairy, often brightly colored covering.

**vel·vet·een** (vĕl′vĭ-tēn′) *n.* [< VELVET.] A velvetlike cotton fabric.

**velvet plant** *n.* MULLEIN 1.

**ven-** *pref. var. of* VENO-.

**ve·na** (vē′nə) *n., pl.* **ve·nae** (vē′nē) [Lat.] *Anat.* A vein.

**ve·na ca·va** (vē′nə kā′və) *n., pl.* **ve·nae ca·vae** (vē′nē kā′vē) [Lat., hollow vein.] Either of the two large veins in air-breathing vertebrates that enter into and return blood to the right atrium of the heart.

**ve·nae** (vē′nē) *n. pl of* VENA.

**ve·nal** (vē′nəl) *adj.* [Lat. *venalis,* for sale < *venum,* sale.] **1. a.** Open or susceptible to bribery. **b.** Capable of betraying one's honor, duty, or scruples for a price : CORRUPTIBLE. **2.** Characterized by dishonest or unscrupulous dealings <a *venal* administration> **3.** Obtainable by purchase or bribery instead of by merit. **—ve′nal·ly** *adv.*

**ve·nal·i·ty** (vē-năl′ĭ-tē) *n., pl.* **-ties. 1.** The quality of being open to bribery or corruption. **2.** The use of an office or a position of trust for dishonest gain.

**ve·nat·ic** (vē-năt′ĭk) *also* **ve·nat·i·cal** (-ĭ-kəl) *adj.* [Lat. *venaticus* < *venari,* to hunt.] **1.** Relating to or used in hunting. **2.** Given to hunting for livelihood or sport.

**ve·na·tion** (vē-nā′shən, vĕ-) *n.* Distribution or arrangement of veins. **—ve·na′tion·al** *adj.*

**vend** (vĕnd) *v.* **vend·ed, vend·ing, vends.** [Fr. *vendre* < OFr. < Lat. *vendere* : *venum,* sale + *dare,* to give.] *—vt.* **1.** To sell. **2.** To offer (e.g., an idea) for public consideration. *—vi.* **1. a.** To sell goods. **b.** To sell by means of a vending machine. **2.** To have a market.

**vend·a·ble** (vĕn′də-bəl) *adj. var. of* VENDIBLE.

**vend·ee** (vĕn-dē′) *n.* A buyer.

**vend·er** *also* **ven·dor** (vĕn′dər) *n.* **1. a.** One that sells or vends. **b.** A sales representative. **2.** A vending machine.

**ven·det·ta** (vĕn-dĕt′ə) *n.* [Ital., revenge < Lat. *vindicta* < *vindicare,* to avenge.] **1.** A bitter blood feud between two families motivated by lust for revenge. **2.** An act or attitude motivated by vengeance.

**vend·i·ble** *also* **vend·a·ble** (vĕn′də-bəl) *adj.* **1.** Capable of being sold or suitable for sale. **2.** Venal. *—n.* Something that can be sold.

**vending machine** *n.* A coin-operated machine that dispenses merchandise.

**ven·dor** (vĕn′dər) *n. var. of* VENDER.

**ven·due** (vĕn′do͞o′, -dyo͞o′, vän′- vĕn-do͞o′, -dyo͞o′, vän-) *n.* [Du. *vendu* < OFr. *vendue* < *vendre,* to sell. —see VEND.] A public sale.

**ve·neer** (və-nîr′) *n.* [G. *Furnier* < *furnieren,* to veneer < OFr. *fournir,* to furnish.] **1.** A thin layer of material, as wood or plastic, bonded to and used for covering a usu. inferior material. **2.** Any of the thin layers glued together in manufacturing plywood. **3.** A misleading or superficial outward show or pretense <a *veneer* of charm> *—vt.* **-neered, -neer·ing, -neers. 1.** To overlay (a surface) with a thin layer of material. **2.** To glue together (layers of wood) in manufacturing plywood. **3.** To conceal (something common or crude) with an attractive but false surface appearance : gloss over. **—ve·neer′er** *n.*

**ven·er·a·ble** (vĕn′ər-ə-bəl) *adj.* [ME < OFr. < Lat. *venerabilis* < *venerari,* to venerate.] **1.** Worthy of respect or reverence by reason of dignity, character, position, or age <a *venerable* minister> **2.** Commanding respect or reverence esp. by religious or historical association <*venerable* remains> **3.** Honored above others. —Used in titles of respect borne by an Anglican archdeacon or by a Roman Catholic who has attained the first degree of sanctity. **—ven′er·a·ble·ness, ven′er·a·bil′i·ty** *n.* **—ven′er·a·bly** *adv.*

**ven·er·ate** (vĕn′ə-rāt′) *vt.* **-at·ed, -at·ing, -ates.** [Lat. *venerari, venerat-,* to venerate.] To regard with respect, reverence, or deference. **—ven′er·a′tion** *n.* **—ven′er·a′tor** *n.*

**ve·ne·re·al** (və-nîr′ē-əl) *adj.* [ME *venerealle* < Lat. *venereus* < *Venus,* Venus, love.] **1.** Of or relating to sexual intercourse. **2. a.** Transmitted by sexual intercourse. **b.** Of or relating to venereal disease. **3.** Of or relating to the genitals.

**venereal disease** *n.* A contagious disease, as syphilis or gonorrhea, contracted through sexual intercourse.

**ve·ne·re·ol·o·gy** (və-nîr′ē-ŏl′ə-jē) *n.* [VENERE(AL) + -LOGY.] Study of venereal diseases. **—ve·ne′re·o·log′i·cal** (-ə-lŏj′ĭ-kəl) *adj.* **—ve·ne′re·ol′o·gist** *n.*

**ven·er·y**[1] (vĕn′ə-rē) *n.* [ME *venerie* < Med. Lat. *veneria* < Lat. *Venus,* Venus, love.] *Archaic.* **1.** Indulgence in or the pursuit of sexual activity. **2.** Sexual intercourse.

**ven·er·y**[2] (vĕn′ə-rē) *n.* [ME < OFr. *venerie* < *vener,* to hunt < Lat. *venari.*] *Archaic.* The act, art, or sport of hunting.

**ven·e·sec·tion** (vĕn′ĭ-sĕk′shən, vē′nĭ-) *n.* [Med. Lat. *venae sectio,* cutting of a vein.] Phlebotomy.

**Ve·ne·tian blind** *also* **ve·ne·tian blind** (və-nē′shən) *n.* [< *Venetian,* of Venice, Italy.] A window blind made of thin horizontal slats that may be raised and lowered and set at a desired angle to control the amount of light admitted.

**venetian blue** *n.* A strong blue to greenish blue.

**venetian red** *n.* A deep to strong reddish brown.

**venge** (vĕnj) *vt.* **venged, veng·ing, veng·es.** [ME *vengen* < OFr. *vengier.* —see VENGEANCE.] *Archaic.* To avenge.

**ven·geance** (vĕn′jəns) *n.* [ME < OFr. < *vengier,* to avenge < Lat. *vindicare.*] The act or motive of punishing another in payment for an injury or wrong he or she has committed : RETRIBUTION. **—with a vengeance. 1.** With great fury or violence. **2.** Excessively <ate and drank *with a vengeance*>

**venge·ful** (vĕnj′fəl) *adj.* **1.** Desiring vengeance : VINDICTIVE. **2.** Indicating or proceeding from an obsession for revenge. **3.** Inflicting or serving to inflict vengeance <*vengeful* gossip> **—venge′ful·ly** *adv.* **—venge′ful·ness** *n.*

**V-en·gine** (vē′ĕn′jən) *n.* An internal-combustion engine with cylinders arranged so that pairs form V shapes.

**veni-** *pref. var. of* VENO-.

**ve·ni·al** (vē′nē-əl, vēn′yəl) *adj.* [ME < OFr. < LLat. *venialis* < Lat. *venia,* forgiveness.] **1.** Easily excused or forgiven : PARDONABLE <a *venial* offense> **2.** *Rom. Cath. Ch.* Minor in nature and calling for only temporal punishment <a *venial* sin> **—ve′ni·al′i·ty** (vē′nē-ăl′ĭ-tē, vēn-yăl′-), **ve′ni·al·ness** *n.* **—ve′ni·al·ly** *adv.*

**ve·ni·punc·ture** (vē′nĭ-pŭngk′chər, vĕn′ĭ-) *n.* Puncture of a vein, as for drawing blood, intravenous feeding, or the administration of medicine.

**ve·ni·re** (və-nī′rē) *n.* [Med. Lat. *venire (facias),* (you should cause) to come, a phrase used in the writ.] *Law.* **1.** *also* **ve·ni·re fa·ci·as** (və-nī′rē fā′shē-əs) A judicial writ ordering a sheriff to summon prospective jurors. **2.** The panel of prospective jurors from which a jury is chosen.

**ve·ni·re·man** (və-nī′rē-mən, -nîr′ē-) *n.* One summoned to jury duty under a venire.

**ven·i·son** (vĕn′ĭ-sən, -zən) *n.* [ME *veneson* < OFr. < Lat. *venatio,* hunting < *venari,* to hunt.] **1.** The flesh of a deer, used for food. **2.** *Archaic.* The flesh of a game animal used for food.

**Venn diagram** (vĕn) *n.* [After John Venn (1824–1923).] A pictorial representation using circles and squares so positioned as to represent an operation in set theory.

**veno-** *or* **veni-** *or* **ven-** *pref.* [< Lat. *vena.*] Vein <*veni*puncture>

**ve·no·gram** (vē′nə-grăm′) *n.* A roentgenogram of a vein or veins.

**ve·nog·ra·phy** (vĭ-nŏg′rə-fē) *n.* Roentgenography of a vein or veins after injection of a radiopaque substance.

**ven·om** (vĕn′əm) *n.* [ME *venim* < OFr. < VLat. **venimen* < Lat. *venenum,* poison.] **1.** A poisonous secretion of an animal, as a snake, spider, or scorpion, usu. transmitted by a bite or sting. **2.** A poison. **3.** Malice : spite.

**ven·om·ous** (vĕn′ə-məs) *adj.* **1.** Secreting and transmitting venom <a *venomous* spider> **2.** Full of or containing venom. **3.** Malicious : spiteful <a *venomous* comment> **—ven′om·ous·ly** *adv.* **—ven′om·ous·ness** *n.*

**ve·nose** (vē′nōs′) *adj.* [Lat. *venosus,* venous < *vena,* vein.] **1.** Having noticeable veins or markings suggesting veins. **2.** Venous.

**ve·nos·i·ty** (vē-nŏs′ĭ-tē) *n.* The quality or state of being venous or venose.

**ve·nous** (vē′nəs) *adj.* [Lat. *venosus* < *vena,* vein.] **1.** Of or relating to a vein or veins. **2.** *Physiol.* Returning to the heart through the great veins. **—ve′nous·ly** *adv.* **—ve′nous·ness** *n.*

**vent** (vĕnt) *n.* [< ME *venten,* to provide with an outlet < OFr. *esventer,* to let out air : *es-,* ex- + *vent,* wind (< Lat. *ventus*).] **1.** A way of escaping or leaving a restricted space : EXIT. **2.** An opening for the passage or escape of a liquid, gas, or vapor. **3.** The small hole at the breech of a gun through which the charge is ignited. **4.** *Zool.* The cloacal or anal excretory opening in animals, esp. in birds, reptiles, amphibians, and fish. *—vt.* **vent·ed, vent·ing, vents. 1.** To give forceful expression or utterance to : EXPRESS <*vented* our resentment to the officials> **2.** To relieve by venting <a device to *vent* excess pressure in the tank> **3.** To discharge through a vent. **4.** To provide with a vent. **—give vent to.** To give utterance or expression to. **—vent′er** *n.*

**vent·age** (vĕn′tĭj) *n.* [< VENT.] A small opening : VENT.

**ven·tail** (vĕn′tāl′) *n.* [ME < OFr. *vantail* < *vent,* wind < Lat. *ventus.*] The lower front part of a medieval helmet, fitting over the neck.

**ven·ter** (vĕn′tər) *n.* [Norman Fr. < Lat.] **1. a.** *Anat.* The abdomen. **b.** The wide swelling part of a muscle. **2.** *Biol.* A swollen structure or part resembling a venter. **3.** A wife or mother who is the source of offspring.

**ven·ti·late** (vĕn′tl-āt′) *vt.* **-lat·ed, -lat·ing, -lates.** [ME *ventilaten* < Lat. *ventilare,* to fan < *ventus,* wind.] **1.** To let fresh air into so as to replace stale air. **2.** To circulate air within so as to freshen <*ventilate* an auditorium> **3.** To provide with a vent or vents. **4.** To expose (a substance) to the circulation of fresh air so as to retard spoilage. **5.** To expose to public discussion or examination. **6.** To aerate or oxygenate (blood). **—ven′ti·la′tion** *n.*

**ven·ti·la·tor** (vĕn′tl-ā′tər) *n.* One that ventilates, esp. a device, as an exhaust fan, that ejects stale air and circulates fresh air. **—ven′ti·la·to′ry** (vĕn′tl-ə-tôr′ē, -tōr′ē) *adj.*

**ven·tral** (vĕn′trəl) *adj.* [Fr. < Lat. *ventralis* < *venter,* belly.] **1.** *Anat.* **a.** Relating to or located on or near the belly : ABDOMINAL. **b.** Relating to the anterior aspect of the human body or the lower surface of the body of an animal. **2.** *Bot.* Of or on the lower or inner surface of an organ. **—ven′tral·ly** *adv.*

**ventral fin** *n. Zool.* A pelvic fin.

**ven·tri·cle** (vĕn′trĭ-kəl) *n.* [ME < OFr. < Lat. *ventriculus,* dim. of *venter,* belly.] A small anatomical cavity or chamber, as of the brain or heart, esp.: **a.** The chamber on the left side of the heart that receives arterial blood from the left atrium and contracts to drive it into the aorta. **b.** The chamber on the right side of the heart that receives venous blood from the right atrium and drives it into the pulmonary artery. **—ven·tric′u·lar** (-trĭk′yə-lər) *adj.*

**ven·tri·cose** (vĕn′trĭ-kōs′) *also* **ven·tri·cous** (-kəs) *adj.* [NLat. *ventricosus* < Lat. *venter,* belly.] Inflated, swollen, or distended. **—ven′tri·cos′i·ty** (-kŏs′ĭ-tē) *n.*

**ven·tric·u·lus** (vĕn-trĭk′yə-ləs) *n., pl.* **-li** (-lī′) [Lat., dim. of *venter,* belly.] A hollow digestive organ, esp. the stomach of an insect or the gizzard of a bird.

**ven·tril·o·quism** (vĕn-trĭl′ə-kwĭz′əm) *also* **ven·tril·o·quy** (-kwē) *n.* [< LLat. *ventriloquus,* speaking from the belly : Lat. *venter,* belly + *loqui,* to speak.] A means of producing vocal sounds so that they seem to come from a source other than the speaker, as from a mechanical or hand-operated dummy. **—ven′tri·lo′qui·al** (vĕn′trə-lō′kwē-əl) *adj.* **—ven′tri·lo′qui·al·ly** *adv.* **—ven·tril′o·quist** (-kwĭst) *n.* **—ven·tril′o·quis′tic** *adj.*

**ven·tril·o·quize** (vĕn-trĭl′ə-kwīz′) *vi.* **-quized, -quiz·ing, -quiz·es.** To practice ventriloquism.

**ven·tril·o·quy** (vĕn-trĭl′ə-kwē) *n. var. of* VENTRILOQUISM.

**ven·ture** (vĕn′chər) *n.* [ME, chance < *aventure,* adventure. —see ADVENTURE.] **1.** A dangerous undertaking. **2.** A daring undertaking. **3.** Something at hazard in a venture : STAKE. *—v.* **-tured, -tur·ing, -tures.** *—vt.* **1.** To expose to danger or risk. **2.** To brave the dangers of <*ventured* the high seas in a flimsy craft> **3.** To express at the risk of denial, criticism, or censure : DARE <*ventured* my opinion> *—vi.* To take a risk or dare. **—ven′tur·er** *n.*

**venture capital** *n.* Capital, such as individual savings or retained corporate earnings, that is invested or is available for investment in the ownership of a new corporate enterprise.

**ven·ture·some** (vĕn′chər-səm) *adj.* **1.** Inclined to venture or to take risks : BOLD. **2.** Involving risk or danger : HAZARDOUS. **—ven′ture·some·ly** *adv.* **—ven′ture·some·ness** *n.*

**ven·tu·ri** (vĕn-tŏŏr′ē) *n.* [After G. B. *Venturi* (1742–1822).] **1.** A short tube with a constricted throat for determining fluid pressures and velocities by measuring differential pressures generated at the throat as a fluid traverses the tube. **2.** A constricted throat in the air passage of a carburetor, causing a reduction in pressure by means of which fuel vapor is drawn out of the carburetor bowl.

**ven·tur·ous** (vĕn′chər-əs) *adj.* **1.** Courageous and daring : ADVENTUROUS. **2.** Hazardous, dangerous, or risky. **—ven′tur·ous·ly** *adv.* **—ven′tur·ous·ness** *n.*

**ven·ue** (vĕn′yōō) *n.* [ME, arrival < OFr. < *venir,* to come < Lat. *venire.*] **1.** *Law.* **a.** The locality where an alleged crime or other cause of legal action occurs. **b.** The locality or political division from which a jury must be called and in which a trial must be held. **c.** The clause within a declaration naming the locality in which the trial is occurring or will occur. **d.** The clause in an affidavit naming the locality where it was made and sworn to. **2.** The locality of a gathering, as for a convention.

**ven·ule** (vĕn′yōōl, vēn′-) *n.* [Lat. *venula,* dim. of *vena,* vein.] A minute vein, as one joining with a capillary or branching from a vein in an insect's wing. **—ven′u·lar** (-yə-lər) *adj.*

**Ve·nus** (vē′nəs) *n.* [ME < OE < Lat. < *venus,* love.] **1.** *Rom. Myth.* The goddess of love and beauty. **2.** The second planet from the sun, having an average radius of 6,114 kilometers or 3,800 miles, a mass 0.816 times that of the earth, and a sidereal period of revolution about the sun of 224.7 days at a mean distance of approx. 108.1 million kilometers or 67.2 million miles.

**Ve·nu·sian** (vĭ-nōō′zhən, -nyōō′-) *adj.* Relating to or typical of the planet Venus.

**Venus's flower basket** *n.* A sponge of the genus *Euplectella,* of deep marine waters, with a cylindrical skeleton of glassy, elaborately interlaced latticework.

**Ve·nus's-fly·trap** (vē′nəs-flī′trăp′, vē′nə-sĭz-) *n.* An insectivorous plant, *Dionaea muscipula* of boggy areas of the southeastern United States, with marginally spined, hinged leaf blades that close and entrap insects.

**Venus's girdle** *n.* A ribbon-shaped marine animal, *Cestum veneris,* with a jellylike bluish-green iridescent body.

**Venus's-hair** (vē′nə-sĭz-hâr′) *n.* A maidenhair fern, *Adiantum capillus-veneris* of moist warm regions, with slender blackish stalks.

**Ve·nus's-look·ing-glass** (vē′nə-sĭz-lŏŏk′ĭng-glăs′) *n.* A plant of the genus *Specularia,* esp. *S. speculum-veneris* native to Europe, and *S. perfoliata* of North America, with small blue or white star-shaped flowers.

**ve·ra·cious** (və-rā′shəs) *adj.* [< Lat. *verax, verac-.*] **1.** Truthful : honest. **2.** Accurate : precise. **—ve·ra′cious·ly** *adv.* **—ve·ra′cious·ness** *n.*

**ve·rac·i·ty** (və-răs′ĭ-tē) *n., pl.* **-ties.** [Med. Lat. *veracitas* < Lat. *verax,* true.] **1.** Adherence to the truth : TRUTHFULNESS. **2.** Conformity to fact : ACCURACY. **3.** Something true.

**ve·ran·dah** *or* **ve·ran·da** (və-răn′də) *n.* [Hindi.] A usu. roofed and often partly enclosed porch or balcony extending along the outside of a building.

**ve·rat·ri·dine** (və-răt′rĭ-dēn′) *n.* [VERATR(INE) + -ID + -INE[2].] A yellowish-white, amorphous powdered alkaloid, $C_{36}H_{51}NO_{11}$, derived from sabadilla seeds and from the rhizome of a species of hellebore, *Veratrum album.*

**ver·a·trine** (vĕr′ə-trēn′, -trĭn) *n.* [Fr. *vératrine* < NLat. *Veratrum,* genus name of a hellebore < Lat. *veratrum,* hellebore.] A poisonous mixture of colorless crystalline alkaloids extracted from sabadilla seeds and once used as a counterirritant.

**verb** (vûrb) *n.* [ME *verbe* < OFr. < Lat. *verbum,* word.] **1. a.** The part of speech that expresses action, existence, or occurrence in most languages. **b.** Any of the words within this part of speech, as *be, walk,* or *believe.* **2.** A phrase or other construction used as a verb.

**ver·bal** (vûr′bəl) *adj.* [OFr. < LLat. *verbalis* < Lat. *verbum,* word.] **1.** Of, relating to, or associated with words <*verbal* orders> **2. a.** Concerned with words instead of with the facts or ideas they represent. **b.** Using or made up of words alone without action <a *verbal* showdown> **3.** Expressed or transmitted in speech : UNWRITTEN <a *verbal* agreement> **4.** Word for word : LITERAL <a *verbal* translation> **5. a.** Relating to, having the nature or function of, or derived from a verb. **b.** Used to form verbs <a *verbal* suffix> **6.** Of or pertaining to competence in the use and comprehension of words <*verbal* skills> *—n.* A verbal noun, adjective, or other word derived from a verb and preserving some of the verb's characteristics. **—ver′bal·ly** *adv.*

**ver·bal·ism** (vûr′bə-lĭz′əm) *n.* **1.** An expression in words. **2.** A phrase or sentence without meaning. **3.** An expression, sentence, or other construction stressing words over content or idea.

**ver·bal·ist** (vûr′bə-lĭst) *n.* **1.** One skilled in the use of words. **2.** One who prefers words over ideas or facts. **—ver′bal·is′tic** *adj.*

**ver·bal·ize** (vûr′bə-līz′) *v.* **-ized, -iz·ing, -iz·es.** *—vt.* **1.** To express in words <tried to *verbalize* my emotions> **2.** To convert (e.g., a noun) to verbal use. *—vi.* **1.** To express oneself in words. **2.** To be verbose. **—ver′bal·i·za′tion** *n.* **—ver′bal·iz′er** *n.*

**verbal noun** *n.* A noun derived from a verb that in some uses retains the verb's characteristics and sense.

**ver·ba·tim** (vər-bā′tĭm) *adj.* [ME < Med. Lat. < Lat. *verbum,* word.] Using precisely the same words : word for word. *—adv.* In precisely the same words.

**ver·be·na** (vər-bē′nə) *n.* [NLat. *Verbena,* genus name < Lat. *verbena,* sing. of *verbenae,* sacred boughs.] **1.** A New World plant of the genus *Verbena,* esp. one of several species raised for their showy variously colored flower clusters. **2.** A plant, as the lemon verbena, similar to or related to the verbena.

**ver·bi·age** (vûr′bē-ĭj, -bĭj) *n.* [Fr. < OFr. *verbier,* to chatter < *verbe,* word < Lat. *verbum.*] **1.** More words than are required for clarity or precision : WORDINESS. **2.** The way in which one expresses oneself in words : DICTION.

**verb·i·fy** (vûr′bə-fī′) *vt.* **-fied, -fy·ing, -fies.** To use (e.g., a noun) as a verb.

**ver·bose** (vər-bōs′) *adj.* [Lat. *verbosus* < *verbum,* word.] Using or having more words than required : WORDY. **—ver·bose′ly** *adv.* **—ver·bose′ness, ver·bos′i·ty** (-bŏs′ĭ-tē) *n.*

**ver·bo·ten** (fər-bōt′n, vər-) *adj.* [G. < OHG *farboten,* p.part. of *farbiotan,* to forbid.] Strictly forbidden.

**ver·dant** (vûr′dnt) *adj.* [OFr. *verdoyant,* pr.part. of *verdoyer,* to become green < *verd,* green < Lat. *viridis* < *virēre,* to be green.] **1. a.** Green with vegetation. **b.** Covered with a green growth. **2.** Green in color. **3.** Inexperienced or unsophisticated. **—ver′dan·cy** *n.* **—ver′dant·ly** *adv.*

**verd antique** *also* **verde antique** (vûrd) *n.* [Obs. Fr., antique green.] **1.** A dull-green mottled or veined serpentine marble used in interior decoration. **2.** Verdigris.

**ver·der·er** *also* **ver·der·or** (vûr′dər-ər) *n.* [AN < OFr. *verdier,* an official in charge of forests < *verd,* green. —see VERDANT.] The official in charge of English royal forests.

**ver·dict** (vûr′dĭkt) *n.* [ME *verdit* < AN, var. of OFr. *veirdit* : *veir,* true (< Lat. *verus*) + *dit,* speech (< Lat. *dictum* < neuter p.part. of *dicere,* to speak).] **1.** The decision arrived at by a jury at the end of a trial. **2.** An expressed conclusion : JUDGMENT.

**ver·di·gris** (vûr′dĭ-grēs′, -grĭs′, -grē′) *n.* [ME *vertegres* < OFr. *vertegrez,* alteration of *vert-de-Grice,* green of Greece.] **1.** A blue or green basic copper acetate used as a paint pigment, fungicide, and insecticide. **2.** A green crust or patina of copper sulfate or copper chloride formed on copper, brass, and bronze because of exposure to air or sea water over a long period.

**ver·din** (vûr′dn) *n.* [Fr., yellowhammer.] A small grayish bird, *Auriparus flaviceps* of the southwestern United States and adjacent Mexico, with a yellowish head and throat.

**ver·di·ter** (vûr′dĭ-tər) *n.* [Alteration of OFr. *verd de terre,* green of earth.] Either of two basic carbonates of copper used as a blue or green pigment.

**ver·dure** (vûr′jər) *n.* [ME < OFr. < *verd,* green. —see VERDANT.] **1. a.** The fresh, vibrant greenness of flourishing vegetation. **b.** Such vegetation itself. **2.** A fresh or flourishing condition <the *verdure* of youth> **—ver′dur·ous** *adj.* **—ver′dur·ous·ness** *n.*

**verge[1]** (vûrj) *n.* [ME < OFr. < Lat. *virga,* rod.] **1.** The extreme edge, rim, or margin : BRINK <the *verge* of a river> **2. a.** An encircling boundary. **b.** The space encircled by such a boundary. **3.** The point beyond which an act, state, or condition is likely to start or happen <on the *verge* of hysterical laughter> **4.** The edge of the tiling that projects over a roof gable. **5.** A rod, wand, or staff carried as a sign of authority or office. **6.** *Obs.* The rod held by a feudal tenant swearing fealty to his lord. **7.** The spindle of a balance wheel in a clock or watch, esp. such a spindle in a clock with vertical escapement. **8.** The male organ of an invertebrate. —*vi.* **verged, verg·ing, verg·es. 1.** To approach the limit or verge <joyfulness *verging* on hysteria> **2.** To make up the verge or limit : BORDER <bicycle paths *verging* on the main streets>

**verge[2]** (vûrj) *vi.* **verged, verg·ing, verg·es.** [Lat. *vergere.*] **1.** To slope or incline. **2.** To be in the process of becoming something else <evening *verging* into night>

**verg·er** (vûr′jər) *n.* **1.** One who carries the verge before a scholastic, legal, or religious dignitary in a procession. **2.** *Chiefly Brit.* One in charge of a church interior.

**ve·rid·i·cal** (və-rĭd′ĭ-kəl) *also* **ve·rid·ic** (-rĭd′ĭk) *adj.* [Lat. *veridicus* : *verus,* true + *dicere,* to say.] Veracious : truthful. **—ve·rid′i·cal′i·ty** (-kăl′ĭ-tē) *n.* **—ve·rid′i·cal·ly** *adv.*

**ver·i·fi·ca·tion** (vĕr′ə-fĭ-kā′shən) *n.* **1.** An act of verifying or the state of being verified. **2. a.** A confirmation of the truth of a theory or fact. **b.** A formal statement of such a confirmation. **3.** *Law.* A short formulaic oath concluding a pleading and affirming that the pleader stands prepared to prove his or her allegations. **—ver′i·fi·ca′tive** *adj.*

**ver·i·fy** (vĕr′ə-fī′) *vt.* **-fied, -fy·ing, -fies.** [ME *verifien* < OFr. *verifier* < Med. Lat. *verificare* : Lat. *verus,* true + Lat. *facere,* to make.] **1.** To prove the truth of by presenting evidence or testimony : SUBSTANTIATE <Two business associates *verified* my story.> **2.** To determine or test the truth or accuracy of, as by comparison, investigation, or reference <made observations to *verify* a suspicion> **3.** *Law.* **a.** To affirm formally or under oath. **b.** To append a verification to (a pleading). **—ver′i·fi′a·ble** *adj.* **—ver′i·fi′er** *n.*

**ver·i·ly** (vĕr′ə-lē) *adv.* [ME *verraily* < *verray,* true. —see VERY.] **1.** In truth : IN FACT. **2.** With confidence : ASSUREDLY.

**ver·i·sim·i·lar** (vĕr′ə-sĭm′ə-lər) *adj.* [Lat. *verisimilis* : *veri,* genitive of *verum,* truth < *verus,* true + *similis,* similar.] Appearing to be real or true. **—ver′i·sim′i·lar·ly** *adv.*

**ver·i·si·mil·i·tude** (vĕr′ə-sĭ-mĭl′ĭ-to͞od′, -tyo͞od′) *n.* [Lat. *verisimilitudo* < *verisimilis,* verisimilar.] **1.** The quality of appearing to be real or true : LIKELIHOOD. **2.** Something appearing to be real or true. **—ver′i·si·mil′i·tu′di·nous** (-to͞od′n-əs, -tyo͞od′-) *adj.*

**ver·ism** (vĕr′ĭz′əm) *n.* [Ital. *verismo* : *vero,* true (< Lat. *verus*) + *-ismo,* -ism.] Realism in literature and art. **—ver′ist** *n.* **—ve·ris′tic** (və-rĭs′tĭk) *adj.*

**ver·i·ta·ble** (vĕr′ĭ-tə-bəl) *adj.* [ME < OFr. < *verite,* truth. —see VERITY.] True : unquestionable <The house was a *veritable* castle.> **—ver′i·ta·ble·ness** *n.* **—ver′i·ta·bly** *adv.*

**ver·i·ty** (vĕr′ĭ-tē) *n., pl.* **-ties.** [ME *verite* < OFr. < Lat. *veritas,* truth < *verus,* true.] **1.** The quality or state of being real, accurate, or correct. **2.** A statement, principle, or belief regarded as established and permanent truth <patriotic *verities*>

**ver·juice** (vûr′jo͞os′) *n.* [ME *verjus* < OFr. *vertjus* : *vert,* green, vert + *jus,* juice.] The acidic juice of sour or unripe fruit, as grapes or crab apples.

**ver·meil** (vûr′məl, -māl′) *n.* [ME *vermayl* < OFr. *vermeil* < LLat. *vermiculus,* a kind of red worm, dim. of *vermis,* worm.] **1.** Vermilion or a similar bright red color. **2.** (vĕr-mā′). Gilded metal, as silver, bronze, or copper. —*adj.* Bright red in color.

**vermi-** *pref.* [< Lat. *vermis,* worm.] Worm <vermicide>

**ver·mi·cel·li** (vûr′mə-chĕl′ē, -sĕl′ē) *n.* [Ital., pl. of *vermicello,* dim. of *verme,* worm < Lat. *vermis.*] Pasta made into long threads thinner than spaghetti.

**ver·mi·cide** (vûr′mĭ-sīd′) *n.* An agent for killing worms. **—ver′mi·cid′al** (-sīd′l) *adj.*

**ver·mic·u·lar** (vər-mĭk′yə-lər) *adj.* [Med. Lat. *vermicularis* < Lat. *vermiculus,* dim. of *vermis,* worm.] **1.** Shaped or moving like a worm. **2.** Having wormlike markings : VERMICULATE. **3.** Caused by or pertaining to worms. **—ver·mic′u·lar·ly** *adv.*

**ver·mic·u·late** (vər-mĭk′yə-lāt′) *vt.* **-lat·ed, -lat·ing, -lates.** [Lat. *vermiculari, vermiculat-* < *vermiculus,* dim. of *vermis,* worm.] To decorate with wavy or winding lines. —*adj.* (-lĭt, -lāt′). **1.** Bearing wormlike wavy lines. **2.** Moving like a worm : WRIGGLING. **3.** Tortuous : sinuous. **4.** Infested with worms : WORM-EATEN.

**ver·mic·u·la·tion** (vər-mĭk′yə-lā′shən) *n.* **1.** Motion like that of a worm, esp. the wavelike contractions of the intestine : PERISTALSIS. **2.** Wormlike marks or carvings, as in mosaic or masonry. **3.** The state of being worm-eaten.

**ver·mic·u·lite** (vər-mĭk′yə-līt′) *n.* [Lat. *vermiculus,* dim. of *vermis,* worm + -ITE[1].] Any of a group of micaceous hydrated silicates of varying composition, related to the chlorites and used as heat insulation and for starting plant seeds and cuttings.

**ver·mi·form** (vûr′mə-fôrm′) *adj.* Wormlike in shape.

**vermiform appendix** *n.* The narrow blind vestigial process of the cecum found in some mammals, including humans.

**vermiform process** *n.* The vermiform appendix.

**ver·mi·fuge** (vûr′mə-fyo͞oj′) *n.* An agent that destroys or expels intestinal worms.

**ver·mil·ion** *also* **ver·mil·lion** (vər-mĭl′yən) *n.* [ME *vermelyon* < OFr. *vermeillon* < *vermeil.* —see VERMEIL.] **1.** A bright red mercuric sulfide used as a pigment. **2.** A vivid red to reddish orange. —*adj.* Of a vivid red to reddish orange. —*vt.* **-ioned, -ion·ing, -ions** *also* **-lioned, -lion·ing, -lions.** To color or dye vermilion.

**ver·min** (vûr′mĭn) *n., pl.* **vermin.** [ME < OFr. < VLat. *_verminum_ < Lat. *vermis,* worm.] **1.** An insect or small animal that is destructive, annoying, or harmful to health, as the cockroach or the rat. **2.** An animal that preys on game, as the fox or the weasel. **3. a.** An offensive or contemptible person. **b.** Offensive or contemptible persons as a group.

**ver·mi·na·tion** (vûr′mə-nā′shən) *n.* **1.** Infestation with vermin or worms. **2.** The breeding of worms, larvae, or vermin.

**ver·min·o·sis** (vûr′mĭ-nō′sĭs) *n., pl.* **-ses** (-sēz′). Infestation with parasitic worms.

**ver·min·ous** (vûr′mə-nəs) *adj.* **1.** Of, relating to, or infested with vermin. **2.** Characteristic of vermin. **—ver′min·ous·ly** *adv.*

**ver·miv·o·rous** (vər-mĭv′ər-əs) *adj.* Feeding on worms.

**ver·mouth** (vər-mo͞oth′) *n.* [Fr. *vermout* < G. *Wermut* < MHG *wermuot,* wormwood < OHG *wermuota.*] A sweet red or dry white wine flavored with aromatic herbs and spices and used mainly as an ingredient in cocktails or as an aperitif.

**ver·nac·u·lar** (vər-năk′yə-lər) *n.* [< Lat. *vernaculus,* native < *verna,* native slave.] **1.** The native language of a country or region, esp. as distinct from literary language. **2.** The substandard or nonstandard daily speech of a country or region. **3.** The idiom of a specific trade or profession <in the political *vernacular*> **4.** An idiomatic word, phrase, or expression. **5.** The common name of a plant or animal. —*adj.* **1.** Native to or commonly spoken by the members of a specific country or region. **2.** Using the native language of a region, esp. as distinct from literary language <*vernacular* poetry> **3.** Relating to, spoken in, or written in the native language or dialect. **4.** Relating to the style of architecture and decoration typical of a particular culture. **5.** Occurring or existing in a specific locality : ENDEMIC <a *vernacular* disease> **6.** Indicating or relating to the common name of a plant or animal. **—ver·nac′u·lar·ly** *adv.*

**ver·nac·u·lar·ism** (vər-năk′yə-lə-rĭz′əm) *n.* A vernacular word, phrase, or expression.

**ver·nal** (vûr′nəl) *adj.* [Lat. *vernalis* < *vernus* < *ver,* spring.] **1.** Of, relating to, or happening in the spring. **2.** Typical of or suggestive of spring. **3.** Fresh and young : YOUTHFUL. **—ver′nal·ly** *adv.*

**vernal equinox** *n.* **1.** The point at which the ecliptic intersects the celestial equator, the sun having a northerly motion. **2.** The moment at which the sun passes through the vernal equinox, approx. Mar. 21, marking the start of spring.

**ver·nal·i·za·tion** (vûr′nə-lĭ-zā′shən) *n.* Subjection of seeds or seedlings to low temperature so as to hasten plant development.

**ver·na·tion** (vər-nā′shən) *n.* [NLat. *vernatio, vernation-* < *vernare,* to flourish.] Arrangement of the folded leaves in a bud.

**Ver·ner's Law** (vûr′nərz, vĕr′-) *n.* [After Karl Adolph *Verner* (1846–1896), its formulator.] A law stating that Proto-Germanic noninitial voiceless fricatives in voiced environments became voiced when the previous syllable was unstressed in Proto-Indo-European.

**ver·ni·er** (vûr′nē-ər) *n.* [After Pierre *Vernier* (1580–1637), its inventor.] **1.** A small movable auxiliary graduated scale attached parallel to a main graduated scale, calibrated to show fractional parts of the subdivisions of the larger scale, and used on certain precision instruments for increasing accuracy in measurement. **2.** An auxiliary device designed to facilitate fine adjustments or measurements on precision instruments. **—ver′ni·er** *adj.*

**vernier caliper** *n.* A measuring instrument consisting of an L-shaped frame with a linear scale along its longer arm and an L-shaped sliding attachment with a vernier scale, for reading directly the dimension of an object represented by the separation between the inner or outer edges of the two shorter arms.

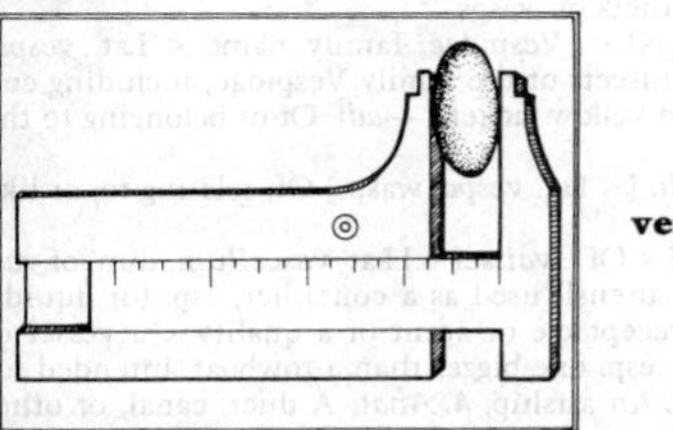
**vernier caliper**

**vernier rocket** *n.* A small rocket engine used primarily for making fine adjustments in velocity and trajectory.

**ver·o·nal** (vĕr'ə-nôl', -nəl) *n.* [Orig. a trademark.] A barbital.

**ve·ron·i·ca**[1] (və-rŏn'ĭ-kə) *n.* [NLat. *Veronica,* genus name.] Any of various plants of the genus *Veronica,* which includes the speedwells.

**ve·ron·i·ca**[2] (və-rŏn'ĭ-kə) *n.* **1.** The representation or image of the face of Jesus, which, in keeping with legend, was impressed on the handkerchief Saint Veronica offered to him on the road to Calvary. **2.** A representation of Jesus' face on a fabric resembling the legendary veronica.

**ve·ron·i·ca**[3] (və-rŏn'ĭ-kə) *n.* [Sp.] A maneuver in bullfighting in which the matador stands immobile while passing the cape slowly in front of the charging bull.

**ver·ru·ca** (və-ro͞o'kə) *n., pl.* **-cae** (-kē) [Lat.] **1.** *Med.* A wart. **2.** *Biol.* A wartlike projection, as on the back of a toad or on some plant leaves.

**ver·ru·cose** (və-ro͞o'kōs') *also* **ver·ru·cous** (-kəs) *adj.* [Lat. *varrucosus* < *verruca,* wart.] Covered with warts or wartlike projections.

**ver·sant** (vûr'sənt) *n.* [Fr. < OFr. < Lat. *versans,* pr.part. of *versare,* to turn frequently. —see VERSATILE.] **1.** The slope of one side of a mountain or mountain range. **2.** The general slope of a region.

**ver·sa·tile** (vûr'sə-təl, -tīl') *adj.* [Fr. < Lat. *versatilis* < *versare,* freq. of *vertere,* to turn.] **1.** Capable of doing many things. **2.** Having many uses or serving various functions <"The most *versatile* of vegetables is the tomato" —Craig Claiborne> **3.** Inconstant or variable : CHANGEABLE <a *versatile* temperament> **4.** *Biol.* Capable of moving freely in all directions, as the antenna of an insect or the loosely joined anther of a flower. **—ver'sa·tile·ly** *adv.* **—ver'sa·til'i·ty** (-tĭl'ĭ-tē), **ver'sa·tile·ness** *n.*

**verse**[1] (vûrs) *n.* [ME *vers,* a line of poetry < OE *fers* and OFr. *vers,* both < Lat. *versus* < p.part. of *vertere,* to turn.] **1.** Writing that follows a metrical pattern : POETRY. **2. a.** One line of poetry. **b.** A section or subdivision of a metrical composition, as a stanza. **3.** Light, often whimsical poetry. **4.** A specific type of metrical composition, as blank verse or free verse. **5.** One of the numbered subdivisions of a chapter in the Bible. *—vt. & vi.* **versed, vers·ing, vers·es.** To versify (something) or to write poetry.

**verse**[2] (vûrs) *vt.* **versed, vers·ing, vers·es.** [Back-formation < *versed,* experienced < Lat. *versatus,* p.part. of *versari,* to occupy oneself.] To make familiar, knowledgeable, or skilled : SCHOOL <*verse* oneself in economics>

**versed cosine** (vûrst) *n.* [VERSED (SINE) + COSINE.] A trigonometric function of an angle equal to one minus the sine of that angle.

**versed sine** *n.* [Transl. of NLat. *sinus versus.*] A trigonometric function of an angle equal to one minus the cosine of that angle.

**ver·si·cle** (vûr'sĭ-kəl) *n.* [ME < OFr. *versicule* < Lat. *versiculus,* dim. of *versus,* verse.] **1.** A short verse. **2.** A short sentence chanted or spoken by a priest and followed by a congregational response.

**ver·si·col·or** (vûr'sĭ-kŭl'ər) *also* **ver·si·col·ored** (-kŭl'ərd) *adj.* [Lat. : *versus,* p.part. of *vertere,* to turn + *color,* color.] **1.** Having a variety of colors : VARIEGATED. **2.** Changing in color : IRIDESCENT.

**ver·si·fy** (vûr'sə-fī') *v.* **-fied, -fy·ing, -fies.** [ME *versifien* < OFr. *versifier* < Lat. *versificare* : *versus,* verse + *facere,* to make.] *—vt.* **1.** To change from prose into metrical form. **2.** To write a poem about. *—vi.* To write verses. **—ver'si·fi·ca'tion** *n.* **—ver'si·fi'er** *n.*

**ver·sine** (vûr'sīn') *n.* [Contraction of VERSED SINE.] A versed sine.

**ver·sion** (vûr'zhən, -shən) *n.* [OFr. < Med. Lat. *versio,* act of turning < Lat. *vertere,* to turn.] **1.** A description, narration, or account told from a particular point of view <two different *versions* of the same accident> **2. a.** A translation from another language. **b.** *often* **Version.** A translation of the entire Bible or a section of it. **3.** A form or variation of an earlier or original model <an electrified *version* of the grandfather clock> **4.** An adaptation of a work of art or literature into another medium or style <the stage *version* of the motion picture> **5.** *Med.* **a.** Manipulation of a fetus in the uterus to bring it into a favorable position for delivery. **b.** A deflection of an organ, as the uterus, from its usual position. **—ver'sion·al** *adj.*

**vers li·bre** (vĕr lē'brə) *n.* [Fr.] Free verse.

**ver·so** (vûr'sō) *n., pl.* **-sos.** [NLat. *verso (folio),* (with the page) turned.] **1.** The left-hand page of a book or the reverse side of a leaf. **2.** The back of a coin or medal.

**verst** (vûrst) *n.* [Fr. *verste* < R. *versta.*] A Russian measure of linear distance equivalent to about two thirds of a mile.

**ver·sus** (vûr'səs) *prep.* [Med. Lat. < Lat., toward < p.part of *vertere,* to turn.] **1.** Against <the Triple Alliance *versus* the Triple Entente> **2.** As an alternative to or in contrast with <partial compromise *versus* total defeat>

**vert** (vûrt) *n.* [ME *verte* < AN < OFr. *vert, verd,* green. —see VERDANT.] **1. a.** Green vegetation that can serve as cover for deer in English forest law. **b.** The right to cut such vegetation. **2.** The color green, esp. in heraldry.

**ver·te·bra** (vûr'tə-brə) *n., pl.* **-brae** (-brē) *or* **-bras.** [Lat. < *vertere,* to turn.] Any of the bones or cartilaginous segments making up the spinal column.

**ver·te·bral** (vûr'tə-brəl, vər-tē'brəl) *adj.* **1.** Of, pertaining to, or characteristic of a vertebra. **2.** Having or consisting of vertebrae. **—ver'te·bral·ly** *adv.*

**vertebral canal** *n.* The spinal canal.

**vertebral column** *n.* The spinal column.

**ver·te·brate** (vûr'tə-brĭt, -brāt') *adj.* [Lat. *vertebratus* < *vertebra,* vertebra.] **1.** Having a backbone or spinal column. **2.** Of or characteristic of a vertebrate or vertebrates. *—n.* A member of the subphylum Vertebrata, a primary division of the phylum Chordata that includes the fishes, amphibians, reptiles, birds, and mammals, all of which have a segmented bony or cartilaginous spinal column.

**ver·tex** (vûr'tĕks') *n., pl.* **-tex·es** *or* **-ti·ces** (-tĭ-sēz') [Lat. < *vertere,* to turn.] **1.** The highest point : APEX. **2.** *Anat.* **a.** The highest point of the skull. **b.** The top of the head. **3.** *Astron.* The highest point reached in the apparent motion of a celestial body. **4. a.** The point at which the sides of an angle intersect. **b.** The point on a triangle opposite to and farthest away from its base. **c.** A point on a polyhedron common to three or more sides. **d.** The fixed point that is one of the three generating characteristics of a conic section.

**ver·ti·cal** (vûr'tĭ-kəl) *adj.* [Fr. or LLat. *verticalis,* both < Lat. *vertex,* highest point.] **1.** Being at right angles to the horizon : extending perpendicularly from a plane. **2.** Relating to or located at the vertex or highest point : directly overhead. **3.** *Anat.* Of or relating to the vertex of the head. **4.** Relating to, made up of, or controlling all the levels or grades in the manufacture and sale of a product. *—n.* **1.** A vertical line, plane, or circle. **2.** A vertical position. **—ver'ti·cal'i·ty** (-kăl'ĭ-tē), **ver'ti·cal·ness** *n.* **—ver'ti·cal·ly** *adv.*

☆ ***syns:*** VERTICAL, PERPENDICULAR, PLUMB, UPRIGHT *adj. core meaning* : being at right angles to the horizon or to level ground <a *vertical* flagpole> ***ant:*** horizontal

**vertical circle** *n.* A great circle on the celestial sphere that passes through the zenith and the nadir and thus is perpendicular to the horizon.

**vertical file** *n.* A collection of articles, as pamphlets, sheets of paper, and mounted photographs, that have been gathered together and arranged for ready reference, as in a library.

**vertical union** *n.* A labor union, the members of which are organized according to the industry for which they work instead of by their particular skill or craft.

**ver·ti·ces** (vûr'tĭ-sēz') *n. var. pl. of* VERTEX.

**ver·ti·cil** (vûr'tĭ-sĭl') *n.* [Lat. *verticillus,* the whorl of a spindle, dim. of *vertex,* highest point.] A circular arrangement, as of flowers or leaves, about a point on an axis : WHORL.

**ver·ti·cil·las·ter** (vûr'tĭ-sə-lăs'tər) *n.* [NLat. < Lat. *verticillus.* —see VERTICIL.] An inflorescence similar to a whorl but actually arising in the axils of opposite leaves. **—ver'ti·cil·las'trate'** (-trāt') *adj.*

**ver·ti·cil·late** (vûr'tĭ-sĭl'ĭt, -āt') *also* **ver·ti·cil·lat·ed** (-sĭl'ā'tĭd) *adj.* Arranged in or forming a whorl or whorls. **—ver'ti·cil'late·ly** *adv.* **—ver'ti·cil·la'tion** *n.*

**ver·tig·i·nous** (vər-tĭj'ə-nəs) *adj.* [Lat. *vertiginosus* < *vertigo,* a whirling < *vertere,* to turn.] **1.** Turning about an axis : REVOLVING. **2.** Affected by vertigo : DIZZY. **3.** Likely to cause vertigo <*vertiginous* altitude> **4.** Liable to rapid change : UNSTABLE. **—ver·tig'i·nous·ly** *adv.* **—ver·tig'i·nous·ness** *n.*

**ver·ti·go** (vûr'tĭ-gō') *n., pl.* **-goes** *or* **-gos.** [Lat. < *vertere,* to turn.] **1.** The sensation of dizziness and the feeling that oneself or one's surroundings are whirling about. **2.** A jumbled, disoriented state of mind.

**ver·tu** (vər-to͞o') *n. var. of* VIRTU.

**ver·vain** (vûr'vān') *n.* [ME *verveine* < OFr. < Lat. *verbena.* —see VERBENA.] Any of several plants of the genus *Verbena,* with slender spikes of small blue, purplish, or white flowers.

**verve** (vûrv) *n.* [Fr. < OFr., fanciful expression < Lat. *verba,* pl. of *verbum,* word.] **1.** Energy and enthusiasm in the expression of ideas and esp. in artistic performance or composition. **2.** Liveliness : vitality. **3.** *Archaic.* Talent : aptitude.

**ver·vet** (vûr'vĭt) *n.* [Fr.] A small, long-tailed African monkey, *Cercopithecus pygerythrus,* with a yellowish-brown or greenish coat.

**ver·y** (vĕr′ē) *adv.* [ME *verray* < OFr. *verai,* true < Lat. *verus.*] **1.** To a high degree : EXTREMELY <*very* sad> **2.** Absolutely : truly <the *very* best we can buy> **3.** Exactly <the *very* same person> ***usage:*** When functioning as an adjective, a form such as *encouraged* or *pleased* may be modified by *very* alone (*a very tired child; a very interested audience*). When a term functions as a past participle in a verb phrase, the appropriate modifier is *very much* (*has been very much praised by critics; was very much enlightened by your explanation*). There are many borderline cases, however, where either modifier may occur, as in *was very (much) mistaken; was very (much) distressed.* —*adj.* **-i·er, -i·est. 1.** Complete : absolute <at the *very* end of my career> **2.** Identical <the *very* complaints you voiced yesterday> **3.** —Used as an intensive to emphasize the importance of the thing described <The *very* mountains shook.> **4. a.** Particularly appropriate or suitable <the *very* item needed for my kitchen> **b.** Precisely as stated <the *very* center of the city> **5.** Mere <The *very* mention of the name was frightening.> **6.** Actual <caught in the *very* act of stealing> **7.** *Archaic.* Real : genuine.

**very high frequency** *n.* A band of radio frequencies between 30 and 300 megahertz.

**very low frequency** *n.* A band of radio frequencies between 3 and 30 kilohertz.

**Ver·y pistol** (vĕr′ē, vîr′ē) *n.* [After Edward W. *Very* (d. 1910), its inventor.] A pistol for firing signal flares.

**ve·si·ca** (və-sī′kə, -sē′-) *n., pl.* **-cae** (-kē, -sē) [Lat.] A bladder, esp. the urinary bladder or the gallbladder. **—ves′i·cal** (vĕs′ĭ-kəl) *adj.*

**ves·i·cant** (vĕs′ĭ-kənt) *n.* A blistering agent, esp. one, as mustard gas, used in chemical warfare. **—ves′i·cant** *adj.*

**ves·i·cate** (vĕs′ĭ-kāt′) *vt.* & *vi.* **-cat·ed, -cat·ing, -cates.** [LLat. *vesicare, vesicat-* < Lat. *vesica,* bladder.] To blister or become blistered. **—ves′i·ca′tion** *n.*

**ves·i·ca·to·ry** (vĕs′ĭ-kə-tôr′ē, -tōr′ē) *n., pl.* **-ries.** A vesicant. **—ves′i·ca·to′ry** *adj.*

**ves·i·cle** (vĕs′ĭ-kəl) *n.* [Fr. *vésicule* < Lat. *vesicula,* dim. of *vesica,* bladder.] **1.** A small bladderlike cell or cavity. **2.** *Anat.* A small sac or bladder, esp. one containing fluid. **3.** *Pathol.* A serum-filled blister that develops beneath or in the skin. **4.** *Geol.* A small cavity or air pocket formed in volcanic rock during solidification.

**ve·sic·u·lar** (vĕ-sĭk′yə-lər, və-) *adj.* **1.** Of or relating to vesicles. **2.** Made up of or containing vesicles. **3.** Having the form of a vesicle. **—ve·sic′u·lar·ly** *adv.*

**ve·sic·u·late** (vĕ-sĭk′yə-lāt′, və-) *vt.* & *vi.* **-lat·ed, -lat·ing, -lates.** To become or make vesicular. —*adj.* (-lĭt, -lāt′). Full of or bearing vesicles : VESICULAR. **—ve·sic′u·la′tion** *n.*

**ves·per** (vĕs′pər) *n.* [Lat., evening star, evening.] **1.** A bell for summoning persons to vespers. **2. Vesper.** The evening star. **3.** *Archaic.* Evening.

▲ word history: *Vesper,* which was borrowed with the meanings "evening" and "evening star," is from Latin *vesper,* which had the same senses. The Latin word is descended from the Indo-European form *wespero–,* which meant "evening, night." The word *Hesperus,* "the evening star," is the Greek descendent of the same Indo-European form. *Wespero–* had a variant form *westo–,* which has descended through Germanic into English as *west,* the direction of sunset.

**ves·per·al** (vĕs′pər-əl) *n.* **1.** A book containing the words and hymns to be used at vespers. **2.** A covering for protecting an altar cloth between services. —*adj.* Of or relating to vesper or vespers.

**ves·pers** also **Ves·pers** (vĕs′pərz) *pl.n.* [OFr. *vespres* < Lat. *vesperas,* acc. pl. of *vespera,* evening < *vesper.*] **1. a.** The sixth of the seven canonical hours. **b.** The time of day reserved for this prayer, in the late afternoon or evening. **2.** A worship service in the late afternoon or evening. **3.** Evening Prayer. **4.** *Rom. Cath. Ch.* A service on Sundays or holy days that includes the office of vespers.

**vesper sparrow** *n.* [From its singing in the evening.] A North American sparrow, *Pooecetes gramineus,* with white markings on its outer tail feathers.

**ves·per·til·i·o·nid** (vĕs′pər-tĭl′ē-ə-nĭd) *n.* [< NLat. *Vespertilionidae,* family name < *Vespertilio,* bat genus < Lat. *vespertilio,* bat < *vesper,* evening.] Any of various widely distributed long-tailed insectivorous bats of the family Vespertilionidae.

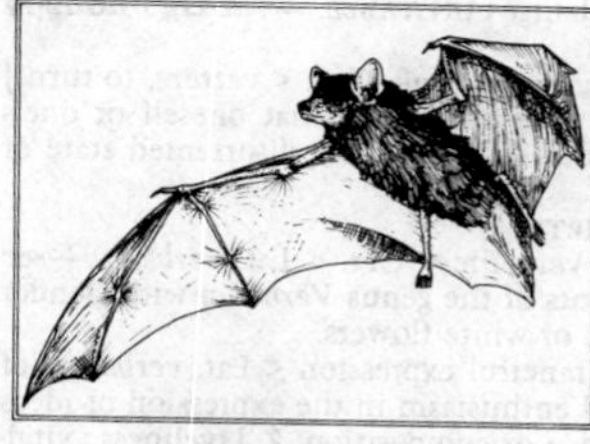

**vespertilionid**
*Over 275 species of greatly varying sizes*

**ves·per·tine** (vĕs′pər-tīn′) also **ves·per·ti·nal** (vĕs′pər-tī′nəl) *adj.* [Lat. *vespertinus* < *vesper,* evening.] **1.** Relating to or occurring in the evening. **2.** *Bot.* Opening or blooming in the evening. **3.** *Zool.* Becoming active in the evening : CREPUSCULAR.

**ves·pi·ar·y** (vĕs′pē-ĕr′ē) *n., pl.* **-ies.** [Lat. *vespa,* wasp + (AP)IARY.] A nest or colony of hornets or wasps.

**ves·pid** (vĕs′pĭd) *n.* [NLat. *Vespidae,* family name < Lat. *vespa,* wasp.] Any of various insects of the family Vespidae, including certain wasps, hornets, and yellow jackets. —*adj.* Of or belonging to the Vespidae.

**ves·pine** (vĕs′pīn′) *adj.* [< Lat. *vespa,* wasp.] Of, relating to, or like a wasp.

**ves·sel** (vĕs′əl) *n.* [ME < OFr. *vaissel* < LLat. *vascellum,* dim. of *vas,* vessel.] **1. a.** A hollow utensil used as a container, esp. for liquids. **b.** One regarded as a receptacle or agent of a quality <a *vessel* of innocence> **2.** A craft, esp. one bigger than a rowboat, intended for navigation on water. **3.** An airship. **4.** *Anat.* A duct, canal, or other tube for containing or circulating a bodily fluid, as a vein or an artery. **5.** *Bot.* One of the tubular conductive structures of woody tissue, composed of cylindrical, often dead cells that are joined end to end.

**vest** (vĕst) *n.* [Fr. *veste* < Ital. < Lat. *vestis,* garment.] **1.** A short sleeveless collarless garment, either open or fastening in front, worn over a blouse or shirt and often under a suit coat or jacket. **2.** A fabric trimming or decoration worn by women to fill in the neckline of a garment, as a coat. **3.** *Chiefly Brit.* An undershirt. **4.** *Archaic.* Clothing : raiment. **5.** *Obs.* An ecclesiastical vestment. —*v.* **vest·ed, vest·ing, vests.** —*vt.* **1.** To clothe or dress with or as if with ecclesiastical vestments. **2.** To place (e.g., ownership) in the possession of <*vest* one's estate in one's child> **3.** To place (e.g., authority) in the control of <*vesting* the President with more autonomy> —*vi.* **1.** To dress oneself, esp. in ecclesiastical vestments. **2.** To be or become legally vested.

**Ves·ta** (vĕs′tə) *n.* [Lat.] **1.** *Rom. Myth.* The goddess of the hearth, worshiped in a temple where the vestal virgins tended the sacred fire. **2.** The third-largest asteroid in the solar system, with a diameter of approx. 386 kilometers or 240 miles. **3. vesta.** A short friction match made of wax or wood.

**ves·tal** (vĕs′təl) *adj.* **1.** Pertaining to or sacred to Vesta. **2.** Relating to or typical of the vestal virgins. **3.** Chaste : pure. —*n.* **1.** A vestal virgin. **2.** A woman who is a virgin. **3.** A nun.

**vestal virgin** *n.* One of the six virgin priestesses who watched over the sacred fire in the temple of Vesta.

**vest·ed** (vĕs′tĭd) *adj.* **1.** *Law.* Settled, complete, or absolute : without contingency. **2.** Dressed or clothed, esp. in ecclesiastical vestments.

**vested interest** *n.* **1.** *Law.* A title or right that can be conveyed to another. **2.** A strong concern for something, as an institution, from which one expects private benefit. **3.** A group with a vested interest.

**vest·ee** (vĕ-stē′) *n.* [< VEST.] A decorative garment worn by women to cover the bosom.

**ves·ti·ar·y** (vĕs′tē-ĕr′ē, -chē-) *adj.* [< ME *vestiarie,* vestry < OFr. < Med. Lat. *vestiarium* < Lat., wardrobe < *vestiarius,* of clothes < *vestis,* garment.] Of or relating to clothes. —*n., pl.* **-ies.** A dressing room, cloakroom, or vestry.

**ves·tib·u·lar** (vĕ-stĭb′yə-lər) *adj.* Of, relating to, or functioning as a vestibule.

**vestibular nerve** *n.* A division of the acoustic nerve.

**ves·ti·bule** (vĕs′tə-byo͞ol′) *n.* [Fr. < Lat. *vestibulum.*] **1.** A small entrance hall or lobby. **2.** An enclosed area at the rear of a passenger car on a railroad train. **3.** *Anat.* A cavity, chamber, or channel that serves as an approach or entrance to another cavity. —*vt.* **-buled, -bul·ing, -bules.** To provide with a vestibule.

**ves·tige** (vĕs′tĭj) *n.* [Fr. < Lat. *vestigium,* footprint.] **1.** A visible trace, evidence, or sign of something once in existence but existing or appearing no more. **2.** *Biol.* A small, degenerate, or rudimentary organ or part existing in an organism as a usu. nonfunctioning remnant of an organ or part completely developed and functional in an earlier generation or developmental stage.

**ves·tig·i·al** (vĕ-stĭj′ē-əl, -stĭj′əl) *adj.* **1.** Of, relating to, or being a vestige. **2.** *Biol.* Existing or persisting as a rudimentary or degenerate structure. **—ves·tig′i·al·ly** *adv.*

**vestigial sideband** *n.* Reduction in bandwidth of a transmitted signal taking place when frequencies to either side of the signal frequency are filtered out.

**vest·ment** (vĕst′mənt) *n.* [ME *vestiment* < OFr. < Lat. *vestimentum* < *vestire,* to clothe < *vestis,* garment.] **1.** A garment, esp. a gown or robe worn as an indication of office or state. **2.** One of the ritual robes worn by members of the clergy, altar boys, or other assistants at services or rites. **—vest·ment′al** (-mĕn′tl) *adj.*

**vest-pock·et** (vĕst′pŏk′ĭt) *adj.* **1.** Intended to fit into a vest pocket <a *vest-pocket* map> **2.** Relatively small : DIMINUTIVE <a *vest-pocket* park>

**ves·try** (vĕs′trē) *n., pl.* **-tries.** [ME *vestrie,* var. of *vestiarie,* vestiary.] **1.** A room in a church where the clergy don their vestments and where these robes and other sacred objects are stored : SACRISTY.

**2.** A church meeting room. **3. a.** A committee of parish or congregation members in the Anglican and Episcopal churches that administers the affairs of the parish or congregation. **b.** A meeting of this group or of the entire congregation. **c.** The place in which such a meeting is held.

**ves·try·man** (vĕs′trē-mən) *n.* A vestry member.

**ves·ture** (vĕs′chər) *n.* [ME, clothes < OFr. < *vestir* < Lat. *vestire,* to clothe < *vestis,* garment.] **1.** Clothing. **2.** Something that cloaks or covers <fields in a *vesture* of mist> —*vt.* **-tured, -tur·ing, -tures.** To clothe.

**ve·su·vi·an** (və-sōō′vē-ən) *n.* [< *Vesuvius,* a volcano in southwestern Italy.] **1.** Idocrase. **2.** A match used esp. for lighting cigars : FUSEE. —*adj.* Characterized by or inclined to abrupt, often violent flare-ups or outbursts <a *vesuvian* temper>

**ve·su·vi·an·ite** (və-sōō′vē-ə-nīt′) *n.* Idocrase.

**vet** (vĕt) *Informal.* —*n.* **1.** A veterinarian. **2.** A veteran. —*v.* **vet·ted, vet·ting, vets.** —*vt.* **1.** To practice veterinary medicine on. **2.** To appraise or examine expertly <*vet* a manuscript> —*vi.* To be or become a veterinarian.

**vetch** (vĕch) *n.* [ME *fecche* < OFr. *veche* < Lat. *vicia.*] Any of various climbing or twining plants of the genus *Vicia,* bearing pinnate leaves and small, usu. purplish flowers.

**vetch·ling** (vĕch′lĭng) *n.* Any of several plants of the genus *Lathyrus,* bearing pinnate leaves, slender tendrils, and small varicolored flowers.

**vet·er·an** (vĕt′ər-ən, vĕt′rən) *n.* [Fr. *vétéran* < Lat. *veteranus* < *vetus,* old.] **1.** One with a long record of service in a particular activity or capacity. **2.** One who has been in the armed forces.

**Veterans Day** *n.* Nov. 11, a holiday commemorating the armistice ending World War I in 1918 and honoring veterans of the armed services.

**vet·er·i·nar·i·an** (vĕt′ər-ə-nâr′ē-ən, vĕt′rə-) *n.* One trained and authorized to treat animals medically.

**vet·er·i·nar·y** (vĕt′ər-ə-nĕr′ē, vĕt′rə-) *adj.* [Lat. *veterinarius,* pertaining to beasts of burden < *veterinus* < *veterinae,* beasts of burden.] Of, relating to, or being the science of the diagnosis and treatment of diseases and injuries of animals, esp. domestic animals. —*n., pl.* **-ies.** A veterinarian.

**veterinary medicine** *n.* The medical science of the diagnosis and treatment of animal diseases and injuries.

**veterinary surgeon** *n.* A veterinarian.

**vet·i·ver** (vĕt′ə-vər) *n.* [Fr. < Tamil *veṭṭivēru.*] **1.** A grass, *Vetiveria zizanioides* of tropical Asia, raised for its aromatic roots that secrete an oil used in perfume manufacture. **2.** The roots of the vetiver.

**vet·i·vert** (vĕt′ə-vûrt′) *n.* [Alteration of VETIVER.] The essential oil of the vetiver.

**ve·to** (vē′tō) *n., pl.* **-toes.** [Lat., I forbid.] **1. a.** The vested power or constitutional right of one branch or department of government, esp. the right of a chief executive, to reject a bill that a legislative body has passed, thus preventing or delaying its enactment into law. **b.** Exercise of this right. **c.** The official document communicating the rejection and the reasons for it. **2.** An authoritative prohibition or rejection of a proposed or intended act. —*vt.* **-toed, -to·ing, -toes. 1.** To prevent (a legislative bill) from becoming law by exercising the power of veto. **2.** To forbid or prevent authoritatively : PROHIBIT. —**ve′to·er** *n.*

☆ ***syns:*** VETO, BLACKBALL, NIX, TURN DOWN *v. core meaning* : to prevent or forbid authoritatively <*vetoed* every proposal for gun control>

**vex** (vĕks) *vt.* **vexed, vex·ing, vex·es.** [ME *vexen* < OFr. *vexer* < Lat. *vexare.*] **1. a.** To irritate or annoy : BOTHER <*vexed* by all the noise> **b.** To bring physical discomfort to <*vexed* by a toothache> **2.** To baffle : puzzle. **3.** To talk about or debate at length <a *vexed* problem> **4.** To toss about : shake up. —**vex′ed·ly** (vĕk′sĭd-lē) *adv.* —**vex′er** *n.* —**vex′ing·ly** *adv.*

**vex·a·tion** (vĕk-sā′shən) *n.* **1.** An act of vexing or the state of being vexed. **2.** A source of annoyance or irritation.

**vex·a·tious** (vĕk-sā′shəs) *adj.* **1.** Causing or creating vexation. **2.** Full of vexation. **3.** Intended to vex. —**vex·a′tious·ly** *adv.* —**vex·a′tious·ness** *n.*

**vex·il·la** (vĕk-sĭl′ə) *n. pl. of* VEXILLUM.

**vex·il·lar·y** (vĕk′sə-lĕr′ē) *n., pl.* **-ies.** [Lat. *vexillarius* < *vexillum,* flag, dim. of *velum,* a covering.] **1.** A member of the oldest class of ancient Roman army veterans who served under a special standard. **2.** A standard-bearer.

**vex·il·late** (vĕk′sə-lĭt, -lāt′) *adj.* Bearing a vexillum.

**vex·il·lum** (vĕk-sĭl′əm) *n., pl.* **-il·la** (-sĭl′ə) [Lat., flag, dim. of *velum,* a covering.] **1.** *Bot.* A usu. enlarged upper petal of certain flowers : STANDARD. **2.** *Zool.* The weblike part of a feather : VANE.

**V format** *n. Computer Sci.* A method of presenting output in such a manner as to start each record with an indication of its length.

**vi·a** (vī′ə, vē′ə) *prep.* [Lat., ablative of *via,* road.] **1.** By way of <went to Harrisburg *via* Philadelphia> **2.** By means of <sent the package *via* parcel post>

**vi·a·ble** (vī′ə-bəl) *adj.* [Fr. < OFr. < *vie,* life < Lat. *vita.*] **1.** Capable of living, as a newborn infant or a fetus arriving at a stage of development that will allow it to live and develop under normal conditions. **2.** Capable of living, developing, or germinating under favorable conditions <*viable* seeds> **3.** Capable of success or ongoing effectiveness : PRACTICABLE <a *viable* production plan> —**vi′a·bil′i·ty** *n.* —**vi′a·bly** *adv.*

**vi·a·duct** (vī′ə-dŭkt′) *n.* [Lat. *via,* road + (AQUA)DUCT.] A series of spans or arches for carrying a road or railroad over a wide valley or over other roads or railroads.

viaduct

**vi·al** (vī′əl) *n.* [ME *viole,* var. of *fiol.* —see PHIAL.] A small container, usu. with a closure, used esp. for liquids. —*vt.* **-aled, -al·ing, -als** *or* **-alled, -al·ling, -als.** To put or keep in or as if in a vial.

**vi·a me·di·a** (vī′ə mē′dē-ə, mĕd′ē-ə, mā′dē-ə, vē′ə) *n.* [Lat.] A middle way.

**vi·and** (vī′ənd) *n.* [ME *viaunde* < OFr. *viande* < VLat. **vivanda,* var. of Lat. *vivenda,* neuter pl. gerund. of *vivere,* to live.] **1. a.** An item of food. **b.** A very choice or delicious dish. **2. viands.** Victuals.

**vi·at·ic** (vī-ăt′ĭk) *also* **vi·at·i·cal** (-ĭ-kəl) *adj.* [Lat. *viaticus* < *via,* road.] Of or relating to traveling, a road, or a way.

**vi·at·i·cum** (vī-ăt′ĭ-kəm, vē-) *n., pl.* **-ca** (-kə) *or* **-cums.** [Lat., traveling provisions < *viaticus,* viatic.] **1.** The Eucharist administered to a dying person or one in danger of death. **2. a.** Supplies for a trip. **b.** An allowance for travel expenses.

**vibes** (vībz) *pl.n.* [Shortened var. of VIBRAPHONE.] **1.** *Informal.* A vibraphone. **2.** *Slang.* VIBRATION 4.

**vi·brac·u·lum** (vī-brăk′yə-ləm) *n., pl.* **-la** (-lə) [NLat. < Lat. *vibrare,* to shake.] One of the whiplike filaments on the surface of certain bryozoan colonies. —**vi·brac′u·lar** *adj.* —**vi·brac′u·loid′** *adj.*

**vi·bra·harp** (vī′brə-härp′) *n.* A vibraphone.

**vi·brant** (vī′brənt) *adj.* **1.** Displaying, marked by, or caused by vibration : VIBRATING. **2.** Throbbing or pulsing with energy or activity <a *vibrant* party> —**vi′brance, vi′bran·cy** *n.* —**vi′brant·ly** *adv.*

**vi·bra·phone** (vī′brə-fōn′) *n.* [VIBRA(TE) + -PHONE.] A musical instrument resembling a marimba but having metal bars and rotating disks in the resonators to make a vibrato. —**vi′bra·phon′ist** *n.*

**vi·brate** (vī′brāt′) *v.* **-brat·ed, -brat·ing, -brates.** [Lat. *vibrare, vibrat-.*] —*vi.* **1.** To move back and forth rapidly. **2.** To produce a sound : RESONATE. **3.** To be moved emotionally : THRILL <*vibrate* with anticipation> **4.** To waver or fluctuate in making choices : VACILLATE. —*vt.* **1.** To cause to tremble or quiver. **2.** To cause to move back and forth rapidly. **3.** To produce (sound) by vibration.

**vi·bra·tile** (vī′brə-tl, -tīl′) *adj.* [Fr. < Lat. *vibrare,* to vibrate.] **1.** Marked by vibration. **2.** Capable of or suitable for vibratory motion. —**vi′bra·til′i·ty** (-tĭl′ĭ-tē) *n.*

**vi·bra·tion** (vī-brā′shən) *n.* **1.** An act of vibrating or the state of being vibrated. **2.** *Physics.* **a.** A rapid linear motion of a particle or of an elastic solid about an equilibrium position. **b.** A periodic process. **3.** A single complete vibrating motion : QUIVER. **4. vibrations.** *Slang.* A distinctive emotional aura or atmosphere that can be instinctively sensed or experienced. —**vi·bra′tion·al** *adj.*

**vi·bra·tive** (vī′brə-tĭv) *adj.* Vibratory.

**vi·bra·to** (və-brä′tō, vē-) *n., pl.* **-tos.** [Ital. < Lat. *vibratus,* p.part. of *vibrare,* to vibrate.] *Mus.* A tremulous or pulsating effect created in an instrumental or vocal tone by barely perceptible minute and rapid variations in pitch.

**vi·bra·tor** (vī′brā′tər) *n.* **1.** Something that vibrates. **2.** An electrically operated device for massage. **3.** An electrical device composed basically of a vibrating conductor interrupting a current.

**vi·bra·to·ry** (vī′brə-tôr′ē, -tōr′ē) *adj.* **1.** Of, marked by, or made up of vibration. **2.** Producing vibration. **3.** Vibrating or able to vibrate.

**vib·ri·o** (vĭb′rē-ō′) *n., pl.* **-os.** [NLat. *Vibrio,* genus name < *vibrare,* to vibrate (from their vibratory motion).] An S-shaped or comma-shaped microorganism of the genus *Vibrio,* esp. *V. comma,* which causes cholera. —**vib′ri·oid′** (-oid′) *adj.*

**vib·ri·o·sis** (vĭb′rē-ō′sĭs) *n., pl.* **-ses** (-sēz′). A disease resulting from vibrios.

**vi·bris·sa** (vī-brĭs′ə, və-) *n., pl.* **-bris·sae** (-brĭs′ē) [Lat. *vibrissae* (pl.) < *vibrare,* to vibrate.] A stiff hair or hairlike projection, as a nostril hair, one of the whiskers of a cat, or one of the modified feathers close to the beak of an insectivorous bird.

**vi·bron·ic** (vī-brŏn′ĭk) *adj.* [VIBR(ATION) + (ELECTR)ONIC.] Of or relating to changes in molecular energy states resulting from vibrational energy.

**vi·bur·num** (vī-bûr′nəm) *n.* [NLat. *Viburnum,* genus name < Lat. *viburnum,* the wayfaring tree.] Any of various shrubs or trees of the genus *Viburnum,* bearing small white flower clusters and berrylike red or black fruit.

**vic·ar** (vĭk′ər) *n.* [ME < OFr. *vicaire* < Lat. *vicarius,* a substitute < *vicarius,* vicarious < *vicis,* change.] **1.** A parish priest in the Church of England who receives a stipend or salary but does not receive the tithes of a parish. **2.** A cleric in the U.S. Episcopal Church in charge of a chapel. **3.** A cleric acting in the place of a rector or bishop in the Anglican Communion generally. **4.** *Rom. Cath. Ch.* A priest acting for or representing another, often higher-ranking cleric. **5.** One who serves as a substitute or agent for another, esp. in the capacity of administrator <the Secretary of State as the President's *vicar* of foreign policy> **—vic′ar·ship′** *n.*

**vic·ar·age** (vĭk′ər-ĭj) *n.* **1.** A vicar's residence. **2.** A vicar's benefice. **3.** A vicar's duties or office.

**vicar apostolic** *n., pl.* **vicars apostolic.** *Rom. Cath. Ch.* **1.** A titular bishop administering a region that is not yet a diocese as a representative of the Holy See. **2.** A titular bishop appointed administrator to a vacant see in which the succession of bishops has been interrupted. **3.** A bishop or archbishop once delegated by the pope to act in his stead in a region.

**vic·ar·ate** (vĭk′ər-ĭt, -ə-rāt′) *n.* A vicariate.

**vicar fo·rane** (fô-rān′, fō-) *n., pl.* **vicars forane.** [Med. Lat. *foranus,* foreign < Lat. *foras,* outside.] *Rom. Cath. Ch.* A priest appointed by a bishop to exercise limited jurisdiction over the clergy in a district of a diocese.

**vicar general** *n., pl.* **vicars general. 1.** *Rom. Cath. Ch.* **a.** A priest acting as deputy to a bishop to help him administer his diocese. **b.** The head of a religious order. **2.** An ecclesiastical official in the Church of England, usu. a layperson, who assists an archbishop or bishop in administrative and judicial duties.

**vi·car·i·al** (vī-kâr′ē-əl, -kăr′-, vĭ-) *adj.* **1.** Of or pertaining to a vicar. **2.** Acting as or holding the position of a vicar. **3.** Serving in the place of another.

**vi·car·i·ate** (vī-kâr′ē-ĭt, -āt′, -kăr′-, vĭ-) *n.* [Med. Lat. *vicariatus* < Lat. *vicarius,* a substitute. —see VICAR.] **1.** The authority or office of a vicar. **2.** The district under a vicar's jurisdiction.

**vi·car·i·ous** (vī-kâr′ē-əs, -kăr′-, vĭ-) *adj.* [Lat. *vicarius.* —see VICAR.] **1.** Endured or done by one person substituting for another <*vicarious* retribution> **2.** Acting in place of someone or something else. **3.** Felt or experienced as if one were taking part in the experience or feelings of another. **4.** *Physiol.* Taking place in or done by a part of the body not usu. associated with a particular function. **—vi·car′i·ous·ly** *adv.* **—vi·car′i·ous·ness** *n.*

**Vicar of Christ** *n. Rom. Cath. Ch.* The pope.

**vice**[1] (vīs) *n.* [ME < OFr. < Lat. *vitium.*] **1. a.** An evil, degrading, or immoral habit or practice. **b.** A serious moral failing. **2.** Wicked or evil habits or conduct : CORRUPTION. **3.** Sexual immorality, esp. prostitution. **4.** A slight personal failing : FOIBLE <the *vice* of sloth> **5.** A flaw or imperfection : DEFECT. **6.** A physical defect or weakness. **7.** Abnormal behavior in a domestic animal. **8. Vice. a.** A character representing generalized or particular vice in English morality plays. **b.** A jester : buffoon.

**vice**[2] (vīs) *n. & v. var. of* VISE.

**vice**[3] (vīs) *n.* [< Lat. *vice,* ablative of *vicis,* change.] One who acts in the place of another : DEPUTY <the *vice*-chairman> *—prep.* **vi·ce** (vī′sē). In place of : REPLACING.

**vice admiral** *n.* A U.S. Navy or Coast Guard flag officer ranking next below an admiral.

**vice-ad·mi·ral·ty** (vīs-ăd′mər-əl-tē) *n., pl.* **-ties.** The office, rank, or command of a vice admiral.

**vice chancellor** *n.* **1.** *Law.* A judge in equity courts ranking below a chancellor. **2.** A university deputy or assistant chancellor. **3.** A deputy or substitute for a head of state or official with the title chancellor. **—vice-chan′cel·lor·ship′** (vīs-chăn′sə-lər-shĭp′, -chăns′lər-) *n.*

**vice consul** *n.* A consular officer who is a subordinate and a deputy of a consul or consul general. **—vice-con′su·lar** (vīs-kŏn′sə-lər) *adj.* **—vice-con′su·late** (-sə-lĭt) *n.* **—vice-con′sul·ship′** (-səl-shĭp′) *n.*

**vice·ge·ren·cy** (vīs-jîr′ən-sē) *n., pl.* **-cies. 1.** The position, function, or authority of a vicegerent. **2.** A district under the jurisdiction of a vicegerent.

**vice·ge·rent** (vīs-jîr′ənt) *n.* [Med. Lat. *vicegerens* : Lat. *vice,* ablative of *vicis,* change + Lat. *gerens,* governing. —see GERENT.] One appointed by a ruler or head of state to act as an administrative deputy. **—vice·ge′ral** (vīs-jîr′əl) *adj.*

**vic·e·nar·y** (vĭs′ə-nĕr′ē) *adj.* [Lat. *vicenarius* < *viceni,* twenty each < *viginti,* twenty.] **1.** Consisting of or relating to 20. **2.** Designating a notation system based on 20.

**vi·cen·ni·al** (vī-sĕn′ē-əl) *adj.* [< LLat. *vicennium,* period of twenty years : Lat. *viciens,* twenty times + Lat. *annus,* year.] **1.** Occurring once every 20 years. **2.** Existing or enduring for 20 years.

**vice president** *n.* **1.** An officer ranking next below a president, usu. empowered to take over the president's duties under conditions such as absence, illness, or death. **2.** A deputy of a president, esp. in a corporation, in charge of a separate department or location <*vice president* of development> **—vice-pres′i·den·cy** (vīs-prĕz′ĭ-dən-sē, -dĕn′-) *n.* **—vice-pres′i·den′tial** (-dĕn′shəl) *adj.*

**vice·re·gal** (vīs-rē′gəl) *adj.* Of or relating to a viceroy, his office, or his responsibility. **—vice·re′gal·ly** *adv.*

**vice regent** *n.* A regent's deputy. **—vice-re′gen·cy** *n.*

**vice·reine** (vīs′rān′) *n.* [Fr. : *vice-,* vice + *reine,* queen < Lat. *regina,* fem. of *rex,* king.] **1.** A viceroy's wife. **2.** A woman who acts as a viceroy.

**vice·roy** (vīs′roi′) *n.* [Fr. : *vice,* vice + *roi,* king < Lat. *rex* < *regere,* to rule.] **1.** A governor of a country, province, or colony, ruling as the representative of a sovereign. **2.** An orange and black North American butterfly, *Limenitis archippus,* similar to but somewhat smaller than the monarch.

**vice·roy·al·ty** (vīs′roi′əl-tē, vīs-roi′-) *n., pl.* **-ties. 1.** The office, authority, or term of service of a viceroy. **2.** A province or district governed by a viceroy.

**vice·roy·ship** (vīs′roi′shĭp′) *n.* Viceroyalty.

**vice squad** *n.* A police division entrusted with the control of vice.

**vi·ce ver·sa** (vī′sə vûr′sə, vīs′) *adv.* [Lat., the position being reversed.] With the meaning or order reversed : CONVERSELY.

**vi·chys·soise** (vĭsh′ē-swäz′, vē′shē-) *n.* [Fr. < fem. of *vichyssois,* of Vichy, a town in France.] A thick creamy potato soup made with chicken stock and flavored with leeks and onions that is usu. served cold.

**Vi·chy water** (vĭsh′ē, vē′shē) *n.* **1.** A naturally effervescent mineral water from the springs at Vichy, France. **2.** A sparkling mineral water similar to Vichy water.

**vic·i·nage** (vĭs′ə-nĭj) *n.* [ME *vesinage* < OFr. *visenage* < Lat. *vicinus,* neighboring. —see VICINITY.] **1. a.** A limited region around a specific area : VICINITY. **b.** A number of places located close to each other and regarded as a whole. **2.** The residents of a specific neighborhood. **3.** The state of living in a neighborhood : PROXIMITY.

**vic·i·nal** (vĭs′ə-nəl) *adj.* [Lat. *vicinalis* < *vicinus,* neighboring. —see VICINITY.] **1.** Of, belonging to, or limited to a specific area or neighborhood : LOCAL. **2.** Indicating a local road instead of a highway. **3.** *Mineral.* Approximating, resembling, or taking the place of a fundamental crystalline form or face. **4.** *Chem.* Designating the consecutive positions of substituted elements or radicals on a benzene ring.

**vi·cin·i·ty** (vĭ-sĭn′ĭ-tē) *n., pl.* **-ties.** [Lat. *vicinitas* < *vicinus,* neighboring < *vicus,* village.] **1.** The state of being near in space or relationship : PROXIMITY <two drugstores in close *vicinity*> **2.** A nearby, surrounding, or adjoining region : NEIGHBORHOOD. **3.** An approximate degree or amount <automobiles priced in the *vicinity* of $12,000>

**vi·cious** (vĭsh′əs) *adj.* [ME < OFr. < Lat. *vitiosus* < *vitium,* vice.] **1.** Having the nature of vice, evil, or immorality : DEPRAVED. **2.** Addicted to vice, immorality, or depravity : EVIL. **3.** Spiteful and malicious <*vicious* rumors> **4.** Having a fault, flaw, or defect <a *vicious* line of reasoning> **5.** Impure : foul. **6.** Marked by violence or ferocity <a *vicious* hurricane> **7.** Savagely aggressive : DANGEROUS <a *vicious* bear> **—vi′cious·ly** *adv.* **—vi′cious·ness** *n.*

**vicious circle** *n.* **1.** A situation in which the solution of one problem in a chain of circumstances leads to a new problem and increases the difficulty of solving the original problem. **2.** A condition in which a disorder or disease gives rise to another that subsequently affects the first. **3.** *Logic.* CIRCLE 10.

**vi·cis·si·tude** (vĭ-sĭs′ĭ-to͞od′, -tyo͞od′) *n.* [OFr. < Lat. *vicissitudo* < *vicissim,* in turn < *vicis,* change.] **1.** *often* **vicissitudes. a.** A change or variation. **b.** The quality of being changeable : MUTABILITY. **2.** One of the abrupt or unexpected changes or shifts often met with in one's life, activities, or surroundings.

**vi·cis·si·tu·di·nar·y** (vĭ-sĭs′ĭ-to͞od′n-ĕr′ē, -tyo͞od′-) *also* **vi·cis·si·tu·di·nous** (-to͞od′n-əs, -tyo͞od′-) *adj.* Marked by, full of, or subject to vicissitudes.

**vic·tim** (vĭk′tĭm) *n.* [Lat. *victima.*] **1.** One harmed or killed by another. **2.** A living creature slain and offered to a deity as a sacrifice or as part of a religious rite. **3.** One harmed by or made to suffer from an act, circumstance, agency, or condition <*victims* of disaster> **4.** One who suffers injury, loss, or death because of a voluntary undertaking <*victims* of their own conniving> **5.** One tricked, swindled, or taken advantage of <the *victim* of a con game>

☆ ***syns:*** VICTIM, CASUALTY, PREY *n. core meaning* : one harmed or killed by another <*victims* of street crime>

**vic·tim·ize** (vĭk′tə-mīz′) *vt.* **-ized, -iz·ing, -iz·es. 1.** To subject to fraud or swindle. **2.** To make a victim of. **—vic′tim·i·za′tion** *n.* **—vic′tim·iz′er** *n.*

**vic·tim·less** (vĭk′tĭm-lĭs) *adj.* Having or involving no victim <a *victimless* crime>

**vic·tim·ol·o·gy** (vĭk′tə-mŏl′ə-jē) *n.* Study of the roles played by victims in crimes against them. **—vic′tim·ol′o·gist** *n.*

**vic·tor** (vĭk′tər) *n.* [ME < Lat. < *vincere*, to conquer.] One who vanquishes or defeats an adversary or opponent.
**vic·to·ri·a** (vĭk-tôr′ē-ə, -tōr′-) *n.* [After Queen *Victoria* of England (1819-1901).] **1.** A low, light four-wheeled carriage for two with a folding top and an elevated driver's seat in front. **2.** A touring car with a folding top usu. covering only the back seat.
**Victoria Cross** *n.* A bronze Maltese cross, Britain's highest military award for conspicuous valor.
**Vic·to·ri·an** (vĭk-tôr′ē-ən, -tōr′-) *adj.* **1.** Of, relating to, or belonging to the period of the reign of Queen Victoria of England <*Victorian* morality> **2.** Displaying qualities, as moral severity or hypocrisy, middle-class stuffiness, and pompous conservatism, that are usu. associated with the time of Queen Victoria. **3.** Being in the highly ornamented massive style of architecture, decor, and furnishings popular in 19th-cent. England. —*n.* One belonging to or displaying features typical of the Victorian period. **—Vic·to′ri·an·i·za′tion** *n.* **—Vic·to′ri·an·ize′** *v.* **(-ized, -iz·ing, -iz·es).**
**Vic·to·ri·an·a** (vĭk-tôr′ē-ăn′ə, -ä′nə, -tōr′-) *n.* Material or a collection of materials of, pertaining to, or typical of the Victorian era.
**Vic·to·ri·an·ism** (vĭk-tôr′ē-ə-nĭz′əm, -tōr′-) *n.* **1.** The quality or state of being Victorian, as in attitude, style, or taste. **2.** Something exhibiting Victorian characteristics.
**vic·to·ri·ous** (vĭk-tôr′ē-əs, -tōr′-) *adj.* [ME < Lat. *victoriosus* < *victoria*, victory.] **1.** Being the winner in a contest or struggle <the *victorious* players> **2.** Typical of or expressing a sense of victory <a *victorious* shout> **—vic·to′ri·ous·ly** *adv.* **—vic·to′ri·ous·ness** *n.*
**vic·to·ry** (vĭk′tə-rē) *n., pl.* **-ries.** [ME < OFr. *victorie* < Lat. *victoria* < *victor*, victor.] **1.** Final and complete defeat of an enemy in a military encounter. **2.** A successful struggle against an opponent or obstacle. **3.** The state of having triumphed.
☆ ***syns:*** VICTORY, CONQUEST, TRIUMPH, WIN *n. core meaning*: a successful struggle against an enemy, opponent, or obstacle <a strategic military *victory*><a *victory* against an opposing ball team><a legal *victory*>
**vic·tress** (vĭk′trĭs) *n.* A woman victor.
**vict·ual** (vĭt′l) *n.* [ME *vitaille* < OFr. < LLat. *victualia*, provisions, pl. of *victualis*, of nourishment < *victus*, nourishment < *vivere*, to live.] **1.** Food suitable for human consumption. **2. victuals.** Food supplies : PROVISIONS. —*v.* **-ualled, -ual·ling, -uals** or **-ualed, -ual·ing, -uals.** —*vt.* To supply with food. —*vi.* **1.** To lay in food supplies. **2.** To eat.
▲ word history: *Victual* was borrowed from Old French *vitaille*, and until the 16th century the spelling of the word conformed to its pronunciation, which rhymes with *little*. During the Renaissance and the renewal of interest in classical languages and literature, scholars pedantically revised the spelling of English words to reflect their origins. *Victual* comes ultimately from Late *victualia*, "provisions." The pronunciation has never reflected this artificial respelling, and the form *vittle*, which adheres more closely to the actual sound of the word, is still sometimes seen.
**vict·ual·er** *also* **vict·ual·ler** (vĭt′l-ər) *n.* **1.** A supplier of victuals : SUTLER. **2.** A supply ship. **3.** *Chiefly Brit.* An innkeeper.
**vi·cu·ña** *also* **vi·cu·na** (vĭ-ko͞on′yə, -ko͞o′nə, -kyo͞o′nə, vī-) *n.* [Sp. < Quechua *wikuña*.] **1.** A llamalike ruminant mammal, *Vicugna vicugna* of the central Andes, with fine silky fleece. **2. a.** The fleece of the vicuña. **b.** Fabric made from this fleece.
**vi·de** (vī′dē, vē′dā′) *imperative v.* [Lat., imper. of *vidēre*, to see.] See. —Used to direct a reader's attention <*Vide* page 64.>
**vi·del·i·cet** (vĭ-dĕl′ĭ-sĕt′, vī-) *adv.* [Lat., clearly : *vidēre*, to see + *licet*, it is permitted.] That is : NAMELY. —Used to introduce examples, lists, or items.
**vid·e·o** (vĭd′ē-ō′) *adj.* [< Lat. *vidēre*, to see.] Of or relating to television, esp. to televised images. —*n.* **1.** The visual portion of a televised broadcast. **2.** Television <a star of stage, screen, and *video*>
**vid·e·o·cas·sette** (vĭd′ē-ō-kə-sĕt′, -kă-) *n.* A videotape recording contained in a cassette.
**vid·e·o·con·fer·ence** (vĭd′ē-ō-kŏn′fər-əns, -frəns) *n.* A teleconference carried on via television. **—vid′e·o·con′fer·enc·ing** *n.*
**vid·e·o·disc** *also* **vid·e·o·disk** (vĭd′ē-ō-dĭsk′) *n.* [Orig. a G. trademark.] A disc recording of sounds and images, as of a motion-picture production, that may be played back on a home television receiver.
**video game** *n.* An electronic or computerized game played by manipulating images on a television or other display screen.
**vid·e·o·gen·ic** (vĭd′ē-ō-jĕn′ĭk) *adj.* [VIDEO + (PHOTO)GENIC.] Appearing to advantage on television : TELEGENIC.
**Vid·e·o·phone** (vĭd′ē-ō-fōn′). A trademark for a telephone equipped for both audio and video transmission.
**Vid·e·o·scan** (vĭd′ē-ō-skăn′). A trademark for a method for machine character recognition in which a video camera records the shapes of the characters to be recognized and then matches the shapes to data held in machine storage.
**vid·e·o·tape** (vĭd′ē-ō-tāp′) *n.* **1.** A wide magnetic tape for recording television images, usu. together with the associated sound, for subsequent playback and broadcasting. **2.** A videotape recording. —*vt.* **-taped, -tap·ing, -tapes.** To make a videotape recording of.
**videotape recorder** *n.* A device for making videotape recordings.
**video terminal** *n.* A computer input-output device utilizing a cathode-ray tube to display data on a screen.

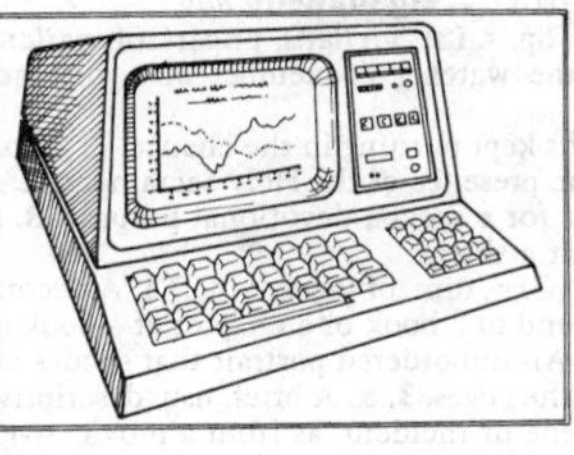
**video terminal**

**vi·dette** (vĭ-dĕt′) *n. var. of* VEDETTE.
**vid·i·con** (vĭd′ĭ-kŏn′) *n.* [VID(EO) + ICON(OSCOPE).] A small television camera tube that forms a charge-density image on a photoconductive surface for subsequent electron-beam scanning.
**vie** (vī) *v.* **vied, vy·ing, vies.** [ME *envien* < OFr. *envier*, to challenge < Lat. *invitare*, to invite.] —*vi.* To strive for victory or superiority : CONTEND. —*vt.* **1.** To offer in competition : MATCH. **2.** To wager.
**Vi·en·na sausage** (vē-ĕn′ə) *n.* [After *Vienna*, Austria.] A small sausage similar to a frankfurter, often served as an hors d'oeuvre.
**Viet·cong** *also* **Viet Cong** (vē-ĕt′kŏng′, -kông′, vyĕt-) *n., pl.* **Vietcong** *also* **Viet Cong.** [Vietnamese, contraction of *Viet Nam Cong Sam*, Vietnamese Communist.] A Vietnamese belonging to or supporting the National Liberation Front of the former nation of South Vietnam. **—Viet′cong, Viet Cong** *adj.*
**Viet·minh** *also* **Viet Minh** (vē-ĕt′mĭn′, vyĕt′-) *n., pl.* **Viet·minh** *also* **Viet Minh.** [Vietnamese, contraction of *Viet Nam Doc Lap Dong Minh Hoi*, Vietnam Federation of Independence.] A member of the Vietnamese army that defeated the Japanese and the French between 1941 and 1954. **—Viet′minh′, Viet Minh** *adj.*
**Viet·nam·ese** (vē-ĕt′nə-mēz′, -mēs′, vyĕt′-) *n.* **1.** A native or inhabitant of Vietnam. **2.** The language of the largest ethnic group in Vietnam and the official language of the nation. **—Viet′nam·ese′** *adj.*
**Viet·nam·ize** (vē-ĕt′nə-mīz′, vyĕt′-) *vt.* **-ized, -iz·ing, -iz·es.** To turn over responsibility for (e.g., military operations) to the Vietnamese <"A policy of *Vietnamizing* the actual fighting" —C.L. Sulzberger> **—Viet′nam·i·za′tion** *n.*
**Vietnam syndrome** (vē′ĕt-näm′, -năm′, vyĕt′-) *n.* A sense of guilt and other complexes regarding the Vietnam War that inhibit similar foreign military involvements by the United States.
**Vietnam War** *n.* A protracted military conflict between the Communist forces of North Vietnam supported by China and the U.S.S.R. and the non-Communist forces of South Vietnam supported by the United States.
**view** (vyo͞o) *n.* [ME *vewe* < OFr. *veue* < *veoir*, to see < Lat. *vidēre*.] **1.** An inspection or examination <used a magnifying glass for a clearer *view*> **2.** A systematic survey : COVERAGE <a *view* of the Victorian novel> **3.** *often* **views.** A specific perception, observation, or interpretation : OPINION <their *views* on labor laws> **4.** The field of vision <Sailboats came into *view*.> **5.** A scene : vista <the *view* from the observation deck> **6.** A picture of a landscape. **7.** A way of showing or seeing something, as from a particular position or angle <an aerial *view* of the fortifications> **8.** An aim : intention. **9.** Expectation : chance <The campaign has no *view* of success.> —*vt.* **viewed, view·ing, views. 1. a.** To see : behold. **b.** To be present at a showing of. **2. a.** To examine : inspect. **b.** To survey or study mentally : CONSIDER. **—in view of.** Taking into account. **—on view.** Being exhibited.
**view·er** (vyo͞o′ər) *n.* **1.** An onlooker or spectator. **2.** An optical device for facilitating the viewing of photographic transparencies by illuminating or magnifying them.
**view finder** *n.* FINDER 2.
**view hal·loo** (vyo͞o′ hə-lo͞o′) *n.* A strident call given during a fox hunt to let the hunters know that a fox has been viewed.
**view·less** (vyo͞o′lĭs) *adj.* **1.** Affording no view. **2.** Lacking or not expressing opinions or views.
**view·point** (vyo͞o′point′) *n.* A point of view.
**view·y** (vyo͞o′ē) *adj.* **-i·er, -i·est. 1.** Displaying extravagant or impractical opinions. **2.** Striking or conspicuous : SHOWY.
**vi·ges·i·mal** (vī-jĕs′ə-məl) *adj.* [< Lat. *vicesimus*, twentieth < *viginti*, twenty.] **1.** Twentieth. **2.** Proceeding or happening in intervals of 20. **3.** Based on or relating to 20.
**vig·il** (vĭj′əl) *n.* [ME *vigile* < OFr. < Lat. *vigilia* < *vigil*, awake.] **1. a.** A watch kept during normal sleeping hours. **b.** A period or act of observing : SURVEILLANCE. **2.** The eve of a religious festival as ob-

served by devotional watching. **3.** *often* **vigils.** Ritual devotions observed on the eve of a holy day.

**vig·i·lance** (vĭj'ə-ləns) *n.* Alert watchfulness.

**vigilance committee** *n.* A volunteer group of citizens that without authority assumes police powers, as pursuing and punishing criminal suspects.

**vig·i·lant** (vĭj'ə-lənt) *adj.* [ME < OFr. < Lat. *vigilans.* —see VIGILANTE.] On the alert : WATCHFUL. **—vig'i·lant·ly** *adv.*

**vig·i·lan·te** (vĭj'ə-lăn'tē) *n.* [Sp. < Lat. *vigilans,* pr.part. of *vigilare,* to be watchful < *vigil,* on the watch.] A member of a vigilance committee.

**vigil light** *n.* **1.** An altar light kept burning in the chancel of Christian churches to symbolize the presence of the Holy Sacrament. **2.** A candle lighted by a worshiper for a special devotional purpose. **3.** A light or candle kept burning at a shrine.

**vi·gnette** (vĭn-yĕt') *n.* [Fr. < OFr., dim. of *vigne,* vine.] **1.** A decorative design at the opening or end of a book or a chapter of a book or along the border of a page. **2.** An unbordered portrait that shades off into the surrounding color at the edges. **3. a.** A brief, usu. descriptive literary sketch. **b.** A short scene or incident, as from a movie. —*vt.* **-gnett·ed, -gnett·ing, -gnettes. 1.** To soften the edges of (a picture) in vignette style. **2.** To describe briefly.

**vi·gnett·er** (vĭn-yĕt'ər) *n.* **1.** A device for printing illustrations and photographs without borders. **2.** *also* **vi·gnett·ist** (-ĭst). A maker of or a specialist in vignettes.

**vig·or** (vĭg'ər) *n.* [ME *vigour* < OFr. < Lat. *vigor* < *vigēre,* to be lively.] **1.** Physical energy or strength <Channel swimmers need *vigor.*> **2.** The capacity for natural growth and survival, as of plants or animals. **3.** Strong feeling : INTENSITY. **4.** Legal effectiveness.

**vig·o·ro·so** (vĭg'ə-rō'sō, -zō, vē'gə-) *adj. & adv.* [Ital. < Med. Lat. *vigorosus* < Lat. *vigor,* vigor.] *Mus.* With emphasis and spirit. —Used as a direction.

**vig·or·ous** (vĭg'ər-əs) *adj.* **1.** Hardy : robust. **2.** Lively : energetic. **—vig'or·ous·ly** *adv.* **—vig'or·ous·ness** *n.*

**vig·our** (vĭg'ər) *n. Chiefly Brit. var. of* VIGOR.

**Vi·king** (vī'kĭng) *n.* [ON *vīkingr.*] One of a seafaring Scandinavian people who raided and plundered settlements on the coasts of northern and western Europe from the 8th through the 10th cent.

**vi·la·yet** (vē'lä-yĕt') *n.* [Turk. *vilāyet* < Ar. *wilāyat,* province < *wālī,* governor.] An administrative division of Turkey.

**vile** (vīl) *adj.* **vil·er, vil·est.** [ME < OFr. *vil* < Lat. *vilis.*] **1.** Disgusting : loathsome <*vile* behavior> **2.** Objectionable or unpleasant <*vile* traffic conditions> **3.** With an abominable taste : UNPALATABLE <*vile* pastry> **4.** Miserably poor : WRETCHED <a *vile* existence in a penal colony> **5.** Depraved. **—vile'ly** *adv.* **—vile'ness** *n.*

**vil·i·fy** (vĭl'ə-fī') *vt.* **-fied, -fy·ing, -fies.** [ME *vilifien* < LLat. *vilificare* : Lat. *vilis,* worthless + Lat. *facere,* to make.] To defame : denigrate. **—vil'i·fi·ca'tion** *n.* **—vil'i·fi'er** *n.*

**vil·i·pend** (vĭl'ə-pĕnd') *vt.* **-pend·ed, -pend·ing, -pends.** [ME *vilipenden* < OFr. *vilipender* < Lat. *vilipendere* : *vilis,* worthless + *pendere,* to consider.] **1.** To regard or treat with contempt : DESPISE. **2.** To abuse or disparage.

**vil·la** (vĭl'ə) *n.* [Ital. < Lat.] **1.** A large and luxurious house in the country. **2.** A Roman country estate with a substantial house. **3.** *Chiefly Brit.* A middle-class suburban house.

**vil·lage** (vĭl'ĭj) *n.* [ME < OFr. < *ville* < Lat. *villa,* country estate.] **1.** A small group of dwellings in a rural area, usu. ranking in size between a hamlet and a town. **2.** An incorporated community smaller in population than a town in some U.S. states. **3.** The inhabitants of a village : VILLAGERS. **—vil'lag·er** *n.*

**vil·lain** (vĭl'ən) *n.* [ME *vilain* < OFr., feudal serf < Med. Lat. *villanus* < Lat. *villa,* country estate.] **1.** A wicked or evil person : SCOUNDREL. **2.** A fictional or dramatic character typically at odds with the hero. **3.** *Obs.* A vile brutish peasant. **4.** (*also* vĭl'ān', vĭ-lān'). *var. of* VILLEIN. **5.** Something said to be the reason for a particular trouble or evil <poverty, the *villain* in the increase of street crime>

▲ word history: The low opinion in which peasants were held by their social superiors is shown by the word *villain. Villain* is ultimately derived from Latin *villa,* "farm," which in Medieval Latin was used to denote a portion of land in the feudal system. *Villanus,* the Medieval Latin noun derived from *villa,* denoted one who worked on such land. *Villanus* appeared in Old French as *vilain,* the direct source of English *villain.* From its adoption into English *villain* meant both "feudal peasant" and "a wicked, depraved person." *Villein,* a variant of *villain,* has only the meaning "peasant."

**vil·lain·age** (vĭl'ə-nĭj) *n. var. of* VILLEINAGE.

**vil·lain·ess** (vĭl'ə-nĭs) *n.* A woman villain.

**vil·lain·ous** (vĭl'ə-nəs) *adj.* **1.** Viciously wicked or criminal. **2.** Highly objectionable : OBNOXIOUS. **—vil'lain·ous·ly** *adv.* **—vil'lain·ous·ness** *n.*

**vil·lain·y** (vĭl'ə-nē) *n., pl.* **-ies. 1.** Viciousness of action or conduct. **2.** Baseness of character or mind. **3.** A vicious or treacherous act.

**vil·la·nelle** (vĭl'ə-nĕl') *n.* [Fr. < Ital. *villanella* < *villanello,* rustic < *villano,* peasant < Med. Lat. *villanus* < Lat. *villa,* country estate.] A 19-line poem of fixed form comprised of five tercets and a final quatrain on two rhymes, with the first and third lines of the first tercet repeated alternately as a refrain ending the succeeding stanzas and united as the final couplet of the quatrain.

**vil·lat·ic** (vĭ-lăt'ĭk) *adj.* [Lat. *villaticus,* of a country estate < *villa,* country estate.] Rural : rustic.

**vil·lein** *also* **vil·lain** (vĭl'ən, -ān', vĭ-lān') *n.* [ME *vilein,* var. of *vilain,* villain.] One of a class of feudal serfs holding the legal status of freemen in their dealings with all persons except their lord.

**vil·lein·age** *also* **vil·lain·age** (vĭl'ə-nĭj) *n.* **1.** The legal status or condition of a villein. **2.** The legal tenure by which a villein held his land.

**vil·li** (vĭl'ī') *n. pl. of* VILLUS.

**vil·li·form** (vĭl'ə-fôrm') *adj.* Having the form of a villus or the appearance of villi.

**vil·lose** (vĭl'ōs') *adj. var. of* VILLOUS.

**vil·los·i·ty** (vĭ-lŏs'ĭ-tē) *n., pl.* **-ties. 1.** The condition of being villous. **2.** A villous surface or coating. **3.** A villus.

**vil·lous** (vĭl'əs) *also* **vil·lose** (-ōs') *adj.* [ME < Lat. *villosus,* hairy < *villus,* shaggy hair.] **1.** Of, relating to, similar to, or covered with villi. **2.** *Bot.* Covered with fine unmatted hairs. **—vil'lous·ly** *adv.*

**vil·lus** (vĭl'əs) *n., pl.* **vil·li** (vĭl'ī') [Lat., shaggy hair.] **1.** *Anat.* A minute projection arising from a mucous membrane. **2.** *Bot.* A fine hairlike epidermal outgrowth.

**vim** (vĭm) *n.* [Lat., accusative of *vis,* force.] Ebullient vitality.

**vin-** *pref. var. of* VINI-.

**vi·na** *also* **vee·na** (vē'nə) *n.* [Hindi *vīṇā* < Skt. *vīṇā.*] A stringed musical instrument of India with a long fretted fingerboard having resonating gourds at each end.

**vi·na·ceous** (vī-nā'shəs, vĭ-) *adj.* [Lat. *vinaceus,* of wine < *vinum,* wine.] Having the color of red wine.

**vin·ai·grette** (vĭn'ĭ-grĕt') *n.* [Fr. < OFr. *vinaigre,* vinegar.] **1.** A small decorative bottle or container with a perforated top for holding an aromatic preparation such as smelling salts. **2.** Vinaigrette sauce.

**vinaigrette sauce** *n.* A cold sauce or dressing consisting of vinegar or lemon juice and oil flavored with finely chopped onions, herbs, and other seasonings.

**vi·nasse** (vĭ-năs', vĭ-) *n.* [Fr. < Lat. *vinacea,* fem. of *vinaceus,* of wine.] The residue in a still after the distillation process.

**vin·blas·tine** (vĭn-blăs'tēn') *n.* [NLat. *Vinca,* periwinkle genus + E. *leuroblast,* a developing leukocyte + -INE.] An alkaloid, $C_{46}H_{58}N_4O_9$, extracted from the Madagascar periwinkle plant, that is used as an antineoplastic drug.

**Vin·cent's angina** (vĭn'sənts) *n.* [After Jean Hyacinthe *Vincent* (1862–1950), its discoverer.] Trench mouth.

**vin·ci·ble** (vĭn'sə-bəl) *adj.* [Lat. *vincibilis* < *vincere,* to conquer.] Capable of being overcome or vanquished. **—vin'ci·bil'i·ty** *n.* **—vin'ci·bly** *adv.*

**vin·cris·tine** (vĭn-krĭs'tēn') *n.* [NLat. *Vinca,* periwinkle genus + Lat. *crista,* crest + -INE.] An alkaloid, $C_{46}H_{56}N_4O_{10}$, extracted from the Madagascar periwinkle plant, that is used as an antineoplastic drug, esp. in treating of leukemia.

**vin·cu·lum** (vĭng'kyə-ləm) *n., pl.* **-lums** *or* **-la** (-lə) [Lat., cord < *vincire,* to tie.] **1.** *Math.* A bar drawn over two or more algebraic terms to show that they are to be considered as a single term. **2.** *Anat.* A ligament that restricts movement of an organ or part. **3.** A bond or tie.

**vin·di·ca·ble** (vĭn'dĭ-kə-bəl) *adj.* That can be vindicated.

**vin·di·cate** (vĭn'dĭ-kāt') *vt.* **-cat·ed, -cat·ing, -cates.** [Lat. *vindicare, vindicat-,* to lay claim to < *vindex,* avenger.] **1.** To clear of accusation, blame, suspicion, or doubt with corroboration or proof. **2.** To justify <*vindicate* one's faith> **3.** To justify or prove the value of, esp. in light of subsequent events. **—vin'di·ca'tor** *n.*

☆ ***syns:*** VINDICATE, ABSOLVE, ACQUIT, CLEAR, EXONERATE *v. core meaning* : to free from a charge of guilt <an accused person who was finally *vindicated* by new evidence>

**vin·di·ca·tion** (vĭn'dĭ-kā'shən) *n.* **1.** An act of vindicating or the state of being vindicated. **2.** The defense, as evidence or argument, that serves to justify a claim or deed.

**vin·di·ca·to·ry** (vĭn'dĭ-kə-tôr'ē, -tōr'ē) *adj.* **1.** Justifying : vindicating. **2.** Exacting retribution : PUNITIVE.

**vin·dic·tive** (vĭn-dĭk'tĭv) *adj.* [< Lat. *vindicta,* vengeance < *vindicare,* to avenge.] **1.** Inclined to seek revenge : REVENGEFUL. **2.** Meant to cause pain or harm : SPITEFUL <*vindictive* remarks> **—vin·dic'tive·ly** *adv.* **—vin·dic'tive·ness** *n.*

☆ ***syns:*** VINDICTIVE, REVENGEFUL, SPITEFUL, VENGEFUL *adj. core meaning* : disposed to seek revenge <a person with a poisonous, *vindictive* mind>

**vine** (vīn) *n.* [ME < OFr. < Lat. *vinea* < fem. of *vineus,* of wine < *vinum,* wine.] **1. a.** A plant with a flexible stem supported by climbing, twining, or creeping along a surface. **b.** The stem of such a plant. **2. a.** A grapevine. **b.** Grapevines as a whole <juice of the *vine*>

▲ word history: The word *vine* has a doublet *wine,* and this pair of words illustrates the history of the Latin sound represented by the letter V. Classical Latin had no *v* sound like the initial sound of *vine;* the letter V represented the sound of the consonant *w.* The Latin word *vinum,* "wine," was borrowed by the ancestral Germanic tongue, probably before the birth of Christ; it appears in Old English as *wine,* "wine," with the original sound of Latin V. A derivative of Latin *vinum, vinea,* "vine," developed into Old French *vine.* The Old French word contained the *v* sound that developed in the later Latin speech, which was the source of the Romance languages.

**vine·dress·er** (vīn′drĕs′ər) *n.* One who prunes and cultivates grapevines.

**vin·e·gar** (vĭn′ĭ-gər) *n.* [ME *vinegre* < OFr. *vinaigre* : *vin,* wine (< Lat. *vinum*) + *aigre,* sour < Lat. *acer.*] An impure dilute solution of acetic acid obtained by fermentation beyond the alcohol stage and used as a condiment and preservative.

**vinegar eel** *n.* A small nematode worm, *Anguillula aceti,* that feeds on the organisms that cause fermentation in vinegar.

**vin·e·gar·roon** (vĭn′ĭ-gə-rōōn′) *also* **vin·e·ga·rone** (-rōn′) *n.* [Mex. Sp. *vinagrón,* aug. of Sp. *vinagre,* vinegar < OFr. *vinaigre.*] A large nonvenomous scorpionlike arachnid, *Mastigoproctus giganteus* of the southern United States and Mexico, that gives off a strong vinegary odor when disturbed.

**vinegarroon**
*Approximately 3 inches long*

**vin·e·gar·y** (vĭn′ĭ-gə-rē, -grē) *also* **vin·e·gar·ish** (-gər-ĭsh, -grĭsh) *adj.* **1.** Having the nature of vinegar : SOUR <a *vinegary* flavor> **2.** Irascible and unpleasant in temperament or speech.

**vin·er·y** (vī′nə-rē) *n., pl.* **-ies.** An area or greenhouse for growing vines.

**vine·yard** (vĭn′yərd) *n.* **1.** Ground planted with cultivated grapevines. **2.** A sphere of spiritual, mental, or physical endeavor.

**vini-** *or* **vino-** *or* **vin-** *pref.* [< Lat. *vinum,* wine.] Wine <*vinic*>

**vi·nic** (vī′nĭk) *adj.* Of, contained in, or derived from wine.

**vin·i·cul·ture** (vĭn′ĭ-kŭl′chər, vī′nĭ-) *n.* Cultivation of grapes : VITICULTURE. **—vin′i·cul′tur·al** *adj.* **—vin′i·cul′tur·ist** *n.*

**vi·no** (vē′nō) *n., pl.* **-nos.** [Ital. and Sp., both < Lat. *vinum.*] Wine.

**vino-** *pref. var. of* VINI-.

**vi·nom·e·ter** (vĭ-nŏm′ĭ-tər, vī-) *n.* A hydrometer for measuring alcohol percentage in a wine.

**vin or·di·naire** (văN′ ôr-dē-nâr′) *n., pl.* **vins or·di·naires** (văN′ ôr-dē-nâr′) [Fr., ordinary wine.] A cheap red table wine.

**vi·nous** (vī′nəs) *adj.* [Lat. *vinosus* < *vinum,* wine.] **1.** Of, relating to, or made with wine. **2.** Affected or produced by the drinking of wine. **3.** Having the color of red wine. **—vi·nos′i·ty** (vī-nŏs′ĭ-tē) *n.* **—vi′nous·ly** *adv.*

**vin·tage** (vĭn′tĭj) *n.* [ME *vyntage,* alteration of *vendage* < OFr. *vendange* < Lat. *vindemia* : *vinum,* grapes + *demere,* to take off.] **1.** The yield of wine or grapes from a specific vineyard or district during one season. **2.** Wine, usu. of high quality, identified as to year and vineyard or district of origin. **3.** The year or place in which a particular wine is bottled. **4. a.** The harvesting of a grape crop. **b.** The early stages of winemaking. **5.** *Informal.* **a.** A group or collection of persons or things with certain characteristics in common. **b.** A year or period of origin <a car of 1942 *vintage*> **c.** Length of existence : AGE. —*adj.* **1.** Of or pertaining to a vintage. **2.** Typified by excellence, maturity, and lasting popularity : CLASSIC. **3.** Old or out-of-date. **4.** Of the best or most distinctive <stories that were *vintage* Saroyan>

**vin·tag·er** (vĭn′tə-jər) *n.* A producer or harvester of wine grapes.

**vintage year** *n.* **1.** The year in which a vintage wine is made. **2.** A year of outstanding success or achievement.

**vint·ner** (vĭnt′nər) *n.* [ME *vineter* < OFr. *vinetier* < Med. Lat. *vinetarius* < Lat. *vinetum,* vineyard < *vinum,* wine.] **1.** A wine merchant. **2.** A maker of wine.

**vin·y** (vī′nē) *adj.* **-i·er, -i·est.** **1.** Of, relating to, or of the nature of vines. **2.** Abounding in or overgrown with vines.

**vi·nyl** (vī′nəl) *n.* [VIN(I)- + -YL.] **1.** The univalent chemical radical $CH_2CH$, obtained from ethylene. **2.** Any of various compounds containing the vinyl radical, typically highly reactive, easily polymerized, and used as basic materials for plastics. **3.** Any of various tough, flexible, and shiny plastics often used for coverings and clothing. **—vi·nyl′ic** (-nĭl′ĭk) *adj.*

**vinyl chloride** *n.* A flammable gas, $CH_2$:CHCl, used as a monomer for polyvinyl chloride.

**vi·ol** (vī′əl) *n.* [OFr. *viole* < OProv. *viola.*] **1.** Any of a family of stringed instruments, principally of the 16th and 17th cent., with a fretted fingerboard, usu. six strings, and a flat back and played with a curved bow. **2.** VIOLA DA GAMBA 1.

**vi·o·la¹** (vē-ō′lə) *n.* [Ital. < OProv., viol.] **1.** A stringed musical instrument of the violin family, somewhat bigger than a violin, tuned a fifth lower, and with a deeper, more sonorous tone. **2.** An organ stop usu. of eight-foot or four-foot pitch yielding stringlike tones. **—vi·o′list** *n.*

**vi·o·la²** (vī-ō′lə, vē-, vī′ə-lə) *n.* [NLat. *Viola,* genus name < Lat. *viola,* the violet.] A plant of the genus *Viola,* which includes the violets and pansies, esp. a variety with flowers similar to violets in size and shape and pansies in coloration.

**vi·o·la·ble** (vī′ə-lə-bəl) *adj.* Capable of being or apt to be violated. **—vi′o·la·bil′i·ty, vi′o·la·ble·ness** *n.* **—vi′o·la·bly** *adv.*

**vi·o·la·ceous** (vī′ə-lā′shəs) *adj.* [NLat. *Violaceae,* family name < Lat. *violaceus,* violet-colored < *viola,* the violet.] **1.** Of or belonging to the family Violaceae, including the violets. **2.** Of the color violet.

**vi·o·la da brac·cio** (vē-ō′lə də brä′chō) *n.* [Ital., viol of the arm.] A stringed instrument of the viol family with approx. the range of the viola.

**vi·o·la da gam·ba** (vē-ō′lə də găm′bə, găm′-) *n.* [Ital., viol of the leg.] **1.** A stringed instrument, the bass of the viol family, with approx. the range of the cello. **2.** An organ stop of eight-foot pitch producing tones sounding similar to those of the viola da gamba.

**vi·o·la d'a·mo·re** (vē-ō′lə dä-môr′ē, -mōr′ē) *n.* [Ital., viol of love.] A stringed instrument, the tenor of the viol family, with six or seven stopped strings and the same number of sympathetic strings that yield a typically silvery tone.

**vi·o·late** (vī′ə-lāt′) *vt.* **-lat·ed, -lat·ing, -lates.** [ME *violaten* < Lat. *violare, violat-* < *vis,* force.] **1.** To break (e.g., a law) intentionally or unintentionally : DISREGARD. **2.** To injure the person or property of, esp. to rape. **3.** To do harm to (property or qualities considered sacred) : DESECRATE. **4.** To disturb rudely or improperly : INTERRUPT <*violated* our seclusion> **—vi′o·la′tive** *adj.* **—vi′o·la′tor** *n.*

**vi·o·la·tion** (vī′ə-lā′shən) *n.* **1.** An act of violating or the state of being violated <*violation* of a peace treaty> **2.** An instance of violating <a speeding *violation*>

**vi·o·lence** (vī′ə-ləns) *n.* **1.** Physical force employed so as to violate, damage, or abuse <acts of *violence*> **2.** An act or instance of violent behavior or action. **3.** Intensity or severity, as in natural phenomena <the *violence* of a tornado> **4.** Abusive or unjust use of power. **5.** Abuse or injury to meaning, content, or intent <do *violence* to a song> **6.** Fervor : vehemence.

**vi·o·lent** (vī′ə-lənt) *adj.* [ME < OFr. < Lat. *violentus* < *vis,* force.] **1.** Characterized or caused by great physical force or rough action <a *violent* onslaught> **2.** Showing or having great emotional force <a *violent* outburst of rage> **3.** Intense : severe <a *violent* headache><a *violent* wind> **4.** Resulting from unexpected force or injury rather than from natural causes <a *violent* death> **5.** Likely to distort or injure meaning, phrasing, or intent. **—vi′o·lent·ly** *adv.*

**vi·o·let** (vī′ə-lĭt) *n.* [ME < OFr. *violete,* dim. of *viole* < Lat. *viola.*] **1. a.** Any of various low-growing plants of the genus *Viola,* bearing spurred, irregular flowers that are purplish-blue and occas. yellow or white. **b.** A similar plant, as the African violet. **2.** Any of a group of colors, reddish blue in hue, that may vary in lightness and saturation; the hue of that portion of the spectrum that may be evoked in the normal observer by radiant energy of wavelengths approx. 420 nanometers.

**vi·o·lin** (vī′ə-lĭn′) *n.* [Ital. *violino,* dim. of *viola,* viola.] A stringed instrument played with a bow, having four strings tuned at intervals of a fifth, an unfretted fingerboard, and a shallower body than the viol and able to produce great flexibility in range, tone, and dynamics. **—vi′o·lin′ist** *n.*

**vi·o·lon·cel·lo** (vē′ə-lən-chĕl′ō, vī′ə-) *n., pl.* **-los.** [Ital., dim. of *violone,* violone.] A cello. **—vi′o·lon·cel′list** *n.*

**vi·o·lo·ne** (vē′ə-lō′nā) *n.* [Ital., aug. of *viola,* viola.] **1.** A 16-foot organ stop producing stringlike tones resembling those of a cello. **2.** A double bass.

**vi·o·my·cin** (vī′ə-mī′sĭn) *n.* [VIO(LET) + -MYCIN.] An antibiotic, $C_{23}H_{36}N_{12}O_8$, produced by the bacterium *Streptomyces puniceus,* that is used in its sulfate form to treat tuberculosis.

**vi·os·ter·ol** (vī-ŏs′tə-rôl′, -rōl′) *n.* [(ULTRA)VIO(LET) + STEROL.] Ultraviolet irradiated ergosterol : VITAMIN $D_2$.

**VIP** (vē′ī-pē′) *n., pl.* **VIPs.** *Informal.* A very important person.

**vi·per** (vī′pər) *n.* [OFr. *vipere* < Lat. *vipera,* snake.] **1.** A venomous Old World snake of the family Viperidae, esp. a common Eurasian species, *Vipera berus.* **2.** A pit viper. **3.** A venomous or presumably venomous snake. **4.** A malicious or treacherous person.

**vi·per·ine** (vī′pə-rīn′) *adj.* Of, similar to, or of the nature of a viper.

**vi·per·ous** (vī′pər-əs) *adj.* **1.** Resembling a viper or venomous snake. **2.** Malicious : venomous. **—vi′per·ous·ly** *adv.*

**viper's bugloss** *n.* A bristly Eurasian plant, *Echium vulgare,* having bright blue flowers.

**vi·ra·gin·i·ty** (vîr′ə-jĭn′ĭ-tē) *n.* [< Lat. *virago, viragin-,* virago.] Masculine mentality and psychology in a woman.

**vi·ra·go** (və-rä′gō, -rā′-, vîr′ə-gō′) *n., pl.* **-goes** *or* **-gos.** [Lat. < *vir,* man.] **1.** A noisy, tyrannizing woman : SCOLD. **2.** A large, strong, and courageous woman. **—vi·rag′i·nous** (və-răj′ə-nəs) *adj.*

**vi·ral** (vīʹrəl) *adj.* Of, relating to, or resulting from a virus. **—viʹral·ly** *adv.*

**vir·e·lay** (vîrʹə-lāʹ) *n.* [ME *virelai* < OFr.] Any of several medieval French verse and song forms, esp. one in which each stanza has two rhymes, the end rhyme recurring as the first rhyme of the following stanza.

**vi·re·mi·a** (vī-rēʹmē-ə) *n.* [VIR(US) + -EMIA.] Presence of viral particles in the blood. **—vi·reʹmic** (-mĭk) *adj.*

**vir·e·o** (vîrʹē-ōʹ) *n., pl.* **-os.** [NLat. *Vireo*, genus name < Lat. *vireo*, a kind of bird < *virēre*, to be green.] Any of various small New World birds of the genus *Vireo*, with grayish or greenish plumage.

**vi·res·cence** (və-rĕsʹəns, vī-) *n.* The state or process of becoming green, esp. the abnormal development of green coloration in plant parts normally not green.

**vi·res·cent** (və-rĕsʹənt, vī-) *adj.* [Lat. *virescens*, p.part. of *virescere*, to become green < *virēre*, to be green.] **1.** Becoming green. **2.** Somewhat green : GREENISH.

**vi·res ma·jo·res** (vīʹrēz mə-jôrʹēz', -jōrʹ-) *n. pl. of* VIS MAJOR.

**vir·ga** (vûrʹgə) *n.* [Lat., twig, virga.] Wisps of precipitation trailing from a cloud but evaporating before reaching the earth.

**vir·gate[1]** (vûrʹgātʹ) *adj.* [Lat. *virgatus*, made of twigs < *virga*, twig.] Shaped like a wand or rod.

**vir·gate[2]** (vûrʹgĭt) *n.* [Med. Lat. *virgata* < *virga*, rod.] An early English measure of land area of varying value, often equivalent to approx. 30 acres.

**vir·gin** (vûrʹjĭn) *n.* [ME < OFr. *virgine* < Lat. *virgo*.] **1.** One who has not experienced sexual intercourse. **2.** A chaste or unmarried woman : MAIDEN. **3.** An unmarried woman who has taken religious vows of chastity. **4. Virgin.** The Virgin Mary. **5.** A female animal that has not mated. **6. Virgin.** Virgo. *—adj.* **1.** Characteristic of or suitable for a virgin : CHASTE. **2.** In a natural or pure state : UNSULLIED <a *virgin* forest> **3.** Unused, uncultivated, or unexplored <*virgin* lands> **4.** Occurring in native or raw form. **5.** Occurring for the first time : INITIAL. **6.** Extracted directly from the first pressing. —Used of vegetable oils.

**vir·gin·al[1]** (vûrʹjə-nəl) *adj.* **1.** Relating to, typical of, or appropriate to a virgin : CHASTE. **2.** Continuing in a state of virginity. **3.** Untouched or unsullied : FRESH. **—virʹgin·al·ly** *adv.*

**vir·gin·al[2]** (vûrʹjə-nəl) *n.* [< VIRGIN, from its being played by young girls.] *often* **virginals.** A small legless rectangular harpsichord popular in the 16th and 17th cent.

**virgin birth** *n.* The theological doctrine that Jesus was miraculously begotten by God and born of Mary, who was a virgin.

**Vir·gin·ia cowslip** (vər-jĭnʹyə, -jĭnʹē-ə) *n.* A plant, *Mertensia virginica* of eastern North America, that bears nodding blue flower clusters.

**Virginia creeper** *n.* A North American climbing vine, *Parthenocissus quinquefolia*, bearing compound leaves with five leaflets and bluish-black berrylike fruit.

**Virginia deer** *n.* The white-tailed deer.

**Virginia fence** *n.* A worm fence.

**Virginia ham** *n.* A lean hickory-smoked dark red ham.

**Virginia reel** *n.* An American country dance in which couples go through various steps together to the instructions of a caller.

**vir·gin·i·ty** (vər-jĭnʹĭ-tē) *n., pl.* **-ties. 1.** The state of being a virgin : CHASTITY. **2.** The state of being pure, unsullied, or untouched.

**Virgin Mary** *n.* The mother of Jesus.

**virgin memory** *n. Computer Sci.* A storage medium that has never been used in a system.

**vir·gin's-bow·er** (vûrʹjĭnz-bouʹər) *n.* A plant of the genus *Clematis*, esp. *C. virginiana* of eastern North America, bearing white flower clusters and plumed seeds.

**virgin wool** *n.* Wool that has not formerly been through a manufacturing process.

**Vir·go** (vûrʹgō, vîrʹ-) *n.* [Lat. < *virgo*, virgin.] **1.** A constellation in the region of the celestial equator. **2. a.** The sixth sign of the zodiac. **b.** One born under this sign.

**vir·gu·late** (vûrʹgyə-lĭt, -lātʹ) *adj.* [< Lat. *virgula*, small rod < *virga*, rod.] Being in the shape of a small rod.

**vir·gule** (vûrʹgyo͞ol) *n.* [Fr., comma < Lat. *virgula*, small rod < *virga*, rod.] A diagonal mark ( / ) used esp. to separate alternatives, as in *and/or*, to represent the word *per*, as in *miles/hour*, and to indicate the ends of verse lines printed continuously, as in *Candy/Is dandy.*

**vir·i·cide** (vīʹrĭ-sīdʹ) *n.* [VIR(US) + -CIDE.] An agent that destroys or inhibits viruses. **—vi'ri·cid'al** (-sīdʹl) *adj.*

**vir·id** (vîrʹĭd) *adj.* [Lat. *viridis* < *virēre*, to be green.] Bright green with or as if with vegetation : VERDANT.

**vir·i·des·cent** (vîrʹĭ-dĕsʹənt) *adj.* [< Lat. *viridescere*, to become green.] Green or somewhat green. **—vir'i·des'cence** *n.*

**vi·rid·i·an** (və-rĭdʹē-ən) *n.* [< Lat. *viridis*, green.] A durable bluish-green pigment.

**vi·rid·i·ty** (və-rĭdʹĭ-tē) *n.* Verdancy : greenness.

**vir·ile** (vîrʹəl, -īlʹ) *adj.* [OFr. *viril* < Lat. *virilis* < *vir*, man.] **1.** Of or having the characteristics of a man : MASCULINE. **2.** Having energy, vigor, or force <*virile* poetry> **3.** Able to perform sexually as a male : POTENT.

**vi·ril·i·ty** (və-rĭlʹĭ-tē) *n.* **1.** The quality or state of being virile. **2.** Masculine vigor : POTENCY.

**vi·ri·on** (vīʹrē-ŏnʹ, vîrʹē-) *n.* [VIR(US) + -ON[1].] A complete virus particle consisting of a definite number of precisely arranged protein molecules enclosing a nucleic acid molecule.

**vi·rol·o·gy** (vī-rŏlʹə-jē) *n.* Study of viruses and viral diseases. **—vi'ro·log'i·cal** (-rə-lŏjʹĭ-kəl), **vi'ro·log'ic** (-ĭk) *adj.* **—vi·rol'o·gist** *n.*

**vir·tu** (vər-to͞oʹ, vîr-) *also* **ver·tu** (vər-to͞oʹ) *n.* [Ital. *virtù* < Lat. *virtus*, excellence. —see VIRTUE.] **1.** A knowledge of, love for, or taste for fine objects of art. **2.** Objects of art, esp. fine antique objets d'art.

**vir·tu·al** (vûrʹcho͞o-əl) *adj.* [ME *virtuall*, effective < Med. Lat. *virtualis* < Lat. *virtus*, excellence. —see VIRTUE.] Existing or resulting in effect or essence though not in actual fact, form, or name <*virtual* elimination of aboriginal cultures> **—vir'tu·al'i·ty** (-ălʹĭ-tē) *n.*

**virtual focus** *n.* The point from which divergent rays of refracted or reflected light seem to have emanated, as from the image of a point in a plane mirror.

**virtual image** *n.* An image from which rays of refracted or reflected light seem to diverge, as from an image observed in a plane mirror.

**vir·tu·al·ly** (vûrʹcho͞o-ə-lē) *adv.* In fact or to all purposes.

**virtual machine** *n.* A computer intended to replicate copies of its entire hardware-software interface so as to develop new software.

**virtual memory** *n.* Computer memory, distinct from a specific machine, that can be used to extend the machine's own memory.

**vir·tue** (vûrʹcho͞o) *n.* [ME *virtu* < OFr. *vertu* < Lat. *virtus*, manliness, excellence < *vir*, man.] **1. a.** Moral excellence and righteousness : GOODNESS. **b.** An example or kind of moral excellence <the *virtue* of generosity> **2.** Chastity, esp. of a girl or woman. **3.** A particularly efficacious, good, or beneficial quality : ADVANTAGE <a plan with the *virtue* of being workable> **4.** Effective force or power <believed in the *virtue* of meditation> **5.** *Obs.* Manly courage : VALOR. **6. Virtues.** An angel of the fifth highest rank in the hierarchy of angels. **—by virtue of.** On the grounds or basis of : by reason of <chosen *by virtue of* outstanding achievement>

☆ ***syns:*** VIRTUE, GOODNESS, MORALITY *n. core meaning* : moral excellence <a leader of great *virtue*> VIRTUE and MORALITY suggest a conforming to standards of what is right and just and to approved codes of behavior; all imply uprightness <*virtue* as its own reward><questioned the *morality* of arms sales to warring countries> GOODNESS often implies inherent qualities of kindness, benevolence, and generosity <the *goodness* and honesty of a local priest>

**vir·tu·o·sa** (vûrʹcho͞o-ōʹsə, -zə) *n.* [Ital., fem. of *virtuoso*, virtuoso.] A woman virtuoso.

**vir·tu·o·si** (vûrʹcho͞o-ōʹsē) *n. var. pl. of* VIRTUOSO.

**vir·tu·os·i·ty** (vûrʹcho͞o-ŏsʹĭ-tē) *n.* The technical skill, fluency, or style displayed by a virtuoso.

**vir·tu·o·so** (vûrʹcho͞o-ōʹsō, -zō) *n., pl.* **-sos** *or* **-si** (-sē) [Ital. < LLat. *virtuosus*, good < Lat. *virtus*, excellence. —see VIRTUE.] **1.** A musician with superb ability, technique, or personal style. **2.** One with outstanding talent or technique in the arts. **3.** One who experiments or investigates in the arts and sciences : SAVANT. **—vir'tu·o'sic** (-ōʹsĭk, -zĭk) *adj.*

**vir·tu·ous** (vûrʹcho͞o-əs) *adj.* **1.** Displaying virtue : RIGHTEOUS. **2.** Endowed with or marked by chastity : PURE. **—vir'tu·ous·ly** *adv.* **—vir'tu·ous·ness** *n.*

**vi·ru·cide** (vīʹrə-sīdʹ) *n.* [VIRU(S) + -CIDE.] A viricide. **—vi'ru·cid'al** (-sīdʹl) *adj.*

**vir·u·lent** (vîrʹyə-lənt, vîrʹə-) *adj.* [ME < Lat. *virulentus* < *virus*, poison.] **1.** Extremely poisonous or pathogenic <a *virulent* disease> **2.** Bitterly hostile or antagonistic : HATEFUL <*virulent* gossip> **3.** Intensely irritating, obnoxious, or harsh. **—vir'u·lence** *n.* **—vir'u·lent·ly** *adv.*

**vi·rus** (vīʹrəs) *n., pl.* **-rus·es.** [Lat., poison.] **1.** Any of various submicroscopic pathogens composed essentially of a core of a single nucleic acid enclosed by a protein coat, able to replicate only within a living cell. **2.** A particular pathogen. **3.** Something that poisons one's soul or mind <the pernicious *virus* of envy>

**vi·sa** (vēʹzə) *n.* [Fr. < Lat., neuter pl. of *visus*, p.part. of *vidēre*, to see.] An official authorization appended to a passport, allowing entry into and travel within a specific country or region. *—vt.* **-saed, -sa·ing, -sas. 1.** To endorse or ratify (a passport). **2.** To give a visa to.

**vis·age** (vĭzʹĭj) *n.* [ME < OFr. < *vis*, face < Lat. *visus*, appearance < p.part. of *vidēre*, to see.] **1.** The face or facial expression of a person : COUNTENANCE. **2.** Aspect : appearance <the *visage* of spring>

**vis·ard** (vĭzʹərd, -ärdʹ) *n. var. of* VIZARD.

**vis-à-vis** (vēʹzə-vēʹ) *n., pl.* **vis-à-vis** (-vēzʹ, -vēʹ) [Fr., face to face.] One of two persons or things opposite or corresponding to the other : COUNTERPART. *—adv.* Face to face. *—prep.* **1.** Opposite to. **2.** Compared with. **3.** In relation to. **—vis'-à-vis'** *adj.*

**Vi·sa·yan** (və-sīʹən) *n.* **1.** A member of the largest native group of the Philippines, found in the Visayan Islands. **2.** The Malay language of the Visayans. **—Vi·sa'yan** *adj.*

**vis·ca·cha** (vĭ-skä′chə) *n.* [Sp. *vizcacha* < Quechua *wiscacha.*] A gregarious burrowing South American rodent of the genera *Lagostomus* or *Lagidium,* related to and similar to the chinchilla.

**vis·cer·a** (vĭs′ər-ə) *pl.n.* [Lat., pl. of *viscus,* flesh.] **1.** The internal organs of the body, esp. those within the abdominal and thoracic cavities. **2.** The intestines.

**vis·cer·al** (vĭs′ər-əl) *adj.* **1.** Relating to, located in, or affecting the viscera. **2.** Perceived in or as if in the viscera : PROFOUND <*visceral* grief> **3.** Instinctive <*visceral* wants> **—vis′cer·al·ly** *adv.*

**vis·cer·o·mo·tor** (vĭs′ər-ə-mō′tər) *adj.* [VISCER(A) + MOTOR.] Causing or related to movements of the viscera.

**vis·cid** (vĭs′ĭd) *adj.* [LLat. *visadus* < Lat. *viscum,* mistletoe, birdlime made from mistletoe berries.] **1.** Thick and adhesive. **2.** Covered with a clammy or sticky coating. **—vis·cid′i·ty** (vĭ-sĭd′ĭ-tē), **vis′cid·ness** *n.* **—vis′cid·ly** *adv.*

**vis·com·e·ter** (vĭ-skŏm′ĭ-tər) *n.* An instrument for measuring viscosity. **—vis′co·met′ric** (vĭs′kə-mĕt′rĭk) *adj.* **—vis·com′e·try** *n.*

**vis·cose** (vĭs′kōs′) *n.* [ME, viscid < LLat. *viscosus* < *viscum,* mistletoe, birdlime made from mistletoe berries.] **1.** A thick golden-brown viscous solution of cellulose xanthate, used in making rayon and cellophane. **2.** Viscose rayon. *—adj.* **1.** Viscous. **2.** Of, pertaining to, or made from viscose.

**viscose rayon** *n.* A rayon manufactured by reconverting cellulose from a soluble xanthate form to tough fibers by washing in acid.

**vis·co·sim·e·ter** (vĭs′kə-sĭm′ĭ-tər) *n.* A viscometer. **—vis·cos′i·met′ric** (vĭ-skŏs′ə-mĕt′rĭk) *adj.*

**vis·cos·i·ty** (vĭ-skŏs′ĭ-tē) *n., pl.* **-ties. 1.** The condition or property of being viscous. **2.** *Physics.* The degree to which a fluid resists flow under an applied force.

**vis·count** (vī′kount′) *n.* [ME < OFr. *visconte* < Med. Lat. *vicecomes, vicecomit-* : *vice-,* vice + *comes,* count.] A peer ranking below an earl and above a baron.

**vis·count·cy** (vī′kount′sē) *n., pl.* **-cies.** The rank, title, or dignity of a viscount.

**vis·count·ess** (vī′koun′tĭs) *n.* A viscount's wife.

**vis·cous** (vĭs′kəs) *adj.* [ME *viscouse* < LLat. *viscosus* < Lat. *viscum,* mistletoe, birdlime made from mistletoe berries.] **1.** Having relatively high resistance to flow. **2.** Having a heavy, gluey quality. **—vis′cous·ly** *adv.* **—vis′cous·ness** *n.*

☆ ***syns:*** VISCOUS, GLUTINOUS, VISCID, VISCOSE *adj. core meaning* : having a heavy, gluey quality <*viscous* motor oil>

**vis·cus** (vĭs′kəs) *n. sing. of* VISCERA.

**vise** *also* **vice** (vīs) *n.* [ME *vice,* spiral staircase < OFr. *vis* < Lat. *vitis,* vine.] A clamping apparatus of metal or wood, usu. two jaws opened or closed by a screw or lever, utilized in carpentry or metalworking for holding a piece in place. *—vt.* **vised, vis·ing, vis·es** *also* **viced, vic·ing, vic·es.** To hold in or as if in a vise.

**Vish·nu** (vĭsh′no͞o) *n.* [Skt. *Viṣṇuḥ.*] The chief Hindu deity worshiped by the Vaishnava and the second member of the trinity including also Brahma and Shiva.

**vis·i·bil·i·ty** (vĭz′ə-bĭl′ĭ-tē) *n., pl.* **-ties. 1.** The fact, state, or degree of being visible. **2.** The greatest distance under given weather conditions to which it is possible to see without the aid of instruments.

**vis·i·ble** (vĭz′ə-bəl) *adj.* [ME < OFr. < Lat. *visibilis* < *vidēre,* to see.] **1.** Capable of being seen <a *visible* speck> **2.** Obvious to the eye <a *visible* improvement in the patient's health> **3.** Apparent : manifest <no *visible* alternatives> **4.** On hand : AVAILABLE <a *visible* supply> **5.** Made or intended to keep important parts in easily accessible view <a *visible* file> **6.** Represented visually, as by symbols. **—vis′i·ble·ness** *n.* **—vis′i·bly** *adv.*

**visible speech** *n.* A system of phonetic notation used as an aid for teaching speech to the deaf and composed of diagrams of the organs of speech in the various positions needed to articulate sounds.

**Vis·i·goth** (vĭz′ĭ-gŏth′) *n.* [LLat. *Visigothi,* the Visigoths.] A member of the western Goths that invaded the Roman Empire in the 4th cent. A.D. and settled in France and Spain, establishing a monarchy that lasted until the early 8th cent. A.D. **—Vis′i·goth′ic** *adj.*

**vi·sion** (vĭzh′ən) *n.* [ME < OFr. < Lat. *visio* < *vidēre,* to see.] **1. a.** The faculty of sight <good *vision*> **b.** Something that is or has been seen. **2.** Unusual capability in discernment or perception : intelligent foresight <a management team of *vision*> **3.** The way in which one sees or conceives of something. **4.** A mental image created by the imagination. **5.** The mystical experience of seeing as if with the eyes the supernatural or a supernatural being. **6.** One of extraordinary beauty. *—vt.* **-sioned, -sion·ing, -sions.** To see in or as if in a vision. **—vi′sion·al** *adj.* **—vi′sion·al·ly** *adv.*

**vi·sion·ar·y** (vĭzh′ə-nĕr′ē) *adj.* **1.** Marked by foresight. **2.** Having the nature of fantasies or dreams. **3.** Marked by or tending to apparitions, prophecies, or revelations. **4.** Not practicable : UTOPIAN <*visionary* schemes> *—n., pl.* **-ies. 1.** One who has visions : SEER. **2.** One given to speculative or impractical ideas : DREAMER.

☆ ***syns:*** VISIONARY, FAR-SIGHTED, PRESCIENT *adj. core meaning* : characterized by foresight <a *visionary* diplomat>

**vis·it** (vĭz′ĭt) *v.* **-it·ed, -it·ing, -its.** [ME *visiten* < OFr. *visiter* < Lat. *visitare,* to go to see, freq. of *visere,* to view, freq. of *vidēre,* to see.] *—vt.* **1.** To go or come to see (a person) : CALL ON <*visit* the grandparents> **2.** To go or come to see (a place), as on a tour <*visit* a museum> **3.** To stay with as a guest. **4.** To go or come to see in a professional or official capacity <a priest *visiting* the parishioners> **5.** To go or come to <*visits* the library on Fridays> **6.** To go or come to in order to aid <*visit* the aged> **7.** To afflict : assail <A plague *visited* the countryside.> **8.** To inflict punishment on or for : AVENGE <The errors of our ancestors were *visited* on us.> *—vi.* **1.** To make a visit. **2.** *Informal.* To converse or chat <enjoyed *visiting* over the back fence> *—n.* **1.** An act or instance of visiting a person, place, or thing. **2.** A stay or sojourn as a guest. **3.** An act of visiting in a professional capacity. **4.** An act of visiting in an official capacity, as an examination or inspection. **—vis′i·tor** *n.*

**vis·it·a·ble** (vĭz′ĭ-tə-bəl) *adj.* **1.** Capable of or appropriate for a visit. **2.** Subject to or permitting official visits, as for inspection.

**vis·i·tant** (vĭz′ĭ-tənt) *n.* [Lat. *visitans, visitant-,* pr.part. of *visitare,* to go to see.] **1.** A visitor : guest. **2.** A supernatural being : GHOST. **3.** A migratory animal or bird that stops in a location for a limited time.

**vis·i·ta·tion** (vĭz′ĭ-tā′shən) *n.* **1.** An act of visiting or an instance of being visited : VISIT. **2.** A visit for the purpose of making an official inspection or examination, as of a bishop to his diocese. **3.** A parent's right to visit a child or have a child as a visitor, as specified in a divorce or separation order. **4. a.** A visit of punishment or affliction or of comfort and blessing considered to be ordained divinely. **b.** A calamitous event or experience. **5.** The arrival or appearance of a supernatural being. **6. Visitation. a.** The visit of the Virgin Mary to her cousin Elizabeth. **b.** *Rom. Cath. Ch.* A religious festival held Jul. 2 in commemoration of this visit. **—vis′i·ta′tion·al** *adj.*

**vis·i·ta·to·ri·al** (vĭz′ĭ-tə-tôr′ē-əl, -tōr′-) *adj.* **1.** Of or relating to an official visitor or visit. **2.** Having the right or power of visitation.

**visiting card** *n.* A calling card.

**visiting fireman** *n. Informal.* **1.** An influential visitor who is lavishly entertained. **2.** A visitor thought to be a free spender and thus heartily welcomed.

**visiting nurse** *n.* A registered nurse employed by a public health agency or hospital to promote community health and esp. to visit sick persons in their homes.

**visiting professor** *n.* A professor on leave invited to lecture or teach at another college or university for a limited time period, as an academic year.

**visiting teacher** *n.* A teacher affiliated with a public school system who visits and instructs sick or handicapped children.

**vis ma·jor** (vĭs mā′jər) *n., pl.* **vi·res ma·jo·res** (vī′rēz mə-jôr′ēz′, -jōr′-) [Lat., greater force.] *Law.* An overwhelming force of nature with unavoidable consequences that under certain circumstances can exempt one from the obligations of a contract.

**vi·sor** *also* **vi·zor** (vī′zər) *n.* [ME *viser* < AN < OFr. *vis,* face < Lat. *visus,* appearance < *vidēre,* to see.] **1.** A piece projecting from the front of a cap to shade the eyes or protect against wind or rain. **2.** A fixed or movable shield against glare over the windshield of a motor vehicle. **3.** The front piece of the helmet of a suit of armor, capable of being raised and lowered and designed for protecting the eyes, nose, and forehead. **4.** A means of concealment or disguise, esp. a mask. *—vt.* **-sored, -sor·ing, -sors** *also* **-zored, -zor·ing, -zors.** To mask or protect with a visor. **—vi′sored** (vī′zərd) *adj.*

**vis·ta** (vĭs′tə) *n.* [Ital. < *visto,* p.part. of *vedere,* to see < Lat. *vidēre.*] **1. a.** A distant view seen through an opening, as between buildings : PROSPECT. **b.** The passage framing the approach to such a scene : AVENUE. **2.** Comprehensive awareness of a series of remembered, present, or anticipated events.

**VISTA** (vĭs′tə) *n.* [V(OLUNTEERS) I(N) S(ERVICE) T(O) A(MERICA).] An organization, sponsored by the U.S. Office of Economic Opportunity, whose volunteer members educate and teach skills to the poor.

**vi·su·al** (vĭzh′o͞o-əl) *adj.* [ME < LLat. *visualis* < Lat. *visus,* sight < *vidēre,* to see.] **1.** Serving, caused by, or relating to the sense of sight. **2.** Capable of being seen by the eye : VISIBLE. **3.** Optical. **4.** Done, maintained, or controlled by using sight alone <*visual* flight> **5.** Having the nature of or creating a mental image. **6.** Of or pertaining to a means of instruction utilizing sight. **—vi′su·al·ly** *adv.*

**visual aid** *n.* Graphic material used in education for imparting instruction by visual means.

**visual field** *n.* The entire area visible to the immobile eyes at a particular moment : FIELD OF VISION.

**vi·su·al·ize** (vĭzh′o͞o-ə-līz′) *v.* **-ized, -iz·ing, -iz·es.** *—vt.* To form a mental image or vision of : ENVISAGE. *—vi.* To form a mental image. **—vi′su·al·i·za′tion** *n.*

**vi·su·al·iz·er** (vĭzh′o͞o-ə-lī′zər) *n.* One who visualizes, esp. one whose mental images are mainly visual.

**visual purple** *n.* A red-light-sensitive pigment of the retina, esp. rhodopsin.

**vi·tal** (vīt′l) *adj.* [ME < OFr. < Lat. *vitalis* < *vita,* life.] **1.** Of or characteristic of life <*vital* force> **2.** Required for the continuation of life : life-sustaining <*vital* functions> **3.** Full of life : ANIMATED. **4.** Giving life or animation : INVIGORATING. **5.** Having immediate importance : ESSENTIAL <"Irrigation was *vital* to early civilization"

—William H. McNeill> **6.** Involved with or recording data pertinent to lives. **7.** Destructive to life. **—vi'tal·ly** *adv.* **—vi'tal·ness** *n.*

**vital capacity** *n.* The amount of air that can be forcibly expelled from the lungs after a full inspiration.

**vi·tal·ism** (vīt'l-ĭz'əm) *n. Philos.* The doctrine that life processes have a unique character radically different from physiochemical phenomena. **—vi'tal·ist** *n.* **—vi'tal·is'tic** *adj.*

**vi·tal·i·ty** (vī-tăl'ĭ-tē) *n., pl.* **-ties. 1.** The characteristic that distinguishes the living from the nonliving. **2.** Capacity to live, grow, or develop. **3.** Physical or intellectual vigor : ENERGY. **4.** Power to survive.

**vi·tal·ize** (vīt'l-īz') *vt.* **-ized, -iz·ing, -iz·es. 1.** To endow with life. **2.** To animate or invigorate. **—vi'tal·i·za'tion** *n.* **—vi'tal·iz'er** *n.*

**vi·tals** (vīt'lz) *pl.n.* **1.** Bodily parts or organs considered as the source of life. **2.** Those elements needed for continued functioning, as of a system.

**vital signs** *pl.n. Med.* A person's pulse rate, temperature, and respiratory rate.

**vital statistics** *pl.n.* Data recording important occasions and dates in human life, as births, deaths, and marriages.

**vi·ta·mer** (vī'tə-mər) *n.* [VITA(MIN) + (ISO)MER.] One of two or more similar chemical compounds capable of fulfilling a specific vitamin function. **—vi'ta·mer'ic** (-mĕr'ĭk) *adj.*

**vi·ta·min** (vī'tə-mĭn) *n.* [G. *Vitamine* : Lat. *vita*, life + *-amine*, amine.] Any of various relatively complex organic substances found in plant and animal tissue and required in small quantities for controlling metabolic processes. **—vi'ta·min'ic** *adj.*

▲ word history: *Vitamin* was borrowed from German *vitamine*, which is a compound of Latin *vita*, "life," and the scientific suffix *-amine*, "amine; an organic compound of nitrogen." It was at first believed that vitamins were based on amino acids. Although this was later found to be untrue, the name *vitamin* was kept.

**vitamin A** *n.* A vitamin or a mixture of vitamins, esp. vitamin $A_1$ or a mixture of vitamins $A_1$ and $A_2$, found primarily in fish liver oils and some yellow and dark-green vegetables, functioning in normal cell growth and development, and responsible in deficiency for roughening and hardening of the skin, night blindness, and deterioration of mucous membranes.

**vitamin $A_1$** *n.* A yellow crystalline compound, $C_{20}H_{30}O$, derived from fish liver oils.

**vitamin $A_2$** *n.* A golden-yellow oil, $C_{20}H_{28}O$, found in pike liver oils and having approx. 40% of the biological activity of vitamin $A_1$.

**vitamin B** *n.* **1.** Vitamin B complex. **2.** A member of the vitamin B complex, esp. thiamine.

**vitamin $B_c$** *n.* Folic acid.

**vitamin $B_1$** *n.* Thiamine.

**vitamin $B_2$** *n.* Riboflavin.

**vitamin $B_6$** *n.* Pyridoxine.

**vitamin $B_{12}$** *n.* A complex, cobalt-containing coordination compound produced in the normal growth of certain microorganisms, occurring in liver, and used in treating pernicious anemia.

**vitamin B complex** *n.* A group of vitamins orig. held to be a single substance, gen. considered to include thiamine, riboflavin, niacin, pantothenic acid, biotin, pyridoxine, folic acid, inositol, and vitamin $B_{12}$, and found mainly in yeast, liver, eggs, and certain vegetables.

**vitamin C** *n.* Ascorbic acid.

**vitamin D** *n.* Any of several chemically similar activated sterols, esp. vitamin $D_2$ or vitamin $D_3$, produced in general by ultraviolet irradiation of sterols, obtained from milk, fish, and eggs, essential for normal bone growth, and used for treating rickets in children and osteomalacia in adults.

**vitamin $D_2$** *n.* A white crystalline compound, $C_{28}H_{44}O$, produced by ultraviolet irradiation of ergosterol.

**vitamin $D_3$** *n.* A colorless crystalline compound, $C_{27}H_{44}O$, with fundamentally the same biological activity as vitamin $D_2$ but significantly more potent in poultry.

**vitamin E** *n.* Any of several chemically related viscous oils, esp. $C_{29}H_{50}O_2$, found mainly in grains and vegetable oils and used in treating sterility and various abnormalities of the muscles, red blood cells, liver, and brain.

**vitamin G** *n.* Riboflavin.

**vitamin H** *n.* Biotin.

**vitamin K** *n.* Any of several natural and synthetic substances required for the promotion of blood clotting and prevention of hemorrhage, occurring naturally in leafy green vegetables, tomatoes, and vegetable oils.

**vitamin P** *n.* A crystalline fraction of citrus juices for treating certain conditions involving hemorrhage into the skin.

**vi·tel·lin** (vĭ-tĕl'ĭn, vī-) *n.* [VITELL(US) + -IN.] A protein in egg yolk.

**vi·tel·line** (vĭ-tĕl'ĭn, -ēn', vī-) *adj.* [VITELL(US) + -INE.] **1.** Of, relating to, or associated with the yolk of an egg. **2.** Having the dull yellow color of an egg yolk. —*n.* Vitellus.

**vi·tel·lus** (vĭ-tĕl'əs, vī-) *n.* [Lat., dim. of *vitulus*, calf.] An egg yolk.

**vi·ti·ate** (vĭsh'ē-āt') *vt.* **-at·ed, -at·ing, -ates.** [Lat. *vitiare, vitiat-* < *vitium*, fault.] **1.** To impair the value or quality of. **2.** To corrupt morally : DEBASE. **3.** To make ineffective : INVALIDATE. **—vi'ti·a·ble** (vĭsh'ē-ə-bəl) *adj.* **—vi'ti·a'tion** *n.* **—vi'ti·a'tor** *n.*

**vit·i·cul·ture** (vĭt'ĭ-kŭl'chər, vī'tĭ-) *n.* [Lat. *vitis*, vine + CULTURE.] Cultivation of grapes. **—vit'i·cul'tur·al** *adj.* **—vit'i·cul'tur·ist** *n.*

**vit·i·li·go** (vĭt'l-ī'gō, -ē'gō) *n.* [NLat. < Lat., tetter.] A skin disorder marked by the appearance of whitish nonpigmented areas with hyperpigmented borders.

**vit·rec·to·my** (vĭ-trĕk'tə-mē) *n., pl.* **-mies.** [VITR(EOUS HUMOR) + -ECTOMY.] Surgical removal of the vitreous humor from the eyeball.

**vit·re·ous** (vĭt'rē-əs) *adj.* [Lat. *vitreus* < *vitrum*, glass.] **1.** Relating to, similar to, or characteristic of glass : GLASSY. **2.** Derived or made from glass. **3.** Of or relating to the vitreous humor. **—vit're·os'i·ty** (-ŏs'ĭ-tē), **vit're·ous·ness** *n.*

**vitreous enamel** *n.* Porcelain enamel.

**vitreous humor** *n.* Clear gelatinous matter that fills the section of the eyeball between the retina and the lens.

**vi·tres·cent** (vĭ-trĕs'ənt) *adj.* [Lat. *vitrum*, glass + -ESCENT.] **1. a.** Apt to turn into glass. **b.** Like or resembling glass. **2.** Capable of being turned into glass. **—vi·tres'cence** *n.*

**vit·ri·fy** (vĭt'rə-fī') *v.* **-fied, -fy·ing, -fies.** [Fr. *vitrifier* < Med. Lat. **vitrificare* : Lat. *vitrum*, glass + Lat. *facere*, to make.] —*vt.* To change or make into glass or a similar substance, esp. through heat fusion. —*vi.* To become vitreous. **—vit'ri·fi'a·bil'i·ty** *n.* **—vit'ri·fi'a·ble** *adj.* **—vit'ri·fi·ca'tion** *n.*

**vit·ri·ol** (vĭt'rē-ōl', -əl) *n.* [ME < Med. Lat. *vitriolum* < LLat. *vitreolum*, neuter of *vitreolus*, of glass < Lat. *vitrum*, glass.] **1.** *Chem.* **a.** Sulfuric acid. **b.** Any of various sulfates of metals, as ferrous sulfate, zinc sulfate, or copper sulfate. **2.** Vituperative expression or feeling. —*vt.* **-oled, -ol·ing, -oles** *or* **-olled, -ol·ling, -ols.** To expose or subject to vitriol.

**vit·ri·ol·ic** (vĭt'rē-ŏl'ĭk) *adj.* **1.** Of, resembling, or obtained from a vitriol. **2.** Bitterly scathing : CAUSTIC <*vitriolic* dispute>

**vit·ta** (vĭt'ə) *n., pl.* **vit·tae** (vĭt'ē) [Lat., ribbon.] **1.** *Biol.* A streak or band of color. **2.** *Bot.* An oil tube in the fruit of certain plants, as the carrot or parsley. **—vit'tate'** (vĭt'āt') *adj.*

**vit·tle** (vĭt'l) *n.* & *v. Nonstandard. var. of* VICTUAL.

**vi·tu·per·ate** (vī-tōō'pə-rāt', -tyōō'-, vĭ-) *vt.* **-at·ed, -at·ing, -ates.** [Lat. *vituperare, vituperat-*.] To berate. **—vi·tu'per·a'tor** *n.*

**vi·tu·per·a·tion** (vī-tōō'pə-rā'shən, -tyōō'-, vĭ-) *n.* **1.** Blame : censure. **2.** Invective : railing.

**vi·tu·per·a·tive** (vī-tōō'pər-ə-tĭv, -tyōō'-, -pə-rā'-, vĭ-) *adj.* Harshly abusive : ACRIMONIOUS. **—vi·tu'per·a·tive·ly** *adv.*

**vi·va** (vē'və, -vä') *interj.* [Ital., long live < *vivere*, to live < Lat.] —Used to express acclamation, salute, or applause.

**vi·va·ce** (vē-vä'chā) *adv.* & *adj.* [Ital. < Lat. *vivax*, vivacious < *vivere*, to live.] *Mus.* In an animated manner. —Used as a direction.

**vi·va·cious** (vĭ-vā'shəs, vī-) *adj.* [< Lat. *vivax* < *vivere*, to live.] Filled with animation and spirit : LIVELY. **—vi·va'cious·ly** *adv.* **—vi·va'cious·ness, vi·vac'i·ty** (-văs'ĭ-tē, vī-) *n.*

**vi·van·dière** (vē'väN-dyâr') *n.* [Fr., fem. of *vivandier* < OFr. < *viande*, food. —see VIAND.] A woman who travels with troops to sell them food, supplies, and liquor.

**vi·var·i·um** (vī-vâr'ē-əm) *n., pl.* **-i·ums** *or* **-i·a** (-ē-ə) [Lat. < neuter of *vivarius*, of living creatures < *vivus*, alive < *vivere*, to live.] A place or enclosure for keeping and raising living animals for research or observation.

**vi·va vo·ce** (vī'və vō'sē) *adv.* & *adj.* [Med. Lat., with the living voice.] By word of mouth.

**vi·ver·rine** (vī-vĕr'ĭn, -īn') *adj.* [< Lat. *viverra*, ferret.] Of or belonging to the family Viverridae, including carnivorous mammals, as the civets and mongooses. —*n.* A member of the Viverridae.

**viv·id** (vĭv'ĭd) *adj.* [Lat. *vividus* < *vivere*, to live.] **1.** Perceived as bright and distinct : BRILLIANT <a *vivid* light on the horizon> **2. a.** Having intensely bright colors <a *vivid* painting> **b.** Very strong <a *vivid* yellow> **3.** Full of the freshness and vigor of immediate experience <a *vivid* description of the trip> **4.** Active in forming lifelike images <a *vivid* imagination> **—viv'id·ly** *adv.* **—viv'id·ness** *n.*

**viv·i·fy** (vĭv'ə-fī') *vt.* **-fied, -fy·ing, -fies.** [OFr. *vivifier* < LLat. *vivificare* : Lat. *vivus*, alive + Lat. *facere*, to make.] **1.** To give or bring life to : ANIMATE. **2.** To make more lively, intense, or striking : ENLIVEN. **—viv'i·fi·ca'tion** *n.* **—viv'i·fi'er** *n.*

**vi·vip·a·rous** (vī-vĭp'ər-əs, vĭ-) *adj.* [Lat. *viviparus* : *vivus*, alive + *parere*, to give birth.] **1.** *Zool.* Giving birth to living offspring that develop inside the body of the mother. **2.** *Bot.* **a.** Germinating or producing seeds that germinate before becoming detached from the parent plant. **b.** Producing bulbils or new plants instead of seed. **—vi'vi·par'i·ty** (vī'və-păr'ĭ-tē, vĭv'ə-) *n.* **—vi·vip'a·rous·ly** *adv.*

**viv·i·sect** (vĭv'ĭ-sĕkt') *v.* **-sect·ed, -sect·ing, -sects.** [Back-formation < VIVISECTION.] —*vt.* To perform vivisection on (a living animal). —*vi.* To practice vivisection. **—viv'i·sec'tor** *n.*

**viv·i·sec·tion** (vĭv′ĭ-sĕk′shən, vĭv′ĭ-sĕk′-) *n.* [Lat. *vivus,* alive + SECTION.] An act of cutting into or dissecting the body of a living animal, esp. for scientific research. **—viv′i·sec′tion·al** *adj.* **—viv′i·sec′tion·al·ly** *adv.* **—viv′i·sec′tion·ist** *n.*

**vix·en** (vĭk′sən) *n.* [ME *fixen.*] **1.** A female fox. **2.** A shrewish woman. **—vix′en·ish** *adj.* **—vix′en·ish·ly** *adv.*

▲ word history: The word *vixen* is one of the few English words that is feminine both in grammatical form and in meaning. *Vixen* is derived from the Old English form *fyxen,* which is a feminine derivative of the noun *fox,* "fox." The spelling of *vixen* with a *v* instead of the expected *f* comes from the pronunciation of the Old English word. The southeastern dialects of Old English regularly changed all *f* sounds at the beginning of words to the closely related *v* sound. The spelling of *fyxen* was unchanged in these dialects, however, because the Old English spelling system had no separate letter for the *v* sound. The letter V was only introduced by French scribes after the Norman Conquest, and in Middle English the dialect form became orthographically distinct. In Middle English the regular standard form *fixen* still survived, but in Modern English the old dialect form *vixen* has replaced it.

**viz·ard** *also* **vis·ard** (vĭz′ərd, -ärd′) *n.* [Alteration of obs. *vizar* < ME *viser.* —see VISOR.] **1.** A visor. **2.** A mask.

**vi·zier** *also* **vi·zir** (vĭ-zîr′) *n.* [Fr. *vizir* < Turk. *vezīr* < Ar. *wazīr.*] A high Moslem governmental officer, esp. in the old Turkish Empire. **—vi·zier′ate** (-ĭt, -āt′) *n.* **—vi·zier′i·al** *adj.*

**vi·zor** (vī′zər) *n.* & *v. var. of* VISOR.

**V-J Day** (vē′jā′) *n.* [V(ICTORY IN) J(APAN) DAY.] Sept. 2, 1945, the official day of victory for the Allies over Japan in World War II.

**V-mail** (vē′māl′) *n.* [< V(ICTORY).] A postal service used during World War II in which letters were reduced photographically for missions overseas, where they were enlarged and delivered.

**V-neck** (vē′nĕk′) *n.* A V-shaped neckline.

**vo·ca·ble** (vō′kə-bəl) *n.* [OFr. < Lat. *vocabulum,* name < *vocare,* to call.] A word regarded only as a sequence of sounds or letters instead of as a unit of meaning. *—adj.* Capable of being voiced or spoken.

**vo·cab·u·lar·y** (vō-kăb′yə-lĕr′ē) *n., pl.* **-ies.** [Med. Lat. *vocabularium* < neuter of *vocabularius,* of words < Lat. *vocabulum,* name < *vocare,* to call.] **1.** A list of words and often phrases, usu. in an alphabetical arrangement and defined or translated : LEXICON. **2.** All the words of a language. **3.** All the words used by, understood by, or at the command of a particular person or group. **4.** A command or reserve of techniques : REPERTOIRE <a dancer's *vocabulary* of movement>

**vo·cal** (vō′kəl) *adj.* [ME < Lat. *vocalis* < *vox,* voice.] **1.** Of or relating to the voice. **2.** Uttered or produced by the voice. **3.** Having a voice and able to emit speech or sound. **4.** Full of voices : RESOUNDING. **5.** Quick to criticize or speak : OUTSPOKEN. **6. a.** Vocalic. **b.** Voiced. *—n.* **1.** A vocal sound. **2.** A popular piece of music for a singer, gen. having instrumental accompaniment. **—vo′cal·ly** *adv.* **—vo′cal·ness** *n.*

**vocal cords** *pl.n.* The lower of two pairs of bands or folds in the larynx that vibrate when drawn together and when air is passed up from the lungs, thereby producing vocal sounds.

**vo·cal·ic** (vō-kăl′ĭk) *adj.* **1.** Containing, characterized by, or made up of vowels. **2.** Of, relating to, or being a vowel. **—vo·cal′i·cal·ly** *adv.*

**vo·cal·ism** (vō′kə-lĭz′əm) *n.* **1.** Use of the voice in talking or singing. **2.** The act, technique, or art of singing. **3.** A vowel or vocalic sound. **4.** A system of vowels, as within a specific language.

**vo·cal·ist** (vō′kə-lĭst) *n.* A singer.

**vo·cal·ize** (vō′kə-līz′) *v.* **-ized, -iz·ing, -iz·es.** *—vt.* **1.** To produce with the voice. **2.** To give voice to : ARTICULATE. **3.** To mark (e.g., a vowelless Hebrew text) with diacritical vowel points. **4. a.** To change (a consonant) into a vowel during articulation. **b.** To voice. *—vi.* **1. a.** To use the voice. **b.** To sing. **2.** To be changed into a vowel. **—vo′cal·i·za′tion** *n.* **—vo′cal·iz′er** *n.*

**vo·ca·tion** (vō-kā′shən) *n.* [ME *vocacioun,* divine call to a religious life < Lat. *vocatio,* a calling < *vocare,* to call.] **1.** A regular occupation or profession, esp. one for which an individual is particularly suited or qualified. **2.** An urge or predisposition to take up a particular type of work, esp. a religious career : CALLING.

**vo·ca·tion·al** (vō-kā′shə-nəl) *adj.* **1.** Of or relating to a vocation. **2.** Relating to, supplying, or undergoing training in a special skill to be followed as a trade. **—vo·ca′tion·al·ly** *adv.*

**vo·ca·tion·al·ism** (vō-kā′shə-nə-lĭz′əm) *n.* Emphasis on vocational training in education. **—vo·ca′tion·al·ist** *n.*

**vocational school** *n.* A school, esp. one on a secondary level, that trains persons with special abilities for qualification in trades.

**voc·a·tive** (vŏk′ə-tĭv) *adj.* [ME *vocatif* < OFr. < Lat. *vocativus* < *vocare,* to call.] **1.** Relating to, typical of, or used in calling. **2.** Relating to or indicating a grammatical case in Latin and certain other languages to designate the person or thing addressed. *—n.* **1.** The vocative case. **2.** A word in the vocative case. **—voc′a·tive·ly** *adv.*

**vo·cif·er·ant** (vō-sĭf′ər-ənt) *adj.* Vociferous.

**vo·cif·er·ate** (vō-sĭf′ə-rāt′) *vi.* & *vt.* **-at·ed, -at·ing, -ates.** [Lat. *vociferari, vociferat-* : *vox,* voice + *ferre,* to bear.] To cry out or utter vehemently, esp. in protest : CLAMOR. **—vo·cif′er·a′tion** *n.* **—vo·cif′er·a′tor** *n.*

**vo·cif·er·ous** (vō-sĭf′ər-əs) *adj.* **1.** Making an outcry : CLAMOROUS. **2.** Characterized by loudness and vehemence <*vociferous* denials> **—vo·cif′er·ous·ly** *adv.* **—vo·cif′er·ous·ness** *n.*

**vod·ka** (vŏd′kə) *n.* [R., dim of *voda,* water.] An alcoholic liquor orig. distilled from fermented wheat mash but now also made from a mash of rye, corn, or potatoes.

**vogue** (vōg) *n.* [Fr. < OItal. *voga* < *vogare,* to row.] **1.** Prevailing fashion, practice, or style <Powdered wigs were once in *vogue.*> **2.** Popular favor or acceptance : POPULARITY. *—adj.* VOGUE 2.

**vogu·ish** (vō′gĭsh) *adj.* **1.** Chic : fashionable. **2.** Temporarily in fashion. **—vogu′ish·ness** *n.*

**voice** (vois) *n.* [ME < OFr. *vois* < Lat. *vox.*] **1. a.** The sound produced by the vocal organs of a vertebrate, esp. by those of a human being. **b.** The ability to produce such sounds. **2.** A sound resembling or reminiscent of vocal utterance. **3.** The specified quality, condition, or timbre of vocal sound <a soft *voice*> **4. a.** A medium or agency of expression <give *voice* to one's feelings> **b.** The right or opportunity to express a choice or opinion. **5.** A verb form showing the relation between the subject and the action expressed by the verb. **6.** Expiration of air through vibrating vocal cords, used in producing vowels and voiced consonants. **7. a.** Musical tone produced by vibrating vocal cords and resonated within the throat and head cavities. **b.** The quality or condition of a person's singing <a tenor in poor *voice*> **c.** A singer <a choir of 30 *voices*> **8.** Any of the melodic parts for a musical composition. *—vt.* **voiced, voic·ing, voic·es. 1.** To give voice to : UTTER. **2.** To pronounce with vibration of the vocal cords. **3.** *Mus.* To regulate the tone of (e.g., the pipes of an organ). **—with one voice.** In unison.

☆ ***syns:*** VOICE, SAY, SAY-SO, SUFFRAGE *n. core meaning* : the right or chance to express an opinion or participate in a decision <The election returns reflect the *voice* of the people.>

**voice box** *n.* The larynx.

**voiced** (voist) *adj.* **1.** Having a voice or a specified type of voice <soft-*voiced*> **2.** Uttered with vibration of the vocal cords, as the consonants *d* and *b.* **—voiced′ness** (voist′nĭs, voi′sĭd-) *n.*

**voice·ful** (vois′fəl) *adj.* Having a voice, esp. a loud one. **—voice′ful·ness** *n.*

**voice·less** (vois′lĭs) *adj.* **1.** Lacking a voice : MUTE. **2.** Uttered with no vibration of the vocal cords, as the consonants *t* and *p.* **—voice′less·ly** *adv.* **—voice′less·ness** *n.*

**voice-o·ver** (vois′ō′vər) *n.* The voice of a film or television narrator who does not appear on camera.

**voice part** *n. Mus.* VOICE 8.

**voice·print** (vois′prĭnt′) *n.* An electronically recorded graphic representation of voice, typically with time plotted on the horizontal axis, frequency on the vertical, and amplitude displayed in a series of contour lines, the configuration characterizing an individual speaker's articulation of a given word.

**voic·er** (voi′sər) *n.* One that voices organ pipes.

**void** (void) *adj.* [ME < OFr. *voide* < VLat. **vocitus,* alteration of Lat. *vacuus* < *vacare,* to be empty.] **1.** Containing no matter : EMPTY. **2.** Unoccupied, as a position : VACANT. **3.** Lacking : devoid <*void* of comprehension> **4.** Useless : ineffective. **5.** Lacking legal force or validity : NULL. *—n.* **1. a.** An empty space. **b.** A vacuum. **2.** An open space or a break in continuity : GAP. **3.** A feeling or state of emptiness, loneliness, or loss. *—v.* **void·ed, void·ing, voids.** *—vt.* **1.** To make void or of no effect : INVALIDATE. **2. a.** To take out (the contents of something) : EMPTY. **b.** To evacuate (body wastes). **3.** To leave : vacate. *—vi.* To evacuate body wastes. **—void′er** *n.*

**void·a·ble** (voi′də-bəl) *adj.* Capable of being voided, esp. capable of being annulled. **—void′a·ble·ness** *n.*

**void·ance** (void′ns) *n.* **1.** An act of voiding, emptying, or evacuating. **2.** The state of being vacant : EMPTINESS.

**void·ed** (voi′dĭd) *adj. Heraldry.* Having the central area left vacant, leaving a narrow border or outline.

**voile** (voil) *n.* [Fr. < Lat. *vela,* neuter pl. of *velum,* covering.] A sheer fabric used esp. for making light dresses and curtains.

**voir dire** (vwär dîr′) *n.* [OFr., to speak the truth.] *Law.* A preliminary examination regarding the competence of a prospective witness or juror.

**voix cé·leste** (vwä′ sā-lĕst′) *n.* [Fr., celestial voice.] An organ stop that produces a gentle tremolo effect.

**Vo·lans** (vō′lănz′) *n.* [Lat. *volans,* pr.part. of *volare,* to fly.] A constellation in the Southern Hemisphere.

**vo·lant** (vō′lənt) *adj.* [Lat. *volans, volant-,* pr.part. of *volare,* to fly.] **1.** Flying or able to fly. **2.** Moving quickly or nimbly : AGILE. **3.** *Heraldry.* Depicted with the wings stretched as in flight.

**Vo·la·pük** (vō′lə-pŏŏk′, vŏl′ə-) *n.* [Volapük, world's speech : *vol,* world (< E. WORLD) + *pük,* speech (< E. SPEECH).] An artificial international language based on English.

**vo·lar** (vō′lər) *adj.* [< Lat. *vola,* sole.] Of or relating to the sole of the foot or the palm of the hand.

**vol·a·tile** (vŏl′ə-tl, -tīl′) *adj.* [Fr. < Lat. *volatilis,* flying < *volare,* to fly.] **1.** Evaporating readily at normal pressures and temperatures. **2.** Capable of being readily vaporized. **3.** Changeable, esp.: **a.** Fickle : inconstant. **b.** Given to violence : EXPLOSIVE. **c.** Flighty : lighthearted. **d.** Fleeting : ephemeral. **4.** VOLANT 1. **—vol′a·tile·ness, vol′a·til′i·ty** (-tĭl′ĭ-tē) *n.*

**volatile oil** *n.* A rapidly evaporating oil, esp. an essential oil, that leaves no stain.

**vol·a·til·ize** (vŏl′ə-tl-īz′) *vi.* & *vt.* **-ized, -iz·ing, -iz·es. 1.** To become or make volatile. **2.** To evaporate or cause to evaporate. **—vol′a·til·iz′a·ble** *adj.* **—vol′a·til·i·za′tion** *n.* **—vol′a·til·iz′er** *n.*

**vol-au-vent** (vô′lō-väN′) *n.* [Fr. : *vol,* flight + *au,* with the + *vent,* wind.] A light pastry shell containing a ragout of fish or meat.

**vol·can·ic** (vŏl-kăn′ĭk, vôl-) *adj.* **1.** Of or suggestive of an erupting volcano. **2.** Produced by or ejected from a volcano. **3.** Powerfully explosive <a *volcanic* disposition> **—vol·can′i·cal·ly** *adv.*

**volcanic glass** *n.* A volcanic igneous rock of vitreous or glassy texture, as obsidian or pitchstone.

**vol·can·ism** (vŏl′kə-nĭz′əm, vôl′-) *also* **vul·can·ism** (vŭl′-) *n.* Volcanic force or activity.

**vol·can·ize** (vŏl′kə-nīz′, vôl′-) *vt.* **-ized, -iz·ing, -iz·es. 1.** To subject to volcanic heat. **2.** To change (something) as a result of volcanic heat. **—vol′can·i·za′tion** *n.*

**vol·ca·no** (vŏl-kā′nō, vôl-) *n., pl.* **-noes** or **-nos.** [Ital. < Lat. *Volcanus,* Vulcan.] **1.** A vent in the earth's crust through which molten lava and gases are ejected. **2.** A mountain formed by the materials ejected from a volcano.

**vol·ca·no·gen·ic** (vŏl′kə-nə-jĕn′ĭk, vôl′-) *adj.* Of volcanic origin.

**vol·ca·nol·o·gy** (vŏl′kə-nŏl′ə-jē, vôl′-) *also* **vul·ca·nol·o·gy** (vŭl′-) *n.* The science concerned with volcanic phenomena. **—vol′ca·no·log′i·cal** (-nə-lŏj′ĭ-kəl) *adj.* **—vol′ca·nol′o·gist** *n.*

**vole**[1] (vōl) *n.* [Short for obs. *volemouse* < Norw. **vollmus* : ON *vǫllr,* field + ON *mūs,* mouse.] Any of various short-tailed rodents of the genus *Microtus* and related genera, similar to rats or mice.

**vole**[2] (vōl) *n.* [Fr. < *voler,* to fly < OFr. < Lat. *volare,* to fly.] A grand slam in a card game.

**vol·i·tant** (vŏl′ĭ-tnt) *adj.* [Lat. *volitans, volitant-,* pr.part. of *volitare,* to fly to and fro, freq. of *volare,* to fly.] **1.** Flying or able to fly. **2.** Moving about rapidly.

**vol·i·ta·tion** (vŏl′ĭ-tā′shən) *n.* **1.** An act of flying : FLIGHT. **2.** Ability to fly. **—vol′i·ta′tion·al** *adj.*

**vo·li·tion** (və-lĭsh′ən) *n.* [Fr. < Med. Lat. *volitio* < Lat. *velle,* to wish.] **1.** An act of willing, choosing, or deciding. **2.** A conscious choice : DECISION. **3.** Power or capability of choosing : WILL. **—vo·li′tion·al** *adj.* **—vo·li′tion·al·ly** *adv.*

**vol·i·tive** (vŏl′ĭ-tĭv) *adj.* **1.** Of, relating to, or arising from the will. **2.** Expressing a wish or permission.

**volks·lied** (fōk′slēt′, fôlk′-) *n., pl.* **-lie·der** (-slē′dər) [G. : *Volk,* people + *Lied,* song.] A folk song.

**vol·ley** (vŏl′ē) *n., pl.* **-leys.** [OFr. *volee* < *voler,* to fly < Lat. *volare.*] **1. a.** Simultaneous discharge of several missiles. **b.** The missiles thus discharged. **2.** A bursting forth <a *volley* of profanity> **3.** A shot, esp. in tennis, made by hitting the ball before it touches the ground. *—v.* **-leyed, -ley·ing, -leys.** *—vt.* **1.** To discharge in or as if in a volley. **2.** To hit (e.g., a tennis ball) before it touches the ground. *—vi.* To be discharged in or as if in a volley. **—vol′ley·er** *n.*

**vol·ley·ball** (vŏl′ē-bôl′) *n.* **1.** A court game in which one team tries to score by grounding a ball on the opposing team's side of a high net. **2.** The large inflated ball used in volleyball.

**vol·plane** (vŏl′plān′, vôl′-) *vi.* **-planed, -plan·ing, -planes.** [Fr. *vol plané,* gliding flight.] To glide toward the earth with the engine shut off. —Used of an aircraft or winged missile. *—n.* The glide of an aircraft.

**Vol·sci** (vôl′skē′, vŏl′sī′) *pl.n.* A people of ancient Italy whose territory the Romans conquered in the 4th cent. B.C.

**Vol·scian** (vŏl′shən, vôl′skē-ən) *adj.* Of or relating to the Volsci or their language. *—n.* The Italic language of the Volsci.

**volt**[1] (vōlt) *n.* [After Count Alessandro *Volta* (1745–1827).] **1.** The International System unit of electric potential and electromotive force, equal to the difference of electric potential between two points on a conducting wire carrying a constant current of one ampere when the power dissipated between the points is one watt. **2.** A unit of electric potential and electromotive force equal to 1.00034 times the International System unit.

**volt**[2] *also* **volte** (vōlt, vôlt) *n.* [Fr. *volte* < OItal. *volta,* turn < *voltare,* to turn, leap. —see VAULT[2].] **1.** A circular movement performed by a horse in manège. **2.** A sudden movement made in evading a thrust in fencing.

**volt·age** (vōl′tĭj) *n.* Electromotive force or potential difference, usu. expressed in volts.

**voltage divider** *n.* A number of resistors in series supplied with taps at some points to make available a variable or fixed fraction of the applied voltage.

**vol·ta·ic** (vŏl-tā′ĭk, vōl-, vôl-) *adj.* [< VOLT.] **1.** Relating to or designating electricity or electric current produced by chemical action : GALVANIC. **2.** Producing electricity by chemical action.

**voltaic battery** *n.* An electric battery consisting of a primary cell or cells.

**voltaic cell** *n.* A primary cell.

**voltaic couple** *n.* Two dissimilar conductors in contact or in the same electrolytic solution, causing a difference of potential between them.

**voltaic pile** *n.* A source of electricity made up of a number of alternating disks of two different metals separated by acid-moistened pads, forming primary cells connected in series.

**vol·ta·ism** (vŏl′tə-ĭz′əm, vōl′-, vôl′-) *n.* [VOLTA(IC) + -ISM.] Galvanism.

**volt·am·me·ter** (vōlt′ăm′mē′tər) *n.* [VOLT-AM(PERE) + -METER.] An instrument for measuring current or potential.

**volt-am·pere** (vōlt′ăm′pîr′) *n.* A unit of electric power equal to the product of one volt and one ampere, equivalent to one watt.

**volte** (vōlt, vôlt) *n. var. of* VOLT[2].

**volte-face** (vôlt-făs′, vôl′tə-) *n.* A reversal, as in policy.

**volt·me·ter** (vōlt′mē′tər) *n.* An instrument, as a galvanometer, for measuring potential differences in volts.

**vol·u·ble** (vŏl′yə-bəl) *adj.* [OFr. < Lat. *volubilis* < *volvere,* to roll.] **1.** Marked by a ready flow of words in speaking : FLUENT. **2.** Turning readily on an axis : ROTATING. **3.** Twining or twisting, as a plant. **—vol′u·bil′i·ty, vol′u·ble·ness** *n.* **—vol′u·bly** *adv.*

**vol·ume** (vŏl′yo͞om, -yəm) *n.* [ME < OFr. < Lat. *volumen,* roll of writing < *volvere,* to roll.] **1.** A collection of printed or written sheets bound together : BOOK. **2.** One of the books included in a complete set. **3.** Written material in a library that has been brought together and catalogued as an individual unit. **4.** A roll of parchment : SCROLL. **5. a.** The size or extent of a three-dimensional object or region of space. **b.** The capacity of such a region or of a specified container. **6.** A large quantity <*volumes* of compliments> **7. a.** The amplitude or loudness of a sound. **b.** A control, as on a radio, for regulating loudness. **—vol′umed** *adj.*

**vol·u·me·ter** (vŏl′yo͞o-mē′tər) *n.* [VOLU(ME) + -METER.] An instrument for measuring the volume of liquids, solids, and gases.

**vol·u·met·ric** (vŏl′yo͞o-mĕt′rĭk) *adj.* [VOLU(ME) + -METRIC.] Of or relating to measurement of volume. **—vol′u·met′ri·cal·ly** *adv.*

**volumetric analysis** *n.* **1.** Quantitative analysis using accurately measured, esp. titrated volumes of standard chemical solutions. **2.** Analysis of a gas by volume.

**vo·lu·mi·nous** (və-lo͞o′mə-nəs) *adj.* [LLat. *voluminosus,* having many folds < Lat. *volumen,* roll of writing < *volvere,* to roll.] **1.** Having great volume, fullness, size, or number. **2. a.** Filling or able to fill volumes. **b.** Prolific in speech or writing. **3.** Having many coils : WINDING. **—vo·lu′mi·nos′i·ty** (-nŏs′ĭ-tē), **vo·lu′mi·nous·ness** *n.* **—vo·lu′mi·nous·ly** *adv.*

**vol·un·ta·rism** (vŏl′ən-tə-rĭz′əm) *n.* Belief in the primacy of will. **—vol′un·ta·rist** *n.* **—vol′un·ta·ris′tic** *adj.*

**vol·un·tar·y** (vŏl′ən-tĕr′ē) *adj.* [ME < Lat. *voluntarius* < *voluntas,* choice < *velle,* to wish.] **1. a.** Arising from one's own free will. **b.** Acting on one's own initiative. **2.** Acting or serving in a designated capacity willingly and with no constraint or guarantee of reward. **3.** Normally controlled by or subject to individual volition. **4.** Capable of exercising will : VOLITIONAL. **5.** Proceeding from impulse : SPONTANEOUS. **6.** *Law.* **a.** Acting or done with no external persuasion or compulsion. **b.** Without legal obligation, payment, or valuable consideration <a *voluntary* conveyance> **c.** Not accidental : INTENTIONAL <*voluntary* manslaughter> *—n., pl.* **-ies. 1.** *Mus.* Solo organ music, occas. improvised, that is played usu. prior to and occas. during or after a church service. **2.** A volunteer. **—vol′un·tar′i·ly** (-târ′ə-lē) *adv.* **—vol′un·tar′i·ness** *n.*

☆ ***syns:*** VOLUNTARY, DELIBERATE, INTENTIONAL, WILLFUL *adj. core meaning* : subject to individual volition <*voluntary* enlistment in the army> VOLUNTARY is the most general; it implies the exercise of free will <a *voluntary* contribution> or of choice <living in *voluntary* exile> DELIBERATE and INTENTIONAL suggest that which is done or said on purpose <a *deliberate* lie><*intentional* insolence> What is WILLFUL is done in accordance with one's own will and often suggests obstinancy <*willful* disobedience> ***ant:*** involuntary

**vol·un·tar·y·ism** (vŏl′ən-tĕr′ē-ĭz′əm) *n.* The principle of dependence on voluntary contributions instead of government funds, as for churches or schools. **—vol′un·tar′y·ist** *n.*

**voluntary muscle** *n.* Muscle normally controlled by individual will.

**vol·un·teer** (vŏl′ən-tîr′) *n.* [Obs. Fr. *voluntaire* < Lat. *voluntarius,* voluntary.] **1.** One who serves or acts of his or her own free will. **2. a.** One who gives help, does a service, or takes an obligation voluntarily. **b.** *Law.* One who holds property under a deed made without valuable consideration. **3.** A cultivated plant growing from self-sown or accidentally dropped seed. *—adj.* **1.** Relating to or composed of volunteers <a *volunteer* army> **2.** Enlisted or serving as a volunteer. **3.** Growing from self-sown or accidentally dropped seed. —Used of a cultivated plant or crop that has reseeded itself. *—v.* **-teered, -teer·ing, -teers.** *—vt.* To give or offer to give on one's own initiative. —*vi.* To enter into or offer to enter into a venture of one's own free will.

**vol·un·teer·ism** (vŏl′ən-tîr′ĭz′əm) *n.* The theory, act, or practice of being a volunteer or of using volunteers in community service work.

**vo·lup·tu·ar·y** (və-lŭp′cho͞o-ĕr′ē) *n., pl.* **-ies.** [LLat. *voluptuarius* < Lat. *voluptarius* < *voluptas,* pleasure.] One whose life is devoted to luxury and sensual pleasures : SENSUALIST. **—vo·lup′tu·ar′y** *adj.*

**vo·lup·tu·ous** (və-lŭp′cho͞o-əs) *adj.* [ME < Lat. *voluptuosus,* full of pleasure < *voluptas,* pleasure.] **1.** Made up of or marked by strong visual and tactile delights <*voluptuous* statuary> **2.** Given over to or frequently indulging in sensual gratifications. **3. a.** Full and appealing in form <a *voluptuous* mouth> **b.** Directed toward or looking forward to sensuous gratification <*voluptuous* thoughts> **c.** Originating in satisfaction of luxurious or sensual desires. **—vo·lup′tu·ous·ly** *adv.* **—vo·lup′tu·ous·ness** *n.*

**vo·lute** (və-lo͞ot′) *n.* [Lat. *voluta* < fem. p.part. of *volvere,* to turn.] **1.** A spiral scroll-like ornament, as that on an Ionic capital. **2.** A spiral or twisted formation, as a whorl on a gastropod shell. **3.** Any of various marine gastropod mollusks of the family Volutidae, with a spiral, often colorfully marked shell. **—vo·lut′ed** (-lo͞o′tĭd) *adj.*

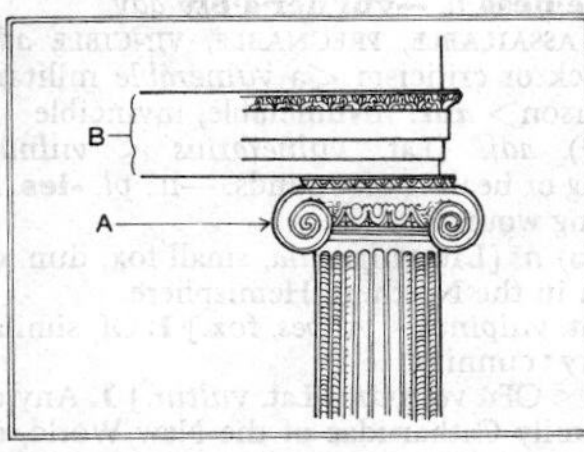


**volute**
*On an Ionic column:*
*A. volute and B. architrave*

**vol·u·tin** (vŏl′yə-tĭn, və-lo͞ot′n) *n.* [G. < NLat. *volutans,* specific epithet of *Spirillum volutans,* a bacterium in which it was first found < Lat., pr.part. of *volutare,* to roll around, freq. of *volvere,* to roll.] A basophilic granular substance occurring in many microorganisms that is thought to be nucleic acid.

**vo·lu·tion** (və-lo͞o′shən) *n.* [< Lat. *volvere, volut-,* to turn.] **1.** A turn or twist about a center : SPIRAL. **2.** *Zool.* A whorl on a spiral shell.

**vol·va** (vŏl′və, vôl′-) *n.* [Lat., a covering.] A cuplike structure surrounding the base of the stalk of certain fungi. **—vol′vate′** (-vāt′) *adj.*

**vol·vox** (vŏl′vŏks′, vôl′-) *n.* [NLat. *Volvox,* genus name < Lat. *volvere,* to turn.] Any of various flagellate protozoans of the genus *Volvox* that form hollow spherical multicellular colonies.

**vol·vu·lus** (vŏl′vyə-ləs, vôl′-) *n.* [NLat. < Lat. *volvere,* to turn.] An intestinal obstruction caused by abnormal twisting.

**vo·mer** (vō′mər) *n.* [Lat., plowshare.] The flat bone in the skull forming the inferior and posterior part of the nasal septum. **—vo′mer·ine′** (-mə-rīn′) *adj.*

**vom·i·ca** (vŏm′ĭ-kə) *n., pl.* **-cae** (-sē′) [Lat., ulcer < *vomere,* to vomit.] **1.** Copious expectoration of putrid matter. **2. a.** An abnormal pus-containing cavity in a lung, induced by degeneration of tissue. **b.** Purulent matter enclosed in such a cavity.

**vom·it** (vŏm′ĭt) *v.* **-it·ed, -it·ing, -its.** [ME *vomiten* < Lat. *vomere.*] —*vi.* **1.** To eject part or all of the contents of the stomach through the mouth, usu. in a series of involuntary spasmodic movements. **2.** To be discharged forcefully and copiously. —*vt.* **1.** To eject from the stomach through the mouth. **2.** To eject or discharge in a gush <a volcano *vomiting* molten lava> —*n.* **1.** An act of ejecting matter from the stomach. **2.** Matter ejected from the stomach. **3.** An emetic. **—vom′it·er** *n.*

**vom·i·tive** (vŏm′ĭ-tĭv) *adj.* Relating to or inducing vomiting. **—vom′i·tive** *n.*

**vom·i·to·ry** (vŏm′ĭ-tôr′ē, -tōr′ē) *adj.* Causing vomiting : VOMITIVE. —*n., pl.* **-ries. 1.** An agent inducing vomiting. **2.** An aperture through which matter is discharged. **3.** A passageway in a Roman amphitheater that leads from the outside wall to the foot of the banked seats.

**vom·i·tu·ri·tion** (vŏm′ĭ-chə-rĭsh′ən, -ĭ-to͞o-) *n.* [VOMIT + (MIC-T)URITION.] Forceful ineffectual attempts to vomit : RETCHING.

**vom·i·tus** (vŏm′ĭ-təs) *n.* [Lat., p.part of *vomere,* to vomit.] VOMIT 2.

**voo·doo** (vo͞o′do͞o) *n., pl.* **-doos.** [Louisiana Fr. *voudou* < Ewe *vòdú.*] **1.** A religious cult marked by a belief in sorcery, fetishes, and rituals in which participants communicate by trance with ancestors, saints, or animistic deities. **2.** A charm, fetish, spell, or curse thought by believers in voodoo to possess magic power. **3.** One who performs rites at a gathering of believers in voodoo. —*vt.* **-dooed, -doo·ing, -doos.** To put under the influence of a voodoo spell : HEX. —*adj. Slang.* Bizarre <*voodoo* economics> **—voo′doo·ism′** (-ĭz′əm) *n.* **—voo′doo·ist** *n.* **—voo′doo·is′tic** *adj.*

**vo·ra·cious** (vô-rā′shəs, və-) *adj.* [< Lat. *vorax, vorac-* < *vorare,* to devour.] **1.** Consuming or eager to consume large quantities of food : RAVENOUS. **2.** Having an insatiable appetite for an activity or pursuit : GREEDY <a *voracious* reader> **—vo·ra′cious·ly** *adv.* **—vo·rac′i·ty** (-răs′ĭ-tē), **vo·ra′cious·ness** *n.*

**vor·la·ge** (fôr′lä′gə, fōr′-) *n.* [G. : *vor,* before + *Lage,* stance.] A posture in skiing in which the skier leans forward from the ankles, usu. without lifting the heels.

**-vorous** *suff.* [Lat. *-vorus* < *vorare,* to devour.] Eating : feeding on <vermi*vorous*>

**vor·tex** (vôr′tĕks′) *n., pl.* **-tex·es** *or* **-ti·ces** (-tĭ-sēz′) [Lat. *vortex, vortic-* < *vertere,* to turn.] **1.** Fluid flow involving rotation about an axis, esp. a whirlpool. **2.** A situation regarded as drawing into its center all that surrounds it <"was swept up in the *vortex* of Hollywood" —*New York Times*>

**vor·ti·cal** (vôr′tĭ-kəl) *adj.* Of, relating to, or suggestive of a vortex : WHIRLING. **—vor′ti·cal·ly** *adv.*

**vor·ti·cel·la** (vôr′tĭ-sĕl′ə) *n., pl.* **-cel·lae** (-sĕl′ē) *or* **-cel·las** [NLat. *Vorticella,* genus name < Lat. *vortex,* vortex.] Any of various bell-shaped, ciliated, stalked protozoans of the genus *Vorticella.*

**vor·ti·ces** (vôr′tĭ-sēz′) *n. var. pl. of* VORTEX.

**vor·ti·cose** (vôr′tĭ-kōs′) *adj.* Vortical.

**vor·tig·i·nous** (vôr-tĭj′ə-nəs) *adj.* [< Lat. *vertigo, vertigin-,* a whirling < *vertere,* to turn.] Vortical.

**vo·ta·ry** (vō′tə-rē) *n., pl.* **-ries.** [< Lat. *votum,* vow < *vovēre,* to vow.] **1.** One bound by vows to live a life of religious service or worship : a monk or a nun. **2.** A fervent devotee to a religion, activity, leader, or ideal.

**vote** (vōt) *n.* [Lat. *votum,* vow < *vovēre,* to vow.] **1. a.** A formal expression of preference for a candidate for office or for a proposed resolution of an issue. **b.** The way by which such a preference is made known, as by a ballot or a raised hand. **2.** The number of votes cast in an election or to resolve an issue <a heavy *vote* in their favor> **3.** A group of voters <the farm *vote*> **4.** The result of an election or referendum. **5.** The right to take part as a voter : SUFFRAGE. —*v.* **vot·ed, vot·ing, votes.** —*vi.* To indicate one's preference by a vote. —*vt.* **1.** To endorse by a vote. **2.** To bring into existence or make available by vote <*vote* new bond issues> **3.** To declare or pronounce by general consent <*voted* the film a failure> **—vote down.** To defeat with a negative vote. **—vote in.** To elect. **—vote out.** To remove from elective office by giving support to the opposition. **—vot′a·ble, vote′a·ble** *adj.* **—vot′er** *n.*

**vote getter** *n.* **1.** A candidate with abilities and qualities that get votes in his or her favor. **2.** A way of drawing votes.

**vote·less** (vōt′lĭs) *adj.* Having no vote, esp. denied a political vote.

**voting machine** *n.* An apparatus that mechanically records and counts votes in polling places.

**vo·tive** (vō′tĭv) *adj.* [Lat. *votivus* < *votum,* vow.] **1.** Dedicated or given in fulfillment of a pledge or vow <a *votive* offering><a *votive* candle> **2.** Expressing a wish, desire, or vow <a *votive* prayer> **—vo′tive·ly** *adv.*

**votive Mass** *n. Rom. Cath. Ch.* A Mass differing from one prescribed for a certain day in that it is celebrated at the direction of authority, because of special circumstances, or by decision of the priest.

**vouch** (vouch) *v.* **vouched, vouch·ing, vouch·es.** [ME *vouchen,* to summon to court < OFr. *voucher* < Lat. *vocare,* to call.] —*vt.* **1.** To substantiate by providing evidence : VERIFY. **2.** *Law.* To summon as a witness to give warranty of title. **3.** *Archaic.* To cite (e.g., an authority) as corroborating evidence for one's statements, opinions, or actions. **4.** *Archaic.* To assert : declare. —*vi.* **1.** To give personal assurance : GUARANTEE. **2.** To furnish supporting evidence : CORROBORATE. —*n. Obs.* A declaration of opinion.

**vouch·er** (vou′chər) *n.* **1.** One who vouches. **2.** A document giving proof that the terms of a transaction have been met.

**vouch·safe** (vouch-sāf′, vouch′sāf′) *vt.* **-safed, -saf·ing, -safes.** [ME *vouchen sauf,* to warrant as safe.] To condescend to grant or bestow (e.g., a privilege) : DEIGN. **—vouch·safe′ment** *n.*

**vous·soir** (vo͞o-swär′) *n.* [Fr. < OFr. *vossoir* < VLat. **volsorium* < **volsus,* var. of Lat. *volutus,* p.part. of *volvere,* to turn.] Any of the wedge-shaped stones that make up the curved portions of an arch or vaulted ceiling.

**vow** (vou) *n.* [ME *vowe* < OFr. < Lat. *votum* < *vovēre,* to vow.] **1.** An earnest promise or pledge that commits one to perform a certain act or behave in a particular way, esp. a solemn promise to live and act according to the prescriptions of a religious body <a monk's *vows*> **2.** A formal assertion or declaration. —*v.* **vowed, vow·ing, vows.** —*vt.* **1.** To promise or pledge solemnly. **2.** To make a pledge or threat to undertake <*vowing* death to the enemy> **3.** To assert or declare formally. —*vi.* To express a promise or pledge : make a vow. **—take vows.** To enter a religious order. **—vow′er** *n.*

**vow·el** (vou′əl) *n.* [ME *vowelle* < OFr. *vouel* < Lat. *vocalis,* sounding < *vox,* voice.] **1.** A speech sound produced by the relatively free passage of breath through the larynx and oral cavity, usu. forming the most prominent and central sound of a syllable. **2.** A letter that represents a vowel, as *a, e, i, o, u,* and sometimes *y* in the English alphabet.

**vowel fracture** *n.* Linguistic breaking.

**vow·el·ize** (vou′ə-līz′) *vt.* **-ized, -iz·ing, -iz·es.** To furnish with vowel points. **—vow′el·i·za′tion** *n.*

**vowel point** *n.* Any of a number of diacritical marks written above or below consonants to designate a preceding or following vowel in languages such as Hebrew and Arabic that are usu. written with no vowel letters.

**vox an·gel·i·ca** (vŏks′ ăn-jĕl′ĭ-kə) *n.* [NLat., angelic voice.] Voix céleste.

**vox hu·ma·na** (vŏks′ hyo͞o-mā′nə, -mä′-) *n.* [Lat., human voice.] An organ reed stop producing tones imitative of the human voice.

**vox pop·u·li** (vŏks′ pŏp′yə-lī′, -lē) *n.* [Lat., voice of the people.] Popular opinion or sentiment.

**voy·age** (voi′ĭj) *n.* [ME < OFr. *veyage* < Lat. *viaticum,* provisions for a journey < *viaticus,* of a journey < *via,* road.] **1.** A long journey, usu. to a foreign or distant country, esp. a journey over an open sea or ocean. **2.** A record or account of a journey of exploration or discovery. *—v.* **-aged, -ag·ing, -ag·es.** *—vi.* To make a voyage. *—vt.* To sail across. **—voy′ag·er** *n.*

**voy·a·geur** (voi′ə-zhûr′, vwä′yä-) *n., pl.* **-geurs** (-zhûr′) [Fr., traveler < *voyage,* journey < OFr. *veyage.* —see VOYAGE.] A woodsman, boatman, or guide, esp. one hired by fur companies to transport furs and supplies between isolated posts in the U.S. and Canadian northwest.

**V-par·ti·cle** (vē′pär′tĭ-kəl) *n.* [From the shape of the track left by its decay product in a cloud chamber.] Any of several subatomic particles with half-lives in the range of $10^{-10}$ to $10^{-6}$ second.

**vroom** (vro͞om) [Imit.] *Slang. —n.* The loud roaring sound made by a motor vehicle, as a race car or motorcycle, accelerated at high speed. *—vi.* **vroomed, vroom·ing, vrooms.** To accelerate a motor vehicle at high speed so as to make such a sound.

**vrouw** *or* **vrow** (frou, frō) *n.* [Du., woman < MDu. *vrouwe.*] A Dutch or Afrikaner woman.

**V-shaped** (vē′shāpt′) *adj.* Shaped like the letter V.

**V sign** *n.* A symbol of victory made by raising the index and middle fingers in a V shape.

**VT fuze** (vē′tē′) *n.* [V(ARIABLE) T(IME) FUZE.] A proximity fuze.

**Vul·can** (vŭl′kən) *n.* [Lat. *Vulcanus, Volcanus.*] *Rom. Myth.* The god of fire and craftsmanship, esp. metalworking.

**vul·ca·ni·an** (vŭl-kā′nē-ən) *adj.* **1.** *Geol.* Of, relating to, or from a volcano or volcanic eruption. **2. Vulcanian. a.** Of or relating to Vulcan. **b.** Of or relating to craftsmanship or metalworking.

**vul·can·ism** (vŭl′kə-nĭz′əm) *n. var. of* VOLCANISM.

**vul·can·ite** (vŭl′kə-nīt′) *n.* A hard rubber made by vulcanization.

**vul·can·ize** (vŭl′kə-nīz′) *vt.* **-ized, -iz·ing, -iz·es.** [< VULCAN.] To increase the strength, resiliency, and freedom from stickiness and odor of (e.g., rubber) by combining with sulfur or other additives in the presence of heat and pressure. **—vul′can·iz′a·ble** *adj.* **—vul′can·i·za′tion** *n.* **—vul′can·iz′er** *n.*

**vul·ca·nol·o·gy** (vŭl′kə-nŏl′ə-jē) *n. var. of* VOLCANOLOGY.

**vul·gar** (vŭl′gər) *adj.* [ME < Lat. *vulgaris* < *vulgus,* the common people.] **1.** Of or associated with the great masses of people : COMMON. **2.** Spoken by or expressed in language used by the common people : VERNACULAR. **3. a.** Deficient in taste, delicacy, or refinement. **b.** Boorish : ill-bred. **c.** Ostentatious in appearance or quality : PRETENTIOUS <a *vulgar* display of wealth> **4.** Indecent : lewd <*vulgar* language> **—vul′gar·ly** *adv.* **—vul′gar·ness** *n.*

**vulgar fraction** *n.* A common fraction.

**vul·gar·i·an** (vŭl-gâr′ē-ən) *n.* A vulgar person.

**vul·gar·ism** (vŭl′gə-rĭz′əm) *n.* **1.** Vulgarity. **2. a.** A vulgar word or phrase. **b.** A word, phrase, or manner of expression used chiefly by uncultivated people.

**vul·gar·i·ty** (vŭl-găr′ĭ-tē) *n., pl.* **-ties. 1.** The quality or state of being vulgar. **2.** Something, as an act or expression, offensive to good taste or propriety.

**vul·gar·ize** (vŭl′gə-rīz′) *vt.* **-ized, -iz·ing, -iz·es. 1.** To make vulgar : DEBASE. **2.** To popularize. **—vul′gar·i·za′tion** *n.*

**Vulgar Latin** *n.* The common speech of the ancient Romans, distinguished from standard literary Latin and the ancestor of the Romance languages.

**vul·gate** (vŭl′gāt′, -gĭt) *n.* [< Lat. *vulgatus,* common < *vulgare,* to make known to all < *vulgus,* the common people.] **1.** The common speech of a people : VERNACULAR. **2.** A widely accepted text or version of a work. **3. Vulgate.** The Latin translation of the Bible made by Saint Jerome at the end of the 4th cent. A.D., now used in a revised form as the Roman Catholic authorized version.

**vul·ner·a·ble** (vŭl′nər-ə-bəl) *adj.* [LLat. *vulnerabilis* < Lat. *vulnerare,* to wound < *vulnus,* wound.] **1.** Susceptible to physical injury. **2.** Susceptible to attack. **3. a.** Subject to criticism or censure : ASSAILABLE. **b.** Liable to yield to temptation or persuasion. **4.** Being in a position to receive greater penalties or bonuses. —Used of the partners of a team that has won one game of a rubber in bridge. **—vul′ner·a·bil′i·ty, vul′ner·a·ble·ness** *n.* **—vul′ner·a·bly** *adv.*

☆ ***syns:*** VULNERABLE, ASSAILABLE, PREGNABLE, VINCIBLE *adj.* *core meaning* : open to attack or criticism <a *vulnerable* military position><a *vulnerable* person> ***ant:*** invulnerable, invincible

**vul·ner·ar·y** (vŭl′nə-rĕr′ē) *adj.* [Lat. *vulnerarius* < *vulnus,* wound.] Used in the treating or healing of wounds. *—n., pl.* **-ies.** A remedy for treating or healing wounds.

**Vul·pec·u·la** (vŭl-pĕk′yə-lə) *n.* [Lat. *vulpecula,* small fox, dim. of *vulpes,* fox.] A constellation in the Northern Hemisphere.

**vul·pine** (vŭl′pīn′) *adj.* [Lat. *vulpinus* < *vulpes,* fox.] **1.** Of, similar to, or typical of a fox. **2.** Foxy : cunning.

**vul·ture** (vŭl′chər) *n.* [ME < OFr. *voltour* < Lat. *vultur.*] **1.** Any of various large birds of the family Cathartidae of the New World, or the family Accipitridae of the Old World, with dark plumage and a naked head and neck and feeding on carrion. **2.** A predatory or rapacious person.

**vulture**
*Wingspan up to 6 feet*

**vul·tur·ine** (vŭl′chə-rīn′) *also* **vul·tur·ous** (-chər-əs) *adj.* **1.** Of, relating to, or typical of a vulture. **2.** Like a vulture : PREDATORY.

**vul·va** (vŭl′və) *n., pl.* **-vae** (-vē′) [Lat., womb, covering.] The external female genitalia, including the labia majora, labia minora, clitoris, and vestibule of the vagina. **—vul′val, vul′var** (-vər, -vär′) *adj.* **—vul′vate′** (-vāt′, -vĭt) *adj.* **—vul′vi·form′** (-və-fôrm′) *adj.*

**vul·vi·tis** (vŭl-vī′tĭs) *n.* Inflammation of the vulva.

**vul·vo·vag·i·ni·tis** (vŭl′vō-văj′ə-nī′tĭs) *n.* Simultaneous inflammation of the vulva and vagina.

**vy·ing** (vī′ĭng) *v. present participle of* VIE.

**w** *or* **W** (dŭb′əl-yo͞o, -yo͞o) *n., pl.* **w's** *or* **W's. 1.** The 23rd letter of the English alphabet. **2.** A speech sound represented by the letter *w.* **3.** The 23rd in a series. **4.** Something shaped like the letter W.

**W** [G. *Wolfram.*] *symbol for* TUNGSTEN.

**wab·ble** (wŏb′əl) *v. & n. var. of* WOBBLE.

**Wac** (wăk) *n.* [Abbr. of *Women's Army Corps.*] A member of the Women's Army Corps of the U.S. Army.

**wack·o** (wăk′ō) *adj. Slang.* Crazy : nuts. **—wack′o** *adv.*

**wack·y** (wăk′ē) *also* **whack·y** (hwăk′ē, wăk′ē) *adj.* **-i·er, -i·est.** [Orig. unknown.] *Slang.* **1.** Irrational or erratic <a *wacky* guy> **2.** Silly <a *wacky* notion> **—wack′i·ly** *adv.* **—wack′i·ness** *n.*

**wad** (wŏd) *n.* [Orig. unknown.] **1.** A small, often rolled or folded mass of soft material used for padding, stuffing, or packing. **2.** A compressed ball, roll, or lump, as of tobacco. **3. a.** A plug, as of cloth or paper, used to hold in a powder charge in a muzzle-loading firearm. **b.** A disk, as of felt or paper, used to keep the powder and shot in place in a shotgun cartridge. **4.** *Informal.* A large amount. **5.** *Informal.* **a.** A large roll of paper money. **b.** A sizable amount of money.

—*v.* **wad·ded, wad·ding, wads.** —*vt.* **1.** To compress into a wad. **2.** To pad, pack, line, or plug with wadding. **3. a.** To hold (shot or powder) in place with a wad. **b.** To insert a wad in (a firearm). —*vi.* To form a wad.

**wad·die** (wŏd'ē) *n. var. of* WADDY[2].

**wad·ding** (wŏd'ĭng) *n.* **1. a.** A wad. **b.** Wads in general. **2.** A soft layer of fibrous cotton or wool for padding or stuffing. **3.** Material for gun wads.

**wad·dle** (wŏd'l) *vi.* **-dled, -dling, -dles.** [Freq. of WADE.] **1.** To walk with short steps that tilt the body from side to side. **2.** To walk heavily and clumsily with a pronounced sway. —*n.* A waddling gait. **—wad'dler** *n.*

**wad·dy[1]** (wŏd'ē) [Native word in Australia.] *Austral.* —*n., pl.* **-dies.** A heavy straight stick or club thrown as a weapon by Australian aborigines. —*vt.* **-died, -dy·ing, -dies.** To strike with a waddy.

**†wad·dy[2]** *also* **wad·die** (wŏd'ē) *n., pl.* **-dies.** [Orig. unknown.] *Western U.S.* **1.** A cowboy. **2.** A cattle rustler.

**wade** (wād) *v.* **wad·ed, wad·ing, wades.** [ME *waden* < OE *wadan.*] —*vi.* **1.** To walk in or through a medium, as water, that hinders normal movement. **2.** To make one's way arduously <*waded* through the angry mob> —*vt.* To cross or pass through by wading <*wade* a stream> **—wade in (*or* into).** To plunge into, begin, or attack resolutely and energetically <*waded into* the assignment> **—wade** *n.*

**wad·er** (wā'dər) *n.* **1.** One that wades. **2.** A long-legged bird that frequents shallow water. **3. waders.** Waterproof hip boots or trousers worn esp. by fishermen or hunters.

**wa·di** *also* **wa·dy** (wä'dē) *n., pl.* **-dis** *also* **-dies.** [Ar. *wādī.*] **1. a.** A valley, gully, or riverbed in northern Africa and southwestern Asia that remains dry except during the rainy season. **b.** A stream that flows through a wadi. **2.** An oasis.

**wading bird** *n.* WADER 2.

**wa·dy** (wä'dē) *n. var. of* WADI.

**Waf** (wăf) *n.* [Abbr. of *Women in the Air Force.*] A member of the Women in the Air Force, formed after World War II.

**wa·fer** (wā'fər) *n.* [ME *wafre* < ONFr. *waufre,* of Germanic orig.] **1.** A small, thin, crisp cake, biscuit, or candy. **2.** A small, thin disk of unleavened bread used in Communion. **3.** A flat tablet of dried flour paste encasing a powdered drug. **4.** A small disk of adhesive material used as a seal for papers. **5.** *Electron.* A small, thin, flat circular disk of a semiconducting material, as pure silicon, that is masked, oxide-coated, doped, and otherwise processed for ultimate separation into numerous individual electronic devices or for packaging as an integrated circuit. —*vt.* **-fered, -fer·ing, -fers. 1.** To fasten together or seal with a wafer. **2.** To prepare in the form of wafers. **3.** *Electron.* To divide into wafers.

**waff** (wăf, wäf) [ME *waffen,* to wave.] *Scot.* —*v.* **waffed, waff·ing, waffs.** —*vi.* To flutter : wave. —*vt.* To cause to flutter or wave. —*n.* **1.** A waving motion. **2.** A gust of air : WAFT.

**waf·fle[1]** (wŏf'əl) *n.* [Du. *wafel.*] A light, crisp batter cake baked in a waffle iron.

**waf·fle[2]** (wŏf'əl) [Prob. freq. of obs. *waff,* to yelp.] *Informal.* —*vi.* **-fled, -fling, -fles.** To speak or write evasively : EQUIVOCATE. —*n.* Evasive or vague expression.

**waffle iron** *n.* An electric appliance having hinged, indented plates that impress a grid pattern into waffle batter as it bakes.

**waft** (wăft, wäft) *v.* **waft·ed, waft·ing, wafts.** [ME **waughten,* to convoy < MDu. or MLG *wachten,* to guard.] —*vt.* **1.** To cause to drift gently and smoothly through the air or over water. **2.** To convey or send floating through the air or over water. —*vi.* To float easily and gently, as on the air : DRIFT. —*n.* **1.** Something, as a scent, carried through the air. **2.** A light breeze. **3.** The act of wafting or waving. **4.** *Naut.* **a.** A flag used for signaling or indicating wind direction. **b.** A signal with a flag.

**waft·age** (wăf'tĭj, wäf'-) *n.* The act or state of being wafted.

**waf·ture** (wăf'chər, wäf'-) *n.* **1.** The act of wafting or waving. **2. a.** Something wafted. **b.** A wavelike motion.

**wag[1]** (wăg) *v.* **wagged, wag·ging, wags.** [ME *waggen.*] —*vi.* **1. a.** To move quickly and repeatedly from side to side, to and fro, or up and down. **b.** To be constantly active. **2.** To walk with a clumsy sway : WADDLE. **3.** *Archaic.* To be on one's way : DEPART. —*vt.* To wag (e.g., the head or a finger) as in playfulness, agreement, admonition, or chatter. —*n.* The act of wagging. **—wag'ger** *n.*

☆ ***syns:*** WAG, SWITCH, WAGGLE, WAVE *v. core meaning* : to move to and fro vigorously and usu. repeatedly <a big dog *wagging* its long tail>

**wag[2]** (wăg) *n.* [Orig. unknown.] A mischievous person.

**wage** (wāj) *n.* [ME < ONFr., of Germanic orig.] **1.** Payment for services to a worker, esp. remuneration on an hourly, daily, or weekly basis or by the piece. **2. wages.** The portion of the national product that represents the aggregate paid for all contributing labor and services as distinguished from the portion retained by management or reinvested in capital goods. **3.** *often* **wages** *(sing. or pl. in number).* A fitting return : RECOMPENSE <the *wages* of sin> —*vt.* **waged, wag·ing, wag·es.** To engage in (a war or campaign).

**wage earner** *n.* **1.** One who works for wages. **2.** One whose earnings support a household.

**wa·ger** (wā'jər) *n.* [ME < AN *wageure* < ONFr. *wagier,* to pledge.] **1. a.** An agreement under which each bettor pledges a certain amount to the other depending on the outcome of an unsettled matter. **b.** The matter bet on : GAMBLE. **2.** Something staked on an unknown outcome : BET. **3.** *Archaic.* A pledge of personal combat to resolve an issue or case. —*v.* **-gered, -ger·ing, -gers.** —*vt.* To risk or stake (an amount of money or a possession) on an unknown outcome : BET. —*vi.* To make a wager. **—wa'ger·er** *n.*

**wage scale** *n.* The scale of wages paid to employees for the various jobs within an industry, factory, or company.

**wage·work·er** (wāj'wûr'kər) *n.* A wage earner.

**wag·ger·y** (wăg'ə-rē) *n., pl.* **-ies. 1.** Waggish spirit or behavior : DROLLERY. **2.** A droll act or remark.

**wag·gish** (wăg'ĭsh) *adj.* Playfully humorous. **—wag'gish·ly** *adv.* **—wag'gish·ness** *n.*

**wag·gle** (wăg'əl) *v.* **-gled, -gling, -gles.** [Freq. of WAG[1].] —*vt.* To move with short, quick motions <*waggle* one's foot> —*vi.* To move shakily : WOBBLE. —*n.* A waggling motion. **—wag'gly** *adj.*

**Wag·ner·i·an** (väg-nîr'ē-ən) *adj.* Of, pertaining to, or typical of Richard Wagner, his music, or his theories. —*n.* **1.** An admirer or disciple of Richard Wagner. **2.** A performer of Wagner's music.

**wag·on** (wăg'ən) *n.* [Du. *wagen* < MDu.] **1.** A four-wheeled, usu. horse-drawn vehicle with a large rectangular body for transporting loads. **2. a.** A light automotive transport or delivery vehicle. **b.** A station wagon. **c.** A police patrol wagon. **3.** A child's low four-wheeled cart hauled by a long handle that governs the direction of the front wheels. **4.** A small table or tray on wheels for serving drinks or food. **5.** *Chiefly Brit.* An open railway freight car. **6. Wagon.** The Big Dipper. —*v.* **-oned, -on·ing, -ons.** —*vt.* To transport by wagon. —*vi.* To transport goods or travel by wagon. **—off the wagon.** *Slang.* No longer abstaining from liquor. **—on the wagon.** *Slang.* Abstaining from liquor.

**wag·on·er** (wăg'ə-nər) *n.* **1.** A wagon driver. **2. Wagoner.** Auriga.

**wag·on·ette** (wăg'ə-nĕt') *n.* A light horse-drawn wagon with two seats facing lengthwise behind the driver's seat.

**wa·gon-lit** (vä'gôN-lē') *n., pl.* **wa·gons-lits** *or* **wa·gon-lits** (vä'gôN-lē') [Fr. : *wagon,* railroad car (< E.) + *lit,* bed (< Lat. *lectus).*] A railroad sleeping car.

**wag·on·load** (wăg'ən-lōd') *n.* The load held by a wagon.

**wagon train** *n.* A line or train of wagons traveling cross-country.

**wag·tail** (wăg'tāl') *n.* A bird of the genus *Motacilla* or related genera, with a long, constantly wagging tail.

**wagtail**
*6½ inches long*

**Wah·ha·bi** *or* **Wa·ha·bi** (wä-hä'bē) *n.* A member of a Moslem sect founded by Abdul Wahhab in the 18th cent., flourishing mainly in Arabia and known for its strict observance of the Koran. **—Wah·ha'bism** (-bĭz'əm) *n.*

**wa·hi·ne** (wä-hē'nē, -nā') *n.* [Hawaiian and Maori.] **1.** A Polynesian woman. **2.** A woman surfer.

**wa·hoo[1]** (wä-hōō', wä'hōō) *n., pl.* **-hoos.** [Dakota *wāhu.*] A shrub or small tree, *Euonymus atropurpureus* of eastern North America, with small purplish flowers and red fruit.

**wa·hoo[2]** (wä-hōō', wä'hōō) *n., pl.* **-hoos.** [Creek *ûhawhu.*] **1.** An elm, *Ulmus alata* of the southeastern United States, bearing twigs with winged, corky edges. **2.** A tree similar to the wahoo.

**wa·hoo[3]** (wä-hōō', wä'hōō) *n., pl.* **wahoo** *or* **-hoos.** [Orig. unknown.] A tropical marine game fish, *Acanthocybium solanderi.*

**†wa·hoo[4]** (wä'hōō') *interj. Chiefly Western U.S.* —Used to express exuberance.

**waif[1]** (wāf) *n.* [ME *waife,* ownerless property < ONFr. *waife,* of Scand. orig.] **1. a.** A homeless person, esp. an orphaned or forsaken child. **b.** An abandoned young animal : STRAY. **2.** Something found and unclaimed, as an object cast up by the sea.

**waif[2]** (wāf) *n.* [Prob. of Scand. orig.] A small signal flag : WAFT.

**wail** (wāl) *v.* **wailed, wail·ing, wails.** [ME *wailen,* of Scand. orig.] —*vi.* **1.** To protest or grieve audibly. **2.** To make a high-pitched, prolonged sound suggestive of a cry <The siren *wailed* through the night.> —*vt. Archaic.* To lament over : BEWAIL. —*n.* **1.** A long, loud, high-pitched cry, as of pain or grief. **2.** A long, loud, high-pitched sound. **—wail'er** *n.* **—wail'ing·ly** *adv.*

**wail·ful** (wāl′fəl) *adj.* **1.** Resembling a wail. **2.** Issuing a sound like a wail. **—wail′ful·ly** *adv.*

**Wailing Wall** *n.* **1.** A wall in the old city of Jerusalem held to be a remnant of the temple of Solomon and revered by Jews as a place of pilgrimage, lamentation, and prayer. **2. wailing wall.** A source of consolation and comfort in times of misfortune.

**wain** (wān) *n.* [ME < OE *wægn.*] **1.** A large open farm wagon. **2. Wain.** The Big Dipper.

**wain·scot** (wān′skət, -skŏt′, -skōt′) *n.* [ME *waynscot* < MDu. *wagenschot.*] **1.** A usu. wood facing or paneling applied to the walls of a room. **2.** The lower part of an interior wall when finished in a material different from that of the upper part. *—vt.* **-scot·ed, -scot·ing, -scots** or **-scot·ted, -scot·ting, -scots.** To line or panel (a room or wall) with wainscot.

**wain·scot·ing** *also* **wain·scot·ting** (wān′skə-tĭng, -skŏt′ĭng, -skō′tĭng) *n.* **1.** A wainscoted wall. **2.** Material for wainscoting.

**wain·wright** (wān′rīt′) *n.* One that builds and repairs wagons.

**waist** (wāst) *n.* [ME *wast.*] **1.** The part of the human trunk between the bottom of the rib cage and the pelvis. **2. a.** The part of a garment encircling the waist. **b.** The upper part of a garment extending from the shoulders to the waistline, esp. the bodice of a woman's dress. **c.** A blouse. **d.** A child's undershirt. **3.** The middle part of an object, esp. when narrower than the rest. **4.** *Naut.* The middle part of the deck of a ship between the forecastle and the quarter-deck.

**waist·band** (wāst′bănd′) *n.* **1.** A garment band encircling and fitting the waist, as on trousers or a skirt. **2.** A sash.

**waist·cloth** (wāst′klôth′, -klŏth′) *n.* A loincloth.

**waist·coat** (wĕs′kĭt, wāst′kōt′) *n.* **1.** *Chiefly Brit.* VEST 1. **2.** A garment once worn by men under a doublet. **—waist′coat′ed** *adj.*

**waist·line** (wāst′līn′) *n.* **1. a.** The natural indentation of the body at the waist. **b.** Measurement of this circumference. **2.** The point or line at which the skirt and bodice of a dress join.

**wait** (wāt) *v.* **wait·ed, wait·ing, waits.** [ME *waiten* < ONFr. *waitier,* to watch, of Germanic orig.] *—vi.* **1. a.** To postpone action or stay in one spot until something anticipated occurs. **b.** To linger until another catches up. **2.** To remain or be in readiness or expectation. **3.** To remain temporarily neglected or postponed <Our vacation has to *wait.*> **4.** To work as a waiter or waitress. *—vt.* **1.** To remain or stay in expectation of : AWAIT <Please *wait* your turn.> **2.** *Informal.* To delay (a meal or event) : POSTPONE <We *waited* dinner.> **3.** To be a waiter or waitress at <*wait* table> **—wait on** (or **upon**). **1.** To be in attendance upon. **2.** To make a formal call upon : VISIT. **3.** To follow as a result. ***usage:*** Although *wait on* occurs in some dialects as the equivalent of *wait for,* this choice of the preposition is not yet a part of standard English. **—wait out.** To defer action until the termination of <*wait out* a storm> **—wait up. 1.** To postpone going to bed in anticipation of something or someone. **2.** *Informal.* To stop or pause so that another can catch up. *—n.* **1.** The act of waiting or the time spent waiting. **2.** *Chiefly Brit.* **a.** One of a group of musicians employed, usu. by a city, to play in parades or public ceremonies. **b.** One of a group of musicians or carolers who perform in the streets at Christmastime. **—lie in wait.** To stay in hiding, awaiting a chance to ambush.

**wait-a-bit** (wāt′ə-bĭt′) *n.* [Transl. of Afr. *wacht-en-bitje.*] Any of several plants with sharp, often hooked thorns.

**wait·er** (wā′tər) *n.* **1.** One who waits on table, as in a restaurant. **2.** A salver or tray.

**wait·ing** (wā′tĭng) *n.* **1.** The act of one that waits. **2.** The period of time spent waiting. **—in waiting.** In attendance.

**waiting game** *n.* The stratagem of deferring action and allowing the passage of time to work in one's favor.

**waiting list** *n.* A list of persons waiting, as for an appointment.

**waiting room** *n.* A room, as in a doctor's office or railroad station, for the use of persons waiting.

**wait·ress** (wā′trĭs) *n.* A woman who waits on table.

**waive** (wāv) *vt.* **waived, waiv·ing, waives.** [ME *weiven,* to abandon < ONFr. *weyver* < *waife,* ownerless property. —see WAIF.] **1.** To give up or relinquish (a right or claim) voluntarily. **2.** To set aside : dispense with <"The original ban on private trading had long since been *waived*" —William L. Schurz> **3.** To put off or postpone for the time.

**waiv·er** (wā′vər) *n.* [AN *weyver* < ONFr. *weyver,* to waive.] **1.** Intentional relinquishment of a right, claim, or privilege. **2.** The document that evidences a waiver.

**Wa·kash·an** (wô′kə-shən, wä-kăsh′ən) *n.* [< Wakashan *waukash,* good.] A family of North American Indian languages spoken by the Nootka and other tribes of Washington and British Columbia. **—Wa′kash·an** *adj.*

**wake¹** (wāk) *v.* **woke** (wōk) *or* **waked** (wākt), **waked** *or* **wok·en** (wō′kən), **wak·ing, wakes.** [ME *waken* < OE *wacian.*] *—vi.* **1. a.** To cease to sleep <*woke* up early> **b.** To be brought into a state of awareness or alertness. **2.** To keep watch or guard, esp. over a corpse. **3.** To be or remain awake. *—vt.* **1.** To rouse from sleep : AWAKEN. **2.** To stir, as from inactivity or dormancy : ROUSE <*wake* old memories> **3.** To make aware of : ALERT <The report *waked* us to the danger.> **4. a.** To keep a vigil over. **b.** To hold a wake over. *—n.* **1. a.** A watch : vigil. **b.** A watch over a corpse before burial, sometimes accompanied by festivity. **2. wakes** *(sing. or pl. in number). Chiefly Brit.* A parish festival held annually, often in honor of a patron saint. **3. wakes** *(sing. or pl. in number). Chiefly Brit.* An annual vacation.

**wake²** (wāk) *n.* [Poss. < MLG < ON *vǫk,* hole in the ice.] **1.** The visible track of turbulence left by something moving through water <the *wake* of an ocean liner> **2.** The track or course left behind something that has passed <The fire left great ruin in its *wake.*> **—in the wake of. 1.** Following directly upon. **2.** In the aftermath of <an invasion *in the wake of* the prime minister's assassination>

**wake·ful** (wāk′fəl) *adj.* **1. a.** Not sleeping or able to sleep. **b.** Without sleep : SLEEPLESS <spent an anxious, *wakeful* night> **2.** Watchful : alert. **—wake′ful·ly** *adv.* **—wake′ful·ness** *n.*

**wake·less** (wāk′lĭs) *adj.* Unbroken <a *wakeless* sleep>

**wak·en** (wā′kən) *v.* **-ened, -en·ing, -ens.** [ME *wakenen* < OE *wæcnan.*] *—vt.* **1.** To rouse from sleep : AWAKE. **2.** To rouse from a quiescent or inactive state : STIR. *—vi.* To wake up. **—wak′en·er** *n.*

**wake-rob·in** (wāk′rŏb′ĭn) *n.* **1.** The trillium. **2.** Any of several plants that bloom early in the spring.

**Wal·den·ses** (wŏl-dĕn′sēz) *pl.n.* [Med. Lat. < Peter *Waldo,* their leader.] A dissenting Christian sect that originated in southern France in the late 12th cent. and adopted Calvinist doctrines in the 16th cent. **—Wal·den′sian** (-chən) *adj.* & *n.*

**Wal·dorf salad** (wôl′dôrf′) *n.* [After the *Waldorf*-Astoria Hotel, New York City, where it was first served.] A salad of diced raw apples, celery, and walnuts mixed with mayonnaise.

**wale** (wāl) *n.* [ME < OE *walu.*] **1.** WELT 3a. **2. a.** One of the parallel ridges or ribs in the surface of some fabrics, as corduroy. **b.** The texture or weave of such a fabric <a narrow *wale*> **3.** *Naut.* **a.** The gunwale. **b.** One of the heavy planks or strakes extending along the sides of a wooden ship. *—vt.* **waled, wal·ing, wales.** To mark (the skin) with wales.

**Wal·hal·la** *n. var. of* VALHALLA.

**walk** (wôk) *v.* **walked, walk·ing, walks.** [ME *walken* < OE *wealcan,* to roll.] *—vi.* **1.** To move over a surface by taking steps at a pace slower than a run. **2.** To go or travel on foot. **3.** To go on foot for exercise or pleasure : STROLL. **4.** To move in a way suggestive of walking. **5.** To conduct oneself or behave in a given manner : LIVE <*walks* in peace> **6.** To roam about in a visible form, as a ghost or specter : APPEAR. **7. a.** *Baseball.* To go to first base after the pitcher has thrown four balls. **b.** *Basketball.* TRAVEL 7. **8.** *Obs.* To be in constant motion. *—vt.* **1.** To go or pass over, on, or through by walking <*walk* the garden paths> **2.** To bring to a given condition by walking <*walked* us to total weariness> **3.** To cause to walk or proceed at a walk <*walk* a bike uphill> **4.** To accompany or escort in walking <*walked* me to the corner> **5.** To traverse on foot in order to survey or measure, as one's property. **6.** To move (a heavy or cumbersome object) in a manner suggestive of walking. **7.** *Baseball.* To allow (a batter) to go to first base by pitching four balls. **—walk away from. 1.** To outdo, outrun, or defeat easily. **2.** To survive (an accident) with minimal injury. **—walk off with. 1.** To win easily or unexpectedly. **2.** To steal. **—walk out. 1.** To go on strike. **2.** To leave suddenly, often as a signal of disapproval. **—walk out on.** *Informal.* To abandon : desert. **—walk over. 1.** *Informal.* To treat badly or contemptuously. **2.** To gain an easy or uncontested victory. **—walk through.** To perform (e.g., a play) in a perfunctory fashion, as at a first rehearsal. *—n.* **1. a.** The act or an instance of walking, esp. a stroll for exercise or pleasure. **b.** The gait of a human being or other biped in which the feet are lifted alternately with one part of a foot always on the ground. **c.** The gait of a quadruped in which at least two feet are always touching the ground, esp. the gait of a horse in which the feet touch the ground in the four-beat sequence of near hind foot, near forefoot, off hind foot, off forefoot. **d.** An astronaut's self-controlled movement in space. **2. a.** The rate at which one walks : PACE. **b.** The manner in which one walks. **3.** Distance covered or to be covered in walking. **4.** A place, as a sidewalk, on which one may walk. **5. a.** *Baseball.* The act or instance of taking first base after four balls have been pitched to the batter. **b.** *Basketball.* The act or an instance of traveling with the ball. **c.** A track event in which contestants compete in walking a specified distance. **6.** An enclosed area set aside for the exercise or pasture of livestock. **7.** An arrangement of or space between trees or shrubs planted in widely spaced rows. **—walk of life.** Occupation or social class. **—walk the plank.** To be executed at sea by walking the length of a plank and falling into the water.

**walk·a·bout** (wôk′ə-bout′) *n.* **1.** A brief retreat to the roaming life of the Australian bush occas. taken by an aborigine as a respite from regular work. **2.** A walking trip.

**walk·a·thon** (wôk′ə-thŏn′) *n.* A usu. long walk organized esp. to raise money for charity.

**walk·a·way** (wôk′ə-wā′) *n.* A contest or victory easily won.

**walk·er** (wô′kər) *n.* **1.** One that walks, esp. a contestant in a footrace. **2.** A frame device used to support a young child learning to walk or a handicapped or convalescent person learning to walk again. **3.** A shoe specially designed for walking comfortably.

**walk·ie-talk·ie** *also* **walk·y-talk·y** (wô′kē-tô′kē) *n., pl.* **-ies.** A battery-powered portable sending and receiving radio set.

**walk-in** (wôk′ĭn′) *adj.* **1.** Large enough to admit entrance <a *walk-in hall* closet> **2.** Located so as to be entered directly from the street <a *walk-in* studio apartment> —*n.* **1.** A room large enough to admit entrance. **2.** An easily won victory, esp. in an election. **3.** One who walks in without having an appointment.

**walking bass** *n. Mus.* A repetitive bass figure of nonsyncopated eighth notes, used in jazz.

**walking catfish** *n.* A catfish, *Clarius batrachus,* able to travel short distances on land between bodies of water.

**walking delegate** *n.* A trade-union official who inspects and confers with the local unions or represents the union in dealings with an employer.

**walking fern** *n.* A North American fern, *Camptosorus rhizophyllus,* with leaflike fronds bearing slender tips that often take root.

**walking leaf** *n.* **1.** A walking fern. **2.** A leaf insect.

**walking papers** *pl.n. Informal.* Notice of discharge or dismissal.

**walking stick** *n.* **1.** A staff or cane used as an aid in walking. **2.** Any of various insects of the family Phasmidae that resemble twigs.

**walk-on** (wôk′ŏn′, -ôn′) *n.* **1.** A minor role in a theatrical production, usu. without speaking lines. **2.** A performer of a walk-on role.

**walk·out** (wôk′out′) *n.* **1.** A labor strike. **2.** The act of leaving or quitting a meeting, company, or organization, esp. as an indication of protest.

**walk·o·ver** (wôk′ō′vər) *n.* **1.** A horse race with only one horse entered, won by the mere formality of walking the length of the track. **2.** A walkaway.

**walk-through** (wôk′thrōō′) *n.* **1.** A brief rehearsal, as of a role or play, performed usu. in an early stage of production. **2.** A television rehearsal during which no cameras are used.

**walk·up** *also* **walk-up** (wôk′ŭp′) *n.* **1.** An apartment house or office building with no elevator. **2.** An apartment or office in a walkup.

**walk·way** (wôk′wā′) *n.* A passage for walking.

**Wal·kyr·ie** (văl-kîr′ē, -kī′rē, văl′kîr′ē, -kī′rē) *n. var. of* VALKYRIE.

**walk·y-talk·y** (wô′ke-tô′kē) *n. var. of* WALKIE-TALKIE.

**wall** (wôl) *n.* [ME < OE *weall* < Lat. *vallum,* palisade < *vallus,* stake.] **1.** An upright structure of building material, as masonry, wood, or plaster, serving to enclose, divide, or protect an area, esp. a vertical construction forming an inner partition or exterior siding of a building. **2.** *often* **walls.** A continuous structure, as of masonry, forming a rampart and built as a defense. **3.** A structure, as of stonework or cement, built to retain a flow of water. **4.** Something resembling a wall <the chest and abdominal *walls*> **5.** The vertical surface of an ocean wave in surfing. **6.** Something virtually impenetrable <a *wall* of mystery> **7.** An extreme or desperate condition or position, as defeat or ruin. —Usu. used in the phrase *to the wall* <were pushed to the *wall* during negotiations><had our backs to the *wall* during the power struggle> —*vt.* **walled, wall·ing, walls.** **1.** To enclose, surround, or fortify with or as if with a wall. **2.** To divide or separate with or as if with a wall. **3.** To enclose within a wall : IMMURE. **4.** To block or close (e.g., an opening or passage) with or as if with a wall. —**off the wall.** *Slang.* Eccentric : crazy. —**up the wall.** *Informal.* Into a state of extreme frustration or distress <Their constant bickering drove me *up the wall.*>

**wal·la·by** (wŏl′ə-bē) *n., pl.* **-bies.** [Native word in Australia.] A marsupial of the genus *Wallabia* or related genera of Australia and adjacent islands, resembling the kangaroos but gen. smaller.

**wal·lah** *also* **wal·la** (wä′lä) *n.* [< Hindi *-wālā,* pertaining to, connected with.] One employed in an occupation or activity <a shipping room *wallah*>

**wal·la·roo** (wŏl′ə-rōō′) *n., pl.* **-roos.** [Native word in Australia.] A kangaroo, *Macropus robustus* or *Osphranter robustus,* of hilly regions of Australia.

**wallaroo**
*Approximately 5 feet high, tail to 4 feet*

**wall·board** (wôl′bôrd′, -bōrd′) *n.* A structural board or sheet of various materials, as gypsum plaster encased in paper or compressed wood fibers and chips, used in construction as a substitute for plaster or wood panels.

**wall creeper** *n.* A long-billed crimson and grayish bird, *Tichodroma muraria* of Old World alpine regions, that seeks food on rocky cliffs or walls.

**wal·let** (wŏl′ĭt) *n.* [ME *walet,* knapsack.] A flat pocket-size folding case, often of leather, for holding paper money, cards, or photographs : BILLFOLD.

**wall·eye** (wôl′ī′) *n.* [Back-formation from WALLEYED.] **1.** An eye in which the cornea is white or opaque. **2.** *Pathol.* **a.** Leukoma of the cornea. **b.** A divergent strabismus. **3.** A North American freshwater food and game fish, *Stizostedium vitreum,* with conspicuous eyes.

**wall·eyed** (wôl′īd′) *adj.* [ME *wawileyed* < ON *vagleygr* : *vagl,* beam + *auga,* eye.] **1.** Having a walleye or walleyes. **2. a.** Having leukoma of the cornea. **b.** Having divergent strabismus. **3. a.** Having large bulging or staring eyes. **b.** *Slang.* Having eyes with greatly distended pupils. **4.** *Slang.* Drunk.

**walleyed pike** *n.* WALLEYE 3.

**wall fern** *n.* A small low-growing fern of the genus *Polypodium,* with creeping stems that form dense mats.

**wall·flow·er** (wôl′flou′ər) *n.* **1. a.** A widely cultivated European plant, *Cheiranthus cheiri,* with fragrant yellow, orange, or brownish flowers. **b.** A similar plant, *Erysimum asperum* of the western United States. **2.** One who does not participate in the activity at a social event esp. because of shyness or unpopularity.

**wall hanging** *n.* A decorative tapestry hung against a wall.

**Wal·loon** (wŏ-lōōn′) *n.* [OFr. *Wallon* < Med. Lat. *Wallo,* of Germanic orig.] **1.** One of a French-speaking people of Celtic descent inhabiting southern and southeastern Belgium and adjacent regions of France. **2.** The dialect of French spoken by the Walloons. —*adj.* Of or relating to the Walloons or their language.

**wal·lop** (wŏl′əp) [ME *walopen,* to gallop < ONFr. *waloper.*] *Informal.* —*v.* **-loped, -lop·ing, -lops.** —*vt.* **1.** To beat soundly : THRASH. **2.** To strike with a hard blow. **3.** To defeat thoroughly. —*vi.* **1.** To move in a rolling, clumsy manner : WADDLE. **2.** To boil noisily. —Used of a liquid. —*n.* **1.** A hard or powerful blow. **2. a.** The ability to strike a wallop. **b.** The capacity to create a forceful effect : IMPACT. —**wal′lop·er** *n.*

**wal·lop·ing** (wŏl′ə-pĭng) *Informal.* —*adj.* **1.** Very large : HUGE <a *walloping* catch of fish> **2.** Very great : REMARKABLE <a *walloping* mistake> —*adv.* To an exaggerated degree <a *walloping* big triumph> —*n.* A sound thrashing or defeat.

**wal·low** (wŏl′ō) *vi.* **-lowed, -low·ing, -lows.** [ME *walowen* < OE *wealwian.*] **1.** To roll the body about clumsily or indolently in water, snow, or mud. **2.** To indulge oneself excessively : REVEL <*wallowing* in self-satisfaction> **3.** To be abundantly supplied with something <*wallowing* in wealth> **4.** To move with difficulty in a clumsy or rolling manner : FLOUNDER. **5.** To swell or surge forth : BILLOW. —*n.* **1.** An act of wallowing. **2. a.** A pool of water or mud where animals go to wallow. **b.** The depression, pool, or pit produced by wallowing animals. **3.** Baseness or degradation. —**wal′low·er** *n.*

**wall·pa·per** (wôl′pā′pər) *n.* Paper printed with designs or colors, used as a decorative wall covering. —*v.* **-pered, -per·ing, -pers.** —*vt.* To cover (e.g., the walls of a room) with wallpaper. —*vi.* To decorate a wall or room with wallpaper.

**wall plate** *n.* **1.** A horizontal timber located along the top of a wall at the level of the eaves and bearing the ends of joists or rafters. **2.** A plate used to attach a device, as a bracket or lamp, to a wall.

**wall plug** *n.* An electric socket, usu. located in a wall, that is connected to and used as a source of electric power.

**wall rock** *n.* The rock that forms the walls of a vein or lode.

**wall rue** *n.* A small delicate fern, *Asplenium ruta-muraria,* growing on rocks or in rocky crevices.

**Wall Street** *n.* [After *Wall Street,* New York City, the main street of the financial district.] The controlling financial interests of the United States. —**Wall′ Street′er** *n.*

**wall system** *n.* A set of modular shelves often with cabinets that can be arranged in various ways along a wall or used as a room divider.

**wall-to-wall** (wôl′tə-wôl′) *adj.* **1.** Covering a floor completely, as carpeting. **2. a.** Present or spreading throughout an entire area <*wall-to-wall* crowds at the resort> **b.** Found everywhere or including everything <*wall-to-wall* extravagance> —*n.* A wall-to-wall carpet.

**wal·nut** (wôl′nŭt′, -nət) *n.* [ME *walnot* < OE *wealhhnutu* : *wealh,* Celt, foreigner + *hnutu,* nut.] **1. a.** A tree of the genus *Juglans,* yielding round, sticky fruit enclosing an edible nut. **b.** The ridged or corrugated nut of the walnut tree. **2.** The hard, dark brown wood of the walnut, used esp. for furniture.

**Wal·pur·gis Night** (väl-pŏŏr′gĭs) *n.* [Partial transl. of G. *Walpurgisnacht* : *Walpurgis,* St. Walpurga + *nacht,* night.] **1.** The eve of May Day, believed in medieval Europe to be the occasion of a witches' Sabbath. **2.** An episode or situation having the quality of nightmarish wildness associated with Walpurgis Night.

**wal·rus** (wôl′rəs, wŏl′-) *n., pl.* **walrus** *or* **-rus·es.** [Du., of Scand orig.] A large marine mammal, *Odobenus rosmarus* of Arctic regions, with tough wrinkled skin and large tusks.

**walrus mustache** *n.* A bushy, drooping mustache.

**waltz** (wôlts) *n.* [G. *Walzer* < MHG *walzen,* to dance < OHG *walzan,* to roll.] **1.** A dance in triple time with a strong accent on the first beat. **2.** The music for a waltz. *—v.* **waltzed, waltz·ing, waltz·es.** *—vi.* **1.** To dance the waltz. **2.** To move lightly and easily : FLOUNCE. **3.** To accomplish a task, chore, or assignment with little effort. *—vt.* **1.** To dance the waltz with. **2.** To lead or force to move briskly and purposefully : MARCH <*waltzed* the culprit into the supervisor's office> **—waltz'er** *n.*

**wam·ble** (wŏm'bəl, wăm'-) *vi.* **-bled, -bling, -bles.** [ME *wamelen,* to feel nausea.] **1.** To move in a rolling, weaving, or wobbling way. **2.** To turn or roll. —Used of the stomach. *—n.* **1.** A wobble or roll. **2.** A stomach upset. **—wam'bling·ly** *adv.* **—wam'bly** *adj.*

**Wam·pa·no·ag** (wăm'pə-nō'ăg') *n., pl.* **Wampanoag** *or* **-ags.** [Natick *Wampan-okhe,* (people of the) eastern land.] **1. a.** A tribe of Indians once inhabiting eastern Rhode Island and adjacent parts of Massachusetts. **b.** A member of this tribe. **2.** The Algonquian language of the Wampanoag. **—Wam'pa·no'ag** *adj.*

**wam·pum** (wŏm'pəm, wôm'-) *n.* [Short for WAMPUMPEAG.] **1.** Small cylindrical beads made from polished shells, once used by North American Indians as currency and jewelry. **2.** *Informal.* Money.

**wam·pum·peag** (wŏm'pəm-pēg', wôm'-) *n.* [Algonquian *wampumpeage,* white strings.] White shell beads used by North American Indians as wampum.

**wan** (wŏn) *adj.* **wan·ner, wan·nest.** [ME, pale, gloomy < OE *wann,* gloomy, dark.] **1.** Unnaturally or unusually pale. **2.** Indicating or suggestive of illness, weariness, or unhappiness : MELANCHOLY <a *wan* little smile> *—vi.* **wanned, wan·ning, wans.** To become pale. **—wan'ly** *adv.* **—wan'ness** *n.*

**wand** (wŏnd) *n.* [ME < ON *vǒndr.*] **1.** A thin supple stick or twig. **2.** A slender rod carried as a symbol of office in a procession : SCEPTER. **3.** A musician's baton. **4.** A stick, baton, or rod used esp. by a magician. **5.** A six-foot by two-foot slat used as an archery target.

**wan·der** (wŏn'dər) *v.* **-dered, -der·ing, -ders.** [ME *wanderen* < OE *wandrian.*] *—vi.* **1.** To move about aimlessly : ROAM. **2.** To go by an indirect route or at no set pace : AMBLE <*wander* toward home> **3.** To proceed in an irregular course or action : MEANDER <a *wandering* stream> **4.** To go astray morally. **5.** To think or express oneself unclearly or incoherently. *—vt.* To wander across or through <*wander* the countryside> *—n.* An act of wandering : STROLL. **—wan'derer** *n.* **—wan'der·ing·ly** *adv.*

**Wandering Jew** *n.* **1.** A legendary Jew condemned to wander until the Day of Judgment for having mocked Christ on the day of Crucifixion. **2. wandering Jew.** A trailing plant native to tropical America, *Tradescantia fluminensis* or *Zebrina pendula,* usu. with variegated foliage and popular as a house plant.

**wan·der·lust** (wŏn'dər-lŭst') *n.* [G. : *wandern,* to wander + *Lust,* desire.] A strong impulse to travel.

**wan·der·oo** (wŏn'də-rōō') *n., pl.* **-oos.** [Singhalese *vandaru,* pl. of *vandurā,* monkey < Skt. *vānaraḥ,* forest dweller < *vanam,* forest.] A monkey, *Macaca silenus* of south-central Asia, with a ruff of gray hair about the face and a glossy black coat.

**wane** (wān) *vi.* **waned, wan·ing, wanes.** [ME *wanen* < OE *wanian.*] **1.** To decrease gradually in size, amount, intensity, or degree : DECLINE. **2.** To show decreasing illuminated area from full moon to new moon. **3.** To near an end. *—n.* **1.** A gradual decrease or decline. **2.** A period or phase of waning, esp. the period of the decrease of the moon's illuminated visible surface. **3.** A defective edge of a board caused by remaining bark or a beveled end. **—on the wane.** In a period of decline : WANING.

**wan·gle** (wăng'gəl) [Orig. unknown.] *Informal. —v.* **-gled, -gling, -gles.** *—vt.* **1.** To make, achieve, or get by contrivance. **2.** To manipulate or juggle, esp. fraudulently. **3.** To extricate (oneself) from difficulty. *—vi.* **1.** To use indirect, tricky, or fraudulent methods. **2.** To extricate oneself by subtle or indirect means, as from difficulty. **—wan'gle** *n.* **—wang'ler** *n.*

☆ ***syns:*** WANGLE, ENGINEER, FINAGLE, FINESSE, WORM *v. core meaning* : to make, achieve, or get through contrivance or guile <*wangled* a better assignment from management>

**wan·i·gan** *also* **wan·ni·gan** (wŏn'ə-gən) *n.* [Ojibwa *wanikkan,* man-made hole.] **1.** A supply chest used in a logging camp. **2.** A shack on wheels or a movable platform, used in a logging camp as a shelter.

**Wan·kel engine** (väng'kəl, wăng'-) *n.* [After Felix *Wankel* (b. 1902).] A rotary internal-combustion engine in which a triangular rotor turning in a specially shaped housing performs the functions allotted to the pistons of a conventional engine, thereby allowing great savings in weight and moving parts.

**wan·ni·gan** (wŏn'ə-gən) *n. var. of* WANIGAN.

**want** (wŏnt, wônt) *v.* **want·ed, want·ing, wants.** [ME *wanten,* to be lacking < ON *vanta.*] *—vt.* **1.** To desire greatly : wish for <*wants* to go to college> **2.** To fail to have : LACK <a manager who *wants* tact> **3.** To need or require <That fence *wants* repair.> **4. a.** To request the presence of. **b.** To seek with intent to capture <The kidnaper is *wanted* by the FBI.> **5. a.** To have a desire for. **b.** To have an inclination toward : LIKE <Say what you *want,* but they are still unqualified for the job.> *—vi.* **1.** To have need. **2.** To be destitute or needy. **3.** To be disposed : WISH <Write to me if you *want.*> **—want in.** *Informal.* To wish to join a project, business, or other enterprise. **—want out.** *Informal.* To wish to leave a project, business, or other enterprise. *—n.* **1.** The quality or condition of lacking a usual or necessary amount. **2.** Pressing need : DESTITUTION <has been living in *want*> **3.** Something needed or desired : NEED <modest *wants*> **4.** A character defect : FAULT.

**want ad** *n. Informal.* A classified advertisement.

**want·ing** (wŏn'tĭng, wôn'-) *adj.* **1.** Lacking : absent. **2.** Not up to standards or expectations. *—prep.* **1.** Without <a car *wanting* fenders> **2.** Minus : less <an hour *wanting* 20 minutes>

**wan·ton** (wŏn'tən) *adj.* [ME *wantowen* : *wan-,* mis- + *towen,* p.part. of *teen,* to bring up < OE *tēon.*] **1.** Immoral or promiscuous : LEWD. **2. a.** Cruel and merciless : INHUMANE. **b.** Marked by malicious cruelty : UNJUST. **3.** Extravagant : excessive <*wanton* spending> **4.** Luxuriant : overabundant <*wanton* tresses> **5.** Playful : unrestrained. **6.** *Obs.* Rebellious. *—v.* **-toned, -ton·ing, -tons.** *—vi.* To act, grow, or move in a wanton way. *—vt.* To waste or squander wantonly. *—n.* **1.** An immoral, lewd, or licentious person, esp. a woman. **2.** One that is playful. **3.** One that is undisciplined or spoiled. **—wan'ton·ly** *adv.* **—wan'ton·ness** *n.*

**wap·en·take** (wŏp'ən-tāk', wăp'-) *n.* [ME < OE *wæpengetæc* < ON *vāpnatak,* act of taking weapons: *vāpn,* weapons + *tak,* act of taking < *taka,* to take.] A historical subdivision of some northern counties in England, corresponding roughly to the hundred.

**wap·i·ti** (wŏp'ĭ-tē) *n., pl.* **wapiti** *or* **-tis.** [Shawnee, white rump.] A large North American deer, *Cervus canadensis.*

**war** (wôr) *n.* [ME *warre* < ONFr. *werre,* of Germanic orig.] **1. a.** A state of open, armed, often prolonged conflict carried on between nations, states, or parties. **b.** The period of such conflict. **2.** A condition of active contention or antagonism <a *war* of ideologies> **3.** The techniques or procedures of war : MILITARY SCIENCE. *—vi.* **warred, war·ring, wars.** **1.** To wage war. **2.** To be in a state of hostility : CONTEND. **—at war.** In an active state of conflict or contention. **—declare war on.** **1.** To state formally the intention to carry on hostilities against. **2.** To state one's intent to suppress or eradicate <*declared war on* poverty>

**war baby** *n.* A child born in wartime.

**war·ble**[1] (wôr'bəl) *v.* **-bled, -bling, -bles.** [ONFr. *werbler* < *werble,* a warbling, of Germanic orig.] *—vt.* To sing (e.g., a note or song) with melodic embellishments, as trills or runs. *—vi.* **1.** To sing with trills, runs, or quavers. **2.** To be sounded in a trilling or quavering way. **3.** To sing. *—n.* **1.** An act of warbling. **2. a.** A warbling sound. **b.** A song, esp. one that is warbled.

**war·ble**[2] (wôr'bəl) *n.* [Prob. of Scand. orig.] **1.** An abscessed swelling under the hide of the back of cattle or other animals, caused by the larva of a warble fly. **2.** The warble fly, esp. in its larval stage. **—war'bled** *adj.*

**warble fly** *n.* A fly of the family Oestridae, whose larvae form warbles within the bodies of cattle and other animals.

**war·bler** (wôr'blər) *n.* **1.** Any of various small New World birds of the family Parulidae, many of which have yellowish plumage or markings. **2.** Any of various small brownish or grayish Old World birds of the subfamily Silviinae.

**war bonnet** *n.* A ceremonial headdress used by some North American Plains Indians, consisting of a cap or band and a trailing extension decorated with erect feathers.

**war bride** *n.* A woman who marries a serviceman during wartime.

**war chest** *n.* **1.** An accumulation of funds to finance a war effort. **2.** A fund reserved for a certain purpose, as a political campaign.

**war club** *n.* A weapon consisting of a weight of stone or iron fixed to a handle, used esp. by American Indians.

**war correspondent** *n.* A journalist, reporter, or commentator assigned to report directly from a combat zone.

**war crime** *n.* A crime, as mistreatment of prisoners of war or genocide, committed during a war and considered to be in violation of the customs of warfare. **—war criminal** *n.*

**war cry** *n.* **1.** A cry uttered by combatants as they attack. **2.** A phrase or slogan used to rally people to a cause.

**ward** (wôrd) *n.* [ME, action of guarding < OE *weard.*] **1.** A division of a town or city for administrative and representative purposes. **2.** A district of some English and Scottish counties corresponding roughly to the hundred or wapentake. **3. a.** A room in a hospital usu. holding six or more patients. **b.** A division in a hospital for the care of a particular group of patients <psychiatric *ward*> **4.** A division of a penal institution, as a jail. **5.** An open court or area of a castle or fortification enclosed by walls. **6. a.** *Law.* A minor or incompetent person placed under the care or protection of a guardian or court. **b.** One under the care or protection of another. **7.** The state of being under guard : CUSTODY. **8.** The act of guarding or protecting someone : GUARDIANSHIP. **9.** A means of protection : DEFENSE. **10.** A defensive movement or attitude, esp. in fencing : GUARD. **11. a.** The projecting ridge of a lock or keyhole that prevents the turning of a key other than the proper one. **b.** The notch cut into a key that corresponds to such a ridge. *—vt.* **ward·ed, ward·ing, wards.** To protect : guard.

**—ward off.** **1.** To turn aside : PARRY <*ward off* an attacker's blows> **2.** To avert <said a prayer to *ward off* calamity>

**-ward** *or* **-wards** *suff.* [ME < OE *-weard.*] **1. a.** In a given direction in time or space <down*ward*> **b.** Toward a given place or position <sky*wards*> **2. a.** Occurring or located in a given direction <left*ward*> **b.** Having a direction toward a given place or position <land*ward*> ***usage:*** The suffixes *-ward* and *-wards* both indicate direction; therefore, it is unnecessary to use the preposition *to* with adverbs incorporating these suffixes. It is correct to write *The ship is sailing westward* (not *to the westward*).

**war dance** *n.* A tribal dance performed before a battle or as a celebration after a victory.

**ward·ed** (wôr′dĭd) *adj.* Having notches or wards. —Used of keys.

**war·den** (wôrd′n) *n.* [ME *wardein* < ONFr. < *warder,* to guard, of Germanic orig.] **1.** The chief administrative official of a prison. **2.** An official, as an air-raid warden, charged with enforcement of certain rules and regulations. **3.** *Chiefly Brit.* **a.** The chief executive official in charge of a market or port. **b.** Any of various crown officers having administrative duties. **4.** The chief executive of a borough in some states. **5.** *Chiefly Brit.* A governing official of some colleges, schools, guilds, or hospitals : TRUSTEE. **6.** A churchwarden.

**war·den·ry** (wôrd′n-rē) *n., pl.* **-ries.** The office, duties, or jurisdiction of a warden.

**ward·er¹** (wôr′dər) *n.* [ME < AN *wardere* < ONFr. *warder,* to guard, of Germanic orig.] **1.** A guard, porter, or watchman of a gate or tower. **2.** *Chiefly Brit.* A prison guard. **—war′der·ship′** *n.*

**ward·er²** (wôr′dər) *n.* [ME, poss. < *warden,* to ward < OE *weardian.*] A baton once used by a commander or ruler to signal orders.

**ward heel·er** (hē′lər) *n. Slang.* A worker for the ward organization of a political machine.

**ward·ress** (wôr′drĭs) *n.* A prison matron.

**ward·robe** (wôr′drōb′) *n.* [ME *warderobe* < ONFr. : *warder,* to keep + *robe,* garment.] **1.** A tall cabinet, closet, or small room designed to hold clothes. **2.** Garments as a whole, esp. all the articles of clothing belonging to one person. **3. a.** The costumes belonging to a theater or theatrical troupe. **b.** The place where theatrical costumes are kept. **4.** The department in charge of wearing apparel, jewelry, and accessories in a royal or noble household.

**ward·room** (wôrd′rōōm′, -rŏŏm′) *n.* **1.** The common recreation area and dining room for the commissioned officers on a warship. **2.** The commissioned officers on a warship.

**-wards** *suff. var. of* -WARD.

**ward·ship** (wôrd′shĭp′) *n.* **1.** The state of being a ward or in the care of a guardian. **2.** Guardianship : custody.

**ware¹** (wâr) *n.* [ME < OE *waru.*] **1.** Articles of the same general kind <kitchen*ware*><copper*ware*> **2.** Pottery or ceramics or a special kind of pottery <earthen*ware*> **3. wares. a.** Articles of commerce : GOODS. **b.** An asset or benefit, as a service or personal accomplishment, regarded as an article of commerce.

**ware²** (wâr) *vt.* **wared, war·ing, wares.** [ME *waren* < OE *warian.*] To beware of. —Used as a command to hunting animals. —*adj. Archaic.* Watchful : wary.

**ware·house** (wâr′hous′) *n.* **1.** A place for the storage of goods or merchandise : STOREHOUSE. **2.** *Chiefly Brit.* A large, usu. wholesale shop. —*vt.* **-housed, -hous·ing, -hous·es.** To place or store in a warehouse, esp. in a bonded or government warehouse.

**ware·room** (wâr′rōōm′, -rŏŏm′) *n.* A room used for the storage or display of wares or goods.

**war·fare** (wôr′fâr′) *n.* [ME : *war,* war + *fare,* journey < OE *faru.*] **1.** The waging of war. **2.** Struggle : conflict.

**war·fa·rin** (wôr′fər-ən) *n.* [After the *W(isconsin) A(lumni) R(esearch) F(oundation)* + (COUM)ARIN.] A colorless crystalline compound, $C_{19}H_{16}O_4$, used to kill rodents and medicinally as an anticoagulant.

**war footing** *n.* Preparedness to wage and maintain a war.

**war game** *n.* A simulated battle in military training maneuvers.

**war hawk** *n.* **1.** A member of the twelfth U.S. Congress (1811–13) who advocated war with Great Britain. **2.** One who advocates war.

**war·head** (wôr′hĕd′) *n.* A part of the armament system in the forward part of a projectile, as a guided missile, torpedo, or bomb, containing the explosive charge.

**war·horse** *also* **war horse** (wôr′hôrs′) *n.* **1.** A horse used in combat : CHARGER. **2.** *Informal.* A person who has been through many battles or struggles. **3.** *Informal.* A dramatic or musical work that has been performed so many times that it has become banal.

**war·like** (wôr′līk′) *adj.* **1.** Having or displaying an eagerness to fight : BELLIGERENT. **2.** Of or relating to war : MARTIAL. **3.** Threatening or indicative of war.

**war·lock** (wôr′lŏk′) *n.* [ME *warloghe* < OE *wærloga,* oath-breaker : *wær,* pledge + *-loga,* liar, < *lēogan,* to lie.] A male witch, sorcerer, wizard, or demon.

▲ word history: The history of *warlock* can be traced to Old English and perhaps to an earlier time. There is, however, a discontinuity between the medieval and modern words in both form and meaning. The Old English meanings of *wælоga,* the ancestor of *warlock,* were "oath-breaker," "wicked person," "damned soul," and "devil," especially and specifically Satan. These meanings persisted until the end of the medieval period The regularly derived modern form of Old English *wæloga* would be *warlow,* which actually does occur in Middle English. During the Middle English period *warlow* developed the meaning "sorcerer, wizard." It had its greatest currency in northern England and Scotland, where it survived well into the modern period. *Warlock* in its current sense, "wizard," acquired new life throughout the English-speaking world through its occurrence in the works of Sir Walter Scott and Robert Burns. The precise reason for the alteration of the second syllable is unknown.

**war·lord** (wôr′lôrd′) *n.* A military commander exercising civil power in a given region, whether in nominal allegiance to the national government or in defiance of it.

**warm** (wôrm) *adj.* **-er, -est.** [ME < OE *wearm.*] **1.** Somewhat hotter than temperate : moderately hot <*warm* weather> **2.** Having the natural heat of living beings. **3.** Preserving or imparting heat <a *warm* sweater> **4.** Having a sensation of unusually high bodily heat, as from hard work or exercise : OVERHEATED. **5.** Enthusiastic : ardent <*warm* approval> **6.** Marked by liveliness, excitement, or disagreement <a *warm* controversy> **7.** Marked by or displaying friendliness or sincerity : CORDIAL <*warm* good wishes> **8.** Loving : passionate <a *warm* hug> **9.** Quick to be aroused : FIERY <a *warm* temperament> **10.** Predominantly red or yellow in color. **11.** Recently made : FRESH <The fugitive's trail was still *warm.*> **12.** Close to discovering, guessing, or finding something, as in certain games. **13.** *Informal.* Uncomfortable because of annoyance or danger <made things *warm* for the suspect> —*v.* **warmed, warm·ing, warms.** —*vt.* **1.** To raise slightly in temperature <*warm* the casserole> **2.** To make ardent or zealous : ENLIVEN. **3.** To fill with pleasant emotions. —*vi.* **1.** To become warm. **2.** To become ardent, enthusiastic, or animated <*warming* to their pet topic> **3.** To become friendly or kindly disposed. **—warm up. 1.** To make or become warm or warmer. **2.** To prepare for an athletic event by exercising or practicing. **3.** To make or become ready for an event or procedure. **4.** To approach a state of confrontation or violence. —*n. Informal.* A warming or heating. **—warm′er** *n.* **—warm′ish** *adj.* **—warm′ly** *adv.* **—warm′ness** *n.*

**warm-blood·ed** (wôrm′blŭd′ĭd) *adj.* **1.** *Zool.* Maintaining a relatively constant and warm body temperature independent of environmental temperature : HOMOIOTHERMOUS. **2.** Ardent : passionate. **—warm′-blood′ed·ness** *n.*

**warmed-o·ver** (wôrmd′ō′vər) *adj. Informal.* **1.** Reheated. **2.** Not new, fresh, or spontaneous : STALE.

**warm front** *n.* A front along which an advancing mass of warm air rises over a mass of cold air.

**warm·heart·ed** (wôrm′här′tĭd) *adj.* Marked by kindness, affection, and generosity : FRIENDLY. **—warm′heart′ed·ly** *adv.* **—warm′heart′ed·ness** *n.*

**warming pan** *n.* A metal pan with a cover and a long handle, designed to hold hot liquids or coals and used to warm a bed.

**war·mon·ger** (wôr′mŭng′gər, -mŏng′-) *n.* One who stirs up or advocates war. **—war′mon′ger·ing** *adj.* & *n.*

**warmth** (wôrmth) *n.* [ME.] **1.** The quality, state, or sensation of producing or having a moderate degree of heat. **2. a.** Kindness and affection <parental *warmth*> **b.** Excitement or intensity, as of love or passion : ARDOR. **3.** The glowing effect produced by using predominantly red or yellow colors.

**warm-up** (wôrm′ŭp′) *n.* An act, procedure, or period of warming up, as before an event or performance.

**warn** (wôrn) *v.* **warned, warn·ing, warns.** [ME *warnen* < OE *warnian.*] —*vt.* **1.** To make aware of potential or probable danger, harm, or evil. **2.** To admonish as to action or manners. **3.** To notify (a person) to go or stay away. **4.** To notify or apprise in advance. —*vi.* To give a warning. **—warn′er** *n.*

☆ ***syns:*** WARN, ALARM, ALERT, CAUTION, FOREWARN *v. core meaning* : to notify (someone) of imminent danger or risk <*warned* them of the thin ice>

**warn·ing** (wôr′nĭng) *n.* **1.** An intimation, threat, or sign of impending danger or evil. **2. a.** Advice to beware. **b.** Counsel to cease an undesirable course of action. **3.** A cautionary or deterrent example. —*adj.* Serving as a warning. **—warn′ing·ly** *adv.*

**war of nerves** *n.* A conflict marked by psychological tactics, as intimidation and threats, intended chiefly to confuse one's enemy and erode morale.

**warp** (wôrp) *v.* **warped, warp·ing, warps.** [ME *werpen* < OE *weorpan,* to throw.] —*vt.* **1.** To turn or twist out of shape. **2.** To turn from a correct, healthy, or true course : PERVERT. **3.** To arrange (strands of yarn or thread) so as to run lengthwise during the process of weaving. **4.** *Naut.* To move (a vessel) by hauling on a line fastened to or around a piling, anchor, or pier. —*vi.* **1.** To become bent or twisted out of shape, as wood. **2.** To deviate from a true, correct, or natural course. **3.** *Naut.* To move a vessel by hauling on a line fastened to or around a piling, anchor, or pier. —*n.* **1.** The state of being twisted or bent out of shape. **2.** A distortion or twist, esp. in a piece of wood. **3.** A mental or moral quirk, aberration, or deviation. **4.** The threads that run lengthwise in a fabric, crossed at right angles by the woof. **5.** *Naut.* A towline used in warping a vessel. **—warp′er** *n.*

**war paint** *n.* **1.** Pigments applied to the face or body by certain tribes, as the Indians of North America, before going to war. **2.** *Informal.* Cosmetics, as lipstick, rouge, or mascara. **3.** *Informal.* Official dress : REGALIA.

**warp and woof** *n.* The underlying structure upon which something is built : BASE.

**war party** *n.* **1.** A band of North American Indians on the attack. **2.** A usu. blatantly patriotic political party supporting a war.

**war·path** (wôr′păth′, -päth′) *n.* **1.** The route taken by a party of North American Indians on the attack. **2.** A hostile course or mood <The boss is on the *warpath* today.>

**war·plane** (wôr′plān′) *n.* A combat aircraft.

**war·rant** (wôr′ənt, wŏr′-) *n.* [ME < ONFr. *warant,* of Germanic orig.] **1.** Authorization, certification, or sanction, as given by a superior. **2.** Justification for an action : GROUNDS. **3.** Something that attests to or guarantees an event or result : PROOF. **4.** An order that serves as authorization for something, esp.: **a.** A voucher authorizing payment or receipt of money. **b.** *Law.* A judicial writ authorizing an officer to execute a judgment or make a search, seizure, or arrest. **c.** A certificate of appointment given to warrant officers. *—vt.* **-rant·ed, -rant·ing, -rants. 1.** To guarantee or attest to the quality, accuracy, or condition of. **2.** To attest to or guarantee the character or reliability of : vouch for. **3. a.** To guarantee (a product). **b.** To guarantee (a purchaser) indemnification against damage or loss. **4.** To guarantee the security or immunity of. **5.** To justify or call for : DESERVE <The problem *warrants* close attention.> **6.** To grant authorization or sanction to (someone). **7.** *Law.* To guarantee clear title to (real property). **—war′rant·a·ble** *adj.* **—war′rant·a·ble·ness** *n.* **—war′rant·a·bly** *adv.* **—war′rant·er** *n.*

**war·ran·tee** (wôr′ən-tē′, wŏr′-) *n. Law.* One to whom a warranty is made.

**warrant officer** *n.* **1.** A military officer, usu. a skilled technician or a helicopter pilot, ranking above a noncommissioned officer and below a commissioned officer and having authority by virtue of a warrant. **2.** A commissioned naval, coast guard, or marine corps officer ranking below an ensign or second lietenant.

**war·ran·tor** (wôr′ən-tər, -tôr′, wŏr′-) *n. Law.* One who makes a warrant or gives a warranty to another.

**war·ran·ty** (wôr′ən-tē, wŏr′-) *n., pl.* **-ties.** [ME *warantie* < ONFr. < *warantir,* to guarantee.] **1.** Official authorization, sanction, or warrant. **2.** Justification for an act or course of action. **3.** *Law.* **a.** An assurance by the seller of property that the goods or property are as represented or will be as promised. **b.** The insured's guarantee that the facts are as stated in reference to an insurance risk or that specified conditions will be fulfilled to keep the contract effective. **c.** A covenant by which the seller of land binds himself or herself and his or her heirs to defend the security of the estate conveyed. **d.** A judicial writ : WARRANT.

**war·ren** (wôr′ən, wŏr′-) *n.* [ME *warenne* < ONFr.] **1. a.** An area where rabbits live in burrows. **b.** A colony of rabbits. **2.** An enclosure for small game animals. **3.** An overcrowded dwelling place.

**war·ren·er** (wôr′ə-nər, wŏr′-) *n.* **1.** One who keeps a rabbit warren. **2.** A gamekeeper.

**war·ri·or** (wôr′ē-ər, wŏr′-) *n.* [ME *werreour* < ONFr. *werreieur* < *werreier,* to make war < *werre,* war.] One engaged or experienced in battle.

**war·saw** (wôr′sô) *n.* [Alteration of Sp. *guasa.*] A large grouper, *Epinephelus nigritus* of warm Atlantic waters.

**war·ship** (wôr′ship′) *n.* A combat ship.

**wart** (wôrt) *n.* [ME < OE *wearte.*] **1. a.** A circumscribed hypertrophy of the outer region of the corium, caused by a virus, covered with a keratinous layer, and occurring usu. on the hands or feet. **b.** A similar protuberance, as on a plant. **2.** One thought to resemble a wart, esp. in smallness or unattractiveness. **3.** An imperfection : defect. **—wart′ed, wart′y** *adj.*

**wart hog** *n.* A wild African tusked hog, *Phacochoerus aethiopicus,* with wartlike protuberances on the face.

**wart hog**
*2½ feet high at shoulder*

**war·time** (wôr′tīm′) *n.* A time or period of war.

**war whoop** *n.* A war cry, esp. of North American Indians.

**war·y** (wâr′ē) *adj.* **-i·er, -i·est.** [ME *ware* < OE *wær.*] **1.** On one's guard : WATCHFUL. **2.** Marked by caution <crossed the busy street with *wary* steps> **—war′i·ly** *adv.* **—war′i·ness** *n.*

**war zone** *n.* A zone in which enemy forces wage war : combat zone.

**was** (wŏz, wŭz; wəz *when unstressed*) [ME < OE *wæs.*] *1st & 3rd person sing. p.t. of* BE.

**†wash** (wŏsh, wôsh) *v.* **washed, wash·ing, washes.** [ME *washen* < OE *wacsan.*] *—v.t.* **1.** To cleanse with water or other liquid, and often with soap, detergent, or bleach, by immersing, dipping, rubbing, or scrubbing. **2.** To soak, rinse out, and remove (dirt or stain) with or as if with water. **3.** To make moist or wet : DRENCH <Tears *washed* the child's face.> **4.** To flow over, against, or past <waves *washing* the dunes> **5.** To carry, erode, remove, or destroy, as topsoil, by the action of moving water. **6.** To cleanse or purify. **7.** To cover or coat with a watery layer of paint or other coloring substance. **8.** *Chem.* **a.** To purify (a gas) by passing through or over a liquid, as to remove soluble matter. **b.** To pass a solvent, as distilled water, through (a precipitate). **9.** To remove particulate constituents from (an ore) by immersion in or agitation with water. **10.** To cause to undergo a swirling action <*washed* the medicine around in the glass> *—vi.* **1.** To wash something in or by means of water or other liquid. **2.** To undergo washing, as a fabric, without fading or other damage. **3.** To be carried away, removed, or drawn by the action of water <The boat *washed* up on the shore.> **4.** To flow, sweep, or beat with a lapping sound <The waves *washed* over the dock.> **5.** *Informal.* To hold up under examination <That alibi just won't *wash.*> **—wash down. 1.** To clean by washing with water from top to bottom, as a wall or car. **2.** To follow the ingestion of (e.g., food) with a drink of liquid <*washed* the pie *down* with milk> **—wash out. 1.** To remove or be removed by washing. **2.** To carry or wear away or be carried or worn away by the action of moving water <The road *washed out* in the wake of the hurricane.> **3. a.** To cause to fade by laundering. **b.** To deplete or become depleted of vitality <felt *washed out* by midafternoon> **c.** To eliminate or be eliminated as unsatisfactory <a basketball player who was *washed out*> **d.** To cause (an event) to be rained out. **—wash up. 1.** To wash one's hands. **2.** *Chiefly Brit.* To wash dishes after a meal. **3.** To burn out : EXHAUST <all *washed up* as a pilot> *—n.* **1.** The act or process of washing or cleansing. **2.** A quantity of articles washed or intended for washing. **3.** Waste liquid : SWILL. **4.** Fermented liquid from which liquor is distilled. **5.** A preparation or product used in washing or coating. **6.** A cosmetic or medicinal liquid, as a mouthwash. **7. a.** A thin layer of water color or India ink spread on a drawing. **b.** A light hue or tint. **8. a.** A rush or surge of water or waves. **b.** The sound of this rush or surge. **9. a.** The removal or erosion of soil by the action of moving water. **b.** A deposit of recently eroded debris. **10. a.** Low or marshy ground washed by tidal waters. **b.** A stretch of shallow water. **11.** *Western U.S.* The dry bed of a stream. **12.** A turbulence in air or water caused by the motion or action of an oar, propeller, jet, or airfoil. *—adj.* **1.** Used for washing. **2.** Capable of being washed : WASHABLE. **—wash one's hands of. 1.** To refuse to accept responsibility for. **2.** To renounce or abandon.

**wash·a·ble** (wŏsh′ə-bəl, wôsh′-) *adj.* Capable of being washed without fading or other damage.

**wash-and-wear** (wŏsh′ən-wâr′, wôsh′-) *adj.* Treated so as to be easily or quickly washed or rinsed clean and to require little or no ironing.

**wash·ba·sin** (wŏsh′bā′sən, wôsh′-) *n.* A washbowl.

**wash·board** (wŏsh′bôrd′, -bōrd′, wôsh′-) *n.* **1. a.** A board having a corrugated surface on which clothes can be rubbed in the process of laundering. **b.** A similar board used as a percussion instrument, as in a jug band. **2.** A board fastened to a wall at the floor : BASEBOARD. **3.** *Naut.* A thin plank fastened to the side of a boat or the sill of a port to keep out water.

**wash·bowl** (wŏsh′bōl′, wôsh′-) *n.* A basin that can be filled with water for use in washing.

**wash·cloth** (wŏsh′klôth′, -klŏth′, wôsh′-) *n.* A small, usu. square cloth of absorbent material for washing oneself.

**wash·day** (wŏsh′dā′, wôsh′-) *n.* A day, often the same day of every week, set aside for doing the household washing.

**washed-out** (wŏsht′out′, wôsht′-) *adj.* **1.** Lacking color or intensity : FADED. **2.** Looking tired or exhausted.

**washed-up** (wŏsht′ŭp′, wôsht′-) *adj.* **1.** No longer successful or needed. **2.** Ready to give up in disgust.

**wash·er** (wŏsh′ər, wôsh′-) *n.* **1.** One that washes. **2.** A small perforated disk placed beneath a nut or at an axle bearing or joint to relieve friction, prevent leakage, or distribute pressure. **3.** A machine for washing clothes. **4.** A dishwasher.

**wash·er·wom·an** (wŏsh′ər-wo͝om′ən, wô′shər-) *also* **wash·wom·an** (wŏsh′wo͝om′ən, wôsh′-) *n.* A woman who washes clothes as a means of livelihood : LAUNDRESS.

**wash·ing** (wŏsh′ĭng, wô′shĭng) *n.* **1.** The act or process of one that washes. **2.** A quantity of articles washed or intended to be washed at one time. **3.** The residue after an ore or other material has been washed. **4.** *often* **washings.** The liquid used to wash something.

**washing machine** *n.* WASHER 3.

ă **pat** ā **pay** âr **care** ä **father** ĕ **pet** ē **be** hw **which** ĭ **pit** ī **tie** îr **pier** ŏ **pot** ō **toe** ô **paw, for** oi **noise** o͝o **took**

**washing soda** *n.* A hydrated sodium carbonate, used as a general cleanser.
**Wash·ing·ton's Birthday** (wŏsh'ĭng-tən z) *n.* Feb. 22, officially observed on the third Monday in February as a legal holiday in most U.S. states in honor of the birthday of George Washington.
**wash·out** (wŏsh'out', wôsh'-) *n.* **1. a.** Erosion of a relatively soft surface, as a roadbed, by a transient stream of water. **b.** A channel produced by washout. **2. a.** A total disappointment or failure. **b.** One who fails to measure up to a standard, esp. one who fails a course of study or training.
**wash·rag** (wŏsh'răg', wôsh'-) *n.* A washcloth.
**wash·room** (wŏsh'ro͞om', -ro͝om', wôsh'-) *n.* A bathroom, rest room, or lavatory, esp. in a public place.
**wash sale** *n.* The illegal buying of stock by a seller's agents to give the impression of an active market.
**wash·stand** (wŏsh'stănd', wôsh'-) *n.* **1.** A stand designed to hold a basin and pitcher of water for washing. **2.** A bathroom sink.
**wash·tub** (wŏsh'tŭb', wôsh'-) *n.* A tub for washing clothes.
**wash·wom·an** (wŏsh'wo͝om'ən, wôsh'-) *n. var. of* WASHERWOMAN.
**wash·y** (wŏsh'ē, wô'shē) *adj.* **-i·er, -i·est. 1.** Diluted : watery. **2.** Lacking strength or intensity. **—wash'i·ness** *n.*
**was·n't** (wŏz'ənt, wŭz'-). Was not.
**wasp** (wŏsp, wôsp) *n.* [ME *waspe* < OE *wæps.*] Any of numerous social or solitary insects, chiefly of the superfamilies Vespoidea and Sphecoidea, having a slender body with a constricted abdomen, membranous wings, and in the females an ovipositor often modified as a sting. **—wasp'like'** *adj.*
**Wasp** *or* **WASP** *n.* [W(HITE) A(NGLO)-S(AXON) P(ROTESTANT).] A white Protestant whose ancestors were Anglo-Saxon. **—Wasp'ish** *adj.* **—Wasp'y** *adj.*
**wasp·ish** (wŏs'pĭsh, wô'spĭsh) *adj.* **1.** Relating to or like a wasp. **2.** Easily irritated. **—wasp'ish·ly** *adv.* **—wasp'ish·ness** *n.*
**wasp waist** *n.* A very slender or tightly corseted waist. **—wasp'-waist'ed** (wŏsp'wās'tĭd, wôsp'-) *adj.*
**wasp·y** (wŏs'pē, wô'spē) *adj.* **-i·er, -i·est.** Characteristic of a wasp.
**was·sail** (wŏs'əl, wŏ-sāl') *n.* [ME *wassayl*, contraction of *wæs hæil*, be healthy < ON *ves heill* : *ves*, imper. sing. of *vera*, to be + *heill*, healthy.] **1. a.** A salutation or toast once given in drinking someone's health or as an expression of good will at a festivity. **b.** The drink used in such toasting, commonly ale or wine spiced with roasted apples and sugar. **2.** A festivity marked by much drinking. —*v.* **-sailed, -sail·ing, -sails.** —*vt.* To drink to the health of : TOAST. —*vi.* To engage in or drink a wassail. **—was'sail·er** *n.*
**Was·ser·mann reaction** (wä'sər-mən) *n.* A complement-fixing reaction to the Wassermann test.
**Wassermann test** *n.* [After August von *Wassermann* (1866–1925), its inventor.] A diagnostic test for syphilis that involves the fixation or inactivation of a complement by an antibody in a blood serum sample.
**wast** (wăst, wŭst) *v.* [WAS + -(E)ST[2].] *Archaic. 2nd person sing. p.t. of* BE.
**wast·age** (wā'stĭj) *n.* **1.** Loss by wear, deterioration, or destruction <"Disease and desertion still caused much greater *wastage* than battle" —Theodore Ropp> **2.** The gradual process of wasting. **3.** Something wasted or lost by wear.
**waste** (wāst) *v.* **wast·ed, wast·ing, wastes.** [ME *wasten* < ONFr. *waster* < Lat. *vastare*, to make empty.] —*vt.* **1.** To use, consume, or expend carelessly or thoughtlessly. **2.** To cause to lose energy, strength, or vigor <Leukemia *wasted* the child's body.> **3.** To fail to take advantage of or use for profit : LOSE <*waste* a golden opportunity> **4. a.** To destroy completely. **b.** *Slang.* To kill : murder. —*vi.* **1.** To lose energy, strength, or vigor. **2.** To pass without being put to use <The hours are *wasting.*> **—waste away.** To grow gradually weaker, thinner, or more feeble. —*n.* **1.** The act of wasting or the state of being wasted. **2.** An uninhabited or uncultivated place or region. **3.** A devastated or destroyed region, town, or building : RUIN. **4. a.** A worthless or useless by-product. **b.** Something, as steam, that escapes without being used. **5.** Garbage : trash. **6.** The undigested residue of food eliminated from the body. —*adj.* **1.** Considered to be or discarded as worthless or useless <*waste* materials> **2.** Used as a conveyance or container for refuse <a *waste* barrel> **3.** Excreted from the body as unusable. **—lay waste.** To ravage : destroy. **—waste one's breath.** To gain nothing by speaking.
☆ ***syns:*** WASTE, CONSUME, DEVOUR, EXPEND, SQUANDER *v. core meaning* : to use up foolishly or needlessly <*wasted* our natural resources><a car that *wastes* gas> ***ant:*** conserve, save
**waste·bas·ket** (wāst'băs'kĭt) *n.* A lidless container for rubbish.
**wast·ed** (wā'stĭd) *adj.* **1.** Not profitably used or maintained. **2.** Needless or superfluous <*wasted* admonitions> **3.** Ravaged : deteriorated. **4.** Physically haggard, as from disease. **5.** *Slang.* STONED 2. **6.** *Archaic.* Elapsed.
**waste·ful** (wāst'fəl) *adj.* Marked by or given to waste : EXTRAVAGANT. **—waste'ful·ly** *adv.* **—waste'ful·ness** *n.*
**waste·land** (wāst'lănd') *n.* **1.** Uncultivated or desolate country. **2.** A place, era, or aspect of life regarded as humanistically, spiritually, or culturally barren.
**waste·pa·per** (wāst'pā'pər) *n.* Discarded paper.
**wast·er** (wā'stər) *n.* **1. a.** One that wastes. **b.** A spendthrift or wastrel. **2.** One that lays waste : DESTROYER.
**wast·ing** (wā'stĭng) *adj.* **1.** Gradually deteriorating : DECLINING. **2.** Sapping the strength, energy, or substance of the body : EMACIATING <a *wasting* illness> **—wast'ing·ly** *adv.*
**wast·rel** (wā'strəl) *n.* [< WASTE.] **1.** One who wastes, esp. one who wastes money. **2.** An idler or loafer.
**wa·tap** (wă-tăp', wä-) *also* **wa·ta·pe** (-tä'pē) *n.* [Cree *watapiy.*] A stringy thread made from the roots of various conifers and used by American Indians in sewing and weaving.
**watch** (wŏch) *v.* **watched, watch·ing, watch·es.** [ME *wachen* < OE *wæccan*, to watch, be awake.] —*vi.* **1.** To look or observe attentively or carefully. **2.** To look and wait expectantly <*watch* for a chance to join in> **3.** To act as a spectator : look on. **4.** To stay awake at night while serving as a guard, sentinel, or watchman. **5.** To keep vigil as a religious or devotional exercise. —*vt.* **1.** To look at or observe continuously or carefully. **2.** To keep a watchful eye on : GUARD. **3.** To observe the course of mentally <*watch* the survey results> **4.** To tend (e.g., a flock). **—watch out.** To be careful or on the alert. **—watch over.** To be in charge of. —*n.* **1.** The act or process of keeping awake or mentally alert. **2.** A part of the night. **3.** A period of close observation, often in order to discover something <a *watch* during the experiment> **4.** A person or group of persons serving, esp. at night, to guard or protect. **5.** The post or period of duty of a guard, sentinel, or watchman. **6.** A small portable timepiece, esp. one worn on the wrist or carried in the pocket. **7. a.** A period of wakefulness, esp. one observed as a religious or devotional vigil. **b.** WAKE[1] 1. **8.** *Naut.* **a.** Any of the periods of time into which the day aboard ship is divided and during which a part of the crew is assigned to duty. **b.** The members of a ship's crew on duty during a specific watch. **c.** A marine chronometer. **9.** A flock of nightingales. **—watch it.** To be careful <*Watch* it when you cross the intersection.> **—watch (one's) step.** To act or go cautiously.
☆ ***syns:*** WATCH, EYE, OBSERVE, SCRUTINIZE *v. core meaning* : to look at attentively or warily <*watched* the magician do the trick><*watched* the prisoners while they were exercising>
**watch cap** *n.* A small woolen cap of dark blue worn for cold-weather duty by naval enlisted personnel.
**watch·case** (wŏch'kās') *n.* The casing for a watch mechanism.
**watch·dog** (wŏch'dôg', -dŏg') *n.* **1.** A dog trained to guard property. **2.** A guardian or protector against waste, loss, or illegal practices.
**watch·er** (wŏch'ər) *n.* **1.** One that watches. **2.** One keeping vigil, as at a sick person's bedside.
**watch·eye** (wŏch'ī') *n.* A walleye, esp. of a dog.
**watch fire** *n.* A fire kept burning at night, as for the use of a watchman or for a signal.
**watch·ful** (wŏch'fəl) *adj.* **1.** Observant or vigilant : ALERT. **2.** *Archaic.* Not sleeping. **—watch'ful·ly** *adv.* **—watch'ful·ness** *n.*
**watch glass** *n.* A shallow glass dish used as a beaker cover or evaporating surface.
**watch·mak·er** (wŏch'mā'kər) *n.* A maker or repairer of watches. **—watch'mak'ing** *n.*
**watch·man** (wŏch'mən) *n.* One who keeps watch or guards.
**watch night** *n.* **1.** New Year's Eve. **2.** A religious service held on New Year's Eve.
**watch·tow·er** (wŏch'tou'ər) *n.* An observation tower upon which a guard or lookout is stationed to keep watch.
**watch·word** (wŏch'wûrd') *n.* **1.** A prearranged reply to a challenge, as from a sentry : PASSWORD. **2.** A rallying cry : SLOGAN.
**wa·ter** (wô'tər, wŏt'ər) *n.* [ME < OE *wæter.*] **1.** A clear, colorless, nearly odorless and tasteless liquid, $H_2O$, the most widely used of all solvents and essential for most plant and animal life. **2.** A form of water, as rain. **3.** A body of water, as a sea, lake, river, or stream. **4.** Any of the liquids passed out of the body, as urine, perspiration, or tears. **5.** The amniotic fluid surrounding the fetus in the uterus. **6.** An aqueous solution of a substance, esp. a gas <ammonia *water*> **7.** A wavy finish or sheen, as of a fabric. **8. a.** Valuation of the assets of a business firm beyond their real value. **b.** Stock issued in excess of paid-in capital. **9. a.** Clarity and luster of a gem. **b.** Degree or quality <of the first *water*> —*v.* **-tered, -ter·ing, -ters.** —*vt.* **1.** To make wet by pouring water upon. **2. a.** To give drinking water to. **b.** To lead (an animal) to drinking water. **3.** To give a wavy finish or sheen to the surface of (silk, linen, or metal). **4.** To increase (the number of shares of stock) without increasing the value of the assets represented. **5.** To irrigate (land). —*vi.* **1.** To produce or discharge fluid, as from the eyes. **2.** To salivate in anticipation of food. **3.** To take on a supply of water, as a ship. **4.** To drink water, as an animal. **—above water.** Out of trouble. **—hold water.** To be logical or consistent <Your explanation doesn't *hold water.*> **—in deep water.** In great difficulty. **—make (one's) mouth water.** To cause to anticipate with relish. **—water down.** To dilute or reduce the strength or effectiveness of. **—wa'ter·er** *n.*
**wa·ter·age** (wô'tər-ĭj, wŏt'ər-) *n. Chiefly Brit.* **1.** Movement of goods or merchandise by water. **2.** The fee paid for waterage.

**water ballet** *n.* The art of synchronized swimming.
**Water Bearer** *n.* Aquarius.
**water bed** *n.* A bed with a mattress made of a tough plastic filled with water.
**water beetle** *n.* Any of various aquatic beetles, esp. of the family Dytiscidae, with a smooth, oval body and flattened hind legs adapted for swimming.
**water bird** *n.* A wading or swimming bird.
**water biscuit** *n.* A biscuit made of flour and water.
**water blister** *n.* A blister having a nonpurulent watery content.
**water bloom** *n.* A growth of algae at or near the surface of a body of water, as a pond.
**water boatman** *n.* Any of various aquatic insects of the family Corixidae, with long oarlike hind legs adapted for swimming.

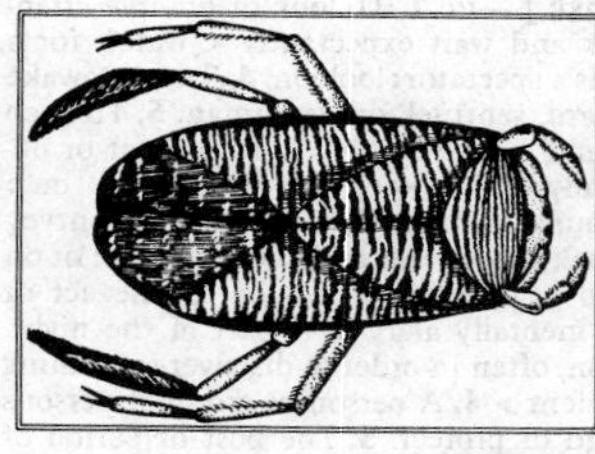
**water boatman**
*Approximately 4½ inches*

**wa·ter·borne** (wô′tər-bôrn′, -bōrn′, wŏt′ər-) *adj.* **1.** Floating on or supported by water : AFLOAT. **2.** Transported by water, as freight. **3.** Transmitted in water, as a disease germ.
**water boy** *n.* One who keeps a group, as a football team, supplied with drinking water.
**wa·ter·buck** (wô′tər-bŭk′, wŏt′ər-) *n.* An African antelope of the genus *Kobus*, having curved, ridged horns and frequenting bodies of water or swamps.
**water buffalo** *n.* A large, often domesticated buffalo, *Bubalus bubalis* of Asia and Africa, with large spreading horns.
**water bug** *n.* Any of various insects of wet places, esp. a large aquatic insect of the family Belostomatidae.
**water cal·trop** (kăl′trəp) *n.* WATER CHESTNUT 1.
**water chestnut** *n.* **1.** A floating aquatic Asian plant, *Trapa natans*, yielding four-pronged, nutlike fruit. **2. a.** A Chinese sedge, *Eleocharis tuberosa*, with an edible corm. **b.** The succulent corm of this plant, used in Oriental cookery.
**water chinquapin** *n.* A North American aquatic plant, *Nelumbo lutea*, related to the lotus and the water lilies and bearing large cup-shaped leaves, large pale yellow flowers, and edible, nutlike seeds.
**water clock** *n.* A time-keeping or time-measuring device based on the motion of running water.
**water closet** *n.* A place with a toilet and often a washbowl.
**water color** *also* **wa·ter·col·or** *or* **wa·ter-col·or** (wô′tər-kŭl′ər, wŏt′ər-) *n.* **1.** A paint composed of a water-soluble pigment. **2.** A work done in water colors. **3.** The art of using water colors. **—wa′ter-col′or** *adj.* **—water colorist** *n.*
**wa·ter-cool** (wô′tər-ko͞ol′, wŏt′ər-) *vt.* **-cooled, -cool·ing, -cools.** To cool (an engine) with water, esp. with circulating water.
**water cooler** *n.* A vessel, device, or apparatus for cooling, storing, and dispensing drinking water.
**wa·ter·course** (wô′tər-kôrs′, -kōrs′, wŏt′ər-) *n.* **1.** A waterway. **2.** The bed or channel of a waterway.
**wa·ter·craft** (wô′tər-krăft′, wŏt′ər-) *n.* **1.** Skill in water-related activities, as swimming or managing boats. **2.** Water vehicles used esp. for cargo transport.
**wa·ter·cress** (wô′tər-krĕs′, wŏt′ər-) *n.* **1.** A Eurasian plant, *Nasturtium officinale*, growing in freshwater ponds and streams and bearing pungent leaves used in salads and as a garnish. **2.** A plant related to or resembling the watercress.
**water cure** *n. Med.* Hydropathy or hydrotherapy.
**water dog** *n.* **1.** A dog at home in water, esp. one trained for hunting waterfowl. **2.** One who is at home in or on the water. **3.** A mud puppy.
**water elm** *n.* The planer tree.
**wa·ter·fall** (wô′tər-fôl′, wŏt′ər-) *n.* A steep descent of water from a height : CASCADE.
**wa·ter·find·er** (wô′tər-fīn′dər, wŏt′ər-) *n.* A dowser.
**water flea** *n.* Any of various small aquatic crustaceans of the order Cladocera that swim with jerking, flealike motions.
**wa·ter·fowl** (wô′tər-foul′, wŏt′ər-) *n., pl.* **waterfowl** *or* **-fowls.** **1.** A swimming bird, as a duck or goose, usu. frequenting freshwater areas. **2.** Swimming game birds as a group.
**wa·ter·front** (wô′tər-frŭnt′, wŏt′ər-) *n.* **1.** Land abutting on a body of water, as a harbor or lake. **2.** The district of a town or city that borders the water, esp. a wharf district where ships dock.
**water gap** *n.* A transverse cleft in a mountain ridge through which a stream flows.
**water gas** *n.* A fuel gas containing approx. 50% carbon monoxide, 40% hydrogen, and small amounts of carbon dioxide and nitrogen, made by passing steam over heated coke.
**water gate** *n.* FLOODGATE 1.
**Wa·ter·gate** (wô′tĕr-gāt′, wŏt′ər-) *n.* [After *Watergate*, a building complex in Washington, D.C., the site of illegal activities that gave rise to such a scandal.] *Informal.* A scandal that involves officials violating public or corporate trust through acts of abuse of power, as perjury or bribery, in order to retain their elective or appointive positions.
**water gauge** *also* **water gage** *n.* An instrument indicating the level of water, as in a boiler, tank, reservoir, or stream.
**water glass** *n.* **1.** A drinking glass or goblet. **2.** A structure, as a tube, having a glass bottom for making observations under water. **3.** Sodium silicate. **4.** A water gauge made of glass. **5.** A clepsydra.
**water gum** *n.* A gum tree, *Nyssa biflora*, growing on swampy land.
**water gun** *n.* A squirt gun.
**water hammer** *n.* **1.** A banging noise heard in a water pipe following an abrupt change of the flow with resulting pressure surges. **2.** A banging noise in steam pipes, caused by steam bubbles entering a cold pipe partially filled with water.
**water hemlock** *n.* A poisonous plant of the genus *Cicuta*, esp. *C. maculata* of marshy areas, with small white flower clusters.
**water hen** *n.* A chickenlike bird of marshy areas, as a coot or rail.
**water hole** *n.* A small natural depression in which water collects, esp. a pool used by animals as a watering place.
**water hyacinth** *n.* A floating aquatic plant native to tropical America, *Eichornia crassipes*, bearing bluish-purple flowers and often forming dense masses in ponds and streams.
**water ice** *n.* A dessert of sweetened, flavored, finely crushed ice.
**watering can** *n.* A watering pot.
**watering hole** *n.* A watering place.
**watering place** *n.* **1.** A place where animals find water. **2.** A health resort featuring water activities or mineral springs : SPA. **3.** A place, as a bar or nightclub, where drinks are served.
**watering pot** *n.* A vessel, often with a spout and a perforated nozzle, for watering plants.
**wa·ter·ish** (wô′tər-ĭsh, wŏt′ər-) *adj.* Watery.
**wa·ter·jack·et** (wô′tər-jăk′ĭt, wŏt′ər-) *vt.* **-et·ed, -et·ing, -ets.** To encase in or provide with a water jacket.
**water jacket** *n.* A casing containing water circulated by a pump, used around a part to be cooled, esp. in water-cooled internal-combustion engines.
**wa·ter·leaf** (wô′tər-lēf′, wŏt′ər-) *n., pl.* **-leafs.** A North American plant of the genus *Hydrophyllum*, with white or purplish flower clusters.
**wa·ter·less** (wô′tər-lĭs, wŏt′ər-) *adj.* **1.** Lacking water : DRY. **2.** Not requiring water, as a cooling system.
**water level** *n.* **1.** The level of the surface of a body of water. **2.** *Geol.* WATER TABLE 2. **3.** The water line of a ship.
**water lily** *n.* **1.** An aquatic plant of the genus *Nymphaea*, with floating leaves and variously colored flowers, esp. *N. odorata*, with fragrant, many-petaled white or pinkish flowers. **2.** A plant related to or resembling the water lily.
**water line** *n.* **1.** *Naut.* **a.** The line on the hull of a ship to which the water surface rises. **b.** Any of several lines marked on the hull of a ship parallel to this line, indicating the depth to which the ship sinks under various loads. **2.** WATERMARK 1.
**wa·ter·log** (wô′tər-lôg′, wŏt′ər-lŏg′) *vt.* **-logged, -log·ging, -logs.** [WATER + *log*, to lie or cause to lie like a log.] To soak or saturate with water, resulting in loss of buoyancy.
**wa·ter-logged** (wô′tər-lôgd′, wŏt′ər-lŏgd′) *adj.* **1.** *Naut.* Heavy and sluggish in the water because of flooding in the hold. **2.** Soaked or saturated with water.
**wa·ter·loo** (wô′tər-lo͞o′, wŏt′ər-) *n., pl.* **-loos.** [After *Waterloo*, Belgium, the town where Napoleon was defeated in 1815.] A disastrous or overwhelming defeat.
**water main** *n.* A principal pipe in a system of pipes for conveying water, esp. one installed underground.
**wa·ter·man** (wô′tər-mən, wŏt′ər-) *n.* A boatman.
**wa·ter·mark** (wô′tər-märk′, wŏt′ər-) *n.* **1.** A mark indicating the height to which water has risen, esp. a line showing the heights of high and low tide. **2. a.** A translucent design impressed on paper during manufacture and visible when the finished paper is held to the light. **b.** The metal pattern producing this design. —*vt.* **-marked, -mark·ing, -marks.** **1.** To mark (paper) with a watermark. **2.** To impress (a pattern or design) as a watermark.
**wa·ter·mel·on** (wô′tər-mĕl′ən, wŏt′ər-) *n.* **1.** A native African vine, *Citrullus vulgaris*, cultivated for its large edible fruit. **2.** The fruit of the watermelon, with a hard green rind and sweet, watery pink or reddish flesh.
**water meter** *n.* An instrument for recording the quantity of water passing through a pipe.

**water milfoil** *n.* An aquatic plant of the genus *Myriophyllum,* with feathery, finely dissected leaves.
**water mill** *n.* A mill with water-driven machinery.
**water moccasin** *n.* A venomous snake, *Agkistrodon piscivorus* or *Ancistrodon piscivorus,* found in lowlands and swampy regions of the southern United States.
**water oak** *n.* Any of various oak trees that grow in wet land.
**water of crystallization** *n.* Water in chemical combination with a crystal and necessary for the maintenance of crystalline properties but capable of being removed by enough heat.
**water of hydration** *n.* Water chemically combined with a substance so that it can be removed, as by heating, without substantially changing the chemical composition of the substance.
**water ouzel** *n.* A small bird of the genus *Cinclus* that feeds along the bottom of swift-moving streams.
**water parting** *n.* WATERSHED 1.
**water pepper** *n.* A marsh plant, *Polygonum hydropiper* or *Persicaria hydropiper,* with reddish stems, clusters of small, greenish flowers, and acrid-tasting leaves.
**water pipe** *n.* **1.** A water conduit. **2.** A hookah.
**water pistol** *n.* A squirt gun.
**water plantain** *n.* An aquatic plant of the genus *Alisma,* with branching clusters of small pinkish or white flowers.
**water polo** *n.* A water sport with two teams, each of which tries to pass a ball into the other's goal.
**wa·ter·pow·er** (wô′tər-pou′ər, wŏt′ər-) *n.* **1. a.** The energy of running or falling water as used for driving machinery, esp. for generating electricity. **b.** A source of such power, as a waterfall. **2.** A water right owned by a mill.
**wa·ter·proof** (wô′tər-pro͞of′, wŏt′ər-) *adj.* **1.** Unaffected by or impenetrable to water. **2.** Made of or treated with rubber, plastic, or a sealing agent to resist water penetration. —*n.* **1.** A waterproof material or fabric. **2.** *Chiefly Brit.* A waterproof garment, as a raincoat. —*vt.* **-proofed, -proof·ing, -proofs.** To make waterproof.
**water purslane** *n.* **1.** An aquatic plant, *Didiplis diandra,* with small greenish flowers. **2.** A marsh plant, *Ludwigia palustris,* with reddish stems and small reddish flowers.
**water rat** *n.* **1.** A semiaquatic rodent, as one of the genus *Hydromis* of Australia and adjacent islands or *Neofiber alleni* of Florida and southern Georgia, resembling the muskrat. **2.** *Slang.* A waterfront thief, ruffian, or habitué.
**wa·ter-re·pel·lent** (wô′tər-rĭ-pĕl′ənt, wŏt′ər-) *adj.* Resistant to water but not entirely waterproof.
**wa·ter-re·sis·tant** (wô′tər-rĭ-zĭs′tənt, wŏt′ər-) *adj.* Water-repellent.
**water right** *n.* **1.** The right to draw water from a particular source, as a lake, irrigation canal, or stream. **2.** The right to navigate on particular waters.
**water sapphire** *n.* A dark-blue cordierite used as a gemstone.
**wa·ter·scape** (wô′tər-skāp′, wŏt′ər-) *n.* A seascape.
**water scorpion** *n.* Any of various aquatic insects of the family Nepidae, bearing a respiratory tube projecting from the posterior part of the abdomen and inflicting a painful sting.
**wa·ter·shed** (wô′tər-shĕd′, wŏt′ər-) *n.* [Prob. transl. of G. *Wasserscheide.*] **1.** A ridge of high land dividing two areas that are drained by different river systems. **2.** The region draining into a river, river system, or body of water. **3.** A critical point serving as a dividing line <reached a *watershed* in the disarmament talks>
**water shield** *n.* **1.** An aquatic plant, *Brasenia schreberi,* with floating oval leaves and purplish flowers. **2.** A plant of the genus *Cabomba,* related to the water shield.
**wa·ter-sick** (wô′tər-sĭk′, wŏt′ər-) *adj.* Not productive due to excessive irrigation. —Used of land.
**wa·ter·side** (wô′tər-sīd′, wŏt′ər-) *n.* Land bordering a body of water : SHORE. —*adj.* **1.** Of, relating to, or located at the waterside. **2.** Living or working along the waterside.
**water skater** *n.* A water strider.
**wa·ter-ski** (wô′tər-skē′, wŏt′ər-) *vi.* **-skied, -ski·ing, -skis.** To ski on water while being towed by a motorboat. —*n. also* **water ski** *pl.* **skis** *or* **ski.** A broad ski used in water-skiing. **—wa′ter-ski′er** *n.*
**water snake** *n.* **1.** A nonvenomous snake of the genus *Natrix,* frequenting freshwater streams and ponds. **2.** Any of various aquatic or semiaquatic snakes.
**wa·ter-soak** (wô′tər-sōk′, wŏt′ər-) *vt.* **-soaked, -soak·ing, -soaks.** To soak or saturate with water.
**water spaniel** *n.* A breed of spaniel marked by a curly water-resistant coat, often used for retrieving waterfowl.
**wa·ter·spout** (wô′tər-spout′, wŏt′ər-) *n.* **1.** A tornado or whirlwind occurring over water and resulting in a whirling column of spray and mist. **2.** A hole or pipe from which water is discharged.
**water sprite** *n.* A sprite or nymph living in or near water.
**water strider** *n.* Any of various insects of the family Gerridae, with long slender legs for moving on the surface of water.
**water supply** *n.* **1.** The water available for an area or community. **2.** The sources and delivery system of a water supply.
**water system** *n.* **1.** A river and its tributaries. **2.** A water supply.
**water table** *n.* **1.** A projecting ledge, molding, or stringcourse along the side of a building, designed to throw off rainwater. **2.** The depth or level below which the ground is saturated with water.
**water thrush** *n.* A brownish New World bird, *Seiurus noveboracensis* or *S. motacilla,* that walks alongside streams or ponds.
**water tiger** *n.* The predacious larva of a diving beetle.
**wa·ter·tight** (wô′tər-tīt′, wŏt′ər-) *adj.* **1.** Made or assembled so that water cannot enter or escape : WATERPROOF. **2.** Having no flaws or loopholes <a *watertight* excuse>
**water tower** *n.* **1.** A standpipe or elevated tank used as a reservoir or for maintaining equal pressure in a water system. **2.** A towerlike fire-fighting apparatus for lifting hoses to the upper levels of a tall structure.
**water turkey** *n.* A blackish New World bird, *Anhinga anhinga* of swampy regions, with a long, slender, flexible neck.
**water vapor** *n.* Water diffused as a vapor in the atmosphere, esp. at a temperature below the boiling point.
**wa·ter·way** (wô′tər-wā′, wŏt′ər-) *n.* A navigable body of water, as a canal, channel, or river.
**wa·ter·weed** (wô′tər-wēd′, wŏt′ər-) *n.* An aquatic plant, esp. of the genus *Anacharis* or *Elodea,* bearing submerged stems with densely crowded, narrow leaves.
**water wheel** *n.* **1.** A wheel propelled by falling or running water used to power machinery. **2.** A wheel with buckets attached to its rim for raising water.
**water wings** *pl.n.* An inflatable device for supporting the body while learning to swim.
**water witch** *n.* One who professes the ability to find underground water esp. by a divining rod.
**wa·ter·works** (wô′tər-wûrks′, wŏt′ər-) *pl.n. (sing. or pl. in number).* **1. a.** The water system, including reservoirs, tanks, buildings, pumps, and pipes, of a town or city. **b.** A single unit, as a pumping station, within such a system. **2.** An exhibition of moving water, as artificial waterfalls or fountains. **3.** *Slang.* Tears.
**wa·ter·y** (wô′tə-rē, wŏt′ə-) *adj.* **-i·er, -i·est. 1.** Filled with, composed of, or containing water : MOIST <*watery* soil> **2.** Resembling or suggestive of water : LIQUID. **3.** Diluted <*watery* tea> **4.** Lacking force : INSIPID <*watery* essays> **5.** Secreting or discharging water, esp. as a symptom of disease. **—wa′ter·i·ness** *n.*
**Wat·son-Crick** (wät′sən-krĭk′) *adj.* Of or pertaining to the Watson-Crick model of DNA.
**Watson-Crick model** *n.* [After James D. *Watson* (b. 1928) and Francis H.C. *Crick* (b. 1916), its devisers.] A structural model of DNA in which the molecule is depicted as a two-stranded helix with each strand composed of alternating links of phosphate and deoxyribose and with the strands connected in a ladderlike fashion by pairs of purine and pyrimidine bases, these in turn connected by hydrogen bonds.
**watt** (wŏt) *n.* [After James *Watt* (1736–1819).] A unit of power in the International System equal to one joule per second.
**watt·age** (wŏt′ĭj) *n.* **1.** An amount of power, esp. electric power, expressed in watts. **2.** The electric power needed by a device.
**watt-hour** (wŏt′our′) *n.* A unit of energy, esp. electrical energy, equal to the energy of one watt acting for one hour and equivalent to 3,600 joules.
**wat·tle** (wŏt′l) *n.* [ME *wattel* < OE *watel.*] **1. a.** Poles intertwined with twigs, reeds, or branches for use in construction, as of fences or walls. **b.** Materials thus used. **2.** A fleshy, often brightly colored fold of skin hanging from the neck or throat, characteristic of certain birds and lizards. **3.** An Australian tree or shrub of the genus *Acacia.* —*vt.* **-tled, -tling, -tles. 1.** To construct from wattle. **2.** To weave into wattle. **—wat′tled** *adj.*
**wat·tle·bird** (wŏt′l-bûrd′) *n.* A bird of the genus *Anthochaera* of Australia and adjacent regions, with pendent wattles on each side of the head.
**watt·me·ter** (wŏt′mē′tər) *n.* An instrument for measuring in watts the power flowing in a circuit.
**Wa·tu·si** (wä-to͞o′sē) *n., pl.* **Watusi** *or* **-sis. 1.** A member of a pastoral people of Rwanda and Burundi in central equatorial Africa, distinguished by their tall stature. **2.** A dance supposedly imitative of Watusi tribal dances. —*vi.* **-sied, -si·ing, -sis.** To dance the Watusi.
**wave** (wāv) *v.* **waved, wav·ing, waves.** [ME *waven* < OE *wafian.*] —*vi.* **1.** To move back and forth or up and down in the air : FLUTTER <flags *waving* in the breeze> **2.** To make a signal with an up-and-down or back-and-forth movement of the hand or of an object in the hand. **3.** To curve or curl, as the hair. —*vt.* **1.** To move back and forth or up and down <*wave* a banner> **2. a.** To move or swing as in giving a signal <*wave* one's arm> **b.** To signal or express by such a movement <*waved* farewell> **3.** To arrange (the hair) into curves or curls. —*n.* **1. a.** A ridge or swell moving along the surface of a large body of water and generated by the action of gravity or the wind. **b.** A small ridge or swell moving across the interface of two fluids and dependent on the surface tension. **2.** *often* **waves.** The sea. **3.** A moving curve or succession of curves in or on a surface : UNDULATION <*waves* of tall grass in the wind> **4. a.** A curve or

succession of curves, as in the hair. **b.** A curved shape, outline, or pattern. **5.** A movement up and down or back and forth <a *wave* of the arm> **6. a.** A sweeping, surging sensation : SURGE <a *wave* of anger> **b.** A movement that sweeps large numbers along with it : CONTAGION <*waves* of tax protests and demonstrations> **c.** A peak of activity <a *wave* of buying> **7.** A widespread, persistent meteorological condition, esp. of temperature <a cold *wave*> **8.** *Physics.* **a.** A disturbance or oscillation propagated from point to point in a medium or in space and described, in general, by mathematical specification of its amplitude, velocity, frequency, and phase. **b.** A graphic representation of the variation of such a disturbance with time. **c.** A single cycle of such a disturbance. **9. a.** A surging movement of large numbers of individuals <a *wave* of immigrants> **b.** A sudden and rapid increase in population density. **—wav'er** *n.*

**Wave** (wāv) *n.* [Abbr. of *Women Accepted for Volunteer Emergency Service.*] A member of the women's reserve of the U.S. Navy, formed during World War II.

**wave·band** (wāv'bănd') *n.* A range of frequencies, esp. of radio frequencies, as those assigned to communication transmissions.

**wave equation** *n.* **1.** A partial differential equation in one, two, or three dimensions whose solution represents the propagation of a wave with constant velocity. **2.** The fundamental equation of wave mechanics.

**wave·form** (wāv'fôrm') *n.* Mathematical representation of a wave, esp. a graph of deviation at a fixed point versus time.

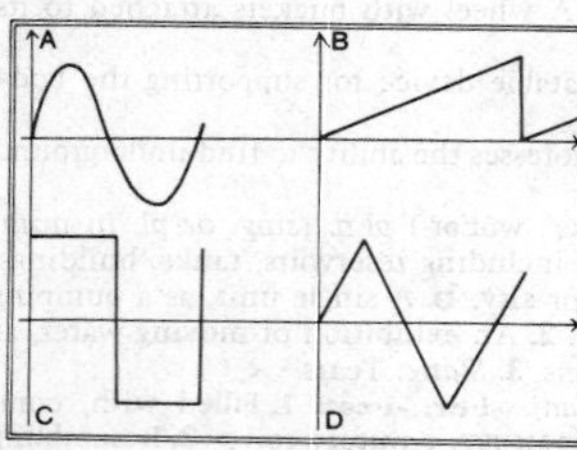


**waveform**
*Single period of common electrical waveforms: A. sine wave, B. saw-toothed wave, C. square wave, D. triangular wave.*

**wave front** *n.* A surface of a propagating wave that is the locus of all points having identical phase, the surface being usu. but not always perpendicular to the direction of propagation.

**wave function** *n.* A mathematical function used in wave mechanics to describe a specified state of a quantum system, the square of the amplitude of the function at a given point being representative of the probability of the system in that state being found at that point.

**wave-guide** (wāv'gīd') *n.* A system of material boundaries in the form of a solid dielectric rod or dielectric-filled tubular conductor capable of guiding high-frequency electromagnetic waves.

**wave·length** (wāv'lĕngth') *n.* **1.** The distance in a periodic wave between two points of corresponding phase in consecutive cycles. **2.** *Informal.* A spontaneous understanding of another person's situation, thoughts, or motivations <on the same *wavelength*>

**wave·let** (wāv'lĭt) *n.* A small wave or ripple.

**wave mechanics** *n. (sing. or pl. in number).* The formulation of quantum mechanics, based on a partial differential equation whose solutions specify the possible dynamic states of an atomic system.

**wave number** *n.* A frequency of a wave divided by its velocity of propagation.

**wave-par·ti·cle duality** (wāv'pär'tĭ-kəl) *n. Physics.* The exhibition of both wavelike and particlelike properties by a single entity, as of both diffraction and linear propagation by light.

**wa·ver** (wā'vər) *vi.* **-vered, -ver·ing, -vers.** [ME *waveren,* to wander.] **1.** To swing or move back and forth : SWAY. **2.** To show indecision or irresolution : VACILLATE <*wavered* over buying the house> **3.** To falter or yield <My courage began to *waver.*> **4.** To flicker or tremble, as light or sound. **—wa'ver** *n.* **—wa'ver·ing·ly** *adv.*

**wave train** *n. Physics.* A succession of similar wave pulses.

**wave trap** *n.* An electronic filtering device designed to exclude unwanted signals or interference from a receiver.

**wav·y** (wā'vē) *adj.* **-i·er, -i·est. 1.** Abounding in, having, or rising in waves. **2.** Proceeding in a wavelike form or motion : SINUOUS. **3.** Having curls, curves, or undulations <*wavy* brown hair> **4.** Typical of or resembling waves. **5.** Unstable. **—wav'i·ly** *adv.* **—wav'i·ness** *n.*

**waw** (väv, vôv) *n. var. of* VAV.

**wax¹** (wăks) *n.* [ME < OE *weax.*] **1. a.** Any of various natural unctuous, viscous, or solid heat-sensitive substances, consisting essentially of high molecular weight hydrocarbons or esters of fatty acids, insoluble in water but soluble in most organic solvents. **b.** A substance secreted by bees : BEESWAX. **c.** A waxy substance in the ears : CERUMEN. **2.** A solid plastic or pliable liquid substance of mineral origin, primarily petroleum, as ozocerite or paraffin, used in paper coating, as insulation, in crayons, and often in medicinal preparations. **3.** A resinous mixture used by shoemakers to wax their thread. **4.** A waxlike substance that is readily molded and impressionable. **5.** A recording for a phonograph. **6.** A preparation containing wax for polishing surfaces, as of floors. —*vt.* **waxed, wax·ing, wax·es.** To treat or coat with wax.

**wax²** (wăks) *vi.* **waxed, wax·ing, wax·es.** [ME *waxen* < OE *weaxan.*] **1.** To increase gradually in size, quantity, strength, or intensity <Indignation *waxed* among the tenants.> **2.** To show a progressively larger light surface, as the moon does in passing from new to full. **3.** To grow or become as specified <The weather *waxed* cool.>

**wax bean** *n.* A variety of string bean with yellow pods.

**wax·ber·ry** (wăks'bĕr'ē) *n.* The waxy fruit of the wax myrtle or the snowberry.

**wax·bill** (wăks'bĭl') *n.* A tropical Old World bird of the genus *Estrilda* or related genera, with a short, often brightly colored waxy beak.

**waxed paper** (wăkst) *n.* Wax paper.

**wax·en** (wăk'sən) *adj.* **1.** Made of or covered with wax. **2.** Like wax, as in being pale or smooth <a *waxen* complexion>

**wax insect** *n.* Any of various scale insects of the family Coccidae that secrete a waxy substance.

**wax moth** *n.* A bee moth.

**wax museum** *n.* An exhibition of life-size wax figures, usu. representing famous individuals.

**wax myrtle** *n.* A shrub, *Myrica cerifera* of the southeastern United States, with evergreen leaves and small berrylike fruit having a waxy coating.

**wax palm** *n.* A palm tree that yields wax, as *Copernica cerifera,* the source of carnauba wax.

**wax paper** *n.* Paper that has been made moistureproof by treatment with wax.

**wax plant** *n.* A tropical Old World vine, *Hoya carnosa,* with waxy pinkish or white flowers.

**wax·wing** (wăks'wĭng') *n.* A bird of the genus *Bombycilla,* with a crested head, brown plumage, and waxy red tips on the secondary wing feathers.

**wax·work** (wăks'wûrk') *n.* **1.** A figure made of wax, esp. a life-size wax representation of a famous person. **2. waxworks** *(sing. or pl. in number).* An exhibition of waxwork. **—wax'work'er** *n.*

**wax·y** (wăk'sē) *adj.* **-i·er, -i·est. 1.** Resembling wax, as: **a.** Pale. **b.** Smooth and lustrous. **c.** Pliable or impressionable. **2.** Consisting of, full of, or covered with wax. **3.** *Pathol.* Containing white insoluble deposits of a waxlike protein in certain portions of the body : AMYLOID.

**†way** (wā) *n.* [ME < OE *weg.*] **1. a.** A road, path, or highway affording passage from one place to another. **b.** An opening affording passage <Show me the *way* into the cellar.> **2.** Room or space to proceed with an action or course of action <Make *way* for the paramedics.> **3.** A course that is or may be used in going from one place to another <Is there any *way* through the jungle?> **4.** Progress or travel along a certain route or in a specific direction <on our *way* south> **5.** A course of conduct or action <did it the hard *way*> **6.** A method of doing something <two *ways* of learning to type> **7.** A usual or habitual manner or mode of being, living, or acting <lived in an aimless *way*> **8.** An individual or personal manner of behaving, acting, or doing <It's not my *way* to hold a grudge.> **9.** Distance <a short *way* off> **10. a.** A specific direction <headed our *way*> **b.** A participant <a four-*way* discussion> **11.** An aspect or feature <I resemble my parents in many *ways.*> **12.** Freedom to do as one wishes : WILL <if we had our *way*> **13.** Talent : facility <has a *way* with children> **14.** *Informal.* A state or condition <in a bad *way* medically> **15.** *Informal.* A district, neighborhood, or area <I was down your *way* yesterday.> **16.** *often* **ways.** A longitudinal strip on a surface that serves to guide a moving machine part. **17. ways** *(sing. or pl. in number). Naut.* The timbered structure on which a ship is built and from which it slides when launched. —*adv. also* **'way.** *Regional.* **1.** At a great distance : FAR. **2.** Away. **—by the way.** Incidentally. **—by way of. 1.** Via : through. **2.** As a means of <did nothing *by way of* recompense> **—go out of one's** *(or* **the)** **way.** To inconvenience oneself by doing more than is required. **—in a way. 1.** Within bounds or with reservations <I like your hairdo, *in a way.*> **2.** From one point of view <*In a way,* it happened for the best.> **—in the way. 1.** In a position to obstruct, impede, or interfere. **2.** In a position to be discovered or utilized <put countless opportunities *in the way*> **—on one's** *(or* **the) way. 1.** In the process of coming, going, or traveling <Spring is *on the way.*> **2.** On the route of one's journey. **—out of the way. 1.** In a position so as not to obstruct, impede, or interfere. **2.** In a remote location. **3.** Of an unusual character. **4.** Improper : wrong.

**way·bill** (wā'bĭl') *n.* A document containing a list of goods and shipping instructions relative to a shipment.

**way·far·er** (wā'fâr'ər) *n.* [ME *weyfarere* : *wey,* way + *faren,* to go.] One who travels, esp. on foot.

**way·far·ing** (wā'fâr'ĭng) *n.* [ME *wayfaringe* < OE *wegfarende* : *weg,* way + *farende,* p.part. of *faran,* to go.] Traveling, esp. on foot. **—way'far'ing** *adj.*

**wayfaring tree** *n.* A shrub, *Viburnum lantana,* with white flower clusters and berries that turn from red to black.

**way·lay** (wā′lā′) *vt.* **-laid** (-lād′), **-lay·ing, -lays. 1.** To lie in wait for and attack from ambush. **2.** To accost unexpectedly. **3.** To intercept or impede the progress or movement of. **—way′lay′er** *n.*

**†ways** (wāz) *n. (sing. in number). Regional.* WAY 9.

**-ways** *suff.* [ME < *weyes,* in such a way < OE *weges,* genitive of *weg,* way.] In a specified way, manner, direction, or position <side*ways*><cross*ways*>

**ways and means** *pl.n.* Means or methods of increasing the financial resources available to a person or group in order to accomplish a specific end.

**way·side** (wā′sīd′) *n.* The side or edge of a road. **—go (***or* **let go) by the wayside.** To postpone or be postponed because of a more urgent or worthy consideration.

**way station** *n.* A station between major stops on a route.

**way·ward** (wā′wərd) *adj.* [ME *awayward,* turned away.] **1.** Stubborn or disobedient : WILLFUL. **2.** Capricious : unpredictable. **—way′ward·ly** *adv.* **—way′ward·ness** *n.*

**we** (wē) *pron.* [ME < OE *wē.*] *(pl. in number).* **1.** —Used to refer to the speaker and another or others. **2.** —Used instead of *I,* esp. by a sovereign or a writer wishing to maintain an impersonal tone. **3.** —Often used to refer to people in general, including the speaker or writer <*We* cannot predict the future.>

**weak** (wēk) *adj.* **-er, -est.** [ME *weike* < ON *veikr.*] **1.** Lacking physical strength, energy, or vigor : FEEBLE. **2.** Likely to fail or break under pressure, stress, or strain <a *weak* link in a chain> **3.** Lacking effectiveness, firmness, or force of will <a *weak* commander> **4.** Lacking the usual, proper, or full strength of a component or ingredient <*weak* tea> **5.** Lacking the capacity to function well or in a normal manner : UNSOUND <*weak* lungs> **6.** Lacking capacity, capability, or skill <a *weak* employee> **7.** Resulting from a lack of persuasiveness : UNCONVINCING <a *weak* excuse> **8.** Lacking authority, influence, or power to rule <a *weak* sovereign> **9.** Lacking power or intensity : FAINT <a *weak* singing voice><*weak* sunlight> **10.** Lacking or deficient in a specified quality or component <was *weak* in spelling> **11.** Designating those verbs in Germanic languages that form a past tense by means of a dental suffix, as *start, started; have, had; bring, brought.* **12.** Unstressed or unaccented, as a syllable. **13.** Being a verse ending in which the stress falls on a word or syllable that is normally unstressed, as a preposition. **14.** Tending downward in price. —Used of the stock market. **—weak′ly** *adv.*

**weak·en** (wē′kən) *vt.* & *vi.* **-ened, -en·ing, -ens.** To make or become weak or weaker. **—weak′en·er** *n.*

**weak·fish** (wēk′fĭsh′) *n., pl.* **weakfish** *or* **-fish·es.** [Obs. Du. *weekvis* : *week,* soft + *vis,* fish.] A marine food and game fish of the genus *Cynoscion,* esp. *C. regalis* of North American Atlantic waters.

**weak force** *n.* Weak interaction.

**weak interaction** *n.* A fundamental interaction between elementary particles that is several orders of magnitude weaker than the electromagnetic interaction and is responsible for some particle decay, nuclear beta decay, and neutrino absorption and emission.

**weak-kneed** (wēk′nēd′) *adj.* Timid : irresolute.

**weak·ling** (wēk′lĭng) *n.* One of weak constitution or character.

**weak·ly** (wēk′lē) *adj.* **-li·er, -li·est.** Sickly. **—weak′li·ness** *n.*

**weak-mind·ed** (wēk′mīn′dĭd) *adj.* **1. a.** Indecisive : irresolute. **b.** Foolish : silly. **2.** Feeble-minded. **—weak′-mind′ed·ness** *n.*

**weak·ness** (wēk′nĭs) *n.* **1. a.** The quality or state of being weak. **b.** An instance or display of being weak. **2.** A personal defect or failing. **3.** A special liking or fondness <a *weakness* for ice cream>

**weak sister** *n. Slang.* A member of a group who is considered weak or incompetent.

**weal**[1] (wēl) *n.* [ME *wele* < OE *wela.*] **1.** Prosperity or happiness. **2.** The welfare of the community : general good.

**weal**[2] (wēl) *n.* [Var. of WALE.] WELT 3a.

**weald** (wēld) *n.* [< *Weald,* a once-forested area in southeastern England < OE *weald,* forest.] *Chiefly Brit.* **1.** A woodland. **2.** An open rolling upland.

**wealth** (wĕlth) *n.* [ME *welthe* < *wele* < OE *wela.*] **1.** Abundance of valuable material possessions or resources : RICHES. **2.** The state of being rich. **3.** A profusion or abundance. **4.** All goods and resources having economic value.

**wealth·y** (wĕl′thē) *adj.* **-i·er, -i·est. 1.** Having wealth. **2.** Richly supplied : ABUNDANT. **—wealth′i·ly** *adv.* **—wealth′i·ness** *n.*

**wean** (wēn) *vt.* **weaned, wean·ing, weans.** [ME *wenen* < OE *wenian.*] **1.** To withhold mother's milk from (the young of a mammal) and substitute other nourishment. **2.** To cause to give up an interest or habit <was *weaned* from gambling>

**wean·ling** (wēn′lĭng) *n.* A recently weaned child or animal.

**weap·on** (wĕp′ən) *n.* [ME *wepen* < OE *wæpen.*] **1. a.** An offensive or defensive combat instrument. **b.** A part of the body, as an animal's claws or horns, used in attack or defense. **2.** A means employed to overcome, persuade, or get the better of another. *—vt.* **-oned, -on·ing, -ons.** To supply with a weapon. **—weap′on·ry** *n.*

**weap·on·eer** (wĕp′ən-îr′) *n.* **1.** One who arms and otherwise prepares a nuclear weapon for release onto a target. **2.** One who devises or designs nuclear weapons.

**wear**[1] (wâr) *v.* **wore** (wôr, wōr), **worn** (wôrn, wōrn), **wear·ing, wears.** [ME *weren* < OE *werian.*] *—vt.* **1.** To put on or have on (e.g., clothes). **2.** To have or carry habitually on one's person <*wear* a watch> **3.** To affect or display : EXHIBIT <*wear* a frown> **4.** To bear, carry, or maintain in a particular manner <used to *wear* my hair short> **5.** To fly or display (colors), as a ship. **6.** To damage, diminish, erode, or use up by long or hard use, as constant exposure or rubbing <*wore* the knees of the pants> **7. a.** To produce by constant use, rubbing, or exposure <*wore* a path through the field> **b.** To bring to a specific state by use or exposure <*wore* the suit to tatters> **8.** To fatigue, weary, or exhaust <Their nagging *wore* our patience.> *—vi.* **1.** To withstand constant or hard use : LAST <a sturdy cotton that *wears* well> **2.** To break down or diminish through use <The highway surface began to *wear.*> **3.** To become by use or attrition <a face *worn* thin by illness> **4.** To pass gradually or tediously <The days *wore* on.> **—wear down.** To break down the resistance of by relentless pressure. **—wear off. 1.** To diminish gradually and vanish <The bad mood *wore off.*> **2.** To become removed from <The wallpaper *wore off* the wall.> **—wear out. 1.** To make or become unusable through heavy use. **2.** To use up : CONSUME <*wore out* their parents' indulgence> **3.** To tire : exhaust <was *worn out* after the hike> **4.** To last through : OUTLAST. *—n.* **1. a.** The act of wearing. **b.** The state of being worn : USE <These clothes have had heavy *wear.*> **2.** Clothing, esp. of a particular kind or for a particular use <casual *wear*> **3.** Gradual damage or diminution resulting from use or age. **4.** The capacity to withstand use : DURABILITY <The car still has plenty of *wear* left.> **—wear stripes.** *Informal.* To do time in prison. **—wear the pants (***or* **trousers).** *Informal.* To exercise controlling authority in a household. **—wear′er** *n.*

**wear**[2] (wâr) *v.* **wore** (wôr, wōr), **worn** (wôrn, wōrn), **wear·ing, wears.** [Orig. unknown.] *Naut. —vt.* To make (a sailing ship) come about with the wind aft. *—vi.* To come about with the stern to windward.

**wear·a·ble** (wâr′ə-bəl) *adj.* **1.** Suitable for wear. **2.** Able to be worn. *—pl.n.* **wearables.** Garments. **—wear′a·bil′i·ty** *n.*

**wear and tear** *n.* Depreciation, damage, or loss resulting from ordinary use or exposure.

**wea·ri·ful** (wîr′ē-fəl) *adj.* Tedious : wearisome. **—wea′ri·ful·ly** *adv.* **—wea′ri·ful·ness** *n.*

**wea·ri·less** (wîr′ē-lĭs) *adj.* Tireless. **—wea′ri·less·ly** *adv.*

**wear·ing** (wâr′ĭng) *adj.* **1.** Designating articles of clothing <*wearing* apparel> **2.** Causing wear or fatigue : EXHAUSTING <a *wearing* day of job hunting> **—wear′ing·ly** *adv.*

**wea·ri·some** (wîr′ē-səm) *adj.* Causing physical or mental fatigue. **—wea′ri·some·ly** *adv.* **—wea′ri·some·ness** *n.*

**wea·ry** (wîr′ē) *adj.* **-ri·er, -ri·est.** [ME *wery* < OE *wērig.*] **1.** Tired : fatigued. **2.** Exhausted of tolerance or patience <*weary* of the incessant criticism> **3.** Causing fatigue : WEARISOME <a *weary* chore> *—vt.* & *vi.* **-ried, -ry·ing, -ries.** To make or become weary. **—wear′i·ly** *adv.* **—wear′i·ness** *n.*

**wea·sand** (wē′zənd) *n.* [ME *wesand.*] The throat or gullet.

**wea·sel** (wē′zəl) *n.* [ME *wesele* < OE *wesle.*] **1.** A long-tailed carnivorous mammal of the genus *Mustela,* with a long, slender body and brownish fur that in many species turns white in winter. **2.** A sneaky, treacherous person. *—vi.* **-seled, -sel·ing, -sels** *also* **-selled, -sel·ling, -sels.** To be evasive : EQUIVOCATE. **—weasel out.** *Informal.* To back out of a situation or commitment in a sneaky or cowardly way.

**weasel word** *n.* [From the weasel's habit of sucking the contents out of an egg without breaking the shell.] An equivocal word used to lessen the force of a statement or evade a direct commitment.

**weath·er** (wĕth′ər) *n.* [ME *weder* < OE.] **1.** The state of the atmosphere at a given time and place, described by specification of variables such as temperature, moisture, wind velocity, and barometric pressure. **2. a.** Unpleasant or destructive atmospheric conditions <protected the new car from the *weather*> **b.** Violent conditions, as high winds and heavy rain on the seas and in the air <The aircraft flew into *weather.*> *—v.* **-ered, -er·ing, -ers.** *—vt.* **1.** To expose to the action of the weather, as for drying, seasoning, or coloring. **2.** To discolor, disintegrate, wear, or otherwise affect adversely by exposure. **3.** To pass through safely : SURVIVE <*weather* a family crisis> **4.** To slope (e.g., a roof) so as to shed water. **5.** *Naut.* To pass to windward of, despite bad weather. *—vi.* **1.** To show the effects of exposure to the weather, as by discoloration or disintegration. **2.** To resist or withstand the effects of weather or adverse conditions. *—adj. Naut.* Of or pertaining to the windward side of a ship. **—under the weather.** *Informal.* Slightly indisposed : UNWELL.

**weather balloon** *n.* A balloon for carrying instruments aloft to gather meteorological data in the atmosphere.

**weath·er-beat·en** (wĕth′ər-bēt′n) *adj.* **1.** Worn by exposure to the weather. **2.** Tanned and leathery from being outdoors.

**weath·er·board** (wĕth′ər-bôrd′, -bōrd′) *n.* Clapboard.
**weath·er·board·ing** (wĕth′ər-bôr′dĭng, -bōr′dĭng) *n.* Weatherboards as a whole : SIDING.
**weath·er-bound** (wĕth′ər-bound′) *adj.* Delayed, halted, or kept indoors by bad weather.
**weather bureau** *n.* A bureau responsible for gathering meteorological data for weather forecasts and study.
**weath·er·cast** (wĕth′ər-kăst′) *n.* A broadcast of weather conditions. **—weath′er·cast′er** *n.*
**weath·er·cock** (wĕth′ər-kŏk′) *n.* **1.** A weather vane, esp. one in the form of a rooster. **2.** One that is very changeable or fickle. *—vi.* **-cocked, -cock·ing, -cocks.** To tend to veer in the direction of the wind. —Used of an aircraft or a missile.
**weath·ered** (wĕth′ərd) *adj.* **1.** Worn, discolored, or warped by or as if by exposure to weather : SEASONED <*weathered* barn board> **2.** Sloped to permit water to run off <a *weathered* masonry joint> **—weathered in.** Having weather conditions that prevent flying.
**weath·er·glass** (wĕth′ər-glăs′) *n.* **1.** A barometer. **2.** Shrewd watchfulness : ALERTNESS.
**weath·er·ing** (wĕth′ər-ĭng) *n.* Any of the mechanical chemical processes by which rocks exposed to the weather decay to soil.
**weath·er·ize** (wĕth′ə-rīz′) *vt.* **-ized, -iz·ing, -iz·es.** To make repairs and improvements on (e.g., one's house) as a protective measure against cold weather.
**weath·er·ly** (wĕth′ər-lē) *adj. Naut.* Capable of sailing close to the wind with little drift to leeward. **—weath′er·li·ness** *n.*
**weath·er·man** (wĕth′ər-măn′) *n.* One who reports weather conditions : METEOROLOGIST.
**weather map** *n.* A map depicting the meteorological conditions over a given geographic area at a given time.
**weath·er·proof** (wĕth′ər-pro͞of′) *adj.* Able to withstand exposure to weather without damage. *—vt.* **-proofed, -proof·ing, -proofs.** To make weatherproof.
**weather ship** *n.* An oceangoing vessel equipped for making meteorological observations.
**weather station** *n.* A station at which meteorological data are gathered, recorded, and released.
**weath·er-strip** (wĕth′ər-strĭp′) *vt.* **-stripped, -strip·ping, -strips.** To fit or equip with weather stripping.
**weather stripping** *n.* A narrow piece of material, as felt, rubber, or metal, installed around doors and windows to protect an interior from external temperature extremes.
**weather vane** *n.* A vane for indicating wind direction.
**weath·er-wise** (wĕth′ər-wīz′) *adj.* Experienced or expert in predicting shifts, as in the weather or public opinion.
**weath·er·worn** (wĕth′ər-wôrn′, -wōrn′) *adj.* Weather-beaten.
**weave** (wēv) *v.* **wove** (wōv), **wo·ven** (wō′vən), **weav·ing, weaves.** [ME *weven* < OE *wefan.*] *—vt.* **1. a.** To make (cloth) by interlacing the threads of the weft and the warp on a loom. **b.** To interlace (yarns) into cloth. **2.** To construct by interlacing or interweaving the materials or components of <*weave* a rug> **3.** To interweave or combine (elements) into a whole <*wove* the events of my life into a novel> **4.** To interject or work in (an element) <*wove* magic into the commonplace> **5.** To spin, as a web. **6.** *p.t.* **weaved.** To move or progress by winding in and out or shuttling from side to side <*weave* one's way through the sunbathers> *—vi.* **1. a.** To engage in weaving an article. **b.** To work at a loom. **2.** *p.t.* **weaved.** To sway or move from side to side. *—n.* The pattern, method of weaving, or construction of a fabric.
**weav·er** (wē′vər) *n.* **1.** One who weaves. **2.** A weaverbird.
**weav·er·bird** (wē′vər-bûrd′) *n.* Any of various chiefly tropical Old World birds of the family Ploceidae, many of which build complex communal nests of intricately woven vegetation.
**weaver's hitch** *n. Naut.* A sheet bend.
**weaver's knot** *n. Naut.* A weaver's hitch.
**web** (wĕb) *n.* [ME < OE.] **1. a.** A textile fabric, esp. one being woven on a loom or in the process of being removed from it. **b.** The structural part of cloth. **2.** An interlacing of materials that forms a latticed or woven structure. **3.** A structure of threadlike filaments spun by spiders or certain insect larvae. **4.** Something intricately constructed, esp. something that traps or entangles <a *web* of intrigue> **5.** A complex network <a *web* of wires and tubes> **6.** A fold of skin or membranous tissue, esp. the membrane connecting the toes of certain water birds. **7.** The vane of a feather. **8.** The surface between the ribs of a ribbed architectural vault. **9.** A metal sheet or plate connecting the heavier sections, ribs, or flanges of a structural element. **10.** A thin metal plate or strip, as the bit of a key or the blade of a saw. **11.** A large continuous roll of paper, as newsprint, in the process of manufacture or as it is fed into a rotary printing press. *—vt.* **webbed, web·bing, webs. 1.** To provide with a web. **2.** To cover or envelop with a web. **3.** To ensnare in a web.
**webbed** (wĕbd) *adj.* Having or connected by a web.
**web·bing** (wĕb′ĭng) *n.* **1.** A strong, gen. narrow, closely woven cotton or nylon fabric used esp. for seat belts, harnesses, or upholstery. **2.** Something forming a web.
**web·by** (wĕb′ē) *adj.* **-bi·er, -bi·est.** Having, resembling, or being a web.
**we·ber** (wĕb′ər) *n.* [After Wilhelm E. *Weber* (1804–1891).] The International System unit of magnetic flux equal to the magnetic flux that in linking a circuit of one turn produces in it an electromotive force of one volt as it is uniformly reduced to zero within one second.
**web-foot·ed** (wĕb′fo͝ot′ĭd) *adj.* Having feet with webbed toes. **—web′foot′** *n.*
**web member** *n.* One of the structural elements connecting the top and bottom flanges of a lattice girder or the outside members of a truss.
**web press** *n.* A printing press that prints on a continuous roll of paper.
**web·ster** (wĕb′stər) *n.* [ME < OE *webbestre,* fem. of *webba,* weaver < *webb,* web.] *Obs.* WEAVER 1.
**web·worm** (wĕb′wûrm′) *n.* Any of various usu. destructive caterpillars that construct webs.
**wed** (wĕd) *v.* **wed·ded, wed** *or* **wed·ded, wed·ding, weds.** [ME *wedden* < OE *weddian.*] *—vt.* **1.** To take as a spouse : MARRY. **2.** To perform the marriage ceremony for. **3.** To join or bind : UNITE. *—vi.* To take a spouse : MARRY.
**we'd** (wēd). **1.** We had. **2.** We should. **3.** We would.
**wed·ding** (wĕd′ĭng) *n.* **1. a.** The act of marrying. **b.** The ceremony or celebration of a marriage. **2.** The anniversary of a marriage. **3.** A close union or association <a *wedding* of theory and practice>
**wedding ring** *n.* **1.** A ring, usu. a gold or platinum band, given by the groom to the bride during the wedding ceremony. **2.** A ring sometimes given by the bride to the groom.
**we·del** (vād′l) *vi.* **-deled, -del·ling, -dels.** [G. < *wedeln,* to fan < *wedel,* fan < OHG *wadal.*] To ski by executing wedelns.
**we·deln** (vād′ln) *n.* A skiing style in which the skier executes a series of short quick parallel turns by moving the back of the skis from side to side at a constant speed.
**wedge** (wĕj) *n.* [ME *wegge* < OE *wecg.*] **1.** A piece of wood or metal tapered for insertion in a narrow crevice and used for splitting, tightening, securing, or levering. **2.** Something having the triangular shape of a wedge <a *wedge* of quiche> **3.** A wedge-shaped formation, as in football or ground warfare. **4.** A tactic, event, policy, or idea that tends to divide or split like a wedge. **5.** *Meteorol.* An elongated, V-shaped region of relatively high atmospheric pressure. **6.** A golf club with a sharply slanted iron face, used to lift the ball, as from sand. **7.** A triangular character in cuneiform writing. *—v.* **wedged, wedg·ing, wedg·es.** *—vt.* **1.** To force apart or split with or as if with a wedge. **2.** To fix in place with a wedge. **3.** To crowd, push, or force into a limited space. *—vi.* To become lodged like a wedge <The heel of my shoe *wedged* fast in the grating.>
**wedg·ies** (wĕj′ēz) *pl.n.* [Orig. a trademark.] Shoes having a wedge-shaped heel joined to a half-sole so as to form a continuous undersurface.
**Wedg·wood** (wĕj′wo͝od′). A trademark for a type of pottery made by Josiah Wedgwood and his successors.
**wed·lock** (wĕd′lŏk′) *n.* [ME *wedlocke* < OE *wedlāc* < *wedd,* pledge.] The state of being married : MATRIMONY.

▲ word history: Wedlock, however indissoluble its bonds are considered to be, has at least no etymological connection with locks of any kind. The element *-lock* is a respelling, perhaps influenced by *lock,* of the Old English suffix *-lāc,* which forms nouns of action expressing the practice or performance of something. It occurs in only a handful of Old English compounds and has not survived except in the word *wedlock.* Old English *wedd* denoted a pledge or a security of any kind, but the compound *wedlāc* seems even in Old English times to have been restricted to the marriage vow.

**Wednes·day** (wĕnz′dē, -dā′) *n.* [ME < OE *Wōdnesdæg,* Woden's day.] The fourth day of the week, after Tuesday and before Thursday.
**wee** (wē) *adj.* **we·er, we·est.** [ME < *we,* a small amount < OE *wæge,* weight.] **1.** Very small : TINY. **2.** Very early <stayed up till the *wee* hours> *—n. Scot.* A short time.
**weed¹** (wēd) *n.* [ME < OE *wēod.*] **1. a.** A plant considered undesirable, unattractive, or troublesome, esp. one growing where it is not wanted, as in a garden. **b.** A rank growth of such plants. **2.** A waterplant, esp. seaweed. **3.** The leaves or stems of a plant as distinguished from the seeds. **4.** *Informal.* **a.** Tobacco. **b.** A cigarette. **5.** *Slang.* Marijuana. **6.** Something detrimental, useless, or worthless, esp. an animal unfit for breeding. *—v.* **weed·ed, weed·ing, weeds.** *—vt.* To remove weeds from <*weed* a row of carrots> *—vi.* To remove weeds from a plot. **—weed out.** To remove or get rid of as unsuitable or unwanted <*weed out* the do-nothings>
**weed²** (wēd) *n.* [ME *wede,* garment < OE *wǣd.*] **1.** A sign of mourning, as a black band worn usu. on the sleeve. **2. weeds.** A widow's mourning clothes. **3.** *often* **weeds.** A garment.
**weed·er** (wē′dər) *n.* A remover of weeds.
**weed·y** (wē′dē) *adj.* **-i·er, -i·est. 1.** Full of or consisting of weeds. **2.** Resembling or typical of a weed. **3.** Of a skinny build. **—weed′i·ly** *adv.* **—weed′i·ness** *n.*

**week** (wēk) *n.* [ME *weke* < OE *wicu.*] **1. a.** A period of seven days <a *week* of travel> **b.** A seven-day calendar period, esp. one starting on Sunday and continuing through Saturday. **2. a.** A week designated by an event or holiday occurring within it <graduation *week*> **b.** A week set aside for the honoring of a cause or institution <National Library *Week*> **3.** The part of a calendar week devoted to work, school, or business. **4. a.** One week from a specified day <I'll meet you Monday *week.*> **b.** One week ago from a specified day <It was Monday *week* that we last saw each other.>

**week·day** (wēk'dā') *n.* **1.** Any day of the week except Sunday. **2.** Any day exclusive of the days of the weekend.

**week·end** (wēk'ĕnd') *n.* The end of the week, esp. the period from Friday evening through Sunday evening. *—vi.* **-end·ed, -end·ing, -ends.** To spend the weekend <*weekending* at the lake>

**week·end·er** (wēk'ĕn'dər) *n.* **1.** One who vacations or visits, esp. habitually, on weekends. **2.** A bag or small suitcase for carrying clothing and personal articles for a weekend.

**week·ends** (wēk'ĕndz') *adv.* On any weekend.

**week·ly** (wēk'lē) *adv.* **1.** Once a week. **2.** Every week. **3.** By the week. *—adj.* **1.** Of or relating to a week. **2.** Occurring once a week or each week. **3.** Computed by the week <a *weekly* salary> *—n., pl.* **-lies.** A publication issued once a week.

**week·night** (wēk'nīt') *n.* A night of the week exclusive of Saturday and Sunday.

**week·nights** (wēk'nīts') *adv.* On any weeknight.

**ween** (wēn) *v.* **weened, ween·ing, weens.** [ME *wenen* < OE *wenan.*] *Archaic.* *—vt.* To think : suppose. *—vi.* To think it possible.

**ween·ie** (wē'nē) *n. Informal.* A wienerwurst.

**wee·ny** (wē'nē) *adj.* **-ni·er, -ni·est.** [Blend of WEE and TINY.] *Informal.* Very small : TINY.

**weep** (wēp) *v.* **wept** (wĕpt), **weep·ing, weeps.** [ME *wepen* < OE *wēpan.*] *—vt.* **1.** To mourn or lament for : BEWAIL. **2.** To shed (tears) as a sign of emotion. **3.** To bring to a specified condition by weeping <*wept* myself to sleep> **4.** To ooze, exude, or let fall drops of liquid. *—vi.* **1.** To express emotion by shedding tears. **2.** To mourn or grieve <*wept* for the dead children> **3.** To emit or run with drops of moisture. *—n. often* **weeps.** A period or fit of weeping.

**weep·er** (wē'pər) *n.* **1.** One that weeps. **2.** A hired mourner. **3.** A badge of mourning once worn by men. **4.** A hole or pipe in a wall to allow water to run off.

**weep·ing** (wē'pĭng) *adj.* **1.** Tearful. **2.** Dropping rain <*weeping* skies> **3.** Having slender, drooping branches.

**weeping willow** *n.* A widely cultivated tree, native to China, *Salix babylonica,* with long, slender, drooping branches and narrow leaves.

**weep·y** (wē'pē) *adj.* **-i·er, -i·est.** Giving to weeping : TEARFUL.

**wee·ver** (wē'vər) *n.* [ONFr. *wivre,* snake < Lat. *vipera.*] Any of several marine fishes of the family Trachinidae, bearing venomous spines.

**wee·vil** (wē'vəl) *n.* [ME *wevel* < OE *wifel.*] Any of numerous beetles, chiefly of the family Curculionidae, that have downward-curving snouts and are destructive to plants and stored plant products. **—wee'vil·y, wee'vil·ly** *adj.*

**weft** (wĕft) *n.* [ME < OE *wefta.*] **1. a.** The horizontal threads interlaced through the warp in a woven fabric : WOOF. **b.** Yarn to be used for the weft. **2.** Woven fabric.

**wei·ge·la** (wī-gē'lə, -jē'lə, wī'jə-lə) *n.* [NLat., genus name, after Christian E. *Weigel* (1748–1831).] A shrub of the genus *Weigela,* esp. *W. florida,* widely cultivated for its pink, white, or red flowers.

**weigh¹** (wā) *v.* **weighed, weigh·ing, weighs.** [ME *weyen* < OE *wegan.*] *—vt.* **1.** To ascertain the weight of by or as if by using a scale or similar instrument. **2.** To measure off an amount equivalent in weight to <*weigh* out two pounds of mushrooms> **3.** To balance in one's mind to determine the worth of : EVALUATE <*weighed* the pros and cons> **4.** *Naut.* To raise (anchor) before sailing. *—vi.* **1.** To have or be of a specific weight. **2.** To have significance or importance : COUNT <The judgment *weighed* heavily against them.> **3.** To be a burden on : OPPRESS <The felony *weighed* on the offender.> **4.** *Naut.* **a.** To raise anchor. **b.** To sail out of port. **—weigh down. 1.** To bend down or overburden. **2.** To burden or oppress. **—weigh in. 1.** To weigh or be weighed before participating in a sports contest. **2.** To have one's baggage weighed. **3.** To enter as a participant <*weighed in* with some nasty remarks> **—weigh (one's) words.** To choose one's words very carefully. **—weigh'er** *n.*

**weigh²** (wā) *n.* [ *var. of* WAY.] *Naut.* Way. —Used in the phrase *under weigh.*

**weight** (wāt) *n.* [ME *wight* < OE *wiht.*] **1.** A measure of the heaviness or mass of an object. **2.** The gravitational force exerted by the earth or another celestial body on an object, equal to the product of the object's mass and the local value of gravitational acceleration. **3. a.** A unit measure of gravitational force <a table of *weights* and measures> **b.** A system of such measures <avoirdupois *weight*><troy *weight*> **4.** The measured heaviness of a particular object <put a two-pound *weight* on the scale> **5.** An object used chiefly to exert a force by virtue of its gravitational attraction to the earth, esp.: **a.** A metallic solid used as a standard of comparison in weighing. **b.** An object used to hold something down. **c.** A counterbalance in a machine. **d.** A heavy object, as a dumbbell, used for exercise or in athletic competition. **6.** *Math.* One of a set of numbers assigned as multipliers to quantities to be averaged to indicate the relative importance of each quantity's contribution to the average. **7.** Burden <the *weight* of myriad duties> **8.** The greatest part : PREPONDERANCE. **9. a.** Influence : authority <The decision carried a lot of *weight.*> **b.** Forceful quality <the *weight* of one's argument> **10.** A classification according to comparative lightness or heaviness. *—vt.* **weight·ed, weight·ing, weights. 1.** To add heaviness or weight to. **2.** To load down : BURDEN. **3.** To treat (fabric) with chemical substances in order to give body or extra weight. **4.** *Math.* To assign a weight or weights to. **—pull (one's) weight.** To do one's job or share. **—throw (one's) weight around.** To make a show of one's importance.

☆ ***syns:*** WEIGHT, BULK, MASS, PREPONDERANCE *n. core meaning* : the greatest part <The *weight* of the evidence points to guilt.>

**weight·less** (wāt'lĭs) *adj.* **1.** Having little or no weight. **2.** Experiencing little or no gravitational force. **—weight'less·ly** *adv.* **—weight'less·ness** *n.*

**weight lifter** *n.* One who lifts heavy weights for exercise or in an athletic competition.

**weight·lift·ing** (wāt'lĭf'tĭng) *n.* The lifting of heavy weights in a prescribed manner as an exercise or in athletic competition.

**weight·y** (wā'tē) *adj.* **-i·er, -i·est. 1.** Having great weight : HEAVY. **2.** Burdensome : oppressive. **3.** Of great consequence : MOMENTOUS <*weighty* matters of life and death> **4.** Influential : efficacious <a *weighty* argument> **5.** Serious : solemn. **6.** Fat. **—weight'i·ly** *adv.* **—weight'i·ness** *n.*

**Wei·mar·an·er** (vī'mä-rä'nər, wī'-) *n.* [G. < *Weimar,* Weimar, Germany.] A large dog orig. bred in Germany, with a smooth grayish coat.

**Weimaraner**
*Approximately 24 inches high at shoulder*

**weir** (wîr) *n.* [ME *wer* < OE.] **1.** A fence or barrier placed in a stream to catch or retain fish. **2.** A dam placed across a river or canal to raise or divert the water.

**weird** (wîrd) *adj.* **-er, -est.** [ME *werde,* having power to control fate < OE *wyrd,* fate.] **1.** Suggestive of or concerned with the supernatural : EERIE. **2.** Of an odd, peculiar, or bizarre character : STRANGE. **3.** Of or relating to fate or the Fates. *—n.* **1. a.** Destiny : fate. **b.** One's assigned lot or fortune : KISMET. **2.** *often* **Weird.** One of the Fates. **—weird'ly** *adv.* **—weird'ness** *n.*

☆ ***syns:*** WEIRD, EERIE, UNCANNY, UNEARTHLY *adj. core meaning* : of a mysteriously strange and usu. frightening nature <a *weird* premonition of disaster>

▲ word history: When Macbeth referred to the witches on the heath as the "weird sisters" he was not being insulting, he was being etymological and innovative at the same time. The phrase *weird sisters* originally referred to the Fates, the three women of classical myth who controlled the destiny of each individual person. The phrase was also extended to include women who prophesied or possessed other attributes of the Fates. *Weird* by itself was a noun that originally meant "fate" or "destiny" pure and simple, and in medieval times the word was used to translate *Parca,* the Latin word for "Fate." This use of *weird* lost currency during the modern period, surviving primarily in the phrase "weird sisters," where it was interpreted as an adjective. The great prestige of Shakespeare preserved this use of *weird,* which was picked up and extended to its current meaning by 19th-century poets and writers. The recent coinages *weirdie* and *weirdo* derive solely from the extended senses "odd" and "strange."

**weird·ie** *also* **weird·y** (wîr'dē) *n., pl.* **-ies.** *Slang.* A weirdo.

**weird·o** (wîr'dō) *n., pl.* **-oes.** *Slang.* One that is strange or bizarre.

**weis·en·hei·mer** (wīz'ən-hī'mər) *n. var. of* WISENHEIMER.

**we·ka** (wē'kə, wā'-) *n.* [Maori.] A flightless bird, *Gallirallus australis* of New Zealand, with brown, mottled plumage.

**welch** (wĕlch) *v. var. of* WELSH.

**wel·come** (wĕl'kəm) *adj.* [ME < OE *wilcuma,* welcome guest.] **1.** Received with pleasure and hospitality into one's company or home <a *welcome* visitor> **2.** Gratifying <a *welcome* break from routine> **3.** Cordially permitted or invited, as to do or enjoy. **4.** Freely granted one's courtesy <"Thank you!" "You're *wel-*

*come!"* > —*n.* **1.** A cordial greeting or hospitable reception. **2.** Willing or glad acceptance. —*vt.* **-comed, -com·ing, -comes. 1.** To greet, receive, or entertain hospitably or cordially. **2.** To receive or accept gladly <*welcomed* the silence> —*interj.* —Used to greet cordially a visitor or recent arrival. **—wel'come·ly** *adv.* **—wel'come·ness** *n.* **—wel'com·er** *n.*

**weld[1]** (wĕld) *v.* **weld·ed, weld·ing, welds.** [Alteration of WELL[1], to weld (obs.).] —*vt.* **1.** To join (metals) by applying heat, sometimes with pressure and sometimes with an intermediate or filler metal having a high melting point. **2.** To bring into close association : UNITE. —*vi.* To be capable of being welded. —*n.* **1.** The union of two metal parts by welding. **2.** The joint formed by welding.

**weld[2]** (wĕld) *also* **wold** (wōld) *n.* [ME *welde.*] **1.** The dyer's rocket. **2.** The yellow dye obtained from the weld.

**weld·ment** (wĕld'mənt) *n.* A unit having an assemblage of pieces welded together.

**wel·fare** (wĕl'fâr') *n.* [ME < *wel faren,* to fare well.] **1. a.** Health, happiness, and general well-being. **b.** Prosperity. **2.** Welfare work. **3.** Provision of economic or social benefits to a certain group of people, esp. aid furnished by the government or private agencies to the disadvantaged or disabled. **—on welfare.** Receiving assistance from the government because of poverty or disability.

**welfare state** *n.* **1.** A social system whereby the state assumes primary responsibility for the welfare of citizens. **2.** A nation that has adopted the welfare system.

**welfare work** *n.* Organized efforts by a community or private agency for the betterment of the poor.

**wel·far·ism** (wĕl'fâr-ĭz'əm) *n.* The set of policies, practices, and social attitudes associated with a welfare state.

**wel·kin** (wĕl'kĭn) *n.* [ME *welken* < OE *wolcen.*] **1.** The vault of heaven : SKY. **2.** The upper air.

**well[1]** (wĕl) *n.* [ME < OE *wælla.*] **1.** A deep hole or shaft dug or drilled to obtain water, oil, gas, or brine. **2.** A container or reservoir, as an inkwell, used to hold a liquid. **3.** A vertical opening passing through the floors of a building, as for stairs or ventilation. **4.** An enclosure in a ship's hold for the pumps. **5.** A cistern with a perforated bottom in the hold of a fishing vessel for storing live fish. **6.** *Chiefly Brit.* The space in a law court where the counsel or solicitor sits. **7.** A spring : fountain. **8.** A source to be drawn upon <*a well* of good advice> —*v.* **welled, well·ing, wells.** —*vi.* **1.** To rise to the surface, ready to flow, as tears. **2.** To rise or surge, as anger or joy, from an inner source. —*vt.* To pour forth.

**well[2]** (wĕl) *adv.* **better** (bĕt'ər), **best** (bĕst) [ME < OE *wel.*] **1.** In a good or proper manner <functioned *well*> **2.** Skillfully or proficiently <reads *well*> **3.** Satisfactorily or sufficiently <dined *well*> **4.** Successfully or effectively <communicates *well* with everyone> **5.** In a comfortable or affluent manner <lived *well* on the generous pension> **6.** Advantageously <married *well*> **7.** Suitably : appropriately. **8.** With reason or propriety : REASONABLY <I can't very *well* refuse.> **9.** Prudently <They would do *well* to keep quiet.> **10.** On close or familiar terms <I knew the author *well.*> **11.** Favorably <spoke *well* of the supervisor> **12.** Thoroughly : completely <likes meat *well* cooked> **13.** Perfectly : clearly <*well* understood the consequences> **14.** Widely : generally <a maxim that is known *well*> **15.** To a suitable or considerable extent or degree <We are *well* satisfied.> **16.** With close and careful attention <heed *well*> **17.** Entirely : fully <*well* worth the money> **18.** Far <*well* before dawn> **19.** In all likelihood <Their accusation may *well* be true.> —*adj.* **1.** In a satisfactory state or circumstances : RIGHT <All is *well* in the family.> **2. a.** In good health. **b.** Cured or healed, as a wound. **3. a.** Advisable : prudent <It would be *well* not to interfere.> **b.** Fortunate : good <It is *well* that they left here early.> —*interj.* **1.** —Used to express surprise. **2.** —Used to introduce a remark or as a filler in a pause during conversation. **—as well. 1.** In addition : ALSO. **2.** With equal or better effect <I might *as well* stay.> **—as well as. 1.** In addition to. ***usage:*** *As well as* is considered redundant in combination with *both.* Therefore a construction such as *both in theory as well as in practice* should be reworded to *both in theory and practice* or *in theory as well as in practice.* **2.** As satisfactorily as <I ran *as well as* you did in the marathon.>

**we'll** (wēl). **1.** We will. **2.** We shall.

**well·a·day** (wĕl'ə-dā') *interj. & n. var. of* WELLAWAY.

**well-ap·point·ed** (wĕl'ə-poin'tĭd) *adj.* Having a full array of suitable furnishings or equipment <a *well-appointed* workshop>

**well·a·way** (wĕl'ə-wā') *also* **well·a·day** (-dā') [ME < OE *wei lā wei.*] *Archaic.* —*interj.* —Used to express lamentation or sorrow. —*n., pl.* **-ways** *also* **-days.** A lamentation.

**well-bal·anced** (wĕl'băl'ənst) *adj.* **1.** Evenly balanced, proportioned, or regulated. **2.** Mentally stable : SOUND.

**well-be·ing** (wĕl'bē'ĭng) *n.* The state of being healthy, happy, or prosperous.

**well-born** (wĕl'bôrn') *adj.* Of good lineage or stock.

**well-bred** (wĕl'brĕd') *adj.* **1.** Of good upbringing : WELL-MANNERED. **2.** Of a good breed. —Used of animals.

**well-de·fined** (wĕl'dĭ-fīnd') *adj.* **1.** Having definite and distinct lines or features <a *well-defined* profile> **2.** Accurately and clearly stated or described <a *well-defined* concept>

**well-dis·posed** (wĕl'dĭs-pōzd') *adj.* Tending to be kindly, friendly, or sympathetic.

**well-done** (wĕl'dŭn') *adj.* **1.** Cooked all the way through. **2.** Satisfactorily accomplished.

**well-fa·vored** (wĕl'fā'vərd) *adj.* Good-looking : attractive.

**well-fed** (wĕl'fĕd') *adj.* **1.** Adequately or properly nourished. **2.** Overfed : fat.

**well-fixed** (wĕl'fĭkst') *adj. Informal.* Financially secure.

**well-found** (wĕl'found') *adj.* Properly furnished or equipped.

**well-found·ed** (wĕl'foun'dĭd) *adj.* Based on sound judgment, reasoning, or evidence <a *well-founded* hypothesis>

**well-groomed** (wĕl'gro͞omd') *adj.* **1.** Attentive to details of dress and personal appearance. **2.** Carefully tended or curried <a *well-groomed* racehorse> **3.** Trim and tidy <a *well-groomed* hedge>

**well-ground·ed** (wĕl'groun'dĭd) *adj.* **1.** Adequately versed in a subject. **2.** Well-founded.

**well-han·dled** (wĕl'hăn'dəld) *adj.* **1.** Managed well. **2.** Showing the signs of much handling <a *well-handled* teddy bear>

**well·head** (wĕl'hĕd') *n.* **1.** The source of a well or stream. **2.** WELLSPRING 2. **3.** The top of a structure built over a well.

**well-heeled** (wĕl'hēld') *adj. Slang.* Well-to-do.

**Wel·ling·ton boot** (wĕl'ĭng-tən) *n.* [After Arthur Wellesley (1769–1852), 1st Duke of *Wellington.*] A boot extending to the top of the knee in front but cut low in back.

**well-in·ten·tioned** (wĕl'ĭn-tĕn'shənd) *adj.* Having or characterized by good intentions <*well-intentioned* suggestions>

**well-knit** (wĕl'nĭt') *adj.* Strongly knit, esp. strongly and firmly constructed <a *well-knit* physique><a *well-knit* play>

**well-known** (wĕl'nōn') *adj.* **1.** Widely known. **2.** Fully known.

**well-man·nered** (wĕl'măn'ərd) *adj.* Courteous : polite.

**well-mean·ing** (wĕl'mē'nĭng) *adj.* Well-intentioned.

**well-meant** (wĕl'mĕnt') *adj.* Well-intentioned.

**well-nigh** (wĕl'nī') *adv.* Nearly : almost <*well-nigh* exhausted>

**well-off** (wĕl'ôf', -ŏf') *adj.* **1.** Being in fortunate circumstances. **2.** Well-to-do.

**well-read** (wĕl'rĕd') *adj.* Knowledgeable through extensive reading.

**well-round·ed** (wĕl'roun'dĭd) *adj.* **1.** Comprehensively developed <a *well-rounded* generalist> **2.** Having a shapely figure.

**well-spo·ken** (wĕl'spō'kən) *adj.* **1.** Selected or expressed with aptness or propriety. **2.** Courteous in speech.

**well·spring** (wĕl'sprĭng') *n.* **1.** The source of a spring or stream. **2.** A source : fountainhead <a *wellspring* of information>

**well-thought-of** (wĕl-thôt'ŭv', -ŏv') *adj.* Respected.

**well-tim·bered** (wĕl'tĭm'bərd) *adj.* **1.** Having a good framework or structure. **2.** Covered with a good growth of timber.

**well-timed** (wĕl'tīmd') *adj.* Occurring or done at an opportune time <a *well-timed* entrance>

**well-to-do** (wĕl'tə-do͞o') *adj.* [From the phrase *to do well.*] Affluent : prosperous.

**well-turned** (wĕl'tûrnd') *adj.* **1.** Expertly turned : SYMMETRIC <a *well-turned* pillar> **2.** Shapely <a *well-turned* ankle> **3.** Concisely or aptly expressed <a *well-turned* metaphor>

**well-wish·er** (wĕl'wĭsh'ər) *n.* One who extends good wishes. **—well'-wish'ing** *adj. & n.*

**well-worn** (wĕl'wôrn', -wōrn') *adj.* **1.** Showing signs of much wear or use. **2.** Repeated too often : HACKNEYED. **3.** Borne in a becoming manner <*well-worn* renown>

**Wels·bach burner** (wĕlz'băk', -bäk'). A trademark for a gauze mantle impregnated with cerium and thorium compounds and used with a gas burner that becomes incandescent when heated, producing light.

**welsh** (wĕlsh, wĕlch) *also* **welch** (wĕlch) *vi.* **welshed, welsh·ing, welsh·es** *also* **welched, welch·ing, welch·es.** [Orig. unknown.] *Slang.* **1.** To swindle a person by not paying a wager or debt. **2.** To fail to fulfill an obligation. **—welsh'er** *n.*

**Welsh** (wĕlsh) *adj.* [ME *Walische* < OE *Wælisc* < *Wealh,* Welshman.] Of or relating to Wales or its people, language, or culture. —*n.* **1.** The natives or inhabitants of Wales. **2.** The Celtic language of Wales.

▲ word history: The Welsh are the descendents of the Celts who lived in Britain before the Roman and Germanic invasions. The Welsh call themselves the *Cymry,* but the invading Anglo-Saxons called them *Wealas,* which in Old English meant "foreigners." *Wealas* became *Wales* in Modern English. *Wælisc,* the Old English adjective derived from *Wealas,* became modern *Welsh.*

**Welsh cor·gi** (kôr'gē) *n.* A short-legged dog orig. bred in Wales, with a long body and a foxlike head.

**Welsh·man** (wĕlsh'mən) *n.* A native or resident of Wales.

**Welsh rabbit** *n.* A dish made of melted cheese, milk or cream, seasonings, and sometimes ale, served hot over toast or crackers.

**Welsh rare·bit** (râr'bĭt) *n.* Welsh rabbit.

**Welsh terrier** *n.* A wire-haired terrier orig. bred in Wales, with a black-and-tan coat.

**welt** (wĕlt) *n.* [ME *welte.*] **1.** A strip of material, as leather, stitched into a shoe between the sole and the upper. **2.** A tape or covered cord sewn into a seam as reinforcement or trimming : WELTING. **3. a.** A ridge or bump raised on the skin by a lash or blow or sometimes by an allergic disorder. **b.** A lash or blow producing such a mark. —*vt.* **welt·ed, welt·ing, welts. 1.** To reinforce or trim with a welt or welting. **2.** To beat severely : FLOG. **3.** To raise welts on.

**Welt·an·schau·ung** (vĕlt′än′shou′o͝ong) *n., pl.* **-ungs** or **-ung·en** (-o͝ong-ən) [G.] A comprehensive world view, esp. from a specified standpoint.

**wel·ter** (wĕl′tər) *vi.* **-tered, -ter·ing, -ters.** [ME *welteren.*] **1.** To wallow, roll, or toss about, as in mud or high seas. **2.** To lie soaked in a liquid, as blood. **3.** To roll and surge, as the sea. —*n.* **1.** Confusion : turmoil. **2.** A confused mass : JUMBLE <a *welter* of odds and ends>

**wel·ter·weight** (wĕl′tər-wāt′) *n.* [Perh. < WELT.] A boxer or wrestler who weighs between 136 and 147 pounds or approx. 62 and 67 kilograms.

**welt·ing** (wĕl′tĭng) *n.* A strip or cord used to welt a seam.

**Welt·schmerz** (vĕlt′shmĕrts′) *n.* [G.] Sadness over the evils of the world, esp. as an expression of romantic pessimism.

**wen**[1] (wĕn) *n.* [ME < OE.] A cyst containing sebaceous matter.

**wen**[2] (wĕn) *n.* [OE.] An Old English runic letter represented by the Modern English *w.*

**wench** (wĕnch) *n.* [ME *wench* < *wenchel* < OE *wencel,* child, girl.] **1.** A young woman or girl, esp. a peasant girl. **2.** A woman servant. **3.** A wanton woman : PROSTITUTE. —*vi.* **wenched, wench·ing, wench·es.** To consort with prostitutes. **—wench′er** *n.*

**wend** (wĕnd) *v.* **wend·ed, wend·ing, wends.** [ME *wenden* < OE *wendan,* to turn.] —*vt.* To proceed on or along : GO <*wend* one's way to town> —*vi.* To go one's way : PROCEED.

**Wend** (wĕnd) *n.* [G. *Wende* < OHG *Winidia.*] One of a Slavic people living in eastern Germany. **—Wend** *adj.*

**Wend·ish** (wĕn′dĭsh) *adj.* Of or relating to the Wends or their language. —*n.* The Slavic language of the Wends.

**went**[1] (wĕnt) *v.* [ME < OE *wende,* p.t. of *wendan,* to go.] *p.t. of* GO[1].

**went**[2] (wĕnt) *v. Archaic. var. p.t. & p.p. of* WEND.

**wen·tle·trap** (wĕnt′l-trăp′) *n.* [Du. *wendeltrappe* : *wendel,* winding < *wenden,* to wind + *trappe,* stairs.] Any of various marine snails of the family Epitoniidae, with a tapering spiral shell having raised longitudinal ridges.

**wept** (wĕpt) *v. p.t. & p.p. of* WEEP.

**were** (wûr) *v.* [ME < OE *wæron.*] **1.** *2nd person sing. p.t. of* BE. **2.** *1st, 2nd, & 3rd person plural p.t. of* BE. **3.** *Past subjunctive of* BE.

**we're** (wîr). We are.

**were·n't** (wûrnt, wûr′ənt). Were not.

**were·wolf** *also* **wer·wolf** (wîr′wo͝olf′, wûr′-, wâr′-) *n.* [ME < OE *werewulf* : *wer,* man + *wulf,* wolf.] A person transformed into a wolf or capable of assuming the form of a wolf at will : LYCANTHROPE.

**wer·geld** (wûr′gĕld′) *also* **wer·gild** or **were·gild** (-gĭld′) *n.* [ME *wargeld* < OE *wergeld* : *wer,* man + *geld,* payment.] A price in Anglo-Saxon and Germanic law, set upon a man's life on the basis of his rank and paid as compensation by the family of a slayer to the kindred or lord of a slain man to free the culprit of further punishment or obligation.

**wer·ner·ite** (wûr′nə-rīt′) *n.* [Fr., after A. G. *Werner* (1750–1817).] *Mineral.* Scapolite.

**wert** (wûrt) *v.* [WERE + *t* as in *shalt.*] *Archaic. 2nd person sing. past indicative & past subjunctive of* BE.

**wer·wolf** (wîr′wo͝olf′, wûr′-, wâr′-) *n. var. of* WEREWOLF.

**wes·kit** (wĕs′kĭt) *n.* [Var. of WAISTCOAT.] A waistcoat : vest.

**Wes·ley·an** (wĕs′lē-ən, wĕz′-) *adj.* Of or pertaining to John or Charles Wesley or to Methodism. —*n.* A Methodist. **—Wes′ley·an·ism** *n.*

**west** (wĕst) *n.* [ME < OE.] **1. a.** The direction opposite to the direction in which the earth rotates on its axis. **b.** One of the four cardinal points on the mariner's compass 90° left of north and 180° from east. **2.** *often* **West.** An area or region lying in a western direction. **3. the West. a.** The part of the earth west of Asia and Asia Minor, esp. Europe and the Western Hemisphere. **b.** The western part of the United States, esp. the region west of the Mississippi River. **c.** The noncommunist countries of Europe and the Americas. —*adj.* **1.** To, toward, of, facing, or in the west. **2.** Coming from or originating in the west, as a wind. **3. West.** Officially designating the western part of a country, continent, or other geographic area <*West Africa*> —*adv.* In, from, or toward the west.

**west·bound** (wĕst′bound′) *adj.* Going toward the west.

**west by north** *n.* The direction or point on the mariner's compass halfway between due west and west-northwest that is 78°45′ west of due north. —*adv. & adj.* Toward or from west by north.

**west by south** *n.* The direction or point on the mariner's compass halfway between due west and west-southwest that is 101°15′ west of due north. —*adv. & adj.* Toward or from west by south.

**west·er** (wĕs′tər) *vi.* **-ered, -er·ing, -ers.** [ME *westeren* < *west,* west.] **1.** To move westward. —Used of the sun, moon, or a star. **2.** To shift to the west. —Used of the wind. —*n.* A wind or storm coming from the west.

**west·er·ly** (wĕs′tər-lē) *adj.* [< obs. *wester,* western < ME < OE *westra* < *west,* west.] **1.** Located toward the west. **2.** From the west. —Used of wind. —*n., pl.* **-lies.** A wind or storm coming from the west. **—west′er·ly** *adv.*

**west·ern** (wĕs′tərn) *adj.* [ME *westeren* < OE *westerne.*] **1.** Located toward, in, or facing the west. **2.** Coming from the west. —Used of wind. **3.** Growing in the west. **4.** *often* **Western.** Of, relating to, or typical of western regions or the West. **5. Western.** Of, relating to, or characteristic of Europe and the Western Hemisphere : OCCIDENTAL <*Western* customs> **6.** *often* **Western.** Of, relating to, or typical of the American West. **7. Western.** Of or relating to the Roman Catholic Church as opposed to the Eastern Orthodox Church. —*n.* **1.** A westerner. **2.** *often* **Western.** A novel, motion picture, or television or radio program about cowboys or frontier life in the American West.

**west·ern·er** (wĕs′tər-nər) *n.* **1.** A native or resident of the west. **2.** *often* **Westerner.** A native or resident of the western United States.

**Western Hemisphere** *n.* The half of the earth that includes North and South America, the surrounding waters, and all neighboring islands.

**west·ern·ize** (wĕs′tər-nīz′) *vt.* **-ized, -iz·ing, -iz·es.** To convert to the customs of Western civilization. **—west′ern·i·za′tion** *n.*

**west·ern·most** (wĕs′tərn-mōst′) *adj.* Farthest west.

**western omelet** *n.* An omelet cooked with diced ham, chopped green pepper, and onion.

**western sandwich** *n.* A sandwich with a western omelet as a filling.

**West Germanic** *n.* A subdivision of the Germanic languages that includes High German, Low German, Yiddish, Dutch, Afrikaans, Flemish, Frisian, and English.

**west·ing** (wĕs′tĭng) *n.* [< WEST.] **1.** *Naut.* **a.** The distance sailed by a ship on a westerly course. **b.** The longitudinal distance from a given meridian on a westward course. **2.** A westward direction or movement.

**west-north·west** (wĕst′nôrth′wĕst′; *Naut.* -nôr′wĕst′) *n.* The direction or point on the mariner's compass halfway between west and northwest that is 67°30′ west of due north. —*adj.* In, facing, or situated toward west-northwest. —*adv.* Toward or from west-northwest.

**West Saxon** *n.* **1.** The dialect of Old English used in southern England that was the chief literary dialect of England before the Norman Conquest. **2.** One of the Saxons inhabiting Wessex during the centuries before the Norman Conquest.

**west-south·west** (wĕst′south′wĕst′; *Naut.* -sou-wĕst′) *n.* The direction or point on the mariner's compass halfway between west and southwest that is 112°30′ west of due north. —*adj.* In, facing, or situated toward west-southwest. —*adv.* Toward or from west-southwest.

**west·ward** (wĕst′wərd) *adj. & adv.* At or toward the west. —*n.* **1.** A direction or point toward the west. **2.** A region situated in or toward the west. **—west′ward·ly** *adj. & adv.* **—west′wards** *adv.*

**wet** (wĕt) *adj.* **wet·ter, wet·test.** [ME < OE *wǣt.*] **1.** Covered or saturated with a liquid, esp. water : MOISTENED. **2.** Not yet firm or dry <*wet* clay><*wet* paint> **3.** Stored or preserved in liquid. **4.** Used or prepared with water or other liquids. **5. a.** Rainy, humid, or foggy <*wet* weather> **b.** Marked by frequent or heavy rainfall or snowfall <a *wet* climate> **6.** *Informal.* Allowing the sale of alcoholic beverages. —*n.* **1.** Something that wets : MOISTURE. **2.** Rainy or snowy weather. **3.** *Informal.* One who supports the legality of the production and sale of alcoholic beverages. —*v.* **wet** or **wet·ted, wet·ting, wets.** —*vt.* **1.** To make wet : DAMPEN. **2.** To make (a bed or one's clothes) wet by urinating. —*vi.* To become wet. **—all wet.** *Slang.* Entirely mistaken. **—wet behind the ears.** Inexperienced : green. **—wet one's whistle.** To take a drink.

**wet blanket** *n. Informal.* One that discourages enjoyment or zeal.

**wet cell** *n.* A primary cell having an electrolyte in the form of a liquid bath.

**weth·er** (wĕth′ər) *n.* [ME < OE.] A gelded male sheep.

**wet·land** (wĕt′lănd′) *n. often* **wetlands.** A lowland area, as a marsh or swamp, that is saturated with moisture, esp. when viewed as the natural habitat of wildlife.

**wet monsoon** *n. Meteorol.* A monsoon.

**wet nurse** *n.* **1.** A woman who suckles another woman's child. **2.** One who treats another with excessive solicitude or care.

**wet-nurse** (wĕt′nûrs′) *vt.* **-nursed, -nurs·ing, -nurs·es. 1.** To serve as wet nurse for. **2.** To treat solicitously.

**wet pack** *n.* A therapeutic pack made of a material, as gauze, that has been moistened in hot or cold water and then wrung out.

**wet suit** *n.* A tight-fitting rubber suit worn by a skin diver in order to retain body heat.

**wetting agent** *n.* A compound that causes a liquid to spread more easily across or penetrate into the surface of a solid by reducing the surface tension of the liquid.

**we've** (wēv). We have.

**whack** (hwăk, wăk) *v.* **whacked, whack·ing, whacks.** [Prob. imit.] —*vt.* To strike with a sharp blow : SLAP. —*vi.* To deal a sharp, resounding blow. —*n.* **1.** A sharp, swift blow. **2.** The sound made by a whack. **—have (or take) a whack at.** *Informal.* To attempt : try. **—out of whack.** *Informal.* Not functioning correctly. **—whacked out.** *Slang.* Insane : crazy.

**whack·ing** (hwăk′ĭng) *Chiefly Brit.* —*adj.* Superlative. —*adv.* Superlatively.

**whack·y** (hwăk′ē, wăk′ē) *adj. var. of* WACKY.

**whale**[1] (hwāl, wāl) *n.* [ME < OE *hwæl.*] **1.** A marine mammal of the order Cetacea, that has a gen. fishlike form with forelimbs modified to form flippers and a tail with horizontal flukes, esp. one of the very large species. **2.** *Informal.* An impressive example <a *whale* of a narrative> —*vi.* **whaled, whal·ing, whales.** To engage in the hunting of whales.

**whale**[2] (hwāl, wāl) *v.* **whaled, whal·ing, whales.** [Orig. unknown.] —*vt.* To strike or hit forcefully and repeatedly : THRASH. —*vi.* To attack vehemently <*whaled* away at the reporters>

**whale·back** (hwāl′băk′, wāl′-) *n.* A steamship with the bow and upper deck rounded so as to shed water.

**whale·boat** (hwāl′bōt′, wāl′-) *n.* **1.** A long rowboat, pointed at both ends and designed to move and turn swiftly, once used in the hunting of whales. **2.** A boat similar to a whaleboat in size and shape.

**whale·bone** (hwāl′bōn′, wāl′-) *n.* **1.** The durable, elastic, hornlike material forming plates or strips in the upper jaw of whalebone whales. **2.** An object made of whalebone, as a corset stay.

**whalebone whale** *n.* The mysticete.

**whale oil** *n.* A yellowish oil obtained from whale blubber, used as a lubricating oil and the manufacture of soap and candles.

**whal·er** (hwā′lər, wā′-) *n.* **1.** One who hunts or processes whales. **2.** A whaling ship. **3.** A whaleboat.

**whale shark** *n.* A large, spotted shark, *Rhincodon typus* of warm marine waters, feeding chiefly on plankton.

**whal·ing** (hwā′lĭng, wāl-) *n.* The business or practice of hunting, killing, and processing whales.

**wham** (hwăm, wăm) *n.* [Imit.] **1.** A forceful, resounding blow. **2.** The sound of a forceful, resounding blow : THUD. —*v.* **whammed, wham·ming, whams.** —*vt.* To strike or smash into with resounding impact. —*vi.* To smash with great force.

**wham·my** (hwăm′ē, wăm′ē) *n., pl.* **-mies.** [Perh. < WHAM.] *Slang.* A supernatural spell capable of subduing an adversary : HEX.

**whang**[1] (hwăng, wăng) *n.* [Var. of ME *thwang,* thong.] *Informal.* **1.** A hide or leather thong or whip. **2. a.** A lashing blow, as of a whip. **b.** The sound of a blow. —*vt.* **whanged, whang·ing, whangs.** *Informal.* **1.** To beat with a thong. **2.** To beat with sharp blows.

**whang**[2] (hwăng, wăng) [Imit.] *Informal.* —*v.* **whanged, whang·ing, whangs.** —*vt.* To strike so as to produce a loud, reverberating noise. —*vi.* To produce a loud, reverberating noise. —*n.* A loud, reverberating noise.

**whang·ee** (hwăng-gē′, wăng′-) *n.* [Chin. (Mandarin) *huang*[2] *li*[2] : *huang*[2], yellow + *li*[2], a kind of bramble.] **1.** A bamboolike Asian grass of the genus *Phyllostachys.* **2.** A walking stick made from the woody stem of the whangee.

**wharf** (hwôrf, wôrf) *n., pl.* **wharves** (hwôrvz, wôrvz) or **wharfs.** [ME *wharfe* < OE *hwearf.*] **1.** A pier where vessels may tie up and load or unload. **2.** *Obs.* A shore or riverbank. —*v.* **wharfed, wharf·ing, wharfs.** —*vt.* **1.** To moor (a vessel) at a wharf. **2.** To take to or store (cargo) on a wharf. **3.** To furnish, equip, or protect with a wharf or wharves. —*vi.* To berth at a wharf.

**wharf·age** (hwôr′fĭj, wôr′fĭj) *n.* **1. a.** Use of a wharf or wharves. **b.** The charges for this use. **2.** Wharves as a whole.

**wharf·in·ger** (hwôr′fĭn-jər, wôr′-) *n.* [Alteration of WHARFAGE + -ER.] The owner or manager of a wharf.

**wharf rat** *n.* **1.** A rat that infests wharves and shipping. **2.** *Slang.* An undesirable person who frequents wharves.

**wharves** (hwôrvz, wôrvz) *n. var. pl. of* WHARF.

**what** (hwŏt, hwŭt, wôt, wŭt; hwət, wət *when unstressed*) *pron.* [ME < OE *hwæt.*] **1. a.** Which thing or which particular one of many <*What* are we having for lunch?><*What* did they do?> **b.** Which kind, character, or designation <*What* is this package?> **c.** One of how much value or significance <*What* is music to a deaf person?> **2. a.** That which : the thing that <Look at *what* I've painted.> **b.** Whatever thing that <Do *what* you will.> **3. a.** *Nonstandard.* Which, who, or that <It's the children *what* get cheated.> **b.** *Informal.* Something <I'll tell you *what.*><Do you know *what?*> —*adj.* **1.** Which one or ones of several or many <*What* company do you work for?><You should know *what* holiday is approaching.> **2.** Whatever <We soon solved *what* problems had arisen.> **3.** *Archaic.* Which degree of : how much. **4.** How great <*What* a spendthrift!> —*adv.* **1.** In what respect : HOW <*What* do you care?> **2.** Which reason : WHY <*What* are you worrying for?> —*conj. Nonstandard.* That <I don't know but *what* I'll go.> —*interj.* **1.** —Used to express surprise, disbelief, or other strong sudden excitement. **2.** *Chiefly Brit.* —Used to express agreement <A fine performance, *what?*> **—and what not.** And other less prominent or unspecified things <needles, thread, buttons, *and what not*> **—what for.** A scolding or strong reprimand <I got *what for* last night.> **—what have you.** What remains and need not be mentioned <dresses, slacks, skirts, and *what have you*> **—what if. 1.** What would occur if : suppose that. **2.** What does it matter. **—what it takes.** The qualities or expertise needed for success. **—what's what.** *Informal.* The fundamentals and details <Tell me *what's what.*> **—what with.** *Informal.* Taking into consideration : because of <*What with* the constant rain, we all caught colds.>

**what·ev·er** (hwŏt-ĕv′ər, hwŭt-wŏt-, wŭt-) *pron.* **1.** Everything or anything that <Do *whatever* makes you happy.> **2.** What amount that : the whole of what <*Whatever* we earn goes into the bank.> **3.** No matter what <*Whatever* happens, be sure to call me.> **4.** *Informal.* Which thing or things <*Whatever* did they say?> —*adj.* **1.** Of any number or kind : ANY <*Whatever* supplies you need will be provided.> **2.** The whole of : all of <We gave *whatever* food remained to the late-comers.> **3.** Of any kind at all <No cyclists *whatever* may leave without helmets.>

**what·not** (hwŏt′nŏt′, hwŭt′-, wŏt′-, wŭt′-) *n.* **1.** A small, trivial, or unspecified object. **2.** A set of light, open shelves for ornaments.

**what·so·ev·er** (hwŏt′sō-ĕv′ər, hwŭt′-, wŏt′-, wŭt′-) *pron.* Whatever. —*adj.* Whatever <no strength *whatsoever*>

**wheal** (hwēl, wēl) *n.* [Var. of WALE.] A small acute swelling on the surface of the skin.

**wheat** (hwēt, wēt) *n.* [ME *whete* < OE *hwǣte.*] **1.** A cereal grass of the genus *Triticum,* esp. *T. aestivum,* widely cultivated in many varieties for its commercially important edible grain. **2.** The grain of a wheat plant, ground to produce flour.

**wheat bread** *n.* A bread made from a blend of white and whole-wheat flours.

**wheat·ear** (hwēt′ĭr′, wēt′-) *n.* [Back-formation from obs. *wheatears,* wheatear, prob. by folk ety. < WHITE + ARSE.] A brown, black, and white bird, *Oenanthe oenanthe,* indigenous to northern regions.

**wheat·en** (hwēt′n, wēt′n) *adj.* Of, relating to, or derived from wheat.

**wheat germ** *n.* The vitamin-rich embryo of the wheat kernel that is separated before milling for use as a cereal or food supplement.

**wheat rust** *n.* **1.** A destructive disease of wheat. **2.** A fungus, as *Puccinia gyraminis,* that causes wheat rust.

**Wheat·stone bridge** (hwēt′stōn′, wēt′-) *also* **Wheat·stone's bridge** (-stōnz′) *n.* [After Sir Charles *Wheatstone* (1802–1875).] An instrument or circuit having four resistors or their equivalent in series, with a galvanometer linking the junction between one pair and the other, used to determine the value of an unknown resistance when the other three resistances are known.

**wheat·worm** (hwēt′wûrm′, wēt′-) *n.* A nematode worm, *Anguina tritici,* parasitic on and destructive to wheat.

**whee** (hwē, wē) *interj.* —Used to express pleasure or zeal.

**whee·dle** (hwēd′l, wēd′l) *v.* **-dled, -dling, -dles.** [Orig. unknown.] —*vt.* **1.** To persuade or attempt to persuade by flattery or guile : CAJOLE. **2.** To obtain through the use of flattery or guile <*wheedled* all my money out of me> —*vi.* To use flattery or guile to achieve one's ends. **—whee′dler** *n.* **—whee′dling·ly** *adv.*

**wheel** (hwēl, wēl) *n.* [ME < OE *hwēol.*] **1.** A solid disk or a rigid circular ring connected by spokes to a hub, designed to turn around an axle passed through the center. **2.** Something resembling a wheel or having a wheel as its principal part or characteristic, as: **a.** An instrument to which a victim was bound for torture during the Middle Ages. **b.** A firework that rotates while burning. **c.** The steering device on a vehicle. **d.** *Informal.* A bicycle. **e.** A spinning wheel. **f.** A water wheel. **g.** A potter's wheel. **h.** A device used in roulette and other games of chance. **3. wheels.** Forces that provide energy, movement, or direction <the *wheels* of business> **4.** The act or process of turning. **5.** A military maneuver to change the direction of movement of a formation, as of troops or ships, in which the formation is maintained while the outer unit describes an arc and the inner unit remains stationary as a pivot. **6. wheels.** *Slang.* A motor vehicle or access thereto. **7.** *Slang.* One with a great deal of power or influence. —*v.* **wheeled, wheel·ing, wheels.** —*vt.* **1.** To roll, move, or transport on a wheel or wheels. **2.** To cause to revolve or rotate around or as if around a central axis. **3.** To provide with a wheel or wheels. —*vi.* **1.** To turn around or as if around a central axis. **2.** To roll or move on or as if on a wheel or wheels. **3.** To fly in a curving or circular course <Seagulls *wheeled* above the waves.> **4.** To turn or whirl around in place : PIVOT <"The boy *wheeled* and the fried eggs leaped from his tray" —Ivan Gold> **5.** To reverse one's opinion or practice. **—at (or behind) the wheel. 1.** Operating the steering mechanism of a vehicle : DRIVING. **2.** In charge. **—wheel and deal.** *Informal.* To engage in the advancement of one's own interests.

**wheel and axle** *n.* A mechanical device, analogous to the lever, having two coaxial wheels of different diameters conjoined so that the effort applied by a cord to the larger wheel in the form of a torque is transmitted as an action by a cord around the circumference of the smaller, yielding a mechanical advantage equal to the ratio of the diameters of the wheels.

**wheel animalcule** *n.* A rotifer.

**wheel·bar·row** (hwēl′băr′ō, wēl′-) *n.* A one- or two-wheeled, handled vehicle used to convey small heavy loads by hand.
**wheel·base** (hwēl′bās′, wēl′-) *n.* The distance from front to rear axle in a motor vehicle, usu. expressed in inches.
**wheel bug** *n.* A large predatory insect, *Arilus cristatus,* with a notched, wheellike projection on the thorax.
**wheel·chair** *also* **wheel chair** (hwēl′châr′, wēl′-) *n.* A chair mounted on large wheels for the use of the sick or disabled.
**wheeled** (hwēld, wēld) *adj.* Having a wheel or wheels.
**wheel·er** (hwē′lər, wē′-) *n.* **1.** One that wheels. **2.** A thing that moves on or is equipped with a wheel or wheels <a two-*wheeler*> **3.** A wheel horse.
**wheel·er-deal·er** (hwē′lər-dē′lər, wē′-) *n. Informal.* One who wheels and deals, esp. in politics or business.
**wheel horse** *n.* **1.** The horse in a team that follows the leader and is harnessed nearest to the front wheels. **2.** A diligent, dependable worker, as in a political organization.
**wheel house** *n.* A pilothouse.
**wheel lock** *n.* A firing mechanism in certain obsolete small arms in which a small wheel produces sparks by revolving against a flint.
**wheel·man** (hwēl′mən, wēl′-) *n.* **1.** A ship's helmsman. **2.** A bicyclist.
**wheels·man** (hwēlz′mən, wēlz′-) *n.* A wheelman.
**wheel·work** (hwēl′wûrk′, wēl′-) *n.* An arrangement of wheels or gears in a mechanical device.
**wheel·wright** (hwēl′rīt′, wēl′-) *n.* One that builds and repairs wheels.
**wheeze** (hwēz, wēz) *v.* **wheezed, wheez·ing, wheez·es.** [ME *whesen,* prob. < ON *hvæsa,* to hiss.] *—vi.* **1.** To breathe with difficulty, producing a hoarse whistling or hissing sound. **2.** To make a sound suggestive of laborious breathing. *—vt.* To produce or utter with a hoarse whistling sound. *—n.* **1.** A wheezing sound. **2.** *Informal.* An old joke. **—wheez′er** *n.* **—wheez′ing·ly** *adv.*
**wheez·y** (hwē′zē, wē′-) *adj.* **-i·er, -i·est. 1.** Given to wheezing. **2.** Producing a wheezing sound <a *wheezy* cough> **—wheez′i·ly** *adv.* **—wheez′i·ness** *n.*
**whelk[1]** (hwĕlk, wĕlk) *n.* [ME *whelke* < OE *weoloc.*] Any of various large, sometimes edible marine snails of the family Buccinidae, with pointed, turreted shells.
**whelk[2]** (hwĕlk, wĕlk) *n.* [ME *whelke* < OE *hwylca* < *hwelian,* to suppurate.] *Pathol.* A swelling, protuberance, or pustule : WHEAL. **—whelk′y** *adj.*
**whelm** (hwĕlm, wĕlm) *vt.* **whelmed, whelm·ing, whelms.** [ME *whelmen,* to turn over.] **1.** To cover with water : SUBMERGE. **2.** To overcome emotionally or mentally : OVERWHELM.
**whelp** (hwĕlp, wĕlp) *n.* [ME *whelpe* < OE *hwelp.*] **1.** A young offspring of a mammal, as a dog or wolf. **2. a.** A child : youth. **b.** An impudent young fellow. **3. a.** A tooth of a sprocket wheel. **b.** A ridge on the barrel of a windlass or capstan. *—v.* **whelped, whelp·ing, whelps.** *—vi.* To give birth to a whelp or whelps. *—vt.* To give birth to (a whelp or whelps).
**when** (hwĕn, wĕn) *adv.* [ME < OE *hwenne.*] **1.** At what time <*When* will we arrive?> **2.** At which time <I know *when* to call it quits.> *—conj.* **1.** At the time that <just before dawn, *when* it's very quiet> **2.** As soon as <called me *when* they got home> **3.** Whenever <*When* we worry too much, we get ill.> **4.** During the time at which : WHILE <*when* I was thinner> **5.** Whereas : although <They gave up *when* they might have succeeded.> **6.** Considering that : IF <How can you be a pianist *when* you won't practice?> *—pron.* What or which time <Since *when* have you been so neat?> *—n.* The time or date <haven't selected the where and *when*> ***usage:*** *When* is often used informally to mean "a situation or event," as in *A dilemma is when you don't know which way to turn.* This usage is best avoided in formal prose.
**when·as** (hwĕn-ăz′, wĕn-) *conj. Archaic.* **1.** When. **2.** Whereas.
**whence** (hwĕns, wĕns) *adv.* [ME *whennes* < *whenne,* whence < OE *hwanon.*] **1.** From what place : from where <*Whence* came this creature?> **2.** From what origin or source <*Whence* comes this lavish display?> *—conj.* **1.** Out of which place : from or out of which. **2.** By reason of which : from which <The cat was snowy white from ears to tail, *whence* the name Snowflake.> ***usage:*** Because *whence* contains the sense of "*from,*" the expression *from whence* is avoided by many because it is considered redundant. Therefore, *Tell us whence* (not *from whence*) *they came* is preferable.
**whence·so·ev·er** (hwĕns′sō-ĕv′ər, wĕns′-) *adv.* From whatever place or source. *—conj.* From any place or source that.
**when·ev·er** (hwĕn-ĕv′ər, wĕn-) *adv.* **1.** At whatever time. **2.** *also* **when ever.** When. *—conj.* **1.** At whatever time that <We can eat *whenever* the guests arrive.> **2.** Every time that <I smile *whenever* I think of that day.>
**when·so·ev·er** (hwĕn′sō-ĕv′ər, wĕn-) *adv.* At whatever time at all : WHENEVER. *—conj.* Whenever.
**where** (hwâr, wâr) *adv.* [ME < OE *hwær.*] **1.** At or in what place <*Where* are the restrooms?> **2.** In what situation or position <*Where* would we be without the loan?> **3.** From what place or source <*Where* did you hear that story?> **4.** To what place : toward what end <*Where* is this discussion leading?> *—conj.* **1.** At what or which place <came to a field *where* wildflowers grew> **2. a.** In a place in which <lives *where* the winter is short> **b.** In any place or situation in which : WHEREVER <*Where* I go, my friend goes too.> **3. a.** To a place in which <You should go *where* you can study.> **b.** To a place or situation in which <won't go *where* I'm not wanted> *—n.* **1.** The place or occasion <knew the when but not the *where* of it> **2.** What place, source, or cause <*Where* are they from?>
**where·a·bouts** (hwâr′ə-bouts′, wâr′-) *adv.* In, at, or near what location <*Whereabouts* can we find it?> *—n. (sing. or pl. in number).* Approximate location <I've just learned their *whereabouts.*>
**where·as** (hwâr-ăz′, wâr-) *conj.* **1.** It being the fact that : INASMUCH AS. **2.** While at the same time. **3.** While on the contrary. *—n.* **1.** An introductory statement to a formal document : PREAMBLE. **2.** A conditional statement.
**where·at** (hwâr-ăt′, wâr-) *conj.* **1.** Toward or at which. **2.** As a result or consequence of : WHEREUPON.
**where·by** (hwâr-bī′, wâr-) *conj.* In accordance with which : by or through which.
**where·fore** (hwâr′fôr′, -fōr′, wâr′-) *adv.* [ME *wherfor* : *wher,* where + *fore,* for.] **1.** For what purpose or reason : WHY. **2.** Therefore. *—n.* A purpose or cause <the whys and *wherefores*>
**where·from** (hwâr′frŏm′, -frŭm′, wâr′-) *conj.* From which.
**where·in** (hwâr-ĭn′, wâr-) *adv.* In what way : HOW <*Wherein* have I erred?> *—conj.* **1.** In which location : WHERE <the city *wherein* my parents live> **2.** During which. **3.** In what way : HOW <told us *wherein* we were mistaken>
**where·in·to** (hwâr-ĭn′to͞o, wâr-) *conj.* Into which.
**where·of** (hwâr-ŏv′, -ŭv′, wâr-) *conj.* **1.** Of what <They know *whereof* they speak.> **2. a.** Of which <ancient art *whereof* many examples have been destroyed> **b.** Of whom.
**where·on** (hwâr-ŏn′, -ôn′, wâr-) *adv. Archaic.* On which or what.
**where·so·ev·er** (hwâr′sō-ĕv′ər, wâr′-) *conj. Archaic.* In, to, or from whatever place at all : WHEREVER.
**where·through** (hwâr′thro͞o′, wâr′-) *conj.* Through, because of, or during which.
**where·to** (hwâr′to͞o′, wâr′-) *adv.* To what place : toward what end. *—conj.* To which.
**where·un·to** (hwâr-ŭn′to͞o, wâr-) *adv. & conj.* Whereto.
**where·up·on** (hwâr′ə-pŏn′, -pôn′) *conj.* **1.** On which. **2.** In close consequence of which <The judge entered the courtroom, *whereupon* we stood up.>
**wher·ev·er** (hwâr-ĕv′ər, wâr-) *adv.* [ME : *wher,* where + *ever,* ever.] **1.** In or to whatever place <Cut the cloth *wherever* indicated.> **2.** *also* **where ever.** Where <*Wherever* did you get those shoes?> *—conj.* In or to whichever place or situation <made friends *wherever* I went>
**where·with** (hwâr′wĭth′, -wĭth′, wâr′-) *adv.* With what or which. *—pron.* The thing or things with which. *—conj.* By means of which.
**where·with·al** (hwâr′wĭth-ôl′, -wĭth-, wâr′-) *conj.* Wherewith. *—pron.* Wherewith. *—n.* The necessary means, esp. financial means <didn't have the *wherewithal* to plan for retirement>
**wher·ry** (hwĕr′ē, wĕr′ē) *n., pl.* **-ries.** [ME *whery.*] **1.** A light swift rowboat built for one person and often used in racing. **2.** A sailing barge used in East Anglia.

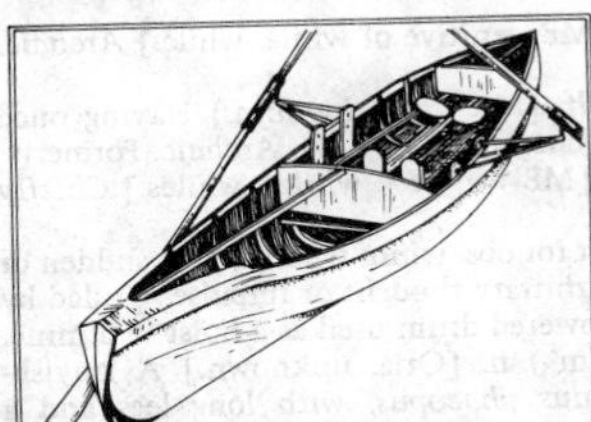

**wherry**
*The rowboat*

**whet** (hwĕt, wĕt) *vt.* **whet·ted, whet·ting, whets.** [ME *whetten* < OE *hwettan.*] **1.** To sharpen or hone <*whet* a knife> **2.** To make more keen : STIMULATE <The sizzling steak *whetted* my appetite.> *—n.* **1.** The act of whetting. **2.** Something that whets. **3.** *Informal.* An appetizer.
**wheth·er** (hwĕth′ər, wĕth′-) *conj.* [ME < OE *hweðer.*] **1.** —Used in indirect questions to introduce one alternative <We must find out *whether* the restaurant is open.> **2.** —Used to introduce alternative possibilities <*Whether* I stay or *whether* I go, I'll need more money.> **3.** Either <won the race, *whether* by luck or skill> *—pron. Obs.* Which. **—whether or no.** Regardless of the prevailing circumstances.
**whet·stone** (hwĕt′stōn′, wĕt′-) *n.* A stone for honing tools.
**whew** (hwo͞o, hwyo͞o) *interj.* —Used to express strong emotion such as amazement or relief.

**whey** (hwā, wā) *n.* [ME < OE *hwaēg.*] The watery part of milk that separates from the curds, as in the process of making cheese. **—whey′ey** *adj.*

**whey-face** (hwā′fās′, wā′-) *n.* One with a pallid face.

**which** (hwĭch, wĭch) *pron.* [ME < OE *hwilc.*] **1.** What particular one or ones <*Which* of these hats is yours?> **2.** The particular one or ones <Give me that *which* is mine.> **3.** The one or ones previously named or implied, esp.: **a.** —Used as a relative pronoun in a clause that provides additional information about the antecedent <my car, *which* is old but reliable> **b.** —Used as a relative pronoun preceded by *that* or a preposition in a clause that defines or restricts the antecedent <that *which* we ordered><the topic on *which* I spoke> **c.** —Used instead of *that* as a relative pronoun in a clause that defines or restricts the antecedent <The sale *which* was held later was better.> **4.** *Archaic.* The person named or implied. **5.** Any of the persons, things, or events named or implied : WHICHEVER <Buy *which* looks the most useful.> **6.** A thing or circumstance that <We stayed late, *which* was foolish.> *—adj.* **1.** What particular one or ones of a number of persons or things <*which* piece of material> **2.** Any one or any number of : WHICHEVER <Take *which* car you please.> **3.** Being the one or ones previously named <The music began, at *which* point the curtain rose.>

**which·ev·er** (hwĭch-ĕv′ər, wĭch-) *pron.* Whatever one or ones. *—adj.* Being any one or any number of a group <See *whichever* films you like.><It's a short trip *whichever* route you take.>

**which·so·ev·er** (hwĭch′sō-ĕv′ər, wĭch′-) *pron. & adj.* Whichever.

**whick·er** (hwĭk′ər, wĭk′-) *vi.* **-ered, -er·ing, -ers.** [Imit.] To whinny. *—n.* A whinny.

**whid·ah** (hwĭd′ə, wĭd′ə) *n. var. of* WHYDAH.

**whiff** (hwĭf, wĭf) *n.* [ME *weffe,* offensive smell.] **1.** A slight, gentle gust of air : WAFT. **2.** A brief, passing odor carried in the air <a *whiff* of cologne> **3.** An inhalation, as of air or smoke <took a *whiff* of the pipe> *—v.* **whiffed, whiff·ing, whiffs.** *—vi.* To be carried in brief gusts : WAFT. *—vt.* **1.** To blow or convey in whiffs. **2.** To inhale through the nose : SNIFF <a horse *whiffing* the air> **—whiff′er** *n.*

**whif·fle** (hwĭf′əl, wĭf′-) *v.* **-fled, -fling, -fles.** [< WHIFF.] *—vi.* **1.** To think or move erratically : VACILLATE. **2.** To blow in fitful gusts : PUFF. **3.** To whistle lightly. *—vt.* To blow, displace, or scatter with gusts of air.

**whif·fle·tree** (hwĭf′əl-trē, wĭf′-) *n.* [Var. of WHIPPLETREE.] The pivoted horizontal crossbar to which the harness traces of a draft animal are attached and which is then attached to a vehicle or an implement.

**Whig** (hwĭg, wĭg) *n.* [Prob. short for *Whiggamore,* a member of a body of 17th-cent. Scottish insurgents.] **1.** A member of an 18th- and 19th-cent. English political party that was opposed to the Tories. **2.** A supporter of the war against England during the American Revolution. **3.** A 19th-cent. American political party formed in opposition to the Democratic Party and favoring high tariffs and a loose interpretation of the Constitution. **—Whig′ger·y** *n.* **—Whig′gish** *adj.* **—Whig′gism** *n.*

**while** (hwīl, wīl) *n.* [ME < OE *hwīl.*] **1.** A period of time <rest a *while*> **2.** The time, effort, or trouble taken in doing something <wasn't worth my *while*> *—conj.* **1.** During the time that : AS LONG AS <We enjoyed it *while* it lasted.> **2.** At the same time that : ALTHOUGH <*While* I like to travel, I'm happiest at home.> **3.** Whereas : and <The dress is satin, *while* the lining is silk.> *—vt.* **whiled, whil·ing, whiles.** To spend (time) idly or pleasantly <*whiled* the hours away>

**whiles** (hwīlz, wīlz) *conj.* [ME, genitive of *while,* while.] *Archaic.* While.

**whi·lom** (hwī′ləm, wī′-) *adj.* [ME < OE *hwīlum.*] Having once been : FORMER <the *whilom* champion> *—adv. Archaic.* Formerly.

**whilst** (hwīlst, wīlst) *conj.* [ME *whylst* < *whiles,* whiles.] *Chiefly Brit.* While.

**whim** (hwĭm, wĭm) *n.* [Short for obs. *whim-wham.*] **1.** A sudden or capricious idea or fancy. **2.** Arbitrary thought or impulse <ruled by *whim*> **3.** A vertical horse-powered drum used as a hoist in a mine.

**whim·brel** (hwĭm′brəl, wĭm′-) *n.* [Orig. unknown.] A grayish-brown wading bird, *Numenius phaeopus,* with long legs and a downward-curving bill.

**whim·per** (hwĭm′pər, wĭm′-) *v.* **-pered, -per·ing, -pers.** [Imit.] *—vi.* **1.** To cry or sob with soft intermittent sounds : WHINE. **2.** To complain. *—vt.* To utter in a whimper. *—n.* A low, broken, sobbing sound : WHINE. **—whim′per·er** *n.* **—whim′per·ing·ly** *adv.*

**whim·sey** (hwĭm′zē, wĭm′-) *var. of* WHIMSY.

**whim·si·cal** (hwĭm′zĭ-kəl, wĭm′-) *adj.* [< WHIMSY.] **1.** Capricious, playful, or fanciful <a *whimsical* idea> **2.** Erratic or unpredictable <a *whimsical* nature> **—whim′si·cal·ly** *adv.*

**whim·si·cal·i·ty** (hwĭm′zĭ-kăl′ə-tē, wĭm′-) *n., pl.* **-ties. 1.** The quality of being whimsical. **2.** A whimsical idea or its expression : CAPRICE.

**whim·sy** *also* **whim·sey** (hwĭm′zē, wĭm′-) *n., pl.* **-sies** *also* **-seys.** [< WHIM.] **1.** An odd or capricious idea : FANCY. **2.** Something quaint, fanciful, or odd.

**whin¹** (hwĭn, wĭn) *n.* [ME *whynne.*] Gorse.

**whin²** (hwĭn, wĭn) *n.* [ME *quin.*] Whinstone.

**whin·chat** (hwĭn′chăt′, wĭn′-) *n.* A brownish Old World bird, *Saxicola rubetra,* frequenting open country.

**whine** (hwīn, wīn) *v.* **whined, whin·ing, whines.** [ME *whinen* < OE *hwīnan,* to make a whizzing sound.] *—vi.* **1.** To utter a long, plaintive, high-pitched sound, as in pain, fear, supplication, or complaint. **2.** To protest or complain in a childish, annoying fashion. **3.** To produce a sustained noise of relatively high pitch <aircraft engines *whining*> *—vt.* To utter with a whine. *—n.* **1.** A whining sound. **2.** The act of whining. **3.** A complaint uttered in a plaintive tone. **—whin′er** *n.* **—whin′ing·ly** *adv.* **—whin′y** *adj.*

**whin·ny** (hwĭn′ē, wĭn′ē) *v.* **-nied, -ny·ing, -nies.** [Prob. imit.] *—vi.* To neigh, as a horse, esp. in a gentle tone. *—vt.* To express in a whinny. *—n., pl.* **-nies.** The sound made in whinnying : NEIGH.

**whin·stone** (hwĭn′stōn′, wĭn′-) *n.* Any of various hard, dark-colored rocks, esp. basalt and chert.

**whip** (hwĭp, wĭp) *v.* **whipped** *or* **whipt** (hwĭpt, wĭpt), **whip·ping, whips.** [ME *wippen.*] *—vt.* **1.** To strike with repeated strokes, as of a strap or rod : LASH. **2. a.** To punish or chastise by repeated striking with a strap or rod : FLOG. **b.** To afflict, castigate, or reprove severely. **3.** To drive, force, or compel by or as if by flogging or lashing. **4.** To strike or affect in a manner similar to whipping or lashing <Stormy winds *whipped* our faces.> **5.** To beat (e.g., cream or eggs) into a froth or foam. **6.** To snatch, pull, or remove in a sudden manner <*whip* off one's hat> **7.** To sew with a loose overcast or overhand stitch. **8.** To wrap or bind (e.g., a rope) with twine to strengthen or prevent fraying. **9.** *Naut.* To hoist by means of a rope passing through an overhead pulley. **10.** *Informal.* To outdo : defeat <Our side can *whip* your side.> *—vi.* **1.** To move in a sudden, quick manner : DART. **2.** To thrash or snap about like a whip <Loose wires *whipped* against the house.> **—whip in.** To keep together, as members of a political party or a pack of hounds. **—whip up. 1.** To arouse : excite <*whip up* the audience><*whip up* indignation> **2.** *Informal.* To prepare quickly <*whip up* a snack> *—n.* **1.** An instrument, either a flexible rod or a flexible thong or lash attached to a handle, used for driving animals or administering punishment. **2.** A whipping or lashing motion or stroke : WHIPLASH. **3.** A blow, wound, or cut made by or as if by whipping. **4.** Something, as an automotive radio antenna, similar to a whip in form or flexibility. **5.** Flexibility, as in the shaft of a golf club. **6.** WHIPPER-IN 1. **7. a.** A member of a legislative body, as the U.S. Congress or the British Parliament, charged with enforcing party discipline and insuring attendance. **b.** A call issued to party members in a lawmaking body to insure attendance at a particular time. **8.** A dessert made of sugar and stiffly beaten egg whites or cream, often with fruit or fruit flavoring <strawberry *whip*> **9.** A windmill arm. **10.** *Naut.* A hoist consisting of a single rope passing through an overhead pulley. **11.** A ride in an amusement park, having small cars that move in a rapid, whipping motion. **—whip into shape.** To bring to a given state or condition, often forcefully. **—whip′per** *n.*

**whip·cord** (hwĭp′kôrd′, wĭp′-) *n.* **1.** A worsted fabric with a distinct diagonal rib. **2.** A strong braided or twisted cord sometimes used in making whiplashes. **3.** Catgut.

**whip hand** *n.* **1.** The hand in which the whip is held. **2.** A dominating position : ADVANTAGE.

**whip·lash** (hwĭp′lăsh′, wĭp′-) *n.* **1.** The lash of a whip. **2.** An injury to the cervical spine caused by an abrupt jerking motion of the head, either backward or forward.

**whiplash injury** *n.* WHIPLASH 2.

**whip·per-in** (hwĭp′ər-ĭn′, wĭp′-) *n., pl.* **whip·pers-in. 1.** One who assists the huntsman in handling a pack of hounds during a foxhunt. **2.** WHIP 7a.

**whip·per·snap·per** (hwĭp′ər-snăp′ər, wĭp′-) *n.* [Alteration of dial. *snippersnapper.*] An insignificant, often impudent person.

**whip·pet** (hwĭp′ĭt, wĭp′-) *n.* [Prob. < WHIP.] A short-haired, swift-running dog orig. bred in England, resembling but smaller than the greyhound.

**whip·ping** (hwĭp′ĭng, wĭp′-) *n.* **1.** The act of one that whips. **2.** A thrashing administered esp. as punishment. **3.** Material, as cord or thread, used to bind or lash parts.

**whipping boy** *n.* **1.** A scapegoat. **2.** A boy formerly raised with a young nobleman and whipped for the latter's misdeeds.

**whip·ple·tree** (hwĭp′əl-trē, wĭp′-) *n.* [Alteration of WHIP + TREE.] A whiffletree.

**whip·poor·will** *also* **whip-poor-will** (hwĭp′ər-wĭl′, wĭp′-, hwĭp′ər-wĭl′, wĭp′-) *n.* [Imit. of its call.] A brownish nocturnal North American bird, *Caprimulgus vociferus.*

**whip·saw** (hwĭp′sô′, wĭp′-) *n.* A narrow two-man crosscut saw. *—vt.* **-sawed** *or* **-sawn** (-sôn′), **-saw·ing, -saws. 1.** To cut with a whipsaw. **2.** To win two bets from (a person) at one time, as in faro. **3.** To defeat or best in two ways at once.

**whip scorpion** *n.* A nonvenomous scorpionlike arachnid of the order Pedipalpi, as the vinegarroon.

**whip snake** *n.* **1.** A slender nonvenomous New World snake of the genus *Masticophis.* **2.** A snake resembling the whip snake.

**whip·stall** (hwĭp′stôl′, wĭp′-) *n.* A usu. intentional stall in which a small aircraft enters a vertical climb, pauses, slips backward momentarily, then drops nose downward.
**whip·stitch** (hwĭp′stĭch′, wĭp′-) *vt.* **-stitched, -stitch·ing, -stitch·es.** To sew with overcast stitches, as in finishing a fabric edge or binding two pieces of fabric together. —*n.* An overcast stitch.
**whipt** (hwĭpt, wĭpt) *v. var. p.t. & p.p. of* WHIP.
**whip·tail** (hwĭp′tāl′) *n.* A New World lizard of the genus *Cnemidophorus,* with a long slender tail.
**whip·worm** (hwĭp′wûrm′, wĭp′-) *n.* A slender whiplike parasitic roundworm, *Trichuris trichiura,* that infests the large intestine.
**whir** (hwûr, wûr) *v.* **whirred, whir·ring, whirs.** [ME *whirren,* of Scand. orig.] —*vi.* To move so as to produce a vibrating or buzzing sound. —*vt.* To cause to make a vibratory sound. —*n.* **1.** A sound of buzzing or vibration. **2.** An excited and noisy activity : BUSTLE.
**whirl** (hwûrl, wûrl) *v.* **whirled, whirl·ing, whirls.** [ME *whirlen* < ON *hvirfla.*] —*vi.* **1.** To revolve rapidly about a center or axis. **2.** To rotate or spin rapidly <a leaf *whirling* to the ground> **3.** To turn quickly, changing direction : WHEEL <I *whirled* around to face my opponent.> **4.** To have the sensation of spinning : REEL. **5.** To move circularly and rapidly in random directions, as the wind. —*vt.* **1.** To cause to rotate or turn rapidly. **2.** To move or drive in a circular or curving course. **3.** To drive at high speed. **4.** *Obs.* To hurl. —*n.* **1.** The act of revolving or rotating rapidly. **2.** Something, as a cloud of dust, that whirls or is whirled. **3.** Tumult : confusion. **4.** A swift succession or round of events <caught up in the social *whirl*> **5.** A state of mental confusion or giddiness : DIZZINESS <My mind is in a *whirl.*> **6.** *Informal.* A short ride or trip. **7.** *Informal.* A brief try <gave skiing a *whirl*> **—whirl′er** *n.*
**whirl·i·gig** (hwûr′lĭ-gĭg′, wûr′-) *n.* [ME *whirlegigge* : *whirlen,* whirl + *-gigg,* something that rotates.] **1.** A toy that spins. **2.** A merry-go-round or carousel. **3.** Something that continuously whirls.
**whirligig beetle** *n.* Any of various beetles of the family Gyrinidae that circle about rapidly on the surface of quiet water.

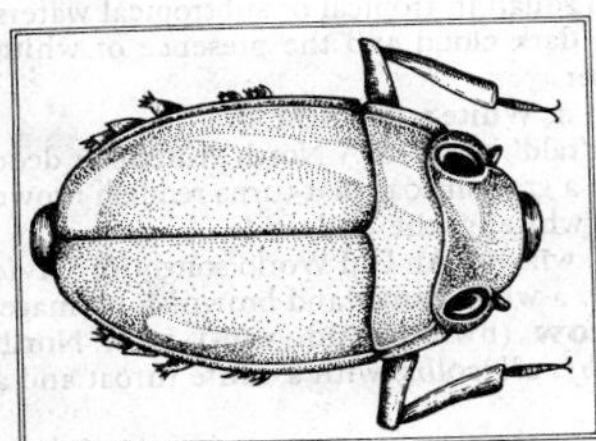
**whirligig beetle**
*1–2 inches long*

**whirl·pool** (hwûrl′po͞ol′, wûrl′-) *n.* **1.** Water in rapid rotating movement, as from the converging of two tides : VORTEX. **2. a.** Turmoil : whirl. **b.** An impelling force into which one may be pulled.
**whirl·wind** (hwûrl′wĭnd′, wûrl′-) *n.* **1. a.** A column of air centered on an area of low atmospheric pressure, rotating violently around a more or less vertical axis and moving forward : TORNADO. **b.** A small momentary current of whirling air over dusty flat land. **2. a.** A tumultuous, confused rush. **b.** A destructive force or thing. —*adj.* Forceful or fast <a *whirlwind* tour of 36 cities>
**whirl·y·bird** (hwûr′lē-bûrd′, wûr′-) *n. Slang.* A helicopter.
**whirr** (hwûr, wûr) *v. & n. Chiefly Brit. var. of* WHIR.
**whisk** (hwĭsk, wĭsk) *v.* **whisked, whisk·ing, whisks.** [ME *wisk,* whisk, of Scand. orig.] —*vt.* **1.** To move or cause to move with quick, light sweeping motions <*whisked* the toddler out of the room> **2.** To whip (eggs or cream). —*vi.* To move lightly, nimbly, and rapidly. —*n.* **1.** A quick, light sweeping motion. **2.** A whiskbroom. **3.** A small bunch, as of twigs or hair, attached to a handle and used in brushing. **4.** A wire kitchen utensil for whipping foodstuffs, as eggs, cream, or potatoes.
**whisk·broom** (hwĭsk′bro͞om′, -bro͝om′, wĭsk′-) *n.* A small short-handled broom used esp. to brush clothes.
**whisk·er** (hwĭs′kər, wĭs′-) *n.* [< WHISK.] **1. a. whiskers.** The unshaven hair on a man's face forming the beard. **b.** A single hair of the beard. **2.** One of the long stiff bristles or hairs growing near the mount of certain animals. **3.** *Informal.* A narrow margin : HAIRSBREADTH <won by a *whisker*> **4.** *Naut.* One of two spars or booms projecting from the side of a bowsprit for spreading the jib or flying-jib guys. **5.** *Chem.* An extremely fine filamentary crystal that can be grown from supersaturated solutions of certain minerals and has extraordinary shear strength and unusual electrical or surface properties. **—whisk′ered, whisk′ery** *adj.*
**whis·key** (hwĭs′kē, wĭs′-) *n., pl.* **-keys.** [Sc. *whiskybae* < Sc. Gael. *uisge beatha,* water of life.] **1.** An alcoholic liquor distilled from grain, as corn, rye, or barley, and containing approx. 40–50% ethyl alcohol by volume. **2.** A drink of whiskey.
**whiskey jack** *n.* [< obs. *whiskyjohn,* by folk ety. < Cree *wiskačan.*] The Canada jay.
**whiskey sour** *n.* A cocktail of whiskey, lemon juice, and sugar.
**whis·ky** (hwĭs′kē, wĭs′-) *n. Scot. var. of* WHISKEY.
**whis·per** (hwĭs′pər, wĭs′-) *n.* [ME *whisperen* < OE *hwisprian.*] **1.** Soft speech produced without full voice. **2.** Something uttered. **3.** A surreptitiously or secretly expressed belief, rumor, or hint <a *whisper* of impropriety> **4.** A low rustling sound <the *whisper* of wind in the tall grass> —*v.* **-pered, -per·ing, -pers.** —*vi.* **1.** To speak softly, without full voice. **2.** To speak quietly or privately, as when imparting gossip, slander, or intrigue. **3.** To make a soft rustling sound. —*vt.* **1.** To utter very softly. **2.** To say or tell privately or secretly. **—whis′per·er** *n.*
**whist** (hwĭst, wĭst) *n.* [Orig. unknown.] A game of cards played with 52 cards by 2 teams of 2 players each.
**whis·tle** (hwĭs′əl, wĭs′-) *v.* **-tled, -tling, -tles.** [ME *whistlen* < OE *hwistlian.*] —*vi.* **1.** To produce a clear musical sound by forcing air through the teeth or through an aperture formed by pursing the lips. **2.** To produce a shrill, sharp musical sound by blowing on or through a device. **3.** To produce a high-pitched sound when moving swiftly through the air <The ball *whistled* past.> **4.** To emit a shrill, sharp, high-pitched cry, as some birds and animals. **5.** To summon by whistling. —*vt.* **1.** To produce by whistling <*whistle* a song> **2.** To summon, signal, or direct by whistling. **3.** To cause to move with a whistling noise. —*n.* **1. a.** A small wind instrument for making whistling sounds by means of the breath. **b.** A device for making whistling sounds by means of forced air or steam <a train *whistle*> **2.** A sound produced by a device or by whistling through the lips. **3.** A whistling sound, as of an animal or projectile. **4.** The act of whistling. **5.** A whistling sound used to summon or command. **—whistle in the dark.** To attempt to keep one's courage up.
**whis·tler** (hwĭs′lər, wĭs′-) *n.* **1.** One that whistles. **2.** A marmot, *Marmota caligata* of the mountains of northwestern North America, with a grayish coat and a shrill, whistling cry. **3.** A bird that produces a whistling sound. **4.** *Physics.* An electromagnetic wave of audio frequency produced by atmospheric disturbances, as lightning, having a decreasing frequency responsible for a whistling sound of descending pitch in detection equipment. **5.** A horse having a respiratory disease marked by wheezing.
**whistle stop** *n.* **1.** A town at which a train stops only if signaled. **2.** A brief appearance of a political candidate in a small town, traditionally on the observation platform of a train.
**whis·tle-stop** (hwĭs′əl-stŏp′, wĭs′-) *vi.* **-stopped, -stop·ping, -stops.** To conduct a political campaign by making brief appearances or speeches in a series of small towns.
**whistling swan** *n.* A North American swan, *Olor columbianus,* that has a black beak marked with yellow at the base.
**whit** (hwĭt, wĭt) *n.* [Alteration of WIGHT.] The least bit <doesn't care a *whit*>
**white** (hwīt, wīt) *n.* [ME < OE *hwīt.*] **1.** An achromatic color of maximum lightness, the complement or antagonist of black, the other extreme of the neutral gray series. **2.** A white or nearly white part, as: **a.** The albumen of an egg. **b.** The white part of an eyeball. **c.** A blank unprinted area, as of an advertisement. **3.** One that is white or nearly white, as: **a. whites.** White trousers or a white outfit <hospital *whites*> **b.** A white wine. **c.** A white pigment. **d.** A white breed of animal. **e.** A Caucasoid. **f.** *often* **whites.** Products of a white color, as flour or sugar. **4. a.** The white or light-colored pieces in chess and checkers. **b.** The player using these pieces. **5. a.** The outermost ring of an archery target. **b.** A hit in this ring. **6. whites.** *Pathol.* Leukorrhea. **7.** A politically ultraconservative or reactionary person. —*adj.* **whit·er, whit·est. 1.** Of the color white : devoid of hue. **2.** Approaching the color white, as: **a.** Weakly colored : PALE <*white* wine> **b.** Pale gray or silvery, as hair. **c.** Bloodless : blanched <a face *white* with fear> **3.** Light or whitish in color or having light or whitish parts. —Used with animal and plant names. **4. a.** Having the comparatively pale complexion typical of Caucasoids. **b.** Of, pertaining to, typical of, or dominated by Caucasians. **c.** *Slang.* Fair or generous : DECENT. **5.** Not written or printed on : BLANK. **6.** Unsullied : pure. **7.** Habited in white <*white* monks> **8.** Accompanied by snow <a *white* Christmas> **9. a.** Incandescent <*white* flames> **b.** Intensely heated : IMPASSIONED <a *white* rage> **10.** Ultraconservative or reactionary. **11.** *Chiefly Brit.* With milk added. —Used of tea or coffee. —*vt.* **whit·ed, whit·ing, whites. 1.** To create or leave blank spaces in (printed or illustrated matter). **2.** *Archaic.* **a.** To whiten : whitewash. **b.** To blanch. **—white′ness** *n.*
**white ant** *n.* A termite.
**white·bait** (hwīt′bāt′, wīt′-) *n.* **1.** The young of various fishes, as the herring, considered a delicacy when fried. **2.** Any of various small edible fishes related to or resembling the whitebait.
**white birch** *n.* A birch tree with white bark, as *Betula pendula* of Europe or the paper birch.
**white blood cell** *n.* A leukocyte.
**white book** *n.* [From its formerly being bound in white.] An official publication of a national government.
**white bryony** *n.* A climbing European vine, *Bryonia dioica,* bearing lobed leaves, greenish-white flowers, and scarlet berries.

**white·cap** (hwīt′kăp′, wīt′-) *n.* A wave with a crest of foam.

**white cedar** *n.* Any of several North American evergreen trees, chiefly of the genus *Chamaecyparis,* with light-colored wood.

**white cell** *n.* A leukocyte.

**white chip** *n.* **1.** A white poker chip of minimal value. **2.** Something of minimal value or worth.

**white cloud** *n.* A small brightly colored freshwater fish, *Tanichthys albonubes,* indigenous to China and popular in home aquariums.

**white clover** *n.* A common Eurasian clover, *Trifolium repens,* bearing rounded white flower heads.

**white-col·lar** (hwīt′kŏl′ər, wīt′-) *adj.* Of or relating to salaried or professional workers, whose jobs usu. do not involve manual labor.

**white corpuscle** *n.* A leukocyte.

**white crappie** *n.* A silvery edible North American sunfish, *Pomoxis annularis.*

**white daisy** *n.* DAISY 1.

**whited sepulcher** *n.* [From the simile applied by Jesus Christ to the scribes and Pharisees in the Gospel according to St. Matthew.] An evil person who pretends to be holy or good **:** HYPOCRITE.

**white dwarf** *n.* A faint, very dense star that has a radius approx. the same as that of the earth.

**white elephant** *n.* **1.** A rare whitish or light-gray form of the Asian elephant, often regarded with special veneration in areas of southeast Asia. **2. a.** A rare and expensive possession that is financially a burden to maintain. **b.** Something of dubious or limited value. **3.** An article, ornament, or household utensil no longer wanted by its owner. **4.** A conspicuous failure.

**white-eye** (hwīt′ī′, wīt′ī′) *n.* A small greenish bird of the genus *Zosterops* of Africa, southern Asia, and the Pacific islands, with a narrow ring of white feathers around each eye.

**white-faced** (hwīt′fāst′, wīt′-) *adj.* **1.** Having a pale face **:** PALLID. **2.** Having a white patch extending from the muzzle to the forehead.

**white feather** *n.* [From the belief that a gamecock with a white feather in its plumage was a poor fighter.] A sign of cowardice. **—show the white feather.** To act like a coward.

**white·fish** (hwīt′fĭsh′, wīt′-) *n., pl.* **whitefish** *or* **-fish·es. 1.** A chiefly North American freshwater food fish of the genus *Coregonus,* with a gen. silvery color. **2.** A fish related to or resembling the whitefish.

**white flag** *n.* A white cloth or flag signaling surrender or truce.

**white·fly** (hwīt′flī′, wīt′-) *n.* Any of various small whitish insects of the family Aleyrodidae, often injurious to plants.

**white-foot·ed mouse** (hwīt′fo͝ot′ĭd, wīt′-) *n.* The deer mouse.

**white fox** *n.* The arctic fox in its winter color phase.

**White Friar** *n.* [From the color of the habit.] A Carmelite.

**white frost** *n.* Hoarfrost.

**white gasoline** *n.* Gasoline containing no tetraethyl lead.

**white gold** *n.* An alloy of gold and nickel, and sometimes palladium or zinc, having a platinumlike color.

**White·hall** (hwīt′hôl′, wīt′-) *n.* [After *Whitehall,* a street in London where most of the departments of government are located.] The British civil-service administration.

**white-head·ed** (hwīt′hĕd′ĭd, wīt′-) *adj.* **1.** Having white hair or plumage on the head, as a bird or animal. **2. a.** White-haired, as from old age. **b.** Flaxen-haired. **3.** *Ir.* FAIR-HAIRED 2.

**white heat** *n.* **1. a.** The temperature of a white-hot substance. **b.** The physical condition of a white-hot substance. **2.** A state of intense emotion or excitement.

**white horse** *n.* A wave capped with foam **:** WHITECAP.

**white-hot** (hwīt′hŏt′, wīt′-) *adj.* **1.** So hot as to glow with a bright white light. **2.** Fervid **:** zealous.

**White House** *n.* [After the *White House,* the official residence of the President of the U.S.] The executive branch of the U.S. government <pressure on Congress from the *White House*>

**white iron pyrites** *n.* Marcasite.

**white lead** *n.* A heavy white poisonous compound of basic lead carbonate, lead silicate, or lead sulfate, used in paint pigments.

**white leather** *also* **whit·leath·er** (hwĭt′lĕth′ər, wĭt′-) *n.* A leather specially treated with salt and alum.

**white lie** *n.* A well-intentioned or diplomatic untruth.

**white list** *n.* [WHITE + (BLACK)LIST.] A list of persons or organizations regarded as worthy of approval or acceptance. **—white′-list′ed** (hwīt′lĭst′ĭd, wīt′-) *adj.*

**white-liv·ered** (hwīt′lĭv′ərd, wīt′-) *adj.* Lily-livered **:** cowardly.

**white magic** *n.* Magic or incantation that is practiced for good purposes or as a counter to evil.

**white mahogany** *n.* PRIMAVERA 2.

**white matter** *n.* White brain and spinal-cord tissue, composed chiefly of myelinated nerve fibers.

**white meat** *n.* Light-colored meat, esp. of poultry.

**white metal** *n.* Any of various whitish alloys, as pewter, containing high percentages of tin or lead.

**white mica** *n.* The mineral muscovite.

**white mulberry** *n.* A native Chinese tree, *Morus alba,* yielding whitish or purplish fruit.

**whit·en** (hwīt′n, wīt′n) *vt. & vi.* **-ened, -en·ing, -ens.** To make or become white. **—whit′en·er** *n.*

**white noise** *n.* Acoustical or electrical noise in which the intensity is the same at all frequencies within a given band.

**white oak** *n.* **1.** A large oak, *Quercus alba* of eastern North America, with heavy, hard, light-colored wood. **2.** ROBLE 1.

**white·out** (hwīt′out′, wīt′-) *n.* A polar weather condition caused by a heavy cloud cover over the snow, in which the light coming from above is approx. equal to the light reflected from below, and which is marked by the absence of shadow, the invisibility of the horizon, and the discernibility of only very dark objects.

**white paper** *n.* **1.** A paper published by a government on a topic. **2.** An investigative television news program on a major issue.

**white pepper** *n.* Peppercorns having their outer black layer removed before grinding.

**white perch** *n.* A small food fish, *Roccus americanus* of the Atlantic coast and freshwater ponds of North America.

**white pine** *n.* **1.** A timber tree, *Pinus strobus* of eastern North America, with needles in clusters of five and durable, easily worked wood. **2.** A pine with needles in clusters of five. **3.** The wood of the white pine.

**white plague** *n.* Tuberculosis of the lungs.

**white poplar** *n.* A native Eurasian tree, *Populus alba,* with leaves having whitish undersides.

**white potato** *n.* POTATO 2.

**white·print** (hwīt′prĭnt′, wīt′-) *n.* A photomechanical copy, usu. of line drawings, in which colored or black lines appear on a white background.

**white room** *n.* A clean room.

**white sauce** *n.* A sauce of butter, flour, and milk, cream, or stock, used esp. as a base for other sauces.

**white slave** *n.* A woman held against her will for purposes of prostitution. **—white slaver** *n.* **—white slavery** *n.*

**white snakeroot** *n.* A poisonous North American plant, *Eupatorium rugosum,* with heart-shaped leaves and flat-topped clusters of small white flowers.

**white squall** *n.* A sudden squall in tropical or subtropical waters, marked by the absence of a dark cloud and the presence of white-capped waves or broken water.

**white·tail** (hwīt′tāl′, wīt′-) *n.* White-tailed deer.

**white-tailed deer** (hwīt′tāld′, wīt′-) *n.* A North American deer, *Odocoileus virginianus,* with a grayish coat that turns reddish brown in summer and a tail that is white on the underside.

**white·throat** (hwīt′thrōt′, wīt′-) *n.* An Old World songbird, *Sylvia communis* or *S. curruca,* with a white throat and brownish plumage.

**white-throat·ed sparrow** (hwīt′thrō′tĭd, wīt′-) *n.* A North American sparrow, *Zonotrichia albicollis,* with a white throat and a distinctive song.

**white tie** *n.* **1.** A white bow tie worn as a part of men's formal evening dress. **2.** Men's formal evening dress. **—white′-tie′** (hwīt′tī′, wīt′-) *adj.*

**white vitriol** *n.* Zinc sulfate.

**white·wall tire** *also* **white·wall** (hwīt′wôl′, wīt′-) *n.* A vehicular tire having a white band on the visible side.

**white walnut** *n.* The butternut.

**white·wash** (hwīt′wŏsh′, -wôsh′, wīt′-) *n.* **1.** A mixture of lime and water, often with whiting, size, or glue added, used to whiten structures, as fences or exterior walls, made of wood, stone, or concrete. **2.** A cosmetic application for whitening the skin. **3.** An act of concealing or glossing over of flaws or failures. **4.** *Informal.* A defeat in a game in which the loser scores no points. —*vt.* **-washed, -wash·ing, -wash·es. 1.** To paint or coat with or as if with whitewash. **2.** To conceal or gloss over (e.g., a flaw). **—white′wash′er** *n.*

**white water** *n.* Turbulent or frothy water, as in rapids. **—white′-wa′ter** (hwīt′wô′tər, -wŏt′ər, wīt′-) *adj.*

**white whale** *n.* A small whale chiefly of northern waters, *Delphinapterus leucas,* that is white when full-grown.

**white·wood** (hwīt′wo͝od′, wīt′-) *n.* The soft, light-colored wood of various trees, as the tulip tree, basswood, or cottonwood.

**whith·er** (hwĭth′ər, wĭth′-) *adv.* [ME < OE *hwider.*] **1.** To what place, result, or condition <*Whither* are we roaming?> **2.** To which specified place or position <lived in the town *whither* fate had led me> **3.** To whatever place, result, or condition <"*Whither* thou goest, I will go" —Ruth 1:16>

**whith·er·so·ev·er** (hwĭth′ər-sō-ĕv′ər, wĭth′-) *adv.* To any place whatsoever.

**whit·ing**[1] (hwī′tĭng, wī′-) *n.* [ME *whityng* < *whiten,* to whiten < *white,* white.] A pure white grade of chalk ground and washed for use in paints, ink, and putty.

**whit·ing**[2] (hwī′tĭng, wī′-) *n.* [ME *whitynge* < MDu. *wijting.*] **1.** A food fish, *Gadus merlangus* of European Atlantic waters, related to the cod. **2.** A marine fish of the genera *Menticirrhus* or *Merluccius,* of North American coastal waters.

**whit·ish** (hwī′tĭsh, wī′-) *adj.* Somewhat white.

**whit·leath·er** (hwĭt′lĕth′ər, wĭt′-) *n. var. of* WHITE LEATHER.

**whit·low** (hwĭt′lō, wĭt′-) *n.* [ME *whitflawe* : *white,* white + *flawe,* flaw.] An inflammation around the nail of a finger or toe.

**Whit·mon·day** *also* **Whit-Mon·day** (hwĭt′mŭn′dē, -dā′, wĭt′-) *or* **Whit·sun-Mon·day** (hwĭt′sən-, wĭt′sən-) *n.* The Monday following Whitsunday.

**Whit·sun** (hwĭt′sən, wĭt′-) *adj.* [ME *whitsone* < *whitsonday,* Whitsunday.] Of, pertaining to, or observed on Whitsunday or at Whitsuntide.

**Whit·sun·day** (hwĭt′sən-dē, -dā′, wĭt′-) *n.* [ME *whitsonday* < OE *hwita sunnandæg,* White Sunday.] PENTECOST 1.

**Whit·sun·tide** *also* **Whit·sun Tide** (hwĭt′sən-tīd′, wĭt′-) *n.* The week beginning with Whitsunday or Pentecost, esp. the first three days of this week.

**whit·tle** (hwĭt′l, wĭt′l) *v.* **-tled, -tling, -tles.** [<ME *whyttel,* knife, var. of *thwitel* < *thwiten,* to whittle < OE *ðwitan.*] *—vt.* **1. a.** To cut small bits or pare shavings from (a piece of wood). **b.** To fashion or shape in this way. **2.** To eliminate or reduce gradually by or as if by whittling with a knife <*whittled* my waist with exercise> *—vi.* **1.** To cut or shape wood with a knife. **2.** To wear oneself or another out by fretting and carping. **—whit′tler** *n.*

**whit·tling** (hwĭt′lĭng, wĭt′-) *n.* A shaving from a piece of wood being whittled.

**whiz** *also* **whizz** (hwĭz, wĭz) *v.* **whizzed, whiz·zing, whiz·zes.** [Imit.] *—vi.* **1.** To make a whirring, buzzing, or hissing sound, as of something rushing through air. **2.** To rush past. *—vt.* To cause to whiz. *—n., pl.* **whiz·zes. 1.** The sound or passage of something that whizzes. **2.** A quick trip. **3.** *Slang.* One who has exceptional skill.

**whiz kid** *n.* A young person who is exceptionally intelligent, innovatively clever, and very successful.

**whizz-bang** (hwĭz′băng′, wĭz′-) *n.* [< *whizzbang,* a shell used in World War I that was heard only an instant before landing and exploding: WHIZZ + BANG.] One that is conspicuous because of noise, speed, or startling effect.

**who** (hōō) *pron.* [ME < OE *hwā.*] **1.** What or which person or persons <*Who* called?> **2.** That. —Used as a relative pronoun to introduce a clause when the antecedent is a human or is understood to be a human <the people *who* live next door><reliable sources *who* were on the scene> **3.** The person or persons that : WHOEVER.

**whoa** (hwō, wō) [ME *whoo,* var. of *ho,* halt!] *interj.* —Used as a command to stop.

**who'd** (hōōd). **1.** Who would. **2.** Who had.

**who·dun·it** (hōō-dŭn′ĭt) *n.* [WHO + DONE + IT.] *Informal.* A mystery story.

**who·ev·er** (hōō-ĕv′ər) *pron.* **1.** No matter who <*Whoever* competes will be honored> **2.** Who <*Whoever* would have believed that incredible story?>

**whole** (hōl) *adj.* [ME *hole,* unharmed < OE *hāl.*] **1.** Containing all components or constituents. **2.** Not divided or disjoined <a *whole* acre of woods> **3. a.** Sound : healthy <a *whole* organism> **b.** Restored : healed <a *whole* person again> **4.** Constituting the full amount, extent, or duration <The children whined the *whole* trip.> **5.** Having the same parents. **6.** *Math.* Not fractional : INTEGRAL. *—n.* **1.** All of the components of a thing. **2.** A complete entity or system. *—adv. Informal.* Wholly : entirely < a *whole* new plan> **—as a whole.** All things considered. **—on the whole.** Considering everything : as a rule. **—whole′ness** *n.*

☆ ***syns:*** WHOLE, ALL, COMPLETE, ENTIRE, GROSS, TOTAL *adj. core meaning* : including every constituent or individual <The *whole* town turned out for the fireworks display.>

**whole blood** *n.* **1.** Blood drawn directly from a living human being and prepared for use in transfusion. **2.** Blood from which no constituent has been removed.

**whole gale** *n.* A wind with a speed of 55 to 63 miles or 88.5 to 103.4 kilometers per hour.

**whole·heart·ed** (hōl′här′tĭd) *adj.* **1.** Marked by total sincerity and enthusiasm. **2.** Marked by earnest commitment. **—whole′heart′ed·ly** *adv.* **—whole′heart′ed·ness** *n.*

**whole hog** *n. Slang.* The whole way or fullest extent.

**whole life insurance** *n.* Insurance providing death protection for the insured's lifetime.

**whole milk** *n.* Milk from which no constituent has been removed.

**whole note** *n. Mus.* A note having, in common time, the value of four beats.

**whole number** *n.* An integer.

**whole·sale** (hōl′sāl′) *n.* [ME *holesale* : *hole,* whole + *sale,* sale.] The sale of goods in large quantities, as for resale by a retailer. *—adj.* **1.** Relating to or engaged in the sale of goods at wholesale. **2.** Sold in large bulk or quantity, usu. at a lower cost. **3.** Made or accomplished extensively and indiscriminately : BLANKET <*wholesale* destruction by fire> *—adv.* **1.** On wholesale terms. **2.** Extensively and indiscriminately. *—v.* **-saled, -sal·ing, -sales.** *—vt.* To sell at wholesale. *—vi.* **1.** To engage in wholesale selling. **2.** To be sold wholesale. **—whole′sal′er** *n.*

**whole·some** (hōl′səm) *adj.* [ME *holsom.*] **1.** Conducive to sound health or well-being : SALUTARY. **2.** Morally or socially salubrious. **3.** Healthy. **—whole′some·ly** *adv.* **—whole′some·ness** *n.*

**whole-wheat** (hōl′hwēt′) *adj.* Made from the entire grain of wheat, including the bran <*whole-wheat* flour>

**who'll** (hōōl). **1.** Who will. **2.** Who shall.

**whol·ly** (hō′lē, hōl′lē) *adv.* **1.** Entirely. **2.** Exclusively.

**whom** (hōōm) *pron.* [ME < OE *hwām,* dative of *hwā,* who.] *The objective case of* WHO.

**whom·ev·er** (hōōm-ĕv′ər) *pron. The objective case of* WHOEVER.

**whom·so·ev·er** (hōōm′sō-ĕv′ər) *pron. The objective case of* WHOSOEVER.

**whoop** (hōōp, hwōōp, wōōp) *n.* [ME *whopen* < OFr. *houpper.*] **1. a.** A cry of excitement or exultation. **b.** A battle cry or hunter's halloo. **2.** A hooting cry, as of a bird. **3.** The paroxysmal gasp characteristic of whooping cough. *—v.* **whooped, whoop·ing, whoops.** *—vi.* **1.** To utter a loud shout or cry. **2.** To utter a hooting cry. **3.** To make the paroxysmal gasp typical of whooping cough. *—vt.* **1.** To utter with a whoop. **2.** To chase, call, urge on, or drive with a whoop. **—whoop it up.** *Slang.* To have a jolly time.

**whoop·ee** (hwōō′pē, wōō′-, hwōō′-, wōō′-) [< WHOOP.] *interj. Slang.* —Used to express jubilance. **—make whoopee.** To celebrate noisily <*made whoopee* after the prom>

**whoop·er** (hōō′pər, hwōō′-, wōō′-) *n.* **1.** One that whoops. **2.** An Old World swan, *Cygnus cygnus* or *Olor cygnus,* with a loud cry.

**whooping cough** *n.* An infectious disease involving catarrh of the respiratory passages and marked by spasms of coughing interspersed with deep, noisy inspiration.

**whooping crane** *n.* A large, long-legged, rare North American bird, *Grus americana,* with black and white plumage and a shrill, trumpeting cry.

**whoops** (hwōōps, wōōps, hwōōps, wōōps) *interj.* —Used to express mild surprise or apology.

**whoosh** (hwōōsh, wōōsh, hwōōsh, wōōsh) *vi.* **whooshed, whoosh·ing, whoosh·es.** [Imit.] **1.** To gush or hurtle rapidly. **2.** To make a sharp sibilant sound. **—woosh** *n.*

**whop** (hwŏp, wŏp) *vt.* **whopped, whop·ping, whops.** [ME *whappen,* var. of *wappen,* to throw violently.] To thrash : defeat. *—n.* A heavy thud or blow. *—adv.* With a whop.

**whop·per** (hwŏp′ər, wŏp′-) *n.* **1.** Something exceptionally big or notable. **2.** A gross untruth.

**whop·ping** (hwŏp′ĭng, wŏp′-) *adj.* & *adv.* —Used as an intensive <a *whopping* error><a *whopping* good story>

**whore** (hôr, hōr) *n.* [ME *hore* < OE *hōre.*] A prostitute. *—vi.* **whored, whor·ing, whores. 1.** To consort with whores. **2.** To behave like a whore. **—whor′ish** *adj.* **—whor′ish·ly** *adv.*

**whore·dom** (hôr′dəm, hōr′-) *n.* [ME *hordom* < ON *hōrdōmr.*] **1.** Fornication : harlotry. **2.** Idolatry.

**whore·house** (hôr′hous′, hōr′-) *n.* A brothel.

**whore·mas·ter** (hôr′măs′tər, hōr′-) *n.* One who consorts with whores.

**whore·mong·er** (hôr′mŭng′gər, -mŏng′gər, hōr′-) *n. Archaic.* A whoremaster.

**whore·son** (hôr′sən, hōr′-) *n.* A bastard. *—adj.* Abominable.

**whorl** (hwôrl, wôrl, hwûrl, wûrl) *n.* [ME *whorle,* prob. var. of *whirle,* whirl < *whirlen,* to whirl < ON *hvirfla.*] **1.** A small flywheel that regulates the speed of a spinning wheel. **2.** *Bot.* An arrangement of three or more parts, as leaves or petals, radiating from a single organ or node. **3.** *Zool.* A single turn or volution of a spiral shell. **4.** One of the circular ridges or convolutions of a fingerprint. **5.** An ornamental device having stylized vine leaves and tendrils. **6.** A coil, curl, or convolution <*whorls* of frost on the windowpane>

**whorled** (hwôrld, wôrld, hwûrld, wûrld) *adj.* **1.** Having or forming a whorl. **2.** Having convolutions, as of vine leaves <"halls, giddy with plush and *whorled* designs in gold" —Djuna Barnes>

**whort** (hwûrt, wûrt) *also* **whor·tle** (hwûrt′l, wûrt′l) *n.* [Var. of dial. *hurt.*] The whortleberry or its fruit.

**whor·tle·ber·ry** (hwûrt′l-bĕr′ē, wûrt′-) *n.* [Var. of dial. *hurtleberry.*] **1.** A small European shrub, *Vaccinium myrtillus,* yielding edible blackish berries. **2.** The fruit of the whortleberry.

**who's** (hōōz). **1.** Who is. **2.** Who has.

**whose** (hōōz) *pron.* [ME *whos* < OE *hwæs.*] (*sing. or pl. in number*). That which belongs to whom <told me *whose* it was> *—adj.* Of or pertaining to whom or which, esp. as the possessor or possessors <a feather bed in *whose* depths I snuggled> ***usage:*** As a possessive form for both persons and things, *whose* is acceptable on all levels. Therefore, its use is entirely correct in a sentence such as *The cabinet, whose doors were decorated with painted panels, was made chiefly of satinwood.*

**who·so·ev·er** (hōō′sō-ĕv′ər) *pron.* Whoever.

**who's who** *or* **Who's Who** *n.* A compilation of short biographical sketches of well-known personages <a *who's who* of composers>

**why** (hwī, wī) *adv.* [ME < OE *hwȳ.*] For what purpose, reason, or cause <*Why* was the meeting canceled?> *—conj.* **1.** The purpose, reason, or cause for which <I know *why* they did it.> **2.** On account of which : for which <"The reason *why* they are called regular is that we can predict what all the other three forms are" —Randolph Quirk> *—n., pl.* **whys. 1.** The cause or intention underlying a given action or situation <the *whys* and wherefores> **2.** A difficult question or problem. —*interj.* —Used to express mild indignation, surprise, or impatience.

**whyd·ah** *also* **whid·ah** (hwĭd′ə, wĭd′ə) *n.* [Prob. alteration of WIDOW (BIRD).] An African bird of the genus *Vidua*, with predominantly black plumage and long tail feathers.

**whydah**
*Over 20 inches long with tail feathers*

**Wich·i·ta** (wĭch′ə-tô′) *n., pl.* **Wichita** *or* **-tas. 1. a.** A confederacy of Indians once living between the Arkansas River and central Texas. **b.** A member of this confederation. **2.** The Caddoan language of the Wichita.

**wick** (wĭk) *n.* [ME *wike* < OE *wēoce.*] **1.** A cord or strand of loosely woven, twisted, or braided fibers, as on a candle or oil lamp, that draws up fuel to the flame by capillary action. **2.** A device similar to a wick that conveys liquid by capillary action.

**wick·ed** (wĭk′ĭd) *adj.* **-er, -est.** [ME, alteration of *wicke*, wicked.] **1.** Morally bad : DEPRAVED <*wicked* lies> **2.** Mischievous or playfully malicious <a *wicked* prank> **3.** Harmful : pernicious <a *wicked* cough> **4.** Obnoxious : offensive <a *wicked* stink> **5.** Excellent or highly skilled : FORMIDABLE <a *wicked* golfer> **—wick′ed·ly** *adv.* **—wick′ed·ness** *n.*

**wick·er** (wĭk′ər) *n.* [ME *wiker*, of Scand. orig.] **1.** A flexible shoot, as of a willow, used in weaving baskets or furniture. **2.** Wickerwork. —*adj.* Constructed, composed of, or covered with wicker.

**wick·er·work** (wĭk′ər-wûrk′) *n.* Woven wicker.

**wick·et** (wĭk′ĭt) *n.* [ME < ONFr. *wiket*, prob. of Germanic orig.] **1.** A small door or gate, esp. one built into or near a larger one. **2.** A small opening or window, often fitted with glass or a grating. **3.** A sluice gate for regulating the amount of water in a millrace or a canal or for emptying a lock. **4.** In cricket: **a.** Either of the two sets of three stumps, topped by bails, that forms the target of the bowler and is defended by the batsman. **b.** A batsman's innings, which may be ended by the ball knocking the bails off the stumps. **c.** Termination of a batsman's innings. **d.** The period during which two batsmen are in together. **e.** The pitch, esp. with respect to wetness or other conditions. **5.** Any of the small, usu. wire arches, through which one tries to direct a croquet ball.

**wick·et·keep·er** (wĭk′ĭt-kē′pər) *n.* The cricket player positioned directly behind the wicket in play.

**wick·i·up** *also* **wik·i·up** (wĭk′ē-ŭp′) *n.* [Fox *wikiyapi*, dwelling.] A frame hut covered with matting, as of bark or brush, used by nomadic North American Indians.

**wic·o·py** (wĭk′ə-pē) *n., pl.* **-pies.** [Cree *wikopiy*, willow bark.] LEATHERWOOD 1.

**Wi·dal test** (vē-däl′) *n.* [After Fernand *Widal* (1862–1929).] A serological test that uses an agglutination reaction to diagnose typhoid fever.

**wid·der·shins** (wĭd′ər-shĭnz′) *adv. var. of* WITHERSHINS.

**wide** (wīd) *adj.* **wid·er, wid·est.** [ME < OE *wīd.*] **1.** Extending over a large area from side to side : BROAD <a *wide* river> **2.** Having a specified extent from side to side <a belt three inches *wide*> **3.** Having great range or scope <a *wide* variety> **4.** Full or ample, as clothing. **5.** Fully open or extended, as the eyes. **6.** Apart or away from the desired goal or point <*wide* of the mark><*wide* of the truth> **7.** LAX 4. —*adv.* **1.** Over a large area : EXTENSIVELY <traveled far and *wide*> **2.** To the full extent : COMPLETELY <opened the gate *wide*> **3.** So as to miss the target : ASTRAY. —*n.* A cricket ball bowled outside of the batsman's reach, counting as a run for the batting team. **—wide′ly** *adv.* **—wide′ness** *n.*

**-wide** *suff.* [< WIDE.] **1.** Extending over a given area or region <city*wide*> **2.** Throughout a given area or region <state*wide*>

**wide-an·gle lens** (wīd′ăng′gəl) *n.* A lens with a relatively short focal length that permits an angle of view wider than approx. 70°.

**wide-a·wake** (wīd′ə-wāk′) *adj.* **1.** Completely awake. **2.** Watchful : alert. **—wide′a·wake′ness** *n.*

**wide-eyed** (wīd′īd′) *adj.* **1.** With the eyes completely opened, as in wonder. **2.** Innocent : credulous.

**wide-mouthed** (wīd′mouthd′, -moutht′) *adj.* **1.** Having a wide mouth <a *wide-mouthed* jar> **2.** With the mouth completely open, as in surprise.

**wid·en** (wīd′n) *vt. & vi.* **-ened, -en·ing, -ens.** To make or become wide or wider. **—wid′en·er** *n.*

**wide-o·pen** (wīd′ō′pən) *adj.* **1.** Opened completely <a *wide-open* gate> **2.** Without laws or law enforcement <a *wide-open* town>

**wide receiver** *n. Football.* A receiver usu. positioned several yards to the side of an offensive formation.

**wide·spread** *also* **wide-spread** (wīd′sprĕd′) *or* **wide-spreading** (-sprĕd′ĭng) *adj.* **1.** Spread or scattered over a considerable extent. **2.** Occurring or accepted widely.

**wid·geon** (wĭj′ən) *n., pl.* **widgeon** *or* **-geons.** [Orig. unknown.] A duck, *Mareca americana* of North America or *M. penelope* of Europe, with brownish plumage.

**wid·get** (wĭj′ət) *n.* [Prob. alteration of GADGET.] A gadget, esp. one that is unnamed or hypothetical.

**wid·ow** (wĭd′ō) *n.* [ME *widewe* < OE *widuwe.*] **1.** A woman whose husband has died and who has not remarried. **2.** An additional hand of cards dealt to the table. **3. a.** An incomplete usu. short line of type, as one ending a paragraph, carried over to the top of the next page or column. **b.** A short line at the bottom of a page or column. —*vt.* **-owed, -ow·ing, -ows.** To make a widow of <was *widowed* a year after the marriage> **—wid′ow·hood′** *n.*

**widow bird** *n.* [From its black plumage.] The whydah.

**wid·ow·er** (wĭd′ō-ər) *n.* [ME *widewer* < *widewe*, widow.] A man whose wife has died. **—wid′ow·er·hood′** *n.*

**widow's mite** (wĭd′ōz) *n.* [From the widow who gave two small coins to the Temple treasury in the Gospel according to St. Mark.] A small contribution made by one who has little.

**widow's peak** *n.* [From the superstition that it is a sign of early widowhood.] A V-shaped point formed by the hair at the middle of the forehead.

**widow's walk** *n.* A railed rooftop gallery on a dwelling, used for observing vessels at sea.

**width** (wĭdth, wĭth) *n.* [< WIDE.] **1.** The quality, state, or fact of being wide. **2.** The measurement of the extent of something from side to side. **3.** Something that has a specified width, esp. a piece of fabric measured from selvage to selvage in sewing.

☆ ***syns:*** WIDTH, BREADTH, BROADNESS, WIDENESS *n. core meaning* : the extent of something from side to side <the *width* of the doorway>

**width·wise** (wĭdth′wīz′, wĭth′-) *adv.* In terms of width.

**wield** (wēld) *vt.* **wield·ed, wield·ing, wields.** [ME *welden* < OE *wieldan.*] **1.** To handle (e.g., a weapon or tool). **2.** To exercise or exert (power or influence). **—wield′a·ble** *adj.* **—wield′er** *n.*

**wield·y** (wēl′dē) *adj.* **-i·er, -i·est.** Easily wielded or managed.

**wie·ner** (wē′nər) *n.* [G., short for *Wienerwurst*, wienerwurst.] A wienerwurst.

**Wie·ner schnit·zel** (vē′nər shnĭt′səl) *n.* [G., Vienna cutlet.] A breaded veal cutlet.

**wie·ner·wurst** (wē′nər-wûrst′, -wŏŏrst′) *n.* [G. : *Wien*, Vienna + *Wurst*, sausage.] A smoked pork or beef sausage, similar to a frankfurter.

**wife** (wīf) *n., pl.* **wives** (wīvz) [ME < OE *wīf.*] **1.** A married woman. **2.** *Archaic.* A woman. **—wife′hood′, wife′dom** *n.* **—wife′ly** *adj.*

**wig** (wĭg) *n.* [Short for PERIWIG.] A headpiece of human or artificial hair worn as personal adornment, part of a costume, or to conceal baldness. —*vt.* **wigged, wig·ging, wigs.** To scold or censure.

**wig·an** (wĭg′ən) *n.* [After *Wigan*, a textile-manufacturing town in England.] A stiff fabric used for interlining.

**wi·geon** (wĭj′ən) *n. Chiefly Brit. var. of* WIDGEON.

**wigged** (wĭgd) *adj.* Wearing a wig.

**wig·gle** (wĭg′əl) *vi. & vt.* **-gled, -gling, -gles.** [ME *wiglen*, prob. < MLG *wiggelen*, to totter.] To move or cause to move with short irregular twisting motions from side to side. —*n.* A wiggling movement or course. **—wig′gly** *adj.*

**wig·gler** (wĭg′lər) *n.* **1.** One that wiggles. **2.** The larva or pupa of a mosquito.

**wight¹** (wīt) *n.* [ME < OE *wiht.*] A human being.

**wight²** (wīt) *adj.* [ME < ON *vīgt*, neuter of *vīgr*, able to fight.] *Archaic.* Brave : valiant.

**wig·wag** (wĭg′wăg′) *v.* **-wagged, -wag·ging, -wags.** [Dial. *wig*, to move + WAG.] —*vt.* **1.** To move back and forth. **2.** To signal by moving (e.g., the hand) back and forth. —*vi.* **1.** To move back and forth : WAG. **2.** To wave the hand or a device in signaling. —*n.* **1.** The act or practice of signaling by wigwagging. **2.** A message relayed by wigwagging. **—wig′wag′ger** *n.*

**wig·wam** (wĭg′wŏm′) *n.* [Abnaki *wikəwam.*] A North American Indian dwelling, usu. having an arched or conical framework overlaid with bark, hides, or mats.

**wik·i·up** (wĭk′ē-ŭp) *n. var. of* WICKIUP.

**wil·co** (wĭl′kō) *interj.* [Short for *will comply.*] —Used esp. in radio signaling to indicate cooperation.

**wild** (wīld) *adj.* **-er, -est.** [ME *wilde* < OE.] **1.** Growing, living, or found in a natural state : not domesticated, cultivated, or tamed. **2.** Not inhabited : DESOLATE. **3.** Uncivilized or barbarous : SAVAGE. **4.** Lacking discipline, restraint, or control : UNRULY <*wild* schoolchildren> **5.** Disorderly: disarranged. **6.** Incoherent or chaotic: FRENZIED <*wild* babbling> **7.** Full of intense, ungovernable emotion <*wild* with anger> **8. a.** Notoriously odd or amusing: OUTLANDISH <a *wild,* eccentric character> **b.** Extravagant : fantastic <a *wild* notion> **9.** Furiously disturbed or turbulent: STORMY. **10.** Reckless: risky. **11.** Random or spontaneous <take a *wild* guess> **12.** Deviating widely: ERRATIC <a *wild* throw> **13.** Having an arbitrary equivalence or value determined by a card holder's needs or choice <Deuces are *wild.*> *—adv.* In a wild way. *—n.* An uninhabited or uncultivated region. **—wild'ly** *adv.* **—wild'ness** *n.*

☆ ***syns:*** WILD, BACKLAND, BUSH, WILDERNESS, WILDNESS *n. core meaning:* an uninhabited region left in its natural state <fought to keep the *wild* from developers>

**wild bergamot** *n.* An aromatic plant, *Monarda fistulosa* of eastern North America, with lilac-purple flower clusters.

**wild boar** *n.* BOAR 2.

**wild carrot** *n.* Queen Anne's lace.

**wild·cat** (wīld'kăt') *n.* **1.** A wild feline of small to medium size, esp. one of the genus *Lynx.* **2.** One who is quick-tempered or fierce. **3.** An oil well drilled in an area not known to yield oil. *—adj.* **1. a.** Unsound or risky, esp. financially. **b.** Issued by a financially irresponsible bank <*wildcat* currency> **c.** Operating or accomplished outside the parameters of standard, ethical business procedures. **2.** Accomplished or operating without official sanction or authority <a *wildcat* strike> *—v.* **-cat·ted, -cat·ting, -cats.** *—vt.* To prospect for (e.g., oil) in a supposedly unproductive area. *—vi.* To prospect in an untapped or questionable area. **—wild'cat'ter** *n.*

**wild celery** *n.* Tape grass.

**wil·de·beest** (wĭl'də-bēst', vĭl'-) *n.* [Obs. Afr. : Du. *wild,* wild + Du. *beest,* beast.] The gnu.

**wil·der** (wĭl'dər) *v.* **-dered, -der·ing, -ders.** [Orig. unknown.] *Archaic. —vt.* **1.** To lead astray: MISLEAD. **2.** To bewilder: perplex. *—vi.* **1.** To lose one's way. **2.** To become bewildered. **—wil'der·ment** *n.*

**wil·der·ness** (wĭl'dər-nĭs) *n.* [ME < *wilddēornes* < *wilddēor,* wild beast.] **1.** An uninhabited region left in its natural condition, esp.: **a.** A large wild tract of land covered with dense vegetation or forests. **b.** An extensive area, as a desert or ocean, that is barren or empty: WASTE. **c.** A piece of land set aside to grow wild. **2.** Something likened to a wild region in its bewildering vastness, perilousness, or unchecked profusion <a *wilderness* of dissenting voices>

**wild-eyed** (wīld'īd') *adj.* **1.** Characterized by a wild expression in the eyes. **2.** Advocating or consisting of extreme social or political measures <*wild-eyed* reactionaries>

**wild·fire** (wīld'fīr') *n.* **1.** A highly flammable material once used in warfare. **2.** A raging fire that travels and spreads rapidly. **3.** Lightning occurring without thunder being heard. **4.** Luminosity that appears at night over swamps or marshes: IGNIS FATUUS. **—like wildfire.** Very quickly and intensely <The rumor spread *like wildfire.*>

**wild·flow·er** *also* **wild flower** (wīld'flou'ər) *n.* **1.** A flowering plant growing in a natural, uncultivated state. **2.** The flower of a wildflower.

**wild·fowl** (wīld'foul') *n., pl.* **wildfowl** or **-fowls.** A wild bird, as a duck, goose, or quail, hunted as game.

**wild geranium** *n.* A North American woodland plant, *Geranium maculatum,* with rose-purple flowers.

**wild ginger** *n.* A North American plant, *Asarum canadense,* with broad leaves, a single brownish flower, and an aromatic root.

**wild-goose chase** (wīld'go͞os') *n.* Hopeless pursuit of an unattainable or imaginary object.

**wild hyacinth** *n.* A plant, *Camassia scilloides* of the central United States, with narrow leaves and pale-blue or white flower clusters.

**wild indigo** *n.* A North American plant of the genus *Baptisia,* esp. *B. tinctoria,* bearing compound leaves with three leaflets and yellow flowers.

**wild·ing** (wīl'dĭng) *n.* [< WILD.] **1.** A plant that grows wild or has escaped from cultivation, esp. a wild apple tree or its fruit. **2.** A wild animal. *—adj.* **1.** Not cultivated : growing wild. **2.** Undomesticated.

**wild·life** (wīld'līf') *n.* Wild animals and vegetation, esp. animals living in a natural, undomesticated state.

**wild lily of the valley** *n.* A woodland plant, *Maianthemum canadense* of eastern North America, with a terminal cluster of small white flowers.

**wild·ling** (wīld'lĭng) *n.* A wild plant or animal, esp. a wild plant transplanted to a cultivated spot.

**wild marjoram** *n.* MARJORAM 2.

**wild mustard** *n.* Charlock.

**wild oat** *n.* **1.** *often* **wild oats.** A native Eurasian grass, *Avena fatua,* related to the cultivated oat. **2. wild oats.** The excesses of youth <sow one's *wild oats*>

**wild olive** *n.* Any of various trees resembling the olive.

**wild pansy** *n.* Heartsease.

**wild pink** *n.* A North American plant of the genus *Silene,* esp. *S. caroliniana,* with pink or white flowers.

**wild pitch** *n. Baseball.* An erratic pitch that the catcher cannot be expected to receive and that enables a runner to advance.

**wild rice** *n.* **1.** A tall aquatic grass, *Zizania aquatica* of northern North America, yielding edible grain. **2.** The grain of the wild rice.

**wild rye** *n.* A grass of the genus *Elymus.*

**wild type** *n.* The characteristic form of an organism as it occurs in nature, as opposed to mutant specimens that may result from selective breeding.

**wild vanilla** *n.* A plant, *Trilisa odoratissima* of the southeastern United States, bearing vanilla-scented leaves.

**Wild West** *n.* The western United States during the period of its settlement, esp. with reference to its lawlessness.

**wild·wood** (wīld'wo͝od') *n.* A wooded area in its natural state.

**wile** (wīl) *n.* [ME *wil.*] **1.** A deceitful stratagem or trick. **2.** A disarming or seductive manner, device, or procedure. **3.** Cunning: trickery. *—vt.* **wiled, wil·ing, wiles. 1.** To lead or influence by means of wiles: ENTICE. **2.** To pass (time) agreeably <*wile* away the hours>

**wil·ful** (wĭl'fəl) *adj. var. of* WILLFUL.

**will**[1] (wĭl) *n.* [ME < OE *willa.*] **1.** The mental faculty by which one deliberately chooses or decides on a course of action: VOLITION. **2.** Exercise of will: CHOICE. **3.** Something desired or decided on by a person of authority or supremacy <It is the ruler's *will* that capital punishment be abolished.> **4.** Deliberate intention or wish <I was held against my *will.*> **5.** Free discretion : INCLINATION. **6.** Bearing or attitude toward others : DISPOSITION <good *will*><ill *will*> **7. a.** The power to arrive at one's own decision and to act on it independently in spite of opposition <a strong *will*> **b.** The collective desire of a given group <the *will* of the citizens> **8. a.** Diligent purposefulness: DETERMINATION <the *will* to excel> **b.** Self-discipline: self-control. **9. a.** A legal declaration of how one wishes one's possessions to be disposed of after one's death. **b.** The document containing this declaration. *—v.* **willed, will·ing, wills.** *—vt.* **1.** To decide on : CHOOSE. **2.** To yearn for : DESIRE. **3.** To decree : order. **4.** To resolve with a forceful will : DETERMINE. **5.** To influence or compel by sheer force of will or by supernatural power <We *willed* the rain to stop.> **6.** To grant in a legal will: BEQUEATH. *—vi.* **1.** To exercise the will. **2.** To decree or make a firm choice. **—at will.** Just as one wishes.

☆ ***syns:*** WILL, CHOICE, DISCRETION, PLEASURE *n. core meaning:* unrestricted freedom to choose <My *will* was to go home early.>

**will**[2] (wĭl) *aux.v.* [ME *willen,* to intend to < OE *willan.*] *p.t.* **would** (wo͝od). —Used to indicate: **1.** Simple futurity <We *will* go tomorrow.> **2.** Likelihood or certainty <You *will* rue this day.> **3.** Willingness <*Will* you lend me your car?> **4.** Requirement or command <You *will* give me a full report.> **5.** Intention <I *will* too quit if I want.> **6.** Customary or habitual action <We *would* go days without speaking.> **7.** Capacity or ability <This siding *will* not rust under any conditions.> **8.** *Informal.* Probability or expectation <That *will* be the delivery person.> *—vt. & vi.* To wish : desire <Stand here if you *will.*><Go where you *will.*>

**willed** (wīld) *adj.* Having a will of a specified kind <firm-*willed*>

**wil·lem·ite** (wĭl'ə-mīt') *n.* [Du. *Willemit,* after *Willem,* William I (1772–1843), king of the Netherlands.] A colorless often fluorescent vitreous to resinous silicate of zinc, $Zn_2SiO_4$, a minor ore of zinc.

**wil·let** (wĭl'ĭt) *n.* [Imit. of its song.] A long-billed New World shore bird, *Catoptrophorus semipalmatus.*

**will·ful** *also* **wil·ful** (wĭl'fəl) *adj.* **1.** Being in accord with one's will : DELIBERATE. **2.** Inclined to impose one's will : OBSTINATE. **—will'ful·ly** *adv.* **—will'ful·ness** *n.*

**wil·lies** (wĭl'ēz) *pl.n.* [Orig. unknown.] *Slang.* Feelings of uneasiness: JITTERS <The old house gave us the *willies.*>

**will·ing** (wĭl'ĭng) *adj.* **1.** Of or resulting from the process of choosing : VOLITIONAL. **2.** Disposed to accept or tolerate : ACQUIESCENT. **3.** Acting or eager to act gladly. **4.** Done, given, accepted, or offered freely and heartily. **—will'ing·ly** *adv.* **—will'ing·ness** *n.*

**wil·li·waw** (wĭl'ē-wô') *n.* [Orig. unknown.] **1.** A strong gust of cold wind blowing seaward from a mountainous coast. **2.** A sudden squall.

**will-o'-the-wisp** (wĭl'ə-thə-wĭsp') *n.* [*Will* (nickname for *William*) + OF + THE + WISP.] **1.** Ignis fatuus. **2.** A delusive goal.

**wil·low** (wĭl'ō) *n.* [ME *wilowe* < OE *welig.*] **1. a.** A deciduous tree or shrub of the genus *Salix,* with usu. narrow leaves, flowers borne in catkins, and strong lightweight wood. **b.** The wood of a willow tree. **2.** Something, as a cricket bat, made from willow. **3.** A textile machine having a spiked drum that revolves inside a chamber fitted internally with spikes, used to open and clean unprocessed cotton or wool. *—vt.* **-lowed, -low·ing, -lows.** To open and clean (textile fibers) with a willow.

**willow herb** *n.* A plant of the genus *Epilobium,* with narrow leaves and pink, purplish, or white flower clusters.

**willow oak** *n.* A timber tree, *Quercus phellos,* of the southern and central United States, with narrow willowlike leaves.

**wil·low·ware** (wĭl'ō-wâr') *n.* Household china decorated with a blue-on-white design depicting a willow tree and often a river.

**wil·low·y** (wĭl'ō-ē) *adj.* **-i·er, -i·est. 1.** Planted with or abounding in willows. **2.** Like a willow tree, esp.: **a.** Flexible : pliant. **b.** Slender and graceful <a *willowy* figure>

**will power** *n.* The ability and strength of mind to carry out one's decisions, wishes, or plans.

**wil·ly-nil·ly** (wĭl′ē-nĭl′ē) *adv.* [Alteration of *will ye, nill ye,* be you willing, be you unwilling. —see NILL.] Whether desired or not. —*adj.* Being or occurring whether desired or not.

**Wil·son's disease** (wĭl′sənz) *n.* [After Samuel A.K. *Wilson* (1877–1937).] A hereditary disease in which the serum level of a glycoprotein needed to metabolize copper is diminished, thus causing accumulation of copper in various bodily organs, as the brain, liver, and kidneys.

**wilt[1]** (wĭlt) *v.* **wilt·ed, wilt·ing, wilts.** [Poss. alteration of dial. *welk* < ME *welken.*] —*vi.* **1.** To become limp or flaccid. **2.** To become less active or energetic : WEAKEN. —*vt.* **1.** To cause to droop or lose freshness. **2.** To deprive of energy or courage : ENERVATE. —*n.* **1.** The act of wilting or the state of being wilted. **2.** Any of various plant diseases marked by slow or rapid collapse of terminal shoots, branches, or entire structures.

☆ ***syns:*** WILT, DROOP, FLAG, SAG *v. core meaning* : to become limp, as from loss of freshness <The flowers *wilted* in the hot sun.>

**wilt[2]** (wĭlt) *aux.v. Archaic. 2nd person sing. present tense of* WILL[2].

**Wil·ton** (wĭl′tən) *n.* [After *Wilton,* England.] A carpet woven on a Jacquard loom and having a velvety surface formed by the cut loops of pile.

**Wilt·shire** (wĭlt′shîr, -shər) *n.* A sheep orig. bred in England, having a long head and pure white fleece.

**wi·ly** (wī′lē) *adj.* **-li·er, -li·est.** Full of wiles : SLY. **—wi′li·ly** *adv.* **—wi′li·ness** *n.*

**wim·ble** (wĭm′bəl) *n.* [ME < AN, prob. < MDu. *wimmel.*] A hand tool for boring holes. —*vt.* **-bled, -bling, -bles.** To bore with a wimble.

**wimp** (wĭmp) *n.* [Orig. unknown.] *Slang.* One who is weak or ineffective. **—wimp′y** *adj.*

**wim·ple** (wĭm′pəl) *n.* [ME *wimpel* < OE.] **1.** A cloth wound around the head, framing the face, and drawn into folds beneath the chin, worn by women in medieval times and as part of the habit of certain orders of nuns. **2. a.** A fold or pleat in cloth. **b.** A ripple, as on the surface of water. **c.** A bend or curve. —*v.* **-pled, -pling, -ples.** —*vt.* **1.** To cover or furnish with a wimple. **2.** To cause to form folds, pleats, or ripples. —*vi.* **1.** To form or lie in folds. **2.** To ripple.

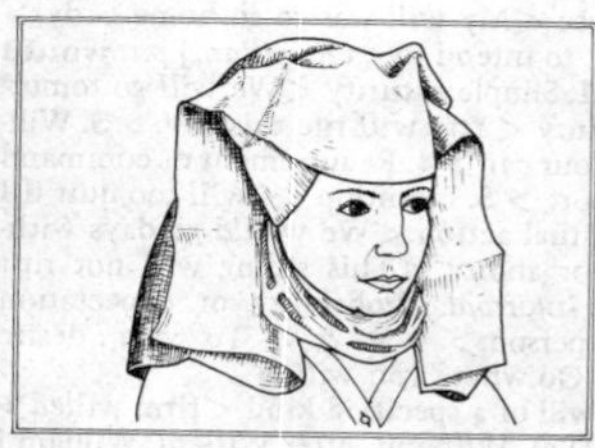

**wimple**

**Wims·hurst machine** (wĭmz′hûrst′) *n.* [After James *Wimshurst* (1832–1903).] An electrostatic generator having oppositely rotating mica or glass disks with metal carriers on which charges are produced by induction, used mainly as a demonstration apparatus.

**win** (wĭn) *v.* **won** (wŭn), **win·ning, wins.** [ME *winnen* < OE *winnan,* to strive.] —*vi.* **1.** To achieve victory over others in a competition or contest. **2.** To achieve success in an effort or venture. —*vt.* **1.** To achieve victory in. **2.** To receive as a prize or reward for performance. **3.** To achieve by effort : EARN <*win* prestige> **4.** To reach with difficulty <The vessel *won* a safe port.> **5.** To take in battle : CAPTURE <*won* the fortress> **6.** To succeed in gaining the favor or support of : prevail upon <Your campaign *won* the public.> **7. a.** To gain the affection or loyalty of. **b.** To appeal successfully to (e.g., someone's sympathy). **c.** To persuade (someone) to marry one. **8.** To make (one's way) with effort. **9. a.** To discover and open (a vein or deposit) in mining. **b.** To extract from a mine. **—win out.** To succeed or prevail. **—win over.** To persuade. **—win through.** To overcome difficulties and attain a desired goal or end. —*n.* **1.** A victory, esp. in a competition. **2.** An amount won or earned. **—win the day.** To be successful.

**wince** (wĭns) *vi.* **winced, winc·ing, winc·es.** [ME *wincen,* to kick.] To start or shrink involuntarily, as in pain or distress : FLINCH. —*n.* A wincing movement or gesture. **—winc′er** *n.*

**winch** (wĭnch) *n.* [ME *winche,* pulley < OE *wince.*] **1.** A stationary motor-driven or hand-powered hoisting machine with a drum around which a rope or chain winds as the load is lifted. **2.** The crank used to give motion to a grindstone or similar device. —*vt.* **winched, winch·ing, winch·es.** To move with or as if with a winch. **—winch′er** *n.*

**Win·ches·ter** (wĭn′chĕs′tər, -chə-stər). A trademark for a shoulder firearm.

**wind[1]** (wĭnd) *n.* [ME < OE.] **1. a.** Moving air, esp. a natural and perceptible current of air parallel to or along the ground. **b.** Moving air where artificially produced, as by a fan. **2.** A movement or current of air blowing from one of the four cardinal points of the compass. **3. a.** Something that disrupts or destroys <an ill *wind*> **b.** *often* **winds.** A tendency : trend <the *winds* of revolution> **4.** *Naut.* The direction from which the wind is blowing. **5.** A current of air carrying an odor, scent, or sound. **6. winds. a.** Orchestral or band wind instruments. **b.** Players of wind instruments. **7.** Gas produced in the body during digestion : FLATULENCE. **8.** Breath, esp. normal or adequate breathing : RESPIRATION. **9. a.** Meaningless utterance : VERBIAGE. **b.** Futile or idle labor or thought. —*vt.* **wind·ed, wind·ing, winds. 1.** To expose to the free movement of air. **2. a.** To catch a scent or trace of. **b.** To pursue by following a scent. **3.** To cause to be short or out of breath. **4.** To afford a recovery of breath. **—get wind of.** To receive hints or intimations of. **—in the wind.** Likely to occur.

**wind[2]** (wīnd) *v.* **wound** (wound), **wind·ing, winds.** [ME *winden* < OE *windan.*] —*vt.* **1.** To wrap (something) around an object or center once or repeatedly. **2.** To wrap or encircle (an object) in a series of coils : ENTWINE. **3.** To set on a curving or twisting course. **4.** To proceed on (one's way) with a curving or twisting course. **5.** To present or introduce in a disguised or devious manner <*wound* a sales pitch into the conversation> **6.** To turn (e.g., a crank) in a series of circular motions. **7.** To coil the spring of (a mechanism) by turning a stem, cord, or similar device <*wind* a clock> **8.** To lift or haul by means of a windlass or winch. —*vi.* **1.** To move in or as if in a bending or coiling course <a stream *winding* through a meadow> **2. a.** To move in or have a spiral or circular course <coils of smoke *winding* from the chimney> **b.** To be coiled or spiraled about something. **3.** To be twisted or whorled into curved forms. **4.** To proceed misleadingly or insidiously in discourse or conduct. **5.** To become wound <a watch that *winds* easily> **—wind down. 1.** To decrease or diminish in energy, intensity, or scope, esp. so as to stop gradually. **2.** *Informal.* To unwind : relax. **—wind up. 1.** *Informal.* To come or bring to a finish : END <*wind up* a business deal> **2.** To put in order <*wound up* their affairs before they left> **3.** *Informal.* To arrive in a place or situation as a result of a given course of action <*wound up* in big trouble> **4.** *Baseball.* To swing back the arm and raise the foot in preparation for pitching the ball. —*n.* **1.** The act of winding. **2.** A single turn, twist, or curve.

☆ ***syns:*** WIND, COIL, CURL, ENTWINE, SNAKE, SPIRAL, TWINE, TWIST, WEAVE *v. core meaning* : to move on a repeatedly curving course <a staircase *winding* up the lighthouse><a vine *winding* around a tree>

**wind[3]** (wīnd, wĭnd) *vt.* **wind·ed** (wīn′dĭd, wĭn′-) *or* **wound** (wound), **wind·ing, winds.** [< WIND[1].] **1.** To sound (a wind instrument) by blowing. **2.** To sound by blowing. **—wind′er** *n.*

**wind·age** (wĭn′dĭj) *n.* **1. a.** The effect of wind on the course of a projectile. **b.** The point or degree at which the wind gauge or sight of a rifle or gun must be adjusted to compensate for the effect of the wind. **2.** The difference in a given firearm between the diameter of the projectile fired and the diameter of the bore of the firearm. **3.** The disturbance of air caused by the passage of a fast-moving object, as a railroad train or missile. **4.** *Naut.* The part of the surface of a ship that is left exposed to the wind.

**wind·bag** (wĭnd′băg′) *n. Slang.* A garrulous person who says little of interest or worth.

**wind·blast** (wĭnd′blăst′) *n.* **1.** A strong gust of wind. **2.** The damaging effect of air friction on a pilot ejected from a high-speed aircraft.

**wind-blown** (wĭnd′blōn′) *adj.* **1.** Blown or dispersed by the wind. **2.** Growing or shaped in a manner determined by the prevailing winds. **3.** Cut short and curled or combed toward the front of the head <a *wind-blown* hairstyle>

**wind-borne** (wĭnd′bôrn′, -bōrn′) *adj.* Carried by the wind.

**wind·break** (wĭnd′brāk′) *n.* A hedge, fence, or row of trees serving to break or lessen the force of the wind.

**Wind·break·er** (wĭnd′brā′kər). A trademark for an outer jacket having close-fitting, often elastic cuffs and waistband.

**wind-bro·ken** (wĭnd′brō′kən) *adj.* Suffering from impairment of respiration, as the heaves. —Used of horses.

**wind·burn** (wĭnd′bûrn′) *n.* A reddened skin irritation caused by exposure to wind. **—wind′burned′** *adj.*

**wind-chill factor** (wĭnd′chĭl′) *n.* The temperature of windless air that would have the same effect on exposed human skin as a given combination of wind speed and air temperature would.

**wind cone** (wĭnd) *n.* A windsock.

**wind·ed** (wĭn′dĭd) *adj.* **1.** Having a given amount of breath or respiratory power <short-*winded*> **2.** Out of breath.

**wind·er** (wīn′dər) *n.* **1.** One that winds, esp. one who winds cloth or materials in a textile factory. **2.** A spool, barrel, or other object around which material is wound. **3.** A device, as a key, for winding up a spring-driven mechanism. **4.** A step of a winding staircase.

**wind·fall** (wĭnd′fôl′) *n.* **1.** Something, as a ripened fruit, blown down by the wind. **2.** A sudden, unexpected piece of good luck or unanticipated personal gain.

**wind·flaw** (wĭnd′flô′) *n.* A sudden blast or gust of wind.

**wind·flow·er** (wĭnd′flou′ər) *n.* ANEMONE 1.

ă **pat** ā **pay** âr **care** ä **father** ĕ **pet** ē **be** hw **which** ĭ **pit**
ī **tie** îr **pier** ŏ **pot** ō **toe** ô **paw, for** oi **noise** ŏŏ **took**

**wind gap** (wĭnd) *n.* A shallow notch or ravine on the side of a deep mountain ridge.
**wind harp** (wĭnd) *n.* An Aeolian harp.
**wind·ing** (wīn'dĭng) *n.* **1. a.** The act of one that winds. **b.** One complete turn of something wound. **2.** Something wound : SPIRAL. **3.** A bend or curve, as of a road. **4.** *Elect.* **a.** Wire wound into a coil. **b.** The manner in which such a coil is wound. **c.** A single loop of such a coil. —*adj.* **1.** Twisting or turning : SINUOUS. **2.** Spiral. **—wind'ing·ly** *adv.*
**winding-sheet** (wīn'dĭng) *n.* A sheet for wrapping a dead body : SHROUD.
**wind instrument** (wĭnd) *n.* A musical instrument, as a clarinet, trumpet, or harmonica, that is sounded by wind, esp. by the breath.
**wind·jam·mer** (wĭnd'jăm'ər) *n.* **1.** A large sailing ship. **2.** A crew member of a sailing ship.
**wind·lass** (wĭnd'ləs) *n.* [ME *wyndlas*, var. of *windas* < ON *vindāss* : *vinda*, to wind + *āss*, pole.] Any of numerous hauling or lifting machines consisting primarily of a drum or cylinder wound with rope and turned by a crank. —*vt.* **-lassed, -lass·ing, -lass·es.** To raise with a windlass.
**win·dle·straw** (wĭn'dəl-strô') *n.* [OE *windelstrēaw* : *windel*, basket (< *windan*, to wind) + *strēaw*, straw.] *Chiefly Brit.* A thin, dried grass stalk.
**wind·mill** (wĭnd'mĭl') *n.* **1.** A mill or other machine powered by a wheel of adjustable blades or slats rotated by the wind. **2.** Something, as a toy pinwheel, similar to a windmill. **3.** An imaginary threat or evil <tilting at *windmills*>
**win·dow** (wĭn'dō) *n.* [ME < ON *vindauga* : *vindr*, wind, air + *auga*, eye.] **1.** An opening in a wall or roof that admits light or air to an enclosure, usu. framed and spanned with glass mounted to permit opening and closing. **2. a.** A framework enclosing a pane of glass : SASH. **b.** A pane of glass, clear plastic, or similar material enclosed in such a framework. **3.** Something held to resemble a window in function or appearance. **4.** CHAFF[1] 4. **5.** A range of electromagnetic frequencies that pass unobstructed through a planetary atmosphere. **6.** A physical space or period of time within which an activity, as a spacecraft launch, must take place to ensure successful completion.
**window box** *n.* **1.** A usu. long narrow box for growing plants, placed on a windowsill. **2.** One of the vertical grooves on the inner sides of a window frame for the weights that counterbalance the sash.
**win·dow-dress·ing** *also* **window dressing** (wĭn'dō-drĕs'ĭng) *n.* **1. a.** Decorative display of retail merchandise in store windows. **b.** Goods and trimmings used in such display. **2.** Something used to improve appearances or create a false favorable impression. **—win'dow-dress'er** *n.*
**window envelope** *n.* An envelope with a transparent panel through which the address on the enclosure is visible.
**win·dow·pane** (wĭn'dō-pān') *n.* A plate of glass in a window.
**window shade** *n.* An opaque fabric mounted to cover or expose a window.
**win·dow-shop** (wĭn'dō-shŏp') *vi.* **-shopped, -shop·ping, -shops.** To look at merchandise in store windows or showcases without making purchases. **—win'dow-shop'per** *n.*
**win·dow·sill** (wĭn'dō-sĭl') *n.* The horizontal ledge at the base of a window opening.
**wind·pipe** (wĭnd'pīp') *n. Anat.* TRACHEA 1.
**wind rose** (wĭnd) *n.* [G. *Windrose*, compass card : *Wind*, wind, air + *Rose*, rose.] Any of a class of meteorological diagrams depicting the distribution of wind direction over a period of time.
**wind·row** (wĭnd'rō') *n.* **1.** A row, as of leaves or snow, heaped up by the wind. **2.** A long row of cut hay or grain left to dry in a field before being bundled. —*vt.* **-rowed, -row·ing, -rows.** To shape or arrange into a windrow. **—wind'row'er** *n.*
**wind·shake** (wĭnd'shāk') *n.* A crack or separation between growth rings in timber, attributed to the straining of tree trunks in high winds.
**wind·shield** (wĭnd'shēld') *n.* **1.** A framed pane of usu. curved glass or other transparent material located at the front of a vehicle to protect the occupants from the wind. **2.** A shield placed to protect an object from the wind.
**wind sleeve** *n.* A windsock.
**wind·sock** (wĭnd'sŏk') *n.* A tapered, open-ended sleeve pivotally attached to a standard that indicates the direction of the wind blowing through it.
**Wind·sor chair** *n.* [After *Windsor*, England.] A wooden chair having a saddle seat, a high spoked back, and outward-slanting legs connected by a crossbar.
**Windsor tie** *n.* A wide silk necktie tied in a loose bow.
**wind sprint** (wĭnd) *n.* A sprint run to develop the breath.
**wind·storm** (wĭnd'stôrm') *n.* A storm with high winds or violent gusts but little or no rain.
**wind·suck·er** (wĭnd'sŭk'ər) *n.* A horse given to swallowing quantities of air.
**wind·swept** (wĭnd'swĕpt') *adj.* Moved by or exposed to the force of the wind <*windswept* English moors>
**wind tee** (wĭnd) *n.* A large weather vane with a horizontal T-shaped wind indicator, found esp. at airfields.
**wind tunnel** (wĭnd) *n.* A chamber through which air is forced at controllable velocities in order to study the aerodynamic flow around and effects on objects, as airfoils or scale models, mounted within.
**wind-up** (wīnd'ŭp') *n.* **1. a.** The act of bringing something to a conclusion. **b.** The concluding part. **2.** *Baseball.* The coordinated movements of a pitcher's arm, body, and legs preparatory to pitching the ball. —*adj.* Having a spring wound by hand for use or operation.
**wind·ward** (wĭnd'wərd) *n.* The direction from which the wind blows. —*adj.* **1.** Of or moving toward the quarter from which the wind blows. **2.** Of or on the side exposed to the wind or to prevailing winds. —*adv.* In a direction from which the wind blows.
**wind·y** (wĭn'dē) *adj.* **-i·er, -i·est. 1.** Marked by or abounding in wind. **2.** Open to the wind. **3.** Resembling wind in swiftness, force, or variability. **4. a.** Marked by lack of substance : EMPTY <a *windy* speech> **b.** Marked by or given to prolonged talk <a *windy* lecturer> **5.** Flatulent. **—wind'i·ly** *adv.* **—wind'i·ness** *n.*
**wine** (wīn) *n.* [ME < OE *wīn*, ult. < Lat. *vinum*.] **1. a.** The fermented juice of any of various kinds of grapes, usu. containing 10–15% alcohol by volume. **b.** The fermented juice of any of various other fruits or plants <dandelion *wine*> **2.** Something that intoxicates or exhilarates. **3.** A dark purplish red. —*v.* **wined, win·ing, wines.** —*vt.* To provide or entertain with drink <*wined* and dined the dignitaries> —*vi.* To drink wine.
**wine cellar** *n.* **1.** A place for storing wine. **2.** A stock of wines.
**wine·glass** (wīn'glăs') *n.* A glass, usu. with a stem, in which wine is served.
**wine·grow·er** (wīn'grō'ər) *n.* One who owns a vineyard and produces wine.
**wine palm** *n.* Any of various palm trees yielding sap or juice from which wine is prepared.
**wine·press** (wīn'prĕs') *also* **wine presser** *n.* A vat in which the juice is pressed from grapes.
**win·er·y** (wī'nə-rē) *n., pl.* **-ies.** A wine-making establishment.
**Wine·sap** (wīn'săp') *n.* A variety of apple bearing fruit with dark red skin.
**wine·skin** (wīn'skĭn') *n.* A bag for holding and dispensing wine, made from the skin of a goat or another animal.
**wing** (wĭng) *n.* [ME *wenge*, of Scand. orig.] **1.** One of a pair of specialized organs of flight, as: **a.** The feather-covered modified forelimb of a bird. **b.** The membranous tissue supported by the elongated digits of the forelimb of a bat. **c.** A reticulated, membranous structure extending from an insect's thorax. **d.** The enlarged pectoral fin of a flying fish. **2.** An organ or structure homologous to or resembling a wing. **3.** *Bot.* **a.** A thin or membranous extension, as of the fruit of the ash or maple. **b.** One of the lateral petals of the flower of a pea or related plant. **4.** *Informal.* A human arm. **5.** An airfoil whose principal function is providing lift, esp. either of two such airfoils symmetrically positioned on each side of the fuselage. **6.** Something resembling a wing in appearance, function, or position relative to a main body. **7.** A means of rapid ascent or flight. **8.** Something that is moved by or that moves against the air, as a weather vane. **9.** *Chiefly Brit.* A car fender. **10.** A folding section, as of a movable partition. **11.** Either of the two side projections on the back of a wing chair. **12. a.** A flat of theatrical scenery projecting onto the stage from the side. **b. wings.** The unseen backstage area on either side of a proscenium stage. **13.** A structure attached to the side of a building. **14.** A section of a large building devoted to a specific purpose <the geriatric *wing* of the hospital> **15.** A group associated with or subordinate to an older or larger organization. **16.** A section of a party, legislature, or community holding distinct, esp. dissenting, political views <the reactionary *wing*> **17.** Either the left or right flank of a military force. **18.** Either of the forward positions played near the sideline, esp. in hockey. **19.** An air force unit larger than a group but smaller than a division or command. **20. wings.** An outspread pair of stylized bird's wings worn as an insignia by qualified pilots. —*v.* **winged, wing·ing, wings.** —*vi.* To move on or as if on wings : FLY. —*vt.* **1.** To furnish with wings. **2.** To feather (an arrow). **3.** To carry or transport by or as if by flying. **4.** To wound superficially, as in the arm. **5.** To furnish with side or subordinate extensions, as an altarpiece or building. **—in the wings.** In the background <a new regime waiting *in the wings*> **—on the wing.** In flight : FLYING. **—take wing.** To fly off. **—under one's wing.** Under one's protection or care. **—wing it.** *Informal.* To ad-lib or improvise.
**wing and wing** *adv. Naut.* With sails extended on both sides.
**wing·back** (wĭng'băk') *n. Football.* **1.** A back positioned on offense behind or outside of an end. **2.** The position of wingback.
**wing·bow** (wĭng'bō') *n.* A mark of color on the bend of the wing in a domestic fowl.
**wing chair** *n.* An armchair with a high back from which project large, enclosing side pieces.
**wing·ding** (wĭng'dĭng') *n.* [Orig. unknown.] *Slang.* A lively or lavish party or celebration.

**winged** (wĭngd, wĭng′ĭd) *adj.* **1.** Having wings or winglike appendages. **2.** Moving on or as if on wings : FLYING. **3.** Elevated : sublime <*winged* thoughts> **4.** Swift : fleet.
**wing·foot·ed** (wĭng′fo͝ot′ĭd) *adj.* Swift : fleet.
**wing·less** (wĭng′lĭs) *adj.* Having rudimentary wings or no wings.
**wing·let** (wĭng′lĭt) *n.* A small or rudimentary wing.
**wing loading** *n.* The gross weight of an airplane divided by the wing area. —Used in stress analysis.
**wing nut** *n.* A nut with winglike projections for thumb and forefinger leverage in turning.
**wing·o·ver** (wĭng′ō′vər) *n.* A flight maneuver or stunt in which an aircraft enters a climbing turn until almost stalled and is allowed to fall while the turn is continued until normal flight is attained in a direction opposite the original heading.
**wing·span** (wĭng′spăn′) *n.* **1.** The linear distance between the extremities of an airfoil. **2.** Wingspread.
**wing·spread** (wĭng′sprĕd′) *n.* The distance between the tips of the wings when fully extended, as of an aircraft, bird, or insect.
**wing tip** *n.* **1.** An often perforated shoe part that covers the toe and extends backward along the sides of the shoe from a point at the center. **2.** A style of shoe having a wing tip.
**wink** (wĭngk) *v.* **winked, wink·ing, winks.** [ME *winken,* to close one's eyes < OE *wincian.*] —*vi.* **1.** To close and open the eyelid of one eye deliberately, as to convey a message, signal, or suggestion. **2.** To close and open the eyelids of both eyes : BLINK. **3.** To shine fitfully : TWINKLE <stars *winking* in the darkness> —*vt.* **1.** To close and open (an eye or the eyes) rapidly. **2.** To signal or express by winking. —*n.* **1.** The act of winking. **2.** The time required for a wink. **3.** A signal or hint conveyed by winking. **4.** A twinkle : gleam. **5.** *Informal.* A brief moment of sleep. **—wink at.** To pretend not to see <The police *winked at* drug deals on the street.> **—wink out. 1.** To come to a close : END. **2.** To cease shining.
**win·kle** (wĭng′kəl) *n.* PERIWINKLE[1].
**win·na·ble** (wĭn′ə-bəl) *adj.* Capable of being won.
**Win·ne·ba·go** (wĭn′ə-bā′gō) *n., pl.* **Winnebago** *or* **-gos** *or* **-goes. 1. a.** A tribe of Indians living in eastern Wisconsin. **b.** A member of this tribe. **2.** The Siouan language of the Winnebago.
**win·ner** (wĭn′ər) *n.* **1.** One that wins, esp. a victor in sports or a successful person. **2.** *Slang.* One of exceptionally superior quality or character.
**winner's circle** *n.* An enclosed area at a racetrack where the winning horse and jockey are brought for awards and publicity.
**win·ning** (wĭn′ĭng) *adj.* **1.** Victorious : successful. **2.** Charming <a child with a *winning* personality> —*n.* **1.** The act of one that wins : VICTORY. **2.** *often* **winnings.** Something won, esp. money. **3.** A section of a mine that has been recently opened or prepared for working. **—win′ning·ly** *adv.* **—win′ning·ness** *n.*
**winning gallery** *n.* An opening below the side penthouse in court tennis.
**winning post** *n.* The post at the end of a racecourse.
**win·now** (wĭn′ō) *v.* **-nowed, -now·ing, -nows.** [ME *wynewer* < OE *windwian* < *wind,* wind.] —*vt.* **1.** To separate the chaff from (grain) by means of an air current. **2.** To blow (chaff) off or away. **3.** To blow away : SCATTER. **4.** To cause to flutter or fly <a breeze *winnowing* the flags> **5.** To examine carefully in order to separate the good from the bad : SIFT. **6.** To separate (a desirable or undesirable part) from something. —*vi.* **1.** To separate grain from chaff. **2.** To separate the good from the bad. —*n.* **1.** A device for winnowing grain. **2.** An act of winnowing. **—win′now·er** *n.*
**win·o** (wī′nō) *n., pl.* **-os.** *Slang.* One habitually drunk on wine.
**win·some** (wĭn′səm) *adj.* [ME *winsum* < OE *wynsum* < *wynn,* joy.] Winning : charming. **—win′some·ly** *adv.* **—win′some·ness** *n.*
**win·ter** (wĭn′tər) *n.* [ME < OE.] **1.** The usu. coldest season of the year, occurring between autumn and spring. **2.** A year as expressed through the recurrence of the winter season. **3.** A period of time marked by coldness, misery, barrenness, or death. —*v.* **-tered, -tering, -ters.** —*vi.* To spend the winter. —*vt.* To lodge, keep, or care for during the winter.
**winter aconite** *n.* A frequently cultivated European plant, *Eranthis hyemalis,* bearing a solitary yellow flower that blooms in winter or early spring.
**win·ter·ber·ry** (wĭn′tər-bĕr′ē) *n.* A North American shrub of the genus *Ilex,* with showy red berries.
**winter cherry** *n.* A frequently cultivated Eurasian plant, *Physalis alkekengi,* with red berries enclosed in inflated papery, orange-red seed cases.
**win·ter-feed** (wĭn′tər-fēd′) *vt.* **-fed** (-fĕd′), **-feed·ing, -feeds.** To feed (livestock) when grazing is not possible.
**win·ter·green** (wĭn′tər-grēn′) *n.* [Transl. of Du. *wintergroen.*] **1. a.** A low-growing plant, *Gaultheria procumbens* of eastern North America, bearing aromatic evergreen leaves, white or pinkish flowers, and spicy, edible red berries. **b.** An oil or flavoring obtained from the wintergreen. **2.** A plant related to the wintergreen, as the pipsissewa.
**win·ter·ize** (wĭn′tə-rīz′) *vt.* **-ized, -iz·ing, -iz·es.** To equip or prepare (e.g., an automobile) for winter weather. **—win′ter·i·za′tion** *n.*
**win·ter·kill** (wĭn′tər-kĭl′) *v.* **-killed, -kill·ing, -kills.** —*vt.* To kill (e.g., plants) by exposure to extremely cold winter weather. —*vi.* To die from exposure to cold winter weather. —Used esp. of plants. —*n.* Death, as of plants, resulting from exposure to winter weather.
**winter melon** *n.* [Transl. of Chin. *dong*[1] *gua*[1].] A melon, *Cucumis melo inodorus,* yielding fruit with sweet, usu. light-colored flesh.
**winter purslane** *n.* A plant, *Montia perfoliata* of western North America, bearing small white flowers and leaves eaten in salads.
**winter savory** *n.* SAVORY[2] 1.
**winter solstice** *n.* *Astron.* SOLSTICE 1.
**winter squash** *n.* A thick-rinded squash, as the acorn squash, that can be stored for long periods.
**win·ter·time** (wĭn′tər-tīm′) *n.* The winter season.
**winter wheat** *n.* Wheat planted in the autumn and harvested the following spring or early summer.
**win·try** (wĭn′trē) *also* **win·ter·y** (wĭn′tə-rē) *adj.* **-tri·er, -tri·est** *also* **-i·er, -i·est. 1.** Belonging to or typical of winter : COLD. **2.** Like winter : CHEERLESS <gave me a *wintry* smile> **—win′tri·ly** *adv.* **—win′tri·ness** *n.*
**win·y** (wī′nē) *adj.* **-i·er, -i·est.** Having the taste or qualities of wine : INTOXICATING.
**winze** (wĭnz) *n.* [Orig. unknown.] An inclined or vertical shaft or passage between levels in a mine.
**wipe** (wīp) *vt.* **wiped, wip·ing, wipes.** [ME *wipen* < OE *wīpian.*] **1.** To subject to light rubbing or friction, as of paper or a cloth, in order to clean or dry. **2.** To remove by rubbing, as dirt or grease. **3.** To rub, move, or pass over something. **4.** *Slang.* To defeat decisively esp. in a sports event. **5.** To form (a joint) in plumbing by spreading solder with a piece of cloth or leather. **—wipe out. 1.** To annihilate : destroy. **2.** *Slang.* To defeat decisively. **3.** *Informal.* To murder. **4.** To lose balance and fall or jump off a surfboard. —*n.* **1.** The act of wiping. **2.** A wiper. **3.** A blow : swipe. **4.** *Informal.* A jeer : gibe.
**wipe·out** (wīp′out′) *n.* **1. a.** The act or an instance of wiping out. **b.** Complete destruction. **2.** *Slang.* A decisive defeat, as in a sports event. **3.** A fall from a surfboard.
**wip·er** (wī′pər) *n.* **1.** One that wipes. **2.** A device designed for wiping, as for a windshield. **3.** A cam that projects from a rotating horizontal shaft to activate another machine part. **4.** *Elect.* A movable electrical contact, as in a rheostat.
**wire** (wīr) *n.* [ME, slender metal rod < OE.] **1.** A usu. pliable metallic strand or rod made in many lengths and diameters, occas. coated with an electrical insulator, and used chiefly for structural support or to conduct electricity. **2.** A group of wire strands bundled or twisted together as a functional unit : CABLE. **3.** Something resembling a wire, as in stiffness or slenderness. **4.** The telegraph service. **5.** A telegram. **6.** An open telephone connection. **7.** The screen on which sheets of paper are formed in a papermaking machine. **8.** A racetrack finish line. **9. wires. a.** The system of wires used to manipulate puppets in a show. **b.** Hidden controlling influences. **10.** *Slang.* A pickpocket. —*v.* **wired, wir·ing, wires.** —*vt.* **1.** To bind, connect, or attach with a wire or wires. **2.** To string (e.g., beads) on wire. **3.** To equip with electrical wires. **4.** To send by telegraph <*wire* condolences> **5.** To send a telegram to. —*vi.* To send a telegram. **—get (in) under the wire.** To arrive or finish in the nick of time. **—pull wires.** To use secret, often underhand means to reach an objective : MANIPULATE.
**wire cloth** *n.* A mesh woven of fine wire.
**wired** (wīrd) *adj. Slang.* **1.** In a fever of excitement : HYPER. **2.** Wearing or equipped with an electronic eavesdropping device <a *wired* conference room>
**wire-draw** (wīr′drô′) *vt.* **-drew** (-dro͞o′), **-drawn** (-drôn′), **-drawing, -draws. 1.** To draw (metal) into wire. **2.** To treat (e.g., a subject) at great length or with excessive detail. **—wire′draw′er** *n.*
**wire gauge** *n.* **1.** A gauge for measuring the diameter of wire, usu. in the form of a disk having variously sized slots in its periphery or a long graduated plate with similar slots along its edge. **2.** A standardized system of wire sizes.
**wire gauze** *n.* A material woven of very fine wires.
**wire glass** *n.* Sheet glass reinforced with wire netting.
**wire·grass** (wīr′grăs′) *n.* A grass with tough, wiry roots or rootstocks, as Bermuda grass.
**wire-haired** (wīr′hârd′) *adj.* Having a coat of very stiff, wiry hair. —Used of breeds of dogs.
**wire·less** (wīr′lĭs) *adj.* Having no wires. —*n.* **1.** A radio telegraph or telephone system. **2.** A message transmitted by wireless telegraph or telephone. **3.** *Chiefly Brit.* Radio. —*v.* **-lessed, -less·ing, -less·es.** —*vt.* To communicate with by wireless. —*vi.* To communicate by wireless.
**wireless telegraphy** *n.* Telegraphy by radio rather than by long-distance transmission lines.
**wireless telephone** *n.* A radiotelephone.
**wire·man** (wīr′mən) *n.* One who works with electric wiring.
**wire netting** *n.* Netting made of woven wire, as for fences.
**Wire·pho·to** (wīr′fō′tō). A trademark for a photograph electrically transmitted over telephone wires.

ă **pat** ā **pay** âr **care** ä **father** ĕ **pet** ē **be** hw **which** ĭ **pit**
ī **tie** îr **pier** ŏ **pot** ō **toe** ô **paw, for** oi **noise** o͝o **took**

**wire·pull·er** (wīr'po͝ol'ər) *n.* **1.** One who pulls wires or strings, as of puppets. **2.** One who uses private influence, subterfuge, or underhand means in order to reach a goal.

**wir·er** (wīr'ər) *n.* **1.** A trapper who uses wire traps to snare game. **2.** One that wires.

**wire recorder** *n.* A forerunner of the tape recorder that recorded sound on a spool of wire rather than on magnetic tape.

**wire rope** *n.* A rope composed of twisted strands of wire.

**wire·tap** (wīr'tăp') *n.* **1.** A hidden listening or recording device connected to a communications circuit. **2.** The act of installing a wiretap. —*v.* **-tapped, -tap·ping, -taps.** —*vt.* **1.** To connect a wiretap to. **2.** To monitor (a telephone call) by means of a wiretap. —*vi.* To install a wiretap or monitor a call with a wiretap. **—wire'tap'per** *n.*

**wire·work** (wīr'wûrk') *n.* **1.** Wire fabric. **2.** Articles made of wire or wire fabric.

**wire·worm** (wīr'wûrm') *n.* **1.** The wirelike larva of various click beetles, causing severe damage by boring into the roots of many kinds of plants. **2.** Any of various millipedes.

**wire-wove** (wīr'wōv') *adj.* **1.** Designating a high grade of writing paper with a smooth finish. **2.** Made of woven wire.

**wir·ing** (wīr'ĭng) *n.* **1.** The act of attaching, connecting, or installing electric wires. **2.** A system of electric wires.

**wir·ra** (wĭr'ə) *interj.* [< Ir. Gael. *a Muire,* O Mary.] *Ir.* —Used to express sorrow.

**wir·y** (wīr'ē) *adj.* **-i·er, -i·est. 1.** Of or pertaining to wire. **2.** Wirelike and kinky, as hair. **3.** Sinewy and lean <a *wiry* physique> **—wir'i·ly** *adv.* **—wir'i·ness** *n.*

**wis·dom** (wĭz'dəm) *n.* [ME < OE *wīsdōm* < *wīs,* wise.] **1.** Understanding of what is true, right, or lasting. **2.** Good judgment : common sense. **3.** Learning : erudition.

**Wisdom of Jesus, the Son of Si·rach** (sī'răk') *n.* —See table at BIBLE.

**Wisdom of So·lo·mon** (sŏl'ə-mən) *n.* —See table at BIBLE.

**wisdom tooth** *n.* [< NLat. *dentes sapientiae,* teeth of wisdom, from their usu. being cut around the age of 20.] One of four molars, the last on each side of both jaws, usu. erupting much later than the others.

**wise**[1] (wīz) *adj.* **wis·er, wis·est.** [ME < OE *wīs.*] **1.** Having wisdom or discernment for what is true, right, or lasting : JUDICIOUS. **2. a.** Displaying common sense : PRUDENT <a *wise* choice> **b.** Shrewd : crafty. **3.** Having great learning : ERUDITE <a *wise* philosopher> **4.** Having knowledge or information : INFORMED <*wise* to my adversary's schemes> **5.** *Slang.* Offensively self-assured : ARROGANT. **—get wise.** *Slang.* **1.** To learn the facts or become aware <*got wise* and invested their money> **2.** To become defiantly insolent <a worker who *got wise* with the foreman> **—wise up.** *Slang.* To make or become aware or sophisticated. **—wise'ly** *adv.*

**wise**[2] (wīz) *n.* [ME < OE *wīse.*] Method or way of doing : FASHION <in no *wise*><in this *wise*><in any *wise*>

**-wise** *suff.* [ME < OE *-wīsan* < *-wīse,* manner.] **1.** In a given manner, direction, or position <clock*wise*> **2.** With reference to : in regard to <dollar*wise*> ***usage:*** The suffix *-wise* has a long history of use with the meaning "in the manner or direction of," as in *clockwise* or *likewise.* Recently *-wise* has come into use as a noun suffix meaning "with relation to," as in *saleswise.* But indiscriminate coinage of terms with this suffix can lead to confusion since the exact nature of the intended relation is not always clear from the context.

**wise·a·cre** (wīz'ā'kər) *n.* [MDu. *wijssegghe*r, soothsayer, alteration of OHG *wīssago,* seer.] *Informal.* An offensively self-assured person.

**wise·crack** (wīz'krăk') *Slang.* —*n.* A flippant, often sardonic remark. —*vi.* **-cracked, -crack·ing, -cracks.** To make or utter a wisecrack. **—wise'crack'er** *n.*

**wise guy** *n. Slang.* A wiseacre.

**wis·en·heim·er** *also* **weis·en·heim·er** (wīz'ən-hī'mər) *n.* [< WISE + G. *-enheimer* (as in G. surnames such as *Oppenheimer*).] *Informal.* A wiseacre.

**wi·sent** (vē'zĕnt') *n.* [G. < OHG *wisunt.*] The European bison, *Bison bonasus.*

**wish** (wĭsh) *n.* [ME *wisshen* < OE *wȳscan.*] **1.** A desire, longing, or strong inclination. **2.** An expression or confession of a desire, longing, or strong inclination. **3.** Something desired or longed for. —*v.* **wished, wish·ing, wish·es.** —*vt.* **1.** To desire or long for : WANT. **2.** To entertain or express wishes for : BID <I *wished* you good night.> **3.** To call upon or invoke <They *wished* you luck.> **4.** To order or entreat <I *wish* you to return at once.> **5.** To impose or force : FOIST <It was a hard task they *wished* on me.> —*vi.* **1.** To have or feel a desire <*wish* for the impossible> **2.** To express a wish. **—wish'er** *n.*

**wish·bone** (wĭsh'bōn') *n.* [From the superstition that when two people pull it apart a wish will be fulfilled for the one who retains the longer piece.] The forked bone, or furcula, anterior to the breastbone of most birds, formed by fusion of the clavicles.

**wish·ful** (wĭsh'fəl) *adj.* Having or expressing a wish. **—wish'ful·ly** *adv.* **—wish'ful·ness** *n.*

**wish fulfillment** *n.* **1.** Gratification of a desire. **2.** *Psychoanal.* Satisfaction of a desire or release of tension by the use of imagination.

**wishful thinking** *n.* Erroneous identification of one's own wishes with reality.

**wish-wash** (wĭsh'wŏsh', -wôsh') *n.* [Redup. of WASH.] *Informal.* A thin watery drink.

**wish·y-wash·y** (wĭsh'ē-wŏsh'ē, -wô'shē) *adj.* **-i·er, -i·est.** [Redup. of *washy* < WASH.] *Informal.* **1.** Watery : thin. **2.** Lacking strength or purpose : INEFFECTUAL <a *wishy-washy* character>

**wisp** (wĭsp) *n.* [ME.] **1.** A small bunch or bundle, as of straw, hair, or grass. **2. a.** One that is thin, frail, or slight. **b.** A thin or faint streak or fragment, as of smoke or clouds. **3.** A fleeting trace or indication : HINT <a *wisp* of a smile> **4.** A flock of birds, esp. of snipe. **5.** Ignis fatuus. —*v.* **wisped, wisp·ing, wisps.** —*vt.* To twist into a wisp. —*vi.* To drift in wisps <clouds *wisping* by> **—wisp'y** *adj.*

**wist** (wĭst) *v. Archaic. p.t. & p.p. of* WIT[2].

**wis·ter·i·a** (wĭ-stîr'ē-ə) *also* **wis·tar·i·a** (wĭ-stâr'ē-ə) *n.* [NLat. *Wisteria,* genus name, after Caspar *Wistar* (1761–1818).] A climbing woody vine of the genus *Wisteria,* with compound leaves and drooping purplish or white flower clusters.

**wist·ful** (wĭst'fəl) *adj.* [< obs. *wistly,* intently.] Full of a pensive or melancholy yearning. **—wist'ful·ly** *adv.* **—wist'ful·ness** *n.*

**wit**[1] (wĭt) *n.* [ME < OE.] **1.** Natural ability to know or perceive : INTELLIGENCE. **2. wits. a.** Keenness of perception or discernment. **b.** Sound mental faculties : SANITY. **3. a.** Ability to perceive and express in an ingeniously humorous way the relationship or similarity between seemingly incongruous or disparate things. **b.** One noted for this ability, esp. one skilled in repartee. **—at (one's) wits' end.** At the limit of one's mental resources. **—have** (or **keep**) **(one's) wits about (one).** To remain alert or calm, esp. in a crisis.

**wit**[2] (wĭt) *v.* **wist** (wĭst), **wit·ting,** *1st & 3rd person present* **wot** (wŏt) [ME < OE *witan.*] *Archaic.* —*vt.* To be or become aware of : LEARN. —*vi.* To know. **—to wit.** That is to say : NAMELY.

**wit·an** (wĭt'ăn) *pl.n.* [OE, pl. of *wita,* councilor.] **1.** The members of the witenagemot in Anglo-Saxon England. **2.** The witenagemot.

**witch** (wĭch) *n.* [ME *wicche* < OE *wicce,* witch, and < OE *wicca,* wizard.] **1.** A woman who practices black magic : SORCERESS. **2.** An ugly, vicious old woman : HAG. **3.** *Informal.* A bewitching young woman or girl. —*vt.* **witched, witch·ing, witch·es. 1.** To bewitch. **2.** To cause, bring, or effect by witchcraft.

**witch·craft** (wĭch'krăft') *n.* **1.** Black magic : sorcery. **2.** A magical or irresistible influence, attraction, or charm.

**witch doctor** *n.* A medicine man or shaman practicing among primitive peoples.

**witch elm** *n. var. of* WYCH ELM.

**witch·er·y** (wĭch'ə-rē) *n., pl.* **-ies. 1.** Witchcraft : sorcery. **2.** Power to fascinate or charm.

**witch·es-broom** (wĭch'ĭz-bro͞om', -bro͝om') *n.* An abnormal brushlike growth of weak, closely clustered shoots or branches on a tree or woody plant caused by fungi or viruses.

**witches' Sabbath** *n.* An orgy of demons, witches, and sorcerers.

**witch grass** *n.* **1.** A North American grass, *Panicum capillare,* with branching purplish panicles. **2.** Couch grass.

**witch hazel** *n.* [From ME *wyche,* WYCH (ELM) + HAZEL.] **1.** A shrub of the genus *Hamamelis,* esp. *H. virginiana* of eastern North America, bearing yellow flowers that bloom in late autumn or winter. **2.** An alcoholic solution containing an extract of the bark and leaves of the witch hazel, applied externally as a mild astringent.

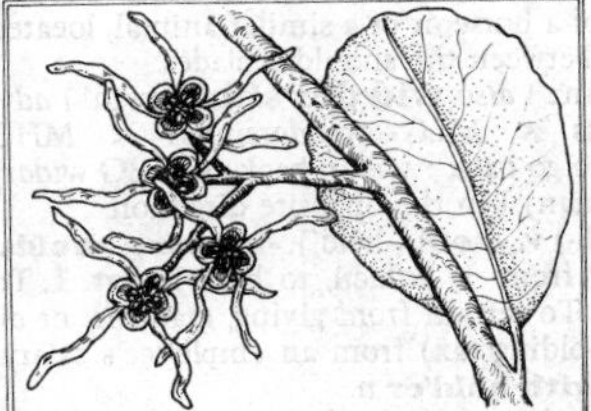

**witch hazel**

**witch-hunt** (wĭch'hŭnt') *n.* A political campaign launched on the pretext of investigating activities subversive to the state. **—witch'-hunt'** *v.* **(-hunt·ed, -hunt·ing, -hunts). —witch'-hunt'er** *n.*

**witch·ing** (wĭch'ĭng) *adj.* **1.** Relating to or suitable for witchcraft. **2.** Having power to charm or enchant : BEWITCHING. —*n.* Witchcraft. **—witch'ing·ly** *adv.*

**witch moth** *n.* [From its nocturnal habits.] A large moth of the genus *Erebus* of the southern United States and tropical America.

**witch of Ag·ne·si** (än-yā'zē) *n.* [After Maria Gaetana *Agnesi* (1718–1799).] A planar cubic curve that is symmetric about the *y*-axis and that approaches the *x*-axis as an asymptote.

**wite** (wīt) *n.* [ME < OE *wīte,* punishment.] *Chiefly Scot.* Blame.

**wit·e·na·ge·mot** (wĭt'n-ə-gə-mōt') *n.* [OE *witena gemōt,* meeting of councilors.] An Anglo-Saxon advisory council to the king, composed of about 100 nobles, prelates, and other officials, convened at intervals to discuss administrative and judicial affairs.

**with** (wĭth, wĭth) *prep.* [ME, with, against, from < OE *wið*.] **1.** As a companion of : ACCOMPANYING <took the dog *with* us> **2.** Next to <Sit *with* them.> **3.** Having as a possession, attribute, or characteristic <people *with* blue eyes> **4. a.** In a manner characterized by <act *with* decision> **b.** In the performance, use, or operation of <problems *with* the plan> **5.** In the charge or keeping of <left the kids *with* a neighbor> **6.** In the opinion or estimation of <Is it O.K. *with* the boss?> **7.** In support of : on the side of <We're *with* you all the way!> **8.** Of the same opinion or belief as <Were they *with* you on that issue?> **9.** In the same group or mixture as : AMONG <planted onions *with* the carrots> **10.** In the membership or employment of <is *with* a major airline> **11.** By the means or agency of <eat *with* chopsticks> **12.** In spite of <*With* all their wealth, they're still unhappy.> **13.** In the same direction as <swim *with* the current> **14.** At the same time as <arose *with* the birds> **15.** In regard to <I am angry *with* you.> **16.** In comparison or contrast to <clothes identical *with* yours> **17.** Having received <*With* your permission, I'll leave.> **18.** And : plus: <had coffee *with* cake> **19.** In opposition to : AGAINST <competitors vying *with* each other> **20.** As a result or consequence of : under the influence of <stiff *with* cold> **21.** To : onto <Link your arm *with* your partner's.> **22.** So as to be free of or separated from <had to part *with* my savings> **23.** In the course of <grows prettier *with* each day> **24.** In proportion to <cheese that improves *with* age> **25.** In relationship to <infatuated *with* a neighbor> **26.** In favorable comparison to : AS WELL AS <We can play tennis *with* the best of them.> **27.** According to the experience or practice of <*With* me, it's a matter of taste.> **28.** —Used as a function word to indicate close association <*With* the jet plane, travel time was cut dramatically.> **—in with.** In league or association with <got *in with* a popular crowd>

**with·al** (wĭth-ôl', wĭth-) *adv.* [ME : *with*, with + *al*, all.] **1.** In addition : BESIDES. **2.** Despite that : NEVERTHELESS. **3.** *Archaic.* Therewith. —*prep. Archaic.* With.

**with·draw** (wĭth-drô', wĭth-) *v.* **-drew** (-dro͞o'), **-drawn** (-drôn'), **-draw·ing, -draws.** [ME *withdrawen* : *with*, away from + *drawen*, to pull.] —*vt.* **1.** To take back or away : REMOVE. **2.** To recall : retract. —*vi.* **1.** To move or draw back : RETIRE. **2.** To remove oneself from activity or a social or emotional involvement.

**with·draw·al** (wĭth-drô'əl, wĭth-) *n.* **1.** The act or process of withdrawing. **2. a.** Termination of the administration of a habit-forming substance <heroin *withdrawal*> **b.** The physiological readjustment that takes place upon such discontinuation <*withdrawal* eased by methadone>

**with·drawn** (wĭth-drôn', wĭth-) *adj.* **1.** Not readily approached : REMOTE. **2.** Socially retiring : MODEST. **3.** Emotionally unresponsive.

**withe** (wĭth, wĭth, wīth) *n.* [ME < OE *wiððe*.] A tough, supple twig, esp. a willow twig, used for binding things together : WITHY.

**with·er** (wĭth'ər) *v.* **-ered, -er·ing, -ers.** [ME *widderen*.] —*vi.* **1.** To dry up or shrivel from or as if from loss of moisture. **2.** To lose freshness : DROOP. —*vt.* **1.** To cause to shrivel or fade. **2.** To render speechless or powerless : STUN.

**with·er·ite** (wĭth'ə-rīt') *n.* [G. *Witherit*, after William Withering (1741–1799).] A white, yellow, or gray vitreous mineral, essentially $BaCO_3$.

**withe rod** *n.* A shrub, *Viburnum cassinoides* of eastern North America, bearing small white flower clusters and bluish-black fruit.

**with·ers** (wĭth'ərz) *pl.n.* [Prob. < obs. *wither-*, against < ME < OE *wiðer-*, from the strain exerted on them when a horse draws a load.] The high point of the back of a horse or of a similar animal, located at the base of the neck and between the shoulder blades.

**with·er·shins** (wĭth'ər-shĭnz') *also* **wid·der·shins** (wĭd'-) *adv.* [Alteration of *widdershins* < MLG *weddersinnes* < MHG *widersinnes* < *widersinner*, to go back : *wider*, back (< OHG *widar*) + *sinner*, to go (< OHG *sinnan*).] In the opposite direction.

**with·hold** (wĭth-hōld', wĭth-) *v.* **-held** (-hĕld'), **-hold·ing, -holds.** [ME *witholden*: *with*, away from + *holden*, to hold.] —*vt.* **1.** To keep in check : RESTRAIN. **2.** To refrain from giving, granting, or allowing. **3.** To deduct (withholding tax) from an employee's salary. —*vi.* To forbear : refrain. **—with·hold'er** *n.*

**withholding tax** *n.* A portion of an employee's wages withheld by the employer as partial payment of the employee's income tax.

**with·in** (wĭth-ĭn', wĭth-) *adv.* [ME *withinne* < OE *wiðinnan* : *wið*, with + *innan*, into < *in*, in.] **1.** In or into the inner part : INSIDE. **2.** Inside the body, mind, heart, or soul : INWARDLY. —*prep.* **1.** In the inner part or parts of : INSIDE <Frustration grew *within* me.> **2.** Inside the limits or extent of in time, degree, or distance <*within* an hour of the city> **3.** Inside the fixed limits of : not beyond <had trouble living *within* my means> **4.** In the scope or sphere of <*within* the legal profession> —*n.* An inner position, place, or area <change from *within*>

**with·in·doors** (wĭth-ĭn'dôrz', -dōrz', wĭth-) *adv.* Indoors.

**with-it** (wĭth'ĭt') *adj. Informal.* Up-to-date : hip.

**†with·out** (wĭth-out', wĭth-) *adv.* [ME *withouten* < OE *wiðūtan* : *wið*, with + *ūtan*, outside of < *ūt*, out.] **1.** On the outside <an imposing house within and *without*> **2.** With something absent or lacking <We had to go *without*.> —*prep.* **1.** Not having : LACKING <a family *without* a dollar><*without* warning> **2.** At, on, to, or toward the outside or exterior of <listening to us *without* the door> —*conj. Regional.* Unless <"You don't know about me *without* you have read a book by the name of *The Adventures of Tom Sawyer*" —Mark Twain>

**with·out·doors** (wĭth-out'dôrz', -dōrz', wĭth-) *adv.* Outside of a house or shelter : OUTDOORS.

**with·stand** (wĭth-stănd', wĭth-) *v.* **-stood** (-stŏŏd'), **-stand·ing, -stands.** [ME *withstanden, withstonden* < OE *wiðstandan* : *wið*, against + *standan*, to stand.] —*vt.* To oppose with force : RESIST. —*vi.* To resist or endure successfully. **—with·stand'er** *n.*

**with·y** (wĭth'ē, wĭth'ē) *adj.* [ME *withye*, flexible twig < OE *wiðig*.] **1.** Made of or as flexible as withes : TOUGH. **2.** Agile and wiry. —*n.*, *pl.* **-ies. 1.** A rope or band made of withes. **2. a.** A long flexible twig, as that of an osier. **b.** A shrub or tree having such twigs.

**wit·less** (wĭt'lĭs) *adj.* Lacking wit or intelligence : STUPID. **—wit'less·ly** *adv.* **—wit'less·ness** *n.*

**wit·loof** (wĭt'lōf') *n.* [Du. : *wit*, white + *loof*, leaf.] ENDIVE 2.

**wit·ness** (wĭt'nĭs) *n.* [ME < OE < *wit*, knowledge.] **1. a.** One who has seen or heard something. **b.** One who gives evidence. **2.** Something that serves as evidence : SIGN. **3.** *Law.* **a.** One who is called upon to testify before a court. **b.** One who is called upon to be present at a transaction in order to attest to what takes place. **c.** One who signs his or her name to a document so as to attest its authenticity. **4.** An attestation to a fact, statement, or event. —*v.* **-nessed, -ness·ing, -ness·es.** —*vt.* **1.** To be present at or have personal knowledge of. **2.** To serve as or furnish evidence of. **3.** To testify to : bear witness. **4.** To be the setting or site of <a church that has *witnessed* many services> **5.** To attest to the legality or authenticity of by signing one's name. —*vi.* To serve as or furnish evidence : TESTIFY. **—wit'ness·er** *n.*

**witness stand** *n.* The place in a courtroom from which a witness presents testimony.

**wit·ti·cism** (wĭt'ĭ-sĭz'əm) *n.* [< WITTY.] A witty saying or remark.

**†wit·ting** (wĭt'ĭng) [Pr.part. of WIT².] *Archaic.* —*adj.* **1.** Aware or conscious. **2.** Done intentionally or with premeditation : DELIBERATE. —*n. Regional.* **1.** Knowledge or awareness : COGNIZANCE. **2.** Information obtained and passed on : NEWS. **—wit'ting·ly** *adv.*

**wit·tol** (wĭt'l) *n.* [ME *wetewold* : *weten*, to know (< OE *witan*) + (*coke*)*wold*, cuckold. —see CUCKOLD.] *Archaic.* A man who tolerates his wife's infidelity.

**wit·ty** (wĭt'ē) *adj.* **-ti·er, -ti·est. 1.** Having or showing wit in speech or writing. **2.** Marked by or having the nature of wit <*witty* repartee> **3.** Quick to discern and express amusing insights. **—wit'ti·ly** *adv.* **—wit'ti·ness** *n.*

**wive** (wīv) *v.* **wived, wiv·ing, wives.** [ME *wiven* < OE *wīfian* < *wīf*, wife.] —*vt.* **1.** To marry (a woman). **2.** To provide a wife for. —*vi.* To marry a woman.

**wi·vern** (wī'vərn) *n. var. of* WYVERN.

**wives¹** (wīvz) *n. pl. of* WIFE.

**wives²** (wīvz) *v. 3rd person sing. present tense of* WIVE.

**wiz** (wĭz) *n.* [Short for WIZARD.] *Informal.* One considered exceptionally gifted or skilled.

**wiz·ard** (wĭz'ərd) *n.* [ME *wysard* < *wys*, wise < OE *wīs*.] **1.** A sorcerer : magician. **2.** One who is extremely skillful or clever <a computer *wizard*> **3.** *Archaic.* A sage. —*adj.* **1.** Of or relating to wizards or wizardry. **2.** *Chiefly Brit.* Excellent.

▲ word history: *Wizard* is a compound word formed from the adjective *wise*, "learned, sensible," and the suffix *-ard*. The word originally meant "a wise man, philosopher." The suffix *-ard*, however, almost always has a pejorative or disparaging sense, as in the words *coward*, *drunkard*, and *sluggard*. *Wizard* was therefore often used contemptuously to mean "a so-called wise man," and from this use it came to mean "sorcerer" and "male witch."

**wiz·ard·ry** (wĭz'ər-drē) *n.* The art, skill, or practice of a wizard.

**wiz·en** (wĭz'ən) *v.* **-ened, -en·ing, -ens.** [ME *wisenen* < OE *wisnian*.] —*vi.* To wither or sear : dry up. —*vt.* To cause to wither or dry up. —*adj.* Shriveled or dried up : WITHERED.

**wiz·ened** (wĭz'ənd) *adj.* Shriveled : wizen.

**wo** (wō) *n. Archaic. var. of* WOE.

**woad** (wōd) *n.* [ME *wode* < OE *wād*.] **1.** An Old World plant, *Isatis tinctoria*, once cultivated for its leaves that yield a blue dye. **2.** The dye obtained from the woad.

**woad·wax·en** (wōd'wăk'sən) *n.* [Alteration of WOODWAXEN.] Dyer's greenweed.

**wob·ble** *also* **wab·ble** (wŏb'əl) [Prob. < LG *wabbeln*.] —*v.* **-bled, -bling, -bles 1.** To move erratically from side to side. **2.** To tremble or quaver, as the voice. **3.** To vacillate in one's feelings or opinions. —*vt.* To cause to wobble. —*n.* **1.** An act or instance of wobbling. **2.** A tremulous, uncertain tone or sound. **—wob'bler** *n.*

**wob·bly** (wŏb'lē) *adj.* **-bli·er, -bli·est.** Tending to wobble.

**Wob·bly** (wŏb'lē) *n.*, *pl.* **-blies.** [Orig. unknown.] *Slang.* A member of the Industrial Workers of the World.

**Wo·den** *also* **Wo·dan** (wōd′n) *n.* [OE *Wōden.*] *Myth.* The chief Teutonic god.
**woe** (wō) *n.* [ME < OE *wā,* woe!] **1.** Deep sorrow : GRIEF. **2.** Misfortune : calamity <political and economic *woes*> —*interj.* —Used to express sorrow or dismay.
**woe·be·gone** (wō′bĭ-gôn′, -gŏn′) *adj.* [ME *wo begon,* beset with woe.] **1.** Mournful or sorrowful. **2.** Being in a sorry state <a *woebegone* old junkyard>
**woe·ful** *also* **wo·ful** (wō′fəl) *adj.* **1.** Afflicted with woe : MOURNFUL. **2.** Causing woe. **3.** Pitiful : deplorable.
**wok** (wŏk) *n.* [Cantonese.] A metal pan with a convex bottom used esp. for frying and steaming in Oriental cooking.
**woke** (wōk) *v. var. p.t. of* WAKE[1].
**wok·en** (wō′kən) *v. var. p.p. of* WAKE[1].
**wold[1]** (wōld) *n.* [ME < OE *weald,* forest.] An unforested rolling plain : MOOR.
**wold[2]** (wōld) *n. var. of* WELD[2].
**wolf** (wo͝olf) *n., pl.* **wolves** (wo͝olvz) [ME < OE *wulf.*] **1. a.** A carnivorous mammal, *Canis lupus* of northern regions or *C. rufus* or *C. niger* of southwestern North America, related to and resembling the dog. **b.** The fur of the wolf. **2.** A mammal related to or like the wolf. **3.** The destructive larva of any of various moths, beetles, or flies. **4. a.** One who is rapacious, predatory, and fierce. **b.** *Slang.* A man given to avid amatory pursuit of women. **5.** *Mus.* **a.** A harshness in some tones of a bowed stringed instrument produced by defective vibration. **b.** Dissonance in some intervals of a keyboard instrument tuned to a system of unequal temperament. —*vt.* **wolfed, wolf·ing, wolfs.** To eat voraciously <"The town's big shots were . . . *wolfing* down the buffet" —Ralph Ellison> **—cry wolf.** To raise a false alarm. **—wolf′ish** *adj.* **—wolf′ish·ly** *adv.*
**wolf·ber·ry** (wo͝olf′bĕr′ē) *n.* A shrub, *Symphoricarpos occidentalis* of western North America, with white berries.
**Wolf Cub** *n. Chiefly Brit.* A Cub Scout.
**wolf dog** *n.* **1.** A dog trained to hunt wolves. **2.** The offspring of a dog and a wolf.
**Wolff·i·an body** (wo͝ol′fē-ən) *n.* [After Kasper Friedrich *Wolff* (1733–1794).] *Biol.* The mesonephros.
**wolf fish** *n.* A northern marine fish of the genus *Anarhichas,* with sharp powerful teeth.
**wolf·hound** (wo͝olf′hound′) *n.* Any of various large dogs trained to hunt wolves or other large game.
**wolf pack** *n.* A group of submarines that attack a single vessel or a convoy.
**wolf·ram** (wo͝ol′frəm) *n.* [G.] Tungsten.
**wolf·ram·ite** (wo͝ol′frə-mīt′) *n.* Any of several red-brown to black minerals with the general formula (Fe,Mn)$WO_4$, a major source of tungsten.
**wolfs·bane** (wo͝olfs′bān′) *n.* The monkshood.
**wol·las·ton·ite** (wo͝ol′ə-stə-nīt′) *n.* [After William H. *Wollaston* (1766–1828).] A mineral, essentially $CaSiO_3$, found in metamorphic rocks and used in ceramics, paints, plastics, and cements.
**Wo·lof** (wō′lŏf′) *n.* A Niger-Congo language of Senegal.
**wol·ver·ine** (wo͝ol′və-rēn′) *n.* [< WOLF.] **1.** A carnivorous mammal, *Gulo gulo* or *G. luscus* of northern regions, with dark fur and a bushy tail. **2. Wolverine.** A native or inhabitant of Michigan.
**wolves** (wo͝olvz) *n. pl. of* WOLF.
**wom·an** (wo͝om′ən) *n., pl.* **wom·en** (wĭm′ĭn) [ME *wimman* < OE *wīfman* : *wīf,* wife + *man,* person.] **1.** An adult female human being. **2.** Women as a group : WOMANKIND. **3.** Feminine quality or aspect : WOMANLINESS. **4.** A maidservant. **5.** MISTRESS 6. **6.** *Informal.* A wife.

▲ word history: Although the word *woman* is now basic to the vocabulary of English, it was not always so, for the word was coined in relatively late Old English times. *Woman* is actually a compound word formed from *wīf,* the inherited Old English word meaning "adult human female," and *man,* which meant in Old English, as it still does today, "human being" in addition to "adult human male." The compound thus literally means "woman-person." During Middle English times *wīfman* became *wimman,* the pronunciation of which varied from dialect to dialect. The irregular plural *women* preserves the irregular plural *men* for *man* used in the original compound; it is not the same plural as the *-en* of *oxen* or *children.* Why the Anglo-Saxons felt a need to create a new word cannot be known, but the coinage was successful and eventually supplanted *wīf,* which now, as *wife,* is almost completely restricted to denoting a married woman.

**wom·an·hood** (wo͝om′ən-ho͝od′) *n.* **1.** The state of being a woman. **2.** Woman's nature. **3.** Womankind.
**wom·an·ish** (wo͝om′ə-nĭsh) *adj.* **1.** Like or typical of a woman. **2.** Effeminate. **—wom′an·ish·ly** *adv.* **—wom′an·ish·ness** *n.*
**wom·an·ize** (wo͝om′ə-nīz′) *v.* **-ized, -iz·ing, -iz·es.** —*vt.* To give feminine characteristics to. —*vi.* To pursue women excessively or illicitly. **—wom′an·iz′er** *n.*
**wom·an·kind** (wo͝om′ən-kīnd′) *n.* Women as a group.

o͞o **boot** ou **out** th **thin** *th* **this** ŭ **cut** ûr **urge** y **young**
yo͞o ab**use** zh vi**sion** ə **a**bout, it**e**m, ed**i**ble, gall**o**p, circ**u**s

**wom·an·ly** (wo͝om′ən-lē) *adj.* **-li·er, -li·est.** Having the becoming qualities of a woman. **—wom′an·li·ness** *n.*
**wom·an·pow·er** (wo͝om′ən-pou′ər) *n.* Power in terms of the women available to a group or required for a task.
**woman suffrage** *n.* The right of women to vote. **—wom′an-suf′fra·gist** *n.*
**womb** (wo͞om) *n.* [ME < OE *wamb.*] **1.** *Anat.* UTERUS 1. **2. a.** A place where something is generated. **b.** A protective and confining organ, receptacle, or area. **3.** *Obs.* The belly.
**wom·bat** (wŏm′băt′) *n.* [Native word in Australia.] An Australian marsupial, *Phascolomis ursinus* or *Lasiorhinus latifrons,* somewhat similar to a small bear.
**wom·en** (wĭm′ĭn) *n. pl. of* WOMAN.
**wom·en·folk** (wĭm′ĭn-fōk′) *also* **wom·en·folks** (-fōks′) *pl.n.* **1.** Womankind. **2.** The women of a family or community.
**women's rights** *pl.n.* **1.** Economic, political, legal, and social rights for women equal to those granted men. **2.** A movement in support of women's rights.
**women's room** *n.* A rest room for women.
**won[1]** (wŭn) *vi.* **wonned, won·ning, wons.** [ME *wonen* < OE *wunian.*] *Archaic.* To dwell : abide.
**won[2]** (wôn) *n., pl.* **won.** [Korean.] —See table at CURRENCY.
**won[3]** (wŭn) *v. p.t. & p.p. of* WIN.
**won·der** (wŭn′dər) *n.* [ME < OE *wundor.*] **1. a.** Something arousing awe, astonishment, surprise, or admiration : MARVEL. **b.** The emotion thus aroused. **2.** A feeling of puzzlement or doubt. **3.** *often* **Wonder.** A monumental human creation regarded with awe, esp. one of seven monuments of the ancient world that appeared on various lists of late antiquity. —*v.* **-dered, -der·ing, -ders.** —*vi.* **1.** To have a feeling of awe or admiration : MARVEL. **2.** To be filled with curiosity or doubt. —*vt.* To have doubts or curiosity about. **—won′der·er** *n.*
**won·der·ful** (wŭn′dər-fəl) *adj.* **1.** Capable of exciting wonder : ASTONISHING. **2.** Excellent : extraordinary. **—won′der·ful·ly** *adv.* **—won′der·ful·ness** *n.*
**won·der·land** (wŭn′dər-lănd′) *n.* **1.** A marvelous imaginary realm. **2.** A marvelous place or scene.
**won·der·ment** (wŭn′dər-mənt) *n.* **1.** Awe, astonishment, or surprise. **2.** Something producing wonder. **3.** Puzzlement or curiosity.
**won·der·work** (wŭn′dər-wûrk′) *n.* A miracle or marvel. **—won′der·work′er** *n.* **—won′der·work′ing** *adj.*
**won·drous** (wŭn′drəs) *adj.* Wonderful. —*adv. Archaic.* To a wonderful or remarkable extent. **—won′drous·ly** *adv.* **—won′drous·ness** *n.*
**won·ky** (wŏng′kē) *adj.* **-ki·er, -ki·est.** [Prob. alteration of dial. *wankle* < ME *wankel* < OE *wancol.*] *Chiefly Brit.* **1.** Shaky : feeble. **2.** Wrong : awry.
**wont** (wônt, wōnt, wŭnt) *adj.* [ME < p.part. of *wonen,* to be used to, dwell. —see WON[1].] **1.** Accustomed or used to. **2.** Apt or likely <is *wont* to be late> —*n.* Usage or custom <It is my *wont* to get up early.> —*v.* **wont, wont** or **wont·ed, wont·ing, wonts.** —*vt.* To make (someone) accustomed to. —*vi.* To be in the habit of.
**won't** (wōnt). Will not.
**wont·ed** (wôn′tĭd, wōn′-, wŭn′-) *adj.* Accustomed : usual <replied with your *wonted* terseness>
**won ton** (wŏn′ tŏn′) *n., pl.* **won tons.** [Cantonese *wan tan.*] **1.** A noodle-dough dumpling filled with spiced minced pork, usu. served in soup. **2.** Soup containing won tons.
**woo** (wo͞o) *v.* **wooed, woo·ing, woos.** [ME *wowen* < OE *wōgian.*] —*vt.* **1.** To seek the affection of with intent to marry. **2. a.** To seek to gain or achieve. **b.** To tempt or invite. **3.** To entreat, solicit, or importune. —*vi.* To court a woman. **—woo′er** *n.*
**wood** (wo͝od) *n.* [ME *wode* < OE *wudu.*] **1. a.** The tough, fibrous cellular substance constituting the xylem of trees and shrubs, lying beneath the bark and composed largely of cellulose and lignin. **b.** This substance, often cut and dried, used for numerous purposes including building material and fuel. **2.** *often* **woods.** A dense growth of trees : FOREST. **3.** An object made of wood, esp.: **a.** A woodwind. **b.** A golf club with a wooden head. —*v.* **wood·ed, wood·ing, woods.** —*vt.* **1.** To fuel with wood. **2.** To cover with trees : FOREST. —*vi.* To gather or be supplied with wood.
**wood alcohol** *n.* Methyl alcohol.
**wood anemone** *n.* A plant, *Anemone quinquefolia* of eastern North America, or *A. nemorosa* of Europe, with deeply divided leaves and a solitary white flower.
**wood betony** *n.* The lousewort.
**wood·bin** (wo͝od′bĭn′) *n.* A box for holding firewood.
**wood·bine** (wo͝od′bīn′) *n.* [ME *wodebinde* < OE *wudubinde* : *wudu,* wood + *bindan,* to bind.] A climbing vine, esp.: **a.** An Old World honeysuckle, *Lonicera periclymenum,* with yellowish flowers. **b.** The Virginia creeper.
**wood·block** (wo͝od′blŏk′) *n.* **1.** A woodcut. **2.** *also* **wood block.** *Mus.* A hollow block of wood struck with a drumstick to produce percussive effects in an orchestra.
**wood·bor·er** (wo͝od′bôr′ər, -bōr′ər) *n.* Any of various insects, insect larvae, or mollusks that bore into wood.
**wood·carv·ing** (wo͝od′kär′vĭng) *n.* **1.** The art of carving in wood. **2.** An object carved from wood. **—wood′carv′er** *n.*

**wood·chat** (wŏŏd'chăt') *n.* An Old World bird, *Lanius senator,* with a reddish crown and black and white plumage.

**wood·chuck** (wŏŏd'chŭk') *n.* [By folk ety. < Cree *oček.*] A common rodent, *Marmota monax* of northern and eastern North America, with grizzled brownish fur and a short-legged, heavy-set body.

**wood coal** *n.* **1.** Charcoal. **2.** Lignite.

**wood·cock** (wŏŏd'kŏk') *n., pl.* **woodcock** or **-cocks.** [ME *wodecok* < OE *wuducocc* : *wudu,* wood + *cocc,* cock.] A game bird, *Scolopax rusticola* of the Old World or *Philohela minor* of North America, with brownish plumage, short legs, and a long bill.

**wood·craft** (wŏŏd'krăft') *n.* **1.** Skill and experience in matters relating to the woods. **2.** The act, process, or art of working with wood.

**wood·cut** (wŏŏd'kŭt') *n.* **1.** A piece of wood on which a design for printing is engraved, esp. in the plane of the grain. **2.** A print made from a woodcut.

**wood·cut·ter** (wŏŏd'kŭt'ər) *n.* A cutter of wood or trees. **—wood'cut'ting** *n.*

**wood duck** *n.* A brightly colored American duck, *Aix sponsa,* that nests in trees.

**wood·ed** (wŏŏd'ĭd) *adj.* Having trees or woods.

**wood·en** (wŏŏd'n) *adj.* **1.** Made of wood. **2.** Stiff and unnatural <a *wooden* performance><a *wooden* smile> **3.** Clumsy and awkward : UNGAINLY. **—wood'en·ly** *adv.* **—wood'en·ness** *n.*

**wood engraving** *n.* **1. a.** A piece of wood on which a design for printing is engraved, usu. on the end grain. **b.** A print from a wood engraving. **2.** The art or process of making wood engravings.

**wood·en·head** (wŏŏd'n-hĕd') *n.* A stupid person : BLOCKHEAD. **—wood'en-head'ed** *adj.*

**wooden Indian** *n.* A wooden effigy of an American Indian brave holding a cluster of cigars and used formerly as the emblem of a tobacconist.

**wood ibis** *n.* A large wading bird of the subfamily Mycteriinae, related to the storks, esp. *Mycteria americana* of the New World.

**wood ibis**
*Wingspan of 5½ feet*

**wood·ie** (wŏŏd'ē) *n. var. of* WOODY.

**wood·land** (wŏŏd'lənd, -lănd') *n.* Land having a cover of trees and shrubs. **—wood'land·er** (-lən-dər) *n.*

**wood·lark** (wŏŏd'lärk') *n.* An Old World songbird, *Lullula arborea,* resembling but smaller than the skylark.

**wood lot** *n.* An area restricted to the growing of forest trees.

**wood louse** *n.* The sow bug.

**wood·man** (wŏŏd'mən) *n.* A woodsman.

**wood·note** (wŏŏd'nōt') *n.* A song or call characteristic of a woodland bird.

**wood nymph** *n.* **1.** A dryad. **2.** A tropical hummingbird of the genera *Thalurania* or *Cyanophaia.* **3.** A butterfly of the family Satyridae, esp. *Cercyonis pegala,* with brownish wings bearing dark eyespots.

**wood·peck·er** (wŏŏd'pĕk'ər) *n.* Any of various birds of the family Picidae, with strong claws and a stiff tail adapted for climbing and clinging to trees and a chisellike bill for drilling through bark and wood.

**wood pigeon** *n.* A large Eurasian pigeon, *Columba palumbus,* with a white band on each wing.

**wood·pile** (wŏŏd'pīl') *n.* A pile of wood, esp. when stacked for use as fuel.

**wood·print** (wŏŏd'prĭnt') *n.* A woodcut.

**wood pulp** *n.* Any of various cellulose pulps ground from wood, chemically processed, and used esp. to make paper.

**wood pussy** *n. Informal.* A skunk.

**wood rat** *n.* PACK RAT 1.

**wood·ruff** (wŏŏd'rəf, -rŭf') *n.* [ME *woderofe* < OE *wudurofe.*] A plant of the genus *Asperula,* esp. the Eurasian variety *A. odorata,* bearing small white flowers and narrow, fragrant leaves used as flavoring and in sachets.

**wood·shed** (wŏŏd'shĕd') *n.* A shed in which firewood is stored.

**woods·man** (wŏŏdz'mən) *n.* One who works or lives in the woods or is skilled in woodcraft : FORESTER.

**wood sorrel** *n.* Any of various plants of the genus *Oxalis,* with compound leaves bearing three leaflets and yellow, white, or pinkish flowers.

**wood spirits** *n.* Methyl alcohol.

**wood sugar** *n.* Xylose.

**woods·y** (wŏŏd'zē) *adj.* **-i·er, -i·est.** Of, pertaining to, typical of, or like the woods.

**wood tar** *n.* A black syruplike viscous fluid that is a by-product of destructive distillation of wood and is used in medicines, pitch, and preservatives.

**wood thrush** *n.* A North American thrush, *Hylocichla mustelina,* with a melodious song.

**wood tick** *n.* A tick of the genus *Dermacentor* that transmits the microorganism that causes Rocky Mountain spotted fever and tularemia in humans.

**wood·turn·ing** (wŏŏd'tûr'nĭng) *n.* The art or process of shaping wood into forms on a lathe. **—wood'turn'er** *n.*

**wood vinegar** *n.* Pyroligneous acid.

**wood·wax·en** (wŏŏd'wăk'sən) *n.* [ME *wodewaxen* < OE *wudu weaxe* : *wudu,* wood + *weaxan,* to grow.] Dyer's greenweed.

**wood·wind** (wŏŏd'wĭnd') *n.* **1.** Any of a group of musical wind instruments, including the bassoons, clarinets, flutes, oboes, and occas. the saxophones. **2. woodwinds.** The section of an orchestra or band composed of woodwind instruments.

**wood·work** (wŏŏd'wûrk') *n.* Objects made of or work done in wood, esp. wooden interior fittings in a house, as moldings, doors, staircases, or windowsills.

**wood·worm** (wŏŏd'wûrm') *n.* A worm or insect larva that bores into wood.

**wood·y** (wŏŏd'ē) *adj.* **-i·er, -i·est. 1.** Forming or made up of wood : LIGNEOUS <*woody* tissue> **2.** Marked by the presence of wood or xylem <*woody* plants> **3.** Characteristic of or like wood <a *woody* odor> **4.** Abounding in trees : WOODED. *—n. also* **wood·ie.** *pl.* **-ies.** A station wagon with exterior wood paneling.

**woof**[1] (wŏŏf, wōōf) *n.* [Alteration of ME *oof* < OE *ōwef* : *ō-,* on + *wefan,* to weave.] **1.** The threads that run crosswise in a woven fabric, at right angles to the warp threads. **2.** The texture of a fabric. **3.** An essential element.

**woof**[2] (wŏŏf) *n.* [Imit.] **1.** The deep, gruff bark of a dog. **2.** A sound similar to a woof.

**woof·er** (wŏŏf'ər) *n.* [< WOOF[2].] A loudspeaker designed to reproduce bass frequencies.

**wool** (wŏŏl) *n.* [ME *wolle* < OE *wull.*] **1.** The dense, soft, often curly hair forming the coat of sheep and certain other mammals, valued as a textile fabric. **2.** A material or garment made of wool. **3.** A filamentous or fibrous covering or substance like the texture of wool.

**wool-clip** (wŏŏl'klĭp') *n.* The annual yield of wool.

**wool·en** *also* **wool·len** (wŏŏl'ən). *—adj.* Of, relating to, or made of wool. *—n. often* **woolens.** Fabric or clothing made from wool.

**wool fat** *n.* Lanolin.

**wool·gath·er·ing** (wŏŏl'găth'ər-ĭng) *n.* Absent-minded indulgence in fanciful daydreams. *—adj.* Indulging in fancies : ABSENT-MINDED. **—wool'gath'er·er** *n.*

**wool·grow·er** (wŏŏl'grō'ər) *n.* One that raises sheep or other animals for the production of wool. **—wool'grow'ing** *n.*

**wool·len** (wŏŏl'ən) *adj.* & *n. var. of* WOOLEN.

†**wool·ly** *also* **wool·y** (wŏŏl'ē) *—adj.* **-li·er, -li·est** *also* **-i·er, -i·est. 1. a.** Of, relating to, or covered with wool. **b.** Resembling wool. **2.** Lacking sharp detail or clarity : BLURRY <*woolly* logic> **3.** Having the characteristics of the rough, gen. lawless atmosphere of frontier America <a wild and *woolly* town> *—n., pl.* **-lies** *also* **-ies. 1.** A garment made of wool, esp. an undergarment. **2.** *Western U.S.* & *Austral.* A sheep. **—wool'li·ness** *n.*

**woolly bear** *n.* The hairy caterpillar of a tiger moth, esp. that of *Isia isabella.*

**wool·pack** (wŏŏl'păk') *n.* **1.** A bag used for packing a bale of wool for shipment. **2.** A cumulus cloud.

**wool·sack** (wŏŏl'săk') *n.* **1.** A sack for wool. **2.** The official seat of the Lord Chancellor in the British House of Lords.

**wool shed** *n.* A building or complex of buildings in which sheep are sheared and wool is prepared for shipment to market.

**wool·skin** (wŏŏl'skĭn') *n.* A sheepskin with the wool still on it.

**wool-sort·er's disease** (wŏŏl'sôr'tərz) *n.* A pulmonary anthrax resulting from inhalation of *Bacillus anthracis* spores in contaminated sheep's wool.

**wool-sta·pler** (wŏŏl'stā'plər) *n.* **1.** A dealer in wool. **2.** One who sorts wool by the quality of the fiber. **—wool'-sta'pling** *adj.* & *n.*

**wool·work** (wŏŏl'wûrk') *n.* Needlework.

**wool·y** (wŏŏl'ē) *adj.* & *n. var. of* WOOLLY.

**wooz·y** (wōō'zē, wŏŏz'ē) *adj.* **-i·er, -i·est.** [Poss. alteration of OOZY.] **1.** Dazed : confused. **2.** Dizzy or queasy. **—wooz'i·ly** *adv.* **—wooz'i·ness** *n.*

**Worces·ter** (wŏŏs'tər) *n.* A trademark for a fine porcelain made in Worcester, England.

**Worces·ter·shire** (wŏŏs'tər-shîr, -shər). A trademark for a piquant sauce of soy, vinegar, and spices.

**word** (wûrd) *n.* [ME < OE.] **1.** A sound or a combination of sounds, or its representation in writing or printing, that symbolizes and communicates a meaning and may consist of a single morpheme or of a combination of morphemes. **2.** An utterance, remark, or comment <a *word* from the sponsor> **3.** *Computer Sci.* A set of bits comprising the smallest unit of addressable memory. **4. words.** A discourse or talk : SPEECH. **5. words.** The text of a vocal musical composition : LYRICS. **6.** An assurance or promise <gave them my *word*> **7. a.** A command or direction <waiting for the president's *word*> **b.** A verbal signal : password or watchword. **8. a.** News <What's the latest *word*?> **b.** Rumor <*Word* has it they eloped.> **9. words.** Hostile or angry remarks made back and forth : QUARREL. **10. Word. a.** The Logos. **b.** The Scriptures or Gospel. —*vt.* **word·ed, word·ing, words.** To express in words. **—by word of mouth.** Orally. **—good word. 1.** A favorable utterance or word <put in a *good word* for the candidate> **2.** Favorable news. **—have no words for.** To be unable to describe or talk about. **—in a word.** In short <*In a word,* I'm broke.> **—in so many words.** Precisely as stated : EXACTLY. **—of few words.** Not conversational or loquacious : LACONIC. **—take (one) at (one's) word.** To be convinced of another's sincerity and act in accordance with the other's statement. **—upon my word.** Indeed : assuredly <*Upon my word,* it is a shame.> **—word for word.** In the same words : VERBATIM.

**word·age** (wûr′dĭj) *n.* **1.** Words as a group. **2.** Use of an excessive number of words : VERBIAGE. **3.** The number of words used, as in a novel. **4.** Wording.

**word blindness** *n.* Alexia. **—word′-blind′** (wûrd′blīnd′) *adj.*

**word·book** (wûrd′bo͝ok′) *n.* A lexicon, vocabulary, or dictionary.

**word deafness** *n.* A form of aphasia in which information in the form of speech is incomprehensible.

**word·ing** (wûr′dĭng) *n.* The act or style of expressing in words : DICTION.

**word·less** (wûrd′lĭs) *adj.* **1.** Having or using no words : UNSPOKEN. **2.** Inarticulate. **—word′less·ly** *adv.* **—word′less·ness** *n.*

**word·mon·ger** (wûrd′mŭng′gər, -mŏng′-) *n.* A pretentious or careless user of language.

**word order** *n.* Syntactic arrangement of words in a sentence, clause, or phrase.

**word play** *n.* **1.** A witty or clever exchange of words : REPARTEE. **2.** A play on words : PUN.

**word processing** *n.* A system of producing typewritten documents, as business letters, by use of automated typewriters and electronic text-editing equipment. **—word processor** *n.*

**word square** *n.* A group of words arranged in a square that read the same vertically and horizontally.

**word·y** (wûr′dē) *adj.* **-i·er, -i·est. 1.** Relating to, consisting of, or having the nature of words : VERBAL. **2.** Expressed in or using more words than necessary. **—word′i·ly** *adv.* **—word′i·ness** *n.*

☆ ***syns:*** WORDY, DIFFUSE, LONGWINDED, PROLIX, VERBOSE *adj. core meaning* : using or containing an excessive number of words <a *wordy* report><a *wordy* explanation>

**wore¹** (wôr, wōr) *v. p.t. of* WEAR¹.

**wore²** (wôr, wōr) *v. p.t. of* WEAR².

**work** (wûrk) *n.* [ME *worke* < OE *weorc.*] **1.** Physical or mental effort or activity directed toward the production or accomplishment of something : LABOR. **2.** Employment : job <out of *work*> **3.** The means by which one earns one's livelihood. **4. a.** Something that one is doing, making, or performing, esp. as a part of one's occupation : a duty or task. **b.** The amount of effort required or done. **5.** Something that has been done, made, or performed as a result of one's occupation, effort, or activity, esp.: **a. works.** The output of an artist or artisan considered or collected as a group. **b. works.** Engineering structures, as bridges or dams. **c.** A piece of needlework or embroidery. **6.** A material or piece being processed in a machine during manufacture. **7.** The area, office, or place where one pursues an occupation. **8. works** *(sing. in number).* A factory, plant, or similar building or system of buildings where a specific type of business or industry is carried on. **9. works.** Mechanism <the *works* of a clock> **10.** The manner or style of working or the quality of treatment <sloppy *work*> **11.** A froth produced during fermentation, as of vinegar or cider. **12.** *Physics.* Transfer of energy from one physical system to another, esp. transfer of energy to a body by application of force, calculated as the line integral between any two points of the scalar product of the force and the body's displacement along the path over which the integral is taken. **13. works.** Moral or righteous acts or deeds <salvation by faith rather than *works*> **14. the works.** *Slang.* The whole : EVERYTHING <ordered pizza with *the works*> —*v.* **worked** *or* **wrought** (rôt), **work·ing, works.** —*vi.* **1.** To exert one's efforts for the purpose of doing or making something : LABOR. **2.** To be employed. **3.** To perform a function or act : OPERATE <The typewriter doesn't *work*.> **4.** To operate effectively or successfully : prove successful <Your solution seems to *work*.> **5.** To be changed into a specified state, esp. gradually or by repeated movement <The masonry *worked* loose.> **6.** To force a passage or way <They *worked* through the jungle to the river.> **7.** To move or contort, as one's mouth, from emotion or pain. **8.** To be handled or processed <Gold *works* very easily.> **9.** To ferment. **10.** *Naut.* To be under strain in heavy seas so that seams loosen and fastenings become slack. **11.** To undergo small motions that result in friction and wear <The gears *work* against each other.> —*vt.* **1.** To cause or effect : bring about <*works* wonders with wood> **2.** To cause to operate or function <*work* a chain saw> **3.** To form or shape : MOLD <*work* copper into jewelry> **4.** To make or decorate by using a needle. **5.** To solve (a problem) by calculation and reasoning. **6.** To handle or manipulate for the purpose of preparing <*work* the flour into the batter> **7.** To achieve (a specified condition) by gradual or repeated effort <*worked* my way to the top> **8.** To arrange : contrive <I *worked* it so that I could go to night school.> **9.** To make productive : CULTIVATE <*work* the land> **10.** To make or force to work or to do work <*worked* the students extra hard> **11.** To excite, rouse, or provoke <*worked* myself into a rage> **12.** To influence or persuade, esp. by underhand means. **13.** *Informal.* To use or employ for one's own ends or purposes <*work* one's business connections> **14.** *Informal.* To practice trickery or deception on : CHEAT. **15.** To function or operate in : COVER <a salesperson who *works* the suburbs> **16.** To ferment (liquors). **—work in. 1.** To put in or introduce : INSERT <*worked in* a plea for money> **2.** To cause to be inserted by repeated, continuous effort. **—work off.** To get rid of : ELIMINATE <*work off* weight by running> **—work on** (or **upon). 1.** To affect <*worked on* our weaknesses> **2.** To attempt to persuade or influence. **—work out. 1.** To exhaust (e.g., a mine or soil). **2.** To accomplish by work or effort. **3.** To find a solution for : SOLVE. **4.** To formulate : develop <*work out* a scheme> **5.** To prove successful, effective, or satisfactory <How did the new route *work out*?> **6.** To perform a series of exercises. **—work over. 1.** To do for a second time : REWORK. **2.** *Slang.* To inflict severe physical damage on : beat up. **—work up. 1.** To arouse the emotions of : EXCITE. **2.** To develop or formulate by mental or physical effort <*worked up* a song-and-dance routine>

☆ ***syns:*** WORK, BUSINESS, EMPLOYMENT, JOB, OCCUPATION *n. core meaning* : what one does to earn a living <found *work* as a ranch hand> WORK, the most general of these terms, can refer to the mere fact of employment or to a specific activity <found *work* in the city> <their *work* as lawyers> A BUSINESS is the activity in which a person engages for a livelihood <went into the shoe *business*> EMPLOYMENT and JOB suggest activity in which a person is hired and paid by another <regular *employment* as a baggage handler><a *job* in a bookstore>, but an OCCUPATION does not necessarily imply being employed by others <Madame Curie's *occupation* as a chemist>

**work·a·ble** (wûr′kə-bəl) *adj.* **1.** Capable of being worked, dealt with, or handled. **2.** Capable of being worked conveniently : FEASIBLE. **—work′a·bil′i·ty, work′a·ble·ness** *n.*

**work·a·day** (wûr′kə-dā′) *adj.* [< ME *werkeday,* workday.] **1.** Relating or appropriate to working days : EVERYDAY. **2.** Mundane : commonplace <"the practical, *workaday* world, of . . . ordinary undistinguished things" —Lionel Trilling>

**work·a·hol·ic** (wûr′kə-hô′lĭk, -hŏl′ĭk) *n.* [WORK + (ALC)OHOLIC.] A compulsive worker. **—work′a·hol′ism** *n.*

**work·bag** (wûrk′băg′) *n.* A bag to hold material on which one is working or implements needed for work.

**work·bench** (wûrk′bĕnch′) *n.* A sturdy bench or table at which manual work is done, as by a machinist, carpenter, or jeweler.

**work·book** (wûrk′bo͝ok′) *n.* **1.** A booklet having problems and exercises in which a student may directly write or calculate. **2.** A manual of operating instructions, as for an appliance or a machine. **3.** A book in which a record is kept of work proposed or accomplished.

**work·box** (wûrk′bŏks′) *n.* A box for implements or materials used in work such as sewing.

**work·day** (wûrk′dā′) *n.* **1.** A day on which work is done. **2.** The part of the day during which one works <a seven-hour *workday*> —*adj.* Workaday.

**work·er** (wûr′kər) *n.* **1.** One that works. **2. a.** One who does manual or industrial labor. **b.** A member of the working class. **3.** A sterile female of certain social insects, as the ant or bee, that performs specialized work.

**workers' compensation** *n.* Payments required by law to be made to an employee injured in the course of work.

**work ethic** *n.* The belief that work is morally good.

**work·fare** (wûrk′fâr′) *n.* Welfare in which recipients of aid are required to perform public service work.

**work·folk** (wûrk′fōk′) *also* **work·folks** (-fōks′) *pl.n.* Laborers, esp. farm laborers.

**work force** *n.* **1.** Those workers employed in a specific capacity : STAFF. **2.** All workers potentially available, as to a nation or project.

**work function** *n.* The amount of work needed to remove an electron from a solid, esp. the work exerted against coulomb forces in removing an electron from just inside to just outside the surface of a metal.

**work hardening** *n.* Increase in strength occas. accompanying plastic deformation of a solid.

**work·horse** (wûrk′hôrs′) *n.* **1.** A horse used for labor. **2.** *Informal.* A tireless worker.

**work·house** (wûrk'hous') *n.* **1.** A prison in which limited sentences, under one year in most systems, are served at manual labor. **2.** *Chiefly Brit.* A poorhouse.

**work·ing** (wûr'kĭng) *adj.* **1.** Relating to or designating one that works : EMPLOYED <a *working* person> **2.** Relating to, used for, or spent in working <*working* hours> **3.** Sufficient or large enough for using or being worked <a *working* knowledge of electricity> **4.** Capable of being used as the basis of further work <a *working* model> **5.** In the process of fermentation. —Used of alcoholic liquors. **6. a.** Functioning, esp. on a reduced scale. **b.** Used as a guide <a *working* diagram>

**working capital** *n.* **1.** The assets of a business that can be applied to its operation. **2.** The current assets of an individual or business as opposed to current liabilities.

**working class** *n.* The part of society whose income is from wages : PROLETARIAT. **—work'ing-class'** *adj.*

**working fluid** *n.* A working substance.

**work·ing·man** (wûr'kĭng-măn') *n.* A man who works for wages, esp. at manual labor.

**working papers** *pl.n.* Legal documents certifying the right of an individual to employment.

**working storage** *n. Computer Sci.* The section of a data storage disk reserved for data to be temporarily stored during the running of a program.

**working substance** *n.* A substance, as a coolant, used to effect a thermodynamic change in a system.

**work·less** (wûrk'lĭs) *adj.* Unemployed.

**work·load** (wûrk'lōd') *n.* **1.** The amount of work assigned to or done by a worker or unit of workers in a given time period. **2.** The capacity of a machine for work in a given time period.

**work·man** (wûrk'mən) *n.* **1.** A man who performs labor. **2.** One who works in a specified way <an innovative *workman*>

**work·man·like** (wûrk'mən-līk') *also* **work·man·ly** (-lē) *adj.* Typical of or befitting a skilled workman or craftsman.

**work·man·ship** (wûrk'mən-shĭp') *n.* **1. a.** The art, skill, or technique of a workman. **b.** The quality of such art, skill, or technique <pewter of fine *workmanship*> **2.** Something produced by a workman. **3.** The product of effort or endeavor.

**workmen's compensation** *n.* Workers' compensation.

**work of art** *n.* **1.** A piece of superior work. **2.** A product of one of the fine arts, as a painting or sculpture.

**work·out** (wûrk'out') *n.* **1.** A period of exercise or practice, esp. in athletics. **2.** An exhausting task.

**work·peo·ple** (wûrk'pē'pəl) *pl.n. Chiefly Brit.* Wage earners.

**work release** *n.* A corrections program in which prisoners are released from confinement every day to work full-time.

**work·room** (wûrk'ro͞om', -ro͝om') *n.* A room where work is done.

**work·shop** (wûrk'shŏp') *n.* **1.** An area, room, or establishment in which manual or industrial work is done. **2.** A group of people who meet regularly for a seminar in a specialized field <a nutrition *workshop*>

**work stoppage** *n.* A protest measure by a group of workers characterized by cessation of work and usu. less serious than a formal strike.

**work·ta·ble** (wûrk'tā'bəl) *n.* A table designed for a specific task or activity, as graphic arts or needlework.

**work·up** (wûrk'ŭp') *n.* A thorough medical diagnostic study.

**work·week** (wûrk'wēk') *n.* The number of hours worked or required to be worked in one week.

**world** (wûrld) *n.* [ME < OE *weorold.*] **1.** The earth. **2.** The universe. **3.** The earth and its inhabitants as a group. **4.** The human race <All the *world* rejoiced.> **5.** Human beings considered as social creatures : the public <the *world's* response to chronic hunger> **6.** *often* **World.** A particular part of the earth <the Occidental *World*> **7.** A particular period in history, including its people, culture, and social order <the Renaissance *world*> **8.** A sphere, realm, or domain including all things relating to or associated with it <the insect *world*><an adolescent's *world*> **9.** A field or sphere of human endeavor <the *world* of serious music> **10.** A specified way of life or state of being <the *world* of the elderly> **11.** Secular life and its concerns <renounced the *world* for a monastic life> **12.** *often* **worlds.** A large amount <has *worlds* of ideas> **13.** A planet or other celestial body <an extraterrestrial *world*> **—for all the world. 1.** For anything or for any reason <I wouldn't quit *for all the world.*> **2.** Precisely : exactly <sounded *for all the world* like an elephant> **—out of this world.** *Informal.* Excellent <champagne that was *out of this world*>

▲ word history: The Old English form of *world,* which was written *weorold,* consisted of two syllables, and this fact gives a clue to its origin. *World* is a compound word, but one formed so long ago that its etymological meaning has become obscured. It was formed in the Common Germanic period, the time before the Germanic language broke up into separate dialects, which scholars put in the millennium before the birth of Christ. *World* is made up of two words whose Old English forms were *wer,* "man," and *ieldo,* "age." The literal meaning would have been "the age of man" or "the period of human life." This meaning persisted in Old English, which used *weorold* to translate Latin *saeculum,* "generation, age," and survives today in the contrast between a life "in the world" and a retired or cloistered existence. The more usual sense of *world* is now the physical globe of Earth and other celestial objects; this meaning was also present in Old English, but it most likely represents an extension of the word's original sense.

**world line** *n.* The path in space-time traveled by an elementary particle for the time and distance it retains its identity.

**world·ling** (wûrld'lĭng) *n.* A worldly person.

**world·ly** (wûrld'lē) *adj.* **-li·er, -li·est. 1.** Of, relating to, or devoted to the temporal world : not religious or spiritual. **2.** Sophisticated or cosmopolitan. *—adv.* In a worldly manner. **—world'li·ness** *n.*

**world·ly-wise** (wûrld'lē-wīz') *adj.* Experienced in the ways of the world : SOPHISTICATED.

**world power** *n.* A political entity whose actions influence or change the course of international events.

**World Series** *n.* The series of professional baseball games played each fall between the championship teams of the American League and National League.

**world's fair** *n.* An exposition featuring international exhibits.

**world-shak·ing** (wûrld'shā'kĭng) *adj.* Of great significance.

**world view** *n.* A Weltanschauung.

**World War I** *n.* A war fought from 1914 to 1918, in which Great Britain, France, Russia, Belgium, Italy, Japan, and the United States defeated Germany, Austria-Hungary, Turkey, and Bulgaria.

**World War II** *n.* A war fought from 1939 to 1945, in which Great Britain, France, the Soviet Union, and the United States defeated Germany, Italy, and Japan.

**world-wea·ry** (wûrld'wîr'ē) *adj.* **-ri·er, -ri·est.** Tired of the world and its pleasures. **—world'-wea'ri·ness** *n.*

**world·wide** (wûrld'wīd') *adj.* Reaching or extending throughout the world : UNIVERSAL. **—world'wide'** *adv.*

**worm** (wûrm) *n.* [ME < OE *wyrm.*] **1.** Any of various invertebrates, as those of the phyla Annelida, Nematoda, or Platyhelminthes, with a long flexible rounded or flattened body, often without obvious appendages. **2.** Any of various insect larvae with soft elongated bodies. **3.** Any of various unrelated animals resembling a worm in habit or appearance, as the shipworm or slowworm. **4.** Something, as a threaded screw, similar to a worm in appearance or movement. **5.** An insidiously tormenting or devouring force <the *worm* of jealousy> **6.** One who is pitiable, contemptible, or weak-willed. **7. worms.** *Pathol.* Intestinal infestation with worms or wormlike parasites. *—v.* **wormed, worm·ing, worms.** *—vt.* **1.** To make (one's way) with or as if with the sinuous crawling motion of a worm. **2.** To elicit by artful or devious means <*wormed* the truth out of the suspect> **3.** To cure of intestinal worms. **4.** *Naut.* To wrap yarn or twine around (rope). *—vi.* **1.** To move in a sinuous way like a worm. **2.** To make one's way by artful or devious means <*wormed* out of a tricky situation>

**worm-eat·en** (wûrm'ēt'n) *adj.* **1.** WORMY 1. **2.** Decayed : rotten. **3.** Antiquated : out-of-date.

**worm fence** *n.* A fence of crossed rails supporting one another and forming a zigzag pattern.

**worm gear** *n.* **1.** A gear having a threaded shaft and a wheel with teeth that mesh into it. **2.** A worm wheel.

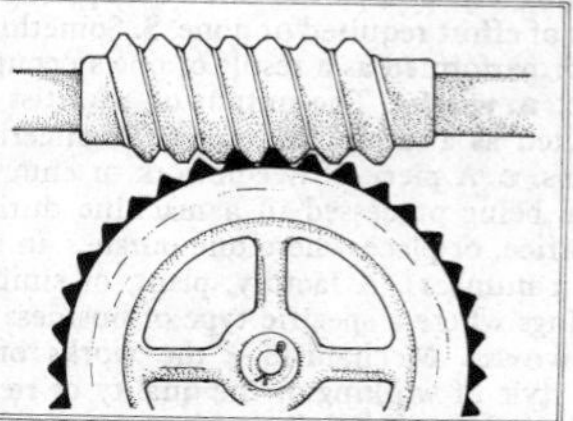

worm gear

**worm·grass** (wûrm'grăs') *n.* [From its use as an anthelmintic.] The pinkroot.

**worm·hole** (wûrm'hōl') *n.* A hole made by a burrowing worm.

**worm screw** *n.* The threaded shaft of a worm gear.

**worm·seed** (wûrm'sēd') *n.* **1.** A tropical American plant, *Chenopodium ambrosioides,* yielding an oil used as an anthelmintic. **2.** Any of several plants used as an anthelmintic.

**worm's-eye view** (wûrmz'ī') *n.* **1.** A close-up view. **2.** A view from a low or inferior position.

**worm wheel** *n.* The toothed wheel of a worm gear.

**worm·wood** (wûrm'wo͝od') *n.* [ME *wormwode,* alteration of *wermode* < OE *wermōd.*] **1.** An aromatic plant of the genus *Artemisia,* esp. the European variety *A. absinthium,* yielding a bitter extract used in making absinthe and in flavoring certain wines. **2.** Something harsh or embittering.

▲ word history: The etymology of *wormwood* is a good example of the process of folk etymology operating at an early period. The earliest Middle English form was *wermode,* regularly descended from Old English *wermōd.* By the 15th century, however, the first syllable was incorrectly interpreted as the word "worm," and the ending *-ode,* which did not make much sense in the new analysis, was reinterpreted as the word *wood.* Old English *wermōd* is apparently a compound of *wer,* "man," and *mōd,* "courage, strength," although it was never used with such a meaning—*wermōd* referred only to the plant. A cognate of *wermōd* entered English by a different route and became *vermouth.*

**worm·y** (wûr′mē) *adj.* **-i·er, -i·est. 1.** Infested with or damaged by worms. **2.** Like a worm. **—worm′i·ness** *n.*

**worn**[1] (wôrn, wōrn) *adj.* [ME, p.part. of *weren,* to wear.] **1.** Affected by use or wear. **2.** Impaired or damaged by use or wear. **3. a.** Exhausted : spent. **b.** Showing fatigue : DRAWN. **4.** Hackneyed : trite.

**worn**[2] (wôrn, wōrn) *v. p.p. of* WEAR[2].

**worn-out** (wôrn′out′, wōrn′-) *adj.* **1.** Worn or used until no longer usable <a *worn-out* sofa> **2.** Thoroughly exhausted : SPENT.

**wor·ri·ment** (wûr′ē-mənt, wŭr′-) *n.* **1.** The act of worrying. **2.** A cause of worry.

**wor·ri·some** (wûr′ē-səm, wŭr′-) *adj.* **1.** Causing worry or anxiety. **2.** Tending to worry : ANXIOUS. **—wor′ri·some·ly** *adv.*

**wor·ry** (wûr′ē, wŭr′ē) *v.* **-ried, -ry·ing, -ries.** [ME *worien,* to strangle < OE *wyrgan.*] *—vi.* **1.** To feel uneasy or troubled. **2.** To pull, bite, or tear at something. **3.** To work under difficulty or hardship : STRUGGLE <*worried* away at the dilemma> *—vt.* **1.** To cause to feel anxious, distressed, or troubled. **2.** To bother : annoy <*worried* me with constant nagging> **3. a.** To grasp and tug at repeatedly <a dog *worrying* a bone> **b.** To touch, press, or handle idly <*worried* the sore paw all day> *—n., pl.* **-ries. 1.** The act of worrying or state of being worried. **2.** A source of nagging concern. **—wor′ri·er** *n.*

**worry beads** *pl.n.* A string of beads that one fingers to keep the hands occupied.

**wor·ry·wart** (wûr′ē-wôrt′, wŭr′-) *n.* One who tends to worry excessively and esp. needlessly.

**worse** (wûrs) *adj.* [ME < OE *wyrsa.*] **1.** More inferior, as in quality, condition, or effect. **2.** More severe or unfavorable. **3.** Further from an ideal or standard. *—n.* Something worse. *—adv.* In a worse way.

**wors·en** (wûr′sən) *vt.* & *vi.* **-ened, -en·ing, -ens.** To make or become worse.

**wors·er** (wûr′sər) *adj.* & *adv. Archaic. var. of* WORSE.

**wor·ship** (wûr′shĭp) *n.* [ME < OE *weorðscipe,* honor : *weorð,* worth + *-scipe,* -ship.] **1. a.** The reverent love and allegiance accorded a deity, idol, or sacred object. **b.** A set of religious forms, as ceremonies or prayers, by which this love is expressed. **2. a.** Ardent, humble devotion. **b.** The object of such devotion. **3.** *often* **Worship.** *Chiefly Brit.* —Used as a title of honor in addressing dignitaries, as magistrates or mayors. *—v.* **-shiped, -ship·ing, -ships** *or* **-shipped, -ship·ping, -ships.** *—vt.* **1.** To honor and love as a deity : VENERATE. **2.** To love or pursue devotedly. *—vi.* **1.** To participate in religious rites of worship. **2.** To perform an act of worship. **—wor′ship·er** *n.*

**wor·ship·ful** (wûr′shĭp-fəl) *adj.* **1.** Given to or showing worship : reverent or adoring. **2.** *Chiefly Brit.* Honorable by virtue of position or rank. —Used in titles of respect. **—wor′ship·ful·ly** *adv.* **—wor′ship·ful·ness** *n.*

**worst** (wûrst) *adj.* [ME < OE *wyrsta.*] **1.** Most inferior, as in quality, condition, or effect. **2.** Most severe or unfavorable. **3.** Furthest from an ideal or standard. *—adv.* In the worst manner or degree. *—vt.* **worst·ed, worst·ing, worsts.** To gain the advantage over : DEFEAT. *—n.* Something that is worst. **—at worst.** Under the most negative foreseeable circumstances. **—get the worst of it.** To suffer a disadvantage or defeat. **—if (the) worst comes to (the) worst.** At the very worst. **—in the worst way.** *Informal.* Very much <wanted *in the worst way* to attend the party>

**wor·sted** (wo͝os′tĭd, wûr′stĭd) *n.* [ME *worthstede,* after *Worthstede,* a village in Norfolk, England.] **1.** Firm-textured, compactly twisted woolen yarn made from long-staple fibers. **2.** Fabric made from worsted. **—wor′sted** *adj.*

**wort** (wûrt, wôrt) *n.* [ME < OE *wyrt,* plant.] **1.** A plant <liver*wort*><milk*wort*> **2.** An infusion of malt fermented to make beer.

**worth**[1] (wûrth) *n.* [ME < OE *weorð.*] **1.** The quality of something that makes it desirable, useful, or valuable <the *worth* of high technology> **2.** Material or market value <a house having a *worth* of $100,000> **3.** The number or amount of something that may be purchased for a specific sum <ten dollars' *worth* of fruit> **4.** Wealth : riches. **5.** The quality within one that renders one deserving of respect <the *worth* of the individual> *—adj.* **1.** Equal in value to something specified <heirlooms *worth* a fortune> **2.** Deserving of : MERITING <a job *worth* doing> **3.** Having wealth or riches amounting to a given figure. **—for all one is worth.** To the utmost of one's powers or ability.

**worth**[2] (wûrth) *vi.* **worthed, worth·ing, worths.** [ME *worthen* < OE *weorðan.*] *Archaic.* To befall : betide.

**worth·less** (wûrth′lĭs) *adj.* **1.** Devoid of worth, use, or value. **2.** Lacking dignity or honor : DESPICABLE. **—worth′less·ly** *adv.* **—worth′less·ness** *n.*

☆ ***syns:*** WORTHLESS, DROSSY, GOOD-FOR-NOTHING, INUTILE, NO-GOOD, NOTHING, VALUELESS *adj. core meaning* : lacking all worth and value <a *worthless* endeavor><bought a lot of *worthless* junk> ***ant:*** valuable

**worth·while** (wûrth′hwīl′, -wīl′) *adj.* Valuable or important enough to justify expenditure of time or effort. **—worth′while′·ness** *n.*

**wor·thy** (wûr′thē) *adj.* **-thi·er, -thi·est. 1.** Having worth, merit, or value. **2.** Honorable : admirable <a *worthy* individual> **3.** Having sufficient worth : DESERVING <*worthy* of respect> *—n., pl.* **-thies. 1.** One esteemed for one's worth, dignity, or importance. **2.** A figure locally renowned or respected. **—wor′thi·ly** *adv.* **—wor′thi·ness** *n.*

**-worthy** *suff.* [< WORTHY.] **1.** Of sufficient worth for <credit*worthy*> **2.** Suitable or safe for <crash*worthy*>

**wot** (wŏt) *v. Archaic. 1st & 3rd person sing. p.t. of* WIT[2].

**Wo·tan** (vō′tän′) *n.* [G.] *Myth.* A Teutonic god identified with Woden.

**would** (wo͝od) *v. p.t. of* WILL[2]

**would-be** (wo͝od′bē′) *adj.* Desiring or pretending to be.

**would·n't** (wo͝od′nt). Would not.

**wouldst** (wo͝odst) *or* **would·est** (wo͝od′ĭst) *v. Archaic. 2nd person sing. p.t. of* WILL[2].

**wound**[1] (wo͞ond) *n.* [ME < OE *wund.*] **1.** An injury, esp. one in which the skin or other external organic surface is torn, pierced, cut, or otherwise broken. **2.** An injury to the feelings. *—vt.* & *vi.* **wound·ed, wound·ing, wounds.** To inflict a wound on or to inflict a wound or wounds.

**wound**[2] (wound) *v. p.t.* & *p.p. of* WIND[2].

**wound**[3] (wound) *v. var. p.t.* & *p.p. of* WIND[3].

**wound·wort** (wo͞ond′wûrt′, -wôrt′) *n.* A plant of the genus *Stachys,* bearing downy leaves once used to treat wounds.

**wove** (wōv) *v. p.t.* & *p.p. of* WEAVE.

**wo·ven** (wō′vən) *v. p.p. of* WEAVE.

**wove paper** *n.* Paper made on a closely woven wire roller or mold and having a faint mesh pattern.

**wow**[1] (wou) *Informal. —interj.* —Used to express wonder or amazement. *—n.* An outstanding success. *—vt.* **wowed, wow·ing, wows.** To have a strong and usu. pleasurable impact on <a pitcher whose performance *wowed* the grandstands>

**wow**[2] (wou) *n.* [Imit.] A slow variation in the pitch of sound reproduced by a phonograph or tape recorder, usu. the result of irregular movement of a mechanical part.

**W particle** *n.* A large elementary particle hypothesized to be responsible for weak interaction.

**wrack**[1] (răk) *n.* [ME < OE *wræc,* punishment.] Severe damage or wreckage <*wrack* and ruin>

**†wrack**[2] (răk) *n.* [ME *wrak* < MDu.] **1. a.** Wreckage, esp. of a ship cast ashore. **b.** *Regional.* Violent destruction of a vehicle or building. **2.** Dried seaweed. **3.** Marine vegetation, esp. kelp. *—vt.* & *vi.* **wracked, wrack·ing, wracks.** To cause the ruin of or to be wrecked.

**wrack**[3] (răk) *v. var. of* RACK[2].

**wraith** (rāth) *n.* [Orig. unknown.] **1.** An apparition of a living person. **2.** The ghost of a dead person.

**wran·gle** (răng′gəl) *v.* **-gled, -gling, -gles.** [ME *wranglen,* prob. of LG orig.] *—vi.* To dispute noisily or angrily : QUARREL. *—vt.* **1.** To win or obtain by argument. **2.** To herd (horses or other livestock). *—n.* **1.** An angry noisy argument. **2.** An act of wrangling.

**wran·gler** (răng′glər) *n.* **1.** One who wrangles. **2.** A cowboy, esp. one who tends saddle horses.

**wrap** (răp) *v.* **wrapped** *or* **wrapt** (răpt), **wrap·ping, wraps.** [ME *wrappen.*] *—vt.* **1.** To arrange or fold about in order to cover or protect something. **2.** To cover, envelop, or encase. **3.** To package, as with paper. **4.** To clasp, fold, or coil about something. **5.** To envelop and obscure, often with the effect of concealing or disguising the nature of <Fog *wrapped* the waterfront.> **6. a.** To suffuse with a particular aura <a mansion *wrapped* in mystery> **b.** To engross <*wrapped* in meditation> *—vi.* **1.** To coil, wind, or twist about or around something. **2.** To put on warm clothing : bundle up. **—wrap up. 1.** To work out and complete the details of <*wrap up* a project> **2.** To encompass in a few words : SUMMARIZE. *—n.* **1.** A garment to be wrapped or folded about one, esp. a robe, cloak, shawl, or coat. **2.** A blanket. **3.** A wrapping or wrapper. **4.** *Computer Sci.* A single turn of metallic magnetic tape in a tape-wound magnetic core.

☆ ***syns:*** WRAP, CLOAK, CLOTHE, ENFOLD, ENSHROUD, ENVELOP, ENWRAP, SHROUD, VEIL *v. core meaning* : to surround and cover completely so as to hide from view <clouds *wrapping* the Alps>

**wrap·a·round** (răp′ə-round′) *n.* **1.** A garment, as a dress or skirt, open to the hem and wrapped around the body before being fastened. **2.** Something that curves and laps over something else. **—wrap′a·round′** *adj.*

**wrap·per** (răp′ər) *n.* **1.** One that wraps. **2.** Material, as paper or foil, in which something is wrapped. **3.** Paper encasing a mailed magazine or newspaper. **4.** A book jacket. **5.** The tobacco leaf covering a cigar. **6.** A loose robe or negligee.

**wrap·ping** (răp′ĭng) *also* **wrap·pings** (-ĭngz) *n.* WRAPPER 2.

**wrapt** (răpt) *v. var. p.t. of* WRAP.

**wrap-up** (răp′ŭp′) *n.* A brief summary, as of the news.

**wrasse** (răs) *n.* [Cornish and Welsh *gwrach.*] Any of various chiefly tropical, often brightly colored marine fishes of the family Labridae.

**wrath** (răth, räth) *n.* [ME < OE *wrǣðð* < *wrāð,* angry.] **1.** Violent, resentful anger : RAGE. **2. a.** A manifestation of rage. **b.** Divine retribution for sin. *—adj. Archaic.* Wrathful.

**wrath·ful** (răth′fəl, räth′-) *adj.* **1.** Very angry. **2.** Proceeding from or expressing wrath. **—wrath′ful·ly** *adv.* **—wrath′ful·ness** *n.*

**wreak** (rēk) *vt.* **wreaked, wreak·ing, wreaks.** [ME *wreken* < OE *wrecan.*] **1.** To inflict (punishment or vengeance) upon another. **2.** To express or gratify (anger, malevolence, or resentment) : VENT. **3.** *Archaic.* To take vengeance for : AVENGE.

**wreath** (rēth) *n., pl.* **wreaths** (rēthz) [ME *wrethe* < OE *wriða.*] **1. a.** A ring or circlet of flowers or leaves worn on the head, used as a decoration or placed as a memorial. **b.** A representation of a wreath, as in woodwork. **2.** A ring or curling form <a *wreath* of cigar smoke> **—wreath′y** *adj.*

**wreathe** (rēth) *v.* **wreathed, wreath·ing, wreathes.** [< WREATH.] *—vt.* **1.** To twist or entwine into a wreath. **2.** To twist or curl into a wreathlike contour or shape. **3.** To crown or decorate with or as if with a wreath. **4.** To curl or coil. **5.** To form a wreath around. *—vi.* **1.** To assume the form of a wreath. **2.** To curl, writhe, or spiral.

**wreck** (rĕk) *n.* [ME *wrek* < AN *wrec,* of Scand. orig.] **1. a.** The act of wrecking or state of being wrecked. **b.** Accidental destruction of a ship. **2.** The stranded hulk of a gravely damaged ship. **3.** The remains of something wrecked or ruined, as by collision. **4.** Fragments of a ship or goods cast ashore by the sea after a shipwreck : WRECKAGE. **5.** One in a shattered, broken-down, or worn-out state <a nervous *wreck*> *—v.* **wrecked, wreck·ing, wrecks.** *—vt.* **1.** To destroy accidentally, as by collision. **2.** To tear down or dismantle. **3.** To bring to a state of ruin. *—vi.* **1.** To suffer destruction, ruin, or shipwreck. **2.** To engage in wrecking or tearing down.

**wreck·age** (rĕk′ĭj) *n.* **1.** WRECK 1a. **2.** The debris of something wrecked.

**wreck·er** (rĕk′ər) *n.* **1. a.** One who wrecks or causes a wreck. **b.** A member of a wrecking or demolition crew. **c.** One who destroys or ruins <a *wrecker* of hopes> **2. a.** One used in recovering or removing a wreck, esp. a truck with a hoist and towing apparatus used in towing wrecked or disabled vehicles. **b.** One that salvages wrecked cargo or parts. **3. a.** One that lures a vessel to destruction, as on a rocky coastline, in order to plunder. **b.** A plunderer.

**wrecking bar** *n.* A small crowbar with a claw at one end and a slight curve at the other end.

**wren** (rĕn) *n.* [ME *wrenne* < OE *wrenna.*] **1.** Any of various small brownish birds of the family Troglodytidae. **2.** Any of various birds similar to the wren.

**Wren** (rĕn) *n. Chiefly Brit.* A member of the Women's Royal Naval Service.

**wrench** (rĕnch) *n.* [< ME *wrenchen,* to twist < OE *wrencan.*] **1.** A sudden, forcible twist or turn. **2.** An injury produced by twisting or straining. **3.** A sudden surge of an emotion, as compassion, sorrow, or anguish. **4. a.** A break or parting that causes emotional distress. **b.** The pain associated with this. **5.** A twisted interpretation or distortion in the original form of something, as a speech. **6.** A hand or power tool with fixed or adjustable jaws for gripping, turning, or twisting an object such as a nut, bolt, or pipe. *—v.* **wrenched, wrench·ing, wrench·es.** *—vt.* **1. a.** To twist or turn suddenly and forcibly. **b.** To twist and sprain <*wrenched* my ankle> **2. a.** To force free by pulling at : YANK. **b.** To pull with a wrench. **3.** To pull at the feelings or emotions of. **4.** To distort or twist the original character or meaning of. *—vi.* To give a wrench, twist, or turn. **—wrench′ing·ly** *adv.*

**wrest** (rĕst) *vt.* **wrest·ed, wrest·ing, wrests.** [ME *wresten* < OE *wrǣstan.*] **1.** To obtain by or as if by pulling with violent twisting movements. **2.** To usurp forcefully <*wrest* power in a coup> **3.** To extract by force, guile, or persistent effort : WRING <*wrested* a confession from the embezzler> **4. a.** To distort or twist the nature or meaning of. **b.** To divert to an improper use : MISAPPLY. *—n.* **1.** The act of wresting. **2.** A small tuning key for the pins of a harp or piano. **—wrest′er** *n.*

**wres·tle** (rĕs′əl) *v.* **-tled, -tling, -tles.** [ME *wrestlen* < OE *wrǣstlian.*] *—vi.* **1.** To contend by grappling and attempting to throw one's opponent to the ground. **2. a.** To contend : struggle <a family *wrestling* with unemployment> **b.** To strive in an effort to master <*wrestle* with one's guilt> *—vt.* **1. a.** To take part in (a wrestling match). **b.** To wrestle with. **2.** To throw (a calf or other animal) for branding. *—n.* **1.** An act of wrestling, esp. a wrestling match. **2.** A struggle. **—wres′tler** *n.*

**wres·tling** (rĕs′lĭng) *n.* A gymnastic exercise or contest between two competitors who attempt to throw each other by grappling.

**wrest pin** *n.* One of the pins to which the strings, esp. of a keyboard stringed instrument, are attached and tuned.

**wretch** (rĕch) *n.* [ME *wrecche* < OE *wrecca.*] **1.** An unfortunate, unhappy, or miserable person. **2.** A wicked or despicable person.

**wretch·ed** (rĕch′ĭd) *adj.* **-er, -est.** [ME *wrecched* < *wrecche,* wretch.] **1.** Living in degradation and misery : MISERABLE <"The *wretched* prisoners huddling in the stinking cages" —George Orwell> **2.** Attended by misery and woes <a *wretched* existence> **3.** Of a poor or mean character : DISMAL <a *wretched* shack> **4.** Contemptible : despicable <*wretched* treatment of animals> **5.** Inferior in performance or quality <*wretched* acting> **6.** Extremely unpleasant : DEPLORABLE <a *wretched* thing to do> **—wretch′ed·ly** *adv.* **—wretch′ed·ness** *n.*

**wrig·gle** (rĭg′əl) *v.* **-gled, -gling, -gles.** [ME *wrigglen* < MLG *wriggeln.*] *—vi.* **1.** To turn or twist the body with sinuous writhing motions : SQUIRM. **2.** To proceed with writhing motions. **3.** To insinuate or extricate oneself by sly or subtle means <*wriggled* out of trouble> *—vt.* **1.** To move with a wriggling motion <*wriggle* a finger> **2.** To make (e.g., one's way) by wriggling < *wriggled* my way into favor> *—n.* The act of wriggling. **—wrig′gly** *adj.*

**wrig·gler** (rĭg′lər) *n.* **1.** One that wriggles. **2.** Mosquito larva.

**wright** (rīt) *n.* [ME < OE *wryhta.*] One who constructs something <play*wright*><wheel*wright*>

**wring** (rĭng) *v.* **wrung** (rŭng), **wring·ing, wrings.** [ME *wringen* < OE *wringan.*] *—vt.* **1.** To compress, as between the rollers of a machine, esp. to extract liquid. **2.** To extract (liquid) by twisting or compressing. **3.** To wrench or twist forcibly or painfully. **4.** To clasp and twist or squeeze, as in distress or pain <*wring* one's hands> **5.** To cause distress to : TORMENT <Your dilemma *wrings* my heart.> **6.** To obtain by applying force or pressure to <*wring* evidence out of a witness> *—vi.* To writhe or squirm, as in pain. *—n.* An act of wringing.

**wring·er** (rĭng′ər) *n.* One that wrings, esp. a device in which laundry is pressed or spun to extract water.

**wrin·kle** (rĭng′kəl) *n.* [ME, back-formation from *wrinkled,* wrinkled, prob. < OE *gewrinclod,* p.part. of *gewrinclian,* to wind.] **1.** A small furrow, ridge, or crease on a normally smooth surface, caused by crumpling, folding, or shrinking. **2.** A line or crease in the skin, as from age. **3.** *Informal.* An ingenious innovation. *—v.* **-kled, -kling, -kles.** *—vt.* **1.** To make a wrinkle or wrinkles in. **2.** To draw up : PUCKER. *—vi.* To form wrinkles. **—wrin′kly** *adj.*

☆ ***syns:*** WRINKLE, ANGLE, GIMMICK, KICKER, TWIST *n. core meaning* : a clever, unexpected new trick or method <a new *wrinkle* in advertising>

**wrist** (rĭst) *n.* [ME < OE.] **1. a.** The junction between the hand and forearm. **b.** *Anat.* The system of bones forming this junction. **2.** The part of a sleeve or glove encircling the wrist.

**wrist·band** (rĭst′bănd′) *n.* A band, as on a long sleeve or on a wrist watch, that encircles the wrist.

**wrist·let** (rĭst′lĭt) *n.* **1.** A band of material worn round the wrist for warmth or additional strength. **2.** A bracelet.

**wrist·lock** (rĭst′lŏk′) *n.* A wrestling hold in which an opponent's wrist is gripped and twisted.

**wrist pin** *n.* A pin that joins a piston to its connecting rod.

**wrist watch** *n.* A watch worn on a band that fastens around the wrist.

**writ**[1] (rĭt) *n.* [ME < OE.] **1.** *Law.* A written court order commanding the party to whom it is addressed to perform or cease performing a specified act. **2.** Writings <the Holy *Writ*>

**writ**[2] (rĭt) *v. Archaic. var. p.t. & p.p. of* WRITE.

**write** (rīt) *v.* **wrote** (rōt), **writ·ten** (rĭt′n), **writ·ing, writes.** [ME *writen* < OE *wrītan.*] *—vt.* **1.** To form (e.g., letters) on a surface with a tool, as a pen or pencil. **2.** To form (e.g., a word) by inscribing letters or symbols on a surface <*write* one's signature> **3.** To compose and set down, esp. in literary or musical form. **4.** To draw up or draft, as a will. **5.** To fill in or cover with writing <*write* a personal check><*write* an exam> **6.** To express <*write* one's ideas> **7.** To communicate, as by correspondence <*write* a letter home> **8.** To underwrite, as an insurance policy. **9.** To mark with the signs of (e.g., a quality) <"Utter dejection was *written* on every face" —Winston Churchill> **10.** To ordain by fate or prophecy. **11.** *Computer Sci.* To record (data) either transiently or permanently in a storage device or on an external medium. *—vi.* **1.** To trace or form letters, words, or symbols on paper or another surface. **2.** To produce written material, as books or articles. **3. a.** To compose and send a letter or letters. **b.** To maintain a correspondence. **—write down. 1.** To reduce in rank, value, or price. **2.** To disparage in writing. **3.** To write in a conspicuously simplified or condescending style. **—write in. 1.** To vote for (a name not listed on a ballot) by insertion. **2.** To insert in a text or document. **—write off. 1.** To reduce the entered value of (an asset) : DEPRECIATE. **2.** To cancel from business accounts as a loss. **3.** To consider as a loss or failure <*wrote* the venture *off*> **—write (one's) own ticket.** To set one's own terms or choose a course that meets one's own desires and requirements. **—write up. 1.** To write a report or description of, as for publication. **2.** To bring (e.g., a journal) up to date. **3.** To overstate the value of (assets). **4.** *Informal.* To write a summons for <*wrote* the speeder *up*>

**write-down** (rīt′doun′) *n.* A reduction of the entered value of an asset.

**write-in** (rīt′ĭn′) *n.* A vote cast by writing in the name of a candidate not on the ballot.

**write-off** (rīt′ôf′, -ŏf′) *n.* **1. a.** A cancellation in account books. **b.** An amount canceled or lost. **2.** A reduction or depreciation of the entered value of an item.

**writ·er** (rī′tər) *n.* One who writes, esp. as a profession : AUTHOR.

**writer's cramp** *n.* A cramp chiefly affecting the muscles of the thumb and two adjacent fingers after prolonged writing.

**write-up** (rīt′ŭp′) *n.* **1.** A published account, review, or notice, esp. a favorable one. **2.** An intentional overevaluation of a corporation's assets.

**writhe** (rīth) *v.* **writhed, writh·ing, writhes.** [ME *writhen* < OE *wrīðan.*] —*vi.* **1.** To twist or squirm, as in pain, struggle, or embarrassment. **2.** To move with a twisting or contorted motion. **3.** To suffer acutely. —*vt.* To cause to twist or squirm : CONTORT. **—writhe** *n.* **—writh′er** *n.*

**writ·ing** (rī′tĭng) *n.* **1.** Written form <Put your request in *writing.*> **2.** Language symbols or characters written or imprinted on a surface. **3.** A written work, esp. a literary composition. **4.** A writer's activity, art, or occupation. **5. Writings.** Hagiographa. **—writing (**or **handwriting) on the wall.** A usu. ominous indication of future events <saw the *writing on the wall* and realized there would soon be heavy layoffs>

**writing paper** *n.* Paper for writing on, esp. in ink.

**writ of election** *n.* A writ ordering that an election be held, esp. a special election to fill a vacancy in an elective office.

**writ of error** *n. Law.* A writ commissioning an appellate court to review the proceedings of another court and correct the judgment given if deemed necessary.

**writ of prohibition** *n. Law.* An order issued by a higher court commanding a lower court to cease from proceeding in a matter not within its jurisdiction.

**writ of summons** *n. Law.* A writ directing one to appear in court to answer a complaint.

**writ·ten** (rĭt′n) *v. p.p. of* WRITE.

**wrong** (rông, rŏng) *adj.* [ME, of Scand. orig.] **1.** Not in conformity with truth or fact : INCORRECT. **2. a.** Contrary to conscience, morality, or law : IMMORAL. **b.** Unfair or unjust. **3.** Not needed, intended, or wanted <rang the *wrong* doorbell> **4.** Not fitting or suitable : INAPPROPRIATE <said the *wrong* thing> **5.** Not in accordance with an established usage, method, or procedure. **6.** Not functioning properly. **7.** Unacceptable or undesirable according to social convention. **8.** Of, pertaining to, or being the side of something that is less finished or opposite to the right, principal, or more prominent side <wore the sweater *wrong* side out> —*adv.* **1.** Mistakenly or erroneously. **2.** Immorally or unjustly. —*n.* **1. a.** An unjust or injurious act. **b.** A breach of ethics or morality. **2. a.** An invasion or violation of another's legal rights. **b.** *Law.* A tort. **3.** The condition of being in error or at fault. —*vt.* **wronged, wrong·ing, wrongs. 1.** To treat unjustly, injuriously, or dishonorably. **2.** To discredit unjustly : MALIGN. **—do (someone) wrong.** *Informal.* To be unfaithful. **—wrong′er** *n.* **—wrong′ly** *adv.*

☆ ***syns:*** WRONG, AGGRIEVE, OPPRESS, OUTRAGE, PERSECUTE *v. core meaning* : to do a wrong to or treat unjustly <You have *wronged* and injured those who trusted you.>

**wrong·do·er** (rông′dōō′ər, rŏng′-) *n.* One who does wrong. **—wrong′do′ing** *n.*

**wrong·ful** (rông′fəl, rŏng′-) *adj.* **1.** Wrong : unjust. **2.** Unlawful. **—wrong′ful·ly** *adv.* **—wrong′ful·ness** *n.*

**wrong-head·ed** (rông′hĕd′ĭd, rŏng′-) *adj.* Persistently misguided in judgment or opinion. **—wrong′-head′ed·ly** *adv.* **—wrong′-head′ed·ness** *n.*

**wrote** (rōt) *v. p.t. of* WRITE.

**wroth** (rôth) *adj.* [ME < OE *wrāð.*] Full of wrath : ANGRY.

**wrought** (rôt) [P.part. of WORK.] *adj.* **1.** Put together <finely *wrought*> **2.** Shaped by hammering with tools. —Used chiefly of metals or metalwork. **3.** Made delicately or elaborately. **—wrought up.** Agitated : excited.

**wrought iron** *n.* An easily welded or forged iron containing approx. 0.2% carbon and total impurities less than approx. 0.5%.

**wrung** (rŭng) *v. p.t. & p.p. of* WRING.

**wry** (rī) *adj.* **wri·er, wri·est** *also* **wry·er, wry·est.** [< ME *wrien,* to turn aside < OE *wrigian,* to move.] **1.** Abnormally twisted or bent to one side : CROOKED <a *wry* neck> **2.** Temporarily twisted in an expression of distaste or displeasure <made a *wry* mouth> **3.** At variance with what is right or proper. **4.** Drily humorous, often with a touch of irony. **—wry′ly** *adv.* **—wry′ness** *n.*

**wry·neck** (rī′nĕk′) *n.* **1.** An Old World bird, *Jynx torquilla* or *J. ruficollia,* capable of twisting its neck into unusual contortions. **2.** *Pathol.* Torticollis.

**wul·fen·ite** (wŏŏl′fə-nīt′) *n.* [G. *Wulfenit,* after Franz X. von *Wulfen* (1728–1805).] A yellow to orange-red mineral, $PbMoO_4$, used as a molybdenum ore.

**wurst** (wûrst, wŏŏrst) *n.* [G. < OHG.] Sausage.

**wu shu** (wōō′ shōō′) *n.* [Chin. (Mandarin) *wu*$^{3}$ *shu*$^{4}$.] The Chinese martial arts.

**Wy·an·dot** *also* **Wy·an·dotte** (wī′ən-dŏt′) *n., pl.* **Wyandot** *or* **-dots** *also* **Wyandotte** *or* **-dottes.** [Wyandot *wădăt,* tribal name.] **1.** An Indian of a tribe in the Huron confederacy. **2.** The Iroquoian language of the Wyandot.

**Wy·an·dotte** (wī′ən-dŏt′) *n.* **1.** A domestic fowl of a breed developed in North America. **2.** *var. of* WYANDOT.

**wych elm** *also* **witch elm** (wĭch) *n.* [< ME *wyche* < OE *wice.*] An Old World elm, *Ulmus glabra,* often planted as a shade tree.

**wye** (wī) *n.* **1.** The letter *y.* **2.** A Y-shaped object.

**wy·vern** *also* **wi·vern** (wī′vərn) *n.* [ME *wyvere,* viper < ONFr. *wivre* < Lat. *vipera.*] *Heraldry.* A two-legged dragon with wings and a barbed and knotted tail.

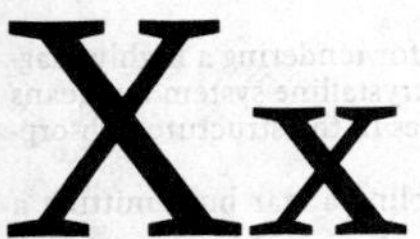

**x** *or* **X** (ĕks) *n., pl.* **x's** *or* **X's. 1.** The 24th letter of the English alphabet. **2.** A speech sound represented by the letter x. **3.** The 24th in a series. **4.** Something shaped like the letter X. **5.** The mark (X) inscribed to represent the signature of an illiterate. **6.** An unknown or unnamed factor, thing, or person. **7. X.** The Roman numeral for ten. —*vt.* **x'd, x'ing, x's. 1.** To mark or sign with an X. **2.** To delete, cancel, or obliterate with a series of x's.

**X** (ĕks) *adj.* Indicating a motion-picture rating of such nature that no one under the age of 17 is to be admitted.

**Xan·a·du** (zăn′ə-dōō′, -dyōō′) *n.* [After *Xanadu,* a place in *Kubla Khan,* a poem by Samuel T. Coleridge (1772–1834).] An idyllic, beautiful place.

☆ ***syns:*** XANADU, SHANGRI-LA, UTOPIA *n. core meaning* : a place of idyllic beauty and contentment <the search for *Xanadu* in song and legend>

**xanth-** *pref. var. of* XANTHO-.

**xan·than gum** (zăn′thən) *n.* A natural gum of high molecular weight produced by culture fermentation of glucose and used as a stabilizer in commercial food preparation.

**xan·thate** (zăn′thāt′) *n.* A salt of a xanthic acid, esp. a simple xanthic acid salt, as of sodium or potassium, used as a flotation collector for copper, silver, and gold.

**xan·thene** (zăn′thēn′) *n.* A yellow crystalline organic compound, $CH_2(C_6H_4)_2O$, that is soluble in ether and used as a fungicide and in organic synthesis.

**xan·thic acid** (zăn′thĭk) *n.* [So called from the yellow color of its salts. —see XANTHO-.] Any of various unstable acids of the form ROC(S)SH, in which R is usu. an alkyl radical.

**xan·thine** (zăn′thēn′, -thĭn) *n.* A yellowish-white purine base, $C_5H_4N_4O_2$, found in blood, urine, and some plants.

**xantho-** *or* **xanth-** *pref.* [NLat. < Gk. *xanthos,* yellow.] **1.** Yellow <*xanthine*> **2.** Xanthic acid <*xanthate*>

**xan·tho·chroid** (zăn′thə-kroid′) *adj.* [< NLat. *xanthochroi,* yellow-haired, fair-skinned people : XANTH(O)- + Gk. *ōkhros,* pale.] Having fair skin and light hair. —*n.* A xanthochroid person.

**xan·tho·ma** (zăn-thō′mə) *n.* A condition marked by the presence of small flat yellowish plaques in the skin, esp. on the eyelids, due to deposits of lipids.

**xan·tho·phyll** (zăn′thə-fĭl′) *n.* [Fr. *xanthophylle* : *xantho-*, xantho- + *-phylle*, -phyll.] A yellow carotenoid pigment, $C_{40}H_{56}O_2$, found with chlorophyll in green plants and in egg yolk.

**xan·thous** (zăn′thəs) *adj.* **1.** Yellow. **2.** Having light-brown or yellowish skin.

**x-ax·is** (ĕks′ăk′sĭs) *n., pl.* **x-ax·es** (-sēz). **1.** The horizontal axis of a two-dimensional Cartesian coordinate system. **2.** One of three axes in a three-dimensional Cartesian coordinate system.

**X-chro·mo·some** (ĕks′krō′mə-sōm′) *n.* The sex chromosome associated with female characteristics, occurring paired in the female and single in the male sex-chromosome pair.

**Xe** *symbol for* XENON.

**xe·bec** (zē′bĕk′) *n.* [Fr. *chebec* < Ar. *shabbāk*.] A small three-masted Mediterranean vessel with both square and triangular sails.

**xebec**

**xen-** *pref. var. of* XENO-.

**xe·ni·a** (zē′nē-ə) *n.* [NLat. < Gk., hospitality < *xenos*, guest, stranger.] *Bot.* The effect on a hybrid plant produced by the transfer of pollen from one strain to the seed of a different strain.

**xeno-** *or* **xen-** *pref.* [NLat. < Gk. *xenos*, stranger.] **1.** Stranger : foreigner <*xeno*phobia> **2.** Strange : foreign : different <*xeno*lith>

**xen·o·blast** (zĕn′ə-blăst′, zē′nə-) *n.* A mineral deposit that has developed during metamorphism without developing crystalline faces.

**xen·o·cryst** (zĕn′ə-krĭst′) *n.* [XENO- + CRYST(AL).] A crystal foreign to the igneous rock in which it occurs.

**xen·o·cur·ren·cy** (zĕn′ō-kûr′ən-sē, -kŭr′-, zē′nō-) *n.* A currency in circulation outside its own country.

**xe·nog·a·my** (zĭ-nŏg′ə-mē) *n. Bot.* Transfer of pollen from one plant to another : CROSS-FERTILIZATION. **—xe·nog′a·mous** *adj.*

**xen·o·gen·e·sis** (zĕn′ə-jĕn′ə-sĭs) *n.* The supposed production of offspring markedly different from and showing no relationship to either of the parents. **—xen′o·ge·net′ic** (-jə-nĕt′ĭk), **xen′o·gen′ic** (-jĕn′ĭk) *adj.*

**xen·o·lith** (zĕn′ə-lĭth′) *n.* A rock fragment foreign to the igneous mass in which it occurs.

**xe·non** (zē′nŏn′) *n.* [< Gk., neuter of *xenos*, stranger.] *Symbol* **Xe** A colorless, odorless, highly unreactive gaseous element found in minute quantities in the atmosphere; atomic number 54; atomic weight 131.30.

**xenon hex·a·flu·o·ride** (hĕk′sə-flōō′ə-rīd′, -flŏŏr′īd′, -flôr′-, -flōr′-) *n.* A highly reactive colorless crystalline compound, $XeF_6$.

**xenon tet·ra·flu·o·ride** (tĕt′rə-flōō′ə-rīd, -flŏŏr′īd′, -flôr′-, -flōr′-) *n.* A colorless crystalline compound, $XeF_4$, derived from fluorine and xenon by heating under pressure and marked by its ease of sublimation in air.

**xen·o·phobe** (zĕn′ə-fōb′) *n.* One who fears or hates strangers or foreigners or anything that is foreign. **—xen′o·pho′bi·a** *n.* **—xen′o·pho′bic** *adj.*

**xer-** *pref. var. of* XERO-.

**xer·ic** (zĕr′ĭk, zîr′-) *adj.* Of, marked by, or adapted to a very dry habitat. **—xer′i·cal·ly** *adv.*

**xero-** *or* **xer-** *pref.* [NLat. < Gk. *xēros*, dry.] Dry : dryness <*xero*derma>

**xe·ro·der·ma** (zîr′ə-dûr′mə) *also* **xe·ro·der·mi·a** (-mē-ə) *n.* Abnormal dryness and roughness of the skin.

**xe·rog·ra·phy** (zĭ-rŏg′rə-fē) *n.* A process for copying printed or pictorial matter in which a negative image formed by a resinous powder on an electrically charged plate is electrically transferred to and thermally fixed as positive on the copy paper. **—xe·rog′raph·er** *n.* **—xe′ro·graph′ic** (zîr′ə-grăf′ĭk) *adj.* **—xe′ro·graph′i·cal·ly** *adv.*

**xe·roph·i·lous** (zĭ-rŏf′ə-ləs) *adj.* Flourishing in or able to withstand a hot dry environment. **—xe·roph′i·ly** *adv.*

**xe·roph·thal·mi·a** (zîr′əf-thăl′mē-ə) *n.* [LLat. < Gk. *xērophthalmia* : *xēros*, dry + *ophthalmia*, ophthalmia < *ophthalmos*, eye.] An abnormally dry and lusterless condition of the eyeball due to a severe deficiency of vitamin A.

**xe·ro·phyte** (zîr′ə-fīt′) *n.* A plant that grows in and is adapted to a moisture-deficient environment. **—xe′ro·phyt′ic** (-fĭt′ĭk) *adj.* **—xe′ro·phyt′i·cal·ly** *adv.*

**xe·ro·sere** (zîr′ə-sîr′) *n.* A sequence of ecological communities beginning in a dry area.

**xe·ro·sis** (zĭ-rō′sĭs) *n.* **1.** Abnormal dryness, as of the skin, eye, or mucous membranes. **2.** Normal sclerosis of aging tissue.

**Xe·rox** (zîr′ŏks). A trademark for a photocopying process or machine using xerography.

**x-height** (ĕks′hīt′) *n.* The height of a lower-case x.

**Xho·sa** *also* **Xo·sa** (kō′sä) *n., pl.* **Xhosa** *or* **-sas** *also* **Xosa** *or* **-sas. 1.** One of a Bantu people of Cape of Good Hope Province, South Africa. **2.** The Bantu language of the Xhosa.

**xi** (zī, sī, ksē) *n.* [Gk. *xei*.] **1.** The 14th letter of the Greek alphabet. —See table at ALPHABET. **2.** *also* **xi particle.** *Physics.* Either of two unstable subatomic particles in the baryon family that have masses 2,572 and 2,585 times that of an electron.

**xiph·i·ster·num** (zĭf′ĭ-stûr′nəm) *n., pl.* **-na** (-nə) [Gk. *xiphos*, sword + STERNUM.] The posterior and smallest of the three divisions of the sternum.

**xiph·oid** (zĭf′oid′) *adj.* [Gk. *xiphoeidēs* : *xiphos*, sword + *eidos*, shape.] **1.** Shaped like a sword. **2.** Of or relating to the xiphisternum. *—n.* The xiphisternum.

**xiph·o·su·ran** (zĭf′ə-sŏŏr′ən) *n.* [<NLat. *Xiphosura*, order name : Gk. *xiphos*, sword + Gk. *oura*, tail.] An arthropod of the order Xiphosura, which includes the horseshoe crab and many extinct forms. *—adj.* Of or belonging to the order Xiphosura.

**X·mas** (krĭs′məs, ĕks′məs) *n.* [< *X*, the Greek letter chi, abbr. of *Khristos*, Christ.] *Informal.* Christmas.

▲ word history: The character X in *Xmas* does not represent the letter X in the Roman alphabet but rather the Greek letter chi. The Greek form of chi is χ. Chi is the first letter of the Greek form of *Christ*, which in Greek is *χριστός*, transliterated *Khristos* or *Christos*. The symbol χ or X has been used as an abreviation for *Christ* since earliest Christian times.

**Xosa** (kō′sä) *n. var. of* XHOSA.

**x-ra·di·a·tion** (ĕks′rā′dē-ā′shən) *n.* **1.** Treatment with or exposure to x-rays. **2.** Radiation composed of x-rays.

**X-rated** (ĕks′rā′tĭd) *adj.* Having the rating X : explicit in the treatment of sex <an *X-rated* movie>

**X rating** *n. Informal.* A classification assigned to a film featuring explicit sex.

**x-ray** *also* **X-ray** (ĕks′rā′) [Transl. of G. *X Strahl*.] **1. a.** A relatively high-energy photon with wavelength in the approximate range from 0.05 angstroms to 100 angstroms. **b.** *often* **x-rays.** A stream of such photons, used for their penetrating power in radiography, radiology, radiotherapy, and research. **2.** A photograph taken with x-rays. *—vt.* **x-rayed, x-ray·ing, x-rays** *also* **X-rayed, X-ray·ing, X-rays. 1.** To irradiate with x-rays. **2.** To photograph with x-rays.

▲ word history: Wilhelm Roentgen, the German scientist who discovered x-rays, gave them the name *x-strahlen*, translated into English as "x-rays." He used *x*, the symbol for an unknown quantity, because he did not completely understand the nature of this kind of radiation.

**x-ray astronomy** *n.* The branch of astronomy that deals with the properties of celestial bodies as indicated by the x-rays they emit.

**x-ray burster** *n.* Any of several celestial phenomena marked by emission of very powerful bursts of x-radiation in cycles lasting from a few seconds to a few minutes.

**x-ray crystallography** *n.* Study of crystal structure by means of x-ray diffraction.

**x-ray diffraction** *n.* Scattering of x-rays by crystal atoms, producing a diffraction pattern that yields data about the structure of the crystal.

**x-ray microscope** *n.* An instrument for rendering a highly magnified image of the atomic structure of a crystalline system by means of the contrasts arising from the differences in the structure's absorption or emission of x-rays.

**x-ray star** *n.* A celestial object resembling a star but emitting a major portion of its radiation in x-rays.

**x-ray therapy** *n.* Radiotherapy with x-rays.

**x-ray tube** *n.* A vacuum tube containing electrodes that accelerate electrons and direct them to a metal anode, where their impacts produce x-rays.

**Xu·thus** (zōō′thəs) *n.* [Lat. < Gk. *Xouthos*.] *Gk. Myth.* The ancestor of the Ionian Greeks.

**xyl-** *pref. var. of* XYLO-.

**xy·lan** (zī′lən) *n.* A yellow gummy pentosan found in plant cell walls and woody tissue and yielding xylose upon hydrolysis.

**xy·lem** (zī′ləm) *n.* [G. < Gk. *xulon*, wood.] The supporting and water-conducting tissue of vascular plants, consisting chiefly of tracheids and vessels.

**xy·lene** (zī-lēn′, zī′lēn′) *n.* **1.** Any of three flammable isomeric hydrocarbons, $C_6H_4(CH_3)_2$, obtained from wood and coal tar. **2.** A mixture of xylene isomers used as a solvent in making lacquers and rubber cement and as an aviation fuel.

**xy·li·dine** (zī′lĭ-dēn′, -dĭn, zĭl′-) *n.* **1.** Any of six toxic isomers, $(CH_3)_2C_6H_3NH_2$, derived from xylene, used primarily as dye intermediates. **2.** Any of various mixtures of xylidine isomers.

**xylo-** *or* **xyl-** *pref.* [< Gk. *xulon,* wood.] **1.** Wood <*xylograph*> **2.** Xylene <*xylidine*>

**xy·lo·graph** (zīʹlə-grăfʹ) *n.* **1.** A wood engraving. **2.** An impression from a wood block. —*vt.* **-graphed, -graph·ing, -graphs.** To print from a wood engraving. **—xy·logʹra·pher** (-lŏgʹrə-fər) *n.*

**xy·log·ra·phy** (zī-lŏgʹrə-fē) *n.* **1.** Wood engraving, esp. of an early period. **2.** The art of printing texts or illustrations, occas. with color, from wood blocks. **—xyʹlo·graphʹic** (-lə-grăfʹĭk), **xyʹlo·graphʹi·cal** (-ĭ-kəl) *adj.* **—xyʹlo·graphʹi·cal·ly** *adv.*

**xy·loid** (zīʹloidʹ) *adj.* Of or resembling wood : LIGNEOUS.

**xy·lol** (zīʹlôlʹ, -lōlʹ) *n.* XYLENE 1.

**xy·loph·a·gous** (zī-lŏfʹə-gəs) *adj.* Feeding on wood.

**xy·lo·phone** (zīʹlə-fōnʹ) *n.* A percussion instrument consisting of a mounted row of wooden bars graduated in length to sound a chromatic scale, played with two small wooden hammers. **—xyʹlo·phonʹist** *n.*

**xy·lose** (zīʹlōsʹ) *n.* A white crystalline aldose sugar, $C_5H_{10}O_5$, used in dyeing and tanning.

**xy·lot·o·my** (zī-lŏtʹə-mē) *n.* Cutting and preparation of sections of wood for microscopic study.

**XY recorder** (eksʹwīʹ) *n.* An output device that sketches the relationship between two variables onto a grid of plane rectangular coordinates.

**xys·ter** (zĭsʹtər) *n.* [Gk. *xustēr,* scraper < *xuein,* to scrape.] A surgical instrument for scraping bones.

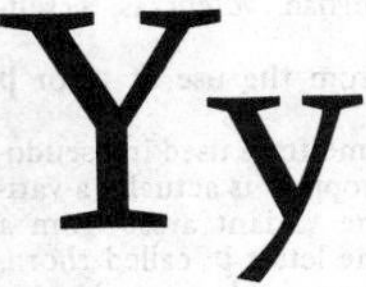

**y** *or* **Y** (wī) *n., pl.* **y's** *or* **Y's. 1.** The 25th letter of the English alphabet. **2.** A speech sound represented by the letter *y.* **3.** The 25th in a series. **4.** Something shaped like the letter Y.

**Y** *symbol for* YTTRIUM.

**-y[1]** *suff.* [ME *-ie, -ey* < OE *-ig.*] **1.** Characterized by : consisting of <*clayey*> **2. a.** Like <*summery*> **b.** To some degree : somewhat : rather <*chilly*> **3.** Tending toward : inclined toward <*sleepy*>

**-y[2]** *suff.* [ME *-ie* < OFr. < Lat. *-ia* and Gk. *-ia,* n. suffixes.] **1.** Condition : state : quality <*jealousy*> **2. a.** Activity <*cookery*> **b.** Instance of a specified action <*entreaty*> **3. a.** Place for an activity <*cannery*> **b.** Result or product of an activity <*laundry*> **4.** Collection : body : group <*soldiery*>

**-y[3]** *suff.* [ME.] **1.** Small one <*doggy*> **2.** Dear one <*sweetie*> **3.** One having to do with or characterized by <*towny*>

**yab·ber** (yăbʹər) [Alteration of JABBER.] *Austral.* —*n.* Jabber. —*vi.* & *vt.* **-bered, -ber·ing, -bers.** To jabber.

**yacht** (yät) *n.* [Obs. Du. *jaghte,* short for *jaghtschip* : *jagen,* to chase + *schip,* ship.] A relatively small and light sailing or mechanically propelled vessel, gen. with smart, graceful lines, used for pleasure cruises or racing. —*vi.* **yacht·ed, yacht·ing, yachts.** To race, sail, or cruise in a yacht.

**yacht club** *n.* A club that promotes and supports yachting.

**yacht·ing** (yätʹĭng) *n.* The sport of sailing or cruising in a yacht.

**yachts·man** (yätsʹmən) *n.* An owner or operator of a yacht. **—yachtsʹman·shipʹ** *n.*

**yack** (yăk) *v.* & *n. var. of* YAK[2].

**yack·e·ty-yak** (yăkʹĭ-tē-yăkʹ) *n.* [Imit.] *Slang.* YAP 2.

**YAG** (yăg) *n.* [Y(TTRIUM) + A(LUMINUM) + G(ARNET).] A hard synthetic yttrium aluminum garnet used in laser technology and as a gemstone.

**ya·gi** (yäʹgē, yăgʹē) *n.* [After Hidetsugu *Yagi* (b. 1888), its inventor.] A directional radio and television antenna consisting of a horizontal conductor with several insulated dipoles parallel to and in the plane of the conductor.

**yah** (yä) *adv.* [Var. of YEA.] *Informal.* Yes.

**ya·hoo** (yäʹhōō, yāʹ-) *n., pl.* **-hoos.** [< *Yahoo,* member of a savage race in *Gulliver's Travels* by Jonathan Swift (1667–1745).] One who is crude, brutish, or stupid. **—yaʹhoo·ism** *n.*

**Yah·weh** (yäʹwā) *also* **Yah·veh** (-vā) *n.* [Heb.] A name for God assumed by modern scholars to be a rendering of the pronunciation of the Tetragrammaton.

**Yah·wist** (yäʹwĭst) *also* **Yah·vist** (-vĭst) *n.* The author of the earliest sources of the Hexateuch, in which God is called Yahweh rather than Elohim. **—Yah·wisʹtic** *adj.*

**yak[1]** (yăk) *n.* [Tibetan *gyag.*] A long-haired bovine mammal, *Bos grunniens,* of Tibet and the mountainous areas of central Asia, where it is often domesticated.

**yak[2]** *also* **yack** (yăk) [Imit.] *vi. Slang.* **yakked, yak·king, yaks** *also* **yacked, yack·ing, yacks.** To chatter persistently and pointlessly. **—yak, yack** *n.*

**ya·ki·to·ri** (yäʹkĭ-tôrʹē) *n.* [J. : *yaki,* roasting + *tori,* bird.] Bite-sized marinated chicken pieces grilled on small bamboo skewers.

**Ya·kut** (yä-kōōtʹ) *n.* **1.** One of a people living in the Yakut region of northeastern Russia. **2.** The Turkic language of the Yakuts.

**y'all** (yôl) *pron. var. of* YOU-ALL.

**†yam** (yăm) *n.* [Port. *inhame,* poss. < Bantu *nyama,* meat, or Bambara *nyana,* wild yam.] **1.** Any of various chiefly tropical vines of the genus *Dioscorea,* many of which have edible tuberous roots. **2.** The starchy root of the yam, used in the tropics as food. **3.** *Southern U.S.* A sweet potato with reddish flesh.

**ya·men** (yäʹmən) *n.* [Chin. (Mandarin) *ya[2] men[2]* : *ya[2],* magistracy (< *ya[2],* tooth, flag with a serrated edge) + *men[2],* gate.] The office or residence of a Chinese imperial official.

**yam·mer** (yămʹər) [Alteration of ME *yomeren,* to lament < OE *gēomrian.*] *Informal.* —*v.* **-mered, -mer·ing, -mers.** —*vi.* **1.** To complain peevishly or whimperingly : WHINE. **2.** To talk rapidly and loudly. —*vt.* To utter or say in a complaining or clamorous tone. **—yamʹmer** *n.* **—yamʹmer·er** *n.*

**yang** *also* **Yang** (yăng) *n.* [Chin. (Mandarin) *yang[2],* sun, light, masculine element.] The active masculine cosmic principle in Chinese dualistic philosophy.

**yank** (yăngk) *v.* **yanked, yank·ing, yanks.** [Orig. unknown.] —*vt. Informal.* To pull or extract suddenly <*yank* a tooth> —*vi.* To pull on something suddenly. —*n.* A sudden vigorous pull : JERK.

**Yank** (yăngk) *n.* [Short for YANKEE.] *Informal.* Yankee.

**Yan·kee** (yăngʹkē) *n.* [Orig. unknown.] **1.** A native or resident of New England. **2.** A native or resident of a Northern state, esp. a Union soldier during the American Civil War. **3.** A native or resident of the United States.

**Yan·kee·dom** (yăngʹkē-dəm) *n.* **1.** The Northern states or New England. **2.** The United States. **3.** Yankees as a group.

**Yankee Doo·dle** (dōōdʹl) *n.* [From the title of a song popular during the Revolutionary War.] A Yankee.

**Yan·kee·ism** (yăngʹkē-ĭzʹəm) *n.* **1.** A Yankee custom or characteristic. **2.** A Yankee peculiarity, as of language or pronunciation.

**yap** (yăp) *v.* **yapped, yap·ping, yaps.** [Imit.] —*vi.* **1.** To bark sharply or shrilly : YELP. **2.** *Slang.* To talk noisily and foolishly. **3.** *Slang.* To talk sharply and abusively : SCOLD. —*vt.* To utter by yapping. —*n.* **1.** A sharp shrill bark : YELP. **2.** *Slang.* Noisy stupid talk. **3.** *Slang.* One who is crude, loud, and stupid. **4.** *Slang.* The human mouth. **—yapʹper** *n.*

**ya·pok** (yə-pŏkʹ) *n.* [After the *Oyapock,* a river in South America.] An aquatic marsupial mammal, *Chironectes minimus* of tropical America, with dense fur, webbed hind feet, and a long tail.

**yapok**
*Approximately 27 inches long*

**Ya·qui** (yäʹkē) *n., pl.* **Yaqui** *or* **-quis. 1. a.** A tribe of North American Indians now living in Sonora, Mexico. **b.** A member of this tribe. **2.** The Uto-Aztecan language of the Yaqui.

**Yar·bor·ough** (yärʹbər-ō, -bər-ə) *n.* [After Charles Anderson Worsley (1809–1897), 2nd Earl of *Yarborough,* said to have bet 1,000 to 1

ōō **boot** ou **out** th **thin** *th* **this** ŭ **cut** ûr **urge** y **young**
yōō **abuse** zh **vision** ə **a**bout, it**e**m, ed**i**ble, gall**o**p, circ**u**s

that such a hand would not occur.] A bridge or whist hand containing no card higher than a nine.

**yard[1]** (yärd) *n.* [ME *yerde,* measuring rod < OE *gerd,* stick.] **1.** The fundamental unit of length in both the U.S. Customary System and the British Imperial System, equal to 0.9144 meter. **2.** *Naut.* A long tapering spar slung at right angles to a mast to support and spread the head of a square sail, lugsail, or lateen.

**yard[2]** (yärd) *n.* [ME < OE *geard,* enclosed area.] **1.** Ground adjacent to, surrounding, or surrounded by a building or group of buildings. **2.** An often enclosed and paved tract of ground used for a specific task, business, or other activity. **3.** An area provided with a system of railroad tracks where trains are made up and cars are switched, stored, or serviced. **4.** A winter pasture for deer and moose in a forest. **5.** An enclosed tract of ground in which animals, as chickens or pigs, are kept. —*v.* **yarded, yard·ing, yards.** —*vt.* To enclose, collect, or put in or as if in a yard. —*vi.* To gather in or as if in a yard.

**yard·age[1]** (yär′dĭj) *n.* **1.** The amount or length of material measured in yards. **2.** Cloth sold by the yard.

**yard·age[2]** (yär′dĭj) *n.* **1.** Use of a livestock yard at a station in the process of transporting cattle by railroad. **2.** The fee paid for yardage.

**yard·arm** (yärd′ärm′) *n. Naut.* Either end of a yard of a square sail.

**yard bird** *n. Slang.* **1. a.** A low-ranking, untrained enlisted person. **b.** An enlisted person confined to base and assigned menial tasks as punishment. **2.** A convict : prisoner.

**yard goods** *pl.n.* Piece goods.

**yard grass** *n.* Any of several weedy grasses of the genus *Eleusine.*

**yard·man** (yärd′mən) *n.* A man employed esp. in a railroad yard.

**yard·mas·ter** (yärd′măs′tər) *n.* A man in charge of a railroad yard.

**yard of ale** *n.* **1.** A slender horn-shaped glass approx. three feet tall and holding about three pints. **2.** The amount of beer or ale contained in a yard of ale.

**yard sale** *n.* A sale of used household belongings held on the front or back lawn of a house.

**yard·stick** (yärd′stĭk′) *n.* **1.** A graduated measuring stick one yard in length. **2.** A test or standard used in measurement, comparison, or judgment : CRITERION.

▲ word history: A yardstick is literally a "stick-stick," for the word *yard* is the descendent of Old English *gerd,* which meant simply "a stick." Since sticks make convenient measuring devices, a stick of a certain length became used as a standard of linear measure. The length of this unit varied over the centuries; the current length of a yard was fixed in the 14th century during the reign of Edward III. The compound *yardstick* is a relatively recent coinage, using the word *yard* in the sense "a unit of length."

**yare** (yâr) [ME < OE *gearo,* ready.] *Archaic.* —*adj.* **1.** Responding easily : MANEUVERABLE. —Used of a vessel. **2.** Bright : lively. **3.** Ready : prepared. —*adv.* Soon : quickly. **—yare′ly** *adv.*

**yar·mul·ke** *also* **yar·mel·ke** (yär′məl-kə, yä′məl-) *n.* [Yiddish < Pol. and Ukranian *yarmulka,* poss. < Turk. *yağmurluk,* raincoat < *yağmur,* rain.] A skullcap worn by Jewish men, esp. those adhering to Orthodox or Conservative tradition.

**yarn** (yärn) *n.* [ME < OE *gearn.*] **1.** A continuous strand of twisted threads of natural or synthetic material, as wool or nylon, used in weaving or knitting. **2.** *Informal.* A long, often elaborate tale of real or fictitious adventures. —*vi.* **yarned, yarn·ing, yarns.** *Informal.* To tell a long complicated story.

**yarn-dyed** (yärn′dīd′) *adj.* Made of yarn dyed before weaving.

**yar·row** (yăr′ō) *n.* [ME *yarow* < OE *gearwe.*] A plant of the genus *Achillea,* esp. *A. millefolium,* native to Eurasia, having finely dissected foliage and flat, usu. white flower clusters.

**yash·mak** *also* **yash·mac** (yäsh-mäk′, yăsh′măk) *n.* [Ar.] A veil worn by Moslem women to cover the face in public.

**yat·a·ghan** *also* **yat·a·gan** (yăt′ə-găn′, -gən) *n.* [Turk. *yatağan.*] A Turkish sword having a double-curved blade and a hilt with an eared pommel and no guard.

**yaup** (yôp) *v.* & *n. var. of* YAWP.

**yau·pon** (yô′pən) *n.* [Catawba *yopun,* dim. of *yop,* tree.] A holly, *Ilex vomitoria* of the southeastern United States, with scarlet fruit and evergreen leaves, once used medicinally.

**yaw** (yô) *v.* **yawed, yaw·ing, yaws.** [Orig. unknown.] —*vi.* **1.** *Naut.* To deviate from the intended course. **2.** To move unsteadily : WEAVE. **3.** To turn about the vertical axis. —Used of an aircraft, spacecraft, or projectile. —*vt.* To cause to yaw. —*n.* **1.** An act of yawing. **2.** Extent of yawing, measured in degrees.

**yawl** (yôl) *n.* [MLG *jolle.*] **1.** A two-masted fore-and-aft-rigged sailing vessel similar to the ketch but having a smaller jigger mast stepped abaft the rudder. **2.** A ship's small boat, manned by oarsmen.

**yawn** (yôn) *v.* **yawned, yawn·ing, yawns.** [ME *yanen* < OE *gēonian.*] —*vi.* **1.** To open the mouth wide with a deep inhalation, usu. involuntarily, from drowsiness, fatigue, or boredom. **2.** To open wide : GAPE <The chasm *yawned* below.> —*vt.* To utter wearily, as if in yawning. **—yawn** *n.* **—yawn′er** *n.*

**yawn·ing** (yô′nĭng) *adj.* Gaping : cavernous. **—yawn′ing·ly** *adv.*

**yawp** *also* **yaup** (yôp) [ME *yolpen,* poss. var. of *yelpen,* to cry aloud. —see YELP.] —*vi.* **yawped, yawp·ing, yawps** *also* **yauped, yaup·ing, yaups.** **1.** To utter a sharp cry. **2.** *Slang.* To talk loudly and stupidly. —*n.* **1.** A yelp. **2.** *Slang.* Loud stupid talk. **—yawp′er** *n.*

**yaws** (yôz) *n.* [Carib.] *(sing. or pl. in number).* An infectious epidemic tropical skin disease caused by a spirochete, *Treponema pertenue,* and marked by multiple red pimples or pustules.

**y-ax·is** (wī′ăk′sĭs) *n., pl.* **y-ax·es** (-sēz). **1.** The vertical axis of a two-dimensional Cartesian coordinate system. **2.** One of three axes in a three-dimensional Cartesian coordinate system.

**Yb** *symbol for* YTTERBIUM.

**Y-chro·mo·some** (wī′krō′mə-sōm′) *n.* The sex chromosome associated with male characteristics, occurring with one X-chromosome in the male sex-chromosome pair.

**y·clept** *also* **y·cleped** (ĭ-klĕpt′, ĭ-klēpt′) *v.* [ME *ycleped* < OE *geclepod,* p.part. of *cleopian, clepian,* to call.] *p.p. of* CLEPE.

▲ word history: The form *yclept* meaning "named, called," preserves an archaic feature of the English verb. In the early Germanic languages, as in modern German, the prefix *ge-* was added to all past participle forms. In Old English, for example, the verb *clepian,* "to call," had a past participle *geclepod,* "called," which is the ancestor of *yclept.* In Old English g before an *e* was pronounced like *y,* and because the prefix *ge-* was unstressed, it was reduced to short *i* before being lost altogether. In Modern English *yclept* is a self-conscious archaism or poetic word.

**ye[1]** (thē) *def. art.* [Alteration of OE *þē,* from the use of *y* for *þ* (thorn) by the early printers.] The.

▲ word history: The word *ye*[1], which is sometimes used in pseudo-archaic phrases such as "ye olde curiosity shoppe," is actually a variant spelling of the definite article *the.* The variant arose from a confusion of the lower-case letter Y with the letter þ, called *thorn,* which was used in medieval times to represent the sounds now spelled with *th.* In early printed English books the common manuscript abbreviations *ye* for *the* and *yt* for *that* were retained. The form *ye* as a living word has now died out.

**ye[2]** (yē) *pron.* [ME < OE *gē.*] YOU 1. —Used esp. in literary or religious contexts and in certain dialects of English.

**yea** (yā) *adv.* [ME < OE *gēa.*] **1.** Yes : aye. **2.** Indeed : truly <They have spoken, *yea,* shouted their reply.> —*n.* **1.** An affirmative statement or vote. **2.** One who votes affirmatively.

**yeah** *also* **yeh** (yĕ′ə, yă′ə, yā′ə) *adv.* [Var. of YEA.] *Informal.* Yes.

**yean** (yēn) *v.* **yeaned, yean·ing, yeans.** [ME *yenen* < OE **geēanian* : *ge-,* verb prefix + *ēanian,* to bear young.] —*vi.* To bear young. —Used of sheep and goats. —*vt.* To give birth to : BEAR.

**yean·ling** (yēn′lĭng) *n.* The young of a sheep or goat : a lamb or kid. —*adj.* Newly born : INFANT.

**year** (yîr) *n.* [ME *yere* < OE *gēar.*] **1. a.** The time period as measured by the Gregorian calendar in which the earth completes a single revolution around the sun, consisting of 365 days, 5 hours, 49 minutes, and 12 seconds of mean solar time divided into 12 months, 52 weeks, and 365 or 366 days, and beginning on Jan. 1 and ending on Dec. 31. **b.** A period approx. equal to a year in other calendars. **2.** Sidereal year. **3.** Tropical year. **4.** A period of approx. the duration of a calendar year <left home a *year* ago> **5.** A period equal to the calendar year but beginning on a different date <a fiscal *year*> **6.** A specific annual period of time, usu. less than a calendar year, devoted to a special activity <the academic *year*> **7. years.** Age, esp. old age <weighed down by *years*> **8. years.** An indefinitely long period of time <It's been *years* since you called.>

**year·book** (yîr′bo͝ok′) *n.* **1.** A documentary, memorial, or historical book published every year, containing data about the previous year. **2.** A yearly record or book published by the graduating class of a school or college.

**year-end** *also* **year·end** (yîr′ĕnd′) *n.* The end of a fiscal year.

**year·ling** (yîr′lĭng) *n.* **1.** An animal that is one year old or has not completed its second year. **2.** A thoroughbred racehorse one year old as reckoned from Jan. 1 of the year in which it was foaled. —*adj.* Being one year old.

**year·long** (yîr′lông′, -lŏng′) *adj.* Lasting one year.

**year·ly** (yîr′lē) *adj.* Occurring or appearing once a year or every year : ANNUAL. —*adv.* Once a year : ANNUALLY. —*n., pl.* **-lies.** A publication issued once a year.

**yearn** (yûrn) *vi.* **yearned, yearn·ing, yearns.** [ME *yernen* < OE *gyrnan.*] **1.** To have a strong desire or longing <*yearn* for solitude> **2.** To feel deep pity, sympathy, or tenderness.

**year-round** (yîr′round′) *adj.* Operating, active, or continuous throughout the year <*year-round* swimming>

**yea-say·er** (yā′sā′ər) *n.* **1.** One who is confidently affirmative in attitude. **2.** One who uncritically agrees : YES MAN.

**yeast** (yēst) *n.* [ME *yeest* < OE *gist.*] **1.** Any of various unicellular fungi of the genus *Saccharomyces* and related genera, reproducing by budding and capable of fermenting carbohydrates. **2.** Froth consisting of yeast cells along with the carbon dioxide they produce in the fermentation process, present in or added to fruit juices and other substances in production of alcoholic beverages. **3.** An either powdered or compressed commercial preparation containing yeast cells and inert material such as meal, used esp. as a leavening agent or as a dietary supplement. **4.** Foam : froth. **5.** An agent of ferment or activity. —*vi.* **yeast·ed, yeast·ing, yeasts. 1.** To ferment. **2.** To froth or foam.

**yeast·y** (yē′stē) *adj.* **-i·er, -i·est. 1.** Of, similar to, or containing yeast. **2.** Marked by ferment or agitation : RESTLESS. **3.** Full of vigor or exuberance. **4.** Frothy : frivolous. **—yeast′i·ly** *adv.* **—yeast′i·ness** *n.*

**yecch** *also* **yech** (yĕKH, yŭKH, yĕk, yŭk) *Slang.* —*interj.* —Used to express strong disgust or contempt. —*n.* Something disgusting. —*adj.* Disgusting : sickening. **—yech′y** *adj.*

**yegg** (yĕg) *n.* [Orig. unknown.] *Slang.* A burglar or safecracker.

**yeh** (yĕ′ə, yă′ə, yā′ə) *adv. var. of* YEAH.

**yell** (yĕl) *v.* **yelled, yell·ing, yells.** [ME *yellen* < OE *giellan.*] —*vi.* To cry out loudly, as in pain, fright, surprise, or enthusiasm. —*vt.* To shout. —*n.* **1.** A loud outcry. **2.** A rhythmic cheer chanted in unison by a group, as at a college football game. **—yell′er** *n.*

☆ ***syns:*** YELL, BAWL, BELLOW, CRY, HOLLER, SHOUT, VOCIFERATE, WHOOP *v. core meaning* : to utter (something) very loudly <The audience *yelled* its disapproval.>

**yel·low** (yĕl′ō) *n.* [ME *yelow* < OE *geolu.*] **1. a.** Any of a group of colors of a hue resembling that of ripe lemons and varying in lightness and saturation; the hue of that portion of the spectrum lying between green and orange; one of the psychological primary hues, evoked in the normal observer by radiant energy of wavelength approx. 580 nanometers; one of the subtractive primaries. **b.** A pigment or dye having this hue. **c.** Something that has this hue. **2.** An egg yolk. **3. yellows.** *(sing. in number).* Any of various plant diseases usu. caused by fungi of the genus *Fusarium* or viruses of the genus *Chlorogenus* and marked by yellowing of the leaves and stunted growth. —*adj.* **-er, -est. 1.** Of the color yellow. **2.** *Slang.* Cowardly. —*vt. & vi.* **-lowed, -low·ing, -lows.** To make or become yellow. **—yel′low·ness** *n.*

**yel·low·bark** (yĕl′ō-bärk′) *n.* Calisaya.

**yel·low-bel·lied** (yĕl′ō-bĕl′ēd) *adj.* **1.** Having a belly yellow or yellowish in color. **2.** *Slang.* Cowardly. **—yel′low-bel′ly** *n.*

**yellow birch** *n.* A North American tree, *Betula lutea,* with yellowish bark and hard light-colored wood used for furniture.

**yel·low·bird** (yĕl′ō-bûrd′) *n.* A yellow or predominantly yellow bird, as the goldfinch or the yellow warbler.

**yellow cake** *n.* The final precipitate formed in milling uranium ore.

**yellow card** *n.* A card raised by a referee in soccer, indicating a player's violation.

**yellow cypress** *n.* The Nootka cypress.

**yel·low-dog contract** (yĕl′ō-dôg′, -dŏg′) *n.* A now illegal employer-employee contract by which the employee agrees not to remain or become a member of a labor union while employed.

**yellow fever** *n.* An acute infectious tropical and subtropical disease caused by a filterable virus vectored by a mosquito of the genus *Aedes* and marked by high fever, jaundice, and dark-colored vomit resulting from hemorrhages.

**yellow flu** *n.* [So called because school buses are usually yellow.] The organized absence of students from school in protest of compulsory busing.

**yel·low·ham·mer** (yĕl′ō-hăm′ər) *n.* [By folk ety. < obs. *yelambre.*] **1.** A common North American woodpecker, *Colaptes auratus,* having yellow wing linings, a red nape, and a black breast band. **2.** A Eurasian bird, *Emberiza citrinella,* having brown and yellow plumage.

**yellowhammer**
*Emberiza citrinella,*
*6½ inches long*

**yel·low·ish** (yĕl′ō-ĭsh) *adj.* Rather yellow. **—yel′low·ish·ness** *n.*

**yellow jack** *n.* **1.** Yellow fever. **2.** *Naut.* A yellow flag hoisted to request pratique or to warn of disease on board. **3.** A silvery and golden food fish, *Caranx bartholomaei,* of western Atlantic and Caribbean waters.

**yellow jacket** *n.* Any of several small wasps of the family *Vespidae,* having yellow and black markings and usu. nesting in the ground.

**yellow journalism** *n.* [From the use of yellow ink in printing "Yellow Kid," a cartoon strip in the *New York World,* a newspaper noted for sensationalism.] Journalism that exploits, distorts, or sensationalizes the news to attract readers.

**yel·low·legs** (yĕl′ō-lĕgz′) *n., pl.* **yellowlegs.** A North American wading bird, *Totanus melanoleucus* or *T. flavipes,* with yellow legs and a long narrow bill.

**yellow ocher** *n.* **1.** A yellow pigment usu. containing limonite. **2.** A moderate orange with yellow overtones.

**yellow pages** *or* **Yellow Pages** *pl.n.* [So called because they are usu. printed on yellow paper.] A section of a telephone directory that lists businesses, services, or products alphabetically according to field.

**yellow peril** *or* **Yellow Peril** *n.* The threatened expansion of the Oriental peoples as magnified in the Western imagination.

**yellow pine** *n.* **1.** A North American evergreen tree having yellowish wood, as *Pinus echinata* of the southeastern United States or the ponderosa pine. **2.** The wood of the yellow pine.

**yellow poplar** *n.* The tulip tree.

**yellow-shafted flicker** (yĕl′ō-shăf′tĭd) *n.* YELLOWHAMMER 1.

**yellow sheet** *n. Slang.* A criminal record : RAP SHEET.

**yellow spot** *n.* The macula lutea.

**yellow streak** *n.* A proneness to cowardice and disloyalty.

**yel·low·tail** (yĕl′ō-tāl′) *n.* **1.** A marine game fish, *Seriola dorsalis* of coastal waters of southern California and Mexico. **2.** A fish other than the yellowtail that has a yellowish tail, as the mademoiselle.

**yel·low·throat** (yĕl′ō-thrōt′) *n.* A small New World bird of the genus *Geothlypis,* esp. *G. trichas,* having a brownish back, a yellow throat, and, in the male, a black facial mask.

**yellow warbler** *n.* A small New World bird, *Dendroica petechia,* with predominantly yellow plumage.

**yel·low·weed** (yĕl′ō-wēd′) *n.* A plant, as the dyer's rocket, having yellow flowers.

**yel·low·wood** (yĕl′ō-wo͝od′) *n.* **1.** A tree, *Cladrastis lutea* of the southeastern United States, with compound leaves, drooping white flower clusters, and yellow wood yielding a yellow dye. **2.** A tree having yellow wood. **3.** The wood of the yellowwood.

**yel·low·y** (yĕl′ō-ē) *adj.* Yellowish.

**yelp** (yĕlp) *v.* **yelped, yelp·ing, yelps.** [ME *yelpen,* to cry aloud < OE *gielpan,* to boast.] —*vi.* **1.** To utter a short, sharp bark or cry. **2.** To cry out sharply, as in pain or surprise. —*vt.* To utter by yelping. —*n.* A short, sharp cry or bark. **—yelp′er** *n.*

☆ ***syns:*** YELP, SQUEAL, YAP, YIP *n. core meaning* : a short, shrill cry <gave a *yelp* of surprise>

**yen[1]** (yĕn) [Cantonese *yan.*] *vi.* **yenned, yen·ning, yens.** To yearn. —*n.* A strong desire or craving.

**yen[2]** (yĕn) *n., pl.* **yen.** [J. *en* < Chin. (Mandarin) *yuan²,* dollar.] —See table at CURRENCY.

**yen·ta** (yĕn′tə) *n.* [Yiddish *yente* < the name *Yente.*] *Slang.* A prying, gossipy woman.

**yeo·man** (yō′mən) *n., pl.* **-men.** [ME *yoman,* perh. contraction of *yong man,* young man.] **1.** An independent farmer, esp. a member of a former class of small freeholding farmers in England. **2.** A yeoman of the guard. **3.** An attendant, servant, or lesser official in a royal or noble household. **4.** A petty officer performing chiefly clerical duties in the U.S. Navy. **5.** An assistant, as of a sheriff. **6.** A diligent, dependable worker. —*adj. also* **yeo·man·ly** (-lē). **1.** Relating to or ranking as a yeoman. **2.** Befitting a yeoman, as in diligence or steadfastness <did a *yeoman* job>

**yeoman of the guard** *n.* A member of a ceremonial guard attending the British sovereign and royal family.

**yeo·man·ry** (yō′mən-rē) *n.* **1.** The class of yeomen : small farmers. **2.** A British volunteer cavalry force organized in 1761 to serve as a home guard and later incorporated into the Territorial Army.

**yep** (yĕp) *adv.* [Alteration of YES.] *Slang.* Yes.

**yer·ba ma·té** (yâr′bə mä-tā′, yûr′bə) *n.* [Am. Sp. : *yerba,* herb (< Lat. *herba*) + *maté,* maté. —see MATÉ.] MATÉ 2.

**Yerk·ish** (yûr′kĭsh) *n.* [After the *Yerkes* Regional Primate Center in Georgia.] An artificial language using geometric forms to represent words that was created for communication between chimpanzees and humans.

**yes** (yĕs) *adv.* [ME < OE *gese.*] It is so : as you say or ask. —Used to express affirmation, agreement, positive confirmation, or consent. —*n., pl.* **yes·es. 1.** An affirmative or consenting reply. **2.** An affirmative vote or voter. —*vt.* **yessed, yes·sing, yes·es.** To give an affirmative reply to.

**ye·shi·va** *or* **ye·shi·vah** (yə-shē′və) *n.* [Heb. *yĕshībhāh* < *yāshābh,* he sat down.] **1.** An Orthodox Jewish institute of learning where students study the Talmud, esp. a rabbinical seminary. **2.** An elementary or secondary school with a curriculum that includes Jewish religion and culture as well as general education.

**yes man** *n. Informal.* One who slavishly agrees with a superior : SYCOPHANT.

**yester-** *pref.* [ME *yister-* < OE *giestran.*] Yesterday <*yester*morning>

**yes·ter·day** (yĕs′tər-dā′, -dē) *n.* [ME < OE *giestran dæg* : *giestran,* yesterday + *dæg,* day.] **1.** The day before the present day. **2.** *often* **yesterdays.** Time in the immediate past. —*adv.* **1.** On the day before the present day. **2.** A short while ago <*Yesterday* we were young.>

**yes·ter·eve·ning** (yĕs′tər-ēv′nĭng) *also* **yes·ter·eve** (-ēv′) *or* **yes·ter·e·ven** (-ē′vən) *n.* The evening of yesterday. **—yes′ter·eve′ning** *adv.*

**yes·ter·morn·ing** (yĕs′tər-môr′nĭng) *also* **yes·ter·morn** (-môrn′) *n.* The morning of yesterday. **—yes′ter·morn′ing** *adv.*
**yes·ter·night** (yĕs′tər-nīt′) *n.* Last night. **—yes′ter·night′** *adv.*
**yes·ter·year** (yĕs′tər-yîr′) *n.* **1.** Last year. **2.** Times past : YORE. **—yes′ter·year′** *adv.*
**yes·treen** (yĕs-trēn′) *n. Scot. var. of* YESTEREVENING.
**yet** (yĕt) *adv.* [ME < OE *gīet.*] **1.** At this time : NOW <Don't leave *yet.*> **2.** Up to the present or some specified time : thus far <The rain has not *yet* started.> **3.** In the time remaining : STILL <There is *yet* a possibility of success.> **4.** In addition : BESIDES <We made *yet* another suggestion.> **5.** Still more : EVEN <a *yet* harder course> **6.** Nevertheless <brave *yet* foolhardy> **7.** At some future time : EVENTUALLY <You may *yet* regret your decision.> *—conj.* And despite this : NEVERTHELESS <They promise to visit, *yet* they never do.> **—as yet.** Up to the present time <no word *as yet*>
**ye·ti** (yĕt′ē) *n.* [Alteration of Tibetan *miti* : *mi*, person + *ti*, a kind of animal.] Abominable snowman.
**yew** (yo͞o) *n.* [ME *ew* < OE *īw.*] **1.** Any of several evergreen trees or shrubs of the genus *Taxus*, of which the flat, dark-green needles and often the scarlet berries are poisonous. **2.** The wood of a yew, esp. the fine-grained, durable wood of an Old World species, *T. baccata*, used in cabinetmaking and for archery bows.
**yé-yé** (yā′yā′) *adj.* [Fr. < E. *yeah, yeah*, a phrase used in rock 'n' roll music.] *Slang.* Of, relating to, or featuring French rock 'n' roll.
**Ygg·dra·sil** *also* **Yg·dra·sil** (ĭg′drə-sĭl, ŭg′-) *n.* [ON.] *Norse Myth.* A great ash tree that holds together earth, heaven, and hell by its roots and branches.
**YHWH** (yä′wā) *also* **YHVH** (yä′vā) *n.* The Hebrew Tetragrammaton representing the name of God.
**Yid·dish** (yĭd′ĭsh) *n.* [Yiddish *Yidish* < MHG *jüdisch (diutsch)*, Jewish (German) < *Jüde*, Jew < OHG *judo* < Lat. *Judaeus.* —see JEW.] A High German language with many borrowings from Hebrew and Slavic that is written in Hebrew characters and spoken chiefly as a vernacular in eastern European Jewish communities and by emigrants from these communities worldwide. **—Yid′dish** *adj.* **—Yid′dish·ism** *n.*
**yield** (yēld) *v.* **yield·ed, yield·ing, yields.** [ME *yielden* < OE *gieldan*, to pay.] *—vt.* **1.** To give forth by a natural process, esp. by cultivation <a vine that *yields* grapes> **2.** To furnish or give in return : PRODUCE <an investment *yielding* high interest> **3.** To surrender (something) in deference or defeat : RELINQUISH <*yielded* the game to the opponent> **4.** To grant or concede as due <*yield* the right of way> *—vi.* **1.** To furnish or give a return : be productive. **2.** To give up : SUBMIT. **3.** To give way to pressure, force, or persuasion <*yielded* to our demands> **4.** To give way to someone or something superior <an incorrect premise that *yielded* to fact> **5.** To break, bend, or give way under physical pressure <The dam *yielded.*> *—n.* **1.** The amount yielded or produced : PRODUCT. **2.** Profit obtained from investment : RETURN. **3.** The energy released by an explosion, esp. by a nuclear explosion, expressed in units of weight of TNT required to produce an equivalent release <a 100-megaton *yield*> **—yield′er** *n.*
☆ ***syns:*** YIELD, BOW, CAPITULATE, FOLD, SUBMIT, SUCCUMB, SURRENDER *v. core meaning* : to give in from or as if from gradual loss of strength <*yielded* to the better player> ***ant:*** withstand
**yield·ing** (yēl′dĭng) *adj.* Inclined to yield or give way : SUBMISSIVE. **—yield′ing·ly** *adv.* **—yield′ing·ness** *n.*
**yin** (yĭn) *n.* [Chin. (Mandarin) *yin*[1], moon, shade, femininity.] The passive female cosmic principle in Chinese dualistic philosophy.
**yip** (yĭp) *n.* [Imit.] A sharp high-pitched bark. *—vi.* **yipped, yip·ping, yips.** To emit sharp high-pitched barks.
**yipe** (yīp) *also* **yipes** (yīps) *interj.* —Used to express surprise, fear, or dismay.
**yip·pee** (yĭp′ē) *interj.* —Used to express joy or elation.
**-yl** *suff.* [Fr. *-yle* < Gk. *hulē*, wood, matter.] A chemical radical <carbon*yl*>
**y·lang-y·lang** (ē′läng-ē′läng) *n.* [Tagalog *ilang-ilang.*] **1.** A tropical Asian tree, *Cananga odorata* or *Canangium odoratum*, having fragrant greenish-yellow flowers yielding an oil used in perfumery. **2.** An oil or perfume obtained from the flowers of the ylang-ylang.
**y·lem** (ī′ləm) *n.* [ME, universal matter < OFr. *ilem* < Med. Lat. *hylem*, accusative of *hyle*, matter < Gk. *hulē.*] A form of matter hypothesized by proponents of the big bang theory to have existed before the formation of the chemical elements.
**yock** (yŏk, yŭk) [Imit.] *Slang.* *—vi.* **yocked, yock·ing, yocks.** To laugh or joke, esp. in a rowdy way. *—n.* A joke : laugh <"it contains a few *yocks*, but the humor . . . never emerges" —*Variety*>
**yod** *also* **yodh** (yŏd, yo͝od) *n.* [Heb *yōdh* < *yādh*, hand.] The tenth letter of the Hebrew alphabet. —See table at ALPHABET.
**yo·del** (yōd′l) *v.* **-deled, -del·ing, -dels** *or* **-delled, -del·ling, -dels.** [G. *jodeln.*] *—vi.* To sing so that the voice fluctuates between the normal chest voice and a falsetto. *—vt.* To sing (a song) in a yodeling manner. **—yo′del** *n.* **—yo′del·er** *n.*
**yodh** (yŏd, yo͝od) *n. var. of* YOD.
**yo·ga** (yō′gə) *n.* [Skt. *yogaḥ*, union, yoking.] **1.** *often* **Yoga.** A Hindu discipline aimed at training the consciousness for a state of perfect spiritual insight and tranquillity. **2.** A system of exercises practiced to promote control of the body and mind.
▲ word history: The word *yoga* is from Sanskrit *yogaḥ*; "union; the uniting of the self with the universe." This word literally meant "a yoking together" and is descended from the Indo-European root *yeug-*, "to join, yoke." *Yeug-* descended into Germanic as *yuk*, whose Old English form is *geoc*, the ancestor of Modern English *yoke.*
**yogh** (yōKH) *n.* [ME, poss. < *yok*, yoke < OE *geoc* (from its shape).] The letter ȝ, representing in Middle English a velar or palatal fricative or the sound of *w* between vowels.
**yo·ghurt** *or* **yo·ghourt** (yō′gərt) *n. vars. of* YOGURT.
**yo·gi** (yō′gē) *n., pl.* **-gis.** [Hindi < Skt. *yogī.*] A practitioner of yoga. **—yo′gic** *adj.*
**yo·gurt** *also* **yo·ghurt** *or* **yo·ghourt** (yō′gərt) *n.* [Turk. *yoğurt.*] A custardlike food prepared from milk curdled by bacteria, esp. *Lactobacillus bulgaricus* and *Streptococcus thermophilus*, often sweetened or flavored with fruit.
**yo·him·bine** (yō-hĭm′bēn′) *n.* [NLat. *yohimbe*, specific epithet of *Corynanthe yohimbe*, species of tree from which it is derived, of Bantu orig.] A poisonous alkaloid, $C_{21}H_{26}N_2O_3$, derived from the bark of a tree, *Corynanthe yohimbe*, and once used as an aphrodisiac, local anesthetic, and mydriatic.
**yoicks** (yoiks) *interj.* —Used as a hunting cry to urge the hounds after the fox.
**yoke** (yōk) *n.* [ME *yok* < OE *geoc.*] **1.** A wooden framework for harnessing together oxen, mules, or other draft animals, consisting of a crossbar with two U-shaped pieces that encircle the necks of the animals. **2.** *pl.* **yoke** *or* **yokes.** A pair of draft animals joined by a yoke or trained to work together. **3.** A frame or crossbar designed to be carried across a person's shoulders with equal loads suspended from each end. **4.** A bar used with a double harness to connect the collar of each horse to the tongue of a wagon or coach. **5.** *Naut.* A crossbar on a ship's rudder to which the steering cables are connected. **6.** A clamp or vise that holds a machine part in place or controls its movement or that holds two such parts together. **7.** A part of a garment that is closely fitted, either around the neck and shoulders or at the hips, and from which an unfitted or gathered part of the garment is hung. **8.** Something that connects or joins together : BOND. **9.** *Electron.* A series of two or more magnetic recording heads fastened securely together for playing or recording on more than one track simultaneously. **10.** An arch made of two upright spears supporting a horizontal spear, under which conquered enemies of ancient Rome were forced to march in submission. **11.** A form or symbol of subjugation or bondage. *—v.* **yoked, yok·ing, yokes.** *—vt.* **1.** To fit or join with a yoke. **2. a.** To harness a draft animal to. **b.** To harness (a draft animal) to something. **3.** To connect, join, or bind together. **4.** To force into bondage or servitude. *—vi.* To become connected, joined, or bound together.

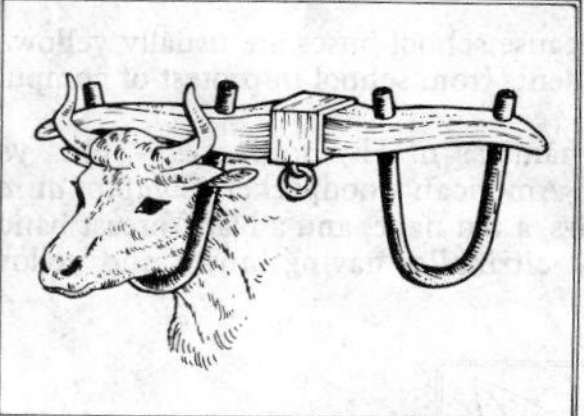
**yoke**

**yo·kel** (yō′kəl) *n.* [Orig. unknown.] A naive or gullible rustic.
**yo·ko·zu·na** (yō′kō-zo͞o-nä′) *n.* [J.] A champion sumo wrestler.
**yolk** (yōk) *n.* [ME *yolke* < OE *geoloca* < *geolu*, yellow.] **1.** The nutritive material of an ovum, consisting chiefly of protein and fat, esp. the yellow, usu. spheroidal mass of the egg of a bird or reptile, surrounded by the albumen. **2.** A greasy substance found in unprocessed sheep's wool. **—yolk′y** *adj.*
▲ word history: The word *yolk*, meaning "the yellow part of an egg," is descended from Old English *geoloca. Geoloca* is a noun derived from the adjective *geolu*, the ancestor of *yellow. Yolk* and *yellow* are related to many other words denoting something bright, shining, or yellow-colored, such as *glass, gloaming, gleam*, and *gold.*
**yolk sac** *n.* A membranous sac attached to the embryo and providing early nourishment in the form of yolk in bony fishes, sharks, reptiles, birds, and primitive mammals, and functioning as the circulatory system of the human embryo before initiation of internal circulation by the pumping of the heart.
**Yom Kip·pur** (yŏm′ kĭp′ər, yōm′ kĭ-po͞or′) *n.* [Heb. *yōm kippūr*, day of atonement.] The holiest Jewish holiday, observed on the tenth day of Tishri, on which fasting and prayer for the atonement of sins are prescribed.
**†yon** (yŏn) *adj. & adv.* [ME < OE *geon.*] Yonder. *—pron. Regional.* That one or those yonder.

**yond** (yŏnd) *adj. & adv.* [ME < OE *geond.*] *Archaic.* Yonder.

**yon·der** (yŏn′dər) *adj.* [ME < *yond,* yond.] Being at an indicated distance, usu. within sight. *—adv.* In or at that indicated place : over there. *—pron.* One that is at an indicated place, usu. within sight.

**yo·ni** (yō′nē) *n.* [Skt. *yoniḥ,* womb, abode, source.] A symbol for the vulva in Indian and Tibetan religion.

**yoo-hoo** (yo͞o′ho͞o) *interj.* —Used to hail persons or to attract attention.

**yore** (yôr, yōr) *n.* [ME < OE *gēara,* long ago < *gēar,* year.] Time long past <days of *yore*>

**York·ist** (yôr′kĭst) *n.* A supporter of the royal house of York against the house of Lancaster during the Wars of the Roses. *—adj.* Of or relating to the royal house that ruled in England from 1461 to 1485.

**York·shire pudding** (yôrk′shîr′, -shər) *n.* A pudding of popover batter made of eggs, flour, and milk and baked in beef drippings.

**Yorkshire terrier** *n.* A toy terrier usu. bred in Yorkshire, England, having a long bluish-gray coat.

**Yo·ru·ba** (yō′ro͞o-bä) *n., pl.* **Yoruba** *or* **-bas. 1.** A member of a West African Negro people living primarily in southwestern Nigeria. **2.** The Kwa language of the Yoruba. **—Yo′ru·ban** *adj.*

**you** (yo͞o) *pron.* [ME < OE *ēow,* dative and accusative of *gē,* ye.] **1.** The one or ones being addressed by the speaker. —Used in all grammatical relations except that of the possessive <*You* ought to try harder.><I'll see *you* next week.> **2.** —Used to indicate an individual of an indefinitely specified group <*You* can't win for losing.>

**†you-all** (yo͞o-ôl′) *also* **y'all** (yôl) *pron. Southeastern U.S.* You. —Used in addressing two or more persons or referring to two or more persons, one of whom is addressed.

**you'd** (yo͞od). **1.** You had. **2.** You would.

**you'll** (yo͞ol, yo͞o′əl, yo͝ol). **1.** You will. **2.** You shall.

**young** (yŭng) *adj.* **-er, -est.** [ME *yong* < OE *geong.*] **1.** Being in the early or undeveloped period of life or growth. **2.** Newly begun or formed <The night is *young.*> **3.** Of or relating to youth or early life <in my *young* years> **4.** Having the vigor or freshness of youth : YOUTHFUL. **5.** Lacking experience : IMMATURE <a *young* outlook> **6.** Being the junior of two people having the same name. **7.** *Geol.* Being of an early stage in a geologic cycle. —Used of bodies of water and land formations. *—n.* **1.** Young persons as a group : YOUTH. **2.** Offspring : brood <seals and their *young*> **—with young.** Pregnant. **—young′ness** *n.*

☆ ***syns:*** YOUNG, ADOLESCENT, IMMATURE, JUVENILE, YOUTHFUL *adj. core meaning* : being between childhood and adulthood. YOUNG and YOUTHFUL are the most general <the *young*—or *youthful*—hero>; both also suggest the freshness and vigor associated with youth <*young* for your age><a *youthful* face> ADOLESCENT, JUVENILE, and IMMATURE stress immaturity <*adolescent* attitudes><*juvenile* behavior><*immature* jokes> ***ant:*** adult

**young·ber·ry** (yŭng′bĕr′ē) *n.* [After B. M. *Young,* 20th-cent. American fruit grower.] **1.** A prickly trailing hybrid between a blackberry and a dewberry, cultivated in the western United States. **2.** The edible, dark-red berry of the youngberry.

**young·ish** (yŭng′ĭsh) *adj.* Rather young.

**young·ling** (yŭng′lĭng) *n.* [ME *yongling* < OE *geongling* < *geong,* young.] **1.** A young person. **2.** A young animal. **3.** A young plant.

**young·ster** (yŭng′stər) *n.* **1.** A young person : YOUTH. **2.** A young animal. **3.** A midshipman of the second-year class in the U.S. Naval Academy.

**Young Turk** *n.* [After the *Young Turks,* a 20th-cent. revolutionary party in Turkey.] A progressive or insurgent member of a political party or other collective enterprise.

**youn·ker** (yŭng′kər) *n.* [Du. *jonker* < MDu. *jonckher,* young nobleman : *jonc,* young + *here,* lord.] **1.** A young man. **2.** A child.

**your** (yo͝or, yôr, yōr; yər *when unstressed) adj.* [ME < OE *ēower,* genitive of *gē,* ye.] **1.** —Used to indicate that the person addressed is the possessor or the agent or recipient of an action <*your* pen><*your* letters><*your* responsibilities> **2.** Of or pertaining to one or oneself <The light switch is right there on *your* left.> **3.** *Informal.* —Used with little or no sense of possession but suggestive of mutual knowledge or experience <I am not one of *your* business school whiz kids.>

**you're** (yo͝or; yər *when unstressed).* You are.

**yours** (yo͝orz, yôrz, yōrz) *pron. (sing. or pl. in number).* [ME *youres,* genitive of *your,* your.] **1.** That or those belonging to you <I can't find my hat, so I'll borrow *yours.*><My books are here and *yours* are over there.> **2.** —Often used with an adverbial modifier in the complimentary close of a letter <*Yours* sincerely,>

**your·self** (yo͝or-sĕlf′, yôr′-, yōr-, yər-) *pron.* **1.** That one identical with you. —Used: **a.** Reflexively as the direct or indirect object of a verb or the object of a preposition <Don't exert *yourself.*><Give *yourself* sufficient time.><Are you mumbling to *yourself?*> **b.** For emphasis <You *yourself* must rely on them.> **c.** In an absolute construction <*Yourself* a victim of fraud, you can certainly understand how we feel.> **2.** Your normal or healthy condition or state <You are not *yourself* today.> **3.** Oneself <Don't hurt *yourself.*>

**your·selves** (yo͝or-sĕlvz′, yōr-, yər-) *pl.pron.* **1.** Those that are identical with you. —Used: **a.** Reflexively as the direct or indirect object of a verb or the object of a preposition <Help *yourselves.*><Have *yourselves* a merry old time.><You should all watch out for *yourselves.*> **b.** For emphasis <You should solve the problem *yourselves.*> **c.** In an absolute construction. **2.** Your normal or healthy condition or state.

**youth** (yo͞oth) *n., pl.* **youths** (yo͞oths, yo͞othz) [ME < OE *geoguð.*] **1.** The quality or state of being young. **2.** An early period of development or existence. **3. a.** The time of life between childhood and maturity. **b.** Young people as a group. **c.** A young person, esp. a young man between childhood and maturity.

**youth·ful** (yo͞oth′fəl) *adj.* **1.** Possessing youth : still young. **2.** Characteristic of youth : FRESH <a *youthful* attitude> **3.** Of or belonging to youth. **4.** Being in an early stage of development : NEW. **5.** *Geol.* Young. **—youth′ful·ly** *adv.* **—youth′ful·ness** *n.*

**youth hostel** *n.* HOSTEL 1.

**you've** (yo͞ov). You have.

**yowl** (youl) *v.* **yowled, yowl·ing, yowls.** [ME *yowlen.*] *—vi.* To utter a long loud mournful cry. *—vt.* To say or utter with a yowl. *—n.* A long loud mournful cry.

**yo-yo** (yō′yō′) *n., pl.* **-yos.** [Orig. a trademark.] **1.** A toy resembling a flattened spool wound in the center with a string by means of which it is spun down from the hand and reeled back up. **2.** *Informal.* One that vacillates. **3.** *Slang.* An incompetent, erratic, or foolish person. *—vi.* **-yoed, -yo·ing, -yos.** *Informal.* To move repeatedly from one position to another : VACILLATE.

**Y·quem** (ē-kĕm′) *n.* [After Chateau d'*Yquem,* an estate in southwestern France where it is made.] A sauterne wine.

**yt·ter·bi·a** (ĭ-tûr′bē-ə) *n.* [NLat. < YTTERBIUM.] Ytterbium oxide.

**yt·ter·bi·um** (ĭ-tûr′bē-əm) *n.* [After *Ytterby,* town in Sweden where it was discovered.] *Symbol* **Yb** A soft bright silvery rare-earth element used as an x-ray source for portable irradiation devices, in some laser materials, and in some special alloys; atomic number 70; atomic weight 173.04. **—yt·ter′bic** (-bĭk) *adj.*

**ytterbium oxide** *n.* A colorless hygroscopic compound, $Yb_2O_3$, used in certain alloys.

**yt·tri·a** (ĭt′rē-ə) *n.* [NLat., after *Ytterby.* —see YTTERBIUM.] Yttrium oxide.

**yt·tri·um** (ĭt′rē-əm) *n.* [< YTTRIA.] *Symbol* **Y** A silvery metallic element used to increase the strength of magnesium and aluminum alloys; atomic number 39; atomic weight 88.905. **—yt′tric** (ĭt′rĭk) *adj.*

**yttrium oxide** *n.* A yellowish powder, $Y_2O_3$, used in optical glasses, ceramics, and color television tubes.

**yu·an** (yü′än′) *n., pl.* **yuan** *or* **yuans.** [Chin. (Mandarin) *yuan*$^2$, dollar.] —See table at CURRENCY.

**Yu·ca·tec** (yo͞o′kə-tĕk′) *n., pl.* **Yucatec** *or* **-tecs. 1.** A member of an Indian people inhabiting the Yucatán Peninsula, Mexico. **2.** The Mayan language of the Yucatec.

**yuc·ca** (yŭk′ə) *n.* [Sp. *yuca,* of American Indian orig.] Any of various chiefly tropical New World plants of the genus *Yucca,* having tall, often rigid stems and a terminal cluster of white flowers.

**Yu·ga** (yo͝og′ə) *also* **Yug** (yo͝og) *n.* [Skt. *yugam,* yoke, pair, era.] One of the four ages constituting the cycle of history in Hinduism.

**Yukon Time** *n.* Time at the 135th meridian west of Greenwich, England, and in the ninth time zone based on it in North America, nine hours earlier than Greenwich time.

**yu·lan** (yo͞o′län, yü′län′) *n.* [Chin. (Mandarin) *yu*$^4$ *lan*$^2$ : *yu*$^4$, jade + *lan*$^2$, orchid.] A tree, *Magnolia denudata,* native to China and often cultivated for its large cup-shaped fragrant white flowers.

**Yule** (yo͞ol) *n.* [ME *yole* < OE *gēol.*] Christmas.

▲ word history: The word *Yule,* in Old English *gēol,* was originally the name of a pagan German religious festival held in midwinter. After the English were converted to Christianity, the name *Yule* was used for the feast of Christmas. The use of native words for Christian terms was encouraged in the Anglo-Saxon church; another example is the use of the name *Easter* for the feast of Christ's resurrection.

**yule log** *n.* A large log traditionally burned in the fireplace at Christmas time.

**Yule·tide** (yo͞ol′tīd′) *n.* The Christmas season.

**Yu·ma** (yo͞o′mə) *n., pl.* **Yuma** *or* **-mas.** [Sp.] **1. a.** A tribe of Indians of southwestern Arizona and the adjacent parts of California and Mexico. **b.** A member of this tribe. **2.** The Yuman language of the Yuma.

**Yu·man** (yo͞o′mən) *n.* A language family comprising the languages of the Yuma and Mohave Indians and other Indian languages of southwestern Arizona and the adjacent parts of California and Mexico. **—Yu′man** *adj.*

**yum·my** (yŭm′ē) *adj.* **-mi·er, -mi·est.** [< *yum,* the sound of smacking the lips.] *Slang.* Delicious : delightful.

**yup·pie** (yŭp′ē) *n.* [Abbr. of *young urban professional* + -IE.] *Informal.* A young, upwardly mobile urban professional person.

**yurt** (yûrt) *n.* [R. *yurta,* of Turkic orig.] A circular domed portable tent used by the nomadic Mongols of Siberia.

**y·wis** (ĭ-wĭs′) *adv. Archaic. var. of* IWIS.

o͞o **boot** ou **out** th **thin** *th* **this** ŭ **cut** ûr **urge** y **young** yo͞o **abuse** zh **vision** ə **about,** **item,** **edible,** **gallop,** **circus**

# Zz

**z** or **Z** (zē) *n., pl.* **z's** or **Z's. 1.** The 26th letter of the English alphabet. **2.** A speech sound represented by the letter z. **3.** The 26th in a series. **4.** Something shaped like the letter Z.

**za·glio·ne** (zä'bəl-yō'nē) *n.* [Ital.] A mixture of egg yolks, sugar, and wine beaten until thick and foamy and served hot or cold as a dessert or as a sauce for puddings or fruit.

**zaf·fer** *also* **zaf·fre** (zăf'ər) *n.* [Ital. *zaffera* < OFr. *safre* < Ar. *ṣufr,* yellow copper.] An impure oxide of cobalt, used to produce a blue color in enamel and in making smalt.

**zaf·tig** or **zof·tig** (zäf'tĭk, -tĭg) *adj.* [Yiddish, juicy < G. *saftig* < *Saft,* juice.] *Slang.* **1.** Full-bosomed. **2.** Comfortably and pleasingly plump.

**Za·greus** (zā'grōōs, -grē-əs) *n.* [Gk.] *Gk. Myth.* The son of Zeus and Persephone who was slain by the Titans and reborn as Dionysus.

**zai·bat·su** (zī'bät-sōō') *n., pl.* **zaibatsu.** [J. : *zai,* wealth (< Chin. *cai²*) + *batsu,* powerful person or family (< Chin. *fa²*).] A powerful family-controlled commercial combine of Japan.

**zai·kai** (zī'kī') *n.* [J. : *zai,* money + *-kai,* community.] The Japanese commercial and financial community.

**zaire** (zîr, zä-îr') *n.* [Port. < Kongo *nzadi,* large river.] —See table at CURRENCY.

**za·mi·a** (zā'mē-ə) *n.* [NLat. *Zamia,* genus name, prob. < Gk. *azainein,* to dry.] Any of various chiefly tropical American cycads of the genus *Zamia,* with a short thick trunk and palmlike terminal leaves.

**zam·in·dar** *also* **zem·in·dar** (zăm'ən-där', zĕm'-, zə-mēn-där') *n.* [Hindi *zamīndār* < Pers. : *zamīn,* earth + *-dār,* holder.] **1.** An official during the Mogul empire in India assigned to collect the land taxes of his district. **2.** A native landholder in British colonial India responsible for collecting and paying to the government the taxes on the land under his jurisdiction.

**zam·in·dar·i** *also* **zem·in·dar·y** (zăm'ən-där'ē, zĕm'-, zə-mēn-) *n., pl.* **-is** *also* **-ies.** [Hindi *zamīndārī* < Pers. *zamīndār,* zamindar.] **1.** The system of tax collection by zamindars. **2.** The area administered by a zamindar.

**za·na·na** (zə-nä'nə) *n. var. of* ZENANA.

**za·ny** (zā'nē) *n., pl.* **-nies.** [Ital. *zani,* buffoon, < *Zanni,* dial. var. of *Gianni,* nickname for *Giovanni,* John, the name of servants who act as clowns in commedia dell'arte.] **1.** A ludicrous, buffoonish character in old comedies who mimics ineptly the tricks of the clown. **2.** A comical person given to extravagant or outlandish behavior. —*adj.* **-ni·er, -ni·est. 1.** Ludicrously comical : CLOWNISH. **2.** Crazy. **—za'ni·ly** *adv.* **—za'ni·ness** *n.*

**zap** (zăp) [Imit.] *Slang.* —*v.* **zapped, zap·ping, zaps.** —*vt.* **1.** To destroy or kill with or as if with a burst of gunfire, flame, or electric current. **2.** To strike, stun, or propel suddenly and forcefully <"His . . . narrative runs marvelously on and on, *zapping* the reader with often surprising and . . . painful glimpses" —*Publishers Weekly*> **3.** To attack (an enemy in warfare) with heavy firepower : strafe or bombard. —*vi.* To move swiftly : ZOOM. —*n.* Something very exciting or interesting. —*interj.* **1.** —Used to imitate the sound made by a gun. **2.** —Used to indicate a sudden occurrence.

**Za·pa·ta mustache** (sä-pä'tä, zə-pä'tə) *n.* [After Emiliano *Zapata* (1880?–1919).] A mustache curving downward on each side.

**za·pa·te·a·do** (zä'pä-tā-ä'dō) *n., pl.* **-dos.** [Sp. < *zapatear,* to tap with the shoe < *zapato,* shoe.] A Spanish flamenco dance marked by rhythmic stamping of the heels.

**Za·po·tec** (zä'pə-tĕk', sä'-) *n.* [Sp. *Zapoteca* < Nahuatl *Tzapoteca.*] Any of a group of related languages spoken in southern Mexico. **—Za'po·tec'** *adj.*

**zap·per** (zăp'ər) *n.* **1. a.** A device used for aiming radiation at a target. **b.** A device for consisting of fluorescent lights and an electrified grid for attracting insects and electrocuting them. **2.** *Slang.* Forceful, pointed criticism.

**zap·py** (zăp'ē) *adj.* **-pi·er, -pi·est.** *Slang.* Zippy.

**za·re·ba** *also* **za·ree·ba** (zə-rē'bə) *n.* [Ar. *zarībah,* pen for cattle, < *zarb,* sheepfold.] **1.** An enclosure of bushes or stakes protecting a campsite or village in northeastern Africa. **2.** A campsite or village protected by a zareba.

**zarf** (zärf) *n.* [Ar. *ẓarf,* container.] A chalicelike holder for a coffee cup, usu. made of ornamental metal, used in the Middle East.

**zas·tru·ga** (zə-strōō'gə) *n. var. of* SASTRUGA.

**zax** (zăks) *n.* [Alteration of ME *sax,* knife < OE *seax.*] A hatchetlike tool for cutting and dressing roofing slates.

**z-ax·is** (zē'ăk'sĭs) *n., pl.* **z-ax·es** (-sēz'). One of three axes in a three-dimensional Cartesian coordinate system.

**za·yin** (zä'yĭn, zī-) *n.* [Heb. < Aram.] The seventh letter of the Hebrew alphabet. —See table at ALPHABET.

**za·zen** (zä'zĕn') *n.* [J. : *za,* to sit down + *zen,* silent meditation. —see ZEN BUDDHISM.] Meditation as practiced in Zen Buddhism.

**zeal** (zēl) *n.* [ME *zele* < LLat. *zelus* < Gk. *zēlos.*] Enthusiastic, diligent devotion in pursuit of a cause, ideal, or goal : FERVOR.

**Zeal·ot** (zĕl'ət) *n.* [LLat. *zelotes* < Gk. *zēlōtēs* < *zēlos,* zeal.] **1.** A member of a fanatical Jewish sect that resisted Roman rule in Palestine during the 1st cent. A.D. **2. zealot. a.** One who is zealous, esp. excessively so. **b.** A fanatically committed person.

**zeal·ot·ry** (zĕl'ə-trē) *n.* Excessive zeal : FANATICISM.

**zeal·ous** (zĕl'əs) *adj.* Filled with or marked by zeal : FERVENT. **—zeal'ous·ly** *adv.* **—zeal'ous·ness** *n.*

**ze·bec** or **ze·beck** (zē'bĕk') *n. vars. of* XEBEC.

**ze·bra** (zē'brə) *n.* [Port. < OSp. *cebro, ecebro,* wild ass, poss. < Lat. *equiferus,* wild horse : *equus,* horse + *ferus,* wild.] **1.** Any of several horselike African mammals of the genus *Equus,* having conspicuous black or dark-brown stripes on a whitish body. **2.** *Slang.* A referee or other official in a football game. **—ze'brine** (zē'brīn') *adj.*

**zebra crossing** *n.* A pedestrian crosswalk marked with broad white stripes.

**zebra finch** *n.* A small Australian bird, *Poephila castanotis,* having black and white striped markings and often kept as a cage bird.

**zebra fish** *n.* A small freshwater tropical fish, *Brachydanio rerio,* of India, having horizontal dark-blue and silvery stripes and often kept in home aquariums.

**ze·bra·wood** (zē'brə-wŏŏd') *n.* **1.** Any of several African or tropical American trees having striped wood. **2.** The wood of the zebrawood, used in cabinetmaking.

**ze·bu** (zē'bōō, -byōō) *n.* [Fr. *zébu.*] A domesticated bovine mammal, *Bos indicus* of Asia and Africa, having a prominent hump on the back and a large dewlap.

**zebu**
*5½ feet high at shoulder*

**Zeb·u·lon** *also* **Zeb·u·lun** (zĕb'yə-lən) *n.* [Heb. *Zĕbhūlōn* < *zĕbhūl,* dwelling < *zābhal,* he dwelled.] **1.** A son of Jacob and Leah in the Old Testament. **2.** A tribe of Israel descended from Zebulon.

**zec·chi·no** (zĕ-kē'nō) *also* **zec·chin** or **zech·in** (zĕk'ĭn) *n., pl.* **-ni** (-nē) or **-nos** *also* **-chins** or **-ins.** [Ital. —see SEQUIN.] SEQUIN 2.

**Zech·a·ri·ah** (zĕk'ə-rī'ə) *n.* [Heb. *Zĕkhar'yah.* —see ZACHARIAS.] —See table at BIBLE.

**zed** (zĕd) *n.* [ME < OFr. *zede* < LLat. *zeta,* zeta < Gk. *zēta.* —see ZETA.] *Chiefly Brit.* The letter z.

**zed·o·ar·y** (zĕd'ō-ĕr'ē) *n., pl.* **-ies.** [ME *zeodoarye* < Med. Lat. *zeodoaria* < Ar. *zadwār* < Pers.] The dried rhizome of a tropical Asian plant, *Curcuma zedoaria,* used as a condiment and stimulant.

**zee** (zē) *n.* The letter z.

**Zee·man effect** (zā'măn') *n.* [After Pieter *Zeeman* (1865–1943).] The splitting of single spectral lines of an emission spectrum into three or more polarized components when the radiation source is in a magnetic field.

**ze·in** (zē'ĭn) *n.* [< NLat. *Zea,* corn genus < Gk. *zeia,* wheat.] A prolamine protein derived from corn and used in manufacturing plastics, coatings, and lacquers.

**Zeit·geist** (tsīt'gīst') *n.* [G. : *Zeit,* time + *Geist,* spirit.] The taste and outlook characteristic of a period : the spirit of the time.

**zek** (zĕk) *n.* [R. < *zaklyuchenny,* prisoner.] An inmate of a Soviet labor camp.

**zem·in·dar** (zăm′ən-där′, zĕm′-, zə-mēn-där′) *n. var. of* ZAMINDAR.

**zem·in·dar·y** (zăm′ən-där′ē, zĕm′-, zə-mēn-) *n. var. of* ZAMINDARI.

**zemst·vo** (zĕmst′vō) *n., pl.* **-vos.** [R. < *zemlya,* land.] An elective council responsible for the local administration of a provincial district in czarist Russia.

**Zen** (zĕn) *n.* Zen Buddhism.

**ze·na·na** *also* **za·na·na** (zə-nä′nə) *n.* [Hindi *zenāna* < Pers. < *zan,* woman.] The part of a house reserved for the women of the household in India and Pakistan .

**Zen Buddhism** *n.* [J. *zen* < Chin. (Mandarin) *chan²,* short for *chan² na⁴,* meditation < Pali *jhānaṃ* < Skt. *dhyānam* < *dhyāti,* he meditates.] A Chinese and Japanese school of Mahayana Buddhism that seeks enlightenment by meditation, self-contemplation, and intuition rather than by the scriptures. **—Zen Buddhist** *n.*

**Zend** (zĕnd) *n.* The Zend-Avesta.

**Zend-A·ves·ta** (zĕn′də-vĕs′tə) *n.* [Pers. *zandavastā* < *Avesta-vazend,* Avesta with an interpretation.] The entire body of sacred writings of the Zoroastrian religion. **—Zend′-A·ves·ta′ic** (-vĕ-stā′ĭk) *adj.*

**ze·ner diode** *or* **Ze·ner diode** (zē′nər) *n.* [After Clarence M. *Zener* (b. 1905).] A silicon semiconductor device used as a voltage regulator because of its ability to conduct heavy currents under reverse bias.

**ze·nith** (zē′nĭth) *n.* [ME *senith* < OFr. *cenith* < OSp. *zenit* < Ar. *samt (arra's),* path (over the head).] **1.** The point on the celestial sphere directly above the observer and diametrically opposite the nadir. **2. a.** The upper region of the sky. **b.** The highest point above the observer's horizon attained by a celestial body. **3.** The highest or culminating point : PEAK <at the *zenith* of one's career>

**ze·o·lite** (zē′ə-līt′) *n.* [Swed. *zeolit* < Gk. *zeein,* to boil (from its swelling and boiling under the blowpipe).] Any of a group of approx. 30 hydrous aluminum silicate minerals or their corresponding synthetic compounds, used chiefly as molecular filters and ion-exchange agents.

**Zeph·a·ni·ah** (zĕf′ə-nī′ə) *n.* [Heb. *Ṣĕphanyāh.*] —See table at BIBLE.

**zeph·yr** (zĕf′ər) *n.* [ME *Zephirus,* Zephyrus < Lat. *Zephyrus.*] **1. a.** The west wind. **b.** A gentle breeze. **2.** A light, soft fabric, yarn, or garment. **3.** Something airy, insubstantial, or transitory.

**zephyr lily** *n.* Any of several plants of the genus *Zephyranthes,* native to tropical America, with grasslike leaves and variously colored flowers.

**Zeph·y·rus** (zĕf′ər-əs) *n.* [Lat. < Gk. *Zephuros.*] *Gk. Myth.* A god personifying the gentle west wind.

**zep·pe·lin** *also* **Zep·pe·lin** (zĕp′ə-lĭn) *n.* [After Count Ferdinand von *Zeppelin* (1838–1917), its inventor.] A rigid airship having a long, cylindrical body supported by internal gas cells.

**ze·ro** (zîr′ō, zē′rō) *n., pl.* **-ros** *or* **-roes.** [Ital. < Arab. *ṣifr,* cipher.] **1.** The numerical symbol "0" : CIPHER. **2.** *Math.* **a.** An element of a set that when added to any other element in the set produces a sum identical with the element to which it is added. **b.** A cardinal number indicating the absence of any or all units under consideration. **c.** An ordinal number indicating an initial point or origin. **d.** An argument at which the value of a function vanishes. **3.** The temperature indicated by the numeral 0 on a thermometer. **4.** A sight setting that enables a firearm to shoot on target. **5.** One having no significance or importance : NONENTITY. **6.** The lowest point <Our hopes fell to *zero.*> **7.** Nothing : nil <Today they learned *zero.*> *—adj.* **1.** Of, relating to, or being zero. **2. a.** Having no measurable or otherwise determinable value. **b.** Absent, inoperative, or irrelevant in specified circumstances <*zero* gravity> **3. a.** Limited by cloud cover to little or no vertical visibility. **b.** Permitting little or no horizontal visibility. *—vt.* **-roed, -ro·ing, -roes.** To adjust (an instrument or device) to zero value. **—zero in. 1.** To aim or concentrate firepower on an exact target location. **2.** To adjust the aim or sight of by repeated firings. **3.** To aim or head for intently : close in <The guests *zeroed in* on the food.>

▲ word history: The word *zero,* "the numeral 0," is not only synonymous with the word *cipher* but is descended from the same Arabic source. The Arabic word *ṣifr* meant "empty" as well as "zero." *Zero* comes from the Italian form of *ṣifr, zefiro,* but *cipher* came into English through Medieval Latin *cifra.* Although *zero* was borrowed as a synonym of *cipher,* the two words have diverged in meaning. *Cipher* has developed the sense of "a code," whereas *zero* is used primarily as the name of the numeral, especially in scientific use.

**ze·ro-base** (zîr′ō-bās′, zē′rō-) *or* **ze·ro-based** (-bāst′) *adj.* Having each expenditure or item justified as to need or cost <"*zero-base* budgeting requires its practitioners to justify every dollar they spend" *—Wall Street Journal*>

**ze·ro-de·fect** (zîr′ō-dē′fĕkt′, -dĭ-fĕkt′, zē′rō-) *adj.* Devoid of flaw or error <a *zero-defect* computer run>

**zero economic growth** *n.* An economic condition marked by negligible increase in a nation's per capita income.

**zero gravity** *n.* A condition of apparent weightlessness that occurs when the centrifugal force on a body exactly counterbalances the gravitational attraction.

**zero growth** *n.* **1.** A policy that inhibits or prevents economic development and expansion. **2.** Zero economic growth. **3.** Zero population growth.

**zero hour** *n.* **1.** The time set for the start of an operation or action, esp. a military offensive. **2.** A critical or decisive point in time.

**ze·ro-point energy** (zîr′ō-point′, zē′rō-) *n.* The irreducible minimum energy possessed by a substance at absolute zero temperature.

**zero population growth** *n.* A rate of population increase limited to the number of live births needed to replace the existing population.

**ze·ro-rate** (zîr′ō-rāt′, zē′rō-) *vt.* **-rat·ed, -rat·ing, -rates.** *Chiefly Brit.* To exempt from paying a value-added tax.

**zero-sum** (zîr′ō-sŭm, zē′rō-) *adj.* Of, pertaining to, or being a situation wherein a gain for one side entails a concomitant loss for the other side.

**zest** (zĕst) *n.* [Obs. Fr., orange or lemon peel.] **1.** Flavor or interest : PIQUANCY. **2.** Spirited enjoyment or enthusiasm : GUSTO. **3.** The outermost part of the rind of an orange or lemon, used as flavoring. *—vt.* **zest·ed, zest·ing, zests.** To give zest, charm, or spirit to. **—zest′·ful** *adj.* **—zest′ful·ly** *adv.* **—zest′ful·ness** *n.*

☆ ***syns:*** ZEST, GUSTO, RELISH *n. core meaning* : spirited enjoyment <ate with *zest*>

**ze·ta** (zā′tə, zē′-) *n.* [Gk. *zēta,* of Phoenician orig.; akin to Heb. *zayin.*] The sixth letter of the Greek alphabet. —See table at ALPHABET.

**Ze·thus** *also* **Ze·thos** (zē′thəs) *n.* [Lat. < Gk. *Zēthos.*] *Gk. Myth.* The son of Zeus and twin brother of Amphion.

**zeug·ma** (zōōg′mə) *n.* [Lat. < Gk., a joining < *zeugnunai,* to yoke.] A figure of speech in which a verb or adjective is applied jointly to two nouns although logically appropriate to only one; e.g., *We changed our minds and our clothes.*

**Zeus** (zōōs) *n.* [Gk.] *Gk. Myth.* The principal god of the Greek pantheon, ruler of the heavens, and father of other gods and mortal heroes.

**zib·e·line** *or* **zib·el·line** (zĭb′ə-lēn′, -lĭn′) *n.* [OFr., sable < OItal. *zibellino,* of Slav. orig.] **1.** A thick, soft fabric of wool and other animal hair, as mohair, having a silky, lustrous nap. **2.** The sable or its fur.

**zib·et** *also* **zib·eth** (zĭb′ĭt) *n.* [Med. Lat. *zibethum* < Ar. *zabād,* civet.] A civet cat, *Viverra zibetha* of southeastern Asia.

**zig·gu·rat** (zĭg′ə-răt′) *n.* [Assyrian *ziqquratu,* summit.] A temple tower of the ancient Assyrians and Babylonians, built in the form of a terraced pyramid of successively receding stories.

**zig·zag** (zĭg′zăg′) *n.* [Fr., prob. < G. *Zickzack.*] **1. a.** A line or course that proceeds by short, sharp turns in alternating directions. **b.** One of a series of short, sharp alternate turns. **2.** Something having the shape of a series of sharp turns, as a design or pattern. *—adj.* Having or moving in a zigzag. *—adv.* In a zigzag manner or pattern. *—v.* **-zagged, -zag·ging, -zags.** *—vi.* To move in or form a zigzag. *—vt.* To cause to move in or form a zigzag.

**zig·zag·ger** (zĭg′zăg′ər) *n.* **1.** One that zigzags. **2.** A sewing-machine attachment for sewing zigzag stitches.

**zilch** (zĭlch) *—n.* [Orig. unknown.] *Slang.* **1.** Zero : nothing. **2.** One who is insignificant : NONENTITY. *—adj.* Amounting to nothing : NIL <"business was *zilch*" *—New York*>

**zill** (zĭll) *n.* [Perh. < Turk. *zil,* cymbals.] One of a pair of small round metal cymbals attached to the fingers and struck together for rhythm and percussion in belly-dancing.

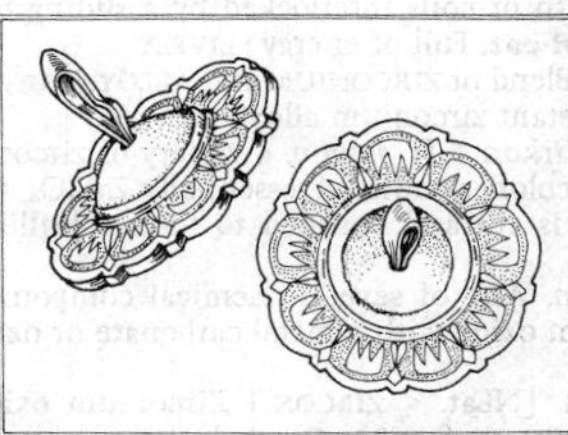

**zill**
*A pair of zills*

**zil·lion** (zĭl′yən) *n.* [Alteration of MILLION.] *Informal.* An extremely large indefinite number.

**Zil·pah** (zĭl′pə) *n.* [Heb. *Zilpāh.*] The servant of Leah who bore Jacob two sons, Gad and Asher.

**zinc** (zĭngk) *n.* [G. *Zink,* poss. < *Zinke,* spike < OHG *zinko.*] *Symbol* **Zn** A bluish-white, lustrous metallic element used to form a wide variety of alloys including brass, bronze, and various solders and in galvanizing iron and other metals; atomic number 30; atomic weight 65.37. *—vt.* **zinced, zinc·ing, zincs** *or* **zincked, zinck·ing, zincs.** To coat or treat with zinc : GALVANIZE.

**zinc·ate** (zĭng′kāt′) *n.* Any of several chemical compounds derived from the reaction of zinc or zinc oxide with certain alkali solutions.

**zinc blende** *n.* Sphalerite.
**zinc·ite** (zĭng′kīt′) *n.* A red to yellow-orange zinc ore, essentially ZnO.
**zinck·en·ite** (zĭng′kə-nīt′) *n. var. of* ZINKENITE.
**zinc·o·graph** (zĭng′kə-grăf′) *n.* **1.** A prepared zinc plate used in zincography. **2.** A print or picture obtained from a zincograph.
**zinc·og·ra·phy** (zĭng-kŏg′rə-fē) *n.* The process of engraving zinc printing plates. **—zinc·og′ra·pher** *n.* **—zinc′o·graph′ic** (zĭng′kə-grăf′ĭk), **zinc′o·graph′i·cal** *adj.*
**zinc ointment** *n.* A salve consisting of about 20% zinc oxide with beeswax or paraffin and petrolatum, used to treat skin disorders.
**zinc oxide** *n.* An amorphous white or yellowish powder, ZnO, used as a pigment, in compounding rubber, in the manufacture of plastics, and in pharmaceuticals and cosmetics.
**zinc sulfate** *n.* A colorless crystalline compound, $ZnSO_4 \cdot 7H_2O$, used as a fungicide, as a preservative for wood and skins, and in medicine as an emetic and astringent.
**zinc white** *n.* Zinc oxide.
**zin·fan·del** *also* **Zin·fan·del** (zĭn′fən-dĕl′) *n.* [Orig. unknown.] A dry red table wine produced in California.
**zing** (zĭng) *n.* [Imit.] **1.** A brief high-pitched humming or buzzing sound, as that made by an object passing by at high speed. **2.** Vitality : vigor. *—v.* **zinged, zing·ing, zings.** *—vi.* **1.** *Informal.* To make a brief high-pitched sound. **2.** *Informal.* To move swiftly with or as if with a zing <The ball *zinged* over the fence.> **3.** To proceed in a lively manner <The party *zinged* along.> *—vt. Informal.* **1.** To attack verbally : criticize sharply <*zing* a play in a review> **2.** ZAP 2.
**zing·er** (zĭng′ər) *n. Informal.* **1.** A witty, often caustic remark. **2.** A sudden, surprising or shocking revelation or turn of events.
**zing·y** (zĭng′ē) *adj.* **-i·er, -i·est.** *Informal.* **1.** Pleasantly stimulating and exciting <"The times are good. The living is easy. The vibes are *zingy*" *—Saturday Review*> **2.** Exceptionally attractive : STRIKING <a *zingy* sports car>
**zink·en·ite** *also* **zinck·en·ite** (zĭng′kə-nīt′) *n.* [G. *Zinkenit*, after J. K. L. *Zinken* (1790–1862).] A steel-gray mineral, essentially $Pb_6Sb_{14}S_{27}$.
**zin·ni·a** (zĭn′ē-ə) *n.* [NLat. *Zinnia*, genus name, after J. G. *Zinn* (1727–1759).] A plant of the genus *Zinnia*, native to tropical America, esp. *Z. elegans*, cultivated for its showy, colorful flowers.
**Zi·on** (zī′ən) *n.* [ME *Sion* < OE < LLat. < Gk. *Seiōn* < Heb. *Ṣīyōn.*] **1. a.** The Jewish people : ISRAEL. **b.** The Jewish homeland as a symbol of Judaism. **2.** A place or religious community regarded as sacredly devoted to God. **3.** An idealized harmonious community : UTOPIA.
**Zi·on·ism** (zī′ə-nĭz′əm) *n.* **1.** A plan or movement of the Jewish people to return from the Diaspora to Palestine. **2.** A movement orig. aimed at the re-establishment of a Jewish national homeland and state in Palestine and now concerned with the development and support of Israel. **—Zi′on·ist** *n.* **—Zi′on·is′tic** *adj.*
**zip** (zĭp) *n.* [Imit.] **1.** A brief, sharp hissing sound. **2.** Energy : vim. **3.** *Slang.* Zero : nothing. *—v.* **zipped, zip·ping, zips.** *—vi.* **1. a.** To move with a sharp hissing sound. **b.** To move or act with a speed that suggests such a sound <The cars *zipped* by on the freeway.> **2.** To act or proceed swiftly and energetically <*zipped* through our chores> **3.** To become fastened or unfastened by a zipper. *—vt.* **1.** To give speed and force to. **2.** To impart life or zest to <*zip* up a room with color> **3.** To fasten or unfasten with a zipper.
**Zip Code** *also* **zip code** *or* **ZIP Code.** A trademark for a system designed to expedite the sorting and delivery of mail by assigning a series of numbers to each delivery area in the United States.
**zip gun** *n.* A crude homemade pistol.
**zip·per** (zĭp′ər) *n.* [< ZIP.] A device for fastening the adjacent edges of an opening, as at the placket of a skirt, consisting of two facing rows of metal or plastic teeth or coils interlocked by a sliding tab.
**zip·py** (zĭp′ē) *adj.* **-pi·er -pi·est.** Full of energy : LIVELY.
**zir·ca·loy** (zûr′kə-loi′) *n.* [Blend of ZIRCONIUM and ALLOY.] Any of several stable, corrosion-resistant zirconium alloys.
**zir·con** (zûr′kŏn′) *n.* [G. *Zirkon* < Fr. *jargon*, a variety of zircon < Ital. *giargone.*] A brown to colorless mineral, essentially $ZrSiO_4$, the transparent form of which is cut and polished to form a brilliant blue-white gem.
**zir·con·ate** (zûr′kə-nāt′) *n.* Any of several chemical compounds formed by heating zirconium oxide with a metal carbonate or oxide in the presence of an acid.
**zir·co·ni·a** (zûr-kō′nē-ə) *n.* [NLat. < ZIRCON.] Zirconium oxide.
**zir·co·ni·um** (zûr-kō′nē-əm) *n. Symbol* **Zr** A lustrous, grayish-white, strong, ductile metallic element used chiefly in ceramic and refractory compounds, as an alloying agent, and in nuclear reactors; atomic number 40; atomic weight 91.22.
**zirconium oxide** *n.* A hard white amorphous powder, $ZrO_2$, derived from zirconium and also found naturally, used chiefly in pigments, abrasives, refractories, and ceramics.
**zit** (zĭt) *n.* [Orig. unknown.] *Slang.* A pimple.
**zith·er** (zĭth′ər, zĭth′-) *also* **zith·ern** (-ərn) *n.* [G. < OHG *zithera* < Lat. *cithara*, cithara < Gk. *kithara.*] A stringed instrument consisting of a flat sound box with 30 to 40 strings, played horizontally with the fingertips and a plectrum on the right thumb. **—zith′er·ist** *n.*
▲ word history: A zither is a stringed musical instrument from Austria. The word *zither* is the German development of Latin *cithara*, which also denoted a kind of stringed instrument. The wo[rd] *cithara* has other descendents in modern English, such as *cittern* a[nd] *guitar*, which denote musical instruments in the same family as th[e] *cithara* and the *zither*.
**zi·ti** (zē′tē) *n., pl.* **ziti.** [Ital. < pl. of *zito*, boy.] Medium-sized tub[u]lar pasta.
**zi·zith** (tsēt-sēt′, tsĭt′sĭs) *pl.n.* [Heb. *ṣīṣīth.*] The tassels or fringes [of] thread on the four corners of prayer shawls worn by Jewish me[n].
**zlo·ty** (zlô′tē) *n., pl.* **zloty** *or* **-tys.** [Pol. *złoty.*] —See table at CU[R]RENCY.
**Zn** *symbol for* ZINC.
**zo-** *pref. var. of* ZOO-.
**zo·a** (zō′ə) *n. var. pl. of* ZOON.
**zo·di·ac** (zō′dē-ăk′) *n.* [ME < OFr. *zodiaque* < Lat. *zodiacus* < G[k.] *zōdiacus (kuklos)*, (circle) of the zodiac < *zōidion*, small represent[ed] figure, dim. of *zōion*, living being.] **1. a.** *Astron.* A band of the cele[s]tial sphere extending about 8° to either side of the ecliptic and e[n]compassing the apparent paths of the principal planets, the moo[n] and the sun. **b.** This band divided into 12 astrological signs, each 3[0°] wide, bearing the name of a constellation for which it was or[ig.] named but with which it no longer coincides owing to the precessi[on] of the equinoxes. **c.** A diagram or figure representing the zodiac. **2.** [A] complete circuit or circle. **—zo·di′a·cal** (-dī′ə-kəl) *adj.*
**zodiacal light** *n.* A faint, hazy cone of light often seen in t[he] west after sunset or in the east before sunrise, apparently caused [by] reflection of sunlight from meteoric particles around the sun.
**zof·tig** (zäf′tĭk, -tĭg) *adj. var. of* ZAFTIG.
**-zoic** *suff.* [< Gk. *zōikos*, of animals < *zōion*, living being.] **1.** P[er]taining to a specified way of animal existence <holozoic> **2.** Of [or] pertaining to a specified geologic era <Archeozoic>
**zoi·site** (zoi′sīt′) *n.* [G. *Zoisit*, after Baron Sigismund *Zois* von Ed[el]stein (1747–1819), its discoverer.] A gray, brown, or pink mine[ral], essentially $Ca_2Al_3(SiO_4)_3(OH)$, used in ornamental stonework.
**zom·bie** *also* **zom·bi** (zŏm′bē) *n.* [Poss. < Kongo *zumbi*, fetis[h].] **1.** A snake god of voodoo cults in West Africa, Haiti, and the sou[th]ern United States. **2. a.** A supernatural power that according to v[oo]doo belief can enter into and reanimate a corpse. **b.** A cor[pse] reanimated to a trancelike state and made to do the bidding o[f a] supernatural power. **3.** One who looks or behaves like an automat[on]. **4.** A tall drink made of various rums, liqueur, and fruit juice.
**zo·nal** (zō′nəl) *also* **zo·na·ry** (-nə-rē) *adj.* **1.** Of or associated w[ith] a zone. **2.** Divided into zones. **—zo′nal·ly** *adv.*
**zo·nate** (zō′nāt′) *also* **zo·nat·ed** (-nā′tĭd) *adj.* Having zone[s]: belted, striped, or ringed.
**zo·na·tion** (zō-nā′shən) *n.* **1.** Arrangement or formation in zo[nes]. **2.** *Ecol.* Distribution of organisms in biogeographic zones.
**zone** (zōn) *n.* [Lat. *zona*, girdle < Gk. *zōnē.*] **1.** An area, region, [or] division distinguished from adjacent parts by a distinctive feature [or] character. **2. a.** Any of the five great regions into which the surf[ace] of the earth is divided according to latitude and prevailing clim[ate], including the Torrid Zone, the North and South Temperate Zo[nes], and the North and South Frigid Zones. **b.** A similar division on ot[her] planets. **3.** *Math.* A portion of the surface of a sphere bounded by [the] intersections of two parallel planes with the sphere. **4.** *Ecol.* An a[rea] marked by distinct physical conditions and populated by comm[uni]ties of certain kinds of organisms. **5.** *Geol.* A region or stratum [dis]tinguished by composition or content. **6.** A section of an area [or] territory set apart for a specific purpose <a war *zone*><a hosp[ital] *zone*> **7.** The total number of railroad stations located in a gi[ven] radius from a shipping point. **8.** *Computer Sci.* A region on a pu[nch] card or on a magnetic tape in which nondigital data are record[ed]. **9.** *Archaic.* A girdle or belt. *—vt.* **zoned, zon·ing, zones. 1.** To [di]vide into zones. **2.** To designate or mark off into zones. **3.** To [sur]round or encircle with or as if with a belt or girdle.
**zone melting** *n.* A purification technique for crystalline s[ub]stances in which a heating system passes slowly over a bar of [the] material to be refined, creating a molten region that carries imp[uri]ties with it across the bar.
**zone refining** *n.* Zone melting.
**zone·time** (zōn′tīm′) *n.* Standard time used at sea according to [the] time zone in which a ship is located.
**zonk** (zôngk, zŏngk) *v.* **zonked, zonk·ing, zonks.** [Orig. [un]known.] *Slang. —vt.* **1.** To stupefy : stun. **2.** To intoxicate or re[nder] senseless with drugs or alcohol <"*zonk* their patients with tranq[uil]izers" *—Psychology Today*> *—vi.* To be intoxicated or sense[less] from drugs or alcohol.
**zoo** (zo͞o) *n., pl.* **zoos.** [Short for ZOOLOGICAL GARDEN.] **1.** A pu[blic] park or large enclosure where live animals are kept for display. [**2.** A] place or situation characterized by rampant confusion or disor[der].
**zoo-** *or* **zo-** *pref.* [Gk. *zōio-* < *zōion*, living being.] **1.** Animal : [ani]mal kingdom <zoography> **2.** Motile <zoospore>
**zo·o·chore** (zō′ə-kôr′, -kōr′) *n.* A plant dispersed by animals.
**zo·o·ge·o·graph·ic region** (zō′ə-jē′ə-grăf′ĭk) *n.* An ecologica[l re]gion marked by the dominance of certain kinds of animal life.

**zo·o·ge·og·ra·phy** (zō′ə-jē-ŏg′rə-fē) *n.* Study of the geographic distribution of animals. **—zo′o·ge·og′ra·pher** *n.* **—zo′o·ge′o·graph′ic** (-ə-grăf′ĭk), **zo′o·ge′o·graph′i·cal** (-ĭ-kəl) *adj.* **—zo′o·ge′o·graph′i·cal·ly** *adv.*

**zo·o·gle·a** (zō′ə-glē′ə) *n., pl.* **-gle·ae** (-glē′ē′) *or* **-gle·as.** [NLat. *Zoogloea*, genus name : ZOO- + Med. Gk. *glia*, gum < Gk. *gloios.*] Any of various bacteria of the genus *Zoogloea*, forming colonies in a jellylike secretion.

**zo·og·ra·phy** (zō-ŏg′rə-fē) *n.* Biological description of animals. **—zo′o·graph′ic** (-ə-grăf′ĭk), **zo′o·graph′ic·al** (-ĭ-kəl) *adj.*

**zo·oid** (zō′oid′) *n.* **1.** *Biol.* An organic cell or organized body that has independent movement within a living organism, esp. a motile gamete such as a spermatozoon. **2.** *Zool.* Any of the usu. microscopic animals forming an aggregate or colony, as of bryozoans. **—zo·oid′al** (-oid′l) *adj.*

**zo·ol·a·try** (zō-ŏl′ə-trē) *n.* Animal worship. **—zo·ol′a·ter** *n.* **—zo·ol′a·trous** *adj.*

**zo·o·log·i·cal** (zō′ə-lŏj′ĭ-kəl) *also* **zo·o·log·ic** (-lŏj′ĭk) *adj.* **1.** Of or relating to animals or animal life. **2.** Of or relating to the science of zoology. **—zo′o·log′i·cal·ly** *adv.*

**zoological garden** *n.* ZOO 1.

**zo·ol·o·gy** (zō-ŏl′ə-jē) *n., pl.* **-gies. 1.** The biological science that deals with animals. **2.** The animal life of a particular area. **3.** The characteristics of an animal group or category <the *zoology* of reptiles> **4.** A treatise on zoology. **—zo·ol′o·gist** *n.*

**zoom** (zo͞om) *v.* **zoomed, zoom·ing, zooms.** [Imit.] *—vi.* **1. a.** To make a loud low-pitched buzzing or humming sound. **b.** To move speedily while making such a sound. **2.** To climb suddenly and sharply in an airplane. **3.** To rise rapidly <Interest rates *zoomed.*> **4. a.** To move rapidly toward or away from a photographic subject. **b.** To simulate such a movement, as by means of a zoom lens. *—vt.* To cause to zoom. **—zoom** *n.*

**zo·om·e·try** (zō-ŏm′ĭ-trē) *n.* Measurement and comparison of the sizes of animals or animal parts, esp. measurement of bulk. **—zo′o·met′ric** (-ə-mĕt′rĭk), **zo′o·met′ri·cal** (-rĭ-kəl) *adj.* **—zo′o·met′ri·cal·ly** *adv.*

**zoom lens** *n.* A camera lens whose focal length can be continuously adjusted, allowing for rapid change in the size of an image without loss of focus.

**zo·o·mor·phism** (zō′ə-môr′fĭz′əm) *n.* **1.** Attribution of animal characteristics or qualities to a deity. **2.** Use of animal forms in symbolism, literature, or graphic representation. **—zo′o·mor′phic** *adj.*

**zo·on** (zō′ŏn′) *n., pl.* **zo·ons** *or* **zo·a** (zō′ə) [NLat. < Gk. *zōion*, living being.] An animal developed from a fertilized egg.

**-zoon** *suff.* [NLat. < Gk. *zōion*, living being.] Animal : independently moving organic unit <spermatozoon>

**zo·on·o·sis** (zō-ŏn′ə-sĭs, zō′ə-nō′sĭs) *n., pl.* **-ses** (-sēz′) [NLat. : ZOO- + Gk. *nosos*, disease.] A disease such as rabies or malaria that can be transmitted from animals to humans.

**zo·oph·a·gous** (zō-ŏf′ə-gəs) *adj.* Feeding on animal matter.

**zo·o·phile** (zō′ə-fīl′) *n.* A lover of animals.

**zo·oph·i·lous** (zō-ŏf′ə-ləs) *adj. Bot.* Pollinated by animals.

**zo·o·pho·bi·a** (zō′ə-fō′bē-ə) *n.* Abnormal fear of animals. **—zo′o·phobe′** *n.* **—zo·oph′o·bous** (zō-ŏf′ə-bəs) *adj.*

**zo·o·phyte** (zō′ə-fīt′) *n.* [Gk. *zōophuton* : *zōion*, animal + *phuton*, plant.] An invertebrate animal, such as a sea anemone or sponge, that remains attached to a surface and superficially resembles a plant. **—zo′o·phyt′ic** (-fĭt′ĭk), **zo′o·phyt′i·cal** *adj.*

**zo·o·plank·ton** (zō′ə-plăngk′tən) *n.* Floating, often microscopic aquatic animals.

**zo·o·plas·ty** (zō′ə-plăs′tē) *n.* Surgical transfer of tissue from a lower animal to a human. **—zo′o·plas′tic** *adj.*

**zo·o·sperm** (zō′ə-spûrm′) *n.* A spermatozoon.

**zo·o·spo·ran·gi·um** (zō′ə-spə-răn′jē-əm) *n., pl.* **-gi·a** (-jē-ə). A sporangium in which zoospores develop.

**zo·o·spore** (zō′ə-spôr′, -spōr′) *n.* A motile, flagellated asexual spore, as of certain algae and fungi. **—zo′o·spor′ic** *adj.*

**zo·os·ter·ol** (zō-ŏs′tə-rôl′, -rōl′) *n.* An animal sterol, as cholesterol.

**zo·o·tech·nics** (zō′ə-tĕk′nĭks) *n. (sing. or pl. in number).* Zootechny.

**zo·o·tech·ny** (zō′ə-tĕk′nē) *n.* [ZOO- + Gk. *tekhnē*, art.] The care, breeding, and improvement of animals : animal husbandry. **—zo′o·tech′ni·cal** *adj.* **—zo′o·tech·ni′cian** (-nĭsh′ən) *n.*

**zo·ot·o·my** (zō-ŏt′ə-mē) *n.* **1.** Dissection of animals other than human beings. **2.** Comparative anatomy.

**zoot suit** (zo͞ot) *n.* [Orig. unknown.] *Slang.* A man's suit popular during the early 1940's, consisting of baggy, tight-cuffed trousers and a long jacket with wide lapels and wide, heavily padded shoulders. **—zoot′-suit′er** *n.*

**zo·ri** (zôr′ē, zōr′ē) *n., pl.* **zori.** [J. *zōri* : *zo*, grass, straw + *-ri*, footwear.] A flat Japanese sandal, usu. made of straw, rubber, or leather, held on the foot by means of a thong that passes between the big toe and the second toe.

**zor·ille** *also* **zor·il** (zôr′ĭl, zŏr′-) *n.* [Fr. < Sp. *zorillo*, skunk, dim. of *zorro*, fox.] An African mammal, *Ictonyx striatus*, resembling the skunk in appearance and defensive action.

**Zo·ro·as·tri·an·ism** (zôr′ō-ăs′trē-ə-nĭz′əm) *n.* A religion founded in Persia by Zoroaster in the 6th cent. B.C. and set forth in the Zend-Avesta, teaching the worship of Ormazd in the context of a universal struggle between the forces of light and darkness. **—Zo′ro·as′tri·an** *adj.* & *n.*

**zos·ter** (zŏs′tər) *n.* [Gk. *zōstēr*, girdle.] **1.** A belt or girdle worn by men in ancient Greece. **2.** Herpes zoster.

**Zou·ave** (zo͞o-äv′) *n.* [Fr. < Berber *Zwāwa*, an Algerian tribe.] **1.** A member of a French infantry unit, once composed of Algerian recruits, noted for their colorful oriental uniforms and precision drilling. **2.** A member of a group patterned after the French Zouaves, esp. a member of such a unit of the Union Army in the Civil War.

**zounds** (zoundz) *interj.* [Shortening and alteration of *by God's wounds.*] —Used to express anger, surprise, or indignation.

**zoy·sia** (zoi′shə, -zhə, -sē-ə, -zē-ə) *n.* [NLat. *Zoysia*, genus name, after Karl von Zois (1756–1800).] Any of several creeping perennial grasses of the genus *Zoysia*, native to Asia and Australia and widely cultivated as a lawn grass.

**Zr** *symbol for* ZIRCONIUM.

**Z score** *n.* A statistical measure of the distance, in standard deviations, of a sample from the mean.

**zuc·chet·to** (zo͞o-kĕt′ō, tso͞o-) *n., pl.* **-tos.** [Ital., dim. of *zucca*, gourd, head < LLat. *cucutia*, gourd.] *Rom. Cath. Ch.* A skullcap worn by the clergy, varying in color with the rank of the wearer.

**zuc·chi·ni** (zo͞o-kē′nē) *n., pl.* **zucchini.** [Ital., pl. of *zucchino*, dim. of *zucca*, gourd < LLat. *cucutia.*] A summer squash having an elongated shape and a smooth dark-green skin.

**Zu·lu** (zo͞o′lo͞o) *n., pl.* **Zulu** *or* **-lus. 1.** A member of a large Bantu nation of southeastern Africa between Natal and Lourenço Marques. **2.** The Bantu language of the Zulu. **—Zu′lu** *adj.*

**Zu·ñi** (zo͞o′nyē, -nē) *n., pl.* **Zuñi** *or* **-ñis. 1. a.** A pueblo-dwelling tribe of Indians of western New Mexico. **b.** A member of this tribe. **2.** The language of the Zuñi.

**Zu·ñi·an** (zo͞on′yē-ən) *n.* A language family consisting only of Zuñi. **—Zu′ñi·an** *adj.*

**zwie·back** (swē′băk′, -bäk′, swī′-, zwē′-, zwī′-) *n.* [G. : *zwie-*, twice + *backen*, to bake.] A usu. sweetened egg bread baked first as a loaf and then cut into slices and toasted.

**Zwing·li·an** (zwĭng′lē-ən, swĭng′-, tsfĭng′-) *adj.* Of or relating to Ulrich Zwingli or his theological teachings, esp. his doctrine that the physical body of Christ is not present in the Eucharist and that the ceremony is merely a symbolic commemoration of Christ's death. *—n.* A follower of Zwingli. **—Zwing′li·an·ism** *n.*

**zwit·ter·i·on** (zwĭt′ər-ī′ən, swĭt′-) *n.* [G. : *Zwitter*, hybrid (< OHG *zwitarn* < *zwi-*, twice) + *ion*, ion < Gk., something that goes, neuter pr.part of *ienai*, to go.] *Physics.* An ion carrying both a positive and a negative charge, thus forming an electrically neutral molecule. **—zwit′ter·i·on′ic** (-ī-ŏn′ĭk) *adj.*

**zy·de·co** (zī′dĭ-kō′) *n.* [Prob. of Fr. orig.] Popular music of southern Louisiana that combines French dance melodies, elements of Caribbean music, and the blues, played by small groups featuring the guitar, the accordion, and a washboard.

**zyg-** *pref. var. of* ZYGO-.

**zyg·a·poph·y·sis** (zĭg′ə-pŏf′ĭ-sĭs, zī′gə-) *n., pl.* **-ses** (-sēz′). One of two usu. paired processes of a vertebra that articulate with corresponding parts of adjacent vertebrae.

**zygo-** *or* **zyg-** *pref.* [NLat. < Gk. *zugon*, yoke.] **1.** Yoke : pair <zygodactyl> **2.** Union <zygospore>

**zy·go·dac·tyl** (zī′gə-dăk′təl) *also* **zy·go·dac·ty·lous** (-tə-ləs) *adj.* Having two toes projecting forward and two backward, as certain birds. *—n.* A zygodactyl bird.

**zy·go·ma** (zī-gō′mə) *n., pl.* **-ma·ta** (-mə-tə) *or* **-mas.** [NLat. < Gk. *zugōma*, bolt < *zugoun*, to join.] **1.** The zygomatic bone. **2.** The zygomatic arch. **3.** The zygomatic process.

**zy·go·mat·ic** (zī′gə-măt′ĭk) *adj.* Of, relating to, or located in the area of the zygoma.

**zygomatic arch** *n.* The bony arch in vertebrates that extends along the side or front of the skull beneath the orbit.

**zygomatic bone** *n.* A small quadrangular bone in vertebrates on the side of the face below the eye, forming, in mammals, part of the orbit and part of the zygomatic arch.

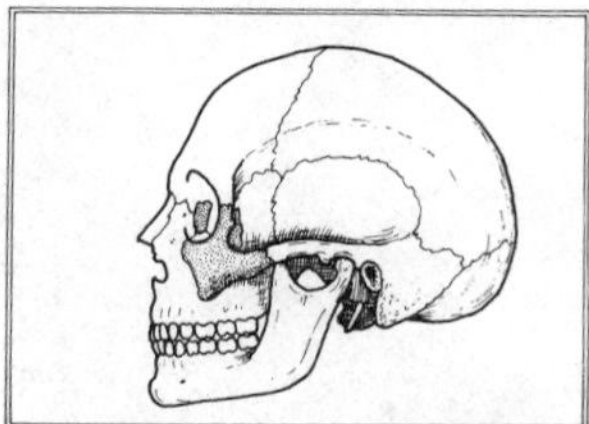

**zygomatic bone**

**zygomatic process** *n.* Any of the three processes that articulate to make up the zygomatic arch.

**zy·go·mor·phic** (zī′gə-môr′fĭk) *also* **zy·go·mor·phous** (-fəs) *adj.* Bilaterally symmetric so as to be capable of being symmetrically divided only along a single longitudinal plane. —Used of organisms or parts. **—zy′go·mor′phism** *n.*

**zy·go·sis** (zī-gō′sĭs) *n., pl.* **-ses** (-sēz′). Union of two gametes to form a zygote : CONJUGATION.

**zy·go·spore** (zī′gə-spôr′, -spōr′) *n.* A thick-walled resting spore formed by conjugation of similar gametes, as in algae or fungi.

**zy·gote** (zī′gōt′) *n.* [< Gk. *zugōtos,* yoked < *zugoun,* to yoke.] **1.** The cell formed by union of two gametes. **2.** The organism that develops from such a cell. **—zy·got′ic** (-gŏt′ĭk) *adj.* **—zy·got′i·cal·ly** *adv.*

**-zygous** *suff.* [Gk. *-zugos,* yoked < *zugon,* yoke.] Having a zygotic constitution of a specified kind <heterozygous>

**zym-** *pref. var. of* ZYMO-.

**zy·mase** (zī′mās′, -māz′) *n.* An enzyme complex that acts in glycolysis, found in yeasts, bacteria, and higher plants and animals.

**-zyme** *suff.* [< Gk. *zumē,* leaven.] Enzyme <lyso*zyme*>

**zymo-** *or* **zym-** *pref.* [NLat. < Gk. *zumē,* leaven.] **1.** Fermentation <*zym*urgy> **2.** Enzyme <*zymo*plastic>

**zy·mo·gen** (zī′mə-jən) *n.* The inactive protein precursor of an enzyme.

**zy·mo·gen·ic** (zī′mə-jĕn′ĭk) *also* **zy·mog·e·nous** (zī-mŏj′ə-nəs) *adj.* **1.** Of or relating to a zymogen. **2.** Capable of causing fermentation. **3.** Enzyme-producing.

**zy·mol·o·gy** (zī-mŏl′ə-jē) *n.* The chemistry of fermentation. **—zy′mo·log′ic** (-mə-lŏj′ĭk), **zy′mo·log′i·cal** *adj.* **—zy·mol′o·gist** *n.*

**zy·mol·y·sis** (zī-mŏl′ĭ-sĭs) *n.* **1.** Enzyme action. **2.** Fermentation. **—zy′mo·lyt′ic** (-mə-lĭt′ĭk) *adj.*

**zy·mo·plas·tic** (zī′mə-plăs′tĭk) *adj.* Producing enzymes.

**zy·mo·scope** (zī′mə-skōp′) *n.* An instrument for determining fermentation efficiency by measuring carbon dioxide produced.

**zy·mo·sis** (zī-mō′sĭs) *n., pl.* **-ses** (-sēz′). [Gk. *zymōsis* < *zumoun,* to leaven < *zumē,* leaven.] **1.** Fermentation. **2.** *Med.* **a.** The process of infection. **b.** An infectious or contagious disease. **—zy·mot′ic** (-mŏt′ĭk) *adj.* **—zy·mot′i·cal·ly** *adv.*

**zy·mur·gy** (zī′mûr′jē) *n.* Technological chemistry that deals with fermentation processes in brewing.

**zyz·zy·va** (zĭz′ə-və) *n.* [NLat. *Zyzzyva,* genus name.] Any of various tropical American weevils of the genus *Zyzzyva,* often destructive to plants.

# Abbreviations

**1.** *also* **a.** are (measurement). **2.** hysics. atto-.

**1.** *also* **a.** *or* **A.** acre. **2.** ammeter. ampere. **4.** area.

**1.** about. **2.** acceleration. **3.** acreage. acting. **5.** adjective. **6.** afternoon. *also* **A.** amateur. **8.** *Lat.* anno (in e year). **9.** *Lat.* annus (year). **.** anonymous. **11.** *also* **A.** answer. **.** *Lat.* ante (before). **13.** anterior.

**1.** academician; academy. **2.** alto. America; American.

angstrom.

**C** airman first class.

**A 1.** Alcoholics Anonymous. **2.** anircraft.

**A.** Associate in Arts.

**AA 1.** Agricultural Adjustment dministration. **2.** antiaircraft artilry.

**and M** ancient and modern.

**and R** artists and repertory.

**ARL** Army Aeromedical Research boratory.

**A.S.** Associate in Applied Scices.

**ATC** automatic air traffic control.

**B** Alberta.

**.** about.

**B. 1.** *or* **a.b.** able-bodied seaman. *Lat.* Artium Baccalaureus (Bacher of Arts).

**b.** abbess; abbey; abbot.

**br.** *or* **abbrev.** abbreviation.

**BCD** accelerated business collecn and delivery.

**BD** all but dissertation.

**BEND** *Computer Sci.* abnormal d of task.

**l.** ablative.

**M** antiballistic missile.

**n** airborne.

**or.** abortion.

**p.** *or* **Abp.** archbishop.

**r.** abridged; abridgment.

**s. 1.** absence; absent. **2.** absolute; solutely. **3.** abstract.

**str.** abstract; abstracted.

**t.** about.

**1.** acre. **2.** air-cooled. **3.** *or* **AC** ernating current.

*Bible.* Acts.

**. 1.** *or* **a/c** air conditioning. *Lat.* ante cibum (*Med.* before eals).

**C. 1.** air corps. **2.** *Lat.* ante Chrism (before Christ).

**c** account; account current.

**ad.** academic; academy.

**cel.** *Mus.* accelerando.

**ct.** account; accountant.

**CCTID** account identifier.

**et.** acetone.

**k.** acknowledge; acknowledgment.

**CLS** advanced cardiac life support stem.

**pt.** acceptance.

**CS** American Community Schools.

**CST** access time.

**CT** American College Test.

**C.T.** Australian Capital Territory.

**tg.** acting.

**v** actual cash value.

**CV** air-cushion vehicle.

**.** adapter.

**D 1.** active duty. **2.** air-dried.

**D.** *Lat.* anno Domini (in the year the Lord). —Usu. used in small pitals <A.D.>

**DC 1.** *also* **a.d.c.** aide-de-camp. Aid to Dependent Children.

**d.** addendum.

**DDDS** automatic direct-distance aling system.

**dn.** addition.

**dnl.** additional.

**f** automatic direction finder.

**DH** antidiuretic hormone.

**int.** *Lat.* ad interim (in the eantime).

**ADIZ** air defense identification zone.

**adj. 1.** adjective. **2.** adjunct. **3.** adjustment. **4.** *also* **Adj.** adjutant.

**adjt.** adjutant.

**ad loc.** *Lat.* ad locum (to *or* at the place).

**adm.** administrative; administrator.

**Adm.** *or* **ADM** admiral; admiralty.

**admin.** administration; administrator.

**ADP** automatic data processing.

**adv. 1.** adverb; adverbial. **2.** *Lat.* adversus (against). **3.** advertisement. **4.** advisory.

**advt.** advertisement.

**AeEng** aeronautical engineer.

**AEF** American Expeditionary Force.

**AEIC** advanced earned income credit.

**aero** aeronautical, aeronautics.

**AF** *also* **A.F. 1.** air force. **2.** audio frequency.

**AFAM** Ancient Free and Accepted Masons.

**AFB** air force base.

**AFDC** Aid to Families with Dependent Children.

**aff** affirmative.

**afft.** affidavit.

**Afg.** Afghanistan.

**Afr.** Africa; African.

**aft.** afternoon.

**A.G.** *also* **AG 1.** adjutant general. **2.** attorney general.

**AGC** advanced graduate certificate.

**agcy.** agency.

**agr.** agricultural; agriculture.

**agric.** agriculture; agriculturist.

**agst.** against.

**agt. 1.** agent. **2.** agreement.

**AH** artificial heart.

**A.h** *or* **a-h** ampere-hour.

**A.H.** *Lat.* **1.** anno Hebraico (in the Hebrew year). **2.** anno Hegirae (in the year of the Hegira).

**ai** airborne intercept.

**a.i.** *Lat.* ad interim (in the meantime).

**AIR** air intercept rocket.

**AK** Alaska.

**a.k.a.** also known as.

**AL** *or* **Ala.** Alabama.

**A.L.A.** Associate in Liberal Arts.

**Alas.** Alaska.

**Alb.** Albania; Albanian.

**alc.** alcohol; alcoholic.

**ALCM** air-launched cruise missile.

**Ald.** alderman.

**alg.** algebra.

**Alg.** Algeria.

**alky.** alkalinity.

**allo.** *Mus.* allegro.

**alpha** alphabetical.

**a.l.s.** *or* **A.L.S.** autograph letter signed.

**alt. 1.** alternate. **2.** altimeter. **3.** altitude.

**Alta.** Alberta.

**ALU** arithmetic logic unit.

**am** *or* **AM** amplitude modulation.

**Am** *Bible.* Amos.

**Am.** America; American.

**A.M. 1.** airmail. **2.** *Lat.* anno mundi (in the year of the world). —Usu. used in small capitals <A.M.> **3.** *also* **a.m.** *Lat.* ante meridiem (before noon). —Usu. used in small capitals <A.M.> **4.** *Lat.* Artium Magister (Master of Arts).

**amb.** *also* **Amb.** ambassador.

**AMC** automatic message counting.

**amdt.** amendment.

**Amer.** America; American.

**Amex** American Stock Exchange.

**AMI** acute myocardial infarction.

**AMM** anti-missile missile.

**Amn** *or* **AMN** airman.

**amp hr.** ampere-hour.

**AMS** auditory memory span.

**amt.** amount.

**amu** *Physics.* atomic mass unit.

**An** *Physics.* actinon.

**AN** *also* **A.N.** Anglo-Norman.

**an.** *Lat.* **1.** ante (before). **2.** anno (in the year).

**anal. 1.** analogous; analogy. **2.** analysis; analytic.

**anat.** anatomical; anatomist; anatomy.

**anc.** ancient.

**and.** *Mus.* andante.

**And.** Andorra.

**Ang.** Angola.

**Angl.** Anglican.

**anhydr.** anhydrous.

**anim.** *Mus.* animato.

**ann. 1.** annals. **2.** annual. **3.** annuity.

**anon.** anonymous.

**ans.** answer.

**ant. 1.** antenna. **2.** antiquarian; antiquity. **3.** antonym.

**Ant.** Antarctica.

**anthrop.** anthropologic; anthropology.

**antiq. 1.** antiquarian; antiquary. **2.** antiquities. **3.** antiquity.

**a/o** account of.

**aor.** aorist.

**AP 1.** airplane. **2.** air police. **3.** American plan. **4.** antipersonnel. **5.** applied physics. **6.** *or* **A.P.** Associated Press.

**ap.** apothecary.

**a.p. 1.** additional premium. **2.** author's proof.

**APB** all points bulletin.

**APC** armored personnel carrier.

**API** air position indicator.

**APL** Adult Performance Level.

**APO** *or* **A.P.O.** Army Post Office.

**Apoc. 1.** Apocalypse. **2.** Apocrypha; Apocryphal.

**app. 1.** apparatus. **2.** appendix. **3.** applied. **4.** appoint; appointed. **5.** apprentice.

**appl.** applied.

**appmt.** appointment.

**approx.** approximate; approximately.

**appt.** appoint; appointment.

**apptd.** appointed.

**APR** annual percentage rate.

**Apr.** April.

**apt. 1.** apartment. **2.** aptitude.

**aq.** aqueous.

**AR 1.** *also* **A/R** accounts receivable. **2.** Arkansas.

**ar.** arrival; arrive.

**Ar. 1.** Arabia; Arabian. **2.** Arabic.

**A.R.** *also* **AR 1.** Airman Recruit. **2.** army regulation.

**Arab. 1.** Arabia; Arabian. **2.** Arabic.

**ARC** American Red Cross.

**arch. 1.** archaic; archaism. **2.** archery. **3.** archipelago. **4.** architect; architectural; architecture.

**Archbp.** archbishop.

**archit.** architecture.

**archt.** architect.

**arg.** argent.

**Arg.** Argentina; Argentine.

**Ariz.** Arizona.

**Ark.** Arkansas.

**ARM** automated route management.

**Arm.** Armenia; Armenian.

**arr. 1.** arranged. **2.** arrival; arrive; arrived.

**art. 1.** article. **2.** artificial. **3.** artillery. **4.** artist.

**arty.** artillery.

**ARV** *Bible.* American (Standard) Revised Version.

**ARVN** Army of the Republic of Vietnam.

**AS 1.** *also* **A.S.** Anglo-Saxon. **2.** antisubmarine.

**As.** Asia; Asian.

**a/s** air speed.

**ASAP** as soon as possible.

**asb.** asbestos.

**ASBC** American Standard Building Code.

**ASE** American Stock Exchange.

**asgd.** assigned.

**asgmt.** assignment.

**ASI** air speed indicator.

**ASK** American simplified keyboard.

**ASL** American Sign Language.

**ASM** assembler.

**asm.** assembly.

**ASR** air-sea rescue.

**assn.** association.

**assoc.** associate; association.

**ASSR** *or* **A.S.S.R.** Autonomous Soviet Socialist Republic.

**asst.** assistant.

**asstd. 1.** assisted. **2.** assorted.

**assy.** assembly.

**Assyr.** Assyrian.

**AST** adiabatic storage test.

**ASTP** Army Specialized Training Program.

**astrol.** astrologer; astrologic; astrology.

**astron.** astronomer; astronomical; astronomy.

**ASV** *Bible.* American Standard Version.

**aT** *Physics.* attotesla.

**At** ampere-turn.

**AT** *also* **a/t** antitank.

**at. 1.** airtight. **2.** atomic.

**atc** around the clock.

**athl.** athlete; athletic; athletics.

**Atl.** Atlantic.

**atm** *Physics.* atmosphere.

**atm.** *or* **atmos.** atmosphere; atmospheric.

**at. no.** *also* **at no** atomic number.

**ATP** *Biochem.* adenosine triphosphate.

**ATR** audio tape recording.

**att. 1.** attached. **2.** attention. **3.** attorney.

**attn.** attention.

**attrib.** attribute; attributive.

**atty.** attorney.

**Atty. Gen.** attorney general.

**ATV** all-terrain vehicle.

**at wt** atomic weight.

**a.u.** *or* **A.u.** angstrom unit.

**A.U.** astronomical unit.

**aud.** audit; auditor.

**aug.** augmentative.

**Aug.** August.

**AUS** Army of the United States.

**Aus.** *or* **Aust. 1.** Australia. **2.** Austria.

**Austl.** Australia; Australian.

**auth. 1.** authentic. **2.** author. **3.** authority. **4.** authorized.

**auto. 1.** automatic. **2.** automotive.

**aux.** auxiliary.

**AV** *or* **A.V. 1.** audio-visual. **2.** *Bible.* Authorized Version.

**av. 1.** *or* **Av.** avenue. **2.** average. **3.** avoirdupois.

**a.v.** *or* **a/v** *Lat.* ad valorem (in proportion to the value).

**AVC** automatic volume control.

**avdp.** avoirdupois.

**ave.** *or* **Ave.** avenue

**AVF** antiviral factor.

**avg.** average.

**avn.** aviation.

**AW 1.** aircraft warning. **2.** Articles of War. **3.** automatic weapon.

**a.w.** all water (transportation).

**A/W** actual weight.

**AWAC** airborne warning and control system.

**ax. 1.** axiom. **2.** axis.

**AYH** American Youth Hostels.

**AZ** Arizona.

**az. 1.** azimuth. **2.** azure.

**Azo.** Azores.

**b** *Physics.* barn.
**B** **1.** baryon number. **2.** bishop (chess).
**b.** or **B. 1.** base. **2.** *Mus.* basso. **3.** bay. **4.** bolivar. **5.** book. **6.** born. **7.** breadth. **8.** brother.
**B. 1.** bachelor. **2.** bacillus. **3.** Baumé scale. **4.** Bible. **5.** British. **6.** brotherhood.
**Ba** *Bible.* Baruch.
**Ba.** Bahamas.
**B.A. 1.** *Lat.* Baccalaureus Artium (Bachelor of Arts). **2.** British Academy. **3.** British Association (for the Advancement of Science).
**Bab.** Babylonia; Babylonian.
**BABS** blind approach beacon system.
**bach.** bachelor.
**bact.** bacteria; bacterial.
**bacteriol.** bacteriologist; bacteriology.
**B.A.E. 1.** Bachelor of Aeronautical Engineering. **2.** Bachelor of Agricultural Engineering. **3.** Bachelor of Architectural Engineering. **4.** Bachelor of Art Education. **5.** Bachelor of Arts in Education.
**B.A.Ed.** Bachelor of Arts in Education.
**B.Ae.E.** Bachelor of Aeronautical Engineering.
**bal.** balance.
**B.A.M. 1.** Bachelor of Applied Mathematics. **2.** Bachelor of Arts in Music.
**B and B** bed-and-breakfast.
**B and E** breaking and entering.
**bankr.** bankruptcy.
**Bap.** or **Bapt.** Baptist.
**BAR** Browning automatic rifle.
**bar. 1.** barometer; barometric. **2.** barrel.
**Barb.** Barbados.
**B.Arch.** Bachelor of Architecture.
**Bart.** baronet.
**B.A.S.** or **B.A.Sc. 1.** Bachelor of Agricultural Science. **2.** Bachelor of Applied Science.
**bat.** battalion.
**Bav.** Bavaria; Bavarian.
**bb** *also* **b.b.** ball bearing.
**BB** B'nai B'rith.
**B.B.A.** Bachelor of Business Administration.
**B.B.Ed.** Bachelor of Business Education.
**bbl** or **bbl.** barrel.
**B.C. 1.** Bachelor of Chemistry. **2.** Bachelor of Commerce. **3.** before Christ. —Usu. used in small capitals <B.C.> **4.** or **BC** British Columbia.
**bcd** or **BCD** *Computer Sci.* binary coded decimal.
**B.C.E. 1.** Bachelor of Chemical Engineering. **2.** Bachelor of Civil Engineering.
**BCG** bacillus Calmette-Guérin (tuberculosis vaccine).
**B.Ch.E.** Bachelor of Chemical Engineering.
**B.C.L. 1.** Bachelor of Canon Law. **2.** Bachelor of Civil Law.
**B.C.S. 1.** Bachelor of Chemical Science. **2.** Bachelor of Commercial Science.
**BCSE** Board of Civil Service Examiners.
**BD 1.** bank draft. **2.** *also* **b/d** bills discounted. **3.** bomb disposal.
**bd. 1.** board. **2.** bound. **3.** bundle.
**B.D.** Bachelor of Divinity.
**b/d 1.** barrels per day. **2.** brought down.
**bd. ft.** board foot.
**bdl** or **bdle.** bundle.
**bdrm.** bedroom.
**bds.** bound in boards.
**B.D.S.** Bachelor of Dental Surgery.
**BDSA** Business and Defense Services Administration.
**BE** Board of Education.
**B.E. 1.** Bachelor of Education. **2.** Bachelor of Engineering.
**B/E 1.** bill of entry. **2.** bill of exchange.
**Bé** Baumé scale.
**BEC** Bureau of Employees' Compensation.
**B.Ed.** Bachelor of Education.
**BEF** British Expeditionary Force.
**bef.** before.
**Bel** or **Belg.** Belgian; Belgium.
**BEM** *also* **B.E.M.** British Empire Medal.
**B.E.M.** Bachelor of Engineering of Mines.
**B.Eng.** Bachelor of Engineering.
**B.Eng.Sci.** Bachelor of Engineering Science.
**bet.** between.
**BeV** *Physics.* billion electron volts.
**bf 1.** board foot. **2.** *also* **b.f.** or **bf.** boldface.
**b.f.** or **B/F** brought forward.
**B.F.A.** Bachelor of Fine Arts.
**bg. 1.** background. **2.** bag.
**B.G.** or **BG** or **BGen** Brigadier General.
**BH** bill of health.
**bhd.** bulkhead.
**BHE** Bureau of Higher Education.
**Bhn.** *Metallurgy.* Brinell hardness number.
**bhp** or **b.hp.** brake horsepower.
**BHT** butylated hydroxytoluene.
**Bhu.** Bhutan.
**BIA** Bureau of Indian Affairs.
**Bib.** Bible; Biblical.
**bibl.** or **Bibl.** Biblical.
**bibliog.** bibliographer; bibliography.
**b.i.d.** *Lat.* bis in die (*Med.* twice a day).
**biog.** biographer; biographical; biography.
**biol.** biological; biologist; biology.
**B.J.** Bachelor of Journalism.
**bk. 1.** bank. **2.** book.
**bkcy.** bankruptcy.
**bkg.** banking.
**bkgd.** background.
**bklr.** black letter.
**bkpg.** bookkeeping.
**bkpt.** bankrupt.
**bks. 1.** barracks. **2.** books.
**bl. 1.** barrel. **2.** black. **3.** blue.
**B.L. 1.** Bachelor of Laws. **2.** Bachelor of Letters; Bachelor of Literature.
**B/L** bill of lading.
**B.L.A.** Bachelor of Liberal Arts.
**bld. 1.** blood. **2.** boldface.
**bldg.** building.
**bldr.** builder.
**B.Lit.** or **B.Litt.** *Lat.* Baccalaureus Litterarum (Bachelor of Literature).
**blk. 1.** black. **2.** block. **3.** bulk.
**BLS** Bureau of Labor Statistics.
**B.L.S.** Bachelor of Library Science.
**blvd.** boulevard.
**BM** basal metabolism.
**bm.** beam.
**b.m. 1.** board measure. **2.** bowel movement.
**B.M. 1.** Bachelor of Medicine. **2.** Bachelor of Music.
**B.M.E. 1.** Bachelor of Mechanical Engineering. **2.** Bachelor of Mining Engineering. **3.** Bachelor of Music Education.
**BMOC** big man on campus.
**BMR** basal metabolic rate.
**B.M.S.** Bachelor of Marine Science.
**B.Mus.** Bachelor of Music.
**bn.** or **Bn. 1.** baron. **2.** battalion.
**B.N.A.** British North America.
**Bngl.** Bangladesh.
**B.O.D.** biochemical oxygen demand.
**Boh.** Bohemia; Bohemian.
**Bol.** Bolivia.
**BOQ** Bachelor Officers' Quarters.
**bor.** borough.
**bot. 1.** botanical; botanist; botany. **2.** bottle. **3.** bottom.
**Bots.** Botswana.
**boul.** boulevard.
**bp** boiling point.
**BP 1.** beautiful people. **2.** or **B/P** bills payable. **3.** blood pressure. **4.** British Pharmacopoeia.
**bp.** bishop.
**B.P. 1.** Bachelor of Pharmacy. **2.** Bachelor of Philosophy.
**bpd** barrels per day.
**B.Pd.** or **B.Pe.** Bachelor of Pedagogy.
**B.P.E.** Bachelor of Physical Education.
**B.Ph.** or **B.Phil.** Bachelor of Philosophy.
**bpi** *Computer Sci.* bits per inch; bytes per inch.
**bpl.** birthplace.
**br. 1.** branch. **2.** brief. **3.** bronze. **4.** brother. **5.** brown.
**Br. 1.** Breton. **2.** Britain; British. **3.** Brother (religious).
**B/R** bills receivable.
**Braz.** Brazil; Brazilian.
**B.R.E.** Bachelor of Religious Education.
**brev.** brevet.
**Br. Gu.** British Guiana.
**Br. Hond.** British Honduras.
**Br. I.** British India.
**brig.** brigade; brigadier.
**Brig. Gen.** brigadier general.
**Brit.** Britain; British.
**bro.** brother.
**bros.** brothers.
**Bru.** Brunei.
**B.S. 1.** Bachelor of Science. **2.** balance sheet. **3.** bill of sale.
**BSA** Boy Scouts of America.
**B.S.A.** Bachelor of Science in Agriculture.
**B.S.A.A.** Bachelor of Science in Applied Arts.
**B.S.Arch.** Bachelor of Science in Architecture.
**B.Sc.** Bachelor of Science.
**B.S.Ec.** Bachelor of Science in Economics.
**B.S.Ed.** Bachelor of Science in Education.
**B.S.E.E.** Bachelor of Science in Electrical Engineering.
**B.S.For.** Bachelor of Science in Forestry.
**B.S.F.S.** Bachelor of Science in Foreign Service.
**bsh.** bushel.
**BSI** British Standards Institution.
**bsk.** basket.
**B.S.N.** Bachelor of Science in Nursing.
**B.S.N.A.** Bachelor of Science in Nursing Administration.
**B.S.N.E.** Bachelor of Science in Nursing Education.
**B.S.Ph.** Bachelor of Science in Pharmacy.
**B.S.P.H.** Bachelor of Science in Public Health.
**Bt.** baronet.
**B.T.** or **B.Th.** Bachelor of Theology.
**btry.** battery.
**Btu** British thermal unit.
**bu. 1.** bureau. **2.** or **bu** bushel.
**bul.** bulletin.
**Bul.** or **Bulg.** Bulgaria; Bulgarian.
**bull.** bulletin.
**bur.** bureau.
**Bur.** Burma; Burmese.
**bus.** business.
**B.V.** Blessed Virgin.
**B.V.M.** Blessed Virgin Mary.
**bvt.** brevet; brevetted.
**BW 1.** biological warfare. **2.** *also* **b/w** black and white.
**B.W.A.** British West Africa.
**B.W.I.** British West Indies.
**bx.** box.
**b.y.** billion years.
**BYO** bring your own.
**BYOB** bring your own bottle (or booze).

**c 1.** *Physics.* candle. **2.** carat. **3.** centi-. **4.** or **C** *Math.* constant. **5.** cubic.
**C 1.** *Elect.* capacitance. **2.** Celsius. **3.** centigrade. **4.** *Physics.* charge conjugation. **5.** coulomb.
**c.** or **C. 1.** capacity. **2.** cape. **3.** carton. **4.** case. **5.** *Baseball.* catcher. **6.** cent. **7.** centime. **8.** century. **9.** chapter. **10.** church. **11.** circa. **12.** *Lat.* congius (*Med.* gallon). **13.** consul. **14.** copy. **15.** copyright. **16.** corps.
**C. 1.** Catholic. **2.** Celtic. **3.** chancellor. **4.** chief. **5.** city. **6.** companion. **7.** Congress. **8.** Conservative. **9.** court.
**ca 1.** centare. **2.** circa.
**CA 1.** California. **2.** *also* **C.A.** chronological age.
**C.A. 1.** Central America. **2.** or **c.a.** chartered accountant.
**c/a** current account.
**CAA** or **C.A.A.** Civil Aeronautics Authority.
**CAB** Civil Aeronautics Board.
**C.A.F.** cost and freight.
**C.A.G.S.** Certificate of Advanced Graduate Study.
**CAI** computer-aided instruction.
**cal** calorie (small).
**Cal** calorie (large).
**cal. 1.** calendar. **2.** caliber.
**Cal.** California.
**calc. 1.** calculation. **2.** calculus.
**Calif.** California.
**Cam.** Cameroon.
**Camb.** Cambodia.
**can. 1.** canceled. **2.** canon. **3.** cant
**Can.** *also* **Canad.** Canada; Can dian.
**canc.** canceled; cancellation.
**C & W** country and western.
**Can. Is.** Canary Islands.
**Cant.** Cantonese.
**CAP** or **C.A.P.** Civil Air Patrol.
**cap. 1.** capacity. **2.** capital (city **3.** capital letter.
**CAPCOM** *Aerospace.* capsule cor municator.
**caps. 1.** capitals (letters). **2.** capsul
**Capt.** or **CAPT** or **CPT** captain.
**car.** carat.
**Card.** Cardinal.
**CARE** Cooperative for America Relief to Everywhere.
**CAT 1.** clear-air turbulence. **2.** co puterized axial tomography.
**cat.** catalogue.
**cath. 1.** cathedral. **2.** cathode.
**CATV** community antenna telev sion.
**caus.** causative.
**cav. 1.** cavalier. **2.** cavalry. **3.** cavi
**CB** or **C.B.** citizens band.
**CBC** complete blood count.
**C.B.D.** cash before delivery.
**CBI** Cumulative Book Index.
**CBW** chemical and biological wa fare.
**cc 1.** carbon copy. **2.** cubic centim ter.
**cc.** chapters.
**CCC 1.** Civilian Conservation Cor **2.** Commodity Credit Corporation.
**CCD** Confraternity of Christia Doctrine.
**CCF** or **C.C.F.** Cooperative Co monwealth Federation of Canada.
**cckw.** counterclockwise.
**CCS** combined chiefs of staff.
**CCTV** closed circuit television.
**CCU** coronary care unit.
**ccw.** counterclockwise.
**cd** *Physics.* candela.
**CD 1.** *also* **C/D** certificate of d posit. **2.** *also* **C.D.** civil defens **3.** *French.* corps diplomatique (dip matic corps).
**cd.** cord.
**c.d.** cash discount.
**CDC** Center for Disease Control.
**Cdr.** or **CDR** commander.
**CDT** or **C.D.T.** Central Daylig Time.
**C.E. 1.** chemical engineer. **2.** civ engineer. **3.** common era. **4.** cu tomer engineer.
**CED** Committee for Economic D velopment.
**CEEB** College Entry Examinati Board.
**CEMF** counter-electromotive forc
**cen. 1.** central. **2.** century.
**Cen. Afr. Rep.** Central Afric Republic.
**cent. 1.** centime. **2.** central. **3.** *L* centum (hundred). **4.** century.
**CEO** *also* **C.E.O.** chief executi officer.
**cert.** certificate; certification; ce fied.
**certif.** certificate.
**CETA** Comprehensive Employme and Training Act.
**cet. par.** *Lat.* ceteris paribus (oth things being equal).
**CF** cystic fibrosis.
**cf. 1.** calfskin. **2.** *Lat.* confer (co pare).
**c.f. 1.** *Baseball.* center field; cen fielder. **2.** or **C.F.** cost and freight.
**C/F** carried forward.
**CFA** *also* **C.F.A.** chartered financ analyst.
**c.f.i.** or **C.F.I.** cost, freight, a insurance.
**cfm** or **c.f.m.** cubic feet per m ute.
**cfs** or **c.f.s.** cubic feet per secor
**cg** centigram.

**g.** 1. center of gravity. 2. *or* **C.G.** onsul general.
**G.** 1. coast guard. 2. commanding eneral.
**gs** *or* **CGS** centimeter-gram-second system of units).
**n** chain (measurement).
**h.** 1. *or* **Ch.** chaplain. 2. chapter. . check. 4. *or* **Ch.** chief. 5. child; hildren. 6. *or* **Ch.** church.
**h.** China; Chinese.
**h.** *or* **C.H.** 1. clearing-house. . courthouse. 3. customhouse.
**han.** channel.
**hanc.** 1. chancellor. 2. chancery.
**hap.** chapter.
**har.** charter.
**hE.** cholinesterase.
**h.E.** chemical engineer.
**hem.** chemical; chemist; chemisry.
**hg.** 1. change. 2. charge.
**hin.** Chinese.
**hl.** chloroform.
**hm.** 1. chairman. 2. checkmate.
**hr** *Bible.* Chronicles.
**hr.** Christ; Christian.
**hron.** 1. chronicle. 2. chronological; chronology.
**hron.** *Bible.* Chronicles.
**hronol.** chronological; chronology.
**i** curie.
**I** cost and insurance.
**IA** Central Intelligence Agency.
**ID** *also* **C.I.D.** Criminal Investigaion Department.
**i.f.** *or* **C.I.F.** cost, insurance, and reight.
**in C** commander in chief.
**ir.** *or* **circ.** circle; circular.
**irc.** 1. circulation. 2. circumference.
**ircum.** circumference.
**it.** 1. citation. 2. cited. 3. citizen.
**iv.** civil; civilian.
**.J.** chief justice.
**k.** 1. cask. 2. check. 3. cook.
**l** centiliter.
**l.** 1. class; classification. 2. clause. . clearance. 4. closet. 5. cloth.
**l.** 1. carload. 2. center line. 3. *or* **.L.** civil law. 4. common law.
**.L.A.** certified laboratory assistant.
**lass.** 1. classic; classical. 2. classifiation; classified; classify.
**lk.** clerk.
**lm.** column.
**lr.** clear.
**LU** *also* **C.L.U.** chartered life unerwriter.
**m** centimeter.
**M** center matched.
**.m.** 1. center of mass. 2. circular ail. 3. court-martial.
**MA** *also* **C.M.A.** certified medical ssistant.
**md.** command.
**mdg.** commanding.
**mdr.** commander.
**ml.** commercial.
**N** centinewton.
**/N** credit note.
**NO** chief of naval operations.
**NS** central nervous system.
**o** *Bible.* Corinthians.
**O** 1. Colorado. 2. *or* **c.o.** commanding officer. 3. *or* **C.O.** conscientious bjector.
**o.** 1. company. 2. county.
**.o.** 1. carried over. 2. cash order.
**/o** *also* **c.o.** care of.
**OD** *or* **C.O.D.** 1. cash on delivery. . collect on delivery.
**oef.** coefficient.
**. of C.** chamber of commerce.
**. of E.** Church of England.
**. of S.** chief of staff.
**og.** cognate.
**ol** *or* **Col.** *Bible.* Colossians.
**ol.** 1. collect; collected; collector. . college; collegiate. 3. colonial; olony. 4. color. 5. column.
**ol.** 1. Colombia. 2. *or* **COL** colonel.
**.** Colorado.
**OLA** cost-of-living adjustment.
**oll.** 1. collateral. 2. collect; collecion; collector. 3. college; collegiate. . colloquial; colloquialism.
**ollat.** collateral.
**olo.** Colorado.

**COM** computer-output microfilm; computer-output microfilmer.
**com.** 1. comedy; comic. 2. comma. 3. commentary. 4. commerce; commercial. 5. *or* **Com.** commissioner. 6. *or* **Com.** committee. 7. common. 8. commune. 9. communication. 10. community.
**Com.** 1. commander. 2. commodore. 3. Communist.
**comb.** 1. combination. 2. combining. 3. combustion.
**comd.** command.
**comdg.** commanding.
**Comdr.** commander.
**Comdt.** commandant.
**coml.** commercial.
**comm.** 1. commerce. 2. commission; commissioner. 3. *also* **Comm.** committee. 4. commonwealth. 5. communication.
**Como.** commodore.
**comp.** 1. companion. 2. comparative; comparison. 3. compensation. 4. compilation; compiled; compiler. 5. complete. 6. compose; composer. 7. composite; composition; compositor. 8. compound. 9. comprehensive. 10. comprising.
**compar.** comparative.
**compd.** compound.
**compt.** compartment.
**Comr.** commissioner.
**con.** 1. concerto. 2. *Law.* conclusion. 3. connection. 4. consolidate; consolidated. 5. *or* **Con.** consul. 6. continued. 7. *Lat.* conjunx (wife).
**Con.** Congo.
**conc.** 1. concentrate. 2. concrete.
**cond.** 1. condition. 2. conductivity. 3. conductor.
**conf.** 1. conference. 2. confidential.
**confed.** confederation.
**cong.** *Lat.* congius (*Med.* gallon).
**Cong.** 1. Congregational. 2. Congress; Congressional.
**conj.** 1. conjugation. 2. conjunction. 3. conjunctive.
**Conn.** Connecticut.
**cons.** 1. consigned; consignment. 2. consonant. 3. *or* **Cons.** constable. 4. constitution; constitutional. 5. construction.
**Cons.** consul.
**consol.** consolidated.
**const.** 1. *or* **Const.** constable. 2. constant. 3. *or* **Const.** constitution. 4. construction.
**constr.** construction.
**cont.** 1. containing. 2. contents. 3. continent. 4. continue; continued. 5. contract. 6. contraction. 7. control.
**contd.** continued.
**contemp.** contemporary.
**contr.** 1. contract. 2. contraction. 3. contralto. 4. control.
**contrib.** contribution; contributor.
**CONUS** Continental United States.
**conv.** 1. convention. 2. convertible.
**coop.** cooperative.
**cop.** copyright.
**Cop.** Coptic.
**cor.** 1. corner. 2. cornet. 3. coroner. 4. corpus. 5. correction. 6. correspondence; correspondent; corresponding.
**Cor.** *Bible.* Corinthians.
**CORE** Congress of Racial Equality.
**corol.** *or* **coroll.** corollary.
**corp.** corporation.
**corr.** 1. correction. 2. correspondence; correspondent.
**correl.** correlative.
**C.O.R.T.** certified operating room technician.
**cos** cosine.
**COS** *or* **C.O.S.** cash on shipment.
**cosec** cosecant.
**cot** cotangent.
**coth** hyperbolic cotangent.
**covers** versed cosine.
**cp** *Physics.* candlepower.
**cP** centipoise.
**CP** 1. chemically pure. 2. command post. 3. Communist Party.
**cp.** compare.
**C.P.** Cape Province.
**CPA** *also* **C.P.A.** certified public accountant.
**CPC** coated paper copier.
**cpd.** compound.

**CPFF** cost plus fixed fee.
**CPI** consumer price index.
**Cpl.** *or* **CPL** corporal.
**cpm** 1. copies per minute. 2. cycles per minute.
**CPO** chief petty officer.
**CPR** cardiopulmonary resuscitation.
**cps** 1. *also* **CPS** characters per second. 2. cycles per second.
**CPS** *also* **C.P.S.** certified professional secretary.
**Cpt.** *or* **CPT** captain.
**CPU** central processing unit.
**CQ** 1. call to quarters. 2. charge of quarters.
**CR** *Psychol.* conditioned reflex; conditioned response.
**cr.** 1. credit; creditor. 2. creek. 3. crescendo. 4. crown.
**C.R.** Costa Rica.
**crit.** critic; critical; criticism.
**CRT** cathode-ray tube.
**C.R.T.T.** certified respiratory therapy technician.
**CS** 1. capital stock. 2. chief of staff. 3. Christian Science; Christian Scientist. 4. civil service. 5. conditioned stimulus.
**cs.** case.
**C.S.A.** Confederate States of America.
**csc** cosecant.
**CSC** civil service commission.
**csch** hyperbolic cosecant.
**CSF** cerebrospinal fluid.
**csk.** 1. cask. 2. countersink.
**CSS** College Scholarship Service.
**CST** 1. *or* **C.S.T.** Central Standard Time. 2. convulsive shock treatment.
**CT** 1. *or* **C.T.** Central Time. 2. *or* **Ct.** Connecticut.
**ct.** 1. cent. 2. certificate. 3. court.
**Ct.** count (title).
**ctf.** certificate.
**ctg.** *or* **ctge.** cartage.
**ctn** cotangent.
**ctn.** carton.
**CTOL** *Aerospace.* conventional takeoff and landing.
**ctr.** 1. center. 2. counter.
**cu.** *or* **cu** cubic.
**cum.** cumulative.
**cur.** 1. currency. 2. current.
**CV** cardiovascular.
**C.V.** Cape Verde.
**CVA** Columbia Valley Authority.
**cvt.** convertible.
**cw** *or* **CW** continuous wave.
**cw.** clockwise.
**CWO** chief warrant officer.
**c.w.o.** cash with order.
**cwt.** hundredweight.
**CY** calendar year.
**cyl.** cylinder.
**CYO** Catholic Youth Organization.
**CZ** *or* **C.Z.** Canal Zone.
**Czech.** Czechoslovakia; Czechoslovakian.

**d** 1. day. 2. deci-. 3. *Physics.* deuteron. 4. dextro-.
**d.** 1. dam. 2. date. 3. daughter. 4. *Lat.* denarius (penny). 5. *or* **D.** deputy. 6. died. 7. *or* **D.** dose. 8. *or* **D.** drachma.
**D** 1. *or* **D.** democrat; democratic. 2. deutrium.
**D.** 1. December. 2. department. 3. *Lat.* Deus (God). 4. diopter. 5. Doctor (in academic degrees). 6. *Lat.* Dominus (Lord). 7. Don (title). 8. duchess. 9. duke. 10. Dutch.
**da** deca-.
**DA** 1. delayed action. 2. deposit account. 3. *also* **D.A.** don't answer.
**Da.** Danish.
**D.A.** 1. *also* **DA** district attorney. 2. Doctor of Arts.
**DAC** Department of the Army Civilian.
**dag** decagram.
**DAGC** *Electron.* delayed automatic gain control.
**DAH** *or* **D.A.H.** Dictionary of American History.
**dam** decameter.
**Dan.** 1. *Bible.* Daniel. 2. Danish.
**D & C** dilatation and curettage.
**DAR** damage assessment routine.

**das** dekastere.
**DASD** *Computer Sci.* direct access storage device.
**dat.** dative.
**dB** decibel.
**DB** *or* **D.B.** daybook.
**d.b.a.** doing business as.
**D.B.A.** Doctor of Business Administration.
**D.B.E.** Dame Commander of the Order of the British Empire.
**d.b.h.** diameter at breast height.
**D.Bib.** Douay Bible.
**dbl.** double.
**dc** *or* **DC** direct current.
**DC** *or* **D.C.** District of Columbia.
**D.C.** 1. *Mus.* da capo. 2. Doctor of Chiropractic.
**D.Ch.E.** Doctor of Chemical Engineering.
**DCI** Director of Central Intelligence.
**D.C.L.** 1. Doctor of Canon Law. 2. Doctor of Civil Law.
**DCM** *also* **D.C.M.** Distinguished Conduct Medal.
**dd.** delivered.
**D.D.** 1. demand draft. 2. dishonorable discharge. 3. *Lat.* Divinitatis Doctor (Doctor of Divinity).
**DDD** Direct Distance Dialing.
**D.D.S.** 1. Doctor of Dental Science. 2. Doctor of Dental Surgery.
**DE** Delaware.
**deb.** debenture.
**dec.** 1. deceased. 2. declaration. 3. declension. 4. declination. 5. decrease.
**Dec.** December.
**decd.** deceased.
**decl.** declension.
**D.Ed.** Doctor of Education.
**def.** 1. defective. 2. defendant. 3. defense. 4. deferred. 5. define. 6. definite. 7. definition.
**deg** *or* **deg.** degree.
**del.** 1. delegate; delegation. 2. delete.
**Del.** Delaware.
**dely.** delivery.
**dem.** demurrage.
**Dem.** Democrat; Democratic.
**demon.** *Gram.* demonstrative.
**Den.** Denmark.
**denom.** denomination.
**dent.** dental; dentist; dentistry.
**dep.** 1. depart; departure. 2. department. 3. deponent. 4. deposed. 5. deposit. 6. depot. 7. deputy.
**Dep.** dependency.
**dept.** 1. department. 2. deputy.
**der.** *or* **deriv.** derivation; derivative.
**Des.** desert.
**det.** 1. detach. 2. detachment. 2. detail.
**Deut.** *Bible.* Deuteronomy.
**dev.** deviation.
**DEW** distant early warning.
**DF** direction finder.
**D.F.** Defender of the Faith.
**D.F.A.** Doctor of Fine Arts.
**DFC** *also* **D.F.C.** Distinguished Flying Cross.
**dft.** draft.
**dg** decigram.
**DH** designated hitter.
**D.H.** Doctor of Humanities.
**D.H.A.** Doctor of Hospital Administration.
**D.H.L.** Doctor of Hebrew Letters; Doctor of Hebrew Literature.
**dia.** diameter.
**diag.** 1. diagonal. 2. diagram.
**dial.** 1. dialect; dialectal. 2. dialectic; dialectical. 3. dialogue.
**diam** diameter.
**dict.** 1. dictation. 2. dictionary.
**diet.** dietetics.
**dif.** *or* **diff.** difference; different.
**dig.** digest.
**dil.** dilute.
**dim.** 1. dimension. 2. diminished. 3. *Mus.* diminuendo. 4. diminutive.
**dimin.** 1. *Mus.* diminuendo. 2. diminutive.
**din.** dinar.
**dipl.** diplomat; diplomatic.
**dir.** director.
**dis.** 1. discount. 2. distance; distant.
**disc.** discount.
**disp.** dispensary.
**diss.** dissertation.
**dissd.** dissolved.

**dist.** 1. distance; distant. 2. district.
**Dist. Atty.** district attorney.
**distr.** distribution; distributor.
**div.** 1. divergence. 2. diversion. 3. divided; division. 4. dividend. 5. divorced.
**dj** dust jacket.
**DJ** disc jockey.
**D.J.** 1. district judge. 2. *Lat.* Doctor Juris (Doctor of Law).
**DJIA** Dow-Jones Industrial Average.
**dk.** 1. dark. 2. deck. 3. dock.
**dkg** dekagram.
**dkl** dekaliter.
**dkm** dekameter.
**dks** dekastere.
**dl** deciliter.
**D/L** demand loan.
**D.Lit.** *or* **D.Litt.** *Lat.* Doctor Litterarum (Doctor of Letters; Doctor of Literature).
**DLO** dead letter office.
**dlr.** dealer.
**D.L.S.** Doctor of Library Science.
**dlvy.** delivery.
**dm** decimeter.
**DM** 1. *Chem.* adamsite. 2. data management. 3. Deutsche mark.
**D.M.A.** Doctor of Musical Arts.
**D.M.D.** *Lat.* Dentariae Medicinae Doctor (Doctor of Dental Medicine).
**D.M.L.** Doctor of Modern Languages.
**DMSO** dimethylsulfoxide.
**DMZ** demilitarized zone.
**Dn** *Bible.* Daniel.
**dn.** down.
**DNB** Dictionary of National Biography.
**DNC** direct numerical control.
**do.** ditto.
**D.O.** 1. Doctor of Optometry. 2. Doctor of Osteopathy.
**DOA** *Med.* dead on arrival.
**DOB** date of birth.
**doc.** document.
**DOD** Department of Defense.
**DOE** Department of Energy.
**dol.** 1. dollar. 2. *Mus.* dolce.
**dom.** 1. domestic. 2. dominant. 3. dominion.
**Dom.** Dominican.
**D.O.M.** *Lat.* Deo Optimo Maximo (to God, the best and the greatest).
**Dom. Rep.** Dominican Republic.
**DOS** disk operating system.
**DOT** Department of Transportation.
**doz.** dozen.
**DP** 1. data processing. 2. dew point. 3. *also* **D.P.** displaced person. 4. *Baseball.* double play.
**DPH** 1. Department of Public Health. 2. *also* **D.P.H.** Doctor of Public Health.
**D.Ph.** *or* **D.Phil.** Doctor of Philosophy.
**dpt.** 1. department. 2. deponent.
**DPT** *Med.* diphtheria, pertussis, tetanus (vaccine).
**DPW** Department of Public Works.
**dr** dram.
**DR** dead reckoning.
**dr.** 1. debit. 2. debtor.
**Dr.** 1. doctor. 2. drive (in street names).
**dram.** dramatic; dramatist.
**dr ap** apothecaries' dram.
**dr avdp** avoirdupois dram.
**dr t** troy dram.
**ds** decistere.
**DS** data set.
**d.s.** 1. *or* **D.S.** *Mus.* dal segno. 2. days after sight. 3. document signed.
**DSC** *also* **D.S.C.** Distinguished Service Cross.
**DSM** *also* **D.S.M.** Distinguished Service Medal.
**DSO** *or* **D.S.O.** Distinguished Service Order.
**d.s.p.** *Lat.* decessit sine prole (died without issue).
**DSRV** deep submergence rescue vehicle.
**DST** *or* **D.S.T.** daylight-saving time.
**Dt** *Bible.* Deuteronomy.
**DT** *or* **D.T.** daylight time.
**d.t.** double time.
**D.T.** Doctor of Theology.
**D.T.'s** delirium tremens.
**Du.** 1. duke (title). 2. Dutch.
**dup.** duplicate.
**D.V.** 1. *Lat.* Deo volente (God willing). 2. *Bible.* Douay Version.
**D.V.M.** Doctor of Veterinary Medicine.
**D.V.S.** Doctor of Veterinary Surgery.
**DW** 1. dead weight. 2. distilled water.
**D/W** dock warrant.
**DWI** driving while intoxicated.
**dwt.** pennyweight.
**dy.** 1. delivery. 2. duty.
**dyn** *Physics.* dyne.
**dz.** dozen.

**e** 1. electron. 2. *or* **e.** *Baseball.* error.
**E** 1. Earth. 2. *also* **E.** *or* **e** *or* **e.** east. 3. *or* **E.** English. 4. excellent.
**e.** *or* **E.** engineer; engineering.
**E.** earl.
**ea.** each.
**E and OE** errors and omissions excepted.
**EbN** east by north.
**EbS** east by south.
**Ec.** Ecuador.
**E.C.** Established Church.
**eccl.** *or* **eccles.** ecclesiastic; ecclesiastical.
**Eccles.** *Bible.* Ecclesiastes.
**ECCS** emergency core cooling system.
**ECG** electrocardiogram.
**ECL** emitter-coupled logic.
**ECM** European Common Market.
**ecol.** ecological; ecology.
**econ.** economics; economist; economy.
**ed.** 1. edition; editor. 2. education.
**E.D.** election district.
**edit.** edition; editor.
**Ed.M.** *Lat.* Educationis Magister (Master of Education).
**EDP** electronic data processing.
**EDT** *or* **E.D.T.** Eastern Daylight Time.
**educ.** education; educational.
**e.e.** errors excepted.
**E.E.** electrical engineer; electrical engineering.
**EEC** European Economic Community.
**EEG** electroencephalogram; electroencephalograph.
**EENT** *or* **E.E.N.T.** eye, ear, nose, and throat.
**EEO** equal employment opportunity.
**EEOC** Equal Employment Opportunity Commission.
**eff.** efficiency.
**EFTS** electronic funds transfer system.
**Eg.** Egypt; Egyptian.
**e.g.** *Lat.* exempli gratia (for example).
**EGD** electrogasdynamics.
**EHF** extremely high frequency.
**EHV** extra high voltage.
**E.I.** East Indian; East Indies.
**EKG** electrocardiogram; electrocardiograph.
**el.** elevation.
**elec.** electric; electrical; electrician; electricity.
**elem.** elementary.
**elev.** elevation.
**ELF** extremely low frequency.
**ELSS** extravehicular life support system.
**EM** 1. electromagnetic. 2. enlisted man.
**E.M.** Engineer of Mines.
**emf** *or* **EMF** electromotive force.
**EMT** 1. electrical-metallic tubing. 2. emergency medical technician. 3. end of magnetic tape.
**emu** electromagnetic unit.
**enc.** *or* **encl.** enclosed; enclosure.
**ency.** *or* **encyc.** *or* **encycl.** encyclopedia.
**ENE** east-northeast.
**eng.** 1. engine. 2. engineer; engineering.
**Eng.** England; English.
**engin.** engineering.
**engr.** 1. engineer. 2. engraved; engraver; engraving.
**enl.** 1. enlarged. 2. enlisted.
**Ens.** *or* **ENS** ensign.
**ENT** *or* **E.N.T.** ear, nose, and throat.
**entom.** entomologic; entomology.
**e.o.** *Lat.* ex officio (by virtue of office).
**EO** executive order.
**e.o.m.** end of month.
**Ep** *Bible.* Ephesians.
**EP** 1. extended play. 2. European plan.
**EPA** Environmental Protection Agency.
**Eph.** *Bible.* Ephesians.
**Epis.** 1. Episcopal; Episcopalian. 2. Epistle.
**Episc.** Episcopal; Episcopalian.
**Epist.** Epistle.
**eq.** 1. equal. 2. equation. 3. equivalent.
**E.Q.** educational quotient.
**Equat. Gui.** Equatorial Guinea.
**equip.** equipment.
**equiv.** equivalency; equivalent.
**ER** emergency room.
**ERA** 1. *Baseball.* earned run average. 2. Equal Rights Amendment.
**ESE** east-southeast.
**Esk.** Eskimo.
**ESL** English as a second language.
**ESOP** employee stock ownership plan.
**ESP** extrasensory perception.
**esp.** especially.
**Esq.** Esquire (title).
**ESR** electron spin resonance.
**Est** *Bible.* Esther.
**EST** *or* **E.S.T.** Eastern Standard Time.
**est.** 1. established. 2. *Law.* estate. 3. estimate.
**esu** electrostatic unit.
**ET** 1. *or* **E.T.** Eastern Time. 2. elapsed time.
**ETA** *or* **e.t.a.** estimated time of arrival.
**et al.** *Lat.* et alii (and others).
**etc.** *Lat.* et cetera (and so forth).
**ETD** *or* **e.t.d.** estimated time of departure.
**Eth.** Ethiopia.
**ETV** educational television.
**etym.** *or* **etymol.** etymological; etymology.
**Eur.** Europe; European.
**EURATOM** European Atomic Energy Community.
**eV** electron volt.
**EVA** extravehicular activity.
**evan.** *or* **evang.** evangelical; evangelist.
**evg.** evening.
**EW** enlisted woman.
**Ex** *or* **Ex.** *Bible.* Exodus.
**ex.** 1. examination. 2. example. 3. except; excepted; exception. 4. exchange. 5. executive. 6. express. 7. extra.
**exam.** examination.
**exc.** 1. excellent. 2. except; exception.
**Exc.** Excellency.
**exch.** 1. exchange. 2. *or* **Exch.** exchequer.
**excl.** 1. exclamation. 2. exclusive.
**exec.** 1. executive. 2. executor.
**Exod.** *Bible.* Exodus.
**exp** *Math.* exponential.
**exp.** 1. expenses. 2. experiment; experimental. 3. expiration; expired. 4. export; exporter. 5. express.
**expt.** experiment.
**exptl.** experimental.
**exr.** executor.
**exrx.** executrix.
**ext.** 1. extension. 2. external; externally. 3. extinct. 4. extra. 5. extract.
**Ezek.** *or* **Ezk** *Bible.* Ezekiel.
**Ezr** *Bible.* Ezra.

**f** 1. *Physics.* femto-. 2. focal length. 3. *or* **F.** *Mus.* forte. 4. function.
**F** 1. Fahrenheit. 2. farad. 3. *or* **F.** fellow (as of a university).
**f.** 1. farthing. 2. *or* **F** *also* **f.** *or* **F.** female. 3. *or* **F.** *Gram.* feminine. 4. *or* **F.** *Metallurgy.* fine. 5. *or* **F.** folio. 6. following. 7. foul. 8. franc.
**F.** 1. February. 2. French. 3. Friday.
**f/** relative aperture of a lens.
**FA** 1. field artillery. 2. *or* **F.A.** fir[e] art. 3. football association.
**f.a.** fire alarm.
**FAA** Federal Aviation Administr[ation].
**f.a.a.** *or* **F.A.A.** free of all averag[e].
**fac.** 1. facsimile. 2. faculty.
**FACA** 1. Fellow of the America[n] College of Anesthesiologists. 2. Fe[l]low of the American College [of] Apothecaries.
**FACC** Fellow of the American Co[l]lege of Cardiologists.
**FACD** Fellow of the American Co[l]lege of Dentists.
**FACOG** Fellow of the America[n] College of Obstetricians and Gyn[e]cologists.
**FACP** Fellow of the American Co[l]lege of Physicians.
**FACR** Fellow of the American Co[l]lege of Radiologists.
**FACS** Fellow of the American Co[l]lege of Surgeons.
**FAD** flavin adenine dinucleotide.
**Fahr.** Fahrenheit.
**FAIA** *or* **F.A.I.A.** Fellow of th[e] American Institute of Architects.
**Falk. Is.** Falkland Islands.
**FAM** Free and Accepted Masons.
**fam.** 1. familiar. 2. family.
**FAO** Food and Agriculture Organiz[a]tion.
**FAQ** fair average quality.
**Far.** Faraday.
**FAS** Foreign Agricultural Service.
**f.a.s.** *also* **F.A.S.** free alongside shi[p].
**fasc.** fascicle.
**fath** *or* **fath.** fathom.
**fb** *also* **f.b.** fullback.
**F.B.** 1. foreign body. 2. freight bi[ll].
**FBA** *or* **F.B.A.** Fellow of the Britis[h] Academy.
**FBI** *also* **F.B.I.** Federal Bureau [of] Investigation.
**fc** foot-candle.
**f.c.** 1. follow copy. 2. font change.
**FCA** Farm Credit Administration.
**FCAP** Fellow of the College [of] American Pathologists.
**fcap.** *or* **fcp.** foolscap.
**FCC** Federal Communications Com[]mission.
**FCPS** Fellow of the College of Phys[i]cians and Surgeons.
**FCS** *or* **F.C.S.** Fellow of the Chemic[al] Society.
**fcy.** fancy.
**FD** 1. fatal dose. 2. fire departmen[t]. 3. focal distance.
**fd.** fjord.
**F.D.** *Lat.* Fidei Defensor (Defend[er] of the Faith).
**FDA** Food and Drug Administratio[n].
**FDIC** Federal Deposit Insuranc[e] Corporation.
**fdn.** foundation.
**fdry.** foundry.
**Feb.** February.
**fec.** *Lat.* fecit (he *or* she made *or* di[d] it).
**fed.** federal; federated; federation.
**fem.** female; feminine.
**FEP** front end processor.
**FEPC** Fair Employment Practice[s] Commission.
**FET** 1. federal excise tax. 2. fiel[d] effect transistor.
**feud.** feudal; feudalism.
**ff** *Mus.* fortissimo.
**ff.** 1. folios. 2. following.
**FG** fine grain.
**f.g.** field goal; field goals.
**fgn.** foreign.
**fgt.** freight.
**FHA** Federal Housing Administra[]tion.
**FHLBB** Federal Home Loan Ban[k] Board.
**fhp** *or* **f.hp.** friction horsepower.
**FICA** Federal Insurance Contribu[]tions Act.
**fict.** 1. fiction. 2. fictitious.
**fid.** fidelity.
**FIFO** first in, first out.
**fig.** 1. figurative; figuratively. 2. fi[g]ure.
**fin.** 1. finance; financial. 2. finish.
**Fin.** Finland; Finnish.
**fl** fluid.
**fL** foot-lambert.

, 1. Florida. 2. focal length. 3. forgn languages.
1. floor. 2. florin. 3. *Lat.* floruit lourished). 4. fluid. 5. flute.
**a.** Florida.
**l.** field.
**dr** fluid dram.
**or.** Florida.
**oz** fluid ounce.
**M** 1. field manual. 2. *or* **F.M.** field arshal. 3. or **fm** frequency modution.
**n.** 1. fathom. 2. from.
**MB** Federal Maritime Board.
**MCS** Federal Mediation and Conliation Service.
**MN** flavin mononucleotide.
. footnote.
**NMA** Federal National Mortgage ssociation.
**)** 1. *or* **F.O.** field officer. 2. field rder. 3. finance officer. 4. *or* **F/O** ight officer. 5. *or* **F.O.** Foreign ffice.
**).b.** *also* **F.O.B.** free on board.
**)BS** fractional orbital bombardent system.
**l.** 1. folio. 2. following.
**r.** 1. foreign. 2. forest; forestry.
**rt.** fortification.
**VD** four-wheel drive.
freezing point.
flash point.
. foolscap.
**).a.** *or* **F.P.A.** free of particular verage.
**PC** 1. Federal Power Commission. fish protein concentrate. Friends Peace Committee.
**m** *or* **f.p.m.** feet per minute.
**PO** fleet post office.
**rf.** fireproof.
**s** *or* **f.p.s.** 1. feet per second. frames per second.
. 1. franc. 2. from.
1. father (clergyman). 2. France; rench. 3. frater. 4. Frau. 5. friar. Friday.
**r.** *Lat.* folio recto (right-hand page).
**RB** Federal Reserve Board.
**RCP** *or* **F.R.C.P.** Fellow of the oyal College of Physicians.
**RCS** *or* **F.R.C.S.** Fellow of the oyal College of Surgeons.
**eq.** 1. frequency. 2. frequentative. frequently.
**R.G.** Federal Republic of Germany.
**RGS** *or* **F.R.G.S.** Fellow of the oyal Geographical Society.
**. Gu.** French Guiana.
**i.** Friday.
**is.** Frisian.
**l.** Fräulein.
**ont.** frontispiece.
**RS** 1. Federal Reserve System. 2. *or* **.R.S.** Fellow of the Royal Society.
**s.** Frisian.
**t.** freight.
**wy.** freeway.
**S** 1. Foreign Service. 2. Forest ervice.
**SA** Federal Security Agency.
**SH** follicle-stimulating hormone.
**SLIC** Federal Savings and Loan nsurance Corporation.
foot.
. fort; fortification.
**TC** Federal Trade Commission.
**-c** foot-candle.
**h.** fathom.
**-lb** foot-pound.
**ır.** furlong.
**ırn.** furnished.
**ıt.** future (in grammar).
**v.** *Lat.* folio verso (on the back of ne page).
**WA** Federal Works Agency.
**WD** front-wheel drive.
**vd.** forward.
**X** foreign exchange.
**Y** fiscal year.
**YI** for your information.
**ZS** *or* **F.Z.S.** Fellow of the Zoological Society.

1. acceleration of gravity. 2. gram.
1. *Physics.* gauss. 2. giga-. 3. *or* **G.** ood. 4. *Physics.* gravitation contant.

**g.** 1. gender. 2. genitive. 3. *or* **G.** gourde. 4. *or* **G.** guilder. 5. *or* **G.** guinea (money). 6. *or* **G.** gulf.
**ga** gauge.
**Ga** *Bible.* Galatians.
**GA** 1. general agent. 2. *also* **G.A.** general assembly. 3. *or* **Ga.** Georgia.
**G.A.** general average.
**gal.** gallon.
**Gal.** *Bible.* Galatians.
**galv.** galvanized.
**Gam.** Gambia.
**GAO** General Accounting Office.
**GAPA** ground-to-air pilotless aircraft.
**GAR** *or* **G.A.R.** Grand Army of the Republic.
**GATT** General Agreement on Tariffs and Trade.
**GAW** guaranteed annual wage.
**gaz.** gazette; gazetteer.
**G.B.** Great Britain.
**GBF** Great Books Foundation.
**GC** gigacycle.
**GCA** ground control approach.
**G.C.B.** Knight of the Grand Cross, Order of the Bath.
**gcd** *or* **g.c.d.** greatest common divisor.
**gcf** *or* **g.c.f.** greatest common factor.
**GCI** ground control intercept.
**GCM** Good Conduct Medal.
**GCT** or **G.c.t.** Greenwich civil time.
**gd.** good.
**G.D.** grand duchy.
**gde.** gourde.
**G.D.R.** German Democratic Republic.
**gds.** goods.
**GED** 1. general educational development. 2. general equivalency diploma.
**GEM** ground-effect machine.
**gen.** 1. gender. 2. general, generally. 3. generator. 4. generic. 5. genitive. 6. genus.
**Gen.** 1. *or* **GEN** general (military rank). 2. *Bible.* Genesis.
**genit.** genitive.
**genl.** general.
**geog.** geographer; geographic; geography.
**geol.** geologic; geologist; geology.
**geom.** geometric; geometry.
**ger.** gerund.
**Ger.** German; Germany.
**GeV** *Physics.* Giga-electron volts.
**GFE** government-furnished equipment.
**GHQ** general headquarters.
**gi** gill (liquid measure).
**GI** 1. gastrointestinal. 2. general issue. 3. *also* **G.I.** Government Issue.
**Gib.** Gibraltar.
**GIGO** *Computer Sci.* garbage in, garbage out.
**Gk.** Greek.
**gl.** gloss.
**GLCM** ground-launched cruise missile.
**gld.** guilder.
**gloss.** glossary.
**gm** gram.
**GM** *or* **G.M.** 1. general manager. 2. grand master. 3. guided missile.
**GMAT** 1. Graduate Management Admissions Test. 2. *or* **G.m.a.t.** Greenwich mean astronomical time.
**GMT** or **G.m.t.** Greenwich mean time.
**GMW** gram-molecular weight.
**Gn** *Bible.* Genesis.
**GNI** gross national income.
**GNP** gross national product.
**GO** general order.
**GOP** *or* **G.O.P.** Grand Old Party.
**Goth.** Gothic.
**gov.** 1. government. 2. *or* **Gov.** governor.
**Gov. Gen.** governor general.
**govt.** government.
**G.P.** *or* **GP** general practitioner.
**GPA** grade-point average.
**g.p.d.** gallons per day.
**g.p.h.** gallons per hour.
**g.p.m.** gallons per minute.
**GPO** 1. general post office. 2. Government Printing Office.
**g.p.s.** gallons per second.

**GPU** *or* **G.P.U.** *Russian.* Gosudarstvennoye Politicheskoye Upravlenie (Government Political Administration).
**GQ** general quarters.
**gr.** 1. grade. 2. grain. 3. gross. 4. group.
**Gr.** Greece; Greek.
**grad.** graduate; graduated.
**gram.** grammar.
**GRAS** generally recognized as safe.
**Grc.** Greece.
**GRE** Graduate Record Examination.
**Grnld.** Greenland.
**gro.** gross.
**gr. wt.** gross weight.
**GS** 1. general staff. 2. ground speed.
**GSA** 1. General Services Administration. 2. Girl Scouts of America.
**GSC** general staff corps.
**GSL** guaranteed student loan.
**GSO** general staff officer.
**GST** *or* **G.s.t.** Greenwich sidereal time.
**GSV** guided space vehicle.
**gt.** 1. gilt. 2. great. 3. *Med.* gutta.
**Gt. Brit.** Great Britain.
**G.T.C.** good till canceled.
**gtd.** guaranteed.
**GTS** gas turbine ship.
**gtt.** *Med.* guttae.
**GU** 1. genitourinary. 2. Guam.
**Guad.** Guadaloupe.
**guar.** guaranteed.
**Guat.** Guatemala.
**Guin.** Guinea.
**Guy.** Guyana.
**gym.** gymnasium; gymnastics.
**gyn.** gynecological; gynecologist; gynecology.
**Gy.Sgt.** gunnery sergeant.

**h** 1. hecto-. 2. *or* **h.** hit. 3. hour. 4. *Physics.* Planck's constant.
**H** 1. *Physics.* Hamiltonian. 2. henry. 3. humidity.
**h.** 1. *or* **H.** harbor. 2. *or* **H.** hard; hardness. 3. or **H.** height. 4. *or* **H.** high. 5. *or* **H.** *Mus.* horn. 6. hundred. 7. *or* **H.** husband.
**ha** 1. hectare. 2. hour angle.
**h.a.** *Lat.* hoc anno (this year).
**Hab** or **Hab.** *Bible.* Habakkuk.
**hab. corp.** habeas corpus.
**Hag.** *Bible.* Haggai.
**Hai.** Haiti.
**hb** *or* **hb.** halfback.
**Hb** hemoglobin.
**H.B.M.** Her *or* His Britannic Majesty.
**HBO** Home Box Office.
**H.C.** 1. hard copy. 2. Holy Communion. 3. House of Commons.
**hcf** *or* **h.c.f.** highest common factor.
**HCL** high cost of living.
**HD** heavy-duty.
**hd.** head.
**hdbk.** handbook.
**hdkf.** handkerchief.
**hdqrs.** headquarters.
**hdwe.** hardware.
**HE** high explosive.
**H.E.** 1. His Eminence. 2. Her or His Excellency.
**Heb** or **Heb.** *Bible.* Hebrews.
**Heb.** Hebrew.
**her.** heraldry.
**herp.** herpetology.
**hex.** hexagon; hexagonal.
**hf** high frequency.
**HF** height finding.
**hf.** half.
**hfs** hyperfine structure.
**hg** 1. hectogram. 2. heliogram.
**Hg** *Bible.* Haggai.
**HG** *also* **H.G.** High German.
**hgb.** hemoglobin.
**HGH** human growth hormone.
**hgt.** height.
**hgwy.** highway.
**H.H.** 1. Her *or* His Highness. 2. His Holiness.
**hhd** hogshead.
**HH.D.** *Lat.* Humanitatum Doctor (Doctor of Humanities).
**HHFA** Housing and Home Finance Agency.
**HHS** Department of Health and Human Services.

**HI** Hawaii.
**H.I.** Hawaiian Islands.
**HIAA** Health Insurance Association of America.
**HID** *Med.* headache, insomnia, depression.
**H.I.H.** Her or His Imperial Highness.
**HII** Health Insurance Institute.
**H.I.M.** Her *or* His Imperial Majesty.
**hist.** historian; historical; history.
**hl** hectoliter.
**H.L.** House of Lords.
**hld.** hold.
**HLF** Heart and Lung Foundation.
**hlt.** halt.
**hm** hectometer.
**H.M.** Her or His Majesty.
**HMAS** *or* **H.M.A.S.** Her *or* His Majesty's Australian Ship.
**HMC** *or* **H.M.C.** Her or His Majesty's Customs.
**HMCS** *or* **H.M.C.S.** Her *or* His Majesty's Canadian Ship.
**HMF** *or* **H.M.F.** Her or His Majesty's Forces.
**HMO** 1. health maintenance organization. 2. heart minute output.
**HMS** *or* **H.M.S.** Her or His Majesty's Ship.
**Ho.** *Bible.* Hosea.
**HO** head office; home office.
**ho.** house.
**Hon.** 1. Honorable (title). 2. *or* **hon.** honorary.
**Hond.** Honduras.
**HOP** high oxygen pressure.
**HOPE** Health Opportunity for People Everywhere.
**hor.** horizontal.
**hort.** horticultural; horticulture.
**Hos.** *Bible.* Hosea.
**hosp.** hospital.
**hp** horsepower.
**HP** high pressure.
**HPF** highest possible frequency.
**HQ** *or* **h.q.** headquarters.
**hr** hour.
**Hr.** Herr.
**h.r.** home run.
**H.R.** 1. home rule. 2. House of Representatives.
**H.R.E.** Holy Roman Emperor; Holy Roman Empire.
**H. Rept.** House report.
**H. Res.** House resolution.
**H.R.H.** Her *or* His Royal Highness.
**hrs** hours.
**HS** *or* **H.S.** high school.
**HSGT** high-speed ground transit.
**H.S.H.** Her *or* His Serene Highness.
**HSL** high-speed launch.
**HST** 1. *or* **H.S.T.** Hawaiian Standard Time. 2. hypersonic transport.
**ht** height.
**HT** 1. halftime. 2. halftone. 3. *or* **H.T.** Hawaiian Time. 4. hydrotherapy.
**Hts.** Heights.
**HUD** *or* **H.U.D.** Housing and Urban Development.
**Hun.** *or* **Hung.** Hungarian; Hungary.
**HV** 1. high velocity. 2. high-voltage.
**hvy.** heavy.
**HW** 1. high water. 2. hot water.
**HWM** high-water mark.
**hwy.** highway.
**hy.** henry.
**hyp.** 1. hypotenuse. 2. hypothesis.
**hypoth.** hypothesis.
**Hz** hertz.

**i** 1. *or* **I** *Elect.* current. 2. *Math.* imaginary unit. 3. interest. 4. intransitive. 5. *or* **I.** island; isle.
**I** isospin.
**IA** *or* **Ia.** Iowa.
**i.a.** *Lat.* in absentia (in absence).
**IAA** indoleacetic acid.
**IADB** Inter-American Defense Board.
**IAEA** International Atomic Energy Agency.
**IALC** instrument approach and landing chart.
**IAP** international airport.
**IAS** indicated air speed.
**IATA** International Air Transport Association.

**ib.** or **ibid.** *Lat.* ibidem (in the same place).
**IBY** International Biological Year.
**IC** integrated circuit.
**ICA** International Cooperation Administration.
**ICAO** International Civil Aviation Organization.
**ICBM** intercontinental ballistic missile.
**ICC** 1. Indian Claims Commission. 2. Interstate Commerce Commission.
**ICE** internal-combustion engine.
**Ice.** or **Icel.** Iceland; Icelandic.
**ICJ** International Court of Justice.
**ICU** intensive care unit.
**ID** 1. or **Id.** Idaho. 2. *also* **I.D.** identification. 3. Intelligence Department. 4. intradermal.
**id.** *Lat.* idem (the same).
**i.d.** inside diameter.
**IDA** International Development Association.
**IDDD** international direct-distance dialing.
**IDP** 1. inosine diphosphate. 2. integrated data processing. 3. international driving permit.
**IE** industrial engineer; industrial engineering.
**i.e.** *Lat.* id est (that is).
**IF** or **i.f.** intermediate frequency.
**IFC** International Finance Corporation.
**IFF** identification, friend or foe.
**IFO** identified flying object.
**Ig** immunoglobulin.
**IG** or **I.G.** inspector general.
**ign.** ignition.
**IGY** International Geophysical Year.
**ihp** or **i.hp.** indicated horsepower.
**IL** Illinois.
**ill.** illustrated; illustration; illustrator.
**Ill.** Illinois.
**illus.** illustrated; illustration; illustrator.
**ILP** or **I.L.P.** Independent Labour Party.
**ILS** instrument landing system.
**IM** intramuscular.
**imdtly** immediately.
**IMF** International Monetary Fund.
**imit.** imitate; imitation.
**immun.** immunity; immunization.
**immunol.** immunology.
**imp.** 1. imperative. 2. imperfect. 3. imperial. 4. import; imported; importer. 5. important. 6. imprimatur.
**in** or **in.** inch.
**IN** Indiana.
**inbd.** inboard.
**inc.** 1. income. 2. incomplete. 3. *also* **Inc.** incorporated. 4. increase.
**incl.** including; inclusive.
**incog.** incognito.
**incr.** 1. increase; 2. incremental.
**IND** investigational new drug.
**ind.** 1. independence; independent. 2. index. 3. indigo. 4. industrial; industry.
**Ind.** 1. India. 2. Indian. 3. Indiana. 4. Indies.
**Ind. E.** industrial engineer.
**indef.** indefinite.
**indic.** 1. indicative (in grammar). 2. indicator.
**indiv.** individual.
**indn.** indication.
**Indon.** Indonesia; Indonesian.
**indus.** industrial; industry.
**inf.** 1. or **Inf.** infantry. 2. inferior. 3. infinitive. 4. influence. 5. information.
**infin.** infinitive.
**infl.** influence; influenced.
**INH** isoniazid.
**inj.** injection.
**INP** International News Photo.
**inq.** inquiry.
**I.N.R.I.** *Lat.* Iesus Nazarenus Rex Iudaeorum (Jesus of Nazareth, King of the Jews).
**INS** Immigration and Naturalization Service.
**ins.** 1. inspector. 2. insulated; insulation. 3. insurance.
**insp.** inspected; inspector.
**inst.** 1. instant. 2. or **Inst.** institute; institution. 3. instrument.
**instr.** 1. instruction; instructor. 2. instrument.
**int.** 1. interest. 2. interior. 3. internal. 4. international. 5. interval.
**inter.** intermediate.
**interj.** interjection.
**interp.** interpreter.
**interrog.** interrogative.
**intl.** international.
**intr.** intransitive.
**intro.** introduction; introductory.
**inv.** 1. invented; invention; inventor. 2. invoice.
**I/O** input/output.
**Ion.** Ionic.
**IP** installment paid.
**IPA** 1. International Phonetic Alphabet. 2. isopropyl alcohol.
**ips** or **i.p.s.** inches per second.
**IQ** or **I.Q.** intelligence quotient.
**i.q.** *Lat.* idem quod (the same as).
**IR** 1. information retrieval. 2. infrared.
**Ir.** Irish.
**IRA** 1. Individual Retirement Account. 2. *also* **I.R.A.** Irish Republican Army.
**IRBM** intermediate range ballistic Missile.
**Ire.** Ireland.
**irid.** iridescent.
**IRO** International Refugee Organization.
**irreg.** irregular; irregularly.
**IRS** Internal Revenue Service.
**Is** *Bible.* Isaiah.
**is.** or **Is.** island.
**Isa.** *Bible.* Isaiah.
**ISBN** International Standard Book Number.
**ISC** interstate commerce.
**isl.** island.
**Isr.** Israel; Israeli.
**ISSN** International Standard Serial Number.
**IST** insulin shock therapy.
**isth.** isthmus.
**ISV** International Scientific Vocabulary.
**It.** Italian; Italy.
**I.T.** inhalation therapist.
**ITA** Initial Teaching Alphabet.
**ital.** italic.
**Ital.** Italian; Italy.
**ITV** instructional television.
**IU** international unit.
**IUD** intrauterine device.
**IV** intravenous; intravenously.
**IW** 1. index word. 2. isotopic weight.
**i.w.** inside width.

**J** 1. current density. 2. joule.
**J.** 1. journal. 2. judge. 3. justice.
**JA** 1. joint account. 2. *also* **J.A.** judge advocate.
**JAG** *also* **J.A.G.** Judge Advocate General.
**Jam.** Jamaica.
**Jan.** January.
**Jav.** Javanese.
**Jb** *Bible.* Job.
**JC** junior college.
**J.C.D.** *Lat.* Juris Canonici Doctor (Doctor of Canon Law).
**JCL** *Computer Sci.* job control language.
**JCS** or **J.C.S.** Joint Chiefs of Staff.
**jct.** junction.
**JD** 1. Justice Department. 2. *also* **J.D.** juvenile delinquent.
**J.D.** *Lat.* Jurum Doctor (Doctor of Laws).
**Jdt** *Bible.* Judith.
**Jer.** *Bible.* Jeremiah.
**JFET** junction field effect transistor.
**jg** junior grade.
**Jg** *Bible.* Judges.
**JIT** job instruction training.
**JJ** 1. judges. 2. justices.
**Jl** *Bible.* Joel.
**Jm** *Bible.* James.
**Jn** *Bible.* John.
**jnr.** junior.
**Jon** or **Jon.** *Bible.* Jonah.
**Jos** *Bible.* Joshua.
**Josh.** *Bible.* Joshua.
**jour.** 1. journal; journalist. 2. journeyman.
**JP** or **J.P.** justice of the peace.
**Jr** *Bible.* Jeremiah.
**jr.** or **Jr.** junior.
**J.S.D.** *Lat.* Juris Scientiae Doctor (Doctor of Juristic Science).
**jt.** joint.
**Judg.** *Bible.* Judges.
**jun.** or **Jun.** junior.
**junc.** junction.
**juv.** juvenile.
**JV** junior varsity.
**jwlr.** jeweler.

**k** 1. karat. 2. kilo-.
**K** 1. kaon. 2. kelvin (temperature unit). 3. Kelvin (temperature scale). 4. kindergarten. 5. king (chess). 6. *Bible.* Kings.
**k.** or **K.** 1. king. 2. knight. 3. kopeck. 4. koruna. 5. krona. 6. krone.
**ka** cathode.
**KB** king's bishop (chess).
**kc** kilocycle.
**K.C.** King's Counsel.
**kcal** kilocalorie.
**kcs** or **kc/s** kilocycles per second.
**KD** 1. kiln-dried. 2. knocked down.
**Ken.** Kentucky.
**keV** kiloelectron volt.
**kg** kilogram.
**kg.** keg.
**K.G.** Knight of the Order of the Garter.
**KGB** or **K.G.B.** *Russian.* Komitět Gosudarstvěnnoi Bezopasnost'i (Committee of State Security).
**KIA** killed in action.
**KKK** or **K.K.K.** Ku Klux Klan.
**km** kilometer.
**kmph** kilometers per hour.
**kmps** kilometers per second.
**kn.** 1. knot. 2. krona. 3. krone.
**Knt** knight.
**K of C** Knights of Columbus.
**Kor.** Korea; Korean.
**KP** 1. king's pawn (chess). 2. kitchen police.
**K.P.** Knights of Pythias.
**KR** king's rook (chess).
**kr.** 1. krona. 2. krone.
**KS** Kansas.
**Kt** knight (chess).
**kt.** karat.
**Kuw.** Kuwait.
**kW** kilowatt.
**kWh** kilowatt-hour.
**KY** or **Ky.** Kentucky.

**l** liter.
**L** 1. lambert. 2. *also* **L.** large.
**l.** 1. *also* **L.** lake. 2. land. 3. late. 4. left. 5. length. 6. line. 7. lira.
**L.** 1. Latin. 2. licentiate (in titles). 3. Linnaean. 4. lodge (society).
**LA** or **La.** Louisiana.
**L.A.** 1. Legislative Assembly. 2. local agent. 3. *also* **LA** Los Angeles.
**lab.** laboratory.
**Lab.** Labrador.
**lam.** laminated.
**Lam.** *Bible.* Lamentations.
**lang.** language.
**lat.** latitude.
**Lat.** 1. Latin. 2. Latvia; Latvian.
**lav.** lavatory.
**LB** Labrador.
**lb.** *Lat.* libra (pound).
**lc** *also* **l.c.** lower case.
**LC** 1. landing craft. 2. or **L.C.** Library of Congress.
**L/C** letter of credit.
**lcd** or **l.c.d.** lowest common denominator.
**LCDR** lieutenant commander.
**LCL** less-than-carload lot.
**lcm** or **l.c.m.** least common multiple.
**LCM** landing craft, mechanized.
**L.Cpl.** or **L/Cpl** lance corporal.
**LCS** landing craft, support.
**LCT** 1. landing craft, tank. 2. local civil time.
**LD** 1. learning disability; learning-disabled. 2. *Med.* lethal dose.
**ld.** 1. lead (in printing). 2. load.
**Ld.** 1. limited. 2. lord (title).
**LDC** less-developed country.
**ldg.** landing.
**lea.** 1. league. 2. leather.
**Leb.** Lebanese; Lebanon.
**lect.** lecture.
**lectr.** lecturer.
**LED** light-emitting diode.
**leg.** 1. legal. 2. legate. 3. *Mus.* legat 4. legislation; legislative; legislatu
**legis.** legislation; legislative; legis ture.
**LEM** lunar excursion module.
**Leso.** Lesotho.
**Lev.** *Bible.* Leviticus.
**lex.** lexicon.
**lf** 1. *also* **l.f.** or **lf.** lightface. 2. lo frequency.
**LG** *also* **L.G.** Low German.
**lg.** or **lge.** large.
**l.h.** *also* **LH** left hand.
**li** link (unit of measurement).
**L.I.** Long Island.
**lib.** 1. liberal. 2. librarian; library.
**Lib.** 1. Liberal. 2. Liberia; Liberian.
**Liech.** Liechtenstein.
**lieut.** lieutenant.
**LIFO** last in, first out.
**lim.** limit.
**lin.** 1. lineal. 2. linear.
**ling.** linguistics.
**liq.** 1. liquid. 2. liquor.
**lit.** 1. liter. 2. literal; literally. 3. lite ary. 4. literature.
**Lit.B.** or **Litt.B.** *Lat.* Litteraru Baccalaureus (Bachelor of Letter Bachelor of Literature).
**Lit.D.** or **Litt.D.** *Lat.* Litteraru Doctor (Doctor of Letters; Doctor Literature).
**lith.** lithograph; lithographic; litho raphy.
**Lith.** Lithuania; Lithuanian.
**litho.** or **lithog.** lithograph; lith graphic; lithography.
**Lk** *Bible.* Luke.
**ll** or **ll.** lines.
**LL.B.** *Lat.* Legum Baccalaureu (Bachelor of Laws).
**LL.D.** *Lat.* Legum Doctor (Doctor Laws).
**LL.M.** *Lat.* Legum Magister (Maste of Laws).
**lm** lumen.
**Lm** *Bible.* Lamentations.
**LM** lunar module.
**LMG** light machine gun.
**LMT** local mean time.
**ln** Napierian logarithm; natura logarithm.
**LNG** liquefied natural gas.
**loc. cit.** *Lat.* loco citato (in th place cited).
**long.** longitude.
**loq.** *Lat.* loquitur (speaks).
**LPG** liquefied petroleum gas.
**LPM** or **lpm** lines per minute.
**LPN** or **L.P.N.** licensed practica nurse.
**LRV** lunar roving vehicle.
**L.S.** *Lat.* locus sigilli (the place of th seal).
**LSAT** Law School Admissions Tes
**LSD** 1. least significant digit. 2. lyse gic acid diethylamide.
**LSS** lifesaving service.
**lt.** light.
**Lt.** or **LT** lieutenant.
**l.t.** or **LT** local time.
**Lt. Col.** or **LTC** lieutenant colone
**Lt. Comdr.** or **LCDR** lieutenan commander.
**ltd.** or **Ltd.** limited.
**Lt. Gen.** or **LTG** lieutenant general
**Lt. Gov.** lieutenant governor.
**LTJG** lieutenant, junior grade.
**LTS** 1. launch telemetry station 2. launch tracking station.
**Luth.** Lutheran.
**Lux.** Luxembourg.
**Lv** *Bible.* Leviticus.
**lv.** 1. leave. 2. livre.
**LW** low water.
**LWM** low-water mark.
**lx** lux.
**LXX** Septuagint.
**lyr.** lyric.
**LZ** landing zone.

**m 1.** or **M a.** em (printing measure). **b.** pica em. **2.** *Physics.* mass. **3.** meter (measure). **4.** milli-. **5.** or **M** *Physics.* modulus.

**M 1.** *Bible.* Maccabees. **2.** *Physics.* Mach number. **3.** mega-. **4.** *Chem.* metal. **5.** middle term (of a syllogism). **6.** *Chem.* molar. **7.** *Physics.* moment. **8.** *Physics.* mutual inductance.

**m. 1.** or **M.** male. **2.** manual. **3.** married. **4.** or **M.** masculine. **5.** or **M.** medium. **6.** or **M.** meridian. **7.** or **M.** *Lat.* meridies (noon). **8.** mile. **9.** month. **10.** morning.

**M. 1.** majesty. **2.** mark (currency). **3.** master (in titles). **4.** medieval. **5.** member (in titles). **6.** mill (currency). **7.** minim (liquid measure). **8.** Monday. **9.** Monsieur.

**mA** milliampere.

**MA 1.** Maritime Administration. **2.** Massachusetts. **3.** *also* **M.A.** mental age. **4.** *also* **M.A.** military academy.

**M.A.** *Lat.* Magister Artium (Master of Arts).

**M.A.B.E.** Master of Agricultural Business and Economics.

**MAC** Municipal Assistance Corporation.

**Maced.** Macedonia; Macedonian.

**mach.** machine; machinery; machinist.

**MACV** Military Assistance Command, Vietnam.

**MAD** mutual assured destruction.

**Mad.** or **Madag.** Madagascar.

**M.A.E. 1.** Master of Aeronautical Engineering. **2.** Master of Art Education. **3.** Master of Arts in Education.

**M.A.Ed.** Master of Arts in Education.

**mag. 1.** magazine. **2.** magnetism. **3.** magneto. **4.** magnitude.

**M.Agr.** Master of Agriculture.

**Maj.** or **MAJ** major.

**Maj. Gen.** or **MG** major general.

**Mal. 1.** *Bible.* Malachi. **2.** Malay; Malayan.

**Mala.** Malaysia.

**M.A.L.S.** Master of Arts in Library Science.

**man.** manual.

**Man.** Manitoba.

**manuf.** or **manufac.** manufacture.

**MAO** monoamine oxidase.

**MAP** modified American plan.

**mar. 1.** maritime. **2.** married.

**Mar.** March.

**March.** marchioness.

**marg.** margin.

**MARS** Military Affiliate Radio System.

**Mart.** Martinique.

**masc.** masculine.

**MASH** Mobile Army Surgical Hospital.

**Mass.** Massachusetts.

**mat.** matinee.

**M.A.T.** Master of Arts in Teaching.

**math.** mathematical; mathematician; mathematics.

**Matt.** *Bible.* Matthew.

**max.** maximum.

**mb** millibar.

**MB** Manitoba.

**M.B.A.** Master of Business Administration.

**MBO** management by objectives.

**mc** millicurie.

**Mc** megacycle.

**MC 1.** Marine Corps. **2.** Medical Corps. **3.** or **M.C.** Member of Congress.

**M.C.** or **m.c.** master of ceremonies.

**MCAT** Medical College Admissions Test.

**mcf** thousand cubic feet.

**M.C.L.** Master of Civil Law.

**MCP** male chauvinist pig.

**M.C.S.** Master of Computer Science.

**MD 1.** or **Md.** Maryland. **2.** medical department. **3.** muscular dystrophy.

**M.D.** *Lat.* Medicinae Doctor (Doctor of Medicine).

**m/d** months after date.

**Mdm.** Madam.

**M.D.S.** Master of Dental Surgery.

**mdse.** merchandise.

**ME 1.** or **Me.** Maine. **2.** *also* **M.E.** Middle English.

**M.E. 1.** mechanical engineer; mechanical engineering. **2.** medical examiner. **3.** military engineer. **4.** mining engineer.

**meas.** measurable; measure.

**mech. 1.** mechanical; mechanics. **2.** mechanism.

**med. 1.** medical; medicine. **2.** medieval. **3.** medium.

**M.Ed.** Master of Education.

**Medit.** Mediterranean.

**Med. Lat.** Medieval Latin.

**mem. 1.** member. **2.** memoir. **3.** memorandum. **4.** memorial.

**mep** or **m.e.p.** mean effective pressure.

**meq.** milliequivalent.

**mer.** meridian.

**met. 1.** metaphor. **2.** metaphysics. **3.** meteorological; meteorology. **4.** metropolitan.

**metal.** or **metall.** metallurgic; metallurgy.

**metaph. 1.** metaphor; metaphoric. **2.** metaphysics.

**meteor.** or **meteorol.** meteorological; meteorology.

**METO** Middle East Treaty Organization.

**mev** or **Mev** million electron volts.

**Mex.** Mexican; Mexico.

**mf** medium frequency.

**mF** millifarad.

**m.f.** *Mus.* mezzo-forte.

**M.F.A.** Master of Fine Arts.

**mfd.** manufactured.

**mfg.** manufacture; manufactured; manufacturing.

**MFN** most-favored nation.

**mfr.** manufacture; manufacturer.

**mg** milligram.

**MG** Major General.

**Mgr. 1.** or **mgr.** manager. **2.** Monseigneur; Monsignor.

**mgt.** management.

**mH** millihenry.

**MH 1.** Medal of Honor. **2.** mental health.

**M.H.A.** Master of Hospital Administration.

**MHD** magnetohydrodynamic.

**M.H.L.** Master of Hebrew Literature.

**MHW** mean high water.

**MHz** megahertz.

**Mi** *Bible.* Micah.

**MI 1.** Michigan. **2.** military intelligence.

**mi. 1.** mile. **2.** mill (monetary unit).

**MIA** missing in action.

**Mic.** *Bible.* Micah.

**Mich.** Michigan.

**MICR** *Computer Sci.* magnetic ink character recognition.

**mid.** middle.

**mil.** military; militia.

**min. 1.** mineralogical; mineralogy. **2.** minimum. **3.** mining. **4.** minor. **5.** or **min** minute.

**Minn.** Minnesota.

**MIPS** million instructions per second.

**misc.** miscellaneous.

**Miss.** Mississippi.

**Mk** *Bible.* Mark.

**mk. 1.** mark. **2.** markka.

**mks** meter-kilogram-second (system of units).

**mksA** meter-kilogram-second-ampere (system of units).

**mkt.** market.

**mktg.** marketing.

**ml** milliliter.

**Ml** *Bible.* Malachi.

**ML** *also* **M.L.** Medieval Latin.

**MLD** minimum lethal dose.

**Mlle.** Mademoiselle.

**Mlles.** Mesdemoiselles.

**M.L.S.** Master of Library Science.

**M.L.T.** Medical Laboratory Technician.

**MLW** mean low water.

**mm** millimeter.

**MM.** Messieurs.

**m.m.** *Lat.* mutatis mutandis (with the necessary changes having been made).

**Mme.** Madame.

**Mmes.** Mesdames.

**mmf** or **m.m.f.** magnetomotive force.

**MMPI** Minnesota Multiphasic Personality Inventory.

**MN 1.** magnetic north. **2.** Minnesota.

**M.N.** Master of Nursing.

**mngr.** manager.

**MO** or **Mo.** Missouri.

**mo.** month.

**m.o.** or **M.O. 1.** mail order. **2.** medical officer. **3.** modus operandi. **4.** *also* **MO** money order.

**mod** *Math.* modulus.

**mod. 1.** moderate. **2.** *Mus.* moderato. **3.** modern.

**modif.** modification.

**MOL** Manned Orbital Laboratory.

**mol.** molecular; molecule.

**mol wt** molecular weight.

**m.o.m.** middle of month.

**mon. 1.** monastery. **2.** monetary.

**Mon.** Monday.

**Mong.** Mongolia; Mongolian.

**Mont.** Montana.

**MOR** middle-of-the-road.

**mor.** morocco (leather).

**Mor.** Moroccan; Morocco.

**morph.** morphological; morphology.

**mos.** months.

**Moz.** Mozambique.

**mp** or **m.p. 1.** melting point. **2.** *Mus.* mezzo-piano.

**MP** or **M.P. 1.** military police; military policeman. **2.** mounted police.

**M.P.** Member of Parliament.

**M.P.A. 1.** Master of Public Administration. **2.** Master of Public Accounting.

**M.Pd.** Master of Pedagogy.

**M.P.E.** Master of Physical Education.

**mpg** or **m.p.g.** miles per gallon.

**mph** or **m.p.h.** miles per hour.

**M.P.H.** Master of Public Health.

**Mr.** Mister.

**MR** map reference.

**mRNA** messenger RNA.

**ms** millisecond.

**MS 1.** Mississippi. **2.** multiple sclerosis.

**ms.** or **MS.** or **ms** manuscript.

**M.S.** or **M.Sc.** *Lat.* Magister Scientiae (Master of Science).

**msec** millisecond.

**MSG 1.** master sergeant. **2.** monosodium glutamate.

**msg.** message.

**Msgr.** Monseigneur; Monsignor.

**M.Sgt.** or **MSG** or **MSGT** master sergeant.

**MSH** melanocyte-stimulating hormone.

**M.S. in L.S.** Master of Science in Library Science.

**m.s.l.** or **M.S.L.** mean sea level.

**mss.** or **MSS.** or **mss** manuscripts.

**MST** or **M.S.T.** Mountain Standard Time.

**M.S.W. 1.** Master of Social Welfare. **2.** Master of Social Work.

**Mt** *Bible.* Matthew.

**MT 1.** or **M.T.** medical technologist. **2.** Montana. **3.** or **M.T.** Mountain Time.

**mt.** or **Mt.** mount; mountain.

**m.t.** or **M.T.** metric ton.

**mtg. 1.** meeting. **2.** mortgage.

**mtge.** mortgage.

**mtn.** mountain.

**mts.** or **Mts.** mountains.

**mun.** or **munic.** municipal; municipality.

**mus. 1.** museum. **2.** music; musical; musician.

**Mus.B.** *Lat.* Musicae Baccalaureus (Bachelor of Music).

**Mus.D.** or **Mus.Dr.** *Lat.* Musicae Doctor (Doctor of Music).

**Mus. M.** *Lat.* Magister Musicae (Master of Music).

**mV** millivolt.

**MV 1.** market value. **2.** mean variation. **3.** megavolt. **4.** motor vessel.

**MVA** Missouri Valley Authority.

**MVD** *Russian.* Ministeyrstvo Vnutreynnikh Deyl (Ministry of Internal Affairs).

**MVP** most valuable player.

**mW** milliwatt.

**MW** megawatt.

**Mx** *Physics.* maxwell.

**mxd.** mixed.

**m.y.** million years.

**myc.** or **mycol.** mycological; mycology.

**myth.** or **mythol.** mythological; mythology.

**n 1.** or **N** en (printing measure). **2.** nano-. **3.** neutron. **4.** *also* **N** or **n-** *Chem.* normal. **5.** *Math.* symbol for indefinite number.

**N 1.** Avogadro number. **2.** knight (chess). **3.** newton. **4.** *also* **N.** or **n** or **n.** north; northern.

**n. 1.** *Lat.* natus (born). **2.** net. **3.** or **N.** noon. **4.** note. **5.** noun. **6.** number.

**N. 1.** Norse. **2.** November.

**Na** *Bible.* Nahum.

**N.A. 1.** Narcotics Anonymous. **2.** National Academician; National Academy. **3.** North America. **4.** not applicable. **5.** not available.

**NAACP** or **N.A.A.C.P.** National Association for the Advancement of Colored People.

**NAB** New American Bible.

**NAD** nicotinamide-adenine dinucleotide.

**NADP** nicotinamide-adenine dinucleotide phosphate.

**Nah.** *Bible.* Nahum.

**NARU** Naval Air Reserve Unit.

**NASA** National Aeronautics and Space Administration.

**NASDAQ** National Association of Securities Dealers Automated Quotations.

**nat. 1.** national. **2.** native. **3.** natural.

**natl.** national.

**NATO** North Atlantic Treaty Organization.

**NATS** Naval Air Transport Service.

**naut.** nautical.

**nav. 1.** naval. **2.** navigable. **3.** navigation.

**Nb** *Bible.* Numbers.

**NB 1.** narrow band. **2.** or **N.B.** New Brunswick.

**n.b.** or **N.B.** nota bene.

**NBA** *also* **N.B.A.** narrow-band allocation.

**NbE** north by east.

**NBS** National Bureau of Standards.

**NbW** north by west.

**NC 1.** no charge. **2.** or **N.C.** North Carolina. **3.** numerical control. **4.** Nurse Corps.

**N.Cal.** New Caledonia.

**NCI** noncoded information.

**NCO** or **N.C.O.** noncommissioned officer.

**NCV** no commercial value.

**ND** or **N.D.** North Dakota.

**n.d.** or **N.D.** no date.

**NDI** no data available.

**N.Dak.** North Dakota.

**NDEA** National Defense Education Act.

**Ne** *Bible.* Nehemiah.

**NE 1.** Nebraska. **2.** or **N.E.** New England. **3.** northeast. **4.** not equal to.

**NEB** New English Bible.

**NEbE** northeast by east.

**NEbN** northeast by north.

**Nebr.** Nebraska.

**NED** or **N.E.D.** New English Dictionary (Oxford).

**neg.** negative.

**Neh.** *Bible.* Nehemiah.

**n.e.i.** not elsewhere included; not elsewhere indicated.

**n.e.m.** not elsewhere mentioned.

**NEP** or **N.E.P.** New Economic Policy.

**Nep.** Nepal.

**n.e.s.** not elsewhere specified.

**NEST** Nuclear Energy Search Team.

**NET** National Educational Television.

**Neth.** Netherlands.

**neur.** or **neurol.** neurological; neurology.

**neut. 1.** neuter. **2.** neutral.

**Nev.** Nevada.

**Newf.** Newfoundland.

**New Hebr.** New Hebrides.

**New M.** New Mexico.

**New Test.** New Testament.

**NF 1.** *also* **N.F.** National Formulary. **2.** neurofibromatosis. **3.** Newfoundland. **4.** nonfiler.

**n/f** no funds.

**Nfld.** Newfoundland.

**NFS** not for sale.

**ng** nanogram.

**NG** *also* **N.G. 1.** National Guard. **2.** no good.

**NGr** *or* **NGr.** New Greek.

**NH** *or* **N.H.** New Hampshire.

**N.Heb.** New Hebrides.

**NHI** National Health Insurance.

**NI** national income.

**Nic.** Nicaragua.

**Nig.** Nigeria.

**NIH** National Institutes of Health.

**N.Ire.** Northern Ireland.

**NIT 1.** National Intelligence Test. **2.** National Invitational Tournament.

**NJ** *or* **N.J.** New Jersey.

**NKVD** *or* **N.K.V.D.** *Russian.* Narodny Kommisariat Vnutrennikh Del (People's Commissariat for Internal Affairs).

**NL** *also* **n.l. 1.** new line. **2.** *also* **N.L.** New Latin.

**n.l.** *Lat.* non licet (not permitted).

**N.L.** *Lat.* non liquet (not clear).

**NLF** National Liberation Front.

**NLRB** *also* **N.L.R.B.** National Labor Relations Board.

**nm 1.** nanometer. **2.** *or* **n.m.** nautical mile. **3.** nuclear magneton.

**NM** *or* **N.M.** New Mexico.

**N.Mex.** New Mexico.

**NNE** north-northeast.

**NNW** north-northwest.

**no.** *or* **No. 1.** north; northern. **2.** number.

**n.o.p.** not otherwise provided (for).

**Nor. 1.** Norman. **2.** north. **3.** Norway; Norwegian.

**NORAD** North American Air Defense Command.

**norm.** normal.

**Norm.** Norman.

**Norw.** Norway; Norwegian.

**nos.** *or* **Nos.** numbers.

**n.o.s.** not otherwise specified.

**Nov.** November.

**NOW 1.** National Organization for Women. **2.** negotiable order of withdrawal.

**NP** neuropsychiatric; neuropsychiatry.

**n.p.** no place.

**N.P. 1.** notary public. **2.** nurse practitioner.

**NPN** nonprotein nitrogen.

**n.p.t.** normal pressure and temperature.

**nr** near.

**NR** no remittance.

**NRA** *also* **N.R.A.** National Recovery Administration.

**NRC 1.** National Research Council. **2.** Nuclear Regulatory Commission.

**ns** *or* **nsec** nanosecond.

**NS 1.** *or* **N.S.** Nova Scotia. **2.** nuclear ship.

**n.s. 1.** new series. **2.** not specified.

**N.S.** New Style.

**n/s** not sufficient.

**NSC** National Security Council.

**NSE** National Stock Exchange.

**NSF** National Science Foundation.

**n.s.f.** *or* **N.S.F.** not sufficient funds.

**N.S.W.** New South Wales.

**NT 1.** net tax. **2.** *also* **N.T.** New Testament. **3.** Northwest Territories. **4.** nurse technicians.

**n.t.p.** *or* **N.T.P.** normal temperature and pressure.

**nt. wt.** net weight.

**n.u.** name unknown.

**num. 1.** number. **2.** numeral.

**Num.** *Bible.* Numbers.

**numis.** *or* **numism.** numismatic; numismatics.

**NV** Nevada.

**NVA** North Vietnamese Army.

**NW** northwest.

**NWbN** northwest by north.

**NWbW** northwest by west.

**n.wt.** net weight.

**N.W.T.** Northwest Territories.

**NY** *or* **N.Y.** New York.

**NYC** *or* **N.Y.C.** New York City.

**NYP** not yet published.

**NYSE** New York Stock Exchange.

**N.Z.** New Zealand.

**O** *or* **O. 1.** ocean. **2.** order.

**o. 1.** *Lat.* octarius (pint). **2.** *or* **O.** octavo.

**O. 1.** October. **2.** Ohio.

**o/a** on or about.

**OAPC** Office of Alien Property Custodian.

**OAS** Organization of American States.

**OAU** Organization for African Unity.

**Ob** *Bible.* Obadiah.

**ob. 1.** *Lat.* obiit (she *or* he died). **2.** *Lat.* obiter (incidentally). **3.** oboe. **4.** obstetric.

**Obad.** *Bible.* Obadiah.

**O.B.E. 1.** Officer of the Order of the British Empire. **2.** *also* **OBE** Order of the British Empire.

**obj. 1.** object; objective (in grammar). **2.** objection.

**obl. 1.** oblique. **2.** oblong.

**obs. 1.** obscure. **2.** observation. **3.** *or* **Obs.** observatory. **4.** obsolete. **5.** obstetric; obstetrician; obstetrics.

**obstet.** obstetric; obstetrics.

**obv.** obverse.

**OC** *also* **O.C. 1.** Office of Censorship. **2.** officer candidate.

**oc.** *or* **Oc.** ocean.

**o.c.** *Lat.* opere citato (in the work cited).

**O.C. 1.** Officer Commanding. **2.** Old Catholic.

**o/c** overcharge.

**OCA** Office of Consumer Affairs.

**OCAS** Organization of Central American States.

**occ. 1.** occident; occidental. **2.** occupation.

**occas.** occasional; occasionally.

**OCD** Office of Civil Defense.

**OCR** optical character recognition.

**OCS** Officer Candidate School.

**oct.** octavo.

**Oct.** October.

**o.d. 1.** *Lat.* oculus dexter (right eye). **2.** olive drab. **3.** on demand. **4.** outside diameter.

**O.D. 1.** Doctor of Optometry. **2.** officer of the day. **3.** *also* **o/d** overdraft. **4.** overdrawn.

**Oe** oersted.

**OE** *also* **O.E.** Old English.

**OECD** Organization for Economic Cooperation and Development.

**OED** *also* **O.E.D.** Oxford English Dictionary.

**OEM** original equipment manufacturer.

**OEO** Office of Economic Opportunity.

**off.** office; officer; official.

**O.F.M.** Order of Friars Minor.

**O.F.S.** Orange Free State.

**OG** *or* **O.G. 1.** officer of the guard. **2.** original gum.

**OGPU** *or* **O.G.P.U.** *Russian.* Ob'edinyonnoye Gosudarstvennoye Politicheskoye Upravlenie (Unified Government Political Administration).

**OH** Ohio.

**OHMS** *or* **O.H.M.S.** On Her *or* His Majesty's Service.

**OIT** Office of International Trade.

**OK** Oklahoma.

**Okla.** Oklahoma.

**Om.** Oman.

**OM.** ostmark.

**O.M.** Order of Merit.

**OMB** Office of Management and Budget.

**ON 1.** *also* **O.N.** Old Norse. **2.** Ontario.

**ONI** Office of Naval Intelligence.

**ONR** Office of Naval Research.

**Ont.** Ontario.

**O.O.D.** officer of the deck.

**op** *or* **OP** *or* **op.** *or* **o.p.** out of print.

**op. 1.** *also* **Op.** operation. **2.** opposite. **3.** *also* **Op.** opus.

**O.P.** Order of Preachers.

**op. cit.** *Lat.* opere citato (in the work cited).

**OPEC** Organization of Petroleum Exporting Countries.

**opp.** opposite.

**opt. 1.** optative. **2.** optical; optician; optics. **3.** optimum. **4.** optional.

**OR** *or* **Or.** Oregon.

**o.r.** owner's risk.

**O.R.** *or* **OR** operating room.

**orch.** orchestra.

**ord. 1.** order. **2.** ordinal. **3.** ordinance. **4.** ordnance.

**ordn.** ordnance.

**Ore.** Oregon.

**org. 1.** organic. **2.** organization; organized.

**orig.** original; originally.

**ornith.** ornithologic; ornithology.

**orth.** orthopedic; orthopedics.

**o.s. 1.** *Lat.* oculus sinister (left eye). **2.** old series. **3.** *or* **o/s** out of stock.

**O.S. 1.** Old Style. **2.** ordinary seaman.

**OSA** *or* **O.S.A.** Order of St. Augustine.

**OSB** *or* **O.S.B.** Order of St. Benedict.

**OSF** *or* **O.S.F.** Order of St. Francis.

**OSHA** U.S. Occupational Safety and Health Administration.

**OSS** Office of Strategic Services.

**OSU** *or* **O.S.U.** Order of St. Ursula.

**OT** *also* **O.T.** Old Testament.

**o.t.** *or* **O.T. 1.** occupational therapy. **2.** overtime.

**OTB** off-track betting.

**OTC** *also* **O.T.C. 1.** Officer in Tactical Command. **2.** Officers' Training Corps.

**otol.** otology.

**OTS** *also* **O.T.S.** Officers' Training School.

**OV** *Aerospace.* orbiter vehicle.

**OWI** Office of War Information.

**Ox.** *or* **Oxf.** Oxford.

**Oxon.** *Lat.* Oxoniensis (of Oxford).

**oz** *also* **oz.** ounce.

**oz ap** apothecaries' ounce.

**oz av** *or* **oz avdp** avoirdupois ounce.

**oz t** troy ounce.

**p 1.** momentum. **2.** *or* **p.** *Mus.* piano (direction). **3.** *Physics.* pico-. **4.** *symbol for* proton.

**P 1.** parental generation. **2.** *Physics.* parity. **3.** pawn (chess). **4.** *Bible.* Peter. **5.** petite. **6.** *Physics.* pressure.

**p. 1.** page. **2.** part. **3.** participle. **4.** past. **5.** penny. **6.** per. **7.** peseta **8.** peso. **9.** pint. **10.** pipe. **11.** pole. **12.** population. **13.** *or* **P.** president. **14.** *or* **P.** prince. **15.** purl.

**P.** priest.

**PA 1.** *or* **Pa.** Pennsylvania. **2.** public-address system.

**p.a.** *Lat.* per annum (by the year).

**P.A. 1.** physician's assistant. **2.** *or* **P/A** power of attorney. **3.** press agent. **4.** prosecuting attorney.

**PABX** *also* **P.A.B.X.** Private Automatic Branch Exchange.

**Pac.** *or* **Pacif.** Pacific.

**Pak.** Pakistan.

**Pal.** Palestine.

**pam.** pamphlet.

**Pan.** Panama.

**P and L** profit and loss.

**par. 1.** paragraph. **2.** parallel. **3.** parenthesis. **4.** parish.

**Par.** *or* **Para.** Paraguay.

**paren.** parenthesis.

**parl.** parliamentary.

**Parl.** Parliament.

**part. 1.** participle. **2.** particular.

**pass. 1.** passage. **2.** passenger. **3.** passive.

**pat.** patent.

**patd.** patented.

**path.** *or* **pathol.** pathological; pathology.

**PAU** *or* **P.A.U.** Pan American Union.

**PAYE** *or* **P.A.Y.E. 1.** pay as you earn. **2.** pay as you enter.

**payt.** payment.

**P.B. 1.** passbook. **2.** prayer book.

**PBI** protein-bound iodine.

**PBS** Public Broadcasting System.

**PBX** *also* **P.B.X.** Private Branc Exchange.

**p.c. 1.** per cent. **2.** *also* **p/c** *or* **P/C** petty cash. **3.** postcard. **4.** *Lat.* post cibum (after meals).

**P.C. 1.** Past Commander. **2.** Police Constable. **3.** Post Commander **4.** Privy Council.

**p/c** *or* **P/C** prices current.

**PCB** polychlorinated biphenyl.

**PCP** phencyclidine.

**pct.** per cent.

**pd.** paid.

**p.d.** *or* **P.D.** per diem.

**P.D. 1.** Police Department. **2.** postal district. **3.** potential difference.

**Pd.B.** *Lat.* Pedagogiae Baccalaureus (Bachelor of Pedagogy).

**Pd.D.** *Lat.* Pedagogiae Doctor (Doctor of Pedagogy).

**Pd.M.** *Lat.* Pedagogiae Magister (Master of Pedagogy).

**PDT** *or* **P.D.T.** Pacific Daylight Time.

**pe** *also* **p.e.** printer's error.

**PE** Prince Edward Island.

**P.E. 1.** physical education. **2.** *Statistics.* probable error. **3.** professional engineer.

**P/E** price/earnings.

**P.E.I.** Prince Edward Island.

**pen.** *or* **Pen.** peninsula.

**Penn.** *or* **Penna.** Pennsylvania.

**per. 1.** period. **2.** person.

**perf. 1.** perfect. **2.** perforated.

**perm.** permanent.

**perp.** perpendicular.

**pers. 1.** person. **2.** personal.

**Pers.** Persia; Persian.

**PERT** program evaluation and review technique.

**pert.** pertaining.

**pet.** petroleum.

**Pet.** *Bible.* Peter.

**petr.** petrology.

**petrog.** petrography.

**petrol.** petrology.

**pf. 1.** pfennig. **2.** preferred.

**Pfc** *or* **Pfc.** *or* **PFC** private first class.

**pfd.** preferred.

**pfg.** pfennig.

**pg.** page.

**Pg.** Portugal; Portuguese.

**P.G. 1.** paying guest. **2.** postgraduate.

**Ph** *Bible.* Philippians.

**PH** *also* **P.H. 1.** Public Health. **2.** Purple Heart.

**ph.** phase.

**PHA** Public Housing Administration.

**phar.** *or* **Phar.** pharmaceutical; pharmacist; pharmacopoeia; pharmacy.

**Phar.B.** *Lat.* Pharmaciae Baccalaureus (Bachelor of Pharmacy).

**Phar.D.** *Lat.* Pharmaciae Doctor (Doctor of Pharmacy).

**pharm.** *or* **Pharm.** pharmaceutical; pharmacist; pharmacopoeia; pharmacy.

**Phar.M.** *Lat.* Pharmaciae Magister (Master of Pharmacy).

**Ph.B.** *Lat.* Philosophiae Baccalaureus (Bachelor of Philosophy).

**Ph.C.** Pharmaceutical Chemist.

**Ph.D.** *Lat.* Philosophiae Doctor (Doctor of Philosophy).

**Ph.G.** graduate in pharmacy.

**phil.** philosopher; philosophical; philosophy.

**Phil. 1.** *Bible.* Philippians. **2.** Philippines.

**Philem.** *Bible.* Philemon.

**Phil. I.** *or* **Phil. Is.** Philippine Islands.

**philol.** philology.

**philos.** philosopher; philosophical; philosophy.

**Phm** *Bible.* Philemon.

**Ph.M.** *Lat.* Philosophiae Magister (Master of Philosophy).

**PHN** public health nurse.

**phon. 1.** phonetic; phonetics. **2.** phonology.

**photog.** photography.

**photom.** photometry.

**phr.** phrase.

**phren.** phrenology.

**PHS** Public Health Service.

**hys.** **1.** physical. **2.** physician. physicist; physics. **4.** physiological; physiology.

**ysiol.** physiological; physiology.

**I.** Philippine Islands.

**nx.** *Lat.* pinxit (she or he painted his).

**zz.** *Mus.* pizzicato.

**K** psychokinesis.

**k.** **1.** pack. **2.** park. **3.** peak. **4.** *or* **pk** eck.

**kg.** *or* **pkge.** package.

**kt.** packet.

**KU** phenylketonuria.

**kwy.** parkway.

**l.** **1.** *or* **Pl.** place. **2.** plate. **3.** plural.

**lat.** **1.** plateau. **2.** platform. **3.** platoon.

**lf.** plaintiff.

**ln.** plain.

**LO** Palestine Liberation Organization.

**LSS** portable life support system.

**lu.** plural.

**m** *also* **p-m** phase modulation.

**M** *or* **P.M.** **1.** past master. **2.** police magistrate. **3.** postmaster. **4.** provost marshal.

**m.** premium.

**m.** *also* **P.M.** post mortem.

**M.** **1.** *also* **p.m.** post meridiem. — Usu. used in small capitals <P.M.> . Prime Minister.

**.M.G.** postmaster general.

**mk.** postmark.

**mt.** payment.

**n.** *or* **P/N** promissory note.

**neum.** pneumatic; pneumatics.

**n.g.** persona non grata.

**o** *or* **p.o.** *Baseball.* putout.

**O** *or* **P.O.** **1.** Personnel Officer. . petty officer. **3.** postal order. *also* **p.o.** post office.

**OE** *or* **P.O.E.** **1.** port of embarkation. **2.** port of entry.

**oet.** poetic; poetical; poetry.

**ol.** political; politician; politics.

**ol.** Poland; Polish.

**olit.** political; politics.

**op.** **1.** popular. **2.** population.

**ort.** Portugal; Portuguese.

**OS** point-of-sale.

**os.** **1.** position. **2.** positive.

**oss.** **1.** possession. **2.** possessive. . possible; possibly.

**ot.** potential.

**OW** *or* **P.O.W.** prisoner of war.

**p** *or* **pp.** *Mus.* pianissimo.

**p.** **1.** pages. **2.** past participle.

**.p.** *or* **P.P.** **1.** parcel post. **2.** parish priest. **3.** past participle. **4.** postpaid.

**PC** plain paper copier.

**pd.** **1.** postpaid. **2.** prepaid.

**ph.** pamphlet.

**.P.S.** *also* **p.p.s.** *Lat.* post postscriptum (additional postscript).

**pt.** precipitate.

**ptn.** precipitation.

**q.** previous question.

**.Q.** *or* **PQ** Province of Quebec.

**r** *Bible.* Proverbs.

**R** *or* **P.R.** **1.** public relations. . Puerto Rico.

**r.** **1.** pair. **2.** present. **3.** price. . printing. **5.** pronoun.

**r.** **1.** priest. **2.** prince. **3.** Provençal.

**.R.** proportional representation.

**rec.** preceding.

**red.** predicate.

**ref.** **1.** preface; prefatory. **2.** preference; preferred. **3.** prefix.

**rem.** premium.

**rep.** **1.** preparation; preparatory; prepare. **2.** preposition.

**repd.** prepared.

**repn.** preparation.

**res.** **1.** present. **2.** president.

**res.** President.

**resb.** *or* **Presby.** Presbyterian.

**ret.** preterit.

**RF** **1.** pulse recurrence frequency. . pulse repetition frequency.

**rf.** proof.

**rim.** **1.** primary. **2.** primitive.

**rin.** **1.** principal. **2.** principle.

**rint.** printing.

**riv.** **1.** private. **2.** privative.

**.r.n.** *Lat.* pro re nata (*Med.* as the situation demands).

**RO** *also* **P.R.O.** public relations officer.

**pro.** professional.

**prob.** **1.** probable; probably. **2.** problem.

**proc.** **1.** proceedings. **2.** process.

**prod.** **1.** produce. **2.** produced. **3.** product; production.

**prof.** **1.** professional. **2.** *also* **Prof.** professor.

**prom.** promontory.

**pron.** **1.** pronominal; pronoun. **2.** pronounced; pronunciation.

**prop.** **1.** proper; properly. **2.** property. **3.** proposition. **4.** proprietary; proprietor.

**propr.** proprietor.

**pros.** prosody.

**Pros. Atty.** prosecuting attorney.

**Prot.** Protestant.

**protec.** protectorate.

**prov.** **1.** province; provincial. **2.** provisional. **3.** provost.

**Prov.** **1.** Provençal. **2.** *Bible.* Proverbs.

**prox.** proximo. .

**Ps** *or* **Ps.** *Bible.* Psalm; Psalms.

**p.s.** passenger steamer.

**P.S.** **1.** permanent secretary. **2.** Police Sergeant. **3.** *also* **p.s.** postscript. **4.** public school.

**PSAT** Preliminary Scholastic Aptitude Test.

**psec** picosecond.

**pseud.** pseudonym.

**psf** *or* **p.s.f.** pounds per square foot.

**psi** *or* **p.s.i.** pounds per square inch.

**PSRO** Professional Standards Review Organization.

**PST** *or* **P.S.T.** Pacific Standard Time.

**psych.** *or* **psychol.** psychological; psychologist; psychology.

**pt.** **1.** part. **2.** payment. **3.** pint. **4.** point. **5.** port. **6.** preterit.

**p.t.** *Lat.* pro tempore (temporarily).

**P.T.** **1.** *also* **PT** Pacific Time. **2.** physical therapy. **3.** physical training. **4.** postal telegraph.

**pta.** peseta.

**ptg.** printing.

**p.t.o.** *or* **PTO** please turn over.

**PTV** **1.** pay television. **2.** public television.

**pty.** proprietary.

**pub.** **1.** public. **2.** publication. **3.** published; publisher.

**publ.** **1.** publication. **2.** published; publisher.

**PV** polyvinyl.

**PVC** polyvinyl chloride.

**pvt.** *or* **Pvt.** *or* **PVT** private.

**PWA** *also* **P.W.A.** Public Works Administration.

**pwr.** power.

**pwt.** pennyweight.

**pxt.** *Lat.* pinxit (he *or* she painted this).

**pyro.** pyrotechnics.

**Q** **1.** queen (chess). **2.** quetzal.

**q.** **1.** quart. **2.** quarter. **3.** quarterly. **4.** *also* **Q.** quarto. **5.** query. **6.** question. **7.** quintal. **8.** quire.

**qb** quarterback.

**QB** queen's bishop (chess).

**Q.B.** Queen's Bench.

**QC** **1.** quality control. **2.** quartermaster corps.

**Q.C.** Queen's Counsel.

**Q.E.D.** *Lat.* quod erat demonstrandum (which was to be demonstrated).

**Q.E.F.** *Lat.* quod erat faciendum (which was to be done).

**QF** quick-firing.

**q.i.d.** *Lat.* quater in die (*Med.* four times a day).

**QKt** queen's knight (chess).

**ql.** quintal.

**Qld.** Queensland.

**qlty.** quality.

**QM** quartermaster.

**Q.M.** *Lat.* quaque mane (every morning).

**QMC** quartermaster corps.

**QMG** Quartermaster General.

**qn.** question.

**Qo** *Bible.* Ecclesiastes.

**QP** queen's pawn (chess).

**q.p.** *or* **q.pl.** *Lat.* quantum placet (as much as you please).

**qq.** questions.

**qq.v.** *Lat.* quae vide (which [things] see).

**QR** queen's rook (chess).

**qr.** **1.** quarter. **2.** quarterly. **3.** quire.

**q.s.** *Lat.* quantum sufficit (as much as suffices).

**QSO** *Astron.* quasi-stellar object.

**QSRS** *Astron.* quasi-stellar radio source.

**QSTOL** quiet short takeoff and landing.

**qt** *or* **qt.** quart.

**qt.** quantity.

**qto.** quarto.

**qty.** quantity.

**qu.** **1.** queen. **2.** query. **3.** question.

**quad.** **1.** quadrangle. **2.** quadrant.

**qual.** qualitative.

**quant.** quantitative.

**quar.** **1.** quarter. **2.** quarterly.

**Que.** Quebec.

**ques.** question.

**quot.** quotation.

**q.v.** *Lat.* quod vide (which see).

**qy.** query.

**r** **1.** *or* **R** radius. **2.** *or* **R** *Elect.* resistance. **3.** *or* **r.** run (in sports).

**R** **1.** *Chem.* gas constant. **2.** *Chem.* radical. **3.** *or* **R.** Réaumur (scale). **4.** response (liturgical). **5.** roentgen (unit of radiation). **6.** rook (chess).

**r.** **1.** *or* **R.** railroad; railway. **2.** range. **3.** rare. **4.** retired. **5.** *or* **R.** right. **6.** *or* **R.** river. **7.** *or* **R.** road. **8.** rod (unit of length). **9.** rouble. **10.** rubber (in card games). **11.** *or* **R.** rupee.

**R.** **1.** rabbi. **2.** rector. **3.** regius. **4.** Republican. **5.** royal.

**Ra.** Range.

**R.A.** **1.** *or* **RADM** rear admiral. **2.** *or* **RA** Regular Army. **3.** *Astron.* right ascension. **4.** Royal Academician; Royal Academy.

**rad.** **1.** radical. **2.** radio. **3.** radius. **4.** radix.

**RADM** rear admiral.

**RAF** *also* **R.A.F.** Royal Air Force.

**RAM** **1.** *Computer Sci.* random-access memory. **2.** *also* **R.A.M.** Royal Academy of Music.

**R & B** rhythm and blues.

**R & D** research and development.

**R and R** rest and recreation.

**RBC** *or* **rbc** red blood cell; red blood (cell) count.

**RBE** *Physics.* relative biological effectiveness.

**rbi** *also* **r.b.i.** run batted in.

**RC** **1.** Red Cross. **2.** Roman Catholic.

**RCAF** *also* **R.C.A.F.** Royal Canadian Air Force.

**R.C.Ch.** Roman Catholic Church.

**RCMP** *also* **R.C.M.P.** Royal Canadian Mounted Police.

**R.C.P.** Royal College of Physicians.

**rcpt.** receipt.

**R.C.S.** Royal College of Surgeons.

**rct.** recruit.

**rd** rod (unit of length).

**RD** **1.** registered dietician. **2.** rural delivery.

**rd.** **1.** *or* **Rd.** road. **2.** round.

**RDA** recommended daily allowance.

**RDF** radio direction finder.

**Re.** rupee.

**R.E.** or **RE** real estate.

**rec.** **1.** receipt. **2.** record; recording. **3.** recreation.

**recd.** *or* **rec'd.** received.

**recip.** reciprocal; reciprocity.

**rect.** **1.** receipt. **2.** rectangle; rectangular. **3.** rectified. **4.** rector; rectory.

**red.** reduced; reduction.

**ref.** **1.** referee. **2.** reference. **3.** referred. **4.** refining. **5.** reformation; reformed. **6.** refunding.

**refl.** **1.** reflection; reflective. **2.** reflex; reflexive.

**reg.** **1.** regent. **2.** regiment. **3.** region. **4.** register; registered. **5.** registrar. **6.** registry. **7.** regular; regularly. **8.** regulation. **9.** regulator.

**regd.** registered.

**regt.** regiment.

**Regt.** regent.

**rel.** **1.** relating. **2.** relative. **3.** released. **4.** religion; religious.

**rem.** remittance.

**REMT** radiological emergency medical team.

**rep.** **1.** repair. **2.** repetition. **3.** report. **4.** reporter. **5.** *or* **Rep.** representative. **6.** reprint. **7.** *or* **Rep.** republic.

**Rep.** Republican.

**r.e.p.** roentgen equivalent, physical.

**repl.** replace; replacement.

**repr.** representing.

**rept.** report.

**Repub.** **1.** republic. **2.** Republican.

**req.** **1.** require; required. **2.** requisition.

**reqd.** required.

**RES** reticuloendothelial system.

**res.** **1.** research. **2.** reserve. **3.** residence; resident; resides. **4.** resolution.

**Res.** **1.** Reservation. **2.** Reservoir.

**resp.** **1.** respective; respectively. **2.** respiration.

**ret.** **1.** retain. **2.** retired. **3.** return.

**rev.** **1.** revenue. **2.** reverse; reversed. **3.** review; reviewed. **4.** revise; revision. **5.** *or* **Rev.** revolution.

**Rev.** **1.** *Bible.* Revelation. **2.** reverend (title).

**Rev. Ver.** *Bible.* Revised Version.

**RF** radio frequency.

**rf.** **1.** reef. **2.** refund.

**r.f.** right field; right fielder.

**RFD** *also* **R.F.D.** rural free delivery.

**r.h.** **1.** relative humidity. **2.** *also* **RH** right hand.

**rhbdr.** rhombohedron.

**rheo.** rheostat.

**rhet.** rhetoric.

**R.H.I.P.** rank has its privileges.

**rhomb.** rhombic.

**rhp** *or* **r.hp.** rated horsepower.

**RI** *or* **R.I.** Rhode Island.

**R.I.P.** *Lat.* requiescat in pace (may she *or* he rest in peace).

**rit.** *Mus.* ritardando.

**riv.** river.

**RJ** road junction.

**R.L.T.** registered laboratory technologist.

**Rm** *Bible.* Romans.

**RM** *also* **Rm.** reichsmark.

**rm.** **1.** ream. **2.** room.

**rms** root mean square.

**RMS** **1.** Railway Mail Service. **2.** *also* **R.M.S.** Royal Mail Service. **3.** *also* **R.M.S.** Royal Mail Steamship.

**RN** *or* **R.N.** **1.** registered nurse. **2.** Royal Navy.

**RNA** ribonucleic acid.

**rnd.** round.

**RNR** *or* **R.N.R.** Royal Naval Reserve.

**ro.** rood (measure).

**rom** *also* **rom.** roman (type).

**ROG** receipt of goods.

**ROM** *Computer Sci.* read-only memory.

**Rom.** **1.** Roman. **2.** Romance (language). **3.** Romania; Romanian. **4.** *Bible.* Romans.

**rot.** rotating; rotation.

**ROTC** Reserve Officers' Training Corps.

**R.Ph.** registered pharmacist.

**rpm** *or* **r.p.m.** revolutions per minute.

**R.P.O.** Railway Post Office.

**RPQ** request for price quotation.

**rps** *or* **r.p.s.** revolutions per second.

**rpt.** **1.** repeat. **2.** report.

**R.Q.** respiratory quotient.

**RR** *also* **R.R.** **1.** railroad. **2.** rural route.

**R.R.** Right Reverend.

**R.R.A.** registered records administrator.

**RRB** Railroad Retirement Board.

**rRNA** ribosomal RNA.

**RS** **1.** recording secretary. **2.** right side. **3.** *also* **R.S.** Royal Society.

**RSFSR** *or* **R.S.F.S.R.** Russian Soviet Federated Socialist Republic.

**RSV** or **R.S.V.** *Bible.* Revised Standard Version.

**R.S.V.P.** *or* **r.s.v.p.** *French.* répondez s'il vous plaît (please reply).

**Rt** *Bible.* Ruth.

**RT** **1.** radio telephone. **2.** room temperature.

**rt.** right.

**rte.** route.

**Rt. Hon.** Right Honorable.

**Rt. Rev.** Right Reverend.

**Rus.** *or* **Russ.** Russia; Russian.

**Rv** *Bible.* Revelations.

**RV** *or* **R.V. 1.** recreational vehicle. **2.** *Aerospace.* reentry vehicle. **3.** *Bible.* Revised Version.

**Rw.** Rwanda.

**R.W. 1.** Right Worshipful. **2.** Right Worthy.

**rwy.** *or* **ry.** railway.

**s 1.** second (unit of time). **2.** second of arc. **3.** siemens. **4.** stere.

**S 1.** *Bible.* Samuel. **2.** *also* **S.** *or* **s** *or* **s.** south; southern. **3.** specialist. **4.** *Physics.* strangeness.

**s. 1.** *or* **S.** school. **2.** *or* **S.** sea. **3.** see. **4.** semi-. **5.** shilling. **6.** singular. **7.** sire. **8.** sister. **9.** small. **10.** *or* **S.** society. **11.** solo. **12.** son. **13.** *or* **S.** soprano. **14.** sou. **15.** stock. **16.** substantive. **17.** surplus.

**S. 1.** Sabbath. **2.** saint. **3.** Saturday. **4.** Saxon. **5.** September. **6.** *Med.* signature. **7.** signor; signore. **8.** Sunday.

**SA** Salvation Army.

**s.a.** *Lat.* sine anno (without date).

**S.A. 1.** South Africa. **2.** South America.

**Sab.** Sabbath.

**SAC** Strategic Air Command.

**SACEUR** Supreme Allied Commander, Europe.

**S.Afr.** South Africa.

**Sal.** El Salvador.

**SALT** Strategic Arms Limitations Talks.

**SAM** surface-to-air-missile.

**Sam.** *Bible.* Samuel.

**s. ap.** apothecaries' scruple.

**SAS** *Chiefly Brit.* Special Air Service.

**SASE** self-addressed stamped envelope.

**Sask.** Saskatchewan.

**SAT** Scholastic Aptitude Test.

**sat.** saturate; saturation.

**Sat.** Saturday.

**satd.** saturated.

**Sau. Ar.** Saudi Arabia.

**S.Austl.** South Australia.

**Sax.** Saxon; Saxony.

**SB** simultaneous broadcast.

**sb.** substantive.

**S.B.** *Lat.* Scientiae Baccalaureus (Bachelor of Science).

**SBA** Small Business Administration.

**SbE** south by east.

**SBN** Standard Book Number.

**SbW** south by west.

**SC 1.** Security Council. **2.** *or* **S.C.** South Carolina.

**sc. 1.** scale. **2.** scene. **3.** science. **4.** *Lat.* scilicet (namely). **5.** scruple (weight). **6.** *Lat.* sculpsit (she or he sculptured [it]).

**Sc.** Scotch; Scottish.

**s.c.** *also* **sc** small capitals.

**S.C.** Supreme Court.

**Scand.** Scandinavia; Scandinavian.

**SCAP** Supreme Commander for the Allied Powers.

**SCC** storage connecting circuit.

**sch.** school.

**sci.** science; scientific.

**Scot.** Scotch; Scotland; Scottish.

**scr.** scruple (unit of weight).

**Script.** Scriptural; Scriptures.

**sct.** scout.

**sctd.** scattered.

**sculp. 1.** *or* **sculpt.** *Lat.* sculpsit (she *or* he sculptured [it]). **2.** sculptor; sculptress; sculpture.

**SD 1.** sight draft. **2.** *or* **S.D.** South Dakota. **3.** special delivery. **4.** standard deviation.

**sd.** sound.

**s.d.** *Lat.* sine die (indefinitely).

**S.Dak.** South Dakota.

**SE 1.** southeast; southeastern. **2.** standard English. **3.** stock exchange. **4.** systems engineer.

**SEATO** Southeast Asia Treaty Organization.

**SEbE** southeast by east.

**SEbS** southeast by south.

**sec 1.** secant. **2.** second. **3.** secondary.

**SEC** Securities and Exchange Commission.

**sec. 1.** secretary. **2.** sector. **3.** *Lat.* secundum (according to).

**sect.** section.

**secy.** secretary.

**sed.** sediment.

**sel.** select; selected.

**SEM** scanning electron microscope.

**sem.** seminary.

**Sem.** Semitic.

**sen.** *or* **Sen. 1.** senate; senator. **2.** senior.

**sep.** separate; separation.

**Sep.** September.

**sepd.** separated.

**Sept.** September.

**seq. 1.** sequel. **2.** *Lat.* sequens (the following).

**seqq.** *Lat.* sequentia (the following [things]).

**ser. 1.** serial. **2.** series. **3.** sermon.

**Serb.** Serbia; Serbian.

**serv. 1.** servant. **2.** service.

**SES** socioeconomic status.

**sess.** session.

**SF** science fiction.

**sf.** *Mus.* sforzando.

**Sfc.** sergeant first class.

**sfz.** *Mus.* sforzando.

**sg** specific gravity.

**Sg** *Bible.* Song of Songs.

**SG** surgeon general.

**S.G.** *or* **SG 1.** sergeant. **2.** solicitor general.

**sgd.** signed.

**Sgt.** *or* **SGT** *or* **SG** sergeant.

**Sgt. Maj.** sergeant major.

**sh. 1.** share. **2.** sheet.

**Shak.** Shakespeare.

**SHAPE** Supreme Headquarters Allied Powers, Europe.

**shf** *or* **SHF** superhigh frequency.

**shp** *or* **s.hp.** shaft horsepower.

**shpt.** shipment.

**shr.** share.

**shtg.** shortage.

**Si** *Bible.* Sirach (Ecclesiasticus).

**SI** *French.* Système Internationale d'Unités (International System of Units).

**Sib.** Siberia; Siberian.

**Sic.** Sicilian; Sicily.

**SIDS** sudden infant death syndrome.

**sig. 1.** signal. **2.** signature. **3.** *or* **Sig.** signor; signore.

**sing.** singular.

**SINS** ships inertial navigational system.

**S.J.** Society of Jesus.

**SJC** supreme judicial court.

**S.J.D.** *Lat.* Scientiae Juridicae Doctor (Doctor of Juridical Science).

**SK** Saskatchewan.

**sk.** sack.

**Skr.** *or* **Skt.** Sanskrit.

**SL 1.** sea level **2.** south latitude.

**sl.** slightly.

**S.L.** Sierra Leone.

**s.l.a.n.** *Lat.* sine loco, anno, vel nomine (without place, year, or name).

**Slav.** Slavic.

**SLBM** submarine-launched ballistic missile.

**sld. 1.** sailed. **2.** sealed. **3.** sold.

**SLIP** symmetric list processor.

**SLV** standard launch vehicle.

**SM 1.** sergeant major. **2.** *or* **S.M.** soldier's medal.

**sm.** small.

**S.M.** Scientiae Major (Master of Science).

**SMN** seaman.

**S.M.Sgt.** *or* **SMSGT** Senior Master Sergeant.

**s.n.** *Lat.* sine nomine (without name).

**SNG** synthetic natural gas.

**so.** *or* **So.** south; southern.

**s.o. 1.** seller's option. **2.** strikeout.

**soc. 1.** social. **2.** socialist. **3.** society.

**SOF** sound on film.

**sol. 1.** solicitor. **2.** soluble. **3.** solution.

**Sol. Is.** Solomon Islands.

**soln.** solution.

**Som.** Somalia.

**SOP** standard operating procedure.

**sop.** soprano.

**soph.** sophomore.

**sou.** *or* **Sou.** south; southern.

**sov.** sovereign (coin).

**Sov. Un.** Soviet Union.

**SP 1.** self-propelled. **2.** shore patrol; shore police.

**sp. 1.** special. **2.** specialist. **3.** species. **4.** specific. **5.** spelling.

**Sp.** Spain; Spanish.

**s.p.** *Lat.* sine prole (without issue).

**Span.** Spanish.

**SPCA** Society for the Prevention of Cruelty to Animals.

**SPCC** Society for the Prevention of Cruelty to Children.

**spec. 1.** special. **2.** specification. **3.** speculation.

**specif.** specifically.

**sp gr** specific gravity.

**sp ht** specific heat.

**SPOT** satellite positioning and tracking.

**spp.** species (plural).

**SPQR** small profits, quick returns.

**S.P.Q.R.** *Lat.* Senatus Populusque Romanus (the Senate and the People of Rome).

**spr.** spring.

**s.p.s.** *Lat.* sine prole supersite (without surviving issue).

**spt.** seaport.

**sq. 1.** squadron. **2.** square.

**sr** steradian.

**Sr. 1.** *or* **sr.** senior. **2.** señor. **3.** sister (religious).

**Sra.** señora.

**SRO 1.** single-room occupancy. **2.** standing room only.

**Srta.** señorita.

**ss.** *Lat.* **1.** *or* **ss** scilicet (namely). **2.** semis (one half).

**S.S. 1.** *or* **SS** steamship. **2.** Sunday school. **3.** sworn statement.

**s/s** same size.

**SSA** Social Security Administration.

**SSE** south-southeast.

**S.Sgt.** *or* **SSG** *or* **SSGT** staff sergeant.

**SSI** Supplemental Security Income.

**ssp.** subspecies.

**SSR** *or* **S.S.R.** Soviet Socialist Republic.

**SSS** Selective Service System.

**SST** supersonic transport.

**SSW** south-southwest.

**ST** standard time.

**st. 1.** stanza. **2.** start. **3.** state. **4.** *or* **St.** statute. **5.** stet. **6.** stitch. **7.** stone. **8.** *or* **St.** strait. **9.** *or* **St.** street. **10.** strophe.

**St.** saint.

**s.t.** short ton.

**sta. 1.** station. **2.** stationary.

**stat. 1.** *Lat.* statim (immediately). **2.** stationary. **3.** statistics. **4.** statuary. **5.** statute.

**stbd.** starboard.

**std.** standard.

**Ste.** *French.* sainte (feminine form of saint).

**steno** *or* **stenog.** *also* **sten.** stenographer; stenography.

**ster.** sterling.

**St. Ex.** stock exchange.

**stg.** sterling.

**stge.** storage.

**stip. 1.** stipend. **2.** stipulation.

**stk.** stock.

**S.T.M.** *Lat.* Sacrae Theologiae Magister (Master of Sacred Theology).

**STOL** short takeoff and landing.

**STP** standard temperature and pressure.

**STR** synchronous transmitter receiver.

**str. 1.** steamer. **2.** *or* **Str.** strait. **3.** stringed.

**stud.** student.

**sub. 1.** subaltern. **2.** substitute. **3.** suburb; suburban.

**subj. 1.** subject. **2.** subjective. **3.** subjunctive.

**subs.** subscription.

**subst. 1.** substantive. **2.** substitute.

**Sud.** Sudan.

**suf.** *or* **suff. 1.** sufficient. **2.** suffix.

**Sun.** Sunday.

**sup. 1.** superior. **2.** superlative (in grammar). **3.** supine (in Latin). **4.** supplement. **5.** supply. **6.** *Lat.* supra (above).

**super. 1.** superintendent. **2.** superior.

**supp.** *or* **suppl.** supplement; supplementary.

**supr.** supreme.

**supt.** *or* **Supt.** superintendent.

**supvr.** supervisor.

**sur. 1.** surface. **2.** surplus.

**Sur.** Surinam.

**surg.** surgeon; surgery; surgical.

**surr.** surrender.

**s.v. 1.** *also* **SV** sailing vessel. **2.** *La* sub verbo; sub voce ([look] und the word).

**svgs.** savings.

**sw** short wave.

**SW** southwest.

**sw.** switch.

**Sw.** Sweden; Swedish.

**SWAT** *also* **S.W.A.T.** Special Wea ons and Tactics Team.

**Swaz.** Swaziland.

**swbd** *or* **swbd.** switchboard.

**SWbS** southwest by south.

**SWbW** southwest by west.

**Swe.** *or* **Swed.** Sweden; Swedish.

**Switz.** Switzerland.

**swp.** swamp.

**sym. 1.** symbol. **2.** symmetric **3.** symphony.

**syn.** synonymous; synonym; sy onymy.

**synd.** syndicate.

**Syr. 1.** Syria; Syrian. **2.** Syriac.

**t 1.** ton. **2.** troy (system of weights

**T 1.** surface tension. **2.** temperatur **3.** *Physics.* tera-. **4.** tesla. **5.** *Math* time reversal.

**t. 1.** tare (weight). **2.** teaspoons teaspoonful. **3.** tempo. **4.** *Lat.* tem pore (in the time of). **5.** *or* **T.** *Mu* tenor. **6.** *Gram.* tense. **7.** termina **8.** *or* **T.** territory. **9.** *or* **T.** time. **10.** c **T.** town; township. **11.** transi **12.** *Gram.* transitive.

**T. 1.** tablespoon; tablespoonfu **2.** Testament. **3.** Tuesday.

**TA** teaching assistant.

**tab.** table.

**TAC** Tactical Air Command.

**tan** tangent.

**Tan.** Tanzania.

**TAS 1.** telephone answering systen **2.** true air speed.

**Tas.** *or* **Tasm.** Tasmania; Tasma nian.

**TAT** Thematic Apperception Test.

**Tb** *Bible.* Tobit.

**TB** *also* **T.B.** tuberculosis.

**t.b. 1.** trial balance. **2.** *also* **T.B** tubercle bacillus.

**TBA** *or* **tba** to be announced.

**tbs.** *or* **tbsp.** tablespoon; table spoonful.

**tchr.** teacher.

**TD 1.** *also* **td** touchdown. **2.** *als* **T.D.** treasury department.

**TDN** *also* **T.D.N.** total digestibl nutrients.

**TDY** temporary duty.

**tech.** technical.

**technol.** technological; technology

**TEFL** teaching English as a foreig language.

**t.e.g.** top edges gilt.

**tel. 1.** telegram. **2.** telegraph; tele graphic. **3.** telephone.

**teleg. 1.** telegram. **2.** telegraph; tele graphic; telegraphy.

**temp. 1.** temperance. **2.** tempera ture. **3.** template. **4.** temporary. **5.** *Lat.* tempore (in the time of).

**ten. 1.** tenor. **2.** *Mus.* tenuto.

**Tenn.** Tennessee.

**ter. 1.** terrace. **2.** territorial; terri tory.

**term. 1.** terminal. **2.** termination.

**terr. 1.** terrace. **2.** territorial; terri tory.

**TESL** teaching English as a second language.

**TESOL** teachers of English to speak ers of other languages.

**test. 1.** testator. **2.** testatrix. **3.** testi mony.

**Test.** Testament.

**Teut.** Teuton; Teutonic.

**Tex.** Texas.

**T.F.** Territorial Force.

**tfr.** transfer.

**TG** transformational grammar.

**t.g.** type genus.

**TGIF** thank God it's Friday.

**Th** *Bible.* Thessalonians.

**Th.** Thursday.
**Thai.** Thailand.
**Th.B.** *Lat.* Theologiae Baccalaureus (Bachelor of Theology).
**THC** tetrahydrocannabinol.
**Th.D.** *Lat.* Theologiae Doctor (Doctor of Theology).
**theat.** theater.
**theol.** theologian; theological; theology.
**therap.** therapeutic; therapeutics.
**Thess.** *Bible.* Thessalonians.
**Th.M.** *Lat.* Theologiae Magister (Master of Theology).
**thp** *or* **t.hp.** thrust horsepower.
**Thurs.** *also* **Thur.** Thursday.
**THz** terahertz.
**t.i.d.** *Lat.* ter in die (*Med.* three times a day).
**Tim.** *Bible.* Timothy.
**tit.** title.
**tk.** truck.
**TKO** technical knockout.
**tkt.** ticket.
**t.l.** *or* **t/l** total loss.
**TLC** tender loving care.
**t.l.o.** total loss only.
**tlr.** tailor.
**Tm** *Bible.* Timothy.
**TM 1.** trademark. **2.** transcendental meditation.
**t.m.** true mean.
**TMV** tobacco mosaic virus.
**TN** Tennessee.
**tn. 1.** ton. **2.** town. **3.** train.
**tng.** training.
**tnpk.** turnpike.
**TNT** trinitrotoluene.
**t.o.** turn over.
**topog.** topographic; topography.
**tp.** township.
**t.p.** title page.
**tpk.** turnpike.
**TQC** total quality control.
**TR** *or* **T-R** transmit-receive.
**tr. 1.** *Gram.* transitive. **2.** translated; translation; translator. **3.** transpose; transposition. **4.** treasurer. **5.** *Law.* trust; trustee.
**trans. 1.** transaction. **2.** *Gram.* transitive. **3.** translated; translation; translator. **4.** transportation. **5.** transpose; transposition. **6.** transverse.
**transl.** translated; translation.
**transp.** transportation.
**trav.** traveler; travels.
**treas.** treasurer; treasury.
**trib.** tributary.
**trig.** *also* **trigon.** trigonometric; trigonometry.
**tripl.** triplicate.
**trit.** triturate.
**tRNA** transfer RNA.
**trop.** tropic; tropical.
**trp.** troop.
**T.S.** *also* **t.s.** *Physics.* tensile strength.
**T.Sgt.** or **TSGT** Technical Sergeant.
**TSH** thyroid-stimulating hormone.
**tsp.** teaspoon; teaspoonful.
**TSS** toxic shock syndrome.
**Tt** *Bible.* Titus.
**TT 1.** telegraphic transfer. **2.** teletypewriter. **3.** transit time. **4.** tuberculin tested.
**Tu.** Tuesday.
**T.U.** trade union.
**Tues.** Tuesday.
**Tun.** Tunisia; Tunisian.
**Tur.** or **Turk.** Turkey; Turkish.
**TVA** Tennessee Valley Authority.
**twp.** township.
**TX** Texas.
**txn.** taxation.
**typ.** typographer; typographical; typography.
**typo.** or **typog.** typographer; typographical; typography.
**typw.** typewriter; typewritten.

**U** *Math.* union.
**u. 1.** *or* **U.** uncle. **2.** unit. **3.** *or* **U.** upper.
**U.** university.
**U.A.E.** United Arab Emirates.
**U.A.R.** United Arab Republic.
**u.c.** *also* **UC** upper case.
**UCMJ** Uniform Code of Military Justice.
**UCS** universal character set.
**UDC** universal decimal classification.
**Ug.** Uganda.
**UGT** urgent (telegram).
**uhf** *or* **UHF** ultrahigh frequency.
**U.K.** United Kingdom.
**ult. 1.** ultimate; ultimately. **2.** ultimo.
**UMT** Universal Military Training.
**UMTS** Universal Military Training Service.
**UN** *or* **U.N.** United Nations.
**unan.** unanimous.
**unb.** *or* **unbd.** unbound.
**UNESCO** United Nations Educational, Scientific, and Cultural Organization.
**UNICEF** United Nations International Children's Emergency Fund.
**Unit.** Unitarian; Unitarianism.
**univ. 1.** universal. **2.** *or* **Univ.** university.
**Univ.** Universalist.
**unm.** unmarried.
**unp.** unpaged.
**UNRRA** United Nations Relief and Rehabilitation Administration.
**UNRWA** United Nations Relief and Works Agency.
**up.** upper.
**UPC** Universal Product Code.
**UPI** *or* **U.P.I.** United Press International.
**UPU** Universal Postal Union.
**Ur.** Uruguay.
**URA** Urban Renewal Administration.
**US** *or* **U.S.** United States.
**u.s.** *Lat.* **1.** ubi supra (where [mentioned] above). **2.** ut supra (as above).
**U.S. 1.** Uncle Sam. **2.** Uniform System (of lens aperture).
**USA** *or* **U.S.A. 1.** United States Army. **2.** United States of America.
**USAF** *also* **U.S.A.F.** United States Air Force.
**USAFA** *also* **U.S.A.F.A.** United States Air Force Academy.
**USAFI** *or* **U.S.A.F.I.** United States Armed Forces Institute.
**USAR** United States Army Reserve.
**USAREUR** United States Army, Europe.
**USASCII** United States of America Standard Code for Information Interchange.
**USASI** United States of America Standards Institute.
**U.S.C.** United States Code.
**U.S.C.A.** United States Code Annotated.
**USCG** *also* **U.S.C.G.** United States Coast Guard.
**USCGA** *also* **U.S.C.G.A.** United States Coast Guard Academy.
**USDA** United States Department of Agriculture.
**USES** United States Employment Service.
**USIA** United States Information Agency.
**U.S.M.** United States Mail.
**USMA** *also* **U.S.M.A.** United States Military Academy.
**USMC** *also* **U.S.M.C.** United States Marine Corps.
**USN** *also* **U.S.N.** United States Navy.
**USNA** *also* **U.S.N.A.** United States Naval Academy.
**USNR** United States Naval Reserve.
**USO** *or* **U.S.O.** United Service Organizations.
**U.S.P.** United States Pharmacopoeia.
**U.S.P.O.** *also* **USPO** United States Post Office.
**U.S.S. 1.** United States Senate. **2.** *also* **USS** United States Ship.
**USSR** *or* **U.S.S.R.** Union of Soviet Socialist Republics.
**usu.** usually.
**usw.** *German.* und so weiter (and so forth).
**UT** Utah.
**ut dict.** *Lat.* ut dictum (*Med.* as directed).
**UV** ultraviolet.
**UW** underwriter.
**ux.** *Lat.* uxor (wife).
**UXB** unexploded bomb.

**V 1.** *Physics.* velocity. **2.** victory. **3.** *Elect.* volt. **4.** volume.
**v. 1.** verb. **2.** verse. **3.** version. **4.** verso. **5.** versus. **6.** *or* **V.** very (in titles). **7.** *or* **V.** vice (in titles). **8.** vide. **9.** *or* **V.** village. **10.** violin. **11.** vocative. **12.** voice. **13.** volume (book). **14.** vowel.
**V. 1.** venerable (in titles). **2.** viscount; viscountess.
**VA** *or* **Va.** Virginia.
**V.A. 1.** *also* **VA** Veterans' Administration. **2.** vicar apostolic.
**VAB** voice answer back.
**vac.** vacuum.
**V.Adm.** or **VADM** vice admiral.
**val. 1.** valley. **2.** valuation; value.
**VAR** visual-aural range.
**var. 1.** variable. **2.** variant. **3.** variation. **4.** variety. **5.** various.
**VAT** value-added tax.
**Vat.** Vatican.
**vb.** verb; verbal.
**VC** *also* **V.C.** Vietcong.
**V.C. 1.** vice chairman. **2.** vice chancellor. **3.** vice consul. **4.** Victoria Cross.
**VD** *also* **V.D.** venereal disease.
**v.d. 1.** vapor density. **2.** various dates.
**VDT** visual display terminal.
**VDU** visual display unit.
**veg.** vegetable.
**vel. 1.** vellum. **2.** velocity.
**Ven.** venerable.
**Venez.** Venezuela.
**ver. 1.** verse. **2.** version.
**vers** versed sine.
**vert.** vertical.
**vet. 1.** veteran. **2.** veterinarian; veterinary.
**veter.** veterinary.
**V.F. 1.** vicar forane. **2.** *also* **VF** video frequency. **3.** *also* **VF** visual field.
**VFD** volunteer fire department.
**VFR** visual flight rules.
**V.G.** vicar general.
**vhf** or **VHF** very high frequency.
**VI** or **V.I.** Virgin Islands.
**v.i.** *Lat.* vide infra (see below).
**V.I.** volume indicator.
**vic. 1.** vicar. **2.** vicinity.
**Viet.** Vietnam; Vietnamese.
**vil.** village.
**VIN** vehicle identification number.
**Vir. Is.** Virgin Islands.
**vis. 1.** visibility. **2.** visual.
**Vis.** *or* **Visct.** viscount; viscountess.
**VISTA** Volunteers in Service to America.
**viz.** *Lat.* videlicet (namely).
**vlf** *or* **VLF** very low frequency.
**V.M.D.** *Lat.* Veterinariae Medicinae Doctor (Doctor of Veterinary Medicine).
**VN** visiting nurse.
**VO** verbal order.
**vo.** verso.
**voc.** vocative.
**vocab.** vocabulary.
**vol. 1.** volcano. **2.** volume. **3.** volunteer.
**VOR 1.** very-high-frequency omnidirectional radio range. **2.** voice-operated relay.
**vou.** voucher.
**VP 1.** variable pitch. **2.** verb phrase. **3.** *or* **V.P.** vice president.
**vs.** versus.
**v.s.** *Lat.* vide supra (see above).
**V.S.** veterinary surgeon.
**vss. 1.** verses. **2.** versions.
**V/STOL** vertical short takeoff and landing.
**VT 1.** vacuum tube. **2.** variable time. **3.** Vermont.
**VTOL** vertical takeoff and landing.
**VTR** videotape recorder; video tape recording.
**VU** volume unit.
**Vul.** Vulgate.
**vulg.** vulgar.
**Vulg.** Vulgate.
**vv.** verses.
**v.v.** vice versa.

**w** *or* **W** *Physics.* work.
**W 1.** *Elect.* watt. **2.** *also* **W.** *or* **w** *or* **w.** west; western.
**w. 1.** week. **2.** weight. **3.** wide. **4.** width. **5.** wife. **6.** with.
**W. 1.** Wednesday. **2.** Welsh.
**WA 1.** Washington. **2.** with average
**W.A.** Western Australia.
**war.** warrant.
**Wash.** Washington.
**WAT** weight, altitude, and temperature.
**WATS** Wide-Area Telecommunications Service.
**W. Aust.** Western Australia.
**Wb** *Physics.* weber.
**w.b. 1.** water ballast. **2.** *also* **W.B.** waybill. **3.** westbound.
**W.B.** Weather Bureau.
**WBC** *or* **wbc** white blood cell; white blood (cell) count.
**WbN** west by north.
**WbS** west by south.
**w.c. 1.** water closet. **2.** without charge.
**WD** or **W.D.** War Department.
**wd. 1.** wood. **2.** word.
**Wed.** Wednesday.
**WEE** western equine encephalitis.
**wf** *or* **w.f.** wrong font.
**w.g.** wire gauge.
**WH** watt-hour.
**wh. 1.** which. **2.** white.
**whf.** wharf.
**WHO** World Health Organization.
**W-hr** watt-hour.
**whs.** warehouse.
**whsle.** wholesale.
**WI** Wisconsin.
**w.i.** when issued (financial stock).
**W.I.** West Indian; West Indies.
**WIA** wounded in action.
**wid** widow; widower.
**Wisd** *Bible.* Wisdom.
**wk. 1.** weak. **2.** week. **3.** work.
**wkly.** weekly.
**WL** or **w.l. 1.** water line. **2.** wavelength.
**wmk.** watermark.
**WN** white noise.
**WNW** west-northwest.
**WO** or **W.O.** warrant officer.
**w/o** without.
**w.o.c.** without compensation.
**WP 1.** weather permitting. **2.** word processing; word processor. **3.** *or* **w/p** without prejudice.
**WPA** Work Projects Administration.
**WPC** watts per candle.
**wpm** *or* **w.p.m.** words per minute.
**wpn.** weapon.
**WPS** word processing secretary.
**WR** or **W.r.** Wassermann reaction.
**Ws** *Bible.* Wisdom.
**WS** working storage.
**WSA** War Shipping Administration.
**WSW** west-southwest.
**WT** withholding tax.
**wt.** weight.
**WV** *or* **W.Va.** West Virginia.
**WVS** Women's Volunteer Service.
**WW I** or **W.W.I** World War I.
**WW II** or **W.W.II** World War II.
**w/w** wall-to-wall.
**WY** or **Wyo.** Wyoming.

**x 1.** *Math.* abscissa. **2.** broken type. **3.** by. **4.** or **X** power of magnification. **5.** or **X** *Math.* **a.** Unknown number. **b.** algebraic variable.
**X 1.** Christ; Christian. **2.** extra. **3.** *Elect.* reactance. **4.** times (multiplied by). **5.** —Used to indicate location, as on a map. **6.** *also* **x** unknown.
**x.** ex.

**XD** *or* **x-div.** ex dividend.
**XI** *or* **x-int.** ex interest.
**XL** **1.** extra large. **2.** extra long.
**Xn** Christian.
**Xnty** Christianity.
**XS** **1.** extra short. **2.** extra small.

**y** *Math.* ordinate.
**Y** **1.** *Elect.* admittance. **2.** *Physics.* hypercharge. **3.** yen (currency). **4. a.** YMCA **b.** YMHA. **c.** YWCA. **d.** YWHA. **5.** yeoman.
**y.** year.
**YA** young adult.
**YB** yearbook.
**yd** yard (measurement).
**yel.** yellow.
**Yem.** Yemen.
**yeo.** yeoman; yeomanry.
**YMCA** *or* **Y.M.C.A.** Young Men's Christian Association.
**YMHA** *or* **Y.M.H.A.** Young Men's Hebrew Association.
**YOB** year of birth.
**yr.** **1.** year. **2.** younger. **3.** your.
**YT** *or* **Y.T.** Yukon Territory.
**Yug.** *or* **Yugo.** Yugoslavia; Yugoslavian.
**YWCA** *or* **Y.W.C.A.** Young Women's Christian Association.
**YWHA** *or* **Y.W.H.A.** Young Women's Hebrew Association.

**Z** **1.** atomic number. **2.** *Elect.* impedance.
**z.** **1.** zero. **2.** zone.
**z.B.** *German.* zum Beispiel (for example).
**Zc** *Bible.* Zechariah.
**Zech.** *Bible.* Zechariah.
**Zeph.** *Bible.* Zephaniah.
**ZI** zone of interior.
**Zl** zloty.
**zool.** zoological; zoology.
**Zp** *Bible.* Zephaniah.
**ZPG** zero population growth.

# Biographical Entries

## A

**Aal·to** (äl'tô), **Alvar.** 1898–1976. Finnish architect & industrial designer.
**Aar·on** (âr'ən, ăr'-). Hebrew high priest. **—Aar·on'ic** (â-rŏn'ĭk, ăr-ŏn'-), **Aar·on'i·cal** *adj.*
**Aaron, Henry Louis ("Hank").** b. 1934. Amer. baseball player.
**Ab·bas·side** *also* **Ab·bas·sid** (ăb'ə-sīd', ə-băs'ĭd'). Dynasty of Moslem caliphs (750–1258).
**Ab·be** (ăb'ē), **Cleveland.** *Father of the Weather Bureau.* 1838–1916. Amer. meteorologist.
**Ab·bey** (ăb'ē), **Edwin Austin.** 1852–1911. Amer. painter & illustrator.
**Ab·bot** (ăb'ət), **Charles Greeley.** 1872–1973. Amer. astrophysicist.
**Ab·bott** (ăb'ət), **Grace.** 1878–1939. Amer. social reformer.
**Abbott,** Sir **John Joseph Caldwell.** 1821–93. Canadian prime minister (1891–92).
**Abbott, Lyman.** 1835–1922. Amer. clergyman, author, & editor.
**Abbott, Robert Sengstacke.** 1868–1940. Amer. civil-rights worker.
**Abbott, William ("Bud").** 1898–1974. Amer. comedian.
**Abd-el-Ka·der** *also* **Abd-al-Ka·dir** (ăb'dəl-kä'dər, -kä-dîr'). 1807?–83. Arab emir & scholar.
**Abd-er-Rah·man Khan** (ăb'dər-rə-män' kän', KHän'). 1844–1901. Afghanistan emir (1880–1901).
**Abd·ul-A·ziz** (ăb'dəl-ə-zēz', äb'dŏŏl-ä-). 1830–76. Turkish sultan (1861–76); deposed.
**Abd·ul Ba·ha** (ăb'dŏŏl bä-hä', äb-dŏŏl'). *Abbas Effendi.* 1844–1921. Persian leader of Bahai sect.
**Abdul Ha·mid** (hä-mēd', -mĭt'). Name of two Turkish sultans, esp. **II,** 1842–1918, ruled 1876–1909; abdicated.
**Abd·ul-Jab·bar** (ăb-dŏŏl'jə-bär'), **Kareem.** b. 1947. Amer. basketball player.
**Abd·ul·lah ibn-Hu·sein** (ăb'dōō-lä' ĭb'n-hōō-sān'). 1882–1951. Emir (1921–46) & king (1946–51) of Jordan; assassinated.
**Abd·ul-Me·djid I** *also* **Abd·ul-Me·jid I** (ăb'dŏŏl-mĕ-jēd', -jĭt'). 1823–61. Turkish sultan (1839–61).
**A·bel** (ā'bəl). Son of Adam & Eve; killed by his brother Cain.
**Abel,** Sir **Frederick Augustus.** 1827–1902. English chemist.
**Ab·e·lard** (ăb'ə-lärd') *also* **A·bé·lard** (ä-bā-lär'), **Peter** or **Pierre.** 1079–1142. French theologian & philosopher; condemned for heresy.
**A·bell** (ā'bəl), **Arunah Sheperdson.** 1806–88. Amer. newspaper publisher.
**Ab·er·crom·by** or **Ab·er·crom·bie** (ăb'ər-krŏm'bē, -krŭm'-), **James.** 1706–81. Scottish-born British army officer.
**Ab·er·nath·y** (ăb'ər-năth'ē), **Ralph David.** b. 1926. Amer. clergyman & civil-rights leader.
**Ab·ing·ton** (ăb'ĭng-tən), **Frances Barton.** 1737–1815. English actress.
**A·bra·ham** (ā'brə-hăm'). Hebrew patriarch.
**A·bra·va·nel** (ä-brä'vä-nĕl'), **Isaac.** 1437–1508. Jewish theologian & scholar.
**A·bruz·zi** (ä-brŏŏt'sē), Duke of. *Prince Luigi Amedeo of Savoy-Aosta.* 1873–1933. Italian explorer & naval officer.
**A·bu-Bakr** (ä'bōō-bä'kər) *also* **A·bu Bekr** (bĕk'ər). 573–634. First Moslem caliph (632–34).
**A·bul Ka·sim** (ä'bŏŏl kä'sĭm, kä-sĭm'). d. 1013? Arab surgeon.
**Ab·zug** (ăb'zŏŏg', -zŭg'), **Bella.** b. 1920. Amer. politician.
**Ach·e·son** (ăch'ĭ-sən), **Dean Gooderham.** 1893–1971. Amer. public official.
**A·ço·ka** (ä-shō'kä). *var. of* ASOKA.
**Ac·ton** (ăk'tən), 1st Baron. *John Emerich Edward Dalberg-Acton.* 1834–1902. English historian.
**Ad·am** (ăd'əm). The first man.
**Adam, Robert** (1728–92) & **James** (1730–94). English architects & designers.
**Ad·ams** (ăd'əmz), **Abigail Smith.** 1744–1818. Amer. letter writer & wife of John Adams.
**Adams, Ansel.** 1902–1984. Amer. photographer.
**Adams, Brooks.** 1848–1927. Amer. historian.
**Adams, Charles Francis.** 1807–86. Amer. public official.
**Adams, Charles Francis, Jr.** 1835–1915. Amer. historian & railroad authority.
**Adams, Franklin Pierce.** *F.P.A.* 1881–1960. Amer. humorist.
**Adams, Hannah.** 1755–1831. Amer. author & compiler.
**Adams, Henry Brooks.** 1838–1918. Amer. historian.
**Adams, Herbert Baxter.** 1850–1901. Amer. historian.
**Adams, James Truslow.** 1878–1949. Amer. historian.
**Adams, John.** 1735–1826. Second U.S. President (1797–1801), diplomat, & political philosopher.
**Adams, John Quincy.** 1767–1848. Sixth U.S. President (1825–29), diplomat, & legislator.
**Adams, Maude Kiskadden.** 1872–1953. Amer. actress.
**Adams, Samuel.** 1722–1803. Amer. Revolutionary leader.
**Adams, Samuel Hopkins.** 1871–1958. Amer. author.
**Adams, Walter Sydney.** 1876–1956. Amer. astronomer.
**Adams, William Taylor.** *Oliver Optic.* 1822–97. Amer. educator and author.
**Ad·am·son** (ăd'əm-sən), **Joy.** 1910–80. Austrian-born naturalist.
**Ad·dams** (ăd'əmz), **Charles Samuel.** b. 1912. Amer. cartoonist.
**Addams, Jane.** 1860–1935. Amer. social reformer & pacifist (Nobel, 1931).
**Ad·di·son** (ăd'ĭ-sən), **Joseph.** 1672–1719. English essayist. **—Ad'di·so'ni·an** *adj.*
**Ade** (ād), **George.** 1866–1944. Amer. humorist.
**Ad·en·au·er** (ăd'n-ou'ər, äd'-), **Konrad.** 1876–1967. West German statesman.
**Ad·ler** (ăd'lər, äd'-), **Alfred.** 1870–1937. Austrian psychiatrist. **—Ad·ler'i·an** (ăd-lîr'ē-ən) *adj.*
**Ad·ler** (ăd'lər), **Cyrus.** 1863–1940. Amer. religious leader & educator.
**Ad·ler** (ăd'lər, äd'-), **Felix.** 1851–1933. German-born Amer. educator & reformer.
**Ad·ler** (ăd'lər), **Mortimer Jerome.** b. 1902. Amer. philosopher & educator.
**Adler, Stella** (b. 1902) & **Luther** (1903–84). Amer. drama educators.
**A·dri·an** (ā'drē-ən). Name of six popes, esp. **IV,** 1100?–59, only English-born pope, reigned 1154–59.
**Adrian, Edgar Douglas.** 1889–1977. English physiologist (Nobel, 1932).
**A·dy** (ô'dē), **Endre.** 1877–1919. Hungarian poet.
**Ael·fric** (ăl'frĭk). *Grammaticus.* 955?–1020? English abbot.
**Aes·chy·lus** (ĕs'kə-ləs, ē'skə-). 525–456 B.C. Greek dramatist. **—Aes'chy·le'an** (-lē'ən) *adj.*
**Ae·sop** (ē'səp, -sŏp'). 6th cent. B.C. Greek fabulist. **—Ae·so'pi·an** (ē-sō'pē-ən), **Ae·sop'ic** (ē-sŏp'ĭk) *adj.*
**Aeth·el·red** (ĕth'əl-rĕd'). ETHELRED.
**A·ga Khan** (ä'gä kän'). Title of the head of the Ismaili Moslems, esp. **III** (1877–1957) & **IV** (b. 1936).
**Ag·as·siz** (ăg'ə-sē), **Alexander.** 1835–1910. Swiss-born Amer. zoologist.
**Agassiz, Elizabeth Cabot Cary.** 1822–1907. Amer. educator.
**Agassiz, (Jean) Louis (Rodolphe).** 1807–73. Swiss-born Amer. naturalist.
**A·gath·o·cles** (ə-găth'ə-klēz'). 361–289 B.C. Sicilian despot & tyrant of Syracuse (317–289).
**A·gee** (ā'jē), **James.** 1910–55. Amer. author & critic.
**A·ges·i·la·us II** (ə-jĕs'ə-lā'əs). 444?–360? B.C. Spartan king (399?–60?).
**Ag·nes** (ăg'nĭs), Saint. d. 304? Christian martyr.
**Ag·new** (ăg'nōō', -nyōō'), **David Hayes.** 1818–92. Amer. surgeon.
**Agnew, Spiro Theodore.** b. 1918. U.S. Vice President (1969–73); resigned.
**Ag·non** (ăg'nôn'), **Shmuel Yosef.** 1888–1970. Polish-born Israeli author (Nobel, 1966).
**A·gric·o·la** (ə-grĭk'ə-lə), **Gnaeus Julius.** A.D. 37–93. Roman soldier & politician.
**A·grip·pa** (ə-grĭp'ə), **Marcus Vipsanius.** 63–12 B.C. Roman soldier & statesman.
**Ag·rip·pi·na** (ăg'rə-pī'nə, -pē'-). *the Elder.* 13 B.C.?–A.D. 33. Roman matron; mother of Caligula.
**Agrippina.** *the Younger.* A.D. 15?–59. Roman empress; mother of Nero.
**A·gui·nal·do** (ä'gē-näl'dō), **Emilio.** 1869–1964. Philippine revolutionary leader.
**Ah·med** (ä'mĕd, ä'mĕt). Name of three Turkish sultans, esp. **III,** 1673–1736, ruled 1703–30; deposed.
**Ai·ken** (ā'kən), **Conrad Potter.** 1889–1973. Amer. author.
**Ai·ley** (ā'lē, ī'lē), **Alvin, Jr.** b. 1931. Amer. choreographer.
**Ains·worth** (ānz'wərth), **William Harrison.** 1805–82. English author.
**A·i·sha** *also* **A·ye·sha** (ä'ē-shə). 611–78. Chief wife of Mohammed.

**Ak·bar** (ăk′bär). *the Great.* 1542–1605. Mogul emperor (1556–1605).

**Ake·ley** (ā′klē), **Carl Ethan.** 1864–1926. Amer. naturalist & sculptor.

**a Kem·pis** (ə kĕm′pĭs, ä), **Thomas.** —See THOMAS A KEMPIS.

**A·khe·na·ton** *or* **A·khe·na·ten** (ä′kə-nät′n, äk′nät′n) *also* **Ikh·na·ton** (ĭk-nät′n). Egyptian pharaoh (1375–58 B.C.) & religious reformer.

**A·ki·ba ben Jo·seph** (ä-kē′bä bĕn jō′zəf, -səf). A.D. 50?–132. Jewish religious leader.

**A·lar·cón** (ä′lär-kôn′), **Pedro Antonio de.** 1833–91. Spanish author.

**Al·ar·ic** (ăl′ər-ĭk). 370–410. Visigoth king & conqueror of Rome (410).

**Alaric II.** d. 507. Visigoth king.

**Al·ba** (ăl′bə). *var. of* ALVA.

**Al·bee** (ôl′bē, ŏl′-, ăl′-), **Edward Franklin.** b. 1928. Amer. playwright.

**Al·be·marle** (ăl′bə-märl), 1st Duke of. George MONCK.

**Al·bé·niz** (äl-bā′nēs′, äl-), **Isaac.** 1860–1909. Spanish composer & pianist.

**Al·bers** (ăl′bərz, ŏl′-), **Josef.** 1888–1976. German-born Amer. painter.

**Al·bert** (ăl′bərt), Prince. 1819–61. Consort of Queen Victoria of England.

**Albert, Carl Bert.** b. 1908. Amer. legislator.

**Albert I.** 1875–1934. Belgian king (1909–34).

**Al·ber·tus Mag·nus** (ăl-bûr′təs măg′nəs), Saint. 1206?–80. German religious philosopher.

**Al·bright** (ôl′brīt, ŏl′-), **Ivan Le Lorraine.** b. 1897. Amer. painter.

**Albright, William Foxwell.** 1891–1971. Amer. archaeologist & educator.

**Al·bu·querque** (ăl′bə-kûr′kē, ăl′bə-kûr′-), **Affonso de.** 1453–1515. Portuguese colonial administrator.

**Al·cae·us** (ăl-sē′əs). fl. 611?–580 B.C. Greek poet.

**Al·ci·bi·a·des** (ăl′sə-bī′ə-dēz′). 450?–404 B.C. Athenian politician & general.

**Al·cott** (ôl′kət, -kŏt, ŏl′-), **Amos Bronson.** 1799–1888. Amer. educator & philosopher.

**Alcott, Louisa May.** 1832–88. Amer. author & reformer.

**Al·cuin** (ăl′kwĭn). 735–804. English prelate & scholar.

**Al·da** (ăl′də, ăl′-), **Frances (Davis).** 1883–1952. New Zealand-born soprano.

**Al·den** (ôl′dən, ŏl′-), **Isabella Macdonald.** 1841–1930. Amer. author.

**Alden, John** (1599?–1687) & **Priscilla Mullins** (b. 1602?). Pilgrim colonists.

**Al·der** (ăl′dər), **Kurt.** 1902–58. German chemist (Nobel, 1950).

**Al·ding·ton** (ôl′dĭng-tən, ŏl′-), **Richard.** 1892–1962. English poet.

**Al·drich** (ôl′drĭch, ŏl′-), **Nelson Wilmarth.** 1841–1915. Amer. financier & legislator.

**Aldrich, Thomas Bailey.** 1836–1907. Amer. author & editor.

**Al·dridge** (ôl′drĭj, ŏl′-), **Ira Frederick.** 1804?–67. Amer. actor.

**Al·drin** (ôl′drĭn, ŏl′-), **Edwin Eugene ("Buzz"), Jr.** b. 1930. Amer. astronaut.

**Al·dus Ma·nu·tius** (ôl′dəs mə-nōō′shəs, -shē′əs, -nyōō′-, ŏl′-). —See MANUTIUS.

**A·lei·chem** (ä-lā′kĕm, -KHĕm), **Shalom** *or* **Sholem.** 1859–1916. Russian-born Amer. Yiddish humorist.

**A·leix·an·dre** (ä′lĕk-sän′drə), **Vicente.** b. 1898. Spanish poet (Nobel, 1977).

**A·le·mán** (ä′lā-män′), **Mateo.** 1547–1610? Spanish-born Mexican author.

**Alemán Val·dés** (väl-dĕs′), **Miguel.** 1902–83. Mexican statesman.

**A·lem·bert** (ăl′əm-bâr′, ä-läN-bĕr′), **Jean Le Rond d′.** 1717–83. French mathematician, scientist, philosopher, & Encyclopedist.

**Al·ex·an·der** (ăl′ĭg-zăn′dər, -zän′-). Name of eight popes, esp.: **a. III.** d. 1181. Reigned 1159–81. **b. VI.** 1431?–1503. Reigned 1491–1503.

**Alexander.** Name of three Russian czars: **a. I.** 1777–1825. Ruled 1801–25. **b. II.** 1818–81. Ruled 1855–81; emancipated serfs (1861). **c. III.** 1845–94. Ruled 1881–94.

**Alexander I** *also* **Alexander O·bre·no·vić** (ō-brĕn′ə-vĭch′). 1876–1903. Serbian king (1889–1903); assassinated.

**Alexander I.** 1888–1934. Yugoslavian king (1921–34); assassinated.

**Alexander III.** *the Great.* 356–323 B.C. Macedonian king (336–323) & conqueror of Greece, Persia, & Egypt. **—Al′ex·an′dri·an** *adj.*

**Alexander Nev·ski** (nĕv′skē, nĕf′-). 1220?–63. Russian saint & national hero.

**Alexander of Tu·nis** (tōō′nĭs, tyōō′-), 1st Earl. *Harold Rupert Leofric George Alexander.* 1891–1969. British field marshal.

**Alexander Se·ve·rus** (sə-vîr′əs), **Marcus Aurelius.** 203?–35. Roman emperor (222–35).

**Al·ex·an·der·son** (ăl′ĭg-zăn′dər-sən), **Ernst Frederick Werner.** 1878–1973. Swedish-born Amer. electrical engineer.

**A·lex·is I Mi·khai·lo·vich** (ä-lĕk′sĭs mĭ-kī′lə-vĭch, mĭ-KHī′-). 1629–76. Russian czar (1645–76) & reformer.

**Alexis Pe·tro·vich** (pĕ-trō′vĭch). 1690–1718. Russian czarevitch; condemned for treason.

**A·lex·i·us Com·ne·nus** (ə-lĕk′sē-əs kŏm-nē′nəs). Name of five Eastern Roman emperors, esp. **I,** 1048–1118, ruled 1081–1118.

**Al·fie·ri** (äl-fē-âr′ē, äl-fyâr′ē), Conte **Vittorio.** 1749–1803. Italian dramatist.

**Al·fon·so** (ăl-fŏn′sō, -zō). Name of six kings of Portugal, esp.: **a. I.** 1112–85. Ruled 1139–85. **b. V.** 1432–81. Ruled 1438–81.

**Alfonso XIII.** 1886–1941. Spanish king (1886–1931, ruled 1902–31); abdicated.

**Al·fred** (ăl′frĭd). *the Great.* 849–99. West Saxon king (871–99), scholar, & lawmaker.

**Alf·vén** (äl-vān′), **Hannes Olof Gösta.** b. 1908. Swedish physicist (Nobel, 1970).

**Al·ger** (ăl′jər), **Cyrus.** 1781–1856. Amer. industrialist.

**Alger, Horatio.** 1832–99. Amer. author.

**A·li** (ä-lē′). 600?–61. Moslem caliph (656–61); assassinated.

**Ali** *or* **Ali Pa·sha** (pä′shä). 1741–1822. Turkish colonial governor; assassinated.

**Ali, Muhammad.** b. 1942. Amer. prizefighter.

**Al·len** (ăl′ən), **Ethan.** 1738–89. Amer. Revolutionary soldier.

**Allen, Florence Ellinwood.** 1884–1966. Amer. jurist.

**Allen, Fred.** 1894–1956. Amer. humorist.

**Allen, Frederick Lewis.** 1890–1954. Amer. editor & historian.

**Allen, Grace Ethel Cecile Rosalie ("Gracie").** 1906–64. Amer. comedienne.

**Allen, Henry Watkins.** 1820–66. Amer. Confederate soldier & public official.

**Allen, John.** 1810–92. Amer. dentistry pioneer.

**Allen, Melvin Israel ("Mel").** b. 1913. Amer. sportscaster.

**Allen, Richard.** 1760–1831. Amer. religious leader.

**Allen, Steve.** b. 1921. Amer. entertainer.

**Allen, William.** 1532–94. English cardinal.

**Allen, Woody.** b. 1935. Amer. actor, writer, & filmmaker.

**Al·len·by** (ăl′ən-bē), 1st Viscount. *Edmund Henry Hynman.* 1861–1936. British field marshal.

**Al·len·de Gos·sens** (ä-yĕn′dā gô′sĕns), **Salvador.** 1908–73. Chilean president (1970–73); died in coup.

**Al·leyn** (ăl′ən, -ĕn′, -ān′), **Edward.** 1566–1626. English actor.

**Al·lou·ez** (ä-lōō-ā′, äl-wā′), **Claude Jean.** 1622–89. French Jesuit missionary.

**All·ston** (ôl′stən, ŏl′-), **Robert Francis Withers.** 1801–64. Amer. agriculturist & politician.

**Allston, Washington.** 1779–1843. Amer. painter.

**Al·ma-Tad·e·ma** (ăl′mə-tăd′ə-mə), Sir **Lawrence.** 1836–1912. Dutch-born English painter.

**Al·mei·da** (äl-mā′də), **Francisco de.** 1450?–1510. Portuguese colonial administrator.

**A·lon·so** (ə-lŏn′zō, ä-lôn′sō), **Alicia.** b. 1921? Cuban ballerina & choreographer.

**Al·sop** (ôl′səp, ŏl′-), **Joseph W., Jr.** (b. 1910) & **Stewart** (1914–74). Amer. journalists.

**Al·ter** (ôl′tər, ŏl′-), **David.** 1807–81. Amer. physicist, physician, & inventor.

**Alt·geld** (ôlt′gĕld, ŏlt′-), **John Peter.** 1847–1902. German-born Amer. politician.

**Alt·man** (ôlt′mən, ŏlt′-), **Benjamin.** 1840–1913. Amer. merchant & art patron.

**Al·va** (ăl′və, äl′vä) *also* **Al·ba** (ăl′bə), Duke of. *Fernando Álvarez de Toledo.* 1508–82. Spanish general & colonial administrator.

**Al·va·ra·do** (ăl′və-rä′dō), **Alonso de** (1490?–1554) & **Pedro de** (1486–1541). Spanish adventurers in the New World.

**Al·va·rez** (ăl′və-rĕz′), **Luis Walter.** b. 1911. Amer. physicist (Nobel, 1968).

**Al·ve·ar** (äl′vā-är′), **Carlos María de.** 1789–1853. Argentine revolutionary soldier & politician.

**A·ma·do** (ə-mä′dōō), **Jorge.** b. 1912. Brazilian author.

**A·ma·ti** (ä-mä′tē). Family of Italian violin makers, esp. **Nicolò** *or* **Nicola** (1596–1684).

**Am·bler** (ăm′blər), **Eric.** b. 1909. English author.

**Am·brose** (ăm′brōz′), Saint. 340?–97. Author, composer, & bishop of Milan (374–97). **—Am·bro′sian** *adj.*

**A·men·ho·tep** (ä′mən-hō′tĕp, ăm′ən-) *also* **Am·e·no·phis** (ăm′ə-nō′fĭs). Name of four Egyptian pharaohs, esp.: **a. III.** Reigned 1411?–1375 B.C. **b. IV.** AKHENATON.

**Ames** (āmz), **Fisher.** 1758–1808. Amer. political leader & author.

**A·min Da·da** (ä-mēn′ dä-dä′), **Idi.** b. 1925? Ugandan dictator (1971–79); deposed.

---

ă **pat** ā **pay** âr **care** ä **father** ĕ **pet** ē **be** hw **which** ĭ **pit**
ī **tie** îr **pier** ŏ **pot** ō **toe** ô **paw, for** oi **noise** ŏŏ **took**

**A·mis** (ā'mĭs), **Kingsley.** b. 1922. English author.
**A·mo·ry** (ā'mə-rē), **Cleveland.** b. 1917. Amer. author & conservationist.
**A·mos** (ā'məs). 8th cent. B.C. Hebrew prophet.
**Am·père** (ăm'pîr, äN-pĕr'), **André Marie.** 1775–1836. French physicist & mathematician.
**A·mund·sen** (ä'mənd-sən, ä'mo͝on-), **Roald.** 1872–1928. Norwegian polar explorer.
**A·nac·re·on** (ə-năk'rē-ən). 572?–488? B.C. Greek poet.
**An·ax·ag·o·ras** (ăn'ăk-săg'ər-əs). 500?–428 B.C. Greek philosopher.
**A·nax·i·man·der** (ə-năk'sə-măn'dər). 611–547 B.C. Greek philosopher & mathematician.
**An·der·sen** (ăn'dər-sən), **Hans Christian.** 1805–75. Danish author.
**An·der·son** (ăn'dər-sən), **Carl David.** b. 1905. Amer. physicist (Nobel, 1936).
**Anderson,** Sir **John.** *Viscount Waverley.* 1882–1958. British politician & public official.
**Anderson, Joseph Reid.** 1813–92. Amer. manufacturer & Confederate general.
**Anderson,** Dame **Judith.** b. 1898. Australian-born actress.
**Anderson, Margaret Caroline.** 1893?–1973. Amer. editor.
**Anderson, Marian.** b. 1902. Amer. contralto.
**Anderson, Mary Antoinette.** 1859–1940. Amer. actress.
**Anderson, Maxwell.** 1888–1959. Amer. playwright.
**Anderson, Philip Warren.** b. 1923. Amer. physicist (Nobel, 1977).
**Anderson, Robert Woodruff.** b. 1917. Amer. dramatist.
**Anderson, Sherwood.** 1876–1941. Amer. author.
**An·dra·da e Sil·va** (ăn-drä'də ē sĭl'və), **José Bonifácio de.** 1763?–1838. Brazilian statesman, scientist, & poet.
**An·drás·sy** (ăn-drăs'ē, ôn'drä-shē), Counts **Gyula** (1823–90) & **Gyula** (1860–1929). Hungarian politicians.
**An·dré** (ăn'drā, ăn'drē), **John.** 1751–80. English soldier; hanged as Revolutionary spy.
**An·dre·a del Sar·to** (ăn-drā'ə dĕl' sär'tō). 1486–1531. Italian painter.
**An·dre·ev** or **An·dre·yev** (ăn-drā'əf, -yəf), **Leonid Nikolaevich.** 1871–1919. Russian author.
**An·drew** (ăn'dro͞o), Saint. One of the 12 Apostles.
**Andrew, John Albion.** 1818–67. Amer. antislavery leader.
**An·drews** (ăn'dro͞oz), **Charles McLean.** 1863–1943. Amer. historian.
**Andrews, Fannie Fern Phillips.** 1867–1950. Canadian-born Amer. reformer.
**Andrews, Roy Chapman.** 1884–1960. Amer. naturalist.
**Andrews, Stephen Pearl.** 1812–86. Amer. abolitionist & reformer.
**Andrews, Thomas.** 1813–85. Irish physicist & chemist.
**An·drić** (ăn'drĭch), **Ivo.** 1892–1975. Yugoslavian author (Nobel, 1961).
**An·dro·pov** (ăn-drŏp'ŏf, -ŏv), **Yuri** 1914–84. Soviet statesman.
**An·dros** (ăn'drŏs, -drəs), Sir **Edmund.** 1637–1714. English colonial administrator.
**An·fin·sen** (ăn'fən-sən), **Christian Boehmer.** b. 1916. Amer. biochemist (Nobel, 1972).
**An·gel·a Me·ri·ci** (ăn'jə-lə mə-rē'chē), Saint. 1474–1540. Italian founder of the Ursuline order.
**An·gel·i·co** (ăn-jĕl'ĭ-kō'), Fra. *Giovanni da Fiesole.* 1387–1455. Italian painter.
**An·gell** (ān'jəl), **James Burrill.** 1829–1916. Amer. educator & diplomat.
**Angell,** Sir **Norman.** 1872–1967. English economist (Nobel, 1933).
**Ång·ström** (ăng'strəm), **Anders Jonas.** 1814–74. Swedish physicist & astronomer.
**An·na I·va·nov·na** (ä'nə ē-vä'nəv-nə). 1693–1740. Russian empress (1730–40).
**Anne** (ăn). 1665–1714. Queen of Great Britain & Ireland (1702–14).
**Anne of Aus·tri·a** (ô'strē-ə). 1601–66. Wife of Louis XIII of France & regent (1643–61).
**Anne of Cleves** (klēvz). 1515–57. English queen as fourth wife of Henry VIII.
**A·nouilh** (ä-no͞o'ē), **Jean.** b. 1910. French dramatist.
**An·selm** (ăn'sĕlm), Saint. 1033–1109. Italian-born English prelate & philosopher.
**An·theil** (ăn'tīl), **George.** 1900–59. Amer. pianist & composer.
**An·tho·ny** (ăn'thə-nē), Saint. c. 250–350. Egyptian ascetic monk.
**Anthony, Susan Brownell.** 1820–1906. Amer. reformer.
**Anthony of Pad·u·a** (păj'o͞o-ə, păd'yo͞o-ə), Saint. 1195–1231. Italian theologian.
**An·tig·o·nus** (ăn-tĭg'ə-nəs). Name of three kings of Macedonia, esp. **I,** 382–301 B.C., one of Alexander III's generals.

o͞o **boot** ou **out** th **thin** *th* **this** ŭ **cut** ûr **urge** y **young**
yo͞o **abuse** zh **vision** ə **about, item, edible, gallop, circus**

**An·ti·o·chus** (ăn-tī'ə-kəs). Name of 13 Seleucid kings of Syria, esp.: **a. III.** *the Great.* c. 241–187 B.C. Ruled 223–187. **b. IV.** *Antiochus Epiphanes.* d. 163 B.C. Ruled 175–163.
**An·tip·a·ter** (ăn-tĭp'ə-tər). 398?–319 B.C. Macedonian general & diplomat.
**An·tis·the·nes** (ăn-tĭs'thə-nēz'). 444?–371? B.C. Greek philosopher.
**An·toine** (ăn-twän', äN-), Père. 1748–1829. Spanish priest in Louisiana.
**An·to·nel·lo da Mes·si·na** (än'tō-nĕl'ō dä mā-sē'nä). 1430?–79. Italian painter.
**An·to·ne·scu** (än'tə-nĕs'ko͞o), **Ion.** 1882–1946. Rumanian general & dictator.
**An·to·ni·nus Pi·us** (ăn'tə-nī'nəs pī'əs). 86–161 A.D. Roman emperor (138–61).
**An·to·ni·on·i** (ăn-tō'nē-ō'nē), **Michelangelo.** b. 1912. Italian filmmaker.
**An·to·ni·us** (ăn-tō'nē-əs), **Marcus.** MARK ANTONY.
**An·za** (än'sə), **Juan Bautista de.** 1735–88? Spanish colonial administrator.
**A·pel·les** (ə-pĕl'ēz). 4th cent. B.C. Greek painter.
**Ap·gar** (ăp'gär), **Virginia.** 1909–74. Amer. physician.
**A·pol·li·naire** (ə-pŏl'ə-nâr'), **Guillaume.** 1880–1918. Polish-born French poet.
**A·pol·lo·dor·us** (ə-pŏl'ō-dôr'əs). *Skiagraphos.* 5th cent. B.C. Greek painter.
**Ap·ol·lo·ni·us of Rhodes** (ăp'ə-lō'nē-əs; rōdz). 3rd–2nd cent. B.C. Greek poet.
**Ap·ple·by** (ăp'əl-bē), **John Francis.** 1840–1917. Amer. inventor.
**Ap·ple·gate** (ăp'əl-gāt'), **Jesse.** 1811–88. Amer. Western surveyor & legislator.
**Ap·ple·seed** (ăp'əl-sēd'), **Johnny.** John CHAPMAN.
**Ap·ple·ton** (ăp'əl-tən), Sir **Edward Victor.** 1892–1965. English physicist (Nobel, 1947).
**A·pra·ksin** *also* **A·prax·in** (ə-prăk'sən), **Fëdor Matveevich.** 1671–1728. Russian admiral.
**Ap·u·lei·us** (ăp'yə-lē'əs), **Lucius.** 2nd cent. A.D. Roman philosopher & satirist.
**A·qui·nas** (ə-kwī'nəs), Saint **Thomas.** 1225?–74. Italian theologian & philosopher.
**Ar·a·fat** (âr'ə-făt'), **Yasir.** b. 1929. Palestinian leader.
**A·ra·gon** (ăr-ə-gôN'), **Louis.** 1897–1982. French author.
**Ar·am** (âr'əm), **Eugene.** 1704–59. English philologist.
**Ar·ber** (är'bər), **Edward.** 1836–1912. English scholar & editor.
**Arber, Werner.** b. 1929. Swiss microbiologist (Nobel, 1978).
**Ar·buth·not** (är-bŭth'nət, är'bəth-nŏt'), **John.** 1667–1735. Scottish physician & author.
**Ar·car·o** (är-kâr'ō), **George Edward ("Eddie").** b. 1916. Amer. jockey.
**Ar·cher** (är'chər), **William.** 1856–1924. Scottish critic, dramatist, & translator.
**Ar·chi·me·des** (är'kə-mē'dēz). 287?–212 B.C. Greek mathematician, engineer, & physicist. **—Ar'chi·me·de'an** *adj.*
**Ar·chi·pen·ko** (är'kə-pĕng'kō), **Alexander Porfirievich.** 1887–1964. Russian-born Amer. sculptor.
**Ar·den** (är'dn), **Elizabeth.** 1891–1966. Canadian-born Amer. businesswoman.
**Ar·drey** (är'drē), **Robert.** 1908–80. Amer. anthropologist.
**A·rendt** (âr'ənt, är'-), **Hannah.** 1906–75. German-born Amer. historian.
**A·re·ti·no** (är'ə-tē'nō), **Pietro.** 1492–1556. Italian author.
**Ar·gall** (är'gôl, -gəl), Sir **Samuel.** 1572?–1626. English colonial administrator.
**Ar·gyll** (är-gīl', är'gīl'), 9th Duke of. *John Douglas Sutherland Campbell.* 1845–1914. English author, politician, & colonial administrator.
**A·ri·os·to** (är'ē-ŏs'tō, -ō'stō, ăr'-), **Lodovico.** 1474–1533. Italian poet.
**Ar·is·tar·chus** (ăr'ĭ-stär'kəs). 220?–150 B.C. Greek grammarian & cirtic.
**Aristarchus of Sa·mos** (sā'mŏs', săm'ŏs'). 3rd cent. B.C. Greek astronomer.
**Ar·is·ti·des** *also* **Ar·is·tei·des** (ăr'ĭ-stī'dēz). *the Just.* 530?–468? B.C. Athenian statesman & general.
**Ar·is·tip·pus** (ăr'ĭ-stĭp'əs). 5th–4th cent. B.C. Greek philosopher.
**Ar·is·toph·a·nes** (ăr'ĭ-stŏf'ə-nēz). 448?–380? B.C. Athenian dramatist.
**Aristophanes of By·zan·ti·um** (bĭ-zăn'shē-əm, -tē-əm). 257?–180? B.C. Greek philologist.
**Ar·is·tot·le** (ăr'ĭ-stŏt'l). 384–322 B.C. Greek philosopher. **—Ar'is·to·te'li·an** (ăr'ĭ-stə-tē'lē-ən, -tēl'yən) *adj.* & *n.*
**A·ri·us** (ə-rī'əs, âr'ē-, ăr'-). 256?–336. Greek theologian; condemned as a heretic. **—Ar·i·an** (âr'ē-ən, ăr'-) *adj.* & *n.*
**Ark·wright** (ärk'rīt'), Sir **Richard.** 1732–92. English inventor & manufacturer.
**Ar·len** (är'lən), **Michael.** 1895–1956. Armenian-born English author.

**Ar·min·i·us** (är-mĭn′ē-əs) *also* **Ar·min** (-mēn′). 17 B.C.–A.D. 21. German hero.
**Arminius, Jacobus.** 1560–1609. Dutch theologian.
**Ar·mour** (är′mər), **Philip Danforth.** 1832–1901. Amer. industrialist.
**Arm·strong** (ärm′strông′), **Edwin Howard.** 1890–1954. Amer. engineer & inventor.
**Armstrong, Hamilton Fish.** 1893–1973. Amer. editor & author.
**Armstrong, John.** 1758–1843. Amer. general, politician, & diplomat.
**Armstrong, Louis ("Satchmo").** 1900–71. Amer. jazz musician.
**Armstrong, Neil Alden.** b. 1930. Amer. astronaut; first to walk on the moon.
**Armstrong, Samuel Chapman.** 1839–93. Amer. educator.
**Armstrong, William George.** 1810–1900. *Baron Armstrong of Cragside.* English inventor & industrialist.
**Arne** (ärn), **Thomas Augustine.** 1710–78. English composer.
**Ar·no** (är′nō), **Peter.** 1904–68. Amer. cartoonist.
**Ar·nold** (är′nəld), **Benedict.** 1741–1801. Amer. Revolutionary general & traitor.
**Arnold, Henry Harley ("Hap").** 1886–1950. Amer. air-force officer.
**Arnold, Matthew.** 1822–88. English poet & critic.
**Arnold, Thomas.** 1795–1842. English educator & historian.
**Arnold, Thurman Wesley.** 1891–1969. Amer. jurist.
**Ar·nold·son** (är′nəld-sən), **Klas Pontus.** 1844–1916. Swedish politician & pacifist (Nobel, 1908).
**Arp** (ärp), **Jean** *or* **Hans.** 1887–1966. French artist.
**Ar·pád** (är′päd). d. 907. Hungarian national hero.
**Ar·rhe·ni·us** (ə-rē′nē-əs, ə-rā′-), **Svante August.** 1859–1927. Swedish physicist & chemist (Nobel, 1903).
**Ar·row** (ăr′ō), **Kenneth Joseph.** b. 1921. Amer. economist (Nobel, 1972).
**Ar·son·val** (är-sôɴ-väl′), **Jacques Arsène d′.** 1851–1940. French physicist.
**Ar·ta·xer·xes** (är′tə-zûrk′sēz′). Name of three Persian kings: **a. I.** d. 424 B.C. Ruled 464–24. **b. II.** d. 359 B.C. Ruled 404–359. **c. III.** d. 338 B.C. Ruled 359–38.
**Ar·te·vel·de** (är′tə-vĕl′də), **Jacob van.** *Brewer of Ghent.* 1290?–1345. Flemish political leader.
**Ar·thur** (är′thər), **Chester Alan.** 1829–86. 21st U.S. President (1881–85).
**Arthur, Timothy Shay.** 1809–85. Amer. journalist & temperance advocate.
**As·bur·y** (ăz′bûr′ē, -bə-rē), **Francis.** 1745–1816. English-born Amer. religious leader.
**Asch** (ăsh), **Sholem** *or* **Shalom.** 1880–1957. Polish-born Amer. Yiddish author.
**As·cham** (ăs′kəm), **Roger.** 1515–68. English scholar & author.
**Ash·burn·er** (ăsh′bûr′nər), **Charles Albert.** 1854–89. Amer. geologist.
**Ashe** (ăsh), **Arthur Robert, Jr.** b. 1943. Amer. tennis player.
**Ash·ley** (ăsh′lē), **William Henry.** 1778?–1838. Amer. fur trader & politician.
**Ash·mun** (ăsh′mən), **Jehudi.** 1794–1828. Amer. colonial agent in Africa.
**Ash·ton** (ăsh′tən), Sir **Frederick.** b. 1906. English choreographer.
**A·shur·ba·ni·pal** (ä′shŏŏr-bä′nə-päl′) *also* **As·sur·ba·ni·pal** (ä′-sŏŏr-). Assyrian king (669–26 B.C.).
**As·i·mov** (ăz′ĭ-mŏf), **Isaac.** b. 1920. Russian-born Amer. author.
**A·so·ka** (ə-sō′kə) *also* **A·ço·ka** (ä-shō′kä). *the Great.* d. 232 B.C. Buddhist king of Magadha (273–32).
**As·pa·sia** (ăs-pā′zhə). fl. c. 440 B.C. Greek courtesan.
**As·quith** (ăs′kwĭth), **Herbert Henry.** *1st Earl of Oxford & Asquith.* 1852–1928. British prime minister (1908–16).
**As·sad** (ä-säd′), **Hafez al-.** b. 1928. Syrian statesman.
**As·ser** (ä′sər), **Tobias Michael Carel.** 1838–1913. Dutch statesman & jurist (Nobel, 1911).
**As·sur·ba·ni·pal** (ä′sŏŏr-bä′nə-päl′). *var. of* ASHURBANIPAL.
**A·staire** (ə-stâr′), **Fred.** b. 1899. Amer. dancer & actor.
**As·ton** (ăs′tən), **Francis William.** 1877–1945. English chemist & physicist (Nobel, 1922).
**As·tor** (ăs′tər), **Caroline Webster Schermerhorn.** 1830–1908. Amer. socialite.
**Astor, John Jacob.** 1763–1848. German-born Amer. fur trader & capitalist.
**Astor, Nancy Witcher Langhorne,** Viscountess. 1879–1964. Amer.-born British politician.
**Astor, William Waldorf.** *1st Viscount Astor of Hever Castle.* 1848–1919. Amer.-born British capitalist.
**As·tu·ri·as** (ə-stŏŏr′ē-əs, ä-stŏŏr′yäs), **Miguel Angel.** 1899–1974. Guatemalan author (Nobel, 1966).
**A·ta·hual·pa** (ä′tə-wäl′pə) *also* **A·ta·ba·li·pa** (ä′tə-bä′lĭ-pä′). 1502?–33. Incan emperor (1525–33); executed.
**At·a·türk** (ăt′ə-tûrk′, ä′tə-), **Kemal.** —See KEMAL ATATÜRK.
**Ath·a·na·sius** (ăth′ə-nā′shəs), Saint. 293–373. Greek patriarch of Alexandria. **—Ath′a·na′sian** *adj.* & *n.*
**Ath·el·stan** (ăth′əl-stăn′). 895–940. English king (924–40).
**Ath·er·ton** (ăth′ər-tən), **Gertrude Franklin Horn.** 1857–1948. Amer. author.
**At·kin·son** (ăt′kĭn-sən), **(Justin) Brooks.** 1894–1984. Amer. critic.
**Atkinson, Edward.** 1827–1905. Amer. economist & industrialist.
**Atkinson, Henry.** 1782–1842. Amer. soldier & explorer.
**At·las** (ăt′ləs), **Charles.** 1894–1972. Italian-born Amer. physical culturist.
**At·tar** (ăt′ər, ə-tär′). 1119–1229? Persian poet & mystic.
**At·ti·la** (ăt′l-ə, ə-tĭl′ə). *Scourge of the Gods.* 406?–53. King of the Huns (433?–53).
**Att·lee** (ăt′lē), **Clement Richard.** 1883–1967. British prime minister (1945–51).
**At·tucks** (ăt′əks), **Crispus.** 1723?–70. Killed in the Boston Massacre.
**At·wat·er** (ăt′wô′tər, -wŏt′ər), **Wilbur Olin.** 1844–1907. Amer. pioneer in agricultural chemistry.
**Au·ber** (ō-bĕr′), **Daniel François Esprit.** 1782–1871. French composer.
**Au·brey** (ô′brē), **John.** 1626–97. English antiquarian.
**Au·chin·closs** (ô′kĭn-klôs, -klŏs), **Louis Stanton.** b. 1917. Amer. author.
**Au·den** (ôd′n), **Wystan Hugh.** 1907–73. English-born Amer. author.
**Au·du·bon** (ô′də-bŏn′, -bən), **John James.** 1785–1851. Haitian-born Amer. ornithologist & artist.
**Au·gus·tine** (ô′gə-stēn′, ô-gŭs′tĭn) *also* **Aus·tin** (ô′stən). *Apostle of the English.* d. 604? Missionary & prelate.
**Augustine,** Saint. 354–430. Church father & philosopher. **—Au′gus·tin′i·an** (ô′gə-stĭn′ē-ən) *adj.* & *n.*
**Au·gus·tus** (ô-gŭs′təs). *Octavian.* 63 B.C.–A.D. 14. First Roman emperor (27 B.C.–A.D. 14). **—Au·gus′tan** *adj.* & *n.*
**Au·rang·zeb** *also* **Au·rung·zeb** *or* **Au·rung·zebe** (ôr′əng-zĕb′). 1618–1707. Hindustani emperor (1658–1707).
**Au·re·lian** (ô-rēl′yən, ô-rē′lē-ən). 212?–75. Roman emperor (270–75).
**Au·ri·ol** (ôr′ē-ôl′, ō-ryôl′), **Vincent.** 1884–1966. French statesman.
**Au·rung·zeb** *or* **Au·rung·zebe** (ôr′əng-zĕb′). *vars. of* AURANGZEB.
**Aus·ten** (ô′stən), **Jane.** 1775–1817. English author.
**Aus·tin** (ô′stən). *var. of* the missionary AUGUSTINE.
**Austin, Alfred.** 1835–1913. English author.
**Austin, John.** 1790–1859. English jurist.
**Austin, Mary Hunter.** 1868–1934. Amer. author.
**Austin, Stephen Fuller.** 1793–1836. Amer. colonizer & political leader.
**Au·try** (ô′trē), **Gene.** b. 1907. Amer. singer & actor.
**Av·en·zo·ar** (ăv′ən-zō′ər). 1090?–1162. Spanish-Arab physician & author.
**A·ver·ro·ës** *or* **A·ver·rho·ës** (ə-vĕr′ō-ēz′, ăv′ə-rō′ēz). 1126–1198. Spanish-Arab physician & philosopher.
**Av·i·cen·na** (ăv′ĭ-sĕn′ə). 980–1037. Arab physician & philosopher.
**A·viz** *also* **A·vis** (ä′vēsh). Dynasty of Portuguese rulers (1385–1850).
**A·vo·ga·dro** (ä′və-gä′drō, äv′ə-), **Amedeo.** 1776–1856. Italian physicist.
**A·von** (ā′vŏn, ăv′ən), Earl of. Anthony EDEN.
**Ax·el·rod** (ăk′səl-räd′), **Julius.** b. 1912. Amer. biochemist (Nobel, 1970).
**Ay·de·lotte** (ād′l-ŏt′), **Frank.** 1880–1956. Amer. educator.
**Ayer** (âr), **Francis Wayland.** 1848–1923. Amer. advertising executive.
**A·ye·sha** (ä′ē-shə). *var. of* AISHA.
**Ayres** (ârz), **Anne.** 1816–96. English-born Amer. religious leader.
**Ayr·ton** (âr′tn), **William Edward.** 1847–1908. English physicist, electrical engineer, & inventor.
**A·za·ña y Dí·ez** (ə-zän′yə ē dē′ĕz, -ĕs), **Manuel.** 1880–1940. Spanish statesman.
**A·ze·glio** (ä-zĕl′yō), Marchese d′. *Massimo Taparelli.* 1798–1866. Italian politician & author.
**A·zue·la** (ə-zwā′lə), **Mariano.** 1873–1952. Mexican author.

# B

**Baa·de** (bä′də), **Walter.** 1893–1960. German-born Amer. astronomer.
**Baal Shem Tov** (bäl′ shĕm′ tōv′, shām′) *also* **Baal Shem Tob** (tōb′). 1700?–60. Jewish religious leader.
**Bab** (bäb, băb), **the.** *Ali Mohammed of Shiraz.* 1819–50. Persian founder of Babism.

**Ba·bar** (bä'bər). *var. of* BABER.
**Bab·bitt** (băb'ĭt), **Irving.** 1865–1933. Amer. humanist & scholar.
**Bab·cock** (băb'kŏk'), **Stephen Moulton.** 1843–1931. Amer. agricultural chemist.
**Ba·ben·berg** (bä'bən-bĕrg'). Franconian dynasty ruling margraviate (976–1156) & duchy (1156–1246) of Austria.
**Ba·ber** *also* **Ba·bar** *or* **Ba·bur** (bä'bər). 1483–1530. Mongol conqueror of India.
**Ba·beuf** *or* **Ba·boeuf** (bä-bœf'), **François Noël** *or* **Émile.** 1760–97. French revolutionary.
**Bab·ing·ton** (băb'ĭng-tən), **Anthony.** 1561–86. English Catholic conspirator; executed for treason.
**Ba·boeuf** (bä-bœf'). *var. of* BABEUF.
**Bab·son** (băb'sən), **Roger Ward.** 1875–1967. Amer. financial statistician.
**Ba·bur** (bä'bər). *var. of* BABER.
**Bach** (bäKH, bäk). Family of German composers & musicians, including **Johann Sebastian** (1685–1750) & his sons **Wilhelm Friedemann** (1710–84), **Karl Philip Emanuel** (1714–88), **Johann Christoph Friedrich** (1732–95), & **Johann Christian** (1735–82).
**Bache** (bāch), **Alexander Dallas.** 1806–67. Amer. physicist.
**Bache, Benjamin Franklin.** 1769–98. Amer. journalist.
**Back·us** (băk'əs), **Isaac.** 1724–1806. Amer. separatist clergyman & historian.
**Ba·con** (bā'kən), **Delia Salter.** 1811–59. Amer. author.
**Bacon, Francis.** *Baron Verulam, Viscount St. Albans.* 1561–1626. English philosopher, essayist, courtier, jurist, & statesman. **—Ba·co'ni·an** *adj. & n.*
**Bacon, Francis.** b. 1910. Irish-born British painter.
**Bacon, Nathaniel.** 1647–76. English-born colonist; led Bacon's Rebellion (1676).
**Bacon, Roger.** *the Admirable Doctor.* 1214?–94. English friar, scientist, & philosopher.
**Ba·den-Pow·ell** (băd'n-pō'əl), Sir **Robert Stephenson Smyth.** 1857–1941. English soldier & founder of the Boy Scouts.
**Ba·do·glio** (bə-dōl'yō), **Pietro.** 1871–1956. Italian general & politician.
**Bae·da** (bē'də). *var. of* BEDE.
**Bae·de·ker** (bā'dĭ-kər), **Karl.** 1801–59. German guidebook publisher.
**Baeke·land** (bāk'lănd'), **Leo Hendrik.** 1863–1944. Belgian-born Amer. chemist.
**Baer** (bâr), **Arthur ("Bugs").** 1886–1969. Amer. journalist.
**Baer, George Frederick.** 1842–1914. Amer. lawyer & railroad magnate.
**Baer, Karl Ernst von.** 1792–1876. Estonian naturalist & pioneer embryologist.
**Bae·yer** (bā'ər, -yər), **(Johann Friedrich Wilhelm) Adolf von.** 1835–1917. German organic chemist (Nobel, 1905).
**Ba·ez** (bī'ĕz', bī-ĕz'), **Joan.** b. 1941. Amer. folk singer.
**Baf·fin** (băf'ĭn), **William.** 1584–1622. English explorer.
**Bage·hot** (băj'ət), **Walter.** 1826–77. English economist, social scientist, & journalist.
**Bag·ley** (băg'lē), **William Chandler.** 1874–1946. Amer. psychologist.
**Bag·nold** (băg'nəld), **Enid.** 1889–1981. English author.
**Ba·gra·tion** (bə-grä'tē-ôn'), Prince **Pëtr Ivanovich.** 1765–1812. Russian general.
**Ba·ha·ul·lah** (bä-hä'o͞o-lä'). 1817–92. Persian founder of Bahai sect.
**Bai·ley** (bā'lē), **Ann.** *White Squaw of the Kanawha.* 1742–1825. English-born Amer. frontier heroine.
**Bailey, Anna Warner.** *Mother Bailey.* 1758–1851. Amer. Revolutionary heroine.
**Bailey, Florence Augusta Merriam.** 1863–1948. Amer. ornithologist.
**Bailey, F(rancis) Lee.** b. 1933. Amer. lawyer.
**Bailey, Gamaliel.** 1807–59. Amer. physician, journalist, & antislavery advocate.
**Bailey, Liberty Hyde.** 1858–1954. Amer. botanist & horticulturist.
**Bailey, Nathan** *or* **Nathaniel.** d. 1742. English lexicographer.
**Bailey, Pearl Mae.** b. 1918. Amer. entertainer.
**Bail·lie** (bā'lē), **Joanna.** 1762–1851. Scottish dramatist.
**Bain** (bān), **Alexander.** 1818–1903. Scottish psychologist.
**Bain·bridge** (bān'brĭj'), **William.** 1774–1833. Amer. naval officer.
**Baird** (bârd), **John Logie.** *Father of Television.* 1888–1946. Scottish inventor.
**Baird, Spencer Fullerton.** 1823–87. Amer. zoologist, naturalist, & ornithologist.
**Bairns·fa·ther** (bârnz'fä'thər), **Bruce.** 1888–1959. English cartoonist & journalist.
**Ba·jer** (bī'ər), **Fredrik.** 1837–1922. Danish pacifist (Nobel, 1908).
**Ba·ker** (bā'kər), **George.** *Father Divine.* 1877–1965. Amer. religious leader.
**Baker, George.** 1915–75. Amer. cartoonist.
**Baker, George Fisher.** 1840–1931. Amer. financier & philanthropist.
**Baker, George Pierce.** 1866–1935. Amer. drama professor.
**Baker, Josephine.** 1906–75. Amer.-born French entertainer.
**Baker, Newton Diehl.** 1871–1937. Amer. politician & public official.
**Baker, Ray Stannard.** 1870–1946. Amer. author & muckraking journalist.
**Baker,** Sir **Samuel White.** 1821–93. English explorer.
**Baker, Sara Josephine.** 1873–1945. Amer. pediatrician & public-health pioneer.
**Bakst** (bäkst), **Léon Nikolaevich.** 1867–1924. Russian painter & scenic designer.
**Ba·ku·nin** (bə-ko͞o'nĭn, bä-), **Mikhail Aleksandrovich.** 1814–76. Russian anarchist.
**Ba·la·ki·rev** (bə-lä'kĭ-rəf, -rəv), **Mili Alekseevich.** 1837–1910. Russian composer.
**Bal·an·chine** (băl'ən-chēn', băl'ən-chēn'), **George.** 1904–83. Russian-born Amer. choreographer.
**Bal·bo** (bäl'bō), Count **Cesar.** 1789–1853. Italian statesman & author.
**Balbo, Italo.** 1896–1940. Italian aviator & politician.
**Bal·bo·a** (băl-bō'ə), **Vasco Núñez de.** 1475–1517. Spanish explorer.
**Bal·bue·na** (bäl-bwā'nä), **Bernardo de.** 1562?–1627. Mexican poet.
**Balch** (bôlch), **Emily Greene.** 1867–1961. Amer. economist & sociologist (Nobel, 1946).
**Bald·win** (bôld'wĭn). Name of five kings of Jerusalem, esp. **I,** 1058–1118, ruled 1108–18.
**Baldwin, Faith.** 1893–1978. Amer. author.
**Baldwin, Henry.** 1780–1844. Amer. legislator & jurist.
**Baldwin, James Arthur.** b. 1924. Amer. author.
**Baldwin, James Mark.** 1861–1934. Amer. psychologist & editor.
**Baldwin, Loammi** (1740–1807) & **Loammi** (1780–1838). Amer. engineers.
**Baldwin, Maria Louise.** 1856–1922. Amer. educator.
**Baldwin, Matthias William.** 1795–1866. Amer. locomotive manufacturer.
**Baldwin, Stanley.** *1st Earl Baldwin of Bewdley.* 1867–1947. British prime minister (1923–24, 1924–29, 1935–37).
**Ba·len·ci·a·ga** (bə-lĕn'sē-ä'gə), **Cristobal.** 1895–1972. Spanish fashion designer.
**Balfe** (bălf), **Michael William.** 1808–70. Irish composer & singer.
**Bal·four** (băl'fo͝or', -fôr', -fōr'), **Arthur James.** *1st Earl of Balfour.* 1848–1930. British prime minister (1902–5) & diplomat.
**Bal·iol** (bāl'yəl), **John de.** 1249–1315. Scottish king (1292–96).
**Ball** (bôl), **John.** *the Mad Priest.* d. 1381. English priest & social agitator; executed.
**Ball, Lucille.** b. 1911. Amer. comedienne.
**Bal·lan·tyne** (băl'ən-tīn'), **James.** 1792–1833. Scottish printer.
**Bal·lou** (bə-lo͞o'), **Adin.** 1803–90. Amer. religious leader & Utopian reformer.
**Bal·lou, Hosea.** 1771–1852. Amer. clergyman & editor.
**Bal·main** (bäl-măN'), **Pierre.** 1914–82. French fashion designer.
**Bal·ti·more** (bôl'tə-môr', -mōr'). Barons. —See CALVERT.
**Baltimore, David.** b. 1938. Amer. microbiologist (Nobel, 1975).
**Bal·zac** (bôl'zăk', băl'-), **Honoré de.** 1799–1850. French author. **—Bal·zac'i·an** *adj.*
**Ban·croft** (băn'krôft'), **George.** 1800–91. Amer. historian & diplomat.
**Bancroft, Hubert Howe.** 1832–1918. Amer. publisher & historian.
**Ban·de·lier** (băn'də-lîr'), **Adolph Francis Alphonse.** 1840–1914. Swiss-born Amer. historian, explorer, archaeologist, & anthropologist.
**Ban·del·lo** (băn-dĕl'ō, bän-), **Matteo.** 1480–1562? Italian prelate & author.
**Bangs** (băngz), **John Kendrick.** 1862–1922. Amer. humorist & editor.
**Bank·head** (băngk'hĕd'), **Tallulah Brockman.** 1903–68. Amer. actress.
**Bankhead, William Brockman.** 1874–1940. Amer. legislator.
**Banks** (băngks), Sir **Joseph.** 1743–1820. English botanist.
**Banks, Nathaniel Prentiss.** 1816–94. Amer. politician & general.
**Ban·ne·ker** (băn'ĭ-kər), **Benjamin.** 1731–1806. Amer. mathematician & astronomer.
**Ban·nis·ter** (băn'ĭ-stər), **Roger.** b. 1929. English physician & runner.
**Ban·ting** (băn'tĭng), Sir **Frederick Grant.** 1891–1941. Canadian physiologist (Nobel, 1923).
**Bar·a** (băr'ə), **Theda.** 1890?–1955. Amer. actress.
**Ba·rab·bas** (bə-răb'əs). Condemned thief whose release was demanded instead of that of Jesus.
**Ba·ra·ka** (bə-rä'kə), **Imamu Amiri.** *LeRoi Jones.* b. 1934. Amer. poet & playwright.

**Ba·ra·nov** (bə-rä'nəf), **Aleksandr Andreevich.** 1746–1819. Russian fur trader & colonial administrator.

**Bá·rá·ny** (bä'rän'yə), **Robert.** 1876–1936. Austrian otologist (Nobel, 1914).

**Bar·ba·ros·sa** (bär'bə-rŏs'ə, -rôs'-). Holy Roman Emperor FREDERICK I.

**Barbarossa, Khair el-Din** (1466?–1546) & **Koruk** (1474–1518). Greek-born Moslem corsairs.

**Bar·ber** (bär'bər), **Samuel.** 1910–81. Amer. composer.

**Bar·bour** (bär'bər), **Philip Pendleton.** 1783–1841. Amer. legislator & jurist.

**Bar·busse** (bär-bo͞os'), **Henri.** 1873–1935. French author & editor.

**Bar·clay** (bär'klē), **Robert.** 1648–90. Scottish Quaker author.

**Bar·clay de Tol·ly** (bär-klī' də tôl'yə), Prince **Mikhail.** 1761–1818. Scottish-born Russian field marshal.

**Bard** (bärd), **John.** 1716–99. Amer. physician; pioneer in dissection.

**Bard, Samuel.** 1742–1821. Amer. physician & educator.

**Bard, William.** 1778–1853. Amer. life-insurance company organizer.

**Bar·deen** (bär-dēn'), **John.** b. 1908. Amer. physicist (Nobel, 1956, 1972).

**Bar·dot** (bär-dō'), **Brigitte.** b. 1935? French actress.

**Bar·en·boim** (băr'ĭn-boim'), **Daniel.** b. 1942. Israeli-born concert pianist.

**Bar·ents** (băr'ənts), **Willem.** 1550?–97. Dutch Arctic explorer.

**Bar·ing** (bâr'ĭng), **Alexander.** *1st Baron Ashburton.* 1774–1848. English financier & statesman.

**Baring, Evelyn.** *1st Earl of Cromer.* 1841–1917. English financier & diplomat.

**Bark·la** (bär'klə), **Charles Glover.** 1877–1944. English physicist (Nobel, 1917).

**Bar·kley** (bär'klē), **Alben William.** 1877–1956. U.S. Vice President (1949–53).

**Bar·low** (bär'lō'), **Joel.** 1754–1812. Amer. poet & diplomat.

**Bar·na·bas** (bär'nə-bəs), **Joses** or **Joseph.** 1st cent. Christian convert & missionary.

**Bar·nard** (bär'nərd, bär-närd'), **Christiaan Neethling.** b. 1923. South African surgeon & pioneer in heart transplants.

**Bar·nard** (bär'nərd), **Edward Emerson.** 1857–1923. Amer. astronomer & pioneer in photography.

**Barnard, Frederick Augustus Porter.** 1809–89. Amer. educator.

**Barnard, George Grey.** 1863–1938. Amer. sculptor & art collector.

**Barnard, Henry.** 1811–1900. Amer. advocate of free public education.

**Barnard, Kate.** 1875–1930. Amer. political reformer.

**Barnes** (bärnz), **Albert Coombs.** 1873–1951. Amer. physician & art collector.

**Barnes, Harry Elmer.** 1889–1968. Amer. sociologist & educator.

**Bar·ne·veldt** or **Bar·ne·veld** (bär'nə-vĕlt'), **Jan van Olden.** 1547–1619. Dutch statesman; beheaded.

**Bar·num** (bär'nəm), **Phineas Taylor ("P.T.").** 1810–91. Amer. showman.

**Ba·ro·ja y Nes·si** (bə-rô'hə ē nĕs'ē), **Pío.** 1872–1956. Spanish author.

**Bar·rès** (bä-rĕs'), **Auguste Maurice.** 1862–1923. French author & politician.

**Bar·rett** (băr'ĭt), **Janie Porter.** 1865–1948. Amer. welfare worker.

**Barrett, Kate Harwood Waller.** 1857–1925. Amer. social worker.

**Bar·rie** (băr'ē), Sir **James Matthew.** 1860–1937. Scottish author.

**Bar·ron** (băr'ən), **Clarence Walker.** 1855–1923? Amer. financial editor.

**Bar·ros** (bär'o͞osh'), **João de.** 1496–1570. Portuguese historian.

**Bar·row** (băr'ō), **Isaac.** 1630–77. English theologian, scholar, & mathematician.

**Bar·ry** (băr'ē), **John.** 1745–1803. Amer. naval commander.

**Barry, Philip.** 1896–1946. Amer. playwright.

**Bar·ry·more** (băr'ĭ-môr', -mŏr'). Family of Amer. actors, including **Maurice Herbert Blythe** (English-born, 1847–1905), **Georgiana Emma Drew** (1854–93), **Lionel Blythe** (1878–1954), **Ethel** (1879–1959), **John Blythe** (1882–1942), **Diana** (1921–60), & **John Drew** (b. 1932).

**Bart** (bär) or **Barth** (bärt), **Jean.** 1651?–1702. French naval hero.

**Barth** (bärth), **John Simmons.** b. 1930. Amer. novelist.

**Barth** (bärt, bärth), **Karl.** 1886–1968. Swiss Protestant theologian.

**Barthes** (bärt), **Roland.** 1915–80. French critic.

**Bar·thol·di** (bär-thŏl'dē, -tôl-dē'), **Frédéric Auguste.** 1834–1904. French sculptor.

**Bar·thol·o·mew** (bär-thŏl'ə-myo͞o'), Saint. One of the 12 Apostles.

**Bart·lett** (bärt'lĭt), **John.** 1820–1905. Amer. publisher & editor.

**Bartlett, John Russell.** 1805–86. Amer. historian & antiquarian.

**Bartlett, Robert Abram ("Captain Bob").** 1875–1946. Amer. Arctic explorer.

**Bar·tók** (bär'tŏk', -tôk'), **Béla.** 1881–1945. Hungarian pianist & composer.

**Bar·to·lom·me·o** (bär-tŏl'ə-mā'ō), Fra. 1475?–1517. Italian painter.

**Bar·ton** (bar'tn), **Bruce.** 1886–1967. Amer. advertising executive, author, & politician.

**Barton, Clara.** 1821–1912. Amer. nurse & founder of the Amer. Red Cross.

**Barton,** Sir **Derek Harold Richard.** b. 1918. English chemist (Nobel, 1969).

**Bar·tram** (bär'trəm), **John** (1699–1777) & **William** (1739–1823). Amer. botanists.

**Ba·ruch** (bə-ro͞ok'), **Bernard Mannes.** 1870–1965. Amer. stock broker, political adviser, & public official.

**Ba·rysh·ni·kov** (bə-rĭsh'nĭ-kôf'), **Mikhail Nikolayavich.** b. 1948. Latvian-born ballet dancer & choreographer.

**Bar·zun** (bär'zŭn), **Jacques Martin.** b. 1907. French-born Amer. educator, author, & historian.

**Bas·com** (băs'kəm), **Florence.** 1862–1945. Amer. geologist.

**Ba·sho** (bä'shō), **Matsuo.** 1644–94. Japanese poet.

**Ba·sie** (bā'sē), **William ("Count").** 1904–1984. Amer. jazz composer & band leader.

**Bas·il** (băz'əl, băs'-, bā'zəl, -səl), Saint. *the Great.* 330?–379? Greek Christian leader.

**Bas·ker·ville** (băs'kər-vĭl'), **John.** 1706–75. English printer & type designer.

**Bas·kin** (băs'kĭn), **Leonard.** b. 1922. Amer. artist.

**Ba·sov** (bä'sôf', -sôv'), **Nikolai Gennadievich.** b. 1922. Russian physicist (Nobel, 1964).

**Bass** (băs), **Sam.** 1851–78. Amer. desperado.

**Bas·sa·no** (bə-sä'nō), **Jacopo.** 1510–92. Venetian genre painter.

**Bate·man** (bāt'mən), **Hezekiah Linthicum.** 1812–75. Amer. actor & theatrical manager.

**Bates** (bāts), **Blanche.** 1873–1941. Amer. actress.

**Bates, Edward.** 1793–1869. Amer. politician & public official.

**Bates, Herbert Ernest ("H.E.").** 1905–74. English author.

**Bates, Katherine Lee.** 1859–1929. Amer. educator & poet.

**Bath·she·ba** (băth-shē'bə, băth'shə-). Second wife of David & mother of Solomon.

**Ba·tis·ta y Zal·di·var** (bə-tēs'tə ē zäl-dē'vär'), **Fulgencio.** 1901–73. Cuban military & political leader; deposed & exiled (1959).

**Bat·ta·ni** (bə-tä'nē), **al-.** 850?–929. Arab astronomer.

**Baude·laire** (bōd-lâr', -lĕr'), **Charles Pierre.** 1821–67. French poet & critic.

**Bau·douin** (bō-dwăN'). b. 1930. Belgian king (since 1951).

**Baum** (bôm, bäm), **Lyman Frank.** 1856–1919. Amer. author.

**Baum** (boum), **Vicki.** 1896–1960. Austrian-born Amer. author.

**Bau·mé** (bō-mā'), **Antoine.** 1728–1804. French chemist.

**Bax·ter** (băk'stər), **Richard.** 1615–91. English nonconformist chaplin & scholar.

**Bay·ard** (bā'ərd, bī'-, bä-yär'), Seigneur de. *Pierre Terrail.* 1473–1524. French military hero.

**Bay·ard** (bī'ərd), **James Asheton** (1767–1815) & **Thomas Francis** (1828–98). Amer. politicians & diplomats.

**Bayes** (bāz), **Nora.** 1880–1928. Amer. vaudeville entertainer.

**Bayle** (bāl, bĕl), **Pierre.** 1647–1706. French philosopher & critic.

**Bay·ley** (bā'lē), **Richard.** 1745–1801. Amer. physician.

**Bay·lor** (bā'lər), **Robert Emmet Bledsoe.** 1793–1873. Amer. politician, lawyer, & clergyman.

**Beach** (bēch), **Alfred Ely.** 1826–96. Amer. inventor & editor.

**Beach, Amy Marcey Cheney.** 1867–1944. Amer. pianist & composer.

**Beach, Frederick Converse.** 1848–1918. Amer. publisher & photographer.

**Beach, Moses Yale** (1800–68) & **Moses Sperry** (1822–92). Amer. publishers & inventors.

**Beach, Sylvia Woodbridge.** 1887–1962. Amer. bookseller & publisher in Paris.

**Bea·cons·field** (bē'kənz-fēld'), 1st Earl of. Benjamin DISRAELI.

**Bea·dle** (bēd'l), **Erastus Flavel.** 1821–94. Amer. publisher.

**Beadle, George Wells.** b. 1903. Amer. biologist (Nobel, 1958).

**Bean** (bēn), **Roy.** 1825?–1903. Amer. frontiersman.

**Beard** (bîrd), **Charles Austin.** 1874–1948. Amer. historian & educator.

**Beard, Daniel Carter ("Dan").** 1850–1941. Amer. author, illustrator, & founder of the Amer. Boy Scouts.

**Beard, Mary Ritter.** 1876–1958. Amer. historian & feminist.

**Beards·ley** (bîrdz'lē), **Aubrey Vincent.** 1872–98. English illustrator.

**Beat·les** (bēt'lz), **the.** Former group of English composers & musicians, including John LENNON, Ringo STARR, Paul MCCARTNEY, & George HARRISON.

**Bea·ton** (bēt'n), **Cecil Walter Hardy.** 1904–80. English photographer, author, & theatrical designer.

**Be·a·trix** (bā'ə-trĭks'). b. 1938. Queen of the Netherlands (since 1980).

**Beat·tie** (bā'tē, bē'-), **James.** 1735–1803. Scottish author & philosopher.

**Beau·fort** (bō′fərt), **Henry.** 1377?–1447. English prelate & statesman.
**Beaufort, Margaret.** *Countess of Richmond & Derby.* 1441–1509. English Lancastrian & patron of education.
**Beau·har·nais** (bō-är-nā′), **Alexandre de.** 1760–94. French politician & military leader.
**Beauharnais, Eugène de.** 1781–1824. French soldier & statesman.
**Beauharnais, Hortense.** 1783–1837. Mother of Napoleon III.
**Beauharnais, Josephine de.** 1763–1814. First wife of Napoleon I; divorced (1809).
**Beau·mar·chais** (bō-mär-shā′), **Pierre Augustin Caron de.** 1732–99. French author.
**Beau·mont** (bō′mŏnt′, -mənt), **Francis.** 1584–1616. English poet & dramatist.
**Beau·mont** (bō′mŏnt′), **William.** 1785–1853. Amer. pioneer surgeon.
**Beau·re·gard** (bō′rĭ-gärd′), **Pierre Gustave Toutant.** 1818–93. Amer. Confederate general.
**Beau·voir** (bō-vwär′), **Simone de.** b. 1908. French author.
**Beaux** (bō), **Cecelia.** 1863–1942. Amer. painter.
**Bea·ver·brook** (bē′vər-brŏŏk′), 1st Baron. *William Maxwell Aitken.* 1879–1964. Canadian-born British publisher & politician.
**Be·bel** (bā′bəl), **(Ferdinand) August.** 1840–1913. German socialist leader.
**Beck·er** (bĕk′ər), **Carl Lotus.** 1873–1945. Amer. historian.
**Becker, George Ferdinand.** 1847–1919. Amer. geologist & physicist.
**Beck·et** (bĕk′ĭt), Saint **Thomas à.** 1118?–70. English Roman Catholic martyr.
**Beck·ett** (bĕk′ĭt), **Samuel.** b. 1906. Irish author (Nobel, 1969).
**Beck·ford** (bĕk′fərd), **William.** 1759?–1844. English author & collector.
**Beck·man** (bĕk′män), **Max.** 1884–1950. German artist.
**Beck·nell** (bĕk′nəl), **William.** 1790?–1832. Amer. frontier explorer.
**Bec·querel** (bĕ-krĕl′). Family of French physicists, including **Antoine César** (1788–1878), **Alexandre Edmond** (1820–91), & **Antoine Henri** (1852–1908; Nobel, 1903).
**Be·da** (bē′də). *var. of* BEDE.
**Bed·does** (bĕd′ōz′), **Thomas Lovell.** 1803–49. English poet & physician.
**Bede** (bēd) *also* **Bae·da** *or* **Be·da** (bē′də). *Venerable Bede.* c. 673–735. English theologian, historian, & scientist.
**Bed·ford** (bĕd′fərd), Duke of. JOHN OF LANCASTER.
**Bee·be** (bē′bē), **Lucius Morris.** 1902–66. Amer. journalist.
**Beebe, (Charles) William.** 1877–1962. Amer. naturalist & explorer.
**Bee·cham** (bē′chəm), Sir **Thomas.** 1879–1961. English conductor.
**Bee·cher** (bē′chər), **Catharine Esther.** 1800–78. Amer. educator & reformer.
**Beecher, Edward.** 1803–95. Amer. clergyman & theologian.
**Beecher, Henry Ward.** 1813–87. Amer. clergyman, editor, & abolitionist.
**Beecher, Lyman.** 1775–1863. Amer. theologian.
**Beer** (bâr), **Wilhelm.** 1797–1850. German banker & astronomer.
**Beer·bohm** (bîr′bōm′), Sir **Max.** 1872–1956. English caricaturist & author.
**Beer·naert** (bâr′närt), **Auguste Marie François.** 1829–1912. Belgian statesman (Nobel, 1909).
**Beers** (bîrz), **Clifford Whittingham.** 1876–1943. Amer. mental-health pioneer.
**Beer·y** (bîr′ē), **Wallace.** 1886?–1949. Amer. actor.
**Bee·tho·ven** (bā′tō′vən), **Ludwig van.** 1770–1827. German composer.
**Be·gin** (bā′gĭn), **Menachem.** b. 1913. Russian-born Israeli statesman (Nobel, 1978).
**Be·han** (bē′ən), **Brendan Francis.** 1923–64. Irish author.
**Beh·ring** (bâr′ĭng, bĕr′-, bā′rĭng). *var. of* BERING.
**Behring, Emil Von.** 1854–1917. German physiologist (Nobel, 1901).
**Behr·man** (bâr′mən), **Samuel Nathaniel.** 1893–1973. Amer. playwright.
**Bei·der·becke** (bī′dər-bĕk′), **Leon Bismark ("Bix").** 1903–31. Amer. jazz composer & cornetist.
**Beis·sel** (bī′səl), **Johann Konrad.** 1690–1768. German religious leader & hymn writer.
**Bé·ké·sy** (bā′kā-shē), **Georg von.** 1899–1972. Hungarian-born Amer. physiologist (Nobel, 1961).
**Be·las·co** (bə-lăs′kō), **David.** 1853?–1931. Amer. playwright & theatrical producer.
**Bel·cher** (bĕl′chər), **Jonathan.** 1682–1757. Amer. merchant & colonial governor.
**Bel·i·sar·i·us** (bĕl′ĭ-sâr′ē-əs). 505?–65. Byzantine general.
**Bell** (bĕl), **Alexander Graham.** 1847–1922. Scottish-born Amer. inventor.
**Bell, (Arthur) Clive (Howard).** 1881–1964. English critic.
**Bell, John.** 1797?–1869. Amer. politician.
**Bel·la·my** (bĕl′ə-mē), **Edward.** 1850–98. Amer. author & utopian socialist.
**Bel·lay** (bə-lā′), **Joachim du.** 1524–60. French poet.
**Bel·lings·hau·sen** (bĕl′ĭngz-hou′zən), **Fabian Gottlieb von.** 1778–1852. Russian naval officer & explorer.
**Bel·li·ni** (bə-lē′nē). Family of Venetian painters, including **Jacopo** (1400?–70?), **Gentile** (1429?–1507), & **Giovanni** (1430?–1516).
**Bellini, Vincenzo.** 1801–35. Italian operatic composer.
**Bel·loc** (bĕl′ŏk′, -ək), **Hilaire.** 1870–1953. French-born English author.
**Bel·low** (bĕl′ō), **Saul.** b. 1915. Canadian-born Amer. novelist (Nobel, 1976).
**Bel·lows** (bĕl′ōz), **Albert Fitch.** 1829–83. Amer. painter.
**Bellows, George Wesley.** 1882–1925. Amer. artist.
**Bel·mont** (bĕl′mŏnt′), **Alva Ertskin Smith Vanderbilt.** 1853–1933. Amer. socialite & suffragist.
**Belmont, August.** 1816–90. German-born Amer. banker, public official, & art collector.
**Bel·mon·te** (bĕl-môn′tā), **Juan.** 1893–1962. Spanish bullfighter.
**Bel·shaz·zar** (bĕl-shăz′ər). Last king of Babylon.
**Be·ly** (byĕ′lē, bĕ′-), **Andrei.** *Boris Nikolaevich Bugaev.* 1880–1934. Russian author.
**Bel·zo·ni** (bĕl-zō′nē), **Giovanni Battista.** 1778–1823. Italian explorer & Egyptologist.
**Be·mel·mans** (bē′məl-mənz, bĕm′əl-), **Ludwig.** 1898–1962. Austrian-born Amer. illustrator & author.
**Be·mis** (bē′mĭs), **Samuel Flagg.** 1891–1973. Amer. historian & educator.
**Ben·a·cer·raf** (bĕn-ăs′ər-əf), **Baruj.** b. 1920. Venezuelan-born Amer. pathologist (Nobel, 1980).
**Be·na·ven·te y Mar·ti·nez** (bĕn′ə-vĕn′tē ē mär-tē′nəs), **Jacinto.** 1866–1954. Spanish playwright (Nobel, 1922).
**Be·na·vi·des** (bĕn′ə-vē′dəs), **Alonzo de.** b. 1580? Spanish missionary in New Mexico.
**Ben Bel·la** (bĕn bĕl′ə), **Ahmed.** b. 1919. Algerian revolutionary statesman; deposed.
**Bench·ley** (bĕnch′lē), **Robert Charles.** 1889–1945. Amer. humorist, critic, & actor.
**Ben·dix** (bĕn′dĭks), **Vincent.** 1882?–1945. Amer. inventor & industrialist.
**Ben·e·dict** (bĕn′ĭ-dĭkt′). Name of 15 popes, esp.: **a. XIV.** 1675–1758. Reigned 1740–58. **b. XV.** 1854–1922. Reigned 1914–22.
**Benedict, Ruth Fulton.** 1887–1948. Amer. anthropologist.
**Benedict of Nur·si·a** (nûr′shē-ə, -shə), Saint. 480?–543. Italian founder of Benedictine order. **—Ben′e·dic′tine** (-tĭn, -tēn′) *adj. & n.*
**Be·neš** (bĕn′ĕsh′), **Eduard.** 1884–1948. Czechoslovakian statesman.
**Be·nét** (bĭ-nā′), **William Rose** (1886–1950) & **Stephen Vincent** (1898–1943). Amer. authors.
**Ben Gur·i·on** (bĕn gŏŏr′ē-ən), **David.** 1886–1973. Polish-born Israeli statesman.
**Ben·ja·min** (bĕn′jə-mən). Youngest son of Rachel and Jacob in the Bible.
**Benjamin, Asher.** 1773–1845. Amer. architect.
**Benjamin, Judah Philip.** 1811–84. British-born Amer. Confederate statesman.
**Ben·nett** (bĕn′ĭt), **(Enoch) Arnold.** 1867–1931. English author.
**Bennett, Edward Herbert.** 1874–1954. English-born Amer. architect.
**Bennett, Floyd.** 1890–1928. Amer. aviator & Arctic explorer.
**Bennett, James Gordon** (1795–1872) & **James Gordon** (1841–1918). Amer. journalists & publishers.
**Bennett, Richard Bedford.** *Viscount Bennett.* 1870–1947. Canadian prime minister (1930–1935).
**Ben·ny** (bĕn′ē), **Jack.** 1894–1974. Amer. comedian.
**Be·noît de Sainte-Maure** (bən-wä′ də săNt-môr′). 12th cent. French trouvère.
**Ben·son** (bĕn′sən), **Arthur Christopher.** 1862–1925. English educator & author.
**Benson, Edward White.** 1829–96. English prelate.
**Benson, Stella.** 1892–1933. English author.
**Bent** (bĕnt), **Charles** (1799–1847) & **William** (1809–69). Amer. pioneers.
**Ben·tham** (bĕn′thəm), **Jeremy.** 1748–1832. English author, reformer, & philosopher. **—Ben′tham·ite′** *n.*
**Ben·tinck** (bĕn′tĭngk), **Lord William Cavendish.** *3rd Duke of Portland.* 1774–1839. English colonial administrator.
**Bentinck, William Henry Cavendish.** 1738–1809. British prime minister (1783, 1807–9).
**Bent·ley** (bĕnt′lē), **Richard.** 1662–1742. English cleric & scholar.
**Ben·ton** (bĕn′tən), **Thomas Hart.** *Old Bullion.* 1782–1858. Amer. legislator.
**Benton, Thomas Hart.** 1889–1975. Amer. artist & writer.

**Benton, William.** 1900–73. Amer. advertising executive, publisher, & public official.
**Benz** (bĕnts), **Karl Friedrich.** 1844–1929. German automobile pioneer.
**Bé·ran·ger** (bā-räN-zhā'), **Pierre Jean de.** 1780–1857. French poet.
**Ber·dya·ev** (bər-dyä'yəf), **Nikolai Aleksandrovich.** 1874–1948. Russian philosopher.
**Ber·en·gar·i·a** (bĕr'ən-gâr'ē-ə). d. 1230? Castilian-born English queen as wife of Richard I.
**Ber·en·son** (bĕr'ĭn-sən), **Bernard** *also* **Bernhard.** 1865–1959. Lithuanian-born Amer. art critic & historian.
**Berg** (bĕrKH), **Alban.** 1885–1935. Austrian composer.
**Berg** (bûrg), **Gertrude.** 1899–1966. Amer. actress & radio, television, and screenwriter.
**Berg, Patricia Jane.** b. 1918. Amer. golfer.
**Berg, Paul.** b. 1926. Amer. chemist (Nobel, 1980).
**Ber·gen** (bûr'gən), **Edgar John.** 1903–78. Amer. ventriloquist.
**Ber·ger** (bĕr'gər), **Hans.** 1873–1941. German psychiatrist.
**Ber·ger** (bûr'gər), **Victor Louis.** 1860–1929. Transylvanian-born Amer. politician & editor.
**Bergh** (bûrg), **Henry.** 1811–88. Amer. founder of the ASPCA.
**Ber·gi·us** (bĕr'gē-əs), **Friedrich.** 1884–1949. German chemist (Nobel, 1931).
**Berg·man** (bûrg'mən), **Ingmar.** b. 1918. Swedish film director.
**Bergman, Ingrid.** 1915–82. Swedish actress.
**Berg·son** (bĕrg'sən, bĕrg-sôN'), **Henri Louis.** 1859–1941. French philosopher & author (Nobel, 1927).
**Berg·strom** (bĕrg'strəm), **Sune Karl.** b. 1916. Swedish biochemist and physician (Nobel, 1982).
**Ber·i·a** (bĕr'ē-ə), **Lavrenti Pavlovich.** 1899–1953. Soviet secret police chief.
**Ber·ing** *also* **Beh·ring** (bâr'ĭng, bĕr'-, bā'rĭng), **Vitus.** 1680–1741. Danish navigator & explorer.
**Berke·ley** (bûrk'lē), **Busby.** 1895–1976. Amer. choreographer & film director.
**Berke·ley** (bärk'lē, bûrk'-), **George.** 1685–1753. English-born Irish prelate & philosopher.
**Berke·ley** (bûrk'lē, bärk'-), Sir **William.** 1606–77. English colonial administrator.
**Berle** (bûrl), **Milton.** b. 1908. Amer. entertainer.
**Ber·le** (bûr'lē), **Adolf Augustus, Jr.** 1895–1971. Amer. educator & diplomat.
**Ber·lich·ing·en** (bĕr'lĭKH-ĭng'ən), **Götz** *or* **Gottfried von.** 1480?–1562. German peasant revolutionary leader.
**Ber·lin** (bûr-lĭn'), **Irving.** b. 1888. Russian-born Amer. songwriter.
**Ber·lin·er** (bûr'l-nər), **Emile.** 1851–1929. German-born Amer. inventor.
**Ber·li·oz** (bĕr'lē-ōz', -ōs'), **(Louis) Hector.** 1803–69. French composer.
**Ber·na·dette of Lourdes** (bûr'nə-dĕt'; lo͝ord, lo͝ordz), Saint. 1844–79. French peasant girl whose visions led to the establishment of the shrine at Lourdes, France.
**Bernadotte** (bûr'nə-dŏt'), Count **Folke.** 1895–1948. Swedish statesman, diplomat, & Red Cross official; assassinated.
**Ber·na·nos** (bĕr-nä-nôs'), **Georges.** 1888–1948. French author.
**Ber·nard** (bĕr-när'), **Claude.** 1813–78. French physiologist.
**Ber·nard of Clair·vaux** (bər-närd', bĕr-när'; klâr-vō'), Saint. 1090–1153. French monastic reformer & political figure.
**Ber·nar·din de Saint-Pierre** (bĕr-nər-dăN' də săN-pyĕr'), **Jacques Henri.** 1737–1814. French author.
**Bern·hardt** (bûrn'härt', bĕr-när'), **Sarah.** *the Divine Sarah.* 1844–1923. French actress.
**Ber·ni·ni** (bər-nē'nē), **Giovanni Lorenzo.** 1598–1680. Italian sculptor, painter, & architect.
**Ber·noul·li** (bər-no͞o'lē). Family of Swiss mathematicians & scientists, including **Jakob** *or* **Jacques** (1654–1705), **Johann** *or* **Jean** (1667–1748), & **Daniel** (1700–82).
**Bern·stein** (bûrn'stīn', -stēn'), **Leonard.** b. 1918. Amer. conductor & composer.
**Bern·storff** (bĕrn'shtôrf), Count **Johann Heinrich von.** 1862–1939. German diplomat.
**Ber·ra** (bĕr'ə), **Lawrence Peter ("Yogi").** b. 1925. Amer. baseball player & manager.
**Ber·ry** (bĕr'ē), **Charles Edward Anderson ("Chuck").** b. 1926. Amer. musician & singer.
**Berry, Martha McChesney.** 1866–1942. Amer. educator.
**Ber·ry·man** (bĕr'ē-mən), **John.** 1914–72. Amer. poet.
**Ber·the·lot** (bĕr-tə-lō'), **Pierre Eugène Marselin.** 1827–1907. French chemist & public official.
**Ber·thier** (bĕr-tyā'), **Louis Alexandre.** *Duc de Valangin, Prince de Neuchâtel & de Wagram.* 1753–1815. French marshal.
**Ber·til·lon** (bûr'tə-lŏn', bĕr-tē-yôN'), **Alphonse.** 1853–1914. French anthropologist & criminologist.
**Ber·ze·li·us** (bər-zē'lē-əs), Baron **Jöns Jakob.** 1779–1848. Swedish chemist.
**Bes·ant** (bĕz'ənt), **Annie Wood.** 1847–1933. English theosophist, philosopher, & political figure in India.
**Bes·sel** (bĕs'əl), **Friedrich Wilhelm.** 1784–1846. Prussian astronomer.
**Bes·se·mer** (bĕs'ə-mər), Sir **Henry.** 1813–98. British inventor & metallurgist.
**Best** (bĕst), **Charles Herbert.** 1899–1978. Amer.-born Canadian physiologist.
**Be·tan·court** (bĕ-tän-ko͞or'), **Romulo.** 1908–81. Venezuelan statesman.
**Be·the** (bā'tə), **Hans Albrecht.** b. 1906. German-born Amer. physicist (Nobel, 1967).
**Beth·mann-Holl·weg** (bĕt'mən-hôl'väg', -män-), **Theobald von.** 1856–1921. German statesman.
**Be·thune** (bə-tho͞on', -thyo͞on'), **Louise Blanchard.** 1856–1913. Amer. architect.
**Bethune, Mary McLeod.** 1875–1955. Amer. educator.
**Bet·je·man** (bĕch'ə-mən), Sir **John.** 1906–84. English poet.
**Bet·ter·ton** (bĕt'ər-tən), **Thomas.** 1635?–1710. English actor.
**Bev·an** (bĕv'ən), **Aneurin.** 1897–1960. English politician.
**Bev·er·idge** (bĕv'ər-ĭj, bĕv'rĭj), **Albert Jeremiah.** 1862–1927. Amer. politician & historian.
**Beveridge,** Sir **William Henry.** *1st Baron Tuggal.* 1879–1963. British economist.
**Bev·er·ley** (bĕv'ər-lē), **Robert.** 1673?–1722. Amer. colonial official & historian.
**Bev·i·er** (bĕv'ē-ā'), **Isabel.** 1860–1942. Amer. home-economics educator.
**Bev·in** (bĕv'ĭn), **Ernest.** 1884–1951. English labor leader & politician.
**Bhu·mi·bol A·dul·ya·dej** (po͞o'mē-pôn' ä-do͞ol'yə-dāt'). b. 1927. Thai king (since 1946).
**Bhut·to** (bo͞o'tō), **Zulfikar Ali.** 1928–79. Pakistani statesman; executed.
**Bi·chat** (bē-shä'), **Marie François Xavier.** 1771–1802. French pioneer anatomist & histologist.
**Bick·er·dyke** (bĭk'ər-dīk'), **Mary Ann Ball.** 1817–1901. Amer. Civil War nurse.
**Bid·dle** (bĭd'l), **Francis.** 1886–1968. French-born Amer. jurist & public official.
**Biddle, George.** 1885–1973. Amer. artist.
**Biddle, James.** 1783–1848. Amer. naval officer & diplomat.
**Biddle, John.** 1615–62. English theologian.
**Biddle, Nicholas.** 1786–1844. Amer. financier & scholar.
**Bid·well** (bĭd'wəl, -wĕl'), **John.** 1819–1900. Amer. Calif. pioneer & politician.
**Bien·ville** (byĕn'vĭl', byăN-vĕl'), Sieur **Jean Baptiste Lemoyne de.** 1680–1768. French colonial administrator.
**Bierce** (bîrs), **Ambrose Gwinett.** 1842–1914? Amer. author.
**Bier·stadt** (bîr'stăt', -shtăt'), **Albert.** 1830–1902. German-born Amer. painter.
**Big·e·low** (bĭg'ə-lō'), **Erastus Brigham.** 1814–79. Amer. inventor & manufacturer.
**Bigelow, Jacob.** 1787–1879. Amer. botanist & physician.
**Bigelow, John.** 1817–1911. Amer. author & diplomat.
**Bigelow, Poultney.** 1855–1954. Amer. author.
**Bi·kel** (bĭ-kĕl'), **Theodore.** b. 1924. Austrian-born actor & folk singer.
**Bil·bo** (bĭl'bō), **Theodore Gilmore.** 1877–1947. Amer. politician.
**Bil·lings** (bĭl'ĭngz), **Frederick.** 1823–90. Amer. businessman & philanthropist.
**Billings, John Shaw.** 1838–1913. Amer. physician & librarian.
**Billings, Josh.** Henry Wheeler SHAW.
**Billings, William.** 1746–1800. Amer. composer.
**Bil·ly the Kid** (bĭl'ē; kĭd). William H. BONNEY.
**Bing** (bĭng), Sir **Rudolf.** b. 1902. Austrian-born opera manager.
**Bing·ham** (bĭng'əm), **Anne Willing.** 1764–1801. Amer. socialite.
**Bingham, George Caleb.** 1811–79. Amer. genre painter.
**Bin·ney** (bĭn'ē), **Horace.** 1780–1875. Amer. lawyer.
**Birds·eye** (bûrd'zī'), **Clarence.** 1886–1956. Amer. inventor.
**Birk·beck** (bûrk'bĕk'), **George.** 1776–1841. English physician & reformer.
**Bir·ken·head** (bûr'kən-hĕd'), 1st Earl of. *Frederick Edwin Smith.* 1872–1930. English statesman.
**Birk·hoff** (bûr'kôf), **George David.** 1884–1944. Amer. mathematician.
**Bir·ney** (bûr'nē), **James Gillespie.** 1792–1857. Amer. politician & abolitionist.
**Bi·ron** (bē'rôn'), **Ernst Johann.** 1690–1772. Russian statesman & regent.
**Bish·op** (bĭsh'əp), **Elizabeth.** 1911–79. Amer. poet.
**Bishop, Hazel Gladys.** b. 1906. Amer. chemist & businesswoman.
**Bis·marck** (bĭz'märk'), Prince **Otto Eduard Leopold von.** *the Iron Chancellor.* 1815–98. Creator & first chancellor of the German Empire (1871–90). **—Bis·marck'i·an** *adj.*

**Bis·sell** (bĭs′əl), **George Henry.** 1821–84. Amer. pioneer oilman.
**Bit·ter** (bĭt′ər), **Karl Theodore Francis.** 1867–1915. Austrian-born Amer. sculptor.
**Bi·zet** (bē-zā′), **(Alexandre César Léopold) Georges.** 1838–75. French composer.
**Björn·son** (byûrn′sən), **Björnstjerne.** 1832–1910. Norwegian author (Nobel, 1903).
**Black** (blăk), **Greene Vardiman.** 1836–1915. Amer. pioneer dentist.
**Black, Hugo La Fayette.** 1886–1971. Amer. jurist.
**Black, Jeremiah Sullivan.** 1810–83. Amer. jurist & public official.
**Black, Joseph.** 1728–99. Scottish chemist.
**Black, Shirley Temple.** b. 1927? Amer. actress & public official.
**Black·beard** (blăk′bîrd′). Edward TEACH.
**Black·ett** (blăk′ĭt), **Patrick Maynard Stuart.** *Baron Blackett.* 1897–1974. English physicist (Nobel, 1948).
**Black Hawk** (blăk′ hôk′). 1767–1838. Amer. Indian leader.
**Black·more** (blăk′môr′, -mōr′), **Richard Doddridge.** 1825–1900. English author.
**Black·mun** (blăk′mən), **Harry Andrew.** b. 1908. Amer. jurist.
**Black·stone** (blăk′stōn′, -stən), Sir **William.** 1723–80. English jurist & educator.
**Black·well** (blăck′wĕl′, -wəl), **Alice Stone.** 1857–1950. Amer. suffragist & reformer.
**Blackwell, Antoinette Louisa Brown.** 1825–1921. Amer. social reformer.
**Blackwell, Elizabeth** (1821–1910) & **Emily** (1826–1910). English-born Amer. physicians & pioneer hospital administrators.
**Black·wood** (blăk′wo͝od′), **William.** 1776–1834. Scottish publisher & editor.
**Blaine** (blān), **James Gillespie.** 1830–93. Amer. politician.
**Blair** (blâr), **Francis Preston** (1791–1876) & **Francis Preston** (1821–75). Amer. politicians.
**Blair, James.** 1655–1743. Scottish-born Amer. clergyman, educator, & colonial official.
**Blair, John.** 1732–1800. Amer. jurist.
**Blair, Montgomery.** 1813–83. Amer. lawyer & public official.
**Blake** (blāk), **James Herbert ("Eubie").** 1883–1983. Amer. pianist and composer.
**Blake, Lillie Devereux.** 1833–1913. Amer. author & suffragist.
**Blake, Robert.** 1599–1657. English admiral.
**Blake, William.** 1757–1827. English mystic, poet, & artist.
**Blake·lock** (blāk′lŏk′), **Ralph Albert.** 1847–1919. Amer. landscape painter.
**Blanc** (bläɴ), **Louis.** 1811–82. French socialist.
**Blanc** (blăngk), **Melvin Jerome ("Mel").** b. 1908. Amer. actor & voice specialist.
**Blan·chard** (blăn′chərd), **Thomas.** 1788–1864. Amer. inventor.
**Bland** (blănd), **James A.** 1854–1911. Amer. composer.
**Blan·ding** (blăn′dĭng), **Sarah Gibson.** b. 1898. Amer. educator & college administrator.
**Blas·co I·bá·ñez** (blä′skō ē-bän′yās), **Vicente.** 1867–1928. Spanish author.
**Blass** (blăs), **Bill.** b. 1922. Amer. fashion designer.
**Blatch** (blăch), **Harriot Eaton Stanton.** 1856–1940. Amer. suffragist.
**Blatch·ford** (blăch′fərd), **Samuel.** 1820–93. Amer. jurist.
**Bla·vat·sky** (blə-văt′skē, -vät′-), Madame **Helena** *or* **Elena Petrovna Hahn.** 1831–91. Russian-born theosophist.
**Blé·riot** (blā′rē-ō, blĕr′ē-ō), **Louis.** 1872–1936. French inventor & aviator.
**Bligh** (blī), **William.** 1754–1817. English naval officer.
**Bliss** (blĭs), **Tasker Howard.** 1853–1930. Amer. army officer & diplomat.
**Blitz·stein** (blĭts′stīn′), **Marc.** 1905–64. Amer. composer.
**Blix·en** (blĕk′sən, blĭk′-), **Karen.** Isak DINESEN.
**Bloch** (blŏk, blôk, blôKH), **Ernest.** 1880–1959. Swiss-born Amer. composer.
**Bloch** (blŏk), **Felix.** 1905–83. Swiss-born Amer. physicist (Nobel, 1952).
**Bloch** (blŏk, blôk, blôKH), **Konrad Emil.** b. 1912. German-born Amer. biochemist (Nobel, 1964).
**Block** (blŏk), **Herbert Lawrence.** *Herblock.* b. 1909. Amer. editorial cartoonist.
**Bloem·ber·gen** (blo͞om′bûr-gən), **Nicolaas.** b. 1920. Dutch-born Amer. physicist (Nobel, 1981).
**Blok** (blôk), **Aleksandr Aleksandrovich.** 1880–1921. Russian poet.
**Bloom·er** (blo͞o′mər), **Amelia Jenks.** 1818–94. Amer. social reformer.
**Bloom·field** (blo͞om′fēld′), **Leonard.** 1887–1949. Amer. linguist.
**Bloor** (blo͝or), **Ella Reeve.** 1862–1951. Amer. labor organizer & political radical.
**Blount** (blŭnt), **William.** 1749–1800. Amer. politician.
**Blow** (blō), **Susan Elizabeth.** 1843–1916. Amer. pioneer in children's education.
**Blü·cher** (blo͞o′kər, -chər), **Gebhard Leberecht von.** *Prince of Wahlstatt.* 1742–1819. Prussian field marshal.
**Blum** (blo͞om), **Léon.** 1872–1950. French author & socialist leader.
**Blum·berg** (blŭm′bərg, blo͞om′-), **Baruch Samuel.** b. 1925. Amer. virologist (Nobel, 1976).
**Blu·men·bach** (blo͞o′mən-bäKH′), **Johann Friedrich.** 1752–1840. German pioneer zoologist & anthropologist.
**Blun·den** (blŭn′dən), **Edmund Charles.** 1896–1974. English poet.
**Bluntsch·li** (blo͝onch′lē), **Johann Kaspar.** 1808–81. Swiss legal scholar & politician.
**Bly** (blī), **Nellie.** Elizabeth Cochrane SEAMAN.
**Bo·ab·dil** (bō′əb-dēl′). d. 1533? Last Moorish king of Granada (1482–83, 1486–92).
**Bo·ad·i·ce·a** (bō′ăd-ĭ-sē′ə) *also* **Bou·dic·ca** (bo͞o-dĭk′ə). d. 62 A.D. Queen of ancient Britain.
**Bo·as** (bō′ăz), **Franz.** 1858–1942. German-born Amer. anthropologist.
**Bo·az** (bō′ăz). Husband of Ruth in the Bible.
**Bo·ba·di·lla** (bō′bə-dēl′yə, -dē′yə), **Francisco de.** d. 1502. Spanish colonial administrator.
**Boc·cac·cio** (bō-kä′chē-ō′, -chō′), **Giovanni.** 1313–75. Italian poet.
**Bod·en·heim** (bōd′n-hīm′), **Maxwell.** 1893?–1954. Amer. author.
**Bod·ley** (bŏd′lē), Sir **Thomas.** 1545–1613. English diplomat & library founder.
**Bo·do·ni** (bō-dō′nē, bə-), **Gianbattista.** 1740–1813. Italian printer & type designer.
**Boeh·me** (bœ′mə) *or* **Boehm** (bœm). *vars. of* BÖHME.
**Bo·e·thi·us** (bō-ē′thē-əs), **Anicius Manlius Severinus.** 480?–524? Roman Christian philosopher.
**Bo·gan** (bō′gən), **Louise.** 1897–1970. Amer. poet.
**Bo·gart** (bō′gärt), **Humphrey DeForest.** 1899–1957. Amer. actor.
**Boh·len** (bō′lĭn), **Charles Eustis.** 1904–74. Amer. diplomat.
**Böh·me** *also* **Boeh·me** (bœ′mə) *or* **Boehm** (bœm), **Jakob.** 1575–1624. German theosophist & mystic.
**Bohr** (bôr, bōr), **Aage Niels.** b. 1922. Danish physicist (Nobel, 1975).
**Bohr, Niels Henrik David.** 1885–1962. Danish physicist (Nobel, 1922).
**Bo·iar·do** (boi-är′dō, bō-yär′dō), **Matteo Maria.** 1434?–94. Italian lyric poet.
**Boi·leau-Des·pré·aux** (bwä′lō-dā′prā-ō′), **Nicolas.** 1636–1711. French critic & poet.
**Bo·i·to** (bō′ē-tō′), **Arrigo.** 1842–1918. Italian composer & librettist.
**Bok** (bŏk), **Edward William.** 1863–1930. Dutch-born Amer. journalist, editor, & pacifist.
**Bol·eyn** (bo͝ol′ĭn, bo͝o-lĭn′), **Anne.** 1507–36. English queen as second wife of Henry VIII; beheaded.
**Bol·ger** (bōl′jər), **Ray.** b. 1904. Amer. dancer & actor.
**Bol·ing·broke** (bŏl′ĭng-bro͝ok′, bo͝ol′-, bō′lĭng-), 1st Viscount. *Henry St. John.* 1678–1751. English statesman & orator.
**Bo·lí·var** (bŏ′lə-vär′, bŏl′ə-, bō-lē′vär), **Simón.** *the Liberator.* 1783–1830. Venezuelan soldier & South American liberator.
**Böll** (bœl), **Henrich.** 1917–85. German author (Nobel, 1972).
**Boltz·mann** (bōlts′män), **Ludwig.** 1844–1906. Austrian physicist.
**Bom·beck** (bŏm′bĕk′), **Erma.** b. 1927. Amer. humorist.
**Bo·na·parte** (bō′nə-pärt′). Corsican family, including: **a. Joseph.** 1768–1844. King of Naples (1806–8) & Spain (1808–13). **b.** NAPOLEON I. **c. Lucien.** *Prince of Canino.* 1775–1840. Politician & diplomat. **d. Louis.** 1778–1846. King of Holland (1806–10); abdicated. **e. Jérôme.** 1784–1860. King of Westphalia (1807).
**Bon·a·ven·ture** (bŏn′ə-vĕn′chər) *also* **Bon·a·ven·tu·ra** (bŏn′ə-vĕn-cho͝or′ə, -to͝or′ə, -tyo͝or′ə), Saint. 1221–74. Italian theologian & philosopher.
**Bond** (bŏnd), **Carrie Jacobs.** 1862–1946. Amer. songwriter & author.
**Bond, Julian.** b. 1940. Amer. politician & civil-rights leader.
**Bond, Thomas.** 1712–84. Amer. physician & hospital founder.
**Bond, William Cranch** (1789–1859) & **George Phillips** (1825-65). Amer. astronomers.
**Bone** (bōn), Sir **Muirhead.** 1876–1953. Scottish artist.
**Bon·fils** (bŏn-fēs′), **Frederick Gilmer.** 1860–1933. Amer. publisher.
**Bon·heur** (bô-nûr′, -nœr′), **Rosa.** 1822–99. French painter.
**Bon·i·face** (bŏn′ə-fās′). Name of nine popes, esp.: **a. I,** Saint. d. 422. Reigned 418–22. **b. VIII.** 1235?–1303. Reigned 1294–1303. **c. IX.** d. 1404. Reigned 1389–1404.
**Boniface,** Saint. 680?–755? English missionary.
**Bon·nard** (bô-när′), **Pierre.** 1867–1947. French painter.
**Bon·ner** (bŏn′ər), **Edmund.** 1500?–69. English prelate.
**Bonner, Robert.** 1824–99. Irish-born Amer. publisher.
**Bon·net** (bô-nā′), **Georges.** 1889–1973. French politician & diplomat.
**Bon·ne·ville** (bŏn′ə-vĭl′), **Benjamin Louis Eulalie de.** 1796–1878. French-born Amer. soldier & explorer.

**Bon·ney** (bŏn′ē), **Mary Lucinda.** 1816–1900. Amer. educator & reformer.
**Bonney, Thérèse.** 1894?–1978. Amer. photographer.
**Bonney, William H.** *Billy the Kid.* 1859–81. Amer. outlaw.
**Bo·non·ci·ni** (bō′nôn-chē′nē) *also* **Buo·non·ci·ni** (bwō′-). Family of Italian composers, including **Giovanni Maria** (1640–78), **Giovanni Battista** (1670?–1750), & **Marcantonio** (1675?–1726).
**Bon·stelle** (bŏn′stĕl), **Jessie.** 1871–1932. Amer. actress & producer.
**Bon·temps** (bôN-täN′), **Arna Wendell.** 1902–73. Amer. author.
**Boole** (bo͞ol), **George.** 1815–64. English mathematician & logician.
**Boone** (bo͞on), **Daniel.** 1734–1820. Amer. frontiersman.
**Booth** (bo͞oth). Family of English & Amer. actors, including **Junius Brutus** (1796–1852); **Edwin Thomas** (1833–93); & **John Wilkes** (1838–65), assassin of Abraham Lincoln, died of gunshot wound.
**Booth.** Family of English and Amer. reformers, including **William** (*General Booth,* 1829–1912), founder of the Salvation Army; **William Bramwell** (1856–1929); **Ballington** (1859–1940), founder of Volunteers of America; **Maud Ballington** (1865–1948); & **Evangeline Cory** (1865–1950).
**Booth, Mary Louise.** 1831–89. Amer. historian & editor.
**Bo·rah** (bôr′ə, bōr′ə), **William Edgar.** 1865–1940. Amer. legislator.
**Bor·den** (bôr′dn), **Gail.** 1801–74. Amer. surveyor & inventor.
**Borden, Lizzie Andrew.** 1860–1927. Amer. accused murderess; acquitted.
**Borden,** Sir **Robert.** 1854–1937. Canadian prime minister (1911–20).
**Bor·det** (bôr-dā′), **Jules Jean Baptiste Vincent.** 1870–1961. Belgian bacteriologist (Nobel, 1919).
**Bo·rel·li** (bō-rĕl′ē), **Giovanni Alfonso.** 1608–79. Italian mathematician, astronomer, & physiologist.
**Borg** (bôrg), **Bjorn.** b. 1956. Swedish tennis player.
**Bor·ge** (bôr′gə), **Victor.** b. 1909. Danish pianist & comedian.
**Bor·ges** (bôr′hās), **Jorge Luis.** b. 1899. Argentinian author.
**Bor·gia** (bôr′jə, -zhə). Influential Italian family, including: **a. Alfonso.** CALIXTUS III. **b. Cesare.** 1475?–1507. Cardinal, diplomat, & soldier. **c. Lucrezia.** *Duchess of Ferrara.* 1480–1519. Patron of learning & the arts. **d. Rodrigo.** ALEXANDER VI.
**Bor·glum** (bôr′gləm), **Gutzon.** 1867–1941. Amer. sculptor.
**Bo·ri** (bôr′ē, bōr′ē), **Lucrezia.** 1887–1960. Spanish-born Amer. lyric soprano.
**Bor·ing** (bōr′ĭng), **Edwin Garrigues.** 1886–1968. Amer. psychologist.
**Bor·laug** (bôr′lôg′), **Norman Ernest.** b. 1914. Amer. agronomist (Nobel, 1970).
**Bor·mann** (bôr′män′), **Martin Ludwig.** b. 1900. German politician; reported dead in 1945.
**Born** (bôrn), **Max.** 1882–1970. German-born physicist (Nobel, 1954).
**Bo·ro·din** (bôr′ə-dēn′, bär′-, bôr′ə-dēn′), **Aleksandr Porfirevich.** 1834–87. Russian composer & chemist.
**Bor·ro·mi·ni** (bŏr′ō-mē′nē), **Francesco.** 1599–1667. Italian artist.
**Bor·row** (bŏr′ō), **George.** 1803–81. English philologist, traveler, & author.
**Bosch** (bŏsh, bôsh), **Carl.** 1874–1940. German chemist (Nobel, 1931).
**Bosch** (bŏsh, bôsh, bŏs, bôs), **Hieronymous.** 1450?–1516. Dutch painter.
**Bos·suet** (bôs-wā′), **Jacques Bénigne.** 1627–1704. French prelate & historian.
**Bos·well** (bŏz′wĕl′, -wəl), **James.** 1740–95. Scottish diarist & biographer.
**Bo·tha** (bō′tə, -tä′), **Louis.** 1862–1919. South African general & statesman.
**Botha, Pieter Willem.** b. 1916. South African statesman.
**Bo·the** (bō′tə), **Walther Wilhelm.** 1891–1957. German physicist (Nobel, 1954).
**Both·well** (bŏth′wĕl′, -wəl, bŏth-), 4th Earl of. *James Hepburn.* 1536?–78. Scottish Protestant nobleman & husband of Mary Queen of Scots.
**Bot·ti·cel·li** (bŏt′ĭ-chĕl′ē), **Sandro.** 1444?–1510. Italian painter.
**Bou·cher** (bo͞o-shā′), **François.** 1703–70. French artist.
**Bou·ci·cault** (bo͞o′sē-kō′), **Dion.** 1820?–90. Irish-born Amer. actor & playwright.
**Bou·dic·ca** (bo͞o-dĭk′ə). *var. of.* BOADICEA.
**Bou·gain·ville** (bo͞o′gən-vĭl′, bo͞o-găN-vēl′), **Louis Antoine de.** 1729–1811. French explorer.
**Bou·lan·ger** (bo͞o-läN-zhā′), **Georges Ernest Jean Marie.** 1837–91. French military & political leader.
**Boulanger, Nadia Juliette.** 1885?–1979. French music teacher.
**Bou·lez** (bo͞o-lĕz′), **Pierre.** b. 1925. French conductor & composer.
**Bour·bon** (bo͝or′bən, bo͝or-bôN′). French royal family ruling in France (1589–1793), Spain (1700–1868, 1874–1931), & Naples & the Two Sicilies (1735–1861).
**Bourbon,** Duc **Charles de.** 1490–1527. French general.
**Bour·geois** (bo͝or-zhwä′), **Léon Victor Auguste.** 1851–1925. French statesman (Nobel, 1920).
**Bour·get** (bo͝or-zhā′), **Paul.** 1852–1935. French author.
**Bourke-White** (bûrk′hwīt′, -wīt′), **Margaret.** 1906–71. Amer. photographer.
**Bourne** (bo͝orn, bôrn, bōrn), **Randolph Silliman.** 1886–1918. Amer. critic & pacifist.
**Bout·well** (bout′wĕl′), **George Sewall.** 1818–1905. Amer. politician.
**Bo·vet** (bō-vā′), **Daniel.** b. 1907. Swiss-born Italian physiologist (Nobel, 1957).
**Bow** (bō), **Clara.** 1905–65. Amer. actress.
**Bow·ditch** (bou′dĭch), **Nathaniel.** 1773–1838. Amer. mathematician & astronomer.
**Bow·doin** (bōd′n), **James.** 1726–90. Amer. merchant & Revolutionary leader.
**Bow·ell** (bō′əl), Sir **Mackenzie.** 1823–1917. English-born Canadian prime minister (1894–96).
**Bow·en** (bō′ən), **Catherine Drinker.** 1897–1973. Amer. author.
**Bowen, Elizabeth Dorothea Cole.** 1899–1973. Irish-born English author.
**Bow·ers** (bou′ərz), **Claude Gernade.** 1878–1958. Amer. journalist, diplomat, & historian.
**Bow·ie** (bo͞o′ē, bō′ē), **James.** 1799–1836. Amer.-born Mexican colonist; died at the Alamo.
**Bow·ker** (bou′kər), **Richard Rogers.** 1848–1933. Amer. author, publisher, & editor.
**Bowles** (bōlz), **Chester Bliss.** b. 1901. Amer. diplomat & author.
**Bowles, Samuel.** 1797–1851. Amer. newspaper publisher.
**Bowles, Samuel.** 1826–78. Amer. newspaper editor & author.
**Bow·man** (bō′mən), **Isaiah.** 1878–1950. Canadian-born Amer. geographer.
**Boyd** (boid), **Belle.** 1844?–1900. Amer. Confederate spy.
**Boy·den** (boid′n), **Seth.** 1788–1870. Amer. inventor & manufacturer.
**Boyden, Uriah Atherton.** 1804–79. Amer. engineer & inventor.
**Boyd Orr** (boid′ ôr′, ōr′), Lord **John.** 1880–1971. English nutritionist (Nobel, 1949).
**Bo·ye** (bô′yə), **Karin.** 1900–41. Swedish author.
**Boy·er** (boi-ā′), **Charles.** 1899–1978. French actor.
**Boyle** (boil), **Kay.** b. 1903. Amer. author.
**Boyle, Robert.** 1627–91. British physicist & chemist.
**Boyl·ston** (boil′stən), **Zabdiel.** 1679–1766. Amer. physician & inoculation pioneer.
**Boze·man** (bōz′mən), **John M.** 1835–67. Amer. explorer.
**Brace** (brās), **Charles Loring.** 1826–90. Amer. social reformer.
**Brack·en·ridge** (brăk′ĭn-rĭj′), **Hugh Henry.** 1748–1816. Scottish-born Amer. author, politician, & judge.
**Brad·bur·y** (brăd′bĕr′ē, -bə-rē), **Ray Douglas.** b. 1920. Amer. science-fiction author.
**Brad·dock** (brăd′ək), **Edward.** 1695–1755. Scottish-born English general in America.
**Brad·ford** (brăd′fərd), **Roark.** 1896–1948. Amer. author.
**Bradford, William.** 1590–1657. English Puritan colonist in America.
**Bradford, William.** 1663–1752. English-born Quaker colonist & pioneer printer.
**Brad·ley** (brăd′lē), **Francis Herbert.** 1846–1924. English philosopher.
**Bradley, Henry.** 1845–1923. English lexicographer & historian.
**Bradley, James.** 1693–1762. English astronomer & educator.
**Bradley, Joseph P.** 1813–92. Amer. jurist.
**Bradley, Lydia Moss.** 1816–1908. Amer. businesswoman & philanthropist.
**Bradley, Milton.** 1836–1911. Amer. game manufacturer.
**Bradley, Omar Nelson.** 1893–1981. Amer. military leader.
**Bradley, Thomas.** b. 1917. Amer. policeman & politician.
**Bradley, William Warren ("Bill").** b. 1943. Amer. basketball player & politician.
**Brad·street** (brăd′strēt′), **Anne Dudley.** 1612–72. English-born colonial poet.
**Bradstreet, Simon.** 1603–97. English colonial administrator.
**Brad·well** (brăd′wĕl), **Myra Colby.** 1831–94. Amer. lawyer, editor, & feminist.
**Bra·dy** (brā′dē), **James Buchanan ("Diamond Jim").** 1856–1917. Amer. financier & philanthropist.
**Brady, Mathew B.** 1823–96. Amer. pioneer photographer.
**Brady, William Aloysius.** 1863–1950. Amer. actor & theatrical manager.
**Bra·gan·za** (brə-găn′zə). Dynasty of Portuguese rulers (1640–1910).
**Bragg** (brăg), **Braxton.** 1817–76. Amer. Confederate soldier.

**Bragg,** Sir **William Henry** (1862–1942) & Sir **William Lawrence** (1890–1971). English physicists (shared Nobel, 1915).
**Brahe** (brä, brä'hē, brä'ə), **Tycho.** 1546–1601. Danish astronomer.
**Brahms** (brämz), **Johannes.** 1833–97. German composer. **—Brahms'i·an** *adj.*
**Braille** (brāl), **Louis.** 1809?–52. French musician, educator, & inventor of writing & printing systems for the blind.
**Bra·man·te** (brə-män'tē, -tā). 1444?–1514. Italian architect.
**Bran·cu·si** (brăn-ko͞o'zē), **Constantin.** 1876–1957. Rumanian-born sculptor.
**Bran·deis** (brăn'dīs', -dīz'), **Louis Dembitz.** 1856–1941. Amer. jurist.
**Bran·do** (brăn'dō), **Marlon.** b. 1924. Amer. actor.
**Brandt** (brănt, bränt), **Willy.** b. 1913. West German statesman (Nobel, 1971).
**Bran·nan** (brăn'ən), **Samuel.** 1819–89. Amer. Calif. pioneer & publisher.
**Brant** (brănt), **Joseph.** 1742–1807. Amer. Indian leader.
**Bran·ting** (brăn'tĭng, brän'-), **(Karl) Hjalmar.** 1860–1925. Swedish statesman & journalist (Nobel, 1921).
**Braque** (bräk, brăk), **Georges.** 1882–1963. French painter.
**Brat·tain** (brăt'n), **Walter Houser.** b. 1902. Amer. physicist (Nobel, 1956).
**Braun** (broun), **Eva.** 1910?–45. German mistress of Adolf Hitler.
**Braun, Karl Ferdinand.** 1850–1918. German physicist (Nobel, 1901).
**Braun, Wernher Magnus Maximilian von.** 1912–77. German-born Amer. rocket engineer.
**Breas·ted** (brĕs'təd, -tĭd), **James Henry.** 1865–1935. Amer. archaeologist & historian.
**Brecht** (brĕkt, brĕKHt), **Bertolt.** 1898–1956. German poet & playwright.
**Breck·in·ridge** (brĕk'ĭn-rĭj'), **John Cabell.** 1821–75. U.S. Vice President (1857–61).
**Breckinridge, Sophonisba Preston.** 1866–1948. Amer. social worker, author, & educator.
**Brel** (brĕl), **Jacques.** 1929–78. Belgian singer & composer.
**Bren·nan** (brĕn'ən), **William Joseph, Jr.** b. 1906. Amer. jurist.
**Brent** (brĕnt), **Margaret.** 1600?–71? English-born colonist & feminist.
**Bresh·kov·sky** (brĕsh-kôf'skē), **Catherine.** 1844–1934. Russian revolutionary & social reformer.
**Bre·ton** (brĭ-tôN'), **André.** 1896–1966. French author & critic.
**Bre·ton·neau** (brĭ-tô-nō'), **Pierre.** 1778–1862. French surgeon.
**Breu·er** (broi'ər), **Marcel Lajos.** 1902–81. Hungarian-born Amer. architect & designer.
**Breu·ghel** (broi'gəl, bro͞o'-, brœ'-). *var. of* BRUEGHEL.
**Brew·er** (bro͞o'ər), **David Josiah.** 1837–1910. Amer. jurist.
**Brew·ster** (bro͞o'stər), **Kingman, Jr.** b. 1919. Amer. educator.
**Brewster, William.** 1567–1644. English Pilgrim colonist.
**Brezh·nev** (brĕzh'nĕf, -nyĕf), **Leonid Ilyich.** 1906–82. Soviet statesman.
**Bri·an Bo·ru** (brī'ən bə-ro͞o', bō-, brēn). 941–1014. Irish king (1002–14).
**Bri·and** (brē-äN'), **Aristide.** 1862–1932. French statesman (Nobel, 1926).
**Brice** (brīs), **Fannie.** 1891–1951. Amer. entertainer.
**Bri·co** (brē'kō), **Antonia.** b. 1902. Dutch-born Amer. conductor & pianist.
**Bridg·er** (brĭj'ər), **James.** 1804–81. Amer. frontiersman & fur trader.
**Bridg·es** (brĭj'ĭz), **Harry.** b. 1900. Amer. labor leader.
**Bridges, Robert Seymour.** 1844–1930. English author.
**Bridg·man** (brĭj'mən), **Laura Dewey.** 1829–89. Amer. educator.
**Bridgman, Percy Williams.** 1882–1961. Amer. physicist (Nobel, 1946).
**Bri·eux** (brē-œ'), **Eugène.** 1858–1932. French playwright.
**Briggs** (brĭgz), **Emily Pomona Edson.** 1830–1910. Amer. journalist.
**Briggs, Henry.** 1561–1630? English mathematician.
**Briggs, Lyman James.** 1874–1963. Amer. physicist.
**Bright** (brīt), **John.** 1811–89. English politician & orator.
**Brill** (brĭl), **Abraham Arden.** 1874–1948. Austrian-born Amer. psychiatrist.
**Bril·lat-Sa·va·rin** (brē-yä'săv'ə-răN', brē-ä'-), **Anthelme.** 1755–1826. French politician & gourmet.
**Brink·ley** (brĭngk'lē), **David.** b. 1920. Amer. broadcast journalist.
**Brin·ton** (brĭn'tən), **Daniel Garrison.** 1837–99. Amer. anthropologist & physician.
**Bris·bane** (brĭz'bān, -bən), **Albert.** 1809–90. Amer. social reformer.
**Brisbane, Arthur.** 1864–1936. Amer. newspaper editor.

o͞o **boot** ou **out** th **thin** *th* **this** ŭ **cut** ûr **urge** y **young** yo͞o **abuse** zh **vision** ə **about**, **item**, **edible**, **gallop**, **circus**

**Bris·tow** (brĭs'tō), **Benjamin Helm.** 1832–96. Amer. lawyer & public official.
**Brit·ten** (brĭt'n), **(Edward) Benjamin.** 1913–76. English composer.
**Bro·gan** (brō'gən), Sir **Denis William.** 1900–74. English political scientist.
**Bro·glie** (brô-glē'), **Louis Victor de.** b. 1892. French physicist (Nobel, 1929).
**Brom·field** (brŏm'fēld'), **Louis.** 1896–1956. Amer. author.
**Bron·të** (brŏn'tē). Family of English novelists, including **Charlotte** (1816–55), **Emily Jane** (1818–48), & **Anne** (1820–49).
**Brook** (bro͝ok), Sir **Alan Francis.** *1st Viscount Alanbrooke.* 1883–1963. British field marshal.
**Brook, Rupert.** 1887–1915. English poet.
**Brook·ings** (bro͝ok'ĭngz), **Robert Somers.** 1850–1932. Amer. businessman & philanthropist.
**Brooks** (bro͝oks), **Gwendolyn Elizabeth.** b. 1917. Amer. author.
**Brooks, Maria Gowen.** 1794?–1845. Amer. poet.
**Brooks, Phillips.** 1835–93. Amer. prelate & author.
**Brooks, Van Wyck.** 1886–1963. Amer. literary historian & critic.
**Broun** (bro͞on), **(Matthew) Heywood (Campbell).** 1888–1939. Amer. journalist.
**Brow·der** (brou'dər), **Earl Russell.** 1891–1973. Amer. socialist leader.
**Brown** (broun), **Alice.** 1856?–1948. Amer. author.
**Brown, Benjamin Gratz.** 1826–85. Amer. politician.
**Brown, Charles Brockden.** 1771–1810. Amer. author & editor.
**Brown, Ford Madox.** 1821–93. English historical painter.
**Brown, Hallie Quinn.** 1850–1945? Amer. educator & lecturer.
**Brown, Helen Gurley.** b. 1922. Amer. editor & author.
**Brown, Henry Billings.** 1836–1913. Amer. jurist.
**Brown, Herbert Charles.** b. 1912. English-born Amer. chemist (Nobel, 1979).
**Brown, Jacob Jennings.** 1775–1828. Amer. military leader.
**Brown, John.** 1800–59. Amer. abolitionist; executed.
**Brown, John Mason.** 1900–69. Amer. drama critic.
**Brown, Martha McClellan.** 1838–1916. Amer. temperance leader.
**Brown, Moses.** 1738–1836. Amer. manufacturer & social reformer.
**Brown, Olympia.** 1835–1926. Amer. clergywoman & suffragist.
**Brown, Robert.** 1773–1858. Scottish botanist.
**Brown, William Wells.** 1815–84. Amer. author & social reformer.
**Browne** (broun), **Charles Farrar.** *Artemus Ward.* 1834–67. Amer. humorist.
**Browne,** Sir **Thomas.** 1605–82. English physician & author.
**Brown·ing** (brou'nĭng), **Elizabeth Barrett** (1806–61) & **Robert** (1812–89). English poets.
**Browning, John Moses.** 1855–1926. Amer. firearms inventor.
**Brown·son** (broun'sən), **Orestes Augustus.** 1803–76. Amer. clergyman, author, & editor.
**Broz** (brŏz, brôz), **Josip.** Marshal TITO.
**Bru·beck** (bro͞o'bĕk), **David Warren.** b. 1920. Amer. pianist and composer.
**Bruce** (bro͞os), **Blanche Kelso.** 1841–98. Amer. politician & public official.
**Bruce,** Sir **David.** 1855–1931. Australian physician & bacteriologist.
**Bruce, David Kirkpatrick Este.** 1898–1977. Amer. diplomat.
**Bruce, Lenny.** 1926–66. Amer. comedian.
**Bruce, Robert the.** ROBERT I of Scotland.
**Bruce,** Viscount **Stanley Melbourne.** 1883–1967. Australian statesman.
**Bruck·ner** (bro͝ok'nər), **Anton.** 1824–96. Austrian composer.
**Brue·ghel** *also* **Brue·gel, Breu·ghel** (broi'gəl, bro͞o'-, brœ'-). Flemish family of painters, including **Pieter** *the Elder* (1525?–69), **Pieter** *the Younger* (1564?–1637), & **Jan** (1568–1625).
**Brue·ning** (bro͞o'nĭng). *var. of* BRÜNING.
**Bruhn** (bro͞on), **Erik.** b. 1928. Danish-born ballet dancer.
**Brulé** (brü-lā'), **Étienne.** 1592?–1632. French explorer.
**Brum·mell** (brŭm'əl), **George Bryan ("Beau").** 1778–1840. English fashionable gentleman.
**Brun·dage** (brŭn'dĭj), **Avery.** 1887–1975. Amer. businessman & sports figure.
**Bru·nel·le·schi** (bro͞o'nə-lĕs'kē), **Filippo.** 1377?–1446. Italian architect.
**Bru·ne·tière** (bro͞o'nə-tyĕr'), **Vincent de Paul Marie Ferdinand.** 1849?–1906. French critic & editor.
**Brü·ning** *also* **Brue·ning** (bro͞o'nĭng), **Heinrich.** 1885–1970. German statesman.
**Bru·no** (bro͞o'nō), **Giordano.** 1548?–1600. Italian philosopher.
**Bruno of Co·logne** (kə-lōn'), Saint. 1030?–1101. German religious writer & founder of Carthusian order.
**Bru·tus** (bro͞o'təs), **Marcus Junius.** 85?–42 B.C. Roman politician, general, & assassin of Julius Caesar.
**Bry·an** (brī'ən), **William Jennings.** 1860–1925. Amer. lawyer & political leader.
**Bry·ant** (brī'ənt), **Paul William ("Bear").** 1913–83. Amer. football coach.

**Bryant, William Cullen.** 1794–1878. Amer. poet & editor.
**Bryce** (brīs), **James.** *Viscount Bryce of Dechmont.* 1838–1922. English statesman, diplomat, & historian.
**Brze·zin·ski** (brĭ-zhĭn'skē), **Zbigniew.** b. 1928. Polish-born Amer. political adviser.
**Bu·ber** (bo͞o'bər), **Martin.** 1878–1965. Austrian-born Judaic scholar & philosopher.
**Buch·an** (bŭk'ən, bŭKH'-), Sir **John.** *1st Baron Tweedsmuir.* 1875–1940. Scottish historian & government official.
**Bu·chan·an** (byo͞o-kăn'ən, bə-), **Franklin.** 1800–74. Amer. Confederate naval officer.
**Buchanan, James.** 1791–1868. 15th U.S. President (1857–61).
**Buch·man** (bo͝ok'mən, bŭk'-), **Frank Nathan Daniel.** 1878–1961. Amer. evangelist.
**Buch·ner** (bo͝ok'nər, bo͝oKH'-), **Eduard.** 1860–1917. German chemist (Nobel, 1907).
**Buck** (bŭk), **Pearl Sydenstricker.** 1892–1973. Amer. author (Nobel, 1938).
**Buck·ing·ham** (bŭk'ĭng-əm, -hăm'), 1st Duke. *George Villiers.* 1592–1628. English political adviser; assassinated.
**Buck·ley** (bŭk'lē), **William Frank, Jr.** b. 1925. Amer. editor & author.
**Buck·ner** (bŭk'nər), **Simon Bolivar.** 1823–1914. Amer. Confederate general & politician.
**Bud·dha** (bo͞o'də, bo͝od'ə). 563?–483? B.C. Indian philosopher & founder of Buddhism. **—Bud'dhist** *adj.* & *n.*
**Budge** (bŭj), **John Donald ("Don").** b. 1915. Amer. tennis player.
**Bu·ell** (byo͞o'əl), **Abel.** 1742–1822. Amer. silversmith, type designer, & engraver.
**Buell, Don Carlos.** 1818–98. Amer. army officer.
**Buf·fa·lo Bill** (bŭf'ə-lō bĭl'). William Frederick CODY.
**Buf·fet** (bo͞o-fā'), **Bernard.** b. 1928. French painter.
**Buf·fon** (bo͞o-fôN'), Comte **Georges Louis Leclerc de.** 1707–88. French naturalist.
**Buis·son** (bwē-sôN'), **Ferdinand.** 1841–1932. French educator (Nobel, 1927).
**Bu·kha·rin** (bo͞o-kär'ĭn), **Nikolai Ivanovich.** 1888–1938. Bolshevik theoretician & revolutionary; executed.
**Bul·finch** (bo͝ol'fĭnch'), **Charles.** 1763–1844. Amer. architect.
**Bulfinch, Thomas.** 1796–1867. Amer. author.
**Bul·ga·nin** (bo͝ol-găn'ĭn), **Nikolai Aleksandrovich.** 1895–1975. Russian military & political leader.
**Bull** (bo͝ol, bo͞ol), **Ole Bornemann.** 1810–80. Norwegian violinist.
**Bul·litt** (bo͝ol'ĭt), **William Christian.** 1891–1967. Amer. diplomat.
**Bü·low** (byo͞o'lō), Prince **Bernhard von.** 1849–1929. German statesman & diplomat.
**Bul·wer** (bo͝ol'wər), **William Henry Lytton Earle.** *Baron Dalling and Bulwer.* 1801–72. English author, politician, & diplomat.
**Bum·bry** (bŭm'brē), **Grace.** b. 1937. Amer. opera singer.
**Bunche** (bŭnch), **Ralph Johnson.** 1904–71. Amer. diplomat (Nobel, 1950).
**Bun·dy** (bŭn'dē), **McGeorge.** b. 1919. Amer. educator & political adviser.
**Bu·ñu·el** (bo͞o-nyo͞o-ĕl'), **Luis.** 1900–83. Spanish-born film director.
**Bu·nin** (bo͞o'nĭn, -nyĭn), **Ivan Alekseevich.** 1870–1953. Russian author (Nobel, 1933).
**Bun·ker** (bŭng'kər), **Ellsworth.** 1894–1984. Amer. diplomat.
**Bun·sen** (bŭn'sən), **Robert Wilhelm.** 1811–99. German chemist.
**Bunt·line** (bŭnt'lĭn, -līn'), **Ned.** Edward Zane Carroll JUDSON.
**Bun·yan** (bŭn'yən), **John.** 1628–88. English preacher & author.
**Buo·non·ci·ni** (bwō'nôn-chē'nē). *var. of* BONONCINI.
**Bur·bage** (bûr'bĭj), **Richard.** 1567?–1619. English actor & theater manager.
**Bur·bank** (bûr'băngk'), **Luther.** 1849–1926. Amer. horticulturist & pioneer plant breeder.
**Burch·field** (bûrch'fēld'), **Charles Ephraim.** 1893–1967. Amer. painter.
**Burck·hardt** (bo͝ork'härt), **Jakob.** 1818–97. Swiss art historian.
**Bur·ger** (bûr'gər), **Warren Earl.** b. 1907. Amer. jurist.
**Bür·ger** (bo͝or'gər), **Gottfried August.** 1747?–94. German poet.
**Bur·gess** (bûr'jĭs), **Anthony.** b. 1917. English author.
**Burgess, (Frank) Gelett.** 1866–1951. Amer. author & illustrator.
**Burgh·ley** or **Bur·leigh** (bûr'lē), 1st Baron. William CECIL.
**Bur·goyne** (bûr-goin', bûr'goin'), **John.** 1722–92. English general & playwright.
**Bur·gun·dy** (bûr'gən-dē). Dynasty of Portuguese rulers (1139–1383).
**Burke** (bûrk), **Billie.** 1886–1970. Amer. actress.
**Burke, Edmund.** 1729–97. British politician, orator, & author.
**Burke** *also* **Burk** (bûrk), **Martha Jane.** *Calamity Jane.* 1852?–1903. Amer. frontier heroine.
**Bur·leigh** (bûr'lē). *var. of* BURGHLEY.
**Burleigh, Harry Thatcher.** 1866–1949. Amer. composer.
**Bur·lin** (bûr'lĭn), **Natalie Curtis.** 1875–1921. Amer. musicologist.
**Bur·lin·game** (bûr'lĭn-găm', -lĭng-), **Anson.** 1820–70. Amer. diplomat.
**Burne-Jones** (bûrn'jōnz'), Sir **Edward Coley.** 1833–98. English pre-Raphaelite painter.
**Bur·net** (bər-nĕt', bûr'nĭt), Sir **Frank Macfarlane.** b. 1899. Australian medical scientist (Nobel, 1960).
**Bur·nett** (bûr-nĕt'), **Carol.** b. 1936. Amer. comedienne.
**Bur·nett** (bûr-nĕt', bûr'nĭt), **Frances Eliza Hodgson.** 1849–1924. English-born Amer. writer.
**Bur·ney** (bûr'nē), **Frances ("Fanny").** 1752–1840. English author.
**Burn·ham** (bûr'nəm), **Daniel Hudson.** 1846–1912. Amer. architect & city planner.
**Burns** (bûrnz), **Arthur Frank.** b. 1904. Austrian-born Amer. economist.
**Burns, George.** b. 1896. Amer. comedian.
**Burns, Robert.** 1759–96. Scottish poet. **—Burns'i·an** *adj.*
**Burn·side** (bûrn'sīd'), **Ambrose Everett.** 1824–81. Amer. politician & general.
**Bur·pee** (bûr'pē), **David.** 1893–1980. Amer. horticulturist.
**Burr** (bûr), **Aaron.** 1756–1836. U.S. Vice President (1801–5), soldier, & adventurer; killed Alexander Hamilton in a duel.
**Bur·ritt** (bûr'ĭt), **Elihu.** 1810–79. Amer. social reformer.
**Bur·roughs** (bûr'ōz), **Edgar Rice.** 1875–1950. Amer. author.
**Burroughs, John.** 1837–1921. Amer. naturalist & author.
**Burroughs, William Seward.** 1855–98. Amer. inventor.
**Bur·rows** (bûr'ōz), **Abe.** 1910–85. Amer. playwright.
**Burt** (bûrt), **William Austin.** 1792–1858. Amer. surveyor & inventor.
**Bur·ton** (bûr'tn), **Harold Hitz.** 1888–1964. Amer. jurist.
**Burton,** Sir **Richard Francis.** 1821–90. English explorer & Orientalist.
**Burton, Robert.** 1577–1640. English clergyman & author.
**Busch** (bo͝osh), **Adolphus.** 1839–1913. German-born Amer. brewer, businessman, & philanthropist.
**Bush** (bo͝osh), **George Herbert Walker.** b. 1924. U.S. Vice President (since 1981).
**Bush, Vannevar.** 1890–1974. Amer. electrical engineer.
**Bush·man** (bo͝osh'mən), **Francis Xavier.** 1883–1966. Amer. actor.
**Bush·nell** (bo͝osh'nəl), **David.** 1742?–1824. Amer. inventor.
**Bushnell, Horace.** 1802–76. Amer. theologian.
**Bu·so·ni** (bo͞o-zō'nē, byo͞o-), **Ferruccio Benvenuto.** 1866–1924. Italian pianist & composer.
**Bu·te·nandt** (bo͞ot'n-änt'), **Adolf Friedrich.** b. 1903. German chemist (Nobel, 1939).
**But·ler** (bŭt'lər), **Benjamin Franklin.** 1818–93. Amer. political & military leader.
**Butler, Joseph.** 1692–1752. English prelate & theologian.
**Butler, Nicholas Murray.** 1862–1947. Amer. educator (Nobel, 1931).
**Butler, Pierce.** 1866–1939. Amer. jurist.
**Butler, Samuel.** 1612–80. English poet.
**Butler, Samuel.** 1835–1902. English novelist.
**But·ter·field** (bŭt'ər-fēld'), **John.** 1801–69. Amer. expressman & financier.
**But·ton** (bŭt'n), **Richard Totten ("Dick").** b. 1929. Amer. figure skater & television producer.
**Bux·te·hu·de** (bo͝ok'stə-ho͞o'də), **Dietrich.** 1637–1707. German composer.
**Byng** (bĭng), **George.** 1663–1733. English admiral.
**Byng, Julian Hedworth George.** 1862–1935. English military leader.
**Byrd** (bûrd), **Richard Evelyn.** 1888–1957. Amer. naval officer & polar explorer.
**Byrd, William.** 1674–1744. Amer. planter, author, & colonial official.
**Byrne** (bûrn), **Jane Margaret.** b. 1934. Amer. public official.
**Byrnes** (bûrnz), **James Francis.** 1879–1972. Amer. politician & jurist.
**By·ron** (bī'rən), **George Gordon.** *6th Baron Byron of Rochdale.* 1788–1824. English poet. **—By·ron'ic** (bī-rŏn'ĭk) *adj.*

# C

**Cab·ell** (kăb'əl), **James Branch.** 1879–1958. Amer. author.
**Ca·be·za de Va·ca** (kə-bā'zə də vä'kə), **Álvar Núñez.** 1490?–1577? Spanish explorer & colonial administrator.
**Ca·ble** (kā'bəl), **George Washington.** 1844–1925. Amer. author.
**Cab·ot** (kăb'ət), **John.** 1450–98. Italian-born explorer.
**Cabot, Sebastian.** 1476?–1557. Italian-born explorer & cartographer.
**Ca·bral** (kə-bräl'), **Pedro Alvares.** 1460?–1526? Portuguese explorer.

**Ca·bri·llo** (kə-brē'lō, -yō, -brēl'yō), **Juan Rodríguez.** d. 1543. Portuguese-born explorer.
**Ca·bri·ni** (kə-brē'nē), Saint **Frances Xavier.** *Mother Cabrini.* 1850–1917. Italian-born Amer. religious leader.
**Cade** (kād), **John ("Jack").** d. 1450. English rebel.
**Cad·il·lac** (kăd'l-ăk'), Sieur **Antoine de la Mothe.** 1656?–1730. French explorer & colonial administrator.
**Caed·mon** (kăd'mən). fl. c. 670. English poet.
**Cae·sar** (sē'zər), **Gaius Julius.** 100–44 B.C. Roman general, statesman, & historian. **—Cae·sar'e·an, Cae·sar'i·an** (sĭ-zâr'ē-ən) *adj.*
**Cage** (kāj), **John Milton, Jr.** b. 1912. Amer. composer.
**Ca·glio·stro** (kăl-yô'strō, kăl-), Count **Alessandro di.** 1743–95. Italian adventurer.
**Cag·ney** (kăg'nē), **James.** b. 1899. Amer. actor.
**Cahan** (kăn), **Abraham.** 1860–1951. Lithuanian-born Amer. Yiddish editor & author.
**Cai·a·phas** (kā'ə-fəs, kī'-), **Joseph.** Jewish high priest (A.D. 18?–36).
**Cain** (kān). Son of Adam & Eve & murderer of Abel.
**Ca·ius** (kā'əs, kī'-). GAIUS.
**Ca·lam·i·ty Jane** (kə-lăm'ĭ-tē jān'). Martha Jane BURKE.
**Cal·der** (kôl'der, kŏl'-), **Alexander.** 1898–1976. Amer. sculptor.
**Cal·de·rón de la Bar·ca** (kăl'də-rôn' dā lə bär'kə), **Pedro.** 1600–81. Spanish author.
**Cald·well** (kôld'wĕl', -dwəl, kŏl'-), **Erskine Preston.** b. 1903. Amer. author.
**Caldwell, Sarah.** b. 1928. Amer. conductor & opera producer.
**Caldwell, (Janet) Taylor.** b. 1900. English-born Amer. author.
**Cal·houn** (kăl-ho͞on'), **John Caldwell.** 1782–1850. U.S. Vice President (1824–32) & political philosopher.
**Ca·lig·u·la** (kə-lĭg'yə-lə). A.D. 12–41. Roman emperor (37–41).
**Ca·lix·tus** (kə-lĭk'stəs). Name of three popes & one antipope, esp. **III,** 1378–1458, reigned 1455–58.
**Cal·kins** (kô'kĭnz), **Mary Whiton.** 1863–1930. Amer. psychologist, philosopher, & educator.
**Cal·la·ghan** (kăl'ə-hən, -hăn'), **James.** b. 1912. British prime minister (1976–79).
**Callaghan, Morley Edward.** b. 1903. Canadian author.
**Cal·las** (kăl'əs, kä'ləs), **Maria Meneghini.** 1923–77. Amer. soprano.
**Ca·lles** (kī'ās, kä'yās), **Plutarco Elías.** 1877–1945. Mexican general & statesman.
**Cal·lim·a·chus** (kə-lĭm'ə-kəs). 5th cent. B.C. Greek sculptor.
**Callimachus.** 3rd cent. B.C. Greek author & scholar.
**Cal·lis·the·nes** (kə-lĭs'thə-nēz'). 360?–328? B.C. Greek philosopher.
**Cal·lo·way** (kăl'ə-wā'), **Cabell ("Cab").** b. 1907. Amer. musician.
**Cal·vert** (kăl'vərt). Family of English colonists & administrators, including **George,** *1st Baron Baltimore* (1580?–1632); **Cecilius,** *2nd Baron Baltimore* (1605–75); **Leonard** (1606–47); & **Charles,** *3rd Baron Baltimore* (1637–1715).
**Cal·vin** (kăl'vĭn), **John.** 1509–64. French-born Swiss Protestant theologian.
**Calvin, Melvin.** b. 1911. Amer. chemist (Nobel, 1961).
**Ca·ma·cho** (kə-mä'chō), **Manuel Ávila.** 1897–1955. Mexican general & statesman.
**Cam·ba·cé·rès** (kän-bä-sā-rĕs'), Duc de. *Jean Jacques Régis.* 1753–1824. French statesman & jurist.
**Cam·by·ses** (kăm-bī'sēz). d. 522 B.C. Persian king (529–522).
**Cam·den** (kăm'dən), **William.** 1551–1623. English historian.
**Cam·er·on** (kăm'ər-ən), **Simon.** 1799–1889. Amer. politician & diplomat.
**Cameron of Loch·iel** (lŏk-ēl', lŏkh-), **Donald.** 1695?–1748. Scottish chieftain & soldier.
**Ca·mo·ëns** (kăm'ō-ənz, kə-mō'-) *also* **Ca·mões** (kə-moinsh'), **Luiz Vaz de.** 1524–80. Portuguese author.
**Camp** (kămp), **Walter Chauncey.** 1859–1925. Amer. football coach & promoter.
**Camp·bell** (kăm'bəl), Sir **Colin.** *Baron Clyde.* 1792–1863. British field marshal.
**Campbell, John.** *4th Earl of Loudoun.* 1705–82. English general in North America.
**Campbell, John Archibald.** 1811–89. Amer. jurist & Confederate official.
**Campbell,** Sir **Malcolm** (1885–1945) & **Donald Malcolm** (1921–67). English automobile & speedboat racers.
**Campbell,** Mrs. **Patrick.** *Beatrice Stella Tanner.* 1867–1940. English actress.
**Campbell, Thomas.** 1777–1844. British author.
**Campbell, Thomas** (1763–1854) & **Alexander** (1788–1866). Irish-born Amer. religious leaders.
**Campbell, William Wallace.** 1862–1938. Amer. astronomer.
**Camp·bell-Ban·ner·man** (kăm'bəl-băn'ər-mən, kăm'əl-), Sir **Henry.** 1836–1908. British prime minister (1905–8).

**Cam·pi** (kăm'pē). Family of Italian painters, including **Galeazzo** (1475?–1536), **Giulio** (1500?–72), **Antonio** (1530?–91), & **Vincenzo** (1532?–91).
**Cam·pi·on** (kăm'pē-ən), **Thomas.** 1567–1620. English poet & songwriter.
**Ca·mus** (kă-mo͞o'), **Albert.** 1913–60. French author (Nobel, 1957).
**Ca·na·let·to** (kăn'ə-lĕt'ō), **Antonio.** 1697–1768. Italian painter.
**Can·by** (kăn'bē), **Henry Seidel.** 1878–1961. Amer. author & editor.
**Cand·ler** (kănd'lər), **Asa Griggs.** 1851–1929. Amer. manufacturer & philanthropist.
**Can·dolle** (kän-dôl'), **Augustin Pyrame de.** 1778–1841. Swiss botanist.
**Ca·net·ti** (kä-nĕt'ē), **Elias.** b. 1905. Bulgarian-born German-language author (Nobel, 1981).
**Can·ning** (kăn'ĭng), Earl **Charles John.** 1812–62. English colonial administrator.
**Canning, George.** 1770–1827. British prime minister (1827).
**Canning,** Sir **Stratford.** *1st Viscount Stratford de Redcliffe.* 1786–1880. English diplomat.
**Can·non** (kăn'ən), **Annie Jump.** 1863–1941. Amer. astronomer.
**Cannon, Joseph Gurney ("Uncle Joe").** 1836–1926. Amer. legislator.
**Ca·no·va** (kə-nō'və), **Antonio.** 1757–1822. Italian sculptor.
**Can·tor** (kăn'tər), **Eddie.** 1892–1964. Amer. comedian.
**Ca·nute** *also* **Cnut** (kə-no͞ot', -nyo͞ot'). *the Great.* 994?–1035. King of England (1016–35), Denmark (1018–35), & Norway (1028–35).
**Canute II** *also* **Cnut II.** HARDECANUTE.
**Ča·pek** (chä'pĕk'), **Karel.** 1890–1938. Czech author.
**Ca·pet** (kā'pĭt, kăp'ĭt, kă-pā'). Dynasty of French kings (987–1328), including **Hugh Capet** (940?–96), ruled 987–96. **—Ca·pe'tian** (kə-pē'shən) *adj.* & *n.*
**Ca·pone** (kə-pōn'), **Alphonse ("Al").** *Scarface.* 1899–1947. Italian-born Amer. gangster.
**Ca·po·te** (kə-pō'tē), **Truman.** 1924–84. Amer. author.
**Capp** (kăp), **Al.** 1909–79. Amer. cartoonist.
**Cap·ra** (kăp'rə), **Frank.** b. 1897. Amer. filmmaker.
**Car·a·cal·la** (kăr'ə-kăl'ə). 188–217. Roman emperor (211–17); assassinated.
**Ca·ra·vag·gio** (kăr'ə-vä'jō, -väzh'ō), **Michelangelo Amerighi** or **Merisa da.** 1565?–1609? Italian painter.
**Car·a·way** (kăr'ə-wā'), **Hattie Ophelia Wyatt.** 1878–1950. Amer. legislator.
**Cár·de·nas** (kär'də-näs'), **Lázaro.** 1895–1970. Mexican soldier & statesman.
**Car·din** (kär-dăn'), **Pierre.** b. 1922. Italian-born French designer.
**Car·do·zo** (kär-dō'zō), **Benjamin Nathan.** 1870–1938. Amer. jurist and author.
**Car·duc·ci** (kär-do͞o'chē), **Giosuè.** 1835–1907. Italian poet (Nobel, 1906).
**Ca·rew** (kə-ro͞o'), **Thomas.** 1595?–1639? English poet.
**Ca·rey** (kâr'ē), **Matthew** (1760–1839) & **Henry Charles** (1793–1879). Amer. publishers & economic theorists.
**Carl XVI Gus·tav** (kärl gŭs'täv, -täf, go͞os'-). b. 1946. Swedish king (since 1973).
**Carle·ton** (kärl'tən), Sir **Guy.** *1st Baron Dorchester.* British soldier & colonial administrator.
**Car·los** (kär'ləs, -lôs), Don. *Count of Molino.* 1788–1855. Spanish pretender to the throne. **—Carl'ist** (kär'lĭst) *adj.* & *n.*
**Carlos de Aus·tri·a** (dĭ ô'strē-ə), Don. 1545–68. Heir to Spanish throne; died mysteriously.
**Car·lo·ta** (kär-lô'tə). 1840–1927. Belgian-born empress of Mexico as wife of Archduke Maximilian of Austria.
**Car·lo·vin·gi·an** (kär'lə-vĭn'jē-ən). *var. of* CAROLINGIAN.
**Carl·son** (kärl'sən), **Chester Floyd.** 1906–68. Amer. inventor.
**Car·lyle** (kär-līl', kär'līl), **Thomas.** 1795–1881. Scottish historian.
**Car·man** (kär'mən), **(William) Bliss.** 1861–1929. Canadian poet.
**Car·mi·chael** (kär'mī-kəl), **Hoagland Howard ("Hoagie").** 1899–1981. Amer. songwriter.
**Car·mo·na** (kär-mō'nə), **Antônio Oscar de Fragoso.** 1869–1951. Portuguese general & statesman.
**Car·ne·gie** (kär'nə-gē, kär-nā'gē, -nĕg'ē), **Andrew.** 1835–1919. Scottish-born Amer. industrialist & philanthropist.
**Car·ne·gie** (kär'nə-gē), **Dale.** 1888–1955. Amer. author & educator.
**Car·not** (kär-nō'), **Lazare Nicolas Marguerite.** 1753–1823. French statesman & military strategist.
**Carnot, Nicolas Léonard Sadi.** 1796–1832. French physicist.
**Carnot, (Marie François) Sadi.** 1837–94. French statesman.
**Car·ol** (kăr'əl). Name of two kings of Rumania, esp. **II,** 1893–1953, ruled 1930–40; abdicated.
**Car·o·lin·gi·an** (kăr'ə-lĭn'jē-ən) *also* **Car·lo·vin·gi·an** (kär'lə-vĭn'-). Dynasty of rulers in France (751–987), Germany (752–911), & Italy (774–961).
**Ca·roth·ers** (kə-rŭth'ərz), **Wallace Hume.** 1896–1937. Amer. chemist & inventor.
**Car·pac·cio** (kär-pä'chō, -chē-ō), **Vittore.** 1460?–1525? Venetian painter.

**Car·ran·za** (kə-răn′zə, -rän′-), **Venustiano.** 1859–1920. Mexican revolutionary statesman.
**Car·rel** (kə-rĕl′, kăr′əl), **Alexis.** 1873–1944. French-born Amer. surgeon & biologist (Nobel, 1912).
**Car·rère** (kə-râr′), **John Merven.** 1858–1911. Amer. architect.
**Car·roll** (kăr′əl), **Anna Ella.** 1815–93. Amer. political pamphleteer & adviser.
**Carroll, Charles.** *Carroll of Carrollton.* 1737–1832. Amer. Revolutionary leader & legislator.
**Carroll, John.** 1735–1815. Amer. religious leader.
**Carroll, Lewis.** Charles Lutwidge DODGSON.
**Car·son** (kär′sən), **Christopher ("Kit").** 1809–68. Amer. frontiersman & Indian agent.
**Carson, Rachel Louise.** 1907–64. Amer. environmentalist.
**Carte** (kärt), **Richard D'Oyly.** 1844–1901. English operetta impresario.
**Car·ter** (kär′tər), **Howard.** 1873–1939. English Egyptologist.
**Carter, James Earl ("Jimmy"), Jr.** b. 1924. 39th U.S. President (1977–81).
**Carter, Samuel Powhatan.** 1819–91. Amer. army & navy officer.
**Car·ter·et** (kär′tər-ət), **John.** *Earl Granville.* 1690–1763. English statesman & diplomat.
**Car·tier** (kär-tyā′, kär′tē-ā′), Sir **George Étienne.** 1814–73. Canadian prime minister (1858–62).
**Cartier, Jacques.** 1491–1557. French explorer.
**Car·tier-Bres·son** (kär-tyā′brĕ-sôN′), **Henri.** b. 1908. French photographer.
**Cart·wright** (kärt′rīt′), **Edmund.** 1743–1823. English clergyman & inventor.
**Car·ty** (kär′tē), **John Joseph.** 1861–1932. Amer. electrical engineer & telephone pioneer.
**Ca·ru·so** (kə-ro͞o′sō, -zō), **Enrico.** 1873–1921. Italian-born operatic tenor.
**Car·ver** (kär′vər), **George Washington.** 1864?–1943. Amer. botanist, agricultural chemist, & educator.
**Carver, John.** 1576?–1621. English-born Pilgrim colonist.
**Carver, Jonathan.** 1710–80. Amer. soldier & explorer.
**Car·y** (kâr′ē), **Alice** (1820–71) & **Phoebe** (1824–71). Amer. authors.
**Cary, Elisabeth Luther.** 1867–1936. Amer. critic.
**Cary, Henry Francis.** 1772–1844. Anglo-Irish poet & translator.
**Cary, (Arthur) Joyce (Lunel).** 1888–1957. Irish-born author.
**Ca·sa·bian·ca** (kä-zä-byäN-kä′), **Louis de.** 1752?–98. French naval officer.
**Ca·sals** (kə-sälz′, -sălz′), **Pablo.** 1876–1973. Spanish-born cellist.
**Ca·sa·no·va de Sein·galt** (kăz′ə-nō′və də săN-gält′, kăs′-), **Giovanni Jacopo.** 1725–98. Italian adventurer & author.
**Case·ment** (kās′mənt), Sir **Roger David.** 1864–1916. British diplomat & Irish rebel; executed for treason.
**Ca·si·mir-Pé·rier** (kăz′ə-mîr-pâr-yā′, -ē-ā′), **Jean Paul Pierre.** 1847–1907. French statesman.
**Cas·lon** (kăz′lən), **William.** 1692–1766. English type designer.
**Cass** (kăs), **Lewis.** 1782–1866. Amer. soldier, politician, & diplomat.
**Cas·satt** (kə-săt′), **Mary Stevenson.** 1845–1926. Amer. painter.
**Cas·sin** (kə-săN′, kä-sēn′), **René.** 1887–1976. French statesman (Nobel, 1968).
**Cas·si·ni** (kə-sē′nē, kä-), **Giovanni Domenico** or **Jean Dominique.** 1625–1712. Italian-born French astronomer.
**Cassini, Oleg.** b. 1913. French-born Amer. fashion designer.
**Cas·si·o·dor·us** (kăs′ē-ə-dôr′əs, -dōr′-), **Flavius Magnus Aurelius.** 6th cent. A.D. Roman statesman & historian.
**Cas·sir·er** (kə-sîr′ər, kä-), **Ernst.** 1874–1945. German philosopher.
**Cas·sius Lon·gi·nus** (kăsh′əs lŏn-jī′nəs), **Gaius.** d. 42 B.C. Roman general & politician.
**Cas·ta·gno** (kä-stä′nyō), **Andrea del.** 1423–57. Florentine painter.
**Cas·ti·glio·ne** (kä′stē-lyō′nā), Count **Baldassare.** 1478–1529. Italian courtier, diplomat, & author.
**Cas·ti·lho** (kəsh-tē′lyo͞o), **Antonio Feliciano de.** 1800–75. Portuguese author & translator.
**Cas·tle** (kăs′əl), **Vernon Blythe** (1887–1918) & **Irene Foote** (1893–1969). English-born dancers.
**Cas·tle·reagh** (kăs′əl-rā′), Viscount. *Robert Stewart, Marquis of Londonderry.* 1769–1822. British statesman.
**Cas·tro** (kăs′trō, kä′strō), **Cipriano.** 1858?–1924. Venezuelan soldier & politician.
**Castro, Fidel.** b. 1927. Cuban revolutionary premier (since 1959).
**Cates·by** (kāts′bē), **Mark.** 1679?–1749. English naturalist.
**Catesby, Robert.** 1573–1605. English conspirator.
**Cath·er** (kăth′ər), **Willa Sibert.** 1876?–1947. Amer. author.
**Cath·e·rine** (kăth′ər-ĭn, kăth′rĭn). Name of two empresses of Russia: **a. I.** 1684?–1727. Wife & successor of Peter the Great; ruled 1725–27. **b. II.** *the Great.* 1729–96. Ruled 1762–96.
**Catherine de Mé·di·cis** (də mĕd′ə-chē′, mā-də-sēs′). 1519–89. Queen of France as wife of Henry II & regent (1560–63).
**Catherine of Ar·a·gon** (ăr′ə-gŏn′). 1485–1536. Queen of England as first wife of Henry VIII.
**Catherine of Bra·gan·za** (brə-găn′zə). 1638–1705. Wife of Charles II of England.
**Catherine of Si·en·a** (sē-ĕn′ə), Saint. 1347–80. Italian religious leader.
**Cat·i·line** (kăt′l-īn′). 108?–62 B.C. Roman politician & conspirator.
**Cat·lin** (kăt′lĭn), **George.** 1796–1872. Amer. artist.
**Ca·to** (kā′tō), **Marcus Porcius.** *the Elder, the Censor.* 234–149 B.C. Roman statesman & general.
**Cato, Marcus Porcius.** *the Younger.* 95–46 B.C. Roman statesman & philosopher.
**Cat·ron** (kăt′rən), **John.** 1786?–1865. Amer. jurist.
**Catt** (kăt), **Carrie (Clinton Lane) Chapman.** 1859–1947. Amer. suffragist.
**Cat·ton** (kăt′n), **(Charles) Bruce.** 1899–1978. Amer. historian, author, & editor.
**Ca·tul·lus** (kə-tŭl′əs), **Gaius Valerius.** 84?–54? B.C. Roman poet.
**Cav·ell** (kăv′əl, kə-vĕl′), **Edith Louisa.** 1865–1915. English nurse; executed.
**Cav·en·dish** (kăv′ən-dĭsh), **Henry.** 1731–1810. English chemist & physicist.
**Cavendish, Spencer Compton.** *Marquis of Compton, 8th Duke of Devonshire.* 1833–1908. English statesman.
**Cavendish, Thomas.** 1555?–92. English navigator.
**Cavendish,** Sir **William.** 1505?–57. English politician.
**Ca·vour** (kə-vo͝or′, kä-vo͝or′), Conte **Camillo Benso di.** 1810–61. Italian political leader.
**Ca·xi·as** (kə-shē′əs), Duque de. *Luiz Alves de Lima e Silva.* 1803–80. Brazilian general & statesman.
**Cax·ton** (kăk′stən), **William.** 1422?–91. First English printer.
**Ceau·ses·cu** (chou-shĕs′ko͞o), **Nicolae.** b. 1918. Rumanian statesman.
**Cec·il** (sĕs′əl), **Robert.** *1st Earl of Salisbury, 1st Viscount Cranborne.* 1563?–1612. English statesman.
**Cecil, (Edgar Algernon) Robert.** *1st Viscount Cecil of Chelwood.* 1864–1958. English statesman (Nobel, 1937).
**Cecil, Robert Arthur Talbot Gascoyne.** *3rd Marquis of Salisbury.* 1830–1903. British prime minister (1885–92, 1895–1902).
**Cecil, William.** *1st Baron Burghley* or *Burleigh.* 1520–98. English statesman.
**Ce·cil·ia** (sĭ-sēl′yə), Saint. 3rd cent. Christian martyr.
**Cel·li·ni** (chə-lē′nē), **Benvenuto.** 1500–71. Italian artist.
**Cel·si·us** (sĕl′sē-əs, -shəs), **Anders.** 1701–44. Swedish astronomer.
**Cen·ci** (chĕn′chē), **Beatrice.** 1577–99. Italian noblewoman; hanged for patricide.
**Cerf** (sûrf), **Bennett Alfred.** 1898–1971. Amer. editor & publisher.
**Cer·van·tes Sa·a·ve·dra** (sər-văn′tēz sä′ə-vā′drə), **Miguel de.** 1547–1616. Spanish author.
**Cé·zanne** (sā-zăn′), **Paul.** 1839–1906. French artist.
**Chad·wick** (chăd′wĭk), **Henry.** 1824–1908. English-born Amer. sportswriter.
**Chadwick,** Sir **James.** 1891–1974. English physicist (Nobel, 1935).
**Cha·fee** (chā′fē), **Zechariah.** 1885–1957. Amer. lawyer & civil libertarian.
**Cha·gall** (shə-gäl′, -găl′), **Marc.** 1887–1985. Russian-born artist.
**Cha·ga·ti** (chăg′ə-tī′). JAGATAI.
**Chain** (chān), **Ernst Boris.** 1906–79. German-born British biochemist (Nobel, 1945).
**Cha·lia·pin** (shä-lyä′pĭn), **Feodor Ivanovich.** 1873–1938. Russian-born French operatic basso.
**Cham·ber·lain** (chām′bər-lĭn). Family of English statesmen, including **Joseph** (1836–1914), Sir **(Joseph) Austen** (1863–1937; Nobel, 1925), & **(Arthur) Neville** (1869–1940; prime minister 1937–40).
**Chamberlain, Owen.** b. 1920. Amer. physicist (Nobel, 1959).
**Cham·ber·lin** (chām′bər-lĭn), **Thomas Chrowder.** 1843–1928. Amer. geologist.
**Cham·bers** (chām′bərz), **Robert.** 1802–71. Scottish publisher & author.
**Chambers, (Jay David) Whittaker.** 1901–61. Amer. journalist & recanted Communist agent.
**Cham·bord** (shäN-bôr′), Comte de. *Henri Charles Ferdinand Marie Dieudonné d'Artois. Duc de Bordeaux.* 1820–83. Bourbon claimant to the French throne.
**Cham·pi·on** (chăm′pē-ən), **Gower.** 1921–80. Amer. dancer & choreographer.
**Cham·plain** (shăm-plān′, shäN-plăN′), **Samuel de.** 1567?–1635. French explorer.
**Cham·pol·lion** (shäN-pô-lyôN′), **Jean François.** 1790–1832. French pioneer Egyptologist.
**Chand·ler** (chănd′lər), **Charles Frederick.** 1836–1925. Amer. chemist & public-health pioneer.
**Chandler, Raymond Thornton.** 1888–1959. Amer. author.

**Chan·dra·gup·ta** (chŭn'drə-go͝op'tə). d. 286? B.C. Indian king (322?–298).
**Chandragupta I.** Indian king (320–30?).
**Chandragupta II.** Indian king (383?–413).
**Chan·dra·se·khar** (chŭn'drə-shä'kər), **Subrahmanyan.** b. 1910. Indian-born Amer. astrophysicist (Nobel, 1983).
**Cha·nel** (shə-nĕl'), **Gabrielle Bonheur ("Coco").** 1883–1971. French fashion designer.
**Cha·ney** (chā'nē), **Lon.** 1883–1930. Amer. actor.
**Chan·ning** (chăn'ĭng), **Edward.** 1856–1931. Amer. historian.
**Channing, William Ellery.** 1780–1842. Amer. religious leader.
**Cha·nute** (shə-no͞ot'), **Octave.** 1832–1910. French-born Amer. engineer & aviation pioneer.
**Chao K'uang-yin** (jou' kwäng'yĭn'). ZHAO KUANGYIN.
**Chap·lin** (chăp'lĭn), Sir **Charles Spencer ("Charlie").** 1889–1977. British-born actor, director, & producer.
**Chap·man** (chăp'mən), **Frank Michler.** 1864–1945. Amer. ornithologist.
**Chapman, George.** 1559?–1634. English author, dramatist, & translator.
**Chapman, John.** *Johnny Appleseed.* 1775?–1845. Amer. pioneer.
**Chapman, John Jay.** 1862–1933. Amer. author.
**Char·cot** (shär-kō'), **Jean Baptiste Étienne Auguste.** 1867–1936. French physician & Antarctic explorer.
**Charcot, Jean Martin.** 1825–93. French neurologist.
**Char·le·magne** (shär'lə-mān'). *Charles I, Charles the Great.* 742–814. King of the Franks (768–814) & emperor of the West (800–14).
**Charles** (chärlz). *Prince of Wales.* b. 1948. English heir apparent.
**Charles.** Name of two kings of England: **a. I.** 1600–49. Ruled 1625–49; beheaded. **b. II.** 1630–85. Ruled 1660–85.
**Charles.** Name of ten kings of France, esp.: **a. IV.** *the Fair.* 1294–1328. Ruled 1322–28. **b. V.** *the Wise.* 1337–80. Ruled 1364–80. **c. VI.** *the Well Beloved.* 1368–1422. Ruled 1380–1422. **d. VII.** 1403–61. Ruled 1422–61. **e. IX.** 1550–74. Ruled 1560–74. **f. X.** 1757–1836. Ruled 1824–30.
**Charles.** Name of seven Holy Roman Emperors, esp.: **a. I.** *the Great.* CHARLEMAGNE. **b. II.** *the Bald.* 823–77. Ruled 875–77; king of France as **Charles I** (840–77). **c. V.** 1500–58. Ruled 1519–56; king of Spain as **Charles I** (1516–56).
**Charles.** Name of 15 kings of Sweden, esp.: **a. XII.** 1682–1718. Ruled 1697–1718. **b. XIV.** *Jean Baptiste Jules Bernadotte.* 1763–1844. Ruled 1818–44.
**Charles, Jacques Alexandre César.** 1746–1823. French physicist & inventor.
**Charles, Ray.** b. 1930. Amer. singer & composer.
**Charles I.** 1887–1922. Austrian emperor (1916–18); deposed.
**Charles Ed·ward Stu·art** (ĕd'wərd sto͞o'ərt, styo͞o'-). —See STUART.
**Charles Lou·is** (lo͞o'ē). 1771–1847. Austrian archduke & military leader.
**Charles Mar·tel** (mär-tĕl'). 689–741. Frankish ruler of Austrasia (715–41).
**Chase** (chās), **Mary Ellen.** 1887–1973. Amer. author & educator.
**Chase, Philander.** 1775–1852. Amer. religious leader.
**Chase, Salmon Portland.** 1808–73. Amer. jurist & politician.
**Chase, Samuel.** 1741–1811. Amer. jurist & Revolutionary War leader.
**Chase, Stuart.** b. 1888. Amer. economist.
**Châ·teau·bri·and** (shä-tō'brē-äN', shăt-ō'-), Vicomte **François René de.** 1768–1848. French political leader, diplomat, & author.
**Chat·ham** (chăt'əm), earls of. —See PITT.
**Chat·ter·ton** (chăt'ər-tn), **Thomas.** 1752–70. English poet.
**Chau·cer** (chô'sər), **Geoffrey.** 1340?–1400. English poet. **—Chau·cer'i·an** (chô-sîr'ē-ən) *adj. & n.*
**Chaun·cy** (chôn'sē, chŏn'-), **Charles.** 1706?–87. Amer. religious leader.
**Chau·temps** (shō-täN'), **Camille.** 1885–1963. French statesman.
**Cha·vannes** (shä-vän'), **Pierre Puvis de.** —See PUVIS DE CHAVANNES.
**Chá·vez** (chä'vĕz'), **Carlos.** 1899–1978. Mexican conductor & composer.
**Chávez, Cesar Estrada.** b. 1927. Amer. labor organizer.
**Cha·yef·sky** (chī-ĕf'skē, chā-), **Paddy.** 1923–81. Amer. playwright & screenwriter.
**Chee·ver** (chē'vər), **John.** 1912–82. Amer. author.
**Che·khov** *also* **Che·kov** (chĕk'ôf, -ŏf, -ŏv), **Anton Pavlovich.** 1860–1904. Russian author. **—Che·kho'vi·an** (chĕ-kō'vē-ən) *adj.*
**Chen** *also* **Ch'ên** (chŭn). Chinese dynasty (557–89).
**Ché·nier** (shā-nyā'), **André Marie de.** 1762–94. French poet; guillotined.
**Chen·nault** (shə-nôlt'), **Claire Lee.** 1890–1958. Amer. air-force officer.
**Che·ops** (kē'ŏps). Egyptian king (2590–2567 B.C.).
**Che·ren·kov** (chə-rĕng'kôf, -kəf), **Pavel Alekseevich.** b. 1904. Soviet physicist (Nobel, 1958).
**Che·ru·bi·ni** (kĕr'ə-bē'nē, kä'ro͞o-), **(Maria) Luigi Carlo Zenobio Salvatore.** 1760–1842. Italian composer.
**Ches·nutt** (chĕs'nŭt'), **Charles Waddell.** 1858–1932. Amer. lawyer & author.
**Ches·ter·field** (chĕs'tər-fēld'), 4th Earl of. *Philip Dormer Stanhope.* 1694–1773. English statesman & author.
**Ches·ter·ton** (chĕs'tər-tən), **Gilbert Keith.** 1874–1936. English author.
**Che·va·lier** (shə-văl'yā, shə-vä-lyā'), **Maurice.** 1888–1972. French actor & singer.
**Ch'i** (chyē). QI.
**Chiang Kai-shek** (chăng' kī'shĕk', jē-äng'). 1887–1975. Chinese military & political leader; exiled in Taiwan (1949–75).
**Ch'ien-lung** (chē-ĕn'lo͝ong'). QIANGLONG.
**Chif·ley** (chĭf'lē), **Joseph Benedict.** 1885–1951. Australian statesman.
**Chi·ka·ma·tsu Mon·za·e·mon** (chē'kä-mät'so͞o môn'-zä-ĕ-môn'). 1653?–1724. Japanese playwright.
**Child** (chīld), **Francis James.** 1825–96. Amer. philologist.
**Child, Julia.** b. 1912. Amer. cookery expert.
**Child, Lydia Maria Francis.** 1802–80. Amer. abolitionist.
**Chil·ders** (chĭl'dərz), **Erskine Hamilton.** 1905–74. Irish statesman.
**Chin** (jĭn). JIN.
**Ch'in** (chĭn). QIN.
**Ch'ing** (chĭng). QING.
**Chip·pen·dale** (chĭp'ən-dāl'), **Thomas.** 1718?–1779. English cabinetmaker.
**Chi·ri·co** (kîr'ĭ-kō', kē'rĭ-), **Giorgio de.** 1888–1978. Italian painter.
**Chis·holm** (chĭz'əm), **Shirley Anita St. Hill.** b. 1924. Amer. politician.
**Chi·sum** (chĭz'əm), **John Simpson.** 1824–84. Amer. cattleman.
**Choate** (chōt), **Joseph Hodges.** 1832–1917. Amer. lawyer & diplomat.
**Choate, Rufus.** 1799–1859. Amer. politician.
**Choi·seul** (shwä-zœl'), Duc **Étienne François de.** 1719–85. French diplomat & statesman.
**Chom·sky** (chŏm'skē), **Noam.** b. 1928. Amer. linguist.
**Cho·pin** (shō'păn'), **Kate O'Flaherty.** 1851–1904. Amer. author.
**Cho·pin** (shō-păn', -păN'), **Frédéric François.** 1810–49. Polish-born French composer.
**Chou** *or* **Chow** (jō). ZHOU.
**Chou En-lai** (jō' ĕn'lī'). ZHOU ENLAI.
**Chou·teau** (sho͞o-tō'). Amer. family of pioneers & fur traders, including **René Auguste** (1749–1829), **Jean Pierre** (1758–1849), **Auguste Pierre** (1786–1838), & **Pierre** (1789–1865).
**Chow** (jō). *var. of* CHOU.
**Chrés·tien de Troyes** *also* **Chré·tien de Troyes** (krā-tyăN dĭ trwä). 12th cent. French trouvère.
**Christ** (krīst). JESUS.
**Chris·tian X** (krĭs'chən). 1870–1947. Danish king (1912–47).
**Chris·tie** (krĭs'tē), Dame **Agatha.** 1891–1976. English mystery writer.
**Chris·ti·na** (krĭ-stē'nə). 1626–89. Swedish queen (1632–54); abdicated.
**Chris·tophe** (krē-stôf'), **Henri.** 1767–1820. Haitian king (1811–20).
**Chris·to·pher** (krĭs'tə-fər), Saint. fl. 3rd cent. Christian martyr.
**Chris·ty** (krĭs'tē), **Edwin P.** 1815–62. Amer. minstrel-show producer.
**Christy, Howard Chandler.** 1873–1952. Amer. artist.
**Chrys·ler** (krīs'lər), **Walter Percy.** 1875–1940. Amer. automobile manufacturer.
**Chry·sos·tom** (krĭs'əs-təm), Saint **John.** 345?–407. Antioch-born Greek church father.
**Church** (chûrch), **Benjamin.** 1734–78? Amer. physician, soldier, & spy.
**Church, Frederick Edwin.** 1826–1900. Amer. painter.
**Chur·chill** (chûr'chĭl', chûrch'hĭl'), **Jennie Jerome.** 1854–1921. Amer. socialite.
**Churchill, John.** *1st Duke of Marlborough.* 1650–1722. English general & statesman.
**Churchill, Randolph Henry Spencer.** 1849–95. English politician.
**Churchill, Winston.** 1871–1947. Amer. author.
**Churchill,** Sir **Winston Leonard Spencer.** 1874–1965. English prime minister (1940–45, 1951–55; Nobel, 1953).
**Chu Teh** (jo͞o' dŭ'). ZHU DE.
**Cia·no** (chä'nō), Conte **Galeazzo.** 1903–44. Italian Fascist statesman; executed.
**Ciar·di** (chär'dē), **John Anthony.** b. 1916. Amer. poet & author.
**Cib·ber** (sĭb'ər), **Colley.** 1671–1757. English author, playwright, & theatrical manager.

**Cic·e·ro** (sĭs'ə-rō'), **Marcus Tullius.** 106–43 B.C. Roman statesman, orator, & philosopher. **—Cic'e·ro'ni·an** *adj.*
**Cid** (sĭd), **the.** *Rodrigo* or *Ruy Díaz de Bivar.* 1040?–99. Spanish soldier & national hero.
**Ci·ma·bu·e** (chē'mə-bōō'ā), **Giovanni.** Late 13th cent. Italian painter.
**Ci·mon** (sī'mən, -mŏn'). 507?–449 B.C. Athenian military & political leader.
**Cin·ci·na·tus** (sĭn'sə-năt'əs, -nā'təs), **Lucius Quinctius.** 519?–439 B.C. Roman general.
**Clair** (klâr), **René.** 1898–1981. French film director.
**Clar·en·don** (klăr'ən-dən), 1st Earl of. *Edward Hyde.* 1609–74. English statesman & historian.
**Clare of As·si·si** (klâr; ə-sē'zē, -sē) *or* **Clar·a** (klâr'ə, klăr'ə), Saint. 1194–1253. Italian nun & religious leader.
**Clark** (klärk), **Alvan.** 1804–87. Amer. astronomer & lens manufacturer.
**Clark, Charles Joseph ("Joe").** b. 1939. Canadian prime minister (1979–80).
**Clark, George Rogers.** 1752–1818. Amer. military leader & frontiersman.
**Clark, James Beauchamp ("Champ").** 1850–1921. Amer. legislative leader.
**Clark, John Bates.** 1847–1938. Amer. economist.
**Clark, Kenneth Bancroft.** b. 1914. Panamanian-born Amer. psychologist & author.
**Clark,** Sir **Kenneth McKenzie.** 1903–83. British art historian.
**Clark, Mark Wayne.** 1896–1984. Amer. army officer.
**Clark, Tom Campbell.** 1899–1977. Amer. jurist.
**Clark, William.** 1770–1838. Amer. Western explorer, military officer, & public official.
**Clarke** (klärk), **Charles Cowden** (1787–1877) & **Mary Victoria Cowden** (1809–98). English Shakespearean scholars.
**Clarke, Helen Archibald.** 1860–1926. Amer. editor & critic.
**Clarke, James Freeman.** 1810–88. Amer. religious reformer.
**Clarke, John Hessin.** 1857–1945. Amer. jurist.
**Claude** (klōd), **Albert.** 1899–1983. Luxembourg-born biologist (Nobel, 1974).
**Clau·del** (klō-dĕl'), **Paul Louis Charles.** 1868–1955. French author & diplomat.
**Clau·di·us** (klô'dē-əs). Name of two Roman emperors: **a. I.** 10 B.C.–A.D. 54. Ruled 41–54. **b. II.** *Gothicus.* 214–70. Ruled 268–70.
**Claudius, Appius.** *Crassus.* 5th cent. B.C. Roman consul.
**Claudius, Appius.** *Caecus.* 4th–3rd cent. B.C. Roman statesman.
**Clau·se·witz** (klou'zə-vĭts), **Karl von.** 1780–1831. Prussian army officer & military theorist.
**Clay** (klā), **Cassius Marcellus.** 1810–1903. Amer. political figure & abolitionist.
**Clay, Cassius Marcellus.** Muhammad ALI.
**Clay, Henry.** 1777–1852. Amer. statesman.
**Clay, Lucius DuBignon.** 1897–1978. Amer. army officer.
**Clay·ton** (klāt'n), **John Middleton.** 1796–1856. Amer. politician & diplomat.
**Cle·an·thes** (klē-ăn'thēz'). 331?–232? B.C. Greek philosopher.
**Cle·ar·chus** (klē-är'kəs). d. 401 B.C. Greek military leader.
**Cleis·the·nes** (klīs'thə-nēz') *or* **Clis·the·nes** (klīs'-). 6th cent. B.C. Greek tyrant.
**Cle·men·ceau** (klĕm'ən-sō', klĕ-mäɴ-sō'), **Georges.** 1841–1929. French statesman.
**Clem·ens** (klĕm'ənz), **Samuel Langhorne.** *Mark Twain.* 1835–1910. Amer. author & humorist.
**Clem·ent** (klĕm'ənt). Name of 14 popes, esp.: **a. I.** Saint. *Clement of Rome.* 1st cent. A.D. Reigned A.D. 88–97. **b. V.** 1264–1314. Reigned 1305–14. **c. VII.** 1478–1534. Reigned 1523–34.
**Clement of Al·ex·an·dri·a** (ăl'ĭg-zăn'drē-ə). 150?–220? Greek Christian theologian.
**Cle·on** (klē'ŏn). d. 422 B.C. Athenian political & military leader.
**Cle·o·pat·ra** (klē'ə-păt'rə, -pä'trə, -pā'-). 69–30 B.C. Egyptian queen.
**Cleve·land** (klēv'lənd), **(Stephen) Grover.** 1837–1908. 22nd & 24th U.S. President (1885–89, 1893–97).
**Cli·burn** (klī'bərn), **Van.** b. 1934. Amer. pianist.
**Clif·ford** (klĭf'ərd), **Clark McAdams.** b. 1906. Amer. lawyer & public official.
**Clifford, Nathan.** 1803–81. Amer. jurist.
**Clift** (klĭft), **Montgomery.** 1920–66. Amer. actor.
**Clin·ton** (klĭn'tən), **DeWitt.** 1769–1828. Amer. politician.
**Clinton, George.** 1686?–1761. English naval officer & colonial administrator.
**Clinton, George.** 1739–1812. U.S. Vice President (1805–12).
**Clinton,** Sir **Henry.** 1738–95. English general in America.
**Clinton, James.** 1733–1812. Amer. military leader.
**Clis·the·nes** (klīs'thə-nēz'). *var. of* CLEISTHENES.
**Clive** (klīv), **Robert.** *Baron Clive of Plassey.* 1725–74. English colonial administrator in India.
**Clough** (klŭf), **Arthur Hugh.** 1819–61. English poet.
**Clo·vis I** (klō'vĭs). 466?–511. Frankish king (481–511).
**Clur·man** (klûr'mən), **Harold.** 1901–80. Amer. theatrical director & critic.
**Cnut** (kə-nōōt', -nyōōt'). *var. of* CANUTE.
**Cobb** (kŏb), **Irwin Shrewsbury.** 1876–1944. Amer. humorist.
**Cobb, Tyrus Raymond ("Ty").** 1886–1961. Amer. baseball player & manager.
**Cob·bett** (kŏb'ĭt), **William.** 1763?–1835. English journalist & social reformer.
**Cob·den** (kŏb'dən), **Richard.** 1804–65. English statesman & reformer.
**Co·ca** (kō'kə), **Imogene.** b. 1914? Amer. comedienne.
**Co·chise** (kō-chēs', -chēz'). 1812?–74. Amer. Apache chief.
**Coch·ran** (kŏk'rən), **Jacqueline.** 1910–80. Amer. aviator & businesswoman.
**Cock·croft** (kŏk'krôft', -krŏft'), Sir **John Douglas.** 1897–1967. English physicist (Nobel, 1951).
**Coc·teau** (kŏk-tō', kôk-), **Jean.** 1891?–1963. French author.
**Co·dy** (kō'dē), **William Frederick.** *Buffalo Bill.* 1846–1917. Amer. frontier scout & showman.
**Coeur de Lion** (kûr' də lē'ən, kœr də lyôɴ'). RICHARD I.
**Cof·fin** (kô'fĭn, kŏf'ĭn), **Levi.** 1789–1877. Amer. abolitionist.
**Coffin, Robert Peter Tristram.** 1892–1955. Amer. author.
**Co·han** (kō'hăn'), **George Michael.** 1878–1942. Amer. singer, songwriter, & playwright.
**Co·hen** (kō'ən), **Morris Raphael.** 1880–1947. Russian-born Amer. educator & philosopher.
**Cohn** (kōn), **Ferdinand Julius.** 1828–98. German botanist & bacteriologist.
**Coke** (kōōk, kōk), Sir **Edward.** 1552–1634. English jurist.
**Col·bert** (kôl-bĕr', kōl-), **Jean Baptiste.** 1619–83. French statesman & financial reformer.
**Col·den** (kōl'dən), **Cadwallader.** 1688–1776. Irish-born Loyalist official, scientist, & philosopher.
**Cole** (kōl), **Nat ("King").** 1919–65. Amer. singer & pianist.
**Cole, Thomas.** 1801–48. Amer. painter.
**Cole·ridge** (kōl'rĭj, kō'lə-rĭj), **Samuel Taylor.** 1772–1834. English poet, critic, & theologian.
**Col·et** (kŏl'ət), **John.** 1467?–1519. English scholar & theologian.
**Co·lette** (kŏl-ĕt', kô-let'), **(Sidonie Gabrielle Claudine).** 1873–1954. French novelist.
**Col·fax** (kōl'făks), **Schuyler.** 1823–85. U.S. Vice President (1869–73).
**Col·gate** (kōl'gāt'), **William.** 1783–1857. Amer. manufacturer & philanthropist.
**Co·li·gny** *or* **Co·li·gni** (kô-lē-nyē', kə-lĭg'nyē), **Gaspard de.** 1519–72. French general & Huguenot leader.
**Col·lier** (kŏl'yər, -ē-ər), **Jeremy.** 1650–1726. English clergyman.
**Collier, John.** 1884–1968. Amer. sociologist & public official.
**Collier, John Payne.** 1789–1883. English lawyer & critic.
**Col·lins** (kŏl'ĭnz), **Michael.** 1890–1922. Irish Sinn Fein leader.
**Collins, (William) Wilkie.** 1824–89. English novelist.
**Collins, William.** 1721–59. English poet.
**Col·man** (kōl'mən), **George.** *Colman the Elder.* 1732–94. English author & theater manager.
**Colman, Ronald.** 1891–1958. English actor.
**Colt** (kōlt), **Samuel.** 1814–62. Amer. inventor & manufacturer.
**Col·um** (kŭl'əm). COLUMBA.
**Col·um** (kŏl'əm), **Padraic** (1881–1972) & **Mary Maguire Gunning** (1887?–1957). Irish-born poets & playwrights.
**Co·lum·ba** (kə-lŭm'bə), Saint. 521–597. Irish missionary.
**Co·lum·bus** (kə-lŭm'bəs), **Christopher.** 1451?–1506. Italian navigator in service of Spain; traditional discoverer of America. **—Co·lum'bi·an** *adj.*
**Co·me·ni·us** (kə-mē'nē-əs), **John Amos.** 1592–1670. Czech educational reformer, scholar, & theologian.
**Co·mines** *also* **Com·mines** *or* **Com·mynes** *or* **Co·mynes** (kô-mēn'), **Phillipe de.** 1447?–1511. French diplomat, political adviser, & historian.
**Com·ma·ger** (kŏm'ə-jər), **Henry Steele.** b. 1902. Amer. historian.
**Com·mines** (kô-mēn'). *var. of* COMINES.
**Com·mo·dus** (kŏm'ə-dəs), **Lucius Aelius Aurelius.** 161–192. Roman emperor (180–92).
**Com·mons** (kŏm'ənz), **John Rogers.** 1862–1945. Amer. political economist.
**Com·mynes** (kô-mēn'). *var. of* COMINES.
**Comp·ton** (kŏmp'tən), **Karl Taylor** (1887–1954) & **Arthur Holly** (1892–1962; Nobel, 1927). Amer. physicists.
**Comp·ton-Bur·nett** (kŏmp'tən-bər-nĕt'), **Ivy.** 1892–1969. English author.
**Com·stock** (kŏm'stŏk', kŭm'-), **Anna Botsford.** 1854–1930. Amer. wood engraver & naturalist.
**Comstock, Anthony.** 1844–1915. Amer. social reformer.

**Comstock, Elizabeth Leslie Rous.** 1815–91. English-born Amer. Quaker leader.
**Comte** (kōNt, kôNt), **(Isidore) Auguste (Marie François).** 1798–1857. French philosopher & mathematician.
**Co·mynes** (kô-mēn'). *var. of* COMINES.
**Co·nant** (kō'nənt), **James Bryant.** 1893–1978. Amer. educator & diplomat.
**Con·boy** (kŏn'boi), **Sara Agnes McLaughlin.** 1870–1928. Amer. labor leader.
**Con·dé** (kôN-dā'), Prince de. *Louis II de Bourbon, the Great Condé.* 1621–87. French general.
**Con·don** (kŏn'dən), **Edward Uhler.** 1902–74. Amer. physicist.
**Con·dor·cet** (kôn-dôr-sĕ'), Marquis de. *Marie Jean Antoine Nicolas Caritat.* 1743–94. French revolutionary, philosopher, & mathematician.
**Con·fu·cius** (kən-fyōō'shəs). 551–479 B.C. Chinese philosopher. **—Con·fu'cian** *adj. & n.*
**Con·greve** (kŏn'grēv', kŏng'-), **William.** 1670–1729. English playwright.
**Conk·ling** (kŏngk'lĭng), **Roscoe.** 1829–88. Amer. politician.
**Con·nel·ly** (kŏn'ə-lē), **Marcus Cook ("Marc").** 1890–1980. Amer. playwright, producer, & director.
**Con·nol·ly** (kŏn'ə-lē), **Maureen Catherine ("Little Mo").** 1934–69. Amer. tennis player.
**Con·nors** (kŏn'ərz), **James Scott ("Jimmy").** b. 1952. Amer. tennis player.
**Con·rad** (kŏn'răd'), **Joseph.** 1857–1924. Polish-born English novelist.
**Con·sta·ble** (kŭn'stə-bəl, kŏn'-), **John.** 1776–1837. English landscape painter.
**Con·stant de Re·becque** (kôN-stäN' də rə-bĕk'), **Benjamin.** 1767–1830. French author & politician.
**Con·stan·tine** (kŏn'stən-tēn', -tīn'). Name of two Greek kings: **a. I.** 1868–1923. Ruled 1913–17, 1920–22; abdicated. **b. II.** b. 1940. Ruled 1964–73.
**Constantine.** Name of two Roman emperors, esp. **I,** *the Great,* 280?–337, ruled 306–37.
**Constantine XI.** 1404?–53. Last Byzantine emperor (1448–53).
**Con·ti** (kōn'tē), **Niccolò de'.** 15th cent. Italian merchant, traveler, & author.
**Con·way** (kŏn'wā), **Thomas.** 1735–1800? Irish-born Amer. Revolutionary general.
**Con·well** (kŏn'wĕl', -wəl), **Russell Herman.** 1843–1925. Amer. clergyman & educator.
**Cook** (kŏŏk), **Frederick Albert.** 1865–1940. Amer. physician & Arctic explorer.
**Cook, James.** *Captain Cook.* 1728–79. English navigator & explorer; murdered in Sandwich Is.
**Cooke** (kŏŏk), **(Alfred) Alistair.** b. 1908. British-born Amer. broadcaster & author.
**Cooke, Jay.** 1821–1905. Amer. financier.
**Cooke, Terence James.** 1921–83. Amer. prelate.
**Coo·ley** (kōō'lē), **Charles Horton.** 1864–1929. Amer. sociologist.
**Cooley, Thomas McIntyre.** 1824–98. Amer. jurist.
**Coo·lidge** (kōō'lĭj), **(John) Calvin.** 1872–1933. 30th U.S. President (1923–29).
**Coolidge, Julian Lowell.** 1873–1954. Amer. mathematician & educator.
**Coo·ney** (kōō'nē), **Joan Ganz.** b. 1929. Amer. television producer.
**Coo·per** (kōō'pər, kŏŏp'ər), **Gary.** 1906–61. Amer. actor.
**Cooper, James Fenimore.** 1789–1851. Amer. novelist.
**Cooper, Leon N.** b. 1930. Amer. physicist (Nobel, 1972).
**Cooper, Peter.** 1791–1883. Amer. manufacturer, inventor, & philanthropist.
**Cooper, Thomas.** 1759–1839. English-born Amer. chemist.
**Co·per·ni·cus** (kō-pûr'nə-kəs, kə-), **Nicolaus.** 1473–1543. Polish astronomer. **—Co·per'ni·can** *adj.*
**Cop·land** (kōp'lənd), **Aaron.** b. 1900. Amer. composer.
**Cop·ley** (kŏp'lē), **John Singleton.** 1738–1815. Amer. portrait painter.
**Cop·pin** (kŏp'ĭn), **Fanny Marion Jackson.** 1837–1913. Amer. educator & missionary.
**Cor·bett** (kôr'bət), **James John ("Gentleman Jim").** 1866–1933. Amer. boxer.
**Cor·bin** (kôr'bĭn), **Margaret Cochran.** 1751–1800. Amer. Revolutionary heroine.
**Cor·co·ran** (kôr'kə-rən, -krən), **Thomas Gardiner.** 1900–81. Amer. public official.
**Corcoran, William Wilson.** 1798–1888. Amer. banker, art collector, & philanthropist.
**Cor·day** (kôr-dā', kôr'dā), **Charlotte.** 1768–93. French revolutionary heroine; guillotined.
**Co·rel·li** (kə-rĕl'ē), **Arcangelo.** 1653–1713. Italian violinist & composer.
**Cor·i** (kôr'ē, kōr'ē), **Carl Ferdinand** (b. Czechoslovakia, 1896–1984) & **Gerty Theresa Radnitz** (1896–1957). Amer. biochemists (shared Nobel, 1947).
**Cor·liss** (kôr'lĭs), **George Henry.** 1817–88. Amer. inventor & manufacturer.
**Cor·mack** (kôr'mək), **Allan MacLeod.** b. 1924. South African-born Amer. physicist (Nobel, 1979).
**Cor·neille** (kôr-nā'), **Pierre.** 1606–84. French dramatist.
**Cor·ne·lia** (kôr-nēl'yə, -nē'lē-ə). 2nd cent. B.C. Roman matron.
**Cor·ne·lius** (kôr-nēl'yəs, -nāl'-, -nā'lē-əs), **Peter von.** 1783–1867. German painter.
**Cor·nell** (kôr-nĕl'), **Ezra.** 1807–74. Amer. businessman & university benefactor.
**Cornell, Katherine.** 1898–1974. Amer. actress.
**Corn·forth** (kôrn'fərth, -fôrth', -fōrth'), **John Warcup.** b. 1917. Australian-born British chemist (Nobel, 1975).
**Corn·wal·lis** (kôrn-wŏl'ĭs, -wô'lĭs), **Charles.** *1st Marquis & 2nd Earl Cornwallis.* 1738–1805. English military & political leader.
**Co·ro·na·do** (kôr'ə-nä'dō, kōr'-), **Francisco Vásquez de.** 1510–54. Spanish explorer & colonial administrator.
**Co·rot** (kô-rō', kə-), **Jean Baptiste Camille.** 1796–1875. French painter.
**Cor·reg·gio** (kə-rĕj'ō, -ē-ō'), **Antonio Allegri da.** 1494–1534. Italian painter.
**Cor·ri·gan** (kôr'ĭ-gən), **Mairead.** b. 1944. Irish peace advocate (Nobel, 1976).
**Cor·tel·you** (kôr'tl-yōō'), **George Bruce.** 1862–1940. Amer. public official.
**Cor·tés** *also* **Cor·tez** (kôr-tĕz', kôr'tĕz'), **Hernando** *or* **Fernando.** 1485–1547. Spanish explorer & conquistador.
**Cor·win** (kôr'wĭn), **Edward Samuel.** 1878–1963. Amer. political scientist & educator.
**Cos·by** (kŏz'bē), **Bill.** b. 1937. Amer. comedian & actor.
**Co·sell** (kō-sĕl'), **Howard.** b. 1920. Amer. sportscaster.
**Cos·grave** (kŏz'grāv'), **William Thomas.** 1880–1965. Irish Sinn Fein leader.
**Cos·ta Ca·bral** (kôs'tə kə-bräl', kôsh'-), **Antônio Bernardo da.** 1803–89. Portuguese statesman.
**Cos·tel·lo** (kŏs-tĕl'ō), **John Aloysius.** 1891–1976. Irish political leader.
**Costello, Lou.** 1908–59. Amer. comedian.
**Cot·ton** (kŏt'n), **Charles.** 1630–87. English poet & translator.
**Cotton, John.** 1584–1652. English-born Amer. clergyman.
**Co·ty** (kō-tē', kô-), **René.** 1882–1962. French statesman.
**Cou·é** (kōō-ā', kwā), **Émile.** 1857–1926. French psychotherapist.
**Coues** (kouz), **Elliott.** 1842–99. Amer. ornithologist, biologist, & editor.
**Cough·lin** (kŏg'lĭn), **Charles Edward.** 1891–1979. Canadian-born Amer. priest & political activist.
**Cou·loumb** (kōō'lŏm', -lōm', kōō-lŏm', -lôN'), **Charles Augustin de.** 1736–1806. French physicist.
**Cou·pe·rin** (kōōp-răN', kōō-pə-), **François.** 1668–1733. French composer & organist.
**Cou·pe·rus** (kōō-pā'rəs, -pĕr'əs), **Louis.** 1863–1923. Dutch novelist.
**Cour·bet** (kōōr-bā'), **Gustave.** 1819–77. French painter.
**Cour·nand** (kōōr-näN'), **André Frédéric.** b. 1895. French-born Amer. physiologist (Nobel, 1956).
**Cour·réges** (kōō-rĕzh'), **André.** b. 1923. French fashion designer.
**Cous·ins** (kŭz'ənz), **Norman.** b. 1915. Amer. editor & writer.
**Cous·teau** (kōō-stō'), **Jacques Yves.** b. 1910. French underwater explorer, film producer, & author.
**Cou·sy** (kōō'zē), **Robert Joseph ("Bob").** b. 1928. Amer. basketball player, coach, & sportscaster.
**Co·var·ru·bias** (kō'və-rōō'bē-əs), **Miguel.** 1904–57. Mexican artist & author.
**Cov·er·dale** (kŭv'ər-dāl'), **Miles.** 1488–1568. English clergyman & Bible translator.
**Cow·ard** (kou'ərd), Sir **Noel Pierce.** 1899–1973. English actor, author, & composer.
**Cowl** (koul), **Jane.** 1884?–1950. Amer. actress.
**Cowles** (koulz), **Gardner.** 1903–85. Amer. publisher.
**Cowles, Henry Chandler.** 1869–1939. Amer. botanist & pioneer ecologist.
**Cow·ley** (kou'lē), **Abraham.** 1618–67. English author.
**Cowley, Malcolm.** b. 1898. Amer. author, editor, & critic.
**Cow·per** (kōō'pər, kou'-, kŏŏp'ər), **William.** 1731–1800. English poet.
**Coxe** (kŏks), **Tench.** 1755–1824. Amer. political economist.
**Cox·ey** (kŏk'sē), **Jacob Sechler.** 1854–1951. Amer. businessman & politician.
**Coz·zens** (kŭz'ənz), **James Gould.** 1903–78. Amer. author.
**Crabbe** (krăb), **George.** 1754–1832. English poet.
**Craig** (crāg), **(Edward) Gordon.** 1872–1966. English theatrical producer, director, & designer.

**Craig·av·on** (krā-găv′ən), 1st Viscount. *James Craig.* 1871–1940. Northern Ireland statesman.
**Crai·gie** (krā′gē), Sir **William Alexander.** 1876–1957. English lexicographer & philologist.
**Cram** (krăm), **Ralph Adams.** 1863–1942. Amer. architect.
**Cra·nach** (krä′näkh′), **Lucas.** 1472–1553. German artist.
**Cran·dall** (krăn′dəl), **Prudence.** 1803–90. Amer. educator & reformer.
**Crane** (krān), **(Harold) Hart.** 1899–1932. Amer. poet.
**Crane, Stephen.** 1871–1900. Amer. author.
**Crane, Walter.** 1845–1915. English artist.
**Crane, Winthrop Murray.** 1853–1920. Amer. manufacturer & legislator.
**Cran·mer** (krăn′mər), **Thomas.** 1489–1556. English prelate & religious reformer.
**Crap·sey** (krăp′sē), **Adelaide.** 1878–1914. Amer. poet.
**Crash·aw** (krăsh′ô), **Richard.** 1613–49. English poet.
**Cras·sus** (krăs′əs), **Marcus Licinius.** 115?–53? B.C. Roman politician & general.
**Cra·ter** (krā′tər), **Joseph Force.** *Judge Crater.* 1889–1937? Amer. jurist; disappeared.
**Craw·ford** (krô′fərd), **Francis Marion.** 1854–1909. Amer. author.
**Crawford, Joan.** 1908–77. Amer. actress.
**Crawford, Thomas.** 1814–57. Amer. sculptor.
**Crawford, William Harris.** 1772–1834. Amer. public official.
**Cra·zy Horse** (krā′zē hôrs′). 1849?–77. Amer. Indian leader.
**Cré·bil·lon** (krā-bē-yôɴ′), **Prosper Jolyot de.** 1674–1762. French poet.
**Cre·mer** (krē′mər), Sir **William Randal.** 1838–1908. English pacifist (Nobel, 1903).
**Crève·coeur** (krĕv-kœr′), **Michel Guillaume Jean de.** 1731?–1813. French agriculturalist, author, & diplomat.
**Crich·ton** (krīt′n), **James.** *the Admirable Crichton.* 1560?–82. Scottish adventurer, linguist, & scholar.
**Crick** (krĭk), **Francis Henry Compton.** b. 1916. British biologist (Nobel, 1962).
**Crile** (krīl), **George Washington.** 1864–1943. Amer. surgeon.
**Cripps** (krĭps), Sir **(Richard) Stafford.** 1889–1952. English statesman.
**Cris·pi** (krĭs′pē, krē′spē), **Francesco.** 1819–1901. Italian statesman.
**Cris·pin** (krĭs′pĭn), Saint. 3rd cent. Christian martyr.
**Crit·ten·den** (krĭt′n-dən), **John Jordan.** 1787–1863. Amer. lawyer & politician.
**Cro·ce** (krō′chā), **Benedetto.** 1866–1952. Italian philosopher, historian, & critic.
**Crock·er** (krŏk′ər), **Charles.** 1822–88. Amer. railroad financier.
**Crock·ett** (krŏk′ĭt), **David ("Davy").** 1786–1836. Amer. politician & frontiersman; died at the Alamo.
**Croe·sus** (krē′səs). d. 546 B.C. Lydian king (560–46).
**Cro·ghan** (krō′gən), **George.** 1720?–82. Irish-born Amer. Indian agent & land speculator.
**Cro·ly** (krō′lē), **Herbert David.** 1869–1930. Amer. author & editor.
**Croly, Jane Cunningham.** 1829–1901. Amer. author, editor, & feminist.
**Cromp·ton** (krŏmp′tən, krŏm′-), **Samuel.** 1753–1927. English inventor.
**Crom·well** (krŏm′wĕl′, -wəl, krŭm′-), **Oliver** (1599–1658) & **Richard** (1626–1712). English military, political, & religious leaders. **—Crom·well′i·an** *adj.*
**Cromwell, Thomas.** *Earl of Essex.* 1485?–1540. English statesman; executed.
**Cro·nin** (krō′nĭn), **Archibald Joseph ("A. J.").** 1896–1981. English physician & novelist.
**Cronin, James W.** b. 1931. Amer. educator & physicist (Nobel, 1980).
**Cron·je** (krôn-yā′), **Piet Arnoldus.** 1840?–1911. Boer general.
**Cron·kite** (krŏn′kīt, krŏng′-), **Walter Leland, Jr.** b. 1916. Amer. broadcast journalist.
**Crook** (kro͝ok), **George.** 1829–90. Amer. general & Indian fighter.
**Crookes** (kro͝oks), Sir **William.** 1832–1919. English chemist & physicist.
**Cros·by** (krôz′bē, krŏz′-), **Frances Jane ("Fanny").** 1820–1915. Amer. hymn writer & poet.
**Crosby, Harry Lillis ("Bing").** 1903–77. Amer. singer & actor.
**Cross** (krôs, krŏs), **Milton.** 1897–1975. Amer. opera commentator.
**Cross, Wilbur Lucius.** 1862–1948. Amer. politician & educator.
**Croth·ers** (krŭth′ərz), **Rachel.** 1878–1958. Amer. playwright, director, & producer.
**Crouse** (krous), **Russel.** 1893–1966. Amer. author.
**Cruik·shank** (kro͝ok′shăngk′), **George.** 1792–1878. English artist.
**Cu·kor** (kyo͞o′kər, -kôr, ko͞o′-), **George.** 1899–1983. Amer. filmmaker.
**Cul·bert·son** (kŭl′bərt-sən), **Ely.** 1891–1955. Amer. contract bridge authority.
**Cul·len** (kŭl′ən), **Countée.** 1903–46. Amer. poet.
**Cul·pep·er** (kŭl′pĕp′ər), Lord **Thomas.** 1635–89. English colonial administrator.
**Cum·mings** (kŭm′ĭngz), **Edward Estlin.** *e. e. cummings.* 1894–1962. Amer. poet.
**Cun·ha** (ko͞o′nyə), **Tristão da.** 1460?–1540. Portuguese navigator & explorer.
**Cun·ning·ham** (kŭn′ĭng-hăm′, -əm), **Allen.** 1784–1842. Scottish author.
**Cunningham, Kate Richards O'Hare.** 1877–1948. Amer. socialist & reformer.
**Cunningham, Merce.** b. 1922? Amer. dancer & choreographer.
**Cu·rie** (kyo͝o-rē′, kyo͝or′ē), **Eve Denise.** b. 1904. French pianist, author, & editor.
**Curie** *also* **Cu·rie-Jo·liot** (-zhô-lyō′), **Irene.** —See JOLIOT-CURIE.
**Curie, Pierre** (1859–1906) & **Marie** (b. Poland, 1867–1934). French chemists & physicists (shared Nobel, 1903).
**Cur·ley** (kûr′lē), **James Michael.** 1874–1958. Amer. politician.
**Cur·ri·er** (kûr′ē-ər, kŭr′-), **Nathaniel.** 1813–88. Amer. lithographer.
**Cur·ry** (kûr′ē, kŭr′ē), **John Steuart.** 1897–1946. Amer. painter.
**Cur·tin** (kûr′tn), **John.** 1885–1945. Australian statesman.
**Cur·tis** (kûr′tĭs), **Benjamin Robbins.** 1809–74. Amer. jurist.
**Curtis, Charles.** 1860–1936. U.S. Vice President (1929–33).
**Curtis, Cyrus Hermann Kotzschmar.** 1850–1933. Amer. publisher.
**Curtis, George Ticknor.** 1812–94. Amer. lawyer & author.
**Curtis, George William.** 1824–92. Amer. journalist & reformer.
**Cur·tiss** (kûr′tĭs), **Glenn Hammond.** 1878–1930. Amer. aviation pioneer.
**Cur·wen** (kûr′wən), **John.** 1816–80. English music educator & publisher.
**Cur·zon** (kûr′zən), **George Nathaniel.** 1859–1925. English colonial administrator.
**Cush·ing** (ko͝osh′ĭng), **Caleb.** 1800–79. Amer. lawyer, politician, & diplomat.
**Cushing, Harvey Williams.** 1869–1939. Amer. brain surgeon.
**Cushing, William.** 1732–1810. Amer. jurist.
**Cushing, William Barker.** 1842–74. Amer. Civil War naval hero.
**Cush·man** (ko͝osh′mən), **Charlotte Saunders.** 1816–76. Amer. actress.
**Cus·ter** (kŭs′tər), **George Armstrong.** 1839–76. Amer. general; killed at Little Bighorn.
**Cut·ler** (kŭt′lər), **Manasseh.** 1742–1823. Amer. clergyman, botanist, & pioneer.
**Cu·vier** (kyo͞o′vē-ā′, ko͞ov-yā′, kü-vyā′), Baron **Georges Léopold Chrétien Frédéric.** 1769–1832. French naturalist.
**Cyn·e·wulf** (kĭn′ə-wo͝olf′) *or* **Cyn·wulf** (kĭn′wo͝olf′). Late 8th cent. Anglo-Saxon poet.
**Cyp·ri·an** (sĭp′rē-ən), Saint. d. 258. Christian prelate & martyr.
**Cy·ra·no de Ber·ge·rac** (sîr′ə-nō də bĕr′zhə-răk′), **Savinien de.** 1619?–55. French satirist & duelist.
**Cyr·il** (sîr′əl), Saint. 827–69. Christian missionary & theologian.
**Cy·rus** (sī′rəs). 600?–529 B.C. *the Great.* Persian king (550–529) & founder of Persian empire.
**Czer·ny** (chĕr′nē), **Karl.** 1791–1857. Austrian pianist & composer.

# D

**Daft** (dăft), **Leo.** 1843–1922. English-born Amer. engineer.
**Da·guerre** (də-gâr′), **Louis Jacques Mandé.** 1787?–1851. French artist & inventor.
**Dahl·gren** (dăl′grən), **John Adolphus Bernard.** 1809–70. Amer. naval officer & inventor.
**Daim·ler** (dīm′lər), **Gottlieb.** 1834–1900. German engineer & inventor.
**Da·kin** (dā′kĭn), **Henry Drysdale.** 1880–1952. English biochemist.
**Da·la·dier** (də-lä′dē-ā′, dăl-ə-dyā′), **Edouard.** 1884–1970. French statesman.
**Dale** (dāl), Sir **Henry Hallett.** 1875–1968. English physiologist (Nobel, 1936).
**Dale,** Sir **Thomas.** d. 1619. English-born colonial administrator.
**Da·lén** (də-lān′), **Nils Gustaf.** 1869–1937. Swedish inventor (Nobel, 1912).
**Da·ley** (dā′lē), **Richard Joseph.** 1902–76. Amer. public official.
**Dal·hou·sie** (dăl-ho͞o′zē, -hou′-), 10th Earl & 1st Marquis of. *James Andrew Broun Ramsay.* 1812–60. English colonial administrator.
**Da·li** (dä′lē), **Salvador.** b. 1904. Spanish artist.
**Dall** (dôl), **Caroline Wells Healey.** 1822–1912. Amer. author & reformer.
**Dal·las** (dăl′əs), **Alexander James.** 1759–1817. West Indian-born Amer. statesman.
**Dallas, George Mifflin.** 1792–1864. U.S. Vice President (1845–49).

**Dal·rym·ple** (dăl-rĭm′pəl, dăl′rĭm-), Sir **James.** *1st Viscount Stair.* 1619–95. Scottish jurist.
**Dalrymple,** Sir **John.** *2nd Earl of Stair.* 1673–1747. Scottish general & diplomat.
**Dal·ton** (dôl′tən), **John.** 1766–1844. English chemist & philosopher.
**Dalton, Robert Hugh.** 1887–1962. English politician.
**Da·ly** (dā′lē), **(John) Augustin.** 1839–99. Amer. playwright & theatrical manager.
**Daly, Marcus.** 1841–1900. Amer. mining financier.
**Dam** (dăm, däm), **(Carl Peter) Henrik.** 1895–1976. Danish biochemist (Nobel, 1943).
**d'Am·boise** (dăN-bwäz′), **Jacques.** b. 1934. Amer. ballet dancer.
**Da·mien de Veus·ter** (dā′mē-ən də vyōōs′tər), **Joseph.** *Father Damien.* 1840–88. Belgian Roman Catholic missionary.
**Dam·pi·er** (dăm′pē-ər), **William.** 1652–1715. English explorer.
**Dam·rosch** (dăm′rŏsh), **Leopold** (1832–85) & **Walter Johannes** (1862–1950). German-born Amer. musicians & composers.
**Dan** (dăn). Fifth son of Jacob in the Old Testament.
**Da·na** (dā′nə), **Charles Anderson.** 1819–97. Amer. newspaper owner & editor.
**Dana, James Dwight.** 1813–95. Amer. geologist & mineralogist.
**Dana, Richard Henry.** 1815–82. Amer. author.
**Dan·dridge** (dăn′drĭj), **Dorothy.** 1923?–65. Amer. actress.
**Dan·iel** (dăn′yəl). Hebrew prophet.
**Daniel, Peter Vivian.** 1784–1860. Amer. jurist.
**Daniel, Samuel.** 1562?–1619. English author.
**Dan·iels** (dăn′yəlz), **Josephus.** 1862–1948. Amer. journalist.
**Da·ni·lo·va** (də-nē′lə-və, -lô-), **Alexandra.** b. 1906. Russian-born Amer. ballerina.
**Dan·nay** (dăn′ā), **Frederic.** With Manfred B. Lee, *Ellery Queen.* 1905–82. Amer. author.
**D'An·nun·zio** (dä-nōōn′tsyō, -tsē-ō′), **Gabriele.** 1863–1938. Italian author.
**Dan·te A·li·ghie·ri** (dän′tā äl′ə-gyâ′rē, dăn′tē). 1265–1321. Italian poet. **—Dan′te·an** *adj.* & *n.* **—Dan·tesque′** (dăn-tĕsk′) *adj.*
**Dan·ton** (däN-tôN′), **Georges Jacques.** 1759–94. French revolutionary leader.
**Dare** (dâr), **Virginia.** 1587–87? First child of English parents born in Amer.
**Da·ri·us** (də-rī′əs). Name of three kings of Persia, esp.: **a. I.** *the Great.* 558?–486? B.C. Ruled 521–486 B.C. **b. III.** 380?–330 B.C. Ruled 336–330 B.C.
**Dar·lan** (där-läN′), **Jean Louis Xavier François.** 1881–1942. French admiral.
**Darn·ley** (därn′lē), Lord. *Henry Stuart* or *Stewart.* 1545–67. Scottish nobleman; second husband of Mary Queen of Scots.
**Dar·row** (dăr′ō), **Clarence Seward.** 1857–1938. Amer. lawyer.
**Dar·win** (där′wĭn), **Charles Robert.** 1809–82. English naturalist. **—Dar·win′i·an** *adj.* & *n.*
**Darwin, Erasmus.** 1731–1802. English physician, scientist, reformer, & poet.
**Dau·bi·gny** (dō-bē-nyē′), **Charles François.** 1817–78. French landscape painter.
**Dau·det** (dō-dā′), **Alphonse.** 1840–97. French author.
**Daudet, Léon.** 1867–1942. French journalist.
**Daugh·er·ty** (dô′ər-tē), **Harry Micajah.** 1860–1941. Amer. lawyer & politician.
**Dau·mier** (dō-myā′, dō′mē-ā′), **Honoré.** 1808–79. French artist.
**Dau·set** (dō-sĕ′, -sā′), **Jean.** b. 1915. French physiologist (Nobel, 1980).
**Dav·e·nant** or **D'Av·e·nant** (dăv′ə-nənt, dăv′nənt), Sir **William.** 1606–68. English dramatist.
**Dav·en·port** (dăv′ən-pôrt′, -pōrt′), **John.** 1597–1670. English Puritan clergyman & colonist.
**Davenport, Thomas.** 1802–51. Amer. inventor.
**Da·vid** (dā′vĭd). 1010?–970? B.C. Second king of Judah & Israel.
**David.** Name of two kings of Scotland, esp. **I,** 1084–1153, ruled 1124–53.
**Da·vid** (dä′vət), **Gerard.** 1450?–1523. Flemish painter.
**Da·vid** (dä-vēd′), **Jacques Louis.** 1748–1825. French painter.
**Da·vid·son** (dā′vĭd-sən), **Jo.** 1883–1952. Amer. sculptor.
**Da·vies** (dā′vēz), **Arthur Bowen.** 1862–1928. Amer. painter.
**Davies, Marion.** 1898?–1961. Amer. actress.
**Dá·vi·la y Pa·di·lla** (dä′və-lə ē pä-dē′ə, -yə), **Agustin.** 1562–1604. Mexican prelate & historian.
**Da·vis** (dā′vĭs), **Angela.** b. 1944. Amer. political activist.
**Davis, Benjamin Oliver.** 1877–1970. Amer. cavalry officer.
**Davis, David.** 1815–86. Amer. politician & jurist.
**Davis, Dwight Filley.** 1879–1945. Amer. public official & tennis player.
**Davis, Jefferson.** 1808–89. Amer. soldier & Confederate statesman.
**Davis, John William.** 1873–1955. Amer. politician & diplomat.
**Davis, Katherine Bement.** 1860–1935. Amer. penologist & social worker.
**Davis, Miles Dewey, Jr.** b. 1926. Amer. musician.
**Davis, Paulina Kellogg Wright.** 1813–76. Amer. suffragist.
**Davis, Rebecca Blaine Harding.** 1831–1910. Amer. author.
**Davis, Richard Harding.** 1864–1916. Amer. journalist.
**Davis, Ruth Elizabeth ("Bette").** b. 1908. Amer. actress.
**Davis, Sammy, Jr.** b. 1925. Amer. entertainer.
**Davis, Stuart.** 1894–1964. Amer. artist.
**Davis, William Morris.** 1850–1934. Amer. geologist & geographer.
**Da·vis·son** (dā′vĭ-sən), **Clinton Joseph.** 1881–1958. Amer. physicist (Nobel, 1937).
**Da·vout** (dä-vōō′), **Louis Nicolas.** 1770–1823. French marshal.
**Da·vy** (dā′vē), Sir **Humphrey.** 1778–1829. English chemist.
**Da·vys** (dā′vĭs), **John.** 1550?–1605. English navigator.
**Dawes** (dôz), **Charles Gates.** 1865–1951. U.S. Vice President (1925–29; Nobel, 1925).
**Dawes, Henry Laurens.** 1816–1903. Amer. politician.
**Day** (dā), **Benjamin Henry.** 1810–89. Amer. printer & journalist.
**Day, Clarence Shepard, Jr.** 1874–1935. Amer. author.
**Day, Dorothy.** 1897–1981. Amer. journalist & reformer.
**Day, Thomas.** 1748–89. English author & philanthropist.
**Day, William Rufus.** 1849–1923. Amer. jurist.
**Da·yan** (dī-än′, dä-yän′), **Moshe.** 1915–81. Israeli political leader.
**Dean** (dēn), **James.** 1931–55. Amer. actor.
**Dean, Jay Hanna** or **Jerome Herman ("Dizzy").** 1911–74. Amer. baseball player & sportscaster.
**Deane** (dēn), **Silas.** 1737–89. Amer. diplomat.
**Dear·born** (dîr′bôrn′, -bərn), **Henry.** 1751–1829. Amer. soldier & politician.
**De Ba·key** (də bā′kē), **Michael Ellis.** b. 1908. Amer. heart surgeon.
**De·bierne** (də-byĕrn′), **André Louis.** 1874–1949. French chemist.
**Deb·o·rah** (dĕb′ər-ə). Hebrew judge & prophetess.
**De Bow** (də bō′), **James Dunwoody Brownson.** 1820–67. Amer. statistician & editor.
**De·breu** (də-brœ′), **Gerard.** b. 1921. French-born Amer. economist (Nobel, 1983).
**Debs** (dĕbz), **Eugene Victor.** 1855–1926. Amer. labor organizer & socialist leader.
**De·bus·sy** (də-byōō′sē), **Claude Achille.** 1862–1918. French composer.
**De·bye** (də-bī′), **Peter Joseph Wilhelm.** 1884–1966. Dutch-born Amer. physicist (Nobel, 1936).
**De·ca·tur** (dĭ-kā′tər), **Stephen.** 1779–1820. Amer. naval officer.
**De·cius** (dē′shəs, -shē-əs). 201–51. Roman emperor (249–51).
**Deck·er** (dĕk′ər). *var. of* DEKKER.
**Dee** (dē), **Ruby.** b. 1924? Amer. actress.
**Dee·ping** (dē′pĭng), **(George) Warwick.** 1877–1950. English novelist.
**Deere** (dîr), **John.** 1804–86. Amer. manufacturer & inventor.
**Def·fand** (dĭ-fäN′), Marquise du. *Marie de Vichy-Chamrond.* 1697–1780. French literary patron.
**De·foe** (dĭ-fō′), **Daniel.** 1660–1731. English author.
**De For·est** (dĭ fôr′ĭst, fŏr′-), **Lee.** *the Father of Radio.* 1873–1961. Amer. inventor.
**De·gas** (də-gä′), **(Hilaire Germain) Edgar.** 1834–1917. French painter.
**De Gaulle** (də gōl′, gôl′), **Charles André Joseph Marie.** 1890–1970. French general & statesman.
**Dek·ker** or **Deck·er** (dĕk′ər), **Thomas.** 1572–1632. English dramatist.
**de Koo·ning** (dĭ kōō′nĭng), **Willem.** b. 1904. Dutch-born Amer. painter.
**De·la·croix** (dĕl′ə-krwä′), **(Ferdinand Victor) Eugène.** 1798–1863. French painter.
**de la Mare** (də lə mâr′), **Walter John.** 1873–1956. English author.
**De·la·mat·er** (də-lăm′ə-tər), **Cornelius Henry.** 1821–1889. Amer. mechanical engineer.
**De Lan·cey** (də lăn′sē), **James.** 1703–60. Amer. colonial official.
**De·land** (də-lănd′), **Margaret.** 1857–1945. Amer. author.
**Del·a·no** (dĕl′ə-nō), **Jane Arminda.** 1862–1919. Amer. nurse.
**De·lan·y** (də-lā′nē), **Martin Robinson.** 1812–85. Amer. physician & social reformer.
**de la Ren·ta** (dā lə rĕn′tə), **Oscar.** b. 1934? Dominican-born Amer. fashion designer.
**De La Rey** (də lə rī′, -rā′), **Jacobus Hercules.** 1847–1914. Boer general & statesman.
**De·la·roche** (dĕl′ə-rōsh′, -rôsh′), **Hippolyte Paul.** 1797–1856. French painter.
**de la Roche** (də lə rōch′, rôsh′), **Mazo.** 1885–1961. Canadian author.
**De Lau·ren·tis** (dē lô-rĕn′təs), **Dino.** b. 1919. Italian filmmaker.
**De·la·vigne** (dĕl′ə-vēn′yə), **(Jean François) Casimir.** 1793–1843. French author.
**De La Warr** (dĕl′ə wâr′), Baron. *Thomas West.* 1577–1618. English-born Amer. colonial administrator.
**Del·brück** (dĕl′brük′, -brōōk′), **Max.** 1906–81. German-born Amer. biologist (Nobel, 1969).

**De·led·da** (dā-lĕd′ə, də-), **Grazia.** 1875–1936. Italian novelist (Nobel, 1926).
**de Les·seps** (də lĕs′ĕps, lĕ-sĕps′), Vicomte **Ferdinand Marie.** 1805–94. French diplomat.
**De·libes** (də-lēb′), **(Clément Philibert) Léo.** 1836–91. French composer.
**De·lius** (dē′lē-əs, dēl′yəs), **Frederick.** 1862–1934. English composer.
**Dell** (dĕl), **Floyd.** 1887–1969. Amer. author.
**del·la Rob·bia** (dĕl′ə rō′bē-ə). Family of Italian sculptors, including **Luca** (1400?–82), **Andrea** (1437–1528), & **Giovanni** (1469–1529?).
**Del·mon·i·co** (dĕl-mŏn′ĭ-kō), **Lorenzo.** 1813–81. Amer. restaurateur.
**De Long** (də lông′), **George Washington.** 1844–81. Amer. Arctic explorer.
**De·lorme** or **de l'Orme** (də-lôrm′), **Philibert.** 1515?–70. French architect.
**De Mille** (də mĭl′), **Agnes George.** b. 1905. Amer. choreographer.
**De Mille, Cecil Blount.** 1881–1959. Amer. movie producer.
**De·moc·ri·tus** (dĭ-mŏk′rĭ-təs). 460?–357 B.C. Greek philosopher.
**Dem·o·rest** (dĕm′ə-rĕst′), **Ellen Louis Curtis.** 1824–98. Amer. businesswoman.
**De Mor·gan** (dĭ môr′gən), **William Frend.** 1836–1917. English artist, author, & inventor.
**De·mos·the·nes** (dĭ-mŏs′thə-nēz′). 385?–322 B.C. Greek orator.
**Demp·sey** (dĕmp′sē), Sister **(Julia) Mary Joseph.** 1856–1939. Amer. hospital administrator.
**Dempsey, William Harrison ("Jack").** 1895–1983. Amer. heavyweight boxer.
**De·muth** (dĭ-mo͞oth′), **Charles.** 1883–1935. Amer. painter.
**Deng Xiao·ping** (dŭng′ shou′pĭng′). b. 1904. Chinese Communist leader.
**De·ni·ker** (dĕ-nē-kĕr′), **Joseph.** 1852–1918. French anthropologist.
**De·nis** or **De·nys** (dĕn′ĭs, də-nē′), Saint. 3rd cent. martyred apostle to the Gauls.
**Den·nett** (dĕn′ĭt), **Mary Coffin Ware.** 1872–1947. Amer. social reformer.
**Den·nie** (dĕn′ē), **Joseph.** 1768–1812. Amer. editor & author.
**Dent** (dĕnt), **Joseph Malaby ("J.M.").** 1849–1926. English publisher.
**De·nys** (dĕn′ĭs, də-nē′). *var. of* DENIS.
**De·pew** (dĭ-pyo͞o′), **Chauncey Mitchell.** 1834–1928. Amer. lawyer, railroad executive, & politician.
**De Quin·cey** (dĭ kwĭn′sē, -zē), **Thomas.** 1785–1859. English essayist.
**De·rain** (də-răN′), **André.** 1880–1954. French artist.
**Der·zha·vin** (dĕr-zhä′vĭn), **Gavriil Romanovish.** 1743–1816. Russian poet.
**De·saix de Vey·goux** (də-sā′ də vā-go͞o′), **Louis Charles Antoine.** 1768–1800. French general.
**De·sargues** (dā-zärg′), **Gérard.** 1593–1662. French army officer & mathematician.
**Des·cartes** (dā-kärt′), **René.** 1596–1650. French mathematician & philosopher. **—Car·te′sian** (kär-tē′zhən) *adj.*
**Des·cha·nel** (dā′shə-nĕl′), **Paul Eugène Louis.** 1856–1922. French statesman & author.
**de Se·ver·sky** (də sə-vĕr′skē), **Alexander Procofieff.** 1894–1974. Russian-born Amer. aeronautical engineer.
**De Si·ca** (də sē′kə), **Vittorio.** 1901–74. Italian filmmaker.
**De Smet** (də smĕt′), **Pierre Jean.** 1801–73. Belgian-born missionary in America.
**Des·mou·lins** (dā-mo͞o-lăN′), **(Lucie Simplice) Camille (Benoît).** 1760–94. French revolutionary; guillotined.
**de So·to** (dĭ sō′tō), **Hernando.** 1496?–1542. Spanish explorer.
**Des·saix** (dĭ-sā′), Comte **Joseph Marie.** 1764–1834. French army officer.
**Des·sa·lines** (dā-sə-lēn′), **Jean Jacques.** 1758–1806. Haitian emperor (1804–06); assassinated.
**De·taille** (dĭ-tī′), **(Jean Baptiste) Édouard.** 1848–1912. French painter.
**De·us Ra·mos** (dĕ′o͞osh rä′mo͞osh), **João de.** 1830–1896. Portuguese poet.
**De Va·le·ra** (dĕv′ə-lĕr′ə, -lîr′ə), **Eamon.** 1882–1975. Amer.-born Irish statesman.
**de Vere** (də vîr′), **Aubrey Thomas.** 1814–1902. Irish author.
**Dev·er·eux** (dĕv′ə-ro͞o′), **Robert.** *2nd Earl of Essex.* 1566–1601. English nobleman & favorite of Elizabeth I.
**De Vin·ne** (də vĭn′ē), **Theodore Low.** 1828–1914. Amer. printer.
**De Vo·to** (də vō′tō), **Bernard Augustine.** 1897–1955. Amer. author & editor.
**De Vries** (də vrēs′), **Hugo.** 1848–1935. Dutch botanist.
**De Vries, Peter.** b. 1910. Amer. author.
**Dew·ar** (do͞o′ər, dyo͞o′-), Sir **James.** 1842–1923. Scottish-born chemist & physicist.
**De Wet** (də wĕt′, vĕt′, vāt′), **Christiaan Rudolph.** 1854–1922. Boer general.
**Dew·ey** (do͞o′ē, dyo͞o′ē), **George.** 1837–1917. Amer. naval officer.
**Dewey, John.** 1859–1952. Amer. philosopher & educator.
**Dewey, Melvil.** 1851–1931. Amer. librarian & founder of decimal system of classification.
**Dewey, Thomas Edmund.** 1902–71. Amer. politician.
**De Witt** (də wĭt′, vĭt′), **Jan.** 1625–72. Dutch statesman; murdered.
**De Wolfe** (də wo͝olf′), **Elsie.** 1865–1950. Amer. actress & interior decorator.
**Dia·ghi·lev** (dē-äg′ə-lĕf′), **Sergei Pavlovich.** 1872–1929. Russian ballet producer.
**Di·an·a** (dī-ăn′ə). *Princess of Wales. Lady Diana Spencer.* b. 1961. British crown princess.
**Di·as** or **Dí·az** (dē′əs, -əsh), **Bartholomeu.** 1450?–1500. Portuguese navigator.
**Dí·az** (dē′äts), **Armando.** 1861–1928. Italian general & statesman.
**Dí·az** (dē′äs, -äz), **(José de la Cruz) Porfirio.** 1830–1915. Mexican general & statesman.
**Dí·az Or·daz** (dē′äs ôr-däz′), **Gustavo.** 1911–79. Mexican statesman.
**Dick** (dĭk), **George Frederick** (1881–1967) & **Gladys Henry.** (1881–1963). Amer. physicians.
**Dick·ens** (dĭk′ĭnz), **Charles John Huffam.** *Boz.* 1812–70. English author. **—Dick·en′si·an** (dĭ-kĕn′zē-ən) *adj.*
**Dick·ey** (dĭk′ē), **James.** b. 1923. Amer. author.
**Dick·in·son** (dĭk′ĭn-sən), **Anna Elizabeth.** 1842–1932. Amer. reformer & lecturer.
**Dickinson, Emily Elizabeth.** 1830–86. Amer. poet.
**Dickinson, John.** 1732–1808. Amer. statesman.
**Di·de·rot** (dē′də-rō′, dē-drō′), **Denis.** 1713–1784. French philosopher, author, & Encyclopedist.
**Did·rik·son** (dĭd′rĭk-sən), **Mildred Ella ("Babe").** 1914–56. Amer. athlete.
**Die·fen·ba·ker** (dē′fən-bā′kər), **John George.** 1895–1979. Canadian prime minister (1957–63).
**Diels** (dēlz, dēls), **Otto Paul Hermann.** 1876–1954. German chemist (Nobel, 1950).
**Di·em** (dē-ĕm′, dyĕm), **Ngo Dinh.** 1901–63. Vietnamese political leader; assassinated.
**Dies** (dīz), **Martin.** 1901–72. Amer. legislator.
**Die·sel** (dē′zəl), **Rudolf.** 1858–1913. German engineer & inventor.
**Die·trich** (dē′trĭk, -trĭKH), **Marlene.** b. 1901? German-born Amer. actress.
**Diez** (dēts), **Friedrich Christian.** 1794–1876. German philologist.
**Dig·by** (dĭg′bē), Sir **Kenelm.** 1603–65. English naval officer, diplomat, & philosopher.
**Diggs** (dĭgz), **Annie LePorte.** 1848–1916. English-born Amer. reformer & politician.
**Dill** (dĭl), Sir **John Greer.** 1881–1944. Irish-born British army officer.
**Dil·lin·ger** (dĭl′ĭn-jər), **John.** 1902–34. Amer. bank robber.
**Dil·lon** (dĭl′ən), **John.** 1851–1927. Irish nationalist political leader.
**Di Mag·gio** (də-mä′zhē-ō, -măj′ē-ō), **Joseph Paul.** b. 1914. Amer. baseball player.
**Di·ne·sen** (dē′nĭ-sən, dĭn′ĭ-), **Isak.** *Baroness Karen Blixen.* 1885–1962. Danish author.
**Din·wid·die** (dĭn-wĭd′ē), **Robert.** 1693–1770. Scottish-born British colonial administrator.
**Di·o·cle·tian** (dī′ə-klē′shən). 245–313. Roman emperor (284–305).
**Di·og·e·nes** (dī-ŏj′ə-nēz′). 412?–323 B.C. Greek philosopher.
**Di·o·ny·si·us** (dī-ə-nĭsh′ē-əs, -nĭsh′əs, -nī′sē-əs). *the Elder* (430?–367 B.C.) & *the Younger* (395?–343? B.C.). Greek tyrants.
**Dionysius of Al·ex·an·dri·a** (ăl-ĭg-zăn′drē-ə), Saint. 190?–264? Christian theologian.
**Dionysius of Hal·i·car·nas·sus** (hăl′ĭ-kär-năs′əs). 1st cent. B.C. Greek historian.
**Dionysius Ex·ig·u·us** (ĕg-zĭg′yə-wəs). 6th cent. monk & scholar.
**Di·or** (dē-ôr′), **Christian.** 1905–57. French fashion designer.
**Di·rac** (dĭ-răk′), **Paul Adrien Maurice.** 1902–84. English mathematician & physicist (Nobel, 1933).
**Dirk·sen** (dûrk′sən), **Everett McKinley.** 1896–1969. Amer. legislative leader.
**Dis·ney** (dĭz′nē), **Walter Elias ("Walt").** 1901–66. Amer. cartoonist, showman, & film producer.
**Dis·rae·li** (dĭz-rā′lē), **Benjamin.** *1st Earl of Beaconsfield. Dizzy.* 1804–81. British prime minister (1868, 1874–80), author, & diplomat.
**Dit·mars** (dĭt′märz′), **Raymond Lee.** 1876–1942. Amer. naturalist & author.
**Dix** (dĭks), **Dorothea Lynde.** 1802–87. Amer. philanthropist, reformer, author, & educator.
**Dix, Dorothy.** Elizabeth Meriwether GILMER.
**Dix, John Adams.** 1798–1879. Amer. politician & diplomat.
**Dix·on** (dĭk′sən), **Jeremiah.** fl. 1763–67. English-born surveyor.
**Dji·las** (jĭl′äs), **Milovan.** b. 1911. Yugoslavian author & political leader.

**Dmow·ski** (də-môf'skē, -môv'-), **Roman.** 1864–1939. Polish nationalist political leader.
**Do·bie** (dō'bē), **James Frank.** 1888–1964. Amer. historian, folklorist, & author.
**Do·brée** (dō'brā'), **Bonamy.** 1891–1974. English literary historian.
**Do·bry·nin** (dō-brē'nĭn), **Anatoly F.** b. 1919. Soviet diplomat.
**Dob·son** (dŏb'sən), **(Henry) Austin.** 1840–1921. English author.
**Dodge** (dŏj), **Grace Hoadley.** 1856–1914. Amer. philanthropist & social worker.
**Dodge, Grenville Mellen.** 1831–1916. Amer. civil engineer & politician.
**Dodge, Josephine Marshall Jewell.** 1855–1928. Amer. antisuffragist & day-care advocate.
**Dodge, Mary Elizabeth Mapes.** 1831–1905. Amer. author.
**Dodg·son** (dŏj'sən), **Charles Lutwidge.** *Lewis Carroll.* 1832–98. English mathematician & author.
**Dods·ley** (dŏdz'lē), **Robert.** 1703–64. English author & editor.
**Doe·nitz** *also* **Dö·nitz** (dœ'nĭts), **Karl.** 1891–1980. German naval officer.
**Do·her·ty** (dō'ər-tē), **Henry Latham.** 1870–1939. Amer. engineer & utilities magnate.
**Doi·sy** (doi'zē), **Edward Adelbert.** b. 1893. Amer. biochemist (Nobel, 1943).
**Dole** (dōl), **Sanford Ballard.** 1844–1926. Amer. jurist & administrator in Hawaii.
**Doll·fuss** (dôl'fo͝os'), **Engelbert.** 1892–1934. Austrian statesman; assassinated.
**Do·magk** (dō'mäk'), **Gerhard.** 1895–1964. German biochemist (Nobel, 1939).
**Do·me·ni·chi·no** (dō-mā'nə-kē'nō). 1581–1641. Italian painter.
**Do·min·go** (də-mĕng'gō), **Placido.** b. 1941. Spanish-born opera singer.
**Dom·i·nic** (dŏm'ə-nĭk), Saint. 1170–1221. Spanish-born founder of Dominican order. **—Do·min'i·can** (də-mĭn'ĭ-kən) *adj. & n.*
**Do·mi·tian** (də-mĭsh'ən). A.D. 51–96. Roman emperor (81–96).
**Don·a·tel·lo** (dŏn'ə-tĕl'ō). 1386?–1466. Italian sculptor.
**Don·i·phan** (dŏn'ĭ-fən), **Alexander William.** 1808–87. Amer. army officer & frontiersman.
**Dö·nitz** (dœ'nĭts). *var. of* DOENITZ.
**Don·i·zet·ti** (dŏn'ĭ-zĕt'ē), **Gaetano.** 1797–1848. Italian composer.
**Donne** (dŭn), **John.** 1572?–1631. English poet.
**Don·nel·ly** (dŏn'ə-lē), **Ignatius.** 1831–1901. Amer. politician & reformer.
**Don·o·van** (dŏn'ə-vən), **William Joseph ("Wild Bill").** 1883–1959. Amer. army officer.
**Doo·lit·tle** (do͞o'lĭt'l), **Hilda.** *H.D.* 1886–1961. Amer. poet.
**Doolittle, James Harold ("Jimmy").** b. 1896. Amer. army officer & aviator.
**Dopp·ler** (dŏp'lər), **Christian Johann.** 1803–53. Austrian physicist & mathematician.
**Do·ra·ti** (də-rä'tē), **Antal.** b. 1906. Hungarian-born Amer. conductor & composer.
**Do·ré** (dô-rā', də-), **(Paul) Gustave.** 1833–83. French artist.
**Do·re·mus** (dô-rē'mŭs), **Sarah Platt Haines.** 1802–77. Amer. social worker & philanthropist.
**Dor·nier** (dôr-nyā'), **Claude.** 1884–1969. German aircraft designer.
**Dorr** (dôr), **Thomas Wilson.** 1805–54. Amer. politician & reformer.
**Dor·set** (dôr'sĭt), 1st Earl of. Thomas SACKVILLE.
**Dos Pas·sos** (dŏs păs'ōs), **John Roderigo.** 1896–1970. Amer. novelist.
**Dos·to·ev·ski** *or* **Dos·to·yev·sky** (dŏs'tə-yef'skē, -toi-, -yĕv'-), **Feodor Mikhailovich.** 1821–81. Russian author. **—Dos'to·ev'ski·an** *adj.*
**Dou** *or* **Dow** *or* **Douw** (dou), **Gerard.** 1613–1675. Dutch genre painter.
**Dou·ble·day** (dŭb'əl-dā'), **Abner.** 1819–93. Amer. army officer & reputed inventor of baseball.
**Doubleday, Frank Nelson.** 1862–1934. Amer. publisher.
**Dough·ty** (dou'tē), **Charles Montagu.** 1843–1926. English traveler & author.
**Doug·las** (dŭg'ləs), **Helen Mary Gahagan.** 1900–80. Amer. actress & politician.
**Douglas,** Sir **John Sholto.** *8th Marquis of Queensberry.* 1844–1900. English nobleman & boxing promoter.
**Douglas, Lloyd Cassel.** 1877–1951. Amer. clergyman & author.
**Douglas, Melvyn.** 1901–81. Amer. actor.
**Douglas, Stephen Arnold.** *the Little Giant.* 1813–61. Amer. legislator.
**Douglas, William Orville.** 1898–1980. Amer. jurist.
**Douglas-Home** (dŭg'ləs-hyo͞om'), Sir **Alexander Frederick.** b. 1903. British prime minister (1963–64).

o͞o **boot** ou **out** th **thin** *th* **this** ŭ **cut** ûr **urge** y **young**
yo͞o **abuse** zh **vision** ə **about,** it**e**m, ed**i**ble, gall**o**p, circ**u**s

**Doug·lass** (dŭg'ləs), **Frederick.** 1817?–95. Amer. abolitionist & journalist.
**Dou·mer** (do͞o-mĕr'), **Paul.** 1857–1932. French statesman; assassinated.
**Dou·mergue** (do͞o-mĕrg'), **Gaston.** 1863–1937. French statesman.
**Dou·vil·lier** (do͞o-vē-lyā'), **Suzanne Théodore Vaillande.** 1778–1826. French-born Amer. dancer & choreographer.
**Douw** *or* **Dow** (dou). *vars. of* DOU.
**Dow** (dou), **Charles Henry.** 1851–1902. Amer. economist & publisher.
**Dow, Herbert Henry.** 1866–1930. Amer. chemist & manufacturer.
**Dow, Neal.** 1804–97. Amer. temperance leader.
**Dow·den** (doud'n), **Edward.** 1843–1913. Irish editor, author, & educator.
**Downes** (dounz), **(Edwin) Olin.** 1886–1955. Amer. music critic.
**Down·ing** (dou'nĭng), **Andrew Jackson.** 1815–52. Amer. landscape architect & horticulturist.
**Dow·son** (dou'sən), **Ernest Christopher.** 1867–1900. English author.
**Doyle** (doil), Sir **Arthur Conan.** 1859–1930. English author.
**D'Oy·ly Carte** (doi'lē kärt'), **Richard.** —See CARTE.
**Drach·mann** (dräk'mən), **Holger Henrik Herholdt.** 1846–1908. Danish author.
**Dra·co** (drā'kō). 7th cent. B.C. Athenian lawgiver. **—Dra·co'ni·an** *adj.*
**Drake** (drāk), **Daniel.** 1785–1852. Amer. physician & pioneer medical educator.
**Drake, Edwin Laurentine.** 1819–80. Amer. oil-industry leader.
**Drake,** Sir **Francis.** 1540?–96. English naval hero & explorer.
**Dra·per** (drā'pər), **Henry.** 1837–82. Amer. pioneer astronomer & photographer.
**Draper, John William.** 1811–82. English-born Amer. chemist & historian.
**Draper, Ruth.** 1884–1956. Amer. monologuist.
**Dray·ton** (drāt'n), **Michael.** 1563–1631. English poet.
**Draper, William Henry.** 1742–79. Amer. Revolutionary leader.
**Drei·ser** (drī'sər, -zər), **Theodore (Herman Albert).** 1871–1945. Amer. author & editor.
**Dress·ler** (drĕs'lər), **Marie.** 1871?–1934. Canadian-born Amer. comedienne.
**Drew** (dro͞o). Amer. family of actors, including **Louisa Lane** (1820–97), **John** (1826–62), & **John** (1853–1927).
**Drew, Daniel.** 1797–1879. Amer. financier.
**Drey·fus** (drī'fəs, drā-), **Alfred.** 1859–1935. French army officer; central figure in the Dreyfus Affair (1894 ff.).
**Driesch** (drēsh), **Hans Adolf Eduard.** 1867–1941. German biologist & philosopher.
**Drink·wa·ter** (drĭngk'wô'tər, -wŏt'ər), **John.** 1882–1937. English author.
**Drum·mond** (drŭm'ənd), **Henry.** 1851–97. Scottish clergyman & author.
**Drummond, William.** 1585–1649. Scottish poet.
**Drummond, William Henry.** 1854–1907. Irish-born Canadian poet & physician.
**Dru·sus** (dro͞o'səs), **Nero Claudius.** 38–9 B.C. Roman general.
**Dry·den** (drīd'n), **John.** 1631–1700. English author.
**Duane** (dwān), **William.** 1760–1835. Amer. journalist & politician.
**Du Bar·ry** (do͞o băr'ē, dyo͞o-, bə-rē'), Comtesse. *Marie Jeanne Bécu.* 1746?–93. Mistress of Louis XV; guillotined.
**Dub·ček** (do͞ob'chĕk), **Alexander.** b. 1921. Czech political leader.
**du Bel·lay** (do͞o bə-lā'), **Joachim.** —See BELLAY.
**Du·bin·sky** (do͞o-bĭn'skē), **David.** b. 1892–1982. Russian-born Amer. labor leader.
**Du·bois** (do͞o-bwä', dyo͞o-), **Eugène.** 1858–1940. Dutch anatomist & paleontologist.
**Dubois, Paul.** 1829–1905. French artist.
**Dubois, Théodore.** 1837–1924. French composer & educator.
**Du Bois** (do͞o bois'), **William Edward Burghardt.** 1868–1963. Amer. sociologist, educator, & author.
**Du·bos** (do͞o-bôs', -bō'), **René Jules.** 1901–82. French-born Amer. bacteriologist.
**Du·buf·fet** (do͞o-bə-fā'), **Jean.** 1901–85. French artist.
**Du Cange** (do͞o käNzh'), Sieur. *Charles du Fresne.* 1610–88. French philologist & historian.
**Du Chail·lu** (do͞o shī'yo͞o, -shäl'-), **Paul Belloni.** 1831–1903. French-born African explorer.
**Du·champ** (do͞o-shäN'), **Marcel.** 1887–1968. French-born modernist painter.
**Du·chesne** (do͞o-shĕn'), **Rose Philippine.** 1769–1852. French-born religious leader in America.
**Du·com·mun** (do͞o'kə-mœN'), **Élie.** 1833–1906. Swiss journalist (Nobel, 1902).
**Dud·ley** (dŭd'lē), **Robert.** *1st Earl of Leicester.* 1532–88. English courtier, politician, & favorite of Elizabeth I.
**Dudley, Thomas** (1576–1653) & **Joseph** (1647–1720). English colonial administrators in America.
**Du·fay** (do͞o-fā'), **Guillaume.** 1400?–74. Flemish composer.

**Duf·fer·in and A·va** (dŭf′ər-ĭn; ä′və), 1st Marquis of. *Frederick Temple Hamilton-Temple Blackwood.* 1826-1902. English diplomat.
**Duff-Gor·don** (dŭf′gôr′dn), Lady **Lucie** *or* **Lucy.** 1821–69. English writer & translator.
**Duf·fy** (dŭf′ē), Sir **Charles Gavan.** 1816–1903. Irish-born writer and political leader in Ireland & Australia.
**Du·fy** (do͞o-fē′), **Raoul.** 1877–1953. French artist.
**Du Gues·clin** (do͞o gĕ-klăN′), **Bertrand.** 1320?–80. French military commander.
**Du·ha·mel** (do͞o′ə-mĕl′, dyo͞o′-), **Georges.** 1844–1966. French author & physician.
**Duke** (do͞ok, dyo͞ok), **Benjamin Newton** (1855–1929) & **James Buchanan** (1856–1925). Amer. tobacco-industry leaders.
**Du·la·ny** (do͞o-lā′nē, də-), **Daniel.** 1722–97. Amer. politician.
**Dul·bec·co** (dŭl-bĕk′ō), **Renato.** b. 1914. Italian-born Amer. virologist (Nobel, 1975).
**Dul·les** (dŭl′ĭs), **Allen Welsh.** 1893–1969. Amer. public official.
**Dulles, John Foster.** 1888–1959. Amer. diplomat & statesman.
**Du·mas** (do͞o-mä′, dyo͞o-), **Alexandre** (*Dumas père;* 1802–70) & **Alexandre** (*Dumas fils;* 1824–95). French authors.
**du Mau·ri·er** (do͞o môr′ē-ā′, dyo͞o), Dame **Daphne.** b. 1907. English author.
**du Maurier, George Louis Palmella Busson.** 1834–96. English illustrator & author.
**du Maurier,** Sir **Gerald.** 1873–1934. English actor & theatrical manager.
**Du·mou·riez** (do͞o-mo͞o′ryā), **Charles François.** 1739–1823. French general.
**Du·nant** (do͞o-näN′), **Jean Henri.** 1828–1910. Swiss philanthropist & founder of the Red Cross (Nobel, 1901).
**Dun·bar** (dŭn′bär), **Paul Laurence.** 1872–1906. Amer. author.
**Dun·bar** (dŭn-bär′), **William.** 1460?–1520. Scottish poet.
**Dun·can** (dŭng′kən), **Isadora.** 1878–1927. Amer. dancer.
**Dun·das** (dŭn-dăs′), **Henry.** *1st Viscount Melville, Baron Dunira.* 1742–1811. English statesman.
**Dun·i·way** (dŭn′ə-wā′), **Abigail Jane Scott.** 1834–1915. Amer. Western pioneer & suffragist.
**Dun·lap** (dŭn′lăp), **William.** 1766–1839. Amer. playwright, theatrical manager, painter, & historian.
**Dun·lop** (dŭn-lŏp′, dŭn′lŏp), **John Boyd.** 1840–1921. Scottish inventor.
**Dun·more** (dŭn-môr′, -mōr′), 4th Earl of. *John Murray.* 1732–1809. English colonial administrator.
**Dunne** (dŭn), **Finley Peter.** 1867–1936. Amer. humorist & journalist.
**Du·nois** (do͞o-nwä′), Comte **Jean de.** 1403?–68. French battle hero & companion of Joan of Arc.
**Dun·sa·ny** (dŭn-sā′nē), 18th Baron. *Edward John Moreton Drax Plunkett.* 1878–1957. Irish author.
**Duns Sco·tus** (dŭnz skō′təs), **John.** *Doctor Subtilis.* 1265?–1308. Scottish Scholastic theologian.
**Dun·stan** (dŭn′stən), Saint. 925?–88. English prelate.
**Dun·ster** (dŭn′stər), **Henry.** 1609–59. English-born Amer. clergyman & educator.
**Du·pleix** (do͞o-plĕks′), Marquis **Joseph-François.** 1697–1763. French colonial administrator.
**Du·ples·sis-Mor·nay** (do͞o-plə-sē′ môr-nā′). —See MORNAY.
**Du Pont** (do͞o-pŏnt′, dyo͞o-, do͞o′pŏnt, dyo͞o′-), **Éleuthère Irénée.** 1771–1834. French-born Amer. industrialist.
**Du Pont, Samuel Francis.** 1803–65. Amer. naval leader.
**Du Pont de Nemours** (də nə-mo͞or′), **Pierre Samuel.** 1739–1817. French-born economist & politician.
**Du·quesne** (do͞o-kān′, dyo͞o-), Marquis **Abraham.** 1610–88. French naval commander.
**Du·rand** (do͞o-rănd′), **Asher Brown.** 1796–1886. Amer. artist.
**Du·rant** (də-rănt′), **Thomas Clark.** 1820–85. Amer. railroad financier.
**Durant, Will(iam James)** (1885–1981) & **Ariel** (1898–1981). Amer. authors.
**Durant, William Crapo.** 1861–1947. Amer. industrialist.
**Du·ran·te** (də-răn′tē), **Jimmy.** 1893–1980. Amer. comedian.
**Dü·rer** (do͝or′ər, dyo͝or′-), **Albrecht.** 1471–1528. German artist.
**Durk·heim** (dûr-kĕm′), **Émile.** 1858–1917. French sociologist & philosopher.
**Du·roc** (do͝or′ŏk, dyo͝or′-), **Géraud Christophe Michel.** *Duc de Friuli.* 1772–1813. French general & diplomat.
**Du·ro·cher** (də-rō′chər, -shər), **Leo Ernest.** b. 1906. Amer. baseball player & manager.
**Dur·rell** (dûr′əl), **Lawrence George.** b. 1912. English author.
**Du·ruy** (do͝or′rwē′), **(Jean) Victor.** 1811–94. French historian & statesman.
**Dur·yea** (do͝or′yā, -ē-ā′), **Charles Edgar** (1861–1938) & **James Frank** (1869–1967). Amer. automobile manufacturers.
**Du·se** (do͞o′zā), **Eleonora.** 1849–1924. Italian actress.
**Dus·tin** (dŭs′tĭn), **Hannah.** 1657–1736? Amer. colonial heroine.
**Du·tra** (do͞o′trə), **Eurico Gaspar.** 1885–1974. Brazilian military & political leader.
**Du·va·lier** (do͞o′väl-yā′), **François** (*Papa Doc;* 1907–71) & **Jean-Claude** (*Baby Doc;* b. 1951). Haitian political leaders.
**Du·vall** (do͞o-väl′), **Gabriel.** 1752–1844. Amer. jurist.
**Du·ve** (do͞o′və), **Christian Marie René Joseph de.** b. 1917. English-born Belgian physiologist (Nobel, 1974).
**du Vi·gneaud** (do͞o vēn′yō, dyo͞o), **Vincent.** 1901–78. Amer. biochemist.
**Dvoř·ák** (də-vôr′zhäk′), **Anton** *or* **Antonín.** 1841–1904. Czech composer.
**Dwig·gins** (dwĭg′ĭnz), **William Addison.** 1880–1956. Amer. type & book designer.
**Dwight** (dwīt), **John Sullivan.** 1813–93. Amer. music editor & critic.
**Dwight, Timothy.** 1752–1817. Amer. clergyman, author, & educator.
**Dwight, Timothy.** 1828–1916. Amer. clergyman & scholar.
**Dyce** (dīs), **Alexander.** 1798–1869. Scottish-born critic & Shakespearean scholar.
**Dy·er** (dī′ər), **John.** 1700?–58. Welsh-born English poet.
**Dyer, Mary.** d. 1660. English-born Amer. Quaker martyr.
**Dy·lan** (dĭl′ən), **Bob.** b. 1941. Amer. musician.
**Dy·ott** (dī′ət), **Thomas W.** 1771–1861. English-born Amer. patent-medicine manufacturer & social reformer.

# E

**Eads** (ēdz), **James Buchanan.** 1820–87. Amer. engineer.
**Ead·wine** *or* **Ead·win** (ĕd′wĭn′). *vars. of* EDWIN.
**Ea·gels** (ē′gəlz), **Jeanne.** 1890–29. Amer. actress.
**Ea·ker** (ā′kər), **Ira Clarence.** b. 1896. Amer. aviator.
**Ea·kins** (ā′kĭnz), **Thomas.** 1844–1916. Amer. artist & educator.
**Eames** (ēmz), **Charles.** 1907–78. Amer. designer.
**Ear·hart** (âr′härt′), **Amelia.** 1897–1937? Amer. aviator; lost on flight over the Pacific.
**Earle** (ûrl), **Alice Morse.** 1851–1911. Amer. antiquarian.
**Earle, Ralph.** 1751–1801. Amer. painter.
**Ear·ly** (ûr′lē), **Jubal Anderson.** 1816–94. Amer. Confederate soldier.
**Earp** (ûrp), **Wyatt.** 1848–1929. Amer. frontier law officer.
**East·man** (ēst′mən), **Charles Alexander.** 1858–1939. Amer. physician & author.
**Eastman, George.** 1854–1932. Amer. inventor, industrialist, & philanthropist.
**Eastman, Max Forrester.** 1883–1969. Amer. editor, author, & translator.
**Ea·ton** (ēt′n), **Cyrus Stephen.** 1883–1979. Amer. industrialist & financier.
**Eaton, Margaret O'Neale** *or* **O'Neill ("Peggy").** 1796–1879. American socialite.
**Eaton, Theophilus.** 1590–1658. English-born Amer. merchant & colonizer.
**Eaton, William.** 1764–1811. Amer. army officer & diplomat.
**E·ban** (ē′bən), **Abba.** b. 1915. South-African-born Israeli political leader.
**E·bert** (ā′bərt), **Friedrich.** 1871–1925. German statesman.
**E·berth** (ā′bərt), **Karl Joseph.** 1835–1926. German bacteriologist & pathologist.
**Ec·cles** (ĕk′əlz), Sir **John Carew.** b. 1903. Australian-born Amer. physiologist (Nobel, 1963).
**Eccles, Marriner Stoddard.** 1890–1977. Amer. economist & public official.
**E·che·ga·ray y Ei·za·guir·re** (ā′chə-gə-rī′ ē ā′sə-gîr′ā, -gwîr′-), **José.** 1832–1916. Spanish mathematician, statesman, & dramatist (Nobel, 1904).
**E·che·ver·rí·a Ál·va·rez** (ā′chə-və-rē′ə äl′və-rĕz′), **Luis.** b. 1922. Mexican statesman.
**Eck** (ĕk), **Johann.** 1486–1543. German theologian.
**Eck·hart** *also* **Eck·art** *or* **Eck·ardt** (ĕk′härt′), **Johannes.** 1260?–1327? German mystic & theologian.
**Ed·ding·ton** (ĕd′ĭng-tən), Sir **Arthur Stanley.** 1882–1944. English mathematician, astronomer, & physicist.
**Ed·dy** (ĕd′ē), **Mary (Morse) Baker.** 1821–1910. Amer. founder of Christian Science.
**Eddy, Nelson.** 1901–67. Amer. singer & actor.
**Ed·el·man** (ĕd′l-mən), **Gerald Maurice.** b. 1929. Amer. biochemist (Nobel, 1972).
**E·den** (ēd′n), Sir **(Robert) Anthony.** *Earl of Avon.* 1897–1977. British prime minister (1955–57).
**E·der·le** (ā′dər-lē), **Gertrude Caroline.** b. 1906. Amer. channel swimmer.
**Edge·worth** (ĕj′wûrth′), **Maria.** 1767–1849. English novelist.

**Ed·in·burgh** (ĕd'n-bûr'ə, -bûrg'), Duke of. Prince PHILIP.
**Ed·i·son** (ĕd'ĭ-sən), **Thomas Alva.** 1847–1931. Amer. inventor.
**Ed·mund II** (ĕd'mənd). 980?–1016. West Saxon king (1016).
**Ed·munds** (ĕd'məndz), **George Franklin.** 1828–1919. Amer. legislator.
**Ed·son** (ĕd'sən), **Katherine Phillips.** 1870–1933. Amer. reformer & public official.
**Ed·ward** (ĕd'wərd). *the Confessor.* 1002–66. West Saxon king (1042–66).
**Edward.** *Prince of Wales, the Black Prince.* 1330–76. English soldier.
**Edward.** Name of eight English kings: **a. I.** 1239–1307. Ruled 1272–1307. **b. II.** 1284–1327. Ruled 1307–27; murdered. **c. III.** 1312–77. Ruled 1327–77. **d. IV.** 1442–83. Ruled 1461–83. **e. V.** 1470–83. Ruled 1483; murdered in the Tower of London. **f. VI.** 1537–53. Ruled 1547–53. **g. VII.** 1841–1910. Ruled 1901–10; also Emperor of India. **h. VIII.** Later *Duke of Windsor.* 1894–1972. Ruled 1936; abdicated. **—Ed·ward'i·an** (ĕd-wôr'dē-ən, -wär'-) *adj. & n.*
**Ed·wards** (ĕd'wərdz), **Jonathan.** 1703–58. Amer. theologian & philosopher.
**Ed·win** or **Ead·wine** or **Ead·win** (ĕd'wĭn'). 585?–633. Northumbrian king (617–33).
**E·gas Mo·niz** (ā-gäs' mô-nēsh'), **Antonio de.** 1874–1955. Portuguese neurologist (Nobel, 1949).
**Eg·bert** (ĕg'bərt). 775?–839. West Saxon king (802–39); first overlord of all the English (829).
**Eg·gle·ston** (ĕg'əl-stən), **Edward.** 1837–1902. Amer. author.
**Eg·mont** (ĕg'mŏnt), Comte **Lamoral d'.** 1522–68. Flemish general & statesman.
**Eh·ren·burg** (ĕr'ən-bŏŏrg', -bŏŏrk'), **Ilya Grigorievich.** 1891–1967. Russian author.
**Ehr·lich** (âr'lĭkн), **Paul.** 1853–1915. German bacteriologist (Nobel, 1908).
**Eif·fel** (ī'fəl, ĭ-fĕl'), **Alexandre Gustave.** 1832–1923. French engineer.
**Ei·gen** (ī'gən), **Manfred.** b. 1927. German chemist (Nobel, 1967).
**Eijk·man** (īk'män', āk'-), **Christiaan.** 1858–1930. Dutch hygienist & pathologist (Nobel, 1929).
**Ein·stein** (īn'stīn'), **Albert.** 1879–1955. German-born Amer. theoretical physicist (Nobel, 1921).
**Eint·ho·ven** (īnt'hō'vən), **Willem.** 1860–1927. Dutch physiologist (Nobel, 1924).
**Ei·sen·how·er** (ī'zən-hou'ər), **Dwight David.** 1890–1969. 34th U.S. President (1953–61) & World War II commander.
**Ei·sen·stein** (ī'zən-stīn'), **Sergei Mikhailovich.** 1898–1948. Soviet filmmaker.
**El·don** (ĕl'dən), 1st Earl of. *John Scott.* 1751–1838. English jurist.
**El·ea·nor of Aq·ui·taine** (ĕl'ə-nər, -nôr'; ăk'wĭ-tān'). 1122?–1204. Queen of France & England.
**Eleanor of Cas·tile** (kə-stēl'). d. 1290. Queen of England as wife of Edward I.
**Eleanor of Pro·vence** (prō-väɴs'). d. 1291. Queen of England as wife of Henry III.
**El·gar** (ĕl'gär', -gər), Sir **Edward.** 1857–1934. English composer.
**El Grec·o** (ĕl grĕk'ō). —See GRECO.
**E·li** (ē'lī). Hebrew judge & teacher of Samuel.
**E·li·jah** (ĭ-lī'jə). 9th cent. B.C. Hebrew prophet.
**El·i·ot** (ĕl'ē-ət), **Charles William.** 1834–1926. Amer. educator & editor.
**Eliot, George.** *Mary Ann Evans.* 1819–80. English novelist.
**Eliot,** Sir **John.** 1592–1632. English statesman.
**Eliot, John.** 1604–90. English missionary in America.
**Eliot, Thomas Stearns ("T.S.").** 1888–1965. Amer.-born English critic & author (Nobel, 1948).
**E·li·sha** (ĭ-lī'shə). 9th cent. B.C. Hebrew prophet.
**E·liz·a·beth** (ĭ-lĭz'ə-bəth). Mother of John the Baptist.
**Elizabeth.** *Queen of Hearts.* 1596–1662. Queen of Bohemia.
**Elizabeth.** 1843–1916. Author & queen of Rumania (1881–1916).
**Elizabeth.** Name of five English queens, esp.: **a. I.** *the Virgin Queen.* 1533–1603. Ruled 1558–1603. **b.** b. 1900. Wife of George VI. **c. II.** b. 1926. Ruled since 1952. **—E·liz'a·be'than** *adj. & n.*
**Elizabeth Pe·trov·na** (pə-trôv'nə). 1709–62. Russian empress (1741–62).
**El·let** (ĕl'ĭt, -ĕt'), **Charles.** 1810–62. Amer. engineer.
**Ellet, Elizabeth Fries Lummis.** 1812?–77. Amer. author, critic, & translator.
**El·ling·ton** (ĕl'ĭng-tən), **Edward Kennedy ("Duke").** 1899–1974. Amer. jazz composer, pianist, & bandleader.
**El·li·ott** (ĕl'ē-ət), **Maxine.** 1871–1940. Amer. actress.
**Elliott, Sara Barnwell.** 1848–1928. Amer. author & suffragist.
**El·lis** (ĕl'ĭs), **Alexander John.** 1814–90. English philologist & mathematician.
**Ellis, (Henry) Havelock.** 1859–1939. English psychologist & author.
**El·li·son** (ĕl'ĭ-sən), **Ralph Waldo.** b. 1914. Amer. author.
**Ells·worth** (ĕlz'wûrth'), **Lincoln.** 1880–1951. Amer. engineer & explorer.
**Ellsworth, Oliver.** 1745–1807. Amer. jurist & political leader.
**El·man** (ĕl'mən), **Mischa.** 1891–1967. Russian-born Amer. violinist.
**El·phin·stone** (ĕl'fĭn-stōn', -stən), **Mountstuart.** 1779–1859. English colonial administrator in India.
**E·ly** (ē'lē), **Richard Theodore.** 1854–1943. Amer. economist.
**El·yot** (ĕl'yət, ĕl'ē-ət), Sir **Thomas.** 1490?–1546. English scholar & diplomat.
**El·y·tis** (ĕl'ē-tēs'), **Odysseus.** b. 1911. Greek poet (Nobel, 1979).
**Em·er·son** (ĕm'ər-sən), **Ralph Waldo.** 1803–82. Amer. author. **—Em'er·so'ni·an** (ĕm'ər-sō'nē-ən) *adj.*
**Em·met** (ĕm'ĭt), **Robert.** 1778–1803. Irish patriot.
**Em·ped·o·cles** (ĕm-pĕd'ə-klēz'). 5th cent. B.C. Greek philosopher.
**En·de·cott** *also* **En·di·cott** (ĕn'dĭ-kət, -kŏt'), **John.** 1589–1665. English-born Amer. colonial governor.
**En·ders** (ĕn'dərz), **John Franklin.** b. 1897. Amer. bacteriologist (Nobel, 1954).
**En·di·cott** (ĕn'dĭ-kət, -kŏt'). *var. of* ENDECOTT.
**E·nes·co** (ə-nĕs'kō), **Georges.** 1881–1955. Rumanian composer.
**En·gels** (ĕng'əlz, -əls), **Friedrich.** 1820–95. German socialist theorist & author.
**En·ver Pa·sha** (ĕn'vĕr pä'shä). 1881?–1922. Turkish soldier & politician.
**E·pam·i·non·das** (ĭ-păm'ə-nän'dəs). 418?–362 B.C. General & statesman of Thebes.
**Ep·ic·te·tus** (ĕp'ĭk-tē'təs). 1st–2nd cent. A.D. Greek philosopher.
**Ep·i·cu·rus** (ĕp'ĭ-kyŏŏr'əs). 342–270 B.C. Greek philosopher. **—Ep'i·cu·re'an** *adj. & n.*
**Ep·stein** (ĕp'stīn'), Sir **Jacob.** 1880–1959. Amer.-born sculptor.
**E·ras·mus** (ĭ-răz'məs), **Desiderius.** 1466?–1536. Dutch Renaissance scholar & theologian.
**E·ras·tus** (ĭ-răs'təs), **Thomas.** 1524–83. German-Swiss theologian & philosopher.
**E·ra·tos·the·nes** (ĕr'ə-tŏs'thə-nēz'). 3rd cent. B.C. Greek mathematician, astronomer, & geographer.
**Er·hard** (ĕr'härt'), **Ludwig.** 1897–1977. German statesman.
**Er·ic** (ĕr'ĭk). *the Red.* 10th cent. Norwegian navigator.
**Er·ic·son** *also* **Er·ics·son** (ĕr'ĭk-sən), **Leif.** fl. c. 1000. Norwegian navigator; discovered Vinland.
**Er·ics·son** (ĕr'ĭk-sən), **John.** 1803–89. Amer. engineer & inventor.
**E·rig·e·na** (ĭ-rĭj'ə-nə), **John Scotus.** 815?–77? Irish-born theologian & philosopher.
**Er·lang·er** (ûr'lăng'ər), **Joseph.** 1874–1965. Amer. physiologist (Nobel, 1944).
**Ernst** (ĕrnst), **Max.** 1891–1976. German-born Amer. surrealist.
**Er·skine** (ûr'skĭn'), **John.** 1509–91. Scottish religious reformer.
**Er·vin** (ûr'vĭn), **Samuel James, Jr.** 1896–1985. Amer. legislator.
**Er·vine** (ûr'vĭn), **St. John Greer.** 1833–1971. Irish author, theatrical manager, & educator.
**Erz·ber·ger** (ĕrts'bĕr'gər), **Matthias.** 1875–1921. German statesman; assassinated.
**Es·a·ki** (ə-sä'kē), **Leo.** b. 1925. Japanese-born physicist (Nobel, 1973).
**E·sar·had·don** (ē'sär-hăd'n). Assyrian king (681–669 B.C.).
**Es·cof·fier** (ĕs-kô-fyā'), **Auguste.** 1847–1935. French chef & author of cookery books.
**E·se·nin** (ĭs-ān'yən, yĭs-), **Sergei Alexsandrovich.** 1895–1925. Russian poet.
**Esh·kol** (ĕsh-kôl'), **Levi.** 1895–1969. Russian-born Israeli statesman.
**Es·par·te·ro** (ĕs'pər-tĕr'ō), **Baldomero.** 1792–1879. Spanish general & statesman.
**Es·po·si·to** (ĕs'pə-zē'tō), **Philip Anthony ("Phil").** b. 1942. Amer. hockey player.
**Es·py** (ĕs'pē), **James Pollard.** 1785–1860. Amer. pioneer meteorologist.
**Es·qui·vel** (ĕs'kē-vĕl'), **Adolfo Perez.** b. 1932. Argentine civil-rights activist (Nobel, 1980).
**Es·sex** (ĕs'ĭks), 2nd Earl of. Robert DEVEREUX.
**Es·taing** (ĕs-tăɴ'), Comte **Charles Hector d'.** 1729–94. French naval commander; guillotined.
**Es·te** (ĕs'tā). Italian princely family (996–1803), including **Isabella d'** (1474–1539), diplomat & patron of the arts.
**Es·ter·ha·zy** (ĕs'tər-hä'zē), **(Marie Charles) Ferdinand Walsin.** 1847–1923. French army officer & forger.
**Es·ther** (ĕs'tər). Persian queen in the Old Testament who saved the Jews from massacre.
**Es·tienne** (ĕs-tyĕn') *or* **É·tienne** (ĕ-tyĕn'). French family of printers, including **Henri** (1460?–1520), **Robert** (1503–59), & **Henri** (1528?–98).
**Es·tour·nelles de Con·stant** (ĕ-stŏŏr-nĕl' də kôn-stäɴ'), Baron **Constant de Rebecque d'.** *Paul Henri Benjamin Balluat.* 1852–1924. French diplomat & pacifist (Nobel, 1909).

**Es·trith** (ĕs′trəth). Danish ruling dynasty (1047–1375).

**Eth·el·bert** (ĕth′əl-bûrt′). 552?–616. Anglo-Saxon king, lawgiver, & Christian convert.

**Eth·el·red** *also* **Aeth·el·red II** (ĕth′əl-rĕd′). *the Unready.* 968?–1016. English king (978–1016).

**Eth·er·ege** (ĕth′ər-ĭj, ĕth′rĭj), Sir **George.** 1635?–91. English dramatist.

**É·tienne** (ĕ-tyĕn′). *var. of* ESTIENNE.

**Euck·en** (oi′kən), **Rudolf Christoph.** 1846–1926. German philosopher (Nobel, 1908).

**Eu·clid** (yo͞o′klĭd). 3rd cent. B.C. Greek mathematician & physicist. **—Eu·clid′e·an, Eu·clid′i·an** *adj.*

**Eu·gene** (yo͞o-jēn′, yo͞o′jēn′). *Prince of Savoy.* 1663–1736. Austrian general.

**Eu·gé·nie** (yo͞o′jə-nē, yo͞o-jē′-). 1826–1920. French empress as wife of Napoleon III.

**Eu·ler** (oi′lər), **Leonhard.** 1707–83. Swiss mathematician.

**Eu·ler-Chel·pin** (oi′lər-kĕl′pĭn), **Hans August Simon von.** 1873–1964. German-born Swedish chemist (Nobel, 1929).

**Eu·rip·i·des** (yo͞o-rĭp′ĭ-dēz′). 480?–406 B.C. Greek dramatist. **—Eu·rip′i·de′an** *adj.*

**Eu·se·bi·us of Caes·a·re·a** (yo͞o-sē′bē-əs; sĕs′ə-rē′ə, sĕz′-). 260?–340? Theologian & church historian.

**Eu·sta·chi·o** (yo͞o-stä′kē-ō), **Bartolommeo.** 1524?–74. Italian anatomist.

**Eu·stis** (yo͞o′stĭs), **Dorothy Leib Harrison Wood.** 1886–1946. Amer. philanthropist.

**E·vald** (ā′vält). *var. of* EWALD.

**Ev·ans** (ĕv′ənz), Sir **Arthur John.** 1851–1941. English archaeologist.

**Evans, Bergen.** 1904–78. Amer. author.

**Evans,** Dame **Edith.** 1888–1976. English-born actress.

**Evans, Elizabeth Glendower.** 1856–1937. Amer. suffragist.

**Evans, George Henry.** 1805–56. Amer. labor & agrarian reformer.

**Evans, Herbert McLean.** 1881–1971. Amer. anatomist & embryologist.

**Evans, John.** 1814–97. Amer. physician, businessman, & philanthropist.

**Evans, Mary Ann.** George ELIOT.

**Evans, Maurice.** b. 1901. English-born actor.

**Evans, Oliver.** 1755–1819. Amer. inventor & steam-engine manufacturer.

**Evans, Walker.** 1903–1975. Amer. photographer.

**Ev·arts** (ĕv′ərts), **William Maxwell.** 1818–1901. Amer. legislator & public official.

**Eve** (ēv). Adam's wife in the Old Testament.

**Eve·lyn** (ēv′lĭn, ēv′-), **John.** 1620–1706. English diarist.

**Ev·er·ett** (ĕv′ər-ĭt, ĕv′rĭt), **Edward.** 1794–1865. Amer. clergyman, orator, educator, & diplomat.

**Ev·ers** (ĕv′ərz), **Charles** (b. 1923) & **Medgar Wiley** (1925–63; assassinated). Amer. civil-rights leaders.

**Ev·ert Lloyd** (ĕv′ərt loid), **Christine Marie.** b. 1954. Amer. tennis player.

**E·wald** *or* **E·vald** (ā′vält), **Johannes.** 1743–81. Danish lyric poet.

**Ew·ell** (yo͞o′əl), **Richard Stoddert.** 1817–72. Amer. Confederate general.

**Eyck** (īk). —See VAN EYCK.

**Ez·e·ki·as** (ĕz′ĭ-kī′əs). HEZEKIAH.

**E·ze·ki·el** (ĭ-zē′kē-əl). 6th cent. B.C. Hebrew prophet.

**Ezekiel, Moses Jacob.** 1844–1917. Amer. artist & musician.

**Ez·ra** (ĕz′rə). 5th cent. B.C. Hebrew scribe & priest.

# F

**Fa·ber** (fā′bər, fä′-), **John Eberhard.** 1822–79. German-born Amer. manufacturer.

**Fa·ber·gé** (făb′ər-zhā′), **Peter Carl.** 1846–1920. Russian designer & jeweler.

**Fa·bi·o·la** (făb-ē-ō′lə). b. 1928. Belgian queen as wife of Baudouin I.

**Fa·bi·us Max·i·mus Ver·ru·co·sus** (fā′bē-əs măk′sə-məs vĕr-yo͞o-kō′səs), **Quintus.** d. 203 B.C. Roman general. **—Fa′bi·an** (fā′bē-ən) *adj. & n.*

**Fa·bre** (fä′brə), **Jean Henri.** 1823–1915. French entomologist.

**Fad·den** (făd′n), Sir **Arthur William.** 1895–1973. Australian statesman.

**Fad·i·man** (făd′ə-mən), **Clifton Paul.** b. 1904. Amer. author & editor.

**Fahd** (făd). *Fahd ibn Abdel Aziz al-Saud.* b. 1922. King of Saudi Arabia (since 1982).

**Fah·ren·heit** (făr′ən-hīt′), **Gabriel Daniel.** 1686–1736. German-born physicist.

**Fair·banks** (fâr′băngks′), **Charles Warren.** 1852–1918. U.S. Vice President (1905–9).

**Fairbanks, Douglas** (1883–1939) & **Douglas Elton, Jr.** (b. 1909). Amer. actors.

**Fair·child** (fâr′chīld′), **Mary Salome Cutler.** 1855–1921. Amer. pioneer librarian.

**Fair·fax** (fâr′făks′), **Thomas.** *3rd Baron Fairfax of Cameron.* 1612–71. English Civil War soldier.

**Fairfax, Thomas.** *6th Baron Fairfax.* 1692–1782. English colonist in America.

**Fai·sal** *also* **Fei·sal** *or* **Fei·sul** (fī′səl). Name of two kings of Iraq: **a. I.** 1885–1933. Ruled 1921–33. **b. II.** 1935–58. Ruled 1939–58; assassinated.

**Faisal** *also* **Feisal** *or* **Feisul.** *Faisal Ibn Abdel Aziz al-Saud.* 1906?–75. Saudi Arabian king (1964–75); assassinated.

**Falk·ner** (fôk′nər). *var. of* FAULKNER.

**Fall** (fôl), **Albert Bacon.** 1861–1944. Amer. politician.

**Fal·lières** (făl-yĕr′), **(Clément) Armand.** 1841–1931. French statesman.

**Fan·euil** (făn′l, făn′yəl), **Peter.** 1700–43. Amer. merchant.

**Far·a·day** (făr′ə-dā′), **Michael.** 1791–1867. English physicist & chemist.

**Far·go** (fär′gō), **William George.** 1818–81. Amer. transportation pioneer.

**Far·ley** (fär′lē), **Harriet.** 1817–1907. Amer. author & editor.

**Farley, James Aloysius.** 1888–1976. Amer. politician.

**Far·man** (fär′mən, fär-mäN′), **Henri** (1873–1934) & **Maurice** (1877–1964). French-born aviation pioneers.

**Far·mer** (fär′mər), **Fannie Merritt.** 1857–1915. Amer. cookbook author.

**Farmer, James Leonard.** b. 1920. Amer. civil-rights leader.

**Farmer, Moses Gerrish.** 1820–93. Amer. electrical pioneer.

**Far·ne·se** (fär-nä′zē), **Alessandro.** 1545–92. Italian general & diplomat.

**Farns·worth** (färnz′wûrth′), **Philo Taylor.** 1906–71. Amer. radio & television pioneer.

**Fa·rouk I** *also* **Fa·ruk I** (fə-ro͞ok′). 1920–65. Egyptian king (1936–52); abdicated.

**Far·quhar** (fär′kwər), **George.** 1678–1707. Irish-born playwright.

**Far·ra·gut** (făr′ə-gət), **David Glasgow.** 1801–70. Amer. admiral.

**Far·rar** (făr′ər), **Frederic William.** 1831–1903. English theologian, educator, & author.

**Far·rar** (fə-rär′), **Geraldine.** 1882–1967. Amer. soprano.

**Far·rar** (făr′ər), **Margaret Petherbridge.** b. 1897. Amer. crossword-puzzle editor.

**Far·rell** (făr′əl), **Eileen.** b. 1920. Amer. soprano.

**Farrell, James Thomas.** 1904–79. Amer. author.

**Farrell, Suzanne.** b. 1945. Amer. ballerina.

**Fa·ruk** (fə-ro͞ok′). *var. of* FAROUK.

**Fass·bind·er** (fäs′bĭn′dər), **Rainer Werner.** 1946–82. German filmmaker.

**Fat·i·ma** (făt′ə-mə). 606–32. Daughter of Mohammed.

**Fat·i·mid** (făt′ə-mĭd′) *also* **Fat·i·mite** (-mīt′). Moslem dynasty of North Africa & Egypt (909–1171).

**Fau·chard** (fō-shär′), **Pierre.** 1678–1761. French dentistry pioneer.

**Faulk·ner** *also* **Falk·ner** (fôk′nər), **William.** 1897–1962. Amer. author (Nobel, 1949).

**Faure** (fôr, fōr), **François Félix.** 1841–99. French statesman.

**Fau·ré** (fō-rā′), **Gabriel Urbain.** 1845–1924. French composer.

**Faust** (foust). *var. of* FUST.

**Fawkes** (fôks), **Guy.** 1570–1606. English conspirator; executed.

**Fech·ner** (fĕk′nər, fĕKH′-), **Gustav Theodor.** 1801–87. German psychologist & physicist.

**Feif·fer** (fī′fər), **Jules.** b. 1929. Amer. cartoonist & playwright.

**Fei·ning·er** (fī′nĭng-ər), **Lyonel Charles Adrian.** 1871–1956. Amer.-born artist.

**Fei·sal** *or* **Fei·sul** (fī′səl). *vars. of* FAISAL.

**Feke** (fēk), **Robert.** 1705?–50? Amer. portrait painter.

**Fel·li·ni** (fə-lē′nē), **Federico.** b. 1920. Italian filmmaker.

**Fell·tham** *also* **Fel·tham** (fĕl′thəm), **Owen.** 1602?–68. English author.

**Fel·ton** (fĕl′tən), **Rebecca Ann Latimer.** 1835–1930. Amer. author, reformer, & legislator.

**Fé·ne·lon** (fā-nə-lôN′), **François de Salignac de la Mothe.** 1651–1715. French prelate & author.

**Feng Yu·xiang** *or* **Feng Yu-hsiang** (fŭng′ yü′shē-äng′). 1880–1948. Chinese military leader.

**Fer·ber** (fûr′bər), **Edna.** 1887–1968. Amer. author.

**Fer·di·nand** (fûr′dn-ănd′). Name of five kings of Castile & León, esp.: **a. I.** *the Great.* d. 1065. Ruled 1037–65. **b. II.** *the Saint.* 1199–1252. Ruled 1230–52. **c. V.** *the Catholic.* 1452–1516. Ruled 1474–1504 with Isabella; also ruled Aragon (1479–1516) as **Ferdinand II** & Naples (1504–16) as **Ferdinand III.**

**Ferdinand.** Name of three Holy Roman Emperors: **a. I.** 1503–64. Ruled 1556–64; also king of Bohemia & Hungary (1526–64). **b. II.** 1578–1637. Ruled 1619–37; also king of Bohemia (1617–19, 1620–37) and Hungary (1618–37). **c. III.** 1608–57. Ruled 1637–57; also king of Hungary (1625–57).

ă **pat** ā **pay** âr **care** ä **father** ĕ **pet** ē **be** hw **which** ĭ **pit**
ī **tie** îr **pier** ŏ **pot** ō **toe** ô **paw, for** oi **noise** o͝o **took**

**Ferdinand.** Name of two Bourbon kings of Spain: **a. VI.** *the Wise.* 1712–59. Ruled 1746–59. **b. VII.** 1784–1833. Ruled 1808 & 1814–33. **c. I.** 1861–1948. Bulgarian king (1908–18).
**Fer·mat** (fĕr-mä'), **Pierre de.** 1601–65. French mathematician.
**Fer·mi** (fĕr'mē), **Enrico.** 1901–54. Italian-born Amer. physicist (Nobel, 1938).
**Fer·nán·dez** (fər-năn'dĕz'), **Juan.** 1536?–1602? Spanish navigator.
**Fer·now** (fûr'nō), **Bernhard Edward.** 1851–1923. Prussian-born Amer. forestry pioneer.
**Fer·ris** (fĕr'ĭs), **George Washington Gale.** 1859–96. Amer. engineer & inventor.
**Fes·sen·den** (fĕs'ən-dən), **Reginald Aubrey.** 1866–1932. Canadian-born inventor & radio pioneer.
**Fessenden, William Pitt.** 1806–69. Amer. politician & financier.
**Fes·tus** (fĕs'təs), **Porcius.** fl. c. 60 A.D. Roman procurator of Judea.
**Feucht·wang·er** (foikt'väng'ər, foiKHt'-), **Lion.** 1884–1958. German-born author.
**Feu·er·bach** (foi'ər-bäKH'), **Ludwig Andreas von.** 1804–72. German philosopher.
**Feuil·let** (fœ-yā'), **Octave.** 1821–90. French author.
**Fewkes** (fyo͞oks), **Jesse Walter.** 1850–1930. Amer. ethnologist & zoologist.
**Feyn·man** (fīn'mən), **Richard Phillips.** b. 1918. Amer. physicist (Nobel, 1965).
**Fi·bi·ger** (fē'bē-gər), **Johannes Andreas Grib.** 1867–1928. Danish pathologist (Nobel, 1926).
**Fich·te** (fĭk'tə, fĭKH'-), **Johann Gottlieb.** 1762–1814. German philosopher.
**Fied·ler** (fēd'lər), **Arthur.** 1894–1979. Amer. musical conductor.
**Field** (fēld), **Cyrus West.** 1819–92. Amer. merchant & financier.
**Field, David Dudley.** 1805–94. Amer. jurist.
**Field, Eugene.** 1850–95. Amer. author.
**Field, Marshall.** 1834–1906. Amer. merchant & philanthropist.
**Field, Stephen Johnson.** 1816–99. Amer. jurist.
**Field·ing** (fēl'dĭng), **Henry.** 1707–54. English author.
**Fields** (fēldz), **Gracie.** 1898–1979. English comedienne.
**Fields, W.C.** 1880–1946. Amer. entertainer.
**Fie·so·le** (fyĕ'zō-lā), **Giovanni Angelica da.** Fra ANGELICO.
**Fi·lene** (fĭ-lēn', fī-), **Edward Albert.** 1860–1937. Amer. merchant.
**Fill·more** (fĭl'môr', -mōr'), **Millard.** 1800–74. 13th U.S. President (1850–53).
**Fil·son** (fĭl'sən), **John.** 1747–88. Amer. explorer & historian.
**Fink** (fĭngk), **Albert.** 1827–97. German-born Amer. railroad engineer.
**Fink, Mike.** 1770?–1822. Amer. frontiersman.
**Fin·lay** (fĭn'lā, fēn-lī'), **Carlos Juan.** 1833–1915. Cuban-born Amer. physician.
**Fin·ney** (fĭn'ē), **Charles Grandison.** 1792–1876. Amer. religious leader & educator.
**Fin·sen** (fĭn'sən), **Niels Ryberg.** 1860–1904. Danish physician (Nobel, 1903).
**Fir·bank** (fûr'băngk'), **Ronald.** 1886–1926. English author.
**Fir·dau·si** (fĭr-dou'sē) *also* **Fir·du·si** (fər-do͞o'-). *Abul Kasim* or *Qasim Mansur.* 940?–1020? Persian epic poet.
**Fire·stone** (fīr'stōn'), **Harvey Samuel.** 1868–1938. Amer. industrialist.
**Fi·scher** (fĭsh'ər), **Emil.** 1852–1919. German chemist (Nobel, 1902).
**Fischer, Ernst Otto.** b. 1918. German chemist (Nobel, 1973).
**Fischer, Hans.** 1881–1945. German chemist (Nobel, 1930).
**Fischer, Robert James ("Bobby").** b. 1943. Amer. chess player.
**Fish** (fĭsh), **Hamilton.** 1808–93. Amer. politician & public official.
**Fish·bein** (fĭsh'bīn'), **Morris.** 1889–1976. Amer. physician, author, & editor.
**Fish·er** (fĭsh'ər), **Andrew.** 1862–1928. Scottish-born Australian statesman.
**Fisher, Clara.** 1811–98. English-born Amer. actress.
**Fisher, Dorothy Canfield.** 1879–1958. Amer. author.
**Fisher, Herbert Albert Laurens.** 1865–1940. English historian, educator, & public official.
**Fisher, Irving.** 1867–1947. Amer. economist.
**Fisher, John Arbuthnot.** *1st Baron Fisher of Kilverstone.* 1841–1920. British naval officer.
**Fisk** (fĭsk), **James.** 1834–72. Amer. railroad financier & speculator.
**Fiske** (fĭsk), **Haley.** 1852–1929. Amer. insurance innovator.
**Fiske, John.** 1842–1901. Amer. historian & philosopher.
**Fiske, Minnie Maddern.** 1865–1932. Amer. actress.
**Fitch** (fĭch), **Asa.** 1809–79. Amer. entomologist.
**Fitch, (William) Clyde.** 1865–1909. Amer. playwright.
**Fitch, John.** 1743–98. Amer. inventor & steamboat pioneer.
**Fitch, Val L.** b. 1923. Amer. physicist (Nobel, 1980).
**Fitz** (fĭts), **Reginald Heber.** 1843–1913. Amer. pathology pioneer.
**Fitz·ger·ald** (fĭts-jĕr'əld), **Ella.** b. 1918. Amer. singer.
**Fitzgerald, F(rancis) Scott (Key).** 1896–1940. Amer. author.

o͞o **boot** ou **out** th **thin** *th* **th**is ŭ **cut** ûr **urge** y **y**oung
yo͞o ab**u**se zh vi**s**ion ə **a**bout, it**e**m, ed**i**ble, gall**o**p, circ**u**s

**Fitz·Ger·ald** (fĭts-jĕr'əld), **Edward.** 1809–83. English poet & translator.
**FitzGerald, George Francis.** 1851–1901. Irish physicist.
**Fitz·gib·bon** (fĭts-gĭb'ən), Sister **Irene.** 1823–96. English-born Amer. hospital pioneer.
**Fitz·her·bert** (fĭts-hûr'bərt), **Marie Anne Smythe.** 1756–1837. First wife of George IV of England; marriage declared invalid.
**Fitz·hugh** (fĭts-hyo͞o'), **George.** 1806–81. Lawyer & pro-slavery advocate.
**Fitz·pat·rick** (fĭts-păt'rĭk), **Thomas.** 1799?–1854. Irish-born Amer. trapper, guide, & Indian agent.
**Fitz·sim·mons** (fĭt-sĭm'ənz), **Robert Prometheus.** 1862–1917. English-born, New Zealand-raised Amer. prizefighter.
**Flagg** (flăg), **James Montgomery.** 1877–1960. Amer. artist & writer.
**Flag·ler** (flăg'lər), **Henry Morrison.** 1830–1913. Amer. capitalist & promoter.
**Flag·stad** (flăg'stăd', flăg'stä'), **Kirsten Marie.** 1895–1962. Norwegian-born operatic soprano.
**Fla·her·ty** (flă'ər-tē), **Robert Joseph.** 1884–1951. Amer. explorer, author, & filmmaker.
**Fla·min·i·us** (flə-mĭn'ē-əs), **Gaius.** d. 217. Roman general & politician.
**Flam·ma·rion** (flə-măr'ē-ŏn'), **Camille.** French astronomer.
**Flan·a·gan** (flăn'ə-gən), **Edward Joseph.** 1886–1948. Amer. clergyman & founder of Boys Town.
**Flan·ner** (flăn'ər), **Janet.** *Genêt.* 1892–1978. Amer. journalist in Paris.
**Flau·bert** (flō-bâr'), **Gustave.** 1821–80. French author. **—Flau·ber'tian** (-shən, -tē-ən) *adj.*
**Flax·man** (flăks'mən), **John.** 1755–1826. English sculptor.
**Flem·ing** (flĕm'ĭng), Sir **Alexander.** 1881–1955. British bacteriologist (Nobel, 1945).
**Fleming, Ian Lancaster.** 1908–64. English author.
**Fleming,** Sir **John Ambrose.** 1849–1945. English electrical engineer & inventor.
**Fleming, Peggy Gale.** b. 1948. Amer. figure skater.
**Fleming, Williamina Paton Stevens.** 1857–1911. Scottish-born Amer. astronomer.
**Fletch·er** (flĕch'ər), **Alice Cunningham.** 1838–1923. Cuban-born Amer. ethnologist.
**Fletcher, John.** 1579–1625. English dramatist.
**Fleu·ry** (flœ-rē'), **André Hercule de.** 1653–1743. French prelate & statesman.
**Flex·ner** (flĕks'nər), **Abraham.** 1866–1959. Amer. educator.
**Flin·ders** (flĭn'dərz), Sir **Matthew.** 1774–1814. English explorer.
**Flint** (flĭnt), **Austin** (1812–86) & **Austin** (1836–1915). Amer. physicians.
**Flint, Timothy.** 1780–1840. Amer. missionary & author.
**Flo·res** (flō'rās, flō'-), **Juan José.** 1800–64. Ecuadorian general & statesman.
**Flo·rey** (flôr'ē, flŏr'ē), Sir **Howard Walter.** 1898–1968. Australian-born English pathologist (Nobel, 1945).
**Flo·ri·o** (flôr'ē-ō', flŏr'-), **John.** 1553?–1625. English lexicographer.
**Flo·ry** (flôr'ē, flŏr'ē), **Paul John.** b. 1910. Amer. chemist (Nobel, 1974).
**Flynn** (flĭn), **Elizabeth Gurley.** 1890–1964. Amer. political radical.
**Flynn, Errol.** 1909–59. Amer. actor.
**Foch** (fôsh, fŏsh), **Ferdinand.** 1851–1929. French army commander.
**Fock·e** (fôk'ə), **Heinrich.** 1890–1980. German aircraft designer & manufacturer.
**Fo·gar·ty** (fō'gər-tē), **Anne.** 1920?–80. Amer. fashion designer.
**Fo·kine** (fô-kēn'), **Michel.** 1880–1942. Russian-born Amer. choreographer.
**Fok·ker** (fŏk'ər, fô'kər), **Anthony Herman Gerard.** 1890–1939. Dutch aircraft designer & manufacturer.
**Fo·ley** (fō'lē), **John Henry.** 1818–74. Irish sculptor.
**Fol·ger** (fōl'jər), **Henry Clay.** 1857–1930. Amer. capitalist & bibliophile.
**Fol·lett** (fŏl'ət), **Mary Parker.** 1868–1933. Amer. author & labor-management counselor.
**Fon·da** (fŏn'də), **Henry.** 1905–82. Amer. actor.
**Fon·tanne** (fŏn-tăn'), **Lynn.** 1887?–1983. English-born Amer. actress.
**Fon·teyn** (fŏn-tān'), Dame **Margot.** b. 1919. English ballerina.
**Foote** (fo͝ot), **Andrew Hull.** 1806–63. Amer. naval officer.
**Foote, Mary Hallock.** 1847–1938. Amer. author & illustrator.
**Foote, Samuel.** 1720–77. English actor & dramatist.
**Forbes** (fôrbz), **Malcolm Stevenson.** b. 1919. Amer. publisher & sportsman.
**Forbes-Rob·ert·son** (fôrbz-rŏb'ərt-sən), Sir **Johnston.** 1853–1937. English actor.
**Force** (fôrs), **Peter.** 1790–1868. Amer. historian & printer.
**Ford** (fôrd, fōrd), **Edward ("Whitey").** b. 1928. Amer. baseball player.
**Ford, Ford Madox.** 1873–1939. English author.

**Ford, Gerald Rudolph.** b. 1913. 38th U.S. President (1974–77).
**Ford, Henry** (1863–1947) & **Henry, Jr.** (b. 1917). Amer. automobile manufacturers.
**Ford, John.** 1586–1639. English dramatist.
**Ford, John.** 1895–1973. Amer. filmmaker.
**Ford, Paul Leicester.** 1865–1902. Amer. author & historian.
**For·es·ter** (fôr′ĭ-stər, fŏr′-), **Cecil Scott ("C. S.").** 1899–1966. English author.
**For·rest** (fôr′ĭst, fŏr′-), **Edwin.** 1806–72. Amer. actor.
**Forrest, Nathan Bedford.** 1821–77. Amer. Confederate general.
**For·res·tal** (fôr′ĭ-stəl, -stôl′, fŏr′-), **James Vincent.** 1892–1949. Amer. banker & public official.
**Forss·mann** (fôrs′män′, -mən, fŏrs′-), **Werner Theodor Otto.** 1904–79. German physician (Nobel, 1956).
**For·ster** (fôr′stər), **Edward Morgan ("E.M.").** 1879–1970. English author.
**For·syth** (fôr-sīth′, fər-, fôr′sīth), **John.** 1780–1841. Amer. politician & diplomat.
**For·tas** (fôr′təs), **Abraham ("Abe").** 1910–82. Amer. jurist; resigned from Supreme Court.
**For·ten** (fôr′tn), **James.** 1766–1842. Amer. businessman & reformer.
**For·tes·cue** (fôr′tĭs-kyo͞o′), Sir **John.** 1394?–1476? English jurist.
**Fos·dick** (fŏz′dĭk), **Harry Emerson.** 1878–1969. Amer. religious leader.
**Fos·se** (fŏs′ē), **Robert Louis ("Bob").** b. 1927. Amer. choreographer & director.
**Fos·ter** (fô′stər, fŏs′tər), **Abigail Kelley ("Abby").** 1810–87. Amer. abolitionist & suffragist.
**Foster, Hannah Webster.** 1759–1840. Amer. author.
**Foster, John Watson.** 1836–1917. Amer. diplomat.
**Foster, Stephen Collins.** 1826–64. Amer. songwriter.
**Foster, William Zebulon.** 1881–1961. Amer. labor leader & radical politician.
**Fou·cault** (fo͞o-kō′), **Jean Bernard Léon.** 1819–68. French physicist.
**Fou·quet** *also* **Fouc·quet** (fo͞o-kā′), **Jean.** 1415?–80. French artist.
**Fouquet** *also* **Foucquet, Nicolas.** *Marquis de Belle-Isle.* 1615–80. French financier & public official.
**Four·dri·nier** (fo͝or-drĭn′ē-ər, fôr-, fōr-), **Henry** (1766–1854) & **Sealy** (d. 1847). English papermakers & inventors.
**Fou·rier** (fo͝or′ē-ā′, fo͞o-ryā′), **François Marie Charles.** 1772–1837. French socialist author.
**Fourier,** Baron **Jean Baptiste Joseph.** 1768–1830. French mathematician & physicist.
**Fow·ler** (fou′lər), **Henry Watson.** 1858–1933. English lexicographer.
**Fowler, William Alfred.** b. 1911. Amer. nuclear physicist (Nobel, 1983).
**Fox** (fŏks), **Charles James.** 1749–1806. English politician & orator.
**Fox, Dixon Ryan.** 1887–1945. Amer. historian.
**Fox, George.** 1624–91. English Quaker religious leader.
**Fox, Gustavus Vasa.** 1821–83. Amer. naval officer.
**Fox, Henry.** *1st Baron Holland.* 1705–74. English statesman.
**Fox, John William, Jr.** 1863–1919. Amer. author.
**Fox, Margaret.** 1833–93. Amer. spiritualist medium.
**Fox, Richard Kyle.** 1846–1922. Irish-born Amer. journalist.
**Foxe** (fŏks), **John.** 1516–87. English martyrologist.
**Foxe** or **Fox** (fŏks), **Richard.** 1448?–1528. English prelate.
**Foy** (foi), **Eddie.** 1856–1928. Amer. entertainer.
**Fra·go·nard** (frăg′ə-när′), **Jean Honoré.** 1732–1806. French artist.
**France** (frăns, fräɴs), **Anatole.** 1844–1924. French author (Nobel, 1921).
**Fran·ce·sca** (frăn-chĕs′kə, frän-), **Piero della.** 1420?–92. Italian painter.
**Francesca da Ri·mi·ni** (də rĭm′ĭ-nē, rē′mə-). d. 1285? Italian noblewoman.
**Fran·cis** (frăn′sĭs). Name of two French kings, esp. **I,** 1494–1547, ruled 1515–47.
**Francis.** Name of two Holy Roman Emperors, esp. **II,** 1768–1835, ruled 1792–1806; also Austrian emperor (1804–35) as **Francis I.**
**Francis of As·si·si** (ə-sē′zē, -sē, ə-sĭs′ē), Saint. 1182?–1226. Italian monk & founder of the Franciscan order. **—Fran·cis′can** (frăn-sĭs′kən) *adj.* & *n.*
**Francis of Sales** (sālz), Saint. 1567–1622. French ecclesiastic.
**Francis Fer·di·nand** (fûr′dn-ănd′). 1863–1914. Austrian archduke; assassinated.
**Francis Jo·seph I** (jō′zəf, -səf). 1830–1916. Austrian emperor (1848–1916).
**Francis Xa·vi·er** (zā′vē-ər, zăv′ē-), Saint. —See XAVIER.
**Franck** (frängk, fräɴk), **César Auguste.** 1822–90. French organist & composer.
**Franck** (frängk), **James.** 1882–1964. German-born Amer. physicist (Nobel, 1925).
**Franck·e** (frăng′kə), **Kuno.** 1855–1930. German-born Amer. historian & educator.
**Fran·co** (frăng′kō, fräng′-), **Francisco.** *El Caudillo.* 1892–1975. Spanish soldier & dictator (1939–75).
**Frank** (frăngk, frängk), **Anne.** 1929–45. Dutch Jewish diarist.
**Frank** (frăngk), **Glenn.** 1887–1940. Amer. editor & educator.
**Frank** (frängk, frăngk), **Ilya Mikhailovich.** b. 1908. Russian physicist (Nobel, 1958).
**Frank·en·thal·er** (frăng′kən-thô′lər, -thŏl′ər), **Helen.** b. 1928. Amer. artist.
**Frank·furt·er** (frăngk′fər-tər), **Felix.** 1882–1965. Austrian-born Amer. jurist.
**Frank·lin** (frăngk′lĭn), **Aretha.** b. 1942. Amer. singer.
**Franklin, Benjamin.** 1706–90. Amer. statesman, diplomat, author, scientist, & printer.
**Franklin,** Sir **John.** 1786–1847. English Arctic explorer.
**Franklin, John Hope.** b. 1915. Amer. historian.
**Franks** (frăngks), Baron **Oliver Shewell.** b. 1905. English educator and diplomat.
**Franz Jo·sef I** (fränts jō′səf, yō′zəf). FRANCIS JOSEPH I.
**Fra·ser** (frā′zər), **James Earle.** 1876–1953. Amer. sculptor.
**Fraser, (John) Malcolm.** b. 1930. Australian statesman.
**Fraser, Peter.** 1884–1950. Scottish-born New Zealand statesman.
**Fraser, Simon.** *12th Baron Lovat.* 1667?–1747. Scottish Jacobite; beheaded.
**Fraser, Simon.** 1776?–1862. Amer.-born Canadian explorer & fur trader.
**Fra·zer** (frā′zər), Sir **James George.** 1854–1941. Scottish anthropologist.
**Fra·zier** (frā′zhər), **Edward Franklin.** 1894–1962. Amer. sociologist.
**Frazier, Joe.** b. 1944. Amer. prizefighter.
**Frazier, Walt.** b. 1945. Amer. basketball player.
**Fred·er·ick** (frĕd′rĭk, -ər-ĭk). Name of nine Danish kings, esp. **IX,** 1899–1972, ruled 1947–72.
**Frederick.** Name of six kings of Denmark & Norway, esp.: **a. II.** 1534–88. Ruled 1559–88. **b. III.** 1609–70. Ruled 1648–70. **c. IV.** 1671–1730. Ruled 1699–1730. **d. V.** 1723–66. Ruled 1746–66.
**Frederick.** Name of three Holy Roman Emperors: **a. I.** *Frederick Barbarossa.* 1123?–90. Ruled 1152–90 (crowned 1155); also king of Germany (1152–90) & Italy (1155–90). **b. II.** 1194–1250. Ruled 1215–50 (crowned 1220); also king of Sicily (1198–1250) as **Frederick I. c. III.** 1415–93. Ruled 1440–93 (crowned 1452); also king of Germany as **Frederick IV.**
**Frederick.** Name of three Prussian kings: **a. I.** 1657–1713. Ruled 1701–13; also elector of Brandenburg (1688–1701) as **Frederick III. b. II.** *the Great.* 1712–86. Ruled 1740–86. **c. III.** 1831–88. Ruled 1888.
**Frederick Wil·liam** (wĭl′yəm). *the Great Elector.* 1620–88. Elector of Brandenburg (1640–88).
**Frederick William.** Name of four Prussian kings: **a. I.** 1688–1740. Ruled 1713–40. **b. II.** 1744–97. Ruled 1786–97. **c. III.** 1770–1840. Ruled 1797–1840. **d. IV.** 1795–1861. Ruled 1840–61.
**Free·man** (frē′mən), **Douglas Southall.** 1886–1953. Amer. historian & editor.
**Freeman, Mary Eleanor Wilkins.** 1852–1930. Amer. author.
**Fre·ling·huy·sen** (frē′lĭng-hī′zən), **Frederick Theodore.** 1817–85. Amer. legislator & public official.
**Fré·mont** (frē′mŏnt′), **John Charles.** 1813–90. Amer. soldier, explorer, & politician.
**French** (frĕnch), **Daniel Chester.** 1850–1931. Amer. sculptor.
**Fre·neau** (frĭ-nō′), **Philip Morin.** 1752–1832. Amer. author.
**Fres·co·bal·di** (frĕs′kə-bäl′dē), **Girolamo.** 1583–1643. Italian composer.
**Fres·nel** (frā-nĕl′), **Augustin Jean.** 1788–1827. French physicist.
**Freud** (froid), **Anna.** 1895–1982. Austrian-born British psychoanalyst.
**Freud, Sigmund.** 1856–1939. Austrian physician & pioneer psychoanalyst. **—Freud′i·an** *adj.* & *n.*
**Frey·berg** (frī′bûrg′), 1st Baron. *Bernard Cyril Freyberg.* 1890–1963. English-born New Zealand soldier & statesman.
**Frey·tag** (frī′täk′, -täg′), **Gustav.** 1816–95. German author.
**Frick** (frĭk), **Henry Clay.** 1849–1919. Amer. industrialist & art patron.
**Fried** (frēd, frēt), **Alfred Herman.** 1864–1921. Austrian pacifist (Nobel, 1911).
**Frie·dan** (frĭ-dăn′), **Betty Naomi Goldstein.** b. 1921. Amer. feminist.
**Fried·man** (frēd′mən), **Esther Pauline** (*Ann Landers*) and **Pauline Esther** (*Abigail Van Buren, Dear Abby*). b. 1918. Amer. advice columnists.
**Friedman, Milton.** b. 1912. Amer. economist (Nobel, 1976).
**Friet·chie** (frĭch′ē), **Barbara Hauer.** 1766–1862. Amer. Civil War heroine.
**Friml** (frĭm′əl), **(Charles) Rudolf.** 1879–1972. Amer. pianist & composer.

**Frisch** (frĭsh), **Karl von.** b. 1886. Austrian-born German zoologist (Nobel, 1973).
**Frisch, Ragnar.** 1895–1973. Norwegian economist (Nobel, 1969).
**Frö·bel** (frœ'bəl). *var. of* FROEBEL.
**Fro·bish·er** (frō'bĭsh-ər), Sir **Martin.** 1535?–94. English navigator.
**Froe·bel** *also* **Frö·bel** (frœ'bəl), **Friedrich Wilhelm August.** 1782–1852. German educator.
**Froh·man** (frō'mən), **Charles.** 1860–1915. Amer. theatrical manager.
**Frois·sart** (froi'särt', frwä-sär'), **Jean.** 1333?–1400. French historian.
**Fromm** (frŏm, frōm), **Erich.** 1900–80. German-born Amer. psychoanalyst.
**Fron·te·nac** (frŏn'tə-năk', frôNt-näk'), Comte **Louis de Buade de.** 1622–98. French colonial administrator.
**Frost** (frôst, frŏst), **Robert Lee.** 1874–1963. Amer. poet.
**Froth·ing·ham** (frŏ*th*'ĭng-hăm', -əm), **Octavius Brooks.** 1822–95. Amer. clergyman & author.
**Froude** (fro͞od), **James Anthony.** 1818–94. English historian.
**Fry** (frī), **Roger Eliot.** 1866–1934. English artist & critic.
**Fu·ad I** (fo͞o-äd'). 1868–1936. Egyptian king (1922–36).
**Fu·en·tes** (fo͞o-ĕn'tās'), **Carlos.** b. 1928. Mexican author.
**Fuer·tes** (fyo͝or'tēz, fyo͞o'ər-), **Louis Agassiz.** 1874–1927. Amer. naturalist & artist.
**Fug·ger** (fo͝og'ər). Family of German bankers, including **Johannes** (1348–1409), **Andreas** (d. 1457), **Jakob I** (d. 1469), **Ulrich** (1441–1510), **Georg** (1453–1506), **Jakob II** (*the Rich;* 1459–1525), **Raymund** (1489–1535), **Anton** (1493–1560), **Hans Jakob** (1516–79), **Ulrich** (1526–84), & **Georg** (d. 1569).
**Fu·ku·i** (fo͞o'ko͞o-ē'), **Kenichi.** b. 1918. Japanese chemist (Nobel, 1981).
**Ful·bright** (fo͝ol'brīt'), **J(ames) William.** b. 1905. Amer. legislator & author.
**Ful·da** (fo͝ol'də), **Ludwig.** 1862–1939. German playwright.
**Ful·ler** (fo͝ol'ər), **Alfred Carl.** 1885–1973. Amer. businessman.
**Fuller, (Richard) Buckminster.** 1895–1983. Amer. architect & inventor.
**Fuller, George.** 1822–84. Amer. painter.
**Fuller, Loie.** 1862–1928. Amer. dancer.
**Fuller, (Sara) Margaret.** 1810–50. Amer. author, critic, & reformer.
**Fuller, Melville Weston.** 1833–1910. Amer. jurist.
**Fuller, Sarah.** 1836–1927. Amer. educator.
**Fuller, Thomas.** 1608–61. English clergyman.
**Ful·ton** (fo͝ol'tn), **Robert.** 1765–1815. Amer. artist, engineer, & inventor.
**Funk** (fŭngk, fo͝ongk), **Casimir.** 1884–1967. Polish-born Amer. biochemist.
**Funk** (fŭngk), **Isaac Kauffman.** 1839–1912. Amer. clergyman, editor & publisher.
**Fun·ston** (fŭn'stən), **Frederick.** 1865–1917. Amer. botanist, explorer, & soldier
**Fur·ness** (fər-nĕs'), **Elizabeth ("Betty").** b. 1916. Amer. consumer advocate.
**Fur·ness** (fûr'nĭs), **Horace Howard** (1833–1912) & **Horace Howard** (1865–1930). Amer. Shakespearean scholars.
**Fur·ni·vall** (fûr'nə-vəl), **Frederick James.** 1825–1910. English scholar & editor.
**Furt·wäng·ler** (fo͝ort'vĕng'lər), **Wilhelm.** 1886–1954. German conductor.
**Fu·sel·i** (fyo͞o'zə-lē'), **Henry.** 1741–1825. Swiss-born English artist.
**Fust** (fo͞ost) *also* **Faust** (foust), **Johann.** 1400?–66? German printer.

# G

**Ga·ble** (gā'bəl), **(William) Clark.** 1901–60. Amer. actor.
**Ga·bo** (gä'bō), **Naum.** 1890–1977. Russian-born Amer. sculptor & designer.
**Ga·bor** (gä'bôr, gə-bôr'), **Dennis.** 1900–79. Hungarian-born British physicist (Nobel, 1971).
**Ga·bo·riau** (gə-bôr'ē-ō), **Émile.** 1835–73. French novelist.
**Gad·da·fi** (gə-dä'fē), **Muammar.** —See QADDAFI.
**Gads·den** (gădz'dən), **James.** 1788–1858. Amer. diplomat, politician, & railroad promoter.
**Gad·ski** (găt'skē), **Johanna.** 1872–1932. German operatic soprano.
**Gág** (gäg), **Wanda Hazel.** 1893–1946. Amer. author & illustrator.
**Ga·ga·rin** (gə-gär'ĭn), **Yuri Alekseyevich.** 1934–68. Soviet cosmonaut; first man in space.
**Gage** (gāj), **Matilda Joslyn.** 1826–98. Amer. feminist.
**Gage, Thomas.** 1721–87. British general & colonial administrator.
**Gail·lard** (gĭl-yärd'), **David Du Bose.** 1859–1913. Amer. army engineer.
**Gaines** (gānz), **Edmund Pendleton.** 1777–1849. Amer. army officer.
**Gains·bor·ough** (gānz'bûr'ō, -bər-ə), **Thomas.** 1727–88. English portrait & landscape painter.
**Gai·ser·ic** (gī'zə-rĭk'). *var. of* GENSERIC.
**Gait·skell** (gāt'skəl), **Hugh Todd Naylor.** 1906–63. British politician.
**Ga·ius** (gā'əs, gī'-). 2nd cent. A.D. Roman jurist.
**Gaj·du·sek** (gī'də-shĕk'), **D(aniel) Carleton.** b. 1923. Amer. virologist (Nobel, 1976).
**Gal·ba** (găl'bə, gôl'-), **Servius Sulpicius.** 5 B.C.?–A.D. 69. Roman emperor (68–69).
**Gal·braith** (găl'brāth'), **John Kenneth.** b. 1908. Canadian-born Amer. economist, diplomat, & author.
**Gale** (gāl), **Zona.** 1874–1938. Amer. novelist & playwright.
**Ga·len** (gā'lən). 130?–201? Greek anatomist, physician, & author.
**Ga·le·ri·us** (gə-lîr'ē-əs). d. 311. Roman emperor (305–11).
**Ga·li·le·o Ga·li·lei** (găl'ə-lē'ō găl'ə-lā'ē', -lā'ō). 1564–1642. Italian astronomer & physicist. **—Gal'i·le'an** *adj.*
**Gal·la·tin** (găl'ə-tĭn), **(Abraham Alfonso) Albert.** 1761–1849. Swiss-born Amer. financier & politician.
**Gal·lau·det** (găl'ə-dĕt'), **Thomas Hopkins.** 1787–1851. Amer. educator.
**Ga·lle·gos Fre·ire** (gä-yā'gōs frā'rā), **Rómulo.** 1884–1969. Venezuelan author.
**Gal·li·co** (găl'ĭ-kō'), **Paul.** 1897–1976. Amer. author.
**Gal·li-Cur·ci** (găl'ĭ-ko͝or'chē), **Amelita.** 1899–1963. Italian operatic soprano.
**Gal·lié·ni** (găl'yā-nē', găl-yā'nē), **Joseph Simon.** 1849–1916. French army officer & colonial administrator.
**Gal·li·e·nus** (găl'ē-ē'nəs, -ā'nəs), **Publius Licinius Valerianus.** d. 268. Roman emperor (253–68).
**Gal·lo·way** (găl'ə-wā'), **Joseph.** 1731?–1803. English Loyalist in America.
**Gal·lup** (găl'əp), **George Horace.** 1901–84. Amer. public-opinion analyst.
**Ga·lois** (găl-wä'), **Évariste.** 1811–32. French mathematician; killed in duel.
**Gals·wor·thy** (gălz'wûr'*th*ē), **John.** 1867–1933. English author (Nobel, 1932).
**Gal·ton** (gôl'tən), Sir **Francis.** 1822–1911. English eugenicist.
**Gal·va·ni** (găl-vä'nē, găl-), **Luigi** *or* **Aloisio.** 1737–98. Italian physicist & physician.
**Gál·vez** (găl'vĕs'), **José de.** 1729–87. Spanish jurist & colonial administrator.
**Gal·way** (gôl'wā), **James.** b. 1939. Irish-born British flutist.
**Ga·ma** (găm'ə, gä'mə), **Vasco da.** 1469?–1524. Portuguese explorer & colonial administrator.
**Ga·mar·ra** (gə-mär'ə), **Agustín.** 1785–1841. Peruvian general & statesman.
**Gam·bet·ta** (găm-bĕt'ə), **Léon.** 1838–82. French political leader.
**Ga·me·lin** (găm-lăN', găm'ə-), **Maurice Gustave.** 1872–1958. French army officer.
**Ga·mow** (gă'mou), **George.** 1904–68. Russian-born Amer. nuclear physicist.
**Gan·dhi** (gän'dē, găn'-), **Indira Nehru.** 1917–84. Indian political leader; assassinated.
**Gandhi, Mohandas Karamchand ("Mahatma").** 1869–1948. Indian nationalist & spiritual leader; assassinated.
**Gan·nett** (găn'ĭt), **Henry.** 1846–1914. Amer. cartographer.
**Gar·a·mond** (găr'ə-mŏnd'), **Claude.** d. 1561. French type designer.
**Ga·rand** (gə-rănd', găr'ənd), **John Cantius.** 1888–1974. Canadian-born Amer. firearms designer.
**Gar·bo** (gär'bō), **Greta.** b. 1905. Swedish-born Amer. actress.
**Gar·cí·a Gu·tiér·rez** (gär-sē'ə go͞o-tyĕr'əs), **Antonio.** 1813–84. Spanish author.
**García Lor·ca** (lôr'kä), **Federico.** 1899–1936. Spanish author.
**García Már·quez** (mär'kəs), **Gabriel.** b. 1928. Colombian-born author (Nobel, 1982).
**García Mo·re·no** (mə-rā'nō), **Gabriel.** 1821–75. Ecuadorian political leader; assassinated.
**García Ro·bles** (rō'bləs), **Alfonso.** b. 1911. Mexican diplomat (Nobel, 1982).
**García y Íñi·guez** (ē ē'nyē-gĕs'), **Calixto.** 1836?–98. Cuban military & political leader.
**Gar·ci·la·so de la Ve·ga** (gär'sə-lä'sō dā lə vā'gə). *El Inca.* 1539?–1616. Peruvian soldier, historian, & translator.
**Gar·den** (gär'dn), **Alexander.** 1730?–91. Scottish-born naturalist & physician.
**Garden, Mary.** 1874?–1967. Scottish-born operatic soprano.
**Gar·den·er** (gär'dn-ər, gärd'nər), **Helen Hamilton.** 1853–1925. Amer. suffragist.
**Gar·di·ner** (gärd'nər, gär'dn-ər), **Samuel Rawson.** 1829–1902. English historian, educator, & editor.
**Gardiner, Stephen.** 1483?–1555. English religious & political leader.

**Gard·ner** (gärd′nər), **Erle Stanley.** 1889–1970. Amer. lawyer & detective novelist.
**Gardner, Isabella Stewart.** 1840–1924. Amer. socialite & art collector.
**Gar·field** (gär′fēld′), **James Abram.** 1831–81. 20th U.S. President (1881); assassinated.
**Garfield, John.** 1913–52. Amer. actor.
**Gar·i·bal·di** (găr′ə-bôl′dē), **Giuseppe.** 1807–82. Italian general & nationalist leader.
**Gar·land** (gär′lənd), **(Hannibal) Hamlin.** 1860–1940. Amer. author.
**Garland, Judy.** 1922–69. Amer. actress & singer.
**Gar·ner** (gär′nər), **John Nance.** 1868–1967. U.S. Vice President (1933–41).
**Gar·nett** (gär′nĭt), **Constance Black.** 1862–1946. English translator.
**Gar·rick** (găr′ĭk), **David.** 1717–79. English actor.
**Gar·ri·son** (găr′ĭ-sən), **Mabel.** 1886–1963. Amer. operatic soprano.
**Garrison, William Lloyd.** 1805–79. Amer. abolitionist.
**Gar·ro·way** (găr′ə-wā′), **Dave.** 1913–82. Amer. radio & television host.
**Gar·vey** (gär′vē), **Marcus (Moziah) Aurelius.** 1887–1940. Jamaican black nationalist active in America; deported.
**Gar·y** (gâr′ē), **Elbert Henry.** 1846–1927. Amer. lawyer & financier.
**Gas·coigne** (găs′koin′), **George.** 1535?–77. English author.
**Gas·kell** (găs′kəl), **Elizabeth Cleghorn Stevenson.** 1810–65. English novelist.
**Gas·ser** (găs′ər), **Herbert Spencer.** 1888–1963. Amer. physiologist (Nobel, 1944).
**Gates** (gāts), **Frederick Taylor.** 1853–1929. Amer. clergyman & philanthropist.
**Gates, Horatio.** 1727?–1806. Amer. Revolutionary general.
**Gates, John Warne ("Bet-you-a-million").** 1855–1911. Amer. speculator & promoter.
**Gat·ling** (găt′lĭng), **Richard Jordan.** 1818–1903. Amer. firearms inventor.
**Gau·dí** (gou′dē), **Antonio.** 1852–1926. Spanish architect.
**Gau·guin** (gō-găN′), **(Eugène Henri) Paul.** 1848–1903. French painter.
**Gauss** (gous), **Karl Friedrich.** 1777–1855. German mathematician & astronomer.
**Gau·tier** (gō-tyā′), **Théophile.** 1811–72. French author.
**Gay** (gā), **John.** 1685–1732. English poet.
**Gay-Lus·sac** (gā′lə-săk′), **Joseph Louis.** 1778–1850. French chemist & physicist.
**Gay·nor** (gā′nər), **Janet.** 1906–84. Amer. actress.
**Gea·ry** (gîr′ē), **John White.** 1819–73. Amer. general & politician.
**Ge·ber** (jē′bər, gā′-). 721–66. Arab scholar & alchemist.
**Ged·des** (gĕd′ēz), **Norman Bel.** 1893–1958. Amer. architect & theatrical & industrial designer.
**Geh·rig** (gĕr′ĭg), **Henry Louis ("Lou").** 1903–42. Amer. baseball player.
**Gei·kie** (gē′kē), Sir **Archibald.** 1835–1924. Scottish geologist.
**Gei·sel** (gī′zəl), **Theodor Seuss.** *Dr. Seuss.* b. 1904. Amer. author & illustrator.
**Gell-Mann** (gĕl′män′), **Murray.** b. 1929. Amer. physicist (Nobel, 1969).
**Ge·net** (zhə-nā′), **Jean.** b. 1910. French author.
**Ge·nêt** (zhə-nā′). Janet FLANNER.
**Genêt, Edmond Charles Edouard.** 1763–1834. French diplomat.
**Gen·ghis Khan** (jĕng′gĭs kän′, gĕng′-). 1162?–1227. Mongol conqueror.
**Gen·ser·ic** (jĕn′sə-rĭk′, gĕn′-) *also* **Gai·seric** (gī′zə-). d. 477. Vandal king (428–77).
**Gen·ti·le da Fa·bri·a·no** (jĕn-tē′lē də fä′brē-ä′nō). 1370?–1427? Italian painter.
**Geof·frey of Mon·mouth** (jĕf′rē; mŏn′məth). 1100?–54? English prelate & chronicler.
**George** (jôrj), Saint. d. c. 303. Christian martyr.
**George.** Name of six kings of Great Britain: **a. I.** 1660–1727. Ruled 1714–27. **b. II.** 1683–1760. Ruled 1727–60. **c. III.** 1738–1820. Ruled 1760–1820. **d. IV.** 1762–1830. Ruled 1820–30. **e. V.** 1865–1936. Ruled 1910–36. **f. VI.** 1895–1952. Ruled 1936–1952. **—Geor′gian** (jôr′jən) *adj. & n.*
**George.** Name of two kings of Greece: **a. I.** 1845–1913. Ruled 1863–1913; assassinated. **b. II.** 1890–1947. Ruled 1922–23, 1935–47.
**George, Henry.** 1839–97. Amer. journalist & reformer.
**George, Stefan.** 1868–1933. German poet.
**Ge·rard** (jə-rärd′, jĕr′ärd′), **Charles.** *1st Baron Gerard of Brandon, 1st Earl of Macclesfield.* 1618?–94. English Royalist commander.
**Ge·rard** (jə-rärd′), **James Watson.** 1867–1951. Amer. diplomat.
**Gé·ri·cault** (zhā-rē-kō′), **Jean Louis André Théodore.** 1791–1824. French painter.
**Ger·man·i·cus Cae·sar** (jər-măn′ĭ-kəs sē′zər). 15 B.C.–A.D. 19. Roman general.
**Gé·rôme** (zhā-rōm′), **Jean Léon.** 1824–1904. French painter.
**Ge·ron·i·mo** (jə-rŏn′ə-mō′). 1829–1909. Amer. Apache leader.
**Ger·ould** (jĕr′əld), **Katherine Elizabeth Fullerton.** 1879–1944. Amer. author.
**Ger·ry** (gĕr′ē), **Elbridge.** 1744–1814. U.S. Vice President (1813–14); died in office.
**Gersh·win** (gûrsh′wĭn), **George.** 1898–1937. Amer. composer.
**Gershwin, Ira.** 1896–1983. Amer. lyricist.
**Ge·sell** (gĭ-zĕl′), **Arnold Lucius.** 1880–1961. Amer. physiologist & pediatrician.
**Ges·ner** (gĕs′nər), **Konrad von.** 1516–65. Swiss encyclopedist & naturalist.
**Get·ty** (gĕt′ē), **George Washington.** 1819–1901. Amer. Union general.
**Getty, J(ean) Paul.** 1892–1976. Amer. oilman.
**Ghi·ber·ti** (gē-bĕr′tē), **Lorenzo.** 1378–1455. Florentine sculptor.
**Ghir·lan·da·jo** *also* **Ghir·lan·da·io** (gēr-lən-dä′yō), **Domenico.** 1449–94. Florentine painter.
**Gia·co·met·ti** (jä-kə-mĕt′ē), **Alberto.** 1901–66. Swiss sculptor & painter.
**Giae·ver** (yā′vər), **Ivar.** b. 1929. Norwegian-born Amer. physicist (Nobel, 1973).
**Gi·an·nin·i** (jē′ə-nĭn′ē, -nē′nē), **Amadeo Peter.** 1870–1949. Amer. banker.
**Gi·auque** (jē-ōk′), **William Francis.** 1895–1982. Canadian-born Amer. chemist (Nobel, 1949).
**Gib·bon** (gĭb′ən), **Edward.** 1737–94. English historian.
**Gib·bons** (gĭb′ənz), **Abigail Hopper.** 1801–93. Amer. social reformer & philanthropist.
**Gibbs** (gĭbz), **Josiah Willard.** 1839–1903. Amer. mathematician & physicist.
**Gibbs, Oliver Wolcott.** 1822–1908. Amer. chemist.
**Gibbs,** Sir **Philip.** 1877–1962. English author & editor.
**Gib·ran** (jə-brän′), **(Gibran) Kahlil.** 1883–1931. Syrian-born Amer. mystic poet & painter.
**Gib·son** (gĭb′sən), **Althea.** b. 1927. Amer. tennis player.
**Gibson, Charles Dana.** 1867–1944. Amer. illustrator.
**Gid·dings** (gĭd′ĭngz), **Joshua Reed.** 1795–1864. Amer. abolitionist & politician.
**Gide** (zhēd), **André.** 1869–1951. French author (Nobel, 1947).
**Gid·e·on** (gĭd′ē-ən). Hebrew judge & hero in the Old Testament.
**Giel·gud** (gĭl′gŏŏd′, gēl′-), Sir **(Arthur) John.** b. 1904. English actor & director.
**Gil·bert** (gĭl′bərt), **Anne Jane Hartley.** 1821–1904. English-born Amer. actress.
**Gilbert, Cass.** 1859–1934. Amer. architect.
**Gilbert, Grove Karl.** 1843–1918. Amer. geologist & surveyor.
**Gilbert, Henry Franklin Belknap.** 1868–1928. Amer. composer.
**Gilbert,** Sir **Humphrey.** 1539?–83. English navigator & soldier; drowned at sea.
**Gilbert, John.** 1897–1936. Amer. actor.
**Gilbert, Walter.** b. 1932. Amer. biologist (Nobel, 1980).
**Gilbert, William.** 1540–1603. English court physician.
**Gilbert,** Sir **William Schwenck.** 1836–1911. English playwright & lyricist.
**Gil·der** (gĭl′dər), **Richard Watson** (1844–1909) & **Jeannette Leonard** (1849–1916). Amer. authors & editors.
**Gil·der·sleeve** (gĭl′dər-slēv′), **Basil Lanneau.** 1831–1924. Amer. philologist.
**Gill** (gĭl), **Theodore Nicholas.** 1837–1914. Amer. zoologist.
**Gil·lett** (jə-lĕt′), **Frederick Huntington.** 1851–1935. Amer. legislator.
**Gil·lette** (jə-lĕt′), **King Camp.** 1855–1932. Amer. inventor & manufacturer.
**Gillette, William Hooker.** 1855?–1937. Amer. actor & playwright.
**Gil·liss** (gĭl′ĭs), **James Melville.** 1811–65. Amer. naval officer & astronomer.
**Gil·man** (gĭl′mən), **Arthur.** 1837–1909. Amer. educator.
**Gilman, Caroline Howard.** 1794–1888. Amer. author.
**Gilman, Charlotte Anna Perkins Stetson.** 1860–1935. Amer. reformer & author.
**Gilman, Daniel Coit.** 1831–1908. Amer. educator.
**Gil·mer** (gĭl′mər), **Elizabeth Meriwether.** *Dorothy Dix.* 1870–1951. Amer. journalist.
**Gil·pin** (gĭl′pĭn), **Charles Sidney.** 1878–1930. Amer. actor.
**Gins·berg** (gĭnz′bərg), **Allen.** b. 1926. Amer. poet.
**Giorgio·ne II** (jôr-jō′nē). 1478?–1511. Venetian painter.
**Giot·to** (jô′tō, jŏt′ō). 1266?–1337. Florentine painter, architect, & sculptor.
**Gi·rard** (zhē-rär′), **Jean Baptiste.** 1765?–1850. Swiss educator.
**Gi·rard** (jə-rärd′), **Stephen.** 1750–1831. French-born Amer. financier & philanthropist.
**Gi·raud** (zhē-rō′), **Henri Honoré.** 1879–1949. French military officer & politician.

**Gi·rau·doux** (zhē-rō-do͞o'), **Jean.** 1882–1944. French author.
**Gir·tin** (gûr'tn), **Thomas.** 1775–1802. English landscape painter.
**Gis·card d'Es·taing** (zhĭ-skär' dĕs-tăng', -täN'), **Valéry.** b. 1926. French political leader.
**Gish** (gĭsh), **Lillian Diana** (b. 1896) & **Dorothy** (1898–1968). Amer. actresses.
**Gis·sing** (gĭs'ĭng), **George Robert.** 1857–1903. English novelist & critic.
**Gist** (gĭst), **Christopher.** 1706?–59. Amer. frontier explorer.
**Gi·ven·chy** (gə-vĭn'chē, zhē-väN-shē'), **Hubert.** b. 1927. French fashion designer.
**Gjel·le·rup** (gĕl'ə-ro͝op'), **Karl.** 1857–1919. Danish author (Nobel, 1917).
**Glack·ens** (glăk'ənz), **William James.** 1870–1938. Amer. artist.
**Glad·den** (glăd'n), **Washington.** 1836–1918. Amer. clergyman & author.
**Glad·stone** (glăd'stōn'), **William Ewart.** 1809–98. British prime minister (four times between 1868 & 1894).
**Gla·ser** (glā'zər), **Donald Arthur.** b. 1926. Amer. physicist (Nobel, 1960).
**Glas·gow** (glăs'kō', -gō', glăz'-), **Ellen Anderson Gholson.** 1874–1945. Amer. novelist.
**Glash·ow** (glăsh'ō), **Sheldon Lee.** b. 1932. Amer. physicist (Nobel, 1979).
**Glas·pell** (glăs'pĕl'), **Susan Keating.** 1882–1948. Amer. author.
**Glass** (glăs), **Carter.** 1858–1946. Amer. legislator.
**Glass, Hugh.** d. 1833. Amer. frontiersman.
**Gla·zu·nov** (glăz'ə-nôf', -nôv'), **Aleksandr Konstantinovich.** 1865–1936. Russian composer.
**Glea·son** (glē'sən), **Herbert John ("Jackie").** b. 1916. Amer. entertainer.
**Glen·dow·er** (glĕn'dou'ər, glĕn-dou'-), **Owen.** 1359?–1416? Welsh rebel.
**Glenn** (glĕn), **John Herschel, Jr.** b. 1921. Amer. astronaut & legislator; first Amer. to orbit in space.
**Glid·den** (glĭd'n), **Charles Jasper.** 1857–1927. Amer. telephone-industry pioneer & sportsman.
**Glidden, Joseph Farwell.** 1813–1906. Amer. farmer, inventor, & manufacturer.
**Glin·ka** (glĭng'kə), **Mikhail Ivanovich.** 1803–57. Russian composer.
**Glov·er** (glŭv'ər), **John.** 1732–97. Amer. Revolutionary general.
**Glover, Sarah Ann.** 1785–1867. English music teacher.
**Gluck** (glo͝ok), **Alma.** 1884?–1938. Rumanian-born Amer. operatic soprano.
**Gluck, Christoph Willibald.** 1714–87. German composer.
**Glyn** (glĭn), **Elinor (Sutherland).** 1864?–1943. English novelist.
**Go·bat** (gō-bä'), **Charles Albert.** 1843–1914. Swiss statesman (Nobel, 1902).
**Go·dard** (gō-där'), **Jean Luc.** b. 1930. French filmmaker.
**God·dard** (gŏd'ərd), **Mary Katherine.** 1738–1816. Amer. printer & publisher.
**Goddard, Robert Hutchings.** 1882–1945. Amer. physicist & rocket pioneer.
**Go·dey** (gō'dē), **Louis Antoine.** 1804–78. Amer. publisher.
**God·frey** (gŏd'frē), **Arthur Michael.** 1903–83. Amer. entertainer.
**Godfrey, Thomas.** 1704–49. Amer. inventor & mathematician.
**Godfrey, Thomas.** 1736–63. Amer. author.
**Godfrey of Bouil·lon** (bo͞o-yôN'). 1061?–1100. French Crusade leader.
**God·kin** (gŏd'kĭn), **Edwin Lawrence.** 1831–1902. Irish-born Amer. journalist.
**Go·dol·phin** (gə-dŏl'fĭn), **Sidney.** *1st Earl of Godolphin.* 1645–1712. English statesman & financier.
**Go·doy** (gō-doi'), **Manuel de.** 1767–1851. Spanish statesman.
**Go·du·nov** (gō'də-nôf', go͝od'ə-, gŏd'ə-), **Boris Fëdorovich.** 1552–1605. Russian czar (1598–1605).
**God·win** (gŏd'wĭn), **William.** 1756–1836. English author & political philosopher.
**God·win-Aus·ten** (gŏd'wĭn-ô'stən), **Henry Haversham.** 1834–1923. English geologist, explorer, & author.
**Goeb·bels** (gœ'bəls), **Joseph Paul.** 1897–1945. German Nazi propaganda minister.
**Goe·ring** (gœ'rĭng, gĕr'ĭng). *var. of* GÖRING.
**Goes** (go͞os), **Hugo van der.** 1440?–82? Flemish painter.
**Goe·thals** (gō'thəlz), **George Washington.** 1858–1928. Amer. army engineer & public official.
**Goe·the** (gœ'tə), **Johann Wolfgang von.** 1749–1832. German poet & dramatist.
**Gogh** (gō, gôKH, KHôKH), **Vincent van.** —See VAN GOGH.
**Go·gol** (gô'gəl, gō'gôl), **Nikolai Vasilievich.** 1809–52. Russian author.
**Gold·berg** (gōld'bərg), **Arthur Joseph.** b. 1908. Amer. jurist and diplomat.
**Goldberg, Reuben Lucius ("Rube").** 1883–1970. Amer. cartoonist.
**Gold·berg·er** (gōld'bər-gər), **Joseph.** 1874–1929. Austrian-born Amer. physician.
**Gol·den** (gōl'dən), **Harry Lewis.** 1902–81. Amer. journalist.
**Gol·den·wei·ser** (gōl'dən-wī'zər, -vī'-), **Alexander.** 1880–1940. Russian-born Amer. educator, anthropologist, & sociologist.
**Gold·ing** (gōl'dĭng), **William Gerald.** b. 1911. English novelist (Nobel, 1983).
**Gold·man** (gōld'mən), **Emma.** 1869–1940. Russian-born Amer. anarchist.
**Gold·mark** (gōld'märk'), **Josephine Clara.** 1877–1950. Amer. investigator of social conditions.
**Gol·do·ni** (gŏl-dō'nē), **Carlo.** 1707–93. Italian dramatist.
**Gold·smith** (gōld'smĭth'), **Oliver.** 1728–74. Irish author.
**Gold·wa·ter** (gōld'wô'tər, -wŏt'ər), **Barry Morris.** b. 1909. Amer. politician.
**Gold·wyn** (gōld'wĭn), **Samuel.** 1882–1974. Polish-born Amer. film producer.
**Gol·gi** (gôl'jē), **Camillo.** 1844–1926. Italian histologist (Nobel, 1906).
**Go·li·ath** (gə-lī'əth). Philistine giant slain by David in the Bible.
**Gó·mez** (gō'mĕz), **Juan Vicente.** 1857?–1935. Venezuelan general & political leader.
**Gom·pers** (gŏm'pərz), **Samuel.** 1850–1924. English-born Amer. labor leader.
**Go·mul·ka** (gō-mo͝ol'kə, -mŭl'-), **Wladyslaw.** 1905–82. Polish political leader.
**Gon·çal·ves Di·as** (gən-säl'vəs dē'əs), **Antônio.** 1823–64. Brazilian poet.
**Gon·court** (gôN-ko͝or'), **Edmond Louis Antoine de** (1822–96) & **Jules Alfred Huot** (1830–70). French authors.
**Gon·do·mar** (gŏn'də-mär'), Count of. *Diego Sarmiento de Acuña.* 1567–1626. Spanish diplomat.
**Gon·za·ga** (gən-zä'gə, -zăg'ə, gän-), Saint **Aloysius.** 1568–91. Italian Jesuit priest.
**Gon·za·les** (gən-zä'lĭs), **Richard Alonzo ("Pancho").** b. 1928. Amer. tennis player.
**Gon·zá·lez** (gən-zä'ləs), **Manuel.** 1833–93. Mexican statesman.
**Gon·za·lo de Cór·do·ba** (gən-zä'lō də kôr'də-bə), **Hernández.** 1453–1515. Spanish general.
**Good·hue** (go͝od'hyo͞o'), **Bertram Grosvenor.** 1869–1924. Amer. architect.
**Good·man** (go͝od'mən), **Benjamin David ("Benny").** b. 1909. Amer. musician & conductor.
**Goodman, Paul.** 1911–72. Amer. author.
**Good·night** (go͝od'nīt'), **Charles.** 1836–1929. Amer. cattleman.
**Good·rich** (go͝od'rĭch'), **Samuel Griswold.** *Peter Parley.* 1793–1860. Amer. author & publisher.
**Good·ridge** (go͝od'rĭj'), **Sarah.** 1788–1853. Amer. miniature painter.
**Good·year** (go͝od'yîr'), **Charles.** 1800–60. Amer. inventor & manufacturer.
**Goo·la·gong Caw·ley** (go͞o'lə-gông kô'lē), **Evonne.** b. 1951. Australian-born tennis player.
**Gor·cha·kov** (gôr'chə-kôf', -kôv'), Prince **Aleksandr Mikhailovich.** 1798–1883. Russian diplomat.
**Gor·din** (gôr'dn), **Jacob.** 1853–1909. Russian-born Amer. Yiddish playwright.
**Gor·don** (gôr'dn), **Charles George.** *Chinese Gordon.* 1833–85. English army officer & colonial administrator.
**Gordon, Charles William.** 1860–1937. Canadian clergyman & missionary.
**Gordon,** Lord **George.** 1751–93. English rebel.
**Gordon, Laura de Force.** 1838–1907. Amer. lawyer & suffragist.
**Go·ren** (gôr'ən), **Charles Henry.** b. 1901. Amer. contract bridge expert.
**Gor·gas** (gôr'gəs), **Josiah.** 1818–83. Amer. Confederate soldier & educator.
**Gorgas, William Crawford.** 1854–1920. Amer. army surgeon.
**Gor·ham** (gôr'əm), **Nathaniel.** 1738–96. Amer. Revolutionary politician & businessman.
**Gö·ring** *also* **Goe·ring** (gœ'rĭng, gĕr'ĭng), **Hermann Wilhelm.** 1893–1946. German Nazi politician.
**Gor·ki** *also* **Gor·ky** (gôr'kē), **Maksim** *also* **Maxim.** 1868–1936. Russian author.
**Gor·ky** (gôr'kē), **Arshile.** 1905?–48. Armenian-born Amer. painter.
**Gor·rie** (gôr'ē), **John.** 1803–55. Amer. physician & inventor.
**Gor·ton** (gôr'tn), **John Grey.** b. 1911. Australian political leader.
**Gos·nold** (gŏz'nōld'), **Bartholomew.** d. 1607. English navigator & Jamestown colonist.
**Gosse** (gŏs), Sir **Edmund William.** 1848–1928. English poet & critic.
**Gott·schalk** (gŏch'ôk', gŏt'shôk'), **Louis Moreau.** 1829–69. Amer. composer & pianist.
**Gou·dy** (gou'dē), **Frederic William.** 1865–1947. Amer. printer & type designer.

**Gould** (go͝old), **Chester.** 1900–85. Amer. cartoonist.
**Gould, Glenn.** 1932–82. Canadian pianist.
**Gould, Jay.** 1836–92. Amer. financier & speculator.
**Gou·nod** (go͞o'nō, go͞o-nō'), **Charles François.** 1818–93. French composer.
**Gour·mont** (go͝or-mô**n**'), **Remy de.** 1858–1915. French novelist & critic.
**Gow·er** (gou'ər, gôr', gōr'), **John.** 1325?–1408. English poet.
**Go·ya y Lu·ci·en·tes** (goi'ə ē lo͞o-sē-ĕn'tēs), **Francisco José de.** 1746–1828. Spanish painter.
**Grac·chus** (grăk'əs), **Tiberius Sempronius** (163–133 B.C.) & **Gaius Sempronius** (153–121 B.C.). *the Gracchi.* Roman statesmen.
**Grace** (grās), Princess. *Grace Patricia Kelly.* 1929–82. Amer. actress & princess of Monaco as wife of Rainier III.
**Grace, William Russell.** 1832–1904. Irish-born Amer. financier, industrialist, & shipping magnate.
**Gra·dy** (grā'dē), **Henry Woodfin.** 1850–89. Amer. editor & orator.
**Gra·ham** (grā'əm), **John.** *1st Viscount Dundee.* 1649?–89. Scottish Jacobite leader.
**Graham, Katherine Meyer.** b. 1917. Amer. newspaper publisher.
**Graham, Martha.** b. 1894. Amer. dancer & choreographer.
**Graham, Sylvester.** 1794–1851. Amer. nutritionist & reformer.
**Graham, Thomas.** 1805–69. Scottish chemist.
**Graham, William Franklin ("Billy").** b. 1918. Amer. evangelist.
**Gra·hame** (grā'əm), **Kenneth.** 1859–1932. English author.
**Gram·mat·i·cus** (grə-măt'ĭ-kəs). AELFRIC.
**Gramme** (grăm), **Zénobe Théophile.** 1826–1901. Belgian electrical inventor.
**Gra·na·dos** (grə-nä'dōs), **Enrique.** 1867–1916. Spanish pianist & composer.
**Gran·di** (grän'dē), Count **(di Mordano) Dino.** b. 1895. Italian politician.
**Grand·ma Mo·ses** (grănd'mä mō'zĭz, -zĭs). Anna Mary Robertson MOSES.
**Grange** (grānj), **Harold Edward ("Red").** b. 1903. Amer. football player.
**Gra·nit** (grä-nēt'), **Ragnar Arthur.** b. 1900. Finnish-born Swedish physiologist (Nobel, 1967).
**Grant** (grănt), **Cary.** b. 1904. English-born Amer. actor.
**Grant, Heber Jedediah.** 1856–1945. Amer. Mormon leader.
**Grant, Ulysses Simpson.** 1822–85. 18th U.S. President (1869–77) & Civil War general.
**Gran·ville-Bar·ker** (grăn'vĭl-bär'kər), **Harley Granville.** 1877–1946. English actor, playwright, & theater manager.
**Grass** (gräs), **Günter Wilhelm.** b. 1927. German writer.
**Grasse** (gräs, gräs), Comte **François Joseph Paul de.** *Marquis de Grasse-Tilly.* 1722?–88. French naval officer.
**Gras·so** (grăs'ō, grä'sō), **Ella.** 1919–81. Amer. public official.
**Gra·tian** (grā'shən, -shē-ən). 359–83. Roman emperor (367–83).
**Grat·tan** (grăt'n), **Henry.** 1746–1820. Irish politician & orator.
**Grau** (grou), **Shirley Ann.** b. 1929. Amer. author.
**Graup·ner** (group'nər), **Johann Christian Gottlieb.** 1767–1836. German-born Amer. musician & composer.
**Graves** (grāvz), **Morris Cole.** b. 1910. Amer. painter.
**Graves, Robert Ranke.** b. 1895. English-born Amer. author & critic.
**Gray** (grā), **Asa.** 1810–88. Amer. botanist.
**Gray, Elisha.** 1835–1910. Amer. inventor.
**Gray, Hanna Holborn.** b. 1930. German-born Amer. educator.
**Gray, Horace.** 1828–1902. Amer. jurist.
**Gray, Robert.** 1755–1806. Amer. trader & explorer.
**Gray, Thomas.** 1716–71. English poet.
**Gra·zia·ni** (grät'sē-ä'nē), **Rodolpho.** 1882–1955. Italian politician & colonial administrator.
**Gre·co** (grĕk'ō), **El.** 1541?–1614? Greek-born Spanish artist, architect, & scholar.
**Gree·ley** (grē'lē), **Horace.** 1811–72. Amer. journalist & politician.
**Gree·ly** (grē'lē), **Adolphus Washington.** 1844–1935. Amer. army officer & Arctic explorer.
**Green** (grēn), **Anna Katherine.** 1846–1935. Amer. author.
**Green, Duff.** 1791–1875. Amer. journalist & politician.
**Green, Henrietta Howland ("Hetty").** 1834–1916. Amer. financier.
**Green, John Richard.** 1837–83. English historian.
**Green, Paul Eliot.** 1894–1981. Amer. author & educator.
**Green, William.** 1873–1952. Amer. labor leader.
**Gree·na·way** (grē'nə-wā'), **Catherine ("Kate").** 1846–1901. English artist & author.
**Greene** (grēn), **Bella Da Costa.** 1883–1950. Amer. librarian & bibliographer.
**Greene, Graham.** b. 1904. English novelist.
**Greene, Nathanael.** 1742–86. Amer. Revolutionary general.
**Greene, Robert.** 1558?–92. English author.
**Green·how** (grē'nou), **Rose O'Neal.** 1815?–1964. Amer. Confederate spy.
**Gree·nough** (grē'nō'), **Horatio.** 1805–52. Amer. sculptor.
**Gregg** (grĕg), **Josiah.** 1806–50. Amer. frontiersman & author.
**Gregg, William.** 1800–67. Amer. cotton manufacturer.
**Greg·o·ry** (grĕg'ə-rē). Name of 16 popes, esp.: **a. I.** Saint. *the Great.* 540?–604. Reigned 590–604. **b. VII.** Saint. 1020?–85. Reigned 1073–85. **c. XIII.** 1502–85. Reigned 1572–85.
**Gregory,** Lady **(Isabella) Augusta Persse.** 1852–1932. Irish playwright.
**Gregory, Cynthia.** b. 1946. Amer. ballerina.
**Gregory of Nys·sa** (nĭs'ə), Saint. 331?–96? Eastern church father.
**Gregory of Tours** (to͝or, to͞or), Saint. 538–94? Frankish prelate.
**Gren·fell** (grĕn'fĕl', -fəl), Sir **Wilfred Thomason.** 1865–1940. English missionary & physician.
**Gren·ville** (grĕn'vĭl', -vəl), **George.** 1712–70. English statesman.
**Gren·ville** (grĕn'vĭl') *or* **Greyn·ville** (grān'-, grĕn'-), Sir **Richard.** 1542?–91. English naval officer.
**Gresh·am** (grĕsh'əm), Sir **Thomas.** 1519?–79. English financier.
**Greuze** (grœz), **Jean Baptiste.** 1725–1805. French painter.
**Gré·vy** (grā-vē'), **(François Paul) Jules.** 1807–91. French statesman.
**Grew** (gro͞o), **Joseph Clark.** 1880–1965. Amer. diplomat.
**Grey** (grā), **Charles.** *2nd Earl Grey.* 1764–1845. British prime minister (1830–34) & parliamentary reformer.
**Grey,** Sir **Edward.** 1862–1933. English statesman.
**Grey,** Lady **Jane.** 1537–54. English queen (1553); executed for treason.
**Grey, Zane.** 1875–1939. Amer. author of Western adventures.
**Greyn·ville** (grān'vĭl', grĕn'-). *var. of* Sir Richard GRENVILLE.
**Grieg** (grēg, grĭg), **Edvard Hagerup.** 1843–1907. Norwegian composer.
**Grier** (grîr), **Robert Cooper.** 1794–1870. Amer. jurist.
**Grier·son** (grîr'sən), Sir **Herbert John Clifford.** 1866–1960. Scottish literary scholar.
**Grieve** (grēv), **Christopher Murray.** Hugh MACDIARMID.
**Grif·fin** (grĭf'ĭn), **Walter Burley.** 1876–1937. Amer. architect.
**Grif·fith** (grĭf'ĭth), **Arthur.** 1872–1922. Irish journalist & Sinn Fein leader.
**Griffith, David Lewelyn Wark ("D.W.").** 1875–1948. Amer. filmmaker.
**Grif·fiths** (grĭf'ĭths), **John Willis.** 1809–82. Amer. naval architect.
**Gri·gnard** (grēn-yär'), **François Auguste Victor.** 1871–1934. French chemist (Nobel, 1912).
**Grill·par·zer** (grĭl'pärt'sər), **Franz.** 1791–1872. Austrian author.
**Grim·ké** (grĭm'kē), **Sarah Moore** (1792–1873) & **Angeline Emily** (1805–79). Amer. feminists & abolitionists.
**Grimm** (grĭm), **Jakob Ludwig Karl** (1785–1863) & **Wilhelm Karl** (1786–1859). German philologists & folklorists.
**Grin·nell** (grə-nĕl'), **George Bird.** 1849–1938. Amer. naturalist & conservationist.
**Gris** (grēs), **Juan.** 1887–1927. Spanish artist.
**Gris·wold** (grĭz'wôld', -wōld'), **Rufus Wilmot.** 1815?–57. Amer. editor & critic.
**Gro·lier de Ser·vières** (grōl'yā də sĕr-vē-âr'), **Jean.** *Viscomte d'Aquisy.* 1479–1565. French bibliophile.
**Gro·my·ko** (grə-mē'kō, grō-), **Andrei Andreevich.** b. 1909. Soviet diplomat.
**Gron·lund** (grŏn'lənd), **Laurence.** 1846–99. Danish-born Amer. lawyer & socialist.
**Groo·te** (grō'tə) *also* **Groot** (grōt), **Gerhard.** 1340–84. Dutch religious reformer.
**Gro·pi·us** (grō'pē-əs), **Walter Adolph.** 1883–1969. German-born Amer. architect.
**Grop·per** (grŏp'ər), **William.** 1897–1977. Amer. artist.
**Gross** (grōs), **Samuel David.** 1805–84. Amer. surgeon & educator.
**Gros·ve·nor** (grōv'nər), **Gilbert Hovey.** 1875–1966. Amer. editor & geographer.
**Grosz** (grōs), **George.** 1893–1959. German-born Amer. artist.
**Grote** (grōt), **George.** 1794–1871. English historian.
**Gro·ti·us** (grō'shē-əs, -shəs), **Hugo.** 1583–1645. Dutch jurist, statesman, & theologian.
**Grou·chy** (gro͞o-shē'), Marquis **Emmanuel de.** 1766–1847. French marshal.
**Grove** (grōv), Sir **George.** 1820–1900. English engineer & musicologist.
**Groves** (grōvz), **Leslie Richard.** 1896–1970. Amer. army officer & director of Manhattan Project.
**Grü·ne·wald** (gro͞o'nə-wôld', -vält'), **Matthias.** 1480?–1530? German painter.
**Gryph·i·us** (grĭf'ē-əs), **Andreas.** 1616–64. German author.
**Guar·nie·ri** (gwär-nyĕr'ē) or **Guar·ne·ri** (-nĕr'ē). Family of Italian violin makers, including **Andrea** (1626–98), **Pietro** (1655–1728), **Giuseppe** (1666–1739), & **Giuseppe Antonio** (1687–1745?).
**Gue·dal·la** (gwĭ-dăl'ə), **Philip.** 1889–1944. English biographer, historian, & essayist.
**Gue·rick·e** (gā'rĭ-kə), **Otto von.** 1602–86. German physicist.

ă **pat** ā **pay** âr **care** ä **father** ĕ **pet** ē **be** hw **which** ĭ **pit** ī **tie** îr **pier** ŏ **pot** ō **toe** ô **paw, for** oi **noise** o͝o **took**

**Gue·rin** (gĕr′ən), **Jules.** 1866–1946. Amer. painter.
**Guesde** (gĕd), **Jules.** 1845–1922. French journalist & socialist.
**Guest** (gĕst), **Edgar Albert.** 1881–1959. English-born Amer. journalist.
**Gue·va·ra** (gā-vär′ə, gə-), **Ernesto ("Che").** 1928–67. Argentine-born revolutionary in Latin America.
**Gug·gen·heim** (go͝og′ən-hīm′), **Daniel.** 1856–1930. Amer. industrialist & philanthropist.
**Guggenheim, Meyer.** 1828–1905. Swiss-born Amer. financier & industrialist.
**Gui·do d'A·rez·zo** (gwē′dō də-rĕt′sō) *or* **A·re·ti·no** (är′ə-tē′nō). 995?–1050? Benedictine monk & music theorist.
**Guil·laume** (gē-ōm′, -yōm′), **Charles Édouard.** 1861–1938. French physicist (Nobel, 1920).
**Guille·min** (gē-măN′), **Roger Charles Louis.** b. 1924. French-born Amer. physicist (Nobel, 1977).
**Guin·an** (gwĭn′ən), **Mary Louise Cecilia ("Tex").** 1884–1933. Amer. actress & hostess.
**Gui·ney** (gī′nē, gwī′-), **Louise Imogen.** 1861–1920. Amer. author.
**Guin·ness** (gĭn′ĭs), Sir **Alec.** b. 1914. English actor.
**Guis·card** (gē-skär′), **Robert.** 1015?–85. Norman leader.
**Guise** (gēz). 2nd Duc. *François de Lorraine.* 1519–63. French general & statesman.
**Guise.** 3rd Duc. *Henri de Lorraine.* 1550–88. French military leader; assassinated.
**Gui·te·ras** (gē-tĕr′əs), **Juan.** 1852–1925. Cuban pathologist & physician.
**Gui·zot** (gē-zō′), **François Pierre Guillaume.** 1787–1874. French historian & diplomat.
**Gull·strand** (gŭl′strănd′), **Allvar.** 1862–1930. Swedish ophthalmologist (Nobel, 1911).
**Gun·nars·son** (gŭn′ər-sən), **Gunnar.** b. 1889. Icelandic author.
**Gun·ter** (gŭn′tər), **Edmund.** 1581–1626. English astronomer, mathematician, & inventor.
**Gun·ther** (gŭn′thər), **John.** 1901–70. Amer. journalist & broadcaster.
**Gus·ta·vus** (gŭs-tā′vəs, -tä′-). Name of six kings of Sweden: **a. I.** 1496–1560. Ruled 1523–60. **b. II.** 1594–1632. Ruled 1611–32. **c. III.** 1746–92. Ruled 1771–92. **d. IV.** 1778–1837. Ruled 1792–1809. **e. V.** 1858–1950. Ruled 1907–50. **f. VI.** 1882–1973. Ruled 1950–73.
**Gu·ten·berg** (go͞ot′n-bûrg′), **Johann** *or* **Johannes.** 1400?–68? German printer & inventor of movable type.
**Guth·rie** (gŭth′rē), **Janet.** b. 1938. Amer. automobile racer.
**Guthrie, Woodrow Wilson ("Woody").** 1912–67. Amer. folk singer & composer.
**Gutz·kow** (go͞ots′kō), **Karl.** 1811–78. German author.
**Guz·mán Blan·co** (go͞os-män′ bläng′kō), **Antonio.** 1828?–99. Venezuelan statesman.
**Gwin·nett** (gwə-nĕt′), **Button.** 1735–77. Amer. Revolutionary patriot.
**Gwyn** *or* **Gwynne** (gwĭn), **Eleanor ("Nell").** 1650?–87. English actress & mistress of Charles II.

# H

**Haa·kon** (hô′kən, -kŏn′). Name of seven kings of Norway, esp. **VII,** 1872–1957, ruled 1905–57.
**Ha·bak·kuk** (hə-băk′ək, hăb′ə-kŭk′). Late 7th cent. B.C. Hebrew prophet.
**Ha·ber** (hä′bər), **Fritz.** 1868–1934. German chemist (Nobel, 1918).
**Habs·burg** (hăps′bûrg′, häps′bo͝ork′). *var. of* HAPSBURG.
**Had·ley** (hăd′lē), **Arthur Twining.** 1856–1930. Amer. educator & economist.
**Hadley, Henry Kimball.** 1871–1937. Amer. composer & conductor.
**Had·ow** (hăd′ō), Sir **(William) Henry.** 1859–1937. English educator & author.
**Ha·dri·an** (hā′drē-ən). A.D. 76–138. Roman emperor (117–38).
**Haeck·el** (hĕk′əl), **Ernst Heinrich.** 1834–1919. German philosopher & naturalist.
**Ha·fiz** (hä-fĭz′, -fēz′). 14th cent. Persian poet, philosopher, & grammarian.
**Ha·gar** (hā′gər, -gär). Egyptian concubine of Abraham in the Old Testament.
**Ha·gen** (hā′gən), **Walter Charles.** 1892–1969. Amer. golfer.
**Hag·er·ty** (hăg′ər-tē), **James C.** 1909–81. Amer. journalist & public official.
**Hag·ga·i** (hăg′ā-ī′). 6th cent. B.C. Hebrew prophet.
**Hag·gard** (hăg′ərd), Sir **(Henry) Rider.** 1856–1925. English novelist.
**Hahn** (hän), **Otto.** 1879–1968. German chemist (Nobel, 1944).
**Hah·ne·mann** (hä′nə-mən), **(Christian Friedrich) Samuel.** 1755–1843. German physician & founder of homeopathy.
**Haig** (hāg), **Alexander Meigs, Jr.** b. 1924. Amer. general & public official.
**Haig, Douglas.** *1st Earl Haig.* Scottish-born British field marshal.
**Hai·le Se·las·sie** (hī′lē sə-lăs′ē, -lä′sē). *Ras Taffari Makonnen.* 1891–1975. Ethiopian emperor (1930–74); exiled (1936–41) & deposed.
**Hak·luyt** (hăk′lo͞ot′), **Richard.** 1552–1616. English geographer.
**Hal·as** (hăl′əs), **George Stanley.** 1895–1983. Amer. football player & coach.
**Hal·dane** (hôl′dān′, -dən), **John Burdon Sanderson.** 1892–1964. English geneticist.
**Haldane, John Scott.** 1860–1936. Scottish-born British scientist.
**Haldane, Richard Burdon.** 1856–1928. Scottish-born British philosopher & statesman.
**Hale** (hāl), **Edward Everett.** 1822–1909. Amer. clergyman & author.
**Hale, George Ellery.** 1868–1938. Amer. astrophysicist & inventor.
**Hale, John Parker.** 1806–73. Amer. politician & diplomat.
**Hale, Lucretia Peabody.** 1820–1900. Amer. author.
**Hale,** Sir **Matthew.** 1609–76. English jurist.
**Hale, Nathan.** 1755–76. Amer. Revolutionary; hanged by the British as a spy.
**Hale, Sarah Josepha Buell.** 1788–1879. Amer. editor & author.
**Hales** (hālz), **Stephen.** 1677–1761. English physiologist & inventor.
**Ha·lé·vy** (äl′ā-vē′, hăl′-), **Jacques Fromental Élie Lévy.** 1799–1862. French composer.
**Halévy, Ludovic.** 1834–1908. French author.
**Ha·ley** (hā′lē), **Alex.** b. 1921. Amer. author.
**Haley, Margaret Angela.** 1861–1939. Amer. educator & labor organizer.
**Haley, William John Clifton, Jr. ("Bill").** 1925–81. Amer. singer & band leader.
**Hal·i·fax** (hăl′ə-făks′), 1st Earl of. *Edward Frederick Lindley Wood.* 1881–1959. English statesman.
**Hall** (hôl), **Abraham Oakey ("O.K.").** 1826–98. Amer. politician & journalist.
**Hall, Asaph.** 1829–1907. Amer. astronomer.
**Hall, Charles Francis.** 1821–71. Amer. Arctic explorer.
**Hall, Charles Martin.** 1863–1914. Amer. chemist & pioneer aluminum manufacturer.
**Hall, Granville Stanley.** 1844–1924. Amer. psychologist.
**Hall,** Sir **James.** 1761–1832. Scottish geologist & chemist.
**Hall, James.** 1793–1868. Amer. author, jurist, & banker.
**Hall, James.** 1811–98. Amer. geologist & paleontologist.
**Hall, James Norman.** 1887–1951. Amer. author.
**Hal·lam** (hăl′əm), **Henry.** 1777–1859. English historian.
**Hallam, Lewis.** 1740–1808. English-born Amer. actor & theatrical manager.
**Hal·leck** (hăl′ək, -ĭk), **Fitz-Greene.** 1790–1867. Amer. poet.
**Halleck, Henry Wager.** 1815–72. Amer. Union general.
**Hal·ley** (hăl′ē, hā′lē), **Edmund.** 1656–1742. English astronomer.
**Hals** (hälz, häls), **Frans.** 1580–1666. Dutch painter.
**Hal·sey** (hôl′sē, -zē, hŏl′-), **William Frederick.** 1882–1959. Amer. naval officer.
**Hal·sted** (hôl′stəd, -stĕd′, hŏl′-), **William Stewart.** 1852–1922. Amer. surgeon.
**Ham** (hăm). Noah's son in the Old Testament.
**Ha·mil·car Bar·ca** (hə-mĭl′kär′ bär′kə, hăm′əl-). 270?–228 B.C. Carthaginian general.
**Ham·ill** (hăm′əl), **Dorothy.** b. 1956. Amer. figure skater.
**Ham·il·ton** (hăm′əl-tən), **Alexander.** 1755?–1804. Amer. statesman & political & economic theorist; killed by Aaron Burr in a duel. **—Ham′il·to′ni·an** *adj.*
**Hamilton, Edith.** 1867–1963. German-born Amer. classicist.
**Hamilton,** Lady **Emma Lyon.** 1765–1815. English socialite & mistress of Lord Nelson.
**Ham·lin** (hăm′lən), **Hannibal.** 1809–91. U.S. Vice President (1861–65).
**Ham·mar·skjöld** (hăm′ər-shəld, -sho͝old′, -shēld′, hä′mər-), **Dag Hjalmar Agné Carl.** 1905–61. Swedish statesman & UN official (Nobel, 1961).
**Ham·mer·stein** (hăm′ər-stīn′, -stēn′), **Oscar.** 1847?–1919. German-born Amer. operatic manager.
**Hammerstein, Oscar, II.** 1895–1960. Amer. lyricist.
**Ham·mett** (hăm′ĭt), **Dashiell.** 1894–1961. Amer. author.
**Ham·mond** (hăm′ənd). Family of Amer. engineers & inventors, including **John Hays** (1855–1936), **John Hays** (1888–1965), & **Laurens** (1895–1973).
**Ham·mu·ra·bi** (hăm′ə-rä′bē). c. 20th or 18th cent. B.C. Babylonian king & lawgiver.
**Hamp·den** (hămp′dən, hăm′-), **John.** 1594–1643. English statesman.
**Hamp·ton** (hămp′tən, hăm′-), **Lionel.** b. 1914. Amer. musician.
**Hampton, Wade** (1752?–1835) & **Wade** (1818–1902). Amer. generals & politicians.

**Ham·sun** (hăm′sən), **Knut.** 1859–1952. Norwegian author (Nobel, 1920).
**Han** (hän). Name of three Chinese dynasties: **Western Han,** 206 B.C.–A.D. 24; **Eastern Han,** A.D. 25–220; & **Later Han,** 947–50.
**Han·cock** (hăn′kŏk′), **John.** 1737–93. Amer. merchant, politician, & Revolutionary leader.
**Hancock, Winfield Scott.** 1824–86. Amer. Civil War general.
**Hand** (hănd), **(Billings) Learned.** 1872–1961. Amer. jurist.
**Han·del** (hăn′dl), **George Frederick.** 1685–1759. German-born English composer.
**Hand·lin** (hănd′lĭn), **Oscar.** b. 1915. Amer. historian & educator.
**Han·dy** (hăn′dē), **William Christopher ("W.C.").** 1873–1958. Amer. musician & composer.
**Han·na** (hăn′ə), **Marcus Alonzo ("Mark").** 1837–1904. Amer. financier & politician.
**Han·nay** (hăn′ā, hăn′ē), **James Owen.** 1865–1950. Irish clergyman & novelist.
**Han·ni·bal** (hăn′ə-bəl). 247?–183 B.C. Carthaginian general.
**Han·no** (hăn′ō). *the Great.* fl. 3rd cent. B.C. Carthaginian political leader.
**Ha·no·taux** (ăn′ə-tō′, ä′nə-), **(Albert Auguste) Gabriel.** 1853–1944. French historian & statesman.
**Han·o·ver** (hăn′ō′vər). English ruling family (1714–1901).
**Han·sard** (hăn′sərd, -särd′), **Luke.** 1752–1828. English parliamentary printer.
**Hans·ber·ry** (hănz′bĕr-ē), **Lorraine.** 1930–65. Amer. playwright.
**Han·sen** (hăn′sən), **Marcus Lee.** 1892–1938. Amer. historian.
**Han·son** (hăn′sən), **Howard Harold.** 1896–1981. Amer. composer & teacher.
**Hanson, John.** 1721–83. Amer. Revolutionary leader.
**Hans·son** (hăn′sən, hän′-), **Per Albin.** 1885–1946. Swedish statesman & journalist.
**Hap·good** (hăp′go͝od′), **Isabel Florence.** 1850–1928. Amer. translator & author.
**Haps·burg** *also* **Habs·burg** (hăps′bûrg′, häps′bo͝ork′). Royal German family ruling Austria (1276–1740) & Spain (1516–1700).
**Har·bach** (här′băk′), **Otto Abels.** 1873–1963. Amer. playwright & librettist.
**Har·de·ca·nute** *or* **Har·di·ca·nute** (här′dĭ-kə-no͞ot′, -nyo͞ot′). 1019?–42. King of England (1040–42) & of Denmark as **Canute II** (1035–42).
**Har·den** (här′dn), Sir **Arthur.** 1865–1940. English biochemist (Nobel, 1929).
**Harden, Maximilian.** 1861–1927. German journalist & critic.
**Har·den·berg** (här′dn-bûrg′, -bĕrk′), Prince **Karl August von.** 1750–1822. Prussian statesman.
**Har·di·ca·nute** (här′dĭ-kə-no͞ot′, -nyo͞ot′). *var. of* HARDECANUTE.
**Har·ding** (här′dĭng), **Chester.** 1792–1866. Amer. portrait painter.
**Harding, Warren Gamaliel.** 1865–1923. 29th U.S. President (1921–23); died in office.
**Hard·wicke** (härd′wĭk′), Sir **Cedric Webster.** 1893–1964. English actor.
**Har·dy** (här′dē), **Oliver.** 1892–1957. Amer. comedian.
**Hardy, Thomas.** 1840–1928. English author.
**Har·greaves** (här′grēvz′), **James.** d. 1778. English inventor.
**Har·ing·ton** *or* **Har·ring·ton** (hăr′ĭng-tən), Sir **John.** 1561–1612. English poet & translator.
**Ha·ri·ri** (hə-rĭr′ē, -rîr′ē), **al-.** 1054–1122. Arabian poet & scholar.
**Hark·ness** (härk′nĭs, -nĕs′), **Anna M. Richardson** (1837–1926) & **Edward Stephen** (1874–1940). Amer. philanthropists.
**Har·lan** (här′lən), **John Marshall** (1833–1911) & **John Marshall** (1899–1971). Amer. jurists.
**Har·ley** (här′lē), **Robert.** *1st Earl of Oxford.* 1661–1724. English statesman & bibliophile.
**Har·low** (här′lō′), **Jean.** 1911–37. Amer. actress.
**Harms·worth** (härmz′wûrth′), **Alfred Charles William** (*Viscount Northcliffe*; Irish-born, 1865–1922) & **Harold Sidney** (*1st Viscount Rothermere*; 1868–1940). English newspaper publishers.
**Har·old** (hăr′əld). Name of two kings of England: **a. I.** d. 1040. Ruled 1035–40. **b. II.** 1022–66. Ruled 1066; killed at the Battle of Hastings.
**Harold.** Name of three kings of Norway, esp. **III,** 1015–66, ruled 1046–66.
**Ha·roun al-Ra·schid** (hə-ro͞on′ äl-rə-shēd′). *var. of* HARUN AL-RASHID.
**Har·per** (här′pər). Family of Amer. printers & publishers, including **James** (1795–1869), **John** (1797–1875), **Joseph Wesley** (1801–70), & **Fletcher** (1806–77).
**Harper, Frances Ellen Watkins.** 1825–1911. Amer. author & reformer.
**Harper, Ida Husted.** 1851–1931. Amer. suffragist & journalist.
**Harper, Robert Goodloe.** 1765–1825. Amer. Federalist politician.
**Harper, William Rainey.** 1856–1906. Amer. educator & scholar.
**Har·ri·man** (hăr′ə-mən), **(William) Averell.** b. 1891. Amer. financier & diplomat.
**Harriman, Edward Henry.** 1848–1909. Amer. railway magnate.
**Harriman, Florence Jaffray.** 1870–1967. Amer. diplomat.
**Har·ring·ton** (hăr′ĭng-tən). *var. of* HARINGTON.
**Harrington, (Edward) Michael.** b. 1928. Amer. author & reformer.
**Har·ris** (hăr′ĭs), **Benjamin.** fl. 1673–1713. English publisher & author.
**Harris, Chapin Aaron.** 1806–60. Amer. pioneer dentist.
**Harris, Frank.** 1854–1931. Irish-born Amer. author.
**Harris, Joel Chandler.** 1848–1908. Amer. author & journalist.
**Harris, Julie.** b. 1925. Amer. actress.
**Harris, Louis ("Lou").** b. 1921. Amer. public-opinion analyst.
**Harris, Patricia Roberts.** b. 1924. Amer. ambassador & educator.
**Harris, Roy Ellsworth.** 1898–1979. Amer. composer.
**Harris, Townsend.** 1804–78. Amer. diplomat.
**Harris, William Torrey.** 1835–1909. Amer. philosopher, editor, & educator.
**Har·ri·son** (hăr′ĭ-sən), **Benjamin.** 1726–91. Amer. Revolutionary statesman.
**Harrison, Benjamin.** 1833–1901. 23rd U.S. President (1889–93).
**Harrison, Elizabeth.** 1849–1927. Amer. pioneer educator.
**Harrison, Frederic.** 1831–1923. English author & philosopher.
**Harrison, George.** b. 1943. English singer & songwriter.
**Harrison, Peter.** 1717–75. English-born Amer. architect & merchant.
**Harrison, Reginald Carey ("Rex").** b. 1908. English actor.
**Harrison, William Henry.** 1773–1841. Ninth U.S. President (1841); died in office.
**Hart** (härt), **Albert Bushnell.** 1854–1943. Amer. historian & editor.
**Hart, Lorenz.** 1895–1943. Amer. lyricist.
**Hart, Moss.** 1904–61. Amer. playwright, librettist, & director.
**Hart,** Sir **Robert.** 1835–1911. English diplomat.
**Hart, William Surrey.** 1870–1946. Amer. actor.
**Harte** (härt), **(Francis) Bret.** 1836–1902. Amer. author.
**Hart·line** (härt′lĭn′), **Haldan Keffer.** 1903–83. Amer. biophysicist (Nobel, 1967).
**Ha·run al-Ra·shid** *or* **Ha·roun al-Ra·schid** (hə-ro͞on′ äl-rə-shēd′) *also* **Harun ar-Ra·shid** (är′-). 764?–809. Caliph of Baghdad (786–809).
**Har·vard** (här′vərd), **John.** 1607–38. Amer. clergyman & philanthropist.
**Har·vey** (här′vē), **Frederick Henry.** 1835–1901. English-born Amer. restaurateur.
**Harvey, George Brinton McClellan.** 1864–1928. Amer. journalist & diplomat.
**Harvey, William.** 1578–1657. English physician, anatomist, & physiologist.
**Harvey, William Hope ("Coin").** 1851–1936. Amer. economist & publicist.
**Has·brouck** (hăz′bro͝ok′), **Lydia Sayer.** 1827–1910. Amer. editor & reformer.
**Has·dru·bal** (hăz′dro͞o′bəl, hăz-dro͞o′-). d. 207 B.C. Carthaginian general.
**Has·sam** (hăs′əm), **(Frederick) Childe.** 1859–1935. Amer. painter.
**Has·sel** (hä′səl), **Odd.** 1897–1981. Norwegian chemist (Nobel, 1969).
**Hass·ler** (hăs′lər), **Ferdinand Rudolf.** 1770–1843. Amer. scientist & coastal surveyor.
**Has·tie** (hā′stē), **William Henry.** 1904–76. Amer. jurist.
**Ha·stings** (hā′stĭngz), 1st Marquis of. *Francis Rawdon-Hastings.* 1754–1826. English general in the Amer. Revolution.
**Hastings, Thomas.** 1860–1929. Amer. architect.
**Hastings, Warren.** 1732–1818. English colonial administrator in India.
**Hatch** (hăch), **William Henry.** 1833–96. Amer. legislator.
**Hath·a·way** (hăth′ə-wā′), **Anne.** 1557?–1623. Wife of William Shakespeare.
**Hat·shep·sut** (hăt-shĕp′so͞ot′) *also* **Hat·shep·set** (-sĕt′). d. c.1481 B.C. Egyptian queen.
**Hauk** *or* **Hauck** (houk), **Minnie.** 1851?–1929. Amer. soprano.
**Haupt·mann** (houpt′män′, houp′-), **Gerhart.** 1862–1946. German author (Nobel, 1912).
**Haus·ho·fer** (hous′hō′fər), **Karl.** 1869–1946. German army officer & geopolitician.
**Hauss·mann** (hous′mən, ōs-män′), Baron **Georges Eugène.** 1809–91. French public official.
**Have·lock** (hăv′lŏk′, -lək), Sir **Henry.** 1795–1857. English general in India.
**Ha·ven** (hā′vən), **Emily Bradley Neal.** 1827–63. Amer. author & editor.
**Hawes** (hôz), **Harriet Ann Boyd.** 1871–1945. Amer. archaeologist.
**Haw·kins** (hô′kĭnz), **Anthony Hope.** 1863–1933. English author.

**Hawkins** *or* **Haw·kyns** (hô′kĭnz), Sir **John.** 1532–95. English naval hero.

**Hawks** (hôks), **Howard Winchester.** 1896–1977. Amer. filmmaker.

**Haw·kyns** (hô′kĭnz). *var. of* Sir John HAWKINS.

**Haw·orth** (hou′ərth, härth), Sir **(Walter) Norman.** 1883–1950. English biochemist (Nobel, 1937).

**Haw·thorne** (hô′thôrn′, hŏth′ôrn′), **Nathaniel.** 1804–64. Amer. author.

**Hay** (hā), **John Milton.** 1838–1905. Amer. diplomat, public official, & author.

**Ha·ya·ka·wa** (hī′ə-kou′ə), **Samuel Ichiye ("S.I.").** b. 1906. Canadian-born Amer. philologist, educator, & legislator.

**Hay·den** (hād′n), **Carl Trumbull.** 1877–1972. Amer. legislator.

**Hayden, Ferdinand Vandeveer.** 1829–87. Amer. geologist & Western explorer.

**Hayden, Melissa.** b. 1923. Canadian-born Amer. ballerina.

**Haydn** (hīd′n), **(Franz) Joseph ("Papa").** 1732–1809. Austrian composer.

**Hay·ek** (hī′ək), **Friedrich August von.** b. 1899. Austrian-born English economist (Nobel, 1974).

**Hayes** (hāz), **Carlton Joseph Huntley.** 1882–1964. Amer. historian & diplomat.

**Hayes, Helen.** b. 1900. Amer. actress.

**Hayes, Isaac Israel.** 1832–81. Amer. physician & Arctic explorer.

**Hayes, Roland.** 1887–1977. Amer. tenor.

**Hayes, Rutherford Birchard.** 1822–93. 19th U.S. President (1877–81).

**Hayne** (hān), **Robert Young.** 1791–1839. Amer. politician & railroad executive.

**Haynes** (hānz), **Elwood.** 1857–1925. Amer. inventor.

**Hays** (hāz), **Arthur Garfield.** 1881–1954. Amer. libertarian lawyer.

**Hays, John Coffee ("Jack").** 1817–83. Amer. frontiersman.

**Hays, William Harrison ("Will").** 1879–1954. Amer. politician & motion-picture executive.

**Hay·wood** (hā′wo͝od′), **William Dudley ("Big Bill").** 1869–1928. Amer. labor leader.

**Ha·zard** (ä-zär′), **Paul Gustave Marie Camille.** 1878–1944. French literary historian.

**Haz·litt** (hăz′lĭt, hāz′-), **William.** 1778–1830. English essayist.

**Head** (hĕd), **Edith.** 1898?–1981. Amer. fashion designer.

**Hea·ly** (hē′lē), **Timothy Michael.** 1855–1931. Irish nationalist politician.

**Hearn** (hûrn), **Lafcadio.** 1850–1904. Greek-born Amer.-Japanese author.

**Hearst** (hûrst), **George.** 1820–91. Amer. mine owner, publisher, & politician.

**Hearst, Phoebe Apperson.** 1842–1919. Amer. philanthropist.

**Hearst, William Randolph.** 1863–1951. Amer. newspaper publisher.

**Heath** (hēth), **Edward Richard George.** b. 1916. British prime minister (1969–74).

**Heat·ter** (hē′tər), **Gabriel.** 1890–1972. Amer. journalist & radio commentator.

**Heav·i·side** (hĕv′ē-sīd′), **Oliver.** 1850–1925. English physicist & electrical theorist.

**Heb·bel** (hĕb′əl), **Friedrich.** 1813–63. German dramatist.

**Hé·bert** (ā-bĕr′), **Jacques René.** 1755–94. French revolutionary & journalist; guillotined.

**Hecht** (hĕkt), **Ben.** 1894–1964. Amer. author.

**Heck·er** (hĕk′ər), **Isaac Thomas.** 1819–88. Amer. priest & founder of Paulist order.

**He·din** (hā-dēn′), **Sven Anders von.** 1865–1952. Swedish explorer & scientist.

**Hef·ner** (hĕf′nər), **Hugh Marston.** b. 1926. Amer. editor & publisher.

**He·gel** (hā′gəl), **Georg Wilhelm Friedrich.** 1770–1831. German philosopher. **—He·ge′li·an** (hā-gā′lē-ən) *adj. & n.*

**Hei·deg·ger** (hī′dĕg′ər, -dĭ-gər), **Martin.** 1889–1976. German philosopher.

**Hei·den·stam** (hād′n-stăm′, -stäm′), **Verner von.** 1859–1940. Swedish author (Nobel, 1916).

**Hei·fetz** (hī′fĭts), **Jascha.** b. 1901. Russian-born Amer. violinist.

**Hei·ne** (hī′nə), **Heinrich.** 1797–1856. German lyric poet & critic.

**Hein·lein** (hīn′līn), **Robert Anson.** b. 1907. Amer. science-fiction author.

**Heinz** (hīnz), **Henry John.** 1844–1919. Amer. manufacturer of prepared foods.

**Hei·sen·berg** (hī′zən-bûrg′, -bĕrk′), **Werner.** 1901–76. German physicist (Nobel, 1932).

**Hei·ser** (hī′zər), **Victor George.** 1873–1972. Amer. physician & public-health pioneer.

**Held** (hĕld), **Anna.** 1865?–1918. French-born Amer. entertainer.

o͞o **boot** ou **out** th **thin** *th* **this** ŭ **cut** ûr **urge** y **young**
yo͞o **abuse** zh **vision** ə **about**, **item**, **edible**, **gallop**, **circus**

**Held, John, Jr.** 1889–1958. Amer. illustrator & author.

**He·li·o·gab·a·lus** (hē′lē-ə-găb′ə-ləs, -lē-ō-). 204–22. Roman emperor (218–22).

**Hel·ler** (hĕl′ər), **Joseph.** b. 1923. Amer. author.

**Hell·man** (hĕl′mən), **Lillian.** b. 1905. Amer. playwright.

**Helm·holtz** (hĕlm′hōlts′), **Hermann Ludwig Ferdinand von.** 1821–94. German physiologist & physicist.

**Hé·lo·ise** (ĕl′ə-wēz′, ā′lə-). 1101?–64? Abelard's beloved.

**Hel·per** (hĕl′pər), **Hinton Rowan.** 1829–1909. Amer. author & antislavery advocate.

**Hel·vé·tius** (hĕl-vā′shəs, -shē-əs, -vē′-), **Claude Adrien.** 1715–71. French philosopher & author.

**He·mans** (hĕm′ənz, hē′mənz), **Felicia Dorothea.** 1793–1835. English poet.

**Hem·en·way** (hĕm′ən-wā′), **Mary Porter Tileston.** 1820–94. Amer. philanthropist.

**Hem·ing** *or* **Hem·minge** (hĕm′ĭng), **John.** 1556?–1630. English actor & Shakespeare editor.

**Hem·ing·way** (hĕm′ĭng-wā′), **Ernest Miller.** 1899?–1961. Amer. author (Nobel, 1954).

**Hem·minge** (hĕm′ĭng). *var. of* HEMING.

**Hench** (hĕnch), **Philip Showalter.** 1896–1965. Amer. physician (Nobel, 1950).

**Hen·der·son** (hĕn′dər-sən), **Arthur.** 1863–1935. Scottish-born British labor leader & statesman (Nobel, 1934).

**Henderson,** Sir **Nevile Meyrick.** 1882–1942. English diplomat.

**Henderson, Richard.** 1735–85. Amer. frontier land developer.

**Hen·dricks** (hĕn′drĭks), **Thomas Andrews.** 1819–85. U.S. Vice President (1885); died in office.

**Hen·ie** (hĕn′ē), **Sonja.** 1912–69. Norwegian-born figure skater.

**Hen·ley** (hĕn′lē), **William Ernest.** 1849–1903. English editor & author.

**Hen·ne·pin** (hĕn′ə-pĭn, ĕn-ə-păN′), **Louis.** 1640?–1701? French-born missionary & explorer in America.

**Hen·ri** (hĕn′rē), **Robert.** 1865–1929. Amer. painter & educator.

**Hen·ry** (hĕn′rē). *the Navigator.* 1394–1460. Portuguese prince.

**Henry.** Name of eight kings of England: **a. I.** 1068–1135. Ruled 1100–35. **b. II.** 1133–1189. Ruled 1154–89. **c. III.** 1207–72. Ruled 1216–72. **d. IV.** 1367–1413. Ruled 1399–1413. **e. V.** 1387–1422. Ruled 1413–22. **f. VI.** 1421–71. Ruled 1422–61, 1470–71. **g. VII.** 1457–1509. Ruled 1485–1509. **h. VIII.** 1491–1547. Ruled 1509–47.

**Henry.** Name of four kings of France: **a. I.** 1008?–1060. Ruled 1031–60. **b. II.** 1519–59. Ruled 1547–59. **c. III.** 1551–89. Ruled 1574–89. **d. IV.** *Henry of Navarre.* 1553–1610. Ruled 1589–1610.

**Henry, Andrew.** 1775?–1833. Amer. frontier explorer & fur trader.

**Henry, Joseph.** 1797–1878. Amer. physicist.

**Henry, Patrick.** 1736–99. Amer. Revolutionary leader & orator.

**Henry IV.** 1050–1106. King of Germany (1056–1106).

**Hens·lowe** (hĕnz′lō), **Philip.** d. 1616. English theatrical manager.

**Hep·burn** (hĕp′bûrn′, -bərn), **Audrey.** b. 1929. Belgian-born actress.

**Hepburn, Katharine.** b. 1909. Amer. actress.

**Hep·ple·white** (hĕp′əl-hwīt′, -wīt′), **George.** d. 1786. English cabinetmaker.

**Her·a·cli·tus** (hĕr′ə-klī′təs). 6th–5th cent. B.C. Greek philosopher. **—Her′a·cli′te·an** *adj.*

**Her·a·cli·us** (hĕr′ə-klī′əs, hĭ-răk′lē-). 575?–641. Byzantine emperor.

**Her·bart** (hĕr′bärt′), **Johann Friedrich.** 1776–1841. German psychologist, philosopher, & educator.

**Her·bert** (hûr′bərt), **George.** 1593–1633. Welsh-born English poet.

**Herbert, Victor.** 1859–1924. Amer. musician, composer, & conductor.

**Herbert, William.** *3rd Earl of Pembroke.* 1580–1630. English statesman & poetry patron.

**Her·block** (hûr′blŏk). Herbert Lawrence BLOCK.

**Her·der** (hĕr′dər), **Johann Gottfried von.** 1744–1803. German philosopher & author.

**He·re·dia** (ā-rād-yä′, -rā′dē-ə, hā-), **José Maria de.** 1842–1905. Cuban-born French poet & translator.

**Her·ford** (hûr′fərd), **Oliver Brooke.** 1863–1935. English author and illustrator.

**Her·ges·hei·mer** (hûr′gəs-hī′mər), **Joseph.** 1880–1954. Amer. novelist.

**Her·ki·mer** (hûr′kə-mər), **Nicholas.** 1728–77. Amer. Revolutionary general.

**Hern·don** (hûrn′dən), **William Henry.** 1818–91. Amer. lawyer & author.

**Herne** (hûrn), **James A.** 1839–1901. Amer. actor & playwright.

**He·ro** (hē′rō, hîr′ō) *or* **He·ron** (hē′rŏn′). 2nd or 3rd cent. A.D. Alexandrian scientist.

**Her·od** (hĕr′əd). *the Great.* 73?–4 B.C. King of Judea (40–4).

**Herod An·ti·pas** (ăn′tĭ-păs′, -pəs). Ruler of Judea & tetrarch in Galilee (4 B.C.–A.D. 40).

**He·rod·o·tus** (hĭ-rŏd′ə-təs). *the Father of History.* 5th cent. B.C. Greek historian.

**He·ron** (hē′rŏn′). *var. of* HERO.

**Her·rick** (hĕr'ĭk), **Myron Timothy.** 1854–1929. Amer. diplomat, businessman, & politician.
**Herrick, Robert.** 1591–1674. English lyric poet.
**Her·ri·ot** (ĕr'ē-ō'), **Edouard.** 1872–1957. French statesman.
**Her·schel** (hûr'shəl). Family of English astronomers, including Sir **William** (German-born; 1738–1822), **Caroline Lucretia** (1750–1848), & Sir **John Frederick William** (1792–1871).
**Her·sey** (hûr'sē), **John Richard.** b. 1914. Amer. author.
**Her·shey** (hûr'shē), **Alfred Day.** b. 1908. Amer. biologist (Nobel, 1969).
**Hershey, Lewis Blaine.** 1893–1977. Amer. general.
**Hershey, Milton Snavely.** 1857–1945. Amer. industrialist & philanthropist.
**Her·sko·vits** (hûr'skə-vĭts'), **Melville Jean.** 1895–1963. Amer. anthropologist.
**Her·ter** (hûr'tər), **Christian Archibald.** 1895–1966. Amer. diplomat.
**Hertz** (hûrts, hĕrts), **Gustav Ludwig.** 1887–1975. German physicist (Nobel, 1925).
**Hertz, Heinrich Rudolf.** 1857–94. German physicist.
**Hert·zog** (hûrt'sŏg', -sôg'), **James Barry Munnik.** 1866–1942. South African statesman.
**Herz·berg** (hûrts'bûrg'), **Gerhard.** b. 1904. German-born Canadian physicist (Nobel, 1971).
**Her·zl** (hĕrt'səl), **Theodor.** 1860–1904. Hungarian-born Austrian founder of Zionism.
**He·si·od** (hē'sē-əd, hĕs'ē-). 8th cent. B.C. Greek poet.
**Hess** (hĕs), Dame **Myra.** 1890–1965. English pianist.
**Hess, (Walter Richard) Rudolf.** b. 1894. German Nazi leader.
**Hess, Victor Franz.** 1883–1964. Austrian-born Amer. physicist (Nobel, 1936).
**Hess, Walter Rudolf.** 1881–1973. Swiss physiologist (Nobel, 1949).
**Hes·se** (hĕs'ə), **Hermann.** 1877–1962. German-born Swiss author (Nobel, 1946).
**He·ve·sy** (hĕv'ə-shē, hə-vĕsh'ē), **George von.** 1885–1966. Hungarian chemist (Nobel, 1943).
**Hew·ish** (hyo͞o'ĭsh), **Antony.** b. 1924. English astronomer (Nobel, 1974).
**Hew·itt** (hyo͞o'ĭt), **Abram Stevens.** 1822–1903. Amer. industrialist, politician, & reformer.
**Hey·er·dahl** (hā'ər-däl', hī'-), **Thor.** b. 1914. Norwegian ethnologist & explorer.
**Hey·mans** (ā-mäɴs'), **Corneille.** 1892–1968. Belgian physiologist (Nobel, 1938).
**Hey·rov·sky** (hā-rôf'skē, -rôv'-), **Jaroslav.** 1890–1967. Czechoslovakian chemist (Nobel, 1959).
**Hey·se** (hī'zə), **Paul von.** 1830–1914. German author (Nobel, 1910).
**Hey·ward** (hā'wərd), **DuBose** 1885–1940. Amer. author.
**Hey·wood** (hā'wo͝od'), **John.** 1497?–1580? English author.
**Heywood, Thomas.** 1574?–1641. English playwright.
**Hez·e·ki·ah** (hĕz'ə-kī'ə). 740?–692? B.C. King of Judah.
**Hich·ens** (hĭch'ənz), **Robert Smythe.** 1864–1950. English novelist & playwright.
**Hick·ok** (hĭk'ŏk'), **James Butler ("Wild Bill").** 1837–76. Amer. frontier scout & marshal.
**Hicks** (hĭks), **Edward.** 1780–1849. Amer. primitive painter.
**Hicks,** Sir **John Richard.** b. 1904. English economist (Nobel, 1972).
**Hi·e·ro I** (hī'ə-rō') or **Hi·e·ron** (-rŏn'). d. 466 B.C. Tyrant of Sicily (478–66 B.C.).
**Hig·gin·son** (hĭg'ən-sən), **Thomas Wentworth Storrow.** 1823–1911. Amer. clergyman, author, & Union soldier.
**High·et** (hī'ĭt), **Gilbert.** 1906–78. Scottish-born Amer. classicist.
**Hil·de·brand** (hĭl'də-brănd'). Pope GREGORY VII.
**Hill** (hĭl), **Ambrose Powell.** 1825–65. Amer. Confederate soldier.
**Hill, Archibald Vivian.** 1886–1977. English physiologist (Nobel, 1922).
**Hill, George Washington.** 1884–1946. Amer. tobacco executive.
**Hill, James Jerome.** 1838–1916. Amer. financier & railway promoter.
**Hill, Joe.** 1872?–1915. Swedish-born Amer. labor organizer.
**Hill,** Sir **Rowland.** 1795–1879. English postal reformer.
**Hil·la·ry** (hĭl'ə-rē), Sir **Edmund Percival.** b. 1919. New Zealand mountaineer, explorer, & author.
**Hil·lel** (hĭl'ĕl). fl. 30 B.C.–A.D. 9. Palestinian rabbi.
**Hil·liard** (hĭl'yərd), **Nicholas.** 1537–1619. English painter.
**Hill·man** (hĭl'mən), **Sidney.** 1887–1946. Amer. labor leader.
**Hill·quit** (hĭl'kwĭt'), **Morris.** 1869–1933. Amer. socialist leader.
**Hil·ton** (hĭl'tən), **Conrad Nicholson.** 1887–1979. Amer. hotel-chain organizer.
**Hilton, James.** 1900–54. English novelist.
**Himm·ler** (hĭm'lər), **Heinrich.** 1900–45. German Nazi SS official.
**Hin·de·mith** (hĭn'də-mĭth, -mĭt), **Paul.** 1895–1963. German violinist & composer.
**Hin·den·burg** (hĭn'dən-bûrg', -bo͝ork'), **Paul von.** 1847–1934. German general & statesman.
**Hine** (hīn), **Lewis Wickes.** 1874–1940. Amer. pioneer photographer.
**Hines** (hīnz), **Duncan.** 1880–1959. Amer. publisher & author of restaurant guides.
**Hines, Earl ("Fatha").** 1905–83. Amer. musician.
**Hin·kle** (hĭng'kəl), **Beatrice Moses Van Geisen.** 1874–1953. Amer. psychiatrist & public-health official.
**Hin·shel·wood** (hĭn'shəl-wo͝od', -chəl-), Sir **Cyril Norman.** 1897–1967. English chemist (Nobel, 1956).
**Hip·par·chus** (hĭ-pär'kəs). d. 6th cent. B.C. Athenian tyrant (527–514).
**Hipparchus.** fl. 130 B.C. Greek astronomer.
**Hip·pi·as** (hĭp'ē-əs). 6th cent. B.C. Athenian ruler.
**Hip·poc·ra·tes** (hĭ-pŏk'rə-tēz'). *the Father of Medicine.* 460?–377? B.C. Greek physician. **—Hip'po·crat'ic** (hĭp'ə-krăt'ĭk) *adj.*
**Hires** (hīrz), **Charles Elmer.** 1851–1937. Amer. soft-drink manufacturer.
**Hi·ro·hi·to** (hîr'ō-hē'tō). b. 1901. Japanese emperor (since 1926).
**Hi·ro·shi·ge** (hîr'ō-shē'gā), **Ando.** 1797–1858. Japanese painter.
**Hirsch·horn** (hûrsh'hôrn'), **Joseph Herman.** 1899–1981. Amer. financier & arts patron.
**Hiss** (hĭs), **Alger.** b. 1904. Amer. public official.
**Hitch·cock** (hĭch'kŏk'), **Alfred Joseph.** 1899–1980. British filmmaker.
**Hitchcock, Edward.** 1793–1864. Amer. geologist & educator.
**Hitchcock, Ethan Allen.** 1835–1909. Amer. manufacturer, diplomat, & public official.
**Hit·ler** (hĭt'lər), **Adolf.** 1889–1945. Austrian-born German Nazi dictator.
**Hit·torf** (hĭt'ôrf'), **Johann Wilhelm.** 1824–1914. German physicist.
**Hoar** (hôr, hōr), **Ebenezer Rockwood.** 1816–95. Amer. jurist.
**Hoar, George Frisbie.** 1826–1904. Amer. legislator.
**Hoard** (hôrd, hōrd), **William Dempster.** 1836–1918. Amer. editor, agricultural pioneer, & politician.
**Hoare** (hôr, hōr), Sir **Samuel John Gurney.** *1st Viscount Templewood.* 1880–1959. English diplomat & public official.
**Ho·ban** (hō'bən), **James.** 1762?–1831. Irish-born Amer. architect.
**Ho·bart** (hō'bärt', -bərt), **Garret Augustus.** 1844–99. U.S. Vice President (1897–99).
**Hob·be·ma** (hŏb'ə-mə), **Meindert.** 1638–1709. Dutch landscape painter.
**Hobbes** (hŏbz), **Thomas.** 1588–1679. English philosopher. **—Hobbes'i·an** *adj.* & *n.*
**Hob·by** (hŏb'ē), **Oveta Culp.** b. 1905. Amer. public official.
**Hob·son** (hŏb'sən), **Richmond Pearson.** 1870–1937. Amer. naval officer, politician, & reformer.
**Hoc·cleve** (hŏk'lēv'), **Thomas.** 1370?–1450? English poet.
**Ho Chi Minh** (hō' chē' mĭn'). 1890–1969. Vietnamese Communist leader.
**Hock·ing** (hŏk'ĭng), **William Ernest.** 1873–1966. Amer. philosopher.
**Hodg·es** (hŏj'ĭz), **Gil.** 1924–72. Amer. baseball player & manager.
**Hodg·kin** (hŏj'kĭn), Sir **Alan Lloyd.** b. 1914. British physiologist (Nobel, 1963).
**Hodgkin, Dorothy Mary Crowfoot.** b. 1910. Egyptian-born British chemist (Nobel, 1964).
**Hoe** (hō), **Richard March.** 1812–86. Amer. inventor & manufacturer.
**Hoe, Robert.** 1784–1833. English-born Amer. manufacturer.
**Ho·fer** (hō'fər), **Andreas.** 1767–1810. Tyrolese patriot; executed.
**Hof·fa** (hŏf'ə), **James Riddle ("Jimmy").** 1913–75? Amer. labor leader; presumed murdered.
**Hoff·man** (hŏf'mən, hôf'-), **Malvina.** 1887–1966. Amer. artist & author.
**Hoff·mann** (hŏf'mən, -män', hôf'-), **August Heinrich.** *Hoffmann von Fallersleben.* 1798–1874. German author, philologist, & literary historian.
**Hoffmann, Ernst Theodor Amadeus (Wilhelm).** 1776–1822. German author, critic, & composer.
**Hoffmann, Roald.** b. 1937. Polish-born Amer. chemist (Nobel, 1981).
**Hof·mann** (hŏf'mən, -män', hôf'-), **August Wilhelm von.** 1818–92. German chemist.
**Hofmann, Hans.** 1880–1966. German-born Amer. artist.
**Hofmann, Josef Casimir.** 1876–1957. Polish-born pianist & composer.
**Hof·manns·thal** (hŏf'məns-täl', hôf'-), **Hugo von.** 1874–1929. Austrian author.
**Hof·stadt·er** (hŏf'stăt'ər), **Richard.** 1916–70. Amer. historian.
**Hofstadter, Robert.** b. 1915. Amer. physicist (Nobel, 1961).
**Ho·gan** (hō'gən), **William Benjamin ("Ben").** b. 1912. Amer. golfer.

**Ho·garth** (hō′gärth′), **William.** 1697–1764. English artist.
**Hogg** (hôg, hŏg), **James.** 1770–1835. Scottish poet.
**Ho·hen·lo·he** (hō′ən-lō′ə). German princely family (12th–19th cent.).
**Ho·hen·stau·fen** (hō′ən-shtou′fən). German princely family, ruling in Germany & Sicily (12th–13th cent.).
**Ho·hen·zol·lern** (hō′ən-zŏl′ərn). German royal family, ruling Brandenburg (1415–1918), Prussia (1701–1918), & Germany (1871–1918).
**Ho·kin·son** (hō′kĭn-sən), **Helen Elna.** 1893–1949. Amer. cartoonist.
**Ho·ku·sai** (hō′ko͞o-sī′, hō′ko͞o-sī′). 1760–1849. Japanese artist.
**Hol·bein** (hōl′bīn, hôl′-), **Hans.** *the Elder.* 1465?–1524. German painter.
**Holbein, Hans.** *the Younger.* 1497?–1543. German-born artist in Switzerland & England.
**Hol·berg** (hôl′bĕrg′), **Ludvig.** 1684–1754. Danish dramatist.
**Hol·brook** (hōl′bro͝ok′), **Josiah.** 1788–1854. Amer. educational reformer.
**Hol·i·day** (hŏl′ĭ-dā′), **Eleanor ("Billie").** *Lady Day.* 1915–59. Amer. singer.
**Hol·in·shed** (hŏl′ən-shĕd′, -ĭnz-hĕd′) *also* **Hol·lings·head** (-ĭngz-hĕd′), **Raphael.** d. 1580? English chronicler.
**Hol·la·day** (hŏl′ə-dā′), **Ben.** 1819–87. Amer. financier.
**Hol·land** (hŏl′ənd), **John Philip.** 1840–1914. Irish-born Amer. inventor & submarine pioneer.
**Holland, Josiah Gilbert.** *Timothy Titcomb.* 1819–81. Amer. author & editor.
**Holland, Sidney George.** 1893–1961. New Zealand statesman.
**Hol·ler·ith** (hŏl′ə-rĭth′), **Herman.** 1860–1929. Amer. inventor.
**Hol·ley** (hŏl′ē), **Alexander Lyman.** 1832–82. Amer. metallurgist, engineer, & author.
**Holley, Marietta.** 1836–1926. Amer. author & feminist.
**Holley, Robert William.** b. 1922. Amer. biochemist (Nobel, 1968).
**Hol·li·day** (hŏl′ĭ-dā′), **Judith Tuvim ("Judy").** 1922–65. Amer. comedienne.
**Hol·lings·head** (hŏl′ĭngz-hĕd′). *var. of* HOLINSHED.
**Holmes** (hōmz, hōlmz), **John Haynes.** 1879–1964. Amer. clergyman & libertarian.
**Holmes, Oliver Wendell.** 1809–94. Amer. physician & author.
**Holmes, Oliver Wendell, Jr.** 1841–1935. Amer. jurist.
**Holt** (hōlt), **Harold Edward.** 1908–67. Australian statesman.
**Holt, Luther Emmett.** 1855–1924. Australian pioneer pediatrician.
**Holt, Winifred.** 1870–1945. Amer. sculptor & philanthropist.
**Hol·yoake** (hŏl′yōk′, hō′lē-ōk′), **Keith Jacka.** b. 1904. New Zealand statesman.
**Ho·mer** (hō′mər). fl. 850? B.C. Greek epic poet. **—Ho·mer′ic** (hō-mĕr′ĭk) *adj.*
**Homer, Louise Dilworth Beatty.** 1871–1947. Amer. opera singer.
**Homer, Winslow.** 1836–1910. Amer. painter.
**Hone** (hōn), **Philip.** 1780–1851. Amer. diarist & public official.
**Hon·eg·ger** (hŏn′ĭ-gər, ŏn′-, ô-nĕ-gĕr′), **Arthur.** 1892–1955. French-born Swiss composer.
**Ho·no·ri·us** (hə-nôr′ē-əs, -nōr′-), **Flavius.** 384–423. Western Roman emperor (395–423).
**Hood** (ho͝od), **John Bell.** 1831–79. Amer. Confederate soldier.
**Hood, Samuel.** 1724–1816. English admiral.
**Hood, Thomas.** 1799–1845. English poet & editor.
**Hook** (ho͝ok), **Sidney.** b. 1902. Amer. philosopher.
**Hooke** (ho͝ok), **Robert.** 1635–1703. English philosopher, inventor, & mathematician.
**Hook·er** (ho͝ok′ər), **Isabella Beecher.** 1822–1907. Amer. philanthropist & feminist.
**Hooker, Joseph.** 1814–79. Amer. Union army officer.
**Hooker,** Sir **Joseph Dalton.** 1817–1911. English botanist & explorer.
**Hooker, Richard.** 1554?–1600. English author & theologian.
**Hooker, Thomas.** 1586?–1647. English-born Amer. colonizer & clergyman.
**Hoo·ton** (ho͞ot′n), **Earnest Albert.** 1887–1954. Amer. anthropologist & educator.
**Hoo·ver** (ho͞o′vər), **Herbert Clark.** 1874–1964. 31st U.S. President (1929–33), relief administrator, & statesman.
**Hoover, J(ohn) Edgar.** 1895–1972. Amer. director of FBI (1924–72).
**Hope** (hōp), **John.** 1868–1936. Amer. educator.
**Hope, Leslie Towne ("Bob").** b. 1903. English-born Amer. entertainer.
**Hop·kins** (hŏp′kĭnz), **Esek.** 1718–1802. Amer. Revolutionary naval officer.

o͞o **boot** ou **out** th **thin** *th* **this** ŭ **cut** ûr **urge** y **young** yo͞o **abuse** zh **vision** ə **a**bout, it**e**m, ed**i**ble, gall**o**p, circ**u**s

**Hopkins,** Sir **Frederick Gowland.** 1861–1947. English biochemist (Nobel, 1929).
**Hopkins, Gerard Manley.** 1844–89. English priest, poet, & artist.
**Hopkins, Johns.** 1795–1873. Amer. financier & philanthropist.
**Hopkins, Mark.** 1802–87. Amer. educator & theologian.
**Hopkins, Samuel.** 1721–1803. Amer. theologian.
**Hop·kin·son** (hŏp′kĭn-sən), **Francis.** 1737–91. Amer. Revolutionary leader & author.
**Hop·pe** (hŏp′ē), **William Frederick.** 1887–1959. Amer. billiards player.
**Hop·per** (hŏp′ər), **(William) DeWolf.** 1858–1935. Amer. musical-comedy actor.
**Hopper, Edward.** 1882–1967. Amer. painter.
**Hopper, Hedda.** 1890–1966. Amer. actress & columnist.
**Hor·ace** (hôr′əs, hŏr′-). 65–8 B.C. Roman poet. **—Ho·ra′tian** (hə-rā′shən) *adj.*
**Hor·na·day** (hôr′nə-dā′), **William Temple.** 1854–1937. Amer. zoologist.
**Horne** (hôrn), **Lena.** b. 1917. Amer. singer.
**Horne, Marilyn.** b. 1934. Amer. operatic soprano.
**Hor·ney** (hôr′nī), **Karen Danielsen.** 1885–1952. German-born Amer. psychoanalyst.
**Horns·by** (hôrnz′bē), **Rogers.** 1896–1963. Amer. baseball player & manager.
**Ho·ro·witz** (hôr′ə-wĭts, hŏr′-), **Vladimir.** b. 1904. Russian-born Amer. pianist.
**Hor·thy** (hôr′tē, hŏr′-), **Miklós von Nagybánya.** 1868–1957. Hungarian admiral & statesman.
**Hor·ton** (hôr′tn), **Edward Everett.** 1887–1970. Amer. actor.
**Hor·wich** (hôr′wĭch), **Frances Rappaport.** b. 1908. Amer. educator.
**Ho·se·a** (hō-zā′ə, -zē′-ə). 8th cent. B.C. Hebrew prophet.
**Hos·mer** (hŏz′mər), **Harriet Goodhue.** 1830–1908. Amer. sculptor.
**Hou·di·ni** (ho͞o-dē′nē), **Harry.** 1874–1926. Amer. magician.
**Hou·don** (ho͞o′dŏn′, o͞o-dôN′), **Jean Antoine.** 1741–1828. French sculptor.
**Hou·dry** (ho͞o′drē, o͞o-drē′), **Eugene Jules.** 1892–1962. French-born Amer. engineer & manufacturer.
**Hough** (hŭf), **Emerson.** 1857–1923. Amer. author.
**Houns·field** (hounz′fēld′), **Godfrey Newbold.** b. 1919. English engineer & inventor (Nobel, 1979).
**House** (hous), **Edward Mandell ("Colonel").** 1858–1938. Amer. diplomat & Presidential adviser.
**House·man** (hous′mən), **John.** b. 1902. Amer. producer, director, & actor.
**Hous·man** (hous′mən), **Alfred Edward.** 1859–1936. English poet & scholar.
**Housman, Laurence.** 1865–1959. English author & illustrator.
**Hous·say** (o͞o-sī′), **Bernardo Alberto.** 1887–1971. Argentine physiologist (Nobel, 1947).
**Hous·ton** (hyo͞o′stən), **Samuel.** 1793–1863. Amer. general & politician.
**Hov·ey** (hŭv′ē), **Richard.** 1864–1900. Amer. poet.
**How·ard** (hou′ərd), **Ada Lydia.** 1829–1907. Amer. educator & college administrator.
**Howard, Catherine.** 1520?–42. Queen of England as fifth wife of Henry VIII (1540–42); executed for adultery.
**Howard, Henry.** *Earl of Surrey.* 1517?–47. English poet & soldier; beheaded for treason.
**Howard, Oliver Otis.** 1830–1909. Amer. Union general.
**Howard, Roy Wilson.** 1883–1964. Amer. journalist & publisher.
**Howard, Sidney Coe.** 1891–1939. Amer. playwright.
**Howe** (hou), **Elias.** 1819–67. Amer. inventor & manufacturer.
**Howe, Irving.** b. 1920. Amer. critic.
**Howe, Julia Ward.** 1819–1910. Amer. author, feminist, & philanthropist.
**Howe, Richard.** *Earl Howe.* 1726–99. English admiral.
**Howe, Samuel Gridley.** 1801–76. Amer. humanitarian & reformer.
**Howe,** Sir **William.** *5th Viscount Howe.* 1729–1814. English general in America.
**How·ells** (hou′əlz), **William Dean.** 1837–1920. Amer. author & editor.
**How·land** (hou′lənd), **Emily.** 1827–1929. Amer. educator, philanthropist, & reformer.
**Hr·dlič·ka** (hûrd′lĭch-kä′), **Aleš.** 1869–1943. Bohemian-born Amer. anthropologist.
**Hrolf** (hrŏlf, rŏlf). ROLLO.
**Hsia** (shē-ä′). XIA.
**Hua Guo·feng** *also* **Hua Kuo·feng** (hwä′ gwō′fŭng′). b. 1920. Chinese Communist leader.
**Huás·car** (wäs′kär′). 1495?–1533. Incan chief.
**Hub·bard** (hŭb′ərd), **Elbert Green.** 1856–1915. Amer. author & publisher.
**Hubbard, Gardiner Greene.** 1822–97. Amer. lawyer & geographer.

**Hub·ble** (hŭb'əl), **Edwin Powell.** 1889–1953. Amer. astronomer.
**Hu·bel** (hyōō'bəl), **David.** b. 1926. Amer. neurobiologist (Nobel, 1981).
**Hud·son** (hŭd'sən), **Henry.** d. 1611. English navigator & Arctic explorer.
**Hudson, Manley Ottmer.** 1886–1960. Amer. jurist & educator.
**Hudson, William Henry.** 1841–1922. English naturalist & author.
**Huer·ta** (wĕr'tə, ōō-ĕr'-), **Victoriano.** 1854–1916. Mexican statesman.
**Hug·gins** (hŭg'ĭnz), **Charles Brenton.** b. 1901. Canadian-born Amer. surgeon (Nobel, 1966).
**Huggins,** Sir **William.** 1824–1910. English astronomer.
**Hugh Ca·pet** (hyōō' kā'pĭt, kăp'ĭt, kă-pā'). —See CAPET.
**Hughes** (hyōōz), **Charles Evans.** 1862–1948. Amer. jurist & statesman.
**Hughes, Howard Robard.** 1905–76. Amer. manufacturer, film producer, & recluse.
**Hughes, (James) Langston.** 1902–67. Amer. author.
**Hughes, Rupert.** 1872–1956. Amer. author.
**Hughes, Thomas.** 1822–96. English reformer & author.
**Hughes, William Morris.** 1864–1952. Australian statesman.
**Hu·go** (hyōō'gō, ü-gō'), **Victor Marie.** 1802–85. French author.
**Hui·zing·a** (hī'zĭng-ə), **Johan.** 1872–1945. Dutch historian.
**Hu·la·gu** (hōō-lä'gōō). 1217–65. Mongol ruler.
**Hull** (hŭl), **Cordell.** 1871–1955. Amer. statesman (Nobel, 1945).
**Hull, Isaac.** 1773–1843. Amer. naval officer.
**Hull, William.** 1753–1825. Amer. general.
**Hum·boldt** (hŭm'bōlt', hōōm'-), Baron **(Friedrich Heinrich) Alexander von.** 1769–1859. German naturalist, author, & statesman.
**Humboldt,** Baron **(Friedrich) Wilhelm (Christian Karl Ferdinand) von.** 1767–1835. German philologist & diplomat.
**Hume** (hyōōm), **David.** 1711–76. Scottish philosopher & historian.
**Hum·per·dinck** (hōōm'pər-dĭngk', hŭm'-), **Engelbert.** 1854–1921. German composer.
**Hum·phrey** (hŭm'frē, hŭmp'-). *Duke of Gloucester & Earl of Pembroke.* 1391–1447. English statesman & collector.
**Humphrey, Doris.** 1895–1958. Amer. dancer & choreographer.
**Humphrey, Hubert Horatio.** 1911–78. U.S. Vice President (1965–69) & legislator.
**Hum·phreys** (hŭm'frēz, hŭmp'-), **Joshua.** 1751–1838. Amer. naval architect.
**Hun·e·ker** (hŭn'ĭ-kər), **James Gibbons.** 1860–1921. Amer. critic.
**Hun·sa·ker** (hŭn'sā'kər), **Jerome Clarke.** 1886–1984. Amer. aeronautical engineer.
**Hunt** (hŭnt), **Haroldson Lafayette ("H.L.").** 1889–1974. Amer. businessman.
**Hunt, Harriot Kezia.** 1805–75. Amer. pioneer physician & reformer.
**Hunt, (William) Holman.** 1827–1910. English painter.
**Hunt, (James Henry) Leigh.** 1784–1859. English author.
**Hunt, Mary Hannah Hanchett.** 1830–1906. Amer. temperance reformer.
**Hunt, Richard Morris.** 1827–95. Amer. architect.
**Hunt, Ward.** 1810–86. Amer. jurist.
**Hunt, William Morris.** 1824–79. Amer. painter.
**Hun·ter** (hŭn'tər), **John.** 1728–93. English anatomist.
**Hunter, Robert Mercer Taliaferro.** 1809–87. Amer. politician & Confederate statesman.
**Hun·ting·ton** (hŭn'tĭng-tən), **Collis Potter.** 1821–1900. Amer. transportation executive.
**Huntington, Ellsworth.** 1876–1947. Amer. geographer & explorer.
**Huntington, Henry Edwards.** 1850–1927. Amer. railroad executive, art collector, & philanthropist.
**Huntington, Samuel.** 1731–96. Amer. Revolutionary leader.
**Hunt·ley** (hŭnt'lē), **Chet.** 1911–74. Amer. broadcast journalist.
**Hu·nya·di** or **Hu·nya·dy** (hōōn'yä-dē, -yô-), **János.** 1387?–1456. Hungarian general.
**Hur·ley** (hûr'lē), **Patrick Jay.** 1883–1963. Amer. public official & diplomat.
**Hur·ok** (hyōōr'ŏk'), **Solomon ("Sol").** 1888–1974. Russian-born Amer. impresario.
**Hurst** (hûrst), Sir **Cecil James Barrington.** 1870–1963. English jurist.
**Hurst, Fannie.** 1889–1968. Amer. author.
**Hur·ston** (hûr'stən), **Zora Neale.** 1901?–60. Amer. author.
**Hus** (hŭs, hōōs). *var. of* HUSS.
**Hu·sain** or **Hu·sayn** (hōō-sān'). *vars. of* HUSSEIN.
**Hu·sein ibn-A·li** (hōō-sān' ĭb'ən-ä-lē'). 1856–1931. King of Hejaz (1916–24) & Arabia (1917–24); abdicated.
**Hu Shi** *also* **Hu Shih** (hōō' shĭr'). 1891–1962. Chinese philosopher & diplomat.
**Huss** or **Hus** (hŭs, hōōs), **John** or **Jan.** 1374–1415. Bohemian religious reformer.
**Hus·sein** or **Hu·sain** or **Hu·sayn ibn Ta·lal** (hōō-sān' ĭb'ən tə-läl'). b. 1935. King of Jordan (since 1953).
**Hus·sey** (hŭs'ē), **Obed.** 1792–1860. Amer. inventor and manufacturer.
**Hus·ted** (hyōō'stĭd'), **Marjorie Child.** *Betty Crocker.* b. 1892? Amer. home-economics executive.
**Hus·ton** (hyōō'stən), **John.** b. 1906. Amer. filmmaker.
**Huston, Walter.** 1884–1950. Amer. actor.
**Hu·szár** (hōōs'är), **Károly.** 1882–1941. Hungarian journalist & politician.
**Hutch·ins** (hŭch'ĭnz), **Robert Maynard.** 1899–1977. Amer. educator.
**Hutchins, Thomas.** 1730–89. Amer. cartographer & engineer.
**Hutch·in·son** (hŭch'ĭn-sən), **Anne.** 1591–1643. English-born Amer. colonist & religious leader.
**Hutchinson, Thomas.** 1711–80. Amer. colonial official.
**Hut·ten** (hōōt'n), **Ulrich von.** 1488–1523. German humanist.
**Hut·ton** (hŭt'n), **Barbara.** 1912–79. Amer. socialite.
**Hux·ley** (hŭks'lē), **Aldous Leonard.** 1894–1963. English author.
**Huxley, Andrew Fielding.** b. 1917. English physiologist (Nobel, 1963).
**Huxley,** Sir **Julian Sorell.** 1887–1975. English biologist & author.
**Huxley, Thomas Henry.** 1825–95. English biologist.
**Hu Yao·bang** *also* **Hu Yao-pang** (hōō' you'băng'). b. 1915. Chinese politician.
**Huy·gens** or **Huy·ghens** (hī'gənz), **Christian.** 1629–95. Dutch physicist & astronomer.
**Huys·mans** (wēs-mäNs'), **Camille.** 1871–1968. Belgian politician & journalist.
**Huysmans, Joris Karl.** 1848–1907. French novelist.
**Hy·att** (hī'ət), **Alpheus.** 1838–1902. Amer. naturalist.
**Hyatt, Anna Vaugh.** 1876–1973. Amer. sculptor.
**Hyatt, John Wesley.** 1837–1920. Amer. inventor.
**Hyde** (hīd), **Douglas.** 1860–1949. Irish nationalist & author.
**Hyde, Edward.** *1st Earl of Clarendon.* 1609–74. English statesman & historian.
**Hy·mans** (hī'mäns', ē-mäNs'), **Paul.** 1865–1941. Belgian diplomat & statesman.

# I

**I·ber·ville** (ē-bər-vēl'), Sieur d'. *Pierre Le Moyne.* 1661–1706. Canadian-born French explorer.
**ibn-Khal·dun** (ĭb'ən-käl-dōōn'). 1332–1406. Arab historian.
**ibn-Sa·ud** (ĭb'ən-sä-ōōd', -soud'), **Abdul Aziz.** 1880–1953. Saudi Arabian king (1932–53).
**Ib·ra·him Pa·sha** (ĭb'rä-hĭm' pä'shə). 1789–1849. Egyptian general & viceroy.
**Ib·sen** (ĭb'sən, ĭp'-), **Henrik.** 1828–1906. Norwegian dramatist. —**Ib·sen'i·an** (-sē'nē-ən, -sĕn'ē-) *adj.*
**Ick·es** (ĭk'ēz, -əs), **Harold LeClair.** 1874–1952. Amer. politician & author.
**Ic·ti·nus** (ĭk-tī'nəs). 5th cent. B.C. Greek architect.
**Ig·na·tius** (ĭg-nā'shəs), Saint. d. c. A.D. 110. Bishop of Antioch; martyred.
**Ignatius of Loy·o·la** (loi-ō'lə), Saint. 1491–1556. Spanish ecclesiastic & founder of the Society of Jesus.
**Ikh·na·ton** (ĭk-nä'tn). AKHENATON.
**In·di·an·a** (ĭn'dē-ăn'ə), **Robert.** b. 1928. Amer. artist.
**In·dy** (ăn'dē, äN-dē'), **(Paul Marie Théodore) Vincent d'.** 1851–1931. French composer.
**Inge** (ĭnj), **William.** 1913–73. Amer. playwright.
**Inge** (ĭng), **William Ralph.** 1860–1954. English clergyman & author.
**In·ger·soll** (ĭng'gər-sôl', -sōl', -səl), **Robert Green.** 1833–99. Amer. politician & lecturer.
**In·gra·ham** (ĭng'grə-hăm', ĭng'grəm), **Prentiss.** 1843–1904. Amer. author & adventurer.
**In·gram** (ĭng'grəm), **Arthur Foley Winnington.** 1858–1946. English prelate & reformer.
**In·gres** (ăN'grə), **Jean Auguste Dominique.** 1780–1867. French painter.
**In·man** (ĭn'mən), **Henry.** 1801–46. Amer. portrait painter.
**In·ness** (ĭn'ĭs), **George.** 1825–94. Amer. landscape painter.
**In·no·cent** (ĭn'ə-sənt). Name of 13 popes, esp.: **a. II.** d. 1143. Reigned 1130–43. **b. III.** 1161–1216. Reigned 1198–1216. **c. IV.** d. 1254. **d. XI.** 1611–89. Reigned 1676–89.
**I·nö·nü** (ĭn'ə-nōō', -nyōō'), **Ismet.** 1884–1973. Turkish statesman.
**In·sull** (ĭn'səl), **Samuel.** 1859–1938. English-born Amer. utilities mogul.
**Io·nes·co** (ē'ə-nĕs'kō, yə-), **Eugène.** b. 1912. Rumanian-born French dramatist.

**I·pa·tieff** (ĭ-pä'tē-ĕf', ĭ-pä'chəf), **Vladimir Nikolaevich.** 1867–1952. Russian-born Amer. chemist.
**Ire·dell** (īr'dĕl'), **James.** 1751–99. Amer. jurist.
**Ire·ton** (īr'tn), **Henry.** 1611–51. English Civil War general.
**I·ri·go·yen** (ĭr'ĭ-gō'yĕn'), **Hipólito.** 1852–1933. Argentine statesman.
**Ir·ving** (ûr'vĭng), Sir **Henry.** 1838–1905. English actor.
**Irving, Washington.** 1783–1859. Amer. writer.
**Ir·win** (ûr'wĭn), **May.** 1862–1938. Amer. comedienne.
**Irwin, Wallace.** 1875–1959. Amer. humorist.
**Irwin, William Henry ("Will").** 1873–1948. Amer. author.
**I·saac** (ī'zək). Hebrew patriarch.
**I·saacs** (ī'zəks), Sir **Isaac Alfred.** 1855–1948. Australian jurist & statesman.
**Is·a·bel·la I** (ĭz'ə-bĕl'ə). *the Catholic.* 1451–1504. Queen of Castile & Aragon.
**I·sa·iah** (ī-zā'ə). 8th cent. B.C. Hebrew prophet.
**Ish·er·wood** (ĭsh'ər-wo͝od'), **Christopher William Bradshaw.** b. 1904. British-born Amer. writer.
**I·shi·i** (ē'shē-ē, ĭsh'ē-ē), Viscount **Kikujiro.** 1866–1945. Japanese diplomat.
**Is·i·dore of Se·ville** (ĭz'ĭ-dôr', -dōr'; sə-vĭl'), Saint. 560?–636. Spanish scholar & ecclesiastic.
**Is·ma·il Pa·sha** (ĭs-mä'ēl pä'shə). 1830–95. Egyptian viceroy (1863–79).
**I·soc·ra·tes** (ī-sŏk'rə-tēz'). 436–338 B.C. Athenian orator & rhetorician.
**I·to** (ē'tō'), Prince **Hirobumi.** 1841–1909. Japanese statesman; assassinated.
**Ito,** Count **Yuko.** 1843–1914. Japanese admiral.
**I·tur·bi** (ĭ-tûr'bē, ē-to͞or'-), **José.** 1895–1980. Spanish-born pianist & conductor.
**I·tur·bi·de** (ē'to͞or-bē'dā), **Agustín de.** 1783–1824. Mexican revolutionary.
**I·van III Va·sil·ie·vich** (ē-vän', ī'vən; və-sĭl'yə-vĭch'). *the Great.* 1440–1505. Grand Duke of Muscovy (1462–1505).
**Ivan IV Vasilievich.** *the Terrible.* 1530–1584. Grand Duke of Muscovy (1533–84) & czar of Russia (1547–84).
**Ives** (īvz), **Charles Edward.** 1874–1954. Amer. composer.
**Ives, Frederick Eugene.** 1856–1937. Amer. inventor.
**Ives, James Merritt.** 1824–95. Amer. lithographer.
**Iz·ard** (ĭz'ərd), **Ralph.** 1742–1804. Amer. Revolutionary leader & diplomat.

# J

**Ja·bir** (jä'bĭr, jä'bər). GEBER.
**Jack·son** (jăk'sən), **Andrew.** *Old Hickory.* 1767–1845. Soldier & seventh U.S. President (1829–37). **—Jack·son'i·an** *adj. & n.*
**Jackson, Charles Thomas.** 1805–80. Amer. physician, chemist, & geologist.
**Jackson, Helen (Maria Fiske) Hunt.** 1830–85. Amer. author.
**Jackson, Howell Edmunds.** 1832–95. Amer. jurist.
**Jackson, Jesse Louis.** b. 1941. Amer. civil-rights leader.
**Jackson, Mahalia.** 1911–72. Amer. singer.
**Jackson, Robert Houghwout.** 1892–1954. Amer. jurist.
**Jackson, Thomas Jonathan ("Stonewall").** 1824–63. Amer. Confederate general.
**Ja·cob** (jā'kəb). Hebrew patriarch.
**Ja·cob** (zhä-kôb'), **François.** b. 1920. French geneticist (Nobel, 1965).
**Ja·co·bi** (jə-kō'bē), **Abraham** (German-born; 1830–1919) & **Mary Corinna Putnam** (English-born; 1842–1906). Amer. physicians.
**Jac·quard** (zhä-kär', jăk'ärd'), **Joseph Marie.** 1752–1834. French inventor.
**Jag·a·tai** (jăg'ə-tī'). d. 1242. Mongol ruler.
**Jag·ger** (jăg'ər), **Michael Philip ("Mick").** b. 1944. English musician.
**Ja·han·gir** (jə-hän-gēr'). 1569–1627. Mongol emperor (1605–27).
**James** (jāmz), Saint. *the Less.* Traditionally regarded as the brother of Jesus.
**James,** Saint. *the Greater.* d. A.D. 44. One of the 12 Apostles; martyred.
**James,** Saint. One of the 12 Apostles.
**James.** Name of six kings of Scotland & Great Britain, esp.: **a. I.** 1566–1625. Ruled 1603–25 in England & 1567–1625 in Scotland as **James VI. b. II.** 1633–1701. Ruled 1685–88; succeeded by his Protestant son-in-law, William of Orange.
**James, Henry.** 1843–1916. Amer. novelist & critic. **—James'i·an** *adj. & n.*
**James, Jesse Woodson.** 1847–82. Amer. outlaw.
**James, William.** 1842–1910. Amer. psychologist & philosopher. **—James'i·an** *adj. & n.*
**Jame·son** (jām'sən, jă'mə-), Sir **Leander Starr.** 1853–1917. Scottish physician & colonial administrator.
**Ja·me·son** (jā'mə-sən), **John Franklin.** 1859–1937. Amer. historian & educator.
**Ja·mi** (jä'mē). 1414–92. Persian poet & mystic.
**Ja·mi·son** (jā'mĭ-sən), **Cecilia Viets Dakin Hamilton.** 1837–1909. Canadian-born Amer. author & painter.
**Ja·ná·ček** (yä'nə-chĕk'), **Leoš.** 1854–1928. Czechoslovakian composer.
**Jan·sen** (jăn'sən, yän'-), **Cornelius.** 1585–1638. Dutch theologian. **—Jan'sen·ist** *adj. & n.* **—Jan'sen·is'tic** *adj.*
**Ja·pheth** (jā'fĭth', jăf'ĭth). Noah's son in the Bible.
**Jaques-Dal·croze** (zhäk'däl-krōz'), **Émile.** 1865–1950. Swiss composer & educator.
**Jar·rell** (jə-rĕl'), **Randall.** 1914–65. Amer. poet.
**Jar·ves** (jär'vəs), **James Jackson.** 1818–88. Amer. critic & art collector.
**Jas·pers** (yäs'pərs), **Karl.** 1883–1969. German psychiatrist & philosopher.
**Jau·rès** (zhō-rĕs'), **Jean Léon.** 1859–1914. French journalist & socialist leader; assassinated.
**Jay** (jā), **John.** 1745–1829. Amer. diplomat & jurist.
**Jeanne d'Arc** (zhäɴ därk'). JOAN OF ARC.
**Jeans** (jēnz), Sir **James Hopwood.** 1877–1946. English astronomer, physicist, & mathematician.
**Jef·fers** (jĕf'ərz), **Robinson.** 1887–1962. Amer. poet.
**Jef·fer·son** (jĕf'ər-sən), **Joseph.** 1829–1905. Amer. actor.
**Jefferson, Thomas.** 1743–1826. Third U.S. President (1801–9), author, scientist, architect, educator, & diplomat. **—Jef'fer·so'ni·an** (-sō'nē-ən) *adj. & n.*
**Jef·frey** (jĕf'rē), **Francis.** *Lord Jeffrey.* 1773–1850. Scottish critic & jurist.
**Jef·fries** (jef'rēz), **John.** 1745–1819. Amer. physician & balloonist.
**Je·hosh·a·phat** (jə-hŏsh'ə-făt', -hŏs'-). 9th cent. B.C. king of Judah.
**Je·hu** (jē'hyo͞o). 9th cent. B.C. Israeli king.
**Jel·li·coe** (jĕl'ĭ-kō'), **John Rushworth.** *1st Earl Jellicoe.* 1859–1935. English naval officer.
**Jen·ghis Khan** *also* **Jen·ghiz Khan** (jĕn'gĭz kän', -gĭs, jĕng'-). GENGHIS KHAN.
**Jen·ner** (jĕn'ər), **Edward.** 1749–1823. English physician.
**Jen·ney** (jĕn'ē), **William Le Baron.** 1832–1907. Amer. inventor & architect.
**Jen·sen** (yĕn'sən), **Johannes Hans Daniel.** 1906–73. German physicist (Nobel, 1963).
**Jen·sen** (yĕn'sən, jĕn'-), **Johannes Vilhelm.** 1873–1950. Danish author (Nobel, 1944).
**Jer·e·mi·ah** (jĕr'ə-mī'ə). 7th–6th cent. B.C. Hebrew prophet.
**Je·ri·tza** (yĕ'rēt-sä', yĕr'ĭt-sə), **Maria.** 1887–1982. Austrian-born operatic soprano.
**Je·rome** (jə-rōm'), Saint. 340?–420. Latin scholar.
**Jer·vis** (jûr'vĭs), **John.** *Earl of St. Vincent.* 1735–1823. English naval officer.
**Jes·per·sen** (yĕs'pər-sən), **(Jens) Otto (Harry).** 1860–1943. Danish philologist.
**Jes·se** (jĕs'ē). King David's father in the Old Testament.
**Jes·sel** (jĕs'əl), **George Albert ("Georgie").** 1898–1981. Amer. actor, film producer, & toastmaster.
**Je·sus** (jē'zŭs) *also* **Je·sus Christ** (krīst). 4? B.C.–A.D. 29? Founder of Christianity.
**Jev·ons** (jĕv'əns), **William Stanley.** 1835–82. English economist & logician.
**Jew·ett** (jo͞o'ĭt), **Sarah Orne.** 1849–1909. Amer. author.
**Jez·e·bel** (jĕz'ə-bĕl'). 9th cent. B.C. queen of Israel.
**Ji·mé·nez** (hē-mā'nəs), **Juan Ramón.** 1881–1958. Spanish poet (Nobel, 1956).
**Jiménez de Cis·ne·ros** (dā sĭs-nĕr'əs), **Francisco.** 1436–1517. Spanish prelate & statesman.
**Jin** (jē-ĭn'). Name of four Chinese dynasties: **Western Jin,** 265–316; **Eastern Jin,** 317–420; **Later Jin,** 936–46; & **Jin,** 1115–1234.
**Jin·nah** (jĭn'ə), **Mohammed Ali.** 1876–1948. Indian nationalist Moslem leader.
**Jo·a·chim** (yō-ä'kĭm, -KHĭm, yō'ə-kĭm', -KHĭm'), **Joseph.** 1831–1907. Hungarian violinist & composer.
**Joan of Arc** (jōn; ärk). 1412–31. French military leader & heroine.
**Job** (jōb). Hebrew patriarch.
**Jo·el** (jō'əl). Hebrew prophet.
**Jof·fre** (zhôf'rə), **Joseph Jacques Césaire.** 1852–1931. French field marshal.
**John** (jŏn), Saint. *the Evangelist.* One of the 12 Apostles & author of the fourth Gospel.
**John,** Saint. *the Baptist.* 5 B.C.–A.D. 30. Baptizer of Jesus.
**John.** Name of 21 popes, esp. **XXIII,** 1881–1963, reigned 1958–63.
**John.** Name of six kings of Portugal, esp. **I,** 1357–1433, ruled 1385–1433.
**John, Augustus Edwin.** 1878–1961. English artist.

**John of Aus·tri·a** (ôs′trē-ə). 1547–78. Spanish general.
**John of Da·mas·cus** (də-măs′kəs). 700–54? Greek theologian.
**John of Gaunt** (gônt, gänt). *Duke of Lancaster.* 1340–99. English nobleman & soldier.
**John of Lack·land** (lăk′lənd). 1167?–1216. English king (1199–1216).
**John of Lan·cas·ter** (lăng′kə-stər, lăn′-). *Duke of Bedford.* 1389–1435. Regent of England & France.
**John of Lei·den** (līd′n). 1509–36. Dutch Anabaptist fanatic.
**John of Salis·bur·y** (solz′bĕr′ē, -brē). d. 1180. English bishop & author.
**John·ny Ap·ple·seed** (jŏn′ē ăp′əl-sēd′). John CHAPMAN.
**John Paul** (jŏn pôl, pōl). Name of two popes: **a. I.** 1912–78. Reigned 1978. **b. II.** b. 1920. Reigned since 1978.
**John III So·bies·ki** (sō-byĕs′kē, sō′bē-ĕs′-). 1629–96. Polish king (1674–96).
**Johns** (jŏnz), **Jasper.** b. 1930. Amer. artist.
**John·son** (jŏn′sən), **Andrew.** 1808–75. 17th U.S. President (1865–69); impeached & acquitted.
**Johnson, Eastman.** 1824–1906. Amer. genre painter.
**John·son** (yōōn′sôn), **Eyvind.** 1904–76. Swedish author (Nobel, 1974).
**John·son** (jŏn′sən), **Hiram Warren.** 1866–1945. Amer. legislator.
**Johnson, Howard Deering.** 1896?–1972. Amer. restaurateur.
**Johnson, Hugh Samuel.** 1882–1942. Amer. public official.
**Johnson, James Weldon.** 1871–1938. Amer. author & educator.
**Johnson, John Arthur ("Jack").** 1878–1946. Amer. prizefighter.
**Johnson, John Harold.** b. 1918. Amer. publisher.
**Johnson, Lyndon Baines.** 1908–73. 36th U.S. President (1963–69).
**Johnson, Osa Helen Leighty.** 1894–1953. Amer. explorer & filmmaker.
**Johnson, Philip Cortelyou.** b. 1906. Amer. architect.
**Johnson, Reverdy.** 1796–1876. Amer. lawyer, politician, & diplomat.
**Johnson, Richard Mentor.** 1780–1850. U.S. Vice President (1837–41) & soldier.
**Johnson, Samuel.** *Dr. Johnson.* 1709–84. English author & lexicographer. **—John·so′ni·an** (jŏn-sō′nē-ən) *adj.* & *n.*
**Johnson, Thomas.** 1732–1819. Amer. jurist.
**Johnson, Tom Loftin.** 1854–1911. Amer. inventor & municipal administrator.
**Johnson,** Sir **William.** 1715–74. British-born Amer. frontiersman & public official.
**Johnson, William.** 1771–1834. Amer. jurist.
**Johnson, William Samuel.** 1727–1819. Amer. Revolutionary leader, jurist, & college administrator.
**John·ston** (jŏn′stən), **Albert Sidney.** 1803–62. Amer. Confederate general.
**Johnston, Annie Fellows.** 1863–1931. Amer. author.
**Johnston, Henrietta.** d. 1729. Irish-born Amer. portrait painter.
**Johnston, Joseph Eggleston.** 1807–91. Amer. Confederate army officer.
**Join·ville** (zhwăN-vēl′), **Jean de.** 1224?–1317. French chronicler.
**Jó·kai** (yô′koi′), **Maurus** *or* **Mór.** 1825–1904. Hungarian author & politician.
**Jo·li·et** *also* **Jol·li·et** (jō′lē-ĕt′, jō′lē-ĕt′, zhô-lyā′), **Louis.** 1645–1700. French-Canadian explorer of America.
**Jo·li·ot-Cu·rie** (zhô-lyō′ kyōō-rē′), **Irène** (1897–1956) & **Frédéric** (1900–58). French physicists (Nobel, 1935).
**Jol·li·et** (jō′lē-ĕt′, jo′lē-ĕt′, zhô-lyā′). *var. of* JOLIET.
**Jol·son** (jōl′sən), **Al.** 1886–1950. Amer. entertainer.
**Jo·nah** (jō′nə). Hebrew prophet.
**Jones** (jōnz), **Henry Arthur.** 1851–1929. English dramatist.
**Jones, Howard Mumford.** 1892–1980. Amer. educator & author.
**Jones, Inigo.** 1573–1652. English architect.
**Jones, Jesse Holman.** 1874–1956. Amer. banker & public official.
**Jones, John Luther ("Casey").** 1864–1900. Amer. locomotive engineer.
**Jones, John Paul.** 1747–92. Scottish-born Amer. naval officer.
**Jones, LeRoi.** Imamu Amiri BARAKA.
**Jones, Mary Harris.** *Mother Jones.* 1830–1930. Irish-born Amer. labor leader.
**Jones, Robert Tyre ("Bobby").** 1902–71. Amer. golfer.
**Jones, Rufus Matthew.** 1863–1948. Amer. Quaker philosopher.
**Jones, Samuel Milton.** 1846–1904. Amer. manufacturer, politician, & reformer.
**Jones, Sybil.** 1808–73. Amer. Quaker preacher.
**Jones, Thomas Hudson.** 1892–1969. Amer. sculptor.
**Jon·son** (jŏn′sən), **Benjamin ("Ben").** 1573–1637. English actor & author.
**Jop·lin** (jŏp′lĭn), **Scott.** 1868–1917. Amer. pianist & composer.
**Jor·dan** (jôr′dn), **David Starr.** 1851–1931. Amer. biologist.
**Jo·seph** (jō′zəf, -səf). Son of Jacob & Rachel in the Bible, sold into slavery in Egypt.
**Joseph.** Husband of Mary, mother of Jesus.
**Joseph,** Chief. 1840?–1904. Nez Percé leader.
**Joseph.** Name of two Holy Roman Emperors, esp. **II,** 1741–90, ruled 1765–90.
**Joseph of Ar·i·ma·the·a** (ăr′ə-mə-thē′ə). Israelite who buried Jesus.
**Jo·sé·phine de Beau·har·nais** (zhō-zā-fēn′ də bō-är-nĕ′). —See BEAUHARNAIS.
**Jo·seph·son** (jō′zəf-sən, -səf-), **Brian David.** b. 1940. English physicist (Nobel, 1973).
**Jo·se·phus** (jō-sē′fəs), **Flavius.** A.D. 37–100? Jewish general & historian.
**Josh·u·a** (jŏsh′ōō-ə). Old Testament Hebrew leader.
**Jo·si·ah** (jō-sī′ə, -zī′ə). King of Judah (638?–607? B.C.).
**Jou·bert** (you-bĕr′, yō-), **Petrus Jacobus.** 1834–1900. Boer general & statesman.
**Jou·haux** (zhōō-ō′), **Léon.** 1879–1954. French politician & labor leader (Nobel, 1951).
**Joule** (jōōl, joul), **James Prescott.** 1818–89. English physicist.
**Jour·dan** (zhōōr-däN′), Comte **Jean Baptiste.** 1762–1833. French marshal.
**Jo·vi·an** (jō′vē-ən). 331?–64. Roman emperor (363–64).
**Jow·ett** (jou′ĭt, jō′-), **Benjamin.** 1817–93. English classical scholar.
**Joyce** (jois), **James.** 1882–1941. Irish author. **—Joyc′e·an** (joi′sē-ən) *adj.*
**Juan Car·los** (wän kär′ləs, -lōs, hwän). b. 1938. Spanish king (since 1975).
**Juan Ma·nuel** (män-wĕl′), Don. 1282–1349. Spanish soldier & author.
**Juá·rez** (hwä′rĕz, -rās), **Benito Pablo.** 1806–72. Mexican statesman.
**Ju·dah** (jōō′də). Hebrew patriarch.
**Ju·das Is·car·i·ot** (jōō′dəs ĭs-kăr′ē-ət). One of the 12 Apostles and betrayer of Jesus.
**Jude** (jōōd), Saint. One of the 12 Apostles.
**Ju·dith** (jōō′dĭth). Jewish biblical heroine.
**Jud·son** (jŭd′sən), **Adoniram.** 1788–1850. Amer. Baptist missionary.
**Judson, Edward Zane Carroll.** *Ned Buntline.* 1823–86. Amer. author.
**Jul·ian** (jōōl′yən). 331–63. Roman emperor (361–63).
**Ju·li·an·a** (jōō′lē-ăn′ə). b. 1909. Queen of the Netherlands (1948–80); abdicated.
**Ju·lius** (jōōl′yəs, -ē-əs). Name of three popes, esp. **II,** 1443–1513, reigned 1503–13.
**Ju·neau** (jōō′nō, jōō-nō′, zhü-nō′), **Solomon Laurent.** 1793–1856. Canadian-born Amer. fur trader & settler.
**Jung** (yōōng), **Carl Gustav.** 1875–1961. Swiss psychologist & psychiatrist.
**Jun·kers** (yōōng′kərz, -kərs), **Hugo.** 1859–1935. German aircraft designer.
**Jusse·rand** (zhüs-räN′), **Jean Jules.** 1855–1932. French scholar & diplomat.
**Jus·tin** (jŭs′tĭn), Saint. A.D. 100?–65. Greek church father.
**Jus·tin·i·an** (jŭ-stĭn′ē-ən). Name of two Byzantine emperors, esp. **I,** *the Great,* 483–565, ruled 527–65.
**Ju·ve·nal** (jōō′və-nəl). A.D. 60?–140? Roman satirist.

# K

**Ka·a·hu·ma·nu** (kä′ə-hōō-mä′nōō). d. 1832. Hawaiian queen regent (1824–32).
**Ká·dar** (kä′där), **János.** b. 1912. Hungarian statesman.
**Kael** (kāl), **Pauline.** b. 1919. Amer. critic.
**Kaf·ka** (käf′kə, kăf′-), **Franz.** 1883-1924. Austrian author. **—Kaf·ka·esque′** (-ĕsk′) *adj.*
**Kahn** (kän), **Herman.** 1922–83. Amer. scientist & futurist.
**Kahn, Louis I.** 1901–74. Estonian-born Amer. architect.
**Kahn, Otto Herman.** 1867–1934. Amer. banker & philanthropist.
**Kai·ser** (kī′zər), **Henry John.** 1882-1967. Amer. industrialist.
**Kalb** (kălb, kälp), **Johann.** *Baron de Kalb.* 1721–80. German general in the Amer. Revolution.
**Ka·li·nin** (kə-lē′nən, -nyən), **Mikhail Ivanovich.** 1875–1946. Russian Communist leader.
**Kal·ten·born** (kôl′tən-bôrn′, kăl′-), **Hans von.** 1878-1965. Amer. news commentator.
**Ka·me·ha·me·ha** (kə-mā′ə-mā′ə). Name of five kings of Hawaii, esp. **I,** *the Great,* 1753?–1819, ruled 1795–1819.
**Ka·me·nev** (kä′mə-nĕf, kăm′ə-), **Lev Borisovich.** 1883-1936. Russian Communist leader.
**Ka·mer·lingh On·nes** (kä′mər-lĭng ô′nəs), **Heike.** 1853–1926. Dutch physicist (Nobel, 1913).
**Kan·din·ski** (kăn-dĭn′skē), **Vasili.** 1866–1944. Russian artist.

**Kane** (kān), **Elisha Kent.** 1820–57. Amer. physician & Arctic explorer.
**Ka·nin** (kā'nĭn), **Garson.** b. 1912. Amer. author.
**Kant** (känt, kănt), **Immanuel.** 1724–1804. German philosopher. **—Kant'i·an** *adj.*
**Kan·tor** (kăn'tər), **MacKinlay.** 1904-77. Amer. author.
**Kan·to·ro·vich** (kăn'tə-rō'vĭch), **Leonid Vitalevich.** b. 1912. Russian economist (Nobel, 1975).
**Ka·pi·tsa** (kä'pyĭt-sə), **Pëtr Leonidovich.** 1894–1984. Russian physicist (Nobel, 1978).
**Kap·lan** (kăp'lən), **Mordecai Menahem.** b. 1881. Amer. Jewish educator.
**Kar·a·george** (kăr'ə-jôrj'). 1776?–1817. Serbian nationalist leader.
**Ka·ra·jan** (kär'ə-yän'), **Herbert von.** b. 1908. Austrian conductor.
**Karl·feldt** (kärl'fĕlt'), **Erik Axel.** 1864-1931. Swedish poet (Nobel, 1931).
**Kar·loff** (kär'lôf', -lŏf'), **Boris.** 1887–1969. English-born Amer. actor.
**Karl Lud·wig** (kärl' lo͞ot'vĭk). CHARLES LOUIS.
**Kár·mán** (kär'män), **Theodor von.** 1881–1963. Hungarian-born Amer. physicist & aeronautical engineer.
**Ká·ro·lyi** (kăr'əl-yē, kär'-), Count **Mihály.** 1875–1955. Hungarian statesman.
**Kar·rer** (kär'ər), **Paul.** 1889–1971. Russian-born Swiss chemist (Nobel, 1937).
**Kar·sa·vi·na** (kär-sä'və-nə), **Tamara.** 1885–1978. Russian ballerina.
**Kast·ler** (käst'lər), **Alfred.** 1902-84. French physicist (Nobel, 1966).
**Katz** (kăts), Sir **Bernard.** b. 1911. German-born English physicist (Nobel, 1970).
**Kauf·man** (kôf'mən), **George Simon.** 1889–1961. Amer. playwright.
**Kau·nitz** (kou'nĭts), Prince **Wenzel Anton von.** *Count of Rietberg.* 1711–94. Austrian statesman.
**Kaut·sky** (kout'skē), **Karl Johann.** 1854–1938. German socialist leader.
**Ka·wa·ba·ta** (kä'wə-bä'tə), **Yasunari.** 1899–1972. Japanese author (Nobel, 1968).
**Kaye** (kā), **Danny.** b. 1913. Amer. entertainer.
**Kaye-Smith** (kā'smĭth'), **Sheila.** 1887–1956. English novelist.
**Ka·zan** (kə-zăn', -zän'), **Elia.** b. 1909. Turkish-born Amer. filmmaker.
**Ka·zan·tza·kis** (kä'zənt-sä'kēs), **Nikos.** 1885–1957. Greek author.
**Kean** (kēn), **Edmund.** 1787–1833. English actor.
**Kear·ny** (kär'nē), **Philip.** 1814–62. Amer. general.
**Kearny, Stephen Watts.** 1794–1848. Amer. army officer.
**Kea·ton** (kēt'n), **Buster.** 1895–1966. Amer. actor.
**Keats** (kēts), **John.** 1795–1821. English poet. **—Keats'i·an** *adj.*
**Ke·ble** (kē'bəl), **John.** 1792–1866. English clergyman & poet.
**Kee·ler** (kē'lər), **Ruby.** b. 1910. Canadian-born dancer & actress.
**Kee·ley** (kē'lē), **Leslie Enraught.** 1834–1900. Amer. physician.
**Keene** (kēn), **Laura.** 1826?–73. English-born actress & theatrical producer.
**Kee·shan** (kē'shən), **Robert James.** *Captain Kangaroo.* b. 1927. Amer. television personality.
**Ke·fau·ver** (kē'fô'vər), **(Carey) Estes.** 1903–63. Amer. legislator.
**Ke·hew** (kē'hyo͞o'), **Mary Morton Kimball.** 1859–1918. Amer. reformer.
**Kei·tel** (kīt'l), **Wilhelm.** 1882–1946. German general; executed.
**Kek·ko·nen** (kĕk'ə-nən, -nĕn'), **Urho Kaleva.** b. 1900. Finnish statesman.
**Kel·land** (kĕl'ənd), **Clarence Budington.** 1881–1964. Amer. author.
**Kel·ler** (kĕl'ər), **Helen Adams.** 1880–1968. Amer. author & lecturer.
**Kel·ley** (kĕl'ē), **Florence.** 1859–1932. Amer. social worker.
**Kelley, Oliver Hudson.** 1826–1913. Amer. farm organizer.
**Kel·logg** (kĕl'ôg', -ŏg'), **Clara Louise.** 1842–1916. Amer. operatic soprano.
**Kellogg, Frank Billings.** 1856–1937. Amer. statesman (Nobel, 1929).
**Kellogg, Louise Phelps.** 1862–1942. Amer. historian.
**Kellogg, Will Keith.** 1860–1951. Amer. cereal manufacturer & philanthropist.
**Kel·ly** (kĕl'ē), **Ellsworth.** b. 1923. Amer. artist.
**Kelly, Emmett.** 1898–1979. Amer. circus clown.
**Kelly, Grace Patricia.** Princess GRACE.
**Kelly, Walter Crawford ("Walt").** 1913–73. Amer. cartoonist & illustrator.
**Kelly, William.** 1811–88. Amer. inventor & manufacturer.
**Kel·vin** (kĕl'vĭn), 1st Baron. *William Thompson.* 1824–1907. Irish-born British mathematician & physicist.
**Ke·mal At·a·türk** (kə-mäl' ăt'ə-tûrk', ä'tə-). 1881–1938. Turkish soldier & statesman.
**Kem·ble** (kĕm'bəl), **Frances Anne ("Fanny").** English-born Amer. actress.
**Kemble, John Philip.** 1757–1823. English actor.
**Kem·pis** (kĕm'pĭs), **Thomas a.** —See THOMAS A KEMPIS.
**Ken** or **Kenn** (kĕn), **Thomas.** 1637–1711. English prelate & hymn writer.
**Ken·dall** (kĕn'dl), **Amos.** 1789–1869. Amer. journalist & public official.
**Kendall, Edward Calvin.** 1886–1972. Amer. biochemist (Nobel, 1962).
**Kendall, (William) Sergeant.** 1869–1938. Amer. artist.
**Ken·drew** (kĕn'dro͞o'), Sir **John Cowdery.** b. 1917. English biologist (Nobel, 1962).
**Kenn** (kĕn). *var. of* KEN.
**Ken·nan** (kĕn'ən), **George Frost.** b. 1904. Amer. diplomat, historian, & author.
**Ken·ne·dy** (kĕn'ĭ-dē), **Joseph Patrick.** 1888–1969. Amer. businessman & diplomat, with his wife **Rose Fitzgerald** (b. 1890) parents of: **a. John Fitzgerald ("Jack").** 1917–63. 35th U.S. President (1961–63); assassinated. **b. Robert Francis ("Bobby").** 1925–68. Legislator & public official; assassinated. **c. Edward Moore ("Ted").** b. 1932. Legislator.
**Ken·nel·ly** (kĕn'ə-lē), **Arthur Edwin.** 1861-1939. Amer. electrical engineer.
**Ken·ny** (kĕn'ē), **Elizabeth.** 1886–1952. Australian-born nursing pioneer.
**Kent** (kĕnt), **James.** 1763–1847. Amer. jurist.
**Kent, Rockwell.** 1882–1971. Amer. artist.
**Ken·yat·ta** (kĕn-yä'tə), **Jomo.** 1893?–1978. Kenyan statesman.
**Ken·yon** (kĕn'yən), **John Samuel.** 1874–1959. Amer. phonetician.
**Ke·o·kuk** (kē'ə-kŭk'). 1790?–1848. Amer. Sauk chief.
**Kep·ler** (kĕp'lər), **Johannes.** 1571–1630. German astronomer & mathematician.
**Kep·pel** (kĕp'əl), **Augustus.** *1st Viscount Keppel.* 1725–86. English naval officer.
**Ke·ren·ski** or **Ke·ren·sky** (kə-rĕn'skē, kĕr'ən-), **Aleksandr Feodorovich.** 1881–1970. Russian revolutionary leader.
**Kern** (kûrn), **Jerome David.** 1885–1945. Amer. composer.
**Ker·ou·ac** (kĕr'o͞o-ăk'), **Jean-Louis ("Jack").** 1922–69. Amer. author.
**Kerr** (kär, kûr), **Jean Collins.** b. 1923. Amer. author.
**Kes·sel·ring** (kĕs'əl-rĭng), **Albert.** 1887–1960. German general.
**Ket·ter·ing** (kĕt'ə-rĭng), **Charles Franklin.** 1876–1958. Amer. electrical engineer & manufacturer.
**Key** (kē), **Francis Scott.** 1779–1843. Amer. lawyer & poet.
**Keyes** (kēz), **Francis Parkinson.** 1885–1970. Amer. author.
**Keynes** (kānz), **John Maynard.** *1st Baron of Tilton.* 1883–1943. English economist. **—Keynes'i·an** *adj.* & *n.*
**Key·ser·ling** (kī'zər-lĭng), Count **Hermann Alexander von.** 1880–1946. Estonian-born German scientist & philosopher.
**Kha·cha·tu·ri·an** (kä'chä-to͝or'ē-ən, kăch'ə-), **Aram Ilich.** 1903-78. Russian composer.
**Kha·da·fy** (kə-dä'fē), **Moammar.** —See QADDAFI.
**Kha·lid** (kä-lēd', KHä-). *Khalid Abdul Aziz al-Saud.* 1913–82. Saudi Arabian king (1975–82).
**Khay·yám** (kī-yäm', -ăm'), **Omar.** —See OMAR KHAYYAM.
**Khe·ra·skov** (kə-räs'kəf), **Mikhail Mateevich.** 1733–1806. Russian poet.
**Kho·mei·ni** (kō-mā'nē, KHō-, hō-), Ayatollah **Ruholla.** b. 1900. Iranian leader.
**Kho·ra·na** (kō-rä'nə), **Har Gobind.** b. 1922. Indian-born Amer. biochemist (Nobel, 1968).
**Khru·shchev** (kro͞osh-chĕf', -chôf', -chôv', -chĕv'), **Nikita Sergeevich.** 1894–1971. Soviet statesman.
**Khu·fu** (ko͞o'fo͞o'). CHEOPS.
**Khwa·riz·mi** (kwär'ĭz-mē), **al-.** 780–850? Arab mathematician.
**Kid** (kĭd). *var. of* KYD.
**Kidd** (kĭd), **Michael.** b. 1917. Amer. choreographer.
**Kidd, William.** *Captain Kidd.* 1645?–1701. Scottish-born English pirate.
**Kie·ran** (kîr'ən), **John Francis.** 1892–1981. Amer. naturalist & journalist.
**Kier·ke·gaard** (kîr'kĭ-gärd', -gôr'), **Sören Aaby.** 1813–55. Danish philosopher & theologian.
**Kie·sing·er** (kē'zĭng-ər), **Kurt Georg.** b. 1904. West German statesman.
**Kil·mer** (kĭl'mər), **(Alfred) Joyce.** 1886–1918. Amer. author.
**Kim Il Sung** (kĭm' ĭl' sŭng', sŏong'). b. 1912? Korean soldier and statesman.
**Kim·mel** (kĭm'əl), **Husband Edward.** 1882–1968. Amer. admiral.
**King** (kĭng), **Billie Jean Moffitt.** b. 1943. Amer. tennis player.
**King, Clarence.** 1842–1901. Amer. geologist & mining engineer.
**King, Coretta Scott.** b. 1927. Amer. civil-rights leader.
**King, Ernest Joseph.** 1878–1968. Amer. naval officer.

**King, Martin Luther, Jr.** 1929–68. Amer. clergyman & civil-rights leader (Nobel, 1964); assassinated.

**King, Richard.** 1825–85. Amer. steamboat captain & rancher.

**King, Rufus.** 1755–1827. Amer. politician & diplomat.

**King, William Lyon Mackenzie.** 1874–1950. Canadian prime minister (1921–26, 1926–30, 1935–48).

**King, William Rufus DeVane.** 1786–1853. U.S. Vice President (1853).

**King·lake** (kĭng′lāk′), **Alexander William.** 1809–91. English historian.

**Kings·ley** (kĭngz′lē), **Charles.** 1819–75. English clergyman & author.

**Kin·kaid** (kĭn-kād′), **Thomas Cassin.** 1888–1972. Amer. admiral.

**Ki·no** (kē′nō), **Eusebio Francisco.** 1645?–1711. Italian-born Jesuit missionary in America.

**Kin·sey** (kĭn′zē), **Alfred Charles.** 1894–1956. Amer. sociologist & biologist.

**Kip·ling** (kĭp′lĭng), **(Joseph) Rudyard.** 1865–1936. English author (Nobel, 1907).

**Kir·by-Smith** (kûr′bē-smĭth′), **Edmund.** 1824–93. Amer. Confederate soldier & educator.

**Kirch·hoff** (kîr′kôf′), **Gustav Robert.** 1824–87. German physicist.

**Kirch·ner** (kîrk′nər, kîrKH′-), **Ernst Ludwig.** 1880–1938. German artist.

**Kirch·wey** (kûrch′wā′), **Freda.** 1893–1976. Amer. editor & publisher.

**Kirk** (kûrk), **Norman.** 1923–74. New Zealand statesman.

**Kirk·land** (kûrk′lənd), **Caroline Matilda Stansbury.** 1801–64. Amer. author.

**Kirkland, Gelsey.** b. 1952. Amer. ballerina.

**Kirk·us** (kûr′kəs), **Virginia.** 1893–1980. Amer. author & critic.

**Ki·rov** (kē′rôf, -rŏv), **Sergei Mironovich.** 1888–1934. Russian revolutionary.

**Kir·sten** (kîr′stən), **Dorothy.** b. 1917. Amer. soprano.

**Kis·sin·ger** (kĭs′ĭn-jər), **Henry Alfred.** b. 1923. German-born Amer. scholar & diplomat (Nobel, 1973).

**Kitche·ner** (kĭch′nər, -ə-nər), **Horatio Herbert.** *1st Earl Kitchener of Khartoum & of Broome.* 1850–1916. Irish-born British soldier & colonial administrator.

**Kit·tredge** (kĭt′rĭj), **George Lyman.** 1860–1941. Amer. educator.

**Klee** (klā), **Paul.** 1879–1940. Swiss artist.

**Klein** (klīn), **Calvin.** b. 1942. Amer. fashion designer.

**Klein, Lawrence R.** b. 1920. Amer. economist (Nobel, 1980).

**Kleist** (klīst), **Heinrich Bernt Wilhelm von.** 1777–1811. German dramatist.

**Klem·per·er** (klĕm′pər-ər), **Otto.** 1885–1973. German composer & conductor.

**Kline** (klīn), **Franz Joseph.** 1919–62. Amer. painter.

**Klop·stock** (klŏp′stŏk′, klôp′shtôk′), **Friedrich Gottlieb.** 1724–1803. German poet.

**Klug** (klŭg, klōōg), **Aaron.** b. 1926. South African-born English biochemist (Nobel, 1982).

**Knel·ler** (nĕl′ər), Sir **Godfrey.** 1646–1723. German-born English painter.

**Knie·vel** (knē′vəl), **Evel.** b. 1938. Amer. daredevil motorcyclist.

**Knight** (nīt), **John S.** 1894–1981. Amer. publisher.

**Knight, Sarah Kemble.** 1666–1727. Amer. educator & diarist.

**Knopf** (knŏpf), **Alfred Abraham.** 1892–1984. Amer. publisher.

**Knox** (nŏks), **Henry.** 1750–1806. Amer. Revolutionary soldier & public official.

**Knox, John.** 1505?–72. Scottish religious reformer.

**Knox, Philander Chase.** 1853–1921. Amer. public official & legislator.

**Knox, Rose Markward.** 1857–1950. Amer. businesswoman.

**Knox, William Franklin ("Frank").** 1874–1944. Amer. soldier, publisher, & public official.

**Knud·sen** (nōōd′sən, kə-nōōd′sən), **William Signius.** 1879–1948. Danish-born Amer. industrialist & public official.

**Knut** (kə-nōōt′, -nyōōt′). CANUTE.

**Koch** (kôk), **Robert.** 1843–1910. German physician & bacteriologist (Nobel, 1905).

**Ko·cher** (kô′kər, -KHər), **Emil Theodor.** 1841–1917. Swiss surgeon (Nobel, 1909).

**Kock** (kôk), **Charles Paul de.** 1794–1871. French author.

**Ko·dál·y** (kō′dī′), **Zoltán.** 1882–1967. Hungarian composer.

**Koest·ler** (kĕst′lər, kĕs′-), **Arthur.** 1905–83. Hungarian-born author.

**Koi·so** (koi′sō, kō′ē-sō′), **Kuniaki.** 1880–1950. Japanese general & public official.

**Ko·kosch·ka** (kə-kôsh′kə), **Oskar.** 1886–1980. Austrian-born artist & author.

**Kol·chak** (kôl′chŏk′), **Aleksandr Vasilievich.** 1875–1920. Russian admiral & counterrevolutionary; executed.

**Kol·lon·tai** (kŏl′ən-tī′), **Aleksandra Mikhailovna.** 1872–1952. Russian revolutionary & author.

**Koll·witz** (kōl′wĭts′, kôl′vĭts′), **Käthe** or **Kaethe Schmidt.** 1867–1945. German artist.

**Kol·tsov** (kōlt-sôf′, -sôv′), **Aleksei Vasilievich.** 1809–42. Russian poet.

**Ko·mu·ra** (kō-mōōr′ä), Marquis **Jutaro.** 1855–1911. Japanese diplomat.

**Kon·dy·les** *or* **Kon·dy·lis** (kôn-dē′ləs, -lēs′), **Georgios.** 1879–1936. Greek general & statesman.

**Ko·nev** (kôn′yĕf′, -yĕv′, -yəf), **Ivan Stepanovich.** 1897–1973. Russian field marshal & revolutionary.

**Ko·no·ye** (kə-nō′ä, -yä), Prince **Fumimaro.** 1891–1945. Japanese statesman.

**Koo** (kōō), **Vi Kyuin Wellington.** b. 1887. Chinese diplomat.

**Koop·mans** (kōōp′mənz), **Tjalling Charles.** b. 1910. Dutch-born Amer. economist (Nobel, 1975).

**Korn·berg** (kôrn′bûrg′), **Arthur.** b. 1918. Amer. biochemist (Nobel, 1959).

**Korn·gold** (kôrn′gōld′, -gōlt′), **Erich Wolfgang.** 1897–1957. Austrian-born Amer. composer & pianist.

**Kor·ni·lov** (kôr-nē′ləf), **Lavr Georgievich.** 1870–1918. Russian general.

**Ko·ro·len·ko** (kôr′ə-lĕng′kō), **Vladimir Galaktionovich.** 1853–1921. Russian author.

**Kor·zyb·ski** (kôr-zĭb′skē, -zĭp′-, kə-zhĭp′-), **Alfred Habdank Skarbek.** 1879–1950. Polish-born Amer. author & scientist.

**Kos·ci·us·ko** (kŏs′ē-ŭs′kō, kôsh-chōōsh′kō), **Thaddeus.** 1746–1817. Polish general & patriot.

**Kos·sel** (kôs′əl), **Albrecht.** 1853–1927. German biochemist (Nobel, 1910).

**Kos·suth** (kŏs′ōōth′, kô′shōōt′), **Ferenc.** 1841–1914. Hungarian politician.

**Kossuth, Lajos.** 1802–94. Hungarian revolutionary patriot & statesman.

**Ko·sy·gin** (kə-sē′gən), **Aleksei Nikolaevich.** 1904–80. Russian statesman.

**Kot·ze·bue** (kŏt′sə-bōō′, kôt′-), **August Friedrich Ferdinand von.** 1761–1819. German dramatist.

**Kou·fax** (kō′făks′), **Sanford ("Sandy").** b. 1935. Amer. baseball player.

**Koun·dou·ri·o·tes** (kōōn-dōōr′ē-ō′tēs), **Pavlos.** 1855–1935. Greek admiral & statesman.

**Kous·se·vitz·ky** (kōō′sə-vĭt′skē), **Sergei Aleksandrovich ("Serge").** 1874–1951. Russian-born Amer. conductor.

**Ko·vacs** (kō′văks′), **Ernie.** 1919–62. Amer. comedian.

**Krafft-E·bing** (krăft′ĕb′ĭng, krăft′ā′bĭng), Baron **Richard von.** 1840–1902. German physician & neurologist.

**Kra·mer** (krā′mər), **Stanley E.** b. 1913. Amer. filmmaker.

**Kraus-Boel·té** (krous′bŏl′tē, -bœl′tə), **Maria.** 1836–1918. German-born Amer. educator.

**Krebs** (krĕbz), Sir **Hans Adolf.** 1900–81. German-born British biochemist (Nobel, 1953).

**Kreh·biel** (krā′bēl′), **Henry Edward.** 1854–1923. Amer. critic.

**Krei·sler** (krī′slər), **Fritz.** 1875–1962. Austrian-born Amer. violinist & arranger.

**Kreps** (krĕps), **Juanita Morris.** b. 1921. Amer. economist & public official.

**Kress** (krĕs), **Samuel Henry.** 1863–1955. Amer. merchant & art patron.

**Krock** (krŏk), **Arthur.** 1886–1974. Amer. journalist.

**Kroe·ber** (krō′bər), **Alfred Louis.** 1876–1960. Amer. anthropologist.

**Krogh** (krôg), **(Schack) August Steenberg.** 1874–1946. Danish physiologist (Nobel, 1920).

**Kroll** (krōl), **Leon.** 1884–1974. Amer. artist.

**Kro·pot·kin** (krə-pŏt′kĭn), Prince **Pëtr Alekseevich.** 1842–1921. Russian scientist & revolutionary.

**Kru·ger** (krōō′gər), **Stephanus Johannes Paulus.** *Oom Paul.* 1825–1904. South African statesman.

**Kru·pa** (krōō′pə), **Gene.** 1909–73. Amer. musician.

**Krupp** (krōōp, krŭp). Family of German steel manufacturers, including **Friedrich** (1787–1826), **Alfred** (1812–87), **Friedrich Alfred** (1854–1902), & **Bertha** (1886–1957) & her husband **Gustav** (1870–1950) & son **Alfred-Felix** (1907–67).

**Krup·ska·ya** (krōōp′skə-yə), **Nadezhda Konstantinovna.** 1869–1939. Russian revolutionary.

**Krutch** (krŭch), **Joseph Wood.** 1893–1970. Amer. educator, critic, & naturalist.

**Ku·bi·tschek** (kōō′bə-chĕk′), **Juscelino.** 1901–76. Brazilian statesman.

**Ku·blai Khan** (kōō′blī kän′) *also* **Ku·bla Khan** (-blə). 1216–94. Mongol emperor.

**Ku·brick** (kōō′brĭk, kyōō′-), **Stanley.** b. 1928. Amer. filmmaker.

**Kuhn** (ko͞on), **Richard.** 1900–67. Austrian chemist (Nobel, 1938; ordered by Nazis to decline).
**Kui·by·shev** (kwē'bĭ-shĕf, ko͞o'ĭ-), **Valerian Vladimirovich.** 1888–1935. Russian Communist leader.
**Kun** (ko͞on), **Béla.** 1885–1937. Hungarian Communist leader.
**Kung** (ko͝ong, go͝ong), Prince. 1833–98. Chinese statesman.
**Kung, H.H.** *K'ung Hsiang-hsi.* 1881–1967. Chinese Nationalist financier.
**Ku·ro·pat·kin** (ko͝or'ə-păt'kĭn, -pät'-), **Aleksei Nikolaevich.** 1848–1925. Russian general.
**Ku·ru·su** (ko͞o-ro͞o'so͞o), **Saburo.** 1888?–1954. Japanese diplomat.
**Kusch** (ko͝osh), **Polykarp.** b. 1911. German-born Amer. physicist (Nobel, 1955).
**Ku·tu·zov** (ko͞o-to͞o'zôf, -zôv), **Mikhail Ilarionovich.** *Prince of Smolensk.* 1745–1813. Russian field marshal.
**Kuz·nets** (ko͝oz'nĕts', kŭz'-), **Simon.** 1901–85. Russian-born Amer. economist (Nobel, 1971).
**Ky** (kē), **Nguyen Cao.** b. 1930. Vietnamese statesman.
**Kyd** or **Kid** (kĭd), **Thomas.** 1558–94. English dramatist.
**Kyn·e·wulf** (kĭn'ə-wo͝olf'). CYNEWULF.

# L

**La Bru·yère** (lä bro͞o-yĕr', brē-), **Jean de.** 1645–96. French moralist author.
**La·chaise** (lə-shāz'), **Gaston.** 1882–1935. French-born Amer. sculptor.
**Ladd-Franklin** (lăd-frăngk'lĭn), **Christine.** 1847–1930. Amer. psychologist & logician.
**Laemm·le** (lĕm'lē), **Carl.** 1867–1939. German-born Amer. film producer.
**Laën·nec** (lä-nĕk'), **René.** 1781–1826. French physician.
**La Farge** (lə färzh', färj'), **John.** 1835–1910. Amer. artist.
**La Farge, Oliver Hazard Perry.** 1901–63. Amer. author.
**La·fa·yette** (lä'fē-ĕt', lăf'ē-), Marquis de. *Marie Joseph Paul Yves Roch Gilbert du Motier de Lafayette.* 1757–1834. French military, political, & revolutionary leader.
**Laf·fite** or **La·fitte** (lə-fēt'), **Jean.** 1780?–1826? French pirate.
**La Flesche** (lä flĕsh', lə), **Susette.** *Bright Eyes.* 1854–1903. Amer. Indian leader.
**La Fol·lette** (lə fŏl'ət), **Robert Marion.** 1855–1925. Amer. politician & reformer.
**La·fon·taine** (lə-fŏn-tān', -fôN-tĕn'), **Henri.** 1854–1943. Belgian politician & pacifist (Nobel, 1913).
**La Fon·taine** (lə fŏn-tān', fôN-tĕn'), **Jean de.** 1621–95. French poet & fabulist.
**La·ger·kvist** (lä'gər-kfĭst', -kwĭst'), **Pär Fabian.** 1891–1974. Swedish author (Nobel, 1951).
**La·ger·löf** (lä'gər-ləv), **Selma Ottiliana Lovisa.** 1858–1940. Swedish novelist (Nobel, 1909).
**La·grange** (lə-grānj', -gräNzh'), Comte **Joseph Louis.** 1736–1813. French mathematician.
**La Guar·di·a** (lə gwär'dē-ə), **Fiorello Henry.** 1882–1947. Amer. politician.
**Lahr** (lär), **Bert.** 1895–1967. Amer. entertainer.
**Lake** (lāk), **Simon.** 1866–1945. Amer. naval architect.
**La·mar** (lə-mär'), **Joseph Rucker.** 1857–1916. Amer. jurist.
**Lamar, Lucius Quintus Cincinnatus.** 1825–93. Amer. jurist & politician.
**Lamar, Mirabeau Buonaparte.** 1798–1859. Amer. politician & diplomat.
**La·marck** (lə-märk'), Chevalier de. *Jean Baptiste Pierre Antoine de Monet.* 1744–1829. French naturalist.
**La·mar·tine** (lä-mär-tēn', lăm'ər-), **Alphonse Marie Louis de Prat de.** 1790–1869. French poet & politician.
**Lamb** (lăm), **Charles.** *Elia.* 1775–1834. English critic & essayist.
**Lamb, Martha Joanna Reade Nash.** 1826–93. Amer. historian.
**Lamb, William.** *2nd Viscount Melbourne.* 1779–1848. British prime minister (1834, 1835–41).
**Lamb, Willis Eugene, Jr.** b. 1913. Amer. physicist (Nobel, 1955).
**Lam·bert** (lăm'bərt), **John.** 1619–83. English Civil War general.
**Lam·masch** (lä'mäsh), **Heinrich.** 1853–1920. Austrian jurist & diplomat.
**La Motte-Fou·qué** (lə môt'fo͞o-kā'), Baron **Friedrich Heinrich Karl.** 1777–1843. German author.
**Lan·cas·ter** (lăng'kə-stər, lăn'-). English royal house (1399–1461); vied against the Plantagenets in the War of the Roses. **—Lan·cas'tri·an** (lăng-kăs'trē-ən) *adj.* & *n.*
**Land** (lănd), **Edwin Herbert.** b. 1909. Amer. inventor.
**Lan·dau** (lăn-dou'), **Lev Davidovich.** 1908–68. Soviet physicist (Nobel, 1962).
**Lan·ders** (lăn'dərz), **Ann.** Esther Pauline FRIEDMAN.
**Lan·dis** (lăn'dĭs), **Kenesaw Mountain.** 1866–1944. Amer. jurist & baseball commissioner.
**Lan·don** (lăn'dən), **Alfred Mossman.** b. 1887. Amer. politician.
**Lan·dor** (lăn'dôr, -dər), **Walter Savage.** 1775–1864. English author.
**Lan·dow·ska** (lăn-dôf'skə, -dôv'-), **Wanda.** 1877–1959. Polish-born musician.
**Land·seer** (lănd'sîr'), Sir **Edwin Henry.** 1802–73. English painter.
**Land·stei·ner** (lănd'stī'nər, länt'-shtī'-), **Karl.** 1868–1943. Austrian-born Amer. pathologist (Nobel, 1930).
**Lane** (lān), **Edward William.** 1801–76. English Egyptologist & Orientalist.
**Lane, James Henry.** 1814–66. Amer. politician & soldier.
**La·ney** (lā'nē), **Lucy Craft.** 1854–1933. Amer. educator.
**Lan·franc** (lăn'frăngk'). 1005?–89. Italian-born English prelate & political adviser.
**Lang** (lăng), **Andrew.** 1844–1912. Scottish author.
**Lang, Cosmo Gordon.** *1st Baron Lang of Lambeth.* 1864–1945. English prelate & reformer.
**Lang, Fritz.** 1890–1976. Austrian filmmaker.
**Lang·dell** (lăng'dəl), **Christopher Columbus.** 1826–1906. Amer. legal educator.
**Lange** (lăng), **Dorothea.** 1895–1965. Amer. photographer.
**Lang·e** (läng'ə), **Christian Louis.** 1869–1938. Norwegian pacifist & historian (Nobel, 1921).
**Lang·er** (lăng'ər), **Susanne Knauth.** 1895–1985. Amer. educator & philosopher.
**Langer, William Leonard.** 1896–1977. Amer. historian.
**Lang·ford** (lăng'fərd), **Nathaniel Pitt.** 1832–1911. Amer. public official, conservationist, & explorer.
**Lang·ley** (lăng'lē), **Samuel Pierpoint.** 1834–1906. Amer. astronomer & aviation pioneer.
**Lang·muir** (lăng'myo͝or'), **Irving.** 1881–1957. Amer. chemist (Nobel, 1932).
**Lang·ston** (lăng'stən), **John Mercer.** 1829–97. Amer. diplomat, politician, & educator.
**Lang·ton** (lăng'tən), **Stephen.** d. 1228? English prelate & author.
**Lang·try** (lăng'trē), **Lillie** or **Lily.** *the Jersey Lily.* 1852–1929. English actress & beauty.
**La·nier** (lə-nîr'), **Sidney.** 1842–81. Amer. author & musician.
**Lan·sing** (lăn'sĭng), **Robert.** 1864–1928. Amer. public official.
**Lan·za** (lăn'zə), **Mario.** 1925–59. Amer. singer & actor.
**Lao-tse** *also* **Lao-tzu** or **Lao-tsze** (lou'dzŭ'). 604?–531? B.C. Chinese philosopher.
**La Pé·rouse** (lä pā-ro͞oz', pə-), Comte de. *Jean François de Galaup.* 1741–88. French explorer.
**La·place** (lə-pläs'), Marquis **Pierre Simon de.** 1749–1827. French mathematician & astronomer.
**Lar·com** (lär'kəm), **Lucy.** 1824–93. Amer. poet.
**Lard·ner** (lärd'nər), **Ringgold Wilmer ("Ring").** 1885–1933. Amer. journalist & author.
**Lar·go Ca·bal·le·ro** (lär'gō kăb'ə-yĕr'ō, -əl-), **Francisco.** 1869–1946. Spanish socialist & labor leader.
**La Roche·fou·cauld** (lä rôsh-fo͞o-kō', -rōsh-), Duc **François de.** 1613–80. French author.
**La·rousse** (lə-ro͞os'), **Pierre Athanase.** 1817–75. French grammarian, lexicographer, & encyclopedist.
**Lar·tet** (lär-tā'), **Edouard Armand Isidore Hippolyte.** 1801–71. French archaeologist & pioneer paleontologist.
**La Salle** (lə săl'), Sieur de. *Robert Cavelier.* 1643–87. French explorer in America.
**Las Ca·sas** (läs kä'səs), **Bartolomé de.** *Apostle of the Indies.* 1474–1566. Spanish missionary & historian.
**Las·ker** (lăs'kər), **Albert Davis.** 1880–1952. Amer. advertising executive, philanthropist, & public official.
**Las·ki** (lăs'kē), **Harold Joseph.** 1893–1950. English political scientist & author.
**Las·ky** (lăs'kē), **Jesse Louis.** 1880–1958. Amer. film producer.
**Las·salle** (lə-săl', -säl'), **Ferdinand.** 1825–64. German socialist.
**La·throp** (lā'thrəp), **Julia Clifford.** 1858–1932. Amer. social reformer.
**Lat·i·mer** (lăt'ə-mər), **Hugh.** 1485–1555. English prelate & religious reformer; executed.
**La Tour** (lə to͝or'), **Georges de.** 1593–1652. French painter.
**La·trobe** (lə-trōb'), **Benjamin Henry.** 1764–1820. English-born Amer. engineer & architect.
**Lat·ti·more** (lăt'ə-môr', -mōr'), **Owen.** b. 1900. Amer. Asian scholar & author.
**Laud** (lôd), **William.** 1573–1645. English prelate; executed for treason.
**Lau·der** (lô'dər), Sir **Harry.** 1870–1950. Scottish singer & songwriter.
**Lau·e** (lou'ə), **Max Theodor Felix von.** 1879–1960. German physicist (Nobel, 1914).
**Laugh·ton** (lôt'n), **Charles.** 1899–1962. English actor.
**Lau·rel** (lôr'əl, lŏr'-), **Arthur Stanley Jefferson ("Stan").** 1890–1965. English-born Amer. comedian.

**Lau·ren·cin** (lô-räN-säN′), **Marie.** 1885–1956. French artist.
**Lau·rens** (lô-räNs′), **Henri.** 1885–1954. French artist.
**Lau·rens** (lôr′ənz, lŏr′-), **Henry.** 1724–92. Amer. Revolutionary statesman & diplomat.
**Lau·ri·er** (lôr′ē-ā′, lŏr′-), Sir **Wilfrid.** 1841–1919. Canadian prime minister (1896–1911).
**La·val** (lə-văl′, -väl′), **Pierre.** 1883–1945. French politician; executed for treason.
**La Val·lière** (lä vəl-yĕr′), Duchesse de. *Françoise Louise de la Baume Le Blanc.* 1644–1710. French mistress of Louis XIV.
**La·ve·ran** (lăv′ə-räN′, läv-räN′), **Charles Louis Alphonse.** 1845–1922. French pathologist (Nobel, 1907).
**La Vé·ren·drye** (lä vā-räN-drē′), Sieur de. *Pierre Gaultier de Varennes.* 1685–1749. French-Canadian explorer.
**La·ver·y** (lā′və-rē, lăv′ə-), Sir **John.** 1856–1941. English painter.
**La·voi·sier** (lə-vwä′zē-ā′, lä-vwä-zyā′), **Antoine Laurent.** 1743–94. French pioneer chemist; guillotined.
**Law** (lô), **(Andrew) Bonar.** 1858–1923. Canadian-born British prime minister (1922–23).
**Law, John.** 1671–1729. Scottish financier & speculator.
**Law, Sallie Chapman Gordon.** 1805–94. Amer. Confederate hospital administrator.
**Law, William.** 1686–1761. English theological author.
**Lawes** (lôz), **Henry.** 1595?–1662. English composer.
**Law·rence** (lôr′əns, lŏr′-), **Abbott.** 1792–1855. Amer. merchant & politician.
**Lawrence, Amos Adams.** 1814–86. Amer. businessman & philanthropist.
**Lawrence, David.** 1888–1973. Amer. journalist.
**Lawrence, David Herbert ("D.H.").** 1885–1930. English novelist.
**Lawrence, Ernest Orlando.** 1901–58. Amer. physicist (Nobel, 1939).
**Lawrence, Gertrude.** 1901–52. English actress.
**Lawrence, James.** 1781–1813. Amer. naval officer.
**Lawrence,** Sir **Thomas.** 1769–1830. English portrait painter.
**Lawrence, Thomas Edward ("T.E.").** *Lawrence of Arabia.* 1888–1935. Welsh-born British soldier, archaeologist, adventurer, & author.
**Law·rie** (lôr′ē, lŏr′ē), **Lee.** 1877–1963. German-born Amer. sculptor.
**Laws** (lôz), **Samuel Spahr.** 1824–1921. Amer. clergyman & inventor.
**Lax·ness** (läks′nĕs′), **Halldór Kiljan.** b. 1902. Icelandic novelist (Nobel, 1955).
**Lay·a·mon** (lā′ə-mən, lī′-). 13th cent. English priest & poet.
**Lay·ard** (lā′ərd, -ärd), Sir **Austen Henry.** 1817–94. English archaeologist & diplomat.
**Laz·a·rus** (lăz′ər-əs). Brother of Mary & Martha in the Bible, believed to have been raised from the dead.
**Lazarus, Emma.** 1849–87. Amer. poet & philanthropist.
**Lea·cock** (lē′kŏk′), **Stephen Butler.** 1869–1944. Canadian economist.
**Leaf** (lēf), **Walter.** 1852–1927. English banker & classical scholar.
**Le·ah** (lē′ə). First wife of Jacob in the Bible.
**Lea·hy** (lā′hē), **William Daniel.** 1875–1959. Amer. naval officer & diplomat.
**Lea·key** (lē′kē), **Louis Seymour Bazett.** 1903–72. English anthropologist.
**Lean** (lēn), **David.** b. 1908. English-born filmmaker.
**Lear** (lîr), **Edward.** 1812–88. English artist & author of nonsense verse.
**Lear, William Powell.** 1902–78. Amer. engineer & manufacturer.
**Leav·en·worth** (lĕv′ən-wərth), **Henry.** 1783–1834. Amer. frontier military officer.
**Leav·itt** (lĕv′ĭt), **Henrietta Swan.** 1868–1921. Amer. astronomer.
**Leavitt, Mary Greenleaf Clement.** 1830–1912. Amer. educator & temperance reformer.
**Le·brun** (lə-brœN′), **Albert.** 1871–1950. French statesman.
**Lebrun, Charles.** 1619–90. French painter.
**le Car·ré** (lə kä-rā′), **John.** *David John Moore Cornwell.* b. 1931. English author.
**Leck·y** (lĕk′ē), **William Edward Hartpole.** 1838–1903. Irish historian.
**Le·conte de Lisle** (lə kôNt′ də lēl′), **Charles Marie.** 1818–94. French poet.
**Le Cor·bu·sier** (lə kôr-bo͞o-zyā′). *Charles Édouard Jenneret.* 1887–1965. Swiss-born architect.
**Led·bet·ter** (lĕd′bĕt′ər), **Huddie ("Leadbelly").** 1888–1949. Amer. musician.
**Led·er·berg** (lĕd′ər-bûrg′, lā′dər-), **Joshua.** b. 1925. Amer. geneticist (Nobel, 1958).
**Le Duc Tho** (lā′ dŭk′ tō′). b. 1911. Vietnamese political leader; declined 1973 Nobel.
**Led·yard** (lĕd′yərd), **John.** 1751–89. Amer. adventurer & traveler.
**Lee** (lē), **Ann.** *Mother Ann.* 1736–84. English religious leader in America.
**Lee, Arthur.** 1740–92. Amer. diplomat & essayist.
**Lee, Charles.** 1731–82. British-born Amer. Revolutionary general.
**Lee, Fitzhugh.** 1835–1905. Amer. Confederate general & politician.
**Lee, Gypsy Rose.** 1914–70. Amer. entertainer & author.
**Lee, Henry ("Lighthorse Harry").** 1756–1818. Amer. Revolutionary statesman & commander.
**Lee, Ivy Ledbetter.** 1877–1934. Amer. public-relations expert.
**Lee, Jason.** 1803–45. Amer. missionary & Western pioneer.
**Lee, Manfred Bennington.** With Frederic Dannay, *Ellery Queen.* 1905–71. Amer. author.
**Lee, Richard Henry.** 1732–94. Amer. Revolutionary leader.
**Lee, Robert Edward.** 1807–70. Amer. Confederate general; surrendered to Ulysses Grant (1865).
**Lee,** Sir **Sidney.** 1859–1926. English biographer & editor.
**Lee, Tsung Dao.** b. 1926. Chinese-born Amer. physicist (Nobel, 1957).
**Leech** (lēch), **Margaret Kernochan.** 1893–1974. Amer. historian.
**Leeu·wen·hoek** *or* **Leu·wen·hoek** (lā′vən-ho͝ok′), **Anton van.** 1632–1723. Dutch microscopy pioneer & naturalist.
**Le Gal·lienne** (lə găl′yən), **Eva.** b. 1899. English-born Amer. actress.
**Le Gallienne, Richard.** 1866–1947. English author.
**Le·gen·dre** (lə-zhän′drə, -zhäN′-), **Adrien Marie.** 1752?–1833. French mathematician.
**Lé·ger** (lā-zhā′), **Alexis Saint-Léger.** *Saint-John Perse.* 1887–1975. French poet & diplomat (Nobel, 1960).
**Léger, Fernand.** 1881–1955. French painter.
**Le·guía y Sal·ce·do** (lə-gē′ə ē′ säl-sā′dō), **Augusto Bernadino.** 1863–1932. Peruvian statesman.
**Le·hár** (lā′här′), **Franz.** 1870–1948. Hungarian composer.
**Leh·man** (lē′mən), **Herbert Henry.** 1878–1963. Amer. banker, politician, & philanthropist.
**Leh·mann** (lā′man), **Lotte.** 1888–1976. German operatic soprano.
**Leib·nitz** *or* **Leib·niz** (līb′nĭts, līp′-), Baron **Gottfried Wilhelm von.** 1646–1716. German philosopher & mathematician.
**Leices·ter** (lĕs′tər), 1st Earl of. Robert DUDLEY.
**Lei·dy** (lī′dē), **Joseph.** 1823–91. Amer. naturalist & anatomist.
**Leigh** (lē), **Vivien.** 1913–67. English actress.
**Leigh·ton** (lāt′n), **Frederick.** *Baron Leighton of Stretton.* 1830–96. English painter & historian.
**Leins·dorf** (līnz′dôrf′, līns′-), **Erich.** b. 1912. Austrian-born Amer. conductor.
**Leis·ler** (lī′slər), **Jacob.** 1640–91. German-born Amer. patriot, merchant, & insurrectionist; executed.
**Leith-Ross** (lēth′rôs′, -rŏs′), Sir **Frederick William.** 1887–1968. English economist & public official.
**Le·land** *or* **Ley·land** (lē′lənd), **John.** 1506?–52. English antiquarian.
**Le·loir** (lā-lwär′), **Luis Federico.** b. 1906. French-born Argentine biochemist (Nobel, 1970).
**Le·ly** (lē′lē), Sir **Peter.** 1618–80. Dutch painter in England.
**Le·maî·tre** (lə-mĕt′rə), Abbé **Georges Édouard.** 1894–1966. Belgian astrophysicist.
**Le May** (lə-mā′), **Curtis Emerson.** b. 1906. Amer. air-force officer.
**Le·nard** (lā′närt′), **Philipp.** 1862–1947. German physicist (Nobel, 1905).
**Len·clos** *or* **L'En·clos** (läN-klō′), **Anne.** 1620–1705. French courtesan.
**L'En·fant** (län-fänt′, läN-fäN′), **Pierre Charles.** 1754–1825. French-born engineer.
**Le·nin** (lĕn′ĭn), **Vladimir Ilich.** 1870–1924. Russian revolutionary leader. **—Len′in·ist′** *adj. & n.*
**Len·non** (lĕn′ən), **John.** 1940–80. English musician & composer; murdered.
**Len·ya** (lān′yə), **Lotte.** 1898–1981. Austrian singer & actress.
**Leo** (lē′ō). Name of 13 popes, esp.: **a. I.** Saint. *the Great.* 390?–461. Reigned 440–61. **b. III.** Saint. 750?–816. Reigned 795–816. **c. X.** 1475–1521. Reigned 1513–21. **d. XIII.** 1810–1903. Reigned 1878–1903.
**Leon·ard** (lĕn′ərd), **William Ellery.** 1876–1944. Amer. poet & educator.
**Le·o·nar·do da Vin·ci** (lē′ə-när′dō də vĭn′chē, lā′-). 1452–1519. Florentine artist, engineer, musician, & scientist.
**Le·on·ca·val·lo** (lā′ōn-kə-vä′lō), **Ruggiero.** 1858–1919. Italian composer.
**Le·on·i·das I** (lē-ŏn′ĭ-dəs). d. 480 B.C. Spartan king (490–80).
**Le·on·tief** (lē-ŏn′tyĕf, -ŏn′-), **Wassily.** b. 1906. Russian-born Amer. economist (Nobel, 1973).
**Le·o·par·di** (lā′ə-pär′dē), Conte **Giacomo.** 1798–1837. Italian poet & philologist.
**Le·o·pold** (lē′ə-pōld′). Name of three kings of Belgium: **a. I.** 1790–1865. Ruled 1831–65. **b. II.** 1835–1909. Ruled 1865–1909. **c. III.** 1901–83. Ruled 1934–51; abdicated.
**Leopold.** Name of two Holy Roman Emperors: **a. I.** 1640–1705. Ruled 1658–1705. **b. II.** 1747–92. Ruled 1790–92.
**Lep·i·dus** (lĕp′ĭ-dəs), **Marcus Aemilius.** d. 13 B.C. Roman triumvir (43–36).

**Ler·mon·tov** (lĕr'mən-tôf', -tôv'), **Mikhail Yurievich.** 1814–41. Russian author.
**Ler·ner** (lûr'nər), **Alan Jay.** b. 1918. Amer. playwright & lyricist.
**Le·sage** (lə-säzh'), **Alain René.** 1668–1747. French author.
**Le·sche·titz·ky** (lĕsh'ə-tĭt'skē), **Theodor.** 1830–1915. Polish pianist, composer, & educator.
**Les·lie** (lĕs'lē, lĕz'-), **Frank.** 1821–80. Amer. publisher.
**Leslie, Miriam Florence Folline.** 1836–1914. Amer. editor & author.
**Les·seps** (lĕs'əps, lē-sĕps'), Vicomte **Ferdinand Marie de.** —See DE LESSEPS.
**Les·sing** (lĕs'ĭng), **Doris.** b. 1919. English author.
**Lessing, Gotthold Ephraim.** 1729–81. German playwright & critic.
**L'Es·trange** (lə-strānj'), Sir **Roger.** 1616–1704. English Royalist pamphleteer.
**Leu·wen·hoek** (lā'vən-hŏŏk'). *var. of* LEEUWENHOEK.
**Leut·ze** (loit'sə), **Emanuel Gottlieb.** 1816–68. German painter.
**Le·vas·seur** (lə-vä-sœr'), **Pierre Émile.** 1828–1911. French economist.
**Le·ven·son** (lĕv'ən-sən), **Sam.** 1911-80. Amer. author & humorist.
**Le·ver** (lē'vər), **Charles James.** 1806–72. Irish novelist.
**Lé·vesque** (lə-vĕk'), **René.** b. 1922. Canadian political leader.
**Le·vi** (lē'vī'). Son of Jacob and Leah in the Bible.
**Le·vi** (lā'vē), **Carlo.** 1902–75. Italian author.
**Le·vine** (lə-vēn'), **Jack.** b. 1915. Amer. painter.
**Le·vy** (lē'vē), **Uriah Phillips.** 1792–1862. Amer. naval officer.
**Lew·es** (lōō'ĭs), **George Henry.** 1817–78. English philosopher & critic.
**Lew·is** (lōō'ĭs), Sir **Arthur.** b. 1915. West-Indies-born British economist (Nobel, 1979).
**Lewis, Cecil Day.** 1904–72. English author.
**Lewis, Clive Staples ("C.S.").** 1898–1963. English novelist.
**Lewis, Isaac Newton.** 1858–1931. Amer. firearms inventor.
**Lewis, John Llewellyn.** 1880–1969. Amer. labor leader.
**Lewis, Matthew Gregory ("Monk").** 1775–1818. English gothic author.
**Lewis, Meriwether.** 1774–1809. Amer. soldier & explorer.
**Lewis, (Harry) Sinclair.** 1885–1951. Amer. novelist (Nobel, 1930).
**Lewis, (Percy) Wyndham.** 1884–1957. English artist.
**Lew·i·sohn** (lōō'ĭ-sən, -zən), **Irene.** 1892–1944. Amer. social worker & theatrical patron.
**Lewisohn, Ludwig.** 1883–1955. German-born Amer. author.
**Ley·land** (lē'lənd). *var. of* LELAND.
**Ley·poldt** (lī'pōlt'), **Frederick.** 1835–84. German-born Amer. editor & publisher.
**Liang** (lē-äng'). Name of two Chinese dynasties (502–57, 907–23).
**Liao** (lē-ou'). Chinese dynasty (916–1125).
**Lib·by** (lĭb'ē), **Willard Frank.** 1908–80. Amer. chemist (Nobel, 1960).
**Lib·er·a·ce** (lĭb'ə-rä'chē), **(Wladziu).** b. 1919. Amer. pianist & entertainer.
**Li Bo** (lē' bō'). *var. of* LI PO.
**Lich·ten·stein** (lĭk'tən-stīn', -stēn'), **Roy.** b. 1923. Amer. painter.
**Li·cin·i·us** (lə-sĭn'ē-əs). 270?–324. Roman emperor (308–24).
**Lid·dell Hart** (lĭd'l härt'), **Basil Henry.** 1895–1970. English military authority.
**Lie** (lē), **Jonas.** 1833–1909. Norwegian author.
**Lie, Jonas.** 1880–1940. Norwegian-born Amer. landscape painter.
**Lie, Trygve Halvden.** 1896–1968. Norwegian statesman & UN secretary-general.
**Lie·ber** (lē'bər), **Francis.** 1800–72. German-born Amer. historian & political economist.
**Lie·big** (lē'bĭg), Baron **Justus von.** 1803–73. German chemist.
**Lieb·knecht** (lēp'knĕkt'), **Karl.** 1871–1919. German Communist leader; assassinated.
**Li·far** (lē-fär', lē'fär), **Serge.** b. 1905. Russian dancer and choreographer.
**Li Hong·zhang** *also* **Li Hung·chang** (lē' hŏŏng'jäng'). 1823–1901. Chinese statesman.
**Lil·ien·thal** (lĭl'yən-thôl'), **David Eli.** 1899–1981. Amer. lawyer & public official.
**Lil·ien·thal** (lĭl'yən-täl', -thôl'), **Otto.** 1848–96. German aeronautical pioneer.
**Li·li·u·o·ka·la·ni** (lə-lē'ə-ō-kə-lä'nē). *Lydia Kamekeha Paki.* 1838–1917. Hawaiian queen (1891–93); deposed.
**Lil·lie** (lĭl'ē), **Beatrice.** *Lady Peel.* b. 1898. Canadian-born comedienne.
**Lil·lo** (lĭl'ō), **George.** 1693?–1739. English playwright.
**Li·món** (lĭ-mōn'), **José Arcadio.** 1908–72. Mexican-born Amer. dancer & choreographer.
**Lin·a·cre** (lĭn'ĭ-kər), **Thomas.** 1460?–1524. English physician & scholar.

ōō **boot** ou **out** th **thin** *th* **this** ŭ **cut** ûr **urge** y **young** yōō **abuse** zh **vision** ə **about**, **item**, **edible**, **gallop**, **circus**

**Lin Biao** (lĭn' bē-ou'). 1907–71. Chinese political leader.
**Lin·coln** (lĭng'kən), **Abraham.** 1809–65. 16th U.S. President (1861–65); assassinated.
**Lincoln, Benjamin.** 1733–1810. Amer. general.
**Lincoln, Mary Todd.** 1818–82. Wife of Abraham Lincoln.
**Lind** (lĭnd), **Jenny.** *Swedish Nightingale.* 1820–87. Swedish singer.
**Lind·bergh** (lĭnd'bûrg', lĭn'-), **Anne Spencer Morrow.** b. 1906. Amer. aviator & author.
**Lindbergh, Charles Augustus.** *Lucky Lindy.* 1902–74. Amer. aviator; first solo transatlantic flight.
**Lind·ley** (lĭnd'lē, lĭn'-), **John.** 1799–1865. English botanist & horticulturist.
**Lind·say** (lĭn'zē), **John Vliet.** b. 1921. Amer. politician.
**Lindsay, (Nicholas) Vachel.** 1879–1931. Amer. poet.
**Lindsay, Howard.** 1889–1968. Amer. playwright & producer.
**Lind·sey** (lĭn'zē), **Benjamin Barr ("Ben").** 1869–1943. Amer. jurist & social reformer.
**Link·la·ter** (lĭngk'lā'tər, -lə-), **Eric.** 1899–1974. English author.
**Link·let·ter** (lĭngk'lĕt'ər), **Art.** b. 1912. Canadian-born entertainer.
**Lin·nae·us** (lĭ-nē'əs, -nā'-), **Carolus.** 1707–78. Swedish botanist & founder of binomial taxonomic classification. **—Lin·nae'an** *adj.*
**Lin Piao** (lĭn' bē-ou', pē-ou'). LIN BIAO.
**Lin Sen** (lĭn' sĕn'). 1867?–1943. Chinese statesman.
**Lin·ton** (lĭn'tən), **Ralph.** 1893–1953. Amer. anthropologist.
**Lin Yu·tang** (lĭn' yōō'täng'). 1895–1976. Chinese-born Amer. philologist.
**Lip·mann** (lĭp'mən), **Fritz Albert.** b. 1899. German-born Amer. biochemist (Nobel, 1953).
**Li Po** (lē' pō', bō') *or* **Li Bo** (bō'). d. 762? Chinese poet.
**Lip·pi** (lĭp'ē), Fra **Filippo** *or* **Lippo** (1406?–69) & **Filippo** *or* **Filippino** (1457?–1504). Florentine painters.
**Lip·pin·cott** (lĭp'ĭn-kŏt', -kət), **Sara Jane Clarke.** 1823–1904. Amer. author.
**Lipp·mann** (lĕp-män'), **Gabriel.** 1845–1921. French physicist (Nobel, 1908).
**Lipp·mann** (lĭp'mən), **Walter.** 1889–1974. Amer. journalist.
**Lip·scomb** (lĭp'skəm), **William Nunn, Jr.** b. 1919. Amer. chemist (Nobel, 1976).
**Lip·ton** (lĭp'tən), Sir **Thomas Johnstone.** 1850–1931. Scottish-born businessman, philanthropist, & sportsman.
**Li·sa** (lē'sə, -sä'), **Manuel.** 1772–1820. Amer. explorer & fur trader.
**Lis·ter** (lĭs'tər), **Joseph.** *1st Baron Lister.* 1827–1912. English founder of antiseptic surgery.
**Liszt** (lĭst), **Franz.** 1811–86. Hungarian pianist & composer.
**Lit·tle·ton** (lĭt'l-tən), Sir **Thomas.** 1407?–81. English jurist.
**Lit·tré** (lĭ-trā'), **Maximilien Paul Émile.** 1801–81. French philosopher & lexicographer.
**Lit·vi·nov** (lĭt-vē'nôf', -nôv', -nəf), **Maxim Maximovich.** 1876–1951. Russian diplomat.
**Liu Shao·qi** *also* **Liu Shao-chi** (lē-ōō' shou'chē'). 1898?–1973. Chinese political leader.
**Liv·er·more** (lĭv'ər-môr', -mōr'), **Mary Ashton Rice.** 1820–1905. Amer. suffragist, reformer, & lecturer.
**Liv·ing·ston** (lĭv'ĭng-stən), **Edward.** 1764–1836. Amer. politician & diplomat.
**Livingston, Henry Brockholst.** 1757–1823. Amer. jurist.
**Livingston, Robert.** 1654–1728. Scottish-born Amer. colonist & public official.
**Livingston, Robert R.** 1746–1813. Amer. Revolutionary leader & diplomat.
**Livingston, William.** 1723–90. Amer. politician.
**Liv·ing·stone** (lĭv'ĭng-stən), **David.** 1813–73. Scottish missionary & African explorer.
**Liv·y** (lĭv'ē), 59 B.C.–A.D. 17. Roman historian.
**Llew·el·lyn** (lōō-ĕl'ĭn), **Richard.** 1906–83. Welsh-born British author.
**Lloyd** (loid), **Harold Clayton.** 1894–1971. Amer. actor.
**Lloyd, Henry Demarest.** 1847–1903. Amer. reformer.
**Lloyd George** (jôrj'), **David.** *1st Earl of Dwyfor.* 1863–1945. British prime minister (1916–22).
**Lo·ba·chev·ski** (lō'bə-chĕf'skē), **Nikolai Ivanovich.** 1793–1856. Russian mathematician.
**Locke** (lŏk), **Alain LeRoy.** 1886–1954. Amer. educator & author.
**Locke, David Ross.** *Petroleum V. Nasby.* 1833–88. Amer. satirist.
**Locke, John.** 1632–1704. English philosopher.
**Lock·hart** (lŏk'ərt, -härt'), **John Gibson.** 1794–1854. Scottish author & editor.
**Lock·wood** (lŏk'wŏŏd'), **Belva Ann Bennett.** 1830–1917. Amer. lawyer & suffragist.
**Lock·yer** (lŏk'yər), Sir **Joseph Norman.** 1836–1920. English astronomer.
**Lodge** (lŏj), **Henry Cabot.** 1850–1924. Amer. politician & author.
**Lodge, Henry Cabot, Jr.** b. 1902. Amer. politician & diplomat.
**Lodge,** Sir **Oliver Joseph.** 1851–1940. English physicist.
**Lodge, Thomas.** 1558?–1625. English author.
**Loeb** (lōb), **Jacques.** 1859–1924. German-born Amer. physiologist.

**Loeb, James Morris.** 1867–1933. German-born Amer. banker & philanthropist.
**Loeb, Sophie Irene Simon.** 1876–1929. Russian-born Amer. journalist & social-welfare advocate.
**Loes·ser** (lĕs′ər), **Frank Henry.** 1910–69. Amer. composer.
**Loewe** (lō), **Frederick.** b. 1904. Austrian-born Amer. composer.
**Loe·wi** (lō′ē), **Otto.** 1873–1961. German-born Amer. pharmacologist (Nobel, 1936).
**Loe·wy** (lō′ē), **Raymond Fernand.** b. 1893. French-born Amer. industrial designer.
**Löff·ler** (lœf′lər, lĕf′-), **Friedrich August Johannes.** 1852–1915. German bacteriologist.
**Lo·gan** (lō′gən), **James.** 1674–1751. Irish-born Amer. colonial politician.
**Logan, James** or **John.** 1725–80. Amer. Indian leader.
**Logan, John Alexander.** 1826–86. Amer. soldier & politician.
**Logan, Joshua.** b. 1908. Amer. producer & director.
**Lo·max** (lō′măks′), **John Avery.** 1867–1948. Amer. folklorist & musicologist.
**Lom·bard** (lŏm′bärd′), **Carole.** 1908–42. Amer. actress.
**Lom·bard** (lŏm′bärd′, -bərd, lŭm′-), **Peter.** 1100?–60? Italian theologian.
**Lom·bar·di** (lŏm-bär′dē, lŭm-), **Vincent Thomas ("Vince").** 1913–70. Amer. football coach.
**Lom·bar·do** (lŏm-bär′dō, lŭm-), **Guy Albert.** 1902–77. Canadian-born Amer. bandleader.
**Lom·bro·so** (lôm-brō′sō), **Cesare.** 1836–1909. Italian criminologist.
**Lon·don** (lŭn′dən), **John Griffith ("Jack").** 1876–1916. Amer. author.
**London, Meyer.** 1871–1926. Amer. labor leader & politician.
**Long** (lông, lŏng), **Crawford Williamson.** 1815–78. Amer. surgeon & pioneer anesthetist.
**Long, Huey Pierce.** 1893–1935. Amer. politician; assassinated.
**Long, Stephen Harriman.** 1784–1864. Amer. railroad engineer & explorer.
**Long·fel·low** (lông′fĕl′ō, lŏng′-), **Henry Wadsworth.** 1807–82. Amer. poet.
**Lon·gi·nus** (lŏn-jī′nəs), **Dionysius Cassius.** 210?–73. Greek philosopher.
**Long·street** (lông′strēt′, lŏng′-), **Augustus Baldwin.** 1790–1870. Amer. clergyman, educator, & author.
**Longstreet, James.** 1821–1904. Amer. Confederate general.
**Long·worth** (lông′wûrth′, lŏng′-), **Alice Roosevelt.** 1884–1980. Amer. socialite & wit.
**Longworth, Nicholas.** 1869–1931. Amer. politician.
**Lönn·rot** (lĕn′rōōt′, -rōt′, lœn′rōt), **Elias.** 1802–84. Finnish scholar & anthologist.
**Lons·dale** (lŏnz′dāl′), **Frederick.** 1881–1954. English playwright.
**Loos** (lōōs), **Anita.** 1893?–1981. Amer. author.
**Ló·pez** (lō′pĕz), **Carlos Antonio** (1790–1862) & **Francisco Solano** (1827–70). Paraguayan political leaders.
**López Ma·te·os** (mə-tā′əs, -ōs), **Adolfo.** 1910–69. Mexican statesman.
**López Por·til·lo** (pôr-tē′ō, -yō), **José.** b. 1920. Mexican statesman.
**Lor·ca** (lôr′kə), **Federico García.** —See GARCÍA LORCA.
**Lord** (lôrd), **Walter.** b. 1917. Amer. author.
**Lo·renz** (lō′rĕnts′), **Konrad Zacharias.** b. 1903. Austrian psychologist (Nobel, 1973).
**Lor·en·zet·ti** (lôr′ən-zĕt′ē), **Ambrogio.** d. 1348? Sienese painter.
**Lor·i·mer** (lôr′ə-mər, lŏr′-), **George Horace.** 1867–1937. Amer. editor.
**Lor·rain** (lō-rān′, lô-răN′), **Claude.** 1600–82. French landscape painter.
**Lor·re** (lôr′ē), **Peter.** 1904–64. Czechoslovakian-born Amer. actor.
**Lot** (lŏt). Abraham's nephew in the Bible, whose wife was turned into a pillar of salt when she looked back as they fled Sodom.
**Lo·thair** (lō-thâr′, -târ′). Name of two Holy Roman Emperors: **a. I.** 795?–855. Ruled 840–55. **b. II.** 1070?–1137. Ruled 1125–37.
**Lo·throp** (lō′thrəp), **Alice Louise Higgins.** 1870–1920. Amer. social worker.
**Lothrop, Harriet Mulford Stone.** *Margaret Sidney.* 1844–1924. Amer. author.
**Lo·ti** (lō-tē′, lô′-), **Pierre.** 1850–1923. French novelist.
**Lou·bet** (lōō-bā′), **Émile.** 1838–1929. French statesman.
**Lou·is** (lōō′ē, lōō-ē′). Name of 18 kings of France, esp.: **a. V.** 966?–87. Ruled 986–87. **b. IX.** *Saint Louis.* 1214–70. Ruled 1226–70. **c. XI.** 1423–83. Ruled 1461–83. **d. XII.** 1462–1515. Ruled 1498–1515. **e. XIII.** 1601–43. Ruled 1610–43. **f. XIV.** *the Sun King.* 1638–1715. Ruled 1643–1715. **g. XV.** 1710–74. Ruled 1715–74. **h. XVI.** 1754–93. Ruled 1774–92; executed. **i. XVII.** 1785–95. Titular king, 1793–95. **j. XVIII.** 1755–1824. Ruled 1814–24.
**Louis.** Name of two Holy Roman Emperors: **a. I.** 778–840. Ruled 814–40. **b. IV.** 1287?–1347. Ruled 1314–47.
**Lou·is** (lōō′ĭs), **Joseph ("Joe").** 1914–81. Amer. prizefighter.
**Lou·is Na·po·le·on** (lōō′ē nə-pō′lē-ən). NAPOLEON III.
**Louis Phi·lippe** (fĭ-lēp′). *the Citizen King.* 1773–1850. French king (1830–48).
**Louns·bur·y** (lounz′bĕr′ē, -bə-rē), **Thomas Raynesford.** 1838–1915. Amer. educator & philologist.
**Lou·ÿs** (lōō-ē′), **Pierre.** 1870–1925. French author.
**Love·joy** (lŭv′joi′), **Elijah Parish.** 1802–37. Amer. clergyman, abolitionist, & journalist.
**Love·lace** (lŭv′lās′), **Richard.** 1618–58. English Cavalier poet.
**Lov·ell** (lŭv′əl), Sir **(Alfred Charles) Bernard.** b. 1913. English astronomer.
**Lov·er** (lŭv′ər), **Samuel.** 1797–1868. Irish author and songwriter.
**Low** (lō), Sir **David Alexander Cecil.** 1891–1963. English political cartoonist.
**Low, Juliette Magill Kinzie Gordon.** 1860–1927. Amer. founder of the Girl Scouts.
**Low, Seth.** 1850–1916. Amer. educator, reformer, & public official.
**Lowe** (lō), **Thaddeus Sobieski Coulincourt.** 1832–1913. Amer. inventor & aeronaut.
**Low·ell** (lō′əl), **Abbot Lawrence.** 1856–1943. Amer. educator.
**Lowell, Amy.** 1874–1925. Amer. poet.
**Lowell, James Russell.** 1819–91. Amer. editor, poet, & diplomat.
**Lowell, Josephine Shaw.** 1843–1905. Amer. philanthropist & reformer.
**Lowell, Percival.** 1855–1916. Amer. astronomer.
**Lowell, Robert Traill Spence, Jr.** 1917–77. Amer. poet.
**Lowes** (lōz), **John Livingston.** 1867–1945. Amer. educator & literary critic.
**Lo·wie** (lō′ē), **Robert Harry.** 1883–1957. Austrian-born Amer. anthropologist.
**Lowndes** (loundz), **William Thomas.** 1798–1843. English bibliographer & bookseller.
**Loy·o·la** (loi-ō′lə), Saint **Ignatius.** —See IGNATIUS OF LOYOLA.
**Lo·zi·er** (lō′zē-ər), **Clemence Sophia Harned.** 1813–88. Amer. physician & feminist.
**Lub·bock** (lŭb′ək), Sir **John.** *1st Baron Avebury.* 1834–1913. English banker, politician, & naturalist.
**Lubbock,** Sir **John William.** 1803–65. English astronomer & mathematician.
**Lu·bitsch** (lōō′bĭch), **Ernst.** 1892–1947. German filmmaker.
**Lu·can** (lōō′kən). A.D. 39–65. Roman poet.
**Luce** (lōōs), **Clare Boothe.** b. 1903. Amer. editor, politician, diplomat, & playwright.
**Luce, Henry Robinson.** 1898–1967. Amer. editor & publisher.
**Lu·cian** (lōō′shən). 2nd cent. A.D. Greek satirist.
**Lu·cre·tius** (lōō-krē′shəs, -shē-əs). 96?–55 B.C. Roman philosopher & poet. **—Lu·cre′tian** *adj.*
**Lu·cul·lus** (lōō-kŭl′əs), **Lucius Licinius.** 110?–57? B.C. Roman general, consul, & patron of the arts. **—Lu·cul′lan** *adj.*
**Lu·den·dorff** (lōōd′n-dôrf′), **Erich Friedrich Wilhelm von.** 1865–1937. German general & political leader.
**Lud·wick** (lŭd′wĭk), **Christopher.** 1720–1801. Amer. Revolutionary patriot, baker, & philanthropist.
**Lu·go·si** (lōō-gō′sē, lə-), **Bela.** 1884–1956. Hungarian-born Amer. actor.
**Lu·han** (lōō′hän′), **Mabel Ganson Dodge.** 1879–1962. Amer. author.
**Luke** (lōōk), Saint. Companion of St. Paul & author of the third Gospel.
**Luks** (lŭks), **George.** 1867–1933. Amer. painter.
**Lul·ly** (lōō-lē′), **Jean Baptiste.** 1632–87. Italian-born French composer.
**Lul·ly** (lŭl′ē), **Raymond.** 1235?–1315. Spanish missionary & scholar.
**Lu·met** (lōō-mĕt′), **Sidney.** b. 1924. Amer. filmmaker.
**Lu·mière** (lōō-myĕr′), **Auguste Marie Louis Nicolas** (1862–1954) & **Louis Jean** (1864–1948). French chemists, inventors, & cinematography pioneers.
**Lu·mum·ba** (lōō-mōōm′bə), **Patrice Emergy.** 1925–61. Congolese (Zaire) statesman; murdered.
**Lun·dy** (lŭn′dē), **Benjamin.** 1789–1839. Amer. abolitionist.
**Lunt** (lŭnt), **Alfred.** 1893–1977. Amer. actor.
**Lu·ri·a** (lōōr′ē-ə), **Salvador Edward.** b. 1912. Italian-born Amer. biologist (Nobel, 1969).
**Lur·ton** (lûr′tn), **Horace Harmon.** 1844–1914. Amer. jurist.
**Lu·ther** (lōō′thər), **Martin.** 1483–1546. German monk & Protestant religious reformer. **—Lu′ther·an** *adj.* & *n.*
**Lu·thu·li** (lōō-tōō′lē, -thōō′-), **Albert John.** 1898–1967. Zulu chieftain & reformer (Nobel, 1960).
**Lux·em·burg** (lŭk′səm-bûrg′, lōōk′səm-bōōrk′), **Rosa.** *Red Rosa.* 1870–1919. German socialist leader; assassinated.
**Lwoff** (lwôf, lə-wôf′), **André Michel.** b. 1902. French microbiologist (Nobel, 1965).
**Ly·cur·gus** (lī-kûr′gəs). 9th cent. B.C. Spartan lawmaker.
**Lyd·gate** (lĭd′gāt′, -gət), **John.** 1370?–1451? English poet.
**Ly·ell** (lī′əl), Sir **Charles.** 1797–1875. English geologist.

**Lyl·y** (lĭl'ē), **John.** 1554?–1606. English novelist & playwright.
**Lynch** (lĭnch), **Charles.** 1736–96. Amer. judge & planter.
**Lynch, John Mary.** b. 1917. Irish political leader.
**Lynd** (lĭnd), **Robert Staughton** (1892–1970) & **Helen** (1897–1982). Amer. sociologists.
**Ly·nen** (lo͞o'nən), **Feodor.** 1911–79. German biochemist (Nobel, 1964).
**Lynn** (lĭn), **Janet.** b. 1953. Amer. figure skater.
**Ly·on** (lī'ən), **Mary Mason.** 1797–1849. Amer. educator.
**Ly·ons** (lī'ənz), **Joseph Aloysius.** 1879–1939. Australian statesman.
**Ly·san·der** (lī-săn'dər). d. 395 B.C. Spartan military leader.
**Ly·sen·ko** (lĭ-sĕng'kō), **Trofim Denisovich.** 1898–1976. Soviet biologist & agronomist.
**Ly·sim·a·chus** (lī-sĭm'ə-kəs). 361?–281 B.C. Macedonian general.
**Ly·sip·pus** (lī-sĭp'əs). 4th cent. B.C. Greek sculptor.
**Lyt·ton** (lĭt'n), 1st Baron. *Edward George Earle Lytton-Bulwer.* 1803–73. English author.
**Lytton,** 1st Earl of. *Edward Robert Lytton Bulwer-Lytton.* 1831–91. English politician & diplomat.
**Lytton,** 2nd Earl of. *Victor Alexander George Robert Lytton.* 1876–1947. English colonial administrator & diplomat.

# M

**Mac·Ar·thur** (mək-är'thər), **Arthur.** 1845–1912. Amer. general.
**MacArthur, Charles.** 1895–1956. Amer. dramatist & screenwriter.
**MacArthur, Douglas.** 1880–1964. Amer. general.
**Ma·cau·lay** (mə-kô'lē), Dame **Rose.** 1881–1958. English author.
**Macaulay, Thomas Babington.** 1800–59. English historian & statesman.
**Mac·beth** (mək-bĕth'). d. 1057. King of Scotland (1040–57).
**Mac·Bride** (mək-brīd'), **Sean.** b. 1904. Irish statesman (Nobel, 1974).
**Mac·ca·bees** (măk'ə-bēz'). Family of Jewish patriots of 2nd & 1st cent. B.C., including **Judas** *or* **Judah Mac·ca·be·us** (măk'ə-bē'əs), d. 160 B.C.
**Mac·Crack·en** (mə-krăk'ən), **Henry Noble.** 1880–1970. Amer. educator.
**Mac·Diar·mid** (mək-dûr'mĭd), **Hugh.** 1892–1978. Scottish poet.
**Mac·don·ald** (mək-dŏn'əld), **Dwight.** 1906–82. Amer. author & editor.
**Macdonald, George.** 1824–1905. Scottish author.
**Macdonald,** Sir **John Alexander.** 1815–91. Canadian prime minister (1867–73, 1878–91).
**Mac·Don·ald** (mək-dŏn'əld), **(James) Ramsay.** 1866–1937. British prime minister (1924, 1929–35).
**Mac·don·ough** (mək-dŏn'ə, -dŭn'ə), **Thomas.** 1783–1825. Amer. naval officer.
**Mac·Dow·ell** (mək-dou'əl), **Edward Alexander.** 1861–1908. Amer. composer.
**Mac·Fad·den** (mək-făd'n), **Bernarr.** 1868–1955. Amer. physical-culture advocate & publisher.
**Mach** (mäk, mäKH), **Ernst.** 1838–1916. Austrian physicist & philosopher.
**Ma·cha·do y Mo·ra·les** (mä-chä'dō ē mə-rä'ləs), **Gerardo.** 1871–1939. Cuban politician.
**Ma·chen** (mā'chən), **John Gresham.** 1881–1937. Amer. theologian.
**Mach·i·a·vel·li** (măk'ē-ə-vĕl'ē), **Niccolò.** 1469–1527. Italian statesman & political theorist. **—Mach'i·a·vel'li·an** *adj. & n.*
**Mac·In·nes** (mə-kĭn'ĭs), **Helen Clark.** b. 1907. Scottish-born Amer. author.
**Mac·I·ver** (mə-kī'vər), **Robert Morrison.** 1882–1970. Scottish-born Amer. sociologist.
**Mack** (măk), **Connie.** 1862–1956. Amer. baseball player & manager.
**Mack·ay** (măk'ē), **Clarence Hungerford.** 1874–1938. Amer. financier & philanthropist.
**Mac·Kaye** (mə-kī'), **Benton.** 1879–1975. Amer. forester & regional planner.
**MacKaye, Percy.** 1875–1956. Amer. poet & playwright.
**MacKaye, (James Morrison) Steele.** 1842–94. Amer. actor, playwright, & producer.
**Mac·ken·zie** (mə-kĕn'zē), Sir **Alexander.** 1764–1820. Scottish explorer.
**Mackenzie, Alexander.** 1822–92. Scottish-born Canadian prime minister (1873–78).
**Mackenzie,** Sir **Compton.** 1883–1972. English author.
**Mackenzie, William Lyon.** 1795–1861. Scottish-born Canadian insurgent.
**Mac·kin·der** (mə-kĭn'dər), Sir **Halford John.** 1861–1947. English geographer.
**Mack·in·tosh** (măk'ĭn-tŏsh'), Sir **James.** 1765–1832. Scottish historian & philosopher.
**Mac·Leish** (mə-klēsh'), **Archibald.** 1892–1982. Amer. poet & dramatist.
**Mac·Len·nan** (mə-klĕn'ən), **Hugh.** b. 1907. Canadian author.
**Mac·leod** (mə-kloud'), **John James Rickard.** 1876–1935. Scottish physiologist (Nobel, 1923).
**Mac·mil·lan** (mək-mĭl'ən), **(Maurice) Harold.** b. 1894. British prime minister (1957–63).
**Mac·Mil·lan** (mək-mĭl'ən), **Donald Baxter.** 1874–1970. Amer. Arctic explorer.
**Mac·Mon·nies** (mək-mŏn'ĭz), **Frederick William.** 1863–1937. Amer. sculptor.
**Mac·Neice** (mək-nēs'), **Louis.** 1907–63. Irish-born British poet & scholar.
**Mac·Neil** (mək-nēl'), **Hermon Atkins.** 1866–1947. Amer. sculptor.
**Ma·con** (mā'kən), **Nathaniel.** 1758–1837. Amer. legislator.
**Mac·pher·son** (mək-fûr'sən), **James.** 1736–96. Scottish poet.
**Mac·rea·dy** (mə-krē'dē), **William Charles.** 1793–1873. English actor.
**Ma·da·ria·ga y Ro·jo** (mä'də-rē-ä'gə ē rō'hō), **Salvador de.** 1886–1978. Spanish diplomat.
**Ma·de·ro** (mə-dîr'ō), **Francisco Indalecio.** 1873–1913. Mexican revolutionary & statesman.
**Mad·i·son** (măd'ĭ-sən), **Dolley Payne Todd.** 1768–1849. Amer. hostess & wife of James Madison.
**Madison, James.** 1751–1836. Fourth U.S. President (1809–17) & political theorist. **—Mad'i·so'ni·an** (-sō'nē-ən) *adj.*
**Mae·ce·nas** (mī-sē'nəs), **Gaius.** 70?–8 B.C. Roman statesman & literature patron.
**Mae·ter·linck** (mā'tər-lĭngk', mĕt'ər-, mä'-), Count **Maurice.** 1862–1949. Belgian poet, dramatist, & naturalist (Nobel, 1911).
**Ma·gel·lan** (mə-jĕl'ən), **Ferdinand.** 1480?–1521. Portuguese navigator; killed while circumnavigating the globe.
**Ma·gi·not** (măzh'ə-nō', măj'-), **André.** 1877–1932. French politician.
**Mag·nes** (măg'nĭs), **Judah Leon.** 1877–1948. Amer. religious leader & educator.
**Ma·gritte** (mə-grēt'), **René.** 1898–1967. Belgian painter.
**Mag·say·say** (mäg-sī'sī'), **Ramón.** 1907–57. Philippine statesman.
**Ma·han** (mə-hăn'), **Alfred Thayer.** 1840–1914. Amer. naval officer & author.
**Mah·ler** (mä'lər), **Gustav.** 1860–1911. Austrian composer & conductor.
**Mah·mud II** (mä-mo͞od'). 1785–1839. Turkish sultan (1803–39).
**Mail·er** (mā'lər), **Norman.** b. 1923. Amer. author.
**Mail·lol** (mä-yôl', -yŏl'), **Aristide.** 1861–1944. French sculptor.
**Mai·mon·i·des** (mī-mŏn'ĭ-dēz'), **Moses.** 1135–1204. Spanish-born Jewish philosopher.
**Maine** (mān), Sir **Henry James Sumner.** 1822–88. English jurist.
**Main·te·non** (măN'tə-nôN', măNt-nôN'), **Marquise de.** *Françoise d'Aubigné.* 1635–1719. French consort of Louis XIV.
**Mait·land** (māt'lənd), **Frederic William.** 1850–1906. English jurist & historian.
**Ma·kar·i·os III** (mə-kăr'ē-əs, -ōs'). 1913–77. Cypriot prelate & statesman.
**Ma·ki·no** (mä-kē'nō), Count **Nobuaki.** 1861–1949. Japanese statesman.
**Mal·a·chi** (măl'ĭ-kī'). 5th cent. B.C. Hebrew prophet.
**Mal·a·mud** (măl'ə-məd), **Bernard.** b. 1914. Amer. author.
**Ma·lan** (mə-lăn', -län'), **Daniel François.** 1874–1959. South African journalist & statesman.
**Mal·bone** (môl'bōn'), **Edward Greene.** 1777–1807. Amer. painter.
**Mal·colm X** (măl'kəm ĕks'). 1925–65. Amer. civil-rights leader; assassinated.
**Male·branche** (măl-bräNsh', măl'ə-, mäl-), **Nicolas de.** 1638–1715. French philosopher.
**Ma·len·kov** (mə-lĕn'kôf', -kôv', măl'ən-kôf', -kôv'), **Georgi Maximilianovich.** b. 1902. Soviet statesman.
**Mal·herbe** (mä-lĕrb'), **François de.** 1555–1628. French poet.
**Mal·i·now·ski** (măl'ə-nôf'skē, -nôv'-, mä'lĭ-), **Bronislaw Kasper.** 1884–1942. Polish-born English anthropologist.
**Mal·lar·mé** (măl'är-mā'), **Stéphane.** 1842–98. French poet.
**Mal·lon** (măl'ən), **Mary.** *Typhoid Mary.* 1870?–1938. Amer. cook & disease carrier.
**Ma·lone** (mə-lōn'), **Edmund** *or* **Edmond.** 1741–1812. Irish scholar & critic.
**Mal·o·ry** (măl'ə-rē), Sir **Thomas.** fl. 1470. English author.
**Mal·pi·ghi** (măl-pē'gē), **Marcello.** 1628–94. Italian anatomist.
**Mal·raux** (măl-rō'), **André.** 1901–76. French author & politician.

**Mal·thus** (măl′thəs, môl′-), **Thomas Robert.** 1766–1834. English economist. **—Mal·thu′sian** (măl-thōō′zhən, môl-) *adj.* & *n.*
**Ma·nas·seh** (mə-năs′ə). 7th cent. B.C. king of Judah.
**Man·de·ville** (măn′də-vĭl′), **Bernard.** 1670?–1733. Dutch-born English physician & satirist.
**Mandeville,** Sir **John.** d. 1372. Pseudonym of unknown compiler of travel books.
**Ma·nes** (mā′nēz). 216?–76? Persian prophet.
**Ma·net** (mə-nā′), **Edouard.** 1832–83. French painter.
**Ma·nil·i·us** (mə-nĭl′ē-əs), **Gaius.** 1st cent. B.C. Roman politician.
**Mann** (măn), **Horace.** 1796–1859. Amer. educator.
**Mann** (măn, män), **Thomas.** 1875–1955. German-born Amer. author (Nobel, 1929).
**Man·ner·heim** (mä′nər-hām′, -hīm′, măn′ər-), Baron **Carl Gustaf Emil von.** 1867–1951. Finnish soldier & statesman.
**Man·nes** (măn′ĭs), **Clara Damrosch.** 1869–1948. Polish-born Amer. musician.
**Mannes, Marya.** b. 1904. Amer. author.
**Man·ning** (măn′ĭng), **Henry Edward.** 1808–92. English religious leader.
**Ma·no·le·te** (mä′nō-lā′tā). 1917–47. Spanish bullfighter.
**Mans·field** (măns′fēld′), **Katherine.** 1888–1923. New Zealand-born British author.
**Mansfield, Richard.** 1854?–1907. English-born Amer. actor.
**Man·son** (măn′sən), Sir **Patrick.** 1844–1922. British parasitologist.
**Man·sur** (măn-sŏŏr′), **al-.** 712?–75. Arab caliph (754–75).
**Man·te·gna** (män-tān′yə), **Andrea.** 1431–1506. Italian painter.
**Man·tle** (măn′tl), **(Robert) Burns.** 1873–1948. Amer. journalist.
**Mantle, Mickey Charles.** b. 1931. Amer. baseball player.
**Ma·nu·ti·us** (mə-nōō′shəs, -shē-əs, -nyōō′-), **Aldus.** 1450–1515. Italian scholar & printer.
**Man·zo·ni** (män-zō′nē), **Alessandro.** 1785–1873. Italian author.
**Mao Ze·dong** (mou′ dzŭ′dŏŏng′) *also* **Mao Tse-tung** (tsə-tŏŏng′). 1893–1976. Chinese Communist leader.
**Ma·rat** (mə-rä′), **Jean Paul.** 1743–93. Swiss-born French revolutionary; assassinated.
**Mar·cel·lus** (mär-sĕl′əs), **Marcus Claudius.** 268?–208 B.C. Roman general.
**March** (märch), **Francis Andrew.** 1825–1911. Amer. philologist & lexicographer.
**Mar·ci·a·no** (mär′sē-ä′nō), **Rocco Francis ("Rocky").** 1924–69. Amer. boxer.
**Mar·co·ni** (mär-cō′nē), Marchese **Guglielmo.** 1874–1937. Italian engineer & inventor (Nobel, 1909).
**Mar·co Po·lo** (mär′kō pō′lō). —See POLO.
**Mar·cos** (mär′kōs, -kəs), **Ferdinand Edralin.** b. 1917. Philippine president (since 1965).
**Marcos, Imelda.** b. 1930. Philippine political leader.
**Mar·cus Au·re·li·us An·to·ni·nus** (mär′kəs ô-rē′lē-əs ăn′tə-nī′nəs). 121–80. Roman emperor & philosopher.
**Mar·cu·se** (mär-kōō′zə), **Herbert.** 1898–1979. German-born Amer. philosopher.
**Mar·gar·et of An·jou** (mär′gə-rət, -grət; ăn-jōō′, äN-zhōō′). 1430–82. Queen of Henry IV of England.
**Margaret of Na·varre** (nə-vär′). 1492–1549. Author and queen of Navarre (1544–49).
**Margaret of Val·ois** (văl-wä′). 1553–1615. Author & queen of Navarre.
**Margaret Rose** (rōz). b. 1930. Princess of Great Britain.
**Mar·gre·the II** (mär-grā′tə). b. 1940. Queen of Denmark (since 1972).
**Ma·ri·a The·re·sa** (mə-rē′ə tə-rā′sə, -zə). 1717–80. Queen of Hungary & Bohemia.
**Ma·rie** (mə-rē′). 1875–1938. Queen of Rumania (1914–27); queen dowager (1927–38).
**Marie An·toi·nette** (ăn′twə-nĕt′). 1755–93. Queen of France (1774–93) as wife of Louis XVI; executed.
**Marie Lou·ise** (lōō-ēz′). 1791–1847. Second wife of Napoleon I.
**Mar·in** (măr′ĭn), **John.** 1872–1953. Amer. painter.
**Ma·ri·net·ti** (măr′ə-nĕt′ē), **Emilio Filippo Tommaso.** 1876–1944. Italian poet.
**Ma·ri·ni** (mə-rē′nē) *or* **Ma·ri·no** (-nō), **Giambattista.** 1569–1625. Italian poet.
**Mar·i·on** (măr′ē-ən, mâr′-), **Francis.** *the Swamp Fox.* 1732?–95. Amer. Revolutionary soldier.
**Ma·ri·tain** (măr′ĭ-tăN′), **Jacques.** 1882–1973. French philosopher & critic.
**Mar·i·us** (mâr′ē-əs), **Gaius.** 155?–86 B.C. Roman general.
**Ma·ri·vaux** (măr′ə-vō′), **Pierre Carlet de Chamblain de.** 1688–1763. French author.
**Mark** (märk), Saint. Author of the second Gospel.
**Mark An·to·ny** (ăn′tə-nē) *or* **Mark An·tho·ny** (ăn′thə-nē). 83?–30 B.C. Roman orator, politician, & soldier.
**Mark·ham** (mär′kəm), **(Charles) Edwin.** 1852–1940. Amer. poet.
**Mar·ko·va** (mär′kə-və, mär-kō′və), Dame **Alicia.** b. 1910. British ballerina.
**Marl·bor·ough** (märl′bər-ə, môl′-), 1st Duke of. John CHURCHILL.
**Mar·lowe** (mär′lō), **Christopher.** 1564–93. English dramatist & poet.
**Marlowe, Julia.** 1866–1950. English-born Amer. actress.
**Mar·mon·tel** (mär-môN-tĕl′), **Jean François.** 1723–99. French author.
**Ma·rot** (mä-rō′), **Clément.** 1495?–1544. French poet.
**Mar·quand** (mär-kwŏnd′), **John Phillips.** 1893–1960. Amer. author.
**Mar·quette** (mär-kĕt′), Père **Jacques.** 1637–75. French missionary & explorer.
**Mar·quis** (mär′kwĭs), **Donald Robert Perry.** 1878–1937. Amer. journalist.
**Mar·ry·at** (măr′ē-ət), **Frederick.** 1792–1848. English naval officer & author.
**Marsh** (märsh), **George Perkins.** 1801–82. Amer. diplomat & linguist.
**Marsh, Ngaio.** 1899–1982. New Zealand author.
**Marsh, Othniel Charles.** 1831–99. Amer. paleontologist.
**Marsh, Reginald.** 1898–1954. Amer. painter.
**Mar·shall** (mär′shəl), **Clara.** 1847–1931. Amer. physician & educator.
**Marshall, George Catlett.** 1880–1959. Amer. soldier, diplomat, & statesman (Nobel, 1953).
**Marshall, James Wilson.** 1810–85. Amer. Western pioneer.
**Marshall, John.** 1755–1835. Amer. jurist & statesman.
**Marshall, Louis.** 1856–1929. Amer. lawyer & religious leader.
**Marshall, Thomas Riley.** 1854–1925. U.S. Vice President (1913–21).
**Marshall, Thurgood.** b. 1908. Amer. jurist.
**Mar·sil·i·us of Pad·u·a** (mär-sĭl′ē-əs, păj′ōō-ə, păd′yōō-ə). 1290?–1343. Italian philosopher.
**Mar·ston** (mär′stən), **John.** 1575?–1634. English dramatist.
**Mar·tel** (mär-tĕl′), **Charles.** —See CHARLES MARTEL.
**Mar·tens** (mär′tnz), **Fedor Fedorovich** *or* **Frédéric Frommhold.** 1845–1909. Russian jurist.
**Mar·tha** (mär′thə). Biblical sister of Lazarus & Mary.
**Mar·tí** (mär-tē′), **José Julian.** 1853–95. Cuban revolutionary leader.
**Mar·tial** (mär′shəl). 1st cent. A.D. Roman epigrammist.
**Mar·tin** (mär′tn). Name of five popes, esp.: **a. I,** Saint. d. 655. Reigned 649–55. **b. V.** 1368–1431. Reigned 1417–31.
**Martin, Archer John Porter.** b. 1910. British chemist (Nobel, 1952).
**Martin, Glenn Luther.** 1886–1955. Amer. airplane manufacturer.
**Martin, Homer Dodge.** 1836–97. Amer. painter.
**Martin, Joseph William, Jr.** 1884-1968. Amer. publisher & politician.
**Martin, Lillien Jane.** 1851–1943. Amer. psychologist.
**Martin, Luther.** 1748?–1826. Amer. Revolutionary leader.
**Martin, Mary.** b. 1913. Amer. actress.
Patron of France.
**Mar·tin Du Gard** (mär-tăN′ dü găr′), **Roger.** 1881–1958. French author (Nobel, 1937).
**Mar·ti·neau** (mär′tə-nō), **Harriet.** 1802–76. English author.
**Martineau, James.** 1805–1900. English theologian & philosopher.
**Mar·ti·ni** (mär-tē′nē), **Simone.** 1283?–1344. Italian painter.
**Mar·tin of Tours** (mär′tn, mär-tăN′; tŏŏr, tōōr), Saint. 315?–99? Patron of France.
**Mar·tins** (mär′tnz), **Peter.** b. 1946. Danish-born ballet dancer and choreographer.
**Mar·tin·son** (mär′tn-sôn′, -tĕn-), **Harry Edmund.** 1904–78. Swedish author (Nobel, 1974).
**Mar·vell** (mär′vəl), **Andrew.** 1621–78. English poet & satirist.
**Marx** (märks). Family of Amer. comedians, including **Leonard ("Chico"),** 1891–1961; **Adolph Arthur ("Harpo"),** 1893–1964; & **Julius ("Groucho"),** 1895–1977.
**Marx, Karl.** 1818–83. German political philosopher & economist. **—Marx′i·an** *adj.* & *n.* **—Marx′ist** *n.*
**Mar·y** (mâr′ē). Mother of Jesus.
**Mary.** Name of two English queens: **a. I.** *also* **Mary Tu·dor** (tōō′dər). *Bloody Mary.* 1516–58. Ruled 1553–58. **b. II.** 1662–94. Ruled jointly with William III (1689–94).
**Mary** *also* **Mary of Teck** (tĕk). 1867–1953. Queen of George V of England.
**Mary Mag·da·lene** (măg′də-lĕn′, -lēn′). Biblical figure cured of evil spirits; also identified with the prostitute who kissed Jesus' feet.
**Mary Queen of Scots** (skŏts) *also* **Mary Stu·art** (stōō′ərt). 1542–87. Queen of Scotland (1542–67); beheaded.
**Ma·sac·cio** (mə-zä′chō, -chē-ō). 1401–28. Italian painter.
**Mas·a·ryk** (măs′ə-rĭk), **Jan Garrigue.** 1886–1948. Czech statesman & diplomat.
**Masaryk, Tomáš Garrigue.** 1850–1937. Czech statesman.
**Mas·ca·gni** (mäs-kän′yē), **Pietro.** 1863–1945. Italian composer.
**Mase·field** (mās′fēld′), **John.** 1878–1967. English author.

ă **pat** ā **pay** âr **care** ä **father** ĕ **pet** ē **be** hw **which** ĭ **pit**
ī **tie** îr **pier** ŏ **pot** ō **toe** ô **paw, for** oi **noise** ŏŏ **took**

**Ma·son** (mā'sən), **Charles.** 1730?–87. English astronomer & surveyor in America.
**Mason, George.** 1725–92. Amer. Revolutionary statesman.
**Mason, James Murray.** 1798–1871. Amer. lawyer & Confederate diplomat.
**Mason, Lowell.** 1792–1872. Amer. musician & composer.
**Mas·sa·soit** (măs'ə-soit'). 1580?–1661. New England Indian leader.
**Mas·se·net** (măs'ə-nā', măs-nā'), **Jules Émile Frédéric.** 1842–1912. French composer.
**Mas·sey** (măs'ē), **Raymond.** 1896–1983. Canadian actor.
**Massey, Vincent.** 1887–1967. Canadian diplomat & statesman.
**Massey, William Ferguson.** 1856–1925. New Zealand statesman.
**Mas·sine** (mä-sēn'), **Léonide.** 1894–1979. Russian-born Amer. choreographer.
**Mas·sin·ger** (măs'ĭn-jər), **Philip.** 1583–1640. English dramatist.
**Mas·son** (măs'ən), **David.** 1822–1907. Scottish author & editor.
**Mas·ters** (măs'tərz), **Edgar Lee.** 1869–1950. Amer. poet.
**Mas·ter·son** (măs'tər-sən), **William Barclay ("Bat").** 1853–1921. Amer. journalist & frontier marshal.
**Ma·ta Ha·ri** (mä'tə här'ē, măt'ə hăr'ē). 1867–1917. German dancer & spy; executed.
**Math·er** (măth'ər), **Increase** (1639–1723) & **Cotton** (1663–1728). Amer. clergymen & authors.
**Mather, Richard.** 1596–1669. English-born Amer. clergyman.
**Mather, Stephen Tyng.** 1867–1930. Amer. conservationist.
**Math·ew·son** (măth'yo͞o-sən), **Christopher ("Christy").** 1880–1925. Amer. baseball player.
**Ma·thi·as** (mə-thī'əs), **Robert Bruce ("Bob").** b. 1930. Amer. Olympic decathlon champion.
**Ma·tisse** (mə-tēs', mä-), **Henri.** 1869–1954. French artist.
**Ma·tsu·o·ka** (măt'sə-wō'kə), **Yosuke.** 1880–1946. Japanese statesman.
**Mat·te·ot·ti** (măt'ē-ŏ'tē), **Giacomo.** 1885–1924. Italian political leader.
**Mat·thew** (măth'yo͞o), Saint. Apostle & author of the first Gospel.
**Mat·thews** (măth'yo͞oz), **(James) Brander.** 1852–1929. Amer. educator & author.
**Matthews, Stanley.** 1824–89. Amer. jurist.
**Mat·ting·ly** (măt'ĭng-lē), **Garrett.** 1900–62. Amer. historian.
**Maugham** (môm), **William Somerset.** 1874–1965. English author.
**Maul·din** (môl'dĭn), **William Henry ("Bill").** 1921–81. Amer. cartoonist.
**Mau·pas·sant** (mō-pə-säN', mō'pə-sänt'), **(Henri René Albert) Guy de.** 1850–93. French author.
**Mau·per·tuis** (mō-pěr-twē'), **Pierre Louis Moreau de.** 1698–1759. French scientist.
**Mau·riac** (môr-yäk', môr'ē-äk'), **François.** 1885–1970. French author (Nobel, 1952).
**Mau·rice** (môr'ĭs, mŏr'-, mô-rēs'). 1521–53. Duke (1541–53) & elector (1547–53) of Saxony.
**Maurice of Nas·sau** (năs'ô). *Prince of Orange.* 1567–1625. Dutch general & statesman.
**Mau·rois** (môr-wä'), **André.** 1885–1967. French author & historian.
**Mau·ry** (môr'ē), **Matthew Fontaine.** 1806–73. Amer. naval officer, meteorologist, & oceanographer.
**Mau·ser** (mou'zər), **Wilhelm** (1834–82) & **Peter Paul** (1838–1914). German inventors.
**Mav·er·ick** (măv'ər-ĭk, măv'rĭk), **Samuel Augustus.** 1803–70. Amer. pioneer & cattle rancher.
**Maw·son** (mô'sən), Sir **Douglas.** 1882–1958. English explorer & geologist.
**Max·im** (măk'sĭm), Sir **Hiram Stevens.** 1840–1916. Amer.-born British inventor.
**Maxim, Hudson.** 1853–1927. Amer. inventor.
**Max·i·mil·ian** (măk'sə-mĭl'yən). 1832–67. Austrian archduke & emperor of Mexico (1864–67); executed.
**Maximilian.** Name of two Holy Roman Emperors: **a. I.** 1459–1519. Ruled 1493–1519. **b. II.** 1527–76. Ruled 1564–76.
**Max·well** (măks'wĕl', -wəl), **Elsa.** 1883–1963. Amer. columnist & hostess.
**Maxwell, James Clerk.** 1831–79. Scottish physicist.
**May** (mā), **Geraldine Pratt.** b. 1895. Amer. air-force officer.
**May,** Sir **Thomas Erskine.** *1st Baron Farnborough.* 1815–86. English jurist.
**Ma·ya·kov·ski** (mä'yə-kôf'skē, -kôv'-, mī'ə-), **Vladimir Vladimirovich.** 1893–1930. Russian poet.
**May·er** (mī'ər), **Julius Robert von.** 1814–78. German physicist & physician.
**May·er** (mā'ər), **Louis Burt.** 1885–1957. Russian-born Amer. motion-picture producer.
**May·er** (mī'ər), **Marie Goeppert.** 1906–72. German-born Amer. physicist (Nobel, 1963).
**May·o** (mā'ō), **Henry Thomas.** 1856–1937. Amer. naval officer.
**Mayo, William James** (1861–1939) & **Charles Horace** (1865–1939). Amer. surgeons & founders of the Mayo Clinic.
**Mays** (māz), **Willie Howard, Jr.** b. 1931. Amer. baseball player.
**Maz·a·rin** (măz'ə-răN'), **Jules.** 1602–61. Italian-born French cardinal & statesman.
**Maz·ze·i** (mät-sā'ē), **Philip.** 1730–1816. Italian-born Amer. physician & merchant.
**Maz·zi·ni** (mät-sē'nē), **Giuseppe.** 1805–72. Italian revolutionary patriot.
**Mc·A·doo** (măk'ə-do͞o'), **William Gibbs.** 1863–1941. Amer. lawyer & politician.
**Mc·Al·lis·ter** (mə-kăl'ĭ-stər), **(Samuel) Ward.** 1827–95. Amer. lawyer & socialite.
**M'Car·thy** (mə-kär'thē), **Justin.** 1830–1912. Irish writer & politician.
**M'Carthy, Justin Huntly.** 1861–1936. Irish author.
**Mc·Au·liffe** (mə-kô'lĭf), **Anthony Clement.** 1898–1975. Amer. soldier.
**Mc·Bride** (mək-brīd'), **Mary Margaret.** 1899–1976. Amer. broadcast journalist.
**Mc·Bur·ney** (mək-bûr'nē), **Charles.** 1845–1913. Amer. surgeon.
**Mc·Car·thy** (mə-kär'thē), **Eugene Joseph.** b. 1916. Amer. politician & poet.
**McCarthy, Joseph Raymond.** 1908–57. Amer. politician.
**McCarthy, Mary Therese.** b. 1912. Amer. author.
**Mc·Cart·ney** (mə-kärt'nē), **(James) Paul.** b. 1942. English musician & composer.
**Mc·Cau·ley** (mə-kô'lē), **Mary Ludwig Hays.** *Molly Pitcher.* 1754–1832. Amer. Revolutionary heroine.
**Mc·Clel·lan** (mə-klĕl'ən), **George Brinton.** 1826–85. Amer. general & politician.
**Mc·Clin·tock** (mə-klĭn'tək, -tŏk'), **Barbara.** b. 1902. Amer. genetic botanist (Nobel, 1983).
**Mc·Clos·key** (mə-klôs-kē), **John.** 1810–85. Amer. religious leader.
**Mc·Cloy** (mə-kloi'), **John Jay.** b. 1895. Amer. banker & public official.
**Mc·Clure** (mə-klo͝or'), **Samuel Sidney.** 1857–1949. Irish-born Amer. editor & publisher.
**Mc·Cor·mack** (mə-kôr'mək, -mĭk), **John.** 1884–1945. Amer. opera singer.
**McCormack, John William.** 1891–1981. Amer. politician.
**Mc·Cor·mick** (mə-kôr'mĭk), **Anne Elizabeth O'Hare.** 1882–1954. English-born Amer. journalist.
**McCormick, Cyrus Hall.** 1809–84. Amer. inventor & manufacturer.
**McCormick, Joseph Medill** (1877–1925) & **Robert Rutherford** (1880–1955). Amer. newspaper publishers.
**Mc·Coy** (mə-koi'), **Joseph Geating.** 1837–1915. Amer. cattleman.
**Mc·Crae** (mə-krā'), **John.** 1872–1918. Canadian physician & poet.
**Mc·Cul·lers** (mə-kŭl'ərz), **Carson Smith.** 1917–67. Amer. author.
**Mc·Cutch·eon** (mə-kŭch'ən), **John Tinney.** 1870–1949. Amer. cartoonist.
**Mc·Dou·gall** (mək-do͞o'gəl), **William.** 1871–1938. English-born Amer. psychologist.
**Mc·Dow·ell** (mək-dou'əl), **Ephraim.** 1771–1830. Amer. surgeon.
**McDowell, Irvin.** 1818–85. Amer. general.
**Mc·Duf·fie** (mək-dŭf'ē), **George.** 1790?–1851. Amer. legislator.
**Mc·Fee** (mək-fē'), **William.** 1881–1966. English author.
**Mc·Gill** (mə-gĭl'), **James.** 1774–1813. Scottish-born Canadian philanthropist.
**Mc·Gil·li·vray** (mə-gĭl'ə-vrā), **Alexander.** 1759?–93. Amer. Indian leader.
**Mc·Gov·ern** (mə-gŭv'ərn), **George Stanley.** b. 1922. Amer. politician.
**Mc·Graw** (mə-grô'), **John Joseph.** 1873–1934. Amer. baseball player & manager.
**Mc·Guf·fey** (mə-gŭf'ē), **William Holmes.** 1800–73. Amer. educator.
**Mc·In·tire** (măk'ĭn-tīr'), **Samuel.** 1757–1811. Amer. architect & craftsman.
**Mc·Kay** (mə-kā'), **Claude.** 1889–1948. Jamaican-born Amer. poet.
**McKay, Donald.** 1810–80. Amer. shipbuilder.
**Mc·Kean** (mə-kēn'), **Thomas.** 1734–1817. Amer. Revolutionary statesman.
**Mc·Ken·na** (mə-kĕn'ə), **Joseph.** 1843–1926. Amer. jurist.
**McKenna, Siobhan.** b. 1923. Irish actress.
**Mc·Kim** (mə-kĭm'), **Charles Follen.** 1847–1909. Amer. architect.
**Mc·Kin·ley** (mə-kĭn'lē), **John.** 1780–1852. Amer. jurist.
**McKinley, William.** 1843–1901. 25th U.S. President (1897–1901); assassinated.
**Mc·Lean** (mə-klēn'), **Alice Throckmorton.** 1886–1968. Amer. volunteer-service organizer.
**McLean, Evalyn Walsh.** 1886–1947. Amer. socialite.
**McLean, John.** 1785–1861. Amer. jurist.

**Mc·Lough·lin** (mək-lôf′lĭn), **John.** 1784–1857. Canadian-born Amer. fur trader.
**Mc·Lu·han** (mə-klo͞o′ən), **(Herbert) Marshall.** 1911–81. Canadian educator.
**Mc·Ma·hon** (mək-mä′ən, -män′), **William.** b. 1908. Australian statesman.
**Mc·Mas·ter** (mək-măs′tər), **John Bach.** 1852–1932. Amer. historian.
**Mc·Mein** (mək-mān′), **Neysa.** 1888–1949. Amer. artist.
**Mc·Mil·lan** (mək-mĭl′ən), **Edwin Mattison.** b. 1907. Amer. physicist & chemist (Nobel, 1951).
**Mc·Na·mar·a** (măk′nə-măr′ə), **Robert Strange.** b. 1916. Amer. public official.
**Mc·Naugh·ton** (mək-nôt′n), **Andrew George Latta.** 1887–1966. Canadian general & statesman.
**Mc·Nutt** (mək-nŭt′), **Paul Vories.** 1891–1955. Amer. lawyer & politician.
**Mc·Pher·son** (mək-fûr′sən), **Aimee Semple.** 1890–1944. Canadian-born Amer. evangelist.
**Mc·Rey·nolds** (mək-rĕn′əldz), **James Clark.** 1862–1946. Amer. jurist.
**Mead** (mēd), **George Herbert.** 1863–1931. Amer. philosopher.
**Mead, Margaret.** 1901–78. Amer. anthropologist.
**Meade** (mēd), **George Gordon.** 1815–72. Amer. Union general.
**Meade, James Edward.** b. 1907. English economist (Nobel, 1977).
**Means** (mēnz), **Gaston Bullock.** 1879–1938. Amer. espionage agent & detective.
**Mea·ny** (mē′nē), **George.** 1894–1980. Amer. labor leader.
**Mears** (mērz), **Helen Farnsworth.** 1872–1916. Amer. sculptor.
**Med·a·war** (mĕd′ə-wər), Sir **Peter Brian.** b. 1915. Brazilian-born British biologist (Nobel, 1960).
**Med·i·ci** (mĕd′ə-chē′). Italian Renaissance family, including: **a. Cosimo de.** 1389–1464. Banker, art patron, & statesman. **b. Giovanni de.** LEO X. **c. Giulio de.** CLEMENT VII. **d. Lorenzo de.** *Lorenzo the Magnificent.* 1449–92. Art patron & statesman. **e. Cosimo I de.** *Cosimo the Great.* 1519–74. Statesman.
**Me·dill** (mə-dĭl′), **Joseph.** 1823–99. Amer. editor & publisher.
**Me·di·na-Si·do·nia** (mə-dē′nə-sə-dōn′yə), 7th Duke of. *Alonso Pérez de Guzmán.* 1550–1615. Spanish naval officer.
**Me·he·met A·li** (mĭ-hĕm′ĕt ä-lē′, mā′mĕt) *also* **Mo·ham·med Ali** (mō-hăm′ĭd). 1769–1849. Viceroy of Egypt (1805–48).
**Meiggs** (mĕgz), **Henry.** 1811–77. Amer. railroad builder.
**Meigh·en** (mē′ən), **Arthur.** 1874–1960. Canadian prime minister (1920–21, 1926).
**Mei·ji** (mā′jē′). MUTSUHITO.
**Mei·kle·john** (mĭk′əl-jŏn′), **Alexander.** 1872–1964. Amer. educator.
**Me·ir** (mī′ər, mā-ēr′), **Golda.** 1898–1978. Russian-born Amer.-Israeli prime minister (1969–74).
**Meis·so·nier** (mā′sən-yā′), **Jean Louis Ernest.** 1815–91. French painter.
**Meit·ner** (mīt′nər), **Lise.** 1878–1968. Austrian-born Swedish physicist.
**Me·lanch·thon** (mə-lăngk′thən), **Philipp.** 1497–1560. German theologian.
**Mel·ba** (mĕl′bə), Dame **Nellie.** 1861–1931. Australian singer.
**Mel·bourne** (mĕl′bərn), 2nd Viscount. *William Lamb.* 1779–1848. British prime minister (1834, 1835–41).
**Mel·chers** (mĕl′chərz), **Gari.** 1860–1932. Amer. painter.
**Mel·chi·or** (mĕl′kē-ôr′), **Lauritz Lebrecht Hommel.** 1890–1973. Danish-born Amer. opera singer.
**Mel·chiz·e·dek** (mĕl-kĭz′ĭ-dĕk′). High priest & king of Salem in the Old Testament.
**Mel·lon** (mĕl′ən), **Andrew William.** 1855–1937. Amer. financier & public official.
**Me·lo·ney** (mə-lō′nē), **Marie Mattingly.** 1878–1943. Amer. journalist.
**Mel·ville** (mĕl′vĭl′), **George Wallace.** 1841–1912. Amer. naval officer & explorer.
**Melville, Herman.** 1819–91. Amer. author.
**Mem·ling** (mĕm′lĭng) *also* **Mem·linc** (-lĭngk), **Hans.** 1430?–95. Flemish painter.
**Me·nan·der** (mə-năn′dər). 4th cent. B.C. Greek dramatist.
**Men·ci·us** (mĕn′shē-əs). 4th cent. B.C. Chinese philosopher.
**Menck·en** (mĕng′kən), **Henry Louis.** 1880–1956. Amer. editor & critic.
**Men·del** (mĕn′dl), **Gregor Johann.** 1822–84. Austrian botanist. **—Men·de′li·an** (mĕn-dē′lē-ən, -dēl′yən) *adj.*
**Men·de·le·ev** (mĕn′də-lā′əf), **Dmitri Ivanovich.** 1834–1907. Russian chemist.
**Men·dels·sohn** (mĕn′dl-sən), **(Jakob Ludwig) Felix.** 1809–47. German composer, pianist, & conductor.
**Mendelssohn, Moses.** 1729–86. German philosopher.
**Men·dès-France** (mäɴ-dĕs-fräɴs′), **Pierre.** 1907–82. French statesman.
**Men·do·za** (mĕn′dō-zə), **Antonio de.** 1485?–1552. Spanish colonial administrator.
**Men·e·lik I** (mĕn′ə-lĭk). 1844–1913. Ethiopian emperor (1889–1913).
**Me·nén·dez de A·vi·lés** (mə-nĕn′dəs dā ä′və-lās′), **Pedro.** 1519–74. Spanish New World colonizer.
**Me·nes** (mē′nēz). fl. 3000 B.C. Egyptian king.
**Meng·ze** (mŭng′dzŭ). MENCIUS.
**Men·ken** (mĕng′kən), **Adah Isaacs.** 1835–68. Amer. actress & poet.
**Men·ning·er** (mĕn′ĭn-jər), **Karl Augustus.** b. 1893. Amer. psychiatrist.
**Me·not·ti** (mə-nŏt′ē), **Gian Carlo.** b. 1911. Italian-born Amer. composer.
**Men·u·hin** (mĕn′yə-wĭn), **Yehudi.** b. 1916. Amer. violinist.
**Men·zies** (mĕn′zēz), Sir **Robert Gordon.** 1894–1978. Australian statesman.
**Mer·ca·tor** (mər-kā′tər), **Gerhardus.** 1512–94. Flemish geographer.
**Mer·cier** (mĕr-syā′, mĕr′sē-ā′), **Désiré Joseph.** 1851–1956. Belgian religious leader.
**Mer·e·dith** (mĕr′ĭ-dĭth), **George.** 1828–1909. English author.
**Meredith, James Howard.** b. 1933. Amer. civil-rights leader.
**Mer·gen·thal·er** (mûr′gən-thô′lər, mĕr′gən-tä′-), **Ottmar.** 1854–99. German-born Amer. inventor.
**Mé·ri·mée** (mĕr′ə-mā′, mā′rə-mā′), **Prosper.** 1803–70. French author & public official.
**Mer·man** (mûr′mən), **Ethel.** 1908–84. Amer. musical-comedy actress.
**Mer·o·vin·gi·an** (mĕr′ə-vĭn′jē-ən, -jən). French ruling dynasty (428–751).
**Mer·rick** (mĕr′ĭk), **David.** b. 1912. Amer. theatrical producer.
**Mer·ritt** (mĕr′ĭt), **Anna Lea.** 1844–1930. Amer. artist.
**Mer·ton** (mûr′tn), **Robert King.** b. 1910. Amer. sociologist.
**Merton, Thomas.** 1915–1968. Amer. clergyman & author.
**Mes·mer** (mĕz′mər, mĕs′-), **Franz** *also* **Friedrich Anton.** 1734–1815. Austrian physician.
**Mes·sa·la Cor·vi·nus** (mə-săl′ə kôr-vī′nəs, mə-sä′lə), **Marcus Valerius.** 1st cent. B.C. Roman general & statesman.
**Mes·sa·li·na** (mĕs′ə-lī′nə), **Valeria.** d. 48 A.D. Roman empress as third wife of Claudius.
**Mes·ser·schmitt** (mĕs′ər-shmĭt′), **Willy.** 1898–1978. German aircraft manufacturer.
**Mes·sier** (mās-yā′, mĕs′ē-ā′), **Charles.** 1730–1817. French astronomer.
**Mes·ta** (mĕs′tə), **Perle.** 1889–1975. Amer. socialite & diplomat.
**Meš·tro·vić** (mĕsh′trə-vĭch′, mĕs′-), **Ivan.** 1883–1962. Yugoslavian-born Amer. sculptor.
**Me·tax·as** (mə-tăk′səs), **Joannes.** 1871–1941. Greek dictator (1936–40).
**Metch·ni·kov** (mĕch′nĭ-kôf′), **Elie.** 1845–1916. Russian zoologist (Nobel, 1908).
**Me·thu·se·lah** (mə-tho͞o′zə-lə). Biblical patriarch said to have lived 969 years.
**Met·ter·nich** (mĕt′ər-nĭk, -nĭKH), Prince **Klemens Wenzel Nepomuk Lothar von.** 1773–1859. Austrian statesman.
**Metz** (mĕts), **Christian.** 1794–1867. Prussian-born Amer. religious leader.
**Mey·er** (mī′ər), **Annie Florance Nathan.** 1867–1951. Amer. educator & author.
**Mey·er·beer** (mī′ər-bîr′), **Giacomo.** 1791–1864. German opera composer.
**Mey·er·hof** (mī′ər-hôf′), **Otto.** 1884–1951. German-born Amer. physiologist (Nobel, 1922).
**Mi·cah** (mī′kə). 8th cent. B.C. Hebrew prophet.
**Mi·chael** (mī′kəl). b. 1921. King of Rumania (1927–30, 1940–47); abdicated.
**Mi·chel·an·ge·lo Buo·nar·ro·ti** (mī′kəl-ăn′jə-lō′ bwôn′ə-rô′tē, mĭk′əl-, mē′kə-lăn′-). 1475–1564. Italian sculptor, painter, architect, & poet.
**Mi·che·let** (mēsh-ə-lā′, mēsh-lā′), **Jules.** 1798–1874. French historian.
**Mi·chel·son** (mī′kəl-sən), **Albert Abraham.** 1852–1931. German-born Amer. physicist (Nobel, 1907).
**Miche·ner** (mĭsh′nər), **Daniel Roland.** b. 1900. Canadian statesman.
**Mich·e·ner** (mĭch′ə-nər, mĭch′nər), **James Albert.** b. 1907. Amer. author.
**Mic·kie·wicz** (mĭts-kyā′vĭch), **Adam.** 1798–1855. Polish poet.
**Mid·dle·ton** (mĭd′l-tən), **Thomas.** 1570?–1627. English dramatist.
**Mies Van Der Ro·he** (mēz′, văn dər rō′ə, rō′, mēs′), **Ludwig.** 1886–1969. German-born Amer. architect.
**Mif·flin** (mĭf′lĭn), **Thomas.** 1744–1800. Amer. Revolutionary soldier & politician.

**Mi·koy·an** (mē'kō-yän'), **Anastas Ivanovich.** 1895–1978. Soviet statesman.
**Miles** (mīlz), **Nelson Appleton.** 1839–1925. Amer. general.
**Mil·haud** (mē-yō'), **Darius.** 1892–1974. French composer.
**Mill** (mĭl), **James.** 1773–1836. Scottish historian & economist.
**Mill, John Stuart.** 1806–73. English philosopher & economist.
**Mil·lais** (mĭ-lā'), Sir **John Everett.** 1829–96. English painter.
**Mil·lay** (mĭ-lā'), **Edna St. Vincent.** 1892–1950. Amer. poet.
**Mil·ler** (mĭl'ər), **Alice Duer.** 1874–1942. Amer. author.
**Miller, Arthur.** b. 1915. Amer. dramatist.
**Miller, Cincinnatus Hiner ("Joaquin").** 1837–1913. Amer. poet.
**Miller, Glenn.** 1909–44. Amer. bandleader.
**Miller, Harriet Mann.** *Olive Thorne Miller.* 1831–1918. Amer. author.
**Miller, Henry John.** 1860–1926. English-born Amer. actor, director, & producer.
**Miller, Henry Valentine.** 1891–1980. Amer. author.
**Miller, Perry Gilbert Eddy.** 1905–63. Amer. historian & critic.
**Miller, Samuel Freeman.** 1816–90. Amer. jurist.
**Miller, William.** 1782–1849. Amer. religious leader.
**Mille·rand** (mēl-räN', -ə-räN'), **Alexandre.** 1859–1943. French statesman.
**Mil·les** (mĭl'əs), **Carl.** 1875–1955. Swedish sculptor.
**Mil·let** (mĭ-lā'), **Jean François.** 1814–75. French painter.
**Mil·li·kan** (mĭl'ĭ-kən), **Robert Andrews.** 1868–1953. Amer. physicist (Nobel, 1923).
**Mills** (mĭlz), **Robert.** 1781–1855. Amer. architect & engineer.
**Mil·man** (mĭl'mən), **Henry Hart.** 1791–1868. English historian & poet.
**Milne** (mĭln), **Alan Alexander ("A.A.").** 1882–1956. English author.
**Milne, John.** 1850–1913. British mining engineer & seismologist.
**Mil·ti·a·des** (mĭl-tī'ə-dēz'). 540?–489 B.C. Athenian general.
**Mil·ton** (mĭl'tən), **John.** 1608–74. English poet.
**Mil·yu·kov** (mĭl'yə-kôf', -kôv'), **Pavel Nikolaevich.** 1859–1943. Russian historian & politician.
**Ming** (mĭng). Chinese dynasty (1368–1644).
**Min·ne·wit** (mĭn'ə-wĭt). *var. of* MINUIT.
**Mi·not** (mī'nət), **George Richards.** 1885–1950. Amer. physician (Nobel, 1934).
**Min·ton** (mĭn'tən), **Sherman.** 1890–1965. Amer. jurist.
**Min·u·it** (mĭn'yo͞o-wĭt) *also* **Min·ne·wit** (-ə-wĭt), **Peter.** 1580–1638. Dutch colonial administrator.
**Mi·ra·beau** (mîr'ə-bō'), Comte de. *Honoré Gabriel Victor Riqueti.* 1749–91. French revolutionist.
**Mi·ró** (mē-rō'), **Joan.** 1893–1983. Spanish artist.
**Mis·tral** (mĭ-sträl'), **Frédéric.** 1830–1914. French Provençal poet (Nobel, 1904).
**Mistral, Gabriela.** 1889–1957. Chilean educator & poet (Nobel, 1945).
**Mitch·ell** (mĭch'əl), **John.** d. 1768. Amer. physician, botanist, & cartographer.
**Mitchell, John.** 1870–1919. Amer. labor leader.
**Mitchell, Lucy Myers Wright.** 1845–88. Persian-born Amer. archaeologist.
**Mitchell, Margaret Julia ("Maggie").** 1832–1918. Amer. actress.
**Mitchell, Margaret Munnerlyn.** 1900–49. Amer. author.
**Mitchell, Maria.** 1818–89. Amer. astronomer.
**Mitchell, Peter Dennis.** b. 1920. British biochemist (Nobel, 1978).
**Mitchell, Silas Weir.** 1829–1914. Amer. physician & author.
**Mitchell, Wesley Clair.** 1874–1948. Amer. economist.
**Mitchell, William ("Billy").** 1879–1936. Amer. soldier & aviation pioneer.
**Mit·ford** (mĭt'fərd), **Mary Russell.** 1787–1855. English author.
**Mitford, William.** 1744–1827. English historian.
**Mith·ri·da·tes VI** (mĭth'rĭ-dā'tēz). *the Great.* 132?–63 B.C. King of Pontus (120–63).
**Mi·tro·pou·los** (mĭ-trŏp'ə-ləs), **Dimitri.** 1896–1960. Greek-born Amer. conductor.
**Mit·ter·rand** (mē'tə-räN', -rănd'), **François Maurice.** b. 1916. French president (since 1981).
**Mix** (mĭks), **Thomas Edwin ("Tom").** 1880–1940. Amer. actor.
**Miz·ner** (mĭz'nər), **Addison.** 1872–1933. Amer. architect & real-estate developer.
**Mo Di** (mō' dē'). 5th–4th cent. B.C. Chinese philosopher.
**Mo·di·glia·ni** (mō-dē'lē-ä'nē, -lyä'nē), **Amedeo.** 1884–1920. Italian artist.
**Mo·djes·ka** (mə-jĕs'kə), **Helena.** 1840–1909. Polish-born Amer. actress.
**Mo·ham·med** (mō-hăm'ĭd) *also* **Mu·ham·mad** (mo͞o-). 570?–632. Arab prophet & founder of Islam.
**Mohammed** *also* **Muhammad.** 1429?–81. Sultan of Turkey (1451–81).
**Mohammed A·li** (ä-lē'). MEHEMET ALI.
**Mo·holy-Nag·y** (mō'hoi-nŏd'yə), **Laszlo** or **Ladislaus.** 1895–1946. Hungarian-born Amer. artist & educator.
**Mois·san** (mwä-säN'), **(Ferdinand Frédéric) Henri.** 1852–1907. French chemist (Nobel, 1906).
**Mo·ley** (mō'lē), **Raymond Charles.** 1886–1975. Amer. journalist.
**Mo·lière** (mōl-yâr'), **Jean Baptiste Poquelin.** 1622–73. French actor & playwright.
**Mol·nár** (mōl'när', mŏl'-), **Ferenc.** 1878–1952. Hungarian author.
**Mo·lo·tov** (mŏl'ə-tôf', -tôv', mōl'-), **Vyacheslav Mikhailovich.** b. 1890. Russian statesman.
**Molt·ke** (môlt'kə), Count **Helmuth von.** 1800–91. Prussian soldier & military strategist.
**Momm·sen** (mŏm'zən), **Theodor.** 1817–1903. German historian (Nobel, 1902).
**Monck** or **Monk** (mŭngk), **George.** *1st Duke of Albemarle.* 1608–70. English general.
**Mon·dale** (mŏn'dāl'), **Walter Frederick.** b. 1928. U.S. Vice President (1977–81).
**Mon·dri·an** (môn'drē-än'), **Piet.** 1872–1944. Dutch painter.
**Mo·net** (mō-nā'), **Claude.** 1840–1926. French painter.
**Mo·ne·ta** (mō-nā'tə), **Ernesto Teodoro.** 1833–1918. Italian journalist & pacifist (Nobel, 1907).
**Mo·niz** (mō-nēz'), **Antonio Caetano de Abreu Freire Egas.** 1874–1955. Portuguese neurologist (Nobel, 1949).
**Monk** (mŭngk). *var. of* MONCK.
**Monk, Maria.** 1816–49. Canadian-born Amer. author.
**Monk, Thelonius Sphere.** 1917–82. Amer. jazz pianist & composer.
**Mon·mouth** (mŏn'məth), Duke of. *James Scott.* 1649–85. English pretender to the throne; beheaded.
**Mon·net** (mō-nā'), **Jean.** 1888–1979. French economist & statesman.
**Mo·nod** (mô-nō'), **Jacques Lucien.** 1910–76. French biochemist (Nobel, 1965).
**Mon·roe** (mən-rō'), **Harriet.** 1860–1936. Amer. editor & poet.
**Monroe, James.** 1758–1831. Fifth U.S. President (1817–25).
**Monroe, Marilyn.** 1926–62. Amer. actress.
**Mon·ta·gna** (mən-tä'nyä), **Bartolommeo.** 1450?–1523. Italian painter.
**Mon·ta·gu** (mŏn'tə-gyo͞o'), **Ashley.** b. 1905. English-born Amer. anthropologist.
**Montagu,** Lady **Mary Wortley.** 1689–1762. English author.
**Mon·taigne** (mŏn-tān'), **Michel Eyquem de.** 1533–92. French essayist.
**Mon·ta·le** (mōn-tä'lā), **Eugenio.** 1896–1981. Italian poet (Nobel, 1975).
**Mont·calm de Saint-Ve·ran** (mŏnt-käm' də săN'vā-räN'), Marquis **Louis Joseph de.** 1712–59. French commander in Canada.
**Mon·tes·quieu** (mŏn'tə-skyo͞o'), Baron **de la Brede et de.** *Charles de Secondat.* 1689–1755. French philosopher & jurist.
**Mon·tes·so·ri** (mŏn'tĭ-sôr'ē, -sōr'ē), **Maria.** 1870–1952. Italian physician & pioneer educator.
**Mon·teux** (môN-to͞o'), **Pierre.** 1875–1964. French-born Amer. conductor.
**Mon·te·ver·di** (mŏn'tə-vâr'dē), **Claudio.** 1567–1643. Italian composer.
**Mon·tez** (mŏn-tĕz'), **Lola.** 1818–61. Irish-born Amer. dancer.
**Mon·te·zu·ma II** (mŏn'tə-zo͞o'mə). 1480?–1520. Last Aztec emperor in Mexico.
**Mont·fort** (mŏnt'fərt), **Simon de.** *Earl of Leicester.* 1208?–65. English soldier & statesman.
**Montfort l'A·mau·ry** (lā'mə-rē), **Simon IV de.** *Earl of Leicester & Comte de Toulouse.* 1160?–1218. French crusader.
**Mont·gol·fi·er** (mŏnt-gŏl'fē-ər, -fē-ā), **Joseph Michel** (1740–1810) & **Jacques Etienne** (1745–89). French aeronautic inventors.
**Mont·gom·er·y** (mənt-gŭm'rē, -gŭm'ə-rē, mŏnt-), Sir **Bernard Law.** *1st Viscount Montgomery of Alamein.* 1887–1976. Irish-born British army officer.
**Mont·mo·ren·cy** (mŏnt'mə-rĕn'sē), Duc **Anne de.** 1493–1567. French marshal.
**Mon·trose** (mŏn-trōz'), 1st Marquis of. *James Graham.* 1612–50. Scottish Covenanter.
**Moo·dy** (mo͞o'dē), **Dwight Lyman.** 1837–99. Amer. evangelist.
**Moody, Helen Wills.** Helen Newington WILLS.
**Moody, William Henry.** 1853–1917. Amer. jurist.
**Moody, William Vaughn.** 1869–1910. Amer. poet & playwright.
**Moon** (mo͞on), **Sun Myung.** b. 1920. South Korean evangelist.
**Moo·ney** (mo͞o'nē), **James.** 1861–1921. Amer. ethnologist.
**Moore** (mo͝or), **Alfred.** 1755–1810. Amer. jurist.
**Moore, Clement Clarke.** 1779–1863. Amer. scholar & poet.
**Moore, George.** 1852–1933. Irish author.
**Moore, George Edward.** 1873–1958. English philosopher.
**Moore, Grace.** 1898–1947. Amer. singer.
**Moore, Henry.** b. 1898. English sculptor.
**Moore, John Bassett.** 1860–1947. Amer. jurist.

**Moore, Marianne Craig.** 1887–1972. Amer. poet.
**Moore, Stanford.** 1913–82. Amer. biochemist (Nobel, 1972).
**Moore, Thomas.** 1779–1852. Irish poet.
**Moore, William Henry** (1848–1923) & **James Hobart** (1852–1916). Amer. financiers.
**Mo·ra·vi·a** (mō-rä'vē-ə), **Alberto.** b. 1907. Italian author.
**More** (môr, mōr), **Hannah.** 1745–1833. English writer & social reformer.
**More, Henry.** 1614–87. English philosopher.
**More, Paul Elmer.** 1864–1937. Amer. philosopher & critic.
**More,** Saint (Sir) **Thomas.** 1478–1535. English statesman & author; beheaded for treason.
**Mor·gan** (môr'gən), **Anne Tracy.** 1873–1952. Amer. philanthropist.
**Morgan, Daniel.** 1736–1802. Amer. Revolutionary soldier.
**Morgan, Helen.** 1900–41. Amer. singer.
**Morgan,** Sir **Henry.** 1635?–88. Welsh buccaneer.
**Morgan, John.** 1735–89. Amer. physician.
**Morgan, John Hunt.** 1825–64. Amer. Confederate soldier.
**Morgan, John Pierpont** (1837–1913) & **John Pierpont, Jr.** (1867–1943). Amer. financiers.
**Morgan, Lewis Henry.** 1818–81. Amer. anthropologist.
**Morgan, Mary Kimball.** 1861–1948. Amer. educator.
**Morgan, Thomas Hunt.** 1866–1945. Amer. biologist (Nobel, 1933).
**Morgan, William.** 1774?–1826. Amer. Freemason; disappeared.
**Mor·gen·thau** (môr'gən-thô'), **Henry.** 1856–1946. German-born Amer. public official & diplomat.
**Morgenthau, Henry, Jr.** 1891–1967. Amer. public official.
**Mor·i·son** (môr'ĭ-sən), **Samuel Eliot.** 1887–1976. Amer. historian.
**Mo·ri·sot** (mô-rē-zō'), **Berthe.** 1841–95. French painter.
**Mor·ley** (môr'lē), **Christopher Darlington.** 1890–1957. Amer. author.
**Morley, Edward Williams.** 1838–1923. Amer. chemist & physicist.
**Morley, John.** *Viscount Morley of Blackburn.* 1838–1923. English statesman & author.
**Mor·nay** (môr-nā'), **Philippe de.** *Duplessis-Mornay.* 1549–1623. French Huguenot leader.
**Mor·phy** (môr'fē), **Paul Charles.** 1837–84. Amer. chess master.
**Mor·rill** (môr'əl), **Justin Smith.** 1810–98. Amer. politician.
**Mor·ris** (môr'ĭs, mŏr'-), **Clara.** 1847–1925. Canadian-born Amer. actress.
**Morris, Esther Hobart McQuigg Slack.** 1814–1902. Amer. suffragist.
**Morris, Gouverneur.** 1752–1816. Amer. diplomat & political leader.
**Morris, Robert.** 1734–1806. Amer. Revolutionary statesman & financier.
**Morris, William.** 1834–96. English poet, artist, & craftsman.
**Mor·ri·son** (môr'ĭ-sən, mŏr'-), Baron **Herbert Stanley.** 1888–1965. British labor leader & politician.
**Morrison, Robert.** 1782–1834. Scottish missionary in China.
**Mor·row** (môr'ō, mŏr'ō), **Dwight Whitney.** 1873–1931. Amer. lawyer, banker, & diplomat.
**Morse** (môrs), **Charles Wyman.** 1856–1933. Amer. speculator & promoter.
**Morse, Jedidiah.** 1761–1826. Amer. religious leader & geographer.
**Morse, Samuel Finley Breese.** 1791–1872. Amer. painter & inventor.
**Mor·ti·mer** (môr'tə-mər), **Roger IV de.** *1st Earl of March.* 1287–1330. Welsh rebel.
**Mor·ton** (môr'tn), **Ferdinand Joseph La Menthe ("Jelly Roll").** 1885–1941. Amer. jazz musician & composer.
**Morton, Julius Sterling.** 1832–1902. Amer. agriculturalist & public official.
**Morton, Levi Parsons.** 1824–1920. U.S. Vice President (1889–93).
**Morton, Oliver Perry.** 1823–77. Amer. legislator.
**Morton, Rogers Clark Ballard.** 1914–79. Amer. public official.
**Morton, Sarah Wentworth Apthorp.** 1759–1846. Amer. poet.
**Morton, Thomas.** d. 1647. English-born Amer. Colonial leader.
**Morton, William Thomas Green.** 1819–68. Amer. dentist & pioneer anesthetist.
**Mos·by** (mōz'bē), **John Singleton.** 1833–1916. Amer. Confederate soldier.
**Mós·cic·ki** (môsh-chēt'skē), **Ignacy.** 1867–1946. Polish chemist & statesman.
**Mose·ley** (mōz'lē), **Henry Gwyn-Jeffreys.** 1887–1915. English physicist.
**Mos·es** (mō'zĭz, -zĭs). Hebrew prophet & lawgiver.
**Moses, Anna Mary Robertson.** *Grandma Moses.* 1860–1961. Amer. painter.
**Moses, Robert.** 1888–1981. Amer. public official.
**Mo·sher** (mō'zhər), **Eliza Maria.** 1846–1928. Amer. physician & educator.
**Mos·ley** (mōz'lē), Sir **Oswald Ernald.** 1896–1981. English politician.
**Moss·bau·er** (mœs'bou'ər, mŏs'-), **Rudolf Ludwig.** b. 1929. German physicist (Nobel, 1961).
**Moth·er·well** (mŭth'ər-wĕl'), **Robert.** b. 1915. Amer. artist.
**Mo Ti** (mō' dē'). MO DI.
**Mot·ley** (mŏt'lē), **John Lothrop.** 1814–77. Amer. historian & diplomat.
**Mo·ton** (mōt'n), **Robert Russa.** 1867–1940. Amer. educator.
**Mott** (mŏt), **John Raleigh.** 1865–1955. Amer. YMCA leader (Nobel, 1946).
**Mott, Lucretia Coffin.** 1793–1880. Amer. social reformer.
**Mott,** Sir **Neville Francis.** b. 1905. British physicist (Nobel, 1977).
**Mot·tel·son** (mŏt'l-sən, -sôn), **Ben Roy.** b. 1926. Amer.-born Danish physicist (Nobel, 1975).
**Mot·teux** (mô-tōō'), **Peter Anthony.** 1663?–1718. French-born English dramatist & translator.
**Moul·ton** (mōl'tən), **Ellen Louise Chandler.** 1835–1908. Amer. author.
**Moulton, Forest Ray.** 1872–1952. Amer. astronomer.
**Moul·trie** (mōōl'trē), **William.** 1730–1805. Amer. Revolutionary soldier.
**Mount** (mount), **William Sidney.** 1807–68. Amer. painter.
**Mount·bat·ten** (mount-băt'n), **Louis.** *1st Earl Mountbatten of Burma.* 1900–79. English naval officer & colonial administrator; assassinated.
**Mow·att** (mou'ət), **Anna Cora Ogden.** 1819–70. French-born Amer. author & actress.
**Mo·zart** (mōt'särt'), **Wolfgang Amadeus.** 1756–91. Austrian composer.
**Mu·bar·ak** (mōō-băr'ək), **Hosni.** b. 1928. Egyptian statesman.
**Mu·ga·be** (mōō-gä'bē), **Robert Gabriel.** b. 1925. Zimbabwean statesman.
**Mu·ham·mad** (mōō-hăm'ĭd). *var. of* MOHAMMED.
**Muhammad, Elijah.** 1897–1975. Amer. religious leader.
**Muhl·en·berg** (myōō'lən-bûrg'), **Frederick Augustus.** 1750–1801. Amer. religious leader & public official.
**Muhlenberg, John Peter Gabriel.** 1746–1807. Amer. religious leader, soldier, & public official.
**Mühl·en·berg** (myōō'lən-bûrg'), **Henry Melchior.** 1711–87. German-born Amer. religious leader.
**Muir** (myŏŏr), **John.** 1838–1914. Scottish-born Amer. naturalist.
**Mul·doon** (mŭl-dōōn'), **Robert David.** b. 1921. New Zealand statesman.
**Mul·ler** (mŭl'ər), **Hermann Joseph.** 1890–1967. Amer. geneticist (Nobel, 1946).
**Mül·ler** (mŭl'ər, myōō'lər), **Friedrich Max.** 1823–1900. German-born English philologist.
**Müller, Johann.** *Regiomontanus.* 1436–76. German mathematician & astronomer.
**Müller, Paul Hermann.** 1899–1965. Swiss chemist (Nobel, 1948).
**Mul·li·kan** (mŭl'ĭ-kən), **Robert Sanderson.** b. 1896. Amer. chemist & physicist (Nobel, 1966).
**Mum·ford** (mŭm'fərd), **Lewis.** b. 1895. Amer. author & critic.
**Munch** (mŏŏngk), **Edvard.** 1863–1944. Norwegian artist.
**Münch·hau·sen** *also* **Mun·chau·sen** (mün**KH**'hou'zən, mŭn'-chou'zən, mŭnch'hou'-), Baron **Karl Friedrich Hieronymus von.** 1720–97. German soldier & anecdote teller.
**Mu·ñoz Ma·rín** (mōō-nyōs' mä-rēn'), **Luis.** 1898–1980. Puerto Rican journalist & statesman.
**Mun·ro** (mən-rō'), **Hector Hugh.** *Saki.* 1870–1916. British author.
**Mun·sey** (mŭn'sē, -zē), **Frank Andrew.** 1854–1925. Amer. publisher.
**Mun·ster·berg** (mōōn'stər-bûrg', myōōn-), **Hugo.** 1863–1916. Prussian-born Amer. psychologist.
**Mu·ra·sa·ki** (mōō'rä-sä'kē), Baroness. 11th cent. Japanese author.
**Mu·rat** (myōō-rä'), **Joachim.** 1767?–1815. French marshal.
**Mur·doch** (mûr'dŏk'), **(Jean) Iris.** b. 1919. Irish-born British author.
**Mur·dock** (mûr'dŏk'), **William.** 1754–1839. Scottish-born British engineer & inventor.
**Mu·rel** (mə-rĕl'). *var. of* MURRELL.
**Mur·free** (mûr'frē), **Mary Noailles.** 1850–1922. Amer. author.
**Mu·ril·lo** (myōō-rĭl'ō), **Bartolomé Esteban.** 1617–82. Spanish painter.
**Mur·phy** (mûr'fē), **Audie.** 1924–71. Amer. soldier & actor.
**Murphy, Frank.** 1890–1949. Amer. jurist.
**Murphy, Michael Charles.** 1861–1913. Amer. athletic coach.
**Murphy, Robert Daniel.** 1894–1978. Amer. diplomat.
**Murphy, William Parry.** b. 1892. Amer. physician (Nobel, 1934).
**Mur·ray** (mûr'ē, mŭr'ē), **(George) Gilbert (Aimé).** 1866–1957. Australian-born British classical scholar.
**Murray,** Sir **James Augustus Henry.** 1837–1915. Scottish-born British philologist & lexicographer.
**Murray, John Courtney.** 1904–67. Amer. theologian.

**Murray, Lindley.** 1745–1826. Amer. grammarian.
**Murray, Philip.** 1886–1952. Scottish-born Amer. labor leader.
**Mur·rell** *or* **Mur·rel** *or* **Mu·rel** (mə-rĕl'), **John A.** 1804?–50? Amer. outlaw.
**Mur·row** (mûr'ō), **Edward Roscoe.** 1908–65. Amer. broadcast journalist.
**Mu·si·al** (myōō'zē-əl), **Stanley Frank ("Stan").** b. 1920. Amer. baseball player.
**Mus·kie** (mŭs'kē), **Edmund Sixtus.** b. 1914. Amer. politician.
**Mus·set** (myōō-sā'), **(Louis Charles) Alfred de.** 1810–57. French poet.
**Mus·so·li·ni** (mōō'sə-lē'nē, mŏŏs'ə-), **Benito.** *Il Duce.* 1883–1945. Italian Fascist dictator (1922–45); assassinated.
**Mus·sorg·sky** (mə-zôrg'skē, mōō-sôrg'skĭ), **Modest Petrovich.** 1835–81. Russian composer.
**Mu·tsu·hi·to** (mōō'tsōō-hē'tō). *Meiji.* 1852–1912. Japanese emperor (1867–1912).
**Muy·bridge** (mī'brĭj'), **Eadweard.** 1830–1904. British-born Amer. motion-picture pioneer.
**Mu·zo·re·wa** (mōōz'ə-rā'wə), **Abel Tendekayi.** b. 1925. Zimbabwean statesman.
**Muz·zey** (mŭz'ē), **David Saville.** 1870–1965. Amer. historian.
**Myr·dal** (mür'däl', mĭr'-), **Alva.** b. 1902. Swedish sociologist & diplomat (Nobel, 1982).
**Myrdal, (Karl) Gunnar.** b. 1898. Swedish economist (Nobel, 1974).
**My·ron** (mī'rən). 5th cent. B.C. Greek sculptor.

# N

**Na·bo·kov** (nə-bô'kəf, nä'bə-kôf'), **Vladimir Vladimirovich.** 1899–1977. Russian-born Amer. author.
**Na·der** (nā'dər), **Ralph.** b. 1934. Amer. consumer advocate.
**Na·hum** (nā'həm, -əm). 7th cent. B.C. Hebrew prophet.
**Nai·du** (nī'dōō), **Sarojini.** 1879–1949. Hindu poet & reformer.
**Nai·smith** (nā'smĭth'), **James.** 1861–1939. Canadian-born Amer. sports educator.
**Na·math** (nā'məth), **Joseph Williams ("Joe").** b. 1943. Amer. football player.
**Na·nak** (nä'nək). 1469–1538. Hindu Sikh religious leader.
**Nan·sen** (năn'sən, nän'-), **Fridtjof.** 1861–1930. Norwegian Arctic explorer, zoologist, & statesman (Nobel, 1922).
**Na·o·mi** (nā-ō'mē). Ruth's mother-in-law in the Bible.
**Naph·ta·li** (năf'tə-lī'). Hebrew patriarch.
**Na·pi·er** (nā'pē-ər, nə-pîr'), Sir **Charles James.** 1782–1853. English general.
**Napier, John.** *Laird of Merchiston.* 1550–1617. Scottish mathematician.
**Napier, Robert Cornelius.** *1st Baron Napier of Magdala.* 1810–90. English general & colonial administrator.
**Na·po·le·on I** (nə-pō'lē-ən, -pōl'yən). *Napoleon Bonaparte.* 1769–1821. French emperor (1804–14); exiled. **—Na·po'le·on'ic** (nə-pō'lē-ŏn'ĭk) *adj.*
**Napoleon II.** *François Charles Joseph Bonaparte.* 1811–32. Titular king of Rome & French emperor.
**Napoleon III.** *Charles Louis Napoleon Bonaparte.* 1808–73. French emperor (1852–70).
**Nar·vá·ez** (när-vä'ās'), **Pánfilo de.** 1478?–1528. Spanish conquistador.
**Nas·by** (năz'bē), **Petroleum V.** David Ross LOCKE.
**Nash** (năsh), **Ogden.** 1902–71. Amer. humorous poet.
**Nash** *or* **Nashe** (năsh), **Thomas.** 1567–1601. English author.
**Nash, Walter.** 1882–1968. English-born New Zealand statesman.
**Na·smyth** (nā'smĭth', năz'mĭth'), **Alexander.** 1758–1840. Scottish portrait painter.
**Nas·ser** (nä'sər, năs'ər), **Gamal Abdel.** 1918–70. Egyptian soldier & statesman.
**Nast** (năst), **Thomas.** 1840–1902. German-born Amer. editorial cartoonist.
**Na·than** (nā'thən). Old Testament prophet.
**Nathan, George Jean.** 1882–1958. Amer. author, editor, & critic.
**Nathan, Maud.** 1862–1946. Amer. reformer.
**Na·than·ael** (nə-thăn'yəl). BARTHOLOMEW.
**Na·thans** (nā'thənz), **Daniel.** b. 1928. Amer. microbiologist (Nobel, 1978).
**Na·tion** (nā'shən), **Carry** *or* **Carrie Amelia Moore.** 1846–1911. Amer. temperance reformer.
**Nat·ta** (nä'tä), **Giulio.** 1903–79. Italian chemist (Nobel, 1963).
**Na·zi·mo·va** (nə-zĭm'ə-və), **Alla.** 1879–1945. Russian-born actress.

**Neb·u·chad·nez·zar II** (nĕb'ə-kəd-nĕz'ər, nĕb'yə-). King of Babylonia (605–562 B.C.).
**Neck·er** (nĕk'ər, nā-kĕr'), **Jacques.** 1732–1804. French financier & statesman.
**Né·el** (nā-ĕl'), **Louis Eugène Félix.** b. 1904. French physicist (Nobel, 1970).
**Nef·er·ti·ti** (nĕf'ər-tē'tē). 14th cent. B.C. Egyptian queen as wife of Akhenaton.
**Ne·he·mi·ah** (nē'hə-mī'ə, nē'ə-). 5th cent. B.C. Hebrew leader.
**Neh·ru** (nā'rōō), Pandits **Motilal** (1861–1931) & **Jawaharlal** (1889–1964). Indian nationalist leaders & statesmen.
**Neil·son** (nēl'sən), **William Allan.** 1869–1946. Scottish-born Amer. educator, author, & lexicographer.
**Nel·son** (nĕl'sən), **Horatio.** *Viscount Nelson.* 1758–1805. English admiral.
**Nelson, Samuel.** 1792–1873. Amer. jurist.
**Ne·pos** (nē'pŏs, nĕp'əs), **Cornelius.** 1st cent. B.C. Roman historian.
**Ne·ri** (nā'rē), San **Filippo de'** *or* Saint **Philip.** 1515–95. Italian ecclesiastic.
**Nernst** (nĕrnst), **Walther Hermann.** 1864–1941. German physicist & chemist (Nobel, 1920).
**Ne·ro** (nîr'ō, nē'rō). A.D. 37–68. Roman emperor (54–68). **—Ne·ro'ni·an** (nĭ-rō'nē-ən) *adj.*
**Ne·ru·da** (nā-rōō'də), **Pablo.** 1904–73. Chilean poet & diplomat (Nobel, 1971).
**Ner·va** (nûr'və), **Marcus Cocceius.** A.D. 30?–98. Roman emperor (96–98).
**Ner·vi** (nĕr'vē), **Pier Luigi.** 1891–1979. Italian architect.
**Nes·to·ri·us** (nĕ-stôr'ē-əs, -stōr'-). d. 451. Syrian-born patriarch of Constantinople.
**Neu·mann** (noi'män'), **John von.** 1903–57. Hungarian-born Amer. mathematician.
**Neu·rath** (noi'rät'), Baron **Konstantin von.** 1873–1956. German diplomat.
**Ne·va·da** (nə-vä'də), **Emma.** 1862–1940. Amer. operatic soprano.
**Nev·el·son** (nĕv'əl-sən), **Louise.** b. 1900. Russian-born Amer. sculptor.
**Nev·ins** (nĕv'ĭnz), **Allan.** 1890–1971. Amer. historian.
**New·bolt** (nōō'bōlt', nyōō'-), Sir **Henry John.** 1862–1938. English author & editor.
**New·comb** (nōō'kəm, nyōō'-), **Josephine Louise Le Monnier.** 1816–1901. Amer. philanthropist.
**Newcomb, Simon.** 1835–1909. Amer. pioneer astronomer.
**New·house** (nōō'hous', nyōō'-), **Samuel I.** 1895–1979. Amer. publisher.
**New·man** (nōō'mən, nyōō'-), **John Henry.** 1801–90. English prelate & theologian.
**New·ton** (nōōt'n, nyōōt'n), Sir **Isaac.** 1642–1727. English mathematician, scientist, & philosopher. **—New·to'ni·an** *adj.*
**Ney** (nī), **Elisabet.** 1833–1907. German-born Amer. sculptor.
**Ney** (nā), **Michel.** *Duc d'Elchingen, Prince de la Moskowa.* 1769–1815. French marshal.
**Nich·o·las** (nĭk'ə-ləs), Saint. 4th cent. prelate.
**Nicholas.** Name of two czars of Russia: **a. I.** 1796–1855. Ruled 1825–55. **b. II.** 1868–1918. Ruled 1894–1918; abdicated & was executed.
**Nicholas of Cu·sa** (kyōō'zə, -sə). 1401?–64. German prelate & philosopher.
**Nich·olls** (nĭk'əlz), **Rhoda Holmes.** 1854–1930. English-born Amer. water-color painter.
**Nich·ols** (nĭk'əlz), **Anne.** 1891–1966. Amer. playwright.
**Nichols, Mike.** b. 1931 German-born Amer. comedian & director.
**Nich·ol·son** (nĭk'əl-sən), **Eliza Jane Poitevent Holbrook.** 1849–96. Amer. author & publisher.
**Nicholson,** Sir **Francis.** 1655–1728. English colonial administrator.
**Ni·ci·as** (nĭsh'ē-əs, nĭs'-). d. 413 B.C. Athenian general & politician; executed.
**Nick·laus** (nĭk'ləs), **Jack William.** b. 1940. Amer. golfer.
**Nic·o·lay** (nĭk'ə-lā'), **John George.** 1832–1901. Amer. author & editor.
**Nic·o·let** (nĭk'ə-lā'), **Jean.** 1598–1642. French explorer.
**Ni·colle** (nē-kôl'), **Charles Jean Henri.** 1866–1936. French physician & bacteriologist (Nobel, 1928).
**Ni·col·let** (nĭk'ə-lā'), **Joseph Nicolas.** 1786–1843. French-born mathematician & explorer.
**Nic·ol·son** (nĭk'əl-sən), Sir **Harold George.** 1886–1968. English diplomat, author, & critic.
**Nie·buhr** (nē'bŏŏr', -bər), **Barthold George.** 1776–1831. German historian, philologist, & diplomat.
**Niebuhr, Reinhold.** 1892–1971. Amer. theologian.
**Niel·sen** (nēl'sən), **Alice.** 1870–1943. Amer. operatic soprano.
**Nielsen, Carl August.** 1865–1931. Danish composer.
**Niem·ce·wicz** (nĕm-sā'vĭch, nē-əm-), **Julian Ursyn.** 1758–1841. Polish author & revolutionary; exiled.

**Nie·mey·er Soa·res Fil·ho** (nē'mī'ər swä'rĕsh fēl'yōō), **Oscar.** b. 1907. Brazilian architect.
**Nie·tzsche** (nē'chə, -chē), **Friedrich Wilhelm.** 1844–1900. German philologist & philosopher. **—Nie'tzsche·an** *adj.* & *n.*
**Night·in·gale** (nīt'n-gāl', nī'tĭng-), **Florence.** *the Lady with the Lamp.* 1820–1910. English nursing pioneer.
**Ni·jin·sky** (nĭ-zhĭn'skē), **Vaslav** *or* **Waslaw.** 1890–1950. Russian-born dancer & choreographer.
**Niles** (nīlz), **Hezekiah.** 1777–1839. Amer. journalist.
**Nils·son** (nĭl'sən), **Birgit.** b. 1918. Swedish operatic soprano.
**Nim·itz** (nĭm'ĭts), **Chester Williams.** 1885–1966. Amer. admiral.
**Nin** (nēn, nĭn), **Anaïs.** 1903–77. French-born Amer. author & diarist.
**Nir·en·berg** (nîr'ĭn-bûrg'), **Marshall Warren.** b. 1927. Amer. biochemist (Nobel, 1968).
**Nit·ti** (nĭt'ē, nē'tē), **Francesco Saverio.** 1868–1953. Italian economist & politician.
**Nix·on** (nĭk'sən), **Richard Milhous.** b. 1913. 37th U.S. President (1969–74); resigned.
**Ni·zer** (nī'zər), **Louis.** b. 1902. English-born Amer. lawyer & author.
**Nkru·mah** (ən-krōō'mə, əng-), **Kwame.** 1909–72. Ghanaian statesman.
**No·ah** (nō'ə). Hebrew patriarch.
**No·bel** (nō-bĕl'), **Alfred Bernhard.** 1833–96. Swedish chemist, engineer, inventor, & philanthropist.
**No·bi·le** (nō'bə-lā), **Umberto.** 1885–1978. Italian soldier, aeronautical engineer, & Arctic explorer.
**No·el-Ba·ker** (nō'əl-bā'kər), **Philip John.** 1889–1982. English statesman & author (Nobel, 1959).
**No·gu·chi** (nō-gōō'chē), **Hideyo.** 1876–1928. Japanese-born Amer. pioneer bacteriologist.
**Noguchi, Isamu.** b. 1904. Amer. sculptor.
**No·mu·ra** (nō-mōōr'ä), **Kichisaburo.** 1887–1964. Japanese naval officer & diplomat.
**Nor·dau** (nôr'dou'), **Max Simon.** 1849–1923. Hungarian-born German author & Zionist leader.
**Nor·den·skjöld** (nōōr'dən-shəld, -shōōld', -shĕld'), Baron **Nils Adolf Erik.** 1832–1901. Swedish Arctic explorer & geologist.
**Nor·di·ca** (nôr'dĭ-kə), **Lillian.** 1859–1914. Amer. operatic soprano.
**No·rell** (nôr-ĕl'), **Norman.** 1900–72. Amer. fashion designer.
**Nor·ris** (nôr'ĭs, nŏr'-). Family of Amer. authors, including **Benjamin Franklin Jr. ("Frank")** (1870–1902), **Kathleen Thompson** (1880–1966), & **Charles Gilman** (1881–1945).
**Norris, George William.** 1861–1944. Amer. legislator.
**Nor·rish** (nôr'ĭsh), **Ronald George Wreyford.** 1897–1978. English chemist (Nobel, 1967).
**North** (nôrth), **Frank Joshua.** 1840–85. Amer. frontiersman.
**North, Frederick.** *2nd Earl of Guilford, "Lord North."* 1732–92. English prime minister (1770–82).
**North,** Sir **Thomas.** 1535?–1601. English translator.
**North·cliffe** (nôrth'klĭf'), Viscount. Alfred Charles William HARMSWORTH.
**Nor·throp** (nôr'thrəp), **John Howard.** b. 1891. Amer. biochemist (Nobel, 1946).
**Nor·ton** (nôr'tn), **Charles Eliot.** 1827–1908. Amer. author, editor, & educator.
**Norton, Thomas.** 1532–84. English lawyer, author, & translator.
**Nos·tra·da·mus** (nŏs'trə-dā'məs, -dä'-, nōs'-). 1503–66. French physician & astrologer.
**Noyes** (noiz), **Alfred.** 1880–1958. English poet.
**Noyes, Arthur Amos.** 1866–1936. Amer. chemist & educator.
**Noyes, Clara Dutton.** 1869–1936. Amer. nurse & educator.
**Noyes, John Humphrey.** 1811–86. Amer. social reformer.
**Nu·re·yev** (nōō-rā'yəf, nōōr'ĭ-yĕv, -ĕf, -ĕv), **Rudolf.** b. 1938. Russian-born Austrian ballet dancer & choreographer.
**Nut·tall** (nŭt'ôl), **Zelia Maria Magdelena.** 1858–1933. Amer. archaeologist.
**Nut·ting** (nŭt'ĭng), **Wallace.** 1861–1941. Amer. clergyman, antiquarian, & painter.
**Nye** (nī), **Edgar Wilson ("Bill").** 1850–96. Amer. humorist.

# O

**Oak·ley** (ōk'lē), **Annie.** 1860–1926. Amer. markswoman.
**Oates** (ōts), **Joyce Carol.** b. 1938. Amer. author.
**Oates, Titus.** 1649–1705. English conspirator.
**O·ba·di·ah** (ō'bə-dī'ə). 6th cent. B.C. Hebrew prophet.
**O·bre·gón** (ō-brā-gôn'), **Álvaro.** 1880–1928. Mexican soldier & statesman.
**O'Ca·sey** (ō-kā'sē), **Sean.** 1880–1964. Irish playwright.
**Oc·cam** (ŏk'əm). *var. of* OCKHAM.
**Oc·cleve** (ŏk'lēv'), **Thomas.** —See HOCCLEVE.
**O·cho·a** (ō-chō'ə), **Severo.** b. 1905. Spanish-born Amer. biochemist (Nobel, 1959).
**Ochs** (ŏks), **Adolph Simon.** 1858–1935. Amer. newspaper publisher.
**Ock·ham** *also* **Oc·cam** (ŏk'əm), **William of.** 1300?–49. English scholastic philosopher.
**O'Con·nor** (ō-kŏn'ər), **Flannery.** 1925–64. Amer. author.
**O'Connor, Thomas Power.** *Tay Pay.* 1845–1929. Irish journalist & politician.
**Oc·ta·vi·an** (ŏk-tā'vē-ən). AUGUSTUS.
**O·dets** (ō-dĕts'), **Clifford.** 1906–63. Amer. playwright.
**O·do·a·cer** (ō'dō-ā'sər) *also* **O·do·va·car** *or* **O·do·va·kar** (-vā'-). 434?–93. Germanic tribal leader.
**Oeh·len·schlä·ger** (œ'lən-shlā-gər), **Adam Gottlob.** 1779–1850. Danish romantic author.
**O'Fao·láin** (ō-fāl'ən, ō-fā'lən), **Seán.** b. 1900. Irish author.
**Of·fen·bach** (ô'fən-bäk'), **Jacques.** 1819–80. French composer.
**O'Fla·her·ty** (ō-flă'hər-tē, -ər-tē), **Liam.** 1896–1984. Irish author.
**Og·den** (ŏg'dən, ŏg'-), **Charles Kay.** 1889–1957. English psychologist & educator.
**Ogden, William Butler.** 1805–77. Amer. railroad executive.
**O·gle·thorpe** (ō'gəl-thôrp'), **James Edward.** 1696–1785. English soldier, philanthropist, & colonizer.
**O'Hair** (ō-hâr'), **Madalyn Murray.** b. 1919. Amer. reformer.
**O'Ha·ra** (ō-hâr'ə, ō-hăr'ə), **John Henry.** 1905–70. Amer. author.
**O. Hen·ry** (ō hĕn'rē). William Sidney PORTER.
**O'Hig·gins** (ō-hĭg'ĭnz), **Bernardo.** 1778–1842. Chilean general & statesman.
**Oh·lin** (ō'lĭn), **Bertil Gottard.** 1899–1979. Swedish economist (Nobel, 1977).
**Ohm** (ōm), **Georg Simon.** 1787–1854. German physicist.
**Ois·trakh** (oi'strək), **David Feodorovich.** 1908–74. Russian violinist.
**O'Keeffe** (ō-kēf'), **Georgia.** b. 1887. Amer. painter.
**O'Kel·ly** (ō-kĕl'ē), **Seán Thomas.** 1883–1966. Irish political leader.
**O·laf** (ō'läf, ō'ləf) *or* **O·lav** (ō'läv, ō'ləv). Name of five Norwegian kings, esp.: **a. I.** *Olaf Tryggvesson.* 969?–1000. Ruled 995?–1000. **b. II.** *Olaf Haraldsson, Saint Olaf.* 995?–1030. Ruled 1015–28. **c. V.** b. 1903. Ruled since 1957.
**Old·cas·tle** (ōld'kăs'əl, -kä'səl), Sir **John.** *Lord Cobham.* d. 1417. English Lollard conspirator; burned for heresy.
**Ol·den·burg** (ōl'dən-bûrg'). Danish ruling house (1448–1523).
**Oldenburg, Claes Thure.** b. 1929. Swedish-born Amer. sculptor.
**Old·field** (ōld'fēld'), **Berna Eli ("Barney").** 1878–1946. Amer. automobile racer.
**Olds** (ōldz), **Ransom Eli.** 1864–1950. Amer. automobile inventor & manufacturer.
**Ol·i·ver** (ŏl'ə-vər), **Joseph ("King").** 1885?–1938. Amer. jazz musician & composer.
**O·liv·i·er** (ō-lĭv'ē-ā'), Sir **Laurence Kerr.** *Baron Olivier of Brighton.* b. 1907. English actor and director.
**Olm·sted** (ŏm'stĕd', -stĭd, ŏm'-), **Frederick Law.** 1822–1903. Amer. landscape architect.
**Ol·ney** (ōl'nē), **Richard.** 1835–1917. Amer. public official.
**O·mar Khay·yám** (ō'mär kī-yäm', -ăm'). 1050–1123. Persian poet, mathematician, & astronomer.
**Om·mi·ad** *also* **O·may·yad** (ō-mī'ăd). Dynasty of Arab caliphs (661–750).
**O·nas·sis** (ō-năs'ĭs, -nä'sĭs), **Aristotle.** 1906–75. Turkish-born Greek financier & shipping magnate.
**Onassis, Jacqueline Lee Bouvier Kennedy.** b. 1929. Amer. socialite & editor.
**O·ña·te** (ôn-yä'tā), **Juan de.** 1549–1624? Spanish explorer & conquistador.
**O'Neill** (ō-nēl'), **Eugene Gladstone.** 1888–1953. Amer. playwright (Nobel, 1936).
**O'Neill, Rose Cecil.** 1874–1944. Amer. artist & author.
**O'Neill, Thomas Philip, Jr. ("Tip").** b. 1912. Amer. legislator.
**On·ions** (ŭn'yənz), **Charles Talbut.** 1873–1965. English philologist & lexicographer.
**On·sa·ger** (ôn'sä'gər), **Lars.** 1903–76. Norwegian-born Amer. chemist (Nobel, 1968).
**Op·pen·heim** (ŏp'ən-hīm'), **E(dward) Phillips.** 1866–1946. English author.
**Op·pen·hei·mer** (ŏp'ən-hī'mər), **J(ulius) Robert.** 1904–67. Amer. physicist.
**Or·ange** (ôr'ĭnj). Dutch ruling family (since 1815).
**Or·cag·na** (ôr-kän'yə), **Andrea.** 1308–68? Florentine artist & architect.
**Or·czy** (ôr'tsē), Baroness **Emmuska.** 1865–1947. Hungarian-born English author.
**O'Reil·ly** (ō-rī'lē), **John Boyle.** 1844–90. Irish-American journalist & poet.
**Orff** (ôrf), **Carl.** 1895–1982. German composer.
**Or·i·gen** (ôr'ĭ-jĕn', -jən, ŏr'-). 185?–254? Greek church father.

**Or·lan·do** (ôr-lăn′dō, -län′-), **Vittorio Emanuele.** 1860–1952. Italian statesman.

**Or·man·dy** (ôr′mən-dē), **Eugene.** 1899–1985. Hungarian-born Amer. conductor.

**Orms·by-Gore** (ôrmz′bē-gôr′, -gōr′), **David.** *5th Baron Harlech.* b. 1918. English diplomat.

**O·roz·co** (ō-rôs′kō), **José Clemente.** 1883–1949. Mexican fresco painter.

**Orr** (ôr), **James Lawrence.** 1822–73. Amer. legislator & diplomat.

**Orr, Robert ("Bobby").** b. 1948. Canadian-born hockey player.

**Or·te·ga y Gas·set** (ôr-tā′gə ē gä-sĕt′), **José.** 1883–1955. Spanish philosopher, author, & politician.

**Or·tiz Ru·bi·o** (ôr-tēz′ ro͞o′bē-ō), **Pascual.** 1877–1963. Mexican statesman.

**Or·well** (ôr′wĕl′, -wəl), **George.** *Eric Blair.* 1903–50. English author.

**Os·born** (ŏz′bərn, -bôrn′, -bōrn′), **(Henry) Fairfield.** 1887–1969. Amer. conservationist.

**Osborn, Henry Fairfield.** 1857–1935. Amer. paleontologist.

**Os·borne** (ŏz′bərn, -bôrn′, -bōrn′), **John James.** b. 1929. English playwright, producer, & actor.

**Osborne,** Sir **Thomas.** *1st Earl of Danby, Marquis of Carmarthen, Duke of Leeds.* 1631–1712. English politician.

**Osborne, Thomas Mott.** 1859–1926. Amer. prison reformer.

**Os·car** (ŏs′kər) *also* **Os·kar** (ôs′kär). Name of two kings of Norway & Sweden, esp. **II,** 1829–1907, ruled Sweden 1872–1907 & Norway 1872–1905.

**Os·ce·o·la** (ŏs′ē-ō′lə, ŏ′sē-). 1804?–38. Amer. Seminole leader.

**Os·kar** (ôs′kär). *var. of* OSCAR.

**Os·ler** (ōs′lər, ŏz′-), Sir **William.** 1849–1919. Canadian-born British physician & educator.

**Os·man** (ŏz′mən, ŏs′-, ōs-män′) *also* **Oth·man** (ŏth′mən, ŏth-män′). Name of three Ottoman rulers, esp. **I,** 1259–1326, emir (1299–1326) & founder of dynasty.

**Os·me·ña** (ŏz-mān′yə, ŏs-), **Sergio.** 1878–1961. Philippine statesman.

**Os·si·etz·ky** (ŏs′ē-ĕt′skē), **Carl von.** 1887–1938. German journalist & pacifist (Nobel, 1935).

**Ost·wald** (ŏst′wôld′, ôst′vält), **Wilhelm.** 1853–1932. German chemist & educator (Nobel, 1909).

**O'Sul·li·van** (ō-sŭl′ə-vən), **Timothy.** 1840–82. Amer. pioneer photographer.

**Os·wald** (ŏz′wôld′), **Lee Harvey.** 1939–63. Amer. alleged Presidential assassin; assassinated.

**Oth·man** (ŏth′mən, ŏth-män′). *var. of* OSMAN.

**O·tho** (ō′thō, ō′tō). *var. of* OTTO.

**O·tis** (ō′tĭs), **Elisha Graves.** 1811–61. Amer. inventor & manufacturer.

**Otis, Harrison Gray.** 1765–1848. Amer. legislator.

**Otis, Harrison Gray.** 1837–1917. Amer. soldier, journalist, & politician.

**Otis, James.** 1725–83. Amer. Revolutionary politician & publicist.

**Ot·ter·bein** (ŏt′ər-bīn′), **Philip William.** 1726–1813. German-born Amer. religious leader.

**Ot·to** (ŏt′ō) *also* **O·tho** (ō′thō, ō′tō). Name of four Holy Roman Emperors, esp. **I,** *the Great,* 912–73, ruled 962–73.

**Ot·way** (ŏt′wā), **Thomas.** 1652–85. English classical poet.

**Oui·da** (wē′də). Marie Louise de la RAMÉE.

**Out·cault** (out′kôlt′), **Richard Felton.** 1863–1925. Amer. cartoonist.

**Ov·id** (ŏv′ĭd). 43 B.C.–A.D. 18. Roman poet.

**Ow·en** (ō′ĭn), Sir **Richard.** 1804–92. English anatomist & paleontologist.

**Owen, Robert.** 1771–1858. Welsh-born British manufacturer, educator, & socialist.

**Owen, Robert Dale.** 1801–77. Scottish-born Amer. social reformer, politician, diplomat, & author.

**Owen, Wilfred.** 1893–1918. English poet.

**Ow·ens** (ō′ĭnz), **Jesse.** 1913–80. Amer. athlete.

**Ox·en·stier·na** *also* **Ox·en·stjer·na** (o͝ok′sən-shĕr′nə, ŏk′-) *or* **Ox·en·stiern** (ŏk′sən-stîrn′, o͝ok′sən-shĕrn), Count **Axel Gustafsson.** 1583–1654. Swedish statesman.

**O·za·wa** (ō-zä′wə), **Seiji.** b. 1935. Japanese-born conductor.

# P

**Paa·si·ki·vi** (pä′sə-kē′vē), **Juho K.** 1870–1956. Finnish statesman.

**Pack·ard** (păk′ərd), **Sophia.** 1824–91. Amer. educator.

**Packard, Vance.** b. 1914. Amer. author.

**Pa·de·rew·ski** (păd′ə-rĕf′skē, -rĕv′-), **Ignace Jan.** 1860–1941. Polish pianist & statesman.

**Pa·ga·ni·ni** (păg′ə-nē′nē, pä′gä-), **Nicolo.** 1782–1840. Italian violinist & composer.

**Page** (pāj), **Thomas Nelson.** 1853–1922. Amer. lawyer, novelist, & diplomat.

**Page, Walter Hines.** 1855–1918. Amer. journalist & diplomat.

**Pag·et** (păj′ĭt), Sir **James.** 1814–99. English surgeon and pathologist.

**Pah·la·vi** (pä′lə-vē), **Mohammed Reza.** 1919–80. Iranian shah (1941–79); deposed.

**Paige** (pāj), **Leroy Robert ("Satchel").** 1906–82. Amer. baseball player.

**Paine** (pān), **Albert Bigelow.** 1861–1937. Amer. author & editor.

**Paine, Robert Treat.** 1731–1814. Amer. Revolutionary leader & jurist.

**Paine, Robert Treat.** 1773–1811. Amer. poet.

**Paine, Thomas.** 1737–1809. English-born Amer. author & Revolutionary leader.

**Pain·le·vé** (păN-lə-vā′), **Paul.** 1863–1933. French mathematician & politician.

**Pa·la·de** (pə-lä′dē), **George Emil.** b. 1912. Russian-born Amer. biologist (Nobel, 1974).

**Pa·le·stri·na** (păl′ĭ-strē′nə), **Giovanni Pierluigi da.** 1526?–94. Italian composer.

**Pa·ley** (pā′lē), **William.** 1743–1805. English theologian & philosopher.

**Paley, William S.** b. 1901. Amer. broadcasting executive.

**Pal·grave** (păl′grāv′, pôl′-), **Francis Turner.** 1824–97. English poet.

**Pal·la·dio** (pə-lä′dē-ō), **Andrea.** 1508–80. Italian architect. **—Pal·la′di·an** *adj.*

**Pal·ma** (päl′mə), **Tomás Estrada.** 1835–1908. Cuban statesman.

**Palm·er** (pä′mər, päl′-), **Alexander Mitchell.** 1872–1936. Amer. lawyer & public official.

**Palmer, Alice Elvira Freeman.** 1855–1902. Amer. educator.

**Palmer, Arnold.** b. 1929. Amer. golfer.

**Palmer, Daniel David.** 1845–1913. Canadian-born Amer. founder of chiropractic.

**Palmer, George Herbert.** 1842–1933. Amer. scholar & educator.

**Palmer, Potter.** 1826–1902. Amer. merchant & real-estate promoter.

**Palm·er·ston** (pä′mər-stən, päl′-), 3rd Viscount. *Henry John Temple.* 1784–1865. British prime minister (1855–58, 1859–65).

**Pa·lo·u** (pə-lō′o͞o), **Francisco.** 1722–1789. Spanish missionary in America.

**Pa·ni·ni** (pä-nē′nē). fl. 350 B.C. Indian Sanskrit grammarian.

**Pank·hurst** (păngk′hûrst′), **Emmeline Goulden.** 1858–1928. English suffragist.

**Pa·nof·sky** (pə-nŏf′skē, -nôv-), **Erwin.** 1892–1968. German-born Amer. art historian.

**Pa·o·li** (pou′lē, pä′ō-), **Pasquale di.** 1725–1807. Corsican patriot.

**Pa·pa·do·pou·los** (păp′ə-dŏ′pə-ləs, pä′pə-), **George.** b. 1919. Greek army officer & statesman.

**Pap·an·dre·ou** (păp′ən-drā′o͞o, pä′pən-), **Andreas George.** b. 1919. Greek statesman.

**Pa·pen** (pä′pən), **Franz von.** 1879–1969. German diplomat.

**Papp** (păp), **Joseph.** b. 1921. Amer. stage producer & director.

**Par·a·cel·sus** (păr′ə-sĕl′səs), **Philippus Aureolus.** 1493–1541. German-Swiss alchemist & physician.

**Pa·ré** (pä-rā′), **Ambroise.** 1517?–90. French pioneer surgeon.

**Pares** (pârz), Sir **Bernard.** 1876–1949. English Slavonic scholar.

**Pa·re·to** (pə-rā′tō), **Vilfredo.** 1848–1923. Italian economist & sociologist.

**Pa·ris** (păr′ĭs), **Matthew.** 1200?–59. English monk & chronicler.

**Park** (pärk), **Mungo.** 1771–1806. Scottish explorer in Africa.

**Park, Robert Ezra.** 1864–1944. Amer. sociologist.

**Park Chung Hee** (pärk′ chŭng′ hē′). 1917–79. South Korean statesman.

**Par·ker** (pär′kər), **Charlie.** 1920–55. Amer. musician & composer.

**Parker, Dorothy Rothschild.** 1893–1967. Amer. author.

**Parker, Francis Wayland.** 1837–1902. Amer. pioneer educator.

**Parker,** Sir **(Horatio) Gilbert (George).** 1862–1932. Canadian author.

**Parker, John.** 1729–75. Amer. Revolutionary leader.

**Parker, Matthew.** 1504–75. English prelate.

**Parker, Theodore.** 1810–60. Amer. abolitionist & clergyman.

**Parkes** (pärks), Sir **Henry.** 1815–96. Australian statesman.

**Par·kin·son** (pär′kĭn-sən), **C(yril) Northcote.** b. 1909. English historian.

**Park·man** (pärk′mən), **Francis.** 1823–93. Amer. historian.

**Parks** (pärks), **Rosa.** b. 1913. Amer. civil-rights leader.

**Par·men·i·des** (pär-mĕn′ĭ-dēz′). 5th cent. B.C. Greek philosopher.

**Par·mi·gia·ni·no** (pär′mĭ-jä-nē′nō) *or* **Par·mi·gia·no** (-jä′nō), **Il.** 1503–40. Italian painter.

**Par·nell** (pär-nĕl′, pär′nəl), **Charles Stewart.** 1846–91. Irish nationalist leader.

**Par·nis** (pär′nĭs), **Mollie.** b. 1905. Amer. fashion designer.

**Parr** (pär), **Catherine.** 1512–48. Queen of England as sixth wife of Henry VIII.

**Par·ring·ton** (păr′ĭng-tən), **Vernon Louis.** 1871–1929. Amer. literary historian & philosopher.
**Par·rish** (păr′ĭsh), **Anne.** 1760–1800. Amer. philanthropist.
**Parrish, Celestia Susannah.** 1853–1918. Amer. educator.
**Parrish, Maxfield Frederick.** 1870–1966. Amer. artist.
**Par·rott** (păr′ət), **Robert Parker.** 1804–77. Amer. army officer & ordnance inventor.
**Par·ry** (păr′ē), Sir **William Edward.** 1790–1855. English navigator & Arctic explorer.
**Par·sons** (pär′sənz), **Louella.** 1893–1972. Amer. newspaper columnist.
**Parsons, Talcott.** b. 1902. Amer. sociologist.
**Parsons, Theophilus.** 1750–1813. Amer. jurist.
**Parsons, Theophilus.** 1797–1882. Amer. legal author.
**Parsons, William.** 1800–67. English astronomer.
**Par·ton** (pär′tn), **James.** 1822–91. Amer. biographer.
**Parton, Sara Payson Willis.** 1811–72. Amer. author.
**Pas·cal** (păs-kăl′, päs-käl′), **Blaise.** 1623–62. French philosopher & mathematician.
**Pa·šić** (pä′shĭch), **Nikola.** 1845?–1926. Serbian & Yugoslavian statesman.
**Pas·sy** (pä-sē′, pä-), **Frédéric.** 1822–1912. French economist & pacifist (Nobel, 1901).
**Passy, Paul Edouard.** 1859–1940. French philologist.
**Pas·ter·nak** (păs′tər-năk′), **Boris Leonidovich.** 1890–1960. Russian author (Nobel, 1958).
**Pas·teur** (păs-tûr′), **Louis.** 1822–95. French chemist. **—Pas·teur′i·an** *adj.*
**Pas·tor** (păs′tər), **Antonio ("Tony").** 1837–1908. Amer. actor & theater manager.
**Pas·tor·i·us** (păs-tôr′ē-əs, -tōr′-), **Francis Daniel.** 1651–1720? German lawyer & colonist in America.
**Patch** (păch), **Sam.** 1807?–29. Amer. daredevil.
**Pa·ter** (pā′tər), **Walter Horatio.** 1839–94. English author.
**Pat·er·son** (păt′ər-sən), **William.** 1745–1806. Irish-born Amer. Revolutionary leader & jurist.
**Pat·more** (păt′môr′, -mōr′), **Coventry Kersey Dighton.** 1823–96. English poet.
**Pa·ton** (pāt′n), **Alan Stewart.** b. 1903. South African author.
**Pa·tri** (pä′trē, păt′rē), **Angelo.** 1877–1965. Italian-born Amer. educator & author.
**Pat·rick** (păt′rĭk), Saint. 389?–461? Patron saint of Ireland.
**Pat·ten** (păt′n), **Gilbert.** *Burt L. Standish.* 1866–1945. Amer. adventure-story author.
**Pat·ter·son** (păt′ər-sən). Family of Amer. newspaper editors & publishers, including **Joseph Medill** (1879–1946), **Eleanor Medill** (1881–1948), & **Alicia** (1906–63).
**Patterson, Floyd.** b. 1935. Amer. prizefighter.
**Patterson, John Henry.** 1844–1922. Amer. salesman & manufacturer.
**Pat·ti** (păt′ē, pä′tē), **Adelina.** 1843–1919. Spanish-born Amer. singer.
**Pat·ti·son** (păt′ĭ-sən), **Mark.** 1813–84. English author & educator.
**Pat·ton** (păt′n), **George Smith, Jr.** 1885–1945. Amer. general.
**Paul** (pôl), Saint. A.D. 5?–67? Apostle to the Gentiles. **—Paul′ine** (-īn) *adj.*
**Paul.** Name of six popes, esp.: **a. III.** 1468–1549. Reigned 1534–49. **b. V.** 1552–1621. Reigned 1605–21. **c. VI.** 1897–1978. Reigned 1963–78.
**Paul, Alice.** 1885–1977. Amer. social reformer.
**Paul I.** 1901–64. Grecian king (1947–64).
**Paul I.** 1754–1801. Russian czar (1796–1801).
**Paul-Bon·cour** (pôl-bôN-ko͝or′), **Joseph.** 1873–1972. French statesman & diplomat.
**Paul·ding** (pôl′dĭng), **Hiram.** 1797–1878. Amer. admiral.
**Paulding, James Kirke.** 1778–1860. Amer. author & public official.
**Pau·li** (pou′lē), **Wolfgang.** 1900–58. Austrian-born Amer. physicist (Nobel, 1945).
**Pau·ling** (pô′lĭng), **Linus Carl.** b. 1901. Amer. chemist (Nobel, 1954, 1962).
**Pau·lus** (pou′ləs), **Julius.** 2nd–3rd cent. A.D. Roman jurist.
**Pau·sa·ni·as** (pô-sā′nē-əs). 2nd cent. A.D. Greek geographer & historian.
**Pav·a·rot·ti** (păv′ə-rŏt′ē, pä′və-), **Luciano.** b. 1935. Italian-born tenor.
**Pav·lov** (păv′lôf′, -lôv′), **Ivar Petrovich.** 1849–1936. Russian physiologist (Nobel, 1904). **—Pav·lo′vi·an** *adj.*
**Pav·lo·va** (păv′lə-və, păv-lō′-, päv′lə-, päv-lō′-), **Anna.** 1885–1931. Russian ballerina.
**Payne** (pān), **John Howard.** 1791–1852. Amer. actor, dramatist, songwriter, & diplomat.
**Pea·bod·y** (pē′bŏd′ē, -bə-dē), **Elizabeth Palmer.** 1804–94. Amer. educator & author.
**Peabody, Endicott.** 1857–1944. Amer. educator.
**Peabody, George.** 1795–1869. Amer. merchant & philanthropist.
**Peabody, Josephine Preston.** 1874–1922. Amer. dramatist.
**Pea·cock** (pē′kŏk′), **Thomas Love.** 1785–1866. English author.
**Peale** (pēl). Family of Amer. painters, including **Charles Willson** (1741–1827), **James** (1749–1831), **Raphael** (1774–1825), **Rembrandt** (1778–1860), **Anna Claypoole** (1791–1878), **Titian Ramsay** (1799–1885), & **Sarah Miriam** (1800–85).
**Peale, Norman Vincent.** b. 1898. Amer. clergyman & author.
**Pear·son** (pîr′sən), **Andrew Russell ("Drew").** 1897–1969. Amer. journalist.
**Pearson, Karl.** 1857–1936. English eugenicist & mathematician.
**Pearson, Lester Bowles.** 1897–1972. Canadian prime minister (1963–68; Nobel, 1957).
**Pea·ry** (pîr′ē), **Robert Edwin.** 1856–1920. Amer. naval officer & Arctic explorer.
**Peck** (pĕk), **Annie Smith.** 1850–1935. Amer. explorer & mountain climber.
**Peck, Gregory.** b. 1916. Amer. actor.
**Peck·ham** (pĕk′əm), **Rufus Wheeler.** 1838–1909. Amer. jurist.
**Pe·cor·a** (pĭ-kôr′ə, -kōr′ə), **Ferdinand.** 1882–1971. Amer. jurist.
**Pe·dro** (pā′drō, -dro͞o). Name of two Brazilian emperors: **a. I.** 1798–1834. Ruled 1822–31; abdicated. **b. II.** 1825–91. Ruled 1831–89 (crowned 1841); forced to abdicate.
**Peel** (pēl), Sir **Robert.** 1788–1850. British prime minister (1834–35, 1841–46).
**Peele** (pēl), **George.** 1556?–96? English author.
**Peerce** (pîrs), **Jan.** 1904–84. Amer. tenor.
**Peg·ler** (pĕg′lər), **Westbrook.** 1894–1969. Amer. journalist.
**Pei** (pā), **Ieoh Ming ("I.M.").** b. 1917. Chinese-born Amer. architect.
**Peirce** (pîrs, pûrs), **Benjamin.** 1809–80. Amer. mathematician & astronomer.
**Peirce, Charles Sanders.** 1839–1914. Amer. pragmatist philosopher, mathematician, & scientist.
**Pei·sis·tra·tus** (pə-sĭs′trə-təs). *var. of* PISISTRATUS.
**Pei·xot·to** (pā-shō′tō), **Ernest Clifford.** 1869–1940. Amer. artist & educator.
**Pe·le** (pā′lā). *Edson Arantes do Nascimento.* b. 1940. Brazilian soccer player.
**Pe·lop·i·das** (pə-lŏp′ĭ-dəs). d. 364 B.C. Theban general.
**Pen·der·gast** (pĕn′dər-găst′), **Thomas Joseph.** 1870–1945. Amer. politician.
**Pen·dle·ton** (pĕn′dl-tən), **Edmund.** 1721–1803. Amer. Revolutionary leader & jurist.
**Penn** (pĕn), **Arthur.** b. 1922. Amer. filmmaker.
**Penn,** Sir **William.** 1621–70. English admiral.
**Penn, William.** 1644–1718. English Quaker colonizer in America.
**Pen·nell** (pĕn′əl, pə-nĕl′), **Elizabeth Robins.** 1855–1936. Amer. author & critic.
**Pennell, Joseph.** 1857–1926. Amer. artist.
**Pen·ney** (pĕn′ē), **James Cash.** 1875–1971. Amer. businessman.
**Pen·rose** (pĕn′rōz′), **Boies.** 1860–1921. Amer. political boss.
**Pen·zi·as** (pĕnt′sē-əs), **Arno Allan.** b. 1933. German-born Amer. physicist (Nobel, 1978).
**Pep·in** (pĕp′ĭn). *the Short.* 714?–68. King of the Franks (751–68).
**Pep·per** (pĕp′ər), **William.** 1843–98. Amer. pioneer physician & educator.
**Pep·per·rell** (pĕp′ər-əl), Sir **William.** 1696–1759. Amer. merchant & colonial official.
**Pepys** (pēps), **Samuel.** 1633–1701? English diarist. **—Pepys′i·an** *adj.*
**Per·cy** (pûr′sē), Sir **Henry.** *Hotspur.* 1364–1403. English soldier; killed in battle.
**Percy, Thomas.** 1729–1811. English prelate, antiquary, & poet.
**Per·el·man** (pĕr′əl-mən), **Sidney Joseph ("S.J.").** 1904–79. Amer. author.
**Pé·rez Gal·dós** (pĕr′əs gäl-dōs′), **Benito.** 1843–1920. Spanish author.
**Pérez Ji·me·nez** (hē-mĕn′ĕs), **Marcos.** b. 1914. Venezuelan soldier & statesman.
**Per·go·le·si** (pĕr′gə-lā′zē), **Giovanni Battista.** 1710–36. Italian composer.
**Per·i·cles** (pĕr′ĭ-klēz′). d. 429 B.C. Athenian statesman & general. **—Per′i·cle′an** (-klē′ən) *adj.*
**Per·kin** (pûr′kĭn), Sir **William Henry.** 1838–1907. English chemist.
**Per·kins** (pûr′kĭnz), **Frances.** 1882–1965. Amer. social reformer & public official.
**Perkins, George Walbridge.** 1862–1920. Amer. financier.
**Perkins, Marlin.** b. 1905. Amer. zoo director.
**Perkins, Maxwell Evarts.** 1884–1946. Amer. editor.
**Perl·man** (pûrl′mən), **Itzhak.** b. 1945. Israeli-born Amer. violinist.
**Pe·rón** (pə-rōn′, pə-). Argentinean popular & political leaders, including **Juan Domingo** (1895–1974), **Maria Eva Duarte de** (*Evita;* 1919–52), & **Isabel** (*Isabelita;* b. 1931).
**Per·rault** (pə-rō′), **Charles.** 1628–1703. French author.

**Per·rin** (pə-răN'), **Jean Baptiste.** 1870–1942. French physicist & chemist (Nobel, 1926).
**Per·rot** (pə-rō'), **Nicolas.** 1644–1717. French explorer & fur trader.
**Per·ry** (pĕr'ē), **Antoinette.** 1888–1946. Amer. actress & director.
**Perry, Bliss.** 1860–1954. Amer. educator, editor, & critic.
**Perry, Matthew Calbraith.** 1794–1858. Amer. naval officer.
**Perry, Oliver Hazard.** 1785–1819. Amer. naval officer.
**Perry, Ralph Barton.** 1876–1957. Amer. philosopher & educator.
**Perse** (pĕrs, pûrs), **Saint-John.** Alexis Saint-Léger LÉGER.
**Per·shing** (pûr'shĭng, -zhĭng), **John Joseph ("Black Jack").** 1860–1948. Amer. army officer.
**Per·sius** (pûr'shəs, -shē'əs). A.D. 34–62. Roman satirist.
**Pe·ru·gi·no** (pĕr'ə-jē'nō), **Il.** 1446–1523? Italian painter.
**Pe·rutz** (pə-rōōts'), **Max Ferdinand.** b. 1914. Austrian-born English biochemist (Nobel, 1962).
**Pe·ruz·zi** (pə-rōōt'sē, pā-), **Baldassare.** 1481–1536. Italian architect & painter.
**Pes·ta·loz·zi** (pĕs'tə-lŏt'sē), **Johann Heinrich.** 1746–1827. Swiss educational reformer.
**Pé·tain** (pā-tăN'), **Henri Philippe.** 1856–1951. French soldier & politician.
**Pe·ter** (pē'tər), Saint. d. A.D. 67? One of the 12 Apostles; martyred.
**Peter.** *the Hermit.* 1050?–1115? French monk & preacher of the First Crusade.
**Peter.** Name of three czars of Russia, esp. **I,** *the Great,* 1672–1725, ruled 1682–1725.
**Peter I Ka·ra·george·vich** (kăr'ə-jôr'jə-vĭch'). 1844–1921. Serbian king (1903–21).
**Peter II.** 1923–70. Yugoslavian king (1934–45).
**Pe·ters** (pā'tərz, -tərs), **Karl.** 1856–1918. German explorer & administrator in Africa.
**Pe·ters** (pē'tərz), **Roberta.** b. 1930. Amer. soprano.
**Pe·ter·son** (pē'tər-sən), **Roger Tory.** b. 1908. Amer. ornithologist & artist.
**Pe·tö·fi** (pĕt'ə-fē), **Sándor.** 1823–49. Hungarian lyric poet & hero.
**Pe·trarch** (pē'trärk', pĕt'rärk') *or* **Pe·trar·ca** (pā-trär'kä), **Francesco.** 1304–74. Italian poet. **—Pe·trarch'an** (pĭ-trär'kən) *adj.*
**Pe·trie** (pē'trē), Sir **(William Matthew) Flinders.** 1853–1942. English Egyptologist.
**Pe·tro·ni·us** (pĭ-trō'nē-əs), **Gaius.** 1st cent. A.D. Roman courtier.
**Pet·ty** (pĕt'ē), Sir **William.** 1623–87. English political economist.
**Pevs·ner** (pĕvz'nər), **Antoine.** 1886–1962. Russian artist.
**Phae·drus** (fē'drəs). 5th cent. B.C. Greek philosopher.
**Phaedrus.** 1st cent. A.D. Roman fabulist.
**Phid·i·as** (fĭd'ē-əs). 5th cent. B.C. Athenian sculptor.
**Phil·ip** (fĭl'ĭp), Saint. One of the 12 Apostles.
**Philip.** d. 1676. Amer. Indian chief.
**Philip,** Prince. *Duke of Edinburgh.* b. 1921. Consort of Elizabeth II of England.
**Philip.** Name of two Dukes of Burgundy, esp. *the Good,* 1396–1467, ruled 1419–67.
**Philip.** Name of six kings of France, esp.: **a. II** *or* **Philip Au·gus·tus** (ô-gŭs'təs). 1165–1223. Ruled 1180–1223. **b. IV.** *the Fair.* 1268–1314. Ruled 1285–1314. **c. VI.** 1293–1350. Ruled 1328–50.
**Philip.** Name of five kings of Macedon, esp. **II,** 382–336 B.C. Ruled 359–36.
**Philip.** Name of five kings of Spain, esp.: **a. I.** *the Handsome.* 1478–1506. Ruled 1504–6; died mysteriously. **b. II.** 1527–98. Ruled 1556–98. **c. V.** 1683–1746. Ruled 1700–46 (abdicated briefly in 1724).
**Phil·ips** (fĭl'ĭps), **Ambrose.** 1675?–1749. English author.
**Phil·lips** (fĭl'ĭps), **Wendell.** 1811–84. Amer. abolitionist.
**Phill·potts** (fĭl'pŏts), **Eden.** 1862–1960. English novelist.
**Phi·lo Ju·dae·us** (fī'lō jōō-dē'əs, -dā'-). 30 B.C.–A.D. 45. Alexandrian Jewish philosopher.
**Phips** (fĭps), Sir **William.** 1651–95. English colonial administrator in America.
**Pho·ci·on** (fō'sē-ŏn'). 402?–317 B.C. Athenian general & statesman.
**Phyfe** (fīf), **Duncan.** 1768?–1854. Scottish-born Amer. cabinetmaker.
**Pi·af** (pē-äf', pē'äf'), **Edith.** 1916–63. French singer.
**Pia·get** (pyä-zhā'), **Jean.** 1896–1980. Swiss psychologist.
**Piaz·zi** (pyät'sē), **Giuseppi.** 1746–1826. Italian astronomer.
**Pi·card** (pē-kär'), **Jean.** 1620–82. French cleric & astronomer.
**Pi·cas·so** (pĭ-kä'sō, -kăs'ō), **Pablo.** 1881–1973. Spanish artist.
**Pic·card** (pē-kär', -kärd'), **Auguste.** 1884–1962. Swiss physicist & aeronaut.
**Piccard, Jean Felix.** 1884–1963. Swiss-born Amer. chemist & aeronautical engineer.
**Pick·er·ing** (pĭk'ər-ĭng), **Edward Charles** (1846–1919) & **William Henry** (1858–1938). Amer. astronomers.
**Pickering, Timothy.** 1745–1829. Amer. soldier, Revolutionary leader, & statesman.

ōō **boot** ou **out** th **thin** *th* **this** ŭ **cut** ûr **urge** y **young**
yōō abuse zh vision ə **a**bout, it**e**m, edib**i**le, gall**o**p, circ**u**s

**Pick·ett** (pĭk'ĭt), **George Edward.** 1825–75. Amer. Confederate general.
**Pick·ford** (pĭk'fərd), **Mary.** 1893–1979. Canadian-born Amer. actress.
**Pi·co del·la Mi·ran·do·la** (pē'kō dĕl'ə mə-răn'də-lə), Count **Giovanni.** 1463–94. Italian philosopher.
**Pidg·eon** (pĭj'ən), **Walter.** 1897–1984. Amer. actor.
**Pieck** (pēk), **Wilhelm.** 1876–1960. German statesman.
**Pierce** (pîrs), **Franklin.** 1804–69. 14th U.S. President (1853–57).
**Pie·ro di Co·si·mo** (pyâr'ō dē kō'zə-mō). 1462–1521. Florentine painter.
**Pike** (pīk), **Albert.** 1809–91. Amer. soldier & Freemason.
**Pike, Mary Hayden Green.** 1824–1908. Amer. novelist & abolitionist.
**Pike, Zebulon Montgomery.** 1779–1813. Amer. army officer & explorer.
**Pi·late** (pī'lət), **Pontius.** Roman governor of Judea (A.D. 26?–36?).
**Pills·bur·y** (pĭlz'bĕr'ē, -bə-rē), **Charles Alfred.** 1842–99. Amer. flour manufacturer.
**Pil·sud·ski** (pĭl-sōōt'skē, -zōōt'-), **Jozef.** 1867–1935. Polish statesman.
**Pin·chot** (pĭn'shō'), **Gifford.** 1865–1946. Amer. conservationist & politician.
**Pinck·ney** (pĭngk'nē). Family of Amer. politicians & diplomats, including **Charles Cotesworth** (1746–1825), **Thomas** (1750–1828), & **Charles** (1757–1824).
**Pin·dar** (pĭn'dər). 522?–443 B.C. Greek poet. **—Pin·dar'ic** (pĭn-dăr'ĭk) *adj.*
**Pi·ne·ro** (pə-nîr'ō), Sir **Arthur Wing.** 1855–1934. English playwright.
**Pin·ker·ton** (pĭngk'ər-tən), **Allan.** 1819–84. Scottish-born Amer. detective.
**Pink·ham** (pĭngk'əm), **Lydia Estes.** 1819–83. Amer. patent-medicine manufacturer.
**Pink·ney** (pĭngk'nē), **William.** 1764–1822. Amer. politician & diplomat.
**Pin·ter** (pĭn'tər), **Harold.** b. 1930. English playwright.
**Pin·tu·ric·chio** (pĭn'tə-rē'kē-ō'). 1454–1513. Italian painter.
**Pin·za** (pĭn'zə), **Ezio.** 1895–1957. Italian-born Amer. basso.
**Pin·zón** (pĭn-zōn'), **Martín Alonso** (1440?–93) & **Vicente Yáñez** (1460?–1524). Spanish navigators.
**Pioz·zi** (pē-ôt'sē), **Hester Lynch.** *Mrs. Thrale.* 1741–1821. English writer.
**Pi·ran·del·lo** (pîr'ən-dĕl'ō), **Luigi.** 1867–1936. Italian author (Nobel, 1934).
**Pi·ra·ne·si** (pîr'ə-nā'zē), **Giambattista.** 1720–78. Italian architect & artist.
**Pire** (pîr, pĕr), **Dominique Georges.** 1910–69. Belgian priest (Nobel, 1958).
**Pi·sa·no** (pĭ-sä'nō, -zä'-), **Andrea.** 1270?–1348. Italian sculptor.
**Pisano, Nicola** *or* **Niccolò** (1220–84) & **Giovanni** (1245–1314). Italian sculptors & architects.
**Pi·sis·tra·tus** (pĭ-sĭs'trə-təs, pī-) *or* **Pei·sis·tra·tus** (pə-). d. 527 B.C. Athenian tyrant (560–27).
**Pis·sar·ro** (pĭ-sär'ō), **Camille.** 1830–1903. French painter.
**Pis·ton** (pĭs'tən), **Walter Hamor.** 1894–1976. Amer. composer & educator.
**Pitch·er** (pĭch'ər), **Molly.** Mary MCCAULEY.
**Pit·kin** (pĭt'kĭn), **Walter Broughton.** 1878–1953. Amer. journalist & educator.
**Pit·man** (pĭt'mən), Sir **Isaac.** 1813–97. English inventor of shorthand.
**Pit·ney** (pĭt'nē), **Mahlon.** 1858–1924. Amer. jurist.
**Pitt** (pĭt), **William.** *1st Earl of Chatham.* 1708–78. English statesman & orator.
**Pitt, William.** *2nd Earl of Chatham.* 1759–1806. British prime minister (1783–1801, 1804–06).
**Pi·us** (pī'əs). Name of 12 popes, esp.: **a. II.** 1405–64. Reigned 1458–64. **b. V.** 1504–72. Reigned 1566–72. **c. VII.** 1742–1823. Reigned 1800–23. **d. IX.** 1792–1878. Reigned 1846–78. **e. X.** 1835–1914. Reigned 1903–14. **f. XI.** 1857–1939. Reigned 1922–39. **g. XII.** 1876–1958. Reigned 1939–58.
**Pi·zar·ro** (pĭ-zär'ō), **Francisco.** 1470?–1541. Spanish explorer & conquistador.
**Planck** (plängk), **Max Karl Ernst Ludwig.** 1858–1947. German physicist (Nobel, 1918).
**Plan·tag·e·net** (plăn-tăj'ə-nĭt). English ruling dynasty (1154–1485).
**Plath** (plăth), **Sylvia.** 1932–63. Amer. author.
**Pla·to** (plā'tō). 427?–347 B.C. Greek philosopher. **—Pla·ton'ic** (plə-tŏn'ĭk) *adj.*
**Platt** (plăt), **Orville Hitchcock.** 1827–1905. Amer. legislator.
**Platt, Thomas Collier.** 1833–1910. Amer. politician.
**Plau·tus** (plô'təs), **Titus Maccius.** 254?–184 B.C. Roman playwright.
**Play·er** (plā'ər), **Gary.** b. 1935. South African-born Amer. golfer.

**Ple·kha·nov** (plĭ-kä′nôf′, -nôv′), **Georgi Valentinovich.** 1857–1918. Russian revolutionary & political philosopher.
**Plimp·ton** (plĭmp′tən), **George Ames.** b. 1927. Amer. author & editor.
**Plim·soll** (plĭm′səl, -sôl′), **Samuel.** 1824–98. English politician & philanthropist.
**Plin·y** (plĭn′ē). *the Elder.* A.D. 23–79. Roman scholar; died observing the eruption of Vesuvius.
**Pliny.** *the Younger.* A.D. 62–113. Roman consul & author.
**Plo·ti·nus** (plō-tī′nəs). A.D. 205?–70. Egyptian-born Roman philosopher.
**Plu·tarch** (plo͞o′tärk′). A.D. 46?–120? Greek biographer & philosopher. **—Plu·tarch′an** (-tär′kən), **Plu·tarch′i·an** (-tär′kē-ən) *adj.*
**Po·ca·hon·tas** (pō′kə-hŏn′təs). 1595?–1617. Amer. Indian princess; rescued John Smith.
**Pod·gor·ny** (pŏd-gôr′nē), **Nikolai Viktorovich.** 1903–83. Russian statesman.
**Poe** (pō), **Edgar Allan.** 1809–49. Amer. author.
**Poin·ca·ré** (pwăN-kä-rā′), **Jules Henri.** 1854–1912. French mathematician & physicist.
**Poincaré, Raymond.** 1860–1934. French statesman.
**Poin·sett** (poin′sĕt′, -sĭt), **Joel Roberts.** 1779–1851. Amer. diplomat.
**Poi·tier** (pwä′ytā), **Sidney.** b. 1927. Amer. actor & director.
**Pole** (pōl), **Reginald.** 1500–58. English prelate.
**Po·li·tian** (pə-lĭsh′ən, pō-). 1454–94. Italian scholar & poet.
**Polk** (pōk), **James Knox.** 1795–1849. 11th U.S. President (1845–49).
**Polk, Leonidas.** 1806–64. Amer. clergyman & Confederate general.
**Pol·lock** (pŏl′ək), **Channing.** 1880–1946. Amer. author.
**Pollock,** Sir **Frederick.** 1845–1937. English jurist & author.
**Pollock, Jackson.** 1912–56. Amer. artist.
**Pollock, Oliver.** 1737?–1823. Irish-born Amer. businessman.
**Po·lo** (pō′lō), **Marco.** 1254?–1325? Venetian traveler.
**Pol Pot** (pŏl pŏt′). b. 1928. Cambodian political leader.
**Po·lyb·i·us** (pə-lĭb′ē-əs). 205?–125? B.C. Greek historian.
**Pol·y·carp** (pŏl′ē-kärp′), Saint. A.D. 69?–155? Christian martyr.
**Pol·y·cli·tus** *or* **Pol·y·clei·tus** (pŏl′ĭ-klī′təs). 5th cent. B.C. Greek sculptor & architect.
**Po·lyc·ra·tes** (pə-lĭk′rə-tēz′). 535?–515 B.C. Tyrant of Samos.
**Pol·yg·no·tus** (pŏl′ĭg-nō′təs). 5th cent. B.C. Greek painter.
**Pom·pa·dour** (pŏm′pə-dôr′, -dōr′, -do͝or′), Marquise de. 1721–64. Mistress of Louis XV of France.
**Pom·pey** (pŏm′pē). *the Great.* 106–48 B.C. Roman general & statesman.
**Pom·pi·dou** (pŏm′pĭ-do͞o′), **Georges Jean Raymond.** 1911–74. French statesman.
**Ponce de Le·ón** (pŏns′ də lē′ən, lē-ōn′), **Juan.** 1460–1521. Spanish explorer.
**Pon·chiel·li** (pông′kē-ĕl′ē), **Amilcare.** 1834–86. Italian composer.
**Pons** (pôNs, pŏnz), **Lily.** 1904–76. French-born Amer. operatic soprano.
**Pon·selle** (pŏn-sĕl′), **Rosa Melba.** 1897–1981. Amer. soprano.
**Pon·ti·ac** (pŏn′tē-ăk′). 1720?–69. Amer. Ottawa chief.
**Pon·tius Pi·late** (pŏn′chəs pī′lət). —See PILATE.
**Pon·top·pi·dan** (pŏn-tŏp′ĭ-dăn′), **Henrik.** 1857–1943. Danish novelist (Nobel, 1917).
**Pool** (po͞ol), **Maria Louise.** 1841–98. Amer. author.
**Poor** (po͝or), **Henry Varnum.** 1812–1905. Amer. historian & economist.
**Pope** (pōp), **Albert Augustus.** 1843–1909. Amer. bicycle manufacturer.
**Pope, Alexander.** 1688–1744. English poet & satirist.
**Pope, John.** 1822–92. Amer. general.
**Por·son** (pôr′sən), **Richard.** 1759–1808. English classical scholar.
**Por·ter** (pôr′tər), **Cole Albert.** 1891?–1964. Amer. composer.
**Porter, David.** 1780–1843. Amer. naval officer.
**Porter, David Dixon.** 1813–91. Amer. naval officer.
**Porter, Eleanor Hodgman.** 1868–1920. Amer. author.
**Porter, Gene Stratton.** 1863?–1924. Amer. author.
**Porter,** Sir **George.** b. 1920. English chemist (Nobel, 1967).
**Porter, Katherine Anne.** 1890–1980. Amer. author.
**Porter, Noah.** 1811–92. Amer. philosopher & lexicographer.
**Porter, Rodney Robert.** b. 1917. English biochemist (Nobel, 1972).
**Porter, Sarah.** 1813–1900. Amer. educator.
**Porter, Sylvia Field.** b. 1913. Amer. economist & journalist.
**Porter, William Sidney.** *O. Henry.* 1862–1910. Amer. author.
**Por·to·lá** (pôr-tō-lä′), **Gaspar de.** 1723?–84? Spanish explorer & colonial administrator.
**Post** (pōst), **Charles William.** 1854–1914. Amer. breakfast-food manufacturer.
**Post, Christian Frederick.** 1710?–85. German-born Amer. missionary.
**Post, Emily Price.** 1872–1960. Amer. etiquette columnist.
**Post, Wiley.** 1899–1935. Amer. aviator.
**Po·tem·kin** (pō-tĕm′kĭn, pə-), **Grigori Aleksandrovich.** 1739–91. Russian army officer & politician.
**Po·tok** (pō′tŏk′), **Chaim.** b. 1929. Amer. author.
**Pot·ter** (pŏt′ər), **Beatrix.** 1866–1943. English author & illustrator.
**Potter, Henry Codman.** 1835–1908. Amer. prelate.
**Potter, Paul.** 1625–54. Dutch painter.
**Pou·lenc** (po͞o-lăNk′), **Francis.** 1899–1963. French composer.
**Pound** (pound), **Ezra Loomis.** 1885–1972. Amer. poet & critic.
**Pound, Roscoe.** 1870–1964. Amer. botanist & jurist.
**Pous·sin** (po͞o-săN′), **Nicolas.** 1594–1665. French painter.
**Pow·der·ly** (pou′dər-lē), **Terence Vincent.** 1849–1924. Amer. labor leader.
**Pow·ell** (pou′əl), **Adam Clayton, Jr.** 1908–72. Amer. clergyman & politician.
**Powell, Anthony.** b. 1905. English author.
**Powell, Cecil Frank.** 1903–69. English physicist (Nobel, 1950).
**Powell, John Wesley.** 1834–1902. Amer. geologist & explorer.
**Powell, Lewis Franklin, Jr.** b. 1907. Amer. jurist.
**Powell, Maud.** 1868–1920. Amer. violinist.
**Pow·er** (pou′ər), **Tyrone.** 1914–58. Amer. actor.
**Pow·ers** (pou′ərz), **Hiram.** 1805–73. Amer. sculptor.
**Pow·ha·tan** (pou′ə-tăn′, pou-hăt′n). 1550?–1618. Amer. Indian chief.
**Pow·ys** (pō′ĭs). Family of English authors, including **John Cowper** (1872–1963), **Theodore Francis** (1875–1953), & **Llewelyn** (1884–1939).
**Pra·do U·gar·te·che** (prä′dō o͞o′gär-tā′chē), **Manuel.** 1889–1967. Peruvian statesman.
**Pra·ja·dhi·pok** (prə-chä′tĭ-pŏk′). 1893–1941. Siamese king (1925–35).
**Pratt** (prăt), **Bela Lyon.** 1867–1917. Amer. sculptor.
**Pratt, Edwin John.** 1883–1964. Canadian poet.
**Prax·it·e·les** (prăk′sĭt′l-ēz′). 400–320 B.C. Greek sculptor.
**Preb·le** (prĕb′əl), **Edward.** 1761–1807. Amer. naval officer.
**Pre·gl** (prā′gəl), **Fritz.** 1869–1930. Austrian chemist (Nobel, 1923).
**Pre·log** (prĕl′ŏg′), **Vladimar.** b. 1906. Yugoslavian-born Swiss chemist (Nobel, 1975).
**Prem·in·ger** (prĕm′ĭn-jər), **Otto Ludwig.** b. 1906. Austrian-born Amer. producer & director.
**Pren·tiss** (prĕn′tĭs), **Elizabeth Payson.** 1818–78. Amer. author.
**Pres·cott** (prĕs′kət, -kŏt′), **Samuel.** 1751–77? Amer. physician & Revolutionary patriot.
**Prescott, William.** 1726–95. Amer. Revolutionary commander.
**Prescott, William Hickling.** 1796–1859. Amer. historian.
**Pres·ley** (prĕs′lē, prĕz′-), **Elvis Aron.** 1935–77. Amer. entertainer.
**Pres·ton** (prĕs′tən), **Ann.** 1813–72. Amer. pioneer physician & educator.
**Pre·to·ri·us** (prĭ-tôr′ē-əs, -tōr′-), **Andries Wilhelmus Jacobus.** 1799–1853. Dutch South African leader.
**Pretorius, Marthinus Wessels.** 1819–1901. South African statesman.
**Pré·vost d'Ex·iles** (prā-vō′ dĕg-zēl′), **Antoine Françoise.** *Abbé Prévost.* 1697–1763. French novelist.
**Price** (prīs), **(Mary) Leontyne.** b. 1927. Amer. operatic soprano.
**Price, Vincent.** b. 1911. Amer. actor.
**Pride** (prīd), **Thomas.** d. 1658. English Parliamentary commander.
**Priest·ley** (prēst′lē), **John Boynton ("J.B.").** 1894–1984. English author.
**Priestley, Joseph.** 1733–1804. English clergyman & chemist.
**Pri·go·gine** (prĭ-gō′zhən, -gô-zhēn′), **Ilya.** b. 1917. Russian-born Belgian chemist (Nobel, 1977).
**Pri·mo de Ri·ve·ra** (prē′mō dā rĭ-vĕr′ə), **José Antonio.** 1903–36. Spanish fascist politician; executed.
**Prince** (prĭns), **Harold.** b. 1928. Amer. stage producer.
**Pri·or** (prī′ər), **Matthew.** 1664–1721. English poet & diplomat.
**Pris·cian** (prĭsh′ən, -ē-ən). fl. A.D. 500. Latin grammarian.
**Pritch·ett** (prĭch′ĭt), **Victor Sawdon ("V.S.").** b. 1900. English author & critic.
**Pro·clus** (prō′kləs, prŏk′ləs). 410?–85. Greek philosopher.
**Pro·co·pi·us** (prə-kō′pē-əs). 6th cent. A.D. Byzantine historian.
**Pro·kho·rov** (prô′KHə-rôf′), **Aleksander Mikailovich.** b. 1916. Russian physicist (Nobel, 1964).
**Pro·kof·iev** (prə-kôf′yəf, -yĕf′, -yĕv′), **Sergei Sergeevich.** 1891–1953. Russian composer.
**Pro·per·tius** (prō-pûr′shəs, -shē-əs), **Sextus.** 50?–15? B.C. Roman poet.
**Pro·tag·o·ras** (prō-tăg′ər-əs). 5th cent. B.C. Greek philosopher. **—Pro·tag′o·re′an** (-ə-rē′ən) *adj.*
**Prou·dhon** (pro͞o-dôN′), **Pierre Joseph.** 1809–65. French utopian socialist.
**Proust** (pro͞ost), **Marcel.** 1871–1922. French author. **—Proust′i·an** *adj.*

**Prynne** (prĭn), **William.** 1600–69. English statesman & pamphleteer.

**Przhe·val·ski** (pûr′zhə-väl′skē, pshə′väl′-, shə′-), **Nikolai Mikhailovoch.** 1839–88. Russian explorer.

**Ptol·e·my** (tŏl′ə-mē). 2nd cent. A.D. Greek astronomer & geographer. **—Ptol′e·ma′ic** (tŏl′ə-mā′ĭk) *adj.*

**Ptolemy.** Name of 14 kings of Egypt, esp. **I,** 367?–283 B.C., reigned 305–285 B.C. **—Ptol′e·ma′ic** (tŏl′ə-mā′ĭk) *adj.*

**Puc·ci** (po͞o′chē), **Emilio.** *Marchese di Barsento.* b. 1914. Italian fashion designer.

**Puc·ci·ni** (po͞o-chē′nē), **Giacomo.** 1858–1924. Italian composer.

**Pu·las·ki** (po͞o-lăs′kē, pə-), **Casimir** or **Kazimierz.** 1748?–79. Polish patriot & general in America.

**Pu·lit·zer** (po͝ol′ĭt-sər, pyo͞o′lĭt-), **Joseph.** 1847–1911. Hungarian-born Amer. journalist & publisher.

**Pull·man** (po͝ol′mən), **George Mortimer.** 1831–97. Amer. industrialist & inventor.

**Pum·pel·ly** (pŭm-pĕl′ē), **Raphael.** 1837–1923. Amer. geologist & explorer.

**Pu·pin** (pyo͞o-pēn′, po͞o-), **Michael Idvorsky.** 1858–1935. Yugoslavian-born Amer. physicist.

**Pur·cell** (pûr-sĕl′), **Edward Mills.** b. 1912. Amer. physicist (Nobel, 1952).

**Pur·cell** (pûr′səl, pûr-sĕl′), **Henry.** 1658?–95. English composer.

**Pur·chas** (pûr′chəs), **Samuel.** 1575?–1626. English clergyman & compiler.

**Pur·kin·je** (pər-kĭn′jē, po͝or′kən-yā′), **Johannes Evangelista.** 1787?–1869. Czechoslovakian physiologist & poet.

**Pu·sey** (pyo͞o′zē), **Edward Bouverie.** 1800–82. English theologian.

**Push·kin** (po͝osh′kĭn, po͝osh′-), **Aleksander Sergeevich.** 1799–1837. Russian author.

**Put·nam** (pŭt′nəm), **Israel.** 1718–90. Amer. general.

**Putnam, Rufus.** 1738–1824. Amer. Revolutionary officer & Western pioneer.

**Pu·vis de Cha·vannes** (pyo͞o-vē′ də shä-văn′, -vēs′), **Pierre.** 1824–98. French artist.

**Pye** (pī), **Henry James.** 1745–1813. English poet.

**Pyle** (pīl), **Ernest Taylor ("Ernie").** 1900–45. Amer. journalist.

**Pyle, Howard.** 1853–1911. Amer. artist & author.

**Pym** (pĭm), **John.** 1584–1643. English Parliamentary leader.

**Py·thag·o·ras** (pĭ-thăg′ər-əs). d. 497 B.C. Greek philosopher & mathematician. **—Py·thag′o·re′an** (-ə-rē′ən) *adj.*

# Q

**Qad·da·fi** (kə-dä′fē), **Muammar.** b. 1943. Libyan political leader.

**Qi** (chē). Name of two Chinese dynasties (479–502, 550–77).

**Qiang·long** (chē-ĕn′lo͝ong′). 1711–99. Chinese emperor (1736–96).

**Qin** (chĭn). Chinese dynasty (221–07 B.C.).

**Qing** (chĭng). Chinese dynasty (1644–1911).

**Qua·dros** (kwä′drōs′), **Janio de Silva.** b. 1917. Brazilian statesman.

**Quan·trill** (kwŏn′trĭl′), **William Clarke.** 1837–65. Amer. Confederate guerrilla leader.

**Quarles** (kwärlz, kwôrlz), **Francis.** 1592–1644. English poet.

**Qua·si·mo·do** (kwä′zē-mō′dō), **Salvatore.** 1901–68. Italian author (Nobel, 1959).

**Quay** (kwā), **Matthew Stanley.** 1833–1904. Amer. politician.

**Queen** (kwēn), **Ellery.** Frederic DANNAY & Manfred B. LEE.

**Queens·ber·ry** (kwēnz′bĕr′ē, -bə-rē), 8th Marquis of. John Sholto DOUGLAS.

**Que·ler** (kwĕl′ər), **Eve.** b. 1936. Amer. conductor.

**Ques·nay** (kā-nā′), **François.** 1694–1774. French physician & political economist.

**Que·zon y Mo·li·na** (kā′sôn′ ē mə-lē′nə), **Manuel Luis.** 1878–1944. Philippine statesman.

**Quid·de** (kvĭd′ə, kfĭd′ə), **Ludwig.** 1858–1941. German politician & pacifist (Nobel, 1927).

**Quil·ler-Couch** (kwĭl′ər-ko͞och′), Sir **Arthur Thomas.** 1863–1944. English author, editor, & educator.

**Quin·cy** (kwĭn′zē, -sē), **Josiah.** 1744–75. Amer. Revolutionary patriot.

**Quincy, Josiah.** 1772–1864. Amer. politician, educator, & historian.

**Quin·te·ro** (kēn-tā′rō), **Serafín Álvarez** (1871–1938) & **Joaquín Álvarez** (1873–1944). Spanish dramatists.

**Quin·til·ian** (kwĭn-tĭl′yən, -ē-ən). A.D. 35?–95? Roman rhetorician.

**Qui·ri·no** (kĭ-rē′nō), **Elpidio.** 1890?–1956. Philippine statesman.

**Quis·ling** (kwĭz′lĭng), **Vidkun Abraham Lauritz.** 1887–1945. Norwegian army officer & political leader; executed for treason.

# R

**Ra·be·lais** (răb′ə-lā′, răb′ə-lā′), **François.** 1494?–1553. French humanist. **—Rab′e·lai′se·an** (-zē-ən, -zhən) *adj.*

**Ra·bi** (rä′bē), **Isidor Isaac.** b. 1898. Austrian-born Amer. physicist (Nobel, 1944).

**Ra·bin** (rä-bēn′), **Itzhak** or **Yitzhak.** b. 1922. Israeli military & political leader.

**Ra·chel** (rā′chəl). Second wife of Jacob in the Old Testament.

**Ra·chel** (rä-shĕl′), **Élisa Félix.** 1820–58. Swiss-born French actress.

**Rach·ma·ni·noff** (rŏk-măn′ə-nôf′, -nôv′, räk-mä′nə-), **Sergei Vasilievich.** 1873–1943. Russian-born composer & pianist.

**Ra·cine** (rə-sēn′, rä-), **Jean Baptiste.** 1639–99. French playwright.

**Rack·ham** (răk′əm), **Arthur.** 1867–1939. English artist.

**Rad·cliffe** (răd′klĭf′), **Ann Ward.** 1764–1823. English novelist.

**Rad·is·son** (rä-dē-sôN′), **Pierre Esprit.** 1636?–1710? French explorer.

**Rae** (rā), **John.** 1813–93. English Arctic explorer.

**Rae·burn** (rā′bərn), Sir **Henry.** 1756–1823. Scottish portrait painter.

**Rae·der** (rā′dər), **Erich.** 1876–1960. German admiral.

**Rae·mae·kers** (rä′mä-kərz, -kərs), **Louis.** 1869–1956. Dutch political cartoonist.

**Raf·fles** (răf′əlz), Sir **Thomas.** 1781–1826. English colonial administrator.

**Ra·fi·nesque** (rä-fē-nĕsk′) *also* **Ra·fi·nesque-Schmaltz** (-shmälts), **Constantine Samuel.** 1783–1840. French-Amer. naturalist.

**Raft** (răft), **George.** 1895–1980. Amer. actor.

**Rag·lan** (răg′lən), 1st Baron. *Fitzroy James Henry Somerset.* 1788–1855. English field marshal.

**Rai·mon·di** (rī-mōn′dē, -mŏn′-), **Marcantonio.** 1475?–1534? Italian engraver.

**Rai·ney** (rā′nē), **Gertrude Malissa Nix Pridgett ("Ma").** 1886–1939. Amer. blues singer.

**Rainey, Henry Thomas.** 1860–1934. Amer. legislator.

**Rai·nier III** (rā-nîr′, rĕ-, rə-, rĕ-nyā′). b. 1923. Prince of Monaco (since 1949).

**Rains** (rānz), **Claude.** 1899–1967. Amer. actor.

**Rain·wat·er** (rān′wô′tər, -wŏt′ər), **L(eo) James.** b. 1917. Amer. physicist (Nobel, 1975).

**Ra·ja·go·pa·la·cha·ria** (rä′jə-gō-pä′lä-chär′yə), **Chakravarti.** 1879–1972. Indian politician.

**Ra·leigh** or **Ra·legh** (rô′lē, rŏl′ē), Sir **Walter.** 1552?–1618. English navigator, courtier, writer, & colonizer; executed.

**Ra·man** (rä′mən), Sir **Chandrasekhara.** 1888–1970. Indian physicist (Nobel, 1930).

**Ra·meau** (rä-mō′), **Jean Philippe.** 1683–1764. French composer.

**Ra·meé** (rə-mā′), **Marie Louise de la.** *Ouida.* 1839–1908. English author.

**Ram·e·ses** (răm′ĭ-sēz′) *or* **Ram·ses** (răm′sēz′). Name of 12 kings of Egypt, esp.: **a. II.** Ruled 1292–1225 B.C. **b. III.** Ruled 1198–1167 B.C.

**Ram·say** (răm′zē), **Allan.** 1686–1758. Scottish poet.

**Ramsay,** Sir **William.** 1852–1916. British chemist (Nobel, 1904).

**Ram·ses** (răm′sēz′). *var. of* RAMESES.

**Ram·sey** (răm′zē), **Arthur Michael.** b. 1904. English prelate.

**Rand** (rănd), **Ayn.** 1905–82. Russian-born Amer. author.

**Rand, Sally.** 1904–79. Amer. entertainer.

**Rand·all** (răn′dl), **Samuel Jackson.** 1828–90. Amer. legislator.

**Ran·dolph** (răn′dŏlf′), **Edmund Jennings.** 1753–1813. Amer. Revolutionary leader & public official.

**Randolph, John.** *Randolph of Roanoke.* 1773–1833. Amer. politician & orator.

**Ran·jit Singh** (rŭn′jĭt sĭng′). *Lion of the Punjab.* 1780–1839. Founder of the Sikh kingdom.

**Ran·ke** (räng′kə), **Leopold von.** 1795–1886. German historian.

**Ran·kin** (răng′kĭn), **Jeannette.** 1880–1973. Amer. feminist & legislator.

**Ran·som** (răn′səm), **John Crowe.** 1888–1974. Amer. educator, author, & editor.

**Raph·a·el** (răf′ē-əl, rā′fē-, rä′-). 1483–1520. Italian artist.

**Rapp** (răp, räp), **George.** 1757–1847. German-born Amer. separatist colonizer.

**Rask** (räsk, răsk), **Rasmus Christian.** 1787–1832. Danish philologist & Orientalist.

**Ras·mus·sen** (răs′mə-sən, räs′mo͝os-ən), **Knud Johan Victor.** 1879–1933. Danish anthologist & Arctic explorer.

**Ras·pu·tin** (răs-pyo͞o′tĭn, -po͞o′-), **Grigori Efimovich.** 1871?–1916. Russian monk influential in czarist court; assassinated.

**Rath·bone** (răth′bōn′), **Basil.** 1892–1967. South-African-born English actor.

**Rath·er** (răth′ər), **Dan.** b. 1931. Amer. broadcast journalist.

**Rat·ti·gan** (răt′ĭ-gən), **Terence.** 1911–78. English playwright.

**Rausch·en·berg** (rou′shən-bûrg′), **Robert.** b. 1925. Amer. painter.

**Rau·schen·busch** (rou'shən-bōōsh'), **Walter.** 1861–1918. Amer. theologian & educator.

**Rausch·ning** (roush'nĭng), **Hermann.** b. 1887. German-born Amer. author.

**Ra·vel** (rə-vĕl', rä-), **Maurice Joseph.** 1875–1937. French composer.

**Raw·lings** (rô'lĭngz), **Marjorie Kinnan.** 1896–1953. Amer. author.

**Raw·lin·son** (rô'lĭn-sən), **George.** 1812–1902. English historian & Orientalist.

**Rawlinson,** Sir **Henry Creswicke.** 1810–95. English diplomat & scholar.

**Ray** (rā), **Charlotte E.** 1850–1911. Amer. lawyer.

**Ray, John.** 1627?–1705. English pioneer naturalist.

**Ray, Man.** 1890–1976. Amer. surrealist painter & photographer.

**Ray·burn** (rā'bûrn'), **Samuel Taliaferro.** 1882–1961. Amer. legislator.

**Ray·leigh** (rā'lē), 3rd Baron. *John William Strutt.* 1842–1919. English physicist (Nobel, 1904).

**Ray·mond** (rā'mənd), **Henry Jarvis.** 1820–69. Amer. journalist & politician.

**Read** (rēd), **George.** 1733–98. Amer. Revolutionary leader, politician, & jurist.

**Read,** Sir **Herbert.** 1893–1968. English art curator, poet, editor, & critic.

**Read, Opie.** 1852–1939. Amer. author & editor.

**Read, Thomas Buchanan.** 1822–72. Amer. poet & painter.

**Reade** (rēd), **Charles.** 1814–84. English author.

**Read·ing** (rĕd'ĭng), 1st Marquis of. *Rufus Daniel Isaacs.* 1860–1935. English politician, diplomat, & colonial administrator.

**Rea·gan** (rā'gən), **Ronald Wilson.** b. 1911. Actor & 40th U.S. President (since 1981).

**Ré·au·mur** (rā'ə-myŏŏr', -ō-), **René Antoine Ferchault de.** 1683–1757. French physicist.

**Re·bec·ca** *also* **Re·bek·ah** (rĭ-bĕk'ə). Wife of Isaac & mother of Jacob & Esau in the Old Testament.

**Ré·ca·mi·er** (rā-kăm'ē-ā, rā-käm-yā'), **Jeanne Françoise Julie Adélaïde Bernard.** 1777–1849. French socialite.

**Red Cloud** (rĕd' kloud'). 1822–1909. Amer. Oglala Sioux Indian leader.

**Red·grave** (rĕd'grāv'). Family of English actors, including Sir **Michael** (1908–85), **Vanessa** (b. 1937), & **Lynn** (b. 1943).

**Red Jack·et** (jăk'ĭt). 1756?–1830. Amer. Seneca Indian leader.

**Red·mond** (rĕd'mənd), **John Edward.** 1851?–1918. Irish nationalist leader.

**Re·don** (rə-dôɴ'), **Odilon.** 1840–1916. French artist.

**Reed** (rēd), **John.** 1887–1920. Amer. journalist.

**Reed, Myrtle.** 1874–1911. Amer. novelist.

**Reed, Stanley Forman.** 1884–1980. Amer. jurist.

**Reed, Thomas Brackett.** 1839–1902. Amer. legislator.

**Reed, Walter.** 1851–1902. Amer. physician & army surgeon.

**Reese** (rēs), **Harold ("Pee Wee").** b. 1919. Amer. baseball player.

**Reese, Lizette Woodworth.** 1856–1935. Amer. poet & educator.

**Reeve** (rēv), **Tapping.** 1744–1823. Amer. jurist & educator.

**Re·gi·o·mon·ta·nus** (rē'jē-ō-mŏn-tā'nəs). Johann MÜLLER.

**Reg·u·lus** (rĕg'yə-ləs), **Marcus Atilius.** d. c. 250 B.C. Roman general & politician.

**Re·han** (rē'ən), **Ada.** 1860–1916. Irish-born Amer. actress.

**Rehn·quist** (rĕn'kwĭst'), **William Hubbs.** b. 1924. Amer. jurist.

**Reich·stein** (rīk'stīn', -shtīn'), **Tadeus.** b. 1897. Polish-born Swiss chemist (Nobel, 1950).

**Reid** (rēd), **Ogden Mills.** 1882–1947. Amer. editor & publisher.

**Reid, Thomas.** 1710–96. Scottish philosopher.

**Reid, Whitelaw.** 1837–1912. Amer. editor & diplomat.

**Rei·nach** (rā-näk', rə-), **Salomon.** 1858–1932. French archaeologist.

**Rei·ner** (rī'nər), **Fritz.** 1888–1963. Hungarian-born Amer. conductor.

**Rein·hardt** (rīn'härt'), **Max.** 1873–1943. German theatrical director & manager.

**Re·marque** (rə-märk'), **Erich Maria.** 1898–1970. German-born Amer. novelist.

**Rem·brandt van Rijn** *or* **van Ryn** (rĕm'brănt' vän rīn', -bränt'). 1606–69. Dutch painter.

**Rem·ing·ton** (rĕm'ĭng-tən), **Eliphalet.** 1793–1861. Amer. firearms manufacturer.

**Remington, Frederic.** 1861–1909. Amer. painter, sculptor, & journalist.

**Rem·sen** (rĕm'sən), **Ira.** 1846–1927. Amer. chemist.

**Re·nan** (rə-näɴ'), **Joseph Ernest.** 1823–92. French philologist, philosopher, & historian.

**Re·nault** (rə-nō'), **Jean Louis.** 1843–1918. French jurist (Nobel, 1907).

**Re·ni** (rā'nē), **Guido.** 1575–1642. Italian painter.

**Ren·ner** (rĕn'ər), **Karl.** 1870–1950. Austrian political leader.

**Ren·oir** (rĕn'wär', rən-wär'), **Jean.** 1894–1979. French filmmaker.

**Renoir, Pierre Auguste.** 1841–1919. French painter.

**Ren·wick** (rĕn'wĭk), **James.** 1818–95. Amer. architect.

**Rep·plier** (rĕp'lîr'), **Agnes.** 1855–1950. Amer. essayist.

**Re·spi·ghi** (rə-spē'gē), **Ottorino.** 1879–1936. Italian composer.

**Res·ton** (rĕs'tən), **James Barrett ("Scotty").** b. 1909. Scottish-born Amer. journalist.

**Retz** (rĕts), Cardinal de. *Jean François Paul de Gondi.* 1614–79. French politician & author.

**Reuch·lin** (roik'lən, roiKH'lĕn, roiKH-lēn'), **Johann.** 1455–1522. German humanist.

**Reu·ter** (roi'tər), Baron **Paul Julius von.** 1816–99. German-born English news-agency founder.

**Reu·ter·dahl** (roi'tər-däl'), **Henry.** 1871–1925. Swedish-born Amer. marine painter.

**Reu·ther** (rōō'thər), **Walter Philip.** 1907–70. Amer. labor leader.

**Re·vere** (rĭ-vîr'), **Paul.** 1735–1818. Amer. silversmith, engraver, & Revolutionary patriot.

**Rev·son** (rĕv'sən), **Charles.** 1906–75. Amer. cosmetics tycoon.

**Rex·roth** (rĕks'rôth'), **Kenneth.** 1905–82. Amer. painter & author.

**Rey·mont** (rā'mŏnt'), **Wladyslaw Stanislaw.** 1867–1925. Polish author (Nobel, 1924).

**Rey·naud** (rā-nō'), **Paul.** 1878–1966. French statesman.

**Reyn·olds** (rĕn'əldz), Sir **Joshua.** 1723–92. English portrait painter.

**Rhee** (rē), **Syngman.** 1875–1965. Korean political leader.

**Rhett** (rĕt), **Robert Barnwell.** 1800–76. Amer. legislator & secessionist.

**Rhine** (rīn), **Joseph Banks.** b. 1895. Amer. psychologist.

**Rhodes** (rōdz), **Cecil John.** 1853–1902. English financier & colonizer.

**Rhodes, James Ford.** 1848–1927. Amer. historian.

**Rhond·da** (rŏn'də, *-thə*), Viscount. *David Alfred Thomas.* 1856–1918. Welsh coal magnate & politician.

**Rib·ben·trop** (rĭb'ən-trŏp', -trōp'), **Joachim von.** 1893–1946. German Nazi diplomat; executed.

**Ri·be·ra** (rē-bĕr'ə), **Jusepe** or **José de.** *Spagnoletta.* 1588–1656? Spanish painter.

**Ri·car·do** (rĭ-kär'dō), **David.** 1772–1823. English economist.

**Rice** (rīs), **Alice Caldwell Hegan.** 1870–1942. Amer. author.

**Rice, Dan.** 1823–1900. Amer. circus clown.

**Rice, Elmer Leopold.** 1892–1967. Amer. author.

**Rice, Grantland.** 1880–1954. Amer. journalist.

**Rice, Thomas Dartmouth.** 1808–60. Amer. minstrel entertainer.

**Rich·ard** (rĭch'ərd). Name of three English kings: **a. I.** *Coeur de Lion, the Lion-Hearted.* 1157–99. Ruled 1189–99. **b. II.** 1367–1400. Ruled 1377–99. **c. III.** 1452–85. Ruled 1483–85.

**Rich·ards** (rĭch'ərdz), **Dickinson Woodruff.** 1895–1973. Amer. physician (Nobel, 1956).

**Richards, Ellen Henrietta Swallow.** 1842–1911. Amer. chemist.

**Richards, Ivor Armstrong ("I.A.").** 1893–1979. English literary critic.

**Richards, Laura Elizabeth Howe.** 1850–1943. Amer. author.

**Richards, Theodore William.** 1868–1928. Amer. chemist (Nobel, 1914).

**Rich·ard·son** (rĭch'ərd-sən), **Henry Hobson.** 1838–86. Amer. architect.

**Richardson,** Sir **Owen Willans.** 1879–1959. English physicist (Nobel, 1928).

**Richardson,** Sir **Ralph David.** 1902–83. English actor.

**Richardson, Samuel.** 1689–1761. English author.

**Ri·che·lieu** (rĭsh'ə-lōō', -lyōō', rē-shə-lyœ'), Duc de. *Armand Jean du Plessis.* 1585–1642. French prelate & political leader.

**Ri·chet** (rē-shā', rĭ-), **Charles Robert.** 1850–1935. French physiologist (Nobel, 1913).

**Rich·ler** (rĭch'lər), **Mordecai.** b. 1928. Canadian author.

**Rich·mond** (rĭch'mənd), **Mary Ellen.** 1861–1928. Amer. pioneer social worker.

**Rich·ter** (rĭk'tər), **Burton.** b. 1931. Amer. physicist (Nobel, 1976).

**Richter, Charles Francis.** b. 1900. Amer. seismologist.

**Rich·ter** (rĭk'tər, rĭKH'-), **Jean Paul Friedrich.** 1763–1825. German satirist.

**Rick·ard** (rĭk'ərd), **George Lewis ("Tex").** 1871–1929. Amer. boxing promoter.

**Rick·en·back·er** (rĭk'ĭn-băk'ər), **Edward Vernon ("Eddie").** 1890–1973. Amer. aviator & businessman.

**Rick·ert** (rĭk'ərt), **Edith.** 1871–1938. Amer. educator & author.

**Rick·ey** (rĭk'ē), **Branch Wesley.** 1881–1965. Amer. baseball executive.

**Rick·o·ver** (rĭk'ō'vər), **Hyman George.** b. 1900. Amer. admiral.

**Rid·dle** (rĭd'l), **Nelson.** b. 1921. Amer. composer, conductor, & arranger.

**Ridg·way** (rĭj'wā'), **Matthew Bunker.** b. 1895. Amer. army officer.

**Rid·ley** (rĭd'lē), **Nicholas.** 1500?–55. English prelate & martyr.

ă pat ā pay âr care ä father ĕ pet ē be hw which ĭ pit
ī tie îr pier ŏ pot ō toe ô paw, for oi noise ŏŏ took

**Ri·el** (rē-ĕl'), **Louis.** 1844–85. French-Canadian rebel; executed.
**Rie·mann** (rē'män'), **Georg Friedrich Bernhard.** 1826–66. German mathematician & educator.
**Ri·en·zi** (rē-ĕn'zē) *or* **Ri·en·zo** (-zō), **Cola di.** 1313?–54. Italian revolutionary leader.
**Ries·man** (rēs'mən), **David, Jr.** b. 1909. Amer. sociologist.
**Riis** (rēs), **Jacob August.** 1849–1914. Danish-born Amer. journalist & reformer.
**Ri·ley** (rī'lē), **James Whitcomb.** 1849–1916. Amer. poet.
**Ril·ke** (rĭl'kə), **Rainer Maria.** 1875–1926. German-Austrian poet.
**Rim·baud** (răm-bō', răN-), **Jean Nicholas Arthur.** 1854–91. French poet.
**Rim·i·ni** (rĭm'ĭ-nē), **Francesca da.** —See FRANCESCA DA RIMINI.
**Rim·mer** (rĭm'ər), **William.** 1816–79. English-born Amer. artist.
**Rim·ski-Kor·sa·kov** *or* **Rim·sky-Kor·sa·kov** (rĭm'skē-kôr'sə-kôf', -kôv'), **Nikolai Andreyevich.** 1844–1908. Russian composer.
**Rin·cón** (rĭng-kōn'), **Antonio del.** 1446–1500. Spanish painter.
**Rine·hart** (rīn'härt'), **Mary Roberts.** 1876–1958. Amer. author.
**Ring·ling** (rĭng'lĭng), **Charles.** 1863–1926. Amer. circus owner.
**Rip·ley** (rĭp'lē), **George.** 1802–80. Amer. minister, scholar, & literary critic.
**Ripley, Robert LeRoy.** 1893–1949. Amer. cartoonist.
**Rit·chard** (rĭ-chärd'), **Cyril.** 1898–1977. Australian-born actor & director.
**Rit·ten·house** (rĭt'n-hous'), **Davis.** 1732–96. Amer. astronomer, mathematician, & public official.
**Rittenhouse, Jessie Belle.** 1869–1948. Amer. poet, anthologist, & critic.
**Rit·ter** (rĭt'ər), **Woodward Maurice ("Tex").** 1905–73. Amer. singer.
**Ri·ve·ra** (rĭ-vĕr'ə), **Diego.** 1886–1957. Mexican artist.
**Rivera y Or·ba·ne·ja** (ē ôr'bə-nā'hä), **Miguel Primo de.** *Marqués de Estella.* 1870–1930. Spanish military & political leader.
**Riv·ers** (rĭv'ərz), **Larry.** b. 1923. Amer. artist.
**Ri·zal** (rĭ-zäl', -säl'), **José.** 1861–96. Philippine national leader.
**Riz·zio** (rĭt'sē-ō'), **David.** 1533?–66. Italian musician & confidant of Mary Queen of Scots.
**Robbe-Gril·let** (rôb-grē-yā'), **Alain.** b. 1922. French author, filmmaker, & agronomist.
**Rob·bins** (rŏb'ĭnz), **Frederick Chapman.** b. 1916. Amer. microbiologist (Nobel, 1954).
**Robbins, Harold.** b. 1916. Amer. author.
**Robbins, Jerome.** b. 1918. Amer. dancer & choreographer.
**Rob·ert** (rŏb'ərt), **Henry Martyn.** 1837–1923. Amer. army officer & parliamentary authority.
**Robert I.** *the Devil.* d. 1035. Duke of Normandy (1028–35).
**Robert I.** *Robert Bruce, the Bruce.* 1274–1329. Scottish king (1306–29).
**Rob·erts** (rŏb'ərts), Sir **Charles George Douglas.** 1860–1943. Canadian author.
**Roberts, Elizabeth Madox.** 1886–1941. Amer. author.
**Roberts, Kenneth.** 1885–1957. Amer. author.
**Roberts, Oral.** b. 1918. Amer. evangelist.
**Roberts, Owen Josephus.** 1875–1955. Amer. jurist.
**Rob·ert·son** (rŏb'ərt-sən), **William.** 1721–93. Scottish historian.
**Robe·son** (rōb'sən), **Paul Bustill.** 1898–1976. Amer. singer & actor.
**Robes·pierre** (rōbz'pîr, -pē-âr', rô-bĕs-pyĕr'), **Maximilien François Marie Isidore de.** 1758–94. French revolutionary; guillotined.
**Rob·in·son** (rŏb'ĭn-sən), **Edward G.** 1893–1973. Amer. actor.
**Robinson, Edwin Arlington.** 1869–1935. Amer. poet.
**Robinson, George Frederick Samuel.** 1827–1909. English politician & colonial administrator.
**Robinson, James Harvey.** 1863–1936. Amer. historian.
**Robinson, John Roosevelt ("Jackie").** 1919–72. Amer. baseball player.
**Robinson, (Esmé Stuart) Lennox.** 1886–1958. Irish author & theatrical manager.
**Robinson, Ray ("Sugar Ray").** b. 1920. Amer. prizefighter.
**Robinson,** Sir **Robert.** 1886–1975. English chemist (Nobel, 1947).
**Robinson, Therese Albertine Louise von Jakob.** 1797–1870. German author.
**Rob·son** (rŏb'sən), **May.** 1858–1942. Amer. actress.
**Ro·cham·beau** (rō'shăm'bō', -shäN-), **Comte de.** *Jean Baptiste Donatien de Vimeure.* 1725–1807. French army officer.
**Rock·e·fel·ler** (rŏk'ə-fĕl'ər). Amer. family including: **a. John Davison.** 1839–1937. Industrialist & philanthropist. **b. John Davison, Jr.** 1874–1960. Philanthropist. **c. Abby Greene Aldrich.** 1874–1948. Philanthropist & art patron. **d. John Davison III.** 1906–78. Philanthropist. **e. Nelson Aldrich.** 1908–79. U.S. Vice President (1974–77). **f. Laurance Spelman.** b. 1910. Conservationist. **g. Winthrop.** 1912–73. Investment manager & politician. **h. David.** b. 1915. Banker.

**Rock·hill** (rŏk'hĭl'), **William Woodville.** 1854–1914. Amer. author & diplomat.
**Rock·ing·ham** (rŏk'ĭng-əm, -həm), 2nd Marquis of. *Charles Watson-Wentworth.* 1730–82. English prime minister (1765–66, 1782).
**Rock·ne** (rŏk'nē), **Knute Kenneth.** 1888–1931. Norwegian-born Amer. football coach.
**Rock·well** (rŏk'wĕl'), **Norman.** 1894–1978. Amer. illustrator.
**Ro·de** (rō'thə), **Helge.** 1870–1937. Danish author.
**Rod·gers** (rŏj'ərz), **John.** 1773–1838. Amer. naval officer.
**Rodgers, Richard.** 1902–79. Amer. composer.
**Ro·din** (rō-dăn', -dăN'), **François Auguste René.** 1840–1917. French sculptor.
**Rod·ney** (rŏd'nē), **George Brydges.** *1st Baron Rodney.* 1719–92. English naval officer.
**Ro·dzin·ski** (rō-jĭn'skē), **Artur.** 1894–1958. Yugoslavian-born Amer. conductor.
**Roeb·ling** (rō'blĭng), **John Augustus** (German-born; 1806–69) & **Washington Augustus** (1837–1926). Amer. engineers.
**Roent·gen** (rĕnt'gən, -jən, rŭnt'-, rĕn'chən, rŭn'-) *or* **Rönt·gen** (rœnt'gən), **Wilhem Konrad.** 1845–1923. German physicist (Nobel, 1901).
**Roe·rich** (rûr'ĭk), **Nicholas Konstantin.** 1874–1947. Russian-born painter & archaeologist.
**Roeth·ke** (rĕt'kē, rĕth'-), **Theodore.** 1908–63. Amer. poet.
**Ro·gers** (rŏj'ərz), **Bruce.** 1870–1957. Amer. book designer & typographer.
**Rogers, Ginger.** b. 1911. Amer. actress.
**Rogers, Harriet Burbank.** 1834–1919. Amer. educator of the deaf.
**Rogers, Henry Huttleston** *or* **Huddleston.** 1840–1909. Amer. business executive.
**Rogers, John.** 1829–1904. Amer. sculptor.
**Rogers, Robert.** 1731–95. Amer. soldier & frontiersman.
**Rogers, Roy.** b. 1912. Amer. singer & actor.
**Rogers, Samuel.** 1763–1855. English poet.
**Rogers, William Penn Adair ("Will").** 1879–1935. Amer. author & actor.
**Ro·get** (rō-zhā', rō'zhā'), **Peter Mark.** 1779–1869. English physician & scholar.
**Ro·kos·sov·ski** (rŏk'ə-sôf'skē, -sôv'-), **Konstantin.** 1896–1968. Russian-Polish army officer & politician.
**Rolf** (rŏlf). *var. of* ROLLO.
**Rolfe** (rŏlf), **John.** 1585–1622. English colonist in America.
**Rol·land** (rô-läN'), **Romain.** 1866–1944. French author (Nobel, 1915).
**Rol·lo** (rŏl'ō). 860?–931? Norse chieftain.
**Röl·vaag** (rōl'väg'), **Ole Edvart.** 1876–1931. Norwegian-born Amer. author.
**Ro·mains** (rō-măN'), **Jules.** 1885–1972. French author.
**Ro·ma·no** (rō-mä'nō), **Giulio.** 1499?–1546. Italian painter.
**Ro·ma·noff** (rō-mä'nəf, rō'mə-nôf', -nŏf). *var. of* ROMANOV.
**Ro·ma·noff** (rō'mə-nôf'), **Michael.** *Prince Mike.* 1892?–1971. Lithuanian-born Amer. impostor & businessman.
**Ro·ma·nov** *also* **Ro·ma·noff** (rō-mä'nəf, rō'mə-nôf', -nŏf). Russian ruling dynasty (1613–1917).
**Rom·bau·er** (rŏm'bou'ər), **Irma von Starkloff.** 1877–1962. Amer. cookery expert.
**Rom·berg** (rŏm'bərg), **Sigmund.** 1887–1951. Hungarian-born Amer. composer.
**Rome** (rōm), **Harold.** b. 1908. Amer. composer.
**Rom·mel** (rŏm'əl), **Erwin.** 1891–1944. German general.
**Rom·ney** (rŏm'nē), **George.** 1734–1802. English painter.
**Romney, George Wilcken.** b. 1907. Amer. businessman & public official.
**Ro·mu·lo** (rŏm'yo͞o-lō'), **Carlos Pena.** b. 1899. Filipino soldier, educator, author, & diplomat.
**Ron·sard** (rôN-sär'), **Pierre de.** 1524–85. French poet.
**Rönt·gen** (rœnt'gən). *var. of* ROENTGEN.
**Roo·ney** (ro͞o'nē), **Mickey.** b. 1920. Amer. actor.
**Roo·se·velt** (rō'zə-vĕlt', rōz'vĕlt', ro͞o'zə-), **(Anna) Eleanor.** 1884–1962. Amer. diplomat, author, & wife of Franklin Delano Roosevelt.
**Roosevelt, Franklin Delano ("FDR").** 1882–1945. 32nd U.S. President (1933–45).
**Roosevelt, Theodore.** 1858–1919. 26th U.S. President (1901–9; Nobel, 1906), soldier, & author.
**Root** (ro͞ot), **Elihu.** 1845–1937. Amer. lawyer & public official (Nobel, 1912).
**Root, John Wellborn.** 1850–91. Amer. architect.
**Ro·rem** (rôr'əm, rōr'-), **Ned.** b. 1923. Amer. composer.
**Ro·sa** (rō'zə), **Salvator.** 1615–73. Italian painter.
**Rose** (rōz), **Billy.** 1899–1966. Amer. showman.
**Rose·ber·y** (rōz'bĕr'ē, -bə-rē), 5th Earl of. *Archibald Philip Primrose.* 1847–1929. British prime minister (1894–95).
**Rose·crans** (rōz'krănz'), **William Starke.** 1819–98. Amer. Union general.
**Ro·sen·bach** (rō'zĭn-băk'), **Abraham Simon Wolf.** 1876–1952. Amer. bibliophile.

**Ro·sen·berg** (rō'zĭn-bûrg', -bĕrg', -bĕrk'), **Alfred.** 1893–1946. German Nazi politician; executed.
**Rosenberg, Anna Marie.** 1902–83. Hungarian-born Amer. public official.
**Rosenberg, Ethel** (1915–53) & **Julius** (1918–53). Amer. spies; executed.
**Ro·sen·wald** (rō'zĭn-wôld'), **Julius.** 1862–1932. Amer. businessman & philanthropist.
**Ross** (rôs), **Betsy Griscom.** 1752–1836. Amer. patriot & legendary maker of first Amer. flag.
**Ross, Harold Wallace.** 1892–1951. Amer. publisher & editor.
**Ross,** Sir **James Clark.** 1800–62. English navigator & polar explorer.
**Ross,** Sir **John.** 1777–1856. English naval officer & Arctic explorer.
**Ross, John.** 1790–1866. Amer. Cherokee Indian leader.
**Ross, Nellie Tayloe.** 1876?–1977. Amer. politician.
**Ross,** Sir **Ronald.** 1857–1932. English physician (Nobel, 1902).
**Ros·sel·li·ni** (rô'sə-lē'nē, rŏs'ə-), **Roberto.** 1906–77. Italian filmmaker.
**Ros·set·ti** (rō-zĕt'ē), **Dante Gabriel** (1828–82) & **Christina Georgina** (1830–94). English pre-Raphaelite poets.
**Ros·si** (rŏs'ē, rôs'ē), **Bruno.** b. 1905. Italian physicist.
**Ros·si·ni** (rō-sē'nē, rə-), **Gioacchino Antonio.** 1792–1868. Italian composer.
**Ros·tand** (rôs-täN'), **Edmond.** 1868–1918. French author.
**Ros·tow** (rŏs'tou'), **Walt Whitman.** b. 1916. Amer. economist.
**Ros·tro·po·vich** (rŏs'trə-pō'vĭch), **Mstislav.** b. 1927. Russian-born musician.
**Roth** (rôth), **Philip Milton.** b. 1933. Amer. novelist.
**Roth·ko** (rŏth'kō), **Mark.** 1903–70. Russian-born Amer. painter.
**Roth·schild** (rŏth'chīld, rŏths'-, rôth'-, rôths'-, rōt'shĭlt'). Family of German bankers, including **Meyer Amschel** (1743–1812), **Salomon** (1774–1855), & **Nathan Meyer** (1777–1836).
**Rou·ault** (ro͞o-ō'), **Georges.** 1871–1958. French artist.
**Rou·get de Lisle** (ro͞o-zhā' də lēl'), **Claude Joseph.** 1760–1836. French soldier & songwriter.
**Rourke** (ro͝ork), **Constance Mayfield.** 1885–1941. Amer. author.
**Rous** (rous), **Francis Peyton.** 1879–1970. Amer. pathologist (Nobel, 1966).
**Rous·seau** (ro͞o-sō'), **Henri.** *Le Douanier Rousseau.* 1844–1910. French painter.
**Rousseau, Jean Jacques.** 1712–78. French author & philosopher.
**Rousseau, Théodore.** 1812–67. French landscape painter.
**Ro·vere** (rō-vîr'), **Richard.** b. 1915. Amer. journalist.
**Rowe** (rō), **Nicholas.** 1674–1718. English author.
**Row·land** (rō'lənd), **Henry Augustus.** 1848–1901. Amer. physicist.
**Row·land·son** (rō'lənd-sən), **Thomas.** 1756–1827. English painter & illustrator.
**Row·ley** (rō'lē), **William.** 1585?–1642? English playwright.
**Row·ling** (rou'lĭng), **Wallace Edward.** b. 1927. New Zealand statesman.
**Row·son** (rou'zən), **Susanna Haswell.** 1762?–1824. English author, actress, & educator.
**Ro·xas y Acu·ña** (rô'häs ē ə-ko͞on'yə), **Manuel.** 1892–1948. Filipino political leader.
**Roy·all** (roi'əl), **Anne Newport.** 1769–1854. Amer. author.
**Royall, Kenneth Claiborne.** 1894–1971. Amer. public official.
**Royce** (rois), **Josiah.** 1855–1916. Amer. educator and philosopher.
**Ro·zelle** (rō-zĕl'), **Alvin Ray ("Pete").** b. 1926. Amer. football commissioner.
**Ru·bens** (ro͞o'bənz), **Peter Paul.** 1577–1640. Flemish painter. **—Ru'ben·esque'** (ro͞o'bə-nĕsk') *adj.*
**Ru·bin·stein** (ro͞o'bĭn-stīn'), **Anton Gregor.** 1829–94. Russian pianist, composer, & educator.
**Rubinstein, Arthur** or **Artur.** 1887–1982. Polish-born Amer. pianist.
**Rubinstein, Helena.** 1871–1965. Polish-born Amer. businesswoman.
**Ru·dolf** (ro͞o'dŏlf'). 1858–89. Austrian crown prince.
**Rudolf I.** 1218–91. Holy Roman Emperor (1273–91) & founder of Hapsburg dynasty.
**Ru·dolph** (ro͞o'dŏlf'), **Wilma Glodean.** b. 1940. Amer. athlete.
**Ruf·fin** (rŭf'ĭn), **Edmund.** 1794–1865. Amer. secessionist, farmer, & publisher.
**Ruis·dael** or **Ruys·dael** (rīz'däl', rīs'-), **Jacob van.** 1628?–82. Dutch painter.
**Ruiz Cor·ti·nes** (ro͞o-ēs' kôr-tē'nəs), **Adolfo.** 1890–1973. Mexican statesman.
**Rum·sey** (rŭm'zē), **James.** 1743–92. Amer. inventor & engineer.
**Rumsey, Mary Harriman.** 1881–1934. Amer. public-welfare pioneer.
**Run·cie** (rŭn'sē), **Robert Alexander Kennedy.** b. 1921. English prelate.
**Rund·stedt** (ro͝on'stĕt', -shtĕt', ro͝ont'-), **Karl Rudolf Gerd von.** 1875–1953. German general.
**Ru·ne·berg** (ro͞o'nə-bûrg'), **Johan Ludvig.** 1804–77. Finnish poet.
**Run·yon** (rŭn'yən), **(Alfred) Damon.** 1884?–1946. Amer. journalist.
**Ru·pert** (ro͞o'pərt), Prince. 1619–82. German-born English military & political leader.
**Ru·rik** (ro͝or'ĭk, ro͞o'rĭk). d. 879. Scandinavian warrior & founder of Russian ruling dynasty (1462–1610).
**Rush** (rŭsh), **Benjamin.** 1745–1813. Amer. physician, politician, author, & educator.
**Rush, Richard.** 1780–1859. Amer. diplomat & public official.
**Rush, William.** 1756–1833. Amer. sculptor & woodcarver.
**Rusk** (rŭsk), **David Dean.** b. 1909. Amer. public official.
**Rus·kin** (rŭs'kĭn), **John.** 1819–1900. English author & critic.
**Rus·sell** (rŭs'əl), **Annie.** 1864–1936. Amer. actress.
**Russell,** Lord **Bertrand Arthur William.** 1872–1970. English pacifist, mathematician, philosopher, & author (Nobel, 1950).
**Russell, Charles Taze.** 1852–1916. Amer. founder of Jehovah's Witnesses.
**Russell, George William.** 1867–1935. Irish author.
**Russell, Henry Norris.** 1877–1957. Amer. astronomer.
**Russell,** Lord **John.** *Viscount Amberly.* 1792–1878. English statesman & political reformer.
**Russell, Lillian.** 1861–1922. Amer. entertainer.
**Russell, Mary Baptist (Katherine).** 1829–98. Irish-born Amer. religious leader.
**Russell, William Felton ("Bill").** b. 1934. Amer. basketball player, coach, & sportscaster.
**Rus·tin** (rŭs'tĭn), **Bayard.** b. 1910. Amer. civil-rights leader.
**Ruth** (ro͞oth). Wife of Boaz in the Bible.
**Ruth, George Herman ("Babe").** 1895–1948. Amer. baseball player.
**Ruth·er·ford** (rŭth'ər-fərd), **Daniel.** 1749–1819. Scottish chemist & physician.
**Rutherford, Ernest.** *1st Baron Rutherford of Nelson.* 1871–1937. New Zealand-born British physicist (Nobel, 1908).
**Rut·ledge** (rŭt'lĭj), **John.** 1739–1800. Amer. jurist & politician.
**Rutledge, Wiley Blount, Jr.** 1894–1949. Amer. jurist.
**Ruys·dael** (rīz'däl', rīs'-). *var. of* RUISDAEL.
**Ru·žič·ka** (ro͞o'zĭch-kə, -zhĭch-), **Leopold.** 1887–1976. Yugoslavian-born Swiss chemist (Nobel, 1939).
**Ry·an** (rī'ən), **Thomas Fortune.** 1851–1928. Amer. financier.
**Ry·der** (rī'dər), **Albert Pinkham.** 1847–1917. Amer. painter.
**Ryle** (rīl), Sir **Martin.** 1918–84. English astronomer & physicist (Nobel, 1974).
**Rys·kind** (rĭs'kĭnd), **Morris.** b. 1895. Amer. playwright.

# S

**Saa·ri·nen** (sär'ə-nən), **Gottlieb Eliel** (1873–1950) & **Eero** (1910–61). Finnish-born Amer. architects.
**Saa·ve·dra La·mas** (sä-vā'drə lä'mas), **Carlos.** 1880?–1959. Argentinean diplomat (Nobel, 1936).
**Sa·ba·tier** (săb'ə-tyā'), **Paul.** 1854–1941. French chemist (Nobel, 1912).
**Sa·ba·ti·ni** (săb'ə-tē'nē, sä'bə-), **Rafael.** 1875–1950. Italian-born English author.
**Sa·bin** (sā'bĭn), **Albert Bruce.** b. 1906. Amer. microbiologist & physician.
**Sabin, Florence Rena.** 1871–1953. Amer. pioneer anatomist.
**Sac·a·ja·we·a** (săk'ə-jə-wē'ə). 1788?–1812. Amer. Indian guide.
**Sac·co** (săk'ō), **Nicola.** 1891–1927. Italian-born Amer. anarchist; executed for murder.
**Sachs** (zäks, säks), **Hans.** 1494–1576. German poet.
**Sachs, Nelly.** 1891–1970. German poet (Nobel, 1966).
**Sack·ville** (săk'vĭl'), **Thomas.** *1st Earl of Dorset, Baron Buckhurst.* 1536–1608. English poet & political adviser.
**Sack·ville-West** (săk'vĭl-wĕst'), **Victoria Mary.** 1892–1962. English author.
**Sa·dat** (sə-dăt', -dät'), **Anwar el-.** 1918–81. Egyptian statesman (Nobel, 1978); assassinated.
**Sade** (säd, săd, săd), Comte **Donatien Alphonse François de.** *Marquis de Sade.* 1740–1814. French author.
**Sad·li·er** (săd'lē-ər), **Mary Anne Madden.** 1820–1903. Irish-born Amer. author.
**Sa·gan** (sā'gən), **Carl.** b. 1934. Amer. astronomer.
**Sa·gan** (sä-gäN'), **Françoise.** b. 1935. French author.
**Sage** (sāj), **Margaret Olivia Slocum.** 1828–1918. Amer. philanthropist.
**Sage, Russell.** 1816–1906. Amer. financier.
**Saint Clair** (sānt klâr', sĭng, sĭn), **Arthur.** 1736–1818. Scottish-born Amer. general.
**Saint-Cyr** (săN-sîr'), Marquis **Laurent de Gouvion.** 1764–1830. French marshal & war minister.
**Saint Denis** (sānt dĕn'ĭs), **Ruth.** 1878–1968. Amer. dancer.

**Sainte-Beuve** (sănt-bœv′), **Charles Augustin.** 1804–69. French poet, critic, & historian.
**Saint-Ex·u·pé·ry** (săN-tĕg-zōō-pā-rē′), **Antoine de.** 1900-44. French author & aviator.
**Saint-Gau·dens** (sānt-gô′dənz), **Augustus.** 1848–1907. Irish-born Amer. sculptor.
**Saint Johns** (sānt jŏnz′, sĭnt), **Adela Rogers.** b. 1894. Amer. journalist.
**Saint-Just** (săN-zhōōst′), **Louis Antoine Léon de.** 1767–94. French revolutionary; guillotined.
**Saint Lau·rent** (săN lô-räN′), **Louis Stephen.** 1882–1973. Canadian prime minister (1948–57).
**Saint Laurent, Yves.** b. 1936. French fashion designer.
**Saint-Saëns** (săN-säNs′), **Charles Camille.** 1835–1921. French composer.
**Saints·bur·y** (sānts′bĕr′ē, -brē), **George Edward Bateman.** 1845–1933. English critic & historian.
**Saint-Si·mon** (săN-sē-mōN′), Duc de. *Louis de Rouvroy.* 1675–1755. French diplomat & author.
**Saint-Simon,** Comte de. *Claude Henri de Rouvroy.* 1760–1825. French philosopher.
**Saint Vrain** (săN vrăN′), **Ceran de Hault de Lassus de.** 1802–70. Amer. pioneer & fur trader.
**Sai·on·ji** (sī-ôn′jē, -ŏn′-), Prince **Kimmochi.** 1849–1940. Japanese statesman.
**Sa·kha·rov** (sä′kə-rôf′, -rôv′), **Andrei Dimitrievich.** b. 1921. Russian physicist & dissident (Nobel, 1975).
**Sa·ki** (sä′kē). Hector Hugh MUNRO.
**Sal·a·din** (săl′ə-dĭn). 1138–93. Sultan of Egypt & Syria.
**Sa·lam** (sä-läm′), **Abdus.** b. 1926. Pakistani physicist (Nobel, 1979).
**Sa·la·zar** (săl′ə-zär′, sä′lə-), **Antonio de Oliveira.** 1889–1970. Portuguese statesman.
**Sal·in·ger** (săl′ĭn-jər), **Jerome David ("J.D.").** b. 1919. Amer. author.
**Salk** (sôlk), **Jonas Edward.** b. 1914. Amer. microbiologist.
**Sal·lust** (săl′əst). 86?–34? B.C. Roman historian & politician.
**Sal·mon** (săm′ən, săl′mən), **Lucy Maynard.** 1853–1927. Amer. historian.
**Sa·lo·me** (sə-lō′mē, săl′ə-mē′). Daughter of Herodias in the Bible.
**Sal·o·mon** (săl′ə-mən), **Haym.** 1740?–85. Polish-born Amer. banker & businessman.
**Sal·ve·mi·ni** (säl-vā′mē-nē), **Gaetano.** 1873–1957. Italian historian.
**Sam·o·set** (săm′ə-sĕt′). d. 1653? Amer. Indian chief & friend of Pilgrim colonists.
**Samp·son** (sămp′sən), **Deborah.** 1760–1827. Amer. Revolutionary soldier.
**Sam·son** (săm′sən). Israelite judge & powerful warrior in the Old Testament.
**Sam·u·el** (săm′yōō-əl). 11th cent. B.C. Hebrew judge & prophet.
**Sam·u·el·son** (săm′yōō-əl-sən, -yōōl-sən), **Paul Anthony.** b. 1915. Amer. economist (Nobel, 1970).
**Sam·u·els·son** (săm′yōō-əl-sən), **Bengt Ingemar.** b. 1934. Swedish physician & biochemist (Nobel, 1982).
**San·chez de Bus·ta·man·te y Sir·vén** (sän′chĕz də bōō-stə-män′tā ē sîr-vĕn′), **Antonio.** 1865–1951. Cuban jurist.
**Sand** (sănd, säNd), **George.** *Amadine Aurore Lucie Dupin, Baroness Dudevant.* 1804–76. French author.
**Sand·burg** (sănd′bûrg′, săn′-), **Carl.** 1878–1967. Amer. poet & biographer.
**San·ders** (săn′dərz), **Harlan ("Colonel").** 1890–1980. Amer. fast-foods businessman.
**San·der·son** (săn′dər-sən), **Sibyl Swift.** 1865–1903. Amer. operatic soprano.
**San·ford** (săn′fərd), **Edward Terry.** 1865–1930. Amer. jurist.
**San·gal·lo** (sän-gä′lō, säng-), **Giuliano da.** 1445–1516. Italian architect, engineer, & sculptor.
**Sang·er** (săng′ər), **Frederick.** b. 1918. English biochemist (Nobel, 1958, 1980).
**Sanger, Margaret Higgins.** 1883–1966. Amer. leader of birth-control movement.
**Sang·ster** (săng′stər), **Margaret Elizabeth Munson.** 1838–1912. Amer. author & editor.
**San Mar·tín** (săn mär-tēn′, sän), **José de.** 1778–1850. Argentine revolutionary leader in South America.
**San·ta An·na** or **San·ta An·a** (săn′tə ăn′ə, săn′tə ä′nə), **Antonio López de.** 1795?–1876. Mexican military & political leader.
**San·tan·der** (săn′tăn-dĕr′), **Francisco de Paula.** 1792–1840. Colombian statesman.
**San·ta·ya·na** (săn′tē-ă′nə, -tə-yä′-, sän′-), **George.** 1863–1952. Spanish-born Amer. educator, philosopher, & poet.
**San·tos-Du·mont** (săn′təs-dōō-mŏnt′, -dyōō-), **Alberto.** 1873–1932. Brazilian pioneer aeronaut.

**Sa·pir** (sə-pîr′), **Edward.** 1884–1939. Amer. linguist & anthropologist.
**Sap·pho** (săf′ō). 7th cent. B.C. Greek poet.
**Sar·ah** (sâr′ə). Wife of Abraham & mother of Isaac in the Old Testament.
**Sar·da·na·pa·lus** (sär′dn-ăp′ə-ləs, -ə-pā′ləs). 9th cent. B.C. Assyrian ruler.
**Sar·dou** (sär-dōō′), **Victorien.** 1831–1908. French playwright.
**Sar·gent** (sär′jənt), **John Singer.** 1856–1925. Amer. painter.
**Sar·gon II** (sär′gŏn′). d. 705 B.C. Assyrian king (722–705).
**Sar·noff** (sär′nôf′), **David.** 1891–1971. Amer. business executive.
**Sa·roy·an** (sə-roi′ən), **William.** 1908–81. Amer. author.
**Sar·ton** (sär′tn), **George Alfred Leon.** 1884–1956. Belgian-born Amer. historian.
**Sar·tre** (sär′trə, särt), **Jean Paul.** 1905–80. French philosopher & author (Nobel, 1964).
**Sas·soon** (sə-sōōn′, să-), **Siegfried Lorraine.** 1886–1967. English poet & biographer.
**Sa·tie** (sä-tē′), **Erik.** 1866–1925. French composer.
**Sa·to** (sä′tō), **Eisaku.** 1901–75. Japanese statesman (Nobel, 1974).
**Sato, Naotake.** 1882–1971. Japanese diplomat.
**Sa·ud** (sä-ōōd′), **Abdul Aziz ibn** (1880–1953; reigned 1932–53) & **Abdul Aziz ibn** (1901?–69; reigned 1953–64). Saudi Arabian kings.
**Saul** (sôl). 11th cent. B.C. Hebrew king.
**Saus·sure** (sō-sōōr′), **Ferdinand de.** 1857–1913. Swiss pioneer linguist.
**Sav·age** (săv′ĭj), **Michael Joseph.** 1872–1940. Australian-born New Zealand labor leader & politician.
**Savage, Richard.** *4th Earl Rivers.* 1697?-1743. English poet.
**Sa·vo·na·ro·la** (săv′ə-nə-rō′lə, sə-vŏn′ə-), **Girolamo.** 1452–98. Italian reformer; executed.
**Sa·voy** (sə-voi′). Ruling dynasty of Sardinia & Italy (1720–1946) & of Spain (1870–73).
**Saxe** (săks), Comte **Hermann Maurice de.** 1696–1750. French military leader.
**Saxe-Co·burg** (săks-kō′bûrg). English ruling house (1901–10).
**Sax·o Gram·mat·i·cus** (săk′sō grə-măt′ĭ-kəs). 1150?–1220? Danish historian.
**Say** (sā), **Thomas.** 1787–1834. Amer. entomologist.
**Say·ers** (sā′ərz), **Dorothy Leigh.** 1893–1957. English author.
**Scal·i·ger** (skăl′ə-jər), **Joseph Justus.** 1540–1609. French scholar.
**Scaliger, Julius Caesar.** 1484–1558. Italian physician.
**Scan·der·beg** (skăn′dər-bĕg′). 1403–68. Albanian national hero.
**Scar·lat·ti** (skär-lä′tē), **Alessandro** (1659–1725) & **(Giuseppe) Domenico** (1685–1767). Italian composers & musicians.
**Scar·ron** (skă-rōN′), **Paul.** 1610–60. French author.
**Schacht** (shäkt, shäKHt), **(Horace Greeley) Hjalmar.** 1877–1970. German banker.
**Schaff** (shăf), **Philip.** 1819–93. Swiss-born Amer. church historian.
**Schal·ly** (shăl′ē), **Andrew Victor.** b. 1926. Polish-born Amer. physiologist (Nobel, 1977).
**Scharn·horst** (shärn′hôrst′), **Gerhard Johann David von.** 1755–1813. Prussian general.
**Schar·wen·ka** (shär-vĕng′kə), **Philipp** (1847–1917) & **Xaver** (1850–1924). German musicians & composers.
**Schaw·low** (shô′lō), **Arthur Leonard.** b. 1921. Amer. physicist (Nobel, 1981).
**Schech·ter** (shĕk′tər), **Solomon.** 1847–1915. Rumanian-born Hebrew scholar.
**Schee·le** (shā′lə), **Karl Wilhelm.** 1742–86. German-born Swedish chemist.
**Schel·ling** (shĕl′ĭng), **Friedrich Wilhelm Joseph von.** 1775–1854. German philosopher.
**Schia·pa·rel·li** (skē-ăp′ə-rĕl′ē, skăp′-), **Elsa.** 1890?–1973. Italian-born fashion designer.
**Schiaparelli, Giovanni Virginio.** 1835–1910. Italian astronomer.
**Schick** (shĭk), **Béla.** 1877–1967. Hungarian-born Amer. pediatrician.
**Schiff** (shĭf), **Dorothy.** b. 1903. Amer. newspaper publisher.
**Schil·ler** (shĭl′ər), **Johann Christoph Friedrich von.** 1759–1805. German poet, playwright, & historian.
**Schip·pers** (shĭp′ərz), **Thomas.** 1930–77. Amer. conductor.
**Schir·mer** (shûr′mər, shîr′-), **Gustav.** 1829–93. German-born Amer. music publisher.
**Schle·gel** (shlā′gəl), **August Wilhelm von.** 1767–1845. German translator, poet, & critic.
**Schlegel, Friedrich.** 1772–1829. German author & philosopher.
**Schlei·cher** (shlī′kər, -KHər), **Kurt von.** 1882–1934. German military & political leader.
**Schlei·er·ma·cher** (shlī′ər-mä′kər, -KHər), **Friedrich Ernst Daniel.** 1768–1834. German philosopher.
**Schles·in·ger** (shlĕs′ĭn-jər), **Arthur Meier** (1888–1965) & **Arthur Meier, Jr.** (b. 1917). Amer. historians & educators.
**Schley** (slī, shlī), **Winfield Scott.** 1839–1911. Amer. naval commander.
**Schlie·mann** (shlē′män′), **Heinrich.** 1822–90. German-born Amer. archaeologist.

**Schmidt** (shmĭt), **Helmut.** b. 1918. West German statesman.
**Schmuck·er** (shmŭk′ər), **Samuel Simon.** 1799–1873. Amer. theologian.
**Schna·bel** (shnä′bəl), **Artur.** 1882–1951. Austrian pianist & composer.
**Schnitz·ler** (shnĭts′lər), **Arthur.** 1862–1931. Austrian physician & author.
**Scho·field** (skō′fēld′), **John McAllister.** 1831–1906. Amer. army officer.
**Schön·berg** (shœn′bûrg, shûrn′-, shœn′bĕrk′), **Arnold.** 1874–1951. Austrian composer.
**School·craft** (skōōl′krăft′), **Henry Rowe.** 1793–1864. Amer. geologist, explorer, & ethnologist.
**Scho·pen·hau·er** (shō′pən-hou′ər), **Arthur.** 1788–1860. German philosopher.
**Schrief·fer** (shrē′fər), **John Robert.** b. 1931. Amer. physicist (Nobel, 1972).
**Schrö·ding·er** (shrœ′dĭng-ər, shrā′-), **Erwin.** 1887–1961. Austrian physicist (Nobel, 1933).
**Schu·bert** (shōō′bərt, -bĕrt′), **Franz Peter.** 1797–1828. Austrian composer.
**Schul·ler** (shōō′lər), **Gunther.** b. 1925. Amer. composer & conductor.
**Schultz** (shōōlts), **Theodore.** b. 1902. Amer. economist (Nobel, 1979).
**Schulz** (shōōlts), **Charles Monroe.** b. 1922. Amer. cartoonist.
**Schu·man** (shōō-män′), **Robert.** 1886–1963. French statesman.
**Schu·man** (shōō′mən), **William Howard.** b. 1910. Amer. composer.
**Schu·mann** (shōō′män′, -mən), **Robert.** 1810–56. German composer.
**Schu·mann-Heink** (shōō′mən-hĭngk′), **Ernestine.** 1861–1936. Amer. contralto.
**Schum·pe·ter** (shōōm′pā-tər), **Joseph Alois.** 1883–1950. Czechoslovakian-born Amer. economist.
**Schur·man** (shōōr′mən, shûr′-), **Jacob Gould.** 1854–1942. Canadian-born Amer. educator & diplomat.
**Schurz** (shōōrts, shûrz), **Carl.** 1829–1906. German-born Amer. general, politician, & editor.
**Schusch·nigg** (shōōsh′nĭk, -nĭg), **Kurt von.** 1897–1977. Austrian statesman.
**Schuy·ler** (skī′lər), **Louisa Lee.** 1837–1926. Amer. social worker.
**Schuyler, Philip John.** 1733–1804. Amer. Revolutionary general & politician.
**Schwab** (shwŏb), **Charles Michael.** 1862–1939. Amer. industrialist.
**Schwann** (shvän), **Theodor.** 1810–82. German physiologist.
**Schweit·zer** (shwĭt′sər, shvĭt′-), **Albert.** 1875–1965. French philosopher, physician, & musician (Nobel, 1952).
**Schwing·er** (shwĭng′ər), **Julian Seymour.** b. 1918. Amer. physicist (Nobel, 1965).
**Scid·more** (sĭd′môr′, -mōr′), **Eliza Ruhamah.** 1856–1928. Amer. travel writer.
**Scip·io** (sĭp′ē-ō′, skĭp′-), **Publius Cornelius.** *the Elder.* 237?–183 B.C. Roman military leader.
**Scipio, Publius Cornelius.** *the Younger.* 185–129 B.C. Roman general & politician.
**Scopes** (skōps), **John Thomas.** 1901–70. Amer. teacher convicted for teaching evolution.
**Scott** (skŏt), **Dred.** 1795?–1858. Amer. slave.
**Scott,** Sir **George Gilbert.** 1811–78. English architect.
**Scott, James Brown.** 1866–1943. Canadian-born Amer. lawyer.
**Scott, Robert Falcon.** 1868–1912. English Antarctic explorer.
**Scott,** Sir **Walter.** 1771–1832. Scottish author.
**Scott, Winfield.** 1786–1866. Amer. general.
**Scot·to** (skŏt′ō), **Renata.** b. 1936? Italian soprano.
**Scri·a·bin** (skrē-ä′bĭn), **Alexander Nikolayevich.** 1872–1915. Russian composer.
**Scribe** (skrēb), **Augustin Eugène.** 1791–1861. French playwright.
**Scripps** (skrĭps). Amer. family of newspaper publishers, including **James Edmund** (1835–1906), **Ellen Browning** (1836–1932), **Edward Wyllis** (1854–1926), & **Robert Paine** (1895–1938).
**Scud·der** (skŭd′ər), **Horace Elisha.** 1838–1902. Amer. author & editor.
**Scudder, Janet.** 1869–1940. Amer. sculptor & painter.
**Scudder, Vida Dutton.** 1861–1954. Amer. educator & author.
**Scu·dé·ry** (skōō′də-rē′), **Madeleine** *or* **Magdeleine de.** 1607–1701. French author.
**Sea·borg** (sē′bôrg′), **Glenn Theodore.** b. 1912. Amer. chemist (Nobel, 1951).
**Sea·bur·y** (sē′bĕr′ē, -bə-rē), **Samuel.** 1729–96. Amer. prelate.
**Sea·man** (sē′mən), **Elizabeth Cochrane.** *Nellie Bly.* 1867–1922. Amer. journalist.
**Sears** (sîrz), **Isaac.** 1730–86. Amer. Revolutionary patriot & businessman.
**Sears, Richard Warren.** 1863–1914. Amer. business executive.
**Se·at·tle** (sē-ăt′l). 1786?–1866. Amer. Indian leader.
**Sedg·wick** (sĕj′wĭk), **Anne Douglas.** 1873–1935. Amer. novelist.
**Sedgwick, Catharine Maria.** 1789–1867. Amer. novelist.
**Sedgwick, Theodore.** 1746–1813. Amer. politician & judge.
**See** (sē), **Thomas Jefferson Jackson.** 1866–1962. Amer. astronomer & mathematician.
**Seeckt** (săkt, zākt), **Hans von.** 1866–1936. German general.
**See·ger** (sē′gər), **Alan.** 1888–1916. Amer. poet.
**Seeger, Peter ("Pete").** b. 1919. Amer. folk singer.
**Se·fe·ri·a·des** (sĕf′ĕr-yä′*th*ēs), **Giorgos Stylianou.** 1900–71. Greek poet & diplomat (Nobel, 1963).
**Se·gal** (sē′gəl), **George.** b. 1924. Amer. sculptor.
**Se·gar** (sē′gär′), **Elzie Crisler.** 1894–1938. Amer. cartoonist.
**Se·go·via** (sĭ-gō′vē-ə), **Andres.** b. 1894. Spanish guitarist.
**Se·grè** (sĭ-grā′, sā-), **Emilio Gino.** b. 1905. Italian-born Amer. physicist (Nobel, 1959).
**Se·ja·nus** (sĭ-jā′nəs), **Lucius Aelius.** d. A.D. 31. Roman courtier.
**Sel·den** (sĕl′dən), **George Baldwin.** 1846–1922. Amer. inventor & patent lawyer.
**Selden, John.** 1584–1654. English antiquary & jurist.
**Sel·des** (sĕl′dĭs), **Gilbert Vivian.** 1893–1970. Amer. author & critic.
**Se·leu·cus** (sĭ-lōō′kəs). Name of six Syrian kings, esp. **I,** 358?–280 B.C., ruled 312–280; assassinated. **—Se·leu′cid** (-sĭd) *adj. & n.*
**Sel·fridge** (sĕl′frĭj), **Harry Gordon.** 1857?–1947. Amer.-born English businessman.
**Se·lig·man** (sĕl′ĭg-mən), **Edwin Robert Anderson.** 1861–1939. Amer. economist.
**Seligman, Joseph.** 1819–80. German-born Amer. financier.
**Sel·in·court** (sĕl′ĭn-kôrt′, -kōrt′), **Hugh de.** 1878–1951. English author.
**Sel·kirk** (sĕl′kûrk′), **Alexander.** 1676–1721. Scottish seaman; reputed original of Robinson Crusoe.
**Selz·nick** (sĕlz′nĭk), **David Oliver.** 1902–65. Amer. film producer.
**Sem·brich** (sĕm′brĭk′, zĕm′brĭKH′), **Marcella** *or* **Marcelline.** 1858–1935. Polish-born Amer. operatic soprano.
**Se·me·nov** (sə-myô′nəf), **Nikolai Nikolayevich.** b. 1896. Soviet chemist (Nobel, 1956).
**Semmes** (sĕmz), **Raphael.** 1809–77. Amer. Confederate naval commander.
**Sem·ple** (sĕm′pəl), **Ellen Churchill.** 1863–1932. Amer. geographer.
**Sen·e·ca** (sĕn′ĭ-kə), **Lucius Annaeus.** *the Younger.* 4 B.C.–A.D. 65. Roman Stoic philosopher, writer, & politician.
**Sen·nach·er·ib** (sĭ-năk′ər-ĭb′). d. 681 B.C. Assyrian king (705–681).
**Sen·nett** (sĕn′ĭt), **Mack.** 1884–1960. Canadian-born Amer. filmmaker.
**Se·quoy·a** or **Se·quoy·ah** (sĭ-kwoi′ə). 1770?–1843. Amer. Cherokee Indian leader.
**Ser·kin** (sûr′kĭn), **Rudolf.** b. 1903. Czechoslovakian-born pianist.
**Ser·ra** (sĕr′ə), **Junípero.** 1713–84. Spanish missionary in Calif.
**Ser·ra·no Su·ñer** (sə-rä′nō sōōn-yĕr′), **Ramón.** b. 1901. Spanish politician.
**Ser·to·ri·us** (sər-tôr′ē-əs, -tōr′-), **Quintus.** d. 72 B.C. Roman general.
**Ser·ve·tus** (sər-vē′təs), **Michael.** 1511–53. Spanish-born theologian & physician; executed for heresy.
**Ser·vice** (sûr′vĭs), **Robert William.** 1874–1958. English-born Canadian author.
**Ses·sions** (sĕsh′ənz) **Roger Huntington.** 1896–1985. Amer. composer.
**Se·ton** (sēt′n), Saint **Elizabeth Ann Bayley.** *Mother Seton.* 1774–1821. Amer. religious leader.
**Seu·rat** (sə-rä′), **Georges Pierre.** 1859–91. French painter.
**Seuss** (sōōs), Doctor. Theodor Seuss GEISEL.
**Sev·a·reid** (sĕv′ə-rĭd′), **Eric.** b. 1912. Amer. broadcast journalist.
**Se·ve·rus** (sə-vîr′əs), **Lucius Septimus.** 146–211. Roman emperor (193–211).
**Se·vier** (sə-vîr′), **John.** 1745–1815. Amer. pioneer, politician, & Indian fighter.
**Sé·vi·gné** (sā-vēn-yā′), Marquise de. *Marie de Rabutin-Chantal.* 1626–96. French author.
**Sew·all** (sōō′əl), **Mary Eliza Wright.** 1844–1920. Amer. educator & reformer.
**Sewall, Samuel.** 1652–1730. English-born Amer. jurist.
**Sew·ard** (sōō′ərd), **William Henry.** 1801–72. Amer. statesman.
**Sew·ell** (sōō′əl), **Anna.** 1820–78. English author.
**Sex·ton** (sĕks′tən), **Anne.** 1928–74. Amer. poet.
**Sey·mour** (sē′môr′, -mōr′), **Horatio.** 1810–86. Amer. politician.
**Seymour, Jane.** 1509?–37. Queen of England as third wife of Henry VIII.
**Sfor·za** (sfôrt′sə), Count **Carlo.** 1873–1952. Italian diplomat & historian.
**Shack·le·ton** (shăk′əl-tən), Sir **Ernest Henry.** 1874–1922. Irish-born British Antarctic explorer.

**Shad·well** (shăd′wəl, -wĕl′), **Thomas.** 1640?–92. English author.
**Sha·fer** (shā′fər), **Helen Almira.** 1839–94. Amer. pioneer educator.
**Shaf·ter** (shăf′tər), **William Rufus.** 1835–1906. Amer. Union general.
**Shaftes·bur·y** (shăfts′bĕr′ē, -bə-rē), 1st Earl of. *Anthony Ashley Cooper.* 1621–83. English statesman.
**Shah Je·han** *or* **Ja·han** (shä′ jə-hän′). 1592?–1666. Mogul emperor (1628–58).
**Shahn** (shän), **Benjamin ("Ben").** 1898–1969. Russian-born Amer. artist.
**Shake·speare** or **Shak·spere** (shāk′spîr), **William.** 1564–1616. English playwright & poet. **—Shake·spear′e·an, Shake·spear′i·an** *adj. & n.*
**Shang** (shăng). Chinese dynasty (1766–1122 B.C.).
**Shan·kar** (shän′kär), **Ravi.** b. 1920. Indian-born musician & composer.
**Sha·pir·o** (shə-pîr′ō), **Karl Jay.** b. 1913. Amer. poet & critic.
**Shap·ley** (shăp′lē), **Harlow.** 1885–1972. Amer. astronomer.
**Shaw** (shô), **Anna Howard.** 1847–1919. English-born Amer. suffragist.
**Shaw, George Bernard.** 1856–1950. Irish-born English playwright (Nobel, 1925). **—Sha′vi·an** (shā′vē-ən) *adj. & n.*
**Shaw, Henry Wheeler.** *John Billings.* 1818–85. Amer. humorist.
**Shaw, Lemuel.** 1781–1861. Amer. jurist.
**Shaw, Robert Lawson.** b. 1916. Amer. director.
**Shawn** (shôn), **Ted.** 1891–1972. Amer. dancer.
**Shays** (shāz), **Daniel.** 1747?–1825. Amer. Revolutionary soldier & insurrectionist.
**Shear·er** (shîr′ər), **Moira.** b. 1926. Scottish-born ballerina.
**Shee·ler** (shē′lər), **Charles.** 1883–1965. Amer. photographer.
**Sheen** (shēn), **Fulton John.** 1895–1979. Amer. prelate.
**Shel·by** (shĕl′bē), **Isaac.** 1750–1826. Amer. pioneer & soldier.
**Shel·ley** (shĕl′ē), **Mary Godwin Wollstonecraft.** 1797–1851. English author.
**Shelley, Percy Bysshe.** 1792–1822. English poet.
**Shen·stone** (shĕn′stən, -stōn′), **William.** 1714–63. English poet.
**Shep·ard** (shĕp′ərd), **Alan Bartlett, Jr.** b. 1923. Amer. astronaut.
**Sher·a·ton** (shĕr′ə-tən), **Thomas.** 1751–1806. English furniture designer.
**Sher·i·dan** (shĕr′ĭ-dən), **Philip Henry.** 1831–88. Amer. Union general.
**Sheridan, Richard Brinsley.** 1751–1816. English playwright & politician.
**Sher·man** (shûr′mən), **James Schoolcraft.** 1855–1912. U.S. Vice President (1909–12).
**Sherman, John.** 1823–1900. Amer. politician.
**Sherman, Roger.** 1721–93. Amer. Revolutionary patriot, politician, & jurist.
**Sherman, Stuart Pratt.** 1881–1926. Amer. critic & educator.
**Sherman, William Tecumseh.** 1820–91. Amer. Union general.
**Sher·riff** (shĕr′ĭf), **Robert Cedric.** 1896–1975. English author.
**Sher·ring·ton** (shĕr′ĭng-tən), Sir **Charles Scott.** 1861–1952. English physiologist (Nobel, 1932).
**Sher·wood** (shûr′wo͝od′), **Robert Emmet.** 1896–1955. Amer. playwright.
**Shev·chen·ko** (shĕf-chĕng′kō), **Taras Grigoryevich.** 1814–61. Ukrainian poet.
**Shi·de·ha·ra** (shē′də-här′ə), Baron **Kijuro.** 1872–1951. Japanese statesman.
**Shi·ge·mit·su** (shē′gə-mĭt′so͞o), **Mamoru.** 1887–1957. Japanese diplomat.
**Shi·ras** (shī′rəs), **George.** 1832–1924. Amer. jurist.
**Shir·er** (shīr′ər), **William Lawrence.** b. 1904. Amer. journalist.
**Shir·ley** (shûr′lē), **James.** 1596–1666. English playwright.
**Shirley, William.** 1694–1771. English colonial administrator.
**Shock·ley** (shŏk′lē), **William Bradford.** b. 1910. English-born Amer. physicist (Nobel, 1956).
**Shoe·ma·ker** (sho͞o′mā′kər), **William Lee ("Willie").** b. 1931. Amer. jockey.
**Sholes** (shōlz), **Christopher Latham.** 1819–90. Amer. journalist & inventor.
**Sho·lo·khov** (shô′lə-kôf′, -kôv′), **Mikhail Aleksandrovich.** 1905–84. Russian novelist (Nobel, 1965).
**Shore** (shôr, shōr), **Dinah.** b. 1917? Amer. entertainer.
**Shos·ta·ko·vich** (shŏs′tə-kō′vĭch, -kô′-, shô′stə-), **Dimitri.** 1906–75. Russian composer.
**Shreve** (shrēv), **Henry Miller.** 1785–1851. Amer. riverboat captain.
**Shu·bert** (sho͞o′bərt), **Lee.** 1875–1953. Amer. producer & theatrical manager.
**Shu Han** (sho͞o′ hän′). Chinese dynasty (221–63).
**Shute** (sho͞ot), **Nevil.** 1899–1960. English author.
**Shver·nik** (shfĕr′nĭk, shvĕr′-), **Nikolai Mikhalovich.** 1888–1970. Soviet political leader.
**Si·be·li·us** (sĭ-bā′lē-əs, -bāl′yəs), **Jean.** 1865–1957. Finnish composer.
**Sib·ley** (sĭb′lē), **Hiram.** 1807–88. Amer. businessman & philanthropist.
**Sick·les** (sĭk′əlz), **Daniel Edgar.** 1825–1914. Amer. Union general & politician.
**Sid·dons** (sĭd′nz), **Sarah.** 1755–1831. English actress.
**Sid·ney** (sĭd′nē), Sir **Philip.** 1554–86. English poet, soldier, & politician.
**Sieg·bahn** (sēg′bän′), **Kai.** b. 1918. Swedish physicist (Nobel, 1981).
**Siegbahn, Karl Manne Georg.** 1886–1978. Swedish physicist (Nobel, 1924).
**Sie·mens** (sē′mənz, zē′-), Sir **William.** 1823–83. German-born English engineer & inventor.
**Sien·kie·wicz** (shĕn-kyā′vĭch), **Henryk.** 1846–1916. Polish author (Nobel, 1905).
**Sie·vers** (sē′vərz, zē′fərs), **Georg Eduard.** 1850–1932. German philologist.
**Sie·yès** (syĕ-yĕs′), **Emmanuel Joseph.** *Abbé Sieyès.* 1748–1836. French political leader.
**Sig·is·mund** (sĭg′ĭs-mənd). 1368–1437. Holy Roman Emperor (1411–37).
**Si·gnac** (sēn-yăk′), **Paul.** 1863–1935. French painter.
**Si·gno·rel·li** (sēn-yō-rĕl′ē), **Luca di Egidio di Ventura de′.** 1441–1523. Italian painter.
**Sig·our·ney** (sĭg′ər-nē), **Lydia Howard Huntley.** 1791–1865. Amer. author.
**Sigs·bee** (sĭgz′bē), **Charles Dwight.** 1845–1923. Amer. naval commander.
**Si·gurds·son** (sĭg′ərd-sən, -ərth-), **Jón.** 1811–79. Icelandic politician & scholar.
**Si·ha·nouk** (sē′ə-no͝ok′), Prince **Norodom.** b. 1922. Cambodian statesman.
**Si·kor·ski** (sĭ-kôr′skē), **Wladyslaw.** 1881–1943. Polish general.
**Si·kor·sky** (sĭ-kôr′skē), **Igor Ivan.** 1889–1972. Russian-born Amer. aviation pioneer.
**Sil·lan·pää** (sĭl′ən-pă′), **Frans Eemil.** 1888–1964. Finnish author (Nobel, 1939).
**Sil·li·man** (sĭl′ə-mən), **Benjamin** (1779–1864) & **Benjamin, Jr.** (1816–85). Amer. science educators & editors.
**Sills** (sĭlz), **Beverly.** b. 1929. Amer. operatic soprano & manager.
**Si·lo·ne** (sĭ-lō′nē), **Ignazio.** 1900–78. Italian novelist.
**Sil·ver·man** (sĭl′vər-mən), **Frederick ("Fred").** b. 1937. Amer. broadcasting executive.
**Silverman, Sime.** 1873–1933. Amer. publisher.
**Si·me·non** (sē-mə-nôN′), **Georges Joseph Christian.** b. 1903. Belgian-born French author.
**Sim·e·on Sty·li·tes** (sĭm′ē-ən stə-lī′tēz), Saint. 390?–459. Syrian ascetic.
**Sim·kho·vitch** (sĭm′kə-vĭch′), **Mary Melinda Kingsbury.** 1867–1951. Amer. settlement worker & reformer.
**Simms** (sĭmz), **Ruth Hanna McCormick.** 1880–1944. Amer. politician.
**Simms, William Gilmore.** 1806–70. Amer. author.
**Si·mon** (sī′mən), 1st Viscount. *John Allesbrook Simon.* 1873–1954. English statesman.
**Simon, Herbert Alexander.** b. 1916. Amer. economist (Nobel, 1978).
**Simon, Neil.** b. 1927. Amer. playwright.
**Simon, Norton.** b. 1907. Amer. businessman.
**Simon, Paul.** b. 1942. Amer. singer & songwriter.
**Si·mon·i·des of Ce·os** (sī-mŏn′ĭ-dēz; sē′ŏs). 6th–5th cent. B.C. Greek poet.
**Si·mo·nov** (sē′mə-nŏf, -nôf), **Konstantin Mikhailovich.** b. 1915. Russian author.
**Simon Ze·lo·tes** (zē-lō′tēz) *or* **Simon the Ca·naan·ite** (kā′nə-nīt′). One of the 12 Apostles.
**Simp·son** (sĭmp′sən, sĭm′-), **Adele.** b. 1903. Amer. fashion designer.
**Simpson, George Gaylord.** 1902–84. Amer. paleontologist.
**Simpson,** Sir **James Young.** 1811–70. Scottish physician.
**Simpson, Jeremiah ("Sockless Jerry").** 1842–1905. Amer. rancher & politician.
**Simpson, Orenthal James ("O.J.").** b. 1947. Amer. football player.
**Sims** (sĭmz), **James Marion.** 1813–83. Amer. pioneer gynecologist.
**Sims, William Sowden.** 1858–1936. Amer. naval commander.
**Si·na·tra** (sə-nä′trə), **Francis Albert ("Frank").** b. 1915. Amer. entertainer.
**Sin·clair** (sĭn-klâr′, sĭng-), **Harry Ford.** 1876–1956. Amer. oil executive.
**Sinclair, May.** 1865–1946. English novelist.
**Sinclair, Upton Beall.** 1878–1968. Amer. author & reformer.
**Sing·er** (sĭng′ər), **Isaac Bashevis.** b. 1904. Polish-born Amer. Yiddish author (Nobel, 1978).
**Singer, Isaac Merritt.** 1811–75. Amer. inventor & manufacturer.
**Si·quei·ros** (sĭ-kā′rōs), **David Alfaro.** 1898?–1974. Mexican mural painter.

**Sis·ley** (sĭs'lē, sĭz'-, sēs-lē'), **Alfred.** 1840?–99. English-born French painter.
**Sis·mon·di** (sĭs-mŏn'dē, sēs-môɴ-dē'), **Jean Charles Léonard Simonde.** 1773–1842. Swiss historian & economist.
**Sit·ter** (sĭt'ər), **Willem de.** 1872–1934. Dutch astronomer.
**Sit·ting Bull** (sĭt'ĭng bo͝ol'). 1834?–90. Amer. Dakota Indian leader.
**Sit·well** (sĭt'wĕl', -wəl). English family of poets & critics, including Sir **George Reresby** (1860–1943), Dame **Edith** (1887–1964), Sir **Osbert** (1892–1969), & **Sacheverell** (b. 1897).
**Skeat** (skēt), **Walter William.** 1835–1912. English philologist.
**Skel·ton** (skĕl'tən), **John.** 1460?–1529. English poet & scholar.
**Skelton, Richard ("Red").** b. 1913. Amer. comedian.
**Skin·ner** (skĭn'ər), **Burrhus Frederick ("B.F.").** b. 1904. Amer. psychologist.
**Skinner, Constance Lindsay.** 1879–1939. Canadian-born Amer. author & historian.
**Skinner, Cornelia Otis.** 1901–79. Amer. actress & author.
**Skinner, Otis.** 1858–1942. Amer. actor.
**Ško·da** (skō'də, shkô'dä), **Emil von.** 1839–1900. Czech engineer & industrialist.
**Skou·ras** (sko͞or'əs), **Spyros Panagiotes.** 1893–1971. Greek-born Amer. film executive.
**Sla·ter** (slā'tər), **Samuel.** 1768–1835. English-born textile pioneer in America.
**Sli·dell** (slī-dĕl', slīd'l), **John.** 1793–1871. Amer. politician & Confederate diplomat.
**Sli·pher** (slī'fər), **Vesto Melvin.** 1875–1969. Amer. astronomer.
**Sloan** (slōn), **Alfred Pritchard.** 1875–1966. Amer. industrialist & philanthropist.
**Sloan, John French.** 1871–1951. Amer. painter.
**Sloat** (slōt), **John Drake.** 1781–1867. Amer. admiral.
**Slo·cum** (slō'kəm), **Henry Warner.** 1827–94. Amer. military & political leader.
**Slo·nim·sky** (slō-nĭm'skē, -nĭmp'-), **Nicolas.** b. 1894. Russian-born Amer. composer & musicologist.
**Smalls** (smôlz), **Robert.** 1839–1915. Amer. Union soldier & politician.
**Sme·ta·na** (smĕt'n-ə), **Bedřich.** 1824–84. Czech composer.
**Smi·bert** (smī'bərt), **John.** 1688–1751. Scottish-born Amer. painter.
**Smig·ly-Rydz** (smĭg'lē-rĭts', -rĭdz', shmĭg'-), **Edward.** 1886–1941? Polish military leader.
**Smith** (smĭth), **Adam.** 1723–90. Scottish political economist & philosopher.
**Smith, Alfred Emanuel.** *the Happy Warrior.* 1873–1944. Amer. politician.
**Smith, Bessie.** 1894?–1937. Amer. blues singer.
**Smith, Charles Henry.** *Bill Arp.* 1826–1903. Amer. author.
**Smith, Francis Marion ("Borax").** 1846–1931. Amer. financier & promoter.
**Smith, Gerritt.** 1797–1874. Amer. reformer & philanthropist.
**Smith, Hamilton Othanel.** b. 1931. Amer. microbiologist (Nobel, 1978).
**Smith, Hannah Whithall.** 1832–1911. Amer. evangelist, author, & reformer.
**Smith, Howard Kingsbury.** b. 1914. Amer. broadcast journalist.
**Smith, Ian.** b. 1919. Zimbabwean political leader.
**Smith, Jedediah Strong.** 1799?–1831. Amer. fur trader & explorer.
**Smith, Jessie Wilcox.** 1863–1935. Amer. artist.
**Smith, John.** 1580?–1631. English adventurer, colonist, explorer, & author.
**Smith, Joseph.** 1805–44. Amer. Mormon religious leader.
**Smith, Julia Evelina** (1792–1886) & **Aby Hadassah** (1797–1878). Amer. suffragists.
**Smith, Kathryn Elizabeth ("Kate").** b. 1910? Amer. entertainer.
**Smith, Margaret Chase.** b. 1897. Amer. politician.
**Smith, Nathan.** 1762–1829. Amer. physician & medical educator.
**Smith, Samuel Francis.** 1808–95. Amer. poet & clergyman.
**Smith, Seba.** *Major Jack Downing.* 1792–1868. Amer. journalist & satirist.
**Smith, Sophia.** 1796–1870. Amer. educational philanthropist.
**Smith, Sydney.** 1771–1845. English clergyman, essayist, & editor.
**Smith, Theobold.** 1859–1934. Amer. pathologist.
**Smith, Walter Bedell.** 1895–1961. Amer. general.
**Smith, Walter Wellesley ("Red").** 1905–82. Amer. sportswriter.
**Smith, William.** 1727–1803. Scottish-born Amer. clergyman & educator.
**Smith, William.** 1769–1839. English geologist.
**Smith·son** (smĭth'sən), **James.** 1765–1829. English chemist, mineralogist, & philanthropist.
**Smo·hal·la** (smō-hăl'ə). 1815?–1907. Amer. Indian religious prophet.
**Smol·lett** (smŏl'ĭt), **Tobias George.** 1721–71. English novelist.
**Smuts** (smŭts, smœts), **Jan Christiaan.** 1870–1950. South African military & political leader.
**Smyth** (smĭth), **Henry DeWolf.** b. 1898. Amer. physicist.
**Snead** (snēd), **Samuel Jackson ("Sam").** b. 1912. Amer. golfer.
**Snell** (snĕl), **George.** b. 1903. Amer. geneticist (Nobel, 1980).
**Snor·ri Stur·lu·son** (snôr'ē stûr'lə-sən). 1178?–1241. Icelandic historian & statesman; assassinated.
**Snow** (snō), **Charles Percy ("C.P.").** 1905–80. English novelist.
**Soar·es** (swär'ĭsh, sə-wär'-), **Mário.** b. 1924. Portuguese statesman.
**So·ci·nus** (sō-sī'nəs), **Faustus.** 1539–1604. Italian theologian.
**Soc·ra·tes** (sŏk'rə-tēz'). 470?–399 B.C. Greek philosopher. **—So·crat'ic** (sō-krăt'ĭk) *adj.*
**Sod·dy** (sŏd'ē), **Frederick.** 1877–1956. English chemist (Nobel, 1921).
**Sö·der·blom** (sœ'dər-blo͞om'), **Nathan.** 1866–1931. Swedish prelate & historian (Nobel, 1930).
**Sol·o·mon** (sŏl'ə-mən). 10th cent. B.C. king of Israel.
**So·lon** (sō'lən, -lŏn'). 638?–559 B.C. Athenian statesman & poet.
**Sol·ti** (sŏl'tē), Sir **Georg.** b. 1912. Hungarian-born English conductor.
**Sol·vay** (sŏl'vā), **Ernest.** 1838–1922. Belgian chemist, industrialist, & philanthropist.
**Sol·zhe·ni·tsyn** (sōl'zhə-nēt'sĭn), **Aleksandr Isayevich.** b. 1918. Russian author (Nobel, 1970).
**Som·er·ville** (sŭm'ər-vĭl'), Sir **James Fownes.** 1882–1949. English naval commander.
**So·mo·za** (sə-mō'zə), **Anastasio Somoza Debayle.** 1925–80. Nicaraguan statesman; assassinated.
**Sond·heim** (sŏnd'hīm'), **Stephen.** b. 1930. Amer. composer & lyricist.
**Song** (so͝ong). Name of three Chinese dynasties: **Northern** (960–1127) & **Southern** (420–79, 1127–1279).
**Son·tag** (sŏn'tăg'), **Susan.** b. 1933. Amer. author & critic.
**Soong** (so͝ong). Prominent Chinese family, including **Charles Jones** (d. 1927), missionary & merchant; **Ai-ling** (1888–1973), wife of H.H. Kung; **Tse-ven** or **Tsu-wen** (1891?–1971), financier; **Ch'ing-ling** (b. 1890), wife of Sun Yat-sen; & **Mei-ling** or **May-ling** (b. 1898), wife of Chiang Kai-shek.
**Soph·o·cles** (sŏf'ə-klēz'). 496?–406 B.C. Greek dramatist. **—Soph'o·cle'an** *adj.*
**Sor·del·lo** (sôr-dĕl'ō). 13th cent. Italian troubadour.
**Sor·el** (sô-rĕl'), **Georges.** 1847–1922. French journalist & philosopher.
**Sor·en·sen** (sûr'ən-sən), **Soren Peter Lauritz.** 1868–1939. Danish chemist.
**Sor·en·sen** (sôr'ĭn-sən), **Theodore Chaikin.** b. 1928. Amer. author & public official.
**Sor·o·kin** (sə-rō'kĭn, sô-), **Pitirim Alexandrovich.** 1889–1968. Russian-born Amer. sociologist.
**So·rol·la y Bas·ti·da** (sə-rôl'yə ē bä-stē'də, -roi'ə), **Joaquín.** 1863–1923. Spanish painter.
**Soth·ern** (sŭth'ərn), **Edward Hugh.** 1859–1933. Amer. actor.
**Sou·lé** (so͞o-lā'), **Pierre.** 1801–70. French-born Amer. politician.
**Soult** (so͞olt), **Nicholas Jean de Dieu.** 1769–1851. French marshal.
**Sou·sa** (so͞o'zə, -sə), **John Philip.** 1854–1932. Amer. bandmaster & composer.
**South** (south), **Robert.** 1634–1716. English clergyman.
**South·amp·ton** (south-hămp'tən, sou-thămp'-), 3rd Earl of. *Henry Wriothesley.* 1573–1624. English courtier, military leader, & patron of Shakespeare.
**Sou·they** (sou'thē, sŭth'ē), **Robert.** 1774–1843. English author.
**South·worth** (south'wûrth'), **Emma Dorothy Eliza Nevitte.** 1819–99. Amer. novelist.
**Sou·tine** (so͞o-tēn'), **Chaim.** 1894–1943. Lithuanian-born painter.
**Spaak** (späk), **Paul Henri Charles.** 1889–1972. Belgian statesman.
**Spaatz** (späts), **Carl.** 1891–1974. Amer. general.
**Spahn** (spän, spôn), **Warren Edward.** b. 1921. Amer. baseball player.
**Spal·ding** (spôl'dĭng), **Albert.** 1888–1953. Amer. violinist & composer.
**Spalding, Albert Goodwill.** 1850–1915. Amer. sports-equipment manufacturer.
**Spark** (spärk), **Muriel Sarah.** b. 1918. Scottish-born author.
**Sparks** (spärks), **Jared.** 1789–1866. Amer. historian.
**Spar·ta·cus** (spär'tə-kəs). d. 71 B.C. Thracian gladiator.
**Speer** (shpîr, spîr), **Albert.** 1905–81. German architect & Nazi politician.
**Spell·man** (spĕl'mən), **Francis Joseph.** 1889–1967. Amer. prelate.
**Spe·mann** (shpā'män'), **Hans.** 1869–1941. German zoologist & physiologist (Nobel, 1935).
**Spen·cer** (spĕn'sər), **Herbert.** 1820–1903. English philosopher. **—Spen·ce'ri·an** (spĕn-sîr'ē-ən) *adj.*
**Spencer, Platt Rogers.** 1800–64. Amer. handwriting expert.
**Spen·der** (spĕn'dər), **Stephen Harold.** b. 1909. English author.
**Speng·ler** (spĕng'lər, shpĕng'-), **Oswald.** 1880–1936. German philosopher.
**Spen·ser** (spĕn'sər), **Edmund.** 1552?–99. English poet. **—Spen·se'ri·an** (spĕn-sîr'ē-ən) *adj.*

**Sper·ry** (spĕr′ē), **Elmer Ambrose.** 1860–1930. Amer. engineer & inventor.
**Sperry, Roger.** b. 1913. Amer. neurobiologist (Nobel, 1981).
**Spiel·berg** (spēl′bûrg′), **Steven.** b. 1947. Amer. film director.
**Spil·lane** (spə-lān′), **Mickey.** b. 1918. Amer. author.
**Spin·garn** (spĭn′gärn′), **Joel Elias.** 1875–1939. Amer. poet & critic.
**Spi·no·za** (spĭ-nō′zə), **Baruch** or **Benedict.** 1632–77. Dutch philosopher & theologian.
**Spit·te·ler** (shpĭt′l-ər, shpĭt′lər, spĭt′-), **Carl.** 1845–1924. Swiss writer & poet (Nobel, 1919).
**Spock** (spŏk), **Benjamin McLane.** b. 1903. Amer. pediatrician, educator, & author.
**Spode** (spōd), **Josiah.** 1754–1827. English potter.
**Spof·ford** (spŏf′ərd), **Harriet Elizabeth Prescott.** 1835–1921. Amer. author.
**Spoo·ner** (spo͞o′nər), **William Archibald.** 1844–1930. English clergyman & educator.
**Spots·wood** (spŏts′wo͝od′), **Alexander.** 1676–1740. English colonial administrator.
**Sprague** (sprăg), **Frank Julian.** 1857–1934. Amer. electrical engineer & inventor.
**Spreck·els** (sprĕk′əlz), **Claus.** 1828–1908. German-born Amer. sugar manufacturer.
**Spru·ance** (spro͞o′əns), **Raymond Ames.** 1886–1969. Amer. naval officer.
**Spy·ri** (shpîr′ē, spîr′ē), **Johanna.** 1827?–1901. Swiss author.
**Squan·to** (skwŏn′tō). d. 1622. Amer. Indian friend of Pilgrims.
**Squibb** (skwĭb), **Edward Robinson.** 1819–1900. Amer. pharmaceutical manufacturer.
**Staël** (stäl), Madame **de.** *Anne Louise Germaine Neckar.* 1766–1817. French novelist, critic, & literary patron.
**Stagg** (stăg), **Amos Alonzo.** 1862–1965. Amer. football coach.
**Stahl·berg** (stôl′bərg, -bĕr′ē), **Kaarlo Juho.** 1865–1952. Finnish statesman.
**Sta·lin** (stä′lĭn, stăl′ĭn), **Joseph.** 1879–1953. Soviet Communist leader. **—Sta′lin·ist** *adj.* & *n.*
**Stan·dish** (stăn′dĭsh′), **Miles** or **Myles.** 1584?–1656. English colonist in America.
**Stan·ford** (stăn′fərd), **Leland.** 1824–93. Amer. financier & politician.
**Stan·is·las I Lesz·czyn·ski** (stăn′ĭ-slôs, -släs; lĕsh-chĭn′skē). 1677–1766. Polish king (1704–09, 1733–35).
**Stan·i·slav·ski** (stăn′ĭ-släv′skē, -släf′-), **Konstantin.** 1863–1938. Russian actor and director.
**Stan·ley** (stăn′lē), **Edward George Geoffrey Smith.** *14th Earl of Derby.* 1799–1869. English political leader.
**Stanley, Francis Edgar** (1849–1918) & **Freelan** (1849–1940). Amer. inventors & automobile manufacturers.
**Stanley,** Sir **Henry Morton.** 1841–1904. Welsh-born journalist & African explorer.
**Stanley, Wendell Meredith.** 1904–71. Amer. biochemist (Nobel, 1946).
**Stan·ton** (stăn′tən), **Edwin McMasters.** 1814–69. Amer. jurist & public official.
**Stanton, Elizabeth Cady.** 1815–1902. Amer. feminist & reformer.
**Stark** (stärk), **Harold Raynsford.** 1880–1972. Amer. naval officer.
**Stark** (stärk, shtärk), **Johannes.** 1874–1957. German physicist (Nobel, 1919).
**Stark** (stärk), **John.** 1728–1822. Amer. Revolutionary general.
**Starr** (stär), **Belle.** 1848–89. Amer. outlaw.
**Starr, Ringo.** *Richard Starkey.* b. 1940. English musician & composer.
**Star·zýn·ski** (stär-zĭn′skē, -zhĭn′-), **Stefan.** 1893–1940? Polish politician & national hero.
**Stas·sen** (stăs′ən), **Harold Edward.** b. 1907. Amer. politician.
**Sta·tius** (stā′shəs, -shē-əs), **Publius Papinus.** A.D. 45?–96? Roman poet.
**Stat·ler** (stăt′lər), **Ellsworth Milton.** 1863–1929. Amer. hotel builder.
**Stau·ding·er** (shtou′dĭng-ər, stou′-), **Hermann.** 1881–1965. German chemist (Nobel, 1953).
**Steed** (stēd), **Henry Wickham.** 1871–1956. English journalist.
**Steele** (stēl), Sir **Richard.** 1672–1729. English politician & author.
**Steen** (stān), **Jan.** 1626–79. Dutch genre painter.
**Ste·fáns·son** (stĕf′ən-sən), **Vilhjálmur.** 1879–1962. Canadian Arctic explorer & ethnologist.
**Stef·fens** (stĕf′ənz), **(Joseph) Lincoln.** 1866–1936. Amer. muckraking journalist.
**Stei·chen** (stī′kən), **Edward Jean.** 1879–1973. Amer. photographer & landscape painter.
**Stein** (stīn), **Gertrude.** 1874–1946. Amer. author.
**Stein, Jules Caesar.** 1896–1981. Amer. entertainment executive.
**Stein, William Howard.** 1911–80. Amer. biochemist (Nobel, 1972).
**Stein·beck** (stīn′bĕk′), **John Ernst.** 1902–68. Amer. novelist (Nobel, 1962).
**Stein·berg** (stīn′bûrg′, -bərg), **Saul.** b. 1914. Rumanian-born Amer. graphic artist & cartoonist.
**Stein·em** (stī′nəm), **Gloria.** b. 1935. Amer. feminist, author, & lecturer.
**Stein·er** (stī′nər, shtī′-), **Rudolf.** 1861–1925. Austrian social philosopher.
**Stein·itz** (stī′nĭts, shtī′-), **William.** 1836–1900. German chess champion.
**Stein·man** (stīn′mən), **David Barnard.** 1886–1960. Amer. civil engineer.
**Stein·metz** (stīn′mĕts′, shtīn′-), **Charles Proteus.** 1865–1923. German-born Amer. electrical engineer & inventor.
**Stein·way** (stīn′wā′), **Henry Englehard.** 1797–1871. German-born Amer. piano manufacturer.
**Stel·la** (stĕl′ə), **Frank Philip.** b. 1936. Amer. painter.
**Sten·dhal** (stĕn-däl′, stăn-, stăN-). *Marie Henri Beyle.* 1783–1842. French novelist & biographer.
**Sten·gel** (stĕng′gəl), **Charles Dillon ("Casey").** 1891–1975. Amer. baseball player & manager.
**Ste·phen** (stē′vən), Saint. 1st cent. Christian martyr.
**Stephen,** Sir **Leslie.** 1832–1904. English author & editor.
**Stephen I.** *Saint Stephen.* 975?–1038. Hungarian king (997–1038).
**Stephen of Blois** (blwä). 1097?–1154. English king (1135–54).
**Ste·phens** (stē′vənz), **Alexander Hamilton.** 1812–83. Amer. political leader & Confederate Vice President (1861–65).
**Stephens, Alice Barber.** 1858–1932. Amer. painter & illustrator.
**Stephens, James.** 1882–1950. Irish author.
**Stephens, Uriah Smith.** 1821–82. Amer. labor leader.
**Ste·phen·son** (stē′vən-sən), **George** (1781–1848) & **Robert** (1803–59). English railway pioneers.
**Stern** (stûrn), **Gladys Bertha.** 1890–1973. English novelist.
**Stern, Isaac.** b. 1920. Russian-born Amer. violinist.
**Stern, Otto.** 1888–1969. German-born Amer. physicist (Nobel, 1943).
**Stern·berg** (stûrn′bûrg′), **George Miller.** 1838–1915. Amer. army physician & bacteriologist.
**Sterne** (stûrn), **Laurence.** 1713–68. English satiric novelist.
**Stet·son** (stĕt′sən), **Augusta Emma Simmons.** 1842–1928. Amer. Christian Science leader.
**Stetson, John Batterson.** 1830–1906. Amer. hat manufacturer.
**Stet·tin·i·us** (stə-tĭn′ē-əs), **Edward Reilly, Jr.** 1900–49. Amer. business executive & public official.
**Steu·ben** (sto͞o′bən, styo͞o′-, sto͞o-bĕn′, styo͞o-, shtoi′bən), Baron **Friedrich Wilhelm Ludolf Gerhard Augustin von.** 1730–94. Prussian-born Amer. Revolutionary military leader.
**Ste·vens** (stē′vənz), **Edwin Augustus.** 1795–1868. Amer. inventor & philanthropist.
**Stevens, George.** 1905–75. Amer. film director.
**Stevens, John** (1749–1838) & **Robert Livingston** (1787–1856). Amer. inventors & engineers.
**Stevens, John Paul.** b. 1920. Amer. jurist.
**Stevens, Risë.** b. 1913. Amer. operatic soprano.
**Stevens, Thaddeus.** 1792–1868. Amer. politician.
**Stevens, Wallace.** 1879–1955. Amer. poet.
**Ste·ven·son** (stē′vən-sən), **Adlai Ewing.** 1835–1914. U.S. Vice President (1893–97).
**Stevenson, Adlai Ewing.** 1900–65. Amer. statesman.
**Stevenson, Andrew.** 1784–1857. Amer. politician.
**Stevenson, Robert Louis Balfour.** 1850–94. Scottish poet & novelist.
**Ste·vin** (stə-vĭn′, -vēn′), **Simon.** 1548–1620. Flemish mathematician.
**Stew·art** (sto͞o′ərt, styo͞o′-), **Alexander Turney.** 1803–76. Irish-born Amer. merchant.
**Stewart, Dugald.** 1753–1828. Scottish philosopher.
**Stewart, James.** b. 1908. Amer. actor.
**Stewart, Potter.** b. 1915. Amer. jurist.
**Stewart, Robert.** Viscount CASTLEREAGH.
**Stewart, William Morris.** 1827–1909. Amer. politician.
**Steyn** (stīn), **Martinus Theunis.** 1857–1916. South African statesman.
**Stie·gel** (stē′gəl, shtē′-), **Henry William.** 1729–85. German-born Amer. iron & glass manufacturer.
**Stieg·litz** (stēg′lĭts), **Alfred.** 1864–1946. Amer. pioneer photographer.
**Stig·ler** (stĭg′lər), **George Joseph.** b. 1911. Amer. economist (Nobel, 1982).
**Stik·ker** (stĭk′ər), **Dirk.** b. 1897. Dutch industrialist & statesman.
**Stiles** (stīlz), **Ezra.** 1727–95. Amer. clergyman & educator.
**Stil·i·cho** (stĭl′ĭ-kō′), **Flavius.** 359?–408. Roman general & statesman; beheaded.
**Still** (stĭl), **Andrew Taylor.** 1828–1917. Amer. pioneer osteopath.

**Stil·well** (stĭl′wĕl′, -wəl), **Joseph Warren.** 1883–1946. Amer. army officer.
**Stim·son** (stĭm′sən), **Henry Lewis.** 1867–1950. Amer. public official.
**Stin·nes** (shtĭn′əs, stĭn′-), **Hugo.** 1870–1924. German industrialist.
**Stock·mar** (stŏk′mär′, shtôk′-), Baron **Christian Friedrich von.** 1787–1863. German physician & adviser to the English crown.
**Stock·ton** (stŏk′tən), **Francis Richard ("Frank").** 1834–1902. Amer. author.
**Stockton, Robert Field.** 1795–1866. Amer. naval commander.
**Stod·dard** (stŏd′ərd), **Richard Henry.** 1825–1903. Amer. poet & critic.
**Stoddard, Solomon.** 1643–1729. Amer. clergyman.
**Stod·dert** (stŏd′ərt), **Benjamin.** 1751–1813. Amer. public official.
**Sto·ker** (stō′kər), **Bram.** 1847–1912. English author.
**Stokes** (stōks), Sir **George Gabriel.** 1819–1903. English mathematician & physicist.
**Stokes, Olivia Egleston Phelps** (1847–1927) & **Caroline Phelps** (1854–1909). Amer. philanthropists.
**Stokes, Rose Harriet Pastor.** 1879–1933. Amer. political activist.
**Sto·kow·ski** (stə-kôv′skē, -kôf′-, -kou′-), **Leopold Antoni Stanislaw.** 1882–1977. English-born Amer. conductor.
**Stone** (stōn), **Edward Durell.** 1902–78. Amer. architect.
**Stone, Harlan Fiske.** 1872–1946. Amer. jurist.
**Stone, Irving.** b. 1903. Amer. author.
**Stone, Isidor Feinstein ("I.F.").** b. 1907. Amer. journalist.
**Stone, Lucy.** 1818–93. Amer. feminist.
**Stone, Melville Elijah.** 1848–1929. Amer. journalist.
**Stopes** (stōps), **Marie Carmichael.** 1880–1958. British paleontologist & birth-control advocate.
**Stor·y** (stôr′ē, stōr′ē), **Joseph.** 1779–1845. Amer. jurist & author.
**Story, William Wetmore.** 1819–95. Amer. poet & sculptor.
**Stout** (stout), **Rex.** 1886–1975. Amer. mystery author.
**Stow** (stō), **John.** 1525?–1605. English historian & antiquarian.
**Stowe** (stō), **Harriet Elizabeth Beecher.** 1811–96. Amer. novelist & reformer.
**Stra·bo** (strā′bō′). 63? B.C.–A.D. 24? Greek geographer.
**Stra·chey** (strā′kē), **(Giles) Lytton.** 1880–1932. English historian, biographer, & critic.
**Strachey, (Evelyn) St. John Loe.** 1901–63. English statesman & author.
**Stra·di·va·ri** (străd′ə-vâr′ē, -vär′ē) *also* **Stra·di·var·i·us** (-vâr′ē-əs), **Antonio.** 1644–1737. Italian violinmaker.
**Straf·ford** (străf′ərd), 1st Earl of. *Thomas Wentworth.* 1593–1641. English statesman; executed.
**Strang** (străng), **James Jesse.** 1813–56. Amer. Mormon leader.
**Stras·berg** (străs′bərg, sträs′-), **Lee.** 1901–82. Austrian-born Amer. theatrical producer, director, & teacher.
**Strat·e·mey·er** (străt′ə-mī′ər), **Edward.** 1862–1930. Amer. author.
**Strath·co·na and Mount Roy·al** (străth-kō′nə; mount roi′əl), 1st Baron. *Donald Alexander Smith.* 1820–1914. English colonial administrator.
**Strat·ton** (străt′n), **Charles Sherwood.** *General Tom Thumb.* 1838–83. Amer. circus performer.
**Stratton, Samuel Wesley.** 1861–1931. Amer. physicist.
**Straus** (strous), **Isidor** (1845–1912) & **Nathan** (1848–1931). German-born Amer. merchants.
**Straus** (strous, shtrous), **Oscar** *or* **Oskar.** 1870–1954. Austrian-born French composer.
**Straus** (strous), **Oscar Solomon.** 1850–1926. Amer. politician & diplomat.
**Strauss** (strous, shtrous). Family of Austrian composers, including **Johann** (1804–49), **Johann,** *the Waltz King* (1825–99), & **Josef** (1827–70).
**Strauss, David Friedrich.** 1808–74. German theologian & author.
**Strauss** (strous), **Levi.** 1829?–1902. Amer. clothing manufacturer.
**Strauss** (strous, shtrous), **Richard.** 1864–1949. German composer.
**Stra·vin·sky** (strə-vĭn′skē), **Igor Fëdorovich.** 1882–1971. Russian-born composer.
**Strei·cher** (strī′kər, -KHər, shtrī′-), **Julius.** 1885–1946. German editor & Nazi leader; executed.
**Strei·sand** (strī′sănd′, -zănd′), **Barbra.** b. 1942. Amer. entertainer.
**Stre·se·mann** (shtrā′zə-män′, strā′-), **Gustav.** 1878–1929. German statesman (Nobel, 1926).
**Strick·land** (strĭk′lənd), **William.** 1787?–1854. Amer. architect, engineer, & graphic artist.
**Strind·berg** (strĭnd′bûrg, strĭn′-, strĭn′bĕr′ē), **(Johan) August.** 1849–1912. Swedish playwright & novelist. **—Strind·berg′i·an** *adj.*
**Stro·heim** (strō′hīm′), **Erich von.** 1885–1957. Austrian-born Amer. actor & director.
**Strong** (strông), **Anna Louise.** 1885–1970. Amer. author.
**Strong, Josiah.** 1847–1916. Amer. clergyman & author.
**Strong, William.** 1808–95. Amer. politician & jurist.
**Stru·en·see** (shtrōō′ən-zā′, strōō′-), Count **Johann Friedrich von.** 1737–72. German-born Danish statesman; executed.
**Strutt** (strŭt), **Joseph.** 1749–1802. English antiquarian.
**Struve** (strōō′və), **Otto.** 1897–1963. Russian-born Amer. astronomer.
**Stu·art** (stōō′ərt, styōō′-). Ruling house of Scotland (1371–1625) & Great Britain (1603–49, 1660–1714).
**Stuart, Charles Edward.** *the Young Pretender.* 1720–88. English prince.
**Stuart, Gilbert Charles.** 1755–1828. Amer. painter.
**Stuart, James Ewell Brown ("Jeb").** 1833–64. Amer. Confederate general.
**Stuart, James Francis Edward.** *the Old Pretender.* 1688–1766. Pretender to English throne.
**Stuart, Ruth McEnery.** 1849–1917. Amer. author.
**Stubbs** (stŭbz), **William.** 1825–1901. English historian.
**Stu·de·ba·ker** (stōō′də-bā′kər, styōō′-), **Clement.** 1831–1901. Amer. manufacturer.
**Stur·gis** (stûr′jĭs), **Russell.** 1836–1909. Amer. architect & author.
**Štur·sa** (shtōōr′sə), **Jan.** 1880–1925. Czech. sculptor.
**Stutz** (stŭts), **Harry Clayton.** 1876–1930. Amer. automobile manufacturer.
**Stuy·ve·sant** (stī′vĭ-sənt), **Peter** *or* **Petrus.** 1592?–1672. Dutch colonial administrator in America.
**Sty·ron** (stī′rən), **William.** b. 1925. Amer. novelist.
**Suá·rez Gon·zá·lez** (swär′əz gən-zäl′əs), **Adolfo.** b. 1932. Spanish politician.
**Suck·ling** (sŭk′lĭng), Sir **John.** 1609?–42? English poet.
**Sucre** (sōō′krā), **Antonio José de.** 1795?–1830. South American revolutionary leader.
**Sue** (sōō), **Eugène.** 1804–57. French novelist.
**Sue·to·ni·us** (swē-tō′nē-əs, sōō′ə-). 2nd cent. A.D. Roman historian.
**Su·gi·ya·ma** (sōō′gē-yä′mə), **Hajime.** 1880–1945. Japanese army officer.
**Su·har·to** (sə-här′tō, sōō-), **Raden.** b. 1921. Indonesian military & political leader.
**Sui** (swā). Chinese dynasty (581–618).
**Su·kar·no** (sōō-kär′nō). 1901–70. Indonesian statesman.
**Su·lei·man I** (sōō′lā-män′, -lə-). 1490?–1566. Turkish sultan (1520–66).
**Sul·la** (sŭl′ə), **Lucius Cornelius.** 138–78 B.C. Roman general & dictator.
**Sul·li·van** (sŭl′ə-vən), **Anne Mansfield.** 1866–1936. Amer. teacher of Helen Keller.
**Sullivan,** Sir **Arthur Seymour.** 1842–1900. English composer of operettas.
**Sullivan, Edward Vincent ("Ed").** 1902–74. Amer. columnist & television host.
**Sullivan, Francis John ("Frank").** 1892–1976. Amer. humorist.
**Sullivan, James Edward.** 1860–1914. Amer. sports promoter.
**Sullivan, John.** 1740–95. Amer. Revolutionary general.
**Sullivan, John Lawrence.** 1858–1918. Amer. boxer.
**Sullivan, John Lawrence.** 1899–1982. Amer. public official.
**Sullivan, Louis Henri** *or* **Henry.** 1856–1924. Amer. architect.
**Sullivan, Timothy Daniel ("Big Tim").** 1862–1913. Amer. politician.
**Sul·ly** (sŭl′ē, sōō-lē′), Duc de. *Maximilien de Béthune.* 1560–1641. French statesman & Huguenot leader.
**Sul·ly** (sŭl′ē), **Thomas.** 1783–1872. English-born Amer. portrait painter.
**Sul·ly-Prud·homme** (sŭl′ē-prōō′dəm, sōō-lē′prōō-dôm′), **René François Armand.** 1839–1907. French poet (Nobel, 1901).
**Sulz·ber·ger** (sŭlz′bûr′gər, sōōlz′-), **Arthur Hays** (1891–1968) & **Arthur Ochs** (b. 1926). Amer. newspaper publishers.
**Sum·ner** (sŭm′nər), **Charles.** 1811–74. Amer. politician.
**Sumner, James Batcheller.** 1887–1955. Amer. biochemist (Nobel, 1946).
**Sumner, William Graham.** 1840–1910. Amer. economist, sociologist, & author.
**Sum·ter** (sŭm′tər), **Thomas.** 1734–1832. Amer. Revolutionary general, politician, & diplomat.
**Sun·day** (sŭn′dē), **William Ashley ("Billy").** 1862–1935. Amer. evangelist.
**Sung** (sŏŏng). SONG.
**Sun Yat-sen** (sōōn′ yät′sĕn′). 1866–1925. Chinese revolutionary leader & statesman.
**Sur·ratt** (sə-răt′), **Mary Eugenia Jenkins.** 1820?–65. Amer. co-conspirator in Abraham Lincoln's assassination; executed.
**Sur·rey** (sûr′ē), Earl of. Henry HOWARD.
**Sur·tees** (sûr′tēz′), **Robert Smith.** 1803–64. English author & editor.
**Su·sann** (sōō-zăn′), **Jacqueline.** 1921–74. Amer. author.
**Suss·kind** (sŭs′kīnd′), **David.** b. 1920. Amer. producer & television host.
**Suth·er·land** (sŭth′ər-lənd), **Earl Wilbur, Jr.** 1915–74. Amer. physiologist (Nobel, 1971).

**Sutherland, George.** 1862–1942. English-born Amer. jurist & politician.
**Sutherland, Joan.** b. 1926. Australian operatic soprano.
**Su·tro** (sōō′trō), **Adolph Heinrich Joseph.** 1830–98. German-born Amer. mining engineer.
**Sut·ter** (sŭt′ər), **John Augustus.** 1803–80. German-born Amer. pioneer in Calif.
**Sutt·ner** (zōōt′nər, sōōt′-), **Bertha von.** 1843–1914. Austrian writer (Nobel, 1905).
**Su·vo·rov** (sōō-vôr′əf, -vär′-), Count **Aleksandr Vasilevich.** 1730?–1800. Russian field marshal.
**Su·zu·ki** (sōō-zōō′kē), **Zenko.** b. 1911 Japanese statesman.
**Sved·berg** (svĕd′bərg, -bĕr′ē), **The** *or* **Theodor.** 1884–1971. Swedish chemist (Nobel, 1926).
**Sver·drup** (svĕr′drəp), **Otto Neumann.** 1855–1930. Norwegian Arctic explorer.
**Swa·dos** (swā′dōs), **Elizabeth.** b. 1957. Amer. writer, composer, & director.
**Swain** (swān), **Clara A.** 1834–1910. Amer. medical missionary.
**Swam·mer·dam** (svä′mər-däm′), **Jan.** 1637–80. Dutch naturalist.
**Swan·son** (swŏn′sən), **Gloria.** 1899–1983. Amer. actress.
**Swayne** (swān), **Noah Haynes.** 1804–84. Amer. jurist.
**Sway·ze** (swā′zē), **John Cameron.** b. 1906. Amer. news commentator.
**Swe·den·borg** (swēd′n-bôrg′), **Emanuel.** 1688–1772. Swedish scientist & theologian.
**Swee·linck** *or* **Swe·linck** (swā′lĭngk, svā′-), **Jan Pieterzoon.** 1562–1621. Dutch composer & organist.
**Sweet** (swēt), **Henry.** 1845–1912. English phoneticist & philologist.
**Swe·linck** (swā′lĭngk, svā′-). *var. of* SWEELINCK.
**Swift** (swĭft), **Gustavus Franklin.** 1839–1903. Amer. meat packer.
**Swift, Jonathan.** *Dean Swift.* 1667–1745. Irish-born English satirist.
**Swin·burne** (swĭn′bûrn′), **Algernon Charles.** 1837–1909. English poet & critic.
**Swin·ner·ton** (swĭn′ər-tən), **Frank Arthur.** 1884–1982. English author.
**Swin·ton** (swĭn′tən), 1st Earl of. *Philip Cunliffe-Lister.* 1884–1972. English statesman.
**Swope** (swōp), **Gerard.** 1872–1957. Amer. business executive & government official.
**Swope, Herbert Bayard.** 1882–1958. Amer. journalist.
**Syl·vis** (sĭl′vĭs), **William.** 1828–69. Amer. labor leader.
**Sy·ming·ton** (sī′mĭng-tən), **(William) Stuart.** b. 1901. Amer. politician & public official.
**Sym·onds** (sĭm′əndz, sī′məndz), **John Addington.** 1840–93. English poet, critic, & literary historian.
**Sy·mons** (sĭm′əndz, sī′mənz), **Arthur.** 1865–1945. English poet & literary critic.
**Synge** (sĭng), **John Millington.** 1871–1909. Irish dramatist.
**Synge, Richard Laurence Millington.** b. 1914. British biochemist (Nobel, 1952).
**Szell** (sĕl, zĕl), **George.** 1897–1970. Hungarian-born Amer. conductor.
**Szent-Györ·gyi** (sānt-jôr′ē, -jôrj′), **Albert.** b. 1893. Hungarian-born Amer. biochemist.
**Szi·ge·ti** (sĭg′ĭ-tē, sĭ-gĕt′ē), **Joseph.** 1892–1973. Hungarian-born Amer. violinist.
**Szi·lard** (zĭl′ərd, zə-lärd′), **Leo.** 1898–1964. Hungarian-born Amer. physicist.
**Szold** (zōld), **Henrietta.** 1860–1945. Amer. Zionist leader.

# T

**Tabb** (tăb), **John Banister.** 1845–1909. Amer. priest & poet.
**Tac·i·tus** (tăs′ĭ-təs), **Publius Cornelius.** A.D. 55?–118? Roman historian & orator.
**Taft** (tăft), **Lorado.** 1860–1936. Amer. sculptor.
**Taft, Robert Alphonso.** 1889–1953. Amer. politician.
**Taft, William Howard.** 1857–1930. 27th U.S. President (1909–13) & jurist.
**Tag·gard** (tăg′ərd), **Genevieve.** 1894–1948. Amer. poet.
**Ta·gore** (tə-gôr′, -gōr′), Sir **Rabindranath.** 1861–1941. Indian poet (Nobel, 1913).
**Taine** (tān, tĕn), **Hippolyte Adolphe.** 1828–93. French philosopher & historian.
**Tal·bot** (tôl′bət, tăl′-), **William Henry Fox.** 1800–77. English inventor & antiquarian.
**Tall·chief** (tôl′chēf′), **Maria.** b. 1925. Amer. dancer.
**Tal·ley·rand-Pé·ri·gord** (tăl′ē-rănd′pĕr′ə-gôr′, tä-lə-räN-pā-rē-gôr′), **Charles Maurice de.** 1754–1838. French statesman & diplomat.

**Tam·er·lane** (tăm′ər-lān′) *or* **Tam·bur·laine** (-bər-). 1336?–1405. Mongol conqueror.
**Tamm** (täm), **Igor Yevgeneevich.** 1895–1971. Russian physicist (Nobel, 1958).
**Ta·na·ka** (tə-nä′kə), **Kakeui.** b. 1918. Japanese statesman.
**Tan·cred** (tăng′krĭd). 1078?–1112. Norman Crusade leader.
**Ta·ney** (tô′nē), **Roger Brooke.** 1777–1864. Amer. jurist.
**Tang** (täng). Name of two Chinese dynasties (618–907, 923–36).
**Tan·guy** (täN-gē′), **Yves.** 1900–55. French-born Amer. painter.
**Tan·ner** (tăn′ər), **Henry Ossawa.** 1859–1937. Amer. painter in Paris.
**Tap·pan** (tăp′ən), **Arthur** (1786–1865) & **Lewis** (1788–1873). Amer. merchants, philanthropists, & antislavery advocates.
**Tar·bell** (tär′bəl), **Ida Minerva.** 1857–1944. Amer. muckraking author & editor.
**Tar·dieu** (tär-dyœ′), **André Pierre Gabriel Amédée.** 1876–1945. French statesman & journalist.
**Tar·king·ton** (tär′kĭng′tən), **(Newton) Booth.** 1869–1946. Amer. author.
**Tas·man** (tăz′mən), **Abel Janszoon.** 1603?–59. Dutch navigator & explorer.
**Tas·so** (tăs′ō, tä′sō), **Torquato.** 1544–95. Italian poet.
**Tate** (tāt), **Allen.** 1899–1979. Amer. poet, critic, editor, & biographer.
**Tate, Nahum.** 1652–1715. English author.
**Ta·ti** (tä-tē′), **Jacques.** *Jacques Tatischeff.* 1908–82. French comedian & filmmaker.
**Ta·tum** (tā′təm), **Art.** 1910–56. Amer. musician.
**Tatum, Edward Lawrie.** 1909–75. Amer. biochemist (Nobel, 1953).
**Taube** (toub), **Henry.** b. 1915. Canadian-born Amer. organic chemist (Nobel, 1983).
**Taus·sig** (tou′sĭg), **Frank William.** 1859–1940. Amer. political economist.
**Taw·ney** (tô′nē), **Richard Henry.** 1880–1962. English economist & educator.
**Tay·lor** (tā′lər), **(James) Bayard.** 1825–78. Amer. journalist & novelist.
**Taylor, Bert Leston.** 1866–1921. Amer. newspaper columnist.
**Taylor, David Watson.** 1864–1940. Amer. naval officer & architect.
**Taylor, (Joseph) Deems.** 1885–1966. Amer. composer & critic.
**Taylor, Edward.** 1645?–1729. English-born Amer. Puritan clergyman & poet.
**Taylor, Edward Thompson.** 1793–1871. Amer. religious leader.
**Taylor, Elizabeth.** b. 1932. English-born Amer. actress.
**Taylor, Frederick Winslow.** 1856–1915. Amer. inventor, engineer, & efficiency expert.
**Taylor, Jeremy.** 1613–67. English bishop & theologian.
**Taylor, John.** *Taylor of Caroline.* 1753–1824. Amer. agriculturalist & political philosopher.
**Taylor, Laurette.** 1884–1946. Amer. actress.
**Taylor, Myron Charles.** 1874–1959. Amer. businessman & diplomat.
**Taylor, Paul.** b. 1930. Amer. choreographer.
**Taylor, Robert.** 1911–69. Amer. actor.
**Taylor, Tom.** 1817–80. English dramatist.
**Taylor, Zachary.** *Old Rough and Ready.* 1784–1850. 12th U.S. President (1849–50) & army officer.
**Tchai·kov·sky** (chī-kôf′skē, -kôv′-), **Peter Ilich.** 1840–93. Russian composer. **—Tchai·kov′sky·an, Tchai·kov′ski·an** *adj.*
**Teach** (tēch), **Edward.** *Blackbeard.* d. 1718. English pirate.
**Teas·dale** (tēz′dāl′), **Sara.** 1884–1933. Amer. poet.
**Te·bal·di** (tə-bäl′dē, tĕ-), **Renata.** b. 1922. Italian-born operatic soprano.
**Te·cum·seh** (tĭ-kŭm′sə, -sē, -kŭmp′-) *or* **Te·cum·tha** (-thə). 1768–1813. Amer. Shawnee chief.
**Ted·der** (tĕd′ər), Sir **Arthur William.** *1st Baron Tedder of Glenguin.* 1890–1967. English air marshal.
**Teil·hard de Char·din** (tā-yär′ də shär-dăN′), **Pierre.** 1881–1955. French priest, paleontologist, & philosopher.
**Teisse·renc de Bort** (tĕs-räN′ də bôr′), **Leon Philippe.** 1855–1913. French meteorologist.
**Te Ka·na·wa** (tĭ kä′nə-wə, kä-nä′wä), **Kiri.** b. 1946. New Zealand-born operatic soprano.
**Te·le·mann** (tā′lə-män′), **Georg Philipp.** 1681–1767. German composer.
**Tel·ler** (tĕl′ər), **Edward.** b. 1908. Hungarian-born Amer. physicist.
**Tem·in** (tĕm′ĭn), **Howard Martin.** b. 1934. Amer. oncologist (Nobel, 1975).
**Tem·ple** (tĕm′pəl), **Frederick.** 1821–1902. English prelate.
**Temple, Shirley.** —See Shirley Temple BLACK.
**Temple,** Sir **William.** 1628–99. English author & statesman.

**Teng Hsiao-ping** (dŭng′ shou′pĭng′). DENG XIAOPING.
**Te·niers** (tə-nîrz′, tän-yā′), **David** (*the Elder*; 1582–1649) & **David** (*the Younger*; 1610–90). Flemish painters.
**Ten·nent** (tĕn′ənt), **William** (1673–1746) & **Gilbert** (1703–64). Irish-born Amer. clergymen.
**Ten·niel** (tĕn′yəl), Sir **John.** 1820–1914. English cartoonist & artist.
**Ten·ny·son** (tĕn′ĭ-sən), **Alfred.** *1st Baron Tennyson; Alfred, Lord Tennyson.* 1809–92. English poet. **—Ten′ny·so′ni·an** (-sō′nē-ən) *adj.*
**Ter·borch** or **Ter Borch** (tər-bôrk′, -bôrKH′), **Gerard.** 1617–81. Dutch painter.
**Ter·ence** (tĕr′əns). 190?–59 B.C. Roman author.
**Te·re·sa** (tə-rē′sə, -zə, -rā′-). *var. of* THERESA.
**Teresa,** Mother. b. 1910. Albanian-born Indian nun (Nobel, 1979).
**Te·resh·ko·va** (tə-rĕsh-kō′və), **Valentina Vladmirovna.** b. 1937. Soviet cosmonaut; 1st woman in space.
**Ter·hune** (tûr-hyo͞on′), **Mary Virginia Hawes** (1830–1922) & **Albert Payson** (1872–1942). Amer. authors.
**Ter·man** (tûr′mən), **Lewis Madison.** 1877–1956. Amer. psychologist.
**Ter·ry** (tĕr′ē), **Alfred Howe.** 1827–90. Amer. Union general.
**Terry, Eli.** 1772–1852. Amer. pioneer clockmaker.
**Terry,** Dame **Ellen Alicia** or **Alice.** 1846–1928. English actress.
**Ter·tul·lian** (tər-tŭl′yən, -tŭl′ē-ən). 160?–230? Carthaginian church father.
**Ter·za·ghi** (tər-zä′gē), **Karl.** 1883–1963. Bohemian-born Amer. engineer & soil expert.
**Tes·la** (tĕs′lə), **Nikola.** 1856–1943. Serbian-born Amer. electrical engineer, physicist, & inventor.
**Tet·zel** or **Te·zel** (tĕt′səl), **Johann.** 1465?–1519. German monk & preacher of indulgences.
**Thack·er·ay** (thăk′ə-rē, thăk′rē), **William Makepeace.** 1811–63. Indian-born English novelist. **—Thack′er·ay·an** *adj.*
**Thal·berg** (thâl′bûrg′), **Irving Grant.** 1899–1936. Amer. movie executive.
**Tha·les** (thā′lēz). 640?–546. B.C. Greek philosopher & geometrician. **—Tha·le′sian** (thā-lē′zhən) *adj.*
**Thant** (thänt, thănt), **U.** 1909–74. Burmese UN secretary-general.
**Tharp** (thärp), **Twyla.** b. 1941. Amer. dancer & choreographer.
**Thatch** (thăch), **Edward.** Edward TEACH.
**Thatch·er** (thăch′ər), **Margaret Hilda.** b. 1925. British prime minister (since 1979).
**Thax·ter** (thăk′stər), **Celia Laighton.** 1835–94. Amer. poet.
**Thay·er** (thā′ər, thâr), **Sylvanus.** 1785–1872. Amer. soldier & educator.
**Thayer, William Roscoe.** 1859–1923. Amer. historian & biographer.
**Thei·ler** (tī′lər), **Max.** 1899–1972. South African-born Amer. microbiologist (Nobel, 1951).
**The·mis·to·cles** (thə-mĭs′tə-klēz′). 527?–460? B.C. Athenian military & political leader.
**The·oc·ri·tus** (thē-ŏk′rĭ-təs). 3rd cent. B.C. Greek poet.
**The·o·do·ra** (thē′ə-dôr′ə, -dōr′-). 508?–48. Byzantine empress as wife of Justinian I.
**The·od·o·ric** (thē-ŏd′ər-ĭk). 454?–526. King of Ostragoth (474–526).
**The·o·do·sius I** (thē′ə-dō′shəs, -shē-əs). *the Great.* 346?–95. Roman emperor (379–95).
**The·o·phras·tus** (thē′ə-frăs′təs). 371–287 B.C. Greek philosopher.
**The·o·rell** (tā′ə-rĕl′), **Axel Hugo Theodor.** 1903–82. Swedish biochemist (Nobel, 1955).
**The·re·sa** or **Te·re·sa** (tə-rē′sə, -zə, -rā′-), Saint. 1515–82. Spanish nun & mystic.
**Thes·pis** (thĕs′pĭs). 6th cent. B.C. Greek poet.
**Thiers** (tē-ĕr′), **Louis Adolphe.** 1797–1877. French statesman & historian.
**Tho·burn** (thō′bûrn′), **Isabella.** 1840–1901. Amer. missionary in India.
**Tho·mas** (tô-mä′), **Ambroise.** 1811–96. French composer.
**Thom·as** (tŏm′əs), Saint. One of the 12 Apostles.
**Thomas, Augustus.** 1857–1934. Amer. editor & playwright.
**Thomas, Dylan Marlais.** 1914–53. Welsh poet.
**Thomas, Edith Matilda.** 1854–1925. Amer. poet.
**Thomas, George Henry.** 1816–70. Amer. Union general.
**Thomas, Isaiah.** 1749–1831. Amer. painter & publisher.
**Thomas, Lowell Jackson.** 1892–1981. Amer. radio commentator & author.
**Thomas, Martha Carey.** 1857–1935. Amer. educator & feminist.
**Thomas, Michael Tilson.** b. 1944. Amer. conductor.
**Thomas, Norman Mattoon.** 1884–1968. Amer. socialist leader.
**Thomas, Seth.** 1785–1859. Amer. clockmaker.
**Thomas, Theodore.** 1835–1905. German-born Amer. violinist & conductor.
**Thomas a Kem·pis** (ə kĕm′pĭs, ä). 1380–1471. German ecclesiastic & writer.
**Thomas A·qui·nas** (ə-kwī′nəs), Saint. —See AQUINAS.
**Thomas of Er·cel·doune** (ûr′səl-do͞on′). 1220?–97? Scottish poet.
**Thomp·son** (tŏmp′sən, tŏm′-), **Benjamin.** *Count Rumford.* 1753–1814. Amer. physicist, Loyalist, & philanthropist.
**Thompson, David.** 1770–1857. Canadian explorer & fur trader.
**Thompson, Dorothy.** 1894–1961. Amer. journalist.
**Thompson, Francis.** 1859–1907. English poet.
**Thompson, James Walter.** 1847–1928. Amer. advertising executive.
**Thompson,** Sir **John Sparrow David.** 1844–94. Canadian prime minister (1892–94).
**Thompson, Smith.** 1768–1843. Amer. jurist & politician.
**Thom·son** (tŏm′sən), **Elihu.** 1853–1937. English-born Amer. electrical engineer & inventor.
**Thomson,** Sir **George Paget.** 1892–1975. English physicist (Nobel, 1937).
**Thomson, James.** 1700–48. Scottish-born British poet.
**Thomson, James.** *the Poet of Despair.* 1834–82. Scottish poet.
**Thomson, John Arthur.** 1861–1933. Scottish biologist.
**Thomson,** Sir **Joseph John.** 1856–1940. English physicist & mathematician (Nobel, 1906).
**Thomson, Virgil Garnett.** b. 1896. Amer. composer & critic.
**Tho·reau** (thə-rō′, thôr′ō), **Henry David.** 1817–62. Amer. essayist & poet. **—Tho·reau′vi·an** *adj.*
**Tho·rez** (tô-rĕz′), **Maurice.** 1900–64. French Communist leader.
**Thor·finn Karl·sef·ni** (thôr′fĭn kärl′sĕv-nē). b. 980? Icelandic navigator & explorer.
**Thorn·dike** (thôrn′dīk′), **Ashley Horace.** 1871–1933. Amer. scholar & educator.
**Thorndike, Edward Lee.** 1874–1949. Amer. educational psychologist.
**Thorndike, Lynn.** 1882–1965. Amer. historian.
**Thorndike,** Dame **Sybil.** 1882–1976. English actress.
**Thorn·ton** (thôrn′tən), **William.** 1759–1828. Amer. architect, inventor, & public official.
**Thorpe** (thôrp), **James Francis.** 1888–1953. Amer. Indian athlete.
**Thorpe, Rose Alnora Hartwick.** 1850–1939. Amer. author.
**Thorpe, Thomas Bangs.** 1815–78. Amer. painter & humorist.
**Thor·vald·sen** or **Thor·wald·sen** (tôr′wôl′sən, thôr′-, to͝or′väl′-), **(Albert) Bertel.** 1768–1844. Danish sculptor.
**Thrale** (thrāl). Mrs. Hester Lynch PIOZZI.
**Thras·y·bu·lus** (thrăs′ə-byo͞o′ləs). d. 389 B.C. Athenian commander & statesman.
**Thu·cyd·i·des** (tho͞o-sĭd′ĭ-dēz′). 471–400 B.C. Greek historian.
**Thumb** (thŭm), General **Tom.** Charles Sherwood STRATTON.
**Thur·ber** (thûr′bər), **James Grover.** 1894–1961. Amer. author & artist.
**Thur·mond** (thûr′mənd), **(James) Strom.** b. 1902. Amer. legislator.
**Thurs·by** (thûrz′bē), **Emma Cecilia.** 1845–1931. Amer. singer & educator.
**Thut·mo·se** (tho͞ot-mō′sə). Name of four kings of Egypt, esp. **III,** ruled 1501–1447 B.C.
**Thwaites** (thwāts), **Reuben Gold.** 1853–1913. Amer. historian, librarian, & editor.
**Thys·sen** (tĭs′ən), **Fritz.** 1873–1951. German industrialist.
**Tib·bett** (tĭb′ĭt), **Lawrence Mervil.** 1896–1960. Amer. baritone.
**Ti·be·ri·us** (tī-bîr′ē-əs). 42 B.C.–A.D. 37. Roman emperor (A.D. 14–37). **—Ti·be′ri·an** *adj.*
**Ti·bul·lus** (tə-bŭl′əs), **Albius.** 54?–18? B.C. Roman poet.
**Tick·nor** (tĭck′nər, -nôr′), **George.** 1791–1871. Amer. author.
**Tieck** (tēk), **Ludwig.** 1773–1853. German poet & critic.
**Tie·po·lo** (tē-ā′pə-lō′, -ĕp′ə-), **Giovanni Battista.** 1696–1770. Italian painter.
**Tif·fa·ny** (tĭf′ə-nē), **Charles Lewis.** 1812–1902. Amer. jeweler & merchant.
**Tiffany, Louis Comfort.** 1848–1933. Amer. painter & decorator.
**Tig·lath-pi·le·ser** (tĭg′lăth-pə-lē′zər, -pī-). Name of three Assyrian kings, esp. **III,** d. 727 B.C., ruled 745?–27.
**Til·den** (tĭl′dən), **Samuel Jones.** 1814–86. Amer. politician & philanthropist.
**Tilden, William Tatem, Jr. ("Big Bill").** 1893–1953. Amer. tennis player.
**Til·lich** (tĭl′ĭk), **Paul Johannes.** 1886–1965. German-born theologian & philosopher.
**Till·man** (tĭl′mən), **Benjamin Ryan.** 1847–1918. Amer. farmer & legislator.
**Till·strom** (tĭl′strəm), **Burr.** b. 1917. Amer. puppeteer.
**Til·ly** (tĭl′ē), Count of. *Johann Tserclass.* 1559–1632. Flemish field marshal.
**Til·you** (tĭl′yo͞o′), **George Cornelius.** 1862–1914. Amer. amusement-park owner & inventor.
**Ti·mo·shen·ko** (tĭm′ə-shĕng′kō), **Semen Konstantinovich.** 1895–1970. Russian army officer.
**Tim·o·thy** (tĭm′ə-thē), Saint. 1st cent. A.D. Christian leader.

**Tim·rod** (tĭm′rŏd′), **Henry.** 1828–67. Amer. Confederate war poet.
**Ti·mur** (tī′mŏŏr) *or* **Ti·mur Lenk** (lĕngk). TAMERLANE.
**Tin·ber·gen** (tĭn′bər-gən, -bĕr′KHən), **Jan.** b. 1903. Dutch economist (Nobel, 1969).
**Tinbergen, Nikolaas.** b. 1903. Dutch-born English ethologist (Nobel, 1973).
**Tin·dal** *or* **Tin·dale** (tĭn′dl). *vars. of* TYNDALE.
**Ting** (tĭng), **Samuel Chao Chung.** b. 1936. Amer. physicist (Nobel, 1976).
**Ting·ley** (tĭng′lē), **Katherine Augusta Westcott.** 1847–1929. Amer. theosophist.
**Tin·to·ret·to** (tĭn′tə-rĕt′ō), **Il.** 1518–94. Italian painter.
**Ti·om·kin** (tē-ŏmp′kĭn, tyŏm′-), **Dimitri.** b. 1899. Russian-born Amer. musician & composer.
**Ti·pu Sa·hib** (tē′pŏŏ sä′ĭb, -ēb, -hĭb) *or* **Tip·poo Sa·hib** (tĭp′ōō). 1751–99. Sultan of Mysore (1782–99).
**Tir·pitz** (tîr′pəts, tûr′-), **Alfred von.** 1849–1930. German admiral.
**Tir·so de Mo·li·na** (tîr′sō dā mə-lē′nə). *Gabriel Téllez.* 1571–1648. Spanish dramatist.
**Ti·se·li·us** (tē-sā′lē-əs), **Arne Wilhelm Kaurin.** 1902–71. Swedish biochemist (Nobel, 1948).
**Titch·e·ner** (tĭch′ə-nər), **Edward Bradford.** 1867–1927. English-born psychologist.
**Ti·tian** (tĭsh′ən). 1477–1576. Italian painter. **—Ti′tian·esque′** *adj.*
**Ti·to** (tē′tō), Marshal. *Josip Broz.* 1892–1980. Yugoslavian statesman.
**Ti·tus** (tī′təs). A.D. 40?–81. Roman emperor (79–81).
**Titus,** Saint. 1st cent. A.D. Christian leader.
**To·bey** (tō′bē), **Mark.** 1890–1976. Amer. artist.
**To·bin** (tō′bĭn, -bən), **James.** b. 1918. Amer. economist (Nobel, 1981).
**Tocque·ville** (tōk′vĭl, tŏk′-, tôk-vēl′), **Alexis Charles Henri Maurice Clérel de.** 1805–59. French statesman, traveler, & historian.
**Todd** (tŏd), Sir **Alexander Robertus.** b. 1907. English chemist (Nobel, 1957).
**Todd, David.** 1855–1939. Amer. astronomer & inventor.
**Todd, Mabel Loomis.** 1856–1932. Amer. author & editor.
**Todd, Thomas.** 1765–1826. Amer. jurist.
**Todt** (tōt), **Fritz.** 1891–1942. German military engineer.
**To·gliat·ti** (tōl-yä′tē), **Palmiro.** 1893–1964. Italian editor & Communist leader.
**To·go** (tō′gō′), Count **Heihachiro.** 1847–1934. Japanese admiral.
**Togo, Shigenori.** 1882–1950. Japanese diplomat & politician.
**To·jo** (tō′jō′), **Hideki** *or* **Eiki.** 1885–1948. Japanese army officer & dictator (1941–44); executed.
**To·klas** (tō′kləs), **Alice B.** 1877–1967. Amer. author.
**Tol·kein** (tŏl′kēn′), **John Ronald Reuel.** 1892–1973. English author & philologist.
**Tol·ler** (tô′lər, tŏl′ər), **Ernst.** 1893–1939. German author & politician.
**Tol·stoy** *or* **Tol·stoi** (tōl′stoi, tŏl′-, tōl-stoi′, tôl-, tōl-), Count **Lev** *or* **Leo Nikolaevich.** 1828–1910. Russian author & philosopher. **—Tol·stoy′an, Tol·stoi′an** *adj.*
**Tom·baugh** (tŏm′bô′), **Clyde William.** b. 1906. Amer. astronomer.
**Tom·ma·si·ni** (tŏm′ə-zē′nē, -sē′-), **Vicenzo.** 1880–1950. Italian composer.
**To·mo·na·ga** (tō′mə-nä′gə), **Shinichiro.** 1906–79. Japanese physician (Nobel, 1965).
**Tomp·kins** (tŏmp′kĭnz, tŏm′-), **Daniel D.** 1774–1825. U.S. Vice President (1817–25).
**Tompkins, Sally Louisa.** 1833–1916. Amer. philanthropist.
**Tone** (tōn), **(Theobold) Wolfe.** 1763–98. Irish revolutionist.
**Ton·ti** *or* **Ton·ty** (tŏn′tē, tōn′-, tôN-tē′), **Henry de.** 1650–1704. Italian explorer & trader in America.
**Tooke** (tŏŏk), **(John) Horne.** 1736–1812. English politician & philologist.
**Toombs** (tōōmz), **Robert Augustus.** 1810–85. Amer. legislator & Confederate statesman.
**Tor·que·ma·da** (tôr′kə-mä′də), **Tomás de.** 1420–98. Spanish grand inquisitor.
**Tor·rey** (tôr′ē, tŏr′ē), **John.** 1796–1873. Amer. botanist & chemist.
**Tor·ri·cel·li** (tôr′ə-chĕl′ē, tŏr′-), **Evangelista.** 1608–47. Italian mathematician & physicist.
**Tor·ri·jos Her·re·ra** (tôr-rē′hôs ə-rĕr′ä), **Omar.** 1929–81. Panamanian general & political leader.
**Tos·ca·ni·ni** (tŏs′kə-nē′nē), **Arturo.** 1867–1957. Italian conductor.
**Tot·le·ben** (tôt′lĕ-bĕn′, -bən, -lə-), Count **Franz Eduard Ivanovich.** 1818–84. Russian military engineer.
**Tou·louse-Lau·trec** (tōō-lōōz′lō-trĕk′, -lə-), **Henri de.** 1864–1901. French painter & lithographer.
**Tour·neur** (tûr′nər), **Cyril.** 1575–1626. English dramatist.

**Tous·saint L'Ou·ver·ture** (tōō-săN′ lōō-vĕr-tür′), **Pierre Dominique.** 1743–1803. Haitian revolutionary & statesman.
**Town** (toun), **Ithiel.** 1784–1844. Amer. architect.
**Townes** (tounz), **Charles Hard.** b. 1915. Amer. physicist (Nobel, 1964).
**Town·send** (toun′zənd), **Francis Everett.** 1867–1960. Amer. physician & social reformer.
**Town·shend** (toun′zənd), **Charles.** 1725–67. English politician.
**Toyn·bee** (toin′bē), **Arnold Joseph.** 1889–1975. English historian & educator.
**Tra·cy** (trā′sē), **Spencer.** 1900–67. Amer. actor.
**Tra·jan** (trā′jən). A.D. 52?–117. Roman emperor (98–117).
**Trask** (trăsk), **Kate Nichols ("Katrina").** 1853–1922. Amer. author & philanthropist.
**Trau·bel** (trou′bəl), **Helen.** 1903–72. Amer. operatic soprano.
**Tra·ven** (trā′vən), **B.** *Berick Traven Torsvan.* 1890–1969. Amer. author.
**Trav·ers** (trăv′ərz), **P(amela) L.** b. 1904. Australian-born English author and actress.
**Trav·is** (trăv′ĭs), **William Barret.** 1809–36. Amer. military leader; killed at the Alamo.
**Tree** (trē), Sir **Herbert Beerbohm.** 1853–1917. English actor & theatrical producer.
**Treitsch·ke** (trīch′kə), **Heinrich Gotthard von.** 1834–96. German historian.
**Trench** (trĕnch), **Richard Chenevix.** 1807–86. English poet & philologist.
**Trev·el·lick** (trĕv′ə-lĭk′), **Richard F.** 1830–95. English-born Amer. labor leader.
**Tre·vel·yan** (trə-vĕl′yən, -vĭl′-), **George McCaulay.** 1876–1962. English historian & biographer.
**Trevelyan,** Sir **George Otto.** 1838–1928. English historian & statesman.
**Tre·vi·no** (trə-vē′no), **Lee.** b. 1939. Amer. golfer.
**Trev·i·thick** (trĕv′ə-thĭk′), **Richard.** 1771–1833. English railroad engineer & inventor.
**Tri·gère** (trĭ-zhâr′), **Pauline.** b. 1912. French-born Amer. fashion designer.
**Tril·ling** (trĭl′ĭng), **Lionel.** 1905–75. Amer. critic & author.
**Trim·ble** (trĭm′bəl), **Robert.** 1777–1828. Amer. jurist.
**Trippe** (trĭp), **Juan.** 1900–81. Amer. aviation pioneer.
**Trist** (trĭst), **Nicholas Philip.** 1800–74. Amer. diplomat & public official.
**Trol·lope** (trŏl′əp), **Anthony.** 1815–82. English novelist.
**Tromp** (trômp, trŏmp), **Maarten** *or* **Martin Harpertszoon.** 1597–1653. Dutch admiral.
**Trot·sky** *or* **Trot·ski** (trŏt′skē, trôt′-), **Leon.** 1879–1940. Russian revolutionary & statesman; assassinated. **—Trot′sky·ist, Trot′sky·ite′** *adj. & n.*
**Troy·on** (trwä-yôN′), **Constant.** 1813–65. French painter.
**Tru·deau** (trōō-dō′, trōō′dō′), **Pierre Elliott.** b. 1919. Canadian prime minister (1968–79 & since 1980).
**Truf·faut** (trōō-fō′), **François.** 1932–84. French filmmaker.
**Tru·ji·llo Mo·li·na** (trōō-hē′ō mō-lē′nə, -yō), **Rafael Leónidas.** 1891–1961. Dominican military & political leader.
**Tru·man** (trōō′mən), **Harry S** 1884–1972. 33rd U.S. President (1945–53).
**Truman, Margaret.** b. 1924. Amer. author & singer.
**Trum·bo** (trŭm′bō), **Dalton.** 1905–76. Amer. screenwriter.
**Trum·bull** (trŭm′bəl), **John.** 1750–1831. Amer. lawyer & poet.
**Trumbull, John.** 1756–1843. Amer. painter.
**Trumbull, Jonathan** (1710–85) & **Jonathan** (1740–1809). Amer. statesmen.
**Truth** (trōōth), **Sojourner.** 1797–1883. Amer. abolitionist.
**Trux·tun** (trŭk′stən), **Thomas.** 1755–1822. Amer. naval officer.
**Tsal·da·res** *or* **Tsal·da·ris** (tsäl-där′ēs, -thär′-), **Panages** *or* **Panagis.** 1868–1936. Greek statesman.
**Tub·man** (tŭb′mən), **Harriet.** 1820–1913. Amer. abolitionist.
**Tubman, William Vacanarat Shadrach.** 1895–1971. Liberian statesman.
**Tuch·man** (tŭck′mən), **Barbara Wertheim.** b. 1912. Amer. historian.
**Tuck·er** (tŭk′ər), **Benjamin Ricketson.** 1854–1939. Amer. anarchist.
**Tucker, Richard.** 1914–75. Amer. tenor.
**Tucker, Sophie.** 1884–1966. Russian-born Amer. entertainer.
**Tu·dor** (tōō′dər, tyōō′-). English ruling family (1485–1603).
**Tudor, Antony.** b. 1909. English choreographer.
**Tug·well** (tŭg′wĕl′, -wəl), **Rexford Guy.** b. 1891. Amer. economist & public official.
**Tul·si Das** (tōōl′sē däs′). 1532–1623. Hindu poet.
**Tun·ney** (tŭn′ē), **James Joseph ("Gene").** 1898–1978. Amer. prizefighter.
**Tup·per** (tŭp′ər), Sir **Charles.** 1821–1915. Canadian prime minister (1896).
**Tu·renne** (tōō-rĕn′), Vicomte de. *Henri de La Tour d'Auvergne.* 1611–75. French military leader.

**Tur·ge·nev** (to͝or-gän'yəf, -gĕn'-), **Ivan Sergeevich.** 1818–83. Russian novelist.

**Tur·got** (to͝or-gō'), **Anne Robert Jacques.** 1727–81. French statesman & economist.

**Tur·ner** (tûr'nər), **Frederick Jackson.** 1861–1932. Amer. historian of the West.

**Turner, Joseph Mallord William.** 1775–1851. English painter.

**Turner, Nat.** 1800–31. Amer. slave leader.

**Tut·ankh·a·men** (to͞o'tăng-kä'mən, -täng-) *or* **Tut·enkh·a·mon** (-tĕng-). fl. c. 1358 B.C. Egyptian pharaoh.

**Tut·wi·ler** (tŭt'wī'lər), **Julia Strudwick.** 1841–1916. Amer. pioneer educator & reformer.

**Twain** (twān), **Mark.** Samuel Langhorne CLEMENS.

**Tweed** (twēd), **William Marcy.** *Boss Tweed.* 1823–78. Amer. politician.

**Ty·ler** (tī'lər), **John.** 1790–1862. 10th U.S. President (1841–45).

**Tyler, Moses Coit.** 1835–1900. Amer. scholar & educator.

**Tyler, Royall.** 1757–1826. Amer. jurist & playwright.

**Tyler, Walter ("Wat").** d. 1381. English peasant revolutionary.

**Tyn·dale** *or* **Tin·dal** *or* **Tin·dale** (tĭn'dl), **William.** 1492–1536. English religious reformer & martyr.

**Tyn·dall** (tĭn'dl), **John.** 1820–93. Irish-born British physicist.

**Tyr·whitt-Wil·son** (tĭr'ĭt-wĭl'sən), **Gerald Hugh.** 1883–1950. English composer & painter.

**Tzu Hsi** (tso͞o' shē'). 1835–1908. Empress dowager of China.

# U

**Uc·cel·lo** (o͞o-chĕl'lō), **Paolo.** 1397–1475? Italian painter.

**U·gar·te** (o͞o-gär'tĕ), **Manuel.** 1874?–1951. Argentine author.

**Uh·land** (o͞o'länt'), **Johann Ludwig.** 1787–1862. German lyric poet.

**U·la·no·va** (o͞o-lä'nə-və), **Galina.** b. 1910. Russian-born ballerina.

**Ul·bricht** (o͝ol'brĭkt, -brĭKHt), **Walter.** 1893–1973. German Communist leader.

**Ul·pi·an** (ŭl'pē-ən). 170?–228. Roman jurist; murdered.

**Um·ber·to** (ŭm-bĕr'tō). Name of two Italian kings: **a. I.** 1844–1900. Ruled 1878–1900. **b. II.** 1904–83. Ruled 1946.

**U·na·mu·no** (o͞o'nə-mo͞o'nō), **Miguel de.** 1864–1936. Spanish philosopher.

**Un·cas** (ŭng'kəs). 1588?–1683? Amer. Mohegan Indian leader.

**Un·der·hill** (ŭn'dər-hĭl'), **John.** 1597?–1672. English-born Amer. colonist, soldier, & public official.

**Un·der·wood** (ŭn'dər-wo͝od'), **Oscar Wilder.** 1862–1929. Amer. legislator.

**Und·set** (o͞on'sĕt'), **Sigrid.** 1882–1949. Danish-born Norwegian novelist (Nobel, 1928).

**Un·ter·mey·er** (ŭn'tər-mī'ər), **Louis.** 1885–1977. Amer. author & editor.

**Un·ter·my·er** (ŭn'tər-mī'ər), **Samuel.** 1858–1940. Amer. lawyer & reformer.

**Up·dike** (ŭp'dīk'), **John Hoyer.** b. 1932. Amer. author.

**Up·john** (ŭp'jŏn'), **Richard.** 1802–1878. English-born Amer. architect.

**Ur·ban** (ûr'bən). Name of eight popes, esp. **II,** 1042?–99, reigned 1088–99.

**U·rey** (yo͝or'ē), **Harold Clayton.** 1893–1981. Amer. chemist (Nobel, 1934).

**Ur·is** (yo͝or'ĭs), **Leon Marcus.** b. 1924. Amer. novelist.

**Ur·quhart** (ûr'kərt, -kärt'), Sir **Thomas.** 1611–1660. Scottish Royalist & author.

**Ur·so** (ûr'sō), **Camilla.** 1842–1902. French-born Amer. violinist.

**Ussh·er** (ŭsh'ər), **James.** 1581–1656. Irish-born prelate & theologian.

**U Thant** (o͞o thänt', thănt). —See THANT.

**U·tril·lo** (yo͞o-trĭl'ō, o͞o-trē-ō'), **Maurice.** 1883–1955. French painter.

# V

**Vail** (vāl), **Alfred Lewis.** 1807–59. Amer. telegraph pioneer.

**Vail, Theodore Newton.** 1845–1920. Amer. communications executive.

**Val·de·mar** (väl'də-mär'). WALDEMAR.

**Val·di·via** (väl-dē'vē-ə), **Pedro de.** 1500?–53. Spanish conqueror of Chile.

**Val·do** (väl'dō, väl'-). WALDO.

**Va·lens** (vā'lənz, -lĕnz'), **Flavius.** 328?–78. Eastern Roman emperor.

**Val·en·tine** (văl'ĭn-tīn'), Saint. 3rd cent. A.D. Christian martyr.

**Val·en·tin·ian** (văl'ĭn-tĭn'ē-ən, -tĭn'yən). Name of three Roman emperors: **a. I.** 321–75. Ruled 364–75. **b. II.** 372?–92. Ruled 383–92; assassinated. **c. III.** 419–55. Ruled 425–55; assassinated.

**Val·en·ti·no** (văl'ĭn-tē'nō), **Rudolf.** 1895–1926. Italian-born Amer. actor.

**Va·le·ra y Al·ca·lá Ga·lia·no** (və-lĕr'ə ē äl'kə-lä' găl'ē-ä'nō), **Juan.** 1824–1905. Spanish author & diplomat.

**Va·le·ri·an** (və-lîr'ē-ən). d. 269? Roman emperor (253–60); overthrown & later killed.

**Va·lé·ry** (văl'ə-rē', və-lā-rē'), **Paul Ambroise.** 1871–1945. French poet.

**Val·lan·dig·ham** (və-lăn'dĭ-gəm), **Clement Laird.** 1820–71. Amer. legislator.

**Val·lee** (văl'ē), **Rudy.** b. 1901. Amer. singer.

**Val·le·jo** (və-lā'ō, -hō), **Mariano Guadalupe.** 1808–90. Mexican-born Calif. pioneer & soldier.

**Va·lois** (väl-wä'). French ruling dynasty (1328–1589).

**Van Al·len** (văn ăl'ən), **James Alfred.** b. 1914. Amer. physicist.

**Van·brugh** (văn'brə, văn-bro͞o'), Sir **John.** 1664–1726. English dramatist & architect.

**Van Bur·en** (văn byo͝or'ən), **Abigail ("Abby").** Pauline Esther FRIEDMAN.

**Van Buren, Martin.** 1782–1862. Eighth U.S. President (1837–41).

**Vance** (văns), **Cyrus Roberts.** b. 1917. Amer. public official.

**Vance, Zebulon Baird.** 1830–94. Amer. politician & soldier.

**Van Cort·landt** (văn kôrt'lənd, -lənt), **Stephanus.** 1643–1700. Amer. colonial merchant & official.

**Van·cou·ver** (văn-ko͞o'vər), **George.** 1757–98. English navigator.

**Van De·man** (văn dē'mən), **Esther Boise.** 1862–1937. Amer. archaeologist.

**Van·den·berg** (văn'dĭn-bûrg'), **Arthur Hendrick.** 1884–1951. Amer. diplomat & politician.

**Van De·poele** (văn' də-po͞ol'), **Charles Joseph.** 1846–92. Amer. electrical inventor.

**Van·der·bilt** (văn'dər-bĭlt'), **Cornelius** (1794–1877), **William Henry** (1821–85), & **Cornelius** (1843–99). Amer. railway promoters & financiers.

**Vanderbilt, George Washington.** 1862–1914. Amer. capitalist & benefactor.

**Vanderbilt, Gloria.** b. 1924. Amer. designer.

**Vanderbilt, Harold Stirling.** 1884–1970. Amer. businessman & sportsman.

**Vanderbilt, William Kissam.** 1849–1920. Amer. capitalist & sportsman.

**Van·der·lyn** (văn'dər-lĭn'), **John.** 1775–1852. Amer. artist.

**van der Ro·he** (văn dər rō'ə). —See MIES VAN DER ROHE.

**Van De·van·ter** (văn' də-văn'tər), **Willis.** 1859–1941. Amer. jurist.

**van Dong·en** (văn dông'ən), **Kees.** 1877–1968. Dutch artist.

**Van Dor·en** (văn dôr'ən, dōr'-), **Carl Clinton.** 1885–1950. Amer. author & critic.

**Van Doren, Mark Albert.** 1894–1972. Amer. poet & critic.

**Van·dyke** *or* **Van Dyck** (văn-dīk'), Sir **Anthony.** 1599–1641. Flemish painter.

**Vane** (vān), Sir **Henry** *or* **Harry.** 1613–62. English colonial administrator & statesman; executed for treason.

**Vane, John Robert.** b. 1927. English pharmacologist (Nobel, 1982).

**van Eyck** (văn īk'), **Jan.** 1370?–1440? Flemish painter.

**van Gogh** (văn gō', gôKH', văn KHôKH'), **Vincent.** 1853–90. Dutch painter.

**Van Heu·sen** (văn hyo͞o'zən), **James ("Jimmy").** b. 1913. Amer. songwriter.

**Van Hise** (văn hīs'), **Charles Richard.** 1857–1918. Amer. geologist & educator.

**Van Rens·se·laer** (văn rĕn'sə-lîr', rĕn'sə-lər), **Killian** *or* **Kiliaen.** 1595–1644. Dutch merchant.

**Van Rensselaer, Mariana Alley Griswold.** 1851–1934. Amer. art critic.

**Van Rensselaer, Stephen.** 1764–1839. Amer. military & political leader.

**Van·sit·tart** (văn-sĭt'ərt), Sir **Robert Bilbert.** *1st Baron Vansittart of Denham.* 1881–1957. English diplomat & author.

**Van Swer·in·gen** (văn swâr'ĭn-jən), **Oris Paxton** (1879–1936) & **Mantis James** (1881–1935). Amer. real-estate developers & railway executives.

**van't Hoff** (vänt hôf', hōf'), **Jacobus Hendricus.** 1852–1911. Dutch chemist (Nobel, 1901).

**Van Vech·ten** (văn vĕk'tən), **Carl.** 1880–1964. Amer. critic.

**Van Vleck** (văn vlĕk'), **John Hasbrouck.** 1899–1980. Amer. physicist (Nobel, 1977).

**Van Zandt** (văn zănt'), **Marie.** 1858–1919. Amer. operatic soprano.

**Van·zet·ti** (văn-zĕt'ē), **Bartolomeo.** 1888–1927. Italian-born Amer. anarchist; executed for murder.

**Va·rèse** (və-răz', -rĕz'), **Edgard.** 1883–1965. French-born Amer. composer.

**Var·gas** (vär′gəs), **Getulio Dornelles.** 1883–1954. Brazilian statesman.
**Var·num** (vär′nəm), **Joseph Bradley.** 1751–1821. Amer. Revolutionary soldier & politician.
**Var·ro** (văr′ō), **Marcus Terentius.** 116–27 B.C. Roman scholar & encyclopedist.
**Va·sa·ri** (və-zär′ē, -sär′ē), **Giorgio.** 1511–74. Italian artist & architect.
**Vas·sar** (văs′ər), **Matthew.** 1792–1868. Amer. brewer & philanthropist.
**Vau·ban** (vō-bäN′), Marquis **Sébastien Le Prestre de.** 1633–1707. French military engineer.
**Vaughan** (vôn), **Henry.** 1622–95. Welsh poet.
**Vaughan, Sarah.** b. 1924. Amer. singer.
**Vaughan Wil·liams** (wĭl′yəmz), **Ralph.** 1872–1958. English composer.
**Vaux** (vôks), **Calvert.** 1824–95. English-born Amer. landscape architect.
**Veb·len** (vĕb′lən), **Oswald.** 1880–1960. Amer. mathematician.
**Veblen, Thorstein Bunde.** 1857–1929. Amer. economist.
**Vega** (vā′gə), **Lope de.** 1562–1635. Spanish author.
**Ve·láz·quez** or **Ve·lás·quez** (və-läs′kəs), **Diego Rodríguez de Silva y.** 1599–1660. Spanish painter.
**Ven·dome** (väN-dōm′), Duc de. *Louis Joseph de Bourbon.* 1654–1712. French general.
**Ve·ni·ze·los** (vĕn′ə-zā′ləs, -zĕl′əs), **Eleutherios.** 1864–1936. Greek statesman.
**Ven·tu·ri** (vĕn-to͝or′ē), **Robert Charles.** b. 1925. Amer. architect.
**Ver·di** (vâr′dē), **Giuseppe.** 1813–1901. Italian composer.
**Ve·re·shcha·gin** (vĕr′əsh-chä′gĭn, vĕr′ə-shä′-), **Vasili Vasilievich.** 1842–1904. Russian painter.
**Ver·gil** (vûr′jəl). *var. of* VIRGIL.
**Ver·laine** (vĕr-lān′, -lĕn′), **Paul.** 1844–1896. French poet.
**Ver·meer** (vər-mîr′, -mâr′), **Jan.** 1632–75. Dutch painter.
**Verne** (vûrn, vĕrn), **Jules.** 1828–1905. French novelist.
**Ver·ner** (vûr′nər, vĕr′-), **Karl Adolph.** 1846–1905. Danish philologist.
**Ver·nier** (vûr′nē-ər, vĕr-nyā′), **Pierre.** 1580–1637. French mathematician.
**Ve·ro·ne·se** (vĕr′ə-nā′sē, -zē), **Paolo.** 1528–88. Italian painter.
**Ver·ra·za·no** or **Verraz·za·no** (vĕr′ə-zä′nō), **Giovanni da.** 1485?–1528? Italian explorer.
**Ver·rett** (və-rĕt′), **Shirley.** b. 1933. Amer. operatic soprano.
**Ver·roc·chio** (və-rō′kē-ō, -rôk′ē-ō), **Andrea del.** 1435–88. Florentine artist.
**Ve·rus** (vîr′əs), **Lucius Aurelius.** 130–69. Roman emperor (161–69).
**Ver·woerd** (fər-vo͝ort′), **Hendrik Frensh.** 1901–66. South African statesman; assassinated.
**Ver·y** (vĕr′ē, vîr′ē), **Jones.** 1813–80. Amer. author.
**Ve·sa·li·us** (vĭ-sā′lē-əs, -zā′-), **Andreas.** 1514–64. Flemish anatomist.
**Ve·sey** (vē′zē), **Denmark.** 1767?–1822. Amer. black insurrectionist.
**Ves·pa·sian** (vĕs-pā′zhən, -zhē-ən). A.D. 9–79. Roman emperor (69–79).
**Ves·puc·ci** (vĕs-po͞o′chē, -pyo͞o′-), **Amerigo.** 1454–1512. Italian navigator & explorer.
**Vick** (vĭk), **James.** 1812–82. English-born Amer. horticulturist.
**Vick·ers** (vĭk′ərz), **Jon.** b. 1926. Canadian-born tenor.
**Vic·tor Em·man·u·el I** (vĭk′tər ĭ-măn′yo͞o-əl). 1759–1824. Sardinian king (1802–21).
**Victor Emmanuel II.** 1820–78. Italian king (1861–78).
**Victor Emmanuel III.** 1869–1947. Italian king (1936–46); abdicated.
**Vic·to·ri·a** (vĭk-tôr′ē-ə, -tōr′-). 1819–1901. British queen (1837–1901) & empress of India (1876–1901). **—Vic·to′ri·an** *adj.* & *n.*
**Vi·da** (vē′də), **Marco Girolamo.** 1480–1566. Italian prelate & poet.
**Vi·dal** (vĭ-däl′), **Gore.** b. 1925. Amer. author.
**Vi·dor** (vĭ-dôr′, vī′dôr′), **King.** 1895–1982. Amer. filmmaker.
**Vi·gée-Le·brun** (vē-zhā′lə-brœN′), **Marie Ann Élisabeth.** 1755–1842. French painter.
**Vi·gno·la** (vēn-yō′lə), **Giacomo da.** 1507–73. Italian architect.
**Vi·gny** (vēn-yē′), Comte **Alfred Victor de.** 1797–1863. French author.
**Vil·la** (vē′ə, vē′yə), **Francisco ("Pancho").** 1877?–1923. Mexican revolutionary leader; assassinated.
**Vil·la-Lo·bos** (vē′lə-lō′bōs), **Heitor.** 1881–1959. Brazilian composer.
**Vil·lard** (vĭ-lär′, -lärd′), **Henry.** 1835–1900. German-born Amer. journalist & railway magnate.
**Villard, Oswald Garrison.** 1872–1949. Amer. journalist & editor.
**Vil·lars** (vĭ-lär′), Duc **Claude Louis Hector de.** 1653–1734. French marshal.
**Vil·lel·la** (və-lĕl′ə), **Edward.** b. 1936. Amer. ballet dancer.
**Vil·liers** (vĭl′ərz, -yərz), **George.** 1st Duke of BUCKINGHAM.
**Vil·lon** (vē-ôN′, -yôN′, -lôN′), **François.** 1431–1462? French poet.
**Villon, Jacques.** 1875–1963. French artist.
**Vin·cent** (vĭn′sənt), **John Heyl.** 1832–1920. Amer. clergyman & educator.
**Vincent de Paul** (də pôl), Saint. 1581–1660. French ecclesiastic.
**Vi·no·gra·doff** (vĭn′ə-grăd′ôf′), Sir **Paul Gavrilovich.** 1854–1925. Russian-born British jurist & historian.
**Vin·son** (vĭn′sən), **Carl** 1884–1981. Amer. legislator.
**Vinson, Frederick Moore.** 1890–1953. Amer. jurist.
**Viol·let-le-Duc** (vē′ə-lā′lə-do͞ok′, -dyo͞ok′), **Eugène Emmanuel.** 1814–79. French architect.
**Vir·chow** (fîr′kō, vîr′-), **Rudolf.** 1821–1902. German pioneer pathologist.
**Vir·gil** *also* **Ver·gil** (vûr′jəl). 70–19 B.C. Roman poet. **—Vir·gil′i·an, Ver·gil′i·an** (vûr-jĭl′ē-ən) *adj.*
**Vir·ta·nen** (vîr′tə-nĕn′), **Artturi Ilmari.** 1895–1973. Finnish biochemist (Nobel, 1945).
**Vis·con·ti** (vĭs-kŏn′tē), **Gian Galeazzo.** 1351?–1402. Milanese ruler (1378–1402).
**Vi·shin·ski** (vĭ-shĭn′skē), **Andrei Yanuarievich.** 1883–1954. Soviet jurist & diplomat.
**Vi·tru·vi·us Pol·li·o** (vĭ-tro͞o′vē-əs pŏl′ē-ō′), **Marcus.** 1st cent. B.C. Roman architect & engineer.
**Vi·val·di** (vĭ-väl′dē, -vôl′-), **Antonio.** 1675?–1741. Italian composer.
**Viz·ca·í·no** (vĭz-kä-ē′nō), **Sebastián.** 1550?–1615. Spanish explorer.
**Vla·minck** (vlə-măNk′), **Maurice de.** 1876–1958. French artist.
**Vo·gler** (fō′glər), **Georg Joseph.** 1749–1814. German organist & composer.
**Vol·stead** (vŏl′stĕd′, vôl′-, vōl′-), **Andrew John.** 1860–1947. Amer. legislator.
**Vol·ta** (vōl′tə, vôl′-), Count **Alessandro.** 1745–1827. Italian physicist & electrical pioneer.
**Vol·taire** (vōl-târ′, vŏl-, vôl-tĕr′). *François Marie Arouet.* 1694–1778. French author.
**Von Eu·ler** (fôn oi′lər), **Ulf Svante.** 1905–83. Swedish physiologist.
**Von·ne·gut** (vŏn′ĭ-gət), **Kurt, Jr.** b. 1922. Amer. author.
**Von Neu·mann** (vŏn noi′män′), **John.** 1903–57. Hungarian-born Amer. mathematician.
**Vo·ro·shi·lov** (vôr′ə-shē′lôf′, -lôv′), **Kliment Efremovich.** 1881–1969. Soviet military & political leader.
**Vor·ster** (fôr′stər), **Balthazar Johannes.** 1915–83. South African statesman.
**Vought** (vôt), **Chance Milton.** 1890–1930. Amer. aircraft designer & manufacturer.
**Voz·ne·sen·ski** (vŏz′nə-sĕn′skē), **Andrei.** b. 1933. Soviet poet.
**Vree·land** (vrē′lənd), **Diana Dalziel.** b. 1903. French-born Amer. editor & fashion expert.
**Vuil·lard** (vwē-yär′), **(Jean) Édouard.** 1868–1940. French painter.

# W

**Waals** (wôlz, väls), **Johannes Diderik van der.** 1837–1923. Dutch physicist (Nobel, 1910).
**Wace** (wās, wäs). 12th cent. Anglo-Norman poet.
**Wad·dell** (wŏ-dĕl′), **James Iredell.** 1824–86. Amer. Confederate naval commander.
**Wade** (wād), **Benjamin Franklin.** 1800–78. Amer. politician.
**Wag·ner** (väg′nər), **(Wilhelm) Richard.** 1813–83. German composer.
**Wag·ner** (wăg′nər), **Robert Ferdinand.** b. 1910. Amer. politician.
**Wag·ner von Jau·regg** (väg′nər fôn you′rĕk′), **Julius.** 1857–1940. Austrian neurologist & psychiatrist (Nobel, 1927).
**Wain·wright** (wān′rīt′), **Jonathan Mayhew.** 1883–1953. Amer. general.
**Wainwright, Richard** (1817–62) & **Richard** (1849–1926). Amer. naval officers.
**Waite** (wāt), **Morrison Remick.** 1816–88. Amer. jurist.
**Waks·man** (wăks′mən), **Selman Abraham.** 1888–1973. Russian-born Amer. microbiologist (Nobel, 1952).
**Wal·cott** (wôl′kət), **Charles Doolite.** 1850–1927. Amer. geologist & paleontologist.
**Wald** (wôld), **George.** b. 1906. Amer. biologist (Nobel, 1967).
**Wald, Lillian D.** 1867–1940. Amer. social reformer.
**Wal·de·mar** (vôl′də-mär′, väl′-) or **Val·de·mar** (väl′-). Name of four Danish kings, esp. **I,** *the Great,* 1132–82, ruled 1157–82.
**Wal·der·see** (väl′dər-zā′, wôl′-), Count **Alfred von.** 1832–1904. German field marshal.
**Wald·heim** (vält′hīm′), **Kurt.** b. 1918. Austrian diplomat & UN secretary-general.

**Wal·do** (wôl′dō, wäl′-) *or* **Val·do** (văl′-, väl′-), **Peter.** 12th cent. French heretic.
**Wa·le·sa** (wä-lĕn′sə, vä-wĕn′sä), **Lech.** b. 1943? Polish labor leader (Nobel, 1983).
**Wal·green** (wôl′grēn′), **Charles Rudolph.** 1873–1939. Amer. pharmacist & businessman.
**Wal·ker** (wô′kər), **Amasa.** 1799–1875. Amer. political economist.
**Walker, David.** 1785–1830. Amer. abolitionist.
**Walker, Francis Amasa.** 1840–97. Amer. statistician & economist.
**Walker, James John ("Jimmy").** 1881–1946. Amer. politician.
**Walker, Joseph Reddeford.** 1798–1876. Amer. frontiersman.
**Walker, Mary Edwards.** 1832–1919. Amer. physician & feminist.
**Walker, Robert John.** 1801–69. Amer. financier & politician.
**Walker, Sarah Breedlove.** 1867–1919. Amer. businesswoman.
**Walker, Thomas.** 1715–94. Amer. physician, explorer, speculator, & politician.
**Walker, William.** 1824–60. Amer. adventurer & South American revolutionary.
**Wal·lace** (wŏl′ĭs), **Alfred Russel.** 1823–1913. English naturalist.
**Wallace, De Witt** (1889–1981) & **Lila Bell Acheson** (1889–1984). Amer. publishers.
**Wallace, George Corley.** b. 1919. Amer. politician.
**Wallace, Henry Agard.** 1888–1965. U.S. Vice President (1941–45).
**Wallace, Irving.** b. 1916. Amer. novelist.
**Wallace, Lewis ("Lew").** 1827–1905. Amer. general, diplomat, & author.
**Wallace, Myron ("Mike").** b. 1918. Amer. broadcast journalist.
**Wallace,** Sir **William.** 1272?–1305. Scottish national hero; executed for treason.
**Wal·lach** (wäl′ək, väl′-), **Otto.** 1847–1931. German chemist (Nobel, 1910).
**Wal·lack** (wŏl′ək), **James William.** 1795?–1864. English-born actor & theatrical manager in America.
**Wallack, Lester.** 1820–88. Amer. actor.
**Wal·len·stein** (wŏl′ən-stīn′), **Albrecht Eusebius Wenzel von.** 1583–1634. Austrian military leader.
**Wal·ler** (wŏl′ər), **Edmund.** 1606–87. English poet.
**Waller, Thomas ("Fats").** 1904–43. Amer. jazz musician.
**Wal·lis** (wŏl′ĭs), **Hal Brent.** b. 1899. Amer. motion-picture producer.
**Wallis, John.** 1616–1703. English mathematician.
**Wal·pole** (wôl′pōl′, wŏl′-), **Horace** *or* **Horatio.** *4th Earl of Orford.* 1717–97. English author.
**Walpole,** Sir **Hugh Seymour.** 1884–1941. New Zealand-born English novelist & critic.
**Walpole,** Sir **Robert.** *1st Earl of Orford.* 1676–1745. English statesman.
**Walsh** (wôlsh), **Thomas James.** 1859–1933. Amer. legislator.
**Wal·ter** (väl′tər), **Bruno.** 1876–1962. German conductor.
**Wal·ter** (wôl′tər), **John.** 1739–1812. English newspaper publisher.
**Wal·ters** (wôl′tərz), **Barbara.** b. 1931. Amer. broadcast journalist.
**Wal·ther** (väl′tər), **Carl Ferdinand Wilhelm.** 1811–87. German-born Amer. educator & clergyman.
**Walther von der Vo·gel·wei·de** (fôn dər fō′gəl-vī′də). 1170?–1230? German poet.
**Wal·ton** (wôl′tən), **Ernest Thomas Sinton.** b. 1903. Irish physicist (Nobel, 1951).
**Walton, Izaak.** 1593–1683. English fisherman & author.
**Walton,** Sir **William Turner.** 1902–83. English composer.
**Wan·a·ma·ker** (wŏn′ə-mā′kər), **John.** 1838–1922. Amer. merchant & public official.
**Wang Jing·wei** *also* **Wang Ching-wei** (wäng′ jē-ĭng′wā′). 1838–1944. Chinese political leader.
**War·beck** (wôr′bĕk′), **Perkin.** 1474–99. Flemish or Walloon pretender to English throne; executed.
**War·burg** (wôr′bərg′, vär′bŏŏrk′), **Otto Heinrich.** 1883–1970. German biochemist (Nobel, 1931).
**Ward** (wôrd), Sir **Adolphus William.** 1837–1924. English historian & editor.
**Ward, Artemus.** 1727–1800. Amer. Revolutionary general.
**Ward, Artemus.** Charles Farrar BROWNE.
**Ward, Barbara.** *Baroness Jackson.* 1914–81. English economist.
**Ward, Elizabeth Stuart Phelps.** 1844–1911. Amer. author.
**Ward, John Quincy Adams.** 1830–1910. Amer. sculptor.
**Ward,** Sir **Joseph George.** 1856–1930. New Zealand statesman.
**Ward, Lester Frank.** 1841–1913. Amer. botanist, geologist, & sociologist.
**Ward, Mary Augusta Arnold.** *Mrs. Humphrey Ward.* 1851–1920. English novelist.
**Ward, (Aaron) Montgomery.** 1843–1913. Amer. mail-order merchant.
**Ward, Nathaniel.** 1578?–1652. English-born clergyman & writer in America.
**War·hol** (wôr′hôl′, -hōl′), **Andy.** b. 1930? Amer. artist.
**War·ing** (wâr′ĭng), **Fred.** 1900–84. Amer. conductor.
**War·ner** (wôr′nər), **Charles Dudley.** 1829–1900. Amer. author & editor.
**Warner, Glenn Scobey ("Pop").** 1871–1954. Amer. football coach.
**Warner, Harry Morris** (1881–1958), **Albert** (1884–1967), **Samuel Louis** (1887–1927), & **Jack.** (b. 1892). Amer. filmmakers.
**Warner, Susan Bogert** (1819–85) & **Anna Bartlett** (1827–1915). Amer. authors.
**War·ren** (wôr′ən, wŏr′-), **Earl.** 1891–1974. Amer. jurist.
**Warren, Gouverneur Kemble.** 1830–82. Amer. army general & engineer.
**Warren, Joseph** (1741–75), **John** (1753–1815), & **John Collins** (1778–1856). Amer. surgeons & physicians.
**Warren, Mercy Otis.** 1728–1814. Amer. author.
**Warren, Robert Penn.** b. 1905. Amer. author.
**Warren, Whitney.** 1864–1943. Amer. architect.
**War·ton** (wôr′tn), **Thomas.** 1728–90. English poet, critic, & scholar.
**War·wick** (wŏr′ĭk), Earl of. *Richard Neville.* 1428–71. English military & political leader.
**Wash·burn** (wŏsh′bûrn′, wôsh′-), **Margaret Floy.** 1871–1939. Amer. psychologist.
**Wash·ing·ton** (wŏsh′ĭng-tən, wôsh′-), **Booker T(aliaferro).** 1856–1915. Amer. educator & author.
**Washington, Bushrod.** 1762–1829. Amer. jurist.
**Washington, George.** 1732–99. First U.S. President (1789–97) & Revolutionary soldier. **—Wash′ing·to′ni·an** (-tō′nē-ən) *adj.*
**Washington, Martha Dandridge Custis.** 1731–1802. Wife of George Washington.
**Was·ser·mann** (wä′sər-mən, vä′-), **August von.** 1866–1925. German bacteriologist.
**Wa·ter·house** (wô′tər-hous′, wŏt′ər-), **Benjamin.** 1754–1846. Amer. physician.
**Wa·ter·man** (wô′tər-mən, wŏt′ər-), **Lewis Edson.** 1837–1901. Amer. inventor & manufacturer.
**Wa·ters** (wô′tərz, wŏt′ərz), **Ethel.** 1896–1977. Amer. actress & singer.
**Wat·son** (wŏt′sən), **Elkanah.** 1758–1842. Amer. banker & agriculturist.
**Watson, James Dewey.** b. 1928. Amer. biologist (Nobel, 1962).
**Watson, Thomas Augustus.** 1854–1934. Amer. telephone pioneer & shipbuilder.
**Watson, Thomas Edward.** 1856–1922. Amer. politician & publisher.
**Watson, Thomas John.** 1874–1956. Amer. businessman.
**Watson,** Sir **(John) William.** 1858–1935. English poet.
**Watt** (wŏt), **James.** 1736–1819. Scottish-born engineer & inventor.
**Wat·teau** (wŏ-tō′, vä-), **Jean Antoine.** 1684–1721. French painter.
**Wat·ter·son** (wô′tər-sən, wŏt′ər-), **Henry.** *Marse Henry.* 1840–1921. Amer. editor & politician.
**Watts** (wŏts), **George Frederick.** 1817–1904. English painter.
**Watts, Isaac.** 1674–1748. English poet, theologian, & hymn writer.
**Watts-Dun·ton** (wŏts′dŭn′tən), **Walter Theodore.** 1832–1914. English author.
**Waugh** (wô), **Alec** (1898–1981) & **Evelyn** (1903–66). English authors.
**Wa·vell** (wā′vəl), 1st Earl of. *Archibald Percival Wavell.* 1883–1950. English army officer.
**Way·land** (wā′lənd), **Francis.** 1796–1865. Amer. clergyman.
**Wayne** (wān), **Anthony ("Mad Anthony").** 1745–96. Amer. Revolutionary general.
**Wayne, James Moore.** 1790–1867. Amer. jurist.
**Wayne, John.** 1907–79. Amer. actor.
**Wea·ver** (wē′vər), **James Baird.** 1833–1912. Amer. Populist leader.
**Weaver, Robert Clifton.** b. 1907. Amer. economist.
**Webb** (wĕb), **Beatrice Potter.** 1858–1943. English socialist & author.
**Webb, Clifton.** 1893–1966. Amer. actor.
**Webb, Sidney James.** *1st Baron Passfield.* 1859–1947. English sociologist & economist.
**We·ber** (vā′bər), **Ernst Heinrich.** 1795–1878. German physiologist & psychologist.
**Weber,** Baron **Karl Maria Friedrich Ernst von.** 1786–1826. German composer & conductor.
**Weber, Max.** 1864–1920. German sociologist & economist.
**We·ber** (wĕb′ər), **Max.** 1881–1961. Russian-born Amer. painter.
**We·ber** (vā′bər), **Wilhelm Eduard.** 1804–91. German sociologist & economist.
**We·bern** (vā′bərn), **Anton von.** 1883–1945. Austrian composer.
**Web·ster** (wĕb′stər), **Alice Jane Chandler ("Jean").** 1876–1916. Amer. author.
**Webster, Daniel.** 1782–1852. Amer. politician, diplomat, & orator.
**Webster, John.** fl. early 17th cent. English dramatist.
**Webster, Noah.** 1758–1843. Amer. lexicographer.

**Wedg·wood** (wĕj'wŏŏd'), **Josiah.** 1730–95. English potter.
**Weed** (wēd), **Thurlow.** 1797–1882. Amer. journalist & politician.
**Weems** (wēmz), **Mason Locke.** 1759–1825. Amer. clergyman & biographer.
**We·ge·ner** (vā'gə-nər), **Alfred Lothar.** 1880–1930. German geophysicist, meteorologist, & explorer.
**Wei** (wā). Name of several Chinese dynasties (220–65, 386–550, & 535–56).
**Weill** (wīl, vīl), **Kurt.** 1900–50. German-born composer.
**Wein·berg** (wīn'bûrg'), **Steven.** b. 1933. Amer. physicist (Nobel, 1979).
**Weir** (wîr). Family of Amer. painters, including **Robert Walter** (1803–99), **John Ferguson** (1841–1926), & **Julian Alden** (1852–1919).
**Weis·man** (vīs'män'), **August.** 1834–1914. German biologist.
**Weiss·mul·ler** (wīs'mŭl'ər, -myŏŏ'lər), **Johnny.** 1903–84. Amer. swimmer & actor.
**Weiz·mann** (vīts'mən', wīts'-, wīz'-), **Chaim.** 1874–1952. Polish-born Israeli chemist & statesman.
**Welch** (wĕlch, wĕlsh), **Joseph Nye.** 1890–1960. Amer. lawyer.
**Welch, William Henry.** 1850–1934. Amer. pathologist & bacteriologist.
**Weld** (wĕld), **Theodore Dwight.** 1803–95. Amer. abolitionist.
**Welk** (wĕlk), **Lawrence.** b. 1903. Amer. musician & bandleader.
**Wel·ler** (wĕl'ər), **Thomas Huckle.** b. 1915. Amer. microbiologist (Nobel, 1954).
**Welles** (wĕlz), **Gideon.** 1802–78. Amer. editor & public official.
**Welles, (George) Orson.** b. 1915. Amer. actor, producer, & director.
**Welles, Sumner.** 1892–1961. Amer. diplomat & journalist.
**Welles·ley** (wĕlz'lē), 1st Marquis. *Richard Colley Wellesley.* 1760–1842. Irish-born British politician & colonial administrator.
**Wel·ling·ton** (wĕl'ĭng-tən). 1st Duke of. *Arthur Wellesley, the Iron Duke.* 1769–1852. Irish-born British military leader & statesman.
**Well·man** (wĕl'mən), **Walter.** 1858–1934. Amer. journalist & Arctic explorer.
**Wells** (wĕlz), **Carolyn.** 1862–1942. Amer. author.
**Wells, David Ames.** 1828–98. Amer. economist.
**Wells, Henry.** 1805–78. Amer. express-company operator.
**Wells, Herbert George ("H.G.").** 1866–1946. English author.
**Wells, Horace.** 1815–48. Amer. dentist & pioneer anesthetist.
**Wells, Mary Georgene.** b. 1928. Amer. businesswoman.
**Wel·ty** (wĕl'tē), **Eudora.** b. 1909. Amer. author.
**Wen·ces·laus** (wĕn'sĭ-slôs') or **Wen·zel** (vĕnt'səl). 1361–1419. Holy Roman emperor (1378–1400; deposed) & king of Germany & Bohemia as **Wenceslaus IV** (1378–1419).
**Wen·dell** (wĕn'dl), **Barrett.** 1855–1921. Amer. educator & author.
**Went·worth** (wĕnt'wûrth'), **Cecile de.** d. 1933. Amer. portrait painter.
**Wentworth, William Charles.** 1793?–1872. Australian statesman.
**Wen·zel** (vĕnt'səl). *var. of* WENCESLAUS.
**Wer·fel** (vĕr'fəl), **Franz.** 1890–1945. Austrian author.
**Wer·ner** (wûr'nər, vĕr'-), **Alfred.** 1866–1919. German-born Swiss chemist (Nobel, 1913).
**Wert·mül·ler** (vĕrt'myŏŏ'lər), **Lina.** b. 1926. Italian filmmaker.
**Wes·ley** (wĕs'lē, wĕz'-), **John.** 1703–91. British founder of Methodism. **—Wes'ley·an** *adj. & n.*
**West** (wĕst), **Benjamin.** 1738–1820. Amer. painter in England.
**West, Jessamyn.** 1902–84. Amer. author.
**West, Mae.** 1892?–1980. Amer. actress.
**West, Nathanael.** 1903–40. Amer. author.
**West,** Dame **Rebecca.** *Cicily Isabel Fairfield.* 1892–1983. English novelist & critic.
**West·cott** (wĕs'kət, wĕst'-), **Edward Noyes.** 1846–98. Amer. banker & author.
**Wes·ter·marck** (wĕs'tər-märk'), **Edward Alexander.** 1862–1939. Finnish anthropologist.
**West·ing·house** (wĕs'tĭng-hous'), **George.** 1846–1914. Amer. engineer & manufacturer.
**West·more·land** (wĕst-môr'lənd, -mōr'-), **William Childs.** b. 1914. Amer. general.
**Wes·ton** (wĕs'tən), **Edward.** 1886–1958. Amer. photographer.
**Wey·den** (wīd'n, vīd'n), **Rogier van der.** 1400?–64. Flemish painter.
**Wey·er·haeu·ser** (wī'ər-hou'zər, wâr'-), **Frederick.** 1834–1914. German-born Amer. lumberman.
**Wey·gand** (vā-gäN'), **(Louis) Maxime.** 1867–1965. French army commander.
**Whar·ton** (hwôr'tn, wôr'-), **Edith Newbold Jones.** 1862–1937. Amer. author.
**Whate·ly** (hwāt'lē, wāt'-), **Richard.** 1787–1863. English theologian & logician.

ŏŏ **boot** ou **out** th **thin** *th* **this** ŭ **cut** ûr **urge** y **young**
yŏŏ ab**u**se zh vi**s**ion ə **a**bout, it**e**m, edib**i**le, gall**o**p, circ**u**s

**Wheat·ley** (hwēt'lē, wēt'-), **Phillis.** 1753?–84. African-born Amer. poet.
**Whea·ton** (hwēt'n, wēt'n), **Henry.** 1785–1848. Amer. diplomat & historian.
**Wheat·stone** (hwēt'stōn', -stən, wēt'-), Sir **Charles.** 1802–75. English physicist & inventor.
**Whee·ler** (hwē'lər, wē'-), **Joseph.** 1836–1906. Amer. Confederate general & politician.
**Wheeler, Wayne Bidwell.** 1869–1927. Amer. lawyer & prohibitionist.
**Wheeler, William Almon.** 1819–87. U.S. Vice President (1877–81).
**Whee·lock** (hwē'lŏk', wē'-), **Eleazar.** 1711–79. Amer. clergyman & educator.
**Whip·ple** (hwĭp'əl, wĭp'-), **George Hoyt.** 1878–1976. Amer. pathologist (Nobel, 1934).
**Whist·ler** (hwĭs'lər, wĭs'-), **James Abbott McNeill.** 1834–1903. Amer. artist in London.
**Whitch·er** (hwĭch'ər, wĭch'-), **Frances Miriam Berry.** *Widow Bedott.* 1811–52. Amer. author.
**White** (hwīt', wīt), **Alma Bridwell.** 1862–1946. Amer. evangelical leader.
**White, Andrew Dickson.** 1832–1918. Amer. editor, historian, & politician.
**White, Byron Raymond.** b. 1917. Amer. jurist & poet.
**White, Canvass.** 1790–1834. Amer. civil engineer.
**White, David.** 1862–1935. Amer. paleobotanist & geologist.
**White, Edward Douglass.** 1845–1921. Amer. jurist & politician.
**White, Ellen Gould Harmon.** 1827–1915. Amer. religious leader.
**White, Elwyn Brooks ("E.B.").** b. 1899. Amer. author.
**White, George Leonard.** 1838–95. Amer. choir leader.
**White, Gilbert.** 1720–93. English naturalist.
**White, John.** d. 1593? English painter & cartographer in America.
**White, Patrick.** b. 1912. Australian author (Nobel, 1973).
**White, Pearl.** 1889–1938. Amer. actress.
**White, Stanford.** 1853–1906. Amer. architect.
**White, Stewart Edward.** 1873–1946. Amer. novelist.
**White, Theodore Harold ("T.H.").** b. 1915. Amer. political journalist.
**White, Walter Francis.** 1893–1955. Amer. author.
**White, William.** 1748–1836. Amer. prelate & author.
**White, William Allen.** 1868–1944. Amer. editor & author.
**White·field** (hwīt'fēld', wīt'-, hwĭt'-, wĭt'-), **George.** 1714–70. English religious leader & orator.
**White·head** (hwīt'hĕd', wīt'-), **Alfred North.** 1861–1947. English mathematician & philosopher.
**Whitehead, William.** 1715–85. English poet.
**White·man** (hwīt'mən, wīt'-), **Paul.** 1891–1967. Amer. jazz musician.
**Whit·lam** (hwĭt'ləm, wĭt'-), **Edward Gough.** b. 1916. Australian statesman.
**Whit·lock** (hwĭt'lŏk', wĭt'-), **Brand.** 1869–1934. Amer. author & diplomat.
**Whit·man** (hwĭt'mən, wĭt'-), **Marcus** (1802–47) & **Narcissa Prentice** (1808–47). Amer. pioneers & frontier missionaries.
**Whitman, Sarah Helen Power.** 1803–78. Amer. poet & critic.
**Whitman, Walter ("Walt").** 1819–92. Amer. poet.
**Whit·ney** (hwĭt'nē, wĭt'-), **Adeline Dutton Train.** 1824–1906. Amer. author.
**Whitney, Anne.** 1821–1915. Amer. sculptor & poet.
**Whitney, Asa.** 1791–1874. Amer. inventor & manufacturer.
**Whitney, Eli.** 1765–1825. Amer. inventor & manufacturer.
**Whitney, Gertrude Vanderbilt.** 1875–1942. Amer. sculptor.
**Whitney, John Hay ("Jock").** 1904–82. Amer. newspaper publisher.
**Whitney, Josiah Dwight.** 1819–96. Amer. geologist.
**Whitney, Mary Watson.** 1847–1921. Amer. astronomer.
**Whitney, William Collins.** 1841–1904. Amer. public official.
**Whitney, William Dwight.** 1827–94. Amer. philologist.
**Whit·ta·ker** (hwĭt'ə-kər, wĭt'-), **Charles Evans.** 1901–73. Amer. jurist.
**Whit·tel·sey** (hwĭt'l-sē, -zē, wĭt'-), **Abigail Goodrich.** 1788–1858. Amer. editor.
**Whit·ti·er** (hwĭt'ē-ər, wĭt'-), **John Greenleaf.** 1807–92. Amer. poet.
**Whit·ting·ton** (hwĭt'ĭng-tən, wĭt'-), **Richard.** 1358?–1423. English merchant & mayor of London.
**Whit·worth** (hwĭt'wûrth', wĭt'-), **Kathrynne Ann.** b. 1939. Amer. golfer.
**Wick·er·sham** (wĭk'ər-shəm), **George Woodward.** 1858–1936. Amer. lawyer.
**Wic·lif** *or* **Wick·liffe** (wĭk'lĭf'). *vars. of* WYCLIFFE.
**Wi·dor** (vē-dôr'), **Charles Marie Jean Albert.** 1845–1937. French organist & composer.
**Wie·land** (vē'länt'), **Christoph Martin.** 1733–1813. German author & translator.
**Wieland, Heinrich.** 1877–1957. German chemist (Nobel, 1927).

**Wien** (vēn), **Wilhelm.** 1864–1928. German physicist (Nobel, 1911).
**Wie·ner** (wē'nər), **Norbert.** 1894–1964. Amer. mathematician.
**Wie·sel** (vē'səl), **Torsten N.** b. 1924. Swedish-born Amer. physiologist (Nobel, 1981).
**Wig·gin** (wĭg'ĭn), **Kate Douglas Smith.** 1856–1923. Amer. author & educator.
**Wig·gins** (wĭg'ĭnz), **Carleton** (1848–1932) & **Guy Carleton** (1883–1962). Amer. painters.
**Wig·gles·worth** (wĭg'əlz-wûrth'), **Michael.** 1631–1705. English-born Amer. clergyman & poet.
**Wig·more** (wĭg'môr', -mōr'), **John Henry.** 1863–1943. Amer. legal educator & author.
**Wig·ner** (wĭg'nər), **Eugene Paul.** b. 1902. Hungarian-born Amer. physicist (Nobel, 1963).
**Wil·ber·force** (wĭl'bər-fôrs', -fōrs'), **William.** 1759–1833. English politician, abolitionist, & philanthropist.
**Wil·bur** (wĭl'bər), **Richard Purdy.** b. 1921. Amer. poet.
**Wil·cox** (wĭl'kŏks'), **Ella Wheeler.** 1850–1919. Amer. author.
**Wilde** (wīld), **Oscar Fingal O'Flahertie Wills.** 1854–1900. Irish poet, playwright, & wit.
**Wil·der** (wīl'dər), **Billy.** b. 1906. Austrian-born Amer. filmmaker.
**Wilder, Burt Green.** 1841–1925. Amer. zoologist & anatomist.
**Wilder, Laura Ingalls.** 1867–1957. Amer. author.
**Wilder, Thornton (Niven).** 1897–1975. Amer. author.
**Wi·ley** (wī'lē), **Harvey Washington.** 1844–1930. Amer. chemist & pure-food reformer.
**Wil·hel·mi·na** (wĭl'ə-mē'nə, vĭl'hĕl-). 1880–1962. Queen of the Netherlands (1890–1948); abdicated.
**Wilkes** (wĭlks), **Charles.** 1798–1877. Amer. naval officer & explorer.
**Wilkes, John.** 1727–97. English political reformer.
**Wil·kins** (wĭl'kĭnz), Sir **George Hubert.** 1888–1958. Australian polar explorer & aviator.
**Wilkins, Maurice Hugh Frederick.** b. 1916. British physicist (Nobel, 1962).
**Wilkins, Roy.** 1901–81. Amer. civil-rights leader.
**Wil·kin·son** (wĭl'kĭn-sən), **Ellen Cicely.** 1891–1947. English labor leader, politician, & suffragist.
**Wilkinson,** Sir **Geoffrey.** b. 1921. English chemist (Nobel, 1973).
**Wilkinson, James.** 1757–1825. Amer. military & political leader.
**Wilkinson, Jemima.** 1752–1819. Amer. religious leader.
**Wil·lard** (wĭl'ərd), **Emma Hart.** 1787–1870. Amer. poet & educator.
**Willard, Frances Elizabeth Caroline.** 1839–98. Amer. author & temperance leader.
**Will·cocks** (wĭl'kŏks'), Sir **William.** 1852–1932. British engineer.
**Wil·liam** (wĭl'yəm). 1882–1951. German crown prince.
**William.** Name of four kings of England: **a. I.** 1027–87. Ruled 1066–87. **b. II.** 1056?–1100. Ruled 1087–1100. **c. III.** 1650–1702. Ruled 1689–1702. **d. IV.** 1765–1837. Ruled 1830–37.
**William.** Name of two kings of Prussia: **a. I.** 1797–1888. Ruled 1861–88. **b. II.** 1859–1941. Ruled 1888–1918.
**William I.** *Prince of Orange.* 1533–84. Dutch stadholder (1579–84).
**William of Malmes·bur·y** (mämz'bĕr'ē, -bə-rē). 1090?–1143? English monk & historian.
**Wil·liams** (wĭl'yəmz), **Bert.** 1876–1922. Amer. entertainer.
**Williams, Edward Bennett.** b. 1920. Amer. lawyer.
**Williams, Eleazar.** 1789?–1858. Amer. missionary.
**Williams, Elizabeth ("Betty").** b. 1943. Irish peace worker (Nobel, 1976).
**Williams, Roger.** 1603?–83. English clergyman in America & founder of Rhode Island.
**Williams, Theodore Samuel ("Ted").** b. 1918. Amer. baseball player.
**Williams, Thomas Lanier ("Tennessee").** 1911–83. Amer. playwright.
**Williams, William Carlos.** 1883–1963. Amer. physician & poet.
**Wil·lis** (wĭl'ĭs), **Nathaniel Parker.** 1806–67. Amer. author.
**Will·kie** (wĭl'kē), **Wendell Lewis.** 1892–1944. Amer. politician.
**Wills** (wĭlz), **Helen Newington.** b. 1906. Amer. tennis player.
**Will·stät·ter** (vĭl'shtĕt'ər, wĭl'shtĕt'-), **Richard.** 1872–1942. German chemist (Nobel, 1915).
**Wil·lys** (wĭl'ĭs), **John North.** 1873–1935. Amer. automobile manufacturer & diplomat.
**Wil·mot** (wĭl'mət, -mŏt'), **David.** 1814–68. Amer. politician.
**Wil·son** (wĭl'sən), **Alexander.** 1766–1813. Scottish-born Amer. ornithologist.
**Wilson, Charles Erwin.** 1890–1961. Amer. automobile executive.
**Wilson, Charles Thomson Rees.** 1869–1959. British physicist (Nobel, 1927).
**Wilson, Edmund.** 1895–1972. Amer. literary critic & author.
**Wilson, (James) Harold.** b. 1916. British prime minister (1964–70, 1974–79).
**Wilson, Harry Leon.** 1867–1939. Amer. novelist.
**Wilson, Henry.** 1812–75. U.S. Vice President (1873–75).
**Wilson, James.** 1742–98. Amer. Revolutionary patriot & jurist.
**Wilson, John.** *Christopher North.* 1785–1854. Scottish author.
**Wilson, Kenneth Geddes.** b. 1936. Amer. physicist (Nobel, 1982).
**Wilson, Meredith.** b. 1902. Amer. composer.
**Wilson, Robert Woodrow.** b. 1936. Amer. physicist & radio astronomer (Nobel, 1978).
**Wilson, William Bauchop.** 1862–1934. Amer. labor leader.
**Wilson, (Thomas) Woodrow.** 1856–1924. 28th U.S. President (1913–21), educator, and author. **—Wil·so·ni·an** (-sō'nē-ən) *adj.*
**Win·chell** (wĭn'chəl), **Walter.** 1897–1972. Amer. journalist.
**Win·ches·ter** (wĭn'chĕs'tər), **Oliver Fisher.** 1810–80. Amer. firearms manufacturer.
**Winck·el·mann** (vĭng'kəl-män'), **Johann Joachim.** 1717–68. German archaeologist & antiquary.
**Win·daus** (vĭn'dous'), **Adolf.** 1876–1959. German chemist (Nobel, 1928).
**Win·dish-Graetz** (vĭn'dĭsh-grĕts'), Prince **Alfred Candidus Ferdinand zu.** 1787–1862. Austrian field marshal.
**Wind·sor** (wĭn'zər). British ruling dynasty (since 1917).
**Windsor,** Duke of. EDWARD VIII.
**Windsor, Wallis Warfield.** *Duchess of Windsor.* b. 1896. Amer. socialite.
**Win·gate** (wĭn'gāt', -gət), **Orde Charles.** 1903–44. English army officer.
**Win·kel·ried** (vĭng'kəl-rēt'), **Arnold von.** 14th cent. Swiss national hero.
**Win·slow** (wĭnz'lō'), **Edward** (English-born; 1595–1655) & **Josiah** (1629?–80). Amer. colonists & administrators.
**Win·sor** (wĭn'zər), **Justin.** 1831–97. Amer. librarian & historian.
**Win·throp** (wĭn'thrəp), **John** (1588–1649), **John** (1606–76), & **John** (1638–1707). English colonial administrators in America.
**Winthrop, John.** 1714–79. Amer. astronomer, mathematician, & physicist.
**Winthrop, Robert Charles.** 1809–94. Amer. legislator & orator.
**Win·ton** (wĭn'tən), **Alexander.** 1860–1932. Scottish-born Amer. automobile manufacturer.
**Wirt** (wûrt), **William.** 1772–1834. Amer. public official & author.
**Wise** (wīz), **Isaac Mayer.** 1819–1900. Bohemian-born Amer. rabbi & editor.
**Wise, John.** 1652–1725. Amer. religious reformer.
**Wise, Stephen Samuel.** 1874–1949. Hungarian-born Amer. religious & civil leader.
**Wise, Thomas James.** 1859–1937. English book collector & forger.
**Wise·man** (wīz'mən), **Nicholas Patrick Stephen.** 1802–65. Spanish-born English prelate & theologian.
**Wiss·ler** (wĭs'lər), **Clark.** 1870–1947. Amer. anthropologist.
**Wis·ter** (wĭs'tər), **Owen.** 1860–1938. Amer. author of Westerns.
**With·er** (wĭth'ər) *or* **With·ers** (-ərz), **George.** 1588–1667. English poet.
**With·er·spoon** (wĭth'ər-spōōn'), **John.** 1723?–94. Scottish-born Amer. clergyman, educator, & Revolutionary patriot.
**Wit·te** (vĭt'ə), Count **Sergei Yulievich.** 1849–1915. Russian statesman.
**Witt·gen·stein** (vĭt'gən-shtīn', -stīn), **Ludwig.** 1889–1951. Austrian-born English philosopher.
**Wit·tig** (vĭt'ĭKH), **Georg.** b. 1897. German chemist (Nobel, 1979).
**Wode·house** (wŏŏd'hous'), **Pelham George ("P.G.").** 1881–1975. English author & humorist.
**Wof·fing·ton** (wŏf'ĭng-tən), **Margaret ("Peg").** 1714?–60. Irish actress.
**Wöh·ler** (wûr'lər, vûr'-, vœ'-), **Friedrich.** 1800–82. German chemist.
**Wol·cott** (wŏŏl'kət). Amer. family of political leaders, including **Roger** (1679–1767), **Oliver** (1726–97), & **Oliver** (1760–1833).
**Wolf** (vôlf). *var. of* Baron Christian von WOLFF.
**Wolf, Friedrich August.** 1759–1824. German classical scholar.
**Wolf, Hugo.** 1860–1903. Austrian songwriter.
**Wolfe** (wŏŏlf), **Charles.** 1791–1823. English poet & clergyman.
**Wolfe, James.** 1727–59. English general in Canada.
**Wolfe, Thomas (Clayton).** 1900–38. Amer. novelist.
**Wolff** *or* **Wolf** (vôlf), Baron **Christian von.** 1679–1754. German mathematician & philosopher.
**Wolff, Kaspar Friedrich.** 1733–94. German pioneer embryologist.
**Wol·fram von Esch·en·bach** (vôl'främ' fôn ĕsh'ən-bäKH'). fl. late 12th cent. German poet.
**Wol·las·ton** (wŏŏl'ə-stən), **William Hyde.** 1766–1828. English chemist & physicist.
**Wolse·ley** (wŏŏlz'lē), 1st Viscount. *Garnet Joseph Wolseley.* 1833–1913. Irish-born British general & colonial administrator.
**Wol·sey** (wŏŏl'zē), **Thomas.** 1475?–1530. English prelate & statesman.
**Wood** (wŏŏd), **Fernando.** 1812–81. Amer. politician.
**Wood, Grant.** 1892–1942. Amer. artist.

**Wood, James Rushmore.** 1813–82. Amer. surgeon & pioneer hospital administrator.
**Wood, Leonard.** 1860–1927. Amer. colonial administrator & military leader.
**Wood, Robert Elkington.** 1879–1969. Amer. general & business executive.
**Wood·ber·ry** (wŏŏd'bĕr'ē, -bə-rē), **George Edward.** 1855–1930. Amer. poet, critic, & educator.
**Wood·bur·y** (wŏŏd'bĕr'ē, -bə-rē), **Helen Laura Sumner.** 1876–1933. Amer. pioneer social economist.
**Woodbury, Levi.** 1789–1851. Amer. jurist & politician.
**Wood·hull** (wŏŏd'hŭl'), **Victoria Clafin** (1838–1927) & **Tennessee** (1846–1923). Amer. publishers & feminists.
**Woods** (wŏŏdz), **William Burnham.** 1824–87. Amer. general & jurist.
**Wood·ward** (wŏŏd'wərd), **C(omer) Vann.** b. 1908. Amer. historian.
**Woodward, Robert Burns.** 1917–79. Amer. chemist (Nobel, 1965).
**Woolf** (wŏŏlf), **(Adeline) Virginia (Stephen).** 1882–1941. English author.
**Wooll·cott** (wŏŏl'kət, -kŏt'), **Alexander.** 1887–1943. Amer. critic & journalist.
**Wool·ley** (wŏŏl'ē), Sir **Charles Leonard.** 1880–1960. English archaeologist.
**Woolley, Mary Emma.** 1863–1947. Amer. educator & reformer.
**Wool·man** (wŏŏl'mən), **John.** 1720–72. Amer. clergyman & abolitionist.
**Wool·sey** (wŏŏl'sē), **Sarah Chauncey.** 1835–1905. Amer. author.
**Woolsey, Theodore Dwight.** 1801–89. Amer. educator & author.
**Wool·son** (wŏŏl'sən), **Constance Fenimore.** 1840–94. Amer. novelist.
**Wool·ton** (wŏŏl'tən), 1st Earl of. *Frederick James Marquis.* 1883–1964. English banker & public official.
**Wool·worth** (wŏŏl'wûrth'), **Frank Winfield.** 1852–1919. Amer. merchant.
**Worces·ter** (wŏŏs'tər), **Dean Conant.** 1866–1924. Amer. zoologist & colonial administrator.
**Worcester, Joseph Emerson.** 1784–1865. Amer. lexicographer.
**Worde** (wôrd), **Wynkyn de.** d. 1534? English printer.
**Words·worth** (wûrdz'wûrth'), **William.** 1770–1850. English poet. **—Words·worth'i·an** *adj.*
**Work** (wûrk), **Henry Clay.** 1832–84. Amer. songwriter.
**Work·man** (wûrk'mən), **Fanny Bullock.** 1859–1925. Amer. traveler, explorer, & author.
**Wot·ton** (wŏŏt'n, wŏt'n), Sir **Henry.** 1568–1639. English diplomat & author.
**Wouk** (wōk), **Herman.** b. 1915. Amer. author.
**Wo·vo·ka** (wō-vō'kə). *Jack Wilson.* 1858?–1932. Amer. Indian mystic.
**Wran·gel** (răng'gəl), Baron **Pëtr Nikolaevich.** 1878–1928. Russian military leader.
**Wren** (rĕn), Sir **Christopher.** 1632–1723. English architect.
**Wright** (rīt), **Carroll Davidson.** 1840–1909. Amer. economist & statistician.
**Wright, Chauncey.** 1830–75. Amer. mathematician & philosopher.
**Wright, Elizur.** 1804–85. Amer. abolitionist, journalist, & public official.
**Wright, Frances ("Fanny").** 1795–1852. Scottish-born Amer. reformer.
**Wright, Frank Lloyd.** 1869–1959. Amer. architect.
**Wright, Harold Bell.** 1872–1944. Amer. author.
**Wright, Henry.** 1835–95. English-born Amer. baseball player & manager.
**Wright, Henry.** 1878–1936. Amer. architect & landscape designer.
**Wright, Joseph.** 1855–1930. English philologist.
**Wright, Patience Lovell.** 1725–86. Amer. sculptor.
**Wright, Richard.** 1908–60. Amer. novelist.
**Wright, Wilbur** (1867–1912) & **Orville** (1871–1948). Amer. aviation pioneers.
**Wright, Willard Huntington.** *S.S. Van Dine.* 1888–1939. Amer. novelist & critic.
**Wrig·ley** (rĭg'lē), **William, Jr.** 1861–1932. Amer. chewing-gum manufacturer.
**Wu** (wōō). Chinese dynasty (222–80).
**Wundt** (vŏŏnt), **Wilhelm.** 1832–1920. German psychologist.
**Wy·att** or **Wy·at** (wī'ət), Sir **Thomas.** 1503–42. English diplomat & poet.
**Wych·er·ley** (wĭch'ər-lē), **William.** 1640?–1716. English playwright.
**Wyc·liffe** *also* **Wick·liffe** or **Wyc·lif** or **Wic·lif** (wĭk'lĭf'), **John.** 1320?–84. English religious reformer.
**Wy·eth** (wī'ĭth), **Newell Convers** (1882–1945) & **Andrew** (b. 1917). Amer. painters.
**Wy·ler** (wī'lər), **William.** 1902–81. Amer. filmmaker.
**Wy·lie** (wī'lē), **Elinor Morton Hoyt.** 1885–1928. Amer. poet & novelist.
**Wylie, Philip Gordon.** 1902–71. Amer. author.
**Wynd·ham** (wĭn'dəm), Sir **Charles.** 1837–1919. English actor.
**Wyndham, George.** 1863–1913. English public official.
**Wynn** (wĭn), **Ed.** 1886–1966. Amer. actor.
**Wy·szyń·ski** (vĭ-shĭn'skē), **Stefan.** 1901–81. Polish prelate.
**Wythe** (wĭth), **George.** 1726–1806. Amer. Revolutionary patriot & jurist.

## X

**Xan·thip·pe** (zăn-thĭp'ē, -tĭp'ē) *or* **Xan·tip·pe** (-tĭp'ē). Proverbially shrewish 5th cent. B.C. wife of Socrates.
**Xa·vi·er** (zā'vē-ər, zăv'ē-), Saint **Francis.** 1506–52. Spanish missionary in the Orient.
**Xe·noph·a·nes** (zə-nŏf'ə-nēz'). 6th cent. B.C. Greek philosopher.
**Xen·o·phon** (zĕn'ə-fən, -fŏn'). 430?–355? B.C. Greek soldier & historian.
**Xer·xes** (zûrk'sēz'). Name of two kings of Persia, esp. **I,** *the Great,* 519?–465 B.C., ruled 486–465.
**Xia** (shē-ä'). First Chinese dynasty (2205–1766 B.C.).

## Y

**Yale** (yāl), **Caroline Ardelia.** 1848–1933. Amer. educator of the deaf.
**Yale, Elihu.** 1649–1721. Colonial-born English merchant & philanthropist.
**Yale, Linus.** 1821–68. Amer. inventor & manufacturer.
**Yal·ow** (yăl'ō), **Rosalyn Sussman.** b. 1921. Amer. physicist (Nobel, 1977).
**Ya·ma·ga·ta** (yä'mə-gä'tə), Prince **Aritomo.** 1838–1922. Japanese soldier & statesman.
**Ya·ma·mo·to** (yä'mə-mō'tō), **Isoroku.** 1884–1943. Japanese naval officer & statesman.
**Ya·ma·ni** (yə-mä'nē), **Ahmed Zaki.** b. 1930. Saudi Arabian oil minister.
**Ya·ma·shi·ta** (yä'mə-shē'tə), **Tomoyuki.** 1885–1946. Japanese general; executed for war crimes.
**Yan·cey** (yăn'sē), **William Lowndes.** 1814–63. Amer. legislator & secessionist.
**Yang Chen Ning** (yäng' jŭn' nĭng'). b. 1922. Chinese-born Amer. physicist (Nobel, 1957).
**Ya·strzem·ski** (yə-strĕm'skē), **Carl.** b. 1939. Amer. baseball player.
**Yeard·ley** (yärd'lē), Sir **George.** 1587–1627. English colonial administrator.
**Yeats** (yāts), **William Butler.** 1865–1939. Irish author, producer, & politician (Nobel, 1923). **—Yeats'i·an** *adj.*
**Yen Hsi-shan** (yĕn' shē'shän'). 1882–1960. Chinese general.
**Yer·by** (yûr'bē), **Frank Garvin.** b. 1916. Amer. novelist.
**Yer·kes** (yûr'kēz), **Charles Tyson.** 1837–1905. Amer. financier.
**Yerkes, Robert Mearns.** 1876–1956. Amer. psychobiologist.
**Yer·sin** (yĕr-săN'), **Alexandre Émile John.** 1863–1943. Swiss bacteriologist.
**Yev·tu·shen·ko** (yĕv'tə-shĕng'kō), **Yevgeny Aleksandrovich.** b. 1933. Soviet poet.
**Yo·nai** (yō'nī'), **Mitsumasa.** 1880–1948. Japanese naval officer & statesman.
**York** (yôrk). English ruling house (1461–85).
**York, Alvin Cullum.** *Sergeant York.* 1887–1964. Amer. World War I hero.
**Yo·shi·hi·to** (yō'shĭ-hē'tō). 1879–1926. Japanese emperor (1912–26).
**You·mans** (yōō'mənz), **Vincent.** 1898–1946. Amer. operetta composer.
**Young** (yŭng), **Andrew Jackson, Jr.** b. 1932. Amer. diplomat & politician.
**Young, Brigham.** 1801–77. Amer. Mormon leader.
**Young, Denton True ("Cy").** 1867–1955. Amer. baseball player.
**Young, Edward.** 1683–1765. English poet.
**Young, Ella Flagg.** 1845–1918. Amer. educator.
**Young, Francis Brett.** 1884–1954. English novelist.
**Young, Murat Bernard ("Chic").** 1901–73. Amer. cartoonist.
**Young, Owen D.** 1874–1962. Amer. corporate executive & public official.
**Young, Thomas.** 1773–1829. English physician, physicist, & Egyptologist.

**Young, Whitney Moore.** 1921–71. Amer. civil-rights leader.
**Young·er** (yŭng′gər), **Thomas Coleman ("Cole").** 1844–1916. Amer. desperado.
**Young·hus·band** (yŭng′hŭz′bənd), Sir **Francis Edward.** 1863–1942. English explorer.
**Yu·an** (yo͞o′än′). Chinese dynasty (1271–1368).
**Yuan Shi·gai** *also* **Yuan Shih-kai** (shē′kī′). 1859–1916. Chinese statesman.
**Yu·ka·wa** (yo͞o-kä′wä), **Hideki.** 1907–81. Japanese physicist (Nobel, 1949).

# Z

**Zagh·lul Pa·sha** (zăg-lo͞ol′ pä′shä), **Saad.** 1860?–1927. Egyptian nationalist leader.
**Za·har·i·as** (zə-hăr′ē-əs), **Mildred Ella Didrikson ("Babe").** —See DIDRIKSON.
**Za·krzew·ska** (zä-kshĕv′skə), **Marie Elizabeth.** 1829–1902. Polish-Amer. physician & pioneer hospital administrator.
**Za·les·ki** (zə-lĕs′kē), **August.** 1883–1972. Polish statesman.
**Za·mo·ra y Tor·res** (zə-môr′ə ē tôr′ĕs), **Niceto Alcalá.** 1877–1949. Spanish politician.
**Zan·gwill** (zăng′gwĭl′, -wĭl′), **Israel.** 1864–1926. English author & Zionist.
**Zan·uck** (zăn′ək), **Darryl Francis.** 1902–79. Amer. motion-picture producer.
**Za·pa·ta** (zə-pä′tə), **Emiliano.** 1877?–1919. Mexican revolutionary.
**Zar·a·thu·stra** (zăr′ə-tho͞o′strə). ZOROASTER.
**Zeb·e·dee** (zĕb′ĭ-dē′). Father of James and John in the New Testament.
**Zech·a·ri·ah** (zĕk′ə-rī′ə). 6th cent. B.C. Hebrew prophet.
**Zed·e·ki·ah** (zĕd′ĭ-kī′ə). King of Judah (597–86 B.C.).
**Zee·man** (zā′män′, -mən), **Pieter.** 1865–1943. Dutch physicist (Nobel, 1902).
**Zef·fi·rel·li** (zĕf′ə-rĕl′ē), **Franco.** b. 1923. Italian filmmaker.
**Zei·sler** (zī′slər), **Fannie Bloomfield.** 1863–1927. Austrian-born Amer. pianist.
**Zeng·er** (zĕng′gər, -ər), **John Peter.** 1697–1746. German-born colonial printer & journalist.
**Ze·no** (zē′nō). 5th cent. B.C. Greek philosopher.
**Zeno.** 342?–270? B.C. Greek Stoic philosopher.
**Zeph·a·ni·ah** (zĕf′ə-nī′ə). 7th cent. B.C. Hebrew prophet.
**Zep·pe·lin** (zĕp′ə-lĭn, zĕp′lĭn, tsĕp′ə-lēn′), Count **Ferdinand von.** 1838–1917. German airship designer & manufacturer.
**Zer·ni·ke** (zâr′nə-kə, zûr′-), **Frits.** 1888–1966. Dutch physicist (Nobel, 1953).
**Zeux·is** (zo͞ok′sĭs). 5th cent. B.C. Greek painter.
**Zhao Kuang·yin** (jou′ kwäng′yĭn′). Chinese emperor (960–76).
**Zhao Zi·yang** (tsē-yäng′). b. 1919. Chinese statesman.
**Zhda·nov** (zhə-dä′nəf), **Andrei Aleksandrovich.** 1896–1948. Soviet soldier & politician.
**Zhou** (jō). Name of several Chinese dynasties (1122–221 B.C., 557–81, & 957–60).
**Zhou En·lai** (ĕn-lī′). 1898–1976. Chinese statesman.
**Zhu De** (jo͞o′ dŭ′). 1886–1976. Chinese Communist leader.
**Zhu·kov** (zho͞o′kəf), **Georgi Konstantinovich.** 1896–1974. Russian army officer.
**Zieg·feld** (zĭg′fēld′, -fĕld′, zēg′-), **Florenz.** 1869–1932. Amer. theatrical producer.
**Zie·gler** (zē′glər, tsē′-), **Karl.** 1898–1973. German chemist (Nobel, 1963).
**Zim·ba·list** (zĭm′bə-lĭst′), **Efrem.** b. 1889. Russian-born Amer. violinist.
**Zim·mer·man** (zĭm′ər-mən, tsĭm′ər-män′), **Arthur.** 1864–1940. German diplomat.
**Zim·mern** (zĭm′ərn), Sir **Alfred.** 1879–1957. English political scientist.
**Zi·nov·iev** (zĭ-nôf′yəf), **Grigori Evseevich.** 1883–1936. Soviet politician.
**Zins·ser** (zĭn′sər), **Hans.** 1878–1940. Amer. bacteriologist & immunologist.
**Zin·zen·dorf** (zĭn′zən-dôrf′, tsĭn′tsən-), Count **Nikolaus Ludwig von.** 1700–60. German Moravian theologian.
**Žiž·ka** (zhĭsh′kə), **Jan.** 1360?–1424. Bohemian Hussite leader.
**Zog I** (zōg). 1895–1961. Albanian king (1928–46); deposed.
**Zo·la** (zō′lə, zō-lä′), **Emile.** 1840–1902. French author.
**Zor·ach** (zō′răk′, -räk′), **William.** 1887–1966. Lithuanian-born Amer. artist.
**Zorn** (sôrn, zôrn), **Anders Leonhard.** 1860–1920. Swedish artist.
**Zo·ro·as·ter** (zōr′ō-ăs′tər, zôr′-). 6th cent. B.C. Persian prophet. **—Zo′ro·as′tri·an** *adj.* & *n.*
**Zor·ri·lla y Mo·ral** (zə-rē′ə ē mə-räl′, -rē′yə), **José.** 1817–93. Spanish author.
**Zsig·mon·dy** (zhĭg′môn-dē), **Richard.** 1865–1929. German chemist (Nobel, 1926).
**Zuk·er·man** (zo͞o′kər-mən), **Pinchas.** b. 1948. Israeli violinist.
**Zweig** (zwīg, swīg, tsvīKH), **Arnold.** 1887–1968. German-born Jewish author.
**Zweig, Stefan.** 1881–1942. Austrian-born English author.
**Zwing·li** (zwĭng′lē, swĭng′-, tsvĭng′-), **Ulrich** or **Huldreich.** 1484–1531. Swiss religious reformer.
**Zwor·y·kin** (zwôr′ĭ-kĭn, zvôr′yə-), **Vladimir Kosma.** 1889–1982. Russian-born Amer. television pioneer.

# Geographic Entries

## A

**Aa·chen** (ä'kən, ä'KHən). City of W West Germany, near the Belgian & Dutch borders. Pop. 242,453.

**Aal·borg** (ôl'bôrg'). *var. of* ÅLBORG.

**Aalst** (älst). City of W central Belgium WNW of Brussels. Pop. 46,659.

**Aa·re** (ä'rə) *or* **Aar** (är). River of central & N Switzerland flowing 183 mi (294.5 km) into the Rhine.

**Aar·hus** (ôr'ho͞os'). *var. of* ÅRHUS.

**A·ba** (ä'bə). City of SE Nigeria N of Port Harcourt. Pop. 177,000.

**A·ba·co and Cays** (ăb'ə-kō'; kēz, kāz). Northernmost islands of the Bahamas.

**Ab·a·dan** (ä'bə-dän', ăb'ə-dăn'). City of SW Iran near the head of the Persian Gulf. Pop. 296,081.

**A·ba·jo** (ä'bə-hō'). Peak, 11,445 ft (3,490.7 m), of SE Utah near the Colo. border in the **Abajo Mts.**

**A·ba·kan** (ä'bə-kän'). **1.** River of central Siberian USSR flowing 350 mi (563.2 km) to the Yenisei. **2.** City of S central Siberian USSR on the Yenisei. Pop. 128,000.

**A·ba·ya** (ə-bī'ə). Lake, 485 sq mi (1,256.2 sq km), of SW Ethiopia.

**Abbe·ville** (ăb-vēl', ăb'ē-vĭl'). City of N France on the Somme. Pop. 25,252.

**Ab·be·ville** (ăb'ē-vĭl'). City of S La. SW of Baton Rouge. Pop. 12,391.

**A·be·o·ku·ta** (ăb'ē-ō-ko͞o'tə). City of SW Nigeria N of Lagos. Pop. 253,000.

**Ab·er·bro·thock** (ăb'ər-brə-thŏk'). *var. of* ARBROATH.

**Ab·er·dare** (ăb'ər-dâr'). Urban district of SW Wales NW of Cardiff. Pop. 38,030.

**Ab·er·deen** (ăb'ər-dēn'). **1.** (*also* ăb'ər-dēn'). Burgh of NE Scotland on the North Sea. Pop. 210,362. **2.** Town of NE Md. ENE of Baltimore. Pop. 11,533. **3.** City of NE S.Dak. NE of Pierre. Pop. 25,956. **4.** City of W Wash. WSW of Tacoma. Pop. 18,739.

**Ab·i·djan** (ăb'ĭ-jän'). Cap. of Ivory Coast, W Africa, in the S part on the Gulf of Guinea. Pop. 685,828.

**Ab·i·lene** (ăb'ə-lēn'). City of W central Tex. WSW of Fort Worth. Pop. 98,315.

**Ab·ing·ton** (ăb'ĭng-tən). Town of E Mass. SSE of Boston. Pop. 13,517.

**Ab·i·tib·i** (ăb'ĭ-tĭb'ē). Lake of E Ont. & SW Que., Canada, source of the **Abitibi R.,** flowing 230 mi (270 km) to the Moose R.

**Ab·kha·zia** (ăb-kä'zhə, -zhē-ə). Region of SW European USSR. **—Ab·kha'zian** (-zhən, -zhē-ən) *adj.* & *n.*

**Å·bo** (ō'bo͞o). TURKU.

**A·bruz·zi** (ä-bro͞ot'sē, ə-) *also* **Abruzzi e Mo·li·se** (ā mô'lə-zā'). Region of central Italy on the Adriatic.

**Ab·sa·ro·ka** (ăb-sär'ə-kə). Range of the Rocky Mts. in NW Wyo. & S Mont., rising to 13,140 ft (4,007.7 m).

**A·bu Dha·bi** (ä'bo͞o dä'bē). Sheikdom & cap. of United Arab Emirates on the Persian Gulf. Pop. 347,000.

**A·bu Qir** *or* **A·bu·kir** (ä'bo͞o-kîr', ăb'o͞o-). Village of N Egypt in the Nile R. delta of the **Bay of Abu Qir,** site of Nelson's victory over a French fleet (1798).

**A·bu Sim·bel** (ä'bo͞o sĭm'bəl). Village of S Egypt on the Nile; site of rock temples dating from c. 1250 B.C. that were raised (1964–66) to avoid flooding from Aswan High Dam.

**A·by·dos** (ə-bī'dŏs). **1.** Ancient town of Asia Minor on the Hellespont. **2.** Ancient city of S Egypt NW of Thebes on the Nile.

**Ab·ys·sin·i·a** (ăb'ĭ-sĭn'ē-ə). ETHIOPIA. **—Ab'ys·sin'i·an** *adj.* & *n.*

**A·ca·di·a** (ə-kā'dē-ə). **1.** Region & former French colony of E Canada, chiefly in N.S. & including N.E., P.E.I., & the coastal area from the St. Lawrence S into Me. **2. National Park.** Scenic area of SE Me. on the Atlantic coast. **—A·ca'di·an** *adj.* & *n.*

**Ac·a·pul·co** (ä'kə-po͞ol'kō, ăk'ə-) *or* **Acapulco de Juá·rez** (də hwär'əs, wär'-). City of S Mexico on the Pacific. Pop. 309,254.

**Ac·ar·na·ni·a** (ăk'ər-nā'nē-ə). Ancient region of W central Greece. **—Ac'ar·na'ni·an** *adj.* & *n.*

**Ac·cad** (ăk'ăd', ä'kăd'). *var. of* AKKAD.

**Ac·cra** (ăk'rə, ə-krä'). Cap. of Ghana in the S part on the Gulf of Guinea. Pop. 564,194.

**Ac·cring·ton** (ăk'rĭng-tən). Borough of NW England N of Manchester. Pop. 36,470.

**A·chae·a** (ə-kē'ə) *also* **A·cha·ia** (ə-kī'ə, ə-kā'ə). Ancient region of S Greece in the N Peloponnesus on the Gulf of Corinth. **—A·chae'an** (ə-kē'ən) *adj.* & *n.*

**Ach·e·lo·us** (ăk'ə-lō'əs). River, 137 mi (220.4 km), of NW Greece.

**Ach·ill** (ăk'ĭl). Island of NW Ireland.

**Ac·o·ma** (ăk'ə-mə). Pueblo of W central N.Mex. W of Albuquerque; regarded as oldest continuously inhabited community in the U.S.

**A·con·ca·gua** (ăk'ən-kä'gwə, äk'-). Mountain, 22,835 ft (6,964.7 m), in the Andes of W Argentina near the Chilean border.

**A·ço·res** (ä-sôr'ĕsh). AZORES.

**A·cre** (ä'krə). **1.** River, c. 400 mi (645 km), of W Brazil. **2.** (*also* ä'kər, ā'kər). AKKO.

**Ac·te** (ăk'tē). Peninsula of NE Greece projecting into the Aegean from SE Macedonia.

**Ac·ti·um** (ăk'shē-əm, -tē-). Promontory and ancient town of W Greece; site of Octavian's victory over Mark Antony and Cleopatra (31 B.C.).

**Ac·ton** (ăk'tən). Town of NE Mass. WNW of Boston. Pop. 17,544.

**A·da** (ā'də). City of S central Okla. SE of Oklahoma City. Pop. 15,902.

**A·dak** (ā'dăk'). Island of W Alas. in the central Aleutians.

**A·da·lia** (äd'l-ē-ä', -l-yä'). ANTALYA.

**A·da·ma·wa Massif** (ăd'ə-mä'wə). Plateau of W central Africa in N central Cameroon & E Nigeria.

**Ad·ams** (ăd'əmz). **1.** Peak, 5,798 ft (1,768.4 m), in N N.H. in the Presidential Range of the White Mts. **2.** Peak, 12,307 ft (3,753.6 m), in SW Wash. in the Cascade Range. **3.** Town of NW Mass. NNE of Pittsfield. Pop. 10,381.

**Ad·am's Bridge** (ăd'əmz). Shoals extending c. 18 mi (30 km) between India and Sri Lanka.

**Adam's Peak.** Sacred mountain, 7,360 ft (2,244.8 m), in S central Sri Lanka.

**A·da·na** (ä'də-nə, ə-dä'nə). City of S Turkey on the Seyhan R. Pop. 475,384.

**A·da·pa·za·ri** (ä'də-pä'zə-rē'). City of NW Turkey E of Istanbul. Pop. 114,130.

**Ad·dis Ab·a·ba** (ăd'ĭs ăb'ə-bə). Cap. of Ethiopia in the center of the country. Pop. 1,196,300.

**Ad·di·son** (ăd'ĭ-sən). Village of NE Ill. W of Chicago. Pop. 28,836.

**Ad·e·laide** (ăd'l-ād'). City of S Australia NW of Melbourne. Metro. area pop. 857,196.

**A·dé·lie Coast** (ə-dā'lē). Region of E Antarctica near Wilkes Land.

**A·den** (äd'n, ād'n). **1.** *also* **Aden Colony.** Former British colony of S Arabia, part of Southern Yemen since 1967. **2.** Former British protectorate of S Arabia between Yemen & Oman, part of Southern Yemen after 1967. **3.** Cap. of Southern Yemen on the NW shore of the **Gulf of Aden,** W arm of the Arabian Sea. Pop. 240,370.

**A·di·ge** (ä'də-jā). River of N Italy flowing c. 225 mi (360 km) to the Adriatic.

**Ad·i·ron·dack Mountains** (ăd'ə-rŏn'dăk') *also* **Ad·i·ron·dacks** (-dăks'). Range of NE N.Y. between the St. Lawrence & Mohawk valleys.

**Ad·mi·ral·ty** (ăd'mər-əl-tē). **1.** Mountains of Antarctica on the N Coast of Victoria Land. **2.** Island of SE Alas. in the Alexander Archipelago SW of Juneau. **3.** Island group of the SW Pacific in the Bismarck Archipelago; part of Papua New Guinea.

**A·do** (ä'dō). City of SW Nigeria NE of Lagos. Pop. 213,000.

**A·dour** (ə-do͝or'). River of SW France flowing 210 mi (337.9 km) to the Bay of Biscay.

**A·do·wa** (ä'də-wə, ăd'ə-). *var. of* ADUWA.

**A·dri·an** (ā'drē-ən). City of SE Mich. SW of Detroit. Pop. 21,186.

**A·dri·a·no·ple** (ā'drē-ə-nō'pəl). EDIRNE.

**A·dri·at·ic Sea** (ā'drē-ăt'ĭk). Arm of the Mediterranean between Italy & the Balkan Peninsula.

**A·du·wa** *or* **A·do·wa** (ä'də-wə, ăd'ə-) *or* **Ad·wa** (ăd'wä). Town of N Ethiopia S of Asmara. Pop. 16,400.

**Ad·zhar·i·a** (ə-jär'ē-ə) *or* **Ad·zhar·i·stan** (-ĭ-stăn'). Region of SE European USSR. **—Ad'zhar'** (ä'jär') *n.* **—Ad·zhar'i·an** *adj.* & *n.*

**Ae·ga·de·an Isles** (ē-gā'dē'ən) *also* **Ae·ga·tes** (-tēz). EGADI IS.

**Ae·ge·an Sea** (ĭ-jē'ən). Arm of the Mediterranean between Greece & Turkey.

**Ae·gi·na** (ĭ-jī'nə). **1.** Greek island in the Saronic Gulf near Athens. **2.** Ancient Greek state on Aegina Is.

**Ae·gos·pot·a·mi** (ē'gəs-pŏt'ə-mī') *or* **Ae·gos·pot·a·mos** (-mŏs'). River & ancient town of S Thrace in present-day W Turkey.

**Ae·o·li·a** (ē-ō'lē-ə). *var. of* AEOLIS.

**Ae·o·li·an Islands** (ē-ō'lē-ən). LIPARI IS.

**Ae·o·lis** (ē'ə-lĭs) *or* **Ae·o·li·a** (ē-ō'lē-ə). Ancient region of the W coast of Asia Minor in present-day Turkey. **—Ae·o'li·an** *adj.* & *n.*

o͞o **boot** ou **out** th **thin** *th* **this** ŭ **cut** ûr **urge** y **young**
yo͞o **abuse** zh **vision** ə **about,** it**e**m, ed**i**ble, gall**o**p, circ**u**s

**Ae·to·li·a** (ē-tō′lē-ə, -tōl′yə). Ancient region of Greece N of the Gulfs of Corinth & Calydon. **—Ae·to′li·an** *adj.* & *n.*

**Af·ghan·i·stan** (ăf-găn′ĭ-stăn′). Country of S central Asia. Cap. Kabul. Pop. 15,540,000. **—Af′ghan′** *adj.* & *n.*

**A·fog·nak** (ə-fôg′năk′, -fŏg-). Island of Alas. NE of Kodiak.

**Af·ri·ca** (ăf′rĭ-kə). Second-largest continent, c. 11,677,240 sq mi (30,244,050 sq km) including nearby islands, S of Europe & between the Atlantic & Indian oceans. **—Af′ri·can** *adj.* & *n.*

**A·ga·na** (ə-gä′nyə). Cap. of Guam on the W coast. Pop. 881.

**Ag·as·siz** (ăg′ə-sē′). Glacial lake of the Pleistocene epoch extending c. 700 mi (1,125 km) over present-day NW Minn., NE N.Dak., S Man., & SW Ont., Canada.

**Ag·a·wam** (ăg′ə-wŏm′). Town of SW Mass. near Springfield. Pop. 26,271.

**A·gen** (ä-zhäɴ′). Town of SW France on the Garonne R. Pop. 33,763.

**Ag·e·nais** (äzh′ə-nā′) *or* **Ag·e·nois** (äzh′ə-nwä′). Ancient region of SW France.

**A·ge·o** (ä′gā-ō′). City of central Honshu, Japan, NNW of Tokyo. Pop. 163,985.

**A·gin·court** (ăj′ĭn-kôrt′, -kōrt′). Village of N France WNW of Arras; scene of Henry V's victory over the French (1415).

**Ag·no** (äg′nō). River, 128 mi (206 km), of NW Luzon, Philippines.

**A·gra** (ä′grə). City of N central India on the Jumna R. SE of New Delhi. Pop. 591,917.

**A·gri·gen·to** (ä′grĭ-jĕn′tō, ăg′rĭ-). City of S Sicily, Italy, on the Mediterranean. Pop. 40,513.

**A·gua·dil·la** (ä′gwə-dē′ə, -yə). Town of NW Puerto Rico. Pop. 22,039.

**A·gua Fri·a** (ä′gwə frē′ə). River, 120 mi (193.1 km), of W Ariz.

**A·guas·ca·lien·tes** (ä′gwəs-kä-lyĕn′tās′). City of central Mexico NE of Guadalajara. Pop. 181,277.

**A·gul·has** (ə-gŭl′əs), **Cape.** Cliffs in South Africa; southernmost point of Africa.

**Ah·ma·da·bad** or **Ah·me·da·bad** (ä′mə-də-bäd′). City of NW India, N of Bombay. Pop. 1,591,832.

**Ah·vaz** *or* **Ah·waz** (ä-wäz′). City of SW Iran SW of Teheran. Pop. 329,006.

**Ah·ven·an·maa** (ä′və-nän-mä′). Archipelago in the Baltic Sea at the entrance to the Gulf of Bothnia between Sweden and Finland.

**Ah·waz** (ä-wäz′). *var. of* AHVAZ.

**Ai·e·a** (ī-ā′ə). City of Oahu, Hawaii near Honolulu. Pop. 12,560.

**Ai·ken** (ā′kən). City of SW S.C. near the Ga. border SW of Columbia. Pop. 14,978.

**Ain·tab** (īn-tăb′). GAZIANTEP.

**Air·drie** (âr′drē). Burgh of S central Scotland, part of Glasgow. Pop. 38,491.

**Aisne** (ān). River of N France flowing 165 mi (265.5 km) to the Oise.

**Aix-en-Pro·vence** (ăk′säɴ-prō-väɴs′, ĕk′-). City of SE France N of Marseilles. Pop. 91,665.

**Aix-la-Cha·pelle** (ăks′lä-shə-pĕl′, ĕks′-). AACHEN.

**Aix-les-Bains** (ăks′lā-băɴ′, ĕks′-). Town of SE France N of Chambéry. Pop. 21,884.

**Ai·zu Wa·ka·mat·su** (ī′zo͞o wä′kə-mä′tso͞o). City of N Honshu, Japan. Pop 108,650.

**A·jac·cio** (ä-yä′chō). Cap. of Corsica on the **Gulf of Ajaccio,** an inlet of the Mediterranean. Pop. 47,065.

**A·jax** (ā′jăks′). **1.** Mountain, 10,900 ft (3,324.5 m), in the S Bitterroot Range on the Mont.-Ida. border. **2.** Town of SE Ont., Canada, on Lake Erie NE of Toronto. Pop. 24,380.

**Aj·man** (ăj-măn′). Sheikdom of E Arabia, one of the United Arab Emirates, on the Persian Gulf. Pop. 4,000.

**Aj·mer** (ŭj-mîr′). City of NW India SW of Delhi. Pop. 262,851.

**A·jodh·ya** (ə-yōd′yə). Pilgrimage village of N India.

**A·ka·shi** (ä-kä′shē). City of SW Honshu, Japan, near Osaka. Pop. 234,905.

**Akh·el·ó·os** (äk′ə-lō′əs). ACHELOUS.

**A·ki·ta** (ä-kē′tə, ä′kĭ-tä′). City of NW Honshu, Japan, on the Sea of Japan. Pop. 261,246.

**Ak·kad** *also* **Ac·cad** (ăk′ăd′, ä′käd′). **1.** Ancient region of N Babylonia. **2.** Cap. of ancient Babylonia in central Mesopotamia.

**Ak·ko** (ä-kō′, ä′kō). City of NW Israel on the Bay of Haifa. Pop. 34,400.

**A·ko·la** (ə-kō′lə). Town of W central India WSW of Amravati. Pop. 168,438.

**Ak·ron** (ăk′rən). City of NE Ohio SSE of Cleveland, Pop. 237,177.

**Ak·sum** or **Ax·um** (äk′so͞om′). Town of N Ethiopia; cap. of ancient Ethiopian empire.

**Ak·tyu·binsk** (äk-tyo͞o′bĭnsk). City of S European USSR WNW of Alma-Ata. Pop. 191,000.

**Al·a·bam·a** (ăl′ə-băm′ə). **1.** River, 315 mi (506.8 km), of S Ala. **2.** State of the S U.S. Cap. Montgomery. Pop. 3,890,061. **—Al′a·ba′mi·an** (-bā′mē-ən), **Al′a·bam′an** *adj.* & *n.*

**A·lai** or **A·lay** (ä′lī′). Mountain range of S Central Asian USSR in the W Tian Shan, rising to c. 19,280 ft (5,880 m).

**Al·a·jue·la** (ä′lə-hwā′lə). City of central Costa Rica W of San José. Pop. 33,122.

**Al·a·kol** (äl′ə-kôl′). Salt lake of S Central Asian USSR near the Chinese border.

**Al·a·mayn** (ăl′ə-mān′), **Al** *var. of* El ALAMEIN.

**Al·a·me·da** (ăl′ə-mē′də). City of W central Calif. near Oakland. Pop. 63,852.

**Al·a·mein** (ăl′ə-mān′), **El,** *or* **Al Al·a·mayn.** Village of N Egypt on the Mediterranean.

**Al·a·mo·gor·do** (ăl′ə-mə-gôr′dō). City of S central N.Mex. SSE of Albuquerque. Pop. 24,024.

**A·land Islands** (ō′länd′). AHVENANMAA.

**A·las·ka** (ə-lăs′kə). **1. Gulf of.** N inlet of the Pacific between the Alaska Peninsula & the Alexander Archipelago. **2.** Mountain range of S central Alas. rising to 20,320 ft (6,197.6 m). **3.** Peninsula of S central Alas. between the Bering Sea & the Pacific. **4.** State of the U.S. in NW North America. Cap. Juneau. Pop. 400,481. **—A·las′kan** *adj.* & *n.*

**A·la-Tau** (ăl′ə-tou′, ä′lə-). Mountain ranges of central Asia in the Tian Shan.

**Al·a·va** (ăl′ə-və). Cape of NW Wash.; westernmost point of conterminous U.S.

**A·lay** (ä′lī′). *var. of* ALAI.

**Al·ba·ce·te** (äl′bə-sä′tē). City of SE Spain WSW of Valencia. Pop. 82,607.

**Al·ba Iu·lia** (äl′bə yo͞o′lyə). Town of W central Rumania on the Mureşul R. Pop. 44,552.

**Al·ba Lon·ga** (ăl′bə lông′gə). City of ancient Latium in central Italy SE of Rome.

**Al·ba·ni·a** (ăl-bā′nē-ə, -bān′yə). **1.** Ancient country of SE Europe in the E Caucasus W of the Caspian Sea. **2.** Republic of SE Europe on the Adriatic. Cap. Tiranë. Pop. 2,590,600. **—Al·ba′ni·an** *adj.* & *n.*

**Al·ba·no** (äl-bä′nō). Lake of central Italy SE of Rome, in an extinct volcanic crater.

**Al·ba·ny** (ôl′bə-nē). **1.** River of W Ont., Canada, flowing 610 mi (981.5 km) to James Bay. **2.** City of W Calif. N of Berkeley. Pop. 15,130. **3.** City of SW Ga. SE of Columbus. Pop. 73,934. **4.** Cap. of N.Y. in the E on the Hudson. Pop. 101,727. **5.** City of NW Ore. S of Salem. Pop. 26,546.

**Al·be·marle** (ăl′bə-märl′). **1. Sound.** Inland body of generally fresh water in NE N.C. **2.** City of S central N.C. ENE of Charlotte. Pop. 15,110.

**Al·ber·ga** (ăl-bûr′gə). Intermittent river, c. 350 mi (565 km), of S central Australia.

**Al·bert** (ăl′bərt), **Lake.** Lake of E central Africa on the Uganda-Zaire border.

**Al·ber·ta** (ăl-bûr′tə). Province of W Canada. Cap. Edmonton. Pop. 2,207,856 **—Al·ber′tan** *adj.* & *n.*

**Albert Lea** (lē). City of S Minn. near the Iowa border S of Minneapolis. Pop. 19,190.

**Albert Nile** (nīl). Name for part of the upper Nile in NW Uganda.

**Albert Ny·an·za** (nī-ăn′zə, nyăn′zə). Lake ALBERT.

**Al·bert·ville** (ăl′bərt-vĭl′). City of NE Ala. NNW of Gadsden. Pop. 12,039.

**Al·bi** (äl-bē′). Town of S France NE of Toulouse. Pop. 43,942.

**Al·bi·on** (ăl′bē-ən). City of S Mich. SW of Detroit. Pop. 11,059.

**Al·borg** *also* **Aal·borg** (ôl′bôrg′). City of N Denmark NNE of Århus. Pop. 154,582.

**Al·bu·quer·que** (ăl′bə-kûr′kē). City of central N.Mex. SW of Santa Fe. Pop. 331,767.

**Al·ca·lá de He·na·res** (ăl′kə-lä′ dā ə-när′əs). Town of central Spain ENE of Madrid. Pop. 59,783.

**Al·ca·mo** (äl′kə-mō′). City of NW Sicily, Italy, SW of Palermo. Pop. 41,448.

**Al·ca·traz** (ăl′kə-trăz′). Island in San Francisco Bay, W Calif.

**Al·co·a** (ăl-kō′ə). City of E Tenn. S of Knoxville. Pop. 7,739.

**Al·coy** (äl-koi′). City of SE Spain N of Alicante. Pop. 61,371.

**Al·dan** (äl-dän′). River of SE Siberian USSR flowing c. 1,400 mi (2,255 km) to the Lena.

**Al·der·ney** (ôl′dər-nē). British island in the English Channel, most northerly of the Channel Is.

**Al·der·shot** (ôl′dər-shŏt′). Borough of S central England SW of London. Pop. 33,750.

**Al·dridge-Brown·hills** (ôl′drĭj-broun′hĭlz′). Urban district of central England. Pop. 89,370.

**A·lek·san·drov** (ăl′ĭk-săn′drəf). ZAPOROZHE.

**Al·en·çon** (ăl-äɴ-sôɴ′). Town of NW France WSW of Paris. Pop. 32,917.

**A·lep·po** (ə-lĕp′ō) *or* **A·lep** (ə-lĕp′). City of NW Syria near the Turkish border. Pop. 639,428.

**A·lès** (ä-lĕs′). City of S France NW of Nîmes. Pop. 33,315.

**Al·es·san·dri·a** (ăl′ə-săn′drē-ə). City of NW Italy ESE of Turin. Pop. 78,644.

**A·leu·tian** (ə-lo͞o'shən). **1. Range.** Mountain chain of SW Alas. rising to 10,200 ft (3,111 m). **2.** *also* **A·leu·tians** (-shənz). Volcanic island chain of SW Alas. curving c. 1,200 mi (1,930 km) W from the Alaska Peninsula. **3. Trench.** Depression, 26,574 ft (8,105.1 m), in floor of the N Pacific S of the Aleutian Is.

**Alexander I.** Island of British Antarctic Territory off W coast of the Antarctic Peninsula.

**Al·ex·an·der Archipelago** (ăl'ĭg-zăn'dər). Group of more than 1,000 islands off SE Alaska.

**Alexander City.** City of E central Ala. SE of Birmingham. Pop. 13,807.

**Al·ex·an·dret·ta** (ăl'ĭg-zăn-drĕt'ə). ISKENDERUN.

**Al·ex·an·dri·a** (ăl'ĭg-zăn'drē-ə). **1.** City of N Egypt on the Mediterranean. Pop. 2,318,655. **2.** City of central La. NW of Baton Rouge. Pop. 51,565. **3.** Independent city of N Va. near Washington, D.C. Pop. 103,217. **—Al·ex·an'dri·an** *adj.* & *n.*

**Al Fay·yam** (ăl' fā'o͞om', -yo͞om'). City of N Egypt on the Nile SSW of Cairo. Pop. 167,081.

**Al·fi·ós** (ăl-fē-ôs', -fyôs'). ALPHEUS.

**Al·föld** (ôl'fəld). Plain of central Hungary, N Yugoslavia, & W Rumania.

**Al·gar·ve** (ăl-gär'və, ăl-). Medieval Moorish kingdom in present-day S coastal Portugal.

**Al·ge·ci·ras** (ăl'jĭ-sîr'əs). City of S Spain on the **Bay of Algeciras** opposite Gibraltar. Pop. 74,754.

**Al·ge·ri·a** (ăl-jîr'ē-ə). Republic of NW Africa on the Mediterranean. Cap. Algiers. Pop. 17,422,000. **—Al·ge'ri·an** *adj.* & *n.*

**Al·giers** (ăl-jîrz'). Cap. of Algeria in the N on the **Bay of Algiers,** an arm of the Mediterranean. Pop. 1,365,400.

**Al·ham·bra** (ăl-hăm'brə). **1.** Hill in Granada, S Spain, with Moorish buildings. **2.** City of S Calif. near Los Angeles. Pop. 64,615.

**A·li·ák·mon** (ăl-yăk'môn, ä'lē-äk'-). River, c. 200 mi (320 km), of N Greece.

**Al·i·can·te** (ăl'ĭ-kăn'tē, ä'lē-kän'tē). City of SE Spain on the Mediterranean S of Valencia. Pop. 177,918.

**Al·ice** (ăl'ĭs, -əs). City of S Tex. W of Corpus Christi. Pop. 20,961.

**A·li·garh** (ăl'ĭ-gär'). City of N central India SE of Delhi. Pop. 252,314.

**Al·i·quip·pa** (ăl'ĭ-kwĭp'ə). Borough of W Pa. on the Ohio NW of Pittsburgh. Pop. 17,094.

**Al Ji·zah** (ăl jē'zə, ĕl). GIZA.

**Alk·maar** (ălk'mär'). Town of NW Netherlands NNW of Amsterdam. Pop. 65,199.

**Al Ku·wait** (ăl ko͞o-wāt'). KUWAIT.

**Al·la·ha·bad** (ăl'ə-hə-băd', ä'lə-hə-bäd'). City of N central India at the junction of the Jumna & Ganges. Pop. 490,622.

**Al·le·ghe·ny** (ăl'ĭ-gā'nē). **1.** River, 325 mi (523km), of N central Pa., W N.Y., & W Pa. flowing to the Monongahela to form the Ohio R. **2.** *also* **Al·le·ghe·nies** (-nēz). W part of the Appalachian Mts. extending from N Pa. to SW Va.

**Al·len Park** (ăl'ən). City of SE Mich. near Detroit. Pop. 34,196.

**Al·len·town** (ăl'ən-toun'). City of E Pa. NNW of Philadelphia. Pop. 103,758.

**Al·li·ance** (ə-lī'əns). City of NE Ohio SW of Youngstown. Pop. 24,315.

**Al·lier** (ă-lyā'). River of central France flowing c. 225 mi (410 km) to the Loire.

**Al·ma** (ăl'mə). City of S central Que., Canada, on the Saguenay R. Pop. 25,638.

**Al·ma-A·ta** (ăl'mə-ä'tə, ăl'mə-ə-tä'). City of SE Central Asian USSR near the Chinese border. Pop. 910,000.

**Al Ma·nam·ah** (ăl' mə-năm'ə). Cap. of Bahrain on the Persian Gulf. Pop. 88,785.

**Al·me·lo** (ăl'mə-lō'). City of E Netherlands. Pop. 62,634.

**Al·me·rí·a** (ăl'mə-rē'ə). City of SE Spain on the Mediterranean. Pop. 104,008.

**Al·or** (ăl'ôr, ä'lôr). Largest of the **Alor Is.** of Indonesia, in the E Lesser Sundas N of Timor in the S Flores Sea.

**A·lost** (ä-lôst'). AALST.

**Al·pe·na** (ăl-pē'nə). City of NE Mich. on an arm of Lake Huron NNE of Saginaw. Pop. 12,214.

**Al·phe·us** (ăl-fē'əs). River of S Greece in the Peloponnesus, flowing c. 70 mi (112 km) to the Ionian Sea.

**Alps** (ălps). Mountain system of S central Europe curving in an arc of c. 500 mi (805 km) from the Riviera on the Mediterranean through N Italy & SE France, Switzerland, S West Germany, & Austria into NW Yugoslavia.

**Al·sace** (ăl-săs', -sās'). Region & former province of E France between the Rhine & the Vosges Mts. **—Al·sa'tian** (-sā'shən) *adj.* & *n.*

**Al·sace-Lor·raine** (ăl'săs-lô-rān', -săs-). Region of NE France comprising Alsace & part of Lorraine.

**Al·sek** (ăl'sĕk'). River of NW Canada & SE Alas. flowing 260 mi (418.3 km) to the Pacific.

**Al·sip** (ôl'sĭp). Village of NE Ill. near Chicago. Pop. 17,134.

**Al·ta Cal·i·for·nia** (ăl'tə kăl'ə-fôr'nyə). UPPER CALIFORNIA.

**Al·tai** or **Al·tay** (ăl'tī'). Mountain system of S Central Asian USSR, W Mongolia, & N China.

**Al·ta·ma·ha** (ôl'tə-mə-hô'). River of SE Ga. flowing 137 mi (220.4 km) into **Altamaha Sound,** an inlet of the Atlantic.

**Al·ta·mi·ra** (ăl'tə-mîr'ə). Caves of N Spain WSW of Santander, containing specimens of Paleolithic art.

**Al·ta·monte Springs** (ăl'tə-mŏnt'). City of E central Fla. N of Orlando. Pop. 22,028.

**Al·ta·mu·ra** (ăl'tə-mo͝or'ə). City of S Italy SSW of Bari. Pop. 44,879.

**Al·tay** (ăl-tī', ăl-, ăl'tī, ăl'-). *var. of* ALTAI.

**Al·ten·burg** (ăl'tən-bûrg', -bo͝ork'). City of S East Germany S of Leipzig. Pop. 51,193.

**Al·ti·pla·no** (ăl-tē-plä'nō). Plateau in the Andes of W Bolivia & S Peru.

**Al·ton** (ôl'tən). City of SW Ill. N of St. Louis, Mo. Pop. 34,171.

**Al·too·na** (ăl-to͞o'nə). City of central Pa. E of Pittsburgh. Pop. 57,078.

**Al·tus** (ăl'təs). City of SW Okla. near the Tex. border SW of Oklahoma City. Pop. 23,101.

**Al U·bay·yid** (ăl' o͞o-bā'ĭd, -yĭd). City of central Sudan SW of Khartoum. Pop. 90,000.

**Al·vin** (ăl'vĭn). City of SE Tex. S of Houston. Pop. 16,515.

**A·ma·ga·sa·ki** (ăm'ə-gə-sä'kē). City of S Honshu, Japan, on Osaka Bay. Pop. 545,783.

**A·mal·fi** (ə-măl'fē). Town of S Italy on the Gulf of Salerno. Pop. 4,205.

**A·ma·mi** (ə-mä'mē). Island group of N Ryukyu Is., Japan, NE of Okinawa between the Philippine Sea & the East China Sea.

**Am·a·ril·lo** (ăm'ə-rĭl'ō, -rĭl'ə). City of N Tex. in the Panhandle N of Lubbock. Pop. 149,230.

**Am·a·zon** (ăm'ə-zŏn', -zən). Second-longest river in the world, flowing c. 3,900 mi (6,275 km) from N Peru across N Brazil to a wide delta on the Atlantic.

**Am·ba·to** (ăm-bä'tō). City of central Ecuador, S of Quito. Pop. 77,955.

**Am·bon** (ăm'bôn) *also* **Am·boi·na** (ăm-boi'nə). Island of E Indonesia in the Moluccas.

**Am·brose Channel** (ăm'brōz'). Dredged channel in SE N.Y. at the entrance to New York harbor.

**Am·chit·ka** (ăm-chĭt'kə). Island of W Alas. in the Rat Is. of the Aleutians.

**A·meri·ca** (ə-mĕr'ĭ-kə). The Western Hemisphere lands, including North America, South America, & Central America. **—A·mer'i·can** *adj.* & *n.*

**American Falls.** Section (167 ft/51 m high) of Niagara Falls within the U.S.

**American Fork.** City of N central Utah S of Salt Lake City. Pop. 12,417.

**American Sa·mo·a** (sə-mō'ə). Unincorporated U.S. territory in the South Pacific comprising the E half of the Samoa Is. chain. Cap. Pago Pago. Pop. 32,395.

**A·meri·cus** (ə-mĕr'ĭ-kəs). City of central SW Ga. SE of Columbus. Pop. 16,120.

**A·mers·foort** (ä'mərz-fôrt', -fôrt', -mərs-). City of central Netherlands NE of Utrecht. Pop. 87,784.

**Ames** (āmz). City of central Iowa N of Des Moines. Pop. 45,775.

**Ames·bur·y** (āmz'bĕr'ē, -bə-rē). Town of NE Mass. NE of Lawrence. Pop. 13,971.

**Am·gun** (ăm-go͞on'). River of SE Far Eastern USSR flowing 490 mi (788.4 km) to the Amur.

**Am·herst** (ăm'ərst, -hərst). **1.** Town of N central N.S., Canada, near the N.B. border. Pop. 10,263. **2.** Town of W central Mass. near Northampton. Pop. 33,229. **3.** City of N Ohio near Lorain. Pop. 10,638.

**Am·i·ens** (ăm'ē-ənz, ä-myăN'). City of N France on the Somme N of Paris. Pop. 129,453.

**A·min·di·vi** (ä'mĭn-dē'vē). Islands of SW India in the Arabian Sea, part of the Laccadive, Minicoy, & Amindivi Is.

**Am·i·rante** (ăm'ə-rănt'). British islands of the W Indian Ocean SW of Seychelles.

**Am·i·ty·ville** (ăm'ĭ-tē-vĭl'). Village of SE N.Y. on the S shore of Long Is. Pop. 9,076.

**Am·man** (ə-män', ə-măn'). Cap. of Jordan in the N central part. Pop. 711,850.

**Am·ne Ma·chin** (ăm'nē mə-jĭn') or **Amne Machin Shan** (shän). Mountains of W central China rising to 23,490 ft (7,164.5 m).

**A·moy** (ä-moi'). City of SE China on the SW shore of **Amoy Is.,** in Formosa Strait W of Taiwan. Pop. 400,000.

**Am·ra·va·ti** (əm-rä'və-tē, ăm'-). Town of central India W of Nagpur. Pop. 193, 800.

**Am·rit·sar** (əm-rĭt'sər). City of NW India NE of Lahore. Pop. 407,628.

**Am·stel·veen** (ăm'stəl-vān'). Town of W Netherlands, a suburb of Amsterdam. Pop. 71,803.

**Am·ster·dam** (ăm'stər-dăm'). **1.** Constitutional cap. of the Netherlands, in the W part on the Ij, an inlet of the Ijsselmeer. Pop. 751,156. **2.** City of E central N.Y. NW of Albany. Pop. 21,872.

**A·mu Dar·ya** (ä'mo͞o där'yə). River of central Asia flowing c. 1,600 mi (2,575 km) along much of the USSR-Afghanistan border to the S Aral Sea.

**A·mund·sen** (ä'mən-sən, ăm'ən-). **1. Sea.** Arm of the S Pacific off the coast of Marie Byrd Land, Antarctica. **2. Gulf.** Inlet of the Arctic Ocean, N.W.T., Canada, opening on the Beaufort Sea.

**A·mur** (ä-mo͝or'). River of NE Asia flowing c. 1,800 mi (2,895 km) along the USSR-Chinese border.

**An·a·con·da** (ăn'ə-kŏn'də). City of SW Mont. WNW of Butte. Pop. 12,518.

**A·na·dyr** (ä'nə-dîr'). River of NE Far Eastern USSR flowing c. 695 mi (1,120 km) into **Anadyr Bay,** an inlet of the Bering Sea.

**An·a·heim** (ăn'ə-hīm'). City of S Calif. SE of Los Angeles. Pop. 221,847.

**A·ná·huac** (ə-nä'wäk'). Plateau of central Mexico.

**An·a·pur·na** (ăn'ə-po͝or'nə, -pûr'-). *var. of* ANNAPURNA.

**An·a·to·li·a** (ăn'ə-tō'lē-ə, -tōl'yə). Asian Turkey, usu. synonymous with Asia Minor. **—An'a·to'li·an** *adj. & n.*

**An·cas·ter** (ăn'kăs'tər). Town of S. Ont., Canada, W of Hamilton. Pop. 14,255.

**An·chor·age** (ăng'kər-ĭj'). City of S Alas. SSW of Fairbanks. Pop. 173,992.

**An·ci·enne-Lo·rette** (ăn'sē-ĕn'lô-rĕt'). Town of S central Que., Canada, W of Quebec. Pop. 11,694.

**An·co·hu·ma** (ăng'kə-ho͞o'mə, -hyo͞o'-). Mountain peak, 21,489 ft (6,550 m), of E Bolivia.

**An·co·na** (ăng-kō'nə, ăn-). City of central Italy on the Adriatic. Pop. 88,427.

**An·da·lu·cí·a** (än'də-lo͞o-sē'ə). ANDALUSIA.

**An·da·lu·sia** (ăn'də-lo͞o'zhə, -zhē-ə). **1.** Region of S Spain on the Mediterranean & the Atlantic. **2.** City of S Ala. S of Montgomery. Pop. 10,415. **—An'da·lu'sian** (-zhən) *adj. & n.*

**An·da·man** (ăn'də-mən, -măn'). **1. Sea.** Part of the Bay of Bengal in SE Asia. **2.** Islands of India S of Burma in the Bay of Bengal.

**An·der·lecht** (än'dər-lĕkt', -lĕKHt'). Commune of central Belgium near Brussels. Pop 95,969.

**An·der·son** (ăn'dər-sən). **1.** River, c. 465 mi (750 km), of NW N.W.T., Canada. **2.** City of E central Ind. NE of Indianapolis. Pop. 64,695. **3.** City of NW S.C. SW of Greenville. Pop. 27,313.

**An·der·son·ville** (ăn'dər-sən-vĭl'). Village of SW Ga.; site of Civil War Confederate prison.

**An·des** (ăn'dēz). Mountain system of W South America extending over 5,000 mi (8,045 km) from Tierra del Fuego to Venezuela & rising at many points to more than 22,000 ft (6,710 m).

**An·di·zhan** (ăn'dĭ-zhăn', än'dĭ-zhän'). City of S Central Asian USSR ESE of Tashkent. Pop. 230,000.

**An·dor·ra** (ăn-dôr'ə, -dŏr-ə). Country of SW Europe between France & Spain in the E Pyrenees. Cap. Andorra la Vella. Pop. 31,000. **—An·dor'ran** *adj. & n.*

**An·do·ver** (ăn'dō'vər, -də-). Town of NE Mass. near Lawrence. Pop. 26,370.

**An·dre·a·nof** (ăn'drē-ăn'əf, -ôf). Islands of SW Alas. in the central Aleutians.

**An·drews** (ăn'dro͞oz). City of W Tex. NW of Midland. Pop. 11,061.

**An·dri·a** (än'drē-ə). City of S Italy WNW of Bari. Pop. 76,405.

**An·dros** (ăn'drəs). **1.** Largest island of the Bahamas, in the W part. **2.** (*also* -drŏs'). Greek island in the Aegean, one of the Cyclades.

**An·dros·cog·gin** (ăn'drə-skŏg'ĭn). River, 157 mi (252.6 km), of NE N.H. & SW Me.

**A·ne·to** (ə-nā'tō), **Pico de.** —See PICO DE ANETO.

**An·ga·ra** (ăn'gə-rä'). River, c. 1,150 mi (1,850 km), of SE Siberian USSR.

**An·garsk** (ăn-gärsk'). City of the SE Siberian USSR on the Angara NW of Irkutsk. Pop. 239,000.

**An·gel Fall** *or* **Falls** (ān'jəl). Highest uninterrupted waterfall in the world (3,212 ft/979.7 m), in SE Venezuela.

**Ang·er·ma·näl·ven** (ông'ər-mə-nĕl'vən). River of E central Sweden flowing c. 280 mi (450 km) to the Gulf of Bothnia.

**An·gers** (äN-zhā'). City of W France near Nantes. Pop. 136,603.

**Ang·kor** (ăng'kôr'). Ruins of Khmer imperial capitals in NW Cambodia.

**An·gle·sey** *or* **An·gle·sea** (ăng'gəl-sē). Island of NW Wales in the Irish Sea.

**An·gle·ton** (ăng'gəl-tən). City of SE Tex. S of Houston. Pop. 13,929.

**An·gli·a** (ăng'glē-ə). **1.** Medieval Latin name for England. **2.** East Anglia. **—An'gli·an** *adj. & n.*

**An·go·la** (ăng-gō'lə, ăn-). Country of SW Africa on the Atlantic. Cap. Luanda. Pop. 7,078,000. **—An·go'lan** *adj. & n.*

**An·gou·lême** (äN-go͞o-lăm', -lĕm'). City of W France NE of Bordeaux. Pop. 46,293.

**An·gou·mois** (äN'go͞o-mwä'). Region & former province of W France in the Charente R. valley.

**An·guil·la** (ăng-gwĭl'ə, ăn-). Island of the British West Indies in the N Leewards.

**An·halt** (än'hält'). Former state of central Germany, part of East Germany since 1945.

**An·hui** *also* **An·hwei** (än'hwā', -wā'). Province of E central China.

**A·ni** (ä'nē). Ancient ruined city of Asia Minor in present-day NE Turkey.

**An·i·ak·chak** (ăn'ē-ăk'chăk'). Volcano, 4,420 ft (1,384.1 m), of SE Alas. in the Aleutian Range.

**An·jou** (ăn'jo͞o', äN-zho͞o'). **1.** Region & former province of W France SE of Brittany in the Loire valley. **2.** Town of S Que., Canada, N of Montreal. Pop. 36,596.

**An·ka·ra** (ăng'kər-ə, äng'-). Cap. of Turkey in the W central part. Pop. 1,701,004.

**An·ke·ny** (ăng'kə-nē). City of central Iowa near Des Moines. Pop. 15,429.

**Ann** (ăn), **Cape.** Peninsula of NE Mass. N of Massachusetts Bay.

**An·na·ba** (ə-nä'bə). City of NE Algeria on the Mediterranean. Pop. 255,900.

**An Na·jaf** (ăn năj'ăf'). City of S central Iraq near the Euphrates. Pop. 128,096.

**An·nam** (ə-năm', ăn'ăm'). Region & former kingdom of central Vietnam on the South China Sea. **—An'na·mese'** (ăn'ə-mēz', -mēs') *adj. & n.*

**An·nap·o·lis** (ə-năp'ə-lĭs). **1.** River of W N.S., Canada, flowing c. 75 mi (120 km) to the **Annapolis Basin,** an arm of the Bay of Fundy. **2.** Cap. of Md. SSE of Baltimore. Pop. 31,740.

**An·na·pur·na** *also* **An·a·pur·na** (ăn'ə-po͝or'nə, -pûr'-). Massif of the Himalayas in N central Nepal rising to 26,502 ft (8,083.1 m) at **Annapurna I.**

**Ann Ar·bor** (ăr'bər). City of SE Mich. W of Detroit. Pop. 107,316.

**An·ne·cy** (ăn'ə-sē', än-sē'). City of SE France in the Alps on **Lake Annecy.** Pop. 53,058.

**An·nis·ton** (ăn'ĭ-stən). City of NE Ala. ENE of Birmingham. Pop. 29,523.

**A·no·ka** (ə-nō'kə). City of E Minn. NNW of Minneapolis. Pop. 15,634.

**An·shan** (än'shän'). City of NE China SSW of Shenyang. Pop. 1,500,000.

**An·so·ni·a** (ăn-sō'nē-ə, -sōn'yə). Town of SW Conn. WNW of New Haven. Pop. 19,039.

**An·ta·kya** (än-tə-kyä') *or* **An·ta·ki·ya** (ăn'tə-kē'yä). ANTIOCH, Turkey.

**An·tal·ya** (än'tl-yä'). City of SW Turkey on the **Gulf of Antalya,** an inlet of the Mediterranean. Pop. 130,774.

**An·ta·na·na·ri·vo** (ăn'tə-năn'ə-rē'vō). Cap. of Madagascar in the E central part. Pop. 451,808.

**Ant·arc·tic** (ănt-ärk'tĭk, -är'tĭk). **1. Ocean.** Waters surrounding Antarctica, the S extensions of the Atlantic, Pacific, & Indian oceans. **2.** Peninsula of W Antarctica extending c. 1,200 mi (1,930 km) N toward South America. **3. Archipelago.** Islands of W Antarctica off the NW coast of the Antarctic peninsula. **4.** Antarctica & surrounding waters.

**Ant·arc·ti·ca** (ănt-ärk'tĭ-kə, -är'tĭ-). Continent chiefly within the Antarctic Circle & asymmetrically centered on the South Pole.

**An·tibes** (äN-tēb'). City of SE France on the Riviera between Nice & Cannes. Pop. 44,226.

**An·ti·cos·ti** (ăn'tĭ-kô'stē, -kŏs'tē). Island of E Que., Canada, at the head of the Gulf of St. Lawrence.

**An·tie·tam** (ăn-tē'təm). Creek in W Md., near Sharpsburg; site of Civil War battle (1862).

**An·ti·go·nish** (ăn'tĭ-gə-nĭsh'). Town of NE N.S., Canada, at the head of **Antigonish Bay** E of New Glasgow. Pop. 5,442.

**An·ti·gua and Bar·bu·da** (ăn-tē'gwə, -gə; bär-bo͞o'də). Country in the N Leewards, comprising the large island of Antigua & the smaller islands of Barbuda & Redonda. Cap. St. John's. Pop. 72,000. **—An·ti'guan** *adj. & n.*

**An·ti-Leb·a·non** (ăn'tē-lĕb'ə-nən). Mountain range on the Syria-Lebanon border.

**An·til·les** (ăn-tĭl'ēz). The West Indies except for the Bahamas, separating the Caribbean Sea from the Atlantic.

**An·ti·och** (ăn'tē-ŏk'). **1.** Ancient town of Phrygia in SW Turkey. **2.** City of S Turkey on the Orontes R. near the Mediterranean. Pop. 77,518. **3.** City of W Calif. ENE of Oakland. Pop. 43,559.

**An·tip·o·des** (ăn-tĭp'ə-dēz'). Rocky islands of the S Pacific SE of New Zealand.

**An·ti·sa·na** (ăn'tĭ-sä'nə). Volcano, 18,885 ft (5,760 m), of N central Ecuador in the Andes SE of Quito.

**An·to·fa·gas·ta** (ăn'tə-fə-gä'stə). City of N Chile on the Pacific. Pop. 157,000.

**An·tung** (än'do͝ong'). DANDONG.

**Ant·werp** (ănt'wûrp', ăn'twərp). City of N Belgium on the Scheldt R. N of Brussels. Pop. 224,543.

**A·nu·ra·dha·pu·ra** (ə-nu-rä'də-po͝or'ə). Town of N central Sri Lanka NNE of Colombo. Pop. 34,836.

**An·vers** (äN-vâr'). ANTWERP.

**An·yang** (än′yäng′). City of E China NNE of Zhengzhou. Pop. 225,000.
**An·zhe·ro-Sud·zhensk** (än-zhĕr′ə-sŏŏd-zhĕnsk′). City of SW Siberian USSR NE of Novosibirsk. Pop. 105,000.
**An·zi·o** (än′zē-ō, än′-). Town of central Italy on the Tyrrhenian Sea. Pop. 14,966.
**Ao·mo·ri** (ou′mə-rē). City of N Honshu, Japan, on **Aomori Bay.** Pop. 264,222.
**Ap·a·lach·ee Bay** (ăp′ə-lăch′ē). Inlet of the Gulf of Mexico in NW Fla.
**Ap·a·lach·i·co·la** (ăp′ə-lăch′ĭ-kō′lə). River of NW Fla. flowing 112 mi (180.2 km) to **Apalachicola Bay,** on the Gulf of Mexico.
**Ap·a·po·ris** (ä′pə-pôr′ēs, -pôr′-). River, c. 500 mi (800 km), of S central Colombia.
**A·pel·doorn** (ăp′əl-dôrn′, -dôrn′, -ä′pəl-). City of E central Netherlands N of Arnhem. Pop. 134,055.
**Ap·en·nines** (ăp′ə-nīnz′). Mountain system extending from NW Italy S to the Strait of Messina.
**A·pi·a** (ə-pē′ə). Cap. of Western Samoa on N Upolu Is. Pop. 32,099.
**A·po** (ä′pō). Volcano, 9,692 ft (2,955.6 m), of S Mindanao, Philippines.
**Ap·pa·la·chi·a** (ăp′ə-lā′chē-ə, -chə, -lăch′ē-ə, -lăch′ə). Region of E U.S. containing the Appalachian Mts.
**Ap·pa·la·chi·an** (ăp′ə-lā′chē-ən, -chən, -lăch′ē-ən, -lăch′ən). **1. Mountains.** *also* **the Ap·pa·la·chi·ans** (-ənz, -chənz). Mountain system of E North America extending c. 1,600 mi (2,575 km) from S Que., Canada, to central Ala. **2. Trail.** Hiking path of E U.S. extending 2,050 mi (3,298.5 km) from Mt. Katahdin in Me. to N Ga.
**Ap·pi·an Way** (ăp′ē-ən). Ancient Roman road between Rome & Brindisi.
**Ap·ple·ton** (ăp′əl-tən). City of E Wis. SW of Green Bay. Pop. 59,032.
**Ap·ple Valley** (ăp′əl). City of E Minn. S of Minneapolis. Pop. 21,818.
**Ap·po·mat·tox Court House National Historical Park** (ăp′ə-măt′əks). Site in S central Va. E of Lynchburg where Robert E. Lee surrendered to Ulysses S. Grant in 1865.
**A·pra Harbor** (ä′prə). Seaport of W Guam in the Mariana Is. of the W Pacific.
**Ap·she·ron** (äp′-shə-rôn′). Peninsula of SW USSR extending into the Caspian Sea.
**A·pu·lia** (ə-pŏŏl′yə, ə-pŏŏ′lē-ə). Region of S Italy.
**A·pu·re** (ə-pŏŏr′ä). River of W central Venezuela originating in Colombia & flowing c. 500 mi (805 km) to the Orinoco.
**A·pu·rí·mac** (ä′pə-rē′mäk′). River of S Peru flowing c. 550 mi (885 km) to the Urubamba & forming the Ucayali.
**A·qa·ba** (ä′kə-bə, äk′ə-). Town of SW Jordan at the head of the **Gulf of Aqaba,** an arm of the Red Sea. Pop. 15,000.
**A·quid·neck** (ə-kwĭd′nĕk′). Island of SE R.I. in Narragansett Bay.
**Aq·ui·taine** (ăk′wĭ-tān′). Historical region of SW France.
**Aq·ui·ta·ni·a** (ăk′wĭ-tā′nē-ə). Roman division of SW Gaul S of the Garonne R. **—Aq′ui·ta′ni·an** *adj. & n.*
**A·ra·bah** *or* **A·ra·ba** (ä′rə-bə). Depression on the Israel-Jordan border from the Dead Sea to the Gulf of Aqaba.
**A·ra·bi·a** (ə-rā′bē-ə). Peninsula of SW Asia between the Red Sea & the Persian Gulf. **—A·ra′bi·an** *adj. & n.*
**Arabian. 1. Sea.** NW part of the Indian Ocean between Arabia & India. **2.** Desert of E Egypt between the Nile Valley & the Red Sea.
**A·ra·ca·ju** (ä′rə-kə-zhŏŏ′). City of E central Brazil near the Atlantic. Pop. 179,512.
**A·rad** (ä-räd′). City of W Rumania on the Mureşul R. Pop. 161,568.
**A·ra·fu·ra Sea** (ä′rə-fŏŏ′rə). Part of the W Pacific between New Guinea & Australia.
**Ar·a·gats** (âr′ə-gäts), **Mount.** Extinct volcano, 13,435 ft (4,097.7 m), of S European USSR in the Caucasus.
**Ar·a·gon** (ăr′ə-gŏn′). Region & former kingdom of NE Spain. **—Ar′a·go·nese′** (ăr′ə-gə-nēz′, -nēs′) *adj. & n.*
**A·ra·guai·a** *or* **A·ra·gua·ya** (ăr′ə-gwī′ə). River, c. 1,300 mi (2,090 km), of central Brazil.
**A·rak** (ə-räk′). City of W central Iran SW of Teheran. Pop. 114,507.
**A·ra·kan** (ăr′ə-kän′, -kăn′). Region of SW Burma.
**A·raks** (ə-räks′). ARAS.
**Ar·al Sea** (ăr′əl). Inland sea of Central Asian USSR E of the Caspian Sea.
**Ar·an** (ăr′ən). Three islands of W Ireland at the entrance to Galway Bay.
**A·ran·sas** (ə-răn′səs, -rănt′-). **1. Bay.** Inlet of the Gulf of Mexico in S Tex. **2. Pass.** Channel of S Tex. between the Gulf of Mexico & the Gulf Intracoastal Waterway near Corpus Christi.
**A·rap·a·hoe** (ə-răp′ə-hō′). Mountain, 13,506 ft (4,119.3 m), of N central Colo. in the Front Range of the Rockies.
**A·ras** (ə-räs′). River rising in NE Turkey & flowing c. 600 mi (965 km) along the Turkey-USSR & USSR-Iran borders.
**A·rau·ca** (ə-rou′kə). River rising in N central Colombia & flowing E c. 500 mi (805 km) to the Orinoco in central Venezuela.
**Ar·au·ca·ni·a** (ə-rou-kä′nē-ə). Region of central Chile. **—Ar′au·ca′ni·an** *adj. & n.*
**A·ra·val·li** (ə-rä′və-lē). Mountain range, c. 300 mi (485 km) long, of NW India.
**Ar·be·la** (är-bē′lə). Ancient Assyrian town in present-day N Iraq.
**Ar·broath** (är-brōth′) *or* **Ab·er·bro·thock** (ăb′ər-brə-thŏk′). Burgh of E central Scotland on the North Sea. Pop. 22,706.
**Ar·buck·le Mountains** (är′bŭk′əl). Range of low hills in S Okla.; site of a national recreation area.
**Ar·ca·di·a** (är-kā′dē-ə). **1.** Region of ancient Greece in the Peloponnesus. **2.** City of S Calif. near Los Angeles. Pop. 45,994.
**Ar·ca·ta** (är-kā′tə). City of NW Calif. NNE of Eureka. Pop. 12,338.
**Arch·an·gel** (ärk′ān′jəl). ARKHANGELSK.
**Ar·ches National Park** (är′chĭz). Area of E Utah with unusual rock formations.
**Arc·tic** (ärk′tĭk, är′tĭk). **1. Ocean.** Waters surrounding the North Pole between North America and Eurasia. **2.** *also* **the Arctic.** Area between the North Pole & the N timberline. **3. Archipelago.** Islands in the Arctic Ocean between North America & Greenland.
**Arctic Red** (rĕd). River of W N.W.T., Canada, flowing c. 310 mi (500 km) to the Mackenzie.
**Ar·de·bil** *also* **Ar·da·bil** (är′də-bēl′). Town of NW Iran near the USSR border. Pop. 147,404.
**Ar·den** (är′dn). District of central England where the **Forest of Arden** was located.
**Ar·dennes** (är-dĕn′). Plateau region of N France, SE Belgium, & Luxembourg, E & S of the Meuse R.
**Ard·more** (ärd′môr, -mōr). City of S Okla. near the Tex. border SSE of Oklahoma City. Pop. 23,689.
**A·re·ci·bo** (är′ə-sē′bō). City of N Puerto Rico on the Atlantic. Pop. 48,779.
**A·re·qui·pa** (är′ə-kē′pə). City of S Peru at the foot of El Misti. Metro. area pop. 304,653.
**A·rez·zo** (ə-rĕt′sō). City of central Italy on the Arno R. Pop. 56,693.
**Ar·gen·teuil** (är-zhəɴ-tœ′yə). City of N France on the Seine near Paris. Pop. 101,542.
**Ar·gen·ti·na** (är′jən-tē′nə). Republic of SE South America. Cap. Buenos Aires. Pop. 27,862,771. **—Ar′gen·tine′** (-tēn′, -tīn′) *adj. & n.* **—Ar′gen·tin′e·an** (-tĭn′ē-ən) *adj. & n.*
**Ar·go·lis** (är′gə-lĭs). **1. Gulf of.** Inlet of the Aegean on the E coast of the Peloponnesus, S Greece. **2.** Ancient region of S Greece in the NE Peloponnesus.
**Ar·gonne** (är-gŏn′, är′gŏn). Region of NE France.
**Ar·gos** (är′gŏs, -gəs). Ancient Greek city in NE Peloponnesus near the head of the Gulf of Argolis.
**Ar·gun** (är-gŏŏn′). River of E central Asia flowing 950 mi (1,528.6 km) along the USSR-China border.
**År·hus** *also* **Aar·hus** (ôr′hŏŏs′). City of central Denmark on **Århus Bay,** an arm of the Kattegat. Pop. 245,941.
**A·ri·nos** (ə-rē′nəs, -nŏŏs′). River, c. 400 mi (645 km), of central Brazil.
**A·ri·pua·nã** (ə-rē′pwä-näɴ′). River of W central Brazil flowing c. 400 mi (645 km) to the Madeira.
**A·ri·us** (ăr′ē-əs, ə-rī′əs). HARI RUD.
**Ar·i·zo·na** (ăr′ĭ-zō′nə). State of the SW U.S. on the Mexican border. Cap. Phoenix. Pop. 2,717,866. **—Ar′i·zo′ni·an** *adj. & n.*
**Ar·ka·del·phi·a** (är′kə-dĕl′fē-ə). City of SW Ark. S of Hot Springs. Pop. 10,028.
**Ar·kan·sas** (är′kən-sô′). **1.** (*also* är-kăn′zəs). River of the S central U.S. rising in central Colo. & flowing c. 1,450 mi (2,335 km) to the Mississippi in SE Ark. **2.** State of the S central U.S. Cap. Little Rock. Pop. 2,285,513. **—Ar·kan′san** (är-kăn′zən) *adj. & n.*
**Arkansas City.** City of S Kans. near the Okla. border SSE of Wichita. Pop. 13,201.
**Ark·han·gelsk** (är-kăn′gĕlsk, -KHän′-). City of NW European USSR on the Northern Dvina. Pop. 385,000.
**Arl·berg** (ärl′bûrg′). Alpine pass in W Austria, 5,946 ft (1,813.5 m) high.
**Arles** (ärlz, ärl). **1.** Medieval kingdom of E & SE France. **2.** City of S central France on the Rhone R. Pop. 37,337.
**Ar·ling·ton** (är′lĭng-tən). **1.** Town of E Mass. near Boston. Pop. 48,219. **2.** City of N Tex. E of Fort Worth. Pop. 160,123. **3.** County & unincorporated city of N Va. near Washington, D.C. Pop. 152,599.
**Arlington Heights.** Village of NE Ill. near Chicago. Pop. 66,116.
**Ar·magh** (är-mä′, är′mä′). Urban district of S Northern Ireland. Pop. 47,500.
**Ar·ma·gnac** (är′mən-yăk′). Region of SW France in Gascony.
**Ar·ma·vir** (är′mə-vîr′). City of SE European USSR on the Kuban R. Pop. 162,000.
**Ar·me·ni·a** (är-mē′nē-ə, -mēn′yə). **1.** Region & former kingdom of Asia Minor in present-day NE Turkey, SE European USSR, & sections of Iranian Azerbaijan. **2.** *also* **Ar·me·ni·an Soviet Socialist Republic** (-mē′nē-ən, -mēn′yən). Constituent republic of SE European USSR in the S Caucasus. **—Ar·me′ni·an** *adj. & n.*
**Ar·men·tières** (är′mən-tîrz′, -tyĕr′). City of N France WNW of Lille. Pop. 23,850.

**Ar·mor·i·ca** (är-môr′ĭ-kə, -mōr′-). Ancient name for NE part of France, esp. Brittany.
**Arn·hem** (ärn′hĕm′, är′nəm). City of E Netherlands on the Rhine. Pop. 126,051.
**Arn·hem Land** (är′nəm). Region of N Australia W of the Gulf of Carpentaria.
**Ar·no** (är′nō). River of central Italy flowing c. 150 mi (240 km) to the Ligurian Sea.
**Ar·nold** (är′nəld). City of E Mo. S of St. Louis. Pop. 19,141.
**A·roe** (ä′rōō). *var. of* ARU.
**A·roos·took** (ə-rōōs′tək, -rōōs′-). River flowing c. 140 mi (225 km) from N Me. to the St. John R. in N.B., Canada.
**Ar·ran** (ăr′ən). Island of W Scotland in the Firth of Clyde.
**Ar·ras** (ăr′əs, ə-räs′). City of N France SSW of Lille. Pop. 45,804.
**Ar·roe** (ä′rōō). *var. of* ARU.
**Ar·roy·o Gran·de** (ə-roi′ō grăn′dē). City of SW Calif. SSE of San Luis Obispo. Pop. 11,290.
**Ar·te·sia** (är-tē′zhə). **1.** City of S Calif. near Los Angeles. Pop. 14,301. **2.** City of SE N.Mex. NNW of Carlsbad. Pop. 10,385.
**Ar·tois** (är-twä′). Region of N France near the English Channel between Picardy & Flanders.
**A·ru** *or* **A·roe** *or* **Ar·roe** (ä′rōō). Islands of E Indonesia, part of the Moluccas, in the Arafura Sea SW of New Guinea.
**A·ru·ba** (ə-rōō′bə). Island of the Netherlands Antilles N of the Venezuela coast.
**A·ru·wi·mi** (är′ə-wē′mē, är′-). River of N Zaire flowing c. 800 mi (1,287 km) to the Congo R.
**Ar·vad·a** (är-văd′ə). City of N central Colo. near Denver. Pop. 84,576.
**A·sa·hi·ga·wa** (ä′sə-hē-gä′wə) *also* **A·sa·hi·ka·wa** (-kä′-). City of W central Hokkaido, Japan. Pop. 320,526.
**A·sa·ma** (ə-sä′mə) *or* **A·sa·ma·ya·ma** (ə-sä′mə-yä′mə). Active volcano, 8,340 ft (2,543.7 m), of central Honshu, Japan.
**A·san·te** (ə-săn′tē). *var. of* ASHANTI.
**As·bes·tos** (ăz-bĕs′təs). Town of SE Que., Canada, N of Sherbrooke. Pop. 9,075.
**As·bur·y Park** (ăz′bĕr′ē, -bə-rē). City of E N.J. on the Atlantic. Pop. 17,015.
**As·ca·lon** (ăs′kə-lŏn′). ASHKELON.
**As·cen·sion** (ə-sĕn′shən). Island in the S Atlantic NW of St. Helena.
**A·schaf·fen·burg** (ä-shä′fən-bərg). City of S central West Germany on the Main R. Pop. 55,398.
**A·sco·li Pi·ce·no** (ä′skə-lē pə-chā′nō). City of central Italy NE of Rome. Pop. 43,041.
**As·cot** (ăs′kət). Village of S central England SW of London; site of Ascot racetrack.
**As·cu·lum** (ăs′kyə-ləm). Ancient Roman town of SE Italy S of present-day Foggia.
**A·shan·ti** (ə-shăn′tē, -shän′-) *or* **A·san·te** (-sän′-). Region & former kingdom of central Ghana, W Africa.
**Ash·bur·ton** (ăsh′bûr′tn, -bər-). River of NW Australia flowing c. 400 mi (645 km) to the Indian Ocean.
**Ash·dod** (ăsh′dŏd′). City of SW Israel on the Mediterranean near the site of ancient **Ashdod,** Philistine city-state. Pop. 40,500.
**Ashe·bor·o** (ăsh′bûr′ō). City of central N.C. S of Greensboro. Pop. 15,252.
**Ashe·ville** (ăsh′vĭl′). City of W N.C. WNW of Charlotte. Pop. 53,281.
**A·shi·ka·ga** (ä′shĭ-kä′gə). City of central Honshu, Japan, N of Tokyo. Pop. 162,359.
**Ash·ke·lon** (ăsh′kə-lŏn′). Ancient city of SW Palestine on the Mediterranean.
**Ash·kha·bad** (ăsh′kə-băd′, -bäd′). City of S Central Asian USSR near the Iranian border. Pop. 312,000.
**Ash·land** (ăsh′lənd). **1.** City of E Ky. on the Ohio-W.Va. border. Pop. 27,064. **2.** City of N central Ohio NE of Mansfield. Pop. 20,326. **3.** City of SW Ore. near Medford & the Calif. border. Pop. 14,943.
**Ash·ley** (ăsh′lē). River of S S.C. flowing 40 mi (64.3 km) into Charleston harbor.
**Ash·ta·bu·la** (ăsh′tə-byōō′lə). City of NE Ohio on Lake Erie. Pop. 23,449.
**Ash·ton-under-Lyne** (ăsh′tən-ŭn′dər-līn′). Borough of NW England E of Manchester. Pop. 48,500.
**Ash·wau·be·non** (ăsh-wô′bə-nən, -wŏb′ə-). Village of NE Wis. near Green Bay. Pop. 14,486.
**A·sia** (ā′zhə, ā′shə). The world's largest continent (17,139,000 sq mi/44,390,010 sq km), occupying the E part of Eurasia & adjacent islands & separated from Europe by the Ural Mts. **—A′sian** *adj.* & *n.*
**Asia Mi·nor** (mī′nər). Peninsula of W Asia between the Black Sea & the Mediterranean.
**As·ma·ra** (ăz-mä′rə). City of N Ethiopia near the Red Sea. Pop. 393,800.
**As·nières-sur-Seine** (ä-nyĕr′sür-sān′). City of N central France, a NW suburb of Paris. Pop. 75,328.
**A·so** (ä′sō′) *or* **A·so-san** (ä′sō-sän′). Volcanic mountain of central Kyushu, Japan.
**As·pen** (ăs′pən). City & ski resort of W central Colo. in the Rockies. Pop. 3,678.
**As·pern** (ăs′pərn, äs′-). Suburb of Vienna, Austria, where Austrians defeated Napoleon in 1809.
**As·sa·teague** (ăs′ə-tēg′). Island along Md. & Va. coasts separating Chincoteague Bay & the Atlantic.
**As·sen** (ä′sən). City of NE Netherlands S of Groningen. Pop. 43,783.
**As·sin·i·boine** (ə-sĭn′ə-boin′). **1.** River of S central Canada flowing 590 mi (949.3 km) from S Sask. to the Red R. at Winnipeg, Man. **2.** Mountain, 11,870 ft (3,620.4 m), in the Canadian Rockies on the B.C.-Alta. border.
**As·si·si** (ə-sē′zē, -sē, ə-sĭs′ē). Town of central Italy ESE of Perugia. Pop. 4,630.
**As·suan** (ăs′wän, äs-wän′, äs-). *var. of* ASWAN.
**As·syr·i·a** (ə-sîr′ē-ə). Ancient empire of W Asia in the upper valley of the Tigris. **—As·syr′i·an** *adj.* & *n.*
**As·ti** (äs′tē). City of NW Italy. Pop. 62,277.
**As·to·ri·a** (ă-stôr′ē-ə, -stōr′-). City of NW Ore. near the mouth of the Columbia. Pop. 9,998.
**As·tra·khan** (ăs′trə-kăn′). City of SE European USSR on the Volga delta. Pop. 461,000.
**As·tu·ri·as** (ăs-tŏŏr′ē-əs, -tyŏŏr′-). Region & former kingdom of NW Spain.
**A·sun·ción** (ə-sōōn′sē-ōn′). Cap. of Paraguay in the S part. Pop. 387,676.
**As·wan** *or* **As·suan** (ăs′wän, äs-wän′, äs-). City of S Egypt on the Nile near the **Aswan High Dam,** completed in 1970. Pop. 144,377.
**As·yut** (äs′ē-ōōt′, ä′sē-). City of E central Egypt on the Nile. Pop. 213,983.
**At·a·ca·ma** (ăt′ə-kăm′ə). Desert of N Chile.
**A·tas·ca·de·ro** (ə-tăs′kə-dâr′ō). City of SW Calif. N of San Luis Obispo. Pop. 15,930.
**At·ba·ra** (ăt′bər-ə). River of NE Africa flowing c. 500 mi (805 km) from NW Ethiopia to the Nile in E Sudan.
**A·tchaf·a·lay·a** (ə-chăf′ə-lī′ə). River of S central La. flowing c. 170 mi (275 km) into **Atchafalaya Bay,** an inlet of the Gulf of Mexico.
**At·chi·son** (ăch′ĭ-sən). City of NE Kans. NW of Kansas City. Pop. 11,407.
**Ath·a·bas·ca** *or* **Ath·a·bas·ka** (ăth′ə-băs′kə). **1.** Lake of NE Alta. & NW Sask., Canada. **2.** River of Alta., Canada, flowing 765 mi (1,230.9 km) to Lake Athabasca. **3. Mount.** Peak, 11,452 ft (3,492.9 m), of W Alta., Canada.
**Ath·ens** (ăth′ənz). **1.** Cap. of Greece in the E central part. Pop. 867,023. **2.** City of N Ala. WNW of Huntsville. Pop. 14,558. **3.** City of NE Ga. ENE of Atlanta. Pop. 42,549. **4.** City of SE Ohio W of Marietta. Pop. 19,743. **5.** City of E Tenn. NE of Chattanooga. Pop. 12,080. **6.** City of E Tex. SE of Dallas. Pop. 10,197.
**Ath·ol** (ăth′ŏl, -ôl). Town of N Mass. W of Fitchburg. Pop. 10,634.
**Ath·os** (ăth′ŏs, ā′thŏs), **Mount.** Peak, c. 6,670 ft (2,035 m), of NE Greece in Macedonia; site of virtually independent monastic community of **Mount Athos.**
**A·ti·tlán** (ä′tē-tlän′). Volcanic lake of SW Guatemala.
**At·ka** (ăt′kə, ät′-). Island of SW Alas. in the Andreanof group of the central Aleutians.
**At·lan·ta** (ăt-lăn′tə). Cap. of Ga. in the NW part. Pop. 425,022.
**At·lan·tic** (ăt-lăn′tĭk). Ocean, c. 31,800,000 sq mi (82,362,000 sq km), extending from the Arctic to the Antarctic between North & South America on the W & Europe & Africa on the E.
**Atlantic City.** City of SE N.J. on the Atlantic. Pop. 40,199.
**Atlantic Provinces.** The E Canadian provinces of N.B., P.E.I., N.S., & Newf.
**At·las** (ăt′ləs). Mountain system of NW Africa extending from SW Morocco to N Tunisia between the Sahara & the Mediterranean, & rising to 13,665 ft (4,167.8 m).
**At·lin** (ăt′lĭn). Lake of NW B.C., Canada.
**A·trak** (ə-trăk′). *var. of* ATREK.
**A·tra·to** (ə-trä′tō). River, c. 375 mi (605 km), of W Colombia.
**A·trek** (ə-trĕk′) *or* **A·trak** (ə-trăk′). River of NE Iran flowing 300 mi (482.7 km), partly along the USSR-Iran border, to the Caspian.
**At·su·gi** (ät-sōō′gē). City of E central Honshu, Japan, near Tokyo. Pop. 108,955.
**At·ta·wa·pis·kat** (ăt′ə-wə-pĭs′kət). River, c. 465 mi (750 km), of N Ont., Canada.
**At·ti·ca** (ăt′ĭ-kə). Ancient region of E central Greece around Athens. **—At′tic** (ăt′ĭk) *adj.*
**At·tle·bor·o** (ăt′l-bûr′ō). City of SE Mass. NE of Providence, R.I. Pop. 34,196.
**At·tu** (ăt′tōō′). Island of SW Alas., westernmost of the Aleutians.
**A·tu·o·na** (ä′tə-wō′nə) *or* **A·tu·a·na** (-wŏn′ə). Town of the Marquesas Is. in the S Pacific in French Polynesia.
**At·wa·ter** (ăt′wô′tər, -wŏt′ər). City of central Calif. WNW of Merced. Pop. 17,530.

**Aube** (ōb). River of NE France flowing 140 mi (225.2 km) to the Seine.

**Au·ber·vil·liers** (ō'bər-vēl-yā'). Town of N central France NE of Paris. Pop. 72,859.

**Au·burn** (ô'bərn). **1.** City of E Ala. NNE of Tuskegee. Pop. 28,471. **2.** City of SE Me. near Lewiston. Pop. 23,128. **3.** Town of S central Mass. S of Worcester. Pop. 14,845. **4.** City of W central N.Y. WSW of Syracuse. Pop. 32,548. **5.** City of W Wash. ENE of Tacoma. Pop. 26,417.

**Auck·land** (ôk'lənd). City of New Zealand on NW North Is. Metro. area pop. 742,786.

**Au·ghra·bies Falls** (ô-grä'bēz). Waterfall, 480 ft (146.4 m), on the Orange R. in South Africa.

**Augs·burg** (ôgz'bûrg', ougz'-). City of S West Germany WNW of Munich. Pop. 249,943.

**Au·gus·ta** (ô-gŭs'tə, ə-gŭs'-). **1.** City of E Ga. on the S.C. border NNW of Savannah. Pop. 47,532. **2.** Cap. of Me. in the SW part NNE of Portland. Pop. 21,819.

**Au·lis** (ô'lĭs). Ancient port of E central Greece in Boeotia on the Gulf of Euboea.

**Aul·nay-sous-Bois** (ō-nā'sōō-bwä'). Town of N central France NE of Paris. Pop. 77,982.

**Au·nis** (ō-nēs'). Region & former province of W France on the Atlantic Ocean.

**Au·rang·a·bad** (ou-rŭng'gə-bäd', -ə-bäd'). Town of W India ENE of Bombay. Pop. 150,483.

**Au·ri·gnac** (ô'rēn-yăk'). Village of S France at the foot of the Pyrenees; site of caves with prehistoric relics.

**Au·ril·lac** (ô'rē-yăk'). Town of S central France SW of Toulouse. Pop. 29,458.

**Au·ro·ra** (ô-rôr'ə, -rōr'-, ə-rôr'ə, -rōr'-). **1.** Town of S Ont., Canada, N of Toronto. Pop. 14,249. **2.** City of N central Colo. near Denver. Pop. 158,588. **3.** City of NE Ill. W of Chicago. Pop. 81,293.

**Au·sa·ble** (ô-sā'bəl). River of NE N.Y. flowing 20 mi (32.1 km) through **Ausable Chasm,** a 2 mi (3.2 km) gorge, to Lake Champlain.

**Ausch·witz** (oush'vĭts'). OŚWIĘCIM.

**Aus·ter·litz** (ô'stər-lĭts', ous'tər-). Town of S Czechoslovakia near site of Napoleon's 1805 defeat of Russian & Austrian armies.

**Aus·tin** (ô'stən, ŏs'tən). **1.** City of SE Minn. SW of Rochester. Pop. 23,020. **2.** Cap. of Tex. in the S central part. Pop. 345,496.

**Aus·tral·a·sia** (ô'strə-lā'zhə, -shə). **1.** Islands of Oceania in the S Pacific, including Australia, New Zealand, New Guinea, & associated islands. **2.** OCEANIA. **—Aus'tral·a'sian** *adj.* & *n.*

**Aus·tra·lia** (ô-strāl'yə). **1.** The world's smallest continent, 2,948,-366 sq mi (7,636,267.9 sq km), SE of Asia between the Pacific & Indian oceans. **2.** *or* **Commonwealth of Australia.** Country comprising the continent of Australia, the island state of Tasmania, two external territories, & several dependencies. Cap. Canberra. Pop. 13,548,448. **—Aus·tra'lian** *adj.* & *n.*

**Australian Alps.** Mountain ranges of SE Australia, the S part of the Eastern Highlands.

**Aus·tra·sia** (ô-strā'zhə, -shə). E portion of the kingdom of the Franks from the 6th to the 8th cent., consisting of parts of E France, W West Germany, & the Netherlands. **—Aus·tra'sian** *adj.* & *n.*

**Aus·tri·a** (ô'strē-ə). Federal republic of central Europe. Cap. Vienna. Pop. 7,507,000. **—Aus'tri·an** *adj.* & *n.*

**Aus·tri·a-Hun·ga·ry** (ô'strē-ə-hŭng'gə-rē). Dual monarchy (1867–1918) of central Europe consisting of Austria, Hungary, Bohemia, & parts of Poland, Rumania, Yugoslavia, & Italy. **—Aus'tro-Hun·gar'i·an** (-gâr'ē-ən) *adj.* & *n.*

**Aus·tro·ne·sia** (ô'strō-nē'zhə, -shə). **1.** Islands of the Pacific, including Indonesia, Melanesia, Micronesia, & Polynesia. **2.** Island area of the S hemisphere extending from Madagascar to Hawaii & Easter Is. **—Aus'tro·ne'sian** *adj.* & *n.*

**Au·vergne** (ō-vĕrn', ō-vûrn'). Region & former province of S central France traversed N to S by the **Auvergne Mts.,** chain of extinct volcanoes.

**Av·a·lon** (ăv'ə-lŏn'). **1.** Peninsula of SE Newf., Canada. **2.** *or* **Isle of Avalon.** District, formerly an island, of SW England.

**A·ve·lla·ne·da** (ăv'ə-zhə-nā'də). City of E Argentina near Buenos Aires. Pop. 337,538.

**A·venches** (ä-väNsh') *or* **A·ven·ti·cum** (ə-vĕn'tĭ-kəm). Commune & ancient city of W Switzerland. Pop. 2,235.

**Av·en·tine** (ăv'ən-tīn', -tēn'). One of the seven hills of ancient Rome.

**A·ver·no** (ä-vĕr'nō) *or* **A·ver·nus** (ə-vûr'nəs). Small crater lake of S Italy W of Naples.

**A·vi·gnon** (ä-vē-nyôN'). City of SE France on the Rhone. Pop. 73,482.

**Á·vi·la** (ä'və-lə). Town of central Spain WNW of Madrid. Pop. 30,958.

**A·vi·lés** (ä'-və-lās'). Town of NW Spain on the Bay of Biscay. Pop. 67,186.

**A·von** (ā'vŏn). **1.** (*also* ā'vən, ăv'ən). *also* **Bris·tol** *or* **Lower Avon** (brĭs'təl). River of SW England flowing 75 mi (120.7 km) to the Severn. **2.** *also* **East Avon.** River of S England flowing 48 mi (77.2 km) to the English Channel. **3.** *also* **Upper Avon.** River of S central England flowing 96 mi (154.5 km) to the Severn. **4.** Town of N central Conn. WNW of Hartford. Pop. 11,201.

**Avon Lake.** City of NE Ohio on Lake Erie. Pop. 13,222.

**A·vranches** (ä-vräNsh'). Town of NW France on the English Channel. Pop. 10,128.

**A·wa·ji** (ə-wä'jē) *or* **A·wa·ji·shi·ma** (ə-wä'jē-shē'mə). Island of Japan in the Inland Sea between SW Honshu and Shikoku.

**A·wash** (ä'wäsh'). HAWASH.

**Ax·el Hei·berg** (ăk'səl hī'bûrg'). Island in the Arctic Ocean, N N.W.T., Canada, W of Ellesmere Is.

**Ax·um** (äk'sōōm'). *var. of* AKSUM.

**A·ya·cu·cho** (ī'ə-kōō'chō). City of S central Peru S of Buenos Aires. Pop. 43,304.

**Ayles·bur·y** (ālz'bĕr'ē, -bə-rē). City of central England NW of London. Pop. 41,420.

**Ayl·mer** (āl'mər). Town of SW Que., Canada, on the Ottawa R. near Hull. Pop. 25,714.

**Ayr** (âr). Burgh of SW Scotland at the mouth of the **Ayr R.** on the Firth of Clyde. Pop. 47,990.

**A·yut·thay·a** (ä-yōōt'ə-yə) *or* **A·yu·dhya** (ä-yōōd'yə). City of S central Thailand on the Chao Phraya R. Pop. 37,213.

**A·zer·bai·jan** (ăz'ər-bī-jän', ä'zər-). **1.** Region of NW Iran. **2.** *also* **Azerbaijan** *or* **A·zer·bai·dzhan Soviet Socialist Republic.** Constituent republic of SE European USSR.

**A·zores** (ā'zôrz, ā'zōrz, ə-zôrz', ə-zōrz'). Islands in the N Atlantic c. 900 mi (1,448 km) W of mainland Portugal. **—A·zor'e·an, A·zor'i·an** *adj.* & *n.*

**A·zov** (ăs'ôf', -ŏf', ā'zôf), **Sea of.** N arm of the Black Sea covering an area of c. 14,000 sq mi (36,260 sq km) in S European USSR.

**A·zu·sa** (ə-zōō'sə). City of S Calif. E of Pasadena. Pop. 29,380.

# B

**Baal·bek** (bäl'bĕk', bā'əl-). Town of Lebanon NE of Beirut. Pop. 15,560.

**Ba·bar** (bä'bär'). Islands of E Indonesia ENE of Timor.

**Bab el Man·deb** (băb' ĕl măn'dəb). Strait, 17 mi (27.4 km) wide, between the Red Sea & the Gulf of Aden.

**Ba·bel·thu·ap** (bä'bəl-tōō'äp'). Island, 120 sq mi (310.8 sq km), of the Palau group in the SW Pacific.

**Ba·bi·a Gó·ra** (bä'bē-ə gōōr'ə). Highest mountain (5,659 ft/1,725.9 m) of the Beskids, in the West Beskids on the border between Poland & Czechoslovakia.

**Ba·bu·yan** (bä'bōō-yän'). Main island of the **Babuyan Is.** in the Philippines N of Luzon.

**Bab·y·lon** (băb'ə-lən, -lŏn'). **1.** Cap. of ancient Babylonia in Mesopotamia on the Euphrates R. **2.** Village of SE N.Y. on S Long Is. Pop. 12,388.

**Bab·y·lo·ni·a** (băb'ə-lō'nē-ə, -lōn'yə). Ancient country of SW Asia in Mesopotamia in the Euphrates valley. **—Bab'y·lo'ni·an** *adj.* & *n.*

**Ba·cau** (bə-kou'). City of E Rumania NNE of Bucharest. Pop. 131,413.

**Back** (băk). River, c. 600 mi (965 km), of N.W.T., Canada.

**Ba·co·lod** (bä-kō'lôd'). City of NW Negros Is., Philippines. Pop. 266,604.

**Bac·tra** (băk'trə). BALKH.

**Bac·tri·a** (băk'trē-ə). Ancient country of SW Asia. **—Bac'tri·an** *adj.* & *n.*

**Ba·da·joz** (bä'də-hōz'). City of SW Spain on the Guadiana R. near the Portugal border. Pop. 80,793.

**Ba·da·lo·na** (bä'də-lō'nə). City of NE Spain on the Mediterranean near Barcelona. Pop. 162,888.

**Ba·den** (bäd'n). Region of SW Germany.

**Ba·den-Ba·den** (bäd'n-bäd'n). City of SW Germany in the Black Forest. Pop. 49,718.

**Bad Go·des·berg** (bät' gō'dĭs-bûrg', -bĕrg'). GODESBERG.

**Bad Hom·burg** (hŏm'bûrg', -bōōrg'). City of central West Germany near Frankfurt. Pop. 51,196.

**Bad Kreuz·nach** (kroits'näk', -näKH'). City of W West Germany. Pop. 42,588.

**Bad·lands National Monument** (băd'lăndz'). Extensive tract of badlands in SW S.Dak.

**Baf·fin Bay** (băf'ĭn). Arm of the Atlantic off NE Canada separating Greenland & **Baffin Is.** (183,810 sq mi/467,068 sq km).

**Ba·fing** (bə-făng'). River of W Guinea & W Mali flowing 350 mi (563.2 km) to the Senegal R.

**Bagh·dad** *or* **Bag·dad** (băg'dăd'). Cap. of Iraq on the Tigris R. Metro. area pop. 1,745,328.

**Ba·gui·o** (bä′gē-ō′). Summer cap. of the Philippines in NW Luzon. Pop. 118,611.

**Ba·ha·ma Islands** (bə-hä′mə, -hā′-) *also* **Ba·ha·mas** (-məz). Island country in the Atlantic SE of Florida. Cap. Nassau. Pop. 223,455. **—Ba·ha′mi·an** (-hā′mē-ən, -hä′-), **Ba·ha′man** (-hä′mən, -hā′-) *adj.* & *n.*

**Ba·ha·wal·pur** (bə-hä′wəl-po͝or′). Region of E central Pakistan between the Sutlej R. & the Indian border.

**Ba·hi·a** (bä-ē′ə, bə-hē′ə). SALVADOR, Brazil.

**Ba·hí·a Blan·ca** (bä-ē′ə bläng′kə, bə-hē′ə bläng′kə). City of SE Argentina on the **Bahía Blanca,** an inlet of the Atlantic. Pop. 182,158.

**Bah·rain** or **Bah·rein** (bä-rān′). Sheikdom & archipelago in the Persian Gulf between Qatar & Saudi Arabia. Cap. Al Manamah. Pop. 358,857. **—Bah·rain′i** *adj.* & *n.*

**Bahr el Gha·zal** (bär′ ĕl′ gə-zäl′, bär′). River of SW Sudan flowing c. 500 mi (805 km) E to Lake No where it joins the Bahr el Jebel.

**Bahr el Jeb·el** (jĕb′əl). River, 594 mi (955.8 km), of S Sudan, a section of the White Nile.

**Bai** (bī). River, c. 350 mi (563 km), of NE China.

**Baie Co·meau** (bā′ kō′mō). Town of E Que., Canada, on the St. Lawrence NNE of Rimouski. Pop. 11,911.

**Bai·kal** or **Bay·kal** (bī-kôl′, -kŏl′). Lake, 12,160 sq mi (31,494 sq km), of SE Siberian USSR.

**Bain·bridge** (bān′brĭj′). **1.** Island in Puget Sound W of Seattle, Wash. **2.** City of SW Ga. near the Fla. border SSE of Columbus. Pop. 10,553.

**Bai·ri·ki** (bī-rē′kē). Cap. of Kiribati on Tarawa in the W central Pacific. Pop. 1,777.

**Ba·ja Cal·i·for·ni·a** (bä′hä kăl′ə-fôr′nyə). LOWER CALIFORNIA.

**Bak·er** (bā′kər). **1.** Lake, c. 1,000 sq mi (2,590 sq km), in N central N.W.T., Canada. **2.** Peak, 12,460 ft (3,800.3 m), of N Colo. **3.** Peak, 10,778 ft (3,287.3 m) of NW Wash. in the Cascades. **4.** U.S.-owned island in the central Pacific near the equator. **5.** City of SE central La. near Baton Rouge. Pop. 12,865.

**Ba·kers·field** (bā′kərz-fēld′). City of S central Calif. NNW of Los Angeles. Pop. 105,611.

**Ba·ku** (bä-ko͞o′). City of SW Central Asian USSR on the Caspian Sea. Pop. 1,022,000.

**Bal·a·kla·va** or **Bal·a·cla·va** (băl′ə-klăv′ə, -klä′və). Section of Sevastopol, SE European USSR, on the Crimean peninsula; site of the Charge of the Light Brigade (1854).

**Bal·a·ton** (băl′ə-tŏn′, bô′lə-tōn′). Lake, 230 sq mi (595.7 sq km), in W central Hungary SW of Budapest.

**Balch Springs** (bôlch). City of NE Tex. near Dallas. Pop. 13,746.

**Bald·win** (bôld′wĭn). Borough of SW Pa. near Pittsburgh. Pop. 24,598.

**Baldwin Park.** City of S Calif. near Los Angeles. Pop. 50,554.

**Bâle** (bäl). BASEL.

**Bal·e·ar·ic** (băl′ē-ăr′ĭk). Archipelago in the W Mediterranean E of Spain.

**Ba·li** (bä′lē). Island, c. 2,220 sq mi (5,700 sq km), of E Indonesia in the Lesser Sundas E of Java. **—Ba′li·nese′** (bä′lĭ-nēz′, -nēs′) *adj.* & *n.*

**Bal·kan** (bôl′kən). **1.** Mountain range extending c. 350 mi (565 km) from E Yugoslavia through central Bulgaria to the Black Sea. **2.** Peninsula of SE Europe bounded by the Black Sea, Sea of Marmara, & the Aegean, Mediterranean, Ionian, & Adriatic seas; occupied by the **Balkan States**: Albania, Bulgaria, continental Greece, SE Rumania, European Turkey, & most of Yugoslavia.

**Bal·kar·i·a** (bôl-kăr′ē-ə). Region of S European USSR.

**Balkh** (bälk). Town of N Afghanistan, cap. of ancient Bactria.

**Bal·khash** (băl-kăsh′, bäl-käsh′). Lake, 6,562 sq mi (16,996 sq km), of SE Central Asian USSR.

**Bal·la·rat** (băl′ə-răt′). City of SE Australia WNW of Melbourne. Metro. area pop. 60,737.

**Ball·win** (bôl′wĭn). City of E Mo. near St. Louis. Pop. 12,750.

**Bal·sas** (bôl′səs, bäl′-). River flowing c. 450 mi (725 km) from E central Mexico to the Pacific.

**Bal·tic** (bôl′tĭk, bŏl′-). **1. Sea.** Arm of the Atlantic in N Europe. **2. States.** Estonia, Latvia, & Lithuania, on the E coast of the Baltic Sea.

**Bal·ti·more** (bôl′tə-môr′, -mōr′, bŏl′-). City of N Md. NE of Washington, D.C. Pop. 786,775.

**Ba·lu·chi·stan** (bə-lo͞o′chĭ-stăn′). Desert region of W Pakistan.

**Ba·ma·ko** (bä′mə-kō′). Cap. of Mali in the SW on the Niger R. Pop. 404,022.

**Bam·berg** (băm′bûrg′, bäm′bĕrk′). City of S West Germany W of Bayreuth. Pop. 74,236.

**Ba·na·na River** (bə-năn′ə). Lagoon in E. Fla. between Cape Canaveral & Merritt Is.

**Ba·na·ras** (bə-när′əs, -ēz). VARANASI.

**Ba·nat** (bə-nät′, bä′nät′). Region of SE central Europe extending across W Rumania, NE Yugoslavia, & S Hungary.

**Ban·bur·y** (băn′bĕr′ē, -bə-rē). Borough of central England SE of Birmingham. Pop. 31,060.

**Ban·da** (băn′də, bän′-). **1. Sea.** Arm of the Pacific Ocean in E Indonesia SE of Sulawesi & N of Timor. **2. Islands.** Archipelago in the Banda Sea in the Moluccas.

**Ban·dar** (bŭn′dər). MASULIPATAM.

**Ban·dar Se·ri Be·ga·wan** (bŭn′dər sĕr′ē bə-gä′wən). Cap. of Brunei. Pop. 36,987.

**Ban·dei·ra** (băn-dĕr′ə). Highest peak (9,482 ft/2,892 m) of Brazil, in the SE.

**Ban·djar·ma·sin** (băn′jər-mä′sĭn, bän′-). BANJARMASIN.

**Ban·dung** (bän′do͝ong′). City of Indonesia in W Java. Pop. 1,201,-730.

**Banff** (bămf). Resort town of SW Alta., Canada, near Lake Louise in **Banff National Park.** Pop. 3,410.

**Ban·ga·lore** (băng′gə-lôr′, -lōr′). City of S central India W of Madras. Pop. 1,540,741.

**Bang·ka** or **Ban·ka** (băng′kə). Island of W Indonesia in the Java Sea SE of Sumatra.

**Bang·kok** (băng′kŏk′, băng-kŏk′). Cap. of Thailand in the SW on the Chao Phraya R. near the Gulf of Siam. Pop. 1,867,297.

**Bang·la·desh** (bäng′glə-dĕsh′, -däsh′, băng′-). Republic of S Asia on the Bay of Bengal between India & Burma. Cap. Dacca. Pop. 87,052,-024.

**Ban·gor** (băng′gôr, -gər). **1.** Borough of E Northern Ireland on Belfast Lough. Pop. 35,260. **2.** City of S Me. on the Penobscot. Pop. 31,643.

**Ban·gui** (bäng-gē′, bäɴ-). Cap. of Central African Republic on the Ubangi R. near the Zaire border. Pop. 279,792.

**Bang·we·u·lu** (băng′wē-o͞o′lo͞o). Lake & swamps of NE Zambia.

**Ba·ni Su·wayf** or **Be·ni Su·ef** (bĕn′ē so͞o-āf′). City of N central Egypt on the Nile. Pop. 118,148.

**Ban·ja Lu·ka** (bä′yə lo͞o′kə). City of W Yugoslavia in Bosnia NW of Sarajevo. Pop. 85,786.

**Ban·jar·ma·sin** (băn′jər-mä′sĭn, bän′-). City of Indonesia in S Borneo. Pop. 281,673.

**Ban·jul** (bän′jo͞ol′). Cap. of Gambia on an island in the Gambia R. Pop. 39,476.

**Ban·ka** (băng′kə). *var. of* BANGKA.

**Banks** (băngks). **1.** Island (c. 26,000 sq mi/67,340 sq km) of NW N.W.T., Canada, in the Arctic Archipelago. **2.** Archipelago in the SW Pacific N of New Hebrides.

**Ban·ning** (băn′ĭng). City of S Calif. NNE of Santa Ana. Pop. 14,020.

**Ban·nock·burn** (băn′ək-bûrn′, băn′ək-bûrn′). Town of central Scotland on the **Ban·nock R.** (băn′ək); site of Robert Bruce's defeat of the English (1314).

**Ban·ská Bys·tri·ca** (bän′skä bĭs′trĭt-sä′). City of E central Czechoslovakia NE of Bratislava. Pop 53,000.

**Ban·try** (băn′trē). Inlet of the Atlantic in SW Ireland.

**Bao·ding** (bou′dĭng′). City of NE China SSW of Beijing. Pop. 350,000.

**Bao·ji** (bou′jē). City of central China W of Xi′an. Pop. 275,000.

**Bao·tou** (bou′tō′). City of N China on the Yellow R. W of Hohhot. Pop. 800,000.

**Ba·ra·cal·do** (băr′ə-käl′dō, bä′rə). City of N Spain, W of Bilbao. Pop. 108,757.

**Ba·ra·co·a** (băr′ə-kō′ə, bä′rə-). City of SE Cuba on the coast near the E end of the island. Pop. 20,926.

**Ba·ra·nof** (băr′ə-nôf′, -nŏf′, bə-rä′nəf). Island off SE Alas. in the Alexander Archipelago.

**Ba·ra·no·vi·chi** (bə-rä′nə-vĭch′ē). City of W European USSR E of Bobruisk. Pop. 131,000.

**Bar·a·tar·i·a Bay** (băr′ə-târ′ē-ə, -tĕr′-). Lagoon of SE La., an inlet of the Gulf of Mexico.

**Bar·ba·dos** (bär-bā′dəs, -dōz′, -dŏs′, -dōs′). Island country of the E West Indies E of the Windward Is. Cap. Bridgetown. Pop. 249,000. **—Bar·ba′di·an** *adj.* & *n.*

**Bar·ba·ry** (bär′bə-rē, -brē). **1.** Region of N Africa on the **Barbary Coast** between the W border of Egypt & the Atlantic. **2. States.** Formerly, Algeria, Tunisia, Tripoli, & sometimes Morocco.

**Bar·ber·ton** (bär′bər-tən). City of NE Ohio near Akron. Pop. 29,751.

**Bar·bi·zon** (bär′bĭ-zŏn′, bär-bē-zôɴ′). Village of N France SSE of Paris.

**Bar·bu·da** (bär-bo͞o′də). Island of Antigua and Barbuda, in the West Indies N of Antigua.

**Bar·ce·lo·na** (bär′sə-lō′nə). **1.** City of NE Spain on the Mediterranean. Pop. 1,741,144. **2.** City of NE Venezuela near the Caribbean coast. Pop. 78,201.

**Ba·reil·ly** *also* **Ba·re·li** (bə-rā′lē). City of N India ESE of Delhi. Pop. 296,248.

**Ba·rents Sea** (băr′ənts, bär′-). Arm of the Arctic Ocean N of Norway & the USSR.

**Bar Harbor** (bär). Town of SE Me. on Mount Desert Is. Pop. 4,124.

**Ba·ri** (bä′rē). City of SE Italy on the Adriatic. Pop. 339,110.

**Ba·ri·sal** (bär′ĭ-sôl′). City of S Bangladesh on the Ganges delta. Pop. 98,127.

**Ba·ri·san** (bä′rē-sän′). Mountain range of W Sumatra, Indonesia, rising to 12,467 ft (3,802.4 m).

**Bar·let·ta** (bär-lĕt′ə). City of S Italy on the Adriatic. Pop. 75,116.

**Bar·na·ul** (bär'nə-ōōl'). City of SW Siberian USSR on the Ob R. SSE of Novosibirsk. Pop. 533,000.
**Bar·ne·gat Bay** (bär'nĭ-găt', -gət). Inlet of the Atlantic between the E coast of N.J. & offshore islands.
**Barns·ley** (bärnz'lē). Borough of N England N of Sheffield. Pop. 74,730.
**Barn·sta·ble** (bärn'stə-bəl). Town of SE Mass. on central Cape Cod. Pop. 30,898.
**Barn·sta·ple** (bärn'stə-pəl). Borough of SW England on **Barnstaple Bay,** an inlet of Bristol Channel. Pop. 17,820.
**Ba·ro·da** (bə-rō'də). City of W central India SE of Ahmadabad. Pop. 466,696.
**Ba·rot·se·land** (bə-rŏt'sē-lănd'). Region W Zambia.
**Bar·qui·si·me·to** (bär'kə-sə-mā'tō). City of NW Venezuela WSW of Caracas. Pop. 330,815.
**Bar·ran·quil·la** (bär'ən-kē'ə, -yə). City of N Colombia on the Magdalena R. near the Caribbean. Pop. 661,009.
**Bar·re** (băr'ē). City of central Vt. SE of Montpelier. Pop. 9,824.
**Barre des E·crins** (bär'dā-zā-krăN'). Peak, 13,461 ft (4,105.6 m), in SE France in the Dauphiné Alps.
**Bar·ren Grounds** or **Barren Lands** (băr'ən). Region of N Canada NW of Hudson Bay & E of the Mackenzie basin.
**Bar·rie** (băr'ē). City of S Ont., Canada, NNW of Toronto. Pop. 34,389.
**Bar·ring·ton** (băr'ĭng-tən). Town of E R.I. near Providence. Pop. 16,174.
**Bar·row** (băr'ō), **Point.** Northernmost point of Alas. on the Arctic Ocean.
**Bar·row-in-Fur·ness** (băr'ō-ĭn-fûr'nĭs). Borough of NW England NW of Manchester. Pop. 73,400.
**Bar·ry** (băr'ē). Borough of S Wales on the Bristol Channel SSW of Cardiff. Pop. 42,780.
**Bar·stow** (bär'stō'). City of SE Calif. NE of Los Angeles. Pop. 17,690.
**Bar·tles·ville** (bär'tlz-vĭl'). City of NE Okla. N of Tulsa. Pop. 34,568.
**Bart·lett** (bärt'lət). **1.** Village of NE Ill. near Chicago. Pop. 13,254. **2.** Town of SW Tenn. near Memphis. Pop. 17,170.
**Bar·tow** (bär'tō'). City of central Fla. E of Tampa. Pop. 14,780.
**Ba·rú** (bä-rōō'). Volcano, 11,070 ft (3,376.6 m), of W Panama near the Costa Rican border.
**Ba·sel** (bä'zəl) *or* **Basle** (bäl). City of N Switzerland on the Rhine. Pop. 199,600.
**Ba·shan** (bā'shən). Ancient region of Palestine NE of the Sea of Galilee.
**Ba·shi** (bä'shē). Channel between the northernmost Philippines & S Taiwan.
**Bash·kir·i·a** (băsh-kîr'ē-ə). Region of E European USSR in the S Urals.
**Ba·si·lan** (bä-sē'län'). **1.** Island & islands of the SW Philippines off SW Mindanao. **2.** City of NW Basilan Is., Philippines. Pop. 171,266.
**Bas·il·don** (băz'əl-dən). Town of E England near London. Pop. 135,720.
**Ba·si·li·ca·ta** (bə-zĭl'ə-kä'tə, -sĭl'-). Region of S Italy forming the instep of the Italian boot.
**Ba·sing·stoke** (bā'zĭng-stōk'). Borough of S central England SSW of Reading. Pop. 60,910.
**Basle** (bäl). *var. of* BASEL.
**Basque Provinces** (băsk). Region of N Spain on the Bay of Biscay.
**Bas·ra** (băs'rə, bŭs'-). City of SE Iraq on the Shatt-al-Arab. Pop. 313,327.
**Bass** (băs). Strait between Tasmania & SE Australia connecting the Indian Ocean & Tasman Sea.
**Bas·sein** (bə-sān'). Town of S Burma on the **Bassein R.** (160 mi/257 km). Pop. 126,045.
**Basse·terre** (băs-târ', bäs-). Cap. of St. Kitts and Nevis on St. Christopher Is. in the Leewards of the West Indies. Pop. 14,725.
**Basse-Terre** (băs-târ', -bäs-). **1.** Island of the French West Indies in the Leewards, the W part of the French department of Guadeloupe. **2.** Cap. of Guadeloupe at the S end of the island. Pop. 15,026.
**Bas·ti·a** (băs'tē-ə, bä-stē'ə). City of NE Corsica, France, on the Tyrrhenian Sea. Pop. 45,387.
**Bas·trop** (băs'trəp). City of NE La. NNE of Monroe. Pop. 15,527.
**Ba·su·to·land** (bə-sōō'tō-lănd'). LESOTHO.
**Ba·taan** (bə-tăn', -tän'). Peninsula of W Luzon, Philippines, between Manila Bay & the South China Sea.
**Ba·tan** (bə-tän'). Northernmost island group of the Philippines, separated from Taiwan by the Bashi Channel.
**Ba·tan·gas** (bə-tăng'gəs, -tăng'-). City of SW Luzon, Philippines, on **Batangas Bay.** Pop. 143,554.
**Ba·ta·vi·a** (bə-tā'vē-ə). **1.** DJAKARTA. **2.** City of NE Ill., W of Chicago. Pop. 12,574. **3.** City of W N.Y. WSW of Rochester. Pop. 16,703.
**Bath** (băth, bäth). **1.** City of SW England ESE of Bristol. Pop. 83,100. **2.** City of SW Me. on the Kennebec near the Atlantic. Pop. 10,246.
**Bath·urst** (băth'ərst). **1.** Cape of NW N.W.T., Canada, extending into an inlet of Beaufort Sea. **2.** Island off N Australia W of Melville Is. **3.** Island in the Arctic Archipelago, N.W.T., Canada; site of the N magnetic pole. **4.** City of N N.B., Canada, on Chaleur Bay NNE of Fredericton. Pop. 16,301. **5.** BANJUL.
**Bat·ley** (băt'lē). Borough of N central England near Leeds. Pop. 41,630.
**Bat·on Rouge** (băt'n rōōzh'). City of SE central La. NW of New Orleans. Pop. 219,486.
**Bat·ter·y** (băt'ə-rē), **The.** Park, 21 acres (8.5 hectares), at S tip of Manhattan Is., New York City.
**Bat·tle** (băt'l). Rural district of SE England; site of the Battle of Hastings (1066). Pop. 4,987.
**Battle Creek.** City of S Mich. E of Kalamazoo. Pop. 35,724.
**Ba·tu·mi** (bə-tōō'mē) *also* **Ba·tum** (-tōōm'). City of SW Central Asian USSR on the Black Sea near the Turkish border. Pop. 123,000.
**Bat Yam** (băt' yăm'). City of W central Israel on the Mediterranean near Tel Aviv-Jaffa. Pop. 124,100.
**Baut·zen** (bout'sən). City of SE East Germany on the Spree R. ENE of Dresden. Pop. 45,851.
**Ba·var·i·a** (bə-vâr'ē-ə). Region & former duchy of S Germany. **—Ba·var'i·an** *adj. & n.*
**Ba·ya·món** (bī'ə-mōn'). Town of NE Puerto Rico near San Juan. Pop. 184,854.
**Bay City** (bā). **1.** City of E Mich. NNW of Detroit. Pop. 41,593. **2.** City of S Tex. SW of Houston. Pop. 17,837.
**Ba·yeux** (bī-ōō', -yōō', bā-, bä-yœ'). Town of N France in Normandy near the English Channel. Pop. 13,381.
**Bay·kal** (bī-kôl', -kōl'). *var. of* BAIKAL.
**Ba·yonne** (bā-ōn', bä-yôn'). Town of SW France near the Bay of Biscay. Pop. 41,281.
**Bay·onne** (bā-yōn'). City of NE N.J. S of Jersey City. Pop. 65,047.
**Bay·reuth** (bī-roit', bī'roit). City of S West Germany NE of Nuremberg. Pop. 67,035.
**Bay·town** (bā'toun'). City of SE Tex. E of Houston. Pop. 56,923.
**Bay Village.** City of NE Ohio near Cleveland. Pop 17,846.
**Beach·y Head** (bē'chē hĕd'). Chalk cliffs, 575 ft (175.4 m) high, on the SE coast of England.
**Bea·con** (bē'kən). City of SE N.Y. on the Hudson S of Poughkeepsie. Pop. 12,937.
**Bea·cons·field** (bē'kənz-fēld'). Town of S Que., Canada, on Montreal Is. SW of Montreal. Pop. 20,417.
**Bear** (bâr). **1.** River of N Utah, SW Wyo., & SE Idaho flowing 350 mi (563.2 km) to Great Salt Lake. **2.** Mountain, 1,284 ft (391.6 m), of SE N.Y. on the Hudson.
**Beard·more** (bîrd'môr', -mōr'). Valley glacier, 260 mi (418.3 km) long, of Antarctica in the Queen Maud Mts.
**Bé·arn** (bā-ärn'). Region & former province of SW France in the Pyrenees.
**Be·as** (bē'äs'). River, 250 mi (402.3 km), of N India.
**Be·at·rice** (bē-ăt'rĭs). City of SE Nebr. S of Lincoln. Pop. 12,891.
**Beau·fort Sea** (bō'fərt). Part of the Arctic Ocean N of Canada & Alas.
**Beau·jo·lais** (bō'zhə-lā'). Region of E central France W of the Saône.
**Beau·mont** (bō'mŏnt'). City of SE Tex. ENE of Houston. Pop.118,102.
**Beau·port** (bō-pôr'). City of S Que., Canada, on the St. Lawrence near Quebec city. Pop. 55,539.
**Beau·vais** (bō-vā'). Town of N France NNW of Paris. Pop. 53,493.
**Bea·ver** (bē'vər). **1.** River of Alta. & Sask., Canada, flowing 305 mi (490.7 km) to the Churchill. **2.** River, 280 mi (450.5 km), of W Colo. & NW Okla.
**Bea·ver·creek** (bē'vər-krēk'). Village of SW Ohio near Dayton. Pop. 31,589.
**Beaver Dam.** City of S central Wis. NE of Madison. Pop. 14,149.
**Beaver Falls.** City of W Pa. NW of Pittsburgh. Pop. 12,525.
**Bea·ver·head** (bē'vər-hĕd'). Mountains on the Idaho-Mont. border in the SE Bitterroot Range.
**Bea·ver·ton** (bē'vər-tən). City of NW Ore. near Portland. Pop. 30,582.
**Bech·u·a·na·land** (bĕch'wän'ə-lănd', bĕch'ə-). **1.** Former British protectorate of S central Africa. **2.** BOTSWANA.
**Beck·ley** (bĕk'lē). City of S W.Va. SE of Charleston. Pop. 20,492.
**Bed·ford** (bĕd'fərd). **1.** Borough of S central England W of Cambridge. Pop. 74,390. **2.** City of S Ind. S of Bloomington. Pop 14,410. **3.** Town of E Mass. NW of Boston. Pop. 13,067. **4.** City of NE Ohio near Cleveland. Pop. 15,056. **5.** City of N Tex. NNE of Fort Worth. Pop. 20,821.
**Bedford Heights.** City of NE Ohio near Cleveland. Pop. 13,214.
**Bed·worth** (bĕd'wərth). Urban district of central England E of Birmingham. Pop. 41,600.
**Bę·dzin** (bĕn'jēn'). Town of SE Poland near Katowice. Pop. 42,787.
**Beech Grove** (bēch). City of central Ind. near Indianapolis. Pop. 13,196.

**Beer·she·ba** (bîr-shē′bə, bĕr-). City of S Israel SW of Jerusalem. Pop. 101,000.

**Bee·ville** (bē′vĭl′). City of S Tex. NNW of Corpus Christi. Pop. 14,574.

**Bei·jing** (bā′jĭng′). Cap. of China in the NE part. Pop. 8,500,000.

**Bei·ra** (bā′rə). City of E central Mozambique on an arm of the Indian Ocean. Metro. area pop. 130,398.

**Bei·rut** (bā-ro͞ot′). Cap. of Lebanon in the W on the Mediterranean. Pop. 474,870.

**Be·jaï·a** (bə-jī′ə). City of N Algeria on the **Gulf of Bejaia,** an arm of the Mediterranean. Pop. 89,500.

**Bé·kés·csa·ba** (bā′kāsh′chô′bô′). City of SE Hungary NE of Szeged. Pop. 67,266.

**Be·la·ya** (bĕl′ə-yə). River of E European USSR flowing c. 880 mi (1,415 km) from the Urals to the Kama R.

**Be·lém** (bə-lĕm′). City of N Brazil on the Pará R. Pop. 934,330.

**Bel·fast** (bĕl′făst′, bĕl-făst′). Cap. of Northern Ireland on **Belfast Lough,** an inlet of the North Channel of the Irish Sea. Pop. 353,700.

**Bel·fort** (bĕl-fôr′). City of E France commanding the **Belfort Gap** between the Vosges & the Jura Mts. Pop. 54,469.

**Bel·gaum** (bĕl-goum′). Town of SE India SSE of Kolhapur. Pop. 192,427.

**Bel·gian Con·go** (bĕl′jən kŏng′gō). ZAIRE.

**Belgian East Af·ri·ca** (ăf′rĭ-kə). RWANDA; BURUNDI.

**Bel·gium** (bĕl′jəm). Constitutional kingdom of NW Europe on the North Sea. Cap. Brussels. Pop. 9,855,110. **—Bel′gian** *adj.* & *n.*

**Bel·go·rod** (bĕl′gə-rŏd′, byĕl′gə-rət). City of S central European USSR on the Donets R. Pop. 240,000.

**Bel·grade** (bĕl′grād′, -gräd′, bĕl-grād′). Cap. of Yugoslavia in the E at the confluence of the Danube & Sava rivers. Pop. 727,945.

**Bel·gra·vi·a** (bĕl-grā′vē-ə). Residential district of SW London, England.

**Be·li·tung** (bə-lē′to͝ong). Island of W Indonesia in the Java Sea between Sumatra & Borneo.

**Be·lize** (bə-lēz′). **1.** Country of Central America on the Caribbean. Cap. Belmopan. Pop. 144,857. **2.** City of E Belize on the Caribbean at the mouth of the **Belize R.** Pop. 39,887.

**Bell** (bĕl). City of S Calif. near Los Angeles. Pop. 25,450.

**Bell·aire** (bĕl-âr′, bə-lâr′). City of S Tex. near Houston. Pop. 14,950.

**Bel·leau Wood** (bĕ-lō′, bĕl′ō). Forested area of N France E of Château-Thierry.

**Belle·fon·taine** (bĕl-foun′tən, -fŏn′-). City of W central Ohio N of Springfield. Pop. 11,888.

**Bellefontaine Neigh·bors** (nā′bərz). City of E Mo. near St. Louis. Pop. 12,082.

**Belle Fourche** (bĕl′ fo͞osh′). River, c. 290 mi (466 km), of NE Wyo. & W S.Dak.

**Belle Glade.** City of SE Fla. on Lake Okeechobee W of West Palm Beach. Pop. 16,535.

**Belle Isle, Strait of.** Channel between SE Labrador & NW Newf., Canada.

**Belle·ville** (bĕl′vĭl′). **1.** City of SE Ont., Canada, near Lake Ontario ENE of Toronto. Pop. 35,311. **2.** City of SW Ill. SE of East St. Louis. Pop. 42,150. **3.** Town of NE N.J. near Newark. Pop. 35,367.

**Belle·vue** (bĕl′vyo͞o′). **1.** City of E Nebr. near Omaha. Pop. 21,813. **2.** Borough of SW Pa. near Pittsburgh. Pop. 10,128. **3.** City of W Wash. near Seattle. Pop. 73,903.

**Bell·flow·er** (bĕl′flou′ər). City of S Calif. near Los Angeles. Pop. 53,441.

**Bell Gar·dens** (gär′dnz). City of S Calif. near Los Angeles. Pop. 34,117.

**Bel·ling·ham** (bĕl′ĭng-hăm′). **1.** Town of S Mass. SE of Worcester. Pop. 14,300. **2.** City of NW Wash. on **Bellingham Bay** near the B.C., Canada, border. Pop. 45,794.

**Bel·lings·hau·sen Sea** (bĕl′ĭngz-hau′zən). Arm of the S Pacific off the coast of Antarctica extending from Alexander I Is. to Thurston Is.

**Bell·mawr** (bĕl′mär′, -môr′). Borough of SW N.J. near Camden. Pop. 13,721.

**Bell·wood** (bĕl′wo͝od′). Village of NE Ill. near Chicago. Pop. 19,811.

**Bel·mont** (bĕl′mŏnt′) **1.** City of W Calif. SSE of San Francisco. Pop. 24,505. **2.** Town of E Mass. near Boston. Pop. 26,100.

**Bel·mo·pan** (bĕl′mō-păn′). Cap. of Belize in the N central part. Pop. 2,932.

**Bel·oeil** (bə-lĭl′). Town of S Que., Canada, on the Richelieu R. near Montreal. Pop. 15,913.

**Be·lo Ho·ri·zon·te** (bā′lō hôr′ĭ-zŏn′tē, bĕl′ō). City of E Brazil N of Rio de Janeiro. Pop. 1,774,712.

**Be·loit** (bə-loit′). City of S Wis. on the Ill. border SSE of Madison. Pop. 35,207.

**Be·lo·rus·sia** (bĕl′ō-rŭsh′ə) *also* **Bye·lo·rus·sia** (bē-ĕl′ō-). **1.** Region of E Europe E of Poland, S of Lithuania & Latvia, & N of the Ukraine. **2.** *also* **Be·lo·rus·sian Soviet Socialist Republic** (-rŭsh′ən). Constituent republic of W central European USSR. **—Be′lo·rus′sian** *adj.* & *n.*

**Be·lo·stok** (bĕl′ə-stôk′). BIALYSTOK.

**Bel·sen** (bĕl′zən). Village of NW West Germany; site of Nazi concentration camp.

**Bel·ton** (bĕl′tən). **1.** City of W Mo. near Kansas City. Pop. 12,708. **2.** City of central Tex. S of Fort Worth. Pop. 10,660.

**Be·lu·kha** (bə-lo͞o′kə). Highest elevation, 15,157 ft (4,662.8 m), of the Altai Mts. in the USSR near Mongolia.

**Bel·vi·dere** (bĕl′vĭ-dîr′). City of N Ill. E of Rockford. Pop. 15,176.

**Be·midj·i** (bə-mĭj′ē). City of NW central Minn. WNW of Duluth. Pop. 10,949.

**Be·na·res** (bə-när′əs, -ēz). VARANASI.

**Ben·brook** (bĕn′bro͝ok′). City of NE Tex. near Fort Worth. Pop. 13,579.

**Bend** (bĕnd). City of central Ore. E of Eugene. Pop. 17,263.

**Ben·di·go** (bĕn′dĭ-gō′). City of SE Australia NNW of Melbourne. Metro. area pop. 50,169.

**Ben·dzin** (bĕn′dēn′). BĘDZIN.

**Be·ne·lux** (bĕn′ə-lŭks′). Tripartite customs union formed in 1947 by Belgium, the Netherlands, & Luxembourg.

**Be·ne·ven·to** (bĕn′ə-vĕn′tō). City of S Italy NE of Naples. Pop. 48,523.

**Ben·gal** (bĕn-gôl′, bĕng-). Region of E India & Bangladesh on the **Bay of Bengal,** arm of the Indian Ocean between Sri Lanka & India on the W & Burma & Thailand on the E. **—Ben′ga·lese′** (bĕn′gə-lēz′, -lēs′, bĕng′-) *adj.* & *n.*

**Beng·bu** (bŭng′bo͞o′). City of E China NW of Nanjing. Pop. 400,000.

**Ben·gha·zi** (bĕn-gä′zē, bĕng-găz′ē). City of NE Libya on the Mediterranean. Pop. 286,943.

**Ben·guel·a** (bĕn-gĕl′ə, -gwĕl′ə). City of W Angola on the Atlantic. Pop. 40,996.

**Be·ni** (bā′nē). River of NW & central Bolivia flowing 994 mi (1,599.3 km) from the Andes to the Mamoré.

**Be·ni·cia** (bə-nē′shə). City of W Calif. NE of Oakland. Pop. 15,376.

**Be·nin** (bə-nĭn′, -nēn′). **1.** Former kingdom of W Africa, now part of Nigeria. **2.** Country of W Africa. Cap. Porto-Novo. Pop. 3,338,240. **3.** City of S Nigeria on the **Benin R.,** flowing c. 100 mi (161 km) into the **Bight of Benin,** an indentation of the Gulf of Guinea. Pop. 136,000.

**Be·ni Su·ef** (bĕn′ē so͞o-āf′). *var. of* BANI SUWAYF.

**Ben Lo·mond** (bĕn lō′mənd). Mountain, 3,192 ft (973.5 m), of S central Scotland on the E shore of Loch Lomond.

**Ben Nev·is** (nē′vĭs, nĕv′ĭs). Highest elevation, 4,406 ft (1,343.8 m), of Great Britain, in the Grampians of W Scotland.

**Ben·ning·ton** (bĕn′ĭng-tən). Town of SW Vt. E of Brattleboro. Pop. 15,815.

**Be·no·ni** (bə-nō′nē′). Town of NE South Africa on the Witwatersrand. Pop. 151,294.

**Ben·sen·ville** (bĕn′sən-vĭl′). Village of NE Ill. WNW of Chicago. Pop. 16,124.

**Ben·ton** (bĕn′tən). City of central Ark. SW of Little Rock. Pop. 17,676.

**Benton Harbor.** City of SW Mich. on Lake Michigan SSW of Grand Rapids. Pop. 14,707.

**Be·nue** (bān′wā) *also* **Bin·ue** (bĭn′wā). River of W Africa flowing c. 670 mi (1,080 km) from Cameroon to the Niger in Nigeria.

**Ben·xi** (bŭn′shē′). City of NE China SSE of Shenyang. Pop. 750,000.

**Bep·pu** (bĕp′po͞o′). City of NE Kyushu, Japan, on **Beppu Bay,** an arm of the Inland Sea. Pop. 133,894.

**Be·rar** (bā-rär′, bə-). Region of W central India.

**Berch·tes·ga·den** (bĕrk′təs-gäd′n, bĕrKH′-). Town of SE West Germany in the Bavarian Alps. Pop. 8,558.

**Be·re·a** (bə-rē′ə). City of NE Ohio near Cleveland. Pop. 19,567.

**Ber·e·ni·ce** (bĕr′ə-nī′sē). Ancient Egyptian city on the Red Sea.

**Be·re·zi·na** (bə-rā′zĭn-ə). River, c. 380 mi (610 km), of E central European USSR.

**Be·rez·ni·ki** (bə-rāz′nĭ-kē′). City of E European USSR on the Kama R. Pop. 185,000.

**Ber·ga·ma** (bĕr-gä′mə, bər-). Town of W Turkey N of Izmir; site of ancient Pergamum. Pop. 29,749.

**Ber·ga·mo** (bĕr′gə-mō′). City of N Italy NE of Milan. Pop. 127,553.

**Ber·gen** (bûr′gən, bĕr′-). City of SW Norway on inlets of the North Sea. Pop. 213,434.

**Ber·gen·field** (bûr′gən-fēld′). Borough of NE N.J. near Hackensack. Pop. 25,568.

**Ber·gen op Zoom** (bĕr′gən ŏp zōm′). Town of SW Netherlands on an estuary of the Scheldt. Pop. 40,770.

**Ber·gisch-Glad·bach** (bĕr′gĭsh-glät′bäk′, -bäKH′). Town of W West Germany near Cologne. Pop. 99,517.

**Be·ring** (bîr′ĭng, bĕr′-). Sea, part of the Pacific, between Siberia & Alas., joined to the Arctic Ocean by the **Bering Strait** (c. 55 mi/90 km wide).

**Berke·ley** (bûrk′lē). **1.** City of W Calif. N of Oakland. Pop. 103,328. **2.** City of E Mo. WNW of St. Louis. Pop. 16,146.

**Berk·ley** (bûrk′lē). City of SE Mich. near Detroit. Pop. 18,637.

**Berk·shire** (bûrk′shîr′, -shər). Range of hills in W Mass.

ă **pat** ā **pay** âr **care** ä **father** ĕ **pet** ē **be** hw **which** ĭ **pit**
ī **tie** îr **pier** ŏ **pot** ō **toe** ô **paw, for** oi **noise** o͝o **took**

**Ber·lin. 1.** (bûr-lĭn'). City of NE East Germany divided since 1945 into **East Berlin,** cap. of East Germany, pop. 1,094,147, & **West Berlin,** part of West Germany, pop. 1,984,837. **2.** (bûr'lĭn). City of N central Conn. near Hartford. Pop. 15,121. **3.** (bûr'lĭn). City of NE N.H. E of Lancaster. Pop. 13,084.
**Ber·me·jo** (bər-mā'hō, bĕr-). River of N Argentina flowing c. 650 mi (1,045 km) to the Paraguay R. at the Paraguay border.
**Ber·mu·da** (bər-myōō'də). British colony in the Atlantic SE of Cape Hatteras, an archipelago of c. 350 islands. Cap. Hamilton. Pop. 67,761. **—Ber·mu'di·an** *adj. & n.*
**Bern** *or* **Berne** (bûrn, bĕrn). Cap. of Switzerland in the NW on the Aare. Pop. 154,700.
**Bern·burg** (bûrn'bûrg', bĕrn'-). City of central East Germany on the Saale R. Pop. 44,428.
**Berne** (bûrn, bĕrn). *var. of* BERN.
**Ber·nese Alps** (bûr-nēz', -nēs'). Range of the Alps in S central Switzerland rising to 14,032 ft (4,279.8 m).
**Ber·ni·ci·a** (bər-nĭsh'ē-ə, -nĭsh'ə, bĕr-). 6th-cent. A.D. Anglian kingdom in present-day NE England.
**Ber·ni·na** (bər-nē'nə, bĕr-). Mountain group of SW Switzerland, part of the Rhaetian Alps on the Swiss-Italian border; highest elevation, **Piz Bernina** (13,287 ft/4,052.5 m).
**Ber·ry** (bĕ-rē'). Former province of central France.
**Ber·wick** (bûr'wĭk). Borough of E central Pa. SE of Wilkes-Barre. Pop. 12,189.
**Ber·wyn** (bûr'wĭn'). City of NE Ill. near Chicago. Pop. 46,849.
**Be·san·çon** (bĭ-zäN-sôN'). City of E France E of Dijon. Pop. 119,803.
**Bes·kids** (bĕs'kĭdz', bĕs-kēdz'). Mountain ranges of the W Carpathians on the Polish-Czechoslovakian border, divided into the **East Beskids** & the **West Beskids.**
**Bes·sa·ra·bi·a** (bĕs'ə-rā'bē-ə). Region of SW European USSR. **—Bes'sa·ra'bi·an** *adj. & n.*
**Bes·se·mer** (bĕs'ə-mər). City of N central Ala. SSW of Birmingham. Pop. 31,729.
**Beth·a·ny** (bĕth'ə-nē). **1.** Village of Biblical Palestine near Jerusalem. **2.** City of central Okla. near Oklahoma City. Pop. 22,130.
**Beth·el** (bĕth'əl). **1.** (*also* bə-thĕl'). Town of Biblical Palestine N of Jerusalem. **2.** Town of SW Conn. NW of Bridgeport. Pop. 16,004.
**Bethel Park.** Borough of SW Pa. near Pittsburgh. Pop. 34,755.
**Be·thes·da** (bə-thĕz'də). City of W central Md. near Washington, D.C. Pop. 78,300.
**Beth·le·hem** (bĕth'lĭ-hĕm', -lē-əm). **1.** Town of the West Bank S of Jerusalem. Pop. 14,439. **2.** City of E Pa. NNW of Philadelphia. Pop. 70,419.
**Beth·sa·i·da** (bĕth-sā'ĭ-də). Town of Biblical Palestine on the NE shore of the Sea of Galilee.
**Bet·si·a·mi·tes** (bĕt'sē-ə-mē'tĕz). River of E Que., Canada, flowing c. 240 mi (385 km) into the St. Lawrence.
**Bet·ten·dorf** (bĕt'n-dôrf'). City of E Iowa near Davenport. Pop. 27,381.
**Bev·er·ley** (bĕv'ər-lē). Borough of NE England NNW of Hull. Pop. 16,920.
**Bev·er·ly** (bĕv'ər-lē). City of NE Mass. near Salem. Pop. 37,655.
**Beverly Hills. 1.** City of S Calif. surrounded by Los Angeles. Pop. 32,367. **2.** Village of SE Mich. near Detroit. Pop. 11,598.
**Bex·hill** (bɛks'hĭl'). Borough of SE England on the English Channel. Pop. 34,680.
**Bex·ley** (bĕks'lē). City of central Ohio surrounded by Columbus. Pop. 13,405.
**Bé·ziers** (bāz-yā'). City of S France SW of Montpellier. Pop. 79,213.
**Bha·gal·pur** (bä'gəl-pŏŏr'). City of NE India on the Ganges. Pop. 172,202.
**Bhat·pa·ra** (bät-pä'rə). City of NE India on the Hooghly R. N of Calcutta. Pop. 204,750.
**Bhav·na·gar** (bou-nŭg'ər). City of W India on the Gulf of Cambay. Pop. 225,358.
**Bhi·ma** (bē'mə). River, c. 400 mi (645 km), of S India.
**Bho·lan** (bō-län'). *var. of* BOLAN.
**Bho·pal** (bō-päl'). City of central India NW of Nagpur. Pop. 298,022.
**Bhu·tan** (bōō-tän', -tăn'). Kingdom of central Asia in the E Himalayas. Cap. Thimbu. Pop. 1,298,000. **—Bhu'tan·ese'** (bōō'tn-ēz', -ēs') *adj. & n.*
**Bi·a·fra** (bē-ăf'rə, -ä'frə). Region of E Nigeria on the **Bight of Biafra,** an inlet of the Gulf of Guinea. **—Bi·a'fran** *adj. & n.*
**Bi·ak** (bē-äk', -yäk'). Largest (948 sq mi/2,455.3 sq km) of the Schouten Is. of Indonesia, off NW New Guinea.
**Bia·ly·stok** (bē-ä'lĭ-stôk'). City of NE Poland near the border of Belorussia. Pop. 166,619.
**Biar·ritz** (bē'ə-rĭts', bē'ə-rĭts'). City of SW France on the Bay of Biscay. Pop. 27,453.
**Bid·de·ford** (bĭd'ə-fərd). City of SW Me. SW of Portland. Pop. 19,638.
**Biel** (bēl, bē'əl). City of NW Switzerland at the NE end of the **Lake of Biel** (15 sq mi/39 sq km). Pop. 63,400.
**Bie·le·feld** (bē'lə-fĕlt'). City of N central West Germany E of Münster. Pop 316,058.
**Biel·la** (bē-ĕl'ə). City of NW Italy WNW of Milan. Pop. 46,453.
**Biel·sko-Bia·la** (bē-ĕl'skô-bē-ä'lə). City of S Poland SW of Kraków. Pop. 105,601.
**Bi·enne** (bē-ĕn'). BIEL.
**Big Bend** (bĭg bĕnd'). **1.** National park of W Tex. in a triangle formed by the Rio Grande. **2.** Portion of the Columbia River in E central Wash.
**Big Black** (blăk). River of Miss. flowing c. 330 mi (530 km) to the Mississippi below Vicksburg.
**Big Blue** (blōō). River of SE Nebr. & NE Kans. flowing c. 300 mi (485 km) to the Kansas R.
**Big·horn** (bĭg'hôrn'). **1.** River flowing 461 mi (741.7 km) from W central Wyo. to the Yellowstone R. in S Mont. **2.** Section of the Rocky Mts. of N Wyo. & S Mont. rising to 13,175 ft (4,018.4 m).
**Big Mud·dy** (mŭd'ē). River of SW Ill. flowing c. 135 mi (217 km) to the Mississippi.
**Big Rapids.** City of W central Mich. on the Muskegon. Pop. 14,361.
**Big Sand·y Creek** (săn'dē). River of central & E Colo. flowing c. 200 mi (320 km) to the Arkansas R.
**Big Sioux** (sōō). River of NE S.Dak. flowing 420 mi (675.8 km) to the Missouri R.
**Big Spring.** City of W Tex. WSW of Abilene. Pop. 24,804.
**Big Sur** (sûr). Rugged resort region of central Calif. coast.
**Big Thick·et** (thĭk'ĭt). Wilderness region of E Tex. NE of Houston.
**Bi·ka** (bē-kä'), **El.** *var. of* Al BIQA.
**Bi·ka·ner** (bē'kə-nîr', -nâr'). City of NW India in the Thar Desert near the Pakistan border. Pop. 188,518.
**Bi·ki·ni** (bĭ-kē'nē). Atoll, c. 2 sq mi (5.2 sq km), in the Marshall Is. of the W central Pacific.
**Bi·lauk·taung** (bē-louk'toung). Mountain range extending c. 250 mi (400 km) along the Thailand-Burma border.
**Bil·ba·o** (bĭl-bä'ō, -bou'). City of N Spain near the Bay of Biscay. Pop. 393,179.
**Bille·ric·a** (bĭl-rĭk'ə, bĕl'ə-). Town of NE Mass. near Lowell. Pop. 36,727.
**Bil·lings** (bĭl'ĭngz). City of S Mont. ESE of Helena. Pop. 66,798.
**Bil·li·ton** (bə-lē'tŏn'). BELITUNG.
**Bi·lox·i** (bə-lŭk'sē, -lŏk'-). City of extreme SE Miss. E of Gulfport. Pop. 49,311.
**Bim·i·ni** (bĭm'ə-nē). Group of small islands in the Straits of Florida, the NW section of the Bahamas.
**Bing·ham·ton** (bĭng'əm-tən). City of S N.Y. near the Pa. border SSE of Syracuse. Pop. 55,860.
**Bin·tan** (bĭn'tän') *also* **Bin·tang** (-täng'). Island of the Riau Archipelago, W Indonesia, off the S tip of the Malay peninsula in the South China Sea.
**Bin·ue** (bĭn'wā). *var. of* BENUE.
**Bí·o-Bí·o** (bē'ō-bē'ō). River of central Chile flowing c. 240 mi (385 km) from the Andes to the Pacific.
**Bi·o·ko** (bē-ō'kō). Island of Equatorial Guinea, W central Africa, in the Gulf of Guinea.
**Bi·qa** (bē-kä'), **Al,** *also* **El Bi·ka** (bē-kä'). Valley of Lebanon and Syria between the Lebanon & Anti-Lebanon ranges.
**Bir·ken·head** (bûr'kən-hĕd'). Borough of W central England at the mouth of the Mersey near Liverpool. Pop. 135,750.
**Bir·ming·ham** (bûr'mĭng-hăm'). **1.** (*also* -əm). City of central England NW of London. Pop. 1,058,800. **2.** City of N central Ala. NE of Tuscaloosa. Pop. 284,413. **3.** City of SE Mich. NW of Detroit. Pop. 21,689.
**Bis·cay** (bĭs'kā), **Bay of.** Arm of the Atlantic indenting the W coast of Europe from NW France to NW Spain.
**Bis·cayne Bay** (bĭs-kān', bĭs'kān'). Inlet of the Atlantic in SE Fla.
**Bi·sce·glie** (bĭ-shāl'yā). City of S Italy on the Adriatic. Pop. 45,014.
**Bish·op Auck·land** (bĭsh'əp ôk'lənd). Urban district of NE England S of Newcastle. Pop. 32,940.
**Bisk** (bĭsk, bēsk). BIYSK.
**Bis·marck** (bĭz'märk'). **1.** Sea in the SW Pacific NE of New Guinea & NW of New Britain. **2.** Archipelago of approx. 200 islands & islets in the SW Pacific NE of New Guinea. **3.** Mountain range of Papua New Guinea on NE New Guinea, rising to 15,400 ft (4,697 m). **4.** Cap. of N.Dak. in the S central part. Pop. 44,485.
**Bis·sau** (bĭ-sou'). Cap. of Guinea-Bissau on an estuary of the Atlantic. Pop. 109,486.
**Bi·thyn·i·a** (bĭ-thĭn'ē-ə). Ancient country of NW Asia Minor in present-day Turkey. **—Bi·thyn'i·an** *adj. & n.*
**Bi·to·la** (bĕt'l-yä'). City of S Yugoslavia near the Greek border. Pop. 64,467.
**Bi·ton·to** (bĭ-tôn'tō). City of S Italy W of Bari. Pop. 39,714.
**Bit·ter** (bĭt'ər). Two lakes, **Great Bitter** & **Little Bitter,** of NE Egypt, crossed & connected by the Suez Canal.
**Bit·ter·root** (bĭt'ər-rōōt', -rŏŏt'). **1.** River, c. 120 mi (195 km), of SW Mont. **2.** Mountain range of the Rockies on the Idaho-Mont. border, rising to 10,961 ft (3,343.1 m).

**Bi·wa** (bē'wä). Lake of S Honshu, Japan, largest (260 sq mi/673.4 sq km) in the country.
**Bi·ysk** (bē'ĭsk, bēsk). City of S central Siberian USSR ESE of Barnaul. Pop. 212,000.
**Bi·zer·te** (bĭ-zûr'tē, bē-zĕrt'). City of N Tunisia on the Mediterranean. Pop. 62,856.
**Black** (blăk). **1.** Sea, c. 159,600 sq mi (413,365 sq km), between Europe & Asia, connected with the Aegean by the Bosporus, the Sea of Marmara, & the Dardanelles. **2.** River of SE Asia flowing c. 500 mi (805 km) from S China to the Red R. in N Vietnam. **3.** River flowing c. 300 mi (485 km) from SE Mo. to NE Ark. **4.** River of N N.Y. flowing c. 120 mi (195 km) to **Black R. Bay,** an inlet of Lake Ontario. **5.** River of central & W Wis. flowing c. 160 mi (260 km) to the Mississippi. **6.** Range of the Blue Ridge in W N.C. **7.** Canyon of the Colorado R. between Ariz. & Nev. **8.** Canyon of the Gunnison R., a national monument in SW Colo.
**Black·burn** (blăk'bûrn'). **1. Mount.** Highest peak (16,523 ft/ 5,039.5 m) of the Wrangell Mts. of S Alas. **2.** City of NW England NNE of Bolton. Pop. 101,670.
**Black·foot** (blăk'fo͝ot'). City of SE Idaho SSW of Idaho Falls. Pop. 10,065.
**Black Forest.** Mountain range of SW Germany between the Rhine & the Neckar.
**Black·heath** (blăk'hēth'). Common, 267 acres (108 hectares), of London, England.
**Black Hills.** Mountains of SW S.Dak. & NE Wyo.
**Black·pool** (blăk'po͞ol'). Borough of NW England on the Irish Sea. Pop. 149,000.
**Blacks·burg** (blăks'bûrg'). Town of SW Va. W of Roanoke. Pop. 30,638.
**Black·town** (blăk'toun'). City of SE Australia near Sydney. Pop. 159,734.
**Black Vol·ta** (vŏl'tə, vōl'-, vôl'-). River of W Africa flowing c. 840 mi (1,350 km) from W Upper Volta to the White Volta in Ghana.
**Black War·rior** (wôr'ē-ər, wŏr'-). River of central Ala. flowing 178 mi (161 km) to the Tombigbee.
**Black·wells Island** (blăk'wĕlz', -wəlz). WELFARE Is.
**Bla·go·vesh·chensk** (blä'gə-vĕsh'chənsk). City of Far Eastern USSR at the confluence of the Amur & Zeya rivers. Pop. 172,000.
**Blaine** (blān). **1.** City of E Minn. N of St. Paul. Pop. 28,558. **2.** City of NW Wash. on the B.C., Canada, border. Pop. 2,263.
**Blanc** (blăngk, blän). **1.** Cape on the coast of Tunisia; northernmost point of Africa. **2. Mont.** —See MONT BLANC.
**Blan·ca** (blăng'kə). Peak, 14,317 ft (4,366.7 m), of S Colo. in the Sangre de Cristo Mts.
**Blan·tyre** (blăn-tīr'). City of S Malawi, SE Africa. Pop. 222,153.
**Blar·ney** (blär'nē). Village of SW Ireland WNW of Cork.
**Blay·don** (blād'n). Urban district of NE England on the Tyne R. Pop. 31,940.
**Bli·da** (blē'də). Town of N Algeria SW of Algiers. Pop. 160,900.
**Block** (blŏk). Island off S R.I. at the E entrance to Long Is. Sound.
**Bloem·fon·tein** (blo͞om'fən-tān', -fŏn-). City of S South Africa ESE of Kimberley. Pop. 149,836.
**Blois** (blwä). Town of central France on the Loire. Pop. 49,134.
**Bloom·field** (blo͞om'fēld'). **1.** Town of N central Conn. near Hartford. Pop. 18,608. **2.** Town of NE N.J. near Newark. Pop. 47,792.
**Bloo·ming·dale** (blo͞o'mĭng-dāl'). Village of NE Ill. WNW of Chicago. Pop. 12,659.
**Bloo·ming·ton** (blo͞o'mĭng-tən). **1.** City of central Ill. ESE of Peoria. Pop. 44,189. **2.** City of S central Ind. SSW of Indianapolis. Pop. 51,646. **3.** City of SE Minn. near Minneapolis. Pop. 81,831.
**Blooms·burg** (blo͞omz'bûrg'). Town of E Pa. on the Susquehanna. Pop. 11,717.
**Blooms·bur·y** (blo͞omz'bĕr'-ē, -bə-rē, -brē). Residential district of N central London, England.
**Blue** (blo͞o). **1.** Mountain range of SE Australia. **2.** Mountains of E Jamaica rising to **Blue Mt. Peak,** c. 7,402 ft (2,257 m). **3.** Mountains of NE Ore. & SE Wash.
**Blue·field** (blo͞o'fēld'). City of S W.Va. SSE of Charleston. Pop. 16,060.
**Blue·grass** (blo͞o'grăs'). Region of central Ky.
**Blue Grot·to** (grŏt'ō). Cave on N coast of Capri, S Italy.
**Blue Island.** City of NE Ill. near Chicago. Pop. 21,855.
**Blue Nile** (nīl). River of NE Africa, chief headstream of the Nile, flowing c. 1,000 mi (1,610 km) from NW Ethiopia into the Sudan, where it merges with the White Nile to form the Nile at Khartoum.
**Blue Ridge.** Mountain range extending from S Pa. to N Ga., part of the Appalachians.
**Blue Springs.** City of W Mo. E of Kansas City. Pop. 25,927.
**Blyth** (blī, blīth). Borough of NE England on the North Sea. Pop. 35,390.
**Blythe·ville** (blī'vəl, blīth'vĭl'). City of NE Ark. N of Memphis, Tenn. Pop. 24,326.
**Bo·bi·gny** (bô'bēn-yē'). City of N central France near Paris. Pop. 43,041.
**Bo·bruisk** (bō-bro͞o'ĭsk). City of W central European USSR SE of Minsk. Pop. 192,000.
**Bo·ca Ra·ton** (bō'kə rə-tōn'). City of SE Fla. S of Palm Beach. Pop. 49,505.
**Bo·chum** (bō'kəm). City of W West Germany in the Ruhr E of Essen. Pop. 414,842.
**Bo·den·see** (bōd'n-zā'). Lake of CONSTANCE.
**Boe·o·tia** (bē-ō'shə, -shē-ə). Ancient region of Greece N of Attica & the Gulf of Corinth. **—Boe·o'tian** *adj.* & *n.*
**Boe·roe** (bo͞or'o͞o). *var. of* BURU.
**Bo·ga·lu·sa** (bō'gə-lo͞o'sə). City of SE La. NNE of New Orleans. Pop. 16,976.
**Bog·nor Re·gis** (bŏg'nər rē'jĭs). Urban district of S central England on the English Channel W of Brighton. Pop. 34,620.
**Bo·gor** (bō'gôr'). City of W Java, Indonesia, S of Djakarta. Pop. 195,882.
**Bo·go·tá** (bō'gə-tä'). Cap. of Colombia in the central part. Pop. 2,696,270.
**Bo Hai** (bō' hī'). Inlet of the Yellow Sea on the NE coast of China W of the Shandong & Liaodong peninsulas.
**Bo·he·mi·a** (bō-hē'mē-ə). Historical region & former kingdom of W Czechoslovakia. **—Bo·he'mi·an** *adj.* & *n.*
**Bohemian Forest.** Mountain range of the N Czechoslovakian-West German border, extending into Austria.
**Bo·hol** (bō-hôl'). Island of the central Philippines SW of Leyte at the N end of the Mindanao Sea.
**Bois de Bou·logne** (bwä' də bo͞o-lōn', -loin'). Park in Paris, France, bordering the suburb of Neuilly-sur-Seine.
**Boi·se** (boi'sē, -zē). **1.** River of SW Idaho flowing c. 160 mi (260 km) to join the Snake at the Ore. border. **2.** Cap. of Idaho in the SW part near the Ore. border. Pop. 102,451.
**Boj·a·dor** (bŏj'ə-dôr'), **Cape.** Headland of NW Africa in the Atlantic on the W central coast of Western Sahara.
**Bo·kha·ra** (bō-kär'ə, -här'-). *var. of* BUKHARA.
**Boks·burg** (bŏks'bûrg'). City of NE South Africa E of Johannesburg. Pop. 106,126.
**Bo·lan** *also* **Bho·lan** (bō-län'). Mountain pass in W Pakistan, c. 60 mi (95 km) long & located at an altitude of 5,880 ft (1,793.4 m).
**Bo·ling·broke** (bō'lĭng-bro͝ok'). Village of NE Ill. SW of Chicago. Pop. 37,261.
**Bo·liv·i·a** (bə-lĭv'ē-ə). Republic of W South America. Caps. Sucre & La Paz. Pop. 5,600,000. **—Bo·liv'i·an** *adj.* & *n.*
**Bo·lo·gna** (bə-lōn'yə). City of N central Italy NE of Florence. Pop. 493,282. **—Bo·lo'gnan, Bo'lo·gnese'** (bō'lə-nēz', -nēs', -lən-yēz', -yēs') *adj.* & *n.*
**Bol·ton** (bōl'tən) *also* **Bol·ton-le-Moors** (-lə-mo͝orz'). Borough of NW England, part of Greater Manchester. Pop. 154,480.
**Bol·za·no** (bōlt-sä'nō, bōl-zä'-). City of N Italy NNW of Venice. Pop. 102,806.
**Bom·bay** (bŏm-bā'). City of W central India on coastal **Bombay Is.** & an adjacent island. Metro. area pop. 5,970,575.
**Bom·o·seen** (bŏm'ə-zēn'). Lake of W Vt. W of Rutland.
**Bo·mu** (bō'mo͞o). River of central Africa flowing c. 500 mi (805 km) from SE Central African Republic & along the boundary with Zaire to the Uele, forming the Ubangi.
**Bon** (bôn), **Cape.** Peninsula of NE Tunisia.
**Bo·na** (bō'nə), **Mount.** Peak, 16,420 ft (5,008 m), of S Alas. at the S end of the Wrangell Mts. near the Canadian border.
**Bo·naire** (bə-nâr'). Island of the Netherlands Antilles in the Leewards off the N coast of Venezuela.
**Bo·nam·pak** (bō-näm'päk). Ruined Mayan city near present-day Tuxtla Gutiérrez in S Mexico.
**Bo·nan·za** (bə-năn'zə). Creek in W Y.T., Canada, flowing c. 20 mi (30 km) to the Klondike near Dawson.
**Bon·a·vis·ta** (bŏn-ə-vĭs'tə). Arm of the Atlantic in E Newf.
**Bône** (bōn). ANNABA.
**Bo·nin** (bō'nĭn). Archipelago of 15 islands in the W Pacific c. 500 mi (804 km) S of Japan.
**Bonn** (bŏn, bôn). Cap. of West Germany, on the Rhine. Pop. 283,711.
**Bon·ne·ville Salt Flats** (bŏn'ə-vĭl'). Plain of NW Utah W of Great Salt Lake, part of the bed of prehistoric **Lake Bonneville.**
**Boone** (bo͞on). **1.** City of central Iowa NNW of Des Moines. Pop. 12,602. **2.** Town of NW N.C. NNW of Lenoir. Pop. 10,191.
**Boones·bor·o** (bo͞onz'bûr'ō). Former settlement of central Ky. on the Kentucky R.
**Boo·thi·a** (bo͞o'thē-ə), **Gulf of.** Inlet of the Arctic Ocean in NE Canada E of **Boothia Peninsula,** northernmost tip of the North American mainland.
**Boo·tle** (bo͞ot'l). Borough of NW England at the mouth of the Mersey R. Pop. 71,160.
**Bo·phu·tha·tswa·na** (bō'po͞o-tät-swä'nə). Autonomous black homeland within South Africa. Cap. Mmabatho. Pop. 1,200,000.
**Bo·ra Bo·ra** (bôr'ə bôr'ə, bōr'ə bōr'ə). Island of French Polynesia in the Leeward group of the Society Is. in the S Pacific.

**Bo·rah** (bôr'ə, bōr'ə), **Mount.** Peak, 12,662 ft (3,861.9 m), of central Idaho.
**Bo·rås** (bo͞o-rōs'). City of SW Sweden E of Göteborg. Pop. 67,537.
**Bor·deaux** (bôr-dō'). City of SW France on the Garonne R. Pop. 220,830.
**Bor·der** (bôr'dər) *also* **Bor·ders** (-dərz). Boundary & adjacent areas between England & Scotland.
**Bor·ger** (bôr'gər). City of N Tex. in the Panhandle NE of Amarillo. Pop. 15,837.
**Bor·ger·hout** (bôr'gər-hout'). City of N Belgium near Antwerp. Pop. 49,002.
**Borgne** (bôrn), **Lake.** Inlet of Mississippi Sound E of New Orleans, La.
**Bor·ne·o** (bôr'nē-ō'). Island, c. 287,000 sq mi (743,300 sq km), of the W Pacific between the Sulu & Java seas SW of the Philippines; divided between Kalimantan & Brunei. **—Bor'ne·an** *adj.* & *n.*
**Born·holm** (bôrn'hōm', -hōlm'). Island group of E Denmark in the Baltic near Sweden.
**Bor·nu** (bôr'no͞o). Former Moslem kingdom of W Africa in present-day NE Nigeria.
**Bo·ro·bu·dur** (bôr'ə-bə-do͝or', bōr'-). Buddhist ruins in central Java, Indonesia.
**Bo·ro·di·no** (bôr'ə-dē'nō, bōr'-). Village of central European USSR W of Moscow; site of French-Russian battle (1812).
**Bos·ni·a** (bŏz'nē-ə). Region of W central Yugoslavia. **—Bos'ni·an** *adj.* & *n.*
**Bos·po·rus** (bŏs'pər-əs) *also* **Bos·pho·rus** (-fər-). Strait separating European & Asian Turkey & joining the Black Sea & the Sea of Marmara.
**Bos·sier City** (bō'zhər). City of NW La. near Shreveport. Pop. 49,969.
**Bos·ton** (bô'stən, bŏs'tən). **1. Mountains.** Ridge of the Ozarks in NW Ark. **2.** Borough of E central England E of Nottingham. Pop. 26,700. **3.** Cap. of Mass. in the E on an arm of Massachusetts Bay. Pop. 562,994. **—Bos·to'ni·an** (bô-stō'nē-ən, bŏs-tō'-) *adj.* & *n.*
**Bos·worth Field** (bŏz'wərth). Site of the final battle in the Wars of the Roses (1485), near Leicester in central England.
**Bot·a·ny Bay** (bŏt'n-ē). Inlet of the Tasman Sea in SE Australia S of Sydney.
**Both·ni·a** (bŏth'nē-ə), **Gulf of.** N arm of the Baltic Sea between Sweden & Finland.
**Bot·swa·na** (bŏt-swä'nə). Republic of S central Africa. Cap. Gaborone. Pop. 819,000.
**Bot·trop** (bŏt'rŏp'). City of W West Germany in the Ruhr NNW of Essen. Pop. 101,495.
**Boua·ké** (bwä'kā). Town of central Ivory Coast. Pop. 173,248.
**Bou·cher·ville** (bo͞o'shər-vĭl', bo͞o'shā-vēl'). Town of S Que., Canada, on the St. Lawrence NE of Montreal. Pop. 25,530.
**Bou·gain·ville** (bo͞o'gən-vĭl', bō'-). Island, c. 3,880 sq mi (10,050 sq km), of Papua New Guinea in the Solomon Is. of the SW Pacific.
**Bou·gie** (bo͞o-zhē'). BEJAIA.
**Boul·der** (bōl'dər). **1.** Former canyon of the Colorado R. between Ariz. & Nev., now inundated by Lake Mead. **2.** City of N central Colo. NW of Denver. Pop. 76,685.
**Bou·logne** (bo͞o-lōn', -loin') *or* **Bou·logne-sur-Mer** (-sûr-mĕr'). City of N France on the English Channel. Pop. 48,309.
**Boulogne-Bil·lan·court** (-bē'äN-ko͞or', -yäN-). City of N central France near Paris. Pop. 103,527.
**Boun·da·ry** (boun'də-rē, -drē). Highest peak, 13,145 ft (4,009.2 m), of Nev., in the SW near the Calif. border.
**Boun·ti·ful** (boun'tĭ-fəl). City of N Utah near Salt Lake City. Pop. 32,877.
**Bour·bon·nais** (bər-bō'nĭs). Village of NE Ill. near Kankakee. Pop. 13,280.
**Bourg-en-Bresse** (bo͝or'käN-brĕs') or **Bourg** (bo͝or). Town of E central France NNE of Lyons. Pop. 40,052.
**Bourges** (bo͝orzh). City of central France SSE of Orléans. Pop. 75,200.
**Bourne** (bôrn, bōrn). Town of SE Mass. on NW Cape Cod. Pop. 13,874.
**Bourne·mouth** (bôrn'məth, bōrn'-, bo͝orn'-). Borough of S central England on an inlet of the English Channel. Pop. 144,100.
**Bou·vet** (bo͞o'vā). Island & Norwegian dependency of the S Atlantic near the Antarctic Circle SSW of the Cape of Good Hope.
**Bow** (bō). River, 315 mi (506.8 km), of S Alta., Canada.
**Bow·er·y** (bou'ə-rē, bou'rē), **the.** Section of Lower Manhattan, New York City.
**Bow·ie** (bo͞o'ē). City of central Md. ENE of Washington, D.C. Pop. 33,695.
**Bowl·ing Green** (bō'lĭng grēn'). **1.** City of S Ky. SSE of Louisville. Pop. 40,450. **2.** City of NW Ohio SSW of Toledo. Pop. 25,728.
**Boyne** (boin). River of E Ireland flowing c. 70 mi (115 km) to the Irish Sea.
**Boyn·ton Beach** (boin'tən). City of SE Fla. N of Boca Raton. Pop. 35,624.
**Boz·ca·a·da** (bōz'jä-ä-dä'). Island of NW Turkey in the Aegean.
**Boze·man** (bōz'mən). City of SW Mont. ESE of Butte. Pop. 21,645.
**Bra·bant** (brə-bănt', -bänt'). Former duchy of the Netherlands divided between the Netherlands & Belgium.
**Brack·nell** (brăk'nəl). Town of S England. Pop. 34,067.
**Bra·den·ton** (brăd'n-tən). City of SW Fla. S of Tampa. Pop. 30,170.
**Brad·ford** (brăd'fərd). **1.** Borough of N central England WSW of Leeds. Pop. 458,900. **2.** City of N Pa. near the N.Y. border ESE of Erie. Pop. 11,211.
**Brad·ley** (brăd'lē). Village of NE Ill. near Kankakee. Pop. 11,008.
**Bra·ga** (brä'gə). City of NW Portugal NNE of Oporto. Pop. 48,735.
**Brah·ma·pu·tra** (brä'mə-po͞o'trə). River of S Asia flowing c. 1,800 mi (2,895 km) from the Himalayas in SW Tibet through NE India & joining with the Ganges to form a delta in central Bangladesh.
**Bră·i·la** (brə-ē'lə). City of SE Rumania on the Danube. Pop. 203,983.
**Brai·nerd** (brā'nərd). City of central Minn. N of St. Cloud. Pop. 11,489.
**Brain·tree** (brān'trē'). Town of E Mass. SSE of Boston. Pop. 36,337.
**Braintree and Bock·ing** (bŏk'ĭng). Urban district of E England W of Colchester. Pop. 26,300.
**Brak·pan** (brăk'păn'). City of NE South Africa S of Johannesburg. Pop. 73,210.
**Bramp·ton** (brămp'tən). City of S Ont., Canada, near Toronto. Pop. 103,459.
**Bran·co** (brăng'kō, -ko͞o). River of N Brazil flowing 350 mi (563 km) to the Río Negro.
**Bran·den·burg** (brăn'dən-bûrg', brän'dən-bo͝org'). **1.** Former duchy of N central Germany around which the kingdom of Prussia developed. **2.** City of central East Germany SW of Berlin. Pop. 94,071.
**Bran·don** (brăn'dən). City of SW Man., Canada, W of Winnipeg. Pop. 34,901.
**Bran·dy·wine** (brăn'dē-wīn'). Creek of SE Pa. & N Del.
**Bran·ford** (brăn'fərd). Town of S Conn. on Long Is. Sound E of New Haven. Pop. 23,363.
**Brant·ford** (brănt'fərd). City of S Ont., Canada, SW of Toronto. Pop. 66,950.
**Bras d'Or Lake** (brä'dôr'). Arm of the Atlantic indenting Cape Breton Is. in SE Canada.
**Bra·sí·lia** (brə-zĭl'yə). Cap. of Brazil in the central plateau NW of Rio de Janeiro. Pop. 1,176,748.
**Bra·şov** (brä-shôv'). City of central Rumania NNW of Ploieşti. Pop. 259,108.
**Bra·ti·sla·va** (brăt'ĭ-slä'və, brä'tĭ-). City of S Czechoslovakia on the Danube near the Austrian & Hungarian borders. Pop. 333,000.
**Bratsk** (brätsk). City of S Siberian USSR NNE of Irkutsk. Pop. 214,000.
**Brat·tle·bor·o** (brăt'l-bûr'ō, -bər-ə). Town of SE Vt. on the Connecticut R. & N.H. border. Pop. 11,886.
**Braw·ley** (brô'lē). City of SE Calif. SE of the Salton Sea. Pop. 14,946.
**Bra·zil** (brə-zĭl'). Republic of E South America. Cap. Brasília. Pop. 119,024,600. **—Bra·zil'i·an** *adj.* & *n.*
**Braz·os** (brăz'əs). River of E N.Mex. & central Tex. flowing c. 950 mi (1,528 km) to the Gulf of Mexico.
**Braz·za·ville** (brăz'ə-vĭl', brä'zə-vēl'). Cap. of Congo on the Congo R. Pop. 298,967.
**Bre·a** (brā'ə). City of S Calif. N of Anaheim. Pop. 27,913.
**Brèche de Ro·land** (brĕsh də rô-läN'). Gorge in the Pyrenees of SW France.
**Brecks·ville** (brĕks'vĭl'). City of NE Ohio near Cleveland. Pop. 10,132.
**Bre·da** (brā-dä'). City of S Netherlands SSE of Dordrecht. Pop. 118,806.
**Breed's Hill** (brēdz). Hill in Charlestown, Mass., near Bunker Hill.
**Bre·men** (brĕm'ən, brā'mən). City of N West Germany on the Weser R. SW of Hamburg. Pop. 572,969.
**Bre·mer·ha·ven** (brĕm'ər-hä'vən, -hā'-). City of N West Germany at the mouth of the Weser near the North Sea. Pop. 143,836.
**Brem·er·ton** (brĕm'ər-tən). City of W central Wash. on an arm of Puget Sound W of Seattle. Pop. 36,208.
**Bren·ham** (brĕn'əm). City of S central Tex. WNW of Houston. Pop. 10,966.
**Bren·ner Pass** (brĕn'ər). Alpine pass, 4,495 ft (1,371 m) high, connecting Innsbruck, Austria, & Bolzano, Italy.
**Brent·wood** (brĕnt'wo͝od'). **1.** Urban district of SE England ENE of London. Pop. 58,690. **2.** Borough of SW Pa. near Pittsburgh. Pop. 11,907.
**Bre·scia** (brĕsh'ə, brā'shə). City of N Italy E of Milan. Pop. 189,092.
**Bres·lau** (brĕs'lou). WROCLAW.
**Brest** (brĕst). **1.** City of NW France on an inlet of the Atlantic. Pop. 163,940. **2.** *also* **Brest Li·tovsk** (lĭ-tôfsk'). City of W European USSR on the Bug R. near the Polish border. Pop. 177,000.
**Bre·ton** (brĕt'n, brĭt'n). Cape of E N.S., Canada, on Cape Breton Is.
**Brew·er** (bro͞o'ər). City of S Me. on the Penobscot R. opposite Bangor. Pop. 9,017.
**Bri·ansk** (brē-änsk'). *var. of* BRYANSK.

**Bri·dal·veil** *also* **Bri·dal Veil** (brīd′l-vāl′). Waterfall, 620 ft (189.1 m) high, in Yosemite National Park, E central Calif.
**Bridge·port** (brĭj′pôrt′, -pōrt′). City of SW Conn. on Long Is. Sound. Pop. 142,546.
**Bridge·ton** (brĭj′tən). **1.** City of E Mo. NW of St. Louis. Pop. 18,445. **2.** City of SW N.J. S of Philadelphia. Pop. 18,795.
**Bridge·town** (brĭj′toun′). Cap. of Barbados, West Indies. Pop. 8,868.
**Bridge·view** (brĭj′vyo͞o′). Village of NE Ill. near Chicago. Pop. 14,155.
**Bridge·wa·ter** (brĭj′wô′tər, -wŏt′ər). Town of E Mass. S of Boston. Pop. 17,202.
**Bridg·wa·ter** (brĭj′wô′tər, -wŏt′ər). Borough of SE England NNE of Taunton. Pop. 26,700.
**Brid·ling·ton** (brĭd′lĭng-tən). Borough of NE England on **Bridlington Bay,** an inlet of the North Sea. Pop. 26,920.
**Brie** (brē). Region of N France E of Paris.
**Bri·enz** (brē-ĕnts′). Town of central Switzerland NE of Interlaken at the NE end of the **Lake of Brienz.** Pop. 2,796.
**Brig·ham City** (brĭg′əm). City of N Utah N of Ogden. Pop. 15,596.
**Brig·house** (brĭg′hous′). Borough of N central England S of Bradford. Pop. 35,320.
**Brigh·ton** (brīt′n). **1.** Borough of SE England on the English Channel S of London. Pop. 156,500. **2.** City of N central Colo. NNE of Denver. Pop. 12,773.
**Brin·di·si** (brĭn′dĭ-zē, brēn′-). City of S Italy on the Adriatic. Pop. 76,612.
**Bris·bane** (brĭz′bən, -bān′). City of E Australia on the **Brisbane R.** (215 mi/345.9 km) above its mouth on Moreton Bay. Pop. 696,740.
**Bris·tol** (brĭs′təl). **1. Bay.** Arm of the Bering Sea in SW Alas. between the mainland & the Alaska Peninsula. **2. Channel.** Inlet of the Atlantic separating Wales from SE England. **3.** City of SW England W of London. Pop. 416,300. **4.** City of central Conn. SW of Hartford. Pop. 57,370. **5.** Borough of SE Pa. NE of Philadelphia. Pop. 10,867. **6.** Town of E R.I. SE of Providence. Pop. 20,128. **7.** City of NE Tenn. & independent city of SW Va. Pop. 23,986 & 19,042.
**Bristol A·von** (ā′vŏn, ā′vən, ăv′ən). —See AVON.
**Brit·ain** (brĭt′n). UNITED KINGDOM.
**Brit·ish A·mer·i·ca** (brĭt′ĭsh ə-mĕr′ĭ-kə) *also* **British North America** (nôrth). Former British possessions in North America N of the U.S.
**British Ant·arc·tic Territory** (ănt-ärk′tĭk, -är′tĭk). British island territory of the S Atlantic & Antarctica.
**British Cam·e·roons** (kăm′ə-ro͞onz′). Former British trust territory of W Africa, divided in 1961 between Nigeria & Cameroon.
**British Co·lum·bi·a** (kə-lŭm′bē-ə). Province of W Canada. Cap. Victoria. Pop 2,716,301.
**British East Af·ri·ca** (ēst ăf′rĭ-kə). Former British territories in E Africa, including Kenya, Uganda, Tanganyika, & Zanzibar.
**British Gui·a·na** (gē-ăn′ə, -ä′nə). GUYANA.
**British Hon·du·ras** (hŏn-do͝or′əs, -dyo͝or′-). The country BELIZE.
**British Isles.** Islands off the NW coast of Europe comprising Great Britain, Ireland, & adjacent smaller islands.
**British Sol·o·mon Islands** (sŏl′ə-mən). Former British protectorate in the Solomon & Santa Cruz Is. of the SW Pacific.
**British So·ma·li·land** (sō-mä′lē-lănd′, sə-). Former British protectorate in E Africa on the Gulf of Aden.
**British To·go·land** (tō′gō-lănd′). Former British protectorate of W Africa, part of present-day Ghana since 1957.
**British Vir·gin Islands** (vûr′jĭn). British colony in the E Caribbean E of the U.S. Virgin Is. Cap. Road Town on Tortola Is. Pop. 11,000.
**British West In·dies** (wĕst ĭn′dēz). Islands of the West Indies that were formerly under British control.
**Brit·ta·ny** (brĭt′n-ē). Region & former province of NW France on a peninsula between the English Channel & the Bay of Biscay.
**Br·no** (bûr′nō). City of central Czechoslovakia SE of Prague. Pop. 335,700.
**Broad** (brôd). River, c. 150 mi (240 km), of N.C. & S.C.
**Broads** (brôdz), **the.** Lowland region of E England along coastal Norfolk & Suffolk.
**Broad·view Heights** (brôd′vyo͞o′). City of NE Ohio near Cleveland. Pop. 10,920.
**Brock·en** (brŏk′ən). Peak of W East Germany, 3,747 ft (1,142.8 m), in the Harz Mts.
**Brock·ton** (brŏk′tən). City of E Mass. S of Boston. Pop. 95,172.
**Brock·ville** (brŏk′vĭl′). City of SE Ont., Canada, on the St. Lawrence S of Ottawa. Pop. 19,903.
**Bro·ken Ar·row** (brō′kən ăr′ō). City of NE Okla. near Tulsa. Pop. 35,761.
**Broms·grove** (brŏmz′grōv′). Urban district of central England SW of Birmingham. Pop. 41,430.
**Bronx** (brŏngks). **1.** River of SE N.Y. flowing c. 20 mi (32 km) through the Bronx into the East R. **2.** *or* **the Bronx.** Borough of New York City, SE N.Y., on the mainland N of Manhattan. Pop. 1,169,-115.
**Brook·field** (bro͝ok′fēld′). **1.** Town of SW Conn. NNE of Danbury. Pop. 12,872. **2.** Village of NE Ill. near Chicago. Pop. 19,395. **3.** City of SE Wis. near Milwaukee. Pop. 34,035.
**Brook·ha·ven** (bro͝ok-hā′vən). City of SW Miss. SSW of Jackson. Pop. 10,800.
**Brook·ings** (bro͝ok′ĭngz). City of E S.Dak. N of Sioux Falls. Pop. 14,951.
**Brook·line** (bro͝ok′līn′). Town of E Mass. near Boston. Pop. 55,062.
**Brook·lyn** (bro͝ok′lĭn). **1.** Borough of New York City, SE N.Y., on W Long Is. Pop. 2,230,936. **2.** City of NE Ohio near Cleveland. Pop. 12,342.
**Brooklyn Center.** City of E Minn. near Minneapolis. Pop. 31,230.
**Brooklyn Park.** City of E Minn. near Minneapolis. Pop. 43,332.
**Brook Park** (bro͝ok). City of NE Ohio near Cleveland. Pop. 26,195.
**Brooks** (bro͝oks). Mountain range in Alas. N of the Arctic Circle, rising to 9,239 ft (2,817.8 m).
**Broom·field** (bro͞om′fēld′). City of N central Colo. near Denver. Pop. 20,730.
**Bros·sard** (brô-sär′, -särd′). Town of S Que., Canada, on the St. Lawrence near Montreal. Pop. 37,641.
**Brown Deer** (broun′ dîr′). Village of SE Wis. near Milwaukee. Pop. 12,921.
**Brown·field** (broun′fēld′). City of NW Tex. SW of Lubbock. Pop. 10,387.
**Browns·ville** (brounz′vĭl′, -vəl). City of S Tex. on the Rio Grande near the Gulf of Mexico. Pop. 84,997.
**Brown·wood** (broun′wo͝od′). City of central Tex. W of Waco. Pop. 19,203.
**Bruges** (bro͞ozh). City of NW Belgium E of Ostend. Pop. 117,220.
**Bru·nei** (bro͞o′nī′). Sultanate of NW Borneo on the South China Sea. Cap. Bandar Seri Begawan. Pop. 212,840.
**Bruns·wick** (brŭnz′wĭk). **1.** Former state of central Germany, chiefly in present-day E West Germany. **2.** City of E West Germany on the Oder R. ESE of Hannover. Pop. 268,519. **3.** City of SE Ga. SSW of Savannah. Pop. 17,605. **4.** Town of SW Me. NE of Portland. Pop. 17,336. **5.** City of NE Ohio near Cleveland. Pop. 27,689.
**Brus·sels** (brŭs′əlz). Cap. of Belgium in the N central part. Metro. area pop. 1,054,970.
**Brut·ti·um** (bro͞ot′ē-əm, brŭt′-). Ancient region of S Italy in the toe of the peninsula.
**Bry·an** (brī′ən). City of E central Tex. NW of Houston. Pop. 44,337.
**Bry·ansk** *also* **Bri·ansk** (brē-änsk′). City of central European USSR SSE of Smolensk. Pop. 394,000.
**Bu·bas·tis** (byo͞o-băs′tĭs). Ancient city of NE Egypt in the Nile delta.
**Bu·ca·ra·man·ga** (bo͞o′kə-rə-mäng′gə). City of N central Colombia. Pop. 291,661.
**Bu·cha·rest** (bo͞o′kə-rĕst′, byo͞o′-). Cap. of Rumania in the SE part. Pop. 1,832,015.
**Bu·chen·wald** (bo͞o′kən-wôld′, -vält′). Village of SW East Germany near Weimar; site of a Nazi concentration camp.
**Buck·ing·ham** (bŭk′ĭng-əm, -hăm′). City of S Que., Canada, NE of Ottawa. Pop. 14,328.
**Bu·co·vi·na** (bo͞o′kə-vē′nə). *var. of* BUKOVINA.
**Bu·cy·rus** (byo͞o-sī′rəs). City of N central Ohio WNW of Mansfield. Pop. 13,433.
**Bu·da·pest** (bo͞o′də-pĕst′, -pĕsht′, byo͞o′-). Cap. of Hungary on the Danube in the N central part. Pop. 2,060,170.
**Bue·na Park** (byo͞o′nə). City of S Calif. WNW of Alameda. Pop. 64,165.
**Bue·na·ven·tu·ra** (bwā′nə-vĕn-to͝or′ə, -tyo͝or′ə, bwĕn′ə-). City of W Colombia on the Pacific. Pop. 115,700.
**Bue·nos Ai·res** (bwā′nəs âr′ēz, ĭr′ĭz, bō′nəs). **1.** Lake of SE Chile & SW Argentina. **2.** Cap. of Argentina in the E part on the Rio de la Plata. Pop. 2,908,000.
**Buf·fa·lo** (bŭf′ə-lō′). City of W N.Y. at the E end of Lake Erie at the Canadian border. Pop. 357,870.
**Buffalo Grove.** Village of NE Ill. NW of Chicago. Pop. 22,230.
**Bug** (bo͞og). **1.** *or* **Western Bug.** River of W European USSR flowing c. 480 mi (770 km) from the Ukraine to the Vistula near Warsaw, Poland. **2.** *or* **Southern Bug.** River of W European USSR flowing c. 490 mi (790 km) through the Ukraine to the Black Sea.
**Bu·gan·da** (bo͞o-găn′də, byo͞o-). Region & former kingdom of E Africa in present-day SE Uganda.
**Bu·jum·bu·ra** (bo͞o′jəm-bo͝or′ə). Cap. of Burundi in the W part on Lake Tanganyika. Pop. 141,040.
**Bu·ka** (bo͞o′kə). Island in the SW Pacific in the N Solomons, part of Papua New Guinea.
**Bu·ka·vu** (bo͞o-kä′vo͞o). City of E Zaire on Lake Kivu. Pop. 182,000.
**Bu·kha·ra** (bo͞o-kär′ə, -här′-) *also* **Bo·kha·ra** (bō-). **1.** Former emirate of central Asia in the Amu Darya basin. **2.** City of S Central Asian USSR W of Samarkand. Pop. 185,000.

**Bu·ko·vi·na** *also* **Bu·co·vi·na** (bo͞o′kə-vē′nə). Historical region of E Europe in W Ukraine & NE Rumania.
**Bu·la·wa·yo** (bo͝ol′ə-wā′ō, -wī′ō). City of SW Zimbabwe. Pop. 359,000.
**Bul·gar·i·a** (bŭl-gâr′ē-ə, bo͝ol-). Republic of SE Europe on the Black Sea. Cap. Sofia. Pop. 8,862,000. **—Bul·gar′i·an** *adj. & n.*
**Bull Run** (bo͝ol′ rŭn′). Small stream of NE Va. SW of Washington, D.C. near Manassas; site of two Civil War battles (1861 & 1862).
**Bun·ker Hill** (bŭng′kər). Height (107 ft/32.6 m) in Charlestown, Boston, Mass.; near site of 1st major Revolutionary War battle (1775).
**Bur·bank** (bûr′băngk′). **1.** City of S Calif. near Los Angeles. Pop. 84,625. **2.** City of NE Ill. near Chicago. Pop. 28,462.
**Bur·gas** (bo͝or-gäs′). City of SE Bulgaria on the Black Sea. Pop. 144,449.
**Bur·gos** (bo͝or′gōs′). City of N Spain SSW of Bilbao. Pop. 118,366.
**Bur·gun·dy** (bûr′gən-dē). Region & former province of E France. **—Bur·gun′di·an** (bər-gŭn′dē-ən) *adj. & n.*
**Bu·rias** (bo͝or′yəs). Island of the Philippines SE of Luzon.
**Burk·bur·nett** (bûrk′bər-nĕt′). City of N Tex. on the Okla. border N of Wichita Falls. Pop. 10,668.
**Bur·le·son** (bûr′lə-sən). City of NE Tex. S of Fort Worth. Pop. 11,734.
**Bur·lin·game** (bûr′lĭn-gām′, -lĭng-). City of W Calif. SSE of San Francisco. Pop. 26,173.
**Bur·ling·ton** (bûr′lĭng-tən). **1.** City of S Ont., Canada, on Lake Ontario near Hamilton. Pop. 104,314. **2.** City of SE Iowa SSE of Cedar Rapids. Pop. 29,529. **3.** Town of NE Mass. NW of Boston. Pop. 23,486. **4.** City of W N.J. NE of Camden. Pop. 10,246. **5.** City of N central N.C. E of Greensboro. Pop. 37,266. **6.** City of NW Vt. on Lake Champlain NW of Montpelier. Pop 37,712.
**Bur·ma** (bûr′mə). Republic of SE Asia on the Bay of Bengal & the Andaman Sea. Cap. Rangoon. Pop. 32,913,000. **—Bur·mese′** (bər-mēz′, -mēs′), **Bur′man** (bûr′mən) *adj. & n.*
**Bur·na·by** (bûr′nə-bē). City of SW B.C., Canada, near Vancouver. Pop. 131,599.
**Burn·ley** (bûrn′lē). Borough of NW England N of Manchester. Pop. 74,300.
**Burns·ville** (bûrnz′vĭl′). City of E Minn. S of Minneapolis. Pop. 35,674.
**Bur·rard** (bə-rärd′). Inlet of the Strait of Georgia, SW B.C., Canada.
**Bur·rill·ville** (bûr′əl-vĭl′). Town of NW R.I. NW of Providence. Pop. 13,164.
**Bur·sa** (bo͝or-sä′, bûr′sə). City of NW Turkey near the Sea of Marmara. Pop. 346,103.
**Bur·ton** (bûr′tn). City of central Mich. WSW of Flint. Pop. 29,976.
**Burton up·on Trent** (ə-pŏn trĕnt′, ə-pôn). Borough of W central England SSW of Derby. Pop. 49,480.
**Bu·ru** *or* **Boe·roe** (bo͝or′o͞o). Island of E Indonesia in the Moluccas W of Ceram.
**Bu·run·di** (bo͞o-ro͞on′dē). Republic of E central Africa NW of Tanzania. Cap. Bujumbura. Pop. 4,021,910. **—Bu·run′di·an** *adj. & n.*
**Bur·y** (bĕr′ē). Borough of NE England, part of Greater Manchester. Pop. 69,550.
**Bury Saint Ed·munds** (sānt ĕd′məndz). Borough of E central England ENE of Cambridge. Pop. 26,800.
**Bu·sto Ar·si·zio** (bo͞o′stō är-sĕt′sē-ō′). City of N Italy NW of Milan. Pop. 72,400.
**Bu·ta·ri·ta·ri** (bo͞o-tär′ē-tär′ē). Atoll of the central Pacific in Kiribati.
**Bute** (byo͞ot). Island of SW Scotland in the Firth of Clyde.
**But·ler** (bŭt′lər). City of W Pa. N of Pittsburgh. Pop. 17,026.
**Butte** (byo͞ot). City of SW Mont. SSW of Helena. Pop. 37,205.
**Bu·tu·an** (bo͞o-to͞o′än). City of NE Mindanao, Philippines. Pop. 172,404.
**Bu·tung** (bo͞o′to͝ong′). Island of central Indonesia off the SE coast of Sulawesi.
**Bu·zău** (bə-zou′, -zō′). City of SE Rumania NE of Bucharest. Pop. 106,738.
**Buz·zards Bay** (bŭz′ərdz). Inlet of the Atlantic in SE Mass.
**Byb·los** (bĭb′ləs, -lŏs′). Ancient city of Phoenicia NNE of present-day Beirut, Lebanon.
**Byd·goszcz** (bĭd′gôsh′, -gôsch′). City of N central Poland NNE of Poznań. Pop. 280,460.
**Bye·lo·rus·sia** (bē-ĕl′ō-rŭsh′ə). *var. of* BELORUSSIA.
**Byrd Land** (bûrd). MARIE BYRD LAND.
**By·tom** (bē′tôm′, bĭ′-). City of SW Poland NNW of Katowice. Pop. 186,993.
**Byz·an·tine Empire** (bĭz′ən-tēn′, -tīn′). E part of the later Roman Empire.
**By·zan·ti·um** (bĭ-zăn′shē-əm, -shəm, -tē-əm). Ancient city of Thrace on the site of present-day Istanbul, Turkey.

# C

**Ca·ba·na·tuan** (kä′bə-nə-twän′). City of central Luzon, Philippines, N of Manila. Pop. 138,297.
**Ca·bin·da** (kə-bĭn′də). Territory of Angola, an exclave on the Atlantic between Congo & Zaire.
**Ca·bot Strait** (kăb′ət). Channel, c. 60 mi (97 km) wide, between SW Newf. & N Cape Breton Is., Canada, connecting the Gulf of St. Lawrence & the Atlantic.
**Cá·ce·res** (kä′sə-rās′). City of W central Spain WSW of Madrid. Pop. 53,108.
**Cache la Pou·dre** (kăsh′ lə po͞o′dər, -drə). River, c. 125 mi (201 km), of N Colo.
**Cad·il·lac** (kăd′l-ăk′). City of NW Mich. NNE of Grand Rapids. Pop. 10,199.
**Cá·diz** (kə-dĭz′, kā′dĭz, kä′-). City of SW Spain NW of Gibraltar on the **Gulf of Cádiz,** an inlet of the Atlantic. Pop. 135,743.
**Cae·li·an** (sē′lē-ən). One of the seven hills of ancient Rome.
**Caen** (käN). City of N France SW of Le Havre. Pop. 116,987.
**Cae·sa·re·a** (sē′zə-rē′ə, sĕs′ə-, -sĕz′ə-). **1.** *also* **Caesarea Pal·es·ti·nae** (păl′ĭ-stī′nē). Ancient seaport & cap. of Roman Palestine S of present-day Haifa, Israel. **2.** *also* **Caesarea Phil·ip·pi** (fĭl′ĭ-pī, fĭ-lĭp′-ī). Ancient city of N Palestine near Mt. Hermon in present-day SW Syria. **3.** *also* **Caesarea Maz·a·ca** (măz′ə-kə). KAYSERI.
**Ca·glia·ri** (käl′yə-rē′). City of Sardinia, Italy, on the S coast on the **Gulf of Cagliari,** an inlet of the Mediterranean. Pop. 211,015.
**Ca·guas** (kä′gwäs′). City of E central Puerto Rico S of San Juan. Pop. 87,218.
**Ca·ha·ba** (kə-hô′bə, -hä′-). River, c. 200 mi (322 km), of central Ala.
**Ca·ho·ki·a** (kə-hō′kē-ə). Village of SW Ill. near East St. Louis & **Cahokia Mounds,** group of 85 prehistoric Indian earthworks. Pop. 18,904.
**Ca·hors** (kä-hôr′, -ôr′). City of S central France N of Toulouse. Pop. 19,288.
**Cai·cos** (kā′kəs). One of the island groups constituting the Turks & Caicos Is. SE of the Bahamas.
**Cairn·gorm** (kârn′gôrm′) *also* **Cairn·gorms** (-gôrmz′). Range of the Grampians in central Scotland.
**Cai·ro. 1.** (kī′rō). Cap. of Egypt on the Nile in the NE part. Pop. 5,084,463. **2.** (kā′rō). Town in S Ill. near the confluence of the Mississippi & Ohio rivers. Pop. 5,931. **—Cai·rene′** (-rēn′) *adj. & n.*
**Ca·ja·mar·ca** (kä′hə-mär′kə). City & ancient Incan center of NW Peru in the Andes. Pop. 37,608.
**Cal·a·bar** (kăl′ə-bär′). City of SE Nigeria on the Gulf of Guinea. Pop. 103,000.
**Ca·la·bri·a** (kə-lā′brē-ə, -lä′). Region of S Italy, a peninsula forming the toe of the Italian boot.
**Ca·lah** (kā′lə). KALAKH.
**Ca·lais** (kă-lā′, kăl′ā). City of N France on the Strait of Dover. Pop. 73,009.
**Ca·la·mi·an** (kä′lə-mē-än′). Islands of the W central Philippines between Mindoro & Palawan.
**Cal·ca·sieu** (kăl′kə-sho͞o′). River of SW La. flowing c. 200 mi (322 km) through **Lake Calcasieu** (c. 15 mi/24 km long) to the Gulf of Mexico.
**Cal·cut·ta** (kăl-kŭt′ə). City of E India on the Ganges delta. Metro. area pop. 7,031,382.
**Cald·well** (kôld′wĕl′, -wəl, kōld′-). City of SW Idaho W of Boise. Pop. 17,699.
**Cal·e·don** (kăl′ĭ-dən). Town of SE Ont., Canada, NW of Toronto. Pop. 22,434.
**Cal·e·do·ni·a** (kăl′ĭ-dō′nē-ə, -dōn′yə). SCOTLAND. **—Cal′e·do′ni·an** *adj. & n.*
**Ca·lex·i·co** (kə-lĕk′sĭ-kō′). City of S Calif. on the Mexican border. Pop. 14,412.
**Cal·ga·ry** (kăl′gə-rē). City of S Alta., Canada, S of Edmonton. Pop. 469,917.
**Ca·li** (kä′lē). City of W Colombia SW of Bogotá. Pop. 898,253.
**Cal·i·cut** (kăl′ĭ-kŭt′). KOZHIKODE.
**Cal·i·for·nia** (kăl′ə-fôr′nyə). **1. Gulf of.** Inlet of the Pacific extending c. 700 mi (1,126 km) between Lower California & the NW Mexican mainland. **2.** State of the W U.S. on the Pacific. Cap. Sacramento. Pop. 23,668,562. **—Cal′i·for′nian** *adj. & n.*
**Cal·la·o** (kə-yä′ō, -you′). City of W central Peru on the Pacific near Lima. Pop. 296,220.
**Cal·ta·nis·set·ta** (käl′tə-nĭ-sĕt′ə). City of central Sicily, Italy, SE of Palermo. Pop. 52,838.
**Cal·u·met** (kăl′yə-mĕt′, -mĭt). Industrial region of NE Ill. & NW Ind. on Lake Michigan SE of Chicago.
**Calumet City.** City of NE Ill. near Chicago. Pop. 39,673.
**Cal·va·ry** (kăl′və-rē, kăl′vrē). Hill outside ancient Jerusalem where Jesus was crucified.
**Cal·y·don** (kăl′ĭ-dŏn′, -dən). **1. Gulf of.** Gulf of PATRAS. **2.** Ancient city of W central Greece N of the Gulf of Patras.
**Cam** (kăm). River, c. 40 mi (64 km), of E central England.
**Ca·ma·güey** (kăm′ə-gwā′). City of E central Cuba. Pop. 216,000.

**Ca·margue** (kə-märg′). Island of SE France, c. 215 sq mi (557 sq km), in the Rhone delta.
**Cam·a·ril·lo** (kăm′ə-rē′ō). City of S Calif. W of Los Angeles. Pop. 37,732.
**Cam·ba·luc** (kăm′bə-lŭk′). KHANBALIK.
**Cam·bay** (kăm-bā′), **Gulf of.** Inlet of the Arabian Sea on the NW coast of India.
**Cam·ber·well** (kăm′bər-wĕl′, -wəl). City of SE Australia near Melbourne. Pop. 89,865.
**Cam·bo·di·a** (kăm-bō′dē-ə). Country of SE Asia between Thailand & Vietnam. Cap. Phnom Penh. Pop. 5,200,000. **—Cam·bo′di·an** *adj.* & *n.*
**Cam·bri·a** (kăm′brē-ə). WALES. **—Cam′bri·an** *adj.* & *n.*
**Cam·bridge** (kām′brĭj). **1.** City of SE Ont., Canada, WNW of Hamilton. Pop. 72,383. **2.** Borough of E central England NNE of London. Pop. 106,400. **3.** City of E Md. on the Eastern Shore SSE of Baltimore. Pop. 11,703. **4.** City of E Mass. near Boston. Pop. 95,322. **5.** City of E Ohio ENE of Zanesville. Pop. 13,573. **—Can′ta·brig′i·an** (kăn′tə-brĭj′ē-ən) *adj.* & *n.*
**Cam·den** (kăm′dən). **1.** City of S Ark. SSW of Little Rock. Pop. 15,342. **2.** City of W N.J. opposite Philadelphia, Pa. Pop. 84,910.
**Cam·e·roon** (kăm′ə-ro͞on′). **1.** Volcano, 13,353 ft (4,072.7 m), in W Cameroon. **2.** *also* **Came·roun** (kăm-ro͞on′). Country of W central Africa E of Nigeria. Cap. Yaoundé. Pop. 8,503,000.
**Cam·e·roons** (kăm′ə-ro͞onz′). Region of W central Africa formerly comprising **British Cameroons** & **French Cameroons** & divided (1960–61) between Cameroon & Nigeria.
**Came·roun** (kăm-ro͞on′). *var. of* CAMEROON.
**Ca·mo·tes** (kə-mō′tās). Sea of central Philippines between Cebu & Leyte.
**Cam·pa·gna di Ro·ma** (kăm-pän′yə dē rō′mə, -păn′). Low-lying region surrounding Rome, Italy.
**Cam·pa·ni·a** (kăm-pā′nē-ə, -pān′yə, -pän′yə). Region of S Italy on the Tyrrhenian Sea.
**Camp·bell** (kăm′bəl). **1.** City of W central Calif. near San Jose. Pop. 27,067. **2.** City of NE Ohio near Youngstown. Pop. 11,619.
**Camp·bell·ton** (kăm′bəl-tən). City of N N.B., Canada, on the Que. border. Pop. 9,282.
**Cam·pe·che** (kăm-pē′chē, kăm-pā′chē). **1. Gulf** *or* **Bay of.** Part of the Gulf of Mexico W of the Yucatán Peninsula. **2.** City of SE Mexico on the W coast of the Yucatán Peninsula. Pop. 69,506.
**Cam·pi·na Gran·de** (kăm′pē-nə grăn′də, -dē). City of extreme E Brazil NW of Recife. Pop. 163,206.
**Cam·pi·nas** (kăm-pē′nəs). City of SE Brazil NNW of São Paulo. Pop. 328,629.
**Cam·po·bel·lo** (kăm′pə-bĕl′ō). Island off SW coast of N.B., Canada.
**Cam·po·for·mi·do** (kăm′pō-fôr′mĭ-dō′) *also* **Cam·po For·mio** (kăm′pō fôr′mē-ō′). Village of NE Italy SW of Udine; site of French-Austrian treaty signing (1797).
**Cam·po Gran·de** (kăm′pō grăn′də, -dē). City of S Brazil NE of the Paraguay border. Pop. 130,792.
**Cam·pos** (kăm′pəs). City of SE Brazil NE of Rio de Janeiro. Pop. 153,310.
**Cam·ranh** *or* **Cam Ranh Bay** (kăm′răn′, -rän′). Inlet of the South China Sea in SE Vietnam.
**Cam·rose** (kăm′rōz′). City of central Alta., Canada, SE of Edmonton. Pop. 10,104.
**Ca·na** (kā′nə). Village of N Palestine near Nazareth.
**Ca·naan** (kā′nən). Ancient region comprising Palestine or the part of it W of the Jordan R.
**Can·a·da** (kăn′ə-də). Country of N North America. Cap. Ottawa. Pop. 24,105,163. **—Ca·na′di·an** (kə-nā′dē-ən) *adj.* & *n.*
**Canadian. 1.** River of S central U.S. flowing 906 mi (1,457.8 km) from NE N.Mex. to the Arkansas R. in E Okla. **2.** Waterfall, c. 160 ft (49 m), forming part of Niagara Falls. **3. Shield.** LAURENTIAN PLATEAU.
**Ca·nal Zone** (kə-năl′). Territory across the Isthmus of Panama formerly administered by the U.S. for the operation of the Panama Canal.
**Can·an·dai·gua** (kăn′ən-dā′gwə). City of W central N.Y. at the N end of **Canandaigua Lake** (15 mi/24.1 km long), one of the Finger Lakes. Pop. 10,419.
**Ca·nar·y** (kə-nâr′ē). Spanish islands off the NW coast of Africa.
**Ca·nav·er·al** (kə-năv′ər-əl, -năv′rəl). Cape of E central Atlantic coast of Fla.
**Can·ber·ra** (kăn′bər-ə, -bĕr′ə). Cap. of Australia in the SE part. Pop. 196,538.
**Can·di·a** (kăn′dē-ə). **1. Sea of.** Sea of CRETE. **2.** CRETE. **3.** City of N Crete, Greece. Pop. 77,506.
**Ca·ney Fork** (kā′nē). River, 144 mi (231.7 km), of central Tenn.
**Can·i·a·pis·cau** (kăn′ē-ə-pĭs′kō). KANIAPISCAU.
**Can·nae** (kăn′ē). Ancient town of SE Italy where Carthaginians under Hannibal defeated the Romans (216 B.C.).
**Cannes** (kăn). Resort city of SE France on the Mediterranean. Pop. 70,226.
**Can·nock** (kăn′ək). Urban district of E central England NNW of Birmingham. Pop. 56,400.
**Can·on City** (kăn′yən). City of S central Colo. WNW of Pueblo. Pop. 13,037.
**Can·ons·burg** (kăn′ənz-bûrg′). Borough of SW Pa. SW of Pittsburgh. Pop. 10,459.
**Ca·no·pus** (kə-nō′pəs). Ancient city of N Egypt E of Alexandria.
**Can·so** (kăn′sō). **1. Strait of.** Channel, 1 mi (1.6 km) wide, between NE N.S. mainland Cape Breton Is., Canada. **2.** Cape of NE extremity of the N.S. mainland, Canada.
**Can·ta·bri·an** (kăn-tā′brē-ən). Mountains of N Spain extending c. 300 mi (483 km) along the coast of the Bay of Biscay.
**Can·ter·bur·y** (kăn′tər-bĕr′ē, -brē, -tə-). **1.** City of SE Australia near Sydney. Pop. 128,710. **2.** City of SE England ESE of London. Pop. 115,600.
**Can·ton** (kăn′tən). **1.** (kăn′tŏn′, kăn′tŏn′). ZHU JIANG. **2.** Coral atoll (3.5 sq mi/9.1 sq km) in the central Pacific, largest of the Phoenix Is., controlled jointly by Great Britain & the U.S. **3.** (kăn′tŏn′, kăn′tŏn′). GUANGZHOU. **4.** City of W central Ill. WSW of Peoria. Pop. 14,626. **5.** Town of E Mass. near Boston. Pop. 18,182. **6.** City of W central Miss. NNE of Jackson. Pop. 11,116. **7.** City of NE Ohio SSE of Akron. Pop. 94,730.
**Can·yon** (kăn′yən). City of N Tex. in the Panhandle S of Amarillo. Pop. 10,724.
**Can·yon·lands National Park** (kăn′yən-lăndz′). Area of SE Utah with deep canyons & erosion-carved land features.
**Cap-de-la-Ma·de·leine** (kăp′də-lä-măd-lān′). City of S Que., Canada, on the St. Lawrence NE of Montreal. Pop. 32,126.
**Cape Bret·on Island** (kāp brĕt′n, brĭt′n). Island, 3,970 sq mi (10,282.3 sq km), forming the NE part of N.S., Canada.
**Cape Cod Bay** (kŏd). S part of Massachusetts Bay W of Cape Cod.
**Cape Cor·al** (kôr′əl, kŏr′-). City of SW Fla. SW of Fort Myers. Pop. 32, 103.
**Cape Fear River** (fîr). River of central & SE N.C. flowing 202 mi (325 km) to the Atlantic.
**Cape Gi·rar·deau** (jə-rär′dō, -rä′-). City of SE Mo. on the Mississippi SSE of St. Louis. Pop. 34,361.
**Cape of Good Hope Province** (go͝od hōp′) *also* **Cape Province.** Province of S South Africa on the Atlantic & Indian oceans.
**Ca·per·na·um** (kə-pûr′nē-əm). City of ancient Palestine on the NW shore of the Sea of Galilee.
**Cape Town** *or* **Cape·town** (kāp′toun′). Legislative cap. of South Africa on the Atlantic in the extreme SW part. Pop. 697,514.
**Cape Verde** (vûrd). Island republic in the N Atlantic W of Senegal. Cap. Praia. Pop. 324,000.
**Cape York Peninsula** (yôrk). Peninsula of NE Australia, c. 450 mi (724 km) long, between the S Pacific & Gulf of Carpentaria.
**Cap-Ha·ï·tien** (kä-pä-ē-syăN′) *or* **Cap Hai·tien** (kăp′ hā′shən). City of N Haiti on the Atlantic. Pop. 46,217.
**Cap·i·to·line** (kăp′ĭ-tə-līn′). One of the seven hills of ancient Rome.
**Cap·i·tol Reef National Park** (kăp′ĭ-tl). Area of S central Utah reserved to protect cliff dwellings & unusual geologic forms.
**Cap·pa·do·cia** (kăp′ə-dō′shə, -shē-ə). Ancient region of Asia Minor in present-day central Turkey.
**Ca·pri** (kə-prē′, kăp′rē, kä′prē). Island of S Italy, c. 5 sq mi (13 sq km), on the S edge of the Bay of Naples.
**Cap·u·a** (kăp′yo͞o-ə). Town of S Italy N of Naples near the site of the ancient Roman city of **Capua,** on the Appian Way. Pop. 13,938.
**Ca·ra·cas** (kə-rä′kəs, -răk′əs). Cap. of Venezuela, in the N part near the Caribbean coast. Pop. 1,035,449.
**Car·bon·dale** (kär′bən-dāl′). **1.** City of S Ill. SSE of East St. Louis. Pop. 27,194. **2.** City of NE Pa. NE of Scranton. Pop. 11,255.
**Car·cas·sonne** (kär′-kə-sôn′, -sŏn′). City of S France SE of Toulouse. Pop. 38,887.
**Car·che·mish** (kär′kə-mĭsh′, kär-kē′mĭsh). Ancient Hittite city in present-day S Turkey on the Euphrates.
**Cár·de·nas** (kär′dn-äs′). City of N Cuba on the Straits of Florida. Pop. 55,209.
**Car·diff** (kär′dĭf). City of SE Wales on Bristol Channel. Pop. 281,500.
**Car·di·gan Bay** (kär′dĭ-gən). Inlet of St. George's Channel in W Wales.
**Car·i·a** (kâr′ē-ə). Ancient region of SW Asia Minor with a coastline on the Aegean.
**Car·ib·be·an** (kăr′ə-bē′ən, kə-rĭb′ē-). Sea of the N Atlantic bounded by the coasts of Central & South America & the West Indies.
**Car·i·boo** (kăr′ə-bo͞o′). Mountains of E B.C., Canada, parallel to & W of the Rockies.
**Ca·rin·thi·a** (kə-rĭn′thē-ə). Region & former duchy of central Europe, in S Austria.
**Car·lisle** (kär-līl′, kär′līl′). **1.** Borough of NW England near the Scottish border. Pop. 99,600. **2.** Borough of S Pa. WSW of Harrisburg. Pop. 18,314.

**Carls·bad** (kärlz′băd′). **1.** Caverns of limestone in **Carlsbad Caverns National Park,** SE N.Mex. **2.** (*also* kärls′bät′). KARLOVY VARY. **3.** City of S Calif. NNW of San Diego. Pop. 35,490. **4.** City of SW N.Mex. on the Pecos R. Pop. 25,496.

**Carls·ru·he** (kärlz′ro͞o′ə). KARLSRUHE.

**Car·mel** (kär-mĕl′). **1.** (kär′məl), **Mount.** Ridge of NW Israel extending c. 15 mi (24 km) to the Mediterranean & rising to 1,800 ft (549 m). **2.** *also* **Car·mel-by-the-Sea** (-bī-thə-sē′). City of W Calif. near Monterey. Pop. 4,707. **3.** City of central Ind. N of Indianapolis. Pop. 18,272.

**Car·ne·gie** (kär′nĭ-gē, kär-nĕg′ē). Borough of SW Pa. near Pittsburgh. Pop. 10,099.

**Car·nic Alps** (kär′nĭk ălps′). Range of the E Alps in S Austria & NE Italy.

**Car·ni·o·la** (kär′nē-ō′lə, kärn-yō′-). Region of NW Yugoslavia NE of Istria. **—Car′ni·o′lan** *adj.* & *n.*

**Ca·ro·li·na. 1.** (kăr′ə-lī′nə). English colony of E North America divided in 1729 into **the Car·o·li·nas** (-nəz), North & South Carolina. **2.** (kär′ə-lē′nə). City of NE Puerto Rico ESE of San Juan. Pop. 147,100.

**Car·o·line** (kăr′ə-līn′, -lĭn). Islands of the W Pacific E of the Philippines, part of the U.S. Trust Territory of the Pacific Is.

**Car·ol Stream** (kăr′əl). Village of NE Ill. W of Chicago. Pop. 15,472.

**Ca·ro·ní** (kär′ə-nē′). River of E Venezuela flowing c. 550 mi (885 km) N to the Orinoco.

**Car·pa·thi·an** (kär-pā′thē-ən). Mountain system of central Europe in E Czechoslovakia, S Poland, W Ukraine, & N & W Rumania.

**Car·pa·thos** (kär′pə-thôs′). KÁRPATHOS.

**Car·pen·tar·i·a** (kär′pən-târ′ē-ə), **Gulf of.** Wide inlet of the Arafura Sea in N Australia.

**Car·pen·ters·ville** (kär′pən-tərz-vĭl′). Village of NE Ill. WNW of Chicago. Pop. 23,272.

**Car·pin·te·ri·a** (kär′pən-tə-rē′ə). City of SW Calif. E of Santa Barbara. Pop. 10,835.

**Car·ran·tuo·hill** (kăr′ən-to͞o′əl). Highest mountain of Ireland, 3,414 ft (1,041.3 m), in the SW part in Macgillicuddy's Reeks.

**Car·ra·ra** (kə-rär′ə). City of N Italy near the Ligurian Sea. Pop. 56,236.

**Car·roll·ton** (kăr′əl-tən). **1.** City of W Ga. WSW of Atlanta. Pop. 14,078. **2.** City of N Tex. near Dallas. Pop. 40,591.

**Car·son** (kär′sən). **1.** River of W Nev. flowing c. 125 mi (201 km) NE into **Carson Sink,** an intermittent lake. **2.** City of S Calif. near Los Angeles. Pop. 81,221.

**Carson City.** Cap. of Nev. in the W part near the Calif. border. Pop. 32,022.

**Car·stensz** (kär′stənz), **Mount.** DJAJA PEAK.

**Car·ta·ge·na** (kär′tə-gā′nə, -jē′-, -hā′-). **1.** City of NW Colombia on the Caribbean. Pop. 292,512. **2.** City of SE Spain on the Mediterranean. Pop. 52,312.

**Car·ter·et** (kär′tə-rĕt′). Borough of NE N.J. S of Elizabeth. Pop. 20,598.

**Car·thage** (kär′thĭj). **1.** Ancient city & state on the N coast of Africa on the Bay of Tunis NE of modern Tunis. **2.** City of SW Mo. NE of Joplin. Pop. 11,104. **—Car′tha·gin′i·an** (kär′thə-jĭn′ē-ən) *adj.* & *n.*

**Car·y** (kăr′ē). Town of central N.C. near Raleigh. Pop. 21,612.

**Cas·a·blan·ca** (kăs′ə-blăng′kə, käz′-). City of NW Morocco, on the Atlantic. Pop. 1,506,373.

**Cas·a Gran·de** (kăs′ə grăn′dē). City of S central Ariz. SSE of Phoenix. Pop. 14,971.

**Cas·cade** (kăs-kād′). Mountain range of NW U.S. extending from NE Calif. through W Ore. & W Wash.

**Cas·co Bay** (kăs′kō). Inlet of the Atlantic in SW Me.

**Ca·ser·ta** (kə-zər′tə). City of S Italy NNE of Naples. Pop. 51,162.

**Cash·mere** (kăsh′mîr′, kăsh-mîr′). KASHMIR.

**Ca·si·quia·re** (kä′sĭ-kyä′rē). River, c. 100 mi (161 km), of S Venezuela linking the Orinoco & Amazon river systems.

**Cas·per** (kăs′pər). City of E central Wyo. NW of Cheyenne. Pop. 51,016.

**Cas·pi·an Sea** (kăs′pē-ən). Salt lake, c. 153,000 sq mi (396,000 sq km), between SE Europe & W Asia.

**Cas·sel** (kăs′əl, kä′səl). KASSEL.

**Cas·sel·ber·ry** (kăs′əl-bĕr′ē). City of E central Fla. NNE of Orlando. Pop. 15,247.

**Cas·tel Gan·dol·fo** (kä-stĕl′ gän-dôl′fō). Town of central Italy SE of Rome. Pop. 2,965.

**Cas·tel·lam·ma·re di Sta·bia** (kä-stĕl′ə-mär′ā dĭ stäb′yə). City of S Italy on the Bay of Naples. Pop. 64,341.

**Cas·tel·lón de la Pla·na** (käs′təl-yōn′ də lä plä′nə). City of E Spain on the Mediterranean NNE of Valencia. Pop. 79,773.

**Cas·tile** (kăs-tēl′). Region & former kingdom of central & N Spain. **—Cas·til′ian** (kă-stĭl′yən, kə-) *adj.* & *n.*

**Cas·ti·lla** (kä-stē′lyä, -yä). CASTILE.

**Cas·tle Peak** (kăs′əl). Mountain, 14,259 ft (4,349 m), in the Elk Mts. of W central Colo.

**Castle Shan·non** (shăn′ən). Borough of SW Pa. near Pittsburgh. Pop. 10,164.

**Cas·tries** (käs′trēz′, -trēs′). Cap. of St. Lucia in the British West Indies. Pop. 42,770.

**Cas·trop-Rau·xel** (käs′trôp-rouk′səl). City of W West Germany in the Ruhr SSW of Münster. Pop. 82,373.

**Cat·a·li·na** (kăt′l-ē′nə). SANTA CATALINA.

**Cat·a·lo·nia** (kăt′l-ōn′yə, -ō′nē-ə). Region of NE Spain bordering on France & the Mediterranean. **—Cat′a·lo′nian** *adj.* & *n.*

**Ca·ta·lu·ña** (kä′tə-lo͞o′nyə). CATALONIA.

**Ca·ta·mar·ca** (kä′tə-mär′kə). City of NW Argentina NW of Córdoba. Pop. 64,410.

**Ca·ta·nia** (kə-tän′yə, -tā′nē-ə). City of E Sicily, Italy, on the E coast of the Ionian Sea. Pop. 403,390.

**Ca·tan·za·ro** (kä′tän-zär′ō, -dzär′ō) City of S Italy near the Ionian Sea. Pop. 52,054.

**Ca·taw·ba** (kə-tô′bə). River, 250 mi (402.3 km), of W N.C. & N S.C.

**Ca·thay** (kă-thā′, kə-). Old name for China.

**Cath·e·rine** (kăth′ə-rĭn, kăth′rĭn), **Mount.** JEBEL KATHERINA.

**Cats·kill** (kăt′skĭl′) *also* **Cats·kills** (-skĭlz). Mountain range in SE N.Y. rising to 4,204 ft (1,282.2 m).

**Cau·ca** (kou′kə). River of NW Colombia flowing c. 600 mi (965 km) N to the Magdalena.

**Cau·ca·sus** (kô′kə-səs) *also* **Cau·ca·sia** (kô-kā′zhə, -shə). Region of SE European USSR between the Black & Caspian seas including the **Caucasus Mountains,** rising to 18,480 ft (5,636.4 m).

**Caucasus In·di·cus** (ĭn′dĭ-kəs). HINDU KUSH.

**Cau·ve·ry** (kô′və-rē). River of S India flowing c. 475 mi (764 km) to the Bay of Bengal.

**Ca·vi·te** (kə-vē′tē). City of SW Luzon, Philippines, on Manila Bay SW of Manila. Pop. 87,813.

**Cawn·pore** (kôn′pôr′, -pōr′). KANPUR.

**Ca·xi·as** (kə-shē′əs). **1.** Town of NE Brazil WNW of Teresina. Pop. 173,082. **2.** *also* **Caxias do Sul** (də so͞ol′). City of S Brazil N of Pôrto Alegre. Pop. 107,487.

**Cay·ce** (kā′sē). City of central S.C. near Columbia. Pop. 11,701.

**Cay·enne** (kī-ĕn′, kā-). Cap. of French Guiana on **Cayenne Is.** in a river mouth near the Atlantic coast. Pop. 30,461.

**Cay·man** (kā-măn′, kā′mən) *also* **Cay·mans** (kā-mănz′, kā′mənz). British-administered island group in the Caribbean NW of Jamaica, including **Grand Cayman, Little Cayman,** & **Cayman Brac.** Cap. Georgetown. Pop. 10,652.

**Cay·u·ga** (kī-o͞o′gə, gyo͞o′-, kā-o͞o′-, -yo͞o′-). Lake of W central N.Y., longest (38 mi/61.1 km) of the Finger Lakes.

**Ce·bu** (sā-bo͞o′). **1.** Island of the central Philippines, one of the Visayans. **2.** City on the E coast of Cebu Is. Pop. 489,208.

**Ce·dar** (sē′dər). River, c. 330 mi (531 km), of SE Minn. & E Iowa.

**Cedar City.** City of SW Utah SSW of Salt Lake City. Pop. 10,972.

**Cedar Falls.** City of NE Iowa near Waterloo. Pop. 36,322.

**Cedar Rapids.** City of E central Iowa WNW of Davenport. Pop. 110,243.

**Ce·la·ya** (sə-lī′ə). City of central Mexico NW of Mexico City. Pop. 79,977.

**Cel·e·bes** (sĕl′ə-bēz′, sə-lē′bēz′). **1.** Sea of the W Pacific between Sulawesi & S Philippines. **2.** SULAWESI.

**Cel·le** (sĕl′ə, tsĕl′ə). City of N West Germany S of Hamburg. Pop. 74,347.

**Cen·ter·ville** (sĕn′tər-vĭl′). City of SW Ohio near Dayton. Pop. 18,886.

**Cen·tral Af·ri·can Republic** (sĕn′trəl ăf′rĭ-kən). Country of central Africa. Cap. Bangui. Pop. 2,284,000.

**Central A·mer·i·ca** (ə-mĕr′ĭ-kə). Region of S North America from S border of Mexico to N border of Colombia. **—Central A·mer′i·can** *adj.* & *n.*

**Central Falls.** City of NE R.I. near Providence. Pop. 16,995.

**Cen·tra·lia** (sĕn-trāl′yə). **1.** City of S Ill. E of East St. Louis. Pop. 15,126. **2.** City of SW Wash. S of Olympia. Pop. 10,809.

**Central Valley.** Valley, c. 450 mi (724 km) long, in central Calif.

**Cen·tre·ville** (sĕn′tər-vĭl′). City of SW Ill. near East St. Louis. Pop. 9,747.

**Ceph·a·lo·ni·a** (sĕf′ə-lō′nē-ə, -lōn′yə). Largest of the Ionian Is. off the W coast of Greece.

**Ce·ram** (sā′räm′). **1.** Sea of the W Pacific among the Moluccas of E Indonesia W of New Guinea. **2.** Island of the Moluccas in E Indonesia.

**Cer·es** (sîr′ēz). City of central Calif. near Modesto. Pop. 13,281.

**Cer·re·do** (sə-rā′dō), **Torre de.** Highest peak (8,687 ft/2,649.5 m) of the Cantabrian Mts. in N Spain.

**Cer·ri·tos** (sə-rē′təs). City of S Calif. SE of Los Angeles. Pop. 52,756.

**Cer·ro de Pas·co** (sĕr′ō də päs′kō). Mountain, 15,100 ft (4,605.5 m), of central Peru.

**Cerro de Pun·ta** (po͞on′tə). Highest mountain (4,400 ft/1,342 m) of Puerto Rico, in the Cordillera Central.

**Cerro Gor·do** (gôr′dō). Mountain pass in S Mexico; site of U.S. victory (1847) in the Mexican War.

**Ce·se·na** (chə-zā′nə). City of N central Italy ENE of Florence. Pop. 49,915.

**Čes·ke Bu·dě·jo·vi·ce** (chěs′kə bo͝od′ə-yô′vət-sə). City of SW Czechoslovakia on the Vltava R. Pop. 80,800.

**Ceu·ta** (sā′o͞o′tə, sě′o͞o-tä). Spanish city of NW Africa on the Strait of Gibraltar. Pop. 60,639.

**Cé·vennes** (sā-věn′). Mountain range of S France, W of the Rhone.

**Cey·lon** (sĭ-lŏn′, sā-). SRI LANKA. **—Cey′lo·nese′** (-nēz′, -nēs′) *adj.* & *n.*

**Chad** (chăd). **1.** Lake of N central Africa, mainly in Chad. **2.** Country of N central Africa. Cap. Ndjamena. Pop. 4,309,000.

**Chaer·o·ne·a** (kĕr′ə-nē′ə, kĭr′-). Ancient city of E Greece.

**Chal·ce·don** (kăl′sĭ-dŏn′, kăl-sēd′n). Ancient Greek city of NW Asia Minor on the Bosporus near present-day Istanbul.

**Chal·cid·i·ce** (kăl-sĭd′ĭ-sē). Peninsula of NE Greece, projecting into the N Aegean with three fingerlike extensions.

**Chal·cis** (kăl′sĭs). Ancient city of SE Greece on the W coast of Euboea.

**Chal·dae·a** *also* **Chal·de·a** (kăl-dē′ə). Ancient region of S Mesopotamia. **—Chal·dae′an, Chal·de′an** *adj.* & *n.*

**Cha·leur Bay** (shə-lo͝or′, -lûr′). Inlet of the Gulf of St. Lawrence between E Que. & N N.B., Canada.

**Cha·lon** (shə-lôN′) *also* **Cha·lon-sur-Saône** (-sûr-sōn′). City of E central France N of Mâcon. Pop. 55,495.

**Cha·lôns** (shä-lôN′) *also* **Cha·lôns-sur-Marne** (-sûr′-märn′). City of NE France on the Marne E of Paris. Pop. 50,870.

**Cham·bal** (chŭm′bəl). River, c. 550 mi (885 km) of W central India.

**Cham·bers·burg** (chăm′bərz-bûrg′). Borough of S Pa. SW of Harrisburg. Pop. 16,174.

**Cham·bé·ry** (shäN-bā-rē′). City of SE France ESE of Lyon. Pop. 52,286.

**Cham·be·zi** (chăm-bē′zĭ). River, c. 300 mi (483 km), of NE Zambia.

**Cham·bly** (shăm′blē, shäN-blē′). City of S Que., Canada, on the Richelieu R. E of Montreal. Pop. 11,815.

**Cham·bord** (shäN-bôr′). Village of N central France; site of Francis I's Renaissance château.

**Cha·mi·zal** (shăm′ĭ-zäl′, chăm′ĭ-säl′). District on N bank of Rio Grande near El Paso, Tex., ceded by U.S. to Mexico in 1963.

**Cham·pagne** (shăm-pān′). Region & former province of NE France.

**Cham·paign** (shăm-pān′). City of E central Ill. adjoining Urbana. Pop. 58,133.

**Cham·pi·gny** (shäN-pē-nyē′) *also* **Cham·pi·gny-sur-Marne** (-sûr-märn′). City of N France near Paris. Pop. 80,189.

**Cham·plain** (shăm-plān′). Lake of NE N.Y., NW Vt., & S Que., Canada.

**Chan·cel·lors·ville** (chăn′sə-lərz-vĭl′, -slərz-). Town of NE Va.; site of Confederate victory (1863).

**Chan·chiang** (jän′jē-äng′). ZHANJIANG.

**Chan·di·garh** (chŭn′dē-gər). City of N India N of Delhi. Pop. 218,743.

**Chan·dler** (chănd′lər). City of S central Ariz. SE of Phoenix. Pop. 29,673.

**Chang·chow** (chăng′jō′). CHANGZHOU.

**Chang·chun** (chăng′cho͞on′). City of NE China in Manchuria SSW of Harbin. Pop. 1,500,000.

**Chang·hua** (jăng′hwä). City of W Taiwan SW of Taipei. Pop. 137,236.

**Chang Jiang** (chăng′ jē-äng′). YANGTZE.

**Chang·sha** (chăng′shä′). City of S China WSW of Shanghai. Pop. 850,000.

**Chang·zhou** (chăng′jō). City of E China on the Grand Canal WNW of Shanghai. Pop. 400,000.

**Chan·kiang** (jän′jē-äng′). ZHANJIANG.

**Chan·nel** (chăn′əl). Islands of Great Britain in the English Channel off the coast of Normandy, France.

**Cha·nute** (shə-no͞ot′). City of SE Kans. E of Wichita. Pop. 10,506.

**Chao Phra·ya** (chou prī′ə). River of Thailand, c. 140 mi (225 km), flowing into the Gulf of Siam.

**Cha·pa·la** (chə-pä′lə). Lake, 408 sq mi (1,057 sq km), of W central Mexico SE of Guadalajara.

**Chap·el Hill** (chăp′əl). Town of N central N.C. WNW of Raleigh. Pop. 32,421.

**Cha·pul·te·pec** (chə-po͝ol′tə-pĕk′). Rocky hill S of Mexico City; site of major Mexican War battles.

**Char·dzhou** (chär-jō′). City of SW Central Asian USSR on the Amu Darya. Pop. 140,000.

**Cha·rente** (shə-ränt′). River of W France flowing c. 220 mi (354 km), to the Bay of Biscay.

**Cha·ri** (shä′rē). SHARI.

**Char·i·ton** (shăr′ĭ-tn). River, c. 280 mi (451 km), of S Iowa & N Mo.

**Char·le·roi** (shär′lə-roi′, shär-lə-rwä′). City of S Belgium S of Brussels. Metro. area pop. 458,000.

**Charles·bourg** (chärlz′bûrg′, shäl-bo͝or′). City of S Que., Canada, near Quebec city. Pop. 63,147.

**Charles·ton** (chärl′stən). **1.** Mountain, 11,919 ft (3,635.3 m), of SE Nev. **2.** City of E Ill. ESE of Decatur. Pop. 19,355. **3.** City of SE S.C. NE of Savannah. Pop. 69,510. **4.** Cap. of W.Va. in the W central part. Pop. 63,968.

**Charles·town** (chärlz′toun′). Former city of E Mass., now the oldest part of Boston.

**Char·le·ville-Mé·zières** (shär′lə-vēl′mā-zhâr′). City of NE France on the Meuse ENE of Paris. Pop. 59,513.

**Char·lotte** (shär′lət). City of S N.C. near the S.C. border SSW of Winston-Salem. Pop. 314,447.

**Charlotte A·ma·lie** (ə-mäl′yə). Cap. of the U.S. Virgin Is. on St. Thomas. Pop. 11,670.

**Char·lottes·ville** (shär′ləts-vĭl′). Independent city of central Va. NW of Richmond. Pop. 45,010.

**Char·lotte·town** (shär′lət-toun′). Cap. of P.E.I., Canada, on the S coast. Pop. 17,063.

**Char·tres** (shärt, shär′trə). City of N France SW of Paris. Pop. 38,574.

**Châ·teau·guay** (shă-tō-gā′, shăt′ə-gā′). Town of S Que., Canada, SW of Montreal. Pop. 36,329.

**Châ·teau·roux** (shă-tō-ro͞o′). City of central France SE of Tours. Pop. 53,166.

**Châ·teau-Thier·ry** (shă-tō-tyĕ-rē′). Town of N France; site of second Battle of the Marne (1918). Pop. 13,379.

**Chat·ham** (chăt′əm). **1.** Island & island group of New Zealand in the SW Pacific E of South Is. **2.** City of SE Ont., Canada, on the Thames R. Pop. 38,685.

**Chat·ta·hoo·chee** (chăt′ə-ho͞o′chē). River, 436 mi (701.5 km), of N Ga.

**Chat·ta·noo·ga** (chăt′ə-no͞o′gə). City of SE Tenn. on the Ga. border SE of Nashville. Pop. 169,565.

**Che·bok·sa·ry** (chĭ-bŏk-sär′ē). City of E central European USSR on the Volga. Pop. 308,000.

**Che·ju** (chĕ′jo͞o′). Island of South Korea separated from the SW coast by **Cheju Strait,** a 50-mi (80-km) channel linking the Yellow Sea & Korea Strait.

**Che·kiang** (jŭj′ē-äng′). ZHENJIANG.

**Che·lan** (shə-lăn′). Lake of N central Wash. in the Cascade Mts.

**Che·liff** (shā-lēf′). River, c. 420 mi (676 km), of N Algeria.

**Chelms·ford** (chĕmz′fərd). **1.** Borough of SE England. Pop. 58,320. **2.** Town of NE Mass. near Lowell. Pop. 31,174.

**Chel·sea** (chĕl′sē). **1.** District & former borough of Greater London, England. **2.** City of E Mass. near Boston. Pop. 25,431.

**Chel·ten·ham** (chĕlt′nəm, chĕl′tən-əm). Borough of W central England. Pop. 75,910.

**Che·lya·binsk** (chĕl-yä′bĭnsk). City of W Siberian USSR S of Sverdlovsk. Pop. 1,030,000.

**Che·lyus·kin** (chĕl-yo͞o′skĭn). Cape of N central Siberian USSR; northernmost point of Asia.

**Chem·nitz** (kĕm′nĭts). KARL-MARX-STADT.

**Che·nab** (chə-näb′). River, 675 mi (1,086.1 km), of N India & E Pakistan.

**Cheng·chow** (jŭng′jō′). ZHENGZHOU.

**Cheng·du** *also* **Cheng·tu** (chŭng′do͞o′). City of central China WNW of Chongqing. Pop. 2,000,000.

**Cher** (shĕr). River, c. 220 mi (354 km), of central France.

**Cher·bourg** (shâr′bo͝org′, shĕr-bo͝or′). City of NW France on the English Channel. Pop. 31,333.

**Che·rem·kho·vo** (chə-rĕm′kə-və, chĕr′əm-kô′və). City of SE Siberian USSR NW of Irkutsk. Pop. 77,000.

**Che·re·po·vets** (chĕr′ə-pə-vĕts′). City of N central European USSR N of Moscow. Pop. 266,000.

**Cher·i·bon** (chĕr′ĭ-bŏn′). TJIREBON.

**Cher·kas·sy** (chər-kăs′ē, -kă′sē). City of S European USSR on the Dnieper. Pop. 91,000.

**Cher·ni·gov** (chər-nē′gəf). City of W central European USSR NNE of Kiev. Pop. 238,000.

**Cher·nov·tsy** (chər-nôft′sē). City of SW European USSR near the Rumanian border. Pop. 219,000.

**Cher·o·kee Strip** (chĕr′ə-kē′ strĭp) or **Out·let** (out′lĕt′, -lĭt). Plot of land, c. 12,000 sq mi (31,080 sq km), in present-day N Okla. purchased by the U.S. in 1891.

**Ches·a·peake** (chĕs′ə-pēk′). **1. Bay.** Inlet of the Atlantic in Va. & Md. **2.** Independent city of SE Va. S of Newport News. Pop. 114,226.

**Chesh·ire** (chĕsh′ər, -ĭr′). Town of S central Conn. W of New Haven. Pop. 21,788.

**Ches·ter** (chĕs′tər). **1.** Borough of W central England SSE of Liverpool. Pop. 117,200. **2.** City of SE Pa. on the Delaware R. near Philadelphia. Pop. 45,794.

**Ches·ter·field** (chĕs′tər-fēld′). **1.** Inlet on NW coast of Hudson Bay extending E into Keewatin District of N.W.T., Canada. **2.** City of N central England S of Sheffield. Pop. 69,480.

**Chev·i·ot** (chĕv′ē-ət, shĭv′-, chē′vē-). Hills extending c. 35 mi (56 km) along the English-Scottish border & rising to **The Cheviot** (2,676 ft/816.2 m).
**Chev·y Chase** (chĕv′ē chās′). Village of W central Md. near Washington, D.C. Pop. 24,000.
**Chey·enne** (shī-ăn′, -ĕn′). **1.** River, c. 290 mi (467 km), of E Wyo. & W S.Dak. **2.** Cap. of Wyo. in the SE part near the Nebr. & Colo. borders. Pop. 47,283.
**Chi·ai** or **Chia-i** (jē-ī′). City of SW Taiwan N of Kaohsiung. Pop. 238,713.
**Chia-ling** (jē-ä′lĭng′). JIALING.
**Chia·mus·su** (jē-ä′mōō′sōō′). JIAMUSI.
**Chiang Mai** (jē-äng′ mī′) *or* **Chieng·mai** (jē-ĕng′mī′). City of NW Thailand on the Ping R. near the Burmese border. Pop. 83,729.
**Chi·an·ti** (kē-än′tē). Range of the Apennines, c. 15 mi (24 km) long, in central Italy.
**Chi·ba** (chē′bə′). City of E central Honshu, Japan, on the NE shore of Tokyo Bay. Pop. 712,488.
**Chi·ca·go** (shĭ-kä′gō, -kô′-). City of NE Ill. on Lake Michigan. Pop. 3,005,072.
**Chicago Heights.** City of NE Ill. S of Chicago. Pop. 37,026.
**Chicago Ridge.** Village of NE Ill. near Chicago. Pop. 13,473.
**Chich·a·gof** (chĭch′ə-gôf′, -gŏf′). Island of SE Alas. in the N Alexander Archipelago.
**Chi·chén It·zá** (chə-chən′ ēt-sä′). Ancient Mayan city of central Yucatán, Mexico.
**Chich·es·ter** (chĭch′ĭ-stər). Borough of S England near the English Channel E of Southampton. Pop. 20,940.
**Chick·a·hom·i·ny** (chĭk′ə-hŏm′ə-nē). River, c. 90 mi (145 km), of E Va.
**Chick·a·sa·whay** (chĭk′ə-sô′wā). River, c. 210 mi (338 km), of SE Miss.
**Chick·a·sha** (chĭk′ə-shā′). City of central Okla. SW of Oklahoma City. Pop. 15,828.
**Chi·cla·yo** (chə-klī′ō). City of NW Peru near the Pacific coast. Pop. 189,685.
**Chi·co** (chē′kō). City of N Calif. N of Sacramento. Pop. 26,601.
**Chic·o·pee** (chĭk′ə-pē). City of SW Mass. near Springfield. Pop. 55,112.
**Chi·cou·ti·mi** (shĭ-kōō′tə-mē). **1.** River, c. 100 mi (161 km), of S Que., Canada. **2.** City of S central Que., Canada, on the Saguenay R. N of Quebec city. Pop. 57,737.
**Chieng·mai** (jē-ĕng′mī′). *var. of* CHIANG MAI.
**Chie·ti** (kyā′tē). City of central Italy near the Adriatic. Pop. 31,895.
**Chi·ga·sa·ki** (chē′gə-sä′kē). City of E central Honshu, Japan, near Yokohama. Pop. 152,023.
**Chi·hua·hua** (chə-wä′wä). City of N Mexico. Pop. 327,313.
**Chil·e** (chĭl′ē). Republic of SW South America with a long Pacific coastline. Cap. Santiago. Pop. 11,198,789. **—Chil′e·an** *adj.* & *n.*
**Chi·lin** (jē′lĭn′). JILIN.
**Chi·llán** (chē-än′, -yän′). City of central Chile ENE of Concepción. Pop. 87,600.
**Chil·li·coth·e** (chĭl′ĭ-kŏth′ē, -kô′thē). City of S central Ohio S of Columbus. Pop. 23,420.
**Chil·li·wack** (chĭl′ə-wăk′). City of SW B.C., Canada, on the Fraser R. E of Vancouver. Pop. 8,634.
**Chi·lo·é** (chē′lə-wā′). Island off S central Chile.
**Chil·pan·cin·go** (chĭl′pən-sĭng′gō) *or* **Chilpancingo de Los Bra·vos** (dā lōs brä′vōs). City of S Mexico SSW of Mexico City. Pop. 36,193.
**Chil·tern** (chĭl′tərn). Hills of S central England NE of the upper Thames.
**Chi·lung** (jē′lŏŏng′). KEELUNG.
**Chim·bo·ra·zo** (chĭm′bə-rä′zō, -rā′-, shĭm′-). Inactive volcano, 20,561 ft (6,271.1 m), in central Ecuador.
**Chim·kent** (chĭm-kĕnt′). City of SW Central Asian USSR N of Tashkent. Pop. 322,000.
**Chin** (chĭn). Hills of W Burma rising to 10,018 ft (3,055 m).
**Chi·na** (chī′nə). **1. Sea.** W part of the Pacific extending along the E coast of Asia from S Japan to the Malay Peninsula & divided by Taiwan into the **East China Sea** & the **South China Sea. 2.** *also* **People's Republic of China.** Country of E Asia. Cap. Beijing. Pop. 958,090,000. **3.** *also* **Republic of China.** TAIWAN.
**Chin·chow** (jĭn′jō′). JINZHOU.
**Chin·co·teague** (shĭng′kə-tēg′, chĭng′-). Narrow bay in NE Va. & SE Md.
**Chin·dwin** (chĭn′dwĭn′). River, c. 550 mi (885 km), of NW Burma.
**Ching·hai** (chĭng′hī′). QINGHAI province.
**Chin·hae** (jĭn′hī′). City of SE South Korea on Korea Strait. Pop. 103,640.
**Chin·ju** (jĭn′jōō′). City of S South Korea W of Pusan. Pop. 154,646.
**Chin·kiang** (jĭn′jē-äng′, chĭn′kē-ăng′). ZHENJIANG.
**Chin·nam·po** (chē′näm′pō′). NAMPO.

ōō **boot** ou **out** th **thin** *th* **this** ŭ **cut** ûr **urge** y **young**
yōō **abuse** zh **vision** ə **about**, **item**, **edible**, **gallop**, **circus**

**Chi·no** (chē′nō). City of S Calif. E of Los Angeles. Pop. 40,165.
**Chiog·gia** (kē-ô′jə). City of NE Italy on an island S of Venice. Pop. 38,200.
**Chi·os** (kī′ŏs′). Island of E Greece in the Aegean off the W coast of Turkey.
**Chip·pe·wa** (chĭp′ə-wô′, -wä′, -wā′). River, c. 180 mi (290 km), of NW Wis.
**Chippewa Falls.** City of W central Wis. near Eau Claire. Pop. 11,845.
**Chis·holm Trail** (chĭz′əm). Former cattle trail from San Antonio, Tex., to Abilene, Kans.
**Chi·ta** (chĭ-tä′). City of SE Siberian USSR E of Irkutsk. Pop. 303,000.
**Chi·tral** (chĭ-träl′). River, c. 300 mi (483 km), of N Pakistan & NE Afghanistan.
**Chit·ta·gong** (chĭt′ə-gông′, -gŏng′). City of SE Bangladesh near the Bay of Bengal. Pop. 889,760.
**Chka·lov** (chə-kä′ləf). ORENBURG.
**Choc·taw·hatch·ee** (chŏk′tə-hăch′ē). River, c. 140 mi (225 km), of S Ala.
**Choi·seul** (shwä-zœl′). One of the Solomon Is. in the SW Pacific SE of Bougainville Is.
**Choi·sy** (shwä-zē′) *or* **Choi·sy-le-Roi** (-zē′lə-rwä′). City of N France SSE of Paris. Pop. 38,629.
**Cho·let** (shō-lā′). City of W France ESE of Nantes. Pop. 49,887.
**Cho·lu·la** (chə-lōō′lə). Town of E central Mexico; ancient Toltec & Aztec center. Pop. 15,399.
**Cho·mo Lha·ri** (chō′mō lär′ē). Peak, 23,997 ft (7,319 m), in the SE Himalayas on the Bhutan-China border.
**Chong·jin** (chông′jĭn′). City of NE North Korea on the Sea of Japan. Pop. 306,000.
**Chong·ju** (chông′jōō′). City of W central South Korea SSE of Seoul. Pop. 192,707.
**Chong·qing** (chông′chē-ĭng′). City of S central China on the Yangtze. Pop. 3,500,000.
**Chon·ju** (chŏn′jōō′). City of SW South Korea S of Seoul. Pop. 311,393.
**Cho O·yu** (chō′ ō-yōō′). Peak, 26,967 ft (8,225 m), in the central Himalayas on the Nepal-China border.
**Cho·rzow** (kô′zhōōf′, -zhōōv′, hô′-). City of S Poland NNW of Katowice. Pop. 151,338.
**Cho·sen** (chō′sĕn′). Old name for Korea.
**Cho·shi** (chō′shē′). City of E central Honshu, Japan, on the Pacific E of Tokyo. Pop. 90,374.
**Cho·ta Nag·pur** (chō′tə näg′pŏŏr). Forested plateau of E India N of the Mahanadi R.
**Chou Shan** (jō′ shän′). ZHOUSHAN.
**Christ·church** (krīst′chûrch′). **1.** Borough of S central England on the English Channel. Pop. 31,610. **2.** City of E South Is., New Zealand, near the Pacific coast. Pop. 171,987.
**Chris·ti·a·ni·a** (krĭs′tē-ăn′ē-ə, -än′-). OSLO.
**Chris·tians·burg** (krĭs′chənz-bûrg′). Town of SW Va. SW of Roanoke. Pop. 10,345.
**Chris·tian·sted** (krĭs′chən-stĕd′). Chief city of St. Croix, U.S. Virgin Is., on the N coast. Pop. 2,846.
**Christ·mas** (krĭs′məs). **1.** Australian-administered island, c. 64 sq mi (166 sq km), in the E Indian Ocean S of W Java. **2.** Largest (c. 222 sq mi/575 sq km) of the Line Is., in the Pacific near the equator & S of Hawaii.
**Chu** (chōō). **1.** ZHU JIANG. **2.** River of S Central Asian USSR flowing c. 600 mi (965 km) E into Issyk-Kul.
**Chu·but** (chə-bōōt′). River of S Argentina flowing c. 500 mi (805 km) E to the Atlantic.
**Chu·chow** (chōō′jō′). ZHUZHOU.
**Chud·sko·ye** (chōōt′skə-yə). PEIPUS.
**Chu·gach** (chōō′găch′, -găsh′). Range of S Alas. extending from Cook Inlet E to the Canadian border & rising to 13,250 ft (4,041.3 m).
**Chuk·chi** (chŭk′chē, chŏŏk′-). **1.** Sea of the Arctic Ocean between NW Alas. & extreme NE USSR. **2.** Peninsula of extreme NE USSR across the Bering Strait from Alas.
**Chu Kiang** (chōō′ jē-äng′). ZHU JIANG.
**Chu·la Vis·ta** (chōō′lə vĭs′tə). City of S Calif. near San Diego. Pop. 83,927.
**Chu·lym** *also* **Chu·lim** (chə-lĭm′). River of S central Siberian USSR flowing c. 700 mi (1,126 km) N & W to the Ob.
**Chun·chon** (chōōn′chôn′). City of N South Korea ENE of Seoul. Pop. 140,530.
**Chung·king** (chŏŏng′kĭng′, jŏŏng′gĭng′). CHONGQING.
**Chur·chill** (chûr′chĭl′). **1.** River of central Sask. & N Man., Canada, flowing c. 1,000 mi (1609 km) NE to Hudson Bay. **2.** River of S Labrador, Canada, flowing 208 mi (334.7 km) over **Churchill Falls** (245 ft/74.7 m high) to Lake Melville.
**Chu·so·va·ya** (chōō′sə-vä′yə). River, c. 460 mi (740 km), of E European USSR.
**Chu·vash** (chōō′väsh′). Region of W European USSR.
**Cic·e·ro** (sĭs′ə-rō′). Town of NE Ill. near Chicago. Pop. 61,232.
**Cien·fue·gos** (sē-ĕn-fwā′gōs). City of S central Cuba on **Cienfuegos Bay,** a narrow-necked inlet of the Caribbean. Pop. 88,000.

**Ci·li·cia** (sĭ-lĭsh′ə). Ancient country of SE Asia Minor along the Mediterranean S of the Taurus Mts. **—Ci·li′cian** *adj. & n.*
**Cilician Gates** (gāts). GÜLEK BOGAZ.
**Cim·ar·ron** (sĭm′ə-rŏn′, -rōn′). River of NE N.Mex., SW Kan., & N Okla. flowing 692 mi (1,113.4 km) E to the Arkansas R.
**Cin·cin·na·ti** (sĭn′sə-năt′ē, -năt′ə). City of extreme SW Ohio on the Ohio R. Pop. 385,457.
**Cinque Ports** (sĭngk′). Group of seaports of SE England (orig. Hastings, Romney, Hythe, Dover, & Sandwich) that formed a maritime & defensive association (11th–19th cent.).
**Cin·tra** (sēn′trə). *var. of* SINTRA.
**Cir·cas·sia** (sər-kăsh′ə, -ē-ə). Region of the USSR on the NE coast of the Black Sea N of the Caucasus. **—Cir·cas′sian** *adj. & n.*
**Cir·cle·ville** (sûr′kəl-vĭl′). City of S central Ohio S of Columbus. Pop. 11,700.
**Cis·al·pine Gaul** (sĭs-ăl′pīn′ gôl′). Part of ancient Gaul S & E of the Alps.
**Cis·cau·ca·sia** (sĭs′kô-kā′zhə, -shə). Part of the Caucasus region N of the main Caucasus range.
**Ci·thae·ron** (sĭ-thē′rən). Mountain, 4,622 ft (1,409.7 m), in SE Greece.
**Ci·tlal·té·petl** (sē′tläl-tā′pĕt-l). ORIZABA.
**Ci·u·dad Bo·lí·var** (sē′o͞o-dăd′ bə-lē′vär, -o͞o-*tha*′). City of E central Venezuela on the Orinoco. Pop. 103,728.
**Ciudad Gua·ya·na** (gwə-yä′nə). City of E Venezuela on the Orinoco. Pop. 143,540.
**Ciudad Juá·rez** (wär′ĕz, -ĕs, -əs, hwär′-). City of N Mexico on the Rio Grande opposite El Paso, Tex. Pop. 424,135.
**Ciudad Tru·jil·lo** (tro͞o-hē′ō, -yō). SANTO DOMINGO.
**Ciudad Vic·to·ri·a** (vĭk-tôr′ē-ə, -tōr′-). City of E central Mexico SSE of Monterrey. Pop. 83,897.
**Ci·vi·ta·vec·chia** (chē′vē-tä-vĕk′yä). City of W central Italy on the Tyrrhenian Sea WNW of Rome. Pop. 41,305.
**Clac·ton** (klăk′tən) *also* **Clac·ton-on-Sea** (-ŏn-sē′, -ôn-). Resort town of SE England on the North Sea. Pop. 39,380.
**Clair·ton** (klâr′tn). City of SW Pa. SSE of Pittsburgh. Pop. 12,188.
**Clare·mont** (klâr′mŏnt′). **1.** City of S Calif. near Pomona. Pop. 30,950. **2.** City of SW N.H. near the Vt. border. Pop. 14,557.
**Clare·more** (klâr′môr′, -mōr′). City of NE Okla. ENE of Tulsa. Pop. 12,085.
**Clark Fork** (klärk). River, c. 360 mi (579 km), of W Mont. & N Idaho.
**Clarks·burg** (klärks′bûrg′). City of N W.Va. SSE of Wheeling. Pop. 22,371.
**Clarks·dale** (klärks′dāl′). City of NW Miss. near the Ark. border. Pop. 21,137.
**Clarks·ville** (klärks′vĭl′). **1.** Town of S Ind. on the Ohio R. opposite Louisville, Ky. Pop. 15,164. **2.** City of NW Tenn. NW of Nashville. Pop. 54,777.
**Claw·son** (klô′sən). City of SE Mich. near Detroit. Pop. 15,103.
**Clay·ton** (klāt′n). City of E Mo. near St. Louis. Pop. 14,219.
**Clear·field** (klîr′fēld′). City of N Utah near Ogden. Pop. 17,982.
**Clear·wa·ter** (klîr′wô′tər, -wŏt′ər). **1.** River, c. 100 mi (161 km), of SW Alta., Canada. **2.** River, c. 130 mi (209 km), of NW Sask. & NE Alta., Canada. **3.** River of N central Idaho flowing c. 190 mi (306 km) W through the **Clearwater Mts.**, a range of the Rockies. **4.** City of W central Fla. W of Tampa. Pop. 85,450.
**Cle·burne** (klē′bərn). City of NE Tex. S of Fort Worth. Pop. 19,218.
**Cler·mont-Fer·rand** (klĕr-mōN′fə-räN′). City of central France W of Lyon. Pop. 153,379.
**Cleve·land** (klēv′lənd). **1. Mount.** Highest peak (10,438 ft/3,183.6 m) in Glacier National Park, NW Mont. **2.** City of NW Miss. near the Ark. border. Pop. 14,524. **3.** City of NE Ohio on Lake Erie. Pop. 573,822. **4.** City of SE Tenn. ENE of Chattanooga. Pop. 26,415. **5.** City of E Tex. NNE of Houston. Pop. 19,218.
**Cleveland Heights.** City of NE Ohio near Cleveland. Pop. 56,438.
**Cleves** (klēvz). City of W West Germany near the Rhine & the Dutch border. Pop. 44,043.
**Cli·chy** (klĭ-shē′). City of N France near Paris. Pop. 47,731.
**Cliff·side Park** (klĭf′sīd′). Borough of NE N.J. opposite New York City. Pop. 21,464.
**Clif·ton** (klĭf′tən). City of NE N.J. near Paterson. Pop. 74,388.
**Clinch** (klĭnch). River of SW Va. & E Tenn. flowing c. 300 mi (483 km) SW to the Tennessee R.
**Cling·mans Dome** (klĭng′mənz dōm′). Highest (6,642 ft/2,025.8 m) of the Great Smoky Mts., on the N.C.-Tenn. border.
**Clin·ton** (klĭn′tən). **1.** Town of S Conn. on Long Is. Sound E of New Haven. Pop. 11,195. **2.** City of E central Iowa NE of Davenport. Pop. 32,828. **3.** Town of E central Mass. NNE of Worcester. Pop. 12,771. **4.** City of W central Miss. WNW of Jackson. Pop. 14,660.
**Clip·per·ton** (klĭp′ər-tən). French island, c. 2 sq mi (5.2 sq km), in the E Pacific SW of Mexico.
**Clo·quet** (klō-kā′). City of NE Minn. W of Duluth. Pop. 11,142.
**Cloud Peak** (kloud). Highest (13,175 ft/4,018.4 m) of the Bighorn Mts. in N Wyo.
**Clo·vis** (klō′vĭs). **1.** City of S central Calif. near Fresno. Pop. 33,021. **2.** City of E N.Mex. near the Tex. border. Pop. 31,194.
**Cluj** (klo͞ozh). City of Rumania NW of Bucharest. Pop. 274,095.
**Clu·ny** (klo͞o′nē, klo͞o-nē′). Town of E central France NNW of Lyon; center of medieval Cluniac order. Pop. 4,335.
**Clyde** (klīd). River of SW Scotland flowing 106 mi (170.6 km) NW to the **Firth of Clyde,** an inlet of the Atlantic.
**Clyde·bank** (klīd′băngk′). Burgh of W central Scotland on the Clyde R. Pop. 47,538.
**Cni·dus** *also* **Cni·dos** (nī′dəs). Ancient Greek city of SW Asia Minor.
**Cnos·sos** *or* **Cnos·sus** (nŏs′əs). *vars. of* KNOSSOS.
**Coast** (cōst). **1. Mountains.** Range in W B.C., Canada, & SE Alas. extending c. 1,000 mi (1,609 km) parallel to the Pacific coast. **2. Ranges.** Series of ranges of extreme W North America along the Pacific Coast from Lower California to SE Alas.
**Coates·ville** (kōts′vĭl′). City of SE Pa. W of Philadelphia. Pop. 10,698.
**Coats Land** (kōts). Region of W Antarctica along the SE shore of the Weddell Sea.
**Co·at·za·co·al·cos** (kō-ät′sə-kō-äl′kəs). City of E Mexico on the Gulf of Campeche. Pop. 69,753.
**Cobh** (kōv). Urban district & resort of S Ireland on Cork Harbor. Pop. 6,076.
**Co·blenz** *also* **Ko·blenz** (kō′blĕnts′). City of W West Germany at the confluence of the Rhine & Moselle rivers. Pop. 118,394.
**Co·bourg** (kō′bûrg′). Town of S Ont., Canada, on Lake Ontario ENE of Toronto. Pop. 11,421.
**Co·burg** (kō′bûrg′) **1.** City of SE Australia near Melbourne. Pop. 58,379. **2.** City of E central West Germany N of Nuremberg. Pop. 46,244.
**Co·cha·bam·ba** (kō′chə-băm′bə). City of E central Bolivia NNW of Sucre. Pop. 204,684.
**Co·chin** (kō′chən). **1.** Region & former state of SW India on the Malabar Coast. **2.** City of SW India on the Malabar Coast; site of Portuguese settlement (1503). Pop. 439,066.
**Cochin Chi·na** (chī′nə). Region of S Indochina comprising the S part of present-day Vietnam.
**Co·chi·nos Bay** (kə-chē′nəs). Bay of PIGS.
**Co·co** (kō′kō). River rising in NW Nicaragua & flowing c. 450 mi (724 km) NE on the Nicaragua-Honduras border to the Caribbean.
**Co·coa** (kō′kō). City of E central Fla. ESE of Orlando. Pop. 16,096.
**Cocoa Beach.** City of E central Fla. on a barrier beach ESE of Cocoa. Pop. 10,926.
**Co·cos** (kō′kəs). Island group in the E Indian Ocean SW of Sumatra, administered by Australia.
**Cod** (kŏd), **Cape.** Peninsula of SE Mass. extending c. 65 mi (105 km) E & N into the Atlantic.
**Coeur d'A·lene** (kôr′ də-lān′). City of N Idaho in the Panhandle E of Spokane, Wash., on **Coeur d'Alene Lake** (c. 60 sq mi/155 sq km). Pop. 20,054.
**Cof·fey·ville** (kô′fē-vĭl′). City of SE Kans. near the Okla. border. Pop. 15,185.
**Co·glians** (kōl-yäns′), **Monte.** KELLERWAND.
**Co·hoes** (kə-hōz′). City of E N.Y. on the Hudson R. near Albany. Pop. 18,144.
**Coim·ba·tore** (koim′bə-tōr′, -tôr′). City of S India SSW of Bangalore. Pop. 356,368.
**Co·im·bra** (kō-ĭm′brə, kwĭm′-). City of central Portugal S of Oporto. Pop. 55,985.
**Col·ches·ter** (kōl′chĕs′tər, -chĭ-stər). **1.** Borough of SE England near the North Sea. Pop. 79,600. **2.** Town of NW Vt. near Burlington. Pop. 12,629.
**Col·chis** (kŏl′kĭs). Ancient region on the Black Sea S of the Caucasus Mts.
**Cold Harbor** (kōld). Locality in E Va. ENE of Richmond, where Confederates defeated Union forces in two battles (1862, 1864).
**Cold·wa·ter** (kōld′wô′tər, -wŏt′ər). River, 220 mi (354 km), of NW Miss.
**Co·li·ma** (kə-lē′mə). City of SW Mexico W of Mexico City. Pop. 58,450.
**Col·lege Park** (kŏl′ĭj). **1.** City of NW Ga. near Atlanta. Pop. 24,632. **2.** City of central Md. near Washington, D.C. Pop. 23,614.
**College Sta·tion** (stā′shən). City of E central Tex. NW of Houston. Pop. 37,272.
**Col·lings·wood** (kŏl′ĭngz-wo͝od′). Borough of SW N.J. near Camden. Pop. 15,838.
**Col·ling·wood** (kŏl′ĭng-wo͝od′). Town of S Ont., Canada, at the S end of Georgian Bay. Pop. 11,114.
**Col·lins·ville** (kŏl′ĭnz-vĭl′). City of SW Ill. near East St. Louis. Pop. 19,613.
**Col·mar** (kōl′mär, kōl-mär′). City of E France between the Vosges Mts. & the Rhine. Pop. 58,585.

**Co·logne** (kə-lōn′). City of W West Germany on the Rhine. Pop. 1,013,771.

**Co·lombes** (kə-lôm′, -lôɴb′). City of N France near Paris. Pop. 83,241.

**Co·lom·bi·a** (kə-lŭm′bē-ə, -lôm′-). Country of NW South America with coastlines on the Pacific & the Caribbean. Cap. Bogotá. Pop. 27,520,000. **—Co·lom′bian** *adj. & n.*

**Co·lom·bo** (kə-lŭm′bō). Cap. of Sri Lanka on the W coast. Metro. area pop. 852,098.

**Co·lón** (kə-lōn′). City of N Panama at the Caribbean entrance to the Panama Canal. Pop. 59,832.

**Co·lo·ni·al Heights** (kə-lō′nē-əl). City of SE Va. S of Richmond. Pop. 16,509.

**Col·o·ra·do** (kŏl′ə-răd′ō, -rä′dō). **1.** River of central Argentina flowing 530 mi (852.8 km) SE to the Atlantic. **2.** River of the SW U.S. flowing 1,450 mi (2,333 km) SW through the **Colorado Plateau** of W Colo., SE Utah, & W Ariz. to the Gulf of California in NW Mexico. **3.** River of Tex. flowing 894 mi (1,438.5 km) SE to the Gulf of Mexico. **4.** Desert in SE Calif. & NW Mexico. **5.** State of the W central U.S. Cap. Denver. Pop. 2,888,834. **—Col′o·ra′dan** *adj. & n.*

**Colorado Springs.** City of central Colo. SSE of Denver. Pop. 215,150.

**Co·los·sae** (kə-lŏs′ē). Ancient city of central Asia Minor. **—Co·los′sian** *adj. & n.*

**Col·ton** (kōl′tən). City of S Calif. near San Bernardino. Pop. 27,419.

**Co·lum·bi·a** (kə-lŭm′bē-ə). **1.** River of SE B.C., Canada, & NW U.S. flowing 1,210 mi (1,945 km) S then W along the Wash.-Ore. border to the Pacific. **2. District of.** —See DISTRICT OF COLUMBIA. **3.** Cape on the N coast of Ellesmere Island; northernmost point of Canada. **4.** City of central Mo. NW of Jefferson City. Pop. 62,061. **5.** Borough of SE Pa. near Lancaster. Pop. 10,466. **6.** Cap. of S.C. in the central part. Pop. 99,296. **7.** City of central Tenn. SSW of Nashville. Pop. 25,767.

**Columbia Heights.** City of E Minn. near Minneapolis. Pop. 20,029.

**Co·lum·bus** (kə-lŭm′bəs). **1.** City of W Ga. on the Ala. border SSW of Atlanta. Pop. 169,441. **2.** City of S central Ind. SSE of Indianapolis. Pop. 30,292. **3.** City of NE Miss. near the Ala. border. Pop. 27,383. **4.** City of E central Nebr. W of Omaha. Pop. 17,328. **5.** Cap. of Ohio in the central part. Pop. 564,871.

**Col·ville** (kōl′vĭl′, kŏl′-). River, c. 320 mi (515 km), of N Alas.

**Col·wyn Bay** (kŏl′wĭn). Borough of N Wales on the Irish Sea. Pop. 25,370.

**Com·ba·hee** (kŭm-bē′, kŭm′bē). River, c. 140 mi (225 km), of S S.C.

**Com·mand·er Islands** (kə-măn′dər). KOMANDORSKIE.

**Com·merce** (kŏm′ərs). City of S Calif. near Los Angeles. Pop. 10,509.

**Commerce City.** City of N central Colo. near Denver. Pop. 16,234.

**Com·mon·wealth of Nations** (kŏm′ən-wĕlth′). Association consisting of the United Kingdom, its dependencies, & many former British colonies.

**Com·mu·nism** (kŏm′yə-nĭz′əm), **Mount.** Highest mountain (24,590 ft/7,500 m) in the USSR, in the Pamirs near the Chinese border.

**Co·mo** (kō′mō). Resort city of N Italy near the Swiss border at the SW end of **Lake Como** (c. 56 sq mi/145 sq km). Pop. 73,257.

**Com·o·rin** (kŏm′ər-ĭn). Cape at the southernmost point of India.

**Com·o·ro** (kŏm′ə-rō′) *or* **Com·o·ros** (-rōz′). Island group off SE Africa between Mozambique & Madagascar; an independent nation excluding the French island of Mayotte. Cap. Moroni. Pop. 292,000.

**Com·piègne** (kōmp-yān′, kôɴ-pyĕn′yə). City of N France NE of Paris. Pop. 37,009.

**Comp·ton** (kŏmp′tən). City of S Calif. near Long Beach. Pop. 81,286.

**Com·stock Lode** (kŏm′stŏk′ lōd). Gold & silver vein discovered in 1859 at Virginia City, W Nev.

**Con·a·kry** (kŏn′ə-krē). Cap. of Guinea, W Africa, in the SW part on the Atlantic. Metro. area pop. 525,671.

**Con·cep·ción** (kən-sĕp′sē-ōn′, -sĕp′shən). City of W central Chile near the Pacific coast SSW of Santiago. Pop. 178,200.

**Concepción del Ur·u·guay** (dĕl ŏŏr′ə-gwī′). City of NE Argentina on the Uruguay R. N of Buenos Aires. Pop. 38,967.

**Con·chos** (kŏn′chəs). River of NW Mexico flowing c. 350 mi (563 km) NE to the Rio Grande.

**Con·cord** (kŏng′kərd). **1.** City of W central Calif. NE of Oakland. Pop. 103,251. **2.** Town of E Mass. WNW of Boston. Pop.16,293. **3.** Cap. of N.H. in the S central part. Pop. 30,400. **4.** (kŏn′kôrd′). City of S central N.C. NE of Charlotte. Pop. 16,942.

**Con·cor·dia** (kən-kôr′dē-ə). City of NE Argentina on the Uruguay R. N of Buenos Aires. Pop. 72,136.

**Con·ey Island** (kō′nē). Resort district of Brooklyn, New York City, on the Atlantic.

**Con·go** (kŏng′gō). **1.** River of central Africa flowing c. 2,900 mi (4,666 km) N, W, & SW through the region of **the Congo** in Zambia & Zaire to the Atlantic. **2.** Republic of W central Africa W of Zaire. Cap. Brazzaville. Pop. 1,537,000. **3. Democratic Republic of the.** ZAIRE. **—Con′go·lese′** (-lēz′, -lēs′) *adj. & n.*

**Con·ne·aut** (kŏn′ē-ôt′). City of extreme NE Ohio on Lake Erie near the Pa. border. Pop. 13,835.

**Con·nect·i·cut** (kə-nĕt′ĭ-kət). **1.** River of NE U.S. flowing 407 mi (654.9 km) from N N.H. S along the Vt.-N.H. border & through Mass. & Conn. to Long Is. Sound. **2.** State of the NE U.S. Cap. Hartford. Pop. 3,107,576.

**Con·nells·ville** (kŏn′əlz-vĭl′). City of SW Pa. SE of Pittsburgh. Pop. 10,319.

**Con·ne·ma·ra** (kŏn′ə-mär′ə). Region of W Ireland on the Atlantic coast.

**Con·ners·ville** (kŏn′ərz-vĭl′). City of E central Ind. E of Indianapolis. Pop. 17,023.

**Con·roe** (kŏn′rō). City of SE Tex. NNW of Houston. Pop. 18,034.

**Con·stance** (kŏn′stəns, -stənts). City of SW West Germany on the **Lake of Constance** (207 sq mi/ 536.1 sq km), on the borders between SW West Germany, N Switzerland, & W Austria. Pop. 70,152.

**Con·stan·ţa** (kən-stän′sə). City of SE Rumania, on the Black Sea. Pop. 279,308.

**Con·stan·tine** (kŏn′stən-tēn′). City of NE Algeria E of Algiers. Pop. 335,100.

**Con·stan·ti·no·ple** (kŏn′stăn-tə-nō′pəl). ISTANBUL.

**Con·way** (kŏn′wā′). **1.** City of central Ark. NNW of Little Rock. Pop. 20,275. **2.** City of E S.C. NE of Charleston. Pop.10,240.

**Cook** (kŏŏk). **1. Inlet.** Inlet of the Pacific in S Alas. W of the Kenai Peninsula. **2. Strait.** Channel separating North & South Is., New Zealand. **3. Mount.** Highest mountain, 12,349 ft (3,766.4 m), of New Zealand, on South Is. in the Southern Alps. **4. Mount.** One of the St. Elias Mts., 13,760 ft (4,196.8 m), on the Alas.-Y.T., Canada, border. **5.** Island group of the S Pacific NE of New Zealand, which holds sovereignty.

**Cooke·ville** (kŏŏk′vĭl′). City of central Tenn. E of Nashville. Pop. 20,350.

**Coo·mas·sie** (kōō-mä′sē, -măs′ē). KUMASI.

**Coon Rapids** (kōōn). City of E Minn. near Minneapolis. Pop. 35,826.

**Coo·per City** (kōō′pər). City of SE Fla. SW of Fort Lauderdale. Pop. 10,140.

**Coo·sa** (kōō′sə). River, 286 mi (460.2 km), of NW Ga. & N Ala.

**Coos Bay** (kōōs). City of SW Ore. SW of Eugene on **Coos Bay,** an inlet of the Pacific. Pop. 14,424.

**Co·pán** (kō-pän′). Ruined Mayan city in W Honduras.

**Co·pen·ha·gen** (kō′pən-hā′gən, -hä′-). Cap. of Denmark in the E on the E coast of Sjaelland. Metro. area pop. 1,327,940.

**Co·pia·pó** (kō′pē-ə-pō′). **1.** Volcano, 19,947 ft (6,083.8 m), in the Andes of N central Chile. **2.** City of N central Chile W of the volcano. Pop. 45,200.

**Cop·per** (kŏp′ər). River, c. 300 mi (483 km), of SE Alas.

**Cop·per·as Cove** (kŏp′ər-əs, -rəs). City of central Tex. SW of Waco. Pop. 19,469.

**Cop·per·mine** (kŏp′ər-mīn′). River of N N.W.T., Canada, flowing c. 525 mi (845 km) N to the Arctic Ocean.

**Co·quil·hat·ville** (kō′kē-ăt′vĭl′, kō-kē′ə-vĭl′). MBANDAKA.

**Co·quim·bo** (kō-kĭm′bō, -kēm′-). City of N central Chile on the Pacific. Pop. 52,700.

**Cor·al** (kôr′əl, kŏr′-). Sea of the SW Pacific between NE Australia, SE New Guinea, & New Hebrides.

**Coral Ga·bles** (gā′bəlz). City of SE Fla. near Miami. Pop. 43,241.

**Coral Springs.** City of SE Fla. near Fort Lauderdale. Pop. 37,349.

**Co·ran·tijn** (kōr′ən-tīn′, kôr′-). COURANTYNE.

**Cor·co·va·do** (kôr′kə-vä′dō). Mountain, 2,310 ft (704.6 m), of SE Brazil overlooking Rio de Janeiro.

**Cor·cy·ra** (kôr-sī′rə). CORFU Is.

**Cor·dele** (kôr-dĕl′, kôr′dĕl′). City of S central Ga. S of Macon. Pop. 10,914.

**Cor·di·lle·ra Cen·tral** (kôr′dl-yâr′ə sĕn-träl′). **1.** Central of three ranges of the Andes in W Colombia. **2.** Range of central Dominican Republic. **3.** Range of the Andes in N central Peru. **4.** Range of N Luzon, Philippines. **5.** Range of S central Puerto Rico.

**Cordillera Mé·ri·da** (mĕr′ĭ-də). Range of W Venezuela.

**Cordillera Oc·ci·den·tal** (ôk′sə-dĕn-täl′). **1.** Range of the W Andes in W Colombia. **2.** Range of the W Andes in Peru along the Pacific coast.

**Cordillera O·rien·tal** (ôr′ē-ĕn-täl′, ōr′-). **1.** Range of the E Andes in central Bolivia. **2.** Range of the E Andes in W Colombia. **3.** Range of the E Andes in SE Peru.

**Cordillera Re·al** (rā-äl′). **1.** Range of the Andes in W Bolivia. **2.** Range of the Andes in Ecuador.

**Cor·dil·le·ras** (kôr′dl-yâr′əz). Entire complex of ranges of W North, Central, & South America from Alas. to Cape Horn.

**Cór·do·ba** (kôr′də-bə, -və). **1.** City of N central Argentina NW of Buenos Aires. Pop. 790,508. **2.** City of S Spain on the Guadalquivir R. Pop. 216,049. **—Cor′do·van** *adj. & n.*

**Cor·en·tyne** (kōr′ən-tīn′, kôr′-). *var. of* COURANTYNE.

**Cor·fu** (kôr-fōō′, kôr′fōō, -fyōō). **1.** One of the Ionian Is. of Greece, 227 sq mi (587.9 sq km), off the NW mainland coast. **2.** City of NW Greece on the E coast of Corfu. Pop. 28,630.

**Cor·inth** (kôr′ĭnth, kŏr′-). **1. Gulf of.** Inlet of the Ionian Sea between the Peloponnesus & central Greece. **2.** Isthmus connecting the Peloponnesus to the rest of Greece, crossed by the **Corinth Canal** (4 mi/6.4 km). **3.** Region of ancient Greece including the Isthmus of Corinth & adjacent NE Peloponnesus. **4.** City of Greece in the NE Peloponnesus on the Gulf of Corinth near the site of the ancient city of **Corinth.** Pop. 20,773. **5.** City of NE Miss. near the Tenn. border. Pop. 13,839. **—Co·rin′thi·an** (kə-rĭn′thē-ən) *adj. & n.*

**Cork** (kôrk). City of S Ireland near the head of **Cork Harbor,** an inlet of the Atlantic. Pop. 128,645.

**Corn Belt** (kôrn bĕlt). Agricultural region of the central U.S. centered in Iowa & Ill.

**Cor·ner Brook** (kôr′nər). City of W central Newf. Canada, WNW of St. John's. Pop. 25,198.

**Corn·ing** (kôr′nĭng). City of S N.Y. near the Pa. border WNW of Elmira. Pop. 12,953.

**Cor·no** (kôr′nō), **Mount** *or* **Monte.** Highest peak (9,560 ft/2,915.8 m) of the Apennines in central Italy.

**Corn·wall** (kôrn′wôl′). **1.** Region of extreme SW England. **2.** City of SE Ont., Canada, on the St. Lawrence & the N.Y. border SW of Ottawa. Pop. 46,121.

**Corn·wal·lis** (kôrn-wŏl′ĭs). Island of N.W.T., Canada, NW of Baffin Is.

**Co·ro** (kôr′ō, kōr′ō). City of NW Venezuela near the Caribbean ENE of Maracaibo. Pop. 68,701.

**Cor·o·man·del Coast** (kôr′ə-măn′dl). Region of SE India.

**Co·ro·na** (kə-rō′nə). City of S Calif. near Riverside. Pop. 37,791.

**Cor·o·na·do** (kôr′ə-nä′dō, kŏr′-). City of S Calif. near San Diego. Pop. 16,859.

**Cor·pus Chris·ti** (kôr′pəs krĭs′tē). City of S Tex. E of San Antonio on **Corpus Christi Bay,** an arm of the Gulf of Mexico. Pop. 231,999.

**Cor·reg·i·dor** (kə-rĕg′ĭ-dôr′, -dōr′). Island, c. 2 sq mi (5 sq km), of N Philippines at the entrance to Manila Bay.

**Cor·ri·en·tes** (kôr′ē-ĕn′tās, kŏr′-). City of NE Argentina across the Paraná from Paraguay. Pop. 136,924.

**Corse** (kôrs). CORSICA.

**Cor·si·ca** (kôr′sĭ-kə). Island of France in the Mediterranean N of Sardinia. **—Cor′si·can** *adj. & n.*

**Cor·si·ca·na** (kôr′sĭ-kăn′ə). City of NE Tex. SSE of Dallas. Pop. 21,712.

**Cort·land** (kôrt′lənd). City of central N.Y. S of Syracuse. Pop. 20,138.

**Co·rum·bá** (kôr′əm-bä′, kōr′-). City of SW Brazil on the Paraguay R. opposite Bolivia. Pop. 48,607.

**Cor·val·lis** (kôr-văl′ĭs). City of W Ore. SSW of Salem. Pop. 40,960.

**Cos** (kŏs, kôs). KOS.

**Co·sen·za** (kō-zĕn′sə, -zĕnt′-). City of S Italy NNE of Reggio di Calabria. Pop. 94,565.

**Co·shoc·ton** (kə-shŏk′tən). City of central Ohio ENE of Columbus. Pop. 13,405.

**Cos·ta Bra·va** (kŏs′tə brä′və, kô′stə, kō′-). NE coast of Spain from Barcelona to the French border.

**Costa Me·sa** (mā′sə). City of S Calif. near Santa Ana. Pop. 82,291.

**Costa Ri·ca** (rē′kə). Country of Central America between Panama & Nicaragua. Cap. San José. Pop. 2,245,000. **—Cos′ta Ri′can** (rē′kən) *adj. & n.*

**Côte d'A·zur** (kōt′ də-zōōr′). Mediterranean coast of SE France.

**Co·ten·tin** (kō-täN-täN′). Peninsula of NW France extending into the English Channel E of the Channel Is.

**Côte-Saint-Luc** (kōt-sănt-lōōk′, -sənt-). City of S Que., Canada, near Montreal. Pop. 25,721.

**Co·to·nou** (kōt′n-ōō′). City of S Benin in W Africa on the Gulf of Guinea. Pop. 178,000.

**Co·to·pax·i** (kō′tə-păk′sē, -pä′hē). Active volcano, 19,347 ft (5,900.8 m), in the Andes of central Ecuador.

**Cots·wold** (kŏts′wōld′, -wəld). Hills of SW England extending c. 50 mi (80 km) NE from Bristol & rising to c. 1,080 ft (330 m).

**Cot·tage Grove** (kŏt′ĭj). City of E Minn. near St. Paul. Pop. 18,994.

**Cott·bus** (kŏt′bəs, -bōōs′). City of SE East Germany near the Polish border. Pop. 94,293.

**Cot·ti·an Alps** (kŏt′ē-ən ălps′). Alpine range between NW Italy & SE France rising to 12,602 ft (3,843.6 m).

**Cot·ton·wood** (kŏt′n-wŏŏd′). River, c. 140 mi (225 km), of SW Minn.

**Coun·cil Bluffs** (koun′səl blŭfs′). City of SW Iowa on the Missouri opposite Omaha, Nebr. Pop. 56,449.

**Coun·try Club Hills** (kŭn′trē klŭb′). City of NE Ill. S of Chicago. Pop. 14,676.

**Cour·an·tyne** *also* **Cor·en·tyne** (kōr′ən-tīn′, kôr′-). River, c. 450 mi (725 km), rising in SE Guyana & forming the Guyana-Surinam border in its lower course.

**Cour·be·voie** (kōōr-bə-vwä′). City of N France on the Seine near Paris. Pop. 54,391.

**Cour·trai** (kōōr-trā′). KORTRIJK.

**Cov·en·try** (kŭv′ĭn-trē). **1.** City of central England ESE of Birmingham. Pop. 336,800. **2.** Town of W central R.I. SW of Providence. Pop. 27,065.

**Co·vi·lhã** (kōō-vĭl-yäN′). Town of E central Portugal NE of Lisbon. Pop. 26,530.

**Co·vi·na** (kō-vē′nə). City of S Calif. E of Los Angeles. Pop. 33,751.

**Cov·ing·ton** (kŭv′ĭng-tən). **1.** City of N central Ga. ESE of Atlanta. Pop. 10,586. **2.** City of extreme N Ky. on the Ohio R. opposite Cincinnati. Pop. 49,013. **3.** Independent city of W Va. N of Roanoke. Pop. 9,063.

**Cow·ans·ville** (kou′ənz-vĭl′). Town of S Que., Canada, SE of Montreal. Pop. 11,902.

**Cowes** (kouz). Town of N coast of the Isle of Wight, S England. Pop. 19,190.

**Cow·litz** (kou′lĭts). River, c. 130 mi (209 km), of SW Wash.

**Crac·ow** (krä′kou, krăk′ou, krā′kou, -kō). KRAKÓW.

**Cra·io·va** (krə-yō′və). City of S Rumania W of Bucharest. Pop. 220,893.

**Cran·brook** (krăn′brŏŏk′). City of SE B.C., Canada, near the Alta. & Idaho borders. Pop. 13,510.

**Cran·ston** (krăn′stən). City of E central R.I. near Providence. Pop. 71,992.

**Cra·ter** (krā′tər). Lake, c. 20 sq mi (52 sq km) & 1,932 ft (589.3 m) deep, of SW Ore. in a volcanic crater in **Crater Lake National Park.**

**Craw·fords·ville** (krô′fərdz-vĭl′). City of W central Ind. S of Lafayette. Pop. 13,325.

**Cré·cy** (krā-sē′, krĕs′ē) *or* **Cré·cy-en-Pon·thieu** (-äN-pōN-tyĕr′). Town of N France NW of Amiens; site of English victory (1346) in the Hundred Years' War.

**Cre·mo·na** (krə-mō′nə). City of N Italy on the Po SE of Milan. Pop. 75,988.

**Crest·wood** (krĕst′wŏŏd′). **1.** Village of NE Ill. near Chicago. Pop. 10,712. **2.** City of E Mo. near St. Louis. Pop. 12,815.

**Crete** (krēt). **1. Sea of.** Part of the S Aegean Sea between Crete & the Cyclades. **2.** Island of SE Greece in the E Mediterranean. **—Cre′tan** *adj. & n.*

**Creve Coeur** (krēv′ kōōr′). City of E Mo. NW of St. Louis. Pop. 12,694.

**Crewe and Nant·wich** (krōō; nănt′wĭch′). Borough of W central England SE of Liverpool. Pop. 98,100.

**Cri·me·a** (krī-mē′ə, krĭ-). Peninsula of S European USSR on the Black Sea & Sea of Azov. **—Cri·me′an** *adj.*

**Cro·a·tia** (krō-ā′shə, -shē-ə). **1.** Region & former kingdom of S Europe along the NE Adriatic coast. **2.** Region of NW Yugoslavia. **—Cro·a′tian** *adj. & n.*

**Croc·o·dile** (krŏk′ə-dīl′). LIMPOPO.

**Crom·well** (krŏm′wĕl′, -wəl). Town of central Conn. S of Hartford. Pop. 10,265.

**Cros·by** (krôz′bē) *or* **Great Cros·by** (grāt). Borough of NW England on Liverpool Bay near Liverpool. Pop. 56,750.

**Cross** (krôs, krŏs). River, c. 300 mi (483 km), of W Cameroon & SE Nigeria.

**Cro·to·ne** (krə-tō′nē). City of S Italy on the Ionian Sea NE of Reggio di Calabria; site of ancient city of **Cro·to·na** (krə-tō′nə). Pop. 44,081.

**Crow·ley** (krou′lē). City of S La. SSW of Baton Rouge. Pop. 16,306.

**Crown Point** (kroun). City of NW Ind. S of Gary. Pop. 16,455.

**Crys·tal** (krĭs′təl). City of E Minn. near Minneapolis. Pop. 25,543.

**Crystal Lake.** City of NE Ill. N of Elgin. Pop. 18,590.

**Ctes·i·phon** (tĕs′ə-fŏn′, tē′sə-). Ancient city of central Iraq on the Tigris SE of Baghdad.

**Cuan·za** (kwăn′zə). River, c. 600 mi (965 km), of W Angola.

**Cu·ba** (kyōō′bə). Island republic in the Caribbean S of Fla. Cap. Havana. Pop. 9,706,369. **—Cu′ban** *adj. & n.*

**Cú·cu·ta** (kōō′kə-tə). City of NE Colombia near the Venezuelan border. Pop. 219,772.

**Cud·a·hy** (kŭd′ə-hē). **1.** City of S Calif. SE of Los Angeles. Pop. 17,984. **2.** City of SE Wis. near Milwaukee. Pop. 19,547.

**Cuen·ca** (kwĕng′kə). City of S central Ecuador SE of Guayaquil. Pop. 104,470.

**Cuer·na·va·ca** (kwĕr-nə-vä′kə). City of S central Mexico in the **Cuernavaca Valley** near Mexico City. Pop. 239,813.

**Cu·ia·bá** (kōō′yə-bä′). City of W central Brazil W of Brasília. Pop. 83,621.

**Cui·to** (kwē′tō). River, c. 400 mi (664 km), of SE Angola.

**Cu·lia·cán** (kōōl′yə-kän′). City of W Mexico on the **Culiacán R.** (c. 175 mi/282 km) WNW of Durango. Pop. 228,001.

**Cull·man** (kŭl′mən). City of N Ala. N of Birmingham. Pop. 13,084.
**Cul·lo·den** (kə-lŏd′n, -lôd′n). Moor in N Scotland E of Inverness; site of defeat of Highland Jacobites by English forces (1746).
**Cul·ver City** (kŭl′vər). City of S Calif. near Los Angeles. Pop. 38,139.
**Cu·mae** (kyōō′mē). Ancient city & Greek colony of S central Italy near present-day Naples.
**Cu·ma·ná** (kōō′mə-nä′). City of NE Venezuela on the Caribbean E of Caracas. Pop. 119,751.
**Cum·ber·land** (kŭm′bər-lənd). **1.** River of S Ky. & N Tenn. flowing c. 690 mi (1,110 km) W to the Ohio. **2.** Falls, 92 ft (28.1 m) high, on the upper Cumberland in SE Ky. **3. Plateau** *or* **Mountains.** SW section of the Appalachians extending along the Va.-Ky. border & into central Tenn. **4. Gap.** Pass, 1,304 ft (397.7 m), through the Cumberland Mts. near the junction of Ky., Va., & Tenn. borders. **5.** Town of SE Ont., Canada, near Ottawa. Pop. 13,541. **6.** City of NW Md. on the W.Va. border. Pop. 25,933. **7.** Town of NE R.I. near Providence. Pop. 27,069.
**Cum·bri·an** (kŭm′brē-ən). Mountains of NW England rising to a height of 3,210 ft (979 m).
**Cu·nax·a** (kyōō-năk′sə). Ancient town of Babylonia NW of Babylon.
**Cu·ne·ne** (kōō-nā′nə). River, c. 750 mi (1,205 km), of SW Angola forming the Angola-Namibia border in its lower course.
**Cu·ne·o** (kōō′nē-ō′). City of NW Italy W of Genoa. Pop. 41,633.
**Cu·per·ti·no** (kōō′pər-tē′nō, kyōō′-). City of W Calif. W of San Jose. Pop. 25,770.
**Cu·ra·çao** (kŏŏr′ə-sō′, -sou′, kyŏŏr′-). Island of the Netherland Antilles in the S Caribbean off the NW coast of Venezuela.
**Cu·ri·có** (kŏŏr′ĭ-kō′). City of central Chile SSW of Santiago. Pop. 41,300.
**Cu·ri·ti·ba** (kŏŏr′ĭ-tē′bə). City of SE Brazil SW of São Paulo. Pop. 1,025,979.
**Cus·co** (kōō′skō). *var. of* CUZCO.
**Cush** (kŭsh, kōōsh). Legendary ancient region of NE Africa often identified with Ethiopia.
**Cutch** (kŭch). KUTCH.
**Cut·tack** (kŭt′ək). City of E India SW of Calcutta. Pop. 194,068.
**Cux·ha·ven** (kŏŏks-hä′fən). City of N West Germany at the mouth of the Elbe R. Pop. 60,353.
**Cuy·a·ho·ga Falls** (kī′ə-hō′gə, kə-hō′-, -hô′-, -hä′-). City of NE Ohio near Akron. Pop. 43,710.
**Cu·yu·ni** (kə-yōō′nē). River, c. 350 mi (563 km), of E Venezuela & NW Guyana.
**Cuz·co** *or* **Cus·co** (kōō′skō). City of S Peru in the Andes ESE of Lima; ancient Inca capital. Pop. 120,881.
**Cyc·la·des** (sĭk′lə-dēz′). Group of islands of SE Greece in the S Aegean.
**Cy·press** (sī′prĭs). City of S Calif. near Long Beach. Pop. 40,391.
**Cy·prus** (sī′prəs). Island republic in the E Mediterranean S of Turkey. Cap. Nicosia. Pop. 629,000. **—Cyp′ri·ot** (sĭp′rē-ət), **Cyp′ri·ote′** (-ōt′, -ət), **Cyp′ri·an** (-rē-ən) *adj. & n.*
**Cyr·e·na·i·ca** (sĭr′ə-nā′ĭ-kə, sī′rə-). Ancient region of NE Libya.
**Cy·re·ne** (sī-rē′nē). Ancient Greek city of Cyrenaica.
**Cy·rus** (sī′rəs). KURA.
**Cy·the·ra** (sĭ-thēr′ə). Island of S Greece in the Mediterranean S of the Peloponnesus.
**Czech·o·slo·va·ki·a** (chĕk′ə-slə-vä′kē-ə, -ō-slō-). Country of central Europe. Cap. Prague. Pop. 15,276,799. **—Czech′o·slo′vak, Czech′o·slo·va′ki·an** *adj. & n.*
**Czę·sto·cho·wa** (chĕn′stə-kō′və). City of S Poland N of Katowice. Pop. 187,613.

# D

**Dą·bro·wa Gór·ni·cza** (dôm-brô′və gŏŏr-nē′chə). City of S Poland near Katowice. Pop. 61,660.
**Dac·ca** (dăk′ə, dä′kə). Cap. of Bangladesh in the E central part. Pop. 1,679,572.
**Da·chau** (dä′kou′). City of SE West Germany near Munich; site of Nazi concentration camp. Pop. 33,207.
**Da·ci·a** (dā′shē-ə, -shə). Ancient region & Roman province corresponding roughly to modern Rumania. **—Da′ci·an** *adj. & n.*
**Da·ho·mey** (də-hō′mē). BENIN.
**Dai·ren** (dī′rĕn′). LÜDA.
**Da·kar** (də-kär′, dăk′är′). Cap. of Senegal in the W part on Cape Verde. Pop. 798,792.
**Da·ko·ta** (də-kō′tə). Former U.S. territory divided in 1889 into **the Da·ko·tas** (-təz), North & South Dakota.
**Da Lat** (dä′ lät′). City of SE Vietnam NE of Ho Chi Minh City. Pop. 105,072.

**Dal·las** (dăl′əs). City of NE Tex. E of Fort Worth. Pop. 904,078.
**Dal·ton** (dôl′tən). City of NW Ga. SE of Chattanooga, Tenn. Pop. 20,743.
**Da·ly City** (dā′lē). City of W Calif. near San Francisco. Pop. 78,519.
**Da·man·hûr** (dăm′ən-hŏŏr′). City of NE Egypt on the Nile delta NW of Cairo. Pop. 188,927.
**Da·mas·cus** (də-măs′kəs). Cap. of Syria in the SW part. Pop. 836,668. **—Dam′a·scene′** (dăm′ə-sēn′) *adj. & n.*
**Dam·a·vand** (dăm′ə-vănd′). *var. of* DEMAVEND.
**Dam·i·et·ta** (dăm′ē-ĕt′ə). City of NE Egypt on the Nile delta NNE of Cairo. Pop. 93,546.
**Da·mo·dar** (dä′mə-där′). River, c. 370 mi (595 km), of NE India.
**Dan** (dăn). River, c. 180 mi (290 km), of S Va. & N N.C.
**Da Nang** (dä′ näng′). City of central Vietnam on the South China Sea. Pop. 492,194.
**Dan·bur·y** (dăn′bĕr′ē, -bə-rē). City of SW Conn. NW of Bridgeport. Pop. 60,470.
**Dan·dong** (dän′dŏŏng′). City of NE China on the Yalu R. opposite North Korea. Pop. 450,000.
**Da·ni·a** (dā′nē-ə). City of SE Fla. near Fort Lauderdale. Pop. 11,811.
**Dan·ube** (dăn′yōōb). River of S central Europe rising in SW West Germany & flowing c. 1,750 mi (2,816 km) SE through Austria, Hungary, Yugoslavia, & Rumania to the Black Sea. **—Dan·u′bi·an** *adj.*
**Dan·vers** (dăn′vərz). Town of NE Mass. near Salem. Pop. 24,100.
**Dan·ville** (dăn′vĭl′). **1.** City of E Ill. ENE of Decatur. Pop. 38,985. **2.** City of central Ky. SSW of Lexington. Pop. 12,942. **3.** Independent city of S Va. near the N.C. border. Pop. 45,642.
**Dan·zig** (dăn′sĭg, dän′-). **1. Gulf of.** Inlet of the S Baltic Sea between N Poland & W European USSR. **2. Free City.** Former state (1919–39) on the Gulf of Danzig surrounding & including Gdańsk. **3.** GDAŃSK.
**Dar·by** (där′bē). Borough of SE Pa. near Philadelphia. Pop. 11,513.
**Dar·da·nelles** (där′dn-ĕlz′). Strait, c. 40 mi (65 km) long & 1–4 mi (1.6–6.4 km) wide, connecting the Aegean with the Sea of Marmara.
**Dar es Sa·laam** (där′ ĕs sə-läm′). Cap. of Tanzania in the E part on the Indian Ocean. Pop. 757,346.
**Dar·i·en** (dâr′ē-ĕn′, dĕr′-, dăr′ē-ən, dĕr′-). **1.** Town of SW Conn. NW of Stamford. Pop. 18,892. **2.** City of NE Ill. W of Chicago. Pop. 14,968.
**Da·ri·én** (dâr′ē-ĕn′, där-yĕn′). **1. Gulf of.** Bay of the Caribbean between NE Panama & NW Colombia. **2.** Region of E Panama.
**Dar·ling** (där′lĭng). **1.** River, c. 1,702 mi (2,739 km), of SE Australia. **2.** Range of hills in SW Australia extending along the coast N & S of Perth.
**Dar·ling·ton** (där′lĭng-tən). Borough of N England S of Newcastle. Pop. 85,120.
**Darm·stadt** (därm′stăt, -shtät′). City of central West Germany near Frankfurt. Pop. 137,018.
**Dart·mouth** (därt′məth). **1.** City of S N.S., Canada, opposite Halifax. Pop. 65,341. **2.** Town of SE Mass. near New Bedford. Pop. 23,966.
**Dar·win** (där′wĭn). City of N Australia on the Timor Sea. Pop. 39,193.
**Dasht-e-Ka·vir** (dăsht′ē-kə-vîr′). Salt desert of N central Iran.
**Da·tong** (dä′tŏŏng′). City of NE China W of Beijing. Pop. 300,000.
**Dau·gav·pils** (dou′gəf-pĭlz′). City of W European USSR SE of Riga. Pop. 116,000.
**Dau·phi·né** (dō-fē-nā′). Range of the W Alps in SE France.
**Da·vao** (dä′-vou′). City of SE Mindanao, Philippines, on **Davao Gulf,** an inlet of the Pacific. Pop. 611,311.
**Dav·en·port** (dăv′ĭn-pôrt′, -pōrt′). City of E Iowa on the Mississippi opposite Moline & Rock Island, Ill. Pop. 103,264.
**Da·vie** (dā′vē). Town of SE Fla. SW of Fort Lauderdale. Pop. 20,877.
**Da·vis** (dā′vĭs). **1.** Strait of the N Atlantic between SE Baffin Is. & SW Greenland. **2.** Mountains of W Tex. SE of El Paso, rising to 8,382 ft (2,556.5 m). **3.** City of N central Calif. W of Sacramento. Pop. 36,640.
**Daw·son Creek** (dô′sən). City of E B.C., Canada, near the Alta. border NE of Prince George. Pop. 10,528.
**Da·xue Shan** (dä′shōō′ shän′). Mountains of S central China SW of Chengdu.
**Day·ton** (dāt′n). City of SW Ohio NNE of Cincinnati. Pop. 203,588.
**Day·to·na Beach** (dā-tō′nə). City of NE Fla. NNE of Orlando. Pop. 54,176.
**Dead Sea** (dĕd). Salt lake, c. 390 sq mi (1,010 sq km), between Israel & Jordan; lowest point on earth (surface 1,292 ft/394 m below sea level).
**Dear·born** (dîr′bôrn′, -bərn). City of SE Mich. W of Detroit. Pop. 90,660.
**Dearborn Heights.** City of SE Mich. near Detroit. Pop. 67,706.
**Death Valley** (dĕth). Desert basin, c. 1,500 sq mi (3,885 sq km), of E Calif. & W Nev., containing the lowest point in the Western Hemisphere (280 ft/85.4 m below sea level) & including **Death Valley National Monument.**
**De·bre·cen** (dĕb′rĭt-sĕn′). City of E Hungary E of Budapest. Pop. 192,484.
**De·ca·tur** (dĭ-kā′tər). **1.** City of N Ala. N of Birmingham. Pop. 42,002. **2.** City of NW Ga. near Atlanta. Pop. 18,404. **3.** City of central Ill. E of Springfield. Pop. 94,081.

**Dec·can** (dĕk′ən, -ăn′). Plateau of S central India between the Eastern & Western Ghats.
**Ded·ham** (dĕd′əm). Town of E Mass. near Boston. Pop. 25,298.
**Deer·field** (dîr′fēld′). Village of NE Ill. near Chicago. Pop. 17,430.
**Deerfield Beach.** City of SE Fla. N of Fort Lauderdale. Pop. 39,193.
**Deer Park** (dîr). City of SE Tex. near Houston. Pop. 22,648.
**De·fi·ance** (dĭ-fī′əns). City of NW Ohio SW of Toledo. Pop. 16,810.
**Deh·ra Dun** (dĕr′ə do͞on′). City of N India NNE of Delhi. Pop. 166,073.
**De Kalb** (dĭ kălb′). City of N Ill. SSE of Rockford. Pop. 33,099.
**De Land** (dĭ lănd′). City of NE Fla. SW of Daytona Beach. Pop. 15,354.
**De·la·no** (də-lā′nō). City of S central Calif. NNW of Bakersfield. Pop. 16,491.
**Del·a·ware** (dĕl′ə-wâr′). **1.** River of NE U.S. rising in SE N.Y. & flowing c. 280 mi (451 km) S to **Delaware Bay,** an inlet of the Atlantic between E Del. & SW N.J. **2.** State of the NE U.S. on the Atlantic. Cap. Dover. Pop. 595,225. **3.** City of central Ohio N of Columbus. Pop. 18,780.
**Del City** (dĕl). City of central Okla. near Oklahoma City. Pop. 28,424.
**Delft** (dĕlft). City of SW central Netherlands SE of The Hague. Pop. 86,103.
**Del·ga·do** (dĕl-gä′dō). Cape on NE coast of Mozambique.
**Del·hi** (dĕl′ē). City of N central India on the Jumna R. Pop. 3,287,-883.
**Del·mar·va** (dĕl-mär′və). Peninsula, c. 180 mi (290 km) long, of the NE U.S. between Chesapeake Bay & the Atlantic.
**Del·men·horst** (dĕl′mĭn-hôrst′). City of N West Germany near Bremen. Pop. 71,488.
**De·los** (dē′lŏs′). Island of SE Greece in the central Cyclades.
**Del·phi** (dĕl′fī′). Ancient town of central Greece near Mt. Parnassus; seat of an oracle of Apollo.
**Del·ray Beach** (dĕl′rā′). City of SE Fla. N of Boca Raton. Pop. 34,325.
**Del Ri·o** (dĕl rē′ō). City of SW Tex. on the Rio Grande W of San Antonio. Pop. 30,034.
**Dem·a·vend** (dĕm′ə-vĕnd′) *or* **Dam·a·vand** (dăm′ə-vănd′). Highest peak (18,934 ft/5,774.9 m) of the Elburz Mts. in N Iran.
**Den Hel·der** (dən hĕl′dər, də). City of NW Netherlands on the North Sea. Pop. 60,421.
**Den·i·son** (dĕn′ĭ-sən). City of N Tex. near the Okla. border NNE of Dallas. Pop. 23,884.
**De·niz·li** (dĕn′ĭz-lē′). City of SW Turkey ESE of Izmir. Pop. 106,902.
**Den·mark** (dĕn′märk′). **1.** Strait, c. 130 mi (209 km) wide, between Greenland & Iceland. **2.** Country of N Europe on Jutland & adjacent islands. Cap. Copenhagen. Pop. 5,124,000. **—Dane** (dān) *n.* **—Dan′·ish** (dā′nĭsh) *adj.* & *n.*
**Den·nis** (dĕn′ĭs). Town of SE Mass. on central Cape Cod. Pop. 12,360.
**Den·ton** (dĕn′tən). City of NE Tex. NNW of Dallas. Pop. 48,063.
**Den·ver** (dĕn′vər). Cap. of Colo. in the N central part. Pop. 491,396.
**De Pere** (dĭ pîr′). City of E Wis. near Green Bay. Pop. 14,892.
**De·pew** (dĭ-pyo͞o′). Village of W N.Y. near Buffalo. Pop. 19,819.
**Der·by** (dûr′bē). **1.** (*also* där′-). City of S central England NNW of London. Pop. 213,700. **2.** City of SW Conn. W of New Haven. Pop. 12,346.
**De Rid·der** (dĭ rĭd′ər). City of S La. SW of Alexandria. Pop. 11,057.
**Der·ry** (dĕr′ē). Town of SE N.H. SE of Manchester. Pop. 18,875.
**De·se·a·do** (dĕs′ē-ä′dō). River, c. 380 mi (611 km), of S Argentina.
**Des·er·et** (dĕz′ə-rĕt′). Area of SW U.S. proposed (1849) by Mormons as a U.S. or independent state.
**Des Moines** (dĭ moin′). Cap. of Iowa in the S central part. Pop. 191,003.
**Des·na** (də-snä′). River, c. 740 mi (1,191 km), of W European USSR.
**De So·to** (dĭ sō′tō). City of NE Tex. near Dallas. Pop. 15,538.
**Des Plaines** (dĕs plānz′). **1.** River, c. 110 mi (177 km), of SW Wis. & NE Ill. **2.** City of NE Ill. near Chicago. Pop. 53,568.
**Des·sau** (dĕs′ou). City of central East Germany N of Leipzig. Pop. 100,820.
**De·troit** (dĭ-troit′). City of SE Mich. on the **Detroit R.** (32 mi/51.5 km) opposite Windsor, Ont., Canada. Pop. 1,203,339.
**Deur·ne** (dûr′nə). City of N Belgium E of Antwerp. Pop. 80,766.
**De·ven·ter** (dā′vən-tər). City of E central Netherlands on the Ijssel R. Pop. 65,557.
**Dev·il's Island** (dĕv′ĭlz). Island in the Caribbean off French Guiana; formerly a penal colony.
**Dev·on** (dĕv′ən). Island of NE N.W.T., Canada, between Baffin & Ellesmere Is.
**Dezh·nev** (dĕzh′nəf, dĕzh′nē-ôf′). Cape of extreme NE Far Eastern USSR on Bering Strait; easternmost point of Asia.
**Dhau·la·gi·ri** (dou′lə-gîr′ē). Peak, 26,810 ft (8,177 m), in the Himalayas of N central Nepal.
**Di·a·man·ti·na** (dī′ə-mən-tē′nə). River, c. 560 mi (901 km), of E central Australia.
**Dia·mond Head** (dī′ə-mənd, dī′mənd). Promontory, 761 ft (232.1 m) high, on SE coast of Oahu, Hawaii.
**Dick·in·son** (dĭk′ĭn-sən). City of SW N.Dak. W of Bismarck. Pop. 15,924.
**Dien Bien Phu** (dyĕn′ byĕn′ fo͞o′). Town of NW Vietnam near the Laos border; site of Vietminh victory over French (1954).
**Di·jon** (dē-zhôɴ′). City of E France SE of Paris. Pop. 149,889.
**Di·nar·ic Alps** (dĭ-năr′ĭk ălps′). Range of W Yugoslavia extending c. 400 mi (645 km) along the Adriatic coast.
**Di·o·mede** (dī′ə-mēd′). Islands in the Bering Strait between Alas. & Siberia, comprising **Little Diomede** (U.S.) & **Big Diomede** (USSR).
**Dis·mal** (dĭz′məl). Swamp of SE Va. & NE N.C.
**Dis·trict of Co·lum·bi·a** (dĭs′trĭkt′; kə-lŭm′bē-ə). Federal district, 69 sq mi (178.7 sq km), of E U.S. on the Potomac R. between Va. & Md.; coextensive with the city of Washington.
**Dix·on** (dĭk′sən). City of N Ill. SSW of Rockford. Pop. 15,659.
**Di·yar·ba·kir** (dĭ-yär′bə-kîr′). City of SE Turkey on the Tigris. Pop. 169,535.
**Dja·ja Peak** (jä′yə). Highest peak of Indonesia, 16,535 ft (5,043.2 m), in central West Irian.
**Dja·kar·ta** (jə-kär′tə). Cap. of Indonesia on the Java Sea coast of NW Java. Pop. 4,576,009.
**Djeb·el Toub·kal** (jĕb′əl to͞ob-käl′). JEBEL TOUBKAL.
**Dji·bou·ti** (jĭ-bo͞o′tē). **1.** Country of E Africa on the Gulf of Aden. Pop. 386,000. **2.** Cap. of Djibouti in the SE part on the Gulf of Aden. Pop. 96,000.
**Djok·ja·kar·ta** (jŏk′yə-kär′tə). City of Indonesia in S central Java. Pop. 342,267.
**Dne·pro·dzer·zhinsk** (nĕp′rō-dər-zhĭnsk′). City of S European USSR on the Dnieper SW of Kharkov. Pop. 253,000.
**Dne·pro·pe·trovsk** (nĕp′rō-pə-trôfsk′). City of S European USSR on the Dnieper SSW of Kharkov. Pop. 1,066,000.
**Dnie·per** (nē′pər). River of W European USSR rising near Smolensk & flowing c. 1,420 mi (2,285 km) S to the NW Black Sea.
**Dnies·ter** (nē′stər). River, c. 850 mi (1,368 km), of SW European USSR.
**Dobbs Fer·ry** (dŏbz′ fer′ē). Village of SE N.Y. on the Hudson near Yonkers. Pop. 10,053.
**Do·be·rai** (dō′bə-rī′). Peninsula of NW West Irian, New Guinea.
**Do·dec·a·nese** (dō-dĕk′ə-nēz′, -nēs′). Islands of SE Greece in the Aegean between Turkey & Crete.
**Dodge City** (dŏj). City of SW Kans. W of Wichita. Pop. 18,001.
**Do·do·na** (də-dō′nə). Ancient city of NW Greece.
**Do·ha** (dō′hə). Cap. of Qatar, SE Arabia, on the Persian Gulf. Pop. 150,000.
**Dol·lard-des-Or·meaux** (dō-yär′dā-zôr-mō′). Town of S Que., Canada, near Montreal. Pop. 36,837.
**Do·lo·mites** (dō′lə-mīts′, dŏl′ə-) *or* **Do·lo·mite Alps** (-mīt′ ălps′). Range of the E Alps in NE Italy.
**Dol·ton** (dōl′tən). Village of NE Ill. S of Chicago. Pop. 24,766.
**Dom·i·ni·ca** (dŏm′ə-nē′kə, də-mĭn′ĭ-kə). Island republic in the E Caribbean between Guadeloupe & Martinique. Cap. Roseau. Pop. 74,089.
**Do·min·i·can Republic** (də-mĭn′ĭ-kən). Republic of the West Indies on the E part of Hispaniola Is. Cap. Santo Domingo. Pop. 5,431,000.
**Don** (dŏn). River of S European USSR flowing 1,222 mi (1,966.2 km) into the NE Sea of Azov.
**Don·cas·ter** (dŏng′kə-stər). Borough of N central England NE of Sheffield. Pop. 81,530.
**Do·nets** (də-nĕts′). **1.** River, c. 650 mi (1,046 km), of S European USSR. **2. Basin.** Industrial region of S European USSR N of the Sea of Azov & W of the Donets R.
**Do·netsk** (də-nĕtsk′). City of S European USSR in the Donets Basin. Pop. 1,021,000.
**Door** (dôr, dōr). Peninsula of E Wis. between Lake Michigan & Green Bay.
**Dor·dogne** (dôr-dōn′, -dōn′yə). River, c. 305 mi (491 km), of SW France.
**Dor·drecht** (dôr′drĕkt′). City of SW Netherlands on the Meuse SE of Rotterdam. Pop. 101,840.
**Dor·is** (dôr′ĭs, dōr′-). Ancient region of central Greece.
**Dor·mont** (dôr′mŏnt′). Borough of SW Pa. near Pittsburgh. Pop. 11,275.
**Dor·set** (dôr′sĭt). Cape of SW Baffin Is., E N.W.T., Canada.
**Dort** (dôrt). DORDRECHT.
**Dort·mund** (dôrt′mənd, -mo͝ont′). City of W West Germany NNE of Cologne. Pop. 630,609.
**Dor·val** (dôr-văl′). Town of S Que., Canada, on S shore of Montreal Is. Pop. 19,131.
**Do·than** (dō′thən). City of SE Ala. near the Fla. border. Pop. 48,750.
**Dou·ai** (do͞o-ā′). Town of N France NE of Amiens. Pop. 43,954.

**Dou·a·la** *or* **Du·a·la** (dōō-ä′lə). City of SW Cameroon on the Bight of Biafra. Pop. 458,246.
**Dou·ay** (dōō-ā′). DOUAI.
**Doug·las** (dŭg′ləs). **1.** City of SE Ariz. on the Mexican border. Pop. 13,058. **2.** City of S central Ga. SSE of Macon. Pop. 10,980.
**Dou·ro** (dôr′ōō, dōr′-). River, c. 475 mi (764 km), of N Spain & N Portugal.
**Do·ver** (dō′vər). **1.** Strait, 21 mi (33.8 km) wide, at E end of English Channel between SE England & N France. **2.** Borough of SE England on the Strait of Dover opposite Calais, France. Pop. 34,160. **3.** Cap. of Del. in the central part. Pop. 23,512. **4.** City of SE N.H. near Portsmouth. Pop. 22,377. **5.** Town of N central N.J. near Morristown. Pop. 14,681. **6.** City of E Ohio S of Akron. Pop. 11,526.
**Dow·ners Grove** (dou′nərz). Village of NE Ill. W of Chicago. Pop. 39,274.
**Dow·ney** (dou′nē). City of S Calif. N of Long Beach. Pop. 82,602.
**Downs** (dounz). Two parallel hill ranges of SE England, **The North Downs** & **The South Downs.**
**Dra·cut** (drā′kət). Town of NE Mass. near Lowell. Pop. 21,249.
**Dra·kens·burg** (drä′kənz-bûrg′). Range of E South Africa, Lesotho, & Swaziland, rising to 11,425 ft (3,484.6 m).
**Drake Passage** (drāk). Strait, c. 500 mi (805 km) wide, between Cape Horn & Antarctica.
**Dram·men** (drä′mən). City of SE Norway WSW of Oslo. Pop. 50,777.
**Dran·cy** (dräN-sē′). City of N central France near Paris. Pop. 64,258.
**Dra·va** *or* **Dra·ve** (drä′və). River, c. 450 mi (725 km), of S Austria & N Yugoslavia.
**Dres·den** (drĕz′dən). City of SE East Germany on the Elbe ESE of Leipzig. Pop. 507,692.
**Drum·mond·ville** (drŭm′ənd-vĭl′, -ən-). City of S Que., Canada, NE of Montreal. Pop. 29,286.
**Dry Tor·tu·gas** (drī tôr-tōō′gəz). Islands of S Fla. W of Key West.
**Du·a·la** (dōō-ä′lə). *var. of* DOUALA.
**Duar·te** (dwär′tē, dōō-är′-). City of S Calif. ENE of Los Angeles. Pop. 16,766.
**Du·bai** (dōō-bī′). City & sheikdom of E United Arab Emirates, E Arabia, on the Persian Gulf. Pop. 60,000.
**Du·bawnt** (dōō-bônt′). River, 580 mi (993.2 km), of SE N.W.T., Canada, flowing through **Dubawnt Lake** (1,654 sq mi/4,284 sq km).
**Dub·lin** (dŭb′lĭn). **1.** Cap. of Ireland in the E central part on the Irish Sea. Pop. 567,866. **2.** Town of W Calif. ESE of Oakland. Pop. 13,641. **3.** City of central Ga. ESE of Macon. Pop. 16,083. **—Dub′lin·er** *n.*
**Du·brov·nik** (dōō′brôv-nĭk′). City of SW Yugoslavia on the Adriatic. Pop. 31,213.
**Du·buque** (də-byōōk′). City of E Iowa on the Mississippi opposite the Ill.-Wis. border. Pop. 62,321.
**Dud·ley** (dŭd′lē). Borough of W central England WNW of Birmingham. Pop. 187,110.
**Due·ro** (dwĕr′ō). DOURO.
**Duis·burg** (dōōs′bûrg′, dōōz′-). City of W West Germany at the confluence of the Rhine & Ruhr rivers. Pop. 591,635.
**Du·luth** (də-lōōth′). City of NE Minn. on Lake Superior opposite Superior, Wis. Pop. 92,811.
**Du·mas** (dōō′məs). City of N Tex. in the Panhandle N of Amarillo. Pop. 12,194.
**Dum·bar·ton** (dŭm-bär′tn). Burgh of W Scotland on the Clyde WNW of Glasgow. Pop. 25,469.
**Dum·fries** (dŭm-frēs′). Burgh of S Scotland SSW of Edinburgh. Pop. 29,259.
**Du·mont** (dōō′mŏnt′, dyōō′-). Borough of NE N.J. near Hackensack. Pop. 18,334.
**Dum·yat** (dōōm-yät′). DAMIETTA.
**Dun·can** (dŭng′kən). City of S Okla. SSE of Oklahoma City. Pop. 22,517.
**Dun·can·ville** (dŭng′kən-vĭl′). City of NE Tex. near Dallas. Pop. 27,781.
**Dun·das** (dŭn′dəs). Town of S Ont., Canada, near Hamilton. Pop. 19,179.
**Dun·dee** (dŭn-dē′). City of E central Scotland on the Firth of Tay. Pop. 194,732.
**Dun·e·din** (dŭn-ēd′n). **1.** City of SE South Is., New Zealand. Pop. 82,546. **2.** City of W central Fla. N of Clearwater. Pop. 30,203.
**Dun·ferm·line** (dŭn-fûrm′lĭn). Burgh of E central Scotland NW of Edinburgh. Pop. 52,098.
**Dun·kirk** (dŭn′kûrk′). **1.** *also* **Dun·kerque** (dœN-kĕrk′). City of N France on the North Sea. Pop. 78,171. **2.** City of W N.Y. on Lake Erie SW of Buffalo. Pop. 15,310.
**Dun Laoghai·re** (dŭn lâr′ə). Borough of E central Ireland on the Irish Sea SE of Dublin. Pop. 53,171.

ōō **boot** ou **out** th **thin** *th* **this** ŭ **cut** ûr **urge** y **young**
yōō **abuse** zh **vision** ə **a**bout, it**e**m, edib**i**le, gall**o**p, circ**u**s

**Dun·more** (dŭn′môr′, -mōr′). Borough of NE Pa. near Scranton. Pop. 16,781.
**Dunn·ville** (dŭn′vĭl′). Town of S Ont., Canada, near Lake Erie SSE of Hamilton. Pop. 11,642.
**Du·que de Ca·xi·as** (dōō′kē də kə-shē′əs). City of SE Brazil on Guanabara Bay NNW of Rio de Janeiro. Pop. 256,582.
**Du·quesne** (dōō-kān′). City of SW Pa. near Pittsburgh. Pop. 10,094.
**Du·ran·go** (dōō-răng′gō). **1.** City of N central Mexico NNW of Guadalajara. Pop. 182,633. **2.** City of SW Colo. near the N.Mex. border. Pop. 11,426.
**Du·rant** (dōō-rănt′). City of S Okla. near the Tex. border SE of Oklahoma City. Pop. 11,972.
**Dur·ban** (dûr′bən). City of E South Africa on the Indian Ocean. Pop. 736,852.
**Dur·ham** (dûr′əm). **1.** Borough of NE England S of Newcastle. Pop. 88,800. **2.** Town of SE N.H. NW of Portsmouth. Pop. 10,652. **3.** City of N central N.C. E of Greensboro. Pop. 100,831.
**Dur·rës** (dōōr′əs). City of W Albania on the Adriatic. Pop. 53,800.
**Du·shan·be** (dōō-shăm′bə, -shäm′-). City of SW Central Asian USSR S of Tashkent. Pop. 494,000.
**Düs·sel·dorf** (dōōs′əl-dôrf′). City of W West Germany on the Rhine NNW of Cologne. Pop. 664,336.
**Dutch East In·dies** (dŭch′ ēst ĭn′dēz). INDONESIA.
**Dutch Gui·a·na** (gē-ăn′ə, -ä′nə, gī-). SURINAM 2.
**Dutch West In·dies** (wĕst ĭn′dēz). NETHERLANDS ANTILLES.
**Dux·bur·y** (dŭks′bĕr′ē, -bə-rē). Town of E Mass. SE of Boston. Pop. 11,807.
**Dvi·na** (də-vē-nä′). **1.** *also* **Northern Dvina.** River, c. 465 mi (748 km), of N European USSR flowing into the White Sea. **2.** *also* **Western Dvina.** River, c. 635 mi (1,022 km), of W European USSR flowing to the Gulf of Riga.
**Dy·ers·burg** (dī′ərz-bûrg′). City of NW Tenn. NNE of Memphis. Pop. 15,856.

# E

**Ea·gan** (ē′gən). City of E Minn. near Minneapolis-St. Paul. Pop. 20,352.
**Ea·gle Pass** (ē′gəl). City of SW Tex. on the Rio Grande WSW of San Antonio. Pop. 21,407.
**Eas·ley** (ēz′lē). City of NW S.C. W of Greenville. Pop. 14,264.
**East** (ēst), **the.** Region of the U.S. E of the Alleghenies & N of the Mason-Dixon Line.
**East An·gli·a** (ăng′glē-ə). Anglo-Saxon kingdom of England in an area now occupied by Norfolk & Sussex.
**East A·von** (ā′vŏn). —See AVON.
**East Ber·lin** (bûr-lĭn′). —See BERLIN.
**East·bourne** (ēst′bôrn′, -bōrn′). Borough of in SE England on the English Channel. Pop. 73,200.
**East Cape.** Northeasternmost point of Asia in Far Eastern USSR on Bering Strait.
**East Chi·ca·go** (shĭ-kä′gō). City of NW Ind. on Lake Michigan near Chicago, Ill. Pop. 39,786.
**East Chi·na Sea** (chī′nə). Arm of the Pacific extending c. 600 mi (965.4 km) between E China & Ryukyu Is.
**East Cleve·land** (klēv′lənd). City of NE Ohio near Cleveland. Pop. 36,957.
**East De·troit** (dĭ-troit′). City of SE Mich. near Detroit. Pop. 38,280.
**East·er Island** (ē′stər). Chilean island in the S Pacific 2,200 mi (3,450 km) W of the mainland; site of ancient massive sculpted heads.
**East·ern Ghats** (ē′stərn gôts). Mountain range of S India.
**Eastern High·lands** (hī′ləndz). Mountain range extending along entire E coast of Australia.
**Eastern Shore.** Sectors of Md. & Va. lying E of Chesapeake Bay.
**East Ger·ma·ny** (jûr′mə-nē). —See GERMANY.
**East Grand Ra·pids** (grănd răp′ĭdz). City of SW Mich. near Grand Rapids. Pop. 10,914.
**East Green·wich** (grĕn′ĭch). Town of central R.I. S of Providence. Pop. 10,211.
**East·hamp·ton** (ēst-hămp′tən). Town of W central Mass. near Northhampton. Pop. 15,580.
**East Hart·ford** (härt′fərd). Town of N central Conn. near Hartford. Pop. 52,563.
**East Ha·ven** (hā′vən). Town of S Conn. on Long Is. Sound E of New Haven. Pop. 25,028.
**East In·dies** (ĭn′dēz). **1.** Historically, the subcontinent of INDIA. **2.** MALAY Archipelago. **3.** Sometimes, SE Asia.
**East·lake** (ēst′lāk′). City of NE Ohio on Lake Erie NE of Cleveland. Pop. 22,104.
**East Lan·sing** (lăn′sĭng). City of S central Mich. near Lansing. Pop. 48,309.
**East Liv·er·pool** (lĭv′ər-pōōl′). City of E Ohio on the W.Va. border S of Youngstown. Pop. 16,687.

**East Lon·don** (lŭn′dən). City of SE South Africa on the Indian Ocean. Pop. 119,727.

**East Long·mea·dow** (lông′mĕd′ō, lŏng′-). Town of SW Mass. near Springfield. Pop. 12,905.

**East Lyme** (līm). Town of SE Conn. WNW of New London. Pop. 13,780.

**East·main** (ēst′măn′). River, c. 510 mi (820 km), of central Que., Canada.

**East Mo·line** (mō-lēn′). City of NW Ill. near Moline. Pop. 20,907.

**East·on** (ē′stən). **1.** Town of SE Mass. near Brockton. Pop. 16,623. **2.** City of E Pa. N of Philadelphia. Pop. 26,027.

**East Or·ange** (ôr′ĭnj, ŏr′-). City of NE N.J. near Newark. Pop. 77,025.

**East Pe·or·i·a** (pē-ôr′ē-ə, -ōr′-). City of N central Ill. opposite Peoria. Pop. 22,385.

**East Prov·i·dence** (prŏv′ĭ-dəns). City of E R.I. near Providence. Pop. 50,980.

**East Prus·sia** (prŭsh′ə). Former province of Prussia divided between Poland & the USSR.

**East Ridge.** City of SE Tenn. near Chattanooga. Pop. 21,236.

**East River.** Narrow strait connecting Upper New York Bay with Long Is. Sound & separating Manhattan Is. from Long Is.

**East Rock·a·way** (rŏk′ə-wā′). Village of SE N.Y. on S Long Is. Pop. 10,917.

**East Saint Lou·is** (sānt lōō′ĭs). City of SW Ill. on the Mississippi opposite St. Louis, Mo. Pop. 55,200.

**East Si·ber·i·an Sea** (sī-bîr′ē-ən). Arm of the Atlantic extending from Wrangel Is. to the New Siberian Is.

**East York** (yôrk′). Borough of metropolitan Toronto, Ont., Canada, on Lake Ontario. Pop. 106,950.

**Ea·ton·town** (ēt′n-toun′). Borough of E central N.J. SE of Red Bank. Pop. 51,509.

**Eb·bw Vale** (ĕb′ōō vāl). Urban district of SE Wales NW of Bristol. Pop. 25,670.

**Eb·ro** (ā′brō). River, c. 575 mi (925 km), of NE Spain flowing to the Mediterranean.

**Ec·ba·ta·na** (ĕk-băt′n-ə). City of ancient Media on the site of present day Hamadan, Iran.

**E·corse** (ē′kôrs′). City of SE Mich. near Detroit. Pop. 14,447.

**Ec·ua·dor** (ĕk′wə-dôr′). Republic of NW South America. Cap. Quito. Pop. 8,354,000. **—Ec′ua·dor′i·an** *adj.* & *n.*

**E·dam** (ē′dəm, ē′dăm′). Town of N central Netherlands on the Ijsselmeer. Pop. 21,507.

**E·de** (ā′də). City of W Nigeria NNE of Ibadan. Pop. 182,000.

**E·den** (ēd′n). City of N N.C. near the Va. border. Pop. 15,672.

**Eden Prairie.** City of E Minn. near Minneapolis. Pop. 16,263.

**E·des·sa** (ĭ-dĕs′ə). Ancient city of Mesopotamia on the site of present-day Urfa, Turkey.

**E·di·na** (ĭ-dī′nə). City of E Minn. near Minneapolis. Pop. 46,073.

**Ed·in·burg** (ĕd′n-bûrg′). City of S Tex. near the Mexican border WNW of Brownsville. Pop. 24,075.

**Ed·in·burgh** (ĕd′n-bûr′ə). Cap. of Scotland in the E on the Firth of Forth. Pop. 470,085.

**E·dir·ne** (ā-dîr′nə). City of NW Turkey NW of Istanbul. Pop. 63,001.

**E·dis·to** (ĕd′ĭ-stō′). River, c. 150 mi (241.4 km), of S S.C.

**E·dith Ca·vell** (ē′dĭth kăv′əl, kə-vĕl′). Mountain, c. 11,033 ft (336.5 m), in the Rockies of SW Alta, Canada.

**Ed·mond** (ĕd′mənd). City of central Okla. N of Oklahoma City. Pop. 34,637.

**Ed·monds** (ĕd′məndz). City of NW Wash. N of Seattle. Pop. 27,526.

**Ed·mon·ton** (ĕd′mən-tən). Capital of Alta., Canada, in the central part N of Calgary. Pop. 461,361.

**Ed·munds·ton** (ĕd′mən-stən). City of NW N.B., Canada, at the Me. border. Pop. 12,710.

**E·dom** (ē′dəm). Ancient country of SW Asia.

**Ed·ward** (ĕd′wərd). Lake in the Great Rift Valley on the Zaire-Uganda border.

**Ed·wards·ville** (ĕd′wərdz-vĭl′). City of SW Ill. NNE of East St. Louis. Pop. 12,460.

**E·fa·te** (ā-fä′tē). Island of central Vanuatu in the New Hebrides.

**Ef·fing·ham** (ĕf′ĭng-hăm′). City of E central Ill. SE of Decatur. Pop. 11,270.

**Eg·a·di** (ĕg′ə-dē). Island group in the Mediterranean W of Sicily.

**E·ger** (ā′gər). City of NE Hungary on the **Eger R.** (193 mi/310.5 km). Pop. 61,283.

**E·gypt** (ē′jĭpt). Republic of NE Africa & SW Asia. Cap. Cairo. Pop. 41,572,000. **—E·gyp′tian** *adj.* & *n.*

**Ei·fel** (ī′fəl). Volcanic plateau of West Germany W of the Rhine R.

**Ei·ger** (ī′gər). Peak, 13,025 ft (3972.6 m), in the Bernese Alps of W central Switzerland.

**Eind·ho·ven** (īnt′hō′vən, ānt′-). City of S Netherlands SE of Rotterdam. Pop. 192,562.

**Ei·sen·ach** (ī′zən-äk′, -äKH′). City of SW East Germany W of Erfurt. Pop. 49,954.

**El Aai·ún** (ĕl′ ī-ōōn′, ī-yōōn′). Town of NW Morocco in Western Sahara. Pop. 24,519.

**E·lam** (ē′ləm). Ancient country of SW Asia in present-day SW Iran.

**El·ba** (ĕl′bə). Italian island, 86 sq mi (222.7 sq km), in the Tyrrhenian Sea.

**El·be** (ĕl′bə, ĕlb). River of Czechoslovakia, East Germany, & West Germany flowing c. 725 mi (1,165 km) to the North Sea.

**El·bert** (ĕl′bərt). Highest (14,431 ft/4,401.5 m) of the U.S. Rocky Mts., in central Colo.

**El·bląg** (ĕl′blông′). City of N Poland ESE of Gdańsk. Pop. 89,835.

**El·brus** (ĕl-brōōz′). Highest mountain, 18,481 ft (5,636.7 m), of Europe, in the Caucasus of SE European USSR.

**El·burz** (ĕl-bōōrz′). Mountain range of N Iran rising to 18,934 ft (5,774.9 km).

**El Ca·jon** (ĕl′ kə-hōn′). City of S Calif. near San Diego. Pop. 10,462.

**El Cam·po** (ĕl kăm′pō). City of SE Tex. SW of Houston. Pop. 10,462.

**El Cen·tro** (ĕl sĕn′trō). City of SE Calif. near the Mexican border. Pop. 23,996.

**El Cer·ri·to** (ĕl′ sə-rē′tō). City of W Calif. near Richmond. Pop. 22,731.

**El·che** (ĕl′chā). City of SE Spain SW of Alicante. Pop. 101,271.

**El Do·ra·do** (ĕl′ də-rä′dō). **1.** City of S Ark. near the La. border SSW of Little Rock. Pop. 23,305. **2.** City of SE Kans. ENE of Wichita. Pop. 10,510.

**E·lec·tric Peak** (ĭ-lĕk′trĭk). Highest peak, 11,155 ft (3,402.3 m), of the Gallatin Range in SW Mont.

**El·e·phan·ti·ne** (ĕl′ə-făn-tī′nē). Island of SE Egypt in the Nile.

**E·leu·sis** (ĭ-lōō′sĭs). Ancient city of Attica, Greece, NW of Athens.

**El Fer·rol** (ĕl′ fə-rôl′) *or* **El Fer·rol del Cau·di·llo** (dĕl′ kou-dē′ō, -yō). City of NW Spain on the Atlantic. Pop. 75,464.

**El·gin** (ĕl′jĭn). City of NE Ill. NW of Chicago. Pop. 63,798.

**El·gon** (ĕl′gŏn′). Extinct volcano, 14,178 ft (4,324.3 m), on the Kenya-Uganda border.

**E·lis** (ē′lĭs). Region & city of ancient Greece in the W Peloponnesus.

**E·lis·a·beth·ville** (ĭ-lĭz′ə-bəth-vĭl′). LUBUMBASHI.

**E·liz·a·beth** (ĭ-lĭz′ə-bəth). City of NE N.J. near Newark. Pop. 106,201.

**Elizabeth City.** City of NE N.C. NW of Raleigh. Pop. 13,784.

**E·liz·a·beth·ton** (ĭ-lĭz′ə-bəth-tən). City of NE Tenn. NE of Knoxville. Pop. 12,431.

**E·liz·a·beth·town** (ĭ-lĭz′ə-bəth-toun′). City of central Ky. S of Louisville. Pop 15,380.

**Elk** (ĕlk). **1.** River, c. 200 mi (320 km), of W Tenn. & Ala. **2.** River, 172 mi (276.7 km), of NW Va. **3.** Range of the Rockies in W central Colo. rising to 14,259 ft (4,349 m).

**Elk Grove Village.** Village of NE Ill. near Chicago. Pop. 28,907.

**Elk·hart** (ĕl′kärt′, ĕlk′härt′). City of N Ind. E of South Bend. Pop. 41,305.

**El·len** (ĕl′ən). Peak, 11,022 ft (3361.8 m), in S Utah.

**El·lens·burg** (ĕl′ĭnz-bûrg′). City of central Wash. N of Yakima. Pop. 11,572.

**Elles·mere** (ĕlz′mîr′). Island, 82,119 sq mi (212,688 sq km), of N.W.T., Canada, in the Arctic Ocean.

**El·lice** (ĕl′ĭs). Group of atolls in the SW Pacific N of Fiji.

**El·lis** (ĕl′ĭs). Island of Upper New York Bay SW of Manhattan.

**Ells·worth** (ĕlz′wûrth′). **1. Mountains.** Range of Antarctica S of Ellsworth Land. **2. Land.** High plateau of Antarctica S of the Antarctic Peninsula.

**El Man·su·ra** (ĕl′ măn-sōōr′ə). City of N Egypt. Pop. 257,866.

**El·mi·ra** (ĕl-mī′rə). City of S N.Y. near the Pa. border W of Binghamton. Pop. 12,710.

**Elm·hurst** (ĕlm′hûrst′). City of NE Ill. near Chicago. Pop. 44,251.

**El Mon·te** (ĕl mŏn′tē). City of S Calif. E of Los Angeles. Pop. 79,494.

**Elm·wood Park** (ĕlm′wŏŏd′). **1.** Village of NE Ill. near Chicago. Pop. 24,016. **2.** Borough of NE N.J. SE of Paterson. Pop. 18,377.

**El O·beid** (ĕl′ ō-bād′). AL UBAYYID.

**El Pas·o** (ĕl păs′ō). City of extreme W Tex. on the Rio Grande. Pop. 425,259.

**El Re·no** (ĕl rē′nō). City of central Okla. W of Oklahoma City. Pop. 15,486.

**El Sal·va·dor** (ĕl săl′və-dôr′). Republic of Central America on the Pacific Ocean. Cap. San Salvador. Pop. 4,813,000. **—El Sal′va·dor′i·an** *adj.* & *n.*

**El Se·gun·do** (ĕl′ sĭ-gŏŏn′dō). City of S Calif., SW of Los Angeles. Pop. 13,752.

**El·si·nore** (ĕl′sə-nôr′, -nōr′). HELSINGØR.

**El·wood** (ĕl′wŏŏd′). City of E central Ind. NNE of Indianapolis. Pop. 10,867.

**E·ly·ri·a** (ĭ-lîr′ē-ə). City of N Ohio WSW of Cleveland. Pop. 57,504.

**Em·bar·ras** *or* **Em·bar·ass** (ăm′brō′). River, 185 mi (297.7 km), of E. Ill. flowing to the Wabash R. in SW Ind.

**Em·den** (ĕm′dən). City of NW West Germany on the Ems R. Pop. 53,509.

**E·mi Kous·si** (ā′mē kōō′sē). Highest peak, 11,204 ft (3,417.2 m), of the Tibesti Massif in NW Chad.

**Em·ma·us** (ə-mā′əs). Borough of E Pa. near Allentown. Pop. 11,001.
**Em·men** (ĕm′ən). City of NE Netherlands near the West German border. Pop. 86,700.
**Em·po·ri·a** (ĕm-pôr′ē-ə, -pōr′-). City of E central Kans. SSW of Topeka. Pop. 25,287.
**Ems** (ĕmz, ĕms). River of NW West Germany flowing c. 208 mi (334.7 km) to the North Sea.
**En·der·by Land** (ĕn′dər-bē). Region of Antarctica between Queen Maud Land & Wilkes Land.
**En·di·cott** (ĕn′dĭ-kət, -kŏt′). Village of S N.Y. near Binghamton. Pop. 14,457.
**En·field** (ĕn′fēld′). Town of N Conn. on the Mass. border. Pop. 42,695.
**En·ga·dine** (ĕng′gə-dēn′). Valley of the Inn R. in E Switzerland.
**En·ga·ño** (ĕn-gän′yō). Cape of Palau Is., Philippines.
**Eng·land** (ĭng′glənd). Part of the United Kingdom, the S part of the island of Great Britain. Pop. 46,220,955. **—Eng′lish** (-glĭsh) *adj. & n.*
**En·gle·wood** (ĕng′gəl-wŏŏd′). **1.** City of N central Colo. near Denver. Pop. 30,021. **2.** City of NE N.J. near New York City. Pop. 23,701. **3.** City of W Ohio near Dayton. Pop. 11,239.
**English Channel.** Arm of the Atlantic Ocean, c. 350 mi (565 km) long, separating France & Great Britain.
**E·nid** (ē′nĭd). City of N central Okla. NNW of Oklahoma City. Pop. 50,363.
**En·i·we·tok** (ĕn′ə-wē′tŏk′, ə-nē′wĭ-). Atoll in Marshall Is., W central Pacific; site of U.S. atomic tests.
**En·nis** (ĕn′ĭs). City of NE Tex. SSE of Dallas. Pop. 12,110.
**En·sche·da** (ĕn′skə-dā′). City of E Netherlands near the West German border. Pop. 141,597.
**En·se·na·da** (ĕn′sə-nä′də). City of NW Mexico on the Pacific. Pop. 77,687.
**En·teb·be** (ĕn-tĕb′ə). City of S Uganda on Lake Victoria. Pop. 21,096.
**En·ter·prise** (ĕn′tər-prīz′). City of SE Ala. NNW of Dothan. Pop. 18,033.
**E·nu·gu** (ā-nōō′gōō). City of SE Nigeria E of the Niger R. Pop. 187,000.
**Eph·e·sus** (ĕf′ĭ-səs). Ancient city of Greek Asia Minor in present-day W Turkey.
**Eph·ra·ta** (ĕf′rə-tə). Borough of SE Pa. NE of Lancaster. Pop. 11,095.
**Ep·i·dau·rus** (ĕp′ĭ-dôr′əs). Ancient city of Greece on the NE shore of the Peloponnesus.
**Ep·som and Ew·ell** (ĕp′səm; yōō′əl). Borough of SE England near London; site of Epsom Downs racetrack. Pop. 70,700.
**E·qua·to·ri·al Guin·ea** (ē′kwə-tôr′ē-əl gĭn′ē, -tōr′-, ĕk′wə-). Republic of W central Africa including islands in the Gulf of Guinea. Cap. Malabo. Pop. 244,000.
**Er·e·bus** (ĕr′ə-bəs). Active volcano, 12,280 ft (3745.4 m), on Ross Is., Antarctica.
**E·re·tri·a** (ĕ-rē′trē-ə). Ancient city of Greece on S coast of Euboea.
**Er·furt** (ĕr′fərt, -fŏŏrt′). City of S central East Germany WSW of Leipzig. Pop. 202,979.
**E·rie** (îr′ē). **1.** One of the Great Lakes, between S Ont. & W N.Y., NW Pa., N Ohio, & SE Mich. **2. Canal.** Former artificial waterway extending c. 360 mi (580 km) across central N.Y. from Albany to Buffalo. **3.** City of NW Pa. on Lake Erie SW of Buffalo, N.Y. Pop. 119,123.
**Er·i·tre·a** (ĕr′ĭ-trē′ə, -trā′ə). Region of N Ethiopia. **—Er′i·tre′an** *adj. & n.*
**Er·lang·en** (ĕr′läng′ən). City of S West Germany near Nuremberg. Pop. 100,671.
**Er·lang·er** (ûr′lăng′gər). City of N Ky. near Covington. Pop. 14,433.
**Er Rif** (ĕr rĭf′). RIF.
**Er·y·man·thos** *or* **Er·y·man·thus** (ĕr′ə-măn′thəs). Mountain range of S Greece in NW Peloponnesus.
**Erz·ge·bir·ge** (ĕrts′gə-bîr′gə). Mountain range extending c. 95 m (155 km) along the border of East Germany & Czechoslovakia.
**Er·zu·rum** (ĕr′zə-rōōm′, ər-). City of E Turkey. Pop. 162,973.
**Es·bjerg** (ĕs′bē-ĕrg′, -ûrg′). City of SW Denmark on the North Sea. Pop. 68,097.
**Es·cam·bi·a** (ə-skăm′bē-ə). River, c. 231 mi (371.7 km), of SE Ala. & N Fla.
**Es·ca·na·ba** (ĕs′kə-nä′bə). City of N Mich. on the Upper Peninsula SSE of Marquette. Pop. 14,355.
**Es·con·di·do** (ĕs′kən-dē′dō). City of S Calif. NE of San Diego. Pop. 62,480.
**Es·dra·e·lon** (ĕz′drə-ē′lən). Fertile plain of N Israel near the Jordan R. valley.
**Es·fa·han** (ĕs′fə-hän′). City of central Iran. Pop. 671,820.
**E·sher** (ē′shər). Urban district of SE England near London. Pop. 63,970.
**Es·kils·tu·na** (ĕs′kĭl-styōō′nə). City of SE Sweden W of Stockholm. Pop. 66,409.

**Es·ki·şe·hir** (ĕs′kĭ-shə-hîr′). City of W central Turkey W of Ankara. Pop. 259,952.
**Es·pí·ri·tu San·to** (ə-spîr′ə-tōō′ săn′tō). Island in the S Pacific, largest & westernmost of the New Hebrides.
**Es·qui·line** (ĕs′kwə-līn′, -lĭn). One of the seven hills of ancient Rome.
**Es·sen** (ĕs′ən). City of W West Germany on the Ruhr R. Pop. 677,568.
**Es·se·qui·bo** (ĕs′ĭ-kē′bō). River of Guyana flowing c. 600 mi (965 km) to the Atlantic.
**Es·sex** (ĕs′ĭks).Town of NW Vt. near Burlington. Pop. 14,392.
**Ess·ling·en** (ĕs′lĭng-ən). City of SW West Germany on the Neckar. Pop. 95,298.
**Es·to·ni·a** (ĕ-stō′nē-ə) *also* **Es·tho·ni·a** (ĕ-stō′-, ĕs-thō′-). **1.** Region & former country of W European USSR. **2.** *also* **Es·to·ni·an Soviet Socialist Republic** (-nē-ən). Constituent republic of W European USSR on the Baltic Sea. **—Es·to′ni·an** *adj. & n.*
**E·thi·o·pi·a** (ē′thē-ō′pē-ə). Country of NE Africa. Cap. Addis Ababa. Pop. 31,065,000. **—E′thi·o′pi·an** *adj. & n.*
**Et·na** (ĕt′nə). Active volcano, 11,122 ft (3,392.2 m), in E Sicily.
**E·to·bi·coke** (ĭ-tō′bĭ-kō′). SW borough of metropolitan Toronto, Ont., Canada, on Lake Ontario. Pop. 297,109.
**E·tru·ri·a** (ĭ-trŏŏr′ē-ə). Ancient country of W central Italy now in Tuscany & parts of Umbria.
**Eu·boe·a** (yōō-bē′ə). Greek island, 1,467 sq mi (3,800 sq km), in the Aegean.
**Eu·clid** (yōō′klĭd). City of NE Ohio near Cleveland. Pop. 59,999.
**Eu·fau·la** (yə-fô′lə). City of SE Ala. ESE of Montgomery. Pop. 12,097.
**Eu·gene** (yōō-jēn′). City of W Ore. S of Salem. Pop. 105,624.
**Eu·less** (yōō′lĭs). City of NE Tex. near Ft. Worth. Pop. 24,002.
**Eu·nice** (yōō′nĭs). City of S central La. W of Baton Rouge. Pop. 12,479.
**Eu·phra·tes** (yōō-frā′tēz). River of SW Asia flowing c. 1,700 mi (2,735 km) from Turkey to the Persian Gulf.
**Eur·a·sia** (yōō-rā′zhə). Land mass comprising the continents of Europe & Asia. **—Eur·a′sian** *adj. & n.*
**Eu·re·ka** (yōō-rē′kə). City of NW Calif. S of Crescent City. Pop. 24,153.
**Eu·rope** (yŏŏr′əp). Sixth-largest continent, extending W from the Dardanelles, Black Sea, & Ural Mts. **—Eur′o·pe′an** (yŏŏr′ə-pē′ən) *adj. & n.*
**E·vans** (ĕv′ənz). Mountain, 14,260 ft (4,349.3 m), in the Front Range of the Rocky Mts. in N central Colo.
**Ev·ans·ton** (ĕv′ən-stən). City of NE Ill. N of Chicago. Pop. 73,706.
**Ev·ans·ville** (ĕv′ənz-vĭl′). City of SW Ind. on the Ohio R. SSW of Indianapolis. Pop. 130,496.
**Ev·er·est** (ĕv′ər-ĭst, ĕv′rĭst). Mountain, 29,028 ft (8,853.5 m), of the central Himalayas on the border of Tibet & Nepal; highest elevation in the world.
**Ev·er·ett** (ĕv′ər-ĭt, ĕv′rĭt). **1.** City of E Mass. near Boston. Pop. 37,195. **2.** City of NW Wash. N of Seattle. Pop. 54,413.
**Ev·er·glades** (ĕv′ər-glādz′). Subtropical swamp area of S Fla. including **Everglades National Park.**
**Ev·er·green Park** (ĕv′ər-grēn′). Village of NE Ill. near Chicago. Pop. 22,260.
**É·vreux** (ā-vrœ′). Town of N France WNW of Paris. Pop. 46,181.
**Ev·voia** (ĕv′yä). EUBOEA.
**E·wab** (ē′wôb). Island group of Indonesia SW of West Irian.
**Ex·cel·si·or Springs** (ĭk-sĕl′sē-ər). City of W Mo. NE of Kansas City. Pop 10,424.
**Ex·e·ter** (ĕk′sĭ-tər). **1.** Borough of SW England on the **Exe R.** (c. 55 mi/90 km) NE of Plymouth. Pop. 93,300. **2.** Town of SE N.H. SW of Portsmouth. Pop. 11,024.
**Ex·moor** (ĕk′smŏŏr′, -smôr′, -smōr′). Moorland plateau of SW England.
**Ex·mouth** (ĕk′smouth′, -sməth). Urban district of SW England on the Exe R. Pop. 26,840.
**Eyre** (âr). **1.** Shallow salt lake of central S Australia. **2.** Peninsula, c. 200 mi (322 km) long, in S Australia.

# F

**Fa·en·za** (fä-ĕn′zə). City of N central Italy SE of Bologna. Pop. 36,241.
**Faer·oe** *or* **Far·oe** (fâr′ō). Islands of Denmark in the N Atlantic between Iceland & the Shetlands.
**Fa·ga·ras** (fə-gə-räsh′). Range of the Transylvanian Alps in S Rumania.
**Fa·ial** *or* **Fa·yal** (fə-yäl′, fī-äl′). Westernmost island of the Azores in the N Atlantic.
**Fair·banks** (fâr′băngks′). City of central Alas. NNE of Anchorage. Pop. 22,645.
**Fair·born** (fâr′bôrn′). City of SW central Ohio near Dayton. Pop. 29,702.

**Fair·fax** (fâr′făks′). Independent city of NE Va. near Washington D.C. Pop. 19,390.

**Fair·field** (fâr′fēld′). **1.** City of SE Australia near Sydney. Pop. 114,603. **2.** City of N central Ala. near Birmingham. Pop. 13,040. **3.** City of W Calif. NNE of Oakland. Pop. 58,099. **4.** Town of SW Conn. on Long Is. Sound. Pop. 54,849. **5.** City of SW Ohio N of Cincinnati. Pop. 30,777.

**Fair·ha·ven** (fâr′hā′vən). Town of SE Mass. near New Bedford. Pop. 15,759.

**Fair Isle** (fâr). Small island at the S tip of the Shetland Is. of N Scotland.

**Fair Lawn** (lôn). Borough of NE N.J. near Paterson. Pop. 32,229.

**Fair·mont** (fâr′mŏnt′). **1.** City of S Minn. near the Iowa border SW of Minneapolis. Pop. 11,506. **2.** City of N W.Va. near the border S of Pittsburgh, Pa. Pop. 23,863.

**Fair Oaks** (ōks). Site, just E of Richmond, Va., of Union victory over the Confederates at the Battle of Seven Pines (1862).

**Fair·view** (fâr′vyōō′). Borough of NE N.J. near Jersey City. Pop. 10,519.

**Fairview Heights.** City of SW Ill. ESE of East St. Louis. Pop. 12,414.

**Fairview Park.** City of NE Ohio near Cleveland. Pop. 19,311.

**Fair·weath·er** (fâr′wĕth′ər). Peak, 15,300 ft (4,666.5 m), on the border between SE Alas. & NW B.C., Canada.

**Faiz·a·bad** (fī′zə-băd′). City of N central India E of Lucknow. Pop. 102,835.

**Fa·jar·do** (fə-här′dō). Town of NE Puerto Rico. Pop. 26,845.

**Fa·ka·ra·va** (fä′kə-rä′və). Atoll of Tuamotu Archipelago, French Polynesia, in the S central Pacific.

**Fal·kirk** (fôl′kûrk′). Burgh of central Scotland W of Edinburgh. Pop. 36,901.

**Falk·land** (fôk′lənd, fôlk′-). Islands of the S Atlantic E of the Strait of Magellan; claimed by Great Britain and Argentina.

**Fall River** (fôl). City of SE Mass. on the R.I. border WNW of New Bedford. Pop. 92,574.

**Fal·mouth** (făl′məth). Town of SE Mass. on SW Cape Cod. Pop. 23,640.

**False Bay** (fôls). Inlet of the Atlantic SW of Cape Town, South Africa.

**Fal·ster** (făl′stər, fôl′-). Island of SE Denmark in the Baltic Sea.

**Fa·ma·gu·sta** (fä′mə-gōō′stə, făm′ə-). City of E Cyprus on the **Bay of Famagusta,** an inlet of the Mediterranean. Pop. 38,960.

**Fan·ning** (făn′ĭng). Island of the central Pacific S of Hawaii, part of Kiribati.

**Far·al·lon** (făr′ə-lŏn′). Islets in the Pacific off San Francisco, Calif., W of the Golden Gate.

**Far East** (fär′ ēst′). SE Asia & the Malay Archipelago. **—Far′ East′ern** *adj.* & *n.*

**Fare·well** (fâr-wĕl′, fâr′wĕl′). Cape at the S tip of Greenland.

**Far·go** (fär′gō). City of E N.Dak. E of Bismarck. Pop. 61,308.

**Far·i·bault** (făr′ə-bō′). City of SE Minn. S of Minneapolis. Pop. 16,241.

**Far·mers Branch** (fär′mərz). Town of NE Tex. near Dallas. Pop. 24,863.

**Far·ming·ton** (fär′mĭng-tən). **1.** Town of central Conn. SW of Hartford. Pop. 16,407. **2.** City of SE Mich. near Detroit. Pop. 11,022. **3.** City of NW N.Mex. SSW of Durango, Colo. Pop. 30,729.

**Farmington Hills.** City of SE Mich. NW of Detroit. Pop. 58,056.

**Far·oe** (fâr′ō). *var. of* FAEROE.

**Far·rukh·a·bad** (fə-rōō′kə-băd′). City of N central India on the Ganges. Pop. 102,768.

**Fars** (färz, färs) *or* **Far·si·stan** (fär′sĭ-stăn′). Historical region of S Iran along the Persian Gulf.

**Fá·ti·ma** (făt′ə-mə). Village & pilgrimage site of W central Portugal NNE of Lisbon.

**Fat·shan** (fät′shän′). FOSHAN.

**Fa·yal** (fə-yäl′, fī-äl′). *var. of* FALAL.

**Fay·ette·ville** (fā′ĭt-vĭl′, -vəl). **1.** City of NW Ark. NNE of Fort Smith. Pop. 36,165. **2.** City of S central N.C. SSW of Raleigh. Pop. 59,507.

**Fear** (fîr), **Cape.** Promontory on an island off SE N.C. at the mouth of the Cape Fear R.

**Feath·er** (fĕth′ər). River of N central Calif. flowing 100 mi (160.9 km) to the Sacramento.

**Fed·er·al District** (fĕd′ər-əl). **1.** District of E Argentina; site of the capital, Buenos Aires. **2.** District of E central Brazil; site of the capital, Brasília. **3.** District of central Mexico; site of the capital, Mexico City. **4.** District of N Venezuela; site of the capital, Caracas.

**Fed·er·at·ed Ma·lay States** (fĕd′ə-rā′tĭd mə-lā′, mā′lā). Former federation of British-protected Malayan states, part of present-day Malaysia.

**Fei·ra de San·ta·na** (fā′rə də săn-tăn′ə, săɴ-). City of E Brazil NNW of Salvador. Pop. 127,105.

**Feld·berg** (fĕlt′bĕrg′). Highest elevation, 4,898 ft (1,493.8 m), in the Black Forest of SW West Germany.

**Fen** (fĕn, fŭn). River of N central China flowing 375 mi (603.4 km) to the Yellow R.

**Feng·tien** (fŭng′tē-ĕn′). SHENYANG.

**Fens** (fĕnz), **the.** Lowland district of E England W & S of the Wash.

**Fer·ga·na** *or* **Fer·gha·na** (fər-gä′nə). City of S Central Asian USSR SW of Andizhan. Pop. 176,000.

**Fer·gus Falls** (fûr′gəs). City of W Minn. SE of Fargo, N.Dak. Pop. 12,519.

**Fer·gu·son** (fûr′gə-sən). City of E Mo. near St Louis. Pop. 24,740.

**Fer·nan·do de No·ro·nha** (fər-năn′dō də nə-rōn′yə). Island group in the Atlantic off the NE coast of Brazil.

**Fer·nan·do Po** (pō′). BIOKO.

**Fern·dale** (fûrn′dāl′). City of SE Mich. near Detroit. Pop. 26,227.

**Fer·ra·ra** (fə-rär′ə). City of N Italy SW of Venice. Pop. 97,507.

**Fer·tile Cres·cent** (fûr′tl krĕs′ənt). Region of the Middle East arching across the N part of the Syrian Desert & extending from the Nile to the Tigris & Euphrates.

**Fez** (fĕz) *also* **Fès** (fĕs). City of N central Morocco NE of Casablanca. Pop. 325,327.

**Fez·zan** (fə-zăn′). Region of SW Libya.

**Fich·tel·ge·birge** (fĭk′təl-gə-bîr′gə). Mountain region of E central West Germany near the East German & Czechoslovak borders.

**Fie·so·le** (fē-ā′zə-lē, -lā′). Resort town of central Italy near Florence. Pop. 3,772.

**Fife** (fīf). Region of E Scotland between the Firths of Forth & Tay.

**Fi·ji** (fē′jē). Island country of the SW Pacific comprising c. 800 islands. Cap. Suva. Pop. 588,068. **—Fi′ji·an** *adj.* & *n.*

**Filch·ner Ice Shelf** (fĭlk′nər). Ice area of Antarctica at the W edge of Coats Land & the head of Weddell Sea.

**Find·lay** (fĭnd′lē). City of NW Ohio S of Toledo. Pop. 35,594.

**Fin·gal's Cave** (fĭng′gəlz). Sea cavern of W Scotland on Staffa Is. in the Inner Hebrides.

**Fin·ger Lakes** (fĭng′gər). Eleven elongated glacial lakes in W central N.Y.

**Fin·is·terre** (fĭn′ĭ-stâr′), **Cape.** Rocky promontory of extreme NW Spain on the Atlantic coast.

**Finke** (fĭngk). River c. 400 mi (645 km), of central Australia.

**Fin·land** (fĭn′lənd). **1. Gulf of.** Arm of the Baltic extending c. 285 mi (460 km) between Finland & the USSR. **2.** Republic of N Europe. Cap. Helsinki. Pop. 4,788,000. **—Finn** (fĭn) *n.* **—Fin′nish** (fĭn′ĭsh) *adj.* & *n.*

**Fin·lay** (fĭn′lē). River of N B.C., Canada, flowing c. 250 mi (400 km) to the Peace R.

**Fin·ster·aar·horn** (fĭn′stər-är′hôrn′). Highest peak, 14,032 ft (4,279.8 m), of the Bernese Alps in S central Switzerland.

**Fiord·land** (fyôrd′lănd′, fyōrd′-). Mountain area of New Zealand extending c. 200 mi (321 km) along the SW coast of South Is.

**Fire Island** (fīr). Barrier island, 32 mi (51.5 km) long, off the S shore of Long Is., SE N.Y., including **Fire Island National Seashore.**

**Fi·ren·ze** (fē-rĕn′tsā). FLORENCE.

**Firth** (fûrth). —See the specific element, e.g., **Firth of Clyde** appears at CLYDE.

**Fish·ers** (fĭsh′ərz). Island off the NE tip of Long Is., SE N.Y.

**Fitch·burg** (fĭch′bûrg′). City of N Mass. N of Worcester. Pop. 39,580.

**Fitz·ger·ald** (fĭts-jĕr′əld). City of S central Ga. ENE of Albany. Pop. 10,187.

**Fi·u·me** (fyōō′mā). RIJEKA.

**Five Forks** (fīv). Crossroads SW of Petersburg in SE Va. where the last important Civil War battle was fought on April 1, 1865.

**Flag·staff** (flăg′stăf′). City of N central Ariz. NE of Prescott. Pop. 34,641.

**Flam·bor·ough Head** (flăm′bûr′ə, -bər-ə). Promontory on coast of NE England.

**Fla·min·i·an Way** (flə-mĭn′ē-ən). Ancient Roman road between Rome & the N central Adriatic coast of Italy.

**Flan·ders** (flăn′dərz). Region of NW Europe including part of N France & W Belgium & bordered by the North Sea. **—Flem′ing** (flĕm′ĭng) *n.* **—Flem′ish** (-ĭsh) *adj.* & *n.*

**Flat·head** (flăt′hĕd′). River flowing c. 240 mi (385 km) from SE B.C., Canada, to the Clark Fork in NW Mont.

**Flat·tery** (flăt′ə-rē), **Cape.** Headland of NW Wash at the entrance to Juan de Fuca Strait.

**Fleet·wood** (flēt′wŏŏd′). Borough of NW England on Morecambe Bay. Pop. 30,070.

**Flens·burg** (flĕnz′bûrg, flĕns′bŏŏrk′). City of N West Germany on **Flensburg Fjord,** an arm of the Baltic at the Danish border. Pop. 93,213.

**Fletsch·horn** (flĕch′hôrn′). Peak, 13,121 ft (4,001.9 m), in the Lepontine Alps of S Switzerland.

**Flin·ders** (flĭn′dərz). **1.** Intermittent river of NE Australia flowing 520 mi (836.7 m) to the Gulf of Carpentaria. **2. Ranges.** Mountain region of S central Australia extending c. 400 mi (644 km) N from Adelaide.

**Flint** (flĭnt). **1.** River of SW Ga. flowing 330 mi (530.9 km) to join the Chattahoochee & form the Apalachicola. **2.** City of SE central Mich. NW of Detroit. Pop. 159,611.

**Flod·den** (flŏd'n). Hill of N England near the Scottish border; site of a battle (1513) in which the English defeated the Scots.

**Flor·al Park** (flôr'əl, flōr'-). Village of SE N.Y. on W Long Is. near Queens. Pop. 16,805.

**Flor·ence** (flôr'əns, flŏr'-). **1.** City of central Italy on the Arno R. Pop. 441,654. **2.** City of NW Ala. WNW of Decatur. Pop. 37,029. **3.** City of N central Ky. SW of Cincinnati, Ohio. Pop. 15,586. **4.** City of NE S.C. NE of Columbia. Pop. 30,062. **—Flor'en·tine'** (-ən-tēn', -tīn') *adj. & n.*

**Flo·res** (flôr'əs, flōr'-). **1.** Sea of E Indonesia between the Java & Banda seas S of Sulawesi. **2.** Island of E Indonesia, one of the Lesser Sundas.

**Flo·ri·a·nó·po·lis** (flôr'ē-ə-nŏp'ə-lĭs, flōr-). City of SE Brazil on an island just off the coast. Pop. 115,665.

**Flor·i·da** (flôr'ĭ-də, flŏr'-). **1. Straits of.** Sea passage between Cuba & the **Florida Keys,** chain of small islands extending SW from Miami to Key West. **2.** Island of SE Solomon Is., SW Pacific. **3.** State of the SE U.S. Cap. Tallahassee. Pop. 9,739,992. **—Flo·rid'i·an** (flə-rĭd'ē-ən), **Flor'i·dan** (flôr'ĭ-dən, flŏr'-) *adj. & n.*

**Flor·is·sant** (flôr'ĭ-sənt, flŏr'-). City of E Mo. near St. Louis. Pop. 55,372.

**Flush·ing** (flŭsh'ĭng). **1.** VLISSINGEN. **2.** Section of New York City in N Queens on W Long Is.

**Fly** (flī). River, c. 650 mi (1,045 km), of Papua New Guinea in SE New Guinea Is.

**Foc·şa·ni** (fôk-shän', -shä'nē). Town of E central Rumania NW of Brăila. Pop. 62,275.

**Fog·gia** (fô'jə). City of S Italy WNW of Barletta. Pop. 136,436.

**Fo·li·gno** (fə-lēn'yō). City of central Italy SE of Perugia. Pop. 26,887.

**Folke·stone** (fōk'stən). Borough of SE England on the Strait of Dover WSW of Dover. Pop. 45,610.

**Fol·som** (fōl'səm). City of central Calif. NE of Sacramento. Pop. 11,003.

**Fond du Lac** (fŏn' də lăk', djə). City of E Wis. on Lake Winnebago E of Sheboygan. Pop. 35, 863.

**Fon·ga·fa·le** (fŏn'gə-fä'lē). Cap. of Tuvalu on Funafuti Is. in the S Pacific. Pop. 2,120.

**Fon·se·ca** (fôn-sā'kə), **Gulf of.** Inlet of the Pacific in W Central America rimmed by El Salvador, Honduras, & Nicaragua.

**Fon·taine·bleau** (fŏn'tĭn-blō'). Town of N France SE of Paris. Pop. 16,436.

**Fon·tan·a** (fŏn-tăn'ə). City of S Calif. near San Bernardino. Pop. 37,109.

**Foo·chow** (fo͞o'jō', -chou'). FUZHOU.

**For·a·ker** (fôr'ə-kər, fŏr'-). Peak, 17,280 ft (5,270.4 m), in the Alaska Range of S central Alas.

**Forbes** (fôrbz). Peak, 11,902 ft (3,630.1 m), in the Rocky Mts., SW Alta., Canada, near the B.C. border.

**For·bid·den City** (fôr-bĭd'n). Walled area of central Beijing, China, containing palaces of former Chinese rulers.

**For·est Grove** (fôr'ĭst, fŏr'-). City of NW Ore. W of Portland. Pop. 11,499.

**Forest Hill.** City of NE Tex. near Fort Worth. Pop. 11,684.

**Forest Hills.** Section of New York City in central Queens on W Long Is.

**Forest Park. 1.** City of NW Ga. near Atlanta. Pop. 18,782. **2.** Village of NE Ill. near Chicago. Pop. 15,177. **3.** City of SW Ohio near Cincinnati. Pop. 18,675.

**For·lì** (fôr-lē'). City of N Italy SE of Bologna. Pop. 83,303.

**For·mo·sa** (fôr-mō'sə). **1.** Strait of the Pacific between Taiwan & China. **2.** TAIWAN.

**For·rest City** (fôr'ĭst, fŏr'-). City of E Ark. ENE of Little Rock. Pop. 13,756.

**For·ta·le·za** (fôr'tl-ā'zə). City of NE Brazil on the Atlantic. Pop. 1,308,859.

**Fort Col·lins** (fôrt kŏl'ĭnz, fōrt). City of N Colo. NW of Greeley. Pop. 64,632.

**Fort-de-France** (fôr-də-fräns'). Cap. of Martinique, French West Indies. Pop. 96,815.

**Fort Dodge** (dŏj). City of central Iowa NNW of Des Moines. Pop. 29,423.

**Fort E·rie** (ĭr'ē). Town of S Ont., Canada, on the Niagara R. opposite Buffalo, N.Y. Pop. 24,031.

**For·tes·cue** (fôr'tĭ-skyo͞o'). River of W Australia flowing 340 mi (547.1 km) to the Indian Ocean.

**Fort George** (jôrj). River of W central Que., Canada, flowing 520 mi (336.6 km) to James Bay.

**Forth** (fôrth, fōrth). River of S central Scotland flowing c. 60 mi (95 km) to the **Firth of Forth,** an inlet of the North Sea extending c. 55 mi (90 km) into SE Scotland.

**Fort-La·my** (fôr'lə-mē'). NDJAMENA.

**Fort Lau·der·dale** (lô'dər-dāl'). City of SE Fla. N of Miami Beach. Pop. 153,256.

**Fort Lee** (lē). Borough of NE N.J. on the Hudson opposite Manhattan. Pop. 32,449.

**Fort Mad·i·son** (măd'ĭ-sən). City of SE Iowa SW of Burlington. Pop. 13,520.

**Fort Mc·Mur·ray** (mək-mûr'ē). City of NE Alta., Canada, on the Athabasca R. Pop. 15,424.

**Fort My·ers** (mī'ərz). City of SW Fla. N of Naples. Pop. 36,638.

**Fort Nel·son** (nĕl'sən). River, 260 mi (416.5 km), of NE B.C., Canada.

**Fort Payne** (pān). City of NE Ala. NE of Gadsden. Pop. 11,485.

**Fort Pierce** (pîrs). City of E central Fla. NNW of Palm Beach. Pop. 33,802.

**Fort Sas·katch·e·wan** (sə-skăch'ə-wän, -wän', săs-kăch'-). Town of central Alta., Canada, NE of Edmonton. Pop. 8,304.

**Fort Smith** (smĭth). City of NW Ark. WNW of Little Rock. Pop. 71,515.

**Fort Thom·as** (tŏm'əs). City of N Ky. near Covington. Pop. 16,012.

**Fort Wal·ton Beach** (wôl'tən). City of NW Fla. E of Pensacola. Pop. 20,829.

**Fort Wayne** (wān). City of NE Ind. NE of Indianapolis. Pop. 172,196.

**Fort Worth** (wûrth). City of NE Tex. W of Dallas. Pop. 385,141.

**Fo·shan** (fō'shän'). City of SE China near Guangzhou. Pop. 125,000.

**Fos·ter City** (fô'stər, fŏs'tər). City of W Calif., near San Francisco. Pop. 23,287.

**Fos·tor·i·a** (fô-stôr'ē-ə, -stōr'-, fŏs-tôr'-, -tōr'-). City of NW Ohio SSE of Toledo. Pop. 15,743.

**Foun·tain Valley** (foun'tən). City of S Calif. SE of Los Angeles. Pop. 55,080.

**Fou·ta Djal·lon** *or* **Fu·ta Jal·lon** (fo͞o'tə jə-lōn'). Mountainous region of NW Guinea, source of the Gambia, Niger, & Senegal rivers.

**Fox** (fŏks). **1.** River, c. 185 mi (298 km), of SE Wis. & NE Ill. **2.** River of central & E Wis. flowing 176 mi (283.2 km) to Green Bay. **3.** Islands of SW Alas. in the E Aleutians.

**Fox·bor·ough** *or* **Fox·bor·o** (fŏks'bûr'ō, -bûr'ə). Town of E Mass. W of Brockton. Pop. 14,148.

**Foxe Basin** (fŏks). Arm of the Atlantic between Melville Peninsula & Baffin Is. in N.W.T., Canada.

**Foyle** (foil). River of W Northern Ireland flowing c. 10 mi (16 km) to **Lough Foyle,** an inlet of the Atlantic.

**Fra·ming·ham** (frā'mĭng-hăm'). Town of E central Mass. WSW of Boston. Pop. 65,113.

**France** (frăns). Republic of W Europe. Cap. Paris. Pop. 53,788,000. **—French** (frĕnch) *adj. & n.*

**Franche-Com·té** (fränsh-kōn-tā'). Region & former province of E France.

**Fran·co·ni·a** (frăng-kō'nē-ə, -kōn'yə, frăn-). Region & former duchy of S West Germany. **—Fran·co'ni·an** *adj. & n.*

**Frank·en** (frăng'kən). FRANCONIA.

**Frank·fort** (frăngk'fərt). **1.** City of W central Ind. NNW of Indianapolis. Pop. 15,168. **2.** Cap. of Ky. in the N central part NW of Lexington. Pop. 25,973.

**Frank·furt** (frănk'fərt, frängk'fo͝ort'). **1.** *or* **Frankfurt am Main** (äm mīn'). City of central West Germany on the Main R. Pop. 636,157. **2.** *or* **Frankfurt an der O·der** (än dər ō'dər). City of E East Germany on the Oder. R. Pop. 70,817.

**Frank·lin** (frăngk'lĭn). **1.** Northernmost district of N.W.T., Canada, composed of the Boothia & Melville peninsulas & the Canadian Arctic Archipelago. **2.** City of S central Ind. SSE of Indianapolis. Pop. 11,563. **3.** Town of SE Mass. near the R.I. border SW of Boston. Pop. 18,217. **4.** City of SW Ohio NNE of Cincinnati. Pop. 10,711. **5.** City of central Tenn. SSW of Nashville. Pop. 12,407. **6.** City of SE Wis. near Milwaukee. Pop. 16,871.

**Franklin D. Roo·se·velt Lake** (rō'zə-vĕlt', ro͞o'-). Reservoir of NE Wash. formed by Grand Coulee Dam on the Columbia R.

**Franklin Park.** Village of NE Ill. near Chicago. Pop. 17,507.

**Franks** (frăngks). Peak, 13,140 ft (4,007.7 m), of the Absaroka Range of the Rocky Mts. in NW Wyo.

**Franz Jo·sef Land** (frănts' jō'zəf, -səf). Archipelago in the Arctic Ocean N of Novaya Zemlya; claimed by the USSR.

**Fra·ser** (frā'zər, -zhər). **1.** River of B.C., Canada, flowing c. 850 mi (1,370 km) from the Rocky Mts. near the B.C.-Alta. boundary to the Strait of Georgia at Vancouver. **2.** City of SE Mich. NNE of Detroit. Pop. 14,560.

**Fred·er·ick** (frĕd'rĭk, -ər-ĭk). City of N Md. W of Baltimore. Pop. 27,557.

**Fred·er·icks·burg** (frĕd'rĭks-bûrg', -ər-ĭks-). Independent city of NE Va. N of Richmond. Pop. 15,322.

**Fred·er·ic·ton** (frĕd'rĭk-tən, -ər-ĭk-). Cap. of N.B., Canada, in the S central part NW of St. John. Pop. 45,248.

**Fred·er·iks·berg** (frĕd'rĭks-bûrg', -ər-ĭks-). City of Denmark near Copenhagen. Pop. 101,874.

**Fre·do·nia** (frĭ-dō'nē-ə, -dōn'yə). Village of W N.Y. near Lake Erie. Pop. 11,126.

**Free·hold** (frē'hōld'). Borough of E central N.J. E of Trenton. Pop. 10,020.
**Free·port** (frē'pôrt', -pōrt'). **1.** City of NW Bahamas on Grand Bahama Is. Pop. 15,546. **2.** City of NW Ill. W of Rockford. Pop. 26,406. **3.** Village of SE N.Y. on SW Long Is. Pop. 38,272. **4.** City of SE Tex. on the Gulf of Mexico S of Houston. Pop. 13,444.
**Free·town** (frē'toun'). Cap. of Sierra Leone in the W part on the Atlantic. Pop. 274,000.
**Frei·berg** (frī'bûrg', -bĕrk'). City of S East Germany ENE of Karl-Marx-Stadt. Pop. 50,815.
**Frei·burg** (frī'bŏŏrg', -bûrg', -bŏŏrk'). **1.** *or* **Freiburg im Breis·gau** (ĭm brīs'gou'). City of SW West Germany near the Rhine at the edge of the Black Forest. Pop. 175,371. **2.** FRIBOURG.
**Fre·man·tle** (frē-măn'tl). City of SW Australia on the Indian Ocean near Perth. Pop. 23,497.
**Fre·mont** (frē'mŏnt'). **1.** City of W Calif. SE of Oakland. Pop. 131,495. **2.** City of E central Nebr. WNW of Omaha. Pop. 23,979. **3.** City of N Ohio SE of Toledo. Pop. 17,834.
**French Broad** (frĕnch brôd'). River of W N.C. & E Tenn. flowing 210 mi (337.9 km) to the Holston to form the Tennessee.
**French E·qua·to·ri·al Af·ri·ca** (ē'kwə-tôr'ē-əl ăf'rĭ-kə, -tōr'-, ĕk'wə-). French federation of W central Africa from 1910 to 1958.
**French Gui·a·na** (gē-ăn'ə, -ä'nə, gī-ăn'ə). French overseas department of NE South America on the Atlantic. Cap. Cayenne. Pop. 64,000.
**French Mo·roc·co** (mə-rŏk'ō). French protectorate over most of present-day Morocco from 1912 to 1956.
**French Pol·y·ne·sia** (pŏl'ə-nē'zhə, -shə). French overseas territory in the S central Pacific including the Society, Marquesas, Tuamotu, Gambier, & Tubuai Is.
**French So·ma·li·land** (sō-mä'lē-lănd'). DJIBOUTI.
**French Su·dan** (sōō-dăn'). MALI.
**Fres·no** (frĕz'nō). City of S central Calif. SE of San Jose. Pop. 218,202.
**Fri·a** (frē'ə). Cape of NW Namibia on the Atlantic.
**Fri·bourg** (frē-bōōr', frī'bûrg'). City of W Switzerland SW of Bern. Pop. 41,600.
**Frid·ley** (frĭd'lē). City of E Minn. near Minneapolis. Pop. 30,228.
**Fridt·jof Nan·sen Land** (frĭt'yôf năn'sən). FRANZ JOSEF LAND.
**Frie·drichs·ha·fen** (frē'drĭks-hä'fən). City of S West Germany on the Lake of Constance. Pop. 51,544.
**Friends·wood** (frĕndz'wŏŏd'). City of SE Tex. SE of Houston. Pop. 10,719.
**Fries·land** (frēz'lənd, -lănd', frēs'-). Region of N Europe on the North Sea between the Scheldt & Weser rivers.
**Fri·o** (frē'ō). River of S Tex. flowing 220 mi (354 km) into the Nueces R.
**Fri·sian** (frĭzh'ən, frē'zhən). Chain of islands in the North Sea off the coasts of the Netherlands, West Germany, & Denmark.
**Fri·u·li** (frē'ə-lē', frē-ōō'lē). Region & former duchy of NE Italy in present-day NE Italy & Slovenia, NW Yugoslavia.
**Fri·u·li-Ve·ne·zia Giu·lia** (frē'ə-lē'və-nĕt'sē-ə jōōl'yə, frē-ōō'lē-). Region of NE Italy bounded by Austria in the N & Yugoslavia in the E.
**Fro·bish·er Bay** (frō'bĭ-shər). Arm of the Atlantic extending c. 150 mi (241 km) into SE Baffin Is., N.W.T., Canada.
**Front Range** (frŭnt). Range of the E Rocky Mts. extending c. 300 mi (482.7 km) in Wyo. & Colo. & rising to 14,274 ft (4,353.6 m).
**Front Roy·al** (roi'əl). Town of N Va. WNW of Alexandria. Pop. 11,126.
**Fro·ward** (frō'wərd, -ərd), **Cape.** Southernmost point of mainland South America, in S Chile on the Strait of Magellan.
**Frun·ze** (frōōn'zə). City of S Central Asian USSR on the Chu R. WSW of Alma-Ata. Pop. 533,000.
**Fu·chou** (fōō'jō', -chou'). FUZHOU.
**Fu·chu** (fōō'chōō'). City of E central Honshu, Japan, near Tokyo. Pop. 182,474.
**Fuer·te·ven·tu·ra** (fōō-ĕr'tē-vĕn-tōōr'ə, fwĕr'-). Second-largest (666 sq mi/1,724.9 sq km) of the Canary Is. off the NW coast of Africa.
**Fu·jai·rah** (fə-jī'rəl). Sheikdom of the United Arab Emirates, E Arabia, on the Gulf of Oman. Pop. 760.
**Fu·ji** (fōō'jē). **1.** *or* **Fu·ji·ya·ma** (fōō'jē-ä'mə, -yä'-) *or* **Fu·ji·no·ya·ma** (-nō-) *or* **Fu·ji-san** (-sän'). Highest peak (12,388 ft/3,778.3 m) in Japan, in central Honshu. **2.** City of central Honshu, Japan, at the foot of Mt. Fuji. Pop. 199,195.
**Fu·jian** (fōō'jē-ăn'). Province of SE China on the East China Sea.
**Fu·ji·no·ya·ma** (fōō'jē-nō-ä'mə, -yä'-). Mt. FUJI.
**Fu·ji·san** (fōō'jē-sän'). Mt. FUJI.
**Fu·ji·sa·wa** (fōō'jē-sä'wə). City of E central Honshu, Japan, near Tokyo. Pop. 265,975.
**Fu·ji·ya·ma** (fōō'jē-ä'mə, -yä'-). Mt. FUJI.
**Fu·kian** (fōō'kē-ĕn'). FUJIAN.
**Fu·ku·i** (fōō-kōō'ē). City of central Honshu, Japan, NNW of Nagoyo. Pop. 231,364.
**Fu·ku·o·ka** (fōō'kə-wō'kə). City of N Kyushu, Japan, on an inlet of the Sea of Japan. Pop. 1,002,201.
**Fu·ku·shi·ma** (fōō'kə-shē'mə). City of NE Honshu, Japan, N of Yokohama. Pop. 246,531.
**Fu·ku·ya·ma** (fōō'kə-yä'mə). City of W Honshu, Japan, near Kure. Pop. 329,714.
**Ful·da** (fōōl'də). City of E central West Germany on the **Fulda R.** (135 mi/217.2 km) near the East German border. Pop. 58,976.
**Ful·ler·ton** (fōōl'ər-tən). City of S Calif. SE of Los Angeles. Pop. 102,034.
**Ful·ton** (fōōl'tən). **1.** City of central Mo. ESE of Columbia. Pop. 11,046. **2.** City of N central N.Y. SSE of Oswego. Pop. 13,312.
**Fu·na·ba·shi** (fōō'nə-bä'shē). City of E central Honshu, Japan, on Tokyo Bay. Pop. 423,101.
**Fun·chal** (fōōn-shäl'). City of SE Madeira Is., Portugal. Pop. 38,340.
**Fun·dy** (fŭn'dē), **Bay of.** Inlet of the Atlantic in SE Canada between N.B. & SW N.S.
**Fur·neaux** (fûr'nō). Island group off NE Tasmania, Australia, in Bass Strait.
**Fürth** (fōōrt, fyōōrt). City of S West Germany W of Nuremberg. Pop. 101,639.
**Fu·san** (fōō'sän'). PUSAN.
**Fu·shun** (fōō'shōōn'). City of NE China E of Shenyang. Pop. 1,700,000.
**Fu·xin** *also* **Fu·sin** (fōō'shĭn'). City of NE China WNW of Shenyang. Pop. 350,000.
**Fu·ta Jal·lon** (fōō'tə jə-lōn'). *var. of* FOUTA DJALLON.
**Fu·tu·na** (fə-tōō'nə). Island & island group of the SW Pacific NE of Fiji, part of the French overseas territory of Wallis and Futuna.
**Fu·zhou** (fōō'jō'). City of SE China on the Min delta. Pop. 900,000.
**Fyn** (fĭn). Island of S central Denmark W of Sjaelland.

# G

**Ga·bès** (gä'bəs, -bĕs'). City of SE Tunisia on the **Gulf of Gabès,** an inlet of the Mediterranean. Pop. 40, 585.
**Ga·bon** (gă-bōN'). Republic of W central Africa. Cap. Libreville. Pop. 551,000.
**Ga·bo·rone** (gä'bə-rōn', -rō'nē). Cap. of Botswana near the South African border. Pop. 21,000.
**Gad·a·ra** (găd'ər-ə). Ancient city of Palestine SE of the Sea of Galilee. **—Gad'a·rene'** (găd'ə-rēn', găd'ə-rēn') *adj.* & *n.*
**Ga·des** (gā'dēz) *or* **Ga·dir** (-dər). CÁDIZ.
**Gads·den** (gădz'dən). **1. Purchase.** Area in extreme S N.Mex. & Ariz., c. 30,000 sq mi (77,700 sq km), purchased (1853) by the U.S. from Mexico. **2.** City of NE Ala. NNW of Anniston. Pop. 47,565.
**Ga·e·ta** (gä-ā'tə). City of W central Italy NW of Naples on the **Gulf of Gaeta,** an inlet of the Tyrrhenian Sea. Pop. 21,973.
**Gaff·ney** (găf'nē). City of NW S.C. near the N.C. border NE of Spartanburg. Pop. 13,453.
**Gaf·sa** (găf'sə). City of W central Tunisia W of Sfax. Pop. 42,225.
**Ga·han·na** (gə-hăn'ə). City of central Ohio near Columbus. Pop. 18,001.
**Gail·lard Cut** (gĭl-yärd', gā'ärd', -lärd'). Excavation, 8 mi (12.8 km) long, through a hill in the Canal Zone, Panama, occupied by the SE section of the Panama Canal.
**Gaines·ville** (gānz'vĭl', -vəl). **1.** City of N Fla. SW of Jacksonville. Pop. 81,371. **2.** City of N central Ga. NE of Atlanta. Pop. 15,280. **3.** Town of NE Tex. NNW of Dallas. Pop. 14,081.
**Gaird·ner** (gârd'nər). Salt lake of S central Australia W of Lake Torrens.
**Gai·thers·burg** (gā'thərz-bûrg'). City of central Md. NNW of Washington, D.C. Pop. 26,424.
**Ga·lá·pa·gos** (gə-lä'pə-gəs, -lăp'ə-). Island group of Ecuador in the Pacific c. 650 mi (1,045 km) W of the mainland.
**Ga·la·ţi** (gä-läts', -lät'sē) *or* **Ga·latz** (gä'läts'). City of E Rumania on the lower Danube. Pop. 252,884.
**Ga·la·tia** (gə-lā'shə, -shē-ə). Ancient country of central Asia Minor in the area around modern Ankara, Turkey. **—Ga·la'tian** *adj.* & *n.*
**Ga·latz** (gä'läts'). GALAŢI.
**Ga·le·ras** (gə-lĕr'əs). Volcano, 13,997 ft (4,269.1 m), in the SW Colombian Andes near the Ecuador border.
**Gales·burg** (gālz'bûrg'). City of NW central Ill. WNW of Peoria. Pop. 35,305.
**Ga·li·cia** (gə-lĭsh'ə, -ē-ə). **1.** Historical region of SE Poland & W Ukraine. **2.** Region & ancient kingdom of NW Spain. **—Ga·li'cian** *adj.* & *n.*
**Gal·i·lee** (găl'ə-lē'). **1. Sea of.** Freshwater lake, 64 sq mi (165.8 sq km), bordered by Israel, Syria, & Jordan. **2.** Region of N Israel. **—Gal'i·le'an** *adj.* & *n.*
**Gal·ion** (găl'yən). City of N central Ohio W of Mansfield. Pop. 12,391.
**Ga·li·tsi·ya** (gə-lēt'sē-ə, -yə). Polish GALICIA.

**Gal·lae·ci·a** (gə-lē′shə, -shē-ə). Spanish GALICIA.
**Gal·la·tin** (găl′ĭ-tən). **1.** River, c. 120 mi (195 km), of NW Wyo. & SW Mont. **2. Range.** Section of the Rocky Mts. in NW Wyo. & SW Mont.rising to 11,155 ft (3,402.2 m). **3.** City of N Tenn. NE of Nashville. Pop. 17,191.
**Galle** (gäl, găl). City of S Sri Lanka on the Indian Ocean. Pop. 72,700.
**Gal·li·a** (găl′ē-ə). GAUL.
**Ga·lli·nas Point** (gə-yē′nəs). Cape of N central Colombia, northernmost point of South America.
**Gal·lip·o·li** (gə-lĭp′ə-lē). Peninsula, c. 50 mi (80 km) long, of W Turkey extending SW between the Aegean & the Dardanelles.
**Gal·lo·way** (găl′ə-wā′). Region of SW Scotland.
**Gal·lup** (găl′əp). City of NW N.Mex. near the Ariz. border WNW of Albuquerque. Pop. 18,161.
**Gal·ves·ton** (găl′vĭ-stən). City of SE Tex. on **Galveston Bay,** an arm of the Gulf of Mexico SE of Houston. Pop. 61,902.
**Gal·way** (gôl′wā′). **1.** Region of W central Ireland. **2.** City of W central Ireland on **Galway Bay,** an inlet of the Atlantic. Pop. 27,726.
**Gam·bi·a** (găm′bē-ə). **1.** River of W Africa flowing 7,700 mi (1,125 km) from N Guinea through SE Senegal & Gambia to the Atlantic. **2.** *or* **The Gambia.** Republic of W Africa on the Atlantic. Cap. Banjul. Pop. 601,000. **—Gam′bi·an** *adj.* & *n.*
**Gam·bier** (găm′bîr′). Islands of the S central Pacific, part of French Polynesia.
**Gan** (găn). River, c. 550 mi (885 km), of SE China.
**Gan·dak** (gŭn′dŭk). River of S Nepal & NE India flowing 420 mi (675.7 km) to the Ganges.
**Gan·der** (găn′dər). Town of NE Newf., Canada. Pop. 9,301.
**Gan·dzha** (gän′jə). KIROVABAD.
**Gan·ges** (găn′jēz′) or **Gan·ga** (gŭng′gə). River of N India & Bangladesh flowing c. 1,560 mi (2,510 km) from the Himalayas to the Bay of Bengal.
**Gan·nett Peak** (găn′ĭt). Mountain, 13,785 ft (4,231.8 m), of the Wind River Range in central Wyo.
**Gan·su** (gän′sōō′). Province of N central China.
**Gao·xiong** (gou′shē-ōōng′). KAOHSIUNG.
**Gar·da** (gär′də). Lake, 143 sq mi (370.4 sq km), of N Italy E of Milan.
**Gar·de·na** (gär-dē′nə). City of S Calif. near Los Angeles. Pop. 45,165.
**Gar·den City** (gär′dn). **1.** City of SW Kans. WNW of Dodge City. Pop. 18,256. **2.** City of SE Mich. near Detroit. Pop. 35,640. **3.** Village of SE N.Y. on W Long Is. Pop. 22,927.
**Garden Grove.** City of S Calif. near Long Beach. Pop. 123,351.
**Garden of the Gods** (gŏdz). Park, 770 acres (311.9 hectares), of central Colo., noted for rock formations.
**Gar·di·ners** (gärd′nərz, gär′dn-ərz). Island, c. 3,000 acres (1,215 hectares), of SE N.Y. in **Gardiners Bay** between two peninsulas of E Long Is.
**Gard·ner** (gärd′nər). City of N Mass. near Fitchburg. Pop. 17,900.
**Gar·field** (gär′fēld′). **1.** Peak, 10,961 ft (3,343.1 m), in the Bitterroot Range of the Rocky Mts. in SW Mont. near the Idaho border. **2.** City of NE N.J. near Passaic. Pop. 26,803.
**Garfield Heights.** City of NE Ohio near Cleveland. Pop. 33,380.
**Ga·ri·glia·no** (gär′əl-yä′nō). River of central Italy flowing 100 mi (160 km) to the Gulf of Gaeta.
**Gar·land** (gär′lənd). City of NE Tex. near Dallas. Pop. 138,857.
**Gar·misch-Par·ten·kir·chen** (gär′mĭsh-pär′tn-kîr′kən). City of S West Germany SW of Munich. Pop. 26,831.
**Gar·mo Peak** (gär′mō). Mt. COMMUNISM.
**Ga·ronne** (gə-rŏn′, -rōn′). River of SW France flowing 402 mi (646.8 km) from the Spanish Pyrenees to the Dordogne to form the Gironde estuary.
**Gar·ri·son Reservoir** (găr′ĭ-sən). Reservoir, 140 mi (225.2 km) long, of W N.Dak., formed in the Missouri R. by **Garrison Dam.**
**Gar·y** (găr′ē, gâr′ē). City of NW Ind. on Lake Michigan. Pop. 151,953.
**Gas·con·ade** (găs′kə-nād′). River, c. 265 mi (426.4 km), of S central Mo.
**Gas·co·ny** (găs′kə-nē). Region & former province of SW France.
**Gash·er·brum** (gŭsh′ər-brōōm′, -brŏŏm′). **1. I.** Peak, 26,470 ft (8,073.3 m), in the Karakoram Range of N Kashmir. **2. II.** Peak, 26,630 ft (8,122.1 m), in the Karakoram Range NW of Gasherbrum I.
**Gas·pé** (găs-pā′). **1.** Peninsula of E Que., Canada, between Chaleur Bay & the mouth of the St. Lawrence. **2.** City of E Que., Canada, on **Gaspé Bay** near the E tip of the Gaspé Peninsula. Pop. 16,842.
**Ga·stein** (gä-stīn′). Valley of central Austria in the N Hohe Tauern.
**Gas·to·ni·a** (găs-tō′nē-ə, -tōn′yə). City of S N.C. near the S.C. border W of Charlotte. Pop. 47,333.
**Gates·head** (gāts′hĕd′). Borough of NE England on the Tyne R. opposite Newcastle. Pop. 91,230.
**Gates of the Arc·tic National Park** (gāts; ärk′tĭk, är′tĭk). Reservation of 7,052,000 acres (2,820,800 hectates) in N central Alas. in the Brooks Range.
**Gath** (găth). Ancient city of Philistia ENE of Gaza.

**Gat·i·neau** (găt′n-ō′). **1.** River of SW Que., Canada, flowing c. 240 mi (385 km) to the Ottawa R. **2.** Town of SW Que., Canada, on the Ottawa R. opposite Ottawa. Pop. 73,479.
**Ga·tún** (gə-tōōn′). Lake of the N Canal Zone, Panama, formed by the **Gatún Dam.**
**Gaul** (gôl). Ancient name for W Europe S & W of the Rhine, W of the Alps, & N of the Pyrenees, comprising approx. modern France & Belgium.
**Ga·var·nie** (găv′ər-nē′). Waterfall, 1,385 ft (422.4 m) high, of SW France in the Pyrenees S of Lourdes & near the **Cirque de Gavarnie,** a natural amphitheater.
**Gäv·le** (yĕv′lə). City of E Sweden on the Gulf of Bothnia. Pop. 67,454.
**Ga·ya** (gə-yä′). City of NE India SSW of Patna. Pop. 179,884.
**Ga·za** (gä′zə, găz′ə, gā′zə). City of SW Asia in the **Gaza Strip,** a Mediterranean coastal area (c. 140 sq mi/370 sq km). Pop. 118,272.
**Ga·zi·an·tep** (gä′zē-än-tĕp′). City of S Asian Turkey N of Aleppo, Syria. Pop. 300,882.
**Gdańsk** (gə-dänsk′, -dănsk′). City of N Poland on the Gulf of Danzig near the mouth of the Vistula R. Pop. 364,285.
**Gdy·ni·a** (gə-dĭn′ē-ə, -dĭn′yə). City of N Poland on the Gulf of Danzig NW of Gdańsk. Pop. 190,125.
**Ge·ba** (gā′bə). Chief river of Guinea-Bissau flowing 200 mi (321.8 km) to the Atlantic.
**Ge·bel Mu·sa** (jĕb′əl mōō′sə). Mountain group of the S Sinai Peninsula between Africa and Asia.
**Gee·long** (jə-lông′). City of SE Australia SW of Melbourne. Metro. area pop. 122,080.
**Ge·la** (jĕ′lə). City of S Sicily, Italy, on the Mediterranean. Pop. 66,845.
**Ge·li·bo·lu** (gĕl′ə-bə-lōō′). GALLIPOLI.
**Gel·sen·kir·chen** (gĕl′zən-kîr′kən). City of W West Germany in the Ruhr NE of Essen. Pop. 322,584.
**Gen·er·al San Mar·tín** (jĕn′ər-əl săn mär-tēn′). City of E Argentina near Buenos Aires. Pop. 360,573.
**Gen·e·see** (jĕn′ĭ-sē′, jĕn′ĭ-sē′). River of N Pa. & W N.Y. flowing 158 mi (254.2 km) to Lake Ontario.
**Ge·ne·va** (jə-nē′və). **1. Lake of.** Lake, 224 sq mi (580.2 sq km), on the Swiss-French border between the Alps & the Jura Mts. **2.** City of SW Switzerland on the Lake of Geneva & bisected by the Rhone. Metro. area pop. 320,200. **3.** City of W central N.Y. on Seneca Lake WSW of Syracuse. Pop. 15,133.
**Genk** (gĕngk). City of NE Belgium ENE of Hasselt. Pop. 57,913.
**Gen·ne·vil·liers** (zhĕn-vēl-yā′). Town of N central France on the Seine R. near Paris. Pop. 50,154.
**Gen·o·a** (jĕn′ə-wə, -ō-ə). City of NW Italy on the Ligurian Sea. Pop. 787,011.
**Gent** (gĕnt). GHENT.
**Gen·tof·te** (gĕn′tŭf′tə). City of E Denmark on Sjaelland Is. near Copenhagen. Pop. 77,744.
**George** (jôrj). **1.** Lake of NE Fla. formed by a widening of the St. Johns R. **2.** Glacial lake of NE N.Y. near Lake Champlain. **3.** River of NE Que., Canada, flowing c. 345 mi (555 km) to Ungava Bay.
**Georges Bank** (jôr′jĭz). Shoal in the Atlantic E of Cape Cod, Mass.
**George·town** (jôrj′toun′). **1.** Cap. of the Cayman Is. on Grand Cayman. Pop. 7,617. **2.** Cap. of Guyana on the Atlantic. Pop. 63,184. **3.** Section of Washington, D.C., in the W. **4.** City of N central Ky. NNW of Lexington. Pop. 10,972. **5.** City of SE S.C. NE of Charleston. Pop. 10,144.
**George Town.** PENANG, Malaysia.
**Geor·gia** (jôr′jə). **1. Strait of.** Channel between mainland B.C. & Vancouver Is., Canada. **2.** Ancient & medieval kingdom coextensive with present-day Georgian SSR. **3.** *also* **Geor·gian Soviet Socialist Republic** (-jən). Constituent republic of SE European USSR in the Caucasus on the Black Sea. **4.** State of the SE U.S. Cap. Atlanta. Pop. 5,464,265. **—Geor′gian** *adj.* & *n.*
**Georgian Bay.** Extension of Lake Huron in SE Ont., Canada.
**Geor·gi·na** (jôr-jē′nə). Intermittent river, c. 700 mi (1,125 km), of N central Australia.
**Ge·ra** (gĕr′ə). City of S East Germany ESE of Jena. Pop. 113,108.
**Ger·la·chov·ka** (gâr′lə-kôf′kə, -kôv′-). Highest peak, 8,737 ft (2,664.8 m), of the Carpathian Mts. in E Czechoslovakia.
**Ger·man Dem·o·crat·ic Republic** (jûr′mən dĕm′ə-krăt′ĭk). —See GERMANY.
**German East Af·ri·ca** (ēst ăf′rĭ-kə). Former German protectorate in E Africa (1885–1922).
**German Fed·er·al Republic** (fĕd′ər-əl). —See GERMANY.
**Ger·ma·ni·a** (jər-mā′nē-ə, -mān′yə). **1.** Ancient region of Europe N of the Danube & E of the Rhine. **2.** Part of the Roman Empire corresponding to present-day NE France & part of Belgium & the Netherlands.
**German New Guin·ea** (nōō′ gĭn′ē, nyōō′). Former German colony in present-day Papua New Guinea.
**Ger·man·town** (jûr′mən-toun′). **1.** Residential section of Philadelphia, Pa.; site of Revolutionary battle (1777). **2.** Town of extreme SW Tenn. E of Memphis. Pop. 20,459. **3.** Village of SE Wis. near Milwaukee. Pop. 10,729.

**Ger·ma·ny** (jûr′mə-nē). Former state of N central Europe bordered on the N by the Baltic & North seas & divided in 1949 into the **German Democratic Republic** (East Germany), cap. East Berlin, pop. 16,737,000; & the **German Federal Republic** (West Germany), cap. Bonn, pop. 61,658,000.

**Ger·mis·ton** (jûr′mĭ-stən). City of NE South Africa on the Witwatersrand. Pop. 221,972.

**Ge·ro·na** (hə-rō′nə). City of NE Spain NE of Barcelona. Pop. 37,095.

**Ge·ta·fe** (hə-tä′fē). Town of central Spain S of Madrid. Pop. 68,680.

**Geth·sem·a·ne** (gĕth-sĕm′ə-nē). Garden E of Jerusalem near the foot of the Mount of Olives.

**Get·tys·burg** (gĕt′ēz-bûrg′). Town of S Pa. ESE of Chambersburg, containing **Gettysburg National Military Park,** site of a Union victory in the Civil War (1863), & of Dwight D Eisenhower's farm, a national historic shrine. Pop. 7,194.

**Ge·zi·ra** (jə-zîr′ə). Region of E central Sudan between the Blue Nile & White Nile.

**Gha·gha·ra** (gä′gə-rə′). GOGRA.

**Gha·na** (gä′nə, găn′ə). **1.** Medieval African kingdom in present-day W Mali. **2.** Republic of W Africa on the Gulf of Guinea. Cap. Accra. Pop. 11,450,000. **—Gha′na·ian, Gha′ni·an** *adj.* & *n.*

**Ghats** (gôts). Two mountain ranges of S India: the **Eastern Ghats,** extending c. 900 mi (1,450 km) along the Bay of Bengal coast, & the **Western Ghats,** extending c. 1,000 mi (1,600 km) along the Arabian Sea coast.

**Gha·zal** (gə-zäl′), **Bahr el.** BAHR EL GHAZAL.

**Ghaz·ni** (gäz′nē). City of E central Afghanistan SW of Kabul. Pop. 30,425.

**Ghaz·ze** (gä′zē) *also* **Ghaz·zah** (gä′zə). GAZA.

**Ghent** (gĕnt). City of W Belgium, NW of Brussels. Pop. 148,860.

**Gi·ant's Causeway** (jī′ənts). Headland on the N coast of Northern Ireland.

**Gi·ba·ra Bay** (hē-bär′ə). Inlet of the Atlantic on N coast of E Cuba; site of the first landing by Columbus in the New World (1492).

**Gib·e·on** (gĭb′ē-ən). Ancient village of Palestine near Jerusalem. **—Gib′e·o·nite** *n.*

**Gib·ral·tar** (jə-brôl′tər). British colony, 2.5 sq mi (6.5 sq km), at the NW end of the **Rock of Gibraltar,** a peninsula on the S central coast of Spain in the **Strait of Gibraltar,** connecting the Mediterranean & the Atlantic between Spain & N Africa. Pop. 29,760.

**Gib·son** (gĭb′sən). Desert of W central Australia bounded by the Great Sandy Desert on the N & Great Victoria Desert on the S.

**Gies·sen** (gē′sən). City of central West Germany N of Frankfurt. Pop. 76,485.

**Gi·fu** (gē′fo͞o′). City of central Honshu, Japan, NNW of Nagoya. Pop. 408,707.

**Gi·jón** (hē-hōn′). City of NW central Spain on the Bay of Biscay. Pop. 159,806.

**Gi·la** (hē′lə). River of SW N. Mex. & S Ariz. flowing 630 mi (1,013.7 km) to the Colorado.

**Gil·bert** (gĭl′bərt). Intermittent river of NE Australia flowing c. 320 mi (515 km) to the Gulf of Carpentaria.

**Gilbert and El·lice Islands** (ĕl′ĭs). Former British colony comprised of atolls in the central Pacific; divided into the independent nations of Kiribati & Tuvalu.

**Gil·bo·a** (gĭl-bō′ə). Hills of NE Israel at SE edge of the Plain of Esdraelon.

**Gil·e·ad** (gĭl′ē-əd). Mountain region of Jordan E of the Jordan R.

**Gil·git** (gĭl′gĭt). **1.** Region of Kashmir in the NW Himalayas. **2.** Town in this region on the **Gilgit R.** (c. 150 mi/241 km), a tributary of the Indus.

**Gil·lette** (jə-lĕt′). City of NE Wyo. SE of Sheridan. Pop. 12,134.

**Gil·ling·ham** (jĭl′ĭng-əm). Borough of SE England N of Maidstone. Pop. 93,900.

**Gil·roy** (gĭl′roi′). City of W Calif. SE of San Jose. Pop. 21,641.

**Gin·za** (gĭn′zə). Shopping & entertainment district of Tokyo, Japan.

**Gi·rard** (jə-rärd′). City of NE Ohio near Youngstown. Pop. 12,517.

**Gi·rar·dot** (hē′rär-dôt′). City of central Colombia on the Magdalena R. Pop. 59,165.

**Gi·re·sun** (jîr′ĭ-so͞on′). City of NE Turkey on the Black Sea. Pop. 38,236.

**Gir·nar** (gîr-när′). Sacred mountain, 3,666 ft (1,118.1 m), of W India.

**Gi·ronde** (jə-rŏnd′, zhē-rōɴd′). Estuary, 45 mi (72.4 km), of SW France formed by the Garonne & Dordogne rivers.

**Giur·giu** (jo͞or′jo͞o′). City of S Rumania, on the Danube. Pop. 53,241.

**Giv·a·ta·yim** (gĭv′ə-tä′yĭm). Town of W central Israel near Tel Aviv-Jaffa. Pop. 48,500.

**Gi·za** (gē′zə). City of N Egypt on the Nile near Cairo; site of Great Pyramids. Pop. 1,246,713.

**Glace Bay** (glās). Town of NW N.S., Canada, on the Atlantic coast of Cape Breton Is. Pop. 21,836.

**Gla·cier** (glā′shər). Mountain, 10,568 ft (3,223.2 m), of NW central Wash. in the Cascades ENE of Everett.

**Glacier Bay National Park.** Mountain & glacier area, 3,878,269 acres (1,551,307 hectares) of SE Alas. near Juneau.

**Glacier National Park.** **1.** Reservation, 1,013,100 acres (410,306 hectares) of NW Mont. straddling the Continental Divide. **2.** Reservation in SE B.C., Canada, in the Selkirk Mts.

**Glad·beck** (glät′bĕk′, glăd′-). City of W West Germany in the Ruhr NNW of Gelsenkirchen. Pop. 80,434.

**Glad·stone** (glăd′stōn′). City of W Mo. surrounded by Kansas City. Pop. 24,990.

**Glå·ma** (glô′mə). GLOMMA.

**Glas·gow** (glăs′kō, -gō). **1.** (*also* glăz′-). City of SW Scotland on the Firth of Clyde. Pop. 880,617. **2.** City of S central Ky. E of Bowling Green. Pop. 12,958. **—Glas·we′gian** (-wē′jən) *adj.* & *n.*

**Glass·bor·o** (glăs′bûr′ə). Borough of SW N.J. S of Camden. Pop. 14,574.

**Glas·ton·bur·y** (glăs′tən-bĕr′ē). City of central Conn. SE of Hartford. Pop. 24,327.

**Glei·witz** (glī′vĭts). GLIWICE.

**Glen·coe.** **1.** (glĕn-kō′). Valley of W Scotland SE of Loch Leven. **2.** (glĕn′kō′). Village of NE Ill. on Lake Michigan near Chicago. Pop. 9,200.

**Glen Cove** (glĕn′ cōv′). City of SE N.Y. on NW Long Is. N of Mineola. Pop. 24,618.

**Glen·dale** (glĕn′dāl′). **1.** City of S central Ariz. near Phoenix. Pop. 96,988. **2.** City of S Calif. near Los Angeles. Pop. 139,060. **3.** City of SE Wis. near Milwaukee. Pop. 13,882.

**Glendale Heights.** Village of NE Ill. near Chicago. Pop. 23,163.

**Glen·do·ra** (glĕn-dôr′ə, -dōr′ə). City of S Calif. ENE of Los Angeles. Pop. 38,654.

**Glen El·lyn** (ĕl′ĭn). Village of NE Ill. near Chicago. Pop. 23,649.

**Glen Rock.** Borough of NE N.J. near Paterson. Pop. 11,497.

**Glens Falls.** City of E N.Y. on the Hudson NNE of Saratoga Springs. Pop. 15,897.

**Glen·view** (glĕn′vyo͞o′). Village of NE Ill. near Chicago. Pop. 30,842.

**Glen·wood** (glĕn′wo͝od′). Village of NE Ill. near Chicago. Pop. 10,538.

**Glit·ter·tind·en** (glĭt′ər-tĭn′ən). Peak, 8,104 ft (2,471.7 m), of S central Norway.

**Gli·wi·ce** (glĭ-vēt′sə). City of SW Poland WNW of Katowice. Pop. 170,912.

**Glom·ma** (glô′mə). River of Norway flowing c. 365 mi (590 km) into the Skagerrak.

**Glos·sop** (glŏs′əp). Borough of central England near Manchester. Pop. 24,820.

**Glouces·ter** (glŏs′tər, glô′stər). **1.** Borough of W central England on the Severn. Pop. 91,600. **2.** City of NE Mass. on the Atlantic NE of Boston. Pop. 27,768.

**Gloucester City.** City of SW N.J. near Camden. Pop. 13,121.

**Glov·ers·ville** (glŭv′ərz-vĭl′). City of E central N.Y. NW of Schenectady. Pop. 17,836.

**Gnos·sus** (nŏs′əs). KNOSSOS.

**Go·a** (gō′ə). Former Portuguese colony (1510–1961) of SW India on the Malabar Coast.

**Goa, Da·man, and Di·u** (də-măn′; dē′o͞o). Union territory of W India on the Arabian Sea, formerly three noncontiguous Portuguese colonies.

**Goat** (gōt). Island of W N.Y. in the Niagara R. dividing Niagara Falls into the American & Canadian falls.

**Go·bi** (gō′bē). Desert of central Asia, c. 500,000 sq mi (1,295,000 sq km), chiefly in Mongolia.

**Go·da·va·ri** (gə-dä′və-rē). River of central India flowing c. 900 mi (1,450 km) from the Western Ghats across the Deccan Plateau to the Bay of Bengal.

**Go·des·berg** (gō′dĭs-bûrg′, -bĕrg′). Section of Bonn, W West Germany, on the Rhine.

**Godt·håb** (gôt′hôb′, gŏt′-). Cap. of Greenland on the **Godthåb Fjord** on the SW coast. Pop. 8,545.

**God·win Aus·ten** (gŏd′wĭn ô′stən, ŏs′tən). Second-highest mountain in the world, 28,250 ft (8,616.3 m), in the Karakoram Range of N Pakistan.

**Goffs·town** (gŏfs′toun′). Township of S N.H. WNW of Manchester. Pop. 11,315.

**Go·ge·bic** (gō-gē′bĭk). Mountain range extending 80 mi (128.7 km) from W Upper Peninsula, N Mich., into N Wis.

**Gog·ra** (gŏg′rə). River of central Asia flowing c. 640 mi (1,030 km) from SW Tibet to the Ganges in E India.

**Goi·â·ni·a** (goi-ăn′ē-ə). City of S central Brazil SE of Brasília. Pop. 362,152.

**Gök·cha** (gœk′chə). SEVAN.

**Go·lan Heights** (gō′län′, -lən). Hill region of NE Israel & SW Syria NE of the Sea of Galilee.

**Gol·con·da** (gŏl-kŏn′də). Ruined city of SE India.

**Gold Coast** (gōld). **1.** Section of coastal W Africa along S shore of Ghana. **2.** Former British colony in S part of Gold Coast, now part of Ghana.

**Gold·en** (gōl′dən). City of N central Colo. W of Denver. Pop. 12,237.
**Golden Gate.** Strait in W central Calif. connecting the Pacific & San Francisco Bay.
**Golden Horn.** Inlet of the Bosporus in European Turkey.
**Golden Valley.** City of E Minn., near Minneapolis. Pop. 22,775.
**Golds·bor·o** (gōldz′bûr′ə). City of E central N.C. SE of Raleigh. Pop. 31,871.
**Gol·go·tha** (gŏl′gə-thə). CALVARY.
**Go·mel** (gō′məl, gô′-). City of W European USSR ESE of Bobruisk. Pop. 383,000.
**Go·me·ra** (gō-mĕr′ə). Island of the Canary group in the Atlantic W of Tenerife Is.
**Go·mor·ra** (gə-môr′ə, -mŏr′ə). Ancient city of Palestine near Sodom.
**Go·na·ïves** (gō′nə-ēv′). City of W Haiti on the NE shore of the **Gulf of Gonaïves,** an arm of the Caribbean. Pop. 29,261.
**Gon·dar** (gŏn′dər, -där′). Town of NW Ethiopia on Lake Tana. Pop. 38,600.
**Good Hope** (go͝od′ hōp′), **Cape of.** Promontory on SW coast of South Africa S of Cape Town.
**Good·win Sands** (go͝od′wĭn). Shoals in the Strait of Dover off the coast of SE England.
**Goose Creek** (go͞os). City of SE S.C. N of Charleston. Pop. 17,811.
**Göp·ping·en** (gœp′ĭng-ən). City of S West Germany ESE of Stuttgart. Pop. 54,365.
**Go·rakh·pur** (gôr′ək-po͝or′, gōr′-). City of N central India E of Lucknow. Pop. 230,911.
**Gor·ham** (gôr′əm). Town of SW Me. W of Portland. Pop. 10,101.
**Go·ri·zi·a** (gə-rēt′sē-ə). City of NE Italy on the Yugoslav border. Pop. 35,912.
**Gor·ki** or **Gor·ky** *also* **Gor·kiy** (gôr′kē). City of E European USSR on the Volga. Pop. 1,344,000.
**Gör·litz** (gûr′lĭts′, gœr′-). City of SE East Germany E of Dresden. Pop. 84,658.
**Gor·lov·ka** (gôr-lôf′kə, -lôv′-). City of S European USSR in the Donets Basin. Pop. 336,000.
**Go·ryn** (gôr′ən). River, 410 mi (660 km), of W European USSR.
**Gor·zów Wiel·ko·pol·ski** (gôr′zo͞of vyĕl′kə-pôl′skē). City of W Poland on the Warta. Pop. 74,267.
**Go·shen** (gō′shən). City of N Ind. ESE of South Bend. Pop. 19,665.
**Gos·lar** (gôs′lär′). City of E West Germany at the foot of the Harz Mts. near the border with East Germany. Pop. 53,957.
**Gos·port** (gŏs′pôrt′, -pōrt′). Borough of S England W of Portsmouth. Pop. 82,300.
**Gö·ta Canal** (yœ′tə). Waterway of S Sweden extending 240 mi (386.2 km) from the Kattegat to the Baltic.
**Gö·te·borg** (yœ′tə-bôr′ē) *or* **Goth·en·burg** (gŏth′ən-bûrg′, gŏt′n-). City of SW Sweden on the Kattegat. Pop. 444,540.
**Go·tha** (gō′tə, -thə). City of SW East Germany W of Erfurt. Pop. 59,243.
**Goth·am** (gŏth′əm). NEW YORK, N.Y. **—Goth′am·ite′** *n.*
**Goth·en·burg** (gŏth′ən-bûrg′, gŏt′n-). *var. of* GÖTEBORG.
**Got·land** (gŏt′lănd′, -lənd). Region of SE Sweden in the Baltic, including **Gotland Is.**
**Go·to·ret·to** (gō′tō-rĕt′ō). Islands of Japan in the East China Sea off W Kyushu.
**Göt·tin·gen** (gœt′ĭng-ən). City of E West Germany ENE of Kassel. Pop. 123,797.
**Gott·wald·ov** (gôt′vəl-dôf′, -dôv′). City of central Czechoslovakia in Moravia. Pop. 84,300.
**Gou·da** (gou′də, go͞o′-). City of W Netherlands NE of Rotterdam. Pop. 56,403.
**Go·ver·na·dor Va·la·da·res** (guv′ər-nə-dôr′ văl′ə-där′əs). City of SE Brazil NE of Belo Horizonte. Pop. 125,174.
**Gov·er·nors Island** (gŭv′ər-nərz). Island of Upper New York Bay S of Manhattan Is., SE N.Y.
**Gow·er** (gou′ər). Peninsula of S Wales.
**Graf·ton** (grăf′tən). Town of S central Mass. near Worcester. Pop. 11,238.
**Gra·ham** (grā′əm). **1. Land.** ANTARCTIC PENINSULA. **2.** Island off NW B.C., Canada, largest of the Queen Charlotte Is.
**Gra·hams·town** (grā′əmz-toun′). City of SE South Africa ENE of Port Elizabeth. Pop. 41,302.
**Gra·ian Alps** (grā′ən ălps′, -yən, grī′ən). N section of the W Alps on the border between SE France & NW Italy.
**Grain Coast** (grān). Historical name of part of the Atlantic coast of W Africa, roughly identical with present-day Liberia.
**Gram·pi·ans** (grăm′pē-ənz), **the.** Mountain range of central Scotland.
**Gra·na·da** (grə-nä′də). **1.** Medieval Moorish kingdom of S Spain. **2.** City of SW Nicaragua at the NW end of Lake Nicaragua. Pop. 34,976. **3.** City of S Spain SE of Córdoba. Pop. 185,799.

**Gran·by** (grăn′bē). City of S Que., Canada, ESE of Montreal. Pop. 37,132.
**Gran Ca·na·ri·a** (grăng′ kä-när′yä). GRAND CANARY.
**Gran Cha·co** (grän′ chä′kō). Lowland plain, c. 250,000 sq mi (647,500 sq km), of central South America, divided among Paraguay, Bolivia, & Argentina.
**Grand** (grănd). **1.** River of S Ont., Canada, flowing c. 165 mi (265 km) to Lake Erie. **2.** River, c. 300 mi (485 km), of SE Iowa & NW Mo. **3.** River of central Mich. flowing 260 mi (418.3 km) to Lake Michigan. **4.** River, 140 mi (225.2 km), of W Mo. **5.** River, 209 mi (336.3 km), of N.Dak. & S.Dak.
**Grand At·las** (ăt′ləs). Mountains of Morocco, highest section of the Atlas Mts. of NW Africa.
**Grand Ba·ha·ma** (bə-hä′mə, -hā′-). Island of the Bahama group in the Atlantic E of West Palm Beach, Fla.
**Grand Banks.** Shoals of the W Atlantic, c. 36,000 sq mi (93,240 sq km), off SE Newf., Canada.
**Grand Canal. 1.** Longest canal in the world, extending c. 1,000 mi (1,610 km) from Tianjin to Hangzhou. **2.** Principal waterway of Venice, Italy.
**Grand Ca·na·ry** (kə-nâr′ē). Principal island of the Canary group in the Atlantic ESE of Tenerife Is.
**Grand Canyon. 1.** Gorge of the Colorado R. in NW Ariz., 217 mi (349.2 km) long, 4–18 mi (6.4–29 km) wide, & c. 1 mi (1.6 km) deep. **2. National Park.** Area of 1,218,375.2 acres (487,350 hectares) in Ariz., including Grand Canyon & Marble Canyon national monuments. **3. of the Ar·kan·sas** (är′kən-sô′). ROYAL GORGE. **4. of the Snake** (snāk). HELLS CANYON.
**Grand Cay·man** (kā-măn′, kā′mən). Largest of the Cayman Is. in the Caribbean NW of Jamaica.
**Grand Cou·lee** (ko͞o′lē). Gorge, c. 30 mi (48 km) long, of N central Wash., carved by the Columbia R.
**Grande Co·more** (gränd kô-môr′). GREAT COMORO.
**Grande Prai·rie** (grănd prâr′ē). City of W Alta., Canada, NW of Edmonton. Pop. 17,626.
**Grande-Terre** (grăn′târ′). Island of E Guadeloupe in the Leeward Is. of the Caribbean.
**Grand Falls.** CHURCHILL FALLS.
**Grand Forks.** City of E N.Dak. N of Fargo. Pop. 43,765.
**Grand Ha·ven** (hā′vən). City of SW Mich. on Lake Michigan WNW of Grand Rapids. Pop. 11,763.
**Grand Island.** City of S Nebr. W of Lincoln. Pop. 33,180.
**Grand Junc·tion** (jŭngk′shən). City of W Colo. near the Utah border. Pop. 28,144.
**Grand Ma·nan** (mə-năn′). Island of S N.B., Canada, in the Bay of Fundy.
**Grand'Mère** (gräN-mĕr′). City of S Que., Canada, NNE of Montreal. Pop. 15,999.
**Grand Me·sa** (mā′sə). Mountain, c. 10,000 ft (3,050 m), of W Colo.
**Grand Prai·rie** (prâr′ē). City of NE Tex. near Dallas. Pop. 71,462.
**Grand Rapids.** City of W central Mich. WNW of Lansing. Pop. 181,843.
**Grand Te·ton** (tē′tŏn′, tēt′n). Highest elevation, 13,766 ft (4,198.6 m), of the Teton Range in **Grand Teton National Park,** NW Wyo.
**Grand Trav·erse Bay** (trăv′ərs). Arm of Lake Michigan in W central Mich.
**Grand Turk** (tûrk). Chief island of the Turks & Caicos Is. in the Atlantic SE of the Bahamas.
**Grand·view** (grănd′vyo͞o′). City of W Mo. S of Kansas City. Pop. 24,502.
**Grand·ville** (grănd′vĭl′). City of SW Mich. near Grand Rapids. Pop. 12,412.
**Grange·mouth** (grānj′məth, -mouth′). Burgh of central Scotland on the Firth of Forth. Pop. 24,430.
**Gran·ite City** (grăn′ĭt). City of SW Ill. near East St. Louis. Pop. 36,815.
**Granite Peak.** Mountain, 12,799 ft (3,903.7 m), of S Mont. NE of Yellowstone National Park.
**Gran Pa·ra·di·so** (grän′ pär′ə-dē′zō). Highest elevation, 13,324. ft (4,063.8 m), of the Graian Alps in NW Italy.
**Gran Sas·so d'I·ta·lia** (grän sä′sō dē-täl′yə). Mountain group of the Apennines in central Italy.
**Grants** (grănts). City of W N.Mex. W of Albuquerque. Pop. 11,451.
**Grants Pass.** City of SW Ore. WNW of Medford. Pop. 14,997.
**Grape·vine** (grāp′vīn′). City of NE Tex. NE of Fort Worth. Pop. 11,801.
**Gras·mere** (grăs′mîr′). Lake of NW England in the Lake District.
**Grasse** (grăs, gräs). Town of SE France, W of Nice. Pop. 24,260.
**Grau·denz** (grou′dĕnts). GRUDZIADZ.
**Graves** (gräv′). Region of SW France in the Garonne valley.
**Graves·end** (grāvz′ĕnd′). Borough of SE England on the Thames near London. Pop. 53,500.
**Grays Harbor** (grāz). Inlet of the Pacific, W Wash.
**Grays Peak.** Highest elevation, 14,274 ft (4,353.5 m), of the Front Range in central Colo.

**Graz** (gräts). City of SE Austria on the Mur SSW of Vienna. Pop. 251,900.
**Great A·ba·co** (grāt ăb′ə-kō′). Largest island of the Abaco and Cays group in the N Bahamas.
**Great Ap·pa·la·chi·an Valley** (ăp′ə-lā′chē-ən, -chən, -lăch′ē-ən, -lăch′ən). Chain of lowlands of the Appalachian Mts. extending from Canada to Ala.
**Great Aus·tra·lian Bight** (ô-strāl′yən). Bay of the Indian Ocean on the S coast of Australia.
**Great Bar·ri·er Reef** (băr′ē-ər). Largest coral reef in the world, c. 1,250 mi (2,010 km) long, off the NE coast of Australia.
**Great Ba·sin** (bā′sĭn). Desert region of the W U.S., 210,000 sq mi (543,900 sq km), comprising most of Nev. & parts of Utah, Calif., Idaho, Wyo., & Ore.
**Great Bear Lake** (bâr). Lake, c. 12,275 sq mi (31,795 sq km), in N central Mackenzie Dist., N.W.T., Canada.
**Great Bend.** City of central Kans. NW of Wichita. Pop. 16,608.
**Great Brit·ain** (brĭt′n). **1.** Island off the W coast of Europe comprising England, Scotland, & Wales. **2.** UNITED KINGDOM.
**Great Com·o·ro** (kŏm′ə-rō′). Largest of the Comoro Is. in the N Mozambique Channel of the Indian Ocean.
**Great Cros·by** (krôz′bē). —See CROSBY.
**Great Di·vid·ing Range** (dĭ-vī′dĭng). Crest line of the Eastern Highlands of Australia.
**Great·er An·til·les** (grā′tər ăn-tĭl′ēz). Island group of the West Indies including Cuba, Jamaica, Hispaniola, & Puerto Rico.
**Greater Lon·don** (lŭn′dən). LONDON, England.
**Greater Sun·das** (sŭn′dəz, so͞on′-) or **Sun·da** (-də). —See SUNDA Is.
**Great Falls. 1.** Waterfall, 35 ft (10.6 m), in the Potomac NW of Washington, D.C., on the Va.-Md. boundary. **2.** City of N central Mont. NNE of Helena. Pop. 56,725.
**Great Glen of Scotland** (skŏt′lənd). Valley of N Scotland extending c. 60 mi (96.5 km) from Moray Firth in the NE to Loch Linnhe in the SW.
**Great In·di·an Desert** (ĭn′dē-ən). THAR.
**Great Ka·by·lia** (kə-bĭ′lē-ə, -bĭl′ē-ə). Mountainous area of N Algeria, E of Algiers.
**Great Kar·roo** (kə-ro͞o′). KARROO.
**Great Lakes.** Group of five freshwater lakes of central North America between the U.S. & Canada, including Lakes Superior, Huron, Erie, Ontario, & Michigan.
**Great Na·ma·qua·land** (nə-mä′kwə-lănd′). Region of S Namibia W of the Kalahari Desert.
**Great Plains.** High grassland region of central North America extending from the Canadian provinces of Alta., Sask., & Man. S into Tex.
**Great Rift Valley** (rĭft). Geologic depression of SW Asia & E Africa extending from the Jordan R. valley to Mozambique.
**Great Saint Ber·nard** (sānt′ bər-närd′). Alpine pass (8,110 ft/ 2,473.6 m) on the Italian-Swiss border.
**Great Salt Lake** (sôlt). Shallow body of salt water, c. 1,000 sq mi (2,590 sq km), of NW Utah between the Wasatch Mts. on the E & **Great Salt Lake Desert** on the W.
**Great Sand Dunes National Monument** (sănd′ do͞onz). Reservation containing large, high sand dunes in S Colo. in the Sangre de Cristo Mts.
**Great San·dy** (săn′dē). Desert of NW Australia N of Gibson Desert.
**Great Slave** (slāv). Lake, c. 10,980 sq mi (28,440 sq km), of S N.W.T., Canada.
**Great Smok·y Mountains** (smō′kē). Part of the Appalachian system on the N.C.-Tenn. border.
**Great Vic·to·ri·a** (vĭk-tôr′ē-ə, -tōr′-). Desert region of SW Australia.
**Great Wall of Chi·na** (wôl; chī′nə). Fortifications, c. 1,500 mi (2,415 km) long, across N China.
**Great Yar·mouth** *also* **Yar·mouth** (yär′məth). Borough of E England on the North Sea NE of London. Pop. 49,410.
**Greece** (grēs). Republic of SE Europe in the S Balkan Peninsula. Cap. Athens. Pop. 9,599,000. **—Gre′cian** (grē′shən) *adj.* **—Greek** (grēk) *adj. & n.*
**Gree·ley** (grē′lē). City of N Colo. NNE of Denver. Pop. 53,006.
**Green** (grēn). **1.** River of Ky. flowing 370 mi (595.3 km) to the Ohio R. near Evansville, Ind. **2.** River flowing 730 mi (1,174.6 km) from W Wyo. through NW Colo. & E Utah to the Colorado R. **3. Mountains.** Range of the Appalachians extending 250 mi (402.3 km) from S Que., Canada, to W Mass.
**Green Bay.** City of E Wis. on **Green Bay,** an arm of Lake Michigan N of Milwaukee. Pop. 87,899.
**Green·belt** (grēn′bĕlt′). City of central Md. near Washington, D.C. Pop. 16,000.
**Green·dale** (grēn′dāl′). Village of SE Wis. near Milwaukee. Pop. 16,928.
**Greene·ville** (grēn′vĭl′, -vəl). Town of NE Tenn. ENE of Knoxville. Pop. 14,097.
**Green·field** (grēn′fēld′). **1.** City of central Ind. E of Indianapolis. Pop. 11,439. **2.** Town of NW Mass. N of Northampton. Pop. 18,436. **3.** City of SE Wis. near Milwaukee. Pop. 31,467.
**Greenfield Park.** Town of S Que., Canada, near Montreal. Pop. 18,430.
**Green·land** (grēn′lənd, -lănd′). **1. Sea.** Section of the S Arctic Ocean off the E coast of Greenland. **2.** Largest island in the world, c. 840,000 sq mi (2,175,600 sq km), part of Denmark, in the North Atlantic off NE Canada. Cap. Godthåb. Pop. 49,719.
**Green River.** City of SW Wyo. WSW of Rock Springs. Pop. 12,807.
**Greens·bor·o** (grēnz′bûr′ə). City of N central N.C. E of Winston-Salem. Pop. 155,642.
**Greens·burg** (grēnz′bûrg′). City of SW Pa. ESE of Pittsburgh. Pop. 17,588.
**Green·ville** (grēn′vĭl′). **1.** City of W Miss. on the Mississippi N of Vicksburg. Pop. 40,613. **2.** City of E N.C. SE of Rocky Mount. Pop. 35,740. **3.** City of W Ohio NW of Dayton. Pop. 12,999. **4.** City of NW S.C. NW of Columbia. Pop. 58,242. **5.** City of NE Tex. NE of Dallas. Pop. 22,161.
**Green·wich. 1.** (grĭn′ĭj, -ĭch, grĕn′-). Borough of Greater London, SE England, on the Thames. Pop. 207,200. **2.** (grĕn′ĭch, grēn′wĭch′, grĭn′wĭch′). Town of SW Conn. on Long Is. Sound near the N.Y. border. Pop. 59,578. **3.** (grĕn′ĭch, -ĭj, grĭn′-). **Village.** Section of Lower Manhattan, New York City.
**Green·wood** (grēn′wo͝od′). **1.** City of central Ind. near Indianapolis. Pop. 19,327. **2.** City of W central Miss. E of Greenville. Pop. 20,115. **3.** City of W S.C. WNW of Columbia. Pop. 21,613.
**Greer** (grîr). City of NW S.C. WNW of Columbia. Pop. 10,525.
**Greifs·wald** (grīfs′vält′). City of N East Germany near the North Sea. Pop. 53,940.
**Gre·na·da** (grə-nā′də). **1.** Island in the Windward Is. of the West Indies, part of the nation of **Grenada,** including the S Grenadines. Cap. St. George's. Pop. 110,000. **2.** City of NW central Miss. NE of Greenville. Pop. 12,641.
**Gren·a·dines** (grĕn′ə-dēnz′). Archipelago in the Windward Is. of the E Caribbean, divided between Grenada & the nation of St. Vincent & the Grenadines.
**Gre·no·ble** (grə-nō′bəl, -nôbl′). City of SE France SSW of Chambéry. Pop. 165,431.
**Gresh·am** (grĕsh′əm). City of NW Ore. near Portland. Pop. 33,005.
**Gret·na** (grĕt′nə). City of SE La. on the Mississippi opposite New Orleans. Pop. 20,615.
**Gretna Green** (grēn). Village of S Scotland on the English border.
**Grey·lock** (grā′lŏk′), **Mount.** Highest peak, 3,491 ft (1,064.8 m), in Mass., in the W in the Berkshires.
**Grif·fin** (grĭf′ĭn). City of W central Ga. SSE of Atlanta. Pop. 20,728.
**Grif·fith** (grĭf′ĭth). Town of NW Ind. S of Hammond. Pop. 17,026.
**Gri·jal·va** (grē-häl′və). River of Central America flowing c. 400 mi (645 km) from SW Guatemala through SE Mexico to the Gulf of Campeche.
**Grims·by** (grĭmz′bē). **1.** Borough of E central England near the mouth of the Humber. Pop. 93,800. **2.** Town of S Ont., Canada, on Lake Ontario ESE of Hamilton. Pop. 15,567.
**Grim·sel** (grĭm′zəl). Pass, 7,159 ft (2,183.5 m) high, of S Switzerland between the Rhone & Aare valleys.
**Grin·del·wald** (grĭn′dl-wôld′, -vält′). Resort village of central Switzerland in the Bernese Alps ESE of Interlaken. Pop. 3,511.
**Gris-Nez** (grē-nā′), **Cape.** Promontory of N France extending into the Strait of Dover.
**Grod·no** (grôd′nō, grŏd′-). City of W European USSR on the Neman. Pop. 195,000.
**Gro·ning·en** (grō′nĭng-ən). City of NE Netherlands near the West German border. Pop. 163,357.
**Grøn·land** (grœn′län′). GREENLAND.
**Groote Ey·landt** (gro͞ot′ ī′lənd). Island of N Australia in the W part of the Gulf of Carpentaria.
**Grosse Pointe Farms** (grōs′ point′). City of SE Mich. near Detroit. Pop. 10,551.
**Grosse Pointe Park.** City of SE Mich. near Detroit. Pop. 13,639.
**Grosse Pointe Wood.** City of SE Mich. near Detroit. Pop. 18,886.
**Gros·se·to** (grō-sā′tō). City of central Italy WNW of Viterbo. Pop. 48,309.
**Gross·glock·ner** (grōs′glŏk′nər). Peak, 12,461 ft (3,800.6 m) high, in S Austria in the Tyrol.
**Gros Ventre** (grō′ vänt′). River, 100 mi (160.9 km), of W central Wyo.
**Grot·on** (grŏt′n). Town of SE Conn. on the Thames opposite New London. Pop. 10,086.
**Grove City** (grōv). City of central Ohio near Columbus. Pop. 16,793.
**Groves** (grōvz). City of SE Tex. near Port Arthur & the La. border. Pop. 17,090.

**Groz·ny** or **Groz·nyy** (grôz′nē, grŏz′-). City of SE European USSR NW of Baku. Pop. 375,000.
**Gru·dziadz** (grōō′jôNts′). City of N central Poland NE of Bydgoszcz. Pop. 75,511.
**Gua·da·la·ja·ra** (gwŏd′l-ə-här′ə). **1.** City of SW Mexico WNW of Mexico City. Pop. 1,478,383. **2.** City of central Spain NE of Madrid. Pop. 30,924.
**Gua·dal·ca·nal** (gwŏd′l-kə-năl′). Volcanic island of the W Pacific, largest of the Solomon Is.
**Gua·dal·qui·vir** (gwŏd′l-kwĭv′ər, -kə-vîr′). River of S Spain flowing c. 350 mi (565 km) to the Gulf of Cádiz.
**Gua·da·lupe** (gwŏd′l-ōōp′, gwŏd′l-ōō′pē). **1.** River, c. 300 mi (485 km), of SE Tex. **2.** Mountain range of S N.Mex. & W Tex. rising to **Guadalupe Peak,** 8,751 ft (2,669 m). **3.** (gwä′tha-lōō′pē). City of NE Mexico E of Monterrey. Pop. 51,899.
**Gua·de·loupe** (gwŏd′l-ōōp′, gwŏd′l-ōōp′). Overseas department of France in the Leeward Is. of the West Indies. Cap. Basse-Terre. Pop. 319,000.
**Gua·dia·na** (gwä′dē-ä′nə). River of S Spain flowing 510 mi (820.6 km) partly along the Spanish-Portuguese border to the Gulf of Cádiz.
**Guam** (gwäm). Island of the W Pacific, largest (209 sq mi/541.3 sq km) of the Mariana Is.; an unincorporated territory of the U.S. Cap. Agana. Pop. 105,821. **—Gua·ma′ni·an** (gwä-mā′nē-ən) *adj. & n.*
**Gua·na·ba·co·a** (gwä′nə-bə-kō′ə). City of W Cuba near Havana. Pop. 69,706.
**Gua·na·ba·ra Bay** (gwä′nə-băr′ə). Inlet of the Atlantic on the SE coast of Brazil.
**Guang·dong** (gwäng′dōōng′). Province of SE China on the South China Sea.
**Guang·xi Zhuang** (gwäng′shē′ jə-wäng′). Autonomous region of S China on the Vietnamese border.
**Guang·zhou** (gwäng′jō′). City of S China on a delta near the South China Sea. Pop. 2,300,000.
**Guan·tá·na·mo** (gwän-tä′nə-mō′). City of SE Cuba N of **Guantánamo Bay,** an inlet of the Caribbean. Pop. 145,000.
**Gua·po·ré** (gwä′pə-rā′). River of central South America flowing c. 750 mi (1,205 km) partly on the Brazil-Bolivia border to the Mamoré.
**Guá·ri·co** (gwär′ĭ-kō′). River, 300 mi (482.7 km), of W Venezuela.
**Gua·te·ma·la** (gwä′tə-mä′lə). **1.** Republic of N Central America. Cap. Guatemala. Pop. 7,262,419. **2.** *also* **Guatemala City.** Capital of Guatemala in the S central part. Pop. 700,538. **—Gua′te·ma′lan** *adj. & n.*
**Gua·via·re** (gwäv-yär′ē). River of central & E Columbia flowing 650 mi (1,045.8 km) to the Orinoco at the Columbia-Venezuela boundary.
**Gua·ya·na** (gwə-yä′nə). GUIANA.
**Gua·ya·quil** (gwī′ə-kēl′, -kĭl). City of W Ecuador near the mouth of the **Guayas R.** (gwī′əs), which drains into the **Gulf of Guayaquil,** an inlet of the Pacific. Pop. 1,022,010.
**Guay·mas** (gwī′məs). City of NW Mexico on the Gulf of California. Pop. 57,492.
**Guay·na·bo** (gwī-nä′bō). City of NE Puerto Rico near San Juan. Pop. 65,091.
**Guelph** (gwĕlf). City of S Ont., Canada, W of Toronto. Pop. 67,538.
**Guer·ni·ca** (gĕr-nē′kə). Town of N central Spain NE of Bilbao. Pop. 12,046.
**Guern·sey** (gûrn′zē). English island of the SW central English Channel in the Channel Is.
**Gui·an·a** (gē-ăn′ə, -ä′nə, gī-ăn′ə). **1.** Region of NE South America, including SE Venezuela, part of N Brazil, & French Guiana, Surinam, & Guyana. **2. Highlands.** Mountainous tableland region of N South America extending from S & SE Venezuela into Guyana & N Brazil. **—Gui·an′an, Gui′a·nese′** (-nēz′, -nēs′) *adj. & n.*
**Gui·enne** or **Guy·enne** (gē-ĕn′). Region & former province of SW central France.
**Guild·ford** (gĭl′fərd). Borough of SE England SW of London. Pop. 58,470.
**Guil·ford** (gĭl′fərd). Town of S Conn. on Long Is. Sound. Pop. 17,375.
**Guilford Court·house** (kôrt′hous′, kōrt′-). Locality in N central N.C. near Greensboro; site of American Revolutionary victory (1781).
**Gui·lin** (gwē′lĭn′). City of SE China NW of Guangzhou. Pop. 225,000.
**Guin·ea** (gĭn′ē). **1. Gulf of.** Large open inlet of the Atlantic formed by the great bend in the W central coast of Africa. **2.** Coastal W Africa from Gambia to Angola. **3.** Republic of W central Africa on the Atlantic. Cap. Conakry. Pop. 5,143,284. **—Guin′e·an** *adj. & n.*
**Guin·ea-Bis·sau** (gĭn′ē-bĭ-sou′). Country of W central Africa on the Atlantic. Cap. Bissau. Pop. 777,214.
**Gui·yang** (gə-wē′yăng′). City of SW China ENE of Kunming. Pop. 1,500,000.
**Gui·zhou** (gwē′jō′). Province of SE China.

**Gu·ja·rat** *also* **Gu·je·rat** (gōō′jə-rät′, gōōj′ə-). Region of W India.
**Guj·ran·wa·la** (gōōj′rən-wä′lə, gōōj′-). City of NE central Pakistan N of Lahore. Pop. 360,419.
**Gü·lek Bo·gaz** (gyōō-lĕk′ bō-gäz′, -äz′). Mountain pass of S Turkey in the Taurus Mts.
**Gulf In·tra·coas·tal Waterway** (gŭlf ĭn′trə-kō′stəl). Inland waterway of bays, canals, & rivers from NW Fla. to Brownsville, Tex.
**Gulf·port** (gŭlf′pôrt′, -pōrt′). **1.** City of W Fla. on Tampa Bay near St. Petersburg. Pop. 11,180. **2.** City of SE Miss. on the Intracoastal Waterway W of Biloxi. Pop. 39,676.
**Gulf States.** States of the S U.S. with coastlines on the Gulf of Mexico: Fla., Ala., Miss., La., & Tex.
**Gulf Stream.** Warm ocean current of the N Atlantic off E North America.
**Gum·ti** (gōōm′tē). River of N India flowing c. 500 mi (804.5 km) to the Ganges.
**Gunn·bjørn** (gōōn′byôrn′). Highest peak, 12,139 ft (3,702.4 m), of Greenland, near the SE coast.
**Gun·ni·son** (gŭn′ĭ-sən). River, 180 mi (289.6 km), of W central Colo.
**Gun·tur** (gōōn-tōōr′). City of SE India NNE of Vijayawada. Pop. 269,991.
**Gu·ryev** (gōōr′yəf). City of SW Central Asian USSR on the Ural R. Pop. 131,000.
**Gü·ters·loh** (gōō′tərz-lō′, gyōō′-). City of W West Germany SSW of Bielefeld. Pop. 77,128.
**Guth·rie** (gŭth′rē). City of central Okla. N of Oklahoma City. Pop. 10,312.
**Guy·a·na** (gī-ăn′ə). Republic of NE South America on the Atlantic. Cap. Georgetown. Pop. 820,000. **—Guy′a·nese′** *adj. & n.*
**Guy·enne** (gē-ĕn′). *var. of* GUIENNE.
**Gwa·li·or** (gwä′lē-ôr′). **1.** Former state of N India. **2.** City of N central India SSE of Agra. Pop. 384,772.
**Győr** (jûr). City of NW Hungary near the Czechoslovak border. Pop. 123,618.

## H

**Haar·lem** (här′ləm). City of W Netherlands near the North Sea W of Amsterdam. Pop. 164,672.
**Haar·lem·mer·meer** (här′lə-mər-mâr′). City of W Netherlands WSW of Amsterdam. Metro. area pop. 72,046.
**Ha·chi·no·he** (hä′chē-nō′hā). City of N Honshu, Japan, on the Pacific. Pop. 224,366.
**Hack·en·sack** (hăk′ən-săk′). City of NE N.J. near Jersey City. Pop. 36,039.
**Ha·da·no** (hä-dä′nō). City of E central Honshu, Japan. Pop. 103,663.
**Had·don·field** (hăd′n-fēld′). Borough of SW N.J. near Camden. Pop. 12,337.
**Ha·dri·an's Wall** (hā′drē-ənz). Ancient Roman wall, 73.5 mi (118.3 km) long, in N England.
**Hae·ju** (hī′jōō). City of SW North Korea on the Yellow Sea S of Pyongyang. Pop. 140,000.
**Ha·fun** (hă-fōōn′). Promontory on Indian Ocean coast of NE Somalia; easternmost point of Africa.
**Ha·gen** (hä′gən). City of W West Germany NE of Cologne. Pop. 229,224.
**Hag·ers·town** (hā′gərz-toun′). City of NW Md. WNW of Baltimore. Pop. 34,132.
**Hague** (hāg), **The.** De facto cap. of the Netherlands in the W part near the North Sea. Pop. 479,369.
**Hai·fa** (hī′fə). City of NW Israel on the Mediterranean. Pop. 227,800.
**Hai·kou** (hī′kou′, -kō′). City of Hainan Is., S China, on Hainan Strait. Pop. 500,000.
**Hai·nan** (hī′nän′). **1.** Strait, c. 30 mi (48 km) wide, of S China between Hainan Is. & Leizhou Peninsula. **2.** Island of S China in the South China Sea.
**Hai·naut** (ā-nō′, hā-). Historical region of SW Belgium & N France.
**Haines City** (hānz). City of central Fla. SSW of Orlando. Pop. 10,799.
**Hai·phong** (hī′fông′). City of NE Vietnam on the Red R. delta near the Gulf of Tonkin. Metro. area pop. 1,279,067.
**Hai·ti** (hā′tē). Country of the West Indies on W part of the island of Hispaniola. Cap. Port-au-Prince. Pop. 5,009,000. **—Hai′tian** (hā′shən, -tē-ən) *adj. & n.*
**Ha·ko·da·te** (hä′kə-dä′tē). City of SW Hokkaido, Japan, on Tsugaru Strait. Pop. 307,453.
**Hal·ber·stadt** (häl′bər-shtät′). City of W East Germany SW of Magdeburg. Pop. 46,669.
**Hal·di·mand** (hôl′də-mənd). Town of SE Ont., Canada, S of Hamilton. Pop. 16,375.
**Ha·le·a·ka·la** (hä′lē-ä′kə-lä′). Mountain, 10,025 ft (3,057.6 m), in **Haleakala National Park,** site of world's largest volcanic crater, 2,720 ft (829.6 m) deep, on E Maui, Hawaii.

**Hal·fa·ya** (hăl-fī'ə). Pass through coastal hills in extreme NW Egypt.
**Hal·i·car·nas·sus** (hăl'ĭ-kär-năs'əs). Ancient Greek city of SW Asia Minor on the Aegean.
**Hal·i·fax** (hăl'ə-făks'). **1.** Cap. of N.S., Canada, on the Atlantic in the S central part. Pop. 117,882. **2.** Borough of central England NE of Manchester. Pop. 88,580.
**Hal·lan·dale** (hăl'ən-dāl'). City of SE Fla. on the Atlantic S of Fort Lauderdale. Pop. 36,517.
**Hal·le** (hä'lə). City of S central E Germany WNW of Leipzig. Pop. 241,425.
**Hal·ma·he·ra** (hăl'mə-hĕr'ə, häl'-). Island of E Indonesia, largest of the Moluccas, between New Guinea & Sulawesi.
**Halm·stad** (hälm'städ'). City of SW Sweden on the Kattegat. Pop. 49,558.
**Halq al Wa·di** (hälk' äl wä'dē). LA GOULETTE.
**Häl·sing·borg** (hĕl'sĭng-bôrg', hĕl'sĭng-bôr'ē). City of NW Sweden on the Oresund. Pop. 80,986.
**Hal·tem·price** (hôl'təm-prīs'). Urban district of NE England near Hull. Pop. 54,850.
**Hal·tom City** (hôl'təm). City of NE Tex. near Fort Worth. Pop. 29,014.
**Hal·ton Hills** (hôl'tən). Town of SE Ont., Canada, near Toronto. Pop. 34,477.
**Ha·ma** *or* **Ha·mah** (hä'mä). City of W Syria on the Orontes R. Pop. 137,421.
**Ham·a·dan** (hăm'ə-dăn', -dän'). City of W Iran WSW of Teheran. Pop. 155,846.
**Ha·mah** (hä'mä). *var. of* HAMA.
**Ha·ma·ma·tsu** (hä'mə-mät'sōō). City of S central Honshu, Japan, near the Pacific WSW of Tokyo. Pop. 468,884.
**Ham·burg** (hăm'bûrg'). **1.** (*also* häm'bŏŏrg', -bŏŏrk'). City of N West Germany on the Elbe. Pop. 1,717,383. **2.** Village of W N.Y. S of Buffalo. Pop. 10,582.
**Ham·den** (hăm'dən). Town of S Conn. N of New Haven. Pop. 51,071.
**Ha·meln** (hä'məln) *or* **Ham·e·lin** (hăm'ə-lĭn, hăm'lĭn). City of N West Germany on the Weser R. SW of Hannover. Pop. 61,066.
**Ham·hung** (häm'hŏŏng'). City of E central North Korea near the Sea of Japan coast. Pop. 484,000.
**Ham·il·ton** (hăm'əl-tən). **1.** Inlet of the N Atlantic in SE Labrador, Newf., Canada. **2. River.** CHURCHILL RIVER, Labrador, Canada. **3.** Cap. of Bermuda on Bermuda Is. Pop. 2,060. **4.** City of S Ont., Canada, at W end of Lake Ontario SW of Toronto. Pop. 312,003. **5.** City of N central North Is., New Zealand, SSE of Auckland. Pop. 87,968. **6.** Burgh of S central Scotland near Glasgow. Pop. 45,495. **7.** City of SW Ohio N of Cincinnati. Pop. 63,189.
**Hamm** (häm, hăm). City of W West Germany SSE of Münster. Pop. 172,210.
**Ham·mer·fest** (hăm'ər-fĕst', hä'mər-). Town of N Norway on an island in the Arctic Ocean; northernmost town of Europe. Pop. 7,610.
**Ham·mond** (hăm'ənd). **1.** City of NW Ind. adjacent to Gary. Pop. 93,714. **2.** City of SE La. E of Baton Rouge. Pop. 15,043.
**Ham·mon·ton** (hăm'ən-tən). Town of S N.J. SE of Camden. Pop. 12,298.
**Hamp·ton** (hămp'tən). **1.** Historic section of London, England. **2.** Town of SE N.H. on the Atlantic SSW of Portsmouth. Pop. 10,493. **3.** Independent city of SE Va. opposite Norfolk on **Hampton Roads,** outlet of the James & Elizabeth rivers into Chesapeake Bay. Pop. 122,617.
**Ham·tramck** (hăm-trăm'ĭk). City of SE Mich. surrounded by Detroit. Pop. 21,300.
**Han** (hän). **1.** River, c. 210 mi (338 km), of S China. **2.** River, c. 700 mi (1,126 km), of central China.
**Han·a·han** (hăn'ə-hăn'). City of SE S.C. near Charleston. Pop. 13,224.
**Ha·nau** (hä'nou'). City of central West Germany on the Main R. E of Frankfurt. Pop. 86,676.
**Han·dan** (hän'dän'). City of E central China SSW of Beijing. Pop. 500,000.
**Han·ford** (hăn'fərd). City of S central Calif. SSE of Fresno. Pop. 20,958.
**Hang·zhou** (häng'jo') *also* **Hang·chow** (hăng'chou', häng'jō'). City of E China at the head of **Hangzhou Bay,** an inlet of the East China Sea. Pop. 1,100,000.
**Han·ka** (häng'kə). KHANKA.
**Han·ni·bal** (hăn'ə-bəl). City of NE Mo. on the Mississippi NW of St. Louis. Pop. 18,811.
**Han·no·ver** *or* **Han·o·ver** (hăn'ō'vər, -ə-vər). City of N West Germany SE of Bremen. Pop. 552,955.
**Ha·noi** (hă-noi', hə-, hä-). Cap. of Vietnam in the N part on the Red R. Metro. area pop. 2,570,905.
**Han·o·ver** (hăn'ō'vər). **1.** (*also* -ə-vər). Former kingdom & province of N West Germany. **2.** (*also* -ə-vər). *var. of* HANNOVER. **3.** Town of E Mass. SE of Boston. Pop. 11,358. **4.** Borough of NE central Pa. near Wilkes-Barre. Pop. 14,890.
**Hanover Park.** Village of NE Ill. near Chicago. Pop. 28,850.
**Han·tan** (hän'dän'). HANDAN.
**Har·a·han** (hăr'ə-hăn'). City of SE La. on the Mississippi near New Orleans. Pop. 11,384.
**Ha·ran** *or* **Har·ran** (hə-rän'). Ancient city of Mesopotamia in present-day SE Turkey.
**Ha·rar** *or* **Har·rar** (hăr'ər). City of E central Ethiopia E of Addis Ababa. Pop. 48,440.
**Har·bin** (här'bĭn, här-bĭn'). City of NE China in Manchuria. Pop. 2,750,000.
**Har·dan·ger·fjord** (här-däng'ər-fyôrd', -fyōrd'). Fjord penetrating 114 mi (183.4 km) from the Atlantic into SW Norway.
**Har·dwar** (här'dwär', hûr'-). City of N India NE of Delhi; Hindu pilgrimage center. Pop. 77,864.
**Har·gei·sa** *or* **Har·gey·sa** (här-gā'sə). City of NW Somalia E Africa. Pop. 40,254.
**Ha·ri Rud** (här'ē rōōd'). River c. 700 mi (1,126 km), of NW Afghanistan, NE Iran, & S Central Asian USSR.
**Har·lem** (här'ləm). **1.** River channel in New York City separating the N end of Manhattan Is. from the Bronx. **2.** Section of New York City in N Manhattan bordering on the Harlem and East rivers.
**Har·lin·gen** (här'lĭn-jən). City of extreme S Tex. NW of Brownsville. Pop. 43,543.
**Har·ney Peak** (här'nē). Highest elevation, 7,242 ft (2,208.8 m), of the Black Hills in SW S.Dak.
**Har·pers Ferry** (här'pərz). Town of extreme E W.Va.; scene of John Brown's rebellion (1859).
**Har·per Woods** (här'pər). City of SE Mich. near Detroit. Pop. 16,361.
**Har·ran** (hə-rän'). *var. of* HARAN.
**Har·rar** (hăr'ər). *var. of* HARAR.
**Har·ris·burg** (hăr'ĭs-bûrg'). Cap. of Pa. in the SE part WNW of Philadelphia. Pop. 53,264.
**Har·ri·son** (hăr'ĭ-sən). **1.** Town of NE N.J. near Newark. Pop. 12,242. **2.** Village of SE N.Y. near Rye. Pop. 23,046.
**Har·ri·son·burg** (hăr'ĭ-sən-bûrg'). Independent city of N central Va. NNW of Charlottesville. Pop. 19,671.
**Har·ro·gate** (hăr'ə-gət, -gāt'). Borough of N central England N of Leeds. Pop. 64,620.
**Hart·ford** (härt'fərd). Cap. of Conn. in the N central part. Pop. 136,392.
**Har·tle·pool** (härt'lē-pōōl', här'tl-). Borough of NE England on the North Sea SSE of Newcastle. Pop. 97,100.
**Har·vard** (här'vərd). **1. Mount.** Peak, 14,414 ft (4,396.3 m), in the Sawatch Range of W central Colo. **2.** Town of NE Mass. NE of Worcester. Pop. 12,170.
**Har·vey** (här'vē). City of NE Ill. near Chicago. Pop. 35,810.
**Harz** (härts). Mountain range of central Germany, c. 60 mi (100 km) long, extending across the West German-East German border NE of Göttingen.
**Has·brouck Heights** (hăz'brŏŏk'). Borough of NE N.J. near Hackensack. Pop. 12,166.
**Has·selt** (häs'əlt). City of NE Belgium, E of Brussels. Pop. 39,663.
**Has·tings** (hā'stĭngz). **1.** Borough of SE England on the Strait of Dover near the site of William the Conqueror's victory over the Saxons (1066). Pop. 74,600. **2.** City of E Minn. SE of St. Paul. Pop. 12,827. **3.** City of S Nebr. S of Grand Island. Pop. 23,045.
**Has·tings-on-Hud·son** (hā'stĭngz-ŏn-hŭd'sən). Village of SE N.Y. on the Hudson NNW of Yonkers. Pop. 8,573.
**Hatch·ie** (hăch'ē). River, c. 180 mi (290 km), of N Miss. & SW Tenn.
**Hat·ter·as** (hăt'ər-əs, hăt'rəs). Long barrier island off the E coast of N.C. between Pamlico Sound & the Atlantic, with **Cape Hatteras** projecting from the SE part.
**Hat·ties·burg** (hăt'ēz-bûrg'). City of SE Miss. SE of Jackson. Pop. 40,829.
**Hau·ra·ki Gulf** (-răk'ē, hou-rä'kē). Inlet of the S Pacific on the N coast of North Is., New Zealand.
**Haute·rive** (ōt-rēv'). Town of E Que., Canada, near Baie Comeau. Pop. 14,724.
**Ha·va·na** (hə-văn'ə). Cap. of Cuba in the NW part on the Gulf of Mexico. Pop. 1,966,435. **—Ha·va'nan** *adj.* & *n.*
**Hav·ant and Wa·ter·loo** (hăv'ənt; wô'tər-lōō', wŏt'ər-). Urban district of S England near the English Channel E of Southhampton. Pop. 112,430.
**Ha·vel** (hä'fəl). River, c. 215 mi (345 km), of N central East Germany & West Berlin.
**Have·lock** (hăv'lŏk'). City of E N.C. SE of Raleigh. Pop. 17,718.
**Hav·er·hill** (hăv'rəl, hā'vər-əl). City of NE Mass. near Lawrence. Pop. 46,865.
**Ha·ví·řov** (hä'və-rôf'). Town of N central Czechoslovakia near Ostrava. Pop. 85,000.
**Hav·re** (hăv'ər). City of N Mont. NE of Great Falls. Pop. 10,891.
**Haw** (hô). River, c. 130 mi (209 km), of N central N.C.

**Ha·wai·i** (hə-wäʹē, -yē, -wīʹē, -yē). **1.** Largest & southernmost island of the state of Hawaii. **2.** State & island group (**Hawaiian Is.**) of the W U.S. in the central Pacific. Cap. Honolulu. Pop. 965,000. —**Ha·waiʹian** *adj. & n.*

**Hawaiian Gardens.** City of S Calif. SE of Los Angeles. Pop. 10,548.

**Hawaii Vol·ca·noes National Park** (vŏl-kāʹnōz). Park on Hawaii Is. containing two active volcanoes.

**Ha·wash** (häʹwäsh′). River, c. 500 mi (805 km), of E Ethiopia.

**Hawke Bay** (hôk). Inlet of the S Pacific & E central coast of North Is., New Zealand.

**Haw·thorne** (hôʹthôrn′). **1.** City of S Calif. near Los Angeles. Pop. 56,447. **2.** Borough of NE N.J. NNE of Paterson. Pop. 18,200.

**Hay** (hā). River, c. 530 mi (885 km), of NE B.C., NW Alta., & S N.W.T., Canada, flowing NE to Great Slave Lake.

**Hayes** (hāz). River, c. 300 mi (483 km), of E Man., Canada, flowing NE to Hudson Bay.

**Hays** (hāz). City of W central Kans. W of Salina. Pop. 16,301.

**Hay·ward** (hāʹwərd). City of W Calif. SE of Oakland. Pop. 94,167.

**Ha·zel Crest** (hāʹzəl). Village of NE Ill. near Chicago. Pop. 13,973.

**Hazel Park.** City of SE Mich. near Detroit. Pop. 10,914.

**Ha·zel·wood** (hāʹzəl-wo͝od′). City of E Mo. near St. Louis. Pop. 12,935.

**Ha·zle·ton** (hāʹzəl-tən). City of E central Pa. SSW of Wilkes-Barre. Pop. 27,318.

**Heard** (hûrd). Australian-claimed island of S Indian Ocean SSE of Kerguelen Is.

**Heart** (härt). River, c. 180 mi (290 km), of SW N.Dak.

**He·bei** (hŭʹbāʹ). Province of NE China.

**Heb·ri·des** (hĕbʹrĭ-dēz′). Islands of W Scotland in the Atlantic, divided into the **Inner Hebrides,** closer to the Scottish mainland, & the **Outer Hebrides,** to the NW. —**Hebʹri·deʹan** *adj. & n.*

**He·bron** (hēʹbrən). City in the West Bank near Jerusalem. Pop. 38,309.

**Hec·ate** (hĕkʹət, -ə-tē). Strait, c. 160 mi (257 km) long & c. 35–80 mi (56–129 km) wide, of W B.C., Canada, separating the Queen Charlotte Is. from coastal islands.

**Heer·len** (hârʹlən). City of SE Netherlands near the West German border. Pop. 71,500.

**He·fei** (hŭʹfāʹ). City of E central China WSW of Nanjing. Pop. 400,000.

**He·gang** (hŭʹgängʹ). City of extreme NE China in Manchuria near the Soviet border. Pop. 350,000.

**Hei·del·berg** (hīdʹl-bûrg′, -bo͝org′). City of SW West Germany on the Neckar R. NNW of Stuttgart. Pop. 129,368.

**Heil·bronn** (hīlʹbrôn′, -brŏn′). City of S West Germany on the Neckar R. N of Stuttgart. Pop. 113,177.

**Hei·long** (hāʹlo͝ongʹ). AMUR.

**Hei·long·jiang** *also* **Hei·lung·kiang** (hāʹlo͝ongʹjē-ängʹ). Province of extreme NE China.

**Hek·la** (hĕkʹlə). Volcano, 4,747 ft (1,447.8 m), of SW Iceland.

**Hel·e·na** (hĕlʹə-nə). Cap. of Mont. in the W central part. Pop. 23,938.

**Hel·go·land** (hĕlʹgō-lănd′). Island of N West Germany, one of the N Frisian Is. in **Helgoland Bay,** an inlet of the North Sea SW of Jutland.

**Hel·i·con** (hĕlʹĭ-kŏn′, -kən). Mountain, 5,736 ft (1,749.5 m), of central Greece.

**He·li·op·o·lis** (hēʹlē-ŏpʹə-lĭs). **1.** Ancient city of N Egypt N of modern Cairo. **2.** BAALBEK.

**Hel·las** (hĕlʹəs). GREECE.

**Hel·les** (hĕlʹəs), **Cape.** Promontory of NW Turkey at the S end of Gallipoli Peninsula.

**Hel·les·pont** (hĕlʹəs-pŏnt′). DARDANELLES.

**Hells Canyon** (hĕlz). Gorge of the Snake R. on the Idaho-Ore. border.

**Hel·mand** (hĕlʹmənd). River, c. 700 mi (1,125 km), flowing SW across Afghanistan.

**Hel·mond** (hĕlʹmônt′). City of SE Netherlands ENE of Eindhoven. Pop. 59,249.

**Helm·stedt** (hĕlmʹshtet′). City of E West Germany at the East German border E of Hannover. Pop. 28,095.

**Hel·sing·borg** (hĕlʹsĭng-bôrg′, hĕlʹsĭng-bôrʹē). HÄLSINGBORG.

**Hel·sing·ør** (hĕlʹsĭng-ûr′). City of E Denmark on the Oresund. Pop. 42,425.

**Hel·sin·ki** (hĕlʹsĭngʹkē, hĕl-sĭngʹ-). Cap. of Finland in the S on the Gulf of Finland. Pop. 502,961.

**Hel·vel·lyn** (hĕl-vĕlʹĭn). Mountain, c. 3,118 ft (951 m), of NW England.

**Hel·ve·tia** (hĕl-vēʹshə, -shē-ə). SWITZERLAND. —**Hel·veʹtian** *adj. & n.*

**Hem·el Hemp·stead** (hĕmʹəl hĕmpʹstĭd). Borough of SE England NW of London. Pop. 71,150.

**Hem·et** (hĕmʹĭt). City of S Calif. E of Santa Ana. Pop. 23,211.

**Hemp·stead** (hĕmpʹstĕd′, -stĭd). Village of SE N.Y. on W Long Is. S of Mineola. Pop. 40,404.

**He·nan** (hŭʹnänʹ). Province of E central China.

**Hen·der·son** (hĕnʹdər-sən). **1.** City of NW Ky. on the Ohio R. S of Evansville, Ill. Pop. 24,834. **2.** City of SE Nev. SE of Las Vegas. Pop. 24,363. **3.** City of N N.C. NNE of Raleigh. Pop. 13,522. **4.** City of NE Tex. ESE of Tyler. Pop. 11,473.

**Hen·der·son·ville** (hĕnʹdər-sən-vĭl′). City of N Tenn. NE of Memphis. Pop. 26,561.

**Heng·e·lo** (hĕngʹə-lō′). City of E Netherlands near the West German border. Pop. 72,281.

**Heng·yang** (hŭngʹyängʹ). City of S central China SSW of Changsha. Pop. 310,000.

**Hen·ley** (hĕnʹlē) *or* **Hen·ley-on-Thames** (-ŏn-tĕmzʹ). Borough of central England W of London. Pop. 11,860.

**Hen·lo·pen** (hĕn-lōʹpən). Cape of SE Del. at S entrance to Delaware Bay.

**Hen·ry** (hĕnʹrē). Cape of SE Va. at entrance to Chesapeake Bay E of Norfolk.

**Her·a·cle·a** (hĕrʹə-klēʹə). Ancient Greek city of S Italy near the Gulf of Taranto.

**He·rat** (hə-rätʹ). City of NW Afghanistan on the Hari Rud. Pop. 163,960.

**Her·cu·la·ne·um** (hûrʹkyə-lāʹnē-əm). Ancient city of S central Italy near Naples; destroyed by eruption of Mt. Vesuvius (A.D. 79).

**Her·e·ford. 1.** (hĕrʹə-fərd). Borough of W central England SW of Birmingham. Pop. 47,800. **2.** (hûrʹfərd). City of NW Tex. in the Panhandle SW of Amarillo. Pop. 15,853.

**Her·ford** (hĕrʹfôrt′). City of N central West Germany WSW of Hannover. Pop. 64,385.

**Her·mon** (hûrʹmən). Highest peak, 9,232 ft (2,815.8 m), of the Anti-Lebanon range on the Syria-Lebanon border.

**Her·mo·sa Beach** (hər-mōʹsə). City of S Calif. SW of Los Angeles. Pop. 18,070.

**Her·mo·sil·lo** (ĕrʹmə-sēʹō, -yō). City of NW Mexico W of Chihuahua. Pop. 232,691.

**Hern·don** (hûrnʹdən). Town of N Va. WNW of Washington, D.C. Pop. 11,449.

**Her·ne** (hĕrʹnə). City of W West Germany NE of Essen. Pop. 190,561.

**Her·rin** (hĕrʹĭn). City of S Ill. SE of East St. Louis. Pop. 10,040.

**Her·ten** (hĕrʹtn). City of W West Germany N of Essen. Pop. 69,400.

**Hert·ford** (härʹfərd, härtʹ-). City of SE England N of London. Pop. 20,760.

**Her·ze·go·vi·na** (hĕrtʹsə-gō-vēʹnə, hûrtʹ-). Region of W central Yugoslavia. —**Herʹze·go·viʹni·an** *adj. & n.*

**Hesse** (hĕs). Region & former grand duchy of central West Germany. —**Hesʹsian** (hĕshʹən) *adj. & n.*

**Hi·a·le·ah** (hīʹə-lēʹə). City of SE Fla. near Miami. Pop. 145,254.

**Hib·bing** (hĭbʹĭng). City of NE Minn. NW of Duluth. Pop. 21,193.

**Hi·ber·ni·a** (hī-bûrʹnē-ə). IRELAND. —**Hi·berʹni·an** *adj. & n.*

**Hick·o·ry** (hĭkʹə-rē, hĭkʹrē). City of W central N.C. NW of Charlotte. Pop. 20,757.

**Hickory Hills.** City of NE Ill. near Chicago. Pop. 13,778.

**Hi·dal·go del Par·ral** (ē-dälʹgō dĕl pə-rälʹ). City of N Mexico S of Chihuahua. Pop. 57,619.

**Hi·er·ap·o·lis** (hī-ə-răpʹō-lĭs). Ancient city of NW Asia Minor.

**Hier·ro** (yĕrʹō). Smallest & westernmost of the Canary Is. of Spain.

**Hi·ga·shi-O·sa·ka** (hē-gäʹshē-ō-säʹkə). City of W central Honshu, Japan, near Osaka. Pop. 524,750.

**High·land** (hīʹlənd). Town of NW Ind. near Gary. Pop. 22,935.

**Highland Park. 1.** City of NE Ill. on Lake Michigan near Chicago. Pop. 30,611. **2.** City of SE Mich. surrounded by Detroit. Pop. 27,909. **3.** Borough of N central N.J. near New Brunswick. Pop. 13,396.

**High·lands** (hīʹləndz, -lənz). Mountainous region of N Scotland N of & including the Grampians.

**High Point** (hī). City of N central N.C. SW of Greensboro. Pop. 64,107.

**High Wyc·ombe** (wĭkʹəm). Borough of S England WNW of London. Pop. 61,190.

**Hil·des·heim** (hĭlʹdəs-hīm′). City of N central West Germany SSE of Hannover. Pop. 105,290.

**Hills·bor·o** (hĭlzʹbûrʹō). City of NW Ore. W of Portland. Pop. 27,664.

**Hills·bor·ough** (hĭlzʹbûrʹō). City of W Calif. S of San Francisco. Pop. 10,451.

**Hills·dale** (hĭlzʹdāl′). Borough of NE N.J. near Hackensack. Pop. 10,495.

**Hi·lo** (hēʹlō). City of Hawaii on E coast of Hawaii Is. Pop. 29,600.

**Hil·ver·sum** (hĭlʹvər-səm). City of central Netherlands SE of Amsterdam. Pop. 94,041.

**Him·a·la·yas** (hĭmʹə-lāʹəz, hĭ-mälʹyəz) *also* **Him·a·la·ya Mountains** (hĭmʹə-lāʹə, hĭ-mälʹyə). Mountain system of S central Asia extending c. 1,500 mi (2,415 km) through Kashmir, N India, S Tibet, Nepal, Sikkim, & Bhutan. —**Himʹa·laʹyan** *adj. & n.*

**Hi·me·ji** (hĭ-mĕjʹē). City of SW Honshu, Japan, WNW of Kobe. Pop. 436,086.

**Hinck·ley** (hĭngk'lē). Urban district of central England E of Birmingham. Pop. 49,310.
**Hin·du Kush** (hĭn'doo koosh'). Mountain range of SW Asia extending c. 500 mi (805 km) W from N Pakistan to NE Afghanistan.
**Hin·du·stan** (hĭn'doo-stăn', -stän', -doo-). **1.** The Indian subcontinent. **2.** N India.
**Hines·ville** (hīnz'vĭl', -vəl). City of SE Ga. SW of Savannah. Pop. 11,309.
**Hing·ham** (hĭng'əm). Town of E Mass. SE of Boston. Pop. 20,339.
**Hins·dale** (hĭnz'dāl'). Village of NE Ill. near Chicago. Pop. 16,726.
**Hip·po** (hĭp'ō) *also* **Hip·po Re·gi·us** (rē'jē-əs). Ancient Numidian city of NW Africa in present-day NE Algeria.
**Hi·ra·tsu·ka** (hĭ-rät'sə-kä'). City of central Honshu on the Pacific WSW of Yokohama. Pop. 195,635.
**Hi·ro·shi·ma** (hîr'ə-shē'mə, hĭ-rō'shə-mə). City of SW Honshu, Japan, on the Inland Sea; destroyed by first atomic bomb used in warfare (1945). Pop. 852,611.
**His·pan·io·la** (hĭs'pən-yō'lə). Island of the West Indies E of Cuba, divided between Haiti & the Dominican Republic.
**Hi·ta·chi** (hĭ-tä'chē). City of E central Honshu, Japan, on the Pacific NE of Tokyo. Pop. 202,383.
**Hitch·in** (hĭch'ĭn). Urban district of SE England NNW of London. Pop. 29,190.
**Hi·va O·a** *also* **Hi·va·o·a** (hē'və-ō'ə). Island of the S Pacific in the SE Marquesas, French Polynesia.
**Hi·was·see** (hī-wŏs'ē). River, c. 150 mi (241 km), of NE Ga., W N.C., & SE Tenn.
**Hjäl·ma·ren** (yĕl'mə-rən). Lake, c. 190 sq mi (492 sq km), of S central Sweden.
**Hka·ka·bo Ra·zi** (kä'kə-bō rä'zē). Highest peak (19,296 ft/5,885.3 m) in Burma, in the extreme N part.
**Ho·bart. 1.** (hō'bärt'). City of SE Tasmania, Australia, on an inlet of the Tasman Sea. Pop. 50,384. **2.** (hō'bərt). City of NW Ind. near Gary. Pop. 22,987.
**Hobbs** (hŏbz). City of SE N.Mex. near the Tex. border SE of Roswell. Pop. 28,794.
**Ho·bo·ken** (hō'bō'kən). **1.** City of N Belgium on the Scheldt near Antwerp. Pop. 33,693. **2.** City of NE N.J. opposite Lower Manhattan. Pop. 42,460.
**Ho Chi Minh City** (hō' chē' mĭn', shē'). City of S Vietnam near the South China Sea. Metro. area pop. 3,419,678.
**Ho·dei·da** (hō-dā'də). City of W central Yemen on the Red Sea. Pop. 80,314.
**Hód·me·ző·vá·sár·hely** (hōd'mə-zər-vä'shər-hā'). City of SE Hungary near the Tisza R. Pop. 54,481.
**Hoek van Hol·land** (hook' vän hô'länt). HOOK OF HOLLAND.
**Hof** (hôf, hōf). City of E central West Germany near the East German & Czechoslovak borders. Pop. 54,357.
**Ho·fei** (hŭ'fā'). HEFEI.
**Hoff·man Estates** (hôf'mən, hŏf'-). Village of NE Ill. near Chicago. Pop. 38,258.
**Ho·fuf** (hoo-foof'). City of E Saudi Arabia E of Riyadh. Pop. 101,271.
**Ho·he Tau·ern** (hō'ə tou'ərn). Range of the E Alps in S Austria, rising to 12,461 ft (3,800.6 m).
**Hoh·hot** (hō'hōt'). City of N China in Inner Mongolia WNW of Beijing. Pop. 700,000.
**Hoi·how** (hoi'hou', hī'hō'). HAIKOU.
**Ho·kang** (hŭ'gäng'). HEGANG.
**Hok·kai·do** (hŏ-kī'dō). Second-largest island of Japan N of Honshu.
**Hol·brook** (hōl'brook'). Town of E Mass. near Brockton. Pop. 11,140.
**Hol·den** (hōl'dən). Town of central Mass. near Worcester. Pop. 13,336.
**Hol·guín** (ôl-gēn'). City of E Cuba NNW of Santiago de Cuba. Pop. 148,000.
**Hol·land** (hŏl'ənd). **1.** NETHERLANDS. **2.** City of SW Mich. SW of Grand Rapids. Pop. 26,281. **—Hol'land·er** *n.*
**Hol·lis·ter** (hŏl'ĭ-stər). City of W Calif. SE of San Jose. Pop. 11,488.
**Hol·lis·ton** (hŏl'ĭ-stən). Town of E central Mass. SW of Boston. Pop. 12,622.
**Hol·ly·wood** (hŏl'ē-wood'). **1.** Community of S Calif., part of Los Angeles. **2.** City of SE Fla. on the Atlantic N of Miami Beach. Pop. 117,188.
**Ho·lon** (hō-lōn'). City of W central Israel near Tel Aviv-Jaffa. Pop. 121,200.
**Hol·stein** (hōl'stīn', -stēn'). Region & former duchy of N West Germany at the base of the Jutland Peninsula.
**Hol·ston** (hōl'stən). River, c. 120 mi (193 km), of NE Tenn.
**Ho·ly Cross** (hō'lē krôs', krŏs'), **Mount of the.** Peak, 13,996 ft (4,268.8 m), in the Sawatch Mts. of W central Colo.
**Hol·yoke** (hōl'yōk'). City of SW Mass. near Springfield. Pop. 44,678.
**Hom·burg** (hŏm'bûrg', -boorg'). City of central West Germany N of Frankfurt. Pop. 41,861.
**Home·stead** (hōm'stĕd'). City of SE Fla. SW of Miami. Pop. 20,668.
**Home·wood** (hōm'wood'). **1.** City of N central Ala. near Birmingham. Pop. 21,271. **2.** Village of NE Ill. near Chicago. Pop. 19,724.
**Homs** (hômz, hôms). City of W central Syria on the Orontes R. Pop. 215,423.
**Ho·nan** (hō'nän'). HENAN.
**Hon·du·ras** (hŏn-door'əs, -dyoor'-). **1. Gulf of.** Inlet of the W Caribbean on the coasts of Belize, Honduras, & Guatemala. **2.** Country of N Central America. Cap. Tegucigalpa. Pop. 3,691,000. **—Hon·du'ran** *adj.* & *n.*
**Hong Kong** *also* **Hong·kong** (hŏng'kŏng', -kông', hông'kông', -kông'). British crown colony, 391 sq mi (1,012.7 sq km), on SE coast of China SE of Guangzhou, including **Hong Kong Is.** (32 sq mi/82.9 sq km) & adjacent areas. Cap. Victoria. Pop. 5,022,000.
**Hong·shui** (hoong'shwā'). River, c. 900 mi (1,448 km), of S China.
**Hong·ze** (hoong'dzŭ'). Lake, c. 65 mi (105 km) long, of E China N of Nanjing.
**Ho·ni·a·ra** (hō'nē-är'ə). Cap. of the Solomon Is. on NW coast of Guadalcanal. Pop. 14,942.
**Hon·o·lu·lu** (hŏn'ə-loo'loo). Cap. of Hawaii on SE coast of Oahu. Pop. 365,048.
**Hon·shu** (hŏn'shoo). Island of Japan in the central part between the Sea of Japan & the Pacific.
**Hood** (hood). **1. Canal.** Narrow arm of W Puget Sound in NW Wash. **2. Mount.** Volcanic peak, 11,235 ft (3,426.7 m), in the Cascade Range of NW Ore.
**Hoogh·ly** (hoo'glē). River, c. 160 mi (257 km), of E India, the W branch of the Ganges on its delta.
**Hook of Hol·land** (hook; hŏl'ənd). Cape & harbor of SW Netherlands on the North Sea WNW of Rotterdam.
**Hoo·sac** (hoo'sək, -sĭk). S range of the Green Mts. in NW Mass. & SW Vt.
**Hoo·ver** (hoo'vər). City of N central Ala. near Birmingham. Pop. 15,064.
**Ho·pat·cong** (hə-păt'kŏn', -kŏng' ). Borough of N central N.J. on **Lake Hopatcong** (c. 7 mi/11 km long) NNW of Morristown. Pop. 15,531.
**Hope** (hōp). City of SW Ark. SW of Hot Springs. Pop. 10,331.
**Ho·pei** *or* **Ho·peh** (hō'bā'). HEBEI.
**Hope·well** (hōp'wĕl'). Independent city of E Va. SSE of Richmond. Pop. 23,397.
**Hop·kins** (hŏp'kĭnz). City of E Minn. near Minneapolis. Pop. 15,336.
**Hop·kins·ville** (hŏp'kĭnz-vĭl'). City of SW Ky. W of Bowling Green. Pop. 27,318.
**Hor·muz** *also* **Hor·moz** (hôr'mŭz', hôr-mooz'). **1.** Strait linking the Persian Gulf with the Gulf of Oman. **2.** Island of S Iran in the Strait of Hormuz.
**Horn** (hôrn), **Cape.** Headland of extreme S Chile in the Tierra del Fuego archipelago; southernmost point of South America.
**Hor·nell** (hôr-nĕl'). City of SW N.Y. WNW of Elmira. Pop. 10,234.
**Hor·sens** (hôr'sənz, -sənts). City of central Denmark at the head of **Horsens Fjord,** an inlet of the Kattegat. Pop. 44,120.
**Horse·shoe Falls** (hôrs'shoo'). CANADIAN FALLS.
**Hor·sham** (hôr'shəm). Urban district of SE England SSW of London. Pop. 26,770.
**Hor·ton** (hôr'tn). River, c. 275 mi (442 km), of W N.W.T., Canada.
**Hos·pi·ta·let** (ŏs-pĭt'l-ĕt', hŏs-). City of NE Spain SW of Barcelona. Pop. 241,978.
**Ho·tan** (hō'tän'). Intermittent river of NW China.
**Hot Springs** (hŏt). City & resort of W central Ark. within **Hot Springs National Park,** WSW of Little Rock. Pop. 35,810.
**Hou·ma** (hō'mə, hoo'-). City of SE La. SW of New Orleans. Pop. 32,602.
**Hou·sa·ton·ic** (hoo'sə-tŏn'ĭk, -zə-). River, 148 mi (238.1 km), of W Mass. & W Conn.
**Hous·ton** (yoo'stən, hyoo'-). City of SE Tex. NW of Galveston. Pop. 1,594,086.
**Hove** (hōv). Borough of SE England on the English Channel W of Brighton. Pop. 72,000.
**How·rah** (hou'rə). City of E India on the Hooghly R. opposite Calcutta. Pop. 737,877.
**Hoy·ers·wer·da** (hoi'ərz-vĕr'də). City of SE East Germany NNE of Dresden. Pop. 64,904.
**Hra·dec Krá·lo·vé** (rä'dĕts krä'lə-və, hrä'-). City of N Czechoslovakia on the Elbe R. E of Prague. Pop. 85,600.
**Hsi·ang** (shē-äng'). XIANG.
**Hsin·chu** (shĭn'choo'). City of NW Taiwan on Formosa Strait SW of Taipei. Pop. 208,038.
**Hsin·kao Shan** (shĭn'gou' shän'). Peak, 13,113 ft (3,999.5 m), of E Taiwan, highest elevation on the island.
**Huai** (hwī). River, c. 680 mi (1,094 km), of E China.
**Huai·nan** (hwī'nän'). City of E central China WNW of Nanjing. Pop. 350,000.
**Hua·lien** (hwä'lē-ĕn'). City of central Taiwan on the Pacific. Pop. 101,010.

**Hua·lla·ga** (wä-yä'gä). River, c. 700 mi (1,126 km), of central Peru.
**Huan·ca·yo** (wäng-kī'ō). City of S central Peru E of Lima. Pop. 115,693.
**Huang** (hwäng). YELLOW R.
**Huang·shi** or **Huang·shih** (hwäng'shē'). City of central China on the Yangtze SE of Wuhan. Pop. 200,000.
**Huas·ca·rán** (wäs'kə-rän'). Extinct volcano, 22,205 ft (6,772.5 m), in the Andes of W central Peru.
**Hu·bei** (ho͞o'bā'). Province of E central China.
**Hu·bli-Dhar·war** (ho͞ob'lē-där-wär'). City of SW India NW of Bangalore. Pop. 379,166.
**Huck·nall** (hŭk'nəl). Urban district of central England NNW of Nottingham. Pop. 27,110.
**Hud·ders·field** (hŭd'ərz-fēld'). Borough of N central England NE of Manchester. Pop. 130,060.
**Hud·son** (hŭd'sən). **1.** Bay of the Atlantic in E central Canada, an inland sea connected to the Atlantic by **Hudson Strait** between S Baffin Is. & N Que. **2.** River, c. 315 mi (505 km), of E N.Y. flowing S to the Atlantic at New York City. **3.** Town of E central Mass. NE of Worcester. Pop. 16,408. **4.** Town of S N.H. near Nashua. Pop. 14,022.
**Hue** (hwā, hyo͞o-ā'). City of central Vietnam near the South China Sea. Pop. 209,043.
**Huel·va** (wĕl'və). City of SW Spain on the Gulf of Cádiz. Pop. 96,689.
**Hu·ey·town** (hyo͞o'ē-toun'). City of N central Ala. near Bessemer. Pop. 13,309.
**Hu·he·hot** (ho͞o'hā-hōt'). HOHHOT.
**Hui·la** (wē'lä). Volcano, c. 18,700 ft (5,703.5 m), in the Cordillera Central of W Colombia.
**Hu·la** *also* **Hu·leh** (ho͞o'lə). Lake of NE Israel N of the Jordan R.
**Hull** (hŭl). **1.** City of SW Que., Canada, opposite Ottawa, Ont. Pop. 61,039. **2.** *also* **Kings·ton-up·on-Hull** (kĭng'stən-ə-pŏn-hŭl', -pôn-). Borough of NE England on the N shore of the Humber. Pop. 276,600.
**Hu·lun Nur** (ho͞o'lo͞on' nŏr'). Lake of NE China in W Manchuria near the Soviet border.
**Hum·ber** (hŭm'bər). Estuary, c. 40 mi (64 km), of the Trent & Ouse rivers in NE England.
**Hum·boldt** (hŭm'bōlt'). **1.** Ocean current of the S Pacific flowing N along coasts of N Chile & Peru. **2.** Bay of the Pacific in NW Calif. **3.** River, c. 290 mi (466.6 km), of N Nev. **4.** Glacier in NW Greenland, c. 60 mi (97 km) wide & 300 ft (91.5 m) high. **5.** City of W central Tenn. NNW of Jackson. Pop. 10,209.
**Hum·phreys Peak** (hŭmp'frēz', hŭm'-). Mountain, 12,633 ft (3,853.1 m), in N Ariz.
**Hu·nan** (ho͞o'nän'). Province of SE central China.
**Hun·ga·ry** (hŭng'gə-rē). Country of central Europe. Cap. Budapest. Pop. 10,709,536. **—Hun·gar'i·an** (-gâr'ē-ən) *adj. & n.*
**Hung·nam** (ho͝ong'näm'). City of central North Korea on the Sea of Japan. Pop. 143,600.
**Hung·shui** (ho͝ong'shwā'). HONGSHUI.
**Hung·tze** (ho͝ong'dzŭ'). HONGZE.
**Hun·ter** (hŭn'tər). River, 287 mi (461.8 km), of SE Australia.
**Hun·ting·ton** (hŭn'tĭng-tən). **1.** City of NE Ind. SW of Fort Wayne. Pop. 16,202. **2.** City of W W.Va. on the Ohio W of Charleston. Pop. 63,684.
**Huntington Beach.** City of S Calif. SE of Long Beach. Pop. 170,505.
**Huntington Park.** City of S Calif. near Los Angeles. Pop. 46,223.
**Hunts·ville** (hŭnts'vĭl'). **1.** Town of SE Ont., Canada, N of Toronto. Pop. 11,123. **2.** City of N central Ala. ENE of Decatur. Pop. 142,513. **3.** City of E central Tex. N of Houston. Pop. 23,936.
**Hu·on Gulf** (hyo͞o'ŏn). Inlet of the Solomon Sea on E coast of New Guinea.
**Hu·peh** or **Hu·pei** (ho͞o'bā'). HUBEI.
**Hu·ron** (hyo͝or'ən, -ŏn'). **1.** Second-largest of the Great Lakes, between SE Ont., Canada, & E Mich. **2.** City of E central S.Dak. NW of Sioux Falls. Pop. 13,000.
**Hurst** (hûrst). City of NE Tex. near Fort Worth. Pop. 31,420.
**Hutch·in·son** (hŭch'ĭn-sən). City of S central Kans. NW of Wichita. Pop. 40,284.
**Huy·ton-with-Ro·by** (hīt'n-wĭ*th*-rō'bē, -wĭth-). Urban district of NW England near Liverpool. Pop. 65,950.
**Hwang** (hwäng). YELLOW R.
**Hy·atts·ville** (hī'əts-vĭl'). City of central Md. near Washington, D.C. Pop. 12,709.
**Hyde Park** (hīd). **1.** Public park in central Greater London, England. **2.** Village of SE N.Y.; birth & burial place of Franklin D. Roosevelt. Pop. 2,805.
**Hy·der·a·bad** (hī'dər-ə-băd', -bäd', hī'drə-). **1.** City of central India ESE of Bombay. Pop. 1,607,396. **2.** City of S Pakistan on the Indus. Pop. 628,310.
**Hy·ères** (ē-âr', yâr). **1.** French island group in the Mediterranean off the SE coast of France. **2.** City of SE France on the Mediterranean. Pop. 29,366.
**Hy·met·tus** (hī-mĕt'əs). Mountain ridge, c. 3,370 ft (1,028 m), in E central Greece near Athens.
**Hyr·ca·ni·a** (hər-kā'nē-ə). Province of ancient Persia on SE shore of the Caspian Sea.

# I

**Ia·şi** (yäsh, yä'shē). City of NE Rumania near the Soviet border. Pop. 262,493.
**I·ba·dan** (ē-bäd'n). City of SW Nigeria NNE of Lagos. Pop. 847,000.
**I·ba·gué** (ē-bä-gā'). City of central Colombia W of Bogotá. Pop. 176,223.
**I·be·ri·a** (ī-bîr'ē-ə). **1.** IBERIAN PENINSULA. **2.** Ancient Spain. **—I·be'ri·an** *adj. & n.*
**Iberian.** Peninsula of SW Europe occupied by Spain & Portugal.
**I·bi·cuí** (ē'bĭ-kwē'). River, c. 300 mi (482.7 km), of S Brazil.
**I·bi·za** *also* **I·vi·za** (ē-vē'zə, -bē'-). Spanish island of the Balearics in the W Mediterranean SW of Majorca.
**I·ca** (ē'kə, ē'kä). City of SW Peru SSE of Lima. Pop. 73,883.
**I·car·i·a** (ī-kâr'ē-ə, ĭ-kâr'-). *var. of* IKARIA.
**Ice·land** (īs'lənd). Island republic in the North Atlantic near the Arctic Circle. Cap. Reykjavík. Pop. 228,785. **—Ice'land·er** *n.* **—Ice·land'ic** *adj.*
**I·chi·ha·ra** (ĭ-chē'här'ə). City of E central Honshu, Japan, on Tokyo Bay opposite Tokyo. Pop. 194,068.
**I·chi·no·mi·ya** (ē'chē-nō'mē-ä', -yä'). City of central Honshu, Japan, NNW of Nagoya. Pop. 238,463.
**I·da** (ī'də). **1.** Mountains of NW Turkey SE of ancient Troy, rising to 5,797 ft (1,768.1 m). **2. Mount.** Highest mountain on Crete (8,058 ft/2,457.7 m), in the central part.
**I·da·ho** (ī'də-hō'). State of the NW U.S. Cap. Boise. Pop. 943,935. **—I'da·ho'an** *adj. & n.*
**Idaho Falls.** City of SE Idaho NNE of Pocatello. Pop. 39,590.
**Ie·per** (yā'pər). City of W Belgium; site of three major World War I battles. Pop. 20,825.
**I·fe** (ē'fā). City of SW Nigeria E of Ibadan. Pop. 176,000.
**I·gua·çú** *also* **I·guas·sú** (ē'gwə-so͞o'). River, c. 380 mi (611 m), of S Brazil flowing W to the Paraná at the Argentina-Paraguay-Brazil border, just above which it forms **Iguaçú Falls** (2.5 mi/4 km wide).
**I·gua·la** (ĭ-gwä'lə). City of S Mexico SSW of Mexico City. Pop. 45,355.
**I·guas·sú** (ē'gwə-so͞o'). *var. of* IGUAÇÚ.
**I·gua·zú** (ē'gwə-so͞o', -zo͞o'). IGUAÇÚ.
**Ijs·sel** or **IJs·sel** (ī'səl). River, c. 70 mi (112.6 km), of E Netherlands flowing out of the Rhine R. N into the Ijsselmeer.
**Ijs·sel·meer** or **IJs·sel·meer** (ī'səl-mâr', -mär'). Dike-enclosed lake, 465 sq mi (1,204 sq km), of NW Netherlands.
**I·ka·ri·a** (ē'kə-rē'ə) *also* **I·car·i·a** (ī-kâr'ē-ə, ĭ-kâr'-). Island of SE Greece in the Aegean W of Samos.
**I·ke·da** (ĭ-kā'də). City of S Honshu, Japan, near Osaka. Pop. 100,268.
**I·le·sha** (ĭ-lĕsh'ə). City of SW Nigeria E of Ibadan. Pop. 224,000.
**I·lhé·us** (ĭl-yā'əs). City of E Brazil on the Atlantic SSW of Salvador. Pop. 58,529.
**I·li** (ē'lē'). River, c. 590 mi (949 km), of NW China & SE Central Asian USSR flowing W & NW into Lake Balkhash.
**Il·i·am·na** (ĭl'ē-ăm'nə). **1.** Lake, c. 1,000 sq mi (2,590 sq km), of SW Alas. **2.** Volcano, 10,016 ft (3,054.8 m), on the N shore of this lake.
**I·li·gan** (ĭ-lē'gän'). City of W central Mindanao, Philippines, on Mindanao Sea. Pop. 165,742.
**Il·kes·ton** (ĭl'kĭ-stən). Borough of central England NW of Nottingham. Pop. 33,690.
**I·llam·pu** (ē-äm'po͞o, ē-yäm'-). Peak, 20,873 ft (6,366.3 m), in the Andes of W Bolivia.
**I·lli·ma·ni** (ē-ə-mä'nē, ē-yə-). Mountain, 21,151 ft (6,451.1 m), in the Andes of W Bolivia.
**Il·li·nois** (ĭl'ə-noi', -noiz'). **1.** River, 273 mi (439.3 km), of N & W Ill. **2.** Waterway, 336 mi (540.6 km), of N Ill. linking Lake Michigan with the Mississippi & including the Chicago, Des Plaines, & Illinois rivers. **3.** State of N central U.S. Cap. Springfield. Pop. 11,418,461. **—Il'li·nois'an** (-noi'ən, -zən) *adj. & n.*
**Il·lyr·i·a** (ĭ-lîr'ē-ə). Ancient region of the NW Balkan Peninsula on the Adriatic coast. **—Il·lyr'i·an** *adj. & n.*
**Il·men** (ĭl'mən). Lake, varying from c. 300 to c. 800 sq mi (780–2,070 sq km), in NW European USSR SSE of Leningrad.
**I·lo·i·lo** (ē'lə-wē'lō). City of central Philippines on SE Panay. Pop. 244,211.
**I·lo·rin** (ē'lə-rēn', ĭ-lôr'ən). City of SW Nigeria NNE of Lagos. Pop. 282,000.
**I·ma·ba·ri** (ē'mə-bär'ē). City of S Japan on the Inland Sea coast of N Shikoku. Pop. 119,726.

**Im·bros** (ĭm′brəs, ēm′vrôs′) *also* **Im·roz** (ĭm-rôz′). Island of NW Turkey in the Aegean off the coast of Gallipoli.
**I·mo·la** (ē′mə-lä′). City of N central Italy SE of Bologna. Pop. 42,111.
**Im·pe·ri·a** (ĭm-pîr′ē-ə, -pĕr′-). City of NW Italy on the Ligurian Sea. Pop. 37,585.
**Im·pe·ri·al Beach** (ĭm-pîr′ē-əl). City of S Calif. on the Mexican border. Pop. 22,689.
**Imperial Valley.** Region of SE Calif. & NE Lower California, Mexico.
**Im·roz** (ĭm-rôz′). *var. of* IMBROS.
**I·na·gua** (ĭ-nä′gwə). Island group of the SE Bahamas, including **Great Inagua** & **Little Inagua.**
**I·na·ri** (ē′nə-rē, ē′när′ē). Lake, c. 500 sq mi (1,295 sq km), of N Finland.
**In·chon** (ĭn′chŏn′). City of NW South Korea on the Yellow Sea. Pop. 800,007.
**In·de·pen·dence** (ĭn′dĭ-pĕn′dəns). **1.** City of SE Kans. SE of Wichita. Pop. 10,598. **2.** City of W Mo. E of Kansas City. Pop. 111,806.
**In·di·a** (ĭn′dē-ə). **1.** Peninsula & subcontinent of S Asia S of the Himalayas, occupied by India, Nepal, Bhutan, Sikkim, Pakistan, & Bangladesh. **2.** Country of S Asia. Cap. New Delhi. Pop. 638,810,051. **—In′di·an** *adj. & n.*
**Indian.** Ocean, c. 28,350,000 sq mi (73,426,500 sq km), extending from S Asia to Antarctica & from E Africa to SE Australia.
**In·di·an·a** (ĭn′dē-ăn′ə). **1.** State of N central U.S. Cap. Indianapolis. Pop. 5,490,179. **2.** Borough of W central Pa. ENE of Pittsburgh. Pop. 16,051. **—In′di·an′i·an** *adj. & n.*
**In·di·an·ap·o·lis** (ĭn′dē-ə-năp′ə-lĭs). Cap. of Ind. in the central part. Pop. 700,807.
**In·di·a·no·la** (ĭn′dē-ə-nō′lə). City of S central Iowa S of Des Moines. Pop. 10,843.
**Indian River.** Lagoon extending c. 120 mi (193 km) along E central Fla. coast.
**Indian Territory.** Former U.S. territory, now part of Okla.
**In·dies** (ĭn′dēz). **1.** EAST INDIES. **2.** WEST INDIES.
**In·di·gir·ka** (ĭn′dĭ-gîr′kə). River, c. 1,113 mi (1,791 km), of NE Siberian USSR flowing N to the E Siberian Sea.
**In·di·o** (ĭn′dē-ō′). City of SE Calif. E of Santa Ana. Pop. 21,611.
**In·do·chi·na** (ĭn′dō-chī′nə). **1.** Peninsula of SE Asia occupied by Vietnam, Laos, Cambodia, Thailand, Burma, & the Malay Peninsula. **2.** Former federation of French colonies & protectorates in SE Asia. **—In′do·chi·nese′** (-nēz′, -nēs′) *adj. & n.*
**In·do·ne·sia** (ĭn′də-nē′zhə, -shə, -dō-). Country of SE Asia in the Malay Archipelago, including Sumatra, Java, Sulawesi, the Moluccas, parts of Borneo, New Guinea, & Timor, & many smaller islands. Cap. Djakarta. Pop. 147,383,075. **—In′do·ne′sian** *adj. & n.*
**Indonesian Ti·mor** (tē′môr′, tĭ-môr′). W part of Timor Is. in the Malay Archipelago.
**In·dore** (ĭn-dôr′, -dōr′). City of W central India NNE of Bombay. Pop. 543,381.
**In·dra·va·ti** (ĭn′drä′və-tē). River, 315 mi (506.8 km), of central India.
**In·dus** (ĭn′dəs). River of S central Asia rising in SW Tibet & flowing c. 1,900 mi (3,057 km) NW through Kashmir & SW through Pakistan to the Arabian Sea.
**In·gle·wood** (ĭng′gəl-wo͝od′). City of S Calif. near Los Angeles. Pop. 94,245.
**In·go·da** (ĭng′gə-də). River, 360 mi (579.2 km), of S Siberian USSR.
**In·gol·stadt** (ĭng′gəl-shtät′). City of S West Germany on the Danube. Pop. 88,500.
**In·gu·lets** (ĭng′gə-lĕts′, -gəl-yĕts′). River, c. 340 mi (547 km), of SW European USSR.
**Ink·ster** (ĭngk′stər). City of SE Mich. near Detroit. Pop. 35,190.
**In·land** (ĭn′lənd). Sea of the Pacific in S Japan extending c. 240 mi (386 km) between Honshu, Shikoku, & Kyushu.
**Inland Passage.** INSIDE PASSAGE.
**Inn** (ĭn). River of E Switzerland, W Austria, & SE West Germany flowing c. 320 mi (515 km) to the Danube.
**In·ner Heb·ri·des** (ĭn′ər hĕb′rĭ-dēz′). —See HEBRIDES.
**Inner Mon·go·li·a** (mŏn-gō′lē-ə, -gōl′yə, mŏng-). Autonomous region of NE China.
**Inns·bruck** (ĭnz′bro͝ok′, ĭns′-). City of SW Austria WSW of Salzburg. Pop. 115,800.
**In·side Passage** (ĭn′sīd′). Natural waterway extending c. 950 mi (1,530 km) along coasts of SE Alas. & W B.C., Canada, through the Alexander Archipelago.
**In·ter·la·ken** (ĭn′tər-lä′kən). Resort town of W central Switzerland SE of Bern. Pop. 4,735.
**In·tra·coas·tal Waterway** (ĭn′trə-kō′stəl). System of navigation channels & canals along the U.S. Atlantic & Gulf coasts.
**In·ver·car·gill** (ĭn′vər-kär′gəl). City of extreme S South Is., New Zealand. Pop. 49,738.
**In·ver Grove Heights** (ĭn′vər grōv). City of E Minn. near St. Paul. Pop. 17,171.
**In·ver·ness** (ĭn′vər-nĕs′). Burgh of N Scotland on the Moray Firth. Pop. 35,801.
**Io·án·ni·na** (yō-ä′nē-nä′). City of NW Greece near the Albanian border. Pop. 40,130.
**I·o·na** (ī-ō′nə). Island of the S Inner Hebrides, NW Scotland.
**I·o·ni·a** (ī-ō′nē-ə). Ancient region of W Asia Minor along the Aegean coast. **—I·o′ni·an** *adj. & n.*
**Ionian. 1.** Sea arm of the Mediterranean between W Greece & S Italy & Sicily. **2.** Islands of W Greece in the Ionian Sea.
**I·o·wa** (ī′ə-wə). **1.** River, c. 329 mi (529.4 km), of N & E Iowa. **2.** State of N central U.S. Cap. Des Moines. Pop. 2,913,387. **—I′o·wan** *adj. & n.*
**Iowa City.** City of E Iowa SSE of Cedar Rapids. Pop. 50,508.
**I·pin** (ē′pĭn′). YIBIN.
**I·poh** (ē′pō). City of W Malay Peninsula, Malaysia, NNW of Kuala Lumpur. Pop. 247,953.
**Ips·wich** (ĭp′swĭch′). **1.** City of E Australia near Brisbane. Pop. 69,242. **2.** Borough of E England near the North Sea NE of London. Pop. 121,500. **3.** Town of NE Mass. NNE of Salem. Pop. 11,158.
**I·qui·que** (ĭ-kē′kē, ĭ-kā′-). City of NW Chile on the Pacific. Pop. 64,500.
**I·qui·tos** (ĭ-kē′tōs). City of NE Peru on the Amazon. Pop. 111,327.
**I·rak** (ĭ-räk′, ĭ-răk′). *var. of* IRAQ.
**I·rá·kli·on** (ĭ-rä′klē-ôn′). CANDIA, Greece.
**I·ran** (ĭ-rän′, ĭ-răn′, ī-răn′). Country of SW Asia. Cap. Teheran. Pop. 37,447,000. **—I·ra′ni·an** (ĭ-rā′nē-ən) *adj. & n.*
**I·ra·pua·to** (îr′ə-pwä′tō). City of central Mexico E of Guadalajara. Pop. 135,596.
**I·raq** *also* **I·rak** (ĭ-räk′, ĭ-răk′). Country of SW Asia. Cap. Baghdad. Pop. 12,767,000. **—I·ra′qi** (ĭ-rä′kē, ĭ-răk′ē) *adj. & n.*
**I·ra·zú** (îr′ə-zo͞o′, -so͞o′). Volcano, c. 11,260 ft (3,434 m), of central Costa Rica.
**Ire·land** (īr′lənd). **1.** Island of the British Isles in the N Atlantic W of Great Britain. **2.** Republic occupying most of Ireland. Cap. Dublin. Pop. 3,440,427. **3.** NORTHERN IRELAND. **—I′rish** (ī′rĭsh) *adj. & n.*
**Irish.** Sea of the N Atlantic between Ireland & Great Britain.
**Ir·kutsk** (îr-ko͞otsk′). City of SE Siberian USSR near Lake Baikal. Pop. 550,000.
**I·ron·de·quoit** (ĭ-rŏn′dĭ-kwoit′, -kwŏt′). Town of W N.Y. W of Rochester. Pop. 57,648.
**I·ron Gate** (ī′ərn). Gorge of the Danube on the Yugoslav-Rumanian border.
**I·ron·ton** (ī′ərn-tən). City of S Ohio on the Ohio R. S of Columbus. Pop. 14,290.
**Ir·ra·wad·dy** (îr′ə-wŏd′ē). Chief river of Burma, flowing c. 1,000 mi (1,609 km) S to the Andaman Sea.
**Ir·tish** *or* **Ir·tysh** (îr-tĭsh′). River of NW China & W Siberian USSR flowing c. 2,650 mi (4,265 km) NW into the Ob.
**I·rún** (ē-ro͞on′). City of N Spain near the Bay of Biscay & the French border. Pop. 38,014.
**Ir·vine. 1.** (ûr′vĭn′). Burgh of SW Scotland SW of Glasgow. Pop. 48,500. **2.** (ûr′vīn′). City of S Calif. SE of Santa Ana. Pop. 62,134.
**Ir·ving** (ûr′vĭng). Town of NE Tex. near Dallas. Pop. 109,943.
**Ir·ving·ton** (ûr′vĭng-tən). Town of NE N.J. near Newark. Pop. 61,493.
**Is·a·bel·a** (ĭz′ə-bĕl′ə). Largest of the Galápagos Is. of Ecuador.
**I·sar** (ē′zär). River, c. 160 mi (257 km), of W Austria & SE West Germany.
**Is·chi·a** (ĭs′kē-ə). Island of S Italy in the Tyrrhenian Sea at the entrance to the Bay of Naples.
**I·se** (ē′sā). **1.** Bay of the Pacific extending c. 15 mi (24 km) into the S central coast of Honshu, Japan. **2.** City of S Honshu, Japan, on Ise Bay. Pop. 104,957.
**I·se·o** (ē-zā′ō). Lake, 24 sq mi (62 sq km), of N Italy ENE of Milan.
**I·sère** (ē-zâr′). River, c. 180 mi (290 km), of SE France.
**I·ser·lohn** (ē′zər-lōn′, ē′zər-lōn′). City of W West Germany NE of Cologne. Pop. 96,174.
**I·se·sa·ki** (ē′sā-sä′kē). City of central Honshu, Japan. Pop. 104,300.
**I·se·yin** (ē′sə-yēn′). City of SW Nigeria NNW of Ibadan. Pop. 115,083.
**Is·fa·han** (ĭs′fə-hän′). ESFAHAN.
**I·shi·ka·ri Bay** (ĭsh′ĭ-kär′ē). Inlet of the Sea of Japan on the W coast of Hokkaido, Japan.
**I·shim** (ĭ-shĭm′). River, c. 1,130 mi (1,818 km), of W Siberian USSR.
**I·shi·no·ma·ki** (ĭsh′ĭ-nō-mä′kē). City of N Honshu, Japan, on the Pacific. Pop. 115,085.
**I·sis** (ī′sĭs). The upper Thames R. in central England.
**Is·kar** *or* **Is·kŭr** (ĭs′kər). River, c. 250 mi (402 km), of NW Bulgaria.
**Is·ken·de·run** (ĭs-kĕn′də-ro͞on′). City of S Turkey on the NE corner of the Mediterranean. Pop. 107,437.
**Is·kŭr** (ĭs′kər). *var. of* ISKAR.
**Is·lam·a·bad** (ĭs-lä′mə-bäd′, ĭz-lăm′ə-băd′). Cap. of Pakistan in the NE part NE of Rawalpindi. Pop. 77,318.
**Ís·land** (ē′slänt′). ICELAND.
**Is·lay** (ī′lā, ī′lə). Island of S Inner Hebrides, W Scotland.

**Isle au Haut** (ī'lə-hō', ē'lə-). Island of S central Me. at the entrance to Penobscot Bay.
**Isle of** (īl). See under final element; e.g., **Isle of Wight** appears at WIGHT.
**Isle Roy·ale** (roi'əl). Island, c. 210 sq mi (544 sq km), of N Mich. in NW Lake Superior, included with adjacent islands in **Isle Royale National Park.**
**I·slip** (ī'slĭp). Town of SE N.Y. on Long Is. Pop. 12,100.
**Is·ma·i·li·a** *also* **Is·ma·i·li·ya** (ĭz'mā-ə-lē'ə). City of NE Egypt on the Suez Canal. Pop. 145,978.
**Is·par·ta** (ĭs'pär-tä'). City of W central Turkey SW of Ankara. Pop. 62,870.
**Is·ra·el** (ĭz'rē-əl). **1.** Ancient kingdom of N Palestine. **2.** Country of SW Asia on the E Mediterranean. Cap. Jerusalem. Pop. 3,878,000. **—Is·rae'li** (ĭz-rā'lē) *adj. & n.*
**Is·sus** (ĭs'əs). Ancient town of SE Asia Minor near modern Iskenderun, Turkey.
**Is·syk-Kul** (ĭs'ĭk-kŭl'). Lake, c. 2,395 sq mi (6,203 sq km), of SE Central Asian USSR in the Tian Shan near the NW Chinese border.
**Is·sy-les-Mou·li·neaux** (ē-sē'lā-mōō'lə-nō'). City of N central France near Paris. Pop. 47,355.
**Is·tan·bul** (ĭs'tăn-bōōl', -tän-, -təm-). Largest city of Turkey, in the NW part on the European side of the Bosporus & the Sea of Marmara. Pop. 2,547,364.
**Is·ter** (ĭs'tər). DANUBE.
**Is·tri·a** (ĭs'trē-ə). Peninsula of NW Yugoslavia projecting into the N Adriatic. **—Is'tri·an** *adj. & n.*
**I·ta·bu·na** (ē'tə-bōō'nə). City of E Brazil SSE of Salvador. Pop. 89,928.
**I·ta·lia** (ē-täl'yə). ITALY.
**It·a·ly** (ĭt'ə-lē). **1.** Peninsula of S Europe projecting c. 600 mi (965 km) into the Mediterranean between the Tyrrhenian & Adriatic seas. **2.** Country of S Europe including the peninsula of Italy, Sardinia, & Sicily. Cap. Rome. Pop. 57,140,000. **—I·tal'ian** (ĭ-tăl'yən) *adj. & n.*
**I·ta·mi** (ē-tä'mē). City of S Honshu, Japan, on Osaka Bay near Osaka. Pop. 171,978.
**I·ta·pe·cu·ru** (ē'tə-pā'kə-rōō'). River, c. 450 mi (724 km), of NE Brazil.
**I·tas·ca** (ī-tăs'kə). Lake, c. 2 sq mi (5.2 sq km), of NW Minn., at source of the Mississippi R.
**I·té·nez** (ē-tā'nəs). GUAPORÉ.
**Ith·a·ca** (ĭth'ə-kə). **1.** Island of W Greece in the Ionian Is. E of Cephalonia. **2.** City of W central N.Y. on Cayuga Lake SSW of Syracuse. Pop. 28,732. **—Ith'a·can** *adj. & n.*
**I·thá·ki** (ĭ-thä'kē). ITHACA.
**I·tim·bi·ri** (ē'tĭm-bîr'ē). River, c. 350 mi (563 km), of N Zaire.
**I·tsu·ku·shi·ma** (ĭt'sōō-kōō'shĭ-mə). Island, 12 sq mi (31 sq km), of SW Japan in the Inland Sea SW of Hiroshima.
**It·u·rae·a** *or* **It·u·re·a** (ĭch'ə-rē'ə). Ancient country of NE Palestine. **—It'u·rae'an, It'u·re'an** *adj. & n.*
**I·tu·rup** (ē'tər-əp). Island of Far Eastern USSR in the Pacific, largest of the Kuriles.
**I·va·no-Fran·kovsk** (ĭ-vä'nō-frăng-kôfsk'). City of extreme SW European USSR in the SW Ukraine. Pop. 150,000.
**I·va·no·vo** (ĭ-vä'nə-və). City of central European USSR NE of Moscow. Pop. 465,000.
**I·vi·za** (ē-vē'sə, -bē'-). *var. of* IBIZA.
**I·vo·ry Coast** (ī'və-rē, īv'rē). Country of W Africa on the Gulf of Guinea. Cap. Abidjan. Pop. 7,920,000.
**I·vry-sur-Seine** (ē-vrē'sōōr-sān', -sĕn'). City of N central France near Paris. Pop. 62,804.
**I·wa·ki** (ī-wä'kē). City of NE Honshu, Japan, on the Pacific. Pop. 330,213.
**I·wa·ku·ni** (ē'wä-kōō'nē). City of SW Honshu, Japan, on the Inland Sea. Pop. 111,069.
**I·wo** (ē'wō). City of SW Nigeria near Ibadan. Pop. 214,000.
**I·wo Ji·ma** (ē'wō jē'mə). Largest (8 sq mi/21 sq km) of the Volcano Is. of Japan in the NW Pacific E of Taiwan.
**Ix·elles** (ēk-sĕl'). City of central Belgium near Brussels. Pop. 86,450.
**Ix·ta·ci·hua·tl** (ēs'tä-sē'wät'l) *also* **Iz·tac·ci·hua·tl** (-täk-). Dormant volcano, 17,342 ft (5,289.3 m), in central Mexico.
**I·za·bal** (ē'zə-băl', ē'sə-băl'). Lake, c. 30 mi (48 km) long & 15 mi (24 km) wide, of E Guatemala.
**I·zal·co** (ĭ-zăl'kō, ē-säl'-). Active volcano, c. 7,828 ft (2,388 m), of W El Salvador.
**I·zhevsk** (ē-zhĕfsk'). City of E central European USSR ENE of Kazan. Pop. 549,000.
**Iz·ma·il** (ĭz'mē-əl, ĭs'mə-ē'əl). City of SW European USSR near the Rumanian border. Pop. 83,000.
**Iz·mir** (ĭz-mîr'). City of W Turkey on the **Gulf of Izmir,** an inlet of the Aegean. Pop. 636,834.

ōō **boot** ou **out** th **thin** *th* **this** ŭ **cut** ûr **urge** y **young**
yōō **abuse** zh **vision** ə **about,** **item,** **edible,** **gallop,** **circus**

**Iz·mit** (ĭz-mĭt'). City of NW Turkey on the **Gulf of Izmit,** E extension of the Sea of Marmara. Pop. 165,483.
**Iz·nik** (ĭz-nĭk'). Lake of NW Turkey E of Sea of Marmara.
**Iz·tac·ci·hua·tl** (ēs'täk-sē'wät'l). *var. of* IXTACIHUATL.
**I·zu·mi** (ĭ-zōō'mē). City of S Honshu, Japan, near Osaka. Pop. 118,237.

# J

**Jab·al·pur** (jŭb'əl-pŏŏr') *or* **Jub·bul·pore** (-pôr', -pōr'). City of central India SSE of Delhi. Pop. 426,224.
**Jack·son** (jăk'sən). **1.** City of S central Mich. W of Detroit. Pop. 39,739. **2.** Cap. of Miss. in the W central part. Pop. 202,895. **3.** City of W Tenn. NE of Memphis. Pop. 49,131.
**Jackson Hole.** Valley of NW Wyo. E of the Teton Range.
**Jack·son·ville** (jăk'sən-vĭl'). **1.** City of central Ark. NE of Little Rock. Pop. 27,589. **2.** City of NE Fla. on the St. Johns R. near the Atlantic & the Ga. border. Pop. 540,898. **3.** City of W central Ill. W of Springfield. Pop. 20,284. **4.** City of E N.C. near the Atlantic NNE of Wilmington. Pop. 17,056. **5.** City of E Tex. SE of Dallas. Pop. 12,264.
**Jacksonville Beach.** City of NE Fla. on the Atlantic near Jacksonville. Pop. 15,462.
**Ja·dot·ville** (zhä-dō-vē'). LIKASI.
**Ja·én** (hä-ān'). City of S Spain NNW of Granada. Pop. 71,145.
**Jaf·fa** (jăf'ə, yäf'ə). Former city of W central Israel, since 1950 a district of Tel Aviv.
**Jaff·na** (jăf'nə). City of extreme N Sri Lanka on Palk Strait. Pop. 112,000.
**Jai·pur** (jī'pŏŏr'). City of NW India SW of Delhi. Pop. 615, 258.
**Ja·kar·ta** (jə-kär'tə). DJAKARTA.
**Ja·la·pa** (hə-lä'pə) *also* **Ja·la·pa En·rí·quez** (ĕn-rē'kəs). City of E central Mexico E of Mexico City. Pop. 161,352.
**Ja·lu·it** (jăl'ə-wĭt). Atoll in the Marshall Is. of the W Pacific.
**Ja·mai·ca** (jə-mā'kə). Island republic in the Caribbean S of Cuba. Cap. Kingston. Pop. 2,161,000. **—Ja·mai'can** *adj. & n.*
**James** (jāmz). **1. Bay.** S arm of Hudson Bay in N.W.T., Canada, between NE Ont. & W Que. **2.** River of E N.Dak. & E S.Dak. flowing 710 mi (1,142.4 km) S to the Missouri. **3.** River of Va. flowing 340 mi (547.1 km) E to Chesapeake Bay.
**James·town** (jāmz'toun'). **1.** Cap. of St. Helena in the S Atlantic. Pop. 1,516. **2.** City of W N.Y. on Chautauqua Lake near the Pa. border. Pop. 35,775. **3.** City of SE N.Dak. E of Bismarck. Pop. 16,280. **4.** Former village of SE Va., first permanent English settlement (1607) in America.
**Jam·mu** (jŭm'ōō). City of N India near the Pakistan border S of Srinagar. Pop. 155,338.
**Jammu and Kash·mir** (kăsh'mîr', kăsh-mîr'). KASHMIR.
**Jam·na·gar** (jäm-nŭg'ər). City of W India on the Gulf of Kutch. Pop. 214,816.
**Jam·shed·pur** (jäm'shĕd-pŏŏr'). City of E India WNW of Calcutta. Pop. 341,576.
**Janes·ville** (jānz'vĭl'). City of S Wis. N of Beloit. Pop. 51,071.
**Jan May·en** (yän mī'ən). Island of Norway, c. 145 sq mi (376 sq km), in the Greenland Sea midway between N Norway & Greenland.
**Ja·pan** (jə-păn'). **1. Sea of.** Part of the Pacific between Japan & the Asian mainland. **2.** Warm ocean current flowing NE from the Philippine Sea past SE Japan into the N Pacific. **3.** Country of Asia on an archipelago off the NE coast. Cap. Tokyo. Pop. 117,057,485. **—Jap'a·nese'** (jăp'ə-nēz', -nēs') *adj. & n.*
**Ja·pu·rá** (zhä'pōō-rä'). River of S Colombia & NW Brazil flowing c. 1,500 mi (2,414 km) SE to the Amazon.
**Jas·per** (jăs'pər). City of N central Ala. NW of Birmingham. Pop. 11,894.
**Jas·sy** (yä'sē). IAŞI.
**Ja·va** (jä'və, jăv'ə). **1.** Sea of the W Pacific between Java & Borneo. **2.** Island of Indonesia SE of Sumatra.
**Ja·va·rí** (zhä'və-rē'). River rising in E Peru & flowing c. 650 mi (1,046 km) along the Peru-Brazil border to the Amazon.
**Jean·nette** (jə-nĕt'). City of SW Pa. ESE of Pittsburgh. Pop. 13,106.
**Jeb·el esh Shar·qi** (jĕb'əl ĕsh shär'kē). ANTI-LEBANON.
**Jebel Kath·e·ri·na** (kăth'ə-rē'nə). Mountain, 8,651 ft (2,638.6 m), of NE Egypt.
**Jebel Mu·sa** (mōō'sə). Mountain, 2,790 ft (851 m), in N Morocco on the Strait of Gibraltar.
**Jebel Toub·kal** (tōōb-käl'). Mountain, 13,671 ft (4,169.7 m), of central Morocco in the Atlas Mts.
**Jed·da** (jĕd'ə). *var. of* JIDDA.
**Jef·fer·son** (jĕf'ər-sən). **1.** River, c. 250 mi (402 km), of SW Mont., a headwater of the Missouri. **2. Mount.** Peak, 10,499 ft (3,202.2 m), in the Cascade Range in NW Ore.
**Jefferson City.** Cap. of Mo. in the central part on the Missouri R. Pop. 33,619.
**Jef·fer·son·town** (jĕf'ər-sən-toun'). City of N Ky. near Louisville. Pop. 15,795.

**Jef·fer·son·ville** (jĕf'ər-sən-vĭl'). City of S Ind. on the Ohio opposite Louisville, Ky. Pop. 21,220.

**Jeh·lam** (jā'ləm). *var. of* JHELUM.

**Je·le·nia Gó·ra** (yā-lĕn'yə go͝or'ə). City of W Poland WSW of Poznań. Pop. 55,720.

**Je·na** (yā'nə). City of S East Germany, SW of Leipzig. Pop. 99,431.

**Jen·nings** (jĕn'ĭngz). **1.** City of SW La. E of Lake Charles. Pop. 12,401. **2.** City of E Mo. near St. Louis. Pop. 17,026.

**Je·qui·tin·hon·ha** (zhə-kēt'n-yōn'yə, -kē'tə-nōn'-). River, c. 500 mi (805 km), of E Brazil.

**Je·rez** (hə-rās') *also* **Je·rez de la Fron·te·ra** (hə-rĕz' də lə frŭn'tĕr'-ə). City of SW Spain NW of Cádiz. Pop. 112,411.

**Jer·i·cho** (jĕr'ĭ-kō'). **1.** Ancient city of Palestine near the NW shore of the Dead Sea. **2.** Town in the West Bank near the site of ancient Jericho. Pop. 5,312.

**Jer·sey** (jûr'zē). Largest (45 sq mi/116.6 sq km) of the Channel Is. in the English Channel.

**Jersey City.** City of NE N.J. on the Hudson opposite Lower Manhattan. Pop. 223,532.

**Je·ru·sa·lem** (jə-ro͞o'sə-ləm, -zə-). Cap. of Israel in the E central part in the West Bank. Pop. 376,000.

**Jer·vis Bay** (jär'vĭs). Inlet of the Tasman Sea on the SE coast of Australia.

**Jez·re·el** (jĕz'rē-ĕl', jĕz'rēl', -rē'əl). Plain of N Israel.

**Jhan·si** (jän'sē). City of N central India SSE of Delhi. Pop. 173,292.

**Jhe·lum** *also* **Jeh·lam** (jā'ləm). River, c. 480 mi (772 km), of N India & NE Pakistan.

**Jia·ling** (jē-ä'lĭng'). River, c. 500 mi (805 km), of central China.

**Jia·mu·si** (jē-ä'mo͞o'sē'). City of NE China, ENE of Harbin. Pop. 275,000.

**Jiang·su** (jē-äng'so͞o'). Province of E China on the Yellow Sea.

**Jiang·xi** (jē-äng'shē'). Province of SE China.

**Jiao·zuo** (jē-ou'dzə-wō'). City of E central China NNW of Zhengzhou. Pop. 300,000.

**Ji·bu·ti** (jĭ-bo͞o'tē). DJIBOUTI.

**Jid·da** (jĭd'ə) *also* **Jed·da** (jĕd'ə). City of W central Saudi Arabia on the Red Sea. Pop. 561,104.

**Ji·lin** (jē'lĭn'). **1.** Province of NE China. **2.** City of NE China E of Changchun. Pop. 1,200,000.

**Ji·long** (jē'lo͝ong'). KEELUNG.

**Ji·nan** (jē'nän'). City of E China on the Yellow R. S of Tianjin. Pop. 1,500,000.

**Jing·de·zhen** (jĭng'dŭ'jŭn'). City of SE China WSW of Shanghai. Pop. 300,000.

**Jin·zhou** (jĭn'jō'). City of NE China ENE of Beijing. Pop. 750,000.

**Ji·xi** (jē'shē'). City of NE China near the Soviet border E of Harbin. Pop. 350,000.

**João Pes·so·a** (zhwouɴ' pə-sō'ə). City of NE Brazil near the Atlantic N of Recife. Pop. 197,398.

**Jodh·pur** (jŏd'pər, -po͝or'). City of W India SW of Delhi. Pop. 317,612.

**Jog·ja·kar·ta** (jŏg'yə-kär'tə). DJOKJAKARTA.

**Jo·han·nes·burg** (jō-hăn'ĭs-bûrg', -hä'nĭs-). Largest city of South Africa, in the NE part. Pop. 654,232.

**John Day** (jŏn' dā'). River, 281 mi (452.1 km), of N Ore.

**John o'Groat's** (ə-grōts'). Point on the NE coast of Scotland, traditionally the N limit of Great Britain.

**John·son City** (jŏn'sən). **1.** Village of S N.Y. near Binghamton. Pop. 17,126. **2.** City of NE Tenn. ENE of Knoxville. Pop. 39,753.

**John·ston** (jŏn'stən). Town of N central R.I. near Providence. Pop. 24,907.

**Johns·town** (jŏnz'toun'). City of SW Pa. E of Pittsburgh. Pop. 35,496.

**Jo·hore Bah·ru** (jə-hôr' bä'ro͞o, -hōr') *also* **Jo·hor Ba·ha·ru** (bə-hä'ro͞o). City of S Malaysia on the S tip of Malay Peninsula opposite Singapore Is. Pop. 136,234.

**Join·vi·le** *or* **Join·vil·le** (zhoiɴ-vē'lē). City of S Brazil NE of Pôrto Alegre. Pop. 77,760.

**Jok·ja·kar·ta** (jŏk'yə-kär'tə). DJOKJAKARTA.

**Jo·li·et** (jō'lē-ĕt', jō'lē-ĕt'). City of NE Ill. SW of Chicago. Pop. 77,956.

**Jo·li·ette** (zhō'lē-ĕt'). City of S Que., Canada, NNE of Montreal. Pop. 18,118.

**Jo·lo** (hō'lō). Island of S Philippines in the Sulu Archipelago.

**Jones·bor·o** (jōnz'bûr'ō, -bûr'ə). City of NE Ark. NE of Little Rock. Pop. 31,419.

**Jön·kö·ping** (yœn'chœ'pĭng). City of S Sweden SW of Stockholm. Pop. 78,650.

**Jon·quière** (zhōɴ'kē-ĕr'). City of S Que., Canada, on the Saguenay R. N of Quebec city. Pop. 60,691.

**Jop·lin** (jŏp'lĭn). City of SW Mo. near the Kans. border WSW of Springfield. Pop. 38,893.

**Jop·pa** (jŏp'ə). JAFFA.

**Jor·dan** (jôr'dn). **1.** River of NE Israel & NW Jordan flowing c. 200 mi (322 km) S through the Sea of Galilee to the Dead Sea. **2.** Country of SW Asia in NW Arabia. Cap. Amman. Pop. 2,152,273. **—Jor·da'ni·an** (jôr-dā'nē-ən) *adj.* & *n.*

**Jos** (jôs). City of central Nigeria S of Kano. Pop. 105,000.

**Jo·tun·hei·men** (yōt'n-hā'mən). Range in S central Norway rising to c. 8,100 ft (2,471 m).

**Juan de Fu·ca** (hwän' də fo͞o'kə, fyo͞o'-, wän'). Strait, c. 100 mi (161 km) long, between NW Wash. & Vancouver Is., B.C., Canada.

**Juan Fer·nán·dez** (fər-năn'dəs). Island group belonging to Chile, in the SE Pacific W of Chile.

**Juá·rez** (wär'ĕz, -ĕs, -əs, hwär'-). CIUDAD JUÁREZ.

**Ju·ba** (jo͞o'bə). River of S Ethiopia & S Somalia flowing c. 1,000 mi (1,609 km) S to the Indian Ocean.

**Jub·bul·pore** (jŭb'əl-pôr', -pōr'). *var. of* JABALPUR.

**Jú·car** (ho͞o'kär'). River, c. 300 mi (483 km), of E Spain.

**Ju·dae·a** (jo͞o-dē'ə, -dā'-). *var. of* JUDEA.

**Ju·dah** (jo͞o'də). Ancient kingdom of S Palestine.

**Ju·de·a** *also* **Ju·dae·a** (jo͞o-dē'ə, -dā'-). Ancient region of S Palestine comprising present-day S Israel & SW Jordan. **—Ju·de'an** *adj.* & *n.*

**Ju·dith** (jo͞o'dĭth). River, 124 mi (199.5 km), of central Mont.

**Ju·go·sla·vi·a** (yo͞o'gō-slä'vē-ə). YUGOSLAVIA.

**Juiz de Fo·ra** (zhə-wēzh' də fôr'ə). City of SE Brazil N of Rio de Janeiro. Pop. 218,832.

**Jul·ian Alps** (jo͞ol'yən ălps'). Range of the E Alps in NW Yugoslavia & NE Italy.

**Jul·lun·dur** (jŭl'ən-dər). City of NW India NNW of Delhi. Pop. 296,106.

**Jum·na** (jŭm'nə). River of N India flowing c. 860 mi (1,384 km) SE to the Ganges.

**Junc·tion City** (jŭngk'shən). City of E central Kans. W of Topeka. Pop. 19,305.

**Jun·di·aí** (zho͞on'dyə-ē'). City of S Brazil NW of São Paulo. Pop. 145,785.

**Ju·neau** (jo͞o'nō'). Cap. of Alas. in the SE Panhandle. Pop. 19,528.

**Jung·frau** (yo͝ong'frou'). Mountain, c. 13,653 ft (4,164 m), in the Bernese Alps in S central Switzerland.

**Ju·ni·at·a** (jo͞o'nē-ăt'ə). River, c. 150 mi (241 km), of S central Pa.

**Ju·nín** (ho͞o-nēn'). **1.** City of E Argentina W of Buenos Aires. Pop. 59,020. **2.** Town of W central Peru NE of Lima, where Bolívar & Sucre defeated the Spaniards (1824). Pop. 8,282.

**Ju·ra** (jo͝or'ə). Range extending c. 200 mi (322 km) along the French-Swiss border.

**Ju·ruá** (zho͝or'ə-wä'). River of E Peru & NW Brazil flowing c. 1,200 mi (1,931 km) NE to the Amazon.

**Ju·rue·na** (zho͝or'ə-wā'nə, zho͝or'wā'-). River, c. 500 mi (805 km), of W central Brazil.

**Jus·tice** (jŭs'tĭs). Village of NE Ill. near Chicago. Pop. 10,552.

**Jut·land** (jŭt'lənd). Peninsula of N Europe comprising mainland Denmark & N West Germany.

**Jyl·land** (yo͞o'lăn). JUTLAND.

**Jy·väs·ky·lä** (yo͞o'və-sko͞o'lə, -skyo͞o'-). City of S central Finland NNE of Helsinki. Pop. 61,209.

# K

**K2** (kā'to͞o'). GODWIN AUSTEN.

**Ka·bul** (kä'bo͝ol, kə-bo͝ol'). **1.** River, c. 300 mi (483 km), of E Afghanistan & N Pakistan. **2.** Cap. of Afghanistan in the E part on the Kabul R. Pop. 905,108.

**Ka·di·yev·ka** (kə-dē'yəf-kə). City of S European USSR in the Ukraine SE of Kharkov. Pop. 108,000.

**Ka·do·ma** (kä-dō'mə). City of S Honshu, Japan, near Osaka. Pop. 143,238.

**Ka·du·na** (kə-do͞o'nə). City of N Nigeria NE of Lagos. Pop. 202,000

**Kae·song** (kā'sông'). City of S North Korea near the South Korean border. Pop. 175,000.

**Ka·fu·e** (kə-fo͞o'ē). River, c. 600 mi (965 km), of central Zambia

**Ka·ge·ra** (kə-gĕr'ə). River of E central Africa flowing c. 250 mi (402 km) along the Rwanda-Tanzania border then E into Lake Victoria

**Ka·go·shi·ma** (kä'gə-shē'mə). City of S Kyushu, Japan, on **Kagoshima Bay,** an inlet of the East China Sea. Pop. 456,827.

**Ka·ho·o·la·we** (kä-hō'ə-lä'wē, -vē). Island, 45 sq mi (116.6 sq km) of Hawaii SW of Maui.

**Kai·e·teur Falls** (kī'ĭ-to͝or'). Waterfall, 741 ft (226 m) high, in the Potaro R. of W central Guyana.

**Kai·feng** (kī'fŭng'). City of E central China SSW of Beijing. Pop 330,000.

**Kai·las** (kī-läs'). Mountain, 22,022 ft (6,714 m), in the Tibetan Himalayas of SW China.

**Kai·lu·a** (kī-lo͞o'ə). City of Hawaii on the E coast of Oahu. Pop 39,700.

**Kair·ouan** (kĕr-wän'). City & Moslem shrine of N Tunisia S of Tunis. Pop. 54,546.

**Kai·sers·lau·tern** (kī'zərs-lou'tərn). City of SW West Germany SW of Frankfurt. Pop. 100,886.

**Ka·ki·na·da** (kä'kə-nä'də). City of SE India on the Bay of Bengal NNE of Madras. Pop. 164,200.
**Ka·ko·ga·wa** (kä'kə-gä'wə). City of S Honshu, Japan, W of Osaka. Pop. 169,293.
**Ka Lae** (kä' lä'ā). Southernmost point of Hawaii Is., Hawaii.
**Ka·la·ha·ri** (käl'ə-här'ē). Desert of S Botswana & E Namibia.
**Ka·lakh** (kä'läKH'). Ancient city of Assyria S of present-day Mosul, Iraq.
**Kal·a·ma·zoo** (käl'ə-mə-zōō'). City of SW Mich. WSW of Lansing. Pop. 79,722.
**Ka·lat** *also* **Khe·lat** (kə-lät'). Region & former state of NW Pakistan.
**Kal·gan** (käl'gän'). ZHANGJIAKOU.
**Kal·goor·lie** (käl-gŏŏr'lē). Town of SW Australia ENE of Perth. Pop. 9,067.
**Ka·li·ma** (kə-lē'mə). City of E Zaire on Lake Tanganyika. Pop. 27,500.
**Ka·li·man·tan** (käl'ə-män'tän', kä'lə-män'tän'). Indonesian part of Borneo.
**Ka·li·nin** (kə-lē'nin). City of central European USSR NW of Moscow. Pop. 412,000.
**Ka·li·nin·grad** (kə-lē'nin-gräd'). City of extreme W European USSR on the Baltic near the Polish border. Pop. 355,000.
**Kal·i·spell** (käl'ĭ-spĕl'). City of NW Mont. near Glacier National Park. Pop. 10,648.
**Ka·lisz** (kä'lĭsh, -lēsh'). City of central Poland W of Łódź. Pop. 81,227.
**Kal·mar** (käl'mär', käl'-). City of SE Sweden on **Kalmar Sound,** an arm of the Baltic between the Swedish mainland & Öland. Pop. 32,049.
**Ka·lu·ga** (kə-lōō'gə). City of central European USSR SW of Moscow. Pop. 265,000.
**Ka·ma** (kä'mə). River of E European USSR flowing 1,262 mi (2,030.6 km) into the Volga.
**Ka·ma·ku·ra** (kä'mə-kŏŏr'ə, käm'ə-). City of SE Honshu, Japan, on the Pacific coast S of Yokohama. Pop. 165,552.
**Kam·chat·ka** (kăm-chăt'kə). Peninsula of Far Eastern USSR extending c. 750 mi (1,207 km) between the Sea of Okhotsk & the Bering Sea.
**Ka·met** (kŭm'āt'). Mountain, 25,447 ft (7,761.3 m), in the NW Himalayas of N India.
**Ka·mi·na** (kə-mē'nə). City of S Zaire ESE of Kinshasa. Pop. 56,300.
**Kam·loops** (kăm'lōōps'). City of S B.C., Canada, NE of Vancouver. Pop. 58,311.
**Kam·pa·la** (käm-pä'lə). Cap. of Uganda in the S part on Lake Victoria. Pop. 478,895.
**Kam·pu·che·a** (kăm'pə-chē'ə, -pōō-). CAMBODIA.
**Kan** (gän). GAN.
**Ka·nan·ga** (kə-näng'gə). City of S central Zaire ESE of Kinshasa. Pop. 428,960.
**Ka·na·ra** (kă'nər-ə). Region of S central India.
**Ka·na·za·wa** (kə-nä'zə-wə, kăn'ə-zä'wə). City of W Honshu, Japan, on the Sea of Japan. Pop. 395,263.
**Kan·chen·jun·ga** (kăn'chən-jŭng'gə, -jŏŏng'-). Mountain, 28,146 ft (8,584.5 m), in the Himalayas on the Sikkim-Nepal border.
**Kan·da·har** (kăn'də-här'). City of SE Afghanistan near the Pakistan border. Pop. 178,409.
**Kan·dy** (kăn'dē). City of central Sri Lanka ENE of Colombo. Pop. 93,602.
**Ka·ne·o·he** (kä'nē-ō'ē, -ō'hā). City of Hawaii on E Oahu on **Kaneohe Bay,** an inlet of the Pacific. Pop. 35,600.
**Kan·ga·roo** (kăng'gə-rōō'). Island, 1,680 sq mi (4,351 sq km), off the S coast of Australia SW of Adelaide.
**Kan·i·a·pis·cau** (kăn'ē-ə-pĭs'kō). River, c. 575 mi (925 km), of N Que., Canada.
**Kan·ka·kee** (kăng'kə-kē'). **1.** River, 225 mi (362 km), of NW Ind. & NE Ill. **2.** City of NE Ill. on the Kankakee R. SSW of Chicago. Pop. 30,141.
**Ka·no** (kä'nō). City of N Nigeria NE of Lagos. Pop. 399,000.
**Kan·pur** (kän'pŏŏr'). City of N India on the Ganges SE of Delhi. Pop. 1,154,388.
**Kan·sas** (kăn'zəs). **1.** River, 169 mi (271.9 km), of NE Kans. **2.** State of the central U.S. Cap. Topeka. Pop. 2,363,208. **—Kan'san** (-zən) *adj.* & *n.*
**Kansas City. 1.** City of NE Kans. on the Missouri adjacent to Kansas City, Mo. Pop. 161,087. **2.** City of W Mo. WNW of St. Louis. Pop. 448,159.
**Kan·su** (gän'sōō'). GANSU.
**Kao·hsiung** (gou'shē-ŏŏng', kou'-). City of SW Taiwan on Formosa Strait. Pop. 1,028,334.
**Kao·lack** (kou'lăk'). City of W Senegal SSE of Dakar. Pop. 106,899.
**Kao·lan** (gou'län'). LANZHOU.
**Ka·pos·vár** (kô'pōsh-vär'). City of SW Hungary SW of Budapest. Pop. 72,330.
**Ka·pu·as** (kä'pōō-äs'). River, 710 mi (1,142.4 km), of W Kalimantan, Borneo.
**Kap·us·ka·sing** (kăp'ə-skā'sĭng). Town of central Ont., Canada, NNE of Sault Ste. Marie. Pop. 12,676.
**Ka·ra** (kär'ə). Sea of the Arctic Ocean between Novaya Zemlya & the Siberian mainland.
**Ka·ra·chi** (kə-rä'chē). City of S Pakistan on the Arabian Sea. Pop. 3,498,634.
**Ka·ra·gan·da** (kăr'ə-gən-dä'). City of NE Central Asian USSR NNE of Tashkent. Pop. 572,000.
**Ka·ra·ko·ram** *also* **Ka·ra·ko·rum** (kăr'ə-kôr'əm, -kōr'-). Range of N Kashmir & SW China.
**Ka·ra·ko·rum** (kăr'ə-kôr'əm, -kōr'-). **1.** *var. of* KARAKORAM. **2.** Ruined ancient Mongol city in central Mongolia.
**Ka·ra Kum** (kăr'ə kōōm'). Desert of SW Central Asian USSR between the Caspian Sea & the Amu Darya.
**Kar·ba·la** (kär'bə-lə). City of central Iraq SSW of Baghdad. Pop. 83,301.
**Ka·re·li·a** (kə-rē'lē-ə, -rēl'yə). Region of NE Europe mainly in the NW USSR between the Gulf of Finland & the White Sea.
**Ka·re·li·an** (kə-rē'lē-ən, -rēl'yən). Isthmus in NW European USSR between Lake Ladoga & the Gulf of Finland.
**Ka·ri·ba** (kə-rē'bə). Lake, c. 165 mi (265.5 km) long, of N Zimbabwe & S Zambia, formed by **Kariba Dam** on the Zambezi R.
**Kar·kheh** (kər-kā'). River, c. 350 mi (563 km), of W Iran & SE Iraq.
**Karl-Marx-Stadt** (kärl-märk'shtät'). City of S East Germany SE of Leipzig. Pop. 303,811.
**Kar·lo·vy Va·ry** (kär'lə-vē vär'ē). City of W Czechoslovakia WNW of Prague. Pop. 43,000.
**Karls·kro·na** (kärls-krōō'nə). City of SE Sweden on the Baltic SSW of Stockholm. Pop. 33,414.
**Karls·ru·he** (kärlz'rōō'ə). City of SW West Germany on the Rhine WNW of Stuttgart. Pop. 280,448.
**Karl·stad** (kärl'städ'). City of SW Sweden on Lake Vänern W of Stockholm. Pop. 51,243.
**Kar·nak** (kär'năk'). Village of E central Egypt on the Nile on part of the site of ancient Thebes.
**Kar·ni·sche Al·pen** (kär'nĭ-shə äl'pən). CARNIC ALPS.
**Kár·pa·thos** (kär'pə-thôs'). Island of SE Greece in the Dodecanese.
**Kar·roo** (kə-rōō'). Semiarid plateau, c. 100,000 sq mi (259,000 sq km), of W South Africa.
**Kars** (kärs). City of NE Turkey near the Soviet border. Pop. 54,892.
**Ka·run** (kə-rōōn'). River of W Iran flowing c. 450 mi (724 km) S into the Shatt-al-Arab.
**Kar·vi·ná** (kär'və-nä'). City of N central Czechoslovakia near the Polish border. Pop. 79,100.
**Ka·sai** (kə-sī'). River of NE Angola & W Zaire flowing c. 1,200 mi (1,931 km) into the Congo.
**Kash·mir** (kăsh'mîr', kăsh-mîr'). Region & former state of N India & NE Pakistan, including in the W part the **Vale of Kashmir,** fertile valley of the Jhelum R.
**Kas·kas·ki·a** (kəs-kăs'kē-ə). River, c. 300 mi (483 km), of S Ill.
**Kas·sa·la** (kăs'ə-lə). City of NE Sudan near the Ethiopian border. Pop. 99,000.
**Kas·sel** (kăs'əl, kä'səl). City of E central West Germany near the East German border. Pop. 205,534.
**Kas·trop-Rau·xel** (käs'trôp-rouk'səl). CASTROP-RAUXEL.
**Ka·su·gai** (kä-sōō'gī). City of central Honshu, Japan, near Nagoya. Pop. 213,857.
**Ka·tah·din** (kə-täd'n), **Mount.** Mountain 5,267 ft (1,606.4 m), in N central Me.
**Ka·tan·ga** (kə-täng'gə, -tăng'-). Region of SE Zaire. **—Kat'an·gese'** (-gēz', -gēs') *adj.* & *n.*
**Ka·thi·a·war** (kä'tē-ə-wär'). Peninsula of W India projecting into the Arabian Sea between the Gulfs of Kutch & Cambay.
**Kat·mai** (kăt'mī'), **Mount.** Active volcano, c. 6,715 ft (2,048 m), in the Aleutian Range of S Alas. at the E end of Alaska Peninsula.
**Kat·man·du** (kăt'măn-dōō', kät'-). Cap. of Nepal in the central part. Pop. 150,402.
**Ka·to·wi·ce** (kä'tə-vēt'sə). City of S Poland WNW of Kraków. Pop. 303,264.
**Ka·tsi·na** (kăt'sĭ-nə). City of N Nigeria NW of Kano. Pop. 109,424.
**Kat·te·gat** (kăt'ĭ-găt'). Strait of the North Sea between SW Sweden & E Jutland, Denmark.
**Ka·tun** (kə-tōōn'). River, c. 415 mi (668 km), of S Siberian USSR.
**Kau·ai** (kou'ī'). Island of Hawaii NW of Oahu.
**Kau·kau·na** (kô-kô'nə). City of E Wis. near Appleton. Pop. 11,310.
**Kau·nas** (kou'nəs, -näs'). City of W European USSR in central Lithuania. Pop. 370,000.
**Ka·vál·la** (kə-văl'ə). City of NE Greece on the Aegean Sea. Pop. 46,234.
**Ka·ve·ri** (kô'və-rē). CAUVERY.
**Ka·vir Desert** (kə-vîr'). DASHT-E-KAVIR.
**Ka·wa·goe** (kə-wä'goi'). City of E central Honshu, Japan, NW of Tokyo. Pop. 225,465.

**Ka·wa·gu·chi** (kä′wə-go͞o′chē). City of SE Honshu, Japan, near Tokyo. Pop. 345,538.

**Ka·wa·ni·shi** (kä′wə-nē′shē). City of S Honshu, Japan, near Osaka. Pop. 115,773.

**Ka·war·tha Lakes** (kə-wôr′thə). Group of 14 lakes in SE Ont., Canada, E of Lake Simcoe.

**Ka·wa·sa·ki** (kä′wə-sä′kē). City of E central Honshu, Japan, on Tokyo Bay near Tokyo. Pop. 1,014,951.

**Kay·se·ri** (kī′zə-rē′). City of central Turkey SE of Ankara. Pop. 207,037.

**Ka·zakh** (kə-zăk′, -zäk′) or **Ka·zakh·stan** (-stăn′, -stän′) *also* **Kazakh Soviet Socialist Republic.** Constituent republic of Central Asian USSR NE of the Caspian Sea.

**Ka·zan** (kə-zăn′). **1.** River, c. 455 mi (732 km), of E N.W.T., Canada. **2.** (*also* -zän′). City of E European USSR on the Volga E of Moscow. Pop. 993,000.

**Ka·zan·lik** (kä′zän-lĭk′) *or* **Ka·zan·lŭk** (-lŭk′). City of central Bulgaria E of Sofia. Pop. 53,607.

**Kaz·bek** (käz-bĕk′), **Mount.** Extinct volcano, 16,541 ft (5,045 m), of S European USSR in the central Caucasus.

**Kaz·da·gi** (käz′dä-gē′). IDA MTS.

**Kaz·vin** (käz-vēn′). City of NW Iran WNW of Teheran. Pop. 138,527.

**Ke·a·la·ke·ku·a Bay** (kā-ä′lə-kə-ko͞o′ə). Inlet of the Pacific on the W coast of Hawaii Is., Hawaii.

**Kear·ney** (kär′nē). City of S central Nebr. WSW of Grand Island. Pop. 21,158.

**Kearns** (kûrnz). City of NW Utah near Salt Lake City. Pop. 17,000.

**Kear·ny** (kär′nē). Town of NE N.J. E of Newark. Pop. 35,735.

**Kecs·ke·mét** (kĕch′kə-māt′). City of central Hungary SE of Budapest. Pop. 91,929.

**Kee·ling Islands** (kē′lĭng). COCOS.

**Kee·lung** (kē′lo͝ong′). City of N Taiwan on the East China Sea. Pop. 342,604.

**Keene** (kēn). City of SW N.H. W of Manchester. Pop. 21,449.

**Kee·wa·tin** (kē-wāt′n). Administrative district, 228,160 sq mi (590,934.4 sq km), of E N.W.T., Canada.

**Ke·fal·li·ní·a** (kĕf′ə-lə-nē′ə). CEPHALONIA.

**Kef·la·vík** (kyĕb′lə-vēk′, kĕf′-). Town of SW Iceland on the Atlantic WSW of Reykjavík. Pop. 5,663.

**Kel·ler·wand** (kĕl′ər-vänt′). Highest (9,220 ft/2,812.1 m) of the Carnic Alps on the Austrian-Italian border.

**Ke·low·na** (kə-lō′nə). City of S B.C., Canada, ENE of Vancouver. Pop. 51,955.

**Kel·so** (kĕl′sō). City of SW Wash. S of Olympia near the Ore. border. Pop. 11,129.

**Ke·me·ro·vo** (kĕm′ər-ə-və, -ə-rō′və, -ər-ə-vō′). City of central Siberian USSR ENE of Novosibirsk. Pop. 471,000.

**Ke·mi·jo·ki** (kĕm′ē-yô′kē). River, c. 345 mi (555 km), of N Finland.

**Kemp·ten** (kĕmp′tən). City of S West Germany SW of Munich. Pop. 56,944.

**Ke·nai** (kē′nī′). Peninsula of S central Alas. between Cook Inlet & the Gulf of Alaska.

**Ken·il·worth** (kĕn′əl-wûrth′). Town of central England SE of Birmingham. Pop. 19,730.

**Ke·ni·tra** (kə-nē′trə). City of N Morocco NE of Rabat. Pop. 139,206.

**Ken·more** (kĕn′môr′, -mōr′). Village of W N.Y. near Buffalo. Pop. 18,474.

**Ken·ne·bec** (kĕn′ə-bĕk′). River, 164 mi (263.9 km), of S Me.

**Ken·ne·dy** (kĕn′ĭ-dē). **1. Mount.** Mountain, 13,095 ft (3,994 m), in the St. Elias Mts. in Y.T., Canada, near the Alas. border. **2. Cape.** Cape Canaveral.

**Ken·ner** (kĕn′ər). City of SE La. on the Mississippi near New Orleans. Pop. 66,382.

**Ken·ne·saw Mountain** (kĕn′ĭ-sô′). Lone peak, 1,809 ft (551.7 m), in NW Ga.; site of Civil War battle (1864).

**Ken·nett** (kĕn′ĭt). City of extreme SE Mo. near the Ark. border. Pop. 10,145.

**Ken·ne·wick** (kĕn′ə-wĭk′). City of S Wash. on the Columbia WNW of Walla Walla. Pop. 34,397.

**Ke·no·ra** (kə-nôr′ə, -nōr′ə). Town of W Ont., Canada, at the N end of the Lake of the Woods. Pop. 10,565.

**Ke·no·sha** (kə-nō′shə). City of extreme SE Wis. on Lake Michigan S of Milwaukee. Pop. 77,685.

**Kent** (kĕnt). **1.** Region & former kingdom of SE England. **2.** City of NE Ohio near Akron. Pop. 26,164. **3.** City of W Wash. near Seattle. Pop. 23,152.

**Ken·tuck·y** (kən-tŭk′ē). **1.** River, c. 250 mi (402 km), of N central Ky. **2.** State of E central U.S. Cap. Frankfort. Pop. 3,661,433. **—Ken·tuck′i·an** *adj. & n.*

**Kent·wood** (kĕnt′wo͝od′). City of W central Mich. near Grand Rapids. Pop. 30,438.

**Ken·ya** (kĕn′yə, kēn′-). **1. Mount.** Extinct volcano, 17,040 ft (5,197.2 m), in central Kenya. **2.** Country of E central Africa. Cap. Nairobi. Pop. 15,327,061. **—Ken′yan** *adj. & n.*

**Ke·o·kuk** (kē′ə-kŭk′). City of extreme SE Iowa on the Mississippi. Pop. 13,536.

**Ke·os** (kē′ŏs′). Island of Greece in the NW Cyclades.

**Kerch** (kĕrch). **1.** Strait connecting the Black Sea & the Sea of Azov in S European USSR. **2.** Peninsula of S European USSR in the E Crimea between the Black Sea & the Sea of Azov. **3.** City of S European USSR on Kerch Strait. Pop. 157,000.

**Ker·gue·len** (kûr′gə-lən, kûr′gə-lĕn′). **1.** Island group in the S Indian Ocean SE of South Africa, administered by France. **2.** Largest island (1,318 sq mi/3,413 sq km) of this group.

**Ke·rin·tji** *or* **Ke·rin·chi** (kə-rĭn′chē). Volcano, 12,467 ft (3,802.4 m), in W central Sumatra, Indonesia.

**Ker·ky·ra** (kĕr′kə-rə). CORFU.

**Ker·kra·de** (kĕr′krä′də). City of SE Netherlands on the West German border. Pop. 46,609.

**Ker·mad·ec** (kər-măd′ək). Island group, c. 13 sq mi (33.7 sq km), of the SW Pacific NE of New Zealand, which administers it.

**Ker·man** (kər-män′, kĕr-). City of E central Iran SE of Teheran. Pop. 140,309.

**Ker·man·shah** (kĕr′män-shä′). City of W Iran WSW of Teheran. Pop. 290,861.

**Kern** (kûrn). River, 155 mi (249.4 km), of S Calif.

**Kerr·ville** (kûr′vĭl′, -vəl). City of SW Tex. NW of San Antonio. Pop. 15,276.

**Ker·u·len** (kĕr′o͞o-lĕn′). River, 785 mi (1,263.1 km), of E Mongolia & NE China.

**Ket·ter·ing** (kĕt′ər-ĭng). City of SW Ohio near Dayton. Pop. 61,186.

**Ket·tle** (kĕt′l). River, c. 160 mi (257 km), of S B.C., Canada, & NE Wash.

**Keu·ka** (kyo͞o′kə, kā-yo͞o′-). One of the Finger Lakes, c. 18 mi (29 km) long, of W central N.Y.

**Kew** (kyo͞o). District of W Greater London, England.

**Ke·wa·nee** (kĭ-wä′nē). City of NW Ill. ESE of Moline. Pop. 14,508.

**Ke·wee·naw** (kē′wə-nô′). Peninsula of NW Mich. extending c. 60 mi (97 km) NE into Lake Superior W of **Keweenaw Bay.**

**Key Lar·go** (kē lär′gō). Island off S Fla., largest of the Florida Keys.

**Key West** (wĕst). City of extreme S Fla. on **Key West Is.,** westernmost of the Florida Keys in the Gulf of Mexico. Pop. 24,292.

**Kha·ba·rovsk** (kə-bär′əfsk). City of SE Far Eastern USSR on the Amur near the Chinese border. Pop. 528,000.

**Khal·ki·dhi·kí** (käl′kyə-*th*ĭ-kē′). CHALCIDICE.

**Khal·kís** (käl-kēs′). CHALCIS.

**Khan·ba·lik** (kän′bə-lēk′). Ancient city of Mongol China on the site of modern Beijing.

**Khan·ka** (käng′kə). Lake, c. 1,700 sq mi (4,403 sq km), between SE Far Eastern USSR & NE China.

**Khar·kov** (kär′kôf′, -kôv′, -kəf). City of S central European USSR E of Kiev. Pop. 1,444,000.

**Khar·toum** *also* **Khar·tum** (kär-to͞om′). Cap. of Sudan in the E central part at the confluence of the Blue Nile & White Nile. Pop. 334,000.

**Khas·ko·vo** (käs′kə-vō′). City of S Bulgaria ESE of Plovdiv. Pop. 82,636.

**Kha·tan·ga** (kə-tăng′gə, -tăng′-). River, c. 715 mi (1,150 km), of N central Siberian USSR.

**Khe·lat** (kə-lät′). *var. of* KALAT.

**Kher·son** (kĕr-sôn′). City of SW European USSR near the Black Sea ENE of Odessa. Pop. 319,000.

**Khing·an** (shĭng′än′). Mountains rising to 5,670 ft (1,729.4 m) in NE China.

**Khí·os** (kē′ŏs′). CHIOS.

**Khi·va** (kē′və). City of SW Central Asian USSR on **Khiva oasis** on the Amu Darya S of the Aral Sea. Pop. 24,139.

**Khmel·nit·sky** (kə-məl-nĭt′skē). City of SW European USSR WSW of Kiev. Pop. 172,000.

**Khmer Republic** (kmĕr). CAMBODIA.

**Kho·per** (kə-pyôr′). River, c. 625 mi (1,006 km), of S European USSR.

**Khor·ram·shahr** (kôr′əm-shär′). City of SW Iran near the Persian Gulf. Pop. 146,709.

**Kho·tan** (kō′tän′). HOTAN.

**Khul·na** (ko͞ol′nə). City of SW Bangladesh near the Ganges delta. Pop. 437,304.

**Khu·zi·stan** (ko͞o′zĭ-stän′, -stăn′). Region of SW Iran at the head of the Persian Gulf.

**Khy·ber** (kī′bər). Pass, c. 3,500 ft (5,632 m), through mountains on the border between W Afghanistan & N Pakistan.

**Kia·ling** (jēä′lĭng′). JIALING.

**Ki·a·mich·i** (kī′ə-mĭsh′ē). River, c. 100 mi (161 km), of SE Okla.

**Kia·mu·sze** (jē-ä′mo͞o′so͞o′). JIAMUSI.

**Kiang·si** (jē-äng′shē′). JIANGXI.

**Kiang·su** (jē-äng′so͞o′). JIANGSU.

**Kick·a·poo** (kĭk'ə-pōō'). River, c. 100 mi (161 km), of SW Wis.
**Kid·der·min·ster** (kĭd'ər-mĭn'stər). Borough of W central England WSW of Birmingham. Pop. 49,960.
**Kiel** (kēl, kē'əl). **1.** Canal, 61 mi (98.1 km) long, of N West Germany connecting the North Sea & the Baltic. **2.** City of N West Germany on **Kiel Bay,** an inlet of the Baltic. Pop. 262,164.
**Kiel·ce** (kē-ĕlt'sä). City of SE Poland S of Warsaw. Pop. 125,952.
**Ki·ev** (kē'ĕf, -ĕv, -əf). City of W European USSR on the Dnieper SW of Moscow. Pop. 2,144,000.
**Ki·ga·li** (kĭ-gä'lē). Cap. of Rwanda in the central part. Pop. 117,749.
**Ki·klá·dhes** (kē-klä'*th*ĭs). CYCLADES.
**Ki·lau·e·a** (kē'lou-ā'ə). Active volcanic crater on Mauna Loa, Hawaii Is., Hawaii.
**Kil·gore** (kĭl'gôr', -gōr'). City of E. Tex. E of Tyler. Pop. 10,968.
**Kil·i·man·ja·ro** (kĭl'ə-mən-jär'ō). Highest mountain in Africa, 19,340 ft (5,898.7 m), in NE Tanzania near the Kenya border.
**Kil·leen** (kĭ-lēn'). City of central Tex. SW of Waco. Pop. 46,296.
**Kil·ling·ly** (kĭl'ĭng-lē). Town of NE Conn. on the R.I. border. Pop. 14,519.
**Kil·mar·nock** (kĭl-mär'nək). Burgh of SW Scotland SSW of Glasgow. Pop. 50,175.
**Kim·ber·ley** (kĭm'bər-lē). City of central South Africa WNW of Bloemfontein. Pop. 105,258.
**Kin·a·ba·lu** (kĭn'ə-bə-lōō'). Mountain, 13,455 ft (4,103.8 m), of N Sabah, Borneo.
**Ki·nesh·ma** (kē'nĭsh-mə). City of central European USSR NE of Moscow. Pop. 101,000.
**King George Mount** (kĭng jôrj'). Mountain, 11,226 ft (3,423.9 m), in the Rockies of SE B.C., Canada, near the Alta. border.
**King George's Falls** (jôr'jĭz). AUGHRABIES FALLS.
**Kings** (kĭngz). **1.** River, c. 125 mi (201 km), of central Calif. rising in headstreams that flow through the gorges of **Kings Canyon National Park** (719 sq mi/1,862.2 sq km) in the Sierra Nevada. **2. Peak.** Highest (13,528 ft/4,126 m) of the Uinta Mts. in NE Utah.
**King's Lynn** (kĭngz' lĭn'). Borough of E England near the Wash. Pop. 29,990.
**Kings·port** (kĭngz'pôrt', -pōrt'). City of NE Tenn. near the Va. border ENE of Knoxville. Pop. 32,027.
**King·ston** (kĭng'stən). **1.** City of S Ont., Canada, on Lake Ontario. Pop 56,032. **2.** Cap. of Jamaica in the SE part on the Caribbean. Pop. 106,791. **3.** City of SE N.Y. on the Hudson SSW of Albany. Pop. 24,481. **4.** Borough of NE central Pa. opposite Wilkes-Barre. Pop. 15,681.
**King·ston-up·on-Hull** (kĭng'stən-ə-pŏn-hŭl', -pôn-). HULL, England.
**Kings·town** (kĭngz'toun'). Cap. of St. Vincent & the Grenadines in the West Indies on the SW coast of St. Vincent Is. Pop. 17,117.
**Kings·ville** (kĭngz'vĭl', -vəl). City of S Tex. SW of Corpus Christi. Pop. 28,808.
**King·teh·chen** (jĭng'dŭ'jŭn'). JINGDEZHEN.
**King Wil·liam** (wĭl'yəm). Island of N.W.T., Canada, in the Arctic Ocean SW of Boothia Peninsula.
**Kin·sha·sa** (kĭn-shä'sə). Cap. of Zaire in the W part on the Congo R. Pop. 1,323,039.
**Kin·ston** (kĭn'stən). City of E central N.C. SE of Raleigh. Pop. 25,234.
**Kin·tyre** (kĭn-tīr'). Peninsula of SW Scotland extending c. 40 mi (64 km) between the Atlantic & the Firth of Clyde.
**Ki·o·ga** *or* **Ky·o·ga** (kē-ō'gə). Lake, c. 1,000 sq mi (2,590 sq km), of central Uganda.
**Kir·ghiz** *or* **Kir·giz Soviet Socialist Republic** (kîr-gēz'). Constituent republic of SE Central Asian USSR bordering on NW China.
**Ki·ri·ba·ti** (kîr'ə-băs'). Island republic of the W central Pacific near the equator. Cap. Bairiki. Pop. 56,213.
**Ki·rin** (kē'rĭn'). JILIN.
**Kirk·cal·dy** (kûr-kô'dē, -kôl'-, -kä'-). Burgh of E Scotland on the Firth of Forth. Pop. 50,207.
**Kirk·land** (kûrk'lənd). City of W central Wash. on Lake Washington near Seattle. Pop. 18,779.
**Kirkland Lake.** Town of E Ont., Canada, NNE of Sudbury. Pop. 13,567.
**Kirk·pat·rick** (kûrk-păt'rĭk), **Mount.** Mountain, 14,856 ft (4,531.1 m), of Antarctica near the S edge of Ross Ice Shelf.
**Kirks·ville** (kûrks'vĭl'). City of N Mo. NW of Hannibal. Pop. 17,167.
**Kir·kuk** (kîr-kōōk'). City of N Iraq SE of Mosul. Pop. 167,413.
**Kirk·wall** (kûrk'wôl', -wəl). Burgh of N Scotland in the Orkney Is. Pop. 4,814.
**Kirk·wood** (kûrk'wŏŏd'). City of E Mo. near St. Louis. Pop. 27,987.
**Ki·rov** (kē'rôf', -rôv', -rəf). City of E central European USSR ENE of Moscow. Pop. 390,000.
**Ki·ro·va·bad** (kĭ-rō'və-băd'). City of S European USSR SE of Tbilisi. Pop. 232,000.
**Ki·ro·vo·grad** (kĭ-rō'və-grăd'). City of SW European USSR SSE of Kiev. Pop. 237,000.
**Kir·yu** (kîr'yōō). City of central Honshu, Japan, NNW of Tokyo. Pop. 134,239.
**Ki·san·ga·ni** (kē'sən-gä'nē). City of N Zaire on the Congo R. Pop. 229,596.
**Ki·sa·ra·zu** (kē'sə-rä'zōō). City of E central Honshu, Japan, across Tokyo Bay from Tokyo. Pop. 96,840.
**Ki·se·levsk** (kĭ-sĕl'yôfsk). City of SW Siberian USSR ESE of Novosibirsk. Pop. 122,000.
**Kish** (kĭsh). Ancient city of Mesopotamia E of present-day Al Hillah, Iraq.
**Ki·shi·nev** (kĭsh'ə-nĕf', -nĕv'). City of SW European USSR near the Rumanian border. Pop. 503,000.
**Ki·shi·wa·da** (kē'shə-wä'də). City of S Honshu, Japan, near Osaka. Pop. 174,952.
**Ki·si** (jē'shē'). JIXI.
**Kis·ka** (kĭs'kə). Island of SW Alas. near the W end of the Aleutians.
**Kis·lo·vodsk** (kĭs'lə-vôtsk'). City of S European USSR in the N Caucasus. Pop. 101,000.
**Kis·sim·mee** (kĭ-sĭm'ē). **1.** River of central Fla. flowing c. 140 mi (225 km) SSE through **Lake Kissimmee** (55 sq mi/142.5 sq km) to Lake Okeechobee. **2.** City of central Fla. S of Orlando. Pop. 15,487.
**Kist·na** (kĭst'nə). River of S India flowing c. 800 mi (1,287 km) E to the Bay of Bengal.
**Ki·ta·kyu·shu** (kē-tä'kyōō-shōō). City of N Kyushu, Japan, on the coast. Pop. 1,058,058.
**Kitch·e·ner** (kĭch'ə-nər, kĭch'nər). City of S Ont., Canada, WSW of Toronto. Pop. 131,870.
**Kí·thi·ra** (kē'thə-rä'). CYTHERA.
**Kit·i·mat** (kĭt'ə-măt'). Town of W B.C., Canada, on an inlet of the Pacific E of Prince Rupert. Pop. 11,791.
**Kit·ta·tin·ny Mountain** (kĭt'ə-tĭn'ē). Ridge of the Appalachians, c. 1,800 ft (549 m), in SE N.Y., NW N.J., & E Pa.
**Kit·ter·y** (kĭt'ə-rē). Town of extreme SW Me. opposite Portsmouth, N.H. Pop. 7,363.
**Kit·ty Hawk** (kĭt'ē hôk'). Village of NE N.C. on a sandy peninsula E of Albemarle Sound; site of Wright Brothers' first successful flight (1903).
**Ki·twe** (kē'twā'). City of N central Zambia near the Zaire border. Pop. 314,794.
**Ki·vu** (kē'vōō). Lake, 1,042 sq mi (2,698.8 sq km), on the Zaire-Rwanda border N of Lake Tanganyika.
**Ki·zil-Ir·mak** *or* **Ki·zil Ir·mak** (kĭ-zĭl'ĭr-mäk'). River of N central Turkey flowing c. 715 mi (1,150 km) to the Black Sea.
**Kjö·len** (chœ'lən). Range of NW Sweden & NE Norway.
**Klad·no** (kläd'nô). City of NW Czechoslovakia WNW of Prague. Pop. 61,200.
**Kla·gen·furt** (klä'gən-fŏŏrt'). City of S Austria SW of Graz. Pop. 74,326.
**Klai·pe·da** (klī'pə-də). City of W European USSR in Lithuania on the Baltic. Pop. 176,000.
**Klam·ath** (klăm'əth). **1.** River flowing c. 263 mi (423 km) from Upper Klamath Lake in SW Ore. through NW Calif. to the Pacific. **2.** Mountains of the Coast Ranges in S Ore. & NW Calif.
**Klamath Falls.** City of S Ore. near the Calif. border ESE of Medford. Pop. 16,661.
**Kle·ve** (klā'və). CLEVES.
**Klon·dike** (klŏn'dīk'). River, c. 90 mi (145 km), of E central Y.T., Canada, flowing through **the Klondike** gold-mining region to the Yukon R.
**Knife** (nīf). River, c. 165 mi (265 km), of W central N.Dak.
**Knos·sos** (nŏs'əs). Ancient city of N Crete near present-day Candia.
**Knox·ville** (nŏks'vĭl', -vəl). City of E Tenn. NE of Chattanooga. Pop. 183,139.
**Ko·be** (kō'bē', -bā'). City of S Japan on Osaka Bay in S Honshu. Pop. 1,360,605.
**Kø·ben·havn** (kœ'bən-houn'). COPENHAGEN.
**Ko·blenz** (kō'blĕnts'). COBLENZ.
**Ko·chi** (kō'chē). City of S Shikoku, Japan. Pop. 280,962.
**Ko·dai·ra** (kō-dīr'ə). City of E central Honshu, Japan, near Tokyo. Pop. 156,181.
**Ko·di·ak** (kō'dē-ăk'). Island of S Alas. in the Gulf of Alaska E of Alaska Peninsula.
**Ko·fu** (kō'fōō). City of central Honshu, Japan, W of Tokyo. Pop. 193,879.
**Ko·ga·nei** (kō'gä-nē'). City of E central, Honshu, Japan, near Tokyo. Pop. 102,714.
**Ko·kand** (kō-kănd'). City of S Central Asian USSR SE of Tashkent. Pop. 153,000.
**Ko·ko·mo** (kō'kə-mō'). City of central Ind. N of Indianapolis. Pop. 47,808.
**Ko·ko Nor** (kō'kō' nôr', nōr'). QINGHAI.
**Ko·la** (kō'lə). Peninsula of NW European USSR projecting E from Scandinavia between the White Sea & Barents Sea.

**Ko·lar Gold Fields** (kō-lär′, kō′lär′). City of S India SE of Bangalore. Pop. 76,112.
**Kol·ding** (kôl′dĭng). City of S central Denmark on the E coast of Jutland. Pop. 41,602.
**Kol·ha·pur** (kō′lə-po͝or′). City of W India SSE of Bombay. Pop. 259,050.
**Kol·mar** (kōl′mär′). COLMAR.
**Köln** (kœln). COLOGNE.
**Ko·lom·na** (kə-lôm′nə). City of central European USSR SE of Moscow. Pop. 147,000.
**Ko·ly·ma** (kə-lē′mə). **1.** River of N Far Eastern USSR flowing c. 1,335 mi (2,148 km) N to the East Siberian Sea. **2.** Range of N Far Eastern USSR extending c. 700 mi (1,126 km) E of the Kolyma R.
**Ko·ma·ki** (kō-mä′kē). City of central Honshu, Japan, near Nagoya. Pop. 101,299.
**Ko·man·dor·skie** *also* **Ko·man·dor·skye** (kŏm′ən-dôr′skē). Island group of Far Eastern USSR in the Bering Sea E of Kamchatka Peninsula.
**Ko·ma·ti** (kə-mä′tē). River, c. 500 mi (805 km), of NE South Africa, N Swaziland, & S Mozambique.
**Ko·ma·tsu** (kō-mät′so͞o). City of W Honshu, Japan, SW of Kanazawa. Pop. 100,273.
**Ko·mo·do** (kə-mō′dō). Island of S central Indonesia between Sumbawa & Flores Is.
**Kom·so·molsk** (kŏm′sə-môlsk′, kŏmp′-). City of S Far Eastern USSR NNE of Vladivostok. Pop. 264,000.
**Ko·na** (kō′nə). Region along W coast of Hawaii Is., Hawaii.
**Kon·gur** (gän′po͝or′), **Mount.** Highest (25,325 ft/7,724.1 m) of the Pamirs in extreme W China.
**Kon·ia** (kôn-yä′). *var. of* KONYA.
**Kö·nigs·berg** (kā′nĭgz-bûrg′, kœ′-). KALININGRAD.
**Kon·stan·ti·nov·ka** (kŏn′stən-tē′nəf-kə). City of S European USSR SSE of Kharkov. Pop. 112,000.
**Kon·stanz** (kôn′stänts′). CONSTANCE.
**Kon·ya** *also* **Kon·ia** (kôn-yä′). City of SW Turkey S of Ankara. Pop. 246,727.
**Ko·o·lau** (kō′ō-lä′o͞o). Range of E Oahu, Hawaii, rising to 3,105 ft (947 m).
**Koo·te·nay** *or* **Koo·te·nai** (ko͞ot′n-ā′). River, 407 mi (654.9 km), flowing from SE B.C., Canada, S through NW Mont., NW through N Idaho, & N into B.C., where it widens to form **Kootenay Lake** (64 mi/103 km long) before joining the Columbia.
**Ko·peysk** (kō-pāsk′). City of W Siberian USSR SSE of Sverdlovsk. Pop. 146,000.
**Kor·do·fan** (kôr′də-făn′). Region of central Sudan.
**Ko·re·a** (kə-rē′ə). **1. Bay.** Inlet of the Yellow Sea between NE China & NW North Korea. **2. Strait.** Channel, c. 110 mi (177 km) wide, between SE South Korea & SW Japan. **3.** Peninsula & former country of E Asia between the Yellow Sea & the Sea of Japan, divided politically since 1948 between **North Korea** (cap. Pyongyang; pop. 17,914,000) & **South Korea** (cap. Seoul; pop. 37,448,836). **—Ko·re′an** *adj. & n.*
**Kó·rin·thos** (kôr′ĭn-thôs′). CORINTH (region).
**Ko·ri·ya·ma** (kôr′ē-ä′mə, -yä′-, kôr′-). City of N Honshu, Japan, N of Tokyo. Pop. 264,628.
**Kö·rös** (kœr′œsh). River, c. 345 mi (555 km), of SE Hungary.
**Kort·rijk** (kôrt′rīk′). City of W Belgium W of Brussels. Pop. 44,961.
**Kos** (kŏs, kôs). Island, 111 sq mi (287.5 sq km), of SE Greece in the N Dodecanese at the entrance to the **Gulf of Kos,** an inlet of the Aegean on the SW coast of Turkey.
**Kos·ci·us·ko** (kŏs′ē-ŭs′kō), **Mount.** Highest mountain (7,316 ft/2,231.4 m) in Australia, in the SE part in the Australian Alps.
**Ko·shi·ga·ya** (kō-shē′gä-yə). City of E central Honshu, Japan, near Tokyo. Pop. 195,917.
**Kosh·tan-Tau** (kôsh′tän-tou′). Mountain, c. 16,880 ft (5,148 m), in the central Caucasus of S European USSR.
**Ko·ši·ce** (kô′shĭt-sā). City of E Czechoslovakia near the Hungarian border. Pop. 169,100.
**Kos·tro·ma** (kŏs′trə-mä′). City of central European USSR on the Volga NE of Moscow. Pop. 255,000.
**Ko·ta** (kō′tə). City of NW India SSW of Delhi. Pop. 212,991.
**Kota Kin·a·ba·lu** (kĭn′ə-bə-lo͞o′). City of E Malaysia on the NW coast of Sabah on Borneo. Pop. 40,939.
**Kott·bus** (kŏt′bəs, -bo͞os′). COTTBUS.
**Kot·ze·bue Sound** (kŏt′sə-byo͞o′). Inlet of the Chukchi Sea in NW Alas. N of Seward Peninsula.
**Kov·no** (kôv′nō). KAUNAS.
**Kov·rov** (kŏv-rôf′, -rôv′). City of central European USSR ENE of Moscow. Pop. 143,000.
**Kow·loon** (kou′lo͞on′). City of Hong Kong colony on **Kowloon Peninsula** opposite Hong Kong Is. on the SE coast of China. Metro. area pop. 2,378,480.
**Koy·u·kuk** (kī′ə-kŭk′). River, c. 500 mi (805 km), of N Alas.
**Ko·zhi·kode** (kō′zhĭ-kōd′). City of SW India on the Arabian Sea SW of Bangalore. Pop. 333,979.
**Kra** (krä). Isthmus, c. 40 mi (64 km) wide, linking Malay Peninsula & the SE Asian mainland.
**Kra·ka·to·a** (krăk′ə-tō′ə) *also* **Kra·ka·tau** (-tou′). Volcanic island of Indonesia between Sumatra & Java.
**Kra·ków** (krä′kou, krăk′ou, krā′kou, -kō). City of S Poland on the Vistula. Pop. 651,300.
**Kra·ma·torsk** (krä′mə-tôrsk′). City of S European USSR SSE of Kharkov. Pop. 178,000.
**Kras·no·dar** (krăs′nə-där′). City of S European USSR in the N Caucasus near the Black Sea. Pop. 560,000.
**Kras·no·yarsk** (krăs′nə-yärsk′). City of S central Siberian USSR on the upper Yenisei. Pop. 796,000.
**Kre·feld** (krā′fĕld′, -fĕlt′). City of W West Germany on the Rhine NNW of Cologne. Pop. 228,463.
**Kre·men·chug** (krĕm′ən-cho͝ok′, -cho͞og′). City of SW European USSR on the Dnieper SE of Kiev. Pop. 210,000.
**Krim** (krĭm). CRIMEA.
**Krish·na** (krĭsh′nə). KISTNA.
**Kris·tian·sand** (krĭs′chən-săn′, -sănd′). City of extreme S Norway on the Skagerrak. Pop. 59,488.
**Kri·voi Rog** *or* **Kri·voy Rog** (krĭv′oi rōg′, rôk′). City of SW European USSR NE of Odessa. Pop. 650,000.
**Kron·shtadt** *also* **Kron·stadt** (krōn′stăt′, krŏn′shtät). City & naval base of NW European USSR on an island in the Gulf of Finland W of Leningrad. Pop. 39,477.
**Kru·ger National Park** (kro͞o′gər). Wildlife preserve, 8,652 sq mi (22,408.7 sq km), in NE South Africa.
**Kru·gers·dorp** (kro͞o′gərz-dôrp′). City of NE South Africa WNW of Johannesburg. Pop. 92,725.
**Krung Thep** (kro͝ong′ tĕp′). BANGKOK.
**Krūs·né Ho·ry** (kro͞osh′nə hôr′ē). ERZGEBIRGE.
**Kua·la Lum·pur** (kwä′lə lo͝om′po͝or′, lŭm′-, ko͞o-ä′lə). Cap. of Malaysia on the SW Malay Peninsula. Pop. 451,977.
**Kuang·chow** (gwäng′jō′). GUANGZHOU.
**Ku·ban** (ko͞o-băn′, -bän′). River of S European USSR flowing c. 570 mi (917 km) N & W to the Sea of Azov.
**Ku·ching** (ko͞o′chĭng). Chief city of Sarawak, Malaysia, in the SW part on the South China Sea. Pop. 63,535.
**Kui·by·shev** *or* **Kuy·by·shev** (kwē′bə-shĕf′, -shĕv′, ko͞o′ē-bə-). City of E central European USSR on the Volga ESE of Moscow. Pop. 1,216,000.
**Ku·ma·mo·to** (ko͞o′mə-mō′tō). City W Kyushu, Japan. Pop. 488,166.
**Ku·ma·si** (ko͞o-mä′sē, -măs′ē). City of S central Ghana NW of Accra. Pop. 260,286.
**Ku·na·shir** (ko͞o′nə-shîr′). Southernmost & largest of the Kurile Is. of Far Eastern USSR.
**Ku·ne·ne** (ko͞o-nā′nə). CUNENE.
**Kun·gur** (ko͞on-go͞or′). KONGUR.
**Kun·lun** (ko͞on′lo͞on′). Range of W China extending from the Kashmir border E along the N edge of Tibet.
**Kun·ming** (ko͞on′mĭng′). City of S China SW of Chongqing. Pop. 1,700,000.
**Kun·san** (go͞on′săn′, ko͞on′-). City of SW South Korea on the Yellow Sea S of Seoul. Pop. 154,780.
**Kuo·pio** (kwô′pē-ô′). City of S central Finland NNE of Helsinki. Pop. 71,684.
**Ku·ra** (kə-rä′, ko͞or′ə). River of NE Turkey & S European USSR flowing c. 940 mi (1,512 km) NE & SE to the Caspian Sea.
**Kurd·i·stan** (ko͞or′dĭ-stăn′, kûr′-). Highland region of SE Turkey, NE Iraq, & NW Iran.
**Ku·re** (ko͞or′ē, kyo͞or′ē). City of SW Honshu, Japan, on the Inland Sea. Pop. 242,655.
**Kur·gan** (ko͞or-găn′, -gän′). City of W Siberian USSR ESE of Sverdlovsk. Pop. 310,000.
**Ku·rile** *or* **Ku·ril** (kyo͞or′ēl′, kyo͞o-rēl′). Island chain of extreme E USSR extending c. 700 mi (1,126 km) in the Pacific between Kamchatka Peninsula & N Hokkaido, Japan. **—Ku·ril′i·an** *adj. & n.*
**Kursk** (ko͞orsk). City of central European USSR SSW of Moscow. Pop. 375,000.
**Ku·ru·me** (ko͞or′ə-mā′). City of NW Kyushu, Japan, SSW of Kitakyushu. Pop. 204,474.
**Kush** (kŭsh, ko͞osh). CUSH.
**Ku·shi·ro** (ko͞osh′ə-rō′, kə-shîr′ō). City of NE Japan in SE Hokkaido on the Pacific. Pop. 206,840.
**Kus·ko·kwim** (kŭs′kə-kwĭm′). River of SW Alas. flowing c. 600 mi (965 km) SW to **Kuskokwim Bay,** an inlet of the Bering Sea.
**Ku·sta·nay** (ko͞os′tə-nī′). City of W Central Asian USSR SE of Sverdlovsk. Pop. 165,000.
**Ku·ta·i·si** (ko͞o-tī′sē). City of S European USSR WNW of Tbilisi. Pop. 194,000.
**Kutch** (kŭch). **1. Gulf of.** Inlet of the Arabian Sea in W India. **2.** RANN OF KUTCH.

**Ku·wait** (kə-wāt′). **1.** Country of the NE Arabian Peninsula at the head of the Persian Gulf. Pop. 1,355,827. Cap. Kuwait. **2.** Cap. of Kuwait in the E central part on the Persian Gulf. Pop. 181,774. **—Ku·wait′i** *adj.* & *n.*

**Kuy·by·shev** (kwē′bə-shĕf′, -shĕv′, ko͞o′ē-bə-). *var. of* KUIBYSHEV.

**Kuz·netsk Ba·sin** (ko͞oz-nĕtsk′). Coal-producing region of W Siberian USSR E of Novosibirsk.

**Kwa·ja·lein** (kwä′jə-lən, -lān′). Atoll, 6.5 sq mi (16.8 sq km), in the Marshall Is. of the W Pacific.

**Kwan·do** (kwän′dō). River, c. 600 mi (965 km), of SE Angola & NE Namibia.

**Kwang·chow** (gwäng′jō′). GUANGZHOU.

**Kwang·cho·wan** (gwäng′jō′wän′). Former coastal territory of SE China on Leizhou Peninsula, leased to France (1898–1946).

**Kwang·ju** (gwäng′jo͞o′). City of SW South Korea S of Seoul. Pop. 607,011.

**Kwang·si Chuang** (gwäng′shē′ jə-wäng′). GUANGXI ZHUANGZU.

**Kwang·tung** (gwäng′do͝ong′, -to͝ong′, kwäng′-). GUANGDONG.

**Kwan·tung** (gwän′do͝ong′, -to͝ong′, kwän′-). Former coastal territory of NE China in S Manchuria, leased to Japan (1905–45).

**Kwan·za** (kwän′zə). CUANZA.

**Kwei·chow** (gwā′jō′, kwā′-). GUIZHOU.

**Kwei·lin** (gwā′lĭn′). GUILIN.

**Kwei·sui** (gwā′swā′). HOHHOT.

**Kwei·yang** (gwā′yäng′). GUIYANG.

**Ky·o·ga** (kē-ō′gə). *var. of* KIOGA.

**Kyong·song** (kē-ông′sông′). SEOUL.

**Kyo·to** (kē-ō′tō). City of S Honshu, Japan, NE of Osaka. Pop. 1,461,-059.

**Kyu·shu** (kē-o͞o′sho͞o). Third-largest island of SW Japan (13,760 sq mi/35,638 sq km).

**Ky·zyl-Kum** (kĭ-zĭl′ko͞om′). Desert of S Central Asian USSR SE of the Aral Sea.

# L

**Laa·land** *or* **Lol·land** (lŏl′ənd). Island of SE Denmark in the Baltic between Sjaelland & N West Germany.

**La Baie** (lä bā′). City of S central Que., Canada, on the Saguenay R. near Chicoutimi. Pop. 20,116.

**La·be** (lä′bə). ELBE.

**Lab·ra·dor** (lăb′rə-dôr′). **1.** Sea of the N Atlantic between NE Canada & SW Greenland. **2.** Ocean current of the NW Atlantic flowing S from Baffin Bay along the coast of Newf. province. **3.** Peninsula of NE Canada between Hudson Bay & the Atlantic, divided between Que. & Newf. **4.** Mainland territory of Newf., Canada, on NE Labrador Peninsula. **—Lab′ra·dor′e·an, Lab′ra·dor′i·an** *adj.* & *n.*

**La·bu·an** (lə-bo͞o′ən). Island of Malaysia off the W coast of Sabah.

**La Ca·na·da-Flint·ridge** (lä′ kən-yä′də-flĭnt′rĭj′). Community of SW Calif. near Pasadena. Pop. 20,652.

**Lac·ca·dive, Mi·ni·coy, and A·min·di·vi Islands** (lăk′ə-dīv′; mĭn′ĭ-koi′; ŭm′ən-dē′vē). Island group in the Arabian Sea off the SW coast of India.

**Lac·e·dae·mon** (lăs′ĭ-dē′mən). SPARTA. **—Lac′e·dae·mo′ni·an** (-də-mō′nē-ən) *adj.* & *n.*

**La·cey** (lā′sē). City of W central Wash. E of Olympia. Pop. 13,940.

**La Chaux-de-Fonds** (lä shō′də-fôN′). City of W Switzerland in the Jura Mts. WNW of Bern. Pop. 38,500.

**La·chine** (lə-shēn′). City of S Que., Canada, on S Montreal Is. Pop. 41,503.

**La·chish** (lā′kĭsh). Ancient city of S Palestine SW of Jerusalem.

**Lach·lan** (lăk′lən). River, c. 922 mi (1,484 km), of SE Australia.

**La·chute** (lə-sho͞ot′). City of S Que., Canada, W of Montreal. Pop. 11,928.

**Lack·a·wan·na** (lăk′ə-wŏn′ə). City of W N.Y. near Buffalo. Pop. 22,701.

**La·co·ni·a** (lə-kō′nē-ə). **1. Gulf of.** Inlet of the Mediterranean on the S coast of Peloponnesus, S Greece. **2.** Ancient region of S Greece on SE Peloponnesus. **3.** City of central N.H. N of Concord. Pop. 15,575.

**La Co·ru·ña** (lä′ kə-ro͞on′yə). City of NW Spain on the Atlantic. Pop. 184,372.

**La Crosse** (lə krôs′, krŏs′). City of W Wis. on the Mississippi NW of Madison. Pop. 48,347.

**La·dakh** (lə-däk′). Region of N India in E Kashmir along the Tibetan border.

**Lad·o·ga** (lăd′ə-gə, lä′də-). Lake, c. 7,000 sq mi (18,130 sq km), of NW European USSR NE of Leningrad.

**La·fay·ette** (lăf′ē-ĕt′, lä′fē-). **1.** City of W Calif. near Oakland. Pop. 20,879. **2.** City of W central Ind. on the Wabash R. Pop. 43,011. **3.** City of S central La. WSW of Baton Rouge. Pop. 81,961.

**La·gash** (lā′găsh). Ancient city of Sumer, S Mesopotamia, in present-day S Iraq.

**La·go·a dos Pa·tos** (lə-gō′ə dəs păt′əs). Tidal lagoon, c. 150 mi (241 km) long, along the coast of S Brazil.

**La·gos** (lā′gŏs′). Cap. of Nigeria in the SW on the Gulf of Guinea. Pop. 1,060,848.

**La Gou·lette** (lä go͞o-lĕt′). City of NE Tunisia on the Mediterranean near Tunis. Pop. 41,912.

**La Grande** (lə grănd′). City of NE Ore. SE of Pendleton. Pop. 11,354.

**La Grange** (lə grānj′). **1.** City of W Ga. near the Ala. border of Columbus. Pop. 24,204. **2.** Village of NE Ill. near Chicago. Pop. 15,681.

**La Grange Park.** Village of NE Ill. near Chicago. Pop. 13,359.

**La Guai·ra** (lə gwīr′ə). City of N Venezuela on the Caribbean NW of Caracas. Pop. 20,344.

**La·gu·na Beach** (lə-go͞o′nə). City of S Calif. SE of Long Beach. Pop. 17,860.

**La Ha·bra** (lə hä′brə). City of S Calif. near Los Angeles. Pop. 45,232.

**La Hague** (lə hāg′), **Cape.** Promontory of NW France at the NW tip of the Cotentin Peninsula on the English Channel.

**La·hon·tan** (lə-hŏn′tən). Extinct Pleistocene lake with remnants surviving in W Nev. & NE Calif.

**La·hore** (lə-hôr′, -hōr′). City of NE Pakistan near the Indian border. Pop. 2,165,372.

**Lah·ti** (lä′tē). City of S central Finland NNE of Helsinki. Pop. 94,864.

**La Jol·la** (lə hoi′ə). Pacific beach district of San Diego, Calif.

**Lake** *or* **Lake of** (lāk) *or* **Loch** (lŏk, lôKH). For names of lakes see the specific element, e.g., for **Lake Erie** see ERIE.

**Lake Charles** (chärlz). City of SW La. E of Beaumont, Tex. Pop. 75,051.

**Lake District.** Scenic area of NW England.

**Lake For·est** (fôr′ĭst, fŏr′-). City of NE Ill. on Lake Michigan near Chicago. Pop. 15,245.

**Lake Hav·a·su City** (hăv′ə-so͞o′). City of W central Ariz. on the Calif. border. Pop. 15,737.

**Lake Jack·son** (jăk′sən). City of SE Tex. SW of Galveston. Pop. 19,102.

**Lake·land** (lāk′lənd). City of central Fla. ENE of Tampa. Pop. 47,406.

**Lake·view** (lāk′vyo͞o′). City of S central Mich. near Battle Creek. Pop. 18,000.

**Lake·ville** (lāk′vĭl′). City of E Minn. S of Minneapolis. Pop. 14,790.

**Lake·wood** (lāk′wo͝od′). **1.** City of S Calif. near Long Beach. Pop. 74,654. **2.** City of N central Colo. near Denver. Pop. 112,848. **3.** Township of E central N.J. SSE of Freehold. Pop. 25,223. **4.** City of NE Ohio near Cleveland. Pop. 61,963.

**Lake Worth** (wûrth). City of SE Fla. on the Atlantic S of West Palm Beach. Pop.27,048.

**Lak·shad·weep** (lək-shäd′wēp′). LACCADIVE, MINICOY, AND AMINDIVI IS.

**La Lí·ne·a** (lä lē′nē-ə). City of SW Spain on the Mediterranean near Gibraltar. Pop. 51,021.

**La Man·cha** (lə män′chə). Region of S central Spain.

**La Marque** (lə märk′). City of SE Tex. near Galveston. Pop. 15,372.

**La·me·sa** (lə-mē′sə). City of NW Tex. S of Lubbock. Pop. 11,790.

**La Me·sa** (lə mā′sə). City of S Calif. near San Diego. Pop. 50,342.

**La·mi·a** (lə-mē′ə). City of E central Greece NW of Athens. Pop. 37,872.

**La Mi·ra·da** (lä′ mə-rä′də). City of S Calif. SE of Los Angeles. Pop. 40,986.

**La·na·i** (lə-nī′). Island of central Hawaii W of Maui.

**Lan·cas·ter** (lăng′kə-stər, -kăs′tər, lăn′-). **1.** Sound, c. 50 mi (80 km) wide, between N Baffin Is. & S Devon Is., N.W.T., Canada. **2.** City of NW England N of Liverpool. Pop. 126,300. **3.** City of S Calif. N of Los Angeles. Pop. 48,027. **4.** Village of W N.Y. near Buffalo. Pop. 13,056. **5.** City of S central Ohio SE of Columbus. Pop. 34,952. **6.** City of SE Pa. W of Philadelphia. Pop. 54,725. **7.** City of NE Tex. S of Dallas. Pop. 14,807.

**Lan·chow** (län′jō′). LANZHOU.

**Land's End** *or* **Lands End** (lăndz′ ĕnd′). Cape of SW England; westernmost extremity of the country.

**Lands·hut** (länts′ho͞ot′). City of SE West Germany on the Isar R. NE of Munich. Pop. 55,858.

**Lang·ley** (lăng′lē). City of S B.C., Canada, near the Wash. border ESE of Vancouver. Pop. 10,123.

**Lans·dale** (lănz′dāl′). Borough of SE Pa. N of Philadelphia. Pop. 16,526.

**Lans·downe** (lănz′doun′). Borough of SE Pa. near Philadelphia. Pop. 11,891.

**Lan·sing** (lăn′sĭng). **1.** Village of NE Ill. near Chicago & the Ind. border. Pop. 29,039. **2.** Cap. of Mich. in the S central part NW of Detroit. Pop. 130,414.

**Lan Tao** *or* **Lan·tao** (län′dou′). Island of Hong Kong W of Hong Kong Is.

**La·nús** (lə-no͞os′). City of E Argentina near Buenos Aires. Pop. 449,824.

**Lan·zhou** (län'jō'). City of central China on the Yellow R. N of Chengdu. Pop. 1,500,000.
**La·od·i·ce·a** (lā-ŏd'ĭ-sē'ə). **1.** Ancient city of W Asia Minor near present-day Denizli, Turkey. **2.** LATAKIA.
**La·os** (lous, lä'ōs', lā'ŏs). Country of SE Asia. Cap. Vientiane. Pop. 3,721,000. **—La·o'tian** (lā-ō'shən) *adj. & n.*
**La Pal·ma** (lə päl'mə). **1.** Island of Spain in the NW Canary Is. **2.** City of S Calif. SE of Los Angeles. Pop. 15,663.
**La Paz** (lə päz', päz', päs'). **1.** Administrative cap. of Bolivia in the W part near Lake Titicaca. Pop. 654,713. **2.** City of NW Mexico on the Gulf of California. Pop. 46,011.
**La Pé·rouse** (lä' pə-ro͞oz'). Strait of the W Pacific between Sakhalin Is., USSR, & N Hokkaido, Japan.
**Lap·land** (lăp'lănd', -lənd). Region of extreme N Europe including N Norway, N Sweden, N Finland, & Kola Peninsula of NW USSR. **—Lap'land·er** *n.*
**La Pla·ta. 1.** (lə plä'tə). **Peak.** Mountain, c. 14,336 ft (4,372 m), in the Sawatch Range of central Colo. **2.** (lä plä'tä). City of E central Argentina SE of Buenos Aires. Pop. 478,666.
**La Porte** (lə pôrt', pōrt'). **1.** City of NW Ind. WSW of South Bend. Pop. 21,796. **2.** City of SE Tex. on Galveston Bay E of Houston. Pop. 14,062.
**Lap·peen·ran·ta** (lăp'ān-răn'tə). City of SE Finland near the USSR border. Pop. 52,682.
**Lap·tev** (lăp'tĕf', -tĕv'). Sea of the Arctic Ocean N of E Siberian USSR between the Taimyr Peninsula & New Siberian Is.
**La Pu·en·te** (lä' po͞o-ĕn'tē). City of S Calif. near Los Angeles. Pop. 30,882.
**L'Aq·ui·la** (lä'kwĭ-lə, läk'wĭ-). City of central Italy NE of Rome. Pop. 66,644.
**Lar·a·mie** (lăr'ə-mē). **1.** River, c. 216 mi (348 km), of N Colo. & SE Wyo. **2.** City of SE Wyo. WNW of Cheyenne. Pop. 24,410.
**Larch** (lärch). River, c. 270 mi (434 km), of N Que., Canada.
**La·re·do** (lə-rā'dō). City of S Tex. on the Rio Grande SSW of San Antonio. Pop. 91,449.
**Lar·go** (lär'gō). City of W Fla. NW of St. Petersburg. Pop. 58,977.
**La·ri·sa** (lär'ĭ-sə) *or* **La·ris·sa** (lə-rĭs'ə). City of E Greece in Thessaly. Pop. 72,336.
**Lark·spur** (lärk'spûr'). City of W Calif. near San Francisco. Pop. 11,064.
**La Ro·chelle** (lä' rə-shĕl'). City of W France on the Bay of Biscay. Pop. 75,367.
**La Salle** (lə săl'). **1.** City of S Que., Canada, on Montreal Is. & the St. Lawrence. Pop. 76,713. **2.** City of N central Ill. SW of Chicago. Pop. 10,307.
**Las·caux** (lă-skō'). Cave in SW France containing Paleolithic paintings.
**Las Cru·ces** (läs kro͞o'sĭs). City of S N.Mex. on the Rio Grande NNE of El Paso, Tex. Pop. 45,086.
**La Se·re·na** (lä' sə-rā'nə). City of N Chile near the Pacific N of Valparaiso. Pop. 61,900.
**Las Pal·mas** (läs päl'məs). Chief city of the Canary Is. of Spain, on the NE coast of Grand Canary Is. Pop. 260,368.
**La Spe·zia** (lä spĕt'sē-ə). City of NW Italy on an arm of the Ligurian Sea. Pop. 121,254.
**Las·sen Peak** *or* **Mount Las·sen** (lăs'ən). Active volcano, 10,453 ft (3,188.2 m), in the Cascade Range in N Calif.
**Las Ve·gas** (läs vā'gəs). **1.** City of SE Nev. near the Calif. & Ariz. borders. Pop. 164,674. **2.** City of N N.Mex. ESE of Santa Fe. Pop. 14,322.
**Lat·a·ki·a** (lăt'ə-kē'ə). City of W Syria on the Mediterranean opposite Cyprus. Pop. 125,716.
**La·ti·na** (lə-tē'nə). City of W central Italy SE of Rome. Pop. 53,003.
**Lat·in A·mer·i·ca** (lăt'n ə-mĕr'ĭ-kə). Countries of the Western Hemisphere S of the U.S. **—Lat'in A·mer'i·can** *n.* **—Lat'in-A·mer'i·can** *adj.*
**Latin Quar·ter** (kwôr'tər). Section of Paris on the S bank of the Seine.
**La·ti·um** (lā'shē-əm, -shəm). Ancient country in W central Italy.
**La·trobe** (lə-trōb'). Borough of SW Pa. ESE of Pittsburgh. Pop. 10,799.
**La Tuque** (lə tyo͝ok'). Town of S Que., Canada, NW of Quebec. Pop. 12,067.
**Lat·vi·a** (lăt'vē-ə). **1.** Region & former country of N Europe on the Baltic. **2.** *also* **Lat·vi·an Soviet Socialist Republic** (-ən). Constituent republic of NW Euopean USSR. **—Lat'vi·an** *adj. & n.*
**Lau·der·dale Lakes** (lô'dər-dāl'). City of SE Fla. near Fort Lauderdale. Pop. 25,426.
**Lau·der·hill** (lô'dər-hĭl'). City of SE Fla. near Fort Lauderdale. Pop. 37,271.
**Lau·rel** (lôr'əl, lŏr'-). **1.** City of central Md. NE of Washington, D.C. Pop. 12,103. **2.** City of SE Miss. NNE of Hattiesburg. Pop. 21,897.
**Lau·rens** (lôr'ənz, lŏr'-). City of NW S.C. S of Spartanburg. Pop. 10,587.
**Lau·ren·tian** (lə-rĕn'chən). **1.** *or* **Lau·ren·tide** (lôr'ən-tīd', lŏr'-). Mountains of S Que., Canada, N of the St. Lawrence & Ottawa rivers. **2. Plateau** *or* **Highlands** *or* **Shield.** Plateau covering E half of Canada.
**Lau·rin·burg** (lôr'ĭn-bûrg', lŏr'-). City of S N.C. SW of Fayetteville. Pop. 11,480.
**Lau·sanne** (lō-zän', -zăn'). City of W Switzerland on the N shore of Lake of Geneva. Pop. 136,100.
**Lau·zon** (lō-zôɴ'). City of S Que., Canada, on the St. Lawrence opposite Quebec city. Pop. 12,663.
**La·val** (lə-văl'). **1.** City of S Que., Canada, on the Ottawa R. near Montreal. Pop. 246,243. **2.** Town of NW France E of Rennes. Pop. 50,734.
**La Verne** (lə vûrn'). City of S Calif. E of Los Angeles. Pop. 23,508.
**Lawn·dale** (lôn'dāl'). City of S Calif. SW of Los Angeles. Pop. 23,460.
**Law·rence** (lôr'əns, lŏr'-). **1.** City of central Ind. near Indianapolis. Pop. 25,591. **2.** City of NE Kans. on the Kansas R. ESE of Topeka. Pop. 52,738. **3.** City of NE Mass. near the N.H. border NNE of Lowell. Pop. 63,175.
**Law·rence·burg** (lôr'əns-bûrg', lŏr'-). City of S Tenn. SSW of Nashville. Pop. 10,175.
**Law·ton** (lôt'n). City of SW Okla. SW of Oklahoma City. Pop. 80,054.
**Lay·san** (lī'sän'). Island of Hawaii in the Leeward group.
**Lay·ton** (lāt'n). City of N Utah S of Ogden. Pop. 22,862.
**Leaf** (lēf). River, c. 180 mi (290 km), of S central & SE Miss.
**League City** (lēg). City of SE Tex. SE of Houston. Pop. 16,578.
**Lea·ming·ton** (lē'mĭng-tən). **1.** *also* **Leamington Spa.** Borough of central England near Warwick. Pop. 44,950. **2.** Town of S Ont., Canada, on Lake Erie SW of Windsor. Pop. 11,169.
**Leav·en·worth** (lĕv'ən-wûrth'). City of NE Kans. NW of Kansas City. Pop. 33,656.
**Lea·wood** (lē'wo͝od'). City of E Kans. near Kansas City. Pop. 13,360.
**Leb·a·non** (lĕb'ə-nən). **1.** Mountains of Lebanon extending c. 100 mi (161 km) parallel to the coast & rising to 10,131 ft (3,090 m). **2.** Country of SW Asia on the Mediterranean N of Israel. Cap. Beirut. Pop. 3,161,000. **3.** City of central Ind. NW of Indianapolis. Pop. 11,456. **4.** City of W N.H. NW of Manchester. Pop. 11,134. **5.** City of W central Ore. E of Corvallis. Pop. 10,413. **6.** City of SE Pa. ENE of Harrisburg. Pop. 25,711. **7.** City of N central Tenn. E of Nashville. Pop. 11,872. **—Leb'a·nese'** (-nēz', -nēs') *adj. & n.*
**Lec·ce** (lā'chē, lĕch'ē). City of extreme SE Italy E of Taranto. Pop. 80,114.
**Lec·co** (lā'kō, lĕk'ō). City of N Italy on Lake Como NNE of Milan. Pop. 53,165.
**Lech** (lĕk). River, c. 175 mi (280 km), of W Austria & S West Germany.
**Le·duc** (lĭ-do͞ok'). Town of central Alta., Canada, S of Edmonton. Pop. 8,576.
**Led·yard** (lĕd'yərd, lĕj'ərd). Town of SE Conn. NNE of Groton. Pop. 13,735.
**Leeds** (lēdz). Borough of N central England NE of Manchester. Pop. 744,500.
**Lees·burg** (lēz'bûrg'). City of N central Fla. NW of Orlando. Pop. 13,191.
**Lee's Summit** (lēz). City of W Mo. SE of Kansas City. Pop. 28,741.
**Leeu·war·den** (lā'vär-dn). City of N Netherlands NE of the Ijsselmeer. Pop. 85,074.
**Lee·ward** (lē'wərd). **1.** Islands of the West Indies in the N Lesser Antilles from Virgin Is. SE to Guadeloupe. **2.** Islands of French Polynesia in the W Society Is. of the S Pacific. **3.** Islands of Hawaii in the central Pacific WNW of the main islands.
**Leg·horn** (lĕg'hôrn'). City of NW Italy on the Ligurian Sea WSW of Florence. Pop. 170,369.
**Le·gna·no** (lā-nyä'nō). City of NW Italy near Milan. Pop. 49,600.
**Leg·ni·ca** (lĕg-nēt'sə). City of SW Poland W of Wroclaw. Pop. 75,843.
**Le Ha·vre** (lə hä'vrə, häv). City of N France on the English Channel. Pop. 216,917.
**Le·high** (lē'hī'). River, c. 103 mi (166 km), of E Pa.
**Leices·ter** (lĕs'tər). City of central England ENE of Birmingham. Pop. 289,400.
**Lei·den** *also* **Ley·den** (līd'n). City of SW Netherlands NE of The Hague. Pop. 99,891.
**Lein·ster** (lĕn'stər). Historical province of SE Ireland.
**Leip·zig** (līp'sĭg, -sĭk). City of S central East Germany SSW of Berlin. Pop. 570,972.
**Leith** (lēth). District of Edinburgh, Scotland, on the Firth of Forth.
**Lei·zhou** (lā'jō'). Peninsula of S China between the Gulf of Tonkin & the South China Sea.
**Lé·man** (lē'mən, lĕm'ən, lə-măn'), **Lake.** Lake of GENEVA.
**Le Mans** (lə mäɴ'). City of NW France WSW of Paris. Pop. 150,289.
**Lem·nos** (lĕm'nŏs, nəs) *also* **Lim·nos** (lēm'nôs'). Island of NE Greece in the Aegean NW of Lesbos.

**Lem·on Grove** (lĕm′ən). City of S Calif. near San Diego. Pop. 20,780.
**Le·na** (lē′nə, lā′-). River of E Siberian USSR flowing c. 2,670 mi (4,296 m) NE & N to the Laptev Sea.
**Le·nex·a** (lə-nĕk′sə). City of E Kans. SSW of Kansas City. Pop. 18,639.
**Len·in** (lĕn′ĭn). Peak, 23,382 ft (7,131.5 m), in the Trans Alai of S Central Asian USSR.
**Le·nin·a·bad** (lĕn′ĭn-ə-băd′). City of S Central Asian USSR on the Syr Darya of Tashkent. Pop. 130,000.
**Le·nin·a·kan** (lĕn′ĭn-ə-kăn′). City of S European USSR near the Turkish border. Pop. 207,000.
**Len·in·grad** (lĕn′ĭn-grăd′). City of NW European USSR on the Gulf of Finland. Pop. 4,073,000.
**Len·insk-Kuz·nets·ki** (lĕn′ĭnsk-ko͝oz-nĕt′skē, -nyĕt′-). City of S central Siberian USSR E of Novosibirsk. Pop. 132,000.
**Le·noir** (lə-nôr′, -nōr′). City of W central N.C. WSW of Winston-Salem. Pop. 13,758.
**Lens** (läns). City of N France SW of Lille. Pop. 39,973.
**Leom·in·ster** (lĕm′ĭn-stər). City of N Mass. near Fitchburg. Pop. 34,508.
**Le·ón** (lā-ōn′). **1.** Region & former kingdom of NW Spain. **2.** City of central Mexico ENE of Guadalajara. Pop. 468,887. **3.** City of W Nicaragua NW of Lake Managua. Pop. 55,625. **4.** City of NW Spain at the foot of the Cantabrian Mts. Pop. 99,702.
**Le·o·ne** (lā-ō′nā), **Monte.** Highest (11,683 ft/3,563.3 m) of the Lepontine Alps, on the Swiss-Italian border.
**Le·o·pold II** (lē′ə-pōld′). Lake, c. 900 sq mi (2,331 sq km), in W central Zaire.
**Le·o·pold·ville** (lē′ə-pōld-vĭl′, lā′-). KINSHASA.
**Le·pan·to** (lĭ-păn′tō), **Gulf of.** Gulf of CORINTH.
**Le·pon·tine Alps** (lĭ-pŏn′tīn′ ălps). Range of the central Alps in S Switzerland & along the Swiss-Italian border.
**Lep·tis Mag·na** (lĕp′tĭs măg′nə). Ancient city of Roman Africa near modern Homs, Libya.
**Lé·ri·da** (lā′rĭ-də, lĕr′ĭ-). City of NE Spain W of Barcelona. Pop. 73,148.
**Ler·ma** (lĕr′mä). River, c. 350 mi (563 km), of central Mexico.
**Ler·wick** (lûr′wĭk, lĕr′ĭk). Burgh of N Scotland on Mainland Is. in the Shetlands. Pop. 6,195.
**Les·bos** (lĕz′bŏs′, -bəs) *also* **Les·vos** (-vôs′). Island of E Greece in the Aegean near the NW coast of Turkey.
**Le·so·tho** (lə-sō′tō). Kingdom of S Africa, an enclave within E central South Africa. Cap. Maseru. Pop. 1,339,000.
**Les·ser An·til·les** (lĕs′ər ăn-tĭl′ēz). Island group of the E West Indies extending in an arc from Curaçao to the Virgin Is.
**Lesser Slave Lake** (slāv). Lake, 461 sq mi (1,194 sq km), in central Alta., Canada.
**Lesser Sun·das** (sŭn′dəz, so͞on-). Island group of S Indonesia from Bali to Timor.
**Les·vos** (lĕz′vôs′). *var. of* LESBOS.
**Letch·worth** (lĕch′wûrth′). Urban district of E central England N of London. Pop. 31,520.
**Leth·bridge** (lĕth′brĭj). City of S Alta., Canada, SSW of Calgary. Pop. 46,752.
**Leu·cas** (lo͞o′kəs) *or* **Lev·kas** (lĕf-käs′). Island of W Greece in the Ionian Is. N of Cephalonia.
**Leuc·tra** (lo͝ok′trə). Village of ancient Greece SW of Thebes; site of Spartan defeat by the Thebans (371 B.C.).
**Leu·ven** (lĕv′ən). LOUVAIN.
**Le·val·lois-Per·ret** (lə-văl-wä′pə-rā′). City of N central France on the Seine near Paris. Pop. 52,460.
**Le·vant** (lə-vănt′). Countries bordering on the E Mediterranean. **—Le′van·tine′** (lĕv′ən-tīn′, -tēn′, lə-văn′-) *adj. & n.*
**Lev·el·land** (lĕv′ə-lănd′). City of NW Tex. W of Lubbock. Pop. 13,809.
**Le·ven** (lē′vən), **Loch. 1.** Arm of Loch Linnhe in W Scotland. **2.** Lake of E Scotland NNW of Edinburgh.
**Le·ver·ku·sen** (lā′vər-ko͞o′zən). City of W West Germany on the Rhine N of Cologne. Pop. 165,947.
**Lé·vis** (lē′vĭs, lā-vē′). City of S Que., Canada, on the St. Lawrence opposite Quebec city. Pop. 17,819.
**Lev·it·town** (lĕv′ĭt-toun′). Urban area of SE N.Y. on W Long Is. Pop. 65,400.
**Lev·kas** (lĕf-käs′). *var. of* LEUCAS.
**Lew·es** (lo͞o′ĭs). **1.** The upper Yukon R. in S Y.T., Canada, above its junction with the Pelly R. **2.** Borough of SE England S of London. Pop. 14,170.
**Lew·is** (lo͞o′ĭs). Range of the Rocky Mts. in NW Mont. rising to 10,448 ft (3,186.6 m).
**Lew·is·ton** (lo͞o′ĭ-stən). **1.** City of NW Idaho SSE of Spokane, Wash. Pop. 27,986. **2.** City of SW Me. opposite Auburn. Pop. 40,481.
**Lew·is·ville** (lo͞o′ĭs-vĭl′, lo͞o′ē-). City of NE Tex., NNW of Dallas. Pop. 24,273.
**Lewis with Har·ris** (hăr′ĭs). Island of NW Scotland, largest (825 sq mi/2,136.8 sq km) & northernmost of the Outer Hebrides.
**Lex·ing·ton** (lĕk′sĭng-tən). **1.** City of N central Ky. ESE of Louisville. Pop. 204,165. **2.** Town of E Mass. near Boston; site of first Revolutionary War battle (1775). Pop. 29,479. **3.** City of central N.C. S of Winston-Salem. Pop. 15,711.
**Ley·den** (līd′n). *var. of* LEIDEN.
**Ley·te** (lā′tē). **1.** Gulf of the W Pacific in the Philippines S of Samar & E of Leyte. **2.** Island of the E central Philippines in the Visayan group N of Mindanao.
**Lha·sa** (lä′sə, lăs′ə). City of SW China, traditional cap. of Tibet. Pop. 175,000.
**Lho·tse** (lōt′sā′). Two peaks of the central Himalayas, 27,923 ft (8,516.5 m) & 27,560 ft (8,405.8 m), on the Nepal-Tibet border.
**Lian·yun·gang** (lē-än′yo͞on′găng′). City of E China near the Yellow Sea SSW of Qingdao. Pop. 300,000.
**Liao** (lē-ou′). River of NE China flowing c. 900 mi (1,448 km) NE & SW to the Gulf of Liaodong.
**Liao·dong** *or* **Liao·tung** (lē-ou′do͝ong′). **1.** N part of the Gulf of Bo Hai in NE China. **2.** Peninsula of NE China projecting SW into the Yellow Sea.
**Liao·ning** (lē-ou′nĭng′). Province of NE China.
**Liao·yang** (lē-ou′yăng′). City of NE China SSW of Shenyang. Pop. 250,000.
**Liao·yu·an** *or* **Liao·yü·an** (lē-ou′yo͞o′ăn′). City of NE China S of Changchun. Pop. 300,000.
**Li·ard** (lē′ərd, lē′ärd′). River, c. 755 mi (1,215 km), of SE Y.T., N B.C., & SW N.W.T., Canada.
**Lib·er·al** (lĭb′ər-əl). City of SW Kans. near the Okla. border SW of Dodge City. Pop. 14,911.
**Li·be·rec** (lĭb′ə-rĕts′). City of NW Czechoslovakia NNE of Prague. Pop. 75,600.
**Li·be·ri·a** (lī-bîr′ē-ə). Country of W Africa on the Gulf of Guinea. Cap. Monrovia. Pop. 1,873,000. **—Li·be′ri·an** *adj. & n.*
**Lib·er·ty** (lĭb′ər-tē). **1.** Island of SE N.Y. in New York Bay; site of the Statue of Liberty. **2.** City of W Mo. NE of Kansas City. Pop. 16,251.
**Lib·er·ty·ville** (lĭb′ər-tē-vĭl′). Village of NE Ill. SW of Waukegan. Pop. 16,520.
**Li·bre·ville** (lē′brə-vĭl′, -vēl′). Cap. of Gabon in the NW on the Gulf of Guinea. Pop. 105,080.
**Lib·y·a** (lĭb′ē-ə). Country of N Africa on the Mediterranean W of Egypt. Cap. Tripoli. Pop. 2,856,000. **—Lib′y·an** *adj. & n.*
**Libyan.** Desert of NE Africa, the NE part of the Sahara.
**Li·can·cá·bur** (lē′kăng-kä′bər). Volcano, 19,455 ft (5,993.8 m) of N Chile near the Bolivian border.
**Li·ca·ta** (lĭ-kä′tə). Town of S Sicily, Italy, on the Mediterranean. Pop. 40,997.
**Lich·field** (lĭch′fēld′). Borough of W central England NNE of Birmingham. Pop. 23,690.
**Lick·ing** (lĭk′ĭng). River of NE Ky. flowing c. 320 mi (515 km) NW to the Ohio R. opposite Cincinnati.
**Li·di·ce** (lĭd′ĭ-sē, -sä′, -ĭt-). Village of NW Czechoslovakia; destroyed by German forces (1942).
**Li·do** (lē′dō). Island in NE Italy separating the lagoon of Venice from the Adriatic.
**Liech·ten·stein** (lĭk′tən-stīn′, -shtīn′). Principality (62 sq mi/161 sq km) in central Europe between Austria & Switzerland. Cap. Vaduz. Pop. 25,220.
**Li·ège** (lē-āzh′, -ĕzh′). City of E Belgium near the Dutch & West German borders. Pop. 145,573.
**Lien·yün·kang** (lē-o͞on′yo͞on′găng′). LIANYUNGANG.
**Lie·pa·ya** *or* **Lie·pa·ja** (lē-ĕp′ə-yə). City of W European USSR on the Baltic. Pop. 108,000.
**Liè·vre** (lē-ĕv′rə). River, c. 200 mi (322 km), of S Que., Canada.
**Lif·fey** (lĭf′ē). River of E Ireland flowing c. 50 mi (80 km) through Dublin to Dublin Bay.
**Light·house Point** (līt′hous′). City of SE Fla. on the Atlantic NE of Pompano Beach. Pop. 11,488.
**Li·gu·ri·a** (lĭ-gyo͝or′ē-ə). Region of NW Italy on the **Ligurian Sea,** an arm of the Mediterranean between NW Italy & Corsica. **—Li·gu′ri·an** *adj. & n.*
**Li·ka·si** (lĭ-kä′sē). City of SE Zaire NW of Lubumbashi. Pop. 146,394.
**Lille** *also* **Lisle** (lēl, lē′əl). City of N France near the Belgian border. Pop. 171,010.
**Li·long·we** (lĭ-lông′wā). Cap. of Malawi in the S central part. Pop. 102,924.
**Li·ma. 1.** (lē′mə). Cap. of Peru in the W central part near the Pacific. Metro. area pop. 2,386,374. **2.** (lī′mə). City of NW Ohio SSW of Toledo. Pop. 47,381.
**Li·mas·sol** (lĭm′ə-sôl′). City of S Cyprus on the Mediterranean. Pop. 79,641.
**Li·may** (lē-mī′). River, c. 250 mi (402 km), of W central Argentina.
**Lim·er·ick** (lĭm′ər-ĭk, lĭm′rĭk). Borough of SW Ireland on the Shannon estuary. Pop. 57,161.
**Lim·nos** (lēm′nôs′). *var. of* LEMNOS.

**Li·moges** (lē-mōzh′). City of W central France NE of Bordeaux. Pop. 136,059.
**Li·món** (lĭ-môn′, -mōn′). City of E Costa Rica on the Caribbean. Pop. 29,621.
**Lim·po·po** (lĭm-pō′pō). River of SE Africa rising near Johannesburg in NE South Africa & flowing c. 1,100 m (1,770 km) in a NE-SE arc to the Indian Ocean in S Mozambique.
**Li·na·res** (lĭ-när′ĭs). City of S Spain ENE of Córdoba. Pop. 45,330.
**Lin·coln** (lĭng′kən). **1. Mount.** Peak, 14,284 ft (4,356.6 m), in the Rocky Mts. of central Colo. **2.** Borough of E England NE of Nottingham. Pop. 73,700. **3.** Town of SE Ont., Canada, on Lake Ontario W of Niagara Falls. Pop. 14,460. **4.** City of central Ill. NNE of Springfield. Pop. 16,327. **5.** Cap. of Nebr. in the SE part SW of Omaha. Pop. 171,932. **6.** Town of NE R.I. near Providence. Pop. 16,949.
**Lincoln Park.** City of SE Mich. near Detroit. Pop. 45,105.
**Lin·coln·wood** (lĭng′kən-wŏŏd′). Village of NE Ill. near Chicago. Pop. 11,921.
**Lin·den** (lĭn′dən). City of NE N.J. near Elizabeth. Pop. 37,836.
**Lin·den·hurst** (lĭn′dən-hûrst′). Village of S N.Y. on S Long Is. near Babylon. Pop. 26,919.
**Lin·den·wold** (lĭn′dən-wōld′). Borough of SW N.J. SE of Camden. Pop. 18,196.
**Lin·des·nes** (lĭn′dĭs-nĕs′). Cape of extreme S Norway.
**Lind·say** (lĭn′zē). Town of S Ont., Canada, NE of Toronto. Pop. 13,062.
**Line** (līn). Islands of the central Pacific in Kiribati S of Hawaii & astride the equator.
**Lin·ga·yen Gulf** (lĭng′gä-yĕn′). Inlet of the South China Sea in W central Luzon, Philippines.
**Ling·ga** (lĭng′gə). Archipelago of Indonesia off the E central coast of Sumatra.
**Lin·guet·ta** (lĭng-gwĕt′tə). Cape of SW Albania.
**Lin·kö·ping** (lĭn′chœ′pĭng′). City of S Sweden SW of Stockholm. Pop. 80,274.
**Linn·he** (lĭn′ē), **Loch.** Inlet of the Firth of Lorne on the W coast of Scotland.
**Linz** (lĭnts). City of N Austria on the Danube. Pop. 205,700.
**Li·ons** (lī′ənz) *also* **Li·on** (-ən), **Gulf of.** Wide inlet of the Mediterranean on the S coast of France.
**Lip·a·ri** (lĭp′ə-rē). **1.** Island group of Italy off the NE coast of Sicily in the Tyrrhenian Sea. **2.** Chief island of this group.
**Li·petsk** (lē′pĕtsk′). City of central European USSR SSE of Moscow. Pop. 396,000.
**Lis·bo·a** (lēzh-vō′ə). LISBON.
**Lis·bon** (lĭz′bən). Cap. of Portugal in the W part on the Tagus R. Pop. 769,410.
**Lis·burne** (lĭz′bûrn′). Cape on the NW coast of Alaska on the Arctic Ocean SW of Point Barrow.
**Li·sieux** (lēz-yœ′). Town of N France S of Le Havre. Pop. 24,972.
**Lisle. 1.** (lēl, lē′əl). *var. of* LILLE. **2.** (līl). Village of NE Ill. W of Chicago. Pop. 13,625.
**Lith·u·a·ni·a** (lĭth′ōō-ā′nē-ə). **1.** Region & former country of N Europe on the Baltic. **2.** *also* **Lith·u·a·ni·an Soviet Socialist Republic** (-ən). Constituent republic of NW European USSR. **—Lith′u·a′ni·an** *adj. & n.*
**Lit·tle A·mer·i·ca** (lĭt′l ə-mĕr′ĭ-kə). U.S. base for explorations in W Antarctica on the Ross Ice Shelf.
**Little Big·horn** (bĭg′hôrn′). River, c. 90 mi (145 km), of W Wyo. & S Mont.
**Little Col·o·ra·do** (kŏl′ə-răd′ō, -rä′dō). River of NE Ariz. flowing c. 315 mi (507 km) NW to the Colorado R.
**Little Fork.** River, c. 132 mi (212 km), of N Minn.
**Little Ka·na·wha** (kə-nô′wə). River, c. 160 mi (257 km), of N W.Va.
**Little Mis·sou·ri** (mĭ-zŏŏr′ē, -zŏŏr′ə). **1.** River, c. 145 mi (233 km), of SW Ark. **2.** River of NE Wyo., SE Mont., NW S.Dak., & W N.Dak., flowing c. 560 mi (901 km) NE to the Missouri.
**Little Pee Dee** (pē′ dē′). River, c. 105 mi (169 km), of S central N.C.
**Little Rock** (rŏk). Cap. of Ark., in the central part. Pop. 153,831.
**Little Sioux** (sōō). River, c. 221 mi (356 km), of SW Minn. & NW Iowa.
**Little Saint Ber·nard** (sānt′ bər-närd′). Mountain pass through the Savoy Alps between Italy & France S of Mont Blanc, rising to 7,180 ft (2,190 m).
**Little Ten·nes·see** (tĕn′ĭ-sē′). River, c. 135 mi (217 km), of NE Ga., SW N.C., & E Tenn.
**Lit·tle·ton** (lĭt′l-tən). City of N central Colo. near Denver. Pop. 28,349.
**Little Wa·bash** (wô′băsh′). River, c. 200 mi (322 km), of E Ill.
**Liv·er·more** (lĭv′ər-môr′, -mōr′). City of W Calif. E of San Leandro. Pop. 48,349.
**Liv·er·pool** (lĭv′ər-pōōl′). Borough of NE England on the Mersey R. near the Irish Sea. Pop. 539,700.
**Liv·ing·stone** (lĭv′ĭng-stən). City of S Zambia on the Zambezi R. & Zimbabwe border. Pop. 71,987.
**Li·vo·ni·a** (lĭ-vō′nē-ə, -vōn′yə). **1.** Region of W European USSR comprising S Latvia & N Estonia. **2.** City of SE Mich. near Detroit. Pop. 104,814. **—Li·vo′ni·an** *adj. & n.*
**Li·vor·no** (lē-vôr′nō). LEGHORN.
**Liz·ard Point** *or* **Head** (lĭz′ərd). Cape of SW England at S tip of **The Lizard,** peninsula extending S into the English Channel; southernmost point of Great Britain.
**Lju·blja·na** (lē-ōō′blē-ä′nə). City of NW Yugoslavia on the Sava R. Pop. 169,064.
**Llan·dud·no** (lăn-dĭd′nō, -dŭd′-). Urban district of NW Wales on the Irish Sea. Pop. 17,700.
**Lla·nel·li** *or* **Lla·nel·ly** (lă-nĕl′ē). Borough of S Wales on an inlet of Bristol Channel. Pop. 25,870.
**LLoyd·min·ster** (loid′mĭn′stər). City on the Alta.-Sask. border, Canada, E of Edmonton. Pop. 10,311.
**Llu·llai·lla·co** (yōō′yī-yä′kō). Volcano, 22,057 ft (6,727.4 m), in the Andes on the N Chile-Argentina border.
**Lo·an·da** (lō-ăn′də). *var. of* LUANDA.
**Lo·an·ge** (lō-ăng′gə). River, c. 425 mi (684 m), of NE Angola & SW Zaire.
**Lo·bi·to** (lō-bē′tō). City of W central Angola on the Atlantic. Pop. 59,528.
**Lo·car·no** (lō-kär′nō). Town of S Switzerland at the N end of Lake Maggiore. Metro. area pop. 39,200.
**Loch** (lŏk, lôKH). —See LAKE.
**Lock·port** (lŏk′pôrt′, -pōrt′). City of W N.Y. NNE of Buffalo. Pop. 24,844.
**Lo·cust Grove** (lō′kəst). Town of SE N.Y. on W Long Is. Pop. 11,648.
**Lodge·pole Creek** (lŏj′pōl′). River, 212 mi (341.1 km), of SE Wyo., SW Nebr., & NE Colo.
**Lo·di** (lō′dī′). **1.** (lō′dē). City of N Italy SE of Milan. Pop. 42,489. **2.** City of central Calif. N of Stockton. Pop. 35,221. **3.** Borough of NE N.J. NE of Passaic. Pop. 23,956.
**Łódź** (lōōj, lŏdz). City of central Poland WSW of Warsaw. Pop. 777,800.
**Lo·fo·ten** (lō′fōt′n). Islands off the NW coast of Norway in the Norwegian Sea.
**Lo·gan** (lō′gən). **1.** Peak, 19,850 ft (6,054.3 m), of the St. Elias Mts. in SW Y.T., Canada, near the Alas. border. **2.** City of N Utah N of Ogden. Pop. 26,844.
**Lo·gans·port** (lō′gənz-pôrt′, -pōrt′). City of N central Ind. NNW of Kokomo. Pop. 17,899.
**Lo·gro·ño** (lə-grōn′yə). City of N Spain on the Ebro R. Pop. 83,117.
**Loir** (lwär, lə-wär′). River, c. 193 mi (311 km), of NW France.
**Loire** (lwär, lə-wär′). Longest river of France, rising in the Cévennes Mts. & flowing c. 630 mi (1,014 km) N & W to the Bay of Biscay.
**Lo·ja** (lô′hä). City of S Ecuador SSE of Guayaquil. Pop. 47,697.
**Lol·land** (lŏl′ənd). *var. of* LAALAND.
**Lo·ma Lin·da** (lō′mə lĭn′də). City of SE Calif. near San Bernardino. Pop. 10,694.
**Lo·ma·mi** (lō-mä′mē). River of Zaire flowing c. 900 mi (1,448 km) N to the Congo R.
**Lo·mas de Za·mo·ra** (lō′mäz də zə-môr′ə, -mōr′ə). City of E Argentina S of Buenos Aires. Pop. 410,806.
**Lom·bard** (lŏm′bärd′). Village of NE Ill. near Chicago. Pop. 37,295.
**Lom·bar·dy** (lŏm′bər-dē). Region of N Italy. **—Lom′bard** *adj. & n.*
**Lom·blen** (lŏm-blĕn′). Island of S central Indonesia in the Lesser Sundas E of Flores Is.
**Lom·bok** (lŏm′bŏk′). Island of S central Indonesia in the Lesser Sundas E of Bali.
**Lo·mé** (lō-mā′). Cap. of Togo in the S part on the Gulf of Guinea. Pop. 148,443.
**Lo·mi·ta** (lō-mē′tə). City of S Calif. near Los Angeles. Pop. 17,191.
**Lo·mond** (lō′mənd), **Loch.** Largest lake in Scotland, 23 mi (37 km) long 1–5 mi (1.6–8 km) wide, in the E central part.
**Lom·poc** (lŏm′pŏk′). City of S Calif. WNW of Santa Barbara. Pop. 26,267.
**Lon·don** (lŭn′dən). **1.** City of SE Ont., Canada, SW of Toronto. Pop. 240,392. **2.** Cap. of the United Kingdom on the Thames R. in SE England; a metropolitan county consisting of the City of London & 32 surrounding boroughs. Metro. area pop. 12,332,900.
**Lon·don·der·ry** (lŭn′dən-dĕr′ē, lŭn′dən-dĕr′ē). **1.** Borough of NW Northern Ireland WNW of Belfast. Pop. 51,200. **2.** Town of SE N.H. near Manchester. Pop. 13,598.
**Long Beach** (lông, lŏng). **1.** City of S Calif. SE of Los Angeles. Pop. 361,334. **2.** City of SE N.Y. of S Long Is. Pop. 34,073.
**Long Branch.** City of E N.J. on the Atlantic N of Asbury Park. Pop. 29,819.
**Long Ea·ton** (ēt′n). Urban district of central England near Nottingham. Pop. 33,560.
**Long Island. 1. Sound.** Arm of the N Atlantic between Long Is. & Conn. **2.** Island, c. 120 mi (193 km) long, of SE N.Y.

**Long·mea·dow** (lông′mĕd′ō, lŏng′-). Town of SW Mass. near Springfield. Pop. 16,301.
**Long·mont** (lông′mŏnt′, lŏng′-). City of N Colo. NNE of Boulder. Pop. 42,942.
**Longs Peak** (lôngz, lŏngz). Peak, 14,255 ft (4,347.8 m), in the Rockies of N. Colo.
**Lon·gueil** (lông-gāl′). City of S Que., Canada, on the St. Lawrence opposite Montreal. Pop. 122,429.
**Long·view** (lông′vyōō′, lŏng′-). **1.** City of E Tex. W of Shreveport, La. Pop. 62,762. **2.** City of SW Wash. on the Columbia near Kelso. Pop. 31,052.
**Long·wood** (lông′wŏŏd′, lŏng′-). City of E central Fla. N of Orlando. Pop. 10,029.
**Look·out** (lŏŏk′out′). **1.** Cape of E N.C. SW of Cape Hatteras. **2. Mountain.** Ridge in SE Tenn.; site of a Civil War Union victory (1863).
**Loop** (lōōp), **the.** Central business district of Chicago, Ill.
**Lop Nur** (lŏp′ nŏŏr′) *or* **Lop Nor** (nôr′). Marshy depression of NW China.
**Lo·rain** (lə-rān′, lô-). City of N Ohio on Lake Erie W of Cleveland. Pop. 75,416.
**Lor·ca** (lôr′kə). City of SE Spain W of Cartagena. Pop. 25,208.
**Lord Howe** (lôrd hou′). Volcanic island of Australia in the Tasman Sea ENE of Sydney.
**Lo·rette·ville** (lə-rĕt′vĭl′). City of S Que., Canada, near Quebec city. Pop. 14,767.
**Lo·rient** (lôr′ē-äɴ′). City of NW France on the Bay of Biscay. Pop. 68,655.
**Lorne** *also* **Lorn** (lôrn), **Firth of.** Inlet of the Atlantic on W coast of Scotland between Mull Is. & the mainland.
**Lor·raine** (lō-rān′, lô-). Region & former province of NE France.
**Los Al·a·mi·tos** (lôs ăl′ə-mē′təs, lŏs). City of S Calif. near Long Beach. Pop. 11,529.
**Los Al·a·mos** (ăl′ə-mōs′). City of N central N.Mex. NW of Santa Fe. Pop. 17,100.
**Los Al·tos** (ăl′təs). City of W Calif. S of Palo Alto. Pop. 25,769.
**Los An·ge·les** (ăn′jə-ləs, -lēz′, ăng′gə-ləs). **1.** (*also* äng′hĕ-lĕs′). City of S central Chile SE of Concepción. Pop. 49,500. **2.** City of S Calif. on the Pacific. Pop. 2,966,763.
**Los Ba·nos** (băn′əs). City of central Calif. NW of Fresno. Pop. 10,341.
**Los Ga·tos** (găt′əs). City of W Calif. near San Jose. Pop. 26,593.
**Lot** (lôt, lŏt). River, c. 300 mi (483 km), of S France.
**Lo·ta** (lō′tə). City of S central Chile on the Pacific SSW of Concepción. Pop. 48,100.
**Lough·bor·ough** (lŭf′bŭr′ə, -bər-ə). Borough of central England SSW of Nottingham. Pop. 49,010.
**Lou·ise** (lōō-ēz′). Lake of SW Alta., Canada, in the Rocky Mts.
**Lou·i·si·ade** (lōō-ē′zē-äd′, -ăd′) Archipelago of the W Pacific SE of New Guinea; part of Papua New Guinea.
**Lou·i·si·an·a** (lōō-ē′zē-ăn′ə, lōō′zē-). **1. Purchase.** Territory of the W U.S. extending from the Mississippi to the Rockies between the Mexican & Canadian borders, purchased (1803) from France. **2.** State of S central U.S. Cap. Baton Rouge. Pop. 4,203,972.
**Lou·is·ville** (lōō′ē-vĭl′, -ə-vəl). City of NW Ky. on the Ohio SW of Cincinnati, Ohio. Pop. 298,451.
**Lourdes** (lŏŏrd, lŏŏrdz). Town of SW France at the foot of the Pyrenees; site of a Catholic shrine. Pop. 17,685.
**Lou·ren·ço Mar·ques** (lə-rĕn′sō mär-kăs′). MAPUTO.
**Lou·vain** (lōō-văɴ′). City of central Belgium E of Brussels. Pop. 30,623.
**Love·land** (lŭv′lənd). City of N Colo. S of Fort Collins. Pop. 30,244.
**Loves Park** (lŭvz). City of N Ill. near Rockford. Pop. 13,192.
**Low** (lō). TUAMOTU.
**Low Countries.** Belgium, the Netherlands, & Luxembourg.
**Low·ell** (lō′əl). City of NE Mass. NW of Boston. Pop. 92,418.
**Low·er A·von** (lō′ər ā′vŏn, ā′vən, ăv′ən). —See AVON.
**Lower Bur·rell** (bûr′əl). City of SW Pa. NE of Pittsburgh. Pop. 13,200.
**Lower Cal·i·for·nia** (kăl′ə-fôr′nyə). Peninsula of NW Mexico extending c. 760 mi (1,223 km) S from the U.S. border.
**Lower Klam·ath** (klăm′əth). Lake of N Calif. connected with Upper Klamath Lake in S Ore.
**Lower Peninsula.** Part of Mich. between Lakes Michigan & Huron.
**Lowes·toft** (lō′stəf, -stəft, -stôft′). Borough of extreme E England on the North Sea. Pop. 53,260.
**Loy·al·ty** (loi′əl-tē). Islands of the SW Pacific NW of New Caledonia.
**Lo·yang** (lō′yăng′). LUOYANG.
**Lu·a·la·ba** (lōō′ə-lä′bə). River, c. 400 mi (644 km), of E Zaire, a headwater of the Congo.
**Lu·an·da** (lōō-ăn′də) *also* **Lo·an·da** (lō-). Cap. of Angola in the W part on the Atlantic. Pop. 475,328.
**Luang Pra·bang** (lwăng′ prə-băng′). City of NW Laos on the Mekong R. Pop. 7,596.
**Lu·ang·wa** (lōō-ăng′wä). River, c. 500 mi (805 km), of E Zambia.
**Lu·an·shya** (lōō-ăn′shä). City of N central Zambia N of Lusaka. Pop. 132,164.
**Lub·bock** (lŭb′ək). City of NW Tex. S of Amarillo. Pop. 173,979.
**Lü·beck** (lōō′bĕk′). City of NE West Germany near the Baltic & the East German border. Pop. 232,270.
**Lu·bi·lash** (lōō-bē′lăsh′). Upper course of the Sankuru R. flowing 285 mi (459 km) through S Zaire.
**Lu·blin** (lōō′blən, -blēn′). City of SE Poland SE of Warsaw. Pop. 235,937.
**Lu·bum·ba·shi** (lōō′bŏŏm-bä′shē). City of SE Zaire near the Zambia border. Pop. 318,000.
**Lu·ca·ni·a** (lōō-kā′nē-ə, -kān′yə). Peak, 17,147 ft (5,229.8 m), of the St. Elias Mts. in SW Y.T., Canada, near the Alas. border.
**Luc·ca** (lōō′kə). City of NW Italy W of Florence. Pop. 54,280.
**Lu·cerne** (lōō-sûrn′). City of central Switzerland on the NW shore of the **Lake of Lucerne** (44 sq mi/114 sq km). Pop. 70,200.
**Lu·chow** (lōō′jō). HEFEI.
**Luck·now** (lŭk′nou). City of N central India ESE of Delhi. Pop. 749,239.
**Lü·da** (lōō′dä′). City of NE China on Korea Bay at the S end of Liaodong Peninsula. Pop. 4,000,000.
**Lü·den·scheid** (lōōd′n-shīt′). City of W West Germany NW of Cologne. Pop. 76,213.
**Lu·dhi·a·na** (lōō′dē-ä′nə). City of N India NNW of Delhi. Pop. 397,850.
**Lud·low** (lŭd′lō). Town of SW Mass. near Springfield. Pop. 18,150.
**Lud·wigs·burg** (lōōd′vĭgz-bûrg′). City of SW West Germany N of Stuttgart. Pop. 83,622.
**Lud·wigs·ha·fen** (lōōd′vĭgz-hä′fən). City of SW West Germany on the Rhine opposite Mannheim. Pop. 170,374.
**Luf·kin** (lŭf′kĭn). City of E Tex. NNE of Houston. Pop. 28,562.
**Lu·ga·no** (lōō-gä′nō). Town of S Switzerland near the Italian border on the Italian-Swiss **Lake of Lugano** (19 sq mi/49 sq km). Pop. 22,280.
**Lu·gansk** (lōō-gänsk′). VOROSHILOVGRAD.
**Lu·go** (lōō′gō). City of NW Spain WSW of Oviedo. Pop. 181,556.
**Lui·chow** (lə-wē′jō′). LEIZHOU.
**Luik** (loik). LIÈGE.
**Lu·le·å** (lōō′lə-ō′). City of NE Sweden on the Gulf of Bothnia. Pop. 42,139.
**Lu·le·älv** (lōō′lə-ôlv′). River, c. 275 mi (442 km), of N Sweden.
**Lum·ber** (lŭm′bər). River, c. 125 mi (201 km), of S central N.C. & NE S.C.
**Lum·ber·ton** (lŭm′bər-tən). City of S N.C. S of Fayetteville. Pop. 18,340.
**Lund** (lŭnd). City of S Sweden NNE of Malmö. Pop. 55,047.
**Lun·dy Isle** (lŭn′dē). Island off the SW coast of England at the mouth of Bristol Channel.
**Lü·ne·burg** (lōō′nə-bûrg′). City of NE West Germany SSE of Hamburg. Pop. 64,586.
**Lü·nen** (lōō′nən). City of NW West Germany ENE of Essen. Pop. 85,685.
**Luo·yang** (lə-wō′yăng′). City of E central China ENE of Xi'an. Pop. 750,000.
**Lu·sa·ka** (lōō-sä′kə). Cap. of Zambia in the S central part. Pop. 538,469.
**Lu·sa·ti·a** (lōō-sā′shē-ə, -shə). Region of central Europe in SE East Germany & SW Poland. **—Lu·sa′tian** *adj.* & *n.*
**Lü·shun** (lōō′shōōn′). City of NE China, now part of Lüda.
**Lu·si·ta·ni·a** (lōō′sĭ-tā′nē-ə). PORTUGAL. **—Lu′si·ta′ni·an** *adj.* & *n.*
**Lü·ta** (lōō′dä′). LÜDA.
**Lu·ton** (lōōt′n). Borough of SE England NNW of London. Pop. 164,500.
**Lutsk** (lōōtsk). City of W European USSR in the W Ukraine NE of Lvov. Pop. 137,000.
**Lux·em·bourg** or **Lux·em·burg** (lŭk′səm-bûrg′). **1.** Country & grand duchy of W Europe bordering on France, Belgium, & West Germany. Cap. Luxembourg. Pop. 364,000. **2.** *also* **Luxembourg City.** Cap. of Luxembourg in the S part. Pop. 78,272.
**Lux·or** (lŭk′sôr′, lōōk′-). City of central Egypt on the Nile on part of the site of ancient Thebes. Pop. 92,748.
**Lu·zon** (lōō-zŏn′). Island of the NW Philippines, largest of the archipelago.
**Lvov** (lə-vôf′, -vôv′). City of W European USSR in the W Ukraine near the Polish border. Pop. 667,000.
**Lya·khov** (lē-ä′kəf). Islands of N Siberian USSR in the S New Siberian group between the Laptev & East Siberian seas.
**Lyc·i·a** (lĭsh′ē-ə, lĭsh′ə). Ancient country & Roman province of SW Asia Minor on the Aegean. **—Lyc′i·an** *adj.* & *n.*
**Lyd·i·a** (lĭd′ē-ə). Ancient country of W central Asia Minor on the Aegean. **—Lyd′i·an** *adj.* & *n.*

**Ly·ell** (lī′əl). Peak, 13,095 ft (3,934 m), of the Sierra Nevada in E central Calif.
**Lym·ing·ton** (lĭm′ĭng-tən). Borough of S England on the Solent opposite the Isle of Wight. Pop. 36,780.
**Lyn·brook** (lĭn′bro͝ok′). Village of SE N.Y. on SW Long Is. near Queens. Pop. 20,431.
**Lynch·burg** (lĭnch′bûrg′). Independent city of SW central Va. ENE of Roanoke. Pop. 66,743.
**Lynd·hurst** (lĭnd′hûrst′). City of NE Ohio near Cleveland. Pop. 18,092.
**Lynn** (lĭn). **1. Canal.** Natural inlet of the Pacific in SE Alas. extending c. 80 mi (129 km) NNW from Juneau. **2.** City of E Mass. near Boston. Pop. 78,471.
**Lynn·field** (lĭn′fēld′). Town of E Mass. near Lynn. Pop. 11,267.
**Lynn·wood** (lĭn′wo͝od′). **1.** City of S Calif. near Los Angeles. Pop. 48,548. **2.** City of W central Wash. near Seattle. Pop. 21,937.
**Ly·on** *or* **Ly·ons** (lē-ôɴ′, lyôɴ). City of E central France at the confluence of the Rhone & Saône rivers. Pop. 454,265.
**Lyth·am Saint Anne's** (lĭth′əm sānt ănz′). Borough of NW England on the Irish Sea N of Liverpool. Pop. 42,120.

# M

**Maas** (mäs). MEUSE.
**Maas·tricht** (mäs′trĭkt′). City of SE Netherlands near the Belgian border. Pop. 111,044.
**Ma·cao** *also* **Ma·cau** (mə-kou′). **1.** Peninsula in the South China Sea W of Hong Kong. **2.** Portuguese overseas province comprising Macao Peninsula & two offshore islands. Cap. Macao (pop. 226,880). Pop. 271,000.
**Ma·ca·pá** (măk′ə-pä′). City of N Brazil on the Amazon. Pop. 51,563.
**Ma·cas·sar** (mə-kăs′ər). *var. of* MAKASSAR.
**Mac·e·do·ni·a** (măs′ĭ-dō′nē-ə, -dōn′yə). **1.** *also* **Mac·e·don** (-dən, -dŏn′). Ancient kingdom N of Greece. **2.** Region of SE Europe including parts of Greece, Bulgaria, & Yugoslavia. **3.** Region of N Greece. **—Mac′e·do′ni·an** *adj. & n.*
**Ma·cei·ó** (măs′ā-ō′). City of NE Brazil on the Atlantic. Pop. 242,867.
**Mac·gil·li·cud·dy's Reeks** (mə-gĭl′ĭ-kŭd′ēz rēks′). Mountain range in SW Ireland rising to c. 3,000 ft (915 m).
**Ma·chi·da** (mə-chē′də). City of E central Honshu, Japan, near Tokyo. Pop. 255,305.
**Ma·chu Pic·chu** (mä′cho͞o pēk′cho͞o). Ancient Incan fortress city in the Peruvian Alps NW of Cuzco.
**Mac·ken·zie** (mə-kĕn′zē). **1.** Range of the N Rocky Mts. in NW Canada. **2.** River of NW Canada flowing c. 1,120 mi (1,800 km) into Beaufort Sea. **3.** District of N.W.T., Canada.
**Mack·i·nac** (măk′ə-nô′). Island of N Mich. in the **Straits of Mackinac,** connecting Lakes Huron & Michigan.
**Ma·comb** (mə-kōm′). City of W Ill. WSW of Peoria. Pop. 19,632.
**Ma·con** (mā′kən). City of central Ga. SE of Atlanta. Pop. 116,860.
**Mâ·con** (mä-kôɴ′). City of E central France N of Lyons. Pop. 39,130.
**Mac·quar·ie** (mə-kwär′ē). River, 590 mi (949.3 km), of New South Wales, Australia.
**Mac·tan** (mäk-tän′). Island of E central Philippines, off the E coast of Cebu Is.
**Mad·a·gas·car** (măd′ə-găs′kər). Island republic in the Indian Ocean off the SE coast of Africa. Cap. Antananarivo. Pop. 8,742,000. **—Mad′a·gas′can** *adj. & n.*
**Ma·dei·ra** (mə-dîr′ə, -dĕr′ə). **1.** River of NW Brazil flowing c. 900 mi (1,450 km) into the Amazon. **2.** Portuguese archipelago in the N Atlantic W of Morocco. **—Ma·dei′ran** *adj. & n.*
**Ma·de·ra** (mə-dĕr′ə). City of central Calif. NW of Fresno. Pop. 21,732.
**Mad·i·son** (măd′ĭ-sən). **1.** River, 183 mi (294.4 km), of NW Mont. **2.** Town of S Conn. on Long Is. Sound. Pop. 14,031. **3.** City of SE Ind. NE of New Albany. Pop. 12,472. **4.** Borough of NE N.J. near Morristown. Pop. 15,357. **5.** Cap. of Wis. in the S central part. Pop. 170,616.
**Madison Heights.** City of SE Mich. near Detroit. Pop. 35,375.
**Mad·i·son·ville** (măd′ĭ-sən-vĭl′). City of W Ky. N of Hopkinsville. Pop. 16,979.
**Ma·dras** (mə-drăs′, -dräs′). City of SE India on the Bay of Bengal. Pop. 2,469,449.
**Ma·dre de Di·os** (mä′drē dā dē-ōs′). River, c. 700 mi (1,125 km), of SE Peru & NW Bolivia.
**Ma·drid** (mə-drĭd′). Cap. of Spain on the central plateau. Pop. 3,146,071.
**Ma·du·ra** (mə-do͝or′ə). Indonesian island separated from NE Java by **Madura Strait.**
**Ma·du·rai** (mä′də-rī′). City of S India SW of Madras. Pop. 549,114.
**Mae·an·der** (mē-ăn′dər). River of W Turkey, the MENDERES.
**Ma·ga·dan** (mä′gə-dăn′, -dän′). City of Far Eastern USSR on the Sea of Okhotsk. Pop. 121,000.
**Ma·ga·dha** (mä′gə-də). Ancient kingdom of NE India.
**Mag·da·le·na** (măg′də-lā′nə). River of Colombia flowing c. 1,000 mi (1,600 km) to the Caribbean.
**Mag·de·burg** (măg′də-bûrg′, mäg′də-bo͝org′). City of W East Germany SW of Berlin. Pop. 276,089.
**Ma·gel·lan** (mə-jĕl′ən), **Strait of.** Channel, c. 350 mi (565 km), separating S South America & Tierra del Fuego.
**Mag·gio·re** (mə-jôr′ē, -jōr′ē). Lake of N Italy & S Switzerland.
**Ma·ghreb** *or* **Ma·grib** (mŭg′rəb). Region of N Africa W of Egypt.
**Mag·na Grae·cia** (măg′nə grē′shə). Ancient Greek colonies of S Italy & Sicily.
**Mag·ni·to·gorsk** (măg-nē′tə-gôrsk′). City of SW Siberian USSR in the Ural Mts. Pop. 406,000.
**Mag·no·lia** (măg-nōl′yə). City of SW Ark. W of El Dorado. Pop. 11,937.
**Ma·gog** (mā′gŏg′). City of S Que., Canada, SW of Sherbrooke. Pop. 13,290.
**Ma·grib** (mŭg′rəb). *var. of* MAGHREB.
**Ma·hal·la el Ku·bra** (mə-hăl′ə ĕl ko͞o′brə). City of Egypt N of Cairo. Pop. 292,853.
**Ma·ha·na·di** (mə-hä′nə-dē). River of central India flowing c. 550 mi (885 km) to the Bay of Bengal.
**Ma·hé** (mä-hā′). Chief island of the Seychelles in the Indian Ocean.
**Ma·hón** (mə-hōn′). Chief town of Minorca. Pop. 17,802.
**Maid·en·head** (mād′n-hĕd′). Borough of S central England W of London. Pop. 48,210.
**Maid·stone** (mād′stən, -stōn′). Borough of SE England ESE of London. Pop. 72,110.
**Mai·kop** (mī-kôp′). City of S European USSR SE of Krasnodar. Pop. 128,000.
**Main** (mīn, mān). River of E West Germany flowing c. 310 mi (500 km) to the Rhine.
**Maine** (mān). **1.** (*also* mĕn). Region of NW France S of Normandy. **2.** State of the NE U.S. Cap. Augusta. Pop. 1,124,660.
**Main·land** (mān′lănd′, -lənd). **1.** Largest of the Orkney Is., N Scotland. **2.** Largest of the Shetland Is., extreme N Scotland.
**Main Line** (mān′ līn′). Suburbs W of Philadelphia, Pa.
**Mainz** (mīnts). City of W central West Germany SW of Frankfort. Pop. 183,880.
**Ma·jor·ca** (mə-jôr′kə, -yôr′-). Spanish island, largest of the Balearics, in the W Mediterranean. **—Ma·jor′can** *adj. & n.*
**Mak·a·lu** (mŭk′ə-lo͞o′). Mountain, c. 27,800 ft (8,480 m), in the Himalayas of NE Nepal.
**Ma·kas·sar** *or* **Ma·cas·sar** (mə-kăs′ər). **1.** Strait between Borneo & Celebes. **2.** City of NW Celebes, Indonesia. Pop. 434,766.
**Ma·ke·yev·ka** (mə-kē′əf-kə). City of S European USSR NE of Donetsk. Pop. 436,000.
**Ma·khach·ka·la** (mə-käch′kə-lä′). City of SE European USSR on the Caspian Sea. Pop. 251,000.
**Ma·kin** (mä′kĭn, mä′-). BUTARITARI.
**Mal·a·bar** (măl′ə-bär′). Region of SW India along the Arabian Sea coast.
**Ma·la·bo** (mä-lä′bō, măl′ə-bō′). Cap. of Equatorial Guinea on Bioko Is. in the Gulf of Guinea. Pop. 37,237.
**Ma·lac·ca** (mə-lăk′ə). Strait between Sumatra & the Malay Peninsula, joining the Andaman & South China seas.
**Má·la·ga** (măl′ə-gə). City of S Spain NE of Gibraltar. Pop. 334,988.
**Mal·a·gas·y Republic** (măl′ə-găs′ē). MADAGASCAR.
**Ma·lang** (mə-läng′). City of E Java, Indonesia. Pop. 422,428.
**Ma·lar** (mā′lär′) *or* **Mä·lar·en** (-ən). Lake of E Sweden extending W from Stockholm.
**Mal·a·spi·na** (măl′ə-spē′nə). Glacier in the St. Elias Mts. of SE Alaska.
**Ma·la·tya** (mä′lə-tyä′). City of E central Turkey in the Taurus Mts. Pop. 154,505.
**Ma·la·wi** (mə-lä′wē). Country of SE Africa. Cap. Lilongwe. Pop. 5,968,000. **—Ma·la′wi·an** *adj. & n.*
**Ma·lay** (mə-lā′, mā′lā). **1.** Archipelago in the Indian & Pacific oceans between Australia & SE Asia. **2.** *also* **Ma·la·ya** (mə-lā′ə, mā-). Peninsula of SE Asia including parts of Malaysia, Thailand, & Burma. **—Ma·la′yan** *adj. & n.*
**Ma·lay·sia** (mə-lā′zhə, -shə). Country of SE Asia consisting of the S Malay Peninsula & N Borneo. Cap. Kuala Lumpur. Pop. 13,435,588. **—Ma·lay′sian** *adj. & n.*
**Mal·den** (môl′dən). City of E Mass. near Boston, Pop. 53,386.
**Mal·dives** (môl′dīvz, măl′-). Island country in the Indian Ocean SW of Sri Lanka. Cap. Male. Pop. 143,046. **—Mal·div′i·an, Mal·di′van** *adj. & n.*
**Ma·le** (mä′lē). Cap. of the Maldives. Pop. 12,000.
**Ma·li** (mä′lē). Country of W Africa. Cap Bamako. Pop. 6,906,000. **—Ma′li·an** *adj. & n.*
**Mal·lor·ca** (mə-yôr′kä, -lyôr′-). MAJORCA.
**Mal·mö** (măl′mō, -mœ). City of S Sweden opposite Copenhagen. Pop. 241,191.

**Mal·ta** (môl'tə). Island country in the Mediterranean S of Sicily, coextensive with the **Malta Is.** Cap. Valletta. Pop. 343,970. **—Mal·tese'** (-tēz', -tēs') *adj. & n.*
**Mal·vern.** **1.** (môl'vərn, mô'-). Hills of W central England rising to 1,395 ft (425.5 m). **2.** (măl'vərn). City of S central Ark. SW of Little Rock. Pop. 10,147.
**Mal·vin·as Islands** (măl-vē'nəs). FALKLAND IS.
**Ma·mar·o·neck** (mə-măr'ə-nĕk'). Village of SE N.Y. near New York City. Pop. 17,616.
**Mam·be·ra·mo** (măm'bə-rä'mō). River, c. 500 mi (805 km), of W New Guinea.
**Ma·mo·ré** (mä-mə-rā'). River, c. 600 mi (965 km), of N Bolivia & Brazil.
**Man** (măn), **Isle of.** British island in the Irish Sea. **—Manx** (măngks) *adj. & n.*
**Ma·na·do** (mə-nä'dō). Town of NE Celebes, Indonesia. Pop. 169,684.
**Ma·na·gua** (mə-näg'wə). **1.** Lake, 390 sq mi (1,014 sq km), of W Nicaragua. **2.** Cap. of Nicaragua on the lake's S shore. Pop. 398,514.
**Ma·nas·sas** (mə-năs'əs). Independent city of N Va. near Washington, D.C.; site of **Manassas National Battlefield Park,** commemorating the Battles of Bull Run (1861, 1862). Pop. 15,438.
**Ma·na·tí** (mä'nə-tē'). City of N Puerto Rico W of San Juan. Pop. 17, 254.
**Ma·naus** (mə-nous'). City of NW Brazil on the Rio Negro. Pop. 248,118.
**Man·ches·ter** (măn'chĕs'tər, -chĭ-stər). **1.** Borough of NE England NE of Liverpool. Pop. 490,000. **2.** Town of central Conn. near Hartford. Pop. 49,761. **3.** City of SE N.H. S of Concord. Pop. 90,936.
**Man·chu·kuo** (măn'chōō'kwō'). Former state of E Asia in Manchuria & E Inner Mongolia.
**Man·chu·ri·a** (măn-chŏŏr'ē-ə). Region of NE China. **—Man·chu'·ri·an** *adj. & n.*
**Man·da·lay** (măn'də-lā'). City of central Burma on the Irrawaddy. Pop. 418,008.
**Man·dan** (măn'dən, -dăn). City of S N.Dak. near Bismarck. Pop. 15,513.
**Man·ga·lore** (măng'gə-lôr'). City of SW India on the Arabian Sea. Pop. 165,174.
**Man·hat·tan** (măn-hăt'n). **1.** City of NE Kans. W of Topeka. Pop. 32,644. **2.** Borough of New York City, SE N.Y., mainly on **Manhattan Is.** (22 sq mi/57.2 sq km). Pop. 1,427,533. **—Man·hat'tan·ite'** (-īt') *n.*
**Manhattan Beach.** City of S Calif. near Los Angeles. Pop. 31,542.
**Ma·nil·a** (mə-nĭl'ə). City of SW Luzon, Philippines, on **Manilla Bay,** an inlet of the South China Sea. Pop. 1,626,249.
**Man·i·to·ba** (măn'ĭ-tō'bə). **1.** Lake, 1,817 sq mi (4,724.2 sq km), of SW Man., Canada. **2.** Province of S central Canada. Cap. Winnipeg. Pop. 1,017,323. **—Man'i·to'ban** *adj. & n.*
**Man·i·tou·lin** (măn'ĭ-tōō'lĭn). Island, c. 80 mi (130 km) long, in N Lake Huron.
**Man·i·to·woc** (măn'ĭ-tə-wŏk'). City of E Wis. on Lake Michigan. Pop. 32,547.
**Ma·ni·za·les** (măn'ĭ-zä'lĭs, -zăl'ĭs). City of W central Colombia. Pop. 199,904.
**Man·ka·to** (măn-kā'tō). City of S Minn. SW of Minneapolis. Pop. 28,651.
**Man·nar** (mə-när'), **Gulf of.** Inlet of the Indian Ocean between S India & Sri Lanka.
**Mann·heim** (măn'hīm', män'-). City of central West Germany on the Rhine. Pop. 314,086.
**Man·re·sa** (män-rā'sə). City of NE Spain NW of Barcelona. Pop. 52,526.
**Mans·field** (mănz'fēld'). **1.** Highest peak, 4,393 ft (1,339.9 m), of the Green Mts. of N central Vt. **2.** Borough of central England N of Nottingham. Pop. 58,450. **3.** Town of NE Conn. N of Willimantic. Pop. 20,634. **4.** Town of SE Mass. SSW of Boston. Pop. 13,453. **5.** City of N central Ohio WSW of Akron. Pop. 53,927.
**Man·te·ca** (măn-tē'kə). City of central Calif. E of Oakland. Pop. 24,925.
**Man·tu·a** (măn'chōō-ə, -tōō-ə). City of N Italy SSW of Verona. Pop. 59,529. **—Man'tu·an** *adj. & n.*
**Ma·nus** (mä'nōōs). Largest of the Admiralty Is. in the SW Pacific.
**Man·ville** (măn'vĭl'). Borough of N central N.J. near New Brunswick. Pop. 11,278.
**Man·za·la** (măn-zä'lə), **Lake.** Lagoon of NE Egypt in the Nile delta.
**Man·za·nil·lo** (măn'zə-nē'ō, -yō). City of SE Cuba. Pop. 82,000.
**Ma·ple Grove** (mā'pəl). City of SE Minn. NW of Minneapolis. Pop. 20,525.
**Maple Heights.** City of NE Ohio near Cleveland. Pop. 29,735.
**Ma·ple·wood** (mā'pəl-wŏŏd'). **1.** City of SE Minn. near St. Paul. Pop. 26,990. **2.** City of E Mo. near St. Louis. Pop. 10,960.



**Ma·pu·to** (mə-pōō'tō). Cap. of Mozambique on the Indian Ocean. Pop. 755,300.
**Ma·quo·ke·ta** (mə-kō'kə-tə). River, c. 130 m (210 km), of E Iowa flowing SE to the Mississippi.
**Ma·ra·cai·bo** (măr'ə-kī'bō, mä'rə-). City of NW Venezuela at the outlet of **Lake Maracaibo,** largest lake of South America, S of the Gulf of Venezuela. Pop. 651,574.
**Ma·ra·cay** (mär'ə-kī'). City of N Venezuela SW of Caracas. Pop. 255,134.
**Mar·ais des Cygnes** (mĕr' də zēn'). River, c. 140 mi (225 km), of E Kans. & W Mo.
**Ma·ra·jó** (mär'ə-zhō'). Island of N Brazil in the Amazon delta.
**Ma·ra·ñón** (mä'rən-yōn'). River flowing c. 1,000 mi (1,610 km) from W central to NE Peru.
**Ma·ras** (mə-räsh'). City of S central Turkey in the Taurus Mts. Pop. 135,782.
**Mar·a·thon** (măr'ə-thŏn'). Plain of ancient Greece NE of Athens; site of Persian defeat (490 B.C.).
**Mar·ble·head** (mär'bəl-hĕd', mär'bəl-hĕd'). Town of NE Mass. NE of Boston. Pop. 20,126.
**Mar·burg** (mär'bûrg', -bŏŏrg'). City of central West Germany N of Frankfurt. Pop. 72,458.
**Marche** (märsh). Region of central France.
**Mar·che** (mär'kā) *or* **the Mar·ches** (-chĭz). Region of E central Italy.
**Mar·cy** (mär'sē). Mountain, 5,344 ft (1,629.9 m), in the Adirondacks, NE N.Y.
**Mar del Pla·ta** (mär' dĕl plä'tə). City of E central Argentina on the Atlantic. Pop. 302,282.
**Ma·ren·go** (mə-rĕng'gō). Village of NW Italy; site of Napoleon's defeat of the Austrians (1800).
**Mar·e·o·tis** (măr'ē-ō'tĭs). Lake of N Egypt in the Nile delta.
**Mar·ga·ri·ta** (mär'gə-rē'tə). Island in the Caribbean off Venezuela.
**Mar·gate** (mär'gāt'). **1.** (*also* -gət). Borough of SE England NE of Canterbury. Pop. 50,290. **2.** City of SE Fla. NW of Fort Lauderdale. Pop. 36,044.
**Mar·i·an·a Islands** (mâr'ē-ăn'ə) *also* **Mar·i·an·as** (-əz). U.S.-administered island group in the W Pacific E of the Philippines.
**Ma·ri·a·nao** (mär'ē-ə-nou'). City of W Cuba near Havana. Pop. 368,747.
**Marianas Trench.** Depression (36,198 ft/11,040.4 m) in the floor of the W Pacific SW of Guam.
**Ma·ri·as** (mə-rī'əs). River of NW Mont. flowing c. 210 mi (338 km) SE to the Missouri.
**Ma·ri·bor** (mär'ĭ-bôr). City of NW Yugoslavia on the Drava. Pop. 94,976.
**Ma·rie Byrd Land** (mə-rē' bûrd') *also* **Byrd Land.** Region of W Antarctica E of Amundsen Sea.
**Mar·i·et·ta** (măr'ē-ĕt'ə, mĕr'-). **1.** City of NW Ga. NW of Atlanta. Pop. 30,805. **2.** City of SE Ohio on the Ohio R. Pop. 16,467.
**Ma·ri·na** (mə-rē'nə). City of W Calif. near Monterey. Pop. 20,647.
**Ma·rin·du·que** (măr'ən-dōō'kā, mär'-). Island of the Philippines S of Luzon.
**Mar·i·nette** (măr'ə-nĕt', mĕr'-). City of NE Wis. on Green Bay & the Mich. border. Pop. 11,965.
**Mar·i·on** (măr'ē-ən, mâr'-). **1.** City of S Ill. W of Harrisburg. Pop. 14,031. **2.** City of E central Ind. NW of Muncie. Pop. 35,874. **3.** City of E central Iowa near Cedar Rapids. Pop. 19,474. **4.** City of central Ohio N of Columbus. Pop. 37,040.
**Mar·i·time Alps** (măr'ĭ-tīm'). Range of the W Alps on the French-Italian border.
**Maritime Provinces** *also* **the Mar·i·times** (măr'ĭ-tīmz'). Canadian provinces of N.S., N.B., & P.E.I.
**Ma·ri·tsa** (mə-rēt'sə). River of W Bulgaria & W Turkey flowing c. 300 mi (485 km) to the Aegean.
**Ma·ri·u·pol** (măr'ē-ōō'pôl). ZHDANOV.
**Mark·ham** (mär'kəm). **1.** Mountain, 14,272 ft (4,353 m), of Victoria Land, Antarctica. **2.** Village of S Ont., Canada, NNE of Toronto. Pop. 56,206. **3.** City of NE Ill. near Chicago. Pop. 15,172.
**Marl** (märl). City of W West Germany, in the Ruhr. Pop. 91,930.
**Marl·bor·ough** *or* **Marl·bo·ro** (märl'bûr'ō, -bər-ə). City of E central Mass. NE of Worcester. Pop. 30,617.
**Mar·ma·ra** (mär'mər-ə), **Sea of.** Sea of NW Turkey connected to the Black Sea & the Aegean through the Bosporus & Dardanelles.
**Mar·mo·la·da** (mär'mə-lä'də). Mountain, 10,964 ft (3,344 m), in the Dolomites of NE Italy.
**Marne** (märn). River, c. 325 mi (523 km), of NE France; scene of battles in World Wars I & II.
**Ma·ro·ni** (mə-rō'nē). River of N South America flowing c. 450 mi (725 km) along the Surinam-French Guiana border to the Atlantic.
**Mar·que·sas Islands** (mär-kā'səz). Archipelago in the South Pacific, part of French Polynesia.
**Mar·quette** (mär-kĕt'). City of NW Mich. on Lake Superior. Pop. 23,288.
**Mar·ra·kesh** *or* **Mar·ra·kech** (mə-rä'kĭsh, măr'ə-kĕsh'). City of W central Morocco near the Atlas Mts. Pop. 332,741.

**Mar·sa·la** (mär-sä'lə). City of W Sicily on the Mediterranean. Pop. 34,150

**Mar·seilles** (mär-sā', -sālz') *also* **Mar·seille** (-sā'). City of SE France on the Mediterranean. Pop. 901,421.

**Mar·shall** (mär'shəl). **1.** Islands in the central Pacific, part of U.S. Trust Territory of the Pacific. **2.** City of SW Minn. WSW of New Ulm. Pop. 11,161. **3.** City of N central Mo. E of Kansas City. Pop. 12,781. **4.** City of E Tex. W of Shreveport, La. Pop. 24,921.

**Mar·shall·town** (mär'shəl-toun'). City of central Iowa NE of Des Moines. Pop. 26,938.

**Marsh·field** (märsh'fēld'). **1.** Town of SE Mass. SE of Boston. Pop. 20,916. **2.** City of central Wis. SW of Wausau. Pop. 18,290.

**Mar·ston Moor** (mär'stən). Site in N England of first Civil War battle (1644).

**Mar·ta·ban** (mär'tə-băn', -bän'), **Gulf of.** Arm of the Andaman Sea off S Burma.

**Mar·tha's Vine·yard** (mär'thəz vĭn'yərd). Island of SE Mass. off Cape Cod.

**Mar·ti·nez** (mär-tē'nəs). City of W Calif. NE of Oakland. Pop. 22,582.

**Mar·ti·nique** (mär'tĭ-nēk', -tn-ēk'). French island & overseas department in the West Indies. Cap. Fort-de-France. Pop. 308,000.

**Mar·tins·burg** (mär'tnz-bûrg'). City of NE W.Va. in the E Panhandle. Pop. 13,063.

**Mar·tins·ville** (mär'tnz-vĭl'). **1.** City of central Ind. SW of Indianapolis. Pop. 11,311. **2.** Independent city of S Va. near the N.C. border. Pop. 18,149.

**Mar·y·land** (mĕr'ə-lənd). State of the E central U.S. Cap. Annapolis. Pop. 4,216,446. **—Mar'y·land·er** *n.*

**Mar·y·ville** (mĕr'ĭ-vəl, -vĭl'). City of E Tenn. S of Knoxville. Pop. 17,480.

**Ma·sa·da** (mə-sä'də, sä'-). Ancient mountaintop fortress in SE Israel.

**Ma·san** (mä'sän'). Port city of SE South Korea. Pop. 371,917.

**Mas·ba·te** (mäs-bä'tē). Philippine island S of Luzon.

**Mas·ca·rene** (măs'kə-rēn'). Islands in the Indian Ocean E of Madagascar.

**Mas·er·u** (măz'ə-rōō'). Cap. of Lesotho in the W part. Pop. 71,500.

**Mash·had** (mə-shăd'). MESHED.

**Ma·son City** (mā'sən). City of N central Iowa NW of Waterloo. Pop. 30,144.

**Ma·son-Dix·on Line** (mā'sən-dĭk'sən). Boundary between Pa. & Md., regarded as the division between free & slave states before the Civil War.

**Mas·sa** (mä'sə). City of N central Italy near the Ligurian Sea. Pop. 56,591.

**Mas·sa·chu·setts** (măs'ə-chōō'sĭts). **1. Bay.** Inlet of the Atlantic off E Mass. **2.** State of the NE U.S. Cap. Boston. Pop. 5,737,037.

**Mas·sa·pe·qua Park** (măs-ə-pē'kwə). Village of SE N.Y. on S Long Is. Pop. 19,779.

**Mas·se·na** (mə-sē'nə). Village of N N.Y. on the St. Lawrence. Pop. 12,851.

**Mas·sif Cen·tral** (mă-sēf' sĕn-träl', săN-). Mountainous plateau of S central France.

**Mas·si·lon** (măs'ə-lən, -lŏn'). City of E central Ohio W of Canton. Pop. 30,557.

**Mas·sive** (măs'ĭv), **Mount.** Peak, 14,418 ft (4,397.5 m), in the Sawatch Mts. of central Colo.

**Ma·su·li·pa·tam** (mə-sōō'lə-pŭt'əm). *also* **Ma·su·li·pat·nam** (-pŭt'nəm). City of E central India on the Bay of Bengal. Pop. 112,612.

**Ma·su·ri·a** (mə-zŏŏr'ē-ə, -sŏŏr'-). Region of NE Poland. **—Ma·su'ri·an** *adj.*

**Mat·a·be·le·land** (măt'ə-bē'lē-lănd'). Region of SW Zimbabwe.

**Ma·ta·di** (mə-tä'dē). City of W Zaire on the Congo R. Pop. 110,436.

**Mat·a·gor·da Bay** (măt'ə-gôr'də). Inlet of the Gulf of Mexico in SE Tex.

**Ma·ta·mo·ros** (măt'ə-môr'əs, -mōr'). City of NE Mexico on the Rio Grande opposite Brownsville, Tex. Pop. 165,124.

**Ma·tane** (mə-tăn'). Town of SE Que., Canada, on Gaspé Peninsula. Pop. 12,726.

**Ma·tan·zas** (mə-tăn'zəs). Port city of W central Cuba. Pop. 94,000.

**Mat·a·pan** (măt'ə-păn'), **Cape.** S tip of the Greek mainland.

**Ma·ta·ró** (mä'tə-rō'). City of NE Spain on the Mediterranean. Pop. 73,129.

**Ma·te·ra** (mə-tĕr'ə). City of S Italy in the Apennines. Pop. 43,026.

**Ma·thu·ra** (mŭt'ər-ə). City of N central India NW of Agra. Pop. 132,028.

**Mat·squi** (măt'skwē). Town of SW B.C., Canada, WSW of Chilliwack. Pop. 31,178.

**Mat·su** (măt'sōō'). Island in Formosa Strait E of SE mainland China.

**Ma·tsu·do** (mät-sōō'dō). City of E central Honshu, Japan, near Tokyo. Pop. 358,139.

**Ma·tsu·e** (mät'sə-wā'). Port city of SW Honshu, Japan. Pop. 127,440.

**Ma·tsu·mo·to** (mät'sə-mō'tō). City of central Honshu, Japan. Pop. 185,595.

**Ma·tsu·ya·ma** (mät'sə-yä'mə). City of NW Shikoku, Japan, on the Inland Sea. Pop. 367,323.

**Mat·tag·a·mi** (mə-tăg'ə-mē). River, 275 mi (442.5 km), of E Ont., Canada.

**Mat·ta·po·ni** (măt'ə-pə-nī'). River, 125 mi (201 km), of E Va.

**Mat·ter·horn** (măt'ər-hôrn', mä'tər-). Mountain, c. 14,685 ft (4,480 m), in the Pennine Alps on the Italian-Swiss border.

**Matte·son** (măt'sən). Village of NE Ill. near Chicago. Pop. 10,223.

**Mat·toon** (mə-tōōn'). City of E central Ill. SE of Decatur. Pop. 19,787.

**Ma·tu·rín** (mä'tə-rēn'). City of NE Venezuela. Pop. 98,188.

**Mau·i** (mou'ē). Island of Hawaii NW of Hawaii Is.

**Mau·mee** (mô-mē', mô'mē). **1.** River of NE Ind. & NW Ohio flowing c. 130 mi (210 km) into Lake Erie. **2.** City of NW Ohio near Toledo. Pop. 15,747.

**Mau·na Ke·a** (mou'nə kā'ə). Dormant volcano, 13,796 ft (4,208 m), in N central Hawaii Is.

**Mauna Lo·a** (lō'ə). Volcano, 13,680 ft (4,172 m), in S central Hawaii Is.

**Mau·re·ta·ni·a** (môr'ĭ-tā'nē-ə, -tān'yə, mär'-). Ancient country of N Africa in present-day Morocco & Algeria. **—Mau're·ta'ni·an** *adj.* & *n.*

**Mau·ri·ta·ni·a** (môr'ĭ-tā'nē-ə, -tān'yə, mär'-). Islamic republic of NW Africa on the Atlantic N of Senegal. Cap. Nouakchott. Pop. 1,634,000. **—Mau'ri·ta'ni·an** *adj.* & *n.*

**Mau·ri·tius** (mô-rĭsh'əs, -ē-əs). Island country in the SW Indian Ocean. Cap. Port Louis. Pop. 959,000. **—Mau·ri'tian** *adj.* & *n.*

**May** (mā), **Cape.** Peninsula of S N.J. between the Atlantic & Delaware Bay.

**Ma·ya·güez** (mī'ə-gwĕz', -gwĕs'). City of W Puerto Rico WSW of San Juan. Pop. 82,703.

**Ma·ya·pán** (mī'ə-pän'). Ruined capital of the Maya in Yucatán, SE Mexico.

**May·fair** (mā'fâr'). District of W London, England.

**May·field** (mā'fēld'). City of SW Ky. S of Paducah. Pop. 10,705.

**Mayfield Heights.** City of NE Ohio near Cleveland. Pop. 21,550.

**Ma·yon** (mä-yōn'), **Mount.** Volcano, 8,070 ft (2,461 m), in SE Luzon, Philippines.

**Ma·yotte** (mä-yôt'). French island of the E Comoros in the Indian Ocean.

**May·wood** (mā'wŏŏd'). **1.** City of S Calif. near Los Angeles. Pop. 21,810. **2.** Village of NE Ill. near Chicago. Pop. 27,998.

**Ma·za·tlán** (mä'sə-tlän'). City of W Mexico on the Pacific. Pop. 147,010.

**Mba·bane** (əm-bə-bän'). Cap. of Swaziland in the NW part. Pop. 23,109.

**Mban·da·ka** (əm-bän-dä'kə). City of W Zaire on the Congo R. Pop. 107,910.

**Mbo·mu** (əm-bō'mōō). BOMU.

**Mbu·ji Ma·yi** (ĕm-bōō'jē mī'yē). City of S central Zaire. Pop. 256,154.

**Mc·Al·es·ter** (mĭ-kăl'ĭ-stər). City of SE Okla. WSW of Muskogee. Pop. 17,255.

**Mc·Al·len** (mĭ-kăl'ən). City of S Tex. on the Rio Grande. Pop. 67,042.

**Mc·Clure Strait** (mĭ-klŏŏr'). Arm of the Beaufort Sea, W N.W.T., Canada.

**Mc·Comb** (mĭ-kōm'). City of SW Miss. ESE of Natchez. Pop. 12,331.

**Mc·Hen·ry** (mĭk-hĕn'rē, mĭ-kĕn'-). City of NE Ill. W of Waukegan. Pop. 10,908.

**Mc·Kees·port** (mĭ-kēz'pôrt'). City of SW Pa. near Pittsburgh. Pop. 31,012.

**Mc·Kin·ley** (mĭ-kĭn'lē), **Mount.** Peak, 20,320 ft (6,198 m), in S central Alas.

**Mc·Kin·ney** (mĭ-kĭn'ē). City of N Tex. NNE of Dallas. Pop. 16,249.

**Mc·Minn·ville** (mĭk-mĭn'vĭl'). **1.** City of NW Ore. SW of Portland. Pop. 14,080. **2.** City of central Tenn. NW of Chattanooga. Pop. 10,683.

**Mc·Mur·do Sound** (mĭk-mûr'dō). Inlet of Antarctica between Ross Is. & Victoria Land.

**Mc·Pher·son** (mĭk-fûr'sən). City of central Kans. NNW of Wichita. Pop. 11,753.

**Mead** (mēd), **Lake.** Reservoir of SE Nev. & NW Ariz. formed by Hoover Dam on the Colorado R.

**Mead·ville** (mēd'vĭl'). City of NW Pa. S of Erie. Pop. 15,544.

**Meaux** (mō). City of N France on the Marne ENE of Paris. Pop. 41,831.

**Mec·ca** (mĕk'ə). City of W Saudi Arabia; birthplace of Mohammed. Pop. 366,801.

**Mech·lin** (mĕk'lĭn) *also* **Me·che·len** (mĕk'ə-lən). City of N central Belgium. Pop. 65,466.

**Meck·len·burg** (mĕk'lən-bûrg'). Region of N East Germany on the Baltic.

**Me·dan** (mā-dän'). City of NE Sumatra, Indonesia. Pop. 635,562.

**Me·del·lín** (mĕd'l-ēn'). City of NW central Colombia NW of Bogotá. Pop. 1,070,924.

**Med·field** (mĕd'fēld'). Town of E Mass. SW of Boston. Pop. 10,220.

**Med·ford** (mĕd'fərd). **1.** City of E Mass. near Boston. Pop. 58,076. **2.** City of SW Ore. W of Klamath Falls. Pop. 39,603.

**Me·di·a** (mē'dē-ə). Ancient country of SW Asia in NW Iran. **—Me'di·an** *adj. & n.*

**Med·i·cine Bow** (mĕd'ĭ-sĭn bō'). **1.** River, c. 120 mi (193 km), of S Wyo. **2. Mountains.** Range of the E Rockies in SE Wyo. & N Colo., including **Medicine Bow Peak** (12,013 ft/3,664 m).

**Medicine Hat.** City of SE Alta., Canada, near the Sask. border. Pop. 32,811.

**Me·di·na** (mĭ-dē'nə). **1.** City of NW Saudi Arabia N of Mecca. Pop. 198,186. **2.** City of NE Ohio WNW of Akron. Pop. 15,268.

**Med·i·ter·ra·ne·an** (mĕd'ĭ-tə-rā'nē-ən). Sea surrounded by Europe, Asia, Asia Minor, the Near East, & Africa, connecting with the Atlantic through the Strait of Gibraltar.

**Mé·doc** (mā-dŏk'). Region of SW France.

**Mee·rut** (mā'rət, mîr'ət). City of N central India NE of Delhi. Pop. 270,993.

**Meg·a·ra** (mĕg'ər-ə). Ancient city of E central Greece, center of **Meg·a·ris** (-əs), district between the Saronic Gulf & Gulf of Corinth.

**Me·gid·do** (mĭ-gĭd'ō). Ancient city of NW Palestine.

**Meis·sen** (mī'sən). City of SE East Germany on the Elbe. Pop. 43,561.

**Mek·nes** (mĕk-nĕs'). City of N Morocco WSW of Fez. Pop. 248,369.

**Me·kong** (mā'kông', -kŏng'). River of SE Asia flowing c. 2,600 mi (4,185 km) from China to the South China Sea through the vast **Mekong delta** in S Vietnam.

**Mel·a·ne·sia** (mĕl'ə-nē'zhə, -shə). Island group in the SW Pacific NE of Australia & S of the equator. **—Mel'a·ne'sian** *adj. & n.*

**Mel·bourne** (mĕl'bərn). **1.** City of SE Australia SW of Canberra. Metro. area pop. 2,479,225. **2.** City of E central Fla. S of Cocoa Beach. Pop. 45,536.

**Me·li·to·pol** (mĕl'ĭ-tô'pəl). City of S European USSR in the Ukraine. Pop. 161,000.

**Mel·rose** (mĕl'rōz'). City of E Mass. near Boston. Pop. 30,005.

**Melrose Park.** Village of NE Ill. near Chicago. Pop. 20,735.

**Mel·ville** (mĕl'vĭl'). **1.** Saltwater lake, c. 1,133 sq mi (2,935 sq km), of SE Labrador, Canada. **2.** Peninsula of E N.W.T., Canada. **3.** Island of N Australia in the Timor Sea. **4.** Island, c. 16,400 sq mi (42,476 sq km), of N N.W.T., Canada, N of Victoria Is.

**Mel·vin·dale** (mĕl'vĭn-dāl'). City of SE Mich. near Detroit. Pop. 12,322.

**Me·mel** (mā'məl). KLAIPEDA.

**Mem·phis** (mĕm'fĭs, mĕmp'-). **1.** Ruined capital of ancient Egypt on the Nile S of Cairo. **2.** City of SW Tenn. on the Mississippi. Pop. 646,356. **—Mem'phi·an, Mem'phite'** (-fīt') *adj. & n.*

**Me·nam** (mə-näm'). CHAO PHRAYA.

**Me·nash·a** (mə-năsh'ə). City of E Wis. N of Oshkosh. Pop. 14,728.

**Men·de·res** (mĕn'də-rĕs'). **1.** River of W Turkey flowing 250 mi (204 km) to the Aegean. **2.** River of NW Turkey flowing c. 60 mi (97 km) to the Dardanelles.

**Men·dip Hills** (mĕn'dĭp'). Range in SW England.

**Men·do·ci·no** (mĕn'də-sē'nō), **Cape.** W extremity of Calif., N of San Francisco.

**Men·do·za** (mĕn-dō'zə). City of W Argentina. Pop. 470,896.

**Men·lo Park** (mĕn'lō). City of W Calif. near Palo Alto. Pop. 25,673.

**Me·nom·i·nee** (mə-nŏm'ə-nē). **1.** River, 118 mi (190 km), of NW Mich. & NE Wis. flowing into Green Bay. **2.** City of N Mich. on the Wis. border. Pop. 10,099.

**Me·nom·o·nee Falls** (mə-nŏm'ə-nē). Village of SE Wis. near Milwaukee. Pop. 27,845.

**Me·nom·o·nie** (mə-nŏm'ə-nē). City of W Wis. WNW of Eau Claire. Pop. 12,769.

**Me·nor·ca** (mə-nôr'kə). MINORCA.

**Men·tor** (mĕn'tər). City of NE Ohio on Lake Erie near Cleveland. Pop. 42,065.

**Meq·uon** (mĕk'wŏn'). City of SE Wis. N of Milwaukee. Pop. 16,193.

**Mer·a·mec** (mĕr'ə-măk'). River of E Mo. flowing 207 mi (333 km) to the Mississippi below St. Louis.

**Mer·ced** (mər-sĕd'). **1.** River, c. 150 mi (240 km), of central Calif. **2.** City of central Calif. NW of Fresno. Pop. 36,499.

**Mer·ce·da·rio** (mĕr'sə-där'ē-ō). Mountain, 22,210 ft (6,774 m), of NW Argentina on the Chilean border.

**Mer·ce·des** (mər-sā'dēz). City of S Tex. WNW of Brownsville. Pop. 11,851.

**Mer·cer Island** (mûr'sər). City of W central Wash., coextensive with **Mercer Is.,** in Lake Washington near Seattle. Pop. 21,522.

**Mer·ci·a** (mûr'shē-ə, -shə). Ancient Anglo-Saxon kingdom of central England. **—Mer'ci·an** *adj. & n.*

**Mé·ri·da** (mĕr'ĭ-də). **1.** City of SE Mexico in the Yucatán. Pop. 233,912. **2.** City of W Venezuela SSE of Maracaibo. Pop. 74,214.

**Mer·i·den** (mĕr'ĭ-dən). City of S central Conn. N of New Haven. Pop. 57,118.

**Me·rid·i·an** (mə-rĭd'ē-ən). City of E Miss. near the Ala. border. Pop. 46,577.

**Mer·o·ë** (mĕr'ə-wē'). Ancient Cushite city of N Sudan on the Nile N of Khartoum.

**Mer·ri·am** (mĕr'ē-əm). City of E Kans. S of Kansas City. Pop. 10,794.

**Mer·rill·ville** (mĕr'əl-vĭl'). Town of NW Ind. near Gary. Pop. 27,677.

**Mer·ri·mack** (mĕr'ə-măk'). **1.** River of S central N.H. & NE Mass. flowing c. 110 mi (175 km) into the Atlantic. **2.** Town of S N.H. S of Manchester. Pop. 15,406.

**Mer·ritt** (mĕr'ĭt). Island of E Fla. between the mainland & Cape Canaveral.

**Mer·sey** (mûr'zē). River of NW England flowing c. 70 mi (113 km) to the Irish Sea at Liverpool.

**Mer·sin** (mĕr-sēn'). City of S Turkey on the Mediterranean. Pop. 152,236.

**Mer·thyr Tyd·fil** (mûr'thər tĭd'vĭl'). Borough of S Wales NW of Cardiff. Pop. 61,500.

**Me·sa** (mā'sə). City of S central Ariz. near Phoenix. Pop. 152,453.

**Me·sa·bi Range** (mə-sä'bē). Low hills in NE Minn. with extensive ore deposits.

**Mesa Verde National Park** (vûrd, vûr'dē). Area of SW Colo. with prehistoric cliff dwellings.

**Me·shed** (mə-shĕd'). City of NE Iran. Pop. 670,180.

**Me·so·lón·gi·on** (mĕs'ə-lông'gē-ôn'). Town of W central Greece on the Gulf of Patras. Pop. 11,614.

**Mes·o·po·ta·mi·a** (mĕs'ə-pə-tā'mē-ə). Ancient country of SW Asia between the Tigris & Euphrates rivers. **—Mes'o·po·ta'mi·an** *adj. & n.*

**Mes·quite** (mə-skēt'). City of N Tex. near Dallas. Pop. 67,053.

**Mes·se·ne** (mĭ-sē'nē). Ancient Greek city in the SW Peloponnesus.

**Mes·se·ni·a** (mĭ-sē'nē-ə, -sēn'yə). Ancient region of SW Greece on the Ionian Sea.

**Mes·si·na** (mĭ-sē'nə). City of NE Sicily on the **Strait of Messina,** a channel separating Sicily from mainland Italy. Pop. 203,937.

**Me·ta** (mā'tə). River, 685 mi (1,102 km), of NE Colombia forming part of the border with Venezuela.

**Met·a·pon·tum** (mĕt'ə-pŏn'təm). Ancient Greek city of SE Italy.

**Me·thu·en** (mə-thōō'ən, -thyōō'-). Town of NE Mass. on the N.H. border near Lawrence. Pop. 36,701.

**Me·tuch·en** (mə-tŭch'ən). Borough of NE N.J. near New Brunswick. Pop. 13,762.

**Metz** (mĕts, mĕs). City of NE France on the Moselle. Pop. 110,939.

**Meuse** (myōōz, mœz). River flowing c. 560 mi (901 km) from NE France through S Belgium & SE Netherlands to the North Sea.

**Mex·i·cal·i** (mĕk'sĭ-kăl'ē). City of NW Mexico near the Calif. border. Pop. 317,228.

**Mex·i·co** (mĕk'sĭ-kō'). **1. Gulf of.** Arm of the Atlantic bordering on E Mexico, SE U.S., & Cuba. **2.** Republic of NW Central America. Cap. Mexico City. Pop. 67,395,826. **3.** City of central Mo. NE of Columbia. Pop. 12,276. **—Mex'i·can** *adj. & n.*

**Mexico City.** Cap. of Mexico at the S end of the central plateau. Pop. 9,377,300.

**Mi·am·i** (mī-ăm'ē). **1.** *also* **Great Miami.** River, c. 160 mi (257 km), of W Ohio flowing c. 160 mi (257 km) into the Ohio R. **2.** City of SE Fla. on Biscayne Bay S of West Palm Beach. Pop. 346,931. **3.** City of extreme NW Ohio WSW of Joplin, Mo. Pop. 14,237.

**Miami Beach.** City of SE Fla. near Miami. Pop. 96,298.

**Mi·am·is·burg** (mī-ăm'ēz-bûrg'). City of SW Ohio S of Dayton. Pop. 15,304.

**Miami Springs.** City of SE Fla. near Miami. Pop. 12,350.

**Mich·i·gan** (mĭsh'ĭ-gən). **1. Lake.** Third-largest of the Great Lakes, between Wis. & Mich. **2.** State of the N U.S. with an Upper Peninsula bordering on Wis. & Lake Superior. Pop. 9,258,344. Cap. Lansing. **—Mich'i·gan'der** *adj. & n.*

**Michigan City.** City of NW Ind. on Lake Michigan. Pop. 36,850.

**Mi·cro·ne·si·a** (mī'krō-nē'zhə, -shə). Islands of the W Pacific E of the Philippines & N of the equator. **—Mi'cro·ne'sian** *adj. & n.*

**Mid·dle At·lan·tic States** (mĭd'l ăt-lăn'tĭk). U.S. states of N.Y., Pa., N.J., Del., & Md.

**Mid·dle·bor·ough** *or* **Mid·dle·boro** (mĭd'l-bûr'ō). Town of SE Mass. N of New Bedford. Pop. 16,404.

**Mid·dle·burg Heights** (mĭd'l-bûrg'). City of NE Ohio near Cleveland. Pop. 16,218.

**Middle East** (ēst). Area of SW Asia & NE Africa. **—Middle East'ern** (ē'stərn) *adj. & n.*

**Mid·dles·bor·ough** *or* **Mid·dles·boro** (mĭd'lz-bûr'ō). City of SE Ky. on the Va. & Tenn. borders. Pop. 12,251.

**Mid·dles·brough** (mĭd'lz-brə). Borough of NE England. Pop. 153,900.

**Mid·dle·sex** (mĭd'l-sĕks'). Borough of NE N.J. NNW of New Brunswick. Pop. 13,480.

**Mid·dle·ton** (mĭd'l-tən). City of S central Wis. near Madison. Pop. 11,779.

**Mid·dle·town** (mĭd'l-toun'). **1.** City of central Conn. near Hartford. Pop. 39,040. **2.** City of SE N.Y. WSW of Newburgh. Pop. 21,454. **3.** City of SW Ohio NNE of Cincinnati. Pop. 43,719. **4.** Borough of SE Pa. near Harrisburg. Pop. 10,122. **5.** Town of SE R.I. near Newport. Pop. 17,216.

**Middle West** (wĕst). Region of the N central U.S. including states in the Upper Mississippi Valley & those bordering on the Great Lakes. **—Middle West'ern** (wĕs'tərn) *adj. & n.* **—Middle West'ern·er** *n.*

**Mid·east** (mĭd-ēst'). MIDDLE EAST. **—Mid·east'ern** *adj.* **—Mid·east'ern·er** *n.*

**Mi·di** (mē-dē'). The S of France.

**Mid·land** (mĭd'lənd). **1.** Town of S Ont., Canada, NW of Toronto. Pop. 11,568. **2.** City of central Mich. W of Bay City. Pop. 37,250. **3.** City of W central Tex. WSW of Abilene. Pop. 70,525.

**Mid·lands** (mĭd'ləndz). The central section of England.

**Mid·lo·thi·an** (mĭd-lō'thē-ən). **1.** Former county of SE Scotland surrounding Edinburgh. **2.** Village of NE Ill. near Chicago. Pop. 14,274.

**Mid·vale** (mĭd'vāl'). City of N Utah near Salt Lake City. Pop. 10,144.

**Mid·way Islands** (mĭd'wā'). Coral atoll & U.S. territory in the central Pacific NW of Honolulu.

**Mid·west** (mĭd-wĕst'). MIDDLE WEST. **—Mid·west'ern** *adj.* **—Mid·west'ern·er** *n.*

**Midwest City.** City of central Okla. near Oklahoma City. Pop. 49,559.

**Mi·lan** (mĭ-lăn', lăn'). City of N Italy. Pop. 1,724,557. **—Mil'a·nese'** (mĭl'ə-nēz', -nēs') *adj. & n.*

**Mi·la·no** (mĭ-lä'nō). MILAN.

**Mi·let·us** (mī-lē'təs). Ancient Ionian city of W Asia Minor.

**Mil·ford** (mĭl'fərd). **1.** City of SW Conn. on Long Is. Sound. Pop. 50,898. **2.** City of S Mass. SE of Worcester. Pop. 23,390.

**Milk** (mĭlk). River, 625 mi (1,006 km), of Alta., Canada., & N Mont.

**Mill·brae** (mĭl'brā'). City of W Calif. near San Francisco. Pop. 20,058.

**Mill·bur·y** (mĭl'bûr'ē). Town of S Mass. near Worcester. Pop. 11,808.

**Mil·ledge·ville** (mĭl'ĭj-vĭl'). City of central Ga. NE of Macon. Pop. 12,176.

**Mill·ing·ton** (mĭl'ĭng-tən). City of SW Tenn. NE of Memphis. Pop. 20,236.

**Mill Valley** (mĭl). City of W Calif. near San Francisco. Pop. 12,967.

**Mill·ville** (mĭl'vĭl'). City of S N.J. W of Atlantic City. Pop. 24,815.

**Mi·los** (mē'lôs) *also* **Mi·lo** (mē'lō, mī'-). Island in the Cylades of SE Greece in the Aegean.

**Mil·pi·tas** (mĭl-pē'təs). City of W Calif. near San Jose. Pop. 37,820.

**Mil·ton** (mĭl'tən). **1.** Town of SE Ont., WSW of Toronto. Pop. 20,756. **2.** Town of E Mass. near Boston. Pop. 25,860.

**Mil·wau·kee** (mĭl-wô'kē). City of SE Wis. on Lake Michigan. Pop. 636,212.

**Mil·wau·kie** (mĭl-wô'kē). City of NW Ore. near Portland. Pop. 17,931.

**Min** (mĭn). **1.** River of SE China flowing c. 350 mi (563 km) to the South China Sea. **2.** River of central China flowing c. 500 mi (805 km) to the Yangtze.

**Minch** (mĭnch). Channel, divided into **North Minch** & **Little Minch,** separating NW Scotland from the Outer Hebrides.

**Min·da·na·o** (mĭn'də-nä'ō, -nou'). Island of S Philippines NE of Borneo, separated from the Visayan Is. by **Mindanao Sea.**

**Min·den** (mĭn'dən). **1.** City of N West Germany on the Weser. Pop. 78,887. **2.** City of NW La. ENE of Shreveport. Pop. 15,074.

**Min·do·ro** (mĭn-dôr'ō, -dōr'ō). Island of W central Philippines S of Luzon.

**Min·e·o·la** (mĭn'ē-ō'lə). Village of SE N.Y. on Long Is. Pop. 20,757.

**Min·er·al Wells** (mĭn'ər-əl). City of N. Tex. W of Fort Worth. Pop. 14,468.

**Mi·nho** (mē'nyōō). River, flowing c. 210 (338 km) from NW Spain to the Atlantic.

**Min·ne·ap·o·lis** (mĭn'ē-ăp'ə-lĭs). City of SE Minn. on the Mississippi adjacent to St. Paul. Pop. 370,951.

**Min·ne·so·ta** (mĭn'ĭ-sō'tə). **1.** River of S Minn., 332 mi (534 km), flowing to the Mississippi near St. Paul. **2.** State of the N U.S. bordering on Ont., Canada, & Lake Superior. Cap. St. Paul. Pop. 4,077,148. **—Min'ne·so'tan** *adj. & n.*

**Min·ne·ton·ka** (mĭn'ĭ-tŏng'kə). City of SE Minn. near Minneapolis. Pop. 38,683.

**Mi·nor·ca** (mĭ-nôr'kə). Spanish island in the Balearics of the W Mediterranean. **—Mi·nor'can** *adj. & n.*

**Mi·not** (mī'nŏt'). City of N N.Dak. N of Bismarck. Pop. 32,843.

**Minsk** (mĭnsk). City of W European USSR SW of Moscow. Pop. 1,262,000.

**Min·ya** (mĭn'yə), **Al.** City of N central Egypt. Pop. 146,423.

**Minya Kon·ka** (kŏng'kə). Peak, 24,900 ft (7,595 m), of central China in the Himalayas.

**Mi·que·lon** (mĭk'ə-lŏn', mēk-lôN'). French island off S Newf. Canada.

**Mir·a·mar** (mîr'ə-mär'). City of SE Fla. S of Fort Lauderdale. Pop. 32,813.

**Mish·a·wa·ka** (mĭsh'ə-wô'kə, -wŏk'ə). City of N. Ind. near South Bend. Pop. 40,224.

**Mis·kolc** (mĭsh'kōlts'). City of NE Hungary NW of Budapest. Pop. 206,727.

**Mis·sion** (mĭsh'ən). **1.** Village of SW B.C., Canada, E of Vancouver. Pop. 14,997. **2.** City of S Tex. WNW of Brownsville. Pop. 22,589.

**Mis·sion·ar·y Ridge** (mĭsh'ə-nĕr'ē). Mountain in SE Tenn. & NW Ga.; site of Civil War battle (1863).

**Mis·sis·sau·ga** (mĭs'ĭ-sô'gə). Town of S Ont., Canada., SW of Toronto. Pop. 250,017.

**Mis·sis·sip·pi** (mĭs'ĭ-sĭp'ē). **1.** River of central U.S. flowing 2,350 mi (3,780 km) to the Gulf of Mexico. **2. Sound.** Arm of the Gulf of Mexico in S La. & S Ala. **3.** State of the S U.S. Cap. Jackson. Pop. 2,520,638. **—Mis'sis·sip'pi·an** *adj. & n.*

**Mis·so·lon·ghi** (mĭs'ə-lông'gē). MESOLÓNGION.

**Mis·sou·la** (mĭ-zōō'lə). City of W Mont. WNW of Helena. Pop. 33,388.

**Mis·sou·ri** (mĭ-zŏŏr'ē, -zŏŏr'ə). **1.** River of the U.S. rising in W Mont. & flowing c. 2,565 mi (4,127 km) to the Mississippi N of St. Louis, Mo. **2.** State of central U.S. Cap. Jefferson City. Pop. 4,917,444. **—Mis·sou'ri·an** *adj. & n.*

**Missouri City.** City of SE Tex. SW of Houston. Pop. 24,533.

**Mis·tas·si·ni** (mĭs'tə-sē'nē). **1.** Lake of central Que. Canada, draining into James Bay. **2.** River of central Que. Canada, flowing c. 200 mi (321 km) into Lake St. John.

**Mis·ti** (mē'stē), **El.** Dormant volcano, 19,098 ft (5,825 m), of S Peru near Arequipa.

**Mitch·ell** (mĭch'əl). **1. Mount.** Peak, 6,684 ft (2,039 m), of W N.C. in the Appalachians. **2.** City of SE S.Dak. WNW of Sioux Falls. Pop. 13,916.

**Mi·ya·za·ki** (mē-yä'zä-kē). City of SE Kyushu, Japan. Pop. 234,347.

**Mo·ab** (mō'ăb). Ancient kingdom of Jordan E of the Dead Sea. **—Mo'a·bite'** (mō'ə-bīt') *adj. & n.*

**Mo·ber·ly** (mō'bər-lē). City of N central Mo. N of Columbia. Pop. 13,418.

**Mo·bile** (mō-bēl', mō'bēl'). City of SW Ala. on **Mobile Bay,** an inlet of the Gulf of Mexico. Pop. 200,452.

**Mo·de·na** (môd'n-ə). City of N Italy WNW of Bologna. Pop. 149,029.

**Mo·des·to** (mō-dĕs'tō). City of central Calif. SE of Stockton. Pop. 106,105.

**Moe·sia** (mē'shə, -shē-ə). Ancient region of SE Europe S of the Danube.

**Mog·a·dish·u** (mŏg'ə-dĭsh'ōō, -dē'shōō). Cap. of Somalia on the Indian Ocean. Pop. 371,000.

**Mo·gi·lev** (mŏg'ə-lĕf', -lĕv'). City of W European USSR on the Dnieper. Pop. 290,000.

**Mo·gol·lon Plateau** (mə-gē-ōn', mō'gə-yōn'). Tableland of E central Ariz.

**Mo·hács** (mō'hăch', -häch'). City of S Hungary on the Danube; site of Ottoman Turk victory (1526) over the Hungarians.

**Mo·ha·ve** (mō-hä'vē). *var. of* MOJAVE.

**Mo·hawk** (mō'hôk'). River of E central N.Y. flowing c. 140 mi (225 km) into the Hudson.

**Mo·hen·jo-Da·ro** (mō-hĕn'jō-där'ō). Ruined prehistoric city of Pakistan in the Indus valley NE of Karachi.

**Mo·ja·ve** *also* **Mo·ha·ve** (mō-hä'vē). **1.** River, c. 100 mi (160 km), of S Calif. **2.** Desert, c. 15,000 sq mi (38,850 sq km), of S Calif. SE of the Sierra Nevada.

**Mok·po** (mŏk'pō'). City of SW South Korea. Pop. 192,958.

**Mol·da·vi·a** (mŏl-dā'vē-ə, -dāv'yə). **1.** Historical region of E Rumania. **2.** *also* **Mol·da·vi·an Soviet Socialist Republic** (-dā'vē-ən, -dāv'yən). Constituent republic of SW European USSR. **—Mol·da'vi·an** *adj. & n.*

**Mo·line** (mō-lēn'). City of NW Ill. near Davenport. Pop. 45,709.

**Mo·li·se** (mô'lĭ-zā'). Region of S central Italy on the Adriatic.

**Mo·lo·kai** (mŏl'ə-kī', mō'lə-). Island of Hawaii between Oahu and Maui.

**Mo·lo·po** (mə-lō'pō). Intermittent river of South Africa flowing c. 600 mi (965 km) W to the Orange R.

**Mo·luc·cas** (mə-lŭk'əz). Islands of E Indonesia between Celebes & New Guinea. **—Mo·luc'can** *adj. & n.*

**Mom·ba·sa** (mŏm-bä'sə). City of SE Kenya mainly on **Mombasa Is.** in the Indian Ocean. Pop. 247,073.

**Mon·a·co** (mŏn'ə-kō', mə-nä'kō). Principality on the Mediterranean, an enclave in SE France. Cap. Monaco *or* Monaco-Ville. Pop. 25,029. **—Mon'a·can** *adj. & n.*

**Mo·nad·nock** (mə-năd'nŏk'). Mountain, 3,165 ft (965 m), of SW N.H.

**Mo·na Passage** (mō′nə). Strait between Puerto Rico & the Dominican Republic connecting the Atlantic & the Caribbean.
**Mön·chen Glad·bach** (mün′kən glät′bäk′). City of W West Germany W of Düsseldorf. Pop. 261,367.
**Monc·ton** (mŭngk′tən). City of SE N.B., Canada, NE of Saint John. Pop. 55,934.
**Mo·nes·sen** (mə-nĕs′ən). City of SW Pa. S of Pittsburgh. Pop. 11,928.
**Mon·go·li·a** (mŏn-gō′lē-ə, -gōl′yə, mŏng-). **1.** Region of E central Asia. **2.** Country of N central Asia between the USSR & China. Cap. Ulan Bator. Pop. 1,594,800. **—Mon·go′li·an** *adj. & n.*
**Mon·mouth** (mŏn′məth). City of W Ill. WSW of Galesburg. Pop. 10,706.
**Mo·non·ga·he·la** (mə-nŏn′gə-hē′lə, -nŏng′-). River, 128 mi (206 km), of N W.Va. & SW Pa.
**Mon·roe** (mən-rō′). **1.** Town of SW Conn. near Bridgeport. Pop. 14,010. **2.** City of N central La. E of Shreveport. Pop. 57,597. **3.** City of SE Mich. SW of Detroit. Pop. 23,531. **4.** City of S N.C. SE of Charlotte. Pop. 12,639. **5.** City of S Wis. SSW of Madison. Pop. 10,027.
**Mon·roe·ville** (mən-rō′vĭl′). Borough of SW Pa. near Pittsburgh. Pop. 30,977.
**Mon·ro·vi·a** (mən-rō′vē-ə). **1.** Cap. of Liberia in the NW part. Pop. 166,507. **2.** City of S Calif. near Los Angeles. Pop. 30,531.
**Mons** (mōɴs). City of SW Belgium near the French border. Pop. 59,362.
**Mon·tan·a** (mŏn-tăn′ə). State of the NW U.S. Cap. Helena. Pop. 786,690. **—Mon·tan′an** *adj. & n.*
**Mon·tau·ban** (mŏn-tō-bäɴ′). City of SW France N of Toulouse. Pop. 35,344.
**Mon·tauk Point** (mŏn′tôk′). E extremity of Long Is. in SE N.Y.
**Mont Blanc** (mônt blăngk, mōɴ bläɴ′). Mountain, 15,771 ft (4,810 m), in the Alps on the French-Italian border.
**Mont·clair** (mŏnt-klâr′). **1.** City of SE Calif. NE of Pomona. Pop. 22,628. **2.** Town of NE N.J. near Newark. Pop. 38,321.
**Mon·te Al·bán** (môn′tĕ äl-bän′). Ruined Zapotec city of SW Mexico.
**Mon·te·bel·lo** (mŏn′tə-bĕl′ō). City of S Calif. near Los Angeles. Pop. 52,929.
**Mon·te Car·lo** (mŏn′tē kär′lō). Town of Monaco on the Riviera. Pop. 11,599.
**Mon·te·go Bay** (mŏn-tē′gō). Town of NW Jamaica on the Caribbean. Pop. 43,521.
**Mon·te·ne·gro** (mŏn′tə-nē′grō). Region of SW Yugoslavia on the Adriatic.
**Mon·te·rey** (mŏn′tə-rā′). City of W Calif. S of San Francisco on **Monterey Bay,** an inlet of the Pacific. Pop. 27,558.
**Monterey Park.** City of S Calif. near Los Angeles. Pop. 54,338.
**Mon·ter·rey** (mŏn′tə-rā′). City of NE Mexico. Pop. 1,006,221.
**Mon·te·vi·de·o** (mŏn′tə-vĭ-dā′ō, -vĭd′ē-ō′). Cap. of Uruguay in the S on the Río de la Plata. Pop. 1,173,254.
**Mont·fer·rat** (mŏnt-fə-răt′). Region of NW Italy S of the Po.
**Mont·gom·er·y** (mŏnt-gŭm′ə-rē, -gŭm′rē). **1.** Cap. of Ala. in the central part. Pop. 176,860. **2.** City of SW Ohio near Cincinnati. Pop. 10,088.
**Mon·ti·cel·lo** (mŏn′tĭ-sĕl′ō). Estate of central Va., home of Thomas Jefferson.
**Mont·lu·çon** (mōɴ-lōō-sōɴ′). City of central France. Pop. 56,337.
**Mont·mag·ny** (mōɴ-mä-nyē′). Town of SE Que., Canada, ENE of Quebec city. Pop. 12,326.
**Mont·mar·tre** (mōɴ-mär′trə). Hill & district of N Paris, France.
**Mont·mo·ren·cy Falls** (mŏnt′mə-rĕn′sē, -rĕnt′sē). Waterfall, 270 ft (82 m), in S Que., Canada, N of Quebec city in the **Montmorency River,** flowing 60 mi (97 km) to the St. Lawrence.
**Mont·par·nasse** (mōɴ-pär-näs′, -năs′). District of S Paris, France, on the left bank of the Seine.
**Mont·pel·ier** (mŏnt-pĕl′yər). Cap. of Vt. in the central part. Pop. 8,241.
**Mont·pel·lier** (mōɴ-pĕl-yā′). City of S France WNW of Marseilles. Pop. 178,136.
**Mon·tre·al** (mŏn′trē-ôl′) *or* **Mont·ré·al** (mōɴ′rā-äl′). City of S Que. Canada, on **Montreal Is.** in the St. Lawrence. Pop. 1,080,546. **—Mon′tre·al′er** *n.*
**Montreal North** (nôrth) *or* **Mont·ré·al-Nord** (mōɴ′rā-äl′nôr). Town of S Que., Canada, near Montreal. Pop. 97,250.
**Mon·treuil** (mōɴ-trœ′yə). Town of N central France near Paris. Pop. 96,441.
**Mon·treux** (mōɴ-trœ′). Resort town of W Switzerland on Lake of Geneva. Pop. 20,421.
**Mont-Saint-Mi·chel** (mōɴ-săɴ-mē-shĕl′). Small island of NW France off the coast of Brittany.
**Mont·ser·rat** (mŏnt′sə-răt′). Island, one of the Leewards, in the British West Indies NW of Guadaloupe.
**Mont·ville** (mŏnt′vĭl′). Town of SE Conn. near New London. Pop. 16,455.
**Mon·za** (mōn′sə, mōnt′sə, mŏn′zə). City of N Italy SE of Milan. Pop. 110,735.
**Moore** (mo͝or, môr). City of central Okla. near Oklahoma City. Pop. 35,063.
**Moor·head** (mo͝or′hĕd′, môr′-). City of NW Minn. near Fargo, N.Dak. Pop. 29,998.
**Moose·head** (mo͞os′hĕd′). Lake, c. 120 sq mi (311 sq km), of W central Me. N of Augusta.
**Moose·jaw** (mo͞os′jô). City of central Sask., Canada, W of Regina. Pop. 32,581.
**Mo·rad·a·bad** (mə-rä′də-bäd′, -răd′ə-băd′). City of N central India. Pop. 258,590.
**Mor·a·ga** (mə-rä′gə). City of W Calif. E of Oakland. Pop. 15.014.
**Mo·ra·va** (môr′ə-və). **1.** River of N Czechoslovakia flowing c. 240 mi (386 km) S to the Danube. **2.** River of E Yugoslavia flowing c. 130 mi (209 km) N to the Danube.
**Mo·ra·vi·a** (mə-rā′vē-ə). Region of central Czechoslovakia. **—Mo·ra′vi·an** *adj. & n.*
**Moravian Gate** *or* **Gap.** Mountain pass of central Europe between the Sudetes & Carpathian Mts.
**Mor·ay Firth** (mûr′ē). Inlet of the North Sea in NE Scotland.
**Mord·vin·i·a** (môrd-vĭn′ē-ə) *also* **Mor·do·vi·a** (môr-dō′vē-ə). Region of E European USSR.
**Mo·reau** (mô′rō, môr′ō). River of NW S.Dak. flowing 250 mi (402 km) E to the Missouri.
**More·cambe and Hey·sham** (môr′kəm, mōr′-; hā′shəm, hē′-). Borough of NW England on **Morecambe Bay,** an inlet of the Irish Sea. Pop. 42,010.
**Mo·re·lia** (mə-rāl′yə). City of SW Mexico WNW of Mexico City. Pop. 199,099.
**More·ton Bay** (môr′tn, mōr′-). Inlet of the Pacific in E Australia.
**Mor·gan City** (môr′gən). City of S La. WSW of New Orleans. Pop. 16,114.
**Morgan Hill.** City of W Calif. SE of San Jose. Pop. 17,060.
**Mor·gan·ton** (môr′gən-tən). City of W N.C. ENE of Asheville. Pop. 13,763.
**Mor·gan·town** (môr′gən-toun′). City of N W.Va. near the Pa. border. Pop. 27,605.
**Mo·ri·o·ka** (môr′ē-ō′kə, mōr′-) City of N Honshu, Japan. Pop. 216,223.
**Mo·roc·co** (mə-rŏk′ō). Kingdom of NW Africa on the Mediterranean & the Atlantic. Cap. Rabat. Pop. 20,242,000. **—Mo·roc′can** *adj. & n.*
**Mo·ro Gulf** (môr′ō, mōr′ō). Inlet of the Celebes Sea in SW Mindanao, Philippines.
**Mo·ro·ni** (mə-rō′nē). Cap. of the Comoros on Great Comoro Is. Pop. 12,000.
**Mor·ris Jes·up** (môr′ĭs jĕs′əp, mŏr′-). Cape of N Greenland on the Arctic Ocean; the northernmost point in the world.
**Mor·ri·son** (môr′ĭ-sən, mŏr′-). **Mount.** Peak, 13,599 ft (4,148 m), of Taiwan.
**Mor·ris·town** (môr′ĭs-toun′, mŏr′-). **1.** Town of N N.J. WNW of Newark. Pop. 16,614. **2.** City of NE Tenn. ENE of Knoxville. Pop. 19,683.
**Mor·ton** (môr′tn). Village of central Ill. SE of Peoria. Pop. 14,178.
**Morton Grove.** Village of NE Ill. near Chicago. Pop. 23,747.
**Mos·cow** (mŏs′kou, -kō). **1.** Cap. of the USSR in the W central European part on the **Moscow R.,** flowing c. 310 mi (499 km) E to the Oka. Pop. 7,831,000. **2.** City of NW Idaho on the Wash. border. Pop. 16,513.
**Mo·selle** (mō-zĕl′). River of NE France & W West Germany flowing 320 mi (515 km) to the Rhine.
**Mos·es Lake** (mō′zĭz). City of E central Wash. SE of Euphrata. Pop. 10,629.
**Mos·kva** (mŏsk-vä′). MOSCOW, USSR.
**Mos·qui·to Coast** (mə-skē′tō). Region of E Nicaragua & NE Honduras.
**Moss Point** (môs). City of SE Miss. E of Gulfport. Pop. 18,998.
**Most** (môst). City of NW Czechoslovakia near the East German border. Pop. 59,400.
**Mos·ta·ga·nem** (mə-stăg′ə-nĕm′). City of NW Algeria on the Mediterranean. Pop. 101,600.
**Mo·sul** (mō-so͞ol′, mō′səl). City of N Iraq on the Tigris. Pop. 315,157.
**Mo·ta·gua** (mō-tä′gwə). River, c. 250 mi (402 km), of S central Guatemala.
**Moth·er·well and Wish·aw** (mŭth′ər-wĕl′, -wəl; wĭsh′ô). Burgh of S central Scotland SE of Glasgow. Pop. 72,991.
**Moul·mein** (mo͞ol-mān′, mōl-). City of S Burma on the Gulf of Martaban. Pop. 171,977.
**Moul·trie** (mōl′trē). City of SW Ga. SE of Albany. Pop. 15,708.
**Mounds View** (moundz′). City of E Minn. near Minneapolis. Pop. 12,593.
**Mounds·ville** (moundz′vĭl′). City of NW W.Va. in the N Panhandle. Pop. 12,419.

**Moun·tain Brook** (moun'tən). City of N central Ala. near Birmingham. Pop. 17,400.
**Mountain View.** City of W Calif. NW of San Jose. Pop. 58,655.
**Mount Ath·os** (mount ăth'ŏs, ā'thŏs). —See ATHOS.
**Mount Clem·ens** (klĕm'ənz). City of SE Mich. NNE of Detroit. Pop. 18,806.
**Mount Des·ert** (dĕz'ərt). Island, c. 100 sq mi (260 sq km), in the Atlantic off the S coast of Me.
**Mount·lake Terrace** (mount'lāk'). City of NW Wash. S of Everett. Pop. 16,534.
**Mount Mc·Kin·ley National Park** (mĭ-kĭn'lē). Scenic area in the Alaska Range in S central Alas.
**Mount Pleas·ant** (plĕz'ənt). **1.** City of central Mich. WNW of Saginaw. Pop. 23,746. **2.** Town of SE S.C. E of Charleston. Pop. 13,838. **3.** City of E Tex. SW of Texarkana. Pop. 11,003.
**Mount Pros·pect** (prŏs'pĕkt'). Village of NE Ill. near Chicago. Pop. 52,634.
**Mount Rai·nier National Park** (rā-nîr', rə-). Scenic area in the Cascade Range in W central Wash.
**Mount Roy·al** (roi'əl). Town of S Que., Canada, near Montreal. Pop. 20,514.
**Mount Rush·more National Memorial** (rŭsh'môr', -mōr'). Mountain in the Black Hills of SW S.Dak. with carved portraits of Washington, Jefferson, Lincoln, & Theodore Roosevelt.
**Mount Ver·non** (vûr'nən). **1.** Estate of NE Va. on the Potomac near Washington, D.C., home of George Washington. **2.** City of S central Ill. SE of East St. Louis. Pop. 16,995. **3.** City of SE N.Y. near the Bronx. Pop. 66,713. **4.** City of central Ohio NE of Columbus. Pop. 14,380. **5.** City of NW Wash. SSE of Bellingham. Pop. 13,009.
**Mourne** (mōrn, môrn). Mountains of SE Northern Ireland rising to 2,796 ft (853 m).
**Mo·zam·bique** (mō'zəm-bēk'). **1.** Arm of the Indian Ocean between Madagascar & Mozambique. **2.** Country of SE Africa. Cap. Maputo. Pop. 12,130,000. **—Mo'zam·bi'can** *adj. & n.*
**Mu·dan·jiang** (mo͞o'dän'jē-äng'). City of NE China. Pop. 400,000.
**Muir Woods National Monument** (myo͝or). Area of W Calif. N of San Francisco with grove of redwood trees.
**Mu·kal·la** (mo͞o-kăl'ə). Town of Southern Yemen on the Gulf of Aden. Pop. 45,000.
**Muk·den** (mo͝ok'dən, mŭk'-, mo͝ok-dĕn'). SHENYANG.
**Mul·ha·cén** (mo͞o'lä-sān', -sĕn'). Mountain, 11,424 ft (3,484 m), of S Spain in the Sierra Nevada.
**Mül·heim** (mo͝ol'hīm, myo͞ol'-). City of W West Germany on the Ruhr R. Pop. 189,259.
**Mul·house** (mə-lo͞oz'). City of E France NW of Basel. Pop. 116,494.
**Mull** (mŭl). Island of the Inner Hebrides of NW Scotland.
**Mul·tan** (mo͞ol-tän'). City of E central Pakistan SW of Lahore. Pop. 542,195.
**Mult·no·mah Falls** (mŭlt-nō'mə). Waterfall, 620 ft (189 m), in a tributary of the Columbia in NW Ore. E of Portland.
**Mün·chen** (mo͞on'KHən). MUNICH.
**Mun·cie** (mŭn'sē). City of E Ind. NE of Indianapolis. Pop. 77,216.
**Mun·de·lein** (mŭn'də-līn'). Village of NE Ill. SW of Waukegan. Pop. 17,053.
**Mun·hall** (mŭn'hôl'). Borough of SW Pa. near Pittsburgh. Pop. 14,532.
**Mu·nich** (myo͞o'nĭk). City of S West Germany near the Bavarian Alps. Pop. 1,314,865.
**Mun·ster** (mŭn'stər). **1.** Province & ancient kingdom of SW Ireland. **2.** Town of NW Ind. on the Ill. border. Pop. 20,671.
**Mün·ster** (mo͞on'stər, mŭn'-, mĭn'-). City of W West Germany NNE of Cologne. Pop 264,546.
**Mur** (mo͝or) *also* **Mu·ra** (mo͝or'ə). River, c. 300 mi (483 km), of central Austria & NW Yugoslavia.
**Mu·rat** (mo͞o-rät'). River of E Turkey flowing 380 mi (611 km) to the Euphrates.
**Mur·chi·son** (mûr'chĭ-sən). **1.** Waterfall in the Victoria Nile, NW Uganda, above Lake Albert. **2.** Intermittent river of W Australia flowing 440 mi (708 km) SW to the Indian Ocean.
**Mur·cia** (mûr'shə, -shē-ə). **1.** Region & former Moorish kingdom of SE Spain on the Mediterranean. **2.** City of SE Spain NNW of Cartagena. Pop. 102,242.
**Mu·re·şul** (mo͝or'ə-so͞ol') *also* **Mu·res** (mo͞o'rĕsh'). River, c. 470 mi (756 km), of N central Rumania & S Hungary.
**Mur·frees·bor·o** (mûr'frēz-bûr'ō). City of central Tenn. SW of Nashville. Pop. 32,845.
**Mur·gab** (mo͝or-gäb'). River, 530 mi (853 km), of NE Afghanistan & S USSR.
**Mur·mansk** (mo͝or-mănsk', -mänsk'). City of NW European USSR on a gulf of Barents Sea. Pop. 381,000.
**Mu·rom** (mo͝or'əm). City of W central European USSR on the Oka. Pop. 111,000.
**Mu·ro·ran** (mo͝or'ə-rän'). City of SW Hokkaido, Japan. Pop. 158,715.
**Mur·ray** (mûr'ē). **1.** River of Australia rising in the Australian Alps & flowing 1,609 mi (2,589 km) W to the Indian Ocean S of Adelaide. **2.** City of SW Ky. near the Tenn. border. Pop. 14,248. **3.** City of N Utah near Salt Lake City. Pop. 25,750.
**Mur·rum·bidg·ee** (mûr'əm-bĭj'ē). River of SE Australia flowing c. 1,050 mi (1,689 km) W to the Murray.
**Mur·rys·ville** (mûr'ēz-vĭl'). Borough of SW Pa. near Pittsburgh. Pop. 16,036.
**Mu·sa·shi·no** (mo͞o-sä'shē-nō'). City of E central Honshu, Japan, near Tokyo. Pop. 139,508.
**Mus·cat** (mŭs'kăt', -kət). Cap. of Oman on the Gulf of Oman. Pop. 7,500.
**Mus·ca·tine** (mŭs'kə-tēn'). City of SE Iowa WSW of Davenport. Pop. 23,467.
**Mus·cle Shoals** (mŭs'əl). City of NW Ala. on the Tennessee R. Pop. 8,911.
**Mus·co·vy** (mŭs'kə-vē). **1.** Principality of Moscow (12th–16th cent.). **2.** Former empire of RUSSIA. **—Mus'co·vite'** *adj. & n.*
**Mu·shin** (mo͞o'shĭn'). City of SW Nigeria near Lagos. Pop. 176,000.
**Mu·si** (mo͞o'sē). River, c. 325 mi (523 km), of S Sumatra, Indonesia.
**Mus·ke·go** (mŭs-kē'gō). City of SE Wis. SW of Milwaukee. Pop. 15,277.
**Mus·ke·gon** (mŭs-kē'gən). **1.** River of W central Mich. flowing 227 mi (365 km) SW to Lake Michigan. **2.** City of SW Mich. NW of Grand Rapids. Pop. 40,823.
**Muskegon Heights.** City of SW Mich. near Muskegon. Pop. 14,611.
**Mus·king·um** (mə-skĭng'əm, -gəm). River of E Ohio flowing 111 mi (179 km) SSE to the Ohio.
**Mus·ko·gee** (mə-skō'gē). City of E Okla. SE of Tulsa. Pop. 40,011.
**Mus·sel·shell** (mŭs'əl-shĕl'). River of central Mont. flowing c. 300 mi (483 km) E & N to the Missouri R.
**Mu·tan·chiang** (mo͞o'dän'jē-äng'). MUDANJIANG.
**Mut·tra** (mŭt'rə). MATHURA.
**Muz·tagh** (mo͞os-tä', -täg'). Mountain, 24,757 ft (7,551 m), in W China near the USSR border.
**Mwe·ru** (mə-wâr'o͞o). Lake, c. 70 mi (113 km) long, in central Africa on the Zaire-Zambia border.
**Myc·a·le** (mĭk'ə-lē). Promontory of W Turkey; site of Greek defeat of the Persian fleet (479 B.C.).
**My·ce·nae** (mī-sē'nē). Ancient Greek city in the NE Peloponnesus.
**Myk·o·nos** (mĭk'ə-nōs', -nəs). Greek island, one of the Cyclades, in the Aegean.
**My·ra** (mī'rə). Ancient Lycian city of S Asia Minor.
**Myr·tle Beach** (mûr'tl). City of E S.C. on the Atlantic. Pop. 18,758.
**My·si·a** (mĭsh'ē-ə). Ancient region of NW Asia Minor. **—My'si·an** *adj. & n.*
**My·sore** (mī-sôr', -sōr'). City of S India. Pop. 355,685.

# N

**Nab·a·tae·a** (năb'ə-tē'ə). Ancient kingdom of Arabia in present-day Jordan. **—Nab·a·tae'an** *adj. & n.*
**Na·blus** (năb'ləs, nä'bləs). City in the West Bank N of Jerusalem. Pop. 41,799.
**Nack·a** (nä'kə). City of E Sweden on the Baltic near Stockholm. Pop. 19,708.
**Nac·og·do·ches** (năk'ə-dō'chĭz). City of E Tex. E of Waco. Pop. 27,149.
**Na·fud** or **Ne·fud** (nə-fo͞od'). Desert area of NW Saudi Arabia.
**Na·ga** (nä'gə) *or* **Na·ga·land** (-lănd'). Hill region on the India-Burma border.
**Na·go·no** (nə-gō'nō). City of central Honshu, Japan, NW of Tokyo. Pop. 306,637.
**Na·ga·o·ka** (nä'gə-ō'kə, nä-gä'ō-kä'). City of central Honshu, Japan, S of Niigata. Pop. 171,742.
**Na·ga·sa·ki** (nä'gə-sä'kē, năg'ə-săk'ē). City of W Kyushu, Japan, on **Nagasaki Bay,** an inlet of the East China Sea. Pop. 450,194.
**Na·go·ya** (nə-goi'ə). City of central Honshu, Japan, at the head of Ise Bay. Pop. 2,079,740.
**Nag·pur** (näg'po͝or'). City of central India NE of Bombay. Pop. 866,076.
**Nagy·vá·rad** (nŏj'vär'äd). ORADEA.
**Na·ha** (nä'hä). City of SW Okinawa, Japan, on the East China Sea. Pop. 295,006.
**Na·huel Hua·pi** (nä-wĕl' wä-pē'). Lake of SW Argentina near the Chilean border.
**Nai·ro·bi** (nī-rō'bē). Cap. of Kenya in the S central part. Pop. 509,286.
**Najd** (năjd). NEJD.
**Na·khod·ka** (nə-kôt'kə). City of Far Eastern USSR E of Vladivostok on the Sea of Japan. Pop. 133,000.
**Nal·chik** (näl'chĭk). City of S European USSR SE of Rostov. Pop. 207,000.

**Na·ma·qua·land** (nə-mä′kwə-lănd′) *or* **Na·ma·land** (nä′-mə-lănd′). Region of SW Africa divided by the Orange R. into **Great Namaqualand** in Namibia and **Little Namaqualand** in South Africa.
**Nam·hoi** (näm′hoi′). FOSHAN.
**Na·mib** (nä′mĭb′). Desert of SW Africa extending c. 800 mi (1,290 km) along the coast of Namibia.
**Na·mib·i·a** (nə-mĭb′ē-ə). Territory of SW Africa currently administered by South Africa. Cap. Windhoek. Pop. 1,200,000. **—Na·mib′i·an** *adj. & n.*
**Nam·oi** (năm′oi′). River, 526 mi (846.3 km), of SE Australia.
**Nam·pa** (năm′pə). City of SW Idaho WSW of Boise. Pop. 25,112.
**Nam·po** (näm′pō′). City of W North Korea on Korea Bay SW of Pyongyang. Pop. 140,000.
**Nam·pu·la** (năm-po͞o′lə). City of NE Mozambique. Pop. 23,072.
**Na-mu Hu** (nä′mo͞o′ ho͞o′) *or* **Nam Tso** (näm′ tsō′). Salt lake, 950 sq mi (2,460.5 sq km), of central Tibet at an altitude of 15,180 ft (4,629.9 m).
**Na·mur** (nə-mo͝or′). City of S central Belgium on the Meuse S of Brussels. Pop. 32,269.
**Nan** (nän). River of W Thailand flowing 350 mi (543.1 km) to the Ping to form the Chao Phraya.
**Na·nai·mo** (nə-nī′mō). City of SW B.C., Canada, on Vancouver Is. & the Strait of Georgia. Pop. 40,336.
**Nan·chang** (nän′chäng′). City of SE China on the Gan R. Pop. 900,000.
**Nan·chong** *also* **Nan·chung** (nän′cho͝ong′). City of central China E of Chengdu. Pop. 275,000.
**Nan·cy** (năn′sē, näN-sē′). City of NE France E of Paris. Pop. 106,906.
**Nan·da De·vi** (nŭn′də dā′vē). Peak, 25,645 ft (7,821.8 m), of N India in the Himalayas.
**Nan·ga Par·bat** (nŭng′gə pûr′bət). Peak, 26,660 ft (8,131.3 m), of NW Kashmir in the Himalayas.
**Nan·jing** (nän′jĭng′) *also* **Nan·king** (năn′kĭng′, nän′-). City of E central China on the Yangtze NW of Shanghai. Pop. 2,000,000.
**Nan Ling** (nän′ lĭng′). Mountain range of S China forming a geographic barrier between central & S China.
**Nan·ning** (nän′nĭng′). City of extreme S China W of Guangzhou. Pop. 375,000.
**Nan Shan** (nän′ shän′). **1.** Mountain range of central China running NW to SE & rising to c. 20,000 ft (6,100 m). **2.** NAN LING.
**Nan·terre** (näN-tĕr′). City of N central France on the Seine. Pop. 94,441.
**Nantes** (nănts, näNt). City of W France on the Loire. Pop. 252,537.
**Nan·ti·coke** (năn′tĭ-kōk′). **1.** City of SE Ont., Canada, on Lake Erie S of Brantford. Pop. 19,489. **2.** City of NE central Pa. near Wilkes-Barre. Pop. 13,044.
**Nan·tong** (nän′to͝ong′). City of E central China on the N side of the Yangtze estuary. Pop. 300,000.
**Nan·tuck·et** (năn-tŭk′ĭt). Island of SE Mass. S of Cape Cod, from which it is separated by **Nantucket Sound,** an arm of the Atlantic.
**Nan·tung** (nän′to͝ong′). NANTONG.
**Nap·a** (năp′ə). City of W Calif. N of Oakland. Pop. 50,879.
**Nap·a·ta** (năp′ə-tə). Ancient city of Nubia on the Nile.
**Na·per·ville** (nā′pər-vĭl′). City of NE Ill. near Chicago. Pop. 42,330.
**Na·pi·er** (nā′pē-ər). City of E central North Is., New Zealand. Pop. 46,994.
**Na·ples** (nā′pəlz). **1.** City of S central Italy on the **Bay of Naples,** an arm of the Tyrrhenian Sea. Pop. 1,214,775. **2.** City of SW Fla. on the Gulf of Mexico S of Fort Myers. Pop. 17,581. **—Ne·a·pol′i·tan** (nē′ə-pŏl′ĭ-tən) *adj. & n.*
**Na·po** (nä′pō). River of N Ecuador & N Peru flowing 550 mi (884.9 km) to the Amazon.
**Na·po·li** (nä′pə-lē). NAPLES, Italy.
**Na·ra** (när′ə). City of S Honshu, Japan, located E of Osaka. Pop. 257,538.
**Na·ra·shi·no** (nä′rä-shē′nô). City of E central Honshu, Japan, near Tokyo. Pop. 117,852.
**Na·ra·yan·ganj** (nə-rä′yən-gŭnj′). City of E central Bangladesh SSE of Dacca. Pop. 270,680.
**Nar·ba·da** (nər-bŭd′ə). River of central India flowing c. 775 mi (1,245 km) to the Gulf of Cambay.
**Nar·bonne** (när-bôn′, -bŭn′). City of S France near the Mediterranean coast. Pop. 36,525.
**Na·rev** (nər-yôf′, -yôv′) *or* **Na·rew** (när′ĕf, -ĕv). River, c. 275 mi (445 km), of W European USSR & NE Poland.
**Nar·ra·gan·sett** (năr′ə-găn′sĭt). Town of S R.I. on **Narragansett Bay,** a deep inlet of the Atlantic SW of Newport. Pop. 12,088.
**Nar·rows** (năr′ōz), **The.** Strait of SE N.Y. between Brooklyn & Staten Is., New York City, & connecting Upper & Lower New York bays.
**Nar·vik** (när′vĭk). City of N Norway, an ice-free port on a fjord opposite Lofoten Is. Pop. 19,582.
**Na·ryn** (nə-rĭn′). River of S Central Asian USSR flowing c. 450 mi (725 km) from the Tian Shan to the Syr Darya.
**Nash·u·a** (năsh′o͞o-ə, -ə-wā′). City of S N.H. S of Manchester. Pop. 67,865.
**Nash·ville** (năsh′vĭl′). Cap. of Tenn. in the central part NW of Chattanooga. Pop. 455,651.
**Nass** (năs). River of W B.C., Canada, flowing 236 mi (380 km) to the Pacific.
**Nas·sau** (năs′ô′). **1.** (*also* nä′sou′). Former duchy of central West Germany N & E of the Main & Rhine rivers. **2.** Cap. of the Bahamas, a port on New Providence. Pop. 105,352.
**Nas·ser** (nä′sər, năs′ər). Lake, c. 1,550 sq mi (4,015 sq km), of SE Egypt & N Sudan formed by the Aswan High Dam on the Nile R.
**Na·tal** (nə-tăl′, -täl′). City of NE Brazil on the Atlantic N of Recife. Pop. 250,787.
**Na·tash·kwan** (nə-tăsh′kwən). River of E Canada flowing 241 mi (387.8 km) from S Labrador across E Que. to the Gulf of St. Lawrence.
**Natch·ez** (năch′ĭz). City of SW Miss. on the Mississippi SSW of Vicksburg. Pop. 22,015.
**Natch·i·toches** (năk′ĭ-tŏsh′). City of NW central La. SE of Shreveport. Pop. 16,664.
**Na·tick** (nā′tĭk). Town of E Mass. WSW of Boston. Pop. 29,461.
**Na·tion·al City** (năsh′ə-nəl, năsh′nəl). City of S Calif. near San Diego. Pop. 48,772.
**Nau·cra·tis** (nô′krə-tĭs). Ancient Egyptian city on the Nile SE of Alexandria.
**Nau·ga·tuck** (nô′gə-tŭk′). Town of SW Conn. S of Waterbury. Pop. 26,456.
**Na·u·ru** (nä-o͞o′ro͞o). Atoll & republic of the central Pacific just S of the equator & W of Kiribati. Cap. Yaren. Pop. 7,254.
**Nav·a·jo** (năv′ə-hō′, nä′və-). Mountain, 10,416 ft (3,176.8 m), of S Utah.
**Na·va·na·gar** (nŭv′ə-nŭg′ər). JAMNAGAR.
**Nav·a·ri·no** (năv′ə-rē′nō). Chilean island S of Tierra del Fuego.
**Na·varre** (nə-vär′). Former kingdom of SW Europe extending from N Spain into France.
**Nav·a·so·ta** (năv′ə-sō′tə). River, c. 130 mi (209.2 km), of E central Tex.
**Na·vy Island** (nā′vē). Island of S Ont., Canada, in the Niagara R. just above Niagara Falls.
**Na·wa** (nä′wä). NAHA.
**Nax·os** (năk′səs, -sōs′). Largest island, c. 160 sq mi (415 sq km), of the Cyclades, in the Aegean.
**Naz·a·reth** (năz′ə-rəth). Town of N Israel SE of Haifa. Pop. 33,300.
**Naze** (nāz), **The. 1.** Headland of SE England on the North Sea. **2.** LINDESNES.
**Ndja·me·na** (ən-jä′mə-nə). Cap. of Chad on the Shari R. Pop. 179,000.
**Ndo·la** (ən-dō′lə). City of N central Zambia near Zaire. Pop. 282,439.
**Neagh** (nā), **Lough.** Lake, 153 sq mi (396.3 sq km), in central Northern Ireland.
**Near** (nîr). Islands of SW Alas. in the W Aleutians.
**Near East** (ēst). **1.** Region that includes nations of the E Mediterranean, the Arabian Peninsula, & sometimes NE Africa. **2.** Formerly, the Balkan Peninsula.
**Ne·bras·ka** (nə-brăs′kə). State of the central U.S. in the Great Plains. Cap. Lincoln. Pop. 1,570,006. **—Ne·bras′kan** *adj. & n.*
**Ne·chak·o** (nə-chăk′ō). River, 287 mi (461.8 km), of central B.C., Canada.
**Nech·es** (nĕch′ĭz). River of E Tex. flowing 416 mi (669.3 km) to Sabine Lake.
**Neck·ar** (nĕk′ər). River of SW West Germany flowing 228 mi (336.9 km) from the Black Forest to the Rhine.
**Ne·der·land** (nē′dər-lănd′). City of SE Tex. near Port Arthur. Pop. 16,855.
**Need·ham** (nē′dəm). Town of E Mass. near Boston. Pop. 27,901.
**Nee·nah** (nē′nə). City of E Wis. on Lake Winnebago NNE of Oshkosh. Pop. 23,272.
**Ne·fud** (nə-fo͞od′). *var. of* NAFUD.
**Ne·gev** (nĕg′ĕv′) *or* **Ne·geb** (-ĕb′). Desert region, c. 5,140 sq mi (13,315 sq km), of S Israel.
**Ne·gro** (nā′grō, nĕg′rō), **Rio** *or* **Río. 1.** River of Argentina flowing c. 400 mi (645 km) to the Atlantic. **2.** River flowing c. 1,400 mi (2,555 km) from E Colombia to the Amazon near Manaus, Brazil. **3.** River flowing c. 500 mi (805 km) from S Brazil to the Uruguay R. in central Uruguay.
**Ne·gros** (nā′grōs, nĕg′rōs). Island, 4,905 sq mi (12,704 sq km), of the Philippines in the Visayan Is. between Panay & Cebu.
**Nei·jiang** *also* **Nei·chiang** (nā′jē-äng′). City of central China SE of Chengdu. Pop. 240,000.
**Nei Mong·gol** (nā′ mŏn′gŏl′, mŏng′-). INNER MONGOLIA.
**Nejd** (nĕjd). Plateau region of central Saudi Arabia.
**Nel·son** (nĕl′sən). **1.** River of Man., Canada, flowing c. 400 mi (645 km) from Lake Winnipeg to Hudson Bay. **2.** Borough of N England NNE of Burnley. Pop. 31,220. **3.** City of N South Is., New Zealand, at the head of Tasman Bay. Pop. 32,793.

**Nem·an** (nĕm′ən). River of W European USSR, flowing c. 580 mi (935 km) to the Baltic.

**Ne·me·a** (nē′mē-ə). Ancient city of Greece W of Corinth. **—Ne′me·an** *adj.* & *n.*

**Ne·mu·nas** (năm′ə-näs′). NEMAN.

**Nen·jiang** *also* **Nen·chiang** (nŭn′jē-äng′). River, 740 mi (1,190.7 km), of NE China.

**Ne·o·sho** (nē-ō′shō, -shə). River of SE Kans. & NE Okla. flowing c. 460 mi (740 km) to the Arkansas R.

**Ne·pal** (nə-pôl′, -päl′, -păl′). Kingdom of central Asia in the Himalayas between India & Tibet. Cap. Katmandu. Pop. 14,179,301. **—Nep′al·ese′** *adj.* & *n.*

**Ness** (nĕs), **Loch.** Lake, 22 mi (35.4 km) long, of N central Scotland.

**Ne·tan·ya** (nə-tän′yə). City of W central Israel on the Mediterranean. Pop. 70,700.

**Neth·er·lands** (nĕth′ər-ləndz). Kingdom of NW Europe on the North Sea. Constitutional cap. Amsterdam; de facto cap. The Hague. Pop. 14,227,000.

**Netherlands An·til·les** (ăn-tĭl′ēz). Autonomous territory of the Netherlands consisting of six islands in the West Indies. Cap. Willemstad. Pop. 246,000.

**Net·til·ling** (nĕch′ə-lĭng). Freshwater lake, 1,956 sq mi (5,066 sq km), of S Baffin Is., N.W.T., Canada.

**Net·tu·no** (nā-to͞o′nō). Town of central Italy on the Tyrrhenian Sea. Pop. 20,927.

**Ne·tza·hual·có·yotl** (nāt-sä′wäl-kō′yōt′l). City of S central Mexico near Mexico City. Pop. 580,436.

**Neu·bran·den·burg** (noi-brän′dən-bo͝org′). City of N East Germany S of Greifswald. Pop. 59,971.

**Neuil·ly-sur-Seine** (nœ-yē-so͞or-sān′, -sĕn′). City of N central France near Paris. Pop. 65,941.

**Neu·mün·ster** (noi-mün′stər). City of N West Germany SSW of Kiel. Pop. 84,777.

**Neun·kir·chen** (noin′kîr-kən). City of W West Germany NE of Saarbrücken. Pop. 54,992.

**Ne·u·quén** (nyo͞o-kān′). River of W central Argentina flowing 365 mi (603.3 km) to the Limay to form the Río Negro.

**Neuse** (no͞os, nyo͞os). River, c. 275 mi (442.5 km), of E N.C.

**Neuss** (nois). City of W West Germany near Düsseldorf. Pop. 148,198.

**Neus·tri·a** (no͞o′strē-ə, nyo͞o′-). W part of the Frankish Merovingian kingdom in the 6th, 7th, & 8th cent. in present-day N France.

**Neu·wied** (noi-vēt′). City of W West Germany on the Rhine. Pop. 60,029.

**Ne·va** (nē′və, nā′-). River of NW European USSR flowing 46 mi (74 km) from Lake Ladoga to the Gulf of Finland.

**Ne·vad·a** (nə-văd′ə, -vä′də). State of the W U.S. Cap. Carson City. Pop. 799,184. **—Ne·vad′an** *aj.* & *n.*

**Ne·vers** (nə-vĕr′). City of central France ESE of Bourges. Pop. 45,122.

**Ne·ves** (nā′vəs). Town of SE Brazil on Guanabara Bay just N of Niterói. Pop. 112,912.

**Ne·vis** (nē′vĭs, nĕv′ĭs). One of the Leeward Is. of the West Indies in the Caribbean.

**New** (no͞o, nyo͞o). River of the SE U.S. flowing c. 320 mi (515 km) from the Blue Ridge in NW N.C. to the Kanawha in S central W.Va.

**New Al·ba·ny** (ôl′bə-nē). City of S Ind. opposite Louisville, Ky. Pop. 37,103.

**New Am·ster·dam** (ăm′stər-dăm′). Settlement established in 1624 by the Dutch on the S end of Manhattan Is. at the mouth of the Hudson; renamed New York by the British (1664).

**New·ark** (no͞o′ərk, nyo͞o′-). **1.** City of W Calif. SSE of Oakland. Pop. 32,126. **2.** City of NW Del. WSW of Wilmington. Pop. 25,247. **3.** City of NE N.J. on **Newark Bay,** an inlet of the Atlantic, opposite Jersey City. Pop. 329,248. **4.** Village of W central N.Y. ESE of Rochester. Pop. 10,017. **5.** City of central Ohio E of Columbus. Pop. 41,200.

**New Bed·ford** (bĕd′fərd). City of SE Mass. on Buzzards Bay. Pop. 98,478.

**New·berg** (no͞o′bûrg′, nyo͞o′-). City of NW Ore. SW of Portland. Pop. 10,394.

**New Ber·lin** (bûr-lĭn′). City of SE Wis. near Milwaukee. Pop. 30,529.

**New Bern** (bûrn). City of E N.C. SE of Raleigh. Pop. 14,557.

**New Braun·fels** (broun′fəlz). City of S central Tex. NE of San Antonio. Pop. 22,402.

**New Brigh·ton** (brīt′n). City of E Minn. near St. Paul. Pop. 23,269.

**New Brit·ain** (brĭt′n). **1.** Volcanic island, c. 14,600 sq mi (37,815 sq km), of Papua New Guinea, largest of the Bismarck Archipelago. **2.** City of central Conn. SSW of Hartford. Pop. 73,840.

**New Bruns·wick** (brŭnz′wĭk). **1.** Province of E Canada. Cap. Fredericton. Pop. 688,926. **2.** City of central N.J. SW of Newark. Pop. 41,442.

**New·burgh** (no͞o′bûrg′, nyo͞o′-). City of SE N.Y. on the Hudson SSW of Poughkeepsie. Pop. 23,438.

**New·bur·y·port** (no͞o′bə-rē-pôrt′, -pōrt′, nyo͞o′-). City of NE Mass. near the Atlantic ENE of Lawrence. Pop. 15,900.

**New Cal·e·do·ni·a** (kăl′ĭ-dō′nē-ə, -dōn′yə). French overseas territory in the SW Pacific including the island of **New Caledonia,** c. 800 mi (1,287.2 km) E of Australia, & several smaller island dependencies. Cap. Nouméa. Pop. 133,233.

**New Ca·naan** (kā′nən). Town of SW Conn. NNE of Stamford. Pop. 17,931.

**New Car·roll·ton** (kăr′əl-tən). City of W Md. near Washington, D.C. Pop. 12,632.

**New Cas·tile** (kăs-tēl′). Region & former kingdom of central Spain.

**New·cas·tle** (no͞o′kăs′əl, nyo͞o′-). **1.** City of SE Australia on the Pacific N of Sydney. Pop. 138,718. **2.** Town of S Ont., Canada, on Lake Ontario near Oshawa. Pop. 31,928. **3.** *or* **New·cas·tle-un·der-Lyme** (-ŭn′dər-līm′). Borough of W central England SSW of Stoke. Pop. 75,940. **4.** *or* **New·cas·tle-up·on-Tyne** (-ə-pŏn-tīn′). Borough of NE England on the Tyne R. opposite Gateshead. Pop. 295,800.

**New Cas·tle** (kăs′əl). **1.** City of E Ind. S of Muncie. Pop. 20,056. **2.** City of W Pa. NNW of Pittsburgh. Pop. 33,621.

**New Del·hi** (dĕl′ē). Cap. of India in the N central part. Pop. 301,801.

**New Eng·land** (ĭng′glənd). **1.** Mountain range & plateau of SE Australia, part of the Great Dividing Range. **2.** Section of the NE U.S. including Me., N.H., Vt., Mass., Conn., & R.I.

**New Fair·field** (fâr′fēld′). Town of SW Conn. near the N.Y. border. Pop. 11,260.

**New·found·land** (no͞o′fən-lənd, -lănd′, -fənd-, nyo͞o′-). Province of E Canada including the island of **Newfoundland** & nearby islands & the mainland area of Labrador with its adjacent islands. Cap. St. John's. Pop. 561,996. **—New′found·land·er** *n.*

**New France** (frăns). French colonial territory in North America including much of SE Canada, the Great Lakes region, & the Mississippi valley.

**New Geor·gia** (jôr′jə). Chief island of the **New Georgia** group, part of the Solomon Is. in the SW Pacific.

**New Glas·gow** (glăs′kō, -gō, glăz′-). Town of N N.S., Canada, NE of Halifax. Pop. 10,672.

**New Gra·na·da** (grə-nä′də). Former Spanish colony of N South America including present-day Colombia, Ecuador, Panama, & Venezuela.

**New Guin·ea** (gĭn′ē). **1.** Island, c. 342,000 sq mi (885,780 sq km), in the SW Pacific N of Australia; the W half is part of Indonesia & the E is part of Papua New Guinea. **2. Trust Territory of.** Former trust territory of Australia consisting of NE New Guinea, the Bismarck Archipelago, & Bougainville in the Solomons. **—New Guin′e·an** *adj.* & *n.*

**New Hamp·shire** (hămp′shər, -shîr′, hăm′-). State of the NE U.S. Cap. Concord. Pop. 920,610. **—New Hamp′shir·ite′** *n.*

**New Har·mo·ny** (här′mə-nē). Village of SW Ind. on the Wabash; site of utopian community led (1825–28) by Robert Owen.

**New Ha·ven** (hā′vən). City of S Conn. on Long Is. Sound. Pop. 126,109.

**New Heb·ri·des** (hĕb′rĭ-dēz′). Island group of the S Pacific E of Australia, forming the republic of Vanuatu.

**New Hope** (hōp). City of E Minn. near Minneapolis. Pop. 23,087.

**New I·be·ri·a** (ī-bîr′ē-ə). City of S La. SW of Baton Rouge. Pop. 32,766.

**New·ing·ton** (no͞o′ĭng-tən, nyo͞o′-). Town of central Conn. near Hartford. Pop. 28,841.

**New Ire·land** (īr′lənd). Volcanic island of the SW Pacific in the Bismarck Archipelago, part of Papua New Guinea.

**New Jer·sey** (jûr′zē). State of the E central U.S. on the Atlantic. Cap. Trenton. Pop. 7,364,158. **—New Jer′sey·ite′** *n.*

**New Ken·sing·ton** (kĕn′zĭng-tən). City of W central Pa. NW of Pittsburgh. Pop. 17,660.

**New Lon·don** (lŭn′dən). City of SE Conn. near Long Is. Sound. Pop. 28,842.

**New·mar·ket** (no͞o′mär′kĭt, nyo͞o′-). Town of S Ont., Canada, N of Toronto. Pop. 24,795.

**New Mex·i·co** (mĕk′sĭ-kō′). State of the SW U.S. on the Mexican border. Cap. Santa Fe. Pop. 1,299,968. **—New Mex′i·can** *adj.* & *n.*

**New Mil·ford** (mĭl′fərd). **1.** Town of W Conn. NNE of Danbury. Pop. 19,420. **2.** Borough of NE N.J. near Hackensack. Pop. 16,876.

**New·nan** (no͞o′nən, nyo͞o′-). City of W Ga. SW of Atlanta. Pop. 11,449.

**New Neth·er·land** (nĕth′ər-lənd). Dutch colony in North America (1624–64) along the Hudson & lower Delaware rivers.

**New Or·leans** (ôr′lē-ənz, ôr′lənz, ôr-lēnz′). City of SE La. between the Mississippi & Lake Pontchartrain. Pop. 557,482.

**New Phil·a·del·phi·a** (fĭl′ə-dĕl′fē-ə). City of NE çentral Ohio S of Canton. Pop. 16,883.

**New·port** (no͞o′pôrt′, -pōrt′, nyo͞o′-). **1.** Borough of the Isle of Wight, S England. Pop. 22,430. **2.** City of N Ky. on the Ohio near Covington. Pop. 21,587. **3.** City of SE R.I. on the Atlantic SSE of Providence. Pop. 21,259. **4.** Borough of SE Wales near the Severn estuary NE of Cardiff. Pop. 110,900.

**Newport Beach.** City of S Calif. S of Santa Ana. Pop. 63,475.

**Newport News.** Independent city of SE Va. on the James R. & Hampton Roads opposite Norfolk. Pop. 144,903.

**New Port Rich·ey** (rĭch′ē). City of W Fla. on the Gulf of Mexico NW of Tampa. Pop. 11,196.

**New Prov·i·dence** (prŏv′ĭ-dəns). **1.** Island of the Bahamas in the West Indies. **2.** Borough of NE N.J. W of Newark. Pop. 12,426.

**New Que·bec** (kwĭ-bĕk′). **1.** Region of N Que., Canada, between Hudson Bay & Labrador N of the Eastmain R. **2. Crater.** Meteoric crater, 3 m (4.8 km) in diameter, of N Que., Canada, on the Ungava Peninsula.

**New Ro·chelle** (rə-shĕl′, rō-). City of SE N.Y. on Long Is. Sound E of Mount Vernon. Pop. 70,794.

**New Si·be·ri·an Islands** (sī-bîr′ē-ən). Archipelago of N Siberian USSR in the Arctic Ocean between the Laptev & East Siberian seas.

**New Smyr·na Beach** (smûr′nə). City of NE Fla. on the Atlantic SSE of Daytona Beach. Pop. 13,557.

**New Spain** (spān). Spanish viceroyalty (1521–1821) including the SW U.S., Mexico, Central America N of Panama, some West Indian islands, & the Philippines.

**New Swe·den** (swēd′n). Swedish colony (1638–55) in North America on the Delaware R., including parts of present-day Pa., N.J., & Del.

**New·ton** (no͞ot′n, nyo͞ot′n). **1.** City of central Iowa ENE of Des Moines. Pop. 15,292. **2.** City of S central Kans. N of Wichita. Pop. 16,332. **3.** City of E Mass. near Boston. Pop. 83,622.

**New·town** (no͞o′toun′, nyo͞o′-). Town of SW Conn. ENE of Danbury. Pop. 19,107.

**New Ulm** (ŭlm′). City of S Minn. WNW of Mankato. Pop. 13,755.

**New West·min·ster** (wĕst-mĭn′stər). City of SW B.C., Canada, on the Fraser R. near Vancouver. Pop. 33,393.

**New York** (yôrk). **1. Bay.** Arm of the Atlantic at the mouth of the Hudson R. in SE N.Y. & NE N.J., divided into **Upper New York Bay** & **Lower New York Bay,** connected by the Narrows. **2.** State of the NE U.S. Cap. Albany. Pop. 17,557,288. **3.** *or* **New York City.** City of SE N.Y. on New York Bay & the mouth of the Hudson R. Pop. 7,071,030.

**New York State Barge Canal.** System of inland waterways in N.Y., 525 mi (845 km) long, connecting the Great Lakes with the Hudson R. & Lake Champlain.

**New Zea·land** (zē′lənd). Island country in the S Pacific SE of Australia. Cap. Wellington. Pop. 3,167,357. **—New Zea′land·er** *n.*

**Ne·ya·ga·wa** (nā′yə-gä′wə). City of SW Honshu, Japan, near Osaka. Pop. 254,311.

**Nga·mi** (əng-gä′mē), **Lake.** Lake of N Botswana.

**Ngau·ru·hoe** (əng-gou′rə-hō′ē). Volcano, 7,515 ft (2,292 m) high, of central North Is., New Zealand.

**Nha Trang** (nyä′ träng′). City of S central Vietnam on the South China Sea. Pop. 216,227.

**Ni·ag·a·ra** (nī-ăg′rə, -ər-ə). River, flowing 34 mi (54.7 km) from Lake Erie to Lake Ontario, forming part of the boundary between W N.Y. & Ont., Canada.

**Niagara Falls. 1.** Falls in the Niagara R. between the cities of Niagara Falls, N.Y., & Niagara Falls, Ont., Canada, divided by Goat Is. into the American Falls & the Canadian Falls. **2.** City of S Ont., Canada, on the Niagara R. opposite Niagara Falls, N.Y. Pop. 69,423. **3.** City of W N.Y. on the Niagara R. NNW of Buffalo. Pop. 71,384.

**Ni·ag·a·ra-on-the-Lake** (nī-ăg′rə-ŏn-thə-lāk′, -ər-ə-). Town of S Ont., Canada, on Lake Ontario at the mouth of the Niagara R. Pop. 12,485.

**Nia·mey** (nē-ä′mā, nyä-mā′). Cap. of Niger in the SW on the Niger R. Pop. 225,314.

**Ni·as** (nē′əs). Volcanic island of Indonesia in the Indian Ocean off W central Sumatra.

**Ni·cae·a** (nī-sē′ə). Ancient city of Bithynia, NW Asia Minor. **—Ni·cae′an** *adj.*

**Nic·a·ra·gua** (nĭk′ə-rä′gwə). **1.** Largest lake, 3,089 sq mi (8,000.5 sq km) of Central America, in SW Nicaragua. **2.** Republic of Central America on the Caribbean Sea & Pacific Ocean. Cap. Managua. Pop. 2,703,000. **—Ni′ca·ra′guan** *adj.* & *n.*

**Nice** (nēs). City of SE France on the Mediterranean. Pop. 331,002.

**Nich·o·las·ville** (nĭk′ə-ləs-vĭl′). City of W central Ky. near Lexington. Pop. 10,400.

**Nick·el Center** (nĭk′əl). Town of S central Ont., Canada, WNW of Lake Nipissing. Pop. 13,157.

**Nic·o·bar Islands** (nĭk′ə-bär′). Indian islands in the Bay of Bengal NW of Sumatra.

o͞o **boot** ou **out** th **thin** *th* **this** ŭ **cut** û**r** **urge** y **young**
yo͞o ab**use** zh vi**sion** ə **a**bout, it**e**m, ed**i**ble, gall**o**p, circ**u**s

**Nic·o·me·di·a** (nĭk′ə-mē′dē-ə). Ancient city of NW Asia Minor near the Bosporus in present-day Turkey.

**Nic·o·si·a** (nĭk′ə-sē′ə). Cap. of Cyprus in the N central part. Pop. 115,718.

**Ni·co·ya** (nĭ-kō′yə), **Gulf of.** Inlet of the Pacific between **Nicoya Peninsula** & the NW mainland of Costa Rica.

**Ni·da·ros** (nē′də-rōs′). TRONDHEIM.

**Nie·men** (nyĕm′ən). NEMAN.

**Ni·ger** (nī′jər). **1.** River of W Africa flowing c. 2,600 mi (4,185 km) from Guinea through Mali, Niger, & Nigeria into the Gulf of Guinea. **2.** Republic of W central Africa. Cap. Niamey. Pop. 5,098,-427.

**Ni·ge·ri·a** (nī-jîr′ē-ə). Republic of W Africa on the Gulf of Guinea. Cap. Lagos. Pop. 82,643,000. **—Ni·ge′ri·an** *adj.* & *n.*

**Ni·hon** (nē′hôn′). JAPAN.

**Ni·i·ga·ta** (nē-gä′tə). City of N Honshu, Japan, on the Sea of Japan. Pop. 423,188.

**Ni·i·ha·ma** (nē′hə-mə). City of N Shikoku, Japan. Pop. 131,712.

**Ni·i·ha·u** (nē′hou′). Island of Hawaii W of Kauai Is.

**Ni·i·za** (nē′zə). City of E central Honshu, Japan, near Tokyo. Pop. 119,991.

**Nij·me·gen** (nī′mā′gən) *or* **Nim·we·gen** (nĭm′vā′gən) *or* **Ni·me·guen** (nī′mā′gən). City of E Netherlands near the West German border. Pop. 148,493.

**Nik·ko** (nĭk′ō). Town & pilgrimage center of central Honshu, Japan, N of Tokyo. Pop. 26,279.

**Ni·ko·la·yev** (nĭk′ə-lä′yəf). Town of S European USSR in the Ukraine at the mouth of the Bug R. Pop. 440,000.

**Ni·ko·pol** (nĭ-kô′pəl). City of S European USSR in the Ukraine on the Dneiper. Pop. 146,000.

**Nile** (nīl). Longest river in the world, flowing c. 4,160 mi (6,695 km) through E Africa from its most remote sources in Burundi to a delta on the Mediterranean in NE Egypt.

**Niles** (nīlz). **1.** Village of NE Ill. near Chicago. Pop. 30,363. **2.** City of SW Mich. N of South Bend, Ind. Pop. 13,115. **3.** City of NE Ohio near Youngstown. Pop. 23,088.

**Ni·me·guen** (nī′mā′gən). *var. of* NIJMEGEN.

**Nîmes** (nēm). City of SE France NE of Montpellier. Pop. 123,914.

**Nim·rud** (nĭm-ro͞od′). Ancient city of Assyria S of present-day Mosul, Iraq.

**Nim·we·gen** (nĭm′vā′gən). *var. of* NIJMEGEN.

**Nin·e·veh** (nĭn′ə-və). Ancient cap. of the Assyrian Empire on the Tigris opposite the site of modern Mosul, Iraq.

**Ning·bo** *also* **Ning·po** (nĭng′bō′). City of E China ESE of Hangzhou on Hangzhou Bay. Pop. 350,000.

**Ning·xia Hui·zu** (nĭng′shē-ä′ hwēd′zo͞o′) *also* **Ning·sia Hui** (hwē′). Autonomous region of N China.

**Ni·o·brar·a** (nī′ə-brâr′ə). River, c. 430 mi (690 km), of E Wyo. & NE Nebr.

**Niort** (nē-ôr). City of W France ENE of La Rochelle. Pop. 59,297.

**Nip·i·gon** (nĭp′ĭ-gŏn′). Lake, c. 1,870 sq mi (4,840 sq km), of central Ont., Canada, N of Lake Superior.

**Nip·is·sing** (nĭp′ĭ-sĭng′). Lake of S Ont., Canada, between the Ottawa R. & Georgian Bay.

**Nip·pon** (nĭ-pŏn′). JAPAN.

**Nip·pur** (nĭ-po͝or′). Ancient city of Babylonia on the Euphrates.

**Niš** *or* **Nish** (nĭsh). City of E Yugoslavia near the Bulgarian border. Pop. 128,231.

**Ni·shi·no·mi·ya** (nĭsh′ə-nō′mē-ä, -yä). City of SW Honshu, Japan, on Osaka Bay. Pop. 400,622.

**Ni·te·rói** (nē′tə-roi′). City of SE Brazil on Guanabara Bay opposite Rio de Janeiro. Pop. 291,970.

**Ni·tra** (nē′trə). City of S central Czechoslovakia on the **Nitra R.,** a tributary of the Danube. Pop. 50,000.

**Ni·u·a·foo** (nē-o͞o′ə-fō′). Island of extreme N Tonga, SW central Pacific.

**Ni·u·e** (nē-o͞o′ā). Island dependency of New Zealand in the S central Pacific E of Tonga.

**Ni·ver·nais** (nĭv′ər-nā′). Region & former province of central France.

**Nizh·ni Nov·go·rod** (nĭzh′nē nŏv′gə-räd′). GORKI.

**Nizhni Ta·gil** (tə-gĭl′). City of E European USSR in the E central Urals. Pop. 398,000.

**No** (nō). Lake of S central Sudan formed by flood waters of the White Nile.

**No·a·tak** (nō-ä′tək). River of NW Alas. flowing c. 400 mi (640 km) to Kotzebue Sound.

**No·be·o·ka** (nō′bē-ō′kə). City of E Kyushu, Japan. Pop. 134,521.

**No·bles·ville** (nō′bəlz-vĭl′). City of central Ind. N of Indianapolis. Pop. 12,056.

**No·gal·es** (nō-găl′ĭs, -gä′lĭs). **1.** City of NW Mexico on the Ariz. border contiguous to Nogales, Ariz. Pop. 14,254. **2.** City of S Ariz. on the Mexican border. Pop. 15,683.

**No·ginsk** (nō-gĭnsk′). City of central European USSR near Moscow. Pop. 120,000.

**Nol·i·chuck·y** (nŏl′ə-chŭk′ē). River, c. 150 mi (240 km), of W N.C. & E Tenn.

**Nome** (nōm). **1.** Cape on the W Alas. coast ESE of the city of Nome on Norton Sound. **2.** Westernmost city of the continental U.S., on Seward Peninsula, W Alas. Pop. 2,301.
**Non·ni** (nŭn′nē′). NENJIANG.
**Noot·ka Sound** (no͝ot′kə, no͞ot′-). Inlet of the Pacific on the W coast of Vancouver Is., SW B.C., Canada.
**Nor·co** (nôr′kō, nŏr′-). City of S Calif. WSW of Riverside. Pop. 21,126.
**Nord·hau·sen** (nôrt′hou′zən). City of W East Germany at the S foot of the Harz Mts. Pop. 44,442.
**Nord·kyn** (nôr′kən, nŏr′-), **Cape.** Northernmost point of the European mainland, in N Norway E of North Cape.
**Nore** (nôr, nōr), **the.** Sandbank in the Thames estuary, SE England.
**Nor·folk** (nôr′fək, -fôk). **1.** Island of the Pacific, a territory of Australia c. 1,035 mi (1,665 km) NE of Sydney. **2.** City of NE Nebr. NW of Omaha. Pop. 19,449. **3.** Independent city of SE Va. on Hampton Roads SE of Richmond. Pop. 266,979.
**Nor·ge** (nôr′gə). NORWAY.
**Nor·i·cum** (nôr′ĭ-kəm, nŏr′-). Province of the Roman Empire, corresponding roughly to modern Austria S of the Danube & W of Vienna.
**No·rilsk** (nə-rēlsk′). Northernmost city in the USSR, in N Siberia. Pop. 180,000.
**Nor·mal** (nôr′məl). Town of central Ill. near Bloomington. Pop. 35,672.
**Nor·man** (nôr′mən). City of central Okla. S of Oklahoma City. Pop. 68,020.
**Nor·man·dy** (nôr′mən-dē). Region & former province of NW France on the English Channel. **—Nor·man** *adj. & n.*
**Nor·ridge** (nôr′ĭj, nŏr′-). Village of NE Ill. near Chicago. Pop. 16,483.
**Nor·ris·town** (nôr′ĭs-toun′, nŏr′-). Borough of SE Pa. NW of Philadelphia. Pop. 34,684.
**Norr·kö·ping** (nôr′chœ′pĭng). City of SE Sweden on an inlet of the Baltic. Pop. 85,244.
**North** (nôrth). **1. Sea.** Arm of the Atlantic NW of central Europe & E of Great Britain. **2. Channel.** Strait, c. 75 mi (120 km) long, between Northern Ireland & Scotland connecting the Irish Sea & the Atlantic. **3. River.** Part of the Hudson R. estuary separating N.J. & New York City. **4.** Cape on N North Is., New Zealand. **5.** Cape of N Norway projecting into the Arctic Ocean. **6.** Island, 44,281 sq mi (114,688 sq km), of N New Zealand. **7. the.** U.S. states N of Md. & the Ohio & Missouri rivers.
**North Ad·ams** (ăd′əmz). City of NW Mass. NNE of Pittsfield. Pop. 18,063.
**North A·mer·i·ca** (ə-mĕr′ĭ-kə). N continent of the Western Hemisphere extending N from the Colombia-Panama border through Central America, the U.S., Canada, & the Arctic Archipelago to the N tip of Greenland. **—North A·mer′i·can** *adj. & n.*
**North·amp·ton** (nôr-thămp′tən, nôrth-hămp′-). **1.** Borough of central England. Pop. 128,290. **2.** City of W central Mass. N of Springfield. Pop. 29,286.
**North An·do·ver** (ăn′dō′vər). Town of NE Mass. near Lawrence. Pop. 20,129.
**North Ar·ling·ton** (är′lĭng-tən). Borough of NE N.J. near Newark. Pop. 16,587.
**North At·tle·bor·o** (ăt′l-bûr′ō, -bər-ə). Town of SE Mass. NNE of Providence, R.I. Pop. 21,095.
**North Au·gus·ta** (ô-gŭs′tə). City of SW S.C. near Augusta, Ga. Pop. 13,593.
**North Bat·tle·ford** (băt′l-fərd). City of W Sask., Canada, on the North Saskatchewan R. NW of Saskatoon. Pop. 13,158.
**North Bay.** City of SE Ont., Canada, WSW of Sudbury. Pop. 51,639.
**North Bell·more** (bĕl′môr′, -mōr′). Town of SE N.Y. on Long Is. near Hempstead. Pop. 23,600.
**North·bor·ough** *or* **North·bor·o** (nôrth′bûr′ō). Town of E central Mass. near Worcester. Pop. 10,568.
**North Bran·ford** (brăn′fərd). Town of S Conn. E of New Haven. Pop. 11,554.
**North·bridge** (nôrth′brĭj′). Town of S central Mass. SSE of Worcester. Pop. 12,246.
**North·brook** (nôrth′bro͝ok′). Village of NE Ill. near Chicago. Pop. 30,735.
**North Ca·na·di·an** (kə-nā′dē-ən). River of N.Mex. & Okla. flowing 760 mi (1,223 km) to the Canadian R. in E Okla.
**North Can·ton** (kăn′tən). City of NE central Ohio near Canton. Pop. 14,228.
**North Car·o·li·na** (kăr′ə-lī′nə). State of the SE U.S. on the Atlantic. Cap. Raleigh. Pop. 5,874,429. **—North Car·o·lin′i·an** (-lĭn′ē-ən) *adj. & n.*
**North Charles·ton** (chärl′stən). City of SE S.C. near Charleston. Pop. 65,630.
**North Chi·ca·go** (shĭ-kä′gō, -kô′-). City of NE Ill. on Lake Michigan. Pop. 38,774.
**North Col·lege Hill** (kŏl′ĭj). City of extreme SW Ohio near Cincinnati. Pop. 10,990.
**North Da·ko·ta** (də-kō′tə). State of N central U.S. Cap. Bismarck. Pop. 652,695. **—North Da·ko′tan** *adj. & n.*
**North·east** (nôrth-ēst′), **the.** Area of the NE U.S. including New England, N.Y., & sometimes Pa. & N.J.
**Northeast Passage.** Water route along the N coast of Europe & Asia between the Atlantic & Pacific.
**North·ern Cook** (nôr′thərn ko͝ok′). Islands of the central Pacific N of the Cook Is., under New Zealand administration.
**Northern Dvi·na** (də-vē-nä′). —See DVINA.
**Northern Hem·i·sphere** (hĕm′ĭ-sfîr′). The half of the earth N of the equator.
**Northern Ire·land** (īr′lənd). Component of the United Kingdom in the NE part of the island of Ireland. Cap. Belfast. Pop. 1,543,000.
**Northern Kar·roo** (kə-ro͞o′). Plateau region of E South Africa.
**Northern Spor·a·des** (spôr′ə-dēz′, spŏr′-). Islands of E Greece in the Aegean.
**North·field** (nôrth′fēld′). City of SE Minn. S of Minneapolis-St. Paul. Pop. 12,562.
**North Fork.** River, c. 100 mi (160 km), of S Mo. & N Ark.
**North Fri·sian Islands** (frĭzh′ən, frē′zhən). Islands in the North Sea off the West German coast.
**North·glenn** (nôrth-glĕn′). City of N central Colo. N of Denver. Pop. 29,847.
**North Ha·ven** (hā′vən). Town of S Conn. NNE of New Haven. Pop. 22,080.
**North High·lands** (hī′ləndz). Town of N central Calif. near Sacramento. Pop. 36,800.
**North Kings·town** (kĭngz′stən). Town of S central R.I. on Narragansett Bay SSW of Providence. Pop. 21,938.
**North Ko·re·a** (kə-rē′ə). —See KOREA.
**North·lake** (nôrth′lāk′). City of NE Ill. near Chicago. Pop. 12,166.
**North Las Ve·gas** (läs vā′gəs). City of S Nev. near Las Vegas. Pop. 42,739.
**North Lau·der·dale** (lô′dər-dāl′). City of SE Fla. NW of Fort Lauderdale. Pop. 18,479.
**North Lit·tle Rock** (lĭt′l rŏk′). City of central Ark. opposite Little Rock. Pop. 64,391.
**North Loup** (lo͞op). River, 212 mi (341.1 km), of N central Nebr.
**North Mi·am·i** (mī-ăm′ē, -ăm′ə). City of SE Fla. near Miami. Pop. 42,566.
**North Miami Beach.** City of SE Fla. on the Atlantic. Pop. 36,481.
**North Minch** (mĭnch). —See MINCH.
**North Olm·sted** (ŭm′stĕd′). City of NE Ohio near Cleveland. Pop. 36,486.
**North Palm Beach** (päm). Village of SE Fla. N of West Palm Beach. Pop. 11,344.
**North Plain·field** (plān′fēld′). Borough of NE central N.J. WSW of Elizabeth. Pop. 19,108.
**North Platte** (plăt). **1.** River of W U.S. flowing c. 680 mi (1,095 km) from N Colo. through SE Wyo. & W central Nebr., joining the South Platte in SW Nebr. to form the Platte. **2.** City of W central Nebr. W of Grand Island. Pop. 24,479.
**North Pole.** N end of the earth's axis of rotation, a point in the Arctic Ocean.
**North·port** (nôrth′pôrt′, -pōrt′). City of W central Ala. near Tuscaloosa. Pop. 14,332.
**North Prov·i·dence** (prŏv′ĭ-dəns). Town of NE R.I. NNE of Providence. Pop. 29,188.
**North Read·ing** (rĕd′ĭng). Town of NE Mass. N of Boston. Pop. 11,455.
**North Rich·land Hills** (rĭch′lənd, -lən). City of NE Tex. near Fort Worth. Pop. 30,592.
**North Ridge·ville** (rĭj′vĭl′). City of NE Ohio WSW of Cleveland. Pop. 21,522.
**North Roy·al·ton** (roi′əl-tən). City of NE Ohio S of Cleveland. Pop. 17,671.
**North Saint Paul** (sānt pôl′). City of E Minn. near St. Paul. Pop. 11,921.
**North Sas·katch·e·wan** (sə-skăch′ə-wən, -wän′, săs-kăch′-). River of central Canada flowing 760 mi (1,222.8 km) from E Alta. to the South Saskatchewan to form the Saskatchewan in central Sask.
**North Slope.** Region of N Alas. N of the Brooks Range.
**North Ton·a·wan·da** (tŏn′ə-wŏn′də). City of W N.Y. N of Buffalo. Pop. 35,760.
**North·um·ber·land Strait** (nôr-thŭm′bər-lənd). Arm of the Gulf of St. Lawrence separating P.E.I. from N.B. & N.S. Canada.
**North·um·bri·a** (nôr-thŭm′brē-ə). Anglo-Saxon kingdom of Britain.
**North Valley Stream.** Town of SE N.Y. on Long Is. Pop. 14,881.
**North Van·cou·ver** (văn-ko͞o′vər). City of SW B.C., Canada, on an inlet of the Strait of Georgia opposite Vancouver. Pop. 31,934.

**North Vi·et·nam** (vē'ət-näm', -năm', vē-ĕt'-, vyĕt'-). Former republic of SE Asia (1954–75).
**North·west** (nôrth-wĕst'), **the. 1.** Formerly, area of the U.S. W of the Mississippi & N of the Missouri R. **2.** U.S. states of Wash., Ore., & Idaho.
**North-West Frontier Province.** Historical region of NW Pakistan on the Afghanistan border.
**Northwest Passage.** Water route from the Atlantic to the Pacific through the Arctic Archipelago of Canada & N of Canada.
**Northwest Territories.** Region of NW Canada including the Arctic Archipelago, the islands in Hudson Bay, & the mainland N of the Canadian provinces.
**Northwest Territory.** Historical U.S. region extending from the Ohio & Mississippi rivers to the Great Lakes.
**North York** (yôrk). Borough of metropolitan Toronto, S Ont., Canada. Pop. 558,398.
**Nor·ton** (nôr'tn). **1. Sound.** Inlet of the Bering Sea, W Alas., S of the Seward Peninsula. **2.** Town of SE Mass. NE of Taunton. Pop. 12,690. **3.** City of NE Ohio near Akron. Pop. 12,242.
**Norton Shores.** City of W Mich. on Lake Michigan. Pop. 22,025.
**Nor·walk** (nôr'wôk'). **1.** City of S Calif. NNE of Long Beach. Pop. 85,232. **2.** City of SW Conn. on Long Is. Sound. Pop. 77,767. **3.** City of N Ohio SSE of Sandusky. Pop. 14,358.
**Nor·way** (nôr'wā'). Kingdom of N Europe in the W part of the Scandinavian peninsula. Cap. Oslo. Pop. 4,092,000. **—Nor·we'gian** (nôr-wē'jən) *adj. & n.*
**Norwegian Sea.** Section of the Atlantic NW of Norway between the Greenland & North seas.
**Nor·wich. 1.** (nŏr'ĭch). Borough of E England W of Great Yarmouth. Pop. 119,200. **2.** (nôr'wĭch', nŏr'-). City of SE Conn. N of New London. Pop. 38,074.
**Nor·wood** (nôr'wo͝od'). **1.** Town of E Mass. SW of Boston. Pop. 29,711. **2.** City of SW Ohio surrounded by Cleveland. Pop. 26,342.
**No·teć** (nô'tĕch'). River of NW Poland flowing c. 270 mi (435 km) to the Warta.
**No·to·gae·a** *or* **No·to·ge·a** (nō'tə-jē'ə). Zoogeographic region including Australia, New Zealand, & the islands of the SW Pacific.
**No·tre Dame Mountains** (nō'trə dăm', däm', nō'tər). Section of the Appalachians extending c. 500 mi (800 km) from the Green Mts. of Vt. into the Gaspé Peninsula, Que., Canada.
**Not·ta·way** (nŏt'ə-wā'). River, c. 140 mi (225 km), of W Que., Canada.
**Not·ting·ham** (nŏt'ĭng-əm). Borough of central England N of Leicester. Pop. 280,300.
**Not·to·way** (nŏt'ə-wā'). River of S Va., flowing 175 mi (281.6 km) to the N.C. border.
**Nouak·chott** (no͞o-äk'shŏt'). Cap. of Mauritania in the W part. Pop. 134,986.
**Nou·mé·a** (no͞o-mā'ə). Cap. of New Caledonia on the island of New Caledonia. Pop. 56,078.
**Nova I·gua·çu** (nô'və ē'gwə-so͞o'). City of SE Brazil near Rio de Janeiro. Pop. 331,457.
**No·va·ra** (nō-vär'ə). City of N Italy W of Milan. Pop. 92,634.
**No·va Sco·tia** (nō'və skō'shə). Province of E Canada. Cap. Halifax. Pop. 837,789. **—No'va Sco'tian** *adj. & n.*
**No·va·to** (nō-vä'tō). City of W Calif. N of San Rafael. Pop. 43,916.
**No·va·ya Zem·lya** (nō'və-yə zĕm'lē-ä'). Archipelago of NW USSR in the Arctic Ocean between the Barents & Kara seas.
**Nov·go·rod** (nŏv'gə-räd'). City of NW European USSR SSE of Leningrad. Pop. 186,000.
**No·vi** (nō'vī'). City of SE Mich. NW of Detroit. Pop. 22,525.
**No·vi Sad** (nō'vē säd'). City of NE Yugoslavia on the Danube. Pop. 143,591.
**No·vo·cher·kassk** (nō'və-chər-kăsk', -käsk'). City of SE European USSR E of Rostov. Pop. 183,000.
**No·vo·kuz·netsk** (nō'vō-ko͞oz-nĕtsk'). City of central Siberian USSR SSE of Leninsk-Kuznetski. Pop. 541,000.
**No·vo·ros·siysk** (nō-və-rə-sēsk'). City of SE European USSR on the Black Sea. Pop. 159,000.
**No·vo·si·birsk** (nō'vō-sə-bîrsk'). City of S Siberian USSR on the Ob R. Pop. 1,312,000.
**No·wy Sącz** (nō'vē sônch'). City of SE Poland SE of Kraków. Pop. 41,103.
**Nu·bi·a** (no͞o'bē-ə, nyo͞o'-). Desert region & ancient kingdom in the Nile valley of S Egypt & N Sudan. **—Nu'bi·an** *adj. & n.*
**Nubian Desert.** Desert of NE Sudan extending E of the Nile to the Red Sea.
**Nu·e·ces** (no͞o-ā'sĭs, nyo͞o-). River of SW Tex. flowing 315 mi (506.8 km) to **Nueces Bay,** an inlet of the Gulf of Mexico near Corpus Christi.
**Nue·vo La·re·do** (no͞o-ā'vō lə-rā'dō). City of NE Mexico across the Rio Grande from Laredo, Tex. Pop. 184,622.
**Nu·ku·a·lo·fa** (no͞o'kə-wə-lô'fə). Cap. of Tonga in the SW Pacific. Pop. 18,356.
**Nu·ku Hi·va** (no͞o'kə hē'və). Volcanic island, 127 sq mi (329 sq km), largest of the Marquesas Is. of French Polynesia.
**Nu·kus** (no͞o-ko͞os'). City of S Central Asian USSR on the Amu Darya. Pop. 109,000.
**Null·ar·bor Plain** (nŭl'ə-bôr', nŭl-är'bər). Region of S central Australia S of the Great Victoria Desert & N of the Great Australian Bight.
**Nu·ma·zu** (no͞o-mä'zo͞o). City of S central Honshu, Japan. Pop. 119,325.
**Num·foor** (no͞om'fôr, -fōr). Island of N West Irian, Indonesia, in the Schouten Is.
**Nu·mid·i·a** (no͞o-mĭd'ē-ə, nyo͞o-). Ancient country of NW Africa corresponding roughly to present-day Algeria. **—Nu·mid'i·an** *adj. & n.*
**Nun·ea·ton** (nə-nēt'n). Borough of central England N of Coventry. Pop. 69,210.
**Nu·ni·vak** (no͞o'nə-văk'). Island off W Alas. in the Bering Sea.
**Nu·rem·berg** (no͝or'əm-bûrg', nyo͝or'-). City of S West Germany NNW of Munich. Pop. 499,060.
**Nu·ri·stan** (no͝or'ĭ-stăn'). Region of NE Afghanistan.
**Nürn·berg** (no͝orn'bĕrg', nyo͝orn'-). NUREMBERG.
**Nu·sa Teng·ga·ra** (no͞o'sə tĕng-gär'ə). Island chain of Indonesia extending from Bali to Timor.
**Nut·ley** (nŭt'lē). Town of NE N.J. near Passaic. Pop. 28,998.
**Ny·as·a** (nī-ăs'ə, nē-). Lake, c. 11,600 sq mi (30,040 sq mi), of SE Africa between Tanzania, Mozambique, & Malawi.
**Ny·as·a·land** (nī-ăs'ə-lănd', nē-). MALAWI.
**Nyir·a·gon·go** (nyîr'ə-gông'gō, gŏng'-). Volcano of E Zaire at the N end of Lake Kivu.
**Nyí·regy·há·za** (nē'rĕj-hä'zô). City of NE Hungary N of Debrecen. Pop. 108,156.

# O

**O·a·hu** (ə-wä'ho͞o). Chief island of Hawaii between Molokai & Kauai.
**Oak Creek** (ōk). City of SE Wis. near Milwaukee. Pop. 16,932.
**Oak·dale** (ōk'dāl'). City of E Minn. near St. Paul. Pop. 12,123.
**Oak Forest.** City of NE Ill. near Chicago. Pop. 26,096.
**Oak Harbor.** City of NW Wash. on Whidbey Is. NW of Everett. Pop. 12,271.
**Oak·land** (ōk'lənd). **1.** City of W Calif. opposite San Francisco. Pop. 339,288. **2.** Borough of NE N.J. near Paterson. Pop. 13,443.
**Oakland Park.** City of SE Fla. on the Atlantic coast. Pop. 21,939.
**Oak Lawn.** Village of NE Ill. near Chicago. Pop. 60,690.
**Oak Park. 1.** Village of NE Ill. near Chicago. Pop. 54,887. **2.** City of SE Mich. near Detroit. Pop. 31,537.
**Oak Ridge.** City of E Tenn. W of Knoxville. Pop. 27,662.
**Oak·ville** (ōk'vĭl'). Town of S Ont., Canada, on Lake Ontario SW of Toronto. Pop. 68,950.
**Oa·xa·ca** (wə-hä'kə). City of S central Mexico S of Orizaba. Pop. 114,948.
**Ob** (ŏb, ôb). River, c. 2,300 mi (3,700 km), of W Siberian USSR flowing to the **Gulf of Ob,** an inlet of the Arctic Ocean.
**O·ber·hau·sen** (ō'bər-hou'zən). City of W West Germany in the Ruhr NW of Essen. Pop. 237,147.
**O·cal·a** (ō-kăl'ə). City of N central Fla. SSE of Gainesville. Pop. 37,170.
**O·cean City** (ō'shən). City of SE N.J. on the Atlantic SW of Atlantic City. Pop. 13,949.
**O·ce·an·i·a** (ō'shē-ăn'ē-ə, -ä'nē-ə). Collective name of islands in the S, W, & central Pacific, usu. including Australia & New Zealand. **—O'ce·an'i·an** *adj. & n.*
**O·cean·side** (ō'shən-sīd'). **1.** City of S Calif. NNW of San Diego. Pop. 76, 698. **2.** City of SE N.Y. on the S shore of Long Is. Pop. 36,400.
**Ocean Springs.** City of extreme SE Miss. near Biloxi. Pop. 14,504.
**Oc·mul·gee** (ōk-mŭl'gē). River, c. 255 mi (410 km), of central Ga.
**O·den·se** (ōd'n-sə). City of S Denmark near the **Odense Fjord,** an arm of the Kattegat. Pop. 168,178.
**O·der** (ō'dər). River of central Europe flowing c. 562 mi (904.3 km) from N central Czechoslovakia through Poland & East Germany to the Baltic Sea.
**O·des·sa** (ō-dĕs'ə). **1.** City of SW Tex. WNW of San Angelo. Pop. 90,027. **2.** City of SW European USSR on **Odessa Bay** of the Black Sea. Pop. 1,046,000.
**O'Fal·lon** (ō-făl'ən). City of SW Ill. E of East St. Louis. Pop. 10,217.
**Of·fen·bach** (ôf'ən-bäk'). City of central West Germany on the Main R. Pop. 115,251.
**Og·bo·mo·sho** (ŏg'bə-mō'shō). City of SW Nigeria N of Ibadan. Pop. 432,000.
**Og·den** (ŏg'dən). City of N Utah N of Salt Lake City. Pop. 64,407.
**Og·dens·burg** (ŏg'dənz-bûrg'). City of N N.Y. on the St. Lawrence R. Pop. 12,375.

**O·gee·chee** (ō-gē'chē). River c. 250 mi (402.3 km), of NE Ga.
**O·hi·o** (ō-hī'ō). **1.** River formed by the confluence of the Allegheny & the Monongahela in W Pa. & flowing 981 mi (1,578.4 km) to the Mississippi in S Ill. **2.** State of the N U.S. in the Great Lakes region. Cap. Columbus. Pop. 10,797,419. **—O·hi'o·an** *adj.* & *n.*
**Oil City** (oil). City of NW Pa. NNE of Pittsburgh. Pop. 13,881.
**Oil Rivers.** Large delta region of the Niger R. in S Nigeria.
**Oise** (wäz). River, 186 mi (299.3 km), of S Belgium & N France.
**O·i·ta** (ō'ī-tä', ō-ē'tä). City of NE Kyushu, Japan, on Beppu Bay. Pop. 320,237.
**O·jos Del Sa·la·do** (ō'hōz dĕl' sə-lä'dō). Peak, 22,539 ft (6,875.4 km), in the Andes on the Argentine-Chilean border.
**O·ka** (ō-kä'). River, c. 925 mi (1,490 km), of central European USSR.
**O·ka·nog·an** (ō'kə-nŏg'ən). River, c. 300 mi (482.7 km), of S B.C., Canada, & N Wash.
**O·ka·van·go** (ō'kə-văng'gō). River of W central Africa flowing c. 1,000 mi (1,610 km) from central Angola to N Botswana.
**O·ka·ya·ma** (ō'kə-yä'mə). City of W Honshu, Japan, on an inlet of the Inland Sea. Pop. 513,471.
**O·ka·za·ki** (ō-kä'zä-kē, ō'kə-zä'kē). City of S Honshu, Japan, NE of Okayama. Pop. 234,510.
**O·kee·cho·bee** (ō'kə-chō'bē). Lake, c. 700 sq mi (1,815 sq km), of SE Fla. N of the Everglades, crossed by the **Okeechobee Waterway,** a manmade & natural water route from the Atlantic to the Gulf of Mexico.
**O·ke·fe·no·kee** (ō'kə-fə-nō'kē). Large swamp of SE Ga. & N Fla.
**O·khotsk** (ō-kŏtsk'), **Sea of.** NW arm of the Pacific W of the Kamchatka Peninsula and Kurile Is.
**O·ki·na·wa** (ō'kĭ-nä'wə, -nou'-). **1.** Island group of the central Ryukyu Is. in the W Pacific SW of Japan. **2.** Largest (454 sq mi/1,175.9 sq km) island of the group.
**O·kla·ho·ma** (ō'klə-hō'mə). State of the SW U.S. Cap. Oklahoma City. Pop. 3,025,266. **—O'kla·ho'man** *adj.* & *n.*
**Oklahoma City.** Cap. of Okla. in the central part of the state. Pop. 403,213.
**Ok·mul·gee** (ōk-mŭl'gē). City of E central Okla. near the Arkansas R. Pop. 16,263.
**Ö·land** (œ'lănd'). Narrow island of SE Sweden in the Baltic Sea.
**O·la·the** (ō-lā'thə). City of E Kans. SW of Kansas City. Pop. 37,258.
**Ol·den·burg** (ōl'dən-bûrg'). City of NW West Germany W of Bremen. Pop. 134,076.
**Old·ham** (ōl'dəm). Borough of NW England near Manchester. Pop. 103,690.
**Ol·du·vai Gorge** (ōl'də-vī'). Gorge in N Tanzania W of Mt. Kilimanjaro; site of early human remains.
**O·le·an** (ō'lē-ăn', ō'lē-ăn'). City of W N.Y. near the Pa. border. Pop. 18,207.
**O·lek·ma** (ō-lĕk'mə). River, c. 820 mi (1,320 km), of SE Siberian USSR.
**O·le·nek** (ōl'ən-yôk'). River of E Siberian USSR flowing c. 1,350 mi (2,175 km) to the Laptev Sea.
**Ol·i·fants** (ŏl'ə-fənts). River, c. 350 mi (563.2 km), of NE South Africa & W Mozambique.
**Ol·ives** (ŏl'ivz), **Mount of,** *also* **Ol·i·vet** (ŏl'ə-vĕt'). Ridge of hills E of Jerusalem.
**O·lo·mouc** (ô'lô-mōts'). City of N central Czechoslovakia on the Morava R. Pop. 82,800.
**Olsz·tyn** (ôl'shtĭn). City of N Poland SE of Gdańsk. Pop. 94,119.
**Olt** (ôlt). River, c. 348 mi (559.9 km), of central Rumania.
**O·lym·pi·a** (ō-lĭm'pē-ə, ə-lĭm'-). **1.** Plain of NW Peleponnesus, Greece; ancient site of the Olympic Games. **2.** Cap. of Wash. on the S end of Puget Sound. Pop. 27,447.
**O·lym·pic** (ō-lĭm'pĭk). **1.** Mountain range of NW Wash., part of the Coast Ranges. **2.** Peninsula of NW Wash. between the Pacific & Puget Sound. **3. National Park.** Large tract of rugged reserved land on the Olympic Peninsula.
**O·lym·pus** (ō-lĭm'pəs, ə-lĭm'-). Mountain range of N Greece near the Aegean coast, rising to 9,570 ft (2,920 m) at **Mount Olympus,** highest point in Greece & home of the mythical Greek gods.
**Om** (ôm). River, c. 450 mi (724 km), of W Siberian USSR.
**O·ma·ha** (ō'mə-hô', -hä'). City of E Nebr. on the Missouri R. Pop. 311,681.
**O·man** (ō-män', ō-măn'). Sultanate of SE Arabian Peninsula on the **Gulf of Oman,** an arm of the Arabian Sea. Cap. Muscat. Pop. 891,000. **—O·man'i** *adj.* & *n.*
**Om·dur·man** (ŏm'dŏŏr-män'). City of central Sudan on the White Nile opposite Khartoum. Pop. 299,000.
**O·mi·ya** (ō-mē'ə). City of E central Honshu, Japan, near Tokyo. Pop. 327,698.
**Om·o·lon** (ŏm'ə-lôn'). River, c. 600 mi (965.4 km), of NE Siberian USSR.
**Omsk** (ômsk, ŏmsk). City of W Siberian USSR at confluence of the Irtysh & Om rivers. Pop. 1,014,000.
**O·mu·ta** (ō'mə-tä'). City of W Kyushu, Japan, NW of Kumamoto. Pop. 165,969.
**O·ne·ga** (ō-nē'gə). Lake, c. 3,800 sq mi (9,840 sq km), of NW European USSR.
**O·nei·da** (ō-nī'də). **1.** Lake, c. 80 sq mi (210 sq km) of central N.Y. NE of Syracuse. **2.** City of central N.Y. E of Syracuse. Pop. 10,810.
**On·e·on·ta** (ō'nē-ŏn'tə). City of central N.Y. WSW of Albany. Pop. 14,933.
**O·nit·sha** (ō-nĭch'ə). City of SE Nigeria on the Niger R. Pop. 220,000.
**On·tar·i·o** (ŏn-târ'ē-ō'). **1.** Smallest of the Great Lakes, between SE Ont., Canada, & NW N.Y. **2.** Province of E central Canada. Cap. Toronto. Pop. 8,551,773. **3.** City of S Calif. E of Los Angeles. Pop. 88,280.
**O·pa-Lock·a** (ō'pə-lŏk'ə). City of SE Fla. near Miami. Pop. 14,460.
**O·pe·li·ka** (ō'pə-lī'kə). City of E Ala. NW of Phenix City. Pop. 22,087.
**Op·e·lou·sas** (ŏp'ə-lōō'səs). City of S central La. W of Baton Rouge. Pop. 18,903.
**O·po·le** (ō-pô'lə). City of S Poland on the Oder R. Pop. 86,500.
**O·por·to** (ō-pôr'tō, -pōr'-). City of NW Portugal near the mouth of the Douro R. Pop. 300,925.
**O·ra·dea** (ô-răd'yä). City of W Rumania near the Hungarian border. Pop. 175,400.
**O·ran** (ō-răn'). City of NW Algeria on the **Gulf of Oran,** an inlet of the Mediterranean. Pop. 491,900.
**Or·ange** (ôr'ĭnj, ŏr'-). **1.** River, c. 1,300 mi (2,090 km), of Lesotho, South Africa, & Namibia. **2.** City of S Calif. NNE of Santa Ana. Pop. 91,788. **3.** Town of SW Conn. near New Haven. Pop. 13,237. **4.** City of NE N.J. near Newark. Pop. 31,136. **5.** City of SE Tex. E of Beaumont. Pop. 83,838.
**Or·ange·burg** (ôr'ĭnj-bûrg', ŏr'-). City of SE central S.C. SSE of Columbia. Pop. 14,933.
**Or·ange·ville** (ôr'ĭnj-vĭl', ŏr'-). Town of S Ont., Canada, WNW of Toronto. Pop. 12,021.
**Or·dos** (ôr'dəs). Sandy desert plateau region of Inner Mongolia in N China.
**Or·dzho·ni·kid·ze** (ôr'jŏn'ĭ-kĭd'zə). City of SE European USSR on the Terek R. Pop. 279,000.
**Ö·re·bro** (œ'rə-brōō'). City of S central Sweden W of Lake Hjälmaren. Pop. 117,877.
**Or·e·gon** (ôr'ĭ-gən, -gŏn', ŏr'-). **1. Territory.** Historical region of NW North America. **2.** State of the NW U.S. in the Pacific Northwest. Cap. Salem. Pop. 2,632,663. **3.** City of NW Ohio near Toledo. Pop. 18,675. **—Or'e·go'ni·an** (-gō'nē-ən) *adj.* & *n.*
**Oregon City.** City of NW Ore. S of Portland. Pop. 14,673.
**Oregon Trail.** Historical overland route to the W U.S. from the Missouri R. to the Oregon Territory.
**O·re·kho·vo Zu·ye·vo** (ôr'ĭ-kôv'ə zōō-yĕv'ō). City of W central European USSR E of Moscow. Pop. 133,000.
**O·rel** (ô-rĕl', ō-rĕl', ôr-yôl'). City of central European USSR on the Oka R. Pop. 305,000.
**O·rem** (ôr'əm, ōr'-). City of N central Utah near Provo. Pop. 52,399.
**O·ren·burg** (ôr'ən-bûrg', ōr'-). City of NW Central Asian USSR on the Ural R. Pop. 459,000.
**O·ren·se** (ô-rĕn'sĕ). City of NW Spain NW of Madrid. Pop. 63,542.
**Ö·re·sund** (œ'rə-sŭn'). Strait between S Sweden & E Denmark connecting the Baltic Sea with the Kattegat.
**O·ril·lia** (ō-rĭl'yə). City of S Ont., Canada, N of Toronto. Pop. 24,412.
**O·ri·no·co** (ôr'ə-nō'kō, ōr'-). River of Venezuela flowing c. 1,500 mi (2,415 km), partly along the Columbia-Venezuela border, to the Atlantic.
**O·ri·za·ba** (ôr'ĭ-ză'bə, ōr'-). **1.** Volcanic peak, c. 18,701 ft (5,703 m), of E central Mexico near the city of Orizaba. **2.** City of E central Mexico W of Veracruz. Pop. 105,150.
**Or·khon** (ôr'kŏn'). River, c. 300 mi (480 km), of N Mongolia.
**Ork·ney Islands** (ôrk'nē) *also* **Ork·neys** (-nēz). Archipelago of c. 70 islands in the Atlantic & the North Sea off the NE coast of Scotland.
**Or·lan·do** (ôr-lăn'dō). City of central Fla. ENE of Tampa. Pop. 128,394.
**Or·land Park** (ôr'lənd). Village of NE Ill. SW of Chicago. Pop. 23,035.
**Or·lé·ans** (ôr-lā-äN'). City of N central France on the Loire S of Paris. Pop. 88,503.
**Or·ly** (ôr-lē', ôr'lē). City of N central France, a suburb SE of Paris. Pop. 26,090.
**Or·mond Beach** (ôr'mənd). City of NE Fla. on the Atlantic. Pop. 21,378.
**Or·muz** (ôr'mŭz', ôr-mōōz'). HORMUZ.
**O·ro·moc·to** (ôr'ə-mŏk'tō, ŏr'-). Town of S central N.B., Canada, on the St. John R. Pop. 10,276.
**O·ron·tes** (ô-rŏn'tēz). River, c. 250 mi (400 km), flowing through Lebanon, Syria, & S Turkey to the Mediterranean.
**Orsk** (ôrsk). City of E European USSR on the Ural R. Pop. 247,000.
**Ort·les** (ôrt'lās). Range of the Alps in N Italy rising to **Ortles** peak, 12,792 ft (3,901.6 km).

**O·sage** (ō-sāj′, ō′sāj′). River c. 360 mi (580 km), of E Kans. & central Mo.

**O·sa·ka** (ō-sä′kə). City of S Honshu, Japan, on **Osaka Bay,** an inlet of the Pacific. Pop. 2,778,987.

**O·sas·co** (o͞o-säs′ko͞o). City of SE Brazil near São Paulo. Pop. 283,303.

**Osh·a·wa** (ŏsh′ə-wä′, -wə). City of SE Ont., Canada, on Lake Winnebago. Pop. 107,023.

**O·shog·bo** (ō-shŏg′bō). City of SW Nigeria NE of Ibadan. Pop. 282,000.

**O·si·jek** (ô′sē-ĕk, -yĕk′). City of N Yugoslavia on the Drava. Pop. 94,989.

**Os·ka·loo·sa** (ŏs′kə-lo͞o′sə). City of SE Iowa SE of Des Moines. Pop. 10,629.

**Os·lo** (ŏz′lō, ŏs′-). Cap. of Norway in the SE at the head of the **Oslofjord,** a deep inlet of the Skagerrak. Pop. 462,732.

**Os·na·brück** (ŏz′nə-bro͝ok′). City of NW West Germany NE of Münster. Pop. 161,671.

**O·sor·no** (ō-sôr′nō). City of S central Chile S of Concepción. Pop. 68,800.

**Os·se·tia** (ō-sē′shə, -shē-ə). Region of the central Caucasus, S European USSR, divided into **North & South Ossetia.**

**Os·si·ning** (ŏs′ə-nĭng′). Village of SE N.Y. on the Hudson R. N of White Plains. Pop. 20,196.

**Ost·end** (ŏs-tĕnd′, ŏs′tĕnd′). City of NW Belgium on the North Sea. Pop. 71,227.

**Os·ter·sund** (œs′tər-so͞ond′). City of central Sweden E of Trondheim, Norway. Pop. 40,056.

**Os·ti·a** (ŏs′tē-ə). Ancient city of E central Italy at the mouth of the Tiber.

**Os·tra·va** (ô′strə-və). City of N central Czechoslovakia near the Oder R. Pop. 293,500.

**Os·we·go** (ŏs-wē′gō). City of N central N.Y. on Lake Ontario NW of Syracuse. Pop. 19,793.

**Oś·wię·cim** (ôsh-vyĕɴt′sēm). City of SE Poland W of Kraków; site (as Auschwitz) of Nazi concentration camp. Pop. 39,600.

**O·ta·ru** (ō-tär′o͞o). City of SW Hokkaido, Japan, WNW of Sapporo. Pop. 184,406.

**O·tsu** (ō′tso͞o). City of S Honshu, Japan, near Lake Biwa. Pop. 191,481.

**Ot·ta·wa** (ŏt′ə-wə, -wä′, -wô′). **1.** River, c. 700 mi (1,125 km), of SE Ont. & S Que., Canada, a major tributary of the St. Lawrence. **2.** Cap. of Canada in SE Ont. at the confluence of the Ottawa R. & Rideau Canal. Pop. 304,462. **3.** City of N central Ill. SW of Aurora. Pop. 18,166. **4.** City of E Kans. SW of Kansas City. Pop. 11,016.

**Ot·to·man Empire** (ŏt′ə-mən). Turkish empire (1299–1919) in SW Asia, NE Africa, & SE Europe.

**Ot·tum·wa** (ə-tŭm′wə, ō-tŭm′-). City of SE Iowa SE of Des Moines. Pop. 27,381.

**Ouach·i·ta** (wŏsh′ĭ-tô′). **1.** River, c. 600 mi (965 km), of SW Ark. & E La. **2.** Low mountain range extending c. 200 mi (320 km) from central Ark. to SE Okla.

**Oua·ga·dou·gou** (wä′gə-do͞o′go͞o). Capital city of Upper Volta in the central part. Pop. 172,661.

**Oudh** (oud). Historical region of N central India.

**Ouj·da** (o͞oj-dä′). City of NE Morocco near the Algerian border. Pop. 175,532.

**Ouse** (o͞oz). **1.** *also* **Great Ouse.** River, c. 155 mi (250 km), of S central England. **2.** River, c. 60 mi (100 km), of NE England.

**Out·er Mon·go·li·a** (out′ər mŏng-gō′lē-ə, maon-). MONGOLIA.

**Ou·tre·mont** (o͞o′trə-mŏnt′, o͞o-trə-môɴ′). City of S Que., Canada, part of Greater Montreal, on Montreal Is. Pop. 27,089.

**O·ver·land** (ō′vər-lənd). City of E Mo. near St. Louis. Pop. 19,620.

**Overland Park.** City of E Kans. near Kansas City. Pop. 81,784.

**O·vie·do** (ō′vē-ā′dō). City of NW Spain near the Cantabrian Mts. Pop. 130,021.

**O·wa·ton·na** (ō′wə-tŏn′ə). City of SE Minn. S of Minneapolis. Pop. 18,632.

**O·wens** (ō′wĭnz). River, c. 120 mi (195 km), of SE Calif.

**O·wens·bor·o** (ō′wĭnz-bûr′ō, -bər-ə). City of W Ky. on the Ohio R. Pop. 54,450.

**O·wen Sound** (ō′wĭn). City of SE Ont., Canada, on **Owen Sound,** an inlet of Georgian Bay. Pop. 19,525.

**Owen Stan·ley** (stăn′lē). Mountain range of SE Papua New Guinea on New Guinea Is., rising to 13,363 ft (4,075.7 m).

**O·wos·so** (ō-wŏs′ō). City of central Mich. W of Flint. Pop. 16,455.

**O·wy·hee** (ō-wī′ē, -hē). River, c. 300 mi (480 km), of SW Idaho, N Nev., & SE Ore.

**Ox·ford** (ŏks′fərd). **1.** Borough of S central England on the Thames. Pop. 117,400. **2.** Town of S Mass. SSW of Worcester. Pop. 11,680. **3.** City of N Miss. SSE of Memphis, Tenn. Pop. 9,882. **4.** Village of SW Ohio NW of Hamilton. Pop. 17,655.

**Ox·nard** (ŏks′närd′). City of S Calif. WNW of Los Angeles. Pop. 108,195.

**O·yo** (ō′yō). City of SW Nigeria SW of Ogbomosho. Pop. 152,000.

**O·zark** (ō′zärk′). **1. Plateau** *also* **the O·zarks** (ō′zärks′). Upland region of S central U.S. extending from NW Ark. to E Okla. **2.** City of SE Ala. NW of Dothan. Pop. 12,721.

**Ozarks, Lake of the.** Manmade lake, 93 sq mi (240.9 sq km), of central Mo., formed in the Osage R. by Bagnell Dam.

# P

**Pa·bia·ni·ce** (pä′byə-nēt′sē). City of central Poland SW of Łódź. Pop. 62,275.

**Pa·chu·ca** (pə-cho͞o′kə) *or* **Pa·chu·ca de So·to** (dĭ sō′tō). City of central Mexico NE of Mexico City. Pop. 83,982.

**Pa·cif·ic** (pə-sĭf′ĭk). Largest & deepest ocean, c. 70,000,000 sq mi (181,300,000 sq km), extending from the W Americas to E Asia & Australia.

**Pa·cif·i·ca** (pə-sĭf′ĭ-kə). City of W Calif. S of San Francisco. Pop. 36,866.

**Pacific Grove.** City of W central Calif. near Monterey. Pop. 15,755.

**Pacific Islands, Trust Territory of the.** U.S.-administered islands of the W Pacific N & NE of New Guinea, including the Carolines, the Marianas (except Guam), & the Marshalls. Pop. 114,773.

**Pacific North·west** (nôrth-wĕst′). Region of the NW U.S. usu. including Wash. & Ore. & sometimes SW B.C., Canada.

**Pa·dang** (pä′däng′). City of W Sumatra, Indonesia, on the Indian Ocean. Pop. 196,339.

**Pa·der·born** (pä′dər-bôrn′). City of N central West Germany NW of Kassel. Pop. 103,705.

**Pa·dre** (pä′drē). Island, c. 115 mi (185 km) long, paralleling the S coast of Tex.

**Pad·u·a** (păj′o͞o-ə, păd′yo͞o-ə). City of NE Italy W of Venice. Pop. 210,950.

**Pa·du·cah** (pə-do͞o′kə, -dyo͞o′-). City of W Ky. on the Ohio R. Pop. 29,315.

**Paes·tum** (pĕs′təm, pēs′-). Ancient city of SW Italy on the Gulf of Salerno.

**Pa·go Pa·go** *also* **Pa·go·pa·go** (păng′ō-păng′ō, păng′gō-păng′gō, päng′gō-päng′gō, pä′gō-pä′gō, pä′gō-pä′gō) *or* **Pan·go Pan·go** (päng′ō päng′ō, păng′gō păng′gō, päng′gō päng′gō). Cap. of American Samoa on the S coast of Tutuila Is. Pop. 3,058.

**Pa·hang** (pə-hăng′). River, c. 285 mi (459 km), of SE Malay Peninsula, Malaysia.

**Paines·ville** (pānz′vĭl′). City of NE Ohio NE of Cleveland. Pop. 16,391.

**Paint·ed Desert** (pān′tĭd). Plateau region of E central Ariz.

**Pais·ley** (pāz′lē). District of W Scotland on the Clyde W of Glasgow. Pop. 94,833.

**Pak·i·stan** (păk′ĭ-stăn′, pä′kĭ-stän′). Country of S Asia. Cap. Islamabad. Pop. 83,782,000. **—Pak′i·stan′i** *adj.* & *n.*

**Pa·lat·i·nate** (pə-lăt′n-ĭt). Either of two historical districts & former states of S Germany: **Lower Palatinate,** in W West Germany between Luxembourg & the Rhine; & **Upper Palatinate,** in E West Germany in NE Bavaria.

**Pal·a·tine** (păl′ə-tīn′). **1.** One of the seven hills of ancient Rome. **2.** Village of NE Ill. NW of Chicago. Pop. 32,166.

**Pa·lat·ka** (pə-lăt′kə). City of NE Fla. S of Jacksonville. Pop. 10,175.

**Pa·lau** (pə-lou′) *or* **Pe·lew** (pə-lo͞o′). Islands within the Caroline group in the W Pacific N of W New Guinea.

**Pa·la·wan** (pə-lä′wən). Island, c. 280 mi (451 km) long, of SW Philippines N of Borneo.

**Pa·lem·bang** (pä′ləm-bäng′). City of SE Sumatra, Indonesia. Pop. 582,961.

**Pa·len·cia** (pə-lĕn′chə, -chē-ə). City of N central Spain NNE of Valladolid. Pop. 58,327.

**Pa·ler·mo** (pə-lûr′mō, -lâr′-). City of NW Sicily, Italy, on the Tyrrhenian Sea. Pop. 556,374.

**Pal·es·tine** (păl′ĭ-stīn′). **1.** Region of SW Asia between the E Mediterranean shore & the Jordan R. **2.** City of NE central Tex. SE of Dallas. Pop. 15,948. **—Pal′es·tin′i·an** (-stĭn′ē-ən) *adj.* & *n.*

**Pal·i·sades** (păl′ĭ-sādz′). Row of cliffs in NE N.J. along the W bank of the Hudson.

**Palisades Park.** Borough of NE N.J. near the Hudson opposite Upper Manhattan. Pop. 13,732.

**Palk** (pôk, pôlk). Strait, 40–85 mi (64–137 km) wide, between India & Sri Lanka.

**Pal·ma** (päl′mä) *also* **Pal·ma de Ma·llor·ca** (də mə-yôr′kə, məl-). City of SW Majorca Is., Spain, on the **Bay of Palma,** an inlet of the Mediterranean. Pop. 191,416.

**Palm Bay** (päm). City of E Fla. on the Atlantic SE of Orlando. Pop. 18,560.

**Palm Beach.** City of SE Fla. on the Atlantic E of Lake Okeechobee. Pop. 9,729.

**Palm Beach Gardens.** City of SE Fla. near West Palm Beach. Pop. 14,407.

**Palm·dale** (päm'dāl').City of S Calif. NE of Los Angeles. Pop. 12,277.

**Palm Desert.** City of SE Calif. E of Los Angeles. Pop. 11,801.

**Pal·mer** (pä'mər). Town of SW Mass. ENE of Springfield. Pop. 11,389.

**Pal·mi·ra** (päl-mîr'ə). City of W Colombia WSW of Bogotá. Pop. 140,481.

**Palm Springs.** City of SE Calif. ESE of Riverside. Pop. 32,271.

**Pal·my·ra** (păl-mī'rə). Ancient city of central Syria NE of Damascus.

**Pal·o Al·to** (păl'ō ăl'tō). City of W Calif. NW of San Jose. Pop. 55,225.

**Pal·o·mar** (păl'ə-mär'), **Mount.** Peak, 6,126 ft (1,868.4 m), of S Calif. NE of San Diego.

**Pa·los Heights** (pā'ləs). City of NE Ill. near Chicago. Pop. 11,096.

**Pal·os Ver·des Estates** (păl'əs vûr'dēz). City of S Calif. SSE of Santa Monica. Pop. 14,376.

**Pa·louse** (pə-lōōs'). River, c. 140 mi (225 km), of NW Idaho & SE Wash.

**Pa·mirs** (pə-mîrz') *also* **Pa·mir** (-mîr'). Mountain region of S central Asia in S Central Asian USSR with extensions in N Afghanistan, N Kashmir, & W China.

**Pam·li·co Sound** (păm'lĭ-kō'). Inlet of the Atlantic, c. 80 mi (129 km) long, between the E coast of N.C. & offshore islands.

**Pam·pa** (păm'pə). City of NW Tex. in the Panhandle NE of Amarillo. Pop. 21,396.

**Pam·plo·na** (păm-plō'nə). City of N Spain SE of San Sebastián. Pop. 175,833.

**Pan·a·ma** (păn'ə-mä', -mô'). **1. Gulf of.** Wide inlet of the Pacific on the S coast of Panama. **2. Isthmus of.** Isthmus, c. 31 mi (50 km) wide, connecting North & South America & separating the Pacific from the Caribbean Sea. **3.** Ship canal, 51 mi (82.1 km) long, across the Isthmus of Panama in the Canal Zone, connecting the Caribbean Sea with the Pacific. **4. Canal Zone.** CANAL ZONE. **5.** Country of SW Central America. Cap. Panama. Pop. 1,830,175. **6.** *also* **Panama City.** Cap. of Panama in the central part on the Gulf of Panama. Pop. 388,638. **—Pan'a·ma'ni·an** (-mā'nē-ən) *adj.* & *n.*

**Panama City.** City of NW Fla. ESE of Pensacola. Pop. 33,346.

**Pan·a·mint** (păn'ə-mĭnt'). Range of SE Calif. near the Nev. border, rising to 11,045 ft (3,368.7 m).

**Pa·nay** (pə-nī'). Island of the central Philippines in the Visayan group NW of Negros.

**Pan·čev·o** (pän'chə-vō'). City of NE Yugoslavia near Belgrade. Pop. 53,979.

**Pan·go Pan·go** (päng'ō päng'ō, păng'gō păng'gō, päng'gō päng'gō). *var. of* PAGO PAGO.

**Pan·mun·jom** (pän'mōōn'jŭm'). Village of NW South Korea; site of Korean War truce signing (1953).

**Pan·no·ni·a** (pə-nō'nē-ə). Ancient Roman province of central Europe including present-day W Hungary & N Yugoslavia.

**Pá·nu·co** (pä'nə-kō'). River, c. 315 mi (507 km), of central Mexico.

**Pao·ki** (bou'jē'). BAOJI.

**Pao·ting** (bou'dĭng'). BAODING.

**Pao·tow** (bou'tō'). BAOTOU.

**Pa·pal States** (pā'pəl). Territories in central Italy ruled by the popes until 1870.

**Pap·u·a New Guin·ea** (păp'yōō-ə nōō' gĭn'ē, nyōō'). Country of the W Pacific comprising the E half of New Guinea, the Bismarck Archipelago, the N Solomons, & adjacent islands. Cap. Port Moresby. Pop. 3,006,799. **—Pap'u·an New Guin'e·an** *n.*

**Pa·rá** (pə-rä'). River, c. 200 mi (320 km), of N Brazil, the SE distributary of the Amazon.

**Par·a·dise** (păr'ə-dīs'). City of N central Calif. N of Sacramento. Pop. 22,571.

**Paradise Valley.** Town of S central Ariz. near Phoenix. Pop. 10,832.

**Par·a·gould** (păr'ə-gōōld'). City of NE Ark. WNW of Blytheville. Pop. 15,214.

**Pa·ra·gua·çu** *or* **Pa·ra·guas·su** (păr'ə-gwə-sōō'). River, c. 300 mi (483 km), of E Brazil.

**Par·a·guay** (păr'ə-gwī', -gwā'). **1.** River of W Brazil & Paraguay flowing c. 1,300 mi (2,092 km) S into the Paraná. **2.** Country of S central South America. Cap. Asunción. Pop. 2,973,000. **—Par'a·guay'an** *adj.* & *n.*

**Pa·ra·í·ba** (păr'ə-ē'bə) *also* **Pa·ra·í·ba do Sul** (də sōōl'). River, c. 650 mi (1,046 km), of SE Brazil.

**Par·a·mar·i·bo** (păr'ə-măr'ə-bō'). Cap. of Surinam on the Suriname R. near the Atlantic. Pop. 102,297.

**Par·a·mount** (păr'ə-mount'). City of S Calif. SE of Los Angeles. Pop. 36,407.

**Pa·ram·us** (pə-răm'əs). Borough of NE N.J. near Paterson. Pop. 26,474.

**Pa·ra·ná** (păr'ə-nä'). **1.** River of central South America rising in E central Brazil & flowing c. 2,040 mi (3,282 km) SW into the Río de la Plata in E Argentina. **2.** City of NE Argentina on the Paraná R. Pop. 127,635.

**Pa·ra·na·guá** (păr'ə-nə-gwä'). City of SE Brazil on the Atlantic. Pop. 51,510.

**Pa·ra·na·í·ba** (păr'ə-nə-ē'bə). River, c. 500 mi (805 km), of S central Brazil.

**Par·du·bi·ce** (pär'dōō-bĭt'sə). City of N central Czechoslovakia on the Elbe R. Pop. 78,500.

**Par·is** (păr'ĭs). **1.** Cap. of France in the N central part on the Seine. Pop. 2,291,554. **2.** City of NW Tenn. WNW of Nashville. Pop. 10,728. **3.** City of NE Tex. NE of Dallas. Pop. 25,498. **—Pa·ri·sian** (pə-rē'zhən, -rĭzh'ən) *adj.* & *n.*

**Park** (pärk). Range of the Rockies in central Colo. & S Wyo. rising to 14,284 ft (4,356.6 m).

**Par·kers·burg** (pär'kərz-bûrg'). City of NW W.Va. on the Ohio R. N of Charleston. Pop. 39,967.

**Park Forest.** Village of NE Ill. near Chicago. Pop. 26,222.

**Park Ridge.** City of NE Ill. near Chicago. Pop. 38,704.

**Par·ma** (pär'mə). **1.** City of N Italy SE of Milan. Pop. 151,967. **2.** City of NE Ohio near Cleveland. Pop. 92,547.

**Parma Heights.** City of NE Ohio near Cleveland. Pop. 23,112.

**Par·na·í·ba** (pär'nə-ē'bə). River, c. 800 mi (1,287 km), of NE Brazil.

**Par·nas·sus** (pär-năs'əs) *also* **Par·nas·sós** (-nä-sôs'). Mountain, c. 8,060 ft (2,460 m), of central Greece N of the Gulf of Corinth.

**Par·ra·mat·ta** (păr'ə-măt'ə). City of SE Australia near Sydney. Pop. 131,659.

**Par·ry** (păr'ē). **1.** Channel through the central Arctic Archipelago, N.W.T., Canada, linking Baffin Bay on the E with Beaufort Sea on the W. **2.** Islands of N N.W.T., Canada, in the Arctic Ocean N of Victoria Is.

**Par·sons** (pär'sənz). City of SE Kans. ESE of Wichita. Pop. 12,898.

**Par·thi·a** (pär'thē-ə). Ancient country of SW Asia corresponding to modern NE Iran. **—Par'thi·an** *adj.* & *n.*

**Pas·a·de·na** (păs'ə-dē'nə). **1.** City of S Calif. near Los Angeles. Pop. 119,374. **2.** City of S Tex. near Houston. Pop. 112,560.

**Pas·ca·gou·la** (păs'kə-gōō'lə). City of extreme SE Miss. E of Biloxi. Pop. 29,318.

**Pas·co** (păs'kō). City of S Wash. on the Columbia near Richland. Pop. 17,944.

**Pas·sa·ic** (pə-sā'ĭk). City of NE N.J. N of Newark. Pop. 52,463.

**Pas·sa·ma·quod·dy Bay** (păs'ə-mə-kwŏd'ē). Arm of the Bay of Fundy between S N.B., Canada, & E Me.

**Pas·sau** (päs'ou'). City of SE West Germany near the Austrian border. Pop. 50,920.

**Pas·ta·za** (pə-stä'zə, -sə). River, c. 400 mi (644 km), of E Ecuador & N Peru.

**Pas·to** (päs'tō). City of SW Colombia near the Ecuador border. Pop. 119,339.

**Pat·chogue** (păch'ôg'). Village of SE N.Y. on S central Long Is. Pop. 11,291.

**Pat·er·son** (păt'ər-sən). City of NE N.J. N of Newark. Pop. 137,970.

**Pat·na** (pŭt'nə). City of NE India on the Ganges. Pop. 473,001.

**Pá·trai** (pä'trā). PATRAS.

**Pa·tras** (pə-trăs', păt'rəs). City of S Greece in NW Peloponnesus on the **Gulf of Patras,** an inlet of the Ionian Sea. Pop. 111,607.

**Pau** (pō). City of SW France in the foothills of the Pyrenees. Pop. 81,560.

**Pa·vi·a** (pə-vē'ə). City of NW Italy S of Milan. Pop. 80,639.

**Paw·tuck·et** (pə-tŭk'ĭt, pô-). City of NE R.I. on the Mass. border near Providence. Pop. 71,204.

**Pay·san·dú** (pī'sän-dōō'). City of W Uruguay on the Uruguay R. Pop. 62,412.

**Pea·bod·y** (pē'bŏd'ē, -bə-dē). City of NE Mass. near Salem. Pop. 45,976.

**Peace** (pēs). River, c. 945 mi (1,521 km), of N B.C. & N Alta., Canada.

**Pearl** (pûrl). **1.** River, 485 mi (780.4 km), of S Miss. forming the Miss.-La. boundary in its lower course. **2.** City of central Miss. near Jackson. Pop. 20,778.

**Pearl City.** Village of Hawaii on Pearl Harbor in S Oahu. Pop. 22,200.

**Pearl Harbor.** Inlet of the Pacific on S coast of Oahu, Hawaii, W of Honolulu.

**Pear·land** (pâr'lənd, -lănd'). City of SE Tex. near Houston. Pop. 13,248.

**Pea·ry Land** (pîr'ē). Peninsula of N Greenland extending into the Arctic Ocean.

**Pe·cho·ra** (pə-chôr'ə, -chōr'ə). River of NE European USSR flowing c. 1,120 mi (1,802 m) N into **Pechora Bay,** SE arm of the Barents Sea.

**Pe·cos** (pā'kəs). **1.** River of E N.Mex. & W Tex. flowing c. 926 mi (1,490 km) SE into the Rio Grande. **2.** City of SW Tex. ESE of El Paso. Pop. 12,855.

**Pécs** (pāch). City of SW Hungary near the Yugoslav border. Pop. 168,788.

**Pee Dee** (pē′ dē′). River, c. 233 mi (375 km), of S central N.C. & NE S.C.
**Peeks·kill** (pēk′skĭl′). City of SE N.Y. on the Hudson N of White Plains. Pop. 18,236.
**Peel** (pēl). River, c. 365 mi (587 km), of N Y.T. & W N.W.T., Canada.
**Pe·gu** (pə-go͞o′). City of S Burma NNE of Rangoon. Pop. 135,000.
**Pei** (bī). BAI.
**Pei·ping** (pā′pĭng′). BEIJING.
**Pei·pus** (pī′pəs). Lake, 1,357 sq mi (3,515 sq km), of NW European USSR SW of Leningrad.
**Pe·ka·long·an** (pə-kä-lông′än′). City of N central Java, Indonesia, on the Java Sea. Pop. 111,537.
**Pe·kin** (pē′kĭn). City of central Ill. S of Peoria. Pop. 33,967.
**Pe·king** (pē′kĭng′). BEIJING.
**Pe·la·gi·an** (pə-lā′jē-ən, -jən) *also* **Pe·la·gie** (pā-lä′jā). Three Italian islands in the Mediterranean between Malta & Tunisia.
**Pe·lée** (pə-lā′). Volcano, c. 4,800 ft (1,464 m), on N Martinique, French West Indies.
**Pe·lew** (pə-lo͞o′). PALAU.
**Pel·ham** (pĕl′əm). Town of SE Ont., Canada, NNW of Welland. Pop. 10,071.
**Pel·la** (pĕl′ə). Ancient city of Greek Macedonia NW of modern Salonika.
**Pel·ly** (pĕl′ē). River, c. 330 mi (531 km), of central Y.T., Canada.
**Pel·o·pon·ne·sus** *also* **Pel·o·pon·ne·sos** (pĕl′ə-pə-nē′səs) *or* **Pel·o·pon·nese** (pĕl′ə-pə-nēz′, -nēs′). Peninsula forming the S part of Greece S of the Gulf of Corinth. **—Pel′o·pon·ne′sian** *adj. & n.*
**Pe·lo·tas** (pə-lō′təs). City of SE Brazil on Lagoa dos Patos. Pop. 150,278.
**Pem·ba** (pĕm′bə). Island of Tanzania in the Indian Ocean N of Zanzibar.
**Pem·broke** (pĕm′bro͝ok′, -brōk′). **1.** City of SE Ont., Canada, on the Ottawa R. NW of Ottawa. Pop. 14,927. **2.** Town of E Mass. SE of Boston. Pop. 13,487.
**Pembroke Pines.** City of SE Fla. SW of Fort Lauderdale. Pop. 35,776.
**Pe·nang** *or* **Pi·nang** (pə-năng′). **1.** Island of Malaysia in the N Strait of Malacca off the W coast of the Malay Peninsula. **2.** City of W Malaysia on Penang Is. Pop. 269,603.
**Pen·dle·ton** (pĕn′dl-tən). City of NE Ore. SW of Walla Walla, Wash. Pop. 14,521.
**Pend O·reille** (pŏn′də-rā′). River, c. 100 mi (161 km), rising in **Pend Oreille Lake** (148 sq mi/383.3 sq km), N Idaho, & flowing W, N, & W into the Columbia in S B.C., Canada.
**Peng·hu** (pŭng′ho͞o′). PESCADORES.
**Peng·pu** (pŭng′po͞o′). BENGBU.
**Pen·ki** (bŭn′chē′). BENXI.
**Pen·nine** (pĕn′īn′). **1. Alps** (ălps). Range of the Alps along the Swiss-Italian border W of Lake Maggiore, rising to 15,203 ft (4,637 m). **2. Chain.** Range of hills extending from S Scotland to central England.
**Penn·syl·va·nia** (pĕn′səl-vān′yə, -vā′nē-ə). State of the E U.S. Cap. Harrisburg. Pop. 11,866,728. **—Penn′syl·va′nian** *adj. & n.*
**Pe·nob·scot** (pə-nŏb′skət, -skŏt′). River of central Me. flowing c. 350 mi (563 km) S into **Penobscot Bay,** an inlet of the Atlantic.
**Pen·sa·co·la** (pĕn′sə-kō′lə). City of extreme NW Fla. ESE of Mobile, Ala. Pop. 57,619.
**Pen·tic·ton** (pĕn-tĭk′tən). City of S B.C., Canada, E of Vancouver. Pop. 21,344.
**Pent·land Firth** (pĕnt′lənd). Channel between NE Scotland & the Orkney Is.
**Pen·za** (pĕn′zə). City of central European USSR NNW of Saratov. Pop. 483,000.
**Pen·zhi·na** (pĕn′zhə-nə). **1.** *also* **Pen·zhin·ska·ya** (pĕn′zhĭn′skə-yə). Bay of the Sea of Okhotsk extending c. 185 mi (298 km) into NE Siberia along NW coast of Kamchatka. **2.** River of NE Asian USSR flowing c. 446 mi (718 km) from the Kolyma Mts. to Penzhina Bay.
**Pe·o·ri·a** (pē-ôr′ē-ə, -ōr′-). **1.** City of S central Ariz. near Phoenix. Pop. 12,251. **2.** City of central Ill. N of Springfield. Pop. 124,160.
**Pe·rei·ra** (pə-rĕr′ə, -rā′rə). City of W central Colombia W of Bogotá. Pop. 174,128.
**Per·ga·mum** (pûr′gə-məm) *also* **Per·ga·mos** (-məs). Ancient Greek city of W Asia Minor on the site of modern Bergama, W Turkey.
**Pé·ri·bon·ca** (pĕr′ə-bŏng′kə). River, c. 280 mi (451 km), of S central Que., Canada.
**Perm** (pĕrm). City of E European USSR on the Kama R. Pop. 999,000.
**Per·nik** (pĕr′nĭk). City of W Bulgaria SW of Sofia. Pop. 87,432.
**Per·pi·gnan** (pĕr-pē-nyäN′). City of S France near the Spanish border & the Mediterranean. Pop. 101,198.
**Per·rys·burg** (pĕr′ēz-bûrg′). City of NW Ohio near Toledo. Pop. 10,215.
**Per·sep·o·lis** (pər-sĕp′ə-lĭs). Ruined city of ancient Persia NE of Shiraz, SW Iran.
**Per·sia** (pûr′zhə, -shə). IRAN. **—Per′sian** *adj. & n.*
**Persian Gulf.** Arm of the Arabian Sea between Arabia & SW Iran.
**Perth** (pûrth). **1.** City of SW Australia near the Indian Ocean. Metro. area pop. 731,275. **2.** Burgh of central Scotland NNW of Edinburgh. Pop. 43,098.
**Perth Am·boy** (ăm′boi′). City of E central N.J. on Raritan Bay opposite Staten Is. Pop. 38,951.
**Pe·ru** (pə-ro͞o′). **1.** Country of W South America on the Pacific. Cap. Lima. Pop. 17,031,221. **2.** City of N Ill. opposite La Salle. Pop. 10,886. **3.** City of N Ind. E of Logansport. Pop. 13,764. **—Pe·ru′vi·an** (-vē-ən) *adj. & n.*
**Pe·ru·gia** (pə-ro͞o′jə, -jē-ə). City of central Italy overlooking the Tiber N of Rome. Pop. 65,975.
**Pe·sa·ro** (pā′zə-rō′). City of central Italy on the Adriatic W of Florence. Pop. 72,104.
**Pes·ca·do·res** (pĕs′kə-dôr′ēz, -ĭs, -dōr′-). Islands of Taiwan in Formosa Strait off the W coast of Taiwan.
**Pes·ca·ra** (pə-skär′ə). City of central Italy on the Adriatic ENE of Rome. Pop. 125,391.
**Pe·sha·war** (pə-shä′wər). City of N Pakistan ESE of Khyber Pass. Pop. 268,366.
**Pe·tah Tiq·wa** *or* **Pe·tah Tiq·va** (pĕt′ə tĭk′və). City of central Israel E of Tel Aviv-Jaffa. Pop. 112,000.
**Pet·a·lu·ma** (pĕt′l-o͞o′mə). City of W Calif. NNW of San Rafael. Pop. 33,834.
**Pe·ter·bor·ough** (pē′tər-bûr′ə, -bər-ə). **1.** City of SE Ont., Canada, NE of Toronto. Pop. 59,683. **2.** Borough of E central England E of Leicester. Pop. 118,900.
**Pe·ters·burg** (pē′tərz-bûrg′). Independent city of SE Va. S of Richmond. Pop. 41,055.
**Pe·ti·tot** (pĕt′ĭ-tō′). River, 295 mi (475 km), of NE B.C. & NW Alta., Canada.
**Pe·tra** (pē′trə). Ancient city of Edom in present-day SW Jordan.
**Pet·ri·fied Forest National Park** (pĕt′rə-fīd′). Section of the Painted Desert in E Ariz. reserved for its petrified trees.
**Pet·ro·pav·lovsk** (pĕt′rə-păv′lôfsk′). **1.** City of N Central Asian USSR W of Novosibirsk. Pop. 207,000. **2.** *also* **Pet·ro·pav·lovsk-Kam·chat·ski** (-kăm-chăt′skē). City of Far Eastern USSR on the Pacific coast of SE Kamchatka. Pop. 215,000.
**Pe·tróp·o·lis** (pə-trŏp′ə-lĭs). City of SE Brazil N of Rio de Janeiro. Pop. 116,080.
**Pet·ro·za·vodsk** (pĕt′rə-zə-vŏtsk′). City of NW European USSR on Lake Onega. Pop. 234,000.
**Pforz·heim** (pfôrts′hīm, pfōrts′-, fôrts′-, fōrts′-). City of SW West Germany WNW of Stuttgart. Pop. 108,635.
**Pha·ros** (fâr′ŏs′). Peninsula, formerly an island, in Alexandria, N Egypt.
**Pharr** (fär). City of extreme S Tex. WNW of Brownsville. Pop. 21,381.
**Phe·nix City** (fē′nĭks). City of E Ala. near Columbus, Ga. Pop. 27,012.
**Phil·a·del·phi·a** (fĭl′ə-dĕl′fē-ə). City of SE Pa. on the Delaware R. Pop. 1,688,210. **—Phil′a·del′phi·an** *adj. & n.*
**Phi·lip·pi** (fə-lĭp′ī′). Ancient town of N central Macedonia, Greece; site of Antony & Octavian's defeat of Brutus & Cassius (42 B.C.).
**Phil·ip·pines** (fĭl′ə-pēnz′). Country of E Asia consisting of the **Philippine Is.,** an archipelago in the W Pacific SE of China. Cap. Manila. Pop. 47,914,017. **—Phil′ip·pine** *adj.*
**Philippine Sea.** Part of the W Pacific E of the Philippines & W of the Marianas.
**Phi·lis·ti·a** (fə-lĭs′tē-ə). Ancient country on the SW coast of Palestine.
**Phil·lips·burg** (fĭl′ĭps-bûrg′). Town of W N.J. on the Delaware R. NE of Bethlehem, Pa. Pop. 16,647.
**Phnom Penh** (pə-nôm′ pĕn′, nŏm′ pĕn′). Cap. of Cambodia in the SW part on the Mekong R. Pop. 300,000.
**Pho·cae·a** (fō-sē′ə). Ancient Ionian Greek city of W Asia Minor on the Aegean N of modern Izmir, Turkey.
**Pho·cis** (fō′sĭs). Region & ancient country of central Greece N of the Gulf of Corinth.
**Phoe·ni·cia** (fĭ-nĭsh′ə, -ē-ə, -nē′shə, -shē-ə). Ancient maritime country of SW Asia consisting of city-states along the E Mediterranean in present-day Syria & Lebanon. **—Phoe·ni′cian** *adj. & n.*
**Phoe·nix** (fē′nĭks). **1.** Islands in the central Pacific N of Samoa. **2.** Cap. of Ariz. in the S central part. Pop. 764,911.
**Pia·cen·za** (pyä-chĕn′sə, pē-ə-). Town of N Italy on the Po R. SE of Milan. Pop. 100,001.
**Pi·at·ra Ne·amt** (pē-ä′trə nē-ămts′). City of NE Rumania E of Cluj. Pop. 84,192.
**Pic·a·yune** (pĭk′ē-o͞on′, -yo͞on′, pĭk′ə-yo͞on′). City of S Miss. near the La. border WNW of Gulfport. Pop. 10,361.
**Pick·er·ing** (pĭk′ə-rĭng). Town of S Ont., Canada, near Lake Ontario NE of Toronto. Pop. 27,879.
**Pi·co Bo·lí·var** (pē′kō bə-lē′vär′). Highest mountain, 16,411 ft (5,003.4 m), in Venezuela, in the W part S of Lake Maracaibo.

**Pico de A·ne·to** (dä ə-nā′tō). Highest peak, 11,168 ft (3,406.2 m), of the Pyrenees, in NE Spain near the French border.

**Pico Ri·ve·ra** (rə-vîr′ə). City of S Calif. E of Los Angeles. Pop. 53,459.

**Pied·mont** (pēd′mŏnt′). **1.** Region of NW Italy. **2.** Plateau region of the E U.S. extending from N.Y. to Ala. between the Appalachians & the Atlantic coastal plain. **3.** City of W Calif. near Oakland. Pop. 10,498. **—Pied′mon·tese′** *adj. & n.*

**Pie·dras Ne·gras** (pē-ā′drəs nā′drəs). City of N Mexico on the Rio Grande N of Monterrey. Pop. 41,033.

**Pierre** (pîr). Cap. of S.Dak. in the central part. Pop. 11,973.

**Pierre·fonds** (pē-ĕr-fôɴ′, pyĕr-). City of S Que., Canada, on Montreal Is. W of Montreal. Pop. 35,402.

**Pie·ter·mar·itz·burg** (pē′tər-măr′ĭts-bûrg′). City of E South Africa WNW of Durban. Pop. 114,882.

**Pigs** (pĭgz), **Bay of.** Small inlet of the Caribbean on the S coast of W Cuba.

**Pikes Peak** (pīks). Mountain, 14,110 ft (4,303.6 m), in the Front Range of central Colo.

**Pi·la** (pē′lə). City of NW Poland N of Poznań. Pop. 43,778.

**Pil·co·ma·yo** (pĭl′kə-mī′ō). River of central South America rising in central Bolivia & flowing c. 1,000 mi (1,609 km) SE along the Argentina-Paraguay border to the Paraguay R.

**Pi·nang** (pə-năng′). *var. of* PENANG.

**Pi·nar del Rí·o** (pə-när′ dĕl rē′ō). City of W Cuba WSW of Havana. Pop. 83,000.

**Pin·dus** (pĭn′dəs). Mountains of NW Greece rising to 8,650 ft (2,638.3 m).

**Pine Bluff** (pīn′ blŭf′). City of S central Ark. SSE of Little Rock. Pop. 56,811.

**Pi·nel·las Park** (pī-nĕl′əs). City of W Fla. near St. Petersburg. Pop. 32,811.

**Pines** (pīnz), **Isle of.** Island in the Caribbean off SW Cuba.

**Pine·ville** (pīn′vĭl′, -vəl). City of central La. on the Red R. opposite Alexandria. Pop. 12,034.

**Ping** (pĭng). River, c. 350 mi (563 km), of NW Thailand.

**Ping·tung** *or* **Ping·dong** (pĭng′dŏŏng′). City of S Taiwan E of Kaohsiung. Pop. 165,360.

**Pi·nole** (pə-nōl′). City of W Calif. near Richmond. Pop. 14,253.

**Pinsk** (pĭnsk). City of W European USSR WNW of Kiev. Pop. 90,000.

**Pio·tr·ków Try·bu·nal·ski** (pyô′tər-kŏŏf′ trĭb′ōō-näl′skē). City of central Poland SSE of Łódź. Pop. 59,683.

**Piq·ua** (pĭk′wā′, -wə). City of SW central Ohio N of Dayton. Pop. 20,480.

**Pi·rae·us** (pī-rē′əs). City of E central Greece on the Saronic Gulf near Athens. Pop. 187,362.

**Pir·ma·sens** (pîr′mə-zĕns′). City of SW West Germany near the French border. Pop. 53,651.

**Pir·na** (pîr′nə). City of SE East Germany on the Elbe R. near the Czech border. Pop. 49,771.

**Pi·sa** (pē′zə, -sä). City of N central Italy on the Arno R. near the Tyrrhenian Sea. Pop. 91,156. **—Pi′san** *adj. & n.*

**Pis·to·ia** (pĭ-stoi′ə, -stō′yə). City of N central Italy NW of Florence. Pop. 55,403.

**Pit** (pĭt). River, c. 200 mi (322 km), of N Calif.

**Pit·cairn** (pĭt′kârn′). British-administered island of the S Pacific ESE of Tahiti.

**Pitts·burg** (pĭts′bûrg′). **1.** City of W Calif. NE of Oakland. Pop. 33,034. **2.** City of SE Kans. near the Mo. border. Pop. 18,770.

**Pitts·burgh** (pĭts′bûrg′). City of SW Pa. at the confluence of the Allegheny & Monongahela rivers, forming the Ohio R. Pop. 423,938.

**Pitts·field** (pĭts′fēld′). City of W Mass. NW of Springfield. Pop. 51,974.

**Piu·ra** (pyŏŏr′ə). City of NW Peru near the Pacific. Pop. 126,702.

**Pla·cen·tia** (plə-sĕn′chə, -chē-ə). **1.** Bay of the Atlantic in SE Newf., Canada. **2.** City of S Calif. near Santa Ana. Pop. 35,041.

**Pla·cid** (plăs′ĭd). Lake of NE N.Y. in the Adirondacks.

**Plain·field** (plān′fēld′). **1.** Town of E Conn. NE of Norwich. Pop. 12,774. **2.** City of NE N.J. SW of Newark. Pop. 45,555.

**Plain·view** (plān′vyōō′). City of NW Tex. S of Amarillo. Pop. 22,187.

**Plain·ville** (plān′vĭl′). Town of central Conn. SW of Hartford. Pop. 16,401.

**Pla·no** (plā′nō′). City of NE Tex. near Dallas. Pop. 72,331.

**Plan·ta·tion** (plăn-tā′shən). City of SE Fla. near Fort Lauderdale. Pop. 48,501.

**Plant City** (plănt). City of W central Fla. E of Tampa. Pop. 19,270.

**Pla·ta** (plä′tə), **Río de la.** —See RÍO DE LA PLATA.

**Pla·tae·a** (plə-tē′ə). Ancient city of central Greece SW of Thebes; site of Greek victory over Persians (479 B.C.).

**Platte** (plăt). River, c. 310 mi (499 km), of S Nebr.

**Platts·burgh** (plăts′bûrg′). City of extreme NE N.Y. on Lake Champlain NW of Burlington, Vt. Pop. 21,057.

**Plau·en** (plou′ən). City of S East Germany near the NW Erzgebirge. Pop. 80,353.

**Pleas·ant Grove** (plĕz′ənt). City of N central Utah near Provo. Pop. 10,669.

**Pleasant Hill.** City of W Calif. NE of Berkeley. Pop. 25,124.

**Pleas·an·ton** (plĕz′ən-tən). City of W Calif. SE of Oakland. Pop. 35,160.

**Pleas·ant·ville** (plĕz′ənt-vĭl′). City of SE N.J. near Atlantic City. Pop. 13,435.

**Plev·en** (plĕv′ən) *or* **Plev·na** (-nə). City of N Bulgaria NE of Sofia. Pop. 107,567.

**Plo·ieș·ti** *or* **Plo·eș·ti** (plô-yĕsht′, -yĕsh′tē). City of S central Rumania N of Bucharest. Pop. 207,009.

**Plov·div** (plôv′dĭf′, -dĭv′). City of S Bulgaria on the Maritsa R. Pop. 300,242.

**Plum** (plŭm). Borough of SW Pa. near Pittsburgh. Pop. 25,390.

**Plym·outh** (plĭm′əth). **1.** Borough of SW England on the English Channel. Pop. 259,100. **2.** Town of W Conn. N of Waterbury. Pop. 10,732. **3.** Town of SE Mass. SE of Boston. Pop. 35,913. **4.** City of SE Minn. near Minneapolis-St. Paul. Pop. 31,615.

**Plzeň** (pŭl′zĕn′, -zĕn′yə). City of W Czechoslovakia WSW of Prague. Pop. 155,000.

**Po** (pō). River of N Italy flowing c. 405 mi (652 km) E to the Adriatic.

**Po·be·da Peak** (pō-bē′də, pə-). Highest (24,406 ft/7,443.8 m) mountain of the Tian Shan on the border between S Central Asian USSR & W China.

**Po·ca·tel·lo** (pō′kə-tĕl′ō, -tĕl′ə). City of SE Idaho near the Snake R. Pop. 46,340.

**Po·co·no** (pō′kə-nō′) *or* **Po·co·nos** (-nōz′). Mountains in NE Pa. rising to c. 1,600 ft (488 m).

**Po·dolsk** (pə-dôlsk′). City of central European USSR S of Moscow. Pop. 202,000.

**Po Hai** (bō′ hī′). BO HAI.

**Po·hang** (pō′hăng′). City of SE South Korea on an inlet of the Sea of Japan. Pop. 134,318.

**Pointe aux Trem·bles** (pwăɴt ō träɴ′blə). City of S Que., Canada, on Montreal Is. N of Montreal. Pop. 35,618.

**Pointe Claire** (point′ klâr′, pwăɴt). City of S Que., Canada, on Montreal Is. WSW of Montreal. Pop. 25,917.

**Pointe-Noire** (pwăɴt-nwär′). City of SW Congo on the Atlantic. Pop. 141,700.

**Point Pleas·ant** (point plĕz′ənt). Borough of E N.J. near the Atlantic ESE of Trenton. Pop. 17,747.

**Poi·tiers** (pwä-tyā′). City of W central France ESE of Nantes. Pop. 78,739.

**Po·land** (pō′lənd). Country of central Europe on the Baltic. Cap. Warsaw. Pop. 35,815,000. **—Po′lish** (pō′lĭsh) *adj. & n.*

**Pol·ta·va** (pəl-tä′və). City of S European USSR in the Ukraine WSW of Kharkov. Pop. 279,000.

**Pol·y·ne·sia** (pŏl′ə-nē′zhə, -shə). Scattered islands of the central & S Pacific roughly between New Zealand, Hawaii, & Easter Is. **—Pol′y·ne′sian** *adj. & n.*

**Pom·er·a·ni·a** (pŏm′ə-rā′nē-ə, -rān′yə). Historical region of N central Europe along the Baltic in present-day NW Poland & NE East Germany. **—Pom′er·a′ni·an** *adj. & n.*

**Po·mo·na** (pə-mō′nə). City of S Calif. near Los Angeles. Pop. 92,742.

**Pom·pa·no Beach** (pŏm′pə-nō′). City of SE Fla. on the Atlantic. Pop. 52,618.

**Pom·pe·ii** (pŏm-pā′, -pā′ē). Ancient city of S Italy near Naples; destroyed by eruption of Mt. Vesuvius (A.D. 79). **—Pom·pe′ian, Pom·pei′ian** *adj. & n.*

**Pomp·ton Lakes** (pŏmp′tən). Borough of NE N.J. NW of Paterson. Pop. 10,660.

**Pon·ca City** (pŏng′kə). City of N Okla. NNE of Oklahoma City. Pop. 26,238.

**Pon·ce** (pôn′sā). City of S Puerto Rico on the Caribbean. Pop. 161,260.

**Pon·di·cher·ry** (pŏn′də-chĕr′ē, -shĕr′ē). City of SE India on the Bay of Bengal SW of Madras. Pop. 90,537.

**Pon·ta Del·ga·da** (pŏn′tə dĕl-gä′də, -găd′ə). Chief city of the Azores on the SW coast of São Miguel Is. Pop. 20,195.

**Pont·char·train** (pŏn′chər-trān′). Lake, c. 630 sq mi (1,632 sq km), of SE La. N of New Orleans.

**Pon·te·ve·dra** (pŏn′tə-vā′drə). City of NW Spain on the Atlantic. Pop. 27,118.

**Pon·ti·ac** (pŏn′tē-ăk′). **1.** City of N central Ill. NE of Bloomington. Pop. 11,227. **2.** City of SE Mich. NW of Detroit. Pop. 76,715.

**Pon·ti·a·nak** (pŏn′tē-ä′näk). City of W Borneo, Indonesia, on the Kapuas R. delta. Pop. 217,555.

**Pon·tus** (pŏn′təs). Ancient country of NE Asia Minor on the S Black Sea coast.

**Pon·ty·pool** (pŏn′tə-pōōl′). Urban district of S Wales NW of Cardiff. Pop. 36,710.

**Poole** (pōōl). Borough of S England on the English Channel WSW of Southampton. Pop. 110,600.

**Poo·na** (pōō′nə). City of W central India SE of Bombay. Pop. 856,105.

**Pop·lar Bluff** (pŏp′lər blŭf′). City of SE Mo. near the Ark. border S of St. Louis. Pop. 17,139.

**Po·po·ca·té·petl** (pō'pə-kăt'ə-pĕt'l). Volcano, 17,887 ft (5,455.5 m), in central Mexico SE of Mexico City.
**Por·cu·pine** (pôr'kyə-pīn'). River, c. 448 mi (721 km), of N Y.T., Canada, & NE Alas.
**Po·ri** (pôr'ē). City of SW Finland near the Gulf of Bothnia NW of Helsinki. Pop. 80,343.
**Por·tage** (pôr'tĭj, pōr'-). **1.** City of NW Ind. on Lake Michigan. Pop. 27,409. **2.** City of SW Mich. S of Kalamazoo. Pop. 38,157.
**Portage la Prai·rie** (lə prâr'ē). City of S Man., Canada, W of Winnipeg. Pop. 12,555.
**Port Al·ber·ni** (pôrt' ăl-bûr'nē, pōrt'). City of SW B.C., Canada, on S central Vancouver Is. Pop. 19,585.
**Port An·ge·les** (ăn'jə-lĭs). City of NW Wash. on Juan de Fuca Strait S of Victoria, B.C., Canada. Pop. 17,311.
**Port A·pra** (ä'prə). APRA HARBOR.
**Port Ar·thur** (är'thər). **1.** LÜSHUN. **2.** City of extreme SE Tex. near the Gulf of Mexico & the La. border. Pop. 61,195.
**Port-au-Prince** (pôrt'ō-prĭns', pōrt'-). Cap. of Haiti in the SW part on the Gulf of Gonaïves. Pop. 306,053.
**Port Ches·ter** (chĕs'tər). Village of SE N.Y. on Long Is. Sound & the Conn. border. Pop. 23,565.
**Port Col·borne** (kōl'bûrn'). City of S Ont., Canada, on Lake Erie at the S end of the Welland Ship Canal. Pop. 20,536.
**Port Co·quit·lam** (kō-kwĭt'ləm). City of SW B.C., Canada, on the Fraser R. E of Vancouver. Pop. 23,926.
**Port E·liz·a·beth** (ĭ-lĭz'ə-bəth). City of SE South Africa on the Indian Ocean. Pop. 392,231.
**Por·ter·ville** (pôr'tər-vĭl', pōr'-). City of S central Calif. N of Bakersfield. Pop. 19,707.
**Port Har·court** (här'kərt). City of SE Nigeria in the Niger delta. Pop. 242,000.
**Port Hope** (hōp). Town of S Ont., Canada, on Lake Ontario ENE of Toronto. Pop. 9,788.
**Port Hue·ne·me** (wī-nē'mē). Town of S Calif. near Oxnard. Pop. 17,803.
**Port Hu·ron** (hyŏŏr'ən). City of SE Mich. on Lake Huron & the St. Clair R. Pop. 33,981.
**Port·land** (pôrt'lənd, pōrt'-). **1.** City of SW Me. S of Lewiston. Pop. 61,572. **2.** City of NW Ore. near the mouth of the Columbia R. Pop. 366,383. **3.** City of S Tex. near Corpus Christi. Pop. 12,023.
**Port La·va·ca** (lə-văk'ə). City of SE Tex. on an inlet of the Gulf of Mexico NE of Corpus Christi. Pop. 10,911.
**Port Lou·is** (lōō'ĭs, lōō'ē, lōō-ē'). Cap. of Mauritius in the NW on the Indian Ocean. Pop. 141,022.
**Port Moo·dy** (mōō'dē). City of SW B.C., Canada, near Vancouver. Pop. 11,649.
**Port Mores·by** (môrz'bē, mōrz'-). Cap. of Papua New Guinea on SE New Guinea Is. Pop. 122,761.
**Port Nech·es** (nā'chĭz). City of extreme SE Tex. near Port Arthur. Pop. 13,944.
**Pôr·to** (pôr'tōō, pōr'-). OPORTO.
**Pôrto A·le·gre** (ə-lĕ'grə). City of SE Brazil at the N end of Lagoa dos Patos. Pop. 1,251,901.
**Port of Spain** (spān). Cap. of Trinidad and Tobago on the NW coast of Trinidad. Pop. 62,680.
**Por·to-No·vo** (pôr'tō-nō'vō, pōr'-). Cap. of Benin in the SE on an inlet of the Gulf of Guinea. Pop. 104,000.
**Port Or·ange** (ôr'ĭnj, ŏr'-). City of NE Fla. near Daytona Beach. Pop. 18,756.
**Port Phil·lip Bay** (fĭl'əp). Inlet of Bass Strait on the SE coast of Australia.
**Port Sa·id** (sä-ēd', sīd). City of NE Egypt at the Mediterranean entrance to the Suez Canal. Pop. 262,620.
**Port Saint Lu·cie** (sānt lōō'sē). City of E Fla. near Fort Pierce. Pop. 14,690.
**Ports·mouth** (pôrt'sməth, pōrt'-). **1.** Borough of S England on the English Channel opposite the Isle of Wight. Pop. 198,500. **2.** City of SE N.H. opposite Kittery, Me. Pop. 26,254. **3.** City of S Ohio on the Ohio R. S of Columbus. Pop. 25,943. **4.** City of SE Va. opposite Norfolk. Pop. 104,577.
**Port Stan·ley** (stăn'lē). STANLEY, Falkland Is.
**Port Su·dan** (sōō-dăn', -dän'). City of NE Sudan on the Red Sea. Pop. 133,000.
**Por·tu·gal** (pôr'chĭ-gəl, pōr'-). Country of SW Europe on the Iberian Peninsula & including Madeira and the Azores. Cap. Lisbon. Pop. 9,933,000. **—Por'tu·gese'** (-gēz', -gēs') *adj.* & *n.*
**Po·sen** (pō'zən). POZNAŃ.
**Po·ta·ro** (pə-tär'ō). River, c. 100 mi (160 km), of central Guyana.
**Potch·ef·stroom** (pŏch'əf-strōōm'). Town of NE South Africa SW of Johannesburg. Pop. 57,443.
**Po·ten·za** (pō-tĕn'sə, -zə). City of S Italy in the Apennines ESE of Naples. Pop. 46,869.

**Po·to·mac** (pə-tō'mək). River of E U.S. rising in NE W.Va. & flowing c. 285 mi (459 km) along the Va.-Md border to Chesapeake Bay.
**Po·to·sí** (pō-tə-sē'). City of S central Bolivia SW of Sucre in the Andes at c. 13,780 ft (4,203 m). Pop. 77,397.
**Pots·dam** (pŏts'dăm'). **1.** City of central East Germany on the Havel R. near Berlin. Pop. 117,236. **2.** Village of N N.Y. E of Ogdensburg. Pop. 10,635.
**Potts·town** (pŏts'toun'). Borough of SE Pa. NW of Philadelphia. Pop. 22,729.
**Potts·ville** (pŏts'vĭl'). City of SE Pa. WNW of Allentown. Pop. 18,195.
**Pough·keep·sie** (pə-kĭp'sē, pō-). City of SE N.Y. on the Hudson N of New York City. Pop. 29,757.
**Pow·der** (pou'dər). **1.** River, c. 150 mi (241 km), of E Ore. **2.** River, c. 486 mi (782 km), of N Wyo. & SE Mont.
**Po·yang** (pō'yäng'). Lake of E China SE of Wuhan.
**Poz·nań** (pŏz'năn', -năn'yə, -nän', -nän'yə, pôz'-). City of W central Poland on the Warta R. Pop. 469,085.
**Poz·zuo·li** (pôt-swô'lē). City of S Italy on the Bay of Naples. Pop. 53,546.
**Prague** (präg). Cap. of Czechoslovakia in the W part on the Vltava R. Pop. 1,162,200.
**Prai·a** (prī'ə). Cap. of Cape Verde Is. on the SE coast of São Tiago Is. Pop. 21,494.
**Prai·rie Village** (prâr'ē). City of E Kans. near Kansas City. Pop. 24,657.
**Pra·to** (prä'tō). City of central Italy NW of Florence. Pop. 108,385.
**Pratt·ville** (prăt'vĭl', -vəl). City of central Ala. NW of Montgomery. Pop. 18,647.
**Pres·cott** (prĕs'kət, -kŏt'). City of central Ariz. NNW of Phoenix. Pop. 20,055.
**Presque Isle** (prĕsk). City of NE Me. near the N.B., Canada, border. Pop. 11,172.
**Pres·ton** (prĕs'tən). Borough of NW England NNE of Liverpool. Pop. 94,760.
**Pre·to·ri·a** (prĭ-tôr'ē-ə, -tōr'-). Administrative cap. of South Africa in the NE part NE of Johannesburg. Pop. 545,450.
**Prib·i·lof** (prĭb'ə-lôf'). Islands of SW Alas. in the Bering Sea.
**Prich·ard** (prĭch'ərd). City of SW Ala. near Mobile. Pop. 39,541.
**Prince Al·bert** (prĭns ăl'bərt). City of central Sask., Canada, on the N Saskatchewan R. Pop. 28,631.
**Prince Ed·ward Island** (ĕd'wərd). Island & province of SE Canada in the S Gulf of St. Lawrence. Cap. Charlottetown. Pop. 121,328.
**Prince George** (jôrj). City of central B.C., Canada, on the upper Fraser R. Pop. 59,929.
**Prince of Wales** (wālz). **1.** Island of extreme SE Alas. in the Alexander Archipelago. **2.** Island of N N.W.T., Canada, in the Arctic Ocean NE of Victoria Is.
**Prince Ru·pert** (rōō'pərt). City of W B.C., Canada, on the Pacific coast near the Alas. border. Pop. 14,754.
**Prince·ton** (prĭn'stən). Borough of central N.J. NNE of Trenton. Pop. 12,035.
**Prince Wil·liam Sound** (wĭl'yəm). Arm of the Gulf of Alaska E of the Kenai Peninsula of S Alas.
**Prin·ci·pe** (prĭn'sə-pā'). Island, 53 sq mi (137 sq km), in the Gulf of Guinea, W Africa, forming part of the republic of São Tomé and Principe.
**Prip·et** (prĭp'ĕt') *or* **Pri·pyat** (-yət). River, c. 440 mi (708 km), of E central European USSR.
**Pro·ko·pyevsk** (prə-kôp'yəfsk). City of E Siberian USSR ESE of Novosibirsk. Pop. 266,000.
**Pros·pect Heights** (prŏs'pĕkt'). City of NE Ill. near Chicago. Pop. 11,808.
**Prov·i·dence** (prŏv'ĭ-dəns, -dĕns'). Cap. of R.I. in the NE on Narragansett Bay. Pop. 156,804.
**Pro·vo** (prō'vō). City of N central Utah SSE of Salt Lake City. Pop. 73,907.
**Prud·hoe Bay** (prōōd'hō, prŭd'-). Inlet of the Arctic Ocean on the N Alas. coast E of the Colville R. delta.
**Prus·sia** (prŭsh'ə). Region & former state of N central Europe including present-day N Germany & N Poland. **—Prus'sian** *adj.* & *n.*
**Pskov** (pə-skôf'). City of NW European USSR SSW of Leningrad. Pop. 176,000.
**Pueb·la** (pōō-ĕb'lə, pwĕb'-). City of E central Mexico ESE of Mexico City. Pop. 465,985.
**Pueb·lo** (pwĕb'lō). City of S central Colo. SSE of Colorado Springs. Pop. 101,686.
**Puer·to Ca·bel·lo** (pwĕr'tə kə-bā'ō, -yō). City of N Venezuela on the Caribbean W of Caracas. Pop. 72,103.
**Puerto la Cruz** (lə krōōz', krōōs'). City of N Venezuela on the Caribbean NE of Barcelona. Pop. 63,276.
**Puerto Montt** (mônt'). City of S central Chile on an inlet of the Pacific. Pop. 62,700.
**Puer·to Ri·co** (pwĕr'tə rē'kō, pôrt'ə, pōrt'ə). Island of the West Indies E of Hispaniola; a self-governing commonwealth of the U.S. Cap. San Juan. Pop. 3,187,076. **—Puerto Ri'can** *adj.* & *n.*
**Pu·get Sound** (pyōō'jĭt). Inlet of the Pacific, c. 100 mi (160 km) long, in NW Wash.

**Pu·la** (po͞oʹlə). City of NW Yugoslavia on the Adriatic. Pop. 47,117.
**Pu·las·ki** (pə-lăsʹkē, po͞o-). Town of SW Va. WSW of Roanoke. Pop. 10,106.
**Pull·man** (po͝olʹmən). City of SE Wash. near the border W of Moscow, Idaho. Pop. 23,579.
**Pun·jab** (pŭn-jäbʹ, -jăbʹ, pŭnʹjäbʹ, -jăbʹ). Region of NW India & NW Pakistan. **—Pun·jaʹbi** *adj. & n.*
**Pun·ta A·re·nas** (po͞onʹtə ə-rĕʹnəs). City of S Chile on the Strait of Magellan. Pop. 61,800.
**Pu·ra·cé** (po͞orʹə-sēʹ). Volcano, c. 15,420 ft (4,703 m), in the Andes of SW Colombia.
**Pur·ga·toire** (pûrʹgə-twärʹ, pûrʹgə-tôrʹē, -tōrʹē). River, c. 186 mi (299 km), of SE Colo.
**Pu·rus** (pə-ro͞osʹ). River of E central Peru & W Brazil flowing c. 2,100 mi (3,379 km) NE into the Amazon.
**Pu·san** (po͞oʹsänʹ). City of extreme SE South Korea on Korea Strait. Pop. 2,453,173.
**Put-in-Bay** (po͝otʹĭn-bāʹ). Bay of W Lake Erie where the U.S. Navy under Perry defeated a British fleet (1813).
**Pu·tu·ma·yo** (po͞oʹtə-mīʹō). River of NW South America rising in SW Colombia & flowing c. 1,000 mi (1,609 km) along the Colombia-Peru border to the Amazon in NW Brazil.
**Puy·al·lup** (pyo͞o-ălʹəp). City of W central Wash. near Tacoma. Pop. 18,251.
**Pyong·yang** (pē-ôngʹ-yängʹ). Cap. of North Korea in the SW part. Pop. 1,250,000.
**Pyr·e·nees** (pĭrʹə-nēzʹ). Mountain range along the French-Spanish border from the Bay of Biscay to the Mediterranean, rising to 11,168 ft (3,406.2 m). **—Pyrʹe·neʹan** *adj.*

# Q

**Qa·tar** (käʹtər). Country of E Arabia on a peninsula in the Persian Gulf. Cap. Doha. Pop. 220,000.
**Qat·ta·ra Depression** (kə-tärʹə). Desert basin of NW Egypt in the Libyan Desert.
**Qe·na** (kēʹnə, käʹ-). City of E central Egypt on the Nile. Pop. 94,013.
**Qing·dao** (chĭngʹdouʹ). City of E China on the Yellow Sea NNW of Shanghai. Pop. 1,900,000.
**Qing·hai** (chĭngʹhīʹ). **1.** Salt lake, 1,625 sq mi (4,210 sq km), of N central China. **2.** Province of W central China.
**Qi·qi·har** (chēʹchēʹhärʹ). City of NE China in Manchuria NW of Harbin. Pop. 1,500,00.
**Qom** (kōm). *var. of* QUM.
**Qu'Ap·pelle** (kwə-pĕlʹ). River, c. 270 mi (434 km), of S Sask. & SW Man., Canada.
**Que·bec** kwĭ-bĕkʹ) *or* **Qué·bec** (kā-). **1.** Province of E Canada. Cap. Quebec. Pop. 6,377,518. **2.** Cap. of Que., Canada, in the S part on the St. Lawrence R. Pop. 177,082. **—Que·beckʹer, Que·becʹer** *n.*
**Queen Charlotte** (kwēn shärʹlət). Islands off the W coast of B.C., Canada, separated from Vancouver Is. to the SE by **Queen Charlotte Sound,** an inlet of the Pacific.
**Queen E·liz·a·beth** (ĭ-lĭzʹə-bəth). Islands of N N.W.T., Canada, in the Arctic Archipelago N of Parry Channel.
**Queen Maud Land** (môd). Region of Antarctica between the Weddell Sea & Enderby Land.
**Queens** (kwēnz). Borough of New York City, SE N.Y., on W Long Is. Pop. 1,891,325.
**Queens·bor·ough-in-Shep·pey** (kwēnzʹbûr-ə-ĭn-shĕpʹē). Borough of SE England on the Isle of Sheppey at the mouth of the Thames. Pop. 31,550.
**Que·moy** (kĭ-moiʹ, kwĭ-, kwēʹmoi). Island off SE China in Formosa Strait, administered by Taiwan.
**Que·ré·ta·ro** (kə-rĕtʹə-rōʹ). City of central Mexico NW of Mexico City. Pop. 146,448.
**Quet·ta** (kwĕtʹə). City of W central Pakistan. Pop. 156,000.
**Que·zal·te·nan·go** (kĕt-sälʹtə-näng'gō). City of SW Guatemala WNW of Lake Atitlán. Pop. 53,021.
**Que·zon City** (kāʹsônʹ, -sōnʹ). City of central Luzon, Philippines, near Manila. Pop. 956,864.
**Quil·mes** (kēlʹmāsʹ). City of E Argentina on the Río de la Plata. Pop. 355,265.
**Quim·per** (kăN-pĕrʹ). City of NW France near the Bay of Biscay SSE of Brest. Pop. 50,856.
**Quin·cy. 1.** (kwĭnʹsē). City of W Ill. on the Mississippi. Pop. 42,352. **2.** (kwĭnʹzē). City of E Mass. SE of Boston. Pop. 84,743.
**Quir·i·nal** (kwĭrʹə-nəl). One of the seven hills of ancient Rome.
**Qui·to** (kēʹtō). Cap. of Ecuador in the N central part. Pop. 599,828.
**Qum** (ko͞om) *or* **Qom** (kōm). City of W central Iran SSW of Teheran. Pop. 246,831.
**Qum·ran** (ko͞om-ränʹ). Ancient village of Palestine on the NW shore of the Dead Sea in present-day NW Israel.
**Qur·net es Sau·da** (ko͞orʹnĭt ĕs souʹdə). Peak, 10,131 (3,090 m), of the Lebanon Mts. in N Lebanon.

# R

**Ra·bat** (rə-bätʹ). Cap. of Morocco in the N part on the Atlantic. Pop. 367,620.
**Rab·bah** (răbʹə) *or* **Rab·bath** (răbʹəth). AMMAN.
**Rac·coon** (ră-ko͞onʹ). River, c. 200 mi (322 km), of N Iowa.
**Race** (rās). Cape at SE end of Newf., Canada, on Avalon Peninsula.
**Ra·ci·bórz** (rät-sēʹbo͞oshʹ). Town of S Poland on the Oder R. near the Czech border. Pop. 40,418.
**Ra·cine** (rə-sēnʹ, rā-). City of SE Wis. on Lake Michigan S of Milwaukee. Pop. 85,725.
**Rad·ford** (rădʹfərd). Independent city of SW Va. WSW of Roanoke. Pop. 13,225.
**Ra·dom** (räʹdôm). City of SE Poland S of Warsaw. Pop. 158,640.
**Rae·ti·a** (rēʹshē-ə, -shə). *var. of* RHAETIA.
**Ra·ges** (rāʹjĭz) *or* **Rha·gae** (-jē). Ancient medieval city of Persia near present-day Teheran, N Iran.
**Ra·gu·sa** (rə-go͞oʹzə). City of SE Sicily, Italy. Pop. 55,751.
**Rah·way** (rôʹwāʹ). City of NE N.J. SW of Elizabeth. Pop. 26,723.
**Ra·ia·te·a** (rīʹə-tāʹə). Volcanic island of the S Pacific, largest of the Leeward group of the Society Is., French Polynesia.
**Rai·nier** (rə-nîrʹ, rā-), **Mount.** Volcanic peak, 14,408 ft (4,394.4 m), of the Cascade Range in W central Wash.
**Rain·y** (rāʹnē). **1.** Lake, c. 345 sq mi (894 sq km), in N Minn. & W Ont., Canada. **2.** River, c. 80 mi (129 km), on the border between N Minn. & SW Ont., Canada.
**Rai·pur** (rīʹpo͞or). City of E central India E of Nagpur. Pop. 174,518.
**Rai·sin** (rāʹzĭn). River, c. 115 mi (185 km), of SE Mich.
**Ra·jah·mun·dry** (räʹjə-mo͞onʹdrē). City of E central India on the Godavari R. Pop. 165,912.
**Raj·kot** (räjʹkōtʹ). City of W central India WSW of Ahmadabad. Pop. 300,612.
**Ra·leigh** (rôʹlē, räʹ-). Cap. of N.C. in the E central part. Pop. 149,771.
**Ra·lik** (räʹlĭk). W chain of the Marshall Is. in the W Pacific.
**Ram·a·po** (rămʹə-pōʹ). Range of the Appalachian Mts. in S N.Y. & N N.J.
**Ra·ma's Bridge** (räʹməz). ADAM'S BRIDGE.
**Ra·mat Gan** (rə-mätʹ gänʹ). City of W central Israel near Tel Aviv-Jaffa. Pop. 120,900.
**Ram·gan·ga** (räm-gŭngʹgə). River, c. 350 mi (563 km), of N India.
**Ram·sey** (rămʹzē). **1.** City of E Minn. near Minneapolis. Pop. 10,093. **2.** Borough of NE N.J. near the N.Y. border N of Paterson. Pop. 12,899.
**Rams·gate** (rămzʹgātʹ, -gĭt). Borough of SE England on the Isle of Thanet. Pop. 40,090.
**Ran·ca·gua** (rän-käʹgwä, räng-). City of central Chile S of Santiago. Pop. 86,500.
**Ran·cho Pal·os Ver·des** (rănʹchō pălʹōs vûrʹdēz, pălʹəs). City of S Calif. near Long Beach. Pop. 35,227.
**Rand** (rănd). WITWATERSRAND.
**Rand·ers** (räʹnərs). City of N central Denmark NNW of Århus. Pop. 58,409.
**Ran·dolph** (rănʹdôlfʹ). Town of E Mass. S of Boston. Pop. 28,218.
**Rand·wick** (răndʹwĭk). City of SE Australia near Sydney. Pop. 119,500.
**Range·ley Lakes** (rānjʹlē). Group of lakes in W Me. & N N.H.
**Ran·goon** (răn-go͞onʹ, răng-). Cap. of Burma in the S central part on the **Rangoon R.** (c. 185 mi/298 km), E distributary of the Irrawaddy. Pop. 1,586,442.
**Ran·noch** (rănʹək), **Loch.** Lake, 9.5 mi (15.3 km) long, of central Scotland in the Grampians.
**Rann of Kutch** *or* **Rann of Cutch** (rŭn; kŭch). Salt marsh, c. 9,000 sq mi (23,310 sq km), of W India & SE Pakistan between the Gulf of Kutch & the Indus delta.
**Ran·toul** (răn-to͞olʹ). Village of E Ill. NNE of Champaign. Pop. 20,161.
**Ra·pa** (räʹpə). Island of the S Pacific in S French Polynesia SSE of Tahiti.
**Ra·pal·lo** (rə-päʹlō). City of NW Italy on the Ligurian Sea. Pop. 22,272.
**Rap·i·dan** (răpʹĭ-dănʹ). River, c. 90 mi (145 km), of N Va.
**Rap·id City** (răpʹĭd). City of SW S.Dak. WSW of Pierre. Pop. 46,492.
**Rap·pa·han·nock** (răpʹə-hănʹək). River, c. 212 mi (341 km), of NE Va.
**Rap·ti** (räpʹtē). River, c. 400 mi (644 km), of Nepal & N India.
**Rar·i·tan Bay** (rărʹĭ-tən). W arm of Lower New York Bay off SE N.Y. & NE N.J.
**Rar·o·ton·ga** (rărʹə-tŏngʹgə). Volcanic island of the S Pacific in the SW Cook Is.
**Ras Da·shan** (räs də-shänʹ). Highest peak, 15,158 ft (4,623.2 m) of Ethiopia, in the N part.

**Rasht** (răsht) *also* **Resht** (rĕsht). City of NW Iran near the Caspian Sea. Pop. 187,203.
**Rat** (răt). Islands of SW Alas. in the W Aleutians.
**Ra·tak** (rä'täk'). Islands of the W Pacific, E chain of the Marshalls.
**Ra·ti·bor** (rä'tē-bôr'). RACIBÓRZ.
**Rat·is·bon** (răt'ĭs-bŏn', -ĭz-). REGENSBURG.
**Ra·ton** (rə-tōn', -to͞on', ră-). Pass, 7,834 ft (2,389.4 m) high, in the Sangre de Cristo Mts. on the Colo.-N.Mex. border.
**Ra·ven·na** (rə-vĕn'ə). **1.** City of N central Italy near the Adriatic NE of Florence. Pop. 75,153. **2.** City of NE Ohio ENE of Akron. Pop. 11,987.
**Ra·vi** (rä'vē). River, 475 mi (764.3 km), of NW India & NE Pakistan.
**Ra·wal·pin·di** (rä'wəl-pĭn'dē, roul-). City of NE Pakistan NNW of Lahore. Pop. 615,392.
**Raw·lins** (rô'lĭnz). City of S Wyo. NE of Laramie. Pop. 11,547.
**Ray** (rā), **Cape.** Promontory of extreme SW Newf., Canada.
**Ray·side-Bal·four** (rā'sīd-băl'fôr', -fər). Town of S central Ont., Canada, near Sudbury. Pop. 16,035.
**Ray·town** (rā'toun'). City of W Mo. near Kansas City. Pop. 31,759.
**Read·ing** (rĕd'ĭng). **1.** Borough of S central England W of London. Pop. 131,200. **2.** Town of E Mass. NNW of Boston. Pop. 22,678. **3.** City of SW Ohio near Cincinnati. Pop. 12,879. **4.** City of SE Pa. NW of Philadelphia. Pop. 78,686.
**Re·ci·fe** (rə-sē'fə). City of NE Brazil on the Atlantic. Pop. 1,204,794.
**Reck·ling·hau·sen** (rĕk'lĭng-hou'zən). City of W West Germany SW of Münster. Pop. 122,437.
**Red** (rĕd). **1.** Sea, c. 1,450 mi (2,333 km) long, between NE Africa & Arabia. **2.** Lake, 451 sq mi (1,168 sq km), of N Minn. **3.** River of S China & N Vietnam flowing c. 730 mi (1,175 km) SE into the Gulf of Tonkin. **4.** River of S central U.S. rising in the Tex. Panhandle & flowing 1,018 mi (1,638 km) E & SE along the Tex.-Okla. border & through SW Ark. & N La. into the Mississippi. **5.** River of N central U.S. & S central Canada flowing c. 310 mi (499 km) N along the Minn.-N.Dak. border into Lake Winnipeg in SE Man., Canada.
**Red Bank** (băngk). **1.** Borough of E central N.J. SE of Perth Amboy. Pop. 12,031. **2.** City of S Tenn. near Chattanooga. Pop. 13,297.
**Red Deer** (dîr). **1.** River, c. 385 mi (620 km), of S Alta. & SW Sask., Canada. **2.** City of S central Alta., Canada, on the Red Deer R. Pop. 41,371.
**Red·ding** (rĕd'ĭng). City of N central Calif. S of Shasta Lake. Pop. 41,995.
**Red·ditch** (rĕd'ĭch). Urban district of central England S of Birmingham. Pop. 44,750.
**Red·lands** (rĕd'ləndz). City of S Calif. near San Bernardino. Pop. 43,619.
**Red·mond** (rĕd'mənd). City of W central Wash. near Seattle. Pop. 23,318.
**Re·don·do Beach** (rĭ-dŏn'dō). City of S Calif. S of Los Angeles. Pop. 57,102.
**Red Wing** (wĭng). City of SE Minn. on the Mississippi SW of St. Paul. Pop. 13,736.
**Red·wood City** (rĕd'wo͝od'). City of W Calif. NW of Palo Alto. Pop. 54,965.
**Redwood National Park.** Reserved area of redwood forests in NW Calif.
**Reed·ley** (rēd'lē). City of central Calif. SE of Fresno. Pop. 11,071.
**Reel·foot** (rēl'fo͝ot'). Lake, 20 mi (32 km) long, of NW Tenn.
**Re·gens·burg** (rā'gənz-bûrg', -bo͝org'). City of SW West Germany on the Danube NNE of Munich. Pop. 131,886.
**Reg·gio** (rĕj'ō, -ē-ō). **1.** *or* **Reggio di Ca·la·bri·a** (dē kə-lä'brē-ə) *also* **Reggio Calabria.** City of extreme S Italy on the Strait of Messina opposite Sicily. Pop. 110,291. **2.** *or* **Reggio nell'E·mi·lia** (nĕl'ə-mēl'yə) *also* **Reggio Emilia.** City of N central Italy WNW of Bologna. Pop. 102,337.
**Re·gi·na** (rĭ-jī'nə). Cap. of Sask., Canada, in the S part. Pop. 149,593.
**Rei·chen·bach** (rī'kən-bäk'). Waterfall, 656 ft (201 m), of S central Switzerland.
**Reids·ville** (rēdz'vĭl', -vəl). City of N N.C. NNE of Greensboro. Pop. 12,492.
**Rei·gate** (rī'gĭt). Borough of S England S of London. Pop. 55,600.
**Reims** *or* **Rheims** (rēmz, răNs). City of NE France ENE of Paris. Pop. 177,320.
**Rein·deer** (rān'dîr'). Lake, 2,467 sq mi (6,390 sq km), of NE Sask. & NW Man., Canada.
**Re·ma·gen** (rā'mä'gən). Town of W West Germany on the Rhine SE of Bonn. Pop. 14,627.
**Rem·scheid** (rĕm'shīt'). City of W West Germany NE of Cologne. Pop. 133,145.
**Rennes** (rĕn). City of NW France in Brittany N of Nantes. Pop. 194,094.
**Re·no** (rē'nō'). City of W Nev. near the Calif. border. Pop. 100,756.
**Rens·se·laer** (rĕn'sə-lîr', rĕn'sə-lər). City of E N.Y. on the Hudson opposite Albany. Pop. 9,047.
**Ren·ton** (rĕn'tən). City of W central Wash. near Seattle. Pop. 30,612.
**Re·pen·ti·gny** (rə-päN-tē-nyē'). Town of S Que., Canada, near Montreal. Pop. 26,698.
**Re·pub·li·can** (rĭ-pŭb'lĭ-kən). River, c. 420 mi (676 km), of E Colo., S Nebr., & N Kans.
**Resht** (rĕsht). *var. of* RASHT.
**Re·sis·ten·cia** (rĕs'ĭ-stĕn'sē-ə). City of NE Argentina on the Paraná R. Pop. 142,848.
**Re·şi·ta** (rĕsh'ĭt-sä'). City of W Rumania in the W Transylvanian Alps. Pop. 90,698.
**Re·thondes** (rə-tôNd'). Village of N France WNW of Reims; site of World War I armistice signing (1918).
**Ré·un·ion** (rē-yo͞on'yən). Island of France in the W Indian Ocean SW of Mauritius.
**Reus** (rē'o͞os). City of NE Spain on the Mediterranean. Pop. 47,240.
**Reut·ling·en** (roit'lĭng-ən). City of SW West Germany S of Stuttgart. Pop. 95,289.
**Re·vere** (rĭ-vîr'). City of E Mass. near Boston. Pop. 42,423.
**Re·vil·la Gi·ge·do** *or* **Re·vil·la·gi·ge·do** (rĭ-vē'ə-hĭ-hā'dō, -yə-). Islands of Mexico in the Pacific S of Lower California.
**Reyes** (rāz), **Point.** Cape on N central Calif. coast WNW of San Francisco.
**Rey·kja·vík** (rā'kyə-vēk', -vĭk'). Cap. of Iceland in the SW part on an inlet of Denmark Strait. Pop. 81,693.
**Rey·nolds·burg** (rĕn'əldz-bûrg'). City of central Ohio near Columbus. Pop. 20,661.
**Rey·no·sa** (rā-nō'sə). City of E Mexico on the Rio Grande. Pop. 181,646.
**Rhae·ti·a** *also* **Rae·ti·a** (rē'shē-ə, -shə). Ancient Roman province including present-day E Switzerland & W Austria. **—Rhae'tian** *adj.* & *n.*
**Rhaetian Alps** (ălps). Range of the central Alps in E Switzerland & W Austria.
**Rha·gae** (ră'jē). *var. of* RAGES.
**Rhe·gi·um** (rē'jē-əm). REGGIO DI CALABRIA.
**Rheims** (rēmz, răNs). *var. of* REIMS.
**Rhein** (rīn). RHINE.
**Rhin** (răN). RHINE.
**Rhine** (rīn). River of W Europe rising in E Switzerland & flowing c. 820 mi (1,319 km) N through W West Germany & the Netherlands to the North Sea.
**Rhine·land** (rīn'lănd', -lənd). Region along the Rhine in West Germany.
**Rhode Island** (rōd). **1.** Island of R.I. in Narragansett Bay. **2.** State of the NE U.S. on the Atlantic. Cap. Providence. Pop. 947,154. **—Rhode Is'land·er** *n.*
**Rhodes** (rōdz). **1.** Island of SE Greece in the Aegean off SW Turkey, largest of the Dodecanese. **2.** City of SE Greece on the N end of Rhodes Is. Pop. 32,092.
**Rho·de·sia** (rō-dē'zhə). **1.** Region of S central Africa comprising Zambia & Zimbabwe. **2.** ZIMBABWE. **—Rho·de'sian** *adj.* & *n.*
**Rhod·o·pe** (rŏd'ə-pē). Mountains of S Bulgaria & NE Greece.
**Rhon·dda** (rŏn'də, -*th*ə, hrŏn'-). Borough of S Wales NW of Cardiff. Pop. 85,400.
**Rhone** *or* **Rhône** (rōn). River of SW Switzerland & SE France flowing c. 505 mi (812 km) W & S to the Mediterranean.
**Rhyl** (rĭl). Urban district of N Wales on Liverpool Bay. Pop. 22,150.
**Ri·al·to** (rē-ăl'tō). **1.** Island of Venice, Italy. **2.** City of S Calif. near San Bernardino. Pop. 35,615.
**Ri·au** (rē'ou). Archipelago of W Indonesia off the SE end of the Malay Peninsula.
**Ri·a·zan** (rē'ə-zăn'). *var. of* RYAZAN.
**Ri·bei·rão Prê·to** (rē'bə-rouN' prā'to͞o). City of SE Brazil N of São Paulo. Pop. 190,897.
**Rich·ard·son** (rĭch'ərd-sən). City of NE Tex. near Dallas. Pop. 72,496.
**Ri·che·lieu** (rĭsh'ə-lo͞o'). River of S Que. flowing c. 210 mi (338 km) N from Lake Champlain to the St. Lawrence.
**Rich·field** (rĭch'fēld'). City of E Minn. near Minneapolis. Pop. 37,851.
**Rich·land** (rĭch'lənd). City of S Wash. on the Columbia ESE of Yakima. Pop. 33,578.
**Rich·mond** (rĭch'mənd). **1.** City of W Calif. NW of Oakland. Pop. 74,676. **2.** City of E Ind. E of Indianapolis. Pop. 41,349. **3.** City of central Ky. SSE of Lexington. Pop. 21,705. **4.** Borough of New York City coextensive with Staten Is. Pop. 352,121. **5.** Cap. of Va. in the E central part. Pop. 219,214.
**Richmond Heights. 1.** City of E Mo. W of St. Louis. Pop. 11,516. **2.** City of NE Ohio near Cleveland. Pop. 10,095.
**Richmond Hill.** City of S central Ont., Canada, N of Toronto. Pop. 34,716.
**Ri·deau** (rĭ-dō'). Canal, 126 mi (202.7 km) long, of S Ont., Canada, connecting the Ottawa R. at Ottawa with Lake Ontario at Kingston.
**Ridge·crest** (rĭj'krĕst'). City of S central Calif. ENE of Bakersfield. Pop. 15,929.

**Ridge·field** (rĭj'fēld'). **1.** Town of SW Conn. near the N.Y. border. Pop. 20,120. **2.** Borough of NE N.J. NNE of Jersey City. Pop. 10,294.
**Ridgefield Park.** Village of NE N.J. near Hackensack. Pop. 12,738.
**Ridge·wood** (rĭj'wo͝od'). Village of NE N.J. near Paterson. Pop. 25,208.
**Rif** *also* **Riff** (rĭf). Range of the Atlas Mts. in NE Morocco along the Mediterranean.
**Ri·ga** (rē'gə). **1. Gulf of.** Inlet of the Baltic off Latvia & Estonia. **2.** City of W European USSR in Latvia on the Gulf of Riga. Pop. 835,000.
**Ri·je·ka** (rē-yĕk'ə). City of NW Yugoslavia on the Adriatic. Pop. 128,883.
**Rijn** (rīn). RHINE.
**Rijs·wijk** (rīs'vīk) *also* **Rys·wick** (rĭz'wĭk). City of W Netherlands near The Hague. Pop. 54,123.
**Ri·mi·ni** (rĭm'ə-nē). City of N central Italy on the Adriatic. Pop. 101,579.
**Ri·mous·ki** (rĭ-mo͞o'skē). City of S Que., Canada, on the St. Lawrence NE of Quebec. Pop. 27,897.
**Ring·wood** (rĭng'wo͝od'). Borough of NE N.J. near the N.Y. border NNW of Paterson. Pop. 12,625.
**Ri·o·bam·ba** (rē'ō-bäm'bə). City of central Ecuador in the Andes S of Quito. Pop. 58,087.
**Ri·o Bran·co** (rē'ō brăng'kō). River, c. 350 mi (565 km), of NW Brazil.
**Río Bra·vo** (brä'vō). RIO GRANDE.
**Rio de Ja·nei·ro** (dā zhə-nâr'ō, dē-). City of SE Brazil on Guanabara Bay. Pop. 5,093,496.
**Río de la Pla·ta** (də lə plä'tə). Estuary of the Paraná & Uruguay rivers on the SE coast of South America extending c. 225 mi (362 km) between Argentina & Uruguay.
**Río de O·ro** (dē ôr'ō, ōr'ō). S part of Western Sahara, NW Africa.
**Río Ga·lle·gos** (gä-yā'gəs, gī-ā'-). Town of extreme S Argentina on the Atlantic N of Strait of Magellan. Pop. 27,833.
**Rio Grande** (grănd', grăn'dē). River, c. 1,885 mi (3,033 km), of SW U.S. rising in S Colo. & flowing SE to the Gulf of Mexico, forming much of the U.S.-Mexican border.
**Rio Gran·de** (grăn'də). **1.** River c. 650 mi (1,046 km), of S Brazil. **2.** City of extreme SE Brazil at the S entrance of Lagoa dos Patos. Pop. 98,863.
**Río Mu·ni** (mo͞o'nē). Mainland part of Equatorial Guinea, W Africa.
**Río Ne·gro** (nā'grō, nĕg'rō). **1.** River of E Colombia & NW Brazil flowing c. 1,400 mi (2,253 km) ESE into the Amazon. **2.** (rē'ô nĕ'grô). River of central Argentina flowing c. 400 mi (644 km) to the Atlantic. **3.** River of S Brazil & central Uruguay flowing c. 500 mi (805 km) WSW to the Uruguay R.
**Rio Roo·se·velt** (rō'zə-vĕlt, rōz'vĕlt). River, c. 400 mi (644 km), of W Brazil.
**Riv·er·dale** (rĭv'ər-dāl'). Village of NE Ill. near Chicago. Pop. 13,233.
**Riv·er Edge** (rĭv'ər). Borough of NE N.J. near Hackensack. Pop. 11,111.
**River Forest.** Village of NE Ill. near Chicago. Pop. 12,392.
**River Grove.** Village of NE Ill. near Chicago. Pop. 10,368.
**River Rouge** (ro͞ozh). City of SE Mich. near Detroit. Pop. 12,912.
**Riv·er·side** (rĭv'ər-sīd'). City of S Calif. NE of Santa Ana. Pop. 170,876.
**Riv·er·view** (rĭv'ər-vyo͞o'). **1.** Town of N.B., Canada, near Moncton. Pop. 14,177. **2.** City of SE Mich. SSW of Detroit. Pop. 14,567.
**Riv·i·er·a** (rĭv'ē-ĕr'ə). Coastal resort area of SE France & NW Italy along the Mediterranean.
**Riviera Beach.** City of SE Fla. on the Atlantic N of West Palm Beach. Pop. 26,596.
**Ri·vière-du-Loup** (rē-vē-ĕr'də-lo͞o'). City of E Que., Canada, on the S shore of the St. Lawrence NE of Quebec city. Pop. 13,103.
**Ri·yadh** (rē-äd', -yäd'). Cap. of Saudi Arabia in the central part. Pop. 666,840.
**Ri·za·i·yeh** (rĭ-zī'ə, -yə). **1.** URMIA. **2.** City of NW Iran WSW of Tabriz. Pop. 163,991.
**Ri·zal** (rĭ-zäl', -säl'). City of central Luzon, Philippines, on Manila Bay S of Manila. Pop. 286,497.
**Road Town** (rōd). Cap. of the British Virgin Is. on Tortola Is. Pop. 2,183.
**Ro·anne** (rō-ăn'). City of central France WNW of Lyon. Pop. 54,999.
**Ro·a·noke** (rō'ə-nōk'). **1.** River, c. 410 mi (660 km), of S Va. & NE N.C. **2.** Island of NE N.C., off the Atlantic coast between Albemarle & Pamlico sounds. **3.** Independent city of SW Va. WSW of Richmond. Pop. 100,427.
**Roanoke Rapids.** City of NE N.C. near the Va. border NE of Raleigh. Pop. 14,702.
**Rob·bins·dale** (rŏb'ĭnz-dāl'). City of E Minn. near Minneapolis. Pop. 14,422.
**Rob·erts** (rŏb'ərts), **Point.** Cape of NW Wash. extending S into the Strait of Georgia from B.C., Canada.
**Rob·er·val** (rŏb'ər-văl'). City of N Que., Canada, on the W shore of Lake St. John. Pop. 8,543.
**Rob·son** (rŏb'sən). Mountain, 12,972 ft (3,956.5 m), of E B.C., Canada; highest elevation of the Canadian Rockies.
**Robs·town** (rŏbz'toun'). City of S Tex. near Corpus Christi. Pop. 12,100.
**Ro·ca** (rō'kə). Cape of W Portugal W of Lisbon; W extremity of Europe.
**Roch·dale** (rŏch'dāl'). Borough of NW England NNE of Manchester. Pop. 93,780.
**Roche·fort** (rôsh-fôr') *or* **Roche·fort-sur-Mer** (-so͞or-mĕr'). City of W central France near the Bay of Biscay NNW of Bordeaux. Pop. 27,264.
**Roch·es·ter** (rŏch'ĭ-stər, -ĕs'tər). **1.** Borough of SE England ESE of London. Pop. 56,030. **2.** City of SE Minn. SE of St. Paul. Pop. 57,855. **3.** City of SE N.H. near Dover. Pop. 21,560. **4.** City of W N.Y. ENE of Buffalo. Pop. 241,741.
**Rock** (rŏk). River, c. 285 mi (459 km), of S Wis. & N Ill.
**Rock Creek Butte.** Mountain, 9,097 ft (2,774.6 m), of NE Ore. in the Blue Mts.
**Rock Falls.** City of NW Ill. on the Rock R. opposite Sterling. Pop. 10,624.
**Rock·ford** (rŏk'fərd). City of N Ill. WNW of Chicago. Pop. 139,712.
**Rock·hamp·ton** (rŏk-hămp'tən). City of E Australia near the Pacific NNW of Brisbane. Pop. 50,132.
**Rock Hill.** City of N S.C. near the border SSW of Charlotte, N.C. Pop. 35,344.
**Rock·ies** (rŏk'ēz). ROCKY Mts.
**Rock Island.** City of NW Ill. adjacent to Moline. Pop. 47,036.
**Rock·land** (rŏk'lənd). Town of E Mass. SSE of Boston. Pop. 15,695.
**Rock·ledge** (rŏk'lĕj', -lĭj). City of E central Fla. near Cape Canaveral. Pop. 11,877.
**Rock Springs.** City of SW Wyo. NE of Salt Lake City, Utah. Pop. 19,458.
**Rock·ville** (rŏk'vĭl', -vəl). City of central Md. NNW of Washington, D.C. Pop. 43,811.
**Rockville Cen·tre** (sĕn'tər). Village of SE N.Y. on SW Long Is. Pop. 25,405.
**Rock·y** (rŏk'ē). Mountain system of N North America extending from N Mexico to NW Alas.
**Rocky Hill.** Town of central Conn. near Hartford. Pop. 14,559.
**Rocky Mount.** City of E central N.C. ENE of Raleigh. Pop. 41,283.
**Rocky Mountain National Park.** Resort area in the Rockies of N Colo.
**Rocky River.** City of NE Ohio near Cleveland. Pop. 21,084.
**Ro·dhos** (rô'*th*ôs'). RHODES.
**Ro·dri·guez** *or* **Ro·dri·gues** (rō-drē'gəs). One of the Mascarene Is. in the W Indian Ocean E of Mauritius.
**Roe·se·la·re** (ro͞o'sə-lär'ə). City of W Belgium WSW of Ghent. Pop. 40,428.
**Rog·ers** (rŏj'ərz). **1. Mount.** Highest peak of Va., 5,729 ft (1,747.3 m), in the SW part. **2.** City of NW Ark. N of Fayetteville. Pop. 17,351.
**Rogue** (rōg). River, c. 200 mi (322 km), of SW Ore.
**Rohn·ert Park** (rō'nərt). City of W Calif. near Santa Rosa. Pop. 22,965.
**Rol·la** (rŏl'ə). City of central Mo. SE of Jefferson City. Pop. 13,303.
**Roll·ing Meadows** (rō'lĭng). City of NE Ill. near Chicago. Pop. 20,167.
**Ro·ma** (rō'mä, -mə). ROME, Italy.
**Ro·ma·gna** (rō-män'yə). Historical region of N central Italy.
**Ro·man** (rō'män'). City of NE Rumania NNE of Bucharest. Pop. 56,466.
**Ro·manche Deep** (rō-mänsh'). Atlantic ocean depth, c. 25,794 ft (7,864 m), at the equator.
**Ro·man Empire** (rō'mən). Empire (27 B.C.–A.D. 395) stretching from Britain to North Africa to the Persian Gulf.
**Ro·ma·ni·a** (rō-mā'nē-ə, -mān'yə). *var. of* RUMANIA.
**Rom·blon** (rŏm-blōn'). **1.** Islands of the central Philippines in the N Visayan Is. in the Sibuyan Sea. **2.** Island of this group, W of Sibuyan Is.
**Rome** (rōm). **1.** ROMAN EMPIRE. **2.** Cap. of Italy in the W central part on the Tiber R. Pop. 2,535,018. **3.** City of NW Ga. NW of Atlanta. Pop. 29,654. **4.** City of central N.Y. WNW of Utica. Pop. 43,826.
**Ro·me·o·ville** (rō'mē-ō-vĭl'). Village of NE Ill. N of Joliet. Pop. 15,519.
**Rom·u·lus** (rŏm'yə-ləs). City of SE Mich. SW of Detroit. Pop. 24,857.
**Ron·ces·valles** (rŏn'səs-vä'yās, -văl'-). Pass, 3,468 ft (1,057.7 m) high, through the W Pyrenees.
**Ronce·vaux** (rōNs-vō'). RONCESVALLES.
**Ron·kon·ko·ma** (rŏng-kŏng'kə-mə, rŏn-kŏn'-). Town of SE N.Y. on central Long Is. Pop. 20,200.

**Ron·ne Ice Shelf** (rō'nə, rŏn'ə). Area of shelf ice, c. 350 mi (563 km) in diameter, in W Antarctica S of the Weddell Sea.
**Roo·de·poort-Ma·rais·burg** (ro͞o'də-po͝ort'mə-rāz'bûrg'). City of NE South Africa W of Johannesburg. Pop. 115,366.
**Ro·rai·ma** (rō-rī'mə). Mountain, 9,094 ft (2,773.7 m), at the junction of the boundaries of Brazil, Guyana, & Venezuela.
**Ro·sa** (rō'zə), **Monte**. Highest (15,203 ft/4,636.9 m) mountain in the Pennine Alps on the Swiss-Italian border.
**Ro·sa·ri·o** (rō-zär'ē-ō', -sär'-). City of E central Argentina on the Paraná NW of Buenos Aires. Pop. 806,942.
**Rose** (rōz). Mountain, 10,778 ft (3,287.3 m), of NW Nev.
**Ro·seau** (rō-zō'). Cap. of Dominica in the Windward Is. of the West Indies, on the SW coast. Pop. 9,968.
**Rose·burg** (rōz'bûrg'). City of SW Ore. SSW of Eugene. Pop. 16,644.
**Ro·selle** (rō-zĕl'). **1.** Village of NE Ill. WNW of Chicago. Pop. 16,948. **2.** Borough of NE N.J. near Elizabeth. Pop. 20,641.
**Roselle Park.** Borough of NE N.J. near Elizabeth. Pop. 13,377.
**Rose·mead** (rōz'mēd'). City of S Calif. near Los Angeles. Pop. 42,604.
**Ro·sen·berg** (rō'zən-bûrg'). City of SE Tex. SW of Houston. Pop. 17,995.
**Ro·set·ta** (rō-zĕt'ə). City of N Egypt on the Nile delta. Pop. 42,962.
**Rose·ville** (rōz'vĭl'). **1.** City of N central Calif. NE of Sacramento. Pop. 24,347. **2.** City of SE Mich. near Detroit. Pop. 54,311. **3.** City of SE Minn. near St. Paul. Pop. 35,820.
**Ros·kil·de** (rŭs'kĭl'ə). City of E Denmark W of Copenhagen. Pop. 44,248.
**Ross** (rôs, rŏs). **1.** Sea of the Pacific in Antarctica S of New Zealand. **2.** Island of Antarctica in the W Ross Sea. **3. Ice Shelf.** Area of shelf ice, c. 400 mi (644 km) in diameter, in Antarctica S of the Ross Sea.
**Ross·bach** (rôs'bäk'). Village of S central East Germany where Frederick II of Prussia defeated the French (1757).
**Ros·tock** (rŏs'tŏk', rôs'tôk'). City of N East Germany near the Baltic. Pop. 210,167.
**Ros·tov** (rə-stôf', -stôv') *also* **Ros·tov-on-Don** (-ŏn-dôn', -dŏn'). City of S European USSR on the Don near the Sea of Azov. Pop. 934,000.
**Ros·well** (rŏz'wĕl', -wəl). **1.** City of NW Ga. near Atlanta. Pop. 23,337. **2.** City of SE N.Mex. SE of Albuquerque. Pop. 39,676.
**Ro·ta** (rō'tə). Island of the W Pacific in the S Marianas N of Guam.
**Roth·er·ham** (rŏ*th*'ər-əm). Borough of N England near Sheffield. Pop. 84,770.
**Ro·to·ru·a** (rō'tə-ro͞o'ə). City of central North Is. New Zealand. Pop. 37,229.
**Rot·ter·dam** (rŏt'ər-dăm'). City of SW Netherlands on the Rhine-Meuse delta SSE of The Hague. Pop. 614,767.
**Ro·tu·ma** (rō-to͞o'mə). Volcanic island in the SW Pacific N of Fiji.
**Rou·baix** (ro͞o-bĕ'). City of N France near Lille. Pop. 109,473.
**Rou·en** (ro͞o-äɴ', -än'). City of N France on the Seine NW of Paris. Pop. 113,536.
**Rou·ma·ni·a** (ro͞o-mā'nē-ə, -mān'yə). *var. of* RUMANIA.
**Round Lake Beach** (round). Village of NE Ill. W of Waukegan. Pop. 12,921.
**Round Rock.** City of central Tex. N of Austin. Pop. 11,812.
**Rous·sil·lon** (ro͞o-sē-yôɴ'). Region & former province of S France bordering on Spain & the Mediterranean.
**Rou·yn** (ro͞o'ĭn, rwăɴ). City of SW Que., Canada, near the Ont. border. Pop. 17,678.
**Rov·no** (rôv'nə). City of W European USSR in the Ukraine NE of Lvov. Pop. 179,000.
**Ro·vu·ma** (ro͞o-vo͞o'mə). *var. of* RUVUMA.
**Roy** (roi). City of N Utah near Ogden. Pop. 19,694.
**Roy·al Gorge** (roi'əl). Canyon formed by the Arkansas R. in S central Colo.
**Royal Oak** (ōk). City of SE Mich. near Detroit. Pop. 70,893.
**Royal Tun·bridge Wells** (tŭn'brĭj wĕlz'). Borough of SE England SW of Maidstone. Pop. 44,800.
**Ru·an·da** (ro͞o-än'də). *var. of* RWANDA.
**Ru·a·pe·hu** (ro͞o'ə-pä'ho͞o). Volcanic peak, 9,175 ft (2,798.4 m), in central North Is., New Zealand.
**Rub al Kha·li** (ro͞ob' ăl kä'lē). Desert in the SE interior of the Arabian Peninsula.
**Ru·bi·con** (ro͞o'bĭ-kŏn'). River, 15 mi (24 km), of N central Italy flowing NE to the Adriatic.
**Ru·dolf** (ro͞o'dŏlf'). Lake, c. 2,500 sq mi (6,475 sq km), of NW Kenya.
**Ru·eil-Mal·mai·son** (ro͞o-ā'măl-mā-zôɴ'). Town of N central France near Paris. Pop. 62,504.
**Ru·fi·ji** (ro͞o-fē'jē). River, c. 375 mi (603 km), of central Tanzania.
**Ru·fisque** (ro͞o-fēsk'). City of W Senegal on the Atlantic E of Dakar. Pop. 54,000.
**Rug·by** (rŭg'bē). Borough of central England ESE of Coventry. Pop. 60,380.
**Rü·gen** (ro͞o'gən). Island, 358 sq mi (927 sq km), of N East Germany, in the Baltic Sea.
**Ruhr** (ro͝or). **1.** River of NW West Germany flowing 145 mi (233.3 km) W to the Rhine. **2.** Industrial region along & N of the Ruhr R.
**Ru·ma·ni·a** (ro͞o-mā'nē-ə, -mān'yə) *also* **Ro·ma·ni·a** (rō-) *or* **Rou·ma·ni·a** (ro͞o-). Country of SE Europe with a short Black Sea coastline. Cap. Bucharest. Pop. 22,048,305. **—Ru·ma'ni·an** *adj. & n.*
**Run·ny·mede** (rŭn'ē-mēd'). Meadow in SE England on the Thames W of London; site of King John's acceptance of the Magna Carta (1215).
**Ru·pert** (ro͞o'pərt). River, c. 380 mi (611 km), of W central Que., Canada.
**Ru·se** (ro͞o'sā). City of NE Bulgaria on the Danube. Pop. 160,351.
**Rush·more** (rŭsh'môr', -mōr'). Mountain, 6,200 ft (1,891 m), in the Black Hills of W S.Dak.; a national memorial with massive carved likenesses of Washington, Jefferson, Lincoln, & Theodore Roosevelt.
**Rüs·sels·heim** (ro͞os'əls-hīm'). City of central West Germany on the Main R. SW of Frankfurt. Pop. 62,067.
**Rus·sia** (rŭsh'ə). **1.** Former empire of E Europe & N Asia superseded by the USSR in 1917. **2.** UNION OF SOVIET SOCIALIST REPUBLICS. **3.** *also* **Rus·sian Soviet Federated Socialist Republic** (-ən). Constituent republic in European, Central Asian, Siberian, & Far Eastern USSR, constituting 75% of the nation's total area & extending from the Baltic Sea to the Pacific Ocean. **—Rus'sian** *adj. & n.*
**Rus·ta·vi** (ro͞o-stä'vē). City of S European USSR in Georgia SE of Tbilisi. Pop. 129,000.
**Rus·ton** (rŭs'tən). City of N La. W of Monroe. Pop. 20,585.
**Ruth·er·ford** (ro͞o*th*'ər-fərd, ro͞oth'-). Borough of NE N.J. near Clifton. Pop. 19,068.
**Ru·the·nia** (ro͞o-thēn'yə, -thē'nē-ə). Region of W European USSR in W Ukraine S of the Carpathians. **—Ru·the'ni·an** *adj. & n.*
**Rut·land** (rŭt'lənd). City of W central Vt. N of Bennington. Pop. 18,436.
**Ru·vu·ma** *also* **Ro·vu·ma** (ro͞o-vo͞o'mə). River of SE Africa rising in N Mozambique & flowing c. 450 mi (724 km) E along the Mozambique-Tanzania border to the Indian Ocean.
**Ru·wen·zo·ri** (ro͞o'ən-zôr'ē, -zōr'ē, -wən-). Mountain range of E central Africa on the Uganda-Zaire border.
**Rwan·da** *also* **Ru·an·da** (ro͞o-än'də). Country of E central Africa S of Uganda. Cap. Kigali. Pop. 4,819,317. **—Rwan'dan** *adj. & n.*
**Rya·zan** (ryä-zän') *also* **Ri·a·zan** (rē-ə-zän'). City of central European USSR SE of Moscow. Pop. 453,000.
**Ry·binsk** (rĭb'ənsk). **1.** Reservoir, c. 2,000 sq mi (5,180 sq km), in N central European USSR on the Volga. **2.** City of N central European USSR on the Volga NNE of Moscow. Pop. 239,000.
**Ryb·nik** (rĭb'nĭk). Town of S Poland WSW of Katowice. Pop. 43,415.
**Rye** (rī). City of SE N.Y. on Long Is. Sound near New York City. Pop. 15,083.
**Rys·wick** (rĭz'wĭk). *var. of* RIJSWIJK.
**Ryu·kyu** (rē-o͞o'ko͞o', -kyo͞o', -yo͞o'-). Islands of SW Japan extending c. 650 mi (1,046 km) between Kyushu, Japan, & Taiwan.
**Rze·szów** (zhĕsh'o͞of'). City of SE Poland E of Kraków. Pop. 82,192.

# S

**Saa·le** (zä'lə, sä'-). River, c. 265 mi (426 km), of E West Germany & SW East Germany.
**Saar** (sär, zär). **1.** River, c. 150 mi (241 km), of NE France & W West Germany. **2.** SAARLAND.
**Saar·brück·en** (zär-bro͝ok'ən, sär-). City of W West Germany on the Saar R. near the French border. Pop. 205,336.
**Saa·re·maa** (sär'ə-mä'). *var. of* SAREMA.
**Saar·land** (sär'lănd', zär'-). Region of W West Germany in the Saar valley, historically contested with France.
**Sa·ba** (sä'bə). Island of the N Netherlands Antilles between St. Martin & St. Eustatius.
**Sa·ba·dell** (sä'bə-bĕl'). City of NE Spain NW of Barcelona. Pop. 148,223.
**Sa·bah** (sä'bä'). Region & state of Malaysia in NE Borneo.
**Sa·ba·lan** (sä'bə-län') *or* **Sa·va·lan** (-və-). Volcanic cone, 15,784 ft (4,814.1 m), of NW Iran.
**Sa·bar·ma·ti** (sä'bər-mŭt'ē). River, c. 220 mi (322 km), of W India.
**Sa·bi** (sä'bē). River, c. 400 mi (644 km), of E Zimbabwe & S Mozambique.
**Sa·bine** (sə-bēn'). River of E Tex. rising NE of Dallas & flowing c. 575 mi (925 km) E & S to the Gulf of Mexico, in its lower course forming the Tex.-La. border & crossing **Sabine Lake** (c. 17 mi/27.4 km long).
**Sa·ble** (sā'bəl). **1. Cape.** Promontory of extreme S N.S., Canada. **2. Cape.** Cape of S Fla.; southernmost extremity of the U.S. mainland. **3.** Island off N.S., Canada, ESE of Halifax.
**Sab·ra·tha** (săb'rə-thə) *also* **Sab·ra·ta** (-tə). Ancient town of Roman Africa in modern Libya W of Tripoli.

**Sa·co** (sô'kō). **1.** River, c. 105 mi (169 km), of E N.H. & SW Me. **2.** City of SW Me. SW of Portland. Pop. 12,921.

**Sac·ra·men·to** (săk'rə-mĕn'tō). **1.** River, c. 380 mi (611 km), of N Calif. **2.** Mountains of S N.Mex. **3.** Cap. of Calif. in the N central part. Pop. 275,741.

**Sa·fa·qis** (sə-fä'kĭs). SFAX.

**Sa·fed Koh** (sə-fĕd' kō'). Mountain range on the Pakistan-Afghanistan border SE of Kabul, rising to 15,620 ft (4,764.1 km).

**Sa·fi** (săf'ē). City of W Morocco on the Atlantic WNW of Marrakesh. Pop. 129,113.

**Sa·fid Rud** (sä-fēd' ro͞od') *also* **Se·fid Rud** (sĕ-). River, c. 450 mi (724 km), of NW Iran.

**Sa·ga** (sä'gä'). City of W Kyushu, Japan, on an inlet of the East China Sea. Pop. 152,258.

**Sa·ga·mi Sea** (sə-gä'mē). Bay of the W Pacific on the SE coast of Honshu, Japan, SW of Tokyo Bay.

**Sa·ga·mi·ha·ra** (sə-gä'mē-här'ə). City of central Honshu, Japan, near Tokyo. Pop. 377,398.

**Sa·gi·naw** (săg'ə-nô'). **1.** Bay of Lake Huron extending into E Mich. **2.** City of E central Mich., NNW of Flint. Pop. 77,508.

**Sa·gua·ro National Monument** (sə-gwär'ō, -wär'ō). Area of saguaro & other desert growth in SE Ariz.

**Sag·ue·nay** (săg'ə-nā'). River, c. 125 mi (201 km), of S Que., Canada.

**Sa·gun·to** (sə-go͞on'tō). City of E Spain NNE of Valencia. Pop. 17,052.

**Sa·har·a** (sə-hâr'ə, -här'ə). Vast desert of N Africa extending from the Atlantic coast to the Nile valley & from the Atlas Mts. S to the Sudan.

**Sa·ha·ran·pur** (sə-här'ən-po͝or'). City of N central India NNE of Delhi. Pop. 225,396.

**Sai·gon** (sī-gŏn'). HO CHI MINH CITY.

**Sai·maa** (sī'mä'). Lake, c. 500 sq mi (1,295 sq km), of SE Finland, largest of the **Saimaa Lakes,** a group of c. 120 connected lakes.

**Saint Al·bans** (sānt ôl'bənz). **1.** Borough of SE England near London. Pop. 123,800. **2.** City of W W.Va. W of Charleston. Pop. 12,402.

**Saint Al·bert** (ăl'bərt). City of central Alta., Canada, NW of Edmonton. Pop. 24,129.

**Saint Ann** (ăn). City of E Mo. near St. Louis. Pop. 15,523.

**Saint Au·gus·tine** (ô'gə-stēn'). City of NE Fla. on the Atlantic SSE of Jacksonville. Pop. 11,985.

**Saint Aus·tell with Fow·ey** (ô'stəl; fō'ē, foi). Borough of SW England on the English Channel W of Plymouth. Pop. 32,710.

**Saint Bar·thol·o·mew** (bär-thŏl'ə-myo͞o') *or* **Saint-Bar·thél·e·my** (săN-bär-tāl-mē'). Island of the French West Indies NNW of Guadeloupe.

**Saint Bru·no de Mon·tar·ville** (bro͞o'nō' də mŏn'tər-vĭl'). Town of S Que., Canada, near Montreal. Pop. 21,272.

**Saint Cath·a·rines** (kăth'ə-rĭnz', kăth'rĭnz). City of S Ont., Canada, on the Welland Ship Canal. Pop. 123,351.

**Saint Charles** (chärlz). **1.** City of NE Ill. W of Chicago. Pop. 17,492. **2.** City of E Mo. on the Missouri near St. Louis. Pop. 37,379.

**Saint Chris·to·pher** (krĭs'tə-fər). Island, 68 sq mi (176 sq km), of the British West Indies WNW of Antigua; part of St. Kitts-Nevis.

**Saint Clair** (klâr). Lake, c. 490 sq mi (1,269 sq km), between SW Ont., Canada, & SE Mich., connected by the **Saint Clair R.** (40 mi/ 64.4 km) with Lake Huron.

**Saint Clair Shores.** City of SE Mich. near Detroit. Pop. 76,210.

**Saint-Cloud** (săN-klo͞o'). Town of N central France near Paris. Pop. 28,052.

**Saint Cloud** (kloud). City of central Minn. NW of Minneapolis. Pop. 42,566.

**Saint Croix** (kroi). **1.** River, c. 164 mi (264 km), of NW Wis. & E Minn. **2.** Largest (81 sq mi/210 sq km) of the U.S. Virgin Is. in the West Indies.

**Saint-De·nis** (săN-də-nē'). **1.** City of N central France near Paris. Pop. 95,808. **2.** Cap. of Réunion on the Indian Ocean. Pop. 80,075.

**Sainte Foy** (săNt fwä'). City of S Que., Canada, near Quebec city. Pop. 71,237.

**Saint E·li·as** (ĭ-lī'əs). **1.** Mountains of the Coast Ranges in SW Y.T., Canada, & E Alas. **2. Mount.** Peak, 18,008 ft (5,492.4 m), in the St. Elias Mts. on the Alas.-Y.T. border.

**Sainte Thé·rèse** (săNt tā-rĕz', sānt' tə-rēz'). City of S Que., Canada, on the St. Lawrence NW of Montreal. Pop. 17,479.

**Saint-É·tienne** (săN-tā-tyĕn'). City of SE France SW of Lyon. Pop. 218,289.

**Saint Eu·stache** (săN tœ-stăsh'). Town of S Que., Canada, W of Montreal. Pop. 21,248.

**Saint Eu·sta·ti·us** (yo͞o-stā'shəs, -shē-əs). Island of the Netherlands Antilles in the Leeward Is. NW of St. Christopher.

**Saint Fran·cis** (frăn'sĭs). **1. Lake.** Expansion of the St. Lawrence R. in SE Ont. & S Que., Canada, SW of Montreal. **2.** River, c. 470 mi (755 km), of SE Mo. & E Ark. **3.** City of SE Wis. near Milwaukee. Pop. 10,066.

**Saint Fran·çois** (săN frăN-swä', sānt' frăn-swä'). River, 165 mi (265.5 km), of S Que., Canada.

**Saint Gall** (gôl', găl'). SANKT GALLEN.

**Saint George** (jôrj). City of SW Utah near the Ariz. border. Pop. 11,350.

**Saint George's** (jôr'jəz). **1. Channel.** Strait, c. 100 mi (161 km) long & 50–95 mi (80–153 km) wide, between W Wales & SE Ireland. **2.** or **Saint George** (jôrj). Cap. of Grenada in the West Indies. Pop. 6,313.

**Saint-Gilles** (săN-zhēl'). City of central Belgium near Brussels. Pop. 55,055.

**Saint Gott·hard** (gŏt'ərd). Mountains of the Lepontine Alps in central Switzerland, crossed by **Saint Gotthard Pass** (6,935 ft/ 2,115.2 m).

**Saint He·le·na** (hə-lē'nə). Island in the S Atlantic W of Angola, a British colony. Cap. Jamestown. Pop. 5,147.

**Saint Hel·ens** (hĕl'ənz) **1. Mount.** Active volcanic peak of the Cascade Range in SW Wash. **2.** Borough of NW England ENE of Liverpool. Pop. 104,890.

**Saint-Hu·bert** (săN-yo͞o-bĕr', sānt'hyo͞o'bərt). Town of S Que., Canada, near Montreal. Pop. 49,706.

**Saint-Hy·a·cinthe** (sānt'hī'ə-sĭnth, săN-tyä-săNt'). City of S Que., Canada, ENE of Montreal. Pop. 37,500.

**Saint-Jean** (săN-zhäN'). City of S Que., Canada, on the Richelieu R. SE of Montreal. Pop. 34,363.

**Saint-Jé·rôme** (săN-zhā-rōm', sānt'jə-rōm'). City of S Que., Canada, NW of Montreal. Pop. 25,175.

**Saint John** (jŏn). **1.** Lake, c. 375 sq mi (971 sq km), of S central Que., Canada. **2.** River, c. 418 mi (673 km), of N Me. & W N.B., Canada. **3.** One of the U.S. Virgin Is. in the West Indies E of St. Thomas. **4.** City of S N.B., Canada, at the mouth of the St. John R. on the Bay of Fundy. Pop. 85,956.

**Saint Johns** (jŏnz). **1.** River, 285 mi (458.6 km), of NE Fla. **2.** SAINT-JEAN.

**Saint John's** (jŏnz). **1.** Cap. of Antigua in the British West Indies, on the N coast. Pop. 21,814. **2.** Cap. of Newf., Canada, on the SE coast of the island. Pop. 86,576.

**Saint Jo·seph** (jō'zəf, -səf). **1.** River, 210 mi (338 km), of SW Mich. & N Ind. **2.** City of NW Mo. on the Missouri R. NNW of Kansas City. Pop. 76,691.

**Saint Kitts** (kĭts). ST. CHRISTOPHER.

**Saint Kitts-Nev·is** (kĭts-nē'vĭs, -nĕv'ĭs). Island country of the British West Indies in the Leeward Is. Cap. Basseterre. Pop. 44,404.

**Saint Lam·bert** (lăm'bərt). City of S Que., Canada, on the St. Lawrence near Montreal. Pop. 20,318.

**Saint Lau·rent** (săN lə-räN', sānt' lô-rĕnt'). City of S Que., Canada, near Montreal. Pop. 64,404.

**Saint Law·rence** (lôr'əns, lŏr'-). **1. Gulf of.** Arm of the NW Atlantic off SE Canada between N.B. & Newf. **2.** River of SE Canada flowing 744 mi (1,197.1 km) NE from Lake Ontario along the Ont.-N.Y. border & through S Que. to the Gulf of St. Lawrence. **3.** Island off W Alas. S of the Bering Strait.

**Saint Lawrence Seaway. 1.** Canal & river route, 182 mi (292.8 km), along the St. Lawrence R. between Lake Ontario & Montreal. **2.** System of rivers, canals, & lakes from the Atlantic through the Great Lakes.

**Saint Lé·o·nard** (săN lā-ō-när', sānt' lĕn'ərd). City of S Que., Canada, N of Montreal. Pop. 78,452.

**Saint Lô** (săN lō', sānt' lō'). Town of NW France W of Caen. Pop. 21,670.

**Saint Lou·is** (lo͞o'ĭs). **1.** (*also* săN lo͞o-ē'). Lake, 57 sq mi (148 sq km), of S Que., Canada, SW of Montreal. **2.** River, c. 160 mi (257 km), of NE Minn. **3.** Independent city of E Mo. on the Mississippi just S of the influx of the Missouri R. Pop. 453,085.

**Saint-Lou·is** (săN-lo͞o-ē'). City of NW Senegal at the mouth of the Senegal R. Pop. 88,404.

**Saint Louis Park.** City of E Minn. near Minneapolis. Pop. 42,931.

**Saint Lu·cia** (lo͞o'shə, lo͞o-sē'ə). Island nation of the West Indies in the Windward Is. S of Martinique. Cap. Castries. Pop. 115,783.

**Saint-Ma·lo** (săN-mə-lō'). Town of NW France on the **Gulf of Saint-Malo,** an inlet of the English Channel. Pop. 43,277.

**Saint Mar·tin** (mär'tn). Island of the West Indies in the W Leeward Is.; divided between France & the Netherlands.

**Saint Mar·ys** (mâr'ēz). **1.** River, c. 175 mi (282 km), rising in SE Ga. & flowing on the Ga.-Fla. border to the Atlantic. **2.** River, 63 mi (101.4 km), flowing from Lake Superior to Lake Huron on the Mich.-Ont. border.

**Saint Mat·thews** (măth'yo͞oz). City of N Ky. near Louisville. Pop. 13,354.

**Saint-Maur-des-Fos·sés** (săN-môr-dā-fō-sā'). City of N central France near Paris. Pop. 80,797.

**Saint Mau·rice** (săN mô-rēs', sānt' môr'ĭs, mŏr'-). River, c. 325 mi (525 km), of S Que., Canada.

**Saint Mo·ritz** (mə-rĭts'). Resort city of SE Switzerland on the Inn R. Pop. 5,699.

**Saint-Na·zaire** (săN-nə-zâr′). City of W central France at the mouth of the Loire R. Pop. 65,228.
**Saint Paul** (pôl). Cap. of Minn. in the E adjacent to Minneapolis. Pop. 270,230.
**Saint Paul's Rocks** (pôlz). Islets of Brazil in the Atlantic NE of Natal.
**Saint Pe·ter** (pē′tər). Lake, 142 sq mi (228 sq km), of S Que., Canada, NE of Montreal.
**Saint Pe·ters** (pē′tərz). City of E Mo. WNW of St. Louis. Pop. 15,700.
**Saint Pe·ters·burg** (pē′tərz-bûrg′). **1.** LENINGRAD. **2.** City of W central Fla. on Tampa Bay. Pop. 236,893.
**Saint Pi·erre** or **Saint-Pi·erre** (sānt′ pîr′, pē-âr′, săN pyĕr′). **1.** Island of the St. Pierre and Miquelon group in the N Atlantic S of Newf., Canada. **2.** Cap. of St. Pierre and Miquelon on St. Pierre Is. Pop. 5,232.
**Saint Pierre and Mi·que·lon** (mĭk′ə-lŏn′, mē-klôN′). French island group, 93 sq mi (241 sq km), in the N Atlantic S of Newf., Canada. Cap. St. Pierre. Pop. 5,840.
**Saint-Quen·tin** (săN-kăN-tăN′, sānt′kwĕn′tən). City of N central France NNE of Paris. Pop. 69,956.
**Saint Si·mons** (sī′mənz). Island of SE Ga. in the Atlantic.
**Saint Tho·mas** (tŏm′əs). **1.** Island of the U.S. Virgin Is. in the West Indies. **2.** City of S Ont., Canada, S of London. Pop. 27,206. **3.** CHARLOTTE AMALIE.
**Saint-Tro·pez** (săN-trô-pā′). Resort town of SE France on the Mediterranean. Pop. 4,484.
**Saint Vin·cent** (vĭn′sənt). **1. Cape.** Promontory at the SW extremity of Portugal. **2.** Island of St. Vincent & the Grenadines in the central Windward Is. of the West Indies.
**Saint Vincent and the Gren·a·dines** (grĕn′ə-dēnz′). Island nation in the central Windward Is. of the West Indies, comprising St. Vincent Is. & the N Grenadines. Cap. Kingstown. Pop. 124,000.
**Sai·pan** (sī-păn′, -pän′, sī′păn). Island of the W Pacific in the S Marianas, part of the U.S. Trust Territory of the Pacific Islands. **—Sai′pa·nese′** (-nēz′, -nēs′) *adj.* & *n.*
**Sa·is** (sā′ĭs). Ancient Egyptian city on the Nile delta.
**Sa·ja·ma** (sə-hä′mə). Mountain, 21,390 ft (6,524 m), in W Bolivia near the Chilean border.
**Sa·kai** (sä′kī′). City of S Honshu, Japan, on Osaka Bay. Pop. 750,688.
**Sa·kar·ya** (sə-kär′yə). River, c. 490 mi (788 km), of NW Turkey.
**Sa·kha·lin** (săk′ə-lēn′, -lən, săk′ə-lēn′). Island of SE Far Eastern USSR in the Sea of Okhotsk N of Hokkaido, Japan.
**Sa·ki·shi·ma** (sä′kĭ-shē′mə). Islands of Japan in the S Ryuku Is. E of Taiwan.
**Sa·kon·net River** (sə-kŏn′ĭt). Inlet of the Atlantic extending into SE R.I.
**Sal·a·do** (sə-lä′dō). **1.** River of N Argentina flowing c. 1,250 mi (2,011 km) SE into the Paraná. **2.** River of W central Argentina flowing c. 750 mi (1,207 km) SSE into the Colorado.
**Sa·la·jar** or **Sa·la·yar** (sə-lä′yär′). Island of central Indonesia in the Flores Sea off SW Sulawesi.
**Sal·a·man·ca** (săl′ə-măng′kə, sä′lə-mäng′kə). **1.** City of central Mexico NE of Mexico City. Pop. 61,039. **2.** City of W central Spain WNW of Madrid. Pop. 125,132.
**Sal·a·mis** (săl′ə-mĭs). **1.** Island of Greece in the Saronic Gulf E of Athens, near which the Greeks defeated the Persians (480 B.C.). **2.** Ancient city of E Cyprus.
**Sa·la·yar** (sə-lä′yär′). *var. of* SALAJAR.
**Sal·can·tay** (säl′kən-tī′). Highest mountain, 20,500 ft (6,252.5 m), in the Cordillera Oriental in SE Peru.
**Sa·lé** (sä-lā′) *also* **Sla** (slä). City of NW Morocco on the Atlantic near Rabat. Pop. 155,557.
**Sa·lem** (sā′ləm). **1.** City of SE India SW of Madras. Pop. 308,716. **2.** City of NE Mass. NE of Boston. Pop. 38,220. **3.** Town of SE N.H. W of Haverhill, Mass. Pop. 24,124. **4.** City of NE Ohio SW of Youngstown. Pop. 12,869. **5.** Cap. of Ore. in the NW part. Pop. 89,233. **6.** Independent city of SW Va. near Roanoke. Pop. 23,958.
**Sa·ler·no** (sə-lûr′nō, -lĕr′-). City of S Italy on the **Gulf of Salerno,** an inlet of the Tyrrhenian Sea. Pop. 146,534.
**Sal·ford** (sôl′fərd). Borough of NW England near Manchester. Pop. 261,100.
**Sa·li·na** (sə-lī′nə). City of central Kans. NNW of Wichita. Pop. 41,843.
**Sa·li·nas** (sə-lē′nəs). **1.** River, c. 150 mi (241 km), of W central Calif. **2.** City of W Calif. near Monterey. Pop. 80,479.
**Salis·bur·y** (sôlz′bĕr′ē, -bə-rē). **1.** Borough of S central England NW of Southampton on the SE edge of **Salisbury Plain,** site of Stonehenge. Pop. 35,460. **2.** City of SE Md. on the Eastern Shore S of Dover, Del. Pop. 16,429. **3.** City of W central N.C. SSW of Winston-Salem. Pop. 22,677. **4.** Cap. of Zimbabwe in the NE part. Pop. 118,500.

**Sal·mon** (săm′ən). River of central Idaho rising in the **Salmon River Mts.** (highest elevation, 10,328 ft/3,150 m) & flowing c. 425 mi (684 km) to the Snake R.
**Sa·lo·ni·ka** (sə-lŏn′ĭ-kə, săl′ə-nē′kə). City of NE Greece on the **Gulf of Salonika,** an arm of the NW Aegean Sea. Pop. 345,799.
**Salt** (sôlt). **1.** River, c. 200 mi (322 km), of S Ariz. **2.** River, c. 200 mi (322 km), of NE Mo.
**Sal·ta** (säl′tə). City of NW Argentina N of Tucumán. Pop. 176,216.
**Sal·ti·llo** (säl-tē′ō, -yō). City of N Mexico N of Mexico City. Pop. 200,712.
**Salt Lake City.** Cap. of Utah in the N part near Great Salt Lake. Pop. 163,033.
**Sal·to** (säl′tō). City of NW Uruguay on the Uruguay R. Pop. 72,940.
**Sal·ton Sea** (sôl′tən). Lake, 370 sq mi (958.3 sq km), of SE Calif.
**Sa·lu·da** (sə-lo͞o′də). River, c. 200 mi (322 km), of W S.C.
**Sal·va·dor** (săl′və-dôr′). **1.** EL SALVADOR. **2.** (*also* säl′və-dôr′). City of E Brazil on the Atlantic. Pop. 1,501,219.
**Sal·ween** (săl′wēn′). River of SE Asia rising in E Tibet & flowing c. 1,750 mi (2,816 km) S through Burma into the Gulf of Martaban.
**Salz·burg** (sôlz′bûrg′, sälz′-, sälz′-). City of W central Austria near the West German border WSW of Linz. Pop. 122,100.
**Salz·git·ter** (zälts′gĭt′ər). City of NE West Germany SE of Hannover. Pop. 117,341.
**Sa·mar** (sä′mär′). Island of E central Philippines in the E Visayan group NE of Leyte.
**Sa·ma·ra** (sə-mär′ə). **1.** River, c. 360 mi (579 km), of SE European USSR. **2.** KUIBYSHEV.
**Sa·mar·i·a** (sə-măr′ē-ə, -mâr′-). **1.** Region of ancient Palestine in present-day NW Jordan. **2.** Ancient N kingdom of Israel. **3.** Ancient city of central Palestine in present-day NW Jordan. **—Sa·mar′i·tan** (-ĭ-tən) *adj.* & *n.*
**Sam·ar·kand** (săm′ər-kănd′). City of S Central Asian USSR SW of Tashkent. Pop. 477,000.
**Sam·ni·um** (săm′nē-əm). Ancient country of central & S Italy. **—Sam′nite** (săm′nīt′) *adj.* & *n.*
**Sa·mo·a** (sə-mō′ə). Island group of the S Pacific ENE of Fiji; divided between **American Samoa** & **Western Samoa. —Sa·mo′an** *adj.* & *n.*
**Sa·mos** (sā′mŏs′). Island of E Greece in the Aegean off W coast of Turkey.
**Sam·o·thrace** (săm′ə-thrās′). Island of NE Greece in the NE Aegean.
**Sam·o·thrá·ki** (sä′mə-thrä′kē). SAMOTHRACE.
**Sam·sun** (säm-so͞on′). City of N Turkey on the Black Sea. Pop. 168,478.
**San** (sän). River, c. 280 mi (451 km), of SE Poland.
**San·a** or **Sa·n'a** *also* **Sa·naa** (sä-nä′, săn′ä). Cap. of Yemen in the central part. Pop. 134,588.
**San An·ge·lo** (săn ăn′jə-lō′). City of SW Tex. SW of Abilene. Pop. 73,240.
**San An·sel·mo** (ăn-sĕl′mō). Town of W Calif. near San Francisco. Pop. 11,927.
**San An·to·ni·o** (ăn-tō′nē-ō′). **1.** River, c. 200 mi (322 km), of S Tex. **2.** Peak, 10,080 ft (3,074.4 m), of the San Gabriel Mts. in S Calif. **3.** City of S central Tex. SSW of Austin. Pop. 785,410.
**San Be·ni·to** (bə-nē′tō). City of extreme S Tex. near Brownsville. Pop. 17,988.
**San Ber·nar·di·no** (bûr′nə-dē′nō, -nər-). **1.** Mountains of S Calif. S of the Mojave Desert. **2.** Pass, 6,770 ft (2,063 m), through the Lepontine Alps in SE Switzerland.
**San Blas** (bläs′), **Gulf of.** Inlet of the Caribbean off N central Panama.
**San Bru·no** (bro͞o′nō). City of W Calif. S of San Francisco. Pop. 35,417.
**San Car·los** (kär′ləs). City of W Calif. SE of San Mateo. Pop. 24,710.
**San Cle·men·te** (klə-mĕn′tē). **1.** Island of S Calif. in SW Santa Barbara Is. S of Santa Catalina. **2.** City of S Calif. SE of Long Beach. Pop. 27,325.
**San Cris·tó·bal** (krĭs-tō′bəl). **1.** Island of Ecuador in the Galápagos Is. **2.** City of extreme W Venezuela near the Colombian border. Pop. 151,717.
**Sanc·ti-Spí·ri·tus** (sängk′tē spîr′ĭ-to͞os′). City of central Cuba WNW of Camagüey. Pop. 83,000.
**San·da·kan** (săn-dä′kən). City of Malaysia in N Borneo on an inlet of the Sulu Sea. Pop. 42,413.
**Sand·hurst** (sănd′hûrst′). Village of S central England SE of Reading.
**San·di·a** (săn-dē′ə). Peak, 10,676 ft (3,256.2 m), of N central N.Mex. NE of Albuquerque.
**San Di·e·go** (dē-ā′gō). City of S Calif. on the Pacific near the Mexican border. Pop. 875,504.
**San Di·mas** (dē′məs). City of S Calif. near Pomona. Pop. 24,014.
**San·dring·ham** (săn′drĭng-əm). Village of E England near the Wash.
**Sand Springs** (sănd). City of NE Okla. near Tulsa. Pop. 13,246.

**San·dus·ky** (sən-dŭs′kē, săn-). **1.** River, c. 150 mi (241 km), of N Ohio. **2.** City of N Ohio on Lake Erie W of Cleveland. Pop. 31,360.
**Sand·wich** (sănd′wĭch′, săn′-). **1.** HAWAIIAN IS. **2.** Borough of SE England, N of Dover. Pop. 4,420.
**Sand·y City** (săn′dē). City of N Utah near Salt Lake City. Pop. 51,022.
**Sandy Hook.** Peninsula of E N.J. at the entrance to Lower New York Bay.
**San Fer·nan·do** (fər-năn′dō). **1.** City of E Argentina NW of Buenos Aires. Pop. 113,249. **2.** City of S Calif. in the **San Fernando Valley,** surrounded by Los Angeles. Pop. 17,731.
**San·ford** (săn′fərd). **1.** City of central Fla. NNE of Orlando. Pop. 23,176. **2.** City of SW Me. W of Biddeford. Pop. 18,020. **3.** City of central N.C. SW of Raleigh. Pop. 14,773.
**San Fran·cis·co** (frən-sĭs′kō). **1. Peaks.** Group of three peaks in N Ariz. N of Flagstaff, consisting of Mt. Humphreys, Mt. Agassiz, & Mt. Fremont. **2.** City of W Calif. on a peninsula between the Pacific & **San Francisco Bay,** an inlet of the Pacific. Pop. 678,974. **—San Fran·cis′can** (-kən) *n.*
**San Ga·bri·el** (gā′brē-əl). **1.** Mountains of S Calif. E & NE of Los Angeles, rising to 10,080 ft (3,074.4 m). **2.** City of S Calif. near Los Angeles. Pop. 30,072.
**San·ga·mon** (săng′gə-mən). River, c. 250 mi (402 km), of central Ill.
**San·gay** (säng-gī′). Active volcano, 17,454 ft (5,323.5 m), in the Andes of E central Ecuador.
**San·ger** (săng′ər). City of S central Calif. E of Fresno. Pop. 12,558.
**San·gi·he** (säng-gē′ə) *also* **Sang·i** (säng′ē). **1.** Islands of N central Indonesia NE of Sulawesi. **2.** Largest (217 sq mi/562 sq km) of the Sangihe Is.
**San Gi·mi·gna·no** (jē′mēn-yä′nō). Town of central Italy NW of Siena. Pop. 2,800.
**San·gre de Cris·to** (săng′grē dē krĭs′tō). Range of the S Rockies extending c. 220 mi (355 km) from S central Colo. into N central N.Mex.
**San·i·bel** (săn′ə-bəl). Island of SW Fla. in the Gulf of Mexico SW of Fort Myers.
**San I·si·dro** (ī-sē′drō). City of E Argentina near Buenos Aires. Pop. 250,008.
**San Ja·cin·to** (jə-sĭn′tō). River, c. 130 mi (209 km), of SE Tex.
**San Joa·quin** (wô-kēn′, wä-). River of central Calif. flowing c. 320 mi (515 km), SW & NW to the Sacramento R.
**San Jo·se** (ə-zā′, ō-, hō-). City of W Calif. SE of San Francisco. Pop. 636,550.
**San Jo·sé** (ə-zā′, ō-, hō-). Cap. of Costa Rica in the central part. Pop. 391,107.
**San Juan** (wän′, hwän′). **1.** River, c. 360 mi (579 km), of SW Colo., NW N.Mex., & SE Utah. **2.** Range of the Rockies in SW Colo. & N N.Mex. **3.** Hill of E Cuba near Santiago de Cuba; site of Spanish-American War battle (1898). **4.** Islands of NW Wash. N of Puget Sound. **5.** City of NW Argentina W of Córdoba. Pop. 217,514. **6.** Cap. of Puerto Rico in the NE part on the Atlantic. Pop. 422,701.
**San Juan Cap·is·tra·no** (kăp′ĭ-strä′nō). City of S Calif. SE of Santa Ana. Pop. 18,959.
**Sankt Gal·len** (zängt gä′lən, zängkt). City of NE Switzerland E of Zurich. Pop. 81,900.
**Sankt Pöl·ten** (pœl′tən). City of N central Austria W of Vienna. Pop. 43,300.
**San·ku·ru** (säng-ko͝or′o͞o). River, c. 750 mi (1,207 km), of S & central Zaire.
**San Le·an·dro** (lē-ăn′drō). City of W Calif. SSE of Oakland. Pop. 63,952.
**San Lu·cas** (lo͞o′kəs). Cape of W Mexico at the S tip of Lower California.
**San Lu·is** (lo͞o′ĭs). Valley of N N.Mex. & S Colo. between the San Juan & Sangre de Cristo ranges.
**San Luis O·bis·po** (ə-bĭs′pō). City of S Calif. NW of Santa Barbara. Pop. 34,252.
**San Lu·is Po·to·sí** (lo͞o-ēs′ pō′tə-sē′). City of central Mexico NE of León. Pop. 271,123.
**San Mar·cos** (mär′kəs). **1.** City of S Calif. NNW of San Diego. Pop. 17,479. **2.** City of S central Tex. NE of San Antonio. Pop. 23,420.
**San Ma·ri·no** (mə-rē′nō). **1.** Republic, 23 sq mi (60 sq km), within N central Italy in the Apennines near the Adriatic. Cap. San Marino. Pop. 19,149. **2.** Cap. of San Marino. Pop. 4,628. **3.** City of S Calif. near Pasadena. Pop. 4,628.
**San Mar·tín** (mär-tēn′). Town of E Argentina near Buenos Aires. Pop. 24,300.
**San Ma·te·o** (mə-tā′ō). City of W Calif. SSE of San Francisco. Pop. 77,561.
**San Mi·guel** (mĭ-gĕl′). City of E El Salvador. Pop. 59,304.
**San Miguel de Tu·cu·mán** (də to͞o′kə-män′). TUCUMÁN.
**San Ni·co·lás de los Gar·zas** (nē′kə-läs′ də lôz gär′səs). City of N Mexico near Monterrey. Pop. 28,803.
**San Pab·lo** (păb′lō). City of W Calif. NNW of Oakland on **San Pablo Bay,** N arm of San Francisco Bay. Pop. 19,750.
**San Pe·dro** (pē′drō). Channel between the S Calif. mainland & Santa Catalina Is.
**San Pe·dro de Ma·co·rís** (pā′drō dā läs kə-lōn′yəs). City of SE Dominican Republic on the Caribbean. Pop. 43,010.
**San Pe·dro Su·la** (pā′drō so͞o′lə). City of NW Honduras. Pop. 150,991.
**San Ra·fael** (rə-fĕl′). City of W Calif. N of San Francisco. Pop. 44,700.
**San Re·mo** (rā′mō, rē′-). City of NW Italy on the Ligurian Sea. Pop. 47,684.
**San Sal·va·dor** (săl′və-dôr′). **1.** Island of the central Bahamas. **2.** Cap. of El Salvador in the W central part. Pop. 337,171.
**San Se·bas·tián** (sə-băs′chĭn). City of N Spain on the Bay of Biscay near the French border. Pop. 159,557.
**San·ta An·a** (săn′tə ăn′ə). **1.** City of W El Salvador NW of San Salvador. Pop. 96,036. **2.** City of S Calif. E of Long Beach. Pop. 203,713.
**Santa Bar·ba·ra** (bär′bər-ə, bär′brə). **1.** Channel between N Santa Barbara Is. & S Calif. coast. **2.** Islands of Calif. in the Pacific off the S coast. **3.** City of S Calif. on the Pacific WNW of Los Angeles. Pop. 74,542.
**Santa Cat·a·li·na** (kăt′l-ē′nə). Island of S Calif. in the S Santa Barbara Is.
**Santa Cla·ra** (klăr′ə, klâr′ə). **1.** City of central Cuba ESE of Havana. Pop. 143,000. **2.** City of W Calif. near San Jose. Pop. 87,746.
**Santa Cruz** (kro͞oz). **1.** River, c. 250 mi (402 km), of S Argentina. **2.** Islands of the W Pacific in the SE Solomon Is. **3.** Island of S Calif. in the N Santa Barbara Is. **4.** City of central Bolivia NE of Sucre. Pop. 254,682. **5.** City of W Calif. on Monterey Bay SSW of San Jose. Pop. 41,483.
**Santa Cruz de Te·ne·ri·fe** (də tĕn′ə-rē′fā, -fē, -rĕf′, -rĭf′). City of the Canary Is. on the NE coast of Tenerife. Pop. 74,910.
**Santa Fe** (fā′). **1.** City of NE Argentina on the Salado R. Pop. 244,655. **2.** Cap. of N.Mex. in the N central part. Pop. 48,899.
**Santa Fe Springs.** City of S Calif. SE of Los Angeles. Pop. 14,559.
**Santa Fe Trail.** 19th cent. wagon & trade route to the SW U.S. extending from Independence, Mo., to Santa Fe, N.Mex.
**Santa Is·a·bel** (ĭz′ə-bĕl′). **1.** Island of the W Pacific in E central Solomon Is. **2.** MALABO.
**Santa Ma·ri·a** (mə-rē′ə). **1.** Active volcano, 12,362 ft (3,770.4 m), of SW Guatemala. **2.** City of S Brazil W of Pôrto Alegre. Pop. 120,667. **3.** City of SW Calif. NW of Santa Barbara. Pop. 39,685.
**Santa Mar·ta** (mär′tə). City of N Colombia on the Caribbean. Pop. 102,484.
**Santa Mon·i·ca** (mŏn′ĭ-kə). City of S Calif. on the Pacific W of Los Angeles. Pop. 88,314.
**San·tan·der** (săn′tän-dĕr′, săn′tăn′-). City of N Spain on the Bay of Biscay. Pop. 130,019.
**Santa Pau·la** (pô′lə). City of S Calif. ENE of Ventura. Pop. 20,552.
**San·ta·rém** (săn′tə-rĕm′). City of N Brazil on the Amazon. Pop. 51,123.
**Santa Ro·sa** (rō′zə). **1.** Island of S Calif. in the NW Santa Barbara Is. **2.** Barrier island of NW Fla. extending c. 50 mi (80 km) along the coast of the Gulf of Mexico. **3.** City of W Calif. N of San Francisco. Pop. 83,205.
**San·tee** (săn-tē′, săn′tē). River, 143 mi (230.1 km), of S.C.
**San·ti·a·go** (săn′tē-ä′gō, săn′-). **1.** Cap. of Chile in the central part. Metro. area pop. 3,448,700. **2.** *also* **Santiago de los Ca·ba·lle·ros** (dā′ lôs kä′bə-lĕr′ōz, -bəl-yĕr′-). City of N Dominican Republic NW of Santo Domingo. Pop. 155,000. **3.** *also* **Santiago de Com·pos·te·la** (də kŏm′pə-stĕl′ə). City of NW Spain S of La Coruña. Pop. 51,620.
**Santiago de Cu·ba** (də kyo͞o′bə). City of SE Cuba on the Caribbean. Pop. 310,000.
**Santiago del Es·te·ro** (dĕl ə-stĕr′ō). City of N central Argentina N of Córdoba. Pop. 105,127.
**San·to An·dré** (săN′to͞o ăn-drā′). City of S Brazil near São Paulo. Pop. 415,025.
**San·to Do·min·go** (săn′tō də-mĭng′gō). Cap. of the Dominican Republic in the S part on the Caribbean. Pop. 673,470.
**San·to·rin** (săn′tə-rēn′, -rĭn′). Island of S Greece in the S Cyclades.
**San·tos** (săn′təs). City of SE Brazil on an offshore island in the Atlantic SE of São Paulo. Pop. 341,317.
**São Ber·nar·do do Cam·po** (souN bər-när′do͞o do͞o kăm′po͞o). City of SE Brazil near São Paulo. Pop. 187,368.
**São Cae·ta·no do Sul** (kī-tä′no͞o do͞o so͞ol′). City of SE Brazil near São Paulo. Pop. 170,675.
**São Fran·cis·co** (frən-sĭs′kō). River of E Brazil flowing c. 1,800 mi (2,896 km) NNE & E to the Atlantic.
**São Gon·ça·lo** (gōN-săl′o͞o). City of SE Brazil on Guanabara Bay opposite Rio de Janeiro. Pop. 161,392.
**São João de Me·ri·ti** (zhwouN′ də mə-rē′tē, -rē-tē′). City of SE Brazil near Rio de Janeiro. Pop. 425,800.
**São Jo·sé do Ri·o Prêt·o** (zho͞o-zā′ do͞o rē′o͞o prĕt′o͞o). City of SE Brazil near São Paulo. Pop. 108,319.
**São José dos Cam·pos** (do͞osh kăm′pəs). City of SE Brazil NNE of São Paulo. Pop. 130,118.

**São Lu·is** (lo͞o-ēs′). City of NE Brazil on an offshore island in the Atlantic ESE of Belém. Pop. 167,529.
**São Ma·nuel** (măn-wĕl′). TELES PIRES.
**São Mi·guel** (mē-gĕl′). Island of the E Azores, largest of the group.
**Saône** (sōn). River, c. 268 mi (431 km), of E central France.
**São Pau·lo** (pou′lo͞o, -lō). City of SE Brazil WSW of Rio de Janeiro. Pop. 8,490,763.
**São Ro·que** (rô′kə). Cape on the NE coast of Brazil N of Natal.
**São Tia·go** (tē-ä′go͞o). Island of S Cape Verde, largest of the group.
**São To·mé** (tə-mā′). **1.** Island, 319 sq mi (826 sq km), in the Gulf of Guinea on the equator, forming part of the republic of São Tomé and Principe. **2.** Cap. of São Tomé and Principe on the SE coast of São Tomé Is. Pop. 7,681.
**São Tomé and Prin·ci·pe** (prēN′sē-pə). Island republic in the Gulf of Guinea, W Africa. Cap. São Tomé. Pop. 85,000.
**São Vi·cen·te** (vē-sāN′tə).City of SE Brazil on the Atlantic near Santos. Pop. 116,075.
**Sap·po·ro** (sə-pôr′ō, -pōr′-). City of SW Hokkaido, Japan, SE of Ishikari Bay. Pop. 1,240,613.
**Sa·pul·pa** (sə-pŭl′pə). City of NE Okla. SSW of Tulsa. Pop. 15,853.
**Saq·qa·ra** (sə-kär′ə). Village of N Egypt near Cairo; site of the oldest pyramid.
**Sar·a·gos·sa** (săr′ə-gŏs′ə). City of NE Spain, on the Ebro R. Pop. 449,319.
**Sa·ra·je·vo** (săr′ə-yĕ-vō′, sâr′ə-yā′-). City of central Yugoslavia SW of Belgrade. Pop. 245,058.
**Sar·a·nac** (săr′ə-năk′). Group of three lakes of NE N.Y. linked by the **Saranac R.** (c. 100 mi/161 km).
**Sa·ransk** (sə-ränsk′, -ränsk). City of central European USSR SSE of Gorki. Pop. 263,000.
**Sar·a·so·ta** (săr′ə-sō′tə). City of SW Fla. on the Gulf of Mexico S of Tampa Bay. Pop. 48,868.
**Sar·a·to·ga** (săr′ə-tō′gə). **1.** Lake of E N.Y. SE of Saratoga Springs. **2.** City of W Calif. near San Jose. Pop. 29,261.
**Saratoga Springs.** City of E N.Y. N of Albany. Pop. 23,906.
**Sa·ra·tov** (sə-rä′təf). City of S European USSR on the Volga R. Pop. 856,000.
**Sa·ra·wak** (sə-rä′wäk, -wäk′). Region & state of Malaysia on NW Borneo.
**Sar·din·i·a** (sär-dĭn′ē-ə, -dĭn′yə). Island of Italy in the Mediterranean S of Corsica. **—Sar·din′i·an** *adj.* & *n.*
**Sar·dis** (sär′dĭs). Ancient city of W Asia Minor NE of modern Izmir, Turkey.
**Sa·re·ma** *or* **Saa·re·maa** (sär′ə-mä′). Island of NW European USSR in the Baltic at the mouth of the Gulf of Riga.
**Sar·gas·so Sea** (sär-găs′ō). Part of the N Atlantic between the West Indies & the Azores.
**Sar·go·dha** (sər-gō′də, -gōd′hə). City of NE Pakistan WNW of Lahore. Pop. 201,407.
**Sark** (särk). One of the Channel Is. in the English Channel E of Guernsey.
**Sar·ma·tia** (sär-mā′shə, -shē-ə). Ancient region of E Europe between the Vistula & the Volga in present-day E Poland & W European USSR. **—Sar·ma′tian** *adj.* & *n.*
**Sar·ni·a** (sär′nē-ə). City of S Ont., Canada, on the St. Clair R. at the S end of Lake Huron. Pop. 55,576.
**Sa·ron·ic Gulf** (sə-rŏn′ĭk). Arm of the Aegean in S Greece between Attica & the Peloponnesus E of Corinth.
**Sa·ros** (sâr′ŏs′, sĕr′-). **Gulf of.** Inlet of the NE Aegean off NW Turkey N of Gallipoli.
**Sarthe** (särt). River, c. 177 mi (285 km), of NW France.
**Sa·se·bo** (sä′sə-bō′). City of W Kyushu, Japan, on the East China Sea. Pop. 250,729.
**Sas·katch·e·wan** (sə-skăch′ə-wən, -wän′, săs-kăch′-). **1.** River, c. 340 mi (550 km), of S central Canada formed by the confluence of the North & South Saskatchewan rivers in central Sask. & flowing E to Lake Winnipeg in Man. **2.** Province of S central Canada. Cap. Regina. Pop. 957,025.
**Sas·ka·toon** (săs′kə-to͞on′). City of central Sask., Canada, NW of Regina. Pop. 133,750.
**Sas·sa·ri** (sä′sə-rē′). City of NW Sardinia, Italy. Pop. 94,312.
**Sa·til·la** (sə-tĭl′ə). River, c. 220 mi (354 km), of SE Ga.
**Sat·pu·ra** (sät′pər-ə). Range of hills in central India extending c. 600 mi (965 km) along the N edge of the Deccan Plateau.
**Sa·tsu·ma** (săt′sə-mä′, săt-so͞o′mə). Peninsula of SW Kyushu, Japan.
**Sa·tu-Ma·re** (sä′to͞o-mär′ə). City of NW Rumania near the Hungarian border. Pop. 108,152.
**Sa·u·di A·ra·bi·a** (sou′dē ə-rā′bē-ə, sä-o͞o′dē). Kingdom comprising most of the Arabian peninsula. Cap. Riyadh. Pop. 8,367,000. **—Sa·u′di, Sa·u′di A·ra′bi·an** *adj.* & *n.*
**Sau·gus** (sô′gəs). Town of NE Mass. near Boston. Pop. 24,746.
**Sauk** (sôk). Village of NE Ill. near Chicago Heights. Pop. 10,906.
**Sault Sainte Ma·rie** (so͞o′ sānt′ mə-rē′). **1. Canals.** Three ship canals, two U.S. & one Canadian, by-passing the rapids on the St. Marys R. between Lakes Superior & Huron. **2.** City of S Ont., Canada, on the St. Marys R. opposite Sault Ste. Marie, Mich. Pop. 81,048. **3.** City of N Mich. on the St. Marys R. N of Detroit. Pop. 14,448.
**Sa·va** (sä′və). River, c. 580 mi (933 km), of N Yugoslavia.
**Sa·vai·i** *or* **Sa·vai′i** (sə-vī′ē). Largest (703 sq mi/1,821 sq km) island of Samoa, in Western Samoa.
**Sa·va·lan** (sä′və-län′). *var. of* SABALAN.
**Sa·van·nah** (sə-văn′ə). **1.** River, c. 314 mi (505 km), forming most of the S.C.-Ga. border. **2.** City of E Ga. near the mouth of the Savannah R. Pop. 141,634.
**Sa·vo** (sä′vō). Island of the W Pacific in the SE Solomon Is.
**Sa·vo·na** (sə-vō′nə). City of NW Italy on the Ligurian Sea WSW of Genoa. Pop. 76,274.
**Sa·voy** (sə-voi′). Region & former duchy of SE France bordering on Switzerland & Italy. **—Sa·voy′ard** (sə-voi′ärd′, săv′ oi-yärd′) *adj.* & *n.*
**Savoy Alps** (ălps). Range of the W Alps in SE France rising to 15,781 ft (4,813.2 m).
**Sa·watch** (sə-wŏch′). Range of the Rockies in central Colo. rising to 14,431 ft (4,398.6 m).
**Saxe** (săks). SAXONY.
**Sax·on·y** (săk′sə-nē). **1.** Former region & duchy of NW Germany. **2.** Former duchy, kingdom, & electorate of central Germany. **—Sax′on** *adj.* & *n.*
**Sa·yan** (sə-yän′). Mountains of S Central Asian & Siberian USSR W of Lake Baikal.
**Say·re·ville** (sā′ər-vĭl′, sâr′-). Borough of E central N.J. SSW of Perth Amboy. Pop. 29,969.
**Sca·fell Pike** (skô′fĕl′). Mountain, 3,210 ft (979 m), of the Cumbrians in NW England; highest peak in England.
**Sca·man·der** (skə-măn′dər). River of NW Turkey, the MENDERES.
**Scan·di·na·vi·a** (skăn′də-nā′vē-ə, -nāv′yə). **1.** Peninsula of N Europe occupied by Norway & Sweden. **2.** Norway, Sweden, & Denmark, & sometimes also Iceland, Finland, & the Faeroe Is. **—Scan′di·na′vi·an** *adj.* & *n.*
**Scap·a Flow** (skăp′ə flō′). Anchorage & British naval base in the Orkney Is. of N Scotland.
**Scar·bor·ough** (skär′bûr′ō, -bər-ə). **1.** E borough of metropolitan Toronto, Ont., Canada, on Lake Ontario. Pop. 387,149. **2.** Borough of NE England on the North Sea N of Hull. Pop. 43,300. **3.** *also* **Scar·bor·o.** Town of SW Me. S of Portland. Pop. 11,347.
**Scars·dale** (skärz′dāl′). City of SE N.Y. near Yonkers. Pop. 17,65[illegible].
**Schaer·beek** (skär′bāk′). City of central Belgium near Bruss[illegible]. Pop. 118,950.
**Schaff·hau·sen** (shäf′hou′zən). Waterfall, 65 ft (19.8 m) high & 377 ft (115 m) wide, in the Rhine N of Zurich, Switzerland.
**Schaum·burg** (shăm′bûrg′). Village of NE Ill. near Chicago. Pop. 52,319.
**Schel·de** (skĕl′də). SCHELDT.
**Scheldt** (skĕlt). River of W Europe rising in N France & flowing c. 270 mi (434 km) NE through W Belgium & SW Netherlands to the North Sea.
**Sche·nec·ta·dy** (skə-nĕk′tə-dē). City of E N.Y. NW of Albany. Pop. 67,972.
**Scher·er·ville** (shîr′ər-vĭl′). Town of NE Ind. near Gary. Pop. 13,209.
**Schie·dam** (skē-däm′). City of SW Netherlands near Rotterdam. Pop. 78,068.
**Schil·ler Park** (shĭl′ər). Village of NE Ill. near Chicago. Pop. 11,458.
**Schles·wig** (shlĕs′wĭg, -wĭk, slĕs′-). **1.** Region & former duchy of N West Germany & S Denmark in S Jutland. **2.** City of N West Germany NW of Kiel. Pop. 30,974.
**Schou·ten** (skout′n). Islands of E Indonesia off the N coast of West Irian.
**Schuyl·kill** (sko͞ol′kĭl′, sko͞o′kəl). River, c. 130 mi (209 km), of SE Pa.
**Schwa·ben** (shfä′bən). SWABIA.
**Schwä·bisch Gmünd** (shfā′bĭsh gə-mo͞ot′). City of S West Germany E of Stuttgart. Pop. 56,422.
**Schwein·furt** (shfīn′fo͝ort′). City of E central West Germany on the Main R. E of Frankfurt. Pop. 56,164.
**Schwe·rin** (shfā-rēn′). City of NW East Germany SW of Rostock. Pop. 104,984.
**Scil·ly** (sĭl′ē). Islands of SW England at the entrance to the English Channel WSW of Land's End.
**Sci·o·to** (sī-ō′tə). River, 237 mi (381.3 km), of Ohio.
**Scit·u·ate** (sĭch′ə-wət). Town of E Mass. SE of Boston. Pop. 17,317.
**Scone** (sko͞on). Village of central Scotland NE of Perth; coronation site of Scottish kings (1150–1488).
**Sco·pus** (skō′pəs). Peak, 2,736 ft (834.5 m), in an Israeli enclave in NW Jordan NNE of Jerusalem.
**Scores·by Sound** (skôrz′bē, skōrz′-). Inlet of the Greenland Sea extending c. 200 mi (322 km) into E Greenland.
**Sco·tia** (skō′shə). SCOTLAND.

**Scot·land** (skŏt′lənd). Constituent country of the United Kingdom of Great Britain & Northern Ireland, comprising N Great Britain and the Hebrides, Shetland Is., & Orkney Is. Cap. Edinburgh. Pop. 5,117,-146. **—Scots** (skŏts) *adj.* **—Scot′tish** (skŏt′ĭsh) *adj. & n.*
**Scotts·bluff** (skŏts′blŭf′). City of W Nebr. near the Wyo. border. Pop. 14,156.
**Scotts·bor·o** (skŏts′bûr′ō). City of NE Ala. E of Huntsville. Pop. 14,752.
**Scotts·dale** (skŏts′dāl′). City of S central Ariz. near Phoenix. Pop. 88,364.
**Scran·ton** (skrăn′tən). City of NW Pa. NW of Wilkes-Barre. Pop. 88,117.
**Scun·thorpe** (skŭn′thôrp′). Borough of NE England SW of Hull. Pop. 68,100.
**Scu·ta·ri** (skōō′tə-rē). **1.** Lake, c. 25 mi (40 km) long, of S Yugoslavia & NW Albania. **2.** SHKODËR.
**Scy·ros** (sī′rəs). *var. of* SKYROS.
**Scyth·i·a** (sĭth′ē-ə, sĭth-). Ancient region of SE Europe & SW Asia between the mouth of the Danube & the Aral Sea. **—Scyth′i·an** *adj. & n.*
**Sea Islands** (sē). Chain of islands off the Atlantic coasts of S.C., Ga., & N Fla.
**Seal Beach** (sēl). City of S Calif. on the Pacific SSE of Los Angeles. Pop. 25,975.
**Sear·cy** (sûr′sē). City of N central Ark. NE of Little Rock. Pop. 13,645.
**Sea·side** (sē′sīd′). City of W Calif. on Monterey Bay. Pop. 36,567.
**Se·at·tle** (sē-ăt′l). City of W central Wash. on Puget Sound & Lake Washington. Pop. 493,846.
**Se·ba·go** (sə-bā′gō). Lake, c. 12 mi (20 km) long, of SW Me.
**Se·bas·to·pol** (sə-băs′tə-pōl′). *var. of* SEVASTOPOL.
**Se·cau·cus** (sĭ-kô′kəs). Town of NE N.J. near Jersey City. Pop. 13,719.
**Se·cun·der·a·bad** (sĭ-kŭn′dər-ə-băd′, -bäd′). City of S central India near Hyderabad. Pop. 250,636.
**Se·da·li·a** (sĭ-dāl′yə). City of central Mo. ESE of Kansas City. Pop. 20,927.
**Se·dan** (sĭ-dăn′). Town of NE France on the Meuse R. Pop. 23,867.
**Sedge·moor** (sĕj′mŏŏr′, -môr′, -mōr′). Marshy tract in SW England where the forces of James II defeated the Duke of Monmouth (1685).
**See·konk** (sē′kŏngk′). Town of SE Mass. on the border near Providence, R.I. Pop. 12,269.
**Se·fid Rud** (sĕ-fēd′ rōōd′). *var. of* SAFID RUD.
**Se·go·vi·a** (sĭ-gō′vē-ə). **1.** COCO. **2.** City of central Spain NW of Madrid. Pop. 41,880.
**Se·guin** (sĭ-gēn′). City of S central Tex. ENE of San Antonio. Pop. 17,854.
**Se·gu·ra** (sā-gŏŏr′ə). River, c. 200 mi (322 km), of SE Spain.
**Seim** *also* **Seym** (sām). River, 460 mi (740 km), of W European USSR.
**Seine** (sān, sĕn). River of N France flowing c. 480 mi (772 km) into the **Bay of the Seine,** an inlet of the English Channel.
**Sek·on·di-Ta·ko·ra·di** (sĕk′ən-dē′tä-kə-rä′dē). City of SW Ghana on the Gulf of Guinea. Pop. 160,868.
**Se·len·ga** (sĕl′ĕng-gä′). River of N Mongolia & S Siberian USSR flowing c. 750 mi (1,207 km) NE into Lake Baikal.
**Se·leu·ci·a** (sĭ-lōō′shē-ə, -shə). Ancient city of Mesopotamia on the Tigris below modern Baghdad.
**Sel·kirk** (sĕl′kûrk′). Range of the Rockies in SE B.C., Canada.
**Sel·ma** (sĕl′mə). **1.** City of S central Ala. W of Montgomery. Pop. 26,684. **2.** City of central Calif. SE of Fresno. Pop. 10,942.
**Se·ma·rang** (sə-mär′äng). City of N Java, Indonesia, on the Java Sea. Pop. 646,590.
**Sem·i·pa·la·tinsk** (sĕm′ē-pə-lä′tĭnsk). City of E Central Asian USSR, on the Irtish R. Pop. 283,000.
**Sen·dai** (sĕn-dī′). City of NE Honshu, Japan, on the Pacific. Pop. 615,473.
**Sen·e·ca** (sĕn′ĭ-kə). Largest (67 sq mi/173 sq km) of the Finger Lakes, in W central N.Y.
**Sen·e·gal** (sĕn′ĭ-gôl′). **1.** River of W Africa rising in W Mali & flowing c. 1,000 mi (1,609 km) NW & W along the Mauritania-Senegal border to the Atlantic. **2.** Country of W Africa on the Atlantic. Cap. Dakar. Pop. 5,508,000. **—Sen′e·ga·lese′** (-gô′lēz′, -lēs′, -gə-) *adj. & n.*
**Sen·e·gam·bi·a** (sĕn′ĭ-găm′bē-ə). Senegal & Gambia.
**Sen·lac** (sĕn′lăk′). Hill in S England near Hastings; site of the Battle of Hastings (1066).
**Sens** (säNs). Town of N central France SE of Paris. Pop. 25,621.
**Seoul** (sōl). Cap. of South Korea in the NW part. Pop. 6,889,502.
**Se·pik** (sā′pĭk). River, c. 700 mi (1,126 km), of N Papua New Guinea.
**Sept-Îles** (sĕt-ēl′). City of E Que., Canada, near the mouth of the St. Lawrence R. Pop. 30,617.
**Se·quoi·a National Park** (sĭ-kwoi′ə). Area in E central Calif. noted for its sequoia forests & mountain scenery.
**Ser·bi·a** (sûr′bē-ə). Region & constituent republic of E Yugoslavia. **—Ser′bi·an** *adj. & n.*
**Ser·en·get·i** (sĕr′ən-gĕt′ē). Plain & wildlife reserve of N Tanzania.
**Ser·pu·khov** (sĕr-pōō′kəf). City of central European USSR S of Moscow. Pop. 140,000.
**Ser·ra da Es·tre·la** (sĕr′rə dä ĕs-trĕl′ə). Range of central Portugal.
**Serra do Mar** (dōō mär′). Coastal range of S Brazil.
**Serra Pa·ca·rai·ma** (păk′ə-rī′mə). SIERRA PACARAIMA.
**Serra Pa·ri·ma** (pə-rē′mə). SIERRA PARIMA.
**Ses·tos** (sĕs′təs). Ancient town of European Turkey at the narrowest point of the Dardanelles.
**Ses·to San Gio·van·ni** (sĕs′tō săn′ jə-vä′nē). City of N Italy near Milan. Pop. 98,151.
**Sète** (sĕt). Town of S France on the Mediterranean. Pop. 39,075.
**Sé·tif** (sā-tēf′). City of NE Algeria ESE of Algiers. Pop. 144,200.
**Se·tú·bal** (sə-tōō′bəl). City of S central Portugal on an inlet of the Atlantic SE of Lisbon. Pop. 49,670.
**Se·van** (sə-văn′). Lake, c. 540 sq mi (1,399 sq km), of S European USSR in the Caucasus.
**Se·vas·to·pol** (sə-văs′tə-pōl′) *also* **Se·bas·to·pol** (-băs′-). City of SW European USSR in the Crimea on the Black Sea. Pop. 301,000.
**Sev·en Hills** (sĕv′ən). City of NE Ohio near Cleveland. Pop. 13,650.
**Sev·ern** (sĕv′ərn). **1.** Inlet of W Chesapeake Bay in central Md. **2.** River of NW Ont., Canada, flowing c. 420 mi (676 km) NE to Hudson Bay. **3.** River of SW Great Britain rising in W Wales & flowing c. 210 mi (338 km) through W England to the Bristol Channel.
**Se·ver·na·ya Zem·lya** (sĕv′ər-nə-yä′ zĕm′lē-ä′). Archipelago of N Siberian USSR in the Arctic Ocean N of Taimyr Peninsula.
**Se·ve·rod·vinsk** (sĕv′ər-əd-vĭnsk′). City of N European USSR on the White Sea W of Arkhangelsk. Pop. 197,000.
**Se·vier** (sə-vîr′). River, c. 280 mi (451 km), of SW Utah.
**Se·vil·la** (sā-vē′yä, -vēl′-). SEVILLE.
**Se·ville** (sə-vĭl′). City of SW Spain on the Guadalquivir R. Pop. 511,447.
**Sè·vres** (sĕv′rə). City of N central France on the Seine SW of Paris. Pop. 21,100.
**Sew·ard** (sōō′ərd). Peninsula of W Alas. projecting c. 200 mi (322 km) into the Bering Sea just below the Arctic Circle.
**Sey·chelles** (sā-shĕl′, -shĕlz′). Island nation in the W Indian Ocean N of Madagascar. Cap. Victoria. Pop. 63,000.
**Sey·han** (sā-hän′). River, c. 320 mi (515 km), of S central Turkey.
**Seym** (sām). *var. of* SEIM.
**Sey·mour** (sē′môr′, -mōr′). **1.** Town of SW Conn. NW of New Haven. Pop. 13,434. **2.** City of SE Ind. S of Columbus. Pop. 15,050.
**Sfax** (sfăks). City of E Tunisia on the Gulf of Gabès. Pop. 171,297.
**Shaan·xi** (shän′shē′). Province of N central China.
**Sha·ba** (shä′bə). Region of SE Zaire.
**Shah·ja·han·pur** (shä′jə-hän′pŏŏr′). City of N central India NW of Lucknow. Pop. 135,604.
**Shah·pur** (shä-pŏŏr′). Ancient city of SW Iran W of Shiraz.
**Sha·ker Heights** (shā′kər). City of NE Ohio near Cleveland. Pop. 32,487.
**Shakh·ty** (shäk′tē). City of S European USSR NE of Rostov. Pop. 209,000.
**Sha·mo·kin** (shə-mō′kĭn). City of E central Pa. NNE of Harrisburg. Pop. 10,357.
**Shan·dong** (shän′dŏŏng′). **1.** Peninsula of E China projecting E between the Gulf of Bo Hai & the Yellow Sea. **2.** Province of E China.
**Shang·hai** (shăng-hī′). City of E China at the mouth of the Yangtze. Metro. area pop. 10,980,000.
**Shang·qiu** (shäng′chē-ōō′) *also* **Shang·kiu** (-kē-ōō′). City of E China ESE of Kaifeng. Pop. 250,000.
**Shan·non** (shăn′ən). River, c. 240 mi (386 km), of W Ireland.
**Shan·si** (shăn′sē′). SHANXI.
**Shan·tar** (shən-tär′). Islands of Far Eastern USSR in the Sea of Okhotsk NW of Sakhalin.
**Shan·tou** (shän′tou′). City of SE China on the South China Sea ENE of Hong Kong. Pop. 400,000.
**Shan·tung** (shăn′tŭng′). SHANDONG.
**Shan·xi** (shän′shē′). Province of NE China.
**Shao·xing** *also* **Shao·hsing** (shou′shĭng′). City of SE China near the East China Sea SW of Shanghai. Pop. 225,000.
**Shao·yang** (shou′yäng′). City of S China SW of Changsha. Pop. 275,000.
**Sha·ri** (shär′ē). River of N central Africa rising in the Central African Republic & flowing c. 1,400 mi (2,253 km) NW through S Chad to Lake Chad.
**Shar·jah** (shär-jä′). **1.** Sheikdom of the United Arab Emirates in E Arabia on the Persian Gulf & the Gulf of Oman. **2.** Chief town of this sheikdom on the Persian Gulf. Pop. 19,200.
**Shark Bay** (shärk). Inlet of the Indian Ocean off W central Australia.
**Shar·on** (shăr′ən). **1.** Town of E Mass. near Brockton. Pop. 13,601. **2.** City of W Pa. on the border NE of Youngstown, Ohio. Pop. 19,057.
**Shar·on·ville** (shăr′ən-vĭl′). City of SW Ohio near Cincinnati. Pop. 10,108.

**Sharps·burg** (shärps′bûrg′). Town of N Maryland; site of Civil War Battle of Antietam (1862).

**Shas·ta** (shăs′tə), **Mount.** Volcanic peak, 14,162 ft (4,319.4 m), of the Cascade Range in N Calif.

**Shatt-al-Ar·ab** *also* **Shatt al Ar·ab** (shăt′ăl-ăr′əb). River, c. 120 mi (193 km), of SE Iraq formed by the confluence of the Tigris & Euphrates rivers.

**Shaw·an·gunk** (shŏng′gəm). Mountains of SE N.Y.

**Sha·win·i·gan** (shə-wĭn′ĭ-gən). City of S Que., Canada, NW of Trois-Rivières. Pop. 24,921.

**Shawinigan-Sud** (shə-wĭn′ĭ-gən-so͞od′). Town of S Que., Canada, on the St. Maurice R. near Shawinigan. Pop. 11,155.

**Shaw·nee** (shô-nē′, shô′nē). **1.** City of E Kans. near Kansas City. Pop. 29,653. **2.** City of central Okla. ESE of Oklahoma City. Pop. 26,506.

**Shcher·ba·kov** (shĕr′bə-kôv′, shchĕr′-). RYBINSK, USSR.

**She·be·li** (shə-bā′lē). WEBBI SHEBELI.

**She·boy·gan** (shə-boi′gən). City of E Wis. on Lake Michigan N of Milwaukee. Pop. 48,085.

**Shef·field** (shĕf′ēld′). **1.** Borough of N central England E of Manchester. Pop. 558,000. **2.** City of NW Ala. near Florence. Pop. 11,903.

**Sheffield Lake.** City of NE Ohio on Lake Erie W of Cleveland. Pop. 10,484.

**Shel·by** (shĕl′bē). City of SW N.C. W of Charlotte. Pop. 15,310.

**Shel·by·ville** (shĕl′bē-vĭl′). **1.** City of central Ind. SE of Indianapolis. Pop. 14,989. **2.** City of central Tenn. SSE of Nashville. Pop. 13,530.

**Shel·i·kof** (shĕl′ĭ-kôf′). Strait of S Alas. between Alaska Peninsula & Kodiak & Afognak Is.

**Shel·ter** (shĕl′tər). Island of SE N.Y. between the two peninsulas of E Long Is.

**Shel·ton** (shĕl′tən). City of SW Conn. NNE of Bridgeport. Pop. 31,314.

**Shen·an·do·ah** (shĕn′ən-dō′ə). **1.** River, c. 150 mi (241 km), of N Va. & NE W.Va. **2. National Park.** Scenic area of N Va. along the crest of the Blue Ridge.

**Shen·si** (shĕn′sē′). SHAANXI.

**Shen·yang** (shŭn′yăng′). City of NE China ENE of Beijing. Pop. 3,750,000.

**Shep·pey** (shĕp′ē), **Isle of.** Island, 30 sq mi (80 sq km), of SE England in the mouth of the Thames.

**Sher·brooke** (shûr′bro͝ok′). City of S Que., Canada, E of Montreal. Pop. 76,804.

**Sher·i·dan** (shĕr′ĭ-dən). City of S Wyo. near the Mont. border. Pop. 15,146.

**Sher·man** (shûr′mən). City of N Tex. near the Okla. border N of Dallas. Pop. 30,413.

**'s Her·to·gen·bosch** (sĕr′tō-gən-bôs′, -gə-). City of S central Netherlands NNW of Eindhoven. Pop. 86,184.

**Sher·wood** (shûr′wo͝od′). **1.** Former royal forest of central England. **2.** City of central Ark. near Little Rock. Pop. 10,474.

**Sherwood Park.** City of central Alta., Canada, near Edmonton. Pop. 26,534.

**Shet·land** (shĕt′lənd) *also* **Shet·lands** (-ləndz). Islands of N Scotland in the Atlantic NE of the Orkney Is.

**Shey·enne** (shī-ĕn′, -ăn′). River, c. 325 mi (523 km), of E N.Dak.

**Shi·be·li** (shĭ-bā′lē). WEBBI SHEBELI.

**Shi·jia·zhuang** *also* **Shih·kia·chwang** (shîr′jē-ä′jə-wăng′). City of NE China SW of Beijing. Pop. 1,500,000.

**Shi·kar·pur** (shĭ-kär′po͝or′). City of S central Pakistan NNE of Karachi. Pop. 70,301.

**Shi·ko·ku** (shĭ-kō′ko͞o). Island of S Japan between SW Honshu & E Kyushu.

**Shil·ka** (shĭl′kə). River, c. 345 mi (555 km), of S Far Eastern USSR.

**Shil·long** (shə-lông′). City of NE India NE of Calcutta. Pop. 87,659.

**Shi·loh** (shī′lō). **1.** Ancient village of central Palestine NW of the Dead Sea. **2.** Locality in SW Tenn.; site of Civil War battle (1862).

**Shi·mi·zu** (shə-mē′zo͞o). City of E central Honshu, Japan, on Suruga Bay. Pop. 243,049.

**Shi·mo·no·se·ki** (shĭm′ə-nō-sĕk′ē). City of extreme SW Honshu, Japan, on Korea Strait. Pop. 266,593.

**Shi·na·no** (shĭ-nä′nō). River of central Honshu, Japan, flowing c. 230 mi (371 km) NNE to the Sea of Japan.

**Shi·nar** (shī′nər, -när′). Ancient country on the lower courses of the Tigris & Euphrates R.

**Shi·raz** (shē-räz′). City of SW Iran SSE of Esfahan. Pop. 416,408.

**Shi·re** (shîr′ā). River, c. 250 mi (402 km), of S Malawi & central Mozambique.

**Shi·shal·din** (shĭ-shăl′dĭn). Volcano, 9,370 ft (2,857.9 m), of SW Alas. on central Unimak Is.

**Shive·ly** (shīv′lē). City of N Ky. near Louisville. Pop. 16,819.

**Shi·zu·o·ka** (shĭz′ə-wō′kə). City of E central Honshu, Japan, on Suruga Bay. Pop. 446,952.

**Shkha·ra** (shə-kä-rä′). Peak, c. 17,064 ft (5,200 m), of the Caucasus in S European USSR.

**Shko·dër** (shkō′dər). City of NW Albania on Lake Scutari. Pop. 55,300.

**Sho·la·pur** (shō′lə-po͝or′). City of W central India ESE of Bombay. Pop. 398,361.

**Shore·view** (shôr′vyo͞o′, shōr′-). City of E Minn. near St. Paul. Pop. 17,300.

**Shore·wood** (shôr′wo͝od′, shōr′-). Village of SE Wis. near Milwaukee. Pop. 14,327.

**Sho·sho·ne** (shə-shō′nē). **1.** River, c. 120 mi (193 km), of NW Wyo. **2.** Falls, 212 ft (64.7 m), in the Snake R. of S Idaho.

**Shreve·port** (shrēv′pôrt′, -pōrt′). City of NW La. near the Tex. border. Pop. 205,815.

**Shrews·bur·y** (shro͞oz′bĕr′ē, -bə-rē). **1.** Borough of W central England on the Severn WNW of Birmingham. Pop. 56,120. **2.** Town of central Mass. near Worcester. Pop. 22,674.

**Shrop·shire** (shrŏp′shîr′, -shər). Region & former county of W England on the Welsh border.

**Shu·ma·gin** (sho͞o′mə-gĭn). Islands of SW Alas. off the SE coast of the Alaska Peninsula.

**Shu·men** (sho͞o′mĕn′). City of NE Bulgaria W of Varna. Pop. 83,525.

**Shu·shan** (sho͞o′shən, -shăn′). SUSA.

**Si** (shē). XI.

**Si·al·kot** (sē-äl′kōt′). City of NE Pakistan near the Kashmir border N of Lahore. Pop. 203,779.

**Si·am** (sī-ăm′). **1. Gulf of.** Arm of the South China Sea between the Malay Peninsula & Indochina. **2.** THAILAND. **—Si′a·mese′** (sī′ə-mēz′, -mēs′) *adj.* & *n.*

**Si·an** (shē′än′). XI'AN.

**Si·ang** (shē-äng′). XIANG.

**Si·ang·tan** (shē-äng′tän′). XIANGTAN.

**Šiau·liai** (shou′lā′). City of W European USSR in Lithuania NNW of Kaunas. Pop. 118,000.

**Si·be·ri·a** (sī-bîr′ē-ə). Region of Asian USSR from the Urals to the Pacific. **—Si·be′ri·an** *adj.* & *n.*

**Si·biu** (sē-byo͞o′). City of central Rumania SSE of Cluj. Pop. 156,854.

**Si·bu·yan** (sē′bo͞o-yän′). Sea in central Philippines bordered by S Luzon, Mindoro, & the Visayans.

**Si·chuan** (sēch′wän′). Province of S central China.

**Si·ci·ly** (sĭs′ə-lē). Island of S Italy W of S end of Italian Peninsula; largest island in the Mediterranean. **—Si·cil′ian** (sĭ-sĭl′yən) *adj.* & *n.*

**Si·cy·on** (sĭsh′ē-ŏn′, sĭs′-). Ancient city of S Greece in the NE Peloponnesus near the Gulf of Corinth.

**Si·di-bel-Ab·bès** (sē′dē-bĕl-ə-bĕs′). City of NW Algeria S of Oran. Pop. 116,000.

**Sid·ley** (sĭd′lē). Mountain, 13,717 ft (4,183.7 m), in Marie Byrd Land, Antarctica.

**Sid·ney** (sĭd′nē). City of W central Ohio WNW of Columbus. Pop. 17,657.

**Si·don** (sīd′n). Ancient city of Phoenicia on the Mediterranean in present-day SW Lebanon.

**Sid·ra** (sĭd′rə). **Gulf of.** Inlet of the Mediterranean off N Libya W of Benghazi.

**Sie·ben·ge·bir·ge** (zē′bən-gə-bîr′gə). Range of hills in W central West Germany along the Rhine S of Bonn.

**Sie·gen** (zē′gən). City of central West Germany E of Cologne. Pop. 116,552.

**Sie·mia·no·wi·ce Śla·skie** (shə-myä′nə-vēt′sə-shlôN′skē-ə). City of S Poland near Katowice. Pop. 67,628.

**Si·en·a** (sē-ĕn′ə). City of central Italy S of Florence. Pop. 56,539. **—Si′e·nese′** (-nēz′, -nēs′) *adj.* & *n.*

**Si·er·ra de Cór·do·ba** (sē-ĕr′ə də kôr′də-bə). Range extending c. 300 mi (483 km) in central Argentina.

**Sierra de Gre·dos** (grā′dōs). Range of W central Spain W of Madrid.

**Sierra de Gua·dar·ra·ma** (gwä′də-rä′mə). Range of central Spain N of Madrid.

**Sierra Le·one** (lē-ōn′). Country of W Africa on the Atlantic coast. Cap. Freetown. Pop. 3,470,000.

**Sierra Ma·dre** (mä′drā). **1.** Mountain system of Mexico comprising three ranges: **a. Sierra Madre del Sur** (dĕl so͞or′), in S Mexico along the Pacific coast. **b. Sierra Madre Oc·ci·den·tal** (ŏk′sə-dĕn′təl), in NW Mexico inland from the Pacific coast. **c. Sierra Madre Ori·en·tal** (ôr′ē-ĕn-täl′, ōr′-), in NE Mexico inland from the Gulf of Mexico coast. **2.** City of S Calif. near Pasadena. Pop. 10,837.

**Sierra Ma·es·tra** (mä-ĕs′trə). Range of SE Cuba.

**Sierra Mo·re·na** (mə-rā′nə). Range of S Spain extending c. 375 mi (603 km) E from the Portuguese border.

**Sierra Ne·va·da** (nə-văd′ə, -vä′də). **1.** Range of S Spain along the Mediterranean coast E of Granada, rising to 11,411 ft (3,480.4 m). **2.** Range of E Calif. extending c. 400 mi (644 km) & rising to 14,494 ft (4,420.7 m).

**Sierra Nevada de Mé·ri·da** (də mĕr′ĭ-də). CORDILLERA DE MÉRIDA.

**Sierra Nevada de San·ta Mar·ta** (də săn′tə mär′tə). Range of N Colombia, rising to 19,020 ft (5,801.1 m).

**Sierra Pa·ca·rai·ma** (păk'ə-rī'mə). Range of SE Venezuela & W Guyana along the Brazilian border.
**Sierra Pa·ri·ma** (pə-rē'mə). Range of S Venezuela along the Brazilian border SW of Sierra Pacaraima.
**Sierra Vis·ta** (vĭs'tə). City of SE Ariz. SE of Tucson. Pop. 25,968.
**Sikes·ton** (sīk'stən). City of SE Mo. WSW of Cairo, Ill. Pop. 17,431.
**Si·kho·te-A·lin** (sē'kə-tā'ə-lēn'). Range of extreme SE Far Eastern USSR extending along the Sea of Japan coast N of Vladivostok.
**Sik·kim** (sĭk'ĭm). State & former semi-independent protectorate of NE India in the E Himalayas between Nepal & Bhutan.
**Si·le·sia** (sī-lē'zhə, -zhē-ə, -shə, -shē-ə, sə-). Region of central Europe in SW Poland & N Czechoslovakia. **—Si·le'sian** *adj. & n.*
**Sil·le·ry** (sĭl'ə-rē, sē'-yə-rē'). City of S Que., Canada, on the St. Lawrence near Quebec city. Pop. 13,580.
**Sil·ver Spring** (sĭl'vər). City of central Md. near Washington, D.C. Pop. 84,300.
**Sil·vret·ta** (sĭlv-rĕt'ə). Range of the Alps in E Switzerland & SW Austria, rising to 11,185 ft (3,411.4 m).
**Sim·coe** (sĭm'kō). **1.** Lake, 539 sq mi (1,396 sq km), of SE Ont., Canada, between Georgian Bay & Lake Ontario. **2.** City of S Ont., Canada, S of Brantford. Pop. 14,189.
**Sim·fer·o·pol** (sĭm'fə-rô'pəl, -rō'-, sĭmp'-). City of S European USSR in the S central Crimea. Pop. 302,000.
**Si·mi Valley** (sē'mē, sĭm'ē). City of S Calif. near Los Angeles. Pop. 77,500.
**Sim·la** (sĭm'lə). Town of N India in the W Himalayas N of Delhi. Pop. 55,368.
**Sim·plon** (sĭm'plŏn'). **1.** Pass, 6590 ft (2,010 m), between the Lepontine & Pennine Alps in S Switzerland near the Italian border. **2.** Tunnel, 12.3 m (19.8 km) long, through Monte Leone on the Italian-Swiss border NE of Simplon Pass.
**Simp·son** (sĭmp'sən). Desert of central Australia.
**Sims·bur·y** (sĭmz'bĕr'ē, -bə-rē). Town of N Conn. NW of Hartford. Pop. 21,161.
**Si·nai** (sī'nī'). **1. Mount.** Mountain, c. 7,500 ft (2,288 m), of NE Egypt on S Sinai Peninsula. **2.** Peninsula of NE Egypt at the N end of the Red Sea.
**Sind** (sĭnd). Region of S Pakistan along the lower Indus R.
**Sin·ga·pore** (sĭng'gə-pôr', -pōr', sĭng'ə-). **1.** Strait off S end of Malay Peninsula between Singapore Is. & Riau Archipelago. **2.** Island, 224 sq mi (580 sq km), off the S end of Malay Peninsula. **3.** Country of SE Asia comprising Singapore Is. & adjacent smaller islands. Cap. Singapore. Pop. 2,413,945. **4.** Cap. of Singapore on Singapore Strait. **—Sin'ga·por'e·an** *adj. & n.*
**Sin·kiang Ui·ghur** (shĭn'jē-äng' wē'gər). XINJIANG UYGUR.
**Si·nop** (sə-nôp'). City of N Turkey on the Black Sea. Pop. 16,098.
**Sint-Ni·klaas** (sĭnt-nē'kläs). City of N Belgium near Antwerp. Pop. 49,214.
**Sin·tra** (sēn'trə, sĭn'-). Town of W Portugal NW of Lisbon. Pop. 15,994.
**Sin·ui·ju** (shĭn'ē-jōō'). City of W North Korea on Korea Bay at the mouth of the Yalu R. Pop. 300,000.
**Si·on** (sē-ōɴ'). Town of SW Switzerland on the Rhone R. Pop. 21,925.
**Sioux City** (sōō). City of W Iowa on the Missouri near the S.Dak.-Nebr. border. Pop. 82,003.
**Sioux Falls.** City of SE S.Dak. near the Minn. border. Pop. 81,343.
**Sip·par** (sĭ-pär'). Ancient city of N Babylonia on the Euphrates near present-day Baghdad.
**Si·ret** (sē-rĕt'). River, c. 280 mi (451 km), of NE Rumania.
**Sí·ros** (sē'rôs'). *var. of* SYROS.
**Sir Sand·ford** (sər săn'fərd), **Mount.** Highest (c. 11,590 ft/3,535 m) of the Selkirk Mts. in SE B.C., Canada.
**Sis·ki·you** (sĭs'kə-yōō'). Range of the Klamath Mts. in N Calif. & SW Ore.
**Sit·ka** (sĭt'kə). Town of SE Alas. on the W coast of Baranof Is. Pop. 7,803.
**Sit·tang** (sĭt'äng'). River of Burma flowing 350 mi (563 km) S to the Gulf of Martaban.
**Sit·ting·bourne and Mil·ton** (sĭt'ĭng-bôrn', -bōrn'; mĭl'tən). Urban district of SE England ESE of London. Pop. 32,830.
**Sit·twe** (sĭt'wē'). City of W Burma on the Bay of Bengal. Pop. 42,329.
**Si·vas** (sĭ-väs', sē-). City of central Turkey E of Ankara. Pop. 149,201.
**Si·vash** (sĭ-väsh'). Lagoon, c. 1,000 sq mi (2,590 sq km), of SW European USSR along the NE coast of the Crimea.
**Si·wa** (sē'wə). Oasis of NW Egypt in the Libyan Desert.
**Si·wa·lik** (sĭ-wä'lĭk). Range of the S Himalayas extending c. 1,050 mi (1,690 km), from SW Kashmir through N India into S Nepal.
**Sjael·land** (shĕl'än'). Largest island of Denmark in the E part between the Kattegat & the Baltic.
**Ska·gen** (skä'gən), **Cape.** The SKAW.
**Skag·er·rak** *also* **Skag·er·ak** (skăg'ə-răk', skä'gə-räk'). Strait, c. 150 mi (241 km) long & 85 mi (137 km) wide, between Norway & Denmark, linking the North Sea & the Kattegat.
**Skag·it** (skăj'ĭt). River, c. 150 mi (241 km), of SW B.C., Canada, & NW Wash.
**Skag·way** (skăg'wā'). Town of SE Alas. at the head of the Lynn Canal NNW of Juneau. Pop. 768.
**Skan·e·at·e·les** (skăn'ē-ăt'ləs, skĭn'-). One of the Finger Lakes, 14 sq mi (36.3 sq km), in central N.Y.
**Skaw** (skô), **The.** Cape on N extremity of Jutland, Denmark, extending into the Skagerrak.
**Skee·na** (skē'nə). River, c. 360 mi (579 km), of W B.C., Canada.
**Skel·lef·te** (shĕ-lĕf'tə). River, c. 255 mi (410 km), of N Sweden.
**Skel·mers·dale and Hol·land** (skĕl'mərz-dāl'; hŏl'ənd). Urban district of NW England NE of Liverpool. Pop. 35,850.
**Skid·daw** (skĭd'ô). Mountain, 3,054 ft (931.5 m), in NW England.
**Ski·en** (shē'ən, shā'-). City of SE Norway SW of Oslo. Pop. 47,105.
**Skik·da** (skēk'də). City of NE Algeria on the Mediterranean. Pop. 107,700.
**Skí·ros** (skē'rôs). *var. of* SKYROS.
**Sko·kie** (skō'kē). Village of NE Ill. near Chicago. Pop. 60,278.
**Skop·lje** (skôp'lā', -yā') *or* **Skop·je** (-yā'). City of SE Yugoslavia in Macedonia on the Vardar R. Pop. 308,117.
**Skunk** (skŭngk). River, 264 mi (424.8 km), of central & SE Iowa.
**Skye** (skī). Island of NW Scotland in the Inner Hebrides.
**Sky·ros** (skī'rəs, -rōs') *also* **Skí·ros** (skē'rôs) *or* **Scy·ros** (sī'rəs). Island of E Greece, largest of the N Sporades, in the Aegean NE of Euboea.
**Sla** (slä). *var. of* SALÉ.
**Slave** (slāv). River, c. 310 mi (499 km), of NE Alta & S N.W.T., Canada.
**Sla·vo·ni·a** (slə-vō'nē-ə, -vōn'yə). Region of N Yugoslavia between the Drava & Sava rivers. **—Sla·vo'ni·an** *adj. & n.*
**Slav·yansk** (slə-vyänsk'). City of S European USSR in the E Ukraine SE of Kharkov. Pop. 140,000.
**Sli·dell** (slī-dĕl'). City of SE La. NE of New Orleans. Pop. 26,718.
**Slide Mountain** (slīd). Highest (4,024 ft/1,227.3 m) of the Catskill Mts. in SE N.Y.
**Sli·go** (slī'gō). City of NW Ireland on the Atlantic. Pop. 16,836.
**Slough** (slou). Borough of SE England near London. Pop. 89,060.
**Slo·vak·i·a** (slō-vä'kē-ə, -văk'ē-ə). Region of E Czechoslovakia. **—Slo·vak'i·an** *adj. & n.*
**Slo·ve·ni·a** (slō-vē'nē-ə, -vēn'yə). Region & constituent republic of NW Yugoslavia. **—Slo'vene', Slo·ve'ni·an** *adj. & n.*
**Smith·field** (smĭth'fēld'). Town of NE R.I. near Providence. Pop. 16,886.
**Smok·y** (smō'kē). River, c. 250 mi (402 km), of W central Alta., Canada.
**Smoky Hill.** River, c. 560 mi (900 km), rising in E Colo. & flowing E across Kans. to the Kansas R.
**Smo·lensk** (smō-lĕnsk'). City of W central European USSR on the Dnieper WSW of Moscow. Pop. 276,000.
**Smyr·na** (smûr'nə). **1.** IZMIR. **2.** City of NW Ga. near Atlanta. Pop. 20,312.
**Snake** (snāk). River of NW U.S. rising in NW Wyo. & flowing 1,038 mi (1,670 km) through S Idaho, along the Ore.-Idaho border, & through SE Wash. into the Columbia.
**Snef·fels** (snĕf'əlz). Peak, 14,143 ft (4,313.6 m), in the San Juan Mts. of SW Colo.
**Sno·qual·mie** (snō-kwŏl'mē). **1.** Falls, 270 ft (82.4 m) high, in the **Snoqualmie R.** (45 mi/72 km) of W central Wash. **2.** Pass, 3,004 ft (916.2 m) high, in the Cascade Range of W central Wash.
**Snow** (snō). Mountains of central New Guinea.
**Snow·don** (snōd'n). Mountain, 3,560 ft (1,085.8 m), of NW Wales.
**Snow·y** (snō'ē). **1.** River, 278 mi (447 km), of SE Australia. **2.** Range of the Australian Alps in SE Australia.
**Sny·der** (snī'dər). City of NW Tex. WNW of Abilene. Pop. 12,705.
**So·bat** (sō'băt'). River, c. 205 mi (330 km), of W Ethiopia & SE Sudan.
**So·chi** (sō'chē). City of S European USSR on the NE shore of the Black Sea. Pop. 287,000.
**So·ci·e·ty** (sə-sī'ĭ-tē). Islands of French Polynesia in the S Pacific E of Samoa.
**So·co·tra** (sə-kō'trə). Island of Southern Yemen in the Indian Ocean at the mouth of the Gulf of Aden.
**Sö·der·täl·je** (sœ'dər-tĕl'yə). City of E Sweden SW of Stockholm. Pop. 58,408.
**Sod·om** (sŏd'əm). City of ancient Palestine.
**So·fi·a** (sō'fē-ə, sō-fē'ə). Cap. of Bulgaria in the W central part. Pop. 965,728.
**Sog·na·fjord** *or* **Sog·ne Fjord** (sông'nə-fyôr). Inlet of the Norwegian Sea in SW Norway.
**So·ho** (sō'hō'). **1.** District of London, England. **2.** District of New York City on Manhattan Is.
**Sois·sons** (swä-sōɴ'). City of N France NW of Paris. Pop. 29,694.
**So·ka** (sō'kə). City of E central Honshu, Japan, near Tokyo. Pop. 167,177.
**So·ko·to** (sō'kə-tō'). City of NW Nigeria WNW of Kano. Pop. 104,000.

**So·lent** (sō′lənt), **The.** Channel, c. 15 mi (24 km) long, between the Isle of Wight & the S English mainland.
**So·li·hull** (sō′lĭ-hŭl′). Borough of central England near Birmingham. Pop. 108,230.
**So·li·mões** (so͞o′lē-moiNsh′). The upper Amazon R. from the Río Negro to the Peruvian border.
**So·ling·en** (zō′lĭng-ən, sō′-). City of W West Germany ESE of Düsseldorf. Pop. 171,180.
**Sol·na** (sōl′nə). City of E Sweden near Stockholm. Pop. 53,992.
**So·lo** (sō′lō). **1.** River, 335 mi (539 km), of Java, Indonesia. **2.** SURAKARTA.
**Sol·o·mon Islands** (sŏl′ə-mən) *also* **Sol·o·mons** (-mənz). **1.** Islands of the W Pacific E of New Guinea, divided between Papua New Guinea & the independent Solomon Is. **2.** Nation comprising the Solomons SE of Bougainville. Cap. Honiara. Pop. 221,000.
**So·lon** (sō′lən). City of NE Ohio SE of Cleveland. Pop. 14,341.
**So·lo·vets·ki** (sŏl′ə-vĕt′skē). Islands of N European USSR in the White Sea.
**Sol·way Firth** (sŏl′wā′). Arm of the Irish Sea separating NW England from SW Scotland.
**So·ma·li·a** (sō-mä′lē-ə, -mäl′yə). Country of extreme E Africa on the Gulf of Aden & the Indian Ocean. Cap. Mogadishu. Pop. 3,645,000. **—So·ma′li·an** *adj. & n.*
**So·ma·li·land** (sō-mä′lē-lănd′, sə-). Region of E Africa including Somalia, Djibouti, & parts of E Ethiopia.
**Som·er·set** (sŭm′ər-sĕt′, -sĭt). **1.** Island of N.W.T., Canada, N of the Boothia Peninsula. **2.** City of S central Ky. S of Lexington. Pop. 10,649. **3.** Town of SE Mass. N of Fall River. Pop. 18,813.
**Som·ers Point** (sŭm′ərz). City of SE N.J. SW of Atlantic City. Pop. 10,330.
**Som·ers·worth** (sŭm′ərz-wûrth′). City of SE N.H. N of Dover. Pop. 10,350.
**Som·er·ville** (sŭm′ər-vĭl′). **1.** City of E Mass. near Boston. Pop. 77,372. **2.** Borough of N central N.J. WNW of New Brunswick. Pop. 11,973.
**Somme** (sŏm, sŭm). River, c. 150 mi (241 km), of N France.
**Song·hai** *also* **Song·hay** (sông′hī′). Ancient empire of W Africa.
**Song·hua** (so͝ong′hwä′). River of Manchuria, NE China, rising near the North Korean border & flowing c. 1,150 mi (1,850 km) NW, E, & NE to the Amur.
**So·no·ra** (sə-nôr′ə, -nōr′-). River, c. 250 mi (402 km) of NW Mexico.
**Soo Canals** (so͞o). SAULT STE. MARIE CANALS.
**Soo·chow** (so͞o′jō′, -chou′). SUZHOU.
**So·pot** (sô′pôt′). City of N Poland on the Gulf of Danzig near Gdańsk. Pop. 47,573.
**Sop·ron** (shō′prōn′). City of NW Hungary near the Austrian border. Pop. 53,930.
**So·rel** (sə-rĕl′). City of S Que., Canada, at the confluence of the St. Lawrence & Richelieu rivers. Pop. 19,666.
**So·ri·a** (sô′ryä). Town of N central Spain W of Saragossa. Pop. 24,744.
**So·ro·ca·ba** (sôr′o͞o-kăb′ə, sōr′-). City of S Brazil W of São Paulo. Pop. 165,990.
**Sor·ren·to** (sə-rĕn′tō). Town of S Italy on the **Sorrento Peninsula,** separating the Bay of Naples from the Gulf of Salerno. Pop. 13,078.
**So·sno·wiec** (sôs-nō′vyĕts). City of S Poland near Katowice. Pop. 144,652.
**Sou·fri·ère** (so͞o′frē-ĕr′). **1.** Volcano, 4,813 ft (1,468 m), on Guadeloupe, French West Indies. **2.** Volcano, 4,048 ft (1,234.6 m), on St. Vincent in St. Vincent and the Grenadines, West Indies.
**Sou·ris** (so͝or′ĭs). River, c. 450 mi (724 km), of SE Sask., N N.Dak., & SW Man.
**Sousse** (so͞os) *also* **Su·sah** or **Su·sa** (so͞o′zə, -sə). City of NE Tunisia on an inlet of the Mediterranean. Pop. 69,530.
**South** (south). **1. the.** Region of the SE U.S. S of Pa. & the Ohio R. & E of the Mississippi. **2. Cape.** KA LAE. **3.** Larger island of New Zealand, SW of North Is.
**South Af·ri·ca** (ăf′rĭ-kə). Republic of S Africa on the Atlantic & Indian oceans. Caps. Pretoria & Cape Town. Pop. 23,771,970. **—South Af′ri·can** *adj. & n.*
**South A·mer·i·ca** (ə-mĕr′ĭ-kə). Continent of the S Western Hemisphere SE of North America between the Atlantic & the Pacific. **—South A·mer′i·can** *adj. & n.*
**South·amp·ton** (south-hămp′tən, sou-thămp′-). **1.** Island of N.W.T., Canada, at the entrance to Hudson Bay. **2.** Borough of S central England on an inlet of the English Channel opposite the Isle of Wight. Pop. 213,700.
**South Bend.** City of N Ind. near the Mich. border. Pop. 109,727.
**South·bridge** (south′brĭj′). Town of S Mass. SW of Worcester. Pop. 16,665.
**South Bur·ling·ton** (bûr′lĭng-tən). City of NW Vt. near Burlington. Pop. 10,679.
**South·bur·y** (south′bĕr′ē, -bə-rē). Town of SW Conn. SW of Waterbury. Pop. 14,156.
**South Car·o·li·na** (kăr′ə-lī′nə). State of the SE U.S. on the Atlantic. Cap. Columbia. Pop. 3,119,208. **—South Car′o·lin′i·an** (-lĭn′ē-ən) *adj. & n.*
**South Charles·ton** (chärl′stən). City of W W.Va. near Charleston. Pop. 15,968.
**South Chi·na Sea** (chī′nə). Arm of the W Pacific bounded by SE China, Taiwan, the Philippines, Borneo, & Vietnam.
**South Da·ko·ta** (də-kō′tə). State of the N central U.S. Cap. Pierre. Pop. 690,178. **—South Da·ko′tan** *adj. & n.*
**South Downs** (dounz). Range of hills in SE England.
**South El Mon·te** (ĕl mŏn′tē). City of S Calif. NNE of Long Beach. Pop. 16,623.
**South·end-on-Sea** (sou′thĕnd-ŏn-sē′, -ôn-). Borough of SE England at the mouth of the Thames. Pop. 159,300.
**South·ern Alps** (sŭth′ərn ălps). Range of W South Is., New Zealand, rising to 12,349 ft (3,766.4 m).
**Southern Yem·en** (yĕm′ən, yā′mən). —See YEMEN.
**South Eu·clid** (yo͞o′klĭd). City of NE Ohio near Cleveland. Pop. 25,713.
**South·field** (south′fēld′). City of SE Mich. near Detroit. Pop. 75,568.
**South·gate** (south′gāt′). City of SE Mich. near Detroit. Pop. 32,058.
**South Gate.** City of S Calif. near Los Angeles. Pop. 66,784.
**South Geor·gia** (jôr′jə). British-administered island of the S Atlantic E of Cape Horn.
**South Hol·land** (hŏl′ənd). Village of NE Ill. near Chicago. Pop. 24,977.
**South Hous·ton** (hyo͞o′stən). City of SW Tex. near Houston. Pop. 13,293.
**South·ing·ton** (sŭth′ĭng-tən). Town of central Conn. NE of Waterbury. Pop. 36,879.
**South Kings·town** (kĭngz′toun′). Town of S R.I. SSW of Providence. Pop. 20,414.
**South Ko·re·a** (kə-rē′ə). —See KOREA.
**South Lake Ta·hoe** (tä′hō). City of E Calif. on Lake Tahoe near the Nev. border. Pop. 20,681.
**South Mi·am·i** (mī-ăm′ē, -ăm′ə). City of SE Fla. near Miami. Pop. 10,884.
**South Mil·wau·kee** (mĭl-wô′kē). City of SE Wis. near Milwaukee. Pop. 21,069.
**South Mountain.** Ridge of S Pa. & W Md. in the Blue Ridge Mts; site of Civil War battle (1862).
**South Na·han·ni** (nə-hăn′ē). River, c. 350 mi (563 km), of SW N.W.T., Canada.
**South Og·den** (ŏg′dən). City of N Utah near Ogden. Pop. 11,366.
**South Or·ange** (ôr′ĭnj, ŏr′-). Village of NE N.J. W of Newark. Pop. 16,971.
**South Ork·ney** (ôrk′nē). British-administered islands in the S Atlantic SE of Cape Horn.
**South Pas·a·de·na** (păs′ə-dē′nə). City of S Calif. near Los Angeles. Pop. 22,681.
**South Pass.** Valley in SW Wyo. at the S end of the Wind River Range.
**South Plain·field** (plān′fēld′). Borough of NE N.J. SW of Elizabeth. Pop. 20,521.
**South Platte** (plăt). River of central & NE Colo. & W central Nebr. flowing c. 450 mi (724 km) E to the North Platte, forming the Platte.
**South Pole.** S end of the earth's axis of rotation, a point in central Antarctica.
**South·port** (south′pôrt′, -pōrt′). Borough of NW England on Liverpool Bay N of Liverpool. Pop. 86,030.
**South Port·land** (pôrt′lənd, pōrt′-). City of SW Me. near Portland. Pop. 22,712.
**South River.** Borough of E central N.J. SW of Perth Amboy. Pop. 14,361.
**South Saint Paul** (sānt pôl′). City of E Minn. near St. Paul. Pop. 21,235.
**South Salt Lake** (sôlt). City of N Utah near Salt Lake City. Pop. 10,561.
**South Sand·wich** (sănd′wĭch′, săn′-). British-administered islands in the S Atlantic ESE of South Georgia Is.
**South San Fran·cis·co** (săn′ frən-sĭs′kō). City of W Calif. on W San Francisco Bay. Pop. 49,393.
**South Sas·katch·e·wan** (sə-skăch′ə-wən, -wän′, săs-kăch′-). River of Canada flowing c. 550 mi (890 km) from S Alta. to central Sask. to join the North Saskatchewan & form the Saskatchewan.
**South Seas. 1.** All seas S of the equator. **2.** The S Pacific.
**South Shet·land** (shĕt′lənd). Islands in the S Atlantic off the Antarctic Peninsula; claimed by United Kingdom, Argentina, & Chile.
**South Shields** (shēldz). Borough of NE England at the mouth of the Tyne R. Pop. 96,900.
**South·west** (south-wĕst′), **the.** Region of the SW U.S. including N.Mex., Ariz., Tex., Calif., Nev., Utah, & Colo.
**South-West Af·ri·ca** (south′wĕst ăf′rĭ-kə). NAMIBIA.

**South Wind·sor** (wĭn′zər). Town of N central Conn. NNE of Hartford. Pop. 17,198.
**So·vi·et Union** (sō′vē-ĕt′, -ĭt, sŏv′ē-). UNION OF SOVIET SOCIALIST REPUBLICS.
**So·we·to** (sə-wē′tō). City of NE South Africa near Johannesburg. Pop. 602,043.
**So·ya** (sō′yä). LA PÉROUSE.
**Spa** (spä). Resort town of E Belgium in the Ardennes. Pop. 9,766.
**Spain** (spān). Country of SW Europe including most of the Iberian Peninsula & the Balearic & Canary Is. Cap. Madrid. Pop. 37,430,000. **—Span′iard** (spăn′yərd) *n.* **—Span′ish** (spăn′ĭsh) *adj.* & *n.*
**Spa·la·to** (spä′lə-tō′). SPLIT.
**Span·dau** (spän′dou′, shpän′-). District of West Berlin, Germany.
**Span·ish** (spăn′ĭsh). River, c. 150 mi (240 km), of S Ont., Canada.
**Spanish Main. 1.** Coast of N South America in colonial times. **2.** Parts of the Caribbean crossed by Spanish shipping in colonial times.
**Spanish Peaks.** Two adjacent mountains, 12,683 ft (3,868.3 m) & 13,623 ft (4,155 m), in S Colo.
**Spanish Sa·ha·ra** (sə-hâr′ə, -hä′rə). WESTERN SAHARA.
**Spanish Town.** Town of SE Jamaica W of Kingston. Pop. 40,731.
**Sparks** (spärks). City of W Nev. E of Reno. Pop. 40,780.
**Spar·ta** (spär′tə). City-state of ancient Greece in the SE Peloponnesus. **—Spar′tan** *adj.* & *n.*
**Spar·tan·burg** (spär′tn-bûrg′). City of NW S.C. NW of Columbia. Pop. 43,968.
**Speed·way** (spēd′wā′). Town of central Ind. near Indianapolis. Pop. 12,641.
**Spen·cer** (spĕn′sər). **1.** City of NW Iowa on the Little Sioux R. NE of Sioux City. Pop. 11,726. **2.** Town of central Mass. W of Worcester. Pop. 10,774.
**Spencer Gulf.** Inlet of the Indian Ocean off S central Australia between the Eyre & Yorke peninsulas.
**Spey** (spā). River, c. 105 mi (169 km), of NE Scotland.
**Spey·er** (spīr, spī′ər, shpīr, shpī′ər). City of SW West Germany on the Rhine R. Pop. 43,663.
**Spice Islands** (spīs). MOLUCCAS.
**Spires** (spīrz). SPEYER.
**Spits·ber·gen** (spĭts′bûr′gən). Norwegian archipelago in the Arctic Ocean N of Norway.
**Split** (splĭt). City of W Yugoslavia on the Dalmatian coast of the Adriatic. Pop. 150,739.
**Spo·kane** (spō-kăn′). City of E Wash. near the Idaho border. Pop. 171,300.
**Spo·le·to** (spə-lā′tō). City of central Italy N of Rome. Pop. 18,013.
**Spor·a·des** (spôr′ə-dēz′, spŏr′-). All the islands of Greece in the Aegean excepting the Cyclades.
**Spot·syl·va·nia** (spŏt′səl-vān′yə). Village of NE Va.; site of a Civil War battle (1864).
**Spree** (sprā, shprā). River, c. 250 mi (402 km), of E East Germany.
**Spring·dale** (sprĭng′dāl′). **1.** City of NW Ark. near Fayetteville. Pop. 23,185. **2.** City of SW Ohio near Cincinnati. Pop. 10,111.
**Spring·field** (sprĭng′fēld′). **1.** Cap. of Ill. in the central part. Pop. 99,637. **2.** City of SW Mass. on the Connecticut R. Pop. 152,319. **3.** City of SW Mo. SE of Kansas City. Pop. 133,116. **4.** Township of NE N.J. W of Newark. Pop. 15,740. **5.** City of W central Ohio W of Columbus. Pop. 72,563. **6.** City of W central Ore. near Eugene. Pop. 41,621. **7.** City of N Tenn. N of Nashville. Pop. 10,814. **8.** Town of SE Vt. near the N.H. border SE of Rutland. Pop. 10,190.
**Springs** (sprĭngz). City of NE South Africa E of Johannesburg. Pop. 142,812.
**Spring Valley** (sprĭng). Village of SE N.Y. near the N.J. border WNW of White Plains. Pop. 20,537.
**Spring·ville** (sprĭng′vĭl′). City of N central Utah near Provo. Pop. 12,101.
**Spuy·ten Duy·vil Creek** (spīt′n dī′vəl). Tidal channel, now a ship canal, in SE N.Y. separating N Manhattan Is. from the mainland.
**Squaw Valley** (skwô). Valley of NE Calif. in the Sierra Nevada.
**Sri Lan·ka** (srē läng′kə). Island nation in the Indian Ocean off SE India. Cap. Colombo. Pop. 14,850,001. **—Sri Lan′kan** *adj.* & *n.*
**Sri·na·gar** (srĭ-nŭg′ər). City of N India, historic cap. of Kashmir, on the Jhelum R. Pop. 403,413.
**St.** For entries beginning with **St.**, see SAINT.
**Staf·fa** (stăf′ə). Island of W Scotland in the Inner Hebrides W of Mull.
**Staf·ford** (stăf′ərd). Borough of W central England NNW of Birmingham. Pop. 54,860.
**Sta·gi·ra** (stə-jī′rə) *also* **Sta·gi·rus** (-rəs). Ancient city of Macedonia, NE Greece; birthplace of Aristotle.
**Staines** (stānz). Urban district of SE England on the Thames near London. Pop. 56,380.
**Sta·lin·grad** (stä′lĭn-grăd′, stăl′ĭn-). VOLGOGRAD.
**Sta·lin Peak** (stä′lĭn, stăl′ĭn). Mt. COMMUNISM.
**Stam·bul** *also* **Stam·boul** (stăm-bo͞ol′). The old section of Istanbul.
**Stam·ford** (stăm′fərd). City of SW Conn. on Long Is. Sound. Pop. 102,453.
**Stan·ley** (stăn′lē). **1.** Series of falls in the Lualaba R. above Kisangani in N central Zaire. **2. Pool.** Lakelike expansion of the Congo R., c. 320 sq mi (829 sq km), on the Zaire-Congo border at Kinshasa & Brazzaville. **3. Mount.** Peak, 16,795 ft (5,122.5 m), in the Ruwenzori Range on the Zaire-Uganda border. **4.** Town of the E Falkland Is. on the Atlantic. Pop. 1,081.
**Stan·ley·ville** (stăn′lē-vĭl′). KISANGANI.
**Stan·o·voi** *or* **Stan·o·voy** (stăn′ə-voi′). Range, c. 450 mi (725 km) long, of SE Far Eastern USSR N of the Amur R.
**Stan·ton** (stăn′tən). City of S Calif. NW of Santa Ana. Pop. 21,144.
**Sta·ra Za·go·ra** (stä′rə zə-gôr′ə). City of central Bulgaria ENE of Plovdiv. Pop. 122,200.
**Stark·ville** (stärk′vĭl′, -vəl). City of E Miss. W of Columbus. Pop. 15,169.
**State Col·lege** (stāt kŏl′ĭj). Borough of central Pa. NE of Altoona. Pop. 36,130.
**Stat·en** (stăt′n). Island in New York Bay, SE N.Y., SW of Manhattan, coextensive with the New York City borough of Richmond.
**States·bor·o** (stāts′bûr-ō). City of E Ga. WNW of Savannah. Pop. 14,866.
**States·ville** (stāts′vĭl′, -vəl). City of W central N.C. N of Charlotte. Pop. 18,622.
**Staun·ton** (stăn′tən). Independent city of W central Va. WNW of Charlottesville. Pop. 21,857.
**Sta·vang·er** (stə-văng′ər). City of SW Norway on an inlet of the North Sea. Pop. 86,639.
**Stav·ro·pol** (stăv-rô′pəl, -rō′-). City of S European USSR SE of Rostov. Pop. 258,000.
**Steele** (stēl), **Mount.** Mountain, 16,644 ft (5,076.4 m), in the St. Elias Mts. of SW Y.T., Canada.
**Steens** (stēnz). Mountains of SE Ore.
**Stel·vi·o** (stĕl′vē-ō). Pass, 9,048 ft (2,759.6 m) high, in the central Alps in N Italy near the Swiss & Austrian borders.
**Ste·phen·ville** (stē′vən-vĭl′). **1.** Town of SW Newf., Canada, on the Gulf of St. Lawrence SW of Corner Brook. Pop. 10,284. **2.** City of N central Tex., SW of Fort Worth. Pop. 11,881.
**Ster·ling** (stûr′lĭng). **1.** City of NE Colo. ENE of Greeley. Pop. 11,385. **2.** City of NW Ill. SW of Rockford. Pop. 16,273.
**Sterling Heights.** City of SE Mich. near Detroit. Pop. 108,999.
**Ster·li·ta·mak** (stĕr′lĭ-tə-măk′). City of E European USSR E of Kuibyshev. Pop. 220,000.
**Stet·tin** (stə-tēn′, shtə-). SZCZECIN.
**Stet·ti·ner Haff** (stə-tē′nər häf′, shtə-) Lagoon on the Baltic coast between NE East Germany & NW Poland.
**Steu·ben·ville** (sto͞o′bĭn-vĭl′, styo͞o′-). City of E Ohio on the Ohio R. near the Pa. border & S of Youngstown. Pop. 26,400.
**Ste·ven·age** (stē′və-nĭj). Urban district of SE England N of London. Pop. 72,600.
**Ste·vens Point** (stē′vĭnz). City of central Wis. S of Wausau. Pop. 22,970.
**Stew·art** (sto͞o′ərt, styo͞o′-). **1.** River, 331 mi (533 km), of central Y.T., Canada. **2.** Island of S New Zealand off the S coast of South Is.
**Sti·kine** (stĭ-kēn′). River, 335 mi (539 km), of NW B.C., Canada, & SE Alas.
**Still·wa·ter** (stĭl′wô′tər, -wŏt′ər). **1.** City of E Minn. ENE of St. Paul. Pop. 12,290. **2.** City of N central Okla. NNE of Oklahoma City. Pop. 38,268.
**Stir·ling** (stûr′lĭng). Borough of central Scotland WNW of Edinburgh. Pop. 29,799.
**Stock·holm** (stŏk′hōlm′, -hōm′). Cap. of Sweden in the E part on the Baltic Sea. Pop. 665,550.
**Stock·port** (stŏk′pôrt′, -pōrt′). City of NW England near Manchester. Pop. 138,350.
**Stock·ton** (stŏk′tən). **1.** STOCKTON-ON-TEES. **2.** City of central Calif. on the San Joaquin R. S of Sacramento. Pop. 149,779.
**Stock·ton-on-Tees** (stŏk′tən-ŏn-tēz′, -ôn-). Borough of NE England near Middlesbrough. Pop. 165,400.
**Stoke-on-Trent** (stōk′ŏn-trĕnt′, -ôn-). Borough of W central England S of Manchester. Pop. 256,200.
**Stol·berg** (stôl′bo͝org′, shtôl′-). City of W West Germany WSW of Cologne. Pop. 57,379.
**Stone·ham** (stō′nəm). Town of E Mass. near Boston. Pop. 21,424.
**Ston·ey Creek** (stō′nē). Town of S Ont., Canada, off Lake Ontario S of Hamilton. Pop. 30,294.
**Ston·ing·ton** (stō′nĭng-tən). Town of SE Conn. on Long Is. Sound. Pop. 16,220.
**Ston·y Point** (stō′nē). Village of SE N.Y. on the Hudson; site of Revolutionary War battle (1779). Pop. 8,270.
**Stor·fjord** (stōr′fyōr). Inlet of the Norwegian Sea in SW Norway.
**Stough·ton** (stōt′n). Town of E Mass. near Brockton. Pop. 26,710.
**Stour** (stour, sto͝or, stōr). River, 40 mi (64.4 km), of SE England emptying into the North Sea in two channels.
**Stow** (stō). City of NE Ohio near Akron. Pop. 25,303.

**Stral·sund** (sträl′zo͝ont′, -so͞ont′, shträl′-). City of N East Germany on the Baltic opposite Rügen Is. Pop. 72,167.
**Stras·bourg** (sträs′bo͝org′, sträz′-). City of NE France near the Rhine R. Pop. 251,520.
**Strat·ford** (străt′fərd). **1.** City of S Ont., Canada, SW of Toronto. Pop. 25,657. **2.** STRATFORD-UPON-AVON. **3.** Town of SW Conn. NE of Bridgeport. Pop. 50,541.
**Strat·ford-up·on-Av·on** (străt′fərd-ə-pŏn-ā′vən). Borough of central England SSE of Birmingham. Pop. 20,080.
**Stream·wood** (strēm′wo͝od′). Village of NE Ill. near Chicago. Pop. 23,456.
**Strea·tor** (strē′tər). City of N central Ill. NE of Peoria. Pop. 14,769.
**Stret·ford** (strĕt′fərd). Borough of NW England near Manchester. Pop. 52,450.
**Strí·mon** (strē-môn′). STRUMA.
**Strom·bo·li** (strŏm′bə-lē). **1.** Island of S Italy in the Lipari Is. off NE Sicily. **2.** Volcano, 3,038 ft (926.6 m), on Stromboli Is.
**Strongs·ville** (strôngz′vĭl′). City of NE Ohio SSW of Cleveland. Pop. 28,577.
**Stru·ma** (stro͞o′mə). River, 216 mi (347.5 km), of W Bulgaria & NE Greece.
**Struth·ers** (strŭ*th*′ərz). City of NE Ohio near Youngstown & the Pa. border. Pop. 13,624.
**Stry·mon** (strī′mŏn′). STRUMA.
**Stry·mon·ic Gulf** (strī-mŏn′ĭk). Inlet of the Aegean off NE Greece E of Chalcidice.
**Stutt·gart. 1.** (stŭt′gärt′). City of E central Ark. ESE of Little Rock. Pop. 10,999. **2.** ( shto͝ot′gärt′, stŭt′-). City of SW West Germany on the Neckar R. Pop. 600,421.
**Styr** (stîr). River, c. 271 mi (436 km), of W European USSR.
**Su·bic Bay** (so͞o′bĭk). Inlet of the South China Sea off W central Luzon, Philippines, W of Manila Bay.
**Su·bo·ti·ca** *also* **Su·bo·ti·tsa** (so͞o′bə-tēt′sə). City of NE Yugoslavia near the Hungarian border. Pop. 89,476.
**Sü·chow** (so͞o′jō′, -chou′, sho͞o′-). XUZHOU.
**Su·cre** (so͞o′krā). Constitutional cap. of Bolivia in the central part SE of La Paz. Pop. 63,625.
**Su·dan** (so͞o-dăn′). **1.** Region of N Africa S of the Sahara & N of the equator. **2.** Country of NE Africa S of Egypt. Cap. Khartoum. Pop. 18,691,000. **—Su′da·nese′** *adj. & n.*
**Sud·bur·y** (sŭd′bĕr′ē, -bə-rē). **1.** City of central Ont., Canada, N of Georgian Bay. Pop. 97,604. **2.** Town of E Mass. W of Boston. Pop. 14,027.
**Su·de·ten** (so͞o-dā′tn). **1.** SUDENTENLAND. **2.** *var. of* SUDETES.
**Su·de·ten·land** (so͞o-dāt′n-lănd′, -länt′). Region of NW Czechoslovakia along the Polish border.
**Su·de·tes** (so͞o-dē′tēz) *also* **Su·de·ten** (so͞o-dāt′n). Mountains extending c. 185 mi (298 km) along the border between NW Czechoslovakia & SW Poland.
**Su·dir·man** (so͞o-dîr′mən). Range of New Guinea in central West Irian.
**Su·ez** (so͞o-ĕz′, so͞o′ĕz′). **1. Gulf of.** N arm of the Red Sea off NE Egypt W of the Sinai Peninsula. **2.** Isthmus of NE Egypt connecting Africa & Asia & traversed by the **Suez Canal** (107 mi/172 km) from the Mediterranean to the Gulf of Suez. **3.** City of NE Egypt at the head of the Gulf of Suez. Pop. 194,001.
**Suf·fern** (sŭf′ərn). Village of SE N.Y. on the N.J. border NW of New York City. Pop. 10,794.
**Suf·folk** (sŭf′ək, -ôk′). Independent city of SE Va. near Portsmouth. Pop. 47,621.
**Sui·sun City** (sə-so͞on′). City of W central Calif. NNE of Oakland & N of **Suisun Bay,** E arm of San Francisco Bay. Pop. 11,087.
**Su·i·ta** (so͞o-ē′tə). City of S Honshu, Japan, near Osaka. Pop. 300,956.
**Su·kho·na** (sə-kô′nə). River, c. 350 mi (563 km), of N European USSR.
**Su·khu·mi** (so͝ok′ə-mē). City of S European USSR on the Black Sea. Pop. 114,000.
**Suk·kur** (so͝ok′ər). City of SE Pakistan on the Indus R. Pop. 158,876.
**Su·la·we·si** (so͞o′lə-wā′sē). Island of central Indonesia on the equator E of Borneo.
**Sul·phur Springs** (sŭl′fər). City of NE Tex. ENE of Dallas. Pop. 12,804.
**Su·lu** (so͞o′lo͞o). **1.** Sea of the W Pacific between the Philippines & N Borneo. **2.** Archipelago of the S Philippines SW of Mindanao.
**Su·ma·tra** (so͞o-mä′trə). Island of W Indonesia in the Indian Ocean W of Borneo & the Malay Peninsula. **—Su·ma′tran** *adj. & n.*
**Sum·ba** (so͞om′bə). Island of S central Indonesia in the Lesser Sundas S of Flores.
**Sum·ba·wa** (so͞om-bä′wə). Island of S central Indonesia in the Lesser Sundas W of Flores.
**Su·mer** (so͞o′mər). Ancient country of Mesopotamia in present-day S Iraq. **—Su·me′ri·an** *adj. & n.*
**Sum·ga·it** (so͞om′gä-ēt′). City of S European USSR on the Caspian Sea. Pop. 196,000.
**Sum·mit** (sŭm′ĭt). **1.** Village of NE Ill. near Chicago. Pop. 10,110. **2.** City of NE N.J. W of Newark. Pop. 21,071.
**Sum·ter** (sŭm′tər). City of central N.C. E of Columbia. Pop. 24,890.
**Su·my** (so͞o′mē). City of W European USSR in the N Ukraine NW of Kharkov. Pop. 228,000.
**Sun** (sŭn). River, c. 130 mi (209 km), of NW Mont.
**Sun·bur·y** (sŭn′bĕr′ē, -bə-rē). City of E central Pa. N of Harrisburg. Pop. 12,292.
**Sun·da** (sŭn′də, so͞on′-). **1.** Strait between Sumatra & Java. **2.** *also* **Sun·das** (-dəz). Islands of the W Malay Archipelago, comprising the **Greater Sundas** (Sumatra, Borneo, Java, & Sulawesi) & the **Lesser Sundas** (from Bali E to Timor).
**Sun·der·land** (sŭn′dər-lənd). Borough of NE England near the North Sea ESE of Newcastle. Pop. 214,820.
**Sunds·vall** (sŭnts′väl′). City of E Sweden on an inlet of the Gulf of Bothnia. Pop. 52,268.
**Sun·ga·ri** (so͞ong′gə-rē′). SONGHUA.
**Sun·ny·vale** (sŭn′ē-vāl′). City of W Calif. near San Jose. Pop. 106,618.
**Sun Prairie.** City of S central Wis. NE of Madison. Pop. 12,931.
**Sun·rise** (sŭn′rīz′). City of SE Fla. near Fort Lauderdale. Pop. 39,681.
**Sun Valley.** Resort town of S central Idaho E of Boise. Pop. 545.
**Su·pe·ri·or** (so͞o-pîr′ē-ər). **1.** Largest & westernmost of the Great Lakes, 31,820 sq mi (82,414 sq km), between the U.S. & Ont., Canada. **2.** City of NW Wis. on Lake Superior opposite Duluth, Minn. Pop. 29,571.
**Su·ra·ba·ya** *or* **Su·ra·ba·ja** (so͝or′ə-bī′ə). City of NE Java, Indonesia, on the Java Sea. Pop. 1,556,255.
**Su·ra·kar·ta** (so͝or′ə-kär′tə). City of central Java, Indonesia, on the Solo R. Pop. 414,285.
**Su·rat** (so͝or′ət, sə-răt′). City of W central India on the Gulf of Cambay. Pop. 471,656.
**Su·ri·ba·chi** (so͝or′ə-bä′chē), **Mount.** Volcanic hill on Iwo Jima in the W Pacific.
**Su·ri·nam** (so͝or′ə-năm′) *also* **Su·ri·na·me** (sü′rē-nä′mə). Country of NE South America on the Atlantic. Cap. Paramaribo. Pop. 352,041. **—Su′ri·nam·ese′** *adj. & n.*
**Su·ri·na·me** (sü′rē-nä′mə) *also* **Su·ri·nam** (so͝or′ə-năm′). River of Surinam flowing c. 400 mi (644 km) N to the Atlantic.
**Sur·rey** (sûr′ē). Region & former county of SE England S of London.
**Su·ru·ga Bay** (so͝or′ə-gə). Inlet of the Pacific on SE coast of Honshu, Japan, SW of Tokyo.
**Su·sa** (so͞o′sə, -zə). Ruined city of SW Iran, cap. of ancient Elam.
**Su·sah** *or* **Su·sa** (so͞o′zə, -sə). *vars. of* SOUSSE.
**Sus·que·han·na** (sŭs′kwə-hăn′ə). River of the NE U.S. rising in central N.Y. & flowing 444 mi (715 km) S through E Pa. & NE Md. to Chesapeake Bay.
**Sus·sex** (sŭs′ĭks). Region & former county of SE England on the English Channel S of London.
**Suth·er·land** (sŭ*th*′ər-lənd). Falls, 1,904 ft (580.7 m) high, on SW South Is., New Zealand.
**Sut·lej** (sŭt′lĕj′). River, c. 900 mi (1,448 km), of SW Tibet, N India, & E Pakistan.
**Sut·ton-in-Ash·field** (sŭt′n-ĭn-ăsh′fēld′). Urban district of central England NNW of Nottingham. Pop. 40,330.
**Su·va** (so͞o′və). Cap. of Fiji on the SE coast of Viti Levu. Pop. 63,628.
**Su·wal·ki** (so͞o-väl′kē). City of NE Poland N of Bialystok. Pop. 25,360.
**Su·wan·nee** (sə-wä′nē). River, c. 240 mi (386 km), of SE Ga. & N Fla.
**Su·won** (so͞o′wŭn′). City of NW South Korea S of Seoul. Pop. 244,145.
**Su·zhou** (so͞o′jō′). City of E China WNW of Shanghai. Pop. 1,300,-000.
**Sval·bard** (sfäl′bär′). Archipelago of Norway comprising Spitsbergen & other islands in the Arctic Ocean N of the Norwegian mainland.
**Sverd·lovsk** (sfĕrd-lôfsk′). City of E European USSR in the central Urals. Pop. 1,211,000.
**Sver·drup** (sfĕr′drəp). Islands of N N.W.T., Canada, in the Arctic Ocean W of Ellesmere Is.
**Swa·bi·a** (swā′bē-ə). Region of SW Germany. **—Swa′bi·an** *adj. & n.*
**Swamp·scott** (swŏmp′skət). Town of NE Mass. NE of Boston. Pop. 13,837.
**Swan** (swän). **1.** River, c. 240 mi (386 km), of SW Australia. **2.** River, c. 110 mi (177 km), of E Sask. & W Man., Canada.
**Swan·sea** (swän′zē, -sē). **1.** Town of SE Mass. near Fall River. Pop. 15,461. **2.** Borough of S Wales on **Swansea Bay,** an inlet of Bristol Channel. Pop. 186,900.
**Swa·tow** (swä′tou′). SHANTOU.
**Swa·zi·land** (swä′zē-lănd′). Country of SE Africa between South Africa & Mozambique. Cap. Mbabane. Pop. 547,000.
**Swe·den** (swēd′n). Country of N Europe on the E Scandinavian peninsula. Cap. Stockholm. Pop. 8,320,000. **—Swede** *n.* **—Swed′ish** *adj. & n.*

**Sweet·wa·ter** (swēt′wô′tər, -wŏt′ər). City of N central Tex. near Abilene. Pop. 12,242.
**Swift Current** (swĭft). City of SW Sask., Canada, W of Regina. Pop. 14,264.
**Swin·don** (swĭn′dən). Borough of S central England ENE of Bristol. Pop. 90,680.
**Swit·zer·land** (swĭt′sər-lənd). Republic of W central Europe. Cap. Bern. Pop. 6,365,960. **—Swiss** (swĭs) *adj. & n.*
**Syb·a·ris** (sĭb′ər-ĭs). Ancient Greek city of S Italy on the Gulf of Taranto.
**Syd·ney** (sĭd′nē). **1.** City of SE Australia on an inlet of the Tasman Sea. Metro. area pop. 2,765,040. **2.** City of N.S., Canada, on Cape Breton Is. Pop. 30,645.
**Syk·tyv·kar** (sĭk-tĭf-kär′). City of NE European USSR ESE of Arkhangelsk. Pop. 171,000.
**Syl·a·cau·ga** (sĭl′ə-kô′gə). City of central Ala. E of Birmingham. Pop. 12,708.
**Syl·va·nia** (sĭl-vān′yə). City of N Ohio on the Mich. border near Toledo. Pop. 15,527.
**Syr·a·cuse** (sîr′ə-kyōōs′, -kyōōz′). **1.** City of SE Sicily, Italy, on the Ionian Sea. Pop. 93,006. **2.** City of central N.Y. ESE of Rochester. Pop. 170,105.
**Syr Dar·ya** (sîr där′yə). River of S Central Asian USSR rising in the Tian Shan & flowing c. 1,380 mi (2,220 km) NW into the Aral Sea.
**Sy·ri·a** (sîr′ē-ə). Country of SW Asia on the E Mediterranean coast. Cap. Damascus. Pop. 8,979,000. **—Syr′i·an** *adj. & n.*
**Syrian.** Desert of N Arabian Peninsula occupying N Saudi Arabia, W Iraq, SE Syria, & E Jordan.
**Sy·ros** (sī′rŏs′) *also* **Si·ros** (sē′rôs′). Island of Greece in the N central Cyclades.
**Syz·ran** (sĭz′rən). City of S central European USSR on the Volga W of Kuibyshev. Pop. 178,000.
**Szcze·cin** (shchĕt′sēn′). City of NW Poland near the mouth of the Oder R. Pop. 337,204.
**Sze·chwan** (sĕch′wän′). SICHUAN.
**Sze·ged** (sĕg′ĕd′). City of S Hungary on the Tisza R. near the Yugoslavian border. Pop. 171,342.
**Szé·kes·fe·hér·vár** (sā′kĕsh-fĕ′ər-vär′). City of central Hungary on the Danube SSW of Budapest. Pop. 103,197.
**Szol·nok** (sōl′nōk′). City of central Hungary ESE of Budapest. Pop. 75,203.
**Szom·bat·hely** (sōm′bôt-hā′). City of W Hungary near the Austrian border. Pop. 82,830.

# T

**Ta·al** (tä-äl′, täl). **1.** Lake, 94 sq mi (243.5 sq km), of SW Luzon, Philippines, S of Manila. **2.** Island in Lake Taal.
**Tab·las** (tä′bləs). Island of central Philippines in the Romblons E of Mindoro.
**Ta·ble** (tā′bəl). **1. Bay.** Inlet of the Atlantic off SW South Africa; harbor of Cape Town. **2.** Flat-topped mountain, 3,550 ft (1,082.8 m), overlooking Cape Town, SW South Africa.
**Ta·briz** (tə-brēz′). City of NW Iran in Azerbaijan E of Lake Urmia. Pop. 598,576.
**Ta·bun Bog·do** (tä′bōōn′ bôg′dō). Mountain, 15,266 ft (4,656.1 m), in the Altai Mts. at the junction of the USSR, China, & Mongolia borders.
**Ta·chi·ka·wa** (tä-chē′kə-wə, tä′chē-kä′wə). City of E central Honshu, Japan, near Tokyo. Pop. 138,129.
**Tac·na** (tăk′nə). Town of S Peru N of Arica, Chile. Pop. 55,752.
**Ta·co·ma** (tə-kō′mə). City of W central Wash. on an arm of Puget Sound S of Seattle. Pop. 158,501.
**Ta·con·ic** (tə-kŏn′ĭk). Range of the Appalachians in SE N.Y. E of the Hudson & in W Mass. & SW Vt.
**Ta·dzhik Soviet Socialist Republic** (tä-jĭk′, -jēk′) *also* **Ta·dzhik·i·stan** (tä-jĭk′ĭ-stăn′, -stän′, -jēk′-). Constituent republic of S Central Asian USSR, bordering on Afghanistan & China.
**Tae·dong** (tă-dōōng′, tī-). River, c. 245 mi (394 km), of SW North Korea.
**Tae·gu** (tă-gōō′, tī-). City of SE South Korea NNW of Pusan. Pop. 1,310,768.
**Tae·jon** (tă-jôn′, -jŏn′, tī-). City of SW South Korea SSE of Seoul. Pop. 506,708.
**Ta·fi·lelt** (tăf′ə-lĕlt′) *or* **Ta·fi·let** (-lĕt′). Oasis, c. 530 sq mi (1,375 sq km), in the Sahara in SE Morocco.
**Tag·an·rog** (tăg′ən-rŏg′). City of S European USSR on the **Gulf of Taganrog,** an arm of the Sea of Azov. Pop. 276,000.
**Ta·gus** (tā′gəs). River of the Iberian Peninsula rising in E central Spain & flowing c. 585 mi (941 km) NW & SW through central Portugal to the Atlantic.
**Ta·hi·ti** (tə-hē′tē). Island of the South Pacific in the Windward group of the Society Is., French Polynesia. **—Ta·hi′tian** *adj. & n.*
**Ta·hoe** (tä′hō). Lake, 193 sq mi (500 sq km), on the Calif.-Nev. border W of Carson City, Nev.
**Tai** (tī). Lake, c. 1,300 sq mi (3,367 sq km), of E central China W of Shanghai.
**Tai·chung** (tī′chōōng′, -jōōng′). City of W central Taiwan SW of Taipei. Pop. 566,525.
**Tai·myr** *also* **Tai·mir** (tī-mîr′). Peninsula of N central Siberian USSR extending N between the Laptev & Kara seas.
**Tai·nan** (tī′nän′). City of SW Taiwan on the South China Sea. Pop. 541,390.
**Tai·na·ron** (tā′nə-rôn′), **Cape.** Cape MATAPAN.
**Tai·pei** *also* **Tai·peh** (tī′pā′, -bā′). Cap. of Taiwan in the N part. Pop. 2,108,193.
**Tai·wan** (tī′wän′). Island off the SE coast of China, constituting with the Pescadores & other smaller islands the Republic of China. Cap. Taipei. Pop. 16,609,961. **—Tai′wan·ese′** *adj. & n.*
**Tai·yu·an** *also* **Tai·yü·an** (tī′yōō-än′). City of NE China SW of Beijing. Pop. 2,725,000.
**Ta·iz** *or* **Ta·′izz** (tă-ĭz′). City of SW Yemen in the SW. Pop. 78,642.
**Tai·zhong** (tī′jōōng′). TAICHUNG.
**Ta·′izz** (tă-ĭz′). *var. of* TAIZ.
**Ta·jo** (tä′hō). TAGUS.
**Ta·ju·mul·co** (tä′hōō-mōōl′kō). Inactive volcano, 13,816 ft (4,213.9 m), in W Guatemala.
**Ta·ka·mat·su** (tä′kə-mät′sōō). City of NE Shikoku, Japan, on the Inland Sea. Pop. 298,999.
**Ta·ka·o·ka** (tə-kou′kə). City of W central Honshu, Japan, near Toyama Bay. Pop. 169,621.
**Ta·ka·ra·zu·ka** (tə-kär′ə-zōō′kə). City of SW Honshu, Japan, near Osaka & Kobe. Pop. 162,624.
**Ta·ka·sa·ki** (tä′kə-sä′kē). City of central Honshu, Japan, NW of Tokyo. Pop. 211,348.
**Ta·ka·tsu·ki** (tə-kät′sōō-kē). City of SW Honshu, Japan, NE of Osaka. Pop. 330,570.
**Tak·ka·kaw** (tăk′ə-kô′). Waterfall, 1,650 ft (503.3 m), in SE B.C., Canada.
**Ta·kli·ma·kan** *also* **Ta·kla·ma·kan** (tä′klə-mə-kän′). Desert of W China between the Tian Shan & Kunlun Mts.
**Ta·ko·ma Park** (tə-kō′mə). City of central Md. near Washington, D.C. Pop. 16,231.
**Ta·la·ud** (tä′lä′ōōd′) *or* **Ta·laur** (-ōōr′). Islands of NE Indonesia NE of Sulawesi.
**Tal·ca** (täl′kə). City of central Chile between Santiago & Concepción. Pop. 94,400.
**Tal·ca·hua·no** (täl′kə-wä′nō, -hwä′-). City of central Chile on the Pacific near Concepción. Pop. 148,300.
**Tal·la·de·ga** (tăl′ə-dē′gə). City of E central Ala. E of Birmingham. Pop. 19,128.
**Tal·la·has·see** (tăl′ə-hăs′ē). Cap. of Fla. in the Panhandle. Pop. 81,548.
**Tal·la·hatch·ie** (tăl′ə-hăch′ē). River, c. 230 mi (371 km), of N Miss.
**Tal·la·poo·sa** (tăl′ə-pōō′sə). River, 268 mi (431.2 km), of NW Ga. & E Ala.
**Tal·linn** *also* **Tal·lin** (tăl′ĭn, tä′lĭn ). City of NW European USSR in Estonia on the Gulf of Finland opposite Helsinki. Pop. 430,000.
**Tall·madge** (tăl′mĭj). City of NE Ohio near Akron. Pop. 15,269.
**Ta·lu·lah** (tə-lōō′lə). City of NE La. WNW of Vicksburg, Miss. Pop. 10,392.
**Ta·man** (tə-män′). Peninsula of S European USSR projecting W between the Sea of Azov & the Black Sea.
**Tam·an·ras·set** (tăm′ən-răs′ĭt). Oasis of S Algeria in the Sahara.
**Tam·a·rac** (tăm′ə-răk′). City of SE Fla. NW of Fort Lauderdale. Pop. 29,142.
**Ta·ma·tave** (tăm′ə-täv′, tä′mə-). City of NE Madagascar on the Indian Ocean. Pop. 77,395.
**Tam·bo·ra** (tăm-bōr′ə, -bôr′ə), **Mount.** Volcano, 9,253 ft (2,822.2 m), on Sumbawa Is., S central Indonesia.
**Tam·bov** (tăm-bôf′, -bôv′). City of central European USSR SE of Moscow. Pop. 270,000.
**Tam·pa** (tăm′pə). City of W central Fla. on **Tampa Bay,** an inlet of the Gulf of Mexico. Pop. 271,523.
**Tam·pe·re** (tăm′pə-rā′, täm′-). City of SW Finland NNW of Helsinki. Pop. 168,118.
**Tam·pi·co** (tăm-pē′kō). City of E central Mexico near the Gulf of Mexico NNE of Mexico City. Pop. 212,188.
**Tam·ri·da** (tăm-rē′də). Chief town of Socotra Is., Southern Yemen.
**Tam·worth** (tăm′wûrth′). Borough of central England NE of Birmingham. Pop. 46,960.
**Ta·na** (tä′nə). **1.** *also* **Tsa·na** (sä′-, tsä′-). Lake, c. 1,400 sq mi (3,626 sq km), of NW Ethiopia. **2.** River, c. 500 mi (805 km), of Kenya. **3.** River, c. 200 mi (322 km), of NE Norway forming part of the Norway-Finland border.
**Tan·a·na** (tăn′ə-nô′). River of Alas. flowing c. 475 mi (764 km) NW to the Yukon R.
**Ta·nan·a·rive** (tə-năn′ə-rēv′). ANTANANARIVO.

**Tan·dil** (tăn-dĭl′). City of E Argentina near Buenos Aires. Pop. 65,876.

**Ta·ne·ga·shi·ma** (tä-nĕg′ə-shē′mə). Island off S Kyushu, Japan.

**Tan·ga** (tăng′gə). City of NE Tanzania on the Indian Ocean. Pop. 103,409.

**Tan·gan·yi·ka** (tăn′gən-yē′kə, tăng′-). **1. Lake.** Lake, c. 12,700 sq mi (32,893 sq km), of E central Africa between Zaire & Tanzania. **2.** Former country of E central Africa that joined with Zanzibar (1964) to form Tanzania. **—Tan′gan·yi′kan** *adj.* & *n.*

**Tan·gier** (tăn-jîr′) *also* **Tan·giers** (-jîrz′). City of N Morocco on the Strait of Gibraltar. Pop. 187,894.

**Tang·shan** (däng′shän′, tăng′-). City of NE China ESE of Beijing. Pop. 1,200,000.

**Ta·nim·bar** (tə-nĭm′bär′, tä-). Islands of SE Indonesia in the S Moluccas.

**Ta·nis** (tā′nĭs). Ancient city of Egypt in the E Nile delta.

**Tan·jore** (tăn-jôr′, -jōr′). THANJAVUR.

**Tan·ta** (tän′tə). City of N Egypt in the Nile delta N of Cairo. Pop. 284,636.

**Tan·tung** (dän′dŏŏng′). DANDONG.

**Tan·za·ni·a** (tăn′zə-nē′ə). Country of E central Africa on the Indian Ocean. Cap. Dar es Salaam. Pop. 17,527,560. **—Tan·za′ni·an** *adj.* & *n.*

**Ta·or·mi·na** (tour-mē′nə). Town of E Sicily, Italy, at the foot of Mt. Etna overlooking the Ionian Sea. Pop. 6,696.

**Taos** (tous). Resort town of N N.Mex. NNE of Santa Fe. Pop. 3,369.

**Ta·pa·chu·la** (tä′pə-chōō′lə). City of SE Mexico near the Guatemalan border. Pop. 60,620.

**Ta·pa·jós** *also* **Ta·pa·joz** (tăp′ə-zhôs′). River, c. 600 mi (965 km), of N Brazil.

**Tap·pan Zee** (tăp′ən zē′). Widening of the Hudson R. in SE N.Y.

**Tap·ti** (tăp′tē). River, 436 mi (702 km), of W central India.

**Ta·qua·ri** (täk′wə-rē′). River, c. 350 mi (563 km), of S central Brazil.

**Tar** (tär). River, 215 mi (346 km), of NE N.C.

**Tar·a** (tăr′ə). Village of E Ireland NW of Dublin; seat of ancient Irish kings.

**Ta·ran·to** (tär′ən-tō′, tə-răn′tō). City of SE Italy on the **Gulf of Taranto,** an arm of the Ionian Sea. Pop. 205,158.

**Ta·ra·wa** (tə-rä′wə, tăr′ə-wä′). Atoll (8 sq mi/21 sq km) of Kiribati in the N Gilbert Is. of the W Pacific.

**Tarbes** (tärb). City of SW France near the Pyrenees WSW of Toulouse. Pop. 54,286.

**Ta·ren·tum** (tə-rĕn′təm). TARANTO.

**Ta·rim** (dä′rēm′, tä′-). River of NW China flowing c. 1,300 mi (2,092 km) E to Lop Nur.

**Tarn** (tärn). River, c. 235 mi (378 km), of S France.

**Tar·nów** (tär′nōōf′). City of SE Poland, E of Kraków. Pop. 85,514.

**Tar·pon Springs** (tär′pən). City of W Fla. on the Gulf of Mexico NW of Tampa. Pop. 13,251.

**Tar·qui·nia** (tär-kwĕn′yə). Town of central Italy WNW of Rome, on site of ancient **Tar·quin·i·i** (-kwĭn′ē-ī′). Pop. 10,300.

**Tar·ra·go·na** (tăr′ə-gō′nə). City of NE Spain on the Mediterranean WSW of Barcelona. Pop. 53,548.

**Tar·ra·sa** (tə-rä′sə). City of NE Spain NNW of Barcelona. Pop. 134,481.

**Tar·ry·town** (tăr′ē-toun′). Village of SE N.Y. N of New York City. Pop. 10,648.

**Tar·shish** (tär′shĭsh). Ancient country on the S coast of Spain.

**Tar·sus** (tär′səs). City of S Turkey near the Mediterranean. Pop. 102,186.

**Tar·ta·ry** (tär′tə-rē) *or* **Ta·ta·ry** (tä′-). Region of E Europe & Asia controlled by the Tartars in the 13th & 14th cent.

**Tar·tu** (tär′tōō). City of NW European USSR in Estonia SE of Tallinn. Pop. 105,000.

**Tash·kent** (tăsh-kĕnt′, -kĕnd′). City of S Central Asian USSR SW of Lake Balkash. Pop. 1,780,000.

**Tas·man** (tăz′mən). **1.** Sea of the S Pacific between Australia & New Zealand. **2. Mount.** Mountain, 11,475 ft (3,449.9 m), of the Southern Alps on W central South Is., New Zealand.

**Tas·ma·ni·a** (tăz-mā′nē-ə, -mān′yə). Island state of SE Australia. **—Tas·ma′ni·an** *adj.* & *n.*

**Ta·tar** (tä′tər). Strait between Sakhalin & the E Asian mainland.

**Ta·ta·ry** (tä′tə-rē). *var. of* TARTARY.

**Ta·tra** (tä′trə) *or* **Ta·tras** (-trəz). Range of the Carpathians along the Czech-Polish border S of Kraków.

**Ta·tung** (dä′tŏŏng′, -dŏŏng′). DATONG.

**Tau·ghan·nock** (tə-găn′ək). Falls, 215 ft (65.6 m), of S central N.Y. NW of Ithaca.

**Taun·ton** (tôn′tən, tŏn′-, tăn′-). **1.** Borough of SW England SW of Bristol. Pop. 37,570. **2.** City of SE Mass. N of Fall River. Pop. 45,001.

**Tau·nus** (tou′nəs). Range of W West Germany extending NE from the Rhine and N of Mainz.



**Tau·po** (tou′pō). Lake, 234 sq mi (606 sq km), of central North Is., New Zealand.

**Tau·ris** (tôr′ĭs). TABRIZ.

**Tau·rus** (tôr′əs). Range of S Turkey extending c. 350 mi (565 km) parallel to the Mediterranean coast.

**Tax·co** (täs′kō). Town of S Mexico SSW of Mexico City. Pop. 27,089.

**Tay** (tā). River of central Scotland rising in the Grampians & flowing c. 118 mi (190 km) E through **Loch Tay** (15 mi/24 km long) to the **Firth of Tay** (25 mi/40.2 km long), an inlet of the North Sea.

**Tay·lor** (tā′lər). **1.** City of SE Mich. near Dearborn. Pop. 77,568. **2.** City of central Tex. NNE of Austin. Pop. 10,619.

**Tay·lor·ville** (tā′lər-vĭl′). City of central Ill. SW of Decatur. Pop. 11,386.

**Tbi·li·si** (tə-bĭl′ĭ-sē). City of S European USSR on the Kura R. Pop. 1,066,000.

**Tea·neck** (tē′nĕk′). Township of NE N.J. ESE of Paterson. Pop. 42,355.

**Tea·pot Dome** (tē′pŏt). Former naval oil reserve site in E central Wyo. N of Casper.

**Tees** (tēz). River, 70 mi (113 km), of NE England.

**Te·gu·ci·gal·pa** (tə-gōō′sə-găl′pə). Cap. of Honduras in the S central part. Pop. 273,894.

**Te·hach·a·pi** (tə-hăch′ə-pē). Range of S Calif. connecting the Sierra Nevada & the Coast Ranges N of Los Angeles.

**Te·he·ran** *or* **Teh·ran** (tā′ə-răn′, -rän′). Cap. of Iran in the N central part. Pop. 4,496,159.

**Te·huan·te·pec** (tə-wän′tə-pĕk′). **1. Isthmus of.** Narrowest part (c. 125 mi/201 km) of S Mexico between the Gulf of Mexico & the Pacific. **2.** Town of S Mexico near the **Gulf of Tehuantepec,** a wide inlet of the Pacific. Pop. 16,179.

**Te·jo** (tā′zhōō). TAGUS.

**Te·jon** (tē-hōn′). Pass, 4,183 ft (1,275.8 m) high, through the Tehachapi Mts. in SW Calif.

**Tel A·viv-Jaf·fa** (tĕl′ə-vēv′-jăf′ə, -yăf′ə). City of Israel in the central part on the Mediterranean. Pop. 343,300.

**Tel·e·scope Peak** (tĕl′ĭ-skōp′). Mountain, c. 11,045 ft (3,368.7 m), in the Panamint Mts. of E Calif.

**Te·les Pi·res** (tĕl′ĭs pîr′ĭs). River, c. 600 mi (965 km), of central Brazil.

**Te·ma** (tē′mə). City of SE Ghana on the Gulf of Guinea. Pop. 60,767.

**Te·meš** (tĕm′ĕsh′). TIMIŞ.

**Te·mes·vár** (tĕm′ĕsh-vär′). TIMIŞOARA.

**Te·mir·tau** (tā′mîr-tou′). City of N Central Asian USSR NW of Karaganda. Pop. 213,000.

**Tem·pe** (tĕm′pē′). **1. Vale of.** Valley of N Greece SE of Mount Olympus. **2.** City of S central Ariz. near Phoenix. Pop. 106,743.

**Tem·ple** (tĕm′pəl). City of E central Tex. S of Fort Worth. Pop. 42,483.

**Temple City.** City of S Calif. near Los Angeles. Pop. 28,972.

**Temple Terrace.** City of W central Fla. near Tampa. Pop. 11,097.

**Te·mu·co** (tā-mōō′kō). City of central Chile S of Concepción. Pop. 110,300.

**Ten·a·fly** (tĕn′ə-flī′). Borough of NE N.J. near the Hudson R. opposite Yonkers, N.Y. Pop. 13,552.

**Ten·er·ife** (tĕn′ə-rē′fē, -fā, -rēf′, -rīf′). Largest of the Canary Is. of Spain, in the Atlantic.

**Ten·gri Khan** (tĕng′grē kän′, tĕng′rē). Mountain, 22,949 ft (6,999.4 m), in the Tian Shan on the border between S Central Asian USSR & NW China.

**Ten·nes·see** (tĕn′ĭ-sē′, tĕn′ĭ-sē′). **1.** River of the SE U.S. rising in E Tenn. & flowing c. 652 mi (1,049 km) through N Ala., W Tenn., & W Ky. to the Ohio R. **2.** State of the SE U.S. Cap. Nashville. Pop. 4,590,-750. **—Ten′nes·se′an** *adj.* & *n.*

**Te·no** (tĕ′nô). TANA R., Norway.

**Te·noch·ti·tlán** (tā-nôch′tē-tlän′). Ancient Aztec cap. on the site of present-day Mexico City.

**Te·nos** (tē′nŏs′). Island of Greece in the N Cyclades SE of Andros.

**Ten·sas** (tĕn′sô′). River, 250 mi (402 km), of NE La.

**Te·o·ti·hua·can** (tā′ə-tē′wä-kän′). Ancient city of Mexico NE of present-day Mexico City.

**Te·pic** (tā-pēk′). City of W Mexico NW of Guadalajara. Pop. 108,924.

**Te·pli·ce** (tĕp′lĭt-sā′). City of NW Czechoslovakia in the Erzgebirge near the East German border. Pop. 52,300.

**Te·quen·da·ma** (tā′kən-dä′mə). Falls, 482 ft (147 m), in central Colombia.

**Ter·cei·ra** (tər-sîr′ə). Island of Portugal in the central Azores.

**Te·rek** (tĕ′rĕk). River, c. 370 mi (595 km), of S European USSR.

**Te·re·si·na** (tĕr′ə-zē′nə). City of NE Brazil on the Parnaíba R. Pop. 181,071.

**Ter·na·te** (tər-nä′tā). Island of W Indonesia in the N Moluccas.

**Ter·ni** (tĕr′nē). City of central Italy N of Rome. Pop. 75,873.

**Ter·no·pol** (tĕr-nō′pəl). City of W European USSR in the W Ukraine WSW of Lvov. Pop. 144,000.

**Terre·bonne** (tĕr′ə-bŏn′). Town of S Que., Canada, N of Montreal. Pop. 11,204.

**Ter·re Haute** (tĕr'ə hōt', hŭt', hôt'). City of W Ind. WSW of Indianapolis. Pop. 61,125.

**Ter·rell** (tĕr'əl). City of NE Tex. E of Dallas. Pop. 13,225.

**Te·ruel** (tĕr'ə-wĕl'). City of E Spain E of Madrid. Pop. 24,856.

**Tes·lin** (tĕz'lĭn). Lake, c. 200 sq mi (322 km), of NW B.C. & S Y.T., Canada.

**Te·ton** (tē'tŏn', tēt'n). **1.** River, 143 mi (230 km), of NW Mont. **2.** *also* **Te·tons** (tē'tŏnz, tēt'nz). Range of the Rockies in NW Wyo. & SE Idaho S of Yellowstone National Park, rising to 13,747 ft (4,192.8 m).

**Te·tuán** (tā-twän', tə-). City of NE Morocco on the Mediterranean. Pop. 139,105.

**Teu·to·bur·ger Wald** (to͞o'tə-bûr'gər wôld, tyo͞o'-). Range of hills in N central West Germany between the upper Ems & the Weser.

**Te·ve·re** (tā'vā-rā). TIBER.

**Tewkes·bur·y** (to͝oks'bĕr'ē, -bə-rē, tyo͝oks'-). Borough of W central England on the Severn R.; site of the final defeat (1471) of the Lancastrians in the Wars of the Roses. Pop. 9,210.

**Tewks·bur·y** (to͝oks'bĕr'ē, -bə-rē, tyo͝oks'-). Town of NE Mass. near Lowell. Pop. 24,635.

**Tex·ar·kan·a** (tĕk'sär-kăn'ə, -sər-). **1.** City of SW Ark. on the Tex. border adjacent to Texarkana, Tex. Pop. 21,338. **2.** City of NE Tex. Pop. 31,271.

**Tex·as** (tĕk'səs). State of the S central U.S. Cap. Austin. Pop. 14,228,-383. **—Tex'an** *adj.* & *n.*

**Texas City.** City of SE Tex. on Galveston Bay. Pop. 41,403.

**Tex·el** (tĕk'səl, tĕs'əl). Island of NW Netherlands in the North Sea in the SW Frisian Is.

**Tha·ban·tsho·nya·na** (tä'băn-chōn-yä'nə). Highest (11,425 ft/ 3,484.6 m) of the Drakensberg Mts. in Lesotho, S Africa.

**Thai·land** (tī'lănd', -lənd). Country of SE Asia on the Gulf of Siam between Burma & Cambodia. Cap. Bangkok. Pop. 46,455,000. **—Thai** *adj.* & *n.*

**Thames** (tĕmz). **1.** River, c. 160 mi (257 km), of SE S Ont., Canada. **2.** River of S England flowing c. 210 mi (338 km) E to a wide estuary on the North Sea.

**Than·et** (thăn'ĭt), **Isle of.** Peninsula of SE England on the North Sea, separated from the mainland by arms of the Stour R.

**Than·ja·vur** (tăn-jôr', -jŏr'). City of SE India on the Cauvery R. delta. Pop. 140,547.

**Thap·sus** (thăp'səs). Ancient city of N Africa on the Mediterranean SE of Carthage in present-day Tunisia.

**Thar** (tär). Desert of NW India & E Pakistan.

**Tha·sos** (thā'sŏs'). Island of NE Greece in the N Aegean E of Chalcidice.

**Thebes** (thēbz). **1.** *also* **The·bae** (thē'bē). Ancient cap. of Upper Egypt on the Nile in present-day central Egypt. **2.** Ancient city of Boeotia in E central Greece NW of Athens.

**The Col·o·ny** (*thə* kŏl'ə-nē). City of NE Tex. near Dallas-Fort Worth. Pop. 11,586.

**The Dalles** (dălz). City of N Ore. on the Columbia E of Portland. Pop. 10,820.

**The·lon** (thē'lŏn'). River, c. 550 mi (885 km), of S central N.W.T., Canada.

**The·ra** (thîr'ə). SANTORIN.

**Ther·mop·y·lae** (thər-mŏp'ə-lē). Pass of E central Greece SE of Lamia; site of Spartan stand against the Persians (480 B.C.).

**Thes·sa·lo·ní·ki** (thĕs'ə-lə-nē'kē) or **Thes·sa·lo·ni·ca** (-nī'kə, -lŏn'ĭ-kə). SALONIKA.

**Thes·sa·ly** (thĕs'ə-lē). Region of E central Greece between the Pindus Mts. & the Aegean. **—Thes·sa'lian, Thes'sa·lo'ni·an** *adj.* & *n.*

**Thet·ford Mines** (thĕt'fərd). City of S Que., Canada, NE of Sherbrooke & S of Quebec city. Pop. 20,874.

**The Vil·lage** (vĭl'ĭj). City of central Okla. near Oklahoma City. Pop. 11,049.

**Thi·bo·daux** (tĭb'ə-dō'). City of SE La. WSW of New Orleans. Pop. 15,810.

**Thim·bu** (thĭm'bo͞o') *also* **Thim·phu** (-po͞o'). Cap. of Bhutan in the W part in the E Himalayas. Pop. 50,000.

**Thí·ra** (thîr'ə). SANTORIN.

**Tho·hoy·an·dou** (tō-hoi'ăn-do͞o'). Cap. of Venda, black enclave in NE South Africa.

**Thom·as·ton** (tŏm'ə-stən). Town of W central Ga. W of Macon. Pop. 18,200.

**Thom·as·ville** (tŏm'əs-vĭl'). **1.** City of S Ga. NNE of Tallahassee, Fla. Pop. 18,463. **2.** City of central N.C. SE of Winston-Salem. Pop. 14,144.

**Thomp·son** (tŏm'sən, tŏmp'-). **1.** River, c. 304 mi (489 km), of S B.C., Canada. **2.** City of N central Man., Canada. Pop. 17,291.

**Thorn** (tôrn). TORUŃ.

**Thorn·ton** (thôrn'tən). City of N central Colo. N of Denver. Pop. 40,343.

**Tho·rold** (thôr'əld, thŏr'-). City of SE Ont., Canada, on the Welland Ship Canal SE of St. Catherines. Pop. 14,944.

**Thou·sand Islands** (thou'zənd). Group of more than 1,500 islands of N N.Y. & SE Ont., Canada, in the St. Lawrence at the outlet of Lake Ontario.

**Thousand Oaks.** City of S Calif. W of Los Angeles. Pop. 77,797.

**Thrace** (thrās). Region & ancient country of the SE Balkan Peninsula N of the Aegean & in ancient times extending as far N as the Danube. **—Thra'cian** *adj.* & *n.*

**Thra·cia** (thrā'shə, -shē-ə). Ancient Thrace.

**Thu·le** (to͞o'lē). Town & U.S. military base of NW Greenland NW of Cape York. Pop. 357.

**Thun** (to͞on), **Lake of.** Lake, c. 18 sq mi (47 sq km), in central Switzerland SE of Bern.

**Thun·der Bay** (thŭn'dər). City of SW Ont., Canada, on **Thunder Bay,** inlet on the NW shore of Lake Superior. Pop. 111,476.

**Thu·ner·see** (to͞o'nər-zā'). Lake of THUN.

**Thu·rin·gi·a** (tho͞o-rĭn'jē-ə, -jə). Region of SW East Germany. **—Thu·rin'gi·an** *adj.* & *n.*

**Thur·rock** (thûr'ək). Urban district of SE England on the Thames E of London. Pop. 127,700.

**Thurs·day** (thûrz'dē). Island of NE Australia in Torres Strait NW of Cape York.

**Thurs·ton** (thûr'stən). Island off W Antarctica.

**Ti·a·hua·na·co** (tē'ə-wə-nä'kō). Site of pre-Incan ruins in W Bolivia S of Lake Titicaca.

**Tian·jin** (tē-än'jĭn'). City of NE China near the Gulf of Bo Hai SE of Beijing. Pop. 7,210,000.

**Tian Shan** (tē-än' shän'). Mountains of central Asia extending c. 1,500 mi (2,414 km) ENE through S Central Asian USSR & NW China.

**Ti·ber** (tī'bər). River of central Italy flowing c. 251 mi (404 km) S & SW through Rome to the Tyrrhenian Sea.

**Ti·be·ri·as** (tī-bîr'ē-əs). **1. Lake.** Sea of GALILEE. **2.** Town of NE Israel on the Sea of Galilee. Pop. 28,300.

**Ti·bes·ti Mas·sif** (tə-bĕs'tē mă-sēf'). Mountains of N Chad in the Sahara, rising to 11,204 ft (3,417.2 m).

**Ti·bet** (tə-bĕt'). Region & former semi-independent theocratic state of SW China. **—Ti·bet'an** *adj.* & *n.*

**Ti·bur** (tī'bər). TIVOLI.

**Ti·bu·ron** (tē'bə-rôn'). Island of NW Mexico in the Gulf of California.

**Ti·ci·no** (tĭ-chē'nō). River, 154 mi (248 km), of S Switzerland & N Italy.

**Ti·con·der·o·ga** (tī'-kŏn-də-rō'gə). Resort village of NE N.Y. N of Lake George. Pop. 2,938.

**Tien Shan** (tē-ĕn' shän'). TIAN SHAN.

**Tien·tsin** (tē-ĕn'sĭn', -ĕnt'-, tĭn'-, tĭnt'-). TIANJIN.

**Ti·er·ra del Fue·go** (tē-ĕr'ə dĕl fo͞o-ā'gō, fyo͞o-). **1.** Archipelago off S South America separated from the mainland by the Strait of Magellan. **2.** Main island of this archipelago divided between Chile & Argentina.

**Tie·tê** (tyə-tā'). River, c. 500 mi (805 km), of S Brazil.

**Tif·fin** (tĭf'ĭn). City of N central Ohio SSW of Toledo. Pop. 19,549.

**Tif·lis** (tĭf'lĭs, tə-flēs'). TBILISI.

**Tif·ton** (tĭf'tən). City of S central Ga. ESE of Albany. Pop. 13,749.

**Ti·gard** (tī'gərd). City of NW Ore. near Portland. Pop. 14,286.

**Ti·gre** (tē'grā). City of E Argentina near Buenos Aires. Pop. 146,451.

**Ti·gris** (tī'grĭs). River of SW Asia rising in E Turkey & flowing c. 1,150 mi (1,850 km) SE through Iraq to the Euphrates R.

**Ti·jua·na** (tē'ə-wä'nə, tē-wä'-). City of extreme NW Mexico on the U.S. border. Pop. 363,154.

**Ti·kal** (tĭ-käl'). Ruined Mayan city of N Guatemala.

**Til·burg** (tĭl'bûrg'). City of S Netherlands near the Belgian border. Pop. 151,513.

**Til·la·mook Bay** (tĭl'ə-mo͝ok'). Inlet of the Pacific off NW Ore.

**Ti·ma·ga·mi** (tə-mä'gə-mē). Lake, 91 sq mi (236 sq km), of S central Ont., Canada, NE of Sudbury.

**Tim·buk·tu** *also* **Tim·buc·too** (tĭm'bŭk-to͞o', tĭm-bŭk'to͞o). City of central Mali near the Niger R. Pop. 20,483.

**Tim·gad** (tĭm'găd'). Ancient Roman city in NE Algeria.

**Ti·miș** (tē'mēsh'). River, c. 270 mi (434 km), of W Rumania & N Yugoslavia.

**Ti·mi·șoa·ra** (tē'mēsh-wär'ə). City of W Rumania near the Yugoslavian border. Pop. 281,320.

**Tim·mins** (tĭm'ĭnz). City of central Ont., Canada, NE of Sault St. Marie. Pop. 44,747.

**Ti·mor** (tē'môr, tē-môr'). **1.** Sea of the Indian Ocean between Timor & Australia. **2.** Island of SE Indonesia, in the Lesser Sundas.

**Tim·pa·no·gos** (tĭm'pə-nō'gəs), **Mount.** Highest (12,008 ft/ 3,662.4 m) of the Wasatch Mts. in N central Utah.

**Ti·ni·an** (tĭn'ē-ăn'). Island of W Pacific in the S Marianas in the U.S. Trust Territory of the Pacific Is.

**Tin·ley Park** (tĭn'lē). Village of NE Ill. near Chicago. Pop. 26,171.

**Tí·nos** (tē'nôs'). TENOS.

**Tin·tag·el Head** (tĭn-tăj'əl). Promontory in SW England in NE Cornwall.

**Tip·pe·ca·noe** (tĭp'ē-kə-no͞o'). River, c. 170 mi (274 km), of N Ind.

**Tip·per·ar·y** (tĭp′ə-râr′ē). Region & town of S central Ireland. Pop. 4,929.
**Ti·ran** (tə-rän′). Strait off the S tip of the Sinai Peninsula, NE Egypt, connecting the Red Sea with the Gulf of Aqaba.
**Ti·ra·në** *also* **Ti·ra·na** (tə-rä′nə). Cap. of Albania in the central part. Pop. 171,300.
**Ti·ras·pol** (tə-räs′pəl). City of SW European USSR on the Dniester R. NW of Odessa. Pop. 139,000.
**Tîr·gu-Mu·reş** (tîr′gōō-mōōr′ĕsh). City of N central Rumania ESE of Cluj. Pop. 129,284.
**Ti·rich Mir** (tîr′ĭch mîr′). Highest elevation (25,263 ft/7,705.2 m) of the Hindu Kush in N Pakistan.
**Ti·rol** (tə-rōl′, tī′rōl′, tĭ-rōl′). *var. of* TYROL.
**Ti·ruch·chi·rap·pal·li** (tîr′ə-chə-rä′pə-lē). City of SE India on the Cauvery R. Pop. 307,400.
**Ti·ryns** (tîr′ĭnz, tī′rĭnz). Ancient city of S Greece in the E Peloponnesus.
**Ti·sza** (tĭs′ô) *also* **Ti·sa** (tē′sə). River of central Europe rising in the Carpathians in W Ukraine & flowing c. 600 mi (965 km) W & S across E Hungary & N Yugoslavia to the Danube.
**Ti·ti·ca·ca** (tĭt′ĭ-kä′kə). Largest freshwater lake (c. 3,200 sq mi/8,288 sq km) in South America, in the Andes on the Bolivia-Peru border.
**Ti·to·grad** (tē′tō-grăd′). City of S Yugoslavia near the Albanian border. Pop. 54,639.
**Ti·tus·ville** (tī′təs-vĭl′). City of E Fla. E of Orlando. Pop. 31,910.
**Tiv·er·ton** (tĭv′ər-tən). Town of E R.I. near the Mass. border SE of Providence. Pop. 13,526.
**Ti·vo·li** (tĭv′ə-lē). City of central Italy ENE of Rome. Pop. 28,393.
**Ti·zi Ou·zou** (tē-zē′ ōō-zōō′). City of N Algeria E of Algiers. Pop. 73,100.
**Tjir·e·bon** (chĭr′ə-bôn′). City of N Java, Indonesia, on the Java Sea. Pop. 178,529.
**Tlal·ne·pan·tla** (tläl′nə-pänt′lä). City of S central Mexico near Mexico City. Pop. 45,575.
**Tla·que·pa·que** (tlä′kā-pä′kā). City of W central Mexico near Guadalajara. Pop. 59,760.
**Tlem·cen** *also* **Tlem·sen** (tlĕm-sĕn′). City of NW Algeria near the Moroccan border. Pop. 109,400.
**To·ba** (tō′bə). Lake, 448 sq mi (1,160 sq km), of N Sumatra.
**To·ba·go** (tə-bā′gō). Island of Trinidad and Tobago in the SE West Indies NE of Trinidad.
**To·bol** (tə-bôl′). River of NE Central Asian USSR flowing c. 1,050 mi (1,689 km) NE to the Irtish R.
**To·bruk** (tō′brōōk′, tō-brōōk′). City of NE Libya on the Mediterranean. Pop. 58,384.
**To·can·tins** (tō′kən-tēNs′). River of central & N Brazil flowing 1,640 mi (2,639 km) N to the Pará R. SW of Belém.
**To·gli·at·ti** *also* **Tol·yat·ti** (tôl-yä′tē). City of E central European USSR on the Volga NW of Kuibyshev. Pop. 502,000.
**To·go** (tō′gō). Country of W Africa on the Gulf of Guinea. Cap. Lomé. Pop. 2,472,000.
**To·ho·pe·kal·i·ga** (tə-hō′pĭ-kăl′ĭ-gə). Lake of central Fla. S of Orlando.
**To·ka·ra** (tō-kär′ə). Islands of Japan in the N Ryukyu group S of Kyushu.
**To·ke·lau** (tō′kə-lou′). Islands of New Zealand in the central Pacific N of Samoa.
**To·ko·ro·za·wa** (tō′kō-rō′zä-wä). City of central Honshu, Japan, near Tokyo. Pop. 196,870.
**To·ku·shi·ma** (tō′kə-shē′mə). City of E Shikoku, Japan, on the Inland Sea. Pop. 239,281.
**To·ky·o** (tō′kē-ō′). Cap. of Japan, in E central Honshu on **Tokyo Bay,** a 50 mi (80 km) inlet of the Pacific. Pop. 8,646,520.
**Tol·bu·khin** (tôl-bōō′kĭn). City of NE Bulgaria N of Varna. Pop. 86,184.
**To·le·do** (tə-lē′dō). **1.** City of central Spain near the Tagus R. SSW of Madrid. Pop. 43,905. **2.** City of NW Ohio on Lake Erie. Pop. 354,635.
**To·li·ma** (tə-lē′mə). Volcanic mountain, 18,438 ft (5,623.6 m), in W central Colombia.
**To·lu·ca** (tə-lōō′kə). City of S central Mexico W of Mexico City. Pop. 136,092.
**Tol·yat·ti** (tôl-yä′tē). *var. of* TOGLIATTI.
**Tom** (tŏm). River, c. 525 mi (845 km), of S Siberian USSR.
**To·ma·ko·mai** (tō-mä′kō-nū′, tō′mə-kō′mī). City of S Hokkaido, Japan, on the Pacific. Pop. 132,477.
**To·ma·szów Ma·zo·wiec·ki** (tô-mä′shōōf mä′zô-vyĕt′skē). City of E central Poland SE of Lódź. Pop. 54,911.
**Tom·big·bee** (tŏm-bĭg′bē). River, c. 400 mi (645 km), of NE Miss. & W Ala.
**Tomsk** (tŏmsk, tômsk, tŏmpsk, tômpsk). City of SW Siberian USSR NE of Novosibirsk. Pop. 421,000.

**Ton·a·wan·da** (tŏn′ə-wŏn′də). City of W N.Y. near Buffalo. Pop. 18,693.
**Ton·bridge** (tŭn′brĭj). Urban district of SE England SE of London. Pop. 31,410.
**Ton·ga** (tŏng′gə). Island nation of the SW Pacific E of Fiji. Cap. Nukualofa. Pop. 90,128.
**Ton·ga·re·va** (tŏng′ə-rĕ′və, -gə-). Island of the central Pacific in the N Cook Is.
**Tong·hua** (tōōng′wä′, -hwä′). City of NE China in Manchuria E of Shenyang. Pop. 275,000.
**Tongue** (tŭng). River, 246 mi (396 km), of N Wyo. & SE Mont.
**Ton·kin** (tŏn′kĭn′, tŏng′-). **1. Gulf of.** NW arm of the South China Sea between N Vietnam & Hainan Is. **2.** Region of N Vietnam.
**Ton·le Sap** (tŏn′lā săp′). Lake, c. 1,100–3,600 sq mi (2,850–9,325 sq km), of central Cambodia.
**Too·el·e** (tōō-ĕl′ē). City of NW Utah SW of Salt Lake City. Pop. 14,335.
**To·pe·ka** (tə-pē′kə). Cap. of Kans. in the NE part W of Kansas City. Pop. 115,266.
**Tor·bay** (tôr-bā′). Borough of SW England on Lyme Bay. Pop. 109,900.
**Tor·cel·lo** (tôr-chĕl′ō). Island of NE Italy in the Lagoon of Venice NE of Venice.
**To·ri·no** (tō-rē′nō). TURIN.
**Tor·ne** (tôr′nə). River of N Sweden rising near the Norwegian border in **Lake Torne** (124 sq mi/321 sq km) & flowing c. 250 mi (402 km) SE to the Gulf of Bothnia, forming the Swedish-Finnish border in its lower course.
**Tor·ni·o** (tôr′nē-ō′). TORNE.
**To·ron·to** (tə-rŏn′tō). Cap. of Ont., Canada, in the S part on Lake Ontario. Pop. 633,318.
**Tor·rance** (tôr′əns, tŏr′-). City of S Calif. S of Los Angeles. Pop. 131,497.
**Tor·re An·nun·zi·a·ta** (tôr′ā ə-nōōn′sē-ä′tə). City of S Italy on the Bay of Naples. Pop. 71,068.
**Torre del Gre·co** (dĕl grĕk′ō). City of S Italy on the Bay of Naples. Pop. 74,752.
**Tor·rens** (tôr′ənz, tŏr′-). Lake, 2,230 sq mi (5,775.7 sq km), of S central Australia.
**Tor·re·ón** (tôr′ē-ōn′). City of N Mexico W of Monterrey. Pop. 244,309.
**Tor·res** (tôr′ĭs). Strait, c. 95 mi (153 km) wide, between New Guinea & Cape York Peninsula of NE Australia.
**Tor·ring·ton** (tôr′ĭng-tən, tŏr′-). City of NW Conn. W of Hartford. Pop. 30,987.
**Tor·to·la** (tôr-tō′lə). Island of the West Indies, largest of the British Virgin Is.
**Tor·tu·ga** (tôr-tōō′gə). Island in the West Indies off N Haiti.
**To·ruń** (tôr′ōōn′, -ōōn′yə). City of N central Poland on the Vistula R. Pop. 129,152.
**To·to·wa** (tō′tə-wə). Borough of NE N.J. near Paterson. Pop. 11,448.
**Tot·to·ri** (tə-tôr′ē, -tōr′ē). City of SW Honshu, Japan, on the Sea of Japan. Pop. 122,312.
**Toub·kal** (tōōb-käl′). JEBEL TOUBKAL.
**Tou·lon** (tōō-lôN′). City of SE France on the Mediterranean ESE of Marseille. Pop. 180,508.
**Tou·louse** (tōō-lōōz′). City of S France on the Garonne SE of Bordeaux. Pop. 371,143.
**Tou·raine** (tōō-rān′, -rĕn′). Region & former province of W central France.
**Tour·coing** (tōōr-kwăN′). City of N France near Lille & the Belgian border. Pop. 102,092.
**Tour·nai** *also* **Tour·nay** (tōōr-nā′). City of SW Belgium on the Scheldt R. Pop. 32,794.
**Tours** (tōōr). City of W central France on the Loire. Pop. 139,560.
**Towns·ville** (tounz′vĭl′). City of NE Australia on the Coral Sea. Pop. 78,653.
**Tow·son** (tou′sən). City of N Md. near Baltimore. Pop. 83,600.
**To·ya·ma** (tō-yä′mə). City of W central Honshu, Japan, on **Toyama Bay,** an inlet of the Sea of Japan. Pop. 290,143.
**To·yo·ha·shi** (tō′yə-hä′shē). City of S central Honshu, Japan, on the Pacific SE of Nagoya. Pop. 284,585.
**To·yo·na·ka** (tō′yō-nä′kə). City of S Honshu, Japan, near Osaka. Pop. 398,384.
**To·yo·ta** (toi-ō′tə). City of S central Honshu, Japan, near Nagoya. Pop. 248,774.
**Trab·zon** (trăb-zŏn′). **1.** TREBIZOND. **2.** City of NE Turkey on the Black Sea. Pop. 97,210.
**Tra·cy** (trā′sē). **1.** Town of S Que., Canada, NE of Montreal. Pop. 12,284. **2.** City of W central Calif. SSW of Stockton. Pop. 18,428.
**Tra·fal·gar** (trə-făl′gər). Cape on the SW coast of Spain NW of the Strait of Gibraltar.
**Tra·lee** (trə-lē′). Urban district of SW Ireland at the head of **Tralee Bay,** an inlet of the Atlantic. Pop. 12,287.
**Trans A·lai** (trăns′ ə-lī′, trănz′). Range of the Pamirs in S Central Asian USSR.

**Trans·al·pine Gaul** (trăns-ăl′pīn′ gôl′, trănz-). Part of ancient Gaul NW of the Alps, including modern France & Belgium.

**Trans·cau·ca·sia** (trăns′kô-kā′zhə, -zhē-ə, trănz′-). Region of S European USSR between the Caucasus Mts. & the borders of Turkey & Iran. **—Trans′cau·ca′sian** *adj. & n.*

**Trans·kei** (trăns-kā′, -kī′). Independent black African homeland in SE South Africa on the Indian Ocean coast. Cap. Umtata. Pop. 2,000,000. **—Trans·kei′an** *adj. & n.*

**Trans·vaal** (trăns-väl′, trănz-). Region & province of NE South Africa.

**Tran·syl·va·nia** (trăn′sĭl-vān′yə, -vā′nē-ə). Region of W Rumania. **—Tran′syl·va′ni·an** *adj. & n.*

**Transylvanian Alps.** S branch of the Carpathian Mts. extending across central Rumania.

**Tra·pa·ni** (trä′pə-nē). City of NW Sicily, Italy, on the Mediterranean. Pop. 90,305.

**Tra·si·me·no** (trä′zə-mā′nō). Lake, 50 sq mi (130 sq km), in central Italy W of Perugia.

**Trav·erse** (trăv′ərs). Lake on border between NE S.Dak. & W Minn.

**Traverse City.** City of N Mich. NNW of Cadillac. Pop. 15,516.

**Treb·i·zond** (trĕb′ĭ-zŏnd′). Byzantine Greek empire (1204–1461) on the S, E, & N coasts of the Black Sea.

**Trent** (trĕnt). **1.** River, c. 150 mi (241 km), of SE Ont., Canada. **2.** River, c. 170 mi (275 km), of central England. **3.** Canal system, 240 mi (386.2 km) long, of SE Ont., Canada, connecting Lake Ontario with Georgian Bay. **4.** *also* **Tren·to** (trĕn′tō). City of N Italy NW of Venice. Pop. 64,272.

**Tren·ton** (trĕn′tən). **1.** Town of SE Ont., Canada, on an inlet of NE Lake Ontario. Pop. 15,465. **2.** City of SE Mich. near Detroit. Pop. 22,762. **3.** Cap. of N.J. in the W central part on the Delaware R. Pop. 92,124.

**Trèves** (trĕv). TRIER.

**Tre·vi·so** (trə-vē′zō). City of NE Italy N of Venice. Pop. 87,447.

**Trier** (trîr). City of W West Germany on the Mosel R. near the Luxembourg border. Pop. 100,338.

**Tri·este** (trē-ĕst′, -ĕs′tē). City of extreme NE Italy on the **Gulf of Trieste,** an inlet of the Gulf of Venice. Pop. 257,259.

**Tri·glav** (trē′gläv′). Peak, 9,392 ft (2,862.6 m), in the Julian Alps in NW Yugoslavia.

**Tri·ko·ra Peak** (trə-kōr′ə). Mountain, 15,518 ft (4,733 m), of E West Irian, Indonesia.

**Tri·na·cri·a** (trə-năk′rē-ə, trī-). SICILY.

**Trin·i·dad** (trĭn′ĭ-dăd′). Island, 1,864 sq mi (4,828 sq km), of Trinidad and Tobago in the Atlantic off NE Venezuela. **—Trin′i·dad′i·an** *adj. & n.*

**Trinidad and To·ba·go** (tə-bā′gō). Country of SE West Indies, consisting of the islands of Trinidad & Tobago. Cap. Port of Spain. Pop. 1,067,108.

**Trin·i·ty** (trĭn′ĭ-tē). River, c. 360 mi (579 km), of E Tex.

**Trip·o·li** (trĭp′ə-lē). **1.** City of NW Lebanon on the Mediterranean. Pop. 127,611. **2.** Cap. of Libya in the NW on the Mediterranean. Pop. 550,438.

**Trip·o·lis** (trĭp′ə-lĭs). TRIPOLITANIA.

**Trip·o·li·ta·ni·a** (trĭp-ŏl′ĭ-tā′nē-ə, -tān′yə, trĭp′ə-lə). Ancient Phoenician colony in present-day NW Libya.

**Tris·tan da Cun·ha** (trĭs′tən də ko͞o′nə). **1.** Islands of the S Atlantic between S Africa & S South America; administered by Great Britain as part of St. Helena. **2.** Chief island of this group.

**Tri·van·drum** (trə-văn′drəm). City of SW India on the Arabian Sea. Pop. 409,627.

**Tro·as** (trō′ăs′). Ancient region of NW Asia Minor surrounding Troy.

**Tro·bri·and** (trō′brē-ănd′). Islands of Papua New Guinea in the Solomon Sea off E New Guinea.

**Trois-Ri·vières** (trwä-rē-vyĕr′). City of S Que., Canada, at the confluence of the St. Lawrence & St. Maurice rivers. Pop. 52,518.

**Trois-Ri·vières-Ouest** (trwä-rē-vyĕr-wĕst′). Town of S Que., Canada, near Trois-Rivières. Pop. 10,564.

**Troll·hät·tan** (trôl′hĕt′n). City of SW Sweden SW of Lake Vänern. Pop. 42,499.

**Trom·be·tas** (trōm-bā′tə). River, c. 470 mi (756 km), of N Brazil.

**Trom·sø** (trŏm′sŏ′). City of N Norway on an offshore island in the Arctic Ocean. Pop. 43,830.

**Trond·heim** (trŏn′hām′). City of central Norway on **Trondheim Fjord,** an 80 mi (129 km) inlet of the Norwegian Sea. Pop. 134,910.

**Tro·o·dos Mountain** (trô′ə-thôs). Highest elevation (6,400 ft/1,952 m) on Cyprus, in the central part.

**Trou·ville** (tro͞o-vēl′) *or* **Trou·ville-sur-Mer** (-so͞or-mĕr′). Town of N France on the English Channel. Pop. 6,618.

**Troy** (troi). **1.** Ancient city of NW Asia Minor near the Dardanelles; site of the Trojan War. **2.** City of SE Ala., SE of Montgomery. Pop. 12,587. **3.** City of SE Mich. near Detroit. Pop. 67,102. **4.** City of E N.Y. on the Hudson near Albany. Pop. 56,638. **5.** City of W Ohio N of Dayton. Pop. 19,086.

**Troyes** (trwä). City of N France on the Seine ESE of Paris. Pop. 71,600.

**Tru·chas Peaks** (tro͞o′chəs). Three peaks (highest, 13,110 ft/3,998.6 m) in N N.Mex. N of Santa Fe.

**Tru·cial O·man** (tro͞o′shəl ō-män′, ō-măn′). UNITED ARAB EMIRATES.

**Truck·ee** (trŭk′ē). River, c. 120 mi (193 km), of E Calif. & W Nev.

**Tru·ji·llo** (tro͞o-hē′ō, -yō). City of NW Peru NW of Lima. Pop. 241,882.

**Trujillo Al·to** (äl′tō′). Town of NE Puerto Rico SE of San Juan. Pop. 41,097.

**Truk** (trŭk, tro͝ok). Islands of the U.S. Trust Territory of the Pacific Is. in the central Carolines.

**Trum·bull** (trŭm′bəl). Town of SW Conn. N of Bridgeport. Pop. 32,989.

**Tru·ro** (tro͝or′ō). Town of central N.S., Canada, NNE of Halifax. Pop. 12,840.

**Tsa·na** (sä′nə, tsä′-). *var. of* Lake TANA.

**Tsang·po** (säng′pô′, tsäng′-). The upper Brahmaputra R. in S Tibet, China.

**Tsa·ri·tsyn** (tsə-rēt′sən). VOLGOGRAD.

**Tse·lin·o·grad** (sə-lĭn′ə-gräd′, -tsə-). City of E Central Asian USSR WSW of Novosibirsk. Pop. 234,000.

**Tsiao·tso** (jēou′jō′). JIAOZUO.

**Tsi·nan** (jē′nän′). JINAN.

**Tsing·hai** (chĭng′hī′). QINGHAI.

**Tsing·tao** (chĭng′dou′, sĭng′tou′, tsĭng′-). QINGDAO.

**Tsi·tsi·har** (chē′chē′här′, sē′sē′-, tsē′tsē′-). QIQIHAR.

**Tsu** (tso͞o). City of S Honshu, Japan, on Ise Bay. Pop. 139,538.

**Tsu·ga·ru** (so͞o-gä′ro͞o, tso͞o-). Strait between Honshu & Hokkaido, N Japan.

**Tsun·yi** (zo͞o′nē′). ZUNYI.

**Tsu·shi·ma** (so͞o-shē′mə, tso͞o-). Islands of SW Japan in Korea Strait between Kyushu & SE South Korea, separated from Kyushu by **Tsushima Strait.**

**Tu·a·mo·tu** (to͞o′ə-mō′to͞o). Archipelago of French Polynesia in the S Pacific E of Tahiti.

**Tü·bing·en** (to͞o′bĭng-ən, tyo͞o′-). City of SW West Germany on the Neckar SSW of Stuttgart. Pop. 71,348.

**Tu·bu·a·i** (to͞ob-wä′ē). Islands of S French Polynesia in the S Pacific S of Tahiti.

**Tuc·son** (to͞o′sŏn′). City of SE Ariz. SSE of Phoenix. Pop. 330,537.

**Tu·cu·mán** (to͞o′kə-män′). City of NW Argentina NNW of Cordoba. Pop. 366,392.

**Tu·ge·la** (to͞o-gā′lə). River, c. 300 mi (483 km), of E South Africa.

**Tu·la** (to͞o′lə). City of central European USSR S of Moscow. Pop. 514,000.

**Tu·la·gi** (to͞o-lä′gē). Island of S central Solomon Is. in the W Pacific.

**Tu·lare** (to͞o-lâr′, -lâr′ē). City of S central Calif. SE of Fresno. Pop. 22,475.

**Tul·la·ho·ma** (tŭl′ə-hō′mə). City of S central Tenn. NW of Chattanooga. Pop. 15,800.

**Tul·sa** (tŭl′sə). City of NE Okla. NE of Oklahoma City. Pop. 360,919.

**Tu·ma·co** (to͞o-mä′kō). City of SW Colombia on the Pacific. Metro. area pop. 38,742.

**Tu·men** (to͞o′mŭn′). River of NE North Korea flowing c. 324 mi (521 km) NE & SE along the Korea-China & Korea-USSR borders to the Sea of Japan.

**Tung·hwa** (to͝ong′wä′, -hwä′). TONGHUA.

**Tun·gus·ka** (to͝on-go͞o′skə, to͝ong-). Any of three rivers of Siberian USSR: **Upper Tunguska,** the lower course of the Angara; **Lower Tunguska,** flowing c. 2,000 mi (3,218 km) N & W to the Yenisei; & **Stony Tunguska,** flowing c. 1,000 mi (1,609 km) WNW to the Yenisei.

**Tu·nis** (to͞o′nĭs, tyo͞o′-). **1.** Former Barbary state on the N coast of Africa. **2.** Cap. of Tunisia in the N part on the Mediterranean. Pop. 550,404.

**Tu·ni·sia** (to͞o-nē′zhə, -shə, -nĭzh′ə, -nĭsh′ə, tyo͞o-). Country of N Africa on the Mediterranean. Cap. Tunis. Pop. 6,367,000. **—Tu·ni′sian** *adj. & n.*

**Tun·ja** (to͞ong′hä). City of central Colombia NE of Bogotá. Pop. 57,620.

**Tu·ol·um·ne** (to͞o-ŏl′ə-mē). River, 155 mi (249 km), of central Calif.

**Tu·pe·lo** (to͞o′pə-lō′, tyo͞o′-). City of NE Miss. NNW of Columbus. Pop. 23,905.

**Tu·pun·ga·to** (to͞o′pən-gä′tō). Mountain, 21,490 ft (6,554.5 m), in the Andes on the Chile-Argentina border.

**Tur·fan** (to͝or′făn′). TURPAN.

**Tu·rin** (to͝or′ĭn, tyo͝or′-). City of NW Italy on the Po WSW of Milan. Pop. 1,181,698.

**Tur·ka·na** (tər-kăn′ə), **Lake.** Lake RUDOLF.

**Tur·key** (tûr′kē). Country of SW Asia & SE Europe between the Mediterranean & the Black Sea. Cap. Ankara. Pop. 45,217,556. **—Tur′kish** (tûr′kĭsh) *adj. & n.*

**Turk·men Soviet Socialist Republic** (tûrk'mən) *also* **Turk·men·i·stan** (tûrk-mĕn'ĭ-stăn', -stän'). Constituent republic of S Central Asian USSR.
**Turks and Cai·cos** (tûrks; kā'kəs, kī'kōs). Island groups of the British West Indies in the Atlantic SE of the Bahamas.
**Tur·ku** (to͝or'ko͞o'). City of SW Finland on the Baltic. Pop. 164,857.
**Tur·lock** (tûr'lŏk'). City of central Calif. SE of Modesto. Pop. 26,291.
**Turn·hout** (tûrn'hout). City of N Belgium near the Dutch border. Pop. 38,007.
**Tur·pan** (to͝or'pän'). Depression (lowest point, 505 ft/154 m below sea level) of NW China at the E end of the Tian Shan.
**Tur·qui·no** (to͞or-kē'nô). Peak, 6,560 ft (2,000.8 m), in the Sierra Maestra of SE Cuba.
**Tus·ca·loo·sa** (tŭs'kə-lo͞o'sə). City of W central Ala. SW of Birmingham. Pop. 75,143.
**Tus·ca·ny** (tŭs'kə-nē'). Region of NW Italy between the N Apennines & the Ligurian & Tyrrhenian seas. **—Tus'can** *adj. & n.*
**Tus·cu·lum** (tŭs'kə-ləm, -kyə-). City of ancient Latium SE of modern Rome, Italy.
**Tus·ke·gee** (tŭs-kē'gē). City of E Ala. E of Montgomery. Pop. 12,716.
**Tus·tin** (tŭs'tĭn). City of S Calif. near Santa Ana. Pop. 32,073.
**Tu·tu·i·la** (to͞o'tə-wē'lə). Largest island of American Samoa in the S Pacific.
**Tu·va·lu** (to͞o-vä'lo͞o). Island nation of the W Pacific N of Fiji. Cap. Funafuti. Pop. 7,349.
**Tux·pan** (to͞os'pän'). City of E central Mexico on the Gulf of Mexico NE of Mexico City. Pop. 33,901.
**Tux·tla Gu·tiér·rez** (to͞os'tlə go͞o-tyĕr'əs). City of SE Mexico near the Isthmus of Tehuantepec. Pop. 66,851.
**Tuz** (to͞oz). Shallow salt lake, c. 625 sq mi (1,619 sq km), of central Turkey.
**Tuz·la** (to͞oz'lä). City of central Yugoslavia NNE of Sarajevo. Pop. 53,836.
**Tweed** (twēd). River, 97 mi (156 km), of SE Scotland forming part of the Scottish-English border.
**Twin Cities** (twĭn). Minneapolis & St. Paul, Minn.
**Twin Falls.** City of S central Idaho in the Snake R. valley. Pop. 26,209.
**Two Rivers** (to͞o). City of E Wis. on Lake Michigan SE of Green Bay. Pop. 13,354.
**Ty·gart** (tī'gərt). River, c. 160 mi (257 km), of E & N W.Va.
**Ty·ler** (tī'lər). City of NE Tex. ESE of Dallas. Pop. 70,508.
**Tyn·dall** (tĭn'dl). Mountain, 14,025 ft (4,277.6 m), in the Sierra Nevada of S Calif.
**Tyne·mouth** (tīn'mouth', -məth). Borough of NE England on the North Sea at the mouth of the **Tyne R.** (80 mi/129 km). Pop. 67,090.
**Tyre** (tīr). Ancient Phoenician city on the E Mediterranean in present-day S Lebanon.
**Ty·ree** (tī-rē'), **Mount.** Peak, 16,290 ft (4,968.5 m), of W Antarctica in the Ellsworth Mts. near the base of the Antarctic Peninsula.
**Ty·rol** *or* **Ti·rol** (tə-rōl', tī'rōl', tī-rōl'). Region of the E Alps in W Austria & N Italy. **—Ty'ro·lese'** *adj. & n.*
**Tyr·rhe·ni·an Sea** (tə-rē'nē-ən). Part of the Mediterranean between the Italian peninsula & the islands of Corsica, Sardinia, & Sicily.
**Tyu·men** (tyo͞o-mĕn'). City of SW Siberian USSR E of Sverdlovsk. Pop. 359,000.
**Tze·kung** (zŭ'go͝ong', dzŭ'-). ZIGONG.
**Tze·po** (zŭ'bō', dzŭ'-). ZIBO.

# U

**Uau·pés** (wou-pĕs'). River of NW South America rising as the **Vaupés** in S central Colombia & flowing c. 500 mi (805 km) ESE through NW Brazil to the Río Negro.
**U·ban·gi** (o͞o-băng'gē, yo͞o-). River of central Africa flowing c. 700 mi (1,126 km) along the NW border of Zaire to the Congo R.
**U·be** (o͞o'bē', -bā'). City of SW Honshu, Japan, on the Inland Sea. Pop. 161,969.
**U·ca·ya·li** (o͞o'kə-yä'lē). River of E Peru flowing c. 1,000 mi (1,609 km) N to the Marañón R. to form the Amazon.
**Uc·cle** (o͝ok'lə). City of central Belgium near Brussels. Pop. 78,909.
**U·dai·pur** *or* **U·day·pur** (o͞o-dī'po͝or, o͞o'dī-po͝or'). City of NW India NE of Ahmadabad. Pop. 161,278.
**Ud·de·val·la** (ŭd'ə-vä'lə). City of SW Sweden on an arm of the Skagerrak. Pop. 32,700.
**U·di·ne** (o͞o'də-nā'). City of NE Italy NE of Venice. Pop. 97,544.
**Ue·le** (wĕl'ē). River, c. 700 mi (1,126 km), of N Zaire.

o͞o **boot** ou **out** th **thin** *th* **this** ŭ **cut** ûr **urge** y **young**
yo͞o ab**u**se zh vi**si**on ə **a**bout, it**e**m, ed**i**ble, gall**o**p, circ**u**s

**U·fa** (o͞o-fä'). **1.** River, c. 600 mi (965 km), of E European USSR in the S Urals. **2.** City of E European USSR in the S Urals at the confluence of the Belaya & Ufa rivers. Pop. 969,000.
**U·gan·da** (yo͞o-găn'də, -gän'-). Country of E central Africa. Cap. Kampala. Pop. 12,630,076. **—U·gan'dan** *adj. & n.*
**U·ga·rit** (o͞o'gə-rēt'). Ancient city of W Syria.
**U·in·ta** (yo͞o-ĭn'tə). Range of the Rockies extending c. 120 mi (195 km) E from NE Utah to SW Wyo.
**U·ji** (o͞o'jē). Town of S Honshu, Japan, near Kyoto. Pop. 133,405.
**Uj·jain** (o͞o'jīn'). City of central India E of Ahmadabad. Pop. 203,278.
**U·ki·ah** (yo͞o-kī'ə). City of NW Calif. NNW of San Francisco. Pop. 12,035.
**Uk·kel** (o͝ok'əl). UCCLE.
**U·kraine** (yo͞o-krān'). **1.** Region of SW European USSR. **2.** *also* **U·krai·ni·an Soviet Socialist Republic** (-krā'nē-ən). Constituent republic of SW European USSR. **—U·krai'ni·an** *adj. & n.*
**U·lan Ba·tor** (o͞o'län bä'tôr'). Cap. of Mongolia in the N central part. Pop. 345,000.
**U·lan-U·de** (o͞o'län-o͞o-dā'). City of S Siberian USSR near Lake Baikal. Pop. 300,000.
**Ulm** (o͞olm). City of S West Germany on the Danube SE of Stuttgart. Pop. 98,237.
**U·lu·a** (o͞o-lo͞o'ə). River, c. 200 mi (322 km), of W Honduras.
**U·lu Dag** (o͞o'lə dä', däg'). Mountain, 8,343 ft (2,544.6 m), of NW Turkey SE of Bursa.
**U·lugh Muz·tagh** (o͞o'lə məz-tä', -täg'). Highest elevation, 25,340 ft (7,728.7 m), of the Kunlun Range in W China.
**Ul·ya·novsk** (o͞ol-yä'nəfsk). City of E central European USSR on the Volga NW of Kuibyshev. Pop. 464,000.
**Um·bri·a** (ŭm'brē-ə). Region of central Italy. **—Um'bri·an** *adj. & n.*
**U·me** (o͞o'mə). River, c. 285 mi (459 km), of N Sweden.
**U·me·å** (o͞o'mə-ô'). City of NE Sweden on an inlet of the Gulf of Bothnia. Pop. 49,715.
**Umm al Qai·wain** (o͞om' ăl kī-wīn'). Sheikdom of the United Arab Emirates in E Arabia on the Persian Gulf. Pop. 2,900.
**Um·nak** (o͞om'năk'). Island of SW Alas. in the E central Aleutians.
**Ump·qua** (ŭmp'kwô'). River, c. 200 mi (322 km), of SW Ore.
**Um·ta·ta** (o͞om-tä'tə). Cap. of the Transkei in the W central part. Pop. 25,216.
**Un·a·las·ka** (ŭn'ə-lăs'kə). Island of SW Alas. in the E Aleutians SW of Unimak Is.
**Un·com·pah·gre** (ŭn'kəm-pä'grē). Peak, 14,306 ft (4,363.3 m), in the San Juan Range of the Rockies in SW Colo.
**Un·ga·va** (ŭn-gā'və, -gä'və, -găv'ə). **1. Bay.** Inlet of Hudson Strait extending c. 200 mi (322 km) into N Que., Canada. **2.** Peninsula of N Que., Canada, between Hudson & Ungava bays.
**U·ni·mak** (yo͞o'nə-măk'). Island of SW Alas. in the E Aleutians SW of Alaska Peninsula.
**Un·ion** (yo͞on'yən). **1.** TOKELAU. **2.** City of NW S.C. NW of Columbia. Pop. 10,523.
**Union City. 1.** City of W Calif. SE of Oakland. Pop. 39,406. **2.** City of NE N.J. near Jersey City & the Hudson R. Pop. 55,593. **3.** City of extreme NW Tenn. near the Ky. border NNW of Jackson. Pop. 10,436.
**Union of Soviet Socialist Republics.** Country of E Europe & N Asia with coastlines on the Baltic & Black seas & the Arctic & Pacific oceans. Cap. Moscow. Pop. 262,436,227.
**Un·ion·town** (yo͞on'yən-toun'). City of SW Pa. near the W.Va. border SSE of Pittsburgh. Pop. 14,510.
**U·nit·ed Ar·ab E·mir·ates** (yo͞o-nī'tĭd ăr'əb ĭ-mîr'ĭts, ĕm'ər-). Country of E Arabia, a federation of seven sheikdoms on the Persian Gulf & the Gulf of Oman. Cap. Abu Dhabi. Pop. 1,040,275.
**United Arab Republic. 1.** EGYPT. **2.** Former union of Egypt & Syria (1958–61).
**United Kingdom** *or* **United Kingdom of Great Brit·ain and Northern Ire·land** (brĭt'n; īr'lənd). Country of W Europe comprising England, Scotland, Wales, & Northern Ireland. Cap. London. Pop. 55,672,000.
**United States** *or* **United States of A·mer·i·ca** (ə-mĕr'ĭ-kə). Country of central & NW North America with coastlines on the Atlantic, Pacific, & Arctic oceans. Cap. Washington, D.C. Pop. 226,504,825.
**U·ni·ver·sal City** (yo͞o'nə-vûr'səl). City of S central Tex. near San Antonio. Pop. 10,720.
**U·ni·ver·si·ty City** (yo͞o'nə-vûr'sĭ-tē). City of E Mo. near St. Louis. Pop. 42,738.
**University Heights.** City of NE Ohio near Cleveland. Pop. 15,401.
**University Park.** City of NE Tex. near Dallas. Pop. 22,254.
**Up·land** (ŭp'lənd). City of S Calif. NE of Los Angeles. Pop. 47,647.
**U·po·lu** (o͞o-pō'lo͞o). Island of Western Samoa in the S Pacific.
**Up·per Ar·ling·ton** (ŭp'ər är'lĭng-tən). City of central Ohio surrounded by Columbus. Pop. 35,648.
**Upper Cal·i·for·ni·a** (kăl'ĭ-fôrn'yə). Spanish possessions along the Pacific coast N of Lower California.
**Upper Klam·ath** (klăm'əth). Lake of S central Ore.

**Upper Peninsula.** N part of Mich. separated from the Lower Peninsula by the Straits of Mackinac.

**Upper Vol·ta** (vōl′tə, vŏl′-, vôl′-). Country of W Africa S of Mali. Cap. Ouagadougou. Pop. 6,908,000. **—Upper Vol′tan** *adj.* & *n.*

**Upp·sa·la** *also* **Up·sa·la** (ŭp′sə-lä′, -sä′lə, ŭp′sä′lə). City of E Sweden NNW of Stockholm. Pop. 101,850.

**Ur** (ûr, o͝or). Ancient city of Sumer, S Mesopotamia, on a site in present-day SE Iraq.

**U·ral** (yo͝or′əl). **1.** River of USSR rising in the S Urals & flowing 1,574 mi (2,533 km) W & S to the Caspian Sea. **2.** *also* **U·rals** (-əlz). Range of the USSR forming the traditional boundary between Europe & Asia, extending c. 1,500 mi (2,400 km) from the Arctic Ocean S to Kazakhstan.

**U·ralsk** (yo͞o-rălsk′). City of NW Central Asian USSR on the Ural R. SSE of Kuibyshev. Pop. 167,000.

**U·ra·ri·coe·ra** *or* **U·ra·ri·cue·ra** (o͞o-rär′ĭ-kwĕr′ə). River, c. 300 mi (483 km), of NW Brazil.

**U·ra·wa** (o͞o-rä′wə). City of E central Honshu, Japan, near Tokyo. Pop. 331,145.

**Ur·ban·a** (ûr-băn′ə). **1.** City of E central Ill. adjoining Champaign. Pop. 35,978. **2.** City of W central Ohio NE of Dayton. Pop. 10,762.

**Ur·ban·dale** (ûr′bən-dāl′). City of central Iowa near Des Moines. Pop. 17,869.

**Ur·fa** (o͝or-fä′). City of SE Turkey near the Syrian border. Pop. 132,934.

**Ur·mi·a** (o͝or′mē-ə). Lake, 1,500-2,300 sq mi (3,885-5,957 sq km), of NW Iran between Tabriz & the Turkish border.

**U·rua·pan** (o͞or-wä′pən). City of W central Mexico W of Mexico City. Pop. 108,124.

**U·ru·bam·ba** (o͞o′ro͞o-bäm′bə). River, c. 450 mi (724 km), of S Peru.

**U·ru·guay** (o͝or′ə-gwī′, -gwā′, yo͝or′-). **1.** River of SE South America rising in S Brazil & flowing c. 1,000 mi (1,609 km) W & S on the Brazil-Argentina border & the Argentina-Uruguay border to the Río de la Plata. **2.** Country of SE South America on the Atlantic & the Río de la Plata. Cap. Montevideo. Pop. 2,899,000. **—U·ru·guay′an** *adj.* & *n.*

**Ü·rüm·qi** *also* **U·rum·chi** (o͞o-ro͞om′chē, o͞or′əm-chē′). City of NW China in the Tian Shan. Pop. 500,000.

**U·se·dom** (o͞o′zə-dôm′). Island in the Baltic Sea at the mouth of the Oder, divided between East Germany & Poland.

**Ush·ant** (ŭsh′ənt). Island of NW France in the Atlantic off W Brittany.

**Us·hua·ia** (o͞o-swä′yə). Town of S Argentina on the S coast of Tierra del Fuego. Pop. 5,373.

**Us·ku·dar** (o͞os′kə-där′). District of Istanbul, Turkey, on the Asian side of the Bosporus.

**Us·pa·lla·ta** (o͞o′spə-yä′tə, -zhä′-). Pass, c. 12,500 ft (3,815 m), through the Andes between Mendoza, Argentina, & Santiago, Chile.

**Us·su·ri** (o͞o-so͞or′ē). River, c. 365 mi (587 km), of SE Far Eastern USSR forming part of the USSR-China border.

**U·sti nad La·bem** (o͞o′stē näd lä-bĕm′). City of NW Czechoslovakia on the Elbe near the East Germany border. Pop. 74,900.

**Ust-Ka·me·no·gorsk** (o͞ost′kə-mĕn′ə-gôrsk′). City of E Central Asian USSR on the Irtish R. S of Novosibirsk. Pop. 274,000.

**Ust·yurt** (o͞o-styo͝ort′). Desert plateau of Central Asian USSR between the Caspian & Aral seas.

**U·su·ma·cin·ta** (o͞o′sə-mə-sĭn′tə). River, c. 600 mi (965 km), of SE Mexico forming part of the Guatemala-Mexico border.

**U·tah** (yo͞o′tô′, -tä′). **1.** Lake, c. 145 sq mi (375 sq km), of N central Utah. **2.** State of the W U.S. Cap. Salt Lake City. Pop. 1,461,037. **—U′tah·an** *adj.* & *n.*

**U·ti·ca** (yo͞o′tĭ-kə). **1.** Ancient city of N Africa on the Mediterranean NW of Carthage. **2.** City of central N.W. E of Syracuse. Pop. 75,632.

**U·trecht** (yo͞o′trĕkt′). City of central Netherlands SSE of Amsterdam. Pop. 250,887.

**U·tsu·no·mi·ya** (o͞ot′sə-nō′mē-ə, -yə). City of central Honshu, Japan, N of Tokyo. Pop. 344,420.

**U·val·de** (yo͞o-văl′dē). City of SW Tex. WSW of San Antonio. Pop. 14,178.

**Ux·bridge** (ŭks′brĭj′). Township of central Ont., Canada, N of Toronto. Pop. 10,936.

**Ux·mal** (o͞oz-mäl′). Ancient ruined Mayan city in Yucatán, SE Mexico.

**Uz·bek Soviet Socialist Republic** (o͞oz′bĕk′, ŭz′-) *also* **Uz·bek·i·stan** (o͞oz-bĕk′ĭ-stăn′, -stän′, ŭz-). Constituent republic of S Central Asian USSR.

# V

**Vaal** (väl). River of South Africa flowing c. 750 mi (1,207 km) SW to the Orange R.

**Vaa·sa** (vä′sə). City of W Finland on the Gulf of Bothnia. Pop. 54,402.

**Vac·a·ville** (văk′ə-vĭl′). City of central Calif. WSW of Sacramento. Pop. 43,367.

**Va·duz** (vä-do͞ots′). Cap. of Liechtenstein in the W part on the Rhine. Pop. 4,614.

**Váh** (vä, väKH). River, c. 245 mi (394 km), of E & central Czechoslovakia.

**Vai·gach** *or* **Vay·gach** (vī-găch′). Island of NE European USSR in the Kara Sea SE of Novaya Zemlya.

**Val-Bé·lair** (văl′bā-lâr′, văl-bā-lĕr′). Town of S Que., Canada, NW of Quebec city. Pop. 10,716.

**Val·dai** *or* **Val·day** (văl-dī′). Hills of NW European USSR between Leningrad & Moscow.

**Val·dez** (văl-dēz′). City of S Alas. on an inlet of Prince William Sound. Pop. 3,079.

**Val·di·vi·a** (văl-dē′vē-ə). City of S central Chile near the Pacific. Pop. 82,300.

**Val d'Or** (văl′ dôr′, väl dôr′). Town of SW Que., Canada, SE of Rouyn. Pop. 19,915.

**Val·dos·ta** (văl-dŏs′tə). City of S Ga. ENE of Tallahassee, Fla. Pop. 37,596.

**Va·lence** (və-läNs′). City of SE France on the Rhone S of Lyon. Pop. 67,101.

**Va·len·ci·a** (və-lĕn′chē-ə, -chə, -sē-ə). **1.** Region & former kingdom of E Spain on the Mediterranean S of Catalonia. **2.** City of E Spain near the **Gulf of Valencia,** a wide inlet of the Mediterranean. Pop. 626,675. **3.** City of N Venezuela WSW of Caracas on the W shore of **Lake Valencia** (125 sq mi/324 sq km). Pop. 367,171.

**Va·len·ci·ennes** (və-lĕn′sē-ĕnz′). City of N France near the Belgian border. Pop. 41,976.

**Va·len·ti·a** (və-lĕn′shē-ə, -shə). Island off SW Ireland.

**Va·let·ta** (və-lĕt′ə). *var. of* VALLETTA.

**Val·la·do·lid** (văl′ə-də-lĭd′, -lē′). City of N central Spain NNW of Madrid. Pop. 227,151.

**Val·le·jo** (və-lā′ō). City of W Calif. on San Pablo Bay N of Oakland. Pop. 80,188.

**Val·let·ta** *also* **Va·let·ta** (və-lĕt′ə). Cap. of Malta on the NE coast. Pop. 14,042.

**Val·ley East** (văl′ē ēst′). Town of central Ont., Canada, near Sudbury. Pop. 19,591.

**Val·ley·field** (văl′ē-fēld′). City of S Que., Canada, on the St. Lawrence SW of Montreal. Pop. 29,716.

**Valley Forge.** Village of SE Pa. on the Schuylkill R.; site of Continental Army winter headquarters (1777–78).

**Valley Stream.** Village of SE N.Y. on SW Long Is. near Queens. Pop. 35,769.

**Va·lois** (văl′wä′, văl-wä′). Region & former duchy of N France.

**Va·lo·na** (və-lō′nə). VLORË.

**Val·pa·rai·so** (văl′pə-rā′zō). **1.** (*also* -rī′-). *also* **Val·pa·ra·í·so** (väl′pä-rä-ē′sō). City of central Chile on the Pacific WNW of Santiago. Pop. 250,400. **2.** City of NW Ind. ESE of Gary. Pop. 22,247.

**Van** (văn). Lake, 1,419 sq mi (3,675 sq km), of E Turkey.

**Van Bu·ren** (byo͝or′ən). City of NW Ark. opposite Fort Smith. Pop. 11,996.

**Van·cou·ver** (văn-ko͞o′vər). **1. Mount.** Mountain, 15,700 ft (4,789 m), in the St. Elias Mts. in SW Y.T., Canada, near the Alas. border. **2.** Island, c. 275 mi (442 km) long, of SW Canada in the Pacific off the SW B.C. mainland. **3.** City of SW B.C., Canada, on the Strait of Georgia opposite Vancouver Is. Pop. 410,188. **4.** City of SW Wash. on the Columbia near Portland, Ore. Pop. 42,834.

**Van·dal·ia** (văn-dāl′yə). City of W central Ohio N of Dayton. Pop. 13,161.

**Van Die·men's Land** (dē′mənz). TASMANIA.

**Vä·nern** (vā′nərn) *or* **Va·ner** (-nər). Lake, c. 2,145 sq mi (5,556 sq km), of SW Sweden.

**Va·nier** (văn′yā′). **1.** City of SE Ont., Canada, on the Ottawa R. Pop. 19,812. **2.** Town of S Que., Canada, adjacent to Quebec city. Pop. 10,683.

**Va·nu·a Le·vu** (və-no͞o′ə lĕv′o͞o). Island of Fiji in the S Pacific NE of Viti Levu.

**Va·nu·a·tu** (vä′no͞o-ä′to͞o). Island republic of the S Pacific E of Australia. Cap. Vila. Pop. 112,596.

**Van Wert** (wûrt). City of W Ohio near the Ind. border SW of Toledo. Pop. 11,035.

**Va·ra·na·si** (və-rä′nə-sē). City of N central India on the Ganges. Pop. 583,856.

**Var·dar** (vär′där′). River, c. 240 mi (386 km), of S Yugoslavia & NE Greece.

**Va·re·se** (və-rā′sē). City of N Italy NW of Milan. Pop. 65,978.

**Var·na** (vär′nə). City of E Bulgaria on the Black Sea. Pop. 251,654.

**Väs·ter·ås** (vĕs′tə-rōs′). City of E Sweden WNW of Stockholm. Pop. 98,858.

**Vat·i·can City** (văt′ĭ-kən). Independent papal state within Rome, Italy. Pop. 728.

**Vat·na·jö·kull** (vät'nä-yœ'-ko͝ol). Glacier, c. 3,150 sq mi (8,160 sq km), in SE Iceland.
**Vät·ter** (vĕt'ər) *or* **Vät·tern** (-ərn). Lake, 733 sq mi (1,898.5 sq km), of S central Sweden SE of Lake Vänern.
**Vaughan** (vôn, vän). Town of SE Ont., Canada, near Toronto. Pop. 17,782.
**Vau·pés** (vou-pās'). —See UAUPÉS.
**Vay·gach** (vī-gäch'). *var. of* VAIGACH.
**Ve·ga Ba·ja** (vā'gə bä'hä). Town of N Puerto Rico WSW of San Juan. Pop. 18,020.
**Ve·ii** (vē'ī). Ancient city of Etruria NW of Rome, Italy.
**Vej·le** (vī'lə). City of central Denmark in W Jutland on **Vejle Fjord,** narrow inlet of the Kattegat. Pop. 43,976.
**Vel·bert** (fĕl'bərt). City of W West Germany in the Ruhr near Essen. Pop. 93,302.
**Ve·li·ki·ye Lu·ki** (və-lē'kē-ə lo͞o'kē, -yə). City of W central European USSR W of Moscow. Pop. 102,000.
**Vel·la La·vel·la** (vĕl'ə lə-vĕl'ə). Island of the W Pacific, in the central Solomons.
**Vel·lore** (və-lôr', -lōr'). Town of SE India W of Madras. Pop. 139,082.
**Ven·da** (vĕn'də). Independent black African homeland in NE South Africa near the Zimbabwe border. Cap. Thohoyandou. Pop. 450,000.
**Ve·ne·ti·a** (və-nē'shē-ə, -shə). Region of NE Italy & NW Yugoslavia.
**Ve·ne·to** (vĕn'ə-tō', vā'nə-). Region of N Italy.
**Ve·ne·zia** (və-nĕt'sē-ə). VENICE, Italy.
**Ven·e·zue·la** (vĕn'ə-zwā'lə, -zwē'-). **1. Gulf of.** Inlet of the Caribbean off NW Venezuela & N Colombia. **2.** Country of N South America on the Caribbean. Cap. Caracas. Pop. 13,913,000. **—Ven'e·zue'lan** *adj.* & *n.*
**Ven·i·am·i·nof Crater** (vĕn-yŏm'ə-nôf'). Active volcano, 8,225 ft (2,509 m), of SW Alas. on Alaska Peninsula.
**Ven·ice** (vĕn'ĭs). **1.** City of NE Italy on islets within a lagoon in the **Gulf of Venice,** a wide inlet of the N Adriatic. Pop. 108,082. **2.** City of SW Fla. S of Sarasota. Pop. 12,153. **—Ve·ne'tian** (və-nē'shən) *adj.* & *n.*
**Ven·lo** (vĕn'lō'). Town of SE Netherlands, near the West German border. Pop. 61,659.
**Ven·ta** (vĕn'tə). River, 217 mi (349 km), of NW European USSR in Lithuania & Latvia.
**Vent·nor City** (vĕnt'nər). City of SE N.J. near Atlantic City. Pop. 11,704.
**Ven·tu·ra** (vĕn-to͝or'ə, -tyo͝or'ə). City of SW Calif. on the Pacific W of Los Angeles. Pop. 490,500.
**Ve·ra·cruz** (vĕr'ə-kro͞oz', -kro͞os') *or* **Ve·ra·cruz Lla·ve** (yä'vā). City of E central Mexico on the Gulf of Mexico E of Puebla. Pop. 255,646.
**Ver·cel·li** (vĕr-chĕl'ē). City of NW Italy WSW of Milan. Pop. 54,934.
**Ver·de. 1.** (vûr'dē, vĕr'-). River, c. 190 mi (306 km), of S Ariz. **2.** (vûrd), **Cape.** Peninsula of W Senegal projecting W into the Atlantic; westernmost point of Africa.
**Ver·di·gris** (vûr'dĭ-grĭs). River, c. 280 mi (451 km), of SE Kans. & NE Okla.
**Ver·dun** (vər-dŭn'). **1.** City of S Que., Canada, on Montreal Is. near Montreal. Pop. 68,013. **2.** City of NE France on the Meuse R.; site of prolonged World War I battle (1916). Pop. 22,889.
**Ve·ree·ni·ging** (fə-rē'nĭ-kĭng'). City of NE South Africa on the Vaal R. S of Johannesburg. Pop. 172,549.
**Ver·kho·yansk** (vĕr'kə-yänsk'). Range of E Siberian USSR parallel to & E of the lower Lena R.
**Ver·mil·ion** (vər-mĭl'yən). City of N Ohio on Lake Erie W of Cleveland. Pop. 11,012.
**Ver·mont** (vər-mŏnt'). State of the NE U.S. Cap. Montpelier. Pop. 511,456. **—Ver·mont'er** *n.*
**Ver·non** (vûr'nən). **1.** City of S B.C., Canada, near the N end of Okanagan Lake. Pop. 17,546. **2.** Town of N Conn. NE of Hartford. Pop. 27,974. **3.** City of N Tex. near the Okla. border. Pop. 12,695.
**Ve·ro Beach** (vîr'ō). City of E Fla. N of Fort Pierce. Pop. 16,176.
**Vé·roi·a** (vâr'yə). City of NE Greece W of Salonika. Pop. 29,528.
**Ve·ro·na** (və-rō'nə). **1.** City of N Italy on the Adige R. W of Venice. Pop. 227,032. **2.** Borough of NE N.J. near Passaic. Pop. 14,166.
**Ver·sailles** (vər-sī', vĕr-). City of N central France near Paris; site of Louis XIV's palace. Pop. 93,359.
**Ves·ta·vi·a Hills** (vĕs-tā'vē-ə). City of N central Ala. near Birmingham. Pop. 15,733.
**Ves·ter·å·len** (vĕs'tə-rô'lən). Archipelago off the NW coast of Norway.
**Ve·su·vi·us** (və-so͞o'vē-əs). Active volcano, 4,190 ft (1,278 m), of Italy on the E shore of the Bay of Naples.
**Vesz·prém** (vĕs'prām). City of W Hungary N of Lake Balaton. Pop. 54,898.
**Vet·lu·ga** (vĕt-lo͞o'gə). River, 528 mi (850 km), of central European USSR.

**Vi·a·reg·gio** (vē'ə-rĕj'ō). City of NW Italy on the Tyrrhenian Sea. Pop. 49,965.
**Vi·cen·te Ló·pez** (və-sĕn'tē lō'pĕz'). City of E central Argentina near Buenos Aires. Pop. 285,178.
**Vi·cen·za** (vĭ-chĕn'sə, -chĕnt'-). City of NE Italy ENE of Verona. Pop. 99,451.
**Vi·cha·da** (vĭ-chä'də). River, c. 400 mi (644 km), of E Colombia.
**Vi·chu·ga** (vĭ-cho͞o'gə). City of central European USSR on the Volga NNW of Gorki. Pop. 52,000.
**Vi·chy** (vĭsh'ē, vē'shē). City & resort of central France; seat of the French government during World War II. Pop. 32,107.
**Vicks·burg** (vĭks'bûrg'). City of W Miss. on the Mississippi W of Jackson. Pop. 25,434.
**Vic·to·ri·a** (vĭk-tôr'ē-ə, -tōr'-). **1. Lake.** *also* **Victoria Ny·an·za** (nē-ăn'zə, nī-). Lake, c. 26,830 sq mi (69,490 sq km), of E central Africa in Uganda, Kenya, & Tanzania. **2.** River, c. 240 mi (386 km), of N Australia. **3.** Falls, c. 420 ft (128 m) high & 1.1 mi (1.7 km) wide, in the Zambesi R. on the Zambia-Zimbabwe border. **4.** Island of N.W.T., Canada, in the Arctic Ocean N of the mainland & S of Parry Channel. **5. Land.** Region of E Antarctica S of New Zealand, bordering the Ross Sea. **6.** Cap. of B.C., Canada, on SE Vancouver Is. & Juan de Fuca Strait. Pop. 62,551. **7.** Cap. of Hong Kong colony on the NW coast of Hong Kong Is. Pop. 1,026,870. **8.** Cap. of the Seychelles on the NE coast of Mahé Is. Pop. 15,559. **9.** City of SE Tex. SE of San Antonio. Pop. 50,695.
**Victoria Nile** (nīl). Section of the Nile, c. 260 mi (418 km), between lakes Victoria & Albert in central Uganda.
**Victoria Ny·an·za** nē-ăn'zə, nī-). Lake VICTORIA.
**Vic·to·ri·a·ville** (vĭk-tôr'ē-ə-vĭl', -tōr'-). Town of S Que., Canada, SE of Trois-Rivières. Pop. 21,825.
**Vic·tor·ville** (vĭk'tər-vĭl'). City of S Calif. N of San Bernardino. Pop. 14,220.
**Vi·da·lia** (vĭ-dāl'yə). City of SE Ga. W of Savannah. Pop. 10,393.
**Vi·din** (vē'dĭn). City of extreme NW Bulgaria on the Danube. Pop. 53,030.
**Vi·dor** (vī'dôr'). City of extreme SE Tex. near Beaumont. Pop. 12,117.
**Vi·en·na** (vē-ĕn'ə). **1.** Cap. of Austria in the NE part on the Danube. Pop. 1,700,000. **2.** Town of NE Va. W of Washington, D.C. Pop. 15,469. **3.** City of NW W.Va. on the Ohio R. near Parkersburg. Pop. 11,618. **—Vi'en·nese'** (vē'ə-nēz', -nēs') *adj.* & *n.*
**Vienne** (vē-ĕn'). River, c. 217 mi (349 km), of central France.
**Vien·tiane** (vyĕn-tyän'). Cap. of Laos in the N central part on the Mekong R. Pop. 132,253.
**Vie·ques** (vē-ā'kĭs). Island, 51 sq mi (132 sq km), off SE Puerto Rico.
**Vier·sen** (fîr'zən). City of W West Germany W of Düsseldorf. Pop. 84,220.
**Viet·nam** (vē-ĕt'näm', -năm', vē'ĕt-, vyĕt'-). Country of SE Asia in E Indochina on the South China Sea. Cap. Hanoi. Pop. 52,741,766. **—Viet'na·mese'** (-nə-mēz', -mēs') *adj.* & *n.*
**Vi·ge·va·no** (vē-jĕv'ə-nō'). City of NW Italy SW of Milan. Pop. 62,855.
**Vi·go** (vē'gō). City of NW Spain on the Atlantic. Pop. 114,526.
**Vii·pu·ri** (vē'pə-rē). VYBORG.
**Vi·ja·ya·wa·da** *also* **Vi·ja·ya·va·da** (vĭj'ə-yə-wä'də, -vä'-). City of SE India near the Kistna R. delta ESE of Hyderabad. Pop. 317,258.
**Vi·la** (vē'lə). Cap. of Vanuatu on Efate Is. in the SW Pacific. Pop. 14,797.
**Vil·lach** (fĭl'äk'). City of S Austria on the Drava R. Pop. 50,993.
**Vi·lla·her·mo·sa** (vē-ə-ĕr-mō'sə, -yə-). City of SE Mexico E of the Isthmus of Tehuantepec. Pop. 133,181.
**Vil·la Park** (vĭl'ə). Village of NE Ill. near Chicago. Pop. 23,185.
**Vi·lla·vi·cen·ci·o** (vē'ə-vĭ-sĕn'sē-ō', -yə-). City of central Colombia SE of Bogotá. Pop. 82,869.
**Vil·leur·banne** (vē'lər-băn', -bän', -yər-). City of SE France near Lyon. Pop. 115,913.
**Vil·ni·us** (vĭl'nē-əs) *or* **Vil·na** (-nə). City of W European USSR in SE Lithuania ESE of Kaunas. Pop. 481,000.
**Vil·yu·i** (vĭl-yo͞o'ē). River of E Siberian USSR flowing c. 1,520 mi (2,446 km) E to the Lena R.
**Vim·i·nal** (vĭm'ə-nəl). One of the seven hills of ancient Rome.
**Vi·ña del Mar** (vēn'yə dĕl mär'). City of central Chile on the Pacific near Valparaiso. Pop. 182,000.
**Vin·cennes. 1.** (văN-sĕn'). City of N central France near Paris. Pop. 44,256. **2.** (vĭn-sĕnz'). City of SW Ind. on the Wabash R. S of Terre Haute. Pop. 20,857.
**Vin·dhya** (vĭn'dyə, -dē-ə). Range of hills in central India, c. 600 mi (965 km) long & rising to c. 3,000 ft (915 m).
**Vine·land** (vīn'lənd). City of S N.J. SSW of Camden. Pop. 53,753.
**Vin·land** (vĭn'lənd). Unidentified coastal region of NE North America visited by early Norse voyagers.
**Vin·ni·tsa** (vĭn'ĭt-sə). City of SW European USSR SW of Kiev. Pop. 314,000.
**Vin·son Mas·sif** (vĭn'sən mă-sēf'). Peak, 16,860 ft (5,142.3 m), in the Ellsworth Mts. of W Antarctica.

**Vir·gin** (vûr′jĭn). **1.** River, c. 200 mi (322 km), of SW Utah & SE Nev. **2.** Islands of the West Indies E of Puerto Rico, divided into the **British Virgin Is.** to the NE & the **Virgin Is. of the United States** to the SW.
**Vir·gin·ia** (vər-jĭn′yə). **1.** State of the E U.S. on Chesapeake Bay & the Atlantic. Cap. Richmond. Pop. 5,346,279. **2.** City of NE Minn. NNW of Duluth. Pop. 11,056. **—Vir·gin′ian** *adj. & n.*
**Virginia Beach.** Independent city of SE Va. on the Atlantic near Norfolk. Pop. 262,199.
**Virgin Islands of the United States.** SW part of the Virgin Is., 133 sq mi (344 sq km), constituting a U.S. territory. Cap. Charlotte Amalie. Pop. 62,468.
**Vi·run·ga** (və-rōōng′gə). Range of central Africa in E Zaire, N Rwanda, & SW Uganda.
**Vis** (vēs). Island of W Yugoslavia off the Dalmatian coast SSW of Split.
**Vi·sa·kha·pat·nam** (vĭ-sä′kə-pŭt′nəm). VISHAKHAPATNAM.
**Vi·sa·lia** (vĭ-sāl′yə). City of S central Calif. SE of Fresno. Pop. 49,729.
**Vi·sa·yan** (və-sī′ən) *also* **Vi·sa·yans** (-ənz). Islands of the central Philippines between Luzon & Mindanao.
**Vis·by** (vĭz′bē). City of SE Sweden on W Gotland Is. on the Baltic. Pop. 19,886.
**Vis·count Mel·ville** (vī′kount mĕl′vĭl′, -vəl). Sound between Victoria & Melville Is. in N N.W.T., Canada.
**Vi·sha·kha·pat·nam** (vĭ-shä′kə-pŭt′nəm). City of E India on the Bay of Bengal. Pop. 352,504.
**Vi·so** (vē′zō), **Mount.** Highest of the Cottian Alps, 12,602 ft (3,843.6 m), in NW Italy near the French border.
**Vis·ta** (vĭs′tə). City of S Calif. N of San Diego. Pop. 35,834.
**Vis·tu·la** (vĭs′chə-lə, vĭsh′-, vĭs′tə-). Longest river of Poland, 678 mi (1,091 km), rising near the Czech border & flowing NE, NW, & N to the Gulf of Danzig.
**Vi·tebsk** (vē′tĕpsk′, -tĕbsk′, vĭ-tĕpsk′, -tĕbsk′). City of W European USSR on the Western Dvina R. NE of Minsk. Pop. 297,000.
**Vi·ter·bo** (vĭ-tĕr′bō). City of central Italy NNW of Rome. Pop. 39,291.
**Vi·ti Le·vu** (vē′tē lĕv′ōō). Largest of the Fiji Is. in the S Pacific.
**Vi·tim** (vĭ-tēm′). River of SE Siberian USSR flowing c. 1,140 mi (1,834 km) N to the Lena R.
**Vi·to·ri·a** (vĭ-tôr′ē-ə, -tōr′-). City of N central Spain SSE of Bilbao. Pop. 124,791.
**Vi·tó·ri·a** (vĭ-tôr′ē-ə, -tōr′-). City of E Brazil on the Atlantic NE of Rio de Janeiro. Pop. 121,978.
**Vi·try-sur-Seine** (vĭ-trē′sōōr-sān′, -sĕn′). City of N central France near Paris. Pop. 87,119.
**Vit·to·ri·a** (vĭ-tôr′ē-ə, -tōr′-). City of SE Sicily, Italy, W of Ragusa. Pop. 43,673.
**Vlaar·ding·en** (vlär′dĭng-ən). City of SW Netherlands near Rotterdam. Pop. 78,311.
**Vla·di·mir** (vlăd′ə-mîr′). City of central European USSR E of Moscow. Pop. 296,000.
**Vlad·i·vos·tok** (vlăd′ə-və-stŏk′, -vŏs′tŏk′). City of extreme SE Far Eastern USSR on the Sea of Japan. Pop. 550,000.
**Vlis·sing·en** (vlĭs′ĭng-ən). City of SW Netherlands on the North Sea. Pop. 43,806.
**Vlo·rë** (vlôr′ə, vlōr′ə) *also* **Vlo·ne** (vlō′nə). City of SW Albania on **Vlorë Bay,** an inlet of the Adriatic. Pop. 50,000.
**Vl·ta·va** (vŭl′tə-və). River, c. 270 mi (434 km), of W Czechoslovakia.
**Vo·gel·kop** (vō′gəl-kŏp′). DOBERAI.
**Vol·ca·no** (vŏl-kā′nō). Islands of Japan in the NW Pacific N of the Marianas.
**Vol·ga** (vŏl′gə, vôl′-, vōl′-). River of European USSR rising in the Valdai Hills NW of Moscow & flowing c. 2,300 mi (3,700 km) E & S to the Caspian Sea.
**Vol·go·grad** (vŏl′gə-grăd′, vôl′-, vōl′-). City of S European USSR on the Volga. Pop. 929,000.
**Vo·log·da** (vô′ləg-də). City of NE central European USSR NNE of Moscow. Pop. 237,000.
**Vo·los** or **Vó·los** (vō′lŏs′). City of E Greece in Thessaly on the **Gulf of Volos,** an inlet of the Aegean Sea. Pop. 51,290.
**Vol·ta** (vŏl′tə, vōl′-, vôl′-). River formed in central Ghana by the confluence of the White Volta & the Black Volta, flowing c. 290 mi (467 km) S through artificial **Lake Volta** (c. 3,275 sq mi/8,482 sq km) to the Gulf of Guinea.
**Volta Re·don·da** (rĭ-dŏn′də). City of E Brazil on the Paraíba R. NW of Rio de Janeiro. Pop. 120,645.
**Vol·tur·no** (vŏl-tōōr′nō, vōl-, vôl-). River, c. 109 mi (175 km), of S Italy.
**Volzh·skiy** (vôlzh′skē). City of S European USSR on the Volga near Volgograd. Pop. 209,000.
**Voor·burg** (vôr′bûrg′, vōr′-). City of SW Netherlands near The Hague. Pop. 45,209.
**Vor·ku·ta** (vôr-kōō′tə). City of extreme NE European USSR above the Arctic Circle. Pop. 100,000.
**Vo·ro·nezh** (və-rô′nĭsh). City of central European USSR on the Don NE of Kharkov. Pop. 783,000.
**Vo·ro·shi·lov·grad** (vôr′ə-shē′ləf-grăd′). City of S central European USSR in the Donets Basin SE of Kharkov. Pop. 463,000.
**Vosges** (vōzh). Mountains of NE France extending c. 120 mi (193 km) along the Rhine.
**Vra·tsa** (vrät′sə). City of NW Bulgaria NNE of Sofia. Pop. 61,265.
**Vyat·ka** (vē-ät′kə, vyät′-). River, c. 850 mi (1,368 km), of E European USSR.
**Vy·borg** (vē′bôrg′). City of NW European USSR NW of Leningrad near the Finnish border. Pop. 76,000.
**Vy·cheg·da** (vĭch′ĭg-də). River, 700 mi (1,126 km), of NE European USSR.

# W

**Waal** (väl). S branch of the lower Rhine in S Netherlands.
**Wa·bash** (wô′băsh′). **1.** River of the E central U.S. rising in W Ohio & flowing c. 475 mi (764 km) W & S across Ind. & on the Ind.-Ill. border to the Ohio R. **2.** City of N central Ind. SW of Fort Wayne. Pop. 12,985.
**Wa·co** (wā′kō). City of E central Tex. S of Dallas-Fort Worth. Pop. 101,261.
**Wad·den·zee** (väd′n-zā′). Inlet of the North Sea off N Netherlands between the Ijsselmeer & the W Frisian Is.
**Wad·ding·ton** (wŏd′ĭng-tən). **Mount.** Peak, 13,260 ft (4,044.3 m), of the Coast Mts. in SW B.C., Canada.
**Wads·worth** (wŏdz′wûrth′). City of NE Ohio W of Akron. Pop. 15,166.
**Wa·gram** (vä′gräm′). Town of NE Austria near Vienna; site of Napoleon's defeat of the Austrians (1809).
**Wa·hi·a·wa** (wä′hē-ə-wä′). City of central Oahu, Hawaii. Pop. 17,598.
**Wai·a·na·e** (wī′ə-nä′ā). Mountains of W Oahu, Hawaii.
**Wai·ka·to** (wī-kä′tō). Longest river of New Zealand, rising in central North Is. & flowing 270 mi (434 km) NW to the Tasman Sea.
**Wai·ki·ki** (wī′kĭ-kē′). Beach & resort district of Honolulu, Hawaii.
**Wai·pa·hu** (wī-pä′hōō). City of S Oahu, Hawaii, on the NW shore of Pearl Harbor. Pop. 29,200.
**Wa·ka·ya·ma** (wä′kə-yä′mə). City of S Honshu, Japan, on the Inland Sea. Pop. 389,717.
**Wake** (wāk). Island of the W Pacific between Hawaii & Guam, belonging to the U.S.
**Wake·field** (wāk′fēld′). Town of E Mass. near Boston. Pop. 24,895.
**Wa·la·chi·a** (wə-lā′kē-ə). *var. of* WALLACHIA.
**Wal·brzych** (väl′b′zhĭk′). City of SW Poland SW of Wroclaw. Pop. 125,048.
**Wal·che·ren** (väl′kə-rən). Region, formerly an island, of SW Netherlands at the mouth of the Scheldt.
**Wal·den** (wôl′dən). **1.** Pond of NE Mass. near Concord; site of Henry David Thoreau's cabin. **2.** Town of central Ont., Canada, near Sudbury. Pop. 10,453.
**Wald·wick** (wôld′wĭk). Borough of NE N.J. NNE of Paterson. Pop. 10,802.
**Wales** (wālz). Principality of SW Great Britain W of England; part of the United Kingdom. Pop. 2,790,462. **—Welsh** (wĕlsh) *adj. & n.*
**Wal·ker** (wô′kər). **1.** Salt lake, c. 105 sq mi (272 sq km), of W Nev. SE of Carson City. **2.** City of W central Mich. near Grand Rapids. Pop. 15,088.
**Wal·lace·burg** (wŏl′əs-bûrg′). Town of SW Ont., Canada, S of Sarnia. Pop. 11,132.
**Wal·la·chi·a** *also* **Wa·la·chi·a** (wə-lā′kē-ə). Region of SE Rumania. **—Wal·la′chi·an** *adj. & n.*
**Wal·la Wal·la** (wŏl′ə wŏl′ə). City of SE Wash. near the Ore. border SSW of Spokane. Pop. 25,618.
**Wal·ling·ford** (wŏl′ĭng-fərd). Town of S Conn. NNE of New Haven. Pop. 37,274.
**Wal·ling·ton** (wŏl′ĭng-tən). Borough of NE N.J. near Passaic. Pop. 10,741.
**Wal·lis and Fu·tu·na** (wŏl′ĭs; fə-tōō′nə). Two island groups of the SW Pacific W of Samoa & NE of Fiji; a French overseas territory. Pop. 12,000.
**Wal·lops** (wŏl′əps). Island in the Atlantic off E Va.
**Wal·low·a** (wä-lou′ə). Mountains of NE Ore.
**Wal·nut Creek** (wôl′nŭt′, -nət). City of W Calif. NE of Oakland. Pop. 53,643.
**Wal·pole** (wôl′pōl′, wŏl′-). Town of E Mass. SW of Boston. Pop. 18,859.
**Wal·sall** (wôl′sôl′, -səl). Borough of W central England near Birmingham. Pop. 182,430.
**Wal·tham** (wôl′thăm′). City of E Mass. near Boston. Pop. 58,200.
**Wal·ton and Wey·bridge** (wôl′tən; wā′brĭj). District of SE England near London. Pop. 51,270.

**Wal·vis Bay** (wôl′vĭs). **1.** District of South Africa, c. 374 sq mi (969 sq km), an exclave in W central Namibia on the Atlantic. **2.** City of Walvis Bay district of W central Namibia on **Walvis Bay,** an inlet of the Atlantic. Pop. 21,725.
**Wan·a·que** (wŏn′ə-kyōō′). Borough of NE N.J. NNE of Paterson. Pop. 10,025.
**Wang·a·nu·i** (wŏng′ə-nōō′ē, -gə-). City of SW North Is., New Zealand, on Cook Strait. Pop. 37,307.
**Wan·tagh** (wŏn′tô′). Town of SE N.Y. on the S shore of Long Is. Pop. 22,300.
**Wap·si·pin·i·con** (wŏp′sə-pĭn′ĭ-kən). River, 255 mi (410 km), of SE Minn. & E Iowa.
**Wa·ran·gal** (wə-rŭng′gəl, wôr′əng-). City of SE India NE of Hyderabad. Pop. 207,520.
**War·bur·ton** (wôr′bûr′tn). River of central Australia flowing 275 mi (442 km) SW of Lake Eyre.
**Ware·ham** (wâr′əm, -hăm′). Town of SE Mass. on Buzzards Bay NE of New Bedford. Pop. 18,457.
**War·ner Rob·ins** (wôr′nər rŏb′ĭnz). City of central Ga. S of Macon. Pop. 39,839.
**War·ren** (wôr′ən, wŏr′-). **1.** City of SE Mich. near Detroit. Pop. 161,134. **2.** City of NE Ohio NW of Youngstown. Pop. 56,629. **3.** Borough of NW Pa. ESE of Erie. Pop. 12,146. **4.** Town of E R.I. NE of Warwick. Pop. 10,640.
**War·rens·burg** (wôr′ĭnz-bûrg′, wŏr′-). City of W central Mo. ESE of Kansas City. Pop. 13,807.
**War·rens·ville Heights** (wôr′ĭnz-vĭl′, wŏr′-). City of NE Ohio near Cleveland. Pop. 16,565.
**War·ring·ton** (wôr′ĭng-tən, wŏr′-). Borough of W central England E of Liverpool. Pop. 65,320.
**War·saw** (wôr′sô′). **1.** Cap. of Poland in the E central part on the Vistula R. Pop. 1,377,100. **2.** City of NE Ind. WNW of Fort Wayne. Pop. 10,647
**War·wick** (wŏr′ĭk, wôr′wĭk). **1.** Borough of central England SE of Birmingham. Pop. 17,870. **2.** City of E central R.I. on Narragansett Bay S of Providence. Pop. 87,123.
**Wa·satch** (wô′săch′). Range of the Rockies extending c. 250 mi (402 km) S from SE Idaho to central Utah.
**Wash** (wŏsh, wôsh), **the.** Inlet of the North Sea off E central England.
**Wash·ing·ton** (wŏsh′ĭng-tən, wôsh′-). **1.** Lake, c. 20 mi (32 km) long & 4 mi (6 km) wide, in W central Wash. E of Seattle. **2. Mount.** Highest (6,288 ft/1,917.8 m) of the White Mts. in N N.H. **3.** Island of NE Wis. in Lake Michigan off the Door Peninsula. **4.** State of the NW U.S. on the Pacific. Cap. Olympia. Pop. 4,130,163. **5.** Cap. of the U.S., coextensive with the District of Columbia. Pop. 637,651. **6.** City of central Ill. E of Peoria, Pop. 10,364. **7.** City of SW Ind. E of Vincennes. Pop. 11,325. **8.** City of SW central Ohio SSW of Columbus. Pop. 12,682. **9.** City of SW Pa. SW of Pittsburgh. Pop. 18,363. **—Wash′ing·to′ni·an** *adj.* & *n.*
**Wash·i·ta** (wŏsh′ĭ-tô′, wôsh′-). River, c. 450 mi (724 km), of NW Tex. & SW Okla.
**Wa·tau·ga** (wŏ-tô′gə). City of NE Tex. near Fort Worth. Pop. 10,284.
**Wat·er·bury** (wô′tər-bĕr′ē, wŏt′ər-). City of W central Conn. NNW of New Haven. Pop. 103,266.
**Wa·ter·ee** (wô′tə-rē, wŏt′ə-). River, c. 145 mi (233 km), of central S.C.
**Wa·ter·ford** (wô′tər-fərd, wŏt′ər-). **1.** Borough of SE Ireland SSW of Dublin. Pop. 31,968. **2.** Town of SE Conn. on Long Is. Sound. Pop. 17,843.
**Wa·ter·loo** (wô′tər-lōō′, wŏt′ər-, wô′tər-lōō′, wŏt′ər-). **1.** Town of central Belgium near Brussels; site of Napoleon's final defeat (1815). **2.** City of SE Ont., Canada, near Kitchener. Pop. 46,623. **3.** City of NE Iowa NW of Cedar Rapids. Pop. 75,985.
**Wa·ter·town** (wô′tər-toun′, wŏt′ər-). **1.** Town of W Conn. near Waterbury. Pop. 19,489. **2.** Town of E Mass. near Boston. Pop. 34,384. **3.** City of N N.Y. N of Syracuse. Pop. 27,861. **4.** City of NE S.Dak. NNW of Sioux Falls. Pop. 15,649. **5.** City of SE Wis. ENE of Madison. Pop. 18,113.
**Wa·ter·ville** (wô′tər-vĭl′, wŏt′ər-). City of S Me. N of Augusta. Pop. 17,779.
**Wa·ter·vliet** (wô′tər-vlēt′, wŏt′ər-). City of E N.Y. on the Hudson near Albany. Pop. 11,354.
**Wat·ford** (wŏt′fərd). Borough of SE England near London. Pop. 77,000.
**Wat·son·ville** (wŏt′sən-vĭl′). City of W Calif. E of Santa Cruz. Pop. 23,543.
**Watts** (wŏts). District of Los Angeles, Calif.
**Wau·ke·gan** (wô-kē′gən). City of NE Ill. on Lake Michigan N of Chicago. Pop. 67,653.
**Wau·ke·sha** (wô′kə-shô′). City of SE Wis. W of Milwaukee. Pop. 50,319.

ōō **b**oo**t** ou **ou**t th **th**in *th* **th**is ŭ **cu**t ûr **ur**ge y **y**oung
yōō ab**u**se zh vi**si**on ə **a**bout, it**e**m, ed**i**ble, gall**o**p, circ**u**s

**Wau·sau** (wô′sô′). City of N central Wis. WNW of Green Bay. Pop. 32,426.
**Wau·wa·to·sa** (wô′wə-tō′sə). City of SE Wis. near Milwaukee. Pop. 51,308.
**Wax·a·hach·ie** (wôk′sə-hăch′ē). City of NE Tex. S of Dallas. Pop. 14,624.
**Way·cross** (wā′krôs′, -krŏs′). City of SE Ga. SW of Savannah. Pop. 19,371.
**Way·land** (wā′lənd). Town of E Mass. W of Boston. Pop. 12,170.
**Wayne** (wān). City of SE Mich. WSW of Detroit. Pop. 21,159.
**Waynes·bor·o** (wānz′bûr′ō, -bər-ə). Independent city of W central Va. W of Charlottesville. Pop. 15,329.
**Wa·zir·i·stan** (wə-zîr′ĭ-stăn′, -stän′). Region of NW Pakistan.
**Weath·er·ford** (wĕ*th*′ər-fərd). City of NE Tex. W of Fort Worth. Pop. 12,049.
**Web·bi She·be·li** *also* **We·bi Shi·be·li** (wā′bē shə-bā′lē). River of NE Africa rising in central Ethiopia & flowing c. 1,200 mi (1,931 km) SE & SW through S Somalia to Indian Ocean coastal swamps.
**We·ber** (wē′bər). River, c. 125 mi (201 4m), of N Utah.
**We·bi Shi·be·li** (wā′bē shə-bā′lē). *var. of* WEBBI SHEBELI.
**Web·ster** (wĕb′stər). Town of S Mass. SSW of Worcester. Pop. 14,480.
**Webster Groves.** City of E Mo. near St. Louis. Pop. 23,097.
**Wed·dell** (wĭ-dĕl′, wĕd′l). Sea of the S Atlantic off W Antarctica E of the Antarctic Peninsula.
**Wee·haw·ken** (wē-hô′kən). Township of NE N.J. on the Hudson opposite New York City. Pop. 13,383.
**Wei** (wā). River of central China flowing c. 450 mi (724 km) E to the Yellow R.
**Wei·fang** (wā′fäng′). City of E China at the base of the Shandong Peninsula. Pop. 260,000.
**Wei·mar** (vī′mär′, wī′-). City of SW East Germany WSW of Leipzig. Pop. 63,144.
**Weir·ton** (wîr′tn). City of N W.Va. in the Panhandle W of Pittsburgh, Pa. Pop. 24,736.
**Weiss·horn** (vīs′hôrn′). Peak, 14,782 ft (4,508.5 m), of the Pennine Alps in S Switzerland.
**Wel·fare** (wĕl′fâr′). Island in the East R. off Manhattan Is., SE N.Y.
**Wel·land** (wĕl′ənd). **1.** Ship canal of SE Ont., Canada, connecting Lake Ontario with Lake Erie & by-passing Niagara Falls. **2.** City of SE Ont., Canada, on the Welland Ship Canal. Pop. 45,047.
**Welles·ley** (wĕlz′lē). Town of E Mass. WSW of Boston. Pop. 27,209.
**Wel·ling·bor·ough** (wĕl′ĭng-bûr′ə, -bər-ə). Urban district of central England ENE of Northampton. Pop. 39,570.
**Wel·ling·ton** (wĕl′ĭng-tən). Cap. of New Zealand on an inlet of Cook Strait in extreme S North Is. Pop. 139,566.
**Wels** (vĕls). City of N Austria SW of Linz. Pop. 47,279.
**Wel·wyn Garden City** (wĕl′ən). Urban district of SE England N of London. Pop. 39,900.
**We·natch·ee** (wə-năch′ē). City of central Wash. NNE of Yakima. Pop. 17,257.
**Wen·zhou** *also* **Wen·chow** (wŭn′jō′, wĕn′-). City of E China near the East China Sea S of Shanghai. Pop. 250,000.
**Wer·ra** (vĕr′ə). River, 181 mi (291 km), of SW East Germany & central West Germany.
**We·sel** (vā′zəl). City of W West Germany on the Rhine NW of Essen. Pop. 56,760.
**We·ser** (vā′zər). River, c. 300 mi (483 km), of central & N West Germany.
**Wes·la·co** (wĕs′lə-kō′). City of extreme S Tex. NW of Brownsville. Pop. 19,331.
**Wes·sex** (wĕs′ĭks). Region & ancient Anglo-Saxon kingdom of S England.
**West** (wĕst), **the. 1.** Former region of the U.S. W of the Alleghenies. **2.** Region of the U.S. W of the Mississippi.
**West Al·lis** (ăl′ĭs). City of SE Wis. near Milwaukee. Pop. 63,982.
**West Bank** (băngk). Disputed territory of SW Asia between Israel & Jordan.
**West Bend** (bĕnd). City of SE Wis. NNW of Milwaukee. Pop. 21,484.
**West·bor·ough** *or* **West·bor·o** (wĕst′bûr′ō, -bər-ə). Town of E central Mass. near Worcester. Pop. 13,619.
**West Bridg·ford** (brĭj′fərd). Urban district of N central England near Nottingham. Pop. 28,340.
**West·brook** (wĕst′brŏŏk′). City of SW Me. near Portland. Pop. 14,976.
**West·bur·y** (wĕst′bĕr′ē, -bə-rē). Village of SE N.Y. on W Long Is. near Mineola. Pop. 13,871.
**West Cald·well** (kôld′wĕl′, -wəl). Borough of NE N.J. W of Passaic. Pop. 11,407.
**West Car·roll·ton** (kăr′əl-tən). City of W central Ohio near Dayton. Pop. 13,148.
**West·ches·ter** (wĕst′chĕs′tər). **1.** Village of NE Ill. near Chicago. Pop. 17,730. **2.** Suburban region & county of SE N.Y.
**West Ches·ter** (chĕs′tər). Borough of SE Pa. W of Philadelphia. Pop. 17,435.

**West Chi·ca·go** (shĭ-kä′gō, -kô′-). City of NE Ill. near Chicago. Pop. 12,550.
**West Co·lum·bi·a** (kə-lŭm′bē-ə). City of central S.C. near Columbia. Pop. 10,409.
**West Co·vi·na** (kō-vē′nə). City of S Calif. E of Los Angeles. Pop. 80,094.
**West Des Moines** (dĭ moin′). City of central Iowa near Des Moines. Pop. 21,894.
**Wes·ter·ly** (wĕs′tər-lē). Town of extreme SW R.I. on the border E of New London, Conn. Pop. 18,580.
**West·ern Dvi·na** (wĕs′tərn də-vē-nä′). —See DVINA.
**Western Ghats** (gôts). Range of S India extending c. 800 mi (1,287 km) along the Arabian Sea coast.
**Western Islands.** HEBRIDES.
**Western Sa·ha·ra** (sə-hâr′ə, -hä′rə). Region of NW Africa on the Atlantic coast, partly annexed (1979) & partly occupied (1979) by Morocco. Pop. 76,425.
**Western Sa·mo·a** (sə-mō′ə). Island nation of the S Pacific comprising the W half of the Samoa Is. Cap. Apia. Pop. 151,983.
**Western Springs.** Village of NE Ill. near Chicago. Pop. 12,876.
**Wes·ter·ville** (wĕs′tər-vĭl′). City of central Ohio near Columbus. Pop. 23,414.
**West Far·go** (fär′gō). City of E N.Dak. near Fargo. Pop. 10,099.
**West·field** (wĕst′fēld′). **1.** City of SW Mass. near Springfield. Pop. 36,465. **2.** Town of NE central N.J. SW of Newark. Pop. 30,447.
**West·ford** (wĕst′fərd). Town of NE Mass. near Lowell. Pop. 13,434.
**West Hart·ford** (härt′fərd). Town of central Conn. near Hartford. Pop. 61,301.
**West Ha·ven** (hā′vən). City of S Conn. near New Haven. Pop. 53,184.
**West He·le·na** (hĕl′ə-nə). City of E Ark. on the Mississippi R. SSW of Memphis, Tenn. Pop. 11,430.
**West In·dies** (ĭn′dēz). Islands between SE North America & N South America, separating the Caribbean Sea & the Atlantic & including the Greater Antilles, the Lesser Antilles, & the Bahamas.
**West Indies Associated States.** Group of former British colonies in the West Indies, including Antigua, Dominica, St. Kitts-Nevis, St. Lucia, & St. Vincent & the Grenadines.
**West I·ri·an** (ĭr′ē-än′). Region & province of Indonesia comprising the W half of New Guinea.
**West Jor·dan** (jôr′dn). City of N Utah S of Salt Lake City. Pop. 26,794.
**West La·fay·ette** (lä′fē-ĕt′, lăf′ē-). City of W Ind. on the Wabash R. opposite Lafayette. Pop. 21,247.
**West·lake** (wĕst′lāk′). City of NE Ohio near Cleveland. Pop. 19,483.
**West·land** (wĕst′lənd). City of NE Mich. near Dearborn. Pop. 84,603.
**West Linn** (lĭn). City of NW Ore. near Oregon City. Pop. 12,956.
**West Mem·phis** (mĕm′fĭs). City of E Ark. on the Mississippi opposite Memphis, Tenn. Pop. 28,198.
**West Mif·flin** (mĭf′lĭn). Borough of SW Pa. near Pittsburgh. Pop. 26,279.
**West·min·ster** (wĕst′mĭn′stər, wĕs′-). **1.** City of S Calif. near Long Beach. Pop. 71,133. **2.** City of N central Colo. near Denver. Pop. 50,211.
**West Mon·roe** (mən-rō′). City of N central La. on the Ouachita R. opposite Monroe. Pop. 14,993.
**West·mont** (wĕst′mŏnt′). Village of NE Ill. near Chicago. Pop. 16,718.
**West·mount** (wĕst′mount′). City of S Que., Canada, on Montreal Is. W of Montreal. Pop. 22,153.
**West New York** (nōō yôrk′, nyōō). Town of NE N.J. on the Hudson opposite Manhattan. Pop. 39,194.
**Wes·ton** (wĕs′tən). Town of E Mass. W of Boston. Pop. 11,169.
**Wes·ton-su·per-Mare** (wĕs′tən-sōō′pər-mâr′). Borough of SW England on the Bristol Channel WSW of Bristol. Pop. 51,960.
**West Or·ange** (ôr′ĭnj, ŏr′-). Town of NE N.J. near Montclair. Pop. 39,510.
**West Palm Beach** (päm). City of SE Fla. opposite Palm Beach & N of Fort Lauderdale. Pop. 62,530.
**West Pat·er·son** (păt′ər-sən). Borough of NE N.J. near Clifton. Pop. 11,293.
**West·pha·lia** (wĕst-fāl′yə, -fā′lē-ə). Region of W West Germany E of the Rhine & centered in the Ruhr district. **—West·pha′lian** *adj.* & *n.*
**West·port** (wĕst′pôrt′, -pōrt′). Town of SW Conn. on Long Is. Sound. Pop. 25,290.
**West Quod·dy Head** (kwŏd′ē). Cape of NE Me. S of Eastport at S entrance to Passamaquoddy Bay.
**West Saint Paul** (sānt pôl′). City of E Minn. near St. Paul. Pop. 18,527.
**West Spits·ber·gen** (spĭts′bûrg′ən). Largest of the Spitsbergen Is. in the Arctic Ocean N of Norway.
**West Spring·field.** (spring′fēld′). Town of SW Mass. near Springfield. Pop. 27,042.
**West U·ni·ver·si·ty Place** (yōō′nə-vûr′sĭ-tē). City of SE Tex. near Houston. Pop. 12,010.
**West Vir·gin·ia** (vər-jĭn′yə). State of the E central U.S. Cap. Charleston. Pop. 1,949,644. **—West Vir·gin′ian** *adj.* & *n.*
**West War·wick** (wôr′ĭk, wôr′wĭk). Town of E central R.I. SSW of Providence. Pop. 27,026.
**West·we·go** (wĕst-wē′gō). City of SE La. on the Mississippi opposite New Orleans. Pop. 12,663.
**West·wood** (wĕst′wŏŏd′). **1.** Town of E Mass. SW of Boston. Pop. 13,212. **2.** Borough of NE N.J. N of Hackensack. Pop. 10,714.
**We·tar** (wē′tär′). Island of SE Indonesia in the S Moluccas off NE Timor.
**Weth·ers·field** (wĕth′ərz-fēld′). Town of central Conn. near Hartford. Pop. 26,013.
**Wet·ter·horn** (vĕt′ər-hôrn′). Peak, c. 12,150 ft (3,706 m), of the Bernese Alps in central Switzerland.
**Wey·mouth** (wā′məth). Town of E Mass. SSE of Boston. Pop. 55,601.
**Weymouth and Port·land** (pôrt′lənd, pōrt′-). Borough of S England on the English Channel WSW of Southampton. Pop. 41,080.
**Whales** (hwālz, wālz), **Bay of.** Inlet of the Ross Sea in the Ross Ice Shelf, W Antarctica.
**Whea·ton** (hwēt′n, wēt′n). **1.** City of NE Ill. near Chicago. Pop. 43,043. **2.** City of central Md. near Washington, D.C. Pop. 73,800.
**Wheat Ridge** (hwēt′, wēt′). City of N central Colo. near Denver. Pop. 30,293.
**Wheel·er Peak** (hwē′lər, wē′-). Mountain, 13,160 ft (4,013.8 m), in N central N.Mex.
**Wheel·ing** (hwē′lĭng, wē′-). **1.** Village of NE Ill. near Chicago. Pop. 23,266. **2.** City of NW W.Va. in the Panhandle on the Ohio R. Pop. 43,070.
**Whid·bey** (hwĭd′bē, wĭd′-). Island of NW Wash. in Puget Sound.
**Whit·by** (hwĭt′bē, wĭt′-). Town of S central Ont., Canada, NE of Toronto. Pop. 28,173.
**White** (hwīt, wīt). **1.** Sea of NW European USSR, an inlet of the Barents Sea. **2.** River of N Ark. & SW Mo. flowing c. 690 mi (1,110 km) SE to the Mississippi. **3.** River of NW Nebr. & S S.Dak. flowing 325 mi (523 km) E to the Missouri. **4.** River, 160 mi (257 km), of NW Colo. & E Utah. **5.** Mountains of the N Appalachians in N N.H. **6.** Pass, 2,888 ft (880.8 m), through the Coast Mts. between SE Alas. & NW B.C., Canada.
**White Bear Lake** (bâr). City of E Minn. near St. Paul. Pop. 22,538.
**White·fish Bay** (hwīt′fĭsh′, wīt′-). Village of SE Wis. near Milwaukee. Pop. 14,930.
**White·hall** (hwīt′hôl′, wīt′-). **1.** City of central Ohio near Columbus. Pop. 21,299. **2.** Borough of SW Pa. near Pittsburgh. Pop. 15,206.
**White·ha·ven** (hwīt′hā′vən, wīt′-). Borough of NW England at the entrance to Solway Firth. Pop. 26,260.
**White·horse** (hwīt′hôrs′, wīt′-). Cap. of Y.T., Canada, in the S part of the Yukon R. Pop. 13,311.
**White Nile** (nīl). Section in the Nile in Sudan from Lake No to Khartoum.
**White Plains.** City of SE N.Y. N of New York City. Pop. 46,999.
**White Rock.** City of SE B.C., Canada, on the Strait of Georgia & the U.S. border. Pop. 12,497.
**White Sands National Monument** (săndz). Area of gypsum sand dunes in S central N.Mex., near site of a bomb & missile testing range.
**White Set·tle·ment** (sĕt′l-mənt). City of NE Tex. near Fort Worth. Pop. 13,508.
**White Vol·ta** (vŏl′tə, vōl′-, vôl′-). River of Upper Volta & N Ghana flowing 550 mi (885 km) S to the Black Volta to form the Volta.
**White·wa·ter** (hwīt′wô′tər, -wŏt′ər, wīt′-). City of SE Wis. WSW of Milwaukee. Pop. 11,520.
**Whit·ney** (hwĭt′nē, wĭt′-). **Mount.** Second-highest peak in the U.S., 14,494 ft (4,420.7 m), in the Sierra Nevada of E central Calif.
**Whit·sta·ble** (hwĭt′stə-bəl, wĭt′-). Urban district of SE England on the North Sea E of London. Pop. 26,980.
**Whit·ti·er** (hwĭt′ē-ər, wĭt′-). City of S Calif. SE of Los Angeles. Pop. 68,872.
**Wich·i·ta** (wĭch′ĭ-tô′). City of S central Kans. SW of Kansas City. Pop. 279,272.
**Wichita Falls.** City of N Tex. near the Okla. border NW of Fort Worth. Pop. 94,201.
**Wick·liffe** (wĭk′ləf). City of NE Ohio on Lake Erie NE of Cleveland. Pop. 16,790.
**Wick·low** (wĭk′lō). Mountains of E Ireland along the coast S of Dublin.
**Wid·nes** (wĭd′nĭs). Borough of W central England ESE of Liverpool. Pop. 58,330.
**Wie·ner·wald** (vē′nər-vält′). Forested range of NE Austria just W of Vienna.

**Wies·ba·den** (vēs′bäd′n, vĭs′-). City of central West Germany on the Rhine W of Frankfurt. Pop. 250,592.

**Wig·an** (wĭg′ən). Borough of NW England NE of Liverpool. Pop. 80,920.

**Wight** (wīt), **Isle of.** Island in the England Channel off S central England.

**Wil·bra·ham** (wĭl′brə-hăm′). Town of SW Mass. near Springfield. Pop. 12,053.

**Wil·helm** (vĭl′hĕlm′), **Mount.** Highest peak (15,400 ft/4,697 m) of Papua New Guinea, in the Bismarck Mts.

**Wil·helms·ha·ven** (vĭl′hĕlmz-hä′fən). City of NW West Germany on the North Sea. Pop. 103,417.

**Wilkes-Bar·re** (wĭlks′băr′ə, -băr′ē). City of NW Pa. NNW of Allentown. Pop. 51,551.

**Wilkes Land** (wĭlks). Region of E Antarctica S of Australia.

**Wil·kins·burg** (wĭl′kĭnz-bûrg′). Borough of SW Pa. near Pittsburgh. Pop. 23,669.

**Wil·lam·ette** (wə-lăm′ĭt). River, 294 mi (473 km), of NW Ore.

**Wil·la·pa** (wĭl′ə-pô′, -pä′). Bay of the Pacific off SW Wash.

**Wil·lem·stad** (vĭl′əm-stät′). Cap. of the Netherland Antilles on the S coast of Curaçao. Pop. 95,000.

**Wil·liams·burg** (wĭl′yəmz-bûrg′). City of SE Va. with restored colonial district, NW of Newport News. Pop. 9,870.

**Wil·liam·son** (wĭl′yəm-sən), **Mount.** Peak, 14,384 ft (4,387.1 m), in the Sierra Nevada of E central Calif.

**Wil·liams·port** (wĭl′yəmz-pôrt′, -pōrt′). City of central Pa. N of Harrisburg. Pop. 33,401.

**Wil·li·man·tic** (wĭl′ə-măn′tĭk). City of E Conn. WNW of Windham. Pop. 14,652.

**Wil·lis·ton** (wĭl′ĭ-stən). City of NW N.Dak. near the Mont. border. Pop. 13,336.

**Will·mar** (wĭl′mär′, -mər). City of SW central Minn. WNW of Minneapolis. Pop. 15,895.

**Wil·lough·by** (wĭl′ə-bē). City of NE Ohio on Lake Erie NE of Cleveland. Pop. 19,329.

**Wil·lo·wick** (wĭl′ə-wĭk′). City of NE Ohio on Lake Erie near Cleveland. Pop. 17,834.

**Wil·mette** (wĭl-mĕt′). Village of NE Ill. on Lake Michigan near Chicago. Pop. 28,229.

**Wil·ming·ton** (wĭl′mĭng-tən). **1.** City of NE Del. SW of Philadelphia, Pa. Pop. 70,195. **2.** Town of E Mass. near Boston. Pop. 17,471. **3.** City of SE N.C. SSE of Raleigh. Pop. 44,000. **4.** City of SW Ohio SE of Dayton. Pop. 10,431.

**Wilms·low** (wĭlmz′lō′, wĭmz′-). Urban district of NW England S of Manchester. Pop. 31,250.

**Wil·son** (wĭl′sən). **1. Mount.** Mountain, 5,710 ft (1,741.6 m), in the San Gabriel Mts. of SW Calif.; site of an observatory. **2.** City of E central N.C. E of Raleigh. Pop. 34,424.

**Wil·ton** (wĭl′tən). Town of SW Conn. N of Norwalk. Pop. 15,351.

**Wilton Manor.** City of SE Fla. near Fort Lauderdale. Pop. 12,742.

**Wim·ble·don** (wĭm′bəl-dən). District of S Greater London, England.

**Win·ches·ter** (wĭn′chĕs′tər, -chĭ-stər). **1.** Borough of S central England WSW of London. Pop. 88,900. **2.** Town of NW Conn. N of Torrington. Pop. 10,841. **3.** City of N central Ky. ESE of Lexington. Pop. 15,216. **4.** Town of E Mass. near Boston. Pop. 20,701. **5.** Independent city of N Va. WNW of Washington, D.C. Pop. 20,217.

**Wind** (wĭnd). River, c. 120 mi (192 km), of W central Wyo.

**Win·der·mere** (wĭn′dər-mîr′). Lake, 10.5 mi (17 km) long, of NW England.

**Wind·ham** (wĭn′dəm). **1.** Town of E Conn. NNW of Norwich. Pop. 21,062. **2.** Town of SW Me. near Portland. Pop. 11,282.

**Wind·hoek** (vĭnt′ho͝ok′). Cap. of Namibia in the central part. Pop. 61,369.

**Wind River Range.** Range of the Rockies in W Wyo.

**Wind·sor** (wĭn′zər). **1.** City of S Ont., Canada, on the Detroit R. opposite Detroit, Mich. Pop. 196,526. **2.** Borough of S central England W of London. Pop. 29,660. **3.** Town of N Conn. near Hartford. Pop. 25,204.

**Windsor Locks.** Town of N Conn. N of Windsor. Pop. 12,190.

**Wind·ward** (wĭnd′wərd). **1. Passage.** Strait, c. 50 mi (80 km) wide, between Cuba & Haiti. **2.** Islands of the SE West Indies, the S group of the Lesser Antilles from Martinique S to Grenada. **3.** Islands of the S Pacific, the E group of the Society Is.

**Win·field** (wĭn′fēld′). City of S Kans. SSE of Wichita. Pop. 10, 736.

**Win·ne·ba·go** (wĭn′ə-bā′gō). Lake, 215 sq mi (557 sq km), of E Wis.

**Win·net·ka** (wə-nĕt′kə). Village of NE Ill. on Lake Michigan NNW of Evanston. Pop. 12,772.

**Win·ni·bi·go·shish** (wĭn′ə-bĭ-gō′shĭsh). Lake, 179 sq mi (464 sq km), of N central Minn.

**Win·ni·peg** (wĭn′ə-pĕg′). **1.** Lake, 9,465 sq mi (24,514 sq km), of S central Man., Canada. **2.** River, c. 200 mi (322 km), of W Ont. & SE Man., Canada. **3.** Cap. of Man., Canada, in the SE part. Pop. 560,874.

**Win·ni·pe·go·sis** (wĭn′ə-pĭ-gō′sĭs), **Lake.** Lake, 2,086 sq mi (5,403 sq km), of W Man., Canada, W of Lake Winnipeg.

**Win·ni·pe·sau·kee** (wĭn′ə-pĭ-sô′kē). Lake, 71 sq mi (184 sq km), of E central N.H.

**Wi·no·na** (wĭ-nō′nə). City of SE Minn. on the Mississippi SE of St. Paul. Pop. 25,075.

**Win·ston-Sa·lem** (wĭn′stən-sā′ləm). City of N central N.C. NNE of Charlotte. Pop. 131,885.

**Win·ter Haven** (wĭn′tər). City of central Fla. E of Lakeland. Pop. 21,119.

**Winter Park.** City of central Fla. near Orlando. Pop. 22,314.

**Win·ter·thur** (vĭn′tər-to͝or′). City of N Switzerland NE of Zurich. Pop. 93,500.

**Win·throp** (wĭn′thrəp). Town of E Mass. near Boston. Pop. 19,294.

**Win·yah** (wĭn′yô′). Bay of the Atlantic off E S.C.

**Wis·con·sin** (wĭs-kŏn′sĭn). **1.** River of Wis. flowing c. 430 mi (692 km) S & W to the Mississippi. **2.** State of the N central U.S. Cap. Madison. Pop. 4,705,335. **—Wis·con′sin·ite′** *n.*

**Wisconsin Rapids.** City of central Wis. S of Wausau. Pop. 17,995.

**Wis·mar** (vĭz′mär′, vĭs′-). City of NW East Germany near the Baltic. Pop. 56,765.

**With·la·coo·chee** (wĭth′lə-ko͞o′chē). **1.** River, c. 160 mi (257 km), of central Fla. **2.** River, c. 115 mi (185 km), of S Ga. & NW Fla.

**Wit·ten** (vĭt′n). City of W Germany on the Ruhr R. Pop. 108,771.

**Wit·ten·berg** (wĭt′n-bûrg′, vĭt′n-bĕrg′). City of central East Germany on the Elbe. Pop. 51,364.

**Wit·wa·ters·rand** (wĭt-wô′tərz-rănd′, -ränd′, -wŏt′ərz-). Gold-rich region of NE South Africa between the Vaal R. & Johannesburg.

**Wlo·cla·wek** (vlôt-slä′vĕk). City of central Poland on the Vistula WNW of Warsaw. Pop. 77,169.

**Wo·burn** (wō′bərn, wo͞o′-). City of E Mass. near Boston. Pop. 36,626.

**Wod·zi·slaw Slą·ski** (vô-jē′släf shlôN′skē). City of S Poland SW of Katowice. Pop. 25,600.

**Wo·king** (wō′kĭng). Urban district of SE England SW of London. Pop. 79,300.

**Wol·cott** (wo͝ol′kət). Town of central Conn. NE of Waterbury. Pop. 13,008.

**Wolds** (wōldz), **The.** Range of chalk hills along the NE coast of England.

**Wolfs·burg** (wo͝olfs′bûrg′, -bo͝org′). City of NE West Germany E of Hannover. Pop. 126,298.

**Wo·lin** (vô′lēn). *var. of* WOLLIN.

**Wol·las·ton** (wo͝ol′ə-stən, wŏl′-). Lake, 796 sq mi (2,062 sq km), in NE Sask., Canada.

**Wol·lin** or **Wo·lin** (vô′lēn). Island of extreme NW Poland in the Baltic near the mouth of the Oder.

**Wol·lon·gong** (wo͝ol′ən-gŏng′, -gông). City of SE Australia on the Tasman Sea SSW of Sydney. Pop. 165,086.

**Wol·ver·hamp·ton** (wo͝ol′vər-hămp′tən, -hăm′-). Borough of W central England NW of Birmingham. Pop. 266,400.

**Won·san** (wŭn′sän′). City of SE North Korea on the Sea of Japan. Pop. 275,000.

**Wood·bridge** (wo͝od′brĭj′). City of NE N.J. SSW of Elizabeth. Pop. 14,200.

**Wood Buf·fa·lo National Park** (wo͝od′ bŭf′ə-lō′). Reserved area of forests & plains in N Alta. & S N.W.T., Canada.

**Wood·burn** (wo͝od′bərn). City of NW Ore. NE of Salem. Pop. 11,196.

**Wood·bur·y** (wo͝od′bĕr′ē, -bə-rē). **1.** City of E Minn. near St. Paul. Pop. 10,297. **2.** City of SW N.J. near Camden. Pop. 10,353.

**Wood Dale.** City of NE Ill. near Chicago. Pop. 11,251.

**Wood·ha·ven** (wo͝od′hā′vən). City of SE Mich. near Detroit. Pop. 10,902.

**Wood·land** (wo͝od′lənd). City of N central Calif. WNW of Sacramento. Pop. 30,235.

**Wood·lark** (wo͝od′lärk′). Island of Papua New Guinea in the Solomon Sea off the SE end of New Guinea.

**Wood·ridge** (wo͝od′rĭj′). Village of NE Ill. near Chicago. Pop. 22,322.

**Wood River.** City of SW Ill. on the Mississippi near St. Louis, Mo. Pop. 12,449.

**Woods** (wo͝odz), **Lake of the.** Lake, 1,485 sq mi (3,846 sq km), of SW Ont., Canada, & N central Minn.

**Wood·stock** (wo͝od′stŏk′). **1.** City of S Ont., Canada, on the Thames R. WSW of Toronto. Pop. 26,779. **2.** City of NE Ill. W of Waukegan. Pop. 11,725.

**Wood·ward** (wo͝od′wərd). City of NW Okla. NW of Oklahoma City. Pop. 13,610.

**Wool·wich** (wo͝ol′ĭj, -ĭch, -wĭch). Township of S central Ont., Canada, N of Kitchener. Pop. 16,197.

**Woon·sock·et** (wo͞on-sŏk′ĭt, wo͝on′sŏk′-). City of N R.I. near the Mass. border NNW of Providence. Pop. 45,914.

**Woo·ster** (wo͝os′tər). City of NE central Ohio NW of Akron. Pop. 19,289.

**Worces·ter** (wo͝os′tər). **1.** Borough of W central England on the Severn SSW of Birmingham. Pop. 73,900. **2.** City of central Mass. W of Boston. Pop. 161,799.

**Work·ing·ton** (wûr′kĭng-tən). Borough of NW England on Solway Firth. Pop. 28,260.
**Work·sop** (wûrk′səp). Borough of N central England ESE of Sheffield. Pop. 36,590.
**Worms** (wûrmz, vôrms). City of S central West Germany on the Rhine NNW of Mannheim. Pop. 75,732.
**Worth** (wûrth). Village of NE Ill. near Chicago. Pop. 11,592.
**Wor·thing** (wûr′*th*ĭng). Borough of SE England on the English Channel SSW of London. Pop. 89,100.
**Wor·thing·ton** (wûr′*th*ĭng-tən). **1.** City of SW Minn. near the Iowa border. Pop. 10,243. **2.** City of central Ohio near Columbus. Pop. 15,016.
**Wound·ed Knee** (wo͞on′dĭd nē′). Creek of SW S.Dak.; site of last major battle of the Indian Wars (1890).
**Wran·gel** (răng′gəl). Island of NE Far Eastern USSR in the Arctic Ocean NW of Bering Strait.
**Wran·gell** (răng′gəl). **1.** Mountains of S Alas. extending c. 100 mi (161 km) from the Copper R. to the Canadian border. **2. Mount.** Volcano, 14,006 ft (4,272 m), in the Wrangell Mts. of S Alas. **3.** Cape of extreme W Alas. on Attu Is. in the Aleutians. **4.** Island of SE Alas. in the Alexander Archipelago NE of Prince of Wales Is.
**Wrath** (răth), **Cape.** Promontory at the NW extremity of the Scottish mainland.
**Wrex·ham** (rĕk′səm). Borough of NE Wales W of Stoke-on-Trent. Pop. 39,530.
**Wro·claw** (vrôt′släf′). City of SW Poland on the Oder R. Pop. 523,318.
**Wu** (wo͞o). River of central China flowing c. 500 mi (805 km) E & N to the Yangtze.
**Wu·han** (wo͞o′hän′). City of E central China on the Yangtze. Pop. 4,250,000.
**Wu·hu** (wo͞o′ho͞o′). City of E central China on the Yangtze SSW of Nanjing. Pop. 300,000.
**Wup·per·tal** (vo͞op′ər-täl′). City of W West Germany ENE of Düsseldorf. Pop. 405,369.
**Würt·tem·berg** (wûr′təm-bûrg′, vîr′təm-bĕrg′). Region & former kingdom of SW West Germany.
**Würz·burg** (wûrts′bûrg′, vîrts′bo͝org′). City of S central West Germany on the Main R. Pop. 112,584.
**Wu·sih** (wo͞o′shē′). CHANGZHOU.
**Wu·tai Shan** (wo͞o′tī′ shän′). Range of NE China extending c. 150 mi (240 km) between Taiyuan & Beijing.
**Wu·xi** (wo͞o′shē′). CHANGZHOU.
**Wy·an·dotte** (wī′ən-dŏt′). **1.** Cave in S Ind. W of New Albany. **2.** City of SE Mich. near Detroit. Pop. 34,006.
**Wye** (wī). River, c. 130 mi (209 km), of E Wales & W England.
**Wy·o·ming** (wī-ō′mĭng). **1.** State of the W U.S. Cap. Cheyenne. Pop. 470,816. **2.** City of W central Mich. near Grand Rapids. Pop. 59,616. **—Wy·o′ming·ite′** *n.*

# X

**Xan·thus** (zăn′thəs). Ancient city of Lycia in present-day SW Turkey.
**Xen·ia** (zēn′yə, zē′nē-ə). City of SW central Ohio ESE of Dayton. Pop. 24,653.
**Xi** (shē). River, c. 300 mi (483 km), of SE China.
**Xi·an** (shē′än′). City of central China SW of Beijing. Pop. 1,900,000.
**Xi·ang** (shē′äng′). River, c. 715 mi (1,150 km), of SE China.
**Xi·ang·tan** (shē-äng′tän′). City of S central China on the Xiang R. SSW of Changsha. Pop. 300,000.
**Xin·gu** (shēng-go͞o′). River of central & N Brazil flowing 1,230 mi (1,979.1 km) to the Amazon.
**Xin·jiang Uy·gur** (shĭn′jē-äng′ wē′gər). Autonomous region of extreme W China.
**Xin·xi·ang** (shĭn′shē-äng′). City of E China SSE of Taiyuan. Pop. 300,000.
**Xi·zang** (shēd′zäng′). TIBET.
**Xu·zhou** (sho͞o′jō′). City of E China NNW of Nanjing. Pop. 1,500,000.

# Y

**Ya·blo·no·vy** (yä′blə-nə-vē′). Mountain range in SE Siberian USSR.
**Yai·zu** (yī′zo͞o′). City of E central Honshu, Japan, near Shizuoka. Pop. 94,102.
**Ya·ki·ma** (yăk′ə-mô′, -mə). **1.** River, 203 mi (326.6 km), of central & SE Wash. **2.** City of S central Wash. SE of Seattle. Pop. 49,286.
**Ya·kut** (yə-ko͞ot′) *also* **Ya·kutsk** (-ko͞otsk′). Region of NE Siberian USSR.
**Ya·kutsk** (yə-ko͞otsk′). City of E Siberian USSR on the Lena R. Pop. 152,000.
**Yal·ta** (yôl′tə). City of SW European USSR in S Crimea on the Black Sea. Pop. 80,000.
**Ya·lu** (yä′lo͞o). River, c. 500 mi (805 km), forming part of the North Korea-China border.
**Ya·lung** (yä′lo͝ong′). River of W China flowing c. 800 mi (1,290 km) to the Yangtze.
**Ya·ma·ga·ta** (yä′mə-gə-tä′). City of N Honshu, Japan, SW of Sendai. Pop. 219,773.
**Ya·mal** (yə-mäl′). Peninsula of NW Siberian USSR extending c. 400 mi (643.6 km) between the Kara Sea & the Gulf of Ob.
**Ya·ma·to** (yä-mä′tō). City of E central Honshu, Japan, near Tokyo. Pop. 145,881.
**Yam·bol** (yäm′bōl′). City of SE Bulgaria E of Stara Zagora. Pop. 75,861.
**Yam·pa** (yăm′pə). River, c. 250 mi (400 km), of NW Colo.
**Ya·na** (yä′nə). River, c. 750 mi (1,206.8 km), of central & N Siberian USSR.
**Yang·chou** (yäng′jō′). YANGZHOU.
**Yang·tze** (yăng′sē, yăngkt′-). Longest river of China & of Asia, flowing c. 3,450 mi (5,550 km) from Tibet to the East China Sea.
**Yang·zhou** (yäng′jō′). City of E central China on the Grand Canal. Pop. 210,000.
**Yank·ton** (yăngk′tən). City of SE S.Dak. on the Missouri R. at the Nebr. border. Pop. 12,011.
**Yao** (you). City of S Honshu, Japan, near Osaka. Pop. 261,639.
**Yaon·dé** *or* **Yaun·dé** (youn-dē′). Cap. of Cameroon in the central part. Pop. 313,706.
**Yap** (yăp, yäp). Island group in W Caroline Is. of the W Pacific.
**Ya·qui** (yä-kē′). River, c. 400 mi (643.6 km), of NW Mexico.
**Yar·kand** (yär-kănd′). River, c. 500 mi (804.5 km), of NW China.
**Yar·mouth** (yär′məth). **1.** GREAT YARMOUTH. **2.** Town of SE Mass. on central Cape Cod. Pop. 18,449.
**Ya·ro·slavl** (yär′ə-slä′vəl). City of E European USSR on the upper Volga R. Pop. 597,000.
**Yaun·dé** (youn-dā′). *var. of* YAOUNDÉ.
**Yazd** (yăzd). City of central Iran SE of Esfahan. Pop. 135,978.
**Ya·zoo** (yə-zo͞o′, yăz′o͞o). River, 188 mi (302.5 km), of W Miss.
**Yazoo City.** City of W central Miss. NNW of Jackson. Pop. 12,426.
**Yea·don** (yād′n). Borough of SE Pa. near Philadelphia. Pop. 11,727.
**Ye·lets** (yə-lĕts′). City of E central USSR E of Orel. Pop. 112,000.
**Yel·low** (yĕl′ō). **1. Sea.** Arm of the Pacific between the Chinese mainland & Korean Peninsula. **2.** River of N China flowing c. 3,000 mi (4,830 km) to the Gulf of Bo Hai.
**Yel·low·knife** (yĕl′ō-nīf′). Capital of N.W.T., Canada, on the N shore of Great Slave Lake. Pop. 8,256.
**Yel·low·stone** (yĕl′ō-stōn′). **1.** River, 671 mi (1,079.6 km) long, of NW Wyo. & Mont. **2. National Park.** Oldest & largest (2,221,773 acres/899,818 hectares) of U.S. national parks, mostly in NW Wyo.
**Yem·en** (yĕm′ən, yā′mən). **1.** *also* **North Yemen.** Country of SW Asia at the SW tip of the Arabian Peninsula. Cap. Sana. Pop. 6,456,189. **2.** *also* **Southern Yemen.** Country of SW Asia at the S edge of the Arabian Peninsula. Cap. Aden. Pop. 1,969,000.
**Yen·a·ki·ye·vo** (yĕn-ə-kē′yə-və). City of SE European USSR in the Ukraine near Donetsk. Pop. 114,000.
**Ye·ni·sei** (yĕn′ĭ-sā′). River of central Siberian USSR flowing c. 2,500 mi (4,025.6 km) to the Kara Sea.
**Yeo·vil** (yō′vĭl′). Borough of SW England on the **Yeo R.** S of Bristol. Pop. 26,180.
**Ye·re·van** (yĕr′ə-vän′). City of SE European USSR S of Tbilisi. Pop. 1,019,000.
**Ye·şil Ir·mak** (yə-shēl′ îr-mäk′). River of N Turkey flowing c. 260 mi (420 km) to the Black Sea.
**Yi·bin** (yē′bĕn′). City of S central China on the Yangtze WSW of Chongqing. Pop. 275,000.
**Yok·kai·chi** (yō-kī′chē). City of W Honshu, Japan, on Ise Bay. Pop. 247,001.
**Yo·ko·ha·ma** (yō′kə-hä′mə). City of SE Honshu, Japan, on the W shore of Tokyo Bay. Pop. 2,621,771.
**Yo·ko·su·ka** (yō′kə-so͞o′kə). City of E central Honshu, Japan, near Tokyo. Pop. 389,557.
**Yo·na·go** (yō-nä′gō). Town of W Honshu, Japan. Pop. 118,332.
**Yon·kers** (yŏng′kərz). City of SE N.Y. N of New York City. Pop. 195,351.
**Yor·ba Lin·da** (yôr′bə lĭn′də). City of S Calif. near Anaheim. Pop. 28,254.
**York** (yôrk). **1. Cape.** Northernmost point of Australia on Torres Strait at the tip of Cape York Peninsula. **2.** Borough of N England on the Ouse R. ENE of Leeds. Pop. 101,900. **3.** W borough of metropolitan Toronto, Ont., Canada, on Lake Ontario. Pop. 141,367. **4.** City of S Pa. near the Md. border SSE of Harrisburg. Pop. 44,619.
**Yorke** (yôrk). Narrow peninsula of S Australia bounded by Spencer Gulf.

**York·ton** (yôrk'tən). City of SE Sask., Canada, NE of Regina. Pop. 14,119.

**York·town** (yôrk'toun'). Village in SE Va.; site of English surrender in Revolutionary War (1781).

**Yo·sem·i·te National Park** (yō-sĕm'ĭ-tē). Rugged area of E central Calif. including **Yosemite Valley** & **Yosemite Falls** (2,425 ft/739.6 m high).

**Yosh·kar-O·la** (yəsh-kär'ə-lä'). City of E central European USSR NW of Kazan. Pop. 201,000.

**Yo·su** (yŭs'ōō'). City of S South Korea on the Korea Strait. Pop. 130,623.

**Youngs·town** (yŭngz'toun'). City of NE Ohio near the Pa. border E of Akron. Pop. 115,436.

**Y·pres** (ē'prə). IEPER.

**Yp·si·lan·ti** (ĭp'sə-lăn'tē). City of SE Mich. WSW of Detroit. Pop. 24,031.

**Yu** (yōō). River of SE China flowing c. 400 mi (644 km) E to the Honshui to form the Xi.

**Yu·an** (yōō'än'). RED R., China.

**Yu·ba City** (yōō'bə). City of N central Calif. NNW of Sacramento. Pop. 18,736.

**Yu·ca·tán** (yōō'kə-tăn', -tän'). Peninsula, mostly in SE Mexico, separating the Caribbean from the Gulf of Mexico.

**Yu·go·sla·vi·a** (yōō'gō-slä'vē-ə). Republic of SE Europe largely in the Balkan Peninsula. Cap. Belgrade. Pop. 22,471,000. **—Yu'go·sla'vi·an** *adj. & n.*

**Yu·kon** (yōō'kŏn'). **1.** River flowing c. 2,000 mi (3,220 km) from S Y.T., Canada, through Alas. to the Bering Sea. **2.** Territory of NW Canada E of Alas. Cap. Whitehorse. Pop. 22,684. **3.** City of central Okla. WNW of Oklahoma City. Pop. 17,112.

**Yu·ma** (yōō'mə). City of SW Ariz. on the Calif. border. Pop. 42,433.

**Yun·nan** (yōō'nän'). Province of S central China.

# Z

**Zaan·dam** (zän-dăm', -däm'). City of W Netherlands near Amsterdam. Pop. 124,795.

**Zab** (zăb). **1.** *also* **Great Zab.** River, c. 265 mi (426.4 km), of SE Turkey & N Iraq. **2.** *also* **Little Zab.** River, c. 250 mi (402.3 km), of NW Iran & N Iraq.

**Zab·rze** (zäb'zhä). City of SW Poland E of Katowice. Pop. 197,214.

**Za·ca·te·cas** (zăk'ə-tā'kəs). City of N central Mexico N of Aguascalientes. Pop. 50,251.

**Za·dar** (zä'där). City of W Yugoslavia in Croatia on the Dalmatian coast. Pop. 43,588.

**Za·gorsk** (zə-gôrsk'). City of central European USSR NE of Moscow. Pop. 107,000.

**Za·greb** (zä'grĕb'). City of NW Yugoslavia on the Sava R. Pop. 561,773.

**Zag·ros** (zăg'rəs). Mountain range of W Iran forming the W & S border of the central Iranian plateau.

**Zaire** (zī'îr, zä-îr'). Republic of W central Africa astride the equator. Cap. Kinshasa. Pop. 28,291.

**Za·ma** (zä'mə). Ancient town in present-day N Tunisia; site of decisive defeat of Hannibal by the Romans (202 B.C.).

**Zam·be·zi** (zăm-bē'zē). River, c. 1,700 mi (2,735 km), of central & S Africa.

**Zam·bi·a** (zăm'bē-ə). Republic of S central Africa. Cap. Lusaka. Pop. 5,679,808. **—Zam'bi·an** *adj. & n.*

**Za·mo·ra** (zə-môr'ə, -mōr'ə). City of NW Spain on the Duero R. Pop. 48,791.

**Zanes·ville** (zānz'vĭl'). City of E central Ohio E of Columbus. Pop. 28,655.

**Zan·jan** (zăn-jăn'). City of NW Iran SW of the Caspian Sea. Pop. 99,967.

**Zan·zi·bar** (zăn'zə-bär'). **1.** Region of E Africa, part of Tanzania. **2.** Island, 641 sq mi (1,660.2 sq km), off the NE coast of Tanzania. **3.** City of Tanzania on the W coast of Zanzibar Is. Pop. 110,669.

**Za·po·ro·zhe** (zä'pə-rô'zhə). City of S European USSR on the Dnieper R. Pop. 781,000.

**Za·ra·go·za** (zăr'ə-gō'zə). SARAGOSSA.

**Za·ri·a** (zä'rē-ə). City of N Nigeria NE of Kaduna. Pop. 224,000.

**Za·wier·ci** (zäv-yĕr'chā). City of S Poland NW of Katowice. Pop. 39,410.

**Zea·land** (zē'lənd). SJAELLAND.

**Zeist** (zīst). City of central Netherlands near Utrecht. Pop. 58,630.

**Zeitz** (tsīts, zīst). City of S East Germany near Leipzig. Pop. 44,582.

**Ze·rav·shan** (zĕr'əf-shän'). River, c. 460 mi (740 km), of S Central Asian USSR.

**Ze·ya** (zā'yə). River of Far Eastern USSR flowing c. 800 mi (1,290 km) to the Amur.

**Zgierz** (zə-gyĕzh'). City of E central Poland near Łódź. Pop. 42,838.

**Zhang·jia·kou** (jäng'jē-ä'kō'). City of NE China near the Great Wall. Pop. 1,100,000.

**Zhan·jiang** (jän'jē-äng'). City of SE China SW of Guangzhou. Pop. 220,000.

**Zhda·nov** (zhdä'nəf). City of S European USSR on the Sea of Azov. Pop. 503,000.

**Zhe·jiang** (jŭj'ē-äng'). Province of E China on the East China Sea.

**Zheng·zhou** (jŭng'jō'). City of E central China SSW of Beijing. Pop. 1,500,000.

**Zhen·jiang** (jŭn'jē-äng'). City of E China on the Grand Canal E of Nanjing. Pop. 250,000.

**Zhi·to·mir** (zhĭ-tô'mîr). City of SW European USSR W of Kiev. Pop. 244,000.

**Zhou·shan** (jou'shän'). Archipelago in the East China Sea at the entrance to Hangzhou Bay, E China.

**Zhu Jiang** (jōō' jē-äng'). River, c. 110 mi (177 km), of SE China flowing into the South China Sea.

**Zhu·zhou** (jōō'jō'). City of S central China on the Xiang R. Pop. 350,000.

**Zi·bo** (zē'bō'). City of E China E of Jinan. Pop. 1,750,000.

**Zie·lo·no Gó·ra** (zhĕ-lô'nə gŏŏr'ə). City of W Poland W of Łódź. Pop. 73,156.

**Zi·gong** (zē'gŏŏng'). City of S central China W of Chongqing. Pop. 350,000.

**Zim·bab·we** (zĭm-bäb'wē, -wā). Republic of S central Africa. Cap. Salisbury. Pop. 7,360,000. **—Zim·bab'we·an** *adj. & n.*

**Zi·on** (zī'ən). City of NE Ill. on Lake Michigan. Pop. 17,861.

**Zla·to·ust** (zlä'tə-ōōst'). City of E European USSR in the S Urals. Pop. 199,000.

**Zon·gul·dak** (zŏŏng'gəl-däk'). City of N Turkey on the Black Sea. Pop. 108,661.

**Zug·spit·ze** (sŏŏk'shpĭt-sə, zŏŏg'spĭt-). Mountain, 9,271 ft (2,965 m), in the Alps of Bavaria S West Germany.

**Zu·lu·land** (zōō'lōō-lănd'). Historical region of NE South Africa.

**Zun·yi** (jŏŏn'yē'). City of S China SSE of Chongging. Pop. 275,000.

**Zu·rich** (zŏŏr'ĭk). City of NE Switzerland at the N tip of **Lake Zurich.** Pop. 401,600.

**Zwick·au** (tsfĭk'ou, zwĭk'-). City of S East Germany S of Leipzig. Pop. 123,069.

**Zwol·le** (zvôl'ə, zwôl'ə). City of N central Netherlands on the Ijssel R. Pop. 77,826.

# Foreign Words and Phrases

**à bas** [Fr.] Down with.
**abeunt studia in mores** [Lat.] Studies turn into habits : zealous pursuits become habits. —Ovid
**ab extra** [Lat.] From outside.
**à bientôt** [Fr.] Goodbye : I'll see you later.
**ab incunabulis** [Lat.] From the cradle : from the earliest stages.
**à bon chat, bon rat** [Fr.] A good cat deserves a good rat : tit for tat.
**à bon marché** [Fr.] At a bargain price : cheap.
**à bouche ouverte** [Fr.] With open mouth : enthusiastically : impetuously.
**ab ovo** [Lat.] From the egg : from the very beginning : tediously. —Horace
**ab ovo usque ad mala** [Lat.] From egg to apples : soup to nuts : from beginning to end. —Horace
**à bras ouverts** [Fr.] With open arms : warmly.
**absit invidia** [Lat.] May there be no envy or malice : no offense intended. —Livy
**absit omen** [Lat.] May it not be an omen. —Used superstitiously after a word or phrase considered of ill omen
**ab uno disce omnes** [Lat.] From one learn to judge all. —Virgil
**ab urbe condita** [Lat.] Dating from the founding of the city (of Rome, 753 B.C.).
**abusus non tollit usum** [Lat.] Abuse does not detract from use : misuse does not invalidate proper use.
**à compte** [Fr.] On account : in partial payment.
**à coup sûr** [Fr.] With sure stroke : confidently.
**acte gratuit** [Fr.] Unwarranted or impulsive act.
**ad arbitrium** [Lat.] Arbitrarily : at will.
**ad calendas Graecas** [Lat.] At the Greek calends (i.e., never, since the Greeks had no calends).
**ad captandum vulgus** [Lat.] To win the crowd : appealing to popular sentiment.
**adeste, fideles** [Lat.] Come, you faithful ones.
**ad extremum** [Lat.] To the extreme : to the very end : at last.
**ad finem** [Lat.] At or toward the end : finally.
**ad majorem Dei gloriam** [Lat.] To the greater glory of God. —St. Gregory the Great, motto of the Society of Jesus
**ad patres** [Lat.] (Gathered) to one's fathers : dead.
**ad referendum** [Lat.] For reference : to be referred to a higher authority.
**à droite** [Fr.] On or toward the right hand.
**ad utrumque paratus** [Lat.] Ready for either (event).
**advocatus diaboli** [Lat.] Devil's advocate.
**aegri somnia vana** [Lat.] A sick man's empty dreams. —Horace
**aequo animo** [Lat.] With equanimity : serenely.
**aere perennius** [Lat.] More enduring than bronze. —Horace
**aetatis** *or* **aetatis suae** [Lat.] At the age of. —Used before a date, esp. on tombstones
**affaire d'honneur** [Fr.] A case of honor : duel.
**à fond** [Fr.] At bottom : thoroughly.
**à gauche** [Fr.] On or toward the left hand.
**age quod agis** [Lat.] Do what you're doing : pay attention to your work.
**à grands frais** [Fr.] At great cost.
**à huis clos** [Fr.] Behind closed doors : in private.
**aide-toi, le ciel t'aidera** [Fr.] Heaven helps those who help themselves. —La Fontaine
**aîné** [Fr.] Elder : senior (masc.).
**aînée** [Fr.] Elder : senior (fem.).
**à l'abandon** [Fr.] Carelessly : in confusion : at random.
**à la belle étoile** [Fr.] Under the open sky : under the stars.
**à la bonne heure** [Fr.] At a good time : splendid : all right.
**à la page** [Fr.] Up to date.
**à la rigueur** [Fr.] Strictly speaking : if absolutely necessary.
**alea jacta est** [Lat.] The die is cast. —Attributed to Julius Caesar
**à l'improviste** [Fr.] Suddenly : unexpectedly.
**aliquando bonus dormitat Homerus** [Lat.] Sometimes good Homer himself nods : even the great make mistakes. —Horace
**aloha oe** [Hawaiian] Love to you : greetings : farewell.
**alter idem** [Lat.] Second self : alter ego. —Cicero
**a maximis ad minima** [Lat.] From the greatest to the least.
**âme damnée** [Fr.] Lost soul : scapegoat.
**amende honorable** [Fr.] Public apology : just restitution.
**à merveille** [Fr.] Wonderfully : to perfection.
**amicus humani generis** [Lat.] Friend of the human race.
**amicus usque ad aras** [Lat.] A friend as far as the altars (i.e., as long as one's beliefs are not concerned) : a friend to the very end.
**ami de cour** [Fr.] Court friend : false or fair-weather friend.
**amor patriae** [Lat.] Love of one's country : patriotism.
**amor vincit omnia** [Lat.] Love conquers all. —Virgil
**ancienne noblesse** [Fr.] Old-time aristocracy : the French aristocracy prior to the Revolution of 1789.
**anguis in herba** [Lat.] Snake in the grass. —Virgil
**animal bipes implume** [Lat.] Two-legged featherless animal : human being. —Attributed to Plato
**anima mundi** [Lat.] World spirit : force governing the physical universe.
**anno aetatis suae** [Lat.] At the age of.
**anno regni** [Lat.] In the year of the reign.
**anno urbis conditae** [Lat.] In the year of the founded city (of Rome, founded 753 B.C.).
**à outrance** [Fr.] To the bitter end : to the death.
**à peu près** [Fr.] Almost : approximately.
**à pied** [Fr.] On foot.
**après nous (***or* **moi) le déluge** [Fr.] After us (or me), the deluge. —Attributed to Marquise de Pompadour
**à propos de bottes** [Fr.] Speaking of boots. —Used to change the subject
**à propos de rien** [Fr.] Apropos of nothing : by the way.
**aqua et igni interdictus** [Lat.] Forbidden the necessities of water and fire : ostracized.
**arbiter elegantiarum** [Lat.] Authority in matters of style, manners, and aesthetics.
**arrière-goût** [Fr.] Aftertaste.
**arrivederci** [Ital.] Goodbye : till we meet again.
**ars amandi** [Lat.] The art of loving.
**ars est celare artem** [Lat.] (True) art is to conceal art.
**ars gratia artis** [Lat.] Art for art's sake.
**ars longa, vita brevis** [Lat.] Art is long, life short.
**a tergo** [Lat.] From or in back : on the back side.
**à tort et à travers** [Fr.] All wrong and crosswise : with no rhyme or reason : at random.
**à trois** [Fr.] Of, for, or among three (persons).
**au bout de son latin** [Fr.] At the end of one's Latin : at the end of one's wits.
**au contraire** [Fr.] On the contrary.
**audentes fortuna juvat** [Lat.] Fortune favors the bold. —Virgil
**audi alteram partem** [Lat.] Listen to the other side of the case.
**au fait** [Fr.] Well-informed : expert : to the point : socially correct.
**au fond** [Fr.] At bottom : basically.
**au grand sérieux** [Fr.] Very seriously : in all seriousness.
**au pays des aveugles les borgnes sont rois** [Fr.] In the land of the blind the one-eyed are kings.
**au pied de la lettre** [Fr.] Literally.
**aurea mediocritas** [Lat.] The golden mean. —Horace
**au reste** [Fr.] Moreover : besides.
**au sérieux** [Fr.] Seriously.
**aussitôt dit, aussitôt fait** [Fr.] No sooner said than done.
**aut Caesar aut nullus (***or* **nihil)** [Lat.] Either Caesar or no one : nothing but the best.
**autre chose** [Fr.] Something else : something different.
**autres temps, autres moeurs** [Fr.] Other times, other customs.
**aut vincere aut mori** [Lat.] Conquer or die.
**aux armes** [Fr.] To arms!
**avanti** [Ital.] Forward : onward!
**ave atque vale** [Lat.] Hail and farewell. —Catullus
**avec plaisir** [Fr.] With pleasure : gladly.
**à votre santé** [Fr.] To your health.

**ballon d'essai** [Fr.] Trial balloon.
**beatae memoriae** [Lat.] Of blessed memory.
**beaux yeux** [Fr.] Beautiful eyes : good looks.
**belle laide** [Fr.] Woman whose unattractiveness is considered appealing.
**bene esse** [Lat.] Well-being : comfort : luxury.
**ben trovato** [Ital.] Well conceived : ingenious : clever invention.
**bien entendu** [Fr.] Well understood : of course.
**bien-pensant** [Fr.] Right-thinkin : holding orthodox or conservativ views.
**bienséance** [Fr.] Propriety : dec rum : good breeding.
**Bildungsroman** [G.] Novel trac ing a character's early life an emotional development.
**bis dat qui cito dat** [Lat.] Wh gives promptly gives twice.
**bonae memoriae** [Lat.] Of happ memory.
**bon appétit** [Fr.] Good appetite enjoy your meal.
**bona vacantia** [Lat.] Goods wit no apparent legal owner.
**bon gré, mal gré** [Fr.] Whethe willing or not : willy-nilly.
**bon jour** [Fr.] Good day : hello.
**bonne à tout faire** [Fr.] Maid o all work.
**bonne chance** [Fr.] Good luck.
**bon soir** [Fr.] Good evening : goo night.
**brutum fulmen** [Lat.] Harmles thunderbolt : idle threat : empt bluster. —Pliny the Elder
**buenas noches** [Sp.] Good nigh
**buenos días** [Sp.] Good day : hello

**cacoethes loquendi** [Lat.] Irre sistible compulsion to talk.
**cacoethes scribendi** [Lat.] Irre sistible compulsion to write. —Juve nal
**cadit quaestio** [Lat.] The ques tion drops : there is nothing left t discuss.
**camino real** [Sp.] Royal road highroad : a direct or certain mean to an end.
**cante hondo (***or* **jondo)** [Sp. Spanish gypsy singing : flamenco.
**caput mortuum** [Lat.] Worthles residue : dregs.
**caramba** [Sp.] Exclamation o surprise, irritation, or pleasure.
**carte d'identité** [Fr.] Identity card.
**causa sine qua non** [Lat.] Indis pensable condition or cause.
**ça va?** [Fr.] How's it going? : every thing all right? —Used esp. as greeting
**caveat lector** [Lat.] Let the reade beware.
**cave canem** [Lat.] Beware of th dog.
**ce n'est que le premier pa qui coûte** [Fr.] The first step i always the hardest. —Attributed t Marquise du Deffand
**c'est à dire** [Fr.] That is to say.
**c'est la guerre** [Fr.] That's war that's what happens in war.
**c'est la vie** [Fr.] That's life : that how it goes.
**c'est plus qu'un crime, c'est une faute** [Fr.] It's worse than a crime, it's a blunder.
**cetera desunt** [Lat.] The rest i missing.
**chacun à son goût** [Fr.] Every one to his own taste.
**chacun à son métier** [Fr. Everyone to his own trade.
**chacun pour soi** [Fr.] Everyone for himself.
**chagrin d'amour** [Fr.] Disap pointment in love.
**châteaux en Espagne** [Fr.] Cas tles in Spain : wishful thinking.

**cherchez la femme** [Fr.] Look for the woman: there's bound to be a woman involved.

**che sarà sarà** [Ital.] What will be will be.

**cheval de bataille** [Fr.] War-horse: habitual or standard argument: favorite topic.

**chose jugée** [Fr.] Something already settled: closed case.

**ci-devant** [Fr.] Former: formerly.

**ci-gît** [Fr.] Here lies. —Used on tombstones

**cogito, ergo sum** [Lat.] I think, therefore I am. —Descartes

**comédie humaine** [Fr.] Comedy of life: the whole variety of human life. —Balzac

**comédie larmoyante** [Fr.] Tearful comedy: sentimental comedy.

**comédie noire** [Fr.] Black comedy.

**comme ci comme ça** [Fr.] So-so: neither good nor bad.

**compte rendu** [Fr.] Report: official review.

**concordia discors** [Lat.] Harmony in discord.

**confessio fidei** [Lat.] Confession of faith.

**contemptus mundi** [Lat.] Contempt for the world.

**contra mundum** [Lat.] Against the world: standing alone.

**corps d'élite** [Fr.] Select group: elite.

**corruptio optimi pessima** [Lat.] The corruption of the best is the worst: when the best is corrupted it becomes the worst.

**coup de foudre** [Fr.] Thunderclap: sudden and surprising event: love at first sight.

**coup de maître** [Fr.] Master-stroke.

**coup d'essai** [Fr.] First attempt.

**coûte que coûte** [Fr.] At all costs.

**credo quia absurdum est** [Lat.] I believe it because it's absurd.

**credo ut intelligam** [Lat.] I believe so that I may understand. —St. Anselm

**cri de cœur** [Fr.] Heartfelt appeal or protest.

**crise de conscience** [Fr.] Crisis of conscience: period of great moral uncertainty.

**crise de nerfs** [Fr.] Crisis of nerves: attack of hysteria.

**cui bono?** [Lat.] Who is the beneficiary?: who will profit by it? —Cicero

**cum grano salis** [Lat.] With a grain of salt.

**d'accord** [Fr.] Agreed: in agreement.

**dame d'honneur** [Fr.] Lady-in-waiting.

**damnant quod no intelligunt** [Lat.] They condemn what they do not understand.

**damnosa hereditas** [Lat.] Accursed inheritance: inheritance that brings more hardship than benefit.

**danke (schön)** [G.] Thank you (very much).

**danse macabre** [Fr.] Dance of death.

**de bene esse** [Lat.] Provisionally: on certain terms.

**de bonne grâce** [Fr.] With good grace: willingly.

**défense de fumer** [Fr.] No smoking.

**défense d'entrer** [Fr.] No admittance.

**de gustibus non est disputandum** [Lat.] There is no arguing in matters of taste.

**de haut en bas** [Fr.] In a condescending or patronizing manner.

**Dei gratia** [Lat.] By the grace of God.

**de integro** [Lat.] Anew: afresh.

**de l'audace, encore de l'audace, et toujours de l'audace** [Fr.] Audacity, more audacity, and ever more audacity. —Georges Jacques Danton

**delenda est Carthago** [Lat.] Carthage must be destroyed: let no obstacle stand in the way. —Cato the Elder

**delineavit** [Lat.] He (or she) drew it.

**de mal en pis** [Fr.] From bad to worse.

**de minimis non curat lex** [Lat.] The law takes no account of trifles.

**de mortuis nil nisi bonum** [Lat.] (Say) nothing but good about the dead.

**de nos jours** [Fr.] Of our time: contemporary. —Used after a noun

**de nouveau** [Fr.] Afresh: all over again.

**deoch-an-doruis** [Sc. Gael.] Parting drink: stirrup-cup.

**Deo favente** [Lat.] With God's favor.

**Deo gratias** [Lat.] Thanks be to God.

**de profundis** [Lat.] From the depths: cry of misery or despair. —*Vulgate (Psalm 129)*

**de règle** [Fr.] Customary: proper: required.

**der Geist der stets verneint** [G.] The spirit that ever denies (i.e., Mephistopheles). —Goethe

**desipere in loco** [Lat.] To be frivolous at the proper time: to relax on occasion. —Horace

**désolé** [Fr.] Extremely sorry.

**Deus misereatur** [Lat.] God have mercy. —*Vulgate (Psalm 66)*

**Deus vobiscum** [Lat.] God be with you.

**Deus vult** [Lat.] God wills it. —Battle cry of the First Crusade

**dialogue des sourds** [Fr.] Dialogue of the deaf: conversation between people who do not listen to each other.

**dies faustus** [Lat.] Lucky day.

**dies infaustus** [Lat.] Unlucky day.

**dies non** [Lat.] Day on which no legal business is transacted.

**Dieu avec nous** [Fr.] God with us.

**Dieu défend le droit** [Fr.] God defends the right.

**Dieu et mon droit** [Fr.] God and my right. —Motto of the British sovereign

**Dieu vous garde** [Fr.] God keep you.

**Ding an sich** [G.] Thing in itself: reality behind appearance. —Kant

**dis aliter visum** [Lat.] The gods deemed otherwise. —Virgil

**disjecta membra** [Lat.] Scattered remains: fragments. —Used esp. of literary works

**divide et impera** [Lat.] Divide and rule.

**docendo discimus** [Lat.] We learn by teaching.

**dolce far niente** [Ital.] Pleasant idleness.

**dolce stil nuovo** [Ital.] Sweet new style. —Dante

**Domine, dirige nos** [Lat.] Lord, guide us. —Motto of the City of London

**Dominus vobiscum** [Lat.] The Lord be with you.

**douceur de vivre** [Fr.] Pleasure of living well.

**Drang nach Osten** [G.] (German) expansionism to the east.

**dulce et decorum est pro patria mori** [Lat.] It is sweet and proper to die for one's country. Horace

**dum vivimus vivamus** [Lat.] Let us live while we have life.

**dux femina facti** [Lat.] A woman was leader of the exploit. —Virgil

**ecce signum** [Lat.] Behold the sign. —Used when introducing evidence or proof

**echt** [G.] Genuine: pure.

**e contrario** [Lat.] On the contrary.

**écrasez l'infâme** [Fr.] Crush the vile thing. —Voltaire

**editio princeps** [Lat.] First printed edition of a book.

**eheu fugaces labuntur anni** [Lat.] Alas, the fleeting years slip by. —Horace

**ein' feste Burg ist unser Gott** [G.] A mighty fortress is our God. —Luther

**embarras de richesses** [Fr.] Embarrassment of riches: abundance which is difficult to appreciate or choose from.

**embarras du choix** [Fr.] Embarrassment of choice: too much to choose from.

**éminence grise** [Fr.] Gray eminence: power behind the throne.

**en ami** [Fr.] In friendship: as a friend.

**en attendant** [Fr.] While waiting.

**en avant** [Fr.] Ahead: move forward!

**en bloc** [Fr.] Wholesale: all together: as one.

**en clair** [Fr.] In ordinary language: not in code.

**en effet** [Fr.] In fact: indeed.

**en face** [Fr.] Facing forward: on facing page of a book.

**en famille** [Fr.] As one of the family: informally: at home.

**enfant chéri** [Fr.] Loved or pampered child.

**enfant gâté** [Fr.] Spoiled child.

**enfants perdus** [Fr.] Lost children: forlorn hope: troops occupying an indefensible position.

**en fête** [Fr.] In a carnival mood: celebratory.

**enfin** [Fr.] In conclusion: in short.

**en garçon** [Fr.] As a bachelor.

**en garde** [Fr.] On guard.

**en grande tenue** [Fr.] In full dress: in full regalia.

**en pantoufles** [Fr.] In slippers: in an informal manner: at ease.

**en plein air** [Fr.] In the open air: outdoors.

**en plein jour** [Fr.] In broad day: openly.

**en rapport** [Fr.] In sympathy or accord: in touch.

**en règle** [Fr.] In order: in proper form.

**en retard** [Fr.] Delayed: late.

**en secondes noces** [Fr.] By or in a second marriage.

**en suite** [Fr.] In series: in succession: connected.

**entre nous** [Fr.] Between ourselves: confidentially.

**eo ipso** [Lat.] By that itself: by that fact.

**épater les bourgeois** [Fr.] Shock the middle classes: disconcert those of conventional tastes or beliefs. —Attributed to Baudelaire

**eppur si muove** [Ital.] And yet it does move. —Attributed to Galileo after being forced to recant his theory that the earth moves around the sun

**Erin go bragh** or **Éire go brách** [Ir. Gael.] Ireland forever.

**errare humanum est** [Lat.] To err is human.

**esprit de l'escalier** [Fr.] Staircase witticism: witty retort that one thinks of after the opportunity has passed. —Diderot

**est modus in rebus** [Lat.] There is a proper measure in things: there is a middle course in everything. —Horace

**et hoc genus omne** [Lat.] And everything of this kind.

**et in Arcadia ego** [Lat.] I too once lived in Arcady: even in Arcady, there am I (i.e., Death). Inscription on tombstones

**et sequentia** [Lat.] And the following. —Used in referring to following pages, chapters, etc.

**et sic de similibus** [Lat.] And so of like things.

**et tu, Brute!** [Lat.] You too, Brutus! —Attributed to Julius Caesar on recognizing his friend Brutus among his assassins

**Ewig-Weibliche** [G.] The eternal feminine. —Goethe

**ex animo** [Lat.] From the heart: genuinely.

**exceptio probat regulam de rebus non exceptis** [Lat.] The exception proves the rule as to things not excepted.

**exceptis excipiendis** [Lat.] With exceptions as appropriate.

**ex gratia** [Lat.] As a favor: not as something owed. —Used of a payment

**ex hypothesi** [Lat.] As a result of the assumptions made: as a matter of course.

**exitus acta probat** [Lat.] The end justifies the means. —Ovid

**ex mero motu** [Lat.] Out of mere impulse: of one's own free will.

**ex more** [Lat.] According to custom.

**ex necessitate rei** [Lat.] From the necessity of the case: necessarily.

**ex nihilo nihil fit** [Lat.] Nothing comes from nothing.

**ex pede Herculem** [Lat.] From the foot of Hercules (one may judge his size): a part bespeaks the whole.

**experto credite** [Lat.] Believe one who knows from experience.

**ex silentio** [Lat.] From lack of evidence to the contrary.

**ex ungue leonem** [Lat.] From the claw of the lion (one may judge its nature): a part bespeaks the whole.

**ex vi termini** [Lat.] From the force of the term.

**ex voto** [Lat.] In fulfillment of a vow.

**facile princeps** [Lat.] Easily first: by far the best.

**facilis descensus Averno** (or **Averni**) [Lat.] The descent to Avernus is easy: the road to hell is smooth. —Virgil

**façon de parler** [Fr.] Manner of speaking: figurative or conventional expression.

**faire suivre** [Fr.] Please forward. —Used as a postal instruction

**faisandé** [Fr.] Racy: risqué.

**faites vos jeux** [Fr.] Place your bets.

**fas est et ab hoste doceri** [Lat.] It is proper to learn even from an enemy. —Ovid

**Fata viam invenient** [Lat.] The Fates will find a way. —Virgil

**faute de mieux** [Fr.] For lack of anything better.

**faux bonhomme** [Fr.] Person whose apparent geniality and good nature is merely feigned.

**faux-naïf** [Fr.] Person pretending to be naive.

**fecit** [Lat.] He (or she) made it.

**femme de chambre** [Fr.] Chambermaid.

**femme savante** [Fr.] Learned and cultured woman: bluestocking.

**festina lente** [Lat.] Make haste slowly: more haste, less speed.

**feux d'artifice** [Fr.] Fireworks: dazzling display, as of wit.

**fiat experimentum in corpore vili** [Lat.] Let the experiment be performed on a worthless body.

**fiat justitia, ruat caelum** [Lat.] Let justice be done even if the heavens fall.

**fiat lux** [Lat.] Let there be light. *Vulgate (Genesis 1:3)*

**Fidei Defensor** [Lat.] Defender of the faith. —Used as a title of English sovereigns

**fidus Achates** [Lat.] Faithful Achates: loyal friend. —Virgil

**fille de chambre** [Fr.] Lady's maid.

**fille de joie** [Fr.] Prostitute.

**fille d'honneur** [Fr.] Maid of honor.

**fils** [Fr.] Son. —Used after a name to distinguish between father and son of the same name

**finem respice** [Lat.] Consider the end.

**finis coronat opus** [Lat.] The end crowns the work.
**fleur de coin** [Fr.] Mint condition of coin.
**floreat** [Lat.] May it flourish. —Used before the name of a school or institution
**floruit** [Lat.] He (or she) flourished. —Used before a date to indicate when a person was most active or productive
**fluctuat nec mergitur** [Lat.] It is tossed by the waves but does not sink. —Motto of Paris
**folie de grandeur (*or* des grandeurs)** [Fr.] Delusion of grandeur.
**fons et origo** [Lat.] Source and origin.
**force de frappe** [Fr.] Strike force. —Used esp. of nuclear forces
**forsan et haec olim meminisse juvabit** [Lat.] Perhaps this too will be pleasant to remember someday. —Virgil
**fortes fortuna juvat** [Lat.] Fortune favors the brave.
**franc-tireur** [Fr.] Sharpshooter : guerrilla.
**frisson** [Fr.] Thrill : shudder : shiver.
**fronti nulla fides** [Lat.] There is no trusting to appearances. —Juvenal
**fuit Ilium** [Lat.] Troy is no more. —Virgil
**furor loquendi** [Lat.] Rage for speaking.
**furor poeticus** [Lat.] Poetic frenzy.
**furor scribendi** [Lat.] Rage for writing.

**Galgenhumor** [G.] Gallows humor.
**garçon d'honneur** [Fr.] Groomsman : best man.
**garde du corps** [Fr.] Bodyguard.
**gardez la foi** [Fr.] Keep the faith.
**gaudeamus igitur** [Lat.] Let us therefore rejoice.
**Gemeinschaft** [G.] Association : community.
**gens d'église** [Fr.] The clergy.
**gens de guerre** [Fr.] The military.
**gens du monde** [Fr.] Worldly people : fashionable society.
**Gesellschaft** [G.] Commercial company.
**gnōthi seauton** [Gk.] Know thyself.
**Götterdämmerung** [G.] Twilight of the gods.
**goût de terroir** [Fr.] Taste of the earth : raciness.
**gracias** [Sp.] Thank you.
**grande dame** [Fr.] Great lady : woman of aristocratic dignity and bearing.
**Grand Guignol** [Fr.] Short horror-play : theatrically macabre.
**grand monde** [Fr.] Fashionable society.
**gravitas** [Lat.] Solemn bearing : sober-mindedness.
**guerre à outrance** [Fr.] All-out war.
**guten Tag** [G.] Good day : hello.

**hasta la vista** [Sp.] See you later : goodbye.
**haute vulgarisation** [Fr.] Popularization of scholarly or intellectual subjects, intended for a general audience.
**haut goût** [Fr.] High flavor : gaminess : slight taint.
**haut monde** [Fr.] High society : the fashionable world.
**heil** [G.] Hail : long live.
**hic et ubique** [Lat.] Here, (there) and everywhere.
**hic jacet** [Lat.] Here lies. —Used in epitaphs
**hic sepultus** [Lat.] Here lies buried. —Used in epitaphs
**hinc illae lacrimae** [Lat.] Hence those tears. —Terence
**hoc age** [Lat.] Do this : attend to your work.
**hoc anno** [Lat.] In this year.
**hoc opus, hic labor est** [Lat.] This is the work, this the labor : here is the real difficulty. —Virgil
**homme d'affaires** [Fr.] Businessman : agent : middleman.
**homme d'esprit** [Fr.] Man of wit.
**homme du monde** [Fr.] Man of the world : sophisticated or worldly-wise man.
**homme moyen sensuel** [Fr.] Average man : man on the street.
**homo sum: humani nil a me alienum puto** [Lat.] I am a man; nothing concerning mankind is alien to me. —Terence
**honi soit qui mal y pense** [Fr.] Shame to him who thinks evil of it. —Motto of the Order of the Garter
**honnête homme** [Fr.] Honest man : decent and upright citizen.
**horribile dictu** [Lat.] Horrible to relate.
**hors concours** [Fr.] Out of the running : not competing for a prize : unexcelled.
**humanum est errare** [Lat.] To err is human.

**ich dien** [G.] I serve. —Motto of the Prince of Wales
**ici on parle français** [Fr.] French spoken here.
**idée reçue** [Fr.] Received idea : doctrine or opinion accepted without questioning.
**id est** [Lat.] That is : that is to say.
**ignorantia juris neminem excusat** [Lat.] Ignorance of the law excuses no one.
**ignotum per ignotius** [Lat.] Explanation that is more obscure than the thing being explained.
**il faut cultiver notre jardin** [Fr.] We must cultivate our garden : let us tend to our own affairs. —Voltaire
**ils ne passeront pas** [Fr.] They shall not pass.
**imperium in imperio** [Lat.] Sovereignty within a sovereignty : absolute authority within the limits of a greater authority.
**incipit** [Lat.] Here begins.
**index librorum prohibitorum** [Lat.] List of books prohibited or censored by the Roman Catholic Church.
**in dubio** [Lat.] In doubt : undetermined.
**in esse** [Lat.] In existence : in actual fact.
**in forma pauperis** [Lat.] Not liable for costs on account of poverty.
**in futuro** [Lat.] In the future.
**in hoc signo vinces** [Lat.] By this sign (of the Cross) shall you conquer. —Motto of Constantine the Great
**in limine** [Lat.] On the threshold : at the outset.
**in omnia paratus** [Lat.] Ready for everything.
**in partibus infidelium** [Lat.] In heathen or heretical areas. —Applied to titular bishop of non-Christian see
**in pectore** [Lat.] In one's heart : secretly : in private.
**in perpetuum** [Lat.] For ever : in perpetuity.
**in posse** [Lat.] In possibility : potentially.
**in praesenti** [Lat.] At the present time : now.
**in rerum natura** [Lat.] In the nature of things.
**in saecula saeculorum** [Lat.] For ever and ever.
**inshallah** [Ar.] If Allah wills it : God willing.
**in statu quo ante** [Lat.] In the same state as before.
**integer vitae scelerisque purus** [Lat.] Blameless of life and free of vice. —Horace
**inter nos** [Lat.] Between ourselves.
**intra muros** [Lat.] Within the walls : within the confines of an institution or policy.
**intra vires** [Lat.] Within the authority or power (of a person or body).
**in usum Delphini** [Lat.] For the use of the Dauphin : expurgated.
**in utero** [Lat.] In the womb.
**in utrumque paratus** [Lat.] Ready for either (outcome). —Virgil
**in vacuo** [Lat.] In a vacuum : without reference to related arguments or circumstances.
**invenit** [Lat.] He (or she) designed it.
**in vino veritas** [Lat.] In wine there is truth : a drunken person always speaks the truth.
**invita Minerva** [Lat.] Minerva unwilling : unblessed by natural talent or inspiration.
**ira furor brevis est** [Lat.] Anger is a brief madness. —Horace

**j'accuse** [Fr.] I accuse : bitter denunciation. —Émile Zola
**jacta alea est (*or* est alea)** [Lat.] The die is cast. —Attributed to Julius Caesar
**j'adoube** [Fr.] I adjust. —Used in chess when adjusting a piece without intending to move it
**janus clausis** [Lat.] Behind closed doors : in secret.
**je maintiendrai** [Fr.] I will maintain. —Motto of the Netherlands
**jeu de mots** [Fr.] Play on words : pun.
**jeu d'esprit** [Fr.] Play of wit : witticism : humorous trifle.
**jeune fille** [Fr.] Young girl.
**jeune premier** [Fr.] Actor playing leading role of young hero or lover.
**jeune première** [Fr.] Actress playing leading role of young heroine or lover.
**Joannes est nomen eius** [Lat.] His name is John. —Motto of Puerto Rico
**jolie-laide** [Fr.] Woman who is considered attractive though not conventionally pretty.
**jour de fête** [Fr.] Holiday : festival : feast day.
**journal intime** [Fr.] Intimate journal : private diary.
**jure divino** [Lat.] By divine law.
**jus canonicum** [Lat.] Canon law.
**jus divinum** [Lat.] Divine law.
**juste milieu** [Fr.] Golden mean : happy medium.
**j'y suis, j'y reste** [Fr.] Here I am and here I stay.

**Kamerad** [G.] Comrade : cry for mercy when surrendering.
**Kinder, Kirche, Küche** [G.] Children, church, kitchen. —Used to express conventional view of a woman's responsibilities
**ktēma es aei** [Gk.] A possession for all time : an artistic work of enduring value. —Thucydides

**la belle dame sans merci** [Fr.] The beautiful, merciless lady.
**laborare est orare** [Lat.] To work is to pray.
**lacrimae rerum** [Lat.] Tears for things : sadness of things : tragedy of life. —Virgil
**laissez-aller** *or* **laisser-aller** [Fr.] Lack of restraint in speech or manner.
**lapsus calami** [Lat.] Slip of the pen.
**lapsus linguae** [Lat.] Slip of the tongue.
**la reine le veut** [Fr.] The queen wishes it.
**l'art pour l'art** [Fr.] Art for art's sake.
**lasciate ogni speranza, voi ch'entrate** [Ital.] Abandon all hope, ye who enter. —Dante
**laudator temporis acti** [Lat.] One who praises times past (and thus criticizes the present). —Horace
**laus Deo** [Lat.] Praise be to God.
**le cœur a ses raisons que la raison ne connaît point** [Fr.] The heart has its reasons that reason knows nothing of. —Pascal
**le roi est mort, vive le roi** [Fr.] The king is dead, long live the king.
**le roi le veut** [Fr.] The king wishes it.
**le roi s'avisera** [Fr.] The king will consider.
**les jeux sont faits** [Fr.] The bets are made : no more stakes may be laid. —Used in roulette
**le style, est l'homme même** [Fr.] The style is the man himself. —Buffon
**l'état, c'est moi** [Fr.] I am the state. —Attributed to Louis XIV
**le tout ensemble** [Fr.] The whole (taken) together : general effect.
**lettre de cachet** [Fr.] Document authorizing summary imprisonment, arrest, or exile.
**lex talionis** [Lat.] Law of retaliation : an eye for an eye.
**liberté, égalité, fraternité** [Fr.] Liberty, equality, fraternity. —Motto of France
**Liederkranz** [G.] Wreath of songs : German singing society.
**littera scripta manet** [Lat.] The written letter abides.
**loco citato** [Lat.] In the place or passage previously cited.
**locus in quo** [Lat.] Place in which : place where a passage occurs.
**locus standi** [Lat.] Recognized position : right to appear in court.
**lucus a non lucendo** [Lat.] Paradoxical explanation : explanation of something by the opposite of what it suggests.
**l'union fait la force** [Fr.] Union makes strength. —Motto of Belgium

**Machtpolitik** [G.] Power politics.
**ma foi!** [Fr.] My faith! : oh my!
**magna est veritas et praevalebit** [Lat.] Truth is great and will prevail.
**magni nominis umbra** [Lat.] The shadow of a great name.
**malade imaginaire** [Fr.] Imaginary invalid : hypochondriac. —Molière
**maladresse** [Fr.] Clumsiness : tactlessness : awkwardness.
**mal du pays** [Fr.] Homesickness.
**mal du siècle** [Fr.] World-weariness : disillusionment at the state of the world.
**malgré lui** [Fr.] In spite of himself.
**malgré tout** [Fr.] In spite of everything.
**malis avibus** [Lat.] Under evil auspices.
**mal vu** [Fr.] Disapproved of : resented.
**mano a mano** [Sp.] Hand in hand : together : alternatingly.
**man spricht Deutsch** [G.] German spoken here.
**mariage de convenance** [Fr.] Marriage of convenience : arranged marriage.
**mauvaise honte** [Fr.] False shame : painful shyness.
**mauvais goût** [Fr.] Bad taste.
**mauvais quart d'heure** [Fr.] Bad quarter-hour : difficult but brief experience.
**mauvais sujet** [Fr.] Rogue : black sheep.
**meden agan** [Gk.] Nothing in excess.
**medio tutissimus ibis** [Lat.] The middle path is the safest. —Ovid
**me judice** [Lat.] I being judge : in my opinion.
**memoria technica** [Lat.] System or device to assist memory.
**mens rea** [Lat.] Criminal intent.

**nens sana in corpore sano** [Lat.] A healthy mind in a healthy body. —Juvenal
**nerci (beaucoup)** [Fr.] Thank you (very much).
**netteur en scène** [Fr.] Designer, director, or stage manager of a theatrical production.
**neum et tuum** [Lat.] Mine and yours. —Used to express principle of rights of property
**nirabile visu** [Lat.] Wonderful to behold.
**niserere nobis** [Lat.] Have mercy on us.
**nissa solemnis** [Lat.] High mass.
**nochree** or **mo chroidhe** [Ir. Gael.] My heart : my darling.
**nœurs** [Fr.] Mores : manners and customs of a place or period.
**nole ruit sua** [Lat.] It collapses from its own massiveness. —Horace
**nonde** [Fr.] World : fashionable society.
**nonstre sacré** [Fr.] Sacred monster : celebrity notorious for his or her unorthodox life or opinions.
**nonumentum aere perennius** [Lat.] A monument more lasting than bronze : immortal work of art or literature. —Horace
**nore suo** [Lat.] In his (or her) own way.
**norituri te salutant (or salutamus)** [Lat.] Those (or we) who are about to die salute you. —Salutation to Emperor by Roman gladiators
**noto perpetuo** [Ital.] Perpetual motion : musical composition with continuous sequence of rapid notes.
**notu proprio** [Lat.] Of one's own accord.
**mudéjar** [Sp.] Spanish architectural style with strong Moorish influence : Moslem living in Christian Spain.
**multum in parvo** [Lat.] Much in little : a great deal in a small space.
**mutato nomine de te fabula narratur** [Lat.] With the name changed the story applies to you. —Horace

**naturam expellas furca, tamen usque recurret** [Lat.] Though you drive nature out with a pitchfork, she will keep coming back. —Horace
**natura non facit saltum** [Lat.] Nature makes no leap : nature takes no shortcuts.
**ne cede malis** [Lat.] Yield not to misfortunes. —Virgil
**nemine contradicente** [Lat.] No one contradicting : unanimously.
**nemine dissentiente** [Lat.] No one dissenting : unanimously.
**nemo me impune lacessit** [Lat.] No one injures me with impunity. —Motto of Scotland and of the Order of the Thistle
**ne quid nimis** [Lat.] Nothing in excess. —Terence
**n'est-ce pas?** [Fr.] Isn't that so?
**nicht wahr?** [G.] Isn't that so?
**nihil ad rem** [Lat.] Beside the point : irrelevant.
**nil admirari** [Lat.] To remain unimpressed : composure of mind. —Horace
**nil desperandum** [Lat.] Never despair. —Horace
**n'importe** [Fr.] No matter : that's all right.
**nolens volens** [Lat.] Whether willing or not : willy-nilly.
**non est (inventus)** [Lat.] He has not (been found) ; legal statement that defendant cannot be found.
**non omnia possumus omnes** [Lat.] We can't all do everything. —Virgil
**non omnis moriar** [Lat.] I shall not wholly die : a part of me will survive death. —Horace
**non sans droict** [OFr.] Not without right. —Motto on Shakespeare's coat of arms
**non sum qualis eram** [Lat.] I am not what I once was. —Horace
**nosce te ipsum** [Lat.] Know thyself.
**nostalgie de la boue** [Fr.] Nostalgia for the mud : longing for esp. sexual degradation.
**nous avons changé tout cela** [Fr.] We have changed all that. —Molière
**nous verrons ce que nous verrons** [Fr.] We shall see what we shall see.
**novus homo** [Lat.] New man : man newly ennobled : parvenu.
**nuit blanche** [Fr.] White night : sleepless night.
**nulla bona** [Lat.] Legal statement that debtor has no goods for seizure.
**nulli secundus** [Lat.] Second to none.
**nyet** [R.] No.

**obiit** [Lat.] He (or she) died. —Used before date of death
**obscurum per obscurius** [Lat.] Explanation that is more obscure than the thing being explained.
**oderint dum metuant** [Lat.] Let them hate, so long as they fear. —Attributed to Lucius Accius
**odi et amo** [Lat.] I hate and I love. —Catullus
**omadhaun** or **amadān** [Ir. Gael.] Fool : madman. —Used as a term of abuse
**omertà** [Ital.] Submission : code of sworn silence and private revenge esp. among the criminal underworld.
**omne ignotum pro magnifico** [Lat.] Everything unknown is assumed to be grand : the unknown always appears more wonderful or imposing than it really is. —Tacitus
**omnia mutantur, nos et mutamur in illis** [Lat.] All things are changing, and we are changing with them.
**omnia vincit amor** [Lat.] Love conquers all. —Virgil
**on dit** [Fr.] It is said : item of gossip.
**onus probandi** [Lat.] Burden of proof.
**opus Dei** [Lat.] The work of God.
**ora pro nobis** [Lat.] Pray for us.
**orare est laborare, laborare est orare** [Lat.] To pray is to work, to work is to pray.
**orbis terrarum** [Lat.] The earth.
**ore rotundo** [Lat.] With rounded mouth : eloquently. —Horace
**Ostpolitik** [G.] Western European country's foreign policy toward a communist Eastern European country.
**O tempora! O mores!** [Lat.] O times! O morals! : what corrupt times we live in! —Cicero
**otium cum dignitate** [Lat.] Leisure with dignity. —Cicero
**où sont les neiges d'antan?** [Fr.] Where are the snows of yesteryear? —Villon

**pallida Mors** [Lat.] Pale Death. —Horace
**panem et circenses** [Lat.] Bread and circuses : benefits provided by a government to pacify the common populace. —Juvenal
**panta rhei** [Gk.] All things are in flux. —Heraclitus
**par avance** [Fr.] In advance : beforehand : by anticipation.
**par avion** [Fr.] By airplane : airmail.
**par exemple** [Fr.] For example.
**Paris vaut bien une messe** [Fr.] Paris is well worth a mass : material gain is worth a sacrifice of principle. —Henry IV of France
**parturient montes, nascetur ridiculus mus** [Lat.] The mountains shall labor and bring forth a ridiculous mouse. —Horace
**pas devant les enfants** [Fr.] Not in front of the children.
**pater patriae** [Lat.] Father of his country.
**paucis verbis** [Lat.] In few words : briefly.
**Pax Britannica** [Lat.] Peace imposed by British rule.
**Pax Romana** [Lat.] Peace imposed by Roman rule.
**pax vobiscum** [Lat.] Peace be with you.
**peine forte et dure** [Fr.] Harsh and severe punishment : torture.
**pendente lite** [Lat.] While a legal suit is pending.
**per angusta ad augusta** [Lat.] Through hardships to greatness.
**père** [Fr.] Father. —Used after a name to distinguish between father and son of the same name
**pereant qui ante nos nostra dixerunt** [Lat.] A plague on those who have proclaimed our bright ideas before us! —Attributed to Aelius Donatus
**pereunt et imputantur** [Lat.] (The hours) pass away and are reckoned against us. —Martial
**perfide Albion** [Fr.] Perfidious England. —Attributed to Napoleon I
**per impossibile** [Lat.] By an impossibility : supposing it were possible, which it isn't.
**per mensem** [Lat.] By the month : monthly payment.
**peu à peu** [Fr.] Little by little.
**peu de chose** [Fr.] Small matter : trifle.
**pièce d'occasion** [Fr.] Musical or literary work composed for a special occasion.
**pièce justificative** [Fr.] Justificatory paper : document serving as evidence.
**pièce montée** [Fr.] Set piece : food arranged in a decorative shape.
**pinxit** [Lat.] He (or she) painted it.
**place aux dames** [Fr.] Make room for the ladies.
**plein air** [Fr.] Open air : representation in painting of natural light or atmospheric conditions.
**pleno jure** [Lat.] With full authority.
**plus ça change, plus c'est la même chose** [Fr.] The more things change, the more they remain the same.
**plus royaliste que le roi** [Fr.] More royalist than the king.
**poeta nascitur, non fit** [Lat.] A poet is born, not made.
**point d'appui** [Fr.] Point of support : fulcrum : basis of argument.
**point de repère** [Fr.] Point of reference.
**pollice verso** [Lat.] Thumbs down. —Gesture by which spectators condemned a defeated Roman gladiator to death
**por favor** [Sp.] Please.
**post hoc, ergo propter hoc** [Lat.] After this, therefore because of it : fallacious argument that because one thing precedes another it necessarily causes it.
**pour acquit** [Fr.] Payment received : paid.
**pour encourager les autres** [Fr.] To encourage the others (by frightening them). —Voltaire
**pour le mérite** [Fr.] For merit.
**pour rire** [Fr.] In jest : laughingly.
**pou stō** [Gk.] Place to stand : base of operation. —Archimedes
**prêt-à-porter** [Fr.] Ready to wear. —Used of clothing
**pro aris et focis** [Lat.] For altars and hearths : for religious and civil liberty.
**pro bono publico** [Lat.] For the public good.
**procul este, profani** [Lat.] Keep your distance, you who are uninitiated. —Virgil
**pro hac vice** [Lat.] For this occasion only.
**pro patria** [Lat.] For one's country.
**pro rege, lege, et grege** [Lat.] For the king, the law, and the people
**pro re nata** [Lat.] For a situatio that has arisen : as needed. —Used medical prescriptions
**pro tanto** [Lat.] To such an extent : so much : so far.
**proxime accessit** [Lat.] Runner-up : honorable mention.
**punica fide** [Lat.] With Punic faith : treacherously.
**pur sang** [Fr.] Pure-blooded : of noble birth : through and through.

**quand même** [Fr.] All the same : even so : despite the consequences.
**quantum mutatus ab illo** [Lat.] How changed from what he once was. —Virgil
**quantum sufficit** [Lat.] As much as necessary. —Used in medical prescriptions
**¿qué pasa?** [Sp.] What's happening? : what's wrong?
**que será será** [Sp.] What will be will be.
**¿qué tal?** [Sp.] How's it going? : how's everything? —Used as a greeting
**¿quién sabe?** [Sp.] Who knows?
**qui facit per alium facit pe se** [Lat.] He who does somethi through another does it throu himself (i.e., shares the responsi ity).
**quis custodiet ipsos c todes?** [Lat.] Who shall guar guardians themselves? —Juven
**qui s'excuse s'accuse** [Fr. excuses himself accuses hims
**quis separabit?** [Lat.] Wh separate us? —Motto of the St. Patrick
**qui va là?** [Fr.] Who goes
**quoad hoc** [Lat.] As to t as this is concerned.
**quod erat demonst** [Lat.] Which was to strated.
**quod erat faciend** Which was to be done.
**quod erat invenie** Which was to be foun
**quod semper, qu quod ab omnibu** (has been held true where, by everyone.
**quod vide** [Lat.] to this.
**quorum pars m** In which I played —Virgil
**quos deus vul dementat** [L god wishes to d mad.
**quot homine** [Lat.] There a there are peo
**quo vadis?** going? —Vul

**raison d'**
**reculer** [Fr.] To leap for better a
**re inf** being plishi
**religi** tity c
**rem** have : yo
**rép** Ple ca
**re**

**rus in urbe** [Lat.] The countryside in the city: rural landscaping in an urban setting. —Martial

**saeva indignatio** [Lat.] Fierce indignation.
**sal Atticum** [Lat.] Attic salt: sharp wit.
**salle à manger** [Fr.] Dining room.
**salto mortale** [Ital.] Deadly leap: full somersault: dangerous or crucial ct.
**lud** [Sp.] Health: to your health. —Ised as a toast
**e or salvete** [Lat.] Welcome.
**a simplicitas** [Lat.] Holy ity: well-meant naiveté. —ed to John Huss
**gue** [Fr.] Seriously: joking
**e** [Fr.] Without a doubt:
[Fr.] Outspokenly: ly.
Without constraint miliar manner.
Without equal:
Without diffi-
**reproche**
above re-
pleasure:
Ised as
him-
:

**silent leges inter arma** [Lat.] The laws are silent in time of war. —Cicero
**s'il vous plaît** [Fr.] If you please: please.
**similia similibus curantur** [Lat.] Like cures like.
**similis simili gaudet** [Lat.] Like takes pleasure in like.
**si monumentum requiris, circumspice** [Lat.] If you seek his monument, look around you. —Epitaph of Sir Christopher Wren in St. Paul's Cathedral, London, which he designed
**siste viator** [Lat.] Stop, traveler. —Used on Roman roadside tombs
**si vis pacem, para bellum** [Lat.] If you want peace, prepare for war.
**solvitur ambulando** [Lat.] It is solved by walking: the problem is solved by practical experiment.
**spalpeen** *or* **spailpin** [Ir. Gael.] Rascal: youngster.
**splendide mendax** [Lat.] Nobly untruthful: lying for an honorable cause. —Horace
**spolia opima** [Lat.] The richest spoils: supreme achievement: arms stripped from the defeated general by the victor.
**status in quo** [Lat.] State in which: the existing state of affairs.
**status quo ante** [Lat.] Previous state of affairs.
**Stimmung** [G.] Mood or tone of an artistic work: atmosphere.
**suaviter in modo, fortiter in re** [Lat.] Soft in manner, strong in deed.
**sub specie aeternitatis** [Lat.] In relation to eternal things: without reference to the changing conditions of everyday life. —Spinoza
**sub verbo (*or* voce)** [Lat.] Under the word. —Used to indicate a oss-referenced word
**ccès de scandale** [Fr.] Success ned through notoriety or scandal.
**estio falsi** [Lat.] Deliberate presentation not involving a ie.
**crimae rerum** [Lat.] e tears for things: life is irgil
at.] In one's own right.
t.] In its rightful place.
[Lat.] By one's own
**ri** [Lat.] Suppres- misrepresentation
place: on the
To each his
eariness
the
se.
**t**

**timeo Danaos et dona ferentes** [Lat.] I fear the Greeks even when they bring gifts. —Virgil
**ton** [Fr.] Fashionable society or style.
**totidem verbis** [Lat.] In so many words: in these exact words.
**toties quoties** [Lat.] As often as: on each occasion: repeatedly.
**totis viribus** [Lat.] With all one's might.
**toto caelo** [Lat.] By the whole extent of the heavens: as far apart as possible: diametrically opposed.
**toujours la politesse** [Fr.] It is always best to be polite.
**toujours perdrix** [Fr.] Always partridge: too much of a good thing.
**tour d'horizon** [Fr.] Circuit of the horizon: general survey.
**tous frais faits** [Fr.] All expenses paid.
**tout à fait** [Fr.] Entirely: altogether: quite.
**tout au contraire** [Fr.] Quite the contrary.
**tout à vous** [Fr.] All yours: at your service.
**tout bien ou rien** [Fr.] Everything done properly or not at all: all or nothing.
**tout comprendre, c'est tout pardonner** [Fr.] To understand everything is to forgive everything. —Attributed to Mme de Staël.
**tout court** [Fr.] Briefly: simply: curtly.
**tout de même** [Fr.] All the same: nevertheless.
**tout de suite** [Fr.] Immediately: all at once.
**tout ensemble** [Fr.] All together: general effect.
**tout est perdu fors (*or* hors) l'honneur** [Fr.] All is lost save honor. —Attributed to Francis I of France
**tout le monde** [Fr.] Everybody: everyone of importance.
**trahison des clercs** [Fr.] Betrayal by the intellectuals (in compromising their own standards). —Julien Benda
**tranche de vie** [Fr.] Slice of life.
**tria juncta in uno** [Lat.] Three joined in one. —Motto of the Order of the Bath
**tristesse** [Fr.] Melancholy: low spirits.
**truditur dies die** [Lat.] Day shoves forth day: one day hurries after another. —Horace
**tuebor** [Lat.] I will defend. —Motto on the Great Seal of Michigan

**Übermensch** [G.] Superman: ideal superior man. —Nietzsche
**ultima ratio regum** [Lat.] The final argument of kings: war.
**ultra vires** [Lat.] Beyond the power or authority of a person or body.
**und so weiter** [G.] And so forth.
**uno animo** [Lat.] With one mind: unanimously.
**uomo universale** [Ital.] Universal man: one of broad education and ability.
**urbi et orbi** [Lat.] To the city (Rome) and to the world. —Used ormerly in papal proclamations
**ile dulci** [Lat.] The useful with pleasurable. —Horace
**fra** [Lat.] As below. —Used in ng to later portion of a text
**sidetis** [Lat.] As you posciple leaving belligerents in of what they have seized.

**ut supra** [Lat.] As above. —Us in referring to earlier portion of text

**vade retro me, Satana** [La Get thee behind me, Satan. —*Vu gate (Mark 8:33)*
**va-et-vient** [Fr.] Coming ar going: traffic: commotion.
**vae victis** [Lat.] Woe to the va quished.
**vale** *or* **valete** [Lat.] Farewell.
**varia lectio** [Lat.] Variant readin
**varium et mutabile semp femina** [Lat.] Woman is ever fickle and changeable thing. —Vir
**vedi Napoli e poi mori** [Ita See Naples and then die.
**veni, vidi, vici** [Lat.] I came, saw, I conquered. —Attributed Julius Caesar
**ventre à terre** [Fr.] Belly to t ground: at full speed.
**verbatim et literatim** [Lat Word for word and letter for lette
**verbum sapienti sat est** [Lat A word to the wise is sufficient.
**vers de société** [Fr.] Society ve : light, witty, topical poetry.
**victor ludorum** [Lat.] Winner the games: athletic champion.
**vieux jeu** [Fr.] Old-fashioned.
**vincit omnia veritas** [Lat Truth conquers all.
**vinculum matrimonii** [Lat Bond of marriage.
**virginibus puerisque** [Lat.] F girls and boys. —Horace
**vis medicatrix naturae** [Lat Healing power of nature.
**vita nuova** [Ital.] New life. - Dante
**vivat regina (*or* rex)** [Lat.] Lor live the queen (or king).
**vive la différence** [Fr.] Long li the difference (between the sexes
**vive la reine (*or* le roi)** [Fr Long live the queen (or king).
**vixere fortes ante Agamem nona** [Lat.] There lived brave me before Agamemnon. —Horace
**vogue la galère** [Fr.] Keep th galley rowing: keep on at all costs. - Rabelais
**voilà** [Fr.] Look! : see! : here it i
**voilà tout** [Fr.] That's all.
**volupté** [Fr.] Pleasure: sensualit
**voulu** [Fr.] Willed: intentional contrived.
**vox et praeterea nihil** [Lat.] voice and nothing more: empt words.
**vox populi, vox Dei** [Lat.] Th voice of the people is the voice c God.

**Wanderjahr** [G.] Years of wande ing: years spent gathering exper ence.
**Weltbild** [G.] Conception of th world.
**Weltpolitik** [G.] World politics foreign policy or diplomacy.
**Wertfreiheit** [G.] Freedom fror value judgments: ethical neutrality
**wie geht's?** [G.] How are thing —Used as a greeting
**Wiegenlied** [G.] Lullaby.
**Wissenschaft** [G.] Science: learr ing.
**wunderbar** [G.] Wonderful: ma velous.
**Wunderkind** [G.] Child prodigy

**zingaro** [Ital.] Gypsy.
**Zollverein** [G.] Customs union.

# MEASUREMENT

## TABLE I. MEASUREMENT UNITS

### Length

| U.S. Customary Unit | U.S. Equivalents | Metric Equivalents |
|---|---|---|
| inch | 0.083 foot | 2.540 centimeters |
| foot | ⅓ yard, 12 inches | 0.305 meter |
| yard | 3 feet, 36 inches | 0.914 meter |
| rod | 5½ yards, 16½ feet | 5.029 meters |
| mile (statute, land) | 1,760 yards, 5,280 feet | 1.609 kilometers |
| mile (nautical, international) | 1.151 statute miles | 1.852 kilometers |

### Area

| U.S. Customary Unit | U.S. Equivalents | Metric Equivalents |
|---|---|---|
| square inch | 0.007 square foot | 6.452 square centimeters |
| square foot | 144 square inches | 929.030 square centimeters |
| square yard | 1,296 square inches, 9 square feet | 0.836 square meters |
| acre | 43,560 square feet, 4,840 square yards | 4,047 square meters |
| square mile | 640 acres | 2.590 square kilometers |

### Weight

| U.S. Customary Unit (Avoirdupois) | U.S. Equivalents | Metric Equivalents |
|---|---|---|
| grain | 0.036 dram, 0.002285 ounce | 64.798 milligrams |
| dram | 27.344 grains, 0.0625 ounce | 1.772 grams |
| ounce | 16 drams, 437.5 grains | 28.350 grams |
| pound | 16 ounces, 7,000 grains | 453.592 grams |
| ton (short) | 2,000 pounds | 0.907 metric ton (1,000 kilograms) |
| ton (long) | 1.12 short tons, 2,240 pounds | 1.016 metric tons |

| Apothecary Weight Unit | U.S. Customary Equivalents | Metric Equivalents |
|---|---|---|
| scruple | 20 grains | 1.296 grams |
| dram | 60 grains | 3.888 grams |
| ounce | 480 grains, 1.097 avoirdupois ounces | 31.103 grams |
| pound | 5,760 grains, 0.823 avoirdupois pound | 373.242 grams |

### Volume or Capacity

| U.S. Customary Unit | U.S. Equivalents | Metric Equivalents |
|---|---|---|
| cubic inch | 0.00058 cubic foot | 16.387 cubic centimeters |
| cubic foot | 1,728 cubic inches | 0.028 cubic meter |
| cubic yard | 27 cubic feet | 0.765 cubic meter |

| U.S. Customary Liquid Measure | U.S. Equivalents | Metric Equivalents |
|---|---|---|
| fluid ounce | 8 fluid drams, 1.804 cubic inches | 29.573 milliliters |
| pint | 16 fluid ounces, 28.875 cubic inches | 0.473 liter |
| quart | 2 pints, 57.75 cubic inches | 0.946 liter |
| gallon | 4 quarts, 231 cubic inches | 3.785 liters |
| barrel | varies from 31 to 42 gallons, established by law or usage | |

| U.S. Customary Dry Measure | U.S. Equivalents | Metric Equivalents |
|---|---|---|
| pint | ½ quart, 33.6 cubic inches | 0.551 liter |
| quart | 2 pints, 67.2 cubic inches | 1.101 liters |
| peck | 8 quarts, 537.605 cubic inches | 8.810 liters |
| bushel | 4 pecks, 2,150.420 cubic inches | 35.239 liters |

| British Imperial Liquid and Dry Measure | U.S. Customary Equivalents | Metric Equivalents |
|---|---|---|
| fluid ounce | 0.961 U.S. fluid ounce, 1.734 cubic inches | 28.413 milliliters |
| pint | 1.032 U.S. dry pints, 1.201 U.S. liquid pints, 34.678 cubic inches | 568.245 milliliters |
| quart | 1.032 U.S. dry quarts, 1.201 U.S. liquid quarts, 69.354 cubic inches | 1.136 liters |
| gallon | 1.201 U.S. gallons, 277.420 cubic inches | 4.546 liters |
| peck | 554.84 cubic inches | 0.009 cubic meter |
| bushel | 1.032 U.S. bushels, 2,219.36 cubic inches | 0.036 cubic meter |

# THE METRIC SYSTEM

### Length

| Unit | Number of Meters | Approximate U.S. Equivalent | Unit | Number of Meters | Approximate U.S. Equivalent |
|---|---|---|---|---|---|
| myriameter | 10,000 | 6.214 miles | meter | 1 | 39.370 inches |
| kilometer | 1,000 | 0.621 mile | decimeter | 0.1 | 3.937 inches |
| hectometer | 100 | 109.361 yards | centimeter | 0.01 | 0.394 inch |
| decameter | 10 | 32.808 feet | millimeter | 0.001 | 0.039 inch |

### Area

| Unit | Number of Square Meters | Approximate U.S. Equivalent | Unit | Number of Square Meters | Approximate U.S. Equivalent |
|---|---|---|---|---|---|
| square kilometer | 1,000,000 | 0.386 square mile | deciare | 10 | 11.960 square yards |
| hectare | 10,000 | 2.477 acres | centare | 1 | 10.764 square feet |
| are | 100 | 119.599 square yards | square centimeter | 0.0001 | 0.115 square inch |

## MEASUREMENT *(continued)*

### Volume

| Unit | Number of Cubic Meters | Approximate U.S. Equivalent | Unit | Number of Cubic Meters | Approximate U.S. Equivalent |
|---|---|---|---|---|---|
| decastere | 10 | 13.079 cubic yards | decistere | 0.10 | 3.532 cubic feet |
| stere | 1 | 1.308 cubic yards | cubic centimeter | 0.000001 | 0.061 cubic inch |

### Capacity

| Unit | Number of Liters | Approximate U.S. Equivalents: Cubic | Dry | Liquid |
|---|---|---|---|---|
| kiloliter | 1,000 | 1.308 cubic yards | | |
| hectoliter | 100 | 3.532 cubic feet | 2.838 bushels | |
| decaliter | 10 | 0.353 cubic foot | 1.135 pecks | 2.642 gallons |
| liter | 1 | 61.024 cubic inches | 0.908 quart | 1.057 quarts |
| deciliter | 0.10 | 6.102 cubic inches | 0.182 pint | 0.211 pint |
| centiliter | 0.01 | 0.610 cubic inch | | 0.338 fluid ounce |
| milliliter | 0.001 | 0.061 cubic inch | | 0.271 fluid dram |

### Mass and Weight

| Unit | Number of Grams | Approximate U.S. Equivalent | Unit | Number of Grams | Approximate U.S. Equivalent |
|---|---|---|---|---|---|
| metric ton | 1,000,000 | 1.102 tons | gram | 1 | 0.035 ounce |
| quintal | 100,000 | 220.462 pounds | decigram | 0.10 | 1.543 grains |
| kilogram | 1,000 | 2.205 pounds | centigram | 0.01 | 0.154 grain |
| hectogram | 100 | 3.527 ounces | milligram | 0.001 | 0.015 grain |
| decagram | 10 | 0.353 ounce | | | |

## METRIC CONVERSION CHART—APPROXIMATIONS

| When You Know | Multiply By | To Find |
|---|---|---|
| **Length** | | |
| millimeters | 0.04 | inches |
| centimeters | 0.39 | inches |
| meters | 3.28 | feet |
| meters | 1.09 | yards |
| kilometers | 0.62 | miles |
| inches | 25.40 | millimeters |
| inches | 2.54 | centimeters |
| feet | 30.48 | centimeters |
| yards | 0.91 | meters |
| miles | 1.61 | kilometers |
| **Area** | | |
| square centimeters | 0.16 | square inches |
| square meters | 1.20 | square yards |
| square kilometers | 0.39 | square miles |
| hectares (10,000m$^2$) | 2.47 | acres |
| square inches | 6.45 | square centimeters |
| square feet | 0.09 | square meters |
| square yards | 0.84 | square meters |
| square miles | 2.60 | square kilometers |
| acres | 0.40 | hectares |
| **Mass and Weight** | | |
| grams | 0.035 | ounce |
| kilograms | 2.21 | pounds |
| tons (100kg) | 1.10 | short tons |
| ounces | 28.35 | grams |
| pounds | 0.45 | kilograms |
| short tons (2000 lb) | 0.91 | tons |
| **Volume** | | |
| milliliters | 0.20 | teaspoons |
| milliliters | 0.06 | tablespoons |
| milliliters | 0.03 | fluid ounces |
| liters | 4.23 | cups |
| liters | 2.12 | pints |

| When You Know | Multiply By | To Find |
|---|---|---|
| **Volume** | | |
| liters | 1.06 | quarts |
| liters | 0.26 | gallons |
| cubic meters | 35.32 | cubic feet |
| cubic meters | 1.35 | cubic yards |
| teaspoons | 4.93 | milliliters |
| tablespoons | 14.78 | milliliters |
| fluid ounces | 29.57 | milliliters |
| cups | 0.24 | liters |
| pints | 0.47 | liters |
| quarts | 0.95 | liters |
| gallons | 3.79 | liters |
| **Volume** | | |
| cubic feet | 0.03 | cubic meters |
| cubic yards | 0.76 | cubic meters |
| **Speed** | | |
| miles per hour | 1.61 | kilometers per hour |
| kilometers per hour | 0.62 | miles per hour |
| **Temperature (exact)** | | |
| Celsius temp. | 9/5, +32 | Fahrenheit temp. |
| Fahrenheit temp. | −32, 5/9 × remainder | Celsius temp. |

Temperatures in degrees Celsius, as in the familiar Fahrenheit system, can only be learned through experience. The following temperatures are ones that are frequently encountered:

| | |
|---|---|
| 0°C | Freezing point of water (32°F) |
| 10°C | A warm winter day (50°F) |
| 20°C | A mild spring day (68°F) |
| 30°C | A hot summer day (86°F) |
| 37°C | Normal body temperature (98.6°F) |
| 40°C | Heat wave conditions (104°F) |
| 100°C | Boiling point of water (212°F) |

**MEASUREMENT *(continued)***

# SCIENTIFIC MEASUREMENT

## TABLE II. SCIENTIFIC UNITS

| Quantity | SI Unit | Symbol | Derivation | Other Units |
|---|---|---|---|---|
| acceleration | meter per second squared | $m/s^2$ | | |
| angular acceleration | radian per second squared | $rad/s^2$ | | |
| angular velocity | radian per second | rad/s | | |
| density | kilogram per cubic meter | $kg/m^3$ | | |
| electric capacitance | farad | F | (A·s/V) | |
| electric charge | coulomb | C | (A·s) | electrostatic unit (esu) = ⅓ $\times 10^{-9}$C |
| electric current | ampere | A | | |
| electric field strength | volt per meter | V/m | | |
| electric resistance | ohm | | (V/A) | |
| energy, work, quantity of heat | joule | J | (N·m) | electronvolt (eV) = 1.60219 $\times 10^{-19}$J<br>calorie (cal) = 4.184 J<br>British thermal unit (Btu) = 1055.87 J<br>erg = $10^{-7}$J<br>foot-pound (ft-lb) = 1.35582 J |
| flux of light | lumen | lm | (cd·sr) | |
| force | newton | N | ($kg \cdot m/s^2$) | dyne (dyn) = $10^{-5}$ N |
| frequency | hertz | Hz | ($s^{-1}$) | formerly cycle per second (cps, c/sec) |
| illumination | lux | lx | ($lm/m^2$) | |
| inductance | henry | H | (V·s/A) | |
| length | meter | m | | angstrom (A) = $10^{-10}$m |
| luminance | candela per square meter | $cd/m^2$ | | |
| magnetic field strength | ampere per meter | A/m | | oersted (Oe) = (1/4) × $10^3$ A/m |
| magnetic flux | weber | Wb | (V/s) | maxwell (Mx) = $10^{-8}$ Wb |
| magnetic flux density | tesla | T | ($Wb/m^2$) | gauss (G) = $10^{-4}$ T |
| magnetomotive force | ampere | A | | |
| mass | kilogram | kg | | |
| power | watt | W | (J/s) | horsepower (hp) = 745.7 W |
| pressure | newton per square meter | $N/m^2$ | | atmosphere (atm) = 1.01325 × $10^5$ $N/m^2$<br>bar = $10^5$ $N/m^2$ |
| velocity | meter per second | m/s | | |
| voltage, potential difference, electromotive force | volt | V | (W/A) | |

# SIGNS AND SYMBOLS

- + plus
- − minus
- ± plus or minus
- ∓ minus or plus
- × multiplied by
- ÷ divided by
- = equal to
- ≠ or ≠ not equal to
- ≈ or ≒ nearly equal to
- ≡ identical with
- ≢ not identical with
- ⇔ equivalent
- ~ difference
- ≅ congruent to
- > greater than
- ≯ not greater than
- < less than
- ≮ not less than
- ≧ or ≥ greater than or equal to
- ≦ or ≤ less than or equal to
- | | absolute value
- ∪ logical sum or union
- ∩ logical product or intersection
- ⊂ is contained in
- ∊ is a member of; permittivity; mean error
- : is to; ratio
- :: as; proportion
- ≐ approaches

- → approaches limit of
- ∝ varies as
- ∥ parallel
- ⊥ perpendicular
- ∠ angle
- ∟ right angle
- △ triangle
- □ square
- ▭ rectangle
- ▱ parallelogram
- ○ circle
- ⌒ arc of circle
- ≟ equilateral
- ≜ equiangular
- √ radical; root; square root
- ∛ cube root
- ∜ fourth root
- Σ sum
- ! or ∟ factorial product
- ∞ infinity
- ∫ integral
- ƒ function
- ∂ or δ differential; variation
- π pi
- ∴ therefore
- ∵ because
- vinculum (above letter)
- ( ) parentheses

- [ ] brackets
- { } braces
- ° degree
- ′ minute
- ″ second
- △ increment
- ω angular frequency; solid angle
- Ω ohm
- μΩ microhm
- MΩ megohm
- Φ magnetic flux
- Ψ dielectric flux; electrostatic flux
- ρ resistivity
- Λ equivalent conductivity
- ℜ reluctance
- → direction of flow
- ⇄ electric current
- ⬡ benzene ring
- → yields
- ⇌ reversible reaction
- ↓ precipitate
- ↑ gas
- ‰ salinity
- ☉ or ☼ sun
- ● or ● new moon
- ☽ first quarter
- ○ or ☺ full moon
- ☾ last quarter
- ☿ Mercury

- ♀ Venus
- ⊖ or ⊕ Earth
- ♂ Mars
- ♃ Jupiter
- ♄ Saturn
- ♅ Uranus
- ♆ Neptune
- ♇ Pluto
- ♈ Aries
- ♉ Taurus
- ♊ Gemini
- ♋ Cancer
- ♌ Leo
- ♍ Virgo
- ♎ Libra
- ♏ Scorpius
- ♐ Sagittarius
- ♑ Capricornus
- ♒ Aquarius
- ♓ Pisces
- ☌ conjunction
- ☍ opposition
- △ trine
- □ quadrature
- ✶ sextile
- ☊ dragon's head, ascending node
- ☋ dragon's tail, descending node
- ● rain
- ✳ snow

- ⊠ snow on ground
- ← floating ice crystals
- ▲ hail
- △ sleet
- ∨ frostwork
- ⊔ hoarfrost
- ≡ fog
- ∞ haze; dust haze
- T thunder
- < sheet lightning
- ⦶ solar corona
- ⊕ solar halo
- ⌜ thunderstorm
- ↘ direction
- ○ or ⊙ or ① annual
- ⊙⊙ or ② biennial
- ♃ perennial
- ♂ or ♂ male
- ♀ female
- □ male (in charts)
- ○ female (in charts)
- ℞ take (from Latin *Recipe*)
- ĀĀ or Ā or āā of each (doctor's prescription)
- ℔ pound
- ℥ ounce
- ʒ dram
- ℈ scruple

- ƒ℥ fluid ounce
- ƒʒ fluid dram
- ♏ minim
- & or & and; ampersand
- ℘ per
- # number
- / virgule; slash; solidus; shilling
- © copyright
- % per cent
- ℅ care of
- ℀ account of
- @ at
- * asterisk
- † dagger
- ‡ double dagger
- § section
- ☞ index
- ´ acute
- ` grave
- ~ tilde
- ˆ circumflex
- ¯ macron
- ˘ breve
- ¨ dieresis
- ¸ cedilla
- ∧ caret